D0803210

The Jepson Manual

FRIENDS OF THE JEPSON HERBARIUM

BENEFACTORS

Special Recognition

Jean and Sherman Chickering
Lawrence R. Heckard
William R. Hewlett
Myrtle R. Wolf

East Bay Chapter, California Native Plant Society

Stanley Smith Horticultural Trust

California Department of Fish and Game

National Science Foundation (BSR-9024888)

Individuals

In Memory of Thomas Hudson Armstrong, Jr.
Glen R. Bailey
David and Ann Bauer
Mary Leo Bowerman
Howard and Phoebe L. Brown
Dale Hollis Clark
Elizabeth Crispin
Katherine Cuneo
Frank W. Ellis, M. D.
Phyllis M. Faber
A. Barlow Ferguson
John R. Gibson
Margaret Hackley Gwinn
Margedant P. Hayakawa
Mr. and Mrs. J. W. Hickling
James C. Hickman
Rob Schonholtz and Jane Hicks
John R. Kiely
Stella May Knouse
In Memory of William W. Lester, Jr.
Greg S. Lyon
Estate of Gerena MacGowan
Lee and Ruth Meadows
Mrs. Kenneth C. Mirov
Diane L. Mitchell
Jane L. and Duane F. Mosier
Charles H. Quibell
Jim and Georgie Robinett
Marjorie G. Schmidt
James R. Shevock
Paul Silva
Sidney Richard Snow
Leslie Wheeler
Thomas J. Zavortink

Organizations

California Botanical Society
California Native Plant Society
  Marin Chapter
  Monterey Bay Chapter
  Sacramento Valley Chapter
  San Luis Obispo Chapter
  State Organization
The George Wright Society

Foundations and Corporations

S. D. Bechtel, Jr. Foundation
Flintridge Foundation
V. T. L. Gardner Botanical Books
Homeland Foundation
Marin Community Foundation
Pacific Gas and Electric Company
David and Lucile Packard Foundation
Santa Cruz Island Foundation
Elvenia J. Slosson Endowment Fund for Ornamental Horticulture
Sunset Publishing Corporation
True North Foundation
The Dean Witter Foundation

Government Agencies

California Energy Commission
U.S. Department of Agriculture
  Forest Service, Pacific Southwest Region
U.S. Department of Interior
  Bureau of Land Management
  Fish and Wildlife Service

THE JEPSON MANUAL PROJECT

Staff:

Project Manager: Dr. Dieter H. Wilken
Assistant Project Manager: Susan M. D'Alcamo
Editor: Dr. James C. Hickman
Scientific Editor: Dr. Thomas J. Rosatti
Production Editor: William J. Stone
Principal Illustrators: Emily E. Reid, Dr. Linda Ann Vorobik
Administrative Assistants: JD Trow, Jeanne A. Visnaw
Principal Investigator: Prof. Lincoln Constance
Principal Consultant: Dr. Lawrence R. Heckard (deceased)

Consultants:

Bobbie Angell (illustrations)
David Comstock (production)
Prof. David J. Keil (Asteraceae)
Karen Klitz (illustrations)
Janet Matias (production consultant)
Dr. Elizabeth Painter (proof-reading)
Michael Paul (illustration plate setup)
Leslie Randall (illustrations)
Dana Wullenwaber (production consultant)
Sarah A. Young (illustrations)

Major Volunteers:

Robert Berman (records)
Sherman Chickering
Maggie Day (Aizoaceae illustrations)
Natalie Hopkins (newsletter editor 1987–88)
Carl Koerper (toxic plants)
Wendy Ellen Kronman (horticultural information)
Katy Martin (reference materials, horticultural information)
Emily E. Reid (illustrations)
Jan Seibert (proof-reading)
Bobby and Ken Shockey
Susan Stanley (illustrations)
Margriet Wetherwax (reference materials)
Karen Wetherell (horticultural information, library and slide curation)
Myrtle R. Wolf
Hisao Yokota (artist, gifts for benefactors)
Sarah A. Young (illustrations)

Editorial Board:

Horticultural Council:

Warren G. Roberts (Chairman), The University Arboretum, University of California, Davis
Frank Chan, Pacific Gas and Electric Company, San Francisco
Barbara Coe, Saratoga Horticultural Foundation, San Martin
Beecher Crampton, Agronomy and Range Science, University of California, Davis (retired)
Steven Darington, Eureka
John Dourley, Rancho Santa Ana Botanic Garden, Claremont (retired)
Dick Dunmire, *Sunset Magazine* (retired) and Los Altos
Stephen Edwards, Regional Parks Botanic Garden, Berkeley
Dara Emery, Santa Barbara Botanic Garden (deceased)
Mike Evans, Tree of Life Nursery, San Juan Capistrano
Ron Gass, Mountain States Nursery, Glendale, Arizona
Brett Hall, Arboretum, University of California, Santa Cruz
Gerda Isenberg, Yerba Buena Nursery, Woodside (represented by Lori Hubbart)
Melanie Baer-Keeley, Theodore Payne Foundation, Sun Valley
Ron Lutsko, Jr., Ron Lutsko and Associates, San Francisco
Elizabeth McClintock, Strybing Arboretum, San Francisco and University of California at Berkeley
Malcolm McLeod, California Polytechnic State University, San Luis Obispo
Alice Mehdy Peden, Santa Fe, New Mexico
Ray and Peg Prag, Forestfarm, Williams, Oregon
Roger Raiche, University of California Botanical Garden, Berkeley
Wayne Roderick, Regional Parks Botanic Garden, Berkeley (retired)
Marjorie Schmidt (Honorary Chairman), Berkeley (deceased)
Jacob Sigg, Strybing Arboretum, San Francisco (retired)
Nevin Smith, Wintergreen Nursery, Watsonville
Margaret J. Williams, Northern Nevada Native Plant Society, Sparks

The Jepson Manual

Higher Plants of California

James C. Hickman, Editor

University of California Press
Berkeley Los Angeles London

University of California Press
Berkeley and Los Angeles, California

University of California Press, Ltd.
London, England

Library of Congress Cataloging-in-Publication Data

The Jepson manual : higher plants of California / James C. Hickman, editor.
p. cm.
Includes index.
ISBN: 0-520-08255-9 (alk. paper)
1. Botany—California. 2. Botany—California—Pictorial works.
3. Plants—Identification. I. Hickman, James C.
QK149.J56 1993
581.9794—dc20 92-21365

Printed in the United States of America

9 8 7 6 5 4 3 2

The paper used in this publication meets the minimum requirements of American National Standard for Information Sciences—Permanence of Paper for Printed Library Materials, ANSI Z39.48-1984.

The Jepson Manual is dedicated to

LAWRENCE R. HECKARD

9 April 1923–26 November 1991

As Curator of The Jepson Herbarium and Library from 1968 until his death, Larry Heckard quietly nourished Jepson's wish that his *Manual* eventually would be updated. Larry's roles in the project have been diverse and crucial—first as originator of the concept, financial planner, and facilitator of the project's beginning; then as Principal Consultant, Chairman of the Editorial Board, Curator of the Jepson Herbarium, major author, and financial benefactor. Without his conviction, energy, and immense knowledge of California plants, this book would not exist.

CONTENTS

SOMETHING · LOST · BEHIND · THE · RANGES · OVER
YONDER · · GO · YOU · THERE · · THE EXPLORER
WILLIS LINN JEPSON

PREFACE

WILLIS LINN JEPSON was born in 1867 near Vacaville in Solano County. By age 13 he had begun cataloguing the plants of the southwestern Sacramento Valley and adjacent mountains and wetlands. He became the first person to receive a doctorate in botany from the newly established University of California. Almost at once he became Assistant Professor, succeeding his mentor, E. L. Greene. He founded the California Botanical Society in 1915, and also helped establish the Save-the-Redwoods League and the Sierra Club.

Jepson published many books that have led thousands to an appreciation of California plants and natural settings. Most beloved is his *Manual of the Flowering Plants of California*, completed in 1925. This book was the first attempt to provide an informative identification manual for all the wild plants of California.

Jepson's ambitious, multi-volume *Flora of California* had not been completed when he died in 1946; his estate was accepted by the University of California so that his work could be continued. The unit he endowed, The Jepson Herbarium and Library, is committed to Jepson's vision of increasing botanical knowledge and enthusiasm among the people of California.

The year before his death, Jepson inscribed a copy of his Manual belonging to his friend Robert E. Burton. The last two sentences are a favorite quote from Kipling that Jepson incorporated into his bookplate. The words he wrote are appropriate as a motto for any who undertake to prepare a new manual for all the wild plants of California.

> The botanist's objective is a furtherance of knowledge of living plants. He wishes by his investigations to discover new facts and establish new principles that will aid in promoting botanical research. If wise, he will never try to produce a work which is perfect, complete and final. Any such work would be a paradox and at cross-purposes with our concepts of living things and our ideas of endless evolution associated with them. Completion, perfection, finality represent an anomaly, a contradiction in the field of biology. The far-seeing botanist, on the contrary, will strive to do work which is inspiring, productive of thought and promoting the soundest progress, so that botanical science will ever advance into new and more fruitful fields. "There is something lost behind the ranges over yonder. Go you there."

ACKNOWLEDGEMENTS

IT SEEMS A MINOR MIRACLE that *The Jepson Manual* is in your hand. We began ten years ago with determination but much naiveté—any number of events could have derailed the project, and some almost did. Somehow, with the cooperation and support of nearly 2000 people, this attempt at a fresh approach to a large plant manual was completed.

Most of those who helped us in innumerable ways cannot be formally acknowledged here, but I would like to name some of the most important.

The stimulus for this book came from Lawrence R. Heckard. He remained devoted to it until his death from AIDS-related complications nine years into the project (see Dedication).

Several others provided critical impetus at the project's beginning. Lincoln Constance was diplomatically and quietly supportive in every way—he has continued his help as the project's Principal Investigator for grant-administration purposes and as our senior adviser. With the encouragement of Myrtle Wolf and Susan D'Alcamo (our first volunteers), Emily E. Reid stepped forward to donate as many illustrations as she could produce for *The Jepson Manual*. This good fortune made it possible to provide new, carefully rendered drawings for about 4000 taxa.

We have been blessed with an extraordinary group of project staff. Despite disagreements, the entire group pulled together, sometimes under difficult circumstances, to make the project work. Lacking any one of the staff, it is unclear how the book could have been finished. However, I want to personally thank Susan D'Alcamo for repeatedly showing us with her infectious enthusiasm what the project could be, Dieter Wilken for accepting the diverse and demanding responsibilities of Project Manager during the two final years of the project and performing admirably under a highly constraining time schedule, Bill Stone for learning how to manage the computer aspects of the project effectively, and Linda Vorobik for her straightforward resolution of many artistic and design issues. All these people made my life much easier.

Nearly 200 botanist-authors have donated their time and expertise to making this book a reality. We cannot adequately thank them for their good work and cooperation—the value in this book is primarily to their joint credit. I express particularly heartful thanks to those authors who understood and supported what we were trying to do with this book, even though it did not conform to their own personal convictions. All authors are listed alphabetically following this section.

Like any project of this magnitude, the beginning was rather amorphous and difficult. Many ideas (some quite short-lived) were tested as widely as possible. A great many people offered important early advice and moral support that helped crystallize the direction of the effort. Among them were Glen and Edna Bailey, Theodore M. Barkley, Mary E. Barkworth, Helen-Mar Beard (deceased), Robert Berman, Edward Burton, Cristalyn M. Cagle (deceased),

Daniel Campbell, Norden Dan Cheatham, David Comstock, Katherine Culligan, Oxford University Press (William Curtis), Lauramay T. Dempster, Frank W. Ellis, Phyllis Faber, Bim and Margaret Flesher, George Hanley, Harvard University Press (Gunder Hefta), Patricia K. Holmgren, Natalie Hopkins, Diane Ikeda, Milton and Ruth Irvine, Jepson Vineyards (Denise Whitaker), Marshall Johnston, David J. Keil, Walter and Irja Knight, Corky Matthews, Sandy Morey, Jo-Ann Ordano, Vernon Oswald, W. Robert Powell, Peter H. Raven, Catherine Rose, Jake and April Ruygt, Suzanne Schettler, James Shevock, Ken and Roberta Shockey, Teresa Sholars, Stanford University Press (William Carver), Veva Stansell, G. Ledyard Stebbins, John L. Strother, Mary Susan Taylor, University of California Press (James H. Clark, Ernest "Chick" Callenbach, and August Frugé), Margaret van Eck, and Virginia L. Waters.

The project was begun with income from W. L. Jepson's endowment that Curator Larry Heckard had saved over the years. Soon, this proved inadequate and Susan M. D'Alcamo agreed to find further financial support. Her success provided viability to the project. She secured the support of many others in publicizing the project and in finding funds. The California Native Plant Society was instrumental in helping us establish The Friends of The Jepson Herbarium. Charlice Danielsen, Suzanne Schettler, and Laurie Kiguchi were especially helpful in the initial stages. Throughout the project, CNPS has been both supportive and cooperative in our fundraising efforts.

Benefactors Myrtle R. Wolf, our first contributor and continuing supporter, and Sherman Chickering, who generously contributed time and effort to obtain other funding, are gratefully acknowledged for their dedication to the project. Others helpful in reaching our financial goals include Eleanor Bade, Wilma Follett, August and Susan Frugé, Virginia L.T. Gardner, Ann Witter Gillette, Milton and Ruth Irvine, Mildred Mathias, Michael D. McCollum, Doug Miller, The Nature Conservancy (Carol Blanton, Robin Cox, Steve Johnson, Nancy Mackinnon, and Steve McCormick), James Shevock, Southern California Botanists (Kathleen Paul and Alan Romspert), Roberta Steiner, Sir George Taylor, UC Development Office (Mike Desler, Anne Garvey and her staff, Sylvia Hansen, Dianne Smith, Julia Voorhies), California state government (Aide Cindy Williams, Senator John Nedjedly, Senator William Campbell, Assemblyman William Baker), and Hisao Yokota (donor of art gifts for benefactors).

Many people provided guidance in areas of special expertise. The following list, like the others, cannot be complete. G. Douglas Barbe (weeds), Rupert Barneby (legumes), Alan Barron (geography), Ken Berg (rare plants), Roxanne Bittman (rare plants), A.I. Chait and John Danielsen (computers), Thomas Duncan (computers), Kent Holsinger (*Clarkia*), Robert Ishi (page design), Harlan Kessel (publication), Carl Koerper (toxic plants), Wendy Kronman (horticultural entries), Bruce MacBryde (rare plants), Elizabeth McClintock (escaped exotics), Richard Moe (computers), Larry E. Morse (rare plants), James Nelson (rare plants), Larry L. Norris (geography), Elizabeth Painter (proofreading), Jim and Georgie Robinett (bulb-bearing plants), Jan Seibert (proofreading), James R. Shevock (geography, rare plants), Mark Skinner (rare plants), John F. Sorenson (mints), UC Botanical Garden (meeting arrangements), Jimmy T. Vale (photography), Margriet Wetherwax (Scrophulariaceae), John Whedon (mailing services), Rick York (rare plants).

JAMES C. HICKMAN

AUTHORS CONTRIBUTING TREATMENTS TO *THE JEPSON MANUAL*

Susan G. Aiken, Canadian Museum of Nature, Ottawa, Ontario
Geraldine A. Allen, University of Victoria, British Columbia, Canada
Kelly W. Allred, New Mexico State University, Las Cruces
Edward Alverson, Oregon State University, Corvallis
Barrett H. Anderson, California Academy of Sciences, San Francisco
Dennis E. Anderson, Humboldt State University, Arcata
Edward F. Anderson, Whitman College, Walla Walla, Washington
Loran C. Anderson, Florida State University, Tallahassee
George Argus, Canadian Museum of Nature, Ottawa, Ontario
Wayne P. Armstrong, Palomar College, San Marcos
Deborah Engle Averett, Ohio State University, Columbus
Tina J. Ayers, Northern Arizona University, Flagstaff
John D. Bacon, University of Texas, Arlington
Susan Bainbridge, Duke University, Durham, North Carolina
John Baird, Northern Kentucky University, Highland Heights
Marc A. Baker, Arizona State University, Tempe
Bruce Baldwin, Duke University, Durham, North Carolina
Theodore M. Barkley, Kansas State University: Contribution 92-168-B from the Kansas Agricultural Experiment Station, Manhattan
Mary E. Barkworth, Utah State University, Logan
Jim A. Bartel, United States Fish and Wildlife Service, Sacramento
David M. Bates, Cornell University, Ithaca, New York
Randall J. Bayer, University of Alberta, Edmonton, Canada
Tania Beliz, University of California at Berkeley
Robert Berman, University of California at Berkeley
John Bleck, University of California at Santa Barbara
David Brandenburg, Dawes Arboretum, Newark, Ohio
Gregory K. Brown, University of Wyoming, Laramie
Richard K. Brummitt, Royal Botanic Gardens, Kew, England
Roy E. Buck, University of California at Berkeley
Gerald D. Carr, University of Hawaii, Honolulu, Hawaii
Robert L. Carr, Eastern Washington State University, Cheney
Kenton L. Chambers, Oregon State University, Corvallis
Anita F. Cholewa, University of Minnesota, St. Paul
Tsan-Iang Chuang, Illinois State University, Normal
Curtis Clark, California Polytechnic State University, Pomona
W. Dennis Clark, Arizona State University, Tempe
Lynn Clark, Iowa State University, Ames
J. Travis Columbus, University of California at Berkeley
Steven A. Conley, Redding
Lincoln Constance, University of California at Berkeley
Raymond Cranfill, University of California at Berkeley
William J. Crins, University of Toronto, Mississauga, Ontario, Canada
Michael Curto, California Polytechnic State University, San Luis Obispo
Susan M. D'Alcamo, University of California at Berkeley
Thomas F. Daniel, California Academy of Sciences, San Francisco
Jerrold I. Davis, Cornell University, Ithaca, New York
W. S. Davis, University of Louisville, Kentucky
Alva G. Day, California Academy of Sciences, San Francisco
Lauramay T. Dempster, University of California at Berkeley
Melinda F. Denton, University of Washington, Seattle
James C. Dice, Natural History Museum, San Diego
Patrick E. Elvander, University of California at Santa Cruz
Barbara J. Ertter, University of California at Berkeley
Frederick Essig, University of South Florida, Tampa
Wayne R. Ferren, University of California at Santa Barbara
Peggy Fiedler, San Francisco State University
Fred R. Ganders, University of British Columbia, Vancouver, Canada
Craig W. Greene, College of the Atlantic, Bar Harbor, Maine
James R. Griffin, University of California at Berkeley
James W. Grimes, New York Botanical Garden, Bronx
Erich Haber, Canadian Museum of Nature, Ottawa, Ontario
Richard R. Halse, Oregon State University, Corvallis
Ronald L. Hartman, University of Wyoming, Laramie
M. J. Harvey, Dalhousie University, Halifax, Canada (retired) and Victoria, British Columbia, Canada
Richard L. Hauke, University of Rhode Island, Kingston (retired) and Atlanta, Georgia
Frank G. Hawksworth, United States Forest Service, Fort Collins, Colorado (retired)
Lawrence R. Heckard, University of California at Berkeley (deceased)
Douglass M. Henderson, University of Idaho, Moscow
James Henrickson, California State University, Los Angeles
James C. Hickman, University of California at Berkeley
Steven R. Hill, Clemson University, Clemson, South Carolina
Peter C. Hoch, Missouri Botanical Garden, St. Louis
Carol A. Hoffman, University of Georgia, Athens
Noel H. Holmgren, New York Botanical Garden, Bronx

Duane Isely, Iowa State University, Ames (retired)
James K. Jarvie, Utah State University, Logan
Judith Jernstedt, University of California at Davis
Dale E. Johnson, Missouri Botanical Garden, St. Louis, and Timber Press, Portland, Oregon
James D. Jokerst, Jones and Stokes Associates, Sacramento
Elaine Joyal, Desert Botanical Garden, Phoenix, Arizona
Steven A. Junak, Santa Barbara Botanic Garden
Glenn Keator, Sebastopol
Jon E. Keeley, Occidental College, Los Angeles
David J. Keil, California Polytechnic State University, San Luis Obispo
Ronald B. Kelley, University of California at Davis
Walter A. Kelley, Mesa College, Grand Junction, Colorado
Daryl L. Koutnik, Los Angeles State and County Arboretum, Arcadia
Arthur C. Kruckeberg, University of Washington, Seattle
John C. La Duke, University of North Dakota, Grand Forks
Meredith A. Lane, University of Kansas, Lawrence
Thomas Lemieux, University of Colorado, Boulder
Harlan Lewis, University of California at Los Angeles (retired)
Richard A. Lis, University of California at Berkeley
R. John Little, University of California at Berkeley
Robert I. Lonard, Pan American University, Edinburg, Texas
Karen Lu, Humboldt State University, Arcata
Joy Mastrogiuseppe, Washington State University, Pullman
Niall F. McCarten, University of California at Berkeley
Elizabeth McClintock, University of California at Berkeley
Katy K. McKinney, Texas A&M University, College Station
Dale W. McNeal, University of the Pacific, Stockton
Michael R. Mesler, Humboldt State University, Arcata
Timothy Messick, Jones and Stokes Associates, Sacramento
Paul Meyers, Pierce College, Los Angeles
Jennifer Milburn, Missouri Botanical Garden
Michael Moore, University of Georgia, Athens
John S. Mooring, University of Santa Clara
Reid Moran, San Diego Natural History Museum (retired) and Sebastopol
James D. Morefield, Rancho Santa Ana Botanic Garden, Claremont
Nancy R. Morin, Missouri Botanical Garden, St. Louis
Michael Nee, New York Botanical Garden, Bronx
Elizabeth Neese, University of California at Berkeley
Guy L. Nesom, University of Texas at Austin
Bryan D. Ness, Pacific Union College, Angwin
Richard Olmstead, University of Colorado, Boulder
Robert Ornduff, University of California at Berkeley
Elizabeth L. Painter, Colorado State University, Fort Collins
Bruce D. Parfitt, Missouri Botanical Garden, St. Louis
Robert W. Patterson, San Francisco State University
Willard W. Payne, Sanibel, Florida
Paul M. Peterson, Smithsonian Institution, Washington, D.C.
C. Thomas Philbrick, Rancho Santa Ana Botanic Garden, Claremont
Duncan M. Porter, Virginia Polytechnic and State University, Blacksburg
A. Michael Powell, Sul Ross State University, Alpine, Texas
Robert A. Price, Indiana University, Bloomington
Barry A. Prigge, University of California at Los Angeles
James S. Pringle, Royal Botanical Gardens, Hamilton, Ontario, Canada
Charles F. Quibell, Sonoma State University, Rohnert Park
Peter H. Raven, Missouri Botanical Garden, St. Louis
Martin Ray, University of California at Berkeley
John R. Reeder, University of Arizona, Tucson
Rhonda Riggins, California Polytechnic State University, San Luis Obispo
Warren Roberts, University of California at Davis
Reed C. Rollins, Harvard University, Cambridge, Massachusetts (retired)
Thomas J. Rosatti, University of California at Berkeley
John O. Sawyer, Jr., Humboldt State University, Arcata
Jeffrey P. Schaffer, University of California at Berkeley
Robert L. Schlising, California State University, Chico
Clifford L. Schmidt, San Jose State University (retired) and Salem, Oregon
Alfred E. Schuyler, Academy of Natural Sciences, Philadelphia, Pennsylvania
John C. Semple, University of Waterloo, Ontario, Canada
James R. Shevock, United States Forest Service, San Francisco
Teresa Sholars, College of the Redwoods, Fort Bragg
Leila M. Shultz, Utah State University, Logan
Beryl B. Simpson, University of Texas at Austin
Mark W. Skinner, California Native Plant Society, Sacramento
Alan R. Smith, University of California at Berkeley
Galen Smith, University of Wisconsin, Whitewater (retired)
James P. Smith, Humboldt State University, Arcata
Robert J. Soreng, Cornell University, Ithaca, New York
Richard Spellenberg, New Mexico State University, Las Cruces
G. Ledyard Stebbins, University of California at Davis
William J. Stone, University of California at Berkeley
John L. Strother, University of California at Berkeley
Scott D. Sundberg, University of Washington, Seattle
Janice C. Swab, St. Mary's College, Raleigh, North Carolina
Fosiée Tahbaz, University of California at Berkeley
Jennifer Talbot, Washington University, St. Louis, Missouri
Dean Taylor, Biosystems Analysis Inc., Santa Cruz
Mary Susan Taylor, Missouri Botanical Garden, St. Louis
W. Carl Taylor, Milwaukee Public Museum, Wisconsin

John Thieret, Northern Kentucky University, Highland Heights
David M. Thompson, Rancho Santa Ana Botanical Garden, Claremont
Robert F. Thorne, Rancho Santa Ana Botanical Garden, Claremont
Steven L. Timbrook, Ganna Walska Lotusland Foundation, Santa Barbara
Gordon C. Tucker, New York State Museum, Albany
John M. Tucker, University of California at Davis
Charles E. Turner, US Department of Agriculture, Albany, California
Staria S. Vanderpool, University of North Dakota, Grand Forks
Barbara Veno, California State University, Los Angeles
Nancy J. Vivrette, Ransom Seed Company, Santa Barbara
Linda Ann Vorobik, University of California at Berkeley
Warren H. Wagner, University of Michigan, Ann Arbor
Warren L. Wagner, Smithsonian Institution, Washington, D.C.
Gary D. Wallace, Los Angeles County Natural History Museum
Michael J. Warnock, Sam Houston State University, Huntsville, Texas
Grady L. Webster, University of California at Davis
Robert Webster, US Department of Agriculture, Beltsville, Maryland
Philip V. Wells, University of Kansas, Lawrence
Thomas L. Wendt, Colegio de Postgraduados, Chapingo, Mexico
Margriet Wetherwax, University of California at Berkeley
R. David Whetstone, Jacksonville State University, Alabama
Sherry Whitmore, Oak Park, Illinois
Delbert Wiens, University of Utah, Salt Lake City
Dieter H. Wilken, Colorado State University, Fort Collins and University of California at Berkeley
Michael P. Williams, University of Washington, Seattle
Dennis W. Woodland, Andrews University, Berrien Springs, Michigan
Thomas Worley, Humboldt State University, Arcata
George Yatskievych, Missouri Botanical Garden, St. Louis

PHILOSOPHY AND HISTORY OF THE JEPSON MANUAL PROJECT

Philosophy

The Jepson Manual Project was based on a virtually revolutionary premise: a single work of this magnitude can be simultaneously accessible to dedicated beginners and indispensable to professional botanists. The audience envisioned for *The Jepson Manual* was very broad, including beginning students, self-taught amateurs, environmental consultants, native-plant gardeners, employees of diverse government agencies, ecologists, and academic systematists.

Our primary purpose was to follow Jepson's lead in making the extraordinarily diverse world of California wild plants available to the widest range of people possible. The central means chosen to achieve this were solicitation of expert volunteer authors to write comprehensive and authoritative treatments, use of accessible rather than arcane language, illustration of the diagnostic features of as many plant taxa as possible, strict comparison of related taxa, judicious use of abbreviations, and close attention to format consistency, readability, and ease of use.

The design and content of *The Jepson Manual* were constrained by both space (a single "portable" volume) and time (a 10-year completion schedule). Because of these constraints, many compromises were necessary that were not easy for either authors or editors. Great conciseness was required, but, where conciseness was seen to sacrifice clarity, clarity was given priority.

The decision to use the best available contributing author for each group brought a diversity of taxonomic philosophies and levels of experience to the project. In an attempt to maintain both accuracy and a reasonably consistent tone, the editors (using representative herbarium specimens, their own experience, and sometimes outside reviews) regularly questioned authors about content. Authors had the final say about recognition of taxa; in this sense, the scientific content of each treatment is the responsibility of its author. However, certain authors could not agree with some editorial practices, especially concerning terminology. For the sake of uniformity, editorial judgement prevailed in these cases.

Treatments had to be made short because of space constraints. This ran counter to the goal of conveying adequately the accrued knowledge of expert authors. The desire to put as much useful information as possible into a small space led to some necessarily painful editing to achieve consistency and conciseness. For authors summarizing a decade or a lifetime of ongoing work, this was often a difficult experience.

Furthermore, intensive study does not necessarily lead to an easier or simpler taxonomic treatment. As taxonomists learn more about the complexities of variation patterns, their best taxonomic assessments automatically change. It may become more difficult to represent what is known in a taxonomic scheme. Scientists continually work toward more integrated perspectives but probably will never be fully successful in the search for simple answers. The more we explore and learn, the more intriguingly complex we find the natural world to be.

Some authors approached certain groups as taxonomic authors for the first time; they confronted a different set of challenges. Almost uniformly, they found that time did not allow a study of the intensity they wished and that more was unknown than they believed when they started. Some treatments in *The Jepson Manual* are simultaneously tentative and the best currently possible. In such situations, there was an attempt to point out clearly what problems remain and to suggest specific further work.

Combining expert authors with a user-oriented philosophy presented a dilemma. Users want and need clear, neat, simple answers to their questions. Experts want to convey their evolving, complex, and dynamic perspectives, which are often neither neat nor simple. The primary goal of The Jepson Manual Project was to find the fine line that best accommodates these diverse wants and needs.

In addition to being authoritative, *The Jepson Manual* attempts to be comprehensive, concise, and accessible. It was dubbed a "people's flora" by some because of the effort devoted to creating a bridge between authors and users — or, more broadly, between the community of plant systematists (who both dispense and are the most sophisticated users of systematic information) and the majority of potential users (who need access to that information but are less sophisticated than those dispensing it). Its users will determine how sturdily this bridge was constructed.

Specific practices developed to achieve comprehensiveness, conciseness, and accessibility are discussed in the chapter "Conventions Used in *The Jepson Manual.*"

History

In 1925, Professor Willis Linn Jepson completed his *Manual of the Flowering Plants of California.* This concise (and slightly misnamed) work was the first written attempt to account for all of the higher plants (ferns, gymnosperms, and flowering plants), both native and alien, growing wild in California. It became a classic, with many devotees more than 65 years after its publication.

Throughout his career, Jepson worked on his more expansive, multi-volumed *Flora of California.* At the time of his death in 1946, even with help from his students (especially Lauramay T. Dempster), treatments for major groups for the *Flora* remained unwritten. He left an endowment to the University of California at Berkeley — the Jepson Research Fund, to be administered by a self-perpetuating committee or Board of Trustees — which was to fund the continuation of his work, particularly the completion of his *Flora* and continued publication and periodic updating of his *Manual.*

The Jepson Herbarium and Library was established by the University in 1950 to continue Jepson's work. Its first Curator was Dr. Rimo Bacigalupi, who immediately began intensive work on Scrophulariaceae, the next major family for the *Flora.* He was assisted for a time by Dr. G. Thomas Robbins, then was joined by Dr. Lawrence R. Heckard in 1960.

During the Jepson Herbarium's early years, several events sidetracked work on both of Jepson's major goals. First, *A California Flora* was published by Professor Philip A. Munz in 1959 (a *Supplement* was added in 1968). Munz's *Flora* was philosophically different from Jepson's *Manual,* more technical, and written more for the professional botanist than the general public. The increased number of taxa treated by Munz reflected the growing knowledge of California's diverse flora. Second, the fourth and final volume of Professor Leroy Abrams' *Illustrated Flora of the Pacific States* was completed in 1959 (four years after Abrams' death) by Roxanna S. Ferris (the first volume had been published in 1923, the same year that Jepson published the early part of his *Manual*). These publications, plus the University of California's emphasis on continuous scholarly publication by Jepson Herbarium staff, greatly slowed activity toward completion of Jepson's *Flora* and updating of his *Manual.*

Following Bacigalupi's retirement in 1974, Heckard became Curator. Although engaged primarily in work on western North American members of *Castilleja*, *Orthocarpus*, and *Cordylanthus* (with Professor Tsan-Iang Chuang of Illinois State University), Heckard preserved the goal of revising and updating Jepson's *Manual.*

In 1980, Dr. James C. Hickman joined Heckard at the Jepson Herbarium to collaborate in a floristic treatment of Snow Mountain in the High North Coast Ranges. In the course of this work, many taxonomic problems became apparent that could not be resolved with available resources; it became evident that the time was right to produce a new manual for the entire state. Heckard believed that the project could be designed for (and should not be allowed to exceed) a ten-year schedule. In late 1982, Hickman accepted the responsibility to produce a new *Manual* by 1992 — in effect, to use the resources of the Jepson Herbarium to prepare an all-new work that would revitalize the ideals of Jepson's 1925 *Manual.*

From the beginning, Heckard was the inspirational force in the design of the project, which emphasized Jepson's determination to make botanical scholarship accessible to the public. He was joined by Professors Lincoln Constance and Thomas Duncan as the initial oversight committee. Hickman sought botanists willing to write treatments on groups in which they had special expertise and was remarkably successful — the number of committed volunteer authors grew to 60 by mid-1984 and finally reached nearly 200. The nature of the project changed from local and modest to multi-authored and authoritative. Coordination and editing became Hickman's primary tasks.

Initially, funding for the project was to have come from Jepson Endowment income. However, this was clearly inadequate for the project that emerged — producing a radically new and better manual for the entire state, a single volume covering an extraordinarily large and complex set of plant taxa. In 1984, Susan D'Alcamo agreed to obtain the increased funding necessary for the project to reach its full potential. To do this, she established the "Friends of The Jepson Herbarium" and secured foundation and agency grants and contracts. A staff of seven and four consultants were hired to handle the complex tasks of planning, coordinating, illustrating, editing, designing, and typesetting. Without the dedicated staff, supported by the financial help of about 1,500 individuals and scores of foundations and agencies, the project could not have succeeded.

The fourth and final *Guide for Contributors* was distributed in mid-1987. It was superseded in mid-1989 by publication of *Introduction to The Jepson Manual*, an 80-page book designed to provide a clear preview of the direction in which the project was headed and to allow comment before final publication of the new Manual.

The larger the number of authors (and the greater the potential authoritativeness of the final product), the greater the coordination and administrative tasks became. The project was reorganized in 1989 with a larger Editorial Board (see p. v) that was broadly representative of the community of users. An equally representative Horticultural Council (see p. v) by then was working to provide concise statements on the garden values of native plants.

In 1990, it was decided that a Project Manager was needed to decrease Hickman's administrative duties and allow him to concentrate on editorial responsibilities. Dieter H. Wilken took a two-year leave of absence from his professorship at Colorado State University to become the administrative head of the project in August 1990, adding another powerful editorial voice to the staff and instituting new policies that assured timely publication of *The Jepson Manual*.

CONVENTIONS USED IN *THE JEPSON MANUAL*

A number of principles and practices were developed to help *The Jepson Manual* achieve the goals of *comprehensiveness*, *conciseness*, and *accessibility* discussed in the chapter on History and Philosophy of The Jepson Manual Project (p. 1). In a work of this sort, precision, conciseness, and clarity are all of utmost importance; however, methods by which they are achieved are often at odds with one another. In an attempt to develop a consistent, workable compromise among them, various assumptions and conventions were adopted for *The Jepson Manual* that may not be found in comparable manuals. To help users adjust to the new style, the most important conventions are outlined below.

This chapter is important for all new users of *The Jepson Manual*, but readers should also familiarize themselves with the chapter on overall philosophy, as well as those detailing *Manual* conventions for abbreviations and symbols, terms, geographic subdivisions, commonness and rarity, and horticultural information.

Beyond the basic structure of keys and descriptions, few conventions were followed absolutely uniformly in *The Jepson Manual*; where clarity and consistency conflicted, clarity was given priority. With nearly 200 authors and a staff of six text and illustration editors, compromises were tailored to circumstances. The recurring issue was "where to draw the line". We believe that this issue, in its many guises, is unsolvable to anyone's full satisfaction; we did our best within the constraints under which the book was produced. We hope, most of all, that treatments in *The Jepson Manual* are usable and helpful.

Overall Design Conventions

Comprehensiveness. All native, validly named taxa are included (no new names are published in *The Jepson Manual*). In addition, all aliens that have become an integral part of the California flora are included. The general policy was not to include (or to note only in passing) waifs or non-reproducing but long-persisting individuals or clones. Including all waifs and possible waifs in California would have added greatly to the length of the book. As in other areas, the desired line was impossible to draw cleanly; strong views of authors about inclusion or exclusion of taxa normally were accommodated.

It is certain that there are fully naturalized alien taxa that might be (but are not) included in *The Jepson Manual*. Most often, this is a result of poor documentation. Alien plants generally are given low priority by collectors. Many botanists do not accept "reports" if they are not documented by specimens. Therefore, there are large gaps in our documentable knowledge of naturalized aliens. Specimens of such plants would make valuable additions to herbaria, especially of those taxa likely to expand their ranges.

Uniformity. Given the diversities of style and training among the nearly 200 contributing authors, achieving consistency among treatments required great effort by both authors and editors. Every attempt has been made to find mutually acceptable compromises where criteria conflict, or where authors have disagreed with practices taken here as standard. As a group, authors have been exceptionally agreeable and cooperative — a central component in the timely completion of the book.

Book Organization and Design. Easy accessibility to information is the primary criterion on which *Manual* organization and design are based. Within each of the four major groups of higher plants (ferns and their allies, gymnosperms, dicots, and monocots), families (and genera within families, species within genera, and infraspecific taxa within species) are all arrayed alphabetically (for an assessment of the patterns of relationships of families, see Appendix II, Classification of California Plant Families). Details of column-layout, type styles, running heads, and order of descriptive features were also evaluated against ease of both accessibility and readability.

Measurements. All linear measurements are given in metric units. Metric conversion scales (centimeters to inches, meters to feet) are provided at the end of the book.

Illustrations. Illustrations provide an important means of identifying plants. In *The Jepson Manual*, they are designed to corroborate understanding of keys and descriptions, as well as to be easily browsed, independently of text. Generally, except for uncommon aliens, each genus is represented, usually by a habit drawing.

Because all taxa could not be illustrated, priority was given to those commonly encountered, those uncommon or threatened enough to warrant monitoring, and those with unusual or difficult diagnostic features. All illustration plates appear on right-handed pages. Except in a few cases, illustrations follow the descriptions of taxa they illustrate and the order is alphabetical.

Index. The index was made as short yet as helpful as possible. All family names, generic names, common names, and names here considered to be synonyms (or to be misapplied to another entity) are indexed in a single alphabetical listing. Because accepted specific and infraspecific names are all ordered alphabetically under the appropriate genus, they can be found easily and are not listed in the index.

Keys and Their Conventions

All Keys.

A key, in our terms, is an artificial, verbal device for identifying a plant. It is a series of paired, mutually exclusive statements that divides a set of objects into progressively smaller subsets until all possibilities but one have been eliminated. Keys are used at all levels, from the separation of groups of families to the separation of varieties of a species.

Keys are made up of numbered, paired, alternative statements about plant features; only one statement of the pair (or "couplet") should be true for a given plant. Couplets are arranged in a series of indented steps, like sequential clues that allow the user to see where any decision leads. Each decision between statements of a couplet eliminates all possibilities included under the rejected statement and leads in turn to the lowest-numbered couplet included under the accepted statement (and thereby to another decision). This process is repeated until the end point is reached, that is, when all inappropriate possibilities have been eliminated and the identity of the plant is revealed.

Much effort was devoted to the development of unambiguous, precisely parallel key statements (avoiding ambiguity had priority over parallelism). All statements have a morphological basis, but inclusion of geographic, habitat, host, and other corroborative information was encouraged. All keys are constructed to differentiate the forms that occur in California in the simplest way; a corollary is that non-naturalized plants from other parts of the world may not key well in *The Jepson Manual*, even though their family or genus is represented.

Complex or variable taxa are keyed more than one way in preference to using cumbersome, synoptic statements. Except in the family key, where the practice would be more confusing than helpful, taxa keyable in more than one way are preceded, in the key itself, by a superscript integer that specifies the number of ways the taxon can be reached in that key. The appearance of a superscript integer in a key may be the first indication that a taxon so marked is variable.

Keys are constructed to emphasize both taxonomically important and readily assessed features. Happily, sometimes a feature has both attributes. More often, the two attributes are at odds and some kind of compromise has been sought. For example, easily assessed features may be given first, with unique but more difficult characters (or those less likely to be present on an average specimen) listed later in the key statement, where, if they are available, they serve to corroborate or correct an impression. In some situations, only a single, more difficult feature (such as fruit type) accurately separates groups, so, by default, it is all that is used. Such situations have been avoided if possible; nevertheless, the taxonomy of many groups (e.g., Apiaceae, Boraginaceae, Ericaceae, *Carex*) relies on features of mature fruit.

Key statements are arrayed so that the inclusively shorter branch is first, allowing the contrasting information to be as physically close as possible. As long as this criterion is satisfied, taxa generally are arrayed alphabetically in keys — that is, in the same relative order in which their descriptions are found.

Sometimes features unique to one part of a couplet (that is, those that cannot be readily compared in the other couplet-half) nevertheless provides useful corroborating evidence for a decision. This happens most often where a single taxon (or small group) is being separated from a large and diverse group of taxa. When such unique information is included, it is set off from the rest of the statement by a dash — indicating that there is no comparable information for the other half of the couplet.

Family Key. The largest and hierarchically most inclusive key, but not the most difficult one to use, is the Key to California Plant Families. Its first portion is a key to groups of families that share taxonomically important or easily seen characters. Users of the family key will do well to become familiar with the kinds of information asked for in the key to groups and practice assessing new material for that information. When one has learned to tell a dicot from a gymnosperm or monocot with ease and to dissect a flower so that its most important parts are revealed, the key to groups will prove easy to use.

Keys are not as formidable as they appear; this is especially true of the Family Key. Although it encompasses about 30 pages and about 700 couplets, the maximum number of decisions between couplet-statements necessary to identify *any* plant to family is 28 (6 to arrive at Group 16, plus 22 within this large group to separate the genus *Tiquilia* of Boraginaceae from members of Gentianaceae). The *average* number of couplets required to identify a plant to family is only 9 or 10. This is true because a single choice generally eliminates many possibilities from consideration.

The family key contains two other important conventions. Variable families are keyed many times, often with the end point a genus or species rather than the whole family (Euphorbiaceae keys in more than 20 ways). Common mistakes and expected misinterpretations of plant features are accounted for in the keys. Thus, even though a technical error may be made, the correct answer can be reached (Euphorbiaceae, with its sometimes flower-mimicking inflorescences, is again an example).

Other Keys. A key to the genera of a family follows each family description (if there is more than one genus in California); a key to species (normally including infraspecific taxa as well) follows each generic description (if there is more than one species in California). For a few very large genera (e.g., *Astragalus*) the key to species has been kept as easy to use as possible; varieties are keyed after the description of their species rather than in the main key to species.

Descriptions and Their Conventions

All Descriptions.

Descriptions emphasize California forms but are constructed to accommodate non-California forms of the taxon insofar as possible. As an ideal, the contents of descriptions are consistently comparable among all members of a taxon and are constructed to focus particular attention on diagnostic features. Characters included in descriptions are treated in the same sequence and with comparable modifiers in the descriptions of all related taxa at the same rank. Key characters are repeated in descriptions. [For large genera like *Astragalus*, *Carex*, and *Eriogonum*, not all key characters can be repeated in every description, so "related taxa" sometimes means only a subset of those at a given rank. Also, where a key separating only two taxa is followed immediately by descriptions of those taxa, key characters are not always repeated in the descriptions.]

Characters are described at the highest rank at which they apply generally and are not standardly repeated in lower level descriptions. This means that the characters common to a family must be known if one is to understand fully the important characters of its included genera. The same applies to included species and to included infraspecific taxa. Where a general state exists but there are exceptions, the abbreviation "gen" is used at the highest possible rank. Then, at lower ranks, only exceptions to the general statement are given.

The means to concisely informative and comparable descriptions are several. These conventions are not difficult to assimilate, but they need to be understood. Also, descriptions must be read carefully; a small difference in form may produce a major difference in correct interpretation.

Descriptions are composed of several statements, most of them cued visually by the all-capital, bold-face abbreviation of a major organ type (stem, leaf, inflorescence, flower, fruit, seed). [In some cases, consistently different, complex forms of inflorescences or flowers are separated into different statements, e.g., staminate catkins, disk flowers.]

Nouns begin both statements and, generally, phrases within statements; they are arrayed from lower to upper on the plant axis and are followed by a specific sequence of adjectives (e.g., position, number, size, shape, color, hair types).

Adjectives always modify the immediately preceding noun. This is very important, because there are not enough types of punctuation to provide an unambiguous hierarchy of descriptors within a complex statement.

First Descriptive Statement. The first statement of a description has the least constrained form of any. It covers any important general aspects of the plant, including growth habit, size and shape, sexuality, general hairiness patterns, etc. Such statements are the only ones that do not begin with bold capitals.

Other Descriptive Statements. Several conventions are followed regarding punctuation within statements of a description. If the entire organ to which the statement pertains is described, the immediately following adjectives are not separated from the main noun by punctuation (e.g., "**LVS** alternate, compound...."). If the main noun is immediately broken into parts that are described sequentially, it is followed by a colon (e.g., "**LF**: petiole 1-3 cm, winged; blade ± 4 cm, round, entire....").

The secondary, noun-initiated phrases of a statement are separated by semicolons; within these secondary phrases, adjectives (and sometimes tertiary nouns) are separated by commas. Where a complex description might be ambiguous, parentheses or other clarifying punctuation may be added — e.g., "**FR** 3–4 mm, ovoid; segments 10–14, margins winged, tips bristly (or segments 8–9, margins rounded, tips glabrous)...." [Note that the tertiary noun "tips" relates to the secondary noun "segments", not to the previous tertiary noun "margins" or the more removed primary noun "FR".]

Articles and conjunctions are used in descriptions only where their absence would create a potentially serious ambiguity. For example, "**LF**: blade lanceolate, margins ciliate, bases, tips tapered...." follows these conventions but raises the question of whether an adjective modifying "bases" has been dropped inadvertently. Normally, therefore, such a construction is altered to "...margins ciliate, bases and tips tapered...".

Familial Descriptions.

Following modern practice, all families are given names based on genera and ending in "-aceae". Changes from some earlier works are Apiaceae (not Umbelliferae), Asteraceae (not Compositae), Brassicaceae (not Cruciferae), Fabaceae (not Leguminosae), Hypericaceae (not Guttiferae), Lamiaceae (not Labiatae), Arecaceae (not Palmae), Poaceae (not Gramineae). A common (English, colloquial) name is given to each family for ease of association. Like other common names, it is printed in Times-Roman small capitals. Authors of family names are not given.

Every family with representatives growing wild in California is described. Family descriptions account for variability throughout the world but focus on key characters rather than ranges of variation. The general form of the description itself is the same as for species.

Approximate numbers are given, on a worldwide basis, for genera and species in each family. Overall range is also summarized. Where appropriate, notes on cultivated, useful, or toxic forms are included. A reference may be given [in brackets] that supplies an appropriate entry into the literature on the family (normally the most recent reference, which cites earlier works), whether it concerns the entire family or only a small part of it. (The same strategy is used for genera and sometimes for species.)

Generic Descriptions.

If a family is made up of only one genus, the family description serves as the generic description. In such cases, the number of species and their overall range are given, together with the derivation of the generic name (in parentheses), an appropriate reference [in brackets], and sometimes additional notes.

If a family is made up of more than one genus, the genera that grow in California are described, adding morphological content to the information given for all genera. If the family is small (e.g., 2–4 genera), descriptions are short and merely diagnostic, particularly if the family is represented in California by only one genus. In large families where many Californian genera must be compared, generic descriptions are sometimes as long as family descriptions. Descriptions of genera containing many species are also long, because citation of general states at generic rank saves much space in specific descriptions.

Genera are assigned a common name (printed in Times-Roman small capitals) only if one is believed to be in common usage. As with families, the general form of the description is the same as for species. Authorship is not given for generic names.

Specific Descriptions.

Descriptions of species are the textual backbone of *The Jepson Manual.* Those conventions of specific descriptions not already covered are described here; except as specified above, they generally apply equally to generic and familial descriptions.

Scientific Names. Binomial scientific names appear in *italics.* Generic names are always capitalized; specific epithets are never capitalized, regardless of their origin. Names of taxa considered native to California are printed in bold-face italic ***Times-Roman***. Names of alien taxa are in italic *Helvetica*, a sans-serif type with a lighter appearance than the names of native taxa.

Authors of Names. Binomials are followed immediately by the name of or abbreviation for the publishing author(s) who validated the name. A scientific name is not strictly complete without the name(s) of the validating author(s) attached. However, the author citation occurs only the primary time a name is encountered — normally at the beginning of its description, but not also where it is mentioned in a note to another description.

In an attempt to standardize the way in which author names are cited, *The Jepson Manual* uses the 1980 *Draft Index of Author Abbreviations compiled at The Herbarium, Royal Botanic Gardens, Kew.* Though international in scope and the most complete source yet released, this reference does not include some current taxonomic authors; moreover, it is out of print. Author names that are not accounted for in the *Draft Index* generally are given here for clarity as one or two initials plus an unabbreviated surname. Mabberley's useful and readily available *The Plant Book* (Cambridge University Press, 1987) includes a list of author abbreviations (pp. 661–706) that is somewhat abridged from the *Draft Index.*

Common Names. In order for them to stand out, common names for species are printed in Times-Roman small capitals. There are two quite different circumstances in which common names for species (and infraspecific taxa) are invoked. The first parallels the situation for genera, in which a taxon is sufficiently common or obvious that there is a colloquial name in common usage. Unless there are mitigating circumstances, common names that are mere English translations or rearrangements of scientific names are omitted (e.g., "Bolander's linanthus" is not used for *Linanthus bolanderi*).

The second reason for including a common name is to provide a cross reference to a taxon that is rare enough to be "sensitive" — it may be legally protected, its abundance may be monitored, or previously unknown populations may be sought. These taxa are often referred to by non-botanists only by their "common names", even though those names may be contrived especially for the purpose. To facilitate communication among all users, *The Jepson Manual* uses the common names accepted in the California Native Plant Society's *Inventory of Rare and Endangered Plants of California*. Names included for this reason nearly always show a structural cross-reference to the *Inventory* (see Commonness and Rarity, p. 29).

Chromosome Numbers. Chromosome reports are presented but, for many reasons, they have not been verified for this work and should not be taken as definitive. There are several problems. When taxonomic ideas change, it is not apparent, without study of an herbarium voucher specimen, what new name an older chromosome count should be connected with. Furthermore, many chromosome-number reports have no known associated voucher specimen; for such reports there is no way to verify the taxon that supplied the chromosome preparation. These problems aside, there are clear examples within the California flora both of chromosomally well studied groups and of groups from which no or few chromosome numbers have ever been reported. Numbers are given here, with cautions, to indicate what seems to be known and what groups at least have been counted. Groups without reported counts may be excellent candidates for chromosomal study.

Habitats and Elevations. Habitats, elevation ranges, and geographic ranges (the last based on plant communities and other aspects of habitat) combine to provide a reasonable prediction for where a plant taxon will be found. Habitats and elevations therefore are given for each taxon at species rank or below.

Plant taxa are highly variable in the specificity of their habitat requirements. Many plant taxa grow in a wide range of plant communities, as long as certain habitat requirements are met. Others share a very specific habitat with a highly predictable group of other taxa. As an ideal, habitat specifications include degree of exposure, soil parent material, soil texture, soil moisture pattern, and constantly associated species or plant communities. Much more remains to be learned about habitat requirements, and it is critical that additional information be made available in a systematic way — on herbarium specimen labels and in databases.

Elevation ranges are given in meters (there is a conversion scale to feet at the back of the book). These ranges have been taken from herbarium specimens and other sources. Due to sampling error, they probably represent a minimum range. To mislead users as little as possible, precise (but not necessarily accurate) elevation ranges were generally rounded down at the bottom and up at the top. Nevertheless, representatives are likely to be found outside the specified elevation range. Specimens from such populations, well labeled, would make excellent additions to herbaria.

Range Descriptions. The four-tiered hierarchical system used to describe geographic ranges in California can be applied both more generally and more precisely than the traditional system based on a more generalized geography and county lines (see Geographic Subdivisions of California, p. 37).

All range specifications assume the restrictions of cited habitats and elevation ranges. At the most general, "CA" means that a taxon occurs in all three floristic provinces in California; there is no major portion of the state where it should not be expected. "CA-FP" means that the taxon should not be expected in the Great Basin or Desert Provinces, but might be found nearly anywhere else in California that has the appropriate habitat and elevation.

A combined habitat and range statement might read "Grasslands, fields; < 500 m. NW". This implies that the taxon might be found along the coast (NCo), in the Klamath Ranges (KR), and in the North Coast Ranges

(NCoR), the three major subdivisions of Northwestern California. However, habitat and elevation exclude it from the higher portions of KR and perhaps entirely from the High North Coast Range district (NCoRH).

Because contributing authors were variably knowledgeable about California geography, an important part of the editorial process involved helping authors use the new system in a uniform way. Like other aspects of scientific content, however, range information was the responsibility of the author.

Some authors included in their citations only regions from which they had seen documentation of occurrence (normally an herbarium specimen). Two conventions were taken to assure that ranges were described as completely as possible. Alien taxa with currently expanding ranges were generally specified as "expected elsewhere" or "expected more widely". For native taxa reported from or expected to be found in a region from which no documentation had been seen, a different wording was used: regions of questionable occurrence were listed, but followed by a "?" (e.g., "s SNH, SNE, w DMoj?."). Documenting specimens from such questionable regions would be valuable.

Synonyms and Misapplied Names. Complete synonymies that show the entire nomenclatural history of a taxon, as it is construed by the current author, cannot be given in a book of this scope. Not all names judged to have been used inappropriately for the taxon at hand are listed; however, those names accepted by Munz (1959 *A California Flora*, as corrected by his 1968 *Supplement*) and by the most recent edition of the California Native Plant Society's *Inventory of Rare and Endangered Plants of California* are all accounted for, either by being recognized and described or by being listed as a synonym, misapplied name, or unclearly differentiated form (see below).

Synonyms (based on either the same or a different type specimen) and misapplied names appear in brackets. The two are differentiated as follows: "[var. *aggregata* Rydb.; *A. rubra* Michaux; *A. coccinea* Rottb. misapplied]". If no species is specified, the one just described is assumed. Included lower-ranking taxa that are not recognized are listed first; other synonyms are then listed alphabetically; misapplied names appear last.

Unclearly Differentiated Forms. Sometimes it is doubtful but uncertain whether a form should be recognized taxonomically. Such forms are normally diagnosed in a note following the synonymy, e.g., "Pls from n SN with reddish hairs have been called *P. rubrotincta* E. Greene". Such a statement reserves judgment about (but calls into question) whether a form should be considered taxonomically distinct or merely part of the variation pattern of the recognized taxon.

Variability and Further Work Needed. Whether unclearly differentiated forms are noted or not, highly variable taxa are so noted in a sentence appended to the description. If particular problems in the variation pattern are known, they are specified; if additional work might result in a more satisfactory taxonomic disposition, it is called for.

Toxicity. Some California plants (e.g., poison hemlock, poison oak, Klamath weed) are seriously toxic, causing the deaths of both humans and livestock. Fuller & McClintock's *Poisonous Plants of California* (University of California Press, 1986) provides an excellent overview of plants that are both major and minor sources of poisoning. Plants that have been toxic to animals or people in California (or are expected to be) are cued with an all-capital "TOXIC...". Usually, some specifics are included, e.g., "TOXIC to livestock but rarely eaten"; "TOXIC: causes severe contact dermatitis"; or "TOXIC: ingestion of a single seed has been fatal to humans".

Weediness. California supports an increasing number of aggressive weeds that were originally native to many parts of the world. Both State and Federal governments have enacted legislation declaring the worst of these (especially those that thrive in agricultural circumstances) as "noxious weeds" that should be eradicated wherever they are found. (G. Douglas Barbe of the California Department of Food and Agriculture prepared computer-generated distribution maps of these taxa.) Listed weeds are specified as "NOXIOUS WEED" in *The Jepson Manual.*

Other non-listed weeds may be seriously aggressive in some circumstances (such as disturbed roadsides and urban gardens) but do not provide an immediate threat to agriculture. Such plants as *Cortaderia jubata* (pampas grass), *Cytisus scoparius* (scotch broom), and *Oxalis corniculata* are also noted as aggressive and pernicious weeds.

Horticultural Value. The final segment of many descriptions of native taxa is a summary of horticultural value and requirements for growth. Such entries are preceded by the symbol ❀. See the chapter, "Horticultural Information in *The Jepson Manual*", p. 31 for more information.

PRONUNCIATION OF SCIENTIFIC NAMES

Much has been written about proposed rules for pronouncing scientific names and about exceptions to the rules. One of the best and most succinct is a small chapter in William T. Stearn's *Botanical Latin* (David & Charles, London, third edition, 1983). The following quotations are taken from pp. 53–54:

"Botanical Latin is essentially a written language, but the scientific names of plants often occur in speech. How they are pronounced really matters little provided they sound pleasant and are understood by all concerned..."

"In English-speaking countries there exist two main systems, the traditional English pronunciation generally used by gardeners and botanists and the 'reformed' or 'restored' academic pronunciation adopted by classical scholars..."

Stearn discusses syllabification, accentuation, the different sounds of vowels and consonants in the two systems, the problem of scientific names derived from proper names, and regional variations in accent. As a classical scholar, he clearly prefers the system of 'restored' academic pronunciation.

Consistent use of the academic system, however, would be impossible for nearly all users of *The Jepson Manual*, including most of its authors and editors. Rather than trying to present a 'correct' system, we wish to emphasize Stearn's point that effective communication is the ultimate criterion for judging pronunciation. Despite a strong inclination by all toward a single 'correct' method of pronunciation, the following points are all true:

1. Classical scholars do not always agree on pronunciation rules.
2. Pronunciation varies strikingly among professional botanists.
3. Individual botanists rarely use any rule of pronunciation consistently.
4. People tend to pronounce names the way they first learned them, regardless of their knowledge of rules.

Although some people avoid saying scientific names out loud at all because they are afraid to mispronounce them, mispronunciation is much less a problem than is commonly feared. We hope the following suggestions will help develop confidence in users of *The Jepson Manual* who feel uncertain about their ability to pronounce scientific names:

1. Divide the word carefully into syllables (it is safest to assume that every vowel belongs to a different syllable).
2. Pronounce each syllable (e.g., "co-to-ne-as-ter", not "cot-on-east-er").
3. Listen to others and practice what sounds good to your ear; conviction is important.
4. Attempt to accent all syllables equally; this is likely to show you where accents fall naturally (some manuals, but not this one, specify accents with stress marks).
5. Develop your own standards for pronouncing common endings like "-aceae", "-iae", "-ensis", etc., and stick to them.
6. Retain pronunciation of proper names used in scientific names (jonesii = "**jones**-eeee", not "jo-**nes**-ee-eye").
7. When someone presumes to correct your pronunciation, a knowing smile is an appropriate response.

W. A. Weber (1986 Madroño 33:235–236) presented three rules intended to help U.S. botanists communicate with those educated in almost any other part of the world. They are worthy of attention and are quoted or summarized (with additions) below:

1. "Try to approach the style of pronunciation of the person you are talking to. This is the cardinal rule."
2. Practice and use the European pronunciation of letters:
   a: "ah", not "ay"
   e: "eh", not "ee"
   i: "ee" or as in "sit", not "eye"

y: as in "sit" or "cynic", not "eye"
-ae: "eye", not "ee" or "ay"
-ii: the vowels may be slightly separated, or merely held longer ("-eeee")
ti, *ci*: generally "tee" or "see", not "she" ("Lo-**ma**-ti-um")
ch: generally as "k", not as in "ouch"
g: as in "go", not as in "gem"

3. In accenting words, retain the sound of word stems and of proper names insofar as possible:
   "A-**den**-o-**phyll**-um", not "Ad-en-**oph**-yl-lum"
   "Chi-ma-**phil**-a", not "Chi-**maph**-il-a"
   "**nutt**-all-ii", not "nu-**tall**-ee-eye"
   "**doug**-las-ii", not "dou-**glas**-ee-eye"

Initially, the practices suggested by Weber will sound strange to the ears of most U.S. citizens, but they are simple and their consistent use would distinctly enhance worldwide communication of spoken scientific names.

GLOSSARY

Illustrations by Dr. Linda Ann Vorobik, Susan Stanley, & Sarah A. Young

The philosophy of a botanical resource is best illustrated by the vocabulary used to differentiate and describe plants. The words chosen represent innumerable decisions about how and to whom to communicate. Many of the decisions made in preparing *The Jepson Manual* are reflected in the glossary.

Beginners will benefit from studying the glossary (especially its illustrations), as a summary of the basic botanical knowledge needed to identify plants. Those with experience in plant identification also need to consult this glossary, because neither keys nor plant descriptions will be consistently interpreted correctly unless the word-usage conventions adopted by *The Jepson Manual* are understood. The same terms are sometimes used in different ways by different botanists and manuals; usages here may be somewhat at odds with those to which some experienced users are accustomed.

The initial major decision regarding *The Jepson Manual* was to design it for an unusually broad community of users — to produce, insofar as possible, a work that would simultaneously satisfy professional needs and invite and allow those with little or no technical botanical training the personal satisfaction of identifying Californian plants. This decision implied a departure from the traditional manual (written, more or less, for trained botanists) and mandated compromises concerning the conventions adopted for use of botanical terms. Two approaches were taken in trying to achieve the best compromises.

First, technical botanical terms used in *The Jepson Manual* were limited to those judged necessary for precise and concise description of diverse plants. Many technical terms were disallowed in favor of more universally understood words and phrases that did not unduly compromise either precision or space considerations. Any descriptive word in the manual not defined in the glossary was judged both to be in common usage and to be used here in a standard manner.

Second, considerable effort was invested in preparing an illustrated glossary of terms as they were used throughout this work. Definitions were structured to be acceptable to most professionals, yet understandable to beginners. A great many people contributed to this effort, and their help is gratefully acknowledged. Nevertheless, they and others will find decisions with which they disagree — whether the result is loss of favored terms or inclusion of unwanted technicality. As in any compromise, our goal was simply "to please most of the people most of the time".

The glossary was structured to facilitate learning the terms. Definitions begin simply, then sometimes have additional information that is more precise and elaborate and which may cite examples. Related terms are cross-referenced. Illustrations are usually clustered by type, to enhance learning related concepts or physically adjacent plant parts. Plurals are given in parentheses if their formation is unusual. Many of the terms are sometimes used, with adjusted spellings and meanings, as parts of speech other than the one given — e.g., the defined noun "tubercle" is also used as the adjective "tubercled".

The page locations of glossary illustrations are given in parentheses before the definition. Some terms are not illustrated in the glossary itself, but supplementary references are given to examples found in illustrations in the main text. These references are to taxa (not pages) and are found at the ends of definitions.

abundant. Very likely to be encountered; nearly always found in appropriate habitats, sometimes forming dense stands. (see common) (see pp. 29, 30 for more information)

achene. (p. 22) Dry, indehiscent, 1-seeded fruit from a 1-chambered ovary, often appearing to be a naked seed.

acuminate. (p. 18) Having a long-tapered, sharp tip, the sides of which are concave. (see acute, awl-like)

acute. (p. 18) Having a short-tapered, sharp tip, the sides of which are convex or straight and converge at less than a right angle. (see acuminate, obtuse)

adherent. Superficially appearing fused to another organ (of like or unlike type) but separable from it, as "perianth adherent to fruit".

adventitious. Arising at unusual times or places. Said of plant structures such as roots on aerial stems.

aggressive. Growing or spreading rapidly, outcompeting other plants, difficult to control. Said especially of weeds. (see weed)

alien. Not native; introduced purposely or accidentally into an area. (see native, naturalized, waif, weed)

alpine. Found above timberline. (see subalpine)

alternate. (p. 17) 1. Arranged singly, often spirally, along an axis — e.g., one leaf per node. (see opposite, whorled)

2. Occurring between structures, or in different ranks, as "stamens alternate petals". (see rank)

angiosperm. Plant that bears true flowers, made up of two major groups, dicots and monocots. (see flower, dicot, monocot)

annual. (p. 14) Completing life cycle (germination through death) in one year or growing season, essentially non-woody. (see biennial, herb, perennial)

annulus (annuli). On the sporangium of most ferns, a row of cells with partly thickened walls that functions in the often catapult-like release of spores.

anther. (p. 21) Pollen-forming portion of a stamen.

appressed. (pp. 15, 23) Pressed against. Said especially of hairs that are parallel or nearly parallel to and often in contact with the surface or axis of origin.

aquatic. Growing under, in, or on water, rooted in bottom sediment or floating. Does not include plants of seeps or wet rocks, but does include those with part of the shoot submerged, even though other parts may be above water. (see emergent) (example, *Nuphar luteum*)

areole. In Cactaceae, a well defined, axillary area bearing one to many spines and generally other, shorter structures. (example, *Ferocactus viridescens*)

aril. Fleshy, corky, or bony appendage arising at or near the point of seed attachment, sometimes completely covering the seed. (example, *Viola purpurea* seed)

armed. General term meaning bearing prickles, spines, or thorns. (see spine)

ascending. (pp. 14, 15) Curving or angling upward from base (generally 30–60° less than vertical or away from axis of attachment). (see decumbent, erect)

asymmetric. (p. 20) Irregular in shape; in no way divisible into identical or mirror-image halves. (see bilateral, biradial, radial)

awl-like. (p. 19) Narrow throughout, but broader at the base and tapered to a sharp tip. (see acuminate)

awn. (pp. 18, 26) 1. Bristle-like appendage or elongation, generally at the tip of a larger structure. 2. Stiff, needle-like pappus element in Asteraceae.

axil. (p. 17) The upper angle between axis and branch or appendage — e.g., between stem and leaf.

axile. (p. 21) Pertaining to an axis, as of a placenta along the central axis in a compound ovary with more than one chamber.

axillary. (pp. 16, 24) Pertaining to or within an axil, especially a leaf axil.

axis (axes). (pp. 21, 26) Line of direction, growth, or extension; structure occupying such a position — e.g., the main stem of a plant or inflorescence, the midrib of a leaf.

banner. (p. 20) Uppermost, often largest petal of many members of Fabaceae.

barbed. (p. 23) Having sharp, normally downward- or backward-pointing projections. Said of an awn, bristle, or other structure.

bark. All tissues outside or covering the wood (hardened xylem tissue) of non-herbaceous plants. Bark patterns are important for identification in many trees and shrubs. (example, *Cupressus forbesii*)

basal. (p. 14) Found at or near the base of a plant or plant part. Especially said of leaves clustered near the ground or of a placenta confined to the base of an ovary.

bell-shaped. (p. 20) Widening more or less abruptly at the base and then generally more gradually above. Generally said of a fused calyx or corolla.

berry. Fleshy, indehiscent fruit in which the seeds are not encased in a stone and are generally more than 1. (see drupe, pome) (example, *Solanum americanum* fruit)

biennial. Completing life cycle (germination through death) in two years or growing seasons (generally flowering only in the second) and non-woody (at least above ground), often with a rosette the first growing season. (see annual, herb, perennial)

bilateral. (p. 20) Divisible into mirror-image halves in only one way. (see asymmetric, biradial, radial)

biradial. (p. 20) Divisible into mirror-image halves in two ways; isobilateral. (see asymmetric, bilateral, radial)

bisexual. (p. 21) Flowers with both fertile stamens and fertile pistils.

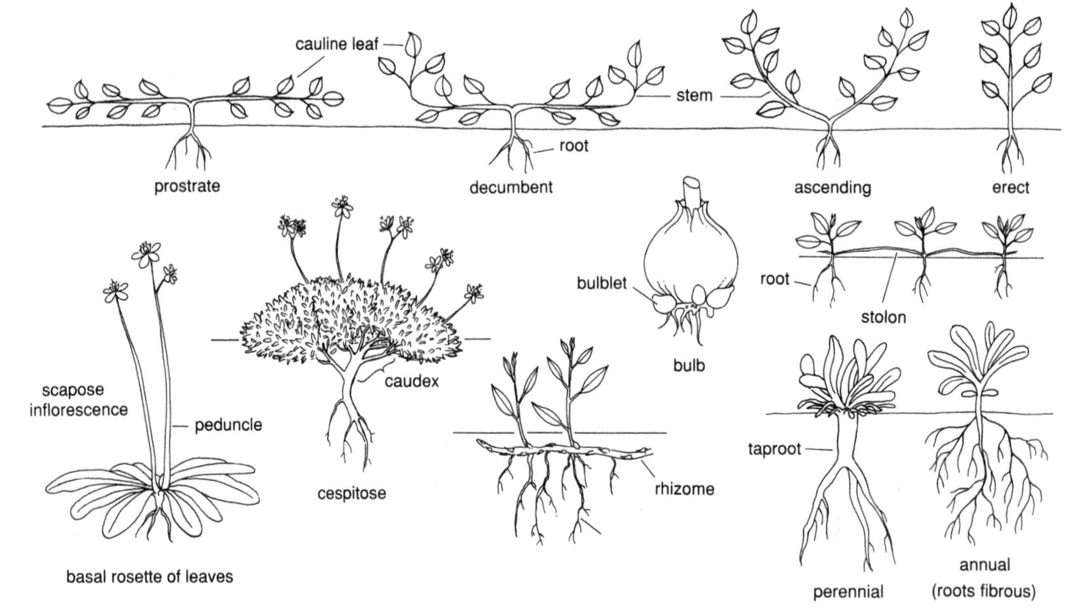

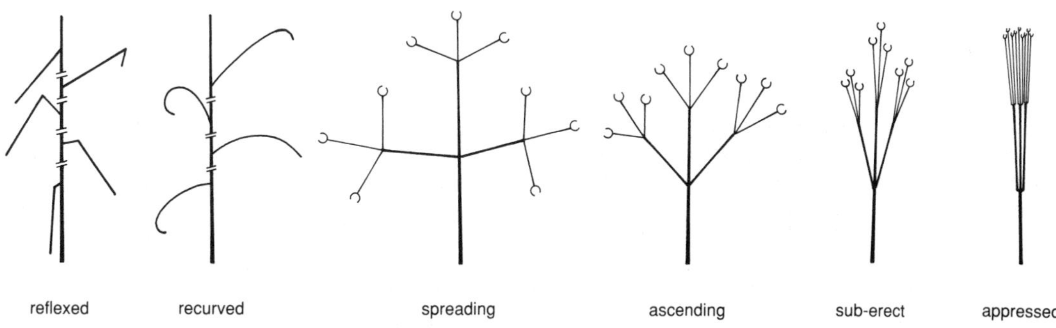

blade. (pp. 17, 21, 24, 26) Expanded portion of a leaf, petal, or other structure, generally flat but sometimes rolled or cylindric.

brackish. Somewhat salty. Said generally of a mixture of marine and fresh water.

bract. (pp. 19, 20) Small, leaf- or scale-like structure associated with an inflorescence or cone. Generally subtends a branch, peduncle, pedicel, flower, or cone scale. (see bractlet)

bractlet. (p. 20) 1. Relatively small, generally secondary bract within an inflorescence. 2. Bract-like structure on a pedicel that often does not directly subtend another structure. (see bract)

bristle. 1. Relatively large, generally stiff, more or less straight hair. (example, *Navarretia nigelliformis*) 2. Fine, generally cylindric pappus element in Asteraceae. (example, *Calycoseris parryi*)

bud. (pp. 16, 20, 24) 1. An incompletely developed shoot or leaf. 2. an unopened flower. Sometimes protected by bud scales or sepals, generally axillary or at a stem tip.

bulb. (p. 14) Short underground stem and the fleshy leaves or leaf bases attached to and surrounding it — e.g., an onion. (see stem)

bulblet. (p. 14) 1. Small bulb generally produced at the base of a bulb. 2. Any small, bulb-like structure that propagates a plant, often in a leaf or bract axil.

bur. Fruit or fruiting inflorescence with awns or bristles, often barbed, that attaches to and is dispersed by a passing animal. (example, *Xanthium strumarium*)

callus. (p. 26) In some Poaceae, enlarged base of floret. Sometimes hairy.

calyx (calyces). (p. 21) Collective term for sepals; outermost or lowermost whorl of flower parts, generally green and enclosing remainder of flower in bud. Sometimes indistinguishable from corolla.

canescent. Covered with dense, fine, generally grayish white hairs. (example, *Phoenicaulis cheiranthoides* leaf)

capsule. (p. 22) Dry, generally many-seeded fruit from compound pistil, nearly always dehiscent (irregularly or by pores, slits, or lines of separation). (see circumscissile, loculicidal, septicidal)

carpel. (p. 21) The basic female structure of flowering plants, in concept derived from a leaf; an evolutionary term rarely used for identification in *The Jepson Manual*. (see pistil)

catkin. (p. 20) Spike of unisexual flowers with inconspicuous perianths, sometimes pendent and often with conspicuous bracts.

caudex (caudices). (p. 14) Short, sometimes woody, more or less vertical stem of a perennial, at or beneath ground level. (see stem)

cauline. (p. 14) Borne on a stem; not basal. Said especially of leaves borne along an above-ground stem.

centimeter. One-hundredth of a meter; 10 millimeters (abbreviation: cm). (see ruler following index)

cespitose. (p. 14) Having a densely clumped, tufted, or cushion-like growth form, with the flowers held above the clump or tuft.

chaff. (p. 25) Dry bracts; in Asteraceae, dry, generally papery or scaly, often persistent bracts on a receptacle.

chamber. (p. 21) Compartment or cavity within an ovary, capsule, or other hollow structure.

ciliate. (p. 23) Having generally straight hairs along the margin or edge.

circumboreal. Found around the world at northern latitudes.

circumscissile. (p. 22) Dehiscent by a transverse line, the top coming off as a lid. Generally said of a capsule. (see loculicidal, septicidal)

claw. (p. 21) Stalk-like base of some free sepals or petals.

cleistogamous. Bud-like, unopening flowers that are generally self-pollinated. (example, *Viola sheltonii*)

clone. Genetically identical individuals resulting from asexual reproduction (fragmentation of rhizomes or stolons, budding, etc.). Often used for an apparent population, the members of which are or were connected — e.g., aspens, cattails, duckweeds.

collar. (p. 26) The back (generally the outer surface) of a grass leaf at the junction of sheath and blade.

column. (p. 21) Structure at the center of an orchid flower formed by fusion of stamen(s) and style.

common. Likely to be encountered. (see abundant, rare, uncommon) (see p. 29 for more information)

compound. (p. 16) 1. Composed of two or more parts, as a *compound leaf* composed of leaflets (see compound leaf) or a *compound pistil* composed of fused or partly fused carpels. 2. Repeating a structural pattern (a compound umbel is an umbel of umbels). (see simple)

compound leaf. (p. 16) A leaf divided into distinct parts. In a *1-compound leaf*, the blade is divided into primary leaflets connected by an axis but no blade material (if there is connecting blade material, the leaf is lobed or dissected); in a *2-compound leaf*, the primary leaflets are so divided into secondary leaflets (if there is connecting blade material, primary leaflets are lobed); etc. (see palmate, pinnate)

compressed. (p. 23) Flattened side-to-side or front-to-back. (see depressed)

concave. Hollowed or indented, as the interior of a curved surface. (see convex)

cone. Reproductive structure composed of an axis, scales, and sometimes bracts. 1. Non-woody structure producing spores (e.g., clubmosses, horsetails) or pollen (e.g., conifers). 2. Generally woody structure producing seeds (e.g., most conifers, alders). (example, *Calocedrus decurrens*)

conic. Having a 3-dimensional shape defined by a wide, more or less round base, the sides evenly tapered to a narrow tip.

continuous. Having parts spaced evenly and without interruption, not clumped. Generally said of inflorescences. (see interrupted)

convex. Rounded outward, as the exterior of a curved surface. (see concave)

cordate. (p. 18) Heart-shaped, as of a leaf. Sometimes said of a leaf base with rounded lobes of which the sides adjacent to the petiole are convex. (see reniform)

corm. Short, thick, unbranched, underground stem often surrounded by dry (not fleshy) leaves or leaf bases. (see bulb, stem) (example, *Muilla maritima*)

corolla. (pp. 21,25) Collective term for petals; whorl of flower parts immediately inside or above calyx, often large and brightly colored. Sometimes indistinguishable from calyx.

cotyledon. Seed-leaf; a modified leaf present in the seed, often functioning for food storage. Persistent in some annuals and of aid in their identification. (example, *Lupinus bicolor*)

crenate. (p. 18) Scalloped — e.g., margins with gen acute sinuses between shallow, rounded teeth.

cylindric. (p. 18) Elongate, with parallel sides and, at any point, round in transverse section.

cyme. (p. 20) Branched inflorescence in which the central or uppermost flower opens before the peripheral or lowermost flowers on any axis. (see panicle)

deciduous. Falling off naturally at the end of a growing period. Generally said of leaves that fall seasonally and all together or of plants that are seasonally leafless. (see evergreen)

decimeter. One-tenth of a meter; 10 centimeters (abbreviation: dm). (see ruler following index)

decumbent. (p. 14) Mostly lying flat on the ground but with tips curving up. (see ascending)

decurrent. (p. 18) Having a wing-like or ridge-like extension beyond the actual or apparent point of attachment. Said especially of a leaf base that seems to continue down its stem.

dehiscent. (p. 22) Splitting open at maturity to release contents. Said especially of fruit or anthers. (see indehiscent)

deltate. (p. 18) More or less equilaterally triangular, with basal corners generally rounded.

dense. Congested or compact. Especially said of disposition of flowers in an inflorescence. (see open)

dentate. (p. 18) Having margins with sharp, relatively coarse teeth pointing outward, not tipward. (see serrate)

depressed. (p. 23) 1. Flattened from above and below. 2. with the center lower than the margins. (see compressed)

dicot. A member of the larger main subgroup of flowering plants; generally having two cotyledons, flower parts in 4's, 5's, or spirals, pinnate or palmate leaf venation, stem veins in rings (but often not all of these) — e.g., poppy, cactus, rose, sunflower.

dioecious. Male and female (or staminate and pistillate) plants separate. Said of a taxon in which individual plants produce either kind of unisexual fertile reproductive structures, but not both. (see monoecious) (example, *Salix reticulata*)

diploid. Having two sets of chromosomes (maternal and paternal), the normal complement in plant cells (except spores, sperm, eggs, some others); 2*n*. (see haploid, *n*, polyploid)

disciform head. In Asteraceae, a head composed of disk flowers and marginal pistillate flowers with minute or missing ligules, superficially similar to discoid head. (see ligulate head, radiate head) (example, *Baccharis pilularis*)

discoid head. (p. 25) In Asteraceae, a head composed entirely of disk flowers. (see disciform head, ligulate head, radiate head)

disk. (p. 21) 1. Fleshy, often nectar-secreting structure near (often surrounding) an ovary base. 2. In Asteraceae, the part of a head made up of disk flowers.

disk flower. (p. 25) In Asteraceae, the generally bisexual (never pistillate), generally radial, ligule-less flower with a 5- (rarely 4-) lobed corolla. Appearing without other flower types (discoid head) or with marginal ray or pistillate flowers (radiate or disciform heads, respectively). (see ligulate flower, ray flower)

dissected. Irregularly, sharply, and deeply cut but not compound. Said especially of leaves. (see compound leaf, lobe) (example, *Geranium dissectum*)

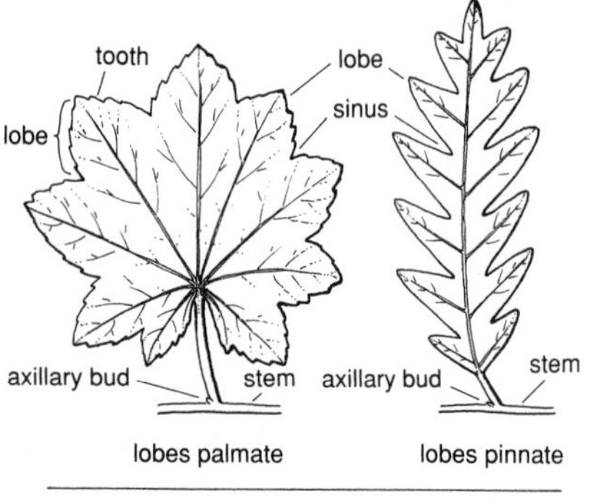

simple leaf

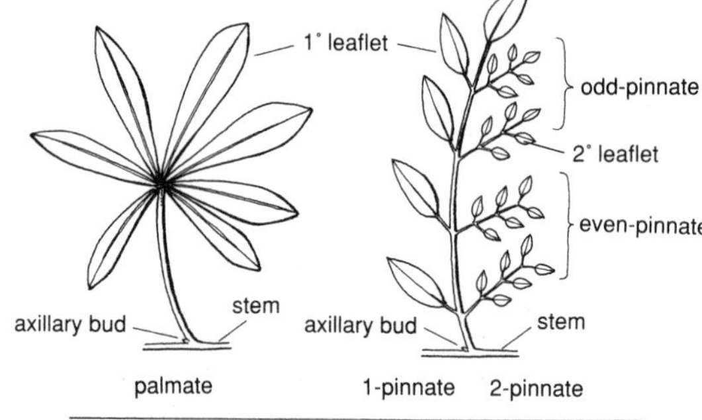

compound leaf

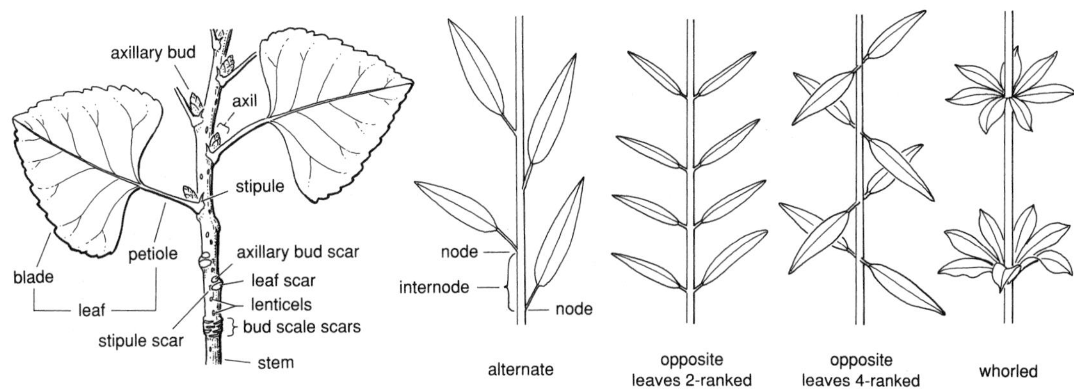

distal. Farther away from the origin or point of attachment, more toward the edge or tip. (see proximal)

drupe. Fleshy or pulpy, indehiscent, superficially berry-like fruit with one seed encased in a hardened stone that is derived from inner ovary tissue (sometimes several seeds are encased separately or together). Ovary outside stone sometimes edible but stone and contents generally inedible. (see berry, nut, pome, stone) (example, *Prunus emarginata* fruit)

elliptic. (p. 18) In the shape of an ellipse (flattened circle). (see oblong)

emergent. 1. Plant normally rooted underwater and extending above the water surface. 2. Any part of a plant normally held above the water surface. (see aquatic) (example, *Menyanthes trifoliata*)

endangered. Survival is in immediate jeopardy. Used only for taxa given such status by law. (see extant, extinct, extirpated, rare, threatened) (see pp. 29, 30 for more information)

endemic. Native to a well defined geographic area and restricted to that area.

entire. (p. 18) Having margins that are continuous and smooth (i.e., without teeth, lobes, etc.).

ephemeral. 1. Lasting a short time. 2. Completing the life cycle (germination through death) or growth cycle in much less than one year, as most desert herbs. Said also of plant parts that are functional for a relatively short time or fall early.

epidermis. Outermost cell layer (or layers) of non-woody plant parts.

epipetalous. (p. 21) Stamens that are partly fused to the petals and therefore appear to arise from them.

erect. (p. 14) Upright; vertically oriented.

evergreen. 1. Never lacking green leaves. 2. Having leaves that remain green and on the plant for more than one season and do not fall all together. (see deciduous)

exceeding. Surpassing another structure tipward. (see exserted)

exserted. (p. 21) Protruded out of surrounding structure(s). (see exceeding, included)

extant. Surviving, in existence; not completely died out or destroyed. (see extinct, extirpated)

extinct. No longer living anywhere; completely died out. (see extant, extirpated)

extirpated. Destroyed or no longer surviving in the area being referred to (may survive outside that area). (see extant, extinct)

exudate. Fluid or solid material discharged from a plant surface that may take on a characteristic color or texture.

fertile. Reproductively functional. Said of a plant or plant part that produces or is associated with the production of functional spores, pollen, ovules, or seeds. (see sterile)

fibrous. (p. 14) 1. Composed of fine or slender structures. 2. Having a root system composed of many roots similar in length and thickness, as in grasses. (see taproot)

filament. (p. 21) Anther-stalk; the often thread-like portion of a stamen.

fleshy. Thick and juicy; succulent. (example, *Cakile maritima*)

floret. (p. 26) In Poaceae, a single flower and its immediately subtending bracts (generally lemma and palea).

follicle. (p. 22) Dry, generally many-seeded fruit from a simple pistil, dehiscent on only one side, along a single suture. A flower may have a simple fruit of 1 follicle or an aggregate fruit of several follicles. (see fruit)

forked. (p. 23) 1. Branching into two parts of about equal size. 2. Hair with branches that do not radiate from a common point. (see stellate)

free. Not fused to other parts; distinct, separate. (see adherent, fused)

free-central. (p. 21) Pertaining to a placenta along the central axis in a compound ovary with only one chamber. (see axile, basal, parietal)

fringed. (p. 23) Having ragged or finely cut margins.

fruit. (p. 22) A ripened ovary and sometimes associated structures. A *simple fruit* develops from one ovary — e.g., cherry, apple, the latter derived largely from the hypanthium; *aggregate* and *multiple fruits* develop from ovaries of one and more than one flower, respectively, held together as a unit — e.g., a strawberry is an aggregate fruit of achenes held together by a juicy, red flower receptacle; a fig is a multiple fruit of achenes surrounded by a fleshy inflorescence receptacle. (see achene, berry, capsule, drupe, follicle, legume, nut, nutlet, pome, utricle)

funnel-shaped. (p. 20) Widening from the base more or less gradually through the throat into an ascending, spreading, or recurved limb. Said usually of a fused calyx or corolla.

fused. (p. 25) United, as the petals together into a corolla tube or stamens onto petals; not free. (see adherent, free)

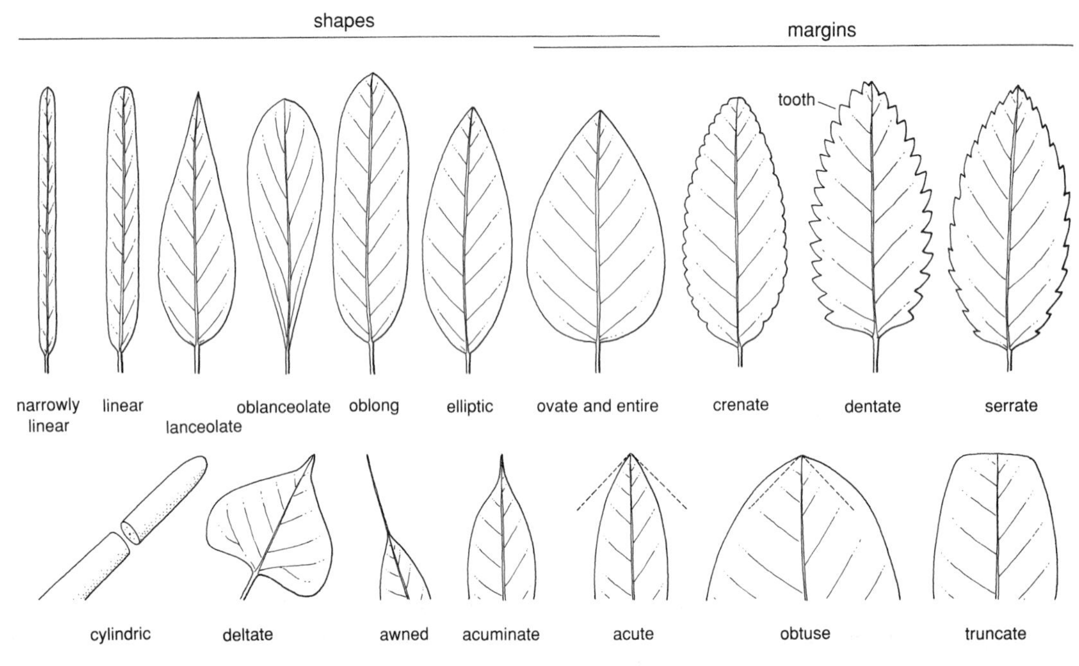

fusiform. (p. 23) Elongate, widest at the middle, tapered to both ends.

gland. A small, often spheric body that exudes a generally sticky substance, on (or embedded in) epidermis or at the tip of a hair. (example, *Ribes bracteosum*)

glaucous. Covered with a generally whitish or bluish, waxy or powdery film that is sometimes easily rubbed off.

glume. (p. 26) In Poaceae, each of generally two sheathing bracts that are the lowermost parts of a spikelet. (see lemma, palea)

granular. (p. 23) Covered with minute bumps. (see papillate, tubercle)

gymnosperm. Woody plant with seeds that are not borne in ovaries but in cones or naked on branches — e.g., pine, sequoia, ephedra, yew.

habit. Characteristic mode of growth; general form or shape of a plant — e.g., cespitose, herb, scapose, shrub.

habitat. Natural setting or abode of a plant, generally specified as a plant community or set of environmental features.

hair. Thread-like epidermal outgrowth. (see puberulent, strigose, trichome)

haploid. Having one set of chromosomes (maternal or paternal), the normal complement in spores, sperm, eggs, and some other cells that are derived from those; *n*. (see diploid, *n*, polyploid)

hastate. (p. 18) Arrowhead-shaped, with two basal lobes oriented more or less perpendicularly to the long axis. (see sagittate)

head. (p. 20) Dense, often spheric inflorescence of sessile or subsessile flowers.

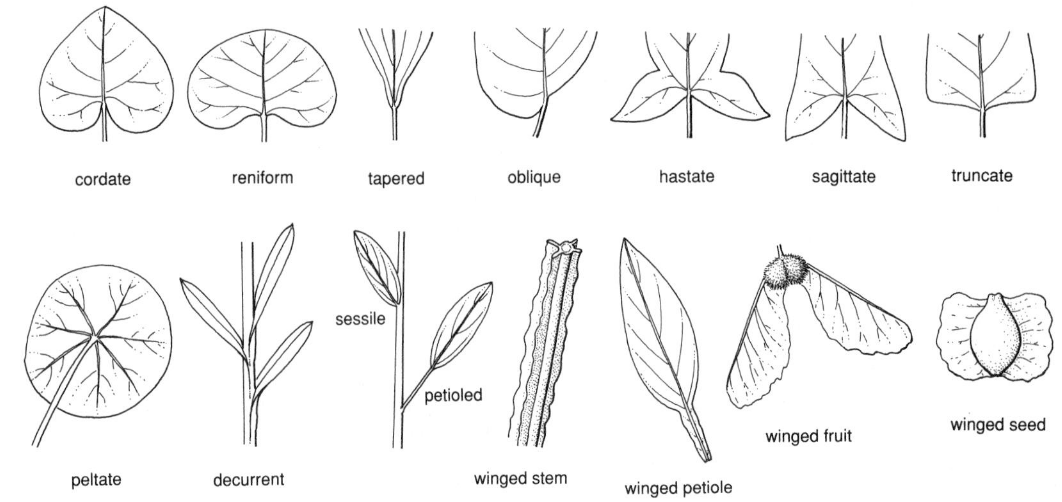

hemispheric. Shaped like a dome or half sphere.

herb. Plant with little or no wood above ground; aboveground parts are of less than one year or growing season duration. All plants called annual, biennial, or perennial in *The Jepson Manual* are herbs. (see annual, biennial, perennial, subshrub)

herbaceous. Lacking wood; having the characteristics of an herb.

herbage. The non-woody, above ground parts of a plant, especially the leaves and young stems taken together.

heterostylous. Having different kinds of style (and stamen) lengths. Said of a species in which individuals produce only one of two or more flower types, each differing in style (and generally stamen) length. (example, *Oxalis pes-caprae* flower)

hypanthium (hypanthia). (p. 21) Structure derived from the fused lower portions of sepals, petals, and stamens and from which these parts seem to arise, the whole generally in the shape of a tube, cup, or plate. An inferior ovary is fused to part or all of a hypanthium. (see inferior ovary)

included. (p. 21) Not protruding out of surrounding structure(s). (see exserted)

indehiscent. (p. 22) Not opening to release contents. Generally said of fruits. (see dehiscent)

interrupted. Having parts spaced unevenly. Generally said of inflorescences in which the axis is elongated between flower clusters. (see continuous)

involucel. (p. 20) A secondary involucre (group of bracts) within an inflorescence — e.g., those subtending the secondary umbels in members of Apiaceae.

involucre. (pp. 20, 25) Group of bracts more or less held together as a unit, subtending a flower, fruit (acorn cup), or inflorescence (the combined phyllaries of a daisy).

keel. (p. 20) 1. Ridge or crease more or less centrally located on the long axis of a structure, generally on the under or outer side. 2. The two lowermost, fused petals of many members of Fabaceae.

lanceolate. (p. 18) Narrowly elongate, widest in the basal half, often tapered to an acute tip.

lateral. Referring to the sides(s) of a structure — e.g., laterally compressed (flattened side-to-side), lateral branch (from "side" of stem).

leaf. (pp. 16, 17, 24, 26) Stem appendage with a structure such as a bud, branch, or flower in its axil, generally green and often composed of a stalk (petiole) and a flat, expanded, photosynthetic area (blade).

leaflet. (pp. 16, 24) One leaf-like unit of a compound leaf, which may be primary, secondary, etc. (see compound leaf)

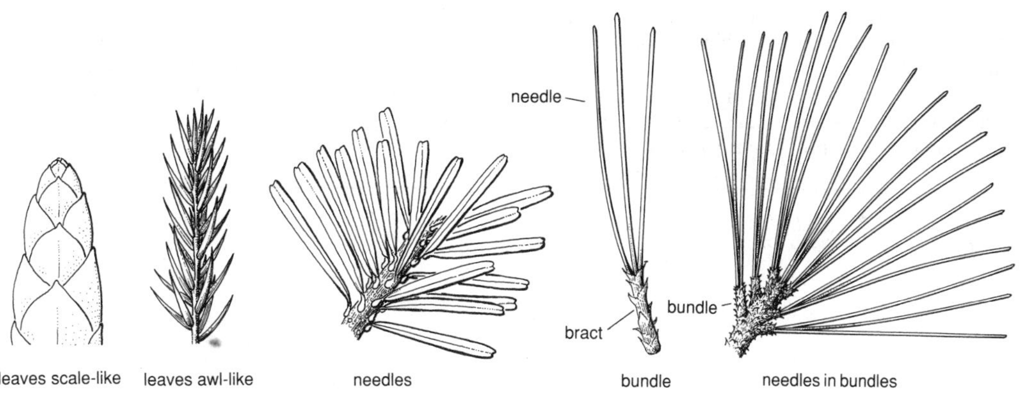

indusium (indusia). (p. 24) In many ferns, a veil- or scale-like outgrowth of the leaf surface or margin that covers a sorus (cluster of sporangia).

inferior ovary. (p. 21) An ovary that appears to be beneath its flower or to bear the sepals, petals, and stamens at or above its summit, owing to its fusion to the hypanthium. (see hypanthium, superior ovary)

inflorescence. (pp. 14, 20, 26) An entire cluster of flowers and associated structures — e.g., axes, bracts, bractlets, pedicels. Often difficult to define as to type and boundaries but generally excluding full-sized foliage leaves.

infraspecific. Below the species level. Said especially of variation within a species, whether taxonomically significant (i.e., characterizing subspecies or varieties) or not.

intergrade. To merge gradually from one extreme to another through a more or less continuous series of intermediates.

intermediate. Between extremes in form and sometimes in other ways.

internode. (p. 17) Stem segment between leaves; segment of an axis between two successive attachment points for appendages. (see node)

legume. (p. 22) 1. In Fabaceae, a dry or somewhat fleshy, one- to many-seeded fruit from a simple pistil, typically dehiscent longitudinally along two sutures and splitting into halves that remain joined at the base, sometimes indehiscent or breaking crosswise into one-seeded segments. 2. A plant with such a fruit.

lemma. (p. 26) In Poaceae, the lower, generally larger of two sheathing bracts that directly subtend a flower; the lowermost part of a floret. (see glume, palea, spikelet)

lenticel. (p. 17) A spongy area (pore), most common on surfaces of twigs or fruits.

lenticular. Lens- or discus-shaped, with both sides convex. (example, *Carex lenticularis* achene)

ligulate flower. (p. 25) In Asteraceae, a bisexual, bilateral flower with the long, outer portion of the corolla (the ligule) 5-lobed. Appears only with other ligulate flowers in a ligulate head. (see disk flower, ray flower)

ligulate head. (p. 25) In Asteraceae, a head composed entirely of ligulate flowers. (see disciform head, discoid head)

ligule. (pp. 25, 26) 1. In Asteraceae, the strap- or blade-like outer portion of the corolla in ligulate and ray flowers. 2. In Poaceae and other grass-like plants, an appendage at the juncture of leaf sheath and blade, generally with a membranous or fringed margin.

limb. (p. 21) In calyces or corollas with fused parts, the expanded, often lobed portion above the tube or throat.

linear. (p. 18) Elongate, with nearly parallel sides, and narrower than oblong.

lip. (p. 20) 1. Upper or lower of two parts in an unequally divided calyx or corolla. 2. In Orchidaceae, generally the largest, lowest, most highly modified perianth part.

lobe. (pp. 16, 21, 24–26) 1. A major expansion or bulge, such as on the margin of a leaf or petal or on the surface of an ovary. 2. The free tips of otherwise fused structures, such as sepals or petals.

loculicidal. (p. 22) A capsule, longitudinally dehiscent through the ovary wall at or near the center of each chamber. (see circumscissile, septicidal)

longitudinal. Pertaining to length or the lengthwise dimension; parallel to the axis. (see transverse)

margin. (pp. 18, 24) The edge, generally of a leaf or perianth part.

membranous. Thin, pliable, sometimes somewhat translucent, sometimes green. (see scarious) (example, *Carex fracta* ligule)

meter. Unit of length in the metric system, equal to 39.4 inches, slightly more than a yard (abbreviation: m). (see ruler following index)

millimeter. One-thousandth of a meter; one-tenth of a centimeter. The smallest unit of size used in *The Jepson Manual* (abbreviation: mm). (see ruler following index)

monocot. The smaller main subgroup of flowering plants; generally having one cotyledon, flower parts in 3's, parallel leaf venation, stem veins scattered (but often not all of these) — e.g., lily, orchid, grass, cat-tail, palm.

monoecious. Male and female (or staminate and pistillate) unisexual structures (flowers) on the same plant. Said of a taxon having only unisexual fertile reproductive structures. (see dioecious) (example, *Alnus rhombifolia*)

montane. Pertaining to mountains; the region between foothills and subalpine.

n. Number of chromosomes in sperm and egg cells. The number in other cells of plants treated in *The Jepson Manual* is generally 2*n*. (see diploid, haploid, polyploid)

native. Occurring naturally in an area, not as either a direct or indirect consequence of human activity; indigenous; not alien.

naturalized. Alien (not native) but reproducing without human fostering. (see native, waif)

nectary. Structure that secretes nectar, often near the base of an ovary or in a perianth spur. Nectar is a nutritive solution consumed by animal visitors that are often pollinators. (example, *Symphoricarpos albus*)

needle. (p. 19) A narrowly linear, often waxy, generally evergreen leaf, especially of conifers.

node. (pp. 17, 26) Position on an axis (generally a stem) from which one or more structures (especially leaves) arise.

nut. A mostly dry, indehiscent fruit in which a single seed is encased in a hard shell that is derived from inner ovary tissue. Ovary tissue outside shell sometimes fleshy (generally inedible) but seed within shell often edible. (see drupe) (example, *Quercus palmeri*)

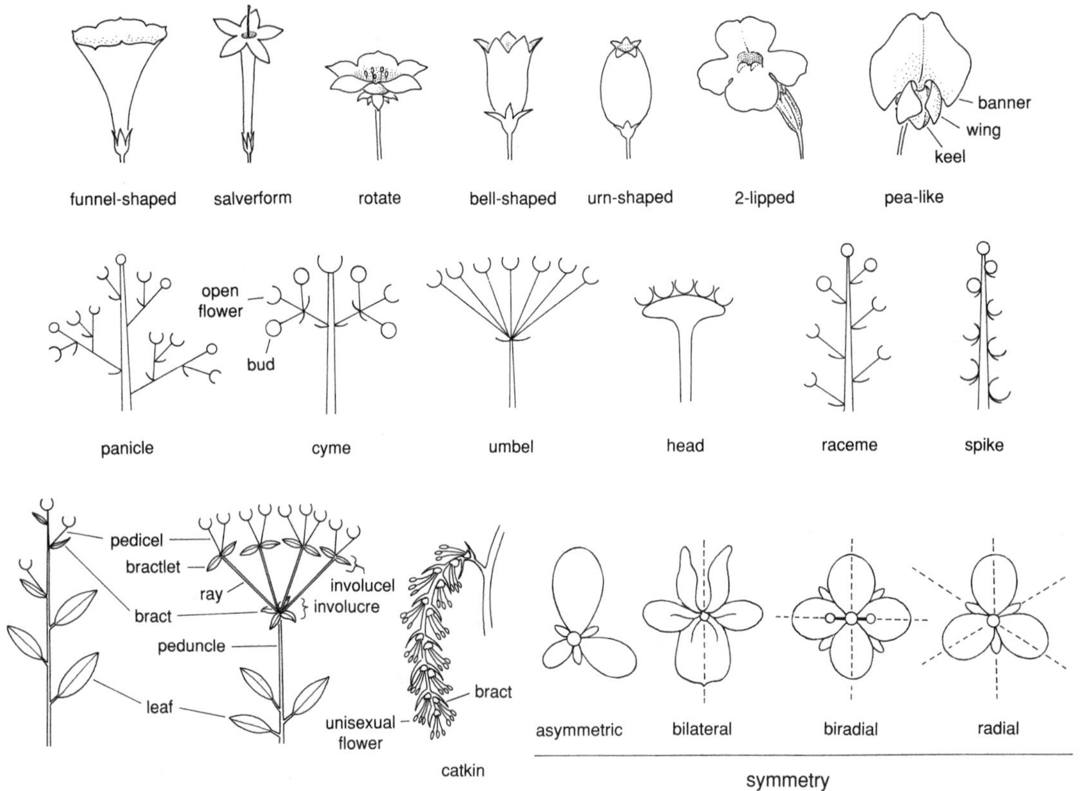

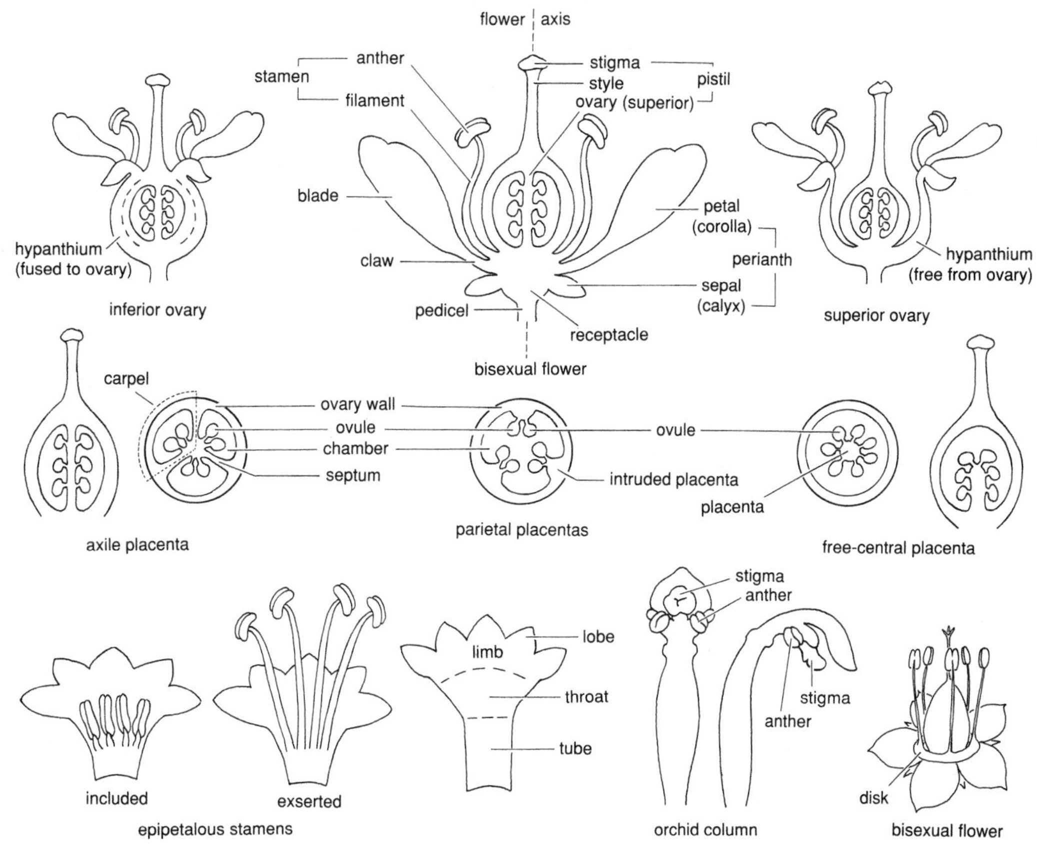

nutlet. Small, dry nut (or nut-like fruit), generally one of several produced by a single flower. (see nut, drupe) (example, *Pectocarya peninsularis*, other Boraginaceae, Lamiaceae)

ob-. (p. 18) A prefix indicating inversion of shape — e.g., lanceolate and oblanceolate leaf blades are widest below and above the middle, respectively.

oblique. (p. 18) Having unequal sides or an asymmetric base.

oblong. (p. 18) Longer than wide, with nearly parallel sides and rounded corners; wider than linear. (see elliptic, linear)

obtuse. (p. 18) Having a short-tapered, blunt tip or base, the sides convex or straight and converging at more than a right angle. (see acute)

open. Uncongested or diffuse. Said especially of the disposition of flowers in an inflorescence. (see dense)

opposite. (p. 17) 1. Located directly across from. 2. Two structures (generally leaves) per node. 3. Superimposed structures that are in the same rank — e.g., "stamens opposite petals". (see alternate)

or. Unless stated otherwise, in *The Jepson Manual*, defined as "one or the other or both"; for example, "leaves toothed or lobed" does not exclude leaves that are both toothed and lobed.

ovary. (pp. 21–23) Ovule-bearing, generally wider, portion of pistil, normally developing into a fruit as ovules become seeds. (see pistil)

ovate. (p. 18) Egg-shaped in two dimensions, widest below the middle, as of a leaf.

ovoid. Egg-shaped in three dimensions, widest below the middle, as of a fruit.

ovule. (p. 21) Structure containing an egg; a seed prior to fertilization.

palea. (p. 26) In Poaceae, the upper and generally smaller of two sheathing bracts subtending a flower, itself generally ensheathed by the lemma. (see glume, lemma, spikelet)

palmate. (p. 16) Radiating from a common point. Generally said of veins, lobes, or leaflets of a leaf.

panicle. (p. 20) Branched inflorescence in which the basal or lateral flowers (or some of them) open before the terminal or central flowers on any axis. (see cyme)

papillate. (p. 23) Bearing small, rounded or conic protuberances (papillae). Said especially of a leaf or fruit surface.

pappus. (p. 25) In Asteraceae, the aggregate of structures such as awns, bristles, or scales arising from the top of the inferior ovary, in the place sepals would be expected.

parasite. A plant that benefits from a physical connection to a host plant of another species and often in time harms the host. *Green parasites* derive water and dissolved substances and often are able to survive without the connection, while *non-green parasites* obtain in addition energy-rich products of photosynthesis and require the connection to survive.

parietal. (p. 21) Pertaining to placentas on the inside surface of the ovary wall in a compound ovary with one or more chambers.

pedicel. (pp. 20, 21) Stalk of an individual flower or fruit. (see peduncle, ray)

peduncle. (pp. 14, 20, 25) Stalk of an entire inflorescence or of a flower or fruit not borne in an inflorescence. (see pedicel, ray)

peltate. (p. 18) With the stalk (of a leaf, scale, or other flat structure) attached toward the middle, not at a margin.

pendent. Drooping, hanging, or suspended from a point of attachment above. (example, *Araujia sericofera* fruit)

perennial. (p. 14) Living more than two years or growing seasons; restricted in *The Jepson Manual* to plants that are essentially non-woody aboveground. (see annual, biennial, herb, subshrub)

perianth. (p. 21) Calyx and corolla collectively, whether or not they are distinguishable.

perianth part. An individual member of a perianth, whether or not calyx and corolla are distinguishable. Normally used when they are not distinguishable.

perigynium. (p. 23) Sac-like structure enclosing the ovary and achene in *Carex*, of diverse form and critical to identification.

persistent. Not falling off; remaining attached. (see deciduous, ephemeral)

pistillate. Having fertile pistils but sterile or missing stamens. Said of flowers, inflorescences, or plants. (example, *Salix laevigata* flower)

placenta. (p. 21) Structure or area to which ovules are attached in an ovary, variously shaped and positioned.

planoconvex. Solid shape, with one side nearly flat, the other rounded. (example, *Carex leporinella* perigynium)

pleated. Having accordion-like folds.

plumose. (p. 23) Plume-like; generally with fine appendages arrayed in three dimensions around an axis, or in tufts held together at the base. Said especially of certain stigmas and pappus elements.

pollination. Placement, in any way, of pollen on a stigma (or other floral surface through which fertilization may be achieved).

polyploid. Having three or more sets of chromosomes; $3n$, $4n$, etc. (see diploid, haploid, n)

pome. In Rosaceae, a fleshy, indehiscent fruit, such as an apple or pear. Derived from a compound, inferior ovary

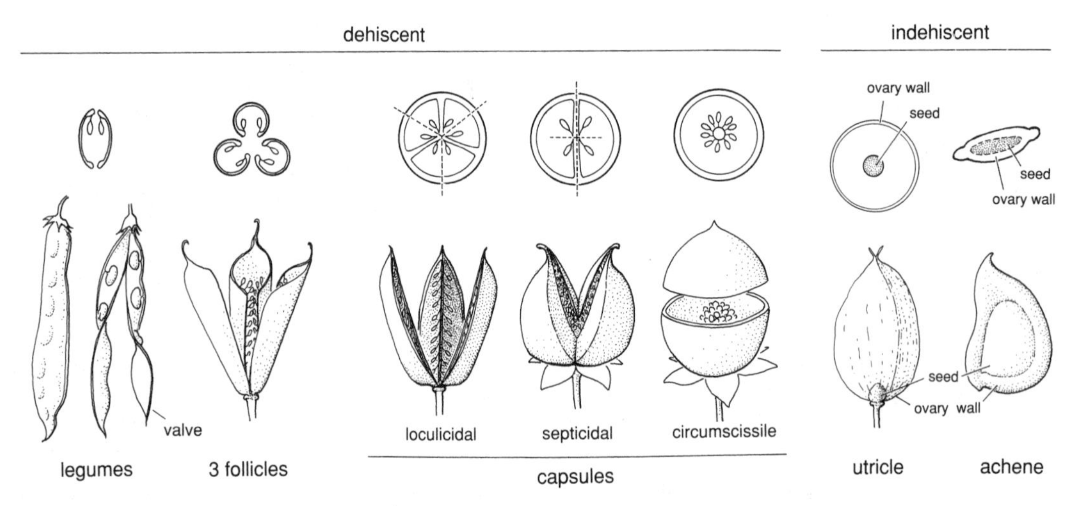

petal. (p. 21) Individual member of the corolla, whether fused or not; often conspicuously colored. (see sepal)

petiole. (pp. 17, 18, 24) Leaf stalk, connecting leaf blade to stem.

phyllary. (p. 25) In Asteraceae, a bract of the involucre that subtends a head.

pinnate. (p. 16) Feather-like, with two rows of structures on opposite sides of an axis. Generally said of veins, lobes, or leaflets arranged in two dimensions along either side of an axis. A leaf is *odd-pinnate* if there is a terminal leaflet, *even-pinnate* if there is not, and either may be *1-pinnate* (blade divided into primary leaflets), *2-pinnate* (primary leaflets divided into secondary leaflets), etc. (see compound leaf)

pistil. (p. 21) Female reproductive structure of a flower, composed of an ovule-containing ovary at the base, one or more pollen-receiving stigmas at the tip, and generally one or more styles between ovary and stigma. A flower may have one or more *simple pistils* (each a single, free carpel with a single ovary chamber, placenta, and stigma) or one *compound pistil* (two or more fused or partially fused carpels, the exact number often equaling the number of ovary lobes, ovary chambers, placentas, styles, or stigmas).

(the core and inner fleshy material) and its surrounding hypanthium (outer fleshy material and skin). (see berry, drupe) (example, *Amelanchier alnifolia* fruit)

prickle. Superficial, sharp-pointed projection, derived from epidermis, bark, etc. (see armed, spine, thorn) (example, *Rosa woodsii* stem)

prostrate. (p. 14) Lying flat on the ground. (see decumbent)

protandrous. Releasing pollen first. Said of a flower (or plant with unisexual flowers) in which pollen release precedes and does not overlap stigma receptivity.

protogynous. Receiving pollen first. Said of a flower (or plant with unisexual flowers) in which stigma receptivity precedes and does not overlap pollen release.

proximal. Closer to the origin or point of attachment (or farther away from the edge or tip). (see distal)

puberulent. (p. 23) Having hairs normally visible only when magnified.

raceme. (p. 20) Unbranched inflorescence of pediceled flowers that open from bottom to top. (see panicle, spike)

radial. (p. 20) Divisible into mirror-image halves in three or more ways. (see asymmetric, bilateral, biradial)

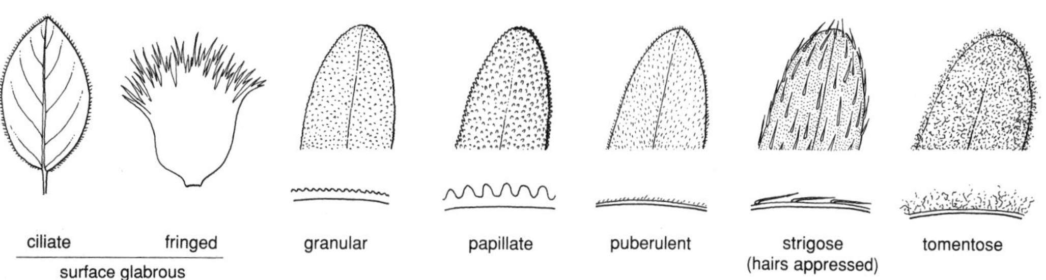

radiate head. (p. 25) In Asteraceae, a head composed of central disk flowers and marginal ray flowers.

rank. (p. 17) 1. A row or column of parts of the same orientation along an axis — e.g., leaves on an erect stem that are arranged in four vertical rows are 4-ranked. (see alternate, opposite) 2. In classification, a level — e.g., family, genus, species. (see taxon)

rare. Extremely unlikely to be encountered, often not present in appropriate habitats and often restricted to a small number of sites. Used only for certain taxa included in *Index of Rare and Endangered Vascular Plants of California* (CNPS), some of which have been accorded such legal status by the State of California. (see endangered, threatened, uncommon) (see pp. 29, 30 for more information)

ray. (p. 20) A primary, radiating axis, as a primary branch in a compound umbel. (see pedicel, peduncle)

ray flower. (p. 25) In Asteraceae, a generally pistillate or sterile, bilateral flower with the long, outer portion of the corolla (ligule) often 3-lobed, appearing on the margin of a head and accompanied by more central disk flowers. (see ligulate flower, disk flower)

receptacle. (pp. 21, 25) 1. In individual flowers, the structure to which flower parts are attached. 2. In heads or head-like inflorescences, especially in Asteraceae, the structure to which flowers or sometimes heads are attached.

recurved. (p. 15) Gradually curved downward or backward.

reduced. Gradually smaller; often narrower, less lobed, etc.

reflexed. (p. 15) Abruptly bent or curved downward or backward.

reniform. (p. 18) Kidney-shaped, as of a leaf; sometimes a leaf base having rounded lobes with the sides adjacent to the petiole clearly concave. (see cordate)

rhizome. (pp. 14, 24, 26) Underground, often elongate, more or less horizontal stem. Distinguished from root by presence of leaves, leaf scars, scales, buds, etc. (see stem)

rib. 1. Ridge, as on a fruit. 2. Raised vein, as on a leaf or perianth part. (example, *Carex hendersonii* perigynium)

root. (pp. 14, 24, 26) Underground structure of a plant, generally branched, without appendages, generally growing into the ground from the base of a stem. Its functions include anchorage, absorption of water and nutrients, and food storage. (see bulb, corm, rhizome, stem)

rosette. (p. 14) A radiating cluster of leaves generally at or near ground level.

rotate. (p. 20) Wheel-shaped, spreading, or saucer-shaped. Said of a fused corolla with a short or nonexistent tube and a spreading limb.

sagittate. (p. 18) Arrowhead-shaped, with two basal lobes oriented nearly parallel to the long axis. (see hastate)

salverform. (p. 20) Having a slender tube and an abruptly spreading, flat limb. Said especially of a fused corolla.

scabrous. Rough to the touch, generally owing to short stiff hairs. (example, *Leersia oryzoides*)

scale. 1. Wide, appressed, membranous, epidermal outgrowth (example, *Cheilanthes covillei*). 2. Structure partially or entirely covering an over-wintering bud (bud scale) (example, *Salix goodingii* bud). 3. In gymnosperms, a woody, seed-bearing structure attached to the cone axis (cone scale) (example, *Pseudotsuga menziesii*). 4. In Asteraceae, a flat, membranous pappus element (example, *Dugaldia hoopesii*). Leaves or bracts may be scale-like in one or more of the preceding ways.

scapose. (p. 14) Pertaining to a plant or an inflorescence having a relatively long peduncle that arises from ground level, often from a rosette, sometimes bearing bracts but without leaves.

scar. (p. 17) Mark left by the natural separation of two structures, as a leaf scar on a stem.

scarious. Thin, dry, pliable, dark-colored or translucent but not green. Often like dry onion peel. (see membranous) (example, *Carex brainerdii* pistillate flower bract)

scree. Relatively unstable, sloping accumulation of small rock fragments, often at a cliff base. (see talus)

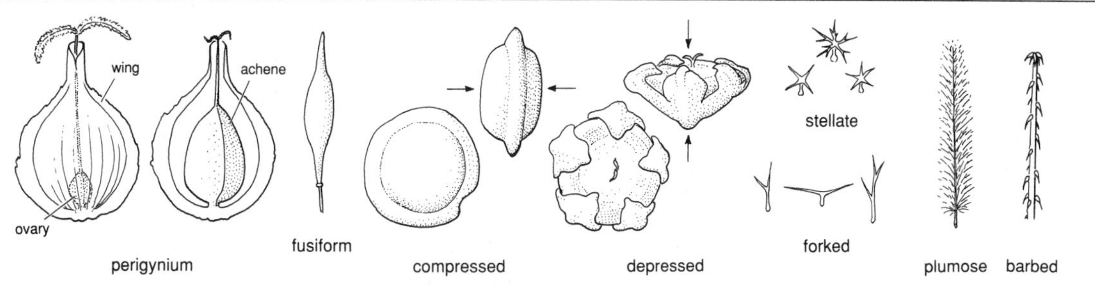

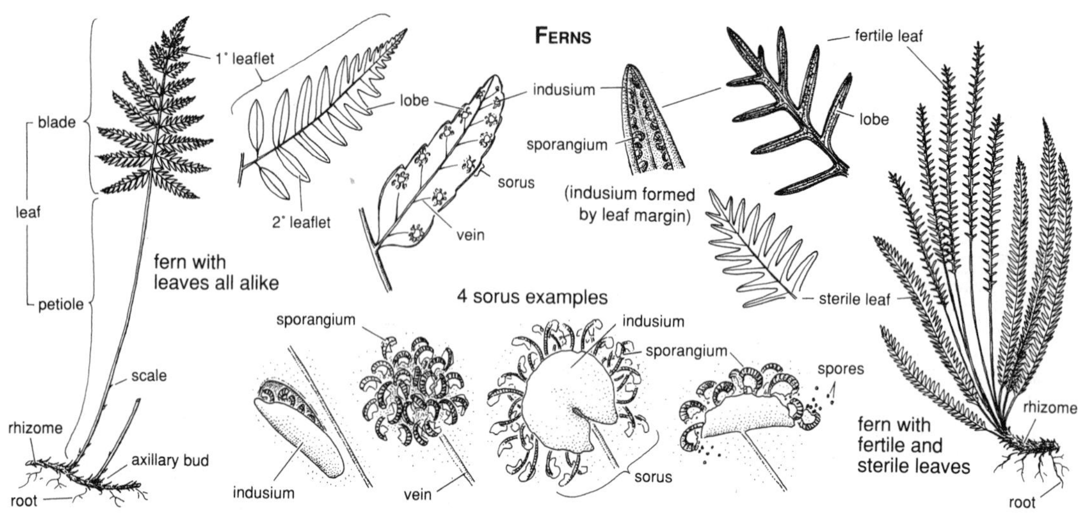

sculpture. Surface ornamentation, often visible only when magnified, as on a seed. (example, *Plagiobothrys glyptocarpus* nutlet)

seed. (p. 22) A fertilized ovule, the earliest product of sexual reproduction in plants. In descriptions, the fully mature form (at full fruit maturation) is assumed unless noted.

segment. One of the repeated components of an organ, such as a perianth, fruit, or leaf (example, *Alcea rosea* fruit). A leaf segment is one of the ultimate or smallest divisions of the blade (not a marginal lobe, tooth, bristle, etc.).

sepal. (p. 21) Individual member of the calyx, whether fused or not, generally green. (see petal)

septicidal. (p. 22) Pertaining to a capsule, dehiscent longitudinally through the ovary wall at or near the center of each septum, such that each resulting valve or segment corresponds to a single chamber. (see circumscissile, loculicidal)

septum (septa). (p. 21) Wall between chambers in a compound ovary.

series. (p. 25) A group of structures of similar size or shape, generally more or less in a whorl — e.g., involucre bracts may be in one or more series.

serpentine. General term for rocks with unusually high concentrations of magnesium and iron or the soils derived from them. Both are characterized by low levels of calcium and other nutrients and high levels of magnesium, iron, and certain toxic metals. Many plant taxa are restricted to or excluded from serpentine.

serrate. (p. 18) Having margins with sharp teeth generally pointing tipward, not outward. (see dentate)

sessile. (p. 18) Without a petiole, peduncle, pedicel, or other kind of stalk.

sheath. (p. 26) Structure that surrounds or partly surrounds another structure, often tubular, as a leaf base in Apiaceae or Poaceae.

shoot. 1. A stem and its appendages collectively. 2. Sometimes used for all aboveground parts of a plant.

shrub. 1. A woody plant of relatively short maximum height. 2. A woody plant much-branched from the base. (see tree, subshrub)

simple. (p. 16) Composed of a single part; undivided; unbranched. (see compound)

sinus. (p. 16) An indentation, as between adjacent lobes of a margin.

sorus (sori). (p. 24) In many ferns, a distinct cluster of sporangia.

spheric. Globe- or ball-shaped; round in three dimensions.

spike. (p. 20) Unbranched inflorescence of sessile flowers, nearly always opening from bottom to top.

spikelet. (p. 26) 1. In Poaceae, the smallest aggregation of florets plus any (generally 2) subtending glumes. 2. In Cyperaceae, the smallest aggregation of flowers (i.e., generally more than 2) and associated bracts.

spine. Sharp-pointed projection, derived from leaf (often vein tip) or other organ, such as ovary wall. Sometimes used for any sharp projection. (see armed, prickle, thorn) (example, *Cirsium arvense* leaf)

sporangium (sporangia). (p. 24) In non-seed plants (fern allies and ferns), a spore-producing organ (some ferns, such as *Marsilea*, bear sporangia in hard cases).

spore. (p. 24) The minute, dispersing, reproductive unit of non-seed plants (fern allies and ferns); one of very many haploid cells dispersed from a diploid parent plant, normally developing into a small haploid plant that produces eggs, sperm, or both, the fusion of which results in new diploid offspring.

spreading. (p. 15) Oriented more or less perpendicularly to the axis of attachment; often, more or less horizontal.

spur. Hollow, often conic, projection or expansion, generally of a perianth part and containing nectar. (example, *Delphinium californicum* flower)

stamen. (p. 21) Male reproductive structure of a flower, typically composed of a stalk-like filament and a terminal, pollen-producing anther. Filaments sometimes partly fuse to the corolla, or to other filaments to form a tube. (see anther, filament, pistil)

staminate. Having fertile stamens but sterile or missing pistils. Said of flowers, inflorescences, or plants. (see pistillate) (example, *Salix laevigata*)

staminode. Sterile stamen, often modified in appearance, sometimes petal-like or elaborate in structure. (example, *Penstemon barnebyi*)

stellate. (p. 23) Star-like. Generally said of a hair with three or more branches radiating from a common point. (see forked)

stem. (pp. 14, 16, 17) Axis or axes of a plant, bearing appendages such as leaves, axillary buds, and flowers.

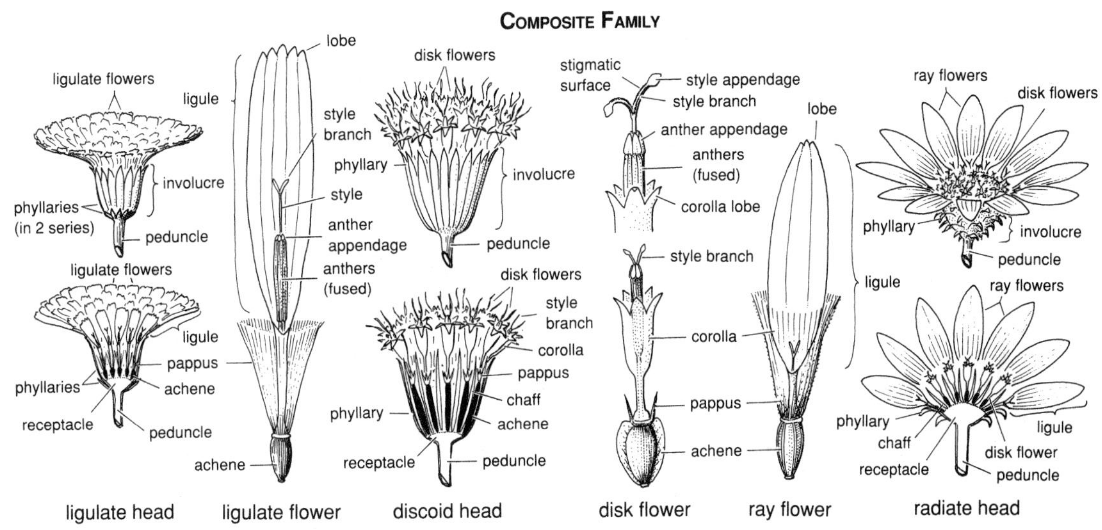

Sometimes below ground. (see bulb, caudex, corm, rhizome, root, stolon, tuber)

sterile. Not reproductively functional. Said of a plant or plant part that does not produce or is not associated with the production of functional spores, pollen, ovules, or seeds. (see fertile)

stigma. (pp. 21, 25, 26) The part of a pistil on which pollen is normally deposited, generally terminal and elevated above the ovary on a style, generally sticky or hairy, sometimes lobed.

stipule. (p. 17) Appendage at base of petiole, generally paired, variable in form but often leaf- or scale-like, sometimes a spine.

stolon. (pp. 14, 26) Runner; a normally thin, elongate stem lying more or less flat on the ground and forming roots as well as erect stems or shoots (i.e., ultimately new plants) at generally widely spaced nodes. (see shoot, stem)

stomate. A minute pore on a leaf or stem through which gases such as carbon dioxide, oxygen, and water vapor pass by diffusion. Features of stomates help identify some plants.

stone. In a drupe, the very hard inner ovary wall and the generally single seed it surrounds; occurring one or more per flower, free or variously fused. (example, *Prunus emarginata* fruit)

stout. Thick, sturdy, not slender.

striate. With fine, longitudinal channels, lines, or ridges.

strigose. (p. 23) With stiff, straight, sharp, appressed hairs.

style. (pp. 21, 25) Stalk-like portion that connects ovary to stigma in many pistils.

sub-. A prefix meaning almost, just below, or somewhat imperfectly.

subalpine. Just below timberline; between montane and alpine.

submersed. 1. A plant normally rooted and remaining underwater. 2. The part of such a plant normally held underwater.

subshrub. A plant with the lower stems woody, the upper stems and twigs not woody (or less so) and dying back seasonally. (see perennial, shrub)

subtend. Occurring immediately below, as sepals subtending petals or leaves subtending axillary buds.

superior ovary. (p. 21) An ovary that is free from the perianth or hypanthium and appears to sit on top of the receptacle. Lower parts of sepals, petals, and stamens (or hypanthium) arise from near ovary base instead of its top. (see inferior ovary)

suture. Groove or line of dehiscence or fusion.

talus. Relatively stable, sloping accumulation of large rock fragments, often at a cliff base. (see scree)

tapered. (p. 18) Gradually (not abruptly) narrower or smaller at base or tip. (see truncate)

taproot. (p. 14) Main, tapered root that generally grows straight down into soil and has smaller, lateral branches.

taxon (taxa). In classification, a group of organisms (such as plants) at any rank — e.g., species, family. (see rank)

tendril. A slender, coiling structure (generally stem, stipule, or leaf tip) by which a climbing plant becomes attached to its support. (example, *Lathyrus latifolius* leaf)

terminal. At the tip of a structure.

ternate. Once or repeatedly lobed or compounded into three parts, as a clover leaf. (example, *Trifolium howellii* leaf)

thorn. Sharp-pointed branch. (see armed, prickle, spine) (example, *Castela emoryi*)

threatened. Survival is in jeopardy but not the extreme jeopardy implied by "endangered". May be used in a general sense, as well as to indicate such status accorded by law (see endangered, rare) (see pp. 29, 30 for more information)

throat. (p. 21) In flowers with fused sepals or petals, the expanded, fused portion above the tube and below the limb.

tomentose. (p. 23) Covered with densely interwoven, generally matted hairs.

tooth (teeth). (pp. 16, 18) A small, pointed projection of a margin. (see dentate, serrate)

transverse. Pertaining to width or the widthwise dimension; perpendicular to the axis. (see longitudinal)

tree. A woody plant of medium to tall maximum height, with generally one relatively massive trunk at the base. (see shrub) (example, *Juniperus occidentalis*)

trichome. Any epidermal outgrowth of a plant. Not used in *The Jepson Manual*. (see hair, scale)

truncate. (p. 18) Abruptly (not gradually) narrower or

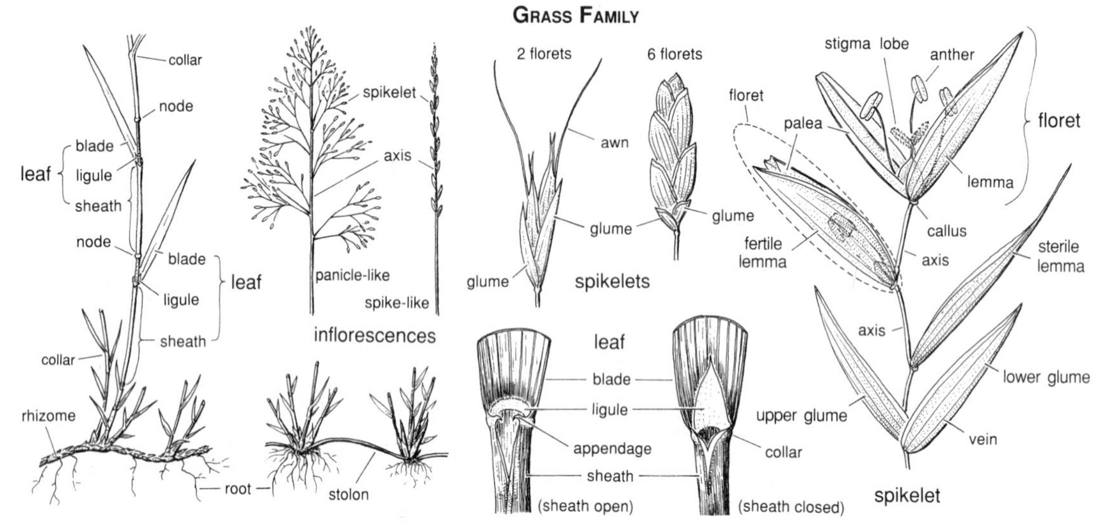

smaller at base or tip, as if cut straight across or nearly so. (see tapered)

tube. (p. 21) In flowers with fused sepals or petals, the more or less cylindric, fused portion at the base.

tuber. Short, thick, fleshy, underground stem for storage (of water, food, or both) and sometimes propagation — e.g., potato. (see stem)

tubercle. Small, wart-like projection. (example, *Plagiobothrys hystriculus* nutlet)

twig. In woody plants, a terminal stem segment, produced during the current or most recent growth period.

twining. Twisting or coiling, normally for the purpose of climbing. Generally said of stems or tendrils that wind around a support. (example, *Araujia sericofera*)

ultimate. Last, most distal, or smallest, as all the tips of a branching stem or the smallest divisions (segments) of a compound leaf.

umbel. (p. 20) Inflorescence in which three to many pedicels radiate from a common point. May be compound, in which case larger inflorescence branches (rays) also radiate from a common point. Characteristic of but not confined to Apiaceae.

uncommon. Unlikely to be encountered and sometimes not present in appropriate habitats. Used in a general sense and (in all capital letters) also for some taxa included in *Inventory of Rare and Endangered Vascular Plants of California* (CNPS). (see common, rare) (see pp. 29, 30 for more information)

unisexual. Having flowers in which either stamens or pistils, but not both, are fertile. (see bisexual, pistillate, staminate)

urn-shaped. (p. 20) Pertaining to a fused calyx or corolla that is gradually or abruptly narrowed toward the tip.

utricle. (p. 22) Mostly dry, generally indehiscent fruit from a generally compound pistil in which a single seed is loosely enclosed by a balloon- or bladder-like ovary wall.

valve. (p. 22) One of the parts into which a capsule or legume splits.

vascular. Pertaining to plant veins or to plants with veins. Only vascular plants are treated in *The Jepson Manual.*

vein. (pp. 24, 26) 1. Tissue specialized for transport of substances within a plant (xylem for water and dissolved substances, phloem for energy-rich, organic compounds). 2. A strand of such tissue, often seen as a bundle in transverse section.

vernal. Pertaining to the spring season.

vestigial. Rudimentary. Said of a structure that is undeveloped, poorly developed, or degenerate and therefore nonfunctional.

vine. A trailing or climbing plant, sometimes attaching to its support by tendrils. (example, *Phaseolus filiformis*)

waif. Alien (not native) and either not reproducing without human intervention or not persisting for more than a few generations and therefore incompletely naturalized. Most known waifs are not treated in *The Jepson Manual.*

weed. A generally alien, generally undesired (sometimes attractive) plant, often adapted to disturbed places, often aggressive, occurring in or near settlements, in fields or gardens, along roadsides, and in relatively undisturbed communities of native plants. Weeds considered legally noxious by the State of California are so noted.

whorl. (p. 17) Group of three or more structures of the same kind (generally leaves or flower parts) at one node.

wing. (pp. 18, 20, 23) 1. Thin, flat extension or appendage of a surface or margin. 2. In many members of Fabaceae, each of two lateral petals.

wiry. Slender, stiff, and tough. Used for stems, etc.

wood. Hard, thickened, vascular tissue (xylem) that develops especially in shrubs, trees, and some vines, generally in concentric rings.

ABBREVIATIONS AND SYMBOLS

The philosophy for choosing abbreviations for *The Jepson Manual* was based on a critical need to conserve space. This need ran counter to that of maximizing ease of understanding. The abbreviations used were selected because they save substantial space, are reasonably unambiguous and self-explanatory, and are relatively easily remembered. They are used throughout *The Jepson Manual*, with the exception of introductory material, such as the glossary. Periods follow abbreviations only where their absence might cause confusion.

Most of the abbreviations are used in describing geographic ranges. About half of the geographic abbreviations were derived especially for the major geographic regions of California. These are preceded by asterisks (*) and are discussed more fully in the chapter "Geographic Subdivisions of California" (p. 37). Other geographic abbreviations are in wide use.

Fewer than a dozen plant features are abbreviated. The most commonly encountered introduce the sentences of a plant description ("st", "lf", "fl", etc). In such positions, they appear in **BOLDFACE CAPITALS** to provide visual guides to the structure of the description. A few other abbreviations (e.g., "lfless") are derived from them. Four special abbreviations are used only in certain illustration captions to save space and one is used only to cross-reference terms in the glossary. Thirteen abbreviations are restricted to summaries of horticultural potential; these summaries follow some descriptions.

Abbreviations

Afr = Africa
AK = Alaska
Am = Americas, western hemisphere
ann = annual
AZ = Arizona

b = back (of an organ, used only in some illustration captions)
B.C. = British Columbia
bien = biennial
br = bract (used only in some illustration captions)

c = central
CA = California
*CA-FP = California Floristic Province
C.Am = Central America
Can = Canada
*CaR = Cascade Range
*CaRF = Cascade Range Foothills
*CaRH = High Cascade Range
*CCo = Central Coast
*ChI = Channel Islands
cm = centimeter
Co. = County
cos. = counties
cult = cultivated
CVS = horticultural entries only: cultivars are available in the trade
*CW = Central Western California

*D = Desert (not abbreviated as a general term)
DFCLT = horticultural entries only: difficult; needs special care in gardens
diam = diameter
dm = decimeter
*DMoj = Mojave Desert
*DMtns = Desert Mountains
DRN = horticultural entries only: requires excellent drainage
DRY = horticultural entries only: intolerant of frequent summer water
*DSon = Sonoran (Colorado) Desert

e = east (ern)
e-c = east-central
e.g. = for example
esp = especially
et al. = and others
etc = and so on
Eur = Europe
exc = except, excluding

f = front (of an organ, used only in some illustration captions)
fl, fls (**FL, FLS**) = flower(s), floral, flowering
fld = flowered
fr (**FR**) = fruit

*GB = Great Basin Floristic Province
gen = generally, usually (also some senses of "mostly")
geog = geographic (-al, -ally)
GRCVR = horticultural entries only: good groundcover
*GV = Great Central Valley

ID = Idaho
i.e. = that is to say
incl = including, included (in)
infl, infls (**INFL, INFLS**) = inflorescence(s)
INV = horticultural entries only: invasive; may displace or overrun other plants

IRR = horticultural entries only: requires moderate summer watering (irrigation)

*KR = Klamath Ranges
KS = Kansas

lf (**LF**) = leaf
lfless = leafless
lflet = leaflet
lvs (**LVS**) = leaves

m = meter
Medit = Mediterranean
Mex = Mexico
mm = millimeter
*MP = Modoc Plateau
MT = Montana
Mtn(s) = Mountain(s) (proper name)
mtn(s) = mountain(s) (not a proper name)

n = north(ern)
n-c = north-central
N.Am = North America
*NCo = North Coast
*NCoR = North Coast Ranges
*NCoRH = High North Coast Ranges
*NCoRI = Inner North Coast Ranges
*NCoRO = Outer North Coast Ranges
NE = Nebraska
ne = northeast(ern)
NM = New Mexico
NV = Nevada
*NW = Northwestern California
nw = northwest(ern)

OK = Oklahoma
OR = Oregon
orn = ornamental

per = perennial herb
peri = perigynium (used only in some *Carex* illustrations)
pl(s) = plant(s)
*PR = Peninsular Ranges

s = south (ern)
S.Am = South America
s-c = south-central
*SCo = South Coast
*SCoR = South Coast Ranges
*SCoRI = Inner South Coast Ranges
*SCoRO = Outer South Coast Ranges
*ScV = Sacramento Valley
SD = South Dakota
se = southeast(ern)
sect(s). = section(s) (abbreviated only as taxonomic rank)
SHD = horticultural entries only: does best in full or part shade
*SnBr = San Bernardino Mountains
*SnFrB = San Francisco Bay Area
*SnGb = San Gabriel Mountains
*SnJt = San Jacinto Mountains
*SnJV = San Joaquin Valley
*SN = Sierra Nevada
*SNE = East of Sierra Nevada (e.g., Mono Valley, Owens Valley)
*SNF = Sierra Nevada Foothills
*SNH = High Sierra Nevada
sp. = species (singular)
spp. = species (plural)
ssp. = subspecies (singular)
sspp. = subspecies (plural)
st, sts, (**ST, STS**) = stem, stems
STBL = horticultural entries only: stabilizer; good for restoring degraded areas
subg. = subgenus, subgenera
subsect(s). = subsection(s) (abbreviated only as taxonomic rank)
SUN = horticultural entries only: does best in ± full sun
*SW = Southwestern California
sw = southwest(ern)

*Teh = Tehachapi Mountain Area
temp = temperate
*TR = Transverse Ranges
trop = tropical
TRY = horticultural entries only: ± untested but worth pursuing
TX = Texas

US = United States
UT = Utah

var(s). = taxonomic variety (taxonomic varieties)
vs = versus

w = west(ern)
WA = Washington (the state)
*W&I = White and Inyo Mountains
w-c = west-central
WET = horticultural entries only: roots need to be in continually moist or wet soil
*WTR = Western Transverse Ranges
*Wrn = Warner Mountains
WY = Wyoming

Symbols

The following symbols are used often. Most are quantitative, referring to number, height, length, width, etc; however, "±" may be qualitative as well, referring to color, fusion, symmetry, etc. Mathematically, the symbol "<" means "up to" (or "approaches as a limit"), so it is generally also equivalent to "=". For example, "ST < 5 m" includes those stems that are exactly 5 m. In cases where equivalency is emphasized, the construction "< or =" is used.

<< much less than (in size); greatly exceeded by
< fewer than (in number); less than or up to (in size); exceeded by
= equal to (generally in size or exsertion)
> more than (in number); greater than or equal to (in size); exceeding
>> much greater than (in size); greatly exceeding
0 none, absent
× multiplication sign, meaning "times" or indicating hybridity
° degree of compoundness, branching, or angle [e.g., 1° (primary), 2° (secondary); 45–60° angle]
❀ precedes a statement of horticultural potential
± more or less, approximately

COMMONNESS AND RARITY

Relatively few plant taxa are truly common; these are of special interest because they help define plant communities by their sheer numbers or conspicuousness. Likewise, relatively few taxa are extremely rare; they, in turn, are of special interest owing simply to their rarity and, often, if they are to survive, to an accompanying need for protection. Because of its dynamic geologic history and current climatic and topographic diversity, California has an unusually high proportion of sensitively rare plants, which receive special attention in *The Jepson Manual.*

Specifying an accurate "likelihood of encounter" for each plant taxon that grows in California is impossible. In *The Jepson Manual*, the most widespread and commonly encountered taxa are termed "abundant". Those that are obvious but somewhat less likely to be encountered or less widespread are considered "common". Together, these two categories make up perhaps 20 percent of the taxa covered in *The Jepson Manual.*

There is no "likelihood of encounter" designation for the large middle group (perhaps 60 percent of the taxa in the manual). This lack of designation indicates that they are neither especially common nor rare.

The "rare" end of the spectrum also makes up about 20 percent of taxa covered in *The Jepson Manual.* "Uncommon" (lower case letters) is comparable with "common" at the high end of the frequency distribution. Taxa so designated are not very likely to be encountered. Nevertheless, they are not rare enough to be considered sensitive or to be included in the CNPS *Inventory* (see below).

A set of specific designations allows correlation with the contents of the California Native Plant Society's *Inventory of Rare and Endangered Vascular Plants of California* (CNPS Special Publication 1, 4th edition); address for copies: CNPS Publications, 909 12th St., Suite 116, Sacramento, CA 95814.

Preparation of edition 5 of the *Inventory* overlapped seriously with final preparations of *The Jepson Manual.* Close communication and correlation were attempted during this process. However, continual flux in taxonomic assessments and legal statuses (as well as in understanding of geographic ranges and rarity) required both projects to make independent decisions that could not be accommodated logistically by the other. Moreover, some rare plants are local forms of highly variable taxa. Authors writing for the *The Jepson Manual* sometimes considered them to be subtle variants not warranting taxonomic recognition; however, contrasting views may be valid and some of these will be recognized in edition 5 of the *Inventory.* For all these reasons, careful readers will find some discrepancies between designations in *The Jepson Manual* and those in the *Inventory.*

The less sensitive taxa included in the *Inventory* on List 3 (more information needed) or List 4 (watch list) are designated "UNCOMMON" (all capital letters) in *The Jepson Manual.* The designation means that the taxon is of special interest because of its potential rarity; the reader is thereby referred to the *Inventory* for more information.

The Jepson Manual restricts application of the word "rare" to those taxa included in or proposed for inclusion in *Inventory* List 1B (rare, threatened, or endangered throughout range) or List 2 (rare, threatened, or endangered in California but more common outside the state). (There are very few exceptions: some taxa found on other *Inventory* lists are legally protected by the State of California.) In *The Jepson Manual*, taxa on *Inventory* List 2 are noted to be "RARE in CA". 475 of the 675 taxa on List 1B are labelled "RARE".

Highlighting is used to note special protection accorded by either California or Federal law. Boldface type summarizes the legal status of the ± 200 taxa of List 1B (and other lists) with such status ("**ENDANGERED** US", "**THREATENED** US", "**ENDANGERED** CA", "**THREATENED** CA", or "**RARE** CA").

Taxa of which living representatives have not been found for many years, despite much effort, make up List 1A of the *Inventory.* They are designated "**PRESUMED EXTINCT**" (or "**PRESUMED EXTIRPATED** in CA", if a population survives outside the state). If extant at all, these are the rarest plants in California. Although without legal status (except very few that are also listed as "Endangered" by the State of California), they are highlighted by boldface designations to draw attention to the fact that they should be sought and protected if found.

The information just given is summarized in the table on the following page.

TABLE 1. RARITY LISTINGS IN *THE JEPSON MANUAL*

CNPS *Inventory* List	Legal Status (as abbreviated in *Inventory*)	*The Jepson Manual*
1A		"**PRESUMED EXTINCT**" "**PRESUMED EXTIRPATED** in CA (extant in...)"
1B	(none)	"RARE"
	CE/FE	"**ENDANGERED** CA, US"
	CE/—	"**ENDANGERED** CA"
	—/FE	"**ENDANGERED** US"
	—/FT	"**THREATENED** US"
	CT/—	"**THREATENED** CA"
	CR/—	"**RARE** CA"
2		"RARE in CA"
3,4		"UNCOMMON"
—		*No designation* (most taxa) or "Uncommon", "Common", or "Abundant".

Legal Status Abbreviations

CE = Endangered status provided by California law
FE = Endangered status provided by United States law
CT = Threatened status provided by California law
FT = Threatened status provided by United States law
CR = Rare status provided by California law

HORTICULTURAL INFORMATION IN *THE JEPSON MANUAL*

James C. Hickman & Warren Roberts

California horticulturists and gardeners are leaders in the use of native plants in gardens. Many of our natives are drought-tolerant and well adapted to the summer-dry, Mediterranean-type climates found in most of our highly populated areas. To support the use of native plants in gardens, *The Jepson Manual* systematically includes horticultural guidelines — the first time such information has been added to a work of this kind.

A statewide Horticultural Advisory Council (see list p. v) gathered information on the uses of native plants in California gardens. The Council designed a format to convey the essence of complex criteria as succinctly as possible, drawing on the expertise and experience painstakingly derived by Council members and by many other enthusiastic native-plant growers. Taxonomic authors are not responsible for the content of horticultural statements.

Native plants useful in gardens come from all climatic regions within California, which has the widest range of climates and growing conditions of any state. Different plants are adapted to each region, and most are restricted in where they can be grown with ease. To help distinguish the areas where plants can be expected to perform well without undue effort, a system of numbered climate zones was adopted. The system was based on an important element of Sunset Publishing Corporation's *Western Garden Book*, by far the best single resource for western gardeners. Sunset Publishing Corporation graciously allowed *The Jepson Manual* to use an adaptation of these zones and an outline map showing their general locations.

We especially encourage the growing of native plants in zones to which they are naturally adapted. Natives so grown are likely to be especially effective and to be "drought-tolerant" — that is, to require a minimum amount of watering.

Interpreting Horticultural Entries

1. Horticultural entries, when present, are found at the end of a species treatment, following the statements of habitat and of elevational and geographic ranges (which are themselves horticulturally useful). Each entry is preceded by the special symbol ❀ and is made up of abbreviations and numbers.
2. Each entry includes capital-letter terms or abbreviations that represent appropriate growing conditions (see list below).
3. Sets of terms or abbreviations are generally followed or preceded by a numerical list of Sunset Publishing Corporations's *Western Garden Book* climate zones in which members of the taxon can be expected to grow if their requirements are met.
4. Zones that are especially appropriate are indicated by bold-face numbers.
5. If a plant has different requirements in different zones, the entry is divided into two or more contrasting or additive parts.

Terms and Abbreviations Used in Horticultural Entries

CVS	Cultivar(s) available in the horticultural trade.
DFCLT	Difficult; needs special care in all zones; has complex requirements.
DRN	Requires excellent drainage. Compacted or other water-holding soils may need to be modified.
DRY	Intolerant of frequent summer water; should not be planted near lawns or other moisture-loving plants.
GRCVR	Good groundcover.
INV	Invasive; used for plants that, once established, tend to outcompete, displace, or overrun others.

IRR	Requires moderate summer watering (irrigation), generally 1–4 times per month depending upon the absorption rate and water retention capacity of the soil.
SHD	Does best in full or part shade; may tolerate morning and winter sun.
STBL	Native plants especially good for stabilizing or restoring disturbed or degraded (including logged or burned) areas, for erosion and slope control, for wildlife food or cover, etc. May be less suitable for general garden use.
SUN	Does best in full or nearly full sun; tolerates summer afternoon sun.
TRY	Insufficiently tested but worth pursuing, especially within its natural range.
WET	Roots need to be in continually moist or wet soil.

Example

The horticultural statement for *Abies bracteata*, bristlecone fir, is "❀DRN:4–6,**15**,**16**,17&IRR:2,3,**18–21**,22–24&SHD:7–9,14". This statement is intended to be read as follows:

"Given well drained soil, bristlecone fir can be grown in zones 4, 5, 6, and 17 but is especially successful in zones 15 and 16; if moderate summer watering is added, it can also be grown in zones 2, 3, 22, 23, and 24 but especially well in zones 18, 19, 20, and 21; if, in addition to well drained soil and summer watering, it is given the protection of partial or full shade, it can also be grown in zones 7, 8, 9, and 14."

Cautions

Although *The Jepson Manual* summarizes garden values for a broad range of native plants, the information presented is general and is intended only as a guide. More detailed guidelines should be sought from experienced gardeners, botanic gardens and arboreta, special classes and seminars, extension agents, nursery managers, and suppliers concerned with growing native plants. Local chapters of the California Native Plant Society are especially valuable sources of information and can help beginning native-plant gardeners in many ways, including recommending and interpreting standard and specialized horticultural library references, books, guides, and journal articles about California plants.

Availability of native plants for gardens is highly variable. Nursery operators, propagators, and seed distributors respond to demand. Asking them to obtain nursery stock of plants you wish to grow will eventually increase their availability. The highly popular and successful sales of native plants by chapters of the California Native Plant Society and by various arboreta and botanic gardens provide additional excellent, diverse sources of garden material propagated from plants already in cultivation. Keep in mind that plants propagated from locally native material will often be better adapted to local conditions than plants that originate from the same species in other parts of its range.

Limitations of space and format reduce exactness in the presentation of horticultural parameters, especially for those plants designated "difficult" ("DFCLT"). Many beautiful and useful plants have exacting requirements regarding the following: 1) seasonality and position of sun and shade; 2) soil factors such as texture and composition, alkalinity or acidity, salinity or other mineral toxicities or deficiencies; 3) the need for fertilizers and "feeding"; 4) biological factors such as mycorrhizal and semi-parasitic associations, mulch, competition, and predation; 5) tolerance of atmospheric conditions such as wind (dry, hot, freezing, salty) and pollution. The importance of these criteria to the successful cultivation of these plants often varies widely in relation to the factors selected for horticultural entries.

We discourage the collection of plants growing in the wild. Plants designated "ENDANGERED", "THREATENED", or "RARE" are afforded legal protection at Federal or State levels. Reduction of the wild population of these taxa for any reason, including horticultural "take", is against the law. No detailed horticultural information is included in *The Jepson Manual* for any taxon protected by either Federal or State law or for any taxon specified as rare in the CNPS ***Inventory*****. Some rare taxa are available in cultivation; they are indicated by the statement "In cult" where horticultural information might be sought. For these taxa, more information on growing requirements is available from nurseries, CNPS chapters, and botanic gardens.**

Under no conditions may readers assume that the inclusion of horticultural information is license or encouragement to remove native plants or plant parts from the wild in any form, including seeds. Neither should they assume that information herein guarantees horticultural success or failure within the designated areas or conditions.

Descriptions of Climate Zones

Many factors combine in a complex interplay to make up a climate. Those factors emphasized in developing Sunset Publishing Corporation's *Western Garden Book* system of climate zones are latitude, elevation, influence of the Pacific Ocean, influence of continental air masses, mountains and hills, and local terrain. The *Western Garden Book* system has 24 horticultural climate zones in the American West; 19 of these are in California.

Adaptations of the original descriptions of these 24 zones are given below, with cross-references to the geographic subdivisions of California used to describe the distributions of plants in the wild. (For more information on those geographic subdivisions, see pp. 37–48.) The nine other western states covered by the climate-zone system are also mentioned where appropriate.

Boundaries of climate zones are arbitrary, because climates vary gradually. Furthermore, the factors used to define climate zones shift back and forth over time, either individually or together. If your garden is well inside the boundaries of a climate zone (see map, p. 35 and, for more detail, those in Sunset Publishing Corporation's *Western Garden Book*), you can assume that the characteristics of the zone hold nearly all the time. If it is near a border, your climate will sometimes resemble that across the line. Very local conditions like fencerows, south-facing walls, slope, or even soil type may produce "microclimates" that are different from the prevailing climate of your region (the same is true of natural communities), perhaps offsetting the effective climate to that of a geographically adjacent zone.

Zones 1–3. Snowy Parts of the West

ZONE 1. Extreme winter cold associated with northern latitudes, continental air masses, or high elevations. Growing season averages 100 days; frosts can occur at any time. High Klamath, Cascade, and Sierra Nevada ranges, high and northern parts of Great Basin, highest desert mountains. [Beyond California, includes most of high-elevation and interior Washington, Oregon, Idaho, Nevada, Utah, and Colorado, and all of Montana and Wyoming.]

ZONE 2. Less cold than Zone 1, but soil freezes in winter. Growing season averages 150 days. Moderate elevations in Klamath, Cascade, and Sierra Nevada ranges (e.g., vicinity of Lake Tahoe), edges of Owens Valley, most desert mountains, high San Bernardino and San Jacinto ranges. [Beyond California, found in moderate-elevation Idaho, Nevada, Arizona, Utah, Colorado, and New Mexico, and interior Washington and Oregon.]

ZONE 3. Mildest high-elevation or cold-interior climates. Growing season averages 160 days. Lowest high-mountain areas, where there are many summer-cabin gardens (e.g., Reno-Markleeville area), much of Transverse and Peninsular ranges. [Beyond California, found in mildest parts of Idaho and northern Nevada and interior Washington and Oregon.]

[Zones 4–6. Pacific Northwest beyond California]

[ZONE 4. Cold-winter parts of western Washington; lower elevations mostly away from coast. Generally colder and wetter than Zone 5.]

[ZONE 5. Northwest coast and Puget Sound marine belt from British Columbia to California border. Cooler and wetter than Zone 17.]

[ZONE 6. Willamette Valley, Oregon; warmer summers and colder winters than Zone 5; wetter (and with cooler summers) than Zone 7.]

Zones 7–9. Great Valley and Surrounding Low Mountains

ZONE 7. Marked seasons of hot summers, moderately cold winters. Lower elevations of Klamath Ranges, Inner North Coast Range, Sierra Nevada Foothills, and much of South Coast Ranges. [Also Rogue River Valley, OR.]

ZONE 8. Cold-air basins (low spots) of Great Valley. Long, very hot summers; cold-air collection in winter may kill sensitive plants; cold just severe enough to satisfy winter dormancy requirements of many other plants; tule fog common in winter.

ZONE 9. Thermal belt zone of Great Valley. Much like Zone 8, but, because cold winter air can flow from Zone 9 to Zone 8, it is safer for cold-sensitive plants; tule fog common in winter.

Zones 10–13. Deserts

[ZONE 10. High deserts beyond California: southern Nevada, southwest Utah, southeast Colorado, Arizona, New Mexico; more rain (especially in summer) and less wind than Zone 11.]

ZONE 11. Medium to high deserts of California and southern Nevada. Wide swings in temperature: cold winters and nights; very hot summers and days; late spring frosts are likely; windy; combination of winter wind and sun may be harmfully desiccating. Owens Valley, all but lowest Mojave Desert, highest parts of Sonoran Desert, western desert edge.

[ZONE 12. Intermediate desert beyond California: confined to Arizona; colder, with more summer rain than Zone 13.]

ZONE 13. Low, essentially subtropical, deserts of California and Arizona. Much warmer winters than Zone 11 (few nights below freezing); summer storms more common than Zone 11. Most of Sonoran Desert [including vicinities of Yuma and Phoenix], lowest parts of eastern Mojave Desert, eastern edge of Peninsular Ranges.

Zone 14. Ocean-influenced Northern and Central California

ZONE 14. Inland areas with ocean or cold air influence. Similar climates with two distinct causes. 1) Marine air influx makes inland areas warmer in winter, cooler in summer. Central and western Great Valley (Modesto, Sacramento areas); eastern and southern valleys of San Francisco Bay area (Orinda, Hollister areas); upper Salinas Valley. 2) Outer Coast Range valley floors and troughs that are colder in winter than surrounding areas because of sinking cold air. Humboldt County to Santa Barbara County (e.g., Napa and Santa Ynez valleys).

Zones 15–17. Coastal Climates of Northern and Central California

ZONE 15. North Coast cold-winter areas (redwood region). Moister, with cooler summers and milder winters than Zone 14; afternoon wind. Much of the moderate-elevation area of the outer North Coast Ranges and Santa Cruz Mountains. South as far as Santa Barbara County, cold-air drainage into valleys near the coast produces the same climatic effect.

ZONE 16. Coast thermal belts. Slopes near the central California coast from which cold air drains; afternoon wind; more heat than Zone 17, warmer winters than Zone 15 (very little frost); one of the finest horticultural climates. Marin County and Berkeley-Oakland Hills, eastern slopes of Santa Cruz Mountains; coastal hills south to Santa Barbara County.

ZONE 17. North Coast marine belt. Tidal saltwater often visible or foghorns audible; winters cool, wet (frost rare); summers cool with frequent fog and wind; immediate shoreline salty, very windy. Beach and coastal-prairie communities plus Sitka-spruce forest in north grading into closed-cone-pine/cypress forests and coastal-sage scrub in south. Oregon to Santa Barbara County.

Zones 18, 19. Interior Valleys of Southwestern California

ZONE 18. Interior basins and hilltops. Cold-air-collecting areas with major influence by continental air masses; frequent frosts. Mid-elevation south slopes of Transverse Ranges, west slopes of Peninsular Ranges.

ZONE 19. Interior thermal belts. Because slopes drain cold air, winters are milder than (but climate otherwise like) Zone 18; frosts rare. Lower slopes of Transverse Ranges, Peninsular Ranges, more interior South Coast (such as near Redlands).

Zones 20, 21. Ocean-influenced Southwestern California

ZONE 20. Ocean-influenced basins and hilltops. Cold-air-collecting valleys near coast with a mixed influence of marine and continental air masses; boundaries move many miles with changing positions of high-pressure areas; summers cooler, moister, frosts less frequent than in Zone 18. South Coast, Transverse Ranges, Peninsular Ranges (e.g., vicinities of Ojai, Burbank, El Monte, Ramona).

Climate Zones of California

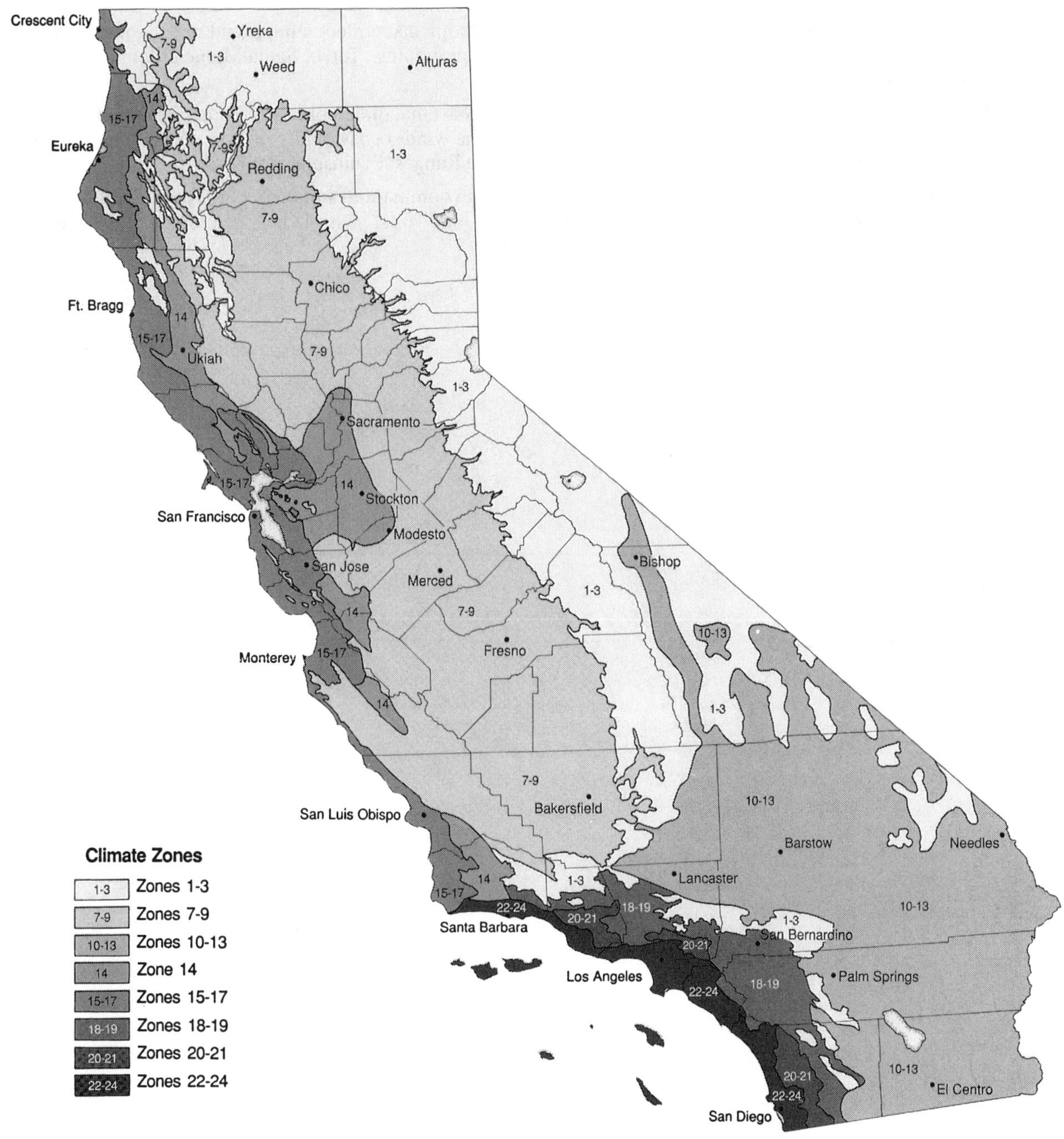

ZONE 21. Ocean-influenced thermal belts. Cold-air-draining slopes; frosts rare. More extensive than (but otherwise like) Zone 20. South Coast, Transverse Ranges, Peninsular Ranges (e.g., vicinities of Santa Paula, Pasadena, Covina, Escondido).

Zones 22–24. Coastal Climates of Southwestern California

ZONE 22. South Coast cold-winter areas. Basins and hilltops that collect winter cold air; marine influence predominates; harmful frosts only in canyons or near their mouths. Top of Santa Monica Mountains, much of Los Angeles/Orange Co. basin.

ZONE 23. South Coast thermal belts. Gentle slopes near coast that drain cold air; summer heat moderated by marine influence; essentially no frost; hot, dry Santa Ana winds occasionally reverse normal marine influence; coastal slopes of South Coast, western Transverse Ranges, Peninsular Ranges.

ZONE 24. South Coast marine belt. Almost complete ocean domination; summers fairly cool (5 degrees Celsius warmer than Zone 17), fog common; winters very mild (frost possible near canyon mouths). Coastal plain only, Point Conception to western Baja California.

GEOGRAPHIC SUBDIVISIONS OF CALIFORNIA

The function of providing geographic ranges in a botanical resource is to help the user predict where plant taxa can be expected to grow. Formulating such predictions is always a challenge — particularly so in California. The state is large (some of it still poorly explored botanically), has great topographic complexity, and contains an unusual diversity of climates and habitats. In an attempt to increase the predictiveness of range descriptions, *The Jepson Manual* departed from earlier manuals in the way geographic ranges are described.

The geographic system used combines features of the natural landscape. The most important features are broadly defined natural vegetation types (and, at a finer scale, more specific plant communities) and geologic, topographic, and climatic variation.

The most useful single resource in creating the system used was the work of Professor A. W. Kuchler, especially his "Natural Vegetation of California", a chapter (p. 909) and detailed map in *Terrestrial Vegetation of California*, edited by Professors Michael G. Barbour and Jack Major (reprinted in 1988 by the California Native Plant Society). Professors Peter Raven & Daniel Axelrod's 1978 *Origins and Relationships of the California Flora* (U California Publ Bot 72) was also important to the development of these ideas. These works are recommended for further information. They contain detailed definitions of plant communities and discuss their composition, architecture, history, and relationships to climate, geology, soils, fire, and human disturbance.

The geographic system of *The Jepson Manual* is organized hierarchically, starting with broadly defined provinces and ending with districts (sub-sub-subdivisions of provinces). Directional modifiers on the geographic units, e.g., "sw NCoRI", are used to improve precision. Such a system allows distribution ranges to be conveyed with clarity, conciseness, and varying levels of precision that suit the state of knowledge or need (see also "Range Descriptions", pp. 9, 10). Combining geographic range descriptions with habitat descriptions and elevation ranges increases the predictiveness of overall range statements.

There are 50 geographic units in this system. Each has a unique abbreviation, used in the text and keys. The abbreviations are as consistent, self-explanatory, and easy to master as they could be made. If a user already knows basic California geography, the units and their abbreviations will prove readily understood. If not, the system (with abbreviations) will be a significant help in learning California geography.

Each of the 50 units is defined and described below. These descriptions will be easier to follow if studied in conjunction with the map and hierarchical outline of subdivisions (inside covers and p. 45).

Emphasis on predictiveness and biological meaning in range descriptions automatically diminishes the importance of county lines, which are the primary alternative for subdividing the state. Counties were added to naturally defined areas when the political information was judged useful. This was often done for rare plants, because county-by-county information is often sought for them. For more details on relationships between geographic subdivisions and counties, see the cross-reference table prepared by Dr. Barbara Ertter (pp. 46-48).

The system of geographic units is four-tiered: provinces, regions, subregions, and districts (see outline, p. 44). There are three provinces at the most inclusive level. All three extend outside of California but the largest, the California Floristic Province, includes most of the state and only small parts of adjacent Oregon, Nevada, and Baja California, Mexico. The other two provinces are Great Basin and Desert.

Each province is subdivided into regions. The California Floristic Province is made up of six regions; in California, each of the other provinces have two. Together, these three provinces and ten regions delineate the broad physiographic and biologic geography of California. It is with them that anyone wishing to understand California plant distributions should begin. They are shown most clearly on the "exploded" inset map (inside covers and p. 45). Like California itself, nearly all of the units are elongate in a more or less north-south direction.

Nine of the ten regions are further divided, into a total of 20 subregions (1–4 subregions per region). Subregions are also defined on the basis of topographic, climatic, and plant-community variations within the region. Seven of the subregions are further divided into districts; variations among them are more localized.

In contrast to the use of biologically arbitrary, often politically determined delimiters, such as county lines, the use of biologically meaningful criteria to define subdivisions results in sometimes frustratingly indefinite or fuzzy lines. Wherever possible, subdivisions are defined on the basis of all three of the main biologically relevant variables: topography, climate, and plant-community types.

There are situations in which these three features do not vary in concert. Where they do not, plant-community type generally takes priority. For example, the grassland-covered, treeless hills that are geologically and topographically lower foothills of the Sierra Nevada support most of the plant communities of the adjacent Great Central Valley; they are considered part of the latter region, up to the point where oak/pine woodlands can sustain themselves.

There are other situations in which plant-community types vary gradually, and there is no basis for drawing a sharp line. Where this occurs, reasonable lines have been sought using a combination of geological and topographic criteria (e.g., the North Fork of the Feather River divides the Cascade Ranges from the Sierra Nevada, with volcanic substrates predominant on the Cascade side) or easy-to-follow, man-made divisions (State Highway 58 through Tehachapi Pass divides the Tehachapi Mountains from the southern Sierra Nevada Foothills).

California Floristic Province (CA-FP)

The largest and most important geographic unit in California is the California Floristic Province (CA-FP). It comprises all of the state west of the dry regions of Great Basin (GB) and Desert (D). CA-FP is equivalent to "cismontane", as used by Jepson, Munz, and others; GB and D together are equivalent to "transmontane". The border between west (CA-FP) and east (GB and D) is thus the main phytogeographic boundary in California.

North of Lake Tahoe, this boundary lies between the Cascade Ranges (CaR) and the Sierra Nevada (SN) (with their predominant montane pine/fir forests) and the Great Basin Province (Modoc Plateau Region, with its predominant juniper savanna and sagebrush steppe). Vegetational, topographic, and geologic boundaries are all indistinct in the north; there are inclusions of sagebrush steppe in the Cascades (especially the large inclusion in Shasta Valley in north-central Siskiyou County) and of montane forest at higher elevations in GB. In CaR, volcanic cones and mountains are more numerous, while GB is generally flatter country, with a greater predominance of lava flows that have been faulted into small mountains with intervening basins.

The boundary runs south from the Oregon border at US Highway 97, along the south side of Lava Beds National Monument, and around Glass Mountain and Black Mountain (barely into Modoc County), curves west again around the Burnt Lava Flow area, and (from the Shasta County border) essentially follows Highways 89, 44, 36 (through Susanville), and 395 south, along the northeastern base of the Diamond Mountains. There is a floristically interesting indentation of the boundary at Sierra Valley (Plumas and Sierra counties). CA-FP extends slightly into Nevada east of Lake Tahoe (e.g., Mount Rose area), with the boundary nearly following Highways 395 and 88.

South of Lake Tahoe, the boundary of CA-FP follows the east slope of the Sierra Nevada, generally defined by an indefinite break between either upper montane (red-fir/lodgepole-pine) forest or Jeffrey-pine forest on the CA-FP side and either juniper/pinyon woodland or sagebrush steppe on the GB side. There is also Jeffrey-pine forest in GB (e.g., Mono Craters area). In some places, highway 395 approximates the boundary; south of Bishop, the boundary lies to the west of that highway, farther up the east slope of the Sierra Nevada.

South of Owens Valley, CA-FP abuts Desert (D), and the boundary lies between the chaparral or pinyon/juniper woodland on the CA-FP side and Joshua-tree or creosote-bush scrub in D; the Little San Bernardino Mountains are considered D. Montane vegetation of the eastern Sierra Nevada, northern Transverse, and eastern Peninsular ranges tends to grade into desert vegetation on the lower slopes of these mountains. Some taxa are limited to this interface, which might be specified as "lower n slope SnBr" or "w edge DSon". In Riverside, San Diego, and southwesternmost Imperial counties, the Santa Rosa, Volcan, Laguna, and Jacumba mountains make up the eastern edge of CA-FP.

Outside California, CA-FP extends north as the Klamath Mountains of southwestern Oregon and south into northwestern Baja California. In California, it is divided into the following six regions, 17 subregions, and 17 districts.

NORTHWESTERN CALIFORNIA (NW).

This region has the most predictable (and wettest) climate in California. The boundary between NW and CaR, the two northernmost regions of CA-FP, is marked geologically by a mostly metamorphic (NW) vs volcanic (CaR) substrate interface and, less clearly, by vegetational differences: oak woodland or montane fir/pine forest with hemlock on the NW side and Sierran montane forest (sugar pine but no hemlock) or sagebrush scrub on the CaR side. This boundary is approximated by Interstate 5 (and the Sacramento River) south to the Great Valley (GV) region at about Redding.

From Redding south to southwestern Solano County, NW abuts GV and the boundary is defined by blue-oak/foothill-pine woodland on the NW side, with prairie (or agricultural land) on the GV side. From southwestern Solano County, the southern boundary of NW jogs westward along a vegetational boundary that excludes salt marsh, coastal prairie, and other maritime communities (see Central Western California), then proceeds through southern Sonoma County to the Pacific Ocean at Bodega Bay. NW is divided into three subregions.

North Coast (NCo). This subregion extends along the Pacific coast the full length of NW, south from the Oregon border to Bodega Bay. It is a variably wide strip that includes truly coastal communities: coastal prairie predominates, along with coastal marsh, closed-cone-pine/cypress forest, and grand-fir/Sitka-spruce forest.

Klamath Ranges (KR). The California portion of this geologically old and distinct subregion is bounded to the north by Oregon and in the northwest corner by the coastal communities of NCo. Its southwestern and southeastern boundaries are with the North Coast Ranges (NCoR). On the west, the boundary has a pronounced geological basis, with the mostly sedimentary Franciscan Complex of NCoR faulted against the older, metamorphic and plutonic rocks of KR. The fault boundary coincides almost exactly with the northwest-flowing Klamath and South Fork of the Trinity rivers.

On the east, the mostly metamorphic KR is bounded by the volcanic Cascade Ranges (CaR), mostly just west of Interstate 5. On the southeast, the boundary excludes the chaparral and pine/oak woodland communities of the Inner North Coast Ranges (NCoRI) in western Shasta and Tehama counties.

There is a gradual change across the KR/NCoR boundary in forest types: KR is characterized more by forests with hemlock, grand fir, or chinquapin, NCoR by forests without hemlock, with noble or red fir rather than grand fir, and with rhododendron rather than chinquapin. There is an abundance of serpentine throughout KR which includes the Marble, Salmon, Scott, Scott Bar, and Siskiyou mountains, the Trinity Alps, and Mount Eddy. The southernmost KR peak above 1500 m is Red Mountain near the Trinity-Shasta-Tehama three-county boundary.

North Coast Ranges (NCoR). NCoR is the largest subregion of NW. It does not support the strictly coastal communities of NCo and is separated from KR by the topography and gradual changes in predominant forest types given immediately above. NCoR is divided into three districts.

Outer North Coast Ranges (NCoRO). The largest district is characterized by redwood, mixed-evergreen, and mixed-hardwood forests and by very high rainfall.

High North Coast Ranges (NCoRH). This district is characterized by montane and subalpine-fir/pine forests, treeless high peaks, heavy snow cover, and floristic similarities to the high Sierra Nevada (SNH). Its major peaks all extend above 1500 m (most above 2000 m), from South Fork Mountain in Humboldt County southeast to the Yolla Bolly Mountains, then south to Goat Mountain (Colusa-Lake county line). Somewhat lower, more western, and more isolated peaks vegetationally similar to Southfork Mountain (e.g., Mount Lassic, Grouse Mountain, Horse Mountain) were included in NCoRO.

Inner North Coast Ranges (NCoRI). This district is characterized by chaparral and pine/oak woodland, with low rainfall and hot, dry summers. It ranges from about Anderson (sw Shasta County) south along the east slope of the NCoR, with a conspicuous westward bulge around the south end of NCoRH to west of the Russian River (from north of Ukiah to Mount St. Helena). NCoR serpentine is widespread, but heavily concentrated in this district.

CASCADE RANGES (CaR).

This volcanic region is bounded to the north by Oregon, to the west by the predominantly metamorphic NW (more or less along Interstate 5) and by NCoRI (along the Sacramento River between Redding and Red Bluff), to the southwest by agricultural land or California prairie of the Great Central Valley (vs chaparral and oak/ pine woodland of CaR Foothills), to the east by juniper savanna of the Great Basin (vs montane fir/pine forests of CaR, see CA-FP above), and to the southeast by the Sierra Nevada (SN).

The boundary between CaR and the Modoc Plateau of the Great Basin is especially unclear vegetationally. A major island of Great Basin communities (sagebrush steppe and juniper savanna) occurs in Shasta Valley (east of Yreka, near the boundary between CaR and KR).

The interface between CaR and the Sierra Nevada is defined geologically by the contact between relatively recent CaR volcanics and the predominant metamorphics (with both granitic intrusions and volcanics) of the northern Sierra Nevada. The contact is located slightly northwest of the canyon of the North Fork of the Feather River, which serves as a reasonable topographic marker. Although the geologic and topographic boundaries nearly coincide, there is no vegetational break in this area; rather, the forests of the Cascade-Sierran axis change gradually with latitude. CaR is divided into two subregions.

Cascade Range Foothills (CaRF). This subregion, in the southwestern part of CaR, is characterized by chaparral and blue-oak/foothill-pine woodland at about 100–500 m elevation. CaRF and adjacent Sierra Nevada Foothills and NCoRI make up a horseshoe of similar foothill communities around the northern and eastern sides of the Great Central Valley.

High Cascade Range (CaRH). This subregion (generally above 500 m) is characterized by ponderosa-pine, montane fir/pine, and lodgepole-pine forests, with treeless alpine communities on Mount Shasta and Lassen Peak.

SIERRA NEVADA (SN).

This primarily metamorphic region abuts volcanic CaR to the north, to the west shares a long border with GV (California prairie on GV side vs foothill communities on SN side), and meets the Southwestern California region (SW) at Tejon Pass. On the east, SN ends at the province boundaries with GB and D. SN is divided into three subregions. The two larger subregions are each further divided north-to-south into three districts. Communities change more or less gradually with latitude in SN; lines between districts were chosen to coincide with areas of more rapid floristic change and with major drainages or drainage boundaries.

Sierra Nevada Foothills (SNF). The lower (upper limit about 500–800 m), mostly narrow, western strip of the Sierra Nevada is adjacent on the west to the Great Central Valley (GV), on the east to High Sierra Nevada (SNH) or Desert (D), and on the south to the Tehachapi Mountains subregion (Teh). Throughout most of its extent, it is characterized by blue-oak/foothill-pine woodlands (vs ponderosa-pine forest of higher elevations in SNH) and is dotted with serpentine. It is divided into three districts from north to south. It is best differentiated from the High Sierra Nevada and Great Central Valley by community type rather than topography or map lines. Lines dividing the three north-south districts are given below.

Northern Sierra Nevada Foothills (n SNF). This district is adjacent to CaRF at its northern boundary (northwest of Oroville). Its arbitrary boundary in the south with c SNF is the Calaveras-Tuolumne county line, which corresponds with the Stanislaus River. Oroville, Auburn, and Placerville are all in or surrounded by n SNF. At about 800 m, Grass Valley is close to the border with n SNH.

Central Sierra Nevada Foothills (c SNF). This district is bounded in the south by the divide between the San Joaquin and Kings river drainages in Fresno County, which is approximated by Highway 168. Sonora, Incline, and Mariposa are all in c SNF.

Southern Sierra Nevada Foothills (s SNF). The boundary between this district and the Tehachapi Mountain subregion to the south is Highway 58, approximating the Tehachapi Creek/Cache Creek divide. The district runs the width of SN at its southern end. Like the Tehachapi Mountains (Teh), s SNF is complex and transitional to SNH, TR, GV, and D.

High Sierra Nevada (SNH). This large subregion, extending from Lassen and Plumas counties in the north to Kern County in the south, is characterized primarily by conifer forest above 500 m (vs chaparral and blue-oak/foothill-pine woodland below 800 m in SNF). Vegetation is complex, including lower montane ponderosa-pine, white-fir, and giant-sequoia forests, upper montane red-fir, Jeffrey-pine, and lodgepole-pine forests, and subalpine mountain-hemlock and whitebark-pine forests, with treeless alpine communities at the highest elevations (± 3000–4400+ m).

The long eastern border of SNH with Great Basin (GB) and Desert (D) (extending more than half the length of the state) is not everywhere easy to define (see CA-FP above). The subregion is divided (like SNF) somewhat arbitrarily into three districts from north to south; vegetational changes are gradual. As with SNF, only the northern and southern dividing lines are given here.

Northern High Sierra Nevada (n SNH). The northern boundary of SNH (within CA-FP) is with CaR, more or less following the orientation of the North Fork of the Feather River from northeastern Butte County to southwestern Lassen County. The southern boundary (with c SNH) follows the Calaveras-Tuolumne, Alpine-Tuolumne, and Alpine-Mono county lines to the border with the Great Basin. Quincy, Downieville, Truckee, and Markleeville are within the boundaries of the district.

Central High Sierra Nevada (c SNH). As for c SNF, the southern boundary (to the west of the Sierran crest) is the divide between the San Joaquin and Kings river drainages. This divide meanders to the south in eastern Fresno County, reaching the crest along the Goddard Divide, near Mount Darwin (4200 m). To the east of the crest, the boundary follows Bishop Creek down to the border with GB at ± 2000 m. Yosemite National Park and Mammoth Lakes are in this district.

Southern High Sierra Nevada (s SNH). In the northern part of the district are California's highest mountains, including 4400+ m Mount Whitney (all but the north tip of Kings Canyon National Park and all of Sequoia National Park are included here). Farther south, the "domelands" northeast of Lake Isabella are notable for their endemism; peaks average ± 3000 m. In the southernmost part of the district, peaks average 2000–2500 m; the convoluted southern boundary (with s SNF, not Teh) is vegetationally defined and relatively indistinct; the higher mountains (Piutes, Kiavahs, Scodies, Breckenridge Mountain) that support yellow or pinyon pines are included here, but not areas of oak/pine woodlands, chaparral, or desert scrub.

NOTE: Districts may also be combined by latitude, without regard to elevation. For example, a distribution in n SNF, n SNH, c SNF, and c SNH is cited "n&c SN".

Tehachapi Mountains (Teh). This small foothill and montane subregion is a floristic melting pot between SN, GV, TR, and D. Elevations rarely exceed 2000 m. Its boundary in the north with s SNF is along Highway 58 through Tehachapi Pass. In the west, it is bounded by the Great Central Valley (marked by prairie communities in the lower foothills vs mixed-woodland communities). Teh ends at the north end of the Western Transverse Ranges district (Tejon Pass on Interstate 5). The eastern boundary is the indistinct CA-FP boundary with Desert (D) (chaparral or pinyon/juniper woodland vs creosote-bush scrub).

GREAT CENTRAL VALLEY (GV).

Now predominantly agricultural, GV once supported grassland (California prairie), marshes, extensive riparian woodlands, and (especially in southern SnJV) islands of valley-oak savanna. It is surrounded by the other regions of CA-FP. On all borders (with NW, CW, SW, SN, and CaR), it ends where oak/pine woodlands or mixed-hardwood forests begin. It is divided into two subregions.

Sacramento Valley (ScV). ScV is the smaller, wetter, northern subregion of GV, extending from Red Bluff in Tehama County to the salt marshes of Suisun Slough in southwest Solano County. The boundary between ScV and the San Joaquin Valley (SnJV) follows the northern borders of Contra Costa and San Joaquin counties. To the west, this line approximately bisects the "delta" of the Sacramento and San Joaquin rivers. To the east, it separates Sacramento and San Joaquin counties (approximating the very low divide between the Cosumnes and Mokelumne river drainages).

San Joaquin Valley (SnJV). SnJv is the larger, southern subregion of GV; its extent is defined under GV and ScV. Hotter and drier than ScV, it supports some desert elements in the south. Some vegetationally similar outliers, such as the Carrizo Plain to the southwest, are included. (This inclusion creates islands of moister habitats above ± 800 m in the Temblor Range and on associated ridges; these islands belong to the Inner South Coast Ranges; see below.)

NOTE: Occurrences restricted to the "delta" of the Sacramento and San Joaquin rivers, but including parts of both ScV and SnJV are cited as "deltaic GV".

CENTRAL WESTERN CALIFORNIA (CW).

This region is bounded by the Pacific Ocean to the west, NCo and NCoRO to the north, GV to the east, and Southwestern California (SW) to the south. The boundary with SW follows the crest of the Santa Ynez Mountains from Point Conception to north of Santa Barbara, where it jogs northeast and east along Mono Creek and beyond, including most of the San Rafael Mountains but excluding Mount Pinos, which is in SW. It is divided into three subregions, one of which comprises two districts; all of these contain many scattered, often small outcrops of serpentine.

Central Coast (CCo). This subregion extends along the Pacific Ocean (and into San Francisco Bay) the entire extent of CW, from Bodega Bay in the north to Point Conception in the south. Like NCo, it is variable in width and supports only truly coastal communities. Coastal-sage scrub predominates toward the south; salt marshes and coastal prairie predominate around San Francisco Bay. The portion around Monterey Bay is notable for its endemism.

San Francisco Bay Area (SnFrB). This subregion occupies the northern one-third of CW, east of CCo. Reasonably defined physiographically, it includes Mount Tamalpais, the Santa Cruz Mountains, and the Diablo Range (including Mount Diablo and Mount Hamilton). The somewhat arbitrary southern boundary follows Highways 156 and 152 from CCo east of Castroville through Hollister and Pacheco Pass to GV near San Luis Reservoir. The subregion is less well defined vegetationally, containing a diversity of community types from wet redwood forest to dry oak/pine woodland and chaparral.

South Coast Ranges (SCoR). This subregion is bounded by CCo to the west, SnFrB to the north, SnJV to the east, and Southwestern California (SW) in the south. All these borders are described above. The subregion is divided into two districts, both of which contain many (generally small) outcrops of serpentine.

Outer South Coast Ranges (SCoRO). The district boundary with SCoRI runs up the Salinas River (approximated by Highway 101) to San Miguel in northern San Luis Obispo County, thence up the Estrella River to the western edge of SnJV near Shandon. The district includes Sierra de Salinas, Santa Lucia Mountains, and San Rafael Mountains to the crest of the Santa Ynez Mountains and Mono Creek. There are small stands of redwood and mixed-hardwood forest in the north near the coast, which give way to southern oak forest farther south. Hotter, more inland slopes support primarily blue-oak/foothill-pine woodland and chaparral.

Inner South Coast Ranges (SCoRI). This district includes the Diablo Range from Pacheco Pass to San Benito Mountain, Gabilan Range, Cholame Hills, and the higher elevations of the Temblor Range and associated ridges (isolated within the southern part of SnJV). It supports a mosaic primarily of summer-dry blue-oak/foothill-pine woodland and chaparral.

SOUTHWESTERN CALIFORNIA (SW).

This region is bounded by the Pacific Ocean (outside the Channel Islands) to the west and by Mexico to the south. It is separated from CW at the crest of the Santa Ynez Mountains and Mono Creek as described above; from GV at the grassland/woodland boundary; from Teh at Tejon Pass along Interstate 5; and from D on the northeast and east as defined under CA-FP above. The region is divided into four subregions and six districts.

South Coast (SCo). This subregion extends along the Pacific Coast from CW (at Point Conception) to Mexico. It is comparable to NCo and CCo of the NW and CW regions, respectively, but is hotter, drier, and extends much farther inland — to San Gorgonio Pass at Banning, the CA-FP/D boundary. Coastal-sage scrub and chaparral communities with many endemics would predominate, but most of the subregion from Santa Barbara to the Mexican border has been urbanized, with great loss of undisturbed habitat.

Channel Islands (ChI). The eight major islands off SCo are floristically similar to SCo, but include enough endemics to justify separate recognition. The subregion is divided into two districts.

Northern Channel Islands (n ChI). This district includes San Miguel, Santa Rosa, Santa Cruz, and Anacapa islands, which are separated from the mainland by the Santa Barbara Channel. Geologically, they are western peaks of the Santa Monica Mountains, which are in the southern part of the Western Transverse Range district (see below).

Southern Channel Islands (s ChI). This district includes Santa Barbara, Santa Catalina, San Clemente, and San Nicolas islands, which are geologically and floristically more isolated and more diverse among themselves than the northern group. They were not so readily colonized from the mainland during glacial times of low sea levels as was the northern group of islands.

Transverse Ranges (TR). This is a subregion of abnormally oriented (east-west) mountains (see California's Geological History and Changing Landscapes, p. 49). TR is separated from the Peninsular Ranges (PR) at San Gorgonio Pass — an essentially four-way boundary (on Interstate 10 at Banning) between the San Bernardino Mountains, San Jacinto Mountains, South Coast, and Desert. The northern boundary with Desert (D) is discussed under CA-FP. TR is characterized at lower elevations by chaparral and at higher elevations by southern oak forest and dry montane forests of white fir or Jeffrey, sugar, or lodgepole pines. Some high peaks are treeless, but apparently none exceed climatic timberline at their latitude. TR is divided into three districts that are progressively higher, hotter, and drier to the east.

Western Transverse Ranges (WTR). This district includes Mount Pinos (2700 m — the highest point), the Santa Ynez, Topatopa, Santa Susanna, Santa Monica, and Liebre mountains, and Sierra Pelona. At the north end of the San Fernando Valley, the topographic boundary with the San Gabriel Mountains (SnGb) follows Interstate 5 north to the Santa Clara River, then east through Soledad Canyon and Soledad Pass to the margin of D south of Palmdale.

San Gabriel Mountains (SnGb). This is a topographically well defined range northeast of Los Angeles. It is separated from the San Bernardino Mountains by the southeast trending Cajon Canyon (occupied by Highways 138 and 15). The highest point, which straddles the Los Angeles-San Bernardino county line, is Mount San Antonio ("Old Baldy", 3070 m), which has a few alpine plants near the summit.

San Bernardino Mountains (SnBr). This is another topographically well defined range. It is bordered by D, SCo and SnGb. The highest point is San Gorgonio Mountain (3500 m), which has the best developed alpine communities south of SN in California. The rapidly developing Big Bear/Lake Baldwin area is notable for its endemism. [The Little San Bernardino Mountains to the east (separated from SnBr by Morongo and Yucca valleys and mostly included in Joshua Tree National Monument) are here considered Desert Mountains (DMtns) because the least desert-like community they support is pinyon/juniper woodland.]

Peninsular Ranges (PR). This subregion includes all of mainland SW exclusive of SCo and south of TR. It includes the Santa Ana, Cuyamaca, Santa Rosa, Laguna, and Jacumba mountains, Mount Palomar, and the separable San Jacinto Mountain district.

San Jacinto Mountains (SnJt). SnJt is the highest range in PR (San Jacinto Peak: 3300 m) and the only one to which many taxa are restricted. A few alpine species grow at the highest elevations.

Great Basin Province (GB)

The Great Basin Province lies to the east of CA-FP in the northern two-thirds of California and abuts D at its southern margin. The boundary with CA-FP is described above; it follows the high eastern margin of CaR and SN. The boundary with D in the north is the transition from sagebrush steppe or pinyon/juniper woodland (GB) to creosote-bush scrub (D). Deep Springs and Fish Lake valleys are in GB; Eureka and Saline valleys are in D. Southward, the mixed vegetation of the Owens Valley is included in GB. (These boundaries make the area covered in Munz's *A Flora of Southern California* identical to SW + Teh + D). Rainfall is low in GB and summers are hot to very hot. GB is divided into two regions and two subregions.

MODOC PLATEAU (MP).

This region (entirely north of Lake Tahoe) is a high plateau (mostly ± 1300–1800 m) in the northeast corner of the state, occupying most of Modoc and Lassen counties. MP is characterized primarily by juniper savanna and sagebrush steppe, but also has extensive areas of ponderosa-pine and Jeffrey-pine forests and some montane pine/fir forest. Substrates are volcanic, with faulted flows predominating over cones (see CaR above).

Warner Mountains (Wrn). This faulted volcanic range (mostly in eastern Modoc County) is the outstanding topographic feature of MP. Eagle Peak, the highest point, exceeds 3000 m. Wrn is recognized as a subregion because it supports a unique flora that includes alpine communities.

EAST OF SIERRA NEVADA (SNE).

This region (entirely south of Lake Tahoe) includes a wide range of elevations, from Owens Lake at 1100 m to White Mountain Peak at 4330 m. SNE (excluding the high White and Inyo mountains separated below) supports primarily a mosaic of sagebrush steppe, pinyon/juniper woodland, and riparian cottonwood-dominated communities. There are also extensive areas of Jeffrey-pine forest in the Mono Craters area, Sierran subalpine- fir/pine forest on Glass Mountain (3400 m), and alpine communities at the top of the Sweetwater Mountains (3550 m). SNE extends along the eastern edge of SN to the southern limit of Owens Valley and the Inyo Mountains, where there is a gradual transition to Mojave Desert creosote-bush scrub. To the east of the junction of the White and Inyo mountains at Westgard Pass lies a low (1500–2000 m) outlier of SNE that includes the Deep Springs and Fish Lake valleys.

White and Inyo Mountains (W&I). These ranges are considered a separate subregion because they support subalpine bristlecone-pine and limber-pine communities as well as unique, treeless, alpine communities. (White Mountain Peak 4330 m; Inyo and Waucoba peaks both ± 3400 m).

Hierarchical Outline of Geographic Subdivisions

CA-FP

California Floristic Province

- **NW** Northwestern California
  - **NCo** North Coast
  - **KR** Klamath Ranges
  - **NCoR** North Coast Ranges
    - **NCoRO** Outer North Coast Ranges
    - **NCoRH** High North Coast Ranges
    - **NCoRI** Inner North Coast Ranges
- **CaR** Cascade Ranges
  - **CaRF** Cascade Range Foothills
  - **CaRH** High Cascade Range
- **SN** Sierra Nevada
  - **SNF** Sierra Nevada Foothills
    - **n SNF** northern Sierra Nevada Foothills
    - **c SNF** central Sierra Nevada Foothills
    - **s SNF** southern Sierra Nevada Foothills
  - **SNH** High Sierra Nevada
    - **n SNH** northern High Sierra Nevada
    - **c SNH** central High Sierra Nevada
    - **s SNH** southern High Sierra Nevada
  - **Teh** Tehachapi Mountains
- **GV** Great Central Valley
  - **ScV** Sacramento Valley
  - **SnJV** San Joaquin Valley
- **CW** Central Western California
  - **CCo** Central Coast
  - **SnFrB** San Francisco Bay Area
  - **SCoR** South Coast Ranges
    - **SCoRO** Outer South Coast Ranges
    - **SCoRI** Inner South Coast Ranges
- **SW** Southwestern California
  - **SCo** South Coast
  - **ChI** Channel Islands
    - **n ChI** northern Channel Islands
    - **s ChI** southern Channel Islands
  - **TR** Transverse Ranges
    - **WTR** Western Transverse Ranges
    - **SnGb** San Gabriel Mountains
    - **SnBr** San Bernardino Mountains
  - **PR** Peninsular Ranges
    - **SnJt** San Jacinto Mountains

GB

Great Basin Province

- **MP** Modoc Plateau
  - **Wrn** Warner Mountains
- **SNE** East of Sierra Nevada
  - **W&I** White and Inyo Mountains

D

Desert Province

- **DMoj** Mojave Desert
  - **DMtns** Desert Mountains
- **DSon** Sonoran Desert (also known as Colorado Desert)

Geographic Subdivisions of California

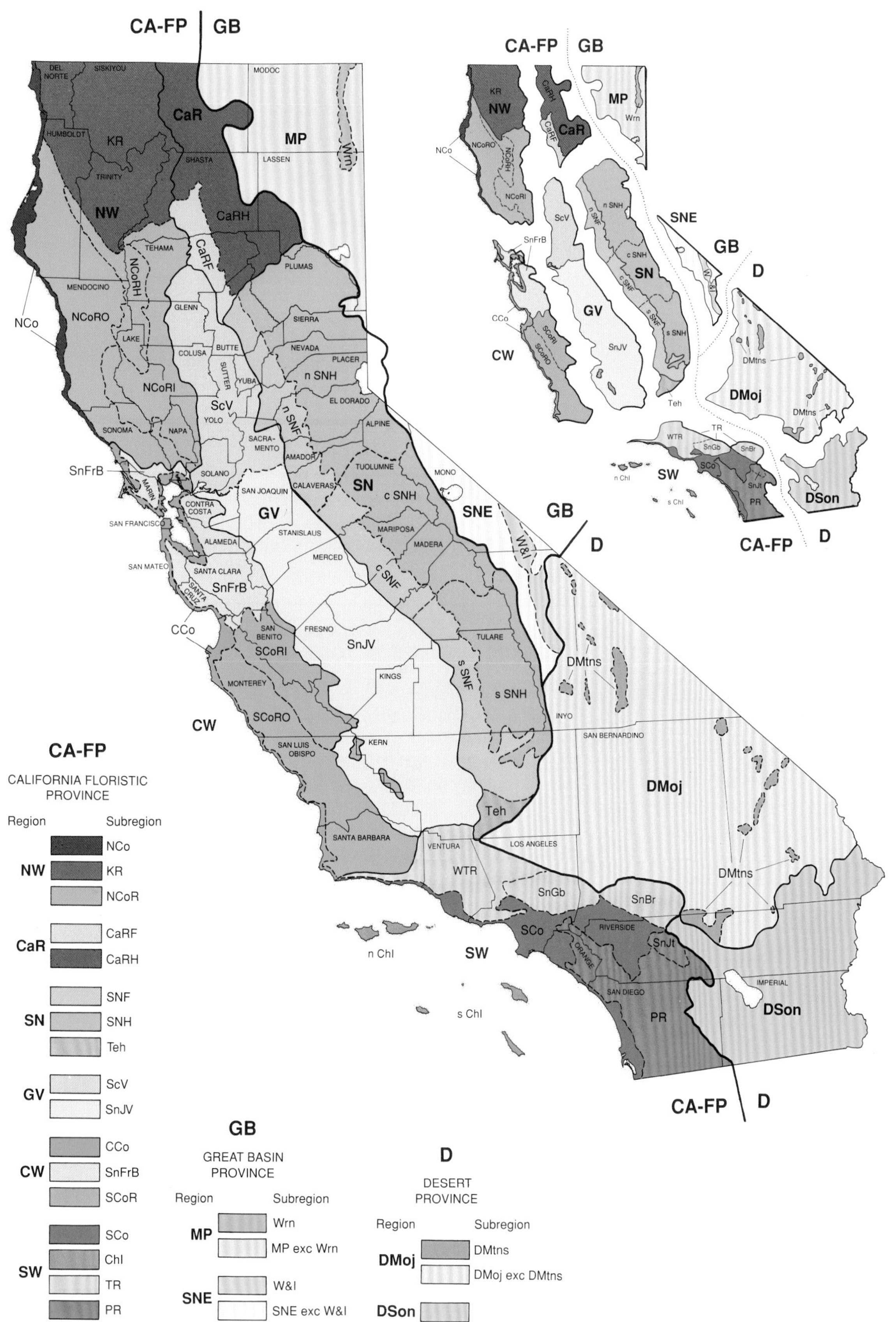

Desert Province (D)

The Desert Province makes up the southeastern portion of California. As discussed above, this province lies east of CA-FP and south of GB. Its climates are unpredictable from year to year. Creosote-bush scrub is characteristic throughout, though there has been much degradation from military activity and other off-road vehicles. Military reservations have also been responsible for preserving some desert communities more or less intact. The province is divided into two regions and one subregion.

MOJAVE DESERT (DMoj).

This region occupies the northern two-thirds of D; it has greater temperature ranges and more diverse elevations than the Sonoran Desert to the south. Mojave creosote-bush scrub is the dominant vegetation type, with saltbush scrub characteristic of alkaline basins.

Desert Mountains (DMtns). Although the entire DMoj is a series of mountains and intervening (often wide) valleys, some ranges reach sufficient elevation to support pinyon/juniper woodland and are therefore recognized as a distinct subregion. They include the Last Chance, Grapevine, Panamint, Coso, Argus, Kingston, Clark, Ivanpah, New York, Providence, Granite, Old Woman, and Little San Bernardino mountains. Four of these (Panamint, Kingston, Clark, and New York mountains) also support white-fir or limber-pine forest at their tops. DMtns have unique elements but also overlap floristically with the pinyon/juniper woodlands of adjacent CA-FP. Some of the eastern ranges support D or GB taxa otherwise unknown in California.

SONORAN DESERT (DSon).

This region lies to the south of DMoj. The physiographic line separating the two desert regions is not always clear, but overall DSon is lower, flatter, warmer, and supports a somewhat different flora. Dominant vegetation is Sonoran creosote-bush scrub.

Cross-Reference Between Geographic Subdivisions and Counties

Prepared by Barbara Ertter

Subdivision:	*Includes parts of these counties:*
CA-FP	
NW	
NCo	Del Norte, Humboldt, Mendocino, Sonoma
KR	Del Norte, Humboldt, Shasta, Siskiyou, Trinity
NCoR	
NCoRO	Humboldt, Lake, Mendocino, Napa, Solano, Sonoma, Trinity
NCoRH	Colusa, Glenn, Humboldt, Lake, Mendocino, Tehama, Trinity
NCoRI	Colusa, Glenn, Lake, Mendocino, Napa, Shasta, Solano, Sonoma, Tehama, Yolo
CaR	
CaRF	Butte, Shasta, Tehama
CaRH	Butte, Lassen, Modoc, Plumas, Shasta, Siskiyou, Tehama
SN	
SNF	
n SNF	Amador, Butte, Calaveras, El Dorado, Nevada, Placer, Sacramento, Yuba
c SNF	Fresno, Madera, Mariposa, Tuolumne
s SNF	Fresno, Kern, Tulare
SNH	
n SNH	Alpine, Amador, Butte, Calaveras, El Dorado, Lassen, Nevada, Placer, Plumas, Sierra, Yuba
c SNH	Fresno, Inyo, Madera, Mariposa, Mono, Tuolumne
s SNH	Fresno, Inyo, Kern, Tulare
Teh	Kern
GV	
ScV	Butte, Colusa, Glenn, Placer, Sacramento, Solano, Sutter, Tehama, Yolo, Yuba

SnJV	Alameda, Contra Costa, Fresno, Kings, Kern, Madera, Mariposa, Merced, Monterey, San Benito, San Joaquin, San Luis Obispo, Santa Barbara, Stanislaus, Tulare
CW	
CCo	Alameda, Contra Costa, Marin, Monterey, Napa, San Francisco, San Luis Obispo, San Mateo, Santa Barbara, Santa Clara, Santa Cruz, Solano, Sonoma
SnFrB	Alameda, Contra Costa, Marin, Monterey, Napa, San Benito, San Mateo, Santa Clara, Santa Cruz, Sonoma
SCoR	
SCoRO	Monterey, San Luis Obispo, Santa Barbara
SCoRI	Fresno, Kern, Kings, Merced, Monterey, San Benito, San Luis Obispo
SW	
SCo	Los Angeles, Orange, Riverside, San Bernardino, San Diego, Santa Barbara, Ventura
ChI	
n ChI	Santa Barbara, Ventura
s ChI	Los Angeles, Ventura
TR	
WTR	Kern, Los Angeles, Santa Barbara, Ventura
SnGb	Los Angeles, San Bernardino
SnBr	Riverside, San Bernardino
PR	Imperial, Los Angeles, Orange, Riverside, San Bernardino, San Diego
SnJt	Riverside
GB	
MP	Lassen, Modoc, Plumas, Shasta, Sierra, Siskiyou
Wrn	Lassen, Modoc
SNE	Alpine, Inyo, Mono
W&I	Inyo, Mono
D	
DMoj	Inyo, Kern, Los Angeles, Riverside, San Bernardino
DMtns	Inyo, Riverside, San Bernardino
DSon	Imperial, Riverside, San Diego

County:	*Includes parts of these subdivisions* (in **CA- FP** unless specified):
Alameda	GV (nw SnJV); CW (n CCo; e SnFrB)
Alpine	SN (n SNH); **GB** (SNE)
Amador	SN (n SNF; n SNH)
Butte	CaR (CaRF; CaRH); SN (n SNF; n SNH); GV (ScV)
Calaveras	SN (n SNF; n SNH)
Colusa	NW (NCoR incl s NCoRH, NCoRI); GV (ScV)
Contra Costa	GV (nw SnJV, deltaic GV); CW (n CCo; e SnFrB)
Del Norte	NW (NCo; KR)
El Dorado	SN (n SNF; n SNH)
Glenn	NW (NCoR incl NCoRH, NCoRI); GV (n ScV)
Fresno	GV (SnJV); CW (SCoR: SCoRI only); SN (c SNF, s SNF, c SNH, s SNH)
Humboldt	NW (n NCo; w KR; NCoR incl n NCoRO, n NCoRH)
Imperial	SW (se PR); **D** (DSon)
Inyo	SN (n&c SNH); **GB** (SNE incl W&I); **D** (n DMoj incl n DMtns)

Kern	SN (s SNF; Teh; s SNH); GV (s SnJV); CW (SCoR: s SCoRI only); SW (TR: n WTR only); **D** (w DMoj)
Kings	GV (SnJV); CW (SCoR: SCoRI only)
Lake	NW (NCoR incl s NCoRO, s NCoRH, NCoRI)
Lassen	CaR (s CaRH); SN (n SNH); **GB** (MP incl s Wrn)
Los Angeles	SW (c SCo; ChI: s ChI only, incl Santa Barbara, Santa Catalina, San Clemente islands; TR incl e WTR, SnGb; nw PR); **D** (w DMoj)
Madera	SN (c SNF; c SNH); GV (e SnJV)
Marin	CW (n CCo; nw SnFRB)
Mariposa	SN (c SNF; c SNH); GV (e SnJV)
Mendocino	NW (c NCo; c NCoR incl c NCoRO, NCoRH, NCoRI)
Merced	GV (w SnJV); CW (se SnFrB; ne SCoRI)
Modoc	CaR (e CaRH); **GB** (MP incl Wrn)
Mono	SN (c SNH); **GB** (SNE incl n W&I)
Monterey	GV (sw SnJV); CW (c CCo; s SnFrB; n SCoRO; w SCoRI)
Napa	NW (NCoR incl s NCoRO, s NCoRI); CW (n CCo; n SnFrB)
Nevada	SN (n SNF; n SNH)
Orange	SW (c SCo; nw PR)
Placer	SN (n SNF; n SNH); GV (e ScV)
Plumas	CaR (s CaRH); SN (n SNH); **GB** (s MP)
Riverside	SW (e SCo; TR: s SnBr only; n PR incl SnJt); **D** (DMoj incl s DMtns; DSon)
Sacramento	SN (n SNF); GV (s ScV, deltaic GV)
San Benito	GV (w SnJV); CW (s SnFrB; n SCoRI)
San Bernardino	SW (e SCo; TR incl e SnGb, SnBr; nw PR); **D** (DMoj incl DMtns; DSon)
San Diego	SW (s SCo; c&s PR); **D** (w DSon)
San Francisco	CW (n CCo; w SnFrB)
San Joaquin	GV (n SnJV, deltaic GV); CW (e SnFrB)
San Luis Obispo	GV (sw SnJV); CW (s CCo; c&s SCoRO; s SCoRI)
San Mateo	CW (n CCo; w SnFrB)
Santa Barbara	GV (sw SnJV); CW (s CCo; s SCoRO); SW (n SCo; ChI: n ChI only, incl San Miguel, Santa Rosa, Santa Cruz islands; TR: w WTR only)
Santa Clara	CW (n CCo; se SnFrB; ne SCoRI)
Santa Cruz	CW (c CCo; sw SnFrB)
Shasta	NW (se KR; NCoR: n NCoRI only); CaR (CaRF, CaRH); **GB** (w MP)
Sierra	SN (n SNH); **GB** (s MP)
Siskiyou	NW (n KR); CaR (n CaRH); **GB** (w MP)
Solano	NW (NCoR, incl s NCoRO, s NCoRI); GV (s ScV, deltaic GV); CW (n CCo)
Sonoma	NW (s NCo; NCoR, incl s NCoRO; s NCoRI); CW (n CCo; n SnFrB)
Stanislaus	GV (n SnJV); CW (e SnFrB)
Sutter	GV (ScV)
Tehama	NW (NCoR incl NCoRH, n NCoRI); CaR (CaRF; s CaRH); GV (n ScV)
Tulare	SN (s SNF; s SNH); GV (s SnJV)
Tuolumne	SN (c SNF; c SNH)
Trinity	NW (KR; NCoR incl n NCoRO, n NCoRH)
Ventura	SW (SCo; ChI incl n ChI Anacapa Island, s ChI San Nicolas Island; TR: WTR only)
Yolo	NW (s NCoRI); GV (ScV)
Yuba	SN (n SNF; n SNH); GV (e ScV)

CALIFORNIA'S GEOLOGICAL HISTORY AND CHANGING LANDSCAPES

Jeffrey P. Schaffer

California has long been a place of dramatic geological change, some of it explosive and wrenching, some gradual and apparently benign. The state we see today is not what always was. Highly complex processes built California into the topographically and climatically diverse place we know. Those processes, together with migration, evolution, and fluctuations in abundance by living plants, have created the patterns of plant distribution and diversity to which this book is devoted.

This essay summarizes the geological history of California since the origin of the flowering plants about 160 million years ago (mya). An understanding of the origins of California's topography is central to understanding the patterns of plant distribution we see today.

Some scientists have tried to deduce past climates and topography from fossil evidence. However, there can be problems with this method because it assumes that past climatic gradients were identical to current ones, that climatic tolerances of plants have not changed with time, that species migrate together, and that plant distributions respond to average environments more than to infrequent, extreme conditions. Therefore, this summary relies more on geological, plate-tectonic, and geophysical evidence than on paleontological evidence.

Particularly because of the relatively recent maturation of the discipline of plate tectonics (the study of the earth's moving lithospheric units), there is much controversy about the details of the geological history of California. The most careful scholars disagree on details of interpretation; their views may change rapidly with new evidence. The ideas in this general review have been gathered from many places that are unreferenced here. Several hundred scientific works could be cited; for those wishing further access to this literature, the following works are reasonable places to begin: Ernst (editor), 1988, *Metamorphism & Crustal Evol of w US*, Prentice- Hall; Bally & Palmer (editors), 1989, *Geol of N.Am: Overview*, Geol Soc Amer, Boulder, Colorado.

Basic Concepts

The earth's surface is made up of lithospheric plates (each composed of crust and underlying upper mantle) that have several possible relative motions. They can spread away from one another, slide past one another, or converge so that they are compressed and wrinkled, or so that one is forced beneath the other — a process called subduction. When subduction occurs, the descending plate tends to melt, the molten rock rising and possibly reaching the surface to cause volcanic eruptions. Subduction has been particularly important in forming the landscapes we now know as California. Today, the Juan de Fuca Plate is being subducted beneath northern California (bringing about volcanic activity in the Cascade Range), while the Pacific Plate is sliding northward past the North American Plate, their surface boundary being the San Andreas fault.

California has not stayed in one place on the earth's surface. During the Late Jurassic, about 160 mya, what we know as Yosemite National Park lay about 650 km northeast of today's Puerto Rico. As the North American Plate grew westward (by plate spreading in the mid-Atlantic Ocean), California traveled with it in a curved line. It reached its northernmost point (near current South Dakota) about 70 mya. By then, all major modern plant groups had evolved and were diversifying.

There are units smaller than plates that are also important in California's geological history. These units are fault-bounded, regional assemblages of rocks with unique geological histories called "terranes". They may break apart to form smaller terranes or become attached (accreted) to continental margins. They may also amalgamate into larger units called "superterranes". Today's California is made up of about 65 terranes.

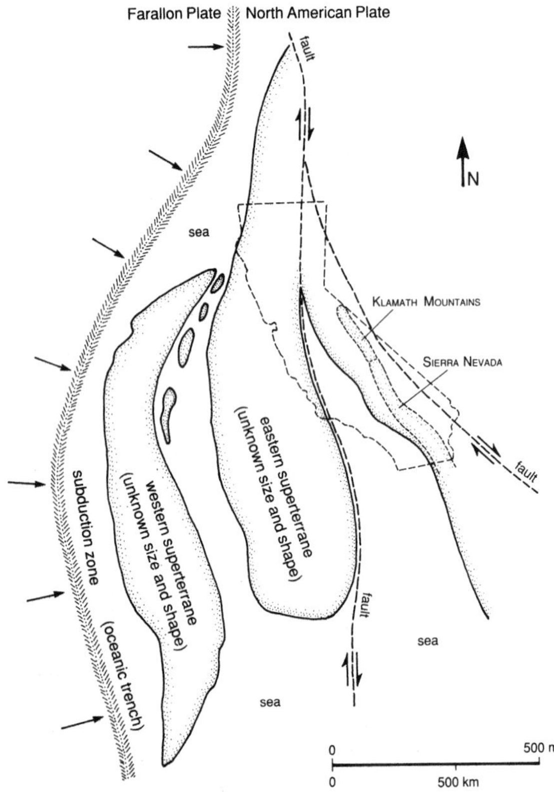

Figure 1. California and two offshore superterranes about 120–100 million years ago (mid-Cretaceous time). Both superterranes were migrating north. The ancestral Klamath and Sierra Nevada ranges were also migrating north with respect to the rest of North America. (The current Transverse and Peninsular ranges lie south of California.)

Mesozoic Era (245 to 65 mya)

(Triassic, Jurassic, and Cretaceous Periods)

In Jurassic time, California (then consisting only of its current eastern part) was mostly a low landscape on the western margin of the westward-growing North American Plate. Late in this geological period, about 160 mya, two above-water superterranes of the subducting oceanic plate to the west compressed against the western landmasses of the North American Plate and initiated a period of mountain building by compressive faulting that lasted well into the Cretaceous Period. Uplift during this time, called the "Nevadan Orogeny", formed an ancestral Sierra Nevada-Klamath Range of unknown height. The overall climate of western North America was then tropical but, if the range were high enough, montane species of some ancient plant groups may have evolved.

During a part of the Cretaceous Period from 120–100 mya, the Sevier Orogeny resulted in the uplift of a mountain range in eastern Nevada and western Utah that rivaled today's Andes in height. Alpine environments probably existed. At this time, the ancestral Sierra Nevada-Klamath Range was capped by volcanoes, associated lava flows, and sediments. Large calderas, similar to the one that makes up most of Yellowstone National Park, developed. Figure 1 shows the region of California as it may have existed during this period.

From Cretaceous time onward, oceanic plates were subducted at northeastward angles to the North American Plate, rather than head-on. A result was that oceanic terranes originating south of California drifted northward — some submerged, some as islands or as adjoining landmasses. These terranes eventually became part of western Canada and Alaska, and the emergent ones may have served as vehicles for the long-range dispersal of some plants.

Terrane accretion actually built most of California. At the start of the Mesozoic Era about 245 mya, California consisted only of a land mass where the Mojave Desert now is. By the close of the era (about 65 mya), most of California had been accreted. This includes the Coast, Transverse, and Peninsular ranges, which were attached by about 24 mya, at the beginning of the Neogene Period, but later would move northward.

Paleogene Period (65 to 24 mya)

(Paleocene, Eocene, and Oligocene Epochs)

Figure 2 shows California at about 60 mya, just before the accretion of terranes that would become the Coast and western Transverse ranges. The Klamath Ranges had already separated from the Sierra Nevada. As their westward migration continued, an ancestral, volcanic Cascade Range developed in their previous position.

This period was a time of relatively low relief, and California probably had significantly fewer habitats than it does today. The Cretaceous Sierra Nevada had been extensively eroded so that its western part lay near sea level; there, gold-bearing gravels were deposited (mostly in the more northern foothills). Fossils from these gravels show that the climate had changed from tropical to temperate. Forests included many genera of trees characteristic of North America today. The coastline, roughly where the Great Central Valley would eventually emerge, varied greatly through time; offshore terranes, including some islands, drifted north along a major fault. These terranes may have served as centers of plant dispersal.

The Paleogene climate was not always mild. The Eocene Epoch (58 to 35 mya) saw many climate changes, including drier climates and one pulse of cold temperatures that led to major worldwide extinctions. Midway through this epoch, Antarctica became covered by ice sheets; consequently, by 24 mya the world climate had cooled significantly. Still, California's climate was milder and more equable than it is today.

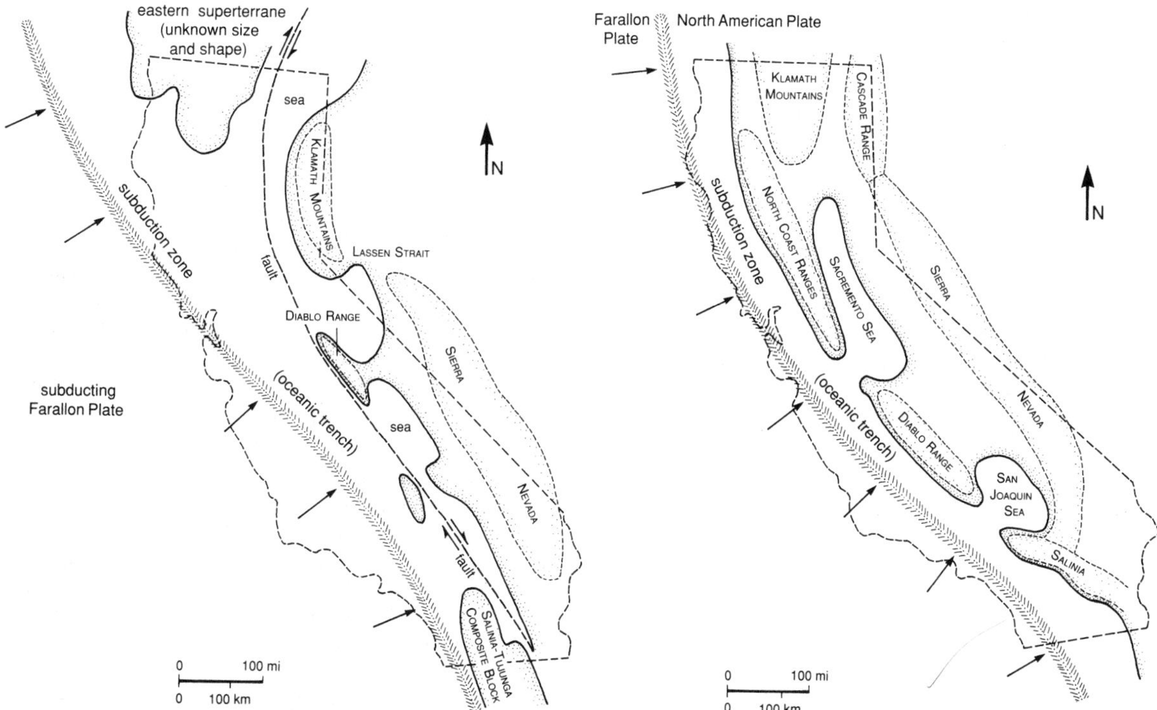

Figure 2. California about 60 million years ago (Paleocene time). The coastline is generalized. By the end of Eocene time (about 35 million years ago), the Klamath Ranges had moved westward close to their current position, and the Salinia-Tujunga Composite Block had become attached to southern California.

Figure 3. California about 30 million years ago (mid-Oligocene time), just before initiation of the San Andreas Fault system. An early Cascade Range had just formed (perhaps similar to today's), and volcanism extended through the northern half of the Sierra Nevada. The low Sierra Nevada was just beginning to rise.

The end of the Paleogene Period marks the inception of the modern California landscape (Figure 3); many changes occurred in close succession. There then existed low, ancestral forms of the mostly complete North Coast Ranges, the Diablo Range, and the Santa Lucia Range. Furthermore, the shallow sea that had covered the site of the Great Central Valley had diminished to two smaller, inland seas, the Sacramento and the San Joaquin. The Tehachapi Mountains, which are a southern extension of the Sierra Nevada, had rotated clockwise to their current position by this time.

Lands east of the Sierra Nevada began to stretch apart and rift; the northern half of the range was blanketed with volcanic flows and ash beds. As north-trending faults developed in the Great Basin (transforming a plateau into a basin-and-range landscape), the Sierra Nevada slowly began to rise.

Neogene Period (24 to 1.6 mya)

(Miocene and Pliocene Epochs)

San Andreas Fault System. Major landscape changes continued. By total subduction of the narrow middle part of the fracturing, shrinking, oceanic Farallon Plate, a precursor to today's San Andreas Fault formed just off the California-Mexico coast, and the Pacific Plate directly abutted the North American Plate for the first time. This fault lengthened with time, causing the remaining pieces of the former Farallon Plate (the Juan de Fuca Plate to the north and the Rivera Plate to the south) to migrate farther apart along an ancestral fault system.

With time (about 16 mya), the fault moved inland by cutting through coastal land, but the San Andreas fault proper, which developed farther inland, did not originate until 12 mya. Its southern extension, which created the Gulf of California, did not originate until 5.5 mya; some parts of this fault system are even younger.

This landward movement of the contact between the North American and Pacific plates transferred land from the North American Plate to the Pacific Plate, which is moving northward. Parallel faults (e.g., Hayward,

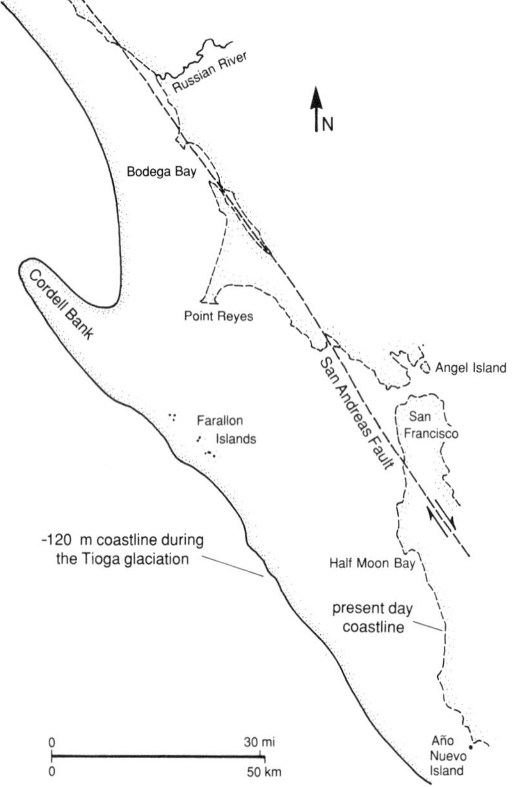

Figure 4. California coastline off San Francisco about 20,000 years ago, at height of last (Tioga) glaciation when sea level was about 120 m lower than today. Land submerged today was then part of a vast coastal plain that extended beyond the Farallon Islands.

Calaveras) also displaced land northward at a lesser rate; the continuing process is slowly deforming California. Local areas of compression and extension around these faults created new lowland habitats — hills, valleys, saline troughs, and new exposures of serpentine rocks. River drainages changed, possibly providing new paths for plant dispersal.

Klamath and eastern mountains, desert. Unrelated to (but simultaneous with) the development of the San Andreas fault system, the Klamath Ranges began another westward migration, which continues today. They experienced several pulses of uplift during the same general period that resulted in massive volcanism in the Cascades and in the northern Sierra Nevada. Like earlier giants, Mount Shasta and Lassen Peak became towering volcanoes.

The rate of uplift of the Sierra Nevada increased. From an elevation of about 1000–1500 m at 25 mya, the highest peaks had reached about 2500 m by 10 mya. Still, there was little rain shadow, because coast ranges were low and lands east of today's Sierran crest had not sunk from down-faulting. By the end of the Neogene, however, the highest Sierran peaks were about 3500 m (they are 4000+ m today). Sporadic cooling and drying episodes at that time (preceding the true ice ages), may have permitted the development of true alpine vegetation in both the Cascades and Sierra Nevada.

East of the Sierra Nevada, a process of extension and subsidence between adjacent faults (beginning 10 to 6 mya) led to the creation of Death Valley, Owens Valley, and others and broke up the area of the MojaveDesert, which formerly was a relatively flat plain perhaps 1000 m in elevation. There, a jumble of small mountain ranges developed. The Mono Basin began to form later (3–4 mya), completing separation of the Sierra Nevada from the White and Inyo Mountains. Just before faulting, the White Mountains were at about their current height of 4300 m, which was probably high enough to support subalpine and alpine communities. As the eastern land mass stretched and sank along faults and the Sierra Nevada continued to rise, the modern Sierran crest came into existence and the rain shadow effect increased. Fault-sunk basins with no natural drainage outlets provided new kinds of wet and saline habitats.

Great Central Valley, southern and western mountains. As the Sierra Nevada rose, erosion increased, cutting today's deep canyons and depositing vast quantities of sediments in the area of the Great Central Valley. The shallow Sacramento Sea filled with sediments a few million years before the deeper San Joaquin Sea. Even then, the San Joaquin basin continued to drain westward into the short but lengthening Salinas trough, which was being transformed from a shallow sea to a coastal valley by terrane transport along the San Andreas fault system.

At the beginning of the Neogene period, the South Coast Ranges, shorter and wider than at present, occupied the space now taken by the western Transverse Ranges. These, in turn, occupied the current space of the northern Peninsular Ranges. All had north-south alignments. As the early San Andreas Fault system moved farther inland in this region, lands to the west sank beneath the sea and the southern California coastline moved eastward.

From 16–12 mya, the still low South Coast and Transverse ranges moved northward along faults, and southern California's major sedimentary basins (such as the Los Angeles Basin) were formed. During the process, the Transverse Ranges were themselves broken up and rotated by a series of faults. Rotation continued after the tearing of Baja California from mainland Mexico 5.5 mya by the southward lengthening of the San Andreas fault system; as their name implies, today the Transverse Ranges have an anomalous east-west orientation.

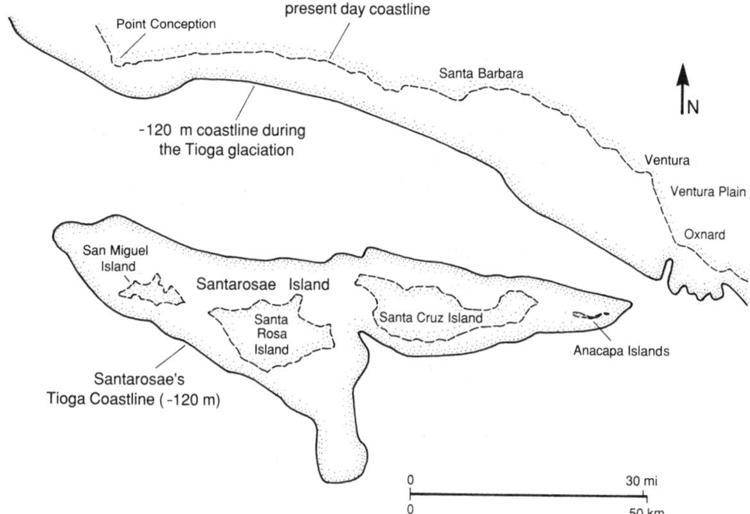

Figure 5. California coastline in the Channel Islands area at the same time as in Figure 4. Today's smaller islands were part of a larger one, Santarosae, which lay only 6 km from the mainland.

By 8 mya, late in the Miocene Epoch, The California Current had become cold enough (and interior high pressure areas strong enough) to permit coastal summer fog banks similar to those of today. The California climate had begun to develop pronounced seasonality and its Mediterranean type climate had begun to form.

A very important event occurred between 5 and 3 mya, during Pliocene time: a minor change in the motion of the Pacific Plate caused it to compress against the North American Plate. This led to some uplift of the formerly low coastal lands to eventually create today's Coast, Transverse, and Peninsular ranges.

Most importantly, the new height of the coastal ranges caught some of the moisture coming off the ocean, and rain shadows developed to their landward and leeward sides. Deprived of their former moisture, southeastern California's woodlands became increasingly desert-like. The Great Central Valley, which had recently lost its status as an inland sea, also slowly became arid.

Ice Age (2.4 million to 13,000 ya)

The period of glaciation commonly called the Ice Age was not restricted to the Pleistocene epoch of our current Quaternary Period (the last 1.6 million years), as is often assumed, but began very late in the Neogene Period. There were some sporadic cooling episodes beginning about 5 mya, but the Ice Age was in full force (with continuous cold-warm cycles) about 2.4 mya, initiating widespread changes for California plant communities. Since the Ice Age began, there have been about 48 cold-warm climate cycles, most lasting about 40,000 years. The most recent cycles, however, have been about 100,000 years long, and each of them has had shorter, minor cycles within the major cycle.

The last major cycle, the Wisconsin glaciation, began about 120,000 years ago; there have been as many as six minor cold-warm cycles within it, the last one the most extreme. That minor cycle (in California, the Tioga glaciation) began about 26,000 years ago and reached a maximum about 20,000 years ago.

The first clear evidence of Sierran glaciation comes from Searles Lake, now a dry lake in the Mojave Desert. The most extensive glaciations started about 2 mya, when the coastal rain shadow was less well developed. During glacial times, alpine plants presumably were much more widespread and abundant than they are today, and moisture-requiring plants were more predominant at lower elevations.

During the last glacial episode, pinyon pines and junipers displaced live oaks and chaparral in the Sierra Nevada, eastern Transverse Ranges, and northern Peninsular Ranges. Large freshwater lakes (now dry or saline basins) were widespread in the Great Basin, Mojave Desert, and Salton Sea trough during most of the Ice Age glaciations. Death Valley's Lake Manly was the largest in California's part of the Great Basin. It covered 1600 km^2 and was 180 m deep. Lake Cahuilla in the Salton Sea trough was even larger. It covered 7800 km^2 but was only about 110 m deep. The total combined surface area of these glacial-period lakes was nearly 30 times that of today's Lake Tahoe. These lakes moderated temperatures, augmented precipitation, and provided extensive riparian habitats in a landscape that today is almost completely devoid of fresh water.

During glacial times, sagebrush scrub probably advanced into montane forest (including the western slopes of the Sierra Nevada) and into eastern basins as well. Despite the cooler temperatures, this advance implies that summers were dry, as they are today. In the higher mountains, conifer zones tended to be displaced lower.

Oaks also retreated downward and probably were abundant in parts of the Great Central Valley. In the cooler, moister coastal ranges and valleys (including the San Francisco Bay Area), oaks were partly replaced by redwood, Douglas-fir, and Monterey pine, all of which extended as far south as the Los Angeles basin. Inland from the coastal fog belt, more drought-tolerant conifers such as incense cedar and ponderosa and sugar pines probably expanded their ranges, replacing oaks.

During glacial times, the vast accumulation of ice on the continents causes the sea level to drop. During the Tioga glaciation, it dropped about 120 m, exposing California's continental shelf and providing greatly increased land area for many coastal communities. In the San Francisco Bay Area, a 50-km wide coastal plain connected the Farallon Islands and Cordell Bank to the mainland (Figure 4). In the south, the northern Channel Islands were connected as one vast island ("Santarosae", Figure 5). Its eastern tip was within 6 km of the mainland.

During Tioga time, many coastal taxa that are rare today likely had significantly larger ranges. These include most cypresses, Santa Lucia fir, and Monterey, Torrey, and Bishop pines. For example, Gowen cypress, which grew on "Santarosae" Island and may have been quite common along the central coast, today is restricted to small patches near the center of the Monterey Peninsula.

Pulses of coast range uplift occurred relatively late in the Ice Age, accompanied by increasing inland aridity and expansion of chaparral communities.

Each species responds independently to environmental changes. The displacements and migrations of the Ice Age produced community types that no longer exist — their component species are now found in different combinations. For example, during Tioga time, sugar pine co-existed with mountain hemlock; shadscale mixed with bristlecone pine. Today these pairs are well separated by elevation.

Current Time and Prospects

Human Influence. At the end of the last glaciation, about 13,000 years ago, humans were an integral part of the landscape throughout North America and had begun to have their own influence on community changes and extinctions (especially of the large animals in the Ice Age fauna).

Modern humans are the most important element of vegetation change in California today. Destruction of habitat by urban and agricultural development, pollution, and atmospheric warming, together with introduction of foreign competitors, grazers, and pathogens have greatly increased rates of change.

Geological Trends. Several geologic trends from California's past are continuing. The most important of these are cold/warm cycles, increasing uplift (and accompanying inland aridity), northward movement of the Pacific Plate, and decreasing volcanism as the Juan de Fuca Plate moves north of California.

Our current epoch, the Holocene, is best considered an interglacial. Even within it, there have been about three periods of slightly cooler weather — "little ice ages" that produced many cirque glaciers in the high Sierra Nevada.

With continued uplift of the Coast and Klamath Ranges, the trend toward increasing aridity inland may continue well into the future. In fact, if the Sierra Nevada becomes drier and more Great Basin plants migrate into it, the continuing rise of the Klamath Ranges may make them a region hospitable to alpine plants.

California's Peninsular and South Coast Ranges, most of which are carried on an oceanic plate (the Pacific Plate), continue to drift northward, possibly carrying some "captive" taxa to their extinction in unsuitable climates.

As the subducting Juan de Fuca Plate migrates northward by lengthening of the San Andreas Fault, the focus of volcanism probably will move northward with it — in a few million years, the current Cascade Range in California will cease its volcanic activities and no longer will produce new soils and new habitats.

These forces will continue in their inexorable way to change the face of California.

CALIFORNIA'S CHANGING CLIMATES AND FLORA

Dieter H. Wilken

The California flora, as treated in *The Jepson Manual*, includes more than 5800 species, of which about 24% are endemic (see Appendix I). These estimates of species diversity underscore the remarkable nature of the flora, because such numbers are extraordinarily high and significantly exceed those of equivalent areas in North America and most of continental Eurasia. Much of this species diversity exists because of California's geological, topographic, and climatic diversity.

Consider a line transecting California, along 39 degrees latitude, between Point Arena on the Pacific coast and Lake Tahoe on the edge of the Great Basin.

At Point Arena, the transect begins on a winter-wet, summer-cool, foggy, coastal headland and passes eastward across the geologically complex North Coast Ranges, where exposed surfaces are rich in serpentine, shales, sandstones, and volcanic rocks like andesite and basalt. On the east side of the Coast Ranges, rainfall decreases as the transect passes across the lower Sacramento Valley, which is characterized by a hotter, drier summer and by soils ranging from rich loams along the Sacramento River to extensive hardpans that permit development of nutrient rich, winter-wet but summer-dry vernal pools.

As the transect enters the Sierra Nevada, it encounters a topography with considerable variability in relief, including steep slopes, cliffs, ridges, peaks, and canyons carved from the intrusive granitic rocks. Winters are much colder, with heavy snowfall, but summer temperatures are moderated by afternoon thunderstorms and high elevations. Lake Tahoe, bordered by 2 fault blocks (Sierra Nevada, Carson Range) and expanded by dams formed by Pliocene volcanos and Pleistocene glaciers, exemplifies the relatively high winter precipitation and low summer evaporation rates of the high montane climate. This climate changes abruptly eastward to one of considerable aridity.

About 300 km (200 miles) to the south, another transect from Mount Whitney to Badwater in Death Valley, interrupted by the intervening Inyo and Panamint Ranges, crosses one of the steepest elevational gradients in North America — about 4500 vertical meters (14,800 feet) in less than 130 km (80 miles). Many other transects serve to supplement these dimensions of California's complex, physical landscape.

The diversity of the California flora is a product of complex geological and climatic changes that have occurred over 65 million years. Changes in species diversity and community organization have been effected by biological processes that respond to geological and climatic change. These include evolution through natural selection, speciation, migration by means of spores, seeds, and fruits, migration as entire populations and communities on mobile geological terranes, and extinction. These dynamic processes are understood partly through observation of ecological succession, partly through studies of dispersal by water, wind, and animals, and partly through modern concepts of speciation within the framework of evolutionary theory.

The history of the flora has been reconstructed almost entirely by fossils in the form of spores, pollen grains, and the mineralized remains of leaves, twigs, seeds, fruits, and cones; only rarely have the fragile, soft tissues of flowers been preserved. Unlike the data of most scientific studies, the fossil record usually represents only a single sample and can never provide a complete understanding of ecosystems present at a particular place and time.

Interpretations of fossils are subject to other sources of error, which have been revealed by studies of modern floras and the deposition of their remains. Some plants produce many more fossilizable parts than others in the same community — species producing cones or fruits that resist decay and avoid consumption by animals may contribute disproportionately to a fossil flora. Deposition and later fossilization can occur at a very local level and may not include common species a few hundred meters distant.

Most fossils were formed in areas of sedimentation along the margins of lakes and flood plains. Such fossil floras often contain a disproportionate number of species favoring wet habitats. Fossils formed in the ash of explosively eruptive volcanos may provide a more representative sample of species, but volcanism has not been continuous over time and space in the last 65 million years. Some geological formations of western North America are particularly rich in the fossil record; others have been subducted beneath geological plates or have been buried beneath more recent sediments and volcanic deposits.

Finally, some uncertainty always exists regarding the true evolutionary relationship between two or more samples of fossils assigned to the same genus or species, especially if such samples are separated by 10 million or more years of geological history. Nevertheless, the fossil records of western North America permit a partial and fascinating interpretation of California's complex floristic history.

Paleogene Period

During the Paleogene, 65 to 24 million years ago (mya), the climate of the emerging western North American continent was very different. Seasonal differences were less pronounced as a result of warmer winters and wetter summers, especially in the interior. The coast supported a dense rainforest, which graded into a tropical savannah southward. The coastal flora was composed of tree ferns, cycads, palms, and evergreen angiosperms whose nearest relatives occur today in southern Mexico and Central America. These members of the flora were eliminated as the climate cooled and became drier over the next 65 mya.

The interior of western North America was forested by a rich mixture of angiosperms and gymnosperms, whose species diversity far exceeded that of any known on the continent today. Among them were the ancestors of many modern California forest species, including alder, birch, fir, hemlock, oak, pine, rhododendron, redwood, giant sequoia, spruce, and tanbark oak. These forests also included many taxa now found elsewhere only in the wetter climates of the northern hemisphere as far away as east Asia. The past and present distribution of such species provide strong evidence for the former coexistence of certain Asian taxa (e.g., dawn redwood and ginkgo) with eastern North American taxa (e.g., beech, sweet gum, persimmon, and sassafras). Some genera, like spicebush (*Calycanthus*) and California nutmeg (*Torreya*), include modern species with a distribution only in California and eastern North America. At least 50% of all modern genera and species in the California flora can be traced with some certainty to the taxa of the Paleogene forests.

The interior forest, which varied locally in species composition and structure, extended along a gradient of decreasing diversity southward and toward the interior. In the vicinity of the present day central and southern Rocky Mountains, the uplands also were forested with species of fir, pine, and spruce, but the lower elevations supported shrublands and woodlands composed of cypress, evergreen oaks, juniper, madrone, barberry, manzanita, mountain mahogany, pinyon pine, sycamore, and sumac, among others. This vegetation extended across what is now the southwestern United States and northern Mexico. Like the northern forests, there existed species of genera not found in the region today and whose descendants may be found in eastern North America and the mountains of central Mexico. About 25% of the genera and 30% of the species in California today may have had their origin in this shrubland and woodland flora of the Paleogene southwest.

To the south and probably intergrading with the tropical savannah of the coast, there occurred a broad belt of subtropical vegetation, including acacia, poinciana, creosote bush, figs, mesquite, palms, and elephant tree, which would contribute the remaining 25% of the genera and another 20% of the species in the modern California flora.

Neogene Period

During the Neogene Period, 24 to 1.6 mya, regional differentiation became evident in association with mountain building, increasingly drier summers, and cooler winters. For example, forests composed partly of redwood were restricted to a broad arc from northern California through Oregon eastward to northern Idaho and Montana and southward to Colorado. South of this arc, in southern Idaho and much of Nevada, redwood is notably absent from the fossil record, but forests including giant sequoia were apparently well developed and common.

The early Neogene redwood forests, including such conifers as spruce, cedar, cypress, and hemlock, were gradually displaced coastward and contributed significantly to the present day vegetation of the northern California coast, the Klamath Mountains, and the North Coast Ranges. Although many genera and species became extinct in this region during the Neogene, the equable climate and relatively high precipitation of northwestern California throughout most of the last 30 million years has contributed much to the diversity of forest communities that now occupy the North Coast Ranges and the Klamath Mountains.

In contrast, the Neogene forests of the interior, characterized partly by giant sequoia, fir, and pine, were to contribute significantly to the montane and subalpine forests of the southern Cascades, the Sierra Nevada, and the mountains of eastern California. As the rainshadow intensified during the latter part of the Neogene, the forests of the Great Basin were displaced westward and upward into the Sierra Nevada. By the end of the Neogene, 1.6 mya, much of the Great Basin to the east was occupied by a woodland composed of drought-tolerant junipers, oaks, and pinyon pines. Conifers such as bristlecone, whitebark, and limber pine became restricted to the higher elevations of the Inyo and White Ranges and the eastern Sierra Nevada.

During the early Neogene, the southern California coast was dominated by a forest that included closed-cone pines, oaks, and palms. The interior, south of the developing southern Sierra Nevada and including much of the modern deserts, was vegetated by a diversity of woodlands. Among their common species were California-lilac, currant, juniper, manzanita, pinyon pine, sumac, and toyon, which together would contribute to the chaparral and oak woodland of modern southern and central California.

Also included were many subtropical genera like acacias, figs, palms, palo verde, smoketrees, and other genera that now occur only in the Sonoran Desert, with its moist summers and frost-free winters. Some of these taxa arrived by means of fruit and seed dispersal; others may have arrived as part of insular floras carried on terranes that originated in the south.

During the Neogene, the woodlands and shrublands began to intermingle more extensively with the northern coniferous forests as California developed into a region of valleys and mountain ranges with a north-south orientation. The broad contact zones between these major floras served to provide new ecological opportunities for adaptive radiation and speciation, especially in such large genera as *Allium, Arctostaphylos, Castilleja, Ceanothus, Cryptantha, Delphinium, Mimulus*, and *Plagiobothrys*.

By the end of the Neogene, forests composed partly of closed-cone pines and cypresses had become dominant along the coast to north of San Francisco Bay and an extensive pine/oak savannah dominated the lower elevations of the Sierra Nevada, the Inner Coast Range foothills, and the valleys of southern California. At the same time, many species became more restricted in distribution or were becoming extinct. For example, the few remaining species of hickory, persimmon, and sweet gum were eliminated from the last summer-moist lowlands of the coastal valleys. Nevertheless, a few genera, like *Magnolia*, persisted in southern California until well into the Pleistocene.

The late Pliocene and Pleistocene glacial episodes and the warm, arid periods that followed (up to the last several thousand years) contributed significantly to the modern composition and structure of California's flora. Changes in the vegetation included the expansion of some species and communities and the segregation of other species into communities with relatively low species diversity.

Northern coastal forest dominated by redwood migrated at least as far south as southern California near Santa Barbara. The Sierran coniferous forest and some associated herbaceous species expanded into the mountains of southern California. At lower elevations, pinyon and juniper woodlands expanded into the lower valleys of the Great Basin and Mojave Desert, only to be replaced later by species derived either from the Great Basin or Sonoran Desert floras. The periodic warm, dry interglacials gradually reduced species diversity at higher elevations and stranded such conifers as bristlecone and limber pine only at the highest, summer-cool elevations. The many glacial-interglacial cycles also contributed to significant fluctuations in community structure, relatively rapid ecological succession, and the opening of many new habitats.

As geographic and elevational ranges of taxa expanded and contracted, many populations became isolated in regional basins and intervening mountains. These events contributed to rapid speciation, with new species arising from ancestral populations in association with rapid climatic change and often catastrophic natural selection.

As a result of the warm and dry interglacials, oak savannah and chaparral expanded their ranges northward into the interior foothills and inland valleys of northern California. Many forest species, like redwood, became extinct along the coast south of Big Sur. The distribution of other species, like the cypresses and closed-cone pines, became fragmented and isolated. Grasslands and shrublands developed on coastal headlands exposed by uplift. Foothill-pine/blue-oak woodland became restricted to a relatively narrow zone in the foothills surrounding the Great Valley. Outside of the Sierra Nevada, a few subalpine and alpine species were restricted to the highest peaks of the Coast, Transverse, Peninsular, and a few desert mountain ranges. Shrublands expanded their ranges, especially in the San Joaquin Valley, the Mojave Desert, and the interior valleys of southern California, while the remaining species-rich woodlands of coastal California became isolated on the Channel Islands.

As humans colonized California, other changes began to take place, especially with the arrival of the first Europeans, who brought with them a culture based largely on cattle grazing and agriculture to the fertile valleys and coastal lowlands. These human activities introduced new species into the California flora, most of which were native to Mediterranean Europe. The oak savannas and perennial grasslands of the Great Valley and adjacent foothills were either cleared or became dominated by an annual grassland composed mostly of naturalized European species. Most of the tule marshes of the delta region were drained and most of the riparian forests along the Sacramento and San Joaquin Rivers in the Great Central Valley disappeared.

Urbanization from San Francisco south to San Diego has reduced and fragmented most of the shrublands, woodlands, and wetlands, especially along the coast. These recent changes in the flora, however, cannot be equated with those of recorded geological history, because they have occurred far more rapidly than any experienced during the last 65 million years. Unfortunately, extinction has become the dominant process of this most recent episode in the history of the California flora.

For those interested in pursuing this topic further, two references are recommended as starting points: Raven & Axelrod, 1978, *Origin and Relationships of the California Flora*, Univ. Calif. Publ. Bot. 72:1–134 and Axelrod, 1976, *History of the Coniferous Forests*, California and Nevada, Univ. Calif. Publ. Bot. 70:1–62.

KEY TO CALIFORNIA PLANT FAMILIES

David J. Keil

Key to Groups

1. Specimens at hand without both stamens and pistils available for examination; fls either not produced at all or, if present, fls unisexual and either staminate or pistillate fls unavailable
 2. Pls aquatic, available specimens reproducing mainly by vegetative means; fls, frs, or spores 0 or not readily apparent .... **Group 1** (p. 60)
 2′ Pls aquatic or terrestrial; fls, frs, seeds, spores, or other specialized reproductive structures present and readily apparent
  3. Pls bearing only spores or pollen; available specimens without ovules, ovaries, seeds, frs, or bulblets
   4. Ferns and their relatives; reproductive spores produced and released directly from sporangia borne on surfaces or in axils of some or all of the lvs or borne in terminal cones; fls and seeds never formed; herbs only [FERNS & FERN-ALLIES] .... **Group 2** (p. 62)
   4′ Angiosperms and gymnosperms in pollen-producing condition only; pollen borne in anthers or pollen cones; staminate fls or other pollen-producing structures present; pistils and other ovule-producing structures 0; herbs, shrubs, and trees
    5. Pls conspicuously woody; tree and shrubs [some ANGIOSPERMS and GYMNOSPERMS] **Group 3** (p. 63)
    5′ Pls herbaceous or woody only at the base; ann and per herbs [some ANGIOSPERMS] ...... **Group 4** (p. 65)
  3′ Pls bearing ovaries, ovules, seeds, frs, or bulblets; ovules borne in ovaries, naked on branches, in cones, or cone-like structures; pollen-producing structures present or 0
   6. Pls herbaceous, sometimes woody at base [some ANGIOSPERMS]
    7. Fls 0, replaced in infls by bulblets that are dispersed in place of seeds or frs .... **Group 5** (p. 66)
    7′ Fls present, unisexual, bearing only ovaries .... **Group 6** (p. 67)
   6′ Pls woody [some ANGIOSPERMS and GYMNOSPERMS]
    8. Ovules not enclosed in ovaries, exposed to air at time of pollination; pollen deposited directly on ovules; stigmas never present; seeds borne between scales of cone or naked on branches [GYMNOSPERMS] .... **Group 7** (p. 68)
    8′ Ovules enclosed in ovaries at time of pollination; pollen deposited on stigmas; seeds borne inside frs derived from ovaries [mostly ANGIOSPERMS] .... **Group 8** (p. 68)

1′ Specimens at hand with both stamens and pistils available for examination; fls bisexual or, if unisexual, both staminate and pistillate fls present
 9. Fl parts (esp perianth) gen in 3's (rarely 1 or in 2's or 4's); principal lf veins gen parallel (rarely pinnate or palmate); vascular bundles of st (in cross-section) scattered or in diffuse circles; sts not increasing in diam through secondary growth exc in some woody Liliaceae; 1st-year taproot gen 0; cotyledon gen 1 [mostly MONOCOTS]
  10. Perianth conspicuous, at least the inner whorl corolla-like in color and texture; infl various .... **Group 9** (p. 70)
  10′ Perianth inconspicuous, the individual perianth parts gen scale-like, bristle-like or 0, never petal-like; infl unit most commonly a spike, spikelet, or head .... **Group 10** (p. 71)
 9′ Fl parts (esp perianth) gen in 4's or 5's (rarely in 3's) or in a spiral with the number of parts indefinite; principal lf veins gen pinnate or palmate (rarely parallel); vascular bundles of young st (in cross-section) gen in a well defined ring; sts gen increasing in diam through secondary growth, often becoming stiffly woody in age; 1st-year taproot gen present; cotyledons gen 2 [mostly DICOTS]
  11. Pistils 2 or more per fl .... **Group 11** (p. 72)
  11′ Pistil 1 per fl
   12. Perianth 0 or in a single whorl (appearing to be sepals or petals but not both)
    13. Pls conspicuously woody; trees or shrubs
     14. Infl, at least the staminate, of catkins or catkin-like spikes .... **Group 12** (p. 73)
     14′ Infl various but not catkin-like .... **Group 13** (p. 73)
    13′ Pls herbaceous (ann or per) or woody only at base
     15. Ovary inferior .... **Group 14** (p. 75)
     15′ Ovary superior .... **Group 15** (p. 76)
   12′ Perianth in 2 or more whorls (gen both sepals and petals) or perianth parts spiraled 2 or more times around floral axis
    16. Petals fused at least at base, forming a ring or a tube, falling as a unit
     17. Ovary superior .... **Group 16** (p. 77)
     17′ Ovary inferior or partly so .... **Group 17** (p. 80)

16′ Petals gen free at least at base, falling singly (in a few families joined and falling in 2's or 3's, but not forming a ring or a tube)
18. Ovary inferior or partly so .... **Group 18** (p. 81)
18′ Ovary superior
19. Stamens > twice as many as petals .... **Group 19** (p. 82)
19′ Stamens twice as many as petals or fewer
20. Lvs compound or so deeply divided as to appear compound .... **Group 20** (p. 83)
20′ Lvs simple, sometimes much reduced
21. Pls conspicuously woody; trees and shrubs .... **Group 21** (p. 84)
21′ Pls herbaceous or woody only at base .... **Group 22** (p. 86)

Group 1: Aquatic Plants in Vegetative Condition

1. Pls of saltwater habitats, either marine or in brackish to saline lakes, ponds, pools, estuaries, etc.
2. Pls not differentiated into lvs or lf-like structures and sts or st-like structures [algal groups, not further treated in this book]
2′ Pls differentiated into lvs or lf-like structures and sts or st-like structures; nodes and internodes readily apparent
3. Pls with a central axis and whorled, cylindric branches [CHAROPHYTA, algal group not further treated here]
3′ Pls with a short to elongate st bearing alternate, opposite, or whorled lvs or lf-like structures with flattened blades
4. Lvs linear to ovate, gen olive-green to reddish or brownish, often with gas-filled floats [PHAEOPHYTA, algal group not further treated here]
4′ Lvs linear, strap-shaped, grass-green, lacking floats [ANGIOSPERMS: MONOCOTS]
5. Pls emergent; lvs and sts able to support themselves out of water
6. Lvs all basal
7. Lf blades wiry, tough or stiffly rigid; internal cross partitions gen present and evident to touch; lvs ± 3-ranged .... **JUNCACEAE**
7′ Lf blades soft, not wiry and tough or stiffly rigid; internal cross partitions 0; lvs ± 2-ranked
8. Lf blades flat, thin .... **POACEAE**
8′ Lf blades cylindric or ± flat, thickish .... **JUNCAGINACEAE** (*Triglochin*)
6′ Lvs basal and cauline
9. Sts ± triangular .... **CYPERACEAE**
9′ Sts round
10. Lvs flat .... **POACEAE**
10′ Lvs cylindric .... **JUNCACEAE**
5′ Pls submersed; lvs and sts weak
11. Lf tips shallowly 2–3-toothed .... **CYMODOCEACEAE**
11′ Lf tips entire
12. Lf blades 1.5 mm wide or more; sts often rather tough; pls of estuaries, intertidal areas, and subtidal oceanic habitats, sometimes on surf-beaten rocks .... **ZOSTERACEAE**
12′ Lf blades gen 1 mm wide or less; sts very slender and delicate; pls gen not in intertidal or subtidal habitats exc sometimes in sheltered bays where protected from wave action
13. Lvs gen alternate; lf bases sheathing; free stipules 0 .... **POTAMOGETONACEAE** (*Ruppia*)
13′ Lvs gen opposite; lf bases not sheathing; free membranous stipules present **ZANNICHELLIACEAE**
1′ Pls of freshwater habitats
14. Pl body consisting of a jointed central axis and whorled branches; lvs, if any, reduced to whorled scales
15. Pl submersed; internodes of central axis solid [algal groups not further treated here]
15′ Pl emergent; internodes of central st obviously hollow .... **EQUISETACEAE**
14′ Pl body various, but not consisting of a jointed central axis and whorled branches; lvs various, but not whorled scales
16. Individual pls gen 1–20 mm in length or diam, free-floating or stranded along shore, sometimes not differentiated into lvs and sts
17. Pl body forked or with forked grooves on upper surface [RICCIACEAE, liverworts not further treated here]
17′ Pl body neither forked nor with forked grooves
18. Pl body differentiated into a short, often branched st covered with tiny, overlapping, scale-like, green or red-purple-tinged lvs .... **AZOLLACEAE**
18′ Pl body spheric to disk-shaped or oblong, not differentiated into sts and lvs .... **LEMNACEAE**
16′ Individual pls often >> 20 mm, anchored or sometimes free-floating, often clearly differentiated into sts and lvs
19. Pls not differentiated into lvs or lf-like structures and sts or st-like structures [various algal groups, not further treated here]
19′ Pls differentiated into lvs or lf-like structures and sts or st-like structures; nodes and internodes readily apparent (or in *Utricularia* producing dissected, lf-like underwater sts)
20. Lvs all scale-like or narrowly linear, gen very thin, entire or minutely toothed, densely overlapping, all alternate; roots never present [MUSCI, true mosses, not further treated here]
20′ Lvs linear to ovate, very thin to thick and ± fleshy, entire to variously toothed, lobed, or compound, gen not densely overlapping exc sometimes in new growth, alternate, opposite, whorled, or sometimes all basal; roots gen present though sometimes few

21. Lvs peltate, sagittate, or deeply notched
22. Lvs peltate
23. Submersed parts gelatinous; lf blades floating, 4–12 cm diam .... **CABOMBACEAE**
23′ Submersed parts not gelatinous; lf blades gen emergent, mostly smaller ..... **APIACEAE** (*Hydrocotyle*)
22′ Lvs sagittate or ± round, the base deeply notched
24. Lf blades pinnately veined; petioles long, stout; rhizome very stout .... **NYMPHAEACEAE**
24′ Lf blades palmately veined; petioles long or short, slender or stout; rhizome, if present, slender
25. Lf blades ovate to sagittate, entire; veinlets passing like rungs of a ladder between prominent main veins .... **ALISMATACEAE**
25′ Lf blades palmately lobed, never sagittate, the lobes often toothed; veinlets forming an irregular network .... **APIACEAE** (*Hydrocotyle*)
21′ Lvs not peltate, sagittate, or deeply notched
26. Pls raft-like, free-floating on water surface, breaking apart into individuals or small clumps; rhizomes short or 0
27. Lf blades light green, velvety-hairy, sessile, lacking floats; blades palmately veined **ARACEAE** (*Pistia*)
27′ Lf blades glossy green, glabrous; petioles often enlarged, forming gas-filled floats; blades finely parallel-veined .... **PONTEDERIACEAE** (*Eichhornia*)
26′ Pls gen anchored in bottom sediments or, if free-floating, most of pl body below water surface, mostly not readily breaking apart; rhizomes present or 0, sometimes elongated
28. Sts gen erect or corm-like; lvs all basal or alternate on an erect, emergent st or borne on a rhizome anchored in bottom sediments
29. Lvs petioled with expanded blades
30. Lvs mostly 1 cm wide or more, often much larger .... **ALISMATACEAE**
30′ Lvs mostly << 1 cm wide .... **SCROPHULARIACEAE** (*Limosella*)
29′ Lvs sessile, linear or reduced to bladeless sheaths
31. Lvs vertically folded and attached "edge-on" to the st, obviously 2-ranked .... **JUNCACEAE** (*Juncus*)
31′ Lvs cylindric, ± angled, with ordinary flat blades, or reduced to bladeless sheaths, not obviously 2-ranked
32. Lvs with flat blades or reduced to bladeless sheaths
33. Lvs bladeless or nearly so, composed mainly of a tubular sheath
34. Sts ± triangular .... **CYPERACEAE**
34′ Sts round, flattened, or several-angled
35. Lf sheaths closed, forming a continuous cylinder around st .... **CYPERACEAE**
35′ Lf sheaths open on one side, with overlapping margins .... **JUNCACEAE**
33′ Lvs with ordinary flat blades and gen also ± tubular sheaths
36. Lf blade rounded on back or flat, lacking a prominent midvein or keel
37. Lf width gen > 5 mm; remnants of old infl a dense spike; inner surface of lf sheaths mucilaginous .... **TYPHACEAE**
37′ Lf width < or = 5 mm; remnants of old infl a raceme; inner surface of lf sheaths not mucilaginous .... **JUNCAGINACEAE** (*Triglochin*)
36′ Lf blade with a well developed midvein or keel
38. Midvein of lf off-center on blade .... **ARACEAE** (*Acorus*)
38′ Midvein of lf central on blade
39. Lf sheaths closed, forming a continuous cylinder around st; st gen triangular; lvs 3-ranked .... **CYPERACEAE**
39′ Lf sheaths open on one side, with overlapping margins; st round; lvs 2-ranked .... **TYPHACEAE** (*Sparganium*)
32′ Lvs cylindric or angled
40. Rhizomes 0; lvs densely tufted
41. A sporangium gen present in axil of each lf, containing either powdery or larger spores; lvs divided lengthwise into 4 hollow chambers .... **ISOETACEAE**
41′ A sporangium never present in lf axil; lvs solid or hollow, but not regularly divided into 4 chambers
42. Pls per; lvs often with internal cross-partitions .... **JUNCACEAE** (*Juncus*)
42′ Pls ann; lvs without internal cross-partitions or partitions obscure
43. Lf bases not sheathing .... **BRASSICACEAE** (*Subularia*)
43′ Lf bases sheathing .... **JUNCAGINACEAE** (*Lilaea*)
40′ Rhizomes present; lvs tufted or borne along rhizome
44. Lvs sharply angled
45. St triangular; lvs 3-ranked .... **CYPERACEAE**
45′ St ± round; lvs 2-ranked .... **TYPHACEAE** (*Sparganium*)
44′ Lvs cylindric
46. Lvs without internal cross-partitions
47. Lvs 2–5 cm .... **SCROPHULARIACEAE** (*Limosella*)
47′ Lvs gen >> 5 cm .... **JUNCAGINACEAE** (*Triglochin*)
46′ Lvs with internal cross-partitions

48. Lvs soft, blunt, mostly 15 cm or shorter, loosely sheathing; cross-partitions often externally prominent .... **APIACEAE** (*Lilaeopsis*)
48′ Lvs gen stiff, wiry, acute, often > 15 cm, tightly sheathing; cross-partitions readily apparent by feel but often externally obscure .... **JUNCACEAE** (*Juncus*)
28′ Sts gen weak, often taking the form of underwater rhizomes or stolons, unable to support pl body outside of water; lvs gen all cauline on submersed sts
49. Lvs (or lf-like branches) compound or very deeply divided
50. Lvs palmately compound — lflets 4, floating or emergent .... **MARSILEACEAE**
50′ Lvs pinnately compound or dissected, repeatedly forked, or irregularly divided
51. Lflets lanceolate or wider; lvs mostly aerial
52. Lflets coarsely serrate or dentate; petioles sheathing .... **APIACEAE**
52′ Lflets entire or teeth rounded; petioles not sheathing .... **BRASSICACEAE** (*Rorippa*)
51′ Lflets or lobes linear, at least on submersed lvs; lvs all or many submersed
53. Underwater parts bearing bladder-like traps that capture aquatic invertebrates and micro-organisms; pls carnivorous .... **LENTIBULARIACEAE** (*Utricularia*)
53′ Underwater parts lacking bladder-like traps; pls not carnivorous
54. Lvs alternate .... **RANUNCULACEAE** (*Ranunculus*)
54′ Lvs opposite or whorled
55. Lvs repeatedly forked .... **CERATOPHYLLACEAE**
55′ Lvs pinnately divided .... **HALORAGACEAE** (*Myriophyllum*)
49′ Lvs simple, the margins entire, toothed, or shallowly lobed
56. Lvs long, narrow, strap-shaped or ribbon-like, often >> 20 cm, at least some basal
57. Lvs all basal .... **ALISMATACEAE**
57′ Lvs basal and cauline .... **TYPHACEAE** (*Sparganium*)
56′ Lvs linear to ovate, gen < 20 cm, gen all cauline
58. Lvs with stipules or sheathing bases; margins entire
59. Principal veins of lf pinnate .... **POLYGONACEAE** (*Polygonum*)
59′ Principal veins of lf parallel or pinnate-parallel or only 1 lf vein apparent
60. Lvs 1.5 mm or more wide; floating lvs sometimes present
61. Roots branched; lvs linear; midvein of lf not clearly differentiated; floating lvs 0 .... **PONTEDERIACEAE** (*Heteranthera*)
61′ Roots gen unbranched; lvs linear to ovate; midvein of lf often differentiated from other veins; floating lvs often present .... **POTAMOGETONACEAE** (*Potamogeton*)
60′ Lvs < 1.5 mm wide; all lvs gen submersed
62. Lf bases sheathing; stipules fused at least in part to lf base .... **POTAMOGETONACEAE**
62′ Lf bases not sheathing; stipules free, axillary, membranous
63. Lvs gen alternate .... **POTAMOGETONACEAE** (*Potamogeton*)
63′ Lvs mostly or all opposite .... **ZANNICHELLIACEAE**
58′ Lvs without stipules or sheathing bases; margins entire, toothed, or shallowly lobed
64. Lvs alternate .... **ONAGRACEAE** (*Ludwigia*)
64′ Lvs opposite or whorled
65. Lvs mostly aerial
66. Lf veins palmate .... **SCROPHULARIACEAE** (*Bacopa, Mimulus*)
66′ Lf veins pinnate
67. Lvs toothed .... **SCROPHULARIACEAE** (*Mimulus, Veronica*)
67′ Lvs entire
68. Lf tip acute and minutely spine-tipped .... **AMARANTHACEAE** (*Alternanthera*)
68′ Lf tip obtuse or acute, not spine-tipped .... **ONAGRACEAE** (*Ludwigia*)
65′ Lvs mostly or all submersed
69. Lvs whorled
70. Lvs entire .... **HIPPURIDACEAE**
70′ Lvs minutely toothed .... **HYDROCHARITACEAE**
69′ Lvs opposite
71. Lvs entire, linear to ovate, not enlarged at base, sometimes with floating upper lvs different from submersed lvs .... **CALLITRICHACEAE**
71′ Lvs minutely toothed to coarsely lobed, gen enlarged at base, all similar in form and submersed .... **HYDROCHARITACEAE** (*Najas*)

Group 2: Ferns and Fern Allies; Seeds Absent, Plants Reproducing by Spores

1. Lvs linear or scale-like, entire or minutely toothed, or (in *Azolla*) 2-lobed; veins 1 or 2
2. Pl floating or stranded on mud; pl body gen < 1 cm .... **AZOLLACEAE**
2′ Pl anchored in soil or growing on other pls; pl body >> 1 cm
3. Lvs basal, linear
4. Pl base corm-like; new lvs erect, never coiled .... **ISOETACEAE**
4′ Pl creeping by rhizomes; new lvs coiled, unrolling as they develop .... **MARSILEACEAE** (*Pilularia*)
3′ Lvs cauline, scale-like

5. Sts hollow, jointed at nodes; lvs whorled, fused at base; sporangia several on undersides of peltate structures that are clustered in terminal cones . . . . **EQUISETACEAE**

5′ Sts solid, not jointed; lvs alternate or opposite; sporangia axillary, 1 per fertile lf, the aggregation of fertile lvs sometimes cone-like

6. Fertile lvs not in well defined ranks, the aggregation appearing round when viewed from above; sporangia all of 1 kind, producing only 1 kind of spore . . . . **LYCOPODIACEAE**

6′ Fertile lvs 4-ranked, the aggregation appearing square when viewed from above; sporangia of 2 kinds, some with 4 large spores and others with many small spores . . . . **SELAGINELLACEAE**

1′ Lvs with well developed blades, not linear or scale-like, often lobed or compound; veins many

7. Pl aquatic or in dried pools; lvs palmately compound; lflets 4 . . . . **MARSILEACEAE** (*Marsilea*)

7′ Pl terrestrial or growing on other pls; lvs simple or 1–4-pinnate; if compound, lflets gen >> 4

8. Sporangia ± 1 mm thick, sessile in clusters on stalk fused to petiole, annulus 0; new lvs never coiled . . . . **OPHIOGLOSSACEAE**

8′ Sporangia << 1 mm thick, slender-stalked, gen in sori on underside of lf, often near margin, annulus present; new lvs coiled, unrolling as they develop

9. Lvs of two dissimilar kinds, the fertile (spore-bearing) blades with greatly narrowed segments

10. Blade deeply pinnately lobed or 1-pinnate; lvs > 30 cm . . . . **BLECHNACEAE** (*Blechnum*)

10′ Blade 2 or more times pinnate or divided; lvs < 20 cm . . . . **PTERIDACEAE**

9′ Lvs all alike or nearly so

11. Sori borne along margin of lf; indusia 0 or sori covered by reflexed lf margin

12. Lflets all entire . . . . **PTERIDACEAE**

12′ Lflets (some or all of them) toothed or lobed

13. Rhizome and base of petiole bearing linear to ovate scales; rhizome short and compact to long-creeping; petiole gen 2.5 mm diam or less, round or nearly so, light brown to black throughout; lvs gen < 80 cm, often much shorter . . . . **PTERIDACEAE**

13′ Rhizome and base of petiole covered with felt-like hairs; rhizome long-creeping; petiole often >> 3 mm diam, strongly grooved, greenish or straw-colored above the base; lvs often > 80 cm, sometimes 2 m or more . . . . **DENNSTAEDTIACEAE**

11′ Sori borne away from margin on underside of lvs or lflets; indusia present or 0

14. Sporangia scattered along veins, not clustered into distinct sori; indusia 0 . . . . **PTERIDACEAE**

14′ Sporangia in distinct sori; indusia sometimes present

15. Indusia 0

16. Blades deeply pinnately lobed . . . . **POLYPODIACEAE**

16′ Blades 2 or more times pinnate or divided . . . . **DRYOPTERIDACEAE** (*Athyrium*)

15′ Indusia present

17. Veins of lf blade forming a network

18. Lvs gen > 1 m; sori elongate, end to end in rows . . . . **BLECHNACEAE** (*Woodwardia*)

18′ Lvs gen < 1 m; sori round, scattered in several series . . . . **DRYOPTERIDACEAE** (*Cyrtomium*)

17′ Veins of lf blade free, not forming a network

19. Blade with 1-celled, needle-like hairs; petiole with 2 vascular strands . . . . **THELYPTERIDACEAE**

19′ Blade without needle-like hairs; petiole with 3 or more vascular strands (exc *Athyrium*)

20. Sori round, with indusium centrally attached . . . . **DRYOPTERIDACEAE**

20′ Sori elongate, with indusium laterally attached

21. Lvs < 30 cm, 1-pinnate or -forked . . . . **ASPLENIACEAE**

21′ Lvs > 30 cm, 2-pinnate or -divided . . . . **DRYOPTERIDACEAE**

Group 3: Woody Gymnosperms and Angiosperms in Pollen-Producing Condition Only; Ovulate Cones or Pistillate or Bisexual Flowers Absent

1. Pls parasitic on sts of woody host pls . . . . **VISCACEAE**

1′ Pls free-living

2. Trees; trunk unbranched, covered with persistent woody lf bases; lvs 1–several m, pinnately compound . . . . **ARECACEAE** (*Phoenix*)

2′ Trees or shrubs; trunk(s) gen branched, gen not covered with persistent lf bases; lvs gen smaller, simple or compound

3. Lvs stiff and sword-like, 0.5–1.5 m; infl a large panicle . . . . **LILIACEAE** (*Nolina*)

3′ Lvs not sword-like, mostly smaller; infls various

4. Lvs opposite or whorled

5. Lvs compound

6. Lvs palmate

7. Lflets 3; infl few-fld, ± flat-topped . . . . **ACERACEAE** (*Acer glabrum*)

7′ Lflets gen 5 or more; infl many-fld, cylindric to conic . . . . **HIPPOCASTANACEAE**

6′ Lvs pinnate

8. Woody vine; sepals gen 4, conspicuous, petal-like . . . . **RANUNCULACEAE** (*Clematis*)

8′ Tree or large shrub; sepals gen 5 or number difficult to determine, very small

9. Fls borne on elongated, thread-like pedicels . . . . **ACERACEAE** (*Acer negundo*)

9′ Fls sessile or borne on short, stout pedicels .... **OLEACEAE** (*Fraxinus*)
5′ Lvs simple
10. Lvs and sometimes sts thick and fleshy — lvs sometimes reduced to fleshy scales
11. Stipules present; petal-like perianth parts 4 .... **BATACEAE**
11′ Stipules 0; petal-like perianth parts 0 .... **CHENOPODIACEAE**
10′ Lvs and sts not fleshy
12. Margins of lvs toothed or lobed
13. Lvs palmately veined and lobed .... **ACERACEAE**
13′ Lvs pinnately veined and toothed .... **EUPHORBIACEAE** (*Tetracoccus*)
12′ Margins of lvs entire
14. Weak-stemmed subshrub
15. Perianth parts in 3's .... **EMPETRACEAE**
15′ Perianth parts in 4's .... **RUBIACEAE** (*Galium*)
14′ Stouter shrub or tree
16. Sts and lvs covered with silvery scales .... **ELAEAGNACEAE** (*Shepherdia*)
16′ Sts and lvs glabrous or variously hairy
17. Fls in catkins or pollen produced in ± catkin-like cones
18. Catkins elongate, pendent; lf blades well developed .... **GARRYACEAE**
18′ Catkins or pollen cones short, not pendent; lvs scale-like
19. Internodes very short; st concealed by persistent, overlapping, green, scale-like lvs; pollen sacs sessile beneath scales of small cone .... **CUPRESSACEAE**
19′ Internodes elongated; st green, bearing dry, scale-like, often early-deciduous lvs; pollen sacs borne on filaments .... **EPHEDRACEAE**
17′ Fls in head-like or raceme-like clusters
20. Lvs thick, leathery, evergreen; fls in peduncled, head-like clusters; sepals well developed .... **SIMMONDSIACEAE**
20′ Lvs thin, deciduous; fls in sessile, axillary clusters or short-peduncled racemes; sepals very small
21. Pl not spiny; bark of young sts reddish; fls in short, axillary racemes .... **EUPHORBIACEAE** (*Tetracoccus*)
21′ Pl spiny; bark smooth, gray; fls sessile, in axillary clusters .... **OLEACEAE** (*Forestiera*)
4′ Lvs alternate
22. Fls not in catkins or spikes; pollen never produced in cones
23. Lvs palmately lobed
24. Fls in terminal, panicles; shrubs or herbs; bark not flaking in large plates; lvs peltate, glabrous .... **EUPHORBIACEAE** (*Ricinus*)
24′ Fls in axillary or terminal clusters of spheric heads; large trees; bark flaking from trunks in large plates; lvs obtuse to cordate at base, densely tomentose when young .... **PLATANACEAE**
23′ Lvs not palmately lobed
25. Sts and lvs densely covered with stellate hairs .... **EUPHORBIACEAE**
25′ Sts and lvs glabrous or hairy but not stellate
26. Stamens many
27. Fls in spheric heads; petals yellow, gen concealed by stamens; lvs simple or 2-pinnate .... **FABACEAE** (*Acacia*)
27′ Fls in racemes; petals white, conspicuous; lvs simple .... **ROSACEAE** (*Oemleria*)
26′ Stamens 12 or fewer
28. Lvs compound
29. Lflets entire or teeth not gland-tipped .... **ANACARDIACEAE**
29′ Lflets with a few, gland-tipped teeth near base .... **SIMAROUBACEAE** (*Ailanthus*)
28′ Lvs simple
30. Fl parts in 3's
31. Low shrubs; fls solitary in lf axils; lvs needle-like .... **EMPETRACEAE**
31′ Woody vines; fls in umbels; lvs ovate .... **LILIACEAE** (*Smilax*)
30′ Fl parts gen in 4's, 5's, or 6's
32. Fls in involucred heads, these gen secondarily clustered .... **ASTERACEAE**
32′ Fls solitary or variously clustered, not in involucred heads
33. Lvs gen grayish, either softly hairy or covered with powdery scales .... **CHENOPODIACEAE**
33′ Lvs green
34. Petals present; infl a many-fld terminal panicle; lvs leathery; pls evergreen **ANACARDIACEAE**
34′ Petals 0; infl a few–many-fld axillary cluster; lvs gen thin; pls evergreen or deciduous
35. Lvs narrow, bases symmetric; shrubs .... **EUPHORBIACEAE** (*Tetracoccus*)
35′ Lvs wide, bases oblique; trees .... **ULMACEAE** (*Celtis*)
22′ Fls in catkins or catkin-like spikes or pollen produced in ± catkin-like cones
36. Lvs compound .... **JUGLANDACEAE**
36′ Lvs simple
37. Shrubs of ± saline desert habitats; lvs thick, often ± fleshy; spikes erect or very short and axillary .... **CHENOPODIACEAE**

37′ Shrubs or trees, gen of moister habitats; lvs mostly thinner; spikes, catkins, or pollen cones erect to pendent
38. Lvs needle-like or scale-like; pls evergreen
39. Pollen sacs borne on filaments .... **TAXACEAE**
39′ Pollen sacs sessile on underside of cone scales
40. Scales with 2 pollen sacs; needle-like lvs 1–30 cm; bark gen separating in flakes or plates **PINACEAE**
40′ Scales with several pollen sacs; needle-like lvs (if present) 1–2.5 cm; bark thick, fibrous, persistent .... **TAXODIACEAE**
38′ Lvs with well developed blades; pls evergreen or deciduous
41. Bractlets of infl well developed and readily visible at time of flowering
42. Catkin-bractlets fringe-margined; trees; petioles long .... **SALICACEAE** (*Populus*)
42′ Catkin-bractlets entire or appearing 3-lobed; shrubs or trees; petioles various
43. Fls inserted individually along catkin axis, each subtended by 1 bractlet (this sometimes early deciduous) .... **SALICACEAE** (*Salix*)
43′ Fls inserted in groups subtended and often somewhat concealed by 1–3 bractlets
44. Staminate infl pendent .... **BETULACEAE**
44′ Staminate infl stiff, erect or spreading
45. Low, brittle shrubs; lvs thin, crenate .... **EUPHORBIACEAE** (*Acalypha*)
45′ Shrubs or trees, not brittle; lvs thick, leathery, entire to sharply serrate .... **FAGACEAE** (*Chrysolepis, Lithocarpus*)
41′ Bractlets of infl 0 or very inconspicuous at time of flowering
46. Evergreen shrubs or trees
47. Main veins of lf parallel or nearly so; fls gen yellow; petals and sepals both present but small .... **FABACEAE** (Mimosoideae)
47′ Main veins of lf clearly pinnate; fls cream-colored or brownish, not clear yellow; petals 0
48. Lvs glabrous or with stellate hairs, not gland-dotted; sepals present; gen of dryland habitats; catkins slender, pendent, gen 2 cm or more .... **FAGACEAE**
48′ Lvs glabrous or with minute, straight hairs, dotted with tiny resin glands; sepals 0; gen of moist, often wetland habitats; catkins short, dense, gen < 2 cm .... **MYRICACEAE**
46′ Deciduous shrubs or trees
49. Sepals small but evident
50. Sepals 5–7; stamens 5–20; veins pinnate; lf bases gen symmetric .... **FAGACEAE**
50′ Sepals 4; stamens 4; veins palmate; lf bases oblique .... **MORACEAE**
49′ Sepals 0
51. Lvs dotted with tiny resin glands; individual fls difficult to distinguish .... **MYRICACEAE**
51′ Lvs not gland-dotted; individual fls clearly distinguishable on close examination ... **SALICACEAE**

Group 4: Herbaceous Monocots and Dicots in Staminate Condition, Producing Only Pollen; Pistillate or Bisexual Flowers Absent

1. Pls fully aquatic, submersed, floating in water, or stranded on mud
2. Pls gen free-floating, the whole pl < 15 mm, often < 5 mm; lvs and sts not differentiated; roots 0–few, unbranched — stamens emerging from tiny flap of tissue .... **LEMNACEAE**
2′ Pls rooted in bottom sediments or free-floating, gen >> 15 mm; lvs and sts clearly differentiated; roots often branched
3. Lvs alternate
4. Lf tips shallowly 2–3-toothed; pls of inland saltwater habitats .... **CYMODOCEACEAE**
4′ Lf tips entire; pls of freshwater or marine habitats
5. Fls in dense, spheric heads, these solitary or in axillary or terminal clusters, not enclosed in lf sheaths; freshwater habitats .... **TYPHACEAE** (*Sparganium*)
5′ Fls in axillary spikes, these gen enclosed and concealed in sheath of subtending lvs; marine habitats .... **ZOSTERACEAE**
3′ Lvs opposite or whorled
6. Lf blades entire, toothed, or shallowly lobed .... **HYDROCHARITACEAE**
6′ Lf blades (at least those of submersed lvs) divided into linear lobes
7. Blades of lvs repeatedly forked .... **CERATOPHYLLACEAE**
7′ Blades of lvs pinnately divided .... **HALORAGACEAE** (*Myriophyllum*)
1′ Pls terrestrial or, if growing in wet places, rooted in place and extending well above water surface
8. Pls parasitic on sts of woody host pls
9. Fls of parasite borne directly on sts of host; remainder of parasite internal within tissues of host; sts and lvs not differentiated .... **RAFFLESIACEAE**
9′ Fls borne on leafy branches of parasite; shoots of parasite external; lvs differentiated though sometimes reduced to scales .... **VISCACEAE**
8′ Pls free-living
10. Lvs opposite or whorled, not all basal

11. Sts or lvs thick and fleshy; infls terminal spikes
12. Lvs with stipules; petal-like perianth parts 4 .... **BATACEAE**
12′ Lvs without stipules; petal-like perianth parts 0 .... **CHENOPODIACEAE**
11′ Sts and lvs of normal texture, not thick and fleshy; infls various
13. Lvs palmately compound or simple and deeply lobed .... **CANNABACEAE**
13′ Lvs simple, entire or toothed
14. Lvs whorled, sessile or nearly so; stinging hairs 0 .... **RUBIACEAE**
14′ Lvs opposite, conspicuously petioled; stinging hairs present .... **URTICACEAE**
10′ Lvs alternate or all basal
15. Lvs stiff and sword-like, 5–15 dm; infl a large panicle; perianth parts 6 .... **LILIACEAE** (*Nolina*)
15′ Lvs not sword-like, often smaller; infls various; perianth parts mostly other than 6
16. Blades of lvs linear or narrowly lanceolate, simple and entire, veins parallel; lf bases sheathing sts; fls in spikelets, these gen in secondary clusters
17. St triangular; nodes not swollen; lf blades often channeled .... **CYPERACEAE**
17′ St round; nodes gen swollen and knot-like; lf blades gen flat .... **POACEAE**
16′ Blades of lvs variously shaped, sometimes toothed, lobed, or compound, veins mostly pinnate or palmate; lf bases often not sheathing sts; fls not in spikelets
18. Lvs all basal; fls with 3 petals and many stamens .... **ALISMATACEAE**
18′ At least some lvs cauline; fls not with both 3 petals and many stamens
19. Pls vines
20. Tendrils 0; perianth not at all petal-like; stamens free .... **CANNABACEAE** (*Humulus*)
20′ Tendrils present; perianth petal-like; stamens fused .... **CUCURBITACEAE**
19′ Pls prostrate to erect herbs
21. Fls in heads or umbels
22. Infl a simple or compound umbel .... **APIACEAE**
22′ Infl of 1 or more heads
23. Heads without involucres .... **CHENOPODIACEAE**
23′ Each head subtended by an involucre
24. Stamens free; petals free or 0 .... **APIACEAE**
24′ Stamens fused by their anthers; petals fused or 0 .... **ASTERACEAE**
21′ Fls in axillary clusters, racemes, or panicles
25. Lvs deeply lobed or compound
26. Lvs palmately compound, the lflets serrate .... **CANNABACEAE** (*Cannabis*)
26′ Lvs pinnately dissected or compound, the lobes or lflets entire to coarsely toothed
27. Lvs simple or 1-pinnate; anthers subsessile, 3–6 mm — sepals 3–9, infl a leafy raceme or panicle .... **DATISCACEAE**
27′ Lvs 2–4-pinnate; anthers exserted on slender filaments
28. Lflets coarsely crenate or shallowly lobed; sepals 4–7; petals 0; anthers linear, 1.5–5 mm; infl an open panicle .... **RANUNCULACEAE** (*Thalictrum*)
28′ Lflets finely serrate; sepals 5; petals 5; anthers oval, < 1 mm; infl a panicle of dense spikes .... **ROSACEAE** (*Aruncus*)
25′ Lvs entire or toothed
29. Blades of lvs with several main veins; perianth parts 6 mm or more .... **LILIACEAE** (*Veratrum*)
29′ Blades of lvs with 1–3 main veins; perianth parts gen 5 mm or less
30. Lvs covered with stellate hairs .... **EUPHORBIACEAE** (*Croton*)
30′ Lvs without stellate hairs
31. Lvs bearing bead-like, sessile hairs or powdery scales .... **CHENOPODIACEAE** (*Atriplex*)
31′ Lvs glabrous
32. Stipules 0; sepals 3–5, in a single series or strongly overlapping .... **AMARANTHACEAE** (*Amaranthus*)
32′ Stipules present, forming a sheath around the st; sepals 6, in two series **POLYGONACEAE** (*Rumex*)

Group 5: Monocots and Dicots; Flowers Replaced by Bulblets that are Dispersed in Place of Seeds and Fruits

1. Lvs linear, sheathing at base
2. Pls with strong odor of onion or garlic; bulblets in heads or umbels .... **LILIACEAE** (*Allium vineale*)
2′ Pls non-scented; bulblets in panicles .... **POACEAE** (*Poa bulbosa*)
1′ Lvs wide, palmately lobed, not sheathing at base
3. Lvs sessile or with short, wide petioles; lf blade entire or sharply toothed ...... **SAXIFRAGACEAE** (*Saxifraga*)
3′ Lvs with slender petioles; lf blade palmately lobed
4. Pls stout, gen > 50 cm; lobes of lvs with sharp teeth .... **RANUNCULACEAE** (*Aconitum columbianum*)
4′ Pls slender, gen < 40 cm; lobes of lvs with rounded teeth .... **SAXIFRAGACEAE** (*Lithophragma*)

Group 6: Herbaceous Monocots and Dicots in Pistillate Condition, Producing Ovules within Ovaries; Staminate or Bisexual Flowers Absent

1. Pls fully aquatic, submersed, floating in water, or stranded on mud
 2. Pls gen free-floating, the whole pl < 15 mm, often < 5 mm; lvs and sts not differentiated; roots 0 or 1–few, unbranched — pistil emerging from tiny flap of tissue .... **LEMNACEAE**
 2′ Pls rooted in bottom sediments or free-floating, gen >> 15 mm; lvs and sts clearly differentiated; roots often branched
  3. Lvs alternate
   4. Lf tips shallowly 2-toothed; pls of inland saltwater habitats .... **CYMODOCEACEAE**
   4′ Lf tips entire; pls of freshwater or marine habitats
    5. Frs borne in umbel-like clusters .... **POTAMOGETONACEAE** (*Ruppia*)
    5′ Frs or fls borne in spikes or heads
     6. Fls in dense, spheric heads, these solitary or in axillary or terminal clusters, not enclosed in lf sheaths; freshwater habitats .... **TYPHACEAE** (*Sparganium*)
     6′ Fls in axillary spikes, these gen enclosed and concealed in sheaths of subtending lvs; marine habitats .... **ZOSTERACEAE**
  3′ Lvs opposite or whorled or all basal
   7. Lvs all basal; fls solitary in lf-axils; style very elongated, often > 5 cm .... **JUNCAGINACEAE** (*Lilaea*)
   7′ Lvs cauline; fls axillary or variously clustered; styles gen << 5 cm (exc in some Hydrocharitaceae)
    8. Lf blades (at least those of submersed lvs) divided into linear lobes
     9. Blades of lvs repeatedly forked .... **CERATOPHYLLACEAE**
     9′ Blades of lvs pinnately forked .... **HALORAGACEAE** (*Myriophyllum*)
    8′ Lf blades entire, toothed, or shallowly lobed
     10. Petals and sepals both 3, evident; perianth and stigmas borne at water surface at end of long, tubular hypanthium; ovary inferior, sessile in lf axil .... **HYDROCHARITACEAE**
     10′ Petals and sepals both 0; stigmas submersed, borne at end of short styles; ovary or ovaries superior, sessile or short-stalked in lf axil
      11. Pistil 1, compound, bearing 2–4 slender stigmas; lvs subentire to finely or coarsely toothed .... **HYDROCHARITACEAE** (*Najas*)
      11′ Pistils 2–10, simple, each with a cup-like stigma; lvs entire .... **ZANNICHELLIACEAE**
1′ Pls terrestrial or, if growing in wet places, rooted in place and extending well above water surface
 12. Pls parasitic on sts of woody host pls
  13. Fls of parasite borne directly on sts of host; remainder of parasite internal within tissues of host; sts and lvs not differentiated .... **RAFFLESIACEAE**
  13′ Fls borne on leafy branches of parasite; shoots of parasite external; lvs differentiated, though sometimes reduced to scales .... **VISCACEAE**
 12′ Pls free-living
  14. Lvs opposite or whorled, not all basal
   15. Sts or lvs thick and fleshy; infls terminal spikes
    16. Stipules present; stigma 1, head-like; ovary chambers 4; ovaries joining into a fleshy multiple fr; perianth 0 .... **BATACEAE**
    16′ Stipules 0; stigmas 2, linear; ovary chamber 1; ovaries maturing as utricles, sometimes surrounded by fleshy bracts and perianth elements .... **CHENOPODIACEAE**
   15′ Sts and lvs of normal texture, not thick and fleshy; infls various
    17. Lvs palmately compound or simple and deeply lobed .... **CANNABACEAE**
    17′ Lvs simple, entire or toothed
     18. Lvs whorled, sessile or nearly so; margin entire or minutely serrate; stinging hairs 0; ovary inferior, 2-lobed .... **RUBIACEAE**
     18′ Lvs opposite, conspicuously petioled; margin serrate, dentate, or crenate; stinging hairs present; ovary superior, unlobed .... **URTICACEAE**
  14′ Lvs alternate or all basal
   19. Lvs stiff and sword-like, 0.5–1.5 m; infl a large panicle; perianth parts 6 .... **LILIACEAE** (*Nolina*)
   19′ Lvs not sword-like, often smaller; infls various; perianth parts mostly other than 6
    20. Blades of lvs linear or narrowly lanceolate, simple and entire; veins parallel; lf bases sheathing st
     21. Lvs all basal; fls solitary in lf axils; styles very elongated, often > 5 cm ..... **JUNCAGINACEAE** (*Lilaea*)
     21′ Lvs basal and cauline or all cauline; fls in spikes or spikelets, these gen in secondary clusters; styles gen << 5 cm
      22. St triangular; nodes not swollen; lf blades often channeled .... **CYPERACEAE**
      22′ St round; nodes gen swollen and knot-like; lf blades gen flat .... **POACEAE**
    20′ Blades of lvs variously shaped, sometimes toothed, lobed, or compound; veins mostly pinnate or palmate; lf bases often not sheathing sts — fls not in spikelets
     23. Lvs all basal; fls with 3 petals and many free pistils .... **ALISMATACEAE**
     23′ At least some lvs cauline; fls not with both 3 petals and many free pistils
      24. Pls vines
       25. Tendrils 0; perianth not at all corolla-like; ovary superior .... **CANNABACEAE** (*Humulus*)
       25′ Tendrils present; perianth corolla-like; ovary inferior .... **CUCURBITACEAE**

24′ Pls prostrate to erect herbs
26. Fls in heads or umbels
27. Infl a simple or compound umbel .... **APIACEAE**
27′ Infl of 1 or more heads
28. Heads without involucres; ovaries superior .... **CHENOPODIACEAE**
28′ Each head subtended by an involucre; ovaries inferior
29. Heads borne in umbels .... **APIACEAE**
29′ Heads borne in cymes, panicles, or racemes .... **ASTERACEAE**
26′ Fls in axillary clusters or in racemes or panicles
30. Lvs deeply lobed or compound
31. Lvs palmately compound, the lflets serrate .... **CANNABACEAE** (*Cannabis*)
31′ Lvs pinnately dissected or compound, the lobes or lflets entire to coarsely toothed
32. Lvs simple or 1-pinnate; ovary inferior — sepals 3–9; infl a leafy raceme or panicle **DATISCACEAE**
32′ Lvs 2–4-pinnate; ovaries 2–many per fl, superior
33. Lflets coarsely crenate or shallowly lobed; sepals 4–7; petals 0; anthers linear, 1.5–5 mm; infl an open panicle .... **RANUNCULACEAE** (*Thalictrum*)
33′ Lflets finely serrate; sepals 5; petals 5; anthers oval, < 1 mm; infl a panicle of dense spikes .... **ROSACEAE** (*Aruncus*)
30′ Lvs entire or toothed
34. Blades of lvs with several main veins; perianth parts 6 mm or more .... **LILIACEAE** (*Veratrum*)
34′ Blades of lvs with 1–3 main veins; perianth parts gen 5 mm or less
35. Lvs densely stellate; ovary chambers gen 3; fr a capsule .... **EUPHORBIACEAE** (*Croton*)
35′ Lvs not stellate; ovary chamber 1; fr an achene or utricle
36. Lvs bearing bead-like, sessile hairs or powdery scales .... **CHENOPODIACEAE** (*Atriplex*)
36′ Lvs glabrous
37. Stipules 0; ovary ovoid, not angled; sepals 3–5, in a single series or strongly overlapping, never adhering to ovary .... **AMARANTHACEAE** (*Amaranthus*)
37′ Stipules present, forming a sheath around the st; ovary triangular; sepals 6, in 2 series, the inner sometimes adherent to the ovary .... **POLYGONACEAE** (*Rumex*)

Group 7: Gymnosperms; Seeds Produced But Not Enclosed in Pistils; Flowers Not Produced

1. Seeds borne in woody cones
2. Subtending bracts present, ± free from cone scales, (sometimes hidden by them); lvs needle-like .... **PINACEAE**
2′ Subtending bracts not evident, fused to cone scales; lvs scale-like, awl-like, or needle-like
3. Lf arrangement opposite or whorled; lvs scale-like or awl-like .... **CUPRESSACEAE**
3′ Lf arrangement alternate; lvs needle-like, awl-like, or scale-like .... **TAXODIACEAE**
1′ Seeds naked, enclosed in a fleshy cone, or surrounded by a cluster of perianth-like scales
4. Sts jointed at nodes, the long, green internodes serving as the main photosynthetic surfaces for the pl; lvs dry, scale-like, often early deciduous, not overlapping; seed(s) surrounded by perianth-like scales; pollen cones with stamen-like structures, the pollen sacs borne on stout filaments .... **EPHEDRACEAE**
4′ Sts not jointed at nodes, the internodes gen very short; lvs green, scale-like, awl-like, or needle-like, persistent, crowded, sometimes overlapping, serving as the main photosynthetic surfaces for the pl; seeds enclosed in fleshy cones or subtended or surrounded by fleshy arils; pollen cones with sessile or stalked pollen sacs
5. Lvs and twigs opposite or whorled; lvs 1–12 mm, scale-like or awl-like; seeds 1–3, enclosed in a fleshy, berry-like cone; pollen cones without stamen-like structures, the pollen sacs sessile on undersurfaces of cone scales .... **CUPRESSACEAE** (*Juniperus*)
5′ Lvs and twigs alternate; lvs 12–70 mm, needle-like; seed 1, ± enclosed by a fleshy, gen colored aril; pollen cones with stamen-like structures, the pollen sacs borne on stout filaments .... **TAXACEAE**

Group 8: Woody Plants, Mostly Angiosperms in Ovule-Producing Condition Only; Staminate or Bisexual Flowers Absent

1. Pls parasitic on sts of woody host pls .... **VISCACEAE**
1′ Pls free-living
2. Trees; trunk unbranched, covered with persistent woody lf bases; lvs to several meters, pinnately compound .... **ARECACEAE** (*Phoenix*)
2′ Trees or shrubs; trunk(s) gen branched, gen not covered with persistent lf bases; lvs gen much smaller, simple or compound
3. Lvs stiff and sword-like, 5–45 dm; infl a large panicle .... **LILIACEAE** (*Nolina*)
3′ Lvs not sword-like, mostly smaller; infls various
4. Lvs opposite or whorled
5. Lvs compound

6. Lvs palmate; lflets 3 .... **ACERACEAE** (*Acer glabrum*)
6′ Lvs pinnate; lflets often > 3
7. Woody vine; sepals gen 4, conspicuous, petal-like; pistils many; styles elongated and plumose in fr .... **RANUNCULACEAE** (*Clematis*)
7′ Tree or large shrub; sepals gen 5 or number difficult to determine, very small; pistil 1; styles never plumose
8. Pistil with 2 styles and 2 short wings; fr 2-winged, often splitting into 2 one-seeded segments .... **ACERACEAE** (*Acer negundo*)
8′ Pistil with 1 style, not winged in fl; fr a 1-winged achene .... **OLEACEAE** (*Fraxinus*)
5′ Lvs simple
9. Lvs and sometimes sts thick and fleshy; lvs sometimes reduced to fleshy scales
10. Stipules present; stigma 1, head-like; ovary chambers 4; ovaries joining into a fleshy, multiple fr; perianth 0 .... **BATACEAE**
10′ Stipules 0; stigmas 2, linear; ovary chamber 1; ovaries maturing as utricles, sometimes surrounded by fleshy bracts and perianth segments .... **CHENOPODIACEAE**
9′ Lvs and sts not fleshy
11. Margins of lvs toothed .... **EUPHORBIACEAE** (*Tetracoccus*)
11′ Margins of lvs entire
12. Weak-stemmed subshrub
13. Perianth parts in 3's, free; ovary superior, unlobed, smooth; NCo .... **EMPETRACEAE**
13′ Perianth parts in 4's, fused; ovary inferior, ± bilobed, smooth to covered with short bristles; widespread .... **RUBIACEAE** (*Galium*)
12′ Stouter shrub or tree
14. Sts and lvs covered with silvery scales .... **ELAEAGNACEAE** (*Shepherdia*)
14′ Sts and lvs glabrous or variously hairy
15. Fls in elongate, pendent catkins .... **GARRYACEAE**
15′ Fls not in catkins
16. Sts green, jointed; lvs thin, dry, scarious, scale-like, often early-deciduous .... **EPHEDRACEAE**
16′ Sts gray or brown or only the very young sts greenish, not jointed; lvs thin or thick, green, not dry and scarious, not scale-like, persistent or deciduous
17. Lvs thick, leathery, evergreen; fls solitary; sepals well developed .... **SIMMONDSIACEAE**
17′ Lvs thin, deciduous; fls in sessile, axillary clusters or short-peduncled racemes; sepals very small
18. Fls subsessile, solitary or few in axils; styles 3–4; ovary 3–4-lobed; fr a capsule .... **EUPHORBIACEAE** (*Tetracoccus*)
18′ Fls pedicelled, in axillary clusters; style 1; ovary unlobed; fr a drupe ... **OLEACEAE** (*Forestiera*)
4′ Lvs alternate
19. Fls in catkins or catkin-like spikes
20. Shrubs of ± saline desert habitats; lvs ± thick and fleshy or very narrow and densely hairy; spikes erect or very short and axillary .... **CHENOPODIACEAE**
20′ Shrubs or trees, gen of moister habitats; lvs thin and wide; spikes or catkins erect or pendent
21. Stipules 0
22. Lvs sessile; blades grayish, entire, not resin-dotted; fls individually enclosed by a pair of appressed bractlets; frs not wax-coated, either hairy or covered with flour-like scales .... **CHENOPODIACEAE**
22′ Lvs petioled; blades green, crenate or serrate, resin-dotted; fls subtended by 2–4 bractlets; frs ± wax-coated .... **MYRICACEAE**
21′ Stipules present, evident at least on new growth, sometimes deciduous but leaving evident scars
23. Styles repeatedly branched into many fine, thread-like divisions; ovary 3-lobed, maturing as a capsule; brittle-stemmed shrub .... **EUPHORBIACEAE** (*Acalypha*)
23′ Styles or style branches 2–4, gen not further divided; ovary gen unlobed, the frs various; sts mostly not brittle
24. Catkins very dense at time of flowering; only stigmas exserted beyond tips of tightly appressed bractlets; at maturity either catkin ± cone-like with ovaries ripening as winged achenes or bractlets expanding into 1 or more tubular involucres, each surrounding a nut **BETULACEAE**
24′ Catkins looser at time of flowering; bractlets, if present, not tightly appressed, the ovaries or calyces gen visible on close inspection; at maturity either catkin becoming a ± open cluster of capsules or a fleshy multiple fr
25. Ovary tightly enwrapped by 4 appressed sepals; catkin bractlets 0; sepals fleshy at maturity, together with ovaries joining as a multiple fr .... **MORACEAE** (*Morus*)
25′ Ovary not enclosed by sepals, the calyx 0 or represented by an entire or slightly lobed rim; fls subtended by bractlets, these sometimes deciduous; each ovary ripening as a capsule containing several-many hair-tufted seeds .... **SALICACEAE**
19′ Fls not in catkins or spikes — ovules never produced in cone-like catkins
26. Lvs palmately lobed
27. Fls in terminal, cymose panicles; shrubs or herbs; bark not flaking in large plates; lvs peltate, glabrous .... **EUPHORBIACEAE** (*Ricinus*)

27′ Fls in axillary or terminal clusters of spheric heads; large trees; bark flaking from trunks in large plates; lvs obtuse to cordate at base, densely tomentose when young .... **PLATANACEAE**
26′ Lvs not palmately lobed
28. Sts and lvs stellate-hairy
29. Ovary clearly superior, 2–4-lobed; fr a capsule; involucre 0; subshrub with entire lvs or desert shrub with crenate lvs .... **EUPHORBIACEAE** (*Bernardia, Croton*)
29′ Ovary inferior but sepals very small and often concealed by bractlets of involucre; fr either an ovoid nut, individually subtended by a cup-like involucre, or a triangular nut, surrounded in groups of 1–3 by a bur-like involucre; shrub or tree with lvs entire, toothed, or pinnately lobed but not crenate .... **FAGACEAE**
28′ Sts and lvs glabrous or hairy but not stellate
30. Lvs compound
31. Petals 0; ovary inferior — fr a nut with a ± dehiscent husk .... **JUGLANDACEAE**
31′ Petals present; ovary or ovaries superior
32. Lflets entire or teeth not gland-tipped; ovary 1, bearing 3 short styles; fr a drupe **ANACARDIACEAE**
32′ Lflets with a few, gland-tipped teeth near base; ovaries 5–6, joined only at style; fr a cluster of winged achenes .... **SIMAROUBACEAE** (*Ailanthus*)
30′ Lvs simple
33. Corolla present; perianth parts in 2 series
34. Fl parts in 3's
35. Low shrubs; fls solitary in lf axils; lvs needle-like .... **EMPETRACEAE**
35′ Woody vines; fls in umbels; lvs ovate .... **LILIACEAE** (*Smilax*)
34′ Fl parts gen in 4's, 5's, or 6's
36. Fls in involucred heads, these gen secondarily clustered; ovary 1 per fl, inferior; calyx modified as a white to brownish pappus .... **ASTERACEAE**
36′ Fls in racemes; ovaries 1–5, superior; calyx green, not modified .... **ROSACEAE**
33′ Corolla 0; perianth parts 0 or in 1 series
37. Lvs needle-like; the apparent ovary actually a naked ovule .... **TAXACEAE**
37′ Lvs linear to ovate or wider but not needle-like; true ovary present
38. Fls subtended or surrounded at base by an involucre of tiny scale-like bractlets; ovary inferior to the minute calyx; fr an ovoid nut, subtended by a cup-like involucre ...... **FAGACEAE** (*Quercus*)
38′ Fls without an involucre; ovary superior or seemingly inferior between a pair of tightly appressed bractlets; fr an achene, utricle, capsule, or drupe
39. Each fl individually enclosed by a pair of tightly appressed bractlets; fr an achene or utricle .... **CHENOPODIACEAE**
39′ Fls not enclosed by paired bractlets; fr a capsule or drupe
40. Lvs narrow; bases symmetric; shrubs .... **EUPHORBIACEAE** (*Tetracoccus*)
40′ Lvs wide; bases oblique; trees .... **ULMACEAE** (*Celtis*)

Group 9: Monocots and some Dicots; Flower Parts in Threes; At Least the Inner Perianth Parts Petal-like

1. Pls aquatic, floating on water surface, stranded on mud, or rooting in bottom sediments; sts often weak and unable to support pls out of water, not extending much above water surface
2. Ovary very deeply lobed or pistils > 1
3. Perianth part gen 1, petal-like; pistils 3–6; frs dehiscent; seeds 2–8 .... **APONOGETONACEAE**
3′ Sepals 3; petals 3; pistils or lobes of ovary 4–many; frs indehiscent; seeds 1–2
4. Lvs not peltate, linear to ovate or sagittate, not gelatinous; fls several–many in open panicles; sepals green; petals white to pink .... **ALISMATACEAE**
4′ Lvs peltate, the lower surfaces and petioles covered with gelatinous material; fls solitary; sepals and petals all similar, reddish purple .... **CABOMBACEAE**
2′ Ovary not or only slightly lobed
5. Stout, free-floating herbs; lvs thick, often with inflated petioles .... **PONTEDERIACEAE** (*Eichhornia*)
5′ Weak-stemmed, submersed herbs; lvs thin, sessile
6. Lvs opposite or whorled; fls unisexual; petals white .... **HYDROCHARITACEAE**
6′ Lvs alternate; fls bisexual; perianth parts pale yellow .... **PONTEDERIACEAE** (*Heteranthera*)
1′ Pls terrestrial or, if growing in water, extending well above water surface; sts able to support pls out of water
7. Woody vines; lvs ovate; sts gen prickly; fls unisexual; dioecious .... **LILIACEAE** (*Smilax*)
7′ Trees, shrubs, or herbs; lvs various; sts gen not prickly; fls bisexual or unisexual; monoecious, dioecious, or with both perfect and unisexual fls
8. Trees, shrubs, or fleshy rosette pls; lvs large, stiff
9. Lvs pinnately compound or palmately lobed, never fleshy; trees; trunk unbranched .... **ARECACEAE**
9′ Lvs simple, entire to coarsely toothed or with marginal fibers, often fleshy; trees, shrubs, or fleshy rosette pls; trunk, if present, branched or unbranched .... **LILIACEAE**
8′ Herbs; lvs gen thinner

10. Ovary inferior or partly so
11. Fertile stamens 1–2, anthers and filaments wholly fused above ovary to stigma and style, forming a column; ovary gen with a half twist; fls bilateral, the lower petal very different from the other two in size, shape, or color .... **ORCHIDACEAE**
11′ Fertile stamens 3 or more, anthers free, filaments free or fused to hypanthium; ovary without half twist; fls radial or bilateral
12. Lvs cordate; stamens 12; perianth parts 3, dark red or brown .... **ARISTOLOCHIACEAE** (*Asarum*)
12′ Lvs linear; stamens 3–6; perianth parts 6, mostly not dark red or brown
13. Stamens 3; lvs 2-ranked, gen folded lengthwise and apparently attached "on edge", only the bottom side of the lf visible .... **IRIDACEAE**
13′ Stamens gen 6; lvs not 2-ranked and folded, top and bottom sides both visible or lvs cylindric **LILIACEAE**
10′ Ovary or ovaries superior
14. Ovary very deeply lobed or pistils > 1 per fl
15. Pl < 10 cm, ann; lvs opposite, < 5 mm; fls 2 per lf pair .... **CRASSULACEAE** (*Crassula*)
15′ Pl gen > 10 cm, per; lvs basal or alternate; infl a raceme, spike, or panicle, fls gen many
16. Pistils or lobes of ovary 6 or more; sepals green; petals white to pink .... **ALISMATACEAE**
16′ Lobes of ovary 3(4); sepals and petals similar, both gen petal-like, the colors various ...... **LILIACEAE**
14′ Ovary unlobed or only shallowly lobed
17. Stamens 5; fl with well developed spur .... **BALSAMINACEAE**
17′ Stamens 6; fl without spur
18. One stamen > other 5, its filament toothed on one side; pls semiaquatic; fls blue; infl a raceme .... **PONTEDERIACEAE** (*Monochoria*)
18′ All stamens similar or 3 large and 3 small, the filaments not toothed; pls gen terrestrial; fls and infl various
19. Outer and inner perianth parts similar, both ± petal-like .... **LILIACEAE**
19′ Outer perianth sepal-like, different from inner
20. Infl an umbel .... **COMMELINACEAE**
20′ Infl various, not an umbel .... **LILIACEAE**

Group 10: Monocots; Perianth Inconspicuous or Absent

1. Lf blade wide; veins pinnate or palmate
2. Herbs with simple, entire or hastate lvs; infl a fleshy spike subtended by a gen showy bract .... **ARACEAE**
2′ Trees with large, pinnately compound or palmately lobed lvs; infl a panicle subtended by a ± woody bract .... **ARECACEAE**
1′ Lf blade linear to narrowly lanceolate (wider if floating) or 0; veins parallel
3. Pls aquatic, submersed or floating
4. Pl < 1 cm, floating or stranded on shore, not differentiated into sts and lvs; roots unbranched or 0 **LEMNACEAE**
4′ Pl >> 1 cm, gen rooted, differentiated into sts and lvs; roots gen present, often branched
5. Lvs opposite or whorled
6. Lvs coarsely to finely toothed, stipules 0 (lf base may have ear-like lobes); pistil 1, compound, with 3–4 stigmas; fr without a stiff beak .... **HYDROCHARITACEAE** (*Najas*)
6′ Lvs entire, stipules free, membranous, entire; pistils > 1, simple, each with 1 stigma; fr with a stiff beak .... **ZANNICHELLIACEAE**
5′ Lvs alternate
7. Infl a dense, spheric head .... **TYPHACEAE** (*Sparganium*)
7′ Infl a spike or fls solitary
8. Pls of marine habitats; stamen 1; pistil 1, compound — infl a flattened spike enwrapped by a sheathing bract .... **ZOSTERACEAE**
8′ Pls of fresh, brackish, or inland saltwater habitats; stamens 2–4; pistils 2–several, simple
9. Lvs gen with 2–3 teeth near tip; fls solitary in axils, unisexual, the pls dioecious; perianth 2 minute protuberances or 0; stamens 2, fused together; pistils 2 .... **CYMODOCEACEAE**
9′ Lvs entire; fls in cylindric spikes, bisexual; perianth segments 4; stamens 2–4, separate; pistils 4 or more .... **POTAMOGETONACEAE**
3′ Pls terrestrial or strongly emergent
10. Fls all unisexual, the infl 1–4 cm diam, either a dense spike or a group of several spheric heads attached to a zigzag main axis .... **TYPHACEAE**
10′ Fls bisexual or, if some or all unisexual, the infls gen < 1 cm diam; infls various
11. Pistils 3(6), free or nearly so, spreading widely in fr .... **SCHEUCHZERIACEAE**
11′ Pistil 1, carpel 1 or carpels 2–6 and strongly fused, sometimes separating in fr
12. Fr a capsule or berry or breaking into segments; ovules 2–many; perianth segments 4–6 in 2 whorls, well developed, all scale-like
13. Infl a panicle or group of head-like clusters or fl solitary; stigmas slender, spreading; fr a capsule .... **JUNCACEAE**

13′ Infl a bractless raceme or a dense spike subtended by a long, flattened bract; fr a gelatinous berry or breaking into segments
14. Infl a dense spike subtended by a long, flattened bract; stigmas strap-shaped, sessile; fr a gelatinous berry .... **ARACEAE** (*Acorus*)
14′ Infl a bractless raceme; stigmas divided into hair-like lobes; fr breaking into segments **JUNCAGINACEAE**
12′ Fr an achene or grain; ovule 1; perianth inconspicuous or 0
15. Lvs cylindric, all basal; pistillate fls axillary, with long, slender styles; bisexual fls in spikes .... **JUNCAGINACEAE** (*Lilaea*)
15′ Lvs flat, cylindric, or scale-like, basal or cauline; infl unit a spikelet of minute fls subtended by scale-like bractlets
16. Sts lfless, the lvs reduced to bladeless basal scales; spikelet 1 .... **CYPERACEAE** (*Eleocharis*)
16′ Sts ± leafy; spikelets > 1
17. Sts gen 3-angled, solid; nodes not swollen; lvs gen 3-ranked; each fl subtended by 1 bract (in *Carex* and *Kobresia* a second, gen hollow, ± flask-shaped bract [perigynium] surrounds each pistillate fl); perianth of bristles or 0; fr an achene .... **CYPERACEAE**
17′ Sts cylindric, the internodes often hollow; nodes swollen, knot-like; lvs 2-ranked or not in obvious ranks; each spikelet gen with 2 bractlets (glumes) at base; each fl gen enclosed by 2 additional bractlets (lemma and palea); perianth of 2 tiny scales or 0; fr a grain .... **POACEAE**

Group 11: Dicots; Pistils Two or More, Simple

1. Perianth 0 or of only one whorl, gen called sepals even when petal-like
2. Hypanthium present .... **ROSACEAE**
2′ Hypanthium 0
3. Trees; lvs palmately lobed; axillary bud covered by petiole base; perianth 0 or reduced to tiny scales; infl a cluster of spheric heads; monoecious; fr a dense head of hairy achenes .... **PLATANACEAE**
3′ Herbs or woody vines; lvs various; axillary bud gen exposed; perianth present, gen well developed; infls various but not spheric heads; fr a cluster of achenes or follicles .... **RANUNCULACEAE**
1′ Perianth of 2 or more whorls or spirals, the outer gen called sepals, the inner called petals
4. Pls aquatic; lvs peltate .... **CABOMBACEAE**
4′ Pls gen terrestrial; lvs not peltate
5. Stamens > 2 times as many as petal-like perianth segments or > 15
6. Pls obviously woody; shrubs or trees
7. Crushed lvs strongly aromatic; lvs opposite .... **CALYCANTHACEAE**
7′ Crushed lvs not aromatic; lvs alternate
8. Lvs glabrous, entire; seeds with arils .... **CROSSOSOMATACEAE**
8′ Lvs hairy, toothed, lobed, or compound; seeds without arils .... **ROSACEAE**
6′ Pls herbaceous or slightly woody at base
9. Hypanthium present .... **ROSACEAE**
9′ Hypanthium 0
10. Sepals enlarging and persisting around frs; stamens maturing from fl center toward edge **PAEONIACEAE**
10′ Sepals deciduous as fl opens or withering as frs mature; stamens maturing from fl edge toward center
11. Sepals 2–3, falling as fl opens, not petal-like; pistils weakly fused in fl, separating in age .... **PAPAVERACEAE** (*Platystemon*)
11′ Sepals 4–many, present in open fls, sometimes petal-like; pistils gen wholly free ... **RANUNCULACEAE**
5′ Stamens 2 times as many as petals or fewer
12. Petals fused together, at least at base
13. Pistils 3 or more; lvs fleshy; sap clear .... **CRASSULACEAE**
13′ Pistils 2; lvs gen not fleshy; sap gen milky
14. Pls prostrate, mat-forming; fls minute, subsessile, axillary, gen concealed by lvs; ovules 1–2 per ovary .... **CONVOLVULACEAE** (*Dichondra*)
14′ Pls erect, prostrate, or twining; fls larger, gen conspicuous, not concealed by lvs; ovules many per ovary
15. Styles 1 or 2; stigmas fused; anthers free from but sometimes lying against style(s) or stigmas .... **APOCYNACEAE**
15′ Styles 2; stigmas fused together and to anthers, the whole unit (style-stigma-anther head) drum-shaped .... **ASCLEPIADACEAE**
12′ Petals free but sometimes attached to a disk-like to tubular hypanthium
16. Lvs and often sts thick and fleshy .... **CRASSULACEAE**
16′ Lvs and sts not very thick and fleshy
17. Hypanthium well developed
18. Ovules gen 1–2 per pistil; lvs simple or compound, gen cauline; stipules gen present; veins often pinnate .... **ROSACEAE**
18′ Ovules gen several–many per pistil; lvs simple, basal or cauline; stipules often 0; veins gen palmate .... **SAXIFRAGACEAE**
17′ Hypanthium 0 or inconspicuous
19. Pls herbaceous

20. Ann; ovule 1 per ovary; lvs pinnately compound or deeply pinnately lobed .... **LIMNANTHACEAE**
20′ Per; ovules several–many per ovary; lvs simple and entire or variously toothed to palmately lobed or palmately compound .... **SAXIFRAGACEAE**
19′ Pls woody
21. Fls solitary in lf axils; fr a cluster of follicles .... **CROSSOSOMATACEAE**
21′ Fls in dense panicles; fr a cluster of winged achenes or ± dry drupes .... **SIMAROUBACEAE**

Group 12: Woody Dicots; Perianth in One Whorl or Absent; Staminate Flowers in Catkins or Catkin-like Spikes

1. Pls parasitic, attached to branches of trees or shrubs, not rooted in soil .... **VISCACEAE**
1′ Pls not parasitic, gen rooted in soil
2. Pls fleshy, ± woody at base; lvs fleshy, scale-like or linear to oblanceolate; catkins erect
3. Stipules present; ovary chambers 4 .... **BATACEAE**
3′ Stipules 0; ovary chamber 1 .... **CHENOPODIACEAE**
2′ Pls non-fleshy shrubs or trees; lvs with well developed flat blades; catkins often drooping
4. Lvs opposite .... **GARRYACEAE**
4′ Lvs alternate
5. Lvs compound .... **JUGLANDACEAE**
5′ Lvs simple
6. Fr a capsule; seeds several–many, each bearing a tuft of hairs; fls subtended by minute glands or fringed bractlets .... **SALICACEAE**
6′ Fr dry or fleshy, indehiscent; seed 1, without a tuft of hairs; bractlets, if any, not fringed
7. Sap milky; sepals of pistillate infl enlarged and fleshy in fr, ripening together with the ovaries as a spheric to ellipsoid or ovoid multiple fr .... **MORACEAE**
7′ Sap clear; sepals 0 or very small, not becoming fleshy; ovaries ripening as individual drupes, nuts, or nutlets
8. Staminate catkins short and densely fld, lacking both bractlets and perianth parts; foliage dotted with minute, bead-like resin glands, sometimes aromatic; stipules 0; fr sometimes ± fleshy, often coated with wax .... **MYRICACEAE**
8′ Staminate catkins elongated, either flexible and pendent or stiff and spike-like; sepals and/or conspicuous bractlets present in staminate catkins; foliage lacking resin glands (exc in some Betulaceae), not noticeably aromatic; stipules present but frequently deciduous; fr dry, not wax-coated
9. Frs many, small, flattened, and often winged, clustered in a cone-like infl; pls deciduous; pistillate fls without sepals .... **BETULACEAE**
9′ Frs 1–3, each a well developed nut, subtended or surrounded by an involucre; pls deciduous or evergreen; pistillate fls with minute sepals — ovary inferior
10. Fr a nut enclosed in a tubular, husk-like involucre; involucral bractlets few, papery; staminate catkins drooping, densely fld, bractlets ± concealing fls; deciduous shrub .... **BETULACEAE** (*Corylus*)
10′ Fr a nut in a scaly cup or 1–3 nuts enclosed in a spiny or prickly bur; involucral bractlets many, scaly or spiny; staminate catkins very slender and drooping or stiff and spike-like, bractlets inconspicuous; deciduous or evergreen trees or shrubs .... **FAGACEAE**

Group 13: Woody Dicots; Perianth in One Whorl or Absent; Inflorescences Not Catkins

1. Infl a dense spike, spheric head, or fls enclosed within a fleshy, hollow receptacle
2. Lvs pinnately compound or modified into linear or oblong, entire, flattened axes without lflets — fr a legume .... **FABACEAE** (Mimosoideae)
2′ Lvs simple, ovate or wider in overall outline, entire or palmately lobed
3. Fls enclosed within a hollow receptacle; the whole infl ripening as an externally smooth, fleshy, multiple fr; petiole base not covering axillary bud; sap milky .... **MORACEAE** (*Ficus*)
3′ Fls in heads or short, dense spikes; pistillate infl ripening as an externally rough, spheric head or spike
4. Petiole bases not expanded or covering axillary buds; stipules membranous, free, early-deciduous; lf-blades entire to crenate or serrate, sometimes irregularly few-lobed, glabrous or sparsely short-hairy; ripened infl a fleshy spike or head; sap milky .... **MORACEAE** (*Maclura, Morus*)
4′ Petiole bases expanded, covering axillary buds; stipules large and green, fused around the st, forming a sheath and conspicuously expanded blades, persistent or tardily deciduous; lf-blades deeply palmately lobed, the lobes entire or coarsely to finely dentate or serrate, densely tomentose when young; ripened infl a dry head of long-hairy achenes; sap clear .... **PLATANACEAE**
1′ Infl not a dense spike, spheric head, nor enclosed in a fleshy receptacle
5. Pls parasitic, attached to branches of trees or shrubs, not terrestrial .... **VISCACEAE**
5′ Pls not obviously parasitic, gen rooted in soil (sometimes root-parasitic)
6. Ovary inferior, partly so, or appearing so
7. Sts and lvs covered with silvery scales .... **ELAEAGNACEAE**
7′ Sts and lvs without silvery scales
8. Infl unit an involucred head (superficially resembling a single fl) .... **ASTERACEAE**

8′ Infl open or fls solitary or paired
9. Trees, evergreen; stamens many; lvs aromatic .... **MYRTACEAE**
9′ Shrubs or woody vines, evergreen or deciduous; stamens 5–12; lvs not aromatic
10. Lvs opposite .... **CAPRIFOLIACEAE** (*Lonicera*)
10′ Lvs alternate
11. Fls bilateral; woody vine .... **ARISTOLOCHIACEAE** (*Aristolochia*)
11′ Fls radial; subshrub .... **SANTALACEAE**
6′ Ovary superior
12. Lvs opposite
13. Fr winged, indehiscent; lvs gen lobed or compound
14. Stigma 1, entire or slightly 2-lobed; fr wings 2 or sometimes 3; lvs palmately lobed or compound or pinnately compound .... **ACERACEAE**
14′ Stigmas 2, elongate; fr wing 1; lvs gen pinnately compound .... **OLEACEAE** (*Fraxinus*)
13′ Fr not winged; lvs simple, entire or toothed
15. Fls bisexual; stamens > 25 .... **ROSACEAE** (*Coleogyne*)
15′ Fls unisexual; stamens gen 4–12
16. Lvs covered with silvery scales .... **ELAEAGNACEAE** (*Shepherdia*)
16′ Lvs glabrous or ± hairy
17. Staminate and pistillate fls in sessile umbels .... **OLEACEAE** (*Forestiera*)
17′ Staminate fls in stalked clusters; pistillate fls solitary
18. Staminate infl head-like, fls subsessile; pistillate fls stalked; fr unlobed .... **SIMMONDSIACEAE**
18′ Staminate infl open, raceme or panicle; staminate and pistillate fls clearly stalked; fr 4–5-lobed .... **EUPHORBIACEAE** (*Tetracoccus*)
12′ Lvs alternate
19. Lvs pinnately compound .... **ANACARDIACEAE**
19′ Lvs simple, sometimes lobed
20. Infl fl-like with a cup-like involucre bearing colored, often petal-like nectaries; each stamen (actually a staminate fl) with a jointed stalk (composed of a filament and a pedicel); pistil (actually pistillate fl) 3-lobed, exserted from involucre; sap milky .... **EUPHORBIACEAE** (*Euphorbia*)
20′ Infls various, but not fl-like; stamens gen > 1 per fl, the filaments not jointed; pistils various; sap mostly clear
21. Fls strongly bilateral; fr an indehiscent, spiny pod .... **KRAMERIACEAE**
21′ Fls radial or nearly so; fr not spiny
22. Anthers opening by pores with hinged lids; lvs strongly aromatic .... **LAURACEAE**
22′ Anthers opening by slits; lvs gen not aromatic
23. Sepals 1.5–4.5 cm, petal-like, yellow or orange; filaments fused into a tube around style; ovules > 3 per chamber .... **STERCULIACEAE** (*Fremontodendron*)
23′ Sepals gen < 1.5 cm, mostly not petal-like, yellow, or orange; filaments not fused around style; ovules 1–2 per chamber
24. Lvs gen > 15 cm, palmately veined and lobed, peltate; ovary prickly; stamens very many, the filaments irregularly joined in clusters .... **EUPHORBIACEAE** (*Ricinus*)
24′ Lvs gen < 15 cm, pinnately veined, gen unlobed, not peltate; ovary without prickles; stamens fewer, not joined in clusters
25. Pls covered with grayish or silvery scales
26. Fls tubular, some or all bisexual; trees or large shrubs; lvs linear-lanceolate .... **ELAEAGNACEAE** (*Elaeagnus*)
26′ Fls not tubular, all unisexual; dioecious shrub or subshrub; lvs various
27. Ovary of pistillate fl enclosed by 2, appressed bractlets; lf surface powder-like, hairs not stellate .... **CHENOPODIACEAE** (*Atriplex*)
27′ Ovary of pistillate fl subtended by 5 sepals; hairs scale-like, stellate **EUPHORBIACEAE** (*Croton*)
25′ Pls without scales
28. Fl parts in 3's
29. Fls solitary, axillary, unisexual; fr a berry .... **EMPETRACEAE**
29′ Fls in involucred umbels, bisexual; fr an achene .... **POLYGONACEAE** (*Eriogonum*)
28′ Fl parts gen in 4's or 5's
30. Stamens as many as and alternating with the sepals .... **RHAMNACEAE**
30′ Stamens opposite the sepals or more numerous
31. Style and stigma 1
32. Style feathery with long, stiff hairs; fr an achene surrounded by a tubular hypanthium; branches stiff .... **ROSACEAE** (*Cercocarpus*)
32′ Style glabrous; fr an exposed drupe; branches pliable .... **THYMELAEACEAE**
31′ Styles or stigmas 2–4
33. Fls bisexual
34. Shrub; lvs mostly narrow, the base symmetric; fr an achene or utricle, not flattened or winged .... **CHENOPODIACEAE**
34′ Tree; lvs wide, flat, the base gen oblique; fr a drupe or round to ovate, winged achene .... **ULMACEAE**

33′ Fls unisexual
35. Large shrubs or small trees; lvs 3–10 cm, green; lf bases oblique; fr a drupe .... **ULMACEAE** (*Celtis*)
35′ Shrubs 2.5 m or less; lvs gen < 3 cm, sometimes gray; lf bases gen symmetric; fr dry
36. Calyx of pistillate fls 0, the ovary enclosed by a pair of tightly appressed bractlets; ovary unlobed; fr indehiscent, seed 1 .... **CHENOPODIACEAE**
36′ Calyx of pistillate fls well developed; ovary 2–3-lobed; fr dehiscent, seeds 2 or more .... **EUPHORBIACEAE**

Group 14: Herbaceous Dicots; Perianth in One Whorl or Absent; Ovary Inferior

1. Pls internal st-parasites on desert shrubs .... **RAFFLESIACEAE**
1′ Pls free-living or root-parasites
2. Perianth parts free or nearly so
3. Infl a simple or compound umbel .... **APIACEAE**
3′ Infl various
4. Pls aquatic, floating or stranded on mud
5. Lvs whorled; blades of submersed lvs pinnately divided into thread-like lobes .... **HALORAGACEAE**
5′ Lvs opposite; blades all entire .... **ONAGRACEAE** (*Ludwigia*)
4′ Pls terrestrial or sometimes growing in damp soil
6. Infl bracts large, petal-like .... **SAURURACEAE**
6′ Infl bracts not at all petal-like
7. Fls solitary, axillary
8. Sepals 3–5, green; sts and lvs covered with bead-like hairs; lvs many, hastate or tapering to a short, winged petiole .... **AIZOACEAE** (*Tetragonia*)
8′ Sepals 3, brownish purple; sts and lvs glabrous to soft-hairy; lvs few, deeply cordate, long-petioled .... **ARISTOLOCHIACEAE** (*Asarum*)
7′ Fls in erect spikes or racemes
9. Lvs all basal, very large, palmately lobed .... **GUNNERACEAE**
9′ Lvs cauline, entire, toothed, or pinnately lobed
10. Style 1, often very short, unbranched, bearing an entire or 2–3-lobed stigma; slender ann; lvs not lobed .... **CAMPANULACEAE**
10′ Styles 3, elongated, each forked; robust per; some lvs gen pinnately lobed .... **DATISCACEAE**
2′ Perianth parts fused
11. Infls of 1–many involucred heads
12. Ovary truly inferior; anthers fused around style or, if free, the fls unisexual; stigma lobes 2 ... **ASTERACEAE**
12′ Ovary actually superior, but appearing inferior (base tightly enwrapped by perianth base); anthers free; stigma unlobed .... **NYCTAGINACEAE**
11′ Infls various but not of involucred heads
13. Lvs alternate or all basal
14. Tendrils present; fls unisexual .... **CUCURBITACEAE**
14′ Tendrils 0; fls bisexual
15. Lvs widely ovate, deeply cordate; fls large, solitary, dark brownish purple; perianth lobes 3 .... **ARISTOLOCHIACEAE**
15′ Lvs linear to narrowly elliptic, acute to rounded; fls smaller, in terminal cymes, greenish white or purplish tinged; perianth lobes 5 .... **SANTALACEAE**
13′ Lvs opposite or whorled
16. Perianth lobes 4 or fewer
17. Stamens 8; lvs with rounded teeth .... **SAXIFRAGACEAE** (*Chrysosplenium*)
17′ Stamens 3 or 4; lvs entire
18. Pls aquatic; perianth green .... **ONAGRACEAE** (*Ludwigia*)
18′ Pls terrestrial; perianth white, yellow, or pale pink .... **RUBIACEAE**
16′ Perianth lobes 5 or more
19. Fls bilateral; ovary truly inferior; lvs of a pair equal .... **VALERIANACEAE**
19′ Fls gen radial or nearly so; ovary actually superior (tightly enwrapped by hardened or winged calyx base); lvs of a pair equal or unequal
20. Calyx green, not at all corolla-like, very inconspicuous; calyx lobes erect, free or nearly so above the level of the ovary, thickened and linear; lvs narrowly linear, needle-like, gen 1–2 cm or shorter, gen those of a pair equal .... **CARYOPHYLLACEAE** (*Scleranthus*)
20′ Calyx corolla-like in color and texture, often very showy; sepals fused well above level of the ovary, the calyx lobes wide and thin; lvs linear to ovate, often > 2 cm, those of a pair often unequal .... **NYCTAGINACEAE**

Group 15: Herbaceous Dicots and Some Monocots; Perianth in One Whorl or Absent; Ovary Superior

1. Lvs alternate or all basal
2. Pls aquatic, floating on water surface — lvs in rosettes, sessile, velvety, palmately veined; fls enclosed by small, sheathing bracts ..... **ARACEAE** (*Pistia*)
2′ Pls terrestrial or rooted in shallow water
3. Infl a ± fleshy spike enclosed by a large, basally sheathing, often petal-like bract; petioles stout ..... **ARACEAE**
3′ Infl various but not a fleshy spike subtended by a sheathing bract; lvs sessile or petioled
4. Lvs with stipules, gen well developed
5. Style 1, unbranched, or stigma sessile
6. Ovule 1; fr an achene ..... **ROSACEAE**
6′ Ovules several–many; fr a capsule ..... **VIOLACEAE**
5′ Styles or style branches 2 or more
7. Lvs palmately compound ..... **CANNABACEAE**
7′ Lvs simple
8. Stipules fused into a sheath around st, sometimes shredded as branches develop ..... **POLYGONACEAE**
8′ Stipules distinct, not fused around st, not shredded as branches develop
9. Fls bisexual, ± 1 mm; style branches 2; sap never milky ..... **CARYOPHYLLACEAE** (*Herniaria*)
9′ Fls unisexual, larger or clustered within fl-like involucres; style branches or styles gen 3, sometimes forked or repeatedly divided; sap often milky ..... **EUPHORBIACEAE**
4′ Lvs without stipules, sometimes reduced to scale-like bracts
10. Perianth parts 6, ± petal-like ..... **POLYGONACEAE**
10′ Perianth parts 5 or fewer, sometimes 0, gen not petal-like
11. Lvs all basal, palmately compound; lflets 3; perianth 0 ..... **BERBERIDACEAE** (*Achlys*)
11′ Lvs cauline, simple; perianth present
12. Infl fl-like with a cup-like involucre bearing colored, often petal-like nectaries; each stamen (actually a staminate fl) with a jointed stalk (composed of a filament and a pedicel); pistil (actually a pistillate fl) 3-lobed, exserted from involucre; sap milky ..... **EUPHORBIACEAE**
12′ Infl various, but not flower-like; stamens gen > 1 per fl, the filament not jointed; pistils various; sap not milky
13. Ovule 1; infls various
14 Herbage ± densely covered with branched hairs
15. Fls bisexual; hairs soft, irregularly branched, branches all short; fr indehiscent ..... **AMARANTHACEAE** (*Tidestromia*)
15′ Fls unisexual; hairs harshly stellate, central branches often long, spreading, bristle-like; fr dehiscent ..... **EUPHORBIACEAE** (*Eremocarpus*)
14′ Herbage glabrous or ± hairy, hairs not branched, not stellate
16. Stigma 1, sessile ..... **URTICACEAE** (*Parietaria*)
16′ Stigmas borne on slender styles
17. Bracts subtending fls dry, scarious; pls neither fleshy nor with powdery or beaded surface; habitats gen not saline ..... **AMARANTHACEAE**
17′ Bracts subtending fls fleshy or herbaceous; pls fleshy or surface powdery or with bead-like hairs; habitats often saline or alkaline ..... **CHENOPODIACEAE**
13′ Ovules 2–many; infl a raceme
18. Pls red- and white-striped, green pigment lacking ..... **ERICACEAE** (*Allotropa*)
18′ Pls green and photosynthetic
19. Sepals 5
20. Fls radial; stamens free; fr a berry, chambers 5–12 ..... **PHYTOLACCACEAE**
20′ Fls bilateral; stamens fused; fr a capsule; chambers 2 ..... **POLYGALACEAE**
19′ Sepals 4
21. Lvs gen >> 3, often narrow; veins pinnate or obscure; fr dry, gen dehiscent ..... **BRASSICACEAE**
21′ Lvs 1–3, gen 2, ovate; veins prominent, parallel; fr a berry ..... **LILIACEAE** (*Maianthemum*)
1′ Lvs opposite or whorled
22. Pls aquatic, weak-stemmed; submersed, floating, or stranded on mud
23. Lvs opposite; ovary lobed or unlobed; seeds 4–many
24. Ovary 2- or 4-lobed, 4-seeded ..... **CALLITRICHACEAE**
24′ Ovary unlobed, many-seeded ..... **LYTHRACEAE** (*Peplis*)
23′ Lvs whorled; ovary unlobed; seed 1
25. Lvs dissected into narrow lobes ..... **CERATOPHYLLACEAE**
25′ Lvs entire ..... **HIPPURIDACEAE**
22′ Pls terrestrial, sometimes in damp soil
26. Ovary chambers 2 or more (doubtful cases should be keyed both ways)
27. Fls unisexual
28. Style 0, stigma sessile; lvs thick, fleshy; sap clear; pls of coastal salt-marsh habitats ..... **BATACEAE**
28′ Styles or stigmas 2–3, free; lvs thin; sap often milky; pls of various habitats but not growing in salt marshes ..... **EUPHORBIACEAE**

27′ Fls bisexual
29. Lvs opposite; fr circumscissile .... **AIZOACEAE**
29′ Lvs appearing whorled; fr splitting lengthwise through chambers .... **MOLLUGINACEAE**
26′ Ovary chamber 1
30. Ovules 3–many; fr a capsule
31. Placentas parietal .... **SAXIFRAGACEAE**
31′ Placentas free-central or basal
32. Capsule circumscissile .... **AIZOACEAE**
32′ Capsule splitting lengthwise, at least at tip
33. Sepals free .... **CARYOPHYLLACEAE**
33′ Sepals fused .... **PRIMULACEAE** (*Glaux*)
30′ Ovule 1; fr an achene, 1-seeded capsule, or utricle
34. Style 1, undivided or stigma 1, sessile
35. Sepals petal-like, fused into a tube; lvs entire or slightly lobed, stinging hairs 0 ...... **NYCTAGINACEAE**
35′ Sepals green, small, inconspicuous; lvs toothed or entire (sometimes with stinging hairs)
36. Herbage glabrous or ± hairy, hairs not branched .... **URTICACEAE**
36′ Herbage densely covered with branched hairs
37. Fls bisexual; hairs soft, irregularly branched, branches all short; fr indehiscent .... **AMARANTHACEAE** (*Tidestromia*)
37′ Fls unisexual; hairs harshly stellate, central branches often long, spreading, bristle-like; fr dehiscent .... **EUPHORBIACEAE** (*Eremocarpus*)
34′ Styles or style branches 2 or more
38. Lvs palmately compound or deeply lobed .... **CANNABACEAE**
38′ Lvs simple, entire or toothed
39. Fr a triangular achene; sepals 6 in two similar to very dissimilar whorls or in a single series .... **POLYGONACEAE**
39′ Fr not triangular; sepals 5 or fewer
40. Lvs with stipules .... **CARYOPHYLLACEAE**
40′ Lvs without stipules
41. Lvs reduced to fleshy scales; infl a fleshy spike .... **CHENOPODIACEAE**
41′ Lvs linear to ovate; infls various
42. Sepals linear, erect, fused below and hardened around ovary; stamens fused to calyx; slender ann, gen 15 cm or less; lvs narrowly linear, 15 mm or less .... **CARYOPHYLLACEAE** (*Scleranthus*)
42′ Sepals erect or spreading, free, not hardened around ovary; stamens free from calyx; pls often larger, ann or per; lvs various
43. Bracts subtending fls dry, scarious; pls neither fleshy nor with powdery or beaded surface; habitats gen not saline .... **AMARANTHACEAE**
43′ Bracts subtending fls lf-like or fleshy; pls often fleshy or surface powdery or with bead-like hairs; habitats often ± saline .... **CHENOPODIACEAE**

Group 16: Dicots; Petals Fused; Ovary Superior

1. Stamens more numerous than corolla lobes
2. Lvs compound
3. Shrubs or trees; lvs 2-pinnate; fr a legume; stamens many, very conspicuous ...... **FABACEAE** (Mimosoideae)
3′ Herbs; lvs once or repeatedly divided (or compounded) into 3 parts; fr a capsule; stamens 10 or fewer
4. Lflets 3; petals 5; stamens 10; fls radial .... **OXALIDACEAE**
4′ Lflets many; petals 4; stamens 6; fls bilateral or biradial .... **PAPAVERACEAE**
2′ Lvs simple, sometimes reduced to linear or scale-like bracts
5. Fl parts in 3's; fr a triangular achene .... **POLYGONACEAE**
5′ Fl parts in 5's; fr a capsule or berry
6. Filaments fused, forming a tube around the style
7. Fls radial; lf veins gen palmate; stamens many; filament tube a continuous cylinder .... **MALVACEAE**
7′ Fls bilateral; lf veins pinnate or nearly parallel; stamens 8; filament tube open on one side **POLYGALACEAE**
6′ Filaments free
8. Fleshy-stemmed, non-photosynthetic herbs; pl entirely non-green — filaments free from corolla .... **ERICACEAE** (Monotropoideae)
8′ Herbs, shrubs, or trees; lvs green and photosynthetic
9. Sts bearing a stout, petiolar spine at each node; corolla tubular, bright red; filaments fused to base of corolla .... **FOUQUIERIACEAE**
9′ Sts spineless; corolla urn-shaped to widely funnel-shaped, variously colored; filaments free from corolla or weakly adherent at base
10. Shrubs or trees; fls often urn-shaped; anthers releasing pollen through pores; ovules several–many; fr a capsule, berry, or drupe .... **ERICACEAE**
10′ Herbs or weak-stemmed subshrubs; fls not urn-shaped; anthers dehiscent by lateral slits; ovule 1; fr an achene tightly enwrapped by hardened or winged base of the corolla-like calyx **NYCTAGINACEAE**

1′ Stamens equal in number to corolla lobes or fewer
11. Fertile stamens fewer than corolla lobes; fls mostly bilateral — 1 or more sterile stamens sometimes present
12. Perianth parts in 6's; fls radial; stigmas and style branches 3 .... **POLYGONACEAE**
12′ Perianth parts in 4's or 5's; fls radial or bilateral; stigmas 1 or 2, the style unbranched or branches 2
13. Pls entirely non-green, ± fleshy .... **OROBANCHACEAE**
13′ Pls green and photosynthetic
14. Pls carnivorous; either aquatic with finely divided, submersed lf-like branches bearing small, hollow traps or terrestrial with lvs all basal, fleshy, and sticky on upper surface with insect-trapping glands; placentas free-central .... **LENTIBULARIACEAE**
14′ Pls not carnivorous; terrestrial or aquatic but lacking finely divided, submersed lf-like branches and fleshy, insect-trapping basal lvs; placentas parietal or axile
15. Ovary deeply 4-lobed; style gen arising from base of ovary lobes — fr 4 nutlets .... **LAMIACEAE**
15′ Ovary entire or ± shallowly lobed; style arising from tip of ovary
16. Lvs covered with silvery scales .... **OLEACEAE** (*Olea*)
16′ Lvs glabrous or variously hairy but not scaly
17. Ovules many per ovary chamber
18. Trees or shrubs
19. Capsule 15–30 cm, linear; trees or large shrubs; lvs long, linear .... **BIGNONIACEAE**
19′ Capsule << 15 cm, variously shaped; gen smaller shrubs; lvs various ...... **SCROPHULARIACEAE**
18′ Herbs
20. Capsule > 5 cm, densely glandular, tipped with a hooked beak several cm long; placentas parietal .... **MARTYNIACEAE**
20′ Capsule < 2 cm, mostly not glandular, beak 0 or < 1 cm; placentas axile .... **SCROPHULARIACEAE**
17′ Ovules 1–4 per ovary chamber
21. The apparent corolla actually a petal-like calyx, narrowed above the ovary; calyx base persistent around ovary, becoming hardened or winged at maturity; stigma, ovary chamber, and ovule 1; fr an achene .... **NYCTAGINACEAE**
21′ True corolla present, not narrowed above ovary; calyx base not persistent around ovary, neither hardened nor winged; stigmas, ovary chambers, and ovules > 1; frs various
22. Ann .... **SCROPHULARIACEAE** (*Collinsia*)
22′ Per herbs or shrubs
23. Lvs toothed, lobed, or compound; fr 2–4 nutlets or a drupe with a ± lobed stone **VERBENACEAE**
23′ Lvs entire; fr a capsule or drupe
24. Lvs gland-dotted; stamens 4; fr fleshy .... **MYOPORACEAE**
24′ Lvs not gland-dotted; stamens 2; fr a capsule
25. Corolla strongly bilateral, red or white; capsule wall thick, stiff .... **ACANTHACEAE**
25′ Corolla radial or nearly so, yellow or white; capsule ± inflated, its wall thin, membranous .... **OLEACEAE** (*Menodora*)
11′ Fertile stamens equal in number to corolla lobes; fls mostly radial
26. Pls entirely non-green, parasitic
27. Pl a thread-like, lfless, twining vine .... **CUSCUTACEAE**
27′ Pl an erect or mound-shaped, fleshy root-parasite with scale-like lvs .... **LENNOACEAE**
26′ Pls green and photosynthetic
28. Ovaries 2, free, sometimes fused by a single style or by a complex stigmatic structure; fr of 1–2 follicles; sap milky
29. Anthers free from stigma or slightly adherent, sometimes forming a cone or a dome over stigma complex; fls solitary or in cymes .... **APOCYNACEAE**
29′ Anthers fused to stigma and top of style, forming a drum-shaped cylindric structure with 5 vertical slits on the sides; fls mostly in umbels .... **ASCLEPIADACEAE**
28′ Ovary 1, sometimes deeply lobed; frs various but not follicles; sap mostly clear
30. Ovary deeply 2–4-lobed, esp in fr; fr breaking apart into 1-seeded nutlets or segments (sometimes only one nutlet maturing)
31. Lvs opposite, at least below; sts 4-angled or round; sts and lvs sometimes aromatic
32. Sts round; lvs gen alternate above or lvs of a pair unequal; sts and lvs not aromatic; infl coiled when young, unrolling in fl or fr, or fls sometimes solitary .... **BORAGINACEAE**
32′ Sts 4-angled; lvs opposite throughout; sts and lvs strongly aromatic; infl never coiled ..... **LAMIACEAE**
31′ Lvs alternate; sts ± round; sts and lvs gen not aromatic
33. Lvs linear to ovate; sts mostly not rooting; infl gen coiled when young, unrolling as it develops, fls seldom concealed by lvs; ovary 2–4-lobed .... **BORAGINACEAE**
33′ Lvs reniform; sts matted, rooting at nodes; fls solitary, concealed by lvs; ovary 2-lobed .... **CONVOLVULACEAE** (*Dichondra*)
30′ Ovary entire or shallowly lobed; fr gen a capsule, berry, drupe, or achene
34. Filaments of stamens fused into a tube around the style; fls large, solitary, yellow or orange; sts and lvs densely stellate .... **STERCULIACEAE** (*Fremontodendron*)
34′ Filaments of stamens free; fls solitary or clustered; sts and lvs mostly not stellate

35. Sepals 2; sts and lvs ± fleshy
36. Fleshy vine with twining sts; fls in slender, many-fld racemes; fr fleshy, indehiscent; 1-seeded .... **BASELLACEAE**
36′ Herb, prostrate to erect; fls solitary or in few-fld clusters; fr dry, dehiscent; seeds 2–many .... **PORTULACACEAE**
35′ Sepals 4 or more; sts and lvs gen not fleshy
37. Perianth parts in 6's or 7's
38. Style and stigma 1; placentas free-central; ovules several–many; fr a capsule .... **PRIMULACEAE**
38′ Styles and stigmas 3; placentas basal; ovule 1; fr a triangular achene .... **POLYGONACEAE**
37′ Perianth parts in 4's or 5's
39. Styles or style-branches 5; sepal tips and subtending bracts wide, obtuse, scarious; lvs all basal .... **PLUMBAGINACEAE** (*Armeria*)
39′ Styles, style-branches, or stigmas gen 1–3; sepal tips and bracts gen not scarious; lvs often cauline
40. The apparent corolla actually a petal-like calyx, narrowed above the ovary; calyx base persistent around ovary, becoming hardened or winged at maturity; stigma, ovary chamber, and ovule 1; fr an achene .... **NYCTAGINACEAE**
40′ True corolla present, calyx gen not narrowed above ovary; calyx base not persistent around ovary, not hardened or winged; stigmas 1–3, ovules mostly > 1; frs various
41. Stamens opposite corolla lobes; placentas free-central .... **PRIMULACEAE**
41′ Stamens alternate with corolla lobes; placentas parietal or axile (rarely basal)
42. Filaments of stamens not fused to corolla, sometimes adhering weakly to base of corolla but readily separable
43. Anthers dehiscent along the sides by slits, not appendaged; petals only slightly fused at base .... **AQUIFOLIACEAE**
43′ Anthers dehiscent at one end by pores, often bearing erect or hooked appendages; petals obviously fused .... **ERICACEAE**
42′ Filaments of stamens definitely fused to corolla tube or throat, not separable without tearing stamen or corolla tissue
44. Infl or branches of infl coiled when young, unrolling and ± open when mature, gen ± 1-sided
45. Stigmas 2, gen not expanded, styles 2 or 1 and ± divided, at least near tip .... **HYDROPHYLLACEAE**
45′ Stigma 1, ± expanded and disk- or head-like, sessile or borne on unidivded style
46. Lvs entire, fleshy, and glabrous or ± bristly; fr breaking into 2 or 4 nutlets .... **BORAGINACEAE** (*Heliotropium*)
46′ Lvs gen toothed or lobed, gen not fleshy, variously hairy; fr capsule **HYDROPHYLLACEAE**
44′ Infl not coiled and unrolling, not 1-sided
47. Calyx lobes gen ± fused by a much thinner, transparent or translucent membrane; stigmas and ovary chambers mostly 3 (less commonly 2 or 1) .... **POLEMONIACEAE**
47′ Calyx lobes without a transparent or translucent marginal membrane; stigmas mostly 2 (less commonly 1); ovary chambers mostly 1–2
48. Lvs palmately compound, long-petioled, glabrous; lflets 3, entire; rhizomed per of bogs, marshes, and lake margins .... **MENYANTHACEAE**
48′ Lvs simple to pinnately compound, often hairy; herbs and shrubs of various habitats
49. Lvs alternate or of unusual arrangement
50. Stigmas 3
51. Twining vines; fls gen large, solitary; lvs wide, palmately veined and often plamately lobed .... **CONVOLVULACEAE** (*Ipomoea*)
51′ Erect or prostrate herbs; fls mostly in clusters; lvs mostly pinnately compound or lobed, if palmately veined then lobes narrow .... **POLEMONIACEAE**
50′ Stigmas 1 or 2
52. Chamber of ovary 1 (doubtful cases should be keyed both ways)
53. Fls 1(–5) per peduncle, gen subtended by a pair of opposite or subopposite bracts appressed against the calyx or borne below the fl(s) on the peduncle; lvs often cordate, sagittate, or hastate; sts often trailing or twining; corolla 2–6 cm .... **CONVOLVULACEAE** (*Calystegia*)
53′ Fls 1–many, not subtended by paired bracts; lvs entire to toothed, lobed, or compound, mostly not cordate, sagittate, or hastate; sts often erect; corolla gen < 2 cm .... **HYDROPHYLLACEAE**
52′ Chambers of ovary 2 or more
54. Styles 2 or style 1 and 2-branched
55. Fls solitary in upper lf axils or on 1–5-fld axillary peduncles; twining vines or trailing herbs; sap often milky .... **CONVOLVULACEAE**
55′ Fls gen in several–many-fld clusters; never twining vines; sap clear .... **HYDROPHYLLACEAE**

54′ Style 1, unbranched
56. Some stamens densely hairy; corolla lobes overlapping in bud; infl a spike or raceme .... **SCROPHULARIACEAE** (*Verbascum*)
56′ Stamens glabrous or nearly so; corolla lobes variously folded in bud; infl variously cyme-like, or fls solitary
57. Lvs shallowly palmately lobed .... **HYDROPHYLLACEAE** (*Romanzoffia*)
57′ Lvs entire or toothed to pinnately lobed .... **SOLANACEAE**
49′ Lvs opposite, whorled, or all basal
58. Corolla lobes 4
59. Shrubs
60. Corolla ± salverform, lobes thin, white to purple; gen dry places .... **BUDDLEJACEAE**
60′ Corolla ± rotate, lobes ± leathery, greenish; coastal salt marshes .... **VERBENACEAE** (Avicennia)
59′ Herbs
61. Corolla of a normal, petal-like texture; infl often a cyme or fls solitary; lvs often cauline .... **GENTIANACEAE**
61′ Corolla dry, scarious; infl a dense spike (rarely fl only 1); lvs in most spp. all basal .... **PLANTAGINACEAE**
58′ Corolla lobes 5
62. Fls 1(–5), infl scapose; sap milky or clear
63. Infl with gen 2 opposite or subopposite bracts, these often appressed against the calyx; sap often milky .... **CONVOLVULACEAE** (*Calystegia*)
63′ Infl bractless; sap clear .... **HYDROPHYLLACEAE** (*Hesperochiron*)
62′ Fls 1–many, borne on a ± leafy st; sap clear
64. Style branches or stigmas 3; fr a capsule .... **POLEMONIACEAE**
64′ Style branches or stigmas 2 or style undivided and stigma 1; fr a capsule or a cluster of 4 nutlets
65. Lvs toothed, lobed, or compound .... **HYDROPHYLLACEAE**
65′ Lvs entire
66. Pls glandular-hairy .... **SOLANACEAE** (*Petunia*)
66′ Pls glabrous or hairy, but not glandular
67. Sts spreading or mat-forming; sts and lvs densely strigose; fls sessile or subsessile subsessile in forks of st; lvs of a pair often unequal ..... **BORAGINACEAE** (*Tiquilia*)
67′ Sts mostly erect or ascending; sts and lvs glabrous or short-hairy; fls axillary or in terminal infls; lvs of a pair equal .... **GENTIANACEAE**

Group 17: Dicots; Petals Fused; Ovary More or Less Inferior

1. Stamens many; spiny, fleshy per or bristly ann; petals many or 5
2. Spiny per with fleshy sts; petals many, overlapping in several series .... **CACTACEAE**
2′ Bristly, slender-stemmed ann; petals 5, in 1 series .... **LOASACEAE**
1′ Stamens 10 or fewer (rarely 20 or more in Styracaceae); herbs, shrubs, or trees but sts gen not fleshy; petals 10 or fewer
3. Tendrils gen present; fls unisexual; monoecious .... **CUCURBITACEAE**
3′ Tendrils 0; fls bisexual or unisexual
4. Ovary actually superior, surrounded by but not fused to the hardened or winged base of the corolla-like calyx; style passing through constriction separating calyx base from petal-like portion and joining to top of ovary — herbs or brittle-stemmed subshrubs, often glandular-hairy; lvs opposite, entire or wavy-margined; fls in heads, racemes, or spikes .... **NYCTAGINACEAE**
4′ Ovary truly inferior, fused to the bases of surrounding perianth parts; style not passing through calyx constriction
5. Fls in dense spikes or in heads of 2–many fls, these often secondarily clustered
6. Infl without an involucre; lvs opposite or whorled; perianth white to pink
7. Shrubs or trees; corolla radial, spur 0; infl spheric, peduncled .... **RUBIACEAE** (*Cephalanthus*)
7′ Ann; corolla ± bilateral, base spurred or swollen; infl head-like or ± cylindric, not peduncled .... **VALERIANACEAE** (*Plectritis*)
6′ Infl subtended by an involucre of 2–many bracts; lf arrangement various; fl color various
8. Calyx of 4–5 green sepals
9. Shrubs, sometimes twining; infl a 2-fld head or a congested or interrupted spike; sepals fused at base, rounded or acute; ovules several; fr a berry .... **CAPRIFOLIACEAE** (*Lonicera*)
9′ Herbs; infl many-fld, very dense; sepals fused into a lobed cup or elongated and bristle-like; ovule 1; fr an achene .... **DIPSACACEAE**
8′ Calyx never green, modified as a white to brownish pappus of 1–many dry scales, bristles, or awns, or sometimes 0
10. Lvs alternate, whorled, or all basal, sometimes opposite at some nodes and alternate at others .... **ASTERACEAE**
10′ Lvs all opposite
11. Anthers fused into a tube surrounding the style, sometimes incl and difficult to observe; bisexual

and pistillate fls often together in the same head — pappus often present; ovule 1 ...... **ASTERACEAE**
11′ Anthers free, gen exserted at least a short distance from corolla; fls gen all bisexual
12. Shrubs; corolla lobes and stamens 5; infl with 2 fls; ovules several ... **CAPRIFOLIACEAE** (*Lonicera*)
12′ Ann or per; corolla lobes 4–5, the stamens 1–4; infl gen with many fls; ovule 1
13. Calyx present as a 4-lobed cup or a pappus of 5 slender awns; stamens gen 4; ovary and fr ± round in X-section, ± angled .... **DIPSACACEAE**
13′ Calyx 0; stamens 3; ovary and fr ± compressed, 2–3-grooved, not angled .... **VALERIANACEAE** (*Valerianella*)
5′ Fls solitary or in few–many-fld cymes or racemes
14. Anthers fused, surrounding style .... **CAMPANULACEAE**
14′ Anthers free
15. Filaments ± free from corolla
16. Herbs; stamens equal in number to corolla lobes; fr a many-seeded capsule .... **CAMPANULACEAE**
16′ Shrubs or trees; stamens twice as many as corolla lobes
17. Lvs glabrous or puberulent; anthers dehiscent at tip by pores; corolla lobes 4–5; fr a berry, gen several–many-seeded .... **ERICACEAE** (*Vaccinium*)
17′ Lvs stellate; anthers dehiscent on sides by slits; corolla lobes 4–10; fr hard, dry, dehiscent, gen 1-seeded .... **STYRACACEAE**
15′ Filaments strongly fused to corolla tube or throat
18. Stamens fewer than corolla lobes
19. Pl a trailing, weak-stemmed shrub with upright, 2-fld peduncles; stamens 4 .... **CAPRIFOLIACEAE** (*Linnaea*)
19′ Pl an erect herb with terminal infl of several–many fls; stamens 1–3 .... **VALERIANACEAE**
18′ Stamens as many as corolla lobes
20. Lvs alternate; infl a raceme or panicle; placentas free-central; fertile stamens opposite corolla lobes; sterile stamens alternate corolla lobes .... **PRIMULACEAE** (*Samolus*)
20′ Lvs opposite or whorled; infls various; placentas axile or apical; fertile stamens alternate corolla lobes; sterile stamens gen 0
21. Shrubs and vines; stipules 0 or, if present, borne on petiole; lvs opposite, simple or compound; pistils 3–5; ovary and fr unlobed, the fr not breaking apart at maturity; corolla lobes gen 5; fls sometimes bilateral .... **CAPRIFOLIACEAE**
21′ Herbs and brittle-stemmed subshrubs (rarely a well developed shrub or small tree); stipules present, well developed, borne on st between bases of adjacent lvs, sometimes very lf-like, thus lvs apparently whorled; lvs simple; pistils 2; ovary and fr gen 2-lobed, gen breaking into 2 indehiscent segments at maturity (rarely a drupe); corolla lobes 4–5; fls radial .... **RUBIACEAE**

Group 18: Dicots; Petals Free; Ovary Inferior

1. Stamens > twice as many as petals or > 15
2. Styles > 1, sometimes partly fused at base
3. Pls herbaceous or woody only at base, gen ± fleshy; petals 4–many
4. Petals many .... **AIZOACEAE**
4′ Petals 4–6 .... **PORTULACACEAE**
3′ Pls shrubs or trees, woody throughout, not fleshy; petals gen 4–5
5. Lvs opposite; fr a capsule .... **PHILADELPHACEAE**
5′ Lvs alternate; fr fleshy, indehiscent .... **ROSACEAE**
2′ Style 1
6. Pls fleshy, spiny; sepals and petals indefinite in number and not sharply differentiated, in spirals **CACTACEAE**
6′ Pls not fleshy, gen not spiny; sepals and petals each 3–7, in whorls, clearly differentiated
7. Herbs or low shrubs with rough, barbed, or stinging hairs .... **LOASACEAE**
7′ Trees or shrubs lacking such hairs
8. Ovary only half inferior, the free portion conic, esp in fr .... **PHILADELPHACEAE** (*Carpenteria*)
8′ Ovary wholly inferior or nearly so, the free portion, if any, rounded
9. Lvs dotted with embedded oil glands, often strongly scented when crushed; ovary chambers gen 2–5, all at 1 level; fls variously colored .... **MYRTACEAE**
9′ Lvs not gland-dotted, not strongly scented; ovary chambers gen 7 or more, some of them at different levels in the ovary; fls bright red-orange .... **PUNICACEAE**
1′ Stamens twice as many as petals or fewer
10. Stamens opposite and equal in number to petals .... **RHAMNACEAE**
10′ Stamens alternate petals or different in number
11. Style 1
12. Pls tendril-bearing vines; monoecious .... **CUCURBITACEAE**
12′ Pls not tendril-bearing; fls gen bisexual
13. Ovules > 1 per chamber
14. Placentas axile .... **ONAGRACEAE**
14′ Placentas parietal
15. Shrubs .... **GROSSULARIACEAE**

15′ Herbs .... **SAXIFRAGACEAE**
13′ Ovule 1 per chamber or per ovary
16. Lvs opposite; sepals and petals 2 or 4 (rarely 5)
17. Fr berry-like; shrubs or trees .... **CORNACEAE**
17′ Fr dry, covered with hooked hairs; herbs .... **ONAGRACEAE** (*Circaea*)
16′ Lvs alternate; sepals and petals 5
18. Pl a sprawling woody vine (flowering sts often erect); lvs smooth, dark green, palmately lobed; infl a cluster of umbels; fr berry-like .... **ARALIACEAE**
18′ Pl a low desert per; lvs gen scabrous, pale green, unlobed; infl ± a raceme, fr a utricle .... **LOASACEAE** (*Petalonyx*)
11′ Styles > 1
19. Seeds gen several–many per chamber
20. Sepals 2 .... **PORTULACACEAE**
20′ Sepals 4–5
21. Herbs .... **SAXIFRAGACEAE**
21′ Shrubs
22. Stamens 4–5 .... **GROSSULARIACEAE**
22′ Stamens 10 .... **PHILADELPHACEAE**
19′ Seeds 1–2 per chamber
23. Fls in umbels or heads
24. Fr a berry; infl a panicle of umbels .... **ARALIACEAE**
24′ Fr dry, splitting in half; infl gen a compound umbel, sometimes a simple umbel or a cluster of heads .... **APIACEAE**
23′ Fls not in umbels or heads
25. Lvs alternate; hypanthium well developed .... **ROSACEAE**
25′ Lvs opposite or whorled; hypanthium very short or 0
26. Petals 2–4; stamens 4–8; fr fleshy or breaking into 1-seeded segments .... **HALORAGACEAE**
26′ Petals 5–6; stamens 10–12; fr a capsule .... **PHILADELPHACEAE**

Group 19: Dicots; Petals Free; Stamens Many; Ovary Superior

1. Ovary chambers 2–many
2. Lvs hollow, tubular, often water-filled .... **SARRACENIACEAE**
2′ Lvs with ordinary flat blades
3. Pls aquatic; lvs large, ovate to round, floating or rising a short distance above the water surface .... **NYMPHAEACEAE**
3′ Pls terrestrial; lvs sometimes small
4. Fls unisexual — ovary gen 3-lobed .... **EUPHORBIACEAE**
4′ Fls bisexual
5. Filaments fused into tube around style; anther chamber 1
6. Fls radial; petals 5; stamens many; filament tube cylindric .... **MALVACEAE**
6′ Fls bilateral; petals 3; stamens 8; filament tube open on one side .... **POLYGALACEAE**
5′ Filaments free or fused at base into groups; anther chambers 2
7. Hypanthium present
8. Ovary unlobed, the lower portion fused to hypanthium; style 1; lvs entire .... **PHILADELPHACEAE** (*Carpenteria*)
8′ Ovary deeply lobed, wholly free from hypanthium; styles > 1; lvs toothed or lobed .... **ROSACEAE**
7′ Hypanthium 0
9. Lvs opposite, entire; petals 5, yellow to orange; sepals persisting as fl opens .... **HYPERICACEAE**
9′ Lvs alternate, toothed or lobed; petals 6, white; sepals deciduous as fl opens **PAPAVERACEAE** (*Romneya*)
1′ Ovary chamber 1
10. Ovules 1–2 per ovary; fr a drupe, berry, follicle, or achene
11. Fls bilateral; lvs compound .... **FABACEAE** (Papilionoideae)
11′ Fls radial; lvs simple
12. Fl parts in 5's .... **ROSACEAE**
12′ Fl parts in 3's
13. Fr a berry or drupe .... **LAURACEAE**
13′ Fr a triangular achene .... **POLYGONACEAE**
10′ Ovules 2–many per ovary; fr a capsule, follicle, legume, or berry
14. Lvs palmately compound, densely covered with stalked glands — fl showy, the parts in 4's **CAPPARACEAE**
14′ Lvs various, glabrous to densely hairy but not glandular
15. Infl a raceme, spike, or head
16. Shrubs or trees; fr a legume .... **FABACEAE** (Mimosoideae)
16′ Herbs; frs various
17. Lvs pinnately compound or deeply palmately lobed; stamens attached all around receptacle; fr a follicle or berry .... **RANUNCULACEAE**

17′ Lvs simple or pinnately lobed; stamens attached on one side of receptacle; fr a capsule, dehiscent at tip .... **RESEDACEAE** (*Reseda*)
15′ Infl a panicle or a cyme or fl solitary
18. Fr a follicle; hypanthium present .... **ROSACEAE**
18′ Fr a capsule; hypanthium gen 0
19. Sepals gen 2–3, sometimes fused into a conical cap, falling as fl opens; sap (at least in roots) gen milky or colored .... **PAPAVERACEAE**
19′ Sepals 2–9, persistent, never fused to tip; sap clear
20. Filaments fused at base into 5 bunches .... **HYPERICACEAE**
20′ Filaments free to base
21. Style 1, sometimes branched at tip; pls not at all fleshy; capsule splitting by valves .... **CISTACEAE**
21′ Styles 3–8, fused only at base; pl gen ± fleshy; capsule circumscissile or splitting by valves .... **PORTULACACEAE**

Group 20: Dicots; Petals Generally Free; Stamens Twice as Many as Petals or Fewer; Leaves Compound or Nearly So; Ovary Superior

1. Lvs 2 or more times compound or divided
2. Shrubs or trees
3. Lflets serrate; fls purple .... **MELIACEAE**
3′ Lflets entire or with a few rounded teeth; fls cream-colored or yellow
4. Ovary unlobed; terminal lflets 0; sts and lvs unscented .... **FABACEAE**
4′ Ovary deeply 4- or 6-lobed; terminal lflets present; sts and lvs strongly scented .... **RUTACEAE** (*Ruta*)
2′ Herbs
5. Ovary very deeply 4–6-lobed; style arising from a deep indentation in middle of ovary
6. Fls yellow; herbage strongly scented .... **RUTACEAE** (*Ruta*)
6′ Fls white to pink; herbage not scented .... **LIMNANTHACEAE** (*Limnanthes*)
5′ Ovary unlobed or only shallowly lobed; style terminal
7. Petals 5
8. Fls solitary on peduncles > 3 cm .... **VIOLACEAE**
8′ Fls several–many in umbels or racemes
9. Petals yellow to orange or reddish; lflets entire; infl a raceme .... **FABACEAE** (*Hoffmanseggia*)
9′ Petals pink to purple; lflets toothed; infl an umbel .... **GERANIACEAE** (*Erodium*)
7′ Petals 4 or 6
10. Petals 6 or more; lvs all basal .... **BERBERIDACEAE** (*Vancouveria*)
10′ Petals 4; lvs basal and cauline
11. Petals all equal; fls radial; infl a bractless raceme .... **BRASSICACEAE**
11′ Petals unequal; fls bilateral or biradial; infl a raceme or a cymose panicle with bracts .... **PAPAVERACEAE** (Fumarioideae)
1′ Lvs once compound
12. Lflets 2 or 3
13. Herbs or subshrubs
14. Fls strongly bilateral
15. Odd petal uppermost in fl; stamens 10, all filaments or 9 of them fused, forming a tube around ovary; pistil 1; fr a legume or indehiscent pod .... **FABACEAE** (Papilionoideae)
15′ Odd petal lowermost in fl; stamens 5, free but tightly appressed against ovary; pistils 3; fr a capsule .... **VIOLACEAE**
14′ Fls radial — stamens free or fused at base
16. Petals 4; stamens 6 .... **BRASSICACEAE**
16′ Petals 5; stamens 10
17. Lflets toothed or lobed; styles thickened and elongated, stiff, esp in fr, persistent as coiled beaks on segments of fr .... **GERANIACEAE** (*Geranium*)
17′ Lflets entire; styles slender, gen not elongated, never forming coiled beaks on segments of fr; fr a capsule
18. Lvs alternate or all basal; stamens ± fused, at least at base; seeds many .... **OXALIDACEAE**
18′ Lvs opposite; stamens free; seeds 1 per chamber .... **ZYGOPHYLLACEAE**
13′ Trees or well developed shrubs
19. Lvs opposite
20. Petals 2 .... **OLEACEAE** (*Fraxinus dipetala*)
20′ Petals 4 or more
21. Lflets coarsely serrate or crenate; styles 2; ovary 2-winged, maturing as a 2-winged fr that splits into 2 one-seeded segments .... **ACERACEAE** (*Acer glabrum*)
21′ Lflets finely serrate; styles 3; ovary not winged, maturing as a bladdery, 3-chambered capsule .... **STAPHYLEACEAE**
19′ Lvs alternate
22. Petals 4 or 6
23. Fls greenish white; lflets gland-dotted; fr a round, winged achene .... **RUTACEAE** (*Ptelea*)

23′ Fls yellow; lflets not gland-dotted; fr a capsule or berry
24. Petals 6; lflets sharply toothed; ovary sessile on receptacle .... **BERBERIDACEAE** (*Berberis*)
24′ Petals 4; lflets entire; ovary borne on a stalk .... **CAPPARACEAE** (*Isomeris*)
22′ Petals 5
25. Fls strongly bilateral; stamens 10, all filaments, or 9 of them fused, forming a tube around ovary .... **FABACEAE** (Papilionoideae)
25′ Fls radial; stamens free .... **ANACARDIACEAE**
12′ Lflets 4 or more
26. Lvs palmately compound
27. Lvs opposite .... **HIPPOCASTANACEAE**
27′ Lvs alternate
28. Petals 4; ovary separated from receptacle by a stalk .... **CAPPARACEAE** (*Cleome*)
28′ Petals 5; ovary sessile on receptacle
29. Shrub or herb, not tendril-bearing; fls bilateral; stamens 10, all filaments or 9 of them fused, forming a tube around ovary .... **FABACEAE** (Papilionoideae)
29′ Woody vine with tendrils; fls radial; stamens 5, free, opposite the petals ..... **VITACEAE** (*Parthenocissus*)
26′ Lvs pinnately compound
30. Petals 2–4
31. Pl a large shrub or small tree; lvs opposite; stamens 2 .... **OLEACEAE** (*Fraxinus dipetala*)
31′ Pl an herb or a small shrub; lvs opposite, alternate, or basal; stamens 6
32. Sepals 2; fls strongly bilateral; 1 petal prolonged, forming a spur ...... **PAPAVERACEAE** (Fumarioideae)
32′ Sepals 3 or 4; fls radial or weakly bilateral; petals not spurred
33. Ovary entire or ± 2-lobed; infl a bractless raceme .... **BRASSICACEAE**
33′ Ovary deeply 3–4-lobed; fls solitary and axillary or infl a leafy-bracted raceme .... **LIMNANTHACEAE**
30′ Petals 5 or more
34. Fls strongly bilateral; stamens gen 10, all filaments or 9 of them fused into a tube around ovary .... **FABACEAE** (Papilionoideae)
34′ Fls radial or weakly bilateral; stamens 5–12, gen free
35. Shrubs or trees
36. Petals 6; lflets spiny-toothed; sts and lvs non-scented .... **BERBERIDACEAE** (*Berberis*)
36′ Petals 5 (rarely 6); lflets entire or toothed but not spiny; sts and lvs often strongly scented
37. Ovary divided into 2–5 flattened lobes; fr a cluster of elongated, winged achenes; lflets with gland-tipped teeth near base .... **SIMAROUBACEAE** (*Ailanthus*)
37′ Ovary entire; fr a drupe or legume; lflets without gland-tipped teeth
38. Fls solitary or few; lflets 5–10 mm, entire — small tree of DSon .... **BURSERACEAE**
38′ Fls in racemes or panicles, often many; lflets mostly > 7 mm, often much longer, toothed or entire
39. Corolla << 1 cm diam, white to greenish or pale yellow; fr a drupe; lflets often toothed .... **ANACARDIACEAE**
39′ Corolla > 1 cm diam, bright yellow; fr a legume; lflets entire .... **FABACEAE** (*Senna*)
35′ Herbs or subshrubs
40. Anther-bearing stamens as many as petals
41. Petals 6; lflets leathery .... **BERBERIDACEAE** (*Berberis*)
41′ Petals 5; lflets thin
42. Lflets entire; infl a spike or raceme .... **FABACEAE** (*Dalea*)
42′ Lflets toothed; infl gen an umbel .... **GERANIACEAE** (*Erodium*)
40′ Anther-bearing stamens twice as many as petals
43. Fls in racemes or panicles; ovary chamber 1
44. Terminal lflet 0; fr a legume .... **FABACEAE** (*Senna*)
44′ Terminal lflet present; fr a capsule .... **RESEDACEAE**
43′ Fls solitary or in pairs; ovary chambers 3–5
45. Stipules 0; ovary deeply lobed at time of pollination; style arising from base of ovary .... **LIMNANTHACEAE**
45′ Stipules present; ovary entire or shallowly lobed at time of pollination; style at tip of ovary .... **ZYGOPHYLLACEAE**

Group 21: Mostly Dicot Shrubs and Trees; Petals Generally Free; Stamens Twice as Many as Petals or Fewer; Leaves Simple; Ovary Superior

1. Lvs opposite or whorled
2. Styles or style branches 2–5, each terminated by a stigma
3. Stamens equal in number to and opposite the petals; petals gen ± cupped; lvs pinnately veined **RHAMNACEAE**
3′ Stamens more numerous than petals; petals flat; lvs palmately veined
4. Tree or large shrub; styles 2(3); ovary 2-winged, maturing as a 2-winged fr that splits into 2 one-seeded segments .... **ACERACEAE**
4′ Small shrub; style 1, elongated, with 5 branches; ovary not winged; fr splitting into 5 segments, each tipped by a spirally coiled beak .... **GERANIACEAE** (*Pelargonium*)

2′ Style 1, unbranched, stigma 1, entire or nearly so
5. Lvs deeply 2-lobed, resin-coated; fls bright yellow; fr densely covered with spreading, white hairs
.... **ZYGOPHYLLACEAE** (*Larrea*)
5′ Lvs not lobed, not resin-coated; fls white or greenish to purplish brown; fr glabrous
6. Lvs whorled at some nodes, alternate at others .... **PITTOSPORACEAE**
6′ Lvs all opposite or some in axillary clusters
7. Lvs toothed, not gland-dotted; sts and lvs not scented; petals greenish to purplish brown **CELASTRACEAE**
7′ Lvs entire, gland-dotted; sts and lvs strongly scented; petals white .... **RUTACEAE** (*Cneoridium*)
1′ Lvs alternate, sometimes reduced to minute scales or so quickly deciduous that the pls are gen lfless
8. Lvs palmately veined, the blade well developed
9. Tendrils 0; pls erect, not climbing
10. Stamens as many as and opposite the petals .... **RHAMNACEAE** (*Ceanothus*)
10′ Stamens twice as many as petals
11. Large shrub or small tree; fls strongly bilateral; umbels sessile; style not elongated, unbranched; ovary chamber 1; fr a flattened legume, not bearing a coiled beak .... **FABACEAE** (*Cercis*)
11′ Small shrub; fls almost radial; umbels peduncled; style 1, elongated, with 5 branches; ovary chambers 5; fr splitting into 5 segments, each tipped by a spirally coiled beak ...... **GERANIACEAE** (*Pelargonium*)
9′ Tendrils present; pl a woody vine, gen trailing or climbing on other pls
12. Lvs entire; fls unisexual; dioecious; petals 3; st covered with spines — infl an umbel ... **LILIACEAE** (*Smilax*)
12′ Lvs toothed or lobed; fls gen bisexual; petals 5 or 6; st not spiny .... **VITACEAE** (*Vitis*)
8′ Lvs pinnately veined, 1-veined, or reduced to bladeless scales
13. Fls strongly bilateral
14. Stamens 4, all inserted on 1 side of ovary, free or nearly so; sepals rose-purple, petal-like, widely spreading; petals 5, the upper 3 similar, stalked, free or 1 fused, the lower 2 reduced to fleshy scales; fr an indehiscent pod bearing slender, barb-tipped spines .... **KRAMERIACEAE**
14′ Stamens 8 or 10, all filaments or 9 of them fused into a tube around the ovary; sepals variously colored, sometimes petal-like; petals 3 or 5, not reduced to fleshy scales; fr a capsule, legume, or spineless, indehiscent pod
15. Sepals all fused, at least at base, forming a cup-shaped to cylindric calyx tube, gen not petal-like; petals 5, the odd petal (banner) uppermost, overlapping the margins of the 2 upper lateral petals; lower 2 lateral petals gen free at base but fused toward the tip, forming a keel that encloses stamens and ovary; fr a legume or indehiscent pod .... **FABACEAE** (Papilionoideae)
15′ Sepals free, petal-like, 2 of them spreading and very different from the other 3; petals 3, the odd petal lowermost, often appendaged, folded and forming a keel enclosing stamens and ovary, a banner petal never present; fr a flattened capsule .... **POLYGALACEAE**
14′ Fls radial or nearly so
16. Stamens more numerous than petals
17. Fls dark blue-purple; sts and lvs gland-dotted, very strongly scented .... **RUTACEAE** (*Thamnosma*)
17′ Fls white to yellow or green; sts and lvs not scented
18. Fls unisexual
19. Ovary unlobed, chamber 1, containing 1 or 2 seeds; hairs 0 or attached at one end; stiffly branched shrub .... **CROSSOSOMATACEAE** (*Glossopetalon*)
19′ Ovary 3-lobed, chambers 3, each 1-seeded; hairs 2-branched, attached in the middle; weak-stemmed subshrub .... **EUPHORBIACEAE** (*Ditaxis*)
18′ Fls bisexual
20. Hypanthium present, well developed, tubular or cup-shaped .... **ROSACEAE**
20′ Hypanthium 0 or small and disk-shaped
21. Anthers opening by small round pores; petals 5 .... **ERICACEAE** (*Ledum*)
21′ Anthers opening along sides by elongated slits; petals 4–6
22. Fls in elongated, terminal, bractless racemes; sepals and petals 4; stamens 6; sts not ending in spines .... **BRASSICACEAE**
22′ Fls solitary or in small, axillary clusters, if in racemes, the fls subtended by bracts; sepals and petals 4–6; stamens 8–12; sts often terminating in stout spines
23. St grooved; ovary sessile on receptacle, chamber 1; lvs linear or narrowly elliptic, ± persistent; fr a follicle .... **CROSSOSOMATACEAE** (*Glossopetalon*)
23′ St smooth, gen lfless most of the year; ovary raised ± 1 mm above receptacle on a short stalk, chambers 2; lvs scale-like, soon deciduous; fr a berry .... **KOEBERLINIACEAE**
16′ Stamens equal in number to petals
24. Lvs all reduced to bladeless scales; twigs very slender, jointed, green .... **TAMARICACEAE**
24′ Lvs with expanded blades, linear to ovate, sometimes early deciduous; twigs not jointed, green or brown
25. Stamens opposite petals; petals gen with cupped blade .... **RHAMNACEAE**
25′ Stamens alternate petals; petals mostly flat
26. Stigmas 3–5, style 1, unbranched, or style branches 3
27. Lvs > 25 mm, petioled, the margin not thickened; style branches or styles 3; fr a drupe
.... **ANACARDIACEAE**
27′ Lvs < 15 mm, sessile or subsessile, the margin thickened; style 1, stigmas 4–5; fr a 1-seeded capsule .... **CELASTRACEAE** (*Mortonia*)

26′ Stigma 1, sessile or borne on an unbranched style
28. Lvs toothed
29. Tree or large shrub; lvs >> 2 cm, glossy green, coarsely dentate with spine-tipped teeth; fr a smooth, red drupe .... **AQUIFOLIACEAE**
29′ Small shrub; lvs < 2 cm, dull green, minutely hairy, finely serrate, dentate, or crenate, the teeth never spine-tipped; fr dry, roughened, breaking into 5 one-seeded pieces **STERCULIACEAE** (*Ayenia*)
28′ Lvs entire
30. Lvs < 2 cm; fr a follicle; low shrub, native in undisturbed, arid and semiarid habitats .... **CROSSOSOMATACEAE** (*Glossopetalon*)
30′ Lvs >> 2 cm; fr a 2- or 3-pistillate capsule with several seeds; large shrub or small tree, escaped from cult in urban areas .... **PITTOSPORACEAE**

Group 22: Mostly Dicot Herbs; Petals Generally Free; Stamens Twice as Many as Petals or Fewer; Leaves Simple; Ovary Superior

1. Fls bilateral
2. Lvs palmately veined
3. Fls 2–many in umbels; ovary chambers and style branches 5; style persistent on fr, forming an elongated beak; beak segments coiled on segments of dry fr .... **GERANIACEAE** (*Pelargonium*)
3′ Fls solitary; ovary chambers 1 or 3, the style unbranched or branches 3; style not persistent as a beak on fr
4. Lvs peltate; sts and lvs strongly scented; uppermost sepal bearing a prominent spur; stamens 10; ovary chambers 3; fr breaking into 3 one-seeded segments .... **TROPAEOLACEAE**
4′ Lvs not peltate; sts and lvs unscented; lowermost petal bearing a pouch or a short to elongated spur; stamens 5; ovary chamber 1; fr a many-seeded capsule .... **VIOLACEAE**
2′ Lvs pinnately veined or veins obscure
5. Lvs with stipules
6. Odd petal uppermost in fl; stamens 10, 9 of them forming a U-shaped tube around ovary; pistil simple; fr a legume or indehiscent pod .... **FABACEAE** (Papilionoideae)
6′ Odd petal lowermost in fl; stamens 5, free but tightly appressed against ovary; pistil compound; fr a capsule .... **VIOLACEAE**
5′ Lvs without stipules
7. Calyx with a protruding, hollow spur; ovary chambers 5; stamens 5 .... **BALSAMINACEAE**
7′ Calyx without a spur; ovary chambers 1–2; stamens mostly > 5
8. Petals 2; stamens 3 .... **RESEDACEAE** (*Oligomeris*)
8′ Petals 3–6; stamens gen 6 or more
9. Filaments fused, forming a U-shaped tube around ovary; petals strongly overlapping .... **POLYGALACEAE**
9′ Filaments free; petals not or scarcely overlapping
10. Petals 4; fls gen not subtended by bractlets .... **BRASSICACEAE**
10′ Petals 5 or 6; each fl subtended by a bractlet .... **RESEDACEAE**
1′ Fls radial
11. Pls without green pigmentation .... **ERICACEAE** (Monotropoideae)
11′ Pls green and photosynthetic
12. Petals 4 or fewer
13. Lvs alternate, sometimes all basal
14. Lvs all basal, covered with stalked, insect-trapping glands .... **DROSERACEAE**
14′ Lvs basal and cauline, without insect-trapping glands
15. Petals 2–3 .... **POLYGONACEAE**
15′ Petals 4
16. Sepals free; stamens gen 6; fls gen not subtended by bractlets .... **BRASSICACEAE**
16′ Sepals fused; stamens 2–3; each fl subtended by a bractlet .... **SAXIFRAGACEAE** (*Tolmiea*)
13′ Lvs opposite or whorled
17. Pls fleshy; dioecious subshrub or per herb; petals present only in staminate fls .... **BATACEAE**
17′ Pls not fleshy; fls bisexual; petals present in all fls
18. Lvs in a single whorl of 3; fl 1, terminal; petals 3 .... **LILIACEAE** (*Trillium*)
18′ Lvs gen in 2 or more pairs or whorls; fls gen 2–many, axillary or in terminal clusters; petals 2–4
19. Sepals half as many as petals, deciduous as fl opens .... **PAPAVERACEAE** (*Meconella*)
19′ Sepals as many as petals, present in opened fls
20. Sepals fused, at least toward base
21. Infl a dense cluster borne on a terminal peduncle .... **PHILADELPHACEAE** (*Whipplea*)
21′ Infl of axillary fls
22. Petals borne on receptacle; style 2–3-branched — calyx cylindric .... **FRANKENIACEAE**
22′ Petals borne on the inner face of a tubular hypanthium; style unbranched .... **LYTHRACEAE**
20′ Sepals free or nearly so
23. Petals 6–20 mm, each bearing a prominent, fringed appendage .... **GENTIANACEAE** (*Frasera*)
23′ Petals < 3 mm, not appendaged

24. Fls pedicelled, terminal or from upper axils, sometimes cymose; lvs without stipules; petals 4; stamens 4 .... **CARYOPHYLLACEAE** (*Sagina*)
24′ Fls sessile in lf axils; lvs with stipules; petals 2–4; stamens 2–4 .... **ELATINACEAE** (*Elatine*)
12′ Petals 5 or more
25. Fls unisexual .... **EUPHORBIACEAE**
25′ Fls bisexual
26. Sepals 2 or 3
27. Lvs not fleshy; sepals deciduous as fls open .... **PAPAVERACEAE** (*Meconella*)
27′ Lvs ± fleshy; sepals persistent on opened fls and fr .... **PORTULACACEAE**
26′ Sepals 5 or more
28. Stamens more numerous than petals
29. Lvs alternate, sometimes all basal
30. Style branches and stigmas 5; ovules 1 per ovary chamber; fr breaking into 5 segments; style elongate, persistent on fr segments as a coiled beak .... **GERANIACEAE**
30′ Styles or style branches 1–3, stigmas 1–3; ovules several–many per ovary chamber; fr a capsule without an elongate beak
31. Stigmas 2–3, sessile or borne on 2–3 styles or style branches .... **SAXIFRAGACEAE**
31′ Stigma 1, sessile or borne on an unbranched style
32. Fls disk-shaped, without a hypanthium; stamens dehiscent by pores .... **ERICACEAE** (Pyroloideae)
32′ Fls with a tubular hypanthium; stamens dehiscent along the sides by slits .... **LYTHRACEAE**
29′ Lvs opposite or whorled
33. Stigma and style 1
34. Fls with a cylindric hypanthium; lvs entire .... **LYTHRACEAE**
34′ Fls without a hypanthium; sepals free to base; lvs toothed
35. Fls sessile or subsessile in lf axils; ann; teeth of lvs gland-tipped .... **ELATINACEAE** (*Bergia*)
35′ Fls pedicelled, borne in a terminal cluster; per; teeth of lvs not gland-tipped .... **ERICACEAE** (*Chimaphila*)
33′ Stigmas 2–6
36. Lvs lobed; opposite lvs a single pair, the remainder all basal; stigmas 3 — ovary chamber 1, placentas parietal .... **SAXIFRAGACEAE** (*Lithophragma*)
36′ Lvs entire or shallowly toothed; opposite lvs gen several–many pairs; stigmas 2–6
37. Ovary chamber 1 (rarely 2–4); placentas free-central (rarely axile); lvs entire **CARYOPHYLLACEAE**
37′ Ovary chambers 5–6; placentas axile; lvs crenate-serrate .... **PHILADELPHACEAE** (*Whipplea*)
28′ Stamens equal in number to petals or fewer
38. Style 1, unbranched
39. Stamens alternate with petals .... **GENTIANACEAE** (*Swertia*)
39′ Stamens opposite petals .... **PRIMULACEAE**
38′ Styles or style branches 2 or more
40. Lvs opposite
41. Sepals free
42. Lvs entire .... **CARYOPHYLLACEAE**
42′ Lvs toothed, lobed, or compound .... **GERANIACEAE**
41′ Sepals fused
43. Lvs linear, acute .... **CARYOPHYLLACEAE**
43′ Lvs oblong to obovate, often curled and appearing narrower, obtuse .... **FRANKENIACEAE**
40′ Lvs alternate, sometimes all basal
44. Lvs linear, entire
45. Lvs cauline; fls solitary or in open clusters .... **LINACEAE**
45′ Lvs all basal; fls in scapose heads .... **PLUMBAGINACEAE**
44′ Lvs wider, often toothed or lobed
46. Lvs covered with long, gland-tipped, insect-trapping hairs .... **DROSERACEAE**
46′ Lvs without insect-trapping hairs
47. Styles, style branches, or stigmas 2–4 .... **SAXIFRAGACEAE**
47′ Styles or style branches 5
48. Style elongating, forming a beak; fr breaking into 5 one-seeded segments, each tipped with a coiled beak segment; fls in umbels .... **GERANIACEAE**
48′ Styles not elongating and forming a beak; fr 1-seeded, not breaking apart, not beaked; fls in panicles .... **PLUMBAGINACEAE**

ASPLENIACEAE SPLEENWORT FAMILY

Alan R. Smith & Thomas Lemieux

Rhizome-scale cells with adjacent walls dark brown to blackish, external clear. **LF**: petiole in transverse section with 1 X-shaped or 2 back-to-back C-shaped vascular strands; segment veins gen free. **SPORANGIA** in linear to oblong sori along veins; indusia linear, opening away from veins; stalk cells in 1 row; spores elliptic, winged. ± 10 genera (limits disputed), 650 spp. (most in *Asplenium*): worldwide, esp trop.

ASPLENIUM SPLEENWORT

Pls in soil or on rocks; rhizome gen short-creeping to erect. **LVS** often tufted, gen glabrous; 1° axis often ± winged; blade simple or 1–many-pinnate, rarely forked; 1° lflets often asymmetric, upward side more developed. **SPORANGIA** in linear sori; indusia persistent, covering sori when young, later reflexed. (Greek: spleen)

1. Lf simple or few-forked, grass-like .... ***A. septentrionale***
1′ Lf 1-pinnate, 1° lflets many
  2. Lf axes greenish or straw-colored above red-brown base .... ***A. trichomanes-ramosum***
  2′ Lf axes dark red- to purple-brown to purple-black ± throughout
    3. Petiole < 0.6 mm wide, narrowly winged; 1° lflets 3–6(7) mm, gen shallowly crenate on upper, outer margins .... ***A. trichomanes*** ssp. ***trichomanes***
    3′ Petiole gen > 0.6 mm wide, unwinged; 1° lflets gen 5–12 mm, margin shallowly lobed ± throughout .... ***A. vespertinum***

A. septentrionale (L.) Hoffm. (p. 93) NORTHERN SPLEENWORT **LVS** simple or 1–2-forked, many, densely clustered, 5–15 cm; petiole 2.5–12 cm, ± 0.5 mm wide, unwinged, red-brown at base, greenish above, dull; blade gen 1–2 mm wide, narrowly linear, sometimes with small, ± linear, gen sterile teeth near or at tip. **SPORANGIA**: sori 5–15 mm, gen 1 per segment but sometimes appearing to have as many as 4, corresponding to gen the same number of teeth per segment. $2n$=144. RARE in CA. Crevices of granite rocks; 2500–3350 m. CaRH, s SNH; to OR, SD (also West Virginia), TX, n Baja CA, Eur, Asia.

A. trichomanes L. ssp. ***trichomanes*** (p. 93) MAIDENHAIR SPLEENWORT **LVS** 1-pinnate, many, clustered, 8–25 cm; petiole 1–3(5+) cm, 0.4–0.6 mm wide, narrowly winged, dark purple-brown, shiny; blade gen 0.5–1.5 cm wide, linear; 1° lflets 20–30(37) pairs, 3–6(7) mm, 3 mm wide, oblong, gen shallowly crenate on upper, outer margins. **SPORANGIA**: sori gen 1–1.5 mm, gen 2–4 pairs per 1° lflet. $2n$=72. RARE in CA. On rocks; 200 m. nw KR (Del Norte Co); widespread in N.Am, Eur, Asia, s temp.

A. trichomanes-ramosum L. GREEN SPLEENWORT **LVS** 1-pinnate, gen few, clustered, 3–20 cm; petiole 2–5 cm, 0.5–1 mm wide, unwinged, red-brown at base, greenish or straw-colored above, dull; blade gen 8–12 mm wide, linear; 1° lflets 8–15 pairs, 3–7 mm, 3–6 mm wide, ± diamond-shaped, gen shallowly crenate on upper, outer margins. **SPORANGIA**: sori gen 1–1.5 mm, gen 1–3 pairs per 1° lflet. $2n$= 72. RARE in CA. On rocks in limestone seams of granite cliffs; 2050 m. n SNH (Sierra Co., e side of Sierra Buttes); to AK, e N.Am, Eur, Asia. [*A. viride* Hudson]

A. vespertinum Maxon (p. 93) **LVS** 1-pinnate, many, clustered, 8–30 cm; petiole 1–5(8) cm, gen 0.5–1 mm wide, unwinged, dark red- to purple-brown, shiny; blade gen 1–2.5 cm wide, linear; 1° lflets 20–30 pairs, gen 5–12 mm, 5 mm wide, oblong, margin shallowly lobed ± throughout. **SPORANGIA**: sori gen 0.5–1.5 mm, gen 2–6 pairs per 1° lflet. Base of overhanging boulders; 500–1000 m. SCo, SnGb, PR; n Baja CA. ❀DRN,SHD:2,3; DFCLT.

AZOLLACEAE MOSQUITO FERN FAMILY

Alan R. Smith

Pl free-floating or stranded on mud, gen 1–5 cm, often fan-shaped; roots pendent from st forks, unbranched. **ST** forked repeatedly or pinnate, thread-like, easily fragmented at joints. **LVS** alternate, in 2 rows, sessile, often overlapped, 0.5–1.5 mm, seemingly paired but actually of 2 roundish to ovate lobes; upper lobe floating or emergent, thick, greenish or reddish, margin whitish; lower lobe submersed, gen slightly larger, thinner, whitish. **SPORANGIA** in seemingly axillary cases of 2 kinds, cases gen in pairs of 1 kind. **MALE SPORANGIUM CASE** 1.2–2 mm diam, spheric; tip dark-pointed; wall transparent; sporangia gen 20–100+, long-stalked; spores 32 or 64, spheric, in gen 3–6 barbed masses. **FEMALE SPORANGIUM CASE** 0.2–0.4 mm diam, hemispheric or spheric; tip obtuse, covered by dark, conic, spongy structures that aid in flotation; wall ± opaque; sporangium 1, sessile; spore 1, spheric. 1 genus, ± 7 spp.: ± worldwide.

AZOLLA

(Greek: dry kill, from pl death in dried habitats) [Perkins et al. 1985 Scanning Electron Microscopy 1985(IV):1719–1734] Used as green manure in rice paddies because of nitrogen-fixing algae in upper lf lobe; spp. identification requires gen female sporangium cases (gen 0 on herbarium specimens), often lf sectioning, compound microscope.

1. Female sporangium case bumpy, each bump often angled, with a tangle of thread-like filaments; papillae on upper side of upper lf lobe 1-celled; male spore mass barbs lacking partitions .... ***A. filiculoides***
1′ Female sporangium case pitted, with few thread-like filaments; papillae on upper side of upper lf lobe often 2-celled; male spore mass barbs with several partitions .... ***A. mexicana***

A. filiculoides Lam. (p. 93) Pl green to reddish, fertile only when ascending. **STS** < 5+ cm: immature prostrate, internodes < 5 mm; mature ascending, internodes < 1 mm. **SPORANGIUM CASES** gen male. Common. Ponds, slow streams, wet ditches; 0–1600 m. ± CA exc D; to WA, AZ, S.Am, Eurasia, Afr; also e US. ❀still water [not saline] or mud:**4–9**,10,**14–24**.

A. mexicana C. Presl MEXICAN MOSQUITO FERN Pl green or blue-green to dark red or red-fringed, gen fertile. **STS** prostrate, gen 1–2 cm; internodes < 1 mm. **SPORANGIUM CASES** often male and female. UNCOMMON. Habitats of *A. filiculoides*; ± 100 m. ScV, possibly n SN; to B.C., c US, S.Am. ❀still water [not saline] or mud:**4–9**,10,**14–17**,18,19,**20–24**.

BLECHNACEAE DEER FERN FAMILY

Alan R. Smith & Thomas Lemieux

Pls in soil; rhizome short-creeping to erect, scaly. **LVS** all alike or of 2 kinds, fertile and sterile; petiole in transverse section with vascular strands in a circle; blade deeply pinnately lobed to 1-pinnate with deeply pinnately lobed 1° lflets, hairs gen 0; veins free or netted. **SPORANGIA** in linear to oblong sori along veins parallel to the nearest midrib; indusium shaped ± like sorus, opening towards nearest midrib; stalk cells in 2–3 rows; spores elliptic, scar linear. ± 9 genera, ± 250 spp.: worldwide, esp trop. New growth often reddish; several spp. cult.

1. Blade deeply pinnately lobed to 1-pinnate with unlobed 1° lflets; lvs of 2 kinds, fertile and sterile; sori 2 per 1° lflet, linear, > 10 × longer than wide, 1 along each side of 1° lflet midrib .... **BLECHNUM**

1′ Blade 1-pinnate with deeply pinnately lobed 1° lflets; lvs all ± alike; sori many per lobe, oblong, 2–4 × longer than wide, end-to-end along each side of lobe midrib, some oblong to linear along 1° lflet midrib as well .... **WOODWARDIA**

BLECHNUM

Rhizome suberect; scales brown. **LVS** all ± alike or of 2 kinds, fertile > sterile, with much narrower 1° lflets, deeply pinnately lobed to 1-pinnate with unlobed 1° lflets; lower 1° lflets reduced or not; veins of sterile lvs free. **SPORANGIA**: sori linear, 2 per 1° lflet, 1 along each side of 1° lflet midrib. > 200 spp.: worldwide, esp trop, subtrop. (Greek: ancient name for ferns)

B. spicant (L.) Smith (p. 93) DEER FERN Rhizome ascending to suberect. **LF**: petiole base blackish, with persistent brownish scales; sterile lvs forming rosette, ± arching, deeply pinnately lobed to 1-pinnate with unlobed 1° lflets, gen < 1 m, 2–10 cm wide, 1° lflets 5–8 mm wide, gen 20–80 pairs, entire to shallowly crenate, lower gradually reduced to semicircular lobes < 5 mm; fertile lvs appearing later than sterile, in center of rosette, ± erect, 1-pinnate, > sterile, 1° lflets ± 2 mm wide, unlobed. $2n$=68. Shaded, moist areas; ± 0–1500 m. NCo, NCoRO, n SNH, SnFrB; to AK, Eur. Locally common; cult. ❀SHD:**4,5**&IRR;**6**,7,14,15,**16,17**,18,19,24.

WOODWARDIA CHAIN FERN

Rhizome prostrate to ascending; scales dense, orange-brown. **LVS** all alike, 1-pinnate with 1° lflets deeply pinnately lobed, the lower gen slightly reduced or not; veins of sterile lvs netted but free at margin. **SPORANGIA**: sori oblong, many per lobe, end-to-end along each side of lobe midrib and 1° lflet midrib. 13 spp.: N.Am, C.Am, e Asia, Eur, gen temp, subtrop. (T.J. Woodward, British phycologist, 1745–1820)

W. fimbriata Smith (p. 93) GIANT CHAIN FERN Rhizome prostrate, short, stout. **LF** gen 1–3 m; petiole gen 5–15 mm wide at base, straw-colored, scales large, orange-brown to straw-colored; 1° lflets gen 15–30 cm, often glandular, lower slightly reduced. **SPORANGIA**: sori gen ± 2–4 mm. $2n$=68. Near streams, springs; ± 0–2300 m. CA-FP (exc GV); OR, NV, AZ, nw Mex. Cult. ❀IRR or WET:**4,5**&SHD;2,3,**6,7**,8,9,**14–17**,18–24.

DENNSTAEDTIACEAE BRACKEN FAMILY

Alan R. Smith & Thomas Lemieux

Pl in soil; rhizome hairy, less often scaly, sometimes from bud near petiole base. **LVS** alike; petiole strongly grooved on upper side, glabrous or hairy, rarely scaly, in X-section vascular strands 1–many in U shape; blade gen 1–5- or more pinnate; axes, blades glabrous or hairy, rarely scaly; veins pinnate from midrib, gen forked beyond, free exc sometimes at margin of fertile segments. **SPORANGIA** at or near margin, gen ± covered by reflexed segment margins (false indusia); true indusium 0, or linear, purse-shaped, opening toward margin or fused with it to form cup; stalk cells in 1–3 rows; spores spheric or elliptic, scar of 3 radiating branches or linear. ± 17 genera, ± 375 spp.: esp trop. Variously defined; *Pteridium* sometimes in its own family.

PTERIDIUM BRACKEN, BRAKE

Rhizome gen deep, long-creeping, branched; scales 0. **LF**: petiole near base blackish, with dense brownish hairs, above straw-colored, ± glabrous; blade 2–4-pinnate, lower surface gen hairy; 1° lflets sometimes with nectaries in axils; veins free exc at margin of fertile segments. **SPORANGIA** marginal, gen continuous exc at sinuses; tips, on connecting veins, ± covered by false indusium (sterile segment margins similarly modified); true indusium inconspicuous or 0; spores spheric. ± 5 spp.: temp, trop. (Greek: small *Pteris*, fern) [Tryon 1941 Contr Gray Herb 134:1–31, 37–67] Often considered (e.g., by Tryon) a single, ± worldwide, highly variable sp., but esp in trop, subtrop, spp. seem distinct.

P. aquilinum (L.) Kuhn var. ***pubescens*** L. Underw. (p. 93) **LF** arched; petiole 10–100 cm; blade gen 15–150 cm, widely-triangular, leathery, gen 3-pinnate below, lower 1° lflets gen longest, ± 45° away from axis; segments or lobes gen 0.5–2 cm, 3–6 mm wide, oblong, round at tip, hairs on lower surface gen dense, straight or ± kinked, clear, sometimes on upper. Pastures, woods, meadows, hillsides, partial to full sun; 0–3200 m. CA-FP (exc GV); to AK, SD, nw Mex; also e Can, ne US. Other vars. in e US, Mex, Eurasia, Afr, Pacific. TOXIC in quantity to livestock and humans; cooking removes some toxins, but carcinogens may remain. ❀**4–6,15–17**, IRR:1–3,**18**,24&SHD:**7**,8–10,**14**,19–23;INV also STBL.

DRYOPTERIDACEAE WOOD FERN FAMILY

Alan R. Smith & Thomas Lemieux

Per, in soil or rock crevices; rhizome gen short-creeping, suberect, or erect, scales large, gen tan to brown, gen 1-colored. **LVS** gen tufted, 5–200+ cm, gen ± alike; petiole gen firm, base gen darker, with 2–many vascular strands; blade 1–4-pinnate, often with scales, hair-like scales, hairs (exc clear, needle-like hairs gen 0), or short-stalked glands on axes, sometimes between veins, veins free to netted; 1° and 2° axes gen grooved on upper side. **SPORANGIA**: sori round, less often oblong or J-shaped, along or at tips of veins; indusia peltate, round-reniform, oblong to linear, J-shaped, hood-like, or cup-like, rarely 0; spores elliptic, winged, ridged, or spiny, scar linear. ± 60 genera, > 1000 spp.: worldwide, esp trop, wooded areas. Sometimes *Woodsia* in Woodsiaceae; *Athyrium, Cystopteris* in Athyriaceae.

1. Indusium cup-like, of many segmented hair- or scale-like fragments or lobes encircling sorus from below, often obscure in age .... **WOODSIA**
1′ Indusium peltate, round-reniform, oblong, J-shaped, or hood-like, ± covering sorus from above or 1 side, rarely 0
 2. Indusium peltate, centrally attached, without a sinus
  3. Veins regularly netted; lf 1-pinnate, teeth without bristle-like tips, sometimes 0 .... **CYRTOMIUM**
  3′ Veins gen free, rarely casually joined; lf 1–3-pinnate, teeth (gen incl bristle-like tips) < 4 mm **POLYSTICHUM**
 2′ Indusium round-reniform and ± centrally attached at a sinus, oblong, J-shaped, hood-like and laterally attached, or 0
  4. Indusium 0, oblong, or J-shaped .... **ATHYRIUM**
  4′ Indusium round-reniform or hood-like
   5. Indusium hood-like; petiole < 1.5 mm wide, base with 2 vascular strands; blade gen 10–20 cm, 5–8 cm wide .... **CYSTOPTERIS**
   5′ Indusium round-reniform; petiole > 1.5 mm wide, base with many vascular strands; blade gen 20–75 cm, 10–20 cm wide .... **DRYOPTERIS**

ATHYRIUM LADY FERN

Rhizome short-creeping to suberect, stout. **LF**: petiole stout, fleshy, easily crushed, straw-colored exc base gen blackened, base scaly, in X-section with 2 crescent-shaped vascular strands; blade gen 2-pinnate or more, ± glabrous, veins free. **SPORANGIA**: sori ± round, ± oblong, or J-shaped; indusium oblong or J-shaped, laterally attached, or 0. ± 100 spp.: gen n temp, esp e Asia. (Greek: doorless, from enclosed sori)

1. Sori gen round, ± near margin; indusium 0 .... ***A. alpestre***
1′ Sori oblong or J-shaped, ± not near margin; indusium oblong or J-shaped .... ***A. filix-femina***

A. alpestre (D.H. Hoppe) T. Moore var. ***americanum*** F.K. Butters (p. 93) **LF**: blade narrow-elliptic or lanceolate, 2–3-pinnate, segments shallowly to deeply pinnately lobed, marginal tooth often reflexed over sorus. $2n$=80. Moist or wet rock crevices, talus, cliffs, boulder bases, streamsides; 1700–3700 m. KR, CaRH, SNH, Wrn; to AK, w Can, MT, Colorado; also e Can, Greenland. [*A. distentifolium* Opiz var. *a* (F.K. Butters) Boivin] Var. *alpestre* in Eurasia. ❀WET:**4,5**&SHD:1–3,**6**,7,15–17;DFCLT.

A. filix-femina (L.) Roth var. ***cyclosorum*** Rupr. (p. 93) **LF**: blade elliptic to lanceolate, 1–2-pinnate, segments deeply pinnately lobed to ± toothed, margin not reflexed over sorus. $2n$=80. Woods, along streams, seepage areas; 0–3200 m. CA (exc GV, PR, D); to AK, w Can, ID, Colorado, n Mex. [var. *californicum* F.K. Butters; var. *sitchense* Rupr.] Highly variable, but named vars. in w N.Am seem indistinct; other vars. worldwide. ❀IRR or WET:**4–6,17** &SHD:1–3,7,14,**15,16**,18–23,**24**.

CYRTOMIUM HOLLY FERN

Rhizome short-creeping or ascending to suberect, stout. **LF**: petiole stout, base firm, scaly, in X-section with many ± round or ovate vascular strands in an arc; blade 1-pinnate, gen thick, leathery, veins regularly netted. **SPORANGIA**: sori round; indusium peltate, not persistent. ± 20 spp.: gen e Asia. (Greek: arch, from pattern of netted veins)

C. falcatum (L.f.) C. Presl Rhizome with old, large, light- to dark-brown, entire to jagged scales. **LF** 30–80 cm; 1° lflets 8–11 cm, often with 1 lobe on distal side of base, teeth 0–10 mm, < 6 mm wide, without bristle-like tips, margin thickened. $2n$=164. Gen moist cliffs, banks, crevices; < 900 m. n SNF, SCo, WTR, SnGb; also se US; native to e Asia; cult as orn.

CYSTOPTERIS FRAGILE FERN

Rhizome gen short-creeping. **LF**: petiole ± fleshy, often with few scales, base in X-section with 2 vascular strands; blade 2–4-pinnate, veins free. **SPORANGIA**: sori round; indusium hood-like, arched over sorus, attached on side away from margin. ± 10 spp.: gen temp. (Greek: bladder fern, from indusium)

C. fragilis (L.) Bernh. (p. 93) Rhizome 2–4 mm diam; scales at tip, lanceolate, brownish, shining, glabrous, entire. **LF** 8–30(37) cm; petiole gen < blade, < 1.5 mm wide, base straw-colored to reddish brown; blade gen 10–24 cm, 5–9 cm wide, ovate-lanceolate, lowest 2–4 1° lflets ± < others. 2*n*=168. Shady, moist rock crevices, meadows, banks, streamsides; 50–3800 m. KR, NCoRO, NCoRH, CaR, n&c SNF, SNH, SnFrB, SCoR, TR, PR, GB, DMtns; worldwide. ❀WET or IRR:**4–6,17**&SHD:1–3,7,14,**15,16**,18–23,**24**.

DRYOPTERIS WOOD FERN

Rhizome short-creeping or ascending to suberect, stout. **LF**: petiole > 1.5 mm wide, firm, more densely scaly than midrib, base in X-section with many round vascular strands in an arc; blade 1–3-pinnate or more, veins free, simple or forked; segments deeply pinnately lobed or not. **SPORANGIA**: sori round; indusium round-reniform, ± centrally attached at a sinus, gen persistent. ± 100 spp.: ± worldwide, esp e Asia. (Greek: oak, fern) Hybrids unknown in CA, frequent in e N.Am. [Montgomery & Paulton 1981 Fiddlehead Forum 8:25–31]

1. Lf ± 3-pinnate, segments deeply pinnately lobed ..... ***D. expansa***
1′ Lf 1–2-pinnate, segments deeply pinnately lobed or not
  2. Longest 1° lflets near base; veins running to teeth tips; locally common, NW, SN, CW, SW ..... ***D. arguta***
  2′ Longest 1° lflets near middle, rarely near base; veins gen ending short of margin; SnBr, W&I ..... ***D. filix-mas***

D. arguta (Kaulf.) Maxon (p. 93) **LF** 30–60(100+) cm, 12–18 (30) cm wide; petiole, midrib minutely glandular; blade lanceolate, 1–2-pinnate, segments deeply pinnately lobed or not, teeth with bristle-like tips or not, veins running to teeth tips; longest 1° lflets near base; scales of 1° lflet midribs lance-ovate to ± linear. 2*n*=82. Locally common. Open, wooded slopes; < 2500 m. NW, SN, CW, SW; to B.C., AZ. ❀DRN:**4–6,17**&SHD:**7,15,16,18**&IRR:2,8,9,**14**, 19–23,**24**.

D. expansa (C. Presl) C.R. Fraser-Jenkins & Jermy (p. 93) **LF** 30–80(100+) cm, 10–30(40) cm wide; petiole gen darker on lower surface, scales gen with a dark central stripe; blade ± lanceolate to widely-deltate, ± 3-pinnate, segments deeply pinnately lobed, teeth with bristle-like tips or not, veins gen ending short of margin; longest 1° lflets near base. 2*n*=82. Shaded, wooded areas, esp banks of streams, creeks; < 500 m. NCo, NCoRO, CCo; to AK, Can, Rocky Mtns. [*D. assimilis* S. Walker; *D. dilatata* (Hoffm.) A. Gray misapplied] ❀**4,5**,IRR,SHD:1,**6,7,14–17**.

D. filix-mas (L.) Schott (p. 93) MALE FERN **LF** ± 40–70(100+) cm, 15–25(30+) cm wide; petiole, midrib nonglandular; blade elliptic, 2-pinnate, segments deeply pinnately lobed or not, teeth ± without bristle-like tips, veins gen ending short of margin; longest 1° lflets near middle, rarely near base; scales of 1° lflet midribs ± linear or hair-like. 2*n*=164. RARE in CA. Granitic cliffs; 2400–3100 m. SnBr, W&I; to B.C., ne N.Am, Eur, Afr. In cult.

POLYSTICHUM SWORD FERN

Rhizome gen suberect to erect, often stout. **LF**: petiole stout, firm, gen densely scaly, base in X-section with many round vascular strands in an arc; blade gen 1–3-pinnate, thin to ± leathery, scaly, veins gen free, rarely casually joined; 1° lflet bases often wider on distal side; teeth gen < 4 mm, gen incl bristle-like tips that are < 2 mm. **SPORANGIA**: sori round; indusium gen peltate, sinus 0. ± 175+ spp.: ± worldwide. (Greek: many rows, from rows of sori on type sp.)

1. Lf gen 1-pinnate; 1° lflets gen simple, ± entire to serrate, in *P. kruckebergii* sometimes 1-lobed
  2. Lf 10–15(25) cm; 1° lflets simple or gen lobed at bases of lowest ..... [2]***P. kruckebergii***
  2′ Lf 10–120(200) cm; 1° lflets simple
    3. Lowest 1° lflets ± deltate, ± 1/2 length of longest; petiole 1/6–1/10 blade length ..... ***P. lonchitis***
    3′ Lowest 1° lflets ovate to lanceolate, ± = to ± 2/3 length of longest; petiole gen 1/5–1/2 blade length
      4. Petiole base scales ovate, ± 3(6) mm wide, those above lowest 1° lflets gen > 1 mm wide, persistent; 1° lflets gen in 1 plane; indusium ciliate ..... ***P. munitum***
      4′ Petiole base scales lanceolate, ± 2–3 mm wide, those above lowest 1° lflets gen < 1 mm wide, falling early; 1° lflets in ± 1 plane or not; indusium ± entire to toothed ..... ***P. imbricans***
        5. Lf 15–80 cm; 1° lflets in ± 1 plane, longest 3–6(10) cm; sori submarginal ..... ssp. ***curtum***
        5′ Lf 15–50 cm; 1° lflets often not in ± 1 plane, longest 2–5 cm; sori ± close to midvein ..... ssp. ***imbricans***
1′ Lf 1–2-pinnate; 1° lflets deeply pinnately lobed to 1-pinnate, at least at bases of lowest
  6. Lflet teeth without bristle-like tips; serpentine ..... ***P. lemmonii***
  6′ Lflet teeth with bristle-like tips; serpentine or not
    7. Lf gen 2- to rarely partly 3-pinnate; 2° lflets abruptly narrowed to attachment < 0.5 mm wide, often with a lobe on distal side of base; main lf blade axis with persistent scales ..... ***P. dudleyi***
    7′ Lf gen 1- to partly 2-pinnate; 2° lflets gradually tapered to attachment < 3 mm wide, lobes 0; main lf blade axis with deciduous scales
      8. Smallest scales on lflet lower surface hair-like; blade gen > 5 cm wide ..... ***P. californicum***
      8′ Smallest scales on lflet lower surface linear, few hair-like; blade gen < 5 cm wide
        9. Lowest 1° lflets ± deltate, longest < 1(–1.5) cm ..... [2]***P. kruckebergii***
        9′ Lowest 1° lflets ± lanceolate, longest 1.5–3 cm ..... ***P. scopulinum***

P. californicum (D. Eaton) Diels (p. 93) **LF** 40–100 cm; petiole < 1/2 blade length, base scales 3–4 mm wide, ovate-lanceolate; blade lanceolate, 1- to partly 2-pinnate; 1° lflets gen 4–7 cm. **SPORANGIA**: indusium ciliate. 2*n*=82,164. Woods, streambanks, to rocky open slopes; 0–800 m. NCoRO, SN (1 site), CCo, SnFrB, SCoRO, SnBr (1 site); to B.C. Probably arises as sterile hybrid between *P. munitum* or *P. imbricans* and *P. dudleyi*. Chromosome doubling (2*n*=164) restores fertility, allows backcrossing to parents. ❀DRN:**4,5**&IRR:**6,17**&SHD:14,**15,16**&WET:**7,18–24**.

P. dudleyi Maxon (p. 93) **LF** 50–100 cm; petiole ± 1/4–1/2 blade length, base scales 3–7(9) mm wide, ovate; blade lanceolate, gen 2- to rarely partly 3-pinnate; 1° lflets ± 5–12 cm. **SPORANGIA**: indusium ciliate. 2*n*=82. Moist forests; < 500 m. NCoRO, CCo, SnFrB, SCoRO. Hybridizes with *P. californicum, P. munitum*. ❀DRN:**4,5**&IRR,SHD:**6**,14,**15–17**&WET:7,18,**19–24**.

P. imbricans (D. Eaton) D.H. Wagner (p. 97) **LF**: petiole gen 1/5–1/2 blade length, base scales ± 2–3 mm wide, lanceolate, those

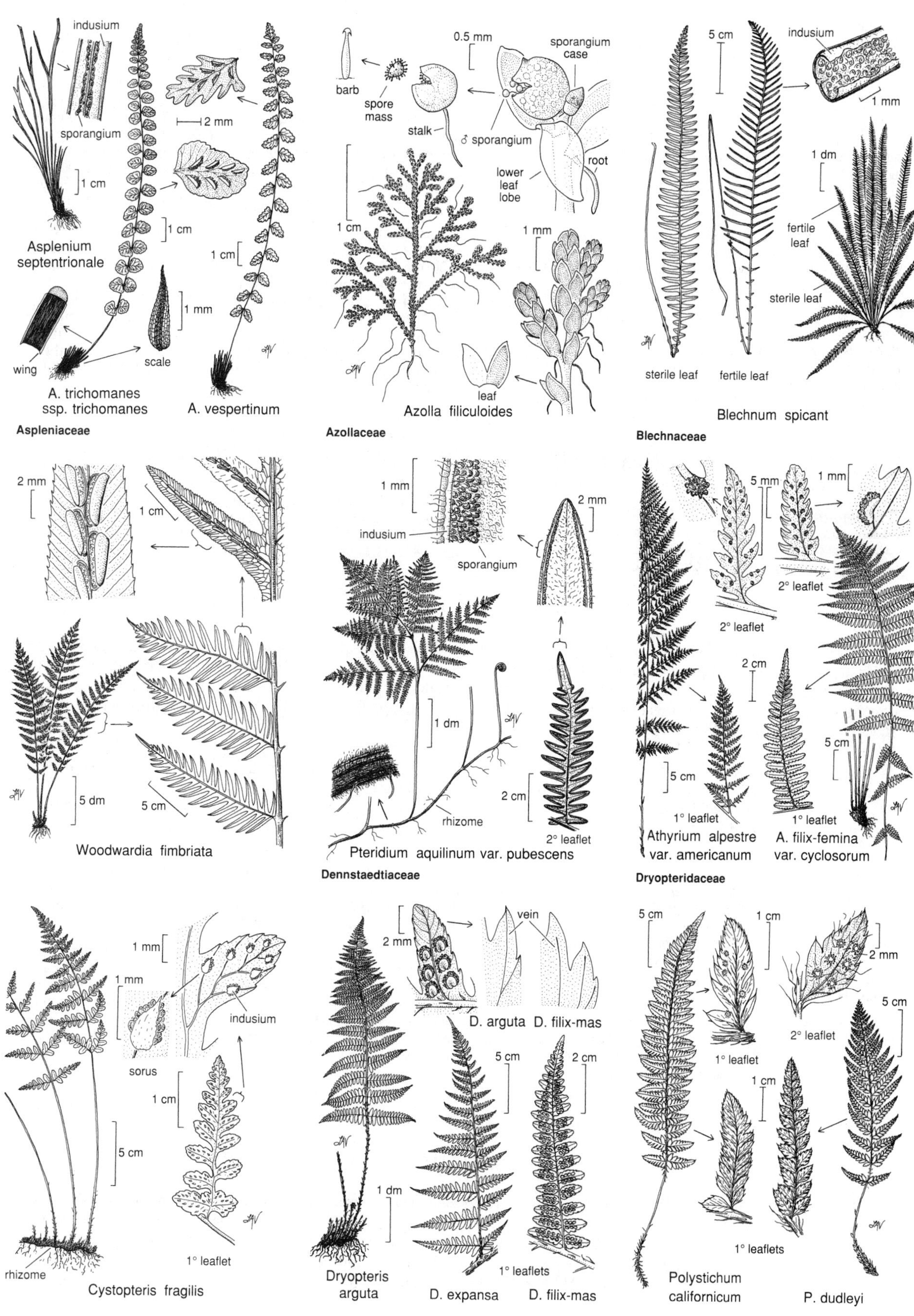
indusium
sporangium
2 mm
1 cm
1 cm
1 cm
1 mm
scale
wing
Asplenium septentrionale
A. trichomanes ssp. trichomanes
A. vespertinum
Aspleniaceae
0.5 mm
barb
spore mass
stalk
sporangium case
♂ sporangium
lower leaf lobe
root
1 cm
1 mm
leaf
Azolla filiculoides
Azollaceae
5 cm
indusium
1 mm
1 dm
fertile leaf
sterile leaf
sterile leaf
fertile leaf
Blechnum spicant
Blechnaceae
2 mm
1 cm
5 dm
5 cm
Woodwardia fimbriata
1 mm
indusium
sporangium
2 mm
1 dm
2 cm
rhizome
2° leaflet
Pteridium aquilinum var. pubescens
Dennstaedtiaceae
5 mm
1 mm
2° leaflet
2° leaflet
2 cm
5 cm
5 cm
1° leaflet
1° leaflet
Athyrium alpestre var. americanum
A. filix-femina var. cyclosorum
Dryopteridaceae
1 mm
1 mm
indusium
sorus
1 cm
5 cm
1° leaflet
rhizome
Cystopteris fragilis
2 mm
vein
D. arguta
D. filix-mas
5 cm
2 cm
1 dm
1° leaflets
Dryopteris arguta
D. expansa
D. filix-mas
5 cm
1 cm
2 mm
5 cm
2° leaflet
1° leaflet
1 cm
1° leaflets
Polystichum californicum
P. dudleyi

above lowest 1° lflets gen < 1 mm wide, falling early; blade narrow-lanceolate to -elliptic, 1-pinnate. 2*n*=82. Shaded or exposed outcrops, banks, slopes, rocky areas; 300–2500 m. KR, NCoR, CaRH, n SNF, SNH, SnFrB, SCoRO, SCo, TR, PR, MP; to B.C. Hybridizes with *P. dudleyi* (called *P. californicum*), *P. lemmonii* (called *P. scopulinum*), *P. munitum*.

ssp. ***curtum*** (Ewan) D.H. Wagner **LF** 15–80 cm; 1° lflets in ± 1 plane, longest 3–6(10) cm. **SPORANGIUM**: sori submarginal; indusium gen toothed. Rocky slopes, crevices, woods; 400–1600 m. n&c SNH, SnFrB, SCoRO, SCo, TR, PR. ❀DRN:4–6&IRR, SHD:14,**15–17**&WET:**7**,18–24.

ssp. ***imbricans*** **LF** 15–50 cm; 1° lflets often not in ± 1 plane, longest 2–5 cm. **SPORANGIUM**: sori ± close to midvein; indusium ± entire to slightly toothed. Habitats of sp.; 300–2500 m. KR, NCoR, CaRH, n SNF, SNH, SnFrB, SCoRO, WTR, MP; to B.C. [*P. munitum* ssp. *i.* (D. Eaton) Munz; *P. munitum* ssp. *nudatum* Ewan] ❀DRN:**4–6**&IRR,SHD:1–3,7,14,**15–17**,18,**19–24**.

P. kruckebergii W. Wagner (p. 97) KRUCKEBERG'S SWORD FERN **LF** 10–15(25) cm; petiole gen ± 1/10 blade length, base scales 2–3 mm wide, ovate to lanceolate; blade linear to very narrow-elliptic, 1- to partly 2-pinnate; 1° lflets gen 6–15 mm, lowest ± deltate, longest < 1(1.5) cm. **SPORANGIA**: indusium entire or minutely toothed. 2*n*=164. UNCOMMON. Rocks, cliffs; 2100– 3200 m. CaRH, n&c SNH, SnBr; to B.C., MT, UT. Probably fertile hybrid between *P. lonchitis*, *P. lemmonii*. ❀DRN:4,5&IRR:1–3,6&SHD: 7,15–17;DFCLT.

P. lemmonii L. Underw. (p. 97) **LF** 15–40 cm; petiole 1/4–1/3 blade length, base scales ± 2 mm wide, gen linear-lanceolate; blade narrow-lanceolate, 2-pinnate; 1° lflets gen 1–3 cm; teeth without bristle-like tips. **SPORANGIA**: indusium entire or minutely toothed. 2*n*=82. Serpentine, rocks, ledges, rocky stream beds; 1200–2600 m. KR, NCoRH, CaRH, n SNH; to B.C. ❀DRN:4,5 &IRR:1–3,6&SHD:7,15–17;DFCLT.

P. lonchitis (L.) Roth (p. 97) HOLLY FERN **LF** gen 10–60 cm; petiole gen 1/6–1/10 blade length, base scales 2–4(5) mm wide, ovate-lanceolate; blade linear, 1-pinnate; 1° lflets gen 1–3(4) cm, lowest ± deltate, ± 1/2 length of longest. **SPORANGIA**: indusium entire or minutely toothed. 2*n*=82. UNCOMMON. Gen shaded, moist or wet, granite or limestone crevices or bluffs; 1800–2600 m. KR, NCoRH (Yolla Bolly Mtns), n SNH; to AK, e Can, Rocky Mtns, Eurasia. ❀DRN:4,5&IRR:1–3,6&SHD:7,15–17;DFCLT.

P. munitum (Kaulf.) C. Presl (p. 97) WESTERN SWORD FERN **LF** gen 50–120(200) cm; petiole gen 1/5–1/2 blade length, base scales ± 3(6) mm wide, ovate, those above lowest 1° lflets gen > 1 mm wide, persistent; blade lanceolate to narrow-elliptic, 1-pinnate; 1° lflets gen in 1 plane, 2–8(14) cm. **SPORANGIA**: indusium ciliate. 2*n*=82. Common. Wooded hillsides, shaded slopes, rarely cliffs, outcrops; < 1600 m. NW, CaR, s SNH, CW, n ChI; to AK, MT; also SD, Guadalupe Island (Mex). Hybrids with *P. dudleyi* are called *P. californicum*. ❀**4,5**, IRR:**6,17** &SHD:**7,14–16** &WET:8,9,18,19, **20–24**;GRCVR;CVS.

P. scopulinum (D. Eaton) Maxon (p. 97) **LF** 10–50 cm; petiole 1/4–1/2 blade length, base scales 1.5–2(3) mm wide, lanceolate to elliptic; blade narrow-lanceolate 1- to partly 2-pinnate; 1° lflets gen 1–3 cm, oblong-lanceolate, lowest ± lanceolate, longest 1.5–3 cm. **SPORANGIA**: indusium entire. 2*n*=164. Serpentine to acidic soils, gen full sun, rock crevices, boulder bases; 400–3200 m. KR, NCoRO, NCoRH, CaRH, n&c SNH, SnBr, SnJt; to B.C., Rocky Mtns, AZ. Probably fertile hybrid between *P. imbricans*, *P. lemmonii*. ❀DRN:**4–6**&IRR,SHD:1–3,7,14–18;DFCLT.

WOODSIA CLIFF FERN

Rhizome gen ascending to suberect, short, with many old petiole bases. **LF** often glandular or hairy; petiole base with 2 vascular strands; blade 1–2-pinnate, segments ± toothed to pinnately lobed, veins free. **SPORANGIA**: sori round, gen not at margins; indusium cup-like, often of many segmented hair- or scale-like fragments or lobes encircling sorus from below, often of crusty, whitish beads, often obscure in age. ± 30 spp.: gen n temp. (J. Woods, Britain, b 1776) [Brown 1964 Beih Nova Hedwigia 16:1–154] ❀TRY.

1. Hairs on lower surface lf axes ± flat, segmented, ± 0.5–1 mm and nonsegmented, glandular, ± 0.1 mm ***W. scopulina***
1′ Hairs on lower surface lf axes 0 or nonsegmented, glandular, ± 0.1 mm
  2. Indusium of segmented hairs .... ***W. oregana***
  2′ Indusium of scale-like fragments or lobes .... ***W. plummerae***

W. oregana D. Eaton (p. 97) **LF** 5–25 cm, 1–3.5 cm wide; blade tip ± acute, nonforked; blade lower surface hairs 0 or ± 0.1 mm, nonsegmented, glandular; 1° lflets 0.5–2.5 cm, 0.3–1.3 cm wide, pinnately lobed to 1-pinnate, margin fine-toothed. **SPORANGIA**: indusium of segmented hairs. 2*n*=76,152. Crevices, rock bases; 900–2800 m. KR, CaRH, n&s SNH, SnBr, PR, MP?, W&I, DMtns; to B.C., e Can, n US, OK, AZ. Sierra Co. citation by Brown a mislabeled specimen.

W. plummerae Lemmon (p. 97) PLUMMER'S WOODSIA **LF** < 25 cm, < 4 cm wide; blade tip often blunt or forked; blade lower surface hairs ± 0.1 mm, nonsegmented, glandular; 1° lflets < 3 cm, < 1.5 cm wide, pinnately lobed to 1-pinnate, margin toothed to shallow-lobed. **SPORANGIA**: indusium of scale-like fragments or lobes ending in hairs or not. 2*n*=152. RARE in CA. Crevices, rock bases; 1600–2000 m. DMtns; to TX, n Mex. San Diego Co. citation by Brown a mislabeled specimen.

W. scopulina D. Eaton (p. 97) **LF** < 36 cm, 5–8 cm wide; blade tip ± acute, nonforked; blade lower surface hairs ± 0.5–1 mm, ± flat, segmented and ± 0.1 mm, nonsegmented, glandular; 1° lflets < 4 cm, 2.5 cm wide, pinnately lobed to 1-pinnate, margin toothed to shallow-lobed. **SPORANGIA**: indusium of narrow scale-like lobes. 2*n*=76,152. Crevices, rock bases; 1400–3500 m. CaRH, SNH, SnBr, W&I; to AK, e Can, w US.

EQUISETACEAE HORSETAIL FAMILY

Richard L. Hauke

Per from rhizome (above ground st ann or per). **ST** gen erect, ridged lengthwise, hollow exc at nodes, sometimes of 2 kinds (sterile, fertile); branches 0 or whorled and alternate lvs, sometimes solid. **LVS** scale-like, whorled, fused into nodal sheath with as many teeth as lvs, gen not green. **SPORANGIA** several on inner surface of peltate scales that are clustered into a terminal cone; spores of 1 kind per sp., spheric, green, unmarked, with 4 strap-like appendages. 1 genus, 15 spp.: worldwide exc Australia, New Zealand. [Hauke 1978 Nova Hedwigia 30:385–455]

EQUISETUM HORSETAIL, SCOURING RUSH

(Latin: horse, bristle, from roots of *E. fluviatile* L.)

1. St regularly branched
 2. Basal internode of branch > subtending nodal sheath ..... [2]***E. arvense***
 2′ Basal internode of branch < subtending nodal sheath
  3. Sheath teeth 5–10 ..... [2]***E. palustre***
  3′ Sheath teeth 14–28 on sterile st, 20–30 on fertile st ..... [2]***E. telmateia*** ssp. ***braunii***
1′ St unbranched or branches scattered
 4. St brown, fleshy, lasting < 1 month irregularly in spring
  5. Sheath teeth 6–14 on sterile st, 6–10 on fertile st ..... [2]***E. arvense***
  5′ Sheath teeth 14–28 on sterile st, 20–30 on fertile st ..... [2]***E. telmateia*** ssp. ***braunii***
 4′ St green, firm, lasting > 1 month
  6. Sheath teeth 5–10, persistent ..... [2]***E. palustre***
  6′ Sheath teeth 10–50, gen deciduous
   7. Cone tip pointed; sheath gen with 2 dark bands, 1 at base, 1 at tip; st per, gen scabrous ***E. hyemale*** ssp. ***affine***
   7′ Cone tip rounded; sheath gen with 1 dark band, at tip; st ann (or per in s), gen not scabrous ..... ***E. laevigatum***

E. arvense L. (p. 97) COMMON HORSETAIL **STS** ann, of 2 kinds. **STERILE ST** 10–60 cm, green; basal internode of branch > subtending sheath; sheath 3–8.5 mm, ± as long as wide, teeth 6–14, 1.5–3.5 mm, dark, often joined but not fused; branch with 3–4 rounded ridges, solid. **FERTILE ST** 11–32 cm, unbranched, fleshy, brown, ephemeral; sheath 5–11 mm, > that of sterile st, teeth 6–10, 3–7.5 mm. Moist, disturbed areas; < 3000 m. CA-FP, MP; N.Am, Eur, Asia. ❀IRR:**1,4–7,15–17**,24&SHD:**2,3,8,9,14**,18–23; INV.

E. hyemale L. ssp. ***affine*** (Engelm.) Calder & R.H. Taylor (p. 97) COMMON SCOURING RUSH **ST** per, of 1 kind, 60–210 cm, green, unbranched; sheath 7–17 mm, ± as long as wide, gen with 2 dark bands, 1 at base, 1 at tip, teeth 22–50, gen deciduous. **SPORANGIA**: cone tip pointed. Streams, moist, sandy or gravelly areas; < 2500 m. CA-FP (exc GV); N.Am (other ssp. in Eur, Asia). [var. *californicum* Milde; var. *robustum* (A. Braun) A.A. Eaton] Intermediates to *E. laevigatum* (aborted spores) are *E.* × *ferrissii* Clute (moist, sandy or gravelly areas; < 3000 m. CA-FP; to B.C., e US. Often mislabeled *E. laevigatum* (that sp. then miscalled *E. kansanum*). ❀IRR:**1–7,14–17,22–24**&SHD:**8,9,18–21**;INV,GRCVR, STBL.

E. laevigatum A. Braun (p. 97) SMOOTH SCOURING RUSH **ST** ann (or per in s), of 1 kind, 30–180 cm, green, unbranched; sheath 6–15 mm, longer than wide, gen with 1 dark band at tip, teeth 10–26, gen deciduous. **SPORANGIA**: cone tip rounded. Moist, sandy or gravelly areas; < 3000 m. CA-FP, MP, DMtns; to B.C., e US. [*E. funstoni* A.A. Eaton; *E. kansanum* J. Schaffner] ❀IRR:**1–7,14–17,22–24**&SHD:**8,9,18–21**;INV.

E. palustre L. (p. 97) MARSH HORSETAIL **ST** ann, of 1 kind, 20–80 cm, green; basal internode of branch < subtending sheath; sheath 4–9 mm, longer than wide, teeth 5–10, 2–5 mm, separate; branch with 4–6 rounded ridges, hollow. RARE in CA. Marshes; < 1000 m. SnFrB (San Francisco); OR to AK, e N.Am. Reported from Lake, San Mateo cos.; scarcity poorly understood; hybrid with *E. telmateia* ssp. *braunii* known from San Mateo Co.

E. telmateia Ehrh. ssp. ***braunii*** (Milde) R.L. Hauke (p. 97) GIANT HORSETAIL **STS** ann, of 2 kinds. **STERILE ST** 30–100 cm, light green; basal internode of branch < subtending sheath; sheath 7–18 mm, ± as long as wide, teeth 14–28, 4–10 mm; branch with 4–5 grooved ridges, solid. **FERTILE ST** 17–45 cm, unbranched, fleshy, brown, ephemeral; sheath 1.5–4 cm, > that of sterile st, teeth 20–30, 5–16 mm. Stream banks, roadside ditches, seepage areas; < 1000 m. NW, CW, SW; along coast to B.C. (other ssp. in Eur, Asia). Sometimes invasive. ❀WET:1,**4–6,15–17**,24&SHD:2,3,**7**,8,9,**14,18**,19–23;INV,STBL.

ISOETACEAE QUILLWORT FAMILY

W. Carl Taylor & Jon E. Keeley

Per, aquatic to terrestrial. **ST** buried, corm-like, 2–3-lobed, corky, brown. **LVS** simple, in grass-like tufts, spirally arranged on st top, erect to spreading, < 25 cm, linear above base. **SPORANGIA** solitary, embedded in wide lf base, < 1 cm, ± covered by a translucent membrane, either male or female; male spores > 10,000, < 0.045 mm, ± bean-shaped, gray or brown in mass; female spores 20–200, 0.2–0.7 mm, spheric, white, ± smooth, ridged, tubercled, or prickly. 1 genus, 150 spp.: worldwide. [Pfeiffer 1922 Ann Missouri Bot Gard 9:79–233]

ISOETES QUILLWORT

(Greek: evergreen, from the habit of some spp.) Perhaps most poorly known pteridophyte genus. Mature female spores, found in decaying lf bases or soil, critical for identification; hand lens for texture when dry, microscope with micrometer for size. Hybrids (spores flattened, highly variable) common between aquatic spp., making them less distinct.

1. Pl terrestrial (or becoming so), of seasonally wet soil, lake margins, temporary streams, vernal pools; gen < 1500 m
 2. Translucent membrane covering < 75% of sporangium ..... ***I. howellii***
 2′ Translucent membrane covering > 75% of sporangium
  3. Pl of wet soil; lf gen > 8 cm, > 1 mm wide at middle, rigid, almost brittle ..... ***I. nuttallii***
  3′ Pl of vernal pools; lf gen < 8 cm, < 1 mm wide at middle, soft, flexible ..... ***I. orcuttii***
1′ Pl underwater in persistent lakes or pools; > 1500 m
 4. Female spore prickly ..... ***I. echinospora***
 4′ Female spore ridged or tubercled
  5. Lf abruptly tapered to tip; female spore 0.3–0.5 mm wide ..... ***I. bolanderi***
  5′ Lf gradually tapered to tip; female spore 0.5–0.7 mm wide ..... ***I. occidentalis***

I. bolanderi Engelm. (p. 97) Pl underwater. **LVS** deciduous, < 20 cm, rigid, not brittle, abruptly tapered to tip, bright green; bases white to brownish. **SPORANGIUM**: membrane covering < 30%; male spores 0.02–0.03 mm, brown in mass; female spores 0.3–0.5 mm diam, ridged, tubercled. 2*n*=22. Persistent lakes, pools; > 1500 m. KR, NCoRH, CaRH, SNH, SnBr, Wrn; to B.C., WY, NM. Hybridizes with *I. echinospora*, *I. occidentalis*. Spores mature late summer. Small pls of c SNH, NV, AZ (lf < 2.5 cm) have been called var. *pygmaea* (Engelm.) Clute.

I. echinospora Durieu (p. 97) Pl underwater. **LVS** ± evergreen, < 20 cm, soft, flexible, gradually tapered to tip, bright green; bases white to brownish. **SPORANGIUM**: membrane covering < 50%; male spores 0.02–0.03 mm, gray in mass; female spores 0.4–0.5 mm diam, prickly. 2*n*=22. Persistent lakes, pools; > 1500 m. KR, n SNH; to AK, e N.Am, Eurasia. [*I. braunii* Durieu; *I. muricata* Durieu] Hybridizes with *I. bolanderi*, *I. occidentalis*. Spores mature late summer.

I. howellii Engelm. (p. 97) Pl becoming terrestrial. **LVS** deciduous, < 30 cm, rigid, not brittle, gradually tapered to tip, bright green; bases brownish to black. **SPORANGIUM**: membrane covering < 75%; male spores 0.025–0.035 mm, brown in mass; female spores 0.3–0.5 mm diam, ridged. 2*n*=22. Vernal pools, lake margins; gen < 1500 m. NCoR, CaRF, SNF, SnFrB, SCoR, SCo, PR; to WA, MT, UT. Spores mature late spring, summer. Small pls of SCo, WA, Baja CA (lf < 10 cm, female spore < 0.42 mm) have been called var. *minima* (A.A. Eaton) Pfeiffer.

I. nuttallii Engelm. (p. 97) Pl terrestrial. **LVS** deciduous, gen > 8 cm, > 1 mm wide at middle, rigid, almost brittle, gradually tapered to tip, light green to gray-green; bases white to brownish, outermost sterile, often surrounded by several black scales. **SPORANGIUM**: membrane covering > 75%; male spores 0.02–0.03 mm, brown in mass; female spores 0.35–0.6 mm diam, ± shiny, ± tubercled. 2*n*=22. Seasonally wet soil, temporary streams; gen < 1500 m. NCoR, CaRF, SN, n SnJV, SnFrB, SCoRO, SCo, PR; to B.C. Spores mature late spring, summer.

I. occidentalis L. Henderson (p. 97) Pl underwater. **LVS** evergreen, < 20 cm, rigid, brittle, gradually tapered to tip, dark green; bases brownish white. **SPORANGIUM**: membrane covering < 50%; male spores 0.035–0.045 mm, gray in mass; female spores 0.5–0.7 mm diam, ridged, tubercled. 2*n*=66. Persistent lakes, pools; > 1500 m. KR, SNH; to B.C., Colorado. [*I. lacustris* var. *paupercula* Engelm.; not *I. lacustris* L.] Hybridizes with *I. bolanderi*, *I. echinospora*. Spores mature late summer.

I. orcuttii A.A. Eaton Pl becoming terrestrial. **LVS** deciduous, gen < 8 cm, < 1 mm wide at middle, soft, flexible, gradually tapered to tip, bright green, bases white to brownish, outermost fertile, often surrounded by several black scales. **SPORANGIUM**: membrane covering > 75%; male spores 0.02–0.025 mm, brown in mass; female spores 0.2–0.4 mm diam, ± shiny, ± smooth. 2*n*=22. Vernal pools; < 1700 m. n SNH, GV, SCo, PR; Baja CA. Spores mature spring.

LYCOPODIACEAE CLUB-MOSS FAMILY

Alan R. Smith & Thomas Lemieux

Pls on ground or other pls, creeping to ± vine-like. **ST**: branches few–many, forked often unequally, also branching laterally or not. **LVS** many, simple, ± alternate, spirally arranged, small, needle- or scale-like, 1-veined, few-toothed to gen entire, those subtending sporangia gen unlike others, sometimes in cones. **CONE** terminal on erect st, erect. **SPORANGIA** solitary in lf axils; spores of 1 kind per pl, scar with 3 radiating branches. 4 genera, ± 400 spp.: worldwide, gen trop. [Øllgaard 1987 Opera Bot 92:153–178]

1. Fertile st unbranched, terminated by 1 cone .... **LYCOPODIELLA**
1′ Fertile st branched, terminated by few–many cones .... **LYCOPODIUM**

LYCOPODIELLA

ST: sterile branched in 1 horizontal plane; fertile erect, unbranched, terminated by 1 cone. ± 40 spp.: gen Am. (Diminutive of *Lycopodium*)

L. inundata (L.) Holub (p. 97) BOG CLUB-MOSS Pl on ground. **ST**: sterile creeping, 5–15(25) cm, 0.5–1 cm wide, incl lvs; fertile gen 1–2, ± 4–6(9) cm. **LVS** ± 3–8 mm, 1 mm wide, those subtending sporangia ± wider at base, ± not bristle-tipped. **CONE** 1.5–3(6) cm, ± 8–10 mm wide. 2*n*=156. RARE in CA. Peat bogs, muddy depressions, pond margins; < 50 m (NCo), ± 1000 m (n SNH). NCo, n SNH (Nevada Co.); to AK, e N.Am, Eur, Asia. [*Lycopodium i.* L.]

LYCOPODIUM

ST: sterile branched in > 1 plane; fertile erect, branched, terminated by few–many cones. ± 40 spp.: worldwide. (Greek: wolf foot, from branch tips)

L. clavatum L. (p. 97) RUNNING-PINE Pl on ground or other pls. **ST**: sterile wide-creeping to ± vine-like, < 0.5 m, 0.5–1.5 cm wide, incl lvs; fertile gen many, ± 10–20 cm, few-much-branched. **LVS** ± 4–7 mm, 0.5–1 mm wide, those subtending sporangia ovate-triangular, bristle-tipped. **CONE** 1–3(5) cm, ± 4–6 mm wide. 2*n*=68. RARE in CA. Moist ground, swamps, rarely on trees; < 200 m. NCo; to AK, MT (also e N.Am), Caribbean, S.Am, Eur, Afr, Asia.

MARSILEACEAE MARSILEA FAMILY

Alan R. Smith & Thomas Lemieux

Pl gen aquatic, gen rooted in, often stranded on mud; rhizome creeping, slender, branched. **LVS** floating, emersed, or out of water, ± alike; blade 1-palmate or 0, << petiole; veins not or repeatedly forked, free or netted. **SPORANGIA** in stalked, spheric or ± flat-ovoid, hard cases of 1 kind, near petiole base. **SPORES** large (female) and small (male), in separate sporangia. 3 genera, ± 70 spp.: esp temp.

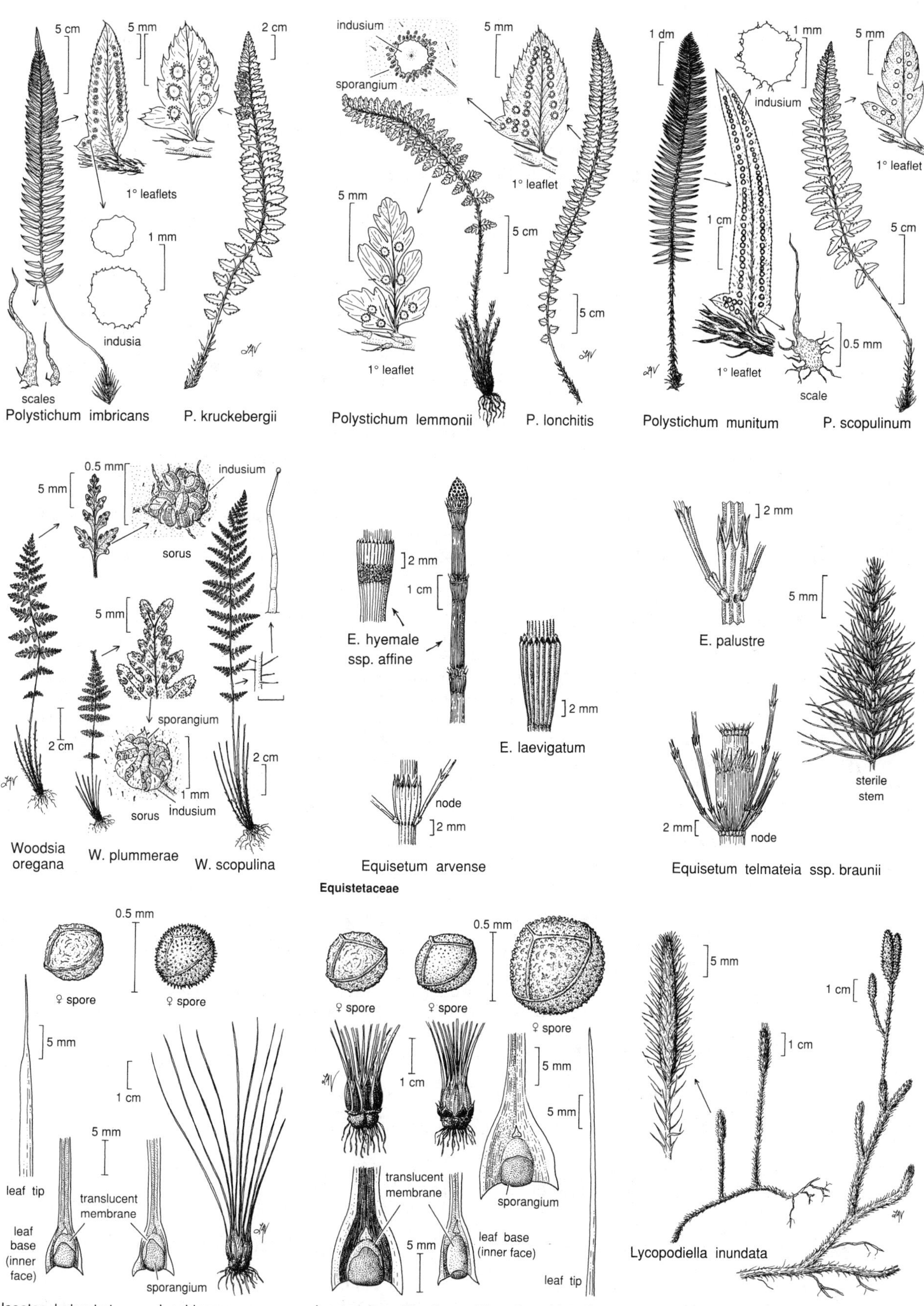

Equistetaceae

Isoetaceae

Lycopodiaceae

1. Lf blade 1-palmate, lflets 4, wedge-shaped, hairy; lf-like that of clover or wood sorrel .... **MARSILEA**
1′ Lf blade 0; lf like that of grass .... **PILULARIA**

MARSILEA

LF like that of clover or wood sorrel; blade 1-palmate, lflets 4, wedge-shaped, hairy . **SPORANGIUM CASE** fused to stalk 0.8–1.7 mm, ± flat-ovoid, hairs long, dense, deciduous or not, teeth 1–2, near base. > 60 spp.: esp temp. (L.F. Marsigli, Italian botanist, 1656–1730) [Johnson 1986 Syst Bot Monogr 11: 1–87]

1. Sporangium case fused to stalk 0.8–1 mm, distal tooth 0 or < 0.4 mm, obtuse .... ***M. oligospora***
1′ Sporangium case fused to stalk 1.1–1.7 mm, distal tooth (0.2)0.4–1.2 mm, acute .... ***M. vestita*** ssp. ***vestita***

M. oligospora Goodd. (p. 103) **LVS**: petioles of floating lvs weak, ± 15 cm, others 3–6 cm; lflet surfaces ± straight, distal margin truncate or convex, faintly fine-crenate. **SPORANGIUM CASE** 5–6 mm, 3–4 mm wide; stalk gen bent near tip. Creek beds, flood basins, vernal pools, etc.; 1400–2000 m. KR, n SNH, MP; to WA, MT, WY. ❀IRR:**4–6**,17,WET:1–3,**7**,14.

M. vestita Hook. & Grev. ssp. ***vestita*** (p. 103) **LVS**: petioles of floating lvs weak, 6–35 cm, others 3–8 cm; lflet sides (or 1) often concave, distal margin convex, ± entire. **SPORANGIUM CASE** 3–8 mm, 3–7 mm wide; stalk not bent. Habitats ± as *M. oligospora*; < 2200 m. KR, NCoRI, CaR, s SNF, SNH, GV, CCo, SnFrB, SCoRO, SCo, WTR, SnBr, PR, MP, DSon; to w&c Can, Mex, also Peru. ❀IRR:**4–7,14–17,24**,WET:1–3,8–13,**18–23**; GRCVR.

PILULARIA

LF grass-like; blade 0. **SPORANGIUM CASE** fused only to stalk tip, spheric, hairy; teeth 0. ± 6 spp.: gen temp. (Latin: little ball, from sporangium case)

P. americana A. Braun (p. 103) **LVS** gen 2–6(11) cm. **SPORANGIUM CASE** ± 2–3 mm diam. 2*n*=20. Vernal pools, mud flats, lake margins, reservoirs, etc.; < 1500 m. NCoRI, CaRF, n&c SNF, n SNH, GV, SCoR, SCo, WTR, PR; OR, Baja CA; also scattered c&se US, S.Am. Poorly collected, often overlooked due to its small, grass-like appearance. ❀TRY.

OPHIOGLOSSACEAE ADDER'S-TONGUE FAMILY

Warren H. Wagner, Jr.

Per, small, fleshy, gen glabrous; caudex gen underground, unbranched; roots glabrous, with bulblets or plantlets or not. **LF** gen 1 per caudex per year, divided into 2 facing parts with a common stalk; sterile part separated from fertile at to well above ground, blade simple to compound, veins free and forked (or netted, with incl veinlets); fertile part bladeless, bearing sporangia, simple to compound. **SPORANGIA** dehiscent into 2 valves, ± 1 mm wide, thick-walled. 3 genera, 70–85 spp.: ± worldwide, gen rare or overlooked. Fern-like pls with many traits of seed pls. Specimens must be carefully spread and pressed for identification; haploid generation underground, fleshy, non-green, associated with fungi.

1. Sterile and fertile lf parts compound (rarely small, simple, entire); sterile lf part gen midribbed, veins free, forked; sporangia not sunken in axis; roots without bulblets or plantlets .... **BOTRYCHIUM**
1′ Sterile and fertile lf parts simple, entire; sterile lf part not midribbed, veins netted with incl veinlets; sporangia sunken in axis; roots often with bulblets or plantlets .... **OPHIOGLOSSUM**

BOTRYCHIUM GRAPE-FERN, MOONWORT

Roots smooth, pale or cork-ridged, dark gray, without bulblets or plantlets. **LF** gen deciduous; bud glabrous or hairy; sterile part gen ± 1–3-pinnate (rarely simple or entire), linear to deltate, segments linear to oblong and midribbed or spoon- to wedge- or fan-shaped and not midribbed, veins free, forked, margins entire to dentate or irregularly cut; fertile part 1–3-pinnate, < to >> sterile. **SPORANGIA** not sunken in axis; stalk 0 or short. 40–50 spp.: gen temp to arctic or alpine. (Greek: bunch of grapes, from clusters of sporangia) [Wagner & Wagner 1983 Amer Fern J 73:53–62] Difficult, needing careful study; most spp. very uncommon, sporadic; good sampling of populations highly desirable in specimens. ❀TRY;DFCLT.

1. Sterile lf part gen > 10 cm; lf bud densely hairy; fertile lf part aborted or well developed; lf leathery, evergreen, each part gen ± 2–3-pinnate .... ***B. multifidum***
1′ Sterile lf part gen < 10 cm; lf bud glabrous; fertile lf part well developed; lf herbaceous, deciduous, each part simple to ± 2-pinnate
  2. Sterile lf part segments oblique-ovate to lanceolate-oblong, midribbed .... ***B. pinnatum***
  2′ Sterile lf part segments gen spoon- to wedge-or fan-shaped, not midribbed
    3. Sterile and fertile lf parts separated near to well below mid-lf; blade tips coarsely lobed to irregularly angled or shallowly rounded
      4. Sterile, fertile lf parts separated near to just below middle of lf; full-sized blades oblong, non-ternate, segments irregularly angled, square or oblong; shady coniferous woods .... ***B. montanum***
      4′ Sterile, fertile lf parts separated well below middle of lf, gen at top of lf sheath; full-sized blades oblong to ovate, ternate, segments wedge- to fan-shaped; open marshes, damp meadows .... ***B. simplex***
    3′ Sterile and fertile lf parts separated near or gen above mid-lf; blade tips finely, regularly, and deeply lobed

5. Lower segments of sterile lf part widely fan-shaped, margins at base meeting at 120–160°
  6. Pl slender, herbaceous-fleshy; sterile lf part gen < 6 cm, < 2 cm wide, segments 3–5 pairs, ± well separated, outer margins gen finely crenate to dentate; fertile lf part 1.3–3 × sterile lf part; marshes, meadows .... ***B. crenulatum***
  6′ Pl robust, fleshy; sterile lf part gen 6–10 cm, 2–4 cm wide, segments 4–9 pairs, gen touching to overlapped, outer margins gen entire; fertile lf part 0.8–2 × sterile; dryish fields, meadows .... ***B. lunaria***
5′ Lower segments of sterile lf part narrowly fan- to wedge-shaped, margins at base meeting at 40–100(120)°
  7. Segments of sterile lf parts distally angled, irregularly toothed, ascending, lower segments wedge-shaped, margins at base meeting at 40–90(120)°, basal often with scattered marginal sporangia .... ***B. ascendens***
  7′ Segments of sterile lf parts distally rounded, ± smooth, not or ± ascending, lower segments ± fan-shaped, margins at base meeting at 50–100(120)°, basal without scattered marginal sporangia ***B. minganense***

B. ascendens W.H. Wagner UPSWEPT MOONWORT **LF**: sterile part separated from fertile near or gen above middle of lf, stalk < 1 cm, blade 1-pinnate, < 6 cm, < 5 cm wide, oblong to oblong-dentate, thin but firm, veiny, yellow-green, segments well separated, ascending, < 5 pairs, obliquely wedge-shaped, not midribbed, lower with margins at base meeting at 40–90(120)°, outer margins sharply and finely dentate to cut, basal segments often with scattered marginal sporangia; fertile part 1–2-pinnate, 1.3–2 × sterile. $2n$=180. RARE. Grassy fields, coniferous woods near streams; 1500–1800 m. s CaRH (near Jonesville, Butte, Tehama cos.); to B.C., NV.

B. crenulatum W.H. Wagner (p. 103) SCALLOPED MOONWORT **LF**: sterile part separated from fertile gen above middle of lf, stalk < 7 mm, blade 1-pinnate, gen < 6 cm, < 2 cm wide, oblong, thin, soft, shiny, yellow-green, segments ± well separated, 3–5 pairs, not midribbed, lower widely fan-shaped, with margins at base meeting at 120–160°, outer margins gen finely crenate to dentate; fertile part 1–2-pinnate, 1.3–3 × sterile. $2n$=90. RARE. Marshes, meadows; 1500–2500 m. NCoRH, CaR, SN, SnGb, SnBr; to WA, UT. Extremely local; often with *B. simplex*.

B. lunaria (L.) Sw. MOONWORT **LF**: sterile part separated from fertile near or gen above middle of lf, stalk < 1.5 mm, blade 1-pinnate, gen 6–10 cm, 2–4 cm wide, oblong, thick, dark green, segments gen touching to overlapped, 4–9 pairs, widely fan-shaped, not midribbed, lower with margins at base meeting at 120–160°, outer margins gen entire; fertile part mostly 1-pinnate, 0.8–2 × sterile. $2n$=90. RARE in CA. Fields, meadows; 3000–3400 m. c SNH (e slope, Tuolumne Co.), n SNE (Mono Co.); n hemisphere, s S.Am. Australia, New Zealand.

B. minganense Victorin MINGAN MOONWORT **LF**: sterile part separated from fertile near or gen above middle of lf, stalk < 2 cm, blade 1-pinnate, < 10 cm, < 2.5 cm wide, linear to oblong, firm, dull, pale green, segments not or ± ascending, < 10 pairs, spoon- to fan- or wedge-shaped, lower ± fan-shaped, not midribbed, lower with margins at base meeting at 50–100(120)°, outer margins subentire to shallowly crenate; fertile part 1-pinnate (2-pinnate in very large pls), 1.5–2.5 × sterile. $2n$=180. RARE in CA. Coniferous forest along streams; 1500–1800 m. CaRH, s SNH; to e N.Am. [*B. lunaria* var. *m.* (Victorin) E.J. Dole]

B. montanum W.H. Wagner WESTERN GOBLIN **LF**: sterile part separated from fertile near to just below middle of lf, stalk 0.2–0.5 × blade, blade lobed to 1-pinnate, < 7 cm, irregularly oblong, gray-green, dull, lobes or segments irregularly angled, square or oblong, not midribbed; fertile part 1-pinnate, 1.5–4.5 × sterile, fleshy. $2n$=90. RARE in CA. Shady coniferous woods; 1500–1800 m. s CaRH (near Jonesville, Butte, Tehama cos.); to B.C., MT.

B. multifidum (S. Gmelin) Rupr. (p. 103) LEATHER GRAPE-FERN Pl often robust, fleshy, leathery, evergreen; caudex < 5 cm, < 1 cm wide; roots 5 mm thick (1 cm from base), encircled by coarse, blackish, corky ridges. **LF**: bud densely hairy; sterile part separated from fertile near ground, stalk gen ± < blade, blade gen ± 2–3-pinnate, < 35 cm wide, segments ovate, oblique, margins entire to shallowly crenate; fertile part 2–3-pinnate or aborted. $2n$=90. Common. Shores, edges of lakes, marshes, among willows; < 2800 m. NCo, CaR, SNH, CCo, MP; to e N.Am., Eur. [sspp. *coulteri* (L. Underw.) R.T. Clausen & *silaifolium* (C. Presl) R.T. Clausen] ❀WET:**4–6,17**&SHD:1–3,7,14,**15,16**.

B. pinnatum H. St. John NORTHWESTERN MOONWORT **LF**: sterile part separated from fertile near middle, stalk very short to 0, blade deeply lobed to 2-pinnate, < 8 cm, < 5 cm wide, oblong to deltate, segments ± ascending, < 7 pairs, oblique-ovate to lanceolate-oblong, shiny, bright green, midribbed, margins entire to minutely crenate; fertile part 2-pinnate, 1–2 × sterile. $2n$=180. RARE in CA. Fields, shrubby slopes; probably ± 2000 m. KR (near Etna Mills, Siskiyou Co.); to AK, Colorado. [*B. boreale* ssp. *obtusilobatum* (Rupr.) R.T. Clausen] *B. lanceolatum* (C. Gmelin) Angström (sterile lf part triangular, fertile lf part branched into 2–4 major axes near base) would key here, might be discovered in mtns of n CA. *B. pumicola* Cov., Oregon moonwort, of Crater Lake region, OR, has not been found in CA.

B. simplex Hitchc. (p. 103) YOSEMITE MOONWORT **LF**: stalk < 1 × blade; sterile part separated from fertile well below middle of lf, gen at top of lf sheath (well above ground in ± young pls), blade simple, deeply lobed, to 2-pinnate, < 12 cm, oblong to ovate, firm, dull green, segments touching to well separated, fan- to wedge-shaped, ± oblique, not midribbed, outer margins entire to slightly crenate; fertile part 1-pinnate, 3–8 × sterile. Uncommon. Open marshes, damp meadows; 2200–3300 m. NCoRH, CaR, SN, SnBr; to e N.Am, Eur, Japan. The w N.Am form probably warrants ssp. or sp. status. w N.Am pls with sterile lf part ternate-pinnate have been called var. *compositum* (Lasch) Milde.

OPHIOGLOSSUM ADDER'S-TONGUE

Roots smooth, pale, gen with bulblets or plantlets. **LF**: sterile part simple, linear to lanceolate or cordate, not midribbed, entire, firm, herbaceous, tip often with small stiff point, veins netted with incl free branched or unbranched veinlets; fertile part gen > sterile, unbranched, slender. **SPORANGIA** in 2 rows, sunken in a linear, long-stalked axis. 20–25 spp.: gen warm temp, trop. (Greek: snake's tongue, from protruding fertile part of lf) ❀TRY;DFCLT.

1. Sterile lf parts 1–2(3) per caudex, blade lanceolate, < 3(–4.3) cm, < 1 cm wide, ± abruptly narrowed to stalk, tip with small, stiff point; fertile lf part 1–2.5 × sterile lf part .... ***O. californicum***
1′ Sterile lf part 1 per caudex, blade oblanceolate to obovate, gen < 10 cm, < 3.5 cm wide, tapered to stalk, tip without point; fertile lf part 2.5–4.5 × sterile lf part .... ***O. pusillum***

O. californicum Prantl (p. 103) CALIFORNIA ADDER'S-TONGUE Caudex < 16 mm, 5 mm wide. **LF**: sterile part separated from fertile at ground, blade gen ± folded, thick, herbaceous, green. **SPORANGIA** 8–15 pairs; rows of sporangia 8–15 mm; sterile axis tip 0.3–1.5 mm. Uncommon. Grassy pastures, chaparral, vernal pool margins; 60–300 m. n&c SNF, CCo, SCo; Mex. [*O. lusitanicum* ssp. *c.* (Prantl) R.T. Clausen] Emerges in early spring during rainy periods; probably under-collected.

O. pusillum Raf. (p. 103) NORTHERN ADDER'S-TONGUE Caudex < 20 mm, 3 mm wide. **LF**: sterile part separated from fertile well above ground, blade flat, thin, herbaceous, pale green. **SPORANGIA**

10–30 pairs; rows of sporangia 10–30 mm; sterile axis tip 1–3 mm. $2n=\pm960$. Very uncommon. Marsh edges, low pastures, grassy roadside ditches; 1000–2000 m. e KR (Siskiyou Co.), n SN (El Dorado Co.); to AK, e N.Am. [*O. vulgatum* L. var. *pseudopodum* (S.F. Blake) Farw.] True *O. v.*, adder's-tongue fern, is unknown in N.Am.

POLYPODIACEAE POLYPODY FAMILY

Sherry Whitmore & Alan R. Smith

Per, on pls, rocks, in rock crevices, less often in soil, humus, or on dunes; rhizome short- to long-creeping, branched, glaucous to not, scaly. **LVS** ± alike or of 2 kinds, fertile and sterile; petiole thin to thick, straw-colored or green to brown or black, gen jointed to persistent knob on rhizome; blade gen simple to 1-pinnate, membranous to fleshy or leathery; veins free to fused, gen forked. **SPORANGIA**: sori round to elongate, rarely linear, gen 1 per areole, in 1–several rows on each side of segment midrib; indusium 0; spores gen ± elliptic, ± smooth to coarse-tubercled or -ridged, scar linear. ± 46 genera, ± 650 spp.: worldwide, esp Old World trop. Numbers of genera, spp. depend on treatment; many spp. cult.

POLYPODIUM POLYPODY

Sherry Whitmore

Rhizome long-creeping; scales lanceolate, ± brownish, 1-colored or often with darker central area or midstripe. **LVS** 0.2–10(20) dm, ± alike or fertile > sterile; petiole glabrous to scaly; blade 1-pinnate to gen deeply pinnately lobed, rarely simple and unlobed, glabrous to hairy, glandular or not, scales on lower surface midrib near base gen lanceolate or linear-lanceolate, gen ± brown; veins free to fused. **SPORANGIA**: sori in 1 row on each side of segment midrib, gen raised, sometimes incl branched or unbranched, glandular hairs, sporangium-like structures, or shriveled sporangia; spores yellow. ± 160 spp.: gen New World, trop, some temp, few boreal. (Latin: many feet, from rhizome) [Whitmore & Smith 1991 Madroño 38:233–248] 50% or more malformed spores indicates hybrid involving 2 or more spp. in CA.

1. Upper surface lf blade midrib glabrous, blade ± membranous to leathery, not fleshy
 2. Lf segment gen < 2.5 cm, gen < 1 cm wide; lf blade ± membranous to ± thick, not leathery, ± firm, not brittle; sori round to ovate, 1–3 mm .... ***P. hesperium***
 2′ Lf segment 2.5–7(10) cm, 0.9–1.8(2.5) cm wide; lf blade thick, leathery, firm, brittle; sori ovate to round, 2–6 mm .... ***P. scouleri***
1′ Upper surface lf blade midrib hairy, if glabrous, blade membranous to fleshy, not leathery, often firm
 3. Veins free; sori gen round .... ***P. glycyrrhiza***
 3′ Veins free and fused; sori gen ovate
  4. Lf blade deltate to ovate, often ± irregular in outline, lower 1–3 segment pairs often = or > those above; sori ± sunken, ± flat, round to gen ovate; NCo, CCo, SCoRO, SW (exc perhaps SnJt) .... ***P. californicum***
  4′ Lf blade oblong-ovate, ± regular in outline, lower 1–3 segment pairs gen < those above; sori raised, ovate; NW (exc NCoRH), CaRF, SN (exc s SNF), ScV, CW .... ***P. calirhiza***

P. californicum Kaulf. CALIFORNIA POLYPODY Rhizome (3)5–10 mm diam, ± glaucous or not, taste bland or acrid; scales ± 1-colored. **LF** summer-deciduous; blade (5)10–25(35) cm, deltate to ovate, membranous to fleshy, often firm, upper surfaces of midribs hairy to ± glabrous, segments serrate, tips obtuse to acute, veins gen 10–50% fused. **SPORANGIA**: sori 1.5–3.5 mm, round to gen ovate, gen ± sunken, flat, sometimes with short, branched, glandular hairs. $2n=74$. Shaded canyons, streambanks, n-facing slopes, roadcuts, cliffs, coastal bluffs, rocks, often granitic or volcanic, humus, not on pls; < 1520 m. NCo, CCo, SCoRO, SW (exc SnJt); Baja CA. Hybrids with *P. hesperium* (SnBr) uncommon, sterile, $2n=111$; see *P. calirhiza, P. scouleri*. ❀DRN,SHD,IRR:6,7,14,**15–17**,18–23,**24**.

P. calirhiza S. Whitmore & A.R. Smith (p. 103) Rhizome 5–10 mm diam, ± glaucous or not, taste acrid and ± sweet; scales 1-colored. **LVS** ± summer-deciduous if conditions dry; blade (5)10–20(40) cm, oblong-ovate, ± membranous to ± firm, upper sides of midribs hairy, segments serrate, tips gen acute to obtuse, veins free to 35(–50)% fused. **SPORANGIA**: sori 1.5–4 mm, gen larger, sometimes fused near segment base, ovate, sometimes with short, many-branched, glandular hairs. $2n=148$. On pls, rocky cliffs or outcrops, roadcuts, often granitic or volcanic, rarely dunes; < 1220 m. NW (exc NCoRH), CaRF, SN (exc s SNF), ScV (Sutter Buttes), CW; to OR, s Mex. Hybrids with *P. glycyrrhiza* sterile, $2n=111$, abundant, of similar geography.

P. glycyrrhiza D. Eaton (p. 103) LICORICE FERN Rhizome 3–6 mm diam, ± glaucous or not, taste sweet-licorice, aftertaste ± bitter; scales 1-colored. **LVS** alive until new lvs formed; blade (5)8–23(38) cm, lanceolate to lanceolate-ovate, membranous to ± firm, upper surfaces of midribs hairy, segments serrate, tips gen acute to elongate-acuminate, veins free. **SPORANGIA**: sori 1–2.5 mm, gen round, sometimes with short, branched, glandular hairs. $2n=74$. Gen near coast, on pls, rocks, moist rocky banks, mossy logs; < 600(1200) m. NCo, KR, NCoRO, NCoRI, n SNF, n&c SNH, CCo, SnFrB; to AK (incl Aleutian Islands), reported from Kamtchatka Peninsula, USSR. Hybrids with *P. scouleri* possible; see *P. calirhiza, P. hesperium*. ❀DRN,SHD:**4–6**&IRR:**7,14–17**, 18–24.

P. hesperium Maxon (p. 103) WESTERN POLYPODY Rhizome 3–6 mm diam, whitish glaucous or not, taste acrid to sweet; scales 1-colored or gen with ± darker central area. **LVS** alive until new lvs formed; blade 2–25 cm, oblong to oblong-ovate, ± membranous to ± thick, ± firm, upper surfaces of midribs glabrous, segments entire to serrate, tips gen obtuse to acute, veins free. **SPORANGIA**: sori 1–2.5 mm, ovate, each with 0–5(10) dark brown or reddish black, shriveled, glandular sporangia. $2n=148$. Rock crevices, talus slopes, under rock ledges; 1400–2980 m. KR, n&c SNH, SnBr, SnJt, W&I, e DMtns (New York Mtns); to B.C., Rocky Mtns, n Mex. [*P. vulgare* L. var. *columbianum* Gilbert] Hybrids with *P. glycyrrhiza* (SNH) uncommon, sterile, $2n=111$; see *P. californicum*. ❀TRY;DFCLT.

P. scouleri Hook. & Grev. (p. 103) LEATHER-LEAF FERN Rhizome 3–12 mm diam, conspicuously whitish glaucous, taste bland; scales with darker midstripe, often partly deciduous in age. **LVS** evergreen 1–several seasons; blade 6–18(50) cm, deltate to oblong-ovate, thick, leathery, firm, brittle, upper sides of midribs glabrous, lower with deltate, dark brown to black-brown scales near base, segments crenate, margins like cartilage, tips gen round or round-obtuse, veins ± often fused. **SPORANGIA**: sori 2–5 mm, ovate to round, often ± merged, sometimes with short, branched, glandular hairs. $2n=74$. Coast, gen in heavy fog-drip or salt-spray zones, on pls, dunes, rocky cliffs, bluffs, mtn ridges, granitic or volcanic rocks, mossy logs, or in soil; 0–610 m. NCo, NCoRO, CCo, SnFrB; to B.C., Baja CA (Guadalupe Island). Pls called *P. scouleri* from n ChI are fertile or sterile hybrids ($2n=111,148$) involving *P. californicum*. ❀DRN,SHD:**4**,**5**&IRR:6,16,**17**,24.

PTERIDACEAE BRAKE FAMILY

Alan R. Smith & Thomas Lemieux

Per, in soil or on or among rocks; rhizome creeping to erect, scaly. **LVS** gen all ± alike (or of 2 kinds, fertile and sterile), gen < 50 cm, often < 25 cm; petiole gen thin, wiry, often dark, in transverse section with vascular strands gen 1–3, less often many in a circle; blade gen pinnate or ± palmate-pinnate (see *Adiantum*), often 2 or more compound, lower surface often with glands, ± powdery exudate, hairs, or scales; segments round, oblong, fan-shaped, or otherwise, veins gen free. **SPORANGIA** in sori or not, marginal, submarginal, or along veins, sometimes covered by recurved, often modified segment margins (false indusia); true indusia 0; spores spheric, sides sometimes flat, scar with 3 radiating branches. ± 40 genera, 500 spp.: worldwide, esp dry areas. Definition of *Cheilanthes* and related genera problematic; traditional limits often untenable.

1. Lvs of 2 dissimilar kinds, fertile more erect, with longer petioles and longer, narrower segments than sterile .... **CRYPTOGRAMMA**
1′ Lvs ± alike (in *Aspidotis densa* fertile and sterile slightly dissimilar)
 2. Lf segment margin gen not recurved, ± unmodified, not covering sporangia; sporangia along veins
  3. Lf 1-pinnate, upper surface glabrous or scaly
   4. Lflet upper surface with stellate scales, margins sometimes shallowly (not deeply) pinnately lobed or dissected .... **ASTROLEPIS**
   4′ Lflet upper surface glabrous, margins ± wavy but not lobed or dissected .... ***Pellaea bridgesii***
  3′ Lf either 1-pinnate with lflets pinnately dissected or lf gen more divided; upper surface of lf glabrous, glandular, or ± covered with exudate, scales 0
   5. Sporangia along veins for outer 1/3–2/3; lf segments narrowed at base .... [2]**ARGYROCHOSMA**
   5′ Sporangia along veins ± throughout (best seen on immature, fertile lf); lf segments not narrowed at base .... **PENTAGRAMMA**
 2′ Lf segment margin gen recurved at least partly, often modified, gen covering sporangia at least partly; sporangia at or near vein tips, thus appearing marginal
  6. Sporangia borne on and covered by highly modified, recurved part of segment margin; segments fan-shaped or oblong, thin-textured .... **ADIANTUM**
  6′ Sporangia borne on unmodified segment surface, gen covered at least partly by recurved part of margin; segments lanceolate, round, or otherwise, gen thick-textured
   7. Lf gen > 40 cm, petiole green to brown, ± thick; alien .... **PTERIS**
   7′ Lf gen < 40 cm, if larger, petiole blackish and wiry (exc in *Pellaea andromedifolia*); native
    8. Lf lower surface with scales, hairs, or glands .... **CHEILANTHES**
    8′ Lf lower surface glabrous or covered with colored exudate
     9. Lf lower surface densely covered with white or yellow exudate, upper sparsely dotted with same .... **NOTHOLAENA**
     9′ Lf without exudate (exc lower surface in *Argyrochosma limitanea*)
      10. Sterile lf segments ± sessile, connected by blade tissue or not, toothed or not; false indusium wide, scarious .... **ASPIDOTIS**
      10′ Sterile lf segments (and fertile) stalked (exc *Pellaea breweri*), not connected by blade tissue, not toothed; false indusium 0 or narrow, scarious or not
       11. Sporangia along veins for outer 1/3–2/3; rhizome scales without darker mid-stripe; false indusium 0 .... [2]**ARGYROCHOSMA**
       11′ Sporangia along veins only at tips; rhizome scales often with darker mid-stripe; false indusium present (exc *P. bridgesii*) .... **PELLAEA**

ADIANTUM

Pl in soil or rock crevices; rhizome short-creeping, scales variously colored. **LF** < ± 1 m; petiole cylindric, gen dark reddish brown to blackish, shiny, ± scaly at base; blade 2–3-pinnate or ± palmate-pinnate (1st division ± palmate, subsequent ones pinnate), segments stalked, fan-shaped or oblong, gen lobed, toothed, or both; axes, blades lacking colored exudate. **SPORANGIA** borne along veins on and covered by highly modified, recurved part of segment margin, appearing to run together at maturity; false indusia ± semi-circular to linear; spores gen smooth, tan. ± 200 spp.: trop, temp. (Greek: unwettable) Widely cult.

1. Lf ± palmate-pinnate .... ***A. aleuticum***
1′ Lf 2–3-pinnate
 2. Lf segment cut or lobed often > 1/4 way to base, margins at base converging at 45–90°, sori (and false indusia) 3–11, rarely 2, gen < 5 mm .... ***A. capillus-veneris***
 2′ Lf segment cut or lobed often < 1/4 way to base, margins at base converging at 90–180°, rarely to 240°, sori (and false indusia) 1–3, rarely to 5, gen > 5 mm .... ***A. jordanii***

A. aleuticum (Rupr.) C.A. Paris (p. 103) FIVE-FINGER FERN **LF** 20–75(100+) cm; petiole reddish brown to blackish; blade ± palmate-pinnate; segments, cut or lobed gen < 1/2 way to midrib, with often > 4 regular lobes, margins at base converging at 45–90°, stalk color extending gradually into base, midvein gen extending part way along 1 margin. **SPORANGIA**: sori (and false indusia) gen 4–6 per segment, gen < 3 mm. 2*n*=58. Shady, moist banks, streamsides, serpentine; < 3400 m. NW, CaR, n&c SNH, SnFrB, SCoRO, n ChI, TR; to AK, w Can, MT, Colorado, NM; also on serpentine, e US & adjacent Can. [*A. pedatum* L. var. *a.* Rupr., var. *subpumilum* W. Wagner, ssp. *calderi* Cody; *A. boreale* Presl; *A.* ×*tracyi* W. Wagner = sterile hybrids with *A. jordanii*] ❀IRR or WET,DRN:**4,5**&SHD:1–3,**6,7**,14,**15–17**,18–24.

A. capillus-veneris L. (p. 103) SOUTHERN MAIDEN-HAIR **LF** gen 20–40 (7–50+) cm; petiole dark brown to blackish; blade 2–3-pinnate; segments cut or lobed often > 1/4 way to base, often with < 4 ± irregular lobes, margins at base converging at 45–90°, stalk color often extending gradually into base, midvein often extending part way along 1 margin. **SPORANGIA**: sori (and false indusia) 3–11, rarely 2 per segment, gen < 5 mm. 2*n*=60. Uncommon (or locally common). Shaded, rocky or moist banks, exposed sites or not; < 2000 m. NCoR, CaRF, n SNF, s SNH, CCo, ScoRO, SW (exc SCo), SNE, D; gen s US, worldwide, esp temp. Widely cult (incl many cultivars); recency of collections, erratic distribution suggest sp. alien in CA. ❀IRR or WET, DRN:4–6&SHD:2,3,**7**,8–12,**13–24**.

A. jordanii C. Mueller (p. 103) CALIFORNIA MAIDEN-HAIR **LF** 20–50 (70+) cm; petiole reddish brown to blackish; blade 2–3-pinnate; segments cut or lobed often < 1/4 way to base, gen with < 4 ± irregular lobes, margins at base converging at 90–180°, rarely to 240°, stalk color often ending ± abruptly at base, midvein forked into ± equal branches, not extending along margin. **SPORANGIA**: sori and false indusia 1–3, rarely to 5 per segment, gen > 5 mm. 2*n*=60. Shaded hillsides, moist woods; < 1200 m. CA-FP (exc uncommon or absent > 1200 m in CaR, SNH); OR. Cult; forms sterile hybrids with *A. aleuticum* (*A.* ×*tracyi* W. Wagner). ❀SHD, DRN,DRY:7,9,14–17;DFCLT.

ARGYROCHOSMA

Pl in soil or rock crevices; rhizome short-creeping, scales linear-lanceolate, tan to ± reddish throughout. **LF** < 40 cm; hairs 0; scales 0; petiole cylindric, dark, glabrous or ± scaly at base; blade 2–5-pinnate, segments stalked, gen < 5 mm, round to oblong, blue- to gray-green, gen thick, veins obscure; axes, blades covered with whitish exudate on lower side or not. **SPORANGIA** along veins for outer 1/3–2/3 of segments; segment margin unmodified, often only slightly recurved; spores tan, coarsely ridged. ±20 spp.: Am. (Greek: silver ornament) Considered closer to *Pellaea* than to *Notholaena*. [Windham 1987 Amer Fern J 77:37–41]. ❀DFCLT.

1. Lf blade without whitish exudate on lower surface, 2–3-pinnate; basal 1° lflets spreading to ± ascending ..... ***A. jonesii***
1′ Lf blade covered with whitish exudate on lower surface, 3–5-pinnate; basal 1° lflets ± strongly ascending ..... ***A. limitanea*** var. ***limitanea***

A. jonesii (Maxon) M.D. Windham **LF** 5–15 cm; petiole dark brown; blade 2–3-pinnate, ± ovate, exudate 0; basal 1° lflets spreading to ± ascending, stalks < 5 mm; 2° lflet stalk 0–1.5 mm; segments gen 2–5 mm. **SPORANGIA** 64-spored. 2*n*=54,108. Gen calcareous rock crevices, cliff bases; 400–1800 m. s SNH, SCoRO, SnGb, W&I, DMtns; to sw UT, AZ. [*Cheilanthes j.* (Maxon) Munz, *Notholaena j.* Maxon]

A. limitanea (Maxon) M.D. Windham var. ***limitanea*** (p. 103) CLOAK FERN **LF** 10–25 cm; petiole dark brown to black; blade 3–5-pinnate, ovate to triangular, covered with whitish exudate on lower surface; basal 1° lflets ± strongly ascending, stalks 5–10 mm; 2° lflet stalk 3–6 mm; segments gen 1.5–3 mm. **SPORANGIA** 32-spored. RARE in CA. In crevices, esp bases of calcareous rocks; 1800 m. e DMtns (New York Mtns, San Bernardino Co.); to UT, NM, nw Mex. [*Cheilanthes l.* (Maxon) Mickel var. *l.*; *Notholaena l.* Maxon] Produces spores asexually; the common name, even more so than others, is applied to ferns in other groups.

ASPIDOTIS

Pl in soil or rock crevices; rhizome short-creeping-decumbent, scales elongate-lanceolate, gen dark, sometimes with very narrow, lighter margin. **LVS** ± alike or fertile and sterile slightly dissimilar; axes grooved, light to dark brown; blade 3–4-pinnate, ovate-triangular to 5-sided, glabrous, upper surface ± glossy; segments long, narrow, with a pointed, sterile tip. **SPORANGIA** in small clusters or continuous along margin; false indusia scarious, irregularly toothed. 4 spp.: gen w N.Am, also e Can, ± dry, mountainous areas. (Greek: shield-bearer, from shield-like false indusia in *A. californica*).

1. Sporangia continuous along both sides of segment midvein; false indusia with many shallow, regular teeth; lf segments linear, ± entire ..... ***A. densa***
1′ Sporangia not continuous along both sides of segment midvein; false indusia with few–many deep, irregular teeth; lf segments lanceolate to triangular, teeth few
  2. False indusia ± as wide as long, free, 1–2(5) per segment, with few coarse, irregular teeth or ± entire ..... ***A. californica***
  2′ False indusia ± as wide as or wider than long, gen fused at bases, 3–5(7) per segment, with many coarse, irregular teeth or lobes ..... ***A. carlotta-halliae***

A. californica (Hook.) Copel. (p. 107) **LF** 3(4)-pinnate, 10–20 (40) cm, thin; lflets pinnately dissected, segments lanceolate to triangular, teeth few. **SPORANGIA**: false indusia ± as wide as long, free, 1–2(5) per segment, with few coarse, irregular teeth or ± entire. 2*n*=60,120. Rock outcrops, crevices; 20–1300 m. NCoR, CaR, SN, GV (Sutter Buttes), CW, SW; Baja CA. ❀SHD,DRN, DRY:7,9,14,**15–17**,18–24;DFCLT

A. carlotta-halliae (W. Wagner & E. Gilbert) Lellinger (p. 107) CARLOTTA HALL'S LACE FERN **LF** 3-pinnate, 8–15(25) cm, leathery; lflets pinnately dissected, segments narrowly lanceolate to triangular, teeth few. **SPORANGIA**: false indusia ± as wide as or wider than long, gen fused at bases, 3–5(7) per segment, with many coarse, irregular teeth or lobes. 2*n*=120. UNCOMMON. Gen serpentine slopes, crevices, outcrops; 100–1400 m. CW. [*Cheilanthes c-h.* W. Wagner & E. Gilbert] A fertile hybrid between *A. californica* and *A. densa*; sometimes backcrosses. ❀SHD,DRN,IRR:7,9,14, **15–17**,18–24;DFCLT.

A. densa (Brackenr.) Lellinger (p. 107) INDIAN'S DREAM **LF** (2)3-pinnate, 15–20(30+) cm, leathery, fertile and sterile slightly dissimilar; lflets sometimes pinnately dissected, segments linear, ± entire. **SPORANGIA**: false indusia continuous along both sides of segment midvein, with many shallow, regular teeth. 2*n*=60. Slopes, crevices, outcrops, esp serpentine; 100–3400 m. NW, CaR, SN, CW; to sw Can, MT, WY, UT; also se Can (Gaspé Peninsula). [*Onychium d.* Brackenr.] ❀DRN,SHD:4–6&IRR:1–3,**7,14–17**,18, 19;DFCLT.

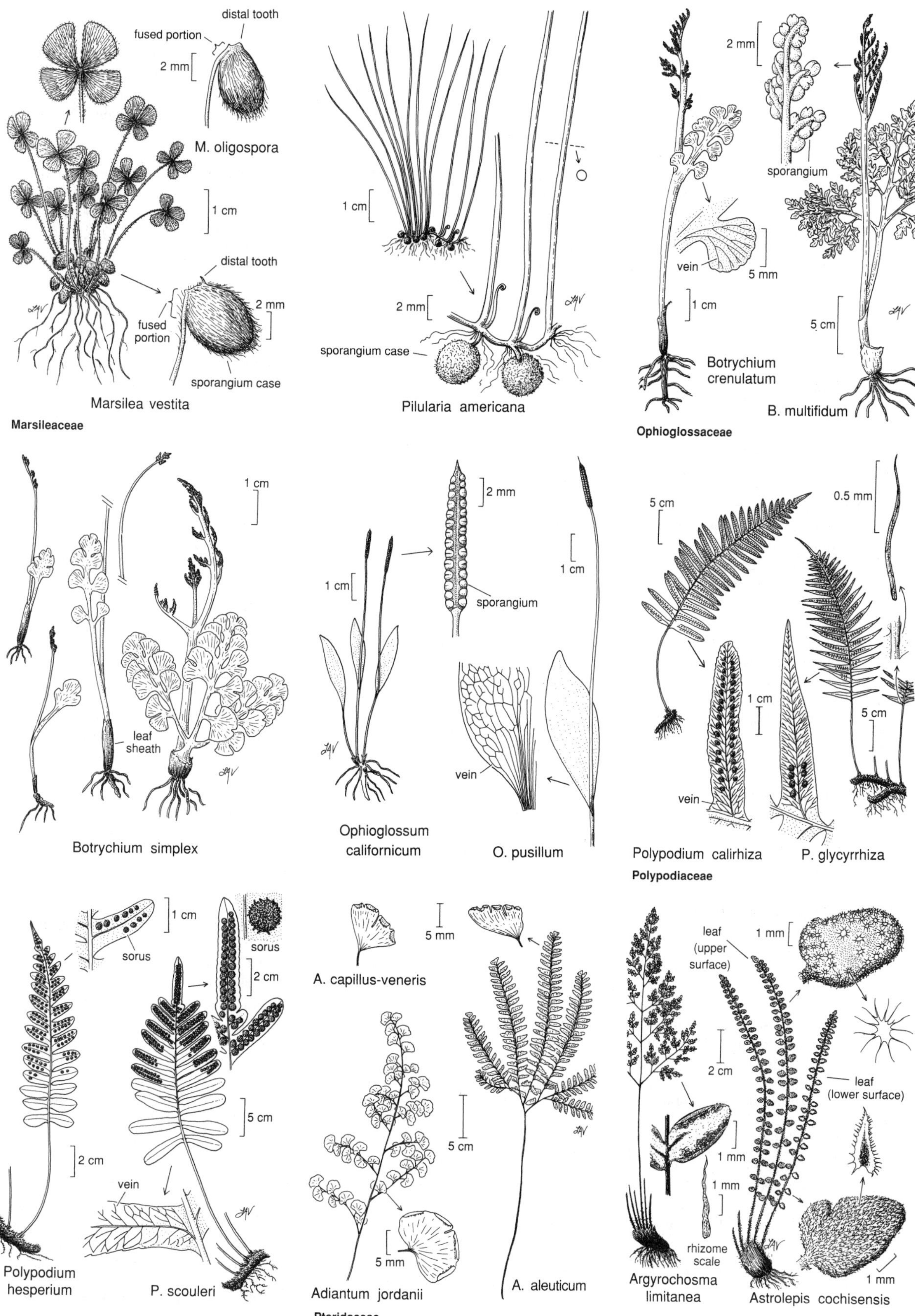

distal tooth
fused portion
2 mm
M. oligospora
1 cm
distal tooth
2 mm
fused portion
sporangium case
Marsilea vestita
Marsileaceae
1 cm
2 mm
sporangium case
Pilularia americana
2 mm
sporangium
vein
5 mm
1 cm
5 cm
Botrychium crenulatum
B. multifidum
Ophioglossaceae
1 cm
leaf sheath
Botrychium simplex
2 mm
1 cm
sporangium
1 cm
vein
Ophioglossum californicum
O. pusillum
5 cm
0.5 mm
1 cm
5 cm
vein
Polypodium calirhiza
P. glycyrrhiza
Polypodiaceae
1 cm
sorus
sorus
2 cm
5 cm
2 cm
vein
Polypodium hesperium
P. scouleri
5 mm
A. capillus-veneris
5 cm
5 mm
Adiantum jordanii
A. aleuticum
Pteridaceae
leaf (upper surface)
1 mm
2 cm
leaf (lower surface)
1 mm
1 mm
rhizome scale
1 mm
Argyrochosma limitanea
Astrolepis cochisensis

ASTROLEPIS

Pl in soil or rock crevices; rhizome ± short-creeping-decumbent, scales gen linear to linear-lanceolate, toothed, pale to reddish brown, older with a darker, irregular central area or not. **LF**: axes gen orange to reddish brown, scaly; blade 1-pinnate, linear, lflets sometimes shallowly but not deeply pinnately lobed or dissected, upper surface with stellate scales. **SPORANGIA** along veins, obscured by dense scales; segment margin unmodified, not recurved. ± 6 spp.: sw US through S. Am. (Greek: star scale) [Benham et al. 1988 Amer J Bot 75(6:2):138]

A. cochisensis (Goodd.) D.M. Benham & M.D. Windham (p. 103) SCALY CLOAK FERN Rhizome short; scales ± 10 mm, 0.1–0.5 mm wide, linear; teeth ± sparse, more pronounced in upper half or not. **LF**: petiole 3–6(10) cm, 1–1.5 mm wide, scales appressed, 0.5 mm, whitish; blade 1-pinnate, 8–15(20) cm, tapered to tip; 1° lflets < 0.5 cm, with jointed stalk and 0–3 pairs of lobes, gen obtuse at tip, upper surface with persistent, stellate scales, lower surface with lanceolate, whitish to tan, abundantly toothed and finely dissected scales covering small (< 0.1 mm) glandular hairs. **SPORANGIA** in a submarginal band when mature, ± visible, erupting through scales. RARE in CA. Limestone slopes, crevices; 900–1800 m. DMtns; to TX, Mex, incl Baja CA. [*Cheilanthes c.* (Goodd.) Mickel; *C. sinuata* (Sw.) Domin var. *c.* (Goodd.) Munz]

CHEILANTHES

Pl in soil or rock crevices; rhizome short- to long-creeping-decumbent, gen many-branched, scales gen linear-lanceolate, pale to dark, with darker mid-stripe or not. **LF** < 75 cm; petiole cylindric, reddish brown to blackish; blade gen 2–3-pinnate, gen oblong to narrowly triangular; segments gen small, ± flat or lower side concave (from recurved margins). **SPORANGIA** along margin, in discrete patches to continuous, partly to completely covered by recurved margin (gen not recurved in *C. cooperae*). 150+ spp.: gen Am, gen dry areas. (Greek: lip fl, from location of sporangia) ❀DFCLT.

1\. Lf blade with scales, with or without hairs, without glands
  2\. Rhizome long-creeping, lvs well separated
    3\. Rhizome scales ovate-lanceolate, sometimes toothed, pale reddish brown or 2-colored with a poorly defined mid-stripe .... ***C. wootonii***
    3′ Rhizome scales linear-lanceolate, ± entire, darkish or 2-colored with a well-defined midstripe .... ***C. clevelandii***
  2′ Rhizome short-creeping, lvs crowded
    4\. Lf blade linear-oblong; lf segments ± round or gen oblong, scales on lower surface deeply dissected, long-ciliate .... ***C. gracillima***
    4′ Lf blade oblong to narrowly triangular; lf segments ± round, scales on lower surface entire to ciliate, gen covering more dissected scales
      5\. Lf segments glabrous on upper side, scales in immediate contact with lower surface dissected but with an obvious and relatively large body, scales of main axes gen > 1 mm wide .... ***C. covillei***
      5′ Lf segments, or at least some, with deeply dissected scales on upper surface, scales in immediate contact with lower surface deeply dissected, with a body only 1–2 cells wide, scales of main axes gen < 1 mm wide .... ***C. intertexta***
1′ Lf blade without scales, with or without hairs, with or without glands
  6\. Rhizome scales with darker mid-stripe
    7\. Hairs on lf segment upper surface very dense, intertwined; darker mid-stripe of rhizome scales thread-like, < 0.1 mm wide at base; 1° lflet stalk 1–2(5) mm; 2° lflet stalk gen brownish on upper, lower surfaces .. ***C. parryi***
    7′ Hairs on lf segment upper surface sparse, gen not intertwined; darker mid-stripe of rhizome scales not thread-like, > 0.1 mm wide at base; 1° lflet stalk ± 0–1 mm; 2° lflet stalk greenish on upper surface, brownish on lower .... ***C. feei***
  6′ Rhizome scales without darker mid-stripe
    8\. Lf surface with ± clear, sticky exudate, gen with sessile or short-stalked glands .... ***C. viscida***
    8′ Lf surface without sticky exudate, with short or long, glandular or non-glandular hairs
      9\. Rhizome scales of 2 kinds, some maroon with a tan, curly, hair-like tip, others tan ± throughout, narrower; lf segment lower surfaces not visible through very dense, tangled, non-glandular hairs; rhizome short- to long-creeping .... ***C. newberryi***
      9′ Rhizome scales of 1 kind, tan to reddish brown ± throughout; lf segment surfaces visible through ± sparse, untangled, glandular and non-glandular hairs; rhizome short-creeping .... ***C. cooperae***

C. clevelandii D. Eaton Rhizome long-creeping, sparsely branched; scales reddish brown or brown, with a darker mid-stripe. **LF** 15–30(40+) cm, 3–5(8+) cm wide, gray-green; petiole < 2(3) mm wide, scales linear-lanceolate, grayish to reddish brown; blade 3–4-pinnate; segments small, ± round, lower surface concave, ± completely covered by scales, non-glandular hairs, upper surface glabrous. **SPORANGIA** gen obscured by recurved segment margin or by scales and hairs. Rocky, exposed areas; 200–1000 m. n ChI, PR; Baja CA. n ChI pls are more robust, have more dissected segment scales.

C. cooperae D. Eaton (p. 107) Rhizome short-creeping; scales lanceolate, tan to reddish brown. **LF** 6–25(32+) cm, 3–6(8+) cm wide, pale green; scales 0; hairs gen spreading, gen glandular; petiole < 2 mm wide; segments gen 1–3 mm, ± oblong, ± flat. **SPORANGIA** submarginal, not obscured by hairs; segment margins unmodified, not (rarely slightly) recurved; spores tan. $2n$=60. Gen in limestone crevices; 100–800 m. SN, SnFrB, TR.

C. covillei Maxon (p. 107) Rhizome short-creeping; scales gen reddish brown, with a darker mid-stripe. **LF** 8–22(30+) cm, 2–4(6) cm wide, dark green; petiole < 2 mm wide, scales linear-lanceolate, whitish to reddish brown; blade 3–4-pinnate; segments small, ± round, upper side glabrous, lower concave, obscured by scales that exceed margin, scales originating from axes, > 2 mm, 1 mm wide, ± entire, covering gen more dissected scales. **SPORANGIA** obscured by recurved segment margin and scales. $2n$=60. Crevices, bases of rocks, sun or shade; 600–2400 m. NCoR, SN, SnFrB, SCoR, TR,

PR, SNE, DMtns; to UT, AZ, Baja CA. Hybridizes uncommonly with *C. newberryi* (*C. ×fibrillosa* L. Underw.) and *C. parryi* (*C. ×parishii* Davenp.) in s CA; also with *C. intertexta*.

C. feei T. Moore (p. 107) Rhizome short-creeping; scales light to reddish brown, gen with darker mid-stripe. **LF** 6–15(18) cm, 1.5–3 cm wide, pale green, scales 0; petiole ± 1 mm wide, hairs < 2 mm, pale or tan with ± orange constrictions; blade gen 3-pinnate; segments small, ± round, lower surface concave, hairs tangled, long, whitish to brownish, ± sparse on upper surface, very dense on lower surface. **SPORANGIA** partly obscured by hairs, less so by segment margin. Gen limestone crevices, slopes, cliffs; 1200–3000 m. W&I, DMtns; to B.C., MT, c US, Mex.

C. gracillima D. Eaton (p. 107) Rhizome short-creeping; scales light brown, with darker (sometimes reddish brown) mid-stripe or not. **LF** 6–18(30) cm, 1–2(3) cm wide, dark green; petiole ± 1 mm wide, scales linear-lanceolate, ciliate at base; blade linear-oblong, gen 3–5 × longer than wide, axes with long, narrow scales; segments small, ± round to gen oblong, lower surface concave, with dense, pale, deeply dissected, long-ciliate scales, upper surface with similar scales or glabrous. **SPORANGIA** on young lvs often entirely obscured by scales and recurved segment margin. Gen granite cliffs, crevices; 400–3200 m. NW, CaR, SN, CW, SNE; to B.C., MT. Hybridizes with *C. intertexta*.

C. intertexta (Maxon) Maxon Rhizome short-creeping; scales pale with darker (reddish brown to blackish) mid-stripe extending ± to margin or not. **LF** 6–14(20) cm, 1.5–3 cm wide, dark green; petiole < 1 mm wide, scales lanceolate, ciliate at base, pale; blade 3-pinnate; segments small, ± round, lower surface concave, with pale to reddish brown scales, scales < 1 mm wide, ciliate at base, covering more deeply dissected scales, barely exceeding segment margin, upper surfaces glabrous or at least some with deeply dissected scales. **SPORANGIA** gen obscured by scales and recurved segment margin. Crevices, bases of rocks; 300–2800 m. NCoRO, NCoRH, SNH, SnFrB, SCoR; NV. Hybridizes with *C. covillei*, *C. gracillima*.

C. newberryi (D. Eaton) Domin Rhizome short- to long-creeping; scales of 2 kinds, some maroon with a tan, curly, hair-like tip, others tan ± throughout, narrower. **LF** 15–20(30) cm, 2–4 cm wide; scales 0; hairs gen very dense, long, tangled or matted, non-glandular, whitish to tan; petiole < 1 mm wide; segments small, ± round, ± flat, upper surface ± gray-green from dense, gen white hairs, lower surface tan from very dense, tan hairs (segment surface not visible). **SPORANGIA** ± visible at segment margin. $2n=60$. Dry, rock outcrops; 100–800 m. SW; Baja CA. Hybridizes uncommonly with *C. covillei* (*C. ×fibrillosa* L. Underw.).

C. parryi (D. Eaton) Domin (p. 107) Rhizome short-creeping, > 6 cm; scales medium brown, most with darker (thread-like) mid-stripe. **LF** 6–15(25) cm, 1–2(3) cm wide; scales 0; petiole < 1 mm wide, hairs short to long, bent, appressed to ± spreading, glandular and not, pale; segments small, ± round, ± flat, hairs long (4+ mm), tangled, gen non-glandular, very dense on both surfaces, making upper surface silver-whitish, lower tan to brown or golden. **SPORANGIA** ± visible through hairs at segment margin; spores blackish. $2n=60$. Limestone, granite crevices, rocks; 100–1500 m. PR, SNE, D; to UT, AZ, Baja CA.

C. viscida Davenp. (p. 107) VISCID LACE FERN Rhizome short-creeping; scales reddish brown, without darker mid-stripe. **LF** 10–15(25) cm, 2(3) cm wide, pale to dull green; petiole < 1 mm wide, scales sparse, at base, gland stalks 0–0.3 mm; segments small, ± oblong, ± flat, upper and lower sides covered with ± clear, sticky exudate from glands. **SPORANGIA** visible at recurved segment margins; spores brownish. UNCOMMON. Limestone, granite crevices, rocks; 100–1600 m. D and adjacent mtns; Baja CA.

C. wootonii Maxon (p. 107) WOOTON'S LACE FERN Rhizome long-creeping; scales light brown, without darker mid-stripe. **LF** 10–20 cm, 2–3 cm wide; petiole 1–2 mm wide; blade 3–4-pinnate; segments small, ± round, lower surface concave, densely covered with ciliate, linear-lanceolate scales, upper side glabrous. **SPORANGIA** gen obscured by dense, overlapping scales. RARE in CA. Rocky outcrops; 1600–1800 m. SNE, DMtns; to TX, Baja CA.

CRYPTOGRAMMA PARSLEY FERN, ROCK-BRAKE

Alan R. Smith, Thomas Lemieux, & Edward Alverson

Pl in rocky places; rhizome creeping-decumbent, scales brown. **LVS** tufted, deciduous or evergreen, of 2 kinds (fertile more erect, with longer petioles, longer, narrower segments than sterile); petioles dark, scaly at base, tan to greenish, glabrous above base; blades 2–4-pinnate, triangular, lanceolate, or elliptic; veins free. **SPORANGIA** along veins, submarginal, appearing to cover surface at maturity; false indusia linear, extending from segment base to tip. Possibly 11 spp.: temp N.Am, S.Am, Eur, Asia. (Greek: hidden line, from protected sori) [Alverson 1989 Amer Fern J 79: 95–102] ❀DFCLT.

1. Sterile lvs thin, translucent when held up to light, glabrous, petiole bases < 1 mm wide, shriveling or deciduous, blades deciduous .... ***C. cascadensis***

1′ Sterile lvs leathery, opaque when held up to light, with very small (0.1 mm), appressed trichomes esp in grooves of axes on upper surface, petiole bases 1–2 mm wide, straw-like, persistent, blades sometimes persistent .... ***C. acrostichoides***

C. acrostichoides R. Br. (p. 107) AMERICAN PARSLEY FERN Rhizome forming small clumps. **LVS**: fertile 10–30 cm; sterile 6–22 cm, blade ovate-lanceolate, ± dark green when mature. $2n=60$. Moist to ± dry rocky slopes, crevices; 1400–3300 m. KR, NCoRH, CaRH, SNH, SnBr, SnJt; w US, w&c Can s to n MI, n MN.

C. cascadensis E. Alverson Rhizome forming small to large clumps. **LVS**: fertile 5–25 cm; sterile 3–20 cm, blade ovate-lanceolate to deltate, green when mature. $2n=60$. Gen ± moist talus slopes, crevices, often granitic or volcanic rock; 2100–3400 m. KR, CaRH, SNH, Wrn; to B.C., MT.

NOTHOLAENA

Pl in soil or often in granite rock crevices; rhizome short-creeping to suberect, scales linear-lanceolate. **LF**: petiole gen cylindric, dark brown to black, glabrous to ± scaly; blade 2–4-pinnate, axes and segments with white to yellow ± powdery exudate on lower, often upper sides, segments gen sessile, sometimes slightly narrower at base. **SPORANGIA** in ± continuous, marginal bands at maturity; segment margin recurved and partly covering sporangia, unmodified; spores finely ridged or granular, often blackish. ± 25 spp.: gen Mex, sw US, few in Caribbean, S.Am. (Greek: false cloak, from lf blade margin not reflexed as it is in *Cheilanthes*) [Tryon 1956 Contr Gray Herb 179:1–106]

N. californica D. Eaton (p. 107) CALIFORNIA CLOAK FERN Rhizome scales rigid, with darker (blackish) midrib extending nearly to margins, finely ciliate. **LF** 3-pinnate, ± 3–13 cm; blade axes brown to black, glabrous or with white to yellow exudate; lowermost 1° lflets each strongly asymmetric (more developed on basal side); segment lower surface covered with white to yellow exudate, hairs 0, scales 0, upper surface sparsely dotted with white to yellow exudate. **SPORANGIA** 32-spored. UNCOMMON. Dry rocky slopes, in rock crevices, under rock ledges; 200–1300 m. s ChI, SnGb, SnBr, PR, DMtns, DSon; AZ, nw Mex. [ssp. *nigrescens* Ewan; *Aleuritopteris cretacea* (Liebm.) Fourn. misapplied] At least 2 forms in CA are chemically distinct: one with pale to bright yellow exudate on lf upper surface; the other with white exudate on lf upper surface. ❀DFCLT.

PELLAEA CLIFF-BRAKE

Pl in soil or rock crevices; rhizome short- to long-creeping, scales overlapping, narrowly linear, light- to reddish or medium-brown, often with darker mid-stripe. **LVS** erect, persistent, < 1 m; petioles ± cylindric, gen dark or reddish brown to blackish, ± shiny, glabrous; blade 1–4-pinnate; segments gen stalked, gen free, linear to rounded, lobed or not, often folded lengthwise when dried; veins gen free. **SPORANGIA** in ± continuous, submarginal bands, among a whitish to yellowish exudate or not; segment margin gen recurved, gen modified; spores tan to light yellow. ± 35 spp.: trop, temp, few in Eur, 0 in Asia. (Greek: dusky, from bluish gray lvs) [Tryon 1957 Ann Missouri Bot Gard 44(2):125–193] Not commonly cult.

1. Lf 1-pinnate, 1° lflets unlobed or deeply 2(3)-lobed
  2. 1° lflet lanceolate-ovate, deeply 2(3)-lobed; fracture lines at base of petiole many; rhizome scales without darker mid-stripe .... ***P. breweri***
  2′ 1° lflet rounded, unlobed; fracture lines at base of petiole 0; rhizome scales with darker mid-stripe .... ***P. bridgesii***
1′ Lf 2–4-pinnate, 1° lflets compound
  3. Lf 2–4-pinnate
    4. Lf segments without a small point at tip, gen 6–15 mm, 3–10 mm wide; petiole ± light brown ***P. andromedifolia***
    4′ Lf segments with a small point at tip, gen 6(8) mm, 2(4) mm wide; petiole dark brown to blackish .... ***P. mucronata*** var. ***mucronata***
  3′ Lf 2-pinnate
    5. Segment 5–15 × longer than wide, > main axis of 1° lflet .... ***P. brachyptera***
    5′ Segment 1–3 × longer than wide, < main axis of 1° lflet
      6. Fertile segment not appearing folded in half, recurved margins not meeting, lower surface therefore visible; DMtns (Providence, New York Mtns) .... ***P. truncata***
      6′ Fertile segment gen appearing folded in half, recurved margins ± meeting, lower surface therefore not visible; c SNH, TR, PR, SNE, DMtns (exc Providence, New York Mtns) .... ***P. mucronata*** var. ***californica***

P. andromedifolia (Kaulf.) Fée (p. 107) COFFEE FERN Rhizome long-creeping, branched, > 20 cm, 0.5 cm wide; scales 2–3 mm, tan to orange-brown, with darker mid-stripe or not. **LVS** ± unclustered, 20–60(80) cm, 10–20(30) cm wide, green to purplish; petiole < ± 3 mm wide, ± light brown; blade 2–4-, gen 3-pinnate, elongate-triangular; segments gen 6–15 mm, 3–10 mm wide, tip ± rounded to obtuse, notched or not. **SPORANGIA** 32- or 64-spored. 2*n*=58,87. Gen rocky or dry areas; 30–1800 m. NCoR, SN, CW, SW; Baja CA. CA pls diploid (assignable to var. *rubens* D. Eaton) or triploid; hairy pls near coast in s CA and on ChI have been called var. *pubescens* D. Eaton. ❀SHD,DRN,DRY:**15–17** &IRR:7,9,**14,18–24.**

P. brachyptera (T. Moore) Baker Rhizome short-creeping, many-branched, compact, > 8 cm, 1 cm wide; scales reddish brown with darker mid-stripe. **LVS** clustered, 15–30(40) cm, 2–3 cm wide, grayish green; petiole 1–2 mm wide; blade 2-pinnate, oblong; segments 5–11(13), < 1(2) cm, linear, 5–15 × longer than wide, with a small point at tip, > main axis of 1° lflet. **SPORANGIA** 64-spored. Rocky crevices, slopes, serpentine or not; 700–2200 m. KR, NCoR, CaR, n SNH; to WA. Hybridizes with *P. mucronata*. ❀DRN, DRY:**4–6**&IRR,SHD:7,14,**15–17**,18,19.

P. breweri D. Eaton (p. 107) Rhizome short-creeping, branched, > 10 cm, 5(7) mm wide; scales very narrowly linear (hair-like), reddish brown, without darker mid-stripe. **LVS** clustered, 8–20(25) cm, 2–3(4) cm wide, pale greenish; petiole < 2 mm wide, fracture lines at base many; blade 1-pinnate, oblong, main axis green at tip; 1° lflets < 2 cm, < 1.5 cm wide, lanceolate-ovate, deeply 2(3)-lobed. **SPORANGIA** 64-spored; spores dark to light brown. 2*n*=58. Gen n-facing granite rock crevices, slopes; 1500–3700 m. KR, SNH, GB, DMtns; to WA, ID, Colorado. ❀DFCLT.

P. bridgesii Hook. (p. 107) Rhizome short-creeping, many-branched, > 15 cm, 0.5 cm wide; scales light to medium brown with darker mid-stripe. **LVS** clustered, 12–25(35) cm, 1.5–2(3.5) cm wide, blue-green; petiole < 1.5 mm wide; blade 1-pinnate, oblong; 1° lflets < 2 cm, < 1.5 cm wide, rounded, unlobed, often folded lengthwise. **SPORANGIA** in marginal bands; segment margin modified but not recurved; spores 64. 2*n*=58. Gen granite rock crevices, slopes; 1200–3200 m. SNH; to OR, ID. Hybrids with *P. mucronata* (*P. ×glaciogena* W. Wagner, A. Reed Smith, & T.R. Pray) are sterile, ± common, c&s SNH, 1500–2400 m, intermediate between parents (see Wagner et al. 1983 Madroño 30:69–83). ❀DFCLT.

P. mucronata (D. Eaton) D. Eaton BIRD'S-FOOT FERN Rhizome short-creeping, branched, > 8 cm, 0.5–1 cm wide; scales brownish with darker mid-stripe. **LVS** ± clustered, greenish to purplish; petiole < 2(3) mm wide; blade 2–3(4)-pinnate, narrowly triangular to oblong; segments 2–6(8) mm, 0.5–2(4) mm wide, linear to oblong, with a small point at tip. **SPORANGIA** 64-spored. 2*n*=58. Rocky or dry areas; 20–3000 m. NCoR, CaRH, SN, GV (Sutter Buttes), CW, SW, SNE, DMtns; Baja CA. Hybrids with *P. bridgesii* (*P. ×glaciogena* W. Wagner, A. Reed Smith, & T.R. Pray) are common, those with *P. brachyptera*, *P. truncata* less so.

var. ***californica*** (Lemmon) Munz & I.M. Johnston (p. 107) **LF** 15–25(33) cm, 2–4(8) cm wide; blade 2-pinnate; 1° lflets often overlapping, ascending; fertile segments gen appearing folded in half, recurved margins ± meeting, lower surface therefore not visible. Habitat of sp.; 1800–3000 m. c SNH, TR, PR, SNE, DMtns. [*P. compacta* (Davenp.) Maxon] ❀DFCLT.

var. ***mucronata*** **LF** 20–40(60) cm, 5–15 cm wide; blade 2–3 (4)-pinnate; 1° lflets not overlapping, gen ± spreading to widely ascending; fertile segments not appearing folded in half, recurved margins not meeting, lower surface therefore visible. Habitat of sp.; 20–2400 m. NCoR, CaR, SN, GV (Sutter Buttes), CW, SW, DMtns; Baja CA. ❀DRN,DRY;2,**7,9,14–18,23,24;**DFCLT.

P. truncata Goodd. (p. 107) CLIFF-BRAKE Rhizome short-creeping, branched, to 8+ cm, 2–3 mm wide; scales brownish with darker mid-stripe. **LVS** clustered, 15–30(36) cm, 4–11 cm wide, olive-green; petiole 1(2) mm wide, ± flat or upper surface grooved; blade 2-pinnate, narrowly triangular to oblong-triangular; 1° lflets gen not overlapping, ± spreading; segments 5–8 mm, 1–4 mm wide, linear to oblong, with a small point at tip, margins wavy-crenate, often (esp sterile) whitish, fertile not appearing folded in half, recurved margins not meeting, lower surface therefore visible. **SPORANGIA** 64-spored. 2*n*=58. RARE in CA. Gen in crevices of or at bases of granite (in CA) or igneous rock; 1200–1900 m. e DMtns (Providence Mtns, New York Mtns); to Colorado, TX, Baja CA. [*P. longimucronata* Hook. misapplied]

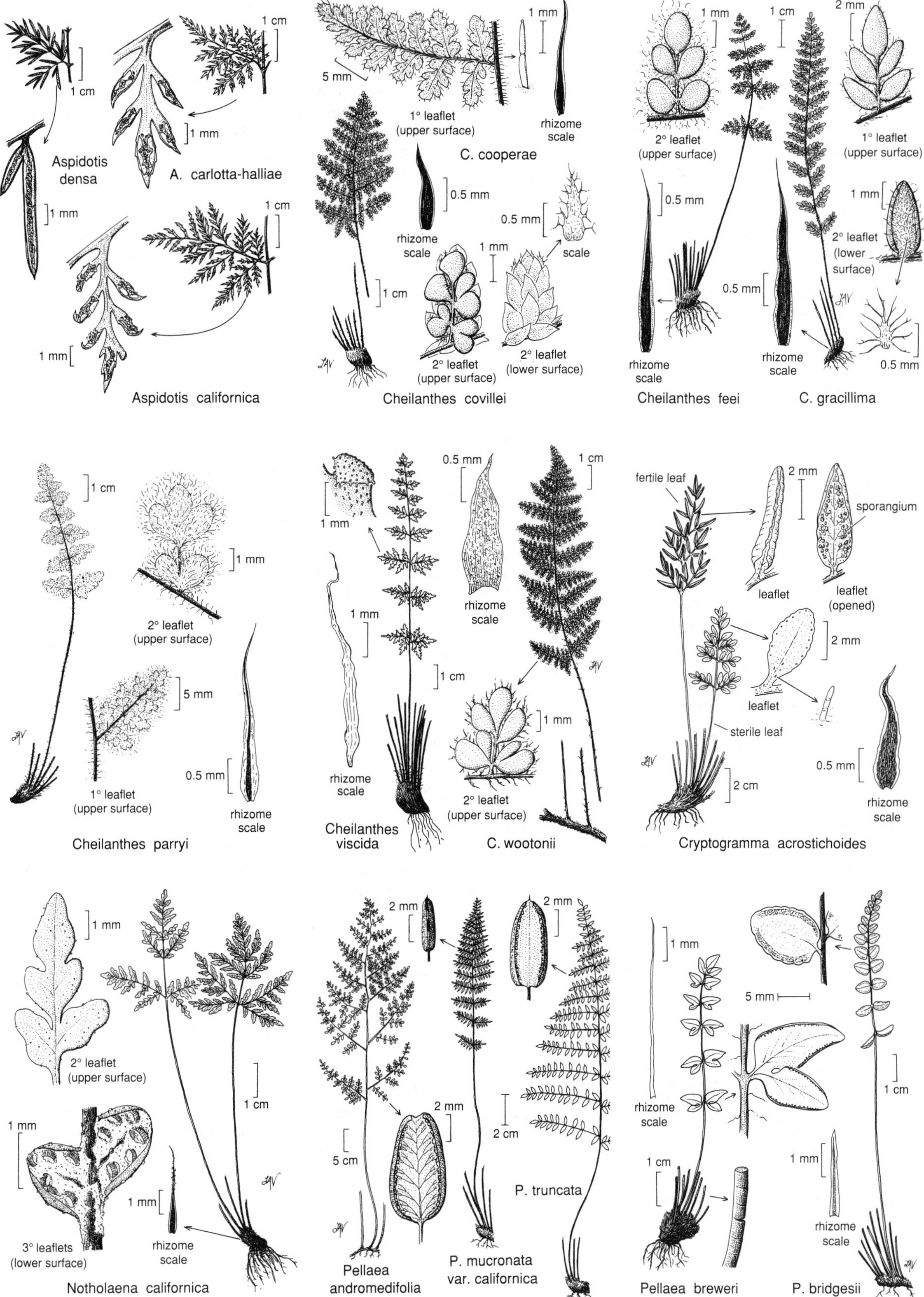
1 cm
1 cm
1 mm
Aspidotis densa
1 mm
A. carlotta-halliae
1 cm
1 mm
Aspidotis californica
5 mm
1 mm
1° leaflet (upper surface)
rhizome scale
C. cooperae
0.5 mm
rhizome scale
0.5 mm
scale
1 mm
1 cm
2° leaflet (upper surface)
2° leaflet (lower surface)
Cheilanthes covillei
1 mm
1 cm
2 mm
2° leaflet (upper surface)
1° leaflet (upper surface)
0.5 mm
1 mm
2° leaflet (lower surface)
0.5 mm
rhizome scale
rhizome scale
0.5 mm
Cheilanthes feei
C. gracillima
1 cm
1 mm
2° leaflet (upper surface)
5 mm
0.5 mm
1° leaflet (upper surface)
rhizome scale
Cheilanthes parryi
1 mm
0.5 mm
1 cm
rhizome scale
1 mm
1 cm
1 mm
rhizome scale
2° leaflet (upper surface)
Cheilanthes viscida
C. wootonii
fertile leaf
2 mm
sporangium
leaflet
leaflet (opened)
2 mm
leaflet
sterile leaf
0.5 mm
2 cm
rhizome scale
Cryptogramma acrostichoides
1 mm
2° leaflet (upper surface)
1 cm
1 mm
3° leaflets (lower surface)
1 mm
rhizome scale
Notholaena californica
2 mm
2 mm
2 mm
5 cm
2 cm
P. truncata
Pellaea andromedifolia
P. mucronata var. californica
1 mm
5 mm
1 cm
rhizome scale
1 cm
1 mm
rhizome scale
Pellaea breweri
P. bridgesii

PENTAGRAMMA GOLDBACK or SILVERBACK FERN

Pl in soil or rock crevices; rhizome short-creeping-decumbent, gen 3–5(8) mm wide, scales linear-lanceolate, with darker mid-stripe. **LF**: petiole 5–20(32) cm, 0.5–2(3) mm wide; blade gen 2–3-pinnate, 2–8(15) cm, triangular or gen 5-sided, with white or yellow exudate on lower surface, with or without exudate on upper surface, main axis shallowly to deeply grooved on upper surface; lowermost 1° lflets each strongly asymmetric (more developed on basal surface); veins free. **SPORANGIA** along veins ± throughout; segment margins unmodified, recurved or not. 2 spp.: w N.Am. A puzzling complex of intergrading chemical, chromosomal, and morphological variants (see Yatskievych et al. 1990 Amer Fern J 80:9–17); for these we prefer the rank of var. but have used ssp. here because combinations at that rank already exist. ❀DFCLT

1. Petiole dark brown to gen blackish; lf blade upper surface with white exudate; SN .... ***P. pallida***
1′ Petiole brown to reddish brown; lf blade upper surface gen without exudate; CA-FP, SNE, DMtns ... ***P. triangularis***
  2. Lf blade upper surface not sticky, gen glabrous .... ssp. ***triangularis***
  2′ Lf blade upper surface sticky or with sparse, minute, yellowish glands
    3. Lf blade upper surface gen with sparse, minute, yellowish glands; 2° lflets on basal surface of lowermost 1° lflets deeply pinnately lobed to nearly 1-pinnate .... ssp. ***maxonii***
    3′ Lf blade upper surface gen sticky; 2° lflets on basal surface of lowermost 1° lflets ± entire to shallowly pinnately lobed .... ssp. ***viscosa***

P. pallida (Weath.) G. Yatskievych, M.D. Windham & E. Wollenweber (p. 113) Rhizome tip, incl scales, covered with white exudate. **LF**: petiole dark brown to gen blackish, with white exudate; blade 2–8(12) cm, pale to olive green, upper surface with white exudate, lower densely covered with white exudate; lower 1° lflets 2–6(8) cm. $2n$=60. Gen in or near granitic soil, rock, boulders; 100–1200 m. SN. [*Pityrogramma triangularis* (Kaulf.) Maxon var. *p.* Weath.]

P. triangularis (Kaulf.) G. Yatskievych, M.D. Windham & E. Wollenweber Rhizome tip, scales without exudate. **LF**: petiole brown to reddish brown, with or without exudate; blade 3–10(18) cm, gen pale to dark green, upper surface gen without exudate. $2n$=60,90,120,150. Common. Gen shaded slopes or rocky areas; < 2300 m. CA-FP, SNE, DMtns; to B.C., NV, Baja CA. [*Pityrogramma t.* (Kaulf.) Maxon]

ssp. ***maxonii*** (Weath.) G. Yatskievych, M.D. Windham & E. Wollenweber **LF**: blade upper surface gen with sparse, minute (0.1 mm), yellowish glands, lower surface often with many yellowish or reddish glands, margins not recurved; lower 1° lflets 2–5(7) cm; upper 1° lflets and 2° lflets on basal surface of lowermost 1° lflets deeply pinnately lobed to nearly 1-pinnate. Gen ± shaded, near rocks, boulders; 300–1400 m. SnBr, SnJt, DMtns; AZ, Baja CA. [*Pityrogramma t.* (Kaulf.) Maxon var. *m.* Weath.]

ssp. ***triangularis*** (p. 113) **LF**: blade upper surface gen glabrous, not sticky, margins not recurved; lower 1° lflets 2–6(11) cm; upper 1° lflets and 2° lflets on basal surface of lowermost 1° lflets deeply pinnately lobed to nearly 1-pinnate. Common. Gen shaded, sometimes rocky or wooded areas; < 2300 m. CA-FP, SNE, DMtns; to B.C., NV, Baja CA. Ssp. *semipallida* (J. Howell) G. Yatskievych, M.D. Windham & E. Wollenweber is chemically distinct, morphologically and geog less distinct, sometimes recognized taxonomically.

ssp. ***viscosa*** (D. Eaton) G. Yatskievych, M.D. Windham & E. Wollenweber **LF**: blade upper surface gen sticky, margins at tips of 2° lflets gen ± recurved; lower 1° lflets 2–5(8) cm; upper 1° lflets and 2° lflets on basal surface of lowermost 1° lflets ± entire to shallowly pinnately lobed. Gen shaded, wooded or grassy slopes; < 800 m. SCo, ChI; Baja CA. [*Pityrogramma t.* (Kaulf.) Maxon var. *v.* (D. Eaton) Weath.] Exudate chemistry unique in genus; incl morphologically intergrading variants.

PTERIS BRAKE

Pl gen in soil; rhizome erect or short- to long-creeping, scaly or hairy. **LVS** gen alike, 1–4-pinnate, erect-arched; petiole, axes grooved on upper surface, grooves from main axis of blade to 1° lflets connected; lowermost 1° lflets strongly asymmetric, 2° lflets on basal surface more developed (exc *P. vittata*). **SPORANGIA** among hair-like structures in continuous, marginal bands; false indusia along segment margins exc at bases, tips, and between lobes, partly covering sporangia, scarious. ± 250–300 spp.: gen trop, subtrop. Popular in cult.

1. Lf blade 2–4-pinnate .... ***P. tremula***
1′ Lf blade 1-pinnate exc basal 1° lflets sometimes deeply lobed to divided
  2. Lf blade 1-pinnate above, lower 1° lflets > others, with 2(3) very deep lobes or 2° lflets; petiole, main blade axis ± glabrous exc base .... ***P. cretica***
  2′ Lf blade 1-pinnate throughout, lower 1° lflets < others; petiole, main blade axis ± scaly throughout ..... ***P. vittata***

P. cretica L. CRETAN BRAKE Rhizome slender, short-creeping. **LF** 15–70(100) cm; petiole gen > blade, ± glabrous exc base; blade olive-green, 1-pinnate above, lower 1° lflets > others, with 2(3) very deep lobes or 2° lflets; 1° lflets 1–5 pairs, terminal > subterminal. $2n$=58,87,116. Disturbed places; < 500 m. Reported from SnFrB, elsewhere in Am; seemingly native to Old World and perhaps s Mex.

P. tremula R.Br. Rhizome stout, short-creeping. **LF** 60–150(200) cm; petiole gen > blade, ± glabrous exc base; blade light green to yellow-green, 2–4-pinnate; 1° lflets many pairs, basal 2–3-pinnate. $2n$=±232. Habitat uncertain; 600 m. SnGb (1 collection); native to Australia, New Zealand, Polynesia.

P. vittata L. Rhizome stout, short-creeping. **LF** 30–75(100) cm; petiole < blade, ± scaly throughout, esp base; blade medium to dull green, 1-pinnate throughout, lower 1° lflets < others; 1° lflets 12–20(30+) pairs, terminal > subterminal. $2n$=116. Moist rock walls, rock crevices, streamsides, seeps, sun or shade; 400–800 m. SnGb; native to se US, Caribbean, S.Am, Eur, Afr, e Asia, Australia, Polynesia.

SELAGINELLACEAE SPIKE-MOSS FAMILY

Dieter H. Wilken

Pls on soil, rocks, or other pls, cespitose, mat-like (± flat), or cushion-like (rounded). **STS** pendent to erect, short to widely spreading, rooting at base or in branch fork; branches intricately intertwined or not. **LVS** many, simple, overlapping, appressed, small, ± scale-like, 1-veined, smooth to grooved on back, sessile to decurrent, gen 4-ranked; fertile lvs at same node of prostrate sts equal or not, if unequal, lvs below st gen appressed, lvs above st ascending to spreading. **CONES** gen terminal; lvs like those on sterile sts or not, gen strongly overlapping, triangular in X-section. **SPORANGIA** 1 per lf axil, spheric to reniform; lower gen with (1–3)4 large, 3-ridged, yellow to orange spores; upper gen with many, small, gen pale-colored spores.

SELAGINELLA SPIKE-MOSS

The only genus. ± 700 spp.: worldwide, gen trop & warm temp. (Latin: small *Selago*, ancient name for some *Lycopodium*). Some cult as groundcover & curiosity (*S. kraussiana, S. lepidophylla*, resurrection plant). Hand lens required to observe lf shape, margin, bristle at tip, cones. ❀TRY;DFCLT.

1. Main sts above ground erect, rooting only at or near st base; rhizome conspicuous at base of erect sts ..... ***S. bigelovii***
1′ Main sts above ground prostrate, spreading or pendent, rooting at branch forks; rhizome 0 (older, buried main sts rhizome-like) or inconspicuous
 2. Lvs on prostrate sts unequal, lvs below wider and > lvs above at the same nodes as viewed at 20×
  3. Mature lf tip acute, not bristly; lf lanceolate; e PR, DSon ..... ***S. eremophila***
  3′ Mature lf tip bristly, bristle 0.5–1.5 mm, lf linear to linear-lanceolate; n of TR
   4. Cone 10–30 mm; lf margins ciliate to minutely dentate in lower half, hairs loosely ascending; n KR, > 1500 m ..... ***S. densa*** var. ***scopulorum***
   4′ Cone 4–8 mm; lf margins ciliate from base to near tip, hairs appressed-ascending; mostly foothills around GV, < 1500 m ..... ***S. hansenii***
 2′ Lvs on prostrate sts equal, lvs below = lvs above as viewed at 20×
  5. Main st segments between branch forks gen > 3 cm (sometimes less in compact pls on exposed site); lateral sts spreading or pendent; lvs weakly appressed — NCo ..... ***S. oregana***
  5′ Main st segments between branch forks gen < 3 cm; lateral sts decumbent to erect; lvs ± strongly appressed
   6. Mature lf acute or abruptly pointed as viewed at 20×, bristle 0 (sometimes bristle-tipped in *S. leucobryoides*)
    7. Lf tip acute; lf margin ciliate throughout, hairs spreading to ascending; s SCo ..... ***S. cinerascens***
    7′ Lf tip abruptly pointed; lf margin ciliate below, minutely dentate above middle; e DMoj ..... ***S. leucobryoides***
   6′ Mature lf bristle-tipped as viewed at 20×
    8. Bristles on distal lvs 0.5–1.5 mm, ± soft, together forming conspicuous tufts at st tip; lvs gray-greeen; sts ± fragile when dry ..... ***S. asprella***
    8′ Bristles on distal lvs gen < 1 mm, ± rigid, together not forming conspicuous tufts; lvs pale to dull yellow-green; sts not fragile when dry
     9. Lf color ± same as st color; lf base sessile; pl ± mat-like, prostrate sts evident, lateral sts decumbent to weakly ascending ..... ***S. wallacei***
     9′ Lf color different from st color; lf base decurrent; pl ± cushion-like, prostrate sts ± obscure, lateral sts crowded, ascending to erect ..... ***S. watsonii***

S. asprella Maxon (p. 113) BLUISH SPIKE-MOSS Pl flat to mat- to cushion-like, open or dense. **STS** strongly intertwined; main sts prostrate to decumbent, 2–6 cm, fragile when dry; lateral branches decumbent to ascending, 1–2(4) cm. **LVS** sessile; upper and lower of main sts equal, 0.5–4 mm, lanceolate to narrowly so, bristle-tipped, bristle < 1.5 mm, soft, together forming conspicuous tufts at st tip, marginal hairs sparse, spreading. **CONE** < 2 cm; lvs lanceolate to ovate; large spores yellow. UNCOMMON. Dry sites, rocks, rocky soils, crevices, coniferous forest; 1600–2700 m. s SNF, TR, PR; Baja CA.

S. bigelovii L. Underw. (p. 113) Pl cespitose. **STS** ascending to erect; main sts clumped, 5–15(20) cm, ± 1 mm wide, rooting at base; lateral branches 1–4 cm, ascending to erect. **LVS** sessile; upper and lower equal, 1–3.5 mm, linear to narrowly lanceolate, bristle-tipped, bristle < 1 mm, ± rigid, marginal hairs spreading to ascending. **CONE** gen < 1 cm; lvs 1.5–2.5 mm, lanceolate to ovate; large spores yellow. Open sites, rocks, crevices, shrubland, woodland; < 2000 m. s NW, s SN, CW, SW; Baja CA. ❀DRN,DRY: **15–17**&SHD:**7**,14,**18**,19;DFCLT.

S. cinerascens Maxon (p. 113) ASHY SPIKE-MOSS Pl ± mat-like. **STS** prostrate to spreading, sometimes ± intertwined; main sts spreading, 5–18 cm; lateral branches < 2 cm, prostrate to ascending. **LVS** gen sessile; upper and lower equal, 1–3 mm, linear to narrowly lanceolate, tip acute to abruptly pointed, not bristly, marginal hairs sparse, spreading. **CONE** < 5 mm; lvs widely ovate; large spores yellow. Uncommon. Dry, open sites or under other pls; < 300 m. s SCo (San Diego Co.); Baja CA.

S. densa Rydb. var. ***scopulorum*** (Maxon) Tryon (p. 113) ROCKY MOUNTAIN SPIKE-MOSS Pl mat- to cushion-like, ± open. **STS** ± spreading to decumbent, 2–8 cm; lateral branches ascending to ± erect, 1–2.5 cm. **LVS**: lower decurrent; upper sessile; upper and lower lvs of main sts unequal (lower > upper), 2–4 mm, linear to narrowly lanceolate, bristle-tipped, bristle < 1 mm, soft, margin sparsely ciliate to minutely dentate, hairs ascending. **CONE** gen 10–30 mm; lvs lanceolate to ovate. UNCOMMON. Open, rocky sites, coniferous forest; 1500–2200 m. n KR; to B.C., ND, n Mex. [*S. engelmannii* Hieron. var. *s.* (Maxon) Reed]

S. eremophila Maxon (p. 113) DESERT SPIKE-MOSS Pl mat-like, dense. **STS** prostrate to spreading; main sts 5–14 cm; lateral branches < 1 cm, ascending. **LVS** decurrent; upper and lower lvs of main sts unequal (lower 1.5–3 mm, upper ± 1.5 mm), ± lanceolate, marginal hairs spreading; young lvs bristle-tipped, bristle < 1 mm, soft, twisted, deciduous; mature tip gen acute. **CONE** 3–10 mm; lvs ovate to deltate; large spores yellow. UNCOMMON. Shaded sites, gravelly soils, crevices, among rocks; < 900 m. e PR, DSon; to TX, Baja CA.

S. hansenii Hieron. (p. 113) Pl mat-like, dense to ± open. **STS** prostrate to spreading; main sts 5–28 cm; lateral branches ascending to erect, 1–2(3) cm. **LVS** sessile; lvs of main sts ± unequal (lower > upper), 1–4 mm, narrowly lanceolate to ± deltate, bristle-tipped, bristle 0.5–1.5 mm, soft, white, together forming conspicuous tufts at st tip, marginal hairs ascending. **CONE** 4–8 mm; lvs ovate to deltate; large spores yellow. Rocky, open to shaded sites, oak woodland, coniferous forest; 250–1500 m. s NCoRI, CaR, SN, ScV (Sutter Buttes), n SCoR.

S. leucobryoides Maxon MOJAVE SPIKE-MOSS Pl cushion-like. **STS** prostrate, strongly intertwined, fragile when dry; main sts 1–2(4) cm; lateral branches ascending to erect, < 1 cm. **LVS** decurrent; upper and lower equal, 1.5–3(4) mm, linear to lanceolate, bristle-tipped, bristle < 0.5 mm, rigid, margin sparsely ciliate to minutely dentate. **CONE** 5–12 mm; lvs lanceolate to ovate; large spores yellow. UNCOMMON. Among rocks, crevices, gen limestone, shrubland, woodland; 600–2300 m. DMtns (Panamint, Kingston, Providence Mtns); sw NV.

S. oregana D. Eaton (p. 113) Pl mat-like, on soil, rocks, or tree trunks. **STS** prostrate, widely spreading to pendent, 10–40(60) cm; lateral branches prostrate or spreading, 1–5+ cm. **LVS** decurrent; upper and lower equal, 2–3 mm, narrowly lanceolate, ± long-tapered, bristle-tipped, bristle << 0.5 mm, ± rigid, margin entire to sparsely ciliate or minutely dentate. **CONE** 10–30 mm, often in pairs at branch tips; lvs lanceolate; large spores yellow. Shaded to open sites, streambanks, coniferous forest; < 300 m. NCo; to WA. ❀DRN:**4,5**&IRR:**6,17**&SHD:7,**15,16**;DFCLT.

S. wallacei Hieron. (p. 113) Pl mat-like (open in moist, shaded sites; dense in open, dry sites). **STS** prostrate; main sts 5–25 cm; lateral branches decumbent to ascending, 1–3(5) cm. **LVS** gen sessile; upper and lower equal, 1.5–3.5 mm, linear-lanceolate to oblong, bristle-tipped, bristle < 1 mm, rigid, margin ciliate to minutely dentate. **CONE** 4–12 mm; lvs ± ovate; large spores yellow to orange. Open to shaded, dry to moist sites, chaparral, woodland; < 1700 m. NW, nnw SNF (Butte Co); to B.C., MT.

S. watsonii L. Underw. (p. 113) Pl cushion-like, ± open to dense. **STS** prostrate to decumbent; main sts 2–12 cm; lateral branches ascending to erect, 1–3(5) cm. **LVS** decurrent; upper and lower equal, 1.5–3.5 mm, linear-lanceolate to oblong, bristle-tipped or acute, bristle << 0.5 mm, rigid, margin entire to sparsely ciliate, hairs ± spreading. **CONE** 10–30 mm; lvs lanceolate to ovate; large spores yellow. Open rocky sites, coniferous forest, alpine; 2300–4100 m. KR (Trinity Alps), SNH, TR, PR, W&I; to OR, MT, UT.

THELYPTERIDACEAE THELYPTERIS FAMILY

Alan R. Smith & Thomas Lemieux

Pls in soil; rhizome short-long-creeping, scales gen hairy, uniformly brownish. **LF**: petiole in transverse section at base with 2 crescent-shaped vascular strands; blade gen with needle-like, clear hairs on axes or between veins; 1° and 2° axes gen grooved on upper surface, grooves not continuous from one axis to next; veins free or lowest pair running together at or near deepest part of sinus. **SPORANGIA** in gen round sori on veins; indusia reniform or round-reniform; stalk cells in 3 rows; spores elliptic, scar linear. 1–30 genera (depending on interpretation), ± 900 spp.: worldwide, esp trop; several cult.

THELYPTERIS

LF: blade 1–2-pinnate; 1° lflets pinnately lobed. [Smith 1971 Univ Calif Publ Bot 59:1–143]

1. Lower 1° lflets << those above, blade widest above base; lowermost veins running to margin above deepest point of sinus .... *T. nevadensis*

1′ Lower 1° lflets ± = or > those above, blade widest at or near base; lowermost veins running to margin at deepest point of sinus .... ***T. puberula*** var. ***sonorensis***

T. nevadensis (Baker) C. Morton (p. 113) Rhizome creeping, 1.5–3 mm wide. **LVS** 0 for several cm, then densely clustered, gen 40–100 cm, 8–15 cm wide; petiole scales ovate, tan, persistent; blade thin, lower surface with many short-stalked or sessile, resinous glands between and on veins, with non-glandular hairs sparse on axes and veins, 0 between veins. **SPORANGIA**: indusia glabrous or sparsely hairy. Springy hillsides, seepage areas; 800–1700 m. KR, NCoRO, n SNH; to B.C. [*Lastrea oregana* (C. Chr.) Copel.] Locally forming large colonies. ❀IRR or WET,DRN:**4–6,17**&SHD:3,7,14,**15,16**;GRCVR(deciduous).

T. puberula (Baker) C. Morton var. ***sonorensis*** A. Reed Smith (p. 113) SONORAN MAIDEN FERN Rhizome creeping, 3–8 mm wide. **LVS** regularly spaced, gen 50–120 cm, 15–30 cm wide; petiole scales lanceolate, brown, not persistent; blade thick, lower surface ± without glands, with non-glandular hairs moderate to dense on axes, veins, and between veins. **SPORANGIA**: indusia densely hairy. UNCOMMON. Along streams, seepage areas; 50–550 m. SCo, WTR, SnGb, SnJt; to AZ, s Mex. [*Lastrea augescens* (Link) J. Smith misapplied] Rarely cult in s CA. Var. *puberula* is found from Mex to Costa Rica. ❀WET:**17,24**&SHD:16,**18–23**; GRCVR.

CUPRESSACEAE CYPRESS FAMILY

Jim A. Bartel

Shrub, tree, evergreen, monoecious or dioecious. **LVS** cauline, opposite and 4-ranked or whorled in 3's and 6-ranked, gen scale-like, decurrent, completely covering young sts. **POLLEN CONE** small, axillary or terminal. **SEED CONE** ± fleshy to woody, gen hard at maturity; scales opposite or whorled. **SEEDS** 1–many per scale, gen angled or winged, gen wind-dispersed. 17 genera, ± 120 spp.: worldwide; all N.Am genera cult. [Elias 1980 Complete Trees N.Am] Juvenile lvs needle- or awl-like, sometimes present in ± mature pls, esp in response to grazing or infection, esp in *Cupressus, Juniperus.*

1. Seed cone ± fleshy, berry-like, scales fused; seeds 1–3 per scale, unwinged; lvs whorled in 3's or not **JUNIPERUS**
1′ Seed cone woody, not berry-like, scales ± free; seeds 2–many per scale, winged; lvs not whorled in 3's
  2. Seed cone ± spheric to widely cylindric, scales peltate, margins abutting; seeds 2–many per scale **CUPRESSUS**
  2′ Seed cone ± oblong, scales not peltate, margins overlapping; seeds 2 per scale
    3. Seed cone ± pendent, scales in 3 pairs, middle pair fertile; lvs opposite, appearing whorled in 4's; exposed part of lf longer than wide; seed wings unequal .... **CALOCEDRUS**
    3′ Seed cone erect to reflexed, scales in 4–6 pairs, middle 2–3 pairs fertile; lvs opposite, ± not appearing whorled in 4's; exposed part of lf ± as long as wide; seed wings equal .... **THUJA**

CALOCEDRUS INCENSE CEDAR

Tree, gen widely conic, monoecious. **ST**: young shoots very flat, in flat clusters. **LVS** opposite, 4-ranked, appearing whorled in 4's, scale-like, closely appressed, overlapping, exposed part longer than wide; lf bases elongated on smallest shoots. **SEED CONE** ± pendent, woody, ± oblong, maturing 1st year; scales in 3 partially overlapping pairs, middle pair fertile. **SEEDS** 2 per fertile scale, flat, unequally 2-winged; cotyledons 2. 3 spp.: w N.Am, e Asia. (Greek: beautiful cedar)

C. decurrens (Torrey) Florin (p. 113) Tree 20–50 m. **ST**: trunk < 6 m diam, tapered from wide base; bark 1–2.5+ cm thick, fibrous, cinnamon-red; lower branches down-curved. **LF** 3–10 mm. **POLLEN CONE** 5–7 mm, oblong, light yellow. **SEED CONE** 18–25 mm, light red-brown. **SEED** 8–12 mm, ± yellow to red-brown, maturing autumn. Common. Mixed-evergreen, yellow-pine forests; 350–2500 m. CA (exc D); OR, w NV, n Baja CA. [*Libocedrus d.* Torrey] ❀1,**2,4–7,15–17**&IRR:**3,8,9**,11,12,**14,18–24**;CVS.

CUPRESSUS CYPRESS

Large shrub, tree, often pyramidal in youth, monoecious. **ST**: young shoots gen cylindric (sometimes 4-angled or flat), gen arrayed in 3-dimensional clusters. **LVS** opposite, 4-ranked, scale-like, closely appressed, overlapping. **POLLEN CONE** gen yellow. **SEED CONE** 6–50 mm, woody, ± spheric to widely cylindric, maturing 1st or 2nd year, often closed > 2 years; scales 6–12, peltate, abutting, shield- or wedge-shaped; projection often present, small, pointed, gen less visible in age. **SEEDS** 2–many per scale, flat, winged; cotyledons 2–5. $2n=22$ for all reports. ± 22 spp.: w N.Am, Eurasia. (Latin: cypress) [Wolf 1948 Aliso 1:1–250]

1. Young shoots in flat clusters; seeds 2–5 per scale, wing prominent; seed cone 6–12 mm, opening in 1–2 years (subg. *Chamaecyparis*)
  2. Bark 15–25 cm thick, fibrous; seed cone scales gen 7–10; lf glands gen visible; lf gen glaucous .... ***C. lawsoniana***
  2′ Bark 1–2 cm thick, scaly; seed cone scales gen 4–6; lf glands gen obscure; lf not glaucous .... ***C. nootkatensis***
1′ Young shoots gen in 3-dimensional clusters (exc sometimes *C. macnabiana*); seeds > 5 per scale, wing ± obscure; seed cone 10–50 mm, opening in 2 or more years or gen closed indefinitely (subg. *Cupressus*)
  3. Outer lf surface gen with conspicuous gland or pit, gen covered with resin that dries clear to white to gray; lf dull to gray-green, often appearing bluish gray
    4. Young shoots in flat clusters; lf pungent-scented; resin copious, sticky when fresh .... ***C. macnabiana***
    4′ Young shoots in 3-dimensional clusters; lf light-scented; resin neither copious nor sticky
      5. Youngest shoots 0.5–1 mm diam; pollen cone 2–3 mm, scales 6–10; seed 3–4 mm; n CA .... ***C. bakeri***
      5′ Youngest shoots 1–2 mm diam; pollen cone (2)3–5 mm, scales 8–16; seed (3)4–8 mm; s CA ..... ***C. arizonica***
        6. Bark smooth, peeling in thin strips or plates; seed cone scale projection 2–4 mm; lf glands gen inactive, sometimes inconspicuous or 0; PR .... ssp. ***arizonica***
        6′ Bark fibrous, not peeling; seed cone scale projection ± 0; lf glands active, conspicuous; s SNH ssp. ***nevadensis***
  3′ Outer lf surface without conspicuous gland or pit, gen not covered with resin (see also 6. above; *C. sargentii* has resin, some glands or pits); lf bright green to dusty green, not grayish
    7. Bark of large branches and trunks cherry-red to mahogany-brown, smooth, polished, not fibrous, peeling in thin plates; small tree gen without dominant terminal shoot; s CA .... ***C. forbesii***
    7′ Bark of large branches and trunks not cherry-red to mahogany-brown, gen fibrous, not peeling in thin plates; small to large tree with dominant terminal shoot; n&c CA
      8. Lf dull, dusty green, glaucous or not; serpentine; NCoR, SnFrB, SCoR .... ***C. sargentii***
      8′ Lf bright or dark green, not glaucous; not on serpentine; NCo, CCo, SnFrB
        9. Crown asymmetric, often open, flat-topped to widely conic; young shoots 1.5–2.5 mm diam; widely planted and naturalized .... ***C. macrocarpa***
        9′ Crown symmetric, often dense, gen pyramidal; young shoots 1–1.5 mm diam; not widely planted and naturalized
          10. Seed 3–5 mm, brown, glaucous or not, attachment scar conspicuous; seed cone gen 15–30 mm, spheric to gen widely cylindric; SnFrB (Santa Cruz Mtns) .... ***C. abramsiana***

10′ Seed 2–4 mm, dark brown to black, not glaucous, attachment scar often inconspicuous; seed cone 10–20(25) mm, ± spheric; NCo, CCo . . . . ***C. goveniana***

11. Tree 5–7 m, terminal shoot not long and whip-like; lf light to yellow-green; CCo . . . . ssp. ***goveniana***

11′ Tree 1–2 m on sterile soil (10–50 m on rich soil), terminal shoot long, whip-like; lf gen dark dull green; NCo . . . . ssp. ***pigmaea***

C. abramsiana C. Wolf (p. 113) SANTA CRUZ CYPRESS Tree 5–7 m. **ST**: bark fibrous, thin, broken in thick vertical strips or plates, gray; young shoots 1–1.5 mm diam, cylindric. **LF** light bright green. **POLLEN CONE** 3–4 mm, 2 mm diam, ± 4-sided; scales 10–16; pollen sacs 4–6 per scale. **SEED CONE** gen 15–30 mm, spheric to gen widely elliptic, brown; scales 8–10, projection 2 mm. **SEED** 3–5 mm, dull brown, glaucous or not, rough; attachment scar conspicuous. **ENDANGERED** US, CA. Yellow-pine, closed-cone-pine/cypress forests; 450–800 m. SnFrB (Santa Cruz Mtns). Threatened by development, agriculture.

C. arizonica E. Greene Tree 5–25 m. **ST**: bark partially peeling in thin strips or plates to fibrous (esp in age), smooth to furrowed, cherry-red to brown to gray; youngest shoots 1–2 mm diam, 4-sided. **LF** dusty gray-green, glaucous or not, appearing blue-green, glandular, often resin-covered. **POLLEN CONE** 2–5 mm, 2 mm diam, ± cylindric to 4-sided; scales 8–16; pollen sacs 3–6 per scale. **SEED CONE** 10–35 mm, spheric to ovoid, often warty, dull gray to brown; scales 4–8, projection 0–4 mm. **SEED** 3–8 mm, light tan to dark brown, slightly warty, often glaucous; attachment scar conspicuous or not. Pinyon/juniper or oak/pine woodland, chaparral; 750–1800 m. s SNH, Teh, PR; AZ, n Baja CA, n&c Mex.

ssp. ***arizonica*** (p. 113) CUYAMACA or ARIZONA CYPRESS Tree 10–16 m. **ST**: bark smooth, thin, peeling in thin strips or plates, cherry-red. **LF**: glands gen inactive, sometimes inconspicuous or 0. **POLLEN CONE**: pollen sacs 3–6 per scale. **SEED CONE** 10–25 mm, spheric, dull gray to brown; scale projection 2–4 mm, often conspicuous, conic. **SEED** 4–8 mm, dark brown, gen not glaucous. $2n=22(23,24)$. RARE in CA. Chaparral; 900–1200 m. PR (Cuyamaca Mtns); AZ, n Baja CA, n&c Mex. [var. *glabra* (Sudw.) Little; var. *revealiana* J. Silba; *C. stephensonii* C. Wolf] Widely planted as windbreak.

ssp. ***nevadensis*** (Abrams) E. Murray (p. 113) PIUTE CYPRESS Tree 5–25 m. **ST**: bark fibrous, not peeling, gray- to red-brown to cherry-red. **LF**: glands active, conspicuous. **POLLEN CONE**: pollen sacs 5–6 per scale. **SEED CONE** 15–35 mm, ovoid, often silver-gray in age; scale projection ± 0. **SEED** 3–6 mm, light brown, gen glaucous. $2n=22(23)$. RARE. Pinyon/juniper or oak/pine woodland, chaparral; 750–1800 m. s SNH (Kern, Tulare cos.). [*C. n.* Abrams]

C. bakeri Jepson (p. 113) BAKER CYPRESS Tree 10–30 m. **ST**: bark partially peeling in plates, ± red, gray when old; youngest shoots 0.5–1 mm diam, cylindric. **LF** gray-green, resin not copious. **POLLEN CONE** 2–3 mm, gen spheric; scales 6–10; sacs 3–5 per scale. **SEED CONE** 10–25 mm, spheric; surface warty, silvery to dull brown; scales 4–8, projection conspicuous. **SEED** 3–4 mm, tan, gen shiny, gen with many wart-like pitch pockets; attachment scar conspicuous. UNCOMMON. Mixed-evergreen forests, open slopes, flats, often serpentine; 1100–1800 m. KR, CaRH, n SNH; sw OR. [ssp. *b.*, Modoc cypress; ssp. *matthewsii* C. Wolf, Siskiyou cypress] ❀DRN:4,5,**6,15–17**&IRR:1,2,**3**,**7**,10,**14**,23.

C. forbesii Jepson (p. 113) TECATE CYPRESS Tree < 10 m, multi-trunked, gen without dominant terminal shoot. **ST**: bark peeling in thin, crisp plates, smooth, polished, cherry-red or mahogany-brown; young shoots 1–1.4 mm diam, cylindric. **LF** rich light green to ± dull green. **POLLEN CONE** 3–4 mm, ± 2 mm diam, ± 4-sided; scales 12–14; pollen sacs 3–5 per scale. **SEED CONE** 20–32 mm, gen spheric, dull brown or gray; scales 6–10, projection < 5 mm. **SEED** 3–6 mm, dark brown, gen with many wart-like pitch pockets; attachment scar conspicuous. $2n=22(23)$. RARE. Chaparral; 450–1500 m. w PR; n Baja CA. Threatened by fire, development.

C. goveniana Gordon Tree gen 5–50 m. **ST**: bark fibrous, smooth, aging rough, brown to gray; young shoots 1–1.5 mm diam, cylindric. **POLLEN CONE** 3–4 mm, 1.5–2 mm diam, ± 4-sided; scales 12–14; pollen sacs 3–6 per scale. **SEED CONE** 10–25 mm, ± spheric, tan to brown to gray, often green beneath surface; scales 6–10, projection inconspicuous to flat, erect to slightly spreading, horn-like. **SEED** 2–4 mm, dark brown to black, not glaucous; attachment scar often inconspicuous. Closed-cone-pine/cypress forests, coastal terraces; < 500 m. c&s NCo, c CCo.

ssp. ***goveniana*** (p. 113) GOWEN CYPRESS Tree 5–7 m, without long, whip-like terminal shoot. **LF** light to yellow-green. **SEED CONE** 10–16 mm, brown to gray-brown. **SEED** dark brown to ± black, dull. RARE. Habitats of sp.; 30–300 m. c CCo (Monterey Peninsula). In cult.

ssp. ***pigmaea*** (Lemmon) J. Bartel PYGMY or MENDOCINO CYPRESS Tree 1–2 m on sterile soil (10–50 m on rich soil), with long whip-like terminal shoot. **LF** gen dark dull green. **SEED CONE** 12–25 mm, tan aging gray. **SEED** ± black, gen shiny. RARE. Habitats of sp.; < 500 m. c&s NCo (Mendocino, Sonoma cos.). [*C. pygmaea* (Lemmon) Sarg.]

C. lawsoniana A. Murray (p. 113) PORT ORFORD CEDAR, LAWSON CYPRESS Tree 20–65 m, pyramidal in youth. **ST**: trunk < 6 m diam; bark 15–25 cm thick, red-brown to tan, fibrous, fire-resistant; young shoots flat, arrayed in flat clusters, held ± horizontally, lower surfaces often paler. **LF** green, gen glaucous; glands gen visible; lateral lf tips gen curved toward st axis. **POLLEN CONE** 2–3 mm, 2–3 mm diam, oblong, pink to red; scales 5–6; pollen sacs 2 per scale. **SEED CONE** 6–10 mm, spheric, red-brown; scales 7–10, projection ± 0. **SEEDS** 2–5 per scale, 2–4 mm, gen with wart-like pitch pockets, light chestnut-brown, gen glandular. UNCOMMON. Coastal conifer, mixed-evergreen, yellow-pine forests, often on serpentine; < 1700 m. NCo, KR, NCoRO; sw OR. [*Chamaecyparis l.* (A. Murray) Parl.] ❀IRR:1,2,**4–6,15–17**,24&SHD3,7,**14**,18–23; CVS.

C. macnabiana A. Murray (p. 113) MCNAB CYPRESS Shrub 3–10 m (rarely tree 10–18 m). **ST**: bark fibrous, gray; young shoots < 1 mm diam, ± 4-sided, arrayed in flat clusters. **LF** blue- to dull gray-green to gray, pungent-scented; resin copious, sticky when fresh. **POLLEN CONE** 2–3 mm, 2 mm diam; scales gen 8; pollen sacs 3–5 per scale. **SEED CONE** 15–25 mm, gen longer than wide, gray to red-brown; scales gen 6, projection prominent, 2–4 mm, conic, pointing upward to incurved. **SEED** 2–5 mm, medium to glaucous brown; attachment scar conspicuous. Dry slopes, flats, chaparral, pine/oak woodlands, often on serpentine; 300–850 m. NCoRI, CaRH, n SNF. ❀4–6,**7**,14,**15**,**16**,17,**18**,19,20,**21**,22–24; CVS.

C. macrocarpa Gordon (p. 119) MONTEREY CYPRESS Tree 20–25 m. **ST**: bark fibrous, rich brown aging ash-gray; young shoots 1.5–2.5 mm diam, cylindric. **LF** rich bright green, not glaucous. **POLLEN CONE** ± 6 mm, 2.5–3 mm diam, ± 4-sided; scales 12–14; pollen sacs 6–10 per scale. **SEED CONE** 25–38 mm, 17–25 mm diam, spheric to elliptic, brown; scales 8–12, projections 2 mm. **SEED** 5–6 mm, 2–3 mm thick, dark brown; attachment scar whitish, conspicuous. RARE. Closed-cone-pine/cypress forests; < 30 m. NCo, CCo (native to Monterey Peninsula; widely planted and naturalized elsewhere). In cult.

C. nootkatensis D. Don (p. 119) YELLOW CYPRESS, ALASKA CEDAR Tree 20–30 m, pyramidal in youth. **ST**: trunk < 6 m diam; bark 1–2 cm thick, gray to purple-brown, fissured, scaly; young shoots ± 4-sided, arrayed in flat, pendent clusters. **LF** green, not glaucous; glands gen obscure; lateral lf tips gen parallel to axis of attachment or diverging. **POLLEN CONE** 2–3 mm, 2–3 mm diam, spheric; scales 4–5; pollen sacs 2 per scale. **SEED CONE** 6–12 mm, spheric, ash-gray; scales gen 4–6, projection ± 0. **SEEDS** 2–5 per scale, 3–4 mm, brown to red-brown. RARE in CA. Cool, moist, forested, well drained mtn slopes; 650–2500 m. KR; to AK. [*Chamaecyparis n.* (D. Don) Spach]

C. sargentii Jepson (p. 119) SARGENT CYPRESS Tree (2)10–20 m. **ST**: bark fibrous, thick, gray or dark brown to ± black; young shoots 1.5–2 mm diam, cylindric to 4-sided. **LF** dull, dusty green to grayish green, glaucous or not. **POLLEN CONE** 3–4 mm, 2 mm

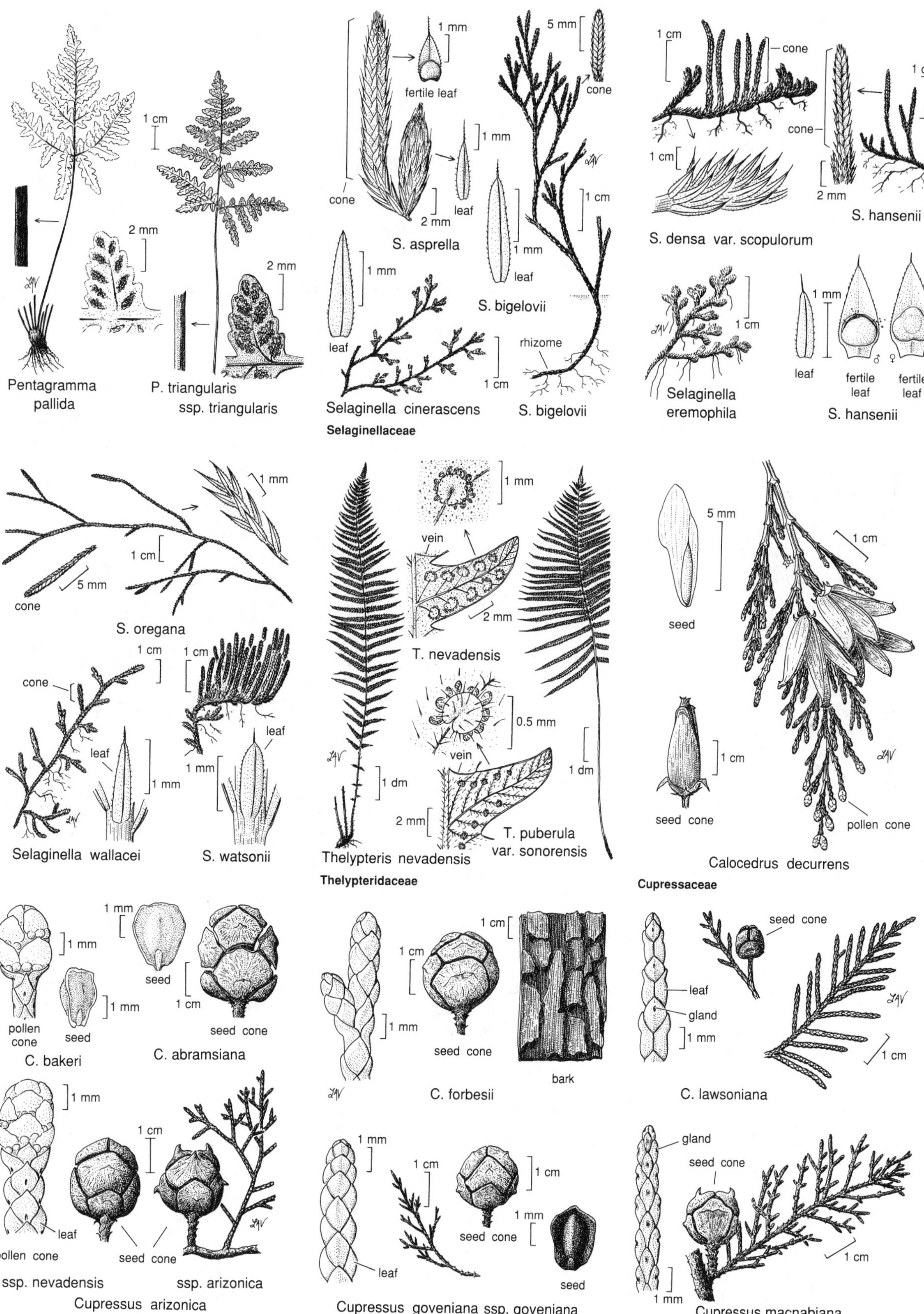

1 cm
2 mm
2 mm
Pentagramma pallida
P. triangularis ssp. triangularis
1 mm
fertile leaf
1 mm
cone
2 mm
leaf
S. asprella
1 mm
leaf
1 cm
Selaginella cinerascens
Selaginellaceae
5 mm
cone
1 cm
1 mm
leaf
S. bigelovii
rhizome
S. bigelovii
1 cm
cone
1 cm
S. densa var. scopulorum
1 cm
cone
2 mm
S. hansenii
1 cm
Selaginella eremophila
1 mm
leaf
fertile leaf
fertile leaf
S. hansenii
1 mm
1 cm
5 mm
cone
S. oregana
1 cm
cone
leaf
1 mm
Selaginella wallacei
1 cm
leaf
1 mm
S. watsonii
1 mm
vein
2 mm
T. nevadensis
0.5 mm
vein
1 dm
2 mm
T. puberula var. sonorensis
1 dm
Thelypteris nevadensis
Thelypteridaceae
5 mm
seed
1 cm
1 cm
seed cone
pollen cone
Calocedrus decurrens
Cupressaceae
1 mm
1 mm
pollen cone
seed
C. bakeri
1 mm
seed
1 cm
seed cone
C. abramsiana
1 cm
1 cm
1 mm
seed cone
bark
C. forbesii
seed cone
leaf
gland
1 mm
1 cm
C. lawsoniana
1 mm
1 cm
leaf
pollen cone
seed cone
ssp. nevadensis
ssp. arizonica
Cupressus arizonica
1 mm
1 cm
1 cm
1 mm
seed cone
leaf
seed
Cupressus goveniana ssp. goveniana
gland
seed cone
1 cm
1 mm
Cupressus macnabiana

diam, 4-sided to cylindric; scales gen 10–12; pollen sacs 3–4 per scale. **SEED CONE** 15–30 mm, spheric, rough-surfaced, dull brown to gray, scales 6–10, projection inconspicuous. **SEED** 5–6 mm, dark brown, gen glaucous; attachment scar gen conspicuous. Closed-cone-pine/cypress, yellow-pine forests, chaparral, serpentine; 200–1000 m. NCoR, SnFrB, SCoR. ❀DRN:3,6,**7**,15,**16**,**17**,**24**&IRR:14,18–23.

JUNIPERUS JUNIPER

Shrub, tree, gen dioecious. **ST**: bark thin, peeling in strips; young shoots 4-angled to cylindric. **LVS** opposite and 4-ranked or whorled in 3's and 6-ranked, scale-like to less often awl- or needle-like. **POLLEN CONE**: pollen sacs 2–6 per scale. **SEED CONE** 5–18 mm, ± fleshy, berry-like, glaucous or not, dry or resinous, sweet, formed by fusion of scales, ± spheric, surrounded at base by minute scale-like bracts, gen maturing 2nd year; scales 3–8, opposite or whorled in 3's. **SEEDS** 1–3(12) per cone, ± flat, unwinged, often not angled, gen animal-dispersed over 2 years; cotyledons 2–6. ± 60 spp.: n hemisphere. (Latin: juniper) [Vasek 1966 Brittonia 18:350–372]

1. Lf needle- or awl-like, sub-erect to spreading but not appressed; pollen, seed cones axillary; shrub < 1 m (sect. *Juniperus*) . . . . ***J. communis***

1′ Lf gen scale-like, closely appressed; pollen, seed cones terminal; tree or shrub > 1 m (sect. *Sabina*)

 2. Seed cone maturing blue-black; tree 5–15 m . . . . ***J. occidentalis***

  3. Bark red-brown; lvs gen whorled in 3's; gen dioecious; NCoRH, SNH, SnGb, SnBr, DMtns . . . . var. ***australis***

  3′ Bark brown; lvs opposite or whorled in 3's; gen ± monoecious; CaRH, MP . . . . var. ***occidentalis***

 2′ Seed cone maturing red-brown; shrub or tree 1–8(10) m

  4. Lf gland obvious; trunks several at base; dioecious; mostly CA-FP . . . . ***J. californica***

  4′ Lf gland obscure; trunk gen 1 at base; monoecious; mostly not CA-FP . . . . ***J. osteosperma***

J. californica Carrière (p. 119) CALIFORNIA JUNIPER Shrub or tree 1–4(10) m, dioecious. **ST**: trunks several at base; bark gray, thin, outer layers persistent. **LVS** gen whorled in 3's, 6-ranked, closely appressed, scale-like; gland obvious. **POLLEN CONE** 2–3 mm, oblong. **SEED CONE** 7–12 mm, spheric to ovoid, bluish maturing red-brown, dry. **SEEDS** 1–3 per cone, 5–7 mm, pointed, angled, brown. Dry slopes, flats, pinyon/juniper woodlands; 50–1500 m. NCoRI, SNF, SCoRI, TR, PR, DMtns; s NV, nw AZ, Baja CA (Cedros, Guadalupe islands). ❀DRN:3,6,**7**,10,**14–16**,17,**18–23**,24&IRR:8,9,11,12.

J. communis L. (p. 119) COMMON JUNIPER Shrub < 1 m, gen dioecious. **ST**: bark red-brown, peeling in papery sheets. **LVS** gen whorled in 3's, 6-ranked, sub-erect to spreading, needle- or awl-like, jointed to decurrent, nongreen base. **POLLEN CONE** 4–5 mm. **SEED CONE** 5–8 mm, ± spheric, red maturing bright blue to black, glaucous, resinous. **SEEDS** 1–3 per cone, 2–5 mm, ovoid, acute, gen 3-angled. 2*n*=22. Rocky or wooded slopes, high-elevation forests; 1900–3400 m. KR; circumboreal. [vars. *jackii* Rehder, *montana* Aiton, *saxatilis* Pallas] ❀DRN:**4–6**,17&IRR:**1**,**2**,7,**15**,16&SHD:**3**,14,18–23.

J. occidentalis Hook. Tree 5–15 m. **ST**: bark brown to red-brown. **LVS** opposite and 4-ranked or whorled in 3's and 6-ranked, closely appressed, scale-like. **POLLEN CONE** 2–3 mm, oblong. **SEED CONE** 5–12 mm, blue-green maturing blue-black, resinous. **SEEDS** 2–3 per cone, 6 mm, ovoid, acute, grooved or pitted. 2*n*=22. Dry slopes, flats, forests, woodlands; 100–3100 m. NCoRH, CaRH, SNH, SnGb, SnBr, MP, DMtns; to WA, ID, w NV.

var. ***australis*** (Vasek) A. Holmgren & N. Holmgren (p. 119) SIERRA JUNIPER Gen dioecious. **ST**: bark red-brown. **LVS** gen whorled in 3's, 6-ranked. **SEED CONE** 5–9 mm. **SEED**: cotyledons 2–4. Exposed, dry, rocky slopes, flats, forests, pinyon/juniper woodlands; 100–3100 m. NCoRH, SNH, SnGb, SnBr, DMtns; w NV. ❀DRN:4,**5**,**6**,15–17,24&IRR:**1–3**,**7**,8–10,**14**,18–23.

var. ***occidentalis*** WESTERN JUNIPER Gen ± monoecious. **ST**: bark brown. **LVS** opposite and 4-ranked or whorled in 3's and 6-ranked. **SEED CONE** 7–12 mm. **SEED**: cotyledons gen 2. Dry slopes, flats, sagebrush, juniper woodlands; 700–2300 m. CaRH, MP; to WA, ID, nw NV. ❀DRN:4,5,**6**,15–17,24&IRR:**1–3**,**7**,8–10,**14**,18–23.

J. osteosperma (Torrey) Little (p. 119) UTAH JUNIPER Tree < 8 m, monoecious. **ST**: trunk gen 1; bark thin, gray-brown aging ash-white. **LVS** gen opposite, 4-ranked, closely appressed, scale-like; gland obscure. **POLLEN CONE** 2–3 mm, cylindric. **SEED CONE** 5–13 mm, spheric, brown maturing red-brown, dry. **SEEDS** 1(2) per cone, 3–4 mm, ovoid, strongly angled. Pinyon/juniper woodlands; 1300–2600 m. SnGb, SnBr, GB, DMtns; to MT, NM. ❀DRN:1,**2**,4–6,15,**16**,22,24&IRR:**3**,**7**,8–12,**14**,**18–21**,23.

THUJA ARBORVITAE

Tree, monoecious. **ST**: trunk gen flared at base; branches spreading to ± pendent, tips often upturned; young shoots 4-sided, in flat clusters, held horizontally. **LVS** opposite, 2-ranked, ± not appearing whorled in 4's, closely appressed, scale-like, exposed part ± as long as wide. **SEED CONE** erect to reflexed, woody, ovoid-oblong, tapered to point, maturing 1st year; scales 4–6 pairs, partly overlapping, sharp-pointed near tip, thin, oblong, acute, leathery, middle 2–3 pairs fertile. **SEEDS** 2 per fertile scale, equally 2-winged; cotyledons 2. 5 spp.: n N.Am, e Asia. (Greek: resinous tree)

T. plicata D. Don (p. 119) WESTERN RED CEDAR, CANOE CEDAR Tree 30–70 m. **ST**: bark 1–2 cm thick, cinnamon-red, fibrous; young shoot upper surface glossy dark green, lower surface gen faintly white-streaked. **POLLEN CONE** 1.5–2 mm, ovate, dark brown. **SEED CONE** 10–19 mm, light brown. **SEED** 4–6 mm, narrow, elliptic, light brown. *n*=11. Coastal conifer forests; < 1800 m. NCo, KR, NCoRO; to AK, MT. ❀**4–6**&IRR:1,2,**7**,**14–24**&SHD:**3**,8,9.

EPHEDRACEAE EPHEDRA FAMILY

James R. Griffin

Shrub or tree-like, densely branched, gen dioecious. **ST** gen erect, < 2 m, jointed; node conspicuous, internode > lf; bark with irregular, longitudinal cracks, gen gray; twigs whorled, grooved, greenish and photosynthetic when young, glaucous, glabrous to scabrous, sometimes thorn-like. **LVS** 2–3 per node, not green; bases ± fused into sheath, thickening with age; tips often ± deciduous. **POLLEN CONES** 1–5 per node, gen short-stalked, gen spheric; bracts gen flexible, lower sterile, upper subtending 2–8 stamen-like structures. **SEED CONES** gen 1–3 per node, sessile or stalk < 5 mm; bracts gen flexible, lower sterile, upper enclosing 1–3 seeds. **SEED** spheric to cylindric, smooth or furrowed, gen angled at top, gen brown. 1 genus, 42 spp.: N.Am, S.Am, Medit, Asia. [Cutler 1939 Ann Mo Bot Gar 26:373–424]

EPHEDRA EPHEDRA, MORMON TEA

The only genus. (Greek: *Equisetum*, for resemblance to those pls)

1. Lvs gen 3 per node
 2. Twig tip thorn-like; lf sheath fibrous, persistent .... ***E. trifurca***
 2′ Twig tip narrowed but not thorn-like; lf sheath not fibrous, not persistent
  3. Seed not or faintly angled; twig yellow-green when young; lf base recurved, thickened when old ***E. californica***
  3′ Seed clearly angled at top; twig gray-green when young; lf base not recurved, not thickened when old ***E. funerea***
1′ Lvs gen 2 per node
 4. Lvs deciduous exc bases gen persistent; seeds gen 2
  5. Twig pale green, glaucous when young; lf base gray; seed cone stalk 1–5 mm .... ***E. nevadensis***
  5′ Twig bright green to yellow-green when young; lf base brown; seed cone sessile or stalk < 5 mm ..... ***E. viridis***
 4′ Lvs persistent; seed gen 1
  6. Seed gen smooth .... ***E. aspera***
  6′ Seed furrowed .... ***E. fasciculata***
   7. Seed < 8 mm .... var. ***clokeyi***
   7′ Seed > 8 mm .... var. ***fasciculata***

E. aspera S. Watson **ST** < 1.5 m; twig greenish when young, aging yellow, tip not thorn-like. **LVS** 2 per node, < 6 mm, gen persistent, sometimes deciduous but leaving thickened bases. **POLLEN CONES** gen 2 per node, < 7 mm. **SEED CONES** gen 2 per node, < 10 mm, ovoid, sessile. **SEED** 1, 5–8 mm, ovoid, gen smooth. Creosote-bush scrub, Joshua-tree woodland; < 1500 m. D; to TX, Mex. [*E. nevadensis* var. *a.* L. Benson] ❀TRY.

E. californica S. Watson (p. 119) DESERT TEA **ST** < 1.5 m; twig yellow-green when young, aging gray-brown, tip not thorn-like. **LVS** 3 per node, < 6 mm, with whitish margins wearing away to leave brown, thickened bases with recurved tips. **POLLEN CONES** 1–3 per node, < 9 mm. **SEED CONES** < 12 mm, ovoid, short-stalked. **SEED** gen 1, < 10 mm, ± spheric, smooth, not or faintly 4-angled. Scattered in arid grassland, chaparral, creosote-bush scrub; < 900 m. s SNF, Teh, w SnJV, SCoR, SW, D; to w AZ, Baja CA. ❀DRN,DRY,SUN:**7–12**,**14**,15–17,**18–21**,22–24.

E. fasciculata Nelson **ST** gen low or prostrate, < 1 m; branch ± flexible; twig pale green when young, aging yellow, tip not thorn-like. **LVS** 2 per node, < 3 mm, persistent; bases slightly thickened, whitish. **POLLEN CONES** gen 2–3 per node, < 8 mm. **SEED CONES** 6–13 mm, sessile or short-stalked. **SEED** gen 1, 6–13 mm, obovoid-ellipsoid, furrowed. Creosote-bush scrub; < 1500 m. D; to AZ, s UT.

var. ***clokeyi*** (Cutler) Clokey **SEED CONE** 6–10 mm, obovoid. **SEED** 5–8 mm. D; to s UT. [*E. clokeyi* Cutler] ❀TRY.

var. ***fasciculata*** (p. 117) **SEED CONE** 6–13 mm, elliptic. **SEED** 8–13 mm. D; to AZ. ❀TRY.

E. funerea Cov. & C. Morton DEATH VALLEY EPHEDRA **ST** < 1.5 m; twig pale gray-green when young, aging gray, tip not thorn-like. **LVS** gen 3 per node, < 5 mm; bases not thickened, persistent, not recurved. **POLLEN CONES** 1–3 per node, 5–8 mm. **SEED CONES** < 15 mm, elliptic, sessile or short-stalked. **SEED** gen 1, sometimes 2–3, 6–9 mm, ovoid, 4-angled, smooth to scabrous. Creosote-bush scrub; < 1700 m. DMoj; to w NV. ❀DRN,DRY, SUN:7,9,**10–12**,14,16,18–23.

E. nevadensis S. Watson (p. 119) **ST** < 1.3 m; twig pale green, glaucous when young, aging yellow to gray, tip not thorn-like. **LVS** 2, sometimes 3 per node, 2–8 mm; bases thickened, persistent, gray. **POLLEN CONES** 1–3 per node, 4–8 mm, elliptic. **SEED CONES** 5–11 mm, spheric; stalk 1–5 mm. **SEEDS** gen 2, 6–9 mm, hemispheric when 2, spheric when 1, smooth. Creosote-bush scrub, Joshua-tree woodland; < 1100 m. s SN, SNE, D; to OR, UT. ❀DRN,DRY,SUN:**3**,**7**,**9–12**,**14**,**18–23**.

E. trifurca Torrey (p. 119) **ST** < 2 m; twig pale green when young, aging yellow to gray-green, rigid, tip thorn-like. **LVS** gen 3 per node, 5–15 mm; tips ± spine-like; sheath fibrous, persistent, aging gray. **POLLEN CONES** 1–4 per node, 6–10 mm. **SEED CONES** 10–14 mm, obovoid, sessile or short-stalked. **SEED** gen 1, sometimes 2–3, 9–15 mm, 4-angled, smooth, light brown. Cresote-bush scrub; < 2100 m. DSon; to TX, Mex. Reportedly uniform, but pls in DMoj may be intermediate to *E. funerea*. ❀DRN, DRY,SUN:**2**,**3**,**7–12**,13,**14**,15,16,**18–23**.

E. viridis Cov. (p. 119) GREEN EPHEDRA **ST** < 1.5 m; twig bright green to yellow-green when young, aging yellow, tip not thorn-like. **LVS** gen 2 per node, < 6 mm; bases thickened, persistent, brown. **POLLEN CONES** 2–5 per node, 5–7 mm. **SEED CONES** 2–6 per node, 6–10 mm, obovoid, sessile or stalk < 5 mm. **SEEDS** 2, 5–8 mm, 3-angled, smooth. $2n$=14. Sagebrush scrub, creosote-bush scrub, pinyon/juniper woodland; 900–2300 m. s SNF, Teh, SnJV, SCoR, WTR, GB; to Colorado. [*E. nevadensis* var. *v.* (Cov.) M.E. Jones] ❀DRN,DRY,SUN:1,**2**,**3**,**7–12**,13,**14–24**.

PINACEAE PINE FAMILY

James R. Griffin

Tree or shrub, monoecious, evergreen. **ST**: young crown conic; twig not grooved, resinous, gen persistent. **LVS** simple, gen alternate, sometimes in bundles or appearing ± 2-ranked, linear or awl-like; bases decurrent, sometimes woody, persistent several years. **POLLEN CONE** gen < 6 cm, not woody, deciduous.**SEED CONE** gen woody; bracts, scales gen persistent; scale not peltate, fused to or free from subtending bract. **SEEDS** 2, on upper side of scale base. 10 genera, 193 spp.: mostly n hemisphere; many of great commercial value, supplying > half the world's timber. [Price 1989 J Arnold Arbor 70:247–305]

1. Lvs gen in bundles of 2–5 (1 in *P. monophylla*), gen 2.5–35 cm; seed cone bract fused to scale, incl, inconspicuous .... **PINUS**
1′ Lvs not in bundles, gen 1–6 cm; seed cone bract free from scale, exserted or incl, conspicuous

2. Lf scar ± flush with twig surface, smooth, round; seed cone erect, scales and bracts deciduous, axis persistent on st .... **ABIES**
2′ Lf scar raised above twig surface on persistent base, sometimes ± smooth and round; seed cone pendent, the whole ± persistent on st
3. Lf scar ± smooth, round, base not woody, projecting slightly above twig surface; seed cone 4–20 cm, bracts exserted, tip 3-toothed .... **PSEUDOTSUGA**
3′ Lf scar gen not smooth, not round, base woody, projecting conspicuously above twig surface; seed cone 1.2–12 cm, bracts incl, entire or fringed
4. Lf tip acute or spine-like, persistent base spreading, peg-like; seed cone 5–12 cm .... **PICEA**
4′ Lf tip gen blunt, persistent base ascending, wedge- or scale-like; seed cone 1.2–7.5 cm .... **TSUGA**

ABIES FIR

ST: young bark smooth, with resin blisters, mature bark thick, deeply furrowed; young branches appearing whorled; twig glabrous or hairy; lf scars smooth, round, flush with surface; bud gen ± spheric, gen < 1 cm, ± resinous. **LVS** 2–9 cm, sessile, twisted at base to become 2-ranked, often curved upward on upper twigs, gen ± flat; upper surface with 2 longitudinal, whitish bands, midrib sometimes depressed; lower surface with or without whitish bands, midrib sometimes ridge-like. **SEED CONE** erect, < 23 cm, maturing 1st season; stalk gen 0; bracts, scales deciduous; bract incl or exserted, free from scale; axis persistent on st. **SEED** with obvious resin deposits on surface; wing < 2.5 cm. $2n=24$ for all reports. 39 spp.: n hemisphere. (Latin: silver fir) [Vasek 1985 Madroño 32:65–77]

1. Lf tip spine-like; bud 1–2.5 cm; seed cone bract exserted 1.5–4.5 cm .... ***A. bracteata***
1′ Lf tip notched, blunt, or acute; bud < 1 cm; seed cone bract incl or exserted < 1 cm
2. Lf upper surface without longitudinal whitish bands
3. Twig lvs 2-ranked, alternating shorter and longer on each side; seed cone 8–15 cm .... ***A. grandis***
3′ Twig lvs not 2-ranked, often crowded on upper side, ± equal throughout; seed cone < 13 cm
4. Lf < 3 cm, dark green; bark ashy gray; twig ± hairy; KR .... ***A. amabilis***
4′ Lf 3–9 cm, blue-green; bark dark gray to ± black; twig glabrous; widespread in CA .... ***A. concolor***
2′ Lf upper surface with 2 faint, longitudinal, whitish bands
5. Lf ± flat; seed cone bract incl .... ***A. lasiocarpa*** var. ***lasiocarpa***
5′ Lf ± 3- or 4-angled (sometimes ± flat on twigs without cones); seed cone bract incl or exserted
6. Seed cone bract incl .... ***A. magnifica*** var. ***magnifica***
6′ Seed cone bract exserted
7. Seed cone bracts slightly exserted, ± reflexed, covering < 25% of cone surface ... ***A. magnifica*** var. ***shastensis***
7′ Seed cone bracts much exserted, strongly reflexed, covering > 90% of cone surface .... ***A. procera***

A. amabilis (Douglas) James Forbes PACIFIC SILVER FIR **ST**: trunk < 75 m, < 2.6 m wide; mature crown steeple-like; bark ± scaly, ashy gray; twig ± hairy; bud resinous. **LVS** not 2-ranked, spreading, < 3 cm, ± equal throughout, dark green; upper surface grooved, without white bands; tip notched or blunt. **SEED CONE** < 15 cm; bract incl. RARE in CA. Subalpine forest; 1700–2140 m. KR (w Siskiyou Co., 2 populations); to B.C.

A. bracteata (D. Don) Nutt. (p. 119) BRISTLECONE FIR **ST**: trunk < 55 m, < 1.3 m wide; mature crown narrow, steeple-like; bark thin, not fire-resistant; branches ± drooping, sometimes to ground; twig glabrous; bud 1–2.5 cm, sharp-pointed, not resinous. **LVS** < 6 cm, dark green, faintly grooved on upper surface; tip sharply spiny. **SEED CONE** < 9 cm; stalk < 15 mm; bract exserted 1.5–4.5 cm, spreading, tip with slender spine. UNCOMMON. Steep, rocky, fire-resistant slopes, gen in canyon-oak phase of mixed-evergreen forest; 210–1600 m. n SCoRO (Santa Lucia Range). [*A. venusta* (Douglas) K. Koch] ❀DRN:4–6,**15**,**16**,17 &IRR:2,3,**18–21**,22–24&SHD:**7**,8,9,**14**.

A. concolor (Gordon & Glend.) Lindley (p. 119) WHITE FIR **ST**: trunk < 61 m, < 2.7 m wide; mature crown rounded; young bark white-gray, old bark gray-brown to ± black, thick, deeply furrowed, with alternate dark and light layers; twig glabrous; bud resinous. **LVS** ± 2-ranked on lower branches, twisted upward on higher branches, 3–9 cm, ± flat; upper surface without white bands; tip gen blunt or acute. **SEED CONE** 6–13 cm; stalk < 5 mm; bract incl. Mixed-conifer to lower red-fir forest; 900–3100 m. KR, NCoR, CaRH, SNH, Teh, TR, PR, MP, DMtns; to c OR, ID, Colorado, n Baja CA. Most CA pls and some Rocky Mtn pls may be called var. *lowiana* (Gordon) Andr. Murray. Relationship of s CA pls (esp from DMtns) to the gen Rocky Mtn var. *concolor* is under study. ❀DRN:1,**4**,5,**6**&IRR:**2**,**3**,**7**,15–17,24&SHD:**10**,**14**,18–23CVS.

A. grandis (Douglas) Lindley GRAND FIR **ST**: trunk < 73 m, < 1.6 m wide; mature crown rounded, with large branches; young bark white-gray, mature bark red-brown, thin; twig hairy; bud resinous. **LVS** 2-ranked, < 5 cm, alternating shorter and longer on each side, shorter toward twig tip; upper surface ± flat, without whitish bands; tip notched or blunt. **SEED CONE** 8–15 cm; stalk < 5 mm; bract incl. Redwood, douglas-fir, mixed-evergreen forests; < 700 m. NCo, NCoR; to B.C., MT. ❀DRN:**4–6**&IRR:2,3,**7**,**15–17**,24 &SHD:14,18–23.

A. lasiocarpa (Hook.) Nutt. var. ***lasiocarpa*** (p. 119) SUBALPINE FIR **ST**: trunk < 30 m, < 1.3 m wide; mature crown very narrow, steeple-like; bark gray, < 5 cm thick; twig hairy several seasons; bud resinous. **LVS** < 3 cm, ± flat; lower surface with whitish bands; tip notched or blunt. **SEED CONE** < 10 cm; bract incl. RARE in CA (1st reported in 1968). Subalpine forest, meadows; 1700–2100 m. KR (w Siskiyou Co., 6 populations); to nw Can, in Rocky Mtns to NM.

A. magnifica Andr. Murray CALIFORNIA RED FIR **ST**: trunk < 57 m, < 2.5 m wide; mature crown ± cylindric, top rounded; young bark gray, with resin blisters, mature bark deeply furrowed, with dark reddish ridges; twig hairy 1st season; bud ± resinous. **LVS** < 3.5 cm, ± 3–4-angled; upper surface with 2 whitish bands; tip notched or blunt. **SEED CONE** < 23 cm; bract incl or ± exserted. mixed-confer to subalpine forests; 1200–2800 m. KR, NCoRH, CaRH, SNH, SNE; s OR (Cascade Mtns), w-c NV.

var. ***magnifica*** Cotyledons 6–13. **SEED CONE**: bract incl. mixed-conifer to subalpine forests; 1200–2600 m. KR, NCoRH, CaRH, n&c SNH, SNE; w-c NV. ❀DRN:**1**,4,5,15&IRR:2,**3**,**6**,16 &SHD:7,14.

var. ***shastensis*** Lemmon (p. 119) SHASTA RED FIR Cotyledons 5–8. **SEED CONE**: bract ± exserted, covering < 25% cone surface. Upper mixed-conifer to subalpine forests; 1350–2800 m. KR, CaRH, s SNH; s OR. May include intermediates to *A. procera* (seed cone bract longer, more reflexed). ❀DRN:**1**,4,5,15&IRR:**2, 3,6**,16&SHD:7,14.

A. procera Rehder (p. 119) NOBLE FIR **ST**: trunk < 85 m, < 2.8 m wide; mature crown cylindric, top rounded; young bark smooth, gray, with resin blisters, mature bark ± reddish brown; twig hairy; bud ± resinous. **LF** < 4 cm, ± 3–4-angled, sometimes ± flat on twigs without cones; upper surface grooved, with 2 whitish bands; tip notched or blunt. **SEED CONE** < 15 cm; bracts ± tapered, exserted, strongly reflexed, covering > 90% cone surface. Upper mixed-conifer to subalpine forests; < 1500 m. n KR, CaRH; to B.C. Intergrades with *A. magnifica* var. *shastensis*. ❀DRN:4,5&IRR:**1–3,6**, 15–17&SHD:7,14.

PICEA SPRUCE

ST: crown conic; young branches appearing whorled, sometimes drooping; bark thin, scaly; twig glabrous or hairy; lf bases spreading, peg-like, < 2 mm, persistent; bud conic to ovoid, resinous. **LVS** often crowded toward upper side of twigs, gen < 3 cm, sessile, ± 4-angled or flat, stiff, often strong-smelling when crushed. **SEED CONE** sometimes conspicuously terminal, pendent, 5–12 cm, maturing 1st season; stalk 0–1 cm; bract incl, free from scale. **SEED**: wing terminal. $2n$=24 for all reports. 34 spp.: n hemisphere. (Latin: pitch) Trunks, esp of *P. sitchensis*, more flared or buttressed at base than other CA conifers.

1. Lf ± 4-angled, with 2 obvious, whitish bands on each surface . . . . ***P. engelmannii***
1′ Lf ± flat, with 2 faint, whitish bands on upper surface, sometimes also lower surface
 2. Lf tip blunt; seed cone scale margin ± entire, bract < 1/4 scale . . . . ***P. breweriana***
 2′ Lf tip acute; seed cone scale margin ± jagged, bract > 1/2 scale . . . . ***P. sitchensis***

P. breweriana S. Watson (p. 119) BREWER SPRUCE **ST**: trunk < 53 m, < 1.3 m wide; branches drooping, sometimes to ground; twig hairy. **LF** < 3 cm, flexible; upper surface ± flat, gen with 2 faint, whitish bands; tip blunt. **SEED CONE** < 12 cm, oblong; scale woody, margin ± entire; bract < 1/4 scale. Uncommon. Cool, moist, forested slopes; 560–2300 m. KR; sw OR. ❀DRN,IRR:2, **3,4**,5,6,15–17.

P. engelmannii Engelm. (p. 119) ENGELMANN SPRUCE **ST**: trunk < 55 m, < 2.4 m wide; branches not drooping; young twigs ± hairy. **LF** < 3 cm, rigid, 4-sided; both surfaces with 2 obvious, whitish bands; tip flattened, acute, not sharp to touch. **SEED CONE** < 7 cm, ovoid-oblong; scale papery, margin ± jagged. RARE in CA. Cool, moist, mixed-conifer, subalpine forests; 1200–2100 m. KR, CaRH; to B.C., in Rocky Mtns to NM. 3 populations known in CA.

P. sitchensis (Bong.) Carrière (p. 119) SITKA SPRUCE **ST**: trunk < 66 m, < 5.1 m wide; branches not drooping; twig glabrous. **LF** 1–3 cm, rigid; upper surface ± flat, with 2 faint, whitish bands, lower rounded, darker green, sometimes with 2 faint, whitish bands; tip acute, sharp to touch. **SEED CONE** < 10 cm, oblong; scale ± papery, margin ± jagged; bract > 1/2 scale. Moist soils near mouths of coastal rivers; < 200 m. NCo; to AK. ❀IRR:2,3,**4,6,15–17**&SHD: **7**,14;CVS.

PINUS PINE

ST: young crown conic, mature crown often rounded or flat; branches ± whorled in young pls; young bark smooth, mature bark furrowed; bud ± conic, gen resinous. **LVS** gen 2.5–35 cm, gen sessile, in bundles of 1–5; bundles solitary in axils of alternate, awl-like bracts, each bundle enclosed at base in a sometimes deciduous sheath of bracts, gen persistent several seasons. **SEED CONES** often whorled, gen maturing and opening 2nd season, sometimes persistent on st; stalk 0 or < 16 cm; bract incl, fused to scale, minute; scale tip reflexed and elongated 3–7 cm, or often with a rounded or angled, often prickled knob < 3 cm. **SEED**: coat hard, sometimes woody. $2n$=24 for all reports. 94 spp.: n hemisphere. (Latin: pine) [Millar & Critchfield 1988 Madroño 35:39–53]

1. Lf-bundle sheath deciduous or mostly so; seed cone scale tip gen without prickle
 2. Lvs gen < 5 per bundle
  3. Lvs gen 4 (3–5) per bundle . . . . ***P. quadrifolia***
  3′ Lvs gen 1–2 per bundle
   4. Lvs gen 2 per bundle . . . . ***P. edulis***
   4′ Lvs gen 1 per bundle . . . . ***P. monophylla***
 2′ Lvs 5 per bundle
  5. Seed cone gen torn apart by animals, gen not on ground intact; seed wings persistent on scale . . . . . ***P. albicaulis***
  5′ Seed cone not torn apart by animals, gen on ground intact; seed wings persistent on scale or not
   6. Seed cone scale thinnest at tip, without prickle
    7. Seed cone stalk < 2 cm; seed wings narrow, gen persistent on scale . . . . ***P. flexilis***
    7′ Seed cone stalk > 2 cm; seed wings broad, deciduous from scale
     8. Seed cone gen > 20 cm; mature bark thick, reddish brown . . . . ***P. lambertiana***
     8′ Seed cone gen < 20 cm; mature bark thin, dark gray . . . . ***P. monticola***
   6′ Seed cone scale thickest at tip, with prickle
    9. Scale knob tip with 2–6 mm prickle; trunks often multiple . . . . ***P. longaeva***
    9′ Scale knob tip with < 1 mm prickle; trunks gen single . . . . ***P. balfouriana***
     10. Lf ± yellow-green; bark reddish brown, in ± square plates; s SN, SNE . . . . ssp. ***austrina***
     10′ Lf ± blue-green; bark gray-brown, in ± narrow ridges; KR, NCoRH . . . . ssp. ***balfouriana***
1′ Lf-bundle sheath persistent; seed cone scale tip with prickle, at least when immature (sometimes 0 at maturity)
 11. Lvs 2 per bundle
  12. Seed cone > 5 cm; lf gen 5–15 cm . . . . ***P. muricata***
  12′ Seed cone < 5 cm; lf 2.5–8.6 cm . . . . ***P. contorta***

13. Seed cone gen closed at maturity, asymmetric .... ssp. ***bolanderi***
13′ Seed cone open at maturity, asymmetric or ± symmetric
14. Seed cone persisting on st many years after opening .... spp. ***contorta***
14′ Seed cone deciduous after opening .... ssp. ***murrayana***
11′ Lvs 3 or 5 per bundle
15. Lvs 5 per bundle .... ***P. torreyana***
15′ Lvs 3 per bundle
16. Basal seed cone scale tips recurved, elongated 3–7 cm; seed cone scale prickles gen worn off by maturity
17. Seed cone yellowish; seed < wing; lf dark green .... ***P. coulteri***
17′ Seed cone brownish; seed > wing; lf gray-green .... ***P. sabiniana***
16′ Basal seed cone scale tips rarely recurved, then elongated < 3 cm; seed cone scales (exc sometimes upper) with a rounded or angled, often prickled knob < 3 cm at maturity
18. Seed cones persisting on sts many years after maturity
19. Seed cone scale tips, exc upper, with an angular knob > 2 cm .... ***P. attenuata***
19′ Seed cone scale tips, exc upper, with a rounded knob < 2 cm .... ***P. radiata***
18′ Seed cones deciduous after maturity
20. Seed cone gen < 9 cm; lf gen 7–12 cm .... ***P. washoensis***
20′ Seed cone gen > 9 cm; lf gen > 15 cm
21. Lower surface of seed cone scale gen not darker than upper; lf grayish blue-green, glaucous; bark odor banana-, pineapple-, or vanilla-like .... ***P. jeffreyi***
21′ Lower surface of seed cone scale gen darker than upper; lf deep yellow-green; bark odor not banana-, pineapple-, or vanilla-like .... ***P. ponderosa***

P. albicaulis Engelm. (p. 123) WHITEBARK PINE **ST** in most exposed places shrubby to prostrate; trunks sometimes multiple, < 26 m, < 1.5 m wide, much wider at base; bark when mature gray-white, smooth, thin; mature crown often deformed by wind. **LVS** 5 per bundle, 3–7 cm, ± curved, dark green, stiff; sheath deciduous. **SEED CONE** sessile, erect, 3.5–9 cm, ovate, purple-brown, gen torn apart (and seeds dispersed) by animals; scale tip knobs angled, prickled. **SEED**: wing persistent on scale. Upper red-fir forest to timberline, esp subalpine forest; < 3700 m. KR, CaRH, SNH, Wrn, SNE; to B.C., WY. ❀DRN,SUN:**1**,6,15–17&IRR:**2,3**,7;DFCLT.

P. attenuata Lemmon (p. 123) KNOBCONE PINE **ST**: trunk < 36 m, < 1.1 m wide; bark gray-brown; mature crown branches many, sometimes with several tops. **LVS** 3 per bundle, 6–16 cm, yellowish green; sheath persistent. **SEED CONE** recurved to reflexed, 6–18 cm, asymmetric, yellow-brown, often remaining closed unless burned, persisting on st many years; stalk < 2 cm, disappearing with st growth; basal scale tip knobs > 2 cm, angled, prickled. **SEED** < wing. Closed-cone-pine forest, chaparral; < 1900 m. NW, CaR, SN, e SnFrB, SCoR, SnBr, PR, MP; sw OR, Baja CA. ❀DRN,SUN: 3,**4–7,14–17,24**&IRR:10,18–23.

P. balfouriana Grev. & Balf. FOXTAIL PINE **ST**: trunk gen single, < 22 m, < 2.6 m wide; mature crown branches short, thick. **LVS** 5 per bundle, 1.5–4 cm, curved, stiff; outer surface ± green, inner white; tip acute, sharp to touch; sheath deciduous. **SEED CONE** pendent, 6–19 cm, ovoid; stalk 7–16 mm; scale tip knobs < 1 mm, angled, with very small prickles. **SEED** < wing. Subalpine forest; 2100–3700 m. KR, NCoRH, s SNH.

ssp. ***austrina*** Mastrogiuseppe & Mastrogiuseppe **ST**: mature bark thick, reddish brown, in ± square plates. **LVS** ± yellow-green, persisting < 30 years. Habitats of sp.; 2700–3700 m. s SNH. In some ways more like *P. longaeva* than ssp. *balfouriana*. ❀DRN:1 &SHD,IRR:2,3,7;DFCLT.

ssp. ***balfouriana*** (p. 123) **ST**: mature bark thin, gray-brown, in narrow ridges. **LVS** ± blue-green, persisting < 15 years. Habitats of sp.; 2100–2500 m. KR, NCoRH. ❀DRN:1,4–6&IRR:2,3,7; DFCLT.

P. contorta Loudon LODGEPOLE PINE **ST**: mature bark scaly, thin; trunk 2–34 m (extremely variable at maturity). **LVS** 2 per bundle, 2.5–8.6 cm; sheath persistent. **SEED CONE** pendent, < 5 cm, brown; stalk ± 0; scale tip knobs < 5 mm, angled, prickled. Coastal to subalpine forest; < 3500 m. CA-FP; to B.C., SD, n Baja CA.

ssp. ***bolanderi*** (Parl.) Critchf. BOLANDER PINE **ST**: trunk gen < 2 m. **SEED CONE** asymmetric, woodier than other sspp., remaining closed, on st for many years. RARE. "Pygmy forest" on coastal terrace soils with claypan or hardpan; < 250 m. NCo (Mendocino Co.) Threatened by development, off-road vehicles. In cult.

ssp. ***contorta*** SHORE PINE **ST**: trunk gen < 15 m. **SEED CONE** asymmetric, persisting on st for many years. Coastal dunes, bluffs lacking claypan, hardpan; < 150 m. NCo; to AK.

ssp. ***murrayana*** (Grev. & Balf.) Critchf. (p. 123) LODGEPOLE PINE **ST**: trunk < 34 m. **SEED CONE** ± symmetric, deciduous within several years. Lodgepole forest, wet meadows, cold places in mixed-conifer forest; < 3500 m. KR, CaRH, SNH, SnGb, SnBr, SnJt, GB; OR, n Baja CA. [*P. murrayana* Grev. & Balf.] ❀SUN: 4–6,15,16&IRR:**1,2,7,17**&SHD:**3**,14.

P. coulteri D. Don (p. 123) COULTER PINE **ST**: trunk < 42 m, 1.5 m wide; young bark dark brown to blackish, with irregular furrows, mature bark forming yellowish plates; mature crown branches thick, lower sometimes reaching ground. **LVS** 3 per bundle, 15–30 cm, stiff; sheath persistent. **SEED CONE** pendent, 19–35 cm, ± ovoid-oblong, yellow-brown, opening slowly during 2nd or 3rd season, deciduous over many years; stalk < 10 cm, persistent with basal scales; scale tips reflexed, elongated 3–5 cm, angled. **SEED** < wing. Chaparral, lower mixed-conifer, mixed-hardwood forests; < 3000 m. CW, SW; n Baja CA. ❀DRN,SUN:**4**,5,**6,7,14–17**,24& IRR:**2,3**,8–10,**18–23**.

P. edulis Engelm. (p. 123) COLORADO PINYON **ST**: trunk < 15 m, < 1.1 m wide; bark shallowly furrowed, red-brown; mature crown ± rounded. **LVS** gen 2 per bundle, 2–6 cm, dark green; sheath deciduous. **SEED CONE** erect, 3–7 cm, ovoid, yellow-brown; stalk ± 0; scale tip knobs < 1 cm, angled, truncate. **SEED**: wing persistent on scale. RARE in CA. Pinyon/juniper woodland; 1300–2700 m. DMtns (New York Mtns); to s WY, w TX, n Mex. Extent of hybridization with *P. monophylla* under study.

P. flexilis James (p. 123) LIMBER PINE **ST**: trunk < 20 m, 2.9 m wide; mature bark dark brown, deeply furrowed, forming rectangular plates; mature crown branches sometimes reaching ground. **LVS** 5 per bundle, 2–9 cm, in dense tufts of bundles at branch ends, stiff, gen curved; sheath deciduous. **SEED CONE** pendent, 4–20 cm, oblong, yellow-brown, opening late 2nd season; stalk < 2 cm; scales thinnest at tips, angled. **SEED**: wing gen persistent on scale. Lodgepole, subalpine, bristlecone forests; < 3700 m. SNH, TR, PR, SNE; to w Can, SD, NM. ❀DRN:**1**,4,6,15–17&IRR:**2,3**,7,10,14, **18**,19.

P. jeffreyi Grev. & Balf. (p. 123) JEFFREY PINE **ST**: trunk < 53 m, 2.3 m wide; mature bark gen red-brown, furrows closely spaced, deep, forming ridges, odor banana-, pineapple-, or vanilla-like; mature crown rounded; bud not resinous, scales light brown, white-hairy. **LVS** 3 per bundle, 15–22 cm, grayish blue-green, glaucous. **SEED CONE** spreading or recurved, gen < 26 cm, ± oblong,

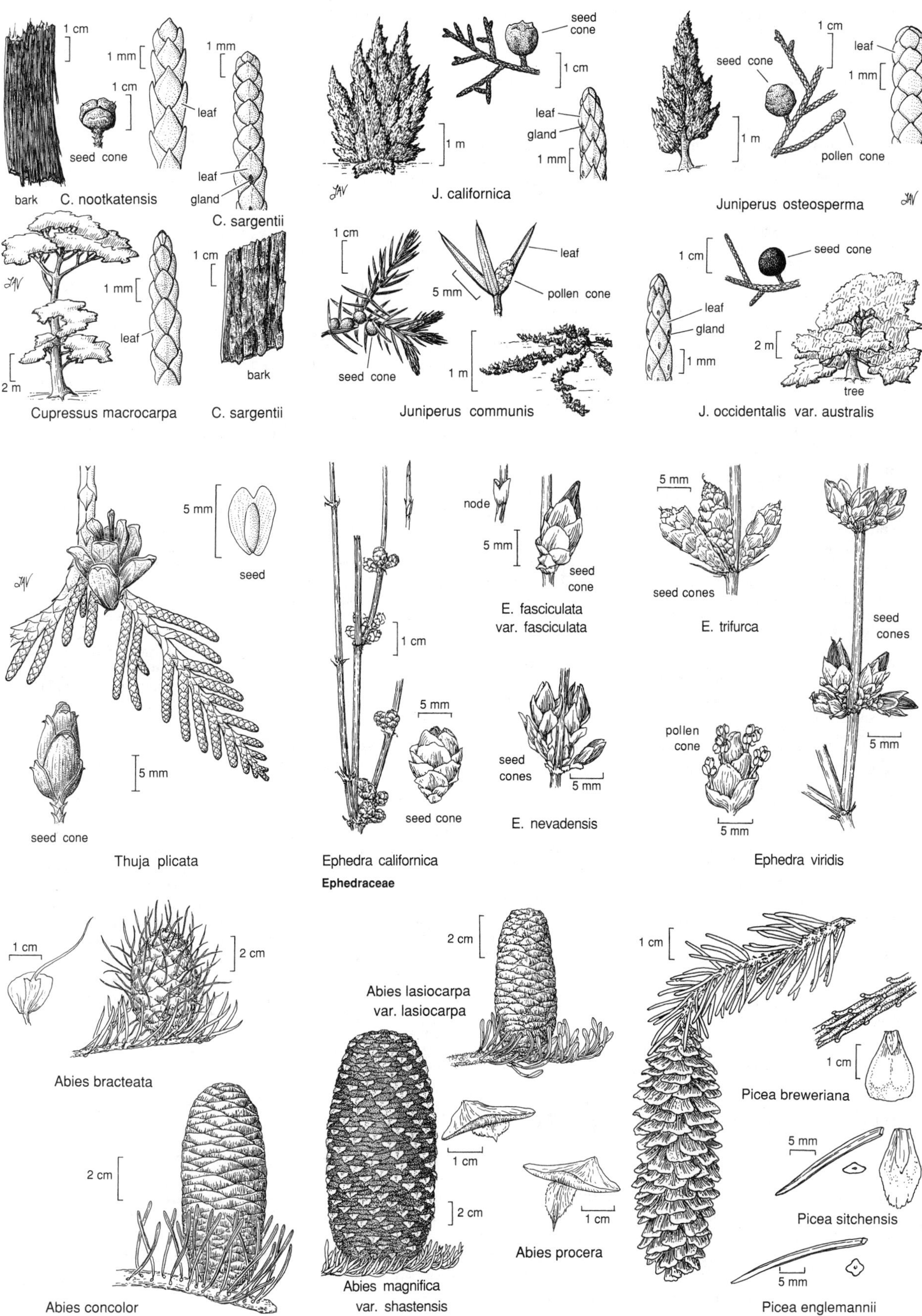
1 cm
1 mm
1 cm
leaf
seed cone
bark
C. nootkatensis
1 mm
leaf
gland
C. sargentii
seed cone
1 cm
1 m
leaf
gland
1 mm
J. californica
1 cm
seed cone
1 m
leaf
1 mm
pollen cone
Juniperus osteosperma
1 mm
leaf
2 m
Cupressus macrocarpa
1 cm
bark
C. sargentii
1 cm
leaf
5 mm
pollen cone
seed cone
1 m
Juniperus communis
1 cm
seed cone
leaf
gland
1 mm
2 m
tree
J. occidentalis var. australis
5 mm
seed
5 mm
seed cone
Thuja plicata
1 cm
5 mm
seed cone
Ephedra californica
Ephedraceae
node
5 mm
seed cone
E. fasciculata var. fasciculata
seed cones
5 mm
E. nevadensis
5 mm
seed cones
E. trifurca
seed cones
5 mm
pollen cone
5 mm
Ephedra viridis
1 cm
2 cm
Abies bracteata
2 cm
Abies concolor
Pinaceae
2 cm
Abies lasiocarpa var. lasiocarpa
1 cm
2 cm
Abies magnifica var. shastensis
1 cm
Abies procera
1 cm
1 cm
Picea breweriana
5 mm
Picea sitchensis
5 mm
Picea englemannii

brown; stalk < 3 cm, persistent with basal scales; scale surfaces gen of similar color, tips reflexed, elongated < 3 cm, with prickled knobs. Upper mixed-conifer, red-fir forests, elsewhere on serpentine; 450–3100 m. KR, NCoR, CaR, SN, SCoRI, TR, PR, GB; sw OR, w NV, n Baja CA. ❀DRN,SUN:1,**4–6,15–17**&IRR:**2,3,7**,8, 9,**14,18,19**.

P. lambertiana Douglas (p. 123) SUGAR PINE **ST**: trunk < 70 m, < 3.3 m wide; mature bark thick, reddish brown, irregularly furrowed into plate-like ridges with deciduous platelets; mature crown flattened, with large, ± horizontal branches. **LVS** 5 per bundle, 5–11 cm, gen stiff, sometimes twisted; sheath deciduous. **SEED CONE** pendent, 15–63 cm, cylindric, yellow-brown; stalk 5–16 cm; scales thinnest at tips. **SEED** < wing. Mixed-conifer, mixed-evergreen forests; < 3200 m. NW, CaR, SN, SW, w GB; OR, n Baja CA. ❀DRN:1,**4–6,15**,16,17&IRR:2,3,7.

P. longaeva D. Bailey (p. 123) WESTERN BRISTLECONE PINE **ST**: trunks often multiple, twisted, strongly tapered upward, < 16 m, < 2 m wide; bark reddish brown; mature crown bushy, irregular. **LVS** 5 per bundle, 1–4 cm, often with white resin spots; sheath deciduous. **POLLEN CONE** when mature deep red. **SEED CONE** ± spreading or pendent, < 5–14 cm, ovoid-oblong, dark red-brown; stalk 0–2 cm; scale tips ± thickened, prickle < 6 mm, slender, reflexed. **SEED** < wing. UNCOMMON. Bristlecone-pine forest; 2200–3700 m. W&I; to e UT. [*P. aristata* Engelm., in part]. ❀DRN:1,4–6,15–17&SHD,IRR:2,3,7,18,19.

P. monophylla Torrey & Frémont (p. 123) SINGLELEAF PINYON PINE **ST**: trunk < 15 m, < 40 cm wide; mature crown much-branched, rounded. **LF** gen 1 per bundle, 2–7 cm, often curved, gray or blue-green; sheath deciduous. **SEED CONE** erect, 3–12 cm, spheric-ovoid, light or reddish brown; stalk < 1 cm; scale tip knobs < 1 cm, angled, truncate. **SEED**: wing persistent on scale. Pinyon/juniper woodland; < 2800 m. c&s SNH, Teh, se SCoRI, TR, PR, SNE, DMtns; to se ID, n Baja CA. [*P. californiarum* D. Bailey] Extent of hybridization with *P. edulis*, *P. quadrifolia* under study. ❀DRN:1,**2,3**,4–6,**7**,10,**14–16**,17,24,**18–23**&IRR:8–12.

P. monticola Douglas (p. 123) WESTERN WHITE PINE **ST**: trunk < 73 m, < 2 m wide; mature bark dark gray, in ± square blocks, relatively thin; mature crown narrowly conic. **LVS** 5 per bundle, 3–10 cm, gen persistent < 4 years, gen straight, flexible, blue-green, glaucous; sheath deciduous. **SEED CONE** pendent, 9–25 cm, cylindric, yellow-brown; stalk 2–5 cm; scales thinnest at tips. **SEED** < wing. Upper mixed-conifer to subalpine forests; 150–3400 m. KR, NCoRH, CaRH, SNH, SNE, MP; to B.C., MT. ❀DRN:1,**4–6** &IRR:**2,3**,7,15,16;DFCLT.

P. muricata D. Don (p. 123) BISHOP PINE **ST**: trunk < 51 m, < 1.2 m wide; bark brown, with rough ridges; mature crown variable, often with large branches. **LVS** gen 2 per bundle, gen 5–15 cm, sometimes twisted, gen green. **SEED CONES** gen whorled, 5–9.7 cm, ovoid, brown, weathering gray, gen closed, persistent many years, either ± perpendicular to st, symmetric, with scale tip knobs < 3 mm and prickled, or reflexed, asymmetric, with basal and middle scale tip knobs < 15 mm, angled, prickled; stalk 0–2 cm. Redwood forest, n coastal coniferous forest, closed-cone-pine forest, oak woodland, chaparral; < 300 m. NCo, CCo, n SnFrB, n ChI; n Baja CA, Cedros Island. [*P. remorata* H. Mason, Santa Cruz Island pine] ❀**5,15–17,24**&IRR:**14**,22,23.

P. ponderosa Laws. (p. 123) PACIFIC PONDEROSA PINE **ST**: trunk < 68 m, 2.2 m wide; branches sometimes lacking in lower half when mature; mature bark gen yellow-brown, furrows shallow, well spaced, forming plates, odor not banana-, pineapple-, or vanilla-like; mature crown short, conic or flat-topped; bud resinous, scales red-brown, dark-hairy. **LVS** gen 3 per bundle, 12–26 cm, < 2 mm thick, not glaucous, deep yellow-green; sheath persistent. **SEED CONE** ± spreading or recurved, gen 9–18 cm, ovoid, dark brown; stalk < 2 cm, persistent with basal scales; scales gen blackish on lower surfaces; knob prickles < 3 mm. **SEED** < wing. mixed-conifer, lower mixed-evergreen forests, gen absent from serpentine; 150–2300 m. CA-FP (exc NCo, GV, SCo); to B.C., MT, Nebraska, n Mex. ❀DRN,SUN:1,**4–6,15,16**,17&IRR:**2,3,7**,8,9,**14, 18**,19.

P. quadrifolia Parl. (p. 123) PARRY PINYON PINE **ST**: trunk < 16 m, 70 cm wide; bark when mature reddish brown. **LVS** gen 4 (3–5) per bundle, 2–5 cm, curved; inner surfaces whitish, outer blue-green; sheath deciduous. **SEED CONE** erect, 4–8 cm, ± spheric, light brown; stalk < 1 cm; scale tip knobs < 15 mm, angled, truncate. **SEED**: wing persistent on scale. Pine forest, pinyon/juniper woodland, chaparral; 1100–1800 m. SnJt, s PR; n Baja CA. ❀DRN: **3**,4–6,**7,15,16**,17,24&IRR:10,**14,18–23**.

P. radiata D. Don (p. 123) MONTEREY PINE **ST**: trunk < 38 m, < 2.1 m wide, when young < 2 m growth per year; mature bark black, deeply grooved; mature crown irregular, round-topped. **LVS** gen 3 per bundle, 6–15 cm, dark green; sheath persistent. **SEED CONE** recurved, 6–15 cm, asymmetric, light brown, opening slowly 2nd season, then persistent < 25 years; stalk < 15 mm; basal scale tip knobs < 2 cm, gen rounded, minutely prickled. **SEED** < wing. RARE. Closed-cone-pine forest, oak woodland; < 1200 m. CCo; islands off Baja CA. Lvs gen 2 per bundle on Cedros, Guadalupe islands. In cult.

P. sabiniana Douglas (p. 123) GRAY or FOOTHILL PINE **ST**: trunk < 38 m, < 2 m wide, often leaning; several major branches after 20–30 years; bark dark gray with irregular furrows, forming yellow plates when very old. **LVS** 3 per bundle, 9–38 cm, gray-green, fragrant; sheath persistent. **SEED CONE** pendent, 10–28 cm, ovate-oblong, brownish, opening slowly 2nd season, then persistent several years; stalk < 7 cm, persistent (with basal cone scales) several years; scale tip reflexed, elongated 3–7 cm, angled. **SEED** > wing. Foothill woodland, n oak woodland, chaparral, infertile soils in mixed-conifer and hardwood forests; 150–1500 m. CA-FP (exc n NW, n CaR, SnJV), w GB, w D. The common name digger pine is pejorative in origin, so best avoided. ❀DRN,SUN:3,**4–7,14–24** &IRR:8–11.

P. torreyana Carrière (p. 123) TORREY PINE **ST**: trunk < 23 m (< 33 m in cult), < 1 m wide; mature crown open, rounded, with many large branches; mature bark with red-brown plates between irregular furrows. **LVS** 5 per bundle, 15–26 cm, stiff, grayish yellow-green to blue-green; sheath persistent. **SEED CONE** spreading to recurved, < 16 cm, widely ovate to spheric, ± symmetric, dark brown, opening slowly 3rd season, then persistent < 15 years; stalk < 4 cm; basal scale tip knobs < 2 cm, pyramidal, minutely prickly. **SEED** < 24 mm, > wing, sometimes remaining in cone < 13 years. RARE. Sandstone, coastal scrub, chaparral; < 150 m. s SCo, n ChI (Santa Rosa Island). [ssp. *insularis* Haller, Santa Rosa Island Torrey pine] Native pls probably total < 9000. Threatened by development. In cult.

P. washoensis H. Mason & Stockw. WASHOE PINE **ST**: trunk < 35 m, < 1.5 m wide; bark yellow-brown, shallowly furrowed; mature crown short, conic or flat-topped. **LVS** 3 per bundle, 7–12 cm, gen > 2 mm thick, dark green; sheath persistent. **SEED CONE** spreading, gen 5–9 cm, ovoid, symmetric, reddish purple to purple-black until mature; stalk > 1 cm, persistent with basal scales; scale tips prickled. UNCOMMON. Subalpine, red-fir forests; < 2700 m. CaRH, n SNH, Wrn; OR, w NV. ❀DRN,SUN:1,**2,4–6,15**&IRR: **3,7**,14,18.

PSEUDOTSUGA DOUGLAS-FIR

ST: young crown conic; branches appearing ± whorled in young pls, smaller ± drooping; young bark smooth, with resin blisters, mature bark thick, deeply furrowed, dark brown; lf scars ± smooth, round, slightly raised above twig surface on persistent lf bases; bud < 15 mm, fusiform. **LVS** persistent < 8 years, 2–4.5 cm, tapered to a short petiole above persistent base, ± spreading, ± flat, with 2 whitish bands on lower surface. **SEED CONE** pendent, 4–20 cm, maturing 1st season, persistent or deciduous; stalk < 2 cm; bract ± exserted, 3-toothed or -lobed at tip, free from scale. **SEED** plus wing < 25 mm. 4 spp.: N.Am, Asia. (Latin & Japanese: false hemlock)

1. Seed cone > 9 cm, lf tip gen acute .... ***P. macrocarpa***
1′ Seed cone < 9 cm; lf tip gen blunt .... ***P. menziesii*** var. ***menziesii***

P. macrocarpa (Vasey) Mayr BIGCONE DOUGLAS-FIR, BIGCONE SPRUCE **ST**: trunk < 44 m, < 2.1 m wide, strongly tapered; mature crown rounded to flat; often with many large, lower branches. **LVS** 2–4.5 cm. **SEED CONE** 9–20 cm, sometimes persistent several years; bracts little exserted exc for central tooth or lobe. Scattered on fire resistant slopes, mixed-evergreen forest; 200–2400 m. s SCoRO, TR, PR. ❀DRN:**3–7,14–17**&IRR:2,8,9,**10**,11,**18**,19,**20–23.**

P. menziesii (Mirbel) Franco var. ***menziesii*** (p. 123) DOUGLAS-FIR **ST**: trunk < 67 m, < 4.4 m wide; mature crown rounded; with large, upper branches. **LVS** 2–4 cm. **SEED CONE** 5–9 cm, deciduous; bracts well exserted. $2n$=26. Widespread in mixed-evergreen, mixed-conifer forests; < 2200 m. KR, NCoR, CaRH, n&c SNH, CCo, SnFrB, n SCoRO; to sw B.C. ❀DRN:**4–6,15–17**&IRR:**2,3,**7&SHD:**14**;CVS.

TSUGA HEMLOCK

ST: crown conic, top slender, gen nodding; branches ± drooping; bark reddish brown; twig hairy or puberulent, lf bases ascending, wedge- or scale-like, persistent; bud ovoid. **LVS** 5–25 mm, tapered to a short petiole above persistent base, spreading, ± 2-ranked or not, ± flat, sometimes grooved on upper surface, ridged on lower. **SEED CONE** pendent, 1.2–7.5 cm, ovoid to oblong, opening 1st season; stalk ± 0; bract incl, free from scale. **SEED** plus wing < 2 cm. $2n$=24 for all reports. 10 spp.: n hemisphere. (Japanese: hemlock)

1. Lvs ± 2-ranked, grooved on upper surface, with 2 whitish bands on lower; seed cone 1.2–2.5 cm ***T. heterophylla***
1′ Lvs not 2-ranked, more crowded above twig, rounded on upper surface, glaucous and with whitish bands on both; seed cone 3–7.5 cm .... ***T. mertensiana***

T. heterophylla (Raf.) Sarg. (p. 127) WESTERN HEMLOCK **ST**: trunk < 50 m, < 2.7 m wide; mature bark thin, reddish brown, with narrow grooves between broad ridges; twig hairy. **LF** 5–20 mm, persistent 4–7 years; tip blunt to rounded. Scattered in n coastal conifer to mixed-evergreen forests; gen < 700 m. NCo, w KR, NCoRO; to AK, MT. ❀IRR,DRN:**4–6,15–17**&SHD:**3,7**,14;CVS.

T. mertensiana (Bong.) Carrière (p. 127) MOUNTAIN HEMLOCK **ST**: trunk < 35 m, 2.2 m wide, prostrate at timberline; mature bark red- or purple-brown, with narrow grooves between narrow ridges; twig puberulent. **LF** 10–25 mm, persistent < 7 years, tip rounded. Subalpine, some scattered in cold areas of red-fir, mixed-conifer forest; 1200–3500 m. KR, NCoRH, CaRH, SNH, MP, n SNE; to AK, MT. ❀DRN,IRR:1,**2**,4,5&SHD:**6**,7,14–17.

TAXACEAE YEW FAMILY

James R. Griffin

Tree or shrub, dioecious, evergreen. **LVS** simple, alternate, not in bundles, gen appearing 2-ranked, < 5 cm, linear, decurrent; tip acute. **POLLEN CONE** with < 32 stamen-like structures. **SEED** solitary at tip of short twig, partly or completely enclosed by subtending aril; coat woody; cotyledons 2. 5 genera, 16 spp.: n hemisphere; some *Taxus* spp. widely used in landscaping. [Florin 1948 Bot Gaz 110:31–39]

1. Aril reddish, open at top, maturing with seed in 2 seasons; lf < 25 mm, spine or bristle tip 0 or < 0.5 mm ... **TAXUS**
1′ Aril green or purplish, closed at top, maturing with seed in 1 season; lf > 25 mm, spine or bristle tip 1–1.5 mm .... **TORREYA**

TAXUS YEW

ST: trunk often of irregular width; branches somewhat drooping; wood not aromatic. **LVS** yellowish green, paler below, not aromatic; tip not sharply spiny. **POLLEN CONE** with a stalked cluster of 4–8 stamen-like structures. **SEED** maturing in 2 seasons; aril reddish, open at top. ± 7 spp.: n hemisphere. (Latin, probably from Greek: bow, for which wood used)

T. brevifolia Nutt. (p. 127) PACIFIC YEW **ST**: trunk < 18 m, < 1.4 m wide; bark shredding, red-brown. **LF** 10–25 mm, ± flexible; spine or bristle tip < 0.5 mm or 0. **SEED** < 10 mm, ovoid; coat smooth. UNCOMMON. Gen dense, mixed-evergreen forest, lower slopes or canyon bottoms; 10–1500 m. NW, CaR, n&c SN, SnFrB; to AK, w MT. ❀IRR,DRN,SHD:1,**2**,3,**7**,14,**15–17**,18–23,**24**;SUN:**4–6**.

TORREYA TORREYA

ST: trunk of regular width; branches horizontal; wood aromatic. **LVS** dark green above, yellowish green, with 2 longitudinal, yellowish grooves below, aromatic; tip sharply spiny. **POLLEN CONE** with 6–8 whorls of 4 stamen-like structures. **SEED** maturing in 1 season; aril green or purplish, closed at top. ± 5 spp.: N.Am, Asia. (John Torrey, New York botanist, 1796–1873)

T. californica Torrey (p. 127) CALIFORNIA NUTMEG **ST**: trunk < 43 m, < 1.5 m wide; bark ± smooth, dark brown. **LF** 25–70 mm, rigid; spine or bristle tip 1–1.5 mm. **SEED** < 50 mm, oblong; coat ± longitudinally grooved. Shady canyons in forest or woodland, sometimes chaparral; 30–2100 m. NCo, NCoR, CaRF, SN, SCoRO. ❀**4–6,17**;IRR,SHD:2,3,7,8,9,**14–16**,18,19,**20,21**,22,24.

TAXODIACEAE BALD CYPRESS FAMILY

James R. Griffin

Tree or shrub, monoecious, evergreen. **ST**: young crown conic, old crown variable; bark fibrous, reddish; twigs not grooved, persistent or deciduous. **LVS** simple, alternate, not in bundles, linear or awl-like (sometimes both on 1 pl), decurrent; tip acute. **SEED CONE** gen woody; scales peltate, fused with bracts, gen persistent. **SEEDS** 2–9 per scale; wings lateral. 10 genera, 16 spp.: N.Am, Asia, Tasmania; many of great commercial value. [Eckenwalder 1976 Madroño 23:237–256] Combination with Cupressaceae supportable.

1. Lvs of 2 kinds, those on sprouting, rapidly growing, or fertile sts awl-like, appressed, not ranked, < 8 mm, others gen linear, spreading, appearing ± 2-ranked, 5–25 mm; seed cone < 35 mm .... **SEQUOIA**
1′ Lvs of 1 kind, awl-like, appressed, ± 4-ranked, < 15 mm; seed cone > 40 mm .... **SEQUOIADENDRON**

SEQUOIA REDWOOD

1 sp.: w N.Am. (Sequoyah, Cherokee chief, ± 1770–1843)

S. sempervirens (D. Don) Endl. (p. 127) Pl gen sprouting vigorously from base if cut, from entire crown if burned. **ST**: trunk < 115 m, < 3.8 m wide; old crown ± cylindric, gen unbranched in lower half; bark < 30 cm thick near base; twigs persistent < 4 years. **LF** remaining green < 3 years, persistent < 4. **SEED CONE** 15–30 mm, ± spheric, maturing in 1 year, persistent < 2. *n*=33. Redwood forest; < 1100 m. NCo, w KR, NCoRO, w NCoRI, CCo, SnFrB, n SCoRO; sw OR. Tallest trees in N.Am. ❀SUN:4,**5**,6,**16,17**&IRR:**7–9,14,15,18-24**:CVS.

SEQUOIADENDRON GIANT SEQUOIA

1 sp.: CA. (Greek: sequoia tree)

S. giganteum (Lindley) Buchholz (p. 127) Pl gen not sprouting. **ST**: trunk < 88 m, < 8 m wide; old crown irregular, with very large branches throughout; bark < 50 cm thick near base; twigs persistent < 20 years. **LF** remaining green < 4 years, persistent < 20. **SEED CONE** 4–9 cm, oblong, maturing in 2 years, persistent < 20. *n*=11. UNCOMMON. Forms overstory in mixed-conifer forest, esp with favorable soil moisture; 825–2700 m. c&s SNH. Most massive trunks in N.Am. ❀DRN,SUN:1,**4–6,17**&IRR:2,3,7,14–16,18–23; CVS.

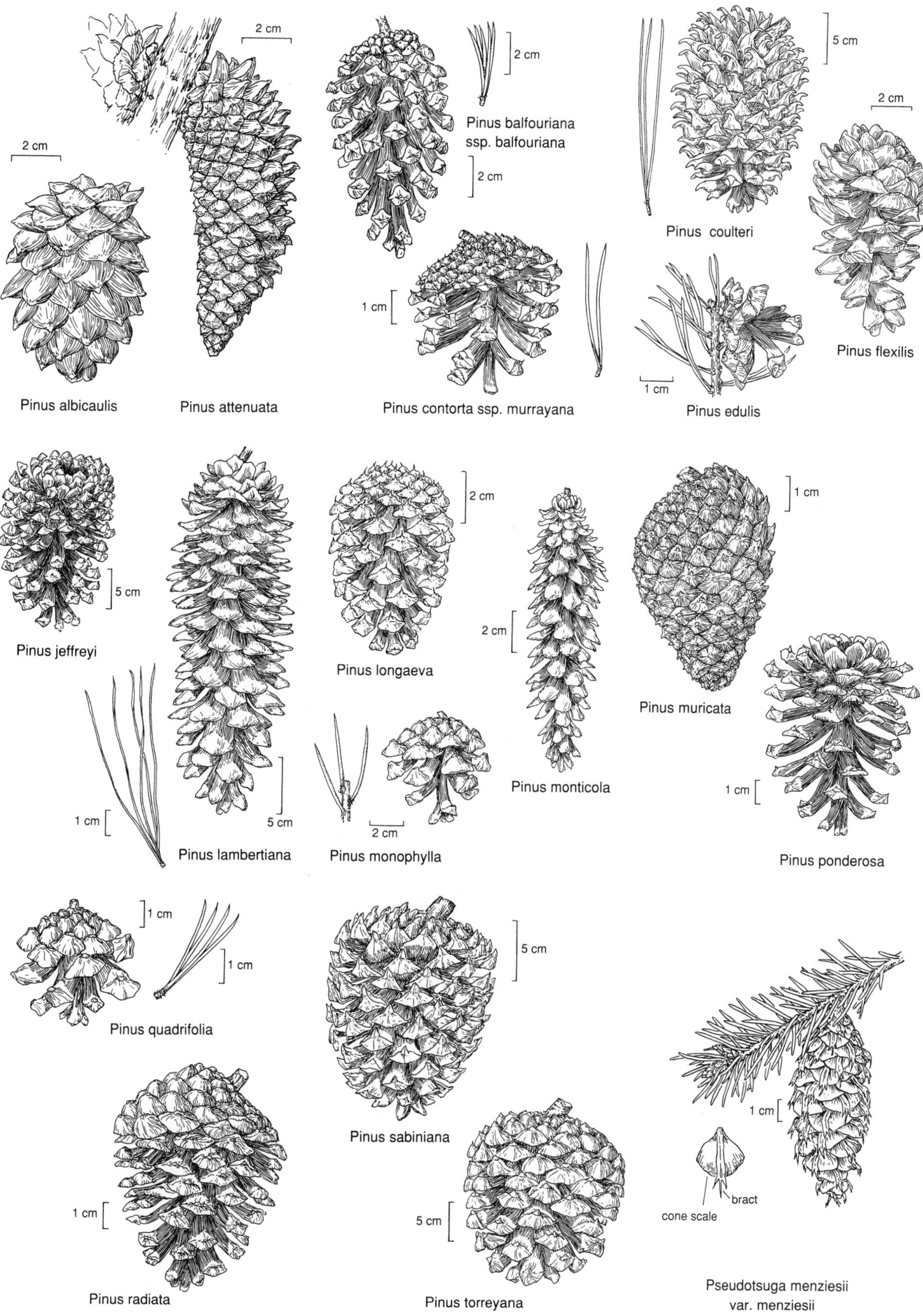
2 cm
2 cm
Pinus albicaulis
Pinus attenuata
2 cm
Pinus balfouriana
ssp. balfouriana
2 cm
1 cm
Pinus contorta ssp. murrayana
5 cm
Pinus coulteri
2 cm
1 cm
Pinus edulis
Pinus flexilis
5 cm
Pinus jeffreyi
1 cm
5 cm
Pinus lambertiana
2 cm
Pinus longaeva
2 cm
Pinus monophylla
2 cm
Pinus monticola
1 cm
Pinus muricata
1 cm
Pinus ponderosa
1 cm
1 cm
Pinus quadrifolia
1 cm
Pinus radiata
5 cm
Pinus sabiniana
5 cm
Pinus torreyana
1 cm
bract
cone scale
Pseudotsuga menziesii
var. menziesii

ACANTHACEAE ACANTHUS FAMILY

Lawrence R. Heckard

Ann, per, shrubs. **LVS** simple, gen opposite; stipules 0. **INFL**: cyme, spike, or raceme, bracted. **FL** bisexual; calyx deeply (3)4–5 lobed; corolla 4–5 lobed, nearly radial to 2-lipped; stamens (2)4, epipetalous, anther sacs sometimes dissimilar in size or placement; ovary superior, chambers 2, ovules 4–many, placentas axile, style long, stigmas 1–2. **FR**: capsule, loculicidal, gen dehiscing explosively. **SEEDS** gen 2 per chamber, ejected by specialized hook-like stalks. 250 genera, 3000 spp.: esp trop; some orn: *Justicia* (Beloperone, shrimp-plant), *Acanthus*, *Thunbergia*. [Daniel 1984 Desert Plants 5:162–179]

1. Corolla resembling pea fl, < 2 cm, tube < lips, white with yellow, maroon-streaked eye on upper lip; anther sacs ± equally inserted on filament, not spurred; fr glabrous .... **CARLOWRIGHTIA**

1′ Corolla long-tubular, > 2 cm, tube > lips, reddish (rarely yellow) throughout; anther sacs unequally inserted on filament, lower sac spurred; fr canescent .... **JUSTICIA**

CARLOWRIGHTIA

Subshrub, shrub. **INFL** gen spike- or panicle-like. **FL**: calyx 5-lobed; corolla appearing 4-lobed, tube < lobes, slender, barely wider at throat, uppermost lobe gen notched (actually 2 lobes ± entirely fused), 2 lateral lobes ascending or reflexed, lowest lobe ± keeled, containing anthers and style; stamens 2, anther sacs ± equally inserted on filament, opening toward uppermost lobe, not spurred. **FR** compressed-ovoid, on flattened stalk. **SEEDS** 4, disk-like, notched at base. 23 spp.: esp sw US, Mex. (Charles Wright 1811–85, Am botanical collector) [Daniel 1988 Brittonia 40: 245–255]

C. arizonica A. Gray (p. 127) ARIZONA CARLOWRIGHTIA **ST** multi-branched, < 1 m, puberulent; hairs erect or recurved. **LVS** variable in size and shape; blade gen < 5 cm, lanceolate to elliptic, sometimes cordate, puberulent; petiole 0–8 mm. **INFL**: spike; fls 1–several per bract. **FL**: calyx 1.5–5 mm, lobes narrow-triangular; corolla 10–18 mm, white with yellow, maroon-streaked eye on upper lip. **FR** 7–11 mm, glabrous. **SEED** 3–4 mm, tubercled; margin finely dentate. $2n=18$. RARE in CA. Rocky slopes; ± 300 m. w DSon (Palm Canyon, Borrego Springs); to TX, C.Am.

JUSTICIA

Ann, per, shrub. **INFL**: gen spike or raceme. **FL**: corolla 2-lipped, tube > lips, wider upward, upper lip notched or 2-lobed, lower lip 3-lobed; stamens 2, gen appressed to upper lip, anther sacs unequally placed on filament, opening toward lower lip, gen at least lower sac spurred. **FR** club-shaped; stalk flattened. **SEEDS** 4, outlined on fr surface. 300 spp.: trop, subtrop. (James Justice, 18th century Scottish horticulturist)

J. californica (Benth.) D. Gibson (p. 127) BELOPERONE, CHUPAROSA Shrub. **ST** < 2 m, gen lfless in fl, canescent. **LF**: petiole < 20 mm; blade 1–6 cm, ovate, triangular, or ± round, puberulent. **INFL**: raceme; bracts < 1 cm, lanceolate-elliptic, falling early; bractlets 2–5 mm, narrow-tapered. **FL**: calyx 5–8 mm, lobes 5, lanceolate; corolla 2–4 cm, puberulent, dull scarlet (yellow), lips 1–2 cm, < tube, lobes of lower lip 1–4 mm. **FR** 1.5–2 cm, canescent. **SEED** 2.5–3.5 mm, ± round, mottled. $2n=28$. Dry sandy or rocky soils, esp washes; < 800 m. w PR (San Diego River e of Wildcat Canyon), DSon; AZ, nw Mex. [*Beloperone c.* Benth.] ❀SUN,DRN:18–22,**23**,24&IRR12,**13**,14;DFCLT < 25° causes damage; CVS.

ACERACEAE MAPLE FAMILY

James R. Shevock

Shrubs, trees, sometimes monoecious, dioecious, or with staminate and bisexual fls. **LVS** opposite, gen simple, gen palmately lobed (rarely pinnate), gen deciduous; stipules 0. **INFL**: panicle, raceme, or umbel-like. **FL** small; perianth gen ± yellowish green; sepals (4)5, free; petals gen 5 (sometimes 0, 4, or 6), free, gen sepal-like; stamens gen 8 (sometimes 5, 10, or 12), gen attached to edge of nectary disk; ovary superior, chambers 2, each 2-ovuled. **FR**: gen pair of achenes, conspicuously winged. **SEED** gen 1 per achene. 2 genera, ± 120 spp.: n temp, trop mtns (*Dipteronia*: 2 spp.: China). Some *Acer* important as timber or orn, often has bright autumn colors. [Ogata 1967 Bull Tokyo U For 63:89–206]

ACER MAPLE

LF simple or pinnately compound. **INFL**: fl clusters drooping, gen appearing before or with emerging lvs. **FR** paired, each with elongate wing ribbed on the proximal side. ± 118 spp.: n hemisphere. (Latin name for maple)

1. Lf pinnately compound, lflets 3–5; petals 0 .... ***A. negundo*** var. ***californicum***

1′ Lf simple, palmately lobed; petals present

2. Lf 10–25 cm wide; infl long, pendent, fls > 30; fr body brownish hairy .... ***A. macrophyllum***

2′ Lf 2–12 cm wide; infl a dense raceme, fls gen < 10; fr body glabrous

3. Lf 5–12 cm wide; fr wings widely spreading (nearly 180°) .... ***A. circinatum***

3′ Lf 2–5 cm wide, fr wings overlapping or spreading < 45° .... ***A. glabrum***

4. Twigs gen whitish; lf gen 1.2–2.8 cm wide .... var. ***diffusum***

4′ Twigs gen reddish; lf gen 3–5 cm wide

5. Fr wings overlapping .... var. ***greenei***
5′ Fr wings spreading ± 45° .... var. ***torreyi***

A. circinatum Pursh (p. 127) VINE MAPLE Shrub or small tree, 1–6 m, often reclining and rooting, vine-like, gen with staminate and bisexual fls. **ST**: twigs slender, glabrous. **LF** 5–12 cm wide, palmately 5–7-lobed to near middle. **INFL** dense, 4–10-fld. **FL**: sepals red-purple; petals greenish white. **FR** reddish when mature; body glabrous; wings widely spreading (nearly 180°). Shaded stream banks; < 1500 m. n&c NW, CaRH, n SN; to AK. ❀IRR: **4–6,17**&SHD:1–3,7,8,9,**14–16**,18–20.

A. glabrum Pursh (p. 127) MOUNTAIN MAPLE Shrub or small tree, 2–6 m, ± erect, dioecious (or staminate pl with some bisexual fls). **LF** 2–4 cm wide, palmately lobed. **INFL** dense, gen < 10-fld. **FL**: petals present. **FR**: body glabrous; wings overlapping or spreading < 45°. Moist to fairly dry, montane, rocky slopes, canyons; 1500–2800 m. CA; to AK, c US, NM. Complex of 6 vars. (var. *glabrum* in US Rocky Mtns, c US).

var. ***diffusum*** (E. Greene) F.J. Smiley **ST**: twigs gen whitish. **LF** gen 1.2–2.8 cm wide; teeth few, blunt. **INFL**: peduncle + pedicel 1–2 cm. **FR**: wings spreading gen 45°. Habitats of sp. SnJt, SNE, DMtns; to UT. ❀DRN,IRR:1–7.

var. ***greenei*** A.C. Keller **ST**: twigs gen reddish. **LF** gen 3–5 cm wide; central lobe largest; teeth few–many, obtuse. **INFL**: peduncle + pedicel 1.7–4 cm. **FR**: wings overlapping. Habitats of sp. s SN (esp Tulare Co). ❀DRN,IRR:1–7.

var. ***torreyi*** (E. Greene) F.J. Smiley **ST**: twigs gen reddish. **LF** gen 3–5 cm wide; central lobe largest; teeth few–many, obtuse. **INFL**: peduncle + pedicel 1.7–4 cm. **FR**: wings spreading ± 45°. Habitats of sp. KR, CaRH, SNH, MP; NV. ❀DRN,IRR:1–7,17 &SHD:15,16;DFCLT.

A. macrophyllum Pursh (p. 127) BIG-LEAF MAPLE Tree 5–30 m. **ST**: twigs coarse, glabrous, greenish to brownish. **LF**: petiole 5–12 cm; blade 10–25 cm wide, lobes 3–5, deep, irregularly few-toothed. **INFL** long, pendent, > 30-fld; staminate and bisexual fls on same raceme. **FL**: petals present. **FR** variable; body brown-hairy; wings spreading < 90°. Common. Streambanks, canyons; < 1500 m; CA-FP (exc GV); to AK. ❀DRN:**4–6**&IRR:1–3,7,8,9, **14–24**.

A. negundo L. var. ***californicum*** (Torrey & A. Gray) Sarg. (p. 133) BOX ELDER Tree 6–20 m, dioecious. **ST**: twigs hairy, greenish. **LF** pinnately compound; petiole 2–8 cm; lflets 3–5, central largest, 5–12 cm, 3–5-lobed, toothed. **INFL** gen appearing before lvs; pistillate pedicels long, drooping. **FL**: petals 0. **FR**: body red, finely hairy, becoming straw-colored; wings widely spreading. Streamsides, bottomlands; < 1800 m. CA-FP. Has been planted widely (esp GV) as orn or street tree. ❀**4–6**;IRR:1–3,**7–9,14–24**; INV;also STBL.

AIZOACEAE FIG-MARIGOLD FAMILY

John Bleck, Wayne R. Ferren Jr., Nancy J. Vivrette

Ann, per, shrub, gen fleshy. **ST** underground or prostrate to erect. **LVS** gen simple, gen cauline, gen opposite; stipule gen 0; blade gen glabrous, often glaucous. **INFL**: cyme or fl solitary. **FL** gen bisexual, radial; hypanthium present; sepals 3–8; petals gen many in several whorls, free or fused at base, linear, sometimes 0; stamens 1–many, free or fused in groups, outer often petal-like; nectary a ring or separate glands; pistil 1, ovary superior to inferior, chambers 1–20, placentas gen parietal, styles 0–20, stigmas 1–20. **FR**: gen capsule, opening by flaps or circumscissile, or berry or nut. **SEEDS** 1–many per chamber, often with aril. 130 genera, 2500 spp.: gen subtrop, esp s Afr; many cult, some waifs in CA (e.g., *Disphyma australe* (Aiton) J. Black: ovary glands convex and minutely crenate, stigmas densely plumose, fr chambers 5, seeds ovate, ± smooth; *Lampranthus* spp.: ovary glands fused, fr chambers 5, seeds pear-shaped, ± black, rough; both genera members of Ruschieae). [Ferren et al. 1981 Madroño 28:80–85] *Glinus*, *Mollugo* are in Molluginaceae.

1. Petals 0
 2. Ovary fully inferior (Tetragonioideae) .... **TETRAGONIA**
 2′ Ovary at least partly superior (Aizooideae)
  3. Styles 3–5; ovary half-superior; lvs of a pair equal .... **SESUVIUM**
  3′ Styles 1–2; ovary fully superior; lvs of a pair unequal
   4. Stamens gen 3; stipules fringed .... **CYPSELEA**
   4′ Stamens 5 or 6; stipules with 2 teeth .... **TRIANTHEMA**
1′ Petals many
 5. Placentas axile (Aptenioideae)
  6. Sepals 4; ovary chambers 4 .... **APTENIA**
  6′ Sepals 5; ovary chambers 5 .... **MESEMBRYANTHEMUM**
 5′ Placentas parietal (Ruschioideae)
  7. Fr fleshy or capsule valves separating; stigmas and ovary chambers 8–20
   8. Basal rosette 0, cauline lvs opposite; fr fleshy, indehiscent (Carpobroteae) .... **CARPOBROTUS**
   8′ Basal rosette present, cauline lvs gen alternate; fr a capsule, valves finally separating (Apatesieae) .... **CONICOSIA**
  7′ Fr a persistent capsule; stigmas and ovary chambers 4–11 (Ruschieae)
   9. Ovary chambers 8–11 .... **MALEPHORA**
   9′ Ovary chambers 4–6
    10. Sepals unequal; lvs often hooked at tip .... **DELOSPERMA**
    10′ Sepals equal; lvs not hooked at tip .... **DROSANTHEMUM**

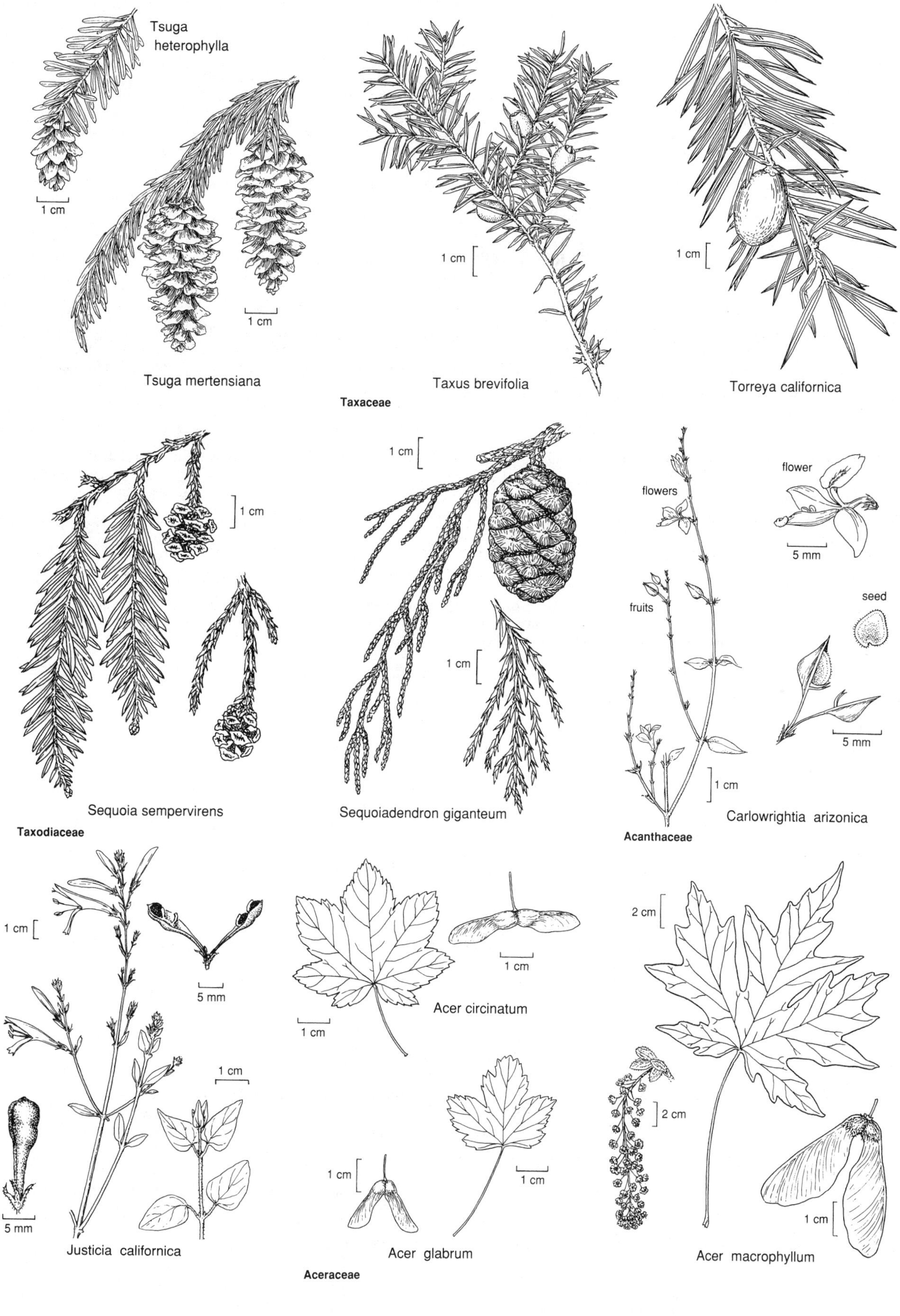

Tsuga heterophylla

1 cm

1 cm

Tsuga mertensiana

1 cm

Taxus brevifolia

Taxaceae

1 cm

Torreya californica

1 cm

Sequoia sempervirens

Taxodiaceae

1 cm

1 cm

Sequoiadendron giganteum

flowers

flower

5 mm

seed

fruits

5 mm

1 cm

Carlowrightia arizonica

Acanthaceae

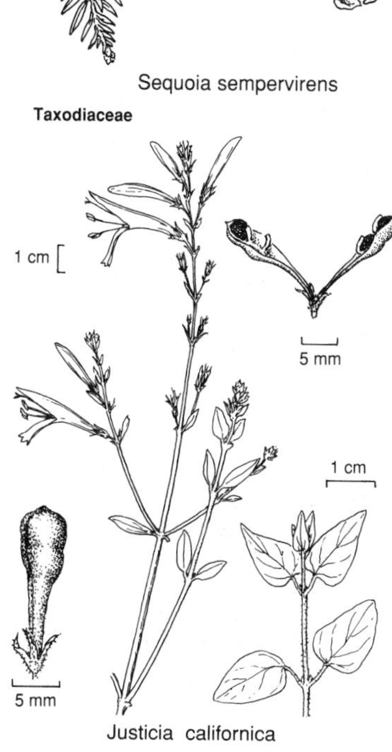

Justicia californica

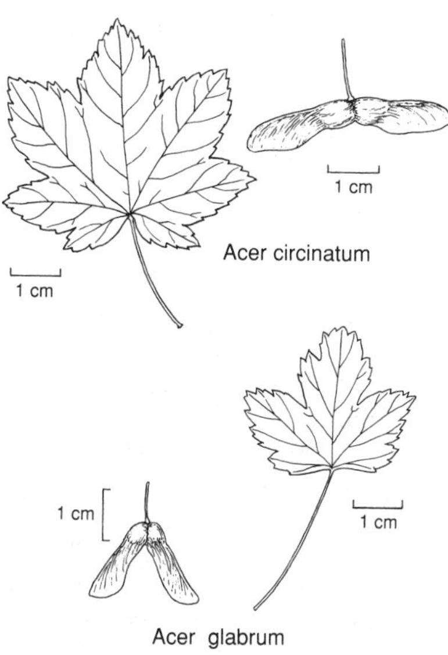

Acer circinatum

Acer glabrum

Aceraceae

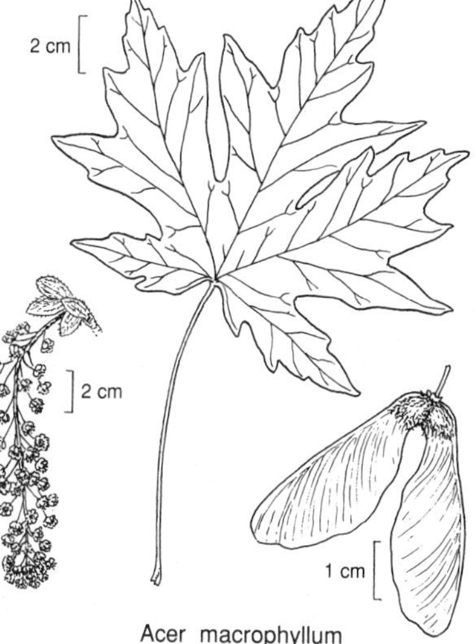

Acer macrophyllum

APTENIA

Nancy J. Vivrette

Per. **ST** prostrate; base woody. **LVS** petioled; blade flat, lanceolate or cordate. **INFL**: fls solitary or whorled, terminal or axillary, sessile or peduncles short; bracts 0. **FL** ± 1 cm diam; hypanthium obconic; sepals 4, unequal, 2 lf-like, 2 tapered; petals fused at base, pink to purple; stamens many, outer sterile and petal-like, inner incurved, ± white; ovary inferior, placentas axile, style 0, stigmas 4. **FR**: capsule; valves 4; valve lids and wings 0. **SEED** flat, tubercled. 2 spp.: s Afr. (Greek: wingless)

A. cordifolia (L.f.) N.E. Br. BABY SUN ROSE **ST** 3–6 dm; nodes widely spaced. **LF** 1–3 cm, cordate, minutely papillate. **INFL**: fl solitary, axillary; peduncle 8–15 mm. **FL**: hypanthium 6–7 mm; calyx lobes ± 5 mm; petals ± 3 mm, purple. **FR** 13–15 mm. *n*=6. Uncommon. Disturbed places, margins of coastal wetlands; < 100 m. CCo, SCo, s ChI; native to s Afr. [*Mesembryanthemum c.* L.f.]

CARPOBROTUS FIG-MARIGOLD

Nancy J. Vivrette

Shrub, smooth, glabrous. **ST** trailing, rooting at nodes, forming mats, < 20 dm; fl-branches ascending. **LF** fleshy, ± triangular in X-section. **INFL**: fl solitary, terminal. **FL** 3–15 cm diam; sepals 5, unequal, smaller 2 or 3 with expanded papery margins; petals free, showy, magenta, pink, or yellow; stamens many, erect; ovary inferior, chambers 8–20, styles 0, stigmas 8–20, sessile, linear, hairy. **FR** berry-like, fleshy, indehiscent. **SEEDS** many. 30 spp.: s Afr, Chile, Australia. (Greek: fr edible)

1. Fl 3–5 cm diam, petals rose-magenta, chambers 8–10; lf 4–7 cm, rounded-triangular in X-section, outer angle smooth; sepal X-section like lf . . . . ***C. chilensis***

1′ Fl 8–10 cm diam, petals pink or yellow aging pink, chambers 10–12; lf 6–10 cm, sharply triangular in X-section, outer angle serrate near tip; sepal X-section like lf . . . . ***C. edulis***

C. chilensis (Molina) N.E. Br. (p. 133) SEA FIG **ST** < 2 m. **LF** 4–7 cm, widest above middle, glaucous. **FL** sessile; sepals 1–2 cm, slightly triangular in X-section, outer angle smooth; petals 1–2.5 cm, rose-magenta. Common. Coastal; < 100 m. NCo, CCo, SCo, n ChI; to OR, Mex; probably native to s Afr. [*Mesembryanthemum c.* Molina; *C. mellei* L. Bolus also applied to material from Chile; *C. aequilaterus* (Haw.) N.E. Br. and *M. a.* Haw. are misapplied to *Carpobrotus*]

C. edulis (L.) N.E. Br. (p. 133) **ST** < 3 m. **LF** 6–10 cm, 10–15 mm diam, widest below middle; outer angle serrate near tip, not glaucous. **FL** peduncled; sepals 3–4 cm, sharply triangular in X-section, outer angle serrate near tip; petals 3–4 cm, pink or yellow aging pink. 2*n*=18. Common. Many coastal habitats, esp on sand; < 100 m. NCo, CCo, SCo, ChI; to Mex; native to s Afr. [*Mesembryanthemum e.* L.] Hybridizes with *C. chilensis*. Extensively planted along highways and for dune stabilization. Invasive.

CONICOSIA

John Bleck

Per, gen short-lived; roots diffuse or tuberous. **ST** underground, vertical; ann shoots prostrate to ascending. **LVS**: basal rosette present; cauline lvs gen alternate, linear, ± triangular in X-section. **INFL**: fl solitary, axillary; bracts 0. **FL** 5–13 cm diam; sepals 5, ± unequal, wider at base, tip cylindric, basal margins of inner 3 papery; petals free, linear, yellow; stamens many, filament base hairy; ovary ± inferior, chambers 10–20, styles 0, stigmas 10–20, thread-like, sessile. **FR**: capsule; conic top opening by 10–20 flaps that do not spread when moistened, finally separating into 10–20 valves. **SEED** spheric, smooth. 10 spp.: s Afr. (Greek: cone-shaped)

C. pugioniformis (L.) N.E. Br. (p. 133) **ST** a single underground caudex, < 30 cm, 1–2 cm diam. **LVS**: cauline scattered, crowded toward st tip, 15–20 cm, 10–12 mm wide, gray-green, glabrous, upper surface grooved or flat. **INFL**: peduncle < 12 cm. **FL** 5–8 cm diam, shiny yellow, lasting several days; odor unpleasant. **SEED** 1.5–1.8 mm. 2*n*=18. Uncommon. Sandy places, esp coastal dunes; < 100 m. CCo; native to s Afr. [*Mesembryanthemum p.* L.; not *M. elongatum* Haw.]

CYPSELEA

Wayne R. Ferren Jr.

Ann, mat-forming, glabrous. **LVS** opposite; stipule sheathing st, scarious, fringed, attached to lower petiole margins. **INFL**: fl solitary in axil; bracts fringed. **FL**: calyx bell-shaped, lobes 4–5, unequal; petals 0; stamens gen 3, alternate calyx lobes; ovary superior, ovoid to round, chamber 1, placenta free-central, styles and stigmas gen 2. **FR**: capsule, circumscissile. **SEEDS** many; aril slender, persistent on placenta. 2 spp.: Caribbean, S.Am. (Greek: beehive)

C. humifusa Turpin (p. 133) **ST** prostrate, slender, diffusely branched from base, forming leafy mats < 10 cm. **LVS** of a pair unequal; petiole ± = blade; larger blades 5–10 mm, elliptic, obtuse; axil of smaller lf with short st, fl, or both. **FL** ± 2 mm diam; calyx lobes ± 1.5 mm, erect, ovate, margin scarious. **FR** ± round, thin-walled. **SEED** ± 0.3 mm, round-reniform, smooth, brown. Uncommon. Seasonally dry margins of wetlands; < 1000 m. NCoR, GV, SnFrB, PR; to NV, se US; native to Caribbean.

DELOSPERMA

John Bleck

Per. **ST** prostrate to erect. **LVS** opposite, sessile, < 7 cm, triangular to round in X-section. **INFL**: cyme or fl solitary; bracts present. **FL** 1.5–8 cm diam; sepals 4–8, gen unequal; petals free; outer stamens sterile, petal-like, inner stamens erect, whitish; nectary glands separate; ovary half-inferior, chambers 5, placentas parietal, stigmas 5. **FR**: capsule, persistent; lids gen 0. **SEED** 0.5–1.5 mm, ± spheric, pale brown, gen with aril. 140 spp.: Afr. (Greek: visible seed)

D. litorale (Kensit) L. Bolus **ST** prostrate, thin, mat-forming, rooting at nodes; nodes widely spaced. **LVS** crowded on short shoots, 2–3 cm, 5–6 mm wide, linear, triangular in X-section; angles white-margined, tip acute, gen reflexed. **INFL**: cyme; fls few; pedicels 2 cm. **FL** 1.5–2.2 cm diam; sepals 5, ± unequal; petals white; inner stamens free, forming cone around stigmas. $2n=36$. Uncommon. Margins of coastal wetlands, bluffs, stabilized dunes; < 35 m. SCo, s ChI; native to s Afr. [*Mesembryanthemum l.* Kensit]

DROSANTHEMUM

John Bleck

Per, shrub, papillate. **ST** prostrate to erect, gen rough. **LVS** 4-ranked, 1–25 cm, 2.5 mm wide, triangular to round in X-section. **INFL**: cyme (fls few) or fl solitary. **FL** < 3.5 cm diam; sepals gen 5; petals in 1 or 2 whorls, free; outer stamens sterile, inner stamens erect; nectaries separate; ovary ± inferior, top flat or convex, chambers 4–6, placentas parietal, stigmas 4–6. **FR**: capsule, persistent; flaps present. **SEED** light brown, rough. 95 spp.: s Afr. (Greek: dew fl).

D. floribundum (Haw.) Schwantes **ST** prostrate, mat-forming, thin; older nodes rooting. **LF** 12–14 mm, 2.5 mm wide, cylindric, slightly curved, wider toward tip, light green. **INFL**: fl solitary; peduncles 2–3 cm. **FL** ± 18 mm diam; sepals ± 3–4 mm, equal, linear, obtuse; petals 8–9 mm. **SEED** 0.5 mm. Uncommon. Coastal habitats; < 35 m. NCo, CCo, SCo, ChI; n Mex; native to s Afr. [*Mesembryanthemum f.* Haw.] *D. hispidum* (L.) Schwantes is an uncommon waif.

MALEPHORA

John Bleck

Per, shrub. **ST** prostrate to erect. **LVS** opposite, < 6 cm, slightly fused at base, triangular to ± cylindric in X-section, smooth, often glaucous. **INFL**: fl solitary. **FL** < 5 cm diam; sepals 4–6; petals free; stamens many in several whorls; nectary a ring; ovary ± inferior, top flat, chambers 8–11, placentas parietal, styles 0, stigmas 8–11, wide, feathery. **FR**: capsule, persistent; valve winged and with lids. **SEED** flat, rough, with tubercles in rows. 15 spp.: s Afr. (Greek: bearing arm-holes, from seed pockets of fr)

M. crocea (Jacq.) Schwantes **ST** prostrate, stout, pale, corky; sometimes rooting at nodes. **LVS** crowded on short shoots, 2.5–6 cm, 6 mm wide, ± triangular in X-section, pale glaucous-green, sometimes reddish. **INFL**: peduncle 1–6 cm. **FL**: calyx 0.8–1.5 cm diam, lobes 4–6, unequal, at least 2 short, acuminate, with translucent margins; petal upper surface gen orange, lower purple; outer stamens sterile, petal-like, inner stamens erect, often hairy at base; ovary chambers gen 8, stigmas gen 8, feathery. **FR**: placental tubercle at outer margin of chamber, 2-lobed with adjacent seed pockets. **SEEDS** many, 1 mm, 0.8 mm wide, lenticular. $2n=27,36$. Common. Margins of wetlands, bluffs along coast; < 50 m. CCo, SCo, s ChI; native to s Afr. [*Mesembryanthemum c.* Jacq.] ± invasive.

MESEMBRYANTHEMUM ICEPLANT

Nancy J. Vivrette

Ann, bien, glabrous, conspicuously papillate. **ST** prostrate to ascending. **LVS** alternate or opposite, cylindric or flat, reddish with age or stress. **INFL**: cyme or fl solitary. **FL**: sepals 4–5, 2 often lf-like; petals free, linear, white; stamens many; ovary inferior, chambers 5–20, styles 5–20. **FR**: capsule, dehiscing when moist. **SEEDS** many, round, compressed, often "D" shaped, with minute tubercles, light or dark reddish brown. 74 spp.: sw Afr, Medit, w Asia. [McVaugh 1974 Taxon 23:820–821] (Greek: afternoon-blooming)

1. Lf flat; fl 10 mm diam .... ***M. crystallinum***
1' Lf cylindric; fl 5 mm diam .... ***M. nodiflorum***

M. crystallinum L. (p. 133) CRYSTALLINE ICEPLANT **ST** trailing, forked, < 1 m. **LF** 2–20 cm, petioled; blade flat, ovate to spoon-shaped, margin wavy, lower ± cordate. **INFL**: cyme or fl axillary, ± sessile. **FL** 7–10 mm diam; hypanthium round, aging red; sepals 5, equal; petals white, aging pink. **FR** coarsely papillate; valves 5. Common. Coastal bluffs, cliffs, disturbed ground; < 100 m. NCo, CCo, SCo, ChI; to AZ, Mex, Medit, S.Am; native to s Afr.

M. nodiflorum L. (p. 133) SLENDER-LEAVED ICEPLANT **ST** prostrate to ascending, branched from base, 15–20 cm. **LF** sessile, 1–2 cm, linear, ± cylindric. **INFL**: cyme or fl axillary; pedicel short. **FL** 4–5 mm diam; hypanthium obconic; sepals 5, equal; petals white, aging yellow. **FR**: valves 5. $2n=36$. Uncommon. Coastal bluffs, margins of saline wetlands; < 100 m. SnFrB, SCo, ChI; to AZ, Mex, Australia; native to s Afr.

SESUVIUM SEA-PURSLANE

Wayne R. Ferren Jr.

Ann, per, shrub, glabrous, gen papillate. **ST** prostrate to erect, forming mats < 2 m diam; nodes sometimes rooting. **LVS** gen < 6 cm; stipule 0; petiole base gen wide with scarious margins; blade linear to ovate, entire. **INFL**: cyme, cluster or fl solitary; bracts 0 or 2. **FL**: hypanthium obconic; calyx lobes 5, gen hooded near tip, reddish within; petals 0; stamens 1–many, often fused at base; ovary half-superior, chambers 2–5, placentas axile, styles 2–5, papillate. **FR**: capsule, circumscissile, ovoid to conic, thin-walled. **SEEDS** many, ± reniform, gen smooth, shiny, black or brown; aril present. 8 spp.: gen trop, subtrop coasts, deserts.

S. verrucosum Raf. (p. 133) WESTERN SEA-PURSLANE Per, branched from base, minutely papillate. **STS** many, < 9 dm; nodes not rooting. **LF** 0.5–4 cm, linear to widely spoon-shaped; base clasping st. **INFL**: fl solitary, axillary, sessile or peduncle short. **FL**: calyx lobes 4–10 mm, margins scarious, hooded or beaked, outer surface papillate; stamens many, filaments fused to midlength, reddish. **FR** 4–5 mm. **SEED** 0.8–1 mm, smooth. Uncommon. Moist or seasonally dry flats, margins of gen saline wetlands; < 1400 m. GV, SCoRO, SCo, WTR, PR, SNE, D; to OR, KS, S.Am.

TETRAGONIA

Nancy J. Vivrette

Ann, per. **ST** prostrate. **LVS** alternate. **INFL**: small cluster or fl solitary, axillary, ± sessile. **FL**: sepals 3–5; petals 0; stamens 1–many; ovary inferior, chambers 3–9, ovule 1, styles 3–9. **FR** nut-like, 8–10 mm diam, angled, with 2–5 horns. **SEED** ± reniform. 50 spp.: s hemisphere. (Greek: four-angled)

T. tetragonioides (Pallas) Kuntze (p. 133) NEW ZEALAND SPINACH Ann, with small crystalline papillae. **STS** many, spreading, 3+ dm. **LF** 2–5 cm; petiole winged; blade triangular-ovate, base truncate, margin entire, ± wavy. **INFL**: cluster; fls 1–3. **FL**: 5 mm; sepals 4–5, spreading. $2n=32$. Common. Sand dunes, bluffs, margins of coastal wetlands; < 100 m. NCo, CCo, SCo; native to s hemisphere. [*T. expansa* Murray]

TRIANTHEMA HORSE-PURSLANE

Wayne R. Ferren Jr.

Ann, per, shrub, branched from base, glabrous, hairy, or papillate, ± fleshy. **ST** gen prostrate. **LVS** ± opposite, unequal; stipules papery; blade linear to round, base tapered, margin entire. **INFL**: fl gen solitary; bracts 2. **FL**: calyx lobes 5, tip pointed; petals 0; stamens 5–10; ovary superior, chambers 1–2, ovules 1–few, placentas basal, style 1–2. **FR**: capsule, papery or leathery, circumscissile; lid winged. **SEED** reniform, rough; aril present. 20 spp.: trop, subtrop, esp Australia. (Greek: 3-fld)

T. portulacastrum L. (p. 133 with *Sesuvium*) Ann. **ST** < 10 dm; young branches with lines of hairs below petioles. **LF**: stipules reduced to 2 teeth on petiole; petiole gen = blade; blade < 4 cm, smaller on twigs, elliptic to ± round, base tapered, tip often notched. **INFL**: fl sessile, ± covered by stipule. **FL**: calyx 3–5 mm, lobes 2.5 mm, lanceolate, purplish within; ovary chamber 1, stigmas 2. **FR** 4–5 mm, cylindric, ± curved; wings of lid 2, prominent, erect. **SEED** 1.5–2 mm, ridged, reddish brown to black. Uncommon. Moist or seasonally dry wetlands, waste places; < 1000 m. SnJV, D; to TX, e N.Am, S.Am., trop Old World.

AMARANTHACEAE AMARANTH FAMILY

James Henrickson

Ann to tree. **ST** prostrate to erect. **LVS** alternate or opposite, simple, gen entire, bract-like upward; stipules 0. **INFL**: cymes, often arrayed ± in spikes or panicles; bracts 1–3 per fl, gen membranous-scarious, tip often short-pointed or spine-like. **FL** gen small, uni- or bisexual, radial; sepals 0–5, fused at base or free, often ± scarious; petals 0; stamens 0–5, opposite sepals (staminodes sometimes alternating), filaments sometimes fused at base; ovary superior, 1-chambered, ovules 1–several, erect or pendent on ± basal stalks, styles 0–3, stigma head-like or 2–3-lobed. **FR**: utricle to circumscissile capsule. **SEED** lenticular to spheric, hard. ± 65 genera, ± 900 spp.: trop, subtrop. [Robertson 1981 J Arnold Arbor 62:267–313]

1. Cauline lvs alternate; fl unisexual; bract midrib thick, margin scarious, tip short-pointed or spine-like .... **AMARANTHUS**
1′ Cauline lvs opposite; fl bisexual; bract scarious throughout, tip gen not short-pointed or spine-like
  2. Herbage hairs short, many-branched; sepals 1–1.5 mm .... **TIDESTROMIA**
  2′ Herbage hairs 0 or simple; sepals 1.5–7 mm
    3. Calyx 3.5–7 mm, sepals free, glabrous or with straight hairs .... **ALTERNANTHERA**
    3′ Calyx 1.5–2.8 mm, sepals fused below, tube densely white-woolly outside .... **GUILLEMINEA**

ALTERNANTHERA

Per in CA. **LVS** opposite. **INFL**: spike or panicle-like; bracts gen persistent. **FL** bisexual; sepals 5, free, inner sometimes < outer, strongly or weakly 3-veined, fleshy to hardened at base, hairs 0 or ± straight; stamens (2–)5, filaments fused at base, gen alternate elongate, entire to branched staminodia; ovule pendent, style 1, short, stigma 0 or 2-lobed. **FR** indehiscent; wall membranous. **SEED** smooth. ± 170 spp.: gen trop Am. (Latin: alternate, from staminodia)

1. Pl terrestrial, ± prostrate; infl a sessile spike; sepals coarsely veined, spine-tipped; lf blade gen 1–3 cm ..... ***A. caracasana***

1′ Pl aquatic, emergent; infl head-like, peduncle 4–9 cm; sepals faintly veined, not spine-tipped; lf blade 4–11 cm ..... ***A. philoxeroides***

A. caracasana Kunth **ST** wavy-hairy, becoming glabrous. **LVS** at a node often unequal; petioles 3–13 mm, narrowly winged, clasping; blades 5–18 mm wide, ± diamond-shaped to round, tip obtuse to acute. **INFL**: spike, 6–7 mm wide, sessile; bracts 3–3.5 mm, entire, whitish, papery, midvein hairy. **FL**: sepals unequal, outer 3.5–4.5 mm, ± widely lanceolate, not keeled, strongly 3-veined below, spine-tipped, inner 2.5 mm, ± linear, keeled, gen 1-veined, whitish, stiff, margins scarious; filaments 0.2–0.4 mm, slightly > staminodia. Gen waste places; < 150 m. SCo; to se US, c Mex, also Afr; native to C.Am, S.Am. [*A. pungens* Kunth; *A. repens* (L.) Kuntze misapplied]

A. philoxeroides (C. Martius) Griseb. (p. 133) ALLIGATOR WEED **ST** glabrous or hairs in 2 lines. **LVS** at a node ± equal; petioles 2–10 mm, narrowly winged, clasping; blades 9–27 mm wide, narrowly obovate, tip acute. **INFL** head-like, 12–18 mm wide, sessile; bracts 1.7–3 mm, keeled, pearly white, glabrous. **FL**: sepals 5–7 mm, ± equal, 2.8–3 mm wide, ± oblong, pearly white; filaments 2–2.5 mm, < staminodia. Uncommon. Shallow water, ditches, ponds; < 200 m. SnJV, SCo; to se US, C.Am; native to S.Am. NOXIOUS WEED.

AMARANTHUS PIGWEED, AMARANTH

Ann, monoecious or dioecious. **LVS** alternate; blade linear to ovate. **INFL**: cymes in dense, spike-like clusters; bract 1, tip gen short-pointed or -spined, gen scarious; bractlets 0–2. **STAMINATE FL**: sepals 3–5, ± equal, ± as bracts; stamens (1)3–5, filaments free; staminodia 0. **PISTILLATE FL**: sepals (3–)5, ± equal, scarious exc midvein, fused at base, falling with fr; ovary compressed-ovoid, styles (2)3, stigmas slender, papillate, ovule 1, erect. **FR** circumscissile or indehiscent, smooth or inflated-wrinkled; walls membranous to spongy-hardened. **SEED** 1, lenticular, smooth, reddish to black. ± 60 spp.: worldwide; some potherbs, some cult for seed. (Greek: unfading, from persistent bracts and sepals) Some spp. (esp *A. cruentus*, *A. powellii*, *A. retroflexus*) hybridize complexly. [Tucker & Sauer Madroño 1958 14:252–261]

1. Dioecious
 2. Pl in hand pistillate
  3. Outermost sepal > others, tip acute, spine-like; bracts >> inner sepals ..... [2]***A. palmeri***
  3′ Outermost sepal slightly > others, obtuse to notched, short-pointed; bracts slightly > inner sepals
   4. Bract midrib keeled; styles gen 3; infl axis and bracts glabrous ..... [2]***A. arenicola***
   4′ Bract midrib moderately thickened, not keeled; styles gen 2; infl axis and bracts sparsely gland-stalked ..... [2]***A. watsonii***
 2′ Pl in hand staminate
  5. Outer sepal spine-tipped, > others; bracts = or > sepals ..... [2]***A. palmeri***
  5′ Outer sepal short-pointed, slightly > others; bracts < or = sepals
   6. Bracts < sepals; infl axis and bracts glabrous ..... [2]***A. arenicola***
   6′ Bracts gen = or > sepals; infl axis and bracts gen sparsely gland-stalked ..... [2]***A. watsonii***

1′ Monoecious (staminate fls sometimes only near infl tip)
 7. Most nodes with 2 rigid spines ..... ***A. spinosus***
 7′ Nodes without spines
  8. Fr indehiscent, gen inflated, thick-walled; pl prostrate ..... ***A. deflexus***
  8′ Fr circumscissile, thin-walled; pl prostrate to erect
   9. Infl clusters axillary, throughout pl, not clearly terminal
    10. Sts ascending to erect, pl spheric; pistillate sepals 3, 0.7–2.2 mm ..... ***A. albus***
    10′ Sts prostrate; pistillate sepals (1)3–5, 1–3.5 mm
     11. Seed 1.2–1.7 mm diam; pistillate sepals 4–5, outer green along midrib, 1.5–3.5 mm ..... ***A. blitoides***
     11′ Seed 0.8–1.1 mm diam; pistillate sepals 1–3, white and scabrous throughout ..... ***A. californicus***
   9′ Infl clearly terminal (sometimes also axillary)
    12. Axillary clusters scattered gen to pl base
     13. Pistillate bracts < sepals; pistillate sepals widely fan-shaped, < or = 2.5 mm wide, margin with small, finger-like projections ..... ***A. fimbriatus***
     13′ Pistillate bracts gen > sepals; pistillate sepals ± spoon-shaped, 0.7–1.5 mm wide, irregularly toothed ..... ***A. torreyi***
    12′ Axillary clusters 0 in lower 1/2 of pl
     14. Inner pistillate sepals spoon-like to obovate, tip obtuse to notched, gen irregularly toothed
      15. Pistillate sepals = fr, ± obovate, 1.8–2.2 mm; pistillate bracts ± = sepals, midrib projecting; infl a long, thick, recurved, terminal spike, sometimes subtended by shorter, lateral spikes ..... ***A. caudatus***

15′ Pistillate sepals > fr, spoon-like, 2.5–3 mm; pistillate bracts 1.5 × sepals, midrib thick; infl many stiff, crowded lateral spikes, terminal not very long .... ***A. retroflexus***

14′ Inner pistillate sepals narrow, sometimes spoon-like, tips acute to short-pointed

16. Pistillate bracts < sepals and fr; infl many lateral, slender, ascending, cylindric spikes .... ***A. cruentus***

16′ Pistillate bracts > sepals and fr; infl various

17. Pistillate bracts gen 2–3.5 mm; staminate sepals and stamens 3–5; infl many short, crowded, lateral spikes .... ***A. hybridus***

17′ Pistillate bracts 4.5–8 mm; staminate sepals and stamens 3–4; infl gen 1 long, terminal spike and few lateral spikes .... ***A. powellii***

A. albus L. (p. 133) TUMBLEWEED Pl ascending to erect, 1–7 dm, bushy, monoecious. **LVS**: cauline petiole 5–50 mm, blade 8–30(70) mm, elliptic to obovate, gen early deciduous; axillary lvs gen smaller, persistent. **INFL**: clusters axillary, throughout pl; bracts 2–3.5 mm, green, spined, margin scarious. **STAMINATE FL**: sepals 3, ± 1 mm; stamens 3. **PISTILLATE FL**: sepals 3, 0.7–2.2 mm, subequal, oblong-lanceolate, tip acute to obtuse, margin scarious. **FR** 1.5–2 mm, circumscissile; surface rough throughout. **SEED** ± 1 mm wide, red-brown. $2n$=32. Common. Weed of waste places, roadsides, fields; < 2200 m. CA; widespread N.Am, to Eurasia; native to trop Am.

A. arenicola I.M. Johnston Pl erect, 4–10(20) dm, bushy, dioecious. **LF**: petiole 5–65 mm; blade 20–75 mm, lanceolate to ± ovate, margins strongly and irregularly curled. **INFL**: terminal spike 20–60 cm, 11–22 mm wide, >> axillary spikes; bracts gen < sepals, scarious, midrib thick, spine-tipped, staminate bracts 1.2–2 mm, pistillate bracts 1.5–2.5 mm. **STAMINATE FL**: sepals 5, 2–3.5 mm; stamens 5. **PISTILLATE FL**: sepals 5, 1.8–2.3 mm, spoon-shaped, tip slightly notched to short-pointed. **FR** ± 2 mm, circumscissile, smooth. **SEED** 0.9–1.2 mm wide, reddish brown. Very uncommon as waif. Sandy soils; < 200 m. ScV, CCo, SCo; native to c&e US.

A. blitoides S. Watson Pl prostrate, 3–7 dm, monoecious. **LVS**: cauline petiole 4–30 mm, blade 5–40 mm, elliptic to widely (ob)ovate, early-deciduous; axillary lvs smaller, persistent. **INFL**: clusters axillary, throughout pl; bracts 1–3.5 mm, < or = sepals, midrib thick, spine-tipped, green, margins scarious esp near base. **STAMINATE FL**: sepals 4–5, 1.5–2.5 mm; stamens (3)4–5. **PISTILLATE FL**: sepals 4–5, 1–3.5 mm, oblong to ovate, outer 2 > others, more reflexed, tips more tapered and spiny. **FR** ± 2.5 mm, circumscissile; lid often rough below. **SEED** 1.3–1.7 mm wide, black; margin acute. Waste places; < 2200 m. CA-FP, D; to WA, e US, c&s Eur. [*A. graecizans* L. misapplied]

A. californicus (Moq.) S. Watson (p. 137) Pl prostrate, monoecious. **ST** 1–5 dm. **LVS**: cauline petiole 1–10(20) mm, blade 3–12 (30) mm, spoon-shaped or obovate, gen early deciduous; axillary lvs much smaller, clustered, persistent. **INFL**: clusters axillary, scattered; bracts 1–2 mm, scarious at margins or ± throughout, often inconspicuous, tip acute, short-spined. **STAMINATE FL**: sepals (2)3, 1–1.5 mm; stamens 3. **PISTILLATE FL**: sepals (1–)3, 1–1.2 mm, subequal, narrowly lanceolate. **FR** 1 mm, circumscissile; lid smooth or wrinkled at base. **SEED** 0.8–1.1 mm wide, red-brown. Seasonally moist flats, lake margins; < 2800 m. CA; to s Can, w TX. ❀TRY:STBL only.

A. caudatus L. QUILETE Pl erect, 3–20+ dm, often reddish, monoecious. **LF**: petiole 25–180 mm; blade 60–200 mm, lanceolate to ovate. **INFL**: panicle of coarse, spike-like clusters, terminal, < 50 cm, ± 15 cm wide; terminal cluster often drooping; spikes at base many, 5–20 cm, ascending; bracts ± = sepals, scarious below middle. **STAMINATE FL**: sepals 5, 2.3–2.7 mm; stamens 5. **PISTILLATE FL**: sepals 5, 1.8–2.2 mm, oblong-obovate to spoon-shaped, margins gen overlapping, acute to obtuse. **FR** ± 1.5 mm, circumscissile; lid inflated at base or not. **SEED** 1.1–1.2 mm wide, dark reddish, black, or whitish. Disturbed places, fields; < 200 m. GV, SnFrB (expected elsewhere); to e US, Asia, Afr; native to w S.Am. Cult for food.

A. cruentus L. Pl erect, 3–20+ dm, often reddish, monoecious. **LF**: petiole 25–180 mm; blade 60–200 mm, lanceolate to ovate. **INFL**: panicle of spike-like clusters, dense, terminal, 30–50 cm, ± 15 cm wide, diamond-shaped; terminal cluster gen erect; bracts ± 2 mm, scarious below middle. **STAMINATE FL**: sepals 4–5, 2–2.4 mm; stamens 5. **PISTILLATE FL**: sepals gen 5, 1.2–1.8 mm, subequal, oblong to elliptic, margins not overlapping in fr, acute to rounded. **FR** 1.5–2.5 mm, circumscissile; style base slender, stigmas 3, erect. **SEED** ± 1 mm wide, dark reddish. Disturbed places, roadsides, fields; < 200 m. GV, SnFrB; to e US, Asia, e Afr; native to S.Am. Widely cult for food.

A. deflexus L. Pl prostrate, 1–4 dm, monoecious, ± soft-crinkled-hairy. **LF**: petiole 6–35 mm, slender; blades 5–50 mm, diamond-shaped to ovate, margins tightly and irregularly curled. **INFL**: clusters terminal, 2–8 cm, 1.5–2.5 cm wide; bracts 1–1.4(2) mm, scarious, midrib thick. **STAMINATE FL**: sepals 2–3, 1.3–1.4 mm; stamens 2–3. **PISTILLATE FL**: sepals 2, 1.5–2.3 mm, < fr, ± oblong, tip short-pointed to -spined. **FR** 2.5–3 mm, indehiscent, ovoid, inflated, smooth or slightly roughened, obscurely 3-veined. **SEED** 1–1.1 mm, smooth, red-black. $2n$=34. Uncommon waif. Disturbed areas; < 650 m. CW, SW; native to Eur. Pls with wrinkled seeds have been called *A. gracilis* Desf.

A. fimbriatus (Torrey) Benth. (p. 137) Pl erect, 4–10 dm, often reddish, monoecious. **LF**: petiole 5–20 mm; blades 20–100 mm, linear to narrowly lanceolate. **INFL**: lower clusters axillary, scattered; terminal clusters spike-like, 10–30 cm, 1–2 cm wide, crowded, leafy; bracts 1–1.8 mm, scarious (exc midrib). **STAMINATE FL**: sepals 5, 1–1.5 mm; stamens 3. **PISTILLATE FL**: sepals 5, 1.5–3.3 mm, < or = 2.5 mm wide, subequal, fan-shaped, tip long-toothed, fringed, reflexed. **FR** ± 1.5 mm, circumscissile; lid strongly inflated-wrinkled near base. **SEED** 0.9–1 mm wide. n=17. Common. Sandy, gravelly slopes, washes, after summer rains; 600–1700 m. D; to s Can, MT, w TX, Baja CA.

A. hybridus L. Pl erect, 3–25 dm, monoecious. **LF**: petiole 10–70 mm; blades 20–150 mm, lower diamond-shaped to ovate, upper ± widely lanceolate. **INFL**: clusters spike-like, 7–12 mm wide, cylindric, crowded; axillary clusters < or = 6 cm; terminal clusters < or = 30 cm; bracts 2–3.5 mm, green, margin scarious below middle, spine-tipped. **STAMINATE FL**: sepals gen 5, 2–3 mm, longest with reflexed spine; stamens 5. **PISTILLATE FL**: sepals 5, 1.5–2.5 mm (outer 1/3 > inner), narrow, spine-tipped, inner short-pointed. **FR** ± 1.5 mm, circumscissile; lid strongly wrinkled at base; style slender, stigmas 3, erect. **SEED** 1.1–1.4 mm wide, reddish brown-black. $2n$=32. Uncommon. Disturbed places; < 300 m. GV, CW, SW; to e US, TX, Mex, S.Am; native to Eurasia, n Afr.

A. palmeri S. Watson (p. 137) Pl erect, 2–20 dm, dioecious. **LF**: petiole 10–90 mm; blade 15–170 mm, lanceolate to ovate. **INFL** spike-like, terminal, solitary or with axillary ones at base, 10–50 cm, 7–15(25) mm wide; bracts ± lanceolate, scarious below, spine-tipped, staminate bracts 2.5–5 mm, pistillate bracts 2.5–6 mm, recurved in fr. **STAMINATE FL**: sepals 5, 2.5–6 mm, outer ± > inner; stamens (3)5. **PISTILLATE FL**: sepals 5, 1.7–3.8 mm, narrowly spoon-shaped, reflexed in fr, tips of inner gen rounded to slightly notched, tips of outer acute to spined. **FR** ± 1.5 mm, circumscissile; lid smooth or rough. **SEED** 0.8–1.4 mm wide, red-brown. Abundant. Roadsides ditches, fields, arroyos; < 1200 m. SnJV, CW, D; to c US, c Mex.

A. powellii S. Watson (p. 137) Pl erect, 3–20 dm, monoecious. **LF**: petiole 10–50 mm; blade 15–90 mm, lower diamond-shaped to ovate, upper ± lanceolate. **INFL**: panicle of spike-like clusters 1–1.8 cm wide; axillary clusters 1–6 cm; terminal clusters < or = 25 cm; bracts 3–6(8) mm, spine-tipped, midrib green, thick, margin scarious in lower 2/3. **STAMINATE FL**: sepals 4–5, 2–3 mm; stamens 3–5. **PISTILLATE FL**: sepals 5, 1.7–3.6 mm, unequal, ± oblong, tip short-pointed. **FR** 2.5–3.6 mm, circumscissile, smooth throughout. **SEED** 0.8–1.4 mm wide, reddish black. Uncommon. Waste places; < 800 m. SNF, GV, CW, SW; to WA, n-c US, TX, Mex, w S.Am.

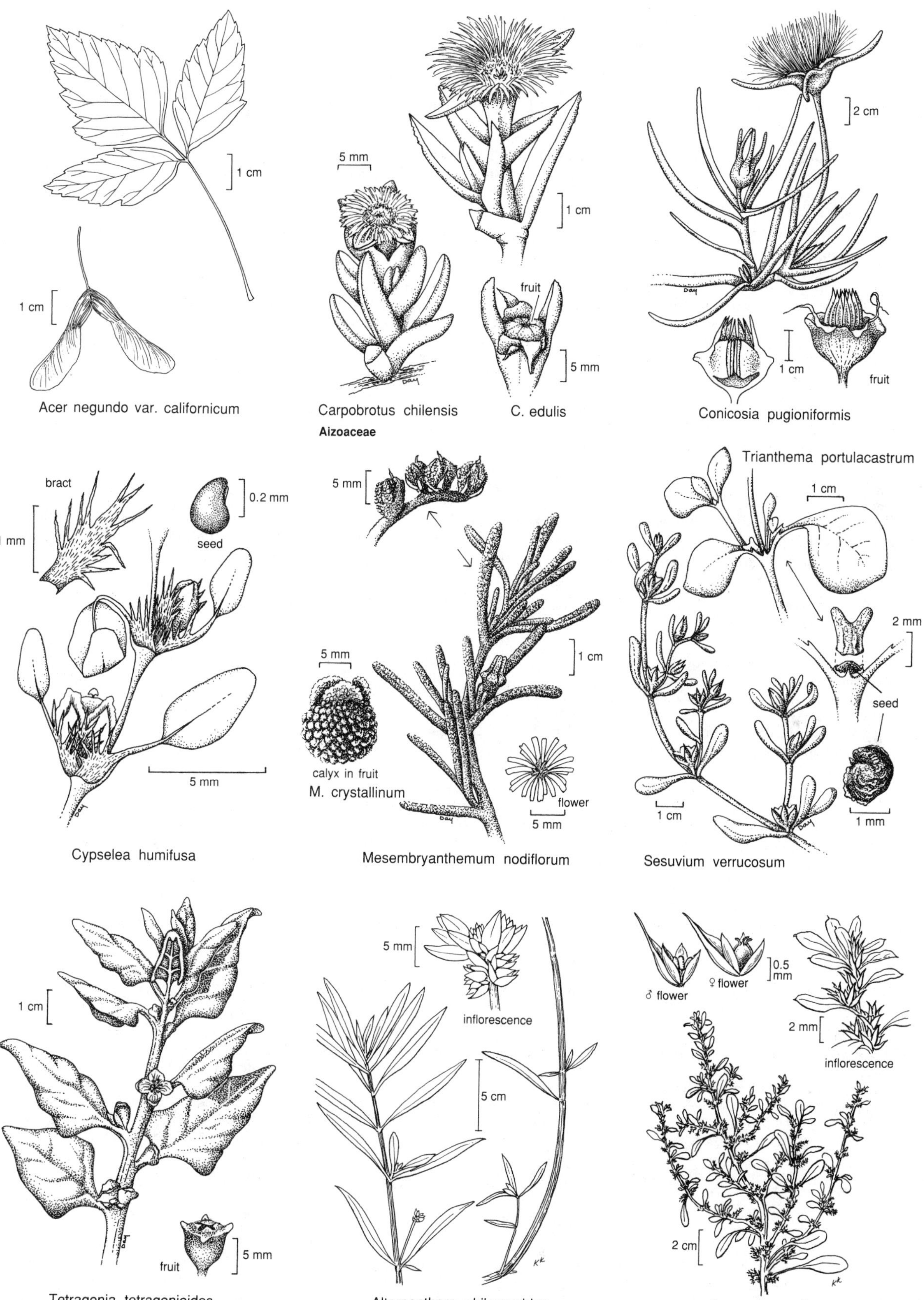

Acer negundo var. californicum

Carpobrotus chilensis

C. edulis

Conicosia pugioniformis

Aizoaceae

Cypselea humifusa

Mesembryanthemum nodiflorum

Sesuvium verrucosum

Tetragonia tetragonioides

Alternanthera philoxeroides

Amaranthus albus

Amaranthaceae

A. retroflexus L. Pl erect, 3–30 dm, monoecious. **LF**: petiole 7–70 mm; blade 20–150 mm, lower diamond-shaped to widely ovate, upper ± lanceolate. **INFL**: gen panicle of spike-like clusters; axillary clusters gen many, 1–5 cm, ± 1–2.5 cm wide; terminal spike < or = 20 cm, 5 cm wide; bracts 2.5–4.5 mm, spined, midrib wide, green, margins scarious in lower 3/4. **STAMINATE FL**: sepals 4–5, 2–3 mm, subequal; stamens (3)4–5. **PISTILLATE FL**: sepals 5, 2–3.2 mm, subequal, oblong to spoon-shaped, becoming reflexed, tip rounded to slightly notched, often abruptly pointed. **FR** ± 1.5 mm, circumscissile; lid smooth or wrinkled near base. **SEED** 1–1.3 mm, black. *n*=16. Common. Wet fields, roadside ditches, waste places; < 2400 m. GV, CW, SW; to se Can, n Mex, also Eurasia; native to trop Am.

A. spinosus L. Pl erect, 2–10 dm, ± spheric, monoecious; spines at lower nodes gen 2, 8–20 mm, slender. **LF**: petiole 12–70 mm; blade 14–80 mm, ± ovate. **INFL** dense; axillary clusters below middle ± spheric, each with 2 bracts elongated, stiff, spiny; terminal clusters spike-like, 3–17 cm, 4–13 mm wide, cylindric; upper bracts 1.5–6 mm, spine-tipped, midrib thick, green, margin scarious below. **STAMINATE FL**: sepals 5, 1.5–2 mm; anthers 5. **PISTILLATE FL**: sepals 5, 1.5–2 mm, oblanceolate to spoon-shaped, tip obtuse to spined. **FR** 1.5–2 mm, irregularly dehiscent; lid smooth throughout or base rough. **SEED** ± 1 mm wide, reddish black. 2*n*=34. Very uncommon as waif. Roadsides, waste places; < 700 m. SCo, w DMoj; to e US, Mex; native to trop Am.

A. torreyi (A. Gray) Benth. (p. 137) Pl erect, 1–7 dm, monoecious. **LF**: petiole 4–16 mm; blade 12–50 mm, oblanceolate to ovate. **INFL**: axillary clusters scattered to st base, crowded above into terminal, leafy spike < or = 10 cm, 7–15 mm wide; bracts 1.3–3.5 mm, midrib green, margins scarious below, sparsely glandular-hairy. **STAMINATE FL**: sepals 5, 2.5–4 mm; stamens 3. **PISTILLATE FL**: sepals 5, 1.4–2.5 mm, narrowly spoon-shaped, tips rounded to slightly notched, jagged. **FR** 1.4–2 mm, circumscissile; lid coarsely wrinkled-inflated near base. **SEED** ± 1 mm wide, black. Uncommon. Sandy flats, arroyos, after late summer rains; 1200–1700 m. DMoj; w AZ.

A. watsonii Standley (p. 137) Pls erect, 1–10 dm, dioecious. **ST** glandular-hairy. **LF**: petiole 3–90 mm; blade 10–80 mm, ± oblong to (ob)ovate. **INFL**: axillary clusters few, scattered below; terminal cluster 2–15 cm, 8–20 mm wide, spike-like; bracts 2.5–4 mm, spine-tipped, midrib ± thick, green, with stalked glands, margins scarious below. **STAMINATE FL**: sepals 5, 1.5–2 mm; stamens 3–5. **PISTILLATE FL**: sepals 5, 1.7–2.2 mm, ± fan-shaped, margin jagged, tip obtuse to slightly notched. **FR** 1.5–2 mm, circumscissile, smooth. **SEED** ± 1 mm wide, reddish black. Very uncommon. Depressions, waste places, after winter rains; < 500 m. s DSon (Imperial Co.); nw Mex.

GUILLEMINEA

Per, prostrate, soft-hairy to tomentose, taprooted. **LVS** opposite, unequal at a node; petiole widely winged, clasping; blade entire. **INFL**: spike, gen in axil of shorter lf, sessile; bracts unequal, white, glabrous, larger ones falling with fr. **FL** bisexual; calyx bell-shaped, lobes 5, ± = tube, glabrous, scarious, tube densely woolly outside; stamens 5, attached to top of calyx tube, filaments fused at base; staminodia 0; ovary compressed-ovoid, style 1, short, stigma ± 2-lobed. **FR** indehiscent; wall membranous. **SEED** lenticular to spheric, smooth. ± 5 spp.: s US to S.Am. (A. Guillemin, French botanist, 1796–1842) [Henrickson 1987 Sida 12:307–337]

G. densa (Willd.) Moq. var. *aggregata* Uline & Bray (p. 137) Pl 2–4 dm wide. **LVS**: basal in rosettes, 10–40 mm, oblong-lanceolate, early deciduous; cauline petiole 1–6 mm, blade 2–16 mm, elliptic to round, acute to obtuse, glabrous or sparsely silky above, long-soft-hairy below; cauline lvs smaller, clustered in axils. **INFL**: clusters of spikes, axillary, < or = 4 mm wide, subtended by and mixed with reduced lvs; bracts 1.5–2.5 mm, glabrous, white-scarious. **FL**: calyx 1.5–2.8 mm, lobes reflexed, erect in age, tip acute to rounded, midvein ending before tip; filament tube 0.1–0.2 mm, free filaments 0.1–0.3 mm. **SEED** ± 0.8 mm, red-brown. Disturbed places; < 100 m. s CCo, SCo; native AZ to TX, S.Am. [*Brayulinea d.* (Willd.) Small]

TIDESTROMIA

Ann to subshrub, ± mounded; hairs short, branched. **LVS** entire, often asymmetric, petioled; lowest alternate to opposite; upper opposite or whorled in 3's, gen reduced, larger falling early. **INFL**: clusters of 1–5 fls, axillary, sessile, subtended and ± enclosed by involucres of 2–3 bract-like lvs that become hardened; bracts scarious. **FL** bisexual; sepals 5, 1–1.5 mm, 1-veined, glabrous inside, sometimes canescent outside, reflexed, inner scarious-margined; stamens 5, filaments fused below into a cup, alternate 5 short staminodes; ovary spheric, style 1, short, stigma head-like or 2-lobed. **FR** indehiscent; wall membranous. **SEED** obovoid, brown. 7 spp.: w N.Am deserts. (Ivar T. Tidestrom, Swedish-born botanist of sw US, 1864–1956)

1\. Ann; young lvs whiter (or grayer) than older lvs; sepals 1.8–2.6 mm; filaments 0.6–1.6 mm .... *T. lanuginosa*
1' Per; young and old lvs ± equally white or gray; sepals 1.3–1.7 mm; filaments < or = 0.5 mm ...... *T. oblongifolia*

T. lanuginosa (Nutt.) Standley Ann 1–9 dm, < or = 20 dm wide. **LVS**: upper << lower; young terminal lvs white-gray, tomentose-canescent, hairs short, much-branched, wearing off, older lvs green above, more whitish below; blades 6–22 mm, (ob)ovate, base wedge-shaped. **FL**: sepals 1.8–2.6 mm; filament tube 0.3–0.4 mm, free filaments 0.6–1.6 mm. Slopes; ± 1200 m. e DMtns (Granite Mtns); to Colorado, TX, Mex.

T. oblongifolia (S. Watson) Standley (p. 137) Per 1–9 dm, < or = 15 dm wide. **LVS**: upper, persistent lvs 2–4(8) mm; blade 10–30(65) mm, narrowly ovate to round, base ± obliquely wedge-shaped, persistently grayish white-canescent. **FL**: sepals 1.3–1.7 mm; filament tube 0.2 mm, free filaments < or = 0.5 mm. Common. Washes, rocky hillsides; < 1200 m. D; NV, AZ, n Baja CA. [ssp. *cryptantha* (S. Watson) Wiggins] ❀TRY:STBL only.

ANACARDIACEAE SUMAC or CASHEW FAMILY

Dieter H. Wilken

Shrub, tree, gen dioecious or fls bisexual and unisexual, ± resinous, sometimes milky, gen aromatic. **LVS** simple or compound, alternate, deciduous or evergreen; stipules 0. **INFL**: raceme or panicle; fls gen many. **FL** gen unisexual, radial; sepals 5, base gen ± fused; petals 5, gen > sepals, free; stamens 5 or 10, reduced and sterile in pistillate fls; ovary superior, vestigial or 0 in staminate fls, subtended by ± lobed, disk-like nectary, chamber gen 1, ovule gen 1, styles 1–3. **FR** drupe-like, glabrous, sticky, or short-hairy; pulp ± resinous, sometimes aromatic. 70+ genera, ± 850 spp.: trop, warm temp; some orn (*Rhus, Schinus*), some cult for fr (*Anacardium*, cashew; *Mangifera*, mango). [Brizicky 1962 J Arnold Arbor 43:359–375] TOXIC: many genera produce contact dermatitis.

1. Trees; lvs compound, lflets 7–15+
 2. Perianth parts 1–7, unequal, bract–like; stamens 4–7 in 1 whorl .... **PISTACIA**
 2′ Sepals 5, petals 5; stamens gen 10, in 2 whorls .... **SCHINUS**
1′ Shrubs, sometimes vine- or tree-like; lvs simple to compound, lflets 3–5
 3. Lvs simple, evergreen; petals white to pinkish
  4. Fr 2–3 mm diam, glabrous; infl branches slender .... **MALOSMA**
  4′ Fr 6–10 mm diam, sticky, puberulent, or short-glandular–hairy; infl branches stout .... **RHUS**
 3′ Lvs deeply lobed to compound, lflets 3–5, deciduous; petals yellow to yellow-green
  5. Infl terminal, branches stiff; fls ± sessile; lflets dull above; fr red .... **RHUS**
  5′ Infl axillary, branches slender, loose, gen arched; fls pedicelled; lflets shiny above; fr creamy white .... **TOXICODENDRON**

MALOSMA

1 sp. (Latin: from odor which resembles that of an apple) [Brizicky 1963 J Arnold Arbor 44:60–80]

M. laurina (Nutt.) Abrams (p. 137) LAUREL SUMAC Shrub, small tree, 2–6 m; fls bisexual or unisexual. **LF** simple, evergreen; petiole 10–40 mm; blade 3–10 cm, 2–4.5 cm wide, elliptic to oblong-lanceolate, ± leathery, ± folded along midrib, tip abruptly pointed, margin entire. **INFL**: branches slender in bud; bractlets < 1.5 mm. **FL**: sepals green, margins entire; petals gen white. **FR** 2–3 mm diam, glabrous, ± white. Slopes, canyons, chaparral; < 1000 m. SW; Baja CA. [*Rhus l.* Nutt.] ❀DRN,DRY:14–16,**17**,19,**20–24**.

PISTACIA PISTACHIO

Shrub, tree, dioecious. **LVS** pinnate; lflets 3–16; lateral gen opposite, membranous, entire to toothed, deciduous. **INFL**: panicle, axillary or terminal, open to dense. **FL**: perianth parts bract-like, gen 1–7, unequal, early deciduous; stamens 4–7, vestigial in pistillate fls; styles 3, fused at base, gen 0 in staminate fl. **FR** spheric to obovoid; pulp fleshy. ± 11 spp.: Medit, e Asia, Mex. (Ancient Arabic or Persian name) [Zohary 1952 Palestine J. Bot 5:187–228] *P. vera*, pistachio, gen with 3 lflets, widely cult for food.

P. atlantica Desf. Tree 3–10 m. **ST**: branches spreading to erect. **LF** odd-pinnate; axis winged; lflets 7–9, tip acute to obtuse. **FR** 6–8 mm, ± obovoid. Flats, roadsides, drainages; < 100 m. ScV, expected elsewhere; native to Medit. Cult for orn, escaping; used as rootstock for *P. vera*. Pls from e ScV and n SNF, with even-pinnate lvs, lf axis cylindric, lflets 10–14, are *P. chinensis* Bunge.

RHUS

Shrub, tree, dioecious or bisexual and pistillate. **LVS** simple or compound, deciduous or evergreen, entire, toothed, or lobed. **INFL**: panicle, terminal on short twigs, open to dense; fls gen sessile. **FL**: stamens 5; styles 3, free or ± fused. **FR** spheric or ± compressed, glabrous or glandular-hairy, gen reddish; pulp thin or thick, ± resinous. ± 150 spp.: warm temp. (Greek: ancient name for sumac) [Brizicky 1963 J Arnold Arbor 44:60–80]

1. Lvs deeply lobed to compound, deciduous .... ***R. trilobata***
1′ Lvs gen simple, evergreen
 2. Lf gen flat, margin entire to toothed, tip ± obtuse; sepals green, glandular-ciliate .... ***R. integrifolia***
 2′ Lf gen folded along midrib, margin entire, tip acute to acuminate; sepal red, margin ciliate, not glandular ***R. ovata***

R. integrifolia (Nutt.) Brewer & S. Watson (p. 137) LEMONADEBERRY Shrub, small tree, 1–8 m, aromatic; fls bisexual or pistillate. **LF** simple, evergreen; petiole 2–7 mm; blade 2.5–6 cm, 2–4 cm wide, widely elliptic to elliptic-lanceolate, ± leathery, ± flat, tip ± obtuse, margin entire to toothed. **INFL**: branches stout in bud; bractlets 2–4 mm. **FL**: sepals green, margins glandular-ciliate; petals white to pinkish. **FR** 7–10 mm diam, glandular-hairy, reddish. Canyons, gen n-facing slopes, chaparral; < 900 m. SW; Baja CA. Cult elsewhere. Hybridizes with *R. ovata*. ❀DRN,DRY:**14**,15, **16**,**17**,19,**20–24**&SHD:8,9.

R. ovata S. Watson (p. 137) SUGAR BUSH Shrub, small tree, 2–10 m; fls bisexual or pistillate. **LF** simple, evergreen; petiole 10–30 mm; blade 3–8 cm, 3–8 cm wide, widely ovate to widely elliptic, ± leathery, folded along midrib, tip ± acute to acuminate, margin entire. **INFL**: branches stout in bud; bractlets < 2 mm. **FL**: sepals red, margins ciliate; petals white to pinkish. **FR** 6–8 mm diam, glandular-hairy, reddish. Canyons, gen s-facing slopes, chaparral; < 1300 m. SW; AZ, Baja CA. Cult elsewhere. Hybridizes with *R. integrifolia*. ❀DRN,DRY:9–12,**14**,15–17,**18–24**.

R. trilobata Torrey & A. Gray (p. 137) SKUNKBRUSH Shrub, 0.5–2.5 m. **LF** deeply lobed to compound, deciduous, thin, flat; petiole 5–15 mm; lobes or lflets gen 3, margins crenate to slightly lobed, lower surfaces tomentose to ± glabrous; terminal lobe or lflet 10–35 mm, ± diamond-shaped; lateral 5–18 mm, gen ovate. **INFL**: branches short, stiff; fl before lvs appear. **FL**: sepals yellow-green to reddish; petals gen yellow. **FR** 5–8 mm diam, sparsely hairy, sticky, gen bright red-orange. Slopes, washes, shrubland; < 2200 m. CA-FP, DMtns; to s Can, c US, n Mex. [var. *anisophylla* (E. Greene) Jepson; var. *malacophylla* (E. Greene) Jepson; var. *quinata* Jepson] Geog variation in w N.Am needs further study. ❀4,5,**6**,**7**, **14**,15–17,22–24;IRR:1–3,8–10,**18–21**;also STBL.

SCHINUS PEPPER TREE

Shrub, tree, dioecious. **LVS** gen odd-pinnate, ± resinous, aromatic; lflets > 5, lateral alternate to opposite, ± leathery, entire to toothed. **INFL**: panicle, axillary or terminal, open to ± dense; pedicels short. **FL**: stamens 10, 2 whorls of 5, reduced and sterile in pistillate fls; styles 3, fused at base. **FR** spheric, leathery, shiny, reddish; pulp resinous to oily, aromatic. ± 25 spp.: trop, warm temp S.Am. (Greek: ancient name for another sp.) [Barkley 1944 Brittonia 5:160–198]

1. Distal branches, twigs drooping; lflets gen > 15, gen < 1 cm wide . . . . ***S. molle***
1′ Distal branches, twigs stiff, spreading to erect; lflets gen 7, gen > 1 cm wide . . . . ***S. terebinthifolius***

S. molle L. PERUVIAN PEPPER TREE Tree 5–18 m, sometimes root-sprouting. **ST**: upper branches gen drooping. **LF** 10–30 cm; lflets gen > 15, sessile, 1–6 cm, 4–8 mm wide, lanceolate to linear-lanceolate, gen entire. **INFL**: pedicel 2–5 mm in fr. **FL** < 3 mm. **FR** 5–8 mm diam, gen pink to red. $2n$=28. Washes, slopes, abandoned fields; < 700 m. SNF, Teh, GV, CW, SW; to TX, Mex; native to S.Am.

S. terebinthifolius Raddi BRAZILIAN PEPPER TREE Tree < 10 m. **ST**: upper branches gen stiff, spreading to erect. **LF** 8–15 cm; lflets gen 7, sessile or short-stalked, 2.5–7 cm, 10–25 mm wide, elliptic to oblong, entire to toothed. **INFL**: pedicel 2–4 mm in fr. **FR** 4–7 mm diam, gen pink to red. Washes, canyons; < 200 m. s SCo; native to S.Am.

TOXICODENDRON POISON OAK, POISON IVY

Shrub, tree, vine, gen dioecious. **LVS** gen ternate or pinnately compound, ± resinous; lflets 3–9, lateral gen opposite, thin to ± leathery, entire, toothed, or lobed. **INFL**: raceme or panicle, axillary, ± open; fls pedicelled. **FL**: stamens 5, sterile or reduced in pistillate fls; styles ± fused, stigmas 3. **FR** gen spheric, becoming papery or leathery, cream-colored to brown; pulp resinous. 6 spp.: Am, e Asia. (Latin: poisonous tree) [Gillis 1971 Rhodora 73:161–237,370–443] TOXIC: resin on lvs, sts, frs causes severe contact dermatitis; one of the most hazardous pls in CA.

T. diversilobum (Torrey & A. Gray) E. Greene (p. 137) WESTERN POISON OAK Shrub (sometimes tree-like) 0.5–4 m or vine < 25 m. **ST**: twigs glabrous to sparsely hairy, gray- to red-brown. **LF**: petiole 1–10 cm; lflets gen 3(5), ± round to oblong, thin to ± leathery, becoming bright red in autumn, base truncate to rounded, tip obtuse to rounded, margin entire, wavy, or slightly lobed, upper surface glabrous, ± shiny, lower sparsely short-hairy; terminal lflet 1–13 cm, 1–8 cm wide; lateral lflets 1–7 cm, 1–6 cm wide. **INFL** drooping, spreading, or erect; pedicels 2–8 mm; bractlets < 1 mm. **FL**: sepals, petals gen ovate, yellow-green. **FR** 1.5–6 mm diam, spheric to slightly compressed, becoming leathery, glabrous to finely bristly, creamy white; pulp white, black-striate. $2n$=30. Canyons, slopes, chaparral, oak woodland; < 1650 m. CA-FP; to B.C., Baja CA. [*Rhus d.* Torrey & A. Gray]

APIACEAE [Umbelliferae] CARROT FAMILY

Lincoln Constance

Ann, bien, per (rarely shrub, tree), often from taproot. **ST** often ± scapose, gen ribbed, hollow. **LVS** basal and gen some cauline, gen alternate; stipules gen 0; petiole base gen sheathing st; blade gen much dissected, sometimes compound. **INFL**: umbel or head, simple or compound, gen peduncled; bracts present (in involucres) or not; bractlets gen present (in involucels). **FLS** many, small, gen bisexual (or some staminate), gen radial (or outer bilateral); calyx 0 or lobes 5, small, atop ovary; petals 5, free, gen ovate or spoon-shaped, gen incurved at tips, gen ± ephemeral; stamens 5; pistil 1, ovary inferior, 2-chambered, gen with a ± conic, persistent projection or platform on top subtending 2 free styles. **FR**: 2 dry, 1-seeded halves that separate from each other but gen remain attached for some time to a central axis; ribs on each half 5, 2 marginal and 3 on back; oil tubes 1–several per interval between ribs. 300 genera, 3,000 spp.: ± worldwide, esp temp; many cult for food or spice (e.g., *Carum*, caraway; *Daucus*; *Petroselinum*); some highly toxic (e.g., *Conium*). Underground structures here called roots, but true nature remains problematic. Mature fr gen critical in identification; shapes gen given in outline, followed by shape in X-section of the 2 fr halves together.

1. Fls in simple umbels or heads, rarely spikes or racemes; fr central axis not an obvious structure
 2. Lf not truly sheathing at base but stipules present, sometimes slightly fused above node when lvs opposite
  3. Herbage hairs stellate; fr not or slightly compressed front-to-back, inflated . . . . **BOWLESIA**
  3′ Herbage hairs simple or 0; fr very compressed side-to-side . . . . **HYDROCOTYLE**
 2′ Lf conspicuously sheathing at base, stipules 0
  4. Fls in open umbels; fr glabrous, with well defined, thickened ribs . . . . **LILAEOPSIS**
  4′ Fls in dense umbels or heads; fr with scales, prickles, or tubercles, without distinct ribs
   5. Fls individually accompanied by a single bractlet, bisexual, sessile . . . . **ERYNGIUM**
   5′ Fls not individually accompanied by a single bractlet, bisexual and staminate (or staminate only) in same umbel or head, staminate often long-pedicelled . . . . **SANICULA**
1′ Fls in compound umbels (these sometimes head-like), sometimes simple umbels as well; fr central axis gen an obvious, separate structure
 6. Ann or bien, from slender taproot or fibrous roots, reproducing solely from seed
  7. Corolla yellow
   8. Lf pinnately dissected into thread-like segments; bractlets 0 . . . . **ANETHUM**
   8′ Lf simple, entire; bractlets lf-like, > fls, frs . . . . **BUPLEURUM**
  7′ Corolla white, greenish white, reddish, or rose-tinged
   9. Fr gen elongate, much longer than wide, beaked or long-tapered to tip
    10. Bractlets entire; fr ribs 0 or obscure, sterile beak << fertile body . . . . **ANTHRISCUS**
    10′ Bractlets lobed or dissected; fr ribs prominent, sterile beak > fertile body . . . . **SCANDIX**

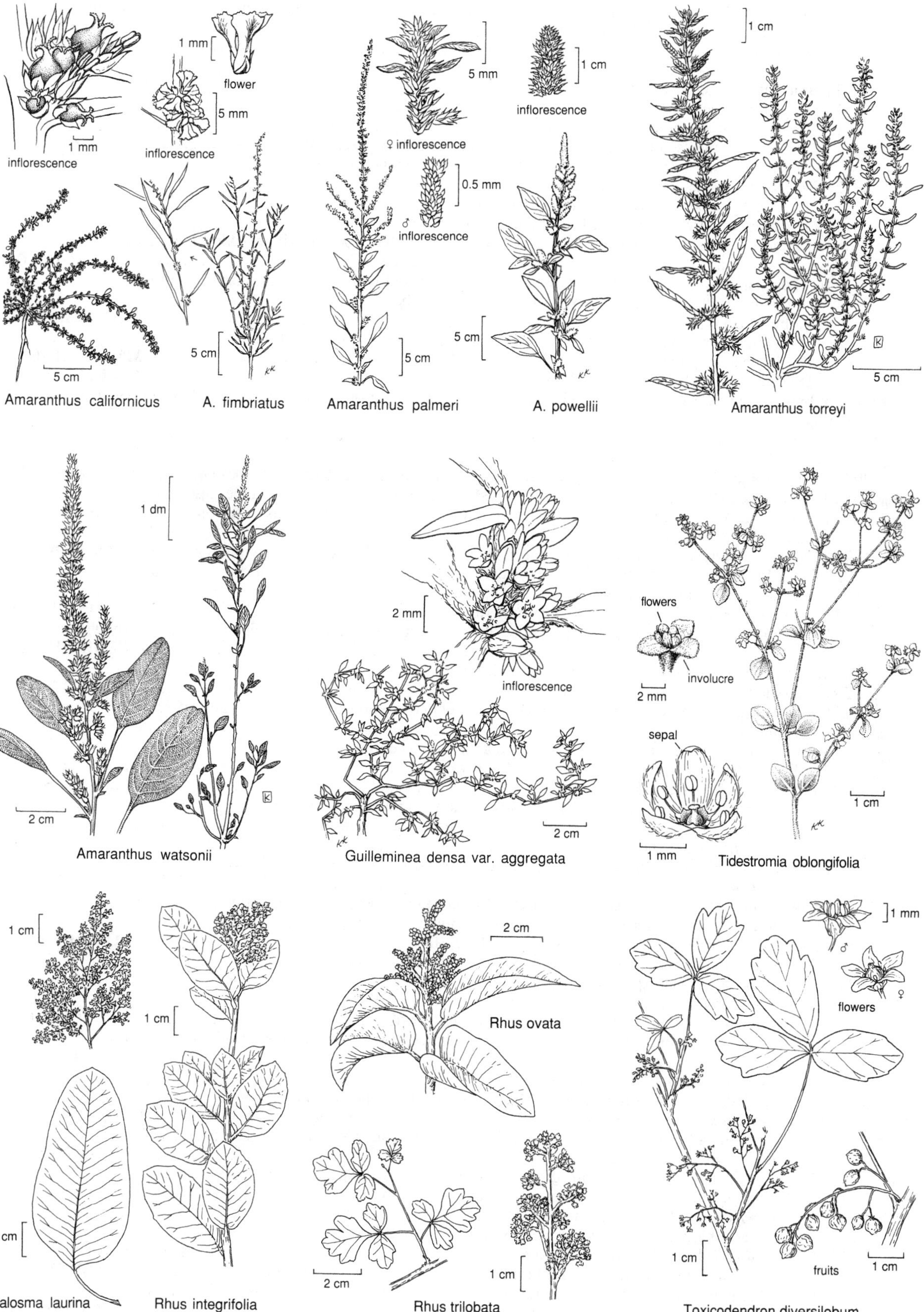

1 mm
flower
5 mm
1 mm
inflorescence
inflorescence
5 cm
5 cm
Amaranthus californicus
A. fimbriatus
5 mm
♀ inflorescence
1 cm
inflorescence
0.5 mm
♂ inflorescence
5 cm
5 cm
Amaranthus palmeri
A. powellii
1 cm
5 cm
Amaranthus torreyi
1 dm
2 cm
Amaranthus watsonii
2 mm
inflorescence
2 cm
Guilleminea densa var. aggregata
flowers
involucre
2 mm
sepal
1 mm
1 cm
Tidestromia oblongifolia
1 cm
1 cm
1 cm
Malosma laurina
Anacardiaceae
Rhus integrifolia
2 cm
Rhus ovata
2 cm
1 cm
Rhus trilobata
1 mm
♂
♀
flowers
1 cm
fruits
1 cm
Toxicodendron diversilobum

9′ Fr ± round to oblong or elliptic, not or only slightly longer than wide, not beaked, not long-tapered to tip
11. Fr ± round, not compressed, halves not separating readily .... **CORIANDRUM**
11′ Fr ovate or cordate to oblong or elliptic, compressed either side-to-side or front-to-back, halves separating readily
12. Bracts lf-like, gen dissected
13. Fr glabrous; herbage glabrous .... **AMMI**
13′ Fr conspicuously bristly; herbage hairy
14. Fr very compressed front-to-back; rays gen many; umbel dense, nest-like in fr .... **DAUCUS**
14′ Fr compressed side-to-side; rays few (1–9); umbel open, not nest-like in fr .... **YABEA**
12′ Bracts 0 or inconspicuous, not lf-like
15. Lvs mostly opposite; rays 2–3; fr elliptic-cordate .... **APIASTRUM**
15′ Lvs mostly alternate; rays 2–14; fr oblong-ovate or ovate
16. Petals very wide, tips narrowed, incurved; fr covered or half covered with prickles; herbage hairy .... **TORILIS**
16′ Petals narrow, tips not narrowed, not incurved; fr covered with prickles, or sharply scabrous, or glabrous; herbage glabrous or ± roughened but not hairy
17. Involucel 0; fr glabrous .... **CICLOSPERMUM**
17′ Involucel present; fr conspicuously bristly or sharply scabrous, at least on ribs
18. Infl exc frs puberulent or minutely hairy; fr oblong-ovate .... **AMMOSELINUM**
18′ Infl exc frs glabrous; fr widely ovate .... **SPERMOLEPIS**
6′ Per or bien, from various, persistent, often swollen, tuberous underground structures, as well as reproducing from seed
19. Pl glabrous, gen in wet or moist soil of marshes, streambanks, etc. or where water stands for long periods; corolla white
20. St gen conspicuously purple-spotted or -streaked; herbage musty-scented; fr oil tubes not apparent [2]**CONIUM**
20′ St green or sometimes purplish but unspotted; herbage not musty-scented; fr oil tubes apparent
21. Sts from well developed single or from clustered tubers or clustered, tuberous or fibrous roots, gen not rooting at lower nodes
22. Calyx lobes forming persistent crown on fr; fr corky; fr central axis not an obvious structure [2]**OENANTHE**
22′ Calyx lobes evident but not forming persistent crown on fr; fr not evidently corky; fr central axis an apparent separate structure
23. Fr very compressed front-to-back, marginal ribs widely thin-winged .... **OXYPOLIS**
23′ Fr slightly or not compressed side-to-side, ribs subequal, thread-like to prominent but not clearly winged .... **PERIDERIDIA**
21′ Sts from stout taproots or rooting at lower nodes or along rhizomes or stolons
24. Calyx lobes forming persistent crown on fr; fr central axis not an obvious structure .... [2]**OENANTHE**
24′ Calyx lobes 0 or minute, gen not forming a persistent crown on fr; fr central axis an apparent separate structure
25. Bracts, bractlets conspicuous; fr halves gen adhering to central axis
26. Fr ribs thread-like, inconspicuous in corky fr wall .... **BERULA**
26′ Fr ribs prominent, corky .... **SIUM**
25′ Bracts, bractlets conspicuous to 0; fr halves not adhering to central fr axis
27. Bien, per; lvs 1-pinnate, lflets lanceolate to ± round; fr axis entire or merely notched at tip; fr not conspicuously corky .... **APIUM**
27′ Per; lvs 1–3-pinnate or 1–3-ternate-pinnate, lflets mostly lanceolate; fr axis divided to base; fr conspicuously corky .... **CICUTA**
19′ Pl glabrous or variously hairy, gen not restricted to wet or moist soil of marshes, streambanks, etc or places where water stands for long periods; corolla white, yellow, or purplish
28. Herbage anise- or licorice-scented; lf segments thread-like .... **FOENICULUM**
28′ Herbage scented or not, but not anise-scented; lf segments wider than thread-like
29. Fr ribs subequal, none expanded into definite wings
30. Bien of disturbed or cult places, alien
31. Sts commonly purple-spotted; cauline lvs like basal; fr ribs prominent, wavy .... [2]**CONIUM**
31′ Sts unspotted; cauline lvs with more, narrower segments than basal; fr ribs thread-like, not wavy .... **PETROSELINUM**
30′ Per of gen undisturbed places, native
32. Pl dwarfed, 2–15 cm, lf segments not sharply serrate; montane or alpine
33. Pl erect or spreading, not cushion-forming, lvs loosely recurved or sometimes spreading at soil surface; umbels not head-like; involucel inconspicuous, not 1-sided, bractlets 0 or free; calyx lobes 0; styles cylindric .... [2]**OROGENIA**
33′ Pls cushion-forming or lvs spreading at soil surface; umbels head-like; involucel conspicuous, 1-sided, bractlets gen partly fused; calyx lobes conspicuous; styles very compressed, tape-like
34. Herbage coarsely hairy to gen tomentose; sterile pedicels many, > fr .... **OREONANA**
34′ Herbage thinly to densely puberulent; sterile pedicels 0–few, < fr .... **PODISTERA**
32′ Pl not dwarfed, 10–120 cm, if < 20 cm, lf segments sharply serrate; gen lower than montane or alpine
35. Fr linear to oblong, 8–22 mm; fr oil tubes obscure; roots licorice-scented .... **OSMORHIZA**
35′ Fr oblong to round, 2–9 mm; fr oil tubes evident, 1–5 per rib-interval; roots not licorice-scented
36. Corolla white; involucel < or > pedicels or gen 0 .... [2]**LIGUSTICUM**
36′ Corolla yellow; involucel present, often > pedicels .... **TAUSCHIA**

29′ Fr ribs unequal, some expanded into definite wings
37. Ribs at margins of fr halves winged, others not or more narrowly so; fr very compressed front-to-back
38. Pl very robust, 1–3 m; outer petals of marginal fls > others; oil tubes extending part way to base of fr .... **HERACLEUM**
38′ Pl not very robust, rarely > 1 m; outer petals of marginal fls = others; oil tubes extending to base of fr
39. Projection atop ovary small, conic, persistent on fr
40. Lf 1–4-pinnately or -ternate-pinnately dissected or compound; corolla white; per **CONIOSELINUM**
40′ Lf 1–2-pinnate; corolla yellow or orange; per, gen bien .... **PASTINACA**
39′ Projection atop ovary 0
41. Marginal fr wings thin or corky, not incurved; fr axis divided to base .... **LOMATIUM**
41′ Marginal fr wings strongly corky, incurved; fr axis represented by a corky flange running length of each fr half along its attachment .... [2]**OROGENIA**
37′ Ribs at margins of fr halves and some others ± winged; fr cylindric to somewhat compressed front-to-back or side-to-side, rarely very much so
42. Pl low, gen spreading, st often 0; rays well developed or not; projection atop ovary 0
43. Pl erect or ± spreading; lf glabrous to scabrous or finely hairy, often glaucous, but lower surface not tomentose; inland — often dry places .... **CYMOPTERUS**
43′ Pl prostrate or ± spreading; lf upper surface green, lower tomentose; seashore .... **GLEHNIA**
42′ Pl erect, st gen present; rays well developed; projection atop ovary conic
44. Pedicels reduced to a disk, 2° umbels head-like .... **SPHENOSCIADIUM**
44′ Pedicels distinct or slightly webbed at base but not reduced to a disk, 2° umbels open, not head-like
45. St base not fibrous; lf segments gen coarse; fr gen very compressed front-to-back, rarely slightly so or cylindric .... **ANGELICA**
45′ St base conspicuously fibrous; lf segments gen fine; fr compressed side-to-side .... [2]**LIGUSTICUM**

AMMI

Ann, bien, taprooted, glabrous. **ST** erect, branched. **LF**: blade oblong to ovate, pinnately or ternately dissected, segments lanceolate to thread-like. **INFL**: umbels compound; bracts, bractlets many, lf-like; rays, pedicels many, spreading-ascending to incurved in fr. **FL**: calyx lobes minute; petals wide, white, tips 2-lobed. **FR** oblong to ovate, subcylindric, compressed side-to-side; ribs subequal, thread-like; oil tubes per rib-interval 1; fr axis entire or divided. **SEED**: face flat. ± 6 spp.: Eurasia, Medit; 2 spp. weedy in CA. (Ancient name used by Dioscorides) TOXIC; rarely eaten by livestock.

1. Infl without disk-like receptacle; rays spreading in fr; cauline lf segments gen lanceolate, finely serrate ***A. majus***
1′ Infl with a disk-like receptacle; rays incurving in fr to form a nest-like umbel; cauline lf segments linear ***A. visnaga***

A. majus L. Pl 2–8 dm. **LF**: petiole 1–5 cm; blade 6–20 cm, oblong, segments 10–15 mm, lanceolate to round; cauline lvs 2-pinnately dissected. **INFL** scabrous; peduncle 8–14 cm; rays 20–60, 2–7 cm, very slender, ± scabrous; pedicels 1–10 mm. **FR** 1.5–2 mm, oblong. $2n=22$. Fields, roadsides, other disturbed places; gen < 1000 m. NCo; native to Eurasia.

A. visnaga L. (p. 143) BISNAGA Pl 2–8 dm. **LF**: petiole ± 1 cm; blade 5–20 cm, triangular-ovate, segments 5–35 mm, thread-like to linear; cauline lvs 1- or 2-pinnately or -ternately dissected. **INFL** glabrous; peduncle 8–14 cm; rays 60–100, 2–5 cm, very slender, glabrous; pedicels 3–13 mm. **FR** 2–2.5 mm, oblong-ovate to ovate. $2n=20$. Disturbed places such as roadsides, railroad tracks; gen < 1000 m. CCo, SCo; native to Eurasia.

AMMOSELINUM

Ann, taprooted. **ST** erect or gen loosely branched, glabrous or roughened. **LF**: petiole entirely sheathing; blade oblong to obovate, ternately or ternately-pinnately dissected, segments linear to spoon-shaped. **INFL**: umbels compound, peduncled or some sessile, puberulent; bracts gen 0; bractlets several, narrow; rays, pedicels few, spreading or spreading-ascending, unequal. **FL**: calyx lobes 0; petals ovate, white, tips obtuse, not narrowed, not incurved. **FR** oblong-ovate, compressed side-to-side; ribs subequal, prominent, conspicuously bristly to glabrous; oil tubes per rib-interval 1–3; fr axis notched at tip. **SEED**: face flat to concave. 4 spp.: 3 N.Am, 1 S.Am. (Greek: sand-parsley)

A. giganteum J. Coulter & Rose DESERT SAND-PARSLEY Pl 1–2 dm. **LF**: petiole 3–8 mm; blade 1.5–2.5 cm, obovate, segments 4–13 mm, linear, glabrous or roughened. **INFL**: peduncles 0–4 cm; rays 4–8, 0–2 cm; pedicels 1–10, 0–8 mm. **FR** 3–5 mm, oblong-ovate; ribs corky, sharply scabrous. $2n=38$. RARE in CA. Heavy soil under shrubs; ± 400 m. DSon (Hayfield Lake, Riverside Co., 1922); AZ, n Mex. Possibly alien in CA.

ANETHUM

1 sp. (Latin: dill)

A. graveolens L. DILL Ann, 0.5–2 m, taprooted; herbage glabrous, glaucous, anise-scented. **ST** erect, branched, hollow. **LVS** mostly cauline; petiole 5–6 cm; blade 12–35 cm, oblong to obovate, pinnately dissected, segments 4–20 mm, thread-like. **INFL**: umbels compound, peduncled; bracts, bractlets 0; rays, pedicels many, slender, subequal, spreading or spreading-ascending. **FL**: calyx lobes 0; petals wide, yellow, tips narrowed. **FR** 2–4.5 mm, ovate, compressed front-to-back, glabrous; ribs unequal, marginal narrowly winged, others thread-like; oil tubes per rib-interval 1; fr axis divided to base. **SEED**: face ± flat. $2n=22$. Disturbed places; gen < 1000 m. SW; native to Medit. Widely but sporadically escaped from cult.

ANGELICA

Per, taprooted. **ST** erect, leafy, hollow. **LVS**: petioles gen inflated; cauline often bladeless; blades compound, rarely dissected, lflets gen wide, distinct, when lf dissected, segments narrow, connected. **INFL**: umbels compound, peduncled; bracts 0; bractlets 0 or many and conspicuous; rays, pedicels many, spreading-ascending or ascending. **FL**: calyx lobes 0 or minute; petals wide, white, pink, red, or purple. **FR** oblong to round, gen very compressed front-to-back (rarely slightly so or cylindric), glabrous to hairy; ribs gen unequal, winged but marginal gen wider than others; oil tubes per rib-interval 1–several, adhering to fr wall or rarely to seed; fr axis divided to base. **SEED**: face flat. 50–60 spp.: temp N.Am, Asia. (Latin: angelic, for cordial & medicinal properties) [DiTomaso 1984 Madroño 31:69–79]

1. Lf 2–3-ternate-pinnately dissected, segments linear or linear-oblong, entire, 2–10 cm .... ***A. lineariloba***
1′ Lf 1–3-pinnate or -ternate-pinnate, lflets lanceolate to widely ovate, gen serrate or dentate, rarely entire, 3–15 cm
 2. Bractlets conspicuous, 5–15 mm; rays, pedicels not webbed at base
  3. 1° lflets gen sharply reflexed; fr very compressed, ovate to round, 3–4 mm, ribs on back much narrower than marginal, thin .... ***A. genuflexa***
  3′ 1° lflets not reflexed; fr cylindric, ± oblong, 4–9 mm, ribs subequal, widely, thickly corky-winged ***A. lucida***
 2′ Bractlets gen 0 or inconspicuous; rays, pedicels webbed at base
  4. Basal lvs 1-ternate-pinnate
   5. Rays subequal, spreading-ascending to reflexed .... ***A. callii***
   5′ Rays unequal, ascending
    6. Lf oblong; lflets entire to remotely serrate; rays 7–14; ovary minutely bristly; fr oblong, 4–5 mm .... ***A. kingii***
    6′ Lf triangular-ovate; lflets sharply serrate; rays 15–50; ovary gen glabrous; fr oblong to ovate, 6–7 mm .... ***A. californica***
  4′ Basal lvs 2–3-ternate-pinnate
   7. Pl glaucous .... ***A. tomentosa***
   7′ Pl bright green at least on lf upper sides
    8. Pl coastal; lf thick, lower side often white-tomentose .... ***A. hendersonii***
    8′ Pl not coastal; lf thin, green, both sides gen glabrous
     9. Petals, ovary glabrous .... ***A. arguta***
     9′ Petals, ovary hairy .... ***A. breweri***

A. arguta Nutt. ANGELICA Pl 1–2 m, gen glabrous. **LF** < 1 m diam, ovate-triangular, 2–3-ternate-pinnate; lflets 6–9 cm, lanceolate to elliptic, acute, sharply serrate. **INFL** gen glabrous; bracts, bractlets gen 0; rays 20–60, 2–10 cm, rays, pedicels webbed at base. **FL**: petals, ovary glabrous. **FR** 8–9 mm, oblong to ovate. 2*n*=22. UNCOMMON. Coniferous forests; 200–2000 m. KR; to B.C., MT, UT. ❀DRN:**4–6,17**&IRR:1,2,7,14,**15,16**.

A. breweri A. Gray Pl 1–2 m, gen hairy. **LF** < 1 m diam, triangular-ovate, 2–3-ternate-pinnate; both surfaces gen glabrous; lflets 6–10 cm, lanceolate, acuminate, serrate. **INFL** hairy; bracts, bractlets gen 0; rays 20–50, 2–10 cm; rays, pedicels webbed at base. **FL**: petals, ovary hairy. **FR** 8–12 mm, oblong to ovate. 2*n*=66. Coniferous forests; 1000–3000 m. CaRH, n&c SNH; NV. ❀TRY.

A. californica Jepson Pl 1–2.5 m, glabrous to sparsely hairy. **LF** 1–12 dm, triangular-ovate, 1-ternate-pinnate; lflets 4–8 cm, lanceolate to oblong, sharply serrate. **INFL** gen glabrous; bracts, bractlets gen 0; rays 15–50, 2–13 cm, unequal, ascending; rays, pedicels webbed at base. **FL**: petals, ovary glabrous to minutely hairy. **FR** 6–7 mm, oblong to ovate. 2*n*=22. Dry slopes; 15–1500 m. NCoR, n SNF. [*A. tomentosa* var. *c.* Jepson] ❀DRN,IRR:**17**&SHD:7,**14–16**.

A. callii Mathias & Constance (p. 143) CALL'S ANGELICA Pl 1–2 m, gen roughened. **LF** 1–4 dm, ovate, 1-ternate-pinnate; lflets 3–13 cm, ovate, acute to obtuse, sharply serrate. **INFL** roughened; bracts, bractlets gen 0; rays 25–50, 2.5–7 cm, subequal, spreading-ascending to reflexed; rays, pedicels webbed at base. **FL**: petals, ovary gen hairy. **FR** 3.5–5 mm, oblong to obovate, moderately compressed front-to-back; marginal ribs thick. 2*n*=22. UNCOMMON. Streambanks in coniferous forests; 1100–2000 m. s SNH (Tulare, Kern cos.). ❀IRR:**1,7**.

A. genuflexa Nutt. Pl 1–2 m, glabrous to hairy. **LF** 1–8 dm, ovate to triangular-ovate, 2–3-ternate-pinnate; 1° lflets gen sharply reflexed, ultimate 4–10 cm, lanceolate to ovate, acute to acuminate, coarsely serrate to irregularly cut. **INFL** hairy; bracts 0; bractlets several, 5–10 mm, conspicuous; rays 25–50, 2–7 cm; rays, pedicels not webbed at base. **FL**: petals, gen ovary glabrous. **FR** 3–4 mm diam, ovate to round. 2*n*=22. Streambanks, wet areas in coniferous forest; 0–1000 m. KR, NCoRO; to AK, e Asia. ❀WET:**4–6,17**.

A. hendersonii J. Coulter & Rose Pl 8–20 dm, often sprawling, tomentose to rarely nearly glabrous. **LF** < 6 dm, triangular-ovate, 2–3-ternate-pinnate; lflets 5–10 cm, lanceolate to oblong, obtuse to acute, doubly serrate to crenate, upper surface green, glabrous, lower often white-tomentose. **INFL** tomentose; bracts, bractlets 0; rays 20–65, 2–8 cm; rays, pedicels webbed at base. **FL**: petals, ovary tomentose. **FR** 6–9 mm, oblong to ovate. 2*n*=22. Coastal bluffs, scrub; 0–100 m. NCo, CCo; to WA. [*A. tomentosa* var. *h.* DiTomaso] ❀SUN:**4,5**&IRR:**15–17**.

A. kingii (S. Watson) J. Coulter & Rose Pl 3–20 dm, glabrous to roughened. **LF** 1.5–4 dm, oblong, 1-ternate-pinnate (rarely ± 2-pinnate); lflets 3–12 cm, ± lanceolate, acute to acuminate, entire to sparsely serrate. **INFL** roughened; bracts, bractlets 0; rays 7–14, 0.5–10 cm, unequal, ascending; rays, pedicels webbed at base. **FL**: petals hairy; ovary minutely bristly. **FR** 4–5 mm, oblong. 2*n*=44. Subalpine streambanks; 1900–3000 m. W&I (White Mtns); to UT. ❀WET:1.

A. lineariloba A. Gray Pl 5–15 dm, nearly glabrous to scabrous. **LF** 1–3.5 dm, triangular-ovate, 2–3-ternate-pinnately dissected; segments 2–10 cm, linear to linear-oblong, acute, entire. **INFL** scabrous; bracts, bractlets 0; rays 20–40, 3–7 cm, subequal; rays, pedicels not webbed at base. **FL**: petals, ovary roughened to becoming glabrous. **FR** 10–13 mm, oblong to wedge-shaped. Rocky open slopes; 2300–3000 m. c&s SNH, SNE, W&I; NV. [var. *culbertsonii* Jepson] ❀TRY.

A. lucida L. (p. 143) Pl 1–1.5 m, glabrous. **LF** 1–3 dm, ovate to triangular-ovate, 2–3-ternate-pinnate; lflets 4–10 cm, widely lanceolate to ovate, acute, serrate to crenate-dentate; both surfaces green, glabrous. **INFL** hairy; bracts 0; bractlets 5–12, 5–15 mm, conspicuous; rays 20–45, 3–8 cm; rays, pedicels not webbed at base. **FL**: petals, ovary glabrous. **FR** 4–9 mm, ± oblong, cylindric; ribs subequal, widely, thickly corky-winged. **SEED** loose in fr wall. 2*n*=28. Coastal bluffs, beaches; 0–50 m. NCo, KR; to AK; coastal e N.Am, e Asia. ❀**4,5,17**;IRR&SHD:**15,16**.

A. tomentosa S. Watson Pl 7–20 dm, glaucous, glabrous to hairy. **LF** < 1 m diam, triangular-ovate, 2–3-ternate-pinnate; lflets 2–12 cm, lanceolate to ovate, acute to acuminate, serrate to entire. **INFL** glabrous to hairy; bracts, bractlets 0; rays 20–60, 2–11 cm;

rays, pedicels webbed at base. **FL**: petals, ovary glabrous to hairy. **FR** 6–10 mm, oblong to ovate. $2n$=22. Gen wooded areas; 660–2400 m. NCoRO, SCoRO; s OR. Appears to intergrade with *A. hendersonii* in NCoRO. ❀**5,IRR:17**&SHD:**15,16**;DFCLT.

ANTHRISCUS

Ann, per, taprooted, glabrous or bristly. **ST** erect, branched. **LVS** mostly cauline; blade oblong to ovate, pinnately or ternately dissected or compound, segments or lflets linear-oblong to ovate. **INFL**: umbels compound, gen peduncled; bracts gen 0; bractlets several, reflexed, entire; rays, pedicels few, spreading. **FLS**: outer sometimes ± bilateral; calyx lobes 0; petals narrow, white. **FR** linear-lanceolate to ovate, cylindric, beaked, smooth or bristly; each half ± cylindric; ribs, oil tubes 0 or obscure; fr axis entire or notched at tip. **SEED**: face grooved. ± 15 spp.: Eurasia, Afr. (Ancient Greek name)

A. caucalis M. Bieb. (p. 143) BUR-CHERVIL Pl 4.5–10 dm. **LF**: petiole 3–8 cm; blade 5–15 cm, oblong-to triangular-ovate, pinnately dissected, segments 1–5 mm, linear-oblong, obtuse. **INFL**: peduncles 0–2 cm; rays 3–6, 1–2.5 cm. **FR** ± 4 mm, ovate; beak << body. $2n$=14. Gen shady places; 0–400 m. w CA-FP; native to Eurasia. [*A. scandicina* (G. Weber) Mansf.]

APIASTRUM

1 sp. (Latin: wild celery)

A. angustifolium Nutt. (p. 143) Ann, 5–50 cm, taprooted. **ST** much-branched throughout. **LVS** mostly cauline, opposite or alternate; petiole 2–4 cm; blade 1–5 cm, round, finely ternately dissected, segments 5–25 mm, linear to oblong. **INFL**: umbels compound, peduncled or not, from axils or opposite upper lvs; bracts, bractlets 0; rays 2–3, pedicels few, both unequal, spreading. **FL**: calyx lobes 0; petals wide, white, tip acute, not incurved; projection atop ovary flat to barely conic. **FR** 1–1.5 mm, elliptic-cordate, compressed side-to-side, papillate-roughened to becoming glabrous; ribs thread-like; oil tubes per rib-interval 1; fr axis divided to base. **SEED**: face concave. $2n$=22. Chaparral, coastal scrub, etc.; 0–1500 m. CA-FP; Baja CA.

APIUM

Ann, bien, per, taprooted or fibrous-rooted from horizontal rhizome. **ST** prostrate to erect, hollow, rooting from lower nodes or not, glabrous. **LF**: blade oblong to obovate, 1-pinnate or ternate-pinnately dissected, lflets paired, lanceolate or ± round (segments linear). **INFL**: umbels compound, peduncled or not; bracts, bractlets conspicuous to 0; rays, pedicels few, spreading-ascending. **FL**: calyx lobes 0 or minute; petals wide, white or greenish white; projection atop ovary sometimes flat. **FR** ovate-oblong to round, compressed side-to-side; ribs subequal, thread-like to obtuse and corky; oil tubes per rib-interval 1; fr axis entire or notched at tip. **SEED**: face flat. $2n$=22. ± 20 spp.: gen s hemisphere (S.Am, s Afr, Australia, New Zealand), also Eurasia, 2 spp. weedy in CA. (Classical name for celery)

1. Pl ± erect, terrestrial; bracts, bractlets 0 .... ***A. graveolens***

1′ Pl prostrate to ascending, aquatic or becoming terrestrial; bracts few or 0, bractlets = or > pedicels ***A. nodiflorum***

A. graveolens L. CELERY Pl 5–15 dm. **ST** not rooting at nodes. **LF**: petiole 0.3–2.5 cm; blade 7–18 cm, oblong to obovate, lflets 2–4.5 cm, ovate to ± round, gen lobed. **INFL**: rays 7–16, 0.7–2.5 cm, subequal; calyx lobes minute. **FR** 1.5–2 mm diam, elliptic to nearly round; fr axis notched at tip. Wet places; gen < 1000 m. CA; temp zones worldwide; native to Eurasia. Cult and naturalized widely.

A. nodiflorum (L.) Lagasca Pl 1.5–10 dm. **ST** rooting at lower nodes. **LF**: petiole 3 dm; blade 10–20 cm, oblong, lflets ± 6 cm, lanceolate to ovate, crenate but terminal 3-lobed. **INFL**: rays 10–20, 1.5–2 cm, unequal; calyx lobes 0. **FR** 1.5–2 mm diam, ovate-oblong; fr axis entire. Wet places; gen < 1000 m. e SnFrB (reported from Niles Canyon, Alameda Co., 1933); native to Eurasia. Sporadically naturalized in temp zones worldwide.

BERULA CUTLEAF WATER-PARSNIP

1 sp. (Latin: water-cress)

B. erecta (Hudson) Cov. (p. 143) Per, < 15 dm, stoloned, glabrous. **ST** ascending or erect, branched, hollow, rooting at lower nodes. **LF**: petiole 4–12 cm, narrowly sheathing; blade 1.5 dm, oblong, 1-pinnate, much-dissected if submersed, lflets 7–12 pairs, 1–8 cm, oblong to ovate, sessile, serrate or lobed. **INFL**: umbels compound, peduncled, terminal or opposite lvs; bracts, bractlets lf-like, lanceolate; rays, pedicels ± unequal, spreading. **FL**: calyx lobes minute, persistent; petals wide, white. **FR** 1.5–2 mm, nearly round, compressed side-to-side, glabrous; ribs thread-like, inconspicuous in corky fr wall; oil tubes apparent but deeply embedded in fr wall; fr axis divided to base, its branches adhering to and falling with fr-halves. $2n$=18. Marshy areas, streams; < 1800 m. CA; Eurasia, Afr. Possibly TOXIC: implicated in some livestock poisonings.

BOWLESIA

Ann, per, taprooted, stellate-hairy. **ST**decumbent to erect, branched. **LVS** opposite to alternate; stipules sometimes slightly fused above node when lvs opposite; blade ovate to round, palmately lobed to compound, rarely subentire. **INFL**: umbels simple, few-fld, peduncled or sessile; bracts gen present; pedicels short to 0. **FL**: calyx lobes gen 0; petals gen wide, yellowish white, green, or purple, obtuse to acutish, tip not narrowed, not incurved. **FR** ovate-oblong to round, cylindric or ± 4-angled or slightly compressed front-to-back, stellate-hairy, prickly, or nearly glabrous; ribs obscure; fr central axis not an obvious structure. **SEED**: face ± flat. 16 spp.: temp S.Am, 1 in CA. (Wm. Bowles, Irish writer on Spanish natural history, 1705–1780) [Mathias & Constance 1965 Univ Calif Publ Bot 38]

B. incana Ruiz Lopez & Pavon (p. 143) Pl 0.5–6 dm. **LVS** gen opposite; petiole 1.5–12 cm; blade 0.5–3 cm, round-reniform, lobes 5–9, ± halfway to base, widely lanceolate to round, gen obtuse. **INFL** axillary, ± sessile; bracts lanceolate. **FL**: calyx lobes minute; petals oblong-ovate, yellowish green. **FR** 1.5–2 mm, ovate to round, inflated. $2n=16$. Shade of trees, rocks, shrubs; 20–1200 m. c&s SNF, CW, SCo, ChI, TR; to TX, S.Am.

BUPLEURUM

Ann, bien, per, shrub, taprooted. **ST** decumbent to erect, branched, gen glaucous. **LVS**: basal petioled; cauline gen sessile, clasping or fused around st at base; blades linear to ovate or obovate, simple, gen parallel-veined, margins entire or minutely serrate. **INFL**: umbels compound; bracts gen present; bractlets gen wide, conspicuous; rays, pedicels few–many, spreading. **FL**: calyx lobes 0; petals narrow to wide, yellow or dark purple, tips narrowed. **FR** oblong to round, slightly compressed side-to-side; ribs subequal, thread-like to narrowly winged; oil tubes per rib-interval gen several; fr axis divided to base. **SEED**: face ± flat. ± 100 spp.: Eurasia, s Afr, n N.Am, 1 reportedly naturalized in CA. (Greek: ox rib)

B. lancifolium Hornem. Pl 0.5–5 dm, glaucous. **LF**: basal 3–10 cm, linear to oblong-lanceolate; cauline lanceolate to widely ovate, base fused around st, margin scarious. **INFL**: peduncles 1–8 cm; bracts 0; bractlets gen 5, > fls and frs, lf-like, ovate to ± round, fused at base; rays 2–3. **FL**: corolla yellow. **FR** 2–3 mm, ovate to round, roughened. $2n=16$. May be naturalized in gardens; gen < 1000 m. CA; e US; native to Medit. Reported as *B. subovatum* Link.

CICLOSPERMUM

Ann, taprooted, glabrous. **ST** decumbent to erect, gen loosely branched. **LVS**: sessile or petioled; blade oblong to triangular-ovate, pinnately or ternate-pinnately dissected or compound, segments or lflets gen thread-like to linear. **INFL**: umbels compound or some simple; bracts, bractlets gen 0; rays few; rays, pedicels spreading. **FL**: calyx lobes 0; petals oblong, white, tips acutish but not incurved. **FR** narrowly ovate to elliptic, compressed side-to-side, glabrous or hairy; ribs subequal, thread-like to prominent and corky; oil tubes per rib-interval 1; fr axis shallowly notched. **SEED**: face flat. 3 spp.: S.Am, 1 ± worldwide aggressive weed. (Greek: circular seed)

C. leptophyllum (Pers.) Britton & E. Wilson (p. 143) Pl 0.5–6 dm. **LF** 3.5–10 cm; petiole 2.5–12 cm, gen 0 on cauline lvs, sheath margin scarious; segments 3–15 mm, thread-like to linear, entire. **INFL** < 2 cm, gen some sessile; rays 1–3; pedicels 6–20, 2–16 mm, spreading. **FR** 1.2–3 mm wide, elliptic to ovate. $2n=14$. Lawns, roadsides; < 150 m. CA; ± worldwide, warm temp. Sporadic in CA. [*Apium l.* (Pers.) Benth.]

CICUTA WATER HEMLOCK

Per, glabrous; rhizome divided internally into chambers, with sap that oxidizes to reddish brown, bearing fibrous or tuberous roots. **ST** erect, hollow. **LF**: blade oblong to triangular-ovate, 1–3-pinnate or ternate-pinnate, lflets linear to ovate-lanceolate, serrate or irregularly cut. **INFL**: umbels compound; bracts gen 0; bractlets gen inconspicuous; rays, pedicels many, spreading. **FL**: calyx lobes minute; petals wide, white, tips narrowed. **FR** ovate to round, slightly compressed side-to-side; ribs low, corky, sometimes unequally spaced; oil tubes per rib-interval 1; fr axis divided to base. **SEED**: face flat or concave. ± 4 spp.: Eurasia, N.Am. (Ancient Latin name) [Mulligan & Munro 1981 Canad J Plant Sci 61: 93–105] More evidence from ripe fr and chromosomes needed to substantiate proposed cryptic spp. TOXIC: both spp. below contain cicutoxin, a virulent poison; many livestock and human deaths recorded. The most lethally toxic native plants.

1. Fr gen round, ribs much wider than intervals between; areas surrounded by veins on lf lower surface coarse, gen some elongate .... ***C. douglasii***

1′ Fr gen ovate, ribs gen as wide as to much narrower than intervals between; areas surrounded by veins on lf lower surface fine, gen rounded or square .... ***C. maculata***

  2. Lf gen 1–2-pinnate; styles < 1 mm; inland .... var. ***angustifolia***

  2′ Lf gen 2-pinnate; styles gen > 1 mm; coastal .... var. ***bolanderi***

C. douglasii (DC.) J. Coulter & Rose (p. 143) Pl 15–30 dm. **LF** 1.5–4.5 dm, narrowly ovate to triangular-ovate, 1–2(3)-pinnate; lflets 1–10(15) cm, linear to widely lanceolate, acute or acuminate, subentire to coarsely serrate, areas surrounded by veins on lower surface coarse, gen some elongate. **INFL**: umbels compound, terminal and lateral; peduncles 2–18 cm; rays 15–30(35), 2–8 cm; pedicels 20–30, 2–10 mm. **FR** 2–4 mm, gen round; ribs much wider than intervals between. $2n=44$. Wet places, often in water; < 2500 m. NCo, CaRH, SNH, CCo, SCo, GB; to B.C., MT.

C. maculata L. Pl 10–15 dm. **LF** 1–4 dm, ovate to triangular-ovate, 1–2-pinnate; lflets 2–10 cm, lanceolate, acute or acuminate, coarsely to sparsely serrate, areas surrounded by veins on lower surface fine, gen rounded or square. **INFL**: umbels compound, terminal and lateral; peduncles 2.5–12 cm; rays 15–30, 2–4.5 cm; pedicels 15–30, 2–10 mm. **FR** 3–4 mm, gen ovate; ribs gen as wide as to much narrower than intervals between. $2n=22$. Wet places; < 2000 m. CA-FP; to AK, c Mex. Vars. distinguishable in CA chiefly by habitat, geog.

var. ***angustifolia*** Hook. **LF** gen 1–2-pinnate. **FL**: styles < 1 mm. Wet meadows, etc.; 1500–2100 m. SnBr, GB. [*C. occidentalis* E. Greene, *C. valida* E. Greene]

var. ***bolanderi*** (S. Watson) Mulligan **LF** gen 2-pinnate. **FL**: styles gen > 1 mm. Coastal wetlands; < 200 m. s ScV (Suisun Marshes), CCo, SCo. [*C. b.* S. Watson]

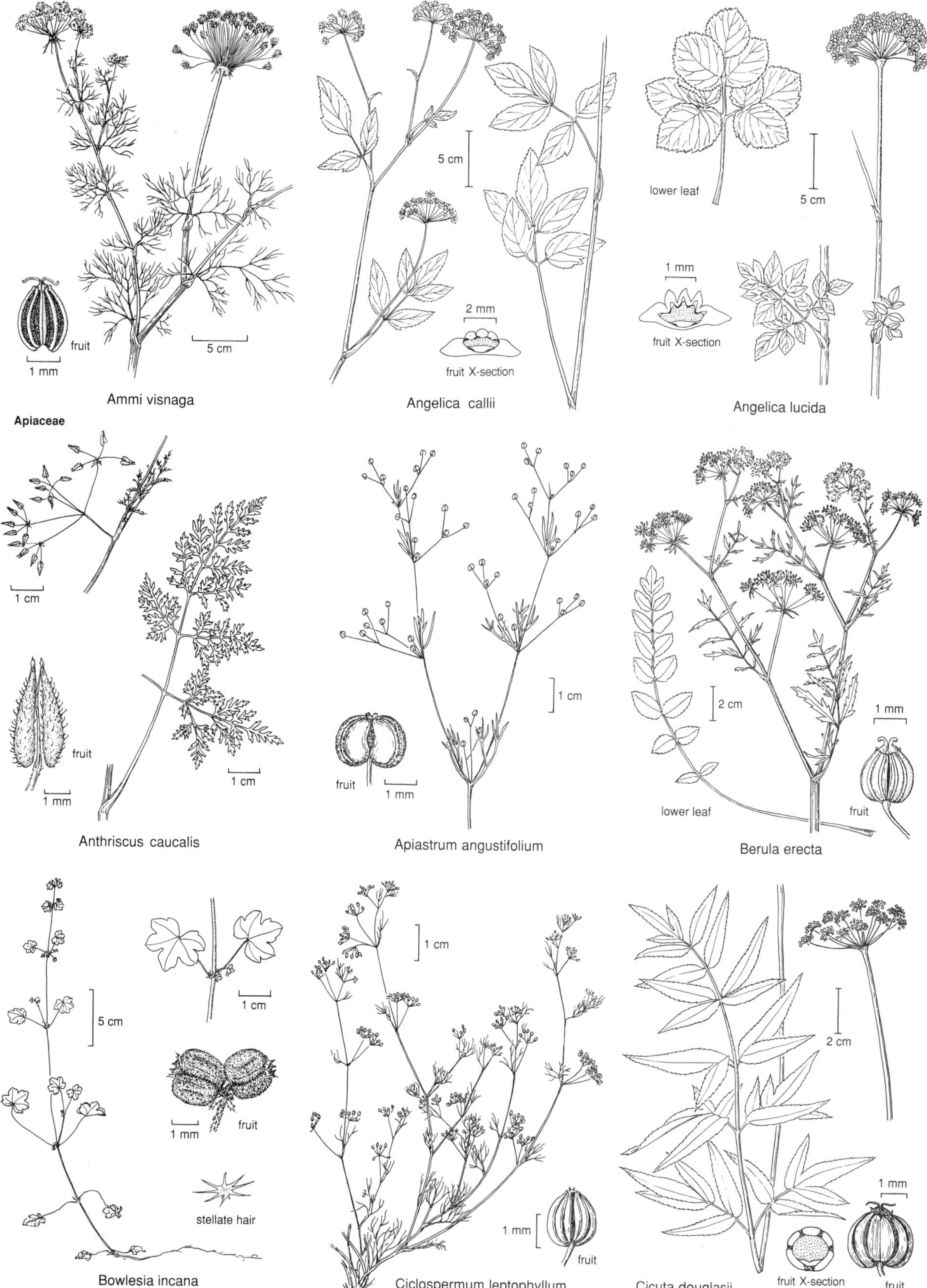
fruit
1 mm
5 cm
Ammi visnaga
5 cm
2 mm
fruit X-section
Angelica callii
lower leaf
5 cm
1 mm
fruit X-section
Angelica lucida
Apiaceae
1 cm
fruit
1 mm
1 cm
Anthriscus caucalis
1 cm
fruit
1 mm
Apiastrum angustifolium
2 cm
1 mm
lower leaf
fruit
Berula erecta
5 cm
1 cm
1 mm
fruit
stellate hair
Bowlesia incana
1 cm
1 mm
fruit
Ciclospermum leptophyllum
2 cm
1 mm
fruit X-section
fruit
Cicuta douglasii

CONIOSELINUM

Per, taprooted or short-rhizomed, glabrous or ± scabrous in infl. **ST** erect, branched. **LF**: blade ovate to widely triangular, 1–4-pinnately or -ternate-pinnately dissected or compound; segments or lflets toothed or lobed to deeply cut; sheaths conspicuously dilated. **INFL**: umbels compound; bracts gen 0; bractlets gen present, ± scarious; rays, pedicels many, spreading-ascending. **FL**: calyx lobes 0 or minute; petals wide, white, tip notched. **FR** oblong to ovate, very compressed front-to-back; marginal ribs widely thin-winged, those on back low, corky, or all ribs winged; oil tubes per rib-interval 1–4; fr axis divided to base. **SEED**: face flat or slightly concave. ± 10 spp.: N.Am, Eurasia. (Combined generic names *Conium* + *Selinum*)

C. pacificum (S. Watson) J. Coulter & Rose (p. 149) Pl 10–40 dm. **LF** triangular-ovate; petiole 3–15 cm; blade 0.5–2 dm, 2–3-pinnate, lflets 1–5 cm, oblong to ovate, coarsely serrate and pinnately cut; cauline lf sheaths inflated. **INFL** terminal and gen some lateral; peduncles 6–20 cm; rays 15–30, 2–4 cm, spreading-ascending; bractlets very narrow, scarious; pedicels 5–10 mm. **FR** 5–8 mm, oblong to oblong-ovate; ribs winged, marginal wider than others. $2n$=44. Ocean bluffs, often in coastal scrub; < 50 m. NCo, CCo; to AK, n Japan. [*C. chinense* (L.) Britton, Sterns, & Poggenburg]

CONIUM POISON HEMLOCK

Bien, taprooted; herbage glabrous, musty-scented. **ST** erect, branched. **LF**: blade ovate to triangular-ovate, pinnately dissected or compound, segments or lflets lanceolate or oblong to ovate, serrate to 1–2-pinnately lobed. **INFL**: umbels compound, terminal and lateral; bracts, bractlets small, few; rays, pedicels ± many, spreading-ascending. **FL**: calyx lobes 0; petals wide, white or yellowish, tips narrowed. **FR** ovate to round, slightly compressed side-to-side; ribs subequal, low; oil tubes per rib-interval 0; fr axis divided to base. **SEED**: face grooved. ± 6 spp.: Eur, s Afr. TOXIC: contains highly toxic alkaloids used in ancient Greece for capital punishment (e.g., Socrates); many human deaths, rarely eaten by livestock. (Greek name used by Dioscorides)

C. maculatum L. (p. 149) Pl 5–30 dm. **ST** gen purple-spotted or -streaked. **LF**: petiole dilated; blade 1.5–3 dm, widely ovate, gen 2-pinnate. **INFL** much branched; peduncles 2–8 cm; bracts 4–6, acuminate; bractlets 5–6, 1.5–2 mm, like bracts, gen ± fused at base, scarious; rays 10–20, 1.5–5 cm. **FR** 2–3 mm wide, ovate; ribs gen wavy. $2n$=22. Common. Moist, esp disturbed places; gen < 1000 m. CA-FP; Am; native to Eur.

CORIANDRUM

Ann, taprooted, glabrous. **ST** erect, branched. **LF**: blade oblong to ovate, ternately or pinnately lobed to pinnately dissected. **INFL**: umbels compound, terminal and lateral; bracts 0; bractlets few, small; rays, pedicels few, spreading-ascending. **FLS**: marginal bilateral; calyx lobes prominent, unequal; petals oblong, white or rosy, marginal 2-lobed, tips of all narrowed; styles elongate. **FR** ± round; halves not separating readily; ribs 1° and 2°, thread-like, in a hard fr-wall; oil tubes 0; fr axis divided to base. **SEED**: face concave. 1–2 spp.: Medit, 1 widely cult. (Greek name for this anciently cult condiment)

C. sativum L. CORIANDER, CILANTRO Pl 2–8 dm, foul-smelling. **LVS**: basal clustered, 3–15 cm, oblong to ovate, ternately or pinnately lobed to 1-pinnate with lflets 1–2 cm, ovate to round, petiole 2–10 cm; upper cauline ovate, pinnately dissected with segments 2–15 mm, thread-like to linear, petiole 0. **INFL**: peduncles 0 or 3–10 cm; bractlets 2–4 mm, linear; rays 2–8, 1–2.5 cm; pedicels 2–5 mm. **FL**: calyx lobes widely lanceolate. **FR** 2.5–5 mm wide. $2n$= 22. Disturbed places, often near gardens; gen < 1000 m. CA-FP; to Mex, trop Am; native to Medit.

CYMOPTERUS

Per, taprooted, gen glabrous. **ST** gen 0 or short. **LVS** mostly basal, membranous to subleathery or fleshy; blade oblong to widely ovate or round, palmately or pinnately lobed to 1–2-pinnately or -ternate-pinnately dissected or compound, segments or lflets linear to obovate, entire to variously lobed, gen spine-tipped. **INFL**: umbels compound, gen terminal, scapose, open to spheric, dense, peduncled; bracts, bractlets conspicuous and scarious (or rarely 0); rays few–many (rays and pedicels sometimes ± 0). **FL**: calyx lobes prominent to 0; petals oblong to obovate, white, yellow, or purple, tips narrowed; projection atop ovary 0. **FR** oblong to ovate, subcylindric to compressed front-to-back; ribs subequal or unequal, marginal and some or all others thin-or corky-winged, or rarely some or all wingless; oil tubes per rib-interval 1–several; fr axis 0 or divided to base. **SEED**: face flat to longitudinally concave or grooved. ± 50 spp.: w N.Am. (Greek: wave wing) [Mathias 1930 Ann Missouri Bot Gard 17:213–476] Generic boundaries fluctuating. Some spp. outside CA are TOXIC to livestock.

1. Rays ± 0 (umbels dense, spheric); bractlets 0 or poorly developed
  2. Lf minutely hairy or roughened; bracts conspicuous, fused below; fr glabrous .... ***C. cinerarius***
  2′ Lf glabrous; bracts 0; fr hairy
    3. Lf round, ternate .... ***C. ripleyi***
    3′ Lf oblong-ovate, ternate–2-pinnately or 2-pinnately dissected
      4. Pl sessile on taproot; fr wings unequal .... ***C. deserticola***
      4′ Pl with lfless stalk above taproot; fr wings subequal .... ***C. globosus***

1′ Rays ± developed (umbels open to ± dense, not spheric); bractlets evident
  5. Pl not woody; corolla purplish; lf simple to 2-pinnately dissected or compound, lobes, segments, or lflets wider than linear

6. Bracts 0; lf round-reniform, ternate .... *C. gilmanii*
6′ Bracts conspicuous, scarious, veiny; lf oblong-ovate, 1–2-pinnate or -ternate
7. Bracts and bractlets greenish or purplish, with many green or purple veins; pedicels < 1 mm *C. multinervatus*
7′ Bracts and bractlets white, with 1–5 green or white veins; pedicels 3–8 mm .... *C. purpurascens*
5′ Pl woody at base; corolla ± yellow or white; lf finely dissected, segments gen ± linear
8. Lf finely hairy; corolla white; umbels ± dense .... *C. aboriginum*
8′ Lf glabrous; corolla ± yellow; umbels open
9. Rays gen unequal; calyx lobes persistent; bractlets gen < mature pedicels .... *C. terebinthinus*
10. Lf blade ± as long as wide; KR, CaRH, SNH .... var. *californicus*
10′ Lf blade much longer than wide; c SNE, W&I .... var. *petraeus*
9′ Rays subequal; calyx lobes 0; bractlets > mature pedicels .... *C. panamintensis*
11. Lf segments not crowded, 3–20 mm, flexible .... var. *acutifolius*
11′ Lf segments ± crowded, 1–5 mm, rigid .... var. *panamintensis*

C. aboriginum M.E. Jones Pl 1–3.5 cm, finely hairy or scabrous, gray-green. **ST** 0. **LF**: petiole 2–13 cm; blade 1–4.5 cm, oblong, ternate–2-pinnately or 3-ternately dissected, segments 2–8 mm, linear. **INFL**: peduncles 8–30 cm, = or > lvs; bracts gen 0; bractlets linear, subscarious, acute; rays 3–10, 4–20 mm; pedicels 3–7 mm. **FL**: corolla white. **FR** 6–11 mm, oblong to ovate; ribs subequal, wings 2 × body in width; oil tubes per rib-interval 2–8. $2n=22$. Rocky slopes; 1500–2500 m. DMtns, W&I; NV.

C. cinerarius A. Gray Pl 7–8 cm, minutely hairy or roughened, glaucous. **ST** 0. **LF**: petiole 3–5 cm; blade 1–2.5 cm, oblong-ovate, 2-pinnately dissected, segments 1–3 mm. **INFL**: peduncles > lvs; bracts conspicuous, scarious-margined, fused below; bractlets obscure; rays, pedicels ± 0. **FL**: corolla white. **FR** 6 mm, narrowly wedge-shaped; ribs subequal, wings < body in width; oil tubes per rib-interval 5–8. Rocky mtn slopes; 2500–3500 m. SNH, W&I; NV.

C. deserticola Brandegee (p. 149) DESERT CYMOPTERUS Pl ± 15 cm, glabrous. **ST** 0. **LF**: petiole 4–10 cm; blade 2–6.5 cm, oblong-ovate, ternate-2-pinnately dissected, segments 1–4 cm, lanceolate, acuminate. **INFL**: peduncles = or > lvs; bracts, gen bractlets 0; rays, pedicels ± 0. **FL**: corolla purple. **FR** 5–7 mm, oblong-ovate to wedge-shaped; ribs unequal, marginal wings < body in width, others inconspicuous; oil tubes per rib-interval 3–5. $2n=22$. RARE. Sandy desert; ± 1500 m. w DMoj.

C. gilmanii C. Morton GILMAN'S CYMOPTERUS **ST** 12–23 cm, glabrous, glaucous; base fibrous. **LF**: petiole 8–18 cm; blade 2.5–4.5 cm, round-reniform, ternate, lflets obovate, deeply lobed, ultimate lobes toothed, spine-tipped. **INFL**: peduncles > lvs; bracts 0; bractlets inconspicuous, lf-like, linear to lanceolate; rays ± 8, 1–2 cm; pedicels 2–5 mm. **FL**: corolla purplish. **FR** 7–8 mm, widely ovate; ribs unequal, marginal and 1 or 2 other wings > body in width. RARE in CA. Limestone, gypsum slopes; 1000–2000 m. ne DMtns; NV.

C. globosus (S. Watson) S. Watson Pl 3–20 cm, with lfless stalk above taproot, glabrous, glaucous. **ST** (above lfless stalk) 0. **LF**: petiole 1–10 cm; blade 1–7 cm, oblong-ovate, ternate–2-pinnately or 2-pinnately dissected, segments 0.5–6 mm, ± indistinct. **INFL**: peduncles = or > lvs; bracts 0; bractlets linear, small; rays, pedicels ± 0. **FL**: corolla white or purple. **FR** 6–11 mm, narrowly wedge-shaped; ribs subequal, wings < body in width; oil tubes per rib-interval gen 1. $2n=22$. Uncommon. Sandy open flats; 1200–2100 m. SNE; NV, UT.

C. multinervatus (J. Coulter & Rose) Tidestrom Pl 4–20 cm, with lfless stalk above taproot or not, glabrous. **ST** (above lfless stalk) 0. **LF**: petiole 2–7 cm; blade 1–8.5 cm, oblong-ovate, 2-pinnately or ternate-pinnately dissected, glaucous, fleshy, segments 0.5–6 mm, ± indistinct. **INFL**: peduncles = or > lvs, 2–14 cm; bracts greenish or purplish, scarious, many-veined, forming a shallow sheath or cup; bractlets like bracts; fertile rays 1–5, 0.5–2.5 cm; pedicels < 1 mm. **FL**: corolla purplish. **FR** 8–17 mm wide, oblong-ovate to ovate; ribs subequal, wings 2–3 × body in width; oil tubes per rib-interval 3–9. Uncommon. Sandy and rocky (limestone?) slopes; 630–1500 m. DMoj; to UT, NM.

C. panamintensis J. Coulter & Rose Pl 0.5–4 dm, glabrous; base woody. **ST** 0. **LF**: petiole 1–10 cm; blade 1–14 cm, oblong-ovate to obovate, ternate–2–3-pinnately dissected, segments 1–5 mm, ± linear, acute, gen ± distinct. **INFL**: peduncles > lvs, 3–25 cm; bracts 0; bractlets linear, acuminate, fused at base; fertile rays 5–15, 1–6.5 cm, subequal; pedicels 4–13 mm. **FL**: corolla greenish yellow. **FR** 6–10 mm, oblong-ovate; ribs subequal, wings = or > body in width; oil tubes per rib-interval 1–5. Uncommon. Rocky slopes, canyon walls; 700–2500 m. DMoj. Vars. questionably distinct.

var. ***acutifolius*** (J. Coulter & Rose) Munz **LF**: segments not crowded, 3–20 mm, flexible. $2n=22$. Rocky canyon walls; 700–1000 m. DMoj.

var. ***panamintensis*** **LF**: segments ± crowded, 1–5 mm, rigid. Rocky slopes; 800–2500 m. DMtns.

C. purpurascens (A. Gray) M.E. Jones Pl 3–15 cm, with lfless stalk above taproot or not, with persistent fibers, glabrous. **ST** (above lfless stalk) 0–15 cm. **LF**: petiole 1–4 cm; blade 1.2–5 cm, oblong-ovate, 1–2-pinnately (rarely ternate-pinnately) dissected, glaucous, fleshy, segments 1–8 mm, ± indistinct. **INFL**: peduncles 1.5–7 cm, = or > lvs; bracts white, with 1–5 green or white veins, fused below; bractlets like bracts; fertile rays 3–5, 4–10 mm; pedicels 3–8 mm. **FL**: corolla purplish. **FR** 8–18 mm wide, widely ovate; ribs subequal, wings 2–3 × body in width; oil tubes per rib-interval 3–4. Shrubby slopes; 1300–2200 m. e DMoj, W&I; to ID, NV, AZ. ❀TRY.

C. ripleyi Barneby RIPLEY'S CYMOPTERUS Pl 10–15 cm, glabrous. **ST** 0. **LF**: petiole 3–10 cm; blade 2–5 cm, round, ternate, lflets wedge-shaped, deeply 3-lobed, lobes again lobed. **INFL**: peduncles > lvs; bracts 0; bractlets small, chaffy; rays, pedicels ± 0. **FL**: corolla purple (rarely white). **FR** 6–7 mm, wedge-shaped to obovate, hairy; ribs unequal, marginal winged, others not; oil tubes minute. $2n=22$. RARE in CA. Sandy soil; 1000–1600 m. s SNH, s SNE, n DMtns; NV. [var. *saniculoides* Barneby]

C. terebinthinus (Hook.) M.E. Jones Pl 1.5–4.5 dm, gray-green, glabrous; base woody. **ST** 0 to very short. **LF**: petiole 2–16 cm; blade 1.5–18 cm, ± ovate, pinnately or ternate-pinnately dissected, segments 1–4 mm, linear, ± rigid, acute. **INFL**: peduncles 1–3.5 cm, gen < lvs; bracts 0; bractlets 2–6 mm, gen linear, acute; rays 3–24, 0.5–8 cm, gen unequal; pedicels 1–8 mm. **FL**: corolla yellow. **FR** 5–10 mm wide, ± ovate; ribs gen subequal, wings often irregularly curled, = or > body in width; oil tubes per rib-interval 3–12. Rocky or sandy slopes; 150–3500 m. n&c CA; WA, OR, to Rocky Mtns. [*Pteryxia t.* (Hook.) J. Coulter & Rose]

var. ***californicus*** (J. Coulter & Rose) Jepson (p. 149) Pl herbage gray-green to bright green. **ST** 0 to very short. **LF**: blades ± as long as wide, appearing ± full from relatively large number and size of segments. $2n=22$. Sand, rocks; 150–3500 m (lowest on serpentine). KR, CaRH, SNH; n NV. [*Pteryxia t.* var. *c.* (J. Coulter & Rose) Mathias]

var. ***petraeus*** (M.E. Jones) Goodrich Pl herbage gray-green. **ST** very short. **LF**: blades much longer than wide, appearing skeleton-like from relatively small number and size of segments. $2n=22$. Rocky alpine slopes; 1800–3400 m. c SNE, W&I; NV, UT. [*Pteryxia p.* (M.E. Jones) J. Coulter & Rose]

DAUCUS

Ann, bien, taprooted, hairy. **ST** decumbent or erect, gen ± branched. **LF**: blade oblong, pinnately dissected, segments linear to lanceolate. **INFL**: umbels compound; bracts, bractlets gen present; bracts conspicuous, gen pinnately lobed; bractlets entire to toothed; rays gen many, spreading, in fr incurving to form a nest-like umbel. **FLS**: outer sometimes ± bilateral; calyx lobes 0 or evident; petals wide, white, tips narrowed, unequally 2-lobed. **FR** oblong to ovate, compressed front-to-back; ribs 10, 1° thread-like and bristly, 2° winged and prickly; oil tubes 1 beneath each 2° ribs; fr axis entire or notched at tip. ± 20 spp.: Am, Eurasia, n Afr, Australia. (Greek: carrot) [Sáenz Laín 1980 Ann Jard Bot Madrid 37:481–533]

1. Bien; bract segments elongate, narrowly linear to ± thread-like; central fl of umbel gen purple; fr widest at middle .... ***D. carota***
1′ Ann; bract segments short, linear to lanceolate; central fl of umbel white; fr widest below middle ...... ***D. pusillus***

D. carota L. CARROT, QUEEN ANNE'S LACE Pl 1.5–12 dm, gen branched. **LF**: petiole 3–10 cm; blade 5–15 cm, segments 2–12 mm, linear to lanceolate, acute, entire or with a few irregular cuts, bristly to glabrous. **INFL**: peduncles 2.5–6 cm, bristles reflexed to spreading; rays 3–7.5 cm; pedicels 3–10 mm. **FR** 3–4 mm, ovate. $2n$=18. Roadsides, disturbed places; 0–1200 m. w CA-FP; to e N.Am; native to Old World. [Small 1978 Canad J Bot 56:248–276] Sporadic in CA; widely naturalized, sometimes hybridizing with cultivars.

D. pusillus Michaux (p. 149) Pl 0.3–9 dm, gen simple or few-branched. **LF**: petiole 4–15 cm; blade 3–10.5 cm, segments 1–5 mm, linear, acute, entire, ± bristly. **INFL**: peduncles 1–4.5 cm, bristles reflexed to spreading; rays 0.4–4 cm; pedicels 2–9 mm. **FR** 3–5 mm, oblong. n=22. Rocky or sandy places; 0–1500 m. CA-FP (esp coastal); to B.C., se US, S.Am.

ERYNGIUM

Bien, per, from taproot, clustered roots, or rhizomes, gen glabrous, gen ± spiny. **ST** creeping to erect, rooting at nodes, branched or not. **LVS** basal and gen also cauline; petioles 0 or present; blades linear to triangular-ovate or round, gen pinnately or palmately lobed or dissected, rarely entire, often sharply toothed or ciliate, net-veined; juvenile lvs linear, segmented. **INFL**: heads 1–many in cymes, racemes, or panicles; bracts in 1 or more series, a single bractlet accompanying each fl; rays, pedicels 0. **FL**: calyx lobes prominent, persistent on fr; petals oblong to ovate, white to blue or purple, tip long; projection atop ovary 0. **FR** obovate to round, not compressed to very compressed front-to-back, densely scaly or tubercled or some surfaces glabrous; ribs 0; oil tubes inconspicuous; fr central axis not an obvious structure. **SEED**: face gen flat. ± 200 spp.: Am, Eurasia, Australia, New Zealand. (Ancient Greek name used by Theophrastus) [Sheikh 1983 Madroño 30:93–101] CA spp. (sect. *Armata*) are gen in vernal pools, polyploid, poorly defined, apparently interbreeding. Basal lvs are described unless stated otherwise.

1. Pl prostrate or decumbent, producing roots, juvenile lvs at nodes; heads in racemes .... ***E. racemosum***
1′ Pl decumbent to erect, not rooting at nodes; heads in cymes
 2. Pl densely, finely hairy throughout; heads 5–7-fld .... ***E. constancei***
 2′ Pl glabrous (rarely sparsely hairy on lvs only); heads > 10-fld
  3. Heads bright blue or purple; bractlets gen regularly 3-toothed at tip .... ***E. articulatum***
  3′ Heads greenish white, sometimes faintly bluish; bractlets entire to deeply pinnately lobed but not regularly 3-toothed at tip
   4. Bracts and bractlets very rigid, margins gen entire, prominently thickened
    5. Infl as a whole yellowish green; lf gen sharply serrate to irregularly cut; coastal .... ***E. armatum***
    5′ Infl as a whole silvery green; lf pinnately sharply lobed; n&c SNF .... ***E. pinnatisectum***
   4′ Bracts and bractlets ± flexible, margin gen sharply toothed, not thickened
    6. Bracts similar to bractlets, spines on outer side dense
     7. Lf blades deeply pinnately lobed, >> petiole; GV, adjacent foothills .... ***E. castrense***
     7′ Lf blades sharply serrate to sharply lobed, > petiole; MP .... ***E. mathiasiae***
    6′ Bracts unlike bractlets, spines on outer side 0 or few
     8. Sepals pinnately sharply lobed or sharply toothed; s GV, c&s SNF .... ***E. spinosepalum***
     8′ Sepals gen entire; NCoR, CaRH, SnFrB, SCoR, SCo, MP
      9. Main st 0, pl branching or spreading directly from basal rosette; basal lvs gen > branches; fr scales subequal; CaRH, MP .... ***E. alismaefolium***
      9′ Main st branching 1–5 cm above basal rosette; basal lvs gen < branches; fr scales gen unequal; NCoR, GV, SnFrB, SCoR, SCo
       10. Petiole < lf blade; fr scales obtuse .... ***E. vaseyi***
       10′ Petiole gen > lf blade; fr scales acuminate .... ***E. aristulatum***
        11. Styles in fr >> calyx; NCoR, SnFrB .... var. ***aristulatum***
        11′ Styles in fr ± = calyx; SnFrB, SCoR, SW
         12. Pl stout, ascending or erect; bractlets sharply toothed .... var. ***hooveri***
         12′ Pl weak, spreading; bractlets entire .... var. ***parishii***

E. alismaefolium E. Greene Pl low, decumbent to ascending, 0.5–3 dm, branching or spreading directly from basal rosette, glabrous. **LF** > branch; petiole short to = blade; blade 6–15 cm, lanceolate to narrowly obovate, sharply serrate and irregularly cut to pinnately lobed. **INFL**: heads 5–12 mm, spheric, in cymes; peduncles 0.5–1.5 cm; bracts gen 7, 7–16 mm, > heads, linear-lanceolate, sharply ciliate; bractlets 4–8 mm. **FL**: sepals 1.5–3 mm, lanceolate, entire; petals oblong, white; styles 3–3.5 mm, 2 × sepals. **FR** 2–2.5 mm, ovate; scales dense, gen subequal, lanceolate, acuminate. $2n$=32. Vernal pools, flooded meadows; 1300–1800 m. CaRH, MP; to OR, ID, NV.

E. aristulatum Jepson Pl gen ascending or erect, 1–9 dm, slender or stout, gen branching loosely from main st, 2–5 cm above basal rosette, gen glabrous. **LF** < branch; petiole 5–27 cm, gen > blade; blade 3–10 cm, lanceolate to oblanceolate, crenate to coarsely sharply serrate, irregularly cut, or lobed. **INFL**: heads 5–12 mm, subspheric, in cymes; peduncles 0.5–1.5 cm; bracts 5–8, 6–27 mm, = to 2 × heads, linear to linear-lanceolate, entire or spine-margined, minutely hairy or roughened; bractlets 5–10 mm. **FL**: sepals 1.7–2.8 mm, lanceolate to ovate, entire; petals oblanceolate, white; styles 1.5–3.5 mm, sometimes purplish. **FR** 1.5–2.5 mm, oblong-ovate; scales dense, unequal, lanceolate to ovate, acuminate, roughened. Abundant. Vernal pools, ditches, etc.; 0–1000 m. NCoR, SnFrB, SCoR, SCo; Baja CA.

var. ***aristulatum*** **FR**: styles >> calyx. 2*n*=32,64. Vernal pools, lake shores, drying lakes, wet depressions; 0–1000 m. NCoR, SnFrB. Most widespread, variable var.; apparently hybridizes with other vars., spp. ❀SUN,IRR:15,**16,17**.

var. ***hooveri*** Y. Sheikh HOOVER'S BUTTON-CELERY Pl stout, ascending to erect. **INFL**: bractlets sharply toothed. **FR**: styles ± = calyx. 2*n*=32. UNCOMMON. Vernal pools, lagunas. s SnFrB, SCoR.

var. ***parishii*** (J. Coulter & Rose) Jepson SAN DIEGO-BUTTON CELERY Pl weak, spreading. **INFL**: bractlets entire. **FR**: styles ± = calyx. 2*n*=32. **ENDANGERED** CA. Vernal pools, marshes; 1–150 m. s SCo, PR (San Diego, Riverside cos.); Baja CA. Now apparently confined to mesas near San Diego, Santa Rosa Mesa.

E. armatum (S. Watson) J. Coulter & Rose Pl low, decumbent to erect, 1–5 dm, gen branching from main st, 2–5 cm above basal rosette, glabrous. **LF** thick; petiole < blade to 0; blade 10–30 cm, oblanceolate, remotely sharply serrate to irregularly cut. **INFL**: heads 5–8 mm, ± spheric, in cymes; peduncles 0–1 cm; bracts 7–8, 1.5–2.3 cm, 1.5–2 × heads, lanceolate, margins gen entire, thickened; bractlets 7.5–10 mm. **FL**: sepals 3–4.5 mm, lanceolate to ovate-lanceolate, gen entire; petals oblanceolate, white or cream-colored (rarely purplish); styles 3.5–4 mm, = or < sepals. **FR** 1.5–2.5 mm, oblong-ovate; scales dense, unequal, ovoid, obtuse. 2*n*=32. Depressions in coastal prairie, bluffs; 0–200 m. NCo, CCo. ❀SUN, IRR:**15,17**.

E. articulatum Hook. Pl stout, erect, 0.3–1.2 m, branching from main st, 2–5 cm above basal rosette, glabrous. **LF**: petiole >> blade; blade 5–10 cm, lanceolate to narrowly ovate, coarsely sharply serrate or irregularly cut. **INFL**: heads 1–2.5 cm, ovoid to subspheric, bright blue or purple, in cymes; peduncles 0.5–2.5 cm; bracts 8–17, 1.5–2.2 cm, 2–3 mm wide, ± = heads, sharply toothed; bractlets 8–10 mm, 1–1.2 mm wide, gen regularly 3-toothed at tip. **FL**: sepals 3–3.5 mm, linear-lanceolate, gen entire; petals oblanceolate, bright blue or purple; styles 2.5–3.5 mm, = or slightly > sepals. **FR** 2.5–4 mm, oblong-ovate; scales dense, lanceolate, acuminate, bearing club-shaped papillae. 2*n*=32. Lake and stream margins, marshes; 0–1800 m. KR, NCoR, CaRH, deltaic ScV (Suisun Marshes), MP; to WA, ID.

E. castrense Jepson Pl ascending to erect, bushy, 2–6 dm, branching profusely from main st 2–5 cm above basal rosette, spiny, glabrous. **LF**: petiole short to 0, << blade; blade 10–30 cm, oblong to lanceolate, deeply pinnately lobed, often more than once so. **INFL**: heads 8–15 mm, ± spheric, in cymes; peduncles 0.5–2.5 cm; bracts 7–9, 1.5–3 cm, 1.5–2 × heads, linear, densely spiny on both surfaces; bractlets 15–20 mm, 2–5 mm wide. **FL**: sepals 3–3.5 mm, lanceolate, margin scarious, often toothed; petals oblanceolate, white or faintly purplish; styles 3.5–4.5 mm, gen > sepals. **FR** 3–3.5 mm, oblong-ovate; scales dense, slightly unequal, lanceolate, acuminate, papillate. 2*n*=32,64. Vernal pools, wet depressions, pond margins; 0–800 m. CaRF, n&c SNF, adjacent GV. [*E. vaseyi* var. *c.* (Jepson) Mathias & Constance]

E. constancei Y. Sheikh (p. 149) LOCH LOMOND BUTTON-CELERY Pl decumbent or ascending, 2–3 dm, slender, branched loosely 1–2 cm above basal rosette, densely minutely hairy throughout. **LF**: petiole 8–12 cm, gen >> blade; blade 3–4 cm, ± lanceolate, ± entire to sharply serrate to sharply lobed. **INFL**: heads 5–7-fld, 3–5 mm, spheric, in cymes; peduncles 5–8 cm; bracts 4–5, 6–7 mm, > heads, linear-lanceolate, spine-margined; bractlets 5–6 mm. **FL**: sepals ± 2 mm, lanceolate, entire; petals oblanceolate, gen white; styles 3–3.5 mm, >> sepals. **FR** 1.6–2.2 mm, ovate; scales dense, unequal, irregularly shaped, roughened. 2*n*=32. **ENDANGERED** CA, US. Vernal pools in meadows; ± 800 m. NCoR (Lake Co.).

E. mathiasiae Y. Sheikh MATHIAS' BUTTON-CELERY Pl stout, ascending to erect, 3–4 dm, branching from main st, 1.5–2 cm above basal rosette, glabrous. **LF**: petiole 6–10 cm, blade; blade 10–17 cm, lanceolate to narrowly obovate, sharply serrate to sharply lobed. **INFL**: heads 8–12 mm, subspheric, in cymes; peduncles 0.8–1.8 cm; bracts 6–7, 13–23 mm, 1.5–2 × heads, linear, densely spiny on both surfaces; bractlets 9–15 mm, 1–1.3 mm wide. **FL**: sepals 3–3.5 mm, lanceolate, gen entire; petals oblong, white; styles 2.5–3 mm, ± = sepals. **FR** 2.5–3 mm, oblong; scales dense, unequal, lanceolate, acuminate to awned, smooth or roughened. 2*n*=64. Roadside ditches, wet depressions; 1000–1500 m. MP.

E. pinnatisectum Jepson (p. 149) TUOLUMNE BUTTON-CELERY Pl erect, 1.5–5 dm, stout, branching from main st 1.5–2.8 cm above basal rosette, glabrous or roughened in infl. **LF**: petiole << blade; blade 10–30 cm, lanceolate; pinnately sharply lobed, lobes gen opposite, white, margins thickened. **INFL**: heads 1.5–2 cm, spheric, in cymes; peduncles 0.2–1 cm; bracts 9–11, 1.5–1.7 cm, 2 × heads, linear-lanceolate, prominently white, margins thickened, gen entire; bractlets 9–13 mm. **FL**: sepals 3.5–4 mm, entire, lanceolate; petals oblanceolate, white; styles 3.5–4 mm, gen = sepals. **FR** 3– 3.5 mm, oblong-ovate; scales dense, unequal, lanceolate, acuminate. 2*n*=32. UNCOMMON. Vernal pools, wet depressions; 250– 450 m. Near boundary of n&c SNF, adjacent GV (Sacramento, Amador, Calaveras, Tuolumne cos.).

E. racemosum Jepson DELTA BUTTON-CELERY Pl prostrate or decumbent, 1–5 dm, slender, branching from main st 1.5–3 cm above basal rosette, producing roots and juvenile lvs at nodes, glabrous. **LF**: petioles > blade; blade 3–5 cm, lanceolate to ± oblong, subentire to sharply serrate or sharply lobed. **INFL**: heads 5–8 mm, spheric-ovoid, in racemes; peduncles 0.5–1 cm; bracts gen 5, 8–12 mm, 2 × heads, linear to narrowly lanceolate, spine-margined; bractlets 6–8 mm. **FL**: sepals 1–1.5 mm, ovate, entire; petals oblong to oblanceolate, white to faintly purplish; styles 2.5 mm, > sepals. **FR** 1.5 mm, ovate; scales dense, unequal, lanceolate, roughened. 2*n*=32. **ENDANGERED** CA. Seasonally flooded clay depressions in riparian scrub; 3–30 m. n SnJV, adjacent SNF. Threatened by agriculture, flood control.

E. spinosepalum Mathias SPINY-SEPALED BUTTON-CELERY Pl erect, 3–7.5 dm, stout, branching from main st 2–5 cm above basal rosette, glabrous. **LF**: petiole 0.5–2 cm, << blade; blade 9–35 cm, oblong to oblanceolate, sharply serrate to pinnately sharply lobed. **INFL**: heads 0.8–2 cm, ovoid to spheric, in cymes; peduncles 0–2 cm; bracts 7–8, 1.5–3 cm, gen > heads, linear-lanceolate, spines 3–4 pairs on margins, gen some on back; bractlets 7–15 mm, 1.5–1.7 mm. **FL**: sepals 3.5–4.5 mm, lanceolate, pinnately sharply lobed or sharply toothed; petals oblong, white; styles 3–4 mm, < sepals. **FR** 2.5–3 mm, oblong-ovate; scales dense, unequal, lanceolate, acuminate, roughened. 2*n*= 32. RARE. Vernal pools, depressions; 100–200 m. e SnJV, adjacent SNF. [*E. vaseyi* var. *globosum* (Jepson) Mathias & Constance] Apparently intergrades with *E. castrense*, possibly *E. vaseyi*.

E. vaseyi J. Coulter & Rose Pl decumbent to ascending, 1.5–5 dm, branching horizontally from main st 1–5 cm above basal rosette, glabrous. **LF**: petiole 1–4 cm, < blade; blade 8–24 cm, lanceolate to oblong, deeply pinnately sharply lobed. **INFL**: heads 8–13 mm, 7–10 mm, subspheric, in cymes; peduncles gen 1–4 cm; bracts 7–8, 1.5–2.5 cm, = to 2 × heads, linear, densely spiny on margins; bractlets 8–15 mm, gen spiny on margins. **FL**: sepals 2–3 mm, lanceolate to ovate, gen entire; petals oblanceolate to oblong, white or cream-colored; styles 2–2.5 mm, = to > sepals. **FR** 2–2.5 mm, ovate; scales dense, unequal, lanceolate, obtuse. 2*n*=32. Vernal pools, (alkaline) depressions; 10–600 m. ScV, SCoRI, SCo. [var. *vallicola* (Jepson) Munz] Perhaps hybridizing historically with *E. castrense*, *E. spinosepalum* in SnJV.

FOENICULUM

1 sp. (Latin: fennel)

F. vulgare Miller (p. 149) FENNEL Per, taprooted, 0.9–2 m, glabrous, glaucous, anise-or licorice-scented. **ST** erect, branched, solid. **LF**: petiole 7–14 cm, conspicuously sheathing; blade 3–4 dm wide, triangular-ovate, pinnately finely dissected, segments 4–40 mm, thread-like. **INFL**: umbels compound, peduncled; bracts, bractlets 0; rays 15–40, unequal, 1–4 cm, spreading-ascending to ascending; pedicels 18–25, 1–10 mm, subequal. **FL**: calyx lobes 0; petals wide, yellow, tips narrowed. **FR** 3.5–4 mm, oblong-ovate, compressed side-to-side, glabrous; ribs subequal, prominent, acute; oil tubes per rib-interval 1; fr axis divided to base. **SEED**: face gen flat. 2*n*=22. Roadsides, waste places; 0–350 m. CA-FP; native to s Eur; widely escaped from cult in w hemisphere. Locally abundant and invasive.

GLEHNIA

1 sp. (P. von Glehn, Russian botanist, 19th century)

G. littoralis (A. Gray) Miq. ssp. ***leiocarpa*** (Mathias) Hultén (p. 149) Per, taprooted, low, prostrate or ± spreading, sparsely puberulent to tomentose. **ST** ± 0. **LF** fleshy; petiole 2.5–15 cm; blade 2.5–15 cm wide, widely ovate, 1–2-ternate or ternate-pinnate, lflets 0.5–5 cm wide, ovate, often 3-lobed, serrate, upper surface green and glabrous or becoming glabrous, lower white-tomentose. **INFL**: umbels compound, peduncled; bracts 0–few; bractlets several, lanceolate, long-acuminate; rays 5–15, 0.5–4.5 cm, stout, hairy; pedicels 0. **FL**: calyx lobes minute; petals widely lanceolate, white, tips narrowed; projection atop ovary 0. **FR** 4–12 mm, oblong to round, compressed front-to-back, glabrous to sparsely tomentose; ribs subequal, conspicuously corky-winged; oil tubes per rib-interval several, large; fr axis divided to base. **SEED**: face concave. 2*n*=22. Ocean beaches; ± 0 m. NCo; to AK. [*G. leiocarpa* Mathias] ❀DRN,IRR,SUN:5,17;DFCLT.

HERACLEUM

Per from taproot or clustered roots. **ST** stout, erect, gen branched, hollow. **LVS**: blades oblong to round, simple to ternately, pinnately, or palmately compound (rarely simple), lflets large, lobed or serrate; uppermost cauline often reduced to enlarged sheaths. **INFL**: umbels compound, large, often sterile at margins; bracts 0–few, often deciduous; bractlets gen present, persistent; rays, pedicels many, spreading-ascending. **FLS**: marginal bilateral, with outer petals > others, 2-lobed; calyx lobes gen 0; petals wide, white, yellowish, or rosy. **FR** oblong-ovate to round or obcordate, very compressed front-to-back; ribs unequal, marginal thin-winged, veined near outer margin, others thread-like; oil tubes per rib-interval 1–2, unequal in length; fr axis divided to base. **SEED**: face flat. ± 80 spp.: Eurasia, e Afr, 1 in N.Am. (Hercules, presumably from large stature of some spp.)

H. lanatum Michaux (p. 155) COW PARSNIP Pl 1–3 m, stout, strong-scented, tomentose. **LF** round to reniform; petioles 1–4 dm, widely sheathing, upper sheaths enlarged, bladeless; blades 2–5 dm wide, ternate, lflets 1–4 dm wide, ovate to round, cordate, coarsely serrate and lobed. **INFL** tomentose or long-hairy; peduncle 5–20 cm; rays 15–30, 5–10 cm, unequal; pedicels 8–20 mm. **FL**: petals obovate, white. **FR** 8–12 mm, obovate to obcordate, ± hairy. 2*n*=22. Moist places, wooded or open; < 2600 m. CA-FP, GB; to AK, e US, AZ. Relationship to some Eurasian taxa unclear; only native sp. in family occurring on both coasts of N.Am. ❀**4,5**,IRR:**6,15–17,24**&SHD:**7**,14,18–23.

HYDROCOTYLE

Per, creeping or sprawling, glabrous or hairy; rhizomes or st rooting at nodes. **LVS** simple; petiole scarious-stipuled, not sheathing; blade ± round, peltate or not, entire to deeply lobed. **INFL**: umbels simple, open or dense; bracts 0 or inconspicuous; pedicels 0–many, spreading. **FL**: calyx lobes 0 or minute; petals obtuse or acute, greenish or yellowish white to purplish, tip not incurved. **FR** elliptic to round, very compressed side-to-side; ribs subequal, thread-like, distinct or not; oil tubes 0 (but individual oil cells in fr wall); fr central axis not an obvious structure. **SEED**: face flat to convex. ± 100 spp.: worldwide, esp s hemisphere. (Greek: water cup, apparently from lf shape)

1. Lf blade round-peltate
  2. Fls long-pedicelled; umbels open ..... ***H. umbellata***
  2′ Fls ± sessile; umbels small, head-like, in an interrupted, simple or forked spike ..... ***H. verticillata***

1′ Lf blade round-reniform, not peltate
  3. Pl ± aquatic; fl, fr pedicelled ..... ***H. ranunculoides***
  3′ Pl terrestrial; fl, fr ± sessile
    4. Pl hairy; fls 10–20 per umbel, subsessile ..... ***H. moschata***
    4′ Pl glabrous; fls 3–10 per umbel, sessile ..... ***H. sibthorpioides***

H. moschata Forster f. Pl low, hairy. **LF**: petiole 1.5–6 cm, slender; blade 1–2.5 cm wide, round-reniform, shallowly 5–7-lobed, minutely crenate, both surfaces hairy. **INFL**: umbels head-like, 10–20-fld; peduncles 1–20 mm, ± hairy; pedicels ± 0. **FR** 1–3 mm, elliptic; ribs acute, evident. 2*n*=48. Reported as lawn weed; ± 100 m. SCo (Brentwood, Los Angeles Co.); native to New Zealand.

H. ranunculoides L.f. Pl fleshy, floating or creeping, glabrous. **LF**: petiole 5–35 cm, stout; blade gen 2–5 cm wide, round-reniform, often wider than long, deeply 3–7-lobed, entire to minutely crenate. **INFL**: umbels dense, 5–10-fld; peduncles 1–5 cm; pedicels short. **FR** 1–3 mm, elliptic to round; ribs obscure. 2*n*=48. Lake margins, pools, etc.; 10–1500 m. Sporadic in CA-FP; to e N.Am, S.Am. ❀WET:**4–7**,8–10,**14–24**;INV.

H. sibthorpioides Lam. Pl low, glabrous. **ST** creeping, thread-like. **LF**: petiole 1–5 cm, thread-like; blade gen 1–2 cm wide, round-reniform, shallowly 5–7-lobed, minutely crenate. **INFL**:

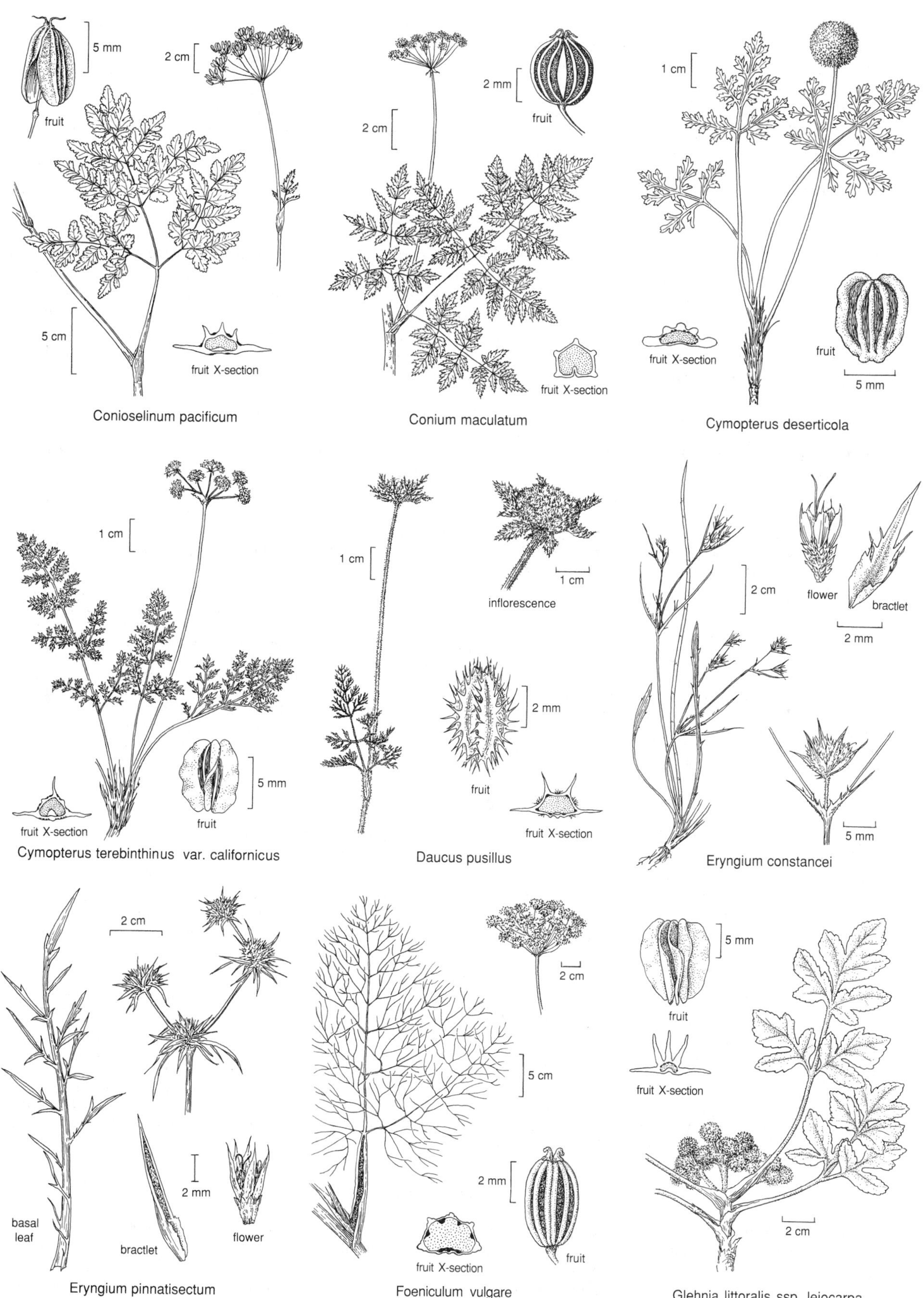
5 mm
fruit
2 cm
5 cm
fruit X-section
Conioselinum pacificum
2 cm
2 mm
fruit
fruit X-section
Conium maculatum
1 cm
fruit X-section
fruit
5 mm
Cymopterus deserticola
1 cm
fruit X-section
fruit
5 mm
Cymopterus terebinthinus var. californicus
1 cm
1 cm
inflorescence
2 mm
fruit
fruit X-section
Daucus pusillus
2 cm
flower
bractlet
2 mm
5 mm
Eryngium constancei
2 cm
basal leaf
bractlet
2 mm
flower
Eryngium pinnatisectum
2 cm
5 cm
2 mm
fruit X-section
fruit
Foeniculum vulgare
5 mm
fruit
fruit X-section
2 cm
Glehnia littoralis ssp. leiocarpa

umbels head-like, 3–10-fld; peduncles 5–15 mm; pedicels 0. **FR** 1–1.5 mm wide, round; ribs acute, evident. $2n=24$. Reported as lawn weed; ± 300 m. SCo (Pasadena, Los Angeles Co.); native to Asia.

H. umbellata L. Pl gen low, glabrous. **ST** creeping or floating. **LF**: petiole 0.5–4.5 cm; blade gen 1–5 cm wide, round-peltate, crenate or very shallowly, crenately, subequally 8–20-lobed. **INFL**: umbel open, 10–60-fld; peduncles 1.5–35 cm; pedicels 2–25 mm, unequal, spreading to reflexed. **FR** 1–2 mm, elliptic; ribs obtuse. $2n=48$. Stream banks, marshy areas; 10–1000 m. CA-FP (esp s); to e N.Am, S.Am. ❀WET:**4–7**,8–10,**14–24**;INV.

H. verticillata Thunb. (p. 155) Pl low, glabrous. **ST** creeping. **LF**: petiole 0.5–25 cm, slender; blade 1–4 cm wide, round-peltate, crenate or very shallowly, crenately, subequally 8–13-lobed. **INFL**: umbels small, 1–15-fld, in a simple or forked spike; peduncles 1.5–20 cm; pedicels 0 or short. **FR** 1–3 mm, elliptic; ribs acute, evident. Swampy ground, lake margins; < 100 m. CA-FP; to e N.Am, S.Am, Hawaii, s Afr. [var. *triradiata* (A. Rich.) Fern.] ❀WET:**5**, 7,14,**15–24**;INV.

LIGUSTICUM

Per, taprooted, glabrous to minutely scabrous. **ST** erect, leafy, gen fibrous at base, gen branched. **LF**: blade oblong to round, ternately or pinnately compound or dissected, lflets oblong to obovate, entire to deeply pinnately lobed, segments linear to oblong. **INFL**: compound umbels; bracts gen 0; bractlets 0 or inconspicuous; rays, pedicels few–many, spreading-ascending. **FL**: calyx lobes minute; petals wide, white (or pinkish). **FR** oblong to elliptic, slightly compressed side-to-side; ribs subequal, thread-like to narrowly winged; oil tubes per rib-interval gen several; fr axis divided to base. **SEED**: face flat to concave. ± 25 spp.: Eurasia, N.Am. (Liguria, Italy, home of the related *Levisticum*, lovage) [Leute 1970 Ann Naturhist Mus Wien 74:457–519] Genus and spp. poorly defined.

1. Herbage and infl minutely scabrous or puberulent; fr ribs thread-like to acute, barely winged .... ***L. apiifolium***
1′ Herbage, infl ± glabrous; fr ribs distinctly but narrowly thin-winged
 2. Cauline lvs 1–3, gen 2–3, ± = basal; rays gen 16–30; lf segments gen wide and obtuse .... ***L. californicum***
 2′ Cauline lvs 0–2, gen 1, << basal; rays gen 5–16; lf segments gen narrow and acute .... ***L. grayi***

L. apiifolium (Nutt.) A. Gray Pl 3–15 dm. **LF**: petiole gen 1–3 dm; blade 0.8–2.5 dm wide, triangular-ovate, ternate-pinnate, lflets 1.5–4.5 cm, ovate, segments obtuse or acute, deeply pinnately lobed, margins minutely scabrous or roughened; cauline lvs ± = basal, gen 2–3, upper subsessile, sometimes paired. **INFL** gen ± puberulent or roughened; peduncles alternate or whorled, 1–3.5 dm; rays 12–23, 2–6 cm, unequal; pedicels 5–10 mm, unequal. **FR** 3–5 mm, oblong; ribs thread-like to acute, ± unwinged; oil tubes per rib-interval 3–6. **SEED**: face concave. $2n=22,44$. Coastal meadows, scrub or woods; < 1800 m. NCo, CCo; to WA. ❀DRN:**5** &IRR:**17**&SHD:**15,16**.

L. californicum J. Coulter & Rose Pl 6–12 dm, glabrous. **LF**: petiole 0.5–4 dm; blade 1–3 dm wide, triangular-ovate, ternate-pinnate, lflets 1.5–4 cm, oblong to ovate, segments gen wide, obtuse, shallowly to deeply pinnately lobed; cauline lvs ± = basal, gen 2–3, upper subsessile, sometimes paired. **INFL**: peduncles gen whorled, terminal 1.5–4 dm, lateral gen staminate; rays gen 16–30, 2–6 cm, unequal; pedicels 5–10 mm, unequal. **FR** 4–6 mm, oblong-ovate; ribs narrowly winged; oil tubes per rib-interval several. **SEED**: face concave. Chaparral, woods, often on serpentine; 15–1500 m. NCo, KR, CaRH, n SNF; sw OR. ❀DFCLT.

L. grayi J. Coulter & Rose (p. 155) Pl 2–8 dm, glabrous. **LF**: petiole 0.2–3 dm; blade 1–2.5 dm, oblong to triangular-ovate, 2-pinnate or ternate-pinnate, lflets 1–4 cm, oblong to ovate, segments gen narrow, acute, deeply pinnately lobed; cauline lvs << basal, gen 1. **INFL** glabrous; peduncles gen whorled, terminal gen 1, 1–3 dm, lateral 1 or more, < terminal, gen staminate; rays 5–18, 1.5–5 cm, unequal; pedicels 5–10 mm, unequal. **FR** 4–5 mm, oblong-ovate; ribs narrowly winged; oil tubes per rib-interval 3–5. **SEED**: face concave. $2n=22,44$. Wet soil of subalpine meadows, coniferous forest; 1400–3300 m. KR, CaRH, SNH, MP; to WA, MT, NV. ❀DFCLT.

LILAEOPSIS

Per, glabrous; rhizomes producing fibrous roots. **ST** prostrate, creeping. **LVS** solitary or tufted, linear to spoon-shaped, cylindric or ± flattened, segmented, entire, without definite blade and petiole, scarious-sheathing at base. **INFL**: umbels simple, open, gen peduncled; bracts several, inconspicuous; pedicels few, spreading to recurved. **FL**: calyx lobes minute; petals wide, white or maroon, short-acuminate, tip not incurved. **FR** ovate to obovate, slightly compressed side-to-side, glabrous; ribs equal or not, 0–all conspicuously spongy-thickened; oil tubes per rib-interval several–many; fr central axis not an obvious structure. **SEED**: face rounded or flat. 13 spp.: Am, Australia, New Zealand, Afr(?). (Greek: like *Lilaea* of Lilaeaceae) [Affolter 1985 Syst Bot Mon 6:1–140]

1. Lf cylindric, gen < 1 mm wide, internal cross-walls of lf 3–8, obscure .... ***L. masonii***
1′ Lf cylindric to flattened, gen > 1 mm wide, internal cross-walls of lf 4–13, clearly evident .... ***L. occidentalis***

L. masonii Mathias & Constance (p. 155) MASON'S LILAEOPSIS **LVS** mostly tufted at tips of vertical branches, 1.5–7.5 cm, 0.4–1.2 mm wide, linear or thread-like. **INFL**: peduncles 2–20 mm; bracts 0.5–1 mm; pedicels 3–8, 1–6 mm. **FR** 1.2–1.6 mm, elliptic or ovate, only marginal ribs rounded, thickened; oil tubes per rib-interval 5–6. $2n=44$. **RARE** CA. Intertidal marshes, streambanks; ± 0 m. s ScV, ne SnFrB. Locally abundant; threatened by development, flood control, agriculture. In cult.

L. occidentalis J. Coulter & Rose **LVS** solitary or tufted, 2.5–30 cm, 0.7–4.5 mm wide, linear or tapered. **INFL**: peduncles 5–65 mm; bracts 0.5–2 mm; pedicels 5–12, 1–12 mm. **FR** 1.3–2.4 mm, elliptic or round, only marginal ribs wide, thickened; oil tubes per rib-interval gen 6–9. $2n=44$. Salty or brackish soil, esp coastal; ± 0 m. NCo, CCo; to B.C. ❀WET,SUN:**4,15–17**;INV.

LOMATIUM

Per from taproot or gen deep-seated tuber, glabrous to tomentose. **ST** 0 or erect, simple or branched; base fibrous (from old lf sheaths) or not. **LF**: blade oblong to triangular-ovate or obovate, ternately, pinnately, or ternate-pinnately dissected or compound, segments or lflets thread-like to wide. **INFL**: umbels compound, peduncled; bracts gen 0; bractlets gen present, 0 to conspicuous; rays, pedicels spreading to erect, often webbed at base. **FL**: calyx lobes gen 0; petals wide, yellow, white, or purple, tips narrowed; projection atop ovary 0. **FR** linear to obovate, very compressed front-to-back; marginal ribs widely to narrowly thin or thick-winged, others thread-like; oil tubes per rib-interval 1–several; fr axis divided to base. **SEED**: face flat to concave. ± 75 spp.: c&s N.Am. (Greek: bordered, from prominent marginal fr wing) [Schlessman 1984 Syst Bot Monogr 4:1–55] Fr wing width expressed as width of 1 wing, not both together.

1. Fr evidently notched at base and tip (wings narrower there); lflets or segments gen large (subg. *Euryptera*)
 2. Lflets coarsely dentate or some 3-lobed (but not pinnately so)
  3. Lf 1–2-ternate; fr wings thickened, >> body in width; s CA .... ***L. lucidum***
  3′ Lf ternate-pinnate or 1–2-pinnate; fr wings thin, = to slightly > body in width; n CA
   4. Pl tall, 1.2–8 dm; fr widely elliptic to round; KR .... ***L. howellii***
   4′ Pl low, 1.2–5 dm; fr widely oblong; NCoR .... ***L. repostum***
 2′ Lflets pinnately lobed and gen also toothed or irregularly cut
  5. Lf blade > petiole; fr 15–18 mm — s ChI .... ***L. insulare***
  5′ Lf blade < to ± = petiole; fr 6–14 mm
   6. Lflet spines weak; fr wings = to > body in width; CCo, SCoR .... ***L. parvifolium***
   6′ Lflet spines strong; fr wings < body in width; SNH, SNE
    7. Mature rays 25–50 mm; fertile pedicels 5–10 mm .... ***L. rigidum***
    7′ Mature rays 1–11 mm; fertile pedicels 0.1–1 mm .... ***L. shevockii***
1′ Fr not evidently notched at base and tip (wings not narrower there); lflets or segments small to large
 8. Pl from a definite, spheric or elongate or irregularly thickened but gen not very deep-seated tuber (if uncertain, try 8′ as well)
  9. Pl puberulent or hairy
   10. Bractlets not scarious, ± fused, reflexing .... [3]***L. macrocarpum***
   10′ Bractlets ± scarious, free or fused, not reflexing .... [5]***L. nevadense***
  9′ Pl glabrous or slightly roughened but not otherwise hairy
   11. Pl short-stemmed, from an elongate, often regularly narrowed tuber; rays suberect; fr linear-oblong .... [3]***L. bicolor*** var. ***leptocarpum***
   11′ Pl truly stless, from a ± spheric to ovoid tuber; rays ± spreading; fr oblong to ovate
    12. Pedicel 5–16 mm; bractlet ± = fl; tuber < 4 cm wide .... ***L. canbyi***
    12′ Pedicel 0–2.5 mm; bractlet 0 or < fl; tuber < 2 cm wide
     13. Corolla white; petiole 3.5–10 cm; rays gen 15–35 mm .... ***L. piperi***
     13′ Corolla yellow; petiole 2–3 cm; rays 1–12 mm .... ***L. stebbinsii***
 8′ Pl from a gen elongated taproot that rarely ends in a deep-seated tuber
  14. Lf dissected or compound, segments or lflets gen few, gen large, gen wide
   15. Peduncle at top conspicuously swollen, inflated .... ***L. nudicaule***
   15′ Peduncle at top not swollen, not inflated, although rays sometimes webbed at base into prominent disk
    16. St 0 or very short, with 0–very few lvs; lf 1–3-pinnate
     17. St base not fibrous; calyx 0 .... ***L. martindalei***
     17′ St base fibrous; calyx prominent (subg. *Cynomarathrum*) .... ***L. parryi***
    16′ St prominent, clearly with lvs; lf 1–2-ternate-pinnate
     18. Herbage glaucous; lflets wedge-shaped to obovate, 10–50 mm wide .... ***L. californicum***
     18′ Herbage green or grayish, not conspicuously glaucous; lflets linear to ovate-lanceolate, 0.5–10 mm wide .... ***L. triternatum***
      19. Ovary and often fr densely puberulent .... var. ***macrocarpum***
      19′ Ovary and fr glabrous .... var. ***triternatum***
  14′ Lf dissected, segments gen many, small, sometimes long but then very narrow (see also *L. parryi*, which may key here)
   20. Ovary definitely puberulent or roughened or hairy, fr hairy to rarely becoming glabrous
    21. Pls gen with a definite st, st lvs 1 or more
     22. Petals conspicuously tomentose .... ***L. dasycarpum***
      23. Lf segments linear; pedicels gen > fr .... ssp. ***dasycarpum***
      23′ Lf segments thread-like; pedicels gen < fr .... ssp. ***tomentosum***
     22′ Petals ± glabrous, not tomentose
      24. Involucel 1-sided, reflexing, not scarious .... [3]***L. macrocarpum***
      24′ Involucel not 1-sided, not reflexing, gen ± scarious
       25. Corolla white; cauline lf sheaths inconspicuous .... [5]***L. nevadense***
       25′ Corolla yellow; cauline lf sheaths conspicuously swollen
        26. Herbage densely puberulent to glabrous, ovaries puberulent; bractlets gen obovate or oblanceolate, often overlapping .... [2]***L. utriculatum***
        26′ Herbage, ovaries gen roughened; bractlets gen lanceolate to oblanceolate, not overlapping .... [2]***L. vaginatum***
    21′ Pls truly stless, st lvs 0
     27. Corolla white, sometimes purple-tinged, anthers often purple
      28. Bractlets puberulent to hairy; lf segments linear to lanceolate .... ***L. ravenii***
      28′ Bractlets gen glabrous; lf segments linear to oblong .... [5]***L. nevadense***
       29. Ovary and often fr densely finely hairy .... var. ***nevadense***
       29′ Ovary and fr ± roughened .... var. ***parishii***

27′ Corolla yellow or purple, anthers gen yellow
30. Petiole > lf blade; corolla gen purple, rarely yellow; fr wings = to > body in width .... ***L. mohavense***
30′ Petiole < lf blade; corolla yellow, rarely purplish; fr wings gen 1/2 body in width ...... ***L. foeniculaceum***
31. Petals minutely ciliate; rays 2–14, 0.5–6 cm .... ssp. ***fimbriatum***
31′ Petals, incl margins, glabrous; rays gen 1, < 5 cm .... ssp. ***inyoense***
20′ Ovary and fr glabrous, sometimes roughened
32. Pl definitely stemmed, st lvs gen 2 or more
33. St base with 1 or more scarious bladeless sheaths; fr pumpkin-seed-like, wings thick, << body in width (subg. *Leptotaenia*) .... ***L. dissectum***
34. Fr subsessile, fertile pedicels gen 1–3 mm, < sterile .... var. ***dissectum***
34′ Fr distinctly pedicelled, fertile pedicels gen 5–15 mm, > sterile .... var. ***multifidum***
33′ St base without conspicuous bladeless sheaths; fr wings thin, < to > body in width
35. Corolla white, cream-colored or purplish
36. Pl glabrous or slightly roughened; bractlets 0–few, inconspicuous .... ***L. martindalei***
36′ Pl densely puberulent to tomentose; bractlets gen conspicuous, sometimes fused into a sheath
37. Involucel 1-sided, reflexing, not scarious .... [3]***L. macrocarpum***
37′ Involucel gen radial, spreading, scarious .... [5]***L. nevadense***
35′ Corolla yellow, sometimes fading or drying white
38. Lvs gen not crowded at st base; fr wings = or > body in width
39. Herbage puberulent to glabrous, ovaries often puberulent — fr gen glabrous .... [2]***L. utriculatum***
39′ Herbage, often ovaries minutely scabrous .... [2]***L. vaginatum***
38′ Lvs crowded at st base; fr wings < body in width
40. Fr linear-oblong; lf segments linear, long, not crowded .... [3]***L. bicolor*** var. ***leptocarpum***
40′ Fr oblong to ovate; lf segments linear to oblong, very short, crowded
41. Lf green, shiny, sheath conspicuous; bractlets free, inconspicuous; KR .... ***L. hallii***
41′ Lf grayish, dull, sheath ± inconspicuous; bractlets fused at base into a conspicuous 1-sided scarious cup; CaRH, n SNH, GB .... ***L. plummerae***
32′ Pl stless or nearly so, cauline lvs 0 or sometimes 1 or more at base of very short st
42. Fr linear to narrowly oblong
43. St base not fibrous; bractlets several, linear .... [3]***L. bicolor*** var. ***leptocarpum***
43′ St base fibrous; bractlets 0 .... ***L. torreyi***
42′ Fr oblong to nearly round
44. Lf segments thread-like or linear, long, not crowded; herbage green, often shiny
45. Bractlets gen > 5, lanceolate to obovate, ± scarious, sometimes 0; pedicels stout — gen < fr .... ***L. caruifolium***
46. Bractlets wide; fr wings ± thin, wide .... var. ***caruifolium***
46′ Bractlets narrow, sometimes 0; fr wings very thick, narrow .... var. ***denticulatum***
45′ Bractlets 0–5, thread-like to linear-lanceolate, not conspicuously scarious; pedicels slender .... ***L. marginatum***
47. Corolla yellow; interior foothills .... var. ***marginatum***
47′ Corolla reddish purple; coast ranges .... var. ***purpureum***
44′ Lf segments lanceolate or oblong to ovate, rarely linear or thread-like, short, often crowded; herbage gen glaucous or grayish
48. Herbage definitely puberulent to tomentose
49. Lf 2–3-pinnately dissected or compound; petiole sheathing in lower half; corolla white; GB .... [5]***L. nevadense***
49′ Lf ternate-pinnately dissected; petiole wholly sheathing; corolla yellow or purple; coast ranges .... ***L. ciliolatum***
50. Corolla gen yellow; fr ribs thick; bractlets scarious-margined .... var. ***ciliolatum***
50′ Corolla purple; fr ribs thin; bractlets scarious ± throughout .... var. ***hooveri***
48′ Herbage glabrous or minutely roughened to sparsely puberulent
51. Pedicels slender, often > fr; bractlets 0
52. St base fibrous; corolla pale yellow .... ***L. congdonii***
52′ St base not fibrous; corolla purplish .... ***L. engelmannii***
51′ Pedicels definitely < fr; bractlets present
53. Lvs obovate, basal sheaths straw-colored; in volcanic soils .... ***L. peckianum***
53′ Lvs oblong to ovate, basal sheaths purplish; on serpentine .... ***L. tracyi***

L. bicolor (S. Watson) J. Coulter & Rose var. ***leptocarpum*** (Torrey & A. Gray) M. Schlessman Pl 2–5 dm, glabrous to minutely scabrous; taproot 0 or elongate; tubers 0 or rarely deep-seated. **ST** 0 or very short. **LF**: petiole 9–14 cm; blade 5–18 cm, widely obovate, ternate-pinnately dissected, segments gen 1–3 mm, linear; cauline lvs 0 or like basal. **INFL** glabrous to minutely scabrous; peduncle 8–25 cm; bractlets several, linear; rays 4–15, 2–13 cm, suberect; pedicels 1–2(5) mm. **FL**: corolla yellow. **FR** 8–17 mm, gen linear-oblong, glabrous; wings < 1/2 body in width; oil tubes per rib-interval several. $2n=22$. Drying adobe, sagebrush slopes; 1000–2250 m. CaRH, MP; to WA, Colorado, AZ. [*L. leptocarpum* Torrey & A. Gray]

L. californicum (Torrey & A. Gray) Mathias & Constance (p. 155) Pl 3–12 dm, glabrous, glaucous; taproot stout, thickened. **ST**: base fibrous. **LF**: petiole 5–25 cm; blade 1–3 dm wide, triangular-ovate, 1–2-ternate-pinnate, lflets 2–5 cm, wedge-shaped to obovate, gen with 3 major lobes, these coarsely toothed or lobed; cauline lvs few, like basal. **INFL** glabrous; peduncle 1.5–3 dm; bractlets 0 or inconspicuous; rays 8–20, 3–8(15) cm, subequal, often forming disk at base; pedicels 4–12 mm, often forming disk at base. **FL**: corolla yellow. **FR** 10–15 mm, oblong-ovate to elliptic, glabrous; wings thickened, < body in width; oil tubes per rib-interval 3–4. $2n=22$. Woodland, brushy slopes; 150–1800 m. KR, NCoR, s SNH, Teh, SCoR, WTR; s OR. ❀DRN:**15–17,24**&SHD:**7,14, 18–23**.

L. canbyi (J. Coulter & Rose) J. Coulter & Rose Pl 7–25 cm, glabrous, glaucous; tuber < 4 cm wide, spheric. **ST** 0. **LF**: petiole 4–6 cm, scarious sheath conspicuous; blade 1–9 cm, oblong to ovate, ternate-2-pinnately or ternate-pinnately dissected, segments 1–5 mm, linear, obtuse. **INFL** glabrous; peduncle 5–10 cm; bractlets ± = fls, linear; rays 5–17, 1–6 cm, ± spreading; pedicels 5–16 mm. **FL**: corolla white. **FR** 6–13 mm, widely oblong; wings thin, very narrow; oil tubes per rib-interval 1–3. 2*n*=22. Barren or rocky places, sagebrush steppe; 2000 m. CaRH, MP; to WA, ID, NV.

L. caruifolium (Hook. & Arn.) J. Coulter & Rose Pl 1.5–4.5 dm, glabrous to finely scabrous or hairy; taproot gen slender. **ST** ± 0. **LF**: petiole 4–7 cm; blade 5–30 cm wide, triangular-ovate to obovate, 1-3-ternately or ternate-2-pinnately dissected, segments 2–60 mm, linear, pointed. **INFL**: peduncle 1–4 dm, erect or spreading; bractlets sessile or not, gen = fls, lanceolate to obovate, entire or toothed, scarious-margined, veined, green to purplish; rays 6–15, 1–12 cm, gen spreading-ascending; pedicels 2–8 mm. **FL**: corolla yellow, rarely purplish. **FR** 6–13 mm, oblong to obovate, glabrous; wings thickish, < body in width. Adobe of wet depressions, vernal pools, open grasslands; 60–600 m. NCoRO (Mendocino Co), NCoRH, CaRF, c&s SNF, ScV, SnFrB, SCoR, ChI. Vars. poorly defined.

var. ***caruifolium*** **INFL**: bractlets obovate, gen overlapping side-to-side. **FR**: wings ± thin, wide. 2*n*=22,44. Adobe of wet depressions, open grassland; 60–600 m. NCoR (Mendocino Co), c&s SNF, SnFrB, SCoR, ChI.

var. ***denticulatum*** Jepson **INFL**: bractlets lanceolate to oblanceolate, not overlapping. **FR** pumpkin-seed-like; wings very thick, narrow. 2*n*=44. Vernal pools, open grassland; 60–500 m. NCoRH, CaRF, ScV. [*L. humile* (J. Coulter & Rose) Mathias & Constance]

L. ciliolatum Jepson Pl 9–30 cm, densely puberulent to tomentose; taproot slender. **ST** 0. **LF**: petiole 2–7 cm, wholly sheathing; blade 3–13 cm, oblong-ovate to triangular-ovate, ternate-pinnately dissected, segments 1–20 mm, linear to ovate. **INFL** glabrous, roughened, or puberulent; peduncle 7–25 cm; bractlets sessile or not, gen = fls, lanceolate to obovate, scarious-margined, veiny; rays 2–14, 0.8–10 cm, unequal; pedicels 2–8 mm. **FL**: corolla yellow or purple. **FR** 7–9 mm, oblong, glabrous; wings thin to thickish, < body in width. Serpentine, ridges, summits, woodland, chaparral; 300–600, 1200–2100 m. NCoR, se SnFrB (Mount Hamilton), SCoRI.

var. ***ciliolatum*** (p. 155) Pl low, stout. **LF**: segments oblong to ovate. **INFL**: fertile rays 2–5, 8–45 cm; bractlets scarious-margined; pedicels 2–4 mm. **FL**: corolla gen yellow, rarely purple. **FR**: ribs thick. 2*n*=22. Serpentine, ridges, summits; 1200–2100 m. NCoRH, NCoRI, se SnFrB (Mount Hamilton), SCoRI.

var. ***hooveri*** Mathias & Constance HOOVER'S LOMATIUM Pl ± erect or spreading, slender. **LF**: segments linear. **INFL**: fertile rays 3–14, 3–10 cm; bractlets scarious ± throughout; pedicels 3–8 mm. **FL**: corolla purple. **FR**: ribs thin. 2*n*=22. UNCOMMON. Serpentine, woodland, chaparral; 300–600 m. s NCoRI.

L. congdonii J. Coulter & Rose (p. 155) CONGDON'S LOMATIUM Pl 18–36 cm, taprooted, glaucous, glabrous, or minutely scabrous. **ST** very short; base fibrous. **LF**: petiole 2–6 cm, wholly scarious-sheathing, persistent; blade 6.5–15 cm, widely oblong, ternate or palmate-pinnately dissected, segments 3–10 mm, linear, tips with sharp points. **INFL** glabrous or roughened; peduncle 12–30 cm; bractlets 0; rays 6–16, 3–13.5 cm, ascending; pedicels 6–15 mm. **FL**: corolla pale yellow. **FR** 7–10 mm, oblong to obovate, glabrous; wings ± 1/2 body in width; oil tube per rib-interval gen 1. 2*n*=22. RARE. Serpentine, woodland; 300–1200 m. c SNF (Tuolumne, Mariposa cos.). Very local.

L. dasycarpum (Torrey & A. Gray) J. Coulter & Rose Pl 1–5 dm, taprooted, gen densely short-hairy to tomentose. **ST** ascending to erect, rarely 0. **LF**: petiole 2.5–12 cm; blade 2–12 cm, oblong to obovate, pinnately or ternate-pinnately dissected, segments 2–6 mm, thread-like to linear; cauline lvs 0 or like basal. **INFL**: peduncle 1–3.5 dm; bractlets linear to narrowly ovate, acute, fused or not; rays 10–21, 1–8.5 cm, spreading; pedicels 5–20 mm. **FL**: corolla greenish white or purplish, conspicuously tomentose. **FR** 8–22 mm, oblong-ovate to round, ± hairy; wings = or > body in width; oil tubes per rib-interval 1–4. Dry ridges; < 1600 m. NCoR, CaRF, SNF, GV, SnFrB, SCoR, PR; Baja CA.

ssp. ***dasycarpum*** **LF**: petiole gen sheathing to middle. **FR**: body sparsely hairy, gen < pedicel, gen < wings in width. 2*n*=22. Rocky (gen serpentine), chaparral, woodland; < 1600 m. NCoR, SnFrB, SCoR, PR; Baja CA. ❀DRN,DRY:**14,18–23**&SUN:**15–17,24**.

ssp. ***tomentosum*** (Benth.) Theob. **LF**: petiole sheathing at base. **FR**: body tomentose, gen > pedicel, gen = wings in width. 2*n*=22. Stony flats, grassland, oak woodland; 25–1500 m. CaRF, SNF, GV. [*L. tomentosum* (Benth.) J. Coulter & Rose]

L. dissectum (Torrey & A. Gray) Mathias & Constance Pl 3–14 dm, glabrous to puberulent or minutely scabrous, ± glaucous; taproot stout, thickened. **ST** rarely 0; base with 1 or more scarious sheaths. **LF**: petiole 3–30 cm; blade 15–35 cm wide, triangular-ovate to round, ternate-pinnately dissected, segments 2–22 mm, linear-oblong; cauline lvs gen few, like basal. **INFL** glabrous; peduncle 1.5–6 dm; bractlets several, > or < fls, linear; rays 10–30, 3–10 cm, spreading; pedicels 1–20 mm. **FL**: corolla maroon-red or yellow. **FR** 12–16 mm, oblong-ovate to elliptic, glabrous, pumpkin-seed-like; wings thick, << body in width; oil tubes obscure. Wooded or brushy slopes; 150–3000 m. CA-FP; to w Can, Baja CA.

var. ***dissectum*** **LF**: segments 2–8 mm, gen 1.5–3 (rarely >> 3) mm wide. **FL**: corolla gen maroon-red, less often yellow. **FR**: pedicels gen 1–3 mm. 2*n*=22. Wooded or brushy slopes; 150–2000 m. NCo, KR, c SNF; to WA, ID. ❀TRY.

var. ***multifidum*** (Torrey & A. Gray) Mathias & Constance **LF**: segments 2–22 mm, 0.5–2 mm wide. **FL**: corolla yellow. **FR**: pedicels gen 5–15 mm. 2*n*=22. Wooded or brushy slopes, often coniferous forest; 600–3000 m. KR, CaRH, SNH, Teh, SCo, SnGb, SnBr, GB; to w Can, Baja CA. ❀TRY.

L. engelmannii Mathias ENGELMANN'S LOMATIUM Pl 1–3 dm, glabrous to sparsely puberulent; taproot slender. **ST** 0. **LF**: petiole 2–10 cm, wholly sheathing; blade 2.5–20 cm, oblong to ovate, ternately or palmate-pinnately dissected, segments 1–15 mm, linear-lanceolate to lanceolate. **INFL** gen glabrous; peduncle 12–20 cm; bractlets 0; rays 2–12 (1–8 fertile), 1–13 cm, spreading or ascending; pedicels 2–12 mm. **FL**: corolla purplish. **FR** 7–14 mm, oblong-ovate, glabrous; wings 1/2 body in width; oil tubes per rib-interval 1–2. 2*n*=22. UNCOMMON. Serpentine slopes in coniferous forest; 1150–2300 m. KR (Siskiyou, Trinity cos.); sw OR.

L. foeniculaceum (Nutt.) J. Coulter & Rose Pl 0.3–3 dm, taprooted, densely puberulent or soft-hairy to tomentose. **ST** 0. **LF**: petiole 1–15 cm, < blade, wholly sheathing; blade gen 2.5–18 cm wide, oblong to obovate, pinnately or ternate-pinnately dissected, segments 1–7 mm, linear to obovate, pointed. **INFL** hairy to nearly glabrous; peduncle 0.3–3 dm; bractlets fused or not, < to = fls, linear to linear-lanceolate, entire or lobed, acute; rays 1–30, gen 1–8 cm, spreading-ascending or spreading; pedicels 1–15 mm. **FL**: corolla yellow, rarely purplish. **FR** 4–12 mm, oblong-ovate, gen hairy; wings gen 1/2 body in width; oil tubes per rib-interval 1–7. Sagebrush, pine woodland, open summits, subalpine scrub; 1600– 3300 m. SNE; to B.C., UT, TX.

ssp. ***fimbriatum*** Theob. Pl 7–30 cm. **LF**: segments 1–5 mm, linear. **INFL**: peduncle 10–30 cm; rays 2–14, 0.5–6 cm. **FL**: petals yellow or purplish, minutely ciliate. 2*n*=22. Sagebrush, pine woodland; 1600–3300 m. SNE; to UT.

ssp. ***inyoense*** (Mathias & Constance) Theob. INYO LOMATIUM Pl 3–12 cm. **LF**: segments 1–2 mm, linear to obovate. **INFL**: peduncle 3–12 cm; rays gen 1, < 5 cm. **FL**: petals pale yellow, glabrous. UNCOMMON. Open summits, subalpine scrub; < 3000 m. W&I (Inyo Mtns); possibly to ID, NV. [*L. inyoense* Mathias & Constance] May be a form induced by high-elevation conditions.

L. hallii (S. Watson) J. Coulter & Rose Pl 2–5 dm; taproot stout; herbage glabrous to minutely roughened. **STS** several; base branched, not fibrous. **LF**: petiole 3–6 cm, sheath conspicuous; blade 5–15 cm, oblong-ovate, gen 3-pinnately dissected, segments

2–4 mm, linear to oblong, pointed. **INFL** roughened or glabrous; peduncle 1.2–3 dm; bractlets several, distinct, linear; rays 8–18, 1.3–5 cm, spreading-ascending; pedicels 4–7 mm. **FL**: corolla yellow. **FR** 5–9 mm, oblong-ovate, glabrous; wings gen 1/2 body in width; oil tubes per rib-interval 2–3. 2*n*=22. Rocky bluffs or slopes; ± 150 m. nw KR (Del Norte Co.); OR. [*L. microcarpum* (J. Coulter & Rose) J. Coulter & Rose, *L. nelsonianum* J.F. Macbr., in part]

L. howellii (S. Watson) Jepson HOWELL'S LOMATIUM Pl 1.2–8 dm, glabrous, glaucous; taproot stout, branched. **ST** short to ± 0; base not fibrous. **LF**: petiole 2–20 cm, sheath short; blade 5–15 dm, triangular-ovate, ternate-pinnate or 2-pinnate, lflets 1–3 cm wide, gen obovate, sharply dentate, often irregularly cut, sometimes 3-lobed; cauline lvs like basal. **INFL**: peduncle 1–4 dm; bractlets several, 1–3 mm, thread-like to lanceolate; rays 8–20, 2.5–6.5 cm, spreading; pedicels 5–10 mm. **FL**: corolla yellow or purplish. **FR** 6–14 mm, widely elliptic to round, glabrous; wings = body in width; oil tubes per rib-interval 2–3. 2*n*=22. UNCOMMON. Serpentine, chaparral, coniferous forest; 150–1000 m. KR (Del Norte, Siskiyou cos.); sw OR.

L. insulare (Eastw.) Munz (p. 155) SAN NICOLAS ISLAND LOMATIUM Pl 1–3.5 dm, glabrous, glaucous; taproot stout, fleshy-thickened. **ST** very short; base fibrous. **LF**: petiole 2–15 cm, < blade, sheath wide; blade 6–24 cm wide, triangular-ovate to round, ternate-pinnate, lflets 2–6 cm wide, gen ovate, gen pinnately lobed, sharply serrate; cauline lvs like basal. **INFL** glabrous; peduncle 1–3 dm; bractlets several, thread-like, often > fls; rays gen 12–20 (4–10 fertile), 3–12 cm, spreading, webbed; pedicels 5–15 mm, webbed. **FL**: corolla yellow. **FR** 15–18 mm, widely elliptic to obovate, glabrous; wings = body in width; oil tubes per rib-interval 2. 2*n*=22. RARE in CA. Sandy soil among rocks; < 800 m. s ChI (San Nicolas Island); Guadalupe Island.

L. lucidum (Torrey & A. Gray) Jepson Pl 1.5–12 dm; taproot slender; herbage glabrous, green, ± fleshy. **ST** very short; base not fibrous. **LF**: petiole 2–12 cm, sheath short; blade 4–12 cm wide, triangular-ovate, 1–2-ternate, lflets 1.5–5 cm, oblong to obovate, coarsely sharply dentate, often 3-lobed; cauline lvs like basal. **INFL** glabrous; peduncle gen 1.5–5 dm; bractlets several, 3–7 mm, linear-lanceolate; rays gen 10–20 (4–18 fertile), 3–8 cm, spreading or spreading-ascending, ± webbed; pedicels 5–12 mm, webbed. **FL**: corolla yellow. **FR** gen 10–15 mm, widely elliptic to round, glabrous; wings thickened, >> body in width; oil tube per rib-interval 1. 2*n*=22. Chaparral, esp on burns; 450–1500 m. SCo, SnGb, SnBr, SnJt; Baja CA.

L. macrocarpum (Torrey & A. Gray) J. Coulter & Rose Pl 1–5 dm; taproot or sometimes tubers slender or basally swollen; herbage gray, gen tomentose to densely short-hairy. **ST** very short; base not fibrous. **LF**: petiole 1.5–7 cm; blade 2.5–15 cm, oblong to obovate, pinnately or ternate-pinnately dissected, segments 1–7 mm, linear to oblong, entire; cauline lvs like basal. **INFL** gen tomentose; peduncle 0.5–3 dm; involucel 1-sided; bractlets several, = to > fls, linear-lanceolate to ovate, acute, ± fused, reflexing, not scarious; rays 5–25, 1–8.5 cm, spreading-ascending; pedicels 1–14 mm. **FL**: corolla white, pale yellow, or purplish; ovary gen hairy. **FR** 9–20 mm, lanceolate or oblong to narrowly elliptic, minutely hairy to nearly glabrous; wings gen < body in width; oil tubes per rib-interval 1–3. 2*n*=22. Rocky (often serpentine) slopes in chaparral or woodland; 150–3000 m. NCo, KR, CaR, SNF, Teh, SnFrB, SCo; to B.C., c Can, ND, UT. [var. *ellipticum* Jepson] ❀TRY.

L. marginatum (Benth.) J. Coulter & Rose Pl 1.5–5 dm, glabrous to minutely scabrous; taproot slender. **ST** ± 0. **LF**: petiole gen 3–8 cm and wholly narrowly sheathing, often purplish; blade 5–20 cm wide, triangular-ovate to obovate, 1–3-ternately or ternate-pinnately dissected, segments 0.5–8 cm, thread-like or linear, pointed. **INFL**: peduncle 1–4 cm; bractlets 0–5, 1–5 mm, thread-like to linear-lanceolate, entire, scarious-margined; rays 6–12, 1.5–15 cm, spreading to ascending; pedicels 2–12 mm. **FL**: corolla yellow or reddish purple. **FR** 8–12 mm, narrowly oblong to obovate, glabrous; wings thin, < body in width; oil tubes obscure. Serpentine slopes, chaparral, woodland; 250–1000 m. KR, NCoR, CaR, ScV, n&c SNF. Vars. poorly defined.

var. ***marginatum*** **INFL**: rays 1.5–10 cm; pedicels 2–6 mm. **FL**: corolla yellow. 2*n*=22. Serpentine slopes, often chaparral; 250–1000 m. KR, CaR, ScV, n&c SNF.

var. ***purpureum*** (Jepson) Jepson **INFL**: rays 2–15 cm; pedicels 5–12 mm. **FL**: corolla reddish purple. 2*n*=22,44. Serpentine slopes, chaparral, woodland; 300–800 m. NCoR.

L. martindalei (J. Coulter & Rose) J. Coulter & Rose COAST RANGE LOMATIUM Pl 1.5–4 dm; taproot stout, often swollen below, carrot-like; herbage glabrous to slightly roughened, ± glaucous, ± fleshy. **ST** 0 or very short; base not fibrous. **LF**: petiole 1.5–5 cm, sheath wide, scarious; blade 2.5–15 cm, oblong to obovate; 1–2-pinnate to ternate-pinnately dissected, lflets or segments 0.8–3 cm, oblong to ovate, often crowded, gen obtuse, dentate, irregularly cut, or pinnately lobed; cauline lvs like basal, gen wholly sheathing. **INFL** gen glabrous; peduncle 0.5–3 dm; bractlets 0 or inconspicuous; rays 3–15, 1–8 cm, ± unequal, spreading to ascending, slightly webbed; pedicels 3–10 mm, slightly webbed. **FL**: corolla yellow. **FR** 8–15 mm, oblong to ± diamond-shaped, glabrous; wings = to < body in width; oil tubes per rib-interval gen 1. 2*n*=22. RARE in CA. Coniferous forest, rocks, meadows, talus, pumice, coastal bluffs; 240–3000 m. KR (Del Norte, Siskiyou cos.); c OR, n WA, B.C. (Vancouver Island).

L. mohavense (J. Coulter & Rose) J. Coulter & Rose (p. 155) Pl 1–4 dm, densely grayish short-hairy; taproot long, often thickened. **ST** ± 0; base fibrous. **LF**: petiole 2–12 cm, > blade; blade 2–10 cm, oblong to ovate, 3–4-pinnately dissected, segments 2–5 mm, linear-oblong to obovate, ± crowded. **INFL** finely hairy; peduncle 8–22 cm; bractlets 8–12, 2–4 mm, linear to linear-lanceolate, acute, free or fused basally, scarious-margined, often obscured by hairs; rays 8–18, 1–5 cm, spreading-ascending, unequal; pedicels 1–10 mm. **FL**: petals yellow or purple, glabrous. **FR** 4.5–11 mm, ovate to round, ± glabrous to densely short-hairy; wings = to > body in width; oil tubes per rib-interval 1–4. 2*n*=22. Desert flats, slopes, summits, scrub, or woodland; 1000–2000 m. SCoR, WTR, D; Baja CA.

L. nevadense (S. Watson) J. Coulter & Rose Pl 1–4.5 dm; taproot slender, sometimes swollen below; herbage grayish, gen finely hairy. **ST** 0 or short. **LF**: petiole 4–6 cm; blade 3.5–10 cm, oblong to obovate, gen 2–3-pinnately dissected, segments gen 2–3 mm, linear or oblong, pointed, often crowded; cauline lvs 0 or like basal. **INFL** finely hairy; peduncle 0.5–3 dm; bractlets 1–10, linear and free to obovate and ± fused (involucel radial to 1-sided), conspicuously scarious or scarious-margined, gen glabrous or nearly so; rays 8–22, 1–2.5 cm, unequal, spreading; pedicels 3–10 mm. **FL**: corolla white to cream-colored. **FR** 6–11 mm, oblong to round or obovate, densely hairy to glabrous; wings < to > body in width; oil tubes per rib-interval 1–9. 2*n*=22. Sagebrush, woodland, desert scrub; 1000–3000 m. CaRH, SNH, SnGB, SnBr, GB, DMtns; to OR, UT, NM, n Mex. Vars. poorly defined.

var. ***nevadense*** **LF** 2–3-pinnately dissected; segments gen 2–3 mm. **FL**: ovary (often fr) densely finely hairy. Sagebrush, woodland; 1500–3000 m. CaRH, n&c SNH, GB, DMtns; to OR, UT, AZ.

var. ***parishii*** (J. Coulter & Rose) Jepson **LF** often 1–2-pinnately dissected; segments < 3 mm. **FL**: ovary (and fr) glabrous or nearly so or ± roughened. Sagebrush, desert scrub, pine woodland; 1000–3000 m. c&s SNH, SnGb, SnBr, GB, DMtns; to NV, NM. [var. *holopterum* Jepson; var. *pseudorientale* Munz]

L. nudicaule (Pursh) J. Coulter & Rose Pl 2.5–7 dm, glabrous, glaucous; taproot long, thickened. **ST** ± 0. **LF**: petiole 0.4–2.5 dm, widely sheathing to middle; blade 9–20 cm, widely ovate, 1–2-ternate-pinnate, lflets 15–90 mm, lanceolate to widely ovate, entire or toothed or lobed, gen near tip. **INFL**: peduncle 1–4 dm, at top conspicuously swollen, often translucent; involucel 0; rays 10–20, 1–20 cm, ascending, ± swollen at top; pedicels 3–15 mm. **FL**: corolla yellow, rarely purple. **FR** 10–14 mm, oblong to elliptic; wings < body in width; oil tubes per rib-interval 1–several. 2*n*=22. Rocky slopes, flats, often pine woodland; 180–2000 m. KR, NCoR, CaRH, n SNF, SnFrB, MP; to B.C., ID, UT.

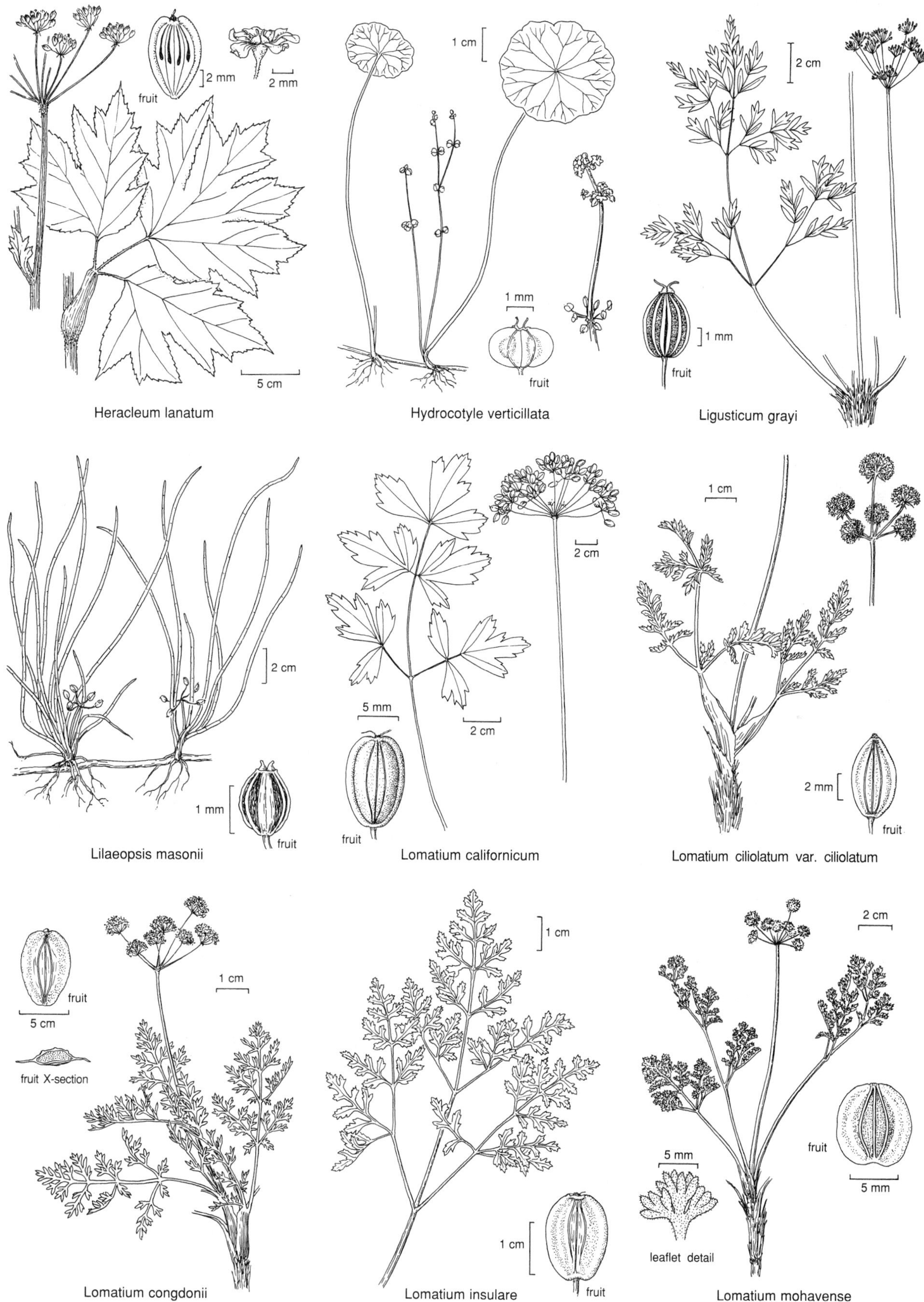

Heracleum lanatum

Hydrocotyle verticillata

Ligusticum grayi

Lilaeopsis masonii

Lomatium californicum

Lomatium ciliolatum var. ciliolatum

Lomatium congdonii

Lomatium insulare

Lomatium mohavense

L. parryi (S. Watson) J.F. Macbr. Pl 2–4 dm, taprooted, glabrous, glaucous. **ST** ± 0; base fibrous. **LF**: petiole 6–10.5 cm, widely sheathing basally; blade 5–20 cm, narrowly oblong, 2–3-pinnate, lflets 2–9 mm, linear, entire, sharp-pointed. **INFL**: peduncle 1–2.5 dm, slightly swollen at top; bractlets 3–8, 3–6 mm, linear, acute, scarious, entire or lobed; rays 8–15, ascending to suberect, 2–4.5 cm, subequal; pedicels 10–17 mm. **FL**: corolla yellow. **FR** 9–15 mm, oblong to ± diamond-shaped; wings = or > body in width; oil tubes per rib-interval 2–3. Rocky slopes, gen in pinyon woodland; 1500–2500 m. DMtns; to UT, AZ.

L. parvifolium (Hook. & Arn.) Jepson SMALL-LEAVED LOMATIUM Pl 1.5–4 dm; taproot slender; herbage glabrous, ± glaucous, ± fleshy. **ST** short. **LF**: petiole 3–15 cm, sheathing basally; blade 3–15 cm wide, oblong- to triangular-ovate, ternate-pinnate, lflets 1–4 cm wide, ovate to obovate, spine-toothed and -tipped (spines weak) or irregularly cut to pinnately lobed. **INFL**: peduncle 1–5 dm; bractlets gen 2–5, 2–8 mm, thread-like to linear-lanceolate, slightly fused basally; rays 8–15, spreading, 0.5–6.5 cm, subequal; pedicels gen 5–12 mm. **FL**: corolla yellow. **FR** 8–14 mm, elliptic to round; wing = to > body in width; oil tubes per rib-interval gen 1. 2*n*=44. UNCOMMON. Pine woods, serpentine outcrops; 70–150 m. CCo, SCoR. [var. *pallidum* (J. Coulter & Rose) Jepson]

L. peckianum Mathias & Constance PECK'S LOMATIUM Pl 1–4.5 dm, glabrous or minutely roughened, glaucous; taproot slender, roughened. **ST** 0. **LF**: petiole < 5 cm, gen wholly scarious-sheathing, straw-colored; blade 5–10 cm wide, obovate, ternately or ternate-pinnately dissected, segments 1–15 mm, linear to oblong, acutish. **INFL**: peduncle 1.2–4 dm, spreading; bractlets gen 5–10, 2–7 mm, linear, ± scarious; rays gen 8–12, 1–6.5 cm, spreading; pedicels 3–6 mm. **FL**: corolla cream-colored to lemon-yellow. **FR** 8–14 mm, oblong-elliptic to elliptic, glabrous; wings < body in width; oil tubes per rib-interval several, obscure. 2*n*=22. RARE in CA. Volcanics, pine-oak woodland; 800–1800 m. KR (Siskiyou Co.); s OR. Locally abundant.

L. piperi J. Coulter & Rose Pl 0.5–2.5 dm, glabrous to minutely scabrous; tuber 0.5–1 cm wide, spheric. **ST** 0. **LF**: petiole 3.5–10 cm, sheathing basally; blade 1.5–5.5 cm wide, triangular-ovate, 2-ternate-pinnately or -ternately dissected, segments 2–40 mm, linear, acute to obtuse; cauline lvs 1–2, gen linear, entire. **INFL** gen glabrous; terminal umbel 1, lateral gen 1–2; peduncle 2–8 cm; bractlets 2–6, 0.5–2.2 mm, < fls, narrowly elliptic, acute; rays 3–10, gen 15–35 mm, spreading-ascending; pedicels 0.5–2.5 mm. **FL**: corolla white; anthers purple. **FR** 4–9 mm, ovate; wings ± 1/2 body in width; oil tubes per rib-interval 1–8. 2*n*=22. Rocky slopes, sagebrush, pine woodland; 1000–1500 m. KR, n SNH, MP; to c OR, WA.

L. plummerae J. Coulter & Rose Pl 1.2–3.5 dm; taproot slender; herbage grayish, dull, ± fleshy; hairs dense, fine, soft to ± 0. **ST** short, lvs crowded at base. **LF**: petiole 3–6 cm, gen wholly scarious-sheathing; blade 5–10 cm, oblong to ovate, ternate-pinnately dissected, segments 3–7 mm, linear to oblong, obtuse or acutish; cauline lvs like basal. **INFL** glabrous or finely soft-hairy; peduncle 0.7–3 dm, spreading-ascending; bractlets 5–10, linear-lanceolate to obovate, gen ± fused into 1-sided, scarious, veiny, irregularly cut cup that is = or > fls; rays 10–25, 0.5–7.5 cm, unequal, spreading-ascending, slightly webbed; pedicels 3–8 mm. **FL**: corolla light yellow. **FR** 9–13 mm, oblong to oblong-ovate, glabrous; wings < body in width; oil tubes per rib-interval 1–several. Rocky places, sagebrush, pine woodland; 1500–2300 m. CaRH, n SNH, GB; w NV. [var. *austiniae* (J. Coulter & Rose) Mathias; var. *sonnei* (J. Coulter & Rose) Jepson] Extremely variable in hairiness.

L. ravenii Mathias & Constance RAVEN'S LOMATIUM Pl 0.5–4 dm; base fibrous; taproot ending in deep-seated tuber; herbage grayish; hairs dense, fine, soft. **ST** 0. **LF**: petiole 2–10 cm, sheathing to middle; blade 4–10 cm, oblong to widely ovate, ternate-pinnately or pinnately dissected, segments 1–4 mm, linear to lanceolate, acute. **INFL** densely puberulent; peduncle 0.5–3 dm; bractlets 5–10, 1–5 mm, linear, puberulent to hairy, ± fused basally, involucel 1-sided; rays 3–18, 1–7 cm, unequal, spreading-ascending; pedicels 2–10 mm. **FL**: corolla white, sometimes purple-tinged; anthers purple; ovary hairs dense, fine. **FR** 6–8 mm, ovate to round, softly hairy to nearly glabrous; wings < body in width; oil tubes per rib-interval 1–2. 2*n*=22. RARE. Flats, slopes, ridges, often ± alkaline soils, sagebrush, pinyon-juniper woodland; 1000–3000 m. MP (e Lassen Co.); to ID, w UT. Abundant in se OR.

L. repostum (Jepson) Mathias NAPA LOMATIUM Pl low, 1.2–5 dm, glabrous, glaucous; taproot slender. **ST** ± 0; lvs clustered at base. **LF**: petiole 1.5–15 cm, sheathing basally; blade 5–15 cm wide, ovate to triangular-ovate, 1–2-ternate or ternate-pinnate, lflets 1–6 cm wide, ovate to widely wedge-shaped, sharply serrate and sometimes sharply lobed; cauline lvs 0 or like basal. **INFL**: peduncle 5–35 cm; bractlets 5–10, 2–8 mm, lanceolate, acuminate, ± fused basally, reflexed; rays 8–20, 3–8 cm, spreading-ascending, webbed; pedicels 3–12 mm. **FL**: corolla greenish yellow or purplish. **FR** 8–12 mm, widely oblong, glabrous; wings = to slightly > body in width; oil tubes per rib-interval 1–3. 2*n*=44. UNCOMMON. Pine-oak woodland, chaparral, often serpentine; 100–800 m. s NCoR. ❀DFCLT.

L. rigidum (M.E. Jones) Jepson STIFF LOMATIUM Pl 1.5–6 dm; taproot massive; herbage glabrous, green. **ST** ± 0; lvs clustered at base. **LF**: petiole 5–15 cm; blade 7–15 cm, oblong to ovate, ternate-pinnate or 2-pinnate, lflets 1–2 cm, ovate, pinnately sharply lobed; cauline lvs 0 or like basal. **INFL**: peduncle 1–5 dm; bractlets 5–8, 3–8 mm, lanceolate, acuminate, ± fused basally, reflexed; rays 10–16, 2.5–5 cm, spreading-ascending, webbed; pedicels 5–10 mm, webbed. **FL**: corolla yellow; calyx lobes < 1 mm, evident. **FR** 6–12 mm, oblong-ovate, glabrous; wings < body in width; oil tubes per rib-interval 3. UNCOMMON. Rocky slopes near streams, sagebrush, pinyon-juniper woodland; 1200–2200 m. SNE (Big Pine, Bishop creeks, Inyo Co.). Vegetatively like *Tauschia parishii.*

L. shevockii R.L. Hartman & Constance (p. 161) OWEN'S PEAK LOMATIUM Pl 4–12 cm, glabrous, very glaucous; taproot elongated. **ST** ± 0; base fibrous. **LF**: petiole 1.5–5 cm; blade 1.5–4 cm, ovate to triangular-ovate, 1-pinnate, lflets gen 3–5, ovate to elliptic, pinnately sharply lobed. **INFL**: peduncle 4–12 cm, spreading; bractlets 3–6, 1–3.5 mm, lanceolate to ovate, scarious-margined, acute or acuminate, gen distinct; rays 5–9, 1–11 mm, spreading; pedicels < 1 mm, webbed. **FL**: calyx lobes < 0.6 mm, evident; corolla purple. **FR** 8–10 mm, widely elliptic to round, glabrous; wings < body in width; oil tubes per rib-interval 3–5. RARE. Rocky slopes, talus, coniferous forest, pine-oak woodland; 2200–2500 m. s SNH (Kern Co.).

L. stebbinsii M. Schlessman & Constance (p. 161) STEBBINS' LOMATIUM Pl 5–15 cm; tuber < 2 cm wide, ovoid to spheric; herbage green, shiny, glabrous. **ST** 0. **LF**: petiole 2–3 cm, sheathing to middle or throughout; blade 2–5 cm, triangular-ovate, ternate-1–2-pinnate, lflets 2–12 mm, linear, acute. **INFL**: peduncle 5–15 cm, spreading; bractlets 0; rays 2–7, 1–12 mm, ascending; pedicels 1–2 mm. **FL**: corolla yellow. **FR** 6–9 mm, narrowly elliptic; wings << body in width; oil tubes per rib-interval 1–4. 2*n*=22. RARE. Gravelly volcanic soil, yellow pine forest; 1250–1700 m. Near border of n&c SNH (Calaveras, Tuolumne cos.)

L. torreyi (J. Coulter & Rose) J. Coulter & Rose Pl 1–3 dm, glabrous to minutely scabrous; base fibrous, with clustered lvs; taproot long. **ST** ± 0. **LF**: petiole 2–5 cm, wholly sheathing; blade 2–15 cm, oblong to ovate, ternate-pinnately dissected, segments 3–10 mm, thread-like to linear, acute. **INFL**: peduncle 0.8–2 dm; bractlets 0; rays 3–9, 1–4 cm, ascending to spreading-ascending; pedicels 1–10 mm. **FL**: corolla yellow. **FR** 10–16 mm, narrowly oblong, glabrous; wings < body in width; oil tubes per rib-interval 1. 2*n*=22. Cracks in granite, pine forests; 1100–3300 m. SNH.

L. tracyi Mathias & Constance (p. 161) TRACY'S LOMATIUM Pl 1–3.5 dm, glabrous to minutely roughened; taproot slender. **ST** ± 0. **LF**: petiole 2.5–8, scarious, gen wholly sheathing; blade 4.5–10 cm, oblong to ovate, ternate-pinnately or 2-pinnately dissected, segments 1–7 mm, thread-like to oblong, acute or obtuse, often overlapping. **INFL**: peduncle 1–2 dm, spreading; bractlets 2–6, 1– 4 mm, linear to oblanceolate, scarious; rays 6–12, 1–6 fertile, 0.5–8 cm, ascending, very unequal; pedicels 1–5 mm. **FL**: corolla yellow. **FR** 6–10 mm, ± oblong-ovate, glabrous; wings << body in width; oil tubes obscure. 2*n*=22. UNCOMMON. Open pine forest, serpentine; 500–1500 m. KR, n NCoR, CaRH.

L. triternatum (Pursh) J. Coulter & Rose Pl 1.5–10 dm, gen finely soft-hairy or puberulent; taproot slender to massive. **ST** prominent. **LF**: petiole 7–20 cm, sheathing ± to middle; blade 7–20 cm, oblong-ovate to triangular-ovate or obovate, 1-2-ternate-pinnate, lflets 1.5–20 cm, gen linear to widely lanceolate, gen entire; cauline lvs 0 or gen wholly sheathing. **INFL**: peduncle gen 1–4.5 dm, spreading to erect; bractlets (0)3–8, 1–5 mm, thread-like to linear-lanceolate, ± scarious; rays 5–20, 2–10 cm, spreading or spreading-ascending, unequal, webbed; pedicels 1–10 mm, webbed. **FL**: corolla yellow; ovary glabrous to densely puberulent. **FR** 6–22 mm, oblong, puberulent or glabrous; wings gen < body in width; oil tubes per rib-interval 1. Uncommon. Sagebrush-juniper, pine woodland or forest, open slopes, meadows, open serpentine ridges, scrub; 200–2000 m. KR, NCoR, GB; to w Can, WY. Variable in GB, adjacent areas. Var. *anomalum* (M.E. Jones) Mathias evidently not in CA. Vars. poorly defined.

var. ***macrocarpum*** (J. Coulter & Rose) Mathias Pl often stout. **FL**: ovary densely puberulent. $2n=44$. Sagebrush-juniper, pine woodland, open slopes, meadows; 200–1500 m. KR, NCoR, GB; to WA, ID, NV. [*L. alatum* J. Coulter & Rose, *L. giganteum* J. Coulter & Rose]

var. ***triternatum*** Pl not very stout. **FL**: ovary glabrous. $2n=22,44$. Open serpentine ridges, scrub, pine forest; 200–2000 m. KR, NCoR, Wrn; to WA, w Can, MT, w NV. [*L. nudicaule* var. *puberulum* Jepson]

L. utriculatum (Torrey & A. Gray) J. Coulter & Rose Pl 1–5 dm, glabrous to densely puberulent; taproot gen slender. **ST** ± leafy. **LF**: petiole 1.5–10 cm, widely sheathing; blade 5–16 cm, oblong to ovate, pinnately or ternate-pinnately dissected, segments 2–25 mm, gen linear; cauline lvs conspicuously wholly sheathing. **INFL**: peduncle 0.5–3 dm; bractlets 3–12, 3–6 mm, gen obovate or oblanceolate, ± scarious, veiny, gen entire; rays 5–20, 2–12 cm, spreading to ascending, unequal; pedicels 2–9 mm, webbed. **FL**: calyx lobes sometimes evident; corolla yellow; ovary puberulent or glabrous. **FR** 5–15 mm, oblong to obovate, glabrous or nearly so; wings thin, = or > body in width; oil tubes per rib-interval 1–4. $2n=22$. Open grassy slopes, meadows, woodland; 50–1550 m. CA-FP; to B.C. [*L. vaseyi* J. Coulter & Rose] Variable. ❀TRY.

L. vaginatum (M.E. Jones) J. Coulter & Rose Pl 1–4.5 dm; taproot stout; herbage green, finely scabrous to nearly glabrous. **ST** leafy. **LF**: petiole 2–12 cm, widely sheathing basally; blade 5–15 cm, oblong-ovate to triangular-ovate, ternate-pinnately or pinnately dissected, segments crowded, 1–5 mm, oblong, obtuse; cauline lvs like basal, sheaths flared. **INFL**: peduncle 2.5–25 cm; bractlets 5–10, 3–7 mm, gen lanceolate to oblanceolate, acute, ± scarious; rays 10–15, 2–7 cm, spreading-ascending, unequal, ± webbed; pedicels 5–15 mm, webbed. **FL**: corolla yellow; ovary gen roughened. **FR** 8–15 mm, oblong to elliptic, gen roughened; wings gen > body in width; oil tubes per rib-interval 1–4. $2n=22$. Sagebrush, grassy slopes, pine woodland; 600–1900 m. KR, NCoR (serpentine), MP; c OR, w NV.

OENANTHE

Per from clustered, fibrous-tuberous roots or rhizomes, glabrous. **ST** gen decumbent or ascending, often rooting at lower nodes. **LF**: blade oblong to triangular-ovate, gen 1–3-pinnate, lflets wide or narrow, gen serrate to pinnately lobed. **INFL**: umbels compound; bracts gen 0 or inconspicuous; bractlets many; rays, pedicels many, spreading or spreading-ascending. **FLS**: outer bisexual or staminate, sometimes bilateral; outer calyx lobes acute, gen prominent, enlarging in fr or deciduous; petals wide, white or reddish, tips narrowed; styles persistent. **FR** oblong-ovate to round, subcylindric or ± compressed front-to-back; ribs low, obtuse, corky; oil tubes per rib-interval gen 1. **SEED**: face flat. ± 30 spp.: N.Am, Eurasia, Afr. (Greek: wine fl)

1. St from clustered fibrous roots bearing small tubers; lflets of uppermost lvs linear; outer fls bilateral, staminate; rays, pedicels thickened . . . . ***O. pimpinelloides***

1′ St from rhizome, rooting at lower nodes; lflets of uppermost lvs ± ovate; fls all radial, bisexual; rays, pedicels unthickened . . . . ***O. sarmentosa***

O. pimpinelloides L. Pl 3–10 dm. **LF**: petiole 3–10 cm; blade 8–12 cm, 2–6 cm wide, 2–3-pinnate, lflets 0.5–2 cm, linear to ovate, irregularly cut to pinnately lobed; upper cauline lvs narrower than basal. **INFL**: peduncle 5–15 cm; bractlets gen 8, 3–5 mm, linear-lanceolate, acute; rays 8–15, 1–2 cm; pedicels of inner bisexual fls 1–4 mm. **FL**: calyx lobes 1–1.5 mm, lanceolate; styles 3–5 mm. **FR** 2.5–3 mm, oblong, ± compressed front-to-back; ribs slender. $2n=22$. Moist places; 1500 m. n NCoRO (Humboldt Co.); native to Eur.

O. sarmentosa J.S. Presl (p. 161) Pl 5–15 dm. **LF**: petiole 1–3.5 dm; blade 1–3 dm, 6–25 cm wide, gen 2-pinnate, lflets 1–6 cm, ± ovate, serrate to lobed; cauline lvs like basal. **INFL**: peduncle 5–13 cm; bractlets many, 4–5 mm, lanceolate, acute; rays 10–20, 1.5–3 cm; pedicels 2–6 mm. **FL**: calyx lobes 0.5–1 mm, lanceolate; styles 2–3 mm. **FR** 2.5–3.5 mm, oblong, subcylindric; ribs wide. $2n=44$. Streams, marshes, ponds, often aquatic; 0–1000 m. NCo, SNF (uncommon), CCo, SCo; to w Can. ❀WET:7,14,**15,16,22–24**&SUN:**4–6,17**;INV.

OREONANA

Per, taprooted, cushion-forming, gen coarsely hairy or tomentose. **ST** 0. **LVS**: lowest bladeless sheaths; upper blades narrowly ovate to round, pinnately or ternately dissected, segments lanceolate or oblong. **INFL**: umbels compound, head-like, spheric or hemispheric; peduncles spreading; bracts 0; involucel 1-sided, bractlets fused to nearly free; rays few–many, spreading to spreading-ascending; fertile pedicels very short, sterile > fr. **FL** bisexual, staminate, or sterile; calyx lobes conspicuous or not; petals spoon-shaped or oblanceolate, gen white or yellow, or becoming purple, early deciduous, tips narrowed; projection atop ovary 0. **FR** ovate to round, slightly compressed side-to-side or cylindric, hairy; ribs subequal, thread-like; oil tubes per rib-interval several; fr axis divided to base. **SEED**: face deeply grooved. 3 spp.: high mtns of CA-FP. (Greek: mountain dwarf) [Shevock & Norris 1981 Fremontia 9:22–25]

1. Herbage white-tomentose; umbels hemispheric; corolla yellow or maroon; anthers yellow; calyx lobes inconspicuous, obscured by hairs . . . . ***O. vestita***

1′ Herbage grayish hairy; umbels spheric; corolla white; anthers purple; calyx lobes conspicuous, stellate-spreading

2. Fls appearing ± with lf blades; bladeless sheaths 0.5–2 cm; umbels 1–3 cm; rays 5–15, 2–8 mm; calyx lobes yellow . . . . ***O. clementis***

2′ Fls appearing before lf blades; bladeless sheaths 3–6 cm; umbels 2.4–4.5 cm; rays 20–35, 5–15 mm; calyx lobes purple . . . . ***O. purpurascens***

O. clementis (M.E. Jones) Jepson Pl 3–8 cm, grayish hairy. **LF**: bladeless sheaths 0.5–2 cm; petiole 2–4 cm; blade 1.5–3.5 cm, ovate, segments 1–3 mm. **INFL**: umbels 1–3 cm diam, spheric, appearing ± with lf blades; peduncles 2–7 cm, spreading; bractlets fused; rays 5–15, 2–8 mm, outer scarious-winged, webbed; sterile pedicels 2–5 mm. **FL** bisexual or staminate; calyx lobes, ± 0.5–1.5 mm, stellate-spreading, yellow; corolla white; anthers purple. **FR** 3–4 mm, ± glabrous to hairy. 2*n*=22. Rocky ridges; 1700–4000 m. s SNH.

O. purpurascens J.R. Shevock & Constance (p. 161) PURPLE MOUNTAIN-PARSLEY Pl 0.8–2.2 cm, grayish hairy. **LF**: bladeless sheaths 3–6 cm; petiole 4–7 cm; blade 5–10 cm, narrowly ovate, segments 1–3 mm, lanceolate or oblong. **INFL**: umbels 2.5–4.5 cm diam, spheric, appearing well before lf blades; peduncles 12–18 cm, spreading; bractlets fused; rays 20–35, 5–15 mm, outer scarious-winged, webbed; sterile pedicels 3–10 mm. **FL** bisexual, staminate, or sterile; calyx lobes 1.5–3 mm, stellate-spreading, purple; corolla white; anthers purple. **FR** 4–5 mm wide, nearly glabrous to hairy. 2*n*=22. RARE. Ridge-tops, on gen metamorphic rocks, red fir or lodgepole pine; 2500–2600 m. s SNH.

O. vestita (S. Watson) Jepson WOOLLY MOUNTAIN-PARSLEY Pl 4–15 cm, white-tomentose. **LF**: bladeless sheaths 1.5–3 cm; petiole 2–10 cm; blade 1.5–5 cm, ovate to round, segments 3–10 mm, oblong. **INFL**: umbels 2–5 cm, hemispheric, appearing ± with lf blades; peduncles 4–12 cm, spreading; bractlets nearly free; rays 10–25, 1–2 cm, not scarious-winged, not webbed; sterile pedicels 10–15 mm. **FL** bisexual or staminate; calyx lobes inconspicuous, obscured by hairs; corolla yellow or maroon; anthers yellow. **FR** 5–6 mm, 3–4 mm wide, tomentose. 2*n*=22. UNCOMMON. Ridge tops; 2410–3500 m. SnGB, SnBr.

OROGENIA

Per, glabrous, ± glaucous; root tuberous. **ST** 0 or very short. **LVS** basal, often recurved or spreading; lowest bladeless, scarious sheaths; blades ovate to triangular-ovate, 1–3-ternately dissected, segments gen linear to lanceolate, elongate. **INFL**: umbels compound, peduncled; bracts 0; bractlets 0–few, minute; rays few, spreading; pedicels few, 0 or very short. **FL**: calyx 0; petals obovate, white, tips narrowed; projection atop ovary 0. **FR** ± ovate, compressed front-to-back; marginal ribs corky-winged, incurved and not very wing-like, others thread-like; oil tubes per rib-interval several; fr axis represented by a corky, rib-like, projection along middle of fr-half. **SEED**: face slightly concave. 2 spp.: w N.Am. (Greek: mountain race)

O. fusiformis S. Watson (p. 161) Pl 5–15 cm; tuber 3–10 mm wide, carrot-like. **LF**: bladeless sheath 2–7 cm, oblong; petiole 3–5 cm; blade 2–8 cm wide, segments 5–60 mm, linear to linear-lanceolate, acute or obtuse. **INFL**: peduncles 1–3, 2–15 cm, recurving; fertile rays 1–8, 5–60 mm; fertile pedicels 2–15, < 1 mm. **FR** 3–4 mm, ovate; ribs on back obscure. Gravelly flats near melting snow; 1500–2000 m. KR, CaRH, n SNH; to c OR.

OSMORHIZA

Per, nearly glabrous to hairy; roots thick, clustered, licorice-scented. **ST** branched, leafy. **LF**: blade oblong to triangular-ovate, 2-pinnate or ternate-pinnate or 2–3-ternate, lflets lanceolate to round. **INFL**: umbels compound; bracts 0; bractlets 0–several and conspicuous; rays, pedicels few, spreading-ascending to spreading. **FL**: calyx lobes 0; petals obovate, white, purple, or greenish yellow (white), tips narrowed; disk sometimes present. **FR** linear to oblong, cylindric to club-shaped, slightly compressed side-to-side, bristly to glabrous; base obtuse or long-tapered into tail, tip tapered into beak or obtuse; ribs thread-like; oil tubes per rib-interval obscure; fr axis divided in upper 1/2. **SEED**: face concave or grooved. ± 10 spp.: Am, e&s Asia. (Greek: sweet root) [Lowry & Jones 1985 Ann Missouri Bot Gard 71:1128–1171]

1. Fr not long-tapered at base, glabrous; lf 2-pinnate (subg. *Glycosma*) .... ***O. occidentalis***
1′ Fr long-tapered at base into tail, bristly; lf 2–3-ternate (subg. *Osmorhiza*)
  2. Involucel conspicuous; corolla greenish yellow .... ***O. brachypoda***
  2′ Involucel 0 or vestigial; corolla white or purple, rarely greenish white
    3. Fr club-like, beakless; rays, pedicels spreading .... ***O. depauperata***
    3′ Fr linear-fusiform or -oblong, beaked; rays, pedicels spreading-ascending
      4. Projection atop ovary conic; disk 0; fr 12–25 mm; corolla white .... ***O. chilensis***
      4′ Projection atop ovary depressed-conic; disk conspicuous; fr 8–15 mm; corolla purple, rarely greenish white .... ***O. purpurea***

O. brachypoda Torrey (p. 161) Pl 3–8 dm, finely hairy. **LF**: petiole 5–20 cm; blade 1–2 dm, ovate to triangular-ovate, 2–3-ternate, lflets 2–8 cm, ovate, serrate to irregularly cut or pinnately lobed. **INFL**: peduncle 9–20 cm; bractlets 2–6, 2–10 mm, linear to lanceolate; rays 2–5, 3.5–12 cm, spreading-ascending; pedicels 1–5 mm. **FL**: corolla greenish yellow; styles gen < 1 mm; disk often conspicuous. **FR** 12–20 mm, oblong-fusiform; tail 1–4 mm; beak slender; ribs bristly. 2*n*=22. Moist ravines, coniferous forests, woodlands; 200–2000 m. c&s SNF, Teh, SnFrB, SCoR, TR, PR; c AZ.

O. chilensis Hook. & Arn. Pl 3–12 dm, nearly glabrous to finely hairy. **LF**: petiole 5–16 cm; blade 4–20 cm wide, widely ovate to obovate, 2-ternate, lflets 2–8 cm, widely lanceolate to round, serrate to irregularly cut or lobed. **INFL**: peduncle 5–25 cm; bractlets gen 0; rays 3–8, 2–12 cm, spreading-ascending; pedicels 4–20 mm. **FL**: corolla white; styles gen < 1 mm; projection atop ovary conic; disk 0. **FR** 12–25 mm, linear-fusiform or -oblong; tail 2–8.5 mm; beak slender; ribs bristly. 2*n*=22. Coniferous forests, woodland, disturbed areas; 100–2800 m. CA-FP; to AK, Alberta, Colorado; also c&e N.Am, s S.Am. ❀STBL.

O. depauperata Philippi Pl 1.5–8 dm, gen sparsely short-hairy to glabrous. **LF**: petiole 3–20 cm; blade 4–12 cm wide, widely ovate to round, 2–3-ternate, lflets 1.5–5 cm, widely lanceolate to ovate, coarsely serrate to deeply cut. **INFL**: peduncle 5–15 cm; bractlets gen 0; rays 2–5, 3–9 cm, spreading; pedicels 8–15 mm. **FL**: corolla white; styles < 0.5 mm; projection atop ovary low- or depressed-conic; disk 0 or inconspicuous. **FR** 10–18 mm, club-like; tail 3–8.5 mm; tip obtuse, beakless; ribs ± bristly. 2*n*=22. Very uncommon. Coniferous forest, aspen woodland; 500–3300 m. Wrn; to AK, MT, Colorado, NM; also in c&e N.Am, s S.Am.

O. occidentalis (Nutt.) Torrey Pl 4–12 dm, glabrous to sparsely fine-hairy. **LF**: petiole 5–25 cm; blade 1–2 dm, oblong to ovate, 2-pinnate, lflets 2–10 cm, oblong-lanceolate to ovate, serrate and gen irregularly cut or lobed. **INFL**: peduncle 6–20 cm; bractlets gen 0; rays 5–12, gen 3–8 cm, ascending to spreading-ascending; pedicels 3–8 mm. **FL**: corolla yellow; styles 0.8–1.4 mm; disk conspicuous. **FR** 12–22 mm, linear-fusiform, not long-tapered at base; tail 0; tip narrowed below; ribs (and intervals) glabrous. 2*n*=22. Coniferous forest, oak woodland; 350–3100 m. KR, NCoR, CaRH, n&c SNH, MP; to w Can, Colorado.

O. purpurea (J. Coulter & Rose) Suksd. Pl 2–6 dm, glabrous to very sparsely fine-hairy. **LF**: petiole 5–12 cm; blade 3–10 cm wide, triangular-ovate to round, 2–3-ternate, lflets 1.5–7 cm, lanceolate to ovate, coarsely serrate to irregularly cut or lobed. **INFL**: peduncle 3–10 cm; bractlets 0; rays 2–6, 3–9.5 cm, spreading-ascending; pedicels 5–25 mm. **FL**: corolla purple or rarely greenish white; styles 0.4–0.8 mm; projection atop ovary depressed-conic; disk conspicuous. **FR** 8–15 mm, linear-fusiform; tail 1–5 mm; tip narrowed below, beaked; ribs bristly at base. $2n$=22. Damp coniferous forest; 150–2200 m. NCo (Del Norte Co.); to AK, MT.

OXYPOLIS

Per, glabrous; tubers clustered. **ST** erect, gen branched. **LF**: blade linear or oblong to triangular-ovate, ternate, pinnate, or simple; main axis segmented, hollow, sometimes bladeless. **INFL**: umbels compound; involucre, involucel extremely variable; bracts, bractlets 0 or 1–several, ± inconspicuous; rays, pedicels few–many, spreading-ascending. **FL**: calyx lobes conspicuous or minute; petals wide, white or purple, tips narrowed. **FR** oblong to obovate, very compressed front-to-back; ribs unequal, marginal widely thin-winged (wings veined on back at inner margin), others thread-like; oil tubes per rib-interval 1; fr axis divided to base. ± 6 spp.: w&e US, Caribbean. (Greek: sharp white)

O. occidentalis J. Coulter & Rose Pl 6–15 dm. **LF**: petiole 1–5 dm; blade 12–30 cm, oblong, 1-pinnate, lflets 5–13, 3.5–9.5 cm, lanceolate to widely ovate, crenate or serrate to irregularly cut; cauline lvs have enlarged petiole, smaller lflets than basal. **INFL**: peduncle 6–30 cm; bracts 0–1(8), 5–20 mm, linear, scarious; bractlets like bracts, < 10 mm; rays 12–24, 2–8.5 cm; pedicels 3–15 mm. **FL**: calyx lobes conspicuous. **FR** 5–6 mm, oblong or ovate. $2n$=36. Bogs, wet meadows, streamsides, often in coniferous forest; 1200–2600 m. CaRF, SN, c SnBr.

PASTINACA PARSNIP

Bien, per, taprooted, nearly glabrous to hairy. **ST** erect, branched. **LF**: blade oblong to triangular-ovate, 1–2-pinnate, lflets oblong to ovate. **INFL**: umbels compound; bracts gen 0; bractlets gen 0; rays 5–20, pedicels many, rays and pedicels spreading-ascending. **FL**: calyx lobes minute; petals wide, yellow or orange, tips narrowed. **FR** oblong to obovate, very compressed front-to-back; ribs unequal, marginal narrowly winged, others thread-like; oil tubes per rib-interval 1, all equal in length; fr axis divided to base. **SEED**: face flat. ± 10 spp.: Eurasia. (Ancient name for parsnip)

P. sativa L. Bien, 0.5–2 m, nearly glabrous to puberulent. **ST** conspicuously angled, grooved. **LF**: petiole 1–1.5 cm; blade 1.5–3 dm, oblong to ovate, 1-pinnate, lflets 5–11, 5–10 cm, oblong to ovate, coarsely serrate and lobed or divided. **INFL**: peduncle 7–15 cm. **FR** 4–6 mm wide, oblong to round; oil tubes visible. $2n$=22. Roadsides, etc.; < 1000 m. CA-FP; to e US; native to Eurasia. Sporadic.

PERIDERIDIA YAMPAH, SQUAWROOT

Per, glabrous, often glaucous; roots tuberous, single or clustered, or fibrous, clustered. **ST** erect, branched. **LF**: blade lanceolate to triangular-ovate, gen 1–2-ternate-pinnate or 1–2-pinnately or ternate-pinnately dissected, lflets or segments gen linear to linear-lanceolate. **INFL**: umbels compound; bracts 0–many, conspicuous and reflexed or not; bractlets several–many, narrow, ± scarious; rays, pedicels few–many, gen spreading-ascending; 2° umbels gen convex on top. **FL**: calyx lobes evident; petals gen obovate, white, tips narrowed. **FR** linear-oblong to round, slightly compressed side-to-side or not at all, glabrous; ribs subequal, thread-like to prominent, not winged; oil tubes per rib-interval 1–several; fr axis divided to base. **SEED**: face flat to grooved. ± 12 spp.: gen w Am. (Greek: around the neck, from involucre) [Chuang & Constance 1969 Univ Calif Publ Bot 55] Roots, basal lvs needed for identification.

1. Roots fibrous or slightly thickened, gen 5–20-clustered; styles gen < 1 mm
 2. Lflets widely lanceolate to ovate, 10–40 mm wide; petals 5–7-veined; fr ribs prominent .... ***P. howellii***
 2′ Lflets linear to lanceolate, 2–8 mm wide; petals 1-veined; fr ribs thread-like .... ***P. kelloggii***
1′ Roots tuberous, single or 2–6-clustered; styles gen 1–2.5 mm
 3. Basal lvs 1–2-ternate or 1–2-pinnate with 1–3 pairs of 1° lflets
  4. Rays in fr unequal, spreading-ascending; oil tubes per rib-interval 1 .... ***P. lemmonii***
  4′ Rays in fr subequal and ascending, or unequal and spreading-ascending; oil tubes per rib-interval 3–4 . ***P. parishii***
   5. Rays gen 12–14, in fr subequal; bractlets ± = pedicels; fr ovate to nearly round, 2.5–3.5 mm ...... ssp. ***latifolia***
   5′ Rays gen 6–11, in fr unequal; bractlets gen < pedicels; fr oblong to ± ovate, 3–5 mm .... ssp. ***parishii***
 3′ Basal lvs ± 1-pinnate with 3–5 pairs of 1° lflets or 2–3-pinnately or ternate-pinnately dissected or compound
  6. Basal lvs ± 1-pinnate with 3–5 pairs of 1° lflets, lower 1° lflets sometimes lobed or ternately dissected .... ***P. gairdneri***
   7. Tuberous roots 2–3-clustered; petals 5- or 7-veined .... ssp. ***borealis***
   7′ Tuberous roots single; petals 1-veined .... ssp. ***gairdneri***
  6′ Basal lvs 2–3-pinnately or ternate-pinnately dissected or compound
   8. Terminal lf segments unlike lateral in size or form
    9. Fr gen not > 6 mm; bractlets wholly scarious, 3–9 mm, acuminate .... ***P. bolanderi*** ssp. ***bolanderi***
    9′ Fr 5–8 mm; bractlets scarious-margined, 1–4 mm, acute
     10. Pl 5–15 dm; tuberous roots cylindric, 5–12 cm; lf segments flattened, 5–10 mm wide; oil tubes per rib-interval 1 .... ***P. californica***
     10′ Pl 3.5–7.5 dm; tuberous roots oblong to ovoid, 1–6 cm; lflets cylindric, < 1 mm wide; oil tubes per rib-interval 3–4 .... [2]***P. pringlei***
   8′ Lf segments or lflets all similar in size or form
    11. Longest rays in fr 5–9 cm

12. 2° umbels gen spheric, gen 40–50-fld; umbels slightly concave on top, rays very unequal; oil tubes per rib-interval 1 .... *P. bacigalupii*
12′ 2° umbels convex on top, gen 20–25-fld; umbels convex on top, rays subequal; oil tubes per rib-interval 3–4 .... [2]*P. pringlei*
11′ Longest rays in fr 1.5–3 cm
13. Bracts 0–2, inconspicuous
14. Umbels strongly concave on top; rays in fr very unequal; fr linear-club-shaped, 5–7 mm ..... *P. leptocarpa*
14′ Umbels convex or flat on top; rays in fr subequal; fr oblong, 3–6 mm .... [2]*P. oregana*
13′ Bracts 6–12, conspicuous
15. Bractlets widely scarious-margined, gen acuminate; oil tubes per rib-interval 2–3 .... *P. bolanderi* ssp. *involucrata*
15′ Bractlets scarious-margined, acute; oil tubes per rib-interval 1 .... [2]*P. oregana*

P. bacigalupii Chuang & Constance (p. 161) BACIGALUPI'S PERIDERIDIA Pl 5–17 dm; roots tuberous, 2–6-clustered, 2–6 cm. **LF**: basal petiole 6–20 cm; basal blade 20–40 cm, ± ovate, gen 2-pinnate, lflets 2–15 cm, linear-lanceolate, gen entire, acute; cauline lvs gen 1-pinnate. **INFL**: peduncle 3–25 cm; bracts 1 or 2, very narrow; bractlets 12–16, 0.5–15 cm, linear-lanceolate, sharply reflexed; rays 4–7, 3–6 cm, unequal, spreading-ascending; pedicels 3–8 mm; 2° umbels gen spheric, 40–50(60)-fld. **FL**: petals 1-veined; styles 0.7–1.2 mm. **FR** 4–6 mm, oblong; ribs thread-like; oil tubes per rib-interval 1. 2*n*=38. UNCOMMON. Chaparral, pine woodland; 450–1000 m. n&c SNF.

P. bolanderi (A. Gray) Nelson & J.F. Macbr. Pl 1.5–9 dm; roots tuberous, single or 2–3-clustered, 1–7 cm. **LF**: basal petiole 2–15 cm; basal blade 10–20 cm, ± ovate, gen 1–2-ternate-pinnately dissected, segments 0.5–6 cm, thread-like to oblong, gen lobed, toothed; cauline lvs ternate-pinnately dissected or 1-ternate. **INFL**: peduncle 2–20 cm; bracts 8–12, 3–12 mm, ± lanceolate, gen acuminate; bractlets 4–10, 3–9 mm, like bracts; rays 9–23, 1–2 cm, subequal, ascending or spreading-ascending; pedicels 2–5 mm; 2° umbels 18–30-fld. **FL**: petals 1-veined; styles 2 mm. **FR** 4–6 mm, oblong; ribs thread-like; oil tubes per rib-interval 2–3. Meadows, scrub, pine forest, blue-oak woodland, summer-dry clay soil; 600–2000 m. NW, SN, GB; to WY, UT. Sspp. quite distinct.

ssp. ***bolanderi*** Pl 1.5–8 dm, green. **LF**: segments 1–4 mm wide, oblong to thread-like, terminal 3–8 cm, lateral 0.5–3 cm. **INFL**: bracts, bractlets wider than linear-lanceolate, deciduous, wholly scarious, margins uneven; 2° umbels 25–30-fld. 2*n*=38. Meadows, scrub, pine forest; 1000–2000 m. KR, CaRH, SNH, MP; to OR, ID, WY, NV. ❀DFCLT.

ssp. ***involucrata*** Chuang & Constance Pl 4–9 dm, glaucous. **LF**: segments 0.6–6 cm, < 1 mm wide, linear, terminal gen like lateral. **INFL**: bracts, bractlets linear-lanceolate, persistent, widely scarious-margined, margins entire; 2° umbels 18–25-fld. 2*n*=38. Blue-oak woodland, in summer-dry clay soil; 600–1000 m. n&c SNF.

P. californica (Torrey) Nelson & J.F. Macbr. Pl 5–15 dm, gen branched; roots tuberous, 2–5-clustered, 5–12 cm, cylindric. **LF**: basal petiole 2–6 cm; basal blade 15–45 cm, ± ovate, 1–2-ternate-pinnately dissected, segments flattened, terminal 3–15 cm, linear-lanceolate, entire, lateral 3–30 mm, 5–10 mm wide, ovate, pinnately lobed; cauline lvs reduced. **INFL**: peduncle 2–20 cm: bracts 5–8, 2–5 mm, ovate-lanceolate, scarious-margined, acuminate; bractlets 10–12, 1–4 mm, ± ovate, scarious-margined; rays gen 8–12, 3–8 cm, subequal, spreading-ascending; pedicels 5–10 mm; 2° umbels 18–28-fld. **FL**: petals 1-veined; styles 1–1.5 mm. **FR** 5–8 mm, oblong; ribs thread-like; oil tubes per rib-interval 1. 2*n*=44. Damp soil along streams; 300–1250 m. c SNF, SnFrB, n SCoR, s SCoR. ❀IRR:**7,14–24**.

P. gairdneri (Hook. & Arn.) Mathias Pl 3–14 dm; roots tuberous, single or 2–3-clustered, 1.5–8 cm, fusiform. **LF**: basal petiole 8–15 cm; basal blade 20–35 cm, oblong to ovate, gen 1-pinnate, lflets 2–12 cm, linear or lanceolate, basal sometimes lobed or ternately dissected; cauline lvs 1–2-pinnate or 1–2-ternate. **INFL**: peduncle 3–20 cm; bracts gen 0; bractlets 8–13, 1–4 mm, linear-lanceolate; rays gen 7–16, 1.5–7 cm, unequal, spreading-ascending; pedicels 4–8 mm; 2° umbels 15–40-fld. **FL**: styles 1–2 mm. **FR** 2.5–3.5 mm, ± round; ribs thread-like; oil tube per rib-interval 1. Moist soil of flats, meadows, streamsides, grasslands, pine groves; 0–3000 m. s NCo, KR, NCoRH, CCo, SCo, Wrn; to sw Can, SD, NM. Sspp. could be considered distinct spp.

ssp. ***borealis*** Chuang & Constance Pl ± rigid; roots tuberous, gen 2–3-clustered, < 10–15 mm wide. **INFL**: rays gen 11–16. **FL**: petals 5- or 7-veined. 2*n*=40,80,120. Gen moist soil of flats, meadows, streamsides; 0–3000 m. KR, NCoRH, Wrn; to sw Can, SD.

ssp. ***gairdneri*** GAIRDNER'S YAMPAH Pl gen flexible; root tuberous, single, 10–15 mm wide. **INFL**: rays gen 7–14. **FL**: petals 1-veined. 2*n*=38. RARE. Coastal flats, grassland, pine groves; 0–350 m. s NCo (Sonoma Co.), CCo (scarce s of Monterey Co.), SCo. In cult.

P. howellii (J. Coulter & Rose) Mathias (p. 161) Pl 8–16 dm; roots fibrous, slightly thickened, > 15-clustered, 4–15 cm. **LF**: basal petiole 10–20 cm; basal blade 30–50 cm, ± lanceolate, 2-pinnate, lflets 2–7 cm, 1–4 cm wide, ± ovate, deeply toothed or lobed; cauline lvs like basal. **INFL**: peduncle 5–25 cm; bracts 8–12, 0.5–2 cm, linear-spoon-shaped, scarious-margined, reflexed; bractlets ± 10, 5–12 mm, lanceolate; rays gen 17–20, 2–7 cm, gen subequal, spreading-ascending; pedicels 5–15 mm; 2° umbels 25–30-fld. **FL**: calyx lobes prominent; petals 5–7-veined; styles < 0.5 mm. **FR** 4–6 mm, oblong; ribs conspicuously corky; oil tubes per rib-interval 1. 2*n*=40. Moist meadows, ravines, stream banks; 300–1500 m. KR, NCoRO, CaR, n&c SNF; OR.

P. kelloggii (A. Gray) Mathias Pl 7–15 dm; roots fibrous, slightly thickened, 5–15-clustered, 3–15 cm. **LF**: basal petiole 3–6 cm, often wholly sheathing; basal blade 15–45 cm wide, lanceolate to triangular-ovate, 1–2-ternate-pinnate, lflets 3–12 cm, 2–8 mm wide, linear to lanceolate, entire; cauline lvs 1-ternate-pinnate. **INFL**: peduncle 5–18 cm; bracts 8–10, 5–15 mm, linear-lanceolate, acuminate, scarious-margined, sharply reflexed; bractlets ± 10, 3–5 mm, linear-lanceolate; rays 9–20, 2–6.5 cm, subequal, spreading-ascending; pedicels 2–5 mm; 2° umbels 23–27-fld. **FL**: petals 1-veined; styles 0.8–1 mm. **FR** 4–5 mm, elliptic-oblong; ribs thread-like; oil tubes per rib-interval 1. 2*n*=40. Open grassland, serpentine outcrops; < 1000 m. NCo, CaRF, n&c SNF, CCo, SnFrB. ❀SUN:**7,14–24**.

P. lemmonii (J. Coulter & Rose) Chuang & Constance Pl 2.5–9 cm; roots tuberous, gen single, 1.5 cm. **LF**: basal petiole 4–15 cm; basal blade 10–30 cm, ± ovate, 1-ternate or 1–2-pinnate with 1–2 pairs of 1° lflets, 1° lflets 3–10 cm, linear-lanceolate, gen entire; cauline lvs 1-ternate. **INFL**: peduncle 3–20 cm; bracts gen 0; bractlets 7–10, 1–2 mm, linear-lanceolate; rays 10–14, 1–3.5 cm, unequal, spreading-ascending; pedicels 3–5 mm; 2° umbels 20–35-fld. **FL**: petals 1-veined; styles 1–1.5 mm. **FR** 3–4.5 mm, oblong, to ± round; ribs thread-like; oil tubes per rib-interval 1. 2*n*=36. Open meadows, coniferous forest edges; 1000–2500 m. SN; se OR, w NV. Grows with similar *P. parishii*, forming masses of "Queen Anne's lace" in late summer.

P. leptocarpa Chuang & Constance NARROW-SEEDED YAMPAH Pl 4.5–7.5 dm; roots tuberous, gen 2–3-clustered, 0.5–3.5 cm. **LF**: basal petiole 1–5 cm; basal blade 10–15 cm, ovate-lanceolate, 2- pinnate, lflets 1–5 cm, narrowly linear, entire, acute; cauline lvs 1- or 2-pinnate. **INFL**: peduncle 5–8 cm; bracts gen 0; bractlets 6–8, 1–4 mm, linear-lanceolate, scarious-margined; rays 12–18, 1–4 cm, very unequal, spreading-ascending or ascending; pedicels 2–4 mm; 2° umbels 20–30-fld. **FL**: petals 1-veined; styles 0.5–1.5 mm. **FR** 5–7 mm, linear-club-shaped; ribs thread-like; oil tubes per rib-interval 1. 2*n*=34. UNCOMMON. Serpentine outcrops in woodland; 600–1500 m. KR. Possibly belongs in *P. oregana*.

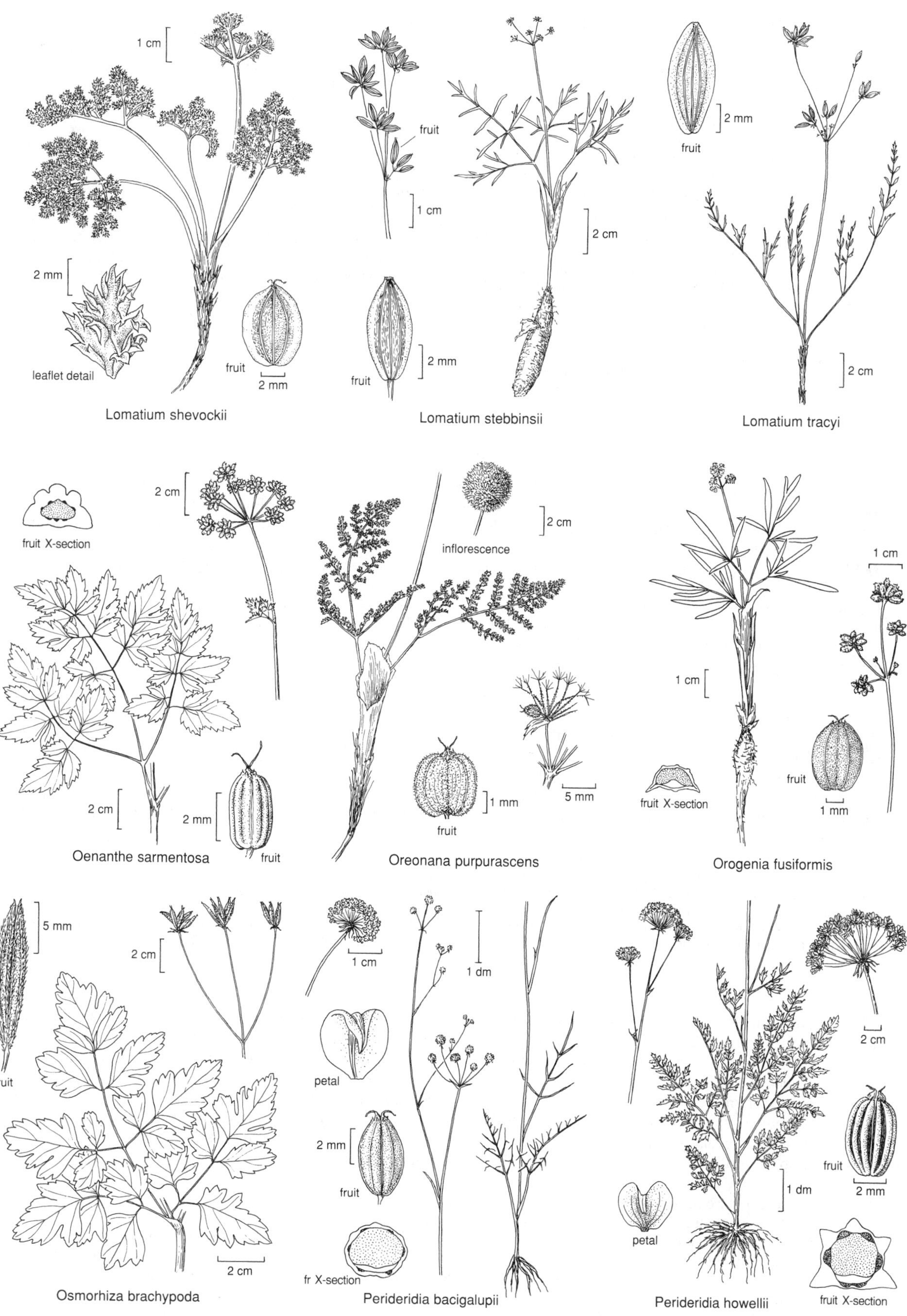
1 cm
2 mm
leaflet detail
fruit
2 mm
Lomatium shevockii
fruit
1 cm
fruit
2 mm
2 cm
Lomatium stebbinsii
2 mm
fruit
2 cm
Lomatium tracyi
fruit X-section
2 cm
2 cm
2 mm
fruit
Oenanthe sarmentosa
inflorescence
2 cm
1 mm
fruit
5 mm
Oreonana purpurascens
1 cm
1 cm
fruit X-section
fruit
1 mm
Orogenia fusiformis
5 mm
2 cm
fruit
2 cm
Osmorhiza brachypoda
1 cm
1 dm
petal
2 mm
fruit
fr X-section
Perideridia bacigalupii
2 cm
fruit
2 mm
1 dm
petal
Perideridia howellii
fruit X-section

P. oregana (S. Watson) Mathias Pl 1–9 dm, green to glaucous; roots tuberous, 2–6-clustered, 0.5–3 cm, fusiform to spheric. **LF**: basal petiole 2–10 cm; basal blade 3–30 cm, ± ovate, 1–2-ternate-pinnate, lflets 0.5–6 cm, linear or oblong, gen entire; cauline lvs 1–2-pinnate. **INFL**: peduncle 3–20 cm; bracts 6–10 (0–2), bristle-like; bractlets 4–8, 2–7 mm, linear-lanceolate; rays 10–29, 1–4.5 cm, subequal, spreading-ascending; pedicels 2–6 mm; 2° umbels 10–29-fld. **FL**: petals 1-veined; styles 1–2 mm. **FR** 3–6 mm, oblong; ribs thread-like; oil tubes per rib-interval 1. $2n$=16,18,20,26. Open flats or slopes, pine-oak woodland; 300–2000 m. KR, NCoR, CaRH, SnFrB, SCoRI, MP; OR. Highly variable.

P. parishii (J. Coulter & Rose) Nelson & J.F. Macbr. Pl 1.5–9 dm, green; roots tuberous, single, 1–2.5 cm, fusiform. **LF**: basal petiole 3–10 cm; basal blade 10–20 cm, ± ovate, gen 1-ternate or 1-pinnate with 1–3 pairs of lflets, lflets 3–15 cm, ± lanceolate, entire; cauline lvs 1-ternate. **INFL**: peduncle 3–20 cm; bracts 0–2, bristle-like; bractlets 3–8, 2–4 mm, linear-lanceolate, scarious-margined; rays 5–20, 1–4.5 cm, subequal or unequal, ascending or spreading-ascending; pedicels 3–4 mm; 2° umbels 13–27-fld. **FL**: petals 1-veined; styles 1–1.5 mm. **FR** 2.5–5 mm wide, oblong to ± round; ribs thread-like; oil tubes per rib-interval 3–4. Moist meadows, open coniferous forests; 2000–3000 m. KR, CaR, SNH, TR, PR; to NV, NM.

ssp. ***latifolia*** (A. Gray) Chuang & Constance **INFL**: umbels flat or convex on top; rays gen 12–14, subequal; bractlets 6–8, ± = pedicels. **FL**: styles 1.5 mm. **FR** 2.5–3.5 mm, ovate to ± round. $2n$=38. Wet meadows, open coniferous forests; 2000–3000 m. KR, CaR, SNH, TR, PR; NV.

ssp. ***parishii*** PARISH'S YAMPAH **INFL**: umbels concave on top; rays gen 6–11, unequal; bractlets 3–5, gen < pedicels. **FL**: styles 1 mm. **FR** 3–5 mm, oblong to ± ovate. $2n$=38. RARE in CA. Damp meadows; 2000–3000 m. SnBr; AZ, NM.

P. pringlei (J. Coulter & Rose) Nelson & J.F. Macbr. PRINGLE'S YAMPAH Pl 3.5–7.5 dm; roots tuberous, 2–4-clustered, 1–6 cm. **LF**: basal petiole 4–15 cm; basal blade 10–30 cm, ± ovate, 2-pinnate, lflets 0.5–8 cm, < 1 mm wide, linear; cauline lvs 1-pinnate. **INFL**: peduncle 3–12 cm; bracts 0–5, bristle-like; bractlets 8–12, 2–4 mm, linear-lanceolate to ovate; rays gen 5–7, 2–8 cm, ± unequal, spreading-ascending; pedicels 5–7 mm; 2° umbels 18–29-gen 20–25-fld. **FL**: petals 1-veined; styles 1 mm. **FR** 5–8 mm, oblong; ribs thread-like; oil tubes per rib-interval 3–4. $2n$=40. UNCOMMON. Grassy slopes, serpentine outcrops; 300–1800 m. Teh, SCoR, WTR.

PETROSELINUM

Ann, bien, taprooted, glabrous. **ST** erect, branched. **LF**: blade oblong to triangular-ovate, 1–3-pinnate or ternate-pinnate, lflets linear-lanceolate to obovate, toothed or lobed. **INFL**: umbels compound; bracts 0–few, very narrow; bractlets several, narrow; rays, pedicels few to ± many, spreading-ascending. **FL**: calyx lobes minute; petals wide, greenish yellow or white. **FR** oblong to ovate, slightly compressed side-to-side; ribs thread-like; oil tubes per rib-interval 1; fr axis divided to base. **SEED**: face flat. 2 spp.: Eurasia (1 cult, sometimes weedy). (Greek: stone wreath; name used by Dioscorides)

P. crispum (Miller) A.W. Hill PARSLEY Pl 3–13 dm. **LF**: basal petiole 1–2 dm; basal blade 3–10 cm, ± ovate, 1–3-pinnate; 1° lflets of lower lvs 2–5 cm, obovate to ovate-lanceolate, toothed, lobed; cauline lvs ± sessile, 1-ternate, lflets gen linear-lanceolate, entire or 3-lobed. **INFL**: peduncle 3–8 cm; rays 10–20, 1–5 cm, subequal; pedicels 2–5 mm. **FR** 2–4 mm ± ovate. $2n$=22. Disturbed places near gardens; < 1000 m. CA-FP; native to Eur.

PODISTERA

Per, taprooted, low, fibrous at base, ± puberulent. **ST** 0. **LVS** basal; blades oblong to widely ovate, 1–2 pinnate, lflets linear to round. **INFL**: umbels compound, dense, head-like; bracts 0 or linear; involucel 1-sided; bractlets several, narrow or wide, gen partly fused; rays few, cylindric or flattened to winged; pedicels like rays. **FL**: calyx lobes conspicuous; petals wide, yellow, purplish, or white, tips narrowed. **FR** oblong-ovate to ovate, slightly compressed side-to-side, glabrous; ribs subequal, thread-like to prominently corky, obtuse; oil tubes per rib-interval 2–several; fr axis not seen. **SEED** compressed front-to-back; face flat to slightly concave. 4 spp.: mtns of w N.Am. (Greek: solid foot, from compact habit)

P. nevadensis (A. Gray) S. Watson (p. 167) SIERRA PODISTERA Pl forming compact cushions 2–5 cm, 2–5 dm diam. **LF**: petiole 3–15 mm, conspicuously white scarious-sheathing; blade 3– 10 mm, oblong to ovate, 1-pinnate, lflets 1–6 mm, linear to lanceolate, entire, pointed. **INFL**: peduncle 5–30 mm; bracts 0; bractlets 2–4 mm, ± = fls and frs, ovate, strongly fused into a cup; rays winged, very short; pedicels 0–few, < fr. **FL**: corolla yellow. **FR** 4–4.5 mm, oblong-ovate; ribs thread-like; oil tubes per rib-interval 3–4. $2n$=22. UNCOMMON. Unglaciated granitic gravel, scree, crevices above timberline; 3000–4000 m. n&c SNH, SnBr, W&I.

SANICULA

Bien, per, rhizomed or tap- or tuberous-rooted, glabrous or minutely scabrous. **ST** gen spreading or erect. **LF**: blade oblong-ovate to obovate, entire to ternately, palmately, subpinnately, or pinnately lobed, dissected, or compound. **INFL**: heads simple, in cymes or racemes, dense, of bisexual and staminate (or only staminate) fls; bracts entire or lobed, < to > heads; bisexual fls pedicelled or not, staminate gen pedicelled. **FL**: calyx lobes prominent, persistent, sometimes fused; petals wide, yellow, red-purple, or greenish white, tips narrowed, often lobed; styles long or short; projection atop ovary 0. **FR** oblong-ovate to round, slightly compressed side-to-side; fr-halves subcylindric, covered with prickles, scales, or tubercles; ribs 0; oil tubes evident or obscure, regularly or irregularly arranged; fr central axis not an obvious structure. **SEED**: face flat or grooved. ± 40 spp.: temp, ± worldwide. (Latin: to heal) [Bell 1954 Univ Calif Publ Bot 27:133–230]

1. Lvs compound, base of main division petiole-like and unwinged
  2. Lvs palmate (sometimes palmate-ternate); ultimate segment of main axis 4–30 mm wide
    3. Main lf divisions often ± 5, central division gen 2–4 cm; styles 2 × calyx; bracts 1–2 mm; fr sometimes clearly stalked above receptacle; common, widespread, variable .... [2]***S. crassicaulis***
    3′ Main lf divisions ± 3, central division gen 4–7 cm; styles ± = calyx; bracts 3–5 mm; fr ± sessile above receptacle; uncommon, coastal .... [2]***S. hoffmannii***

2′ Lvs somewhat pinnate (first division often ternate, others pinnate); ultimate segment of main axis 1–5 mm wide
4. Fr with many clearly hooked bristles
5. Staminate fls 4–6 per umbel, inconspicuous; st leafy, ± erect .... ***S. bipinnata***
5′ Staminate fls 7–12 per umbel, conspicuous; st ± lfless, branches often spreading widely from near base .... ***S. graveolens***
4′ Fr with ± rounded tubercles, hooked bristles 0 or few
6. Pl from a taproot that is rarely tuber-like; staminate fls slightly > fr, not very conspicuous — NCo, CaRF .... ***S. tracyi***
6′ Pl from a distinct, spheric or irregular tuber; staminate fls often >> fr and very conspicuous
7. Corolla straw- or salmon-colored; fr 2.5–3 mm, most tubercles nipple-like or with a bristle; SnFrB ***S. saxatilis***
7′ Corolla yellow; fr 1.5–2 mm, tubercles unarmed; widespread, variable .... ***S. tuberosa***
1′ Lvs simple, entire to deeply lobed or cut, axis of main division with at least a narrow, toothed wing throughout
8. Bracts yellow, >> heads; pl prostrate; lvs bright yellowish green when fls open .... ***S. arctopoides***
8′ Bracts greenish, inconspicuous, < or = heads; pl spreading to erect; lvs green when fls open
9. Lf margins not or barely toothed, some basal lvs entire .... ***S. maritima***
9′ Lf margins sharply toothed, all lvs ± lobed
10. Lvs palmately lobed or dissected; main divisions 3–5
11. Main lf division narrowed to an almost petiole-like base .... [2]***S. hoffmannii***
11′ Main lf division often 10 mm or more wide at base
12. Outline of lf margin ± rounded, teeth gen ± 1 mm; pl stout, 24–120 cm, gen branched well above base; common, widespread, variable .... [2]***S. crassicaulis***
12′ Outline of lf margin sharply angled, teeth gen ± 2 mm; pl slender, 5–30 cm, gen branched near base; NCo, CCo .... ***S. laciniata***
10′ Lvs somewhat pinnately lobed or dissected; main divisions gen 7 or more, each gen deeply cut
13. Lvs appearing almost palmate, with some pinnate divisions — corolla yellow; SW .... ***S. arguta***
13′ Lvs clearly pinnate, with some ternate divisions
14. Corolla purple or yellow; frs 3–8 per head; staminate fls inconspicuous, < fr; widespread ... ***S. bipinnatifida***
14′ Corolla yellow; frs 1–5 per head; staminate fls conspicuous, > fr; KR (Del Norte Co.) .... ***S. peckiana***

S. arctopoides Hook. & Arn. FOOTSTEPS OF SPRING, YELLOW MATS Pl prostrate, 10–30 cm wide, taprooted. **LF** simple, ± palmately or ternately dissected, bright yellowish green; blade 2–6.5 cm, triangular-ovate to round, lobes coarsely toothed to lobed. **INFL**: peduncle 3–21 cm; bracts 8–17, fused at base, 5–18 mm, >> heads, oblanceolate, entire or 3-lobed, yellow; pedicel of bisexual fl ± 2 mm, of staminate 3–4 mm. **FLS**: bisexual 10–12; staminate 10–13; calyx lobes 1–2 mm, ovate, acute, fused below middle; corolla yellow; styles 2–3 × calyx lobes. **FR** 2–5 mm, obovate to nearly round, smooth or with stout, curved, inflated, bulbous-based prickles above, prickleless tubercles below. **SEED**: face concave. $2n$=16. Open coastal bluffs, headlands, dunes; 0–250 m. NCo, CCo; to B.C. Fls late winter, early spring. ❀DFCLT.

S. arguta J. Coulter & Rose Pl 10–50 cm, minutely scabrous; taproot turnip-shaped. **LF** simple, somewhat pinnately (but appearing almost palmately) dissected, green; blade 3–10 cm, triangular-ovate, main divisions gen ± 7, some divisions pinnate, margins sharply toothed. **INFL**: peduncle 2.5–11 cm; bracts ± 10, 2–4 mm, < heads, lanceolate, entire or 3-lobed; pedicel of bisexual fl short, of staminate 2–4 mm. **FLS**: bisexual 5–10; staminate ± 10; calyx lobes ovate, obtuse, fused below middle; corolla yellow; styles 2–3 × calyx lobes. **FR** 4–6 mm, obovate; prickles stout, curved, inflated, bulbous-based above, little developed below. **SEED**: face concave to flat. $2n$=16. Open grassy or rocky slopes; 0–600 m. SW; Baja CA. [*S. simulans* Hoover] ❀DRN,DRY,SUN:**15–17, 22–24**.

S. bipinnata Hook. & Arn. POISON SANICLE Pl 12–60 cm; taproot ± swollen. **ST** leafy. **LF** compound, ternate then 1–2-pinnate, green; blade 3–10 cm, ± ovate, lflets well separated, stalked, margins entire to lobed, ultimate segment of main axis 1–5 mm wide. **INFL**: peduncle 1.5–9 cm; bracts 6–8, 1–2 mm, linear to lanceolate, entire, < heads; pedicel of bisexual fl 0, of staminate 1–1.5 mm. **FLS**: bisexual 3–10; staminate 4–6, inconspicuous; calyx lobes ovate, acute, fused at base; corolla yellow; styles 2–3 × calyx lobes. **FR** 2–3 mm wide, ovate to nearly round, covered with bulbous tubercles, each gen with short curved prickles. **SEED**: face grooved. $2n$=16. Open grassland or pine-oak woodlands; 20–1000 m. NCoR, SNF, ScV, SnFrB, SCoR, n&c SW. Possibly TOXIC, but no record of poisonings exist. ❀DFCLT.

S. bipinnatifida Hook. PURPLE SANICLE, SHOE BUTTONS Pl 12–60 cm, taprooted. **LF** simple, 1–2-pinnately dissected, green, glaucous, or purplish; blade 4–19 cm, oblong-ovate to ± round, main divisions gen 7 or more, gen narrow, coarsely toothed. **INFL**: peduncle 0.5–16 cm; bracts 6–8, 2.5 mm, lanceolate, slightly fused at base, < heads; pedicel of bisexual fl 0, of staminate 2 mm, < fr. **FLS**: bisexual 8–10; staminate 10–12; calyx lobes 0.8–1 mm, widely lanceolate, acute, slightly fused at base; corolla purple or yellow; styles 2 × calyx lobes. **FRS** 3–8 per head, 3–6 mm, ovate to round, with stout, curved, inflated, bulbous-based prickles. **SEED**: face concave. $2n$=16. Open grassland, often serpentine, or pine-oak woodland; 20–1200 m. NW, SNF, CW, SW; to B.C., Baja CA. ❀SUN:4,5,**6,7**,8,9,**14–24**.

S. crassicaulis DC. Pl 24–120 cm, stout, taprooted. **LF** gen simple, palmately lobed, green; blade 3–12 cm, gen ± rounded-cordate, lobes 3–5, obovate, gen ± deeply cut, margins finely sharply serrate, teeth gen ± 1 mm, central lobe often > 10 mm wide at base, its ultimate segment 4–20 mm wide. **INFL**: peduncle 0.7–8 cm; bracts ± 5, 1–2 mm, narrowly lanceolate, < heads; pedicels short. **FLS**: bisexual 3–8; staminate 3–5; calyx lobes 0.5–0.7 mm, ± lanceolate, acute, fused at base; corolla yellow; styles 2 × calyx lobes. **FR** 2–5 mm, nearly round, sometimes clearly stalked, with stout, curved prickles throughout or these little developed below, all bulbous-based. **SEED**: face grooved. $2n$=32,48,64. Open slopes, ravines, woodlands; 0–1000 m. NW, CW, SW, SNF; to B.C., Baja CA; s S.Am. Highly variable. ❀DRN:4,5,**6,17**&SHD:**7,14–16,18–24**;INV.

S. graveolens DC. Pl low-spreading to erect, 5–45 cm, slender, taprooted (or roots tuberous). **LF** compound, ternate then 1–2-pinnate, green or purplish; blade 1.5–4 cm, ± ovate, lflets stalked, margins crenate to lobed, ultimate segment of main axis 1–5 mm wide. **INFL**: peduncle 0.5–8 cm; bracts 6–10, ± 1 mm, linear to ovate, fused at base, < heads; pedicel of bisexual fl 0, of staminate 2–2.5 mm. **FLS**: bisexual 3–5; staminate 7–12; calyx lobes 0.5–1 mm, ovate, acute, ± fused; corolla yellow; styles 3 × calyx lobes. **FR** 3–5 mm, ovate to nearly round, prickles stout, curved, inflated, bulbous-based, lower sometimes poorly developed. **SEED**: face concave. $2n$=16. Open forest or rocky slopes, sometimes serpentine; 600–2600 m. NW, SNH; to B.C., MT; s S.Am.

S. hoffmannii (Munz) C. Bell HOFFMANN'S SANICLE Pl 30–90 cm, stout, taprooted. **LF** gen compound, palmate, bluish green; blade 4.5–13.5 cm, triangular, lflets (or deepest lobes) gen 3, obovate, ± cut, margins irregularly serrate, central lflet gen 4–7 cm, 4–30 mm wide, narrowed to a petiole-like base. **INFL**: peduncle 1.5–12 cm; bracts 5–7, 3–5 mm, lanceolate, < heads; pedicels 0 or short. **FLS**: bisexual 4–10; staminate 3–5; calyx lobes 1–2.3 mm, widely lanceolate, acute, fused at base; corolla greenish yellow; styles ± =

calyx lobes. **FR** 3–5 mm, ovate to obovate, with stout, curved prickles throughout, these poorly developed below. **SEED**: face flat. 2*n*=16. UNCOMMON. Shrubby coastal hills, pine woodland; 80–300 m. CCo (Santa Cruz Co.), SCo, n ChI. ❀DFCLT.

S. laciniata Hook. & Arn. Pl 5–30 cm, slender, taprooted. **ST** gen branched near base. **LF** simple, palmately or ± ternately lobed, green; blade 1.5–4 cm, ovate to round, lobes 0–5, margins sharply angled, teeth gen ± 2 mm, central lobe often > 10 mm wide at base. **INFL**: peduncle 1.5–8 cm; bracts ± 10, lanceolate to ovate, < heads; pedicel of bisexual fl 0, of staminate 1.5 mm. **FLS**: bisexual 4–8; staminate 10–12; calyx lobes 0.5–0.8 mm, ovate, acute, fused below middle; corolla yellow; styles 2 × calyx lobes. **FR** 2–4 mm wide, ovate to nearly round; prickles slender, curved, slightly bulbous-based above, poorly developed below. **SEED**: face grooved. 2*n*=16. Coastal, open or brushy slopes, woodland; 30–900 m. NCo, CCo; s OR. ❀DFCLT.

S. maritima S. Watson (p. 167) ADOBE SANICLE Pl 8–40 cm, taprooted. **LF** simple, entire to somewhat pinnately lobed or dissected, green or yellowish green; blade 3–8 cm, ovate-cordate to obovate, lobes or segments 0, 3, or 5, obtuse, margins entire to barely toothed. **INFL**: peduncle 1.5–12 cm; bracts ± 10, 1–4 mm, lanceolate, pointed, < heads; pedicel of bisexual fl 0, of staminate 2.5–3 mm. **FLS**: bisexual 3–8; staminate 10–12; calyx lobes 1.5–2 mm, ovate-lanceolate, obtuse, fused at base; corolla yellow; styles 3 × calyx lobes. **FR** ± 5 mm, obovate; prickles stout, curved, inflated, bulbous-based above, nearly 0 below. **SEED**: face concave. 2*n*=16. **RARE** CA. Coastal, grassy, open wet meadows, ravines; ± 150 m. SnFrB (apparently extirpated), CCo (1 site, San Luis Obispo Co. In cult.

S. peckiana J.F. Macbr. PECK'S SANICLE Pl 16–38 cm, glabrous. **LF** simple, ± 1–2-pinnately dissected, green or ± glaucous; blade 5–10 cm, oblong-ovate, main divisions gen 7 or more, margins sharply toothed. **INFL**: peduncle 2–75 mm; bracts ± 6, ± 1 mm, ovate-lanceolate, acute, fused at base, < heads; pedicel of bisexual fl 0, of staminate 2.5–5 mm, > fr. **FLS**: bisexual 1–5; staminate 5–18; calyx lobes 0.5–0.8 mm, ± lanceolate, acute, fused below middle; corolla yellow; styles 3–4 × calyx lobes. **FRS** 1–5 per head, 1.5–3 mm wide, ovate to round, covered by bulbous tubercles, only those near top with short curved prickles, others unarmed. **SEED**: face concave. 2*n*=16. UNCOMMON. Serpentine, chaparral, woodland; 150–800 m. nw KR (Del Norte Co.); sw OR.

S. saxatilis E. Greene ROCK SANICLE Pl 10–25 cm, stout; tuber 2–3.5 cm wide, ± spheric, deep-seated. **LF** compound, ternate then 1–2-pinnate, purplish or glaucous; blade 3–9 cm, triangular, lflets stalked, margins finely serrate. **INFL**: peduncle 0.5–4 cm; bracts 3–5, widely lanceolate, entire or 3-lobed, fused at base, < heads; pedicel of bisexual fl 0, of staminate 3–8 mm, >> fr. **FLS**: bisexual 3–5; staminate 6–12; calyx lobes ± 0.5 mm, triangular-ovate, acute; corolla pale salmon to straw-colored; styles ± 3 × calyx lobes. **FR** 2.5–3 mm wide, ovate to nearly round, tubercles ± rounded, most nipple-like or with a bristle. **SEED**: face ± plane. 2*n*=16. **RARE** CA. Rocky ridges or talus, chaparral, woodland; 900–1100 m. e&s SnFrB (Mount Diablo, Contra Costa Co.; Mount Hamilton, Santa Clara Co.). In cult.

S. tracyi Shan & Constance (p. 167) TRACY'S SANICLE Pl 35–60 cm, slender, taprooted. **LF** compound, ternate then 1–2-pinnate, green or purplish; blade 2.5–3.5 cm, ± ovate, lflets irregularly toothed or lobed, margins serrate. **INFL**: peduncle 0.5–4 cm; bracts 5–8, lanceolate or ovate, fused below middle, < heads; pedicel of bisexual fl 0, of staminate 2–4 mm, slightly > fr. **FLS**: bisexual 2–3; staminate 4–7; calyx lobes 0.3–0.5 mm, ovate, acute, fused below middle; corolla yellow; styles 3 × calyx lobes. **FR** 2–3 mm wide, obovate or nearly round; tubercles inflated, unarmed or a few at top prickled. **SEED**: face concave. 2*n*=16. RARE. Openings in coniferous forest, woodland; 100–1000 m. NCo (Humboldt, Trinity cos.), CaRF (Butte Co.).

S. tuberosa Torrey Pl 5–80 cm, slender; tuber 5–15 mm wide, spheric. **LF** compound, ternate then 1–2-pinnate, green, glaucous, or purplish; blade 2–13 cm, triangular to ovate, lflets entire to deeply pinnately lobed. **INFL**: peduncles 0.5–8.5 cm; bracts 6–10, 1–3 mm, oblong to lanceolate, fused, < heads; pedicel of bisexual fl 0, of staminate 2–7 mm, often >> fr. **FLS**: bisexual 3–5; staminate 4–10; calyx lobes 0.3–0.5 mm, ovate, acute, fused below middle; corolla yellow; styles ± 3 × calyx lobes. **FR** 1.5–2 mm, ovate to nearly round, with rounded, unarmed tubercles. **SEED**: face flat. 2*n*=16. Open gravelly meadows, chaparral, woodland, pine forest; 150–2700 m. NCoR, CaR, SN, SCoR, WTR, PR; Baja CA. Variable. ❀DFCLT.

SCANDIX

Ann, taprooted, ± hairy. **ST** spreading or erect, branched. **LF**: blade oblong to ovate, pinnately dissected, segments thread-like to linear. **INFL**: umbels compound, gen opposite a lf; bracts gen 0; bractlets few, entire, lobed, or dissected, conspicuous; rays, pedicels few. **FLS** slightly bilateral; calyx lobes gen 0; petals white, slightly notched, narrowed at tips, outer > others. **FR** gen ± bristly; body linear to oblong, slightly compressed side-to-side; beak > body, sterile, very compressed front-to-back or cylindric; ribs thread-like; oil tubes inconspicuous; fr axis entire or divided at tip. **SEED**: face grooved. 15–20 spp.: Medit. (Greek: chervil)

S. pecten-veneris L. (p. 167) VENUS' NEEDLE Pl 1.5–5 dm. **LF**: petiole gen 2–10 cm; blade 2–10 cm, 1–5 cm wide, segments 1.5 mm, linear. **INFL**: peduncle 1–6 cm; bractlets 2–5, 5–10 mm, linear-lanceolate to obovate, ciliate; rays 1–3, gen 1–2(5) cm, ascending; pedicels 2–3 mm, stout. **FR** bristly; body 6–15 mm, 1–2 mm wide; beak 2–7 cm, very compressed front-to-back, bristly-ciliate; ribs low, wide, rounded. 2*n*=16,26. Grassy slopes, roadsides; 15–1000 m. CA-FP, esp NCo, SnFrB; native to Medit. Sporadic.

SIUM

Per, glabrous; roots clustered, fibrous or ± tuberous. **ST** erect or ascending, branched. **LF**: blade oblong to ovate, 1-pinnate, sometimes 2-pinnate when submerged, lflets distinct, serrate, irregularly cut, or pinnately lobed. **INFL**: umbels compound, gen opposite a lf; bracts, bractlets lf-like, often reflexed, conspicuous; rays, pedicels many, spreading-ascending. **FL** sometimes ± bilateral; calyx lobes 0 or minute; petals wide, white, narrowed at tips, outer slightly > others. **FR** ovate to round, slightly compressed side-to-side; ribs prominent, subequal, corky; oil tubes per rib-interval 1–3; fr axis entire or divided to base, adhering to fr halves or not. **SEED**: face flat. 2*n*=12. ± 10 spp.: N.Am., Eurasia, Afr. (Greek: for some aquatic member of family)

S. suave Walter (p. 167) Pl 6–12 dm, stout. **LF**: petiole 1–8 dm, segmented; blade 6–25 cm, 7–18 cm wide, lflets 1–4 cm, linear or lanceolate, serrate or irregularly cut. **INFL**: peduncle 4–10 cm; bracts 6–10, 3–15 mm, linear or lanceolate, acute, entire or irregularly cut, reflexed; bractlets 4–8, 1–3 mm, linear-lanceolate; rays 10–20, 1.5–3 cm, slender, subequal; pedicels 3–5 mm. **FR** 2–3 mm wide. 2*n*=12. Wet soil of swamps, marshes, streambanks; < 2000 m. SnFrB (Suisun Marshes), GB; to B.C., e N.Am, e Asia. ❀TRY.

SPERMOLEPIS

Ann, taprooted, glabrous. **ST** gen spreading, branched. **LF**: blade oblong to ovate, ± ternate-pinnately dissected, segments thread-like to linear. **INFL**: umbels compound, terminal and lateral, peduncled or not; bracts 0; bractlets few, narrow; rays, pedicels few, suberect, gen ± spreading. **FL**: calyx lobes 0; petals oblong to ovate, tips not narrowed, not incurved. **FR** ovate, slightly compressed side-to-side, smooth, tubercled, or short-bristly; ribs low, thread-like; oil tubes per rib-interval 1–3; fr axis divided at tip. **SEED**: face grooved. 5 spp.: s US, Hawaii, s S.Am. (Greek: seed scale, from tubercled or bristly fr)

S. echinata (DC.) A.A. Heller Pl low, spreading, 5–40 cm. **LF** ovate; petiole 3–20 mm; blade 7–25 mm wide, segments 2–18 mm, thread-like. **INFL**: peduncles 1–5 cm; bractlets few, thread-like to linear, entire or toothed; rays 5–14, 1–15 mm, gen ± ascending, very unequal; pedicels gen < 7 mm, those of central fl of each 2° umbel gen 0. **FR** 1.5–2 mm wide, widely ovate; ribs prominent, short-bristly. $2n$=16,20. Very uncommon. Rocky slopes, sandy flats; 60–1500 m. DSon (Borrego Valley); to se US, n Mex.

SPHENOSCIADIUM

1 sp. (Greek: wedge umbrella, from umbel)

S. capitellatum A. Gray (p. 167) SWAMP WHITE HEADS, RANGER'S BUTTONS Per, ± scabrous; root tuberous. **ST** erect, 5–18 dm, gen branched, leafy. **LF**: petiole 1–4 dm; blade 1–4 dm, oblong to ovate, 1–2-pinnate or ternate then pinnate, lflets 1–12 cm, gen ± lanceolate, acute, sparsely toothed to irregularly cut or pinnately lobed; cauline lf sheaths conspicuously enlarged. **INFL**: umbels compound, tomentose; peduncle 7–40 cm; bracts 0; bractlets many, linear, bristle-like; rays 4–18, 1.5–10 cm, ascending to reflexed; pedicels reduced to a disk; 2° umbels head-like, spheric. **FL**: calyx lobes 0; petals obovate, white or purplish, tips narrowed; styles slender. **FR** 5–8 mm, wedge-shaped-obovate, very compressed front-to-back, tomentose; ribs unequally winged, marginal wider than others; oil tubes per rib-interval 1; fr axis divided to base. **SEED**: face ± flat. $2n$=22. Wet meadows, streamsides, lakeshores; 900–3000 m. NCoRI, CaR, SNH, SW, GB; to OR, ID, NV, Baja CA. TOXIC to livestock, but rarely eaten. ❀WET:1,2,6,7,14–16; DFCLT.

TAUSCHIA

Per, taprooted or roots tuberous, glabrous to hairy. **ST** 0 or low. **LF**: blade oblong to obovate, 1–2-pinnate or -ternate, lflets wide, margins entire to pinnately lobed. **INFL**: umbels compound, terminal; peduncle gen > lf; bracts gen 0; involucel 1-sided; bractlets inconspicuous to lf-like; rays, pedicels few–many, ascending to reflexed, gen few fertile. **FL**: calyx lobes evident to 0; petals wide, yellow, tips narrowed; styles short to slender; projection atop ovary inconspicuous. **FR** oblong to round, ± compressed side-to-side, glabrous; ribs prominent to thread-like, subequal, unwinged; oil tubes per rib-interval several; fr axis gen divided ± to base. **SEED**: face gen grooved or concave. (I.F. Tausch, Czech botanist, 1793–1848) ± 35 spp.: w N.Am., C.Am, n S.Am.

1. Peduncle gen < lf; bractlets very large, >> umbels, wide, sharply pinnately lobed, lf-like .... *T. howellii*
1′ Peduncle > lf; bractlets small, < to slightly > umbels, narrow, entire or nearly so, not lf-like
  2. Fr oblong to narrowly elliptic, much longer than wide, ribs prominent, acute, ± wing-like; lf ± leathery
    3. Lf 1-pinnate; rays unequal, spreading-ascending .... *T. arguta*
    3′ Lf 2-pinnate; rays subequal, spreading and reflexed .... *T. parishii*
  2′ Fr nearly round, ± as wide as long, ribs low, obtuse, thread-like; lf not leathery
    4. Pl glabrous, smooth; rays 5–12; pedicel 1–3 mm .... *T. glauca*
    4′ Pl minutely roughened; rays 10–30; pedicel 2–15 mm
      5. Several bractlets gen < fls, frs; fr 4–7 mm, 4–5 mm wide .... *T. hartwegii*
      5′ All bractlets gen < fls, frs; fr 3–5 mm, 4–6 mm wide .... *T. kelloggii*

T. arguta (Torrey & A. Gray) J.F. Macbr. (p. 167) Pl 3–7 dm, glabrous. **ST** gen > lvs. **LF**: petiole 6–20 cm; blade 8–16 cm, oblong to ovate, gen 1-pinnate, lflets 3–8 cm, oblong to ovate, sharply dentate. **INFL**: peduncle 17–45 cm; bractlets several, 2–10 cm, linear to lanceolate, entire or few-lobed; rays 12–25, 2–12 cm, unequal, spreading-ascending; pedicels 3–9 mm. **FL**: calyx lobes evident; corolla yellow; styles slender. **FR** 6–9 mm, oblong; ribs acute, very prominent; oil tubes per rib-interval 3–5; fr axis divided to base. $2n$=22. Chaparral, woodland; 100–1500 m. SCoRO, SW; Baja CA.

T. glauca (J. Coulter & Rose) Mathias & Constance GLAUCOUS TAUSCHIA Pl 2–4 dm, glabrous. **ST** 0 or gen < lvs. **LF**: petiole 2–12 cm; blade 6–12 cm, ovate to round, 2-ternate or ternate-pinnate, lflets 10–17 mm, ovate to round, coarsely serrate or lobed. **INFL**: peduncle 2–4 dm; bractlets several, 1–7 mm, lanceolate; rays 5–12, 1–6 cm, very unequal; pedicels 1–3 mm. **FL**: calyx lobes minute; corolla yellow; styles slender. **FR** 2–3 mm, nearly round; ribs thread-like; oil tubes per rib-interval 2–3; fr axis divided to base or nearly so. $2n$=22. UNCOMMON. Gravelly, often serpentine flats in coniferous forest; 80–1700 m. NCo; s OR.

T. hartwegii (A. Gray) J.F. Macbr. Pl 3–10 dm, minutely scabrous. **ST** 0. **LF**: petiole 5–25 cm; blade 12–24 cm, oblong to widely ovate, 1–2-ternate-pinnate, lflets 25–60 mm, oblong to ovate, coarsely serrate, often lobed. **INFL**: peduncle 2.5–8 dm; bractlets several, 5–12 mm, ± lanceolate, entire, reflexed, some slightly > fls, frs; rays 10–30, 2–12 cm, unequal; pedicels 2–7 mm. **FL**: calyx lobes minute; corolla yellow; styles slender. **FR** 4–7 mm, nearly round; ribs thread-like; oil tubes per rib-interval 3–5; fr axis divided to base. $2n$=22. Chaparral, pine-oak woodland; 300–1500 m. n SNF, SnFrB, SCoRO.

T. howellii (J. Coulter & Rose) J.F. Macbr. (p. 167) HOWELL'S TAUSCHIA Pl 5–8 cm, glabrous. **ST** short but > lvs. **LF**: petiole 2–3 cm; blade 1.5–3 cm, ovate, 1-pinnate or -ternate, lflets 5–15 mm, oblong to ovate, irregularly sharply toothed. **INFL**: peduncle < 2 cm, gen < lf; bractlets several, 1–2 cm, divided and sharply toothed like lvs, >> umbels; rays 3–5, 8–16 mm, unequal, or some 2° umbels sessile; pedicels < 5 mm. **FL**: calyx lobes prominent; corolla yellow; styles slender. **FR** 2–4 mm, oblong; ribs thread-like; oil tubes per rib-interval several; fr axis divided nearly to base. $2n$=22. RARE. Granitic gravel, ridge tops, *Abies* forest; 2000–2500 m. KR (Salmon Mtns); s OR.

T. kelloggii (A. Gray) J.F. Macbr. Pl 2–7 dm, minutely roughened. **ST** gen 0. **LF**: petiole 5–15 cm; blade 8–20 cm, ovate to round, 1–3-ternate or -ternate-pinnate, lflets 15–35 mm, oblong to ovate, coarsely serrate, often irregularly cut or lobed. **INFL**: peduncle 2–5 dm; bractlets few, 3–8 mm, linear, entire, < fls, frs; rays 10–20, 2–12 cm, unequal; pedicels 3–15 mm. **FL**: calyx lobes minute; corolla yellow; styles slender. **FR** 3–5 mm, nearly round; ribs thread-like; oil tubes per rib-interval 2–3; fr axis divided nearly to base. 2*n*=22. Scrub or chaparral, woodland or coniferous forest; 50–1700 m. NCo, s CaRF, n SNF, n CCo; s OR.

T. parishii (J. Coulter & Rose) J.F. Macbr. Pl 1–4 dm, glabrous, ± glaucous. **ST** 0. **LF**: petiole 5–15 cm; blade 8–15 cm, oblong to ovate, 2-pinnate, lflets 15–40 mm, oblong to ovate, sharply serrate to pinnately lobed. **INFL**: peduncle 10–30 cm; bractlets few, 5–12 mm, linear, entire; rays 12–18, 3–6 cm, subequal, spreading and reflexed; pedicels 2–7 mm. **FL**: calyx lobes evident; styles slender. **FR** 5–8 mm, oblong to narrowly elliptic; ribs narrow, prominent, acute; oil tubes per rib-interval 4–5; fr axis divided in upper 2/3. 2*n*=22. Rocky or sandy soil, pine woodland; 1200–2400 m. Teh, SnGb, SnBr, SnJt, SNE.

TORILIS

Ann, taprooted, hairy or bristly. **ST** spreading or erect, branched. **LF**: blade lanceolate to triangular, 1-pinnately dissected, segments narrow. **INFL**: umbels compound, terminal or opposite a lf, peduncled or sessile; bracts gen 0 or small; bractlets several, thread-like to linear; rays 0–few, spreading-ascending; pedicels 0 or short. **FL** gen slightly bilateral; calyx lobes 0 or evident; petals obcordate, white or reddish, tips narrowed, outer petals ± > others; styles short. **FR** oblong to ovate, slightly compressed side-to-side; 1° ribs thread-like, prickly, 2° ribs densely prickly or tubercled; oil tubes per interval between 1° ribs 1; fr axis divided in upper 1/2. **SEED**: face grooved. 10–15 spp.: Eurasia. (Name used by Adanson in 1763, meaning obscure)

1. Pl erect; peduncle > lf; umbel open, not head-like; outer, inner fr-halves equally prickly .... ***T. arvensis***
1′ Pl spreading; peduncle < lf; umbel dense, head-like; outer fr-half prickly, inner only tubercled .... ***T. nodosa***

T. arvensis (Hudson) Link (p. 167) Pl 3–10 dm, slender. **LF**: petiole 2–8 cm; blade 5–12 cm, ± ovate, 2–3-pinnate, lflets 5–60 mm, lanceolate to ovate, regularly pinnately cut; upper cauline lvs gen 1-pinnate. **INFL**: peduncle 2–12 cm; bracts 0–2; bractlets several, 2–4 mm, linear; rays 2–10, ± equal; pedicels 1–4 mm. **FR** 3–5 mm, oblong-ovate; prickles spreading, gen = fr in width. 2*n*=12. Disturbed places; 40–1600 m. CA-FP (esp NW, CW, n SNF); native to c&s Eur. Rapidly spreading; waifs in CA seem to be ssp. *purpurea* (Ten.) Hayek, incl pls sometimes referred to *T. japonica* (Houtt.) DC.

T. nodosa (L.) Gaertner Pl 1–5 dm. **LF**: petiole 3–10 cm; blade 3–9 cm, oblong to elliptic, pinnately dissected, segments 3–8 mm, linear to lanceolate, gen entire; cauline lvs like basal. **INFL**: peduncle 0–2.5 cm; bracts gen 0; bractlets 2–4, 1.5–2 mm, linear; rays, pedicels very short to 0. **FR** 2–3 mm, ovate; prickles gen ascending, < or > fr in width. 2*n*=22,24. Disturbed places; < 1500 m. n&c CA-FP, s ChI; native to Eurasia, esp Medit.

YABEA

1 sp. (H. Yabe, Japanese botanist, 1876–1931)

Y. microcarpa (Hook. & Arn.) Koso-Polj. (p. 173) Ann, slender, ± hairy, taprooted. **ST** erect, 3–40 cm. **LF**: petiole 2.5–3.5 cm; blade 2–6 cm, oblong to ovate, pinnately dissected, segments 2–8 mm, thread-like to linear. **INFL**: umbels gen compound; peduncle 2–10 cm; bracts 2–5, lf-like, pinnately lobed to compound, 1–5 cm; bractlets 1–5, 1–10 mm, entire or some often 3-lobed at tip or pinnately lobed; rays 1–9, 1–8 cm, unequal, erect or ascending; pedicels 2–9, < 15 mm, erect. **FL** minute, slightly bilateral; calyx lobes evident; petals obovate, white, tips narrowed; styles very short. **FR** oblong, compressed side-to-side; 1° ribs bristly, alternating with prickly wings; oil tubes per interval between 1° ribs 1; fr axis divided in upper 1/4. **SEED**: face deeply grooved. 2*n*=12. Grassy slopes, dunes, chaparral, woodland; 0–1000 m. CA-FP, esp coast, SNF, SnGb, SnBr, SnJt; to WA, ID, AZ, Baja CA. [*Caucalis m.* Hook. & Arn.]

APOCYNACEAE DOGBANE FAMILY

Lauramay T. Dempster

Per (sometimes ann, shrub, vine, tree); sap milky. **LVS** simple, entire, opposite, alternate, or subwhorled; stipules 0 or small. **INFL**: cyme; fls 1–many, axillary or terminal. **FL** bisexual, radial; perianth parts overlapping, at least in bud; sepals 5, fused at base, persistent; petals 5, fused in ± basal half; stamens 5, attached to corolla tube or throat, alternate lobes; ovaries 2, ± superior, gen free, styles and stigmas fused. **FR**: gen 2 follicles. **SEEDS** many, often with tuft of silky hairs. ± 150–200 genera, 1000–2000 spp.: esp trop; many orn (*Oleander*; *Plumeria*, frangipani); some alkaloids highly toxic, some used in medicine. [Rosatti 1989 J Arnold Arbor 70:307–401]

1. Stamens attached near base of corolla tube, apparently below level of stigma; seed with tuft of long hairs
   2. Lvs, sts, roots not fleshy; fl < 8 mm; infl many-fld .... **APOCYNUM**
   2′ Lvs, sts, roots fleshy; fl > 15 mm, infl few-fld .... **CYCLADENIA**
1′ Stamens attached near top of corolla tube, near level of stigma; seed glabrous
   3. Lvs alternate to subwhorled; pl erect .... **AMSONIA**
   3′ Lvs opposite; pl erect or sprawling
      4. Corolla pink, tube ± cylindric; st erect .... **CATHARANTHUS**
      4′ Corolla blue, tube funnel-shaped; st sprawling .... **VINCA**

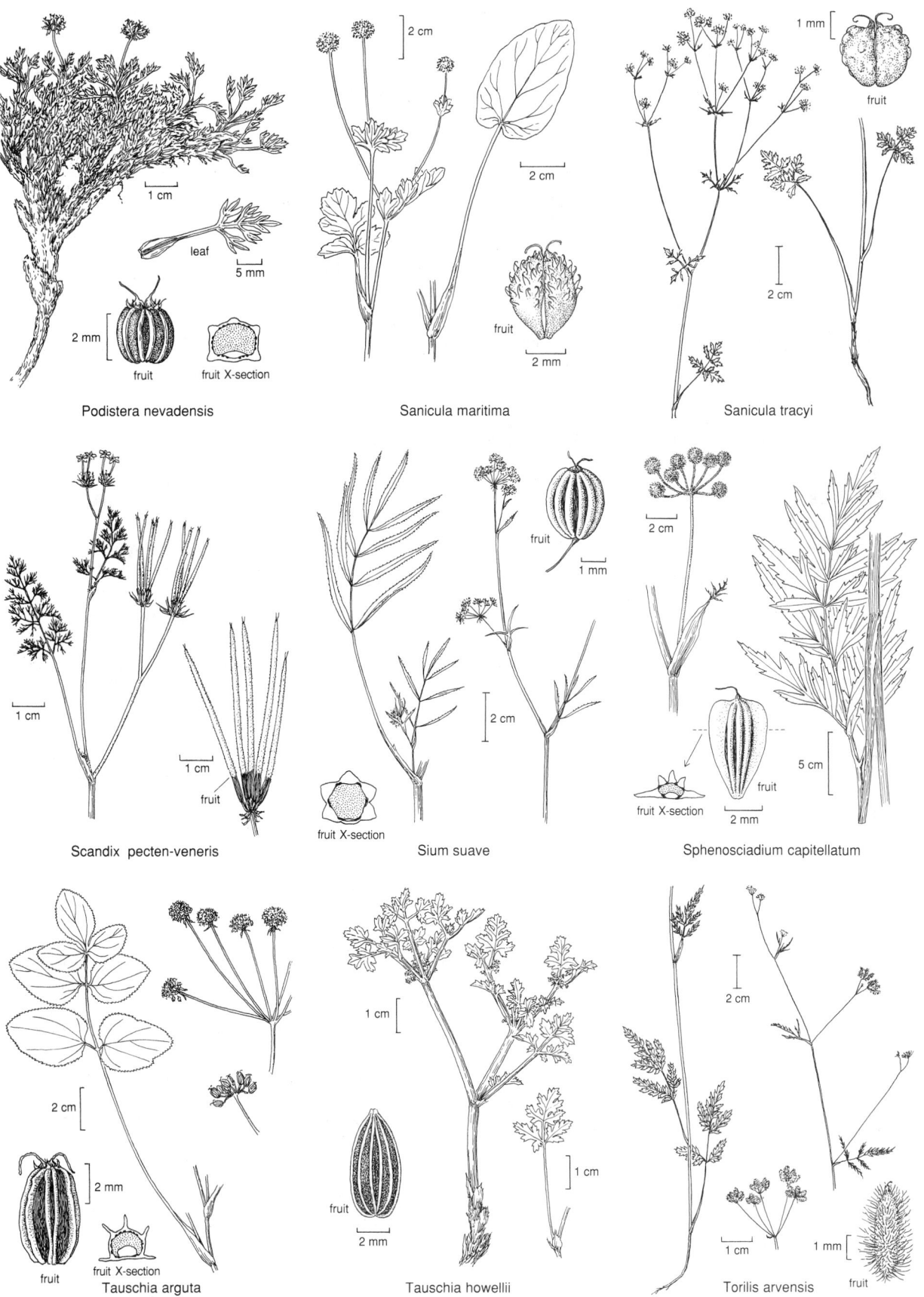

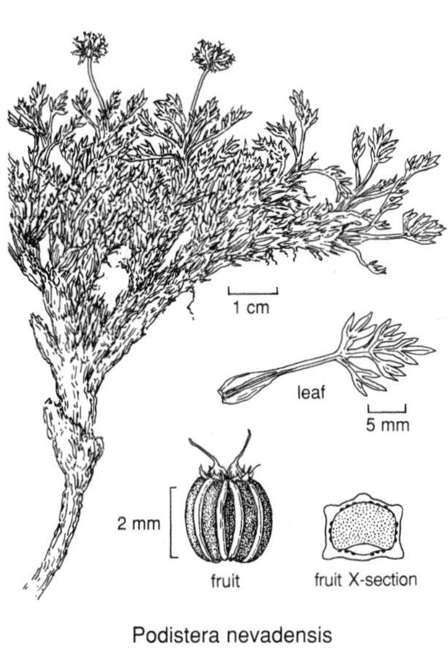

Podistera nevadensis

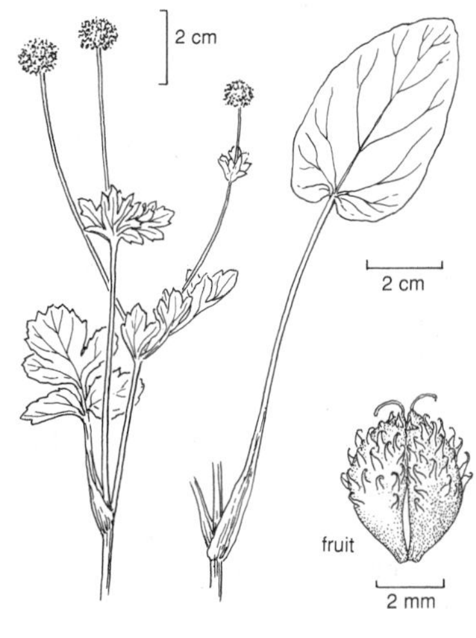

Sanicula maritima

1 mm

fruit

2 cm

Sanicula tracyi

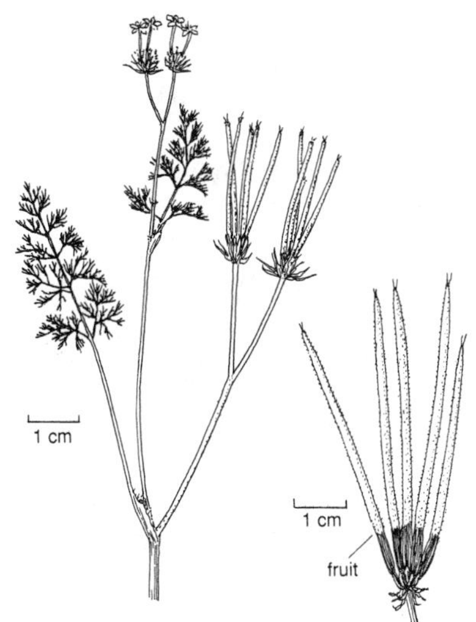

Scandix pecten-veneris

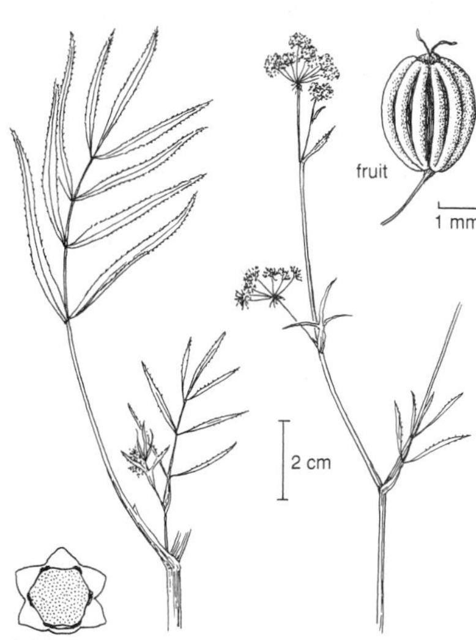

Sium suave

2 cm

5 cm

fruit

fruit X-section

2 mm

Sphenosciadium capitellatum

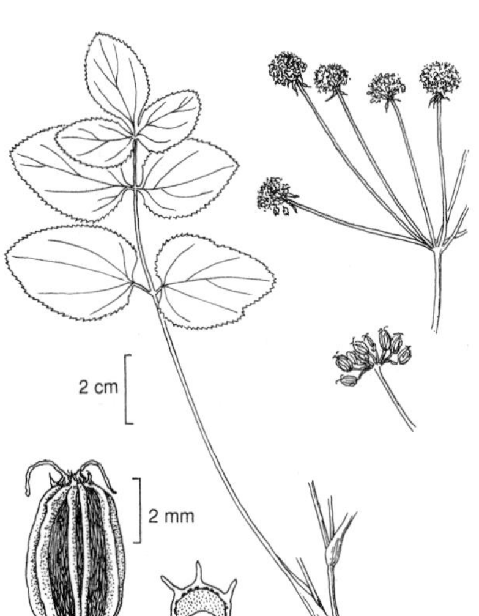

Tauschia arguta

1 cm

1 cm

fruit

2 mm

Tauschia howellii

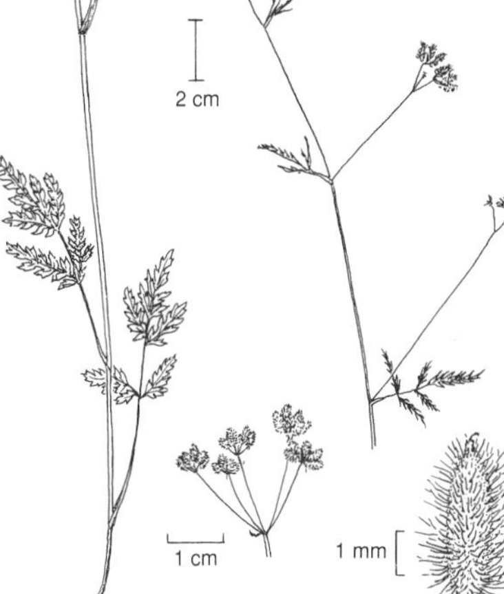

Torilis arvensis

AMSONIA

Pl erect. **ST** ± woody. **LVS** alternate to subwhorled. **INFL**: fls several–many. **FL**: corolla salverform; anthers free from each other and stigma; nectary 0 or low ring around ovaries; style ± thread-like, stigma skirted at base. **SEED** glabrous. 5–25 spp.: N.Am, Japan. (Probably for Charles Amson, Virginian physician, 18th century) [McLaughlin 1982 Ann Missouri Bot Gard 69:336–350]

A. tomentosa Torrey & Frémont (p. 173) Pl glabrous or grayish tomentose. **STS** several–many from woody crown, 16–36 cm. **LF** 2–4 cm; petiole short or 0; blade ovate-lanceolate. **FL**: calyx lobes erect, thread-like above base; corolla whitish, blue, or greenish, tube ± 15 mm, inflated above middle, narrower just below spreading lobes; style spheric just below stigma. **FR** 3–8 cm, narrowed between seeds, often breaking into 1-seeded segments. Desert plains, canyons; 300–1800 m. SnBr (n slope), s DMtns, DSon; to UT. [*A. brevifolia* A. Gray] Hairiness is probably a simple, genetically recessive trait. ❀TRY.

APOCYNUM DOGBANE, INDIAN HEMP

ST ascending to erect. **LVS** opposite. **INFL**: fls several–many. **FL** small; corolla cylindric to bell-shaped, 5-lobed, with 5 triangular appendages alternate stamens; stamens attached at base of tube, filaments short, wide, anthers forming adherent cone around stigma, each partly sterile, sharply sagittate; nectaries 5, free, around and < ovaries; style ± 0, stigma massive, ovoid, obscurely 2-lobed. **FR** slender, cylindric, pointed. **SEED** with tuft of long hairs. ± 7 spp.: N.Am. (Greek: away from, dog, from ancient use as dog poison). The 2 CA spp. hybridize extensively; many hybrid forms have been named.

1. Pl 1.6–3 dm; lf drooping to spreading, ovate to round, dark green above, pale below; corolla reddish purple or pink to white, sometimes with pink stripes, lobes spreading to recurved; calyx << corolla tube .... ***A. androsaemifolium***

1′ Pl 3–12 dm; lf ascending, lanceolate to narrowly ovate, yellowish green above, below; corolla greenish or white, lobes ± erect; calyx ± = corolla tube .... ***A. cannabinum***

A. androsaemifolium L. (p. 173) BITTER DOGBANE **ST** diffusely branched. **LF**: petiole << blade; blade 4–6 cm, base gen round or cordate, tip round or obtuse to ± acute. **FL**: corolla 4–8 mm, ± bell-shaped. **FR** 7–11 cm, pendent to erect. Open slopes, rocky places, with conifers, chaparral; 200–2500 m. KR, NCoRH, NCoRI, CaRH, SNH, SnFrB, SCoRI, SnBr, SnJt, MP; to e N.Am, Can. [var. *glabrum* Macoun; *A. medium* E. Greene var. *floribundum* (E. Greene) Woodson; *A. pumilum* (A. Gray) E. Greene] ❀DRN: **1**,4–6,15,17&SHD:**2,3**,16&IRR:**7**,14,18,20,21;STBL,INV.

A. cannabinum L. (p. 173) INDIAN HEMP **ST** stout, ± stiffly erect, branched near top. **LF**: petiole << blade; blade 5–8 cm, base tapered to cordate, sometimes clasping st, tip obtuse to acute. **FL**: corolla 2.5–5 mm, cylindric to urn-shaped. **FR** 6–9 cm, ± pendent. Moist places near streams, springs, etc., or weed in orchards; < 2000 m. KR, NCoR, CaRH, SNH, Teh, SnJV, SnFrB, SCoR, TR, PR, GB; to B.C., e N.Am. [var. *glaberrimum* A. DC.; *A. sibiricum* Jacq. var. *salignum* (E. Greene) Fern.] ❀WET:**1-3**,4,5,**6–9**,10,**14–21**;STBL,INV; important traditional source of fiber.

CATHARANTHUS

Pl ± puberulent. **LVS** ± opposite. **INFL**: fls gen solitary in axils. **FL**: calyx lobes long, slender; corolla pink, tube ± cylindric, lobes asymmetric; stamens attached near top of corolla tube, filaments ± straight, anthers held above, free from stigma, each completely fertile; nectaries 2, alternate and gen > ovaries; style thread-like, stigma skirted at base. **SEED** glabrous. 3–7 spp.: Madagascar, India. (Greek: pure fl) [Taylor & Farnsworth 1975 The *Catharanthus* alkaloids]

C. roseus (L.) G. Don MADAGASCAR PERIWINKLE **ST** erect, 30–60 cm. **LF**: petiole gen < 1 cm; blade ± elliptic. **FL**: corolla 3–5 cm wide, pink. **FR** ± straight. Desert spring; 160 m. DSon (Borrego Springs, San Diego Co.); native to Madagascar. [*Vinca rosea* L.] Widely cult, naturalized in warm temp to trop regions. Alkaloids used to treat childhood leukemia, Hodgkin's disease, other cancers.

CYCLADENIA

Pl fleshy (incl large root), glabrous to tomentose. **ST** ± erect. **LVS** opposite. **INFL**: fls 2–6 near tips of axillary peduncles. **FL**: calyx lobes slender; corolla funnel-shaped with 5 rounded appendages behind anthers; stamens appearing attached at base of corolla tube but filaments epipetalous up to stigma level, anthers forming cone around but free from stigma, each partly sterile, sharply sagittate; nectary disk 5-lobed, around and < ovaries; style thread-like, stigma skirted at base. **SEED** with tuft of long hairs. 1 sp.: CA. (Greek: ring gland, from nectary)

C. humilis Benth. Pl 6–12 cm, gen glabrous, glaucous. **LF**: pairs 2–5; lf < 9 cm; petiole < to > blade; blade ovate or rounded. **FL**: calyx lobes narrowly triangular; corolla 15–20 mm, rose-purple, lobes obovate or round, margins wavy; filaments hairy. **FR** 3–5 cm. Sandy flats to talus slopes in open pine forest, chaparral; 1200–2800 m. KR, NCoRH, CaRH, n SNH, SCoRO, SnGb, SNE. ❀TRY.

1. Outside of corolla gen glabrous; inside glabrous exc densely hairy around stigma; n CA .... var. ***humilis***

1′ Outside of corolla soft-hairy; inside sparsely hairy exc densely so around stigma; SCoRO, SnGb, SNE .... var. ***venusta***

var. ***humilis*** (p. 173) Pl gen glabrous, glaucous, sometimes densely tomentose throughout. Loose gravel or sand, talus slopes, often with ponderosa pine; 1200–2800 m. KR, NCoRH, CaRH, n SNH (Plumas, Sierra cos.). [var. *tomentosa* (A. Gray) A. Gray] Tomentose and glabrous pls occur together.

var. ***venusta*** (Eastw.) Munz Pl glabrous, glaucous, but infl hairy. Talus, loose gravel, dry ground in light shade of pines, chaparral; 1550–2500 m. SCoRO (Santa Lucia Range), SnGb, SNE. Pls of Inyo Co. have narrower, darker red corolla lobes.

VINCA PERIWINKLE

Pl ± glabrous. **LVS** ± opposite. **INFL**: fls solitary in axils. **FL**: calyx lobes long, slender; corolla blue, tube funnel-shaped, lobes asymmetric; stamens attached near top of corolla tube, filaments sharply bent near base, anthers held around but free from stigma, each partly sterile; nectaries 2, alternate and gen < ovaries; style thread-like, stigma skirted at base. **SEED** glabrous. 6–7 spp.: s Eur, n Afr, sw Asia. (Latin: possibly, to bind or conquer) [Taylor & Farnsworth 1973 The *Vinca* alkaloids]

V. major L. GREATER PERIWINKLE Pl sprawling. **ST** arching, rooting at tips. **LF**: petiole gen < 1 cm; blade ± 7 cm, ovate. **FL**: corolla 3–5 cm wide, purplish blue, rarely white. **FR** curved, gen not produced. Sheltered places, esp along streams; 2–200 m. NCoRO, SnFrB, s SCoRO, SCo; to s US; native to Eur. Cult as orn.

AQUIFOLIACEAE HOLLY FAMILY

Elizabeth McClintock

Shrub, tree, evergreen or deciduous, dioecious or nearly so. **LVS** gen alternate, simple, often spiny, toothed, or lobed. **INFL**: gen cyme, gen axillary. **FL** gen unisexual, radial, small; sepals gen 4, gen fused at base; petals 4, fused at base or free, white or green; stamens gen 4, gen alternate petals; ovary superior, chambers often 4, gen 1-ovuled, style terminal, 0 or short, stigma head-like or lobed. **FR**: drupe, gen red (purple to black or yellow to whitish); stones often 4. **SEEDS** often 4. 4 genera, 300–400 spp.: trop, temp. [Brizicky 1964 J Arnold Arbor 45:227–234]

ILEX HOLLY

LVS entire, spiny, or toothed, petioled; upper surface gen shiny. **FL**: sepals persistent in fr; petals oblong or obovate, fused at base. **FR** round, pulpy; stones 4, walls hard. **SEED** 1 per stone. 300–400 spp.: esp trop. (Latin: name for Medit holly-oak, *Quercus ilex* L.)

I. aquifolium L. ENGLISH HOLLY Shrub, small tree, evergreen, gen dioecious. **ST** erect; branches many; branchlets minute-hairy. **LVS** 2.5–6 cm, ovate or oblong-ovate, entire to lobed (on 1 pl), often with widely spaced, stiff, spine-like teeth. **FL**: petals dull white. **FR** ± 7–8 mm wide, gen red, smooth. Cool, wooded areas; < 200 m. NCoRO, SnFrB, expected elsewhere; native to Eur, w Asia. Cult since ancient time for showy fr.

ARALIACEAE GINSENG FAMILY

Elizabeth McClintock

Shrub, woody vine, rarely per, tree; juvenile pl often unlike fl pl. **ST** gen branched. **LVS** simple or compound, gen alternate; stipules ± fused to ± sheathing petiole base or 0. **INFL**: umbels solitary to panicled; bracts deciduous or not. **FL** gen bisexual, gen radial, gen < 5 mm; sepals gen 5, fused at base, inconspicuous, persistent; petals gen 5, free, ± white to green, deciduous; stamens gen 5, gen alternate petals; ovary inferior, chambers 1–15, 1-ovuled, styles as many as chambers, free or fused, persistent. **FR**: berry or drupe. 60–70 genera, 700 spp.: esp trop, subtrop; medicinal (e.g. *Panax*, ginseng; *Aralia*, sarsaparilla), orn (e.g., *Aralia, Fatsia, Hedera, Polyscias*). [Graham 1966 J Arnold Arbor 47:126–136]

1. Lf pinnately compound; erect per .... **ARALIA**
1′ Lf simple, lobed to ± entire; woody vine .... **HEDERA**

ARALIA SPIKENARD

Per, shrub, small tree. **ST** erect. **LVS** 1–3-ternate or -pinnate, deciduous; stipule ± fused to petiole base. **INFL**: umbels in clusters of 2–few or in spreading panicles. **FR**: berry. **SEEDS** 3–5. ± 30 spp.: N.Am, Asia, Malaysia. (Latinized from old French Canadian, Aralie)

A. californica S. Watson (p. 173) ELK CLOVER Per; roots large; juice milky. **ST** 2–3 m, stout. **LF** 1–3-pinnate, 1–2 m, ± glabrous; petiole < 3 dm; lflets 15–30 cm, ovate to oblong, serrate, base subcordate. **INFL** 30–45 cm. **FR** ± 5 mm, spheric, black. 2*n*=48. Moist shade, canyons, streamsides; < 2000 m. CA-FP; OR. ❀IRR or WET:4,**5**&SHD:3,**7**,8,9,14,15,**16**,**17**,18–23.

HEDERA IVY

Woody vine. **ST**: juvenile sts climbing by aerial rootlets; fl sts fewer, nonclimbing. **LVS** simple, evergreen; those on juvenile sts lobed; those on fl sts ± entire; stipules 0; hairs stellate. **FR**: berry, black. ± 6 spp.: Eurasia. (Latin: sacred pl of Bacchus, god of wine)

H. helix L. ENGLISH IVY **LVS**: on juvenile sts palmately 3–5-lobed, < 35 cm; on fl sts < 15 cm, ovate to ± diamond-shaped; petiole, lower surfaces hairy to ± glabrous. **FR** ± 5 mm. 2*n*=48. Disturbed areas; 0–1000 m. CA-FP; native to Eur, widely cult in mild-winter regions, sometimes spreading aggressively. TOXIC: juice can cause contact dermatitis; berries and lvs toxic when eaten.

ARISTOLOCHIACEAE PIPEVINE FAMILY

Michael R. Mesler & Karen Lu

Per from rhizome, woody vine, shrub, aromatic. **ST** branched, sometimes nearly all underground. **LVS** simple, basal, cauline, or arising singly from rhizome, alternate; blade gen cordate, entire. **INFL**: fl gen solitary, axillary or terminal. **FL** bisexual, radial or bilateral; sepals 3, free or fused; petals gen 0; stamens gen 6 or 12, free or fused to style; pistil gen 1, ovary gen inferior or partly so, chambers gen 6. **FR**: gen capsule. **SEEDS** many. 10 genera, 600 spp.: mainly trop, warm temp; some cult (*Aristolochia, Asarum*). [Gregory 1956 Amer J Bot 43:110–122]

1. Woody vine; fl bilateral .... **ARISTOLOCHIA**
1′ Per from rhizome; fl radial .... **ASARUM**

ARISTOLOCHIA PIPEVINE, BIRTHWORT

Per, woody vine. **ST** gen climbing. **LVS** cauline. **INFL**: fl axillary. **FL** bilateral, often foul smelling; sepals fused into a gen curved tube, deciduous, lobes 1–3; stamens 6, fused to style. **FR**: capsule. **SEED** flat. 500 spp.: gen trop, warm temp. (Greek: best birth, from use as medication in childbirth)

A. californica Torrey (p. 173) Pl soft-hairy. **ST** < 5 m, twining. **LVS** deciduous; blade 3–15 cm, ovate-cordate to sagittate. **FL**: fragrance metallic; calyx tube 2–4 cm, U-shaped, ± green or light brown, lined with a pink to red pad of thickened tissue, veins purple, lobes 3. **FR**: winged capsule. $2n$=28. Streamsides, forest, chaparral; < 700 m. NCoR, CaRF, n&c SNF, ScV, SnFrB, n SCoRO. Pollinated by fungus gnats. ❀5,6,**15–17**,;SHD:**14**&IRR:**7–9**,10,**18–23**,24.

ASARUM WILD-GINGER

Per from horizontal rhizome at soil surface or deep, ± vertical rhizome, spreading or clumped, gingery-aromatic. **LVS** basal or arising singly from rhizome, gen evergreen; blade cordate to reniform. **INFL**: fl terminai, at ground level. **FL** radial, gen dark colored; sepals 3, persistent, adherent into a tube or fused; stamens 12, free from style. **FR**: fleshy capsule. **SEED** with fleshy appendage, dispersed by ants. 90 spp.: n temp. (Greek: derivation unknown)

1. Lf blade marbled white along major veins; rhizomes deep, ± vertical, pls clumped; sterile anther tips > pollen sacs
  2. Inner surface of calyx tube white with several red stripes that are covered by bands of white hairs; lf margin hairs curved toward lf tip; anther tips pale .... ***A. hartwegii***
  2′ Inner surface of calyx tube dark red-maroon with scattered dark hairs; lf margin hairs ± perpendicular-spreading; anther tips dark .... ***A. marmoratum***
1′ Lf blade uniformly green; rhizomes near soil surface, horizontal, pl gen spreading; sterile anther tips < pollen sacs
  3. Calyx lobes long-tapered, spreading in fl .... ***A. caudatum***
  3′ Calyx lobes obtuse to acute or abruptly soft-pointed, strongly reflexed in fl .... ***A. lemmonii***

A. caudatum Lindley (p. 173) Pl gen spreading, forming a loose mat; rhizome near soil surface, horizontal. **LF**: blade uniformly green. **INFL**: peduncle ± horizontal. **FL**: calyx tube inner surface white with single median red stripe, lobes 2–9 cm, ± spreading from tube in fl, long-tapered, maroon (rarely ± green); anther tip < pollen sac, dark. $2n$=26. Moist places in forests; 0–2200 m. KR, NCoR, SnFrB; to B.C., MT. A form with short, reflexed, tapered calyx lobes from KR (McCloud River) is separable from *A. lemmonii* by lobe taper.❀IRR:**4,5,17**&SHD1,2,**6,7**,14,**15,16**,18–23,**24**;GRCVR.

A. hartwegii S. Watson (p. 173) Pl closely clumped; rhizome deep, ± vertical. **LF**: blade marbled white along major veins, margin hairs margin curved toward tip. **INFL**: peduncle ± erect. **FL**: calyx tube inner surface white with several red stripes covered with bands of white hairs, lobes gen spreading from tube in fl, tapered, maroon; anther tip > pollen sac, pale. $2n$=26. Dry rocky slopes in open forest: 150–2200 m. KR, CaRH, SNH; sw OR. ❀DRN&SHD: **4**,5,**6**,17;IRR:1–3,**7**,14,**15,16;**DFCLT.

A. lemmonii S. Watson (p. 173) Pl gen spreading, forming loose mat; rhizome near soil surface, horizontal. **LF**: blade uniformly green. **INFL**: peduncle bent downward in fl. **FL**: calyx tube inner surface white, gen not striped (rarely with a partial median stripe), lobes 0.8–1.5 cm, strongly reflexed in fl, appressed to tube, obtuse, acute, or abruptly soft-pointed, red; anther tip < pollen sac. $2n$=26. Shady wet places; 1100–1900 m. SNH. KR (McCloud River) pls with tapered calyx lobes are *A. caudatum*. ❀IRR,DRN&SHD:1, 2,4,**5**,6,7,14,15,**16,17**;GRCVR.

A. marmoratum Piper (p. 173) MARBLED WILD-GINGER **LF**: blade gen marbled (rarely uniformly green), margin hairs ± perpendicular-spreading. **INFL**: peduncle ± erect. **FL**: calyx tube inner surface dark red with scattered dark hairs, lobes ± erect in fl, olive-brown; anther tip > pollen sac, dark. UNCOMMON. Moist forests, exposed rocky slopes; 200–1800 m. KR (n Del Norte, Siskiyou cos.); sw OR. ❀TRY

ASCLEPIADACEAE MILKWEED FAMILY

Carol A. Hoffman

Ann, per, shrub, vine; sap milky. **LVS** simple, gen opposite or whorled; stipules 0 or small. **INFL**: cyme, terminal or axillary, umbel- or raceme-like, or fl solitary. **FL** bisexual, radial; sepals 5, gen reflexed; petals 5, gen reflexed or spreading; stamens 5, fused to form filament column and anther head, gen with 5 elaborate appendages on outside of filament column, pollen removed in pairs of massive sacs; ovaries 2, superior, free, style tips gen fused into massive pistil head surrounded by anther head. **FR**: follicle (1 ovary gen aborts). **SEEDS** many, ± flat, with tuft of silky hairs. 50–250 genera, 2000–3000 spp.: esp trop, subtrop S.Am, s Afr; orn (*Asclepias, Hoya, Stapelia*). [Rosatti 1989 J Arnold Arbor 70:307–401;443–514] Cardiac glycosides produced by some; used as arrow poisons, in medicine to control heart contraction, and by some insects for defense.

1. St not twining; widespread . . . . **ASCLEPIAS**
1′ St twining in CA; gen D
2. Filament column without appendages . . . . **CYNANCHUM**
2′ Filament column with appendages
3. Filament-column appendages fused margin-to-margin into a lobed, cup- or plate-like structure around anther head . . . . **MATELEA**
3′ Filament-column appendages free
4. Filament-column appendages solid, with out-curved margins; ring of tissue at base of corolla 0; weed in citrus groves . . . . **ARAUJIA**
4′ Filament-column appendages hollow, spheric; ring of tissue at base of corolla present; D, native . . . . **SARCOSTEMMA**

ARAUJIA BLADDER-FLOWER

Per. **ST** twining. **LVS** opposite; blade cordate, hastate, or ovate. **INFL** raceme- or panicle-like. **FL**: sepals large, lf-like, ± erect; corolla ± erect (> stamens, pistils), ring of tissue at base 0; filament-column appendages free, attached to base of filament column and base of corolla, without projections, solid (margins curved out but not nearly meeting on side away from column). **FR** pendent, gen ovoid, with coarse longitudinal grooves. 5 spp.: S.Am. (P.A. de Matos Araujo, probably a Brazilian or Portuguese collector, born probably latter 1700's) [Spellman & Gunn 1976 Castanea 41:139–148]

A. sericofera Brot. (p. 173) **ST** < 12 m, soft-tomentose when young. **LF**: petiole > 1 cm; blade 5–12 cm, upper surface glabrous, lower gen densely puberulent. **FL**: corolla 2–3 cm, bell- or funnel-shaped, white; pistil head with 2 erect, elongate lobes. **FR** 10–12 cm. Citrus groves, disturbed places; 100–400 m. CA-FP; native to S.Am. **NOXIOUS WEED CA.**

ASCLEPIAS MILKWEED

Per, ann, shrub. **ST** prostrate to erect. **LVS** alternate, opposite, or whorled; blade narrowly linear to ovate or cordate. **INFL**: umbel-like, gen terminal and in (esp upper) axils. **FL**: ring of tissue at base of corolla 0; filament-column appendages (hoods) free, sometimes elevated above corolla base, each often with an elongate projection (horn) attached to inside, solid, margins curved in and meeting or nearly meeting on side adjacent to column but not fused; top of pistil head flat or conic. **FR** gen erect (pedicel gen pendent), narrowly ovoid, smooth or with tubercles. 100 spp.: Am. (Asklepios, ancient Greek physician) [Woodson 1954 Ann Missouri Bot Gard 41:1–211]

1. Lvs ephemeral (st gen lfless), narrowly linear
2. Hoods < anther head . . . . ***A. albicans***
2′ Hoods >> anther head . . . . ***A. subulata***
1′ Lvs persistent, narrowly linear to broadly ovate
3. Lf 1 per node (alternate or clustered)
4. Lvs alternate to clustered, lanceolate; hoods not elevated above corolla base, subreflexed to spreading at base, ± = anther head . . . . ***A. asperula***
4′ Lvs alternate, linear; hoods elevated above corolla base, ± erect at base, > anther head . . . . ***A. linaria***
3′ Lvs > 1 per node (opposite or whorled)
5. Horn 0 or incl in hood
6. Hoods > or ± = anther head
7. Corolla purple; lf sessile, blade base clasping st . . . . ***A. cordifolia***
7′ Corolla greenish yellow; lf with short petiole, sometimes sessile, blade base rarely clasping st ***A. cryptoceras***
6′ Hoods < anther head
8. St decumbent to erect; pl densely white-hairy; tops of hoods above base of anther head (hoods, anther head overlapping) . . . . ***A. californica***
8′ St prostrate; pl ± hairy, purplish or green (appears glabrous at arm's length); tops of hoods below base of anther head (hoods, anther head not overlapping) . . . . ***A. solanoana***
5′ Horn exserted from hood (length sometimes < hood)
9. Horn > hood
10. Corolla red, hoods ± orange; lvs opposite, axillary lf clusters 0; filament column ± 2 mm; uncommon garden escape . . . . ***A. curassavica***
10′ Corolla and hoods greenish to pale yellow; lvs opposite or whorled, often with axillary clusters of small lvs; filament column ± 1 mm; common native . . . . ***A. fascicularis***
9′ Horn < to ± = hood
11. Hoods > anther head by at least 1/2 their length
12. Corolla greenish white; hoods gen not elevated above corolla base, ± = exserted horns, ± erect, tips rounded . . . . ***A. nyctaginifolia***
12′ Corolla rose-purple; hoods slightly elevated above corolla base, >> exserted horns, ± ascending, tips acute . . . . ***A. speciosa***
11′ Hoods not > anther head by at least 1/2 their length

13. Axillary infls ± sessile .... ***A. vestita***
13′ Axillary infls peduncled
14. Lvs opposite, sessile, blade bases often (shallowly) cordate; hoods slightly > anther head .... ***A. erosa***
14′ Lvs opposite or whorled, sessile to short-petioled, blade bases rarely cordate; hoods slightly < anther head .... ***A. eriocarpa***

A. albicans S. Watson WHITE-STEMMED or WAX MILKWEED Shrub, ± hairy, waxy. **ST** ± erect. **LVS** whorled in 3's, ephemeral; blade narrowly linear. **INFL** gen terminal. **FL**: corolla reflexed, greenish white, sometimes tinged brown or pink; hoods elevated above corolla base, < anther head, yellowish; horns > hoods. Dry washes, gravelly slopes; 200–1100 m. D; AZ, Baja CA.

A. asperula (Decne.) Woodson ssp. ***asperula*** (p. 173) ANTELOPE HORNS Per, ± hairy. **ST** decumbent to ascending. **LVS** alternate to clustered in 3's, persistent; petiole short; blade narrowly lanceolate. **INFL** terminal. **FL**: corolla spreading to ascending, greenish white; hoods not elevated above corolla base, ± = anther head, purplish; horns incl in hoods. Dry, open, rocky places; 1500–2000 m. DMoj; sw U.S., Mex. [*A. capricornu* ssp. *occidentalis* Woodson] ❀TRY.

A. californica E. Greene (p. 191) CALIFORNIA or ROUND-HOODED MILKWEED Per, densely hairy. **ST** decumbent to ± ascending. **LVS** opposite, persistent; petiole 0-short; blade ovate. **FL**: corolla reflexed, purplish; hoods elevated above corolla base, < anther head, dark purple; horns 0-minute. Flats, grassy or brushy hillsides; 200–2100 m. c&s SNF, CW, SW, D; n Baja CA. Pls from c&s SNF, CW with hoods elevated above corolla base have been called ssp. *greenei* Woodson. ❀TRY.

A. cordifolia (Benth.) Jepson PURPLE MILKWEED Per, ± glabrous. **ST** ascending. **LVS** opposite, persistent, sessile; blade cordate, base clasping st. **FL**: corolla spreading to reflexed, dark red-purple; hoods slightly elevated above corolla base, > anther head, purple; horns 0. Rocky slopes, talus, woods, chaparral, lava flows; 50–2000 m. NW, CaR, SN; OR, NV. ❀TRY.

A. cryptoceras S. Watson HUMBOLDT MOUNTAINS MILKWEED Per, ± glabrous. **ST** prostrate to decumbent. **LVS** opposite, persistent, sometimes sessile; blade ovate to nearly round, sometimes cordate, base rarely clasping st. **INFL** gen terminal. **FL**: corolla reflexed, greenish yellow; hoods not elevated above corolla base, ± = anther head, pinkish tan; horns 0 or incl in hoods. **FR**: pedicel erect. Sandy or gravelly slopes, canyon bottoms, arid plains; 1400–1700 m. SNE (Mono Co.); to WA, WY, Colorado, AZ. If the dubious sspp. are recognized, CA pls (with hoods slightly < anther head) are ssp. *davisii* (Woodson) Woodson. Ssp. *cryptoceras* (hoods gen slightly > anther head) has been reported from SNE (Mono Co.), is listed as UNCOMMON. Even if recognized, its occurrence in CA must be doubted. ❀TRY.

A. curassavica L. Ann, ± hairy. **ST** erect. **LVS** opposite, persistent; petiole short; blade lanceolate. **FL**: corolla reflexed, bright crimson, rarely yellow or white; hoods elevated above corolla base, ± = anther head, gen ± orange; horns > hoods. **FR**: pedicel erect. Moist, disturbed sites; < 2000 m. SW; to trop, subtrop; native to C.Am or S.Am. Cult as orn; possibly a waif in N.Am.

A. eriocarpa Benth. KOTOLO or INDIAN MILKWEED Per, very hairy (sometimes becoming less so). **ST** erect. **LVS** opposite or in whorls of 3–4, persistent; petiole 0 to short; blade lanceolate, elliptic, or ovate, base tapered or obtuse, rarely cordate. **FL**: corolla reflexed to ascending, cream-colored, sometimes tinged pink; hoods slightly elevated above corolla base, < to ± = anther head, cream, sometimes tinged purple; horns exserted, ± = hoods. Dry, barren areas; 200–1900 m. CA; NV, n Baja CA. ❀DRN,DRY,&SUN:**7–9**,10, **14,15**,16,17,**18**,19–24.

A. erosa Torrey (p. 191) DESERT MILKWEED Per, glabrous to very hairy. **ST** ascending to erect. **LVS** opposite, persistent, sessile; blade lanceolate, elliptic, or ovate, base tapered to cordate, sometimes clasping st. **FL**: corolla reflexed to spreading, pale cream-colored or greenish white; hoods elevated above corolla base, > anther head, cream-colored, yellow, or reddish; horns exserted, < to ± = hoods. Dry slopes, washes; 150–1900 m. SW, D; AZ, Baja CA.

A. fascicularis Decne. NARROW-LEAF MILKWEED Per, gen glabrous. **ST** ascending to erect. **LVS** in whorl of 3–5, often with axillary clusters of small lvs, persistent; petiole short; blade narrowly lanceolate, base tapered. **FL**: corolla reflexed, greenish white, sometimes tinged purple; hoods elevated above corolla base, gen < anther head, greenish white; horns gen > hoods (and anther head). **FR**: pedicel erect. Dry ground, valleys, foothills; 50–2200 m. CA (exc NCo, CCo, SCo, D); to WA, UT, Baja CA. [not *A. mexicana* Cav.] ❀DRN,DRY,&SUN:**3,7–9**,10,**14,15**,16,17,**18**,19–24.

A. linaria Cav. Per or shrub, ± hairy. **ST** erect. **LVS** alternate, persistent, sessile; blade linear, resembling pine needle. **FL**: corolla reflexed, inner surface greenish white, outer surface pinkish or purplish; hoods gen elevated above corolla base, > anther head, greenish white; horns exserted, < hoods. Uncommon. Open woodland, limestone ridges, rocky hills, canyons, arroyos, dry abandoned pastures; 1000–1400 m. D; AZ, Mex.

A. nyctaginifolia A. Gray MOJAVE MILKWEED Per, hairy. **ST** decumbent to ascending. **LVS** opposite, persistent, petioled; blade lanceolate, elliptic, or ovate. **FL**: corolla reflexed, greenish white, outer surface sometimes purple; hoods not to slightly elevated above corolla base, >> anther head, greenish white; horns exserted, ± = hoods. Arroyos, dry slopes; 1000–1700 m. DMoj; to NM.

A. solanoana Woodson SOLANOA, PROSTRATE MILKWEED Per, ± hairy, purplish. **ST** prostrate, ± flat. **LVS** opposite, persistent, petioled; blade reniform, cordate, or ovate, base sometimes clasping st. **INFL** spheric. **FL**: corolla reflexed to ascending, purple; hoods slightly elevated above corolla base, < anther head, brownish yellow; horns 0. UNCOMMON. Serpentine outcrops; 700–1600 m. NCoR. ❀TRY.

A. speciosa Torrey GREEK or SHOWY MILKWEED Per, hairy. **ST** ascending to erect. **LVS** opposite, persistent; petiole short; blade elliptic to ovate, base rarely cordate and clasping st. **FL**: corolla reflexed, rose-purple; hoods slightly elevated above corolla base, >> anther head, pink, aging yellow; horns exserted, << hoods, ± = anther head. Many habitats incl fields, roadsides; 0–1900 m. CA; to B.C., c Can, TX. [*A. giffordii* Eastw. (see Gilmartin 1980 Bull Torrey Bot Club 104:496–505)] ❀SUN&DRN:1,**2,3**,4,5,**6,10,14, 15**,16,17&IRR:**7–9,14,15,18**,19–24;STBL,INV.

A. subulata Decne. RUSH MILKWEED, AJAMETE Per, gen glabrous exc infl. **ST** erect. **LVS** opposite, ephemeral, sessile; blade narrowly linear. **FL**: corolla reflexed, yellowish white; hoods slightly elevated above corolla base, >> anther head, yellowish white; horns exserted, = or slightly < hoods, >> anther head. **FR** pendent. Arroyos, washes; < 700 m. D; NV, AZ, n Mex.

A. vestita Hook. & Arn. WOOLLY MILKWEED Per, gen densely hairy, sometimes becoming ± glabrous. **ST** ascending. **LVS** opposite, persistent; petiole gen short; blade lanceolate, elliptic, or ovate. **FL**: corolla reflexed, yellowish white or purplish; hoods slightly elevated above corolla base, ± = anther head, yellowish white, sometimes with vertical, brown stripe; horns exserted, ± = hoods. Dry plains, brushy flats, hillsides, desert canyons; 50–1350 m. GV, CW, TR, DMoj. Pls from CW, TR, DMoj with purple corollas have been called var. *parishii* Jepson.

CYNANCHUM VINE MILKWEED

Per, shrub. **ST** twining (elsewhere sometimes prostrate to erect). **LVS** opposite; blade linear to ovate. **INFL** axillary, umbel- or raceme-like. **FL**: corolla ± erect, ring of tissue at base 0; filament-column appendages 0 (elsewhere gen free); pistil head flat, conic, or with 2 lobes on top. **FR** gen erect, fusiform to narrowly ovoid. ± 200 spp.: temp, esp trop. (Greek: dog strangle, from ancient supposition of or use as dog poison) [Sundell 1981 Evol Monogr 5:1–63]

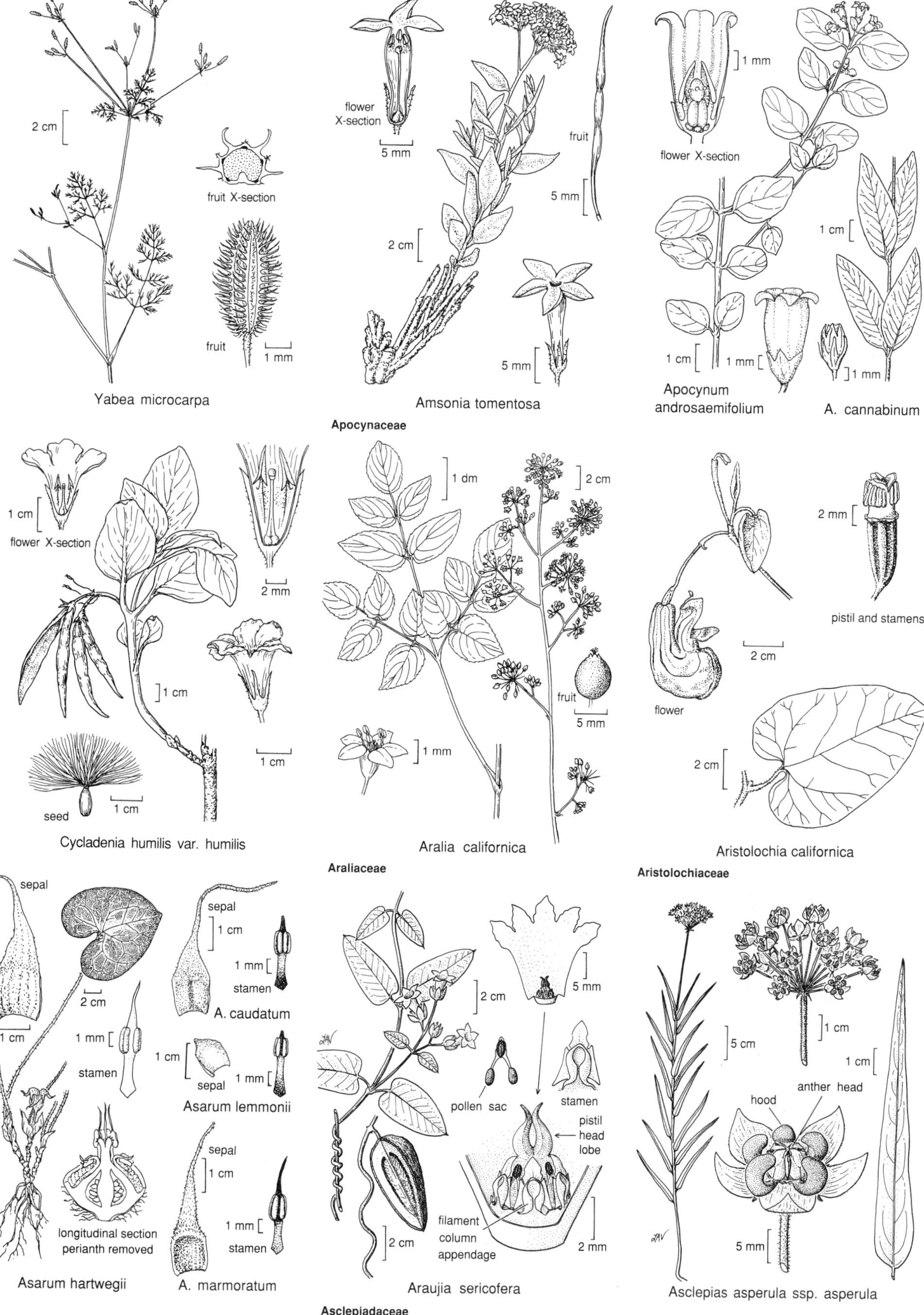
2 cm
fruit X-section
fruit
1 mm
Yabea microcarpa
flower X-section
5 mm
fruit
5 mm
2 cm
5 mm
Amsonia tomentosa
1 mm
flower X-section
1 cm
1 cm
1 mm
1 mm
Apocynum androsaemifolium
A. cannabinum
Apocynaceae
1 cm
flower X-section
2 mm
1 cm
1 cm
seed
1 cm
Cycladenia humilis var. humilis
1 dm
2 cm
fruit
5 mm
1 mm
Aralia californica
Araliaceae
2 mm
pistil and stamens
2 cm
flower
2 cm
Aristolochia californica
Aristolochiaceae
sepal
2 cm
1 cm
1 mm
stamen
longitudinal section perianth removed
Asarum hartwegii
sepal
1 cm
1 mm
stamen
A. caudatum
1 cm
sepal
1 mm
Asarum lemmonii
sepal
1 cm
1 mm
stamen
A. marmoratum
2 cm
5 mm
pollen sac
stamen
pistil head lobe
filament
column
appendage
2 cm
2 mm
Araujia sericofera
Asclepiadaceae
5 cm
1 cm
1 cm
hood
anther head
5 mm
Asclepias asperula ssp. asperula

C. utahense (Engelm.) Woodson (p. 191) UTAH CYNANCHUM Per. **ST** slender, much branched, < 1 m. **LF**: blade 1.5–4 cm, linear, becoming reflexed. **INFL** umbel-like. **FL**: corolla 1.5–3 mm, bell-shaped, lobes incurved, hood-like, yellow, becoming orange. **FR** 4–6 cm, with fine longitudinal grooves. UNCOMMON. Dry, sandy or gravelly areas; < 1000 m. DMoj; to UT, AZ.

MATELEA

Per, shrub. **ST** twining (elsewhere sometimes prostrate to erect). **LVS** opposite; blade often ± cordate. **INFL**: fls 1–2, axillary (elsewhere various). **FL**: corolla ± spreading, ring of tissue at base 0; filament-column appendages gen fused margin-to-margin into 5-lobed, cup- or plate-like structure around anther head, fused to base of filament column, each lobe with a vertical, flap-like ridge; pistil head flat. **FR** erect or pendent, fusiform to ± ovoid, smooth, gen tubercled. ± 200 spp.: trop, warm temp Am. [Stevens 1976 Diss Abstr B 37(2):587]

M. parvifolia (Torrey) Woodson (p. 191) SPEARLEAF, TALAYOTE Per. **ST** slender, much branched, < 0.5 m. **LF**: blade 0.5–2 cm, cordate-sagittate. **FL**: corolla greenish or purple, each sinus with acute, turned out tooth. **FR** ± 7 cm, with fine longitudinal grooves. RARE in CA. Dry rocky areas; 700–1000 m. D; to NV, TX, Baja CA. Populations widely scattered.

SARCOSTEMMA CLIMBING MILKWEED

Per, shrub. **ST** twining (elsewhere sometimes prostrate to ± erect). **LVS** opposite; blade often linear to narrowly lanceolate or hastate. **INFL** axillary, often umbel-like. **FL**: corolla gen ± spreading to ± erect, with ring of tissue at base; filament-column appendages free, ± spheric, attached to base of filament column, projections hollow; pistil head ± conic, 2-lobed, or both. **FR** gen erect, narrowly fusiform to narrowly ovoid, with fine longitudinal grooves. ± 34 spp.: N.Am, Afr to Australia. (Greek: fleshy crown, from sac-like filament-column appendages) [Holm 1950 Ann Missouri Bot Gard 37:477–560]

1. Pl sparsely hairy; corolla ± purple; filament-column appendages free from ring of tissue at base of corolla .... ***S. cynanchoides*** ssp. ***hartwegii***
1′ Pl densely hairy; corolla ± greenish white; filament-column appendages fused to ring of tissue at base of corolla .... ***S. hirtellum***

S. cynanchoides Decne. ssp. ***hartwegii*** (Vail) R. Holm (p. 191) CLIMBING MILKWEED Pl green; hairs gen sparse, ± appressed. **LF**: blade base gen hastate or truncate. **FL**: corolla pink to purple, or lobes white with purple streak; filament-column appendages free from ring of tissue at base of corolla. **FR** gen 1. Dry, sandy, rocky arroyos or plains, in ditches near cult; 30–1600 m. D; to UT, AZ, Mex. ❀TRY.

S. hirtellum (A. Gray) R. Holm TRAILING TOWNULA Pl gray-green; hairs gen dense, short, erect. **LF**: blade base variously tapered. **FL**: corolla white to greenish white; filament-column appendages fused to ring of tissue at base of corolla. **FR** gen 2, often spreading. Hard desert pavement, washes; 150–1200 m. D; NV, AZ. ❀TRY.

ASTERACEAE [Compositae] SUNFLOWER FAMILY

David J. Keil, Family Editor and author, except as specified

Ann to tree. **LVS** basal or cauline, alternate to whorled, simple to compound. **INFL**: 1° infl a head, each resembling a fl, 1–many, gen arrayed in cymes, gen subtended by ± calyx-like involucre; fls 1–many per head. **FLS** bisexual, unisexual, or sterile, ± small, of several types; calyx 0 or modified into pappus of bristles, scales, or awns, which is gen persistent in fr; corolla radial or bilateral (rarely 0), lobes gen (0)4–5; stamens 4–5, anthers gen fused into cylinder around style, often appendaged at tips, bases, or both, filaments gen free, gen attached to corolla near throat; pistil 1, ovary inferior, 1-chambered, 1-seeded, style 1, branches 2, gen hair-tufted at tip, stigmas 2, gen on inside of style branches. **FR**: achene, cylindric to ovoid, gen deciduous with pappus attached. ± 1300 genera, 21,000 spp. (largest family of dicots): worldwide. The largest family in CA. See glossary p. 25 for illustrations of general family characteristics.

Key to Groups

1. Ligules (strap-shaped, petal-like corollas) 0
  2. Heads disciform [fls of 2 kinds (pistillate or sterile and bisexual or staminate), in same or different heads] **Group 1**
  2′ Heads discoid [fls bisexual, gen all fertile (sometimes outermost enlarged, ± bilateral)]
    3. Receptacle bearing chaff (scale-like bractlets) or stiff hairs among fls
      4. Receptacle chaffy .... **Group 2**
      4′ Receptacle bristly .... **Group 3**
    3′ Receptacle ± naked (sometimes bearing minute scales or short hairs among fls)
      5. Pappus 0 or only a low crown .... **Group 4**
      5′ Pappus well developed
        6. Pappus of bristles (sometimes with an additional series of shorter bristles or scales) .... **Group 5**
        6′ Pappus of flat, ± membranous scales or stiff, ± needle-like awns .... **Group 6**
1′ Ligules present
  7. Heads ligulate or of 2-lipped disk fls; sap milky or clear
    8. Corollas readily withering, 1-lipped; ligules 5-lobed; sap gen milky .... **Group 7**
    8′ Corollas gen not withering, 2-lipped; ligules 3-lobed (second lip shorter, recurved); sap clear
      9. Fls whitish to pink or purple; stout per .... **ACOURTIA**
      9′ Fls yellow; shrub .... **TRIXIS**

7′ Heads radiate; fls of 2 kinds (ray fls pistillate or sterile; disk fls gen bisexual); sap gen clear
10. Receptacle bearing chaff scales among disk fls (sometimes only in a ring between ray and disk fls)
11. Phyllaries in 1 series, each subtending a ray fl .... **Group 8**
11′ Phyllaries in 2+ series, not all subtending ray fls .... **Group 9**
10′ Receptacle naked or bearing minute scales or hairs (rarely stiff bristles) among fls
12. Pappus 0 or only a low crown .... **Group 10**
12′ Pappus well developed on ray or disk frs (or both)
13. Pappus of flat, ± membranous scales or stiff, ± needle-like awns .... **Group 11**
13′ Pappus of bristles (sometimes with an additional series of shorter bristles or scales)
14. Ligules white to purple .... **Group 12**
14′ Ligules yellow to orange or red .... **Group 13**

Group 1: Heads disciform; fls of 2 kinds

1. Pistillate and staminate fls in different heads
2. Pistillate and staminate heads on same pl (monoecious)
3. Subshrubs or shrubs
4. Lvs linear to ovate, toothed or lobed; bracts of pistillate heads spiny or knob-like, sometimes not evident .... [2]**AMBROSIA**
4′ Lvs thread-like or with thread-like lobes; bracts of pistillate heads flat, scarious .... **HYMENOCLEA**
3′ Herbs
5. Sts armed with 3-branched spines .... ***Xanthium spinosum***
5′ Sts unarmed
6. Staminate heads in long terminal infl; burs gen 3–10 mm .... [2]**AMBROSIA**
6′ Staminate heads congested; burs gen > 20 mm .... ***Xanthium strumarium***
2′ Pistillate and staminate heads on different pls (dioecious)
7. Shrubs or erect herbs with sticky lvs, not woolly .... **BACCHARIS**
7′ Herbs, sometimes prostrate, not sticky; pls gen ± woolly
8. Sts from rhizomes; lf rosettes 0; cauline lvs long, all exc uppermost ± equal; pappus bristles free [2]**ANAPHALIS**
8′ Sts from lf rosette or thick, often branched caudex; cauline lvs gen ± reduced, at least uppermost short, 1–3 cm; pappus bristles fused at base .... **ANTENNARIA**
1′ Pistillate or sterile fls in same heads with staminate or bisexual fls
9. Outer fls without corollas; heads not embedded in wool
10. Heads long-peduncled; outer frs stalked .... **COTULA**
10′ Heads sessile or short-peduncled; frs sessile
11. Per to shrubs; fr not winged .... [2] **IVA**
11′ Ann; fr winged
12. Pl erect; heads in racemes or panicles; fr not spine-tipped; D .... **DICORIA**
12′ Pl low, often prostrate; heads sessile at st forks; fr spine-tipped; CA-FP .... **SOLIVA**
9′ Outer fls with (sometimes narrowly cylindric) corollas; heads sometimes embedded in wool
13. Outer fls sterile (corollas sometimes expanded, ± bilateral); heads not very small, not embedded in wool
14. Fertile fls 1–2 per head .... **CRUPINA**
14′ Fertile fls many per head
15. Lvs spiny .... **CNICUS**
15′ Lvs not spiny
16. Pappus of awns with reflexed barbs; phyllaries in 2 series, unarmed .... **BIDENS**
16′ Pappus of smooth or minutely rough bristles; phyllaries in 3+ series, often fringed or spine-tipped .... **CENTAUREA**
13′ Outer fls pistillate (corollas often narrowly cylindric); heads sometimes very small, embedded in wool
17. Phyllaries (or outermost bracts of head) papery, membranous or scarious, sometimes green only below middle or in narrow, central band; heads often embedded in wool
18. Phyllaries gen many, overlapping, inner > fls; receptacle naked; pistillate fls not subtended by bracts
19. Pappus 0 or a minute crown .... [3]**ARTEMISIA**
19′ Pappus of bristles
20. Herbage short-appressed-hairy, glandular, or silky; upper half of outer phyllaries leathery, scarious margin 0 or narrow .... [2]**PLUCHEA**
20′ Herbage ± long-hairy (lvs gen densely woolly below); phyllaries thin, membranous, margin scarious above middle or throughout, gen wide
21. Pl from creeping rhizome .... [2]**ANAPHALIS**
21′ Pl from taproot or caudex .... **GNAPHALIUM**
18′ Phyllaries 0 or 6–8, not overlapping, < fls; outer receptacle chaffy; some or all pistillate fls individually subtended by bracts
22. Disk fls bisexual, pappus gen of > 12 bristles, ± exserted; inner pistillate fls with pappus .... **FILAGO**
22′ Disk fls staminate, pappus 0 or 1–12 bristles, incl; pistillate pappus 0
23. Chaff scales subtending disk fls (disk chaff) enlarged, rigid, ± spreading, very different from scales subtending pistillate fls (pistillate chaff)
24. Disk chaff scale tips gen spine-like, strongly hooked inward; pistillate chaff scales closed, strongly 3-veined, tips gen strongly scarious-winged; receptacle not bristly, gen widest at tip .... **ANCISTROCARPHUS**

24′ Disk chaff scale tips erect or spreading, ± flat to folded, not spine-like, obtuse; pistillate chaff scales open, concave, veins obscure, tips not or barely scarious-winged; receptacle bristly, widest at base .... **HESPEREVAX**

23′ Disk chaff 0 or scales ± gradually reduced, scarious, erect

25. Lvs gen opposite; chaff scales net-veined, inner surface with scarious wing that is hidden in head; pappus 0 .... **PSILOCARPHUS**

25′ Lvs alternate or seeming whorled; chaff scales parallel-veined, scarious wing visible in head; pappus gen of 1–12 bristles

26. Phyllaries 4–6, equal, scarious, obovate, rounded, abruptly different from chaff scales; style on inner edge of fr; receptacle length ± 1–2 × width .... **MICROPUS**

26′ Phyllaries 0 or 1–4, unequal, vestigial or ± like chaff scales; style ± at fr tip; receptacle length ± 3–8 × width .... **STYLOCLINE**

17′ Phyllaries green gen throughout, sometimes scarious-margined; heads gen not embedded in wool

27. Pappus well developed

28. Pappus of flattened scales or barbed awns

29. Lvs gen opposite throughout .... **LASTHENIA**

29′ Lvs mostly alternate

30. Pl dotted with sessile resin glands; fr 3–4-angled .... [2]**AMBLYOPAPPUS**

30′ Pl densely woolly; fr compressed .... **LEMBERTIA**

28′ Pappus of bristles

31. Main phyllaries in 1 series, equal .... **ERECHTITES**

31′ Main phyllaries in 2+ series

32. Sts thread-like; pistillate fls 1–5 .... [3]**PENTACHAETA**

32′ Sts stouter; pistillate fls several–many

33. Pappus bristles 1–2; lvs palmately lobed or toothed .... ***Perityle emoryi***

33′ Pappus bristles gen many; lvs entire to pinnately lobed

34. Heads ± flat, button-like .... **ERIGERON**

34′ Heads gen cylindric or bell-shaped

35. Ann .... **CONYZA**

35′ Per or shrub .... [2]**PLUCHEA**

27′ Pappus 0 or reduced to a minute crown

36. Phyllary margins widely scarious or transparent

37. Delicate ann; pistillate corollas thread-like .... [3]**PENTACHAETA**

37′ Per or stout ann; pistillate corollas short, ± wide

38. Heads many in wide, flat-topped clusters

39. Lvs simple, entire or with 2–5 ± linear lobes; subshrub .... ***Sphaeromeria cana***

39′ Lvs pinnate; lflets serrate to 1–2 × divided; per from rhizome .... **TANACETUM**

38′ Heads 1–few in small ± flat-topped clusters or few–many in spikes, racemes, or panicles

40. Heads in spikes, racemes, or panicles .... [3]**ARTEMISIA**

40′ Heads 1 or few in ± flat-topped clusters .... **SPHAEROMERIA**

36′ Phyllary margins gen not scarious or transparent

41. Ovary of disk fls much reduced; style tip truncate or tack-shaped

42. Lvs widely triangular, gen ± basal or cauline below middle, green above, white below; inf open; corollas white; fr glandular .... **ADENOCAULON**

42′ Lvs linear to narrowly ovate, surfaces ± alike, gen green to gray on one or both surfaces; infl ± dense; corollas ± yellowish white; fr not glandular

43. Anthers fused .... [3]**ARTEMISIA**

43′ Anthers ± free .... [2] **IVA**

41′ Ovary of disk fl well developed; style tip ± branched

44. Lvs wide, palmately lobed or toothed .... ***Perityle emoryi***

44′ Lvs linear

45. Pl glandular, ± strongly scented

46. Pistillate corollas 0.5–1 mm .... [2]**AMBLYOPAPPUS**

46′ Pistillate corollas gen > 1 mm .... **MADIA**

45′ Pl not glandular, not strongly scented

47. Lvs opposite .... ***Lasthenia microglossa***

47′ Lvs alternate .... [3]**PENTACHAETA**

Group 2: Heads discoid; receptacle chaffy

1. Lf margins spiny

2. Heads many-fld .... **CARTHAMUS**

2′ Heads 1-fld, grouped in spheric 2° heads .... **ECHINOPS**

1′ Lf margins not spiny

3. Pappus of bristles or of stout, needle-like awns

4. Pappus of bristles

5. Corollas yellow; fr many per head; shrub .... **BEBBIA**

5′ Corollas purple; fr 1–2 per head; herb .... **CRUPINA**

4′ Pappus of awns

6. Lvs pinnately compound, lflets lanceolate to ovate .... **BIDENS**
6′ Lvs linear or dissected into linear lobes .... **THELESPERMA**
3′ Pappus 0 or of thin, membranous scales
7. Ovary and fr strongly flattened, long-ciliate
8. Shrub, strigose .... ***Encelia frutescens***
8′ Per, sticky-glandular .... ***Geraea viscida***
7′ Ovary and fr ± thick, not ciliate
9. Ann; outer phyllaries tapered to long, bristle-like tips .... ***Chaenactis carphoclinia***
9′ Bien to shrub; phyllaries obtuse to acute
10. Lf ± cylindric, lobes many, small .... **SANTOLINA**
10′ Lf flat, entire to few-lobed
11. Lf lobes long, slender .... ***Artemisia palmeri***
11′ Lf lobes 0 or wide
12. Receptacle strongly conic or columnar; corollas dark brown .... ***Rudbeckia occidentalis***
12′ Receptacle flat to convex; corollas yellow or orange
13. Lvs ± linear .... **EASTWOODIA**
13′ Lvs ovate to triangular .... ***Wyethia invenusta***

Group 3: Heads discoid; receptacle bristly

1. Lvs not spiny
2. Phyllaries spine-tipped
3. Phyllary spines hooked at tip .... **ARCTIUM**
3′ Phyllary spines straight .... **CENTAUREA**
2′ Phyllaries not spine-tipped
4. Phyllary tips (at least inner) prominently expanded, ± fringed with short spines or irregular teeth **CENTAUREA**
4′ Phyllary tips not expanded or fringed
5. Lower lvs linear to oblanceolate, often lobed or divided; attachment scar of fr lateral .... **ACROPTILON**
5′ Lower lvs lanceolate to triangular-ovate, serrate; attachment scar of fr basal .... **SAUSSUREA**
1′ Lvs spiny
6. Corollas yellow to orange-red
7. Pappus of many, unequal, narrow scales in several series (sometimes 0), gen 0 on outer frs; fls all fertile; fr 4-angled .... **CARTHAMUS**
7′ Pappus of 20 stiff bristles or awns in 2 series, on all frs; outer fls sterile (corolla 3-lobed; ovary vestigial); fr cylindric, 20-ribbed .... **CNICUS**
6′ Corollas white to blue, red, or purple
8. Pappus of long-plumose bristles
9. Largest lvs toothed to deeply lobed; involucre body 1–6 cm diam; receptacle gen not fleshy; phyllaries linear to ovate .... **CIRSIUM**
9′ Largest lvs often ± compound; involucre body 3–15 cm diam; receptacle fleshy; phyllaries ovate .... **CYNARA**
8′ Pappus of ± rough or barbed bristles or slender scales
10. Sts spiny-winged .... **CARDUUS**
10′ Sts not spiny-winged
11. Lvs not blotched along veins; heads not long-peduncled; pappus of free, slender scales ***Carthamus leucocalos***
11′ Lvs white-blotched along veins; heads long-peduncled; pappus of many long bristles, deciduous in a ring .... **SILYBUM**

Group 4: Heads discoid; receptacle naked; pappus 0 or reduced to crown

1. Fls 1–7 per head; heads often in tight, head-like 2° clusters
2. Lvs coarsely 1–2-pinnately lobed, spiny-margined .... **ECHINOPS**
2′ Lvs entire or toothed
3. Shrubs; 1° lvs ± persistent as spines
4. Heads 1-fld, clustered in 2° heads; phyllaries ± 15 in several overlapping series .... **HECASTOCLEIS**
4′ Heads 5–6-fld, short-peduncled in terminal clusters; phyllaries 5–6 in 1 series .... ***Tetradymia comosa***
3′ Herbs; lvs never forming spines
5. Herbage glabrous .... **FLAVERIA**
5′ Herbage hairy or glandular
6. Pl 1–2.5 cm, tufted, ± woolly .... ***Eriophyllum mohavense***
6′ Pl 15–80+ cm, erect, glandular, ± bristly .... ***Madia glomerata***
1′ Fls 10–many; heads not in 2° heads
7. Fr compressed
8. Ann; lower lvs 2–4 × dissected .... ***Chaenactis artemisiifolia***
8′ Per or subshrub; lvs entire to lobed
9. Phyllaries in 1 series, weakly fused; lvs long-acuminate .... **PERICOME**
9′ Phyllaries in 2–3 series, ± equal, free; lvs not long-acuminate .... **PERITYLE**
7′ Fr not compressed
10. Receptacle flat-topped or weakly convex; pl woolly, ± tufted
11. Shrub; heads many; phyllaries in several overlapping series; corollas not yellow .... **ARTEMISIA**

11′ Low, tufted ann; heads 1–few; phyllaries in 1–2 subequal series; corollas yellow .... *Eriophyllum pringlei*
10′ Receptacle conic to spheric; pls glabrous or puberulent, not tufted
12. Lvs 2–3 × dissected, not decurrent; phyllary margins scarious .... CHAMOMILLA
12′ Lvs ± entire, decurrent; phyllary margins not scarious .... HELENIUM

Group 5: Heads discoid; receptacle naked; pappus of bristles

1. Lvs and phyllaries dotted or streaked with embedded, translucent oil glands, otherwise glabrous; odor strong, unpleasant .... POROPHYLLUM
1′ Lvs and phyllaries without oil glands (sometimes stalked- or sessile-glandular); odor 0 or not very unpleasant
2. Shrubs or subshrubs
3. Corollas white or cream-colored to dull purple or greenish
4. Ovaries very reduced, sterile; style incl .... BACCHARIS
4′ Ovaries forming frs; style often exserted
5. Fr 10-ribbed .... [2]BRICKELLIA
5′ Fr 5-ribbed
6. Lvs narrowly linear, blade not evident, entire .... *Chrysothamnus albidus*
6′ Lvs petioled, blade obvious, gen toothed
7. Blade gen = or > petiole; pappus of bristles only .... [2]AGERATINA
7′ Blade gen << petiole; pappus of bristles and short scales .... PLEUROCORONIS
3′ Corollas yellow to orange
8. Pappus bristles 1–2
9. Lf blade 3–12 cm, petiole 1.5–5 cm, phyllaries in 1 series, weakly fused .... [2]PERICOME
9′ Lf blade gen < 2.5 cm, petiole 0.1–0.6 cm; phyllaries in 2 equal series, free .... PERITYLE
8′ Pappus bristles gen many
10. Heads ± spheric; phyllaries elliptic or widely ovate .... *Acamptopappus sphaerocephalus*
10′ Heads cylindric to bell-shaped; phyllaries linear to narrowly ovate
11. Phyllaries in 1–2 series, ± equal
12. Phyllaries 9–18; pl glabrous, resinous-glandular; fls 12–21 .... PEUCEPHYLLUM
12′ Phyllaries 4–7; pl ± woolly; fls 4–8 .... TETRADYMIA
11′ Phyllaries in several graduated series, unequal
13. Lvs (at least upper) scale-like
14. Heads solitary; fr appressed-hairy .... *Machaeranthera carnosa*
14′ Heads in panicle-like clusters; fr ± glabrous .... *Lepidospartum squamatum*
13′ Lvs linear to oblong or obovate
15. Lvs toothed
16. Corolla throat expanded gradually above tube .... HAZARDIA
16′ Corolla throat expanded abruptly above tube .... [2]ISOCOMA
15′ Lvs entire
17. Phyllaries in 5 ± distinct vertical ranks .... [2]CHRYSOTHAMNUS
17′ Phyllaries not in distinct vertical ranks
18. Sts ± tomentose
19. Phyllaries acuminate
20. Lvs glabrous to ± tomentose .... *Chrysothamnus parryi*
20′ Lvs stalked-glandular .... *Ericameria discoidea*
19′ Phyllaries obtuse to acute
21. Sts loosely tomentose; pappus 3–5 mm .... *Isocoma menziesii*
21′ Sts closely tomentose exc glabrous ribs; pappus 8–11 mm .... *Lepidospartum latisquamum*
18′ Sts glabrous or ± hairy
22. Corolla throat expanded abruptly above tube .... [2]ISOCOMA
22′ Corolla throat expanded gradually above tube
23. Fls 9–70
24. Dwarf alpine shrubs; gen > 3200 m .... *Chrysothamnus viscidiflorus*
24′ Non-alpine shrubs; < 2900 m .... ERICAMERIA
23′ Fls 2–8
25. Phyllaries not grading into lvs, tips erect .... [2]CHRYSOTHAMNUS
25′ Phyllaries grading into upper lvs, tips spreading to recurved .... *Ericameria ophitidis*
2′ Herbs
26. Corollas white to purple
27. Lvs and phyllaries spiny .... ONOPORDUM
27′ Lvs and phyllaries not spiny
28. Pappus bristles 2–6
29. Lvs linear, gen entire; involucre cylindric to narrowly bell-shaped; DSon .... MALPERIA
29′ Lvs oblong to ovate, toothed; involucre widely bell-shaped to hemispheric; wet places, GV, SW .... TRICHOCORONIS
28′ Pappus bristles gen >> 6
30. Phyllaries in 1–2 series, ± equal
31. Fls 10–many; per; pappus bristles simple .... [2]AGERATINA
31′ Fls 2(3); ann; pappus bristles plumose .... DIMERESIA

30′ Phyllaries in several graduated series, unequal
32. Ann; outer fls often bilateral, ± ray-like .... [2]**LESSINGIA**
32′ Per; fls all radial
33. Herbage glabrous, sticky; lvs entire or finely toothed; involucre 3.5–5 mm — corollas white .... ***Baccharis douglasii***
33′ Herbage puberulent to loosely tomentose; lvs sharply toothed; involucre 10–15 mm
34. Corollas cream-colored .... [2]**BRICKELLIA**
34′ Corollas gen ± purple .... **SAUSSUREA**
26′ Corollas yellow to orange
35. Phyllaries in 2–several series, unequal, often strongly graduated
36. Pl low, mounded, densely gray-woolly, strongly scented; lvs often wider than long; phyllaries in 2 series, outer phyllaries wide, tips spreading .... [2]**PSATHYROTES**
36′ Pl spreading to erect, not densely woolly, scented or not; lvs gen longer than wide; phyllaries in 3–several series
37. Pappus bristles flat, readily deciduous; involucre gummy .... **GRINDELIA**
37′ Pappus bristles ± cylindric, ± persistent; involucre gen not gummy
38. Ann; outer fls often bilateral, ± ray-like .... [2]**LESSINGIA**
38′ Per; fls radial
39. Pl glabrous; fls 4–7 .... ***Chrysothamnus gramineus***
39′ Pl ± hairy, sometimes glandular; fls 14–60
40. Phyllary tips spreading
41. Lvs lanceolate to triangular; corollas creamy yellow .... ***Brickellia grandiflora***
41′ Lvs linear to ± obovate; corollas bright yellow .... ***Machaeranthera canescens***
40′ Phyllary tips gen appressed (lowest sometimes spreading)
42. Lvs linear to narrowly oblong; disk wide, button-like .... [2]**ERIGERON**
42′ Lvs linear to ovate; disk ± narrow
43. Hairs smooth; various habitats .... **ASTER**
43′ Hairs minutely knobby (at 20×); gen dry streambanks, gravel bars .... ***Heterotheca oregona***
35′ Phyllaries in 1–2 series, ± equal
44. Pappus bristles ± plumose
45. Lvs ± basal .... **RAILLARDELLA**
45′ Some or all lvs cauline
46. Phyllaries in 2 series; pappus sub-plumose .... [2]**ARNICA**
46′ Phyllaries in 1 series; pappus plumose .... **RAILLARDIOPSIS**
44′ Pappus bristles smooth or barbed
47. Cauline lvs opposite (uppermost sometimes alternate)
48. Pappus bristles many; phyllaries in 2 series, free .... [2]**ARNICA**
48′ Pappus bristles 1–2; phyllaries in 1 series, fused at base .... [2]**PERICOME**
47′ Cauline lvs alternate
49. Ann or short-lived per from slender taproot
50. Main phyllaries in 1 series, outer, if present, << inner .... [2]**SENECIO**
50′ Main phyllaries in 2 series, ± equal
51. Pappus bristles free; heads short-peduncled; lvs entire or blunt-toothed .... [2]**PSATHYROTES**
51′ Pappus bristles fused at base in 5 groups; heads long-peduncled; lvs sharply toothed or lobed .... **TRICHOPTILIUM**
49′ Per, often from caudex or rhizome
52. Lvs palmately veined and lobed
53. Erect herb .... **CACALIOPSIS**
53′ Twining vine .... ***Senecio mikanioides***
52′ Lvs pinnately veined (or lateral veins obscure), entire to pinnately lobed
54. Pappus double, outer pappus of minute bristles or scales .... [2]**ERIGERON**
54′ Pappus single
55. Lvs densely white-tomentose below; corollas creamy yellow; phyllaries not black-tipped .... **LUINA**
55′ Lvs glabrous to loosely tomentose below; corollas yellow; phyllaries often black-tipped .... [2]**SENECIO**

Group 6: Heads discoid; receptacle naked; pappus of scales or awns

1. Heads ± sessile, sometimes in dense 2° clusters
2. Heads 1-fld, in dense 2° heads; pl ± spiny
3. Herb, 10–20 dm, stout, erect; lvs large, 1–2-pinnately lobed, spiny-margined; 2° heads spheric .... **ECHINOPS**
3′ Shrub, 4–7 dm; lvs small, spine-tipped, margins weakly spiny; 2° heads not spheric, ± concealed by veiny bracts .... **HECASTOCLEIS**
2′ Heads 3–9-fld, solitary or clustered; pl not spiny
4. Lf tips gen 3-lobed; pappus not falling in a ring .... **ERIOPHYLLUM**
4′ Lf tips entire; pappus falling in a ring .... **OROCHAENACTIS**
1′ Heads peduncled
5. Phyllaries in 2–many overlapping series, outer << inner
6. Pappus of short scales alternating with longer awns or bristles
7. Ann; lvs linear, gen entire .... **MALPERIA**

7′ Per or subshrub; lvs diamond-shaped, gen few-toothed .... **PLEUROCORONIS**
6′ Pappus of readily deciduous awns or of scales that are often ± dissected into bristles
8. Shrubs; phyllary tips appressed .... **ACAMPTOPAPPUS**
8′ Herbs; phyllary tips often spreading to recurved
9. Involucre gummy; pappus awns readily deciduous .... **GRINDELIA**
9′ Involucre hairy or glandular; pappus scales often ± dissected into bristles .... **LESSINGIA**
5′ Phyllaries in 1–3 series, subequal, weakly overlapping or not, outer ± = inner
10. Heads ± spheric; lvs decurrent .... **HELENIUM**
10′ Heads cylindric to hemispheric; lvs not decurrent
11. Phyllary margins thin, ± scarious, often brownish to purple
12. Pappus scales bristle-tipped; corollas cream-colored to white .... **HYMENOTHRIX**
12′ Pappus scales obtuse to acute; corollas gen yellow
13. Per; pappus scales 12–22 .... **HYMENOPAPPUS**
13′ Ann; pappus scales gen 8 .... **SCHKUHRIA**
11′ Phyllary margins not evidently scarious, gen ± green
14. Corollas white to purple
15. Fr 3–9 mm; pappus scale midrib 0; lvs often toothed or lobed .... **CHAENACTIS**
15′ Fr 10–15 mm; pappus scale midrib well developed; lvs entire .... **PALAFOXIA**
14′ Corollas yellow
16. Outer corollas bilateral, much enlarged, ray-like; pappus scales entire or toothed *Chaenactis glabriuscula*
16′ Outer corollas radial, not enlarged; pappus scales plumose or dissected into bristles
17. Lf teeth and lobes blunt; involucre densely glandular; pappus scales plumose .... *Layia discoidea*
17′ Lf teeth and loves sharp; involucre not glandular; pappus scales dissected into bristles **TRICHOPTILIUM**

Group 7: Heads ligulate; ligules 5-lobed; sap gen milky

1. Receptacle with long chaff scales among fls
2. Lvs mostly basal, not spiny; chaff scales narrow, not enclosing frs .... **HYPOCHAERIS**
2′ Lvs mostly cauline, spiny, thistle-like; chaff scales wide, enclosing frs .... **SCOLYMUS**
1′ Receptacle gen naked (sometimes short bristly or minutely scaly)
3. Pappus 0
4. Heads solitary; st lfless .... **PHALACROSERIS**
4′ Heads few–many in cyme-like infl; st often leafy
5. Outer fr long, curved, spreading .... **RHAGADIOLUS**
5′ Outer fr short, straight, enclosed in involucre until released
6. Lvs mostly basal; corollas white; D .... **ATRICHOSERIS**
6′ Lvs mostly cauline; corollas yellow; CA-FP .... **LAPSANA**
3′ Pappus present
7. Corollas blue
8. Pappus of short scales .... **CICHORIUM**
8′ Pappus of long bristles .... **[2]LACTUCA**
7′ Corollas yellow or white to purple
9. Pappus of scales, stiff awns, or awn-tipped scales, not plumose
10. Shrub; s ChI (San Clemente Island) .... *Stephanomeria blairii*
10′ Herbs; widespread
11. Outer frs enfolded by phyllaries, pappus scales short, deltate; inner frs with pappus scales long, narrow .... **HEDYPNOIS**
11′ Outer frs gen not enfolded by phyllaries; pappus of outer and inner frs ± alike
12. Corollas pink to whitish; sts much-branched — lvs short, linear, entire .... **CHAETADELPHA**
12′ Corollas cream-colored to orange; st branches 0–few
13. Outer phyllaries long, linear; pappus scales stiff, bristle-like; uncommon alien .... **TOLPIS**
13′ Outer phyllaries < or = inner, gen wide (if linear, short); pappus scales few–many, bristle-tipped; native
14. Per, scapose; rosette lvs entire, narrow, long-tapered; heads erect; pappus scales 10–30, narrowly lanceolate, silvery, bristle-tip not plumose .... **NOTHOCALAIS**
14′ Ann or per, scapose or not; lvs entire to lobed; pappus scales 1–many, silvery to blackish, bristle-tip smooth, barbed, or plumose
15. Ann; head erect; involucre glabrous, outer phyllaries not < 1/4 inner; pappus scales 5, lanceolate, silvery bristle short, smooth, from notched scale tip .... **UROPAPPUS**
15′ Ann or per; head ± nodding in bud; involucre glabrous to hairy, outer phyllaries often < 1/4 inner; pappus scales 1–35, variously colored, tip entire to unevenly cut, bristle from tip smooth to plumose
16. Per from fleshy taproot(s); ligules >> involucre; pappus scales 5–35 .... **MICROSERIS**
16′ Ann; ligules ± = involucre; pappus scales 5 (often < 5 in *Microseris douglasii*)
17′ Head gen nodding in bud; pappus scales gen deltate to ovate (if lanceolate, tapered to bristle); lvs basal .... **MICROSERIS**
17. Head not strongly nodding; pappus scales narrowly lanceolate, tip irregularly cut; cauline lvs sometimes present .... **STEBBINSOSERIS**
9′ Pappus of simple bristles or plumose
18. Pappus of plumose bristles or awns

19. Pappus bristles wide at base (scale-like), silvery .... *Microseris nutans*
19′ Pappus of bristle or awns, slender or thick at base, not widened or silvery
20. Pappus bristles very unequal
21. Pappus bristles much longer on 1 side of all frs; phyllary margins papery, much wider than midrib .... ANISOCOMA
21′ Outer fr pappus of short scales, inner fr pappus of long and short bristles; phyllary margins not wide and papery .... [2]LEONTODON
20′ Pappus bristles ± equal
22. Lvs linear to lanceolate, grass-like, entire; involucre 2.5–5 cm
23. Per; phyllary series several; fr beak 0 .... SCORZONERA
23′ Bien; phyllary series 1; fr beak > body .... TRAGOPOGON
22′ Lvs lanceolate to elliptic (or scale-like), larger lvs ± toothed or lobed; involucre 0.5–2 cm
24. Corollas white to pink or lavender; native
25. Fr beaked; corollas white or cream-colored, sometimes red-veined .... RAFINESQUIA
25′ Fr beak 0; corollas pink to pale lavender .... STEPHANOMERIA
24′ Corollas yellow; alien
26. Lvs in basal rosette; heads solitary on scapose peduncles — outer phyllaries inconspicuous .... [2]LEONTODON
26′ Lvs cauline; heads 1–many, not on scapose peduncles
27. Phyllaries in 2 series, outer wide, free, inner narrow; common weed .... PICRIS
27′ Phyllaries in 1 series, wide, bases slightly fused; uncommon .... UROSPERMUM
18′ Pappus of many, smooth or barbed, stiff or soft bristles
28. Fr ± compressed
29. Fr beaked (beak sometimes short, thick) .... [2]LACTUCA
29′ Fr beak 0 .... SONCHUS
28′ Frs not compressed
30. St much-branched, thorny; corollas white to pink .... *Stephanomeria spinosa*
30′ St little-branched, not thorny; corollas yellow, white, or lavender
31. Fr widely cylindric or base ± narrowed, not beaked
32. Involucre narrow; fls 3–5 .... PRENANTHELLA
32′ Involucre wider; fls many
33. Some or all pappus bristles falling from fr — long glandless hairs 0 .... [2]MALACOTHRIX
33′ Pappus bristles persistent
34. Pl long-nonglandular-hairy esp below; heads few or several; sts gen branched .... HIERACIUM
34′ Pl glabrous or puberulent; heads solitary on scapose peduncles
35. Ann; pappus bristle 5 or fewer
35′ Per; pappus bristles > 20
36. Pappus bristles brownish, barbed .... *Microseris borealis*
36′ Pappus bristles white, smooth .... [2]*Nothocalais alpestris*
31′ Fr tapered at both ends, often beaked
37. Fr tip with ring of small scales at base of slender beak .... CHONDRILLA
37′ Fr tip without ring of scales, fr beaked or not
38. Pappus bristles ± all deciduous
39. Fr widely cylindric or tapered at both ends, not beaked .... [2]MALACOTHRIX
39′ Fr short-beaked
40. Fr tapered to beak; lf margins not hard .... CALYCOSERIS
40′ Fr abruptly beaked; lf margins white, hard .... GLYPTOPLEURA
38′ Pappus bristles persistent
41. St branched, leafy (lvs often reduced); beak present or 0 .... CREPIS
41′ Sts unbranched, lfless (exc some ann *Agoseris*)
42. Phyllaries in 2 series; fr beak long, slender .... TARAXACUM
42′ Phyllaries equal or in several ± graded series; fr beak 0 or short to long, slender to stout
43. Pls gen ± hairy; outer phyllaries gen < inner, often striped purplish, speckled or not .... AGOSERIS
43′ Pls ± glabrous (or petioles ciliate); outer phyllaries ± = inner, finely speckled purple — KR, n SNH .... [2]*Nothocalais alpestris*

Group 8: Heads radiate; phyllaries in 1 series, subtending ray fls; receptacle chaffy (at least between ray and disk fls)

1. Phyllary margins ± flat, not clasping or enclosing ray ovaries
2. Pappus 2.5–4 mm, of plumose bristles .... *Layia hieracioides*
2′ Pappus 0 or < 1 mm, of scales or awns
3. Lvs woolly; ray corollas thin, deciduous; receptacle minutely chaffy in center .... *Eriophyllum ambiguum*
3′ Lvs scabrous; ray corollas leathery, persistent; receptacle chaffy throughout, scales awn-tipped SANVITALIA
1′ Phyllary margin folded, ± clasping or enclosing ray ovaries
4. Disk pappus 0
5. Phyllary margins not adjacent or overlapping, ± clasping fr, but not completely enclosing fr
6. Style of disk fl hairy below minutely notched tip .... [2]BLEPHARIPAPPUS

6′ Style of disk fl glabrous below tapered branches
7. Upper lvs and phyllaries tipped by sessile, open-pit glands .... **HOLOCARPHA**
7′ Upper lvs and or phyllary glands 0 or stalked
8. Upper lvs and phyllaries without tack-shaped glands; lower lvs often toothed or lobed ...... [2]**HEMIZONIA**
8′ Upper lvs and phyllaries with tack-shaped glands; lower lvs gen entire
9. Ray fr 10-ribbed, hairy .... [2]**BLEPHARIZONIA**
9′ Ray fr 3-angled, gen glabrous .... [2]**CALYCADENIA**
5′ Phyllary margins adjacent or overlapping, completely enclosing fr
10. Per from long fleshy rhizome .... [2]**HOLOZONIA**
10′ Ann
11. Ray fr compressed side-to-side, 3-angled, or obovoid; phyllary margins folded around angled or rounded ovary face; lower lvs gen entire .... [2]**MADIA**
11′ Ray fr compressed front-to-back; phyllary margins folded across flat ovary face; lower lvs gen toothed or lobed
12. Disk fls 1–2 .... ***Madia minima***
12′ Disk fls gen 6–many
13. Disk fls 6, staminate; heads gen closing by mid-day .... **LAGOPHYLLA**
13′ Disk fls 12–100+, gen fruiting; heads open throughout day .... [3]**LAYIA**
4′ Disk pappus present
14. Disk pappus bristles or scales plumose
15. Per
16. Lvs mostly basal .... **RAILLARDELLA**
16′ Lvs cauline
17. Rays 8–15 .... [2]**MADIA**
17′ Rays 1–3 .... ***Raillardiopsis scabrida***
15′ Ann
18. Ray fr compressed front-to-back, pappus 0; phyllary margins folded across flat ovary face, enfolding fr .... [3]**LAYIA**
18′ Ray fr cylindric to obconic, pappus present; phyllary margins sometimes folded around ovary, partly enclosing fr
19. Style of disk fl hairy below minute notch .... [2]**BLEPHARIPAPPUS**
19′ Style of disk fl glabrous below tapered branches
20. Pl 30–180 cm; infl gen with tack-shaped glands; phyllaries clasping, not enclosing fr [2]**BLEPHARIZONIA**
20′ Pl 5–25 cm; tack-shaped glands 0; phyllaries completely enclosing fr .... ***Madia stebbinsii***
14′ Disk pappus bristles or scales not plumose, sometimes minutely fringed
21. Phyllary margins clasping ovary, not overlapping, partly enclosing fr
22. Upper lvs and or phyllaries tipped by tack-shaped glands; ligules gen deeply and palmately lobed, lobes spreading .... [2]**CALYCADENIA**
22′ Upper lvs and phyllaries without tack-shaped glands; ligules gen ± shallowly 2–3-lobed, lobes ± parallel (spreading in *Osmadenia*)
23. Lvs gen toothed or lobed; disk pappus scales not long-acuminate or bristle-tipped .... [2]**HEMIZONIA**
23′ Lvs entire; some disk pappus scales long-acuminate or bristle-tipped — ligule lobes spreading **OSMADENIA**
21′ Phyllary margins overlapping around ovary, completely enclosing fr
24. Ray fr compressed side-to-side, club-shaped, or ± triangular with angle toward center of head
24′ Ray fr ± cylindric or compressed front-to-back
25. Per from fleshy rhizome .... [2]**HOLOZONIA**
25′ Ann
26. Pappus of 10 blunt, shiny, white scales in 2 series, inner 6–11 mm; fr 10-ribbed, ± cylindric .... **ACHYRACHAENA**
26′ Pappus of bristles or scales < 5 mm; fr not ribbed .... [3]**LAYIA**

Group 9: Heads radiate; phyllary series 2+; receptacle chaffy ± throughout

1. Chaff scales folded around disk ovaries
2. Ray fls fruiting (style present, ovary well developed)
3. Ann; disk fr compressed, winged .... ***Verbesina encelioides***
3′ Per; disk fr gen 4-angled
4. Lvs mostly ± basal, cauline 0 or few; pappus 0 .... **BALSAMORHIZA**
4′ Lvs basal and cauline; pappus of scales or 0 .... **WYETHIA**
2′ Ray fls sterile (ovary vestigial, style gen 0)
5. Fr weakly compressed or not, margin not thin
6. Receptacle widely cylindric to conic .... **RUDBECKIA**
6′ Receptacle flat to convex
7. Pappus 0 .... **HELIOMERIS**
7′ Pappus present
8. Herbs; pappus scales gen readily deciduous .... **HELIANTHUS**
8′ Shrubs; pappus scales often persistent .... **VIGUIERA**
5′ Fr strongly compressed, margin ± thin
9. Fr margin ciliate

10. Subshrubs or shrubs .... **ENCELIA**
10′ Ann or per (sometimes with woody caudex)
11. Lvs all basal or subbasal, gray or silvery, entire .... [2]**ENCELIOPSIS**
11′ Lvs mostly cauline, green, entire or toothed .... ***Geraea canescens***
9′ Fr margin not ciliate
12. Lvs all basal or subbasal, silvery or gray; fr margin corky .... [2]**ENCELIOPSIS**
12′ Some or all lvs cauline, green; fr margin thin or winged
13. Per; lvs basal and cauline, entire .... **HELIANTHELLA**
13′ Subshrub; lvs cauline, some often toothed .... ***Verbesina dissita***
1′ Chaff scales flat, linear to ovate, not folded around disk ovaries
14. Ray corollas white to purple
15. Disk corollas white
16. Lvs finely dissected, aromatic; heads many in flat-topped clusters; ligules wide .... **ACHILLEA**
16′ Lvs entire or toothed, not aromatic; heads solitary or in small cymes; ligules narrowly linear .... **ECLIPTA**
15′ Disk corollas yellow
17. Lvs entire or toothed
18. Lvs opposite, ± ovate .... **GALINSOGA**
18′ Lvs alternate, linear .... [2]**RIGIOPAPPUS**
17′ Lvs pinnately lobed to compound
19. Phyllaries in 2 very different series (inner ± membranous)
20. Lvs pinnate; lflets flat< serrate; ray fls short, inconspicuous .... ***Bidens pilosa***
20′ Lvs 2-pinnately dissected, segments narrowly linear, entire; ray fls long, conspicuous .... ***Cosmos bipinnata***
19′ Phyllaries graduated in several series, unequal, margins scarious or transparent
21. Base of corolla narrow .... **ANTHEMIS**
21′ Base of corolla wide, enclosing fr tip .... **CHAMAEMELUM**
14′ Ray corollas yellow to orange-red (sometimes multi-colored)
22. Phyllaries ± alike, in 2+ series
23. Phyllaries graduated in several series, unequal, margins scarious or transparent — pappus 0 ***Anthemis tinctoria***
23′ Phyllaries not strongly graduated, ± subequal, margins not scarious or transparent
24. Lvs oblong to ovate, entire to pinnately lobed; pappus scales awn-tipped .... **GAILLARDIA**
24′ Lvs linear, entire; pappus awns stiff, tapered .... [2]**RIGIOPAPPUS**
22′ Phyllaries in 2 very different series (inner ± membranous)
25. Receptacle convex or conic; pappus 0 .... **GUIZOTIA**
25′ Receptacle flat; pappus often present
26. Pappus 0
27. Fr < 7 mm, ± compressed, not beaked .... [2]**COREOPSIS**
27′ Fr 16–28 mm, subcylindric, ± 4-angled, often beaked .... [2]***Cosmos sulphurea***
26′ Pappus present
28. Pappus of flat scales .... [2]**COREOPSIS**
28′ Pappus of barbed awns
29. Fr subcylindric, ± 4-angled, often beaked .... [2]***Cosmos sulphurea***
29′ Fr ± compressed, ± wedge-shaped, not beaked
30. Inner phyllaries free; lflets wide .... **BIDENS**
30′ Inner phyllaries fused partway; lflets thread-like .... **THELESPERMA**

Group 10: Heads radiate; receptacle naked; pappus 0 or low crown

1. Lvs opposite, at least below
2. Upper lvs alternate
3. Ray corollas < 1 mm .... [2]**AMBLYOPAPPUS**
3′ Ray corollas 2–20+ mm
4. Ligules white
5. Ligules not red-veined below; PR, e DSon .... [2]***Eriophyllum wallacei***
5′ Ligules red-veined below; SCoRI, TR, SnJt .... [2]***Syntrichopappus lemmonii***
4′ Ligules yellow
6. Ray corollas without small lobe opposite ligule .... [2]**ERIOPHYLLUM**
6′ Ray corollas with small lobe opposite ligule .... [2]**MONOLOPIA**
2′ Upper lvs opposite
7. Phyllaries in 2+ series, widely overlapping; per, fleshy, coastal .... **JAUMEA**
7′ Phyllaries in 1–2 series, barely overlapping; ann or per, gen not fleshy, coastal or interior
8. Pl white-hairy and sessile-glandular; ligules persistent; disk fls staminate .... **WHITNEYA**
8′ Pl glabrous or ± hairy; ligules deciduous; disk fls gen fruiting
9. Ray fl gen 1 per head; ligules < 1 mm .... **FLAVERIA**
9′ Ray fls 4–many; ligules gen >> 1 mm
10. Lf margins without embedded oil glands, base not bristly-ciliate; herbage gen not scented; spring fl .... **LASTHENIA**
10′ Lf margins with embedded oil glands, base bristly-ciliate; herbage strongly scented; summer/autumn fl .... **PECTIS**
1′ Lvs alternate througout

11. Ligules 1 mm or less
12. Shrub .... *Artemisia bigelovii*
12′ Herb
13. Lvs finely dissected; phyllaries unequal, overlapping in several series, scarious-margined .... [2]**TANACETUM**
13′ Lvs entire to ± lobed; phyllaries ± equal, in 1–3 series, margins barely scarious or not
14. Phyllaries in 1 series; lower lvs often toothed or lobed
15. Pl glabrous or sticky-glandular; fr 1.5–2 mm .... [2]**AMBLYOPAPPUS**
15′ Pl ± woolly; fr ± 3 mm .... **LEMBERTIA**
14′ Phyllaries in 2–3 series; lvs linear, entire
16. Ligules white; inner phyllaries flat .... [2]**PENTACHAETA**
16′ Ligules pale yellow, becoming purplish; inner phyllaries ± enfolding ray frs .... **RIGIOPAPPUS**
11′ Ligules > 1 mm
17. Ray corollas white to purple (sometimes multi-colored)
18. Phyllaries unequal, in 2+ series, gen conspicuously overlapping, scarious-margined
19. Subshrub — disk fr 1–3-winged .... **ARGYRANTHEMUM**
19′ Herbs
20. Ann; ligules multi-colored; ray fr 2-winged .... *Chrysanthemum carinatum*
20′ Per; ligules white; ray fr unwinged
21. Heads solitary on long peduncles; lvs serrate or pinnately lobed .... **LEUCANTHEMUM**
21′ Heads gen many on short peduncles; lvs 2–3-pinnately dissected .... *Tanacetum parthenium*
18′ Phyllaries ± equal, in 1–2 series, gen little overlapping, gen not scarious-margined
22. Ligules 25–40 mm; coarse per .... **OSTEOSPERMUM**
22′ Ligules < 10 mm; ann or slender per
23. Lvs widely ovate, coarsely dentate or palmately lobed, abruptly petioled .... *Perityle emoryi*
23′ Lvs linear to obovate, entire to pinnately lobed (tip sometimes 3-lobed), sessile or tapered to petiole
24. Lower lvs pinnately lobed .... [2]**BLENNOSPERMA**
24′ Lvs entire
25. Per; lawn weed .... **BELLIS**
25′ Ann; dry habitats
26. Pl tomentose or becoming glabrous
27. Ligules not red-veined below; PR, w DSon .... [2]*Eriophyllum wallacei*
27′ Ligules red-veined below; SCoRI, TR, SnJt .... [2]*Syntrichopappus lemmonii*
26′ Pl short-hairy
28. Pl prostrate .... **MONOPTILON**
28′ Pl ± erect .... [2]**PENTACHAETA**
17′ Ligules yellow to orange
29. Phyllaries unequal, overlapping in several series
30. Phyllary margins conspicuously scarious
31. Pl sticky-glandular .... **HETERANTHEMIS**
31′ Pl not glandular
32. Heads solitary or in open, leafy cymes; outer frs 3-angled .... **CHRYSANTHEMUM**
32′ Heads in dense, flat-topped clusters; frs cylindric .... [2]**TANACETUM**
30′ Phyllary margins not or barely scarious
33. Lvs widely ovate, serrate, puberulent; phyllaries wide, outer spreading or reflexed .... **VENEGASIA**
33′ Lvs oblong to obovate, pinnately lobed, tomentose at least below; phyllaries narrow, appressed or tips spreading
34. Ligules yellow throughout .... **ARCTOTHECA**
34′ Ligules purple at base .... **VENIDIUM**
29′ Phyllaries ± equal, in 1–3 series, slightly overlapping or not
35. Ray fr prickly, knobby, or papillate on back; phyllary margins gen narrowly scarious
36. Involucre 5–8 mm .... [2]**BLENNOSPERMA**
36′ Involucre 10–20 mm
37. Ray fr strongly curved (crescent-shaped or almost loop-like); disk fls staminate .... **CALENDULA**
37′ Ray fr 3-angled, not curved; disk fls fruiting, fr flat .... **DIMORPHOTHECA**
35′ Ray fr smooth on back; phyllary margins gen not scarious
38. Heads ± spheric; phyllaries spreading or reflexed .... **HELENIUM**
38′ Heads cylindric to hemispheric; phyllaries erect or ascending
39. Pl green, glabrous to strigose, glandular or puberulent
40. Phyllaries in 2–3 series; pl glandular .... **BAHIA**
40′ Phyllaries in 1 series; pl ± glabrous .... *Blennosperma nanum*
39′ Pl ± gray-tomentose
41. Ligules when dry persistent, papery, reflexed .... **BAILEYA**
41′ Ligules when dry withering, deciduous
42. Ray corollas with small lobe opposite ligule .... [2]**MONOLOPIA**
42′ Ray corollas without small lobe opposite ligule
43. Fr 4–5-angled .... [2]**ERIOPHYLLUM**
43′ Fr ± compressed .... **PSEUDOBAHIA**

Group 11: Heads radiate; receptacle gen naked; pappus of scales or awns

1. Lvs and phyllaries dotted or streaked with embedded, translucent oil glands — herbage strongly scented
2. Ligules white, sometimes purple-tinged .... **NICOLLETIA**
2′ Ligules yellow to red
3. Phyllaries in 1 series
4. Phyllaries free, falling with ray frs; disk pappus gen of bristles .... **PECTIS**
4′ Phyllaries fused, persistent; disk pappus of scales .... **TAGETES**
3′ Phyllaries in 2–3 series — pappus scales ± divided into bristles or awns
5. Heads ± sessile .... **DYSSODIA**
5′ Heads evidently peduncled
6. Lvs toothed or divided into ± flat segments; involucre 10–20 mm .... **ADENOPHYLLUM**
6′ Lvs divided into stiff, needle-like segments; involucre 5–6 mm .... **THYMOPHYLLA**
1′ Lvs and phyllaries without embedded oil glands (sometimes glandular-hairy or gland-dotted)
7. Ligules white to purple
8. Per
9. Receptacle bristly — pappus scales wide, overlapping .... **ARCTOTIS**
9′ Receptacle naked
10. Lvs compound; pappus scales short .... ***Tanacetum parthenium***
10′ Lvs simple; pappus scales long, bristle-like .... **TOWNSENDIA**
8′ Ann
11. Lvs wide, palmately lobed or toothed .... ***Perityle emoryi***
11′ Lvs narrow, entire
12. Pl tomentose
13. Ligules 1.5–2.5 mm; pappus scales 2 .... [2]**EATONELLA**
13′ Ligules 3–7 mm; pappus scales ± 10 .... ***Eriophyllum lanosum***
12′ Pl puberulent or short-stiff-hairy
14. Pl prostrate; ray corollas 5–12 mm .... **MONOPTILON**
14′ Pl erect; ray corollas < 2 mm .... [2]**RIGIOPAPPUS**
7′ Ligules yellow to red (sometimes multi-colored)
15. Receptacle bristly .... **GAILLARDIA**
15′ Receptacle smooth or minutely hairy
16. Phyllaries in several series, strongly graduated
17. Lvs tomentose below; sts creeping, herbaceous .... **ARCTOTHECA**
17′ Lvs glabrous to short-hairy; sts erect, woody below
18. Phyllary tips spreading to recurved; involucre gummy — pappus of 2–8 slender deciduous awns .... **GRINDELIA**
18′ Phyllary tips ± appressed; involucre not or barely gummy
19. Ann; fr winged; lvs toothed or lobed .... **HETERANTHEMIS**
19′ Per to shrub; fr not winged; lvs entire
20. Heads ± spheric; disk fls 30–80 .... **ACAMPTOPAPPUS**
20′ Heads narrow; disk fls 1–13
21. Lvs elliptic to obovate; disk pappus scales twisted, bristle-like .... **AMPHIPAPPUS**
21′ Lvs linear; disk pappus scales straight, wide, flat .... **GUTIERREZIA**
16′ Phyllaries in 1–3 series, ± equal, not strongly graduated
22. Lvs basal
23. Heads 3.5–8 cm diam; ray base gen dark-spotted; fr hairs concealing pappus; sap milky .... **GAZANIA**
23′ Heads gen < 3 cm diam; ray base not dark-spotted; fr hairs not concealing pappus; sap clear
24. Pl ± densely glandular-hairy .... ***Hulsea vestita***
24′ Pl not glandular-hairy, often dotted with sessile resin glands .... ***Hymenoxys acaulis***
22′ Some or all lvs cauline
25. Lvs gen opposite throughout .... **LASTHENIA**
25′ Lvs gen alternate (sometimes opposite below)
26. Old ray corollas papery, persistent when dry .... **PSILOSTROPHE**
26′ Old ray corollas withering, deciduous
27. Pappus of rigid, tapered awns; lvs linear, entire .... [2] **RIGIOPAPPUS**
27′ Pappus of ± thin scales; lvs various, entire to lobed
28. Phyllaries in 1 series
29. Ligules < 1 mm
30. Pl dotted with sessile resin glands; fr 3–4-angled .... **AMBLYOPAPPUS**
30′ Pl densely woolly; fr compressed .... **LEMBERTIA**
29′ Ligules 1.5–10+ mm
31. Disk fr ± compressed; ann .... [2]**EATONELLA**
31′ Disk fr (exc sometimes outermost) ± cylindric, club-shaped, or 3–4-angled; ann to shrub .... **ERIOPHYLLUM**
28′ Phyllaries in 2–3 ± equal series
32. Lvs deeply linear-lobed or 1–3 × dissected
33. Pl glandular; lvs 1–3 × pinnately dissected .... **BAHIA**
33′ Pl not glandular; lvs dissected into linear lobes .... **HYMENOXYS**

32′ Lvs entire to shallowly lobed
34. Pappus scales 4, in 2 unequal pairs; pl glandular-hairy .... **HULSEA**
34′ Pappus scales 5–10, equal or not; pl gen not glandular
35. Lvs not thread-like or decurrent; disk hemispheric; phyllaries gen spreading (or reflexed in age) .... **DUGALDIA**
35′ Lvs thread-like or decurrent; disk ± spheric; phyllaries gen ± reflexed .... **HELENIUM**

Group 12: Heads radiate; ligules white to purple; receptacle naked; pappus of bristles

1. Lvs palmately lobed or dentate
2. Ann; lvs cauline; pappus bristle 1; D .... ***Perityle emoryi***
2′ Per; lvs basal, cauline lvs reduced to sheathing scales; pappus bristles many; moist forests .... **PETASITES**
1′ Lvs entire to pinnately lobed
3. Ligules inconspicuous, barely surpassing disk or not
4. Phyllaries without orange resin-filled veins
5. Pappus single; phyllaries often ± thickened near tip .... [4]**ASTER**
5′ Pappus double, outer of short bristles or scales; phyllaries not thickened near tip
6. Disk wide, flat; disk fls many; fr not beaked .... [4]**ERIGERON**
6′ Disk narrow; disk fls 15–25; fr short-beaked .... [2]**TRACYINA**
4′ Phyllaries with 1–3 orange resin-filled veins
7. Resin-filled phyllary veins gen 3
8. Subshrub; branches often thorny .... [3]**CHLORACANTHA**
8′ Herb; branches not thorny .... **TRIMORPHA**
7′ Resin-filled phyllary vein 1
9. Disk narrow, disk fls few; pappus single .... **CONYZA**
9′ Disk wide, flat, disk fls many; pappus double .... [4]**ERIGERON**
3′ Ligules conspicuous, gen much exceeding disk
10. Pappus bristles alternating with well-developed scales — pl ill-scented .... **NICOLLETIA**
10′ Pappus of bristles only or scales very short
11. Pappus bristles < 10
12. Pappus double, of 5 bristles alternating with 5 narrow scales or of 1 plumose-tipped bristle and a low crown .... **MONOPTILON**
12′ Pappus single, of 3–5 bristles
13. Fr beakless .... **PENTACHAETA**
13′ Fr short-beaked .... [2]**TRACYINA**
11′ Pappus bristles gen > 20
14. Bristles flattened, barbed to subplumose .... **TOWNSENDIA**
14′ Bristles cylindric, smooth to barbed
15. Main phyllaries ± equal in 1–3 series
16. Pappus ± 1 mm, bristles fused at base; pl woolly .... **SYNTRICHOPAPPUS**
16′ Pappus gen >> 1 mm, bristles free; pl gen not woolly
17. Pappus gen double, outer of short bristles or scales, inner of soft bristles; rays gen < 2 mm wide .... [4]**ERIGERON**
17′ Pappus single; rays often > 2 mm wide
18. Main phyllaries subequal, in 2–3 series .... [4]**ASTER**
18′ Main phyllaries equal, in 1 series (often with very short outer phyllaries) .... **SENECIO**
15′ Main phyllaries unequal, overlapping in 2–several series
19. Shrubs or subshrubs
20. Phyllary veins orange, resin-filled; pl often thorny .... [3]**CHLORACANTHA**
20′ Phyllary vein not orange or resin-filled; pl unarmed
21. Ray fls sterile (ovary vestigial); pappus brown or reddish .... [3]**LESSINGIA**
21′ Ray fls fruiting; pappus gen ± white
22. Ray fls 15–40 mm; lvs often sharply toothed
23. Heads 15–20 mm diam, gen in open cymes .... ***Machaeranthera canescens***
23′ Heads 35–70+ mm diam, gen solitary .... [2]**XYLORHIZA**
22′ Ray fls 5–10 mm; lvs entire
24. Slender subshrub, ± strigose, ± glandular .... [2]**CHAETOPAPPA**
24′ Stiff shrub, glabrous, sticky-resinous .... ***Ericameria gilmanii***
19′ Herbs
25. Ray fls sterile (style 0, ovary ± vestigial)
26. Disk style branch tips densely tufted with stiff yellow hairs; pappus brown or reddish .... [2]**LESSINGIA**
26′ Disk style branches without stiff yellow hairs; pappus ± white .... [2]**MACHAERANTHERA**
25′ Ray fls fruiting, style present
27. Pappus double, outer of short bristles or scales, inner of long bristles
28. Resin-filled phyllary veins gen 3; sts often thorny .... [3]**CHLORACANTHA**
28′ Resin-filled phyllary veins 0–1; sts unarmed
29. Rays often wider than narrowly linear, 8–16 .... ***Aster scopulorum***
29′ Rays narrowly linear, often very many .... [4]**ERIGERON**
27′ Pappus single, of ± equal bristles

30. Phyllaries ± herbaceous throughout (or margin scarious to ± whitish)
 31. Lvs gen 3–20 cm; pls often > 20 cm; style appendages lanceolate to awl-shaped, acute ..... [4]**ASTER**
 31′ Lvs gen 1–10 mm; pls gen < 15 cm; style appendages oblong to ovate, obtuse .... [2]**CHAETOPAPPA**
30′ Phyllary tips ± herbaceous, bases white to straw-colored
 32. Per, gen from rhizome; lvs entire to toothed; phyllary tips gen appressed .... [4]**ASTER**
 32′ Ann to per from taproot; lvs gen toothed, teeth sometimes bristle-tipped; phyllary tips often spreading
  33. Heads gen < 3 cm diam, often in cyme-like clusters .... [2]**MACHAERANTHERA**
  33′ Heads 3.5–6 cm diam, solitary .... [2]**XYLORHIZA**

Group 13: Heads radiate; ligules yellow to red; receptacle naked; pappus of bristles

1. Lvs opposite at least below
 2. Lvs and phyllaries dotted or streaked with embedded, translucent oil glands, very strongly scented
  3. Lvs linear, entire, base bristly-ciliate; phyllaries 8, in 1 series .... **PECTIS**
  3′ Lvs sharply toothed to pinnately dissected; phyllaries in 2 series
   4. Subshrubs; peduncles 2–15 cm .... [2]**ADENOPHYLLUM**
   4′ Ann; peduncles 1–5 mm .... **DYSSODIA**
 2′ Lvs and phyllaries without embedded oil glands, not or faintly scented
  5. Trailing fleshy per; pappus bristles 1–5; coastal saline habitats .... **JAUMEA**
  5′ Erect ann or per, not fleshy; pappus bristles many; inland
   6. Per; involucre gen 8–20 mm; pappus >> 2 mm, free .... **ARNICA**
   6′ Ann; involucre 5–7 mm; pappus 1–2 mm, fused at base .... [2]**SYNTRICHIPAPPUS**
1′ Lvs alternate througout
 7. Ann or bien
  8. Ray pappus 0
   9. Phyllaries tipped by a tack-shaped gland; disk fls staminate .... ***Lessingia occidentalis***
   9′ Phyllaries without tack-shaped glands
    10. Phyllaries in 1 series, fused at base .... [2]**CROCIDIUM**
    10′ Phyllaries in 2–several series, free .... [2]**HETEROTHECA**
  8′ Ray (and disk) frs with pappus
   11. Main phyllaries ± equal, in 1–3 series
    12. Fr compressed .... [2]**PENTACHAETA**
    12′ Fr not compressed
     13. Rays many; pappus double, outer of short scales, inner of bristles .... [2]**PULICARIA**
     13′ Rays 5–13; pappus single
      14. Receptacle conic; phyllaries fused at base .... [2]**CROCIDIUM**
      14′ Receptacle flat; phyllaries free
       15. Phyllaries gen ± 13, not enfolding ray frs .... [3]**SENECIO**
       15′ Phyllaries gen 5, ± enfolding ray frs .... [3]**SYNTRICHOPAPPUS**
   11′ Main phyllaries unequal, in 2–several series
    16. Pappus bristles flat, deciduous; involucre gummy .... [3]**GRINDELIA**
    16′ Pappus bristles ± cylindric, gen persistent; involucre not gummy
     17. Fr short-beaked; phyllaries deciduous when head dry .... **TRACYINA**
     17′ Fr not beaked; phyllaries persistent on old heads
      18. Lvs ± entire .... [2]**PENTACHAETA**
      18′ Lvs serrate to 1–2-pinnately lobed
       19. Pl 3–25 cm; disk corollas 4.5–5.5 mm .... ***Machaeranthera gracilis***
       19′ Pl 50–150 cm; disk corollas ± 7 mm .... **PRIONOPSIS**
 7′ Per to shrubs
  20. Shrubs or subshrubs
   21. Outer phyllaries with an embedded, translucent oil gland near tip; pls pungently ill-scented .... [2]**ADENOPHYLLUM**
   21′ Outer phyllaries without oil glands; pls scented or not
    22. Main phyllaries in 1 series, equal (often a few, gen much shorter, outer phyllaries present) ...... [3]**SENECIO**
    22′ Main phyllaries in 2–7 series, often graduated, often unequal
     23. Pappus bristles flattened; phyllaries ± ovate
      24. Phyllary tips recurved to coiled; involucre gummy; ray fls 20–60 .... [3]**GRINDELIA**
      24′ Phyllary tips ± appressed; involucre ± resinous but not gummy; ray fls 1–14
       25. Rays 5–14, >> involucre; disk fls 30–80, fruiting .... ***Acamptopappus shockleyi***
       25′ Rays 1–2, barely > involucre; disk fls 3–7, staminate .... **AMPHIPAPPUS**
     23′ Pappus bristles ± cylindric; phyllaries gen linear to narrowly lanceolate
      26. Woody sts ± prostrate, pl cushion-forming .... [2]**STENOTUS**
      26′ Woody sts gen erect
       27. Phyllaries ± graduated in 2–4 series, bases loose to tightly appressed, tips weakly thickened, or not, green or straw-colored, never tomentose; lvs entire .... **ERICAMERIA**
       27′ Phyllaries strongly graduated in 5–7 series, bases tightly appressed, straw-colored, tips clearly thickened, green (sometimes phyllaries densely tomentose); lvs often toothed
        28. Upper lvs gen not much reduced .... **HAZARDIA**

28′ Upper lvs reduced to scale-like bracts .... *Machaeranthera pinnatifida*
20′ Herbs
29. Main phyllaries in 1 series, equal (often with few, very short, outer phyllaries) .... [3]SENECIO
29′ Phyllaries in 2+ series, subequal to strongly graduated
30. Disk pappus double, outer of very short bristles or scales
31. St simple, erect; phyllaries ± equal — fr 2-ribbed .... ERIGERON
31′ St gen branched, prostrate to erect; phyllaries gen ± unequal
32. Ligules 3–12 mm; disk fr 1.5–4 mm, ± flat; pappus 3–7 mm .... [2]HETEROTHECA
32′ Ligules 1.5–2 mm; disk fr ± 1 mm, 5-ribbed; pappus 2–3 mm .... [2]PULICARIA
30′ Disk pappus single
33. Pappus bristles flat, deciduous; involucre strongly gummy, esp in bud .... [3]GRINDELIA
33′ Pappus bristles ± cylindric, gen persistent; involucre not strongly gummy
34. Lvs toothed or lobed
35. Heads gen small, in racemes or panicles, often clustered on 1 side of branches .... [2]SOLIDAGO
35′ Heads gen not small, not in 1-sided clusters
36. Pl from stout taproot; basal rosette well developed .... [2]PYRROCOMA
36′ Pl from slender taproot or caudex; rosette 0 or poorly developed
37. Fr 2–3 mm, obconic; lf teeth or lobes bristle- or minutely spine-tipped; upper lvs scale-like .... MACHAERANTHERA
37′ Fr 3–10 mm, ± cylindric or compressed; lf margin not bristle- or minutely spine-tipped; upper lvs little reduced
38. Fr 5–10 mm, glabrous; phyllaries in 4–5 series, strongly graduated .... *Hazardia whitneyi*
38′ Fr 3–5 mm, hairy; phyllaries in 3–4 series, subequal to weakly graduated .... TONESTUS
34′ Lvs entire
39. Phyllaries in vertical ranks; disk fls staminate .... PETRADORIA
39′ Phyllaries not in vertical ranks; disk fls fruiting
40. Heads small, in clusters at branch tips; infl leafy, panicle-like; st erect, from rhizome; fr ± 1 mm .... EUTHAMIA
40′ Heads 1–several, small to large; infl various; st from taproot, caudex, or rhizome; fr gen >> 1 mm
41. Pl from stout taproot — basal rosette well developed .... [2]PYRROCOMA
41′ Pl from ± branched caudex or rhizome
42. Pls densely long-stalked-glandular .... *Tonestus lyallii*
42′ Pls glandless or short-stalked-glandular
43. Ligules 1–6 mm; cauline lvs distributed well up st .... [2]SOLIDAGO
43′ Ligules 7–12 mm; cauline lvs gen only at st base .... [2]STENOTUS

ACAMPTOPAPPUS GOLDENHEAD

Meredith A. Lane

Subshrubs, appearing glabrous. **STS** decumbent, widely spreading or erect, striate; old growth gray, bark sometimes shreddy with age; new growth white below, tips green. **LVS** simple, sometimes in axillary clusters below, spreading-ascending to appressed-erect, linear to narrowly oblanceolate, gen minutely spine-tipped, pale green to light gray-green. **INFL**: heads radiate or discoid, 1–many in rounded to ± flat-topped clusters, very small in bud, expanding rapidly in fr; involucres widely bell-shaped to nearly spheric; phyllaries 20–24 in 2–3 series, ovate to ovate-elliptic, bases cream-yellow, tips green, margins scarious; receptacle deeply pitted, with projections between fls but not chaffy. **RAY FLS** present or 0; ligules yellow. **DISK FLS** many; corollas funnel-shaped, yellow, sinuses deep, lobes spreading to reflexed, style-branch appendages lanceolate. **FR** densely long-hairy, hairs white, bronze, or brownish; pappus of 20–25 wide, stiff, widely spreading bristles, slightly > fr. 2 spp.: sw US. (Greek: unbending pappus) [Lane 1988 Madroño 35:247–265]

1. Ray fls present; involucres bell-shaped to hemispheric .... *A. shockleyi*
1′ Ray fls 0; involucres hemispheric to spheric .... *A. sphaerocephalus*
2. Sts and lvs rough-puberulent .... var. *hirtellus*
2′ Sts and lvs glabrous (or lf margins sparsely rough-puberulent) .... var. *sphaerocephalus*

A. shockleyi A. Gray (p. 191) **STS** decumbent to ascending, gen < 4 dm, gen minutely hairy to scabrous. **LVS** < 2.5 cm, < 4 mm wide, gen oblanceolate, densely, minutely hairy to scabrous. **INFL**: heads few, terminal; involucre bell-shaped to hemispheric. **RAY FLS** 5–14; corollas < 2 cm, ligules < 6 mm wide. **DISK FLS** 30–80; corollas 2–5 mm. **FR** < 5 mm. $2n$=18. Slopes and ridges; 500–2000 m. SNE, DMoj; s NV. ❀TRY.

A. sphaerocephalus (A. Gray) A. Gray (p. 191) **STS** much-branched, ascending to erect, gen < 1 m. **LVS** < 1.5 cm, < 3 mm wide, gen linear to oblanceolate. **INFL**: heads very many, borne singly or clustered; involucre hemispheric to spheric. **RAY FLS** 0. **DISK FLS** 13–27; corollas 2–3.5 mm. **FR** < 3 mm. $2n$=18. Gravelly or rocky soils on flats or slopes in deserts to juniper woodlands; 60–2200 m. Teh, SnGb, SnBr, PR, SNE, D; to UT, s NV, AZ. Vars. sometimes occur together and intergrade.

var. ***hirtellus*** S.F. Blake Herbage ± densely rough-puberulent. **STS** gen < 6 dm. Habitats of sp.; < 1600 m. Teh, SnGb, SnBr, SNE, DMoj; AZ, s NV.

var. ***sphaerocephalus*** Herbage glabrous (or lf margins sparsely rough-puberulent). **STS** gen < 1 m. Habitats and range (exc SNE) of sp.; < 2200 m.

ACHILLEA YARROW, MILFOIL

Per, strongly scented. **LVS** alternate, simple to 3-pinnately dissected, ± hairy, ± reduced upward. **INFL**: heads gen radiate, many, in flat-topped clusters; involucre bell-shaped or ovoid; phyllaries graded in 3–4 unequal series, ovate, obtuse; margins membranous; receptacle flat to rounded; chaff scales narrow, transparent. **RAY FLS** few; ligules short, round, white, pink, or yellow. **DISK FLS** ± many; corollas short, white to purple or yellow. **FR** oblong to obovate, compressed, thick-margined, glabrous; pappus 0. ± 85 spp.: N.Am, Eurasia, n Afr. (Greek: Achilles of ancient mythology) [Tyrl 1975 Brittonia 27:187–196]

A. millefolium L. (p. 191) Pl 10–200 cm. **LVS** very finely 3-pinnately divided; cauline lvs ± clasping. **INFL**: phyllaries 4–9 mm. **RAY FLS** gen 3–8; ligules 2.5–4 mm, ovate to round, white to pink. **DISK FLS** 15–40; corollas 2–3 mm, white to pink. **FR** ± 2 mm. $2n$=36,45,54,63,72. Many habitats; < 3500 m. CA (exc D); circumboreal. [*A. borealis* Bong. sspp. *arenicola* (A.A. Heller) Keck, *californica* (Pollard) Keck; *A. lanulosa* Nutt. sspp. *alpicola* (Rydb.) Keck, *lanulosa*; *A. m.* vars. *alpicola* (Rydb.) Garrett, *arenicola* (A.A. Heller) Nobs, *borealis* (Bong.) Farwell, *californica* (Pollard) Jepson, *gigantea* (Pollard) Nobs, *lanulosa* (Nutt.) Piper, *littoralis* Nobs, *pacifica* (Rydb.) G.N. Jones, *puberula* (Rydb.) Nobs] Highly variable polyploid complex; lf size and hairiness esp variable. ❀SUN,DRN:**4–6,17**&IRR:**1-3,7–9,14–16,18–24**& part SHD:**10**,11–13;GRNCVR;CVS; rather INV; also STBL.

ACHYRACHAENA BLOW-WIVES

1 sp. (Greek & Latin: chaffy achene, from fr)

A. mollis Schauer (p. 191) Ann 2.5–40 cm, ± soft-hairy. **ST** erect, gen few-branched. **LVS** simple, sessile, ± sheathing; lower opposite and fused around st; upper alternate; blades 2–12 cm, linear, obtuse, entire or few-toothed. **INFL** soft-hairy, ± glandular; heads radiate, 1–few; peduncles (0.5)3–11 cm, bractless; involucre 1–2 cm, cylindric to ellipsoid or narrowly bell-shaped; phyllaries 3–8 in 1 series, each tightly enwrapping ovary of a ray fl; chaff scales ± equal, in 1 involucre-like series, gen slightly < phyllaries, flat, widely membrane-margined. **RAY FLS** 3–8; corollas 10–12 mm, yellow to red; ligules 5–6 mm, inconspicuous, slightly exceeding phyllaries. **DISK FLS** 6–35; corollas 6–10 mm, yellow to red, tube = throat, lobes 0.5–0.8 mm; style branches tapered, short-bristly. **FR**: ray achenes ± cylindric, smooth, black, pappus 0; disk achenes 4.5–8.5 mm, 10-ribbed, scabrous, black, pappus of 10 oblong, blunt, shining, white scales in 2 series, 3–6 mm and 6–11 mm, scales widely spreading in fr. $2n$=16. Common. Grassy areas; < 900 m. CA-FP; s OR, n Baja CA. ❀SUN,DRN:**7–9**,10,11,**14–24**.

ACOURTIA

Per. **STS** erect or spreading. **LVS** simple, alternate, sessile, pinnately veined. **INFL**: heads discoid but often appearing radiate, solitary to many in panicles; involucre cylindric to bell-shaped; phyllaries graduated in several series; receptacle naked. **FLS** few–many; corollas white to pink or purple, strongly 2-lipped, inner lip deeply 2-lobed, recurved or coiled, outer lip entire to shallowly 3-lobed, spreading, ligule-like; anther tips lance-oblong, basal appendages stiff, tail-like; style tips rounded to truncate. **FR** cylindric, ribbed, glandular; pappus of many bristles. ±40 spp.: N.Am. (Mrs. A'Court, English amateur botanist) [Reveal & King 1973 Phytologia 27:228–232]

A. microcephala DC. (p. 191) **STS** several from woody caudex, 6–16 dm, very leafy, gen branched only in upper half. **LVS** 2.5–15 cm, widely ovate to elliptic or oblong; base truncate to widely clasping; tip acute to widely obtuse, finely dentate; surfaces densely glandular. **INFL**: heads many; peduncles 1–10 mm; involucre 5–10 mm diam; phyllaries 7–10 mm, linear to lanceolate, acute or obtuse, glandular. **FLS** 10–20; corollas 8–11 mm, white to pink-purple. **FR** 1.5–4 mm; pappus 7–10 mm. $2n$=54. Shrubby and wooded slopes, esp after fires; < 1550 m. SCoR, SW; nw Baja CA. [*Perezia m.* (DC.) A. Gray] ❀STBL.

ACROPTILON RUSSIAN KNAPWEED

David J. Keil & Charles E. Turner

1 sp.: Eurasia. (Greek: feather-tipped, from pappus bristles)

A. repens (L.) DC. (p. 195) RUSSIAN KNAPWEED Per 3–10 dm, from dark rhizome. **STS** erect; branches ascending, ± cobwebby-tomentose. **LVS** below middle 4–10 cm, oblong, 1–2-pinnately lobed, above middle 1–3 cm, linear to narrowly lanceolate, entire to toothed; blades glabrous to ± tomentose. **INFL**: heads discoid in ± flat-topped or panicle-like clusters, leafy; involucre 10–14 mm, ± ovoid; phyllaries in several unequal series, entire, ± soft-hairy, inner narrower, tips widely scarious; receptacle bristly; fls ± 30. **FL**: corolla ± 15 mm, white to blue, tube very slender, abruptly wider, lobes linear; anther bases short-tailed, tips oblong; style with minutely hairy, swollen node above, papillate above node, branches very short, tips triangular. **FR** 3–4 mm, obovoid, slightly compressed, glabrous; pappus bristles many, 6–8 mm, ± deciduous, barbed below, short-plumose above. $2n$=26. Fields, roadsides, cult ground; < 1900 m. CA (exc wettest NW, driest GB, D); N.Am; native to c Asia. [*Centaurea r.* L.] NOXIOUS WEED.

ADENOCAULON

Ann, per. **STS** slender; base ± leafy. **LVS** alternate, reduced upward; lower lvs long-petioled; blades wide, thin. **INFL**: heads small, disciform, in open panicle-like cymes; involucre inconspicuous; phyllaries in 1–2 series, ± equal, reflexed in fr, deciduous; receptacle convex, naked. **PISTILLATE FLS** few in 1 series; corolla deeply 4–5-lobed, white, soon deciduous. **DISK FLS** few, staminate, ± persistent; corolla cylindric, white; anther bases sagittate, tips narrowly triangular; style undivided. **FR** club-shaped to obovoid, weakly veined, prominently stalked-glandular; pappus 0. 6 spp.: w N.Am, C.Am, s S.Am, e Asia. (Greek: gland st)

A. bicolor Hook. (p. 195) TRAIL PLANT Per 3–10 dm, gen ± erect, openly branched, ± tomentose below, stalked-glandular above. **LVS** at base; petiole ± winged; blade 3–25 cm, ± triangular to ovate, base cordate or hastate, subentire to shallowly lobed, white-tomentose below, glabrous above. **INFL**: phyllaries 1–2.5 mm, ovate. **PISTILLATE FLS** 3–7; corollas 0.5–1 mm. **DISK FLS** 3–10; corollas 1.5–2 mm. **FR** 5–8 mm, club-shaped. $2n$=46. Gen in shade, woods, forests; < 2000 m. NW, CaR, SN, nw CW; to B.C., n-c US.

ADENOPHYLLUM

Ann, per, subshrubs. **STS** erect. **LVS** simple or pinnate, opposite or alternate, dotted with embedded oil glands. **INFL**: heads radiate or discoid, solitary or in few-headed, often leafy-bracted, often open cymes; peduncles slender, with lf-like or scale-like bracts; involucre bell-shaped to cylindric; phyllaries in 3 series, outer << others, free, 2 inner ± equal, fused or free, gland-dotted; receptacle flat, bearing short, fringed scales. **RAY FLS** 0–few; corollas yellow to red. **DISK FLS** few–many; corollas yellow to orange; style tips tapered. **FR** obconic, ribbed; pappus of scales dissected into slender bristles. 10 spp.: sw US, n Mex. (Greek: gland-leaf) [Strother 1986 Sida 11:371–378]

1. Lvs simple, ovate to oblanceolate, coarsely toothed or shallowly lobed; involucres 15–18 mm; pappus scales 15–20 .... ***A. cooperi***
1′ Lvs deeply pinnately divided into 3–5 linear lobes; involucres 10–15 mm; pappus scales 8–12 ***A. porophylloides***

A. cooperi (A. Gray) Strother (p. 195) Subshrub, glabrous or short-rough-hairy; odor unpleasant. **STS** many, 3.5–6 dm, gen much-branched. **LVS** gen alternate, sessile, 8–25 cm, ovate to oblanceolate, coarsely toothed or shallowly lobed; blade with 1–2 glands near base, 1 near tip. **INFL**: heads radiate (rarely discoid); peduncles 6–15 cm, swollen beneath heads, bracts lf-like; involucre cylindric; outer phyllaries 12–22, 5–8 mm, linear, long-acuminate, each with 1 central gland; inner phyllaries ± 20, 15–18 mm, linear, fused below, gland-dotted. **RAY FLS** (0)7–13; ligule 8–9 mm, yellow to red-orange. **DISK FLS** many; corollas yellow 8–10 mm. **FR** 5–7 mm, hairy; pappus scales 15–20, 8–10 mm, each dissected into 5–9 bristles. $2n$=26. Dry, sandy slopes and washes; 600–1550 m. DMoj; s NV, nw AZ. [*Dyssodia c.* A. Gray]

A. porophylloides (A. Gray) Strother (p. 195) Subshrub, glabrous; odor unpleasant. **STS** many, 2–6 dm, gen much-branched. **LVS** opposite below, alternate above, 1.5–4 cm, deeply divided; lobes linear, entire to sharply serrate, each with 1 gland at base, 1 near tip. **INFL**: heads radiate (discoid); peduncles 2–8 cm, naked or with 1–5 narrow bracts; involucre cylindric; outer phyllaries 12–16, 3–8 mm, erect or recurved, each with a central gland; inner phyllaries 12–20, 10–15 mm, lanceolate, fused below, gland-dotted. **RAY FLS** (0)8–12; ligules 2–4 mm, yellow to red-orange. **DISK FLS** many; corollas 7–8 mm, yellow-orange. **FR** 5 mm, sparsely hairy; pappus scales 8–12, 7–8 mm, each dissected into 7–11 bristles. $2n$=26. Dry, rocky hillsides, washes; 200–1460 m. s DMoj, DSon; to AZ, n Mex. [*Dyssodia p.* A. Gray]

AGERATINA

A. Michael Powell

Per, shrubs. **LVS** gen opposite; blade elliptic to triangular, margin entire to lobed. **INFL**: heads discoid, solitary or in ± flat-topped cymes; phyllaries subequal, in 1–2(3) series; receptacle flat to conic, naked. **FLS** 10–60; corollas ± white or blue to pink-tinged, cylindric (or throat wider). **FR** 5-angled, gen 5-ribbed; pappus of 5–40 slender scabrous bristles, often easily detached. (Latin: resembling *Ageratum*) [King & Robinson 1987 Monogr Syst Bot Missouri Bot Garden 22:428–436]

1. Heads gen solitary (rarely 2–3) .... ***A. shastensis***
1′ Heads clustered
 2. Lvs gen alternate; heads 8–10 mm .... ***A. occidentalis***
 2′ Lvs opposite; heads 6–8 mm
  3. Lf blade gen 40–100 mm; petiole gen 20–40 mm .... ***A. adenophora***
  3′ Lf blade gen 15–55 mm; petiole gen 5–10 mm .... ***A. herbacea***

A. adenophora (Sprengel) R. King & H. Robinson Per; base woody. **ST** gen 5–15 dm, erect, ± purple, glandular-hairy. **LVS** opposite; blade deltoid-ovate, serrate, purple below, glandular-puberulent, esp below. **INFL**: heads ± 6.5 mm, clustered; phyllaries glandular-puberulent. **FL**: corollas white, pink-tinged. **FR** 1.7–2 mm. $3n$=51(n=17). Disturbed places; gen < 300 m. CCo, SnFrB, SCoRO, SCo; native of Mex, widely naturalized. [*Eupatorium a.* Sprengel] Reproduces by asexual seed; cult as orn; may be seriously invasive in mild coastal situations.

A. herbacea (A. Gray) R. King & H. Robinson Per; caudex woody. **ST** 4.5–7 dm, erect or spreading, green, puberulent. **LVS** gen opposite; blade gen triangular to ± cordate, yellowish to light or grayish green, glabrous to puberulent. **INFL**: heads 6–8 mm, in dense clusters; phyllaries puberulent. **FL**: corolla white. **FR** 2–3 mm. $2n$=34. Common. Rocky pinyon/juniper woodland; 1600–2200 m. e DMtns (Clark, New York, Providence mtns); to Colorado, NM, w TX, n Mex. [*Eupatorium h.* (A. Gray) E. Greene] ❀TRY.

A. occidentalis (Hook.) R. King & H. Robinson (p. 195) Per; caudex woody, ± rhizomatous. **ST** 1.5–7 dm, erect or ascending, ± green or purple, puberulent. **LVS** gen alternate; blade ± triangular, ± serrate, glandular. **INFL**: heads 8–10 mm, clustered; phyllaries puberulent. **FL**: corollas ± white to blue. **FR** 3–3.5 mm. $2n$=34. Rocks; 2100–3700 m. Common. NW, CaRH, SN, Wrn; to WA, ID, UT. [*Eupatorium o.* Hook.] Pls from SN are smaller than those from NCoR. ❀DRN,SHD:**2**,**7**,8–11,14–21;DFCLT.

A. shastensis (D. Taylor & Stebb.) R. King & H. Robinson (p. 195) SHASTA EUPATORIUM Per; caudex woody. **ST** 1.3–4.5 dm, erect or ascending, glandular-hairy. **LVS** gen alternate; blade ± deltoid-ovate, serrate, ± hairy. **INFL**: heads gen 1(–3), ± 1.2 cm; phyllaries glabrous to glandular-hairy. **FL**: corollas white. **FR** 3–5.5 mm. $2n$=34. UNCOMMON. Limestone cliffs; 400–1800 m. CaR. [*Eupatorium s.* D. Taylor & Stebb.] ❀TRY.

AGOSERIS

Kenton L. Chambers

Ann or per, gen scapose; sap milky. **LVS** ± basal, narrow, entire to pinnately lobed. **INFL**: head 1, ligulate, erect, 8–60 mm; phyllaries subequal or overlapping in 2–4 series, inner often elongated in fr; receptacle flat, naked. **FLS** gen many; ligules = to >> involucre, yellow or red-orange, soon withering, outer ± reddish below. **FR** linear to fusiform, ± 10-ribbed; beak often = or >> body; pappus of many fine, simple, white bristles. ± 10 spp: Am. (Greek: goat chicory) [Q. Jones 1954 PhD Harvard U.] Like *Taraxacum* but closely related to *Nothocalais*. Hybrids and polyploidy complicate variation in some spp.

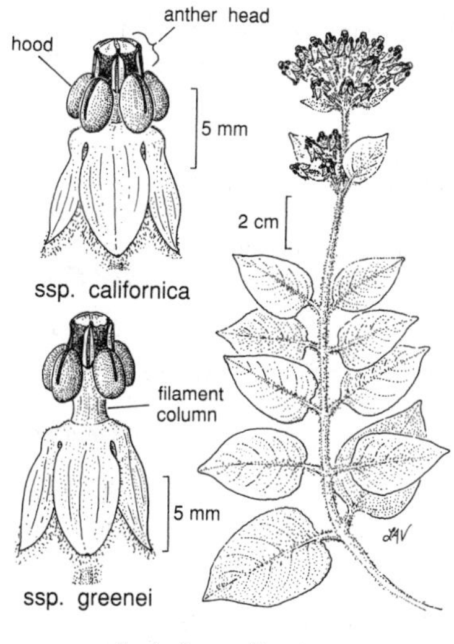

hood
anther head
5 mm
ssp. californica
2 cm
filament column
5 mm
ssp. greenei

Asclepias californica

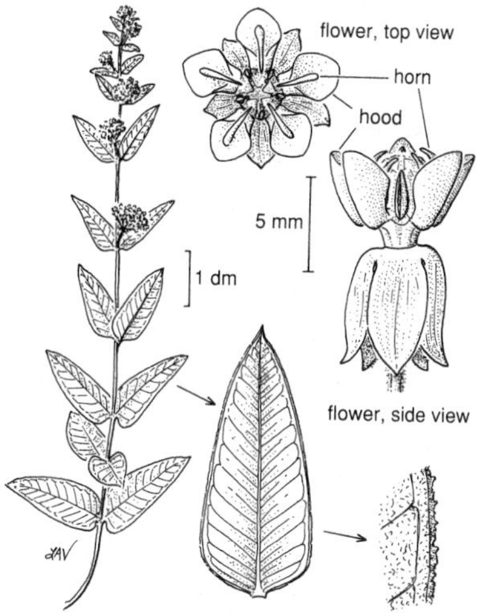

flower, top view
horn
hood
5 mm
1 dm
flower, side view

Asclepias erosa

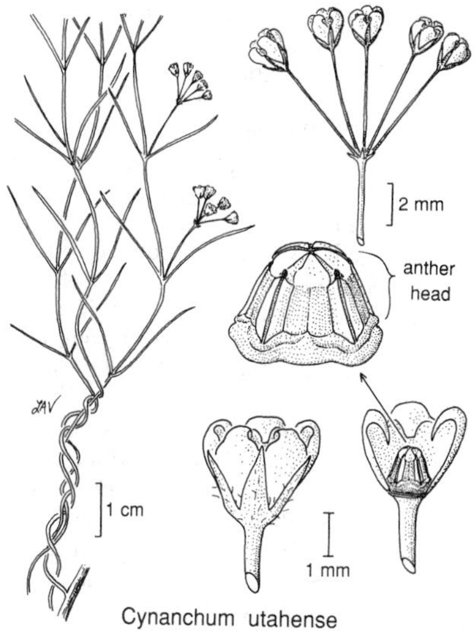

2 mm
anther head
1 cm
1 mm

Cynanchum utahense

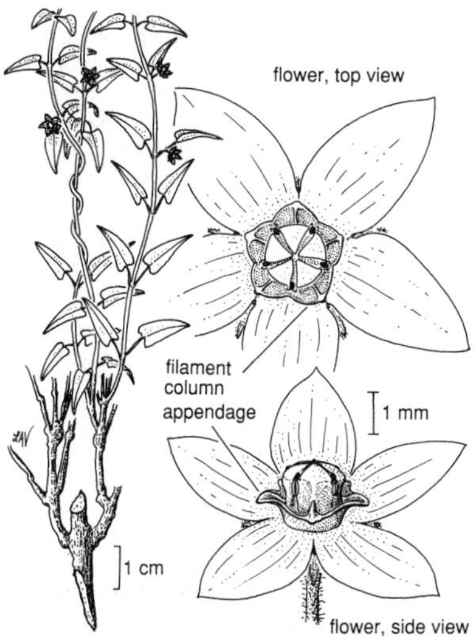

flower, top view
filament column appendage
1 mm
1 cm
flower, side view

Matelea parvifolia

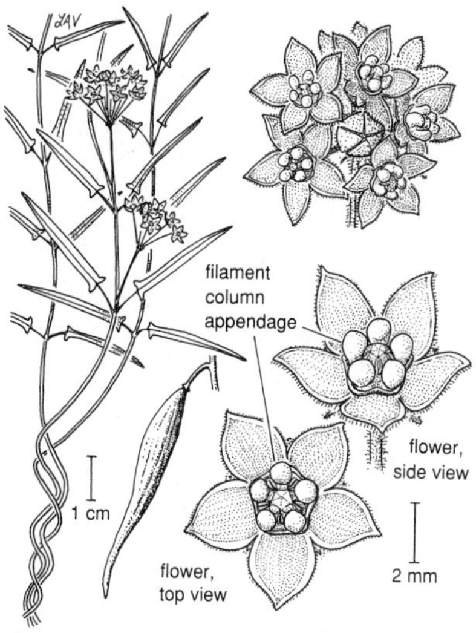

filament column appendage
flower, side view
1 cm
2 mm
flower, top view

Sarcostemma cynanchoides ssp. hartwegii

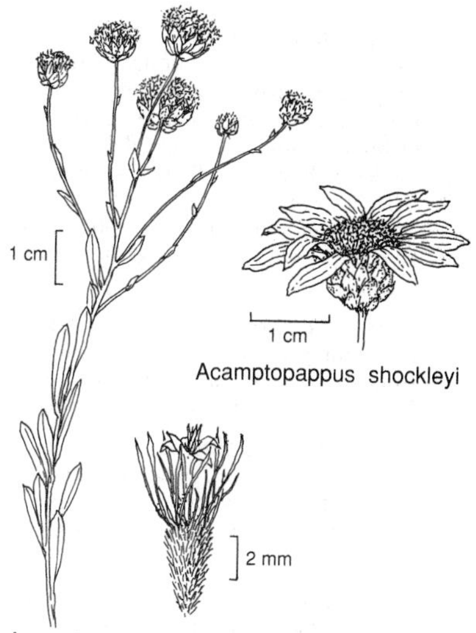

1 cm
1 cm

Acamptopappus shockleyi

2 mm

Acamptopappus sphaerocephalus

Asteraceae

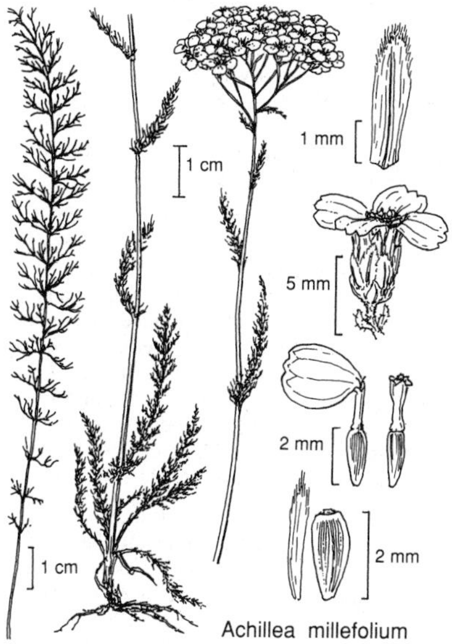

1 mm
1 cm
5 mm
2 mm
1 cm
2 mm

Achillea millefolium

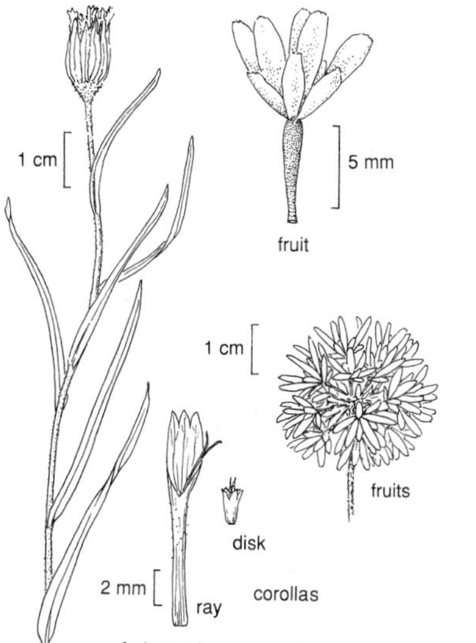

1 cm
5 mm
fruit
1 cm
fruits
disk
2 mm
ray
corollas

Achyrachaena mollis

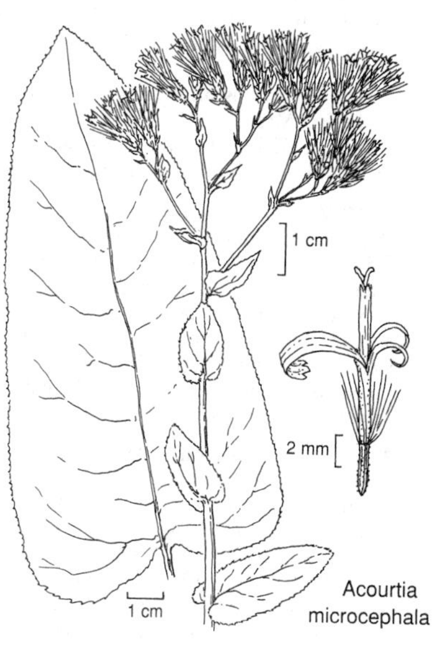

1 cm
2 mm
1 cm

Acourtia microcephala

1. Ann from slender taproot; involucre 10–25 mm in fr; outer frs sometimes wavy-ribbed or inflated (and ribless) . . . . . *A. heterophylla*
1′ Per from stout taproot and caudex; involucre 15–60 mm in fr; outer frs gen straight-ribbed
2. Corollas orange or brick-red, drying purplish — KR, NCoRH, MP, CaRH, SNH . . . . . *A. aurantiaca*
2′ Corollas yellow (often reddish on back), drying pinkish
3. Fr ± flat-topped, tapered abruptly to beak; beak >> fr; lf lobes gen angled toward lf base — widespread . . . . . *A. retrorsa*
3′ Fr tapered gradually to beak; beak sometimes < fr; lf lobes gen 0, spreading, or angled toward lf tip
4. Beak > 2 × fr body; corolla ± = involucre — widespread . . . . . *A. grandiflora*
4′ Beak < 2 × fr body; corolla > involucre
5. Involucre 8–21 mm; lf tip obtuse to acute . . . . . *A. apargioides*
6. Lf blade narrow throughout, gen deeply lobed; ± inland NCo, NCoRO, CW, n ChI < 500 m . var. *apargioides*
6′ Lf blade much wider toward tip, gen obtuse or rounded, entire to few-lobed; coastal NCo, CCo . . . . . var. *eastwoodiae*
5′ Involucre often > 21 mm in fr; lf tip acute to gen long-tapered — KR, CaR, SN, GB, > 500 m
7. Pl 30–65 cm; involucre 25–35 mm in fr; beak = or > fr body — SNH . . . . . *A. elata*
7′ Pl 3–50 cm; involucre 10–30 mm in fr; beak < fr body . . . . . *A. glauca*
8. Pl 20–50 cm; infl glabrous; lvs entire or few-toothed, ± glabrous; MP, W&I . . . . . var. *glauca*
8′ Pl 3–25 cm; upper infl axis (and sometimes phyllaries) hairy; KR, CaR, SN, GB
9. Infl hairs nonglandular, opaque; outer phyllaries lanceolate, often abruptly long-tapered; lvs long-tapered, irregular lobes often angled toward base . . . . . var. *laciniata*
9′ Some infl hairs glandular, translucent; outer phyllaries ± ovate, evenly tapered; lvs acute, ± entire . . . . . var. *monticola*

A. apargioides (Less.) E. Greene Per 10–45 cm, decumbent to erect. **LVS** variable; tip often obtuse, glabrous to soft- or rough-hairy. **INFL**: base of head tomentose; involucre 8–20 mm; phyllaries overlapping in 3–4 series, outer oblong-lanceolate, < 1/2 inner, glabrous to tomentose, glandular or not. **FLS** many, > involucre; ligules yellow. **FR**: body 3–6 mm, fusiform; ribs narrow, gen straight; beak 2 × body. Gen coastal dunes, grassland, open woodland, in sandy or clay soil; < 500 m. w NW, CW, n ChI; to WA. Hybridizes with *A. heterophylla*. Vars. weak, ± ecologically defined.

var. ***apargioides*** (p. 195) **LVS** gen erect, 6–30 cm, gen narrow throughout, deeply lobed, acute. $2n=18$. Fields, slopes away from ocean; < 500 m. NCo, CCo, SnFrB (less common in NCoRO, SCoRO, n ChI); OR. [*A. hirsuta* (Hook.) E. Greene]

var. ***eastwoodiae*** (Fedde) Munz (p. 195) **LVS** erect or spreading, 2–15 cm, widened above, obtuse to rounded, entire to irregularly lobed below wide tip. $2n=36$. Coastal dunes, beaches, bluffs; < 50 m. NCo, CCo; to WA. [ssp. *maritima* (Sheldon) Q. Jones]

A. aurantiaca (Hook.) E. Greene (p. 195) Per 10–50 cm. **LVS** gen linear to (widely) oblanceolate, (acute to) long-tapered, entire to irregularly lobed in lower 2/3, (sub)glabrous. **INFL**: base of head tomentose; involucre 12–30 mm; outer phyllaries gen narrowly oblong-lanceolate, long-tapered, < inner phyllaries in fr, glabrous to soft-hairy, nonglandular. **FLS** many, = to > involucre; ligules orange or brick-red, drying purplish or dark pink. **FR**: body 4–9 mm, fusiform; ribs often minutely hairy; beak sometimes < body. $2n=18,36$. Meadows, shrubland, streamsides; 1500–3500 m. KR, NCoRH, CaRH, SNH, MP; to AK, w Can, SD, NM. Reported yellow-fld pls from outside CA much like *A. glauca*. ❀TRY.

A. elata (Nutt.) E. Greene Per 30–65 cm, stout, ± tomentose. **LVS** ± oblanceolate, acute, entire to few-lobed, (sub)glabrous. **INFL**: involucre 25–35 mm; phyllaries overlapping in 3–4 series, outer oblong-lanceolate, acute, ciliate, soft-hairy, gen ± glandular. **FLS** many, > involucre; ligules yellow, drying pinkish. **FR**: body 7–10 mm; ribs sometimes minutely hairy; beak sometimes > body. Meadows, shrubby slopes; 1600–3200 m. SNH; to WA. Problematic sp. (type supposedly from Willamette Valley, OR — either incorrect or extirpated). Much like *A. aurantiaca*.

A. glauca (Pursh) Raf. Per 3–50 cm, decumbent to erect, gen ± tomentose. **LVS** linear to (ob)lanceolate, acute to long-tapered, entire to lobed. **INFL**: involucre 10–25 mm; phyllaries overlapping in several series, outer ovate and acute to lanceolate and long-tapered, glabrous to ciliate or soft-hairy. **FLS** many, > involucre; ligules yellow. **FR**: body 4–10 mm, linear-fusiform; ribs straight, smooth; beak < body. Sagebrush scrub, coniferous forest, meadows, alpine slopes; 1400–3800 m. CaRH, SNH, GB; to w Can, n-c US, NM.

var. ***glauca*** (p. 195) Pl 20–50 cm, glabrous. **LVS** linear to oblanceolate, long-tapered, entire or few-toothed, (sub)glabrous, glaucous. **INFL**: phyllaries lanceolate, long-tapered, often marked reddish, glabrous. $2n=36$. Meadows, open coniferous woods; 1400–2500 m. MP, W&I; range of sp. outside CA. Robust, hairy, n MP pls suggestive of var. *agrestis* (Osterh.) Q. Jones of Rocky Mtns.

var. ***laciniata*** (D. Eaton) F.J. Smiley (p. 195) Pl 3–25 cm, ± soft-white-opaque-hairy (esp below head), nonglandular. **LVS** ± narrowly lanceolate, long-tapered, gen pinnately (and downwardly) lobed. **INFL**: outer phyllaries lanceolate, ± long-tapered, gen glabrous, red-marked. $2n=18$. Dry, open sagebrush scrub, coniferous woods, alpine slopes; 1400–3800 m. CaRH, SNH, GB; to ID, MT, WY, NM. Intergrades with var. *monticola*.

var. ***monticola*** (E. Greene) Q. Jones (p. 195) **STS** 3–40 cm, ± soft-yellow-translucent-hairy (esp below head), ± glandular. **LVS** gen oblanceolate, acute, glabrous, entire to short-lobed. **INFL**: outer phyllaries ± ovate, ± acute, gen hairy and glandular, often red-marked. $2n=18,36$. Moist meadows, streambanks, coniferous woods, alpine slopes; 1500–3800 m. KR, CaRH, SNH, GB; to WA, NV. [var. *dasycephala* (Torrey & A. Gray) Jepson misapplied]

A. grandiflora (Nutt.) E. Greene (p. 195) Per 25–85 cm, gen ± soft-white-opaque-hairy (esp base of head), nonglandular. **LVS** linear to oblanceolate, acute to long-tapered, entire to deeply (± upwardly) lobed. **INFL**: involucre 15–40 mm; outer phyllaries lanceolate to ovate, often red-tinged, ± soft-hairy, tips prominent; inner phyllaries much elongated in fr. **FLS** many, = involucre; ligules yellow. **FR**: body 3.5–7 mm, fusiform; ribs straight, smooth; beak >> body. $2n=18$. Grassland, shrubland, woodland; < 2000 m. CA-FP (exc SnJV, MP); to B.C., ID, NV. ❀TRY.

A. heterophylla (Nutt.) E. Greene (p. 195) Ann 5–40 cm, ± white-hairy (exc densely yellow- to reddish hairy and glandular at base of head). **LVS** gen oblanceolate, acute, entire to few-toothed, less often narrower and lobed, gen softly or roughly white-hairy. **INFL**: involucre 8–25 mm; outer phyllaries oblong-lanceolate, gen ciliate and glandular-hairy on back, often reddish tinged or striped, tip tapered, green; inner phyllaries much elongated in fr. **FLS** 5–many, = to >> involucre; ligules yellow. **FR**: body 2–5 mm, variable, often wavy-ribbed, sometimes inflated; beak >> body. $2n=18,36$. Many open habitats; < 2000 m. CA (exc SNE, D); to B.C., ID, NM, Mex. Variable; small-fld races self-fertilizing.

A. retrorsa (Benth.) E. Greene (p. 195) Per 10–50 cm, stout, gen soft-white-opaque-hairy (esp lvs, base of head), nonglandular. **LVS** linear to oblanceolate; lobes tapered, mostly angled downward. **INFL**: involucre 20–60 mm; outer phyllaries oblong-lanceolate, acute, ± white-tomentose, gen ± red-marked, inner phyllaries linear, much elongated in fr. **FLS** many, > involucre; ligules yellow. **FR**:

body 4–7 mm, flat-topped; beak 15–25 mm, >> body. $2n$=18. Dry shrubland, oak woodland, coniferous forest; 400–2800 m. CA-FP (exc coast, GV), MP; to WA, NV. ❀TRY.

AMBLYOPAPPUS

1 sp. (Greek & Latin: blunt pappus)

A. pusillus Hook. & Arn. (p. 195) Ann 3–40 cm, glabrous or glandular, sticky; odor strong. **ST** erect, simple or much-branched. **LVS** simple, alternate or lowermost opposite, 1–4 cm, linear to narrowly oblanceolate, entire or pinnately lobed ± fleshy. **INFL**: heads disciform or minutely radiate, many in a round- or flat-topped, leafy cyme; peduncles gen 1–6 mm; involucre 3–6 mm diam, bell-shaped; phyllaries 3–8, in 1 series, equal, free, overlapping, 2–5 mm, obovate; receptacle short-conic, naked. **PISTILLATE FLS** 1 per phyllary; corollas 0.5–1 mm, tubular, yellow, ligules minute, 2–3-lobed. **DISK FLS** 2–30; corollas 0.5–1 mm, yellow; style branches very short, tip triangular. **FR** 1.5–2 mm, obconic, 3–4 angled, black; hairs short, ascending; pappus a crown of 7–12 scales, each 0.2–0.5 mm, oblong, blunt, white or purplish. $2n$=16. Coastal dunes, beaches, headlands; < 50 m. s CCo, SCo, ChI; Baja CA; also w S.Am. ❀STBL.

AMBROSIA RAGWEED, BUR-SAGE

Willard W. Payne

Ann to shrub, monoecious. **LVS** often opposite below, gen alternate above, gen petioled, hairy, glandular. **INFL**: staminate heads gen many in ± spikes or racemes, phyllaries fused into shallow cup; pistillate heads gen clustered below staminate, gen spiny, bur-like; involucre ± 0; receptacle chaffy; chaff scales spirally arrayed, fused below, tips gen becoming spiny; each pistillate fl in separate chamber. **STAMINATE FLS** ± many; corolla yellow or translucent; anthers free; style unbranched. **PISTILLATE FLS** 1–5; corolla 0; style branches long. **FR** enclosed in bur; pappus 0. (Greek: early name for aromatic plants; the mythic food of the gods) [Payne 1976 Pl Syst Evol 125:169–178] Wind-blown pollen often highly allergenic.

1. Subshrub or shrub; sts persisting 2+ seasons
  2. Spines of fr heads straight, flat or round in X-section
    3. Cauline lvs ± compound; fr heads (sub)glabrous ..... ***A. dumosa***
    3' Cauline lvs few-toothed to shallowly lobes; fr heads long-soft-silky ..... ***A. eriocentra***
  2' Spines of fr heads hooked, round in X-section
    4. Cauline lvs sessile, clasping ..... ***A. ilicifolia***
    4' Cauline lvs petioled
      5. Lf blades long-triangular, 15+ cm; fr heads long, body ± glabrous or gland-dotted ..... ***A. ambrosioides***
      5' Lf blades round, < 10 cm; fr heads round, body white-woolly ..... ***A. chenopodiifolia***
1' Ann or per
  6. Ann from slender taproot
    7. Cauline lvs mostly opposite, palmately lobed ..... ***A. trifida***
    7' Cauline lvs mostly alternate, pinnately lobed and divided
      8. Spines of fr heads many, scattered, sharply pointed; distal staminate phyllary tips ± black-lined ..... ***A. acanthicarpa***
      8' Spines of fr heads few, below beak, blunt; staminate phyllary tips uniformly green ..... ***A. artemisiifolia***
  6' Per from rhizome-like roots or caudex
    9. Fr head spines many, scattered, or in several rows
      10. Fr heads 5–10 mm; spines straight ..... ***A. chamissonis***
      10' Fr heads 2–5 mm; spines hooked ..... ***A. confertiflora***
    9' Fr heads spines 0–few, vestigial, mostly on upper half
      11. Lvs simple or sparsely 1-pinnate, lobes toothed; pls gen 4–20 dm ..... ***A. psilostachya***
      11' Lvs 3–4-pinnately divided into many tiny segments; pls gen 1–3.5 dm ..... ***A. pumila***

A. acanthicarpa Hook. (p. 195) ANNUAL BUR-SAGE Ann 4–15 dm, from slender taproot. **ST** gray-green, ± stiffly strigose-bristly. **LVS**: petioles winged; blade pinnately divided, < 8 cm, 7 cm wide. **INFL**: staminate heads 2–5 mm diam, involucre lobes 3–9, tips of longest 3 ± black-lined along midveins; pistillate heads 1-fld. **FR**: body of bur 5–7 mm, ovoid, gen ± golden, glabrous or puberulent; spines 0–30, scattered, flat, straight, sharp. $2n$=36. Sandy plains, disturbed sites, many communities; < 2200 m. NCoR, SN, SW, GB, D; to WA, MT, WY, TX, nw Mex. [*Franseria a.* (Hook.) Cov.]

A. ambrosioides (Cav.) Payne Shrub < 2.5 m, gen dark. **STS** coarsely long-hairy or bristly. **LVS**: blade 2–20 cm, ± lanceolate, coarsely toothed, dark green, sticky, fragrant. **INFL**: staminate heads 6–8 mm diam, involucre lobes 7–12; pistillate heads 3–5-fld. **FR**: bur 10–15 mm, elliptic, ± brown, finely glandular; spines gen > 50, scattered, ± cylindric, hooked. $2n$=36. Coastal scrub: < 100 m. s SCo (near San Diego); AZ, nw Mex. [*Franseria a.* Cav.] Perhaps not currently in CA.

A. artemisiifolia L. (p. 195) COMMON RAGWEED Ann < 7 dm, much-branched. **ST** green, red- or black-marked, weakly hairy. **LVS** opposite below; cauline lvs 3–12 cm, widely ovate, gen 2–3-pinnately parted, ± hairy. **INFL**: staminate heads 2–5 mm diam, involucre asymmetric, lobes 5–10, green; pistillate heads 1-fld. **FR**: bur 2–4 mm, widely obconic, green to brown, ± puberulent; spines 4–12, blunt, vestigial, ± in 1 whorl below beak. $2n$=36. Uncommon. Disturbed sites; < 650 m. NW, e ScV, SCo, SNE; native to e US.

A. chamissonis (Less.) E. Greene (p. 195) BEACH-BUR Per < 4 m, sprawling from caudex or taproot. **STS** brown to gray-canescent, harshly to silky-hairy, ± striate. **LVS** opposite below, variable; blade 2–5 cm, oblanceolate to widely triangular, toothed to 3-pinnately lobed, canescent. **INFL**: staminate heads 4–8 mm diam, involucre lobes ± 10, ± black-lined above; pistillate heads 1-fld. **FR**: bur 5–10 mm, ovoid, ± brown, puberulent to tomentose; spines 10–20+, scattered, ± cylindric, straight, sharp, bases wide. $2n$=32,36. Beaches, dunes; < 25 m. NCo, CCo, SCo, ChI; to B.C., Baja CA; also w S.Am. [*Franseria c.* Less. incl ssp. *bipinnatisecta* (Less.) Wiggins & Stockw.] Lf forms intergrade; pls with pinnately compound lvs have been called *A. bipinnatifida* (Nutt.) E. Greene. ❀DRN,SUN:4,5,**17**&IRR:**14–16**,19–21,**22–24**.

A. chenopodiifolia (Benth.) Payne (p. 195) SAN DIEGO BUR-SAGE Shrub < 3.5 m, much-branched. **STS** slender, tough, green to red-brown, tomentose or becoming glabrous. **LVS**: blades 1–3 cm, ± ovate, entire to palmately 3-lobed, white-hairy. **INFL**: staminate heads few, 4–8 mm diam, involucre 5–8-lobed, ± veiny above; pistillate heads 2–3-fld. **FR**: bur 5–7 mm, ± spheric, body densely woolly; spines 12–25, scattered, slender, hooked. 2*n*=36. RARE in CA. Coastal scrub; < 150 m. s SCo (sw San Diego Co.); Baja CA. [*Franseria c.* Benth.]

A. confertiflora DC. Per 3–18 dm, from rhizome-like roots. **STS** green to brown, bristly, ± puberulent. **LVS**: petioles with lobed wings; blade < 16 cm, ± ovate, 2–4-pinnate. **INFL**: staminate heads 3–9 mm diam, involucre lobes 5–9, green above; pistillate heads 1(2)-fld. **FR**: bur 2–5 mm, 1.5–5 mm diam, ovoid, base ± tapered; spines few (sometimes vestigial), scattered, short, cylindric, hooked, pitted at base on upper surface. 2*n*=72,108. Disturbed areas; < 350 m. SNF, SnJV, SnFrB, SCo; to c US, n Mex. [*Franseria c.* (DC.) Rydb.] Often forms large colonies.

A. dumosa (A. Gray) Payne (p. 199) BURRO-WEED Shrub 2–9 dm, much-branched. **STS** softly gray-white canescent. **LVS** gen ± clustered on short branches, (sub)sessile; blade 0.5–4 cm, ± ovate, 1–3-pinnate, canescent. **INFL**: staminate heads 3–5 mm diam, involucre lobes 5–8; pistillate heads 2-fld. **FR**: bur 5–9 mm, spheric, golden to purple or brown, puberulent; spines 12–35, scattered, flat, straight, sharp. 2*n*=36,72,108,126,144. Creosote-bush scrub; < 1600 m. W&I, D; to sw UT, AZ, nw Mex. [*Franseria d.* A. Gray] ❀TRY.

A. eriocentra (A. Gray) Payne (p. 199) WOOLLY BUR-SAGE Shrub 3–18 dm, ± spheric. **STS** gray-brown, ± woolly, becoming glabrous. **LVS**: petioles winged; blades 1–9 cm, ± lanceolate, coarsely toothed or pinnately lobed, ± rolled under, ± green above, **INFL**: staminate heads few, 5–7 mm diam, involucre lobes 5–8; pistillate heads gen l-fld. **FR**: bur 8–11 mm, greenish brown, densely long-soft-hairy; spines 12–20, gen near middle, straight, flat, stout, sharp, tips ± hair-tufted. 2*n*=36. Dry washes and slopes; 800–1700 m. e DMoj, DMtns; to sw UT, nw AZ.

A. ilicifolia (A. Gray) Payne (p. 199) Shrub < ± 1 m, matted. **STS** greenish when young, densely glandular and stiffly short-hairy; older bark gray-white. **LVS** 2–10 cm, ovate to round, leathery, brittle, dark gray-green, sticky, veiny below; teeth long, spine-tipped; sessile, ± clasping. **INFL**: staminate heads 10–15 mm diam, involucre lobes 10–15, lanceolate, spine-tipped; pistillate heads gen 2-fld. **FR**: bur < 20 mm, spheric, ± brown, sticky; spines 20–70, scattered, < 6 mm, ± cylindric, strongly hooked. 2*n*=36. Sandy washes, rocky canyons, creosote-bush scrub; < 500 m. SnGb, SnBr, s D; w AZ; Baja CA. [*Franseria i.* A. Gray]

A. psilostachya DC. (p. 199) WESTERN RAGWEED Per 5–20 dm, from rhizome-like roots, little-branched. **STS** ± straw-colored, soft-hairy to bristly. **LVS** opposite at least below, 4–12 cm, lanceolate to ovate, gen 1–2-pinnately lobed, hairy; lobes entire to toothed; upper lvs often sessile. **INFL**: staminate heads 2–5 mm diam, involucre lobes 3–5, obscure; pistillate heads 1-fld. **FR**: bur < 6 mm, obovoid, greenish brown, puberulent; spines 0–7 below beak, blunt or vestigial. 2*n*=36,72,108,144. Common. Roadsides, dry fields; < 1000 m. CA-FP; to ID, e N.Am, TX, n Mex.

A. pumila (Nutt.) A. Gray (p. 199) SAN DIEGO AMBROSIA Per < 5 dm, from rhizome-like roots, gen few-branched. **STS** green to straw-colored, densely short-hairy. **LVS**: petioles with lobed wings; blades 3–13 cm, lanceolate to widely ovate, 2–4-pinnate, softly gray-white-hairy. **INFL**: staminate heads 3–5 mm diam, involucre lobes 5–8, obscure, widely obtuse; pistillate heads 1-fld. **FR**: burs < 2 mm, obovoid, ± brown, puberulent; spines (0)1–5, vestigial, scattered. 2*n*=36. RARE in CA. Disturbed sites; < 150 m. s SCo (San Diego Co.); Baja CA.

A. trifida L. (p. 199) GIANT RAGWEED Ann < 2 m, coarse. **ST** thick, fleshy, black-lined, soft-hairy to ± bristly; base ± woody. **LVS** opposite; blades 6–35 cm, lanceolate to widely ovate, lobes palmate, (0)3–5, serrate, teeth black-lined below, green, short-rough-hairy. **INFL**: staminate heads 2–5 mm diam, involucre lobes ± 10, obscure, 3 longest dark-lined; pistillate heads 1-fld. **FR**: burs 6–12 mm, obovoid, green or straw-colored, glabrous or ± short-rough-hairy; spines 5–8, vestigial, clustered around blunt beak. 2*n*=24. Dry, disturbed places; gen < 100 m. n SnJV, n CW, SW, MP; native to c&e US. NOXIOUS WEED, but gen not long persisting.

AMPHIPAPPUS CHAFF-BUSH

Meredith A. Lane

1 sp. (Greek: double pappus) [Porter 1943 Amer J Bot 30:481–483]

A. fremontii Torrey & A. Gray (p. 199) Shrub, glabrous. **STS** much-branched, widely spreading to ascending, gen < 6 dm, striate, smooth, white above, grayish below; lfless sts spiny. **LVS** short-petioled, gen < 2 cm, obovate or elliptic, entire, sometimes ± thick, light yellow- or gray-green. **INFL**: heads radiate, in crowded, flat-topped clusters gen 3–5 cm wide; involucres cylindric, ± 5 mm, < 3 mm wide; phyllaries 7–12, ovate, whitish or pale green. **RAY FLS** 1–2, barely > involucre, tips 2- or 3-toothed, yellow. **DISK FLS** 3–7, staminate; corollas narrowly funnel-shaped, sinuses deep, lobes reflexed, style-branch appendages lanceolate to rounded. **FR** of ray fls < or = 3 mm, hairy; pappus of 15–20 stout bristles fused at base, gen 1 mm; ovary of disk fls < or = 1 mm, glabrous; pappus of 25 flattened, twisted bristles gen < or = 3 mm. 2*n*=18. Rocky or gravelly flats, slopes, canyons; < 1600 m. c&e DMoj; to UT, NV, AZ. Vars. not known to intergrade.

1. Lvs glabrous .... var. ***fremontii***
1′ Lvs scabrous .... var. ***spinosus***

var. ***fremontii*** **LVS** glabrous, sometimes gummy. Habitats of sp. ne DMoj; NV.

var. ***spinosus*** (Nelson) C.L. Porter **LVS** minutely scabrous on surfaces and margins. Habitats of sp. e-c DMoj; to UT, NV, AZ. [ssp. *s.* (Nelson) Keck] ❀TRY.

ANAPHALIS PEARLY EVERLASTING

G. Ledyard Stebbins

Per, white-woolly, dioecious (or pistillate heads with few staminate fls), rhizomed. **ST** gen erect (cespitose), leafy. **LVS** basal, alternate, linear to lanceolate, entire, not reduced upward. **INFL**: heads discoid or disciform, small, many in terminal panicles, staminate and pistillate heads similar; involucre hemispheric; phyllaries many, free, overlapping in several unequal series; receptacles naked. **DISK FLS** functionally staminate; corolla narrowly tubular, yellowish; anther bases tailed, tips ± ovate; style branches wider at tip, truncate, ovary vestigial, pappus of fine bristles. **PISTILLATE FLS**: corolla slender, ± yellow, lobes minute. **FR** cylindric or obconic; pappus of many fine, minutely barbed bristles. 60–100 spp.: N.Am, Asia. (Greek: name of some everlasting)

A. margaritacea (L.) Benth. & Hook. (p. 199) **ST** 2–12 dm, erect; rhizome slender. **LVS** 3–10 cm, sessile, linear to lanceolate, entire, green or gray above, white-tomentose below. **INFL**: involucre 5–6 mm; phyllaries ovate, pearly white. **DISK FL**: corolla 4–5 mm. **PISTILLATE FL**: corolla 4–6 mm. **FR** 0.5–1 mm, papillate; pappus = corolla. 2*n*=28. Woods, roadsides, disturbed places; < 3200 m. NW, CaRH, SNH, CW, SnBr, MP; to AK, e N.Am, Eurasia. Locally common. ❀**4–6,15–17**&IRR:1–3,7,8,**9**,10,11, **14,18–24**; aromatic.

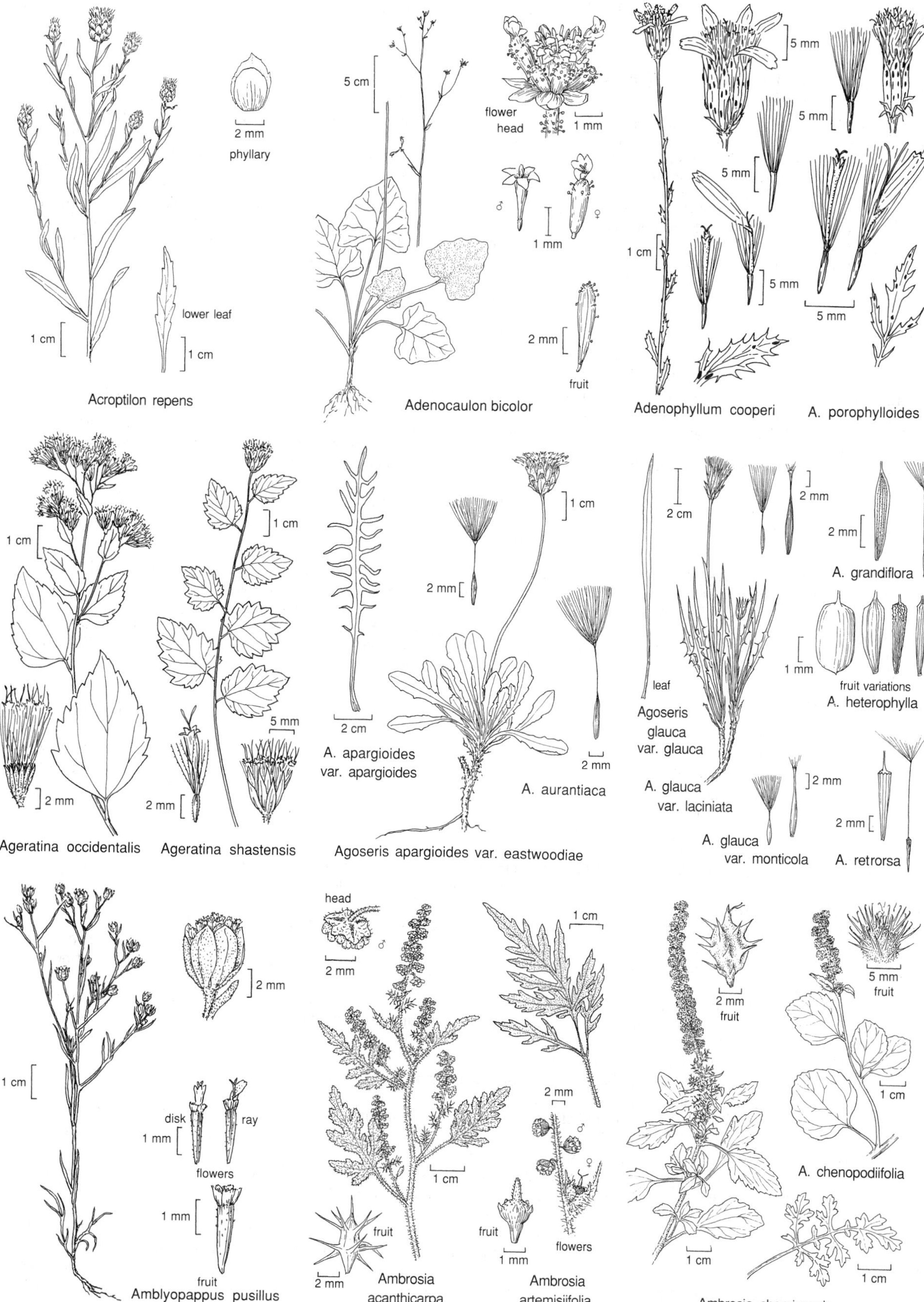
phyllary
2 mm
lower leaf
1 cm
1 cm
Acroptilon repens
5 cm
flower head
1 mm
1 mm
2 mm
fruit
Adenocaulon bicolor
5 mm
5 mm
5 mm
1 cm
5 mm
5 mm
Adenophyllum cooperi
A. porophylloides
1 cm
1 cm
2 mm
5 mm
2 mm
Ageratina occidentalis
Ageratina shastensis
2 cm
A. apargioides var. apargioides
1 cm
2 mm
2 mm
A. aurantiaca
Agoseris apargioides var. eastwoodiae
2 cm
2 mm
2 mm
A. grandiflora
1 mm
fruit variations
A. heterophylla
leaf
Agoseris glauca var. glauca
A. glauca var. laciniata
2 mm
A. glauca var. monticola
2 mm
A. retrorsa
2 mm
1 cm
disk
ray
1 mm
flowers
1 mm
fruit
Amblyopappus pusillus
head
2 mm
1 cm
fruit
2 mm
Ambrosia acanthicarpa
1 cm
2 mm
fruit
1 mm
flowers
Ambrosia artemisiifolia
2 mm
fruit
5 mm
fruit
1 cm
A. chenopodiifolia
1 cm
1 cm
Ambrosia chamissonis

ANCISTROCARPHUS WOOLLY FISHHOOKS

James D. Morefield

1 sp. (perhaps another undescribed). (Greek: fishhook chaff) Often placed in *Stylocline*, but more closely related to *Hesperevax* or Eur *Cymbolaena*.

A. filagineus A. Gray (p. 199) Ann, ± gray, cobwebby to tomentose. **STS** 1–several from base, spreading to erect, < 15 cm, ± forked. **LVS** simple, alternate, ± sessile, < 30 mm, linear-oblanceolate to ± elliptic, entire. **INFL**: heads disciform, ± sessile in leafy-bracted groups of 2–5, < 9 mm; phyllaries 0 or 3–6, vestigial, ± equal, membranous; receptacle 1–2 × longer than wide, expanded at tip, chaffy; outer chaff scales phyllary-like, each enclosing a pistillate fl, falling with a fr, woolly, strongly 3-veined, body hard, tip scarious-winged; innermost chaff scales each subtending a disk fl, 2.7–4.1 mm (> outer), open, concave, persistent, thinly woolly, in fr rigid, enlarged, spreading, tip hooked strongly inward, forming a spine. **PISTILLATE FLS** 5–10 in 1–2 series; corollas tubular. **DISK FLS** staminate, 3–5, 1–1.5 mm, 4–6-lobed. **FR** 1.4–2 mm, 0.6–0.9 mm wide, obovoid, compressed front-to-back, smooth, dull, black-banded near base; pappus 0. Bare or grassy, often serpentine slopes, road beds, vernally moist places; < 1900 m. CA-FP (exc coast), MP, sw DMoj; to OR, ID, NV, nw Baja CA. [*Stylocline f.* (A. Gray) A. Gray incl var. *depressa* Jepson]

ANISOCOMA

G. Ledyard Stebbins

1 sp.: sw US, n Mex. (Greek: unequal pappus)

A. acaulis Torrey & A. Gray (p. 199) Ann; sap milky. **LVS** all basal, 3–5 cm, pinnately lobed; lobes toothed, ± hairy. **INFL**: head ligulate, solitary, 2–3 cm; peduncle 5–20 cm, glabrous; phyllaries papery-transparent on margins, often with reddish tips and dots, in several series, outer short, oblong, blunt, inner linear, pointed; receptacle naked. **FLS** many; ligules pale yellow, ± ephemeral. **FR** 10–15 veined; pappus of plumose bristles in 1 series, white, those on inner side of achene < on outer. $2n$=18. Sandy washes, dry slopes; 600–2400 m. s SN, Teh, SnJV, SCoRO, TR, PR, SNE, D; NV, AZ, Baja CA. ❀TRY.

ANTENNARIA PUSSY-TOES

G. Ledyard Stebbins & Randall J. Bayer

Per, often matted, dioecious; staminate pls gen present. **LVS** alternate, entire, gen ± tomentose. **INFL**: heads discoid or disciform, 1 or in cyme-like clusters; phyllaries many in several series, papery or membranous (staminate wider, more conspicuous); receptacle naked. **STAMINATE FLS** 2–5 mm; corollas white, yellow, or red; pappus bristle tips gen enlarged. **PISTILLATE FLS** 2–10 mm; corollas barely lobed, white, yellow or red. **FR** 0.5–3.5 mm, ± elliptic; pappus bristles many, soft, weakly barbed. ± 40 spp.: Am, n Eurasia. (Latin: antenna) [Bayer 1990 Can J Bot 68:1389–1397 & Madroño 37:171–183] Many races reproduce by asexual seeds, their populations entirely pistillate pls.

1. Heads gen 1(–3) per fl st
 2. St 5–12 cm; lvs thick, green above, tips notched ..... ***A. suffrutescens***
 2′ St < 4 cm; lvs thin, gray-tomentose above, tips acute
  3. Pl without stolons ..... ***A. dimorpha***
  3′ Pl bearing long, slender, ± lfless stolons ..... ***A. flagellaris***
1′ Heads 3–many per fl st
 4. Pl without stolons, not forming mats
  5. Lvs cauline, ± equal; lower involucre densely tomentose ..... ***A. geyeri***
  5′ Lvs basal and cauline, reduced upward; involucre ± glabrous
   6. Lower lvs 7–20 mm wide; sts 18–40 cm; phyllaries white or cream-colored, unequal ..... ***A. argentea***
   6′ Lower lvs 2–5 mm wide; sts 10–25 cm; phyllaries straw-colored or pale, ± equal ..... ***A. luzuloides***
 4′ Pl with stolons, forming mats
  7. Rosette lvs elliptic, green above; infl raceme- or panicle-like, main axis 3–11 cm; peduncles 10–30 mm; infl finely glandular ..... ***A. racemosa***
  7′ Rosette lvs oblanceolate to spoon-shaped; infl cyme-like, main axis ± 0; peduncles < 5 mm; infl glandular or not
   8. Rosette lvs green or very lightly tomentose above, 1–3-veined
    9. Rosette lvs > 6 mm wide, 1–3-veined; cauline lvs 20–40 mm; infl not glandular; NW ***A. howellii*** ssp. ***howellii***
    9′ Rosette lvs < or = 6 mm wide, 1-veined; cauline lvs 7–16 mm; infl gen glandular; SnBr ..... ***A. marginata***
   8′ Rosette lvs gray-tomentose above, 1-veined
    10. Phyllaries dark brown to black; sts 3–13 cm
     11. Lowest cauline lf > 11 mm; herbage gen not glandular; pistillate corolla > 3 mm in fr; staminate corolla gen > 2.7 mm in fl ..... ***A. media***
     11′ Lowest cauline lf < or = 11 mm; herbage gen glandular; pistillate corolla < or = 3 mm in fr; staminate corolla gen < or = 2.8 mm in fl ..... ***A. pulchella***
    10′ Phyllaries white to rose or pale brown; sts 6–40 mm
     12. Phyllaries white-tipped with a prominent blackish brown spot near base of scarious part ..... ***A. corymbosa***
     12′ Phyllaries white, rose, straw-colored or (paler) brown, not dark-spotted near base
      13. Stolons slightly woody, ascending; phyllaries pale yellow to pale brown, tips blunt; staminate pls present — n SNH ..... ***A. umbrinella***
      13′ Stolons not woody, horizontal or ascending; phyllaries various colors, tips acute; staminate pls ± 0 ..... ***A. rosea***
       14. Longest lf of fl rosettes < 20 mm; phyllaries sometimes brown ..... ssp. ***confinis***
       14′ Longest lf of fl rosettes > 20 mm; phyllaries gen not brown ..... ssp. ***rosea***

A. argentea Benth. (p. 199) **STS** 18–40 cm, from branched caudex; stolons 0. **LVS**: basal 20–50 mm, oblanceolate to elliptic, 1-3-veined, ± gray-tomentose; cauline 15–45 mm, lanceolate. **INFL**: heads 10–75; involucres 4–5 mm, ± glabrous; phyllaries wide, acute, whitish. **STAMINATE FLS**: corollas 2.5–3.5 mm. **PISTILLATE FLS**: corollas 3–4 mm. **FR** 1–1.5 mm, glandular; pappus 3–4 mm. 2*n*=28. Dry coniferous forest; 800–2000 m. KR, NCoRH, CaRH, MP; OR. ❀TRY.

A. corymbosa E. Nelson (p. 199) **STS** 6–15 cm, slender; stolons horizontal, 1–10 cm. **LVS**: basal 18–22 mm, spoon-shaped, 1-veined, thin, ± gray-tomentose; cauline 8–13 mm. **INFL**: heads 3–7; involucres 4–5.3 mm, densely hairy below; phyllaries wide, tips blunt (staminate) or acute (pistillate), white, distinctly spotted blackish brown near base. **STAMINATE FLS**: corollas 2–3.2 mm. **PISTILLATE FLS**: corollas 2.5–3.5 mm. **FR** 0.5–1 mm, slightly papillate; pappus 3.5–4.5 mm. 2*n*=28. Moist meadows, streamsides; 1900–3200 m. SNH; to OR, sw Can, MT, WY, Colorado. ❀TRY.

A. dimorpha (Nutt.) Torrey & A. Gray (p. 199) **STS** < 4 cm, cespitose from much-branched caudex; stolons 0. **LVS** basal, 8–11 mm, linear to narrowly spoon-shaped, 1-veined, ± gray-tomentose. **INFL**: head 1; involucres 6–8 mm (staminate) or 10–11 mm (pistillate), base hairy; phyllaries narrow, acute-acuminate, dingy brown. **STAMINATE FLS**: corollas 3–5 mm; pappus bristle tips slender. **PISTILLATE FLS**: corollas 8–10 mm. **FR** 2–3.5 mm, hairy; pappus 10–12 mm. 2*n*=28,56. Dry places; 800–2400 m. KR, NCoRH, SNH, TR, GB; to sw Can, MT, WY, NE.

A. flagellaris (A. Gray) A. Gray STOLONIFEROUS PUSSY-TOES **STS** < 1.5 cm, from slender caudex; stolons 3–10 cm, slender, lfless exc tip. **LVS** basal, 16–18 mm, linear-oblanceolate, 1-veined, ± gray-tomentose. **INFL**: head 1; involucres 6–7 mm (staminate) or 7–9 mm (pistillate), base hairy; phyllaries wide, tips acute, brown to blackish or whitish. **STAMINATE FLS**: corollas 3–4.5 mm; pappus bristle tips slender. **PISTILLATE FLS**: corollas 5–7 mm. **FR** 2–3 mm, papillate; pappus 6–8 mm. 2*n*=28. UNCOMMON. Seasonally moist sagebrush scrub; 1750 m. MP; to WA, ID, WY, NV.

A. geyeri A. Gray (p. 199) **STS** 3–14 cm; base woody, branched; stolons 0. **LVS** cauline, 11–35 mm, linear-lanceolate, 1-veined, ± gray-tomentose. **INFL**: heads 3–25; involucres 6–8 mm, densely tomentose below; phyllaries wide, upper part red to pink, light brown, or white, tips blunt (staminate) or acute (pistillate). **STAMINATE FLS**: corollas 3–4.5 mm; pappus bristle tips slender. **PISTILLATE FLS**: corollas 5–6 mm. **FR** 2–2.5 mm, hairy, papillate; pappus 6–7 mm. 2*n*=28. Dry, open woods, shrubland; 900–2400 m. KR, NCoRH, n SNH, MP; to WA, NV. ❀DRN,SUN:(1–3,7, 15–17);DFCLT.

A. howellii E. Greene ssp. ***howellii*** Pls pistillate. **STS** 15–30 cm; stolons 1–4 cm. **LVS**: basal 25–40 mm, oblanceolate to spoon-shaped, 1–3-veined, green above, tomentose below; cauline 20–40 mm, linear. **INFL**: heads 5–12; involucres 6–7.5 mm, base hairy; phyllaries narrow, acute, upper part light brown to white. **FLS**: corollas 5–6 mm. **FR** 1.5–2 mm, papillate; pappus 6–8 mm. 2*n*=56,84. Open, pine woods, rocky slopes; 1300–2000 m. KR; to w Can, MT, SD. [*A. neglecta* E. Greene var. *h.* Cronq.; *A. neodioica* E. Greene ssp. *h.* (E. Greene) R. Bayer] ❀DRN,IRR:**1–3,6,7,14–16**&SUN:**4,5,17**.

A. luzuloides Torrey & A. Gray **STS** 7–25 cm, from partly woody caudex; stolons 0. **LVS**: basal 18–45 mm, linear to spoon-shaped, 1–3-veined, ± gray-tomentose; cauline 12–45 mm. **INFL**: heads 8–30; involucres 4–5 mm, narrow, glabrous; phyllaries narrow, acute, pale brown to whitish. **STAMINATE FLS**: corollas 2–3.5 mm. **PISTILLATE FLS**: corollas 2.5–4 mm. **FR** 1–2 mm, papillate; pappus 2.5–4 mm. 2*n*=28. Dry, open slopes, sagebrush scrub; 1300–1900 m. n SNH, MP; to B.C., MT, Colorado, NV. [*A. microcephala* A. Gray]

A. marginata E. Greene Pls pistillate (in CA). **STS** 15–18 cm; stolons horizontal, 2–7 cm. **LVS**: basal 15–20 mm, spoon-shaped, 1-veined, green above with white-woolly margins, gray-tomentose below; cauline 7–16 mm, linear. **INFL**: heads 5–8; involucre 5–7 mm, sparsely hairy at base; phyllaries wide, tips acuminate, upper part white. **FLS**: corollas 4.5–6.5 mm. **FR** 0.8–2 mm, papillate; pappus 5.5–8.5 mm. 2*n*=28,56,84,112,140. Uncommon. Dry woods; ± 2150 m. SnBr (s Fork Santa Ana River); to Colorado, NM.

A. media E. Greene (p. 199) **STS** 5–13 cm; stolons many, matted. **LVS**: basal 6–19 mm, ± linear to spoon-shaped, 1-veined, densely woolly; cauline 5–20 mm, linear. **INFL**: heads 2–7; involucres 4–8 mm, base woolly; phyllaries narrow and acute (pistillate) or wide and blunt (staminate), upper part dark brownish black. **STAMINATE FLS**: corollas 2.8–4.5 mm. **PISTILLATE FLS**: corollas 3–4.5 mm. **FR** 0.6–1.6 mm, papillate or smooth; pappus 4–5.5 mm. 2*n*=56,98,112. High meadows, snow basins, ridges; 1800–3900 m. KR, NCoRH, CaRH, SNH, SnBr, W&I; to sw Can, MT, Colorado, NM. [*A. alpina* (L.) Gaertner var. *m.* (E. Greene) Jepson] ❀DRN,IRR,SUN:1–3,7,14–17;DFCLT.

A. pulchella E. Greene BEAUTIFUL PUSSY-TOES **STS** 3–12 cm; stolons many, matted. **LVS**: basal 6–12 mm, ± linear to spoon-shaped, 1-veined, densely woolly, often purple-glandular; cauline 3–11 mm, linear. **INFL**: heads 4–6; involucres 3.5–5 mm, base woolly; phyllaries narrow and acute (pistillate) or wide and blunt (staminate), upper part gen dark brownish black (to whitish at very tip). **STAMINATE FLS**: corollas 1.9–2.8. **PISTILLATE FLS**: corollas 2–3 mm. **FR** 0.7–1.3 mm, papillate or smooth; pappus 2.5–3.5. 2*n*=28. UNCOMMON. High meadows, snow basins, ridges; 2800–3700 m. SNH. [*A. alpina* (L.) Gaertner var. *scabra* Jepson]

A. racemosa Hook. **STS** 12–50 cm, glandular above; stolons horizontal, 3–8 cm. **LVS**: basal 30–100 mm, elliptic, 3-veined, green above, tomentose below; cauline 10–30 mm. **INFL**: heads 3–12; cluster ± raceme-like; peduncles 10–30 mm; involucres 4–9 mm (staminate < pistillate), wide, blunt or acute, finely glandular, upper part white to light brown. **STAMINATE FLS**: corollas 3–4 mm; pappus bristle tips slender. **PISTILLATE FLS**: corollas 3.5–5.5 mm. **FR** 1–1.5 mm, barely papillate; pappus 4.5–7 mm. 2*n*=28. Montane forest; 1200–2000 m. KR; to sw Can, SD, WY.

A. rosea E. Greene Pls gen all pistillate. **STS** 9–40 cm; stolons short, horizontal or ascending. **LVS**: basal 8–40 mm, spoon- to wedge-shaped, 1-veined, ± gray-tomentose; cauline 6–36 mm. **INFL**: heads 3–16; involucre 3.5–7.5 mm, base hairy; phyllaries wide, acute, tips white, rose, yellowish, or brownish. **PISTILLATE FLS**: corollas 2.5–5.5 mm. **FR** 0.7–1.8 mm, papillate or smooth; pappus 3.5–6.5 mm. 2*n*=42,56,70. Woods, meadow edges, rock barrens, dry ridges; 1200–3700 m. KR, CaRH, SNH, SnGb, SnBr, SnJt, GB; to AK, e Can, NM. [*A. microphylla* Rydb. misapplied, in part] Highly variable; sspp. intergrade extensively.

ssp. ***confinis*** (E. Greene) R. Bayer (p. 199) **STS** 9–25 cm; stolons decumbent, 15–45 mm. **LVS**: basal 8–20 mm; cauline 6–20 mm. **INFL**: heads 4–11; involucre 4–6.5 mm; phyllaries often brown. **FL**: corollas 2.5–4 mm. **FR**: pappus 3.5–5 mm. Habitat and range of sp.

ssp. ***rosea*** **STS** 10–40 cm; stolons decumbent, 20–70 mm. **LVS**: basal 20–40 mm; cauline 8–36 mm. **INFL**: heads 6–20; involucre 5–8 mm; phyllaries not brown. **FL**: corollas 3–4.5 mm. **FR**: pappus 4–6 mm. Habitats and range of sp. ❀DRN,IRR,SUN:1–3,6,7,15–17;DFCLT.

A. suffrutescens E. Greene EVERGREEN EVERLASTING **STS** many, 5–12 cm, densely tufted, woody at base; stolons 0. **LVS** dense, 5–12 mm, spoon-shaped, evergreen, 1-veined, green above, tomentose below; tips notched. **INFL**: head 1; involucres 5–9 mm (staminate) or 10–15 mm (pistillate), woolly and finely glandular at base; phyllaries wide, blunt or acute, greenish yellow below, upper part white. **STAMINATE FLS**: corollas 4–5 mm. **PISTILLATE FLS**: corollas 5–8 mm. **FR** 1–2 mm, papillate; pappus 7–9 mm. 2*n*=28. UNCOMMON. Dry, open coniferous woods, serpentine barrens; 500–1600 m. KR, NCoRH; OR. ❀DRN,IRR,SUN:1,5–7,15–17;DFCLT.

A. umbrinella Rydb. **STS** 7–16 cm; base ± woody; stolons or sterile shoots ascending, slightly woody. **LVS**: basal 10–17 mm, spoon- to wedge-shaped, 1-veined, ± gray-tomentose; cauline 8–18 mm. **INFL**: heads 3–8; involucres 3–6.5 mm, base woolly; phyllaries narrow and acute (pistillate) or wide and blunt (staminate), upper part whitish to pale brownish. **STAMINATE FLS**: corollas

2.5–3.5 mm. **PISTILLATE FLS**: corollas 2.5–3.5 mm. **FR** 0.5–1.2 mm, glabrous; pappus 3–5 mm. 2*n*=28,56. Uncommon. Dry sagebrush scrub, open yellow-pine forest; 1800–2000 m. n SNH (Plumas Co.); to sw Can, MT, Colorado, NV.

ANTHEMIS DOG-FENNEL, CHAMOMILE

Elizabeth McClintock

Ann, per, gen aromatic. **STS** erect, gen branched. **LVS** alternate, 1–3-pinnately lobed. **INFL**: heads gen radiate, solitary; phyllaries overlapping in several series, margins scarious; receptacle conic, chaffy; chaff scales narrowly linear. **RAY FLS** < 25, fertile or sterile; ligules white or yellow. **DISK FLS** many; corollas yellow; anther tips ovate, bases rounded or ± cordate; style branches truncate with shrub-like tips. **FR** cylindric, ribbed or angled; pappus 0 or a short crown. ± 200 spp.: Eur, w Asia, n Afr. (Greek for chamomile, *Anthemis nobilis*, see *Chamaemelum n.*) [Ross-Craig 1960–1963 Drawings Brit Pls 16: plates 4,5]

1. Per; ligules yellow; lvs 1-pinnately lobed .... ***A. tinctoria***
1' Ann; ligules white; lvs 2–3-pinnately lobed
 2. Herbage not scented; receptacle chaffy throughout; fr smooth .... ***A. arvensis***
 2' Herbage ill-smelling; receptacle chaffy only in upper half; fr tubercled along ribs .... ***A. cotula***

A. arvensis L. CORN CHAMOMILE Ann; herbage gen not scented. **STS** gen > 20 cm, erect or lowermost decumbent, sometimes rooting at nodes, gen branched from base, ± hairy. **LVS**: blades 2–5 cm, 2-pinnately lobed. **INFL**: heads radiate, 2–3 cm diam, gen solitary; peduncles long; phyllaries > 5 mm, green, margins brown, scarious; receptacle conic, chaffy throughout; chaff persistent, scales lanceolate, tapered upward to a sharp awn, ± same length as disk fls. **RAY FLS** 15–20, pistillate; ligules 10–15 mm, white, reflexed. **FR** ± 2 mm, smooth, 5–10-ribbed; pappus a short crown. 2*n*=18. Uncommon. Escape from cult in disturbed areas, roadsides, fields; > 1000 m. KR, NCoRO, c SNH, expected more widely; native to Eur.

A. cotula L. (p. 199) MAYWEED, STINKWEED, DOG-FENNEL Ann; herbage ill-scented. **STS** < 24 cm, erect, gen lightly hairy. **LVS** < 6 cm, 2–3-pinnately divided; segments linear. **INFL**: heads radiate, 1–2.5 cm diam; phyllaries ± 2.5 mm, green, margins scarious, pale brown; receptacle conic, chaffy in upper half, chaff persistent, 2–3 mm. **RAY FLS** sterile, 10–15; ligules 6–9 mm, white, reflexed in age. **FR** 1–2 mm; 10-ribbed; ribs with tubercles; pappus a short crown. 2*n*=18. Abundant. Weed in disturbed areas, fields, roadsides, coastal dune areas, chaparral, oak woodland; < 2000 m. NW, c SN, GV, CW, SCo, TR, PR; native to Eur.

A. tinctoria L. GOLDEN MARGUERITE Per; herbage not scented. **STS** erect, < 30 cm, hairy, sparingly branched. **LVS** 2–5 cm, pinnately lobed; segments toothed or shallowly lobed. **INFL**: heads radiate, 2.5 cm diam; phyllaries 3 mm, green, margins scarious; receptacle hemispheric, chaffy throughout, chaff persistent, scales 2 mm, linear. **RAY FLS** 10–20; ligules 7–15 mm, yellow. **FR** ± angled, 2 mm; pappus a short crown. 2*n*=18. Uncommon. Escape from cult in disturbed areas, roadsides; < 2000 m. NCoRI, CaRF, n SNH; native to s Eur, w Asia. Fl heads yield a yellow dye.

ARCTIUM BURDOCK

David J. Keil & Charles E. Turner

Bien. **STS** 1–several, branched above. **LVS** basal and cauline, alternate, long-petioled, widely ovate; base deeply cordate; margin entire or toothed, gradually reduced upward in CA. **INFL**: heads discoid, in leafy-bracted clusters; involucre ± spheric; phyllaries overlapping in many series, ± linear in CA, bases appressed, tips stiffly radiating, hooked-spiny; receptacle bristly. **FLS** many; corolla pink to ± purple, lobes narrowly triangular; anther base tailed, tips ovate, acute to obtuse; style branched just above hairy ring, branches oblong, obtuse. **FR** ± compressed, rough or ribbed, glabrous, attachment basal; pappus several series of rough bristles, readily deciduous. 10 spp.: Eurasia. (Greek: bear)

1. Heads gen 25–40 mm diam, gen long-peduncled in ± flat-topped clusters; inner phyllaries gen green, margins minutely hairy; pappus 2–6 mm .... ***A. lappa***
1' Heads gen 10–25 mm diam, sessile to short-peduncled in raceme-like clusters; inner phyllaries gen purple-tinged, margins minutely serrate; pappus ± 2 mm .... ***A. minus***

A. lappa L. Pl < 25 dm. **ST**: branches ascending. **LF**: blade < 40+ cm, green, nearly glabrous above, thinly gray-tomentose below. **INFL**: heads gen ± long-peduncled in ± rounded or flat-topped clusters; involucre 25–40 mm diam; inner phyllaries gen green, margins minutely spreading- or reflexed-hairy. **FL**: corolla 9–12 mm. **FR** 6–7 mm, ± brown, sometimes dark-spotted; pappus 2–6 mm. 2*n*=32. Uncommon. Disturbed places; < 500 m. CA-FP; native to Eur.

A. minus (Hill) Bernh. (p. 199) Pl < 15 dm. **ST**: branches stiffly ascending. **LF**: blade < 40+ cm, green, puberulent or soft-hairy above, becoming glabrous, ± gray-tomentose below. **INFL**: heads sessile to short-peduncled in raceme- or panicle-like clusters; involucre 10–25 mm diam; inner phyllaries gen purple-tinged, margins minutely serrate. **FL**: corolla 8–10 mm. **FR** 4–7.5 mm, dark brown or dark-spotted; pappus ± 2 mm. 2*n*=32,36. Disturbed places; < 1200 m. NW; native to Eur.

ARCTOTHECA

Elizabeth McClintock

Ann, per. **STS** creeping or decumbent, spreading by stolons or rhizomes. **LVS** gen pinnately lobed; basal in loose rosettes; cauline alternate. **INFL**: heads radiate, solitary, long-peduncled; phyllaries overlapping in 3–6 series; receptacle flat, naked. **RAY FLS** gen sterile; ligules yellow. **DISK FLS** many; corolla funnel-shaped; stamen tips ovate-triangular, bases minutely sagittate; style slender below, thickened above a minutely hairy node, branches very short. **FR** 3–5-ribbed, glabrous or woolly; pappus of scales or 0. ± 6 spp.: s Afr, Australia. (Greek: bear container)

A. calendula (L.) Levyns CAPEWEED **STS** creeping to decumbent, ribbed, soft, ± juicy, tomentose. **LVS** 5–25 cm, pinnately lobed, or upper nearly entire; upper surface finely cobwebby; lower surface densely white-woolly. **INFL**: heads ± 5 cm diam; peduncles scapose, 15–20 cm; outer phyllaries green, woolly, tips reflexed. **RAY FLS** < 20; ligules yellow above, greenish or purplish below. **DISK FLS** many, purplish or yellow. **FR** ± 3 mm, covered with woolly hairs; pappus scales ± obscured by woolly hairs. 2*n*=18. Disturbed, urban, coastal habitats; < 250 m. NCo, SnFrB; native to s Afr. A vegetatively reproducing race is cult, sometimes escapes; a seed-producing race in highly invasive and NOXIOUS, as yet uncommon in CA.

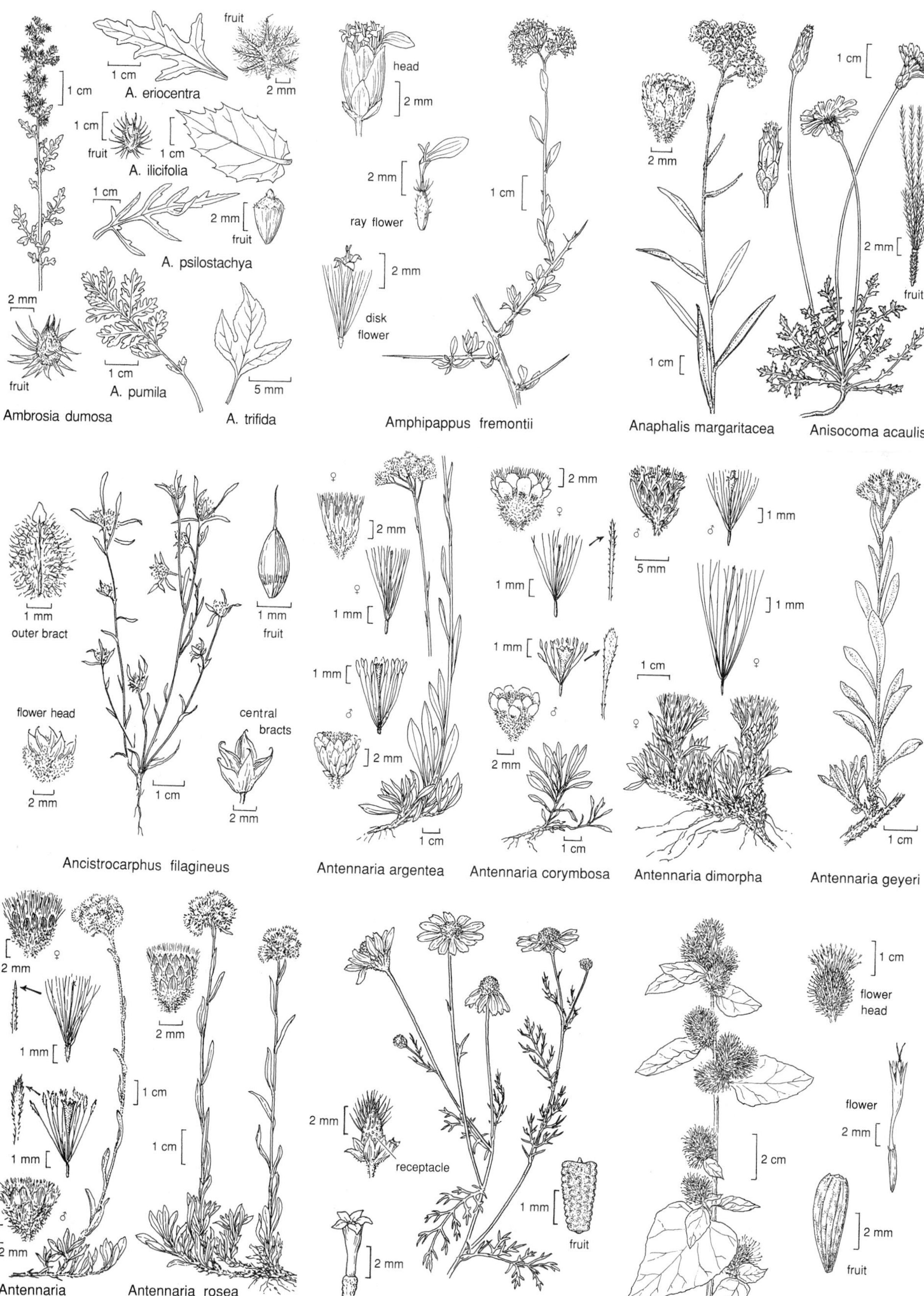
fruit
1 cm
A. eriocentra
2 mm
1 cm
1 cm
fruit
1 cm
A. ilicifolia
1 cm
2 mm
fruit
A. psilostachya
2 mm
fruit
1 cm
A. pumila
5 mm
Ambrosia dumosa
A. trifida
head
2 mm
2 mm
ray flower
2 mm
disk flower
1 cm
Amphipappus fremontii
2 mm
1 cm
1 cm
2 mm
fruit
Anaphalis margaritacea
Anisocoma acaulis
1 mm
outer bract
1 mm
fruit
flower head
2 mm
1 cm
central bracts
2 mm
Ancistrocarphus filagineus
2 mm
1 mm
1 mm
2 mm
1 cm
Antennaria argentea
2 mm
1 mm
1 mm
2 mm
1 cm
Antennaria corymbosa
1 mm
5 mm
1 mm
1 cm
Antennaria dimorpha
1 cm
Antennaria geyeri
2 mm
1 mm
1 mm
2 mm
1 cm
Antennaria media
2 mm
1 cm
Antennaria rosea ssp. confinis
2 mm
receptacle
2 mm
1 mm
fruit
Anthemis cotula
1 cm
flower head
flower
2 mm
2 cm
2 mm
fruit
Arctium minus

ARCTOTIS AFRICAN DAISY

Elizabeth McClintock

Per. **STS** erect or decumbent, branched. **LVS** alternate, entire to pinnately lobed. **INFL**: heads radiate, solitary, long-peduncled; phyllaries overlapping in several series; receptacle flat, pitted, gen bristly. **RAY FLS** 15–25; ligules yellow. **DISK FLS** many; corolla purple or brown; stamen tips ovate-triangular, bases minutely sagittate; style slender below, thickened above a minutely hairy node, branches very short. **FR** obovoid, gen hairy; pappus of 2 rows of overlapping scales. ± 100 spp.: s Afr. (Greek: bear's ear, from shape of pappus scales) [Norlindh 1964 Svensk Bot Tidskr 58:194–203]

A. stoechadifolia P. Bergius BLUE-EYED AFRICAN DAISY **STS** stout, 7.5–10 dm, densely white-woolly. **LVS**: blade 1–4 cm, obovate to oblong, gen fiddle-shaped, pinnately lobed, lobes entire or irregularly toothed, white-woolly when young, less so later. **INFL**: heads < 8 cm diam; peduncles gen > lvs. **RAY FLS** ± 20; ligules < 3 cm, white above, violet below. **DISK FLS**: corollas violet. **FR** 2–3 mm, obovoid, 3–5-ridged, with 2–3 cavities on one side; base with many straight hairs > achene; inner row of pappus scales > achene and its hairs, outer scales < achene. $2n=18$. Uncommon. Escape from cult in urban coastal areas, waste ground, roadsides; < 300 m. SCo; native to s Afr. [var. *grandis* (Thunb.) Less.]

ARGYRANTHEMUM

Elizabeth McClintock

Subshrubs. **STS** branched. **LVS** alternate, gen variously lobed. **INFL**: heads radiate, solitary, gen standing above lvs; phyllaries overlapping, in 3–4 series, green, margins scarious; receptacle convex to conic, naked. **RAY FLS** white, yellow, or pink. **DISK FLS** many; corollas yellow; anther tips ovate, bases rounded or ± cordate; style branches truncate with shrub-like tips. **FR**: ray achenes 3-angled, 2–3-winged, pappus a crown; disk achenes 4-angled or compressed, wings 1–2 or 0, pappus a crown. 22 spp.: Canary Islands, Madeira. [Humphries 1976 Bull Brit Mus Bot 5:147–240] *A. frutescens* (Paris daisy) is widely cult.

A. foeniculaceum (Willd.) Schultz-Bip. **STS** gen < 1 m, robust, much-branched. **LVS** < 10 cm, obovate in outline, deeply 2- or 3-pinnately lobed, glaucous; primary lobes ± 8, segments linear. **INFL**: heads ± 5 cm diam; phyllaries with jagged, scarious margins, midrib ± fleshy. **RAY FLS** ± 20; ligules 15–22 mm, white. **DISK FLS** many; corollas 4 mm. **FR**: ray achenes 5–6 mm, 3-angled, 3-winged; disk achenes < 3 mm, compressed, wing 1, small, pappus a short, toothed crown. $2n=18$. Uncommon. Escape from cult in disturbed, urban, coastal areas; < 200 m. SnFrB, SCo; native to Canary Islands. [*Chrysanthemum f.* (Willd.) Desf.; *C. anethifolium* Willd.]

ARNICA

Theodore M. Barkley

Per gen from long, naked rhizome. **LVS**: basal 0 or gen withered by fl; cauline opposite. **INFL** ± flat-topped; heads radiate or discoid, 1–many; involucre hemispheric to obconic; phyllaries gen in 2 ± equal series; receptacle ± flat, naked. **RAY FLS** (0)6–21; ligules (orange-) yellow. **DISK FLS** many; corolla gen soft-hairy, colored like ligules; anther bases entire or slightly sagittate, tips triangular; style branches flat, tips truncate, very short, hair-tufted. **FR** ± cylindric, 5–10-veined; pappus of many barbed to subplumose bristles, white to red-brown. ± 27 spp.: N.Am, Eurasia. (Latin or Greek: ancient name) [Downie & Denford 1988 Rhodora 90:245–275] Diploid spp. sexual; polyploid spp. gen form seeds asexually.

1. Heads discoid (some marginal fls may have ± expanded corollas)
  2. Lvs widely sessile
    3. Lf toothed, net-veined between 3–5 prominent veins; fr hairy, nonglandular .... ***A. venosa***
    3′ Lf entire, 3–5-veined, not otherwise strongly veined; fr stalked-glandular .... ***A. viscosa***
  2′ Lower and middle lvs (wing-) petioled
    4. Phyllaries very narrow; young heads nodding .... [2]***A. parryi***
    4′ Phyllaries wide; young heads erect
      5. Lowermost lvs ± cordate; petiole weakly winged; fr both hairy and stalked-glandular .... ***A. discoidea***
      5′ Lowermost lvs ± spoon-shaped; petiole weakly defined, widely winged; fr stalked-glandular, gen otherwise glabrous .... ***A. spathulata***

1′ Heads radiate (pls rarely discoid)
  6. Cauline lvs gen 5–12 pairs, ± equal or gradually reduced upward; heads 4–12
    7. Phyllaries gen ± obtuse, inner face of tip hair-tufted; tube of disk corolla gen 3–4.5 mm .... ***A. chamissonis*** ssp. ***foliosa***
    7′ Phyllaries ± acute, tip no more hairy than body; tube of disk corolla gen 2–3 mm
      8. Sts 1–few from rhizome; lvs toothed .... ***A. amplexicaulis***
      8′ Sts clustered on short caudex; lvs ± entire .... ***A. longifolia***
  6′ Cauline lvs gen 2–4(5) pairs, often strongly reduced upward; heads gen < 5
    9. Pappus (yellow-) brown, bristles subplumose — rhizome short, freely rooting
      10. Rays < 15 mm; lower herbage ± long-spreading-hairy .... [2]***A. parryi***
      10′ Rays gen 15–25 mm; lower herbage becoming glabrous to hairy or glandular
        11. Heads ± narrowly obconic; cauline lvs elliptic to deltate, middle gen largest .... ***A. diversifolia***
        11′ Heads ± hemispheric; cauline lvs variable, lowest gen largest .... ***A. mollis***
    9′ Pappus ± white, bristles short-barbed (if bristles subplumose, rhizome long)
      12. Lf blade narrow (length ± 3–6+ × > width); basal lvs gen densely clustered; fr < 6 mm
        13. Lower axils with tufts of long, brown, woolly hairs; disk corollas spreading-hairy, stalked-glandular .... ***A. fulgens***

13′ Lower axils glabrous or sparsely white-hairy; disk corollas stalked-glandular only .... ***A. sororia***

12′ Lf blade gen ± wide (length gen < 2.5 × width); basal lvs sometimes persistent, not densely clustered; fr 6–8.5 mm

14. Fr glabrous toward base, gen hairy above — lvs ± toothed; involucre with 0–few long, prominent hairs

15. Pappus weakly subplumose; heads solitary; cauline lvs petioled; pls of serpentine soils .... ***A. cernua***

15′ Pappus short-barbed; heads 1–3(+); cauline lvs ± sessile; pls not of serpentine soils .... ***A. latifolia***

14′ Fr short-hairy (or glandular) ± to base

16. Phyllaries ± obtuse, with hair-tuft on inner face near tip; heads 3–7; lvs ovate to elliptic, lightly tomentose — uncommon .... ***A. tomentella***

16′ Phyllaries ± acute, hair-tuft near tip gen 0; heads 1–3 (or 3+ in pls with cordate lvs); lvs either cordate or glandular

17. Lower lvs gen toothed, ± strongly cordate (exc stunted alpine pls); involucre (esp base) gen long-white-hairy; pappus short-barbed .... ***A. cordifolia***

17′ Lower lvs ± entire, gen rounded (subcordate); long involucre hairs 0–few; pappus ± subplumose .... ***A. nevadensis***

A. amplexicaulis Nutt. (p. 207) Pl 3–8 dm from short rhizome or caudex, gen hairy and glandular, esp upward. **STS** 1–few, sometimes branched above middle. **LVS**: basal petioled; cauline 5–10 pairs, ± sessile, 4–12 cm, ± elliptic to ovate, ± toothed. **INFL**: heads radiate, (3)4–9+; involucre 9–15 mm, bell-shaped or widely obconic; phyllaries acute, tip hairier. **RAY FLS** 8–14; ligules 1–2 cm. **FR** 4–6 mm, sparsely hairy, sometimes glandular; pappus subplumose, brownish. 2*n*=38,57,76. Moist, open woodlands, streambanks; 2200–3500 m. KR, CaRH, SNH, SNE; to s AK, MT. Gen asexual. ❀TRY.

A. cernua Howell (p. 207) SERPENTINE ARNICA Pl 1–3 dm, (sub)glabrous. **STS** 1(–3), gen unbranched. **LVS**: cauline gen 3–4 pairs, petioled, uppermost smaller, subsessile, blade 1–5+ cm, (ob)ovate to subcordate, subentire to weakly lobed, thickish. **INFL**: heads radiate, 1(–4); involucre 12–18 mm, obconic to bell-shaped; phyllaries acute to acuminate, short-hairy, often glandular, long-ciliate. **RAY FLS** 7–10; ligules < 3 cm. **FR** 6–8 mm, ± forked-hairy above; pappus short-barbed or subplumose, white. 2*n*=38. UNCOMMON. Serpentine soils, coniferous forest; 500–1500 m. KR; sw OR. ❀DRN,IRR,SHD:5,7,15–17;DFCLT.

A. chamissonis Less. ssp. ***foliosa*** (Nutt.) Maguire (p. 207) Pl 3–8 dm, ± hairy, gen glandular upward. **ST** 1, branched above. **LVS**: cauline 5–10 pairs, sessile or short-petioled, 5–20(30) cm, blade (ob)lanceolate, (sub)entire, upper ± reduced. **INFL**: heads radiate, 5–10; involucre 8–15 mm, bell-shaped; phyllaries ± obtuse, inner face gen hair-tufted near tip. **RAY FLS** gen 13; ligules 1–2 cm. **FR** 4–5 mm, ± glabrous to short-hairy, glandular; pappus short-barbed or weakly subplumose, dirty white to straw-colored. 2*n*=38,57,106–108. Damp meadows, rocky places, coniferous forest; 1800–3500 m. KR, CaR, SN, SnBr, Wrn, W&I; to B.C., MT, NM. [vars. *bernardina* (E. Greene) Maguire, *incana* (A. Gray) Hultén, *jepsoniana* Maguire] Gen asexual. ❀IRR,SHD:1–3,**4–7**,10,14,**15–18**,19–24.

A. cordifolia Hook. Pl 1–5 dm, unevenly glandular and short-hairy; rhizome tip ± scaly. **STS** 1–few, loose, gen unbranched. **LVS**: basal often persistent on sterile rosettes; cauline 2–4 pairs, petioled, reduced upward, blades 3–11 cm, cordate, gen shallowly toothed. **INFL**: heads radiate, 1–5; involucre 15–20(30) mm, obconic to bell-shaped; phyllaries acute to acuminate, ± long-white-hairy, glandular or not. **RAY FLS** (6)10–15; ligules < 3 cm. **DISK FLS**: corolla soft-hairy, glandular. **FR** 6–10 mm, short-forked-hairy or glandular; pappus short-barbed, whitish. 2*n*=38,57,76,95,114,152, ±198. High meadows, coniferous forest; 1200–3000 m. KR, NCoR, CaRH, n&c SNH, e SnFrB, MP; to WY, n-c N.Am, NM. Gen asexual. Stunted alpine pls with ovate lvs, glandular fr have been called var. *pumila* (Rydb.) Maguire. ❀DRN,IRR:**1**,2,3,**4–7**,14,**15–17**,18–24.

A. discoidea Benth. Pl 2–6 dm, short-glandular, gen ± long-hairy. **ST** gen 1, branched or not. **LVS**: basal often in prominent sterile rosettes; cauline 3–7 pairs, lower petioled, blade 2–12 cm, lanceolate to ovate (subcordate), ± toothed, reduced upward. **INFL**: heads discoid, 3–10(30); involucre 10–17 mm, obconic to subhemispheric; phyllaries acute to acuminate, stalked-glandular, long-hairy. **DISK FLS**: corolla ± soft-hairy, glandular. **FR** 6–8 mm, glandular and forked-hairy above; pappus strongly short-barbed, white or straw-colored. 2*n*=38,76. Chaparral, foothill woodland; 500–1500 m. NW, CaR, n&c SN, CW, WTR, w PR; to WA, w NV. [vars. *alata* (Rydb.) Cronq. & *eradiata* (A. Gray) Cronq.] Sexual or not. Highly variable; pls with expanded outer disk corollas resemble *A. cordifolia*. ❀SHD,DRY:1–3,6,7,14–24;DFCLT.

A. diversifolia E. Greene Pl 1–4 dm from short rhizome, ± glandular-hairy. **STS** 1–several, unbranched. **LVS**: cauline (2)3–4 pairs, short-petioled, middle largest, blade 4–8 cm, narrowly elliptic to deltate, irregularly toothed. **INFL**: heads radiate, 1–5; involucre 10–14 mm, ± obconic; phyllaries acute, soft-hairy or short-glandular. **RAY FLS** ± 8 or 13; ligules 1–2 cm. **FR** 5–6.5 mm; pappus subplumose, straw-colored to brownish. 2*n*=38,57,76,108–114. Uncommon. Grassy or wet, rocky sites, coniferous forest; 1800–3400 m. KR, CaRH, SNH; to AK, MT, Colorado. Gen asexual.

A. fulgens Pursh HILLSIDE ARNICA Pl 1–6 dm, from short, densely scaly, branched rhizome, stalked-glandular, gen hairy, esp upward; axils brown-woolly-tufted. **STS** 1–several, unbranched. **LVS**: basal persistent, petioled, blade 3–12 cm, narrowly elliptic to oblanceolate, ± entire; cauline 2–4 pairs, reduced upward. **INFL**: heads radiate, 1(3); involucre 10–15 mm, widely hemispheric; phyllaries gen ± obtuse, glandular, hairy on both faces, ± ciliate. **DISK FLS**: corolla hairy, stalked-glandular. **RAY FLS** ± 13 or 21; ligules 1–3 cm. **FR** 4–6 mm, forked-hairy; pappus short-barbed, white or pale straw-colored. 2*n*=38. RARE. Open, damp depressions in sagebrush scrub or grassland; 1800–2700 m. n SNH (e slope), MP; to B.C., c N.Am, Colorado.

A. latifolia Bong. Pl 1–5 dm from ± scaly caudex and long rhizomes, becoming glabrous to unevenly soft-hairy and glandular. **STS** 1–several, gen unbranched. **LVS**: basal sometimes in sterile rosettes, petioled; cauline gen 2–4 pairs, ± subsessile, middle gen largest, blade 2–10(15) cm, lanceolate to subcordate, gen toothed. **INFL**: heads radiate, 1–3(5); involucre 7–18 mm, obconic to subhemispheric; phyllaries acute to acuminate, gen glandular, sometimes sparsely long-hairy. **RAY FLS** 8–11(13); ligules 1–3 cm. **DISK FLS**: corolla sparsely soft-hairy. **FR** 6–8 mm, short-glandular or short- ± forked-hairy above; pappus short-barbed, white. 2*n*=38,76,112. Meadows, open coniferous forest; 1800–2400 m. KR, c SNH; to AK. Sexual or not. ❀TRY.

A. longifolia D. Eaton Pl 3–6 dm from caudex-like rhizome, scabrous, short-hairy upward, often ± sticky. **STS** clustered, often forming large patches. **LVS**: basal few, < cauline; cauline 5–7 pairs, lowest pairs often fused around st, blade 5–12 cm, ± lanceolate, ± entire, upper ± reduced. **INFL**: heads radiate, 3–20; involucre 7–10 mm, bell-shaped or widely obconic; phyllaries acute, glandular, ± long-hairy, (esp tip). **RAY FLS** 8–13; ligules 1–2 cm. **FR** 4–6 mm, subglabrous to glandular and hairy; pappus short-barbed or subplumose, red- to yellow-brown. 2*n*=38,±50. Gen wet meadows or open coniferous forest; 1800–3500 m. KR, s NCoRH, CaRH, SNH, MP; to w Can, MT, Colorado. [ssp. *myriadenia* (Piper) Maguire] Gen asexual in CA. ❀IRR,DRN,SHD:**1**,2–5,**6**,7,14,**15–17**,18–24.

A. mollis Hook. Pl 2–6 dm from short rhizome or loose caudex, ± hairy, glandular. **STS** 1–several, little-branched. **LVS**: cauline 3–5 pairs, (sub)sessile, lowest gen largest, blade 4–20 cm, (ob)lanceolate to (ob)ovate, entire to unevenly dentate. **INFL**: heads radiate, 1–3(7); involucre 10–16 mm, hemispheric to spreading-bell-shaped; phyllaries acute, sparsely soft-hairy. **RAY FLS** 12–18; ligules 1–3 cm. **FR** 6–7 mm, forked-hairy, gen glandular; pappus subplumose, (yellow-) brown. $2n$=38,57,76,152. Meadows, streambanks in subalpine zone; 2500–3500 m. KR, CaRH, SNH; to WA, w Can, MT, Colorado. Sexual or not. ❀TRY.

A. nevadensis A. Gray Pl 1–3 dm, glandular, short- (or long-) hairy; rhizome tip scaly. **STS** 1–several, loose, gen unbranched. **LVS**: basal sometimes in sterile rosettes; cauline 2–3 pairs, short-petioled, lower largest, blade 3–8 cm, elliptic to widely ovate, (sub)entire, base rounded to subcordate. **INFL**: heads radiate, 1–3; involucre 10–18 mm, ± obconic; phyllaries acute to acuminate, stalked-glandular, base often ± long-hairy. **RAY FLS** 6–11(15); ligules gen 1–2 cm. **DISK FLS**: corolla glandular. **FR** 7–8.5 mm, glandular or short-forked-hairy; pappus short-barbed to subplumose, white to straw-colored. $2n$=38,76. Coniferous forest, meadows; 2000–3000 m. KR, CaRH, n&c SNH; to WA, NV. Sexual or not. Alpine forms intergrade with *A. cordifolia*.

A. parryi A. Gray Pl 1–6 dm from short rhizome; hairs various. **ST** 1, gen unbranched. **LVS**: basal ± sessile, ± ovate, < cauline; cauline 3–4 pairs, lower pairs ± crowded, petioled, blade 5–20 cm, lanceolate to ± ovate, (sub)entire, upper pairs sessile, reduced. **INFL**: heads gen radiate, (1)3–9, gen nodding in bud; involucre 15–20 mm, ± obconic; phyllaries acute, sparsely soft-hairy. **RAY FLS** (0)6–10; ligules < 1.5 cm. **FR** 4–6 mm, glabrous to unevenly hairy and glandular; pappus subplumose, brownish yellow. $2n$=38,57,76,±97. Slopes in coniferous forest; 600–2400 m. s NCoRH, n&c SNH; to nw Can, MT, Colorado. CA pls gen radiate, have been called ssp. *sonnei* (E. Greene) Maguire.

A. sororia E. Greene (p. 207) TWIN ARNICA Pl 1–5 dm from short, sparsely scaly rhizome, stalked-glandular, unevenly hairy upward; axils often sparsely white-hairy. **STS** 1–several, branched or not. **LVS**: cauline 3–6 pairs, lower petioled, blades 4–14 cm, ± oblanceolate, (sub)entire; upper lvs sessile, reduced. **INFL**: heads radiate, 1–5; involucre 10–15 mm, widely hemispheric; phyllaries acute, stalked-glandular, hairy, ciliate, inner face glabrous. **DISK FLS**: corollas stalked-glandular. **RAY FLS** 8–17; ligules 1–3 cm. **FR** 3–5.5 mm, hairy, sometimes sparsely glandular; pappus short-barbed, white. $2n$=38. RARE. Open sagebrush scrub; 1400–1800 m. GB; to w Can, WY. Similar to *S. fulgens*.

A. spathulata E. Greene (p. 207) KLAMATH ARNICA Pl 1–5 dm, gen ± stalked-glandular and spreading-hairy. **STS** 1–few, loose, often branched above. **LVS**: basal gen prominent, < 15 cm, ± oblanceolate, ± toothed, petiole obscure; cauline 3–5 pairs, lowest like basal, reduced upward. **INFL**: heads discoid, 3–10(20); involucre obconic to narrowly bell-shaped; phyllaries obtuse to acute, glandular, coarsely spreading-hairy. **DISK FLS**: corolla sparsely soft-hairy and glandular below. **FR** 5–10 mm, sparsely stalked-glandular; pappus short-barbed, ± white. $2n$=38,76. UNCOMMON. Open, dry, disturbed oak/conifer woodland, gen on serpentine; 1000–1800 m; KR; sw OR. [ssp. *eastwoodiae* (Rydb.) Maguire] Sexual or not.

A. tomentella E. Greene Pl 2–5 dm, glandular, hairy. **ST** 1, unbranched. **LVS**: cauline gen 3–4 pairs, larger short-petioled, blade 2–7 cm, elliptic to ovate, subentire to toothed, sparsely tomentose. **INFL**: heads radiate, (3)5–7; involucre gen 10–13 mm, obconic; phyllaries ± obtuse, sparsely hairy, hair-tufted on inner face near tip. **RAY FLS** 9–11; ligules 1–2 cm. **FR** 6–7 mm, short-forked-hairy, sparsely stalked-glandular; pappus strongly short-barbed or subplumose, white to pale straw-colored. Uncommon. Open coniferous forest; 1500–2200 m. KR, CaRH, n SNH (Truckee River), s SNH (Middle Tule River); s OR. Perhaps asexual form of *A. nevadensis*.

A. venosa H.M. Hall (p. 207) SHASTA COUNTY ARNICA Pl < 5 dm from branched, scaly caudex, unevenly hairy, stalked-glandular. **STS** 0–few-branched. **LVS**: basal 0; cauline 6–10 pairs, (sub)sessile, middle largest, blade 3–7 cm, lanceolate to ± ovate, unevenly toothed, strongly 3–5-veined, strongly net-veined below; upper lvs often bract-like. **INFL**: heads discoid, 1; involucre 15–20(+) mm, ± obconic; phyllaries obtuse to acute, hairy, stalked-glandular. **DISK FLS**: corolla densely soft-hairy below. **FR** 6–8 mm, densely forked-hairy; pappus short-barbed, white. $2n$=38. UNCOMMON. Open, often disturbed oak/pine woodland; 400–1400 m. KR. Some pls seem intermediate to *A. discoidea*.

A. viscosa A. Gray (p. 207) MOUNT SHASTA ARNICA Pl 2–5 dm from woody, scaly caudex, densely stalked-glandular, sparsely hairy upward. **STS** 1–several, much-branched. **LVS**: basal 0; cauline 5–10 pairs (fewer on branches), sessile, blade 2–4(5+) cm, ovate-oblong, ± entire, weakly 3–5-veined. **INFL**: heads discoid, gen 15–20; involucre 8–15 mm, ± obconic; phyllaries acute, stalked-glandular, short-hairy. **DISK FLS**: corolla glandular. **FR** 5–6.5 mm, stalked-glandular; pappus short-barbed to subplumose, gen white (brownish). $2n$=38. UNCOMMON. Open, rocky, subalpine to alpine sites; 2000–2500 m. KR, nw CaR; s OR.

ARTEMISIA SAGEBRUSH

Leila M. Shultz

Ann to shrubs, gen aromatic. **LVS** entire to ± lobed, glabrous to densely hairy; hairs glandular (resin-filled) or T-shaped, hollow. **INFL**: gen panicle; heads gen discoid or disciform, in racemes or panicles; involucre ovoid to hemispheric, gen concealing fls; phyllaries in several series, margins scarious; receptacle conic, gen naked. **PISTILLATE FLS** 0–many; corollas gen < 2 mm. **DISK FLS** 4–many, gen forming frs, sometimes staminate; corollas < 2 mm, pale yellow; anther tips acute to awl-shaped; style branches flat, fringed or blunt (sometimes simple, tack-shaped in staminate fls). **FR** < 2 mm, obovoid or fusiform, ribbed or smooth, glabrous, hairy, or resinous; pappus gen 0 or minute crown. ± 300 spp.: esp n hemisphere. (Greek: Artemis, goddess of the hunt, and noted herbalist, Queen of Anatolia) [Keck 1946 Proc Calif Acad Sci (4)25:421–468; Shultz 1983 PhD thesis Claremont Graduate School]

1. Herbs, base sometimes woody; axillary lf clusters gen 0
 2. Lvs gen entire, < 5 mm wide, anise- or tarragon-scented .... ***A. dracunculus***
 2′ Some lvs lobed, gen > 5 mm wide; odor, if any, not anise- or tarragon-scented
  3. Rhizome present
   4′ Lvs gen 1–8 cm wide, ± glabrous above, glabrous or hairy below; sts 3–25 dm
    5. Lowest lvs 2-pinnately divided, lateral lobes acute; lvs gen glabrous above, tomentose (or becoming glabrous) below; strongly lemon-scented; subalpine to alpine .... ***A. michauxiana***
    5′ Lvs entire to pinnately lobed, lateral lobes obtuse to acute, green and glabrous above, densely tomentose below; ± scented (not lemony); coastal and inland valleys
     6. Involucre bell-shaped, 2–4 mm diam; phyllaries gray-hairy, ovate; fls > 15; lvs green to gray-green above .... ***A. douglasiana***
     6′ Involucre narrow, < 2 mm diam; phyllaries yellowish, glabrous to sparsely hairy, oblong to ± ovate; fls < 15; lvs dark green above .... ***A. suksdorfii***
   4. Lvs gen < 1 cm wide, densely gray-hairy (or becoming ± glabrous above); sts gen < 8 dm ..... ***A. ludoviciana***
    7. Infl open; lvs gen 1–2 cm .... ssp. ***albula***
    7′ Infl narrow; lvs gen 2–11 cm
     8. Lvs entire to toothed or shallowly lobed .... ssp. ***ludoviciana***

8′ Lvs deeply 1–2-divided into narrow lobes
9. Involucre 4–5 mm, 4–7 mm diam; lvs white-tomentose .... ssp. ***candicans***
9′ Involucre 2.5–3.5 mm, 2–3.5 mm diam; lvs upper surface often green .... ssp. ***incompta***
3′ Rhizome 0
10. St glabrous, simple, basal lvs 0
11. Per; st wand-like, > 8 dm; lvs coarsely lobed; receptacle chaffy — s SCo, PR .... ***A. palmeri***
11′ Ann or bien; st erect, < 8 dm; lvs finely divided; receptacle naked
12. Heads nodding; ann, aromatic; lvs 2–3-pinnately divided .... ***A. annua***
12′ Heads erect; bien, unscented; lvs 2-pinnately divided .... ***A. biennis***
10′ St hairy, branched from base, from basal lf cluster
13. Lvs densely hairy; phyllaries densely gray-hairy, margins widely scarious; coastal .... ***A. pycnocephala***
13′ Lvs glabrous or ± hairy; phyllaries sparsely hairy, margins dark or scarious; montane
14. Bien or per from taproot; infl narrow, spike-like, phyllary margins scarious ..... ***A. campestris*** ssp. ***borealis***
14′ Per from branched caudex; infl open, phyllary margins black .... ***A. norvegica*** ssp. ***saxatilis***
1′ Subshrubs or shrubs; axillary lf clusters present
15. Pl spiny, compact (hairy); infl short, heads ± hidden by lvs; lvs palmately 3–5-divided; fl spring/early summer .... ***A. spinescens***
15′ Pl unarmed, compact to tall; infl long, stalked, gen ± exceeding lvs, heads not hidden; lf lobes 0–5; fl late summer/autumn
16. Lvs gen > 2 cm, linear, often ± thread-like or lobes < 3 mm wide, gen > 1 cm
17. Pl 6–18 dm; lf segments gen < 1 mm wide, margins curled under; mainland, some ChI .... ***A. californica***
17′ Pl gen < 6 dm; lf segments 1–3 mm wide, flat; s ChI .... ***A. nesiotica***
16′ Lvs gen < 2 cm, oblong or wedge-shaped, gen > 3 mm wide, entire to shallowly 3-lobed at tip, lobes < 1 cm
18. Heads nodding; lvs not clustered, entire or teeth 3, sharply pointed — W&I, DMoj .... ***A. bigelovii***
18′ Heads erect; lvs often in axillary clusters, entire or teeth 3, rounded to acute
19. Lvs gen linear, entire, winter-deciduous; heads gen > 5 mm wide .... ***A. cana*** ssp. ***bolanderi***
19′ Lvs gen wedge-shaped, 3-toothed, persistent in winter; heads gen < 5 mm wide
20. Pl gen < 4 dm; infl narrow, gen < 3 cm wide
21. Shrub with evident trunk at base .... ***A. tridentata*** ssp. ***wyomingensis***
21′ Shrub branched at base, trunk not evident
22. Lvs of fl-sts entire; heads < 3 mm diam, solitary on slender peduncle; phyllaries shiny-resinous, straw-colored .... ***A. nova***
22′ Lvs of fl-sts lobed; heads gen > 3 mm diam, sessile, 2–4 per cluster; phyllaries gray-hairy ***A. arbuscula***
23. Lvs of fl-st shallowly lobed; heads 3–4 mm diam .... ssp. ***arbuscula***
23′ Lvs of fl-st deeply lobed; heads 2–3 mm diam .... ssp. ***thermopola***
20′ Pl gen > 4 dm; infl > 3 cm wide
24. Pl sticky-resinous; heads > 3 mm diam; meadows, c&s SNH, SnGb, SnBr .... ***A. rothrockii***
24′ Pl not sticky; heads gen < 3 mm diam (4.5–6 mm in *A. spiciformis*); widespread in mtns, dry valleys
25. Lvs entire to irregularly 3–6-lobed, partly deciduous; pl sprouting from roots; high mtn meadows .... ***A. spiciformis***
25′ Lvs gen regularly 3-lobed, wedge-shaped, evergreen; pl not root-sprouting; widespread .... ***A. tridentata***
26. Infl narrow, branches gen erect .... ssp. ***vaseyana***
26′ Infl wide, branches erect to spreading or drooping
27. Infl branches drooping; fr hairy .... ssp. ***parishii***
27′ Infl branches erect to spreading; fr glandular .... ssp. ***tridentata***

A. annua L. SWEET ANNIE Ann 3–20 dm, glabrous, green, sweet-scented. **ST** 1, erect. **LVS** 2–5 cm, triangular, 2–3-pinnately divided. **INFL** wide, much-branched; heads discoid, 1.5–2 mm diam, nodding; phyllaries lanceolate, green, margins narrowly scarious. **PISTILLATE FLS** 0. **DISK FLS** 18–24. **FR** < 0.2 mm, glabrous. $2n$=18. Uncommon. Disturbed places; < 1000 m. GV; native to Eur. Cult for oils.

A. arbuscula Nutt. LOW SAGEBRUSH Shrub < 3 dm, mounded, gray, evergreen. **STS** much-branched. **LVS** 3–9 mm, wedge-shaped, 3-lobed, hairy, gray-green. **INFL** spike-like; heads discoid, 2–4 mm diam, gen 2–4 per cluster, ± sessile; phyllaries ovate, densely hairy, margins transparent. **PISTILLATE FLS** 0. **DISK FLS** 4–6. **FR** 0.7–0.8 mm, light brown, finely resinous. $2n$=18,36. Clay soils, valleys, slopes; 1500–3800 m. NCoRH, SNH, MP, W&I; to WA, MT, WY, UT.

ssp. ***arbuscula*** **LVS** of fl-sts shallowly 3-lobed. **INFL**: heads 3–4 mm diam. $2n$=18, 36. Habitats and range of sp. ❀SUN:1,**2**, **17**&IRR,DRN:**3**,**7**,8–10,**14–16**,**18–24**.

ssp. ***thermopola*** Beetle **LVS** of fl-sts deeply 3-lobed. **INFL**: heads 2–3 mm diam. Uncommon. Igneous rock; 2200–2500 m. NCoRH, n&c SNH.

A. biennis Willd. (p. 207) Ann or bien gen 3–20 dm, glabrous, green, unscented. **ST** 1, erect, finely striate, often reddish. **LVS** 4–13 cm, widely lanceolate, 2-pinnately divided ± to midrib; lobes sharply toothed. **INFL** ± dense, leafy; heads 2–4 mm diam, erect, 1–several in tight groups; phyllaries widely elliptic to obovate, green, margins widely scarious. **PISTILLATE FLS** 0. **DISK FLS** 15–40. **FR** 0.2–0.9 mm, 4–5-veined, glabrous. $2n$=18. Disturbed moist sites; < 2200 m. NCoRO, SnJV, CW, SCo, WTR, SnBr, GB; N.Am; native to Eur.

A. bigelovii A. Gray BIGELOW SAGEBRUSH Shrub 4–6 dm, rounded, branched from base. **STS** many, slender, curved, silvery-canescent. **LVS** 0.5–3 cm, entire to sharply 3-toothed, densely hairy. **INFL** 6–23 cm, 1–4 cm wide; heads 2–2.5 mm diam, nodding; phyllaries 12–15, ovate, obtuse, densely hairy, margins narrowly scarious. **PISTILLATE FLS** 0–2; ligules < 1 mm. **DISK FLS** 2–3; corollas < 1.5 mm. **FR** elliptic in outline, 5-ribbed, glabrous. $2n$=18,36. Sandy, often limestone soils; 1300–1900 m. W&I, DMtns; to Colorado, TX. [*A. petrophila* Wooton & Standley] ❀TRY.

A. californica Less. (p. 207) CALIFORNIA SAGEBRUSH Shrub (2)15–25 dm, rounded, branched from base. **STS** slender, flexible, wand-like, glabrous or canescent. **LVS** 1–10 cm, pinnately divided; lobes 2–4, < 5 cm, < 1 mm wide, thread-like, ± hairy, light green to gray; margins curled under. **INFL** leafy, narrow, sparse; heads < 5 mm diam, nodding in fr; phyllaries widely (ob)ovate, sparsely canescent, margins wide, scarious. **PISTILLATE FLS** 6–10. **DISK FLS** 15–30. **FR** 0.8–1.5 mm, resinous; pappus a minute crown. $2n$=18. Coastal scrub, chaparral, dry foothills, esp near coast; < 800 m. CW, SW; n Baja CA. ❀SUN:**15–17**&IRR:7–9,**14,18–24**;CVS: GRNCVR.

A. campestris L. ssp. ***borealis*** (Pallas) H.M. Hall & Clements Bien or per 1–4 dm, mounded, light green, unscented. **STS** erect, ± silky, reddish. **LVS** mostly basal, 2–7 cm, 2–3 cm wide, 2–3-pinnately divided into linear lobes, ± glabrous to densely silky; cauline widely spaced, 3–5 cm, entire to pinnately lobed. **INFL** spike-like, < 17 cm, 2–3 cm wide; heads < 5 mm diam, gen erect; phyllaries ovate to widely elliptic, glabrous to loosely tomentose, margins widely scarious, midrib prominent. **PISTILLATE FLS** 10–25. **DISK FLS** 15–30, staminate. **FR** < or = 1 mm, glabrous. $2n$=18, 36. Meadows; ± 2200 m. KR; to AK, Can, n Eurasia. ❀TRY.

A. cana Pursh ssp. ***bolanderi*** (A. Gray) G. Ward (p. 207) SILVER SAGEBRUSH Shrub < 9 dm, from woody trunk. **STS** white-felty. **LVS** (1.5)3–4 cm, linear to narrowly lanceolate, gen entire, winter-deciduous. **INFL** 12–20 cm, 1–2 cm wide, leafy; heads gen sessile, 2–3 per cluster, gen > 5 mm diam, erect; phyllaries elliptic to ovate, outer acute, densely hairy, inner obtuse, membranous, margins scarious. **PISTILLATE FLS** 0. **DISK FLS** 8–16. **FR** < 1.2 mm, resinous. $2n$=18,36. Gravelly soils, meadows, streambanks; 1600–3300 m. CaRH, s SNF, SNH, GB; to OR, NV. [*A. tridentata* ssp. *b.* (A. Gray) H.M. Hall & Clements] ❀IRR,SUN,DRN:**1–3**,**7**,14, **15–17**,18–21.

A. douglasiana Besser (p. 207) MUGWORT Per 5–25 dm, from rhizome. **STS** many, erect, brown to gray-green. **LVS** evenly spaced, 1–11(15) cm, narrowly elliptic to widely oblanceolate, entire or coarsely 3–5-lobed near tip, sparsely tomentose above, densely white-tomentose below. **INFL** 10–30 cm, 3–9 cm wide, leafy; branches widely spreading; heads 2–4 mm diam, bell-shaped, gen nodding; phyllaries ± widely (ob)ovate, gray-tomentose, margins wide, transparent. **PISTILLATE FLS** 6–9. **DISK FLS** 9–25, staminate. **FR** < 1 mm, glabrous. $2n$=18,36,54. Common. Open to shady places, often in drainages; < 2200 m. CA-FP; to WA, ID, Baja CA. ❀**4–6**,**17**&IRR:1–3,**7–9**,10,**14–16**,**18–24**; may be INV; STBL.

A. dracunculus L. TARRAGON Per 5–15 dm, from rhizome and woody caudex, gen glabrous, odorless or tarragon-scented. **STS** many, stiff, erect, brown. **LVS** basal and cauline, 1–7 cm, linear, entire or with few linear lobes, bright green, glabrous (exc D). **INFL** 15–45 cm, leafy; heads 2–3.5 mm diam, gen nodding; phyllaries widely ovate, glabrous, light brown, membranous, margins widely transparent. **PISTILLATE FLS** 14–25. **DISK FLS** 8–20, staminate. **FR** 0.5–0.8 mm, glabrous. $2n$=18. Common. Meadows, disturbed sites; < 3400 m. CA; to AK, n Can, n-c US, n Mex, Eurasia. ❀DRN,SUN:4,**5**,**6**;IRR:1,**2**,3,7,8,9,**14–24**; [CVS non CA].

A. ludoviciana Nutt. SILVER WORMWOOD Per 3–10 dm, from rhizome. **STS** many, simple, gray- to white-tomentose. **LVS** 1–11 cm, linear to narrowly elliptic, entire to deeply lobed, densely tomentose. **INFL** 5–30+ cm, narrow, open to dense; heads < 7 mm diam, gen nodding; phyllaries lanceolate to (ob)ovate, densely tomentose, margins narrowly transparent. **PISTILLATE FLS** 5–12. **DISK FLS** 6–30. **FR** < 0.5 mm, glabrous. $2n$=36,54. Common. Gen dry, sandy to rocky soils; < 3500 m. NCoRO, SN, SnJV, SW (exc ChI), GB, D; to WA, e Can, TX, n Mex.

ssp. ***albula*** (Wooton) Keck **LVS** gen 1–2 cm, linear to obovate, entire to ± serrate or lobed, ± persistently white-tomentose on both surfaces, often ± curled under. **INFL** open; involucre ± 3 mm. **DISK FLS** 8–13. Dry sandy soils, shrubland, forest; < 1400 m. PR, D; to Colorado, w TX, nw Mex. ❀SUN:**4–6**&IRR:2,3,**7–10**,11,**14–24**;CVS;INV;STBL.

ssp. ***candicans*** (Rydb.) Keck **LVS**: lower gen 5–10 cm, deeply divided into narrow, entire to toothed lobes; upper entire to divided, gray-white-tomentose both surfaces, margins flat. **INFL** narrow; involucre 4–5 mm. **DISK FLS** 17–42. Dry woodland, shrubland; < 2400 m. n SNH, MP; to WA, MT, UT. ❀TRY.

ssp. ***incompta*** (Nutt.) Keck **LVS**: lower gen 2–8 cm, 1–2-divided into lance-linear lobes; upper entire to divided, white-tomentose below, becoming ± glabrous above, ± flat. **INFL** narrow; involucre 2.5–3.5 mm. **DISK FLS** 15–30. $2n$=36. Shrubland, woodland, coniferous forest; < 3500 m. SN, TR, PR, GB, DMtns; to OR, MT, Colorado. ❀STBL.

ssp. ***ludoviciana*** **LVS**: lower gen 3–11 cm, entire to few-toothed ± near tip, white-tomentose on both surfaces or becoming ± glabrous above, flat. **INFL** narrow; involucre 3–4 mm. **DISK FLS** 6–20. $2n$=36. Rocky soils, shrubland, coniferous forest; < 2600 m. NCoRO, SN, SnJV, SW (exc ChI), GB, DMoj; range of sp. outside CA. ❀STBL.

A. michauxiana Besser LEMON SAGEWORT Per 3–10 dm, from rhizome, lemon-scented. **STS** many, unbranched, green. **LVS** 1.5–11 cm, linear to narrowly elliptic, 1–2-pinnately divided, green, glabrous and dotted with yellow glands above, white-tomentose or becoming ± glabrous below. **INFL** 8–15 cm, narrow; heads < 5.5 mm diam, nodding; phyllaries elliptic to widely (ob)ovate, often purplish and dotted with yellow glands, gen ± glabrous, margins widely scarious. **PISTILLATE FLS** 9–12. **DISK FLS** 15–35. **FR** 0.5–1 mm, glabrous. $2n$=18,36. Subalpine to alpine scree, talus, drainages; 3000–3500 m. W&I; to B.C., MT, Colorado. ❀TRY; DFCLT.

A. nesiotica Raven ISLAND SAGEBRUSH Shrub < 6 dm, rounded. **STS** slender, soft, wand-like from brittle woody base, ± canescent. **LVS** 1–10 cm, blunt, pinnately divided; lobes few, 1–3 mm wide, flat, ± canescent. **INFL** narrow, sparse, leafy; heads ± 5 mm diam, ± nodding; phyllaries widely ovate, canescent, margins widely scarious. **PISTILLATE FLS** 10–15. **DISK FLS** 20–40. **FR** 1-1.5 mm, resinous; pappus present. Rocky slopes; gen < 100 m. s ChI. [*A. californica* var. *insulare* (Rydb.) Munz] Distinct from *A. californica*. ❀TRY.

A. norvegica Fries ssp. ***saxatilis*** H.M. Hall & Clements BOREAL SAGEWORT Per 4–6 dm, from branched caudex, mildly fragrant. **STS** erect, loosely tomentose. **LVS**: basal 6–15 cm; cauline < 10 cm; 1–2-pinnately divided; lobes 1–2 mm wide. **INFL** open; peduncles 0.3–10 cm, slender; heads ± 9–11 mm diam, nodding; phyllaries widely (ob)ovate, sparsely hairy, margins dark, scarious. **PISTILLATE FLS** 6–12. **DISK FLS** 30–80. **FR** glabrous. Rocky slopes; 2300–3800 m. KR, SNH, SnJt, W&I; to AK, n Can, MT, Colorado. ❀TRY.

A. nova Nelson BLACK SAGEBRUSH Shrub 1–3 dm, loosely branched from short trunk. **STS** canescent, becoming ± glabrous. **LVS** 0.5–2 cm, wedge-shaped, gen 3-toothed at tip (gen entire on fl-sts), evergreen. **INFL** slender; branches erect; heads 2–3 mm diam; phyllaries elliptic to ovate, inner gen glabrous, shiny-resinous, straw-colored, margins transparent. **PISTILLATE FLS** 0. **DISK FLS** 3–6. **FR** 0.8–1 mm, ribbed, glabrous or resinous. $2n$=18,36. Shallow rocky soils in desert valleys, dry slopes; < 2300 m. CaRH, SnJt, GB, DMtns; to OR, MT, NM. [*A. arbuscula* ssp. *n.* (Nelson) G. Ward] Often confused with *A. arbuscula*. ❀DRN:1, **2**,**3**&IRR:**7**,**15–21**.

A. palmeri A. Gray PALMER SAGEWORT Bien or per 8–30 dm, from ± woody base, strongly scented. **STS** brittle, wand-like, glabrous. **LVS** cauline, 3.5–12(15) cm, deeply and coarsely pinnately 3–7-lobed, glabrous to sparsely hairy above, gray-green-canescent below. **INFL** < 30 cm, open; heads 2.5–3.5 mm diam, gen nodding; phyllaries ± widely ovate, glabrous to sparsely hairy, membranous, margins ± transparent; receptacle chaffy. **PISTILLATE FLS** 0. **DISK FLS** 8–30. **FR** 1–1.2 mm, smooth to shiny-glandular. $2n$=18. RARE. Moist drainages, sandy soil; < 600 m. SCo, PR; Baja CA. Threatened by development. In cult.

A. pycnocephala DC. (p. 207) COASTAL SAGEWORT Per forming mounds 3–7 dm, densely white-gray tomentose throughout, faintly aromatic. **STS** > 1 m, decumbent, densely leafy. **LVS**: lower cauline 3–8 cm, finely 2–3-pinnately divided, lobes 1–2 mm wide; upper 1–2-divided. **INFL** spike-like, narrow, leafy; heads 3–4.5 mm diam, erect, sessile, clustered; phyllaries ovate, densely soft-hairy, margins widely scarious. **PISTILLATE FLS** 8–15. **DISK FLS** 8–25, staminate. **FR** < 1.5 mm, glabrous. $2n$=18. Rocky or sandy soils, coastal strand; < 200 m. NCo, CCo; OR. ❀DRN,SUN:4,**5**, **17**&IRR:7–9,14,**15**,**16**,19–21,**22–24**;CVS.

A. rothrockii A. Gray ROTHROCK SAGEBRUSH Shrub 2–5 dm from narrow trunk, sticky-resinous, dark green throughout (sts sometimes ± white-hairy), pungently aromatic. **LVS** 1–1.5(2) cm, (0)3-lobed (gen entire on fl-sts), evergreen, canescent, sometimes becoming ± glabrous. **INFL** narrow; heads 3–5 mm diam, erect, sessile to short-peduncled; phyllaries ± ovate, sparsely hairy, straw-colored, shiny, margins wide, scarious. **PISTILLATE FLS** 0. **DISK FLS** gen 10–16. **FR** 0.8–2 mm, smooth, resinous; pappus present on outer frs. $2n$=18,36,54. Clay soils, meadows; 2000–3100 m. c&s SNH, SnGb, SnBr. ❀IRR,SUN,DRN:**1**,**2**,3,**7**,14,**15**,**16**,17–21.

A. spiciformis Osterh. SNOWFIELD SAGEBRUSH Shrub 3–8 dm, gray-tomentose, widely branched, root-sprouting. **LVS** 2.5–5.5 cm, entire to irregularly 3–6-toothed (gen entire on fl-sts), turning yellow, ± winter-deciduous. **INFL** 5–20 cm, narrow, leafy; heads 4.5–6 mm diam, erect; phyllaries oblanceolate to obovate, ± hairy, margins brownish scarious. **PISTILLATE FLS** 0. **DISK FLS** 8–23. **FR** ± 1 mm, glabrous. $2n$=18,36,54. Common. Moist open slopes to rocky meadows; 2100–3700 m. c&s SNH, W&I; to WY, Colorado. Often confused with *A. rothrockii*; perhaps hybrid of *A. cana* & *A. tridentata*. ❀TRY.

A. spinescens D. Eaton (p. 207) BUDSAGE Shrub < 3 dm, stout, mound-like, pungently aromatic. **STS** canescent; old branches forming thorns. **LVS** < 2 cm, round, palmately 3–5-divided into narrow segments, densely soft-hairy. **INFL** spike-like, ± surrounded by lvs; heads < 5 mm diam; phyllaries few, widely obovate, densely soft-hairy, margins obscurely scarious. **PISTILLATE FLS** 2–8. **DISK FLS** 2–15, densely soft-hairy, staminate; corolla widely bell-shaped; style unbranched. **FR** < 0.5 mm, lightly veined, long-hairy. $2n$=18,36. Clay or gravelly, often saline soils, shrubland; 900–1600 m. SnBr, SNE, DMoj; to OR, MT, NM.

A. suksdorfii Piper COASTAL MUGWORT Per 5–20 dm from rhizome. **STS** many, erect, brown. **LVS** 5–10(15) cm, sessile, lanceolate, coarsely and irregularly lobed, dark green above, white-tomentose below. **INFL** 17–30 cm, 4–5 cm wide; branches erect, dense, gen not leafy; heads 1.5–2 mm diam, gen erect; phyllaries oblong to ± ovate, shiny, (sub)glabrous, straw-colored to yellow-green, margins scarious. **PISTILLATE FLS** 3–7. **DISK FLS** 2–8. **FR** < 0.8 mm, glabrous. $2n$=18. Coastal drainages, roadsides; < 300 m. NCo, s ChI; to B.C. Hybridizes with *A. douglasiana*. ❀SUN,IRR: **4**,**5**,6,7,14,**15–17**,19–21,**22–24**;also STBL.

A. tridentata Nutt. (p. 207) Shrub < 30 dm, from thick trunk, gray-hairy. **STS** gen glabrous. **LVS** 1–3(6) cm, gen wedge-shaped, gen 3(0–5)-toothed at tip, often in axillary clusters, persistent, gray-green, densely hairy. **INFL**: heads 2–2.5 mm diam, gen erect; phyllaries oblanceolate to widely obovate, densely tomentose, margins ± transparent. **PISTILLATE FLS** 0. **DISK FLS** 4–6. **FR** 1–2 mm, glandular or hairy. $2n$=18,36. Common. Dry soils, valleys, slopes; 300–3000+ m. CaRH, SNH, SCoRI, s SnJV, SCo, TR, GB, DMoj; to WA, n-c US, NM.

ssp. ***parishii*** (A. Gray) H.M. Hall & Clements Pl 10–30 dm. **LVS** gen 3–6 cm, linear to narrowly wedge-shaped. **INFL** < 30 cm, wide, often surrounded by lvs; branches gen drooping. **FR** hairy. Uncommon. Dry sandy soils; 300-2000 m. SCoRI, SCo, WTR, W&I, DMoj.

ssp. ***tridentata*** BIG SAGEBRUSH Pl < 20 dm. **LVS** gen 1.2–4 cm, narrowly wedge-shaped. **INFL** < 30 cm, 5–15 cm wide, often surrounded by lvs; branches gen spreading. **FR** glandular. $2n$= 18,36. Valleys, benches, sandy to coarse gravelly soils; 800–1900 m. CaRH, s SnJV, SCoRI, WTR, GB; to WA, MT, NM. ❀DRN, SUN:1,**2**,**3**,6,**7–10**,11,**14–16**,17,**18–21**,22–24;may be INV.

ssp. ***vaseyana*** (Rydb.) Beetle MOUNTAIN SAGEBRUSH Pl gen < 10 dm. **LVS** gen 2–3 cm, wedge-shaped. **INFL** < 30 cm, narrow, exposed; branches erect, exceeding vegetative branches. **FR** glandular. $2n$=18,36. Common. Dry slopes; 1800–3000+ m. CaRH, SNH, TR, n SNE, W&I; to ID, ND, Colorado, NM. ❀TRY.

ssp. ***wyomingensis*** Beetle & A. Young WYOMING SAGEBRUSH Pl gen < 4 dm. **LVS** gen < 1.2 cm, wedge- to fan-shaped. **INFL** gen < 6 cm, narrow, often surrounded by vegetative branches; branches short, stiffly spreading, persistent, giving pl a twiggy appearance. **FR** glandular. $2n$=36. Uncommon. Valleys, slopes; < 2200 m. MP; to ID, WY, NM.

ASTER ASTER

Geraldine A. Allen

Ann or per from caudex or rhizome. **ST** gen erect, 1–20 dm. **LVS** basal, cauline, or both, alternate, gen entire; basal gen petioled. **INFL**: heads gen radiate, solitary or in a cyme or panicle; involucre obconic to hemispheric; phyllaries in 2–6 series, outer gen < inner, free, at least inner with pale, papery margins; receptacle ± flat, naked. **RAY FLS** 0–many; corolla violet to pink or white. **DISK FLS** many; corolla and anthers gen yellow, tube gen < throat; anther tips ± triangular; style branches flat on inner face, base ± warty, tip acute, hairy. **FR** gen rounded, ± ribbed, ± brown; pappus of bristles, white to brownish. ± 250 ssp.: N.Am, Eurasia, Afr. (Greek: star) [Allen 1984 Syst Bot 9:175–191]

1. Ann, taprooted; rays < 8 mm, gen > disk fls by 2 mm or less, pink to violet
 2. Phyllaries obtuse to rounded; lvs oblanceolate to elliptic, gen obtuse .... ***A. frondosus***
 2′ Phyllaries acute to acuminate; lvs linear to oblanceolate, acute .... ***A. subulatus*** var. ***ligulatus***
1′ Per, gen from rhizome or caudex; rays (if any) > 8 mm, conspicuous, white to pink or violet
 3. Heads solitary; cauline lvs 0 or < 1/3 basal lvs
  4. Involucre and st glabrous to ± hairy, not glandular .... ***A. alpigenus*** var. ***andersonii***
  4′ Involucre and upper st glandular .... ***A. peirsonii***
 3′ Heads gen > 1 per st; cauline lvs well developed, > 1/3 basal lvs (or basal lvs 0)
  5. Outer pappus bristles < 2 mm, << inner; pl 12 cm or less; lvs cauline, < 15 mm .... ***A. scopulorum***
  5′ Outer pappus bristles gen ± = inner; pl > 12 cm or basal lvs present and >> 15 mm
   6. Rays 0–7 per head
    7. Smallest infl branches very short (heads clustered); ligules white .... ***A. oregonensis***
    7′ Smallest infl branches not very short (heads not clustered); ligules 0 or pale violet
     8. Phyllaries linear to lanceolate, gen acuminate; 1500–2700 m .... ***A. breweri***
     8′ Phyllaries oblong to ovate, acute; 600–2000 m .... ***A. brickellioides***
   6′ Rays 8 or more per head
    9. Lvs sharply toothed, 10 cm or less, gen 2–4 × longer than wide; tube of disk corollas > throat ***A. radulinus***
    9′ Lvs gen entire, or > 10 cm, or > 4 × longer than wide; tube of disk corollas < throat
     10. Involucre and peduncles ± glandular
      11. Lvs linear to lanceolate, < 10 mm wide
       12. Lvs ± similar in size; pl with both glandular and nonglandular hairs .... ***A. campestris***

12′ Lvs smaller upward; pl glabrous except for glandular involucre and upper st .... ***A. pauciflorus***
11′ Lvs narrowly elliptic to obovate, largest gen 10 mm wide or more
13. Basal lvs > cauline, present at fl .... ***A. integrifolius***
13′ Basal lvs ± = cauline, 0 at fl .... [2]***A. ledophyllus***
10′ Involucre glabrous to ± hairy, not glandular
14. Phyllaries ± keeled, with ± prominent green midvein, gen pale and papery almost to ± purple tip; rays gen few; basal lvs 0 in fl
15. Lvs 4–10 cm, green on both sides, ± glabrous; rays gen white .... ***A. engelmannii***
15′ Lvs < 6 cm, pale green and ± tomentose beneath; rays 0 or light purple .... [2]***A. ledophyllus***
14′ Phyllaries not keeled, green throughout or papery at base, the tips not purple; rays many; basal lvs present or 0 in fl
16. Outer phyllaries obtuse to rounded, < inner ones, ± pale-margined at base, with the green area at tip gen < 2.5 × longer than wide; pl gen densely and uniformly strigose, at least beneath heads
17. Infl open, branches gen > 5 cm; lvs glabrous to ± strigose, gen lacking short leafy shoots in axils; sts 2–6 dm .... ***A. ascendens***
17′ Infl narrow, branches gen < 5 cm; lvs strigose, gen with short leafy shoots in axils; sts 4–10 dm .... ***A. bernardinus***
16′ Outer phyllaries acute to obtuse, < or = inner ones, green throughout, or ± pale-margined at base with green area at tip gen > 2.5 × longer than wide; pl glabrous or with hairs gen ± in lines
18. Lvs rough-hairy on both surfaces, elliptic to obovate, cauline > basal; phyllaries acute ...... ***A. greatae***
18′ Lvs glabrous to ± soft-hairy, gen ± linear to narrowly elliptic (if wider, basal > cauline); phyllaries acute to obtuse
19. Basal or lower cauline lvs largest; basal lvs gen present at fl; st gen 1–6 dm; 1200–3200 m
20. Lvs < 7 × longer than wide; outermost phyllaries gen green throughout, slightly < to > inner .... ***A. foliaceus***
21. Cauline lvs clasping; outer phyllaries = or > inner .... var. ***lyallii***
21′ Cauline lvs not clasping; outer phyllaries < or = inner .... var. ***parryi***
20′ Lvs 7+ × longer than wide; outermost phyllaries pale-margined at base, < inner ..... ***A. occidentalis***
22. Lvs lanceolate to narrowly elliptic .... var. ***occidentalis***
22′ Lvs ± linear .... var. ***yosemitanus***
19′ Lvs ± equal or middle cauline lvs largest; basal lvs gen 0 at fl; st gen > 4 dm; < 2000 m (following spp. commonly intergrade, incl many polyploids)
23. Outer phyllaries pale-margined < 1/2 length, gen < 3 × longer than wide
24. Ligules violet; infl gen open, lateral branches gen 10+ cm; outer phyllaries gen < inner, obtuse .... ***A. chilensis***
24′ Ligules gen pink to white; infl gen narrow, lateral branches gen < 10 cm; phyllaries gen ± subequal, outer acute to obtuse .... ***A. eatonii***
23′ Outer phyllaries pale-margined > 1/2 length, gen 3+ × longer than wide
25. Sts hairy in lines; lvs lanceolate to narrowly elliptic; lvs of infl well developed .... ***A. lanceolatus*** ssp. ***hesperius***
25′ Sts glabrous; lvs linear to narrowly lanceolate; lvs of infl small, bract-like .... ***A. lentus***

A. alpigenus (Torrey & A. Gray) A. Gray var. ***andersonii*** (A. Gray) M. Peck (p. 207) Per; caudex taprooted. **STS** decumbent to ± erect, 1–4 dm, ± hairy above. **LVS** mainly basal, sessile, 3–25 cm, linear to narrowly elliptic, acute, ± glabrous; cauline 0 or < 1/3 as long. **INFL**: heads solitary; phyllaries subequal or outer < inner, lanceolate to oblong, acute, green to ± purple-tinged, pale-margined at base. **RAY FLS** many; corollas gen 10–16 mm, violet to light purple. **FR** ± hairy throughout. 2*n*=18,36. Meadows; 1500–3700 m. KR, NCoR, CaR, SN, SnJt, Wrn, W&I; OR, NV. Var. *alpigenus* farther n. ❀**4–6**;IRR:1,15–17&SHD;2,3,7,14,18;DFCLT.

A. ascendens Lindley (p. 207) Per; rhizomes elongate. **STS** 2–6 dm, strigose at least above. **LVS** basal and cauline, smaller upward, 4–15 cm, oblong to oblanceolate, acute, glabrous to strigose. **INFL**: heads in an open cyme; phyllaries oblong to narrowly obovate, obtuse (outer) to ± acute (inner), green at tip, pale-margined at base. **RAY FLS** many; corollas 8–15 mm, violet. **FR** hairy. 2*n*=26,52. Meadows, disturbed places; 500–2000 m. SN (e slope), Teh, SnGb, SnBr, GB; to w Can, Colorado, n AZ. [spelling corrected from *A. adscendens*] ❀TRY.

A. bernardinus H.M. Hall Per; rhizomes long. **STS** 4–10 dm, strigose ± throughout. **LVS** basal and cauline (basal 0 at fl), 4–12 cm, narrowly oblong to oblanceolate, acute, strigose; smaller lvs tufted in axils. **INFL**: heads in a narrow cyme; phyllaries ± oblong, obtuse (outer) to ± acute (inner), green at tip, pale-margined at base. **RAY FLS** many; corollas 8–12 mm, ± white to pale violet. **FR** hairy. 2*n*=36. Grassland, disturbed places; < 1500 m. SnGb, SnBr, PR. ❀TRY.

A. breweri (A. Gray) Semple Per; caudex woody. **STS** 1–9 dm, ± glandular to ± tomentose. **LVS** cauline, 2–5 cm, lanceolate to ovate, acute, entire to ± toothed, ± glabrous to tomentose or glandular-hairy. **INFL**: heads discoid, in a cyme; phyllaries ± keeled, linear to narrowly lanceolate, acute to acuminate; base ± pale, midvein and tip green, ± tomentose to glandular-hairy. **RAY FLS** 0. **FR** ± hairy. 2*n*=18. Subalpine meadows, open woods; 1500–2700 m. KR, CaRH, SNH, SnBr. [*Chrysopsis b.* A. Gray]

A. brickellioides E. Greene BRICKELLBUSH ASTER Per; caudex woody. **STS** 3–8 dm, ± minutely tomentose. **LVS** cauline; lowest scale-like, 3–6 cm, oblong to elliptic, gen obtuse, ± tomentose beneath. **INFL**: heads gen discoid, in a cyme; phyllaries ± keeled, oblong to narrowly ovate, acute, base ± white, midvein green, tip green to purple; ± tomentose, ± ciliate. **RAY FLS** gen 0; corollas (if present) pale violet. **FR** ± hairy. UNCOMMON. Dry woods, rocky places; 600–2000 m. KR, NCoRI; s OR. [var. *glabratus* E. Greene]

A. campestris Nutt. Per; rhizomes elongate. **STS** 1–5 dm, glandular and gen strigose. **LVS** cauline, 2–5 cm, linear to narrowly oblong, acute to obtuse, glabrous to ± short-hairy. **INFL**: heads in a narrow cyme; phyllaries linear to lanceolate, acute, green, ± pale-margined at base, ± glandular. **RAY FLS** many; corollas 6–10 mm, violet. **FR** hairy. 2*n*=10. Dry meadows; 1800–2700 m. n SN (e slope), MP; to e WA, UT. ❀TRY.

A. chilensis Nees (p. 207) Per; rhizomes elongate. **STS** 4–10 dm, hairy above. **LVS** basal and cauline (basal 0 at fl), ± sessile, ±

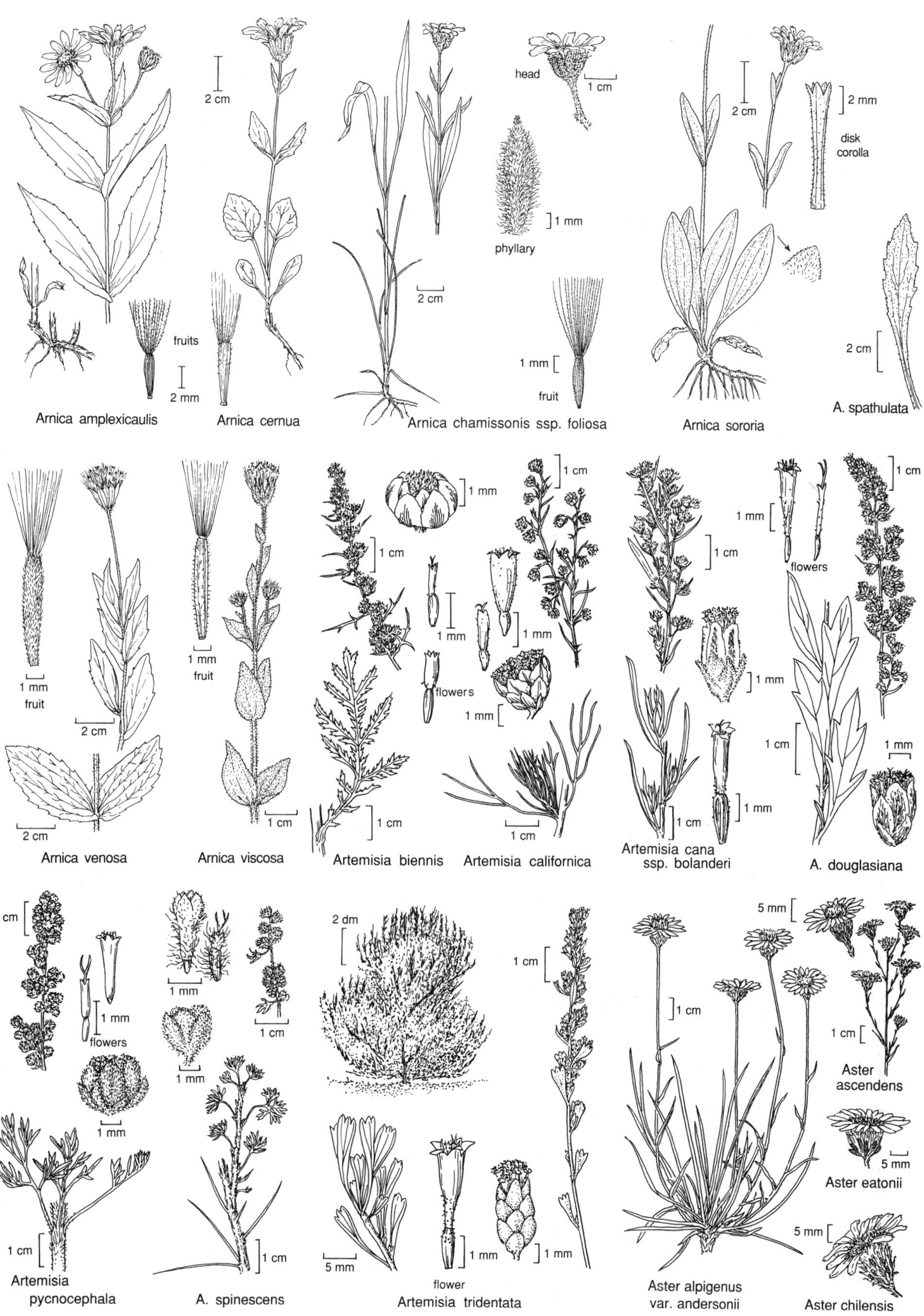

2 cm
fruits
2 mm
Arnica amplexicaulis
Arnica cernua
head
1 cm
phyllary
1 mm
2 cm
1 mm
fruit
Arnica chamissonis ssp. foliosa
2 cm
2 mm
disk
corolla
Arnica sororia
2 cm
A. spathulata
1 mm
fruit
2 cm
2 cm
Arnica venosa
1 mm
fruit
1 cm
Arnica viscosa
1 mm
1 cm
1 mm
flowers
1 cm
Artemisia biennis
1 cm
1 mm
1 mm
1 cm
Artemisia californica
1 cm
1 mm
1 cm
1 mm
Artemisia cana
ssp. bolanderi
1 mm
flowers
1 cm
1 cm
1 mm
A. douglasiana
1 cm
1 mm
flowers
1 mm
1 cm
Artemisia
pycnocephala
1 mm
1 mm
1 cm
1 cm
A. spinescens
2 dm
1 cm
5 mm
1 mm
1 mm
flower
Artemisia tridentata
1 cm
Aster alpigenus
var. andersonii
5 mm
1 cm
Aster
ascendens
5 mm
Aster eatonii
5 mm
Aster chilensis

oblanceolate, acute, entire to finely serrate, ± hairy. **INFL**: heads in a cyme; phyllaries oblong to ovate, outer green ± to base, obtuse, inner pale-margined, ± acute. **RAY FLS** many; corollas 8–12 mm, violet. **FR** hairy. 2*n*=48,64. Grasslands, salt marshes, disturbed places; < 500 m. w&c NW, w&c CW, n SCo, n ChI; to B.C. ❀**4–6**&IRR:1–3,**7–10**,11,**14–24**;also STBL;INV.

A. eatonii (A. Gray) Howell (p. 207) Per; rhizomes short. **STS** 4–10 dm, ± hairy above. **LVS** basal and cauline (basal 0 at fl), sessile, 5–15 cm, narrowly lanceolate, acute, gen entire, glabrous to ± hairy. **INFL**: heads in a narrow, many-headed cyme; phyllaries ± subequal, oblong to ovate, inner pale-margined at base, acute, outer ± green throughout, obtuse to acute. **RAY FLS** many; corollas 6–12 mm, white to pink or pink-purple. **FR** hairy. 2*n*=16,32,48,64. Wet places; 500–2000 m. CaR, SN, GB; to B.C., Colorado. ❀TRY.

A. engelmannii (Eaton) A. Gray Per; caudex woody. **STS** 5–15 dm, ± hairy. **LVS** cauline with lowest scale-like, 4–10 cm, elliptic to ovate, ± acute, gen entire, ± glabrous. **INFL**: heads in a cyme; phyllaries ± keeled, oblong to ovate, acute, pale with green midvein and gen purple-margined tip, ciliate. **RAY FLS** gen 8 or 13; corollas 15–22 mm, white, ± becoming pink. **FR** ± hairy. 2*n*=18. Uncommon. Meadows, open woods; 1800–2000 m. KR, NCoRH; to w Can, Colorado, NV. ❀TRY.

A. foliaceus Lindley Per; rhizomes long. **STS** 1–6 dm, ± hairy above. **LVS** basal and cauline, 6–15 cm, elliptic to obovate, acute to obtuse, ± entire, gen glabrous. **INFL**: heads in 1–few–headed cyme; phyllaries oblong to ovate, acute, outer ± lf-like and green throughout, inner pale-margined. **RAY FLS** many; corollas 12–25 mm, violet to purple. **FR** gen hairy. 2*n*=16,32,48,64,80,96. Open woods to alpine meadows; 1400–3200 m. NW, CaR, SN, SnBr, SnJT, Wrn; to AK, MT, NM. Highly variable.

var. ***lyallii*** (A. Gray) Cronq. **STS** 3–6 dm. **LVS** gen 5–15 cm; cauline lvs clasping. **INFL**: outer phyllaries > inner. 2*n*=16,32. Habitat of sp.; 1500–2200 m. KR, CaR, c SN (Tuolumne Co.); s OR, also n ID. ❀TRY.

var. ***parryi*** (Eaton) A. Gray **STS** 1–4 dm. **LVS** 2–12 cm, not clasping. **INFL**: outer phyllaries < inner. 2*n*=16,32,64. Habitat of sp.; 1400–3200 m. Range of sp. in CA; to MT, NM. 1-headed alpine pls have been called var. *apricus* A. Gray. ❀TRY.

A. frondosus (Nutt.) Torrey & A. Gray Ann. **STS** decumbent to erect, 2–6 dm, ± glabrous. **LVS** basal and cauline, sessile, gen 2–5 cm, elliptic to obovate, ± obtuse, glabrous to finely ciliate. **INFL**: heads in a narrow cyme; phyllaries oblanceolate to obovate, obtuse to rounded, green, inner ± pale-margined at base. **RAY FLS** many; corollas < 8 mm, pink-purple, slightly > disk corollas. **FR** ± hairy. 2*n*=14. Marshes, lake edges, often alkaline; 700–2200 m. CaR, SN, TR, PR, GB; to B.C., WY, AZ.

A. greatae Parish GREATA'S ASTER Per; rhizomes elongate. **STS** ascending to erect, 5–12 dm, sparsely hairy. **LVS** basal and cauline (basal gen 0 at fl), 6–15 cm, elliptic to obovate, ± clasping, acute, entire to serrate, hairy. **INFL**: heads in an open cyme, gen leafy-bracted; phyllaries gen ± subequal, lanceolate to oblong, acute, green, ± pale-margined at base, ± spreading. **RAY FLS** many; corollas 8–15 mm, pale violet. **FR** hairy. 2*n*=16. UNCOMMON. Damp places in canyons; 800–1500 m. SnGb (s slope). Formerly misspelled *A. greatai*.

A. integrifolius Nutt. (p. 211) Per; rhizomes short, ± woody. **STS** 2–7 dm, glandular above. **LVS** basal and cauline, 6–22 cm (smaller upward), glabrous to ± hairy; basal narrowly elliptic to obovate, acute, abruptly pointed, or rounded; cauline lanceolate to oblong, acute. **INFL**: heads in a cyme; phyllaries oblong, acute, glandular, outer green ± to base, inner green at tip, ± purple-tinged. **RAY FLS** 10–20; corollas 10–15 mm, violet to purple. **DISK FLS**: corollas yellow to ± purple. **FR** short-hairy. 2*n*=18. Dry meadows; 1600–3200 m. CaR, SN; to WA, MT, Colorado. ❀TRY.

A. lanceolatus Willd. ssp. ***hesperius*** (A. Gray) Semple & J. Chmielewski (p. 211) Per; rhizomes long. **STS** 6–16 dm; hairs ± in lines throughout. **LVS** basal and cauline (basal 0 at fl), 8–16 cm, lanceolate, acute, entire to ± serrate, ± hairy. **INFL**: heads in an open, leafy-bracted cyme; phyllaries linear to oblong, acute to acuminate, pale-margined at base to > 1/2 length. **RAY FLS** many; corollas 6–12 mm, ± white to violet. **FR** hairy. 2*n*=64. Wet places; < 1500 m. SCo, PR, SNE; to Alberta, n-c US. [*A. h.* A. Gray] ❀TRY.

A. ledophyllus A. Gray (p. 211) Per; caudex woody. **STS** 3–8 dm, ± hairy. **LVS** cauline, 2–6 cm (lowest scale-like), elliptic to oblong, acute to obtuse, gen entire; upper surface ± glabrous; lower surface ± tomentose. **INFL**: heads in a cyme; phyllaries ± keeled, oblong to ovate, acute, pale at base, ± purple at tip, with green midvein, ciliate. **RAY FLS** gen 8, 13, or 21; corollas 10–15 mm, violet to purple. **FR** ± hairy. 2*n*=18. Meadows, open woods; 1300–1900 m. KR, NCoR; to n WA. Intergrades with *A. engelmannii* (Eaton) A. Gray. ❀TRY.

A. lentus E. Greene (p. 211) SUISUN MARSH ASTER Per; rhizomes long. **STS** 4–15 dm, ± glabrous. **LVS** basal and cauline (basal gen 0 at fl), small and bract-like in infl, sessile, 5–15 cm, linear to narrowly lanceolate, acute, gen glabrous. **INFL**: heads at branch tips, in an open cyme; phyllaries linear to oblong, acute to ± obtuse, ± pale-margined at base. **RAY FLS** many; corollas 8–14 mm, violet. **FR** hairy. 2*n*=64. **ENDANGERED** US. Marshes; < 150 m. s ScV, CCo, SnFrB. [*A. chilensis* Nees vars. *l.* (E. Greene) Jepson & *sonomensis* (E. Greene) Jepson] Grades into *A. chilensis*. Threatened by habitat loss. In cult.

A. occidentalis (Nutt.) Torrey & A. Gray Per; rhizomes long. **STS** 2–8 dm, glabrous to ± hairy. **LVS** basal and cauline, 5–15 cm, ± smaller upward, ± linear to narrowly elliptic, acute, ± entire, ± glabrous. **INFL**: heads in an open cyme; phyllaries linear to oblong, acute or outer ± obtuse, green at tip, ± pale-margined at base. **RAY FLS** many; corollas 8–12 mm, violet. **FR** hairy. 2*n*=16,32, 48,64. Meadows; 1200–2800 m. NW, CaR, SN, TR, SnJt, GB; to B.C., Colorado, Baja CA. Variable.

var. ***occidentalis*** (p. 211) **STS** 2–8 dm. **LVS** lanceolate to narrowly elliptic. 2*n*=16,32,48,64. Habitats and range of sp. [vars. *delectabilis* (H.M. Hall) Ferris & *parishii* (A. Gray) Ferris] Larger pls have been called var. *intermedius* A. Gray; smaller forms intergrade with *A. foliaceus* var. *parryi*. ❀**4–6**&IRR:1–3,**7**,14,**15–17**, 18.

var. ***yosemitanus*** (A. Gray) Cronq. WESTERN BOG ASTER **STS** 3–8 dm. **LVS** ± linear. 2*n*=16,32. UNCOMMON. Meadows; 1200–2100 m. KR, CaR, SN; s OR. [*A. paludicola* Piper]

A. oregonensis (Nutt.) Cronq. (p. 211) Per; caudex woody, fibrous-rooted. **STS** ± erect, gen 4–10 dm, glabrous to sparsely hairy. **LVS** mostly cauline; lowermost withering early, gen 3–8 cm, oblanceolate to elliptic, acute, ± hairy. **INFL**: heads in a flat-topped cyme; peduncles of individual heads very short, thus heads in clusters; phyllaries ± keeled, oblong, acute, base ± white, tip green, outer ± spreading. **RAY FLS** gen ± 5; corollas 4–7 mm, white. **DISK FLS**: anthers ± purple. **FR** appressed-hairy. 2*n*=18. Open woods; 800–2200 m. KR, NCoR (Humboldt Co.), CaR, SN; to w WA. [ssp. *californicus* (Durand) Keck] ❀TRY.

A. pauciflorus Nutt. Per; rhizomes long. **STS** 3–12 dm, glabrous at base, glandular above. **LVS** basal and cauline, smaller upward (bract-like in infl), sessile, 6–10 cm, ± linear, acute to acuminate, ± glabrous. **INFL**: heads in an open cyme with heads at ends of branches; phyllaries ± oblong, acute, ± green, ± glandular. **RAY FLS** many; corollas 5–8 mm, ± white to pale purple. **FR** hairy. 2*n*= 18. Uncommon. Damp alkaline places; 200–700 m. n DMoj (Inyo Co); to c Can, TX, Mex.

A. peirsonii Sharsm. Per; caudex taprooted. **STS** gen decumbent, 2–8 cm, glabrous to ± glandular. **LVS** mainly basal, sessile, 1.5–6 cm, linear, acute, ± glabrous; cauline 0 or << basal, often bract-like. **INFL**: heads solitary; phyllaries ± subequal, ± oblong, acute, green to ± purple, ± pale-margined. **RAY FLS** many; corollas gen 12–16 mm, violet to purple. **FR** glabrous at base, ± hairy toward tip. Alpine meadows, gravel slopes; 3000–3800 m. c&s SNH. ❀TRY;DFCLT.

A. radulinus A. Gray (p. 211) Per; rhizomes ± woody, fibrous-rooted. **STS** 2–7 dm, gen hairy. **LVS** mainly cauline, 4–10 cm, widely elliptic to obovate, firm, ± acute, sharply serrate, scabrous at least below. **INFL**: heads in a flat-topped cyme or panicle; phyllaries oblong, ± acute, base ± white with dark midvein, tip green, often purple-margined. **RAY FLS** 10–15; corollas 8–13 mm, white to pale violet. **DISK FLS**: corolla tube > throat. **FR** ± hairy; pappus ± minutely barbed. 2*n*=18. Dry woods; 150–1500 m. NW, CaR, n&c SN, SCoR, n ChI; to B.C. ❀DRN:**4–6**&IRR:1,2,7,8–10,**14–18**,19–24;also STBL;INV.

A. scopulorum A. Gray (p. 211) Per; caudex fibrous-rooted. **STS** ascending to erect, 4–12 cm, ± hairy. **LVS** cauline, crowded on lower 2/3 of st, 0.3–1.2 cm, narrowly oblanceolate to elliptic, firm, ± abruptly pointed, densely short-hairy. **INFL**: heads solitary; phyllaries lanceolate to oblong, acute to acuminate, green to ± purple, inner pale-margined below. **RAY FLS** gen 8–16; corollas 7–12 mm, violet to purple. **FR** hairy; pappus outer bristles ± 1 mm, << inner. $2n=18$. Dry, rocky places; 1500–3000 m. Wrn, W&I; to e OR, WY, NV. ❀TRY;DFCLT.

A. subulatus Michaux var. ***ligulatus*** Shinn. (p. 211) Ann. **STS** erect, 2–8 dm, glabrous. **LVS** basal and cauline, sessile, gen 3–6 cm, linear to oblanceolate, acute, glabrous. **INFL**: heads in an open cyme; phyllaries linear to lanceolate, acute to long-tapered, green, ± pale-margined. **RAY FLS** many; corollas gen < 7 mm, pink to violet, barely > disk corollas. **FR** ± hairy. $2n=10$. Wet places, often alkaline; < 200 m. GV, CW, SW; to c&se US, Mex. [*A. exilis* Elliott]

ATRICHOSERIS TOBACCO-WEED, GRAVEL-GHOST

G. Ledyard Stebbins

1 sp.: sw US. (Greek: chicory-like pl without pappus)

A. platyphylla A. Gray (p. 211) Ann, 2–18 dm, glabrous; sap milky. **ST** erect. **LVS**: basal and lowest cauline gen flat against soil, sessile or tapering to a short, winged petiole, 3–10 cm, obovate, obtuse, finely dentate, gray-green, often purple-tinged, esp below; cauline few, reduced to inconspicuous, triangular scales. **INFL**: heads ligulate, 1.5–2.5 cm diam, few–many in an open scapose cyme; involucre 6–10 mm; phyllaries in 2–4 series, outer short, triangular to lanceolate, inner 10–15, ± equal; receptacle naked. **FLS** fragrant; ligules white, ± ephemeral. **FR** 4–4.5 mm with 5 thick corky ribs, whitish; pappus 0. $2n=18$. Desert valleys and washes; < 1400 m. D; to UT, AZ. ❀TRY.

BACCHARIS

Scott Sundberg

Per to shrub, dioecious, sometimes aromatic, often ± sticky-resinous. **STS** erect, channeled. **LVS** cauline, alternate, simple, reduced to bracts above. **INFL**: heads discoid and disciform, borne in terminal or lateral racemes, panicles or cymes; phyllaries overlapping in several series; receptacle naked or chaffy. **DISK FLS** gen many, functionally staminate; corollas white to pink-tinged; ovary much reduced; pappus of bristles < involucre. **PISTILLATE FLS** gen many; corollas thread-like, ± whitish. **FR** ± cylindric, 4–10-ribbed; pappus of many bristles > involucre. 250–400 spp.: Am. (Latin: Bacchus, god of wine) [Boldt 1989 *Baccharis* TX Agric Exp Sta, College Station]

1. Sts hairy or stalked-glandular
 2. Involucre of pistillate heads < 6 mm; pistillate fls < 18, corollas < 2.8 mm; fr pappus < 6.5 mm; involucre of staminate heads < 5.2 mm; sts finely glandular-puberulent; lvs 0.7–3 mm wide, entire; D, SnBr, SnJt .... ***B. brachyphylla***
 2′ Involucre of pistillate heads > 6 mm; pistillate fls > 20, corolla > 3.5 mm; fr pappus > 7 mm; involucre of staminate heads > 5 mm; sts with fine curled hairs or stalked glands; lvs 1–12 mm wide, entire to toothed; CCo, SCoRO, SCo, WTR, n ChI .... ***B. plummerae***
  3. Sts with stalked glands; lvs 1–3 mm wide, ± linear, principal vein gen 1(3), entire to toothed .... ssp. ***glabrata***
  3′ Sts with fine curled hairs; lvs 4–12 mm wide, oblanceolate, principal veins 3, margin toothed ssp. ***plummerae***
1′ Sts glabrous
 4. Per or subshrub; lf entire or finely toothed; frs ± glandular and hairy at tip; pistillate fls 80–150 per head .... ***B. douglasii***
 4′ Shrub (sts definitely woody); lf entire or coarsely toothed; frs gen glabrous; pistillate fls gen < 50 per head
  5. Fr ribs 5; pistillate fls > 50 per head; lvs lanceolate, largest gen > 5 cm, persistent at fl .... ***B. salicifolia***
  5′ Fr ribs 8–10; pistillate fls < 43 per head; lf shape various, < 5 cm or not lanceolate, persistent or deciduous at fl
   6. Pls without many erect branches; bracts not << st lvs; lvs coarsely toothed, gen persistent at fl, principal veins 3
    7. Bracts gen linear, entire; lvs light green; involucre of pistillate heads 6–8 mm; fr pappus 8–12 mm . ***B. emoryi***
    7′ Bracts obovate, sometimes toothed; lvs dark green; involucre of pistillate heads 3.5–5 mm; fr pappus 5.5–9 mm .... ***B. pilularis***
   6′ Pls with many erect branches; bracts gen << st lvs; lvs gen entire, gen 0 at fl, principal vein 1
    8. Larger lvs widely oblanceolate or wider; lateral heads sessile; receptacles conic to convex, gen with chaff; pistillate involucres < 4.5 mm; staminate involucres < 4 mm; staminate corollas < 3.3 mm, pappus < 3 mm; fr < 1.5 mm, pappus gen < 3 mm .... ***B. sergiloides***
    8′ Larger lvs linear to narrowly oblanceolate; lateral heads gen pedicelled; receptacles concave to convex, chaff 0; pistillate involucres > 5.3 mm; staminate involucres > 4 mm; staminate corollas > 4 mm, pappus > 3 mm; fr > 2 mm, pappus > 4 mm
     9. Pistillate involucres cylindric to bell-shaped, base glabrous; phyllaries lanceolate to ovate, tip acute to rounded, glabrous; widespread PR, D .... ***B. sarothroides***
     9′ Pistillate involucres funnel-shaped, base glandular-puberulent; phyllaries narrowly tapered, gen puberulent; s SCo .... ***B. vanessae***

B. brachyphylla A. Gray Shrub < 1 m, glandular. **STS** much-branched, wand-like. **LVS** sessile, < 17 mm, linear-lanceolate, entire; principal vein 1. **INFL**: heads in open panicles; involucre funnel- or bell-shaped, of staminate heads 3.8–5.2 mm, of pistillate heads 4.5–6 mm; phyllaries in 4–5 series, lanceolate, glandular-puberulent, acute to long-tapered; receptacle convex, honeycombed, puberulent, chaff 0. **STAMINATE FLS** (8)12–18(29); corollas 3.3–4.2 mm; pappus 3–4.4 mm. **PISTILLATE FLS** 8–18; corollas 2–2.8 mm. **FR** 2–3 mm, glandular-puberulent; ribs 5; pappus 4.5–6.5 mm. $2n=18$. Canyon bottoms, dry washes; 300–1200 m. SnBR, SnJt, D; to TX, n Mex. ❀TRY;STBL.

B. douglasii DC. (p. 211) MARSH BACCHARIS Per 1–2 m, from rhizome, glabrous, sticky. **STS** gen 1 from base; branches 0–few, ascending. **LVS**: petioles < 10 mm, winged; blades < 130 mm, lanceolate, entire or finely toothed, principal veins 1–3. **INFL**: heads in flat-topped clusters; involucre hemispheric, of staminate heads 3.5–5 mm, of pistillate heads 3.8–4.8(6) mm; phyllaries in 2–3 series, linear to lanceolate, narrow, mostly equal, long-tapered, glabrous, gen sticky; receptacle domed to convex, smooth or sometimes honeycombed, glandular, chaff 0 or 1 series of inconspicuous scales. **STAMINATE FLS** 26–40; corollas 3.5–4.3 mm; pappus 3–4(5) mm. **PISTILLATE FLS** 80–150; corollas 1.7–3 mm. **FR** 0.6–1.5 mm, ± glandular, hairy at tip; ribs 4–5; pappus 2.6–4(7) mm. 2*n*=18. Moist salt marshes, stream edges; 0–750 m. w NW, SNF, GV, CW, SCo. ❀STBL;INV.

B. emoryi A. Gray (p. 211) Shrub < 4 m, glabrous, gen sticky. **STS** erect; branches ascending, sometimes lfless at fl. **LVS** sessile or with winged petiole, 35–70 mm, oblanceolate, 0–8-toothed or -lobed; base wedge-shaped; principal veins 3. **INFL**: heads in clusters of 3–5 in a panicle; involucre cylindric to bell-shaped, of staminate heads 4–6 mm, of pistillate heads 6–8 mm; phyllaries in 5–7 series, linear-lanceolate, glabrous, gen sticky, tip acute; receptacle flat, honeycombed, gen hairy (sometimes long-bristly), chaff 0. **STAMINATE FLS** 17–38; corollas (3.2)4–5 mm; pappus 3–5 mm. **PISTILLATE FLS** 20–40; corollas 3.5–4 mm. **FR** 1.2–2 mm, glabrous; ribs 10; pappus 8–12 mm. Sandy edges of rivers and washes, salt marshes; 0–600 m. SCoRI, SCo, WTR, PR, D; to UT, TX, n Mex. ❀TRY;STBL.

B. pilularis DC. (p. 211) CHAPARRAL BROOM, COYOTE BRUSH Shrub < 3 m, glabrous, gen sticky. **STS** prostrate to erect; branches spreading or ascending. **LVS** barely petioled, 8–55 mm, oblanceolate to obovate, entire to toothed; principal veins 3. **INFL**: heads in a leafy panicle; involucre hemispheric to bell-shaped, of staminate heads 3.2–5 mm, of pistillate heads 3.5–5 mm; phyllaries in 4–6 series, lanceolate to ovate, glabrous, tip rounded to acute; receptacle convex to conic, honeycombed, chaff 0. **STAMINATE FLS** (13)20–34; corollas 3–4.2 mm; pappus 3–4.2 mm. **PISTILLATE FLS** 19–43; corollas 2.5–3.5(4.5) mm. **FR** 1–2 mm, glabrous; ribs 8–10; pappus 5.5–9 mm. 2*n*=18. Coastal bluffs to oak woodland, sometimes on serpentine; 0–750(1500) m. NW, SNF, w CW, SCo, ChI, WTR, PR; OR, n Mex. [var. *consanguinea* (DC.) Kuntze] Variable; erect pls are gen mixed (and intergrade completely) with prostrate pls. ❀DRN:**5,15–17**&IRR:7–9,**14**,18,**19–24**;male CVS incl. GRCVR.

B. plummerae A. Gray Subshrub or shrub, < 2 m, hairy or glandular (at least infl). **STS**: branches many, slender, wand-like, ascending. **LVS** sessile, 8–45 mm, linear to oblanceolate, entire to finely bristle-toothed; base wedge-shaped; principal veins 1–3. **INFL**: heads in a panicle; involucre bell-shaped, of staminate heads 5–6.5 mm, of pistillate heads 6–8.5 mm; phyllaries in 5–6 series, linear-lanceolate, hairy, glandular, tip acute to long-tapered; receptacle convex, honeycombed, puberulent. **STAMINATE FLS** 19–26; corollas 4–7 mm; pappus 3.5–4.5 mm. **PISTILLATE FLS** 20–30; corollas 3.5–5 mm. **FR** 2.7–3.6 mm, puberulent; ribs 5; pappus 7–8.5 mm. Rocky slopes, woodlands; 0–425 m. CCo, SCoRO, SCo, n ChI, WTR.

ssp. ***glabrata*** Hoover (p. 211) SAN SIMEON BACCHARIS **STS** with stalked glands. **LVS** 8–35 mm, linear to narrowly oblanceolate, entire or toothed; principal vein gen 1(3). 2*n*=18. UNCOMMON. Shrubby slopes; < 100 m. c CCo (nw San Luis Obispo Co.).

ssp. ***plummerae*** (p. 211) PLUMMER'S BACCHARIS **STS** with fine curled hairs. **LVS** 20–45 mm, oblanceolate, toothed; principal veins 3. 2*n*=18. UNCOMMON. Habitats and range of sp. ❀DRN, SUN:5,**15–17,22–24**&IRR:7,**14,19–21**.

B. salicifolia (Ruiz Lopez & Pavon) Pers. (p. 211) MULE FAT, SEEP-WILLOW, WATER-WALLY Shrub < 4 m, glabrous to minutely puberulent, often ± sticky. **STS**: main sts gen 1–few; branches few–many, short, spreading or ascending. **LVS**: petioles winged; blades < 150 mm, lanceolate, entire to toothed, principal veins 1–3. **INFL**: heads in a pyramid-shaped to rounded panicle; involucre hemispheric, of staminate heads (3)4–6 mm, of pistillate heads 3–6 mm; phyllaries in 4–5 series, awl-shaped to lanceolate, irregularly toothed, gen tinged red, glabrous (exc ciliate margins), tip obtuse to long-tapered; receptacle flat to convex, smooth, glabrous to tomentose, chaff 0. **STAMINATE FLS** (10)17–48; corollas (3)4–6 mm, pappus (3)3.6–5 mm. **PISTILLATE FLS** 50–150; corollas 2.2–3.5 mm. **FR** 0.8–1.3 mm, glabrous; ribs 5; pappus 4.2–6 mm. 2*n*=18. Canyon bottoms, moist streamsides, irrigation ditches, often forming thickets; < 1250 m. NW, CaRF, SNF, GV, Teh, CW, SW, D; to TX, Mex, S.Am. Summer forms (infl terminal, lvs mostly toothed) formerly separated from winter forms (infls lateral, lvs entire). [*B. glutinosa* Pers.; *B. viminea* DC.] ❀STBL.

B. sarothroides A. Gray (p. 211) BROOM BACCHARIS Shrub < 4 m, glabrous, sticky. **STS**: branches dense, erect, often lfless at fl. **LVS** sessile, < 35 mm, linear-oblanceolate, thick, entire, ± rolled under; principal vein 1. **INFL**: heads in a panicle; involucre cylindric to hemispheric, of staminate heads 4.2–5.2 mm, of pistillate heads 5.3–7 mm; phyllaries in 5–6 series, lanceolate to ovate, hard, glabrous, gen sticky, tip rounded to acute, receptacle flat to concave, honeycombed, chaff 0. **STAMINATE FLS** 18–35; corollas 4.2–5 mm; pappus 3–4.5 mm. **PISTILLATE FLS** 19–31; corollas 2.5–3.5 mm. **FR** 2–2.6 mm, glabrous; ribs 10; pappus 7–10.8 mm. 2*n*=18. Gravelly and sandy washes, roadsides; < 850 m. PR, D; to TX, n Mex. *B. emoryi* without st lvs may key here. ❀SUN, DRN:15–18,**19–24**&IRR:**7–14**;male CV.

B. sergiloides A. Gray SQUAW WATERWEED, DESERT BACCHARIS Shrub < 2 m, glabrous, gen sticky. **STS**: branches gen many, erect, often lfless at fl. **LVS** barely petioled, < 35 mm, oblanceolate to obovate, 0–4-toothed; principal vein 1. **INFL**: heads in a panicle; involucre funnel- to bell-shaped, of staminate heads 2.2–4 mm, of pistillate heads 3–4.5 mm; phyllaries in 4–6 series, lanceolate, hard, glabrous, gen sticky, tip rounded to acute; receptacle conic to convex, gen smooth, glabrous to puberulent, chaff sometimes 0 in staminate pls. **STAMINATE FLS** (11)24–33; corollas 2.2–3.3 mm; pappus 2–3 mm. **PISTILLATE FLS** 15–30; corollas 1.6–2.7 mm. **FR** 1–1.5 mm, glabrous; ribs 10; pappus 1.7–3.1 mm. 2*n*=18. Gravelly or sandy stream beds; 600–1575 m. Teh, PR, D; NV, AZ, n Mex. ❀TRY.

B. vanessae Beauch. (p. 211) ENCINITAS BACCHARIS Shrub < 2 m from root crown, gen glabrous, sticky. **STS**: branches dense, erect, often lfless at fl, glabrous (exc glandular-puberulent below heads). **LVS** sessile, 1–45 mm, linear, entire; principal veins 1. **INFL**: heads in a raceme or open panicle, or few-headed clusters; involucre funnel-shaped, 5.5–6.8 mm, phyllaries in 4 series, narrowly tapered, thick, rounded on back, gen glandular-puberulent, tip acute to long-tapered; receptacle ± convex, honeycombed, puberulent, chaff 0. **STAMINATE FLS** 15–22; corollas 4 mm; pappus 4 mm. **PISTILLATE FLS** ± 25; corollas ± 2.5 mm. **FR** 2–3 mm, puberulent; pappus 7–10 mm. 2*n*=18. **ENDANGERED** CA. Chaparral in Torrey-pine forest, on sandstone; 60–335 m. s SCo (San Diego Co). Threatened by development.

BAHIA

Ann, bien, per, subshrubs. **LVS** simple or ternately dissected, opposite or alternate, sessile or petioled, dotted with resin glands. **INFL**: heads radiate, in few–many-headed cymes; involucre hemispheric or bell-shaped; phyllaries in 1–3 series, free, ± equal, reflexed in age; receptacle flat or slightly rounded, naked. **RAY FLS** 5–20; ligules yellow or white. **DISK FLS** gen many; corollas yellow, radial or outer bilateral; style tips obtuse. **FR** 4-sided, narrowly obpyramidal; pappus scales (0)8–15, lanceolate to obovate. 12 spp.: sw US, Mex, S.Am. (J.F. Bahi, botany professor, Barcelona) [Ellison 1964 Rhodora 66:67–86,177–215,281–311]

B. dissecta (A. Gray) Britton (p. 211) Ann, bien. **ST** gen 1, erect, 2–12 dm, openly branched above, glandular. **LVS** alternate; petiole 1–6 cm or upper lvs sessile; blade 1–3.5 cm, 1–several × ternately divided into linear to oblong segments 1–5 mm wide, strigose or hairs short, spreading. **INFL** many-headed, panicle-like; peduncles 0.7–7 cm; involucre 7–10 mm diam, hemispheric to widely bell-shaped; phyllaries 12–24 in 2–3 series, 4.5–6 mm, oblanceolate or lanceolate, acuminate, soft-hairy, glandular. **RAY FLS** 10–20;

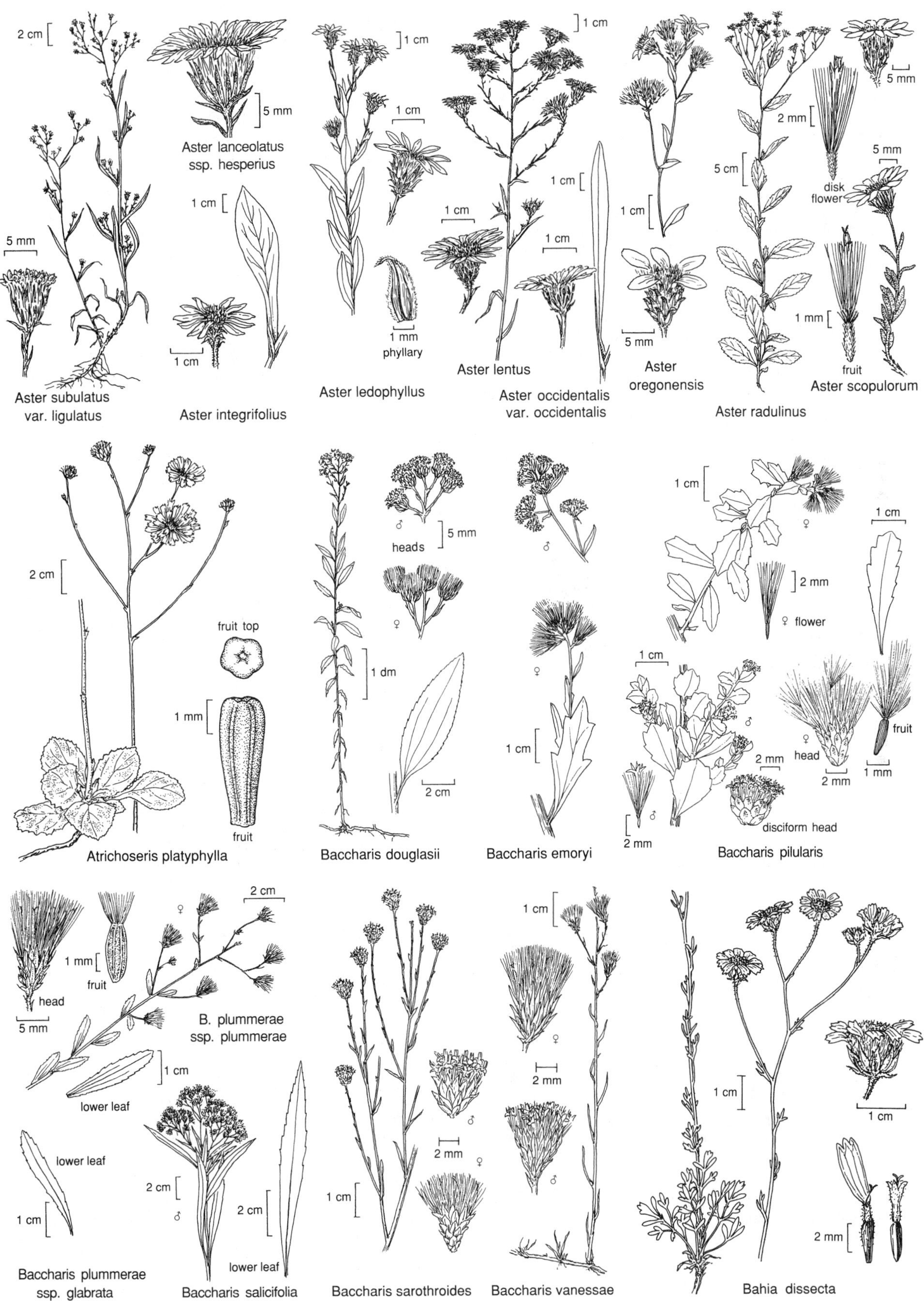

2 cm
5 mm
Aster subulatus var. ligulatus
5 mm
Aster lanceolatus ssp. hesperius
1 cm
1 cm
Aster integrifolius
1 cm
1 cm
1 mm
phyllary
Aster ledophyllus
1 cm
1 cm
Aster lentus
1 cm
1 cm
Aster occidentalis var. occidentalis
1 cm
5 mm
Aster oregonensis
5 cm
Aster radulinus
5 mm
2 mm
disk flower
5 mm
1 mm
fruit
Aster scopulorum
2 cm
fruit top
1 mm
fruit
Atrichoseris platyphylla
♂ heads
5 mm
♀
1 dm
2 cm
Baccharis douglasii
♂
♀
1 cm
Baccharis emoryi
1 cm
♀
2 mm
♀ flower
1 cm
1 cm
♂
♀ head
fruit
2 mm
1 mm
2 mm
♂
2 mm
disciform head
Baccharis pilularis
head
5 mm
1 mm
fruit
♀
2 cm
B. plummerae ssp. plummerae
1 cm
lower leaf
lower leaf
1 cm
Baccharis plummerae ssp. glabrata
2 cm
♂
2 cm
lower leaf
Baccharis salicifolia
♂
2 mm
♀
1 cm
Baccharis sarothroides
1 cm
♀
2 mm
♂
Baccharis vanessae
1 cm
1 cm
2 mm
Bahia dissecta

ligules 5–7.5 mm, yellow. **DISK FLS**: corollas radial or bilateral, 2.5–4.5 mm; stamens often abortive. **FR** 3–4.5 mm, dark brown or black, glabrous to short-rough-hairy; pappus 0 (rarely scales 1–13, 1.5–3 mm, lanceolate). $2n=36$. Dry, open, forest slopes; 1800–2650 m. SnBr, e PR (Santa Rosa Mtns); to WY, Colorado, TX, n Mex. Triploid, reproducing asexually.

BAILEYA DESERT-MARIGOLD

Ann, per, ± tomentose throughout. **STS** 1–many from base. **LVS** basal and alternate, simple, entire to deeply lobed, ± reduced upward, petioled or upper sessile, tomentose and glandular. **INFL**: heads radiate, solitary or in few-headed cymes; peduncles short to very long; involucre cylindric to bell-shaped or hemispheric; phyllaries in 1–2 ± equal series, linear-lanceolate; receptacle flat to slightly rounded, naked or with scattered, narrow chaff. **RAY FLS** 4–many; ligules ovate, 3-lobed, sessile on ovary, yellow, drying cream-colored, papery, reflexed and persistent on fr when dry. **DISK FLS** 8–many; corollas yellow, gland-dotted, lobes triangular, long-hairy; anther tips triangular; style tips short-triangular. **FR** linear to club-shaped, cylindric or ± angled, short-rough hairy and gland-dotted to glabrous; pappus 0. 4 spp.: sw US, Mex. (J.W. Bailey, Am microscopist, born 1811) [Brown 1973 PhD dissertation, AZ State Univ]

1. Ray fls 4–8, pale yellow; involucre gen 4–6 mm diam; heads 2–3 .... ***B. pauciradiata***
1′ Ray fls 15–many, bright yellow; involucre 10+ mm diam; heads solitary
  2. Ligules 10–20 mm, widely linear or oblong, prominently 3-lobed; fr cylindric or barely angled; ribs ± equal; peduncles gen 10–20+ cm, ± scape-like .... ***B. multiradiata*** var. ***multiradiata***
  2′ Ligules 6–10 mm, widely elliptic to obovate, shallowly 3-lobed to truncate; fr ribbed and angled; ribs of angles most prominent; peduncles gen 3–10 cm, gen leafy-bracted .... ***B. pleniradiata***

B. multiradiata A. Gray var. ***multiradiata*** (p. 217) Ann, per, canescent-tomentose. **STS** 2–5 dm, gen branched only at base. **LVS** mostly basal and on lower sts, these 2–10 cm; petioles winged; blades 1–3-pinnately divided, lobes linear to ovate; mid-st lvs 0 or reduced to linear, entire bracts. **INFL**: heads solitary, showy; peduncles 1–3 dm, ± naked; involucre 10–25 mm diam, hemispheric; phyllaries 5–8 mm. **RAY FLS** 50–60 in > 1 series; ligule 10–20 mm, widely linear, bright yellow; lobes prominent, lanceolate to ovate. **DISK FLS** many; corollas 3–4 mm. **FR** 2.5–4 mm, cylindric, not or only slightly angled, ± equally ribbed. $2n=32$. Desert roadsides, flats, washes, hillsides; 600–1600 m. DMoj; to NV, AZ, NM, n Mex. ❀SUN,DRN,DRY:7–9,**10–13**,14,**18–21**,22,23.

B. pauciradiata A. Gray (p. 217) Ann (rarely per), ± loosely tomentose. **STS** 1–5 dm, gen branched above base. **LVS** mostly cauline; basal soon withering; cauline lvs 4–14 cm, not markedly reduced upward, linear-oblong to lanceolate or oblanceolate, entire or divided into linear lobes. **INFL**: heads 2–3 in cymes; peduncles 2–5 cm; involucre 4–6 mm diam, cylindric to bell-shaped; phyllaries 8–10, 4–6 mm. **RAY FLS** 4–8; ligules 4–10 mm, obovate, shallowly 3-lobed, pale yellow. **DISK FLS** 8–20; corolla 2.5–3 mm. **FR** 4–5 mm, narrowly club-shaped, evenly ribbed. $2n=32$. Sandy desert soils, esp dunes; < 1100 m. D; w AZ, n Mex.

B. pleniradiata A. Gray (p. 217) Ann, canescent-tomentose. **STS** 1–5 dm, branched mostly in basal half. **LVS** mostly cauline; basal gen withering by time of fl; basal and lower cauline similar, 2–8 cm, petioled; blade 1–3 × divided into oblong to ovate lobes; upper cauline sessile, simple, linear to narrowly oblanceolate, gen entire. **INFL**: heads solitary; peduncles 2–10 cm; involucre 7–10 mm diam, hemispheric; phyllaries 20–30, 4–6 mm. **RAY FLS** 20–60, in 2 or more series; ligules 6–10 mm, obovate, ± entire to shallowly 3-lobed, bright yellow. **DISK FLS** many; corollas 3 mm. **FR** 3–4 mm, cylindric, distinctly angled; ribs of angles most prominent. $2n=32$. Desert roadsides, sandy soils; < 1500 m. D; to UT, AZ, n Mex. ❀TRY.

BALSAMORHIZA BALSAM-ROOT

Per from fleshy taproot; caudices 1–many. **STS** erect. **LVS** ± basal, few cauline, long-petioled; blade entire to 1–3-pinnately lobed. **INFL**: heads 1–few, raditate; peduncles long, bracts 0–few; involucre hemispheric to bell-shaped; phyllaries in 2–4 series; receptacle flat; chaff folded around frs. **RAY FLS** showy; ligules yellow. **DISK FLS** many; corollas yellow, tube short, throat cylindric to narrowly bell-shaped; style branches tapered. **FR** oblong, 3–4-angled; pappus 0. ± 12 spp.: w N.Am. (Greek: balsam root, from sticky sap of taproot) [Weber 1982 Phytologia 50:357–359] Hybrids common. ❀TRY.

1. Lvs ± sagittate or hastate, entire to coarsely dentate
  2. Lvs thinly hairy; phyllaries loosely soft-hairy .... ***B. deltoidea***
  2′ Lvs densely canescent-tomentose, at least when young; phyllary bases permanently tomentose .... ***B. sagittata***
1′ Lvs not sagittate or hastate, coarsely toothed to compound
  3. Lvs coarsely serrate but gen not deeply lobed .... ***B. serrata***
  3′ Lvs gen deeply lobed to compound
    4. Sts densely glandular (and gen also nonglandular) .... ***B. hirsuta***
    4′ Sts strigose to densely tomentose
      5. Outer phyllaries lanceolate, 2–4 mm wide, acuminate; herbage densely tomentose .... ***B. hookeri*** var. ***lanata***
      5′ Outer phyllaries ovate-lanceolate, gen > 5 mm wide; herbage finely strigose, green to densely silvery-canescent
        6. Outer phyllaries 20–40 mm, >> disk; lvs thinly strigose, green or lightly canescent .... ***B. macrolepis*** var. ***macrolepis***
        6′ Outer phyllaries 13–24 mm, gen = or slightly > disk; lvs gen densely strigose, lightly canescent to silvery
          7. Lvs gray (gen not silvery), lobes gen again divided; e CaR, MP .... ***B. macrolepis*** var. ***platylepis***
          7′ Lvs silvery, lobes gen entire; e KR .... ***B. sericea***

B. deltoidea Nutt. **STS** 2–9 dm, densely glandular, sparsely long-hairy. **LVS**: basal 20–60 cm, blade widely triangular, entire to coarsely dentate, scabrous to stiff-hairy, base cordate and ± hastate; cauline 2–several, alternate or opposite, oblanceolate, entire. **INFL**: heads 1–few; outer phyllaries 10–40 mm, 3–9 mm wide, oblong-lanceolate, obtuse to acute, glandular and ± long-hairy, often ciliate. **RAY FLS**: ligules 2–3 cm. **DISK FLS**: corollas 5–7 mm. **FR** 7–8 mm. Common. Grassy slopes, open forests, shrubby areas; 300–2400 m. NW, CW, WTR; to B.C.

B. hirsuta Nutt. **STS** 2–5 dm, densely glandular-puberulent, sparsely long-hairy. **LVS**: basal 15–30 cm, blade lanceolate to elliptic, pinnately lobed, further divided into linear-oblong segments 2–3 mm wide, both surfaces green, strigose, glandular-puberulent;

cauline gen 2 near base, opposite. **INFL**: head 1; outer phyllaries 15–22 mm, 4–5 mm wide, ovate-lanceolate, long-acuminate, ciliate, minutely glandular, tips often spreading or reflexed. **RAY FLS**: ligules 2–4 cm. **DISK FLS**: corollas 7–8 mm. **FR** 6–7 mm. Sagebrush scrub; 1450–1800 m. MP; to WA, nw NV. [*B. hookeri* (Hook.) Nutt. var. *neglecta* (W. Sharp) Cronq.]

B. hookeri Nutt. var. ***lanata*** W. Sharp **STS** 1–3 dm, densely tomentose. **LVS**: basal 10–30 cm, blade lanceolate to widely ovate, pinnately lobed, further divided into linear-oblong segments 2–3 mm wide, both surfaces whitish tomentose; cauline gen 2 near base, opposite. **INFL**: head 1; outer phyllaries 11–17 mm, 2–4 mm wide, oblong-lanceolate, acute, tomentose. **RAY FLS**: ligules 2–2.5 cm. **DISK FLS**: corollas 7–8 mm. **FR** 5–6 mm. Open woods, grassy slopes; 800–1050 m. n CaR (Shasta Valley, Siskiyou Co.).

B. macrolepis W. Sharp (p. 217) **STS** 2–6 dm. **LVS**: basal gen 20–50 cm, blade lanceolate to elliptic, pinnately divided into linear to ovate segments, finely strigose, green to silvery canescent; cauline often 2–3 near base. **INFL**: head 1; outer phyllaries 13–40 mm, gen 5–10 mm wide, ovate-lanceolate, tip wide, flat, obtuse to acute, appressed to spreading, finely strigose, minutely glandular. **RAY FLS**: ligules 2–3 cm. **DISK FLS**: corollas 7–10 mm. **FR** 5–8 mm. Dry slopes, valleys; < 1800 m. e CaR, SNF, ScV, SnFrB, MP; nw NV.

var. ***macrolepis*** **STS** 2–6 dm, finely hairy, gen also glandular. **LVS** 20–40 cm; blade green to canescent. **INFL**: outer phyllaries 20–40 mm, gen 6–10 mm wide, finely strigose, minutely glandular. **FR** 7–8 mm. Open grassy slopes, valleys; gen < 1400 m. SNF, ScV, SnFrB.

var. ***platylepis*** (W. Sharp) Ferris **STS** 0.5–4 dm, ± densely hairy, becoming ± glabrous. **LVS** 10–30 cm; blade ± gray. **INFL**: outer phyllaries 13–24 mm, 4–8 mm wide, densely strigose-canescent. **FR** 5–7 mm. Dry slopes, valleys; 1300–1800 m. e CaR, MP; nw NV.

B. sagittata (Pursh) Nutt. (p. 217) **STS** 2–6 dm, ± short-tomentose, minutely glandular. **LVS**: basal 20–50 cm, blade widely triangular, entire, acute or obtuse, base cordate and ± hastate, upper surface soft-hairy, lower surface short-tomentose to finely strigose, ± canescent; cauline gen several, linear to oblanceolate. **INFL**: heads 1–few; outer phyllaries 10–25 mm, 4–9 mm wide, oblong-lanceolate to ovate, obtuse to acute, ± tomentose. **RAY FLS**: ligules 2.5–4 cm. **DISK FLS**: corollas 6–8 mm. **FR** 7–9 mm. Open forest, scrub; 1400–2600 m. SNH, GB; to B.C., Rocky Mtns.

B. sericea W.A. Weber SILKY BALSAM-ROOT **STS** 2–4 dm. **LVS**: basal 10–30 cm, blade lanceolate to elliptic, pinnately lobed (lobes gen entire), finely silvery canescent; cauline gen 2–3 near base. **INFL**: head 1; outer phyllaries 15–20 mm, gen 7–8 mm wide, widely ovate, tip wide, flat, obtuse to acute, appressed to spreading, finely strigose, minutely glandular. **RAY FLS**: ligules 2–3 cm. **DISK FLS**: corollas 7–10 mm. **FR** 5–8 mm. UNCOMMON. e KR (Siskiyou, Trinity cos.); sw OR. Perhaps not distinct from *B. macrolepis* var. *platylepis*.

B. serrata Nelson & J.F. Macbr. (p. 217) **STS** 1–3 dm, glandular-puberulent, thinly tomentose to soft-hairy. **LVS**: basal 8–20 cm, blade lanceolate to widely ovate, serrate or irregularly pinnately lobed, lobes wide, serrate, green, strigose or scabrous; cauline gen 2 near base, opposite. **INFL**: head 1; outer phyllaries 12–22 mm, 2–3 mm wide, linear-lanceolate, long-acuminate, tomentose, tips often spreading or reflexed. **RAY FLS**: ligules 2–3 cm. **DISK FLS**: corollas 7–8 mm. **FR** 5–8 mm. Sagebrush scrub; 1400–1500 m. n MP; to WA, nw NV.

BEBBIA SWEETBUSH

Subshrubs, shrubs, ± strongly scented. **STS** many, slender, from thick, woody root-crown, short-lived, very brittle, often lfless. **LVS** simple, opposite (or upper alternate), sessile or petioled; blades linear to triangular, entire to dentate or irregularly lobed. **INFL**: heads discoid, solitary or in open, rounded cymes; peduncles slender; involucre cylindric to narrowly bell-shaped; phyllaries graduated in several series; receptacle rounded, chaffy, scales folded around frs. **FLS** many; corollas yellow; anther tips ovate, acute; style tips tapered, acute. **FR** club-shaped, compressed, 3-angled, brown to black; hairs ascending, white; pappus of 15–30 subplumose bristles. 2 spp.: sw US, nw Mex. (M.S. Bebb, Am botanist, 1833–1895) [Whalen 1977 Madroño 24:112–123]

B. juncea (Benth.) E. Greene var. ***aspera*** E. Greene (p. 217) **STS** 5–15 dm, much-branched, forming a rounded bush < or = 3 m diam, glabrous or short-bristly. **LVS** 1–3(9) cm, linear and entire or with few, sharp, pinnate lobes, drought-deciduous. **INFL**: heads few; peduncles 1.5–6 cm; involucre 4–15 mm diam; phyllaries 1–7 mm, lanceolate to linear, acute. **FLS**: corollas 6.5–10 mm. **FR** 2–3.5 mm; pappus 6–10 mm. $2n$=18. Common. Dry, rocky slopes, desert plains, washes; < 1500 m. SW, SNE, D; to s NV, TX, nw Mex. ❀DRN:7–11,**12**,**13**,14,18–24;bright green sts.

BELLIS

Ann, per. **LVS** simple, basal or restricted to lower st, petioled. **INFL**: heads radiate, solitary on slender, ± naked peduncles; involucre hemispheric; phyllaries in 2 equal series; receptacle conic, naked. **RAY FLS** many; ligules white to pink or purple. **DISK FLS** many; corollas yellow; style tips flat, triangular. **FR** flat; pappus 0. 7 spp.: Eur, Medit. (Latin: pretty)

B. perennis L. (p. 217) ENGLISH DAISY Per; roots fibrous. **LVS** 2–10 cm; blades oblanceolate or obovate, tapered to winged petiole, tip obtuse, margins entire to crenate or serrate, surfaces loosely soft-hairy. **INFL**: peduncle 2–25 cm, strigose or hairs soft, spreading; phyllaries 3–6 mm, ovate, obtuse, hairs soft, spreading. **RAY FLS**: ligule 7–10 mm, narrow, white to purple. **DISK FLS**: corollas ± 2 mm. **FR** 1–1.5 mm, light brown, glabrous. $2n$=18. Lawns, damp, grassy areas; gen < 200 m. NW, CW, SW; native to Eur.

BIDENS STICKTIGHT, SPANISH-NEEDLES, BEGGAR'S TICK

Ann, per, shrubs. **STS** prostrate to erect. **LVS** simple or pinnate, gen opposite, sessile or petioled. **INFL**: heads radiate or discoid, gen few in CA; involucre cylindric to bell-shaped; phyllaries in 2 dissimilar series, outer gen ± lf-like in texture, inner thinner, with transparent or scarious margins; receptacle chaffy; chaff scales narrow, flat. **RAY FLS** 0 or few; ligules yellow or white. **DISK FLS** gen many; corollas yellow, radial (or outermost white, bilateral). **FR** narrowly club-shaped, thick or compressed front-to-back; pappus 0 or awns 1–several, gen barbed. ±230 spp.: worldwide. (Latin: 2 teeth) [Sherff & Alexander 1955 N.Am Flora 2(2):70–129]

1. Lvs simple
 2. Lvs petioled .... ***B. tripartita***
 2′ Lvs sessile
  3. Fr with ± thickened margin, prominently ribbed; tips of chaff scales yellow; ligules 0 or < 15 mm .... ***B. cernua*** var. ***cernua***

3′ Fr flat, without thick margin or prominent ribs; tips of chaff scales reddish; ligules 15–30 mm .... ***B. laevis***
1′ Lvs pinnately compound
4. Fr narrow, ± cylindric .... ***B. pilosa*** var. ***pilosa***
4′ Fr wedge-shaped, flat
5. Outer phyllaries 5–8; inner phyllaries = disk; fr ± black .... ***B. frondosa***
5′ Outer phyllaries 10–16; inner phyllaries < disk; fr brown or dark green .... ***B. vulgata***

B. cernua L. var. ***cernua*** NODDING BUR-MARIGOLD Ann. **ST** 1–9 dm, erect, glabrous to short-rough-hairy. **LVS** simple, sessile; bases fused around st; blades 4–20 cm, linear-lanceolate to lanceolate, acuminate, serrate, glabrous. **INFL**: heads radiate or discoid, erect in fl, nodding in fr; peduncles 1–7 cm; involucre 1–2 cm diam, hemispheric; outer phyllaries 5–8, 1–3 cm, linear-lanceolate, spreading; inner phyllaries 8–10 mm, ovate-lanceolate; chaff scales 6–8 mm, yellowish. **RAY FLS** 0 or 6–8; ligules 8–15 mm, yellow. **DISK FLS**: corollas 3–4 mm, yellow. **FR** 5–7 mm, narrowly wedge-shaped, 4-angled, flat; angles thick, barbed; pappus awns gen 4, 2–3 mm. $2n=24$. Freshwater wetlands; < 2000 m. KR, NCoR, CaR, n SN, SnFrB, MP; widespread n hemisphere. Much like *B. laevis*. ❀TRY.

B. frondosa L. (p. 217) STICKTIGHT Ann, ± glabrous. **STS** 5–12 dm, square. **LVS** compound, petioled; lflets 2–8 cm, lanceolate, gen acuminate, serrate. **INFL**: heads radiate or discoid, erect; peduncles 2–10 cm; involucre ± 1 cm diam, hemispheric; outer phyllaries 5–8, 1–5 cm, ± linear, ciliate; inner phyllaries 5–7 mm, ovate; chaff scales 5–7 mm, brownish. **RAY FLS** 0–few; corollas 2–3.5 mm. **DISK FLS**: corollas ± 2 mm, orange. **FR** 6–10 mm, blackish, narrowly wedge-shaped, flat, ± glabrous to stiffly hairy; pappus awns gen 2, 3–4.5 mm. $2n=48$. Uncommon. Damp soil, esp disturbed sites; gen < 1600 m. NCoR, GV, SW, GB; to e N.Am.

B. laevis (L.) Britton, Sterns & Pogg. (p. 217) BUR-MARIGOLD Ann, per, gen ± glabrous. **STS** 2–25 dm, ± decumbent to erect, ± cylindric. **LVS** simple, sessile; bases sometimes fused around st; blades 5–15 cm, ± lanceolate, acute to acuminate, serrate. **INFL**: heads radiate, erect in fl, often nodding in fr; peduncle 2–10 cm; involucre 1–2 cm diam, bell-shaped; outer phyllaries 6–8, 1–2 mm, linear-lanceolate, sparsely ciliate; inner phyllaries 8–16 mm, obovate; chaff scales reddish tipped. **RAY FLS** 7–8; ligules 1.5–3 cm, yellow. **DISK FLS**: corollas 4–6 mm, yellow. **FR** narrowly wedge-shaped, flat or 3–4-angled; angles thin, barbed; pappus awns 2–4, 3–5 mm. $2n=22$. Freshwater wetlands; gen < 300 m. GV, SCoR, SW, DMoj (Mojave River); to e N.Am. Much like *B. cernua*. ❀WET,SUN:7,**8**,**9**,10,11,**12**,**14–16**,**18–24**.

B. pilosa L. var. *pilosa* (p. 217) COMMON BEGGAR-TICKS, SPANISH-NEEDLES Ann, glabrous or ± soft-hairy. **STS** erect, 3–18 dm, square. **LVS** compound, petioled; axis sometimes winged; lflets gen 3–5, 2–6 cm, lanceolate to ovate, acute, serrate, base often asymmetric. **INFL**: heads discoid or radiate, few–many, erect; peduncles 1–9 cm; involucre 7–8 mm diam in fl, hemispheric; outer phyllaries 7–9, 4–5 mm, linear; inner phyllaries 4–7 mm, lanceolate; chaff scales linear, acuminate. **RAY FLS** 0 or vestigial; corolla white. **DISK FLS** alike; corolla ± 2 mm, radial, yellow (or not alike, with outer corollas 2–3 mm, white, ± bilateral). **FR** 4–16 mm, narrowly club-shaped or slightly flat, 4-angled, black, short-rough-hairy; pappus awns 2–4, 2–4 mm, ± yellow. $2n=72$. Disturbed sites; gen < 750 m. SCoR, SW; subtrop, trop worldwide.

B. tripartita L. Ann, glabrous. **STS** erect, 4–15 dm. **LVS** simple, petioled; blades < or = 20 cm, lanceolate, acuminate, serrate. **INFL**: heads gen discoid, erect; peduncles 1–6 cm; involucre 7–20 mm diam (exc leafy tips of outer phyllaries), hemispheric; outer phyllaries 4–5, gen 1–3 cm, ± linear, spreading; inner phyllaries 7–8 mm, ovate; chaff scales 8–10 mm, black or greenish with fine black lines. **RAY FLS** gen 0. **DISK FLS**: corollas ± 3 mm. **FR** wedge-shaped, ± flat; outer achenes 3.5–4 mm, 2–3-angled; inner achenes 5–8 mm, 4-angled; barbs of angles ascending below, reflexed above; pappus awns gen 4, 1–3.6 mm. $2n=48, 72$. Freshwater wetlands; 50–1600 m. c SN, SnJV; native to e N.Am, Eur. [*B. connata* Muhlenb. var. *petiolata* (Nutt.) Farw.]

B. vulgata E. Greene Ann, ± puberulent. **STS** erect, 3–15 dm, 4-angled. **LVS** compound, petioled; lflets 2–8 cm, lanceolate, acuminate, serrate. **INFL**: heads radiate; peduncles 2–23 mm; involucre bell-shaped; outer phyllaries 10–16, 1–2 cm, ± linear, ciliate; inner phyllaries 7–9 mm, ovate-lanceolate. **RAY FLS** few; ligules 2.5–3.5 mm. **DISK FLS**: corollas ± 3 mm, yellow. **FR** 6–12 mm wedge-shaped, yellowish or olive-brown, flat; margin with stiff, ascending hairs or reflexed barbs; faces glabrous or short-rough-hairy; pappus awns 2, 3–4 mm. $2n=24$. Uncommon. Freshwater wetlands; < 300 m. NCoR, ScV; to WA, native to e N.Am.

BLENNOSPERMA

Robert Ornduff

Ann < 30 cm, monoecious, ± glabrous. **ST** branched, erect. **LVS** cauline, alternate, < 15 cm, entire or lobed. **INFL**: heads radiate, borne in cymes; phyllaries in 1 series, thin, margins clear; receptacle flat, naked. **RAY FLS** 6–13, pistillate, (pistillate fls without corollas sometimes also present); ligules ± yellow (or 0). **DISK FLS** 2–100, staminate; corolla yellow; anthers obcordate, pollen yellow or white; style undivided, tip head-like. **FR** 2–4.5 mm, obovoid, gen papillate (if so, gray, sticky when wet); pappus 0. 3 spp.: CA & Chile. (Greek: slimy seed) [Ornduff 1964 Brittonia 16:289–295]

1. Stigma of ray fl dark red; upper lvs 1–3-lobed .... ***B. bakeri***
1′ Stigma of ray fl yellow; most lvs with 3–15 pinnate lobes .... ***B. nanum***
2. Pollen gen white; fr < 3 mm; disk fls 20–70; gen inland .... var. ***nanum***
2′ Pollen yellow; fr > 3 mm; disk fls 60–100; coastal .... var. ***robustum***

B. bakeri Heiser (p. 217) BAKER'S BLENNOSPERMA **LVS** 5–15 cm, linear; lower entire; upper 1–3-lobed, lobes 1–3 cm. **INFL**: involucre 6–8 mm. **RAY FLS**: ligules 5–7 mm, yellow; stigma dark red; pistillate fls without corollas 0. **DISK FLS**: stigma and pollen white. **FR** 3–4 mm, strongly 4–6-angled, papillate. $2n=18$. **ENDANGERED** CA,US. Vernal pools, wet grasslands; < 100 m. NCoR, ne SnFrB (s Sonoma Co.). In cult.

B. nanum (Hook.) S.F. Blake **LVS** 0.5–6 cm, linear, pinnately lobed; lobes 3–15, short, linear. **INFL**: involucre 5–6 mm. **RAY FLS**: ligules 3–7 mm, yellow or rarely white, ± purple on back; stigma yellow; pistillate fls without corollas present. **DISK FLS**: stigmas and pollen white or yellow. **FR** 2.5–4.5 mm, smooth or ridged. Grassland, scrub, woods, gen wet, open areas; < 1600 m. NCoR, SNF, GV, CW, SCo, ChI.

var. ***nanum*** (p. 217) **DISK FLS** 20–70; pollen gen white. **FR** 2.5–3 mm, ridged, papillate. $2n=14$. Common (very uncommon in s). Wet open areas, woods, grassland; < 1600 m. NCoR, SNF, GV, CW, SCo, ChI. Fls early spring. ❀SUN:6,**7–9**,10,**14–17**,18,**19–24**.

var. ***robustum*** J. Howell POINT REYES BLENNOSPERMA **DISK FLS** 60–100; pollen yellow. **FR** 3–4.5 mm, ridged, gen papillate. $2n=14$. **RARE** CA. Grassy places in shrubs; < 100 m. n CCo (Point Reyes peninsula, Marin Co.). Fls late spring. Some populations on Point Reyes peninsula are intermediate to var. *nanum* in fr length, pollen color. In cult.

BLEPHARIPAPPUS

Bruce G. Baldwin & Susan J. Bainbridge

1 sp. (Greek: eyelash pappus)

B. scaber Hook. (p. 217) Ann 7–40 cm, slender-branched, scabrous to short-hairy, glandular above. **LVS** ± alternate, sessile, gen ascending, < 35 mm, linear; margin gen rolled under. **INFL**: heads 1 or in leafy cymes, radiate; involucre obconic to hemispheric; phyllaries partly surrounding ray fr, 3–8 mm, glandular; outer receptacle scales phyllary-like, tips of inner green, hairy. **RAY FLS** 2–8; ligules 2–8 mm, white, purple-veined below; style branches long, glabrous. **DISK FLS** 6–25(60); corollas 2–3.5 mm, white; anthers purple, tips acute; style barely branched, bristly. **FR** 2–3.5 mm, ± obconic, black, gen ± white-hairy; pappus (0 or) gen < 26 flattened, plumose bristles. 2*n*=16. Yellow-pine forest, juniper woodland; 1000–3000 m. KR, CaR, n SN, MP; to WA, ID, NV. [ssp. *laevis* (A. Gray) Keck] ❀TRY.

BLEPHARIZONIA BIG TARWEED

1 sp. (Greek: eyelash girdle, from frs weakly held by phyllaries)

B. plumosa (Kellogg) E. Greene (p. 221) Ann 3–18 dm, strongly scented. **ST** gen 1, erect, much-branched, soft-hairy to bristly, puberulent, ± glandular. **LVS** basal and lower cauline 4–15 cm, crowded; blades linear, entire or serrate, often withering in fl; cauline alternate, gen subtending smaller lf clusters, base 1–3 cm wide; clustered lvs < cauline, margins with tack-shaped glands. **INFL** ± glandular; heads radiate, axillary or terminal, in a narrow to open panicle; bracts gen glandular-ciliate; involucre 5–7.5 mm, hemispheric; phyllaries 1 per ray fl, elliptic, hairy and often glandular, base closely appressed to ovary, tip free, acute; chaff scales between ray and disk fls, = phyllaries, flat. **RAY FLS** 5–13; corolla 6–9 mm; ligule 5–8 mm, white, red-veined. **DISK FLS** 10–35; corolla 4–5.5 mm, white, throat > tube, lobes 0.5 mm; anthers brown; style branches tapered, short-bristly. **FR** 3.5–4 mm, club-shaped, 10-ribbed, densely puberulent, ± dark; ray pappus 0 or minute crown, sometimes of narrow, plumose scales < 1 mm; disk pappus 1.5–3 mm, scales 15–20, narrow, plumose, sometimes 0 or reduced to crown. 2*n*=28. Dry grassy areas; < 1100 m. w SnJV, SnFrB, SCoR.

1. Disk pappus 1.5–3 mm; herbage sparsely glandular below heads; older pls with many heads lateral on infl branches ..... ssp. ***plumosa***

1′ Disk pappus 0–1 mm; herbage densely glandular; older pls with heads mostly terminal on slender wand-like, bracted peduncles ..... ssp. ***viscida***

ssp. ***plumosa*** Pl gray-green, sparsely glandular below heads. **INFL**: branches of older pls spike-like or raceme-like; many heads lateral on infl branches; involucre canescent. **FR**: disk pappus 1.5–3 mm. Dry hills, plains; < 400 m. nw SnJV (n of Alameda Co.), e SnFrB.

ssp. ***viscida*** Keck Pl yellow-green, densely glandular. **INFL**: branches of older pls wand-like; heads mostly terminal; involucre ± bristly. **FR**: disk pappus 0–1 mm. Foothill woodland, chaparral, grassland; < 1100 m. w SnJV (Alameda Co. s), e SnFrB, SCoRI.

BRICKELLIA BRICKELLBUSH

A. Michael Powell

Per from ± woody caudex or shrub (rarely ann). **LVS** alternate or opposite, simple, veiny, gen resinous-dotted. **INFL**: heads discoid, gen clustered; involucre cylindric to bell-shaped; phyllaries overlapping, strongly nerved; receptacle gen flat, chaff 0. **FLS**: corollas cylindric, ± white or tinged red; anther bases rounded or slightly cordate, tips ovate; style branches long, club-shaped, tips rounded. **FR** 10-ribbed, gen cylindric, hairy; pappus gen of many, gen scabrous bristles. (John Brickell, early botanist in Georgia) [King & Robinson 1987 Monogr Syst Bot Missouri Bot Gard 22:220–224]

1. Per or subshrubs from woody caudex
 2. Per; root ± tuber-like; lvs deltate to ± cordate, 3–12 cm, nonglandular ..... ***B. grandiflora***
 2′ Subshrubs; lvs linear to ovate, 1–3 cm, glandular
  3. Lvs 10–20 mm wide, ovate; fr ± 7 mm ..... ***B. greenei***
  3′ Lvs < 8 mm wide, linear to ± ovate; fr 4–6 mm ..... ***B. oblongifolia*** var. ***linifolia***
1′ Shrub — 2–20 dm
 4. Pl gen 10 dm or more
  5. Heads 8–18-fld
   6. Heads 12–14 mm; phyllaries ± glabrous ..... ***B. californica***
   6′ Heads 8–10 mm; phyllaries puberulent ..... ***B. desertorum***
  5′ Heads 3–7-fld
   7. Lvs linear; phyllaries 10–12 ..... ***B. longifolia***
   7′ Lvs lanceolate to ovate; phyllaries ± 20
    8. Lvs gen serrate ..... ***B. knappiana***
    8′ Lvs ± entire ..... ***B. multiflora***
 4′ Pl gen 2–6 dm
  9. Phyllary tips spreading or recurved
   10. Lvs white-woolly ..... ***B. nevinii***
   10′ Lvs ± green
    11. Sts glandular-hairy ..... ***B. microphylla***
    11′ Sts tomentose-puberulent ..... ***B. watsonii***
  9′ Phyllary tips erect
   12. Pls white-tomentose; heads ± 24 mm ..... ***B. incana***
   12′ Pls green; heads gen 12–15 mm
    13. Outer phyllaries ovate ..... ***B. atractyloides***
    13′ Outer phyllaries ± linear to widely lanceolate

14. Lvs oblong to spoon-shaped, not leathery, entire; outer phyllaries linear-oblong, acute, ± 3 mm, puberulent to minutely tomentose .... ***B. frutescens***
14′ Lvs ovate, leathery, gen sharply toothed; outer phyllaries widely lanceolate, long-tapered, ± 10 mm, gen minutely scabrous .... ***B. arguta***
15. Outer phyllaries ± entire .... var. ***arguta***
15′ Outer phyllaries clearly dentate .... var. ***odontolepis***

B. arguta Robinson (p. 221) Shrub 2–4 dm. **STS** much-branched, zigzag, glandular-puberulent. **LVS** alternate, subsessile, 1–2 cm, ovate, leathery, prominently toothed or entire, scabrous, gen glandular-puberulent. **INFL**: heads 13–15 mm, solitary; involucre bell-shaped; phyllaries 26–33, overlapping to subequal; outer lance-ovate, ± entire to dentate, inner linear, 4–13-nerved, greenish, gen minutely scabrous, tips erect. **FLS** 40–55. **FR** 4 mm. Rocky places; < 1500 m. D; n Baja CA. Perhaps same as *B. atractyloides*.

var. ***arguta*** **INFL**: outer phyllaries ± entire. $2n=18$. Common. Habitats and range of sp. ❀DRN,SUN:7–16,18–21;STBL.

var. ***odontolepis*** Robinson **INFL**: outer phyllaries clearly dentate. Uncommon. Habitats of sp. DSon.

B. atractyloides A. Gray Shrub 1.7–3.5 dm. **STS** much-branched, gen glandular-puberulent. **LVS** alternate, gen short-petioled, 1.5–3 cm, ± lanceolate to ovate, leathery, clearly toothed, scabrous, gen glandular-puberulent. **INFL**: heads 12–14 mm, solitary; involucre bell-shaped; phyllaries 24–32, overlapping to subequal, outer ± ovate, innermost linear, 4–15-veined, greenish to tan, minutely scabrous, tips gen erect. **FLS** 50–65. **FR** 4–5 mm. $2n=18$. Uncommon. Rocks, dry slopes, washes; 600–1400 m. DMoj; NV, AZ, UT. Intergrading with *B. arguta* in Inyo Co.

B. californica (Torrey & A. Gray) A. Gray (p. 221) Shrub 5–20 dm. **STS** many, branched, puberulent to weakly tomentose. **LVS** alternate, short-petioled, 1–6 cm, triangular-ovate, ± serrate, gland-dotted, puberulent, reduced upward. **INFL**: heads 12–14 mm, in a leafy panicle with small clusters on short lateral branchlets; involucre cylindric; phyllaries 21–35, overlapping, 3–5-veined, green or ± purple, ± glabrous, outer ± lanceolate to ovate, inner linear to oblong, tips rounded to ± acute, erect. **FLS** 8–18. **FR** 3 mm. $2n=18$. Common. Diverse dry habitats; < 2700 m. NW, CaRF, SN, SnJV, CW, SW, PR, D; to ID, Colorado, TX, n Mex. Variable. ❀DRN, SUN:1–3,**7**,8–13,**14–16,18–24**;STBL.

B. desertorum Cov. Shrub 8–15 dm. **STS** intricately branched, puberulent to short-tomentose. **LVS** opposite or alternate, short-petioled, 0.5–1.2 cm, ovate, ± serrate, minutely tomentose. **INFL**: heads 8–10 mm, in small clusters on short lateral branchlets; involucre cylindric; phyllaries ± 21, overlapping, linear to oblong, 3–5-veined, ± green, purple, or yellow, ± minutely tomentose, tips erect. **FLS** 8–12. **FR** 2–3 mm. $2n=18$. Common. Rocky places, desert scrub; 200–1400 m. D; NV, AZ. ❀TRY;STBL.

B. frutescens A. Gray (p. 221) Shrub 3–6 dm, aromatic. **STS** slender, rigid, spreading, ± short-tomentose. **LVS** alternate, short-petioled, 0.5–2 cm, oblong to spoon-shaped, puberulent, entire. **INFL**: heads 13–15 mm, gen 1(3), terminal; involucre cylindric; phyllaries ± 21, overlapping, linear-oblong, 4-veined, green or purplish-tinged, ± minutely tomentose, tips erect. **FLS** 20–30. **FR** 3–4.8 mm. Common. Rocky slopes, desert scrub; 600–1200 m. DSon; NV, Baja CA. ❀TRY;STBL.

B. grandiflora (Hook.) Nutt. (p. 221) Per 3–7 dm; root ± tuber-like. **STS** simple, puberulent. **LVS** gen opposite; petiole 5–30 mm; blade 3–12 cm, lanceolate to triangular-cordate, crenate to dentate, gland-dotted, puberulent. **INFL**: heads 12–15 mm, ± loosely clustered, nodding; involucre widely cylindric to bell-shaped; phyllaries 30–40, loosely overlapping, outer gen lanceolate, puberulent, greenish, tips long-tapered, gen reflexed or curled, inner ± lanceolate, 5–7-veined. **FLS** 20–40. **FR** ± 4 mm. $2n=18$. Common. Among rocks, shaded forest; 1500–2700 m. NCoRH, CaRH, c SNF, s SNH; to WA, MT, WY, NE, TX, Mex. ❀TRY;STBL.

B. greenei A. Gray (p. 221) Per 2–5 dm, leafy. **STS** many, ± simple, glandular-hairy, sticky. **LVS** alternate, short-petioled; blades 1.5–3 cm, ovate, gen serrate, glandular, sticky. **INFL**: heads 17–20 mm, gen solitary (in cymes), subtended by bracts; involucre widely cylindric; phyllaries ± 40, overlapping, linear, 2–4-veined, greenish or straw-colored, glandular-puberulent, tips rounded to acute, erect. **FLS** ± 60. **FR** ± 7 mm. $2n=18$. Common. Among rocks; 900–2700 m. NW (exc NCoRO), CaRH, n SNH; OR. ❀TRY;STBL.

B. incana A. Gray (p. 221) Shrub 4–10 dm, ± spheric. **STS** many, simple or branched, white-tomentose. **LVS** alternate, sessile or short-petioled, 1–3 cm, ovate, entire to minutely serrate, tomentose. **INFL**: heads ± 24 mm, solitary; involucre bell-shaped, tomentose; phyllaries ± 40, overlapping, ± gray (greenish to purple beneath hairs), outer oblong-ovate, inner linear-lanceolate, 5–9-veined, hidden by dense hairs, tips erect. **FLS** ± 60. **FR** ± 1 cm. $2n=18$. Uncommon. Sandy washes, flats; < 1700 m. D; NV. ❀DRN, DRY:2,3,7–12,**14**,18–24;DFCLT.

B. knappiana Drew (p. 221) KNAPP'S BRICKELLIA Shrub 10–20 dm. **STS** slender, willow-like, ± sticky, puberulent. **LVS** alternate; petioles 4–5 mm; blades 2.5–3.5 cm, lanceolate or narrowly ovate, gen serrate, gland-dotted. **INFL**: heads ± 7 mm, in a panicle-like cluster with most heads ± lateral; involucre cylindric; phyllaries ± 20, overlapping, linear-elliptic, 4-veined, greenish, puberulent, tips erect. **FLS** 5–7. **FR** 2.5 mm. RARE. Desert scrub; 800–1700 m. n&e DMtns (Panamint, n Kingston mtns). Origin possibly *B. multiflora* × *B. californica* or *B. desertorum*; needs further study.

B. longifolia S. Watson (p. 221) Shrub 10–15 dm. **STS** much-branched, ± glabrous. **LVS** alternate, ± sessile, 3–10 cm, linear, entire, appearing varnished, subglabrous to puberulent. **INFL**: heads ± 5 mm, in small raceme-like clusters; involucre cylindric; phyllaries 10–12, overlapping, 4-veined, outer ovate, inner ± lanceolate, greenish, glabrous or minutely puberulent, tips erect. **FLS** 3–5. **FR** 1.8 mm. Uncommon. Washes; 1000–1700 m. n&w DMoj; to UT, AZ. Indistinct from *B. multiflora*.

B. microphylla (Nutt.) A. Gray Shrub 3–6 dm. **STS** erect, branched from base, glandular-hairy. **LVS** green, short-petioled; blade 0.7–2 cm, ovate to round, entire to serrate-lobed, glandular. **INFL**: heads 10–12 mm, 1–3 per panicle on short branches; involucre cylindric to narrowly bell-shaped; phyllaries ± 45, overlapping, outer obovate, inner linear-oblong, 3–5-veined, greenish, glandular, tips darker, recurved or spreading. **FLS** ± 22. **FR** 4–4.5 mm. $2n=18$. Uncommon. Among rocks; 1000–2700 m. s SNF, SnGb, Wrn, SNE, DMoj; to ID, NV.

B. multiflora Kellogg Shrub 10–20 dm. **STS** erect, branched. **LVS** alternate, short-petioled; blade 3–8 cm, lanceolate, ± entire, gummy, gland-dotted, sparsely bristly. **INFL**: heads ± 7 mm, in small raceme-like clusters; involucre cylindric; phyllaries ± 20, overlapping, lanceolate to ovate, 4-veined, greenish, glabrous or minutely puberulent, tips erect. **FLS** 3–5. **FR** 1.7–2.2 mm. Uncommon. Rocks, washes; 600–2400 m. DMoj; NV.

B. nevinii A. Gray (p. 221) NEVIN'S BRICKELLBUSH Shrub 3–5 dm. **STS** erect, branched, dense, white-tomentose. **LVS** alternate, ± sessile, 0.6–1.8 cm, ovate to ± cordate, dentate-serrate, tomentose. **INFL**: heads ± 1.5 cm, few in panicle-like clusters on short branches; involucre subcylindric; phyllaries ± 30, overlapping, linear-oblong, 3–6-veined, whitish woolly, tips spreading or recurved. **FLS** ± 23. **FR** ± 4 mm. $2n=18$. UNCOMMON. Desert scrub; 300–1900 m. SCoRI, WTR, SnGb, DMoj; NV. Related to *B. microphylla*, *B. watsonii*. ❀DRN,DRY:7–12,14,18–24;DFCLT.

B. oblongifolia Nutt. var. ***linifolia*** (D. Eaton) Robinson (p. 221) Subshrub 2–4 dm, rounded. **STS** much-branched, gray-hairy, gen glandular. **LVS** alternate, ± sessile, 1–3 cm, linear to ± ovate, 0–2-toothed, ± densely puberulent, gen glandular. **INFL**: heads 15–17 mm, gen solitary; involucre cylindric to bell-shaped; phyllaries 25–35, overlapping, outer shorter, lance-oblong, inner linear, gen 4-veined, green, glandular-puberulent, tips erect. **FLS** 35–45. **FR** 4–6 mm. $2n=18$. Common. Desert, woodland; < 2800 m. SnJt, SNE, D; to B.C., MT, UT, Colorado. Variable; 2 other vars. ❀TRY;STBL.

1 cm
1 cm
involucre
1 cm
2 mm
fruit
Baileya multiradiata var. multiradiata
1 cm
5 mm
2 mm
5 mm
2 mm
2 cm
2 mm
Baileya pauciradiata
B. pleniradiata
5 cm
2 mm
chaff
fruit
Balsamorhiza sagittata
B. macrolepis
2 cm
2 cm
Balsamorhiza serrata
1 cm
5 mm
1 cm
fruit
chaff
2 mm
Bebbia juncea var. aspera
1 cm
flower
1 mm
Bellis perennis
1 cm
1 cm
5 mm
1 cm
chaff
fruit
Bidens frondosa
5 mm
1 cm
5 mm
Bidens pilosa var. pilosa
2 mm
fruit
chaff
1 cm
Bidens laevis
1 cm
leaf
1 mm
fruit
Blennosperma bakeri
1 cm
1 mm
fruit
B. nanum var. nanum
flower head
5 mm
phyllary
2 mm
2 mm
2 cm
2 mm
fruit
disk
ray
flowers
Blepharipappus scaber

B. watsonii Robinson Shrub or subshrub 2–3 dm, aromatic. **STS** much-branched, tomentose-puberulent. **LVS** alternate, short-petioled; blades 5–15 mm, ovate, light green, > tomentose, ± dentate. **INFL**: heads 9–11 mm, in panicle-like clusters with 1–3 heads at branchlet ends; involucre cylindric; phyllaries ± 25, ± spreading, ± ovate, 3–5-veined, glandular, tips green. **FLS** 15–18. **FR** 3.5–4.5 mm. 2*n*=18. Uncommon. Among rocks, on rock walls, gen in woodland; 1500–2400 m. DMoj; to UT.

CACALIOPSIS

1 sp. (Greek: like *Cacalia*, a relative of *Senecio*) [Strother 1978 N.Am Fl II 10:160–173]

C. nardosmia (A. Gray) A. Gray (p. 221) Per 1.5–9 dm, rhizomed. **STS** gen erect. **LVS** alternate, reduced upward; petioles 7–30 cm; blades 7–35 cm, widely reniform, palmately lobed and further divided, ± tomentose below, ± glabrous above. **INFL**: heads gen 3–11 in raceme-like to ± flat-topped cyme, discoid; involucre gen 12–16 mm, ± bell-shaped; phyllaries in 1–2 equal series, linear to lanceolate; receptacle flat, naked. **FLS** 20–50; corollas gen 5–7 mm, ± yellow, lobes lanceolate, spreading; anther bases entire, tips narrowly lanceolate; style branches ± short-bristly, tips flattened, conic. **FR** 6–8.5 mm, cylindric, 12–15-veined, glabrous, brown; pappus of many fine bristles 10–18 mm, white. Meadows, open forest, sometimes serpentine; 200–1900 m. KR, NCoR; to s B.C. ❀DRN:4–6&SHD,IRR:7,15–17;DFCLT.

CALENDULA

Elizabeth McClintock

Ann, per, gen glandular-hairy. **STS** erect, branched. **LVS** simple, alternate. **INFL**: heads radiate, solitary, peduncled; phyllaries in 2–3 ± equal series, linear, margins narrowly scarious; receptacle flat, naked. **RAY FLS** many; ligules yellow or orange. **DISK FLS** many, staminate; corollas yellow to brown; anthers with ovate or triangular-ovate tips, sagittate at base, short-tailed; style with ring of hairs just below tip, very shallowly divided. **FR** incurved, rough or prickly on back; pappus 0. ± 15 spp.: Eur, n Afr, w Asia. (Latin: calendar, for long fl season)

1. Heads gen < 4 cm diam, nodding at maturity; fr strongly incurved, almost a circle; lvs ± lanceolate ***C. arvensis***
1′ Heads gen > 5 cm diam, erect at maturity; fr gen boat-shaped; lvs oblong to obovate .... ***C. officinalis***

C. arvensis L. FIELD-MARIGOLD Ann, finely glandular-hairy. **STS** slender, < 15 cm. **LVS** petioled; blade < 7 cm, ± lanceolate, thinnish, becoming sessile upward. **INFL**: head < 4 cm diam, nodding at maturity. **RAY FLS**: ligules yellow. **DISK FLS**: corollas ± 3 mm. **FR** 3 mm, strongly incurved, almost in a circle, with many prickles on back; inner prickles blunt; outer prickles sharp. 2*n*=18,36,44. Uncommon. Escape from cult in disturbed, urban, coastal areas; < 200 m. NCo, CCo, SCoRO, expected more widely; native to c Eur, Medit.

C. officinalis L. POT-MARIGOLD Ann, finely hairy. **STS** slender to ± coarse, sparingly branched. **LVS** sessile, < 15 cm, oblong to obovate, thickish; base gen clasping. **INFL**: heads 4–10 cm in diam, erect at maturity, closing at night. **RAY FLS**: ligules pale yellow to orange. **DISK FLS**: corollas ± 6 mm. **FR** 3 mm, incurved, boat-shaped, with many sharp prickles on back. 2*n*=14,32. Uncommon. Escape from cult in disturbed, urban, coastal areas; < 500 m. CCo, SCoRO; native orgin unknown.

CALYCADENIA

Robert L. Carr & Gerald D. Carr

Ann, gen erect, with tack- or saucer-shaped glands (and gen more normal glandular hairs), gen aromatic. **LVS** gen alternate, ± sessile, ± linear, entire, gen reduced upward; lower often deciduous by fl. **INFL**: heads radiate, variously clustered; peduncle bracts gen with 1+ tack-like glands; each phyllary partly enfolding a ray fr; chaff scales between ray and disk fls, fused, forming a cup, or ± free. **RAY FLS** 1–6; ligules 3–13 mm, 3-lobed, white to rose or yellow, sometimes with red spot near base. **DISK FLS** 1–many; corollas 2–9 mm, colored ± like ligules. **FR**: ray fr ± 3-angled, inner surface ± flat, outer surface rounded to ± angled, pappus 0; disk fr angled, base tapered, pappus 0 or gen of ± lanceolate (and sometime wider) scales. ± 9 spp.: esp CA (OR, NV). (Greek: cup gland)

1. Peduncle bracts tipped by 1 tack-like gland; chaff scales gen without tack-like glands; pl often > 1.5 dm
  2. Corollas yellow; ray frs decidedly rough-wrinkled .... ***C. truncata***
  2′ Corollas white; ray frs smooth (exc some *C. hooveri*)
    3. Heads 2–4-fld, inconspicuous; branchlets many, ± thread-like, flexible; ray fr glabrous .... ***C. hooveri***
    3′ Heads 6–many-fld, conspicuous; branchlets 0–many, rigid; ray fr hairy
      4. Peduncle bracts ± cylindric distally, tip truncate to strongly concave .... ***C. spicata***
      4′ Peduncle bracts flat, tip rounded to truncate (terminal gland sometimes 0) .... ***C. villosa***
1′ Peduncle bracts, chaff scales, or both gen with 1 tack-like gland (or in pls < 1.5 dm of Butte Co., peduncle bracts with 1 tack-like gland and chaff scales without such glands)
  5. Lvs opposite throughout; heads appearing whorled .... ***C. oppositifolia***
  5′ Lvs gen alternate at least above; heads not appearing whorled
    6. Central ligule lobe << and narrower than lateral lobes, widest at base .... ***C. multiglandulosa***
    6′ Central ligule lobe < or = lateral lobes, widest near middle
      7. St ± densely soft-hairy; peduncle bracts oblanceolate, flexible, tack-like glands small, ± sessile, ± marginal; chaff scale tack-like glands gen 0; ray fr rough-wrinkled .... ***C. mollis***
      7′ St short-stiff-hairy (and sparsely soft-hairy); peduncle bracts club-shaped or acuminate, rigid, tack-like glands stalked, on margins and surfaces; chaff scales gen with tack-like glands; ray fr smooth
        8. Ray fls (2)3–6; disk fls 6–20; branches gen few, ± rigid, main axis gen obvious; receptacle cup width = length; corollas sometimes yellow .... ***C. fremontii***
        8′ Ray fls gen 1(2); disk fls 2–5; branches many, flexible, often ± zigzag and without obvious axis; receptacle cup width ± 1/2 length; corollas not yellow .... ***C. pauciflora***

C. fremontii A. Gray **ST** 1–10 dm, ± finely strigose (some hairs longer); branches gen few, ± rigid. **LVS**: lower 3–8 cm. **INFL**: heads 1–several per node; peduncle bracts 4–10 mm, rigid, tack-like glands gen > 1 (1 in rare n SNF pls); phyllaries and chaff scales 4–7 mm, outer surfaces finely scabrous, gen glandular, sparsely long-hairy, tack-like glands 1–several (exc some n SNF pls). **RAY FLS** 2–6; ligules white to reddish or yellow, lobes ± equal. **DISK FLS** 6–20. **FR**: ray fr smooth, gen ± glabrous; disk pappus scales ± 10, alternately long and short. $2n=12$. Common. Open, dry meadows, hillsides, gravelly outwashes; 100–1400 m. NW, CaR, n SNF, ScV; OR. [*C. ciliosa* E. Greene; *C. elegans* E. Greene] Highly variable but not easily divisible. Intergrades with *C. pauciflora*.

C. hooveri G. Carr (p. 221) HOOVER'S CALYCADENIA **ST** 1–6 dm; branchlets many above, flexible, minutely scabrous, glandular. **LVS**: lower many, 1–8 cm. **INFL** ± spike-like or terminal on branchlets; heads 1–4 per node; peduncle bracts 1–5 mm, tack-like gland gen 1; phyllaries 2.5–3.5 mm, outer surface minutely glandular, tack-like gland 0–1; chaff scales 3–5 mm, outer surface glandular, tack-like gland 0. **RAY FLS** 0–2; ligules white, gen deeply lobed, middle lobe smallest. **DISK FLS** 1–2. **FR**: ray fr smooth or rough-wrinkled, ± glabrous; disk pappus scales 6–13, lanceolate, acuminate. $2n=14$. RARE. Rocky, exposed places; < 300 m. n&c SNF.

C. mollis A. Gray (p. 221) **ST** 3–9 dm, gen simple, often zigzag-curved, long-soft-hairy. **LVS** 2–8 cm, often longest at mid-st. **INFL**: heads few–many in dense cymes; peduncle bracts 3–9 mm, flat, oblanceolate, tack-like glands gen several, marginal; phyllaries and chaff scales 4–5 mm, outer surfaces soft-spreading-hairy, densely glandular (tack-like glands gen 0). **RAY FLS** 1–4; ligules yellow, white or rose, lobes ± equal. **DISK FLS** 2–10. **FR**: ray fr gen rough-wrinkled, ± glabrous; disk pappus scales ± 8, gen lanceolate, acuminate (or 1–3 shorter, ± rounded). $2n=14$. Common. Open, dry meadows, fields; 150–1500 m. c&s SNF, SnJV.

C. multiglandulosa DC. (p. 221) **ST** 1–7 dm, ± simple (often with small, many-bracted branchlets ± throughout), gen hairy (and often glandular) esp above. **LVS**: lower often persistent, 3–8 cm. **INFL** sometimes dense, esp above; peduncle bracts many, 4–20 mm, often reddish, tack-like glands (0) gen many; phyllaries and chaff scales 4–10 mm, outer surfaces often long-hairy, glandular, and reddish, tack-like glands (0)1–many. **RAY FLS** 2–6; ligules yellow or white to rose; middle lobe << lateral lobes, widest at base. **DISK FLS** 4–20. **FR**: ray fr smooth, glabrous to ± appressed-hairy; disk pappus scales ± 11, lanceolate, acuminate (2–3 shorter, blunt). $2n=12$. Common. Gen dry, open valleys, hillsides, rocky ridges; 75–1000 m. NCoR, SN, GV, SnFrB, SCoRI; w NV. [sspp. *bicolor* (E. Greene) Keck, *cephalotes* (DC.) Keck, *robusta* Keck; *C. hispida* (E. Greene) E. Greene incl ssp. *reducta* Keck] Variable.

C. oppositifolia (E. Greene) E. Greene (p. 221) BUTTE COUNTY CALYCADENIA **ST** < 3 dm, strigose and sparsely ± spreading-hairy; branches 0–few. **LVS** opposite, 1–5 cm, little reduced upward. **INFL** dense; heads appearing whorled; peduncle bracts 4–10 mm, tack-like glands (0)1–several; phyllaries and chaff scales 4–7 mm, outer surfaces minutely scabrous, often sparsely long-hairy, often with terminal tack-like gland. **RAY FLS** 2–4; ligules white to reddish, lobes ± equal. **DISK FLS** 4–20. **FR**: ray fr gen smooth, glabrous; disk pappus scales ± 8, gen lanceolate, acuminate (often 2–4 shorter, blunt). $2n=14$. UNCOMMON. Open, dry meadows, hillsides; < 900 m. CaRF, n SNF.

C. pauciflora A. Gray **ST** 1–5 dm, finely strigose (some hairs longer); branches gen many, slender, ± divergent, ± zigzag, fan-shaped, without obvious main axis. **LVS**: lower 1–5 cm. **INFL**: heads 1–4 per node; peduncle bracts 2–3 mm, tack-like glands 1–5; phyllaries and chaff scales 4–6 mm, outer surfaces finely scabrous, sparsely long-hairy, tack-like glands many. **RAY FLS** 1(2); ligules white to reddish, lobes ± equal. **DISK FLS** 2–5. **FR**: ray fr smooth, glabrous; disk pappus scales ± 10, alternately long and short. $2n=10,12$. Common. Open, dry, gen rocky meadows, hillsides; 100–1000 m. NCoR. Intergrades with *C. fremontii*.

C. spicata (E. Greene) E. Greene (p. 221) **ST** 2–6 dm, densely hairy and glandular above; branches 0 or above middle. **LVS** 2–5 cm, often longest at mid-st. **INFL**: heads 1–several per node; peduncle bracts many, 3–7 mm, ± concealing heads, cylindric distally, tip gen concave, tack-like gland 1; phyllaries and chaff scales 6–7 mm, outer surfaces often long-hairy, tack-like glands often many, small. **RAY FLS** 1–5; ligules white, fading reddish, lobes ± equal. **DISK FLS** 4–11. **FR**: ray fr smooth, ± densely long-hairy; disk pappus scales ± 9, lanceolate, acuminate. $2n=8$. Common. Dry, open hillsides; < 1400 m. SNF, GV.

C. truncata DC. (p. 221) ROSIN WEED **ST** 2–12 dm, gen reddish, gen glabrous to ± scabrous above; branches 0–many above middle. **LVS**: lower 2–10 cm. **INFL**: heads gen 1(few) per node; peduncle bracts 2–12 mm, tack-like gland 1; phyllaries and chaff scales 5–10 mm, outer surfaces glabrous or sparsely stiff-appressed- or long-straight-hairy, esp above, tack-like glands gen 0. **RAY FLS** 3–6; ligules yellow, lobes variable. **DISK FLS** 3–25; anthers sometimes purple. **FR**: ray fr rough-wrinkled, glabrous; disk pappus scales 0 or 7–12, short, often blunt, toothed. $2n=14$. Common. Dry, open hillsides, rocky ridges, talus; < 1500 m. NW, CaR, SNF, GV, CW; OR. [sspp. *microcephala* Keck & *scabrella* (E. Drew) Keck]

C. villosa DC. (p. 221) DWARF CALYCADENIA **ST** < 4 dm, ± densely long-hairy; branches 0 or ascending from base. **LVS**: lower gen very many, often persistent, 2–5 cm. **INFL**: heads 1–3 per node; peduncle bracts very many, 3–6 mm, ± concealing heads, deeply grooved, tack-like gland 1; phyllaries and chaff scales 5–6 mm, outer surfaces ± glabrous, tack-like glands 0. **RAY FLS** 1–4; ligules white to pinkish, central lobe much narrower than lateral lobes, widest at base. **DISK FLS** 5–15. **FR**: ray fr smooth, ± densely long-hairy; disk pappus scales ± 10, lanceolate, acuminate. $2n=14$. RARE. Dry, rocky hills, ridges; < 1100 m. c&s SCoRO.

CALYCOSERIS

G. Ledyard Stebbins

Ann; sap milky. **ST** 1–3 dm, branched, glabrous below, with tack-shaped glands above. **LVS**: basal and lower cauline petioled, 1–2-pinnately divided into long, linear lobes, glabrous; cauline alternate, gradually or abruptly reduced, upper sessile. **INFL**: heads ligulate, showy, solitary or in open, few-headed cymes; phyllaries in 2 series, scarious-margined, outer short, wide, inner many, linear; receptacle minutely bristly, otherwise naked. **FLS** many; ligules yellow or white, ± ephemeral. **FR** 5–6-ribbed, tapered to short beak; pappus white, of many slender bristles that fall together. 2 spp.: sw US, nw Mex. (Greek: cup-like chicory)

1. Ligules yellow; fr smooth or nearly so . . . . ***C. parryi***
1′ Ligules white; fr roughened on ribs . . . . ***C. wrightii***

C. parryi A. Gray (p. 221) **ST**: glands dark colored. **LF** 3–12 cm. **INFL**: heads 2–4 cm diam; involucre 10–15 mm. **FL**: ligules yellow. **FR** 7–9 mm, smooth or nearly so; pappus 6–8 mm. $2n=14$. Sandy to gravelly soils, washes, slopes; < 2000 m. Teh, PR, SNE, D; to UT, AZ, Baja CA.

C. wrightii A. Gray **ST**: glands pale. **LF** 4–10 cm. **INFL**: heads 3–4 cm diam; involucre 10–15 mm. **FL**: ligules white, often purple-tinged, esp below. **FR** 5–8 mm, roughened on ribs; pappus 5–6.5 mm. $2n=14$. Desert washes, gravelly slopes, desert plains; 150–1150 m. W&I, D; to UT, TX.

CARDUUS PLUMELESS THISTLE

David J. Keil & Charles E. Turner

Ann to per. **STS** erect. **LVS** alternate, reduced upward, decurrent as spiny wings, spiny-dentate and pinnately lobed, glabrous to tomentose; basal tapered to winged petiole; cauline sessile. **INFL**: heads discoid, 1–20 at branch tips; involucre cylindric to spheric; phyllaries overlapping in several series, spine-tipped; receptacle flat, whitish bristly. **FLS**: corollas white to pink or purple, tube long, slender, throat abrupt, short, lobes linear; anther bases short-sagittate, tips oblong; style with slightly swollen node, cylindrical above node, branches very short. **FR** ovoid, slightly compressed, glabrous; base slightly angled; pappus of many flat, minutely barbed, persistent bristles. ± 90 spp.: Eurasia, e Afr. (Latin: ancient name) [Howell 1959 Leaflets W Bot 9:17–29]

1. Phyllaries gen > 2 mm wide; heads gen solitary, conspicuously peduncled; involucres 2–7 cm diam ..... ***C. nutans***
1′ Phyllaries gen < 2 mm wide; heads 1–several, often clustered at branch tips, short-peduncled or sessile; involucres 1–3 cm diam
  2. Involucres spheric or hemispheric .... ***C. acanthoides***
  2′ Involucres cylindric or narrowly elliptic
    3. Lower lvs 4–10-lobed; heads gen 2–5 per cluster; phyllaries not scarious-margined, ± persistently, tomentose, tips scabrous .... ***C. pycnocephalus***
    3′ Lower lvs 12–20-lobed; heads 5–20 per cluster; phyllaries scarious-margined, glabrous to sparsely tomentose, tips not or barely scabrous, sometimes short-ciliate .... ***C. tenuiflorus***

C. acanthoides L. (p. 227) PLUMELESS THISTLE Bien. **STS** 3–15 dm, ± glabrous to loosely woolly, strongly spiny-winged. **LVS**: basal 10–20 cm, spiny-toothed or pinnately lobed, gen sparsely hairy. **INFL**: heads 1–5 in clusters; involucre 1–2.5 cm diam, spheric; phyllaries narrowly lanceolate, tips appressed to spreading. **FLS**: corollas 14–20 mm, purple; tube 7–10 mm. **FR** 2.5–3 mm, golden to brown; pappus 11–13 mm. $2n$=16,20,22. Roadsides, pastures, waste areas; < 1300 m. NCoRI, n SN, SnFrB, n SCoRO, MP; N.Am.; native to Eur. NOXIOUS WEED. Scattered.

C. nutans L. (p. 227) MUSK THISTLE Bien. **STS** 4–15 dm, glabrous to woolly, narrowly spiny-winged. **LVS**: basal 10–40 cm, 1–2-pinnately lobed; cauline glabrous or sparsely hairy. **INFL**: heads gen solitary, often nodding; involucre 2–7 cm diam, spheric; phyllaries spreading, lanceolate to ovate, spine-tipped. **FLS**: corollas 20–25 mm, purple; tube 12–14 mm, throat 4–5 mm, lobes 4–6 mm. **FR** 4–5 mm, golden to brown; pappus 13–25 mm. $2n$=16. Roadsides, pastures, waste areas; 100–1200 m. KR, CaR, n SNH, MP (also SCo, DMoj); N.Am.; native to Eur. [var. *leiophyllus* (Petrovic) Arènes] NOXIOUS WEED.

C. pycnocephalus L. (p. 227) ITALIAN THISTLE Ann or bien. **STS** 2–20 dm, glabrous or slightly woolly, narrowly spine-winged. **LVS**: basal 10–15 cm, 4–10-lobed; cauline ± tomentose. **INFL**: heads 2–5 per cluster, sessile or short-peduncled; involucre 1–2 cm diam, cylindric to elliptic; phyllary bases loosely tomentose, margin not scarious, tips ascending, linear-lanceolate, spiny, scabrous. **FLS**: corollas 10–14 mm, pink to purple; tube 5–8 mm, throat 2–3 mm, lobes 4–5 mm. **FR** 4–6 mm, golden to brown; veins 20; pappus 10–15 mm. $2n$=62–64. Roadsides, pastures, waste areas; < 1000 m. s NCo, s NCoR, SNF, CW; native to Medit.

C. tenuiflorus Curtis Ann or bien. **STS** 2–20 dm, glabrous or slightly woolly, widely spiny-winged. **LVS**: basal 10–15 cm, 12–20-lobed; cauline ± tomentose. **INFL**: heads 5–20 per cluster, sessile or short-peduncled; involucre 1–2 cm diam, cylindric to elliptic; phyllary margin scarious, tips ascending, linear-lanceolate, spined, glabrous or ciliate. **FLS**: corollas 10–14 mm, pink to purple; tube 4–6 mm, throat 2–3 mm, lobes 4–5 mm. **FR** 4–5 mm, brown; veins 10–13; pappus 10–15 mm. $2n$=±54,62–64. Roadsides, pastures, waste areas; < 1000 m. NCo, NCoR, SNF, CW, SW; native to c Eur. Perhaps not different from *C. pycnocephalus*.

CARTHAMUS DISTAFF THISTLE

David J. Keil & Charles E. Turner

Ann in CA. **ST** gen erect, leafy, branched above or throughout. **LVS** alternate, gen pinnately lobed, ± spiny; basal often 0 by fl; cauline gen clasping, gen spreading to recurved, lanceolate to ovate, rigid. **INFL**: heads discoid, solitary; involucre ± urn-shaped; outer phyllaries ± lf-like, inner with spiny appendages in CA; receptacle convex to conic, chaffy. **FLS**: corolla tube very slender, throat abruptly expanded, lobes linear; filaments gen densely hairy, anther bases short-tailed, tips oblong; style tip with minutely hairy node and terminal segment minutely papillate, barely notched. **FR** oblong to obpyramidal, ± 4-angled, glabrous, attached at side; outer frs gen ± roughened, pappus gen 0; inner frs smooth, pappus 0 or of many, narrow, gen unequal scales. 14 spp.: Medit. (Arabic: frs from fl color)

1. Corollas pale purple; cauline lvs ± deeply lobed .... ***C. leucocaulos***
1′ Corollas yellow to red; cauline lvs toothed to deeply lobed
  2. Cauline lvs dentate, weakly spiny; corollas yellow to red; fr white .... ***C. tinctorius***
  2′ Cauline lvs ± deeply lobed, very spiny; corollas yellow; fr brown
    3. Outer phyllaries gen 2 × inner; herbage ± sparsely hairy; st white .... ***C. baeticus***
    3′ Outer phyllaries gen < 1.5 × inner; herbage ± cobwebby; st straw-colored .... ***C. lanatus***

C. baeticus (Boiss. & Reuter) Nyman (p. 227) SMOOTH DISTAFF THISTLE Pl 4–10 dm, ± sparsely hairy. **ST** white. **LVS** very spiny. **INFL**: involucre 20–25 mm, gen becoming glabrous; outer phyllaries 35–55 mm, recurved, gen 2 × inner, tip appendages prominently spiny. **FLS**: corollas 25–35 mm, yellow. **FR** 4–6 mm, brown; pappus 8–10 mm. $2n$=64. Disturbed ground; < 500 m. c SNF, c GV, SnFrB, s SCoR, SW, w DMoj; native to Medit. NOXIOUS WEED.

C. lanatus L. (p. 227) WOOLLY DISTAFF THISTLE Pl 4–10 dm, ± densely glandular, loosely cobwebby to ± woolly. **ST** straw-colored. **LVS** very spiny. **INFL**: involucre 25–35 mm, gen ± tomentose; outer phyllaries 35–50 mm, ascending or ± spreading, gen < 1.5 × inner, tip appendages prominently spiny. **FLS**: corollas 25–35 mm, yellow. **FR** 4–6 mm, brown; pappus 10–13 mm. $2n$=44. Disturbed ground; < 1100 m. ± NW, c SN, ± CW; native to Medit.

C. leucocaulos Sibth. & Smith (p. 227) WHITESTEM DISTAFF THISTLE Pl 3–10 dm, (becoming) ± glabrous. **ST** gen shiny white. **LVS**: cauline lobes 2–3 pairs, long, spine-tipped. **INFL**: involucre 10–13 mm, ± glabrous; outer phyllaries reflexed, 2.5–3 × inner, very shiny, tip appendages prominently spiny. **FLS**: corollas 13–17 mm, pale purple. **FR** 3–5 mm, brown; pappus 5–7 mm. $2n$=20. Grassland; < 200 m. s NCoR (expected more widely); native to Medit. NOXIOUS WEED.

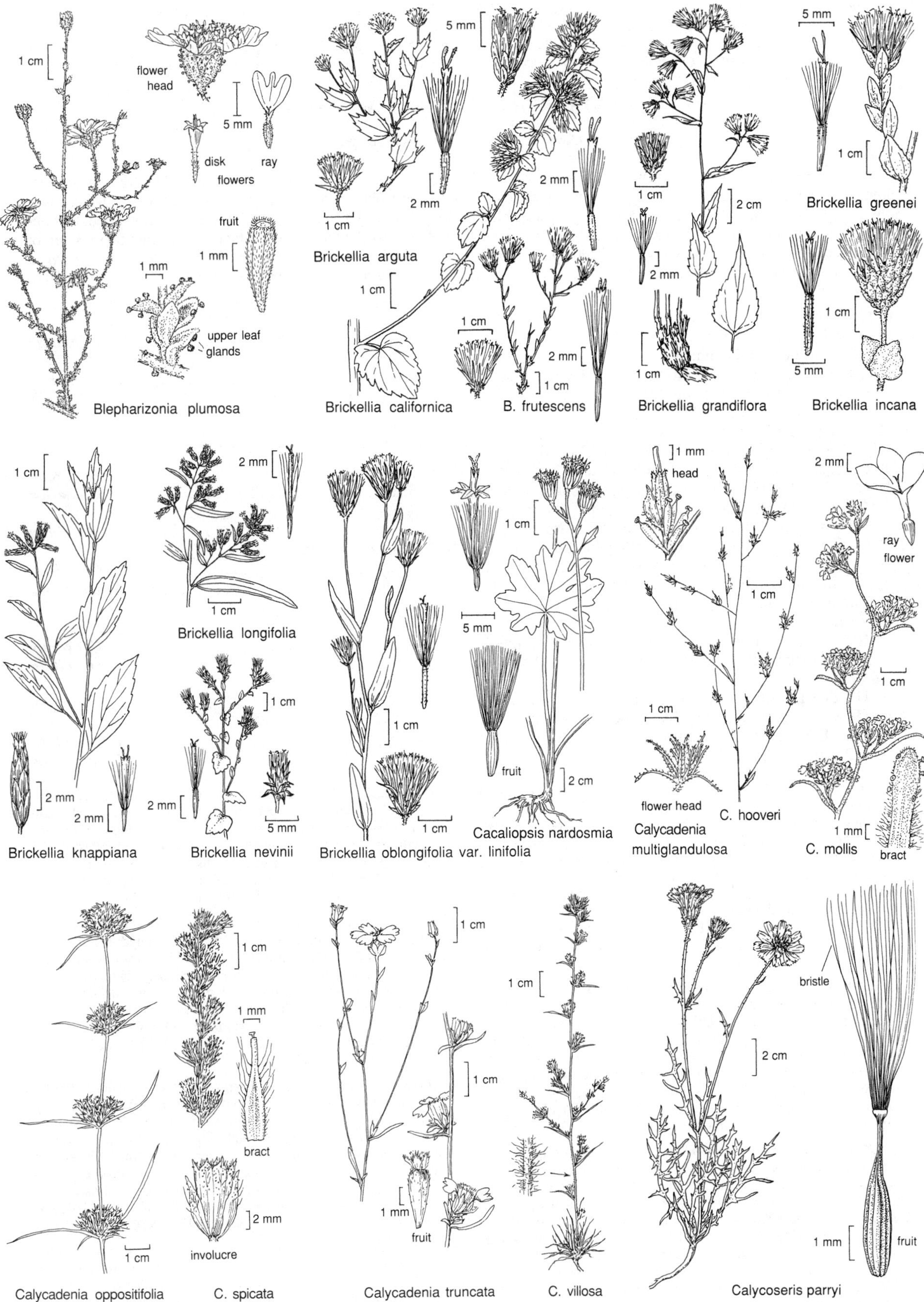
1 cm
flower head
5 mm
disk
ray
flowers
fruit
1 mm
1 mm
upper leaf glands
Blepharizonia plumosa
5 mm
2 mm
1 cm
Brickellia arguta
1 cm
2 mm
1 cm
2 mm
1 cm
Brickellia californica
B. frutescens
1 cm
2 cm
2 mm
1 cm
Brickellia grandiflora
5 mm
1 cm
Brickellia greenei
1 cm
5 mm
Brickellia incana
1 cm
2 mm
2 mm
Brickellia knappiana
2 mm
1 cm
Brickellia longifolia
1 cm
2 mm
5 mm
Brickellia nevinii
1 cm
5 mm
1 cm
1 cm
Brickellia oblongifolia var. linifolia
fruit
2 cm
Cacaliopsis nardosmia
1 mm
head
1 cm
1 cm
flower head
Calycadenia multiglandulosa
C. hooveri
2 mm
ray flower
1 cm
1 mm
C. mollis
bract
1 cm
1 mm
bract
2 mm
involucre
1 cm
Calycadenia oppositifolia
C. spicata
1 cm
1 cm
1 mm
fruit
Calycadenia truncata
1 cm
C. villosa
bristle
2 cm
1 mm
fruit
Calycoseris parryi

C. tinctorius L. SAFFLOWER Pl 3–10+ dm, ± glabrous. **ST** ± straw-colored. **LVS**: cauline dentate, teeth minutely spine-tipped, shiny green. **INFL** ± flat-topped; involucre 2–4 cm diam, ± glabrous; outer phyllaries spreading to reflexed, 1.5–2 × inner, tip appendages minutely spiny. **FLS**: corollas 20–30 mm, yellow to red. **FR** 7–9 mm, white; pappus 0 or 1–4 mm. 2*n*=24. Disturbed places, roadsides; < 1000 m. GV and margins (esp SnJV); native to Medit. Waif from cult for oil, dye, birdseed.

CENTAUREA KNAPWEED, STAR-THISTLE

David J. Keil & Charles E. Turner

Ann to per, ± branched. **LVS** alternate; lower gen deeply 1–2-lobed, segments gen narrow; upper reduced. **INFL**: heads discoid (sterile outer fls sometimes ± ray-like); involucre cylindric to hemispheric; phyllaries many, graded, gen ± ovate, scarious-margined, tip appendages fringed to spiny; receptacle flat, long-bristly. **FLS**: inner fruiting; anther bases tailed, tips oblong; style top minutely hairy, tips minutely branched. **FR** ± barrel-shaped, ± compressed, attached ± at side; pappus gen of stiff, unequal bristles or narrow scales. ± 500 spp.: esp Eurasia, n Afr (± 2 N.Am); some cult (waifs may incl *C. cineraria* L., *C. eriophora* L., *C. jacea* L., *C. moschata* L., *C. muricata* L., *C. salmantica* L.). (Greek: ancient name) Many NOXIOUS or invasive weeds.

1. Corollas yellow
  2. Corollas gen > 25 mm .... ***C. sulphurea***
  2′ Corollas gen 10–20 mm
    3. Central spines of main phyllaries 5–10 mm .... ***C. melitensis***
    3′ Central spines of main phyllaries 10–25 mm .... ***C. solstitialis***
1′ Corollas white to pink, blue, or purple
  4. Main phyllaries ± spine-tipped
    5. Central spines of main phyllaries 10–25 mm
      6. Young lvs ± gray-tomentose; involucre body 6–8 mm diam .... ***C. calcitrapa***
      6′ Young lvs minutely bristly, green; involucre body 8–14 mm diam .... ***C. iberica***
    5′ Central spines of main phyllaries gen < 10 mm
      7. Outer corollas ± ray-like, 20–30 mm, purple .... [2]***C. diluta***
      7′ Outer corollas not or barely enlarged, 7–13 mm, white to pink or pale purple
        8. Involucre 10–13 mm; corollas 12–13 mm .... ***C. diffusa***
        8′ Involucre 7–8 mm; corollas 7–9 mm .... ***C. squarrosa***
  4′ Main phyllaries not spine-tipped
    9. Ann
      10. Herbage ± tomentose; inner corollas 10–15 mm .... ***C. cyanus***
      10′ Herbage (sub)glabrous; inner corollas ± 20 mm .... [2]***C. diluta***
    9′ Per
      11. Fls all fruiting, corollas ± equal .... ***C. nigra***
      11′ Outer fls sterile, ± ray-like; inner fls fruiting
        12. Involucres 10–13 mm .... ***C. maculosa***
        12′ Involucres 15–18 mm .... ***C. ×pratensis***

C. calcitrapa L. (p. 227) PURPLE STAR-THISTLE Ann to per 2–10 dm, often ± mounded, puberulent to ± tomentose. **LVS** becoming ± glabrous, resin-dotted; lower 10–20 cm, ± deeply 1-3-lobed. **INFL**: heads few; involucre 15–20 mm, ovoid; main phyllaries greenish or straw-colored, appendages spiny-fringed at base, tip spine 10–25 mm, stout. **FLS** 25–40; corollas 15–24 mm, equal, purple. **FR** 2.5–3.5 mm, white or brown-streaked, glabrous; pappus 0. 2*n*=20. Disturbed places; gen < 1000 m. NW, s CaRF, SNF, GV, CW, SW; native to s Eur. NOXIOUS WEED. *C. ×pouzinii* DC. (*C. c.* × *C. aspera* L.), with shorter spines and pappus, is established in e SCo (Riverside Co.).

C. cyanus L. (p. 227) BACHELOR'S BUTTON, CORNFLOWER Ann 2–10 dm, loosely tomentose. **LVS**: lower 3–10 cm, entire or few-lobed; cauline little reduced. **INFL**: heads ± few; involucre 12–16 mm, bell-shaped; main phyllaries green, appendages fringed with slender, white to black teeth. **FLS** 25–35; corollas gen blue (white to purple), sterile corollas 20–25 mm, ray-like. **FR** 4–5 mm, straw-colored (bluish), finely hairy; pappus bristles 2–3 mm. 2*n*=24. Disturbed grassland, open woods; gen < 2000 m. CA-FP, MP; native to s Eur. Escaped from cult.

C. diffusa Lam. DIFFUSE KNAPWEED Ann to per 2–8 dm, much-branched, puberulent, ± gray-tomentose. **LVS**: lower 10–20 cm, ± deeply 2-lobed. **INFL** panicle-like; involucre 10–13 mm, cylindric to narrowly ovoid; main phyllaries pale green, prominently parallel-veined, appendages fringed with slender, straw-colored spines, tip spine 1–3 mm. **FLS** few; corollas 12–13 mm, equal, white to pink or pale purple, sterile corollas slender. **FR** ± 2.5 mm, dark brown; pappus 0 or scales < 1 mm, white. 2*n*=18. Fields, roadsides; < 2300 m. NW, CaR, n&c SN, SCoR, MP, W&I; native to se Eur. NOXIOUS WEED.

C. diluta Aiton Ann < 13 dm, ± thinly hairy. **LVS**: lower 10–15 cm, coarsely lobed. **INFL** open; involucre 15–18 mm, ovoid; main phyllaries greenish or straw-colored, appendages fringed with slender teeth, tip spine 0–5 mm, slender. **FLS** many; corollas 20–30 mm, pink-purple, sterile corollas ray-like. **FR** ± 3.5 mm; pappus bristles ± 4.5 mm, white. 2*n*=20. Disturbed places; < 100 m. w CW, SCo; native to sw Eur. Escaped from cult.

C. iberica Spreng. IBERIAN STAR-THISTLE Per 5–10 dm, ± loosely tomentose, becoming ± glabrous. **LVS** resin-dotted; lower 10–20 cm, coarsely lobed. **INFL** leafy; involucre 15–18 mm, ovoid; main phyllaries greenish or straw-colored, appendage short-spined at base, tip spine 10–25 mm, stout. **FLS** many; corollas 15–20 mm, ± equal, rose-pink or whitish, sterile corollas slender. **FR** ± 3 mm, white or marked brown; pappus bristles ± 1 mm, white. 2*n*=16,20. Disturbed places; < 1000 m. s NCoRI, n&c SNF, c GV, CW, s SCo, PR; native to se Eur. NOXIOUS WEED.

C. maculosa Lam. (p. 227) SPOTTED KNAPWEED Bien 3–10 dm, ± gray-tomentose. **LVS** resin-dotted; lower 10–15 cm, ± deeply 1-2-lobed. **INFL** open; heads gen many; involucre 10–13 mm, ovoid; phyllaries pale green or pink-tinged, prominently parallel-veined, appendages fringed with slender, dark teeth. **FLS** 30–40; corollas 12–25 mm, white to pink or purple, sterile corollas ± slender. **FR** 3–3.5 mm, ± pale brown, finely hairy; pappus bristles 1–2 mm, white. 2*n*=18,36. Disturbed areas; < 2000 m. NW, CaR, SN, n ScV, n CW, MP, n SNE, s PR; native to Eur. NOXIOUS WEED.

C. melitensis L. TOCALOTE Ann 1–10 dm, ± gray-hairy. **LVS** resin-dotted, ± scabrous; lower 2–15 cm, entire to lobed, gen 0 at fl; cauline long-decurrent. **INFL**: heads 1–few; involucre 10–15 mm, ovoid, ± cobwebby or becoming glabrous; main phyllaries ± straw-colored, appendage purplish, base spine-fringed, tip spine 5–10 mm, slender. **FLS** many; corollas 10–12 mm, ± equal, yellow, sterile corollas slender. **FR** ± 2.5 mm, ± light brown, finely hairy; pappus bristles 2.5–3 mm, white. 2*n*=18,24,36. Disturbed fields, open woods; < 2200 m. CA-FP, D (uncommon); native to s Eur. Invasive.

C. nigra L. Per 3–10 dm, ± soft-crinkly-hairy. **LVS**: lower 5–25 cm, entire to irregularly lobed. **INFL** ± flat-topped; heads few; involucre 15–18 mm, hemispheric; main phyllaries brown to black, appendages conspicuously fringed with long, tapered, minutely bristly teeth. **FLS** many, fruiting; corollas 15–18 mm, purple. **FR** 2.5–3 mm, tan, finely hairy; pappus bristles (0 or) gen < 1 mm, blackish. 2*n*=22,44. Disturbed places; < 500 m. NW, SnFrB; native to Eur.

C. ×pratensis Thuill. Per 3–10 dm, ± soft-crinkly-hairy, becoming ± glabrous. **LVS**: lower 5–25 cm, entire to irregularly lobed. **INFL** ± flat-topped; heads few; involucre 15–18 mm, hemispheric; main phyllaries ± brown, appendages fringed with tapered teeth. **FLS** many; corollas 15–22 mm, purple, sterile corollas ± ray-like. **FR** 2.5–3 mm, tan, finely hairy; pappus scales ± 0.5 mm. 2*n*=44. NW, s SnFrB; native to Eur. Variable; ± stable hybrid (*C. nigra* × *C. jacea* — the latter a waif in NW).

C. solstitialis L. (p. 227) YELLOW STAR-THISTLE Ann 1–10 dm, ± rounded, gray-tomentose. **LVS** ± scabrous-bristly; lower 5–15 cm, 1–2-lobed, gen 0 at fl; cauline long-decurrent. **INFL** open; heads 1–many; involucre 13–17 mm, ovoid; main phyllaries pale green to straw-colored or brown, appendages palmately spiny, tip spine 10–25 mm, stout. **FLS** many; corollas 13–20 mm, ± equal, yellow, sterile corollas slender. **FR** 2–3 mm, glabrous; outer fr dark brown (pappus 0); inner fr ± mottled light brown (pappus bristles 2–4 mm, fine, white). 2*n*=16. Pastures, roadsides, disturbed grassland or woodland; < 1300 m. CA-FP; native to s Eur. Invasive, esp in pastures. Cumulatively TOXIC to horses.

C. squarrosa Willd. (p. 227) Per 2–5 dm, much-branched, scabrous, ± loosely tomentose. **LVS**: lower 10–15 cm, deeply 1–2-lobed, gen 0 at fl. **INFL** panicle-like; heads ± many; involucre 7–8 mm, ± cylindric; main phyllaries pale green to straw-colored (or tinged purple), appendages fringed with slender, straw-colored spines, tip-spine 1–3 mm. **FLS** few; corollas 7–9 mm, equal (or sterile corollas ± larger), pink or pale purple. **FR** 2.5–3.5 mm, ± light brown; pappus bristles (0)2–2.5 mm, white. 2*n*=18. Disturbed places; < 1400 m. KR, CaR, MP; native to Asia. NOXIOUS WEED. [*C. virgata* Lam. var. *s.* (Willd.) Boiss.]

C. sulphurea Willd. SICILIAN STAR-THISTLE Ann 1–10 dm, ± soft-crinkly-hairy. **LVS**: basal 10–15 cm, lobed; cauline ± spiny-serrate. **INFL**: heads 1–few; involucre 2–3 cm, ovoid; main phyllaries greenish or straw-colored to purple-black, appendages sometimes reflexed, palmately spiny, tip spine 10–25 mm, stout. **FLS** many; corollas ± 30 mm, ± equal, yellow, sterile corollas slender. **FR** ± 5 mm; pappus bristles 6–7 mm, dark. 2*n*=24. Disturbed places; < 300 m. c SNF, c GV, s SnFrB; native to sw Eur. NOXIOUS WEED.

CHAENACTIS PINCUSHION

James D. Morefield

Ann to subshrubs, gen ± hairy. **LVS** alternate or basal, gen petioled, reduced upward, entire and linear or gen elliptic to ovate or obovate and 1–4-pinnately lobed; 1° lobes longest near middle or base of blade. **INFL**: heads discoid (but outer fls often enlarged, ± ray-like), 1–many per st, gen in terminal cymes; peduncle gen hairy like phyllary bases; involucre gen < 15 mm diam, cylindric to obconic or hemispheric; phyllaries in 1–2 ± equal series, gen linear to lanceolate, tips gen ± flat, gen ± green; receptacle flat to rounded, gen naked. **FLS** 10–many; corollas radial (outer, if enlarged, ± bilateral), gen white to pinkish or yellow, gen opening in daytime; anthers gen exserted. **FR** club-shaped, gen not compressed, stiffly hairy; pappus 0 or of 4–20 fringed scales in 1–few series. 18 spp.: w N.Am. (Greek: gaping ray, from enlarged outer corollas of some) [Mooring 1980 Amer J Bot 67:1304–1309] Spp. of sect. *Chaenactis* hybridize.

1. Per (bien?) to subshrubs, rarely fl 1st year; pappus scales (6)8–20 per fr in indistinct series, ± equal; outer corollas radial, not strongly enlarged (sect. *Macrocarphus*)
 2. Pls not scapose or cespitose or matted; heads (1)3–many per st; gen below subalpine
  3. Gen per; hairs near st base thin, cobwebby, gray, thinning with age; basal rosette ± persistent; longest lf lobes near middle of blade, tips curled .... ***C. douglasii*** var. ***douglasii***
  3′ Subshrubs; hairs near st base dense, felt-like, white, persistent; basal rosette 0 or withering before fl; longest lf lobes near base of blade, tips ± flat
   4. Longest phyllaries 10–13 mm, thinly tomentose, not or obscurely glandular; PR .... ***C. parishii***
   4′ Longest phyllaries 14–17 mm, strongly glandular-hairy, not or scarcely tomentose; KR, CaR .. ***C. suffrutescens***
 2′ Pls ± scapose, cespitose to ± matted; heads 1(–3) per st; gen subalpine to alpine
  5. Phyllaries ± tomentose, glandless .... ***C. alpigena***
  5′ Phyllaries glandular-hairy
   6. Lf blades ± ovate, longest lobe near base, tips flat .... ***C. nevadensis***
   6′ Lf blades linear to elliptic, longest lobe near middle, tips curled to twisted
    7. Largest lf blades ± widely elliptic, lobes 3–7 pairs, adjacent but not densely crowded .... ***C. douglasii*** var. ***alpina***
    7′ Largest lf blades linear to narrowly elliptic, lobes gen > 9 pairs, densely crowded .... ***C. santolinoides***
1′ Ann; pappus scales 0 or 4–8 in 1–2 distinct series, outer series (if present) gen << inner; outer corollas often bilateral and strongly enlarged
 8. Lower st and lvs whitish scaly- or granular-puberulent, appearing ± granular, not cobwebby; largest lf blades (2)3–4-pinnately lobed, longest lobe near base (sect. *Acarphaea*)
  9. Tips of longest phyllaries sharp but not acuminate or needle- like, ± flat, green; lf tips flat; fr compressed, pappus 0; TR, PR .... ***C. artemisiifolia***
  9′ Tips of longest phyllaries acuminate, needle-like, cylindric, gen reddish; lf tips ± cylindric; fr not compressed, at least the inner with pappus; SNE, D .... ***C. carphoclinia***
   10. Lvs basal and cauline, < 7 cm; bases of petioles not strongly enlarged .... var. ***carphoclinia***
   10′ Lvs gen basal, longest 7–10 cm; bases of petioles strongly enlarged .... var. ***peirsonii***
 8′ Lower st and lvs ± cobwebby or glabrous; largest lf blades entire or 1–2-pinnately lobed, longest lobe gen near middle (sect. *Chaenactis*)
  11. Corollas white to pinkish, rarely pale yellow
   12. Corollas 9–11 mm, ± 2 × fr, anthers ± incl .... ***C. macrantha***

12′ Corollas < 9 mm, ± = fr, anthers exserted
13. Longest phyllaries 10–18 mm, bases glabrous, tips densely puberulent, glands 0; pappus scales in 2 series; corollas radial; peduncles not glandular .... ***C. xantiana***
13′ Longest phyllaries < 10(–12) mm, ± uniformly glandular- hairy or cobwebby or glabrous; pappus scales in (0)1 series (rarely 2 in *C. stevioides*); outer corollas bilateral, expanded; peduncles ± glandular-hairy near heads
14. St base glabrous at fl time; lvs entire or 1-pinnately lobed, lobes 1–2(5) pairs, well separated; pappus of inner fls > buds, tips visible; outer phyllaries sharp, becoming glabrous .... ***C. fremontii***
14′ St base at least thinly cobwebby at fl time; lvs 1–2-pinnately lobed, 1° lobes gen 4–8 pairs, often crowded; pappus < buds, tips hidden; outer phyllaries gen blunt, persistently glandular-hairy or cobwebby .... ***C. stevioides***
11′ Corollas ± bright to deep yellow .... ***C. glabriuscula***
15. Pappus scales in 2 series, scales of outer series (1)3–4, << inner .... var. ***heterocarpha***
15′ Pappus scales in 1 series
16. Pls ± scapose, ± densely white-woolly .... var. ***lanosa***
16′ Pls not scapose, gen thinly gray-cobwebby
17. Basal lvs strongly fleshy, 2-pinnately lobed; sts ± spreading; SCo .... var. ***orcuttiana***
17′ Basal lvs not strongly fleshy, gen 1-pinnately lobed; sts ± erect; not SCo
18. Longest phyllaries 5–7 mm, < 2 mm wide, gen ± gray- tomentose at least on midrib, often glandular-hairy; longest pappus < 0.75 × corolla .... var. ***glabriuscula***
18′ Longest phyllaries 7–9 mm, 2–3 mm wide, green, ± glabrous; longest pappus ± = corolla .... var. ***megacephala***

C. alpigena Sharsm. (p. 227) Per, scapose, cespitose to ± matted. **STS** gen several, erect to prostrate, < 10 cm, tomentose near base, hairs gen thinning with age. **LVS** < 4 cm, densely tomentose, rarely becoming glabrous, not fleshy; basal rosette persistent; largest blades linear to obovate, curled and 1–2-pinnately lobed (s pls) to flat and ± palmately lobed or entire (n pls), 1° lobes 2–7 pairs, ± crowded, longest near middle. **INFL**: heads 1 per st; peduncles < 10 cm; involucre obconic to cylindric, ± tomentose, glandless; longest phyllaries 9–14 mm, tips erect, ± rigid, blunt. **FLS**: corollas radial, 5.5–8 mm, ± equal, white to pinkish. **FR** 5–8 mm; pappus scales 8–20 in indistinct series, ± equal, longest 5–8.5 mm. Sand or gravel; 2300–4000 m. c&s SNH, W&I (n White Mtns.); w NV.

C. artemisiifolia (A. Gray) A. Gray (p. 227) Ann, branched above middle, robust. **ST** gen 1, erect, < 10(20) dm, whitish scaly-puberulent below and on lvs, hairs thinning with age. **LVS** < 15(–20) cm, not fleshy; basal rosette withering; largest blades (2)3–4-pinnately lobed, 1° lobes 5–10 pairs, ± crowded, longest near base, tips flat. **INFL**: heads gen several per st; peduncles < 7 cm; involucre hemispheric, densely wavy-hairy; longest phyllaries 7–10 mm, tips erect, ± rigid, sharp. **FLS**: corollas radial, 5–7 mm, ± equal, white to pinkish. **FR** 4–7 mm, compressed; pappus 0. $2n$=16. Open slopes, disturbed areas; < 1600 m. TR, PR; n Baja CA. Much like *C. lacera* E. Greene of Baja CA.

C. carphoclinia A. Gray PEBBLE PINCUSHION Ann, branched above middle. **ST** gen 1, ± erect, < 7 dm, whitish granular-puberulent below and gen on lvs; hairs thinning with age. **LVS** < 11 cm, gen not fleshy; basal rosette withering; largest blades (2)3–4-pinnately lobed, 1° lobes 2–7(10) pairs, ± crowded, longest near base, tips ± cylindric. **INFL**: heads gen several per st; peduncles < 7 cm; involucre cylindric to hemispheric, glandular- or wavy-hairy; longest phyllaries 7–10 mm, tips (at least of inner) erect, rigid, acuminate, needle-like, cylindric, gen reddish; receptacle gen ± chaffy, chaff resembling phyllaries. **FLS**: corollas radial, 4–6 mm, white to pinkish, outer spreading but barely enlarged. **FR** 3–4 mm; pappus scales in 1 series, scales of outer fr (0–)4, unequal, < inner, scales of inner fr 4, equal, longest 3–5 mm. Gen open rocks or gravel; < 1900 m. SNE, D; to UT, AZ, nw Mex. Often mistaken for *C. stevioides* but closer to *C. artemisiifolia*.

var. ***carphoclinia*** (p. 227) **ST** < 4 dm. **LVS** basal and cauline, all < 7 cm; petiole narrow, soft, base not strongly enlarged. $2n$=16. Habitats and range of sp. [var. *attenuata* (A. Gray) M.E. Jones]

var. ***peirsonii*** (Jepson) Munz (p. 227) PEIRSON PINCUSHION **ST** gen 4–6 dm. **LVS** gen basal, longest 7–10 cm; petiole stout, hard, base strongly enlarged. RARE. Habitats of sp; < 500 m. e PR (e Santa Rosa Mtns), adjacent w DSon?.

C. douglasii (Hook.) Hook. & Arn. Per, bien(?), sometimes fl lst year. **STS** 1–several, erect to spreading, < 50 cm, gen thinly grayish cobwebby; hairs thinning with age. **LVS** < 15 cm, gen cobwebby to ± tomentose, not fleshy; basal rosette ± persistent; largest blades gen 2-pinnately lobed, 1° lobes 3–7 pairs, ± crowded, longest near middle, tips curled. **INFL**: heads 1–many per st; peduncles < 10 cm; involucre obconic to hemispheric, gen glandular-hairy; longest phyllaries 9–14 mm, tips gen erect, ± rigid, blunt. **FLS**: corollas radial, 5–8 mm, white to pinkish, outer somewhat enlarged. **FR** 5–8 mm; pappus scales 8–20 in indistinct series, ± equal, longest gen 3–6 mm. Dry, open areas, alpine crevices; 1000–3500 m. NW, CaR, SN, GB, n DMtns; to B.C., Colorado, AZ.

var. ***alpina*** A. Gray ALPINE DUSTY MAIDENS Per, scapose, cespitose to ± matted. **STS** < 10 cm. **LVS** < 7 cm, sometimes becoming glabrous. **INFL**: heads 1(–2) per st; longest phyllaries 9–12 mm. $2n$=12. RARE in CA. Open, subalpine to alpine gravel and crevices; 3000–3400 m. n SNH (Alpine, El Dorado cos., esp Freel Peak area), n DMtns (Panamint Mtns); to ID, MT, Colorado. [*C. a.* (A. Gray) M.E. Jones] Intergrades downslope with var. *douglasii*; more uniform outside CA (CA pls are ± atypical).

var. ***douglasii*** (p. 227) DUSTY MAIDENS Per, bien(?), sometimes fl lst year, gen branched below middle. **STS** < 50 cm. **LVS** < 15 cm. **INFL**: heads (1)3–many per st in a ± leafy cyme; longest phyllaries 11–14 mm. $2n$=12,24,36 (extra chromosomes also reported). Dry, open areas; 1000–3500 m. NW, CaR, SN, GB, n DMtns; to B.C., MT, Colorado, n AZ. [vars. *achilleifolia* (Hook. & Arn.) A. Gray, *montana* M.E. Jones, *rubricaulis* (Rydb.) Ferris; *C. ramosa* Stockw.] More distinctive variants occur outside CA. ❀DRN:1,15,16&IRR:2,3,7,18;DFCLT.

C. fremontii A. Gray (p. 227) DESERT PINCUSHION Ann, branched below middle, sometimes ± scapose. **STS** 1–many, ± erect, < 40 cm, becoming glabrous before fl time exc near heads. **LVS** < 9 cm, gen glabrous, ± fleshy; basal rosette withering; largest blades entire or 1-pinnately lobed, lobes 1–2(5) pairs, well separated, longest near middle, tips cylindric. **INFL**: heads 1–few per st; peduncles < 11 cm, gen glandular-hairy near heads; involucre obconic to hemispheric, ± truncate at base, becoming glabrous; longest phyllaries 8–10(12) mm, tips erect, rigid, sharp. **FLS**: corollas white to pinkish, outer bilateral, greatly enlarged, inner radial, 5–7 mm. **FR** (3)6–8 mm; pappus scales in 1 series, smaller and unequal on outer fr, on inner fr 4(5), equal, longest gen 6–8.5 mm, > buds, tips visible. $2n$=10. Open sand or gravel; < 1600 m. Teh, s SnJV, SCoRI, more common in s SNE, D; to sw UT, w AZ, n Baja CA. Scattered. ❀TRY.

C. glabriuscula DC. YELLOW PINCUSHION Ann. **STS** gen 1–few, erect to spreading, < 50 cm; hairs thinning with age. **LVS** < 11 cm, ± cobwebby, fleshy or not; largest entire or 1–2-pinnately lobed, 1° lobes 1–7 pairs, longest near blade middle, tips flat to curled or cylindric. **INFL**: heads 1–several per st; peduncle < 20 cm; involucre widely cylindric to obconic or hemispheric, gen ± tomentose or glandular-hairy; longest phyllaries 4.5–9 mm, tips erect, ± rigid, gen blunt. **FLS**: corollas ± bright to deep yellow, outer bilateral, greatly enlarged, inner radial, 4–8 mm. **FR** 3–9 mm; pappus scales (1)4–8 in 1–2 series, scales of outer fr gen < inner, unequal, scales of inner fr gen equal, longest 1–8 mm. 2*n*=12. Gen dry open places, sometimes dunes or serpentine; < 1600 m. CA-FP; n Baja CA. Highly variable; some forms like *C. stevioides* exc fl color.

var. ***glabriuscula*** Pls branched above or below middle, thinly grayish cobwebby. **STS** < 50 cm, ± erect. **LVS** < 9 cm, gen not fleshy; basal rosette withering; largest blades 1(2)-pinnately lobed, 1° lobes 2–7 pairs, well separated to ± crowded, tips flat to ± cylindric. **INFL**: heads gen several per st; peduncles < 5 cm; involucre obconic to hemispheric; longest phyllaries 5–7 mm, < 2 mm wide, gen ± grayish tomentose at least on midrib. **FLS**: inner corollas 4–6 mm. **FR** 3–5.5 mm; pappus scales 4 in 1 series, longest 2–4 mm, sometimes 0 on outer fr. 2*n*=12. Open slopes, sandy places; < 1600 m. s SNF, Teh, CW, SW, w edge D; n Baja CA. [vars. *curta* (A. Gray) Jepson & *tenuifolia* (Nutt.) H.M. Hall] ❀TRY.

var. ***heterocarpha*** (A. Gray) H.M. Hall Pls branched above or below middle, thinly grayish cobwebby. **STS** < 50 cm, spreading to erect. **LVS** < 5 cm, gen not fleshy; basal rosette withering or ± persistent; largest blades gen 2-pinnately lobed, 1° lobes 2–7 pairs, ± crowded, tips gen curled. **INFL**: heads few per st; peduncles < 10 cm; involucre hemispheric to widely cylindric; longest phyllaries 6–9 mm, 1.5–2.5 mm wide, glabrous and green to ± grayish tomentose. **FLS**: inner corollas 4.5–7 mm. **FR** 6–9 mm; pappus scales (5)7–8 in 2 series, scales of outer series (1)3–4, << inner, scales of inner series 4, longest gen 2–7 mm. 2*n*=12. Open (often serpentine) slopes, ridges; < 1600 m. NCoR, SNF, GV, CW, most common n. [var. *gracilenta* (E. Greene) Keck; *C. tanacetifolia* A. Gray] Possibly best considered a separate sp.; like *C. nevii* A. Gray (fr compressed, pappus 0 or vestigial; OR). ❀DRY,SUN,DRN:**7**,8,9,**14–16**,**18**,19–24.

var. ***lanosa*** (DC.) H.M. Hall Pls ± scapose, ± densely white-woolly. **STS** < 10 cm, decumbent to erect. **LVS** < 11 cm, gen not fleshy; basal rosette ± persistent; largest blades entire or 1-pinnately lobed, lobes 1–2(5) pairs, well separated, tips flat to cylindric. **INFL**: heads 1 per st; peduncles < 20 cm; involucre obconic to hemispheric; longest phyllaries 6–8 mm, < 2 mm wide, whitish tomentose. **FLS**: inner corollas 5–6.5 mm. **FR** 4–6 mm; pappus scales 4 in 1 series, longest 4–6 mm. 2*n*=12. Loose sand; < 700 m. SCoRI, SCo, TR, nw edge DSon. [var. *denudata* (Nutt.) Munz] ❀DRN,DRY,SUN:**7**,**8**,**9**,**11**,**14–17**,**19–24**.

var. ***megacephala*** A. Gray (p. 227) Pls branched gen below middle, gen becoming glabrous. **STS** < 50 cm, ± erect. **LVS** < 9 cm, gen not fleshy; basal rosette withering; largest blades gen 1-pinnately lobed, lobes 2–7 pairs, ± well separated, tips flat. **INFL**: heads 1–few per st; peduncles < 20 cm; involucre widely cylindric to hemispheric; longest phyllaries 7–9 mm, 2–3 mm wide, green, ± glabrous. **FLS**: inner corollas 5–8 mm. **FR** 5–8.5 mm; pappus scales 4–5 in 1 series, longest 5–8 mm. Dry, often sandy slopes; < 1600 m. s SNF, Teh, SnJV, CW, n SW. ± intermediate between vars. *glabriuscula* & *heterocarpha*. ❀TRY.

var. ***orcuttiana*** (E. Greene) H.M. Hall Pls branched above and below middle, thinly grayish cobwebby. **STS** < 30 cm, ± spreading. **LVS** < 7 cm; lowest strongly fleshy; basal rosette withering; largest blades gen 2-pinnately lobed, 1° lobes 2–7 pairs, ± crowded, tips flat to ± curled. **INFL**: heads few per st; peduncles < 4.5 cm; involucre hemispheric; longest phyllaries 4.5–6.5 mm, < 2 mm wide, greenish, gen glandular-hairy. **FLS**: inner corollas 4.5–5.5 mm. **FR** 4–5 mm; pappus scales 4 in 1 series, longest 1–2.5 mm. 2*n*=12. Coastal dunes, bluffs; < 100 m. SCo; nw Baja CA. ❀TRY.

C. macrantha D. Eaton (p. 227) MOJAVE PINCUSHION Ann, branched below middle. **STS** 1–few, erect, < 35 cm, thinly tomentose; hairs thinning with age. **LVS** < 6 cm, not fleshy; basal rosette withering; largest blades 1-pinnately lobed, lobes gen 2–5 pairs, ± well separated, longest near middle to ± near base, tips flat. **INFL**: heads 1–few per st; peduncles < 9 cm; involucre widely cylindric, ± tomentose, glands 0; longest phyllaries 12–18 mm, tips gen recurved, soft, blunt. **FLS**: corollas all radial, 9–11 mm, ± equal, white to pinkish, gen opening at night; anthers ± incl. **FR** 5–6 mm; pappus scales 8 in 2 series, longest 5–7 mm, outer << inner. 2*n*=12. Open (often calcareous) sand or gravel; 1000–2000 m. GB, n DMoj; to se OR, sw ID, w UT, n AZ.

C. nevadensis (Kellogg) A. Gray (p. 227) Per, ± scapose, cespitose to ± matted. **STS** several, spreading to erect, < 10 cm, tomentose near base; hairs gen thinning with age. **LVS** < 5 cm, loosely tomentose, not fleshy; basal rosette persistent; largest blades 1–2-pinnately lobed, 1° lobes 2–4 pairs, ± crowded, longest near base, tips flat. **INFL**: heads 1(–2) per st; peduncles < 10 cm; involucre obconic to cylindric, glandular-hairy; longest phyllaries 9–12 mm, tips erect, ± rigid, blunt. **FLS**: corollas all radial, 6–7 mm, ± equal, white to pinkish. **FR** 5.5–7.5 mm; pappus scales 10–16 in indistinct series, ± equal, longest 3–4.5 mm. 2*n*=12. Sand or gravel; 1900–3200 m. CaRH, n SNH. ❀TRY.

C. parishii A. Gray (p. 227) PARISH'S CHAENACTIS Subshrub, branched below middle. **STS** several, erect, 20–60 cm; hairs near base dense, felt-like, white, persistent. **LVS** < 6 cm, finely tomentose, not fleshy; basal rosette 0 or withering; largest blades 1-pinnately lobed, lobes gen 2–5 pairs, ± well separated, longest near base, tips ± flat. **INFL**: heads 1–few per st; peduncles < 20 cm; involucre obconic, thinly tomentose, not or obscurely glandular; longest phyllaries 10–13 mm, tips erect, ± rigid, blunt. **FLS**: corollas all radial, 7–8.5 mm, ± equal, white to pinkish. **FR** 6–8.5 mm; pappus scales 10–16 in indistinct series, ± equal, longest 6–8 mm. 2*n*=12. UNCOMMON. Open but vegetated slopes; 1300–2500 m. PR; n Baja CA.

C. santolinoides E. Greene (p. 231) Per, scapose, cespitose to ± matted. **STS** several, erect to ± spreading, < 10 cm, tomentose near base; hairs gen thinning with age. **LVS** < 12 cm, loosely tomentose, not fleshy; basal rosette persistent; largest blades linear to narrowly elliptic, 1–2-pinnately lobed, 1° lobes (7)10–many pairs, crowded to ± densely packed, longest near middle, tips curled to twisted. **INFL**: heads 1(–3) per st; peduncles < 25 cm; involucre obconic to cylindric, glandular-hairy; longest phyllaries 8–13 mm, tips erect, ± rigid, gen blunt. **FLS**: corollas all radial, 5–7 mm, ± equal, white to pinkish. **FR** 4–6 mm; pappus scales 10–16 in indistinct series, ± equal, longest 3–4.5 mm. 2*n*=12. Open sand or gravel; 1500–2800 m. s-most SN, TR. Closely related to *C. douglasii* var. *alpina*. ❀TRY.

C. stevioides Hook. & Arn. (p. 231) DESERT PINCUSHION Ann, branched above or below, sometimes ± scapose. **STS** 1–many, erect, < 45 cm, cobwebby (at least thinly so near base); hairs thinning with age. **LVS** < 11 cm, ± cobwebby, gen not fleshy; basal rosette gen withering; largest blades 1–2-pinnately lobed, lobes 4–8 pairs, crowded to well separated, longest near middle, tips gen curled. **INFL**: heads gen several per st; peduncles < 11 cm; involucre obconic to hemispheric, glandular-hairy or ± cobwebby; longest phyllaries 5.5–8(10) mm, tips erect, ± rigid, gen blunt. **FLS**: corollas white to pink, rarely pale yellow, outer bilateral, greatly enlarged, inner radial, 4.5–6.5 mm. **FR** 4–6.5 mm; pappus scales 4–5 in 1(–2) series, scales of outer fr < inner, unequal, scales of inner fr equal to unequal, longest 1.5–6 mm. 2*n*=10. Common. Open flats, slopes; < 2100 m. SCoRI, SNE, D; to se OR, sw ID, w Colorado, sw NM, nw Mex. [var. *brachypappa* (A. Gray) H.M. Hall; *C. mexicana* Stockw.] Highly variable, but not clearly divisible. SCoRI pls resemble *C. gillespiei* Stockw. of c AZ (st ± glabrous, lf lobes elongate). See also *C. carphoclinia*. ❀TRY.

C. suffrutescens A. Gray (p. 231) SHASTA CHAENACTIS Subshrub, branched below middle. **STS** several, erect, 10–60 cm; hairs near base dense, felt-like, white, persistent. **LVS** < 9 cm, tomentose, not fleshy; basal rosette 0 or withering; largest blades 1–2-pinnately lobed, lobes 2–5 pairs, ± well separated, longest near base, tips flat. **INFL**: heads 1–few per st; peduncles 5–15 cm; involucre cylindric, strongly glandular-hairy, not or scarcely tomentose; longest phyllaries 14–17 mm, tips gen erect, ± rigid, gen blunt. **FLS**: corollas white to pinkish, all radial, 8.5–10 mm, ± equal. **FR** 7–9 mm; pappus scales 10–16 in indistinct series, ± equal, longest 7–9 mm. 2*n*=12. RARE. Dry, open areas; 800–2100 m. KR, CaR. In cult.

C. xantiana A. Gray (p. 231) Ann, branched above middle. **STS** 1–several, erect, < 50 cm; base gen becoming glabrous before fl time. **LVS** < 7 cm, gen glabrous, ± fleshy; basal rosette withering; largest blades entire or 1-pinnately lobed, lobes 1–2(5) pairs, well separated, longest near middle, tips cylindric. **INFL**: heads few–many per st; peduncles < 6 cm; involucre widely obconic, base (at least of outer phyllaries) glabrous, tips densely puberulent, glandless; longest phyllaries 10–18 mm, gen recurved, tips soft, blunt. **FLS**: corollas white to pinkish, all radial, 7–9 mm, outer somewhat enlarged. **FR** 5–9 mm; pappus scales 8 in 2 series, outer << inner, longest 5–9 mm. $2n$=14. Sand; 300–2500 m. s SN, SCoR, TR, GB, w DMoj; se OR, w NV, nw AZ. ❀DRN,SUN,DRY:1–3,7,8–12,**14–16,18–24**.

CHAETADELPHA

G. Ledyard Stebbins

1 sp.: w US. (Greek: fused bristles) [Tomb 1972 Madroño 21:459–462]

C. wheeleri A. Gray (p. 231) Per from branched caudex; sap milky; herbage glabrous. **STS** 1–4 dm, much-branched. **LVS** all cauline, alternate, 1–5 cm, linear to narrowly lanceolate (or scale-like), acute, entire. **INFL**: heads ligulate, solitary on branchlets 2–5(7) cm; involucre 11–15 mm, narrowly cylindric; phyllaries in 2 series, outer 7–10 short, wide, inner 5 long, linear-oblong; receptacle naked. **FLS** 5 per head; ligules 6–12 mm, pale lavender-white, readily withering. **FR** 8–12 mm, columnar, 4–5-ridged, glabrous; pappus double, of 5 stiff awns and gen many bristles, all fused at base, yellowish or tan. $2n$=18. Sand dunes, desert scrub; 900–1900 m. W&I, n DMoj (Eureka Valley); se OR, w NV. ❀TRY.

CHAETOPAPPA

Geraldine A. Allen

Ann to subshrub. **LVS** alternate, linear to oblanceolate, entire. **INFL**: heads radiate, solitary; phyllaries in 2–6 series, outer < inner, margins translucent, papery; receptacle ± flat, chaff 0. **RAY FLS** 5–24; ligules coiled at maturity, white to blue or pink-purple. **DISK FLS** 5–many; corollas yellow; anther tips triangular; style tips triangular. **FR** gen linear-fusiform, round or ± compressed; pappus 0 or of scales, bristles, or both. ± 10 spp.: sw US, n Mex. (Greek: bristle-pappus) [Nesom 1988 Phytologia 64:448–456]

C. ericoides (Torrey) G. Nesom (p. 231) Per; caudex ± woody. **STS** ascending to erect, 5–15 cm, gen branched, strigose, ± glandular. **LVS** basal and cauline (basal 0 at fl), 4–12 mm, linear to obovate, obtuse to abruptly pointed, ± glandular, bristly-hairy below. **INFL**: phyllaries lanceolate to oblong, acute to acuminate, green, ± purple-tipped. **RAY FLS** gen 12–21; corolla 5–10 mm, white to ± pink. **FR** ± rounded, brown, hairy; pappus of minutely barbed bristles, ± white. $2n$=16,32. Dry slopes; 1300–2900 m. SN (e slope), W&I, DMtns; to NE, TX, n Mex. [*Leucelene e.* (Torrey) E. Greene] ❀TRY.

CHAMAEMELUM

Elizabeth McClintock

Ann, per, glabrous or hairy. **STS** gen branched, erect or prostrate, < 60 cm. **LVS** alternate, 1–3-pinnately lobed. **INFL**: heads gen radiate, solitary; phyllaries in 2 or more series, scarious-margined; receptacle rounded to conic, chaffy. **RAY FLS** < 25, gen pistillate; ligules gen white. **DISK FLS** many; corolla yellow, tube swollen at base, enclosing tip of fr; anther tips ovate, bases rounded or ± cordate; style branches truncate with shrub-like tips. **FR** slightly compressed, faintly striate; pappus 0. 4 spp.: Eur. (Derivation unknown) [Ross-Craig 1960–1963 Drawings Brit Pls 16: plate 6]

1. Ann; receptacle chaff tips, phyllary tips and margins brown .... ***C. fuscatum***
1′ Per; chaff of receptacle and phyllaries transparent throughout .... ***C. nobile***

C. fuscatum (Brot.) Carv. Vasc. Ann, aromatic. **STS** < 30 cm, simple or sparingly branched. **LVS** ± 1–4 cm, 1–2-pinnately lobed; lobes linear; lower petioled; upper sessile. **INFL**: heads ± 3 cm diam; phyllaries 3–4 mm, tips and margins dark brown, becoming reflexed; receptacle hemispheric, chaff scales ± 3 mm, transparent. **RAY FLS** ± 12; ligules 10–15 mm, white. **DISK FLS**: corollas 7–10 mm. **FR** ± 1 mm, all same size, elliptic. $2n$=18. Uncommon. Escape from cult in disturbed areas, esp vineyards; < 100 m. s NCoRO (Sonoma Co.); native to Medit. [*Anthemis f.* Brot.]

C. nobile (L.) All. CHAMOMILE, RUSSIAN CHAMOMILE Per, pleasantly aromatic. **STS** < 15 cm, creeping, forming mats, ± hairy. **LVS** < 5 cm, 2–3-pinnately lobed, sessile; lobes linear. **INFLS**: heads ± 1 cm diam; phyllaries 3–4 mm, scarious; receptacle hemispheric, chaff scales ± 5 mm, transparent. **RAY FLS** ± 20; ligules ± 10 mm, white, rarely 0. **DISK FLS**: corollas ± 10 mm. **FR** ± 1 mm, ± cylindric; ray achenes slightly > disk achenes. $2n$=18. Uncommon. Escape from cult in disturbed, coastal areas; < 100 m. NCo, expected more widely; native to w Eur. [*Anthemis n.* L.] TOXIC: fl heads used medicinally, but tea may produce anaphylactic shock.

CHAMOMILLA

Elizabeth McClintock

Ann, sometimes aromatic. **STS** branched, erect or decumbent. **LVS** alternate, irregularly 2–3-pinnately lobed; segments linear; petiole short or 0. **INFL**: heads discoid, solitary or 2–3; receptacle conic, naked; phyllaries in 2–3 unequal series, margins scarious. **FLS** many, yellow, tubular, 4-lobed, narrowed above; anthers very small, tips ovate, bases rounded or ± cordate; style short, branches truncate with shrub-like tips. **FR** cylindric, sometimes gelatinous when wet, ribbed; pappus a narrow crown or 0. ± 5 spp.: Eur, N.Am. (Derivation of name not known) [Moe 1977 Dissertation, Univ CA, Berkeley]

1. Tip of achene and pappus crown 2-lobed; pls < 70 cm; peduncles < 6.5 cm; undisturbed wet places ***C. occidentalis***
1′ Tip of achene entire, pappus crown narrow; pls < 50 cm; peduncles < 1.5 cm; disturbed places ...... ***C. suaveolens***

C. occidentalis (E. Greene) Rydb. Pls 15–45(70) cm; herbage not strongly scented. **STS** often branched only above. **LVS** sessile, < 7 cm, glabrous. **INFL**: heads gen < 1.5 cm diam, ± conic to spheric, remaining intact at maturity. **FLS**: corollas 1–2 mm. **FR** angled,

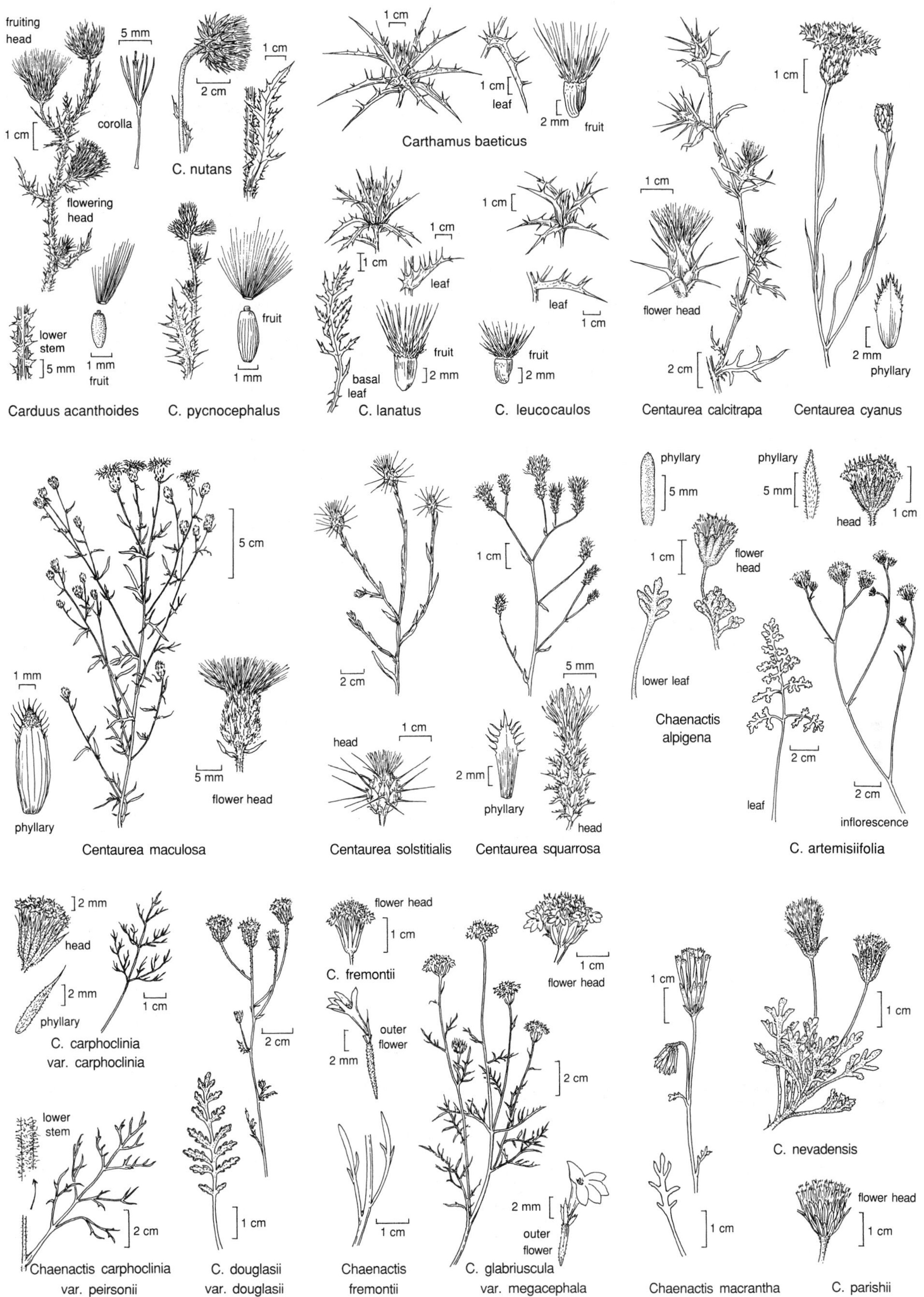
fruiting head
5 mm
corolla
1 cm
flowering head
lower stem
5 mm
1 mm
fruit
Carduus acanthoides
2 cm
1 cm
C. nutans
fruit
1 mm
C. pycnocephalus
1 cm
Carthamus baeticus
1 cm
leaf
2 mm
fruit
1 cm
1 cm
leaf
basal leaf
fruit
2 mm
C. lanatus
1 cm
leaf
1 cm
fruit
2 mm
C. leucocaulos
1 cm
flower head
2 cm
Centaurea calcitrapa
1 cm
2 mm
phyllary
Centaurea cyanus
5 cm
1 mm
phyllary
5 mm
flower head
Centaurea maculosa
2 cm
head
1 cm
Centaurea solstitialis
1 cm
5 mm
2 mm
phyllary
head
Centaurea squarrosa
phyllary
5 mm
1 cm
flower head
lower leaf
Chaenactis alpigena
phyllary
5 mm
head
1 cm
2 cm
leaf
2 cm
inflorescence
C. artemisiifolia
2 mm
head
2 mm
phyllary
1 cm
C. carphoclinia var. carphoclinia
lower stem
2 cm
Chaenactis carphoclinia var. peirsonii
2 cm
1 cm
C. douglasii var. douglasii
flower head
1 cm
C. fremontii
outer flower
2 mm
1 cm
Chaenactis fremontii
1 cm
flower head
2 cm
2 mm
outer flower
C. glabriuscula var. megacephala
1 cm
1 cm
Chaenactis macrantha
1 cm
C. nevadensis
flower head
1 cm
C. parishii

gelatinous when wet; tip and pappus crown 2-lobed, with wide brown gland extending from tip of each lobe to ± middle of achene. Common. Undisturbed alkali flats, vernal pools, edges of salt marshes; < 2400 m. NCoRO, CaRH, SNH, SnJV, SnFrB, SCoRO, SCo. [*Matricaria o.* E. Greene] ❀SUN:5,**7–9,14–24**;used as substitue for chamomile.

C. suaveolens (Pursh) Rydb. (p. 231) PINEAPPLE WEED, RAYLESS CHAMOMILE Pls (1)10–30(50) cm; herbage sweet-scented. **STS** gen branched from base. **LVS** < 5 cm, glabrous, sessile. **INFL**: heads gen ± 1 cm diam, conic, shattering at maturity. **FLS**: corollas 1–2 mm. **FR** 3–5-veined, gelatinous when wet; tip truncate; pappus a narrow crown, with linear, brown glands extending nearly length of achene. $2n$=18. Abundant. Disturbed sites, sand bars, river banks, footpaths, roadsides, grazed land; gen < 800(2400) m. KR, NCoR, CaRH, SN, GV, w CW, SCo; native to nw N.Am, ne Asia. [*Matricaria matricarioides* (Less.) Porter]

CHLORACANTHA

Guy L. Nesom

1 sp.: sw US, Mex, C.Am. (Latin: green thorns) [Nesom et al. 1991 Phytologia 70:371–381]

C. spinosa (Benth.) G. Nesom var. ***spinosa*** (p. 231) Subshrub from stout rhizome, ± glabrous. **STS** erect, 5–15(25) dm; branches sometimes thorn-like. **LVS** alternate, 10–50 cm, oblanceolate, gen entire, 1-veined, early deciduous. **INFL**: heads radiate, few–many in loose clusters; phyllaries in (3)4–5 series, inner 4.5–7.5 mm, ± lanceolate, veins (1)3(5), parallel, orange-resinous, tips gen rounded, margins transparent; receptacle naked. **RAY FLS** 10–33; corollas 4–8(11) mm, white, ligules 1–2 mm wide, coiled at maturity. **DISK FLS** many; corollas 3–6 mm, yellow; style appendages ± deltate. **FR** 1.5–3.5 mm, slightly compressed, glabrous; veins 5(6), whitish to golden-brown; pappus of many barbed bristles in ± 2 series, outer gen << inner. $2n$=18. Seeps, moist streamsides, ditches, sometimes saline or drier areas; < 1250 m. PR, DSon; to UT, s-c US, Mex. [*Aster s.* Benth.]

CHONDRILLA

G. Ledyard Stebbins

Bien, per; sap milky. **STS** much-branched. **LVS** basal and cauline, entire to pinnately lobed, gen much-reduced above. **INFL**: heads ligulate, many, peduncled or lateral subsessile; involucre cylindric; phyllaries in 2 series, outer much reduced; receptacle naked. **FLS** many; ligules yellow, readily withering. **FR** oblong, many-ribbed, beaked, or beakless; pappus of many, slender, white bristles. 25 spp.: Eurasia. (Greek: endive or chicory)

C. juncea L. SKELETON WEED **STS** 4–10 dm, glabrous or bristly at base. **LVS**: basal and lower cauline wing-petioled, oblong to obovate, shallowly lobed, lobes pointed, often reflexed; upper cauline linear, entire. **INFL**: heads subsessile, mostly in interrupted spike-like clusters; involucre 9–12 mm; phyllaries linear-lanceolate, glabrous or sparsely tomentose. **FLS** 7–12; corollas 12–18 mm. **FR** 8–12 mm, glabrous; body with pointed tubercles near tip; beak slender; pappus 5–6 mm. $2n$=14+1B,30. Uncommon. Disturbed ground; 0–600 m. NCo, n SN, GV, SnFrB, SCoRO, n SCo; native to s Eur. NOXIOUS WEED.

CHRYSANTHEMUM

Elizabeth McClintock

Ann. **STS** erect. **LVS** alternate, toothed or pinnately lobed. **INFL**: heads radiate, peduncled; phyllaries in 2–3 series, overlapping, scarious-margined; receptacle flat to convex or conic, naked. **RAY FLS**: ligules gen yellow or white. **DISK FLS** many; corollas yellow; anther tips ovate, bases rounded or ± cordate; style branches truncate with shrub-like tips. **FR**: ray achenes 3-angled, wings present or 0; disk achenes ± cylindric, pappus 0 or a narrow crown. ± 5 spp.: Medit. (Greek: gold flower) [Haw 1986 The Garden 111(11):525–528; Humphries 1976 Bull Brit Mus Bot. 5:147–240] See also *Argyranthemum, Leucanthemum, Tanaretum.*

1. Fl heads multicolored, ray fls red-purple, white, or yellow, disk fls purple; fr 2-winged .... ***C. carinatum***
1' Fl heads yellow; fr not winged
 2. Lvs 2-pinnately lobed, segments toothed .... ***C. coronarium***
 2' Lvs simple, toothed or pinnately lobed .... ***C. segetum***

C. carinatum Schousboe **STS** branched, < 30 cm. **LVS** 2-pinnately lobed, ± fleshy, glabrous. **INFL**: heads 5 cm diam; phyllaries gen in 3 series, 8–10 mm, ± keeled. **RAY FLS** ± 20; ligules white or yellow with basal red or purple markings forming ring around disk fls. **DISK FLS**: corollas purple. **FR** 3–4 mm, flat, 2-winged, ± regularly ribbed or ribs 0; pappus a narrow crown. $2n$=18. Uncommon. Escape from cult in urban waste ground; < 100 m. CCo, SCo; native to Morocco.

C. coronarium L. GARLAND or CROWN DAISY **STS** stiffly branched from base, > 10 dm. **LVS** < 6 cm, obovate, sessile, 2-pinnately lobed almost to midrib. **INFL**: heads solitary, < 4 cm diam; phyllaries gen in 3 series, obovate, not keeled, margins transparent. **RAY FLS** < 15; ligules yellow. **DISK FLS**: corollas 14–15 mm, yellow. **FR**: ray achenes < 3 mm, 3-angled, winged; disk achenes ± 2.5 mm, angled and ribbed, pappus 0. $2n$=18. Common escape from cult in urban coastal areas, vacant lots, roadsides, waste ground; < 200 m. CCo, SCo, expected more widely; native to s Eur.

C. segetum L. (p. 231) CORN CHRYSANTHEMUM **STS** < 80 cm, simple or branched from near base. **LVS**: blades of middle and lower lvs 3.5–6 cm, pinnately lobed, lobes gen not reaching midrib; upper lvs < 3.5 cm, oblong to obovate, irregularly toothed or subentire, sessile. **INFL**: heads < 3 cm diam; phyllaries gen in 3 series, margins widely scarious. **RAY FLS** ± 20; ligules < 15 mm. **DISK FLS** many; corollas 3–4 mm. **FR**: ray achenes ± 2 mm, ± compressed, lateral wings 2; disk achenes ± cylindric, 10-ribbed, pappus 0. $2n$=18. Common. Escape from cult in urban coastal areas, fields; < 200 m. NCo, NCoRO, n CW, expected more widely; native to Eur, sw Asia.

CHRYSOTHAMNUS RABBITBRUSH

Loran C. Anderson

Per to shrub. **STS** erect, often densely clustered. **LVS** alternate, sessile, entire. **INFL**: heads discoid, arrayed in ± dense cymes, peduncled or subsessile; involucre gen cylindric; phyllaries gen in 3–5 series (± 5 vertical ranks), free, overlapping, keeled; receptacle naked. **FLS** 2–20 (often 5) per head; corollas gen yellow, lobes 0.5–3 mm, gen spreading; style branches long, slender, gen exserted. **FR** narrowly cylindric, ± 5-ridged, gen light brown; pappus of many white to brownish bristles. 16 spp.: sw Can to n Mex. (Greek: golden shrub) Closely related to (perhaps part of) *Ericameria*.

1. Sts glabrous, puberulent, or gland-dotted
2. Herbage gland-dotted; corollas yellow
3. Involucres 6–8 mm, obconic; phyllaries weakly vertically ranked, tips not swollen .... ***C. paniculatus***
3′ Involucres 7–9.5 mm, cylindric; phyllaries strongly vertically ranked, tips swollen, green .... ***C. teretifolius***
2′ Sts (and gen lvs) not gland-dotted; corollas gen yellow (white)
4. Corollas white; lvs gland-dotted .... ***C. albidus***
4′ Corollas yellow; lvs not gland-dotted
5. Fr glabrous; lvs flat, rigid, persistent
6. Involucres gen 10–13 mm; phyllaries strongly keeled .... ***C. depressus***
6′ Involucres gen 12–15 mm; phyllaries not keeled .... ***C. gramineus***
5′ Fr (gen densely) hairy; lvs gen twisted
7. Heads overtopped by lvs; fls 2–3(4) per head .... ***C. humilis***
7′ Heads not overtopped by lvs; fls 3+ per head
8. Phyllaries acuminate to bristle-tipped; lvs < 1 mm wide .... ***C. greenei***
8′ Phyllaries obtuse to acute; lvs 1–10 mm wide .... ***C. viscidiflorus***
9. Upper sts (and gen lvs) hairy
10. Sts bristly near infl; lvs > 2 mm wide, ± rough-hairy .... ssp. ***lanceolatus***
10′ Herbage finely puberulent; lvs 1–2(4) mm wide, glabrous .... ssp. ***puberulus***
9′ Sts glabrous; lvs glabrous (exc gen ciliate)
11. Lvs ± 1 mm wide; fls 3–4(5) per head; involucre ± obconic .... ssp. ***axillaris***
11′ Lvs 1–10 mm wide; fls gen 4+ per head; involucre cylindric .... ssp. ***viscidiflorus***
1′ Sts tomentose to felted
12. Phyllaries long-acuminate, weakly keeled; infl gen raceme-like; heads 5–18-fld .... ***C. parryi***
13. Lvs gen > 3 cm, > 5 mm wide .... ssp. ***latior***
13′ Lvs gen < 3 cm, 4 mm wide or less
14. Heads 10–18-fld .... ssp. ***imulus***
14′ Heads 5–10-fld
15. Lvs oblanceolate, glands short-stalked .... ssp. ***asper***
15′ Lvs linear to oblanceolate, stalked glands 0
16. Heads 1–3 per infl; corolla 8–9 mm .... ssp. ***monocephalus***
16′ Heads many per infl; corolla > 9 mm
17. Phyllaries tips recurved; corolla 10–11 mm .... ssp. ***nevadensis***
17′ Phyllaries straight; corolla 11–12.5 mm .... ssp. ***vulcanicus***
12′ Phyllaries obtuse to acute, strongly keeled; infl gen panicle-like; heads 5-fld .... ***C. nauseosus***
18. Involucre subglabrous to tomentose; sts gen whitish; lvs gen dark green to grayish white
19. Fr glabrous exc white-hair-tufted at top .... ssp. ***washoensis***
19′ Fr densely hairy
20. Involucre 10–13 mm .... [2]ssp. ***bernardinus***
20′ Involucre 7–10(11) mm
21. Corolla lobes 1–2 mm; style appendage > stigma .... ssp. ***albicaulis***
21′ Corolla lobes 0.5–1 mm; style appendage < stigma .... ssp. ***hololeucus***
18′ Involucre glabrous (outer phyllaries sometimes ciliate or scaly-margined); sts greenish or brownish; lvs greenish yellow or 0
22. Fr glabrous .... ssp. ***leiospermus***
22′ Fr hairy
23. Lvs 1–3 mm wide, 1–5-veined; involucre > 10 mm .... [2]ssp. ***bernardinus***
23′ Lvs 1 mm wide or less, 1-nerved; involucre < 10 mm
24. Phyllaries abruptly pointed, recurved .... ssp. ***ceruminosus***
24′ Phyllaries acute, erect
25. Involucres 6–10 mm; sts gen leafy .... ssp. ***consimilis***
25′ Involucres 8.5–12 mm; sts often lfless .... ssp. ***mohavensis***

C. albidus (A. Gray) E. Greene (p. 231) Shrub 3–12 dm. **STS** brittle, glabrous. **LVS** 2–3.5 cm, linear, ± cylindric, very resinous. **INFL**: heads in cymes; involucres 7.4–9.5 mm, ± cylindric; phyllaries ± lanceolate, in poorly defined ranks, barely keeled, straw-colored, resinous, tips abruptly pointed, green. **FLS** 5–6; corollas 6–8 mm, white, lobes gen recurved; style appendage > stigma. **FR** 4–5 mm, hairy; pappus slightly < corolla. $2n$=18. Uncommon. Saline or alkaline soils; 300–1300 m. SNE, n DMoj; to w UT.

C. depressus Nutt. (p. 231) Subshrub 1–2 dm; caudex much-branched. **STS** unbranched below infl, brittle, rough, ± gray-hairy. **LVS** 2–3 cm, oblanceolate, rough-hairy. **INFL**: heads in small, dense cymes; involucre 10–15 mm, ± cylindric; phyllaries in well defined vertical ranks, strongly keeled, acute to acuminate, straw-colored or purplish. **FLS** 5–6; corollas 8–11 mm. **FR** 5–6.5 mm, gen glabrous (sparsely glandular); pappus ± = corolla. $2n$=18. Rocky crevices; 1000–2100 m. DMtns; to Colorado, NM.

C. gramineus H.M. Hall (p. 231) Per 2–5 dm. **STS** ± glabrous. **LVS** < 9 cm, ± lanceolate, firm, glabrous exc finely rough-hairy margins. **INFL**: heads few; involucre 11–17.5 mm, cylindric; phyllaries weakly vertically ranked, oblong to ovate, firm, not keeled, tips irregularly toothed or notched, abruptly pointed. **FLS** 4–7; corollas 9.5–11.5 mm; style appendage > stigma. **FR** 7–9 mm, glabrous; pappus ± = corolla. $2n=18$. Uncommon. Pinyon/juniper woodland, bristlecone-pine forest; 2200–2900 m. s W&I, n DMtns (Panamint Mtns); s NV. [*Petradoria discoidea* L. Anderson]

C. greenei (A. Gray) E. Greene Shrub 1–2 dm. **STS** gen glabrous. **LVS** 1–2 cm, thread-like, straight or twisted. **INFL**: heads in dense cymes; involucre 5–8 mm, cylindric; phyllaries widely lanceolate, clearly vertically ranked, weakly keeled, tips (at least outer) gen ascending, long-tapered, sticky. **FLS** 4–5; corollas 4–5 mm; style appendage < stigma. **FR** 3–3.5 mm, hairy; pappus < corolla. $2n=18$. Uncommon. Sandy washes; 1340–1830 m. SNE (Fish Lake Valley), n DMtns (Cottonwood Mtns); to Colorado, NM.

C. humilis E. Greene (p. 231) Subshrub 1–2 dm, ± bristly. **STS** leafy. **LVS** 2–3 cm, ± oblanceolate; axillary lf clusters gen 0. **INFL**: heads in small, dense cymes gen overtopped by lvs; involucre 8–10 mm, cylindric to obconic; phyllaries weakly keeled and ranked, outermost often green, bristly, lf-like, inner ovate-oblong, straw-colored, not very sticky, tips irregularly toothed. **FLS** 2–3(4); corollas (5)7–8 mm, pale yellow, lobes straight; style ± incl, appendage gen > stigma. **FR** 4.5–5.5 mm, subglabrous to sparsely strigose; pappus ± = corolla. $2n=18$. Uncommon. Sagebrush grassland; 1400–3040 m. ne SNH, MP; to WA, ID, NV. [*C. viscidiflorus* ssp. *h.* (E. Greene) H.M. Hall & Clements]

C. nauseosus (Pallas) Britton RUBBER RABBITBRUSH Shrub < 28 dm, ± tomentose; odor strong. **STS** whitish to green, ± flexible, very leafy or naked at fl. **LVS** 1–7 cm, thread-like to narrowly (ob)lanceolate. **INFL** dense, flat-topped or rounded, panicle-like; involucre 6–14.5 mm, cylindric; phyllaries ± lanceolate to ovate, in vertical ranks, ± strongly keeled, firm, obtuse to acute. **FLS** gen 5; corollas 6–12 mm; style appendage gen > stigma. **FR** 3–8 mm, gen hairy; pappus gen = corolla. $2n=18$. Diverse habitats; 50–3300 m. NW, CaR, SN, SCoR, TR, PR, GB, DMoj; to B.C., MT, Colorado, Baja CA. Highly variable; 22 sspp. in w US.

ssp. ***albicaulis*** (Nutt.) H.M. Hall & Clements (p. 231) Pl 2–10 dm, gray-white to dark green. **LVS** 3–6 cm, ± narrowly lanceolate. **INFL**: involucres 9–14 mm, ± tomentose. **FLS**: corollas gen 9–11 mm. Common. Diverse dry habitats; 50–3300 m. NW, CaR, SN; to B.C., MT, Colorado. Pls with large, wide lvs have been called var. *macrophyllus* Howell or *C. californicus* E. Greene. Intergrades with sspp. *consimilis* & *hololeucus*. Stabilized hybrids with *Ericameria discoidea* [*C. parryi* ssp. *bolanderi* (A. Gray) H.M. Hall & Clements] occur in c SNH (sw Mono Co.). ❀DRN,SUN, DRY:1–3,7,8,9,**10**,11–16,**18–21**,22–24.

ssp. ***bernardinus*** (H.M. Hall) H.M. Hall & Clements Pl 5–15 dm, white-tomentose to green. **LVS** 2.5–5 cm, ± narrowly lanceolate. **INFL**: involucres 10–14 mm, glabrous to hairy. **FLS**: corollas 10–12 mm. Open yellow-pine forest; 1200–2900 m. s SN, Teh, TR, n PR. Much like robust ssp. *albicaulis*. Sometimes also confused with sspp. *consimilis* & *mohavensis*. ❀TRY.

ssp. ***ceruminosus*** (Durand & Hilg.) H.M. Hall & Clements (p. 231) Pl 6–15 dm, gray- or yellow-green. **STS** ± lfless. **LVS** (if any) 1–3 cm, thread-like, curved. **INFL**: involucres 7–8.2 mm, glabrous, sticky; phyllary tips abruptly narrowed, recurved. **FLS**: corollas 6–7 mm. Gravelly arroyos; 700–1700 m. DMoj. Locally common.

ssp. ***consimilis*** (E. Greene) H.M. Hall & Clements Pl 5–25 dm. **STS** leafy, yellowish green. **LVS** 2–6 mm, ± thread-like. **INFL** ± narrow; involucres 6–10 mm, glabrous; phyllary midribs brownish. **FLS**: corollas 6–7 mm. Common. Gen alkaline soils; 1000–2900 m. SN (e slope) Teh, s SCoRI, SnBr, PR, GB; to OR, MT, WY, NM, Baja CA. Sometimes intergrades with sspp. *albicaulis* (gen higher elevation) and *mohavensis*. Robust pls from s SNE have been called ssp. *viridulus* (H.M. Hall) H.M. Hall & Clements. ❀TRY.

ssp. ***hololeucus*** (A. Gray) H.M. Hall & Clements Pl 3–25 dm, white to gray- or yellow-green. **LVS** 3–7 cm (shorter in infl), ± narrowly lanceolate, sometimes reflexed. **INFL**: involucres gen 8–9 mm, straw-colored, less hairy inward; phyllary midribs brownish. **FLS**: corollas 8–9.5 mm, lobes erect or bent inward. Common. Well-drained granitic or limestone soils in scrub or woodland; 150–2500 m. SNH, Teh, SCoRO, WTR, SNE, DMtns; to s OR, UT, n AZ. Yellow-green pls have been called ssp. *gnaphalodes* (E. Greene) H.M. Hall & Clements. Ssp. *viscosus* Keck is hybrid, ssp. *h.* ×*Ericameria cuneata*. ❀TRY.

ssp. ***leiospermus*** (A. Gray) H.M. Hall & Clements (p. 231) Pl gen 3–6 dm. **STS** gen lfless, yellowish. **LVS** (if any) 1–3 cm, thread-like. **INFL**: involucres 8–9.5 mm, gen straw-colored (purplish), glabrous. **FLS**: corollas 6–7.5 mm, lobes erect or bent inward. **FR** glabrous. Common. Dry sand, gravel, rocky crevices; 700–2400 m. DMtns, SNE; to UT, n AZ. ❀TRY.

ssp. ***mohavensis*** (E. Greene) H.M. Hall & Clements Pl 5–28 dm. **STS** few-branched, often ± lfless. **LVS** (if any) 1.5–3 cm, thread-like. **INFL**: involucres 8.5–12 mm, straw-colored, glabrous. **FLS**: corollas 7–10.5 mm. Common. Dry scrub; 400–2400 m. SCoRO, TR, DMoj; s NV. Sometimes intergrades with sspp. *consimilis* & *hololeucus*. ❀DRN,SUN,DRY:1–3,7–10,**11**,12–16,18–21.

ssp. ***washoensis*** L. Anderson (p. 231) Pl 3–10 dm. **STS** gen whitish. **LVS** ± 3 cm, ± narrowly oblanceolate. **INFL**: involucre 10–12 mm, sparsely long-hairy and short-tomentose. **FLS**: corollas 7.5–9 mm, lobes nearly erect, sparsely long-hairy. **FR** glabrous exc white hair-tuft near top. Uncommon. Dry juniper or pinyon grassland; 1500–1700 m. MP; nw NV.

C. paniculatus (A. Gray) H.M. Hall (p. 237) BLACK-STEM Shrub 5–10(18) dm. **STS** erect, clustered, gland-dotted, gen ± black-banded (from fungal or insect attack). **LVS** 1.5–3.5 cm, thread-like, light green, resinous. **INFL**: heads in elongate cymes; involucre 6–8 mm, obconic; phyllaries oblong to ovate, weakly keeled, weakly 5-ranked, obtuse, resinous. **FLS**: corollas 5.5–7 mm; style appendage ± = stigma. **FR** 3.5–4 mm, hairy; pappus ± = corolla. $2n= 18$. Common. Gravelly washes; 400–1600 m. D; to sw UT, nw AZ. ❀TRY.

C. parryi (A. Gray) E. Greene Subshrub or shrub, 1–9 dm. **STS** prostrate to erect, white to green, tomentose. **LVS** 1–7 cm, gen < 4 mm wide, thread-like to ± oblanceolate, ± glabrous to tomentose. **INFL**: heads gen ± many, in long or rounded cymes; involucre 10–18 mm, widely cylindric; phyllaries ± lanceolate, weakly 5-ranked, keeled, ± membranous, tips green. **FLS** 5–18; corollas 8–12.5 mm; style appendage > stigma. **FR** 4–7 mm, hairy; pappus gen < corolla. $2n=18$. Dry places, gen open forest; 700–3700 m. KR, NCoR, CaR, SNH, SnBr, GB, DMtns; to s OR, Colorado, AZ. 12 sspp. in w US.

ssp. ***asper*** (E. Greene) H.M. Hall & Clements **LVS** 1.5–4 cm, straight or curved, gray or green, short-glandular. **INFL** spike-like or branched, ± dense; phyllaries straight or spreading, whitish, straw-colored, or purplish. **FLS** 5–10. Dry forest to alpine barrens, often in pumice or gravel; 1900–3300 m. c&s SNH, SnBr, SNE, DMtns; NV.

ssp. ***imulus*** H.M. Hall & Clements **LVS** 1–2 cm, spreading to recurved, green; glands sessile. **INFL**: cyme, dense; phyllaries strongly recurved, yellowish green, ± resinous. **FLS** 10–18. Uncommon. Dry flats, open yellow-pine forest; ± 2200 m. SnBr (Bear Valley). Perhaps only an extreme form of ssp. *asper*.

ssp. ***latior*** H.M. Hall & Clements (p. 237) **STS** green. **LVS** 4.5–7 cm, 6–14 mm wide, ± oblanceolate, abruptly pointed, green, flat; glands 0. **INFL**: cyme, dense; phyllaries long-tapered, straw-colored. **FLS** 5. Open pine/fir forest; 700–1900 m. KR, NCoR, CaR. ❀TRY.

ssp. ***monocephalus*** (Nelson & Kenn.) H.M. Hall & Clements **LVS** 1–2.5 cm, green; glands 0. **INFL**: heads gen 1(–3), gen overtopped by lvs; phyllaries erect or spreading, straw-colored. **FLS** 5–8. Common. Open subalpine forest, talus, alpine barrens; 2800–3700 m. c SNH, n SNE (Sweetwater Mtns); w NV. Probably derived from ssp. *nevadensis*.

ssp. ***nevadensis*** (A. Gray) H.M. Hall & Clements (p. 237) **STS** stout. **LVS** 2–4 cm, green; glands 0. **INFL** raceme-like; phyllaries long-tapered, spreading or reflexed. **FLS** 5–6. Scrub, open yellow-pine forest, rarely on serpentine; 1100–2700 m. n&c SNH, GB; to s OR, sw UT, n AZ. Intergrades with ssp. *monocephalus* s of Lake Tahoe. ❀TRY.

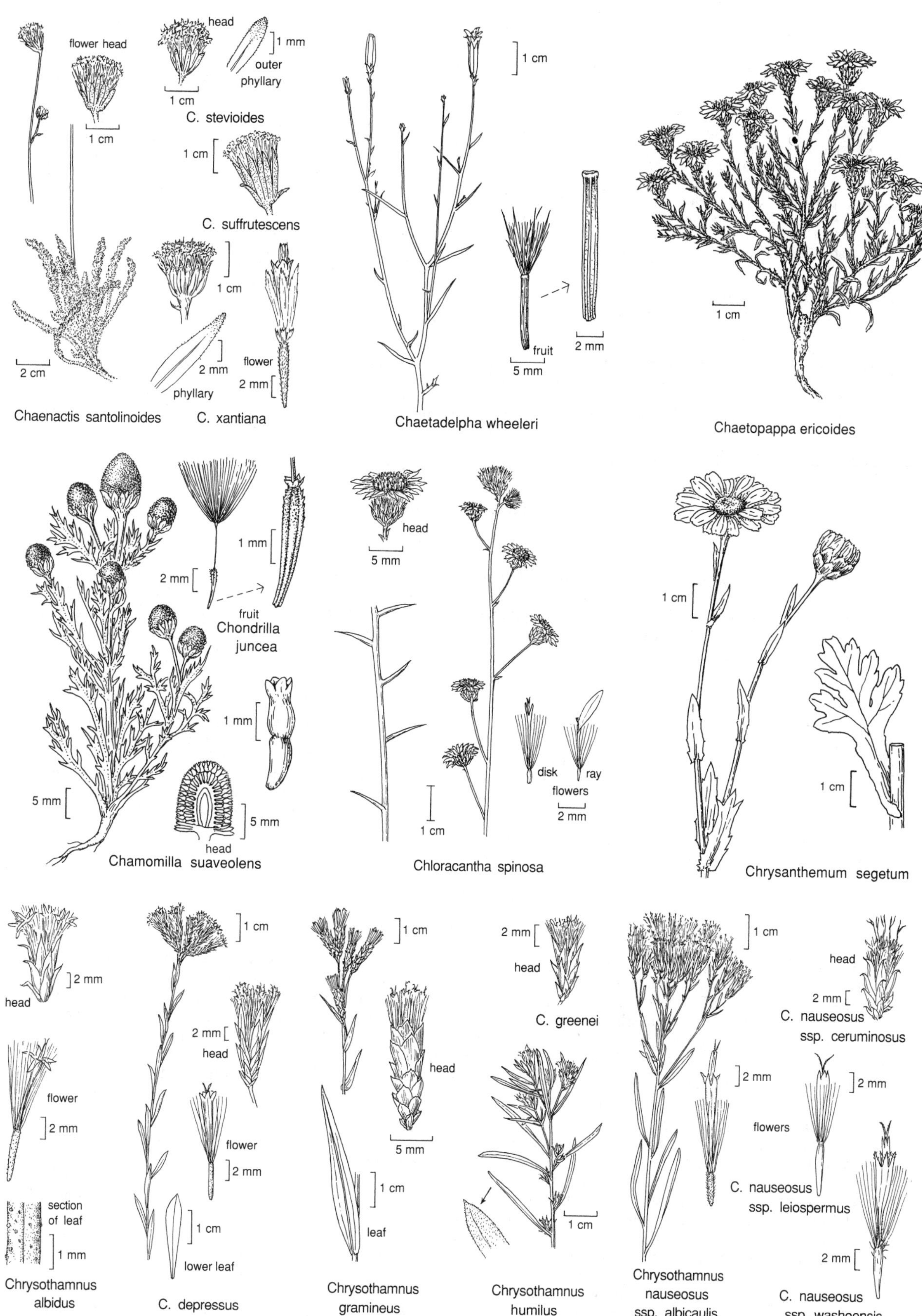

flower head
1 cm
2 cm
head
1 mm
outer phyllary
1 cm
C. stevioides
1 cm
C. suffrutescens
1 cm
2 mm
phyllary
flower
2 mm
Chaenactis santolinoides
C. xantiana
1 cm
fruit
5 mm
2 mm
Chaetadelpha wheeleri
1 cm
Chaetopappa ericoides
1 mm
2 mm
fruit
Chondrilla juncea
1 mm
5 mm
5 mm
head
Chamomilla suaveolens
head
5 mm
1 cm
disk
ray
flowers
2 mm
Chloracantha spinosa
1 cm
1 cm
Chrysanthemum segetum
2 mm
head
flower
2 mm
section of leaf
1 mm
Chrysothamnus albidus
1 cm
2 mm
head
flower
2 mm
1 cm
lower leaf
C. depressus
1 cm
head
5 mm
1 cm
leaf
Chrysothamnus gramineus
2 mm
head
C. greenei
1 cm
Chrysothamnus humilus
1 cm
2 mm
Chrysothamnus nauseosus ssp. albicaulis
head
2 mm
C. nauseosus ssp. ceruminosus
2 mm
flowers
C. nauseosus ssp. leiospermus
2 mm
C. nauseosus ssp. washoensis

ssp. ***vulcanicus*** (E. Greene) H.M. Hall & Clements **STS** slender. **LVS** 3–5 cm, green; glands 0. **INFL**: cymes, dense, much-branched; phyllaries long-tapered, gen straight, straw-colored. **FLS** 6–7. Uncommon. Rocky slopes, sagebrush flats; 1400–3200 m. c&s SNH.

C. teretifolius (Durand & Hilg.) H.M. Hall (p. 237) Shrub 2–12 dm. **STS** much-branched, brittle, green, gland-dotted. **LVS** 1–3.5 cm, thread-like, subcylindric, dark green, resinous. **INFL**: heads scattered or in long, dense cymes; involucre 7–9.5 mm, narrowly cylindric; phyllaries strongly ranked, straw-colored, ± keeled, tips enlarged, green. **FLS** 5–7; corollas 6.5–8 mm; style appendage < stigma. **FR** 4–5 mm, hairy; pappus slightly < corolla. 2*n*=18. Rocky flats, slopes; 600–2400 m. Teh, TR, SnJt, SNE, DMoj; s NV, nw AZ.

C. viscidiflorus (Hook.) Nutt. YELLOW RABBITBRUSH Shrub 1–15 dm; **STS** gen erect, brittle, white (greener upward). **LVS** 1–7.5 cm, 1–10 mm wide, thread-like to oblong, flat or twisted, (gray-) green, ± sticky. **INFL**: heads in dense, flat-topped or rounded cymes; involucre 5–10 mm, gen cylindric; phyllaries gen ± lanceolate, in ± 5 vertical ranks, keeled, yellow-green, ± sticky, tips obtuse to acute. **FLS** 3–13; corolla 3.5–7.5 mm; style exserted, appendage gen >> stigma. **FR** 3–5 mm, hairy; pappus ± = corolla. 2*n*=18, 36,54. Sagebrush, pinyon/juniper woodland; 900–4000 m. KR, CaR, SN, TR, GB, DMtns; to s Can, MT, NM. Highly variable; 5 sspp. in w US.

ssp. ***axillaris*** (Keck) L. Anderson **STS** glabrous. **LVS** 1–3 cm, thread-like, glabrous (exc ± ciliate). **INFL**: involucre ± obconic. **FLS** 3–5; corolla 3.5–4.5 mm. 2*n*=18. Uncommon. Gravelly washes; 1300–2000 m. s W&I, ne DMtns; to UT, n AZ. [*C. a.* Keck]

ssp. ***lanceolatus*** (Nutt.) H.M. Hall & Clements **STS** bristly near infl. **LVS** 2–4.5 cm, > 2 mm wide, ± lanceolate, rough-hairy below. **FLS** gen 5; corolla 5.5–6 mm. 2*n*=18,36. Uncommon. Juniper/sagebrush savanna; 1200–2500 m. c SNH, MP; to s Can, MT, NM. Intergrades with ssp. *viscidiflorus* esp in n GB. Common outside CA.

ssp. ***puberulus*** (D.C. Eaton) H.M. Hall & Clements (p. 237) Pl gray-green, densely puberulent. **LVS** 1–3 cm, gen 1–2 mm wide, thread-like to ± oblanceolate. **FLS** 4–7; corolla 4.5–5.5 mm. 2*n*=18,36 (tetraploids larger, lower elevations). Sagebrush, pinyon/juniper, subalpine slopes; 1500–3000 m. SN (e slope), SnBr, GB, DMtns; to s ID, UT, n AZ.

ssp. ***viscidiflorus*** (p. 237) **STS** glabrous. **LVS** 0.5–7.5 cm, linear to ± lanceolate, glabrous (exc ± ciliate). **FLS** 4–13; corolla 4–7.5 mm. 2*n*=18,36,54. Common. Sagebrush, pinyon/juniper, alpine talus; 900–4000 m. KR, CaR, SN, n WTR, SnBr, GB, s DMtns; to WA, MT, NM. Intergrading forms have been called sspp. *latifolius* (D.C. Eaton) H.M. Hall & Clements (wide lvs); *pumilus* (Nutt.) H.M. Hall & Clements (small pls); *stenophyllus* (A. Gray) H.M. Hall & Clements (narrow lvs). ❀TRY.

CICHORIUM CHICORY

G. Ledyard Stebbins

Ann, bien, per; sap milky. **STS** 3–10 dm, branched. **LVS** basal and cauline, reduced upward; lower 1–2 dm, toothed or pinnately lobed, wing-petioled; middle sessile, sometimes clasping; upper greatly reduced. **INFL**: heads ligulate, showy, terminal and axillary, lateral sessile; involucre ± cylindric; phyllaries in 2 series, hardened in basal half, outer short, spreading, inner elongate, erect; receptacle naked. **FLS** many; ligules blue to purple. **FR** oblong, glabrous, 5-angled; pappus of minute blunt scales. 8 spp.: Eur, Medit, Afr. (Old Arabic name)

1. Ann or bien; peduncles of terminal heads thickened; uncommon garden escape ..... ***C. endiva***
1′ Per from deep taproot; peduncles of terminal heads not thickened; common roadside weed ..... ***C. intybus***

C. endiva L. ENDIVE Ann, bien. **STS** 6–9 gm, glabrous. **LVS** elliptic, lobed. **INFL**: peduncles of terminal heads strongly thickened. **FLS**: ligules 11–13 mm, purple. **FR** 1.5–2.5 mm; pappus scales < 0.5 mm. 2*n*=18,36. Uncommon. Disturbed places near gardens; gen < 200 m. SnFrB, SCoRI (Salinas Valley); native to Eur. Cult as salad green.

C. intybus L. (p. 237) Per < 1 m, from deep, woody taproot; herbage glabrous to short-bristly, esp near base. **STS** erect; branches stout, ascending. **LVS**: lower oblong to elliptic in outline, subentire to coarsely pinnately lobed. **INFL**: peduncles of terminal heads not thickened. **FLS**: ligules blue (pink, white). **FR** 1.5–2.5 mm; pappus scales < 0.5 mm. 2*n*=18. Common. Roadsides, waste places; < 1500 m. CA-FP; widespread in N.Am; native to Eur. Roasted roots used as coffee flavoring or substitute.

CIRSIUM THISTLE

David J. Keil & Charles E. Turner

Ann to per (sometimes short-lived, dying after fl once). **STS** gen erect. **LVS**: lower gen tapered or petioled, often wavy-margined, gen pinnately lobed, ± dentate, lobes and teeth spine-tipped, margin gen spiny-ciliate, glabrous to tomentose; upper gen sessile, ± reduced. **INFL**: heads discoid, 1–many; involucre cylindric to spheric; phyllaries many, graduated in several series, outer spine-tipped; receptacle flat, long-bristly. **FLS** gen many; corollas ± bilateral, white to red or purple, tube long, slender, lobes linear; anther bases sharply sagittate, tips oblong; style tip with slightly swollen node, appendage (above node) long, cylindric, branches very short. **FR** ovoid, glabrous; scar slightly angled; pappus bristles many, plumose, ± persistent or falling in ring. ± 200 spp.: N.Am, Eurasia. (Greek: thistle) Taxa difficult, incompletely differentiated, hybridize.

1. Fls unisexual (dioecious); heads gen 1–1.5 cm ..... ***C. arvense***
1′ Fls bisexual; heads gen > 1.5 cm
  2. Lf surfaces harshly bristly above; sts spiny-winged ..... ***C. vulgare***
  2′ Lf surfaces glabrous to tomentose above; sts wingless or ± short-winged
    3. Heads nodding; lf surface ± glandular-puberulent above ..... ***C. fontinale***
      4. Fr 4–5 mm; phyllary spines 1–2 mm; sw SnFrB ..... var. ***fontinale***
      4′ Fr ± 3.5–4 mm; phyllary spines 2–6 mm; e SnFrB, c SCoRO
        5. Cauline lvs with larger spines 10–18 mm; most outer phyllaries with 0 marginal spines; e SnFrB var. ***campylon***
        5′ Cauline lvs with larger spines 4–7 mm; many outer phyllaries with marginal spines; c SCoRO var. ***obispoense***
    3′ Heads gen erect; lf surface not glandular above
      6. Outer phyllary margins spiny-fringed or with expanded, scarious appendages
        7. Corollas 30–45 mm, gen red to rose-purple; inner phyllaries long, straight, entire, purple-tipped [3]***C. andersonii***

7′ Corollas 16–30 mm, white to purple; inner phyllaries often ± crinkled, sometimes with expanded, scarious appendages, gen not purple-tipped
8. Heads gen not closely subtended by clustered leafy bracts, each often subtended by 1 lf
9. Phyllary midribs conspicuously sticky-glandular; outer phyllary margins minutely spiny-ciliate .... [2]***C. hydrophilum***
10. Heads gen 2.5–3 cm; fr ± 5 mm; deltaic GV (Suisun Marsh) .... var. ***hydrophilum***
10′ Heads gen 3–3.5 cm; fr 4–5 mm; n SnFrB (Mount Tamalpais) .... var. ***vaseyi***
9′ Phyllary midribs not glandular (or inner inconspicuously glandular); outer phyllary margins gen ± conspicuously fringed or spiny-ciliate
11. Bien, stout, gen 10–30 dm; st base 2–10 cm diam, hollow; middle and inner phyllaries entire; SnJV .... [2]***C. crassicaule***
11′ Per, slender, gen 1–11 dm; st base gen < 2 cm diam, not hollow; middle and inner phyllaries often with fringed appendages; NW, n CW .... [4]***C. remotifolium***
8′ Heads gen closely subtended by clustered, ± leafy bracts
12. Lvs closely gray-white-tomentose (felt-like) on both surfaces, very prominently wavy-margined; corollas white or pale yellow — dunes, headlands, s CCo .... ***C. rhothophilum***
12′ Lvs glabrous to cobwebby-tomentose above, sometimes white-tomentose below, flat to wavy-margined; corollas white to purple
13. Phyllaries conspicuously cobwebby-tomentose
14. Upper lvs ± stiff, strongly spiny, spines stout, often 10–15 mm; most outer phyllaries spiny-margined; s NCo, n CCo .... ***C. andrewsii***
14′ Upper lvs thin, ± weakly spiny, spines slender, gen < 8 mm; few outer phyllaries spiny-margined; NCo, NCoRO, CC, SnFrB, SCo .... [2]***C. brevistylum***
13′ Phyllaries ± glabrous or very thinly cobwebby
15. Pl 10–30 dm; most outer phyllaries spiny-ciliate, also minutely rough-ciliate; SnJV .... [2]***C. crassicaule***
15′ Pl gen 1–10 dm; most outer phyllaries entire or minutely rough-ciliate; s CCo .... [4]***C. loncholepis***
6′ Outer phyllary margins gen entire (bracts subtending head gen spiny)
16. Middle and outer phyllaries gen spreading to reflexed
17. Main phyllary spine tips gen 7–20+ mm
18. Per; heads sessile or on short leafy peduncles, sometimes in ± congested cyme-like clusters
19. Phyllaries densely, persistently cobwebby; CCo .... [3]***C. occidentale*** var. ***compactum***
19′ Phyllaries thinly cobwebby or becoming glabrous
20. Corollas bright pink to red; corolla tube < throat; s SNH, W&I .... [2]***C. arizonicum***
20′ Corollas rose-purple; corolla tube > 2 × throat; DMtns .... [2]***C. nidulum***
18′ Bien; heads gen long-peduncled, often in ± open cyme-like clusters
21. Corolla tube abruptly expanded to throat; corollas gen white to pale purple (sometimes darker purple to reddish)
22. Cauline lvs with main spines gen 8–15 mm; e DMoj .... ***C. neomexicanum***
22′ Cauline lvs with main spines gen < 8 mm; SN, SCoR, SW .... [2]***C. occidentale*** var. ***californicum***
21′ Corolla tube gradually expanded to throat; corollas gen rich purple to bright pink or red (sometimes white to pale purple) .... [2]***C. occidentale***
23. Pls compact, rounded, mound-like, gen < 5 dm — CCo .... [3] var. ***compactum***
23′ Pls erect, often >> 5 dm
24. Head length gen ± = width; phyllaries gen very narrow, needle-like, radiating; corollas gen purple; mostly coastal, CW, SW .... [2] var. ***occidentale***
24′ Heads gen longer than wide; phyllaries gen clearly flattened, widened toward base, spreading to reflexed; corollas gen red or bright pink, sometimes purple; gen interior mtns, n&c CA vars. *candidissimum*, *venustum* (see 33., 33′)
17′ Main phyllary spine tips gen 1–6 mm
25. Corollas white to pale purple
26. Bien; heads gen in open cyme-like clusters; SN, c&s SCoR, SW .... [2]***C. occidentale*** var. ***californicum***
26′ Per; heads gen in ± congested flat-topped or raceme-like clusters; gen n CA
27. Some phyllaries gen fringed or spiny-toothed .... [4]***C. remotifolium***
27′ All phyllaries entire
28. Sts, lvs tomentose; jointed-multicellular hairs 0; KR .... [2]***C. ciliolatum***
28′ Sts, lvs with hairs of 2 kinds, tomentose and jointed, multicellular hairs, esp along midveins (lf lower surface); widespread .... [3]***C. cymosum***
25′ Corollas deep purple to bright pink or red
29. Corolla tube abruptly expanded to throat; per
30. Corollas 30–45 mm, gen red to rose-purple; inner phyllaries long, straight, entire .... [3]***C. andersonii***
30′ Corollas 16–25 mm, white to purple; inner phyllaries gen ± crinkled, sometimes with expanded, scarious appendages .... [4]***C. remotifolium***
29′ Corolla tube gradually expanded to throat; bien .... [2]***C. occidentale***
31. Pls compact, rounded, mound-like, gen < 5 dm — CCo .... [3]var. ***compactum***
31′ Pls erect, often >> 5 dm
32. Head length gen ± = width; corollas gen purple; mostly coastal, CW, SW .... [2]var. ***occidentale***
32′ Heads gen longer than wide; corollas gen red or bright pink; gen interior mtns, n&c CA
33. Pl densely white-tomentose; phyllaries persistently white-tomentose (exc spine tip); outer phyllaries gen very long, spreading to reflexed; KR, NCoR, CaR, MP, n SN .... var. ***candidissimum***

33′ Pl gray-tomentose to ± green (rarely densely white-tomentose); phyllaries ± cobwebby to tomentose or becoming ± glabrous; outer phyllaries short to long, ascending to spreading or reflexed; NCoR, s SN, and W&I southward .... var. ***venustum***

16′ Middle and gen outer phyllaries appressed or stiffly ascending (spine tip sometimes abruptly spreading)

34. Lvs all basal or crowded on very short, densely leafy st; heads ± sessile, subtended by rosette lvs

35. Per from creeping rootstock; gen ± dry habitats — NCo, NCoRO, CCo, SnFrB, n SCoRO [3]***C. quercetorum***

35′ Bien or short-lived per that fls once, taprooted; gen ± wet habitats

36. Pls ± glabrous; main lf spines gen 7–10 mm; s CCo .... [4]***C. loncholepis***

36′ Pls gen ± tomentose (esp lower surface of lvs); main lf spines gen 2–7 mm; gen interior mtns .... [3]***C. scariosum***

34′ Lvs basal and cauline; heads sessile or peduncled, clearly raised above rosette lvs

37. Phyllary spine tips 10–20+ mm

38. Corolla rose-purple to red, lobes = or > tube

39. Corolla bright pink to red; corolla tube < throat; s SNH, W&I .... [2]***C. arizonicum***

39′ Corolla rose-purple; corolla tube > 2 × throat; DMtns .... [2]***C. nidulum***

38′ Corolla dull white to pale purple, lobes < tube

40. Pl sparsely cobwebby-tomentose, becoming glabrous; heads closely subtended by well developed lvs — s CCo .... [4]***C. loncholepis***

40′ Pl persistently tomentose; heads gen not closely subtended by lvs

41. Corollas 20–27 mm; bien from taproot .... [3]***C. canovirens***

41′ Corollas 29–37 mm; per from creeping rootstock .... [2]***C. ochrocentrum***

37′ Phyllary spine tips gen < 10 mm

42. Corollas gen 25–50 mm

43. Corollas bright purple to red

44. Phyllaries in 3–4 series, outer linear-lanceolate, inner much longer, straight, red- or purple-tipped; corollas red or rose-purple .... [3]***C. andersonii***

44′ Phyllaries strongly graduated in 5–10 series, outer lanceolate to ovate, inner often twisted or crinkled at tip; corollas purple

45. Bien or short-lived per, taprooted — gen interior mtns .... [3]***C. scariosum***

45′ Per from creeping rootstock

46. Phyllaries gen ovate, midribs not glandular; NCo, NCoRO, CCo, SnFrB, SCoRO [3]***C. quercetorum***

46′ Phyllaries gen lanceolate, midribs with sticky-glandular zone; weed, s CaR, SnFrB, SCoRO, SCo, MP .... [2]***C. undulatum***

43′ Corollas white to pink or lavender

47. Bien or short-lived per, taprooted — gen ± wet soil of meadows

48. Pls ± glabrous; main lf spines gen 7–10 mm; s CCo .... [4]***C. loncholepis***

48′ Pls gen ± tomentose (esp lower surface of lvs); main lf spines gen 2–7 mm; gen interior mtns .... [3]***C. scariosum***

47′ Per from creeping rootstock

49. Phyllary spine tips gen 5–10 mm; lvs gen < 3 cm wide, deeply divided into many narrow, spiny lobes .... [2]***C. ochrocentrum***

49′ Phyllary spine tips gen 1–5 mm; lvs often > 3 cm wide, variously divided, variously spiny

50. Phyllaries ± persistently tomentose .... [3]***C. cymosum***

50′ Phyllaries gen becoming glabrous

51. Phyllaries gen ovate, midribs not glandular; NCo, NCoRO, CCo, SnFrB, SCoRO [3]***C. quercetorum***

51′ Phyllaries gen lanceolate, midribs with sticky-glandular zone; weed, s CaR, SnFrB, SCoRO, SCo, MP .... [2]***C. undulatum***

42′ Corollas gen 16–24 mm

52. Phyllary midribs gen without a sticky-glandular area

53. Heads gen conspicuously peduncled .... [4]***C. remotifolium***

53′ Heads gen ± sessile or short-peduncled

54. Phyllaries connected by tangled cobwebby hairs; involucres 2.5–3.5 cm; heads closely subtended by ± leafy bracts; NCo, NCoRO, CCo, SnFrB, SCo .... [2]***C. brevistylum***

54′ Phyllaries ± tomentose or becoming glabrous; involucres 1.5–2.5 cm; heads often not closely subtended by leafy bracts; SNH (e slope), n GB .... [3]***C. canovirens***

52′ Some phyllary midribs with a ± sticky-glandular area or raised translucent gland

55. Phyllary bodies (at least outer) ascending or loosely appressed, spine tip gen erect or ascending; heads commonly medium to large, in cyme- or raceme-like clusters

56. Hairs all cobwebby, often dense and persistent on lvs .... [2]***C. ciliolatum***

56′ Hairs of 2 kinds, cobwebby and jointed, multicellular, gen ± loose and irregularly deciduous from lvs in age .... [3]***C. cymosum***

55′ Phyllary bodies tightly appressed, spine tip ascending to abruptly spreading; heads commonly small, ± ovoid, in panicle-like clusters

57. Lvs (both surfaces) thinly tomentose, becoming ± glabrous above

58. Hairs of 2 kinds, cobwebby and jointed, multicellular, gen ± loose and irregularly deciduous from lvs in age; SNH (e slope), n GB .... [3]***C. canovirens***

58′ Hairs all cobwebby, often dense and persistent on lvs; deltaic GV, n SnFrB .... [2]***C. hydrophilum*** vars. (see 10., 10′)

57′ Lvs (both surfaces) densely tomentose, esp below
59. Phyllaries greenish or straw-colored; corollas white to pale purple; SNE, DMoj ...... ***C. mohavense***
59′ Phyllaries gen blackish near tips; corollas gen dark purple (rarely white) .... ***C. douglasii***
60. Upper cauline lvs with larger spines mostly < 7 mm; KR, n NCoR, CaR, n SN, MP .... var. ***breweri***
60′ Upper cauline lvs with larger spines gen 7–20 mm; NCo, NCoR, SnFrB, n CCo .... var. ***douglasii***

C. andersonii (A. Gray) Jepson (p. 237) Per, often appearing bien, (1.5)4–7(10) dm; caudices erect, taprooted, from rootstock. **STS** 1–several, simple or branches ascending, ± glabrous to puberulent and or tomentose. **LVS** gen green above, ± persistently gray-tomentose below; lower 8–35 cm, petiole spiny-winged, coarsely dentate or shallowly to deeply 1–2 × lobed, main spines 1–5 mm; middle ± clasping; upper gen much reduced, linear-oblong, often spinier than lower. **INFL**: heads 1–few; peduncles < 20 cm (lateral heads sometimes ± sessile); involucre 3–5 cm, 2–4 cm diam, widely cylindric to narrowly bell-shaped, closely subtended by 1–several bracts, loosely tomentose or glabrous; phyllaries strongly graduated, ascending, ± lanceolate, long-acuminate, middle and outer entire or spiny-ciliate, tip spines 1–3 mm, inner tips entire, long, flat, purple. **FLS**: corollas 30–45 mm, red to reddish purple, tube 10–20 mm, throat 10–16 mm, lobes 9–11 mm. **FR** 6–7 mm, compressed, brown; pappus 25–40 mm. 2*n*=32,64. Open places, woodland, forest; (760)1400–3150 m. KR, CaR, SN; to se ID?, NV.

C. andrewsii A. Gray (p. 237) FRANCISCAN THISTLE Bien (or short-lived per) 6–20 dm. **STS** gen much-branched below, ± fleshy, thinly cobwebby, soon becoming glabrous. **LVS** thinly cobwebby above, becoming glabrous, gray-cobwebby-tomentose below; lower 3–7.5 dm, gen 1–2 × lobed, main spines 2–9 mm; middle and upper smaller, clasping, bases wide, spiny-margined, ear-like; uppermost much reduced, gen very spiny, spines < 15 mm. **INFL**: heads 1–few in loose cymes, closely subtended by upper lvs, sessile or peduncles < 7 cm; involucres 1.5–3 cm, 1.5–5 cm diam, ovoid to hemispheric or bell-shaped, densely cobwebby; phyllaries ± linear, tips widely spreading, spines 5–15 mm, outer and middle spiny-ciliate, inner with tips flat, straight. **FLS**: corollas 17–24 mm, dark reddish purple, tube 8–11 mm, throat 3.5–6 mm, lobes 5–7 mm. **FR** 4–5 mm, dark brown, ± compressed; pappus ± 15 mm. 2*n*=32. UNCOMMON. Bluffs, ravines, seeps, sometimes on serpentine; < 100 m. s NCo, n CCo. ❀TRY,DFCLT.

C. arizonicum (A. Gray) Petrak ARIZONA THISTLE Per 2–10 dm; taproot ± woody. **ST** gen 1, few-branched above, thinly tomentose or ± glabrous. **LVS** ± tomentose (esp below), becoming ± glabrous; lower 1–2 dm, tapered to spiny-winged petioles, oblong-obovate, ± lobed, lobes gen further lobed or toothed, main spines 5–15 mm; middle and upper not strongly reduced, clasping or short-decurrent. **INFL**: heads gen few, short-peduncled, ± closely subtended by uppermost lvs; involucres 3–4 cm, 1.5–2 cm diam when fresh, cylindric or narrowly ovoid, sparsely tomentose; phyllaries linear or linear-lanceolate, entire, gen erect or ascending, outer and middle tipped with spines 10–20+ mm, inner with tips flat or short-spined, straight, often red or purple, puberulent. **FLS** exserted; corollas 30–34 mm, red, tube 7–9 mm, throat 11–14 mm, lobes 12–13 mm. **FR** 4–6 mm, ± compressed, shiny brown; pappus 24–28 mm. 2*n*=32. Open forests, sagebrush scrub; 2400–3700 m. s SNH, W&I; to UT, NM. [*C. eatonii* (A. Gray) Robinson & *C. nidulum* (M.E. Jones) Petrak misapplied] Variable, needs further study; close to *C. nidulum*. ❀TRY.

C. arvense (L.) Scop. (p. 237) CANADA THISTLE Per 5–10 dm, dioecious; rootstock creeping; herbage green. **STS** colonial, very leafy. **LVS** 5–20 cm, mostly cauline, gradually reduced upward, sessile, tapered at base, sometimes decurrent as spiny wings, subentire to coarsely dentate or 1–2 × lobed; main spines 3–5 mm. **INFL**: heads several–many, cymes tight to ± open, rounded or flat-topped; peduncles 0–4 cm; involucre hemispheric to ovoid, 1–2 cm, 1–2 cm diam, gen ± purplish, ± tomentose when young; outer phyllaries ovate, tipped by spines ± 1 mm, inner lanceolate, tips flat, membranous. **FLS**: corollas gen purplish, sometimes white or pink. **STAMINATE FL**: corollas 12–13 mm, > pappus, tube 8 mm, throat 1–1.5 mm, lobes 3–4 mm. **PISTILLATE FL**: corollas 14–20 mm, gen < pappus, tube 10–16 mm, throat ± 1 mm, lobes 2–3 mm. **FR** 2–3 mm; pappus 13–23 mm. 2*n*=34,36. Disturbed places; < 1800 m. CA-FP (exc SnJV, s SN), SNE, w DMoj; N.Am.; native to Eur. NOXIOUS WEED.

C. brevistylum Cronq. (p. 237) INDIAN THISTLE Ann to bien (or short-lived per) 3–20 dm. **STS** gen 1, gen branched above middle, ± hairy with jointed hairs, often cobwebby below heads. **LVS** sparsely hairy (hairs jointed) above, gray-tomentose below, sometimes becoming ± glabrous; lower 15–25 cm, tapered to spiny-winged petioles, gen shallowly lobed and dentate, main spines 3–7 mm; cauline gradually reduced, sometimes ± clasping or short-decurrent. **INFL**: heads 1–several at st tips, sometimes also in upper axils, ± sessile, closely subtended by bract-like lvs; involucres 2.5–3.5 cm, 2.5–4 cm diam, hemispheric to ± bell-shaped, loosely cobwebby; phyllaries in ± 5 series, linear, not strongly graduated, outermost sometimes spiny-ciliate near base, tip spines 3–5 mm, inner with tips straight, flat. **FLS**: corollas 20–24 mm, white to purple, tube 10–14 mm, throat 4–5 mm, lobes 3–5 mm. **FR** 3–4 mm, ± brown; pappus 10–20 mm. 2*n*=34. Moist places; < 500 m. NCo, NCoRO, CCo, SnFrB, SCo; to WA, BC.

C. canovirens Rydb. GRAY-GREEN THISTLE Bien or short-lived per 3–10 dm, from taproot. **ST** gen 1, gen simple below, ± cobwebby and soft-hairy (hairs jointed, multicellular). **LVS** ± tomentose (esp lower surface), sometimes becoming ± glabrous; lower ± petioled, 1–2(3) dm, oblong or narrowly oblanceolate, gen lobed, main spines 5–10 mm; middle and upper smaller, decurrent as spiny wings, toothed or shallowly lobed. **INFL**: heads 1–many in flat-topped to raceme-like clusters, sessile or peduncles gen 1–10 cm; involucres 1.5–2.5 cm, 1.5–3 cm diam, ovoid to hemispheric, ± loosely tomentose; phyllaries linear-lanceolate to ovate, ascending to spreading, midribs of outer and middle often with narrow glandular area, spines 5–10 mm, inner with tips flat, straight. **FLS**: corollas 20–27 mm, dull white to pale purple, tube 7–10 mm, throat 7–10 mm, lobes 4–7 mm. **FR** 4–6 mm, dark brown, ± compressed; pappus 15–25 mm. 2*n*=34. Shrubby areas, open forests, roadsides; 1600–3400 m. e SNH, MP, n SNE. [*C. utahense* Petrak misapplied] Needs further study; closely related to *C. subniveum* Rydb.

C. ciliolatum (L. Henderson) J. Howell (p. 237) ASHLAND THISTLE Per 6–20 dm; rootstock branched. **STS** gen simple below, ± cobwebby to densely white-tomentose. **LVS** gray-tomentose above, ± white below: rosettes from rootstock branches; lowest 1–2.5 dm, wing-petioled, entire to shallowly lobed, spines 0 or 1–2 mm; cauline well developed, gradually reduced, short-decurrent as spiny wings, shallowly to deeply lobed, weakly spiny, uppermost bract-like. **INFL**: heads few–many in ± raceme- or panicle-like clusters; peduncles gen 1–10 cm; involucres ± 2 cm, 2–5 cm diam, ovoid to hemispheric, thinly tomentose; phyllaries ± linear, tips ascending to spreading, midribs of outer and middle with sticky glandular area, spine tips 1–3 mm, inner with tips flat, straight. **FLS**: corollas 17–25 mm, dull white to lavender, tube 7–11 mm, throat 5–7 mm, lobes 5–7 mm. **FR** 3.5–4.5 mm, straw-colored to brown, thick; pappus ± 15 mm. **ENDANGERED** CA. Open woodland; 1000–1400 m. KR; s OR.

C. crassicaule (E. Greene) Jepson (p. 237) Ann or bien 1–3 m. **STS** gen 1, gen simple below, openly branched above, thinly cobwebby, soft-hairy (hairs jointed), at least below; base < 10 cm diam, hollow. **LVS** thinly cobwebby-tomentose above, gray-tomentose below, sometimes midvein with jointed hairs; lower 1.5–7 dm, petioles spiny-lobed, winged, blades elliptic to widely oblanceolate, 1–2 × lobed, lobes sometimes dentate, main spines 3–8 mm; middle and upper smaller, narrower, sessile, spiny-margined, bases wide-clasping; upper much reduced, gen very spiny, spines < ± 10 mm. **INFL**: heads in loose to crowded cymes (sometimes on short axillary branches), closely subtended by uppermost lvs; peduncles 0–15 cm; involucres 1.5–3 cm, 1.5–3 cm diam, ovoid to ± bell-shaped, ± glabrous; phyllaries ± linear, at least outer irregularly spiny-fringed, outer and middle ascending or ± spreading, spines 3–5 mm, inner with tips erect, ± twisted. **FLS**: corollas 19–26 mm, pale rose-purple (white), tube 9–12 mm, throat 4–6 mm, lobes 5–9 mm. **FR** 5–5.5 mm, dark brown, ± compressed; pappus 1.5–2 cm. 2*n*=32. RARE. Freshwater marshes; < 100 m. SnJV.

C. cymosum (E. Greene) Jepson PEREGRINE THISTLE Per 2.5–6(>10) dm; rootstock branched. **STS** gen simple below, ± cobwebby, soft-hairy (hairs jointed). **LVS** ± cobwebby on both surfaces or becoming glabrous above, soft-hairy (hairs jointed), esp on midrib below; rosettes from rootstock branches; lower 1–2(5) dm, petioled, gen 1–2 × lobed, main spines 3–4 mm, middle and upper clasping or short-decurrent, upper much reduced, gen very spiny. **INFL**: heads few–many in ± flat-topped cymes (sometimes raceme- or panicle-like), sessile or peduncles gen 1–10 cm; involucres 1.5–3 cm, 1.5–5 cm diam, ovoid to hemispheric, ± tomentose; phyllaries ± linear, midribs of outer and middle gen with a narrow sticky glandular area, tips ascending to spreading, spines 2–4 mm, inner with tips flat, straight. **FLS**: corollas 20–33 mm, dull white, tube 10–15 mm, throat 8–12 mm, lobes 5–8 mm. **FR** 7–8 mm, dark brown to black, ± flattened; pappus 15–25 mm. $2n$=30, 34. Shrubland, woodland, open forest, meadows, sometimes on serpentine; 100–2100 m. KR, NCoR, CaR, n&s SN, SnFrB, n SCoRI, MP; s OR. Variable. ❀STBL.

C. douglasii DC. SWAMP THISTLE Bien (or short-lived per) 7–25 dm, gray-tomentose with appressed felt-like hairs. **STS** 1–several, gen much-branched. **LVS** tomentose (both surfaces) or becoming ± greenish above; basal 3–11 dm, petiole spiny-toothed, blade coarsely dentate to deeply 1–2 × lobed and dentate, main spines 3–10 mm; cauline smaller, expanded to ear-like, decurrent bases, shallowly to deeply lobed, spines often 7–12+ mm. **INFL**: heads several–many, clusters ± leafy, panicle-like; peduncles 1–4 cm; involucres 1.5–3 cm, 2–4.5 cm diam, ovoid to hemispheric; phyllaries often with dark purple patch near tip, tightly appressed, outer phyllaries ovate, inner lanceolate, outer and middle tipped by spreading spines 1–7 mm, inner with tips flat, acuminate, dark purple. **FLS**: corollas 18–21 mm, ± dark purple or white, tube 8–9 mm, throat 5–6 mm, ± = lobes. **FR** 5–6 mm, dark-brown to black, gen not flattened; pappus 15–20 mm. Wet places; < 2200 m. NCo, KR, NCoR, CaR, n SN, SnFrB, n SCoR, MP; s OR. Closely related to *C. mohavense* & *C. hydrophilum*.

var. ***breweri*** (A. Gray) Keil & C. Turner (p. 237) **STS** < 2.5 m. **LVS**: cauline gen shallowly lobed or not, spines gen < 7 mm. **INFL**: heads gen 2–3 cm. $2n$=30,34. Habitat of sp.; 1000–2200 m. KR, n NCoR, CaR, n SN, MP; s OR. [var. *canescens* (Petrak) J. Howell; *C. breweri* A. Gray] ❀TRY;DFCLT.

var. ***douglasii*** **STS** < 1.5 m. **LVS**: cauline gen deeply lobed, spines gen 7–20 mm. **INFL**: heads gen 2.5–3.5 cm. Habitat of sp.; < 1000 m. NCo, NCoR, n CCo, SnFrB, n SCoR. ❀TRY;DFCLT.

C. fontinale E. Greene Per 5–22 dm, thinly to densely tomentose, glandular. **STS** 1–several. **LVS** gen very strongly wavy; lower 1–7 dm, petioles spiny-lobed or toothed, blades shallowly to deeply 1–2 × lobed and dentate, main spines 1–10 mm, abrupt; cauline gradually reduced, well-distributed, ± lobed, upper clasping (bases expanded, ear-like), sometimes short-decurrent, spines often 10–12 mm. **INFL**: heads ± grouped in panicle-like cluster, ± nodding, closely subtended by uppermost lvs; peduncles 0–7 cm; involucres hemispheric or bell-shaped, green to purple, 1.5–3 cm, 2–4.5 cm diam; phyllaries lanceolate to ovate, spreading to reflexed, spines 1–5 mm. **FLS**: corollas 20–22 mm, white to pink or lavender. **FR** 4–5 mm, ± brown; pappus 12–15 mm. Serpentine seeps and streams; < 750 m. CW.

var. ***campylon*** (H. Sharsm.) Keil & C. Turner (p. 237) HAMILTON THISTLE **STS** < 2 m, green. **LVS** densely tomentose; glandular hairs concealed. **INFL**: heads strongly, permanently nodding; outer phyllaries 20–30 mm, green, strongly recurved, channeled, widest below middle, upper 1/2 tapered to spine 3–5 mm. **FLS**: corolla tube 5–6 mm, throat 10 mm, lobes 4–5 mm; style branches 4–6 mm. **FR** glabrous. UNCOMMON. Habitat of sp.; 300–750 m. e SnFrB.

var. ***fontinale*** (p. 237) FOUNTAIN THISTLE **STS** < 1.3 m, reddish. **LVS** thinly tomentose, clearly glandular. **INFL**: heads ± nodding in fl, gen erect in fr; outer phyllaries 15–20 mm, reddish, moderately recurved, flat to ± channeled, widest above middle, abruptly tipped by 1–2 mm spine. **FLS**: corolla tube 10 mm, throat 5–6 mm, lobes 5–6 mm; style branches 3–4 mm. **FR** glabrous. $2n$=34 + 1. RARE. Habitat of sp.; ± 120 m. sw SnFrB (San Mateo Co.). In cult.

var. ***obispoense*** J. Howell CHORRO CREEK BOG THISTLE **STS** < 2 m, green to purplish. **LVS** densely tomentose, esp lower surface; glandular hairs concealed. **INFL**: heads ± nodding in fl, gen erect in fr; outer phyllaries 15–20 mm, green to dark purple, strongly recurved, ± channeled, widest below middle, upper 1/2 tapered to spine 1–4 mm. **FLS**: corolla tube 7 mm, throat 8 mm, lobes 5 mm; style branches 4–6 mm. **FR** minutely scabrous above. RARE. Habitat of sp.; < 300 m. c SCoRO (San Luis Obispo Co.).

C. hydrophilum (E. Greene) Jepson Bien (or short-lived per) 1–2.1 m. **STS** ± cobwebby, becoming glabrous. **LVS** thinly cobwebby-tomentose (both surfaces) or becoming glabrous above; lower 3–9 dm, petiole spine-margined, blade gen ± lobed, lobes gen with 2–4(6) wide 2° lobes or coarse teeth, main spines 2–9 mm; middle and upper smaller, clasping or short-decurrent with ear-like bases, uppermost much reduced, gen very spiny, lobes narrower. **INFL**: heads solitary or in small groups in ± open cymes or panicle-like clusters; peduncles 0–6 cm; involucres 1.5–2.5 cm, 1.5–3 cm diam, ovoid to bell-shaped, thinly cobwebby, becoming glabrous; outer and middle ovate to ovate-lanceolate, minutely spiny-ciliate, appressed, midribs with narrow sticky glandular area, spines 1–3 mm, spreading, inner linear-lanceolate or oblong, tips flat, ± twisted. **FLS**: corollas 18–23 mm, pale rose-purple, tube 8–10 mm, throat 5–6 mm, lobes 5–7 mm. **FR** 4–5 mm, dark brown to black, ± thick; pappus ± 15 mm. Wet soils; < 450 m. Deltaic GV, n SnFrB. Related to *C. douglasii* AND *C. mohavense*.

var. ***hydrophilum*** (p. 237) SUISUN THISTLE **STS** gen branched above middle. **INFL**: heads gen 2.5–3 cm. **FR** ± 5 mm, oblong, tapered near base. $2n$=32. RARE. Salt marsh; ± 0 m. Deltaic GV (Suisun Marsh, Solano Co.).

var. ***vaseyi*** (A. Gray) J. Howell MOUNT TAMALPAIS THISTLE **STS** often branched below middle. **INFL**: heads gen 3–3.5 cm. **FR** 4–5 mm, ± oblong or elliptic. $2n$=32. RARE. Serpentine seeps; 300–450 m. n SnFrB (Mount Tamalpais). [*C. vaseyi* (A. Gray) Jepson]

C. loncholepis Petrak (p. 237) LA GRACIOSA THISTLE Bien (or short-lived per) 1–10+ dm. **STS** 0 or 1–several from base, spreading (pls low, mound-like) or erect, often ± fleshy, ridged, glabrous. **LVS** strongly wavy-margined, loosely tomentose, soon becoming glabrous; lower 10–30 cm, spiny-petioled, gen deeply lobed, lobes ± ovate, gen with radiating 2° lobes or teeth, main spines 7–10 mm; middle and upper not much reduced, sessile or petioled, decurrent as short, spiny wings. **INFL**: heads in tight clusters at st tips, ± sessile or short-peduncled, often closely subtended by several lvs; involucres 2–3.5 cm, 2–4 cm diam, ovoid to bell-shaped, glabrous; phyllaries lanceolate to linear-lanceolate, minutely ciliate or spiny-ciliate, outer tipped by ascending spines 2–5 mm, inner with tips straight, flat, linear. **FLS**: corollas 25–30 mm, ± white, purplish, tube 13–17 mm, throat 5–9 mm, lobes 5–7 mm; anthers purple. **FR** 3–4 mm, oblong-obovate, ± compressed; pappus 15–25 mm. $2n$=34. RARE. Wetlands in dunes; < 50 m. s CCo (s San Luis Obispo, n Santa Barbara cos.). Related to *C. scariosum*.

C. mohavense (E. Greene) Petrak MOJAVE THISTLE Bien (or short-lived per) 5–25 dm. **STS** gen 1, gen simple below, openly branched above, ± white-tomentose. **LVS** ± densely tomentose, lighter below; lower 1.5–6 dm, spiny-petioled, elliptic to oblanceolate, toothed to deeply lobed, lobes gen rigidly spreading, simple or with 1–2 pairs of coarse teeth or 2° lobes, main spines 3–20 mm; middle and upper smaller, narrower, decurrent as spiny wings, upper much reduced, often long-acuminate, gen very spiny, spines 4–25 mm. **INFL**: heads in loose to crowded cyme- or panicle-like clusters (sometimes on short axillary branches); peduncles leafy, 0–10 cm; involucres 1.5–2.5 cm, 1.5–2 cm diam, ± ovoid, ± loosely tomentose, becoming glabrous; phyllaries strongly graduated (outer ovate, inner oblong), entire, tightly appressed, midribs of middle often with glandular area, spines 3–7 mm, ascending or spreading, inner with erect, ± twisted flat tips. **FLS**: corollas 16–25 mm, white to lavender or pink, tube 7–11 mm, throat 4–7 mm, lobes 4–8 mm. **FR** 3.5–6 mm, straw-colored to brown, gen not compressed; pappus ± 15 mm. $2n$=30,32. Damp soil around springs, canyons, streams, ditches; 400–2800 m. SNE, DMoj; NV. Related to *C. douglasii* and *C. hydrophilum*.

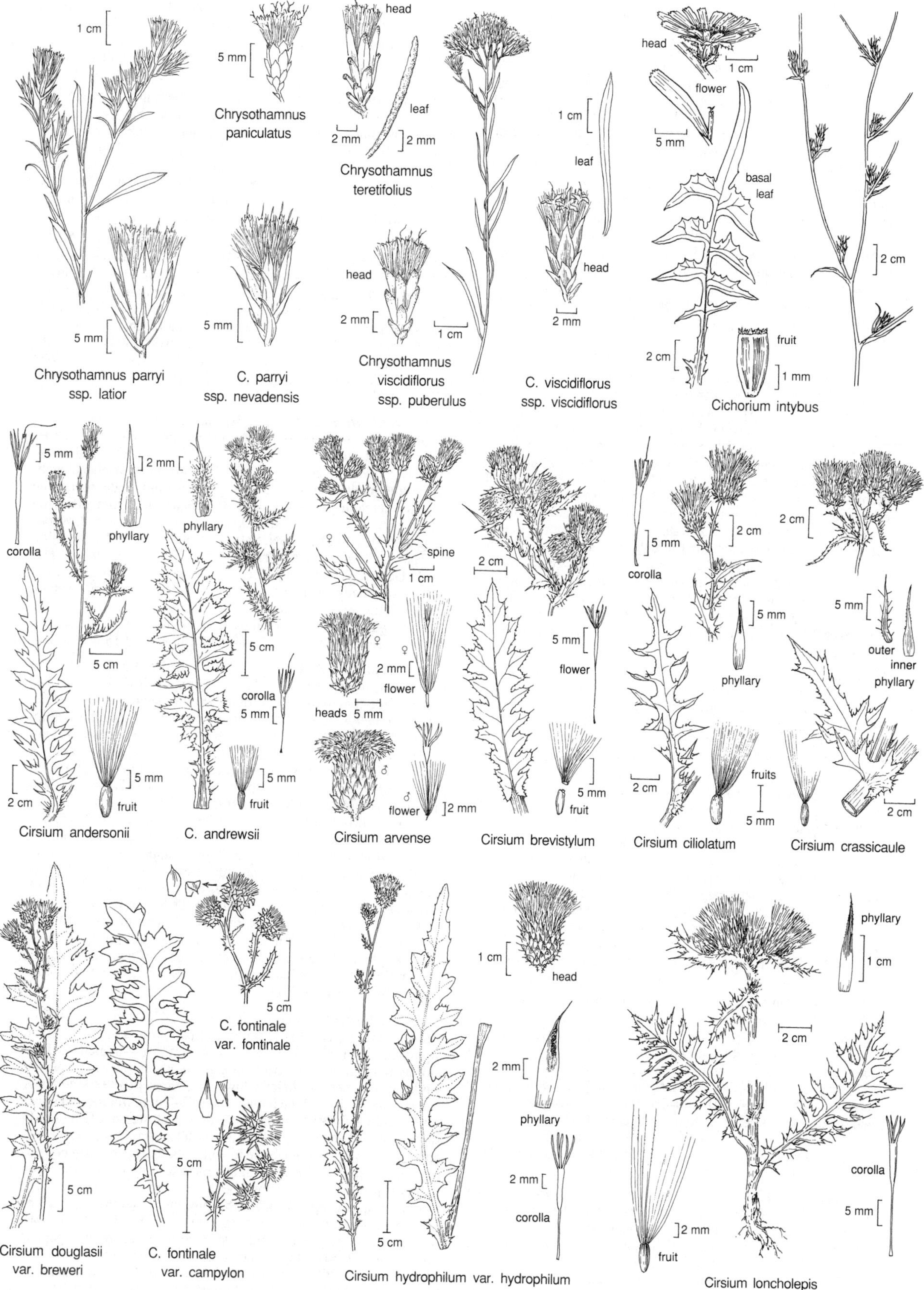

1 cm
5 mm
Chrysothamnus parryi ssp. latior
5 mm
Chrysothamnus paniculatus
5 mm
C. parryi ssp. nevadensis
head
leaf
2 mm
2 mm
Chrysothamnus teretifolius
head
2 mm
1 cm
Chrysothamnus viscidiflorus ssp. puberulus
1 cm
leaf
head
2 mm
C. viscidiflorus ssp. viscidiflorus
head
1 cm
flower
5 mm
basal leaf
2 cm
fruit
1 mm
2 cm
Cichorium intybus
5 mm
corolla
2 mm
phyllary
5 cm
2 cm
fruit
5 mm
Cirsium andersonii
phyllary
5 cm
corolla
5 mm
5 mm
fruit
C. andrewsii
♀
spine
1 cm
♀
♀
2 mm
flower
heads 5 mm
♂
♂
flower
2 mm
Cirsium arvense
2 cm
5 mm
flower
5 mm
fruit
Cirsium brevistylum
5 mm
corolla
2 cm
5 mm
phyllary
2 cm
fruits
5 mm
Cirsium ciliolatum
2 cm
5 mm
outer
inner
phyllary
2 cm
Cirsium crassicaule
5 cm
Cirsium douglasii var. breweri
5 cm
C. fontinale var. fontinale
5 cm
C. fontinale var. campylon
1 cm
head
2 mm
phyllary
2 mm
corolla
5 cm
Cirsium hydrophilum var. hydrophilum
phyllary
1 cm
2 cm
2 mm
fruit
corolla
5 mm
Cirsium loncholepis

C. neomexicanum A. Gray DESERT THISTLE Bien (or short-lived per) 4–29 dm. **STS** gen 1, gen simple below; branches few above, ascending, ± white-cobwebby-tomentose, puberulent. **LVS** ± persistently gray-tomentose (both surfaces), lighter below; lower 6–35 cm, petioled or tapered to spiny-winged base, oblong-elliptic to oblanceolate, ± lobed, lobes gen rigidly spreading, simple or with 2–4 coarse teeth or 2° lobes, main spines 5–15 mm; middle and upper gen smaller, narrower, decurrent as spiny wings, uppermost well separated, much reduced, ± bract-like, sometimes barely a cluster of long spines. **INFL**: heads 1–few in open cyme-like clusters (sometimes on short axillary branches); peduncles 2.5–30 cm, leafy; involucres 2–2.5 cm, 2.5–5 cm diam, hemispheric or bell-shaped, ± loosely tomentose, sometimes glabrous; phyllaries linear-lanceolate, entire, sometimes midveins with glandular area, outer and middle spreading to reflexed, spines 4–15 mm, inner with tips ± erect, flat. **FLS**: corollas 18–27 mm, white to pale lavender or pink, tube 8–14 mm, throat 4–7 mm, lobes 5–9 mm. **FR** 5–6 mm, dark brown, ± flattened; pappus 15–20 mm. $2n$=30. Canyons, slopes, roadsides; 800–2100 m. e DMoj; to Colorado, NM. Closely related to *C. occidentale*.

C. nidulum (M.E. Jones) Petrak Per 2–10 dm; taproot woody. **STS** gen 1, few-branched above, thinly tomentose, sometimes ± glabrous. **LVS** ± tomentose (esp below), becoming ± glabrous; lower 1–3 dm, tapered to spiny-winged petioles, oblong-elliptic, long-acute, divided ± to midvein, lobes with 2–4 narrow 2° lobes or coarse teeth, main spines 10–30 mm; middle and upper sessile, not strongly reduced, clasping or short-decurrent, exceedingly spiny. **INFL**: heads gen few, short-peduncled, ± closely subtended by uppermost lvs; involucres ± 3 cm, 1.5–2 cm diam when fresh, cylindric or narrowly ovoid, sparsely tomentose or becoming glabrous; phyllary bodies lanceolate to ± ovate, entire, tip spreading to erect, spines 20–25 mm, inner with tips straight, flat or short-spiny, often red or purple, puberulent. **FLS** exserted; corollas 28–30 mm, ± rose-purple, tube 11–14 mm, throat 4–6 mm, lobes 12–13 mm. **FR** 5–6 mm, ± flattened, shiny brown; pappus 18–20 mm. Uncommon. Pinyon/juniper woodland; 1500–2300 m. e DMtns (Clark, New York Mtns). The most spiny thistle in CA. Related to *C. arizonicum*.

C. occidentale (Nutt.) Jepson Bien 1–30 dm, erect or low, mound-like. **STS** gen 1, branched above (near base in dwarf pls), ± tomentose. **LVS** ± densely gray- or whitish tomentose, esp below; lower 1–4 dm, petioles spiny-winged, blade oblanceolate, lobed 1/2+ to midvein, lobes widely triangular, dentate or further lobed, main spines 1–10 mm; upper gradually reduced, sessile, ± clasping or short-decurrent, linear or oblong, often entire, often spinier than lower, uppermost bract-like. **INFL**: heads 1–several in loose to tight cluster (barely raised above rosette in dwarf pls); peduncles 1–30 cm; involucres 1.5–5 cm, 1.5–8 cm diam, ovoid to spheric; phyllaries ± equal to strongly graduated, linear or linear-lanceolate, straight, ascending and appressed to widely radiating, often connected side-to-side by conspicuous cobwebby hairs, spines 3–10+ mm, inner with tips flat, straight. **FLS**: corollas 18–40 mm, white to purple or red, tube 8–18 mm, throat 5–7 mm, lobes 5–10 mm. **FR** 5–6 mm, shiny, ± brown; pappus 15–30 mm. $2n$=30. Many habitats; < 3600 m. CA-FP, MP, W&I, w DMoj; to s OR, sw ID, w NV. Variable; ± distinctive, intergrading races often treated as sp.

var. ***californicum*** (A. Gray) Keil & C. Turner CALIFORNIA THISTLE Pls gen 5–20 dm, erect. **LVS** green to gray above, gray below. **INFL**: heads short- to long-peduncled in ± open clusters, well elevated above lower lvs; involucre 1.5–5 cm diam, subglabrous to densely cobwebby; middle phyllary tips gen < 1 cm, sometimes much longer, 1–3 mm wide, appressed to loosely spreading or ascending, sometimes twisted. **FLS**: corollas 18–35 mm, white to purple or rose (rarely red). $2n$=28,30. Often disturbed places, woodland, open forest; < 2300 m. SN, c&s SCoR, SW. [*C. c.* A. Gray; *C. c.* var. *bernardinum* (E. Greene) Petrak] Intergrading with var. *venustum* in s SN & s SCoR; pls intermediate to var. *candidissimum* in n SN.

var. ***candidissimum*** (E. Greene) J.F. Macbr. (p. 243) SNOWY THISTLE Pls gen 4–20 dm, erect or ± bushy. **LVS** densely white-tomentose (both surfaces). **INFL**: heads gen long-peduncled, sometimes in tight clusters at ends of peduncles, well elevated above lower lvs or branches spreading; involucre gen 2–6 cm diam, densely white-tomentose; middle phyllary tips gen 1.5–3 cm, gen 2–3 mm wide, rigidly spreading or reflexed. **FLS**: corollas 26–40 mm, gen red (rarely white or pink). $2n$=30–33,60. Disturbed places, shrubby areas, open woodland; < 1900 m. KR, NCoR, CaR, n SN, MP; to s OR, sw ID, w NV. [*C. pastoris* J. Howell] Intergrades with var. *venustum* in s KR, NCoR, c SNH, with var. *occidentale* in s NCo. ❀SUN:2,15,16&IRR:**3**,**7**,8–10,**14**,**18–21**

var. ***compactum*** Hoover COMPACT COBWEBBY THISTLE Pls gen 0.5–3 dm, low, mound-like. **INFL**: heads short-peduncled, closely subtended by basal and large cauline lvs; involucre 5–8 cm diam, densely cobwebby; middle phyllary tips gen 1–2 cm, gen 1–2 mm wide, straight, ± spreading. **FLS**: corollas 25–30 mm, dark rose-purple. $2n$=30. RARE. Bluffs; < 50 m. CCo (n San Luis Obispo, Monterey cos., formerly San Francisco). Some inland pls suggest weak separation from var. *occidentale*. In cult.

var. ***occidentale*** (p. 243) COBWEBBY THISTLE Pls gen 3–15+ dm, erect. **INFL**: heads gen long-peduncled, sometimes in tight clusters at ends of peduncles, well elevated above lower lvs; involucre gen 4–5 cm diam, ± densely cobwebby; middle phyllary tips gen 1–2 cm, gen 1–2 mm wide, straight or curved upward, ascending to spreading. **FLS**: corollas gen 25–35 mm, gen ± purple. $2n$= 30. Common. Grassland, coastal dunes, oak woodland, scrub, often in disturbed places; gen < 200 m. CW, SW. [*C. coulteri* Harvey & A. Gray] Intergrades with var. *venustum* in SnFrB. ❀DRN,SUN:5,14,**15–17**,21–24.

var. ***venustum*** (E. Greene) Jepson (p. 243) VENUS THISTLE Pls gen 5–30 dm, erect. **INFL**: heads gen long-peduncled, sometimes in tight clusters at ends of peduncles, well elevated above lower lvs; involucre 2–6 cm diam, subglabrous to densely cobwebby; middle phyllary tips 5–20+ mm, gen 2–3 mm wide, ascending to rigidly spreading or reflexed. **FLS**: corollas 23–35 mm, gen ± red (white, pink, purple). $2n$=30. Disturbed places, grassland, woodland; < 3600 m. NCoR, s SN, SnFrB, SCoR, WTR, W&I, w DMoj. [*C. coulteri* Harvey & A. Gray misapplied; *C. proteanum* J. Howell] Gen more inland than var. *occidentale* but some pls from SnFrB not readily separable. Intergrades with var. *candidissimum* in KR, NCoR, & SNH, with var. *californicum* in s&c SN & SCoR. Pls with pale corolla sometimes difficult to separate from var. *californicum*. ❀DRN,SUN:**7**,**14–16**,17&IRR:1,2,**3**,**18–21**,22–24.

C. ochrocentrum A. Gray (p. 243) YELLOWSPINE THISTLE Per 2.5–10 dm; roots creeping. **STS** gen simple below, few-branched above, white-tomentose. **LVS** thinly gray-tomentose above, white-tomentose below; lower 10–25 cm, tapered to spiny petioles, elliptic to oblanceolate, deeply lobed, lobes gen rigidly spreading, simple or with 2–4 ± narrow 2° lobes or coarse teeth, main spines 3–10 mm; middle and upper gradually reduced, decurrent as spiny-margined wings, gen very spiny, spines 5–15 mm. **INFL**: heads 1–few in cymes; peduncles 0–10 cm, leafy; involucres 2.5–3.5 cm, 2–3.5 cm diam, ± ovoid to bell-shaped, ± loosely tomentose, becoming glabrous; phyllaries strongly graduated (outer ovate, inner oblong), minutely roughened or toothed, tightly appressed, midribs of middle phyllaries often with glandular area, spines 5–12 mm, stout, spreading to reflexed, inner with tips erect or recurved, ± twisted, flat, sometimes ± expanded and fringed. **FLS**: corollas 29–37 mm, white to pale lavender or pink, tube 14–17 mm, throat 7–10 mm, lobes 8–11 mm. **FR** 7–8 mm, light brown, ± thick; pappus 25–30 mm. $2n$=15,16,17. Disturbed places, fields; < 1700 m. n NCo, KR, e SN, n ChI, TR, PR, MP; native to c US. Hybridizes with *C. undulatum*. NOXIOUS WEED.

C. quercetorum (A. Gray) Jepson (p. 243) BROWNIE THISTLE Per (appearing bien) 0.5–7(9) dm, compact, low, sometimes erect, taprooted from rootstock. **STS** 1–several, few-branched. **LVS** loosely tomentose above, becoming green, ± glabrous, gen ± persistently gray-tomentose below; lower < 35 cm, petioles smooth or spiny, blades ± 1–3 × lobed, main spines 2–7 mm; middle and upper well developed. **INFL**: heads 1–several, ± clustered, often closely subtended by upper lvs; peduncles 0–3(12) cm; involucres 2.5–5 cm, 2–6 cm diam, hemispheric to widely bell-shaped, soon becoming glabrous; phyllaries strongly graduated in 5–10 series, gen tightly appressed, outer and middle ovate to lanceolate, gen rounded or obtuse, tipped by spines 1–2(5) mm, inner tips thin, flat. **FLS**: corollas 25–30 mm, white to purple, tube 10–15 mm, throat 7–10 mm, lobes 5–8 mm. **FR** 5–6.5 mm, ± compressed, brown; pappus 20–40 mm. $2n$=16. Open places, grassland, woodland; < 350 m. c&s NCo, NCoRO, n&c CCo, SnFrB, n SCoRO. [var. *walkerianum* (Petrak) J. Howell; *C. w.* Petrak; var. *xerolepis* Petrak]

C. remotifolium (Hook.) DC. (p. 243) Bien or per 3–15 dm, ± tomentose (at least when young), taprooted. **STS** 1–several, often much-branched. **LVS** ± persistently tomentose below, gen becoming ± glabrous above; lower 1.5–5 dm, petioles spiny, coarsely dentate or ± 1–2 × lobed and dentate, main spines 1–5 mm; cauline gradually reduced, well separated or crowded, often ± clasping, uppermost bract-like. **INFL**: heads 1–several, ± clustered; peduncles 0–12 cm; involucres (1)2–3 cm, (1)1.5–4.5 cm diam, hemispheric or bell-shaped, ± persistently tomentose or becoming glabrous, sometimes cobwebby; phyllaries linear to widely oblong, outer entire or spiny-ciliate, with spine tips 1–3 mm, middle and inner tips flat, tapered, entire, or spiny-ciliate and sometimes widened into spiny-fringed or irregularly toothed appendages. **FLS**: corollas (16)22–25 mm, white to purple, tube (6)8–12 mm, throat (6)8–10 mm, lobes 4–6 mm. **FR** 5–7 mm, compressed, dark; pappus 15–21 mm. $2n$=34. Grassy areas, openings in woodlands, forests, sometimes on serpentine; < 2000 m. NW, n CCo, SnFrB; to WA, s WY, Colorado. [var. *odontolepis* Petrak; *C. acanthodontum* S.F. Blake; *C. amblylepis* Petrak; *C. callilepis* (E. Greene) Jepson, *C. c.* var. *pseudocarlinioides* (Petrak) J. Howell]

C. rhothophilum S.F. Blake (p. 243) SURF THISTLE Bien (or short-lived per) 1–10 dm, ± fleshy, bush-like or low, mounded, gray-tomentose throughout; hairs appressed, felt-like. **STS** 1–several. **LVS** gen very strongly wavy; lower 10–25 cm, wing-petioled, entire or ± widely lobed, lobes entire or few-toothed, main spines 1–4 mm; cauline gradually reduced, well-distributed, upper sessile, clasping, ± lobed, often spinier than lower, spines < 8 mm. **INFL**: heads 1–several, ± clustered, closely subtended by uppermost lvs; peduncles 0–7 cm; involucres 3–4 cm, 4–6 cm diam, hemispheric or bell-shaped; phyllaries linear-lanceolate, persistently tomentose, outer and middle spiny-ciliate, spine tips 2–5 mm. **FLS**: corollas 20–25 mm, white to pale yellow, tube 11–13 mm, throat 5–6 mm, lobes 5–8 mm; anthers brown. **FR** 5–7 mm, light-brown to black, thick or ± compressed; pappus 15–20 mm. $2n$=34. RARE. Dunes, bluffs; < 20 m. s CCo (s San Luis Obispo and n Santa Barbara cos.).

C. scariosum Nutt. (p. 243) ELK THISTLE Bien (or short-lived per) 0.5–10+ dm. **STS** often 0 or very short, sometimes erect, often ± fleshy, ridged, glabrous to loosely tomentose, sometimes coarse-hairy. **LVS** often basal, glabrous to loosely tomentose above, glabrous to densely tomentose below; lower 1–4 dm, tapered or spiny-petioled, oblong to oblanceolate, subentire to deeply lobed, sometimes with 2° lobes or teeth, main spines gen 2–7 mm; cauline 0 or well distributed, ± petioled. **INFL**: heads gen ± sessile in basal lf rosette or in clusters at st tips, gen closely subtended by lvs; involucres 2–4.5 cm, 2–5 cm diam, ovoid to bell-shaped, glabrous; phyllaries linear-lanceolate to ovate, entire to minutely toothed, outer tipped by ascending spines 2–5 mm, inner with tips entire or toothed, flat or crinkled. **FLS**: corollas 23–32 mm, ± white to purple, tube 10–16 mm, throat 6–10 mm, lobes 5–6 mm. **FR** 3–5 mm, oblong-obovate, ± compressed; pappus 15–30 mm. $2n$=34. Moist places, meadows; (400)700–3400 m. KR, CaR, SN, SCoR (very uncommon), TR, PR, GB; to B.C., MT, Colorado, NM, Baja CA. [*C. drummondii* Torrey & A. Gray & *C. foliosum* (Hook.) DC., misapplied; *C. tioganum* Congdon] Extremely variable, needing study; dwarf and tall pls sometimes occur together. Pls from SNE & SnBr, with sts 0, heads small, corollas purple, have been called *C. congdonii* R. Moore & Frankton.

C. undulatum (Nutt.) Spreng. (p. 243) WAVYLEAF THISTLE Per 2.5–10 dm, from creeping (sometimes tuber-like) roots. **STS** gen simple below, few-branched above, white-tomentose. **LVS** gray-tomentose above, white- below; lower 15–30 cm, tapered to spiny petioles, elliptic to oblanceolate, ± shallowly lobed, lobes simple or with 2–4 narrow to wide 2° lobes or coarse teeth, main spines 2–4 mm, yellow; middle and upper gradually reduced, decurrent as spiny-margined wings, often spinier than lower, spines 2–10 mm. **INFL**: heads 1–few in cymes; peduncles 0–15 cm, ± leafy; involucres 2–5 cm, 2–5 cm diam, ± ovoid to bell-shaped, ± loosely tomentose on phyllary margins, becoming glabrous; phyllaries strongly graduated (outer ovate, inner lance-oblong), entire or minutely roughened, tightly appressed, midribs of outer and middle glandular area, spines 2–5 mm, spreading to reflexed, inner with tips flat, erect or recurved, ± twisted. **FLS**: corollas 25–50 mm, white to pale lavender or pink, tube 11–25 mm, throat 7–11 mm, lobes 7–15 mm. **FR** 4–7 mm, light brown, ± compressed; pappus 25–40 mm. $2n$=24,26. Disturbed places; < 1600 m. s CaR, SnFrB, s SCoRO, SCo, MP; native to c US. Hybridizes with *C. ochrocentrum*. NOXIOUS WEED.

C. vulgare (Savi) Ten. (p. 243) BULL THISTLE Bien 3–20 dm. **STS** gen 1, ± openly branched above middle, loosely tomentose, often glandular-hairy. **LVS** harshly bristly above, sometimes ± tomentose when young, ± tomentose below; main veins prominently raised on lower surface, ± glandular; lower 10–40 cm, sessile or wing-petioled, shallowly to deeply 1–2 × lobed; cauline gradually reduced, long-decurrent as spiny wings, gen spinier than lower, main lobes gen rigidly spreading, spine-margined, otherwise entire, tip prolonged, main spines < 15 mm. **INFL**: heads 1–several, ± clustered, closely subtended by bract-like uppermost lvs; peduncles 1–6 cm; involucres 3–4 cm, 2–4 cm diam, hemispheric or bell-shaped; phyllaries graduated in 5–10 series, tips linear to linear-lanceolate, spreading to reflexed, spines 1–5 mm. **FLS**: corollas 25–35 mm, purple, tube 18–25 mm, throat 5–6 mm, lobes 5–6 mm. **FR** 3.5–4.5 mm, light-brown or tan; pappus 20–30 mm. $2n$=68. Common. Disturbed areas; < 2300 m. CA-FP, GB; N.Am; native to Eur.

CNICUS BLESSED THISTLE

David J. Keil & Charles E. Turner

1 sp.: Eur. (Latin: from Greek name for safflower)

C. benedictus L. Ann < 6 dm. **STS** gen branched throughout, gen reddish, ± loosely tomentose. **LVS** ± cauline, alternate, sessile, short decurrent or tapered to winged petiole, 6–25 cm, (ob)lanceolate, spine-toothed or pinnately lobed, strongly veined, sparsely hairy, gland-dotted. **INFL**: heads disciform, sessile among lf-like bracts; involucre 2–4 cm, ± spheric; outer phyllaries ovate, base appressed, spine tips spreading, inner phyllaries lanceolate, spine tips pinnately divided; receptacle bristly. **STERILE FLS** few, marginal; corollas very slender, lobes 3, linear. **DISK FLS** many; corollas ± 2 cm, yellow, lobes linear; anther bases sagittate, tips long-appendaged; style tip minutely hair-ringed just below rounded lobes. **FR** ± 8 mm, cylindric, ribbed, attached laterally, glabrous; top with 10-toothed rim; pappus 2 series of awns, outer ± 10 mm, smooth or rough, inner 2–5 mm, rough from short, spreading hairs. $2n$=22. Roadsides, fields, waste places; < 800 m. NCoR, GV, CW, SCo, w DMoj; native to Eur. Scattered.

CONYZA

Ann or per. **STS** gen erect, leafy. **LVS** alternate, linear to ± (ob)lanceolate, entire to pinnately dissected, obtuse to acute. **INFL**: heads gen disciform (or minutely radiate), gen many in raceme- or panicle-like clusters; phyllaries in 2–3 equal or unequal series, free, linear to lanceolate, often narrowly scarious-margined, reflexed in age; receptacle naked. **PISTILLATE FLS** gen many; corollas white, pink, or cream-colored, narrowly cylindric. **DISK FLS** ± few; corollas gen yellow, lobes short-triangular; style tips lanceolate, incl or short-exserted. **FR** elliptic, compressed, puberulent; pappus of bristles. ± 50 spp.: esp trop. (Greek: flea; name of Pliny & Dioscorides for a fleabane)

1. Midvein of phyllary brown, resin-filled
 2. Heads disciform (pistillate fls liguleless) .... ***C. bilboana***

2′ Heads radiate; ligules present (gen < 1 mm) .... ***C. canadensis***
1′ Midvein of phyllary green to purple, not resin-filled
3. Lvs clasping, regularly dentate or spreading-lobed; herbage densely glandular; pistillate fl styles long-exserted .... ***C. coulteri***
3′ Lvs not clasping, gen entire (or lower irregularly serrate to ascending-lobed); herbage gray-hairy, not or barely glandular; pistillate fl styles ± = corollas
4. Central st gen overtopped by lateral branches; pressed heads gen ± 1 cm wide; dry, reflexed phyllaries whitish or dull brown inside .... ***C. bonariensis***
4′ Central st > lateral branches (unless pl damaged); pressed heads gen < 1 cm wide; dry, reflexed phyllaries reddish brown inside .... ***C. floribunda***

C. bilboana E.J. Remy (p. 243) Ann 30–100 cm. **ST** gen simple below, much-branched above, coarsely hairy below, strigose above. **LVS** 1–8 cm, entire or shallowly dentate, glabrous to scabrous or strigose. **INFL**: peduncles 0–5 mm; fresh involucres 3–4 mm diam; phyllaries 1–3.5 mm, glabrous, midvein brown, resin-filled. **PISTILLATE FLS** 20–30; corollas ± 2 mm; ligules ± 0; style incl. **DISK FLS** 5–8; corollas 2–2.5 mm. **FR** 1–1.2 mm; pistillate fr straw-colored; disk fr red-margined; pappus 2–2.5 mm. Disturbed, urban sites; < 100 m. SnFrB; S.Am. Perhaps = *C. canadensis*.

C. bonariensis (L.) Cronq. (p. 243) Ann 20–120 cm, ± gray-hairy. **STS** 1–several from base, strigose and long-soft-hairy. **LVS** 1–9 cm, entire to shallowly lobed. **INFL**: lateral clusters often overtopping central; peduncles gen 1–4 cm; fresh involucres 5–7 mm diam; phyllaries 2–6 mm, often purple-tipped, densely soft-hairy, whitish or dull brown inside when dry and reflexed, midvein green to purple, not resin-filled. **PISTILLATE FLS** very many; corollas 3–4 mm; ligule ± 0. **DISK FLS** 10–20; corollas 3.5–4 mm, greenish yellow. **FR** ± 1.5 mm; pappus 3–4 mm, gen ± reddish in age. $2n$=54. Disturbed, gen urban sites; gen < 1000 m. CW, SW; to e N.Am; native to S.Am.

C. canadensis (L.) Cronq. (p. 243) HORSEWEED Ann 1–20(30) dm. **ST** gen simple below, much-branched above, strigose or stiff-spreading-hairy. **LVS** 1–10 cm, entire to shallowly few-lobed, glabrous to strigose, often ciliate. **INFL**: heads obscurely radiate; lateral clusters not overtopping central; fresh involucre gen 2.5–4 cm diam; phyllaries 1–3.5 mm, glabrous to strigose, midvein brown, resin-filled. **PISTILLATE FLS** 20–40; corollas 2.5–3 mm, ligule < 1 mm. **DISK FLS** 7–13; corollas 2.5–3 mm. **FR** ± 1.5 mm; pappus 2.5–3 mm, gen whitish in age. $2n$=18,36,54. Common. Waste ground; gen < 2000 m. CA; ± worldwide.

C. coulteri A. Gray (p. 243) Ann 2–10(20+) dm, densely soft-hairy and glandular. **ST** gen simple below, much-branched above. **LVS** 2–8 cm, sessile, ± clasping, gen coarsely dentate or shallowly spreading-lobed. **INFL**: lateral clusters gen not overtopping central; peduncles 3–10 mm; fresh involucres 4–5 mm diam; phyllaries 2–4 mm, densely glandular, midvein ± green, not resin-filled. **PISTILLATE FLS** many; corollas 2 mm, very slender, cream-colored; style long-exserted, branches unequal. **DISK FLS** 5–15; corollas 3–3.5 mm. **FR** 1 mm; pappus 3–4 mm, gen white in age. $2n$=18. Disturbed places; gen < 1000 m. s SNF, SnJV, CW, SW, D; to Colorado, TX, Mex.

C. floribunda Kunth Ann 1–20 dm. **ST** gen 1, ± bristly and soft-hairy. **LVS** 1–7 cm, linear, gen entire (or lower serrate to lobed), densely strigose or canescent. **INFL**: heads disciform or minutely radiate; lateral clusters gen not overtopping central; peduncles gen 5–20 mm; fresh involucres 5–7 mm diam; phyllaries 3–4.5 mm, finely canescent, tip green, midvein ± green, not resin-filled. **PISTILLATE FLS** many; ligules ± 0. **DISK FLS** ± 10; corollas 3.5–4 mm. **FR** ± 1 mm; pappus 3–4 mm, gen ± yellowish in age. Disturbed places; < 100 m. NW, ScV, SW; widely naturalized; native to trop Am. Sometimes considered to = *C. bonariensis*.

COREOPSIS TICKSEED

Ann, per, shrubs. **STS** slender to stout and fleshy. **LVS** simple to several × pinnately dissected, basal or cauline, opposite or less commonly alternate, sessile or petioled. **INFL**: heads radiate, solitary or in few–many-headed cymes; peduncles short to long; involucre hemispheric or bell-shaped; phyllaries in 2 series, outer ± spreading, thick, green, inner thin, membranous; receptacle flat to rounded, chaffy; scales flat, scarious. **RAY FLS** fertile or sterile; ligules gen yellow, showy. **DISK FLS** many; corollas 4–5-lobed, yellow; style tips truncate to long-tapered. **FR**: ray and disk achenes alike or different, gen compressed front-to-back, often winged; pappus 0 or of 2 awns or scales. ± 114 spp.: Am, Afr. (Greek: bedbug-like, from fr) [Smith 1984 Sida 10:276–289]

1. Lvs opposite; ray fls sterile
2. Per; ligules yellow throughout; st nodes gen 2–5 .... ***C. lanceolata***
2′ Ann; ligules gen with reddish brown spot at base; st nodes gen > 5
3. Disk corollas 4-lobed; outer phyllaries gen 2–3 mm, << inner phyllaries .... ***C. tinctoria***
3′ Disk corollas 5-lobed; outer phyllaries 5–8 mm, ± = inner phyllaries .... ***C. wrightii***
1′ Lvs alternate or all basal; ray fls fertile
4. Per; sts stout, 3–20 dm; ligules oblong to oblanceolate (much longer than wide); heads (incl ligules) gen 4–10 cm diam
5. Pl with an erect fleshy trunk to ± 2 m; sts solid; heads several–many per cyme; peduncles 6–20 cm .. ***C. gigantea***
5′ Pl without trunk; sts hollow; heads gen 2–4 per cyme; peduncles 15–50 cm .... ***C. maritima***
4′ Ann; sts slender, gen 1–4 dm; ligules ovate to obovate; heads 1–7 cm diam
6. Disk achenes ciliate, pappus scales 2; ray achenes glabrous, pappus scales gen 2
7. Outer phyllaries ovate; pappus scales 2.5–5 mm .... ***C. calliopsidea***
7′ Outer phyllaries linear-oblong; pappus scales < or = 3 mm
8. Pappus scales 1.7–2.8 mm; chaff scales and disk achenes fused at base, falling together; ligules spreading horizontally in fully opened heads .... ***C. bigelovii***
8′ Pappus scales 0.9–1.3 mm; chaff scales free from disk achenes falling separately; ligules reflexed in fully opened heads .... ***C. hamiltonii***
6′ Disk and ray achenes similar, never ciliate; pappus 0
9. Lvs flat, 1–2 × pinnately divided into oblanceolate lobes 1–3 mm wide; outer phyllary bases with gland-tipped teeth .... ***C. stillmanii***
9′ Lvs or their lobes linear, thread-like, 0.3–1.3 mm wide; phyllary teeth 0

10. Fr tan to dull brown or reddish, gen blotched with darker spots at least when young, puberulent, wings thick, corky .... *C. californica* var. *californica*

10′ Fr dark brown, shiny, ± glabrous, wings thin .... *C. douglasii*

C. bigelovii (A. Gray) H.M. Hall (p. 243) Ann gen 1–3 dm, glabrous. **STS** 1–many, erect. **LVS** basal (or few cauline, alternate); petiole 1–5 mm; blade 2–8 cm, 1–2-pinnately divided into linear segments 1–2 mm wide, grooved above. **INFL**: heads solitary, ± scapose; involucre cylindric, base truncate; outer phyllaries 4–7, 5–12 mm, linear; inner phyllaries 6–8, 6–10 mm, ovate, acute, margin scarious; chaff scales gen 5–8 mm, lanceolate to oblanceolate, fused to base of disk achenes. **RAY FLS** 5–10, fertile; ligules 5–25 mm, obovate, spreading, yellow. **DISK FLS** 20–50; corollas ± 4 mm, yellow. **FR**: ray achenes 3–5 mm, oblong to obovate, brown or splotched with tan, rough, glabrous, wing narrow, pappus 0; disk achenes 4–6 mm, oblong to oblanceolate, dark brown or splotched with tan, shiny, outer face glabrous, inner face with central row of hairs, margins ciliate, hairs 1–1.5 mm; pappus scales 1.7–2.8 mm, lanceolate. 2*n*=24. Open woodlands, grasslands, deserts; 150–1500 m. SCoRI, TR, Teh, s SNF, DMoj, n DSon. ❀DRN,SUN:**8,9**,10,11,**12,13,18–24**&DRY:7,14–16.

C. californica (Nutt.) H. Sharsm. var. ***californica*** (p. 243) Ann gen 5–30 cm, glabrous. **STS** 1–many, erect. **LVS** all basal or few cauline, gen erect, 2–10 cm, 0.5 mm wide, linear, thread-like, entire or lobes 1–2, short, linear, ± cylindric, tip obtuse, red. **INFL**: heads solitary, ± scapose; involucre widely cylindric, base rounded; outer phyllaries 2–7, 4–7 mm, narrowly lanceolate, hairs at base yellow or red, glandular, tip red; inner phyllaries 5–8, 6–10 mm, widely lanceolate, acute, margin narrowly scarious; chaff scales 4–5.5 mm, linear to oblanceolate, free from disk achenes. **RAY FLS** 5–12, fertile; ligules 5–15 mm, obovate, yellow. **DISK FLS** 10–30; corollas 3.5–5 mm, yellow. **FRS** alike, 2.5–4.3 mm, obovate, rusty-tan to light brown, often red- or black-spotted near margin, puberulent (hairs club-shaped), wing irregularly thickened; pappus 0 (or scales 2). 2*n*=24. Desert plains, washes; 30–600 m. s SnJV, s SCoRI, TR, D; Baja CA. ❀TRY.

C. calliopsidea (DC.) A. Gray (p. 243) Ann gen 1–4(6+) dm, glabrous. **STS** 1–many, erect, simple or few-branched. **LVS** basal and alternate; petiole 1–5 mm; blade 1–5 cm, 1–2-pinnately divided into linear segments 0.5–2 mm wide, grooved above; upper lvs sometimes simple. **INFL**: heads solitary; involucre bell-shaped; outer phyllaries 4–6, 3–8 mm, triangular-ovate, fused at base; inner phyllaries gen 8, 8–10 mm, ovate, acute, margin narrowly scarious; chaff scales 6–7 mm, lanceolate to oblanceolate, fused to base of disk achene. **RAY FLS** gen 8, fertile; ligules 10–35 mm, obovate, yellow. **DISK FLS** 15–50; corollas ± 5 mm, yellow. **FR**: ray achenes 5–6 mm, ovate, tan or brown, glabrous, wing smooth, flat, pappus 0; disk achenes 6–7 mm, linear to oblanceolate, outer face dark brown, shiny, glabrous, inner face covered with white hairs, margin ciliate, hairs 2–3 mm; pappus scales 2.5–5 mm, lanceolate. 2*n*=24. Deserts, dry grassy areas; 200–1100 m. SCoRI, TR, w DMoj. ❀TRY.

C. douglasii (DC.) H.M. Hall (p. 243) Ann gen 5–25 cm, glabrous, glaucous. **STS** 1–few, erect. **LVS** basal and alternate on lower st, 2–8 cm, 1 mm wide, linear, entire (or lobes 1–2, short, linear), ± fleshy, grooved above. **INFL**: heads solitary, ± scapose; involucre widely cylindric, base rounded; outer phyllaries 2–7, 4–7 mm, narrowly lanceolate; inner phyllaries 5–8, 6–10 mm, obovate, acute; chaff scales 4–5 mm, linear, free from disk achenes. **RAY FLS** 5–8, fertile; ligules 5–8 mm, ovate, yellow. **DISK FLS** 10–30; corollas 4–4.5 mm, yellow. **FRS** alike, 2.5–5 mm, obovate, dark brown, shiny, glabrous; wing ± yellow; pappus 0. 2*n*=24. Dry, rocky slopes; 150–600 m. SCoRI, n WTR. ❀TRY.

C. gigantea (Kellogg) H.M. Hall (p. 247) GIANT COREOPSIS Shrub, glabrous. **ST**: trunk gen 1–2 m, 4–10 cm diam, erect, fleshy, few-branched. **LVS** alternate, tightly clustered at st tips; petiole 3–7 cm; blade 3–25 cm, 3–4-pinnate into fleshy, linear segments 1–5 cm, 0.5–1.5 mm wide. **INFL**: heads many in cymes; peduncles 6–20 cm, leafy-bracted; involucre bell-shaped; outer phyllaries 5–12, 5–20 mm, lanceolate to oblong; inner phyllaries 10–15, 10–15 mm, oblong-ovate, obtuse to acute; chaff scales 8–10 mm, linear, free from disk achenes. **RAY FLS** 10–16, fertile; ligules 20–30 mm, elliptic to oblong, yellow. **DISK FLS** many; corollas 6–6.5 mm, yellow. **FRS** alike, 5–6 mm, oblong to obovate, dark brown, ± glabrous; wing narrow, thin; pappus 0. 2*n*=24. Shrubby hillsides, coastal dunes, seabluffs; 15–60 m. s CCo, n&c SCo, ChI. ❀DRN, SUN,DRY:14,**15–17**,19,**20–24**;summer dormant.

C. hamiltonii (Elmer) H. Sharsm. (p. 247) MOUNT HAMILTON COREOPSIS Ann 5–25 cm, glabrous. **STS** 1–many, erect. **LVS** basal (or few cauline, alternate); petiole 1–3 cm; blade 5–20 mm, 1–2-pinnately divided into linear segments 1 mm wide, grooved above. **INFL**: heads solitary, ± scapose; involucre cylindric, base truncate; outer phyllaries 4–7, 3–6 mm, linear; inner phyllaries 6–8, 5–8 mm, ovate, acute, margin scarious; chaff scales gen 5–6 mm, linear, free from disk achenes. **RAY FLS** 5–8, fertile; ligules 3–8 mm, oblong to obovate, reflexed, yellow. **DISK FLS** 20–30; corollas ± 3–4.5 mm, yellow. **FR**: ray achenes 5 mm, obovate, brown or tan splotched with brown, smooth, glabrous, shiny; wing narrow; pappus 0; disk achenes 5–6 mm, obovate, tan splotched with brown, shiny, both surfaces covered with ascending hairs, margins ciliate, hairs 1–1.5 mm; pappus scales ± 1 mm, obovate. 2*n*=24. RARE. Dry exposed slopes; 600–1300 m. e SnFrB (Diablo Range). In cult.

C. lanceolata L. GARDEN COREOPSIS Per 3–6 dm from branched rootstock, glabrous to ± hairy. **STS** 1–few, erect, simple or few-branched. **LVS** basal and opposite; petiole 5–15 cm or upper lvs sessile; blade 5–15 cm, simple and oblanceolate or pinnate with 3–7 linear to oblanceolate lflets. **INFL**: heads few–many in leafy-bracted cymes; peduncle 15–30 cm; involucre hemispheric; outer phyllaries 8–10, 5–10 mm, narrowly lanceolate; inner phyllaries 8–12 mm, lanceolate to ovate, obtuse or acute, margins scarious; chaff scales 4–6 mm, lanceolate or ovate, free from disk achenes. **RAY FLS** gen 8, sterile; ligules 1.5–3 cm, oblanceolate to obovate, gen 4-lobed, yellow. **DISK FLS** many; corollas ± 4 mm, 5-lobed, yellow. **FR** 2.5–3 mm, round or depressed; surfaces black, rough; wing wide, thin; pappus scales 2, < or = 1 mm. 2*n*=26. Disturbed places; < 500 m. CCo, SCo; native to e US. Cult as orn.

C. maritima (Nutt.) Hook. f. SEA-DAHLIA Per 3–8 dm, from fleshy taproot, glabrous. **STS** few–many, stout, hollow, much-branched. **LVS** alternate, fleshy; petiole 2–15 cm; blade 2–3-pinnately divided into linear lobes 1.5–50 mm, 2–3 mm wide. **INFL**: heads 2–4 in cymes; peduncles 2–5 dm; involucre bell-shaped; outer phyllaries 6–10, 10–25 mm, oblong to obovate; inner phyllaries 10–15, 12–15 mm, widely lanceolate to ovate, acute; chaff scales 8–12 mm, linear to oblanceolate, free from disk achenes. **RAY FLS** 15–20, fertile; ligules 25–40 cm, oblong, yellow. **DISK FLS** many; corollas 6–6.5 mm, 5-lobed, yellow. **FRS** alike, 6–7 mm, oblong to obovate, dark brown, smooth or minutely rough; wing narrow, thin; pappus 0. 2*n*=24. RARE in CA. Seabluffs; < 20 m. s SCo (San Diego Co.); Baja CA. In cult.

C. stillmanii (A. Gray) S.F. Blake Ann 5–30 cm, glabrous. **STS** 1–few, erect. **LVS** basal and alternate on lower st; petiole 1–5 cm; blade 0.5–5 cm, 1–2-pinnate, lobes 1–3 mm wide, flat, oblanceolate, terminal often widest. **INFL**: heads solitary, ± scapose; involucre widely cylindric, base rounded; outer phyllaries 4–8, 3–10 mm, linear or narrowly oblanceolate, green or reddish; inner phyllaries 5–10, 5–10 mm, ovate, acute, narrowly scarious-margined; chaff scales 5–6 mm, lanceolate, free from disk achenes. **RAY FLS** 5–8, fertile; ligules 5–8 mm, ovate, yellow. **DISK FLS** 10–40; corollas 3.5–4 mm, yellow. **FRS** alike, 2.5–5 mm, obovate, flat, dark brown, smooth, not shiny, glabrous or hairs few, short; wing ± yellow; pappus 0 or awns 1–2, < or = 1 mm. 2*n*=24. Grassy slopes; 30–900 m. NCoRI, SNF, GV, n SCoRI. ❀DRN,SUN:7,**8,9**,11,**14–17,19–24**.

C. tinctoria Nutt. CALLIOPSIS Ann 3–12 dm, glabrous. **ST** gen 1, erect, much-branched above. **LVS** opposite; petiole 1–2 cm; blade 3–12 cm, lower 1–2 × divided into linear to narrowly lanceolate, entire lobes, upper simple, linear. **INFL**: heads many in open, leafy-bracted cymes; peduncles 4–10 cm; involucre bell-shaped; outer phyllaries 7–8, 1–3 mm, triangular; inner phyllaries 4.5–8 mm, ovate, obtuse or acute; chaff scales 4–5 mm, linear, red-orange, free from disk achenes. **RAY FLS** gen 8; ligules 6–20 mm, yellow (or base red-brown). **DISK FLS** many; corollas ± 3 mm, 4-lobed, reddish. **FR** 1.5–2.8 mm, linear-oblong to widely elliptic, black, smooth or minutely rough; wing 0 to wide; pappus 0. 2*n*=12,13.

Disturbed places; < 1000 m. CA-FP; native nw US to e US. [*C. atkinsoniana* Lindley] Cult as orn.

C. wrightii (A. Gray) H. Parker Ann, 2–5 dm, ± glabrous. **ST** gen 1, erect, much-branched above. **LVS** opposite; petiole 1–5 cm; blade 5–7 cm, 2–3-pinnately lobed, lobes of lower lvs linear to oblanceolate, lobes of upper lvs thread-like. **INFL**: heads many in open, leafy-bracted cymes; peduncles 5–15 cm; involucre bell-shaped; outer phyllaries 7–8, 5–8 mm, linear; inner phyllaries 6–9 mm, ovate, obtuse or acute; chaff scales 4–5 mm, linear, red-orange, free from disk achenes. **RAY FLS** gen 8; ligules 12–20 mm, gen 3–4-lobed, yellow, base red-brown. **DISK FLS** many; corollas ± 3 mm, 5-lobed, reddish. **FR** 1.5–2 mm, obovate, scarcely flat, black, rough, wing twisted to 1 side of achene; pappus 0. $2n=26$. Disturbed places; < 350 m. s SCo (San Diego Co.); native to TX, OK. [*C. basalis* (C.F.E. Otto & D. Dietr.) S.F. Blake var. *w.* (A. Gray) S.F. Blake]

COSMOS

Ann, per. **STS** gen erect. **LVS** simple to 1–3 × pinnately divided, opposite. **INFL**: heads radiate, 1–few in cymes; peduncles long; involucre hemispheric; phyllaries in 2 series, fused at base, outer gen ± lf-like in texture, inner thinner, wider; receptacle ± flat, chaffy, scales ± flat, entire. **RAY FLS** sterile; ligules yellow to white, pink, or purple. **DISK FLS** many; corollas yellow; style tips thickened, acute. **FR** ± cylindric, gen 4-angled, often beaked or much narrowed toward tip; pappus 0 or of 2–8 stiff, barbed awns. ± 26 spp.: trop, subtrop Am, esp Mex. (Greek: ornament) [Sherff & Alexander 1955 N.Am Flora 2(2):130–146]

1. Ray fls white to pink or purple; lf lobes gen < or = 1.5 mm wide .... ***C. bipinnatus***
1′ Ray fls yellow to red-orange; lf lobes gen 2–5 mm wide .... ***C. sulphureus***

C. bipinnatus Cav. GARDEN COSMOS Ann, glabrous or puberulent. **STS** gen erect, 3–20 dm. **LVS** sessile or short-petioled; blade 6–11 cm, 1–2 × divided into linear segments < or = 1.5 mm wide. **INFL**: peduncles 1–2 dm; involucre 7–15 mm diam; outer phyllaries gen 8, 9–13 mm, lanceolate; inner phyllaries ovate-lanceolate. **RAY FLS** gen ± 8; ligules 1–3 cm, white to pink or purple. **DISK FLS**: corollas 5–7 mm. **FR** 7–16 mm, black; pappus awns 0–3, 1–1.5 mm. $2n=24$. Uncommon. Disturbed places; < 1000 m. CA-FP; native to trop Am. Cult as orn.

C. sulphureus Cav. Ann, ± stiffly hairy. **STS** gen erect, 4–21 dm. **LVS** petioled; blade 5–12 cm, 2–3 × divided into linear to lanceolate segments 2–5 mm wide. **INFL**: peduncles 1–2 dm; involucre 6–10 mm diam; outer phyllaries gen 8, 5–7 mm, linear, tapered; inner phyllaries 9–13 mm, oblong-lanceolate. **RAY FLS** gen ± 8; ligules 2–3 cm, yellow to red-orange. **DISK FLS**: corollas 6–7 mm. **FR** 16–28 mm; pappus awns 0–3, 4–7 mm. $2n=24$. Uncommon. Disturbed places; < 1000 m. CA-FP; native to Mex. Cult as orn.

COTULA

Elizabeth McClintock

Ann, per, ± aromatic. **STS** low, erect or creeping, gen branched. **LVS** alternate, entire, lobed, or dissected. **INFL**: heads disciform, solitary; phyllaries in 2 or more series, narrowly scarious-margined; receptacle flat or convex, naked or roughened by persistent fl stalks. **PISTILLATE FLS** in 1–several series, borne on slender stalks; corolla 0; ovary flattened. **DISK FLS** many, very small, bisexual or functionally staminate, sessile or short-stalked; corolla tubular, 4-lobed, yellow; style and stamens incl; anther tips ovate, bases rounded or ± cordate; style branches truncate with shrub-like tips. **FR** of pistillate fls compressed front-to-back, winged; fr of disk fls unwinged or wing narrow; pappus 0. ± 80 spp.: mostly in s hemisphere. (Greek: "small cup," for sheathing lf bases of some spp.)

1. Lvs 2–3-pinnately dissected into many fine lobes, often short-petioled, not sheathing; ann; sparsely hairy; sts very slender .... ***C. australis***
1′ Lvs entire to coarsely 1-pinnately lobed, sessile, sheathing; per; glabrous, ± fleshy; sts stout ...... ***C. coronopifolia***

C. australis (Sieber) Hook. f. Ann. **STS** < 20 cm, very slender, branched from base, sparsely hairy. **LVS** 2–6 cm, 2–3-pinnately divided into linear segments, petioled or upper sessile. **INFL**: heads 3–6 mm diam; phyllaries in 2 series, margins scarious, tips brownish. **PISTILLATE FLS** in 2–3 series. **DISK FLS** few–many, sessile or subsessile; corolla ± 1 mm, pale yellow. **FR**: marginal achenes ± 1 mm, widely winged, slightly < stalks; inner achenes < 1 mm, not or scarcely winged, ± sessile. $2n=36,40$. Common. Weed, esp in urban coastal areas; < 250 m. NCo, CCo, SCo; native to Australia.

C. coronopifolia L. (p. 247) BRASS-BUTTONS Per. **STS** gen < 50 cm, ± fleshy, glabrous, decumbent, rooting from nodes. **LVS** 2–7 cm, linear to lanceolate or oblong, entire or irregularly toothed or lobed into widely linear to lanceolate divisions; base forming a fused sheath around st. **INFL**: heads 6–15 mm diam; phyllaries in 3 ± unequal series. **PISTILLATE FLS** in 1 series. **DISK FLS** many, subsessile; corollas ± 1 mm, bright yellow. **FR**: marginal achenes 1.5–2 mm, widely winged, ± = stalks; inner achenes 1–1.5 mm, not or scarcely winged, nearly sessile. $2n=20$. Common. Saline and freshwater marshes along coast; < 300 m. NCo, CCo, SCoR, SCo; native to s Afr.

CREPIS HAWKSBEARD

G. Ledyard Stebbins

Ann, bien, per from taproot; sap milky. **STS** erect, < 8 dm. **LVS** basal or cauline, entire to pinnately lobed. **INFL**: heads ligulate, clustered in cymes; phyllaries in 2 distinct series; receptacle naked. **FLS** 5–60; ligules yellow. **FR** tapered at both ends, sometimes beaked; pappus of many soft, hair-like bristles. ± 200 spp.: esp n hemisphere. (Greek: sandal, for unknown reason) Sexual forms of native spp. are distinct but (exc *C. nana, C. runcinata*) connected by many asexually reproducing forms of hybrid origin that obscure boundaries. Asexual forms are all placed in the same sp. as sexual forms, exc for the asexual group described under the name *C. intermedia*, for which no key is attempted. ❀TRY.

1. Lvs oblanceolate to obovate, entire or few toothed; dwarf alpine pls .... ***C. nana***
1′ Lvs toothed to pinnately lobed; taller pls of lowland, montane, or subalpine regions
 2. Fr tapered to a distinct slender beak; alien ann or short-lived per
  3. Beak very slender, ± 2 × fr body; per, esp of lawns .... ***C. bursifolia***

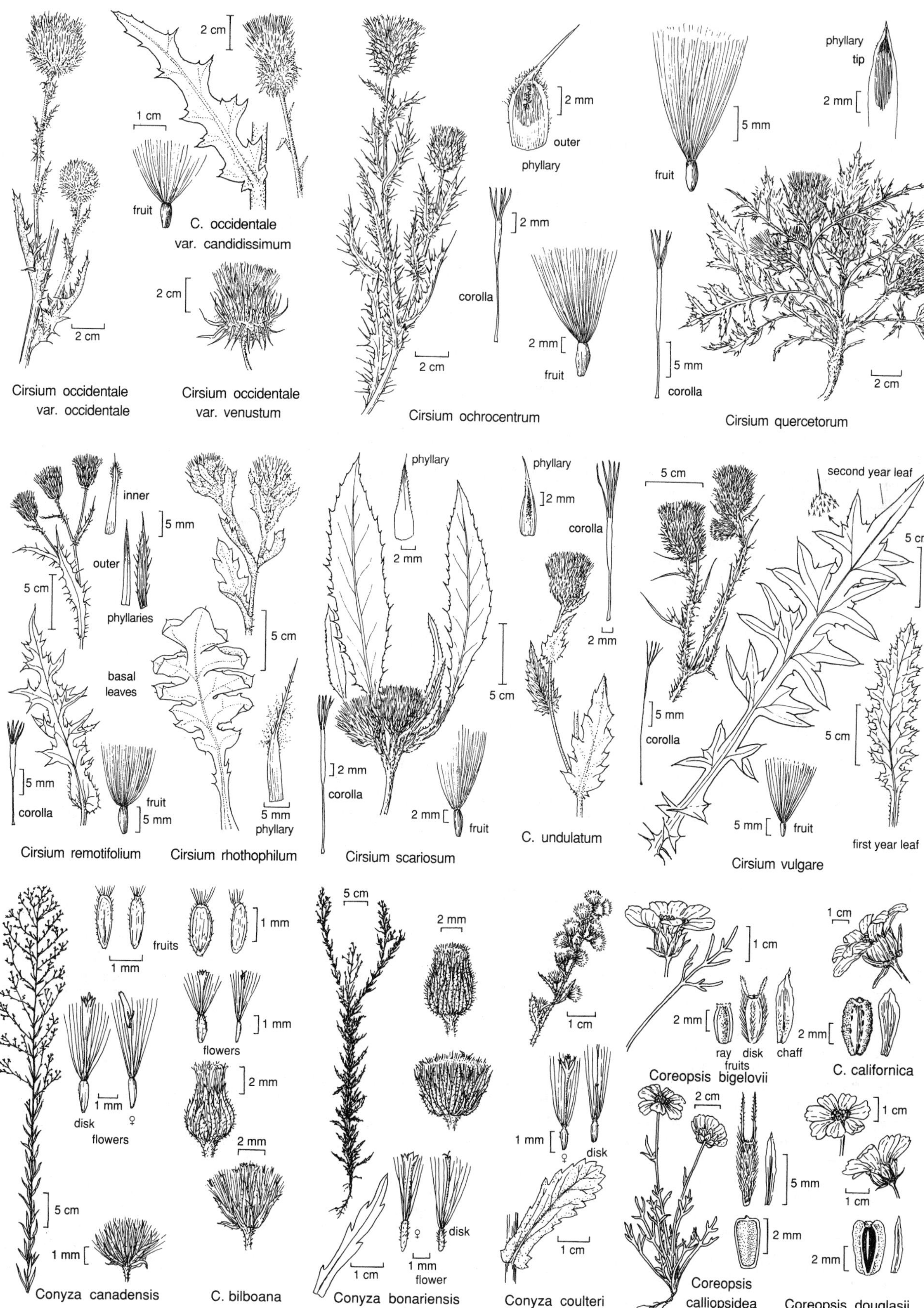
2 cm
1 cm
fruit
C. occidentale
var. candidissimum
2 cm
2 cm
Cirsium occidentale
var. occidentale
Cirsium occidentale
var. venustum
2 mm
outer
phyllary
2 mm
corolla
2 cm
2 mm
fruit
Cirsium ochrocentrum
5 mm
fruit
phyllary
tip
2 mm
5 mm
corolla
2 cm
Cirsium quercetorum
inner
5 mm
outer
phyllaries
5 cm
basal
leaves
5 cm
5 mm
corolla
fruit
5 mm
5 mm
phyllary
Cirsium remotifolium
Cirsium rhothophilum
phyllary
2 mm
5 cm
2 mm
corolla
2 mm
fruit
Cirsium scariosum
phyllary
2 mm
corolla
2 mm
C. undulatum
5 cm
second year leaf
5 cm
5 mm
corolla
5 cm
5 mm
fruit
first year leaf
Cirsium vulgare
fruits
1 mm
1 mm
1 mm
flowers
1 mm
disk
♀
flowers
2 mm
2 mm
5 cm
1 mm
Conyza canadensis
C. bilboana
5 cm
2 mm
1 cm
♀
disk
1 mm
flower
Conyza bonariensis
1 cm
1 mm
♀
disk
1 cm
Conyza coulteri
1 cm
2 mm
ray
disk
fruits
chaff
Coreopsis bigelovii
1 cm
2 mm
C. californica
2 cm
5 mm
2 mm
Coreopsis
calliopsidea
1 cm
1 cm
2 mm
Coreopsis douglasii

3′ Beak slightly < or = fr body; ann or bien of waste places
4. Involucres and sts bearing many, coarse yellow bristles .... ***C. setosa***
4′ Involucres and sts with few blackish bristles, also lightly tomentose .... ***C. vesicaria*** ssp. ***taraxacifolia***
2′ Fr cylindric, tapered at tip or with a short, stout beak
5. Involucre 5–8 mm; ann weed of lowlands .... ***C. capillaris***
5′ Involucre 8–24 mm; native per of montane to subalpine regions
6. Lvs glabrous, nearly all basal; pls of moist places or alkaline flats .... ***C. runcinata***
6′ Lvs gray-tomentose or hairy, some cauline; pls of montane slopes or dry bench lands
7. Lvs green, larger ones lobed, teeth conspicuously white-tipped .... ssp. ***andersonii***
7′ Lvs glaucous, not lobed, teeth not conspicuously white-tipped .... ssp. ***hallii***
8. Herbage densely covered with long gland-tipped hairs; involucre 18–24 mm; outer phyllaries linear-lanceolate .... ***C. monticola***
8′ Herbage tomentose, often with a few glandless or short-glandular hairs; involucres 9–20 mm; outer phyllaries linear, lanceolate, or ovate
9. Involucre narrow; inner phyllaries 5–8; fls 5–10 (see also *C. intermedia*)
10. Phyllaries glabrous or lightly tomentose; cauline lvs 1–3-pinnately lobed .... ***C. acuminata***
10′ Phyllaries conspicuously tomentose on margins, less so on midrib; cauline lvs reduced, mostly toothed or sparsely short-lobed .... ***C. pleurocarpa***
9′ Involucre narrow to wide; inner phyllaries 7–15; fls 8–60
11. Phyllaries conspicuously glandular-hairy .... ***C. bakeri***
11′ Phyllaries without glandular hairs (see also(see also *C. intermedia*)
12. Phyllaries with scattered to many dark, spreading hairs .... ***C. modocensis***
12′ Phyllaries soft-hairy to gray-tomentose, without dark, spreading hairs .... ***C. occidentalis***

C. acuminata Nutt. (p. 247) Per from deep taproot. **STS** 2–7 dm, branched, tomentose. **LVS**: basal 12–40 cm, tomentose, gray-green, lobes narrowly triangular, acute, sometimes with short secondary lobes; cauline similar, smaller. **INFL**: heads many in a compound cyme; involucre 9–10 mm, cylindric to obconic; outer phyllaries narrowly triangular; inner phyllaries 5–8, lanceolate, smooth and shining or lightly tomentose. **FLS** gen 5–6(10). **FR** 6.5–8 mm, tapered to both ends, ± 12-ribbed, pale yellow, ± brown, or dusky white. $2n$=22,33,44,55,88. Open, rocky places; 1200–3300 m. KR, NCoRH, CaRH, SNH, Teh, TR, GB; to WA, MT, WY, Colorado.

C. bakeri E. Greene Per. **STS** 0.8–3 dm, few-branched, dark green, glandular-hairy. **LVS**: basal 8–20 cm, elliptic, lobes lanceolate to narrowly elliptic, dark green with red-purple midveins, glandular-hairy; cauline similar, smaller. **INFL**: heads 10–22 in cyme; peduncles long; involucre 11–20 mm, bell-shaped; outer phyllaries lanceolate, up to half as long as inner; inner phyllaries 10–14, lanceolate, lightly tomentose and glandular-bristly. **FLS** 8–60. **FR** 6–10 mm, tapered to both ends, ± 13-ribbed, dark brown to ± yellow; pappus dusky- or yellowish white. $2n$=22,33,44,55. Dry mtn slopes and flats; 1000–1900 m. KR, CaRH, n&c SNH, MP. [ssp. *cusickii* (Eastw.) Babc. & Stebb.]

C. bursifolia L. Per. **STS** 0.5–3.5 dm, several from base, branched, short-tomentose. **LVS**: basal and lower cauline 5–25 cm, oblanceolate, glabrous or puberulent, lateral lobes lanceolate, terminal lobe ovate, margins ± dentate; cauline gen few, ± reduced, ± linear. **INFL**: heads 2–14; involucre 9–11 mm, cylindric; outer phyllaries unequal, << inner; inner phyllaries 8–10, linear to linear-lanceolate, canescent, ± yellow-bristly. **FLS** 30–60. **FR** 6–7.5 mm, 10-ribbed, pale brown; beak thread-like, 2 × body; pappus white. $2n$=8. Lawn weed; < 100 m. SnFrB; native to s Eur.

C. capillaris (L.) Wallr. (p. 247) Ann, bien with short taproot. **STS** 0.2–9 dm, branched, hairy near base. **LVS**: basal and lower cauline 3–30 cm, glabrous or hairy on veins beneath, lobes lanceolate. **INFL**: heads many; involucre 5–8 mm, cylindric; outer phyllaries linear, << inner; inner 8–16, lanceolate, glandular-bristly. **FLS** 20–60. **FR** 1.5–2.5 mm, 10-ribbed, beakless, brownish; pappus white. $2n$=6. Weed in pastures, waste places; 0–1300 m. NW, CaR, n&c SN, SnFrB, SCo; native to Eur.

C. intermedia A. Gray Per from stout taproot. **STS** 3–7 dm, stout, branched near middle or above, tomentose. **LVS**: basal 15–40 cm, tomentose, lobes narrowly triangular, often bearing small 2° lobes; cauline similar, smaller. **INFL**: heads many; involucre 10–16 mm, cylindric to bell-shaped; outer phyllaries small, lanceolate-deltate; inner phyllaries 7–10, densely and evenly tomentose, sometimes glandular. **FLS** 7–12. **FR** 5.5–9 mm, narrowed at both ends, 10–12-ribbed, yellow, buff, or brown; pappus dusky to white. $2n$=33,44,55,88. Dry slopes, open forest; 1400–3300 m. KR, NCoRH, CaRH, NCoRH, n&c SN, GB; to WA, w Can, WY, Colorado. Complex series of asexually reproducing forms, probably of hybrid origin, combining characters of *C. acuminata, C. modocensis, C. occidentalis, & C. pleurocarpa.*

C. modocensis E. Greene Per from deep taproot. **STS** 1–4.5 dm, few-branched, bearing straight, spreading, glandless bristles at least below. **LVS**: basal 7–25 cm, deeply lobed, lightly tomentose and with some gen blackish bristles, lobes toothed and secondarily lobed. **INFL**: heads 1–10, involucre 11–21 mm, cylindric; outer phyllaries lanceolate, acute; inner phyllaries 8–18, lanceolate, acute, bristly. **FLS** 10–60. **FR** 7–12 mm, deep reddish brown to greenish black; beak 0 or short; pappus dusty to white. $2n$=22,33, 44,55,66,88. Sagebrush scrub, open, dry mtn slopes; 1100–2900 m. CaRH, n&c SNH, SnBr, MP; to B.C., MT, WY, Colorado. [ssp. *subacaulis* (Kellogg) Babc. & Stebb.]

C. monticola Cov. (p. 247) Per from deep taproot. **STS** 1–3 dm, leafy, simple or branched. **LVS** 10–25 cm; toothed or lobed, lobes toothed, densely glandular-bristly. **INFL**: heads 4–20 in cymose clusters; involucre 18–24 mm; outer phyllaries linear-lanceolate, 1/2–2/3 length of inner; inner phyllaries 7–12, linear to narrowly lanceolate, densely covered with long, gland tipped hairs. **FLS** 16–20. **FR** 5.5–9 mm, narrowed at both ends, strongly ribbed, reddish brown; pappus white. $2n$=22,33,44,55, 77,88. Dry, open forests; 700–2400 m. KR, NCoRH, CaRH, n SNH, se SnFrB (Mount Hamilton); OR.

C. nana A. Richards (p. 247) Per from taproot or creeping rhizome, dwarf, purplish green. **STS** many, 2–7 cm, much-branched, in dense clumps. **LVS** mostly basal, glabrous, oblanceolate to obovate or elliptic, entire or shallowly lobed. **INFL**: heads in clusters of 2–4 among lvs; involucre 10–13 mm; outer phyllaries lanceolate or ovate, acute; inner phyllaries 10, oblong, obtuse, short-hairy at tip. **FLS** 9–12. **FR** 4–6 mm, golden brown; tip acuminate or obtuse and beak short; pappus < or = fr. $2n$=14. Stony or gravelly scree; 2600–3700 m. c&s SNH, SnGb, SNE (Sweetwater Mtns), n DMtns (Panamint Mtns); to AK, MT, WY, Colorado, ne N.Am, Asia. [ssp. *ramosa* Babc.]

C. occidentalis Nutt. (p. 247) Per from deep taproot; herbage densely gray-tomentose, without spreading, glandless or glandular hairs. **STS** 1.5–4 dm, branched from base or middle. **LVS** 10–30 cm, toothed to deeply lobed. **INFL**: heads 10–30, clustered in cymes; involucre 11–19 mm; outer phyllaries linear to deltate; inner phyllaries lanceolate, acute, tomentose and sometimes with short, gland-tipped hairs. **FLS** 9–40. **FR** 6–10 mm, tapered to both ends, beakless, strongly 10–18-ribbed, light to dark brown; pappus dusky to white. $2n$=22,33,44,55,66,77,88. Dry terraces, mtn slopes; 800–2700 m. KR, NCoRH, CaRH, SNH, Teh, se SnFrB (Mount Hamilton), WTR, SnBr, GB; to w Can, MT, WY, NM. Pls called sspp. *conjuncta* (Jepson) Babc. & Stebb., *costata* (A. Gray) Babc.

& Stebb., and *pumila* (Rydb.) Babc. & Stebb. intergrade extensively. Highly variable; incl genes from most or all of the dry-land montane spp.

C. pleurocarpa A. Gray (p. 247) Per from taproot. **STS** 1.5–6 dm, gen branched near base, tomentose to ± glabrous. **LVS**: basal 7–28 mm, lanceolate, entire or with widely separated, triangular lobes; cauline lvs much reduced, gen not lobed. **INFL**: heads 7–40 in cymes; involucre 8.5–17 mm; outer phyllaries small, lanceolate to deltate; inner phyllaries 5–8, small, lanceolate to deltate, thin and densely hairy near margins, thicker and nearly glabrous along midrib. **FLS** 5–8. **FR** 5–8 mm, with 10 prominent ribs, beakless, brown; pappus dusky to white. 2*n*=22,33,44, 55,77,88. Dry, lightly wooded or open slopes, often on serpentine; 400–1900 m. KR, NCoRO, NCoRH, CaRH, n SNH; to WA.

C. runcinata Torrey & A. Gray Per from taproot. **STS** 2.5–8 dm, ± lfless, glabrous; branches 0. **LVS**: basal 7–27 cm, oblanceolate or elliptic, pinnately lobed, minutely dentate, or entire, glabrous; cauline 0–few, much reduced. **INFL**: heads (in CA sspp.) 3–20 in open cyme; involucre 8–21 mm; outer phyllaries linear to lanceolate; inner phyllaries 10–16, narrowly to widely lanceolate, glandular-hairy. **FLS** 20–50. **FR** 3.5–7.5 mm, 10–13-ribbed light to dark-brown; tip acuminate or beak short; pappus white. 2*n*=22. Moist depressions, streambanks; 1250–1500 m. SNE; to WA, s-c Can, n-c US.

ssp. ***andersonii*** (A. Gray) Babc. & Stebb. **LVS** green, strongly and closely dentate, larger definitely lobed, teeth conspicuously white-tipped. **FR**: beak definite, short. 2*n*=22. Uncommon. Moist, ± alkaline valley bottoms; ± 1500 m. s MP (Sierra Valley, se Plumas Co.).

ssp. ***hallii*** Babc. & Stebb. (p. 247) HALL'S MEADOW HAWKSBEARD **LVS** glaucous, winged at base, closely dentate. **FR** tapered or beak short. 2*n*=22. RARE in CA. Moist, alkaline valley bottoms; 1250–1450 m. SNE (Benton, s Mono Co. to Bishop, n Inyo Co.); NV. Declining from grazing, habitat drainage.

C. setosa Haller f. Ann. **STS** 2–8 dm; hairs spreading. **LVS**: basal few, 0.6–3 dm, lower half lobed, upper half with widely spaced teeth, finely hairy; cauline sessile, narrower. **INFL**: heads 5–30 in cymose cluster; involucre 8–11 mm; outer phyllaries linear; inner phyllaries 12–16, with stiff, spreading, glandless bristles. **FLS** 20–25. **FR** 3–5 mm, slender-beaked, 10-ribbed, yellowish brown; pappus white, bristles in 1 series, deciduous. 2*n*=8. Uncommon. Disturbed places; 50–300 m. NCoRO; native to Eur.

C. vesicaria L. ssp. *taraxacifolia* (Thuill.) Thell. Ann, bien. **STS** 0.3–8 dm, much-branched, lightly tomentose, hairy below. **LVS** 4–10 cm, lobed below nearly glabrous, larger, terminal segment toothed. **INFL**: heads many in compound cymes; involucre 8–12 mm; outer phyllaries lanceolate; inner phyllaries 9–13, lanceolate, often with short, glandular hairs. **FLS** many. **FR** 6–8 mm incl slender beak (± 1/3 body length), 10-ribbed, pale brown; pappus white. 2*n*=8,16. Uncommon. Disturbed places; 0–300 m. NCoRO, SnFrB, SCo; native to Eur.

CROCIDIUM SPRING GOLD

1 sp. (Greek: little loose tufts, from hair in lf axils) [Morton 1978 N.Am Fl II 10:147]

C. multicaule Hook. (p. 247) Ann 5–30 cm. **STS** 1–few from base, gen unbranched, glabrous to ± tomentose (esp lower axils). **LVS** alternate, 10–25 mm, linear to obovate, ± fleshy, entire or few-toothed, reduced upward. **INFL**: head 1, radiate, long-peduncled; involucre ± bell-shaped; phyllaries 5–13 in 1 series, fused at base, ovate or elliptic, thin, veiny; receptacle conic, naked. **RAY FLS** 5–13; ligules 4–10 mm, yellow. **DISK FLS** many; corollas 2–3 mm, yellow; anther bases rounded, tips narrowly triangular; style branches flattened, tips triangular, minutely papillate. **FR** 1.5–2.5 mm, elliptic, ribbed, brown, puberulent, gelatinous when wet; pappus of barbed bristles 1–3 mm, soon deciduous (often 0 on ray fr). 2*n*=18. Sandy soils, grassland, open woodland; gen < 1500 m. NCoR, SNF, e SnFrB, SCoR (very uncommon), MP; to B.C. ❀TRY.

CRUPINA

David J. Keil & Charles E. Turner

Ann. **STS** erect, openly branched above. **LVS**: basal entire or toothed; cauline alternate, pinnately lobed. **INFL**: heads disciform; involucre cylindric to ovoid; phyllaries in several overlapping series, oblong-lanceolate, acute; receptacle chaffy. **FLS** 3–15; outer sterile, inner 1–2 fertile; corolla purple, tube slender, gradually wider, lobes linear; anther bases short-tailed, tips narrowly triangular; style with minutely hairy cylindric node and papillate terminal segment, branches short, triangular. **FR** cylindric or ± compressed; base puberulent, soft-hairy upward; pappus (on fertile fr only) of small scales and stiff bristles. 2 spp.: Eur. (Latin: ancient name)

C. vulgaris Cass. (p. 247) BEARDED CREEPER **STS** gen 2–6 dm, leafy up to branches. **LVS**: basal sessile or petioled, oblong to obovate, scabrous, gen 0 at fl; cauline sessile, 1–3.5 cm, lobes linear, minutely toothed. **INFL**: peduncles 1–3 cm; involucre 8–20 mm. **FLS** 3–5; corollas ± 14 mm. **FR** gen 1 per head, 3–4 mm, barrel-shaped, attached at base; pappus scales < 1 mm, bristles 5–7 mm, blackish brown. 2*n*=30. Grassy places; ± 250 m. s NCoR (Sonoma Co.), MP; native to s Eur. NOXIOUS WEED.

CYNARA

David J. Keil & Charles E. Turner

Per, 5–25 dm and ± tomentose in CA. **STS** gen erect, leafy, branched, stout. **LVS** basal and cauline, alternate, 3–10 dm in CA, 1–2-pinnately lobed to divided, ± spiny. **INFL**: heads discoid, large, 1–few per cyme; involucre ovoid or hemispheric; phyllaries in many overlapping series, gen ovate, leathery, entire, glabrous, tips gen triangular; receptacle flat, fleshy, bristly. **FLS** many; corollas blue-purple in CA, tube very slender, throat widened abruptly, lobes linear; anther bases long-sagittate, tips oblong; style appendage long, cylindric, minutely papillate, tip barely notched. **FR** cylindric to obconic, ± 4-angled or ± compressed, glabrous, attached at base; pappus of many stiff bristles in several series, white or brownish, plumose below, fused and falling together. 10 spp.: Medit, Canary Islands. (Greek: dog, from tooth-like phyllaries)

1. Lf very spiny; phyllaries clearly spine-tipped .... ***C. cardunculus***
1′ Lf spines 0 or weak; phyllaries spines 0 or minute .... ***C. scolymus***

C. cardunculus L. (p. 247) CARDOON, ARTICHOKE THISTLE Pls 5–25 dm. **LVS** very spiny; upper surface loosely cobwebby; lower surface densely gray-tomentose. **INFL**: involucre 3–6 cm, 4–7 cm diam (not incl tips), narrowed above; phyllaries spine-tipped. **FLS**: corollas ± 5 cm. **FR** 6–8 mm; pappus 2.5–4 cm. 2*n*=34. Disturbed places; < 500 m. CA-FP; native to Medit. Cult as orn or for edible root and petioles. NOXIOUS WEED.

C. scolymus L. (p. 247) ARTICHOKE Pls 8–20 dm, bushy. **LVS** unarmed or minutely spiny; surfaces ± densely gray-tomentose. **INFL**: involucre 6–15 cm, 7–15 cm diam (not incl tips), little narrowed above; phyllaries sometimes notched, spines 0 or weak. **FLS**: corollas 4–5 cm. **FR** 5–6 mm; pappus 3–5.5 cm. $2n$=34. Disturbed places; < 500 m. CA-FP; native to s Eur. Cult for edible phyllary bases and receptacles. Intergrades with (probably derived from) *C. cardunculus*.

DICORIA

Willard W. Payne

Ann, taprooted, much-branched, gen ± white-hairy. **LVS** petioled, gen opposite below, alternate above. **INFL**: heads disciform, many; involucre cup- to bell-shaped; phyllaries few in 1 series, small, free; receptacle chaffy. **PISTILLATE FLS** 0–2; corolla 0; style branches long. **STAMINATE FLS** 5–20; corolla greenish to dull purple; anthers free, exserted, filaments fused, attached at corolla tube base. **FR** compressed, winged, falling with chaff scale; pappus 0. 3–4 spp.: sw US, nw Mex. (Greek: 2 bugs, from 2-fr heads)

D. canescens A. Gray (p. 247) Pl 3–9 dm, gen white-hairy. **LVS** gen ± densely canescent; lower 3–5 cm, lanceolate to triangular-ovate, toothed, gen 3-veined; upper reduced, ± round, ± entire. **INFL**: phyllaries gen 3–5, 2–5 mm, reflexed in age. **PISTILLATE FLS** gen 2. **STAMINATE FLS**: corolla ± 3 mm. **FR** 3–6 mm, ± keeled, glabrous, wings toothed. $2n$=36. Sandy soil; < 1300 m. D; to s UT, w AZ, nw Mex. Highly variable; features vary ± independently. [sspp. *clarkiae* (Kenn.) Keck & *hispidula* (Rydb.) Keck] ❀TRY;STBL.

DIMERESIA

1 sp.: w N.Am. (Greek: 2 numbers, from 2-fld head) [Morton 1978 N.Am Fl II 2:47–49]

D. howellii A. Gray (p. 251) DOUBLET Ann gen < 4 cm, tufted, cobwebby at base, ± glandular above. **LVS** opposite, 1–3 cm, short-petioled, oblanceolate to ovate, entire, clustered around heads. **INFL**: heads discoid, gen in sessile clusters; involucre 4–6 mm, narrowly bell-shaped; phyllaries 2–3, each subtending a fl, ± fused at base, obovate, tip rounded, back convex; receptacle naked. **FLS** 2–3; corollas 4–6 mm, tube and throat ± purple, lobes spreading, white to purple; anther bases sagittate, tips ± triangular; style branches ± flat, ± papillate, tip slightly expanded, rounded. **FR** ± 3 mm, cylindric, ribbed, glabrous; pappus bristles ± 20, 3–5 mm, plumose, fused at base. $2n$=14. UNCOMMON. Dry volcanic soils; 1600–2300 m. MP; to sw ID, nw NV. ❀TRY;DFCLT.

DIMORPHOTHECA CAPE-MARIGOLD, AFRICAN DAISY

Elizabeth McClintock

Ann. **STS** erect, gen branched. **LVS** alternate, simple, entire to pinnately lobed. **INFL**: heads radiate, solitary; peduncles long; phyllaries in 1 series, linear to lanceolate; receptacle flat or nearly so, naked. **RAY FLS** < 15; corollas yellow, blue, or variously colored. **DISK FLS** many; corolla with very short tube and long throat, gen yellow; anthers with ovate or triangular-ovate tips, sagittate at base, short-tailed; style branches very short, appendages widely obtuse. **FR**: ray achenes 3-angled, with knobby tubercles; disk achenes smooth, obovate, flattened, winged, pappus 0. ± 7 spp.: s Afr. (Greek: 2 forms of frs) [Norlindh 1943 Studies in Calenduleae]

D. sinuata DC. Pls glandular-hairy. **STS** 10–30 cm, simple or sparingly branched from base. **LVS** < 10 cm; lower tapered to petiole-like base; upper sessile, blade oblong to oblanceolate, entire to coarsely dentate, upper smaller, sometimes linear. **INFL**: heads 3–7 cm diam; involucre ± bell-shaped; phyllaries 10–15 mm, linear-lanceolate, acuminate, narrowly scarious-margined. **RAY FLS**: ligules 2–2.5 cm, orange to yellow, sometimes violet at base or tip. **DISK FLS**: corollas 4.5–5.5 mm, yellow or orange, often purple-tipped. **FR**: ray achenes 4–5 mm; disk achenes 6–7 mm. $2n$=18. Uncommon. Escape from cult, roadsides, disturbed places, sometimes sown in wildflower mixes; < 1000 m. SnJV, SCoRO, SCo, PR; native to s Afr.

DUGALDIA

Per, glabrous to densely hairy. **STS** erect. **LVS** simple, basal and cauline, alternate; lower wing-petioled; upper sessile; blade 3-veined. **INFL**: heads large, radiate, 2–15 in ± flat-topped cymes; peduncles short to long; involucre disc-like or widely bell-shaped; phyllaries in 2 series; receptacle ovoid to spheric, naked. **RAY FLS** 14–35; ligules yellow to orange, elongated, 3–4-lobed. **DISK FLS** many; corollas yellow; tips of style branches truncate, shrub-like. **FR** oblong, 4-angled; hairs straight; pappus of 5–7 lanceolate scales. 3 spp.: w US to C.Am. (Dugald Stewart, Scotland, 1753–1828) [Bierner 1974 Brittonia 26:385–392]

D. hoopesii (A. Gray) Rydb. (p. 251) **STS** 3–9 dm, glabrous to ± hairy. **LVS** entire, obtuse to acute; lower cauline and basal oblong to oblanceolate; middle oblong to elliptic; upper linear to lanceolate. **INFL**: peduncles 3–16 cm, white-tomentose below heads; disk 12–17 mm, 19–26 mm diam; phyllaries in 2 series, outer lanceolate to ovate, ± fused, ± hairy, inner smaller, elliptic, acuminate, free, glabrous. **RAY FLS** 14–26; ligules 1.5–3.5 cm, yellow to orange. **DISK FLS**: corollas 4–5.5 mm. **FR** 3.5–4.5 mm; pappus 3–4 mm. $2n$=30. Meadows; 1500–3000 m. KR, SN, Wrn; to OR, WY, NM. [*Helenium h.* A. Gray] ❀IRR or WET,SUN:**1**,**2**,3,6,7,15,16,18.

DYSSODIA

Ann, per. **STS** spreading to erect. **LVS** gen pinnately divided or compound, alternate or opposite, sessile or petioled, gland-dotted. **INFL**: heads radiate or discoid, solitary to many in cymes or secondary heads; peduncle 0 to long; bracts 0–few, scattered; involucre widely obconic to hemispheric; phyllaries in 2–3 series, free, outer (when present) linear, inner wider, overlapping; receptacle flat to convex, naked. **RAY FLS** (0)5–8; corollas yellow to orange. **DISK FLS** gen many; corollas yellow to orange; style tips truncate to conic. **FR** obconic or obpyramidal; pappus of scales dissected into bristles. 4 spp.: N.Am., S.Am. (Greek: disagreeable odor) [Strother 1986 Sida 11:371–378]

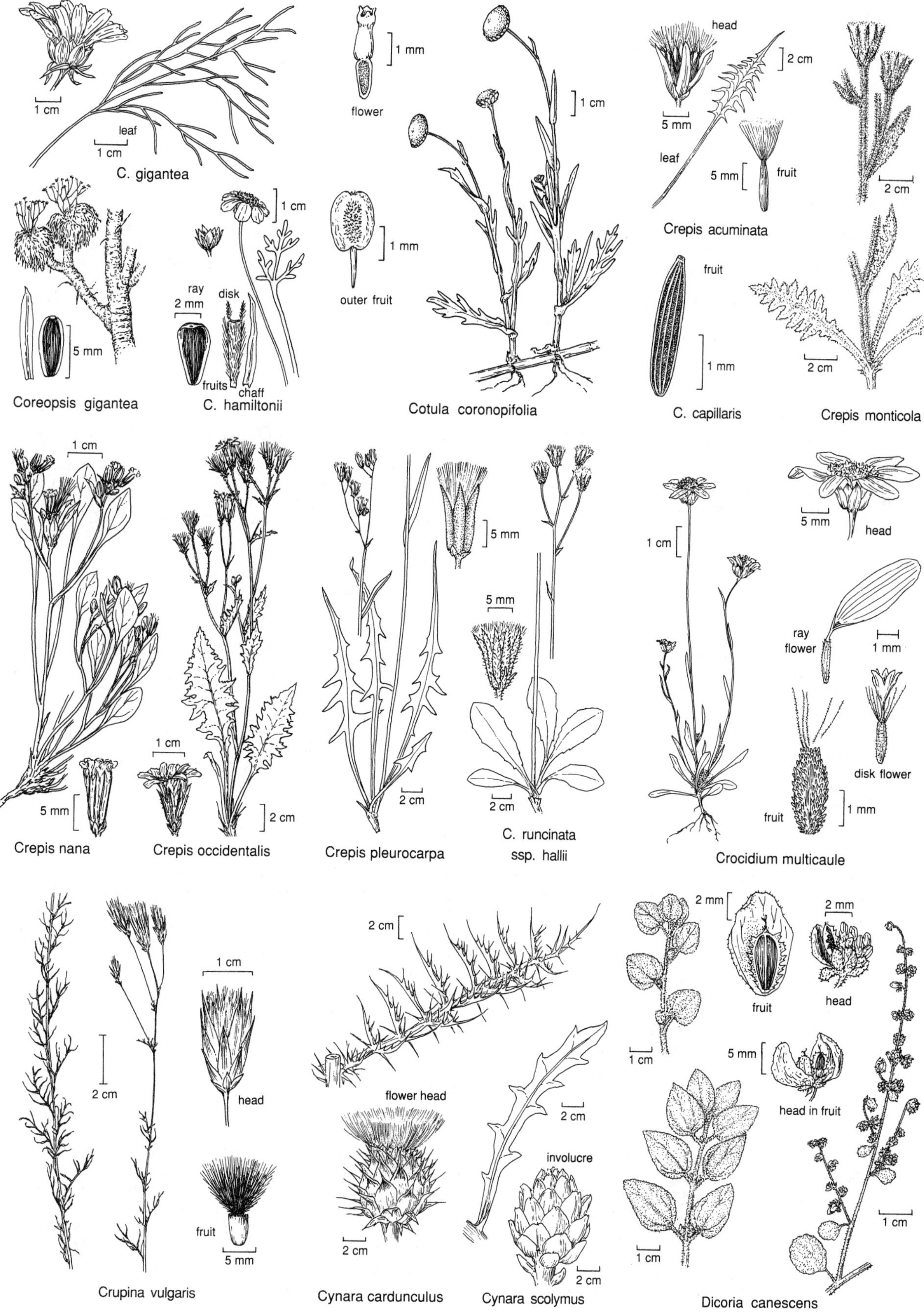
1 cm
leaf
1 cm
C. gigantea
5 mm
Coreopsis gigantea
ray
2 mm
disk
fruits
chaff
1 cm
C. hamiltonii
1 mm
flower
1 mm
outer fruit
1 cm
Cotula coronopifolia
head
2 cm
5 mm
leaf
5 mm
fruit
Crepis acuminata
2 cm
fruit
1 mm
C. capillaris
2 cm
Crepis monticola
1 cm
5 mm
Crepis nana
1 cm
2 cm
Crepis occidentalis
5 mm
2 cm
Crepis pleurocarpa
5 mm
2 cm
C. runcinata
ssp. hallii
1 cm
5 mm
head
ray
flower
1 mm
disk flower
fruit
1 mm
Crocidium multicaule
2 cm
1 cm
head
fruit
5 mm
Crupina vulgaris
2 cm
flower head
2 cm
Cynara cardunculus
2 cm
involucre
2 cm
Cynara scolymus
2 mm
fruit
2 mm
head
1 cm
5 mm
head in fruit
1 cm
1 cm
Dicoria canescens

D. papposa (Vent.) A. Hitchc. (p. 251) Ann gen 1–4 dm, glabrous or sparsely puberulent. **LVS** opposite below, often alternate above, 2–3 cm, divided into 11–15 narrow lobes, these sometimes again divided. **INFL**: heads radiate, 1–few in dense cymes; peduncles 1–5 mm; involucre 4–10 mm diam, bell-shaped; outer phyllaries 4–9, linear, inner phyllaries gen 8, scarious-margined, gland-dotted. **RAY FLS** 8 or fewer; ligules 1.5–2 mm. **DISK FLS** 25–30; corollas ± 3 mm, pale yellow. **FR** 3–3.5 mm, hairy; pappus 2–3 mm. $2n$=26. Roadsides, disturbed fields; < 350 m. e SCo (sw San Bernardino Co.); widespread weed in Am; native to c&e N.Am, Mex; also in w S.Am.

EASTWOODIA YELLOW MOCK ASTER

Meredith A. Lane

1 sp. (Alice Eastwood, western American botanist, 1859–1953)

E. elegans Brandegee (p. 251) Shrub, appearing glabrous. **STS** < or = 1 m, ± striate; new growth yellow to tan; bark of older growth gray-white, shredding with age. **LVS** alternate, sometimes in axillary clusters below, < or = 4 cm, < or = 4 mm wide, ± linear, entire, minutely scabrous. **INFL**: heads discoid, solitary or in open panicles; involucre bell-shaped, < 2 cm, < or = 2 cm wide; phyllaries in 2–4 series, narrow, whitish, tips acuminate, green; receptacle hemispheric, chaffy, scales readily deciduous. **FLS** 30–40; corollas < or = 6 mm, cylindric to funnel-shaped, yellow; style branch appendages lanceolate. **FR** < or = 2 mm, narrowly obconic, obscurely 3–4-angled, hairy, esp on angles; pappus of 5–8 scales, > fr. $2n$=18. Banks and hillsides; 60–1250 m. Teh, s SnJV, se SnFrB, SCoRI. ❀TRY;DFCLT.

EATONELLA

1 sp.: w US. (Daniel C. Eaton, American botanist, 1834–1895)

E. nivea (D.C. Eaton) A. Gray (p. 251) Ann 1–4 cm, densely tomentose. **STS** congested. **LVS** basal and alternate, very crowded, gen 1 cm or less, linear to obovate, entire. **INFL**: heads radiate, solitary, terminal, subsessile, in fl ± concealed by lvs; peduncle in fr elongating to ± 1 cm; involucre 4–5 mm diam, bell-shaped; phyllaries 8–12 in 1 series, 4–5 mm, linear-oblong, free, in fr reflexed; receptacle flat, naked. **RAY FLS** 1 per phyllary, inconspicuous; ligules 1.5–2.5 mm, light yellow, often drying ± purple. **DISK FLS** many; corollas 2 mm, 4–5-lobed, yellow. **FR** ± 3 mm, oblanceolate, flattened; surfaces glabrous, black and shiny; margins ciliate with long, white hairs; pappus of 2 fringed scales 1–2 mm. $2n$=19. Sandy soil; 1350–2900 m. GB; se OR, w NV. ❀TRY.

ECHINOPS GLOBE THISTLE

David J. Keil & Charles E. Turner

Per, gen ± tomentose. **LVS** basal and cauline, alternate, spiny, dentate to 1–3-pinnately lobed. **INFL**: heads discoid, many, 1-fld, sessile in spheric 2° heads; 1° involucre elliptic, subtended by bristles; phyllaries in several overlapping series; receptacle obconic, chaffy; 2° heads peduncled; 2° involucre bracts reflexed. **FLS** 1 per 1° head; corollas white to blue; anther bases sharply tailed, tips narrowly triangular, acute; style with ring of minute hairs just below oblong branches. **FR** ± cylindric, 4-angled, hairy; pappus of many, short scales. ± 100 spp.: Medit, Eurasia. (Greek: hedgehog-like)

E. sphaerocephalus L. Pls 1–2 m. **STS** simple to much branched, glandular and ± tomentose. **LVS** oblong-elliptic; base clasping; upper surface green, glandular; lower surface cobwebby-tomentose; lobes lanceolate to triangular. **INFL**: 1° involucre 15–25 mm, 1° phyllary tips expanded, fringed, spine-tipped, ± glandular; 2° heads 3–6 cm diam. **FLS**: corolla tube long, throat very short, lobes ± linear. **FR** 7–8 mm, densely appressed-hairy; pappus scales 1–1.5 mm, ciliate, ± fused. $2n$=32. Disturbed places; ± 1300 m. MP; native to Eurasia. Sometimes cult.

ECLIPTA

Ann. **STS** prostrate to erect, often rooting below, simple to much-branched. **LVS** simple, opposite, sessile. **INFL**: heads radiate, small, solitary or in small cymes; peduncles short or 0; involucre hemispheric; phyllaries in 1–2 ± equal series, free, ovate; receptacle rounded, chaffy, scales narrowly linear, bristle-like. **RAY FLS** many; corollas white; ligules short, narrowly linear. **DISK FLS** many; corollas white; style branches very short, incl; anthers brown, incl. **FR** 4-angled, ± flat, brown, glabrous; pappus a crown of minute bristles or 0. 4 spp.: esp trop. (Greek: deficient, from absence of pappus)

E. prostrata (L.) L. (p. 251) Pl ± strigose throughout. **STS** 1–10 dm. **LVS**: blades linear, lanceolate, or narrowly elliptic, entire or short-toothed, tip acute. **INFL**: peduncles 0–15 mm; involucre 4–10 mm diam; phyllaries 4–5 mm, acute. **RAY FLS**: corolla 1.5–3 mm. **DISK FLS**: corolla 1.5–2 mm. **FR** 1.7–2.2 mm, obovate, smooth or ± warty; pappus < or = 0.2 mm. $2n$=12. Damp places; < 300 m. GV, SCoR, SW, DSon. [*E. alba* (L.) Hassk.] A weed in all continents. Source of dark dye, medicine against roundworm parasites.

ENCELIA

Curtis Clark

Shrubs. **STS** gen many from base. **LVS** alternate, gen drought-deciduous, simple, petioled, entire or rarely toothed. **INFL**: heads radiate or discoid, solitary or in cyme-like panicles; peduncles gen long; involucre hemispheric; phyllaries in 2–3 series, free; receptacle chaffy, scales folded around frs and falling with them. **RAY FLS** sterile; style 0; ligules yellow. **DISK FLS** many; corollas yellow or brown-purple, tube slender, throat abruptly expanded, lobes triangular; anther tips ovate, ± acute; style tips triangular. **FR** strongly compressed, obovate or wedge-shaped; edges long-ciliate; faces glabrous or short-hairy; pappus of 2 narrow scales or 0. 13 spp.: w N.Am, w S.Am. (Christopher Encel, 16th century) Commonly hybridizing, esp in disturbed areas; *E. farinosa* × *E. frutescens* is common; *E. farinosa* × *E. californica*, *E. farinosa* × *E. actonii*, *E. actonii* × *E. frutescens*, *E. frutescens* × *E. virginensis*, *E. farinosa* × *Geraea canescens* have been reported.

1. Heads in panicles, always radiate; lvs ± hairy, with curled hairs
2. Heads in tight panicles; ray corollas well developed; lvs densely hairy, without strigose hairs .... ***E. farinosa***
2′ Heads in loose panicles; ray corollas often short, few, and deeply lobed; lvs moderately hairy, often with some strigose hairs .... ***E. farinosa × E. frutescens***
1′ Heads solitary, rayed or rayless; lvs glabrous or hairy
3. Disk corollas brown; lvs glabrous .... ***E. californica***
3′ Disk corollas yellow or orange; lvs glabrous or hairy
4. Rays 0; lvs strigose but not canescent .... ***E. frutescens***
4′ Rays present; lvs canescent
5. Lvs silvery canescent, with no strigose hairs; rays gen > 14, with ligules > 10 mm and shallowly toothed; sw SnJV and adjacent WTR (Cuyama Valley), w D and adjacent CA-FP, DMtns .... ***E. actoni***
5′ Lvs with some strigose hairs intermixed with a softer canescence; rays gen < 21, with ligules < 15 mm and deeply toothed; e DMoj .... ***E. virginensis***

E. actoni Elmer (p. 251) Shrub 5–15 dm, with many slender branches from base. **STS** branched below; young sts hairy; older sts with fissured bark. **LVS** scattered along sts; petioles 6–12 mm; blades 2.5–4 cm, ovate to deltate, acute, silvery green, canescent. **INFL**: heads radiate, solitary; peduncles canescent; involucre 8–14 mm; phyllaries ovate. **RAY FLS** 15–25; ligules 10–25 mm. **DISK FLS**: corollas 5–6 mm, yellow. **FR** 5–7 mm; pappus gen 0. $2n$=36. Open areas, rocky slopes, roadsides; 800–1500 m. sw SnJV and adjacent WTR (Cuyama Valley), w D and adjacent CA-FP, DMtns; sw NV, n Baja CA. [*E. virginensis* Nelson ssp. *a.* (Elmer) Keck] ❀DRY,SUN,DRN:7,**8**,**9**,10,**11**,12,13,**14**,15,16,18,**19–21**,22–24.

E. californica Nutt. (p. 251) Shrub 5–15 dm, with many slender branches from base. **STS** branched below; young sts glabrous; older sts with smooth or roughened bark. **LVS** scattered along sts; petioles 5–25 mm; blades 3–6 cm, diamond-shaped or narrowly ovate, acute, green, glabrous. **INFL**: heads radiate, solitary; peduncles hairy; involucre 10–12 mm; phyllaries lanceolate. **RAY FLS** 15–25; ligules 15–35 mm. **DISK FLS**: corollas 5–6 mm, brown-purple. **FR** 5–7 mm; pappus 0. $2n$=36. Coastal scrub; < 600 m. SCo, s CCo, w PR, WTR; n Baja CA. ❀DRN:7,**14**,15,16,18,**19–24**;IRR:**8**,**9**,10–12;also STBL.

E. farinosa Torrey & A. Gray (p. 251) BRITTLEBUSH, INCIENSO Shrub 3–15 dm, from 1 or several trunks; sap fragrant. **STS** much-branched above; young sts tomentose; older sts with smooth bark. **LVS** clustered near st tips; petioles 10–20 mm; blades 2–7 cm, ovate to lanceolate, obtuse or acute, silver- or gray-tomentose. **INFL**: heads radiate, 3–9 in panicles; peduncles ± yellow, glabrous exc just below heads; involucre 4–10 mm; phyllaries lanceolate. **RAY FLS** 11–21; ligules 8–12 mm. **DISK FLS**: corollas 5–6 mm, yellow or brown-purple. **FR** 3–6 mm; pappus 0. $2n$=36. Coastal scrub, stony desert hillsides; < 1000 m. e SCo and adjacent PR, D; to sw UT, AZ, nw Mex. Dried resin used as incense. [var. *phenicodonta* (S.F. Blake) I.M. Johnston] ❀SUN,DRN:**8–14**,15,16,18,**19–24**.

E. farinosa × E. frutescens Shrub 5–12 dm with few–many branches from 1–several short trunks. **STS** branched along their length; young sts hairy; older with rough bark. **LVS** scattered along sts; petioles 5–15 mm; blades 1–5 cm, elliptic, lanceolate, or narrowly ovate, obtuse, gray-green, lightly tomentose with some strigose hairs. **INFL**: heads radiate or sometimes discoid, 2–5 in loose panicles; peduncles glabrous or strigose; involucre 5–12 mm; phyllaries lanceolate. **RAY FLS** 1–11; ligules 5–20 mm, deeply 3-lobed. **DISK FLS**: corollas 5–6 mm, yellow. **FR** 5–8 mm; pappus 2 slender scales or 0. $2n$=36. Roadsides, waste places, desert washes, flats; < 800 m. D; w AZ, n Baja CA. Hybrids and backcrosses commonly found with parent spp. Many herbarium specimens of "*E. virginensis*" are these.

E. frutescens (A. Gray) A. Gray (p. 251) Shrub 5–15 dm, with many slender branches from 1–several short trunks. **STS** branched below; young sts glabrous; older sts with fissured bark. **LVS** scattered along sts; petioles 2–7 mm; blades 1–2.5 cm, elliptic or narrowly ovate, obtuse, green, strigose. **INFL**: heads discoid, solitary; peduncles strigose; involucre 6–12 mm; phyllaries lanceolate. **RAY FLS** 0. **DISK FLS**: corollas 5–6 mm, yellow. **FR** 6–9 mm; pappus 2 slender scales or 0. $2n$=36. Desert washes, flats, slopes, roadsides; < 800 m. D; s NV, w AZ, Baja CA. ❀SUN,DRN:**8**,**9**,10–13,**14**,15,16,18,**19–24**.

E. virginensis Nelson Shrub 5–15 dm, with many slender branches from base. **STS** branched below; young sts hairy; older sts with fissured bark. **LVS** scattered along sts; petioles 2–7 mm; blades 1.2–2.5 cm, narrowly ovate to deltate, acute or obtuse, gray-green, lightly canescent, with some strigose hairs. **INFL**: heads radiate, solitary; peduncles canescent; involucre 9–13 mm; phyllaries narrow ovate. **RAY FLS** 11–21; ligules 8–15 mm. **DISK FLS**: corollas 5–6 mm, yellow. **FR** 5–8 mm; pappus gen 0. $2n$=36. Desert flats, rocky slopes, roadsides; 500–1500 m. e DMoj; to sw UT, nw AZ. ❀SUN,DRN:2,**7–9**,10,**11**,12,13,**14**,15,16,**18–24**.

ENCELIOPSIS

Curtis Clark

Per from stout caudex, subscapose. **STS** densely leafy at base, lfless above. **LVS** basal and closely alternate, simple, petioled or sessile, entire, 3-veined. **INFL**: heads radiate or discoid, solitary; peduncles long; involucre hemispheric; phyllaries in 2–3 series, free; receptacle chaffy, scales folded around frs and falling with them. **RAY FLS** sterile; style 0; ligules yellow. **DISK FLS** many; corollas yellow, tube slender, throat abruptly expanded, lobes triangular; anther tips ovate, ± acute; style tips triangular. **FR** strongly compressed, wedge-shaped; edges ± white, corky, glabrous or long-ciliate; faces black, glabrous or ± hairy; pappus of 2 narrow awns and a crown of shorter scales. 3 spp.: w N.Am. (Greek: like *Encelia*)

1. Petioles winged, wings merging with blades, blades diamond-shaped or widely elliptic, herbage silvery ***E. covillei***
1′ Petioles not or barely winged, blades ovate, herbage dull gray .... ***E. nudicaulis***

E. covillei (Nelson) S.F. Blake (p. 251) PANAMINT DAISY Pl 1.5–8(10+) dm; herbage silvery-canescent, hairs fine, ± appressed. **STS** woody at base. **LVS**: petioles winged, wings merging with blades; blades 4–10 cm, 2–8 cm wide, diamond-shaped or widely elliptic, 3-veined. **INFL**: heads radiate, 9–13 cm diam; peduncles 3–10 dm, gray-puberulent; involucre 1.8–3 cm; phyllaries in 3 series, lanceolate to ovate, acuminate, densely gray-puberulent. **RAY FLS** 20–35; ligules 3–5 cm. **FR** ± 10 mm, 6.5 mm wide, glabrous or puberulent; pappus awns ± 1 mm, smooth. $2n$=36. RARE. Stony hillsides, canyons; 400–1250 m. n DMtns (w side Panamint Mtns). [*E. argophylla* (D.C. Eaton) Nelson var. *grandiflora* (M.E. Jones) Jepson]

E. nudicaulis (A. Gray) Nelson (p. 251) NAKED-STEMMED DAISY Pl 1–4 dm; herbage dull gray, hairs short, ± spreading. **STS** woody at base. **LVS**: petioles not or only slightly winged; blades 2–6 cm, 2–6 cm wide, ovate, 3-veined. **INFL**: heads radiate, 4–9 cm diam; peduncles 1.5–4.5 dm, gray-puberulent; involucre 1–2 cm; phyllaries in 3 series, narrowly lanceolate from ovate base, acute, densely gray-puberulent. **RAY FLS** ± 21; ligules 2–4 cm. **FR** ± 9 mm, 3.5 mm wide, silky-hairy; pappus awns 1–1.5 mm, smooth. $2n$=36. UNCOMMON. Stony hillsides and canyons; 950–2000 m. W&I, DMtns; to ID, UT, n AZ. ❀TRY;DFCLT.

ERECHTITES FIREWEED

Theodore M. Barkley

Ann or weak per from taproot, often ill-smelling. **LVS** alternate, subentire to pinnately dissected, sometimes clasping. **INFL**: heads gen many, disciform; involucre cylindric to urn-shaped, phyllaries equal in 1 series (or with few smaller bractlets), green or black-tipped; receptacle flat, naked. **PISTILLATE FLS** many; corolla narrowly cylindric, ± light yellow. **DISK FLS** many, sometimes staminate; corollas ± light yellow; anther bases rounded, tips short-triangular; style tips ± hairy, gen papillate. **FR** cylindric, ribbed; pappus of many fine bristles, white in CA. ±25 spp.: Am, Australasia (or 5 spp.: Am). (Greek: name of a *Senecio* sp.) [Barkley & Cronquist 1978 N.Am Fl II 10:139–142]

1. Phyllaries 10–17 mm; receptacle ± wide; head urn-shaped or obconic .... ***E. hieracifolia*** var. ***hieracifolia***
1′ Phyllaries 4–8 mm; receptacle narrow; head cylindric
 2. Lvs pinnately lobed; phyllaries ± 13 .... ***E. glomerata***
 2′ Lvs toothed but not lobed; phyllaries ± 8 .... ***E. minima***

E. glomerata (Poiret) DC. (p. 259) Pl 6–20 dm, lightly gray-hairy, becoming ± glabrous. **LVS** (2)7–15 cm; blade narrowly lanceolate, pinnately lobed, lobes gen sharp-toothed. **INFL** ± flat-topped; involucre cylindric; phyllaries ± 13, 4–7 mm; receptacle narrow. **FR** gen < 2 mm; pappus persistent. $2n$=60. Woods near coast; < 500 m. NCo, NCoRO, CCo, SCoRO; OR; native to Australasia. [*E. arguta* (A. Richards) DC.]

E. hieracifolia (L.) DC. var. *hieracifolia* Pl 8–15+ dm, glabrous or sparsely hairy. **LVS** 6–20 cm; blade widely lanceolate, serrate to lobed, lower withering, uppermost bract-like, clasping. **INFL** ± flat-topped; heads few–many; involucre urn-shaped or obconic; phyllaries 7–21, 10–17 mm; receptacle 5–8 mm diam. **FR** 2.2–3 mm, 10–12-ribbed; pappus deciduous. $2n$=40. Uncommon. Disturbed sites near coast; < 500 m. NCo, n CCo; native to e N.Am, Caribbean.

E. minima (Poiret) DC. (p. 259) Pl 6–20 dm (sometimes weak per), unevenly hairy, becoming ± glabrous. **LVS** 7–20 cm; blade ± lanceolate, sharply but irregularly toothed. **INFL** loose, ± flat-topped; involucre cylindric; phyllaries ± 8, 4–8 mm; receptacle narrow. **FR** gen < 2 mm; pappus persistent. $2n$=60. Grassland, woodland, coastal scrub; gen < 500 m. NCo, NCoRO, CCo, SCoRO; OR; native to Australasia. [*E. prenanthoides* (A. Richards) DC.]

ERICAMERIA GOLDENBUSH

Gregory K. Brown & David J. Keil

Shrubs < 50 dm, resinous, gen gland-dotted. **LVS** < 10 cm, thread-like to wedge-shaped, entire. **INFL** various; heads radiate or discoid; involucre 3–14 mm, obconic to hemispheric; phyllaries in 2–6 series, ± lanceolate to ovate, gen resinous, tips erect to recurved, obtuse to acuminate or tailed, midrib often thickened with a resin gland. **RAY FLS** 0–30; corollas 2–12 mm, gen yellow. **DISK FLS** 4–70+; corollas 3–11 mm, yellow. **FR** 2–8 mm, ribbed; pappus white to brown. ±27 spp.: w N.Am. [Nesom 1990 Phytologia 68:144–155] Gen fls summer/autumn. Some spp. hybridize with *Chrysothamnus nauseosus*.

1. Lvs narrowly oblong to obovate, widest gen > 2.5 mm wide
 2. Involucres gen 3–10 mm wide
  3. Pl 15–50 dm; lvs 20–60 mm, linear to narrowly elliptic; heads discoid; fr ribs 9–12 .... ***E. parishii***
  3′ Pl < 12 dm; lvs 5–20 mm, ± wedge-shaped to widely obovate; heads radiate or discoid; fr ribs 4 (fr 4-angled) .... ***E. cuneata***
   4. Disk fls 36–70; c PR .... [2]var. ***macrocephala***
   4′ Disk fls 7–32; widespread
    5. Lvs wedge-shaped, sessile, largest 3–18 mm .... var. ***cuneata***
    5′ Lvs oblanceolate, blunt, gen distinctly petioled, largest 9–25 mm .... var. ***spathulata***
 2′ Involucre gen 5–15 mm wide
  6. Heads discoid
   7. Branches glabrous .... [2]***E. cuneata*** var. ***macrocephala***
   7′ Branches white-tomentose .... ***E. discoidea***
  6′ Heads radiate
   8. Heads 1–few in cymes
    9. Lvs 6–12 mm; involucres 7–9 mm; outer phyllaries parchment-like, only tips green; disk fls 15–18 .... ***E. gilmanii***
    9′ Lvs 15–40 mm; involucre 8–15 mm; outer phyllaries green, lf-like; disk fls 13–40 .... ***E. suffruticosa***
   8′ Heads many in crowded racemes, panicles, or cymes
    10. Sts 2–6 dm; lvs 2–7 cm; involucre 8–12 mm wide; phyllaries green-tipped; disk fls 4–13 .... ***E. bloomeri***
    10′ Sts 1–2.5 dm; lvs 1.5–3 cm; involucre 12–15 mm wide; outer phyllaries green, lf-like; disk fls 7–20 .... ***E. greenei***
1′ Lvs linear to thread-like and cylindric, widest < 2.5 mm wide
 11. Involucre 6–18 mm wide, obconic to hemispheric
  12. Involucre 8–14 mm
   13. Heads radiate; lvs 10–55 mm .... ***E. linearifolia***
   13′ Heads discoid; lvs < 15 mm .... ***E. ophitidis***
  12′ Involucre (late-season in *E. pinifolia*) 5–7 mm
   14. Heads radiate or discoid; disk fls 20–25, corollas 8–10 mm; phyllaries scarious, resinous, pale yellow .... ***E. fasciculata***
   14′ Heads radiate; disk fls 12–18, corollas 6–7.5 mm; phyllaries woolly-ciliate, green-tipped .... ***E. pinifolia***
 11′ Involucre 3–6 mm wide, ± narrowly obconic
  15. Lvs not gland-dotted, glabrous, ± sticky .... ***E. nana***

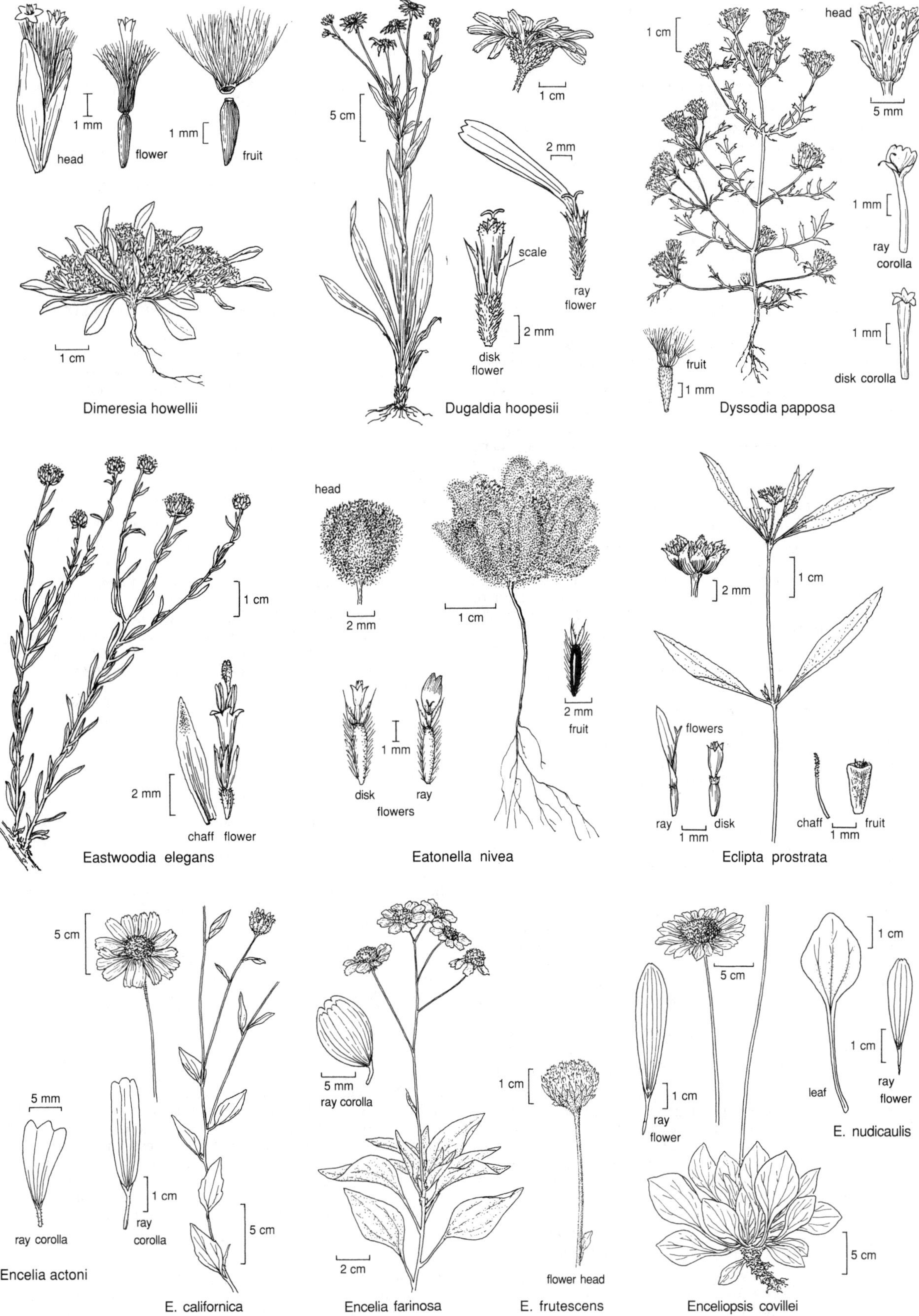
head
1 mm
flower
1 mm
fruit
1 cm
Dimeresia howellii
5 cm
1 cm
2 mm
scale
ray flower
2 mm
disk flower
Dugaldia hoopesii
1 cm
head
5 mm
1 mm
ray corolla
1 mm
disk corolla
fruit
1 mm
Dyssodia papposa
1 cm
2 mm
chaff
flower
Eastwoodia elegans
head
2 mm
1 cm
2 mm
fruit
1 mm
disk
ray
flowers
Eatonella nivea
2 mm
1 cm
flowers
ray
disk
1 mm
chaff
fruit
1 mm
Eclipta prostrata
5 cm
5 mm
ray corolla
1 cm
ray corolla
5 cm
Encelia actoni
E. californica
5 mm
ray corolla
2 cm
Encelia farinosa
1 cm
flower head
E. frutescens
5 cm
1 cm
ray flower
5 cm
Enceliopsis covillei
1 cm
1 cm
leaf
ray flower
E. nudicaulis

15′ Lvs gland-dotted, resinous, glabrous or puberulent
16. Young herbage glabrous; lvs 10–60 mm; pl < 3 m
17. Rays fls 3–11 . . . . *E. laricifolia*
17′ Ray fls 0
18. Lvs 30–60 mm; disk fls 18–25 . . . . *E. arborescens*
18′ Lvs 10–25 mm; disk fls 9–14 . . . . *E. brachylepis*
16′ Young herbage puberulent; lvs 3–40 mm; pl < 1 m
19. Involucres 4–5 mm; disk corollas 3–5 mm; ray fls (0)1–2 . . . . *E. cooperi*
19′ Involucres 5–9 mm; disk corollas 5–8 mm; ray fls 1–8
20. Lvs 4–12 mm; involucre 5–6 mm wide, outer phyllaries acuminate; fr < 3 mm, glabrous to hairy, ribs 8–10; pappus ± = disk corolla, white to tan . . . . *E. ericoides*
20′ Lvs 6–40 mm; involucre 3.5–4.5 mm wide, outer phyllaries obtuse to acute; fr 3–4 mm, hairy, ribs 4–7; pappus > disk corolla, brown . . . . *E. palmeri*
21. Sts 5–15 dm; lvs 5–16 mm . . . . var. *pachylepis*
21′ Sts 1–4 dm; lvs 20–40 mm . . . . var. *palmeri*

E. arborescens (A. Gray) E. Greene (p. 259) GOLDEN-FLEECE Pl < 30 dm, glabrous, strongly gland-dotted. **LF** 30–60 mm, thread-like to linear, acute. **INFL**: heads discoid in rounded cymes at branch tips; involucre < or = 4.5 mm, 4–4.5 mm diam, obconic; phyllaries 20–25 in 3–4 series, lanceolate, acute, (sub)glabrous, midrib brown. **DISK FLS** 18–25; corollas 4.7–5.5 mm. **FR** 2 mm, 5-angled, densely appressed-hairy; pappus > disk corollas, dull white. $2n$=18. Woodland, open forest, chaparral, esp after fire; gen < 1200 m. KR, NCoR, SNF, s SNH (< 2900 m), SCoR, WTR. [*Haplopappus a.* (A. Gray) H.M. Hall] Closely related to *E. parishii*. ❀DRN,DRY,SUN:7–9,14–16,19–24;DFCLT.

E. bloomeri (A. Gray) J.F. Macbr. (p. 259) Pl 2–6 dm, glabrous to barely woolly or glandular. **LF** 20–70 mm, thread-like to ± oblanceolate, acute or acuminate. **INFL**: heads radiate; involucre 8–11 mm, 8–12 mm diam, narrowly bell-shaped; phyllaries loosely overlapping in 3–6 series, rigid, linear to lanceolate, resinous, tips recurved, margins scarious, woolly-ciliate, all but innermost abruptly narrowed to linear green tips. **RAY FLS** 1–5; corollas 8–12 mm. **DISK FLS** 4–13; corollas 7–11 mm, yellow. **FR** 5.5–7.5 mm, 5-angled, brown, ± glabrous to hairy; pappus ± = disk corolla, light brown. $2n$=18. Open, gravelly, coniferous forest; 1000–2900 m. CaR, SN, MP; to WA, w NV. [*Haplopappus b.* A. Gray] ❀TRY.

E. brachylepis (A. Gray) H.M. Hall BOUNDARY GOLDENBUSH Pl 10–20 dm, ± glabrous, gland-dotted. **LVS** crowded, 10–25 mm, thread-like, obtuse to ± acute, curled under. **INFL**: heads discoid, in raceme-like or narrowly panicle-like clusters at ends of leafy branchlets; peduncles 5–20 mm; involucre 4.5–5.5 mm, 4–6 mm diam, obconic; phyllaries 16–22 in 3–4 series, linear to ovate, obtuse to acute, midrib brown, resinous, margin white-scarious, minutely ciliate. **DISK FLS** 9–14; corollas 6–7 mm. **FR** < or = 5 mm, densely soft-hairy; pappus 6–7 mm, ± white. $2n$=36. Chaparral; < 1400 m. s PR, s SCo; n Baja CA. [*Haplopappus propinquus* S.F. Blake]

E. cooperi (A. Gray) H.M. Hall var. ***cooperi*** (p. 259) Pl 3–6 dm, puberulent, gland-dotted. **LF** 3–15 mm, ± linear, acute. **INFL**: heads gen radiate, in open cymes; involucre 4–5 mm, 3–4 mm diam, narrowly bell-shaped; phyllaries 9–15 in 3–4 series, 2.5–4 mm, oblong to ovate, obtuse or outer acute. **RAY FLS** 0–2; corollas 4–9 mm. **DISK FLS** 4–12; corollas 3–5 mm. **FR** 3–3.5 mm, obconic to subcylindric, softly silky-hairy, veins 10–12, thin; pappus 3–4.5 mm, white. $2n$=18. Rocky slopes, valleys, in creosote-bush scrub, Joshua-tree woodland; 800–2000 m. DMoj; s NV. [*Haplopappus c.* (A. Gray) H.M. Hall] Fls spring. Hybridizes with *E. linearifolia*. ❀TRY;DFCLT.

E. cuneata (A. Gray) McClatchie Pl 1–10 dm, glabrous, ± gland-dotted. **LF** 2–25 mm, ± oblanceolate or obovate, obtuse. **INFL**: heads radiate or discoid in small compact cymes; involucre 6–12 mm, 4–14 mm diam, obconic; phyllaries 20–30 in 4–6 series, lanceolate to obovate, glabrous, sometimes resinous. **RAY FLS** 0–3; corollas < 5 mm. **DISK FLS** 7–70; corollas ± 5.5 mm. **FR** 2.5–3 mm, 5-ribbed, silky-hairy; pappus < corolla, sparse, brown. Outcrops, slopes, cliffs; 100–2800 m. SN, SCoR, TR, PR, SNE, D; s NV, AZ, nw Mex. [*Haplopappus c.* A. Gray]

var. ***cuneata*** **LVS**: largest 3–14(18) mm, 2–9(12) mm wide, wedge-shaped, sessile. **INFL**: heads radiate or discoid, 8–11 mm, 5–7 mm diam. **DISK FLS** 12–33. $2n$=18. Granite outcrops; 1000–2800 m. SN, WTR, SnGb, PR, SNE. ❀DRN,DRY,SUN:1–3,**7**,14,**15,16,18**,19–24;GRNCVR.

var. ***macrocephala*** Urb. LAGUNA MOUNTAIN GOLDENBUSH **LVS**: largest gen 12–20 mm, 6–14 mm wide, ± obovate; base petiole-like. **INFL**: heads discoid, 9–12 mm, 6–10 mm diam. **DISK FLS** 36–70. $2n$=18. RARE. Rock crevices; ± 1850 m. c PR (Laguna Mtns, San Diego Co.).

var. ***spathulata*** (A. Gray) H.M. Hall (p. 259) **LVS**: largest (9)12–25 mm, 4–16 mm wide, ± obovate; base petiole-like; tip gen widely obtuse or notched. **INFL**: heads gen discoid, 8–11 mm, 5–7 mm diam. **DISK FLS** 7–15. $2n$=18. Rock outcrops; 100–1900 m. SCoRI, Teh, D; s NV, AZ, nw Mex. Intergrades with var. *cuneata* in Kern Co. [*Haplopappus c.* var. *s.* (A. Gray) S.F. Blake] ❀TRY; DFCLT.

E. discoidea (Nutt.) G. Nesom Pl 1–4 dm. **ST** densely white-tomentose. **LF** 10–30 mm, oblong to oblanceolate, sessile, obtuse to acute, stalked-glandular. **INFL**: heads discoid, 1–few in terminal clusters; involucre 9–13 mm, 8–12 mm diam, obconic to bell-shaped; phyllaries in 2–3 series, lanceolate, acuminate, scarious, grading into upper lvs. **DISK FLS** 10–26; corollas 9–11 mm. **FR** 5–6 mm, narrowly obconic, hairy; pappus brownish. $2n$=18. Rocky slopes; 2700–3700 m. SNH, Wrn, n SNE (Sweetwater Mtns); to OR, ID, WY, Colorado, UT. [*Haplopappus macronema* A. Gray] Stabilized hybrids with *Chrysothamnus nauseosus* ssp. *albicaulis* have been called *C. parryi* (A. Gray) E. Greene ssp. *bolanderi* (A. Gray) H.M. Hall & Clements.

E. ericoides (Less.) Jepson (p. 259) Pl < 10 dm, ± lightly puberulent. **LVS** crowded, 4–12 mm, subcylindric, grooved on back, ± obtuse, gen with fan-shaped axillary clusters. **INFL**: heads radiate, many in panicle-like clusters; involucre 5–7 mm, 5–6 mm diam, obconic; phyllaries 16–22 in 3–4 series, outer lanceolate, acuminate, inner narrowly oblong, acute, tomentose-ciliate, midrib a thread-like gland. **RAY FLS** 2–6; corollas 4–5.5 mm. **DISK FLS** 8–14; corollas 6–7 mm. **FR** < or = 3 mm, cylindric, 8–10-ribbed, striate, glabrous to ± silvery hairy; pappus = corolla, white to tan. $2n$=18. Dunes, inland sandy soils; < 500 m. s NCo, CCo, SCoRO, c SCo. [*Haplopappus e.* (Less.) Hook. & Arn. incl ssp. *blakei* C. Wolf] Perhaps 1 variable sp. with *E. palmeri*, *E. pinifolia*. Hybridizes with *Chrysothamnus nauseosus*. ❀DRN,SUN:**17**&IRR:14,**15,16**,22–24.

E. fasciculata (Eastw.) J.F. Macbr. (p. 259) EASTWOOD'S GOLDENBUSH Pl < 5 dm, densely leafy, glabrous to sparsely puberulent. **LF** 10–20 mm, subcylindric, reflexed with age, ± obtuse. **INFL**: heads gen radiate, 1–few in cymes; involucre 6–7 mm, 6–8 mm diam, obconic; phyllaries 22–26 in 4–5 series, scarious, pale yellow, resinous, outer lanceolate or ovate, subacuminate, inner oblong, acute. **RAY FLS** (0)1–2; corollas 5–6.5 mm. **DISK FLS** 20–25; corollas 8–10 mm. **FR** < 3.5 mm, subcylindric, densely long-silky-hairy; pappus > corolla, dull white to reddish brown. $2n$=18. RARE. Dunes, coastal chaparral, closed-cone pine forest; < 100 m. c CCo (n Monterey Co.). [*Haplopappus eastwoodiae* H.M. Hall] In cult.

E. gilmanii (S.F. Blake) G. Nesom (p. 259) GILMAN'S ERICAMERIA Pl 2–4 dm, aromatic, glabrous. **LF** 6–12 mm, oblanceolate, obtuse, often folded. **INFL**: heads radiate, 1–few in cymes; involucre 7–9 mm, ± 5 mm diam, narrowly bell-shaped; phyllaries ± 25 in 4–6 series, outer widely lanceolate, tip often thickened, green, recurved, inner phyllaries oblong, parchment-like, resinous. **RAY FLS** 4–6; corollas 8–10 mm, white or pale yellow. **DISK FLS** 15–18; corollas 7–7.5 mm. **FR** 3–4 mm, ± cylindric, 5-ribbed, silky-hairy; pappus < 6.5 mm, whitish. $2n$=18. RARE. Open coniferous forests, gen on limestone; 2100–3400 m. W&I, n DMtns (Panamint Mtns). [*Haplopappus g.* S.F. Blake]

E. greenei (A. Gray) G. Nesom Pl 1–2.5 dm, glabrous to ± sticky-glandular, puberulent or white-tomentose above. **LF** 15–30 mm, ± oblanceolate, obtuse. **INFL**: heads radiate, gen in crowded, terminal clusters; involucre 8–12 mm, 12–15 mm diam, obconic; phyllaries linear to lanceolate, acute, glandular and sticky to minutely tomentose, all but innermost with lf-like tips. **RAY FLS** 2–6; corollas 7–10 mm. **DISK FLS** 7–20; corollas 9–9.5 mm. **FR** 6–7 mm, narrowly oblong, ± glabrous to densely appressed-soft-hairy; pappus = disk corollas, white to light brown. Rocky areas in open coniferous forest; 1500–2200 m. KR, NCoRH, CaRH, n SNH, Wrn; to WA, ID. [*Haplopappus g.* A. Gray] ❀TRY;DFCLT.

E. laricifolia (A. Gray) Shinn. TURPENTINE-BRUSH Pl 3–10 dm, glabrous, aromatic. **LF** 10–30 mm, gen subcylindric, ± acute. **INFL**: heads radiate, in cymes; involucre 3–5 mm, 3–5 mm diam, obconic; phyllaries 12–20 in 3–4 series, ± linear, acute, glabrous, midrib a brownish to yellowish gland. **RAY FLS** 3–11; corollas 8–11 mm. **DISK FLS** 10–18; corollas 5.5–6.5 mm. **FR** 3.5–4 mm, narrowly obconic, obscurely 4-ribbed, densely white-soft-hairy; pappus 5–8 mm, tan. $2n$=18. Rocky canyons, pinyon/juniper woodland, creosote-bush scrub; 1000–2000 m. DMtns; to TX, n Mex. [*Haplopappus l.* A. Gray] ❀TRY;DFCLT.

E. linearifolia (DC.) Urb. & J. Wussow (p. 259) INTERIOR GOLDENBUSH Pl 4–15 dm, glabrous to ± puberulent. **LF** 10–55 mm, linear, acute; base narrowed. **INFL**: head radiate, 1, on ± lfless peduncles; involucre 8–14 mm, 10–18 mm diam, hemispheric; phyllaries in 2–3 series, linear to lanceolate, acuminate, stalked-glandular, center green, margin cut-ciliate, scabrous. **RAY FLS** 13–18; corollas 9–20 mm. **DISK FLS** many; corollas 6–10 mm. **FR** 4–5 mm, compressed, 6–8-veined, densely silky-hairy; pappus 5.5–7 mm, white. $2n$=18. Dry slopes, valleys; < 2000 m. s SNF, ScV (Sutter Buttes), s SnJV, e CW, WTR, DMoj, w DSon; NV, sw UT, w AZ. [*Haplopappus l.* DC.] Fls spring. Hybridizes with *E. cooperi.* ❀DRN,DRY,SUN:2,3,7,9–12,14–16,18–24;DFCLT.

E. nana Nutt. Pl 1–5 dm, glabrous. **LF** 10–15 mm, linear to narrowly oblanceolate, gen curved, acute, not gland-dotted, ± sticky. **INFL**: heads radiate, in dense leafy cymes; involucre 5.5–7.5 mm, 3–4 mm diam, obconic; phyllaries 20–30 in 4–5 series, lanceolate, acute to acuminate, glabrous. **RAY FLS** 1–7; corollas 2–3 mm. **DISK FLS** 4–8; corollas 4.5–6.5 mm. **FR** 4.5–5.5 mm, cylindric, faintly 5-angled, glabrous to densely hairy; pappus = disk corollas, light brown. $2n$=18. Rocky soils, cliffs; 2100–2800 m. c SNE; to WA, ID, UT. [*Haplopappus n.* (Nutt.) D. Eaton]

E. ophitidis (J. Howell) G. Nesom SERPENTINE GOLDENBUSH Pl < 3 dm, (sub)glabrous. **LF** 5–15 mm, linear, recurved, abruptly pointed. **INFL**: head discoid, 1; involucre 12–14 mm, 8–10 mm diam, subcylindric to obconic; phyllaries ± 25 in 5–6 series, outer lf-like, inner parchment-like, tips green, recurved. **DISK FLS** 5–6; corollas ± 10 mm. **FR** ± 7 mm, 5-angled, appressed-hairy; pappus ± 10 mm. UNCOMMON. Open coniferous forests, gen on serpentine; ± 1600 m. n NCoR (s Trinity, sw Shasta, w Tehama cos). [*Haplopappus o.* (J. Howell) Keck] ❀DRN,SUN:1,6,17&IRR:2,7,14–16;DFCLT.

E. palmeri (A. Gray) H.M. Hall (p. 259) Pl 5–40 dm, glabrous to puberulent, gland-dotted when young. **LF** < 40 mm, linear, acute. **INFL**: heads radiate, many; involucre 5–8.5 mm, 3.5–4.5 mm diam, obconic to ± cylindric; phyllaries 16–24 in 4–5 series, < disk fls, oblong, tips greenish, oblong or acute, margins narrow, white, ciliate. **RAY FLS** 1–8; corollas 4–6 mm. **DISK FLS** 5–20; corollas 5–8 mm. **FR** 3–4 mm, subcylindric, 4–7-angled, hairy; pappus of disk fls > corollas, brown. Plains, foothills; < 800 m. SCo, w DSon; n Baja CA. [*Haplopappus p.* A. Gray] See *E. ericoides.*

var. ***pachylepis*** (H.M. Hall) G. Nesom Pl 5–15 dm, very leafy, puberulent. **LF** 5–16 mm, thread-like. **INFL**: involucre 6–7 mm. **RAY FLS** 1–6. Coastal scrub, disturbed chaparral; < 800 m. ChI, n SCo, w DSon. [*Haplopappus palmeri* A. Gray var. *p.* H.M. Hall] ❀TRY.

var. ***palmeri*** PALMER'S GOLDENBUSH Pl 1–4 dm, stout. **LF** 20–40 mm, often curved. **INFL**: involucre 5–6.5 mm. **RAY FLS** 4–10. $2n$=18. RARE. Coastal scrub; < 600 m. s SCo (s San Diego Co.); n Baja CA. Threatened by urban development.

E. parishii (E. Greene) H.M. Hall var. ***parishii*** Pl 15–50 dm, very leafy, glabrous. **LF** 20–60 mm, ± elliptic, acute, leathery. **INFL**: heads discoid, in dense, terminal cymes; involucre 5–6.5 mm, 3–4 mm diam, obconic; phyllaries ± lanceolate, acute, whitish, midrib thick, brown. **DISK FLS** 8–18; corollas 5–6 mm. **FR** 2–2.5 mm, obconic, minutely hairy, veins 12 or fewer; pappus > disk corollas, copious, dull white to brown. $2n$=18. Dry slopes, chaparral, open forest, esp after fires; 400–2200 m. SnGb, SnBr, PR. [*Haplopappus p.* (E. Greene) S.F. Blake] Var. *peninsularis* (Moran) G. Nesom in n Baja CA. Sometimes treated as ssp. of *E. arborescens.* ❀TRY.

E. pinifolia (A. Gray) H.M. Hall PINE-BUSH Pl 6–25 dm, ± glabrous. **LF** 10–40 mm, thread-like, subacute; short lvs clustered in axils. **INFL**: heads radiate; head in spring 1, large; heads in autumn many, smaller; involucre 5–10 mm, 5–15 mm diam, obconic to hemispheric; phyllaries 20–26 in ± 4 series, lanceolate-acuminate to oblong, woolly-ciliate, tips green, outer often tailed, inner acute, margin narrow. **RAY FLS** 15–30 in spring, 5–10 in autumn; corollas 3.5–5 mm. **DISK FLS** 12–many, corollas 6–7.5 mm. **FR** 4.5–5 mm, subcylindric, striate, lightly hairy; pappus > disk corollas, reddish or tan. $2n$=18. Chaparral, oak woodland, scrub away from coast; < 1700 m. WTR, PR, w DSon. [*Haplopappus p.* (A. Gray) H.M. Hall] See *E. ericoides.* ❀DRN,DRY,SUN:7,14–17,19–21.

E. suffruticosa (Nutt.) G. Nesom Pl 1.5–4 dm, glandular-puberulent. **LF** 15–40 mm, ± oblanceolate, acute. **INFL**: heads radiate, 1–few in small cymes; involucre 8.5–15 mm, 10–15 mm diam, bell-shaped; phyllaries lanceolate, outer green and leafy throughout, inner green above, acuminate, glandular. **RAY FLS** 1–6; corollas 7–12 mm. **DISK FLS** 13–40; corollas 8.5–10.5 mm. **FR** 5.5–6.5 mm, narrowly obconic, angled, hairy; pappus < disk corollas, white to pale yellow. $2n$=18. Coniferous forest, alpine places; 2400–3700 m. SNH, n SNE (Sweetwater Mtns), W&I; to OR, MT, WY, AZ. [*Haplopappus s.* (Nutt.) A. Gray]

ERIGERON FLEABANE DAISY

Guy L. Nesom

Ann to per (subshrub). **STS** gen erect. **LVS** alternate, gen entire. **INFL**: heads gen radiate, 1–many in loose, panicle-like or flat-topped clusters; involucre hemispheric; phyllaries narrowly lanceolate, in 2–several equal to strongly graded series; receptacle flat to steeply conic, naked, smooth to shallowly pitted. **RAY FLS** (0) gen 10–many; ligules gen white, pink, or blue (yellow). **DISK FLS** many; corollas gen narrowly funnel-shaped, yellow; style tips 0.1–0.8 mm, ± triangular. **FR** 0.5–3 mm, gen ± oblong, compressed to ± cylindric, gen 2-ribbed, gen sparsely hairy; pappus (0) gen of 6–50 longer, inner bristles and shorter outer bristles, narrow scales, or short crown. ± 375 spp.: wordwide. (Greek: early old age) [Nesom 1992 Phytologia 72:157–208]

1. Heads discoid or disciform; pistillate fls 0 or << involucre
 2. Pistillate fls present, corollas inconspicuous
  3. Heads 2–many; sts gen branched near mid-st or below
   4. Per — pappus bristles 12–17 . . . . *E. aphanactis* var. ***aphanactis***

4′ Ann or bien
5. Pappus bristles 15–20 .... *E. calvus*
5′ Pappus bristles 6–9(12) .... [3]*E. divergens*
3′ Heads solitary; sts unbranched
6. Ann or bien; pappus bristles 6–9(12) .... [3]*E. divergens*
6′ Per; pappus bristles 12–25
7. Lvs 2–3-ternately dissected .... [2]*E. compositus*
7′ Lvs entire
8. Sts minutely glandular; disk corollas gen much wider upward; fr base ringed whitish; outer pappus of narrow scales or obvious bristles .... *E. aphanactis* var. *congestus*
8′ Sts glandless; disk corollas narrowly funnel-shaped; fr base ringed yellowish; outer pappus of minute bristles .... *E. austiniae*
2′ Pistillate fls 0
9. Basal lvs present at fl; cauline lvs much reduced or 0 upward
10. Caudex simple .... *E. lassenianus* var. *deficiens*
10′ Caudex branched
11. Sts ascending, 15–40 cm, sometimes spreading-hairy; caudex branches slender, rhizome-like; basal lvs oblanceolate to spoon-shaped .... *E. supplex*
11′ Sts erect, 5–15 cm, sometimes strigose; caudex branches stout; basal lvs ± linear .... *E. bloomeri*
12. Herbage and phyllaries strigose .... var. *bloomeri*
12′ Herbage and phyllaries ± glabrous .... var. *nudatus*
9′ Basal lvs 0 at fl; cauline lvs gen evenly sized and spaced
13. Glandless st hairs spreading
14. St hairs dense, short, stiff, slightly reflexed, grayish .... *E. inornatus* var. *keilii*
14′ St hairs sparse to dense, long, ± crinkly-spreading, white
15. Sts 30–90 cm, erect; lvs 20–40 mm; heads 12–15 mm diam .... [2]*E. biolettii*
15′ Sts 5–20(30) cm, decumbent or ascending; lvs 7–25 mm; heads 7–12 mm diam
16. Inner phyllaries 3.5–5 mm .... *E. miser*
16′ Inner phyllaries 5.5–7 mm .... *E. petrophilus*
17. Herbage ± nonglandular; hairs stiff, straight or curved .... var. *viscidulus*
17′ Herbage densely glandular; nonglandular hairs gen loose, often crinkly
18. Phyllaries distinctly expanded and purplish near tip .... var. *petrophilus*
18′ Phyllary tips like lower portions .... var. *sierrensis*
13′ Glandless st hairs 0-sparse, appressed
19. Phyllaries glabrous .... *E. inornatus*
20. Sts 10–20 cm, decumbent to ascending; hairs below middle spreading to reflexed; lower and middle lf blades and margins stiffly spreading-hairy; inner phyllaries 5–6.5 mm .... var. *calidipetris*
20′ Sts 30–90 cm, ascending; hairs below middle appressed; lower and middle lf blades and margins gen short-strigose; inner phyllaries 4.5–5.5 mm .... var. *inornatus*
19′ Phyllaries glandular (sometimes sparsely nonglandular)
21. Herbage glandular; lvs narrowly oblanceolate .... [2]*E. biolettii*
21′ Herbage nonglandular; lvs linear to oblanceolate
22. Lvs ± oblanceolate; sts long-ascending from woody taproot .... *E. inornatus* var. *inornatus*
22′ Lvs linear; sts from caudex or thin, rhizome-like branches
23. Sts 30–90 cm, many, from woody caudex .... *E. angustatus*
23′ Sts 8–20(30) cm, arising singly from slender, rhizome-like caudex branches .... *E. reductus*
24. Phyllary tips gen purplish (variable n CCo); pappus bristles (38)46–61 .... var. *angustatus*
24′ Phyllary tips greenish; pappus bristles 20–30 .... var. *reductus*
1′ Heads radiate; pistillate fls with obvious ligules
25. Ann from shallow fibrous roots; ray pappus 0
26. St hairs spreading; lvs gen coarsely toothed .... *E. annuus*
26′ St hairs appressed; lvs gen entire .... *E. strigosus*
25′ Per from woody taproot, branched caudex, or rhizome; ray pappus present
27. Lvs 1–4-ternately dissected
28. Lvs 2–4-ternately dissected; ligules < 1 mm wide; caudex branches thick, ascending .... [2]*E. compositus*
28′ Lvs deeply 3-lobed at tip; ligules 1–2 mm wide; caudex branches thin, rhizome-like .... *E. vagus*
27′ Lvs entire to shallowly toothed
29. Lf bases hardened, white-shiny, expanded, enclosing lower st
30. Caudex branches thick, lf bases persistent; ligules yellow .... *E. linearis*
30′ Caudex branches slender, rhizome-like, lf bases not persistent; ligules white to blue
31. Heads 13–18 mm diam; basal lvs gen 2–5 mm wide; pappus bristles 25–40; phyllary hairs spreading; > 2100 m .... *E. barbellulatus*
31′ Heads 6–11 mm diam; basal lvs gen 0.5–1 mm wide; pappus bristles 20–30; phyllary hairs appressed; 1000–2100 m .... *E. elegantulus*
29′ Lf bases ± thin, not hardened, barely or not expanded
32. Fr ribs 4–8; herbage gen ± silvery, hairs dense, appressed
33. Fr ribs 6–8; basal lvs tufted, persistent .... *E. argentatus*
33′ Fr ribs 4(6); basal lvs gen 0 at fl
34. Outer pappus of narrow, conspicuous scales; sts silvery-white .... *E. parishii*

34′ Outer pappus inconspicuous; sts gray-green, hairs ± sparse .... ***E. utahensis***
32′ Fr ribs gen 2 (if more, herbage not silvery)
35. Phyllaries strongly graded; lvs entire, not clasping, basal 0, cauline gen evenly sized and spaced; infls arising near st tips (sect. *Pycnophyllum*)
36. Sts minutely glandular .... ***E. aequifolius***
36′ Sts not glandular
37. St hairs dense, crinkly; fr glabrous .... ***E. blochmaniae***
37′ Sts hairs 0 to dense, straight, stiffly spreading to ± reflexed or appressed; fr sparsely strigose
38. St hairs dense, stiffly spreading, gen slightly reflexed .... ***E. breweri***
39. Phyllaries densely glandular, glandless hairs 0 or very sparse
40. Sts 7–15 cm, prostrate to decumbent; lvs 5–12 mm .... var. ***jacinteus***
40′ Sts gen > 20 cm, ascending to erect; lvs gen > 15 mm
41. Herbage hairs 0.1–0.3 mm; inner phyllary tips green; sts from rhizome-like bases ...... var. ***breweri***
41′ Herbage hairs 0.5–1 mm; inner phyllary tips widely, white-thick-margined; sts from woody basal offsets and roots .... var. ***klamathensis***
39′ Phyllaries not or barely glandular, glandless hairs prominent
42. Sts 20–30 cm, wiry, brittle; phyllary hairs long, thick-based, stiffly spreading, translucent (also some glandular) .... var. ***porphyreticus***
42′ Sts (30)40–75 cm, not wiry or brittle; phyllary hairs short, white, ± appressed, glandless
43. Phyllary hairs thin-based, ± appressed, slightly less dense on inner phyllaries .... var. ***bisanctus***
43′ Phyllary hairs thick-based, ascending, much less dense on middle and inner phyllaries .. var. ***covillei***
38′ St hairs ± 0 or sparse, appressed
44. Sts prostrate to decumbent, from slender rhizomes .... ***E. elmeri***
44′ Sts ascending to erect, from woody taproot
45. Ray fls 9–13; ciliate hairs thin-based; phyllaries minutely nonglandular, densely sessile-glandular .... ***E. serpentinus***
45′ Ray fls 15–50; ciliate hairs thick-based; phyllaries glandular or nonglandular
46. Phyllaries ± nonglandular, 0.5–0.8 mm wide, margins gen thick or narrowly scarious
47. Lvs 2–4(5) cm, oriented variously; inner phyllaries 3.2–4.5 mm .... ***E. foliosus*** var. ***foliosus***
47′ Lvs (3)4–6 cm, often ± oriented to 1 side of st; inner phyllaries (4)5–6 mm
48. Sts 30–85 cm; axillary lf-tufts 0; lvs 35–80 mm, 1–2(4) mm wide; inner phyllaries not strongly scarious; ray fls 26–50 .... ***E. foliosus*** var. ***hartwegii***
48′ Sts 15–20 cm; axillary lf-tufts present; lvs 25–45 mm, (2)5–8 mm wide; inner phyllary margins strongly scarious; ray fls 18–22 .... ***E. mariposanus***
46′ Phyllaries minutely but prominently glandular, 0.8–1 mm wide, margins gen widely scarious .... part of ***E. foliosus***
49. Phyllary hairs all glandular; lvs gen 1–2 mm wide .... var. ***confinis***
49′ Phyllary hairs ± dense, glandless; lvs gen 2–4 mm wide
50. Lf blades ± strigose; phyllary midvein raised, orange-resinous; ray corollas 7–10 mm .... var. ***franciscensis***
50′ Lf blades ± glabrous (ciliate); phyllary midvein gen not raised or orange-resinous; ray corollas 10–15 mm .... var. ***mendocinus***
35′ Phyllaries gen ± equal; lvs entire to lobed, sometimes clasping, basal sometimes present, cauline often strongly reduced upwards; infls arising near mid-st
51. Cauline lvs ± clasping; pls fibrous-rooted, gen from obvious lateral rhizomes
52. Glandless phyllary hairs with black cross-walls .... ***E. coulteri***
52′ Glandless phyllary hairs with clear cross-walls
53. Bien or short-lived per, gen not rhizomed; ray fls 150–400 .... ***E. philadelphicus***
53′ Per; ray fls 30–165
54. Lvs thick, fleshy, entire or toothed; coastal habitats .... ***E. glaucus***
54′ Lvs ± thin, entire; inland habitats
55. Phyllary hairs dense, all glandular
56. Sts glabrous; ray fls 20–30(45) .... ***E. cervinus***
56′ Sts ± appressed-hairy; ray fls 30–105 .... ***E. peregrinus***
57. Herbage strigose .... var. ***callianthemus***
57′ Herbage spreading-hairy .... var. ***hirsutus***
55′ Phyllary hairs prominent, glandless (sometimes also glandular)
58. Sts from thin, often stolon-like caudex branches with few fibrous roots .... ***E. sanctarum***
58′ Sts from short, thick, strongly fibrous-rooted rhizomes
59. Sts 2–20(30) cm, hairs spreading; phyllaries blackish purple; pappus bristles 12–16(20) .... [2]***E. algidus***
59′ Sts 30–100 cm, hairs appressed; phyllaries greenish; pappus bristles 20–30 .... ***E. aliceae***
51′ Cauline lvs not at all clasping; pls taprooted
60. Sts spreading-hairy
61. Ann from slender taproot; caudex gen simple; sts branched; heads many
62. Sts and phyllaries densely, evenly puberulent, hairs < 0.5 mm .... ***E. divergens***
62′ Sts and phyllaries glandular-puberulent, also sparsely spreading-hairy (hairs 0.5–2 mm) ***E. lobatus***
61′ Per from rhizome or stout taproot and branched caudex; sts gen unbranched; head 1
63. Lf blades clearly petioled

64. Basal lvs 20–70 mm; cauline lvs much reduced; sts 4–30 cm; phyllaries blackish-purple; ray fls (30)50–125, ligules 7–13 mm . . . . . [2]***E. algidus***
64′ Basal lvs 5–25 mm; cauline lvs 0; sts 1–4 cm; phyllaries green; ray fls 15–40, ligules 4–6 mm . . . . . ***E. uncialis*** var. ***uncialis***
63′ Lvs linear or gradually tapered to base, petiole indistinct
65. Margins of lower lvs gen hairy but not stiffly spreading-ciliate; > 2200 m
66. Basal lvs 20–80 mm; sts ± nonglandular; phyllaries green; ray fls 25–55, ligules 6–11 mm . . . . . ***E. clokeyi***
66′ Basal lvs 6–35 mm; sts densely glandular; phyllaries blackish purple; ray fls 20–37, ligules 4–7(10) mm . . . . . ***E. pygmaeus***
65′ Margins of lower lvs stiffly long-spreading-ciliate; 1200–1800 m
67. Disk corollas (sub)glabrous; pappus bristles 12–22; outer pappus of bristles or narrow scales; ray fls (40)55–115, ligules 8–12 mm . . . . . ***E. pumilus*** var. ***intermedius***
67′ Disk corollas sharply scabrous; pappus bristles 5–15; outer pappus of wide scales; rays 40–60, ligules 7–9 mm . . . . . ***E. concinnus***
68. Sts leafy; heads many . . . . . var. ***concinnus***
68′ Sts ± lfless; head 1 . . . . . var. ***condensatus***
60′ Sts glabrous or appressed-hairy
69. Ray fls 75–125; disk corolla throats wide . . . . . ***E. multiceps***
69′ Ray fls 14–80; disk corollas cylindric, throat ± 0
70. Lvs few-toothed near tip, basal 0, only upper cauline reduced; axillary lf-tufts present; ray fls 45–80 . . . . . ***E. karvinskianus***
70′ Lvs entire, basal gen persistent, cauline reduced; axillary lf-tufts 0; ray fls 14–45
71. Basal lvs linear
72. Sts 2–8 cm, lfless; st hairs straight, appressed; basal lvs gen 1–3 cm; fr ribs densely ciliate, faces glabrous . . . . . ***E. compactus***
72′ Sts 15–30 cm, leafy below; st hairs curved up, ± loose; basal lvs gen 2–7 cm; fr ribs and faces sparsely hairy . . . . . ***E. filifolius*** var. ***filifolius***
71′ Basal lvs narrowly oblanceolate to spoon-shaped
73. Pls cespitose; caudex stout, branched; basal lvs spoon-shaped, petioled; cauline lvs abruptly reduced . . . . . ***E. tener***
73′ Pls not cespitose, caudex gen slender; basal lvs narrowly oblanceolate, petiole indistinct; cauline lvs gradually reduced
74. Lvs gen 1-veined . . . . . ***E. lassenianus*** var. ***lassenianus***
74′ Lvs gen 3-veined
75. Sts 25–55 cm; heads (1)2–4; basal lvs gen 0 at fl . . . . . ***E. decumbens*** var. ***robustior***
75′ Sts 4–15(30) cm; heads 1–2(7); basal lvs persistent at fl . . . . . ***E. eatonii***
76. Heads (14)17–23 mm diam; longest phyllaries 7–11 mm; disk corollas 4.4–6.8 mm . . . . . var. ***nevadincola***
76′ Heads 8–16 mm diam; longest phyllaries 5–8 mm; disk corollas 3–5 mm
77. Sts 10–23 cm; disk corollas 3–4 mm; pappus bristles 16–20, 3–3.5 mm . . . . . var. ***plantagineus***
77′ Sts 4–12(21) cm; disk corollas 3.5–5 mm; pappus bristles 18–30, 3.5–5 mm . . . . . var. ***sonnei***

E. aequifolius H.M. Hall (p. 259) HALL'S DAISY Per 10–20 cm, from woody roots and slender-branched caudex, ascending to erect, little-branched, short-loose-spreading-hairy, minutely sessile-glandular. **LVS** cauline, 6–20 mm, narrowly elliptic to oblanceolate, evenly sized and spaced. **INFL**: head 1, short-peduncled, 7–10 mm diam; phyllaries strongly graded in 3–5 series, minutely glandular (or base sparsely nonglandular). **RAY FLS** 14–30; corollas 5–8 mm, ligules drying blue, weakly coiled. **FR**: pappus bristles 20–35. $2n$=18. RARE. Rock ledges, crevices; 1500–2100 m. s SNH.

E. algidus Jepson (p. 259) Per 2–30 cm, from (sub)simple caudex and fibrous roots, unbranched, sparsely and loosely spreading-hairy and glandular. **LVS**: basal 2–7 cm, oblanceolate to spoon-shaped; cauline much reduced. **INFL**: head 1, 8–16 mm diam; phyllaries ± equal, blackish purple, tips spreading to reflexed. **RAY FLS** 30–125; corollas 7–13 mm, ligules blue or pink to white, coiled. **FR**: pappus bristles 12–20. Alpine meadows, talus; 2600–3700 m. SNH, SNE; w NV. [*E. petiolaris* E. Greene] ❀DRN, SUN:**1**&IRR:**2**,3,15–18;DFCLT.

E. aliceae Howell (p. 259) Per 30–100 cm, from short rhizome, few-branched near mid-st, sparsely strigose. **LVS**: basal petioled, blade oblanceolate, spreading-hairy; mid-cauline 5–15 cm, lanceolate, gen ± clasping. **INFL**: heads 1–7, long-peduncled, 12–20 mm diam; phyllaries ± equal, loose-spreading-hairy, densely sessile-glandular. **RAY FLS** 45–80; corollas 10–15 mm, ligules white, often drying bluish, tips coiled slightly. **FR**: pappus bristles 20–30. Meadows, openings in woods; 1300–2200 m. NW; to WA. ❀TRY; DFCLT.

E. angustatus E. Greene (p. 259) Per 30–90 cm, from stout, woody taproot, many-stemmed, few-branched above, (sub)glabrous. **LVS** cauline, 1–6 cm, thread-like to linear, evenly sized and spaced. **INFL**: heads discoid; disk 8–12 mm diam; phyllaries ± equal, densely sessile-glandular, glandless hairs 0, tip gen purple. **FR**: pappus bristles 26–38. Serpentine areas, gen in shrubby vegetation; < ± 150 m. s NCoRO.

E. annuus (L.) Pers. Ann 50–120 cm, from fibrous roots, sparsely spreading- or appressed-hairy. **LVS**: basal gen petioled, blades 4–16 cm, elliptic to obovate, coarsely serrate; cauline gen slightly reduced upward. **INFL** flat-topped; heads many, 5–10 mm diam; phyllaries ± equal, spreading-hairy, sparsely glandular. **RAY FLS** 80–120; corollas 5–8 mm, ligules white, not coiled or reflexed. **FR**: ray pappus bristles 0; disk pappus bristles 10–15. $2n$=27. Disturbed places; < 2000 m. KR, n&c SNH; worldwide weed; native to e US. Asexual triploid. Intergrades with *E. strigosus*.

E. aphanactis (A. Gray) E. Greene (p. 259) Per 8–25 cm, from taproot and short-branched caudex, often densely cespitose, stiffly spreading-hairy, sessile-glandular. **LVS**: basal 4–7 cm, sometimes long-petioled, linear-oblanceolate, gradually reduced upwards. **INFL** flat-topped; heads disciform, 1–many, 8–12 mm diam; phyllaries ± equal. **PISTILLATE FLS** many; ligules 0 or < involucre. **DISK FLS**: corollas gen abruptly inflated to throat. **FR**: pappus bristles 12–17. Sagebrush or juniper scrub; 1300–2600 m. SnBr, SNE, DMtns; to OR, Colorado, NM.

var. ***aphanactis*** **STS** gen branched in lower 1/2. **INFL**: heads gen 2–many. $2n$=18. Habitats and ± range of sp.

var. ***congestus*** (E. Greene) Cronq. **STS** unbranched. **INFL**: head 1. Habitats of sp.; 1800–2600 m. SnBr; to UT. Geog separate from var. *a.*

E. argentatus A. Gray (p. 259) Per 10–40 cm, from woody taproot and short-branched caudex, often cespitose, densely silvery-hairy. **LVS**: basal many, erect, gen 2–5 cm, narrowly oblanceolate; cauline scattered on lower 2/3 of st. **INFL**: head 1, 12–22 mm diam; phyllaries ± equal. **RAY FLS** 25–48; corollas 10–16 mm, ligules blue, coiled. **FR** 6–8-ribbed, densely hairy; pappus bristles 25–40. 2*n*=18. Rocky slopes, pinyon/juniper woodland; 2000–2300 m. W&I; to UT, w AZ. ❀TRY;DFCLT.

E. austiniae E. Greene Per 3–12 cm, from woody taproot and short-branched caudex, cespitose, sparsely short-spreading-hairy. **LVS** ± basal, erect, 2–8 cm, linear to oblanceolate. **INFL**: head disciform, 1; disk 10–15 mm diam; phyllaries ± equal. **PISTILLATE FLS** many; ligules ± 0 (yellow if apparent). **FR**: pappus bristles 15–25. 2*n*=18. Crevices, rocky slopes, sagebrush scrub; 1200–1700 m. MP; to se OR, sw ID, n NV. [*E. chrysopsidis* ssp. *a.* (E. Greene) Cronq.]

E. barbellulatus E. Greene (p. 259) Per 5–15 cm, from slender-branched caudex (or taproot), ± appressed-short-hairy. **STS** ascending, unbranched, lower 1/4 gen leafy, white-shiny or purplish. **LVS** gen basal, 2–5 cm, narrowly oblanceolate; bases wider, hard, white-shiny, enclosing lower st. **INFL**: head 1, 13–18 mm diam; phyllaries ± equal, sparsely and finely spreading-hairy. **RAY FLS** 15–35; corollas 7–15 mm, ligules white to blue, drying bluish, not coiled or reflexed. **FR**: pappus bristles 25–40. 2*n*=18. Gravelly or rocky slopes, sagebrush/pine to subalpine forest; 2100–3300 m. CaRH, n&c SNH. ❀TRY;DFCLT.

E. biolettii E. Greene (p. 259) Per 30–90 cm, from woody caudex, gen ascending, branched above, sparsely spreading-hairy, densely glandular. **LVS** cauline, 2–4 cm, narrowly oblanceolate, evenly sized and spaced. **INFL** flat-topped; heads discoid, 12–15 mm diam; phyllaries strongly graded in 3–5 series, densely glandular, tips of inner gen purple. **FR**: pappus bristles 22–38. Dry slopes, rocks, ledges along rivers; < 1100 m. NCoRO. [*E. inornatus* var. *b.* (E. Greene) Jepson] Intergrades with *E. inornatus*.

E. blochmaniae E. Greene (p. 259) BLOCHMAN'S LEAFY DAISY Per 40–80 cm, from woody caudex (or rhizome), gen ascending, branched above, densely and minutely curled- or bent-hairy. **LVS** cauline, 1–3 cm, linear to narrowly oblanceolate, evenly sized and spaced. **INFL** flat-topped; heads 9–14 mm diam; phyllaries strongly graded in 3–5 series, densely short-stiff-spreading-hairy, densely sessile-glandular. **RAY FLS** 45–72; corollas 8–11 mm, ligules white to blue, weakly coiled. **FR** glabrous; pappus bristles 21–36. 2*n*=18. RARE. Sand dunes and hills; < 30 m. s CCo. [*E. foliosus* var. *b.* (E. Greene) H.M. Hall] Threatened by coastal development. In cult.

E. bloomeri A. Gray Per 4–15 cm, from taproot and short-branched caudex, cespitose, ± glabrous to densely white-strigose. **LVS**: basal 2–7 cm, ± linear; cauline lvs on lower st. **INFL**: head discoid, 1, 7–20 mm diam; phyllaries ± equal, hairy, sparsely sessile-glandular. **FR**: pappus bristles 25–40. Rocky slopes, lava beds, meadows; 600–2300 m. KR, CaR, n SNH, GB; to WA, ID, NV.

var. ***bloomeri*** Herbage strigose. 2*n*=18. Habitats of sp.; 800–2000 m. Range of sp. (exc KR). [var. *pubens* Keck] ❀DRN, SUN:1&IRR:2,7,14–17;DFCLT.

var. ***nudatus*** (A. Gray) Cronq. WALDO DAISY Herbage ± glabrous. RARE in CA. Serpentine slopes, rocky ridges; 600–2300 m. KR; sw OR.

E. breweri A. Gray Per 7–75 cm, from woody roots and slender-branched caudex, gen not wiry or brittle, gen densely short-spreading-hairy; glands gen 0. **LVS** cauline, 5–40 mm, linear to oblanceolate, evenly sized and spaced. **INFL**: heads gen radiate, 8–15 mm diam; phyllaries strongly graded in 3–5 series, glandular or not, tip gen ± like body. **RAY FLS** 12–45, 4–7 mm, ligules white to pink or drying blue, weakly coiled. **FR**: pappus bristles 22–46. Many habitats; 300–3100 m. KR, SN, TR, SnJt, SNE, DMoj; OR, w NV.

var. ***bisanctus*** G. Nesom (p. 259) **ST** 30–75 cm; base thicker. **LF** 20–35 mm; hairs < 0.1 mm. **INFL**: phyllary hairs short, white, thin-based, ascending, barely fewer from outer to inner phyllaries, nonglandular. Open, dry slopes and washes; 300–1600 m. SnGb, SnBr.

var. ***breweri*** (p. 259) **ST** 12–60 cm, from ± slender rhizomes, gen ascending. **LF** 10–40 mm; hairs 0.1–0.3 mm. **INFL**: phyllaries narrow-margined, green-tipped, densely glandular. 2*n*=18. Open, rocky habitats; 1200–3100 m. n&s SNH, SnBr; c-w NV. Many pls of s SNH are tall, more erect, lack rhizomes, have narrower lvs.

var. ***covillei*** (E. Greene) G. Nesom **ST** 30–75 cm; base thicker. **LF** 15–30 mm; hairs 0.2–0.4 mm. **INFL**: phyllary hairs nonglandular, short, white, thick-based, ascending, much sparser inward. Open, rocky sagebrush, chaparral, juniper scrub; ± 1000–1900 m. s SNH (e slope), SnGb, SnBr, SnJt, DMoj. [*E. foliosus* var. *c.* (E. Greene) Compton]

var. ***jacinteus*** (H.M. Hall) Cronq. **ST** 7–15 cm, prostrate to decumbent. **LF** 5–12 mm; hairs 0.2–0.4 mm. **INFL**: phyllaries densely glandular. Open, rocky slopes and crests; ± 2700–2900 m. SnBr, SnGb, SnJt.

var. ***klamathensis*** G. Nesom (p. 259) **ST** gen 10–25 cm, gen ascending, sparsely hairy; base wiry. **LF** 12–30 mm; hairs 0.5–1 mm. **INFL**: phyllaries densely glandular. Open, rocky slopes, crevices; ± 1600–2200 m. KR; s-c OR.

var. ***porphyreticus*** (M.E. Jones) Cronq. **ST** 20–30 cm, wiry, brittle. **LF** 5–30 mm; hairs 0.1–0.2 mm. **INFL**: phyllaries thick-margined, hairs glandular and not, long, thick-based, stiffly spreading, translucent. 2*n*=18. Open, rocky sagebrush to yellow-pine forest; 1200–2600 m. SnBr, SNE, DMoj; sw NV. Some pls in SnBr have hairs like var. *breweri*.

E. calvus Cov. BALD DAISY Bien or short-lived per 10–14 cm, from taproot, ± long-spreading-hairy; base much-branched. **LVS**: basal 3–5 cm, spoon-shaped; cauline abruptly reduced. **INFL**: heads disciform or obscurely radiate, 1–few, 13–14 mm diam; phyllaries ± equal, sessile-glandular, hairs thick-based. **RAY FLS** 0–many, < phyllaries. **FR**: pappus bristles 15–20. RARE. Sagebrush and desert scrub; ± 1200 m. s SNE (w base Inyo Mtns). Closely related to *E. divergens*; also confused with *E. aphanactis*.

E. cervinus E. Greene (p. 263) SISKIYOU DAISY Per 15–30 cm, from short rhizomes, 0–3-branched above middle, glabrous or sparsely sessile-glandular. **LVS**: basal 4–12 cm, oblanceolate to narrowly spoon-shaped; cauline slightly clasping, little reduced. **INFL**: heads 1–4, short-peduncled, 10–14 mm diam; phyllaries ± equal, sessile-glandular. **RAY FLS** 20–45; corollas 7–10 mm, ligules white, coiled. **FR**: pappus bristles 12–15. UNCOMMON. Open, rocky slopes, meadows, pine to fir woods; 900–1900 m. KR; sw OR. [*E. delicatus* Cronq., Del Norte daisy] ❀DRN:1,5,6 &IRR:7,15–17.

E. clokeyi Cronq. (p. 263) Per 5–20 cm, from stout taproot and (sub)simple caudex, ascending to erect, unbranched; hairs short, stiffly spreading to reflexed, glandless. **LVS**: basal 2–8 cm, ± oblanceolate; cauline not clasping, gen strongly reduced by mid-st. **INFL**: head 1, 8–12 mm diam; phyllaries ± equal, sessile-glandular, hairs short, stiffly spreading to reflexed. **RAY FLS** 25–55; corollas 6–11 mm, ligules white to blue, reflexed. **FR**: pappus bristles 13–22. 2*n*=18. Sagebrush scrub to alpine talus; 2200–3400 m. s SNH, SNE; to UT. ❀TRY;DFCLT.

E. compactus S.F. Blake (p. 263) Per 2–8 cm, from taproot and short-branched caudex, cespitose, unbranched, short-white-appressed-hairy. **LVS** basal, 5–25 mm, linear. **INFL**: head 1, 10–14 mm diam; phyllaries ± equal. **RAY FLS** 15–32; corollas 7–11 mm, ligules white, sometimes striped lilac, coiled. **FR**: ribs densely ciliate; faces glabrous; pappus bristles 30–40. Rocky slopes, pinyon/juniper woodland; 1800–2300 m. W&I; to UT. ❀TRY;DFCLT.

E. compositus Pursh (p. 263) Per 3–15 cm, from stout taproot and short-branched caudex, cespitose, unbranched, densely sessile-glandular, sometimes also sparsely soft-bristly. **LVS**: gen basal, 1–5 cm, oblanceolate to spoon-shaped, (1)2–3-ternately divided; cauline lvs (if any) much reduced, gen entire. **INFL**: head ± radiate or disciform, 1; disk 10–18 mm diam; phyllaries ± equal. **RAY FLS**

(0) ± 30–60; corollas 1–5 mm, ligules gen < styles in CA, blue to pinkish or white, weakly coiled. **FR**: pappus bristles 12–15. 2*n*=18, 27,36,45,54,63. Rocky slopes, crevices, talus; 2000–4300 m. SNH, Wrn, SNE; to AK, e Can, Colorado, AZ. [vars. *discoideus* A. Gray & *glabratus* Macoun] ❀DRN,IRR:1,2,**3**,6,**7**,**14–16**,**18**.

E. concinnus (Hook. & Arn.) Torrey & A. Gray Per 6–16 cm, from woody taproot and thick, short-branched caudex, cespitose, 1–4-branched below mid-st; hairs spreading or reflexed, soft-bristly and minutely stalked-glandular. **LVS**: basal erect, crowded, 2–6 cm, ± linear to oblanceolate; cauline 0 or gradually reduced upward. **INFL** flat-topped; heads 1–many, 7–11 mm diam; phyllaries ± equal. **RAY FLS** 40–60; corollas 7–9 mm, ligules blue to white or pink, reflexed or weakly coiled. **DISK FLS**: corollas ± widely funnel-shaped. **FR**: pappus bristle 7–15, outer series of prominent, wide scales 0.2–0.5 mm. Sandy to rocky slopes, crevices; 1200–1800 m. DMtns; to ID, WY, Colorado, NM.

var. ***concinnus*** **ST** leafy. **INFL**: heads many. 2*n*=18. Habitats and range of sp. [*E. pumilus* ssp. *concinnoides* Cronq.]

var. ***condensatus*** D. Eaton **ST** ± lfless. **INFL**: head 1. Habitats and range of sp. Scattered variant. [Nesom 1983 Sida 10:159–166]

E. coulteri Porter (p. 263) Per 20–70 cm, from thin rhizomes, 0–3-branched above mid-st, glabrous below, sparsely hairy above. **LVS**: basal gen 5–12 cm, ± oblanceolate, entire or with 2–6 pairs of shallow teeth; cauline oblanceolate to elliptic-ovate, ± clasping, little reduced. **INFL**: heads 1–4, 10–16 mm diam; phyllaries ± equal, sessile-glandular, nonglandular hairs with black crosswalls and bases. **RAY FLS** 45–140; corollas 11–20 mm, ligules white, coiled. **FR**: pappus bristles 20–25. 2*n*=18. Streambanks, wet meadows, coniferous forest; ± 1900–3400 m. SNH, Wrn; to OR, WY, NM. ❀DRN,IRR,SHD:**1,2,6,7,14–16**,17,**18**,19–21.

E. decumbens Nutt. var. ***robustior*** (Cronq.) Cronq. (p. 263) ROBUST DAISY Per 25–55 cm, from taproot and simple caudex, gen decumbent, few-branched near mid-st, gen purplish, sparsely ± appressed-hairy. **LVS**: basal 8–17 cm, linear to narrowly oblanceolate, 3-veined; cauline gradually reduced upwards. **INFL**: heads (1)2–4, long-peduncled, 12–18 mm diam; phyllaries ± equal, ± densely shaggy-hairy. **RAY FLS** 21–36; corollas 9–19 mm, ligules bluish, slightly coiled. **FR**: pappus bristles 14–20. UNCOMMON. Grassy openings, meadows, sometimes on serpentine; 200–500 m. NCoRO. Var. *decumbens* in w OR.

E. divergens Torrey & A. Gray Ann 10–45 cm, from slender taproot, branched near mid-st, ± densely reflexed- to spreading-hairy, minutely sessile-glandular near and on heads. **LVS**: basal and lower cauline 2–6 cm, ± obovate, entire to lobed, gradually reduced upward, not clasping. **INFL** often flat-topped; heads gen radiate, gen many, 7–11 mm diam; phyllaries ± equal. **RAY FLS** gen 75–150; corollas (0)5–10 mm, white to purple. **DISK FLS**: corollas abruptly wider at throat. **FR**: pappus bristles 6–9(12), outer series short bristles or narrow scales. 2*n*=18,27,36. Desert scrub to yellow-pine forest; 500–2600 m. s SN, SnGb, SnBr, SnJt, GB, D; to B.C., TX, nw Mex. Variable. ❀DRN,SUN:1,**2,3,7,10**,12,**14**,15, 16,**22–24**&IRR:**8,9**,11,**18–21**.

E. eatonii A. Gray Per 4–33 cm, from taproot and (sub)simple caudex, prostrate to erect, 0–few-branched below mid-st, ± ascending-hairy. **LVS** linear to narrowly oblanceolate, 3-veined; basal gen present at fl; middle cauline 1–3(5) cm, strigose. **INFL**: heads 1(4), long-peduncled, 8–23 mm diam; phyllaries ± equal. **RAY FLS** 15–39; corollas 7–15 mm, ligules white, gen bluish or pinkish beneath, weakly coiled. **FR**: pappus bristles 16–30. Open grassland, rocky flats, gen in sagebrush or pinyon/juniper scrub; 1000–2900 m. CaR, n&c SNH, GB; to OR, WY, Colorado, AZ. CA vars. intergrade; other vars. in Rocky Mtns.

var. ***nevadincola*** (S.F. Blake) G. Nesom **STS** 14–33 cm. **INFL**: head 14–23 mm diam; longest phyllaries 7–11 mm. **DISK FLS**: corollas 4.4–6.8 mm. **FR**: pappus bristles 20–26, 4–5 mm. Habitats of sp.; 1400–2900 m. MP; n NV. [*E. n.* S.F. Blake]

var. ***plantagineus*** (E. Greene) Cronq. **STS** 10–23 cm. **INFL**: head 8–16 mm diam; longest phyllaries 5–8 mm. **DISK FLS**: corollas 3–4 mm. **FR**: pappus bristles 16–20, 3–3.5 mm. 2*n*=18. Open grassy or sagebrush scrub, gen on volcanic rock; 1000–2500 m. CaR, MP; sc OR.

var. ***sonnei*** (E. Greene) G. Nesom (p. 263) **STS** 4–21 cm. **INFL**: head 8–16 mm diam; longest phyllaries 5–8 mm. **DISK FLS**: corollas 3.5–5 mm. **FR**: pappus bristles 18–30, 3.5–5 mm. 2*n*=18. Rocky grassland or sagebrush scrub; 1800–2800 m. n&c SNH, SNE; c-w NV. [*E. s.* E. Greene]

E. elegantulus E. Greene (p. 263) VOLCANIC DAISY Per 3–15 cm, from taproot and slender-branched caudex, unbranched, sparsely and minutely white-appressed-hairy. **LVS** ± basal, 15–50 mm, linear to narrowly oblanceolate; base wider, hard, white-shiny, enclosing lower st. **INFL**: head 1, 6–11 mm diam; phyllaries ± equal. **RAY FLS** 15–25; corollas 6–9 mm, ligules blue or pink, weakly coiled. **FR**: pappus bristles 20–30. 2*n*=27. UNCOMMON. Sagebrush scrub, juniper woodland; 1000–2200 m. MP; e OR. Asexual triploid.

E. elmeri (E. Greene) E. Greene (p. 263) Per 6–20 cm, from woody roots and rhizomes, prostrate to ± erect, wiry, gen unbranched, sparsely and minutely appressed-thin-based-hairy. **LVS** cauline, 5–20 mm, ± linear to narrowly oblanceolate, entire. **INFL**: heads 1(–3), 7–10 mm diam; phyllaries strongly graded in 3–5 series, minutely glandular, inner often purple-tipped. **RAY FLS** 12–21; corollas 6–9 mm, ligules drying white to pink or blue, weakly coiled. **FR**: pappus bristles 18–26. Rock ledges, crevices, talus; 1300–3300 m. c SNH. [*E. breweri* var. *e.* (E. Greene) Jepson]

E. filifolius Nutt. var. ***filifolius*** (p. 263) Per 15–30 cm, from taproot and slender-branched caudex, 0–few-branched near mid-st, ± densely and minutely white-upcurved-hairy. **LVS**: basal 2–7 cm, thread-like to linear; cauline restricted to lower st or not. **INFL**: heads 1–4, 6–11 mm diam; phyllaries ± equal, spreading-hairy, minutely sessile-glandular. **RAY FLS** 20–45; corollas 4–10 mm, ligules gen blue (white to pink), weakly coiled. **FR**: pappus bristles 20–30. 2*n*=18,36. Sagebrush scrub, juniper to yellow-pine woods; 1200–2000 m. MP; to B.C., MT, UT. Var. *robustior* M.E. Peck in OR, WA. ❀TRY.

E. foliosus Nutt. Per 20–100 cm, from woody roots and often slender-branched caudex, gen ascending, branched above, sparsely appressed-hairy or becoming glabrous. **LVS** cauline, gen well distributed around st, 10–65 mm, thread-like to widely oblanceolate, evenly sized and spaced. **INFL** often flat-topped; heads 1–several, 10–16 mm diam; phyllaries strongly graded in 3–5 series, sometimes glandular. **RAY FLS** 15–60; corollas 5–15 mm, ligules gen blue, weakly coiled. **FR**: pappus bristles 20–34. Many habitats; < 2900 m. CA-FP (exc GV); OR, n Baja CA. Vars. *confinis* and *mendocinus* intergrade.

var. ***confinis*** (Howell) Jepson **LVS** 2–6 cm, ± linear, sparsely strigose. **INFL**: phyllaries 0.8–1 mm wide, inner 4–5 mm, minutely but obviously glandular, midvein ± prominent, margins gen widely scarious. **RAY FLS**: corollas 6–12 mm. Rocky sites, chaparral to fir woods, often on serpentine; < 2200 m. KR; sw OR. ❀DRN, SUN:1,6,17&IRR:2,7,14–16.

var. ***foliosus*** (p. 263) **LVS** 2–5 cm, thread-like to widely oblanceolate, glabrous to sparsely strigose. **INFL**: phyllaries 0.5–0.8 mm wide, inner 3.2–4.5 mm, barely glandular, midvein raised, orange-resinous, margins gen thick (or narrowly scarious). **RAY FLS**: corollas 6–11 mm. 2*n*=18. Open, rocky grassland or chaparral to pine-fir forest; < 2900 m. s SN, SCoRO, TR, PR; n Baja CA. [var. *stenophyllus* (Nutt.) A. Gray] ❀DRN,SUN:1,2,7,14–24.

var. ***franciscensis*** G. Nesom **LVS** 2–4 cm, gen 2–4 mm wide, ± strigose. **INFL**: phyllaries 0.8–1 mm wide, inner 4–6 mm, minutely but obviously glandular, midvein raised, orange-resinous, margins widely scarious. **RAY FLS**: corollas 7–10 mm. Grassy dunes, chaparral, oak woodlands; < 800 m. SnFrB, n SCoR (scattered). ❀TRY.

var. ***hartwegii*** (E. Greene) Jepson **LVS** 3–6 cm, 1–2 mm wide, ± oriented to 1 side of st, sparsely strigose. **INFL**: phyllaries 0.5–0.8 mm wide, inner 4–6 mm, ± densely nonglandular-spreading-hairy, midvein ± prominent, margins gen thick (or narrowly scarious). **RAY FLS**: corollas 7–12 mm. Rocky riverbanks, oak woodland; 100–600 m. n&c SNF. ❀DRN,IRR,SUN:7,9,14–16, 19–21.

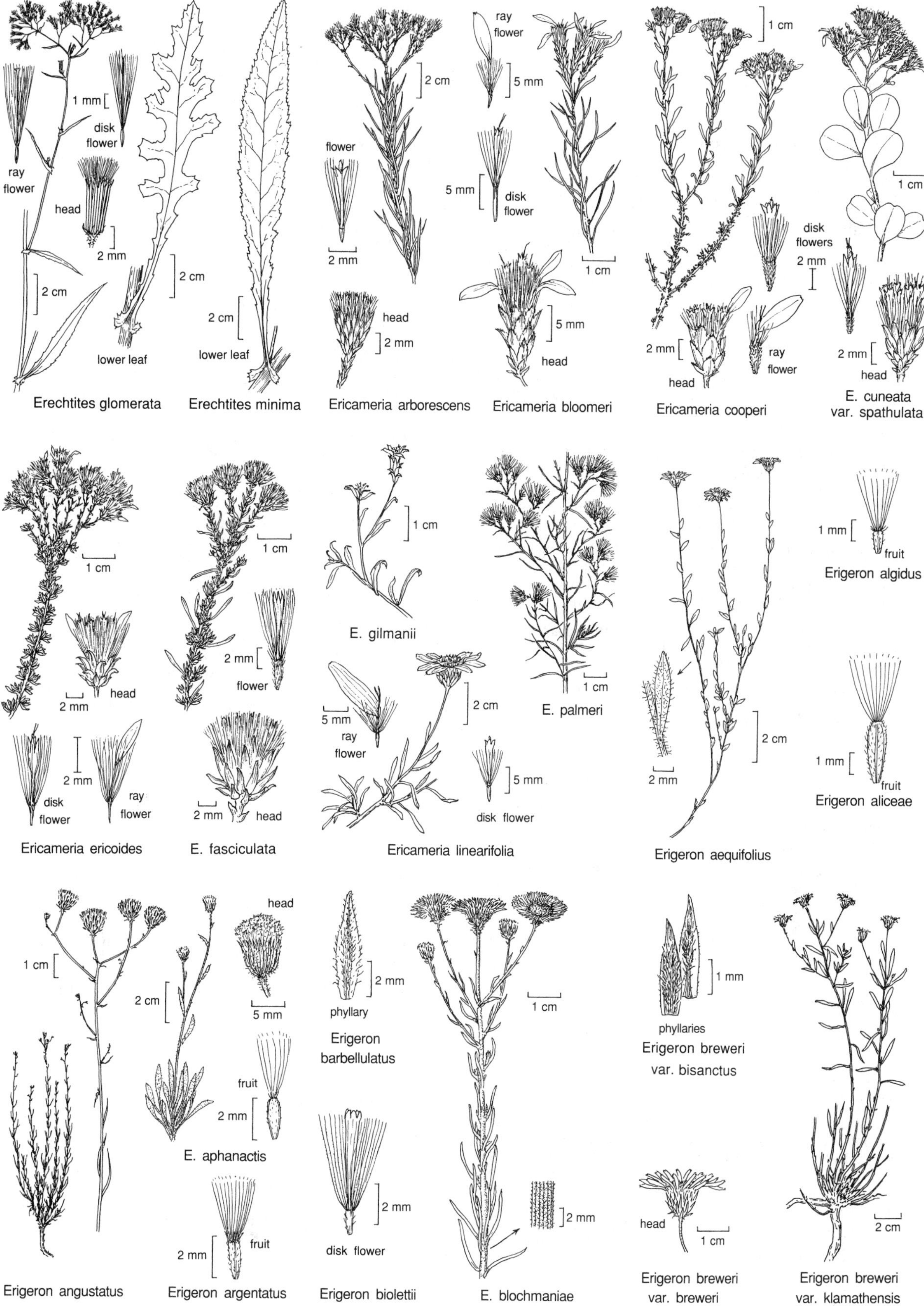
ray flower
1 mm
disk flower
head
2 mm
2 cm
2 cm
lower leaf
2 cm
lower leaf
Erechtites glomerata
Erechtites minima
flower
2 mm
2 cm
head
2 mm
Ericameria arborescens
ray flower
5 mm
5 mm
disk flower
1 cm
5 mm
head
Ericameria bloomeri
1 cm
disk flowers
2 mm
2 mm
head
ray flower
Ericameria cooperi
1 cm
2 mm
head
E. cuneata var. spathulata
1 cm
2 mm
head
2 mm
disk flower
ray flower
Ericameria ericoides
1 cm
2 mm
flower
2 mm
head
E. fasciculata
1 cm
E. gilmanii
5 mm
ray flower
2 cm
5 mm
disk flower
Ericameria linearifolia
1 cm
E. palmeri
2 mm
2 cm
Erigeron aequifolius
1 mm
fruit
Erigeron algidus
1 mm
fruit
Erigeron aliceae
1 cm
Erigeron angustatus
2 cm
head
5 mm
fruit
2 mm
E. aphanactis
2 mm
fruit
Erigeron argentatus
2 mm
phyllary
Erigeron barbellulatus
2 mm
disk flower
Erigeron biolettii
1 cm
2 mm
E. blochmaniae
1 mm
phyllaries
Erigeron breweri var. bisanctus
head
1 cm
Erigeron breweri var. breweri
2 cm
Erigeron breweri var. klamathensis

var. ***mendocinus*** E. Greene **LVS** 2–5 cm, gen 2–4 mm wide, ± glabrous (exc ciliate). **INFL**: phyllaries 0.8–1 mm wide, inner 4–5.5 mm, minutely but prominently glandular, midvein ± indistinct, margins gen widely scarious. **RAY FLS**: corollas 10–15 mm. River bars, banks, ledges, dry slopes; < 800 m. NCoRO. [*E. m.* E. Greene] ❀TRY.

E. glaucus Ker-Gawler (p. 263) SEASIDE DAISY Per 5–30 cm, from thick rhizomes and offsets, ± decumbent, gen branched near mid-st, glabrous or glandular to densely spreading-hairy. **LVS** thick, 2–13 cm, widely obovate to spoon-shaped, entire or with 2–4 pairs of shallow teeth; cauline gen little reduced, sometimes ± clasping. **INFL**: heads 1–15, 15–35 mm diam; phyllaries ± equal. **RAY FLS** 80–165; corollas 8–15 mm, ligules white to purple or blue, coiled. **FR** 2–6-ribbed; pappus bristles 20–30. 2*n*=18. Coastal bluffs, dunes, beaches; < 20 m. NCo, CCo, SCoRO, n ChI; OR. ❀DRN:**5**&IRR:**17,22–24**&SHD:9,**14–16**,18–21;CVS.

E. inornatus A. Gray Per 1–9 cm, from woody taproot, decumbent to ascending, branched near st-tips, gen ± spreading-short-hairy below, (sub)glabrous above. **LVS** cauline, 1–6 cm, narrowly oblanceolate to oblong, glabrous or stiff-hairy, gen ciliate. **INFL** flat-topped; heads discoid, 5–many; disk 7–12 mm diam; phyllaries strongly graded in 3–5 series, sometimes sparsely glandular. **FR**: pappus bristles 28–60. Chaparral to pine/fir woods, lava beds; gen 150–2300 m. NW, CaR, SN, MP; to WA, nw NV.

var. ***calidipetris*** G. Nesom (p. 263) **STS** 10–20 cm, decumbent to ascending, spreading- or reflexed-hairy below middle. **LVS**: middle and lower stiffly spreading-hairy and -ciliate. **INFL**: inner phyllaries 5–6.5 mm. Loose sand, lava beds, depression edges, to pine/fir woods; 1100–1400 m. CaR, MP.

var. ***inornatus*** (p. 263) **STS** 30–90 cm, ascending, appressed-hairy below. **LVS**: middle and lower gen short-strigose. **INFL**: inner phyllaries 4.5–5.5 mm. 2*n*=18. Habitats and range (exc MP) of sp. Hairs variable in c SN (e slope), w NV.

var. ***keilii*** G. Nesom **STS** 40–65 cm, erect, densely hairy. **LVS** short-strigose. **INFL**: inner phyllaries 4.5–6 mm. Dry slopes, meadows, in coniferous forest; ± 1800–2200 m. s SN (Tulare Co.).

E. karvinskianus DC. Per 50–100 cm, from woody roots, sprawling to erect, sparsely strigose or becoming glabrous. **LVS**: basal 0 at fl; cauline gen 1–5 cm (not reduced), elliptic to obovate, weakly 3-veined, entire or with 1–2 pairs of acute teeth or shallow lobes near tip; lf-tufts present in axils. **INFL**: heads 1–many, long-peduncled, 7–10 mm diam; phyllaries ± equal. **RAY FLS** 45–80; corollas 5–8 mm, ligules white, drying white to pink, slightly coiled. **FR**: pappus bristles 15–27. 2*n*=18,27,36. Shaded rock walls, moist disturbed habitats; < 1100 m. c SNF, CCo, SnFrB; native Mex to S.Am; cult.

E. lassenianus E. Greene (p. 263) Per 9–35 cm, from taproot and simple caudex, decumbent to erect, gen few-branched near mid-st, purplish, appressed-to ± spreading-hairy. **LVS**: basal 5–15 cm, linear to spoon-shaped, gen 1-veined; cauline little reduced. **INFL**: heads radiate or disciform, 1–8, long-peduncled, 6–12 mm diam; phyllaries ± equal, sessile-glandular. **RAY FLS** (0)14–36; corollas 5–8 mm, ligules white to pink or lavender, slightly coiled. **FR**: pappus bristles 12–24. Coniferous forest; 600–2000 m. KR, CaRH, n SNH. ❀TRY.

var. ***deficiens*** Cronq. **RAY FLS** 0. Habitat of sp.; ± 1200–1400 m. n SNH (Plumas Co.).

var. ***lassenianus*** **RAY FLS** well developed. 2*n*=18. Habitat and range of sp. [*E. flexuosus* Cronq.]

E. linearis (Hook.) Piper (p. 263) Per 5–20 cm, from taproot and thick-branched caudex, cespitose, sparsely short-white-appressed-hairy. **LVS**: bases persistent on caudex; basal 2–9 cm, ± linear, bases wider, hard, white-shiny, enclosing lower st; cauline 0 on upper 1/3–3/4 of st. **INFL**: head 1, 8–12 mm diam; phyllaries ± equal. **RAY FLS** 25–38; corollas 4–8 mm, ligules yellow, weakly coiled. **FR**: pappus bristles 10–20. 2*n*=18,27,36,45. Grassland, sagebrush scrub, open, rocky slopes; 1300–3100 m. n SNH, MP; to B.C., MT, NV. ❀TRY;DFCLT.

E. lobatus Nelson Ann gen 15–50 cm, from slender taproot, branched at base, glandular-puberulent. **LVS**: basal 5–10 cm, ± obovate, pinnately 4–8-lobed; cauline gradually reduced upwards. **INFL**: heads radiate, 6–10 mm diam; phyllaries glandular-puberulent. **RAY FLS** 85–110; corollas 6–8 mm, drying blue. **FR**: pappus bristles 11–12. Sandy soil, creosote-bush scrub; ± 550 m. DMoj (e San Bernardino Co.); NV, AZ, n Mex

E. mariposanus Congdon Per 15–28 cm, from short-, slender-branched caudex, decumbent to ascending, 0–few-branched near st tips, sparsely appressed-hairy or becoming glabrous. **LVS** cauline (gen with axillary lf-tufts), 25–45 mm, oblanceolate, reduced upward, ± loosely strigose, not glandular. **INFL**: heads gen 1–3, 8–12 mm diam; phyllaries strongly graded in 3–5 series, outer sparsely strigose, midrib orange-resinous, inner 4–5 mm, margins widely scarious. **RAY FLS** 18–22; corollas 7–9 mm, ligules ± blue, weakly coiled. **FR**: pappus bristles 28–32. Foothill woodland; 600–800 m. SNF.

E. miser A. Gray (p. 263) STARVED DAISY Per 5–25 cm, many-stemmed, from woody caudex, decumbent to ascending, gen unbranched, densely long-spreading-hairy. **LVS** cauline, 7–16 mm, evenly sized and spaced, narrowly oblanceolate. **INFL**: heads discoid, 1–few, long-peduncled, 7–14 mm diam; phyllaries strongly graded in 3–5 series, densely sessile-glandular. **FR**: pappus bristles 12–28. 2*n*=18. UNCOMMON. Rocky sites; 1900–2300 m. n SNH.

E. multiceps E. Greene (p. 263) KERN RIVER DAISY Per 12–20 cm, from taproot and (sub)simple caudex, decumbent to ascending, few-branched ± mid-st, ± ascending-hairy. **LVS**: basal persistent, 2–5 cm, oblanceolate to spoon-shaped, entire or few-toothed; cauline reduced. **INFL**: heads few, 7–10 mm diam; phyllaries ± equal, spreading-hairy, sessile-glandular. **RAY FLS** 75–125; corollas 5–8 mm, ligules white or purplish, not reflexed or coiled. **DISK FLS**: corollas somewhat wider at throat. **FR**: pappus bristles 5–8. RARE. Riverbanks, sandy flats, meadows in pine or aspen woodlands; 1500–2500 m. s SNH, SnBr. Indistinct from *E. divergens*.

E. parishii A. Gray (p. 263) PARISH'S DAISY Per 10–35 cm, from thick taproot and branched caudex, 0–few-branched near mid-st, silvery-hairy, esp above. **LVS**: basal 3–6 cm, ± linear, often 0 by fl; cauline ± reduced. **INFL**: heads 1–10, 10–15 mm diam; phyllaries ± equal, sessile-glandular. **RAY FLS** 30–55; corollas 6–13 mm, ligules bluish to pink or white, coiled. **FR** 4-ribbed, ± hairy; pappus bristles 18–26. RARE. Rocky blackbush or creosote-bush scrub to pinyon/juniper woodland, often on limestone; 800-2000 m. n SnBr (Cushenbury Canyon). Threatened by limestone mining. In cult.

E. peregrinus (Pursh) E. Greene Per 8–45 cm, from short rhizome, branched on upper sts, glabrous or strigose above. **LVS**: basal 5–20 cm, oblanceolate to spoon-shaped, glabrous or sparsely spreading-hairy; cauline ± reduced upward, gen lanceolate to ovate, subclasping. **INFL** flat-topped; heads 1–4, 10–21 mm diam; phyllaries ± equal, with long-acuminate, loosely spreading tips, densely stalked-glandular. **RAY FLS** 30–105; corollas 8–15 mm, ligules white to purple, coiled. **FR** 4–7-ribbed; pappus bristles 20–30. Clearings, talus, alpine meadows; 1300–3400 m. KR, CaRH, SNH, Wrn, SNE; to AK, MT, Colorado, NM, e Asia. ❀TRY;DFCLT.

var. ***callianthemus*** (E. Greene) Cronq. (p. 263) Herbage strigose. 2*n*=18. Habitats and range of sp. (exc n Can to e Asia). [var. *angustifolius* (A. Gray) Cronq.]

var. ***hirsutus*** Cronq. Herbage spreading-hairy. Habitats of sp.; 2200–3200 m. c&s SNH, SNE.

E. petrophilus E. Greene Per 10–30 cm, many-stemmed, from woody root and rhizome or caudex, decumbent to ascending, 0–few-branched near st tips, ± long-loose-spreading-hairy, gen densely glandular near heads. **LVS** cauline, 10–25 mm, narrowly oblong to oblanceolate, evenly sized and spaced, gen proximally long-ciliate. **INFL** dense; heads discoid, 10–15 mm diam; phyllaries strongly graded in 3–5 series, densely sessile-glandular. **FR**: pappus bristles 22–35. Rocky foothills to montane forest, sometimes on serpentine; 300–2700 m. KR, NCoRO, n SNF, SnFrB, SCoRO; sw OR.

var. ***petrophilus*** Herbage densely glandular-hairy; nonglandular hairs gen loose, often crinkly. **INFL**: phyllary tips obviously wider, purplish. $2n$=18. Habitats of sp.; 500–2100 m. NCoRO, SnFrB, SCoRO. ❀DRN,IRR:1,2,4–6,**7**,14–17.

var. ***sierrensis*** G. Nesom Herbage densely glandular-hairy; nonglandular hairs gen loose, often crinkly. **INFL**: phyllary tips like body. Habitats of sp.; 300–1900 m. n SNF.

var. ***viscidulus*** (A. Gray) G. Nesom (p. 263) Herbage ± nonglandular; hairs stiff, straight or curved. **INFL**: phyllary tips like body. Habitats of sp.; 1500–2700 m. KR; sw OR. [*E. inornatus* var. *v.* A. Gray] Intergrades with *E. reductus*.

E. philadelphicus L. (p. 265) Bien or short-lived per 25–80 cm, from fibrous roots and often short rhizome, branched ± near st tips, wider below heads, sparsely loose-spreading-hairy. **LVS**: basal 8–15 cm, oblong-obovate to spoon-shaped, gen coarsely toothed; cauline lanceolate to ovate, clasping, little reduced. **INFL** flat-topped; heads 1–many, 6–15 mm diam; phyllaries ± equal, bases often ± fused. **RAY FLS** ± 150–400; corollas 6–9 mm, ligules white or pinkish, coiled. **FR**: pappus bristles 20–30. $2n$=18. Streamsides, other moist habitats; < 1200 m. CA (exc GV, DMtns); to e US. Scattered. ❀WET or IRR:1,2,**14,18–24**&SUN:**3–7**,8–10,**15–17**;INV.

E. pumilus Nutt. var. ***intermedius*** Cronq. (p. 265) Per 8–35 cm, from woody taproot and thick-branched caudex, cespitose, 0–4-branched below mid-st, stiffly spreading- or reflexed-hairy, minutely stalked-glandular. **LVS**: basal erect, crowded, 2–8 cm, linear to oblanceolate; cauline gradually reduced. **INFL** flat-topped; heads 9–14 mm diam; phyllaries ± equal. **RAY FLS** 40–115; corollas 8–12 mm, ligules blue to white or pink, reflexed, sometimes weakly coiled. **DISK FLS**: corollas abruptly wider at throat. **FR**: pappus bristles 12–16, outer series inconspicuous scales. $2n$=18, 36. Open slopes, meadows; 1200–1800 m. CaR, n SNH, MP, W&I; to B.C., MT, WY, UT. Var. *pumilus* on e side Rocky Mtns. ❀TRY.

E. pygmaeus (A. Gray) E. Greene (p. 265) Per 1–6 cm, from taproot and short-branched caudex, gen cespitose, unbranched, ± spreading-hairy, densely sessile-glandular. **LVS** gen basal, < 4 cm, linear to narrowly oblanceolate; cauline (if any) below mid-st. **INFL**: head 1, 6–15 mm diam; phyllaries ± equal, purple-black. **RAY FLS** 20–37; corollas 4–10 mm, ligules blue or purple, rarely white, not coiled or reflexed. **FR**: pappus bristles 15–25. $2n$=18. Rocky sites, subalpine forest to alpine talus; 2900–4100 m. c SNH, SNE; w-c NV. ❀TRY;DFCLT.

E. reductus (Cronq.) G. Nesom Per 8–30 cm, from woody roots and slender-branched caudex, ascending to erect, 0–few-branched near st tips, (sub)glabrous. **LVS**: basal 0; cauline gen 1–4 cm, thread-like to linear, evenly sized and spaced, ascending-ciliate. **INFL**: heads discoid, 1–2(5+), 8–10 mm diam; phyllaries strongly graded in 3–5 series, densely and minutely glandular. **FR**: pappus bristles 20–61. Rocky sites, sometimes on serpentine, often in pine or oak woodland; 600–2400 m. KR, NCoRO, NCoRH, n SNH, SnFrB.

var. ***angustatus*** (A. Gray) G. Nesom (p. 265) **INFL**: phyllary tips purplish (variable in n SnFrB). **FR**: pappus bristles 38–61. Habitats of sp.; 600–1400 m. NCoRO, NCoRH, SnFrB. [*E. inornatus* var. *a.* A. Gray] ❀IRR,SUN:1,7,15,16.

var. ***reductus*** CALIFORNIA RAYLESS DAISY **INFL**: phyllary tips greenish. **FR**: pappus bristles 20–30. UNCOMMON. Crevices, other open, rocky sites; 700–2400 m. KR, n SNH. [*E. inornatus* var. *r.* Cronq.]

E. sanctarum S. Watson (p. 265) SAINTS' DAISY Per 5–35 cm, from slender-branched caudex, ascending, 0–few-branched near mid-st, sparsely short-spreading-hairy. **LVS**: basal 2–5 cm, oblanceolate to spoon-shaped; cauline lanceolate, barely < basal, sometimes subclasping, mostly below mid-st. **INFL**: heads 1(–3), 12–17 mm diam; phyllaries ± equal, hairs long, spreading, ± flat. **RAY FLS** 45–90; corollas 7–13 mm, ligules blue, weakly coiled. **FR**: pappus bristles 18–25. UNCOMMON. Coastal scrub or woodland; < 300 m. s CCo, s SCoRO, n ChI. ❀DFCLT.

E. serpentinus G. Nesom (p. 265) SERPENTINE DAISY Per 30–45 cm, many-stemmed, from woody taproot, few-branched near st tips, gen glabrous. **LVS** cauline, 2–4 cm, linear to thread-like, evenly sized and spaced, ± ciliate. **INFL** loose; heads radiate, several; disk 9–12 mm diam; phyllaries strongly graded in 3–5 series, minutely sessile-glandular. **RAY FLS** 9–13; corollas 7–8 mm, ligules drying blue, weakly coiled. **FR**: pappus bristles 26–32. RARE. Serpentine shrubland; 400–600 m. NCoRO (nw Sonoma Co.). Like *E. angustatus*.

E. ***strigosus*** Willd. (p. 265) Ann or bien 30–80 cm, from fibrous roots, branched above mid-st, ± glabrous to sparsely hairy. **LVS**: basal gen petioled, blades 3–6 cm, elliptic to obovate; cauline gen reduced upwards, strigose. **INFL** flat-topped; heads many, 5–10 mm diam; phyllaries ± equal, sparsely spreading-hairy. **RAY FLS** ± 60–110; corollas 5–6 mm, ligules white, not coiled or reflexed. **FR**: ray pappus 0; disk pappus bristles 10–15. $2n$=18,27,36. Disturbed sites; < 1100 m. CA-FP; w US; native to e US. Scattered weed, producing asexual seeds.

E. supplex A. Gray (p. 265) SUPPLE DAISY Per 15–40 cm, from taproot and slender-branched caudex, ascending, simple, sparsely spreading-hairy. **LVS**: basal 4–8 cm, elliptic to oblanceolate, ± sessile; cauline gen gradually reduced, not clasping, ciliate. **INFL**: head discoid, 1, 14–20 mm diam; phyllaries ± equal, sessile-glandular. **FR**: pappus bristles 17–30. RARE. Coastal areas, bluffs; < 50 m. n&c NCo. Threatened by coastal development.

E. tener A. Gray Per 2–15 cm, from taproot and branched caudex, cespitose, gen simple, short-white-appressed-hairy, glandless. **LVS**: basal 1–8 cm, long-petioled, oblanceolate to elliptic; cauline much reduced. **INFL**: head 1(2), 5–12 mm diam; phyllaries ± equal, spreading-hairy, sessile-glandular. **RAY FLS** 18–40; corollas 4–8 mm, ligules bluish, not coiled or reflexed. **FR**: pappus bristles 15–30. Crevices or ledges from sagebrush scrub to yellow-pine forest; 2300–3400 m. KR, SNH, MP; to se OR, WY, UT.

E. uncialis S.F. Blake var. ***uncialis*** (p. 265) LIMESTONE DAISY Per 1–4 cm, from taproot and slender-branched caudex, simple, spreading-hairy. **LVS** basal, 5–25 mm, long-petioled, widely elliptic to obovate. **INFL**: head 1, 6–11 mm diam; phyllaries ± equal, densely sessile-glandular. **RAY FLS** 15–40; corollas 4–6 mm, ligules white to pinkish, not coiled or reflexed. **FR**: pappus bristles 13–22. UNCOMMON. Limestone crevices from sagebrush scrub to subalpine forest; 2100–2900 m. W&I, DMoj; c NV. Var. *conjugans* S.F. Blake in s NV.

E. utahensis A. Gray (p. 265) Per 10–50 cm, from thick, peeling taproot and branched caudex, 0–few-branched near mid-st, densely silvery-hairy. **LVS**: basal 6–8 cm, narrowly oblanceolate, gen 0 by fl; cauline ± reduced. **INFL**: heads 1–4(10), 10–15 mm diam; phyllaries ± equal, densely white-strigose, minutely glandular. **RAY FLS** 16–28; corollas 12–15 mm, ligules bluish, coiled. **FR** 4(6)-ribbed; pappus bristles 20–35. $2n$=18. Limestone slopes; ± 1500 m. e DMtns (Providence Mtns); to Colorado, n AZ.

E. vagus Payson (p. 265) Per 2–5 cm, from (taproot and) branched, spreading caudex, simple, spreading-hairy, sometimes glandular. **LVS** gen 1–3 cm, oblanceolate to spoon-shaped, with 3 long-obovate lobes at tip; cauline gen 0 or much reduced. **INFL**: head 1, 8–16 mm diam; phyllaries ± equal, purple at least distally. **RAY FLS** 25–35; corollas 4–7 mm, ligules white to pink, sometimes with midstripe, coiled. **FR**: pappus bristles 16–20. $2n$=18. Alpine talus; 3300–4400 m. c SNH, W&I; to se OR, Colorado.

ERIOPHYLLUM WOOLLY SUNFLOWER

John S. Mooring & Dale E. Johnson

Ann to shrubs, ± woolly. **LVS** gen alternate, entire to nearly compound. **INFL**: heads 1–many, gen radiate; cluster often ± flat-topped; involucre obconic to hemispheric; phyllaries in 1 series, free or ± fused; receptacle flat to columnar, gen naked. **RAY FLS** gen ± 1 per phyllary; ligules entire to lobed, gen yellow (white). **DISK FLS** 10–300; corolla yellow; anther tips ovate, deltate or awl-shaped. **FR** 4-angled or flattened in outer fls, gen club-shaped in inner fls; pappus 0–15 ± jagged or fringed scales. 14 spp.: w N.Am. (Greek: woolly lf) [Mooring 1991 Madroño 38:213–226] Ann spp. by Dale E. Johnson.

1. Ann
2. Peduncles < 5 mm
3. Heads radiate .... ***E. multicaule***
3′ Heads discoid
4. Lf weakly rolled under, tips sharp-pointed; phyllaries 3–4 .... ***E. mohavense***
4′ Lf strongly rolled under, tips rounded; phyllaries 6–8 .... ***E. pringlei***
2′ Peduncles > 5 mm
5. Anther tips awl-like; disk corolla lobes glandular; ligules white or yellow
6. Pappus of alternating long and short scales; fr 2.5–4.5 mm; ligules white .... ***E. lanosum***
6′ Pappus 0 or of ± = scales; fr ± 2 mm; ligules cream-colored to yellow .... ***E. wallacei***
5′ Anther tips deltate; disk corolla lobes nonglandular; ligules yellow
7. Pappus 0 or of minute scales; anther tips glabrous; s SN, s SNE, D .... ***E. ambiguum***
8. Disk corolla lobes glabrous; pappus of minute scales .... var. ***ambiguum***
8′ Disk corolla lobes with 1-celled hairs; pappus gen ± 0 .... var. ***paleaceum***
7′ Pappus of scales > 0.5 mm; anther tips glandular; c SN (Mariposa Co.)
9. Branches open, spreading; ligules > 2 mm; < 1900 m .... ***E. congdonii***
9′ Branches strictly ascending; ligules < 2 mm; > 1800 m .... ***E. nubigenum***
1′ Subshrub or shrub
10. Heads (incl ligules) < 15 mm diam, gen ± densely clustered
11. Ray fls (0)6–9; lvs 3–20 cm; phyllaries 8–12; coastal
12. Lvs 8–20 cm; ligules ± 2 mm; pappus scales > 1 mm; s ChI .... ***E. nevinii***
12′ Lvs 3–7 cm; ligules (0)3–5 mm; pappus scales < 1 mm; mainland, n ChI .... ***E. staechadifolium***
11′ Ray fls (0)4–6; lvs 1–7 cm; phyllaries 4–8; esp inland
13. Heads gen 10–30+, dense; ligules 2–5 mm; ray fls (0)4–6; frs gen < 35; common .... ***E. confertiflorum*** var. ***confertiflorum***
13′ Heads 1–10, loose; ligules gen 5–10 mm; ray fls gen 5–9; frs gen 30–75; possibly local derivatives of *E. confertiflorum* × *E. lanatum*
14. Subshrub; lf lobes ± triangular, deep; phyllaries barely overlapping — cw SnFrB .... ***E. latilobum***
14′ Shrub; lf lobes ± linear, almost to midrib; phyllaries clearly overlapping
15. Lvs ± obovate; peduncles < 3 cm; ray fls (0)4–6; ligules 4–5 mm; c SNF .... ***E. confertiflorum*** var. ***tanacetiflorum***
15′ Lvs ovate; peduncles 4–14 cm; ray fls gen 6–8; ligules 6–10 mm; e CW .... ***E. jepsonii***
10′ Heads (incl ligules) 15–45 mm diam, 1 or loosely clustered
16. Ray fls (0 or 4) gen 5–9; ligules gen 5–10 mm; involucres 4–7 mm; possibly derivatives of *E. confertiflorum* × *E. lanatum* (see 13′)
16′ Ray fls (0) gen 8–13; ligules 6–20 mm; involucres 5–12 mm; widespread complex .... ***E. lanatum***
17. Tube of disk fls glabrous; rare, Teh, SCoRO .... var. ***hallii***
17′ Tube of disk fls very glandular or hairy; widespread
18. Ray fls gen 8(5–10); phyllaries 5–10, strongly keeled; lvs gen entire to 3-lobed, margins flat .... var. ***integrifolium***
18′ Ray fls gen 9–13; phyllaries gen 9–13, keeled or flat; lvs entire to nearly compound, margins flat or rolled under
19. Lvs densely woolly above and below, entire or coarsely toothed (esp above middle), margins flat
20. Fr hairy; ligules 7–10 mm; KR .... var. ***lanceolatum***
20′ Fr glabrous; ligules 6–7 mm; s SN, SnBr .... var. ***obovatum***
19′ Lvs less hairy (often green) above, entire to nearly compound, margins rolled under
21. Lvs narrowly oblanceolate or wider, serrate to sharply lobed, glabrous above; pappus 0 or minute; peduncles 30–100 mm; ligules golden-yellow
22. Lvs often diamond-shaped, lower sharply 3–5-lobed, woolly-tufted below; coastal ... var. ***arachnoideum***
22′ Lvs often obovate, coarsely serrate or lobed distally, silky-woolly below; SN .... var. ***croceum***
21′ Lvs linear to ovate, entire to 2-compound, often woolly-tufted above; pappus 0.5–1.5 mm; peduncles 30–300 mm; ligules yellow
23. Lvs 1–3 cm, lobed to ± 2-pinnately compound; peduncles 30–100 mm, slender; involucres 6–8 mm; ligules 6–9 mm .... var. ***achillaeoides***
23′ Lvs 3–8 cm, entire to pinnately lobed; peduncles 100–300 mm, gen swollen beneath heads; involucres 8–10 mm; ligules 10–20 mm .... var. ***grandiflorum***

E. ambiguum (A. Gray) A. Gray Ann 5–30 cm, decumbent to ascending. **LF** < 4 cm, oblong-oblanceolate, entire to shallowly lobed. **INFL**: head 1; peduncle 1–8 cm; involucre 3–6 mm, obconic to hemispheric; phyllaries 6–10, acuminate; receptacle conic. **RAY FLS** 6–10; ligules 2–10 mm. **DISK FLS** many; corollas 1.3–3 mm, tube and throat hairy; anther tips deltate, smooth. **FR** 2.2–3 mm, ± strigose; pappus 0 or < 0.5 mm. $2n$=14. Woodland, desert; < 2800 m. s SN, s SNE, Teh, D; s NV.

var. ***ambiguum*** (p. 265) **LF** pinnately lobed. **INFL**: involucre 4–5.5 mm; phyllaries sometimes fused; receptacle naked. **DISK FLS**: corolla lobes glabrous. **FR**: pappus of 6–10 irregularly toothed scales 0.2–0.5 mm. Open chaparral, foothill woodland; 200–1900 m. s SNF, Teh. ❀TRY.

var. ***paleaceum*** (Brandegee) Ferris (p. 265) **LF** entire or 3-lobed near tip. **INFL**: involucre 5–7 mm; phyllaries free; receptacle tip sometimes scaly. **DISK FLS**: corolla lobes with 1-celled hairs. **FR**: pappus 0 or scales entire, < 0.2 mm. Desert scrub or woodland; 100–2800 m. s SN, Teh, D; s NV. ❀TRY.

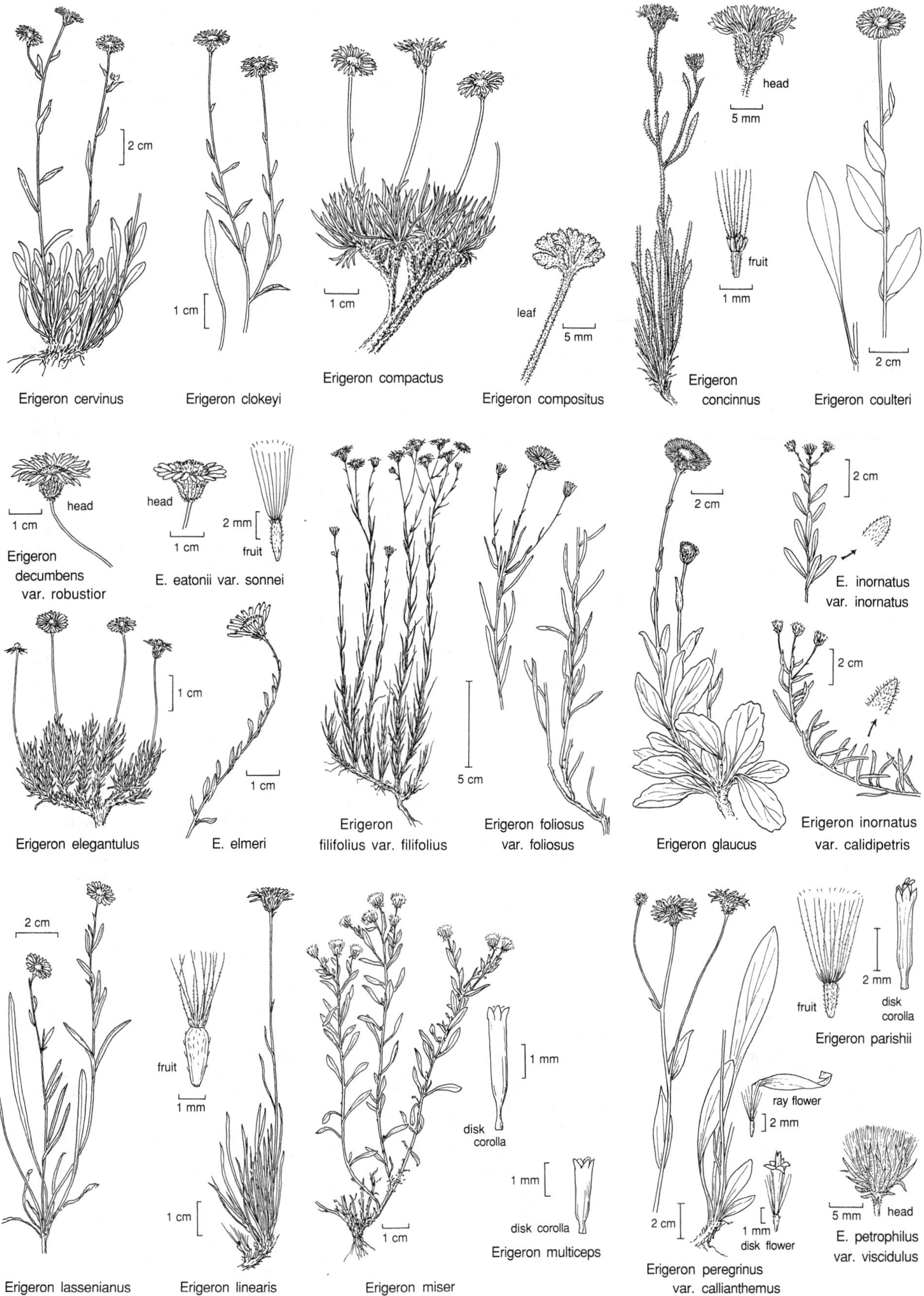
2 cm
Erigeron cervinus
1 cm
Erigeron clokeyi
1 cm
Erigeron compactus
leaf
5 mm
Erigeron compositus
head
5 mm
fruit
1 mm
Erigeron concinnus
2 cm
Erigeron coulteri
head
1 cm
Erigeron decumbens var. robustior
head
1 cm
2 mm
fruit
E. eatonii var. sonnei
1 cm
Erigeron elegantulus
1 cm
E. elmeri
5 cm
Erigeron filifolius var. filifolius
Erigeron foliosus var. foliosus
2 cm
Erigeron glaucus
2 cm
E. inornatus var. inornatus
2 cm
Erigeron inornatus var. calidipetris
2 cm
Erigeron lassenianus
fruit
1 mm
1 cm
Erigeron linearis
1 cm
Erigeron miser
1 mm
disk corolla
1 mm
disk corolla
Erigeron multiceps
2 cm
ray flower
2 mm
1 mm
disk flower
Erigeron peregrinus var. callianthemus
fruit
2 mm
disk corolla
Erigeron parishii
5 mm
head
E. petrophilus var. viscidulus

E. confertiflorum (DC.) A. Gray (p. 265) GOLDEN-YARROW Subshrub or shrub gen 2–7 dm. **LF** 1–5 cm, ± obovate, deeply 3–5-lobed to nearly 2-pinnately compound, rolled under, becoming ± glabrous above. **INFL**: heads 3–30+; peduncles 0–25 cm; involucres 3–7 mm, bell-shaped, tomentose; phyllaries 4–7, obtuse, keeled, strongly overlapping, ± free; receptacle ± convex. **RAY FLS** (0)4–6; ligules 2–5 mm. **DISK FLS** 10–75; corollas 2–4 mm, puberulent to glandular. **FR** 2–4 mm; pappus gen < 1 mm. Many dry habitats; < 3000 m. NCoR, SN, CW, SW; Baja CA. Highly variable, intergrading complex. Hybridizes with *E. lanatum*.

var. ***confertiflorum*** Pl gen persistently tomentose. **INFL**: heads gen 10–30+; clusters dense; peduncles gen < 10 mm; involucres 3–5 mm. **RAY FLS** sometimes 0. **DISK FLS** gen 10–35; corollas 2–3 mm. **FR** 2–3 mm; pappus scales 5–14, ± equal. $2n$=16,32,48,64. Habitats and range of sp. [var. *discoideum* E. Greene; var. *laxiflorum* A. Gray] ❀DRN,SUN:5,**15–17**&IRR:**7**, 8,9,**14,18–24**;also STBL.

var. ***tanacetiflorum*** (E. Greene) Jepson Pl sometimes becoming glabrous. **INFL**: heads 3–10; clusters gen loose; peduncles 0–25 mm; involucres 5–7 mm. **DISK FLS** 35–75; corollas 3.5–4 mm. **FR** 3.5–4 mm; pappus scales ± 8, unequal. $2n$=64. Oak woodland; 600–800 m. c SNF. Possible hybrid derivative of var. *c.* × *E. lanatum*. ❀TRY.

E. congdonii Brandegee (p. 265) CONGDON'S WOOLLY SUNFLOWER Ann 1–3 dm; branches spreading. **LF** 1–4 cm, oblanceolate, entire or lobed near tip. **INFL**: head 1; peduncles 3–10 cm; involucres 5–8 mm, bell-shaped; phyllaries 8–10, acute, free; receptacle flat to conic. **RAY FLS** 8–10; ligules 3–5 mm. **DISK FLS** many; corollas 2–3 mm, glabrous; anther tips deltate, glandular. **FR** 2.5–3 mm, strigose; pappus scales 1–2 mm, unequal. $2n$=14. **RARE** CA. Rocky, open foothill woodland, pine forest; 500–1900 m. c SNF (Mariposa Co.). [*E. nubigenum* var. *c.* (Brandegee) Constance]

E. jepsonii E. Greene (p. 265) JEPSON'S WOOLLY SUNFLOWER Subshrub 5–8 dm. **LF** 3–6 cm, ovate, pinnately 5–7-lobed (lobes linear, obtuse), rolled under, woolly-tufted, becoming glabrous above. **INFL**: heads 1–5; peduncles 4–14 cm; involucres 4–7 mm, widely bell-shaped; phyllaries 5–8, acute, strongly overlapping, keeled, free; receptacle convex to low-conic. **RAY FLS** 6–8; ligules 6–10 mm. **DISK FLS** gen 35–50; corollas 3–5 mm, glandular-puberulent or bristly. **FR** 3–4.5 mm; pappus scales gen 8, ± 1–1.5 mm, unequal. $2n$=64. UNCOMMON. Dry oak woodland; 200–1000 m. e CW. Possible derivative of *E. confertiflorum* × *E. lanatum*. ❀DRN,DRY,SUN:7,8,9,**14–17**,18–24.

E. lanatum (Pursh) James Forbes Gen ± subshrub 1–10 dm. **LF** 1–8 cm, linear to ovate, entire to ± 2-pinnately compound, gen becoming glabrous above. **INFL**: heads 1–5+; peduncles 3–30 cm; involucres 5–12 mm, bell-shaped to hemispheric; phyllaries 5–15; receptacle ± flat to ± conic. **RAY FLS** (0) gen 8–13; ligules 6–20 mm, oblong to elliptic. **DISK FLS** 20–300; corollas 2.5–5 mm, tube gen glandular. **FR** gen variable, 2–5 mm, glandular or hairy; pappus scales 0 or 6–12, 0–2 mm, translucent. Many (gen dry) habitats; < 4000 m. CA (exc SnJV, D); to B.C., MT, WY, NV. Polyploid pillar complex of intergrading races; key is to modal populations; some vars. hybridize with *E. confertiflorum*.

var. ***achillaeoides*** (DC.) Jepson Ann to per. **LF** 1–3 cm, gen ± 1–2-compound. **INFL**: peduncles 3–10 cm, gen slender; involucres 6–8 mm. **RAY FLS**: ligules 6–9 mm. **DISK FLS**: corollas 2.5–3 mm. **FR** 2.2–3 mm; pappus gen < 1 mm. $2n$=16,32. Dry, often rocky sites; < 1300 m. KR, NCoR, CaRH, n&c SNF, c SNH, w CW, MP; OR. Intergrades with vars. *arachnoideum*, *grandiflorum* in CA. Rayless NCoRI pls with $2n$=32 have been called var. *aphanactis* J. Howell. ❀DRN,SUN,DRY:**7–9,14–24**.

var. ***arachnoideum*** (Fischer & Avé-Lall.) Jepson (p. 265) Per, sometimes stoloned. **LF** 2–5 cm, ± diamond-shaped, thin: lower sharply 3–5-lobed, woolly-tufted below. **INFL**: peduncles 3–10 cm; involucres 8–11 mm. **RAY FLS**: ligules 8–10 mm, golden-yellow. **DISK FLS**: corollas 3–4 mm. **FR** 2–4 mm, gen glabrous; pappus 0 or minute. $2n$=16,32. ± moist places; < 400 m. NCo, NCoRO, SnFrB, SCoRO?. ❀DRN,SUN:**5**,6,**17**&IRR:7–9,**14–16**,18, **19–24**.

var. ***croceum*** (E. Greene) Jepson Per, often stoloned. **LF** 2–5 cm, oblanceolate to obovate, coarsely serrate or lobed distally, silky-woolly below. **INFL**: peduncles 3–8 cm; involucres 5–8 mm. **RAY FLS**: ligules 8–10 mm, golden-yellow. **DISK FLS**: corollas 3–4 mm. **FR** 2–3 mm; pappus 0 or minute. $2n$=16,32. Gen under conifers; 1300–2000 m. SNH. Like var. *arachnoideum*. ❀TRY.

var. ***grandiflorum*** (A. Gray) Jepson **LF** 3–8 cm, gen entire to toothed. **INFL**: peduncles 1–3 dm, gen swollen below head; involucres 8–10 mm. **RAY FLS**: ligules 10–20 mm. **DISK FLS**: corollas 4 mm. **FR** 3–4 mm; longest pappus scales gen > 1 mm. $2n$= 16,32,48,64. Dry, gen rocky sites; < 1700 m. KR, NCoRI, CaR, n&c SN, ScV. ❀TRY.

var. ***hallii*** Constance FORT TEJON WOOLLY SUNFLOWER **LF** 2.5–5 cm, gen opposite below, ovate, thin, pinnately lobed. **INFL**: peduncles 5–12 cm; involucres 8–10 mm. **RAY FLS** 8–9; ligules 10–13 mm. **DISK FLS**: corollas 4–5 mm, tube glabrous. **FR** 4–5 mm. $2n$=16. RARE. Dry sites; 1200–1500 m. s Teh (near Fort Tejon), se SCoRO (Sierra Madre Mtns). Threatened by grazing.

var. ***integrifolium*** (Hook.) F.J. Smiley OREGON SUNSHINE **LF** 1–4 cm, wedge-shaped to obovate, gen entire (to 5-lobed at tip); margins flat. **INFL**: head 1; peduncle 3–10 cm; involucres 6–8 mm; phyllaries gen 8 (5–10), strongly keeled. **RAY FLS** gen 8(5–10); ligules 6–10 mm. **DISK FLS**: corollas ± 4 mm. **FR** 3–4 mm; pappus < 2 mm. $2n$=16,32,48,64. Cold, dry sites; 1400–3500 m. KR?, CaRH, n&c SNH, GB; to WA, WY, NV. [var. *monoense* (Rydb.) Jepson] Variable and complex. ❀TRY;DFCLT.

var. ***lanceolatum*** (Howell) Jepson **LF** 2–4 cm, lanceolate to ovate, thick, entire to coarsely serrate, ± flat, ± woolly above. **INFL**: head 1; peduncle 3–10(15) cm, slender; involucre 8–12 mm; phyllaries 10–15, ± flat. **RAY FLS** 10–15; ligules 7–10 mm. **DISK FLS**: corollas 3–4 mm. **FR** 2–3 mm, hairy; pappus < 1 mm. $2n$=16, 32. Dry, rocky oak/conifer forest; 200–2200 m. KR; OR. ❀TRY.

var. ***obovatum*** (E. Greene) H.M. Hall (p. 265) WOOLLY SUNFLOWER **LF** 1–5 cm, entire to few-toothed distally. **INFL**: head 1; peduncle 3–10(15) cm, often ± swollen below head; involucre 7–10 mm. **RAY FLS**: ligules 6–7 mm. **DISK FLS**: corollas 3–4 mm. **FR** 2.5–3 mm, ± glabrous; pappus < 1 mm. $2n$=16. UNCOMMON. Open conifer forest; 1300–2500 m. s SNH, SnBr. Like var. *lanceolatum*.

E. lanosum (A. Gray) A. Gray (p. 267) Ann 1–15 cm, decumbent-ascending, often reddish, sparsely woolly. **LF** 5–20 mm, linear-oblanceolate, entire or lobed at tip. **INFL**: head 1; peduncles 1–5 cm; involucre 5–7 mm, ± cylindric; phyllaries 8–10, acuminate, free; receptacle conic. **RAY FLS** 8–10; ligules 3–7 mm, oblong, white, sometimes red-veined. **DISK FLS** many; corollas 2–3 mm, glabrous; anther tips awl-like, glabrous. **FR** 2.5–4.5 mm, linear or narrowly club-like, glabrous to minutely strigose; pappus scales 0.5–2.5 mm, gen very unequal, longest awned. $2n$=8. Desert scrub; < 1400 m. D; to sw UT, AZ, nw Mex. ❀TRY.

E. latilobum Rydb. (p. 267) SAN MATEO WOOLLY SUNFLOWER Subshrub 2–5 dm, becoming ± glabrous. **LF** 2–6 cm, diamond-shaped to ± obovate, thin, deeply triangular-lobed, glabrous above. **INFL**: heads 1–10; peduncles 1–8 cm; involucres 4–7 mm, widely bell-shaped; phyllaries 6–10, acute, barely overlapping, free; receptacle flat (exc conic in center). **RAY FLS** 6–13; ligules gen 6–10 mm. **DISK FLS** 40–70; corollas 3–4 mm, glandular. **FR** 3–4 mm; angles gen strigose; pappus 0.3–1 mm, disk scales > ray scales. $2n$=32. RARE. Gen oak woodland; 100–150 m. cw SnFrB (San Mateo Co.). Probable derivative of *E. lanatum* var. *arachnoideum* × *E. confertiflorum*. Threatened by development. In cult.

E. mohavense (I.M. Johnston) Jepson (p. 267) BARSTOW WOOLLY SUNFLOWER Ann 1–2.5 cm, tufted, spreading, loosely white-woolly. **LVS** 3–10 mm, spoon- to wedge-shaped, entire or 2–3-lobed (lobes pointed). **INFL**: head 1, discoid, ± sessile; involucre 3–4 mm, ± cylindric; phyllaries 3–4, acute, free; receptacle ± pointed-columnar, flanges 3, protruding between frs, spine-tipped. **RAY FLS** 0. **DISK FLS** ± 3; corollas ± 2 mm, throat minutely puberulent; anther tips narrowly deltate. **FR** 2–2.5 mm, narrowly obconic, strigose; pappus ± 1.5 mm. RARE. Creosote-bush scrub; 500–800 m. c DMoj (wc San Bernardino Co.).

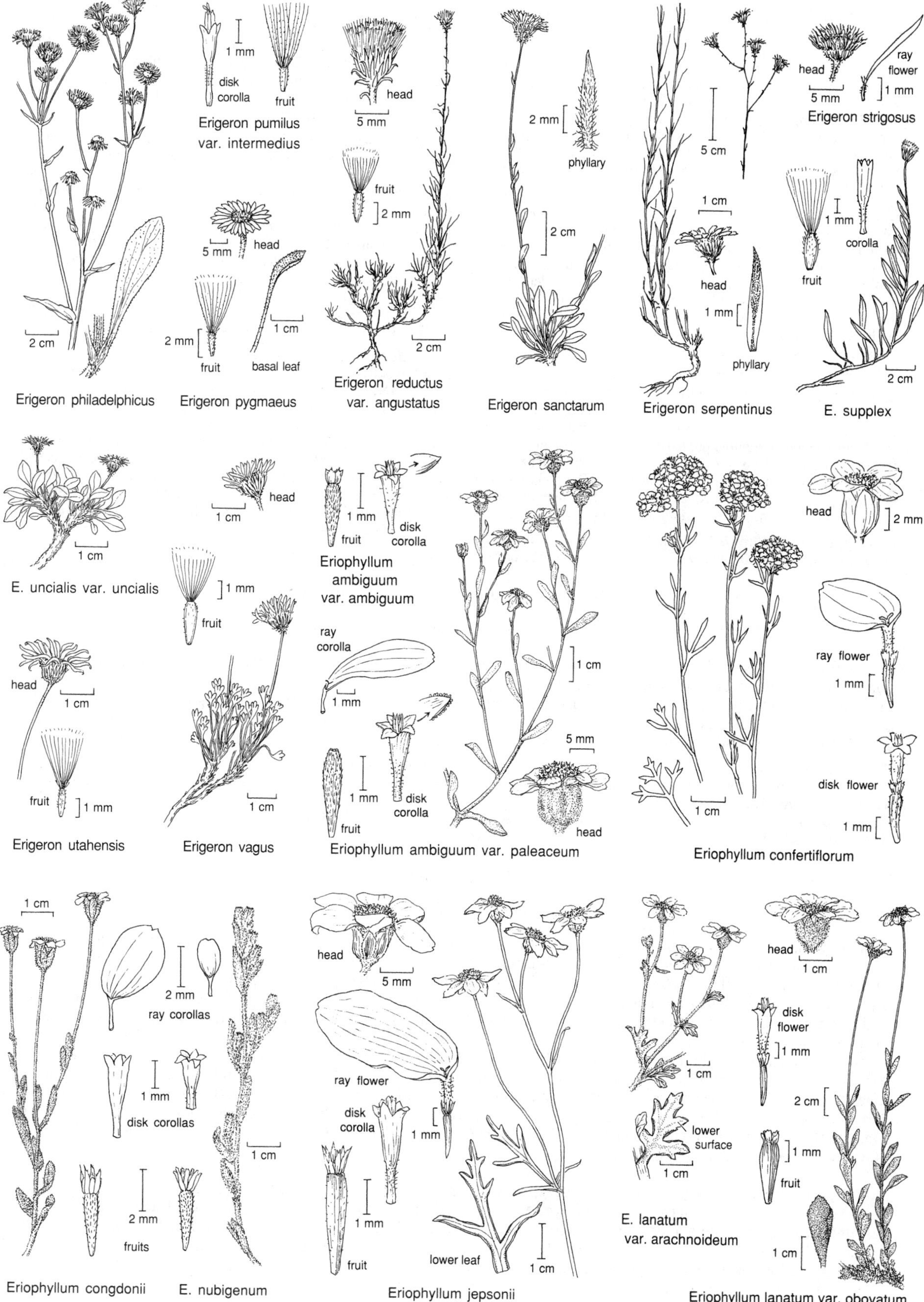
1 mm
disk corolla
fruit
Erigeron pumilus var. intermedius
head
5 mm
fruit
2 mm
2 mm
phyllary
2 cm
5 cm
1 cm
head
1 mm
phyllary
head
5 mm
ray flower
1 mm
Erigeron strigosus
1 mm
corolla
fruit
2 cm
5 mm
head
2 mm
fruit
1 cm
basal leaf
2 cm
2 cm
Erigeron philadelphicus
Erigeron pygmaeus
Erigeron reductus var. angustatus
Erigeron sanctarum
Erigeron serpentinus
E. supplex
1 cm
E. uncialis var. uncialis
head
1 cm
1 mm
fruit
1 mm
fruit
disk corolla
Eriophyllum ambiguum var. ambiguum
ray corolla
1 mm
1 cm
1 mm
disk corolla
fruit
5 mm
head
head
1 cm
fruit
1 mm
1 cm
Erigeron utahensis
Erigeron vagus
Eriophyllum ambiguum var. paleaceum
head
2 mm
ray flower
1 mm
disk flower
1 mm
1 cm
Eriophyllum confertiflorum
1 cm
2 mm
ray corollas
1 mm
disk corollas
1 cm
2 mm
fruits
Eriophyllum congdonii
E. nubigenum
head
5 mm
ray flower
disk corolla
1 mm
1 mm
fruit
lower leaf
1 cm
Eriophyllum jepsonii
1 cm
lower surface
1 cm
E. lanatum var. arachnoideum
head
1 cm
disk flower
1 mm
2 cm
1 mm
fruit
1 cm
Eriophyllum lanatum var. obovatum

E. multicaule (DC.) A. Gray (p. 267) Ann 2–15 cm, decumbent-ascending, green to purple, ± fleshy, often becoming glabrous. **LF** ± 1 cm, wedge-shaped, 2–3-lobed, ± woolly. **INFL**: heads ± sessile in leafy clusters at branch tips; involucre 3–4 mm, bell-shaped to hemispheric; phyllaries 5–7, acute, free; receptacle convex. **RAY FLS** 5–7; ligules ± 2 mm. **DISK FLS** 13–25; corollas ± 2 mm, throat minutely puberulent; anther tips deltate. **FR** ± 2 mm, narrowly club-shaped, glabrous to minutely strigose; pappus ± 1 mm. $2n$=14,18. Open coastal scrub, chapparal; < 1600 m. sw SnJV (e San Luis Obispo, ne Santa Barbara cos.), SCoR, SCo, SnGb, PR.

E. nevinii A. Gray (p. 267) NEVIN'S ERIOPHYLLUM Subshrub 3–10 dm, thick, ascending. **LF** 8–20 cm, widely ovate, deeply 1–2-pinnately lobed (lobes 9–25+, oblong), rolled under, white-tomentose. **INFL**: heads 10–20; peduncles < 1 cm; involucres 5–7 mm, narrowly bell-shaped; phyllaries 8–12, obtuse, keeled, barely overlapping, free; receptacle low-convex. **RAY FLS** 6–8; ligules 2 mm. **DISK FLS**: corollas 3–4 mm, glandular-hairy. **FR** 3–4 mm; angles bristly; pappus < 2 mm, very unequal. $2n$=32,34. RARE. Coastal bluffs; < 35 m. s ChI (Santa Barbara, Santa Catalina, San Clemente Islands). Threatened by goats. In cult.

E. nubigenum A. Gray (p. 265) YOSEMITE WOOLLY SUNFLOWER Ann 5–15 cm, gray-woolly; branches strictly ascending. **LF** 1–2 cm, oblanceolate, entire. **INFL**: heads 1 or clustered; peduncles ± 1 cm; involucre 5–6 mm, cylindric; phyllaries 4–6, acute, free; receptacle flat to conic. **RAY FLS** 4–6; ligules ± 1 mm. **DISK FLS** 3–16; corollas ± 2 mm; anther tips deltate, glandular. **FR** 2.5–3 mm, strigose; pappus 0.5–1.5 mm, unequal. $2n$=14. RARE. Open, gravelly forest; 1800–2500 m. c SNH (Mariposa Co.).

E. pringlei A. Gray (p. 267) Ann 1–8 cm, ± tufted-spreading, white-woolly. **LF** 3–10 mm, wedge-shaped, gen 3-lobed, very woolly; margins rolled under. **INFL**: heads discoid, ± sessile in leafy clusters at branch tips; involucre 3–6 mm, hemispheric; phyllaries 6–8, acuminate, free; receptacle convex. **RAY FLS** 0. **DISK FLS** 10–25; corollas ± 2 mm, minutely glandular; anther tips deltate. **FR** 1.5–2 mm, strigose; pappus ± 1 mm. $2n$=14+. Chaparral, sagebrush or desert scrub or woodlands; 300–2200 m. s SNF, Teh, s SCoRO, SCoRI, TR, SNE, D; to s NV, AZ. ❀TRY.

E. staechadifolium Lagasca (p. 267) SEASIDE WOOLLY SUNFLOWER Subshrub 3–15 dm, much-branched, becoming glabrous. **LF** 3–7 cm, lanceolate to ovate, entire to ± 1–2-pinnately compound, becoming glabrous above; margins rolled under. **INFL**: heads 5–15+; peduncles < 1 cm; involucres 5–7 mm, bell-shaped; phyllaries 8–11, obtuse or acute, keeled, barely overlapping, free; receptacle convex. **RAY FLS** (0)6–9; ligules 3–5 mm. **DISK FLS**: corollas 4 mm, glandular. **FR** 3–4 mm, linear-oblong, ± glandular and bristly; pappus often ± 1 mm, unequal, obtuse. $2n$=30. Dunes, coastal scrub; < 100 m. NCo, CCo, n ChI; s OR. [var. *artemisiifolium* (Less.) J.F. Macbr.; var. *depressum* E. Greene] ❀DRN:**5,17**&IRR:7,**14–16,22–24**&SHD:8,9,**19–21**;GRCVR;also STBL.

E. wallacei (A. Gray) A. Gray (p. 267) Ann 1–15 cm, often tufted, woolly. **LF** 7–20 mm, spoon-shaped to obovate, entire or 3-lobed. **INFL**: head 1; peduncles 1–3 cm; involucre 5–7 mm, bell-shaped; phyllaries 5–10, acute, free; receptacle hemispheric. **RAY FLS** 5–10; ligules 3–4 mm, sometimes cream-white. **DISK FLS** many; corollas 2–3 mm, throat minutely puberulent; anther tips awl-like, glabrous. **FR** ± 2 mm, narrowly club-shaped, glabrous or minutely strigose; pappus (0)0.4–0.8 mm. $2n$=10+. Chaparral, sagebrush or desert scrub or woodlands; < 2400 m. e SnFrB, SnGb, SnBr, PR, SNE, D; to sw UT, nw AZ, n Baja CA. [var. *rubellum* (A. Gray) A. Gray] ❀TRY.

EUTHAMIA GRASS-LEAVED GOLDENRODS

John C. Semple

Per from rhizome, ascending to erect, branched above. **LVS** alternate, sessile, linear-lanceolate, 3–5-veined, entire, resin-dotted; margins finely scabrous. **INFL** dense, sometimes flat-topped; heads radiate, subsessile; involucre ± ovoid; phyllaries in graded series, midrib gen ± swollen, translucent; receptable convex, naked, pitted. **RAY FLS**: ligules yellow. **DISK FLS**: corollas yellow, ± glabrous; style branches finely papillate, appendages narrowly triangular. **FR** fusiform; pappus bristles 25–45, long, in 1 whorl. ± 8 spp.: N.Am. (Greek: well-crowded, from dense infl) [Sieren 1981 Rhodora 83: 551–579]

E. occidentalis Nutt. (p. 267) WESTERN GOLDENROD **STS** < 2 m, smooth, sometimes ± white. **LVS** < 10 cm, < 6 mm wide; lower deciduous; middle largest. **INFL** large, panicle-like, ± resinous; branches ascending; involucre 3–5 mm; phyllaries in 3–4 series. **RAY FLS** 15–25; ligules 1.5–2.5 mm. **DISK FLS** 6–15; corollas 3–4 mm. **FR** 1 mm, strigose. $2n$=18. Ditches, marshes, streambanks, meadows; < 2300 m. CA (exc D); to w Can, n-c US, NM, n Baja CA. [*Solidago o.* (Nutt.) Torrey & A. Gray] ❀IRR:1,**2,3**,4,5,**6,7**,8–10,**14–24**;INV.

FILAGO HERBA IMPIA

James D. Morefield

Ann, grayish, cobwebby to tomentose. **STS** gen ± evenly leafy below, ± lfless between upper forks. **LVS** simple, alternate or seeming whorled, ± sessile, entire. **INFL**: heads disciform, ± sessile, gen in groups of 2–10(20), ± ovoid to conic until mature; bracts lf-like; phyllaries ± 0; receptacle gen < 2 × longer than wide, gen expanded at tip, chaffy; chaff scales gen 10–20, ± phyllary-like, each subtending a pistillate fl, gen evenly curved inward; outer scales each ± folded around a fl, gen falling with a fr, ± woolly, back gen rounded, tip gen narrowly obtuse to acute, ± scarious-winged; innermost chaff scales gen > outer, open, boat-shaped, persistent, ± glabrous, ± rigid throughout, gen spreading at maturity. **PISTILLATE FLS** in (3)4–8 series, all or outer subtended by chaff scales; corollas tubular. **DISK FLS** bisexual, not subtended by chaff scales; corolla lobes 4–5. **FR** ± obovoid, gen ± compressed side-to-side; outer fr enfolded by chaff scale, gen erect, straight, smooth, shiny, pappus 0; inner fr not enfolded by chaff scale, slightly < outer, rougher or papillate, dull, pappus gen of 16–30 bristles, ± deciduous, gen cohering in a ring. ± 25 spp.: Eur, n Afr, sw Asia, sw N.Am. (Latin: with threads, from woolly hairs) [Wagenitz 1976 Sida 6:221–223] Characters may be unreliable in dwarf pls. Subg. *Oglifa* sometimes treated as genus *Logfia* in Eur. 3 Eur. aliens occur near CA: *F. arvensis, minima, vulgaris.*

1\. Longest heads 5–7 mm, yellowish, in dense, spheric groups of 12–20, largest groups 11–15 mm; receptacle 5–15 × as long as wide, narrowly cylindric; middle chaff scales acuminate, ± awned, persistent, obscurely winged; very uncommon alien (subg. *Filago*) .... ***F. pyramidata*** var. ***pyramidata***

1' Longest heads 3–4.5 mm, grayish, in loose ± hemispheric groups of 1–10, all groups < 11 mm; receptacle 0.5–2 × as long as wide, expanded at tip; middle chaff scales narrowly obtuse to acute, each falling with a fr, scarious-winged (subg. *Oglifa*)

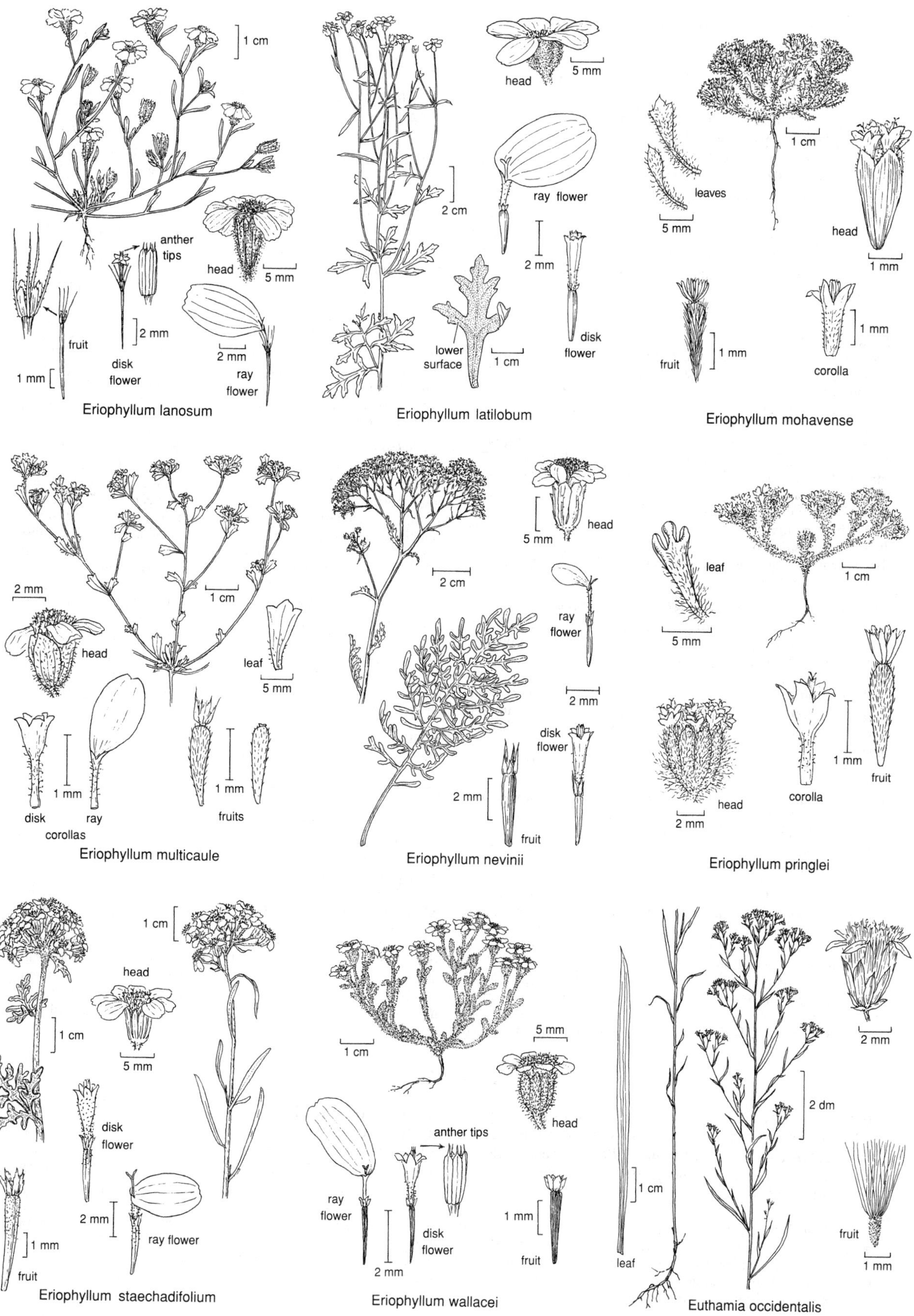
1 cm
anther tips
head
5 mm
fruit
disk flower
2 mm
1 mm
ray flower
Eriophyllum lanosum
head
5 mm
ray flower
2 cm
2 mm
lower surface
1 cm
disk flower
Eriophyllum latilobum
1 cm
leaves
5 mm
head
1 mm
fruit
corolla
Eriophyllum mohavense
2 mm
1 cm
head
leaf
5 mm
disk
ray
corollas
fruits
Eriophyllum multicaule
head
5 mm
2 cm
ray flower
2 mm
disk flower
fruit
Eriophyllum nevinii
leaf
1 cm
5 mm
head
corolla
fruit
Eriophyllum pringlei
1 cm
head
5 mm
disk flower
2 mm
ray flower
1 mm
fruit
Eriophyllum staechadifolium
5 mm
1 cm
head
anther tips
ray flower
disk flower
2 mm
1 mm
fruit
Eriophyllum wallacei
2 mm
2 dm
1 cm
leaf
fruit
1 mm
Euthamia occidentalis

2. Lvs gen awl-like, ± stiff; outer chaff scales bent abruptly inward above fr, tightly closed, body very hard; longest chaff scale 3.3–4.1 mm . . . . *F. gallica*

2′ Lvs gen elliptic to obovate or linear, flexible; outer chaff scales evenly curved above fr, loosely closed, body soft to firm; longest chaff scale 2.1–3.3 mm

3. Fls inside innermost chaff scales 4–12, 0–2 pistillate; heads gen in groups of 4–10, restricted to forks and tips of branches, subtending lvs ± linear, gen 2 × heads or longer; sts gen lfless between lower forks, ± glabrous, purplish to black . . . . *F. arizonica*

3′ Fls inside innermost chaff scales 12–40, 9–33 pistillate; heads gen in groups of 1–4, not restricted to forks and tips of branches, subtending lvs elliptic to obovate, gen < 1.5 × heads; sts ± evenly leafy between lower forks, cobwebby, grayish to greenish

4. Pappus bristles of disk fr 17–23, falling in rings; inner fr gen papillate; lobes of disk fl gen 4, gen reddish purple; sts gen erect, not forked, central axis dominant; lvs gen oblanceolate, acute . . . . *F. californica*

4′ Pappus bristles of disk fr 11–15, falling singly or in 2s; inner fr gen smooth; lobes of disk fl gen 5, gen yellowish to brownish; sts gen spreading, forked (no dominant central axis); lvs gen elliptic to obovate, obtuse . . . . *F. depressa*

F. arizonica A. Gray (p. 273) **STS** gen several from base, spreading, forked, < 20 cm, gen lfless between lower forks, purplish to black, ± glabrous; central st gen 0 or not dominant. **LVS** < 25 mm, linear to narrowly oblanceolate, acute, flexible, grayish to green, cobwebby; uppermost gen 2 × heads or longer. **INFL**: heads in loose, ± hemispheric groups of 4–10, restricted to forks and tips of branches, longest ± 4 mm, ± 3 mm wide, grayish, largest groups 8–10 mm diam; chaff scales in 5 vertical ranks, longest 2.2–2.7 mm, body gen ± hard. **PISTILLATE FLS** in 3–4 series; each fl subtended by a chaff scale. **DISK FLS** 4–10; corollas 1.2–1.7 mm, lobes 5, gen brownish to yellowish. **FR**: outer fr ascending, 0.9–1 mm, ± bent, compressed front-to-back; inner fr densely and minutely papillate, pappus bristles 17–23, 1.3–2 mm, falling in a ring. *n*=14. Locally or seasonally moist, gen clay soils; 0–800 m. SCo, s ChI, PR, w-most DSon; s-c AZ, nw Mex.

F. californica Nutt. (p. 273) **STS** 1–several from base, ± erect, not forked, < 55 cm, grayish to green, cobwebby; central axis dominant. **LVS** < 20 mm, gen oblanceolate, acute, flexible, grayish to green, cobwebby; uppermost gen < 1.5 × heads. **INFL**: heads solitary or in loose, ± hemispheric groups of 2–4, not restricted to forks and tips of branches, longest 3.5–4.5 mm, 2.5–3 mm wide, grayish, largest groups 5–9 mm diam; chaff scales in spiral ranks, longest 2.7–3.3 mm, body gen ± hard. **PISTILLATE FLS** in 4–8 series; the inner 9–30 fls not subtended by chaff scales. **DISK FLS** 4–7; corollas 1.9–2.8 mm, lobes 4, gen bright reddish purple. **FR**: outer fr erect, 0.9–1 mm, straight, compressed side-to-side; inner fr gen sparsely papillate, pappus bristles 17–23, 1.9–3 mm, falling in a ring. *n*=14. Common, ± weedy. Bare, rocky, or grassy places, drainages; 0–1800 m. CA-FP (most common s), SNE, D (esp DMtns); to sw UT, w TX, nw Mex. Dwarf pls like *F. depressa* exc adherent pappus bristles, 4-lobed disk corollas.

F. depressa A. Gray (p. 273) **STS** gen several from base, ± spreading, forked, < 11 cm, grayish to whitish, tomentose; central axis not dominant. **LVS** < 11 mm, elliptic to obovate, obtuse, rarely acute, flexible, grayish to whitish, tomentose; uppermost ± equaling heads. **INFL**: heads in loose, ± hemispheric groups of 2–5, not restricted to forks and tips of branches; longest 3–4 mm, ± 2–2.5 mm wide, grayish to whitish, largest groups 5–9 mm diam; chaff scales in spiral ranks, longest 2.1–3.1 mm, body (exc central vein) soft. **PISTILLATE FLS** in 4–7 series; inner 4–20 fls not subtended by chaff scales. **DISK FLS** 3–5; corollas 1.3–2 mm, lobes 5, gen brownish to yellowish. **FR**: outer fr erect, 0.7–0.9 mm, straight, compressed side-to-side; inner fr smooth (rarely sparsely papillate), dull; pappus bristles 11–15, 1.3–2.4 mm, falling singly or in 2's. Sandy washes or open alluvium; 0–1500 m. s SNE, D (rare s SnJV margin, SCo); s NV, se AZ, nw Mex. See *F. californica*.

F. gallica L. (p. 273) **STS** 1–few from base, erect, forked, < 50 cm, grayish to green, cobwebby; central axis gen not dominant. **LVS** < 30(–40) mm, gen awl-like, acute, ± stiff, grayish to green, cobwebby; uppermost gen 2 × heads or longer. **INFL**: heads in loose, ± hemispheric groups of 3–10, restricted to forks and tips of branches; longest 3.5–4.5 mm, 2–3 mm wide, grayish, largest groups 8–10 mm diam; chaff scales in 5 ± vertical ranks, longest 3.3–4.1 mm; outer chaff scales bent abruptly inward above fr, tightly closed, body very hard. **PISTILLATE FLS** in 4–8 series; inner 9–30 fls not subtended by chaff scales. **DISK FLS** 3–5; corollas 2.2–3 mm, lobes 4, gen brownish to yellowish. **FR**: outer fr ascending, 0.9–1 mm, ± bent, compressed side-to-side; inner fr sparsely papillate; pappus bristles 19–30, 2.2–3 mm, falling in a ring. 2*n*=28. Common, weedy. Bare or grassy places; < 1400 m. CA-FP (where possibly naturalized); also in nw Baja CA, Hawaii, Chile; native to Medit.

F. pyramidata L. var. *pyramidata* **ST** erect, forked, < 35 cm, grayish to green, cobwebby. **LVS** < 30 mm, gen widely oblanceolate, obtuse, flexible, grayish to green, cobwebby to tomentose; uppermost gen < 2 × heads. **INFL**: heads in dense, spheric groups of 12–20, restricted to forks and tips of branches; longest 5–7 mm, 3–4 mm wide, yellowish, largest groups 11–15 mm; receptacle 5–15 × as long as wide, narrowly ± cylindric; chaff scales gen 22–40 in 5 vertical ranks, longest 4.5–6 mm, body ± soft; outer chaff scales ± persistent, sharply keeled, tip acuminate, ± awned, obscurely winged; innermost chaff scales < outer, ascending at maturity. **PISTILLATE FLS** in 4–8 series; each fl subtended by a chaff scale. **DISK FLS** 5–11; corollas 2–3 mm, lobes 4, gen brownish to yellowish. **FR**: outer ± ascending, 0.7–1 mm, straight, compressed side-to-side, finely papillate, dull; inner fr sparsely papillate, pappus bristles (15)17–21, 2–2.8 mm, falling in a ring. 2*n*=28. Uncommon weed, possibly a waif. Disturbed places; 100–500 m. NCoR, n&c SnFrB; native to Medit. [*F. vulgaris* Lam. misapplied]

FLAVERIA

Ann, per, shrubs. **STS** prostrate to erect. **LVS** simple, opposite, sessile or petioled. **INFL**: heads radiate or discoid, borne in stalked or sessile, open to very condensed, sometimes head-like cymes; peduncles 0 or slender; involucre ± cylindric; phyllaries 2–5. **RAY FLS** 0 or 1; corollas yellow or cream-colored; ligules very small. **DISK FLS** 1–15; corollas yellow; style tips flattened, obtuse. **FR** 10-ribbed, ± flattened, glabrous, shining; pappus 0 (rarely of 2–4 scales). 21 spp.: N.Am, S.Am, Australia. (Latin: yellow)

F. trinervia (Sprengel) C. Mohr Ann, often rounded. **STS** 15–80 cm. **LVS** 3–15 cm, lanceolate, oblanceolate, elliptic or ovate, petioled or upper sessile; bases often fused around st; tip acute to obtuse; margin dentate or serrate; surfaces glabrous. **INFL**: heads in dense sessile, head-like clusters at forks of st; involucres < 1 mm diam; phyllaries gen 2, 4–4.5 mm, oblong, obtuse. **RAY FLS** gen 1; ligule 0.5–1 mm, creamy yellow. **DISK FLS**: corollas 2–2.5 mm. **FR** 2–2.6 mm, black. 2*n*=36. Moist soil in waste places, cult areas; < 200 m. e SCo; native to AZ to se US, S.Am.

GAILLARDIA INDIAN-BLANKET

Ann or per, gen rough-hairy. **LVS** alternate, reduced upward; blades oblong to ± ovate in CA, entire to pinnately lobed. **INFL**: heads gen radiate, 1 or in leafy cyme; peduncles slender, bracts 0–few; involucre hemispheric; phyllaries in 2–3 series, gen reflexed in fr; receptacle rounded to spheric, chaffy or bristly. **RAY FLS** 0–many, sometimes sterile; ligules showy, yellow to red, purple or multicolored, gen deeply 3-lobed. **DISK FLS** many; corollas yellow to red or purple, gen long-hairy; style tips long, tapered. **FR** obpyramidal, gen hairy; pappus of 5–10 scales, often awn-tipped. ± 28 spp.: Am. (Gaillard de Merentonneau, French botanist, 1700's)

1. Per; ligules yellow or only base purple .... ***G. aristata***
1′ Ann; ligules red or purple, sometimes yellow-tipped or multicolored .... ***G. pulchella***

G. aristata Pursh (p. 273) Per 2–7 dm, gen erect. **LVS**: lower 5–15 cm. **INFL**: heads gen 1–2; peduncles 8–20 cm; phyllaries 9–11 mm, acuminate; receptacle bristles clearly > frs. **RAY FLS** gen (6–) 13(–16); ligules 10–35 mm, yellow or base purple. **DISK FLS**: corollas 7–9 mm, purple or brown, densely purplish woolly. **FR** 3–4 mm, densely hairy; pappus 7–10 mm, awn ± 2 × body. $2n$=34,36, 68,72. Uncommon. Open sites, grassland; < 2000 m. e CaR, SN, SW; to B.C., ND, Colorado; native to OR, B.C. Sometimes cult.

G. pulchella Foug. Ann 2–4 dm. **LVS**: lower 4–8 cm. **INFL**: heads gen several; peduncles 5–15 cm; phyllaries ± 10 mm; receptacle bristles ± = frs. **RAY FLS** 10–many; ligules 12–20 mm, purple or red, sometimes yellow-tipped or multicolored. **DISK FLS**: corollas ± 7 mm, red to purple. **FR** ± 2 mm, hairy; pappus 5–8 mm, awn ± = body. $2n$=34,36. Disturbed, urban areas; < 1000 m. SnFrB, SW; native to TX. Cult forms variable; orn hybrids with *G. aristata* are *G.* ×*grandiflora* Van Houtte.

GALINSOGA

Ann. **STS** gen erect. **LVS** simple, opposite, petioled; blade 3-veined. **INFL**: heads radiate or discoid, gen small, in leafy-bracted cymes; peduncles slender; involucre bell-shaped; phyllaries in 2 series, free; receptacle conic, chaffy, scales of 2 kinds, outer fused in groups of 2–3 together with a phyllary around a ray fl, inner narrower, subtending disk fls. **RAY FLS** (0)5–8, ligule short, white. **DISK FLS** 8–50; corolla yellow; style tips acute. **FR** obconic, round to ± angled; pappus of fringed scales or 0. ± 15 spp.: Am trop; some widespread weeds. (D. Mariano Martinez de Galinsoga, Spanish physician, 18th century) [Canne 1977 Rhodora 79:319–389]

1. Phyllaries and chaff persistent after frs fall; outer phyllaries 2–4, margins scarious; inner chaff scales deeply 3-lobed; ligules gen < or = 1.5 mm; pappus scales of disk achenes 15–20, obtuse to acute ..... ***G. parviflora***
1′ Phyllaries and chaff deciduous with fr; outer phyllaries 1–2, margins herbaceous; inner chaff scales entire or shallowly 2–3-lobed; ligules < or = 2.5 mm; pappus scales of disk achenes 0–20, acute to bristle-tipped .... ***G. quadriradiata***

G. parviflora Cav. var. *parviflora* (p. 273) Ann, glabrous or sparsely soft-hairy, sometimes also glandular. **ST** 10–60 cm, simple or much-branched. **LF**: petiole < or = 2.5 cm; blade 1–11 cm, ± ovate, acute, finely dentate to coarsely serrate. **INFL**: heads 2–6 mm diam; cymes round- to ± flat-topped; peduncles 1–40 mm; phyllaries and chaff persistent; outer phyllaries 2–4, 1.2–2.2 mm, 0.6–1.5 mm wide, margins scarious; inner phyllaries gen 5, 2.5–3.5 mm, 1.3–2.6 mm wide; inner chaff scales deeply 3-lobed. **RAY FLS** gen 5; ligules 1–1.5 mm. **DISK FLS** 8–50; corollas 1.3–1.8 mm. **FR** 1.2–2.5 mm, glabrous or strigose; pappus of ray achenes of 5–8 unequal scales < or = 1 mm; pappus of disk achenes of 15–20 obtuse to acute scales < or = 2 mm. $2n$=16. Gardens, fields; gen < 1000 m. SnFrB, SW, SNE; worldwide; native to S.Am.

G. quadriradiata Ruiz Lopez & Pavon Ann, ± soft-hairy, sometimes also glandular. **ST** 10–60 cm, simple or much-branched. **LF**: petiole 2–6 cm; blade 1.5–9.5 cm, ± ovate, acute or obtuse, finely to coarsely serrate. **INFL**: heads 2–10 mm diam; cymes round- to ± flat-topped; peduncles 0–5 mm; phyllaries and chaff deciduous; outer phyllaries gen 1–2, 1–3 mm, 0.5–2 mm wide, margins herbaceous; inner phyllaries gen 5, 2.5–4 mm, 1.5–2.7 mm wide; inner chaff scales entire or shallowly 2–3-lobed. **RAY FLS** gen 5; ligule 1–2.8 mm. **DISK FLS** 15–65; corolla 1–2 mm. **FR** 1–1.8 mm, glabrous or strigose; pappus of ray achenes 0 or of 8–20 scales 0.2–1.4 mm; pappus of disk achenes 0 or of (few)15–20 bristle-tipped scales < or = 1.5 mm. $2n$=32,48,64. Gardens, fields; < 200 m. SnFrB; ± worldwide; native to Mex.

GAZANIA

Elizabeth McClintock

Per from stolons or rhizomes; juice milky. **LVS** all basal, ± entire to dentate. **INFL**: heads radiate, large, showy, solitary, scapose; peduncles long; phyllaries in 2 or more series, fused in basal half; receptacle flat or convex, pitted, naked. **RAY FLS** yellow or orange, variously marked, sterile, closing at night. **DISK FLS** many; corollas variously colored; stamen tips ovate-triangular, bases minutely sagittate; style slender below, thickened above a minutely hairy node, branches very short. **FR** obovate, covered with long hairs; pappus of slender scales ± hidden by hairs of achene. ± 16 spp.: esp s Afr. (Theodorus of Gaza, died 1478, translator of works of Theophrastus) [Hutchinson 1934 Curtis's Bot Mag Tab 9354]

G. linearis (Thunb.) Druce **STS** branching from base, spreading along ground, ± mat-like. **LVS** in loose basal rosettes; petioles long, winged, gradually expanding upward; blades linear to lanceolate, entire or irregularly dentate, glabrous or nearly so above, white-wooly below. **INFL**: heads 3.5–8 cm diam; phyllaries ± pouched at base. **RAY FLS** ± 20–21; ligules 4–5 cm, yellow or orange, gen with dark spot at base. **DISK FLS**: corollas ± 8 mm, reddish orange. **FR** 1–2 mm; pappus scales 7–8, 3–4 mm, hidden by the longer hairs of fr. Uncommon. Escape from cult in urban coastal areas, roadsides, waste places; < 200 m. CCo, SnFrB, SCo; native to s Afr. [*G. longiscapa* DC.]

GERAEA

Curtis Clark

Ann, per. **STS** erect; branches ascending. **LVS** basal and alternate, simple, sessile or petioled, entire or toothed, 3-veined from base. **INFL**: heads radiate or discoid, solitary or in few-headed panicles; peduncles ± elongated; involucre hemispheric; phyllaries in 2–3 series, free; receptacle chaffy, scales folded around frs and falling with them. **RAY FLS** sterile; style 0; ligules yellow. **DISK FLS** many; corollas yellow, tube slender, throat gradually expanded, lobes triangular; anther tips ovate, ± acute;

style tips triangular. **FR** strongly compressed, narrowly wedge-shaped; edges ± white, long-ciliate; faces black, ± hairy; pappus of 2 narrow awns. 2 spp.: sw US, nw Mex. (Greek: old, from white-haired involucre)

1. Rays present; ann from taproot; lvs sessile or with winged petioles, ± canescent, bases not ear-like; phyllaries narrow, acute, ciliate . . . . **G. canescens**

1′ Rays 0; per from an underground caudex; lvs sessile, glandular-puberulent, bases ear-like; phyllaries wide, obtuse, glandular . . . . **G. viscida**

G. canescens A. Gray (p. 273) DESERT-SUNFLOWER Ann, taprooted; herbage bristly or soft-hairy. **STS** 1–8 dm, simple to openly much-branched. **LVS** 1–10 cm, sessile above, wing-petioled below; blade lanceolate or ovate to elliptic or oblanceolate, green or ± canescent, tip acute, base tapering to wing, margin entire or dentate. **INFL**: heads radiate, solitary or few–many in panicles; involucre 7–12 mm; phyllaries narrowly lanceolate, acute, green, ciliate. **RAY FLS** 10–21; ligules 1–2 cm. **DISK FLS**: corollas 4–5 mm. **FR** 6–7 mm; pappus awns 3–4 mm. $2n=36$. Sandy desert soils; < 1300 m. D; to sw UT, w AZ, n Mex. Sometimes hybridizes with *Encelia farinosa*. ❀TRY.

G. viscida (A. Gray) S.F. Blake (p. 273) STICKY GERAEA Per from underground caudex; herbage densely glandular-puberulent and ± bristly. **STS** several from caudex, 3–10 cm, simple or few-branched. **LVS** 3–9 cm, sessile; blade ovate to oblong, green, tip obtuse, base with ear-like basal lobes, margin entire or dentate. **INFL**: heads discoid, solitary or several in ± flat-topped clusters; involucre 10–15 mm; phyllaries narrowly lance-oblong, obtuse, green, densely glandular. **RAY FLS** 0. **DISK FLS**: corollas 6–8 mm. **FR** 7–10 mm; pappus awns 3–5 mm. $2n=36$. RARE in CA. Openings in chaparral; 450–1700 m. s PR (s San Diego Co.); nw Baja CA.

GLYPTOPLEURA

G. Ledyard Stebbins

1 sp.: w N.Am. (Greek: carved side, from sculptured fr)

G. marginata D. Eaton (p. 273) Ann, taprooted, forming small tufts; sap milky. **STS** many, semi-prostrate, 2–5 cm. **LVS** 2–5 cm, lobed; lobes rounded, toothed, margins whitish, hard. **INFL**: heads ligulate, gen 1 (2–3 in cymes); involucre 10–12 mm, cylindric; phyllaries in 2 series, outer few, linear, inner 7–12, equal, narrow; receptacle naked. **FLS** 7–16; ligules cream-colored to pale yellow. **FR** 4 mm, oblong, often curved, obtusely 5-angled; ribs alternating with 5 rows of pits, abruptly short-beaked; pappus ± 8 mm, of many white bristles, outer falling separately, inner persistent. $2n=18$. Local on sandy flats; 600–2100 m. s SNH, DMoj; to OR, UT, AZ. [*G. setulosa* A. Gray] ❀TRY.

GNAPHALIUM CUDWEED, EVERLASTING

G. Ledyard Stebbins

Ann or per, gen ± woolly or tomentose. **LVS** alternate, sessile, entire. **INFL**: heads disciform, many, small, ± sessile in clusters; involucres ± cylindric to spheric, often bell-shaped when pressed; phyllaries graded in several series, transparent to opaque at tips or scarious ± throughout; receptacle flat, naked. **PISTILLATE FLS** many, in several series; corollas very slender, minutely lobed, cream-colored to pale yellow or tip reddish. **DISK FLS** few; corollas ± cylindric to funnel-shaped, whitish to purplish; anther bases short-tailed; style branches wider at tip, truncate. **FR** < 1 mm, oblong; pappus of many fine bristles, bases sometimes fused. ± 120 spp.: worldwide. (Greek: lock of wool)

1. Lvs green at least above, often glandular, sometimes sparsely tomentose
  2. Lvs green, glandular on both surfaces, sometimes sparsely tomentose
    3. Larger cauline lvs 10–15 mm wide, upper green at fl; heads spheric; phyllaries gen white . . . . . . **G. californicum**
    3′ Larger cauline lvs 5–7 mm wide, all gen withered at fl; heads narrower than high; phyllaries dull white to greenish or pinkish . . . . . . **G. ramosissimum**
  2′ Lvs green, glabrous or glandular above, densely tomentose below
    4. Heads densely clustered in leafy-bracted axils; phyllaries brown
      5. Basal lvs many more than cauline; per; leafy bracts short . . . . . . **G. collinum**
      5′ Basal lvs 0 or fewer than cauline; ann; leafy bracts long . . . . . . **G. japonicum**
    4′ Heads in open to dense, panicle-like clusters; leafy bracts 0; phyllaries white or pale yellow
      6. Pl from rhizome; lf surface not glandular above; herbage not scented; pistillate corollas 4–5 mm . . . . . . ***Anaphalis margaritacea***
      6′ Pl from taproot; lf surface densely glandular-puberulent above; herbage gen ± strongly scented pistillate corollas 3–4 mm
        7. Cauline lvs 4–16 mm wide, clasping; heads 5.5–6 mm . . . . . . **G. bicolor**
        7′ Cauline lvs 2–4 mm wide, not clasping; heads 6.5–7 mm . . . . . . **G. leucocephalum**

1′ Lvs gray, densely tomentose on both surfaces
  8. Per (bien); basal lf-tufts present at fl . . . . . . **G. canescens**
    9. Heads 4–5 mm; pistillate corollas 2.5–3 mm; lvs of sterile shoots narrowly spoon-shaped . . . . . . ssp. ***thermale***
    9′ Heads 5–6 mm; pistillate corollas 3–4 mm; lvs of sterile shoots linear to oblanceolate
      10. Lvs narrowly lanceolate, cauline ascending to erect . . . . . . ssp. ***beneolens***
      10′ Lvs ± oblong to spoon-shaped, cauline spreading
        11. Middle phyllaries ± 3 × > wide, tips ± acute . . . . . . ssp. ***canescens***
        11′ Middle phyllaries ± 2 × > wide, tips obtuse or rounded . . . . . . ssp. ***microcephalum***
  8′ Ann or bien; basal lf-tufts 0 at fl
    12. Hairs closely appressed, rarely loose; phyllaries brown to purple . . . . . . **G. purpureum**
    12′ Hairs not appressed, ± woolly, often loose and tufted; phyllaries white (at least tips)
      13. St gen < 20 cm; phyllaries brown, tips short, whitish; fls 40–60 per head . . . . . . **G. palustre**

13′ St 30–60 cm in healthy pls; phyllary upper 1/2 white or pale yellow, transparent or opaque; fls 50–120 per head

14. Heads 3–4.2 mm; pistillate corollas 1.5–2 mm; pappus bristles falling in clusters .... ***G. luteo-album***

14′ Heads 3.8–5.2 mm; pistillate corollas 1.8–2.5 mm; pappus bristles falling singly .... ***G. stramineum***

G. bicolor Bioletti (p. 273) Per 20–120 cm, sweet- or pungently scented. **STS** woody at base, variably branched, densely gray-white-tomentose. **LVS** 2–8 cm, (ob)lanceolate, sessile, green and glandular above, gray-white-tomentose below; base widely clasping. **INFL** dense; involucre 5–6 mm, ovoid to ± bell-shaped; phyllaries white or pale yellow, shiny, tips obtuse to acute, outer glabrous (exc loosely tomentose base). **FLS** 25–50 per head; pistillate corollas 3–3.5 mm: **FR**: pappus bristles free. Dry slopes, coastal dunes; < 600 m. c&s SNF, CCo, SCoR, SCo, SnJt, ChI; Baja CA. ❀SUN:**4,5,15–17,24**&IRR:**7**,8,**9,14,19–23**;fragrant lvs & sts.

G. californicum DC. (p. 273) Ann or bien 20–85 cm, scented, glandular, sometimes sparsely gray-woolly. **STS** branched above. **LVS** 2–15 cm, linear to (ob)lanceolate, sessile, short-decurrent, green and glandular on both surfaces. **INFL** often wider than high; involucre 5.5–6 mm, ± spheric; phyllaries white to straw-colored or pink-tinged, outer glabrous or base loosely tomentose. **FLS** 50–75; pistillate corollas 3–4 mm. **FR**: pappus bristles free. $2n$=28. Dry, open or wooded hills; < 1800 m. CA-FP (exc GV); OR, Baja CA. ❀SUN,DRN:4–6,**7,15–17,22–24**&IRR:8–10,**14,18–21**.

G. canescens DC. Bien or short-lived per 20–110 cm, much-branched above, scented or not, gray-tomentose throughout; basal lf-tufts gen present. **LVS** 15–80 mm, linear to (ob)lanceolate, sometimes decurrent. **INFL** panicle-like; heads many; involucre 4.5–6 mm, ovoid; phyllaries white or pale straw-colored, outer ± tomentose (esp base). **FLS** 25–45; pistillate corollas 2.5–4 mm. **FR**: pappus bristles free. Many habitats; < 2500 m. CA (exc DSon); to B.C., WY, n Mex.

ssp. ***beneolens*** (Davidson) Stebb. & Keil Pl 40–110 cm, scented. **LVS** 20–55 mm, linear (basal tufts) to narrowly lanceolate; cauline ascending to erect, upper decurrent. **INFL** open. **FLS** 30–45; pistillate corollas 2.8–4 mm. $2n$=14. Open slopes; < 1200 m. NCo, NCoRO, n&c SNF, CCo, SnFrB, SCoRO, SCo, n ChI, SnBr, PR. [*G. b.* Davidson] ❀TRY.

ssp. ***canescens*** DC. Pl 25–70 cm, barely scented or not. **LVS** 20–55 mm, narrowly oblanceolate; cauline spreading, gen not decurrent. **INFL** open. **FLS** 30–45; pistillate corollas ± 3.5 mm. $2n$= 14. Canyons, rocky slopes; < 2000 m. DMtns; to NM, n Mex.

ssp. ***microcephalum*** (Nutt.) Stebb. & Keil Pl 35–100 cm, ± unscented. **LVS** 15–60 mm, (ob)lanceolate to spoon-shaped; cauline spreading, not decurrent. **INFL** narrow or open. **FLS** 35–50; pistillate corollas 3–3.5 mm. $2n$=28. Open slopes, chaparral; < 900 m. CCo, SCoRO, SCo, n ChI, SnBr, PR; n Baja CA. [*G. m.* Nutt.] ❀TRY.

ssp. ***thermale*** (E. Nelson) Stebb. & Keil (p. 273) Pl 20–70 cm, ± scented. **LVS** 15–80 mm, linear to ± spoon-shaped (basal tufts); cauline ascending, upper decurrent. **INFL**: branches short, ascending. **FLS** 20–35; pistillate corollas 3–3.5 mm. Dry woods, roadsides; 1000–2500 m. KR, CaRH, SNH, SnBr, SnJt, MP, W&I; to B.C., WY, Colorado.

G. collinum Labill. Per 5–40 cm, from creeping rhizome and leafy stolons, unscented. **STS** few-branched, densely tomentose; internodes long. **LVS**: basal 2–10 cm, lanceolate to oblong-spoon-shaped, petioled, green above, densely white-tomentose below; cauline 2–8, 15–40 mm, linear-lanceolate, not decurrent. **INFL** dense, leafy-bracted, spheric; involucre 4–4.5 mm, cylindric; phyllaries brown, bases loosely tomentose. **FLS** 20–30; pistillate corollas ± 3 mm. **FR**: pappus bristles falling separately. Moist, grassy places, open woodland; < 800 m. NCo, NCoRO; native to Australia.

G. japonicum Thunb. Ann 5–75 cm, unscented. **STS** simple or branched from base (little branched above), tomentose. **LVS** cauline, 10–70 mm, oblong to spoon-shaped, petioled, glabrous above, densely white-tomentose below. **INFL** dense, spheric; involucre ± 4 mm, cylindric; phyllaries smooth and brown above, transparent, base white-woolly. **FLS** ± 20; pistillate corollas 3–3.5 mm. **FR**: pappus bristles falling separately. $2n$=28. Disturbed places; < 700 m. NCo, NCoRI, s SnJV; native to Australasia.

G. leucocephalum A. Gray Per 50–80 cm, ± sweet-scented. **STS** ± white-woolly. **LVS** many, 25–65 mm, linear to lanceolate, short-decurrent, green and glandular above, white-tomentose below. **INFL** much-branched; involucre 6–7 mm, ovoid; phyllaries white, smooth and opaque above, tomentose below. **FLS** 50–100; pistillate corollas ± 4 mm. **FR**: pappus bristles free. $2n$=28. Dry, sandy creek bottoms; < 150 m. SCo; to TX, nw Mex. ❀TRY.

G. luteo-album L. (p. 273) Ann 15–60 cm, gen white- (yellow-) woolly, unscented. **STS** 1–several from base, branched below or above, densely leafy below. **LVS** 10–60 mm, linear to spoon-shaped, short-decurrent. **INFL**: heads many in spheric clusters; involucre 3–4.5 mm, widely cylindric to ovoid; phyllaries smooth, transparent, whitish to yellowish above, ± glabrous. **FLS** 40–100; pistillate corollas 1.5–2 mm. **FR**: pappus bristles falling together or in clusters. $2n$=14. Fields, waste places; < 2100 m. NCoRO, c&s SN, GV, w CW, SW, SNE, D; widespread weed; native to Eurasia.

G. palustre Nutt. (p. 273) Ann 1–30 cm, unscented, ± tomentose throughout. **STS** gen branched at base, leafy. **LVS** 4–30 mm, oblong to spoon-shaped, clasping or short-decurrent. **INFL**: clusters small, terminal or axillary; involucre 3–3.5 mm, ± cylindric to ± ovoid; phyllaries smooth and brown with whitish tips above, tomentose below. **FLS** 40–60; pistillate corollas 1.5–2 mm. **FR**: pappus bristles free. $2n$=14. Common. Moist places; < 2700 m. Most of CA (exc MP); to w Can, MT, NM.

G. purpureum L. (p. 273) Ann or bien 6–60 cm, unscented. **STS** branched or not. **LVS**: basal 16–125 mm, oblanceolate to spoon-shaped, sometimes purplish, becoming ± glabrous above, densely appressed-tomentose below; cauline oblanceolate, short-decurrent. **INFL** dense, spike-like, or interrupted and leafy-bracted; involucre 4.5–5.5 mm, ovoid or urn-shaped; phyllaries brown or purplish above, base glabrous or loosely tomentose. **FLS** ± 60; pistillate corollas ± 2.5 mm. **FR**: pappus bristles fused at base, falling together. $2n$=14,28. Disturbed or waste places; < 1200 m. NCo, NCoRO, c SNF, SnJV, w CW, ChI; to e US. [*G. peregrinum* Fern.]

G. ramosissimum Nutt. Bien 50–120 cm, sweet-scented. **STS** 1–several from caudex. **LVS** 20–85 mm, narrowly lanceolate, sessile, decurrent, greenish, becoming glabrous, gen withering at fl. **INFL** large, much-branched; involucre 5–6 mm, cylindric or narrowly ovoid; phyllaries whitish to greenish or pink, opaque, base loosely tomentose. **FLS** 40–50; pistillate corollas 3–3.5 mm. **FR**: pappus bristles free. $2n$=28. Dry, open slopes; < 600 m. w NW, w CW, SCo, n ChI, SnGb, PR.

G. stramineum Kunth Ann or bien 8–70 cm, unscented, ± tomentose throughout. **STS** l–several from caudex. **LVS** 10–70 mm, lanceolate to oblong or narrowly spoon-shaped, decurrent. **INFL**: heads in dense terminal clusters; involucre 3.8–5.5 mm, ovoid; phyllaries transparent to opaque, white to straw-colored in upper part, base glabrous or loosely tomentose. **FLS** 65–110; pistillate corollas 1.8–2.5 mm. **FR**: pappus bristles free. $2n$=28. Moist, disturbed places; < 2000 m. CA; to B.C., TX, Mex. [*G. chilense* Sprengel]

GRINDELIA GUMPLANT

Meredith A. Lane

Bien to subshrub from taproot or woody caudex, glabrous to tomentose or glandular-sticky. **LVS** entire to pinnately lobed, gen clasping, gland-dotted. **INFL**: heads gen radiate, 1–many; involucres obconic to hemispheric, gen gummy; phyllaries in 4–10 series, bases gen tough, tips green; receptacle flat to convex, naked, ± pitted. **RAY FLS** 0–many; ligules yellow.

DISK FLS: corollas yellow; style appendages linear to lanceolate, gen = or > stigmatic portion. **FR** cylindric or swollen-obconic, shiny white to ± brown, glabrous, smooth to ridged; pappus of 1–6 awns ± < disk corollas, gen < 0.2 mm wide, gen U-shaped in X-section, gen entire, deciduous. ± 80 spp.: c&w N.Am, S.Am. (D.H. Grindel, 1776–1836, Latvian botanist) Hybrids common.

1. Pappus awns > 0.3 mm wide at base, minutely serrate, V-shaped in X-section; lvs ± fleshy; coastal ...... ***G. stricta***
 2. Subshrubs, gen erect; sts gen ± red-brown throughout; cauline lvs narrower near middle, tips gen acute; salt marshes, SnFrB .... var. ***angustifolia***
 2′ Per, prostrate to erect; caudex woody; upper sts gen greenish; cauline lvs widest at middle, tips truncate to acute
  3. Pls prostrate to decumbent; lf tips truncate to rounded; dunes, sea bluffs .... var. ***platyphylla***
  3′ Pls decumbent to erect; lf tips gen rounded to acute; sloughs, bluffs — NCo .... var. ***stricta***
1′ Pappus awns < 0.3 mm wide at base, entire, U-shaped in X-section or flat; lvs gen not fleshy; inland
 4. Outer phyllaries long-acuminate, gen coiled 360°
  5. Lvs dentate, teeth rounded, tips swollen — weed .... ***G. squarrosa*** var. ***serrulata***
  5′ Lvs entire or serrate, teeth pointed
   6. Sts whitish, gen appearing varnished; involucres 17–25 mm wide; cauline lvs 3–5 × longer than wide; sw CA .... ***G. camporum*** var. ***bracteosum***
   6′ Sts yellow-brown to coppery-red, gen not appearing varnished; involucres 13–18 mm wide; cauline lvs gen 5–8 × longer than wide; ne CA .... ***G. nana***
 4′ Outer phyllaries acute to acuminate, erect to reflexed (not tightly coiled)
  7. Involucres obconic; phyllaries ascending to erect; heads in panicles — e DMoj .... ***G. fraxino-pratensis***
  7′ Involucres bell-shaped or hemispheric; phyllaries erect or recurved; heads in cymes
   8. Sts gen whitish, appearing varnished; lvs gen yellow-green, cauline widest below tip, gen abruptly reduced upwards; fr gen white to golden-brown — involucres gen not subtended by phyllary-like bracts .... ***G. camporum*** var. ***camporum***
   8′ Sts gen reddish, gen not appearing varnished; lvs ± gray-green or purplish, cauline gen widest at base, little reduced; fr gen red- or gray-brown .... ***G. hirsutula***
    9. Phyllaries acuminate, weakly ascending or reflexed; pappus gen ± = disk corollas .... var. ***davyi***
    9′ Phyllaries acute, erect or ascending; pappus gen ± 1/2 disk corollas
     10. Involucres gen < 12 mm diam — PR, w DSon .... var. ***hallii***
     10′ Involucres gen > 12 mm diam
      11. Fr gen reddish, smooth or narrowly 2–3-ribbed, top gen truncate; esp NCoR, CW .... var. ***hirsutula***
      11′ Fr gen golden or grayish, gen deeply ridged, top gen flanged or knobby; bluffs, hills, CCo .. var. ***maritima***

G. camporum E. Greene Per 6–25 dm, ± erect, ± glabrous. **STS** appearing white-varnished. **LVS**: cauline 2–3 cm, lanceolate to ovate, entire to serrate, stiff, light yellow- to gray-green. **INFL**: head 1; involucre 17–25 mm diam, hemispheric; phyllaries 6–7 series, acuminate, strongly reflexed to coiled 360°, glabrous. **RAY FLS** (0)25–39; ligules 8–11 mm. **DISK FLS** many; corolla throat ± narrow. **FR** 2–5 mm, gen 2–3-angled, white to golden-brown; tops gen flanged; pappus awns 2–6. 2*n*=24. Sandy or saline bottomlands, fields, roadsides; < 1400 m. NCoRI, GV, SnFrB, SCoRO, SW; Baja CA.

var. ***bracteosum*** (J. Howell) M.A. Lane (p. 273) Pl 6–12 dm, few-branched above. **INFL**: head gen subtended by phyllary-like bracts. **RAY FLS** 0 or 25–27. Clay or sandy roadsides, streambanks, dry washes; 150–1400 m. sw GV, s SCoRO, SCo, WTR, PR; Baja CA. [*G. robusta* Nutt. incl var. *b.* (J. Howell) Keck and part of var. *r.*] Possibly derived from var. *c.* × *G. hirsutula* var. *h.* and *G. squarrosa* var. *serrulata*.

var. ***camporum*** (p. 273) Pl < 25 dm, much-branched from base. **INFL**: head gen not subtended by bracts. **RAY FLS** gen 32–39. Habitats of sp. NCoRI, GV, SnFrB, SCoRO, SW. [var. *parviflora* Steyerm., Great Valley gumplant; *G. procera* E. Greene; *G. robusta* Nutt. var. *r.* in part] Hybrids: *G.* ×*paludosa* E. Greene (× *G. stricta* var. *angustifolia*); *G.* ×*latifolia* Kellogg in part & *G.* ×*robusta* in part (× *G. stricta* var. *platyphylla*). ❀DRN,SUN:**7–9**,**14–16**,17,18,**19–21**,22–24.

G. fraxino-pratensis Rev. & Beatley ASH MEADOWS GUMPLANT Per 5–12 dm, erect, branched above, tan to reddish, glabrous, resinous. **LVS**: cauline 1–7 cm, gen oblanceolate to oblong, entire to serrate toward tip, acute, gen dark green, densely gland-dotted, bases narrowly clasping. **INFL**: heads 1–4; involucres 5–10 mm diam, ± obconic; phyllaries 4–5 series, ± erect. **RAY FLS** ± 13; corollas 4–4.5(7) mm. **DISK FLS** ± 15; throat ± narrow. **FR** 2.5–4 mm, golden-brown, oblong; top gen truncate; pappus awns 2, ± 3/4 disk corolla. 2*n*=24. **THREATENED** US. Wet clay of meadows, woodland borders; ± 700 m. e DMoj (Ash Meadows, Inyo Co.); NV. Threatened by water diversion.

G. hirsutula Hook. & Arn. Per 2–15 dm, erect, few-branched above, green to red-purple or -brown, glabrous to tomentose. **LF** 1–10 cm, oblong to lanceolate, entire to lobed (less so upward) yellow-, red- or gray-green. **INFL**: heads often subtended by phyllary-like bracts; involucres 7–32 mm diam, hemispheric to bell-shaped; phyllaries 4–5 series, gen lanceolate-acute, outer erect to reflexed. **RAY FLS** 10–50+; ligules 8–20 mm. **DISK FLS** many; throat narrow. **FR** 2.5–5.5 mm, golden-to red-brown, smooth to ridged; pappus awns flat. 2*n*=12,24. Sandy, clay, or serpentine slopes or roadsides; < 1700 m. NCoR, n&c SNF, ScV, CW, WTR, PR, DSon; to B.C.?. Vars. (exc *hallii*) intergrade.

var. ***davyi*** (Jepson) M.A. Lane (p. 279) Pl 3–6 dm, ± coppery. **LF** yellowish gray-green. **INFL**: heads gen not subtended by bracts; involucre 12–20 mm diam; phyllaries erect to barely recurved, ± linear-acuminate. **RAY FLS** gen 20–40; ligules 12–18 mm. **FR** 3.5–5.5 mm; top gen flanged or knobby. 2*n*=12. Habitats of sp. NCoR, n&c SNF, ScV. Possible derivative of var. *hirsutula* × *G. camporum*. ❀DRN,SUN:7,8,**9**,**14–17**,**19–24**.

var. ***hallii*** (Steyerm.) M.A. Lane (p. 279) SAN DIEGO GUMPLANT Pl 2–6 dm, gen white to coppery. **LF** light yellowish green. **INFL**: heads gen not subtended by bracts; involucre gen 8–12 mm diam; phyllaries erect to ± recurved. **RAY FLS** gen 12–20; ligules 8–9 mm. **FR** 4.5–5 mm. 2*n*=6. RARE. Meadows, dry slopes, open pine/oak woodlands; 800–1700 m. PR, w DSon (San Diego Co.). [*G. hallii* Steyerm.] In cult.

var. ***hirsutula*** (p. 279) Pl 2–12 dm, green to deep red-brown, puberulent to tomentose. **LF** yellow-, red-, or gray-green. **INFL**: heads gen subtended by bracts; involucre gen 12–23 mm diam; phyllaries erect to ± spreading. **RAY FLS** gen 20–60; ligules 14–20 mm. **FR** 4–5 mm, gen ± red-brown; top gen truncate. 2*n*=12,24. Sandy, clay, or serpentine slopes; < 900 m. CW, WTR; to B.C.? [ssp. *rubricaulis* (DC.) Keck; *G. humilis* Hook. & Arn., marsh gumplant] Hybrids: *G.* ×*latifolia* Kellogg in part & *G.* ×*robusta* Nutt. in part (× *G. camporum* & *G. stricta* var. *platyphylla*). ❀DRN,SUN:5,14,**15–17**,19–21,**22–24**.

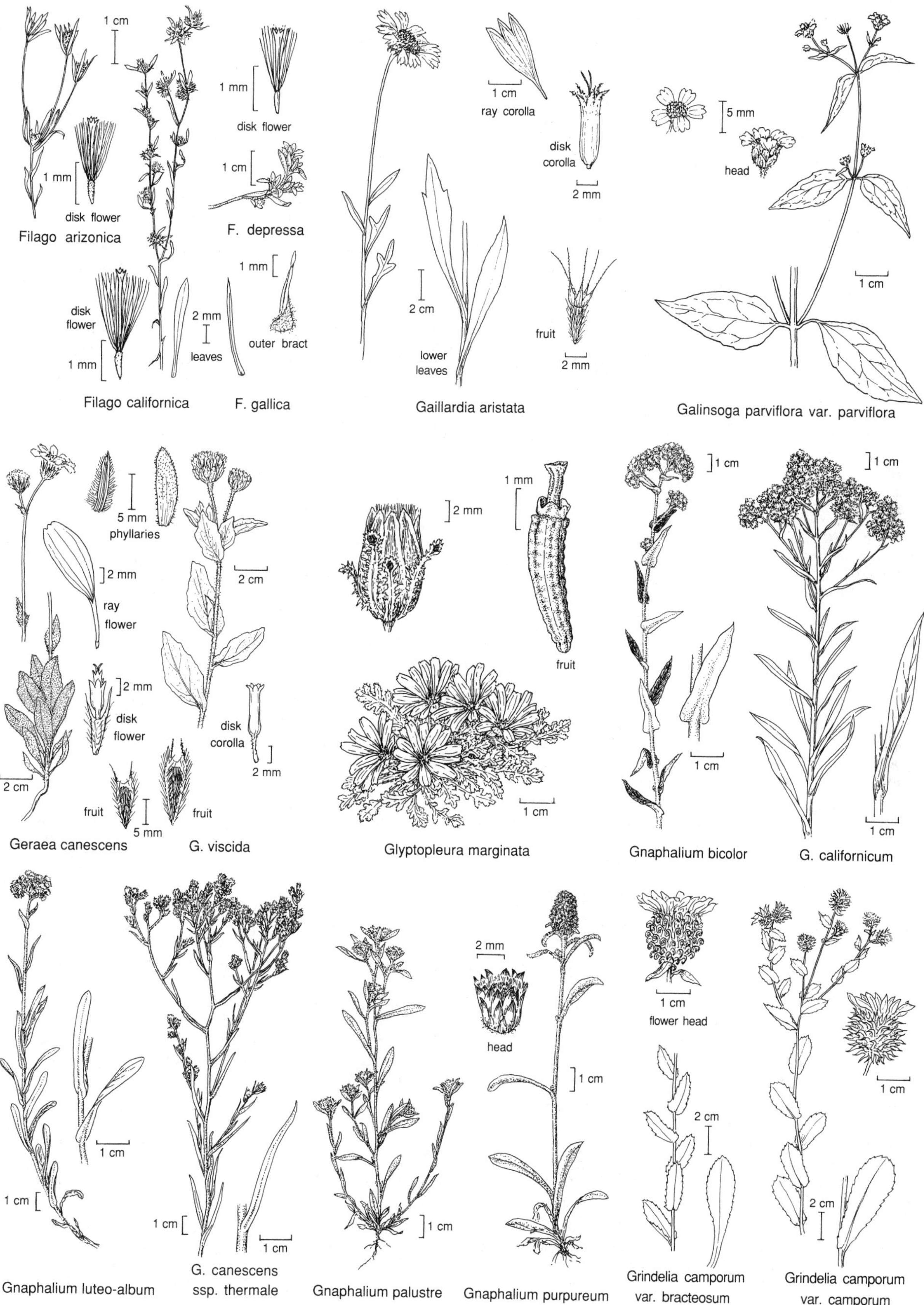
1 cm
1 mm
disk flower
1 mm
disk flower
Filago arizonica
1 cm
F. depressa
1 mm
disk flower
1 mm
outer bract
2 mm
leaves
1 mm
disk flower
Filago californica
F. gallica
1 cm
ray corolla
disk corolla
2 mm
2 cm
lower leaves
fruit
2 mm
Gaillardia aristata
5 mm
head
1 cm
Galinsoga parviflora var. parviflora
5 mm
phyllaries
2 mm
ray flower
2 cm
2 mm
disk flower
disk corolla
2 mm
2 cm
fruit
5 mm
fruit
Geraea canescens
G. viscida
2 mm
1 mm
fruit
1 cm
Glyptopleura marginata
1 cm
1 cm
Gnaphalium bicolor
1 cm
1 cm
G. californicum
1 cm
1 cm
Gnaphalium luteo-album
1 cm
1 cm
G. canescens ssp. thermale
1 cm
Gnaphalium palustre
2 mm
head
1 cm
Gnaphalium purpureum
1 cm
flower head
2 cm
Grindelia camporum var. bracteosum
1 cm
2 cm
Grindelia camporum var. camporum

var. ***maritima*** (E. Greene) M.A. Lane (p. 279) SAN FRANCISCO GUMPLANT Pl 3–5 dm, often leaning, red-brown to -purple, ± glabrous. **LF** gen gray-green. **INFL**: heads ± subtended by bracts; involucres gen 12–15(25) mm diam; phyllaries erect to ± spreading. **RAY FLS** 10–40; ligules 11–15 mm. **FR** 3.5–4.2 mm, golden- to gray-brown, deeply ridged; top gen flanged or knobby; pappus awns 4–5. 2*n*=24. RARE. Sandy or serpentine slopes, sea bluffs; < 400 m. n CCo (San Francisco, San Mateo cos.). [*G. m.* (E. Greene) Steyerm.] Possible derivative of var. *hirsutula* × *G. stricta* var. *platyphylla*. In cult.

G. nana Nutt. (p. 279) Per 1–12 dm, decumbent to erect, branched from base, ± coppery, gen not appearing varnished, ± glabrous. **LF** 3–9 cm, (ob)lanceolate, gen entire, yellow- to gray-green; base gen tapered. **INFL**: heads sometimes subtended by bracts; involucres 7– 18 mm diam, bell-shaped; phyllaries 5–7 series, linear-lanceolate, outer coiled 360°. **RAY FLS** 11–28; ligules 5–11 mm. **DISK FLS** gen 20–50; corolla throat abruptly wider. **FR** 3.5–4 mm, light brown, smooth; top flanged; pappus awns 2. 2*n*= 12. Dry, sandy hills, fields, roadsides; 100–1800 m. CaR, MP, (introduced elsewhere); to WA, MT, ID. Hybrids with *G. hirsutula* var. *davyi* have been called var. *altissima* Steyerm. ❀SUN,DRN, IRR:1,2,**6**,**7**,14, **15–17**,18–24.

G. squarrosa (Pursh) Dunal var. *serrulata* (Rydb.) Steyerm. (p. 279) Bien 1–6 dm, decumbent to erect, branched, white to yellowish, glabrous. **LF** 1.5–7 cm, oblong to ovate, dentate (teeth with rounded, swollen tips), gray-green, glabrous. **INFL**: heads sometimes subtended by bracts; involucres 12–20 mm diam, bell-shaped; phyllaries 5–6 series, outer coiled 360°. **RAY FLS** 0 or 24–36; corollas 8–10 mm. **DISK FLS** gen > 120; corolla throat abruptly wider. **FR** 2.3–3 mm, light brown to yellowish, smooth to 2–3-ribbed; top truncate; pappus awns 2–3(6). 2*n*=12,24. Disturbed roadsides, streamsides; < 1000 m. SN (e slope), c ScV, SCo, SNE, DMoj; native from WY to NM. TOXIC: concentrates selenium.

G. stricta DC. Per or subshrub 2–15 dm, prostrate to erect, green to deep purplish red, glabrous to tomentose. **LF** 1–15 cm, oblong to lanceolate, ± fleshy, green or red-veined. **INFL**: heads 1–many, gen not subtended by bracts; involucres 10–55 mm diam, hemispheric; phyllaries 4–6 series, spreading to slightly recurved. **RAY FLS** gen 20–60; ligules 12–25 mm. **DISK FLS** many; corolla throat abruptly wider than tube. **FR** 3.5–7 mm, whitish or gray- to red-brown, ridged, top knobby; pappus awns 2–6, becoming reflexed to coiled, > 0.3 mm wide, V-shaped in X-section, minutely serrate. Tidal flats, marshes, dunes, seabluffs; < 200 m. NCo, CCo, SCo, ChI; to AK. Vars. intergrade.

var. ***angustifolia*** (A. Gray) M.A. Lane (p. 279) MARSH GUMPLANT Subshrub 10–15 dm, erect, glabrous. **LVS**: cauline narrower near middle, tips gen acute. **INFL**: heads gen not subtended by bracts. **RAY FLS** 16–56; ligules 12–17 mm. **FR** 5–7 mm. 2*n*= 12,24. UNCOMMON. Tidal areas; < 10 m. n CCo. [*G. humilis* Hook. & Arn. misapplied] Hybrid: *G.* ×*paludosa* E. Greene (× *G. camporum*).

var. ***platyphylla*** (E. Greene) M.A. Lane (p. 279) Per 1–2 dm, ± prostrate, glabrous to sparsely tomentose. **LVS**: cauline gen widest at truncate to rounded tips. **INFL**: head subtended by bracts or not. **RAY FLS** gen 20–60; ligules 12–20 mm. **FR** 3.5–5 mm. 2*n*=12,24. Windswept coastal bluffs, dunes; < 200 m. NCo, CCo, SCo, ChI. [ssp. *venulosa* (Jepson) Keck] Hybrids: *G.* ×*latifolia* Kellogg in part & *G.* ×*robusta* Nutt. in part (× *G. hirsutula*); *G.* ×*robusta* in part (× *G. camporum*). ❀DRN:5,**17**&IRR:**15**,**16**,**24** &SHD:7–9,**14**,19–21,**22**,**23**;GRNCVR.

var. ***stricta*** (p. 279) Per < 10 dm, decumbent to erect, glabrous or sparsely tomentose, esp below heads. **LVS**: basal long-tapered, larger; cauline gen widest below rounded to acute tips. **INFL**: heads subtended or not by phyllary-like bracts. **RAY FLS** 30–60; ligules 13–25 mm. **FR** 3.5–7 mm. 2*n*=24. Sloughs, salt marshes, seabluffs; < 500 m. NCo; to AK. [ssp. *blakei* (Steyerm.) Keck, Humboldt Bay gumplant] Variable with distance from salt water. May hybridize with *G. nana* OR to B.C. ❀DRN,SUN:**4**,**5**,**17** &IRR:**15**,**16**,**24**&SHD:7–9,**14**,**19–23**;some forms GRNCVR.

GUIZOTIA

Ann, per. **STS** erect. **LVS** simple; lower opposite; upper sometimes alternate, sessile. **INFL**: heads radiate, borne in a terminal, leafy-bracted cyme; peduncles slender; involucre bell-shaped; phyllaries few, outer ± lf-like in texture, inner membranous; receptacle conic, chaffy; chaff scales flat. **RAY FLS** few; corollas yellow; ligules wide, conspicuously 3-lobed. **DISK FLS** many; corollas yellow; style tips tapering. **FR** 4-angled, ± flattened, glabrous, shining; pappus 0. 6 spp.: Afr. (Pierre Guizot, 1787–1874; French politician, historian)

G. abyssinica (L.f.) Cass. RAMTILLA **STS** 3–15 dm. **LVS** 7–16 cm, linear to lance-ovate, clasping; tip acute to obtuse; margin serrate; surfaces sparsely puberulent to scabrous. **INFL**: peduncles 3–5 cm; involucre 1–1.5 cm diam; outer phyllaries 5, ± 1 cm, ovate or obovate, obtuse. **RAY FLS** 8–13; ligule 1.5–2 cm. **DISK FLS**: corollas 4–5 mm. **FR** ± 4 mm, brown or black. 2*n*=30. Waste places; < 200 m. SnFrB, SCo; native to Afr. Cult as orn, oil crop, and birdseed.

GUTIERREZIA SNAKEWEED, MATCHWEED

Meredith A. Lane

Ann to subshrubs, appearing glabrous. **STS** single and branching above or branching from base, ascending, < 1.5 m, ± striate, gen fibrous, gummy, minutely scabrous, yellow to tan or gray. **LVS** alternate, sometimes in axillary clusters, entire, gland-dotted, sometimes gummy, glabrous or minutely scabrous, dark gray-green. **INFL**: heads radiate, solitary or in short-peduncled clusters; involucres narrowly to widely obconic; phyllaries in 3–4 series, whitish yellow, tips green; receptacle naked, minutely hairy. **RAY FLS** 1–13; corollas yellow. **DISK FLS** 1–13 (in CA spp.); corollas yellow, club- or narrowly funnel-shaped, lobes short, recurved; style appendages lanceolate. **FR** narrowly obconic, light tan, hairy; hairs appressed, white; pappus of 1–2 series of finely toothed, white or yellowish scales gen 1/2 fr length (in CA spp.) or much reduced. 25 spp.: 10 w N.Am, 15 S.Am. (Gutierrez, surname of a noble Spanish family) [Lane 1985 Syst Bot 10:7–28] TOXIC to livestock, fresh or dried in hay.

1. Involucres cylindric; phyllaries 4–6 .... ***G. microcephala***
1′ Involucres obconic or bell-shaped; phyllaries 12 or more
  2. Ray fls 4–13; disk fls 4–13; total fls 8–20; heads gen solitary .... ***G. californica***
  2′ Ray fls 2–8; disk fls 2–9; total fls 8–14; heads gen in clusters of 2–5 .... ***G. sarothrae***

G. californica (DC.) Torrey & A. Gray (p. 279) CALIFORNIA MATCHWEED Subshrub 2–6 dm. **STS** sprawling or erect, sometimes reddish. **LVS** ± linear. **INFL**: heads 8–20-fld, solitary or in groups of 2–3; peduncles gen > 1.5 mm; involucres gen bell-shaped, sometimes obconic, gen < or = 6.5 mm, < 3.5 mm diam; phyllaries gen 9–21 in 3 series. **RAY FLS** 4–13; corollas 2.5–7.2 mm. **DISK FLS** 4–13, fertile; corollas 2.3–4.2 mm. **FR** 1–2.8 mm. 2*n*=16,24. Grasslands, slopes, outcrops, sometimes on serpentine; 100–300 m. NCoRI, SnJV, CW, SCo, WTR, SnGb, PR; nw Baja

CA. [*G. bracteata* Abrams] Variable in involucre shape, arrangement of heads. ❀DRN,DRY,SUN:7,**8**,**9**,11–13,**14–16**,17,**19–24**.

G. microcephala (DC.) A. Gray (p. 279) STICKY SNAKEWEED Subshrub 2–6 cm, much-branched, often nearly spheric. **STS** brown below, yellow or green above. **LVS** linear to thread-like. **INFL**: heads 1–3-fld, in groups of 5–6, sessile; involucres gen < 3.2 mm, < 1.2 mm diam, cylindric; phyllaries 4–6 in 2 series. **RAY FLS** 1–2; corollas 2.1–3.5 mm. **DISK FLS** 1–2, functionally staminate; corollas 2.2–3.3 mm. **FR** 1–1.5 mm. 2*n*=8,16,24,32. Grasslands, sand dunes; 1800–2500 m. SCo, SnBr, PR, D; to Colorado, c Mex. ❀DRN,DRY,SUN:**8–11**,**14–24**.

G. sarothrae (Pursh) Britton & Rusby (p. 279) BROOM SNAKEWEED, MATCHWEED Subshrub 1–6 dm. **STS** sprawling or upright, brown below, green or tan above. **LVS** lance-linear if single, thread-like if clustered. **INFL**: heads 6–14-fld, in clusters of 5 or fewer, on peduncles < 1.5 mm or sometimes sessile; involucre gen < or = 4.5 mm, < or = 2.5 mm diam, narrowly obconic; phyllaries 8–21 in 2–3 series. **RAY FLS** 2–8; corollas 3–5.4 mm. **DISK FLS** 2–9, fertile; corollas 2.3–3.5 mm. **FR** 0.9–1.6 mm. 2*n*=8,16, 32. Grasslands, deserts, montane areas; 50–2900 m. SCo, WTR, PR, D; to WA, s-c Can, c US, n Mex. Intergrades with *G. californica* in c&s SCo, Baja CA. ❀DRN,DRY,SUN:**7–11**,**14–16**,17,**18–24**.

HAZARDIA

Gregory K. Brown & W. Dennis Clark

Per or shrub, gen resinous. **STS** < 2.5 m, leafy. **LVS** gen sessile or short-petioled. **INFL**: heads radiate or discoid; involucres obconic to bell-shaped; phyllaries in several overlapping series, linear to oblanceolate, gen recurved. **RAY FLS** 0–25; corollas < 9 mm, yellow. **DISK FLS** 4–60; corollas 4–10 mm, gradually flared from middle. **FR** 1–10 mm, 4–5-angled; pappus of 20–60 bristles, 2.5–12 mm, white to reddish brown. 13 spp.: w N.Am. (Barclay Hazard, 19th Century CA botanist) [Clark 1979 Madroño 26:105–127]

1. Per or subshrub from woody root-crown ..... ***H. whitneyi***
 2. Heads discoid; disk achenes < 10 mm ..... var. ***discoidea***
 2′ Heads radiate; disk achenes gen 5–8 mm ..... var. ***whitneyi***
1′ Shrubs; sts arising from woody trunk
 3. Heads radiate
  4. Ray fls >> disk fls; lf margin entire ..... ***H. orcuttii***
  4′ Ray fls < disk fls; lf margin gen subentire to toothed
   5. Lvs soft-hairy to scabrous; pl 2–8 dm ..... ***H. brickellioides***
   5′ Lvs densely tomentose, at least below; pl 6–25 dm
    6. Lvs thin, becoming glabrous above; phyllary tips loosely woolly-tufted; disk corollas 5–8 mm ..... ***H. cana***
    6′ Lvs thick, densely tomentose above; phyllaries gen densely woolly throughout; disk corollas 8–10 mm ..... ***H. detonsa***
 3′ Heads discoid
  7. Fls 4–8; lvs 15–25 mm, 7–12 mm wide ..... ***H. stenolepis***
  7′ Fls 9–30; lvs 15–50 mm, 10–20 mm wide ..... ***H. squarrosa***
   8. Herbage gen sparsely hairy, gen not resinous; involucres 8–12 mm; fls 9–16, corollas 9–10 mm; pappus white to reddish brown ..... var. ***grindelioides***
   8′ Herbage glabrous to sparsely hairy, gen resinous; involucres 11–15 mm; fls 18–30, corollas 10–11 mm; pappus tan
    9. Sts glabrous to scabrous; phyllaries smooth, resinous, ± obtuse, abruptly pointed, ± erect ..... var. ***obtusa***
    9′ Sts sparsely hairy above or becoming glabrous; phyllaries glandular, obtuse to acute, recurved .. var. ***squarrosa***

H. brickellioides (S.F. Blake) W. Clark (p. 279) Shrub 2–8 dm, yellow-glandular, hairy to scabrous. **LVS** 10–35 mm, elliptic to obovate, leathery, teeth (0)2–8, teeth and tip spiny. **INFL**: heads radiate; involucre 4–5 mm wide, cylindric to obconic; phyllaries 15–25, 3–7 mm, lanceolate, acute, recurved (or inner erect), bristly-glandular. **RAY FLS** 5–8; tube 4–5 mm; ligules 2–4 mm. **DISK FLS** 8–12; corollas 6–8 mm. **FR** 2–3 mm, hairy; veins 5, white; pappus 5–7 mm, white to brownish. 2*n*=12. Limestone outcrops, cliffs; 700–2100 m. DMoj. [*Haplopappus b.* S.F. Blake] Locally common.

H. cana (A. Gray) E. Greene (p. 279) SAN CLEMENTE ISLAND HAZARDIA Shrub 6–20 dm, ± woolly-tomentose. **LVS** 4–12 cm, oblanceolate, thin, subentire to finely serrate, obtuse; upper surface becoming glabrous. **INFL**: heads radiate; involucre 5–8 mm wide, obconic; phyllaries 25–35, 2–7 mm, oblong, acute, tips of outer woolly-tufted, inner ± glabrous. **FLS** sometimes becoming red-purple. **RAY FLS** 6–14; tube 4–5 mm, ligules < 1.8 mm. **DISK FLS** 15–25; corollas 5–8 mm. **FR** 3–4 mm, 4-ribbed, canescent; pappus 4–7 mm, brown or reddish-brown. 2*n*=10. RARE. Coastal bluffs, scrub; 200–500 m. s ChI; Guadalupe Island, Mex. [*Haplopappus c.* (A. Gray) S.F. Blake] In cult.

H. detonsa (E. Greene) E. Greene (p. 279) ISLAND HAZARDIA Shrub 6–25 dm, ± densely woolly-tomentose. **LVS** 4–14 cm, (ob)ovate, thick, obtuse, ± serrate. **INFL**: heads radiate; involucre 10–13 mm wide, ± bell-shaped; phyllaries 30–50, 4–10 mm, oblong, acute, ± woolly. **FLS** sometimes becoming red-purple. **RAY FLS** 6–14; ligules < 2.2 mm, tube 5–6 mm. **DISK FLS** 30–40; corollas 8–10 mm. **FR** 3–4 mm, 4-ribbed, hairy; pappus 6–9 mm, (reddish) brown. 2*n*=10. UNCOMMON. Open rocky hillsides, canyon walls; < 300 m. n ChI. ❀DRN,DRY:9,**14**,19–21&SUN:5, **15–17**,**22–24**.

H. orcuttii (A. Gray) E. Greene (p. 279) ORCUTT'S HAZARDIA Shrub 5–10 dm, glabrous, ± resinous. **LVS** 2–5 cm, (ob)lanceolate to narrowly obovate, gen abruptly pointed, ± entire. **INFL**: heads radiate; involucre 4–6 mm wide, obconic; phyllaries 30–40, 4–6 mm, linear, acute to obtuse, recurved, resinous. **RAY FLS** 8–12; ligules 2–3 mm (>> disk fls), tube 3–4 mm. **DISK FLS** 10–20; corollas 5–7 mm. **FR** 3–4.5 mm, faintly ribbed, sparsely strigose; pappus 4–5 mm, brownish. 2*n*=10. RARE. Chaparral, coastal scrub; < 200 m. s SCo (San Diego Co.); nw Baja CA. Threatened by development.

H. squarrosa (Hook. & Arn.) E. Greene (p. 279) SAW-TOOTHED GOLDENBUSH Shrub 3–23 dm, glabrous to sparsely hairy, often ± resinous. **LVS** 1.5–5 cm, leathery or stiffly papery, oblong to obovate, obtuse, toothed. **INFL**: heads discoid; involucre 7–10 mm wide, obconic; phyllaries 30–60, 3–10 mm, oblong to lanceolate. **DISK FLS** 9–30; corollas 9–11 mm, tinged red. **FR** 5–8 mm, 5-angled, glabrous; pappus 7–12 mm, white to reddish brown. 2*n*=10. Shrubland; < 1300 m. s SnJV, CW, SW, n Baja CA. [*Haplopappus s.* Hook. & Arn.]

var. ***grindelioides*** (DC.) W. Clark Pl gen sparsely tomentose, (esp near heads, on lf upper surface, phyllary margins), gen not resinous. **INFL**: involucre 8–12 mm; phyllaries gen recurved. **FLS** 9–16; corollas 9–10 mm. **FR**: pappus white to reddish brown.

Foothills, coastal mtns; 100–1300 m. s SCoRO, SW; n Baja CA. [*Haplopappus s.* ssp. *g.* (DC.) Keck] ❀TRY;STBL.

var. ***obtusa*** (E. Greene) Jepson Pl ± glabrous, resinous. **INFL**: involucre 11–15 mm; phyllaries ± erect, subtruncate to obtuse, abruptly pointed. **FLS** 18–30; corollas 10–11 mm. **FR**: pappus tan. Dry canyons; 600–1200 m. s SnJV, WTR. [*Haplopappus s.* ssp. *o.* (E. Greene) H.M. Hall]

var. ***squarrosa*** Pl ± glabrous (exc sometimes sts), resinous. **INFL**: involucre 11–15 mm; phyllaries recurved, glandular, obtuse to acute. **FLS** 18–30; corollas 10–11 mm. **FR**: pappus tan. Foothills, coastal mtns; < 700 m. SCoRO. [*Haplopappus s.* Hook. & Arn. ssp. *s.*] ❀DRY,DRN,SUN:7,**14–17**,18–24.

H. stenolepis (H.M. Hall) Hoover (p. 279) Shrub 3–10 dm. **STS** scabrous. **LVS** 15–25 mm, oblong to obovate, leathery, obtuse, abruptly pointed, toothed, glabrous, resinous. **INFL**: heads discoid; involucre 3–6 mm wide, narrowly obconic; phyllaries 20–30, 3–23 mm, linear, acute, glabrous. **DISK FLS** 4–8; corollas 9–11 mm. **FR** 5–8 mm, 5-angled, glabrous; pappus 7–12 mm, red-brown. $2n=10$. Serpentine or loose shale; 150–1200 m. SCoR. [*Haplopappus squarrosus* Hook. & Arn. ssp. *st.* H.M. Hall] Locally common.

H. whitneyi (A. Gray) E. Greene (p. 279) Per or subshrub 2–5 dm, glabrous to scabrous-glandular, resinous. **LVS** 25–50 mm, widely oblong to oblanceolate, acute, serrate. **INFL**: heads radiate or discoid; involucre 8–12 mm wide, bell-shaped; phyllaries grading into lvs, 5–12 mm, linear-lanceolate, acute, sometimes recurved. **RAY FLS** 0 or 5–18. **DISK FLS** 15–30. **FR** 5–10 mm, glabrous, 5-angled; pappus 7–10 mm, brownish. $2n=8$. Open coniferous forest; 1000–3500 m. NW, CaR, SN; OR. [*Haplopappus w.* A. Gray]

var. ***discoidea*** (J. Howell) W. Clark **RAY FLS** 0. **FR** < 10 mm. Habitat of sp.; 1000–2500 m. NW; OR. [*Haplopappus w.* ssp. *d.* (J. Howell) Keck]

var. ***whitneyi*** **RAY FLS** 5–18. **FR** 5–8 mm. Habitat of sp.; 1200–3500 m. SN. [*Haplopappus w.* ssp. *w.*]

HECASTOCLEIS

1 sp. (Greek: each enclosed, from 1-fld heads)

H. shockleyi A. Gray (p. 279) Shrub. **STS** 4–7 dm, stiff, much-branched, glandular-puberulent or becoming glabrous exc for tufts of soft hair in axils of persistent lf bases. **LVS** simple, alternate, some also clustered in axils of older sts; primary lvs 1–3 cm, linear to linear-lanceolate, sessile, spine-tipped, sparsely spiny-dentate, ± persistent as spines when dry; clustered axillary lvs narrower, obtuse to acute, gen not toothed. **INFL**: heads 1-fld, sessile, in dense, head-like clusters surrounded by involucre of persistent, ovate, spiny-toothed, net-veined bracts 1–2 cm; true involucre narrowly cylindric; phyllaries ± 15 in several unequal series, 4–10 mm, linear, acuminate, loosely soft-hairy; receptacle naked. **FL** 1 per head; corolla pink to reddish purple in bud, greenish white in fl, lobes linear, equal; filaments inserted near base of corolla, anthers purple, exserted, bases with stiff, bristle-like tails, tips short-triangular; style tips very shallowly lobed. **FR** cylindric, glabrous; pappus a crown of fringed scales < or = 1 mm. $2n=16$. Dry, rocky slopes; 1200–2200 m. SNE, DMtns; w NV.

HEDYPNOIS

G. Ledyard Stebbins

Ann; sap milky. **STS** branched, spreading. **LVS** basal and cauline, alternate, entire to dentate or pinnately lobed. **INFL**: heads ligulate, small; phyllaries in 2 series, outer small, inner equal, hardened, enveloping outer frs; receptacle naked. **FLS** many; ligules yellow, readily withering. **FR** cylindric, not beaked, often strongly incurved; pappus of long or short scales, sometimes mixed with bristles. 3 spp.: Atlantic islands, Medit. (Greek: name of Pliny for a kind of wild endive)

H. cretica (L.) Dum.-Cours. Herbage finely bristly; hairs minutely forked or barbed at tip. **STS** 0.5–4 dm. **LVS** 5–18 cm, gen oblong to oblanceolate; lower tapered to base. **INFL**: heads solitary or 1–several in open cymose clusters; peduncles gen thickened; involucre 8–10 mm; phyllaries linear, strongly incurved, glabrous to densely bristly. **FR** 5–7 mm, minutely scabrous; pappus of outer achenes a low crown of fused scales; pappus of inner achenes short scales + elongated, bristle-tipped scales. $2n=8,11,12,13,14,15,16,18$. Locally naturalized weed of gardens, pastures, roadsides, etc; 0–750 m. NCoRO, SNF, SnJV, SnFrB, SCoRO, SCo; native to Medit.

HELENIUM SNEEZEWEED

Ann, per, glabrous to ± hairy. **STS** gen erect, often winged, gen glabrous below, ± short-hairy above. **LVS** simple, alternate, gland-dotted. **INFL**: heads radiate or discoid; involucre disk-shaped; phyllaries in 1–3 series, free or ± fused, subequal or outer longer; receptacle strongly convex to spheric or conic, naked. **RAY FLS** fertile or sterile; corollas yellow; ligule fan-shaped, gen strongly 3-lobed. **DISK FLS** many; corollas 4–5-lobed, yellow to red or purplish brown; anther appendages short-triangular; style branches ± truncate. **FR** obconic, ribbed, ± hairy; pappus of 5–10 scales or 0. ± 35 spp.: N.Am, S.Am. (Helen of Troy) [Bierner 1972 Brittonia 24:331–355]

1. Lvs very narrowly linear, thread-like, not decurrent; ann ..... ***H. amarum***
1′ Lvs wider, with flat blade, decurrent; ann or per
  2. Lf margins ± toothed ..... ***H. autumnale***
    3. Ligules 1.5–2.5 cm; herbage sparsely puberulent or becoming glabrous ..... var. ***grandiflorum***
    3′ Ligules ± 1 cm; herbage densely puberulent and glandular ..... var. ***montanum***
  2′ Lf margins entire
    4. Heads several–many; ligules 4–10 mm (sometimes rays 0); disk corollas gen 4-lobed, 2–2.7 mm; pappus < or = 1 mm ..... ***H. puberulum***
    4′ Heads 1–few; ligules 13–30 mm; disk corollas 5-lobed, 3–5 mm; pappus 1.3–4.5 mm
      5. Peduncles sparsely to moderately hairy; pappus gen 1.3–2.7 mm; widespread but gen not NCo ***H. bigelovii***
      5′ Peduncles gen densely hairy; pappus gen 3–4.5 mm; NCo ..... ***H. bolanderi***

H. amarum (Raf.) Rock Ann. **ST** 2–7 dm; branches gen many. **LVS** cauline, very many, narrowly linear to thread-like, not decurrent, entire, glabrous. **INFL**: heads several–many in ± flat-topped cymes; peduncles 4–10 cm, glabrous or sparsely hairy; receptacle 2–3 mm diam. **RAY FLS** 7–10; ligules 6–12 mm. **DISK FLS**: corollas 2.5–3 mm, 5-lobed, yellow throughout. **FR** 1 mm, with long, straight hairs on ribs; pappus scales 1–2 mm, awn-tipped. $2n=30$. Disturbed areas; < 100 m. SnJV, SnFrB; native to se US.

H. autumnale L. Per. **ST** 2–10 dm; branches few–many. **LVS** mostly cauline; lower often with short, winged petioles; upper sessile, decurrent; blades oblanceolate to elliptic, obtuse to acute, gen ± toothed, glabrous or sparsely hairy. **INFL**: heads several–many in a ± flat-topped cyme; peduncles 3–10 cm, short-hairy; receptacle 3–5 mm diam. **RAY FLS** 10–20; ligules 10–25 mm. **DISK FLS**: corollas 5-lobed, 3–3.5 mm, yellow throughout. **FR** 1.5–2 mm, sparsely hairy; pappus scales awn-tipped, 1–1.5 mm. $2n$=32,34,36. Wet places; < 2000 m. NCo, KR, CaR, MP; to e N.Am. Highly variable.

var. ***grandiflorum*** (Nutt.) Torrey & A. Gray Pl gen 4–12 dm, sparsely puberulent or becoming glabrous. **RAY FLS**: ligules 15–25 mm. Wet places; < 500 m. n NCo, w KR; to B.C. ❀IRR or WET,SUN:**4**,**5**,15,16,**17**.

var. ***montanum*** (Nutt.) Fern. Pl gen 1.5–6 dm, densely puberulent and glandular. **RAY FLS**: ligules ± 10 mm. Wet places; 500–2000 m. e KR, MP; to B.C., WY, Colorado, NM.

H. bigelovii A. Gray (p. 279) Per. **STS** 3–13 dm; branches 0–few. **LVS** basal and cauline, entire, glabrous or ± short-hairy; basal with short, winged petioles; cauline sessile, clasping, decurrent, lower and middle lanceolate to oblong-elliptic, upper linear. **INFL**: heads 1–few; peduncles 10–30 cm; receptacle 4–8 mm diam. **RAY FLS** 14–20; ligules 13–25 mm. **DISK FLS**: corollas 3–5 mm, 5-lobed, tube and throat yellow, lobes yellow to red, brown, or purple. **FR** 1.8–2.4 mm, ± long-hairy; pappus scales 1.3–2.7 mm, awn-tipped. $2n$=32. Wet meadows, marshes, bogs; ± 0–3000+ m. KR, NCoR, CaR, SN, SCoR, TR, PR. ❀IRR or WET,SUN:1–3,**4–6**, 7–9,14,**15–17**,18–24.

H. bolanderi A. Gray Per. **STS** 2–14 dm, glabrous below, ± short-hairy above; branches 0–few. **LVS** basal and cauline, entire, glabrous or ± short-hairy; basal and lower cauline oblanceolate to obovate, tips rounded; middle and upper cauline sessile or wing-petioled, oblanceolate or elliptic to ovate, clasping, decurrent. **INFL**: heads gen 1(–3); peduncles 10–30 cm, gen tomentose; receptacle 4–8 mm diam. **RAY FLS** 15–30; ligules 15–30 mm. **DISK FLS**: corollas 4–5 mm, 5-lobed, yellow, lobes yellow to red, brown, or purple. **FR** 1.5–2.5 mm, ± long-hairy; pappus scales 3–4.5 mm, awn-tipped. $2n$=32. Wet soil; < 100 m. NCo; OR. ❀IRR or WET, SUN:**5**,14,**15–17**,24.

H. puberulum DC. (p. 279) Ann, per. **STS** 5–16 dm; branches several–many. **LVS** basal and cauline, entire, glabrous or ± short-hairy; basal oblanceolate, wing-petioled; cauline gen sessile, decurrent, lower and middle lanceolate to oblong-elliptic, upper narrower. **INFL**: heads gen several–many; peduncles 9–17 cm, ± short-hairy; receptacle 5–10 mm diam. **RAY FLS** 0 or 13–15; ligules gen 4–10 mm. **DISK FLS**: corollas 2–2.7 mm, 4–5-lobed, yellow, lobes yellow to brown or purple. **FR** 1.2–1.9 mm, ± long-hairy; pappus scales 0.4–1 mm, short-awned. $2n$=58. Streamsides, marshes, other wetlands; 0–1200 m. NCoR, SNF, SnJV, SCoR, WTR, w PR; Baja CA. ❀STBL.

HELIANTHELLA

Per from taproot and caudex. **STS** 1–several, erect. **LVS** basal and cauline, alternate or opposite; petioles short to long; blades gen 3-veined, linear to ± ovate, entire, ± scabrous. **INFL**: heads radiate, 1–few, in cymes; peduncles long, with 0–few lf-like bracts; involucre hemispheric; phyllaries in 2–3 series, free; receptacle flat to ± convex, chaff scales folded around ovary, entire, obtuse, tip hairy. **RAY FLS** sterile; corolla yellow; style 0. **DISK FLS** many; corolla and anthers yellow or purple; style tips short-triangular. **FR** thick or compressed; pappus 0 or 2 awns 1–2 mm (sometimes also crown of low scales). (Latin: diminutive of *Helianthus*) [Weber 1952 Amer Midl Naturalist 48:1–35]

1. Outer phyllaries gen enlarged, lf-like, incurved; involucre 2.5–4 mm diam; fr thick .... ***H. castanea***
1′ Outer phyllaries seldom enlarged or lf-like, not incurved; involucre gen 1.5–2 cm diam; fr compressed .... ***H. californica***
 2. Pappus 0 .... var. ***californica***
 2′ Pappus present
  3. Disk gen 1.5–2 cm diam; pappus gen of 2 awns, 1–2 mm, and crown of short scales .... var. ***nevadensis***
  3′ Disk gen 1–1.5 cm diam; pappus gen of 2 awns, 1 mm .... var. ***shastensis***

H. californica A. Gray **STS** 1–6 dm, glabrous to coarsely hairy. **LVS**: blades 0.5–4 cm wide, linear to oblanceolate. **INFL**: peduncles 10–30 cm, sometimes ± scapose, ± short-hairy; involucre gen 1–2 cm diam; phyllaries seldom lf-like, gen 1–2 cm, short-ciliate. **RAY FLS** 9–21; ligules gen 1–2 cm. **DISK FLS**: corollas 4–5 mm, yellow; anthers yellow to purple. **FR** 6–8 mm, obovate, thin, glabrous; pappus 0 or 2 awns or narrow scales, sometimes with crown of short scales. Grassy sites, woodland, forest; < 2600 m. KR, NCoR, CaR, SN, s PR; s OR.

var. ***californica*** **STS** gen 3–6 dm. **LVS** 1–4 cm wide. **INFL**: heads 1–few; disk gen 1.5–2 cm diam. **FR**: pappus 0. $2n$=30. Open, grassy sites; < 1000 m. NCoR. ❀SUN,DRN:4,5,**6**,**14–17**, 18,22–24&IRR:1–3,**7**,**9**,19–21.

var. ***nevadensis*** (E. Greene) Jepson **STS** gen 3–6 dm. **LVS** gen 1.5–4 cm wide. **INFL**: heads 1–few; disk gen 1.5–2 cm diam. **FR**: pappus of 2 awns or narrow scales 1–2 mm and crown of short scales. Habitats of sp.; 250–2600 m. NCoRI, CaR, SN, s PR; s OR. ❀TRY.

var. ***shastensis*** W.A. Weber **STS** gen 1–3 dm. **LVS** gen 0.5–1(4) cm wide, often linear; cauline few. **INFL**: head gen 1; disk gen 1–1.5 cm diam. **FR**: pappus of 2 awns or narrow scales 1 mm. Open forest; 1000–2600 m. KR, CaR. ❀TRY.

H. castanea E. Greene (p. 279) DIABLO HELIANTHELLA **STS** 1–5 dm, glabrous to coarsely hairy. **LVS**: cauline few; petioles long; blades 2–6 cm wide, narrowly to widely elliptic. **INFL**: head gen 1; peduncle 7–20 cm, stout, ± rough-hairy, often with 1–few bracts near tip; involucre 2.5–4 cm diam; outer phyllaries gen lf-like, 3–10 cm, 7–20 mm wide, curving up around head; inner phyllaries 2–2.5 cm, coarsely ciliate. **RAY FLS** 13–21; ligules 1–3 cm. **DISK FLS**: corollas 6–7 mm, yellow; anthers yellow. **FR** 8–10 mm, obovate, glabrous; center thick; edges thin; pappus awns 0 or 2, < 1 mm. $2n$=30. RARE. Open, grassy sites; 200–1300 m. n SnFrB. In cult.

HELIANTHUS SUNFLOWER

Ann or per. **STS** gen erect. **LVS** opposite or alternate, gen reduced upward, often 3-veined from near base, gen rough-hairy. **INFL**: heads radiate, solitary or in cymes; involucre bell-shaped to hemispheric; phyllaries in 1–3 gen ± equal series, free; receptacle flat to rounded; chaff scales 0–3-lobed. **RAY FLS** 10–many, sterile; ligules yellow. **DISK FLS** many; corollas yellow to red or purple, tube short, throat base often swollen, lobes triangular; style appendages triangular. **FR** oblanceolate to obovate, ± compressed; sides rounded; pappus gen of 2 deciduous, lanceolate to ovate scales (sometimes also 1–several shorter scales). 67 spp.: Am. (Greek: sun fl) [Heiser 1969 Mem Torrey Bot Club 22(3):1–218]

1. Ann; petioles ± long, well developed
 2. Phyllaries gen > 4 mm wide, abruptly acuminate; lf base cordate .... ***H. annuus***

2′ Phyllaries gen < 4 mm wide, acute or gradually acuminate; lf base gen truncate to wedge-shaped
3. Central chaff scales tipped with stiff white hairs
4. Herbage and phyllaries densely gray-canescent .... ***H. niveus*** ssp. ***canescens***
4′ Herbage and phyllaries strigose, green .... ***H. petiolaris*** ssp. ***petiolaris***
3′ Central chaff scales glabrous or very finely appressed-hairy
5. Chaff scales awn-tipped, >> disk fls, glabrous; lvs green .... ***H. bolanderi***
5′ Chaff scales acute, = or slightly > disk fls, rough-hairy; lvs densely white-hairy ...... [2]***H. niveus*** ssp. ***tephrodes***
1′ Per; lvs sessile (or petiole < 3 cm)
6. Disk corolla lobes red to purple
7. Lvs blue-green, ± glabrous, margins often crinkled; roots horizontal, rhizome-like; ligules gen < or = 1 cm; phyllaries < disk .... ***H. ciliaris***
7′ Lvs green or densely white-hairy, margins gen flat; pl taprooted; ligules gen >> 1 cm; phyllaries = or > disk
8. Lvs green, short-rough-hairy; phyllaries not > disk .... [2]***H. gracilentus***
8′ Lvs densely white-hairy; phyllaries = or slightly > disk .... [2]***H. niveus*** ssp. ***tephrodes***
6′ Disk corolla lobes yellow
9. Phyllaries not > disk, obtuse to acute .... [2]***H. gracilentus***
9′ Phyllaries > disk, acuminate
10. Phyllaries gen 12–18; most lvs opposite, gen < 12 cm; pls tap-rooted .... ***H. cusickii***
10′ Phyllaries gen > 25; most lvs alternate, often > 15 cm: pls rhizomed
11. Lf blades folded along midvein; infl gen raceme-like .... ***H. maximilianii***
11′ Lf blades ± flat; infl flat-topped or rounded
12. Phyllaries gen 3+ mm wide, folded back at maturity; st gen glabrous .... ***H. californicus***
12′ Phyllaries gen < 3 mm wide, ± erect at maturity; st gen ± hairy .... ***H. nuttallii***
13. Phyllaries sparsely hairy; lf blade short-rough-hairy below .... ssp. ***nuttallii***
13′ Phyllaries densely hairy; lf blade finely tomentose below .... ssp. ***parishii***

H. annuus L. Ann < 3 m. **STS** ± rough-hairy. **LVS** long-petioled; blade 10–40 cm, widely lanceolate to widely ovate, base gen ± cordate, tip obtuse to acute, margin serrate. **INFL**: heads gen few–many; peduncles 2–20 cm; involucre 1.5–3(20) cm diam; phyllaries 15–25 mm, lanceolate to widely ovate, abruptly acuminate, glabrous to rough-hairy, gen ciliate; chaff scales deeply 3-lobed, middle lobe long-acuminate. **RAY FLS** 15–many; ligule gen > 2.5 cm. **DISK FLS**: corollas 5–8 mm, lobes red to purple or yellow. **FR** 3–15 mm; pappus scales 2–3.5 mm. $2n$=34. Disturbed areas, shrubland, many other habitats; < 1900 m. CA; to e N.Am. Highly variable; hybridizes with several other spp. [ssp. *jaegeri* (Heiser) Heiser; ssp. *lenticularis* (Douglas) Cockerell; var. *macrocarpus* (DC.) Cockerell] Head 1, very large in cult forms. ❀SUN: 2–5,**6,17**&IRR:**7–16,18–24**.

H. bolanderi A. Gray (p. 285) Ann < 1.5 m. **STS** rough-hairy. **LVS** petioled; blades 3–15 cm, narrowly lanceolate to ovate, base truncate to wedge-shaped, tip gen acute, margin entire to serrate. **INFL**: heads 1–several; peduncles 3–13 cm; involucre 1.5–2.5 cm diam; phyllaries 8–25 mm, oblong to lanceolate, long-acuminate, gen >> disk, hairs long, soft to stiff; chaff scales 3-toothed, awn-tipped, >> disk fls, tip glabrous. **RAY FLS** 10–17; ligules 1–2 cm. **DISK FLS**: corollas 4–6 mm, lobes yellow or red-purple. **FR** 3–4.5 mm; pappus scales 2–3 mm. $2n$=34. Grassy, often disturbed places; gen < 1200 m. KR, NCoR, CaR, SNF, GV, SCoR; s OR. Slender pls on serpentine have been called *H. exilis* A. Gray, serpentine sunflower. ❀SUN,DRN:4–6,**17**&IRR:**7–16**,18,**19–24**.

H. californicus DC. (p. 285) Per 15–35 dm; roots thick, woody; rhizome short. **STS** glabrous, glaucous, grooved. **LVS** gen alternate; petiole 0–3 cm; blade 10–20 cm, gen lanceolate, entire or few-toothed. **INFL**: heads several; peduncles 3–15 cm; involucre 1–2.5 cm diam; phyllaries 10–25 mm, gen >> disk, widely lanceolate, bent back in fr, margin glabrous or rough-ciliate; chaff scales 10–11 mm, sharply 3-lobed, middle lobe acute, short-rough-hairy. **RAY FLS** 12–21; ligules 2–3 cm. **DISK FLS**: corollas 6–8 mm, lobes yellow. **FR** ± 5 mm; pappus scales 3–4 mm. $2n$=102. Dry, rocky soils or marshes, streambanks; gen < 1600 m. s NCoR, c SNF, ScV, n CW, SCo, PR; n Baja CA. ❀SUN:4–6,**17**&IRR:**7–16**,18,**19–24**;INV.

H. ciliaris DC. BLUEWEED Per 4–7 dm, from rhizome-like, horizontal roots. **STS** ± glabrous, glaucous. **LVS** gen opposite, sessile, 3–7.5 cm, oblong or lanceolate, acute, entire to shallowly lobed, wavy, glabrous, sparsely short-rough-hairy, glaucous, sometimes ciliate. **INFL**: heads few; peduncles 3–13 cm; involucre 12–25 mm diam; phyllaries gen unequal, 3–8 mm, < disk, oblong to ovate, tip obtuse or abruptly pointed to acute, ciliate, (sub)glabrous; chaff scales 8–10 mm, entire or 3-toothed, tip hairy. **RAY FLS** (0)10–18; ligules ± 1 cm. **DISK FLS**: corollas 4–5 mm, lobes red. **FR** ± 3 mm; pappus scales ± 2 mm. Irrigated fields, roadsides; gen < 300 m. $2n$=68,102. ScV, SnFrB, SCo, w D; native to sc US, n Mex. **NOXIOUS WEED**.

H. cusickii A. Gray Per < 12 dm, from stout, ± fleshy taproot. **STS** glabrous to sparsely long-hairy. **LVS** opposite (or upper alternate), ± sessile; blade 6–15 cm, ± lanceolate, acute, entire, glabrous or scabrous. **INFL**: heads few; peduncles 5–15 cm; involucre 1.2–2.8 cm diam; phyllaries 12–18, 12–25 mm, ± lanceolate, gen > disk, strigose or stiff-spreading-hairy; chaff scales 10–13 mm, entire or 3-toothed, sparsely hairy. **RAY FLS** 12–21; ligules 2–4 cm. **DISK FLS**: corollas 5–7 mm, lobes yellow. **FR** 4–5 mm; pappus scales 3–4 mm. $2n$=34. Dry slopes, open woods; 1200–2000 m. e CaR, MP; to WA, ID, w NV. ❀DRN:1–3,**7,14–16**&SUN:4,5,**6,17**.

H. gracilentus A. Gray (p. 285) Per 6–20 dm, from stout taproot. **STS** rough-hairy. **LVS** opposite (or upper alternate); petioles 0–3 cm; blade 5–11 cm, ± lanceolate, acute, entire or few-toothed. **INFL**: heads few; peduncles 4–30 cm; involucre 1.3–2 cm diam; phyllaries gen unequal, 5–8 mm, ± lanceolate, < or = disk, glabrous to short-rough-hairy, ciliate; chaff scales 8–9 mm, entire or 3-toothed, tip short-ciliate. **RAY FLS** 13–21; ligules 15–25 mm. **DISK FLS**: corollas 4–5 mm, lobes yellow or red. **FR** 3–4 mm; pappus scales ± 3 mm. $2n$=34. Dry slopes, esp after fires; < 1800 m. s CW, SW; n Baja CA. ❀SUN,DRN:3,**7,14–16**,17,**18–24**.

H. maximilianii Schrader Per 5–30 dm, from well developed rhizome. **STS** ± scabrous. **LVS** alternate; petiole 0–2 cm; blade 10–30 cm, lanceolate, folded along midvein, acute, entire, scabrous. **INFL** spike-like or raceme-like clusters; peduncles 2–6 cm; involucre 1.6–2.8 cm diam; phyllaries 15–25 mm, linear-lanceolate, long-acuminate, >> disk, canescent, ciliate; chaff scales 7–8 mm, acute to acuminate, entire or weakly 3-toothed, tip hairy. **RAY FLS** 12–21; ligule 2.5–3.5 cm. **DISK FLS**: corollas 5–7 mm, lobes yellow. **FR** 3–4 mm; pappus scales 2–3 mm. $2n$=34. Disturbed places; < 100 m. s SnJV (Fresno Co.); native to c&e N.Am. Cult as orn.

H. niveus (Benth.) Brandegee Ann or per < 15 dm, from taproot. **LVS** gen alternate, ± long-petioled; blade lanceolate to ovate, wedge-shaped to ± cordate at base, obtuse to acute, densely canescent to long-silky. **INFL**: heads 1–few; involucre 8–28 mm diam; phyllaries 8–12 mm, ± lanceolate, acute, canescent to soft-hairy, ± = disk; chaff scales 10–11 mm, entire to deeply 3-lobed. **RAY FLS** 13–21; ligules 12–25 mm. **DISK FLS**: corollas 5–7 mm, lobes red to dark purple in CA. **FR** 3–8 mm; pappus scales 2–3 mm (gen also several shorter scales). $2n$=34. Open, sandy places; gen < 300. SCo, DSon; to TX, n Mex.

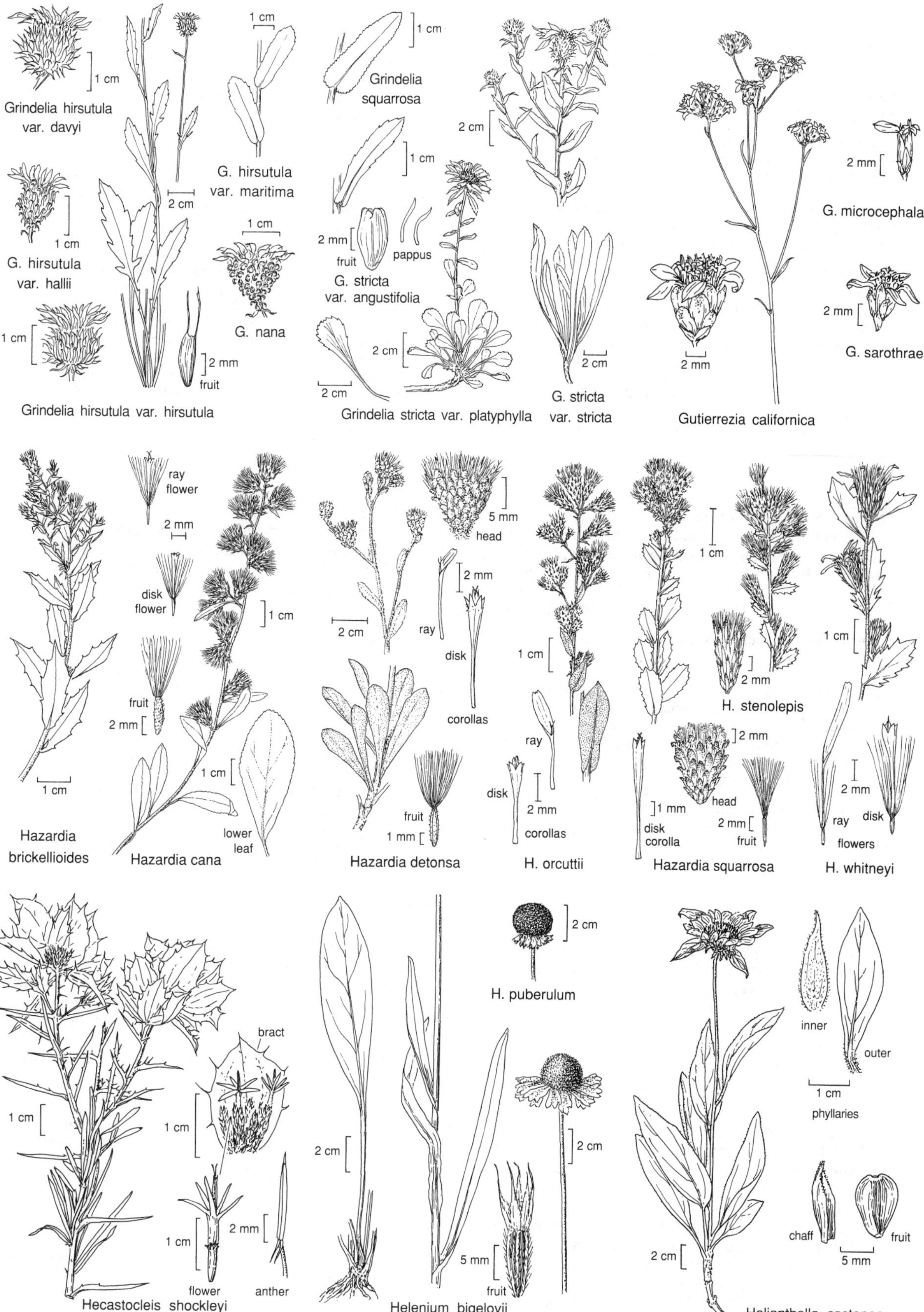
1 cm
Grindelia hirsutula var. davyi
1 cm
G. hirsutula var. hallii
1 cm
2 cm
1 cm
G. hirsutula var. maritima
1 cm
G. nana
2 mm
fruit
Grindelia hirsutula var. hirsutula
1 cm
Grindelia squarrosa
1 cm
2 mm
fruit
pappus
G. stricta var. angustifolia
2 cm
2 cm
2 cm
Grindelia stricta var. platyphylla
2 cm
G. stricta var. stricta
2 mm
G. microcephala
2 mm
2 mm
G. sarothrae
Gutierrezia californica
ray flower
2 mm
disk flower
fruit
2 mm
1 cm
1 cm
1 cm
Hazardia brickellioides
Hazardia cana
lower leaf
5 mm
head
2 mm
2 cm
ray
disk
corollas
fruit
1 mm
Hazardia detonsa
1 cm
ray
disk
2 mm
corollas
H. orcuttii
1 cm
2 mm
H. stenolepis
2 mm
1 mm
head
disk corolla
2 mm
fruit
Hazardia squarrosa
1 cm
2 mm
ray
disk
flowers
H. whitneyi
1 cm
bract
1 cm
1 cm
2 mm
flower
anther
Hecastocleis shockleyi
2 cm
2 cm
H. puberulum
2 cm
5 mm
fruit
Helenium bigelovii
inner
outer
1 cm
phyllaries
2 cm
chaff
fruit
5 mm
Helianthella castanea

ssp. ***canescens*** (A. Gray) Heiser Ann or per. **STS** strigose and stiff-spreading-hairy. **LVS**: blades 2–12 cm, lanceolate to ovate, obtuse, densely stiff-hairy. **INFL**: phyllaries gen slightly > disk; central chaff scales gen stiff-white-hairy. **FR** 3–4 mm, short-hairy. Habitats and range of sp. [*H. petiolaris* Nutt. var. *c.* A. Gray] ❀DRN,SUN:**15,16**,17,&IRR:**8–14,19–24**.

ssp. ***tephrodes*** (A. Gray) Heiser (p. 285) ALGODONES DUNES SUNFLOWER Gen per, ± shrubby. **STS** soft-white-appressed-hairy. **LVS**: blades gen 3–7 cm, triangular-ovate, densely soft-hairy. **INFL**: phyllaries not > disk; central chaff scales glabrous or fine-appressed-hairy. **FR** 4–8 mm, densely long-hairy. $2n$=34. **ENDANGERED** CA. Sand dunes; gen < 100 m. s DSon (Imperial Co.); sw AZ, n Mex. [*H. tephrodes* A. Gray]

H. nuttallii Torrey & A. Gray Per 5–40 dm, from clustered, tuber-like roots; rhizome short. **STS** glabrous or hairy. **LVS** alternate or opposite, subsessile; blade 10–20 cm, narrowly lanceolate to ovate, acute to acuminate, entire or serrate, glabrous to hairy. **INFL**: heads few–many; peduncles 1–18 cm; involucre 1–2 cm diam; phyllaries ± erect, 8–16 mm, gen < 3 mm wide, ± linear, = or slightly > disk, glabrous or tomentose; chaff scales 8–12 mm, entire or 3-toothed, acute, short-rough-hairy. **RAY FLS** 12–20; ligules 15–25 mm. **DISK FLS**: corollas 5–6 mm, lobes yellow. **FR** 3–4 mm; pappus scales 3–4 mm (sometimes also with shorter scales). Damp meadows, marshes; < 500 (c-w SW) or 1200–2500 m. SnGb, SnBr, GB; to B.C., e Can, NM.

ssp. ***nuttallii*** (p. 285) **STS** glabrous to scabrous. **LVS**: blade scabrous above, stiffly hairy below. **INFL**: peduncles glabrous; phyllaries glabrous or strigose. $2n$=34. Damp meadows; 1200–2500 m. SnGb, SnBr, GB; to B.C., e Can, NM. ❀IRR,SUN:1–5, **6**,7–10,**14–24**.

ssp. ***parishii*** (A. Gray) Heiser (p. 285) LOS ANGELES SUNFLOWER **STS** glabrous to tomentose. **LVS**: blade rough-hairy to densely tomentose above, ± finely tomentose below. **INFL**: peduncles and phyllaries tomentose. **PRESUMED EXTINCT**. Marshes; < 500 m. c-w SW. Last seen in 1937.

H. petiolaris Nutt. ssp. *petiolaris* Ann 4–20 dm. **STS** ± glabrous, rough-hairy, or strigose. **LVS** long-petioled; blades 4–15 cm, lanceolate to triangular, widely wedge-shaped to truncate at base, acute, entire to serrate, strigose. **INFL**: heads few–many; peduncles 5–40 cm; involucre 1–2.5 cm diam; phyllaries 9–12 mm, lanceolate to ovate, short-rough-hairy, margin glabrous or ciliate; chaff scales 9–10 mm, sharply 3-lobed, middle lobe slightly > disk fls, central scale tips densely ciliate or bristly. **RAY FLS** 15–30; ligules 2.5–3 cm. **DISK FLS**: corollas 5–7 mm, lobes reddish purple. **FR** 3.5–4.5 mm; pappus scales 2–3 mm. $2n$=34. Disturbed habitats; gen < 450 m. SnFrB, SW; native to c US.

HELIOMERIS GOLDEN-EYE

Ann, per. **STS** 1–many from base, slender, spreading or erect. **LVS** simple, opposite or upper alternate, linear to narrowly elliptic, entire. **INFL**: heads radiate, solitary or in few-headed cymes; peduncles slender, bracts 0–few, linear; involucre hemispheric, appearing disk-shaped when pressed; phyllaries in 2–3 ± equal series, linear; receptacle conic, chaffy, scales linear-lanceolate. **RAY FLS** sterile; corollas yellow-orange; ligules showy, oblong, entire or nearly so. **DISK FLS** many; corollas yellow-orange; style tips triangular. **FR** oblanceolate, ± 4-angled, flat, glabrous; pappus 0. 6 spp.: N.Am. (Greek: sun part, from showy heads) [Yates & Heiser 1979 Proc Indiana Acad Sci 88:364–372]

1. Ann; lvs conspicuously ciliate .... ***H. hispida***
1′ Per; lvs not ciliate .... ***H. multiflora*** var. ***nevadensis***

H. hispida (A. Gray) Cockerell Ann from slender taproot. **STS** 1.5–6 dm, erect, simple to much-branched, strigose; hairs stiff, ascending. **LVS** 2–10 cm, linear to linear-lanceolate, stiffly hairy; lower surface resin-dotted. **INFL**: peduncles 5–8 cm; involucre 12–15 mm diam; phyllaries 5.5–10 mm, lanceolate, streaked with 3 vein-like glands; hairs stiff, spreading or ascending. **RAY FLS** 10–15; ligules ± 7 mm, ovate. **DISK FLS**: corollas 2–3 mm; anthers dark brown. **FR** 1.5–2.5 mm. $2n$=16. Damp, disturbed areas; < 100 m. SW; native to UT, AZ, NM, n Mex. [*Viguiera ciliata* (Robinson & Greenman) S.F. Blake]

H. multiflora Nutt. var. ***nevadensis*** (Nelson) Yates (p. 285) Per from branched, woody rootstock. **STS** 3–9 dm, slender, erect or spreading, glabrous or finely strigose or puberulent. **LVS** 2–6 cm, linear to oblong or narrowly elliptic, ± stiffly hairy near base; tip obtuse to acute; surfaces strigose. **INFL**: peduncles 5–15 cm; involucre 7–13 mm diam; phyllaries 4–7 mm, linear, strigose. **RAY FLS** 8–15; ligules 15–20 mm, oblong to ovate. **DISK FLS**: corollas 3 mm; anthers yellow. **FR** 2 mm. $2n$=16. Dry, rocky slopes, upland valleys; 1200–2400 m. SNE, DMtns; to UT, AZ. [*Viguiera m.* (Nutt.) S.F. Blake var. *n.* (Nelson) S.F. Blake] Other vars. widespread in mtns of w N.Am. ❀TRY.

HEMIZONIA TARPLANT, TARWEED

Ann to shrub, gen glandular, aromatic. **STS** gen branched above middle or throughout. **LVS** gen cauline (some also basal and cauline, gen alternate, gen linear to (ob)lanceolate, entire to pinnately lobed, gen not spine-tipped; lower gen toothed to lobed; upper gen entire. **INFL**: heads radiate, gen 1–many in open cymes; involucre gen hemispheric; phyllaries gen linear to lanceolate, half-enclosing ray frs; chaff scales gen in 1 ring between ray and disk fls (scattered). **RAY FLS** 3– many; ligules gen 3-lobed, white to yellow. **DISK FLS** 3–many, staminate or fruiting; corollas white to yellow, becoming red; anther tips ovate; style branches long, tips bristly. **FR**: ray achenes ± 3-angled; pappus 0; disk achenes cylindric or obconic, pappus 0 or scales gen linear to lanceolate. ± 25 spp.: CA, OR, w AZ, n Baja CA. (Greek: half girdle, from sheathing phyllaries) [Tanowitz 1982 Syst Bot 7:314–339; Venkatesh 1958 Amer J Bot 45:77–84]

1. Subshrub or shrub
  2. Anthers very dark; phyllaries barely keeled; older lf axils woolly-tufted; ChI .... ***H. clementina***
  2′ Anthers yellow; phyllaries strongly keeled; older lf axils not woolly-tufted; s WTR .... ***H. minthornii***
1′ Ann or per
  3. Lvs rigidly spine-tipped .... ***H. pungens***
    4. Disk pappus 0
      5. Chaff scales obtuse to acute, not spine-tipped; SCo, PR .... ssp. ***laevis***
      5′ Chaff scales spine-tipped; widespread
        6. Lvs and bracts scabrous — SCoR, SnJV, SW .... ssp. ***pungens***
        6′ Lvs and bracts glabrous
          7. Heads 7–10 mm diam, 5–40 per dense cluster; < 50 m; s NCoR, c GV, n CW .... ssp. ***maritima***
          7′ Heads gen 4–6 mm diam, scattered; < 500 m; CaRF, ScV .... ssp. ***septentrionalis***

4′ Disk pappus present
8. Chaff scales obtuse, long-soft-hairy .... ***H. fitchii***
8′ Chaff scales obtuse to acute, ± resinous-thickened, not long-soft-hairy .... ***H. parryi***
9. Herbage glandular and long-hairy
10. Anthers very dark; involucre 4–5.5 mm; ligules ± 2 mm; SCo .... ssp. ***australis***
10′ Anthers yellow; involucre 5–10 mm; ligules 3–6 mm; s NCoR, n&c CW .... ssp. ***parryi***
9′ Herbage not glandular, variously hairy
11. Herbage not puberulent; involucre 4.5–8 mm, gen overtopped by subtending bracts; ligules 2.5–3 mm; c&s CW .... ssp. ***congdonii***
11′ Herbage rough-puberulent; involucre 3.5–5 mm, barely overtopped by subtending bracts or gen not; ligules ± 2 mm; c-w GV .... ssp. ***rudis***
3′ Lvs not spine-tipped
12. Ray achenes beakless; ligules white to yellow; chaff scales scattered throughout, ± fused at base — NCoR, ScV, CW .... ***H. congesta***
13. Corollas ± yellow .... ssp. ***congesta***
13′ Corollas white
14. Phyllary tips gen >> body
15. Heads gen solitary, overtopped by upper lvs; herbage densely glandular; NCoRI, n ScV .... ssp. ***calyculata***
15′ Heads gen ± clustered, gen little overtopped; herbage soft-hairy, barely glandular or not; s NCoR, SnFrB .... ssp. ***leucocephala***
14′ Phyllary tips gen < body
16. Heads gen ± sessile along infl branches, in clusters .... ssp. ***clevelandii***
16′ Heads gen peduncled, ± terminal on infl branches
17. Lvs green to silky-canescent or tomentose, gen ± sticky-glandular (esp upward and in autumn); fls spring to autumn; NCoRI, c ScV, CW .... ssp. ***luzulifolia***
17′ Lvs green, puberulent, barely glandular or not; fls in spring; NCoR .... ssp. ***tracyi***
12′ Ray achenes with short, off-center beak; ligules yellow; chaff scales in 1 ring between ray and disk fls, free or ± fused
18. Anthers yellow; upper lvs lanceolate; sts gen ± white
19. Disk pappus 0; disk fls 18–60
20. Sts solid; herbage ± soft-bristly; involucre glands dense, stalked; disk fls 18–25, sometimes fruiting, corolla glabrous; ray achenes ± 2.5 mm; basal lvs deeply toothed; w DMoj (e Kern Co.) .... ***H. arida***
20′ Sts hollow; herbage becoming ± glabrous; involucre glands sessile; disk fls 28–60, staminate, corolla base minutely glandular; ray achenes 3.5–4 mm; basal lvs (sub)entire; SCoRI .... ***H. halliana***
19′ Disk pappus present; disk fls 6–21
21. Ray fls 8–12; disk fls 10–21; phyllaries lanceolate .... ***H. pallida***
21′ Ray fls 5; disk fls 6; phyllaries ± linear
22. Infl open, panicle-like; pappus scales ± linear, entire or ± fringed; basal lvs clearly toothed; SnJV, s SNF, e SnFrB, SW .... ***H. kelloggii***
22′ Infl dense; pappus scales lanceolate to rectangular, deeply cut; basal lvs subentire; s SnBr .... ***H. mohavensis***
18′ Anthers very dark; upper lvs ± linear; sts reddish
23. Pappus ± 0; ray fls 18–35; disk fls > 27, staminate; coast .... ***H. corymbosa***
24. Infl open, panicle-like; sts gen branched above; ray fls 18–24(28); disk fls 28–35(40); involucre 6–9.5 mm diam; c&s NCo, n CCo, SnFrB .... ssp. ***corymbosa***
24′ Infl dense; sts often branched near base; ray fls 20–35; disk fls 40–70; involucre 9–13.5 mm diam; n&c CCo .... ssp. ***macrocephala***
23′ Pappus well developed; ray fls gen 3–18(20); disk fls 3–31, sometimes fruiting; coast or interior
25. Ray fls 3–5; phyllaries strongly keeled; lower lvs abruptly pointed
26. Herbage hairs not bulb-based; pls bright green; infl ± flat-topped to panicle-like; lower lvs gen weakly lobed; phyllaries not scabrous .... ***H. pentactis***
26′ Herbage hairs bulb-based; pls gray-green; infl open, panicle-like; lower lvs deeply 1–2-pinnately lobed; phyllaries scabrous on keel
27. Ray fls 5; disk fls 6; heads gen 2–3 in clusters, subsessile; c SCoR .... ***H. fasciculata***
27′ Ray fls 3; disk fls 3; heads gen solitary, slender-peduncled; CaRF, ne&c CW .... ***H. lobbii***
25′ Ray fls 8–20; phyllaries keeled or rounded; lower lvs acuminate
28. Disk fls gen fruiting; disk corollas hairy; pappus scales oblong or elliptic
29. Ray fls 13–20; ligules narrowly obovate; disk fls 24–31; basal lvs (sub)entire; infl ± flat-topped; peduncles 5–15 mm; herbage soft-hairy; pappus scales barely fringed, reddish brown; s PR .... ***H. floribunda***
29′ Ray fls 8; ligules widely obovate; disk fls 8–13; basal lvs deeply lobed; infl panicle-like; peduncles 1–4 cm; herbage bristly; pappus scales fringed, white; SCo, sw PR .... ***H. paniculata***
28′ Disk fls gen staminate; disk corollas (sub)glabrous; pappus scales ± linear (narrowly lanceolate)
30. Phyllaries keeled, with short-stalked and large, sessile flat glands near margin; phyllary tips acuminate; herbage soft-bristly .... ***H. conjugens***
30′ Phyllaries not keeled, densely glandular (glands small, sessile, rounded); phyllary tips abruptly pointed; herbage stiff-bristly .... ***H. increscens***
31. Pl gray-green, soft-hairy; ray fls 13; disk fls 18–31; involucre 8–10 mm diam; infl rounded or flat-topped .... ssp. ***villosa***

31′ Pl deep green, stiff-bristly; ray fls 8–13; disk fls 14–30; involucre 5–9 mm diam; infl panicle-like
32. Ray fls gen 8(10); disk fls 18–30; involucre 5–7.5 mm diam .... ssp. ***foliosa***
32′ Ray fls gen (8)13; disk fls 14–20; involucre 6–9 mm diam .... ssp. ***increscens***

H. arida Keck (p. 285) RED ROCK TARPLANT Ann 2–10 dm. **STS** erect; branches ascending, ± bristly, ± sparsely glandular. **LVS**: lower 4–8 cm, linear-oblanceolate, dentate, glabrous; upper widely lanceolate, entire, ± bristly. **INFL** ± flat-topped; heads short-peduncled; involucre 4.5–6.5 mm; phyllaries soft-bristly, densely glandular. **RAY FLS** 5–10; ligules 5–9 mm, pale yellow. **DISK FLS** 18–25, gen staminate; corollas and anthers yellow. **FR** ± 2.5 mm; beaked; disk pappus ± 0. 2*n*=24. **RARE** CA. Clay soil of washes; 300–950 m. w DMoj (Red Rock Canyon, e Kern Co.). Threatened by vehicles.

H. clementina Brandegee (p. 285) ISLAND TARPLANT Subshrub or shrub 3–8 dm. **STS** many from base, decumbent to erect, densely leafy, soft-hairy or puberulent, minutely glandular; lower axils often woolly-tufted. **LVS**: lower opposite, 3–8 cm, sparsely dentate, gen deciduous; upper alternate, gen with lf-clusters in axils, entire. **INFL** ± dense; peduncles short; involucre 5–8 mm, sometimes bell-shaped; phyllaries obtuse; chaff scales in 2 series (outer ± fused, inner free), acute or obtuse, hairy. **RAY FLS** ± 13; ligules 4.5–6.5 mm, yellow. **DISK FLS** 18–30, staminate; corollas yellow; anthers brown or black. **FR** 2–3 mm; beak curved; back rough; disk pappus scales 7–15, 1–3 mm, linear. 2*n*=24. UNCOMMON. Dry slopes, bluffs, rocky places; < 200 m. n ChI (Anacapa Island), s ChI.

H. congesta DC. HAYFIELD TARWEED Ann 1–8 dm. **ST** simple below, branched above. **LVS** gen linear to narrowly elliptic, entire to sparsely dentate; lower 5–18 cm; upper entire, gen densely glandular; lf clusters in upper axils gen 0. **INFL**: heads solitary or in small groups, sessile to long-peduncled; involucre 3.5–12 mm; phyllaries long-soft-hairy and densely stalked-glandular; chaff scales scattered, fused at base, falling with disk fls, tips blunt, ± hairy. **RAY FLS** 5–13; ligules 5–11 mm, gen white, gen red- or purple-striped below, central lobe narrow. **DISK FLS** 5–60, staminate; corollas white to yellow; anthers appearing black. **FR** 2–3 mm; beak 0; disk pappus 0. Common. Grassland, fallow fields; < 1400 m. NCoR, w&c GV, CW; sw OR. ± interfertile races; intergradation compounded by seasonal changes in form.

ssp. ***calyculata*** Babc. & H.M. Hall (p. 285) MENDOCINO TARPLANT **ST** ± short-soft-hairy and glandular. **LVS**: lower rough-puberulent, glandular. **INFL**: heads in open cymes, short-peduncled; involucre 6–12 mm; phyllary tips gen > body. 2*n*=28. UNCOMMON. Grassland, open woods and forests; < 1400 m. NCoRI, n ScV. [*H. ca.* (Babc. & H.M. Hall) Keck]

ssp. ***clevelandii*** (E. Greene) Babc. & H.M. Hall (p. 285) **ST** soft-hairy (esp below) and glandular (esp above). **LVS**: lower densely soft-hairy, glandular. **INFL**: heads (exc young pls) sessile or short-peduncled along infl branches, often closely subtended by bracts; involucre 4–7 mm; phyllary tips gen < body. 2*n*=28. Grassland, open woods and forests; < 1400 m. NCoR; sw OR. [*H. cl.* E. Greene]

ssp. ***congesta*** **ST** long-soft-hairy and glandular. **LVS**: lower dark green to silvery-canescent. **INFL** open; heads peduncled; involucre 5–6.5 mm; phyllary tips < or = body. **RAY FLS**: ligules yellow (rarely white-tipped). 2*n*=28. Coastal grassland, sometimes serpentine; gen < 300 m. s NCoR, n CW. [*H. lutescens* (E. Greene) Keck, *H. multicaulis* Hook. & Arn. incl ssp. *vernalis* Keck] Intergrades with sspp. *leucocephala*, *luzulifolia*; dried pls may not be separable. ❀SUN,IRR:**15–17**,24.

ssp. ***leucocephala*** (B.D. Tanowitz) Keil (p. 285) HAYFIELD TARPLANT **ST** ± soft-hairy, obscurely glandular above. **LVS**: lower densely soft–hairy; upper gen glandless. **INFL**: heads solitary or in small groups, short- to long-peduncled; involucre 6–9 mm; phyllaries elliptic with inconspicuous stalked-glands, tips = to >> body. 2*n*=28. UNCOMMON. Grassy valleys and hills, often in fallow fields; gen < 200 m. NCoR, SnFrB. [*H. l.* Tanowitz; *H. co.* ssp. *co.* misapplied] Intergrades with ssp. *co.*; dried pls may not be separable.

ssp. ***luzulifolia*** (DC.) Babc. & H.M. Hall (p. 285) **ST** tomentose or ± silky below, glandular above or throughout. **LVS**: lower silky-canescent or puberulent to tomentose. **INFL** open; heads peduncled; involucre 3.5–6.5 mm; phyllary tips gen << body. 2*n*=28. Common. Grassland, fallow fields; < 1000 m. NCoRI, c GV, CW. [*H. l.* DC. incl ssp. *rudis* (Benth.) Keck] Intergrades with ssp. *co.*; dried pls may not be separable.

ssp. ***tracyi*** Babc. & H.M. Hall TRACY'S TARPLANT **ST** ± soft-hairy and puberulent. **LVS**: lower green, finely strigose to puberulent; upper gen glandless. **INFL** open; heads peduncled; involucre 5–8 mm, phyllary tips gen < body. 2*n*=28. UNCOMMON. Grassland, fallow fields; < 1200 m. n&c NCoR. [*H. t.* (Babc. & H.M. Hall) Keck]

H. conjugens Keck (p. 285) OTAY TARPLANT Ann 1–4 dm, short-bristly below, glandular above. **STS** branched ± above middle. **LVS**: lower 2–4.5 cm, oblanceolate, lobed; middle entire to toothed; upper appressed to pedicels. **INFL**: heads 1–many in dense clusters, subsessile to short-peduncled; involucre 5–7.5 mm; phyllaries with flat, sessile glands. **RAY FLS** 8–10; ligules 3–6 mm, yellow. **DISK FLS** 13–21, fertile or staminate; corollas yellow; anthers black. **FR** 2.5–3 mm, beaked; disk pappus scales 6–9, white. 2*n*=24. **ENDANGERED** CA. Clay soil, coastal scrub; < 300 m. s PR (s San Diego Co.); n Baja CA. Threatened by development.

H. corymbosa (DC.) Torrey & A. Gray Ann 2–10 dm. **STS** bristly below, bristly or puberulent above. **LVS**: lower lanceolate to widely oblanceolate, lobed; upper crowded below infl, entire, soft–hairy, glandular. **INFL** open or dense; involucre spheric, 5–9.5 mm; phyllaries densely stalked-glandular, tips gen = to >> body. **RAY FLS** 18–35; ligule 4–7 mm, bright yellow. **DISK FLS** 24–70, staminate; corollas yellow; anthers black. **FR** ± 2.5 mm, beaked; disk pappus ± 0. Coastal grassland; < 300 m. s NCo, n CW.

ssp. ***corymbosa*** **STS** gen erect, gen branched above. **LVS**: lower 3–9 cm; uppermost 3–8 mm. **INFL** open; heads gen peduncled; involucre 6–9.5 mm diam. **RAY FLS** 18–24. **DISK FLS** 24–40(48). 2*n*=20. Habitats of sp. c NCo, SnFrB, n&c CCo.

ssp. ***macrocephala*** (Nutt.) Keck (p. 285) **STS** gen decumbent, gen branched near base. **LVS**: lower 3.5–12 cm; uppermost 6–15 mm. **INFL** dense; heads subsessile or short-peduncled; involucre 9–13.5 mm diam. **RAY FLS** 21–32. **DISK FLS** 40–70. 2*n*=20. Habitats of sp. c CCo.

H. fasciculata (DC.) Torrey & A. Gray (p. 285) Ann 0.5–10 dm. **STS** ± glabrous to sparsely short-bristly. **LVS**: lower 3.5–15 cm, dentate to deeply lobed, bristly; upper linear, entire, appressed to st. **INFL** open to dense; heads peduncled; involucre 4.5–5.5 mm, narrowly bell-shaped; phyllaries short-bristly, with sessile glands. **RAY FLS** 5; ligule 6–14 mm, deep yellow. **DISK FLS** 6, gen staminate; corollas yellow; anthers black. **FR** 2.5–3 mm, beaked; disk pappus scales 5–12, white. 2*n*=24. Common. Coastal grassland, woodland; < 900 m. s CCo, s SCoRO, SW; to c Baja CA. [*H. ramosissima* Benth.] ❀SUN:**15–17,22–24**.

H. fitchii A. Gray (p. 285) Ann, 1–5 dm, ill-scented. **STS** densely long-straight-hairy, glandular-puberulent. **LVS**: lower gen present at fl, often dense, 5–15 cm, entire or deeply divided with linear spine-tipped segments, gen densely long-hairy; upper gen with axillary lf clusters, linear, soft-hairy, glands black. **INFL** open or dense; heads subsessile or short-peduncled; involucre 5–10 mm, gen overtopped by bracts; phyllaries linear, keeled, densely soft-hairy, spine-tipped; chaff scales scattered, tips blunt, soft-hairy. **RAY FLS** many; ligules 3–4 mm, 2-lobed, yellow. **DISK FLS** many, gen staminate; corollas yellow; anthers yellow to black. **FR** 2–3 mm, beaked; disk pappus scales 8–12. 2*n*=26. Fields, open woodland; 30–1000 m. NW, CaRF, n&c SNF, n&c GV, SCoR, n ChI (Santa Cruz Island), n SnBr; sw OR.

H. floribunda A. Gray (p. 285) TECATE TARPLANT Ann 3–10 dm, soft-hairy below, glandular above. **LVS**: lower 3–5 cm, oblanceolate, toothed; upper linear-lanceolate, entire. **INFL**: heads short-peduncled; involucre 4.5–6 mm; phyllaries lanceolate. **RAY FLS** 13–20; ligule 4–7 mm, deep yellow. **DISK FLS** 24–31, gen fertile;

corollas yellow; anthers black. **FR** 2–2.5 mm, beaked; disk pappus scales 6–10, red-brown-flecked. 2*n*=26. RARE. Dry valleys, foothills; 300–700 m. s PR (s San Diego Co.); n Baja CA.

H. halliana Keck (p. 285) HALL'S TARPLANT Ann 2–12 dm. **STS** hollow, erect; branches ascending, ± glabrous below, soft-hairy and sticky-glandular above. **LVS** linear–lanceolate, glabrous; lower 4–8 cm, entire to lobed, margin scabrous; upper entire. **INFL**: heads in ± flat-topped cymes, peduncled; involucre 5–7 mm; phyllaries soft-hairy, sticky glandular. **RAY FLS** 10–14; ligule 6–10 mm, pale yellow. **DISK FLS** 28–60, staminate; corollas and anthers yellow. **FR** 3.5–4 mm, beaked; disk pappus ± 0. 2*n*=20. UNCOMMON. Clay soil of flood plains; 300–950 m. c SCoRI.

H. increscens (Keck) B.D. Tanowitz Ann 3–9 dm. **STS** erect, bristly. **LVS** gen deep green, gen bristly; lower 1–3 cm wide; upper gen glandular-puberulent. **INFL**: heads subsessile to short-peduncled; involucre 5–7.5 mm. **RAY FLS** 8–13; ligule 5–8 mm, deep yellow. **DISK FLS** 14–30, gen staminate; corollas yellow; anthers black. **FR** 2.5–3 mm, beaked; disk pappus scales 5–14, white. 2*n*=24. Common. Coastal foothills, valleys; < 500 m. CW, n SCo, n ChI.

ssp. ***foliosa*** (Hoover) B.D. Tanowitz **STS** gen branched below middle. **LVS**: lower 5–9 cm. **INFL**: involucre 5–7.5 mm diam, glandular-puberulent. **RAY FLS** gen 8(10). **DISK FLS** 14–20. **FR**: disk ovaries sometimes fertile. 2*n*=24. Dry foothills, interior valleys; 300–500 m. s SCoR.

ssp. ***increscens*** (p. 289) **STS** gen branched above middle. **LVS**: lower 8–13 cm. **INFL**: involucre 6–9 mm diam, glandular-puberulent. **RAY FLS** 8(13). **DISK FLS** 14–20. **FR**: disk ovaries sometimes fertile. 2*n*=24. Coastal grassland, foothills; 0–300 m. CCo, SnFrB (uncommon), SCoR, n SCo, n ChI. [*H. paniculata* A. Gray ssp. *i.* Keck]

ssp. ***villosa*** B.D. Tanowitz GAVIOTA TARPLANT **STS** gen branched near base. **LVS** gray-green, soft-hairy; lower 5–8.5 cm; upper glandular. **INFL**: involucre 8–10 mm diam, densely soft-hairy, glandular. **RAY FLS** 13. **DISK FLS** 18–31. **FR**: disk ovaries gen sterile. 2*n*=24. **ENDANGERED** CA. Coastal fields, bluffs; < 50 m. n SCo (w Santa Barbara Co.). Threatened by development.

H. kelloggii E. Greene (p. 289) Ann 1–10 dm. **STS** soft-hairy to bristly below, bristly and glandular-puberulent above. **LVS**: bristly; lower 3–9 cm, oblong-oblanceolate, dentate to lobed; upper linear. **INFL** open; heads long-peduncled; involucre 4.5–5 mm; phyllaries soft-hairy to densely glandular. **RAY FLS** 5; ligule 4–8 mm, pale yellow. **DISK FLS** 6, gen staminate; corollas and anthers yellow. **FR** 2.5–3 mm, beaked; disk pappus scales 6–12. 2*n*=18. Common. Open areas; < 700 m. s SNF, SnJV, e SnFrB, SW, alien in s NCoRI; n Baja CA. Like *H. pentactis*.

H. lobbii E. Greene (p. 289) Ann 0.5–6 dm. **STS** bristly. **LVS**: lower 3–7 cm, lanceolate to widely oblanceolate, deeply 1–2-pinnately lobed, bristly; upper appressed to st, glandular and with hairs bulb-based. **INFL**: heads peduncled; involucre 3.5–5 mm, narrowly cylindric; phyllaries densely hairy, hairs glandular or bulb-based. **RAY FLS** 3; ligule 3.5–5 mm, lemon yellow. **DISK FLS** 3, gen staminate; corollas yellow; anthers black. **FR** 1.5–2 mm, beaked; disk pappus scales 4–12, white. 2*n*=22. Interior valleys, foothills; < 700 m. CaRF, e SnFrB, SCoRI.

H. minthornii Jepson (p. 289) SANTA SUSANA TARPLANT Shrub, 6–10 dm, < 3 dm diam, rounded, sticky, fragrant. **STS** many from base, spreading to erect, densely leafy, puberulent to densely short-bristly, minutely glandular. **LVS** linear, entire; lower 2–3 cm, often deciduous; upper many, gen with axillary clusters, thick, glandular, short-bristly. **INFL**: heads 1 or few per cluster; peduncles densely bracted; involucre 5.5–7.5 mm, ± bell-shaped; phyllary tips acute, short-bristly, glandular; chaff scales scattered, free. **RAY FLS** gen 8; ligules 5.5–6.5 mm, yellow. **DISK FLS** 18–23, staminate; corolla and anthers yellow. **FR** 2.5–3 mm, short-beaked; disk pappus scales 8–12, 1.2–2.6 mm, linear. 2*n*=24. **RARE** CA. Chaparral; 300–500 m. s WTR (Santa Susana, Santa Monica mtns). Threatened by development.

H. mohavensis Keck MOJAVE TARPLANT Ann 1.5–3 dm. **STS** erect, soft-hairy and sticky-glandular. **LVS**: lower 3–5 cm, oblanceolate, (sub)entire, bristly; upper oblong-oblanceolate, glandular-puberulent. **INFL**: heads sessile, clustered; involucre 4.5–6 mm; phyllaries soft-hairy to densely glandular. **RAY FLS** 5; ligule ± 5 mm, yellow. **DISK FLS** 6, staminate; corollas and anthers yellow. **FR** 2.5–3 mm, beaked; disk pappus scales 5–9, ± fused. 2*n*=22. **PRESUMED EXTINCT**; **ENDANGERED** CA; 900–1300 m. s SnBr (2 sites; last seen in 1933).

H. pallida Keck (p. 289) KERN TARPLANT Ann 1–6 dm. **STS** erect, bristly below, short-bristly above. **LVS**: soft-hairy or bristly; lower 5–10 cm, dentate to lobed; upper linear. **INFL**: heads peduncled; involucre 4.5–6.5 mm; phyllaries bristly, densely stalked-glandular. **RAY FLS** 8–12; ligule 8–10 mm, pale yellow. **DISK FLS** 10–21, gen staminate; corollas and anthers yellow. **FR** ± 2.5 mm, beaked; disk pappus scales 4–9, white. 2*n*=18. UNCOMMON. Interior valleys, foothills; < 700 m. s SNF, s SnJV, s SCoRI, w DMoj (Kern Co.).

H. paniculata A. Gray (p. 289) Ann 1–8 dm, bristly below, glandular above. **LVS**: basal 1–10 cm, linear to oblanceolate, deeply toothed or lobed; upper linear, entire. **INFL**: heads solitary, long-peduncled; involucre 5–7 mm; phyllaries densely glandular. **RAY FLS** 8–13; ligule 3.5–6 mm, deep yellow. **DISK FLS** 8–13, fertile or staminate; corollas yellow; anthers black. **FR** 2.5–3 mm, beaked; disk pappus scales (4)6–12, white. 2*n*=24. Dry foothills, mesas; gen < 300 m. SCo, sw PR; n Baja CA.

H. parryi E. Greene Ann, 1–7 dm, mildly scented. **STS** prostrate to spreading, or erect, long-straight-hairy, often puberulent with pale yellow glands. **LVS** gen puberulent, glandular or not, soft-hairy or bristly; lower gen 0 at fl, 5–20 cm, deeply 1–2-divided; upper linear, entire or few-toothed, spine-tipped, gen with axillary lf clusters. **INFL** open or dense; heads (sub)sessile; involucre 3.5–10 mm, often overtopped by subtending bracts; phyllaries keeled, spine-tipped; chaff scales scattered, tips obtuse to acute, ± resinous-thickened. **RAY FLS** 9–30+; ligule 2–6 mm, 2-lobed, yellow or becoming red. **DISK FLS** many, gen staminate; corollas yellow; anthers yellow to black. **FR** ± 2 mm, beaked; disk pappus scales 3–5. Seasonally wet (often saline or alkaline) grassland near coast; < 200 m. sw NCoR, w CW, w SW; nw Baja CA.

ssp. ***australis*** Keck (p. 289) SOUTHERN TARPLANT Pls gen erect, densely glandular, esp above. **INFL**: involucre 4–5.5 mm, barely overtopped or not. **RAY FLS**: ligules ± 2 mm, often aging yellow-orange. **DISK FLS**: anthers brown or black. 2*n*=22. RARE. Seasonally moist (saline) grassland; < 200 m. SCo; n Baja CA. [*H. a.* (Keck) Keck]

ssp. ***congdonii*** (Robinson & Greenman) Keck (p. 289) CONGDON'S TARPLANT Pls prostrate to erect, not puberulent or glandular. **INFL**: involucre 4.5–8 mm, gen overtopped. **RAY FLS**: ligules 2.5–3 mm, remaining yellow. **DISK FLS**: anthers yellow. 2*n*=24. RARE. Grassland; < 100 m. c&s CW.

ssp. ***parryi*** Pls prostrate to erect, ± densely glandular-puberulent, esp above; glands subsessile. **INFL**: involucre 5–10 mm, gen overtopped. **RAY FLS**: ligules 3–6 mm, remaining yellow. **DISK FLS**: anthers yellow. 2*n*=24. Coastal grassland, salt (or alkaline) marshes; gen < 100 m. s NCoR, s ScV, n&c CW.

ssp. ***rudis*** (E. Greene) Keck (p. 289) Pls gen erect, rough-puberulent, gen nonglandular. **INFL**: involucre 3.5–5 mm, barely overtopped or not. **RAY FLS**: ligules ± 2 mm, often aging yellow-orange. **DISK FLS**: anthers yellow. 2*n*=22. Grassy areas; gen < 100 m. c-w GV.

H. pentactis (Keck) Keck SALINAS RIVER TARPLANT Ann 0.5–5 dm. **STS** ± bristly. **LVS**: lower 3–10 cm, 1–2 × lobed, bristly; upper linear, appressed to sts, glandular with few bulb-based hairs. **INFL** open, sometimes dense; heads peduncled; involucre 4.5–5.5 mm, cylindric; phyllaries densely glandular, some hairs bulb-based. **RAY FLS** 5; ligule ± 4.5 mm, lemon yellow. **DISK FLS** 6, gen staminate; corollas yellow; anthers black. **FR** 1.5–2 mm, beaked; disk pappus scales 5–11, white. 2*n*=22. UNCOMMON. Interior foothills, valleys; < 800 m. CW. Like *H. kelloggii*.

H. pungens (Hook. & Arn.) Torrey & A. Gray COMMON SPIKEWEED Ann 1–12 dm. **STS**: branches gen stiff, bristly. **LVS**: lower 5–15 cm, linear-lanceolate, deeply 2 × divided; upper linear, spine-tipped, glabrous to scabrous, margin stiff-ciliate, gen with axillary lf clusters. **INFL** sometimes dense; heads ± clustered; involucre 3–6 mm, gen overtopped by upper lvs; phyllaries keeled, spine-tipped, scabrous; chaff scales scattered, gen spine-tipped. **RAY FLS** many; ligule 3–5 mm, 2-lobed, yellow. **DISK FLS** many, gen staminate; corollas and anthers yellow. **FR** ± 2 mm, beaked; disk pappus 0. $2n$=18. Grassland, depressions, marshes; < 500 m. CaRF, GV, SCoR, alien in SW; to WA.

ssp. ***laevis*** Keck (p. 289) SMOOTH TARPLANT **LVS** and bracts not scabrous. **INFL**: heads gen 4–6 mm diam, solitary or few in loose clusters along infl branches; chaff scales not spine-tipped. $2n$=18. RARE. Grassland; < 400 m. SCo, PR. [*H. l.* (Keck) Keck] Threatened by flood-control, development.

ssp. ***maritima*** (E. Greene) Keck **LVS** and bracts not scabrous. **INFL**: heads 7–10 mm diam, 5–40 per dense cluster. Salt marshes, low sites; < 50 m. s NCoR, GV, n CW.

ssp. ***pungens*** (p. 289) **LVS** and bracts scabrous. **INFL**: heads 4–10 mm diam, 1–few in clusters along infl branches. Low (saline or alkaline) grassland; < 500 m. SnJV, SCoR, alien in SW; OR.

ssp. ***septentrionalis*** Keck Pl gen erect, 5–10 dm. **STS** loosely branched. **LVS** and bracts not scabrous. **INFL**: heads gen 4–6 mm diam, scattered. Grassland; < 500 m. CaRF, ScV; to WA.

HESPEREVAX

James D. Morefield

Ann < 18 cm, ± green to gray-woolly. **LVS** alternate or seeming whorled; base hard, yellowish; blade oblanceolate to ± round, entire. **INFL**: heads disciform, sessile, 1–many above (sometimes scattered below), leafy-bracted, < 5 mm; phyllaries 0; receptacle conic, ± acuminate, short-shiny-bristly; chaff scales phyllary-like, subtending fls, flat to concave, persistent, rigid, outer acute, ± glabrous, inner obtuse, finely woolly inside. **PISTILLATE FLS** in 1–5 series, cylindric. **DISK FLS** staminate, 2–6(12), ± bilateral, 3–5-lobed. **FR** obovoid, compressed front-to-back, ± angled, smooth, dull, ± black-banded near base; pappus 0. 3 spp.: CA, s OR. (Greek: western *Evax*) [Morefield 1992 Syst Bot 17:293–310] Segregated from *Evax*; perhaps closest to *Ancistrocarphus*. *Evax multicaulis* DC. and *E. prolifera* of AZ & NM not adequately documented from CA.

1. Upper heads in loose groups of 3–5 subtended by 1–4 lvs, length 2+ × width; inner chaff scales 0.3–0.4 × head, tip incl, ± erect; lower heads ± = upper ..... ***H. sparsiflora***
2. Longest lvs 6–12(14) mm, ± round, densely woolly ..... var. ***brevifolia***
2′ Longest lvs (10)13–30 mm, ± obovate, cobwebby ..... var. ***sparsiflora***

1′ Upper heads in dense groups of 1–40 subtended by 6–20 lvs, length < 1.5 × width; inner chaff scales 0.5–1 × head, tip exserted, spreading; lower heads 0 or < upper
3. Longest lvs gen 33–90 mm; petiole > 2 × blade; heads in groups of 10–40, largest group > 9 mm wide; chaff scales in 5 vertical ranks ..... ***H. caulescens***
3′ Longest lvs < 32 mm; petiole < 1.5 × blade; head gen 1, sometimes 2–40 in group < 8 mm wide; chaff scales in spiral ranks ..... ***H. acaulis***
4. St ± erect; longest lvs 9–32 mm, 2–5 mm wide; inner chaff scales 2.5–3.2 mm ..... var. ***robustior***
4′ St ± prostrate or 0; longest lvs gen 4–12 mm, < 2 mm wide; inner chaff scales 1.6–2.4 mm
5. Lvs ± sessile, oblanceolate, acute, longest gen 7–12 mm ..... var. ***acaulis***
5′ Lvs petioled, ± round, obtuse, longest gen 4–7 mm ..... var. ***ambusticola***

H. acaulis (Kellogg) E. Greene **ST** 0–8 cm. **LVS** ± crowded near st tips, longest 4–32 mm, < 5 mm wide; petiole < 1.5 × blade. **INFL**: head gen 1, 2–4 mm, 1.5–3.5 mm wide, sometimes in dense groups of 2–8, groups < 8 mm wide, subtended by gen 6–12 lvs; lower heads 0 or < upper; receptacle < 2 × longer than wide; chaff in spiral ranks, inner scales 1.6–3.2 mm, tip exserted, spreading over outer scales. **PISTILLATE FLS** in 2–5 series. **DISK FLS** 0.6–1 mm. **FR** 0.6–1.6 mm. Sandy slopes, flats, sometimes burns; 100–1300 m. CA-FP. [*Evax a.* (Kellogg) E. Greene (1883); *E. a.* E. Greene (1894) misapplied] Vars. intergrade.

var. ***acaulis*** (p. 289) DWARF EVAX **STS** 0 or ± prostrate, < 5 cm. **LVS** ± sessile, longest gen 7–12 mm, < 2 mm wide, oblanceolate, acute. **INFL**: upper heads 2.5–3 mm, 2–2.5 mm wide; inner chaff scales 1.9–2.4 mm. **PISTILLATE FLS** in gen 3–4 series. **FR** gen 0.6–0.8 mm. UNCOMMON. Dry or vernally moist places; 100–1000 m. SNF, e GV.

var. ***ambusticola*** J. Morefield (p. 289) FIRE EVAX **STS** ± prostrate, < 5 cm. **LVS** petioled; longest gen 4–7 mm, 1–2 mm wide, obtuse. **INFL**: upper heads ± 2–2.5 mm, 1.5–2 mm wide; inner chaff scales 1.6–2 mm. **PISTILLATE FLS** in gen 2–3 series. **FR** gen 0.8–1 mm. UNCOMMON. Bare, gen burned slopes; 500–1300 m. s NCoRI, c-s SNF, e SnFrB, SCoRO, PR.

var. ***robustior*** J. Morefield (p. 289) **STS** ± erect, < 8 cm. **LVS** gen ± sessile; longest 9–32 mm, 2–5 mm wide, oblanceolate to obovate, obtuse to acute. **INFL**: upper heads gen 3–4 mm, 2.5–3.5 mm wide; inner chaff scales 2.5–3.2 mm. **PISTILLATE FLS** in gen 3–5 series. **FR** gen 1–1.6 mm. Dry or vernally moist places; 100–1100 m. NW, c-n SNF, ScV margin, CW; sw OR. [*Evax sparsiflora* var. *brevifolia* (A. Gray) Jepson misapplied]

H. caulescens (Benth.) A. Gray (p. 289) HOGWALLOW STARFISH **STS** 0–few, erect, < 18 cm. **LVS** ± crowded near st tips, long-petioled, longest gen 33–90 mm, 7–20 mm wide. **INFL**: upper heads in dense groups of 10–40 subtended by gen 10–20 lvs, 3–5 mm, 2.5–4 mm wide, largest group > 9 mm wide; lower heads 0; receptacle < 2 × longer than wide; chaff in 5 vertical ranks, inner scales 3–4 mm, tip exserted, spreading over outer scales. **PISTILLATE FLS** in 2–4 series. **DISK FLS** 1.1–1.6 mm. **FR** gen 1.5–2 mm. Dry mud of vernal pools and flats; 0–500 m. s SNF, n&c GV, sw PR. [*Evax c.* (Benth.) A. Gray; *E. acaulis* E. Greene (1894); *E. a.* (Kellogg) E. Greene (1883) misapplied] Tall, narrow-lvd pls have been called *E. involucrata* E. Greene. Becoming uncommon.

H. sparsiflora (A. Gray) E. Greene **STS** gen several, ascending to erect, < 18 cm. **LVS** ± equal, ± evenly spaced, petioled; longest 6–30 mm, 2–10 mm wide. **INFL**: upper heads in loose groups of 3–5 subtended by gen 1–4 lvs and mixed with lvs, 3–4.5 mm, 1.5–2 mm wide, groups < 5 mm wide; lower heads ± = upper; receptacle 2+ × longer than wide; chaff in spiral ranks, inner scales 1.1–1.8 mm, tip incl, ± erect. **PISTILLATE FLS** in 1–3 series. **DISK FLS** 0.8–1.1 mm. **FR** gen 1–1.7 mm. Sandy soils or serpentine; < 900 m. NCo, s NCoR, deltaic GV, CW, ChI, sw PR; sw OR. Vars. distinctive; some intermediates may occur in SnFrB.

var. ***brevifolia*** (A. Gray) J. Morefield (p. 289) SHORT-LEAVED EVAX **STS** < 10 cm. **LVS**: longest gen 6–12 mm; blade ± round, obtuse, greenish, densely woolly. **INFL**: longest heads 3–3.7 mm. UNCOMMON. Sandy bluffs and flats; 0–200 m. NCo, n-most CCo; sw OR. [*Evax s.* var. *b.* (A. Gray) Jepson, in part]

var. ***sparsiflora*** (p. 289) **STS** < 18 cm. **LVS**: longest gen 13–30 mm; blade ± obovate, obtuse to acute, grayish to greenish, cobwebby. **INFL**: longest heads 3.6–4.5 mm. Common. Open, gen serpentine soil; 10–900 m. s NCoR, deltaic GV, CW, ChI, sw PR. [*Evax s.* (A. Gray) Jepson] Santa Rosa Island pls are uniformly stless.

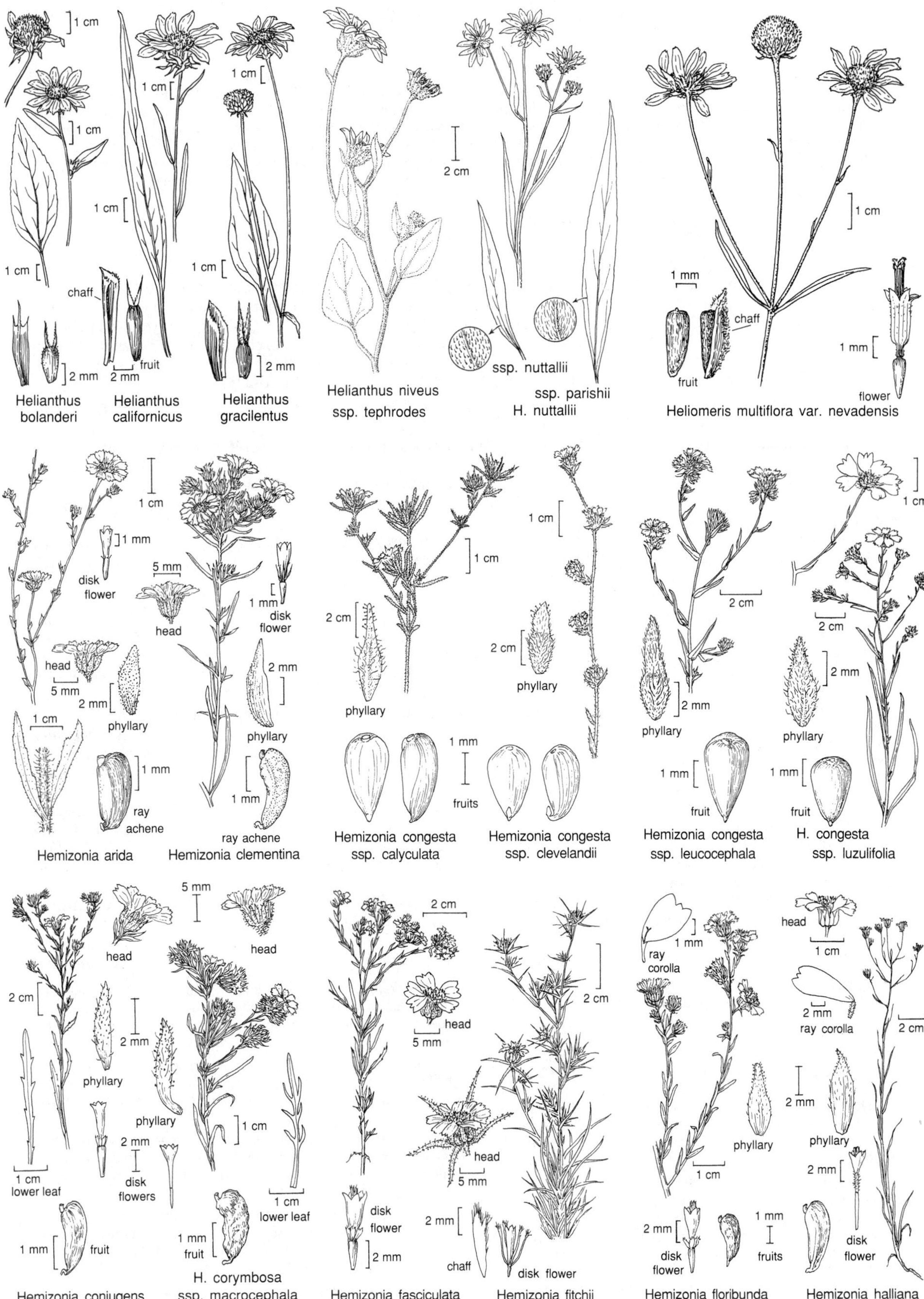

1 cm
2 mm
chaff
fruit
2 mm
Helianthus bolanderi
Helianthus californicus
Helianthus gracilentus
2 cm
Helianthus niveus ssp. tephrodes
ssp. nuttallii
ssp. parishii
H. nuttallii
1 mm
chaff
fruit
flower
Heliomeris multiflora var. nevadensis
disk flower
head
5 mm
phyllary
ray achene
Hemizonia arida
Hemizonia clementina
Hemizonia congesta ssp. calyculata
Hemizonia congesta ssp. clevelandii
fruits
Hemizonia congesta ssp. leucocephala
H. congesta ssp. luzulifolia
fruit
lower leaf
disk flowers
Hemizonia conjugens
H. corymbosa ssp. macrocephala
disk flower
chaff
Hemizonia fasciculata
Hemizonia fitchii
ray corolla
Hemizonia floribunda
Hemizonia halliana

HETERANTHEMIS

1 sp. (Greek: other *Anthemis*, from similarity to that genus) [Talavera 1987 Fl Vasc Andalucía Occ 3:65–66]

H. viscidehirta Schott Ann 20–80 cm. **STS** erect, few-branched above, sticky-glandular. **LVS** alternate, finely glandular-puberulent; lower 4–6 cm, oblanceolate, coarsely pinnately lobed and toothed; upper shorter, oblong, toothed or shallowly lobed, clasping. **INFL**: heads radiate, peduncled, solitary or in few-headed cymes; involucre widely hemispheric; phyllaries in 2–3 series, outer 2–6 mm, ± triangular, inner 6–10 mm, oblong to widely obovate, margins and tip widely scarious; receptacle conic, naked. **RAY FLS** 13–21; ligule 2–2.5 cm, yellow. **DISK FLS** many; corolla yellow. **FR** glabrous; pappus 0; ray achenes ± 4 mm, triangular, expanded to thick disk at tip, 3-winged, wings tipped by pappus-like spines; disk achenes ± 4 mm, compressed, 1–2-winged. Uncommon. Sandy soil near coast; < 50 m. CCo; native to sw Eur, n Afr. Originally reported as *Chrysanthemum segetum* L.

HETEROTHECA GOLDENASTER, TELEGRAPH WEED

John C. Semple

Ann to per, taprooted, branched above, strigose-bristly; hairs minutely knobby. **LVS** gen ± cauline, alternate; lower oblanceolate to ovate, base or petiole ± spreading-hairy; upper reduced, glandular. **INFL** ± flat-topped; heads gen radiate; involucre ± bell-shaped; phyllaries in 3–5 graded series; receptacle naked, pitted. **RAY FLS** (0)10–30(40); ligules yellow. **DISK FLS** many; corollas yellow; style branches finely papillate, appendage narrowly triangular. **FR** obconic; ray fr ± 3-angled, pappus 0 or of bristles; disk fr compressed, outer pappus of narrow scales < 1 mm, inner of 30–45 bristles 3–7 mm. ± 30 spp.: esp w N.Am. (Greek: different cases, from ray and disk frs) [Semple 1990 Brittonia 42:221–228]

1. Ray fls 0; outer pappus bristles very few — gen dry streambeds (sect. *Ammodia*) .... ***H. oregona***
 2. Lvs on branches ± (sub)glabrous, ± densely glandular
  3. Lvs on branches gen > 13 mm, > 3 mm wide; outer phyllaries lanceolate-deltate; NCoR, CaR, n&c SCoR .... var. ***oregona***
  3′ Lvs on branches gen < 13 mm, < 3 mm wide, ± scabrous; outer phyllaries ovate-deltate; s SnFrB, SCoR .... var. ***scaberrima***
 2′ Lvs on branches ± densely hairy, sparsely glandular
  4. Sts gen branched below middle; lvs long-soft-hairy, appearing gray-green .... var. ***compacta***
  4′ Sts gen loosely branched above middle; lvs short-bristly, appearing green .... var. ***rudis***
1′ Ray fls present; outer pappus bristles gen many
 5. Ann to short-lived per; ray frs ± glabrous; ray pappus 0 (sect. *Heterotheca*)
  6. Upper lvs not clasping; ± CA-FP, common, weedy .... ***H. grandiflora***
  6′ Upper lvs clasping; SCo, DSon, uncommon .... ***H. psammophila***
 5′ Per; ray frs hairy; ray pappus present (sect. *Phyllotheca*)
  7. Corolla lobes ± sparsely hairy; lf margins often wavy, hairs stiff or soft; Teh, CW, SW .... ***H. sessiliflora***
   8. Upper lf margins flat to weakly wavy, hairs ± 2 mm
    9. Lower st sparsely woolly; upper lvs oblanceolate, little reduced, flat, strigose to long-woolly; NCo, n CCo .... ssp. ***bolanderi***
    9′ Lower st bristly; upper lvs elliptic to lanceolate, gen much reduced, sometimes wavy-margined, strigose and bristly; inland c&s CA .... ssp. ***echioides***
   8′ Upper lf margins distinctly wavy, hairs ± 1 mm
    10. Heads not subtended by lf-like bracts; upper lvs whitish, stiff; TR and PR > 200 m .... ssp. ***fastigiata***
    10′ Heads subtended by large lf-like bracts; upper lvs green, not stiff; SCo < 10 m .... ssp. ***sessiliflora***
  7′ Corolla lobes (sub)glabrous; lf margins flat, hairs stiff; CaR, SN, MP, sw DMtns .... ***H. villosa***
   11. Herbage ± densely strigose, gen glandless (exc infl) — KR, CaR .... var. ***villosa***
   11′ Herbage ± sparsely hairy, ± glandular
    12. Mid-st lvs (ob)lanceolate, ± sparsely strigose and glandular — SN, MP .... var. ***hispida***
    12′ Mid-st lvs narrowly triangular-lanceolate, ± scabrous to bristly, ± densely glandular
     13. Sts 2–5 dm; mid-st lvs ascending, flat; sw DMtns .... var. ***scabra***
     13′ Sts 2–13 dm; mid-st lvs very stiff, often reflexed, ± inrolled; s SNF .... var. ***shevockii***

H. grandiflora Nutt. (p. 289) TELEGRAPH WEED Ann to short-lived per < 20 dm, erect, ± densely bristly, glandular. **LVS**: lower lvs petioled, basal lobes ear-like, clasping; middle lvs densely appressed-hairy, lanceolate; upper lvs sessile, ascending, less hairy and more glandular upward. **INFL** ± panicle-like, densely glandular; involucre 6–9 mm; phyllaries in 4–6 series. **RAY FLS** 25–40; ligules 5–8 mm. **DISK FLS** 30–75; corollas 4–6 mm. **FR** 2–5 mm: rays frs ± glabrous, pappus 0; disk frs strigose, outer pappus 0.2–0.7 mm, inner 3–5 mm. 2*n*=18. Disturbed areas, dry streams, sand dunes; gen < 300 mm. s NW, SNF, s Teh, GV, CW; nw Mex; introduced into AZ, UT. ❀SUN:7,8,**9**,10,**14–17**,18,**19–24**;INV.

H. oregona (Nutt.) Shin. (p. 289) Per 2–10 dm, ascending to erect, ± bristly, glandular. **LVS**: lower (ob)lanceolate, 0 by fl; upper sessile, wider. **INFL**: heads discoid; involucre 7.5–14 mm; phyllaries in 5–7 series, outer 1–1.5 mm, gen lanceolate-deltate. **DISK FLS** 14–60; corollas 7.5–14 mm. **FR** 2–5 mm, sparsely strigose; outer pappus bristles few, < 0.5 mm, inner 30–40, ± = corolla. 2*n*=18. Seasonally dry streambeds; < 1000 m. NW, CaR, n SNF, CW; to WA. Vars. intergrade. [*Chrysopsis o.* (Nutt.) A. Gray]

var. ***compacta*** (Keck) Semple **STS** ± densely branched below middle. **LVS** of branches long-soft-hairy and ± densely bristly, sparsely glandular, appearing gray-green. **INFL** ± dense. Habitats and ± range of sp. in CA. [*C. o.* var. *c.* Keck] ❀TRY.

var. ***oregona*** **STS** ± sparsely branched above middle. **LVS** of branches gen > 13 mm, > 3 mm wide, sparsely hairy, ± densely glandular. **INFL** ± open. Habitats and ± range (exc SNF, to WA) of sp.

var. ***rudis*** (E. Greene) Semple **STS** ± sparsely branched above middle. **LVS** of branches ± densely short-bristly, appearing green, sparsely glandular. **INFL** ± open. Habitats and ± range of sp. in CA. [*C. o.* var. *r.* (E. Greene) Jepson]

var. ***scaberrima*** (A. Gray) Semple **STS** ± sparsely branched above middle. **LVS** < 13 mm, < 3 mm wide, ± scabrous, glandular. **INFL** much-branched; outer phyllaries ovate-deltate. Habitats and elevations of sp. ± CW. [*C. o.* var. *s.* A. Gray]

H. psammophila B. Wagenkn. CAMPHOR WEED Ann 3–15 dm, ascending to erect, ± bristly-scabrous, glandular (densely so above);

large pls much-branched. **LVS**: lower wing-petioled, basal ear-like lobes clasping; middle ± sessile, ovate, clasping; upper spreading. **INFL**: involucre 6–12 mm; phyllaries in 4–6 series. **RAY FLS** 20–30; ligules 4–10 mm. **DISK FLS** 40–75; corollas 4– 6 mm. **FR** 2–4 mm; ray frs (sub)glabrous, pappus 0; disk frs densely strigose, outer pappus 6–9 mm, inner 0.2–0.6 mm. 2*n*=18. Uncommon. Disturbed sandy soils; < 300 m. SCo, e DSon; to TX, nw Mex. [*H. subaxillaris* (Lam.) Britton & Rusby misapplied] Probably naturalized in SCo.

H. sessiliflora (Nutt.) Shinn. Per 1–13 dm, ± bristly to woolly, ± glandular above. **LVS**: lower tapered; upper ± sessile, 1–5 cm. **INFL**: involucre 8–14 mm, gen not subtended by lf-like bracts, ± hairy and glandular. **RAY FLS** 3–30; ligules 3–10 mm. **DISK FLS** 20–50; corollas 3–10 mm; lobes gen sparsely hairy. Dunes, grassland, oak woodland; < 2700 m. ± c&s CA-FP; Baja CA. Highly variable, esp in CW; sspp. ± merge where ranges overlap.

ssp. ***bolanderi*** (A. Gray) Semple (p. 289) Pl sparsely woolly below. **LVS** flat, little reduced upward; hairs ± 2 mm. **INFL**: heads gen few. **DISK FLS** 30–50; corolla lobes (sub)glabrous. 2*n*=18. Dunes, headlands; < 150 m. c&s NCo, n CCo. [*Chrysopsis villosa* (Pursh) Nutt. var. *b.* (A. Gray) Jepson] ❀SUN:**5**&IRR:7,**14–17**, **19–24**.

ssp. ***echioides*** (Benth.) Semple (p. 289) Pl bristly below. **LVS** flat (or ± wavy in Teh, TR), much reduced upward; hairs ± 2 mm. **INFL**: heads many in tall pls. **DISK FLS** 30–50. 2*n*=18. Grasslands, oak woodlands; < 1100 m. c&s SNF, Teh, SnJV, CW, SCo, TR, PR. [*Chrysopsis villosa* var. *e.* (Benth.) A. Gray] Sparsely glandular pls from serpentine in SnFrB with long-hairy disk corolla lobes have been called var. *bolanderoides* Semple; densely glandular pls from s SnFrB, n SCo with ± glabrous disk corolla lobes have been called var. *camphorata* (Eastw.) Semple [*Chrysopsis villosa* (Pursh) Nutt. var. *c.* Eastw.] ❀TRY.

ssp. ***fastigiata*** (E. Greene) Semple (p. 289) **LVS** wavy-margined; upper stiff, ± whitish, hairs gen ± 1 mm. **INFL**: heads few–many. **DISK FLS** 20–40. 2*n*=18. Woodland; < 2700 m. TR, c PR, SnJt. [*Chrysopsis villosa* (Pursh) Nutt. var. *f.* (E. Greene) H.M. Hall] PR pls (more glandular, less hairy) have been called var. *sanjacintensis* Semple. ❀TRY.

ssp. ***sessiliflora*** (p. 289) **LVS** wavy-margined; upper not stiff, green, hairs gen ± 1 mm. **INFL**: heads few, subtended by lf-like bracts. **DISK FLS** 30–50; corolla lobes (sub)glabrous. Beaches; < 10 m. s CCo, SCo; Baja CA. [*Chrysopsis villosa* (Pursh) Nutt. var. *s.* (Nutt.) A. Gray] ❀TRY.

H. villosa (Pursh) Shinn. Per 1–13 dm, decumbent to erect, ± bristly-strigose, sometimes densely glandular above. **LVS** ± (ob)lanceolate, gen flat; hairs gen < 1 mm; lower tapered to base; upper tapered to subclasping. **INFL**: heads 1–many. **RAY FLS** 7–30; ligules 4–12 mm. **DISK FLS** 30–80; corollas 4–8 mm, lobes (sub)glabrous (any hairs < 0.2 mm), glands few. **FR** 2–4 mm. Crevices, open sand and gravel, lava flows; 400–3100 m. KR, CaR, n&c SNH, s SNF, MP, sw DMtns; w N.Am. Highly variable; local forms often ± distinct.

var. ***hispida*** (Hook.) V. Harms **LVS**: mid-st lvs (ob)lanceolate, flat, ± moderately strigose (not bristly) and glandular. **RAY FLS** 7–21; ligules 4–10 mm. **DISK FLS** 15–70; corollas 4–8 mm. **FR** 2–3 mm. 2*n*=18,36. Lava flows, rocky sites; 600–3100 m. CaRH, n&c SNH, MP. [*Chrysopsis villosa* (Pursh) Nutt. var. *h.* (Hook.) D. Eaton]

var. ***scabra*** (Eastw.) Semple Pl < 5 dm. **LVS**: mid-st lvs narrowly triangular, flat, ± sparsely scabrous to bristly, ± densely glandular (esp later growth). **RAY FLS** 9–14; ligules 5–6 mm. **DISK FLS** 30–50; corollas 5–7 mm. **FR** 2–3 mm. 2*n*=18,36. Uncommon. Rock crevices; 1200–1300 m. sw DMtns (Little San Bernardino Mtns); to UT, n AZ. [*Chrysopsis v.* (Pursh) Nutt. var. *hispida* (Hook.) D. Eaton misapplied, in part] Possibly alien in CA.

var. ***shevockii*** Semple (p. 289) SHEVOCK'S HAIRY GOLDENASTER Pl < 13 dm, ± scabrous to bristly, ± densely glandular. **LVS**: mid-st lvs narrowly triangular, very stiff, often reflexed, ± inrolled (esp large pls), ± sparsely scabrous to bristly, ± densely glandular. **RAY FLS** 9–18; ligules 5–12 mm. **DISK FLS** 30–80; corollas 5–8 mm. **FR** 3–4 mm. 2*n*=36. RARE. Ditches, crevices, shallow sand; 400–800 m. s SNF (Kern River Canyon). [*Chrysopsis v.* var. *hispida* (Hook.) D. Eaton misapplied, in part]

var. ***villosa*** (p. 289) Herbage ± densely strigose, gen glandless. **LVS** ± (ob)lanceolate, flat. **INFL** densely glandular. **RAY FLS** 9–30; ligules 5–9 mm. **DISK FLS** 30–80; corollas 4–8 mm. **FR** 2–3 mm. Uncommon. Rocky places, crevices; 500–2000 m. KR, CaR; to B.C., e Can, c US, TX.

HIERACIUM HAWKWEED

G. Ledyard Stebbins

Per; sap milky; herbage gen long-hairy. **STS** erect, 1–10 dm. **LVS** basal or cauline, alternate. **INFL**: heads ligulate, few–many in cymes or panicles; involucre cylindric; phyllaries in 2–4 series of different lengths; receptacle naked. **FLS** few–many; ligules yellow, white, or orange, readily withering. **FR** cylindric, slender; pappus of many slender bristles, brittle, dull white, tawny, or brownish. ± 250 spp.: ± worldwide. (Greek: hawk) Many reproduce only by asexual seeds.

1. Fls orange or white
 2. Fls white; stolons 0; native per of forests . . . . . ***H. albiflorum***
 2′ Fls orange; stolons present; weed . . . . . ***H. aurantiacum***
1′ Fls yellow
 3. Lvs glabrous on one or both surfaces
  4. Lvs glabrous on lower surface, sparsely hairy on upper; phyllaries glabrous or with a few scattered hairs; fls 8–20 . . . . . ***H. bolanderi***
  4′ Lvs glabrous on both surfaces; phyllaries lightly tomentose and with short, glandular hairs; fls 20–35 . ***H. gracile***
 3′ Lvs ± hairy on both surfaces
  5. Lvs and sts grayish with short, stellate hairs and sometimes longer hairs; frs 5–6 mm . . . . . ***H. greenei***
  5′ Lvs and sts with long, unbranched hairs; frs 2–3 mm
   6. Sts mostly 0.8–3 dm; lvs densely hairy; fls 6–15 . . . . . ***H. horridum***
   6′ Sts mostly 3–10 dm; lvs sparsely to moderately hairy; fls 14–50
    7. At least some lvs with prominent, widely spaced teeth; c&s CA . . . . . ***H. argutum***
    7′ Lvs all entire; n-most CA . . . . . ***H. scouleri***

H. albiflorum Hook. (p. 293) Stolons 0. **STS** 4–8 dm; lower part densely hairy. **LVS** mostly basal, 0.8–1.5 dm, oblong to oblanceolate, entire or few-toothed, coarsely long-hairy; cauline gen restricted to lower part of st, smaller, upper part of st nearly or quite lfless. **INFL**: heads few–many, in open cymes or panicles; involucres 9–10 mm; phyllaries glabrous or glandular. **FLS** gen 15–30; ligules white. **FR** 2–3 mm, red-brown; pappus dull white or tawny. 2*n*=18. Dry forests; 0–2900 m. CA-FP, Wrn; to AK, Colorado. ❀DRN,SHD:1–3,**7–9**,10,**14–24**.

H. argutum Nutt. (p. 293) Stolons 0. **STS** 3–10 dm, densely hairy below, glabrous above. **LVS** mostly basal, 0.8–1.6 dm, oblong-lanceolate; margins with widely spaced teeth, coarsely long-hairy; cauline gen restricted to lower part of st, smaller, upper part of st nearly or quite lfless. **INFL**: heads many in open panicles; involucres 7–10 mm; phyllaries with short, gland-tipped dark or yellowish hairs. **FLS** 14–22; ligules yellow. **FR** 2–3 mm, dark brown or black; pappus tawny or dull brown. 2*n*=18. Dry slopes, often in woodland; 0–1530 m. c SNF, SCoRO, SCo, n ChI, WTR, SnGb. [var. *parishii* (A. Gray) Jepson] ❀DRN,**15–17**&SHD:**7–9,14,18–24**.

H. aurantiacum L. Stolons and lf rosettes many. **STS** 2–7 dm, densely hairy. **LVS** all or mostly basal, 0.5–2 dm, oblanceolate to narrowly elliptic, entire or obscurely toothed, densely soft-hairy. **INFL** scapose; heads in dense to loose cymes; involucres 5–9 mm; phyllaries glandular and black-hairy. **FLS** 32–45; ligules orange-red, often drying ± purple. **FR** 1.5–2 mm; pappus white. 2*n*= 30,36,45,54. Invasive weed of lawns, disturbed places; ± 300 m. n SNF (Grass Valley, Nevada Co.), expected more widely; native to Eur; noxious weed in ne US.

H. bolanderi A. Gray (p. 293) Stolons 0. **STS** 1–3 dm, simple or few-branched, glabrous or puberulent. **LVS** mostly basal, 2–7 cm, widely oblanceolate, entire or few-toothed; upper surface hairy; lower surface gen glabrous. **INFL**: heads few in open or dense clusters; involucres 8–9 mm; phyllaries ± glabrous. **FLS** 8–20; ligules yellow. **FR** dark brown, 4 mm; pappus tawny. Dry forest; 300–2700 m. KR, NCoR; OR. ❀DRN:**15–17**&SHD:7,8,9,**14**,18–24.

H. gracile Hook. Stolons 0. **STS** 1–several from basal rosette, 1–3 dm, puberulent. **LVS** all or mostly basal, 2–8 cm, oblong-lanceolate, entire or nearly so, glabrous or sparsely hairy. **INFL**: heads few, in raceme-like or cymose clusters; involucres 7–8 mm; phyllaries with black, mostly glandular hairs. **FLS** 20–35; ligules yellow. **FR** 2 mm; pappus tawny. 2*n*=18. Moist forests, rocky open places, meadows; 1650–3550 m. KR, CaRH, SNH; to BC, MT, WY, Colorado.

H. greenei A. Gray Stolons 0. **STS** 2–4 dm, densely stellate-to-mentose and coarsely long-hairy below. **LVS**: basal and lower cauline 5–10 cm, oblong-oblanceolate, nearly entire or with blunt teeth, densely covered with stellate hairs and long, simple hairs; upper lvs much reduced. **INFL**: heads several–many, in open cymes or panicles; involucre 8–10 mm, narrow. **FLS** 8–15; ligules yellow. **FR** 5–6 mm, 10-ribbed; pappus deep tawny. 2*n*=18. Dry slopes in montane forest; 900–2750 m. KR, NCoRH, CaRH; OR. ❀TRY.

H. horridum Fries (p. 293) Stolons 0; herbage densely long-hairy; hairs whitish or brownish. **STS** 1–several from taproot, 1–3.5 dm, much-branched. **LVS** mostly cauline, 3–10 cm, oblong, entire. **INFL**: heads many, in open cymes or panicles; involucre 6–9 mm, narrow; phyllaries glandular and long-hairy. **FLS** 6–15; ligules yellow. **FR** 3 mm; pappus brownish. 2*n*=18. Rocky places, crevices; 1350–3300 m. KR, CaRH, SNH, SnJt, GB; OR. ❀TRY.

H. scouleri Hook. Stolons 0. **STS** 3–7 dm, simple, ± glabrous to bristly. **LVS**: gen basal and lower cauline, 10–20 cm, lanceolate to oblanceolate, entire, long-hairy. **INFL** glabrous to hairy or glandular; heads many, in cymes; involucre 8–10 mm, glandular-hairy. **FLS** 15–50; ligules yellow. **FR** 3 mm; pappus dull brown. 2*n*=18. Open woods, shrubby places; 450–2250 m. KR, NCoRH, CaRH, n SNH, GB; to B.C., MT. Pls with naked sts and many, dark, glandular hairs have been called var. *nudicaule* (A. Gray) Cronq. [*H. cynoglossoides* Arv.-Touv. misapplied to CA pls] ❀DRN:4,5,**6,15–17**&SHD:1,2,**7**,8–10,**14**,18–24.

HOLOCARPHA

Ann, densely glandular, strongly scented. **LVS** alternate or lower opposite, linear to oblanceolate, entire to lobed; upper reduced, often with axillary lf clusters, tipped by open-pit glands. **INFL**: heads radiate, in flat-topped to panicle-like clusters; phyllaries half-enclosing ray ovaries, falling with frs, back with stalked, open-pit glands, tip cylindric, gland-tipped; receptacle chaffy. **RAY FLS** 3–16; ligules 3-lobed (middle narrowest), yellow. **DISK FLS** few–many, most sterile; corolla yellow; anther tips widely triangular; style branches tapered. **FR**: pappus 0; ray achenes ± 3-angled, short-beaked, dark; disk achenes ± club-shaped. 4 spp.: CA. (Greek: wholly chaffy, from receptacle) [Palmer 1982 PhD Univ CA Davis] ± identical forms often intersterile.

1. Anthers yellow; herbage ± puberulent above
 2. Herbage densely puberulent above; involucre ± spheric; glands on phyllaries 25–50, very slender ..... ***H. heermannii***
 2′ Herbage glabrous or ± sparsely puberulent above; involucre obconic; glands on phyllaries 5–15(20), stout ..... ***H. obconica***

1′ Anthers black; herbage gen not puberulent above
 3. Disk fls 40–90; ray fls 8–16; infl clusters dense, heads large; coastal ..... ***H. macradenia***
 3′ Disk fls 9–25; ray fls 3–7; infl raceme-like, heads small; interior ..... ***H. virgata***
  4. Branchlets slender, gracefully curved; peduncles long ..... ssp. ***elongata***
  4′ Branchlets stiffly ascending; peduncles 0 to short ..... ssp. ***virgata***

H. heermannii (E. Greene) Keck (p. 293) **STS** 2–12 dm; branches glandular throughout, hairy below, densely gray-puberulent above. **LVS** canescent; lower 3–10 cm; upper densely glandular. **INFL** raceme- to panicle-like; peduncle bracts 2–5 mm, overlapping, appressed to ± recurved, flat, gland-tipped; involucres 5–6 mm, ± hemispheric; phyllaries with 25–50 very slender gland-tipped projections, puberulent, sticky; receptacle 1–2 mm diam, rounded. **RAY FLS** 3–10; ligules 5–6.5 mm. **DISK FLS** 9–22; corollas 3.5–5 mm; anthers yellow. **FR**: ray achenes 2.5–4 mm. 2*n*= 12. Grassy slopes; < 1300 m. SNF, Teh, SnJV, e SnFrB, SCoR, n WTR.

H. macradenia (DC.) E. Greene (p. 293) SANTA CRUZ TARPLANT **STS** 1–5 dm; branches few, stiffly spreading, bristly, glandular. **LVS**: lower < 12 cm, widely linear, slender-toothed; upper bract-like, rolled under, ± bristly; uppermost gland-tipped. **INFL** dense or heads ± sessile in upper axils; peduncle bracts 5–10 mm, upper appressed to involucre base, flat; involucres 5.5–8 mm, ± spheric; phyllaries with ± 25 long, cylindric, gland-tipped projections, puberulent; receptacle 2.5–3 mm diam, conic. **RAY FLS** 8–16; ligules 4.5–5.5 mm. **DISK FLS** 40–90; corollas 4–4.5 mm; anthers black. **FR**: ray achenes 2.5–3 mm. 2*n*=8. **ENDANGERED** CA. Grassy coastal terraces; < 100 m. n CCo (n&c Monterey Bay), sw SnFrB; once more widespread in SnFrB. Threatened by development, agriculture.

H. obconica (J. Clausen & Keck) Keck (p. 293) **STS** 1–12 dm; branches stiffly ascending, overtopping main st, bristly below, glabrous or puberulent above, resinous. **LVS**: lower 6–15 cm, linear, bristly, resinous; upper well spaced. **INFL** open, ± flat-topped; peduncle bracts 1–3 mm, overlapping, ascending to spreading, ± cylindric, gen gland-tipped; involucres 4–5 mm, ± obconic; phyllaries with 5–20 stout, gland-tipped projections, glabrous or minutely glandular; receptacle 2–2.5 mm diam, flat or rounded. **RAY FLS** 4–9; ligules 4–6 mm. **DISK FLS** 11–21; corollas 2.5–4 mm; anthers yellow. **FR**: ray achenes 2.5–3.5 mm. 2*n*=12. Grassland; < 500 m. s SNF, nw SnJV, ne CW. Highly variable; densely leafy, late-fl pls from nw SnJV, e SnFrB have been called ssp. *autumnalis* Keck.

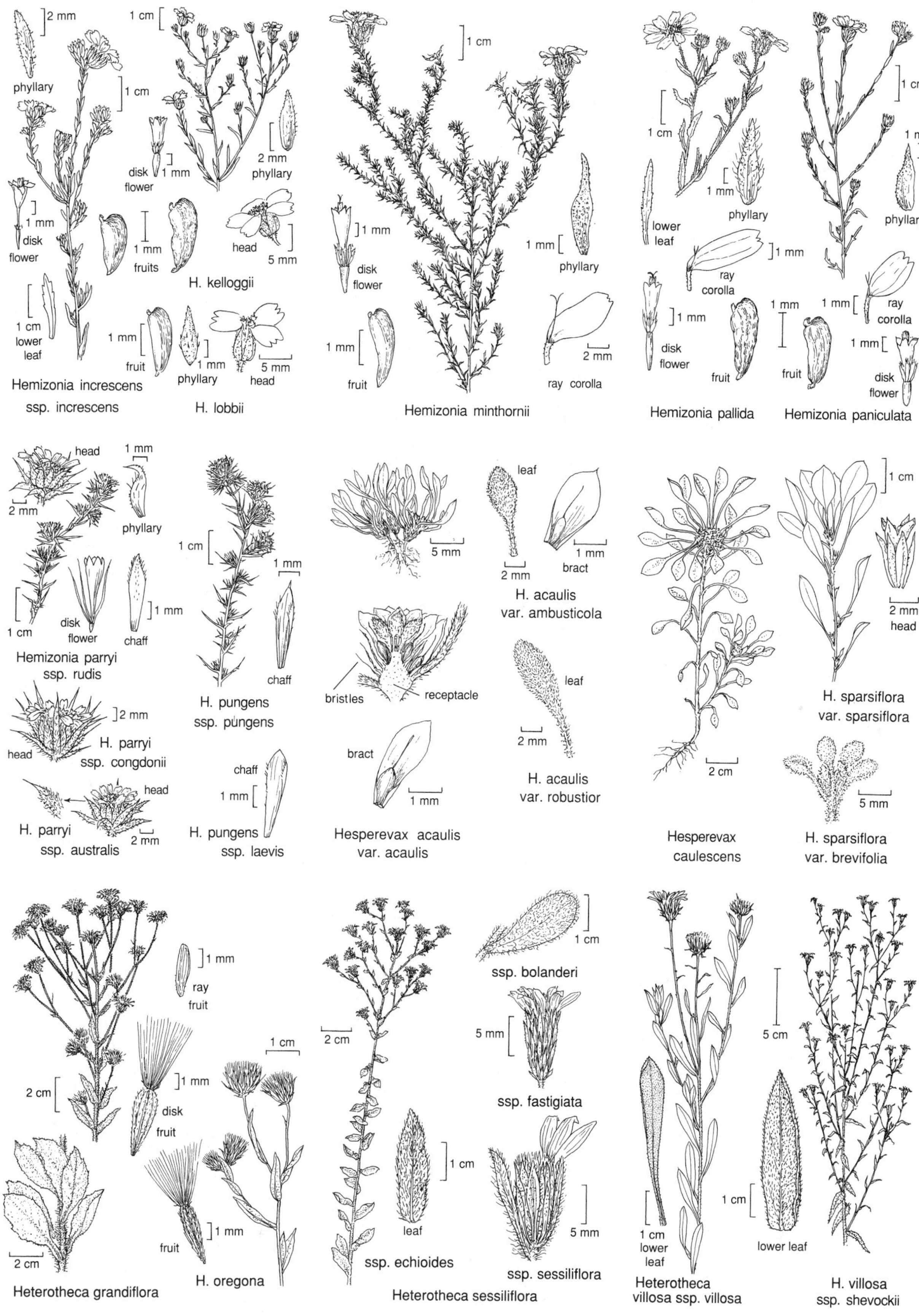

phyllary
2 mm
1 cm
1 mm
disk flower
1 cm
lower leaf
Hemizonia increscens ssp. increscens
disk flower
2 mm
phyllary
fruits
head
5 mm
H. kelloggii
fruit
phyllary
head
H. lobbii
disk flower
fruit
phyllary
ray corolla
Hemizonia minthornii
lower leaf
phyllary
ray corolla
disk flower
fruit
Hemizonia pallida
Hemizonia paniculata
head
Hemizonia parryi ssp. rudis
chaff
H. parryi ssp. congdonii
H. parryi ssp. australis
H. pungens ssp. pungens
H. pungens ssp. laevis
leaf
bract
H. acaulis var. ambusticola
bristles
receptacle
H. acaulis var. robustior
Hesperevax acaulis var. acaulis
Hesperevax caulescens
H. sparsiflora var. sparsiflora
H. sparsiflora var. brevifolia
ray fruit
disk fruit
Heterotheca grandiflora
H. oregona
ssp. bolanderi
ssp. fastigiata
ssp. echioides
ssp. sessiliflora
Heterotheca sessiliflora
Heterotheca villosa ssp. villosa
lower leaf
H. villosa ssp. shevockii

H. virgata (A. Gray) Keck **STS** 2–12 dm; branches gen not overtopping main st, soft-hairy and glandular below, canescent or short-bristly above, resinous. **LVS**: lower 6–15 cm, linear, bristly, resinous; upper crowded. **INFL** spike- or raceme-like; heads sessile or short-peduncled; peduncle bracts 2–5 mm, overlapping, appressed to involucre base, spreading to ± recurved, ± cylindric, gen gland-tipped; involucres 5–6 mm, ± obconic; phyllaries with 5–20 stout, gland-tipped projections, otherwise glabrous; receptacle 1–2 mm diam, nearly flat. **RAY FLS** 3–7; ligules 4–6 mm. **DISK FLS** 9–25; corollas 3.5–4.5 mm; anthers black. **FR**: ray achenes 2.4–3.5 mm. $2n$=8. Grassland; < 800 m. NCoRI, n&c SNF, GV, ne CW, c&s SCo.

ssp. ***elongata*** Keck GRACEFUL TARPLANT **STS** slender; branches many, gracefully curved. **INFL**: peduncles < 15 cm, openly bracted. UNCOMMON. Grassland; < 600 m. c&s SCo.

ssp. ***virgata*** (p. 293) **STS** stout, rigid; branches few–many, stiffly ascending. **INFL**: peduncles short, densely bracted. Habitats and range of sp. (exc c&s SCo).

HOLOZONIA

Bruce G. Baldwin & Susan J. Bainbridge

1 sp. (Greek: whole girdle) [Thompson 1983 PhD thesis Univ CA Davis]

H. filipes (Hook. & Arn.) E. Greene (p. 293) Per 3–15 dm, from rhizomes, slender-branched, ± white-hairy and glandular above. **LVS** < 10 cm, linear to lanceolate; glands tack-shaped (esp margins); lower lvs opposite, fused around st; upper lvs alternate, sessile. **INFL**: heads radiate in leafy cyme; involucre obconic; peduncles < 6 cm; phyllaries 3–5 mm, each enfolding a ray fr, long-white-hairy, often glandular; receptacle convex; chaff scales 8–13, between ray and disk fls, hairy, edges fused. **RAY FLS** 6–10, fruiting; ligules 2.5–5 mm, deeply linear-lobed, white, purplish below, sparsely hairy; style branches long, glabrous. **DISK FLS** 9–28, staminate; corollas 3–4.5 mm, white, hairy; anthers purple, acute; style branches long, short-bristly. **FR** 3–5 mm, narrowly club-shaped, compressed front-to-back, glabrous; inner surface ribbed; ray pappus 0; disk pappus bristles 0–5. $2n$=28. Dry beds and borders of lakes, streams. NCoR, CaRF, n&c SNF, GV (edge), SnFrB, c SCoRO.

HULSEA

Dieter H. Wilken

Ann to per. **STS** 1–5, 1–15 dm, ± hairy, glandular. **LVS** alternate; petioles gen ciliate; blades gen ± oblanceolate, entire to lobed, ± reduced upward. **INFL** raceme-like; heads radiate, 1–many; bracts ± narrowly lanceolate; involucre hemispheric to obconic; phyllaries many, in 2–3 series, linear to obovate, green, ± glandular; receptacle naked. **RAY FLS**: ligules yellow to red. **DISK FLS** many; corollas 5–9 mm, yellow to orange, gen glabrous. **FR** 4–10 mm, linear to club-like, black, ± hairy; pappus scales gen 2 pairs, 4–10 mm, gen deeply cut, gen translucent. ± 8 spp.: w US. (G.W. Hulse, US Army surgeon, botanist, 1807–1883) [Wilken 1977 Madroño 24:48–55] Self-sterile.

1. Lower lvs gen glandular, gen barely hairy (exc woolly in *H. nana*)
  2. St lfless, gen < 4 dm; heads 1–2; lvs abruptly smaller above rosette; gen alpine or subalpine
    3. Phyllaries narrowly oblong, long-tapered; ray fls 25–60 .... ***H. algida***
    3′ Phyllaries oblong to obovate, acuminate; ray fls 12–30 .... [2]***H. nana***
  2′ St leafy, gen > 4 dm; heads 3–many; lvs gradually smaller upward; montane
    4. Ligules < 2 mm wide; red .... ***H. heterochroma***
    4′ Ligules > 2 mm wide; yellow
      5. Ray fls 10–23, corolla tube hairs glandular and non-glandular .... ***H. brevifolia***
      5′ Ray fls 20–35, corolla tube hairs all glandular .... ***H. mexicana***
1′ Lower lvs ± woolly or long-soft-wavy-hairy
  6. Lvs gradually smaller upward; heads 3–4, upper peduncles < 5 cm .... ***H. californica***
  6′ Lvs abruptly smaller above rosette; heads gen 1 per st (or peduncles gen > 5 cm)
    7. Lvs gen oblanceolate, sparsely glandular; n CA .... [2]***H. nana***
    7′ Lvs gen spoon-shaped, woolly or long-soft-wavy-hairy; s CA .... ***H. vestita***
      8. Basal lf blade entire to weakly scalloped, gradually tapered; petiole gen < blade
        9. St lfless; bracts densely woolly .... ssp. ***vestita***
        9′ St ± leafy in lower 1/3; bracts glandular to barely woolly
          10. Petioles green; st gen > 5 dm; ligules gen yellow to orange; n PR .... ssp. ***callicarpha***
          10′ Petioles ± reddish; st gen < 5 dm; ligules ± red-tinged; WTR, SnGb .... ssp. ***gabrielensis***
      8′ Basal lf blades scalloped to lobed, abruptly tapered, petiole gen = or > blade
        11. Ligules yellow, 12–18 mm .... ssp. ***inyoensis***
        11′ Ligules red to orange, 5–8 mm
          12. Hairs of lower lf surface glandular and nonglandular; phyllaries gen green, tinged red; montane .. ssp. ***parryi***
          12′ Hairs of lower lf surface gen glandular only; phyllaries deep-red purple; alpine .... ssp. ***pygmaea***

H. algida A. Gray (p. 293) Per < 4 dm, ± long-soft-hairy, ± glandular. **LVS**: basal < 10 cm, ± coarsely toothed. **INFL**: heads 10–20 mm, 10–25 mm wide, phyllaries 8–15 mm, 1–3 mm wide, narrowly oblong, long-tapered. **RAY FLS** 25–60; corollas 10–15 mm, 2–4 mm wide, yellow, puberulent. **FR** 6–10 mm, sparsely hairy; pappus scales < 1.5 mm, ± equal. $2n$=38. Subalpine to alpine talus; 3000–4000 m. SNH, W&I; to OR, MT, NV.

H. brevifolia A. Gray (p. 293) SHORT-LEAVED HULSEA Per 3–6 dm, ± glandular. **LVS**: basal gen 5–6 cm, coarsely toothed; petiole long-ciliate. **INFL**: heads gen 10 mm, < 20 mm wide; phyllaries 8– 10 mm, 1–3 mm wide, lanceolate, acuminate. **RAY FLS** 10–23; corollas 10–12 mm, 2–4 mm wide, yellow, short-hairy. **FR** 6–8 mm, sparsely hairy; pappus scales 1–3 mm, unequal, tinged red. $2n$=38. UNCOMMON. Gravelly soils, montane forest; 1500–2700 m. c&s SNH.

H. californica Torrey & A. Gray (p. 293) SAN DIEGO SUNFLOWER Bien 4–12 dm, ± woolly, ± glandular. **LVS**: basal 6–10 cm, entire. **INFL**: heads ± 20 mm, 20 mm wide; phyllaries 9–14 mm, linear-lanceolate, long-acuminate. **RAY FLS** 22–40; corollas 15–20 mm, 3–5 mm wide, yellow, puberulent. **FR** 4–6 mm, silky-hairy; pappus scales < 2 mm, ± equal. $2n$=38. RARE. Open sites; 1000–2000 m. s PR (Laguna, Cuyamaca Mtns, San Diego Co.).

H. heterochroma A. Gray (p. 293) Ann or per < 15 dm, ± densely glandular. **LVS**: basal 10–20 cm, coarsely toothed, green. **INFL**: heads gen 20 mm, 15 mm wide; phyllaries 10–14 mm, linear to lanceolate, long-acuminate. **RAY FLS** 30–60; corollas 6–10 mm, linear, red, hairy. **FR** 6–8 mm, very hairy; pappus scales 1–3 mm, unequal. 2*n*=38. Open sites, recent burns; 300–2500 m. SN, SCoRO, TR, n PR; to UT. Hybridizes with *H. vestita* sspp. *callicarpha* & *parryi*. ❀DRN,SUN:1–7,14–24;DFCLT.

H. mexicana Rydb. MEXICAN HULSEA Ann or bien < 10 dm, ± soft-hairy and glandular. **LVS**: basal 5–15 cm, coarsely toothed. **INFL**: heads gen 20 mm, 20 mm wide; phyllaries 8–12 mm, linear-lanceolate, narrowly acuminate. **RAY FLS** 20–35; corollas 12–18 mm, 3–5 mm wide. **FR** 4–6 mm, sparsely hairy; pappus scales 1–2 mm, ± equal. 2*n*=38. RARE in CA. Burns, disturbed sites; 1200 m. se PR (se San Diego Co.); Baja CA.

H. nana A. Gray (p. 293) Per < 2 dm, ± long-hairy and glandular. **LVS**: basal 2–6 cm, gen lobed. **INFL**: head gen 1 per st, < 15 mm, < 12 mm wide; phyllaries 8–12 mm, oblong to obovate, acuminate, sometimes woolly. **RAY FLS** 12–30; corollas 8–12 mm, 2– 5 mm wide, yellow, puberulent. **FR** 6–8 mm, sparsely hairy; pappus scales 1–2 mm, ± equal. 2*n*=38. Volcanic talus; 2400–3000 m. CaRH, MP; to WA.

H. vestita A. Gray Per gen 1–10 dm. **STS** gen leafy in lower 1/3–1/2. **LVS**: basal < 8 cm, 1–3 cm wide, spoon-shaped, entire to lobed, woolly above; cauline few. **INFL**: heads < 15 mm, < 12 mm wide; bracts lanceolate to ovate, glandular, ± long-soft-hairy; phyllaries 8–11 mm, oblong to obovate, acuminate, hairy. **RAY FLS** 9–32; corollas 5–18 mm, 2–5 mm wide, yellow to red, puberulent. **FR** 5–7 mm, moderately hairy; pappus 1–2 mm, gen ± equal. Open gravel, talus slopes; 1300–3900 m. c&s SNH, TR, PR, SNE, DMtns; NV. Sspp. gen geog separated.

ssp. ***callicarpha*** (H.M. Hall) Wilken Pl gen > 5 dm. **LVS**: petiole gen < blade, green; basal lvs entire to weakly scalloped. **INFL**: bracts sometimes barely woolly; phyllary tips green or red-tinged. **FLS** yellow to orange; ray corollas 6–10 mm. 2*n*=38. Montane chaparral, pine forest; 1300–2500 m. n PR. [*H. c.* (H.M. Hall) Rydb.] Intergrades with *H. californica*.

ssp. ***gabrielensis*** Wilken Pl gen < 5 dm. **LVS**: petioles gen < blade, ± red; basal lvs ± scalloped. **INFL**: bracts sometimes barely woolly; phyllary tips red-tinged. **RAY FLS**: corollas 6–8 mm, red-tinged. **DISK FLS**: corollas yellow to orange. 2*n*=38. Montane forest; 1500–2500 m. e WTR, SnGb.

ssp. ***inyoensis*** (Keck) Wilken INYO HULSEA Pl < 7 dm. **LVS**: petioles gen > blade, green; basal lvs lobed. **INFL**: bracts barely woolly; phyllary tips ± green. **FLS** yellow; ray corollas 12–18 mm. 2*n*=38. RARE in CA. Rocky pinyon/juniper woodland; 1700–3000 m. n DMtns; w NV. [*H. i.* (Keck) Munz]

ssp. ***parryi*** (A. Gray) Wilken (p. 293) Pl gen 2–6 dm. **LVS**: petiole gen > blade, ± green; basal lvs deeply lobed, glandular and non-glandular below. **INFL**: bracts sparsely woolly; phyllary tips red-tinged. **FLS** orange to red; ray corollas 5–7 mm, ± red. 2*n*=38. Sagebrush to fir forest ; 2000–2500 m. SnBr, sw DMtns. [*H. p.* A. Gray]

ssp. ***pygmaea*** (A. Gray) Wilken Pl < 1 dm. **STS** ± lfless. **LVS**: petiole gen = blade, ± green or purplish; basal lvs lobed, entirely glandular below. **INFL**: bracts sparsely woolly; phyllary tips reddish or purple. **RAY FLS**: corollas 5–8 mm, red. **DISK FLS**: corollas orange. 2*n*=38. Subalpine forest, alpine barrens; 3200–3900 m. s SNH, SnBr.

ssp. ***vestita*** (p. 293) Pl 1–4 dm. **STS** sometimes ± lfless. **LVS**: petiole = blade, green; basal lvs entire, woolly. **INFL**: bracts densely woolly; phyllary tips green or red-tinged. **FLS** gen yellow; ray corollas 5–9 mm. 2*n*=38. Montane sagebrush to fir forest; 2400–3000 m. c&s SNH, n SNE.

HYMENOCLEA WINGED RAGWEED

Willard W. Payne

Subshrub to small tree, monoecious. **STS** slender. **LVS** alternate, entire or narrowly lobed, white-hairy above; margins curled upward. **INFL**: staminate heads many, in spike- or panicle-like clusters; phyllaries 4–8, fused into shallow cup, greenish; receptacle chaffy; pistillate heads ± spheric, beaked, in fr winged, 1-fld, ± bur-like, phyllaries 0–few, chaff scales ± fused, free tips winged, papery. **STAMINATE FLS** 4–18; corolla translucent; filaments and anthers free; style unbranched. **PISTILLATE FL**: corolla 0; style branches long; pappus 0. **FR** enclosed in bur, dispersed by wind or water. 2–4 spp.: sw US, n Mex. (Greek: membrane-enclosed, from bur) [Peterson & Payne 1974 Brittonia 26:397] Like *Ambrosia, Xanthium*.

1. Wings of bur < 1.5 mm wide, in 1 whorl; fls autumn .... ***H. monogyra***
1′ Wings of bur 2–8 mm wide, often scattered; fls spring .... ***H. salsola***
 2. Wings ± 7, in ± 1 central whorl .... var. ***pentalepis***
 2′ Wings ± 10–13, scattered
  3. Wings appressed, enclosing body of bur .... var. ***patula***
  3′ Wings spreading perpendicular to body of bur .... var. ***salsola***

H. monogyra A. Gray (p. 293) Shrub or small tree 1–4 m, much-branched above. **STS** straw-colored to dark brown; old bark gray-brown. **INFL**: fr heads 2–4 mm, fusiform; wings 7–12, 1.5–2 mm, < 1.5 mm wide, oblanceolate, in 1 central whorl. Washes, dry riverbeds; < 500 m. SW, SNE, DSon; to w NV, TX, n Mex. Fls autumn.

H. salsola A. Gray (p. 293) BURROBRUSH Subshrub < 2 m, branched throughout. **STS** pale straw-colored. **INFL**: fr heads 2.8–6.3 mm, widely fusiform; wings 5–19, 2–7 mm, 1.5–8 mm wide, often reniform, in 1 whorl or scattered. Dry flats, washes, fans; < 1800 m. s SnJV, s SCoRI, SW, SNE, D; to sw UT, AZ, nw Mex. Fls spring. Vars. intergrade.

var. ***patula*** (Nelson) Peterson & Payne **INFL**: bracts subtending bur gen several, widely ovate, gen green, hairy. **FR**: bur wings (10–)13(–19), scattered, appressed, base pitted. Habitats of sp. s SnJV, SCoRI, SCo, WTR, DMoj; to sw UT, w AZ. [var. *salsola* misapplied]

var. ***pentalepis*** (Rydb.) L. Benson **INFL**: bract subtending bur gen 1, obovate, transparent. **FR**: bur wings (5–)7(–13), whorled at center, spreading. Habitats of sp. SW, D; s NV, sw AZ, nw Mex.

var. ***salsola*** **INFL**: bracts subtending bur gen several, ovate, green or transparent, hairy. **FR**: bur wings (5–)10(–16), scattered, spreading. Habitats of sp. s SnJV, s SCoRI, SW, SNE, D; to sw UT, w AZ, Baja CA.

HYMENOPAPPUS

Bien, per from taproot bearing 1–several caudices. **STS** erect. **LVS** simple to 2-pinnately dissected; basal and alternate, reduced upward, gland-dotted. **INFL**: heads radiate or discoid, in few–many-headed panicles; involucre obconic to widely bell-shaped; phyllaries in 2–3 series, ± equal, scarious-margined; receptacle flat or rounded, gen naked. **RAY FLS** 0 (in CA) or 8; ligules white. **DISK FLS** 10–many; corollas yellow, white, or reddish purple, tube slender, throat abruptly enlarged, lobes triangular, reflexed; style branches obtuse. **FR** obpyramidal, 4-angled; pappus 0 or of many thin, transparent, linear-oblong to ovate, obtuse scales. 10 spp.: N.Am. (Greek: membranous pappus) [Turner 1956 Rhodora 58:163–186, 208–242, 250–269, 295–308]

H. filifolius Hook. Per 0.5–10 dm, ± glabrous to densely tomentose. **LVS**: basal 3–20 cm, 2-pinnately dissected into 2–50 mm linear or thread-like segments, minutely gland-dotted; cauline 0 or few, gen much reduced upwards. **INFL**: heads discoid, 1–many; peduncles 0.5–16 cm; principal phyllaries 3–14 mm, 2–5 mm wide, margin white or yellowish scarious for 1–4 mm from acute to obtuse tip; fls 10–70 per head. **FLS**: corollas 2–7 mm, yellow or white. **FR** 3–7 mm, densely short-hairy; pappus scales 12–22, gen linear-oblong. Many ± dry habitats, sometimes on limestone; ± 1000–3000 m. TR, PR, W&I, DMtns; to WA, SD, AZ, n Mex.

1. Corollas 3–4 mm; anthers 2–3 mm; fr 4.5–5.5 mm .... var. *nanus*
1′ Corollas 4–7 mm; anthers 3–4 mm; fr 4–7 mm
  2. Fls white; peduncles 8–16 cm; cauline lvs 2–7; hairs on fr 0.5–1 mm .... var. *eriopodus*
  2′ Fls gen yellow; peduncles 2–12 cm; cauline lvs 0–6; hairs on fr 1–2 mm
    3. Cauline lvs (1)2–7; tips of basal lvs 5–30 mm; phyllaries 8-12(14) mm .... var. *megacephalus*
    3′ Cauline lvs 0–3; tips of basal lvs 3–15 mm; phyllaries 6–10 mm .... var. *lugens*

var. ***eriopodus*** (Nelson) B. Turner Pl 4–8 dm, ± tomentose below, becoming glabrous above. **LVS**: basal 10–20 cm, divisions 10–20 mm, 0.4–1 mm wide, thread-like; cauline 2–7, gen glossy green. **INFL**: heads 3–8; peduncles 8–16 cm; main phyllaries 7–10 mm, 2–4 mm wide. **FLS**: corollas 4–5 mm, ± white; anthers 3–4 mm. **FR** 5.5–6 mm, evenly hairy; hairs 0.5–1.5 mm. Limestone soil, with pines, junipers; 1600–1700 m. e DMtns (Clark, New York mtns); s NV, sw UT.

var. ***lugens*** (E. Greene) Jepson (p. 297) Pl 2–6 dm, densely gray-tomentose to ± glabrous exc woolly axils. **LVS**: basal 5–14 cm, divisions 3–10 mm, 1–3 mm wide, flat, linear; cauline 0–3. **INFL**: heads 3–8; peduncles 2–12 cm; main phyllaries 6–10 mm, 3–4 mm wide. **FLS**: corollas 4–6 mm, gen yellow; anthers 3–4 mm. **FR** 4–6 mm, evenly hairy; hairs gen 1–1.5 mm. $2n$=34,68. Forests, woodlands; 1300–2700 m. TR, PR; to s UT, AZ, Baja CA.

var. ***megacephalus*** B. Turner Pl 3–7 dm, tomentose throughout. **LVS**: basal 8–20(30) cm, divisions 8–30 mm, 1–2 mm wide, flat, linear; cauline 2–6. **INFL**: heads 3–14; peduncles 2–10 cm; main phyllaries 8–14 mm. **FLS**: corollas 4–7 mm, yellow; anthers 3–4 mm. **FR** 5–7 mm, evenly hairy; hairs 1–2 mm. Sandy or gravelly desert soils in valleys, washes; ± 1000–1500 m. e DMtns (Providence Mtns); to w Colorado, AZ.

var. ***nanus*** (Rydb.) B. Turner Pl 0.5–5 dm, ± evenly, sparsely tomentose throughout. **LVS**: basal 2–12 cm, divisions 5–15 mm, 0.5–1 mm, thread-like; cauline 0–3. **INFL**: heads 1–6; peduncles 3–15 cm; main phyllaries 6–9 mm. **FLS**: corollas 3–4 mm, pale yellow; anthers 2–3 mm. **FR** 4.5–5.5 mm, evenly hairy; hairs gen 0.2–1 mm. $2n$=34. Limestone soil, often with juniper; 1500–3000 m. W&I (Inyo Mtns); also c&e NV to w UT, nw AZ.

HYMENOTHRIX

Ann, bien, per, taprooted. **STS** 1–several from base, erect, simple or much-branched above. **LVS** basal and cauline, alternate; lower petioled, pinnately or ternately dissected into narrow lobes; upper reduced, sessile. **INFL**: heads radiate or discoid, in many-headed, ± flat-topped clusters; peduncles slender; involucre obconic to hemispheric; phyllaries in 1–2 ± equal series, free; receptacle rounded, naked. **RAY FLS** 0 (in CA) or few; ligules yellow. **DISK FLS** few–many; corollas white to yellow; anther tips triangular; style tips narrowly triangular. **FR** club-shaped, 4–5-angled, ± hairy; pappus of 10– 15 bristle-tipped scales. 5 spp.: sw US, Mex. (Greek: membranous bristle, from pappus) [Turner 1962 Brittonia 14:101– 119]

1. Corollas cream-colored, outer often larger than inner; anthers yellow or light brown .... ***H. loomisii***
1′ Corollas white, equal; anthers dark purple .... ***H. wrightii***

H. loomisii S.F. Blake Ann, bien, or short-lived per, < or = 1.5 m. **LVS** dissected into 1–2 mm wide linear lobes, glabrous or short-hairy. **INFL**: peduncles 1–7 cm, bracts 0 or scattered, linear; involucre obconic; phyllaries 3–5 mm, linear, obtuse to acute, tips often cream-colored, glabrous to gland-dotted or puberulent. **FLS**: corollas 5–7 mm, cream-colored, outer often larger than inner, ± bilateral; anthers 2–3 mm, yellow to light brown. **FR** 3–5 mm; pappus scales 3–7 mm. $2n$=48. Disturbed sites; < 1000 m. SNF, SW; native to AZ.

H. wrightii A. Gray (p. 297) Bien, per < or = 1.5 m, glabrous to densely glandular or soft-hairy. **LVS** densely crowded on lower st; lobes gen 1 mm wide or less; margins rolled under. **INFL**: peduncle 1–7 cm; bracts scattered, scale-like; involucre hemispheric; phyllaries 3–8 mm, oblanceolate to obovate, obtuse, tips gen purple, glabrous or glandular. **FLS**: corollas 5–8 mm, outer and inner equal, white or purple-tinged, lobes linear, spreading; anthers ± 3 mm, dark purple. **FR** 3–6 mm; pappus scales 3–7 mm. $2n$=24. Open slopes; ± 1500 m. s PR (Cuyamaca, Laguna mtns); to NM, n Mex. ❀TRY.

HYMENOXYS

Ann, bien, per. **STS** erect, simple to much-branched. **LVS** simple, basal or alternate, sessile or petioled, entire or divided 1 or more times into linear lobes, dotted with sunken resin-glands. **INFL**: heads gen radiate, solitary or in few–many-headed cymes; involucre hemispheric; phyllaries in 2–3 similar or dissimilar series; receptacle flat to rounded, naked. **RAY FLS** few–many; corolla yellow, fan-shaped, 3–5-lobed. **DISK FLS** many; corolla yellow. **FR** obpyramidal, gen 5-angled, hairy; pappus of ± 5 membranous, often awn-tipped scales. 28 spp.: w N.Am, S.Am. (Greek: sharp membrane, from pappus) ❀TRY.

1. Lvs entire, gen all basal; head solitary, scapose .... ***H. acaulis*** var. ***arizonica***
1′ Lvs divided into linear lobes, basal and cauline; heads gen several–many in a cyme
  2. Involucre diam gen 15 mm or more .... ***H. lemmonii***
  2′ Involucre diam gen 10 mm or less
    3. Bien or short-lived per; basal lvs gen present at time of fl; involucre diam gen 9–10 mm .... ***H. cooperi***
    3′ Ann; basal lvs 0–few at time of fl; involucre diam 5–7 mm .... ***H. odorata***

H. acaulis (Pursh) K. Parker var. ***arizonica*** (E. Greene) K. Parker (p. 297) Per 8–20 cm from branched caudex. **STS** unbranched. **LVS** gen all basal, 2–6 cm, linear to narrowly oblanceolate, entire, soft-hairy, esp at base, becoming ± glabrous in age. **INFL**: heads solitary, scapose; involucre tomentose; outer phyllaries = inner, 5–6 mm, oblanceolate to obovate, free. **RAY FLS** 8–13; ligule 10–15 mm, bright yellow, fading to cream-colored and ± persistent on fr. **DISK FLS**: corolla 3.5–4 mm. **FR** 3–4 mm; pappus scales 5–7,

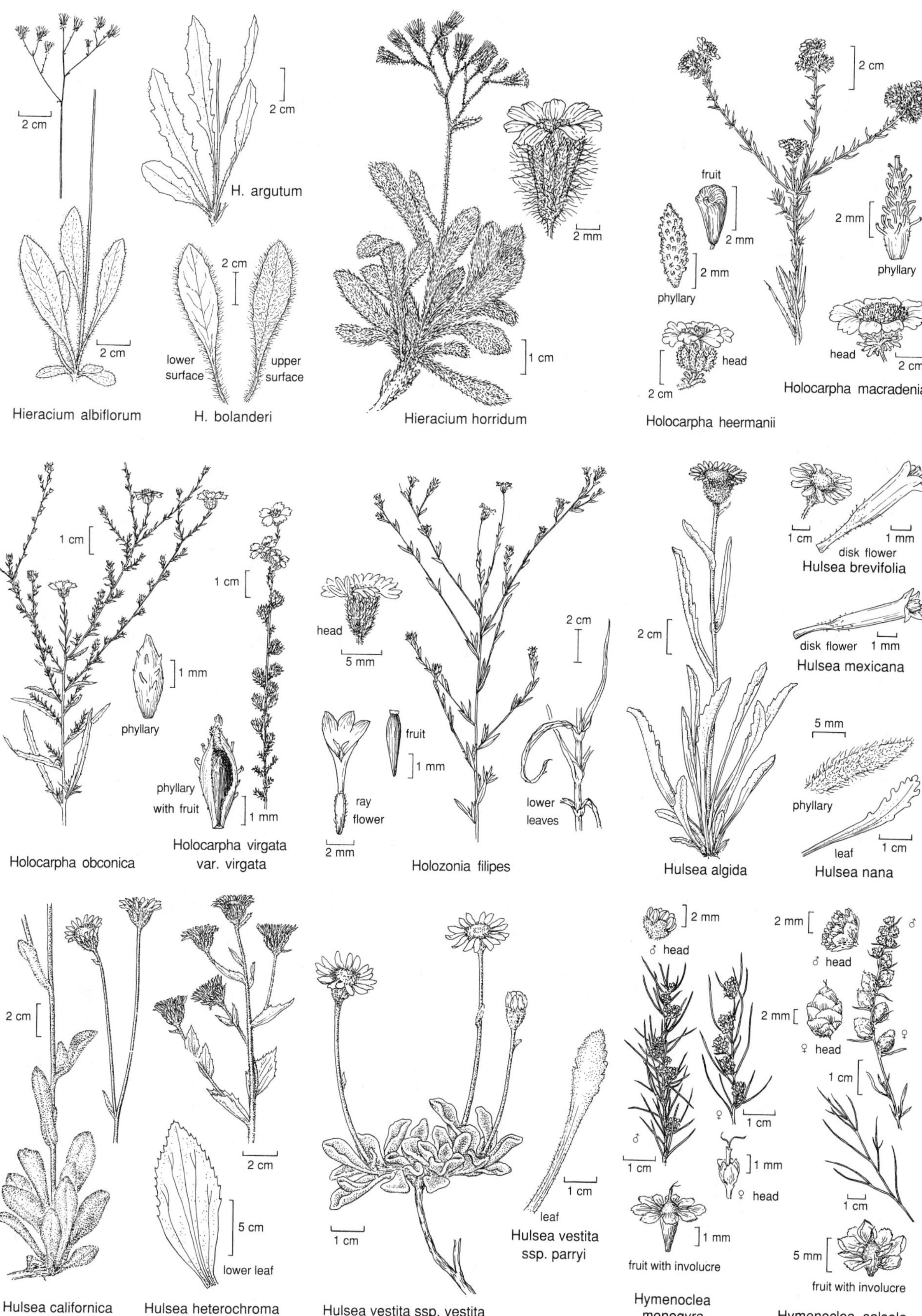

2 cm
2 cm
H. argutum
2 cm
2 cm
lower surface
upper surface
Hieracium albiflorum
H. bolanderi
2 mm
1 cm
Hieracium horridum
2 cm
fruit
2 mm
2 mm
phyllary
2 mm
phyllary
head
2 cm
head
2 cm
Holocarpha macradenia
Holocarpha heermanii
1 cm
1 cm
1 mm
phyllary
phyllary with fruit
1 mm
Holocarpha obconica
Holocarpha virgata var. virgata
head
5 mm
2 cm
fruit
1 mm
ray flower
2 mm
lower leaves
Holozonia filipes
2 cm
Hulsea algida
1 cm
1 mm
disk flower
Hulsea brevifolia
disk flower
1 mm
Hulsea mexicana
5 mm
phyllary
leaf
1 cm
Hulsea nana
2 cm
2 cm
5 cm
lower leaf
Hulsea californica
Hulsea heterochroma
1 cm
1 cm
leaf
Hulsea vestita ssp. parryi
Hulsea vestita ssp. vestita
2 mm
♂ head
2 mm
♂ head
2 mm
♀ head
♂
♀
1 cm
1 cm
♀
♂
1 cm
1 mm
♀ head
1 mm
fruit with involucre
1 cm
5 mm
fruit with involucre
Hymenoclea monogyra
Hymenoclea salsola

2–3 mm, obovate, entire or awn-tipped. $2n=30,60$. Dry, rocky slopes; 1400–2800 m. SnJt, DMtns; to WY, Colorado, AZ.

H. cooperi (A. Gray) Cockerell Bien or short-lived per 15–90 cm, lightly canescent with soft, white, jointed hairs, ± glabrous in age. **ST** gen branched above middle. **LVS** basal and cauline, 4–9 cm, divided into 1–2 mm wide linear lobes. **INFL**: heads in a ± flat-topped cyme; peduncles > 10 cm; involucre gen 9–10 mm diam; outer phyllaries fused at base, 5–7 mm, < or = inner series. **RAY FLS** 8–14; ligules 10–13 mm. **DISK FLS**: corolla 3.5–4.5 mm. **FR** ± 3 mm; pappus scales, 1–2 mm, ovate, entire or short-awn-tipped. $2n=30$. Dry, rocky slopes; 1200–2000 m. DMtns; NV, AZ.

H. lemmonii (E. Greene) Cockerell (p. 297) Bien or per 4–50 cm, persistently tomentose or glabrous in age. **STS** often > 1, gen branched above middle. **LVS** basal and cauline, 2–9 cm, divided into 1–5 mm wide linear lobes. **INFL**: heads in a ± flat-topped cyme; peduncles gen < 8 cm; involucre gen 15 mm diam or more; outer phyllaries fused at base, 5–7 mm, < or = inner series. **RAY FLS** 8–14; ligules 10–15 mm. **DISK FLS**: corolla 3.5–4.5 mm. **FR** ± 3 mm; pappus scales, 1–2 mm, lanceolate to ovate, entire. $2n=30$. Meadows or dry, rocky slopes; 700–3500 m. e KR, GB; se OR, s ID, NV, UT. The dwarf, subalpine form has been called *H. cooperi* (A. Gray) Cockerell var. *canescens* (D. Eaton) K. Parker.

H. odorata DC. Ann 15–60 cm, soft-hairy, ± glabrous in age. **STS** branched above. **LVS**: basal gen 0 at time of fl; cauline 1–5 cm, divided into linear lobes 1 mm or less wide. **INFL**: heads in flat-topped cyme; peduncles 2–15 cm; involucre gen 5–7 mm diam; outer phyllaries 3–5 mm, < inner phyllaries, thickened and fused at base. **RAY FLS** ± 8; ligules gen 1 cm or less, yellow, fading to cream-colored and ± persistent on fr. **DISK FLS**: corolla 3–4 mm. **FR** ± 2 mm; pappus scales 2 mm, lanceolate. $2n=11,12,14,15$. Sandy flats near Colorado River; < 150 m. e DSon; to UT, TX, n Mex. TOXIC range pl outside CA.

HYPOCHAERIS CAT'S-EAR

G. Ledyard Stebbins

Ann, per; sap milky. **STS** erect, 1–8 dm. **LVS** basal, entire or lobed. **INFL**: heads ligulate, 1–few, in ± flat-topped cyme; phyllaries few, overlapping in a graduated series; receptacle chaffy with thin scales. **FLS** gen many; ligules yellow. **FR** long-beaked or outer cylindric, beakless; pappus of stiff, plumose bristles, or outer shorter and barbed. ± 50 spp.: Eurasia, n Afr, S.Am. (Greek: name from Theophrastus)

1. Ann, glabrous; outer fr cylindric, beakless, inner beaked .... ***H. glabra***
1′ Per, rough-hairy; fr all beaked .... ***H. radicata***

H. glabra L. (p. 297) SMOOTH CAT'S-EAR Ann, glabrous. **STS** 1–many from slender taproot, 1–4 dm. **LVS** 2–10 cm, entire or shallowly lobed. **INFL**: involucre 12–16 mm. **FLS**: ligules short, inconspicuous. **FR**: outer achenes 4 mm, cylindric, beakless, pappus short and spreading; inner achenes longer, beak = body, pappus ± 9 mm, tawny or dull white. $2n=10,12$. Common. Weed of disturbed places; < 1200 m. CA-FP; native to Eur.

H. radicata L. ROUGH CAT'S-EAR Per, rough-hairy. **STS** several from fleshy per root, 4–8 dm. **LVS** 6–14 cm, toothed or lobed. **INFL**: involucre 12–16 mm. **FLS**: ligules large, conspicuous. **FR**: achene body 3–4 mm, < or = slender beak; pappus 9–10 mm, tawny or dull white. $2n=8$. Abundant. Pernicious weed of lawns, disturbed places, esp n CA and near coast; < 500 m. NW, CaRF, n SNF, ScV, CW, SCo; native to Eur.

ISOCOMA GOLDENBUSH

Meredith A. Lane

Subshrubs, glabrous to scabrous or hairy. **STS** prostrate or decumbent to ascending or erect, < 3 m, ± striate below, minutely scabrous, yellow-white or gray to red-brown. **LVS** alternate, sometimes clustered in axils, entire or toothed, gland-dotted, sometimes gummy, glabrous, minutely scabrous, or tomentose, light to dark gray-green. **INFL**: heads discoid, in loose to tight clusters, these borne in flat-topped or ± spheric cymes; involucres obconic; phyllaries yellow-white below, texture cartilage-like, tips green; receptacles flat, naked. **FLS** yellow; tubes narrowly cylindric, abruptly expanded into a larger cylindric throat; sinuses shallow, lobes erect; style branch appendages triangular. **FR** narrowly obconic, light tan, silky-hairy; hairs white, yellow, tan or light red-tan; pappus of 1–2 series of white, yellowish, or red-tan bristles ± 2 × fr. ± 10 spp.: sw N.Am, Mex. (Greek: equal hair-tuft, from fls) [Nesom 1991 Phytologia 70:69–114]

1. Phyllaries oblong, tips blunt to rounded, green < 1/4 total length of phyllary, often swollen and wart-like .... ***I. acradenia***
  2. Lvs conspicuously reduced upward, grading into bracts below heads, these grading into phyllaries; fls 10–17 .... var. ***bracteosa***
  2′ Lvs uniform throughout, or only slightly reduced upward, not grading into bracts below heads; fls 6–14
    3. Lvs narrowly oblanceolate, entire; involucres 4–5 mm .... var. ***acradenia***
    3′ Lvs narrowly obovate, with long-tapering base, to obovate, with 4–6 teeth per side; involucres 5–7 mm .... var. ***eremophila***

1′ Phyllaries lanceolate, tips acute to acuminate, green 1/3–1/2 total length of phyllary, flat
  4. Involucres 4.5–7 mm, 2.5–5 mm diam; upper sts yellow-white .... ***I. arguta***
  4′ Involucres 8–10 mm, 5–8 mm diam; upper sts gray, green, tan, or red .... ***I. menziesii***
    5. Pl prostrate; lvs widely obovate with a long-tapering base, margins toothed, fleshy; exposed areas on beaches, dunes, sea-cliffs .... var. ***sedoides***
    5′ Pl decumbent to erect; lvs linear, (ob)ovate or narrowly spoon-shaped, not fleshy; inland, or protected areas on coast
      6. Sts scabrous to tomentose; lvs gen not clustered in axils, green to gray-green, not closely spaced; st readily visible .... var. ***menziesii***
      6′ Sts glabrous to scabrous; lvs clustered, dark green, often so closely spaced that the st is obscured .... var. ***vernonioides***

I. acradenia (E. Greene) E. Greene Pl < 1.3 m, rounded or open. **STS** erect or ascending, branched from ground or above, glabrous or minutely scabrous, yellow-white, varnished, shining, becoming yellow-tan or gray with age. **LVS** 1.5–6 cm, 1.5–15 mm wide, linear, obovate or spoon-shaped, entire or toothed, gland-dotted, glabrous or minutely scabrous, gen light gray-green. **INFL**: heads in loose to tight clusters of 4–5; involucre 4–5 mm, 4–5 mm diam; phyllaries 22–36 in 3–6 series, oblong, tips blunt, rounded, or acute, green or tan to 1/4 total length of phyllary, swollen by glandular exudate below surface, appearing wart-like. **FLS** 6–17. **FR** 2–3.5 mm; pappus 3–5.5 mm, white-yellow, bristles unequal. $2n$=12. Sandy or clay soils in alkaline or gypsum flats or slopes; < 1300 m. Teh, SnJV, SCoR, SnBr, D; NV, AZ, Baja CA. [*Haplopappus a.* (E. Greene) S.F. Blake] Vars. intergrade somewhat.

var. ***acradenia*** (p. 297) **STS** < 0.8 m. **LVS** < 5 cm, not much-reduced above, entire. **INFL**: involucre not closely subtended by bracts; phyllaries 22–28 in 3–4 series, tips blunt, sometimes abruptly soft-pointed. **FLS** 6–12. Alkaline soils; < 1000 m. SnJV, SCoR, SnBr, DMoj.

var. ***bracteosa*** (E. Greene) G. Nesom (p. 297) **STS** < 0.9 m. **LVS** < 5 cm, much-reduced above, grading into bracts, sometimes toothed. **INFL**: involucre closely subtended by bracts grading into phyllaries; phyllaries 25–36 in 4–6 series, tips blunt to acute (inner ones), abruptly soft-pointed. **FLS** 10–17. Sandy, alkaline soils; < 900 m. SnJV, SCoRI, nw DMoj. [*Haplopappus a.* ssp. *b.* (E. Greene) H.M. Hall]

var. ***eremophila*** (E. Greene) G. Nesom (p. 297) **STS** < 1.3 m. **LVS** < 6 cm, not much-reduced above; margins with 4–6 shallow to deep, abruptly soft-pointed teeth per side. **INFL**: involucre not closely subtended by bracts; phyllaries < 28 in 3–5 series, tips widely rounded. **FLS** 7–14. Alkali or gypsum silt on flats or slopes; < 1300 m. Teh, SCoRI, D; AZ, NV, Baja CA. [*Haplopappus a.* ssp. *e.* (E. Greene) H.M. Hall]

I. arguta E. Greene (p. 297) CARQUINEZ GOLDENBUSH Pl < 0.5 m, < 0.5 m diam. **STS** erect to slightly decumbent, branched at base, scabrous, sometimes with scattered, long hairs. **LVS** not clustered in axils, not fleshy, < 2 cm, linear to oblong, glabrous, light gray-green. **INFL**: involucres 4.5–7 mm, 2.5–5 mm diam; phyllaries 20–24 in 3–4 series. **FLS** 12–18. RARE. Alkaline soils, flats, lower hills; < 20 m. Deltaic ScV (Suisun Slough). [*Haplopappus venetus* (Kunth) S.F. Blake ssp. *a.* (E. Greene) Keck] Perhaps a hybrid between *I. menziesii* and *I. acradenia*.

I. menziesii (Hook. & Arn.) G. Nesom Pl < 2 m, mat-forming, rounded or open. **STS** prostrate to erect, branched from base or rarely above, glabrous, minutely scabrous, or tomentose, sometimes with stalked glands, yellow-tan, gray, gray-green, or red-brown. **LVS** 0.7–4.5 cm, 5–15 mm wide, (ob)ovate to widely spoon-shaped, entire or toothed, gland-dotted or sometimes with stalked glands, glabrous, minutely scabrous, or tomentose, gray-green. **INFL**: heads in loose to tight clusters of 4–10, these arranged variously; involucres 4.5–10 mm, 2.5–8 mm diam; phyllaries 20–40 in 3–6 series, lanceolate, tips acute, abruptly soft-pointed, green to 1/3–1/2 total length of phyllary, flat. **FLS** 12–22. **FR** < 3.5 mm; pappus 3–5 mm, white to tan-white. $2n$=12. Sandy soils, coastal and inland; < 1200 m. sw ScV, CW, SCo, ChI, WTR, PR; Baja CA. [*I. veneta* (Kunth) E. Greene; *Haplopappus v.* (Kunth.) S.F. Blake] Here treated as 3 strongly intergrading vars.

var. ***menziesii*** (p. 297) Pls < 2 m, 1.5–2.3 m diam. **STS** erect, decumbent or prostrate, branched at base, glabrous, scabrous, or tomentose. **LVS** clustered in axils or not, not fleshy, < 4.5 cm, linear to spoon-shaped, glabrous, scabrous, or tomentose. **INFL**: involucres 8–10 mm, 5–8 mm diam; phyllaries 20–32 in 3–6 series. **FLS** 12–22. Soils from sandstone or granite, landward side of dunes, hillsides, arroyos; < 1200 m. SCo, s ChI, PR; Baja CA. [ssp. *oxyphyllus* (E. Greene) H.M. Hall, *H. venetus.* ssp. *furfuraceus* (E. Greene) H.M. Hall]

var. ***sedoides*** (E. Greene) G. Nesom (p. 297) Pls < 0.5 m, 4–12 m diam. **STS** prostrate, branched at base, glabrous or scabrous. **LVS** gen not clustered in axils, fleshy, < 4 cm, oblong to widely spoon-shaped, glabrous, scabrous or sometimes tomentose. **INFL**: involucres 8–10 mm, 5–8 mm diam; phyllaries 24–40 in 3–6 series. **FLS** 12–22. Exposed places on beaches, headlands, sea-cliffs; < 70 m. CCo, SCo, n ChI; Baja CA. [*H. venetus.* var. *s.* (E. Greene) Munz] ❀SUN,DRN,IRR:5,14,**15–17**,19–21,**22–24**;GRCVR.

var. ***vernonioides*** (Nutt.) G. Nesom (p. 297) Pls < 1.2 m. **STS** erect to decumbent, branched at base, glabrous or scabrous. **LVS** clustered in axils, often so dense as to obscure the st, not fleshy, < 4.5 cm, obovate, ovate, or oblong, glabrous or scabrous. **INFL**: involucres 8–10 mm, 5–8 mm diam; phyllaries 20–40 in 3–6 series. **FLS** 12–22. Protected sites on dunes, lagoon shores, marshes; < 400 m. CW, sw ScV, SCo, s ChI; n Baja CA. [*H. venetus* ssp. *veronioides* (Nutt.) H.M. Hall] ❀SUN,DRN,IRR:5,14,**15–24**.

IVA WORMWOOD

Willard W. Payne

Ann to shrub. **LVS** gen opposite below, ± alternate above. **INFL**: heads many, disciform; involucre cup-shaped, shallow; phyllaries few, free or fused; receptacle chaffy. **PISTILLATE FLS** 0–6; corolla 0 or cylindric. **STAMINATE FLS** 5–20; corolla translucent; anthers free, yellow; style tip truncate or tack-shaped. **FR** obovate, ± compressed; pappus 0. (Latin: from mint *Ajuga iva*, with similar aroma) [Jackson 1960 Univ KS Sci Bull 41:793–876]

1. Ann .... ***I. nevadensis***
1′ Per or shrub
  2. Phyllaries ± fused .... ***I. axillaris*** ssp. ***robustior***
  2′ Phyllaries free
    3. Lower lvs pinnately lobed, tips sharp; fr densely long-soft-hairy .... ***I. acerosa***
    3′ Lvs simple, tips blunt; fr glabrous .... ***I. hayesiana***

I. acerosa (Nutt.) R. Jackson (p. 297) Shrub 1–2 m, sometimes appearing rush-like. **STS** ± green. **LVS** 5–15 cm, ± thick, ± linear, entire to pinnately linear-lobed, finely canescent. **INFL** panicle-like; head erect; involucre 3–4 mm; phyllaries free; chaff scales ± cut, tips ± wider. **PISTILLATE FLS**: corolla 0. **FR** plump, densely long-soft-hairy. $2n$=36. Saline soils; < 700 m. n DMoj; to Colorado, NM. [*Oxytenia a.* Nutt.; Bolick 1983 Adv Clad 2:125–141]

I. axillaris Pursh ssp. ***robustior*** (Hook.) Bassett (p. 297) POVERTY WEED Per 1–6 dm, from rhizome. **STS** green to brown. **LVS** 1–4 cm, linear to obovate, entire, subsessile, ± red-gland-dotted. **INFL** raceme-like; heads short-peduncled, nodding; involucre 5–6 mm; phyllaries ± fused; chaff scales linear. **PISTILLATE FLS**: corolla vestigial. **FR** plump, glabrous. $2n$=36,54. Common. Many (saline) habitats; < 2500 m. CA; to B.C., MT, c US, TX.

I. hayesiana A. Gray (p. 297) SAN DIEGO MARSH-ELDER Per or subshrub < 1 m. **STS** ± brown. **LVS** 3–6 cm, ± oblanceolate, fleshy, entire, 3-veined from base, ± gland-dotted, gen short-appressed-hairy, reduced upward. **INFL** raceme-like; heads short-peduncled, nodding; involucre 5–6 mm; phyllaries free, (ob)ovate; chaff scales linear. **PISTILLATE FLS**: corolla 0. **FR** plump, glabrous. RARE. Alkaline flats; < 500 m. s SCo (sw San Diego Co.); Baja CA. In cult.

I. nevadensis M.E. Jones (p. 297) Ann < 3 dm, from stout taproot, ill-scented. **STS** yellow. **LVS** 8–18 mm, 2-pinnately dissected (segments ± obtuse), ± glandular and white-strigose. **INFL**: small cymes, leafy-bracted; heads erect; involucre 2–3 mm; phyllaries gen 3, free outer chaff scales wide, inner linear. **PISTILLATEFLS**: corolla short. **FR** glabrous; inner face flat or concave. Alkaline, sandy plains; 1000–2000 m. nw W&I; w NV. [Bolick 1983 Advances in Cladistics 2:125–141]

JAUMEA

Per from stolons and slender rhizomes; herbage glabrous. **STS** prostrate to ascending, rooting below. **LVS** simple, opposite, sessile; blades narrow, entire, fleshy. **INFL**: heads radiate or discoid, gen solitary (–3); peduncles bractless; involucre cylindric or ovoid; phyllaries overlapping in 3–5 series, tightly appressed, unequal, inner longest; receptacle conic, naked. **RAY FLS** 0–few; ligules yellow. **DISK FLS** many; corollas yellow; anther tips narrowly triangular; style tips short-triangular. **FR** ± cylindric, angled, glabrous; pappus 0 or a crown of short bristles. 2 spp.: w coasts of N.Am, s S.Am. (J.H. Jaume St. Hilaire, French botanist, 1772–1845)

J. carnosa (Less.) A. Gray (p. 301) **STS** long, weak. **LVS** gen 1.5–5 cm, linear to narrowly oblong-oblanceolate; bases fused 1–2 mm around st, tips ± obtuse. **INFL**: heads radiate, 1.2–2 cm (incl fls); peduncles 1–2 cm, enlarged below head; involucre 5–12 mm diam; phyllaries ovate, gen ± purple, tips rounded. **RAY FLS**: ligules 3–5 mm, narrow. **DISK FLS**: corollas 6–7 mm. **FR** 2–3 mm, brown; pappus 0 or bristles 1–5, < or = 0.5 mm. $2n=38$. Coastal salt marshes, bases of sea cliffs; 0–5 m. NCo, CCo, SCo; to B.C., n Baja CA. ❀WET,SUN:**4,5**,8,9,14,15,**16,17**,20,21,**22-24**;GRNCVR;stbl.

LACTUCA LETTUCE

G. Ledyard Stebbins

Ann to per; sap milky. **STS** erect, 0.5–1.5+ m. **LVS** basal and cauline, alternate, entire to pinnately lobed. **INFL**: heads ligulate, many, in panicles; involucre cylindric; phyllaries in 2–several series; receptacle naked. **FLS** few–many; ligules yellow or cream-colored to blue, readily withering. **FR** flattened, short- or long-beaked; pappus of many bristles, falling separately. ± 100 spp.: ± worldwide temp. (Latin: milky)

1. Fr beak short, thick; bien or per
  2. Bien from short taproot; heads in fl 1–1.5 cm wide; NCo, NCoRO, SnFrB; alien ..... ***L. biennis***
  2′ Per from long, deep rhizome; heads in fl 2–3 cm wide; CaRH, GB; native ..... ***L. tatarica***

1′ Fr beak slender, = or > body; ann or bien
  3. Fr with 1 median rib on each face
    4. Heads 10–15 mm; fls 13–25 ..... ***L. canadensis***
    4′ Heads 15–22 mm; fls 20–56 ..... ***L. ludoviciana***
  3′ Fr with several veins on each face
    5. Bien; basal rosette lvs toothed or shallowly lobed; cauline lvs widely clasping ..... ***L. virosa***
    5′ Ann; basal lvs few; cauline lvs with narrow, acute, basal lobes
      6. Sts arching upward, 5–10 dm; lvs linear-lanceolate with narrow lobes, margins entire or few-toothed; peduncles often appressed to axis; fls 8–12 ..... ***L. saligna***
      6′ Sts erect, 6–15 dm; lvs oblong-elliptic in outline, margins prickly-toothed; panicle branches widely spreading; fls 14–20 ..... ***L. serriola***

L. biennis (Moench) Fern. (p. 301) Bien from taproot, 0.5–4 m. **LVS** many, ovate to elliptic in outline, irregularly lobed and coarsely toothed; base clasping. **INFL** large, dense, compound; heads in fl 10–15 mm wide; involucre in fr 8–10 mm. **FLS** > 15; corolla pale blue to cream-colored. **FR** 5–7 mm, oblong, thickish, mottled, ± 3-veined on each face, unwinged, tapered to a very short beak; pappus brownish. $2n=34$. Locally established; gen < 500 m. NCo, NCoRO, SnFrB; native to e US.

L. canadensis L. Mostly bien from taproot, 0.5–2.5 m. **LVS** many, elliptic in outline, lobed; lobes linear, curved, mostly entire; bases acute, clasping. **INFL** open, heads in fl 5–10 mm wide; involucre in fr 10–15 mm. **FLS** 13–25; corolla pale yellow. **FR** 5–6 mm, elliptic, blackish, rough, with 1 rib on each face, unwinged; beak pale, ± = body; pappus 5–7 mm, white. $2n=34$. Forests and waste places; < 1050 m. KR, n SN; native to e N.Am.

L. ludoviciana (Nutt.) DC. Bien 1–2 m. **LVS** lanceolate in outline, lobed; lobes widely lanceolate, dentate. **INFL** open, heads in fl 4–5 mm wide; involucre in fr 15–22 mm. **FLS** 20–56; corolla blue to ± yellow. **FR** 3–4 mm, elliptic, mottled gray and black, transversely roughened, 1-ribbed on each face, unwinged; beak slender, ± = body; pappus 7–12 mm, white. $2n=34$. Waste places; < 300 m. CaRF, SCo; native to c US.

L. saligna L. (p. 301) Ann from taproot, 0.5–1(2) m. **STS** decumbent or erect below, glabrous or bristly below. **LVS** few–many, linear to lanceolate, entire or with linear lobes, clasping, glabrous or prickly-bristly on midvein. **INFL** narrow, spike-like; branches and peduncles often appressed; heads in fl 4–5 cm wide; involucre in fr 10–17 mm. **FLS** 8–12; corolla pale yellow. **FR** 7–9 mm (incl beak), brown, dark-mottled, rough, glabrous, several-veined on each face, unwinged; beak = or > body; pappus ± 6 mm, white. $2n=18$. Waste, mostly urban places; < 750 m. NCoR, s ScV, CCo, SnFrB, SCoRO; native to Eur.

L. serriola L. (p. 301) PRICKLY LETTUCE Ann from taproot, 0.5–1.5 m. **STS** erect, prickly-bristly. **LVS** few–many, oblanceolate to obovate, dentate to coarsely lobed, clasping, prickly-bristly on midvein. **INFL** open; branches often widely spreading; heads in fl 4–6 mm wide; involucre in fr 10–12 mm. **FLS** 14–20; corolla pale yellow. **FR** 6–7 mm (incl beak), light to dark brown, rough-hairy, several-veined on each face, unwinged, beak = or > body; pappus 4–5 mm, white. $2n=18$. Abundant. Weed of disturbed places; < 2000 m. CA; native to Eur. [var. *integra* Gren. & Godron]

L. tatarica (L.) C. Meyer ssp. ***pulchella*** (Pursh) Stebb. (p. 301) Per from extensive rhizomes, 3–10 dm. **STS** erect, glabrous. **LVS** many, 5–15 cm, linear to elliptic, entire to lobed. **INFL**: heads in fl 2–3 cm wide; involucre in fr 1.4–1.8 mm. **FLS** 10–15; corolla bright blue, conspicuous. **FR** 5–6 mm (incl beak), pale brown, glabrous, several-veined on each face, unwinged; beak very short; pappus 8–11 mm, white. $2n=18$. Dry to moist alluvial valleys; 1150–2000 m. CaRH, GB; to AK, e Can, c US, NM. ❀TRY.

L. virosa L. (p. 301) Bien from taproot, 0.6–2 m. **STS** erect, arising from a rosette, glabrous or bristly below. **LVS**: basal obovate, toothed or shallowly lobed; cauline many, upper widely clasping, prickly-bristly on midvein. **INFL**: cyme, compound; heads many, in fl 3–4 cm wide; involucre in fr 8–12 mm. **FLS** 15–20; corolla pale yellow. **FR** many-ribbed, rough; body dark, wing-margined; beak white, ± = body; pappus 8 mm, white. $2n=18$. Disturbed, shrubby and wooded slopes; < 300 m. SnFrB, SCoRO; native to Eur.

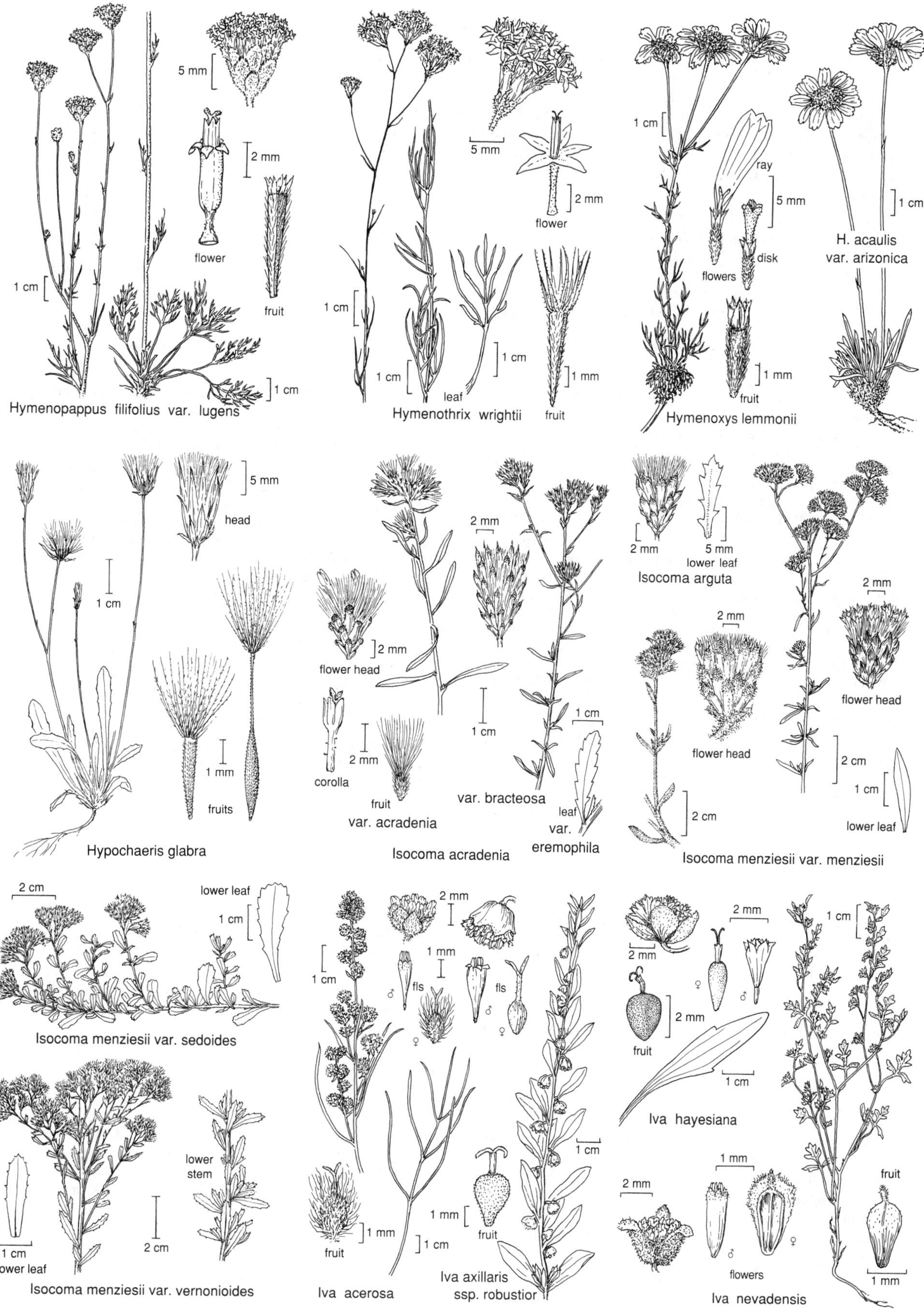
5 mm
2 mm
flower
fruit
1 cm
1 cm
Hymenopappus filifolius var. lugens
5 mm
2 mm
flower
1 cm
1 cm
1 cm
leaf
1 mm
fruit
Hymenothrix wrightii
1 cm
ray
5 mm
disk
flowers
1 mm
fruit
Hymenoxys lemmonii
1 cm
H. acaulis
var. arizonica
5 mm
head
1 cm
1 mm
fruits
Hypochaeris glabra
2 mm
2 mm
flower head
1 cm
2 mm
corolla
fruit
var. acradenia
var. bracteosa
1 cm
leaf
var.
eremophila
Isocoma acradenia
2 mm
5 mm
lower leaf
Isocoma arguta
2 mm
2 mm
flower head
flower head
2 cm
2 cm
1 cm
lower leaf
Isocoma menziesii var. menziesii
2 cm
lower leaf
1 cm
Isocoma menziesii var. sedoides
lower
stem
2 cm
1 cm
lower leaf
Isocoma menziesii var. vernonioides
2 mm
1 mm
1 cm
fls
fls
♂
♂
♀
♀
1 mm
fruit
1 cm
1 mm
fruit
1 cm
Iva acerosa
Iva axillaris
ssp. robustior
2 mm
2 mm
♀
♂
2 mm
fruit
1 cm
Iva hayesiana
1 cm
1 mm
2 mm
♂
♀
flowers
fruit
1 mm
Iva nevadensis

LAGOPHYLLA

Ann 1–15 dm, green to canescent, gen self-sterile. **LVS** alternate (or lower opposite), linear to oblanceolate, entire to toothed; lower soon deciduous; upper ± bract-like. **INFL** spike- to panicle-like; heads radiate, opening in evening, gen closing by ± midday, ± short-peduncled; phyllaries fully folded around ray ovaries, falling with frs, tips flat, erect or spreading; receptacle flat, short-hairy; chaff scales 5 in ring between ray and disk fls, weakly fused. **RAY FLS** 5; ligule 3-lobed, yellow, red-veined, often aging purplish. **DISK FLS** 6, staminate; corolla yellow; anthers black, tips widely triangular; ovaries slender, style branches undivided. **FR**: pappus 0; ray achenes compressed front-to-back. 4 spp.: CA. (Greek: hare lf, from copious lf hairs) [Thompson 1983 PhD Univ CA Davis]

1. Ligules 3–5.5 mm, pale yellow .... ***L. ramosissima***
  2. Heads in dense, leafy-bracted clusters; branches 0–few, short, stout .... ssp. ***congesta***
  2′ Heads in open clusters or few at st tips; branches gen many, long, slender .... ssp. ***ramosissima***

1′ Ligules 8–13 mm, bright yellow
  3. Infl spike- to panicle-like .... ***L. glandulosa***
  3′ Infl open, gen forking cymes
    4. Fr obovate, dull, many-veined, midvein obscure; phyllaries short-curved-ciliate; SNF, e SnJV, SCoRI .... ***L. dichotoma***
    4′ Fr oblanceolate, glossy, midvein evident; phyllaries long-straight-ciliate; NCoRI, n SNF (El Dorado Co.) .... ***L. minor***

L. dichotoma Benth. (p. 301) **STS** repeatedly forked, strigose, dark purple-brown. **LVS** 2–5.5 cm, toothed (esp basal); upper finely ciliate. **INFL** nearly glandless; heads in st forks; involucre hemispheric; phyllaries 6–8 mm, short-hairy, curved-ciliate, tips narrowly triangular, ± = body. **RAY FLS**: ligules 8–13 mm, bright yellow. **DISK FLS**: corollas 3–4 mm. **FR** 3–3.5 mm, obovate, finely striate, dull, dark or mottled; midvein obscure. $2n=14$. Grassland, woodland; < 800 m. n SCoRI, e SnJV, w SNF. Fls ± early summer.

L. glandulosa A. Gray (p. 301) **STS** 0–many-branched, ± sparsely hairy, becoming ± glabrous, straw-colored or purple-tinged. **LVS**: lower 3–12 cm, toothed, soon deciduous; middle entire, deciduous, with axillary lf clusters; upper densely canescent, ± glandular. **INFL** spike- to panicle-like, glandular; involucre obconic; phyllaries 5–7 mm, short-hairy, long-ciliate, tip narrowly triangular, < or = body. **RAY FLS**: ligules 8–13 mm, bright yellow. **DISK FLS**: corollas 4–4.5 mm. **FR** 2.7–3.9 mm, narrowly obovate, dark brown, smooth; midvein evident. $2n=14$. Grassland, chaparral, woodland; < 850 m. c NCoRI, CaRF, SNF, n GV. [ssp. *serrata* (E. Greene) Keck] Fls ± late summer.

L. minor (Keck) Keck (p. 301) **STS** repeatedly forked, often spreading, strigose, dark purple-brown. **LVS** 2–5.5 cm, linear; basal toothed; cauline entire, minutely strigose or short-rough-hairy; upper finely long-ciliate. **INFL** ± nonglandular; heads borne in st forks; involucre hemispheric; phyllaries 4–5 mm, acuminate, soft-hairy, very-long-ciliate. **RAY FLS**: ligules 8–13 mm, bright yellow. **DISK FLS**: corollas 3–3.5 mm. **FR** 2.1–2.7 mm, oblanceolate, glossy black; midvein evident. $2n=14$. Foothill woodland, often on serpentine; < 700 m. NCoRI, n SNF (El Dorado Co.). Fls spring.

L. ramosissima Nutt. **STS** 0–many-branched, soft-hairy, often becoming glabrous. **LVS**: lower 3–12 cm, toothed, early deciduous; middle entire, deciduous, with axillary lf clusters; upper bract-like, densely canescent to silvery-hairy, long-ciliate, uppermost glandular. **INFL** head- to panicle-like, ± glandular; heads gen closing early morning; involucre obconic; phyllaries 4.4–7.5 mm, lanceolate, short-hairy, densely long-ciliate, tip < body. **RAY FLS**: ligules 3–5.5 mm, pale yellow. **DISK FLS**: corollas 3.5–4 mm. **FR** 2.5–4 mm, narrowly obovate, dull, ± black; midvein obscure. $2n=14$. Many dry habitats; < 1600 m. CA-FP, MP; to WA, ID, MT, n NV. Fls spring to autumn. Self-fertile.

ssp. ***congesta*** (E. Greene) Keck (p. 301) **STS** gen stout; branches 0–few, short, densely leafy. **INFL** dense, leafy-bracted; clusters head-like. Habitats of sp.; < 1400 m. NCoRI, SNF, sw ScV, e SnFrB. [*L. c.* E. Greene]

ssp. ***ramosissima*** (p. 301) **STS** gen slender; branches often many, long, slender, gen sparsely leafy. **INFL**: heads solitary or few per cluster. Habitats and range of sp.

LAPSANA NIPPLEWORT

G. Ledyard Stebbins

Ann or per; sap milky. **ST** gen 1 from base. **LVS** alternate, toothed or lobed. **INFL** panicle-like; heads ligulate; peduncles slender, subtended by scale-like bracts; involucre subcylindric; phyllaries in 2 series, outer scale-like, inner much longer, equal; receptacle naked. **FLS** 6–15; ligules yellow, soon withering. **FR** oblong; tip rounded; pappus 0. 9 spp.: Eurasia. (Greek: name used by Dioscorides)

L. communis L. Ann, ± soft-hairy or becoming glabrous. **ST** erect, 3–8 dm, leafy. **LVS** 2–5 cm, ovate, thin, petioled or upper ± sessile, dentate. **INFL**: involucre 4–5 mm; main phyllaries 8. **FR** 2.5–9 mm, outer >> inner, 20–30-veined, glabrous. $2n=12,14,16$. Uncommon. Shady places; < 1400 m. NW, n SNF, n CW; native to Eur.

LASTHENIA GOLDFIELDS

Robert Ornduff

Ann, per, glabrous or hairy. **ST** gen branched, gen erect, < 60 cm. **LVS** opposite, < 20 cm, entire to pinnately cut. **INFL**: heads radiate, solitary or in cymes; phyllaries in 1 or 2 series, free or partly fused; receptacle narrowly conic to hemispheric, naked, smooth, pitted, or rough. **RAY FLS** 4–21; ligules gen yellow. **DISK FLS** gen many; corollas gen 5-lobed, gen yellow; anther tips acuminate to triangular; style tips triangular or round, gen hair-tufted. **FR** < 5 mm, cylindric to obovoid, black or gray; pappus of awns, scales, or 0. 17 spp.: w N.Am, Chile. (Greek: female pupil of Plato) [Ornduff 1966 Univ Calif Publ Bot 40:1–92] Gen self-incompatible (cross-pollinated).

1. Phyllaries partly fused
  2. Phyllaries fused < 1/2 their length; lvs entire to pinnately cut .... ***L. conjugens***
  2′ Phyllaries fused > 2/3 their length; lvs entire

3. Pappus present ..... ***L. glaberrima***
3′ Pappus 0
4. Fr strongly flattened, margin fringed with blunt hairs ..... ***L. chrysantha***
4′ Fr round or barely flattened, glabrous or surface hairy
5. Fr hairs short, curved ..... ***L. ferrisiae***
5′ Fr glabrous or hairs wart-like ..... ***L. glabrata***
6. Fr hairs wart-like ..... ssp. ***coulteri***
6′ Fr glabrous ..... ssp. ***glabrata***
1′ Phyllaries free
7. Receptacle narrowly conic; phyllaries gen 4–6
8. Rays < 2 mm; involucre cylindric; disk corollas gen 4-lobed ..... ***L. microglossa***
8′ Rays > 2 mm; involucre obconic or bell-shaped; disk corollas gen 5-lobed
9. Sts coarse, hairy; anther tips triangular ..... ***L. debilis***
9′ Sts fine, wiry, glabrous below; anther tips awl-shaped ..... ***L. leptalea***
7′ Receptacle conic or ± spheric; phyllaries gen > 6
10. All lvs ± entire; corollas dark red in alkali solution
11. Ann; coastal or inland ..... ***L. californica***
11′ Per; coastal ..... ***L. macrantha***
12. St erect, simple to few-branched; roots fleshy; lvs gen < 2 mm wide ..... ssp. ***bakeri***
12′ St ± decumbent, gen branched; roots fibrous; lvs gen > 2 mm wide ..... ssp. ***macrantha***
10′ Middle cauline lvs gen pinnately divided; corollas yellow in alkali solution
13. Pappus of awns tapered to tip; involucre spheric; phyllaries persistent ..... ***L. platycarpha***
13′ Pappus elements alike, of 2 kinds, or 0; involucre ± spheric; phyllaries ultimately deciduous
14. Pappus 0 or elements of 1 kind (if of 2 kinds, awns 0); pl gen glandular ..... ***L. coronaria***
14′ Pappus 0 or elements of 2 kinds; glands 0
15. Fr > 2 mm
16. Pappus awns > 4; ray corollas < 3 mm ..... ***L. maritima***
16′ Pappus awns 2–3; ray corollas > 4 mm ..... ***L. minor***
15′ Fr < 1.5 mm
17. Pappus gen of 1 long awn and many short scales ..... ***L. burkei***
17′ Pappus gen of 3 or more long awns mixed with several short scales ..... ***L. fremontii***

L. burkei (E. Greene) E. Greene (p. 301) BURKE'S GOLDFIELDS Ann < 30 cm. **ST** simple or freely branched, hairy. **LVS** < 5 cm, linear, entire or pinnately lobed, glabrous to ± hairy. **INFL**: involucre 4–6 mm, dome-shaped or obconic; phyllaries 7–16, free, hairy; receptacle dome-shaped or conic, glabrous or hairy. **RAY FLS** 8–13; ligules < 6 mm. **DISK FLS** many; anther tips linear to ± ovate; style tips triangular. **FR** < 1.5 mm, club-shaped, hairy; pappus gen of 1 long awn and many short scales. $2n$=12. **ENDANGERED** CA,US. Vernal pools, wet meadows; < 500 m. s NCoRI (s Mendocino, s Lake and ne Sonoma cos.). In cult.

L. californica Lindley (p. 301) Ann < 40 cm. **ST** simple or freely branched, ± hairy. **LVS** 0.8–7 cm, linear to oblanceolate, entire, hairy, ± fleshy in coastal forms. **INFL**: involucre 5–10 mm, bell-shaped or hemispheric; phyllaries 4–13, free, hairy; receptacle conic, rough, glabrous. **RAY FLS** 6–13; ligules 5–10 mm. **DISK FLS** gen many; anther tips triangular; style tips triangular. **FR** < 3 mm, linear to ± club-shaped, glabrous or hairy; pappus of 1–7 narrow awns, wider awned scales, or 0. $2n$=16,32. Abundant. Many habitats; < 1500 m. CA-FP, w DMoj; sw OR, AZ, Mex. Highly variable; needs further study. [*L. chrysostoma* (Fischer & C. Meyer) E. Greene] ❀SUN:**7–10**,11,12,**14–24**.

L. chrysantha (A. Gray) E. Greene Ann < 28 cm. **ST** simple or freely branched, glabrous or ± hairy. **LVS** < 8 cm, linear, entire, ± glabrous. **INFL**: involucre 5–7 mm, hemispheric; phyllaries 8–14, fused, ± glabrous; receptacle conic, warty, ± glabrous. **RAY FLS** 6–10; ligules 6–7 mm. **DISK FLS** many; anther tips ovate or triangular; style tips triangular. **FR** < 3 mm, obovoid, flattened, black; margin fringed with blunt, gen curved hairs < 0.4 mm; pappus 0. $2n$=14. Vernal pools, wet saline flats; < 100 m. s ScV, SnJV. ❀TRY.

L. conjugens E. Greene (p. 301) CONTRA COSTA GOLDFIELDS Ann < 40 cm. **ST** simple or freely branched, ± glabrous. **LVS** < 8 cm, linear, entire or pinnately lobed, glabrous. **INFL**: involucre 6–10 mm, hemispheric or obconic, hairy; phyllaries 12–18, fused < 1/2 their length; receptacle dome-shaped or obconic, glabrous or hairy. **RAY FLS** 6–13; ligules 5–10 mm. **DISK FLS** many; anther tips linear to ± ovate; style tips triangular. **FR** < 1.5 mm, club-shaped, glabrous; pappus 0. $2n$=12. **ENDANGERED** CA. Vernal pools; < 100 m. Formerly NCo, s ScV, SnFrB, SCo; now ± deltaic ScV (Napa, Solano cos.).

L. coronaria (Nutt.) Ornd. (p. 301) Ann < 40 cm; herbage sweetly scented. **ST** simple or much-branched; hairs short, glandular, or long, non-glandular, or mixed. **LVS** 1.5–6 cm, linear, entire or pinnately lobed or cut, gen glandular-hairy. **INFL**: involucre 4–7 mm, obconic to hemispheric; phyllaries 6–14, free, gen glandular-hairy; receptacle conic, hairy. **RAY FLS** 6–15; ligules 3–10 mm. **DISK FLS** many; anther tips elliptic; style tips triangular or dome-shaped. **FR** < 2.5 mm, linear to narrowly club-shaped, hairy; pappus gen a mixture of lanceolate or oblong, truncate scales or 0, gen different in ray and disk frs. $2n$=8,10. UNCOMMON. Sunny, open places; < 700 m. SCo, PR, w D; nw Baja CA, Guadalupe Island, Mex. ❀TRY.

L. debilis (A. Gray) Ornd. (p. 301) Ann < 30 cm. **ST** simple or branched, ascending, hairy. **LVS** 1–8 cm, ± entire (rarely 1–2-toothed), linear to linear-oblong, hairy. **INFL**: involucre 5–7 mm, obconic to bell-shaped; phyllaries 5, free, slightly hairy; receptacle narrowly conic, with small ridges and grooves, glabrous. **RAY FLS** 5–10; ligules 3–5 mm, yellow to white. **DISK FLS** many; corolla yellow to white; anther tips triangular with 1–4 wart-like glands on face; style tips triangular with wide tuft of hairs. **FR** < 3 mm, ± linear, hairy; pappus 0 or scales < 4, lanceolate, white or brown. $2n$=8. ± shaded, moist slopes; < 500 m. c&s SNF, Teh, SCoRI.

L. ferrisiae Ornd. (p. 301) Ann < 40 cm. **ST** erect, simple or branched, glabrous or slightly hairy. **LVS** 1–8 cm, linear, entire, glabrous. **INFL**: involucre 5–10 mm, hemispheric; phyllaries 6–14, fused, glabrous; receptacle conic, papillate, glabrous. **RAY FLS** 6–13; ligules 6–10 mm. **DISK FLS** many; anther tips ovate or triangular; style tips triangular, hair-tufted. **FR** < 2.5 mm, ± flattened, obovate to oblong, sparsely to densely short-hairy and papillate; pappus 0. $2n$=14. Vernal pools or wet saline flats; < 700 m. ScV (2 stations), SnJV. Variable; believed to be derived from hybridization between *L. glabrata* and *L. chrysantha*. ❀TRY.

L. fremontii (A. Gray) E. Greene Ann < 35 cm. **ST** erect, simple or branched from base, ± hairy. **LVS** 1–6 cm, linear and entire, or with 1–3 pairs of linear lobes, glabrous to sparsely hairy. **INFL**: involucre 4–6 mm, dome-shaped or obconic; phyllaries 8–16, free, hairy; receptacle dome-shaped, glabrous or hairy. **RAY FLS** 6–13; ligules 5–7 mm. **DISK FLS** many; anther tips linear to narrowly ovate; style tips triangular, with a short tuft of hairs. **FR** < 1.5 mm, club-shaped, gen hairy; pappus of < 5 narrow awns intermixed with short scales, rarely of narrow awns only, or rarely 0. $2n$=14. Vernal pools, wet meadows; < 700 m. GV. ❀SUN:7,**8,9,14–17,19–24**.

L. glaberrima A.DC. Ann < 35 cm. **ST** sprawling or erect, simple or freely branched, glabrous. **LVS** 3–10 cm, linear, entire, glabrous. **INFL**: involucre 5–7 mm, hemispheric or bell-shaped; phyllaries 5–10, fused, hairy; receptacle conic, papillate, glabrous. **RAY FLS** 6–13; ligules very short, pale yellow. **DISK FLS** many; corollas gen 4-lobed, pale yellow or greenish; anther tips obovate or oblong, blunt; style tips short, glabrous. **FR** < 4 mm, ± linear, hairy; pappus of < 10 narrowly tapered or elliptic scales. $2n$=10. Vernal pools, wet places; < 900 m. NCoR, ScV, n SnJV, CW, MP (rare); to sw WA. Gen self-pollinated.

L. glabrata Lindley Ann < 60 cm. **ST** erect, simple or branched, glabrous or slightly hairy. **LVS** 4–15 cm, linear or awl-shaped, entire, glabrous. **INFL**: involucre 5–10 mm, hemispheric; phyllaries 10–14, fused, glabrous; receptacle conic, papillate, glabrous or sparsely hairy. **RAY FLS** 7–15; ligules 4–14 mm. **DISK FLS** many; anther tips ovate or triangular; style tips triangular, hair-tufted. **FR** < 3.5 mm, club-shaped or ovoid, glabrous or papillate; pappus 0. $2n$=14. Saline places, vernal pools; < 1000 m. NCoRI, Teh, ScV, n SnJV, SnFrB, SCoRO, SCo, n ChI (Santa Rosa Island), PR, w DMoj.

ssp. ***coulteri*** (A. Gray) Ornd. (p. 301) COULTER GOLDFIELDS **FR** warty-hairy. $2n$=14. RARE. Habitats of sp. Teh (1 station), s SCoRO, SCo, n ChI (Santa Rosa Island), PR, w DMoj. [var. *c.* A. Gray]

ssp. ***glabrata*** (p. 301) **FR** glabrous. $2n$=14. Habitats of sp.; < 550 m. NCoRI, ScV, n SnJV, SnFrB, n SCoRO. [var. *californica* Jepson] ❀SUN:7,**8,9,14–17,19–24**.

L. leptalea (A. Gray) Ornd. (p. 301) SALINAS VALLEY GOLDFIELDS Ann < 15 cm, erect. **ST** simple or branched, glabrous below. **LVS** 3–20 mm, linear, entire, sparsely hairy. **INFL**: involucre 4–6 mm, obconic to bell-shaped; phyllaries gen 4–6, hairy at tips; receptacle narrowly conic, glabrous. **RAY FLS** 6–9; ligules 2.5–5 mm. **DISK FLS** many; anther tips awl-shaped; style tips ± triangular, long hair-tufted. **FR** < 2 mm, narrowly club-shaped, sparsely hairy; pappus of < 4 awns, each narrowly tapered, white or yellowish (or 0 in some fls). $2n$=16. UNCOMMON. Open areas in woods; < 500 m. SCoR (Monterey, San Luis Obispo cos.).

L. macrantha (A. Gray) E. Greene (p. 301) Per (rarely ann) < 40 cm. **ST** simple or branched, hairy. **LVS** 2–21 cm, linear to oblong, entire, glabrous to densely hairy. **INFL**: involucre 9–14 mm, bell-shaped to depressed-hemispheric; phyllaries 13–16, free, hairy; receptacle conic, rough, gen glabrous. **RAY FLS** 8–16; ligules 5–16 mm. **DISK FLS** many; anther tips triangular to sublanceolate; style tips triangular, hair-tufted. **FR** < 4 mm, linear to narrowly club-shaped, gen glabrous; pappus of < 4 narrow awns, or 0. Grasslands, dunes, woods; < 500 m. NCo, CCo. Sspp. intergrade.

ssp. ***bakeri*** (J. Howell) Ornd. Roots fleshy, clustered. **ST** erect, simple or few-branched, < 40 cm. **LVS** gen < 2 mm wide; basal clustered. $2n$=48. Grasslands, woods, near coast; < 500 m. c&s NCo (Mendocino, Sonoma cos.). [*Baeria m.* var. *b.* (J. Howell) Keck] ❀TRY.

ssp. ***macrantha*** Roots fibrous, not clustered. **ST** gen decumbent, gen branched at base, < 40 cm. **LVS** > 2 mm wide. $2n$=48. Grasslands, dunes along immediate coast; < 500 m. NCo, CCo (2 stations). ❀DRN,SUN,IRR:5,14,**15–17**,22–24.

L. maritima (A. Gray), M.C. Vasey MARITIME GOLDFIELDS Ann < 25 cm. **ST** prostrate or decumbent, branched; nodes gen hairy. **LVS** 1–9 cm, narrowly to widely strap-shaped, blunt, fleshy, entire or variously lobed, glabrous. **INFL**: involucre 4–7 mm, hemispheric; phyllaries 6–14, free, hairy on margins and midribs; receptacle conic, glabrous, rough. **RAY FLS** 7–12; ligules gen < 3 mm. **DISK FLS** many; anther tips ± oblong, obtuse; style tips triangular or dome-shaped, often not hair-tufted. **FR** < 3.2 mm, hairy, linear to narrowly club-shaped; pappus of < 9 brownish awns and fewer narrow scales, or 0. $2n$=8. UNCOMMON. Seabird nesting and roosting areas, gen offshore rocks and islands; < 100 m. NCo, CCo; to sw B.C. [*L. minor* ssp. *maritima* (A. Gray) Ornd.] Mostly self-pollinated. ❀TRY.

L. microglossa (A.DC.) E. Greene (p. 301) Ann < 25 cm. **ST** sprawling or erect, simple or much-branched, hairy. **LVS** 1.5–8 cm, linear or awl-shaped, ± entire, hairy. **INFL**: involucre 6–8.5 mm, cylindric to narrowly obconic; phyllaries ± 4, hairy; receptacle narrowly conic, glabrous. **RAY FLS** ± 4; ligules < 1 mm or sometimes 0. **DISK FLS** few; corolla gen 4-lobed; anther tips narrowly tapered; style tips lanceolate, glabrous. **FR** < 5 mm, ± linear, hairy; pappus of < 4 scales, each lanceolate, yellowish or white, awn-tipped, or 0. $2n$=24. Shaded slopes; < 1000 m. NCoRI, ScV (1 collection), SnFrB, SCoR, TR, PR, DMoj. Mostly self-pollinated.

L. minor (A.DC.) Ornd. Ann < 35 cm. **ST** erect, simple or much-branched, sparsely to densely woolly. **LVS** 2–12 cm, linear, entire or irregularly toothed (teeth < 1.5 cm), glabrous to soft-hairy. **INFL**: involucre 4–6 mm, hemispheric, phyllaries 7–14, free, hairy; receptacle conic, glabrous, rough. **RAY FLS** < 13; ligules 4–8 mm. **DISK FLS** many; anther tips ovate or elliptic; style tips triangular or dome-shaped, hair-tufted. **FR** < 2.6 mm, narrowly club-shaped, glabrous or hairy; pappus of 2–3 narrowly tapered to lanceolate, brown or white scales intermixed with ± truncate, fringed shorter scales, or 0. $2n$=8. Grasslands; < 700 m. NCo, s ScV, SnJV, CCo. ❀DRN,SUN:5,**8,9,14–17**,19–21,**22–24**.

L. platycarpha (A. Gray) E. Greene (p. 301) Ann < 30 cm. **ST** erect, simple or branched from base, ± glabrous or hairy. **LVS** 1–6 cm, entire (or pinnate, lobes few, linear), glabrous or hairy. **INFL**: involucre 6–8 mm, obconic; phyllaries 6–9, free, glabrous or hairy; receptacle conic, rough, glabrous or sparsely hairy. **RAY FLS** 6–13; ligules 7–8 mm. **DISK FLS** many; anther tips triangular; style tips triangular, hair-tufted. **FR** < 3.5 mm, narrowly club-shaped, hairy; pappus of < 6 scales, each lanceolate to ovate, white or yellowish. $2n$=8. Alkali flats; < 120 m. ScV, n SnJV, SnFrB (very uncommon). ❀TRY.

LAYIA

Bruce G. Baldwin & Susan J. Bainbridge

Ann, gen ascending to erect, often black-glandular, gen ± purplish or brownish. **LVS** gen linear to (ob)lanceolate, sessile, gen pinnately lobed, reduced upward; lower lvs opposite. **INFL**: heads gen radiate; involucre obconic to urn-shaped; phyllaries gen folded around ray fr, falling with fr, gen ± hairy; receptacle flat to slightly convex, chaff scales free. **RAY FLS** (0)3–27; ligules white (often aging pinkish) to yellow, tubes hairy. **DISK FLS** 5–many; corollas yellow, puberulent; anther tips acute, long-tapered; style branches long, bristly. **FR** gen 2–5 mm, gen club-shaped, black; ray fr compressed back to front, gen ± glabrous, pappus 0; disk fr ± straight, gen ± hairy, pappus various. 14 spp.: w N.Am. (George T. Lay, early 19th century English pl collector) [Kyhos et al. 1990 Ann Missouri Bot Gard 77:84–95]

1. Pappus 0
  2. Pl nonglandular; disk fls each subtended by chaff scale — ligules yellow with white tips . . . [2]***L. chrysanthemoides***
  2′ Pl glandular; chaff scales in 1 involucre-like series between ray and disk fls
    3. Basal lvs gen minutely dentate to serrate — ligules white to cream-colored . . . . [2]***L. heterotricha***
    3′ Basal lvs conspicuously lobed
      4. Phyllaries bulged out at base (involucre ± urn-shaped); st gen purple-streaked
        5. Ray fls 8–18 in 1 series, ligules yellow or white-tipped; disk fr 3–4 mm; pl scented . . . . [2]***L. gaillardioides***

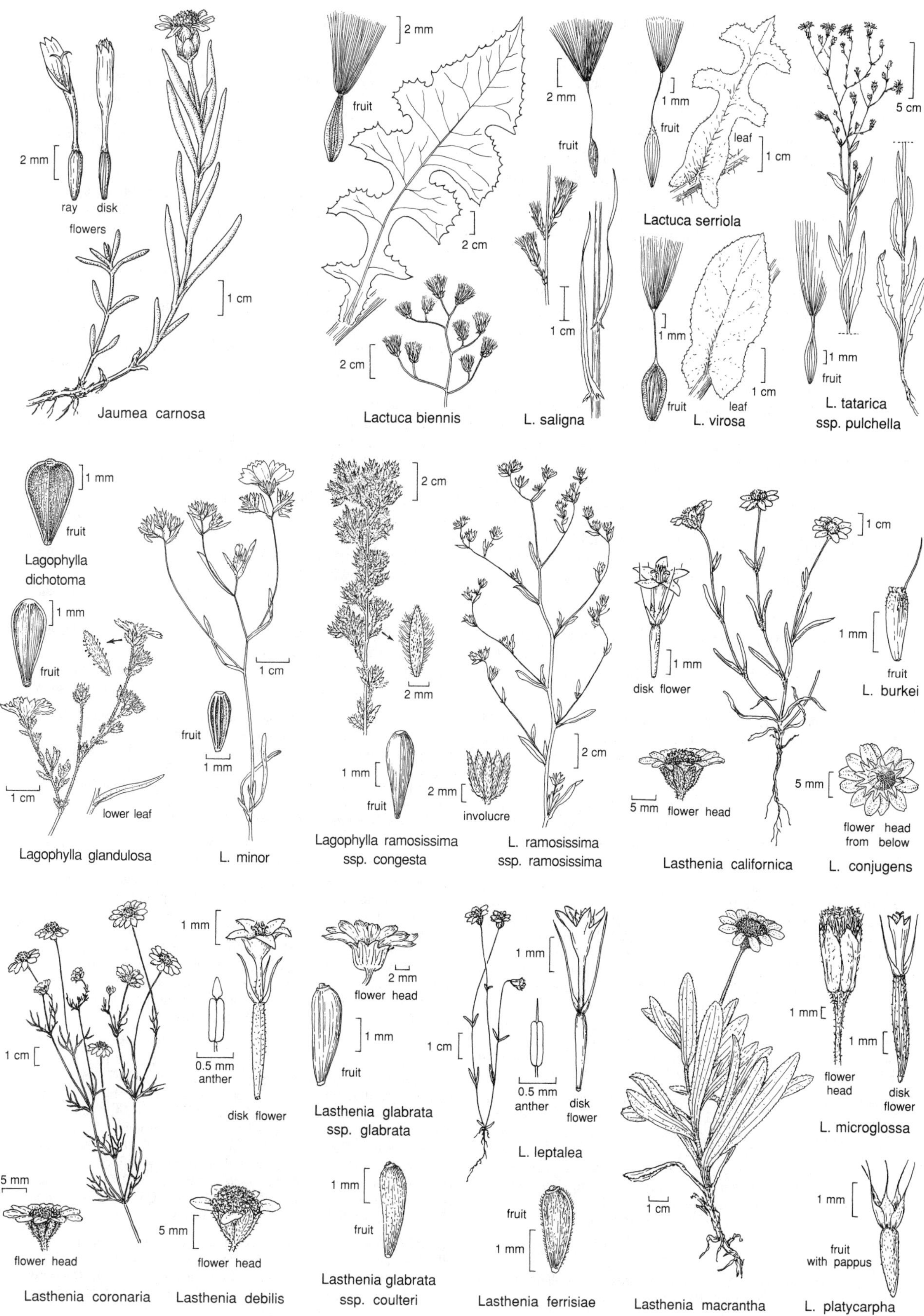
2 mm
ray disk
flowers
1 cm
Jaumea carnosa
2 mm
fruit
2 cm
2 cm
Lactuca biennis
2 mm
fruit
1 cm
L. saligna
1 mm
fruit
leaf
1 cm
Lactuca serriola
1 mm
1 cm
fruit
leaf
L. virosa
5 cm
1 mm
fruit
L. tatarica
ssp. pulchella
1 mm
fruit
Lagophylla
dichotoma
1 mm
fruit
1 cm
lower leaf
Lagophylla glandulosa
1 cm
fruit
1 mm
L. minor
2 cm
2 mm
1 mm
fruit
Lagophylla ramosissima
ssp. congesta
2 mm
involucre
2 cm
L. ramosissima
ssp. ramosissima
1 mm
disk flower
5 mm
flower head
1 cm
Lasthenia californica
1 mm
fruit
L. burkei
5 mm
flower head
from below
L. conjugens
1 mm
1 cm
0.5 mm
anther
disk flower
5 mm
flower head
Lasthenia coronaria
5 mm
flower head
Lasthenia debilis
2 mm
flower head
1 mm
fruit
Lasthenia glabrata
ssp. glabrata
1 mm
fruit
Lasthenia glabrata
ssp. coulteri
1 mm
1 cm
0.5 mm
anther
disk
flower
L. leptalea
fruit
1 mm
Lasthenia ferrisiae
1 cm
Lasthenia macrantha
1 mm
flower
head
1 mm
disk
flower
L. microglossa
1 mm
fruit
with pappus
L. platycarpha

5′ Ray fls 13–27 in 2 series, ligules yellow, white-tipped; disk fr 2.5–3 mm; pl not scented .... [2]***L. jonesii***
4′ Phyllaries not bulged out; st gen not purple-streaked
6. Pl scented; phyllary tip gen < base; ligules white or yellow; anthers yellow .... [2]***L. pentachaeta***
6′ Pl not scented; phyllary tip gen > base; ligules yellow, gen white-tipped; anthers gen purple (sometimes yellow in SW) .... [2]***L. platyglossa***
1′ Pappus present
7. Heads discoid .... ***L. discoidea***
7′ Heads radiate (ligules sometimes inconspicuous)
8. Disk fls each subtended by chaff scale; pl non-glandular
9. Pappus of gen very unequal, awl-shaped awns or bristles .... [2]***L. chrysanthemoides***
9′ Pappus of ± equal, lanceolate scales .... ***L. fremontii***
8′ Chaff scales in 1 series between ray and disk fls; pl glandular
10. Pappus of lanceolate or elliptic scales
11. Ligules white; anthers yellow; ray frs sparsely hairy .... ***L. leucopappa***
11′ Ligules yellow, white-tipped; anthers purple; ray frs glabrous or sparsely hairy
12. Ray fls 13–27 in 2 series; phyllaries bulged out at base (involucre ± urn-shaped); st gen purple-streaked; pappus < 2 mm; ray fr shiny; c CCo, c SCoRO .... [2]***L. jonesii***
12′ Ray fls 6–15 in 1 series; phyllaries not bulged out; st not purple-streaked; pappus 2–3.5 mm; ray fr dull; SnJV .... ***L. munzii***
10′ Pappus of plumose awns or plumose to scabrous bristles
13. Pappus bristles plumose, densely woolly on inner surface, or scabrous
14. Pappus of 10–15 flattened, linear, plumose awns; ligules white or yellow .... ***L. glandulosa***
14′ Pappus of 14–32 densely woolly or scabrous bristles; ligules yellow or white-tipped
15. Involucre gen ± hemispheric; phyllary tips gen > base; pappus scabrous (exc SW); ligules yellow, gen white-tipped; anthers gen purple (sometimes yellow in SW) .... [2]***L. platyglossa***
15′ Involucre narrower; phyllary tips gen < base; pappus plumose below with inner wool; ligules and anthers yellow .... ***L. septentrionalis***
13′ Pappus bristles plumose, ± not woolly on inner surface
16. Pappus readily deciduous; lvs entire to minutely toothed; ligules white to cream-colored [2]***L. heterotricha***
16′ Pappus persistent; basal lvs lobed; ligules white, yellow, or yellow with white tips
17. Anthers yellow .... [2]***L. pentachaeta***
17′ Anthers purple
18. St not purple-streaked; ligules white, 1.5–3.5 mm; ray fr narrowly sickle-shaped .... ***L. carnosa***
18′ St purple-streaked; ligules yellow or white-tipped, 1–7 mm; ray fr club-shaped
19. Ligules 3.5–7 mm, yellow or white-tipped; pappus bristles 15–24 .... [2]***L. gaillardioides***
19′ Ligules 1–4 mm, yellow; pappus bristles 10–16 .... ***L. hieracioides***

L. carnosa (Nutt.) Torrey & A. Gray (p. 307) BEACH LAYIA Pl 2–18 cm, prostrate to erect, glandular, not scented. **LVS** 3–45 mm, oblong to ovate, fleshy; lower lvs gen lobed < 1/2 to midvein. **INFL**: peduncles < 3 cm; phyllaries 4–8 mm, bases strongly overlapping; chaff scales between ray and disk fls. **RAY FLS** 4–10; ligules 1.5–3.5 mm, white. **DISK FLS** 5–45; corollas 2–4 mm; anthers purple. **FR**: ray fr curved, ± hairy; disk pappus bristles 24–32, 2.5–3.5 mm, white to brownish (esp at base), ± long-plumose below, scabrous above. $2n$=16. RARE. Coastal dunes; < 60 m. n NCo, CCo.

L. chrysanthemoides (DC.) A. Gray (p. 307) Pls 4–53 cm, nonglandular, not scented. **LVS** < 11 cm, linear to (ob)lanceolate, sometimes ± fleshy, gen scabrous-ciliate; lower lvs often lobed ± to midvein. **INFL**: peduncles < 10 cm; phyllaries 4–12 mm, margins gen papillate-scabrous, bases interlocked by cottony hairs; chaff scales each subtending a disk fl. **RAY FLS** 6–16; ligules 3–18 mm, yellow, white-tipped. **DISK FLS** 28–100+; corollas 3–5 mm; anthers purple. **FR**: disk pappus of (0)2–18 awns or bristles, < 3.5 mm, unequal, awl-shaped, whitish, scabrous. $2n$=14. Grassy or open heavy soil; < 800 m. NCo, NCoR, GV, CW. [ssp. *maritima* Keck, maritime layia] ❀SUN:7,**8,9,14–17,19–24**.

L. discoidea Keck (p. 307) RAYLESS LAYIA Pl 3–20 cm, glandular, not scented. **LVS** 2–35 mm, linear to (ob)lanceolate, thin; lowest lvs lobed < 1/2 to midvein. **INFL**: heads discoid; peduncles < 4 cm; involucre 0 (chaff scales involucre-like, in 1 series). **RAY FLS** 0. **DISK FLS** 5–35; corollas 2.5–4 mm; anthers yellow. **FR**: disk pappus of 8–15 irregular scales, gen < 1.5 mm, often unevenly notched or cut, brownish, long-plumose. $2n$=16. RARE. Serpentine soils; 900–1500 m. SCoRI (San Benito Co.). Close to *L. glandulosa*.

L. fremontii (Torrey & A. Gray) A. Gray (p. 307) Pl < 40 cm, not glandular, not scented. **LVS** < 7 cm, linear to (ob)lanceolate, ± fleshy; lower lvs < 30-lobed. **INFL**: peduncles < 9 cm; phyllaries 4–11 mm, ± papillate-scabrous, bases interlocked by cottony hairs; chaff scales each subtending a disk fl. **RAY FLS** 3–15; ligules 5–18 mm, yellow, white-tipped. **DISK FLS** 4–100, 3.5–5.5 mm; anthers purple. **FR**: disk fr densely hairy, pappus of 9–12 scales, 2–5 mm, white, glabrous, tip long-tapered. $2n$=14. Grassy slopes in heavy soil; < 600 m. c NCoR (Mendocino Co.), CaRF, SNF, GV. Close to *L. chrysanthemoides*. ❀SUN:7,**8,9,14–17,19–24**.

L. gaillardioides (Hook. & Arn.) DC. (p. 307) Pl 9–100 cm, often stout, glandular, scented. **LVS** < 10 cm, linear to (ob)lanceolate; lower lvs serrate or lobed < 1/2 lobed to midvein. **INFL**: peduncles < 7 cm; involucre gen ± urn-shaped; phyllaries 4–8 mm, ± bulged out at base, bases interlocked by cottony hairs; chaff-scales between ray and disk fls. **RAY FLS** 6–18; ligules 3.5–15 mm, yellow or white-tipped. **DISK FLS** 14–86; corollas 3–5 mm; anthers purple. **FR**: disk pappus of (0)15–24 bristles, slightly flattened at base, 1–4 mm, whitish to red-brown, plumose below or throughout, scabrous above. $2n$=16. Open, wooded, or shrubby slopes; < 1200 m. NCo, NCoR, CW. ❀SUN:7,14,**15,16**,17.

L. glandulosa (Hook.) Hook. & Arn. (p. 307) WHITE LAYIA Pl 4–60 cm, glandular, sometimes spicy-scented. **LVS** < 10 cm, linear to obovate, thin, gen ± irregularly lobed below. **INFL**: peduncles 0–7 cm; phyllaries 4–11 mm, bases interlocked by cobwebby hairs; chaff scales between ray and disk fls. **RAY FLS** 3–14; ligules 3–22 mm, white to yellow. **DISK FLS** 17–105; corollas 3.5–6.5 mm; anthers yellow. **FR** gen ± hairy; disk pappus of 10–15 awns, 2–5 mm, linear, flat, white, plumose below, often woolly on inner surface, scabrous above. $2n$=16. Open, sandy soils, < 2700 m. CaRH, SNH, Teh, SnJV, CW, SW, GB, D; to WA, ID, UT, NM. [ssp. *lutea* Keck] ❀DRN,SUN:1–3,**7–10**,11,12,**14–24**.

L. heterotricha (DC.) Hook. & Arn. (p. 307) PALE-YELLOW LAYIA Pl 13–90 cm, gen stout, glandular, apple- or banana-scented. **LVS** < 12 cm, elliptic to ovate, fleshy, often clasping, entire to minutely toothed. **INFL**: peduncles 0–7 cm; phyllaries 7–12 mm, bases overlapping; chaff scales between ray and disk fls. **RAY FLS** 7–13; ligules 5–24 mm, white to cream-colored. **DISK FLS** ± 43–91; corollas 4–7mm; anthers yellow. **FR** 4–6.5 mm; disk pappus of (0)14–20 deciduous bristles, 3–6 mm, ± long-plumose below, scabrous above. 2*n*=16. RARE. Open, clay soils; < 1600 m. s Teh, w SnJV, SCoR, n WTR.

L. hieracioides (DC.) Hook. & Arn. (p. 307) Pl 5–130 cm, ± stout, glandular, sweetly or pungently scented. **LVS** < 15 cm, linear to (ob)lanceolate, thin, sessile to ± clasping; lower lvs toothed or ± irregularly and deeply lobed. **INFL**: peduncles < 6 cm; involucre ± narrow, elliptic to ± urn-shaped; phyllaries 4–9 mm, sometimes flat and not enfolding ray fr, base gen bulged out; chaff scales between ray and disk fls. **RAY FLS** 6–16; ligules 1–4 mm, yellow. **DISK FLS** 9–80; corollas 2.5–4.5 mm; anthers purple. **FR** 3–6 mm; disk pappus of 10–16 bristles, 2.5–4 mm, whitish to red-brown, plumose below, scabrous above. 2*n*=16,32. Open or disturbed light soil; < 1200 m. CW, WTR. [*L. paniculata* Keck] Close to *L. gaillardioides*.

L. jonesii A. Gray (p. 307) JONES' LAYIA Pl 7–55 cm, glandular, not scented. **LVS** < 7 cm, linear to (ob)lanceolate, ± fleshy; lower lvs lobed > 1/2 to midvein. **INFL**: peduncles 1–8 cm; involucre widely urn-shaped; phyllaries 4–8 mm, ± scabrous, bases bulged out, interlocked by cobwebby hairs; chaff scales between ray and disk fls. **RAY FLS** 13–27; ligules 5–10 mm, yellow, white-tipped. **DISK FLS** 35–100; corollas 3–5 mm; anthers purple. **FR**: ray fr shiny; disk pappus of (0)8–14 scales, 0.5–2 mm, whitish, ciliate, base sparsely long-hairy. 2*n*=14. RARE. Open serpentine or clay slopes; < 400 m. c CCo, c SCoRO (San Luis Obispo Co.).

L. leucopappa Keck (p. 307) COMANCHE POINT LAYIA Pl 8–60 cm, straw-colored, glandular. **LVS** < 5 cm, oblong to oblanceolate, fleshy, ± glaucous, gen scabrous-ciliate; lower lvs ± lobed. **INFL**: peduncles 1–12 cm; involucre ± urn-shaped; phyllaries 3.5–8 mm, bases often ± bulged out, interlocked by cottony hairs; chaff scales between ray and disk fls. **RAY FLS** 6–15; ligules 3–12 mm, white to cream-colored. **DISK FLS** 20–100; corollas 2.5–5 mm; anthers yellow. **FR**: disk fr densely white-hairy, pappus of 10–13 scales, 2–3.5 mm, white, ciliate, base sparsely long-hairy. 2*n*=14. RARE. Open slopes in heavy soil; 150–350 m. w Teh (Kern Co.). Threatened by development.

L. munzii Keck (p. 307) MUNZ' TIDY-TIPS Pl 1–50 cm, decumbent to erect, glandular, not scented. **LVS** < 6 cm, linear to oblanceolate; lower lvs lobed > 1/2 to midvein. **INFL**: peduncles 1–10 cm; involucre ± urn-shaped; phyllaries 5–9 mm, ± scabrous, bases often slightly bulged out, interlocked by cobwebby hairs; chaff scales between ray and disk fls. **RAY FLS** 6–15; ligules 3–14 mm, yellow, white-tipped. **DISK FLS** 16–108; corollas 3.5–5 mm; anthers purple. **FR**: ray fr ± dull; disk pappus of 9–12 scales, 2–3.5 mm, whitish, scabrous, base sparsely long-hairy. 2*n*=14. UNCOMMON. Alkaline clay soils; < 700 m. SnJV. Close to *L. jonesii*, *L. leucopappa*. ❀SUN:7,**8,9,14**,15–17,**19–24**.

L. pentachaeta A. Gray Pl 5–100 cm, ± stout, glandular, with acrid or lemon-like odor. **LVS** < 11 cm, linear to (ob)lanceolate, thin; lower lvs ± irregularly 1–2 × lobed. **INFL**: peduncles < 6 cm; phyllaries 5–12 mm, bases interlocked by hairs; chaff scales between ray & disk fls. **RAY FLS** 4–14; ligules 3–26 mm, white or yellow. **DISK FLS** 12–125; corollas 3–6 mm; anthers yellow. **FR**: disk pappus of (0)5–22 bristles, 1.5–3.5 mm, fine to flattened, whitish, gen plumose below, gen not woolly on inner surface. 2*n*=16. Light soils; < 1800 m. SNF, n SNH, Teh, SnJV, SCoRI. Sspp. hybridize in s SNF. Close to *L. glandulosa*.

ssp. ***albida*** Keck (p. 307) **FL**: ligules white. Habitats and range of sp. (exc n&c SN). ❀SUN,DRN:**7–9,14–24**.

ssp. ***pentachaeta*** **FL**: ligules yellow. Habitats and range of sp. (exc Teh, rare in SCoRI). ❀SUN,DRN:**7–9,18–24**.

L. platyglossa (Fischer & C. Meyer) A. Gray (p. 307) TIDY-TIPS Pl 3–70 cm, decumbent to erect, glandular, not scented. **LVS** 4–100 mm, linear to (ob)lanceolate, sometimes fleshy; lower lvs lobed ± 1/2 to midvein. **INFL**: peduncles < 13 cm; phyllaries 4–18 mm, bases interlocked by cottony hairs; chaff scales between ray and disk fls. **RAY FLS** 5–18; ligules 3–20 mm, yellow, gen white-tipped. **DISK FLS** 6–124; corollas 3.5–6 mm; anthers gen purple. **FR** 2.5–7 mm; ray fr dull; disk pappus of (0)14–32 bristles, 1–6 mm, whitish, gen scabrous throughout, not woolly on inner surface (in SW, rarely short-plumose below and woolly inside). 2*n*=14. Common. Many habitats; < 2000 m. NW, GV, CW, SW. [ssp. *campestris* Keck; *L. ziegleri* Munz, Ziegler's tidy-tips] ❀SUN:1–6,**7–9,14–23**,24.

L. septentrionalis Keck (p. 307) COLUSA LAYIA Pl 6–35 cm, glandular, not scented. **LVS** 4–70 mm, linear to (ob)lanceolate, thin; lower lvs gen lobed > 1/2 to midvein. **INFL**: peduncles < 8 cm; involucre ± narrow elliptic to bell-shaped; phyllaries 5–12 mm, bases interlocked by cottony hairs; chaff scales between ray and disk fls. **RAY FLS** 5–9; ligules 4–15 mm, yellow. **DISK FLS** 10–67; corollas 5–8 mm; anthers yellow. **FR** 3.5–7.5 mm; disk pappus of 16–22 bristles, 4–7 mm, white, densely plumose below, scabrous above, densely woolly on inner surface. 2*n*=16. UNCOMMON. Serpentine or sandy soils; 100–800 m. c&s NCoRI, ScV (Sutter Buttes). ❀SUN:**7,8**,9,**14**,15–18,**19–24**.

LEMBERTIA

Dale E. Johnson

1 sp. (John Baptist Lembert, CA, 1840–1896) [Johnson 1991 Novon 1:119–124]

L. congdonii (A. Gray) E. Greene (p. 307) SAN JOAQUIN WOOLLY THREADS Ann 5–30 cm, loosely woolly. **STS** decumbent-ascending. **LVS** alternate, 1–4.5 cm, narrowly oblong, entire to shallowly wavy-dentate. **INFL**: heads clustered at branch tips, appearing discoid; involucre ± 4.5 mm, hemispheric; phyllaries 4–7 in 1 series, acute, gen black-hairy near tips, becoming reflexed; receptacle ± domed, naked. **RAY FLS** 1 per phyllary; corolla with small lobe opposite ligule; ligules ± 0.5 mm, 3-lobed, yellow. **DISK FLS** 1–1.5 mm, 4-lobed, bell-shaped, yellow; lobes glandular. **FR** ± 3 mm; ray achenes 3-angled; disk achenes 2-angled, fringed, compressed front-to-back; pappus of 2–7 scales, ± 1 mm. 2*n*=20,22. RARE. Sandy grassland, alkali sink; 90–700 m. sw SnJV. [*Eatonella c.* A. Gray]

LEONTODON HAWKBIT

G. Ledyard Stebbins

Ann to per; sap milky. **STS** 1–many from base. **LVS** all basal, simple or few-branched. **INFL**: heads ligulate, 1–few per scape; phyllaries in 1–several series; receptacle naked. **FLS** many; ligules yellow, readily withering. **FR** ± cylindric, narrowed at summit or beaked; pappus of narrow scales and or stiff, plumose bristles. ± 45 spp.: Old World. (Greek: lion tooth)

L. taraxacoides (Villars) Mérat (p. 307) **STS** many, 1–3 dm, unbranched. **INFL**: heads solitary; involucre 7–9 mm; principal phyllaries subequal, ± basal bractlets. **FR**: outer achenes smooth, pappus a crown of short scales; inner achenes roughened, pappus of short and long bristles with expanded bases. 2*n*=8. Uncommon. Weed in disturbed ground; < 1000 m. NW, n SN, SnJV, CCo, SnFrB; native to Eur. [*L. leysseri* (Wallr.) G. Beck]

1. Ann (rarely bien); beak of inner fr 2–3 mm ssp. ***longirostris***
1′ Per (rarely bien); beak of inner fr ± 1 mm ssp. ***taraxacoides***

ssp. *longirostris* Finch & Sell Per (rarely bien). **FR**: beak of inner achenes ± 1 mm. Habitats of sp. NCo; native to Eur.

ssp. *taraxacoides* Ann (rarely bien). **FR**: beak of inner achenes 2–3 mm. Habitats of sp. NW, n SN, SnJV, CCo, SnFrB; native to Eur. Records scattered.

LEPIDOSPARTUM SCALE-BROOM

Theodore M. Barkley

Shrubs or small trees gen < 3 m, broom-like. **LVS** alternate, entire, thread- to needle- or scale-like. **INFL**: heads discoid, in panicle-like clusters at branch tips; involucre cylindric to obconic; phyllaries overlapping; receptacle naked. **RAY FLS** 0. **DISK FLS** 4–17; corollas yellow, tube long, throat abruptly wider, lobes long; anther bases sagittate to tailed, tips ± lanceolate; style branches long, tips conic or hair-tufted. **FR** ± fusiform; pappus many bristles in 3–4 series. 3 spp.: w N.Am. (Greek: scale-broom) [Strother 1978 N.Am Fl II 10:171–173]

1. Lvs of fl-sts 20–30 mm, thread- or needle-like; fls gen 5; fr densely hairy .... ***L. latisquamum***
1′ Lvs of fl-sts 2–3 mm, scale-like; fls 9–17; fr ± glabrous .... ***L. squamatum***

L. latisquamum S. Watson (p. 307) Pl narrow. **STS** striate; ribs glabrous, grooves felted-tomentose. **LVS** of fl-sts gen 20–30 mm, thread- or needle-like, short-hairy or becoming glabrous. **INFL**: heads 3–5; main inner phyllaries 3–5, 6–8 mm, outer grading into subtending bracts. **FLS** gen 5; corollas pale yellow. **FR** 5–6.5 mm, 5-veined, ± long-white-hairy between veins; pappus bristles < 11 mm, white to brownish. $2n=60$. Sandy or gravelly pine/juniper woodlands, open scrubland; 1400–1500 m. SnGb (n slope), W&I, DMtns; to UT. ❀DRN,SUN:**7**,8–11,14,**19–21**,22–24.

L. squamatum (A. Gray) A. Gray (p. 307) Pl spreading, round-topped, woolly, soon becoming glabrous. **LVS** of fl-sts 2–3 mm, scale-like, appressed; axils often woolly-tufted. **INFL**: heads 1–5; main inner phyllaries 7–23, 4–7 mm, outer grading into subtending scale-like bracts. **FLS** gen 9–17; corollas yellow. **FR** 3.5–5 mm, 10–15-veined, ± glabrous; pappus bristles 5–8 mm, whitish brown. $2n=\pm 90$. Sandy or gravelly washes, stream terraces; < 1800 m. SNF, SCoRI, SW, D; Baja Ca. [var. *palmeri* (A. Gray) L. Wheeler] Apparently TOXIC but unpalatable. ❀DRN,SUN:**7–9**,10,11,**14–16,18–21**,22–24.

LESSINGIA

Meredith A. Lane

Ann, per, subshrubs, decumbent to erect, taprooted. **STS** simple or branched from base or above. **LVS** simple, entire to pinnately lobed; basal petioled, ovate to spoon-shaped; cauline reduced upward, sessile, (ob)lanceolate to (ob)ovate. **INFL**: heads radiate or discoid, terminal, solitary or clustered; involucres hemispheric, bell-shaped, obconic, or cylindric; phyllaries in 4–9 series, thin to tough but flexible, tips green or tinged purplish; receptacle concave, naked, shallowly pitted. **RAY FLS** present or 0, fertile or sterile; corollas yellow or white to purple. **DISK FLS** gen fertile (exc *L. occidentalis*); corollas yellow, violet, purple, pink, or white, funnel-shaped to cylindric, marginal ones in rayless heads often enlarged and bilateral, deeply lobed on inner side, lobes spreading from head center; style appendages flat, triangular, awl-shaped, or cusped. **FR** obconic, mottled purple-brown; hairs dense, appressed, silky; pappus 0 in ray fls, in disk fls of many bristles, these free or fused at base, or fused throughout into awns, white, tan, or red-brown. 14 spp.: CA; NV, AZ, n Baja CA. (C.F. Lessing, 1809–1862, German specialist in Asteraceae) Incl CA sp. previously treated in *Benitoa* and *Corethrogyne*.

1. Ray fls present
 2. Ligules yellow .... ***L. occidentalis***
 2′ Ligules purple, pink, or white .... ***L. filaginifolia***
  3. Heads 1–5 per st; involucres ± as long as wide; disk corollas 6–8 mm .... var. ***californica***
  3′ Heads 3–20+ per st; involucres ± 2 × as long as wide; disk corollas 4–6 mm ...... var. ***filaginifolia***
1′ Ray fls 0
 4. Corollas yellow (exc sometime pink marginal fls of *L. tenuis*)
  5. Pls without glands, or only the phyllaries with sparse, nail-shaped glands .... ***L. germanorum***
  5′ Pls with nail- and or tack-shaped glands
   6. Margins of cauline lvs without glands; tips of inner phyllaries gen purple or purple-tinged .... ***L. tenuis***
   6′ Margins of cauline lvs glandular; tips of inner phyllaries gen green
    7. Phyllaries with glands on margins and back of tips; pls gen rigidly much-branched; cauline lvs much reduced upward .... ***L. glandulifera***
     8. Sts prostrate, herbage densely tomentose .... var. ***tomentosa***
     8′ Sts erect, herbage glabrous or thinly tomentose
      9. Sts light tan .... var. ***glandulifera***
      9′ Sts dark red-brown .... var. ***pectinata***
    7′ Phyllaries with glands on margins only; pls gen flexible, few-branched; cauline lvs only slightly reduced upwards .... ***L. lemmonii***
     10. Phyllaries tomentose .... var. ***peirsonii***
     10′ Phyllaries puberulent
      11. Cauline lvs not clasping base of involucre .... var. ***lemmonii***
      11′ Cauline lvs clasping base of involucre .... var. ***ramulosissima***
 4′ Corollas white, purple, or pink
  12. Phyllaries glabrous or puberulent
   13. Involucres widely obconic to bell-shaped; basal lvs many, present at fl .... ***L. ramulosa***
   13′ Involucres narrowly obconic; basal lvs few, gen withering before fl
    14. Corollas funnel-shaped; style-branch appendages 0.5–0.9 mm, without abrupt point .... ***L. nemaclada***
    14′ Corollas tubular; style-branch appendages < 0.3 mm, with evident abrupt point .... ***L. micradenia***

15. Margins of cauline lvs without glands .... var. ***glabrata***
15′ Margins of cauline lvs with glands .... var. ***micradenia***
12′ Phyllaries woolly or tomentose
16. Tips of inner phyllaries white .... ***L. nana***
16′ Tips of inner phyllaries green, scarious, or purplish
17. Herbage without glands
18. Pappus < fr; style appendages short-triangular, 0.2–0.4 mm; basal lvs gen fallen before fl, glabrous if present .... ***L. arachnoidea***
18′ Pappus = or > than fr; style appendages tapered, 0.6–0.8 mm; basal lvs present at fl, tomentose .... ***L. hololeuca***
17′ Herbage with glands
19. Fls 6–25 per head; lvs and sts glabrous or hairy; cauline lvs with glandular dots only; branches slender, flexible, spreading .... ***L. leptoclada***
19′ Fls 3–6 per head; lvs and sts persistently tomentose, at least in part; cauline lvs with sessile and stalked glands; branches stout, wand-like, ascending .... ***L. virgata***

L. arachnoidea E. Greene (p. 307) CRYSTAL SPRINGS LESSINGIA Ann; herbage without glands, puberulent below, tomentose above. **STS** erect, 3–8 dm. **LVS** < 11 cm, lanceolate to awl-shaped, entire to toothed, reduced upwards to 0.2–4 cm. **INFL**: heads discoid, solitary, terminal; involucres 5–8 mm, narrowly obconic; phyllaries lanceolate, acute, tomentose (sometimes only on margins), sessile or tack-shaped glands 0 or rare; tips purplish tinged, white or tannish white diffusing along margins and midveins of inner phyllaries. **RAY FLS** 0. **DISK FLS** (3)8–18; corollas funnel shaped, pale to deep lavender; style-branches 0.7–1.5 mm, appendages 0.2–0.4 mm, with or without abrupt point. **FR** 2–3 mm; pappus very short, bristles many, free or fused at base in groups of 2–3, white or tannish white. 2*n*=10. RARE. Open serpentine barrens; < 200 m. sw SnFrB (San Mateo Co., near Crystal Springs Reservoir). [*L. hololeuca* E. Greene var. *a.* (E. Greene) J. Howell]

L. filaginifolia (Hook. & Arn.) M.A. Lane CALIFORNIA-ASTER Per, subshrubs; herbage gen densely white-tomentose, sometimes mostly glabrous or glandular above. **STS** erect to decumbent or trailing, < 1 m. **LVS** < 7 cm, linear to (ob)ovate, often toothed, reduced and sessile upwards. **INFL**: heads radiate, solitary or tightly clustered, terminal; involucres 6–13 mm, obconic or bell-shaped to hemispheric; phyllaries linear to narrowly lanceolate, tips often purplish, becoming reflexed when frs mature. **RAY FLS** 10–43, sterile; ligules purple, pink, or white. **DISK FLS** 12–120+; corollas tubular, yellow. **FR** < 5 mm; pappus 3–8 mm, bristles free, red-brown. 2*n*=10. Widespread in coastal scrub, oak woodlands, grasslands; < 2600 m. NCo, KR, NCoRO, s SN, SnJV, CW, SW; sw OR, n Baja CA. Highly variable sp.; further division requires additional research. Vars. intergrade.

var. ***californica*** (DC.) M.A. Lane (p. 307) **STS** gen decumbent to ascending, gen unbranched above, < 6 dm. **LVS** obovate, often long-tapered, toothed. **INFL**: heads gen solitary; involucres 5–10 mm, as long as wide at fl; phyllaries in 3–5 series. **DISK FLS** 30–120+; corollas 6–8 mm. Coastal scrub, oak woodlands, grasslands; < 1000 m. NCo, KR, NCoRO, CW; sw OR. [*Corethrogyne c.* DC, incl. vars. *lyonii* S.F. Blake & *obovata* (Benth.) Kuntze] ❀SUN,IRR:**14,16**.

var. ***filaginifolia*** (p. 307) **STS** gen erect, often branched above, < 1 m. **LVS** narrowly long-tapered obovate to linear, entire to toothed. **INFL**: heads 1–20+ per st; involucres 6–13 mm, ± 2 × as long as wide at fl; phyllaries in 4–9 series. **DISK FLS** 12–40; corollas 4–6 mm. Coastal scrub, oak woodlands, grasslands; < 2600 m. s SN, SnJV, CW, SW; n Baja CA. [*Corethrogyne f.* (Hook. & Arn.) Nutt., incl. vars. *bernardina* (Abrams) H.M. Hall, *brevicula* (E. Greene) Canby, *hamiltonensis* Keck, *incana* (Nutt.) Canby, *latifolia* H.M. Hall, *linifolia* H.M. Hall, *peirsonii* Canby, *pinetorum* I.M. Johnston., *robusta* E. Greene, *sessilis* (E. Greene) Canby, *virgata* (Benth.) A. Gray, *viscidula* (E. Greene) Keck; *C. leucophylla* Jepson] ❀DRN,SUN:**7**,8,**9**,**14–24**.

L. germanorum Cham. (p. 311) SAN FRANCISCO LESSINGIA Ann; herbage without glands, loosely grayish tomentose, becoming glabrous with age. **STS** 0.5–3 dm, decumbent, rarely erect, reddish brown. **LVS**: basal deciduous, < 5 cm, oblanceolate to long-tapered obovate, pinnately lobed, lobes toothed; cauline 0.5–3 cm, oblong to oblanceolate, entire to pinnately lobed. **INFL**: heads discoid, solitary or rarely tightly clustered, terminal; involucres 4–8 mm, bell-shaped; phyllaries lanceolate, tips with abrupt point, recurved, margins of outer phyllaries often lobed, of inner entire. **RAY FLS** 0. **DISK FLS** 20–40; corollas funnel-shaped to tubular, deep lemon-yellow with reddish brown band in throat; style branches 0.8–2 mm, appendages 0.3–0.4 mm, with or without short, abrupt point. **FR** 1–3 mm; pappus bristles free or fused at base in groups of 2–4, white or tannish white. 2*n*=10. RARE. Apparently restricted to sandy soils; < 100 m. SnFrB. Vars. of J. Howell given other allocations here.

L. glandulifera A. Gray Ann; herbage with glands, glabrous, puberulent, or tomentose, gen becoming glabrous with age. **STS** depressed or erect, 0.5–8 dm. **LVS**: basal deciduous, < 11 cm, oblanceolate, toothed to pinnately lobed, lobes toothed; cauline oblong to ovate, entire, toothed or pinnately lobed, upper with tack-shaped glands. **INFL**: heads discoid, solitary, terminal; involucres 5–8 mm, widely obconic; phyllaries with dense nail- and tack-shaped glands, hairy, puberulent, or glabrous. **RAY FLS** 0. **DISK FLS** 15–30; corollas funnel-shaped, limbs of marginal ones expanded, yellow with brown band in throat; style branches 1–2.5 mm, appendages 0.2–0.6 mm, with or without abrupt point. **FR** 2–3 mm; pappus bristles many in 1–2 series, free to base, white or tannish white. Sandy soils, coastal, desert washes, pine forests; < 1700 m. SNF, SnJV, CW, SCo, TR, PR; n Baja CA. Vars. intergrade.

var. ***glandulifera*** (p. 311) **STS** erect, light tan or whitish; herbage glabrous or partly persistently thin-tomentose. **LVS**: cauline entire or pinnately lobed. 2*n*=10. Sandy soils, coastal, desert washes, pine forests; < 1700 m. SNF, SnJV, SnFrB, SCoR, SCo, TR, PR; n Baja CA. [*L. germanorum* var. *gl.* (A. Gray) J. Howell]

var. ***pectinata*** (E. Greene) Jepson **STS** erect, dark brown; herbage glabrous or thin-tomentose. **LVS**: cauline pinnately lobed, lobes very narrow, with abrupt points. 2*n*=10. Sandy soils near the coast; < 200 m. CCo, SnFrB, SCoRO. [*L. germanorum* var. *p.* (E. Greene) J. Howell]

var. ***tomentosa*** (E. Greene) Ferris WARNER SPRINGS LESSINGIA **STS** prostrate, white or tannish white; herbage densely, persistently white-tomentose. **LVS**: cauline entire. RARE. Sandy soil; 200–300 m. e PR (Warner's Ranch, San Diego Co.). [*L. germanorum* var. *t.* (E. Greene) J. Howell]

L. hololeuca E. Greene (p. 311) WOOLLY-HEADED LESSINGIA Ann; herbage without glands, tomentose, becoming glabrous with age. **STS** decumbent or erect, 0.5–4 dm. **LVS**: basal persistent, < 13 cm, linear to long-tapered obovate, entire to lobed; cauline 0.2–3 cm, reduced upwards, lanceolate to awl-shaped, tip with abrupt point, entire. **INFL**: heads discoid, solitary or tightly clustered, terminal or axillary; involucres 0.5–1.3 cm, widely or rarely narrowly obconic; phyllaries lanceolate, woolly, tips purplish tinged, margins and midveins of inner ones ± white or tannish white. **RAY FLS** 0. **DISK FLS** 10–20; corollas funnel-shaped, pink to lavender; style branches 1.5–2.5 mm, appendages 0.6–0.8 mm, awl-shaped. **FR** 3–5 mm; pappus bristles many in 2 unequal series, free to base, white or tannish white. 2*n*=10. UNCOMMON. Fields, roadside ditches; < 300 m. s NCoR, s ScV, n SnFrB.

L. lemmonii A. Gray Ann; herbage grayish tomentose, sometimes becoming glabrous with age. **STS** decumbent or erect, 0.3–4 dm. **LVS**: basal deciduous, < 6 cm, oblanceolate or long-tapered obovate, entire to pinnately lobed; cauline < 2 cm, awl-shaped or linear to obovate, entire or tip with few teeth. **INFL**: heads discoid, solitary, terminal; involucres 4–6 mm, narrowly obconic to bell-shaped; phyllaries oblong, obtuse or acute, puberulent or tomentose, margins with large sessile glands. **RAY FLS** 0. **DISK FLS** 10–25; corollas funnel-shaped to tubular, marginal corollas funnel-shaped in small heads, yellow, with or without white band in throat; style branches 1–2 mm, appendages 0.7–1.3 mm, with long, abrupt point. **FR** 1.5–3.5 mm; pappus bristles many, free to base, white or tannish white. 2*n*=10. Sandy soils; 200–1850 m. TR, W&I, DMoj; w NV, nw AZ. Vars. intergrade.

var. ***lemmonii*** Herbage tomentose, partly glabrous with age. **LVS**: cauline fewer above, not clasping base of involucre. **INFL**: phyllaries puberulent. Dry, sandy soil; < 1900 m. TR, W&I, DMoj; nw AZ. [*L. germanorum* Cham. var. *l.* (A. Gray) J. Howell]

var. ***peirsonii*** (J. Howell) Ferris Herbage densely and persistently tomentose. **LVS**: cauline few, clasping base of involucre. **INFL**: phyllaries densely and persistently tomentose. Open slopes in sandy soil; 300–1850 m. e WTR, w DMoj; w NV. [*L. germanorum* Cham. var. *p.* J. Howell]

var. ***ramulosissima*** (Nelson) Ferris Herbage tomentose, becoming less so with age. **LVS**: cauline many, clasping base of involucre. **INFL**: phyllaries puberulent. Very dry, sandy soil; 800–1200 m. TR, W&I, DMoj; w NV. Sometimes forms tumbleweeds.

L. leptoclada A. Gray Ann; herbage gland-dotted, glabrous to ± tomentose. **STS** erect, 0.3–9 dm; lateral branches > primary axis. **LVS**: basal < 8 cm, deciduous, toothed to lobed; cauline 0.5–5 cm, reduced upward, awl-shaped, lanceolate or ovate, entire or toothed. **INFL**: heads discoid, solitary or in clusters of 2–5, terminal or in axils; involucres 5–10 mm, hemispheric or widely obconic; phyllaries oblong, persistently tomentose. **RAY FLS** 0. **DISK FLS** 6–25; corollas pale to deep lavender-blue, drying darker, marginal corollas funnel-shaped or palmately lobed and enlarged; style branches 0.8–2 mm, appendages 0.3–0.6 mm, with or without abrupt point. **FR** 3–4 mm; pappus bristles many in 1–2 series, free or ± fused at base, white or tannish white. 2*n*=10. Open slopes and valleys; 400–1800 m. SN. ❀SUN,DRN:7,8,**9,14–24**.

L. micradenia E. Greene (p. 311) Ann; herbage thinly tomentose, persistently so on upper surface of cauline lvs. **STS** ± erect 0.5–6 dm; branches spreading, > main st. **LVS**: basal deciduous, < 6 cm, entire, toothed or lobed; cauline 0.2–2 cm, linear, awl-shaped, or lanceolate, tip with abrupt point, mostly entire, with or without glands. **INFL**: heads discoid, solitary or clustered, terminal or rarely sessile in axils; involucres 4–6 mm, narrowly cylindric-obconic; phyllaries lanceolate, gen with glands, tips purplish. **RAY FLS** 0. **DISK FLS** 3–10; corollas tubular, marginal ones slightly palmately lobed, white to lavender; style branches 0.7–1 mm, appendages 0.2–0.3 mm, with evident abrupt point. **FR** 2–4 mm; pappus bristles ± fused into 5 awns, white or tannish white. 2*n*=10. Thin, gravelly soils of serpentine outcrops and roadcuts; < 500 m. SnFrB.

var. ***glabrata*** (Keck) Ferris SMOOTH LESSINGIA **LVS**: cauline without glands on margins. **INFL**: phyllaries glabrous, with or without terminal and marginal sessile glands. **FLS** 3–5 per head. RARE. Serpentine outcrops and gravelly roadcuts; < 300 m. s SnFrB (Santa Clara Co.). [*L. ramulosa* A. Gray var. *g.* Keck]

var. ***micradenia*** TAMALPAIS LESSINGIA **LVS**: cauline with tack-shaped marginal glands. **INFL**: phyllaries puberulent, with tack-shaped glands. **FLS** 5–10 per head. RARE. Thin, gravelly soils of serpentine outcrops and roadcuts; 100–500 m. n SnFrB (Mount Tamalpais, Marin Co.). [*L. ramulosa* A. Gray var. *m.* (E. Greene) J. Howell]

L. nana A. Gray (p. 311) Ann, prostrate; herbage with sessile glands, densely woolly, persistently so at least on upper surfaces of cauline lvs. **STS** decumbent, 0.2–0.3 dm (rarely pls upright, sts erect, < 2.5 dm). **LVS**: basal deciduous, < 6 cm, tips with abrupt point, entire to toothed or lobed; cauline 0.5–3 cm, lanceolate to awl-shaped, tips with abrupt point, entire to toothed, frequently intergrading with phyllaries. **INFL**: heads discoid, solitary or clustered, terminal, or sessile in rosette or axils; involucres 7–10 mm, narrowly obconic; outer phyllaries green, woolly; inner phyllaries > corollas, white-scarious, with tough, flexible, gristly abrupt point. **RAY FLS** 0. **DISK FLS** 10–20; corollas tubular, rose to purple; style branches 1–2 mm, appendages 0.2–0.4 mm, gen with abrupt point. **FR** 2–3 mm; pappus bristles many in 2 series, free to base, red, purplish, or tan. 2*n*=10. Open places on plains or slopes; 200–900 m. CaRF, SNF, GV. ❀TRY.

L. nemaclada E. Greene Ann; herbage with sessile and nail-shaped glands, persistently tomentose on upper surface of cauline lvs. **STS** erect, 0.5–6 dm; branches > main st. **LVS**: basal deciduous, < 7 cm, toothed; cauline 0.2–2 cm, reduced upwards, lanceolate, entire. **INFL**: heads discoid, solitary, or rarely in 3–4-headed clusters; involucres 5–6 mm, very narrowly obconic; phyllaries lanceolate, glabrous, with sessile and nail-shaped glands. **RAY FLS** 0. **DISK FLS** 3–10; corollas funnel-shaped, white, pale pink, or lavender; style branches 0.8–2 mm, appendages 0.45–0.9 mm, without abrupt point. **FR** 3–4 mm; pappus of many bristles, somewhat fused at base, or of 5 awns, white or tannish white. 2*n*=10,12. Open fields, gravelly roadcuts and borders of woodlands; 200–1100 m. NW, CaR, SN, GV, SnFrB, SCoRI. [var. *albiflora* (Eastw.) J. Howell] ❀DRN,SUN:**7–9,14–24**.

L. occidentalis (H.M. Hall) M.A. Lane (p. 311) BENITOA Ann; herbage with tack- and nail-shaped glands, scantily tomentose when young. **STS** erect, stout, 2–11 dm, red-green. **LVS** 4–8 cm, linear-lanceolate to oblanceolate; tips acute, margins entire to wavy, reduced upwards. **INFL**: heads radiate, in an open panicle; involucres 8–10 mm, narrowly obconic to cylindric; phyllaries linear, long-tapered, surfaces glandular, tips straight, each with a single, very large nail-shaped gland. **RAY FLS** 5–8, fertile; ligules yellow. **DISK FLS** 9–12, functionally staminate; corollas 4.6–5.2 mm, narrowly funnel-shaped, yellow, red-tinged; style branches 1.5 mm. **FR** 3.5–4 mm; pappus of 5–8 slender, deciduous, white bristles. 2*n*=10. UNCOMMON. Apparently restricted to serpentine outcrops; 450–850 m. SCoRI (areas of Parkfield, se Monterey Co., and New Idria, se San Benito Co.). [*Benitoa o.* (H.M. Hall) Keck]

L. ramulosa A. Gray Ann; herbage with dense, small, nail- and tack-shaped glands, tomentose. **STS** erect, 2–5 dm. **LVS**: basal persistent, < 9 cm, entire to shallowly or deeply toothed; cauline 0.5–1.5 cm, ovate, entire to toothed. **INFL**: heads discoid, solitary, sessile, axillary; involucres 5–7 mm, widely obconic to bell-shaped; phyllaries lanceolate, with few large, nail-shaped glands, tips and margins purplish tinged. **RAY FLS** 0. **DISK FLS** 5–15; corollas funnel-shaped, the marginal scarcely enlarged, rose to pale pink, drying purple; style branches 0.8–1.5 mm, appendages 0.2–0.3 mm, without abrupt point. **FR** 3–4 mm; pappus bristles many, free to base or fused into 5 awns, white or tannish white. 2*n*=10. Roadsides and open hills; < 1000 m. NCoR, n SnFrB.

L. tenuis (A. Gray) Cov. (p. 311) SPRING LESSINGIA Ann; herbage gray-green to reddish; hairs appressed on upper or both lf surfaces. **STS** ± erect, 0.2–1.5 dm. **LVS**: basal deciduous, < 4 cm, entire to deeply and irregularly lobed, lobes entire or toothed; cauline 0.3–1.5 cm, entire, lobed or irregularly toothed. **INFL**: heads discoid, solitary, terminal; involucres 4–7 mm, obconic to bell-shaped; phyllaries acute, tips recurved, sparsely tomentose, inner ones with nail-shaped glands, purplish. **RAY FLS** 0. **DISK FLS** 10–25; corollas funnel-shaped, limb of marginal ones enlarged, yellow with purple band in throat, marginal ones rose-white or tannish white or yellow diffused with purple; style branches 0.5–0.8 mm, appendages 0.1–0.2 mm, without abrupt point. **FR** 2–3 mm; pappus bristles many, free to base, or reduced to awns, white or tannish white. 2*n*=10. UNCOMMON. Dry, open slopes; 300–2150 m. SnFrB, SCoR, WTR. [*L. germanorum* Cham. vars. *t.* (A. Gray) J. Howell and *parvula* (E. Greene) J. Howell]

L. virgata A. Gray (p. 311) Ann; herbage with sessile and nail-shaped glands, woolly-tomentose. **STS** erect, 0.5–6 dm. **LVS**: basal deciduous, < 6 cm, shallowly to very deeply toothed or shallowly lobed; cauline appressed upward, 7–10 mm, oblong-ovate to ovate, entire. **INFL**: heads discoid, solitary, sessile, axillary; involucres 5–7 mm, very narrowly obconic to cylindric; phyllaries lanceolate, outer ones green- or purple-tipped, with glands, loosely woolly, inner ones scarious, margins with straight hairs. **RAY FLS** 0. **DISK FLS** 3–6; corollas tubular, white to lavender; style branches 1–1.2 mm, appendages 0.3–0.8 mm, with abrupt point or slender-tapering overall. **FR** 2–2.5 mm; pappus bristles many, ± fused at base, white or tannish white. 2*n*=10. Dry plains and foothills; < 350 m. CaRF, n&c SNF, e GV.

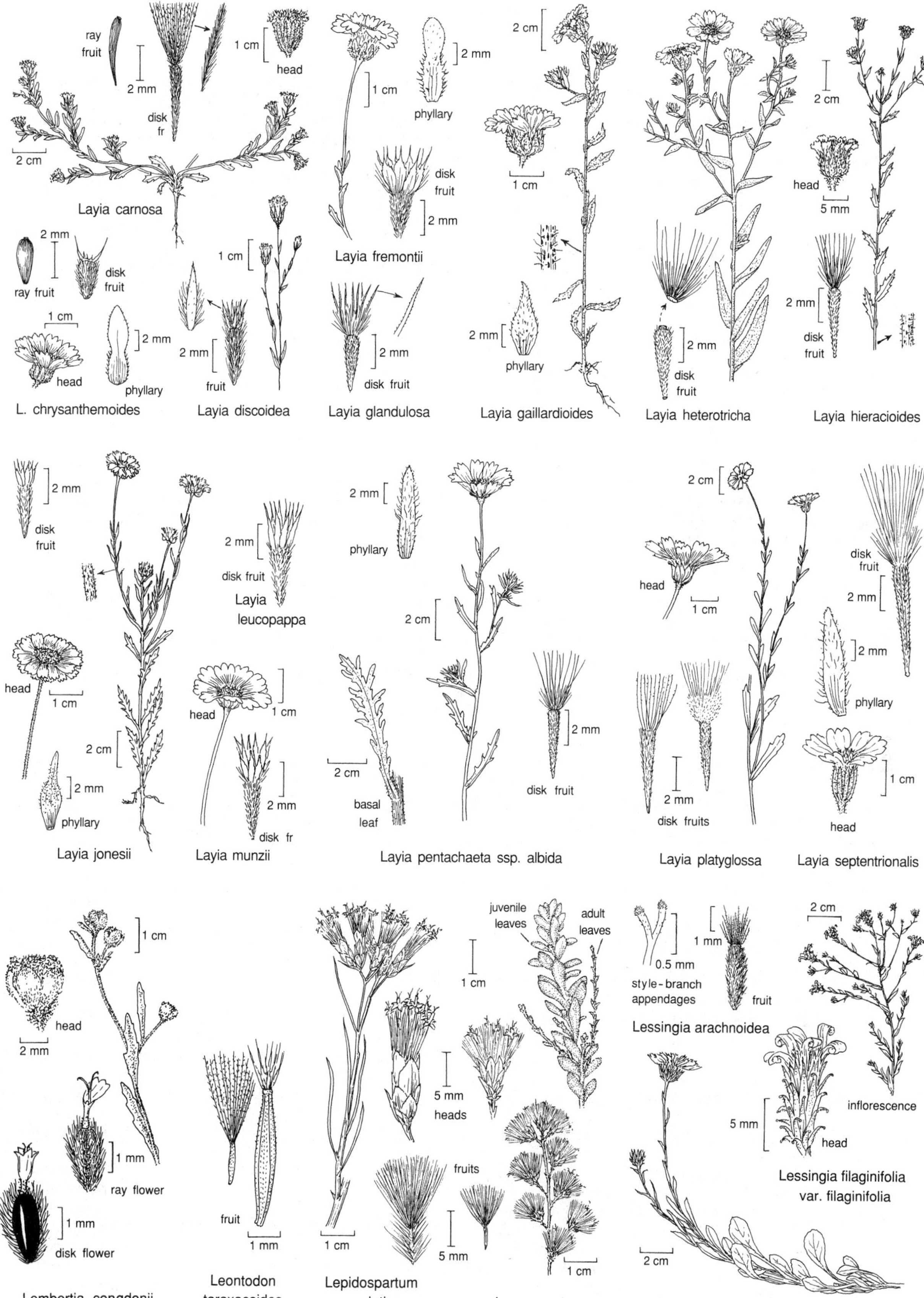
ray fruit
2 mm
disk fr
1 cm
head
2 cm
Layia carnosa
2 mm
ray fruit
disk fruit
1 cm
head
2 mm
phyllary
L. chrysanthemoides
1 cm
2 mm
fruit
Layia discoidea
1 cm
2 mm
phyllary
disk fruit
2 mm
Layia fremontii
2 mm
disk fruit
Layia glandulosa
2 cm
1 cm
2 mm
phyllary
Layia gaillardioides
2 mm
disk fruit
Layia heterotricha
2 cm
head
5 mm
2 mm
disk fruit
Layia hieracioides
2 mm
disk fruit
head
1 cm
2 cm
2 mm
phyllary
Layia jonesii
2 mm
disk fruit
Layia leucopappa
head
1 cm
2 mm
disk fr
Layia munzii
2 mm
phyllary
2 cm
2 cm
basal leaf
2 mm
disk fruit
Layia pentachaeta ssp. albida
2 cm
head
1 cm
2 mm
disk fruits
Layia platyglossa
disk fruit
2 mm
2 mm
phyllary
1 cm
head
Layia septentrionalis
1 cm
head
2 mm
1 mm
ray flower
1 mm
disk flower
Lembertia congdonii
fruit
1 mm
Leontodon taraxacoides
1 cm
5 mm
heads
fruits
1 cm
5 mm
Lepidospartum latisquamum
juvenile leaves
adult leaves
1 cm
L. squamatum
0.5 mm
style-branch appendages
1 mm
fruit
Lessingia arachnoidea
2 cm
inflorescence
5 mm
head
Lessingia filaginifolia var. filaginifolia
2 cm
Lessingia filaginifolia var. californica

LEUCANTHEMUM

Elizabeth McClintock

Per. **STS** simple or branched. **LVS** alternate, entire to pinnately lobed or toothed. **INFL**: heads radiate, gen long-peduncled, gen solitary; phyllaries in 2–3 series, margins scarious; receptacle gen convex to conic, chaff 0. **RAY FLS** ± 20(0), pistillate; ligules gen white. **DISK FLS** many; corollas yellow; anther tips ovate, bases rounded or ± cordate; style branches truncate with shrub-like tips. **FR** gen 10-ribbed; pappus 0 or a narrow crown. ± 20 spp.: Eur, n Afr. (Greek: white flower) [Humphries 1976 Bull Brit Mus Bot 5:147–240]

1. Heads 7–9 cm diam; lvs ± regularly toothed; middle st lvs > 5 cm .... ***L. maximum***
1′ Heads 3–7 cm diam; lvs irregularly pinnately lobed ± halfway to midrib; middle st lvs < 5 cm .... ***L. vulgare***

L. maximum (Ramond) DC. **STS** from rhizome, 3–7 dm, stiff, robust, gen unbranched, ± glabrous. **LVS** serrate; basal lf blades < 25 cm, oblanceolate, petioles winged; cauline lvs sessile, lanceolate, becoming shorter upward. **INFL**: heads gen 7–9 cm diam. **RAY FLS** 20–30; ligules 3–4 cm. **DISK FLS**: corollas ± 3 mm. **FR** 2–3 mm, 10-ribbed. $2n$=90,108. Uncommon. Escape from cult in disturbed areas, spreading locally; < 200 m. NCo, CCo, SCoRO, WTR; native to Eur. [*Chrysanthemum m.* Ramond]. Presumed to be one of the parents of Shasta daisy (*C.* ×*superbum*). Much like *L. vulgare*.

L. vulgare Lam. (p. 311) OX-EYE DAISY **STS** from a creeping rootstock, 2–5 dm, erect, stout, sparingly branched, ± glabrous. **LVS** pinnately lobed or toothed; lowermost and middle blades < 12 cm, obovate to spoon-shaped, petiole = blade, wingless; upper cauline lvs shorter, oblong, sessile. **INFL**: heads 3–7 cm diam. **RAY FLS** ± 22(0); ligules 1–3.5 mm, white. **DISK FLS**: corollas ± 2.5 mm. **FR** ± 2 mm, 10-ribbed. $2n$=18,36,54,72,90. Common. Escape from cult in pastures, disturbed mtn meadows, roadsides, fields; < 2000 m. KR, NCoRO, n&c SNH, SnFrB, WTR, PR; native to Eur. [*Chrysanthemum leucanthemum* L.]

LUINA

Per from branched caudex. **STS** slender, gray-tomentose. **LVS** cauline, alternate, lanceolate to ovate, entire or toothed. **INFL**: heads discoid, gen in ± flat-topped clusters; involucre obconic; phyllaries in 1–2 equal series; receptacle flat or convex, naked. **DISK FLS** gen 12–25; corollas ± yellow, tube > throat; anther bases short-tailed, tips narrowly lanceolate; style branches rounded-truncate. **FR** ± cylindric; pappus of many bristles in 1–2 series. 2 spp.: nw N.Am. (Anagram of *Inula*) [Strother 1978 N.Am Fl II 10:160–173]

L. hypoleuca Benth. (p. 311) Pl 1.5–6 dm. **LVS** ± sessile, 2–6.5 cm, white-tomentose below, glabrous above. **INFL**: heads gen 4–12; peduncles 1–5 cm; phyllaries 5–9 mm, lanceolate to narrowly ovate, gen tomentose, sometimes glandular-puberulent. **FLS** 11–15(23); corollas gen 8–10 mm, creamy yellow, tube 3–4 mm, throat slightly expanded, lobes 1–2 mm, erect. **FR** 3–4 mm, ± 9-veined, glabrous or strigose; pappus 8–10 mm. Rocky places, cliffs, sometimes on serpentine; < 2100 m. KR, NCoR, sw SnFrB; to B.C. ❀TRY;DFCLT.

MACHAERANTHERA

David J. Keil & Gregory K. Brown

Ann to subshrubs. **STS** from taproot or ± branched caudex. **LVS** simple, alternate, entire to pinnately dissected; teeth or lobes often ± bristle-tipped. **INFL**: heads radiate or discoid, solitary or cymosely clustered; involucre bell-shaped, hemispheric, or obconic; phyllaries in 2–several series of unequal length, basal portion straw-colored to purplish, tips green; receptacle convex, naked or with short, triangular scales (not chaff). **RAY FLS** 8–many; corollas yellow, white, pink, blue, or purple. **DISK FLS** 10–many; corollas yellow; style tips triangular to linear, acute. **FR** linear to club-shaped or obovoid, smooth or several–many-ribbed, glabrous to densely hairy; pappus of many unequal bristles (ray pappus sometimes 0). ± 35 spp.: temp w N.Am. (Greek: sword-like anthers) [Hartman 1990 Phytologia 68:439–465]

1. Rays 0
2. Fl-sts ± hairy or glandular .... ***M. canescens*** var. ***shastensis***
2′ Fl-sts glabrous, glaucous .... ***M. carnosa***
1′ Rays present
3. Ray corollas yellow
4. Ann .... ***M. gracilis***
4′ Per
5. Bracts of upper peduncles overlapping, grading into phyllaries; sts nearly lfless above base .... ***M. juncea***
5′ Bracts of upper peduncles few, well separated, not grading into phyllaries; sts gen leafy, at least in lower half .... ***M. pinnatifida*** var. ***gooddingii***
3′ Ray corollas white to purple
6. Ray achenes without pappus .... ***M. arida***
6′ Ray achenes with pappus
7. Lvs 1–2-pinnately dissected .... ***M. tanacetifolia***
7′ Lvs entire or toothed
8. Phyllary tips narrowly awl-shaped, gen hairy throughout .... ***M. asteroides***
9. Involucre hemispheric; phyllary tips 3–6 mm; lvs of mid-st minutely serrate .... var. ***asteroides***
9′ Involucre obconic; phyllary tips 1–3 mm; lvs of mid-st entire or nearly so .... var. ***lagunensis***
8′ Phyllary tips acute to awl-shaped, gen ± glandular .... ***M. canescens***
10. Heads gen 14–20 mm; subshrub or long-lived per; e PR (Santa Rosa Mtns) .... var. ***ziegleri***
10′ Heads gen 6–12 mm; ann to short-lived per; widespread
11. Sts glandular, but not canescent-puberulent .... var. ***leucanthemifolia***
11′ Sts either canescent-puberulent throughout, without glandular hairs, or sts canescent and glandular

12. Sts stiffly erect, < 60 cm; branches many, stiff, spreading; heads many .... var. ***incana***
12′ Sts various, gen smaller; branches gen fewer, more flexible; heads few–many
13. Ray fls with styles, fertile, well developed; involucre gen 8–12 mm; phyllaries gen in 5–10 series .... var. ***canescens***
13′ Ray fls without styles, sterile, often ± reduced; involucre 6–9 mm; phyllaries gen in 3–5 series .... var. ***shastensis***

M. arida B. Turner & D. Horne (p. 311) SILVER LAKE DAISY Ann. **STS** branched from base, 5–30 cm, glandular with nonglandular hairs interspersed. **LVS** gen sessile, 1–30 mm, 1–10 mm wide, oblong, toothed to pinnately lobed; lobes and teeth bristle-tipped; upper lvs reduced, appressed. **INFL**: heads radiate, solitary or in cymes; involucre 3–6 mm, 5–10 mm wide, hemispheric; phyllaries in 2–3 series, oblong to oblanceolate, glandular. **RAY FLS** 25–35, < or = 7 mm, white to lavender. **DISK FLS** 28–45, 2.5–3 mm. **FR** 1.4–1.9 mm, hairy; ray pappus 0 or obscure; disk pappus of many bristles, 2–3 mm, white. $2n=10$. Uncommon. Riverbanks, sandy, alkaline flats, roadsides; 30–300 m. D; s NV, s AZ, n Mex. [*Psilactis coulteri* A. Gray misapplied]

M. asteroides (Torrey) E. Greene Bien, per < 10 dm, gen canescent-puberulent and nonglandular. **STS** 1–several from base, gen branched above and ± bushy. **LVS** gen 3–10 cm, gen 5–25 mm wide, lanceolate to oblanceolate, irregularly dentate to minutely serrate or subentire; lower tapered; upper clasping. **INFL**: heads radiate; phyllaries gen in 5–12 series, tips short-triangular to elongate, acuminate, spreading to bent backward, puberulent. **RAY FLS** many; corollas blue-purple; ligules 1–2 cm. **DISK FLS** many; corollas 5.5–8 mm. **FR** 2.5–3.5 mm, narrowly obovate, weakly curved and ± flattened with 5–7 ribs on each face, glabrous or ± silky; pappus 6–8 mm. Chaparral, woodland, shrubland; ± 100 m; 800–2400 m. PR, DSon; to s NV, sw NM, n Mex. Perhaps not a different sp. from *M. canescens*.

var. ***asteroides*** (p. 311) **LVS**: middle cauline lvs 6–15(25) mm wide, minutely serrate. **INFL**: involucre hemispheric; phyllary tips 3–6 mm, narrowly acuminate. $2n=8$. Shrubland; ± 100 m. e DSon (near Colorado River); to sw NM, n Mex.

var. ***lagunensis*** (Keck) B. Turner (p. 311) LAGUNA MOUNTAINS ASTER **LVS**: middle cauline lvs gen 2–5 mm wide, entire or obscurely toothed. **INFL**: involucre widely obconic to ± hemispheric; phyllary tips 1–3 mm, acute to short-acuminate. **RARE** CA. Chaparral, woodland; 800–2400 m. PR (Laguna Mtns, San Diego Co.); n Baja CA. [*M. lagunensis* Keck]

M. canescens (Pursh) A. Gray HOARY-ASTER Ann to per < 12 dm, gen canescent-puberulent and often glandular. **STS** 1–several from base, gen branched above and ± bushy. **LVS** gen 3–10 cm, gen 2–6 mm wide, linear to obovate, subentire to dentate or minutely serrate; lower tapered; upper sometimes clasping. **INFL**: heads radiate; phyllaries gen in 3–10 series, tips short-triangular to elongate, acuminate, spreading to bent backward, gen ± glandular or glabrous. **RAY FLS** many (0 in var. *shastensis*); corollas blue-purple; ligules 1–2 cm. **DISK FLS** many; corollas 5.5–8 mm. **FR** 2.5–3.5 mm, narrowly obovate, weakly curved and ± flattened with 5–7 ribs on each face, glabrous or ± silky; pappus 6–8 mm. Common. Many habitats; 300–3400 m. KR, CaR, SN, TR, PR, GB, DMtns; w N.Am. Variable; perhaps not a different sp. from *M. asteroides*.

var. ***canescens*** (p. 311) Ann to short-lived per. **STS** 1–5 dm, spreading to erect; branches ascending to loosely spreading. **INFL**: heads radiate, 6–12(14) mm, 10–15 mm wide (when pressed); involucres 8–12 mm. **RAY FLS** present, fertile (rarely reduced or 0); style well developed. Open montane habitats; 2000–3000 m. CaR, SN, TR; to WA, c Can, Colorado, AZ. [*M. shastensis* A. Gray vars. *glossophylla* (Piper) Cronq. & Keck & *montana* (E. Greene) Cronq. & Keck; *M. tephrodes* (A. Gray) E. Greene] ❀DRN,SUN:1–3,**15**,**16**.

var. ***incana*** (Lindley) A. Gray Ann to short-lived per. **STS** < 6 dm, stiffly erect; branches stiffly spreading. **INFL**: heads radiate, 6–12(14) mm, 10–15 mm wide (when pressed); involucres 7–9(12) mm. **RAY FLS** present, fertile; style well developed. Dry fields, sandy streamsides; 300–1100 m. KR; to B.C., ID.

var. ***leucanthemifolia*** (E. Greene) Welsh Ann to short-lived per. **STS** 1–5 dm, spreading or erect; branches ascending to loosely spreading. **INFL**: heads radiate; 6–12(14) mm, 10–15 mm wide (when pressed); involucres 5–9 mm. **RAY FLS** present, fertile; style well developed. Desert scrub; 1000–2000 m. e PR, DMtns; to e OR, UT. [*M. l.* (E. Greene) E. Greene]

var. ***shastensis*** (A. Gray) B. Turner Ann to short-lived per. **STS** 1–5 dm, spreading or erect; branches loosely spreading to ascending. **INFL**: heads discoid or radiate; 6–12(14) mm, 10–15 mm wide (when pressed); involucres 6–9 mm. **RAY FLS** 0 or reduced, often sterile; style 0. Various montane habitats; 1500–3400 m. KR, CaR, n SN, MP; s OR, w NV. [*M. s.* A. Gray, incl var. *eradiata* (A. Gray) Cronq. & Keck] ❀DRN,SUN:1–3,**6**,**7**,**15**,**16**.

var. ***ziegleri*** (Munz) B. Turner (p. 311) ZIEGLER'S ASTER Long-lived per or subshrub. **STS** 1–5 dm, spreading or erect; branches loosely spreading to ascending. **INFL**: heads radiate, (12)14–20 mm, 15–20 mm wide (when pressed); involucres 14–15 mm. **RAY FLS** present, fertile; style well developed. UNCOMMON. Dry coniferous forests; 1400–2470 m. PR (Santa Rosa Mtns). [ssp. *z.* Munz]

M. carnosa (A. Gray) G. Nesom (p. 311) SHRUBBY ALKALI ASTER Subshrub 5–9 dm, much-branched, ± glabrous, glaucous. **LVS**: lower 1–2 cm, linear, entire, fleshy; upper reduced to appressed, awl-shaped scales 1–4 mm. **INFL**: heads discoid, solitary at branch tips; involucre 5.5–7 mm; phyllaries gen in 4–5 series, with tips acute or acuminate, appressed or outer spreading. **RAY FLS** 0. **DISK FLS** 12–18; corollas ± 6 mm. **FR** 2–3 mm, subcylindric, many-ribbed, appressed-hairy; pappus ± 6 mm. Alkaline soils; 100–1600 m. SnJV, SNE, DMoj; s NV, w&s AZ. [*Aster intricatus* (A. Gray) S.F. Blake]

M. gracilis (Nutt.) Shinn. (p. 317) Ann. **STS** erect, 3–25 cm, leafy, branched at or above base, bristly throughout. **LVS** 1–3 cm, 3–7 mm wide; lower oblanceolate, elliptic, or oblong in outline, 1–2-pinnately lobed; upper linear, reduced, lobes and teeth bristle-tipped. **INFL**: heads radiate, solitary or in cymes; involucre 6–7 mm, 7–12 mm wide, hemispheric; phyllaries in 4–6 series, linear-lanceolate, bristle-tipped, hairy. **RAY FLS** 16–18; ligules 7–12 mm, yellow. **DISK FLS** 44–65; corollas 4.5–5.5 mm. **FR** 2.2–2.8 mm, canescent; pappus < or = 5 mm, bristles unequal, slightly wider at base, white to reddish brown. $2n=4,8$. Sandy or rocky places; < 1500 m. D; to Colorado, TX, n Mex. [*Haplopappus g.* (Nutt.) A. Gray; *H. ravenii* R. Jackson]

M. juncea (E. Greene) Shinn. (p. 317) RUSH-LIKE BRISTLEWEED Per. **STS** spreading to erect, 4–10 dm, sometimes woody at base, sparsely strigose, slightly glandular near heads, nearly lfless above base. **LVS** 1–2 cm, < or = 4 mm wide, gen linear, pinnately lobed or serrate; lobes and teeth bristle-tipped; upper lvs reduced and gen entire, glabrous or sometimes tomentose. **INFL**: heads radiate, solitary or in cymes; peduncles long, with overlapping bracts grading into phyllaries; involucre 5–8 mm, 10–12 mm wide, hemispheric; phyllaries in 5–6 series, linear, bristle-tipped, glandular. **RAY FLS** 15–25; ligules 5–6 mm, yellow. **DISK FLS** 25–40; corollas 5–6.5 mm. **FR** 2.5–3 mm, hairy; pappus ± = disk corolla, bristles unequal, tan. UNCOMMON. Dry hillsides; < 1000 m. PR; s AZ, n Mex. [*Haplopappus j.* E. Greene]

M. pinnatifida (Hook.) Shinn. var. ***gooddingii*** (Nelson) B. Turner & R. Hartman Per. **STS** erect or ascending, 2–6 dm, gen leafy, at least in lower half, ± glandular-puberulent to canescent. **LVS** 2–5 cm, pinnately lobed; lobes linear, bristle-tipped; upper reduced, entire. **INFL**: heads radiate, solitary at tips of long branches; involucre 6–9 mm, 10–18 mm wide, hemispheric; phyllaries linear-lanceolate, tip with short bristle, greenish, prominently glandular-puberulent and scabrous. **RAY FLS** 30–45; ligules 6–16 mm, yellow. **DISK FLS** many; corollas 4–6 mm. **FR** 2–3 mm, appressed-hairy; pappus of many unequal bristles 3–5 mm, tan. Rocky places; < 625 m. D; NV, AZ, nw MEX. [*Haplopappus g.* (Nelson) Munz & I.M. Johnston] ❀TRY.

M. tanacetifolia (Kunth) Nees Ann, bien 1–7 dm, puberulent to densely glandular. **STS** 1–several from base, gen branched above and ± bushy. **LVS** gen 3–12 cm, 1–2-pinnately dissected. **INFL**: heads radiate; phyllaries gen in 3–5 series, tips elongate, acuminate, spreading to bent backward. **RAY FLS** many; corollas blue-purple; ligules 1–2 cm. **DISK FLS** many; corollas 5–7 mm. **FR** 3–4 mm, narrowly obovate, ± flattened; ribs 4–6 on each face, silky; pappus 4–6 mm. $2n$=8. Uncommon. Desert scrub, pinyon/juniper woodland; ± 1700 m. e DMtns (New York Mtns.); to MT, SD, TX, n-c Mex.

MADIA TARWEED

Ann or per, gen densely glandular, aromatic. **STS** 1–several, gen simple below, ± branched above. **LVS** gen opposite below, alternate above, gen linear to lanceolate, entire to slightly toothed. **INFL**: heads gen radiate, gen peduncled, few–many; phyllaries gen 1–20, free, enclosing (and falling with) ray achenes; receptacle ± flat, gen glabrous; chaff scales gen ± fused, in ring between ray and disk fls. **RAY FLS** gen 1–20, sometimes minute; ligules 2–3-lobed, gen yellow. **DISK FLS** 1–many, sometimes staminate; corollas yellow or maroon; anther tips triangular-ovate; style tips linear to oblong, acute, bristly. **FR** club-shaped or obovoid; ray achenes compressed, thickened, or 3-angled (1 angle toward center of head), ridged, sometimes beaked; pappus 0 or of short scales; disk achenes ± symmetric; pappus 0 or of 4–10 scales or bristles. 21 spp.: w N.Am, sw S.Am. (Chilean name) [Nelson & Nelson 1980 Brittonia 32:323–325]

1. Pappus present; heads open at midday; ann or per
 2. Per; lvs opposite well up st; rosette lvs widest
  3. Phyllaries 10–12 mm; anthers black; disk fls fertile; ray achenes beakless .... ***M. bolanderi***
  3′ Phyllaries 5–6 mm; anthers yellow; disk fls staminate; ray achenes often short-beaked .... ***M. madioides***
 2′ Ann; lvs mostly alternate; rosette lvs narrow
  4. Ray achenes bowed outward, short-beaked; disk fls staminate
   5. Anthers yellow; KR .... ***M. doris-nilesiae***
   5′ Anthers brown to black; SN
    6. Ray fls 15–20, corolla 5–9 mm; disk pappus purplish .... ***M. rammi***
    6′ Ray fls 2–8, corolla 2.5–3 mm; disk pappus not purplish .... ***M. yosemitana***
  4′ Ray achenes club-shaped (slightly bowed outward or not), beakless; some or all disk fls fertile — s NCoR, n SnFrB
   7. Disk pappus scales 0.2–0.5 mm .... ***M. hallii***
   7′ Disk pappus scales 1.2–3.7 mm
    8. Heads nodding in bud and fr; disk fls fertile; pappus scales fringed, not plumose .... ***M. nutans***
    8′ Heads gen erect; central disk fls staminate; pappus scales plumose .... ***M. stebbinsii***
1′ Pappus 0; heads often closing by midday; ann
 9. Disk fls 1–2, corollas glabrous; phyllary tips not flat; sts thread-like
  10. Ray achenes compressed side-to-side, sharply angled; involucre prominently yellow-glandular ..... ***M. exigua***
  10′ Ray achenes compressed front-to-back, not angled; involucre minutely black-glandular .... ***M. minima***
 9′ Disk fls gen 2–many, corollas puberulent; phyllary tips ± flat; sts not thread-like
  11. Ray achenes with short, off-center beak; chaff scales deciduous; heads open at midday .... ***M. radiata***
  11′ Ray achenes with beak minute or 0; chaff scales persistent; heads gen closing by midday
   12. Disk fls staminate, not forming frs; receptacle hairy; rays fls often very conspicuous
    13. Ray achenes barely compressed, ± triangular with wide backs; ligules inconspicuous
     14. Ligules greenish yellow; anthers black; herbage lemon-scented .... ***M. citriodora***
     14′ Ligules bright yellow; anthers yellow; herbage not lemon-scented .... ***M. elegans*** ssp. ***wheeleri***
    13′ Ray achenes strongly compressed, with narrow backs; ligules gen conspicuous .... ***M. elegans***
     15. Basal rosette well developed; sts stout, very leafy below, heavily glandular; phyllary tips = or > body; fl summer to autumn .... ssp. ***densifolia***
     15′ Basal rosette weakly developed or 0; sts slender to ± stout, lvs crowded or well separated, moderately to heavily glandular; phyllary tips gen = body; fl spring or summer
      16. Herbage soft-hairy below, ± densely glandular above; gen > 1000 m; fl summer .... ssp. ***elegans***
      16′ Herbage sparsely hairy to densely bristly, barely glandular or not below infl; gen < 1000 m; fl spring .... ssp. ***vernalis***
   12′ Disk fls fertile, forming frs; receptacle glabrous; ray fls gen inconspicuous
    17. Ray fls 0–3; ray achenes widest at middle, ends truncate .... ***M. glomerata***
    17′ Ray fls 3–9(14); ray achenes widest toward top, tapered to base
     18. Frs obovoid, plump, black, shiny .... ***M. anomala***
     18′ Frs compressed side-to-side, dull
      19. Pls strongly glandular throughout; sts stout; involucre 7–15 mm .... ***M. sativa***
      19′ Pls gen glandular only above middle; sts often slender; involucre gen < or = 10 mm
       20. Heads subsessile or short-peduncled, in dense spike- or raceme-like clusters
        21. Herbage dark green; involucre 6–10 mm; widespread .... ***M. gracilis***
        21′ Herbage yellow-green; involucre 6–7 mm; SNF .... ***M. subspicata***
       20′ Heads slender-peduncled, in open raceme- or panicle-like clusters
        22. Branches > main st; cauline lvs widest at base; herbage densely ± appressed-soft-hairy .. ***M. citrigracilis***
        22′ Branches < main st; cauline lvs narrow at base; herbage coarsely spreading-hairy .... ***M. gracilis***

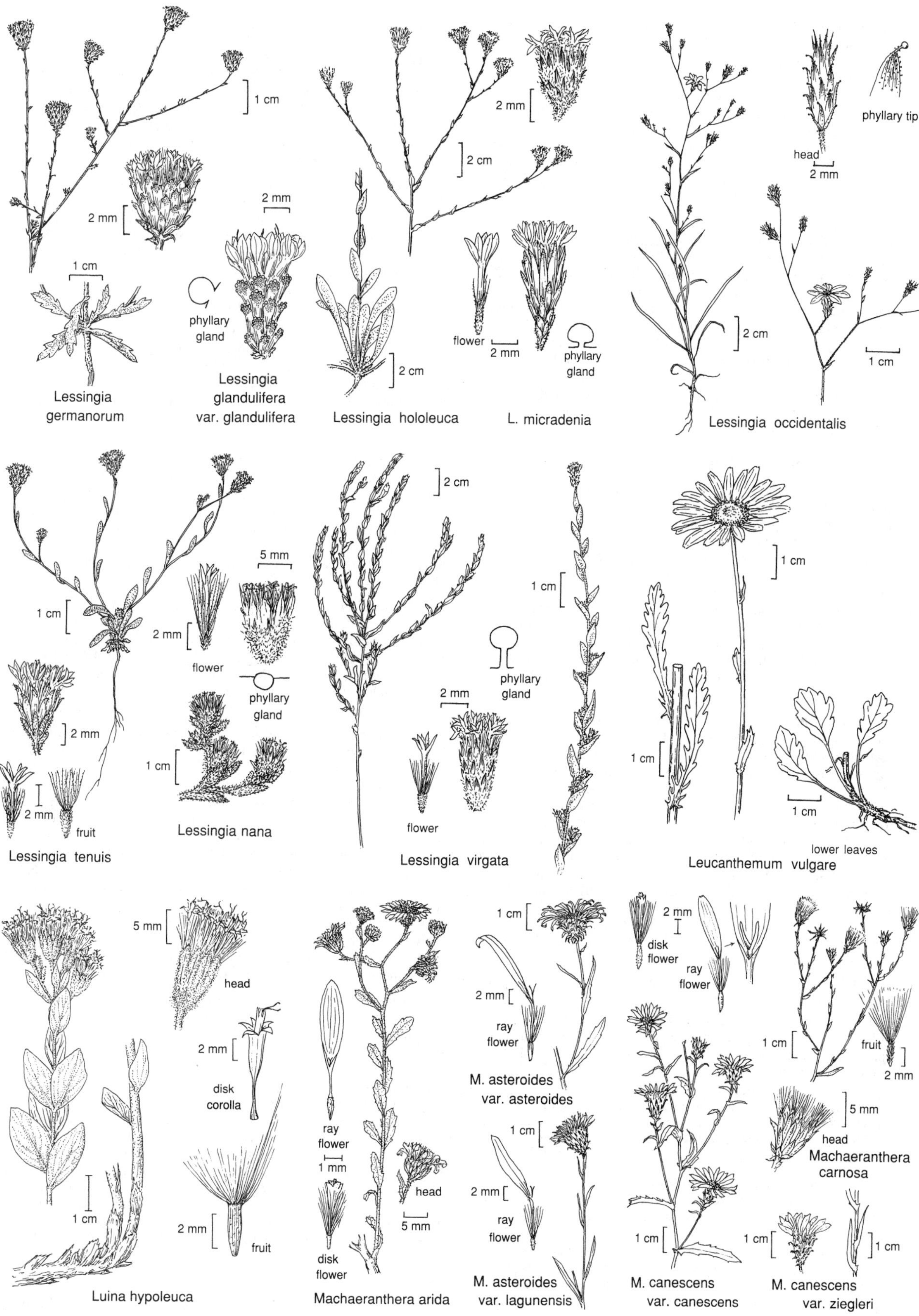
1 cm
2 mm
1 cm
Lessingia germanorum
2 mm
phyllary gland
Lessingia glandulifera var. glandulifera
2 mm
2 cm
2 cm
Lessingia hololeuca
flower
2 mm
phyllary gland
L. micradenia
phyllary tip
head
2 mm
2 cm
1 cm
Lessingia occidentalis
1 cm
5 mm
2 mm
flower
phyllary gland
2 mm
2 mm
fruit
Lessingia tenuis
1 cm
Lessingia nana
2 cm
1 cm
phyllary gland
2 mm
flower
Lessingia virgata
1 cm
1 cm
1 cm
lower leaves
Leucanthemum vulgare
5 mm
head
2 mm
disk corolla
1 cm
2 mm
fruit
Luina hypoleuca
ray flower
1 mm
disk flower
head
5 mm
Machaeranthera arida
1 cm
2 mm
ray flower
M. asteroides var. asteroides
1 cm
2 mm
ray flower
M. asteroides var. lagunensis
2 mm
disk flower
ray flower
1 cm
M. canescens var. canescens
1 cm
fruit
2 mm
5 mm
head
Machaeranthera carnosa
1 cm
1 cm
M. canescens var. ziegleri

M. anomala E. Greene (p. 317) Ann 2–5 dm. **STS** bristly below; upper branches puberulent, ± loosely bristly, glandular; glands golden to brown. **LVS** 3–7 cm, soft-spreading- to appressed-hairy, ± glandular above. **INFL**: heads few–many, in ± open flat-topped to rounded clusters; peduncles gen 1–3 cm, bracted; involucre 6–7 mm, spheric; phyllaries acute, stalked-glandular, sticky-puberulent, tips flat; chaff scales fused only near base. **RAY FLS** 3–7; corolla tubes 2–2.5 mm, ligules 3–3.5 mm. **DISK FLS** 3–6, fertile; corollas 3.5–4 mm; anthers black. **FRS** alike, 3–4 mm, obovoid, not flat, glabrous, black, shiny; beak 0; pappus 0. $2n$=32. Grassy slopes; < 500 m. NCoR, s ScV, SnFrB. Interfertile with *M. gracilis*.

M. bolanderi (A. Gray) A. Gray (p. 317) Per 5–12 dm, strongly scented. **STS** ± bristly below; upper branches densely stalked-glandular; glands dark brown to black. **LVS** 10–30 cm, opposite well up st, fused at base, bristly; upper alternate, often much reduced. **INFL**: heads few, long-peduncled, in rounded or flat-topped clusters; involucre 9–13 mm, bell-shaped or hemispheric; phyllaries densely glandular, ± bristly, tips acuminate; chaff scales free or fused only near base. **RAY FLS** 8–12, showy; corolla tubes ± 3 mm, ligules 7–13 mm. **DISK FLS** 30–65, fertile; corollas 5–7 mm; anthers black. **FR**: ray achenes 5–7 mm, club-shaped, moderately curved, ± compressed side-to-side, obscurely 5-veined, dark brown or black, ± glabrous, beak ± 0, pappus 0 or minute crown; disk achenes 5–8 mm, narrowly club-shaped, short-hairy, brown; pappus scales 5–10, 1–5 mm, linear-lanceolate, ciliate. $2n$=12. Meadows, streambanks; 1000–2600 m. KR, NCoRH, CaRH, SNH; s OR.

M. citrigracilis Keck Ann 2.5–5 dm. **STS** gen few-branched throughout; branches ascending, often overtopping main st, bristly below, soft-hairy, stalked-glandular above. **LVS** 4–8 cm, widest at base, finely soft-hairy, sometimes ± bristly; upper glandular. **INFL**: heads in raceme- or panicle-like clusters; involucre 6–8 mm, obovoid; phyllaries ± bristly, densely stalked-glandular, margins ciliate, tips flat; chaff scales fused > 1/2 length. **RAY FLS** 5(–14); corolla tubes ± 1.5 mm, ligules 5–8 mm, pale yellow. **DISK FLS** 3–10(30), fertile; corollas 2.5–3 mm; anthers black. **FRS** alike, ± 4 mm, glabrous, moderately compressed side-to-side, black or dark brown; beak ± 0; pappus 0. $2n$=48. Open areas, shrubland, forests; 1400–2000 m. e CaRH, n SNH (n Plumas Co.), MP. Polyploid derivative of *M. citriodora* × *M. gracilis*.

M. citriodora E. Greene LEMON-SCENTED TARWEED Ann 1.5–7 dm, lemon-scented. **STS** soft-hairy to ± bristly, esp below, stalked-glandular above; glands black. **LVS** 2–9 cm, sometimes ± serrate, finely appressed- to spreading-hairy, also stalked-glandular above. **INFL**: heads in open, rounded to ± flat-topped clusters; lateral branches gen overtopping main st; involucre 6–8 mm, widely ovoid or urn-shaped; phyllaries soft-hairy or ± bristly, barely glandular or not, tips flat; receptacle hairy; chaff scales strongly fused. **RAY FLS** gen 5–12; corolla tubes 0.5–1 mm, ligules 4–10 mm, greenish yellow. **DISK FLS** 15–50, staminate; corollas ± 3 mm; anthers dark. **FR**: ray achenes 3.3–4.5 mm, obovoid, ± triangular in X-section, back rounded, glabrous, dark brown or black, sometimes mottled, beak ± 0, pappus 0; disk ovaries slender, pappus 0. $2n$=16. Open dry slopes; 400–1600 m. KR, NCoRI, n SNF, MP; to WA. Widely scattered. ❀TRY.

M. doris-nilesiae T.W. Nelson & J.P. Nelson (p. 317) NILE'S MADIA Ann 9–26 cm. **STS** simple or few-branched below, sparsely bristly; upper branches minutely stalked-glandular; glands black. **LVS** 1–4 cm, minutely dentate, sparsely bristly. **INFL**: heads long-peduncled, in open cymes; involucre 3–5 mm, ovoid; phyllaries long-soft hairy, tips ± flat; chaff scales ± free. **RAY FLS** 4–8; corolla tubes 0.5–1 mm, ligules 2.5–5.5 mm, bright yellow. **DISK FLS** 8–20, staminate; corollas 2.5–3.5 mm; anthers yellow. **FR**: ray achenes 2–3 mm, strongly bowed out, black, beaked, pappus scales < 0.3 mm; disk pappus scales 7–10, 0.25–0.9 mm, fringed. $2n$=18. RARE. Serpentine; 850–1400 m. s KR (Trinity Co.).

M. elegans Lindley COMMON MADIA Ann 1–25 dm, strongly scented. **STS** simple to branched throughout, often very leafy below, soft-hairy below, sparsely to densely stalked-glandular esp above; glands yellow to black. **LVS** 2–20 cm, linear to widely lanceolate, entire or ± serrate, soft-hairy to bristly, ± glandular. **INFL**: heads in open, rounded to ± flat-topped cymes (sometimes panicle-like); involucre 4.5–12 mm, bell-shaped to hemispheric; phyllaries ± bristly, glandular or not, tips flat; receptacle hairy; chaff scales strongly fused. **RAY FLS** 5–21; corolla tubes 0.5–1 mm, ligules (2.5)4–20 mm, fan-shaped, deeply lobed, yellow, base gen maroon-spotted. **DISK FLS** 25–50+, staminate; corollas 2.5–5 mm, yellow or maroon; anthers yellow or black. **FR**: ray achenes 2.5–5 mm, compressed side-to-side or ± 3-angled, glabrous, black or dark brown, sometimes mottled, beak 0, pappus 0; disk ovaries slender, much reduced, pappus 0. Grassland, open forest; < 3350 m. CA-FP, GB; to n OR, Baja CA. Highly variable; intermediates blur distinctions among extremes.

ssp. ***densifolia*** (E. Greene) Keck (p. 317) Pl 2.5–25 dm. **LVS**: basal rosetted; cauline densely overlapping, densely soft-hairy or bristly, upper (sometimes lower) strongly glandular, hairy. **INFL**: involucre 6–12 mm; phyllary tips often > body. **RAY FLS** (5)12–20; ligules gen 10–20 mm. **DISK FLS** 15–many; corollas yellow; anthers purple-black. **FR**: ray achenes compressed. $2n$=16. Grassy slopes, valleys; gen < 1000 m. CA-FP, GB; to OR. Fl summer-autumn. ❀SUN:2,3,5,6,**7–9,14–24**.

ssp. ***elegans*** Pl 2–9 dm. **LVS**: basal few, rosette small; lower cauline ± crowded, soft-hairy, upper ± strongly glandular. **INFL**: involucre 4–10 mm; phyllary tips < or = body. **RAY FLS** 8–16; ligules 6–15 mm. **DISK FLS** few–many; corollas yellow or maroon; anthers purple-black. **FR**: ray achenes compressed. $2n$=16. Meadows, dry slopes, open forests; 900–2500 m. CA-FP (exc GV), SNE (uncommon); OR. Fl summer. ❀TRY.

ssp. ***vernalis*** Keck Pl 3–8 dm. **LVS**: basal few (rosette poorly developed or 0); cauline linear to (ob)lanceolate, scattered, soft-hairy, glandless to glandular. **INFL**: involucre 8–11 mm; phyllary tips ± = body. **RAY FLS** gen 12–14; ligules 10–20 mm. **DISK FLS** gen many; corollas yellow or maroon; anthers black. **FR**: ray achenes 4.5–5 mm, compressed. $2n$=16 Grassy sites, valleys, foothills; < 1000 (1800) m. CA-FP; OR. Fl spring. ❀SUN:5,6,**7–9, 14–24**.

ssp. ***wheeleri*** (A. Gray) Keck Pl 1–4.5 dm. **LVS**: basal few (rosette poorly developed or 0); cauline linear, scattered, soft-hairy, slightly glandular or not. **INFL**: involucre 4.5–5.5 mm; phyllary tips < body. **RAY FLS** 5–13; ligules (2.5)4–5(10) mm. **DISK FLS** 15–50+; corollas yellow; anthers yellow. **FR**: ray achenes 2.5–3.5 mm, 3-angled, barely compressed. $2n$=16. Dry forest slopes; 1500–3350 m. s SN, WTR, SnBr, SnJt; n Baja CA. Fl summer.

M. exigua (Smith) A. Gray (p. 317) THREADSTEM MADIA Ann 1–50 cm, aromatic. **STS** slender, simple below to branched throughout, ± bristly or soft-hairy, esp below, gen stalked-glandular. **LVS** 1–4 cm, strigose to ± bristly. **INFL**: heads in open, ± flat-topped to panicle-like clusters; peduncles 1–4 cm, lfless, thread-like, stiff; involucre 2.5–5 mm, top-shaped; phyllaries densely stalked-glandular (glands yellow), tips not flat; chaff scales strongly fused into ± urn-shaped tube. **RAY FLS** gen 5 or 8, inconspicuous; corolla tubes 0.6–1 mm, ligules 0.7–1 mm, pale yellow. **DISK FLS** 1(2), fertile; corollas 1.3–1.8 mm; anthers yellow. **FR**: ray achenes 1.8–2.9 mm, strongly bowed out, compressed side-to-side, back rounded, sharp-angled toward center of head, glabrous, dark brown or black, beaked, pappus 0; disk achene 1.5–2.7 mm, obovoid, weakly compressed, glabrous, pappus 0. $2n$=32. Grasslands, open woodlands; < 2500 m. CA-FP; to B.C., MT, n Baja CA.

M. glomerata Hook. (p. 317) MOUNTAIN TARWEED Ann 1.5–8 (12) dm, strongly ill-scented. **STS** simple below or with stiff ascending branches, very leafy, soft-hairy to ± bristly, esp below, stalked-glandular above; glands yellow. **LVS** 2–10 cm, mostly cauline, often with axillary clusters, gen ascending, loosely strigose, often bristly-ciliate; upper glandular. **INFL**: heads discoid or inconspicuously radiate, ± sessile to short-peduncled, in dense cymes or panicle-like clusters; involucre (incl chaff) 5.5–9 mm, narrowly ovoid or ellipsoid, sometimes curved; phyllaries 0–3, soft-hairy or ± bristly, ± glandular, tips flat; chaff scales 1–few, free, ± like phyllaries. **RAY FLS** 0–3; corolla tubes 1–3.5 mm, ligules 1.5–3 mm, greenish yellow or purple-tinged. **DISK FLS** 1–5(12), fertile; corollas 3–4.5 mm; anthers black. **FRS** alike, 4–6 mm, oblanceolate, compressed side-to-side, glabrous, black; beak 0; pappus 0. $2n$=28. Forest openings; 1050–2700 m. CaR, c&n SNH, SnBr, GB; to AK, SD, Colorado, NM. Naturalized in NCo, n SCoR.

M. gracilis (Smith) Keck (p. 317) SLENDER TARWEED Ann 1–10 dm, ± fragrant. **STS**: branches ascending, gen not overtopping main st, ± bristly to long-soft-hairy below, gen glandular above. **LVS** 3–10 cm, soft-hairy, stalked-glandular. **INFL**: heads ± sessile to long-peduncled, in raceme- or panicle-like clusters; involucre 6–10 mm, depressed-spheric to urn-shaped; phyllaries ± densely stalked-glandular, tips flat; chaff scales fused > 1/2 length, easily separated. **RAY FLS** 3–9; corolla tubes 2–3 mm, ligules 1.5–8 mm, lemon-yellow. **DISK FLS** 2–12, fertile; corollas 3–5 mm; anthers black. **FRS** alike, 2.8–5 mm, obovate, flat, ± bowed outward, glabrous, often mottled; beak ± 0; pappus 0. $2n$=32,48. Grassy areas, woodlands, open forests, many habitats; < 2400 m. CA-FP, MP; to B.C., MT, UT, n Baja CA; also s S.Am. [sspp. *collina* Keck & *pilosa* Keck] Highly variable; ± interfertile with, sometimes hybridizing with *M. anomala, M. citriodora,* & *M. sativa.*

M. hallii Keck (p. 317) HALL'S MADIA Ann 5–18 cm. **STS** simple or branches stiffly spreading; lower gen very leafy, densely bristly, branches nearly lfless (exc bases, tips), densely stalked-glandular; glands black or dark brown. **LVS** 0.5–3 cm, mostly crowded along lower sts and at branching points, bristly, thick-edged. **INFL**: heads solitary or in rounded to flat-topped clusters; peduncles mostly 1–5 cm, stiff, densely glandular; involucre 4–5.2 mm, obovoid; phyllaries acute, soft-hairy, glandular or not, tips not flat; chaff scales free. **RAY FLS** 3–6; corolla tubes ± 1 mm, ligules 2–5 mm, bright yellow, base sometimes red-tinged. **DISK FLS** 8–20, gen fertile; corollas 2.5–3.5 mm; anthers yellow. **FRS** alike, 2.8–3.2 mm, club-shaped, slightly curved, slightly compressed; beak 0; pappus forming low crown, scales 0.2–0.5 mm, minutely fringed, sometimes purplish. $2n$=18. RARE. Serpentine barrens in open chaparral; ± 500 m. s NCoRI.

M. madioides (Nutt.) E. Greene (p. 317) Per 1.5–7.5 dm. **STS** ± bristly below, densely stalked-glandular above; glands golden to dark brown. **LVS** 6–12 cm, opposite well up st, fused at base, linear to oblanceolate, entire or ± serrate, bristly-strigose or soft-hairy; upper alternate, uppermost often much reduced. **INFL**: heads few, long-peduncled, in open raceme-like or flat-topped cymes; involucre 4–6 mm, spheric to urn-shaped; phyllaries densely stalked-glandular, tip short, not flat; chaff scales sometimes fused only near base. **RAY FLS** 8–15; corolla tubes 1–1.5 mm, ligules 4–10 mm. **DISK FLS** 10–30, staminate; corollas 4–5 mm; anthers yellow to brown. **FR**: ray achenes 3–5 mm, oblanceolate to obovate, moderately bowed, strongly compressed side-to-side, obscurely 5-veined, beaked or not, pappus 0 or a minute crown; disk ovaries 3–4 mm, hairy, pappus scales 5–8, 0.2–1 mm, fringed. $2n$=14. Forests; < 1300 m. NW, n SNH, SnFrB, SCoRO; to s B.C. ❀DRN,IRR:**4**& SHD:1,**5**,**6**,7,14,**15–17**.

M. minima (A. Gray) Keck (p. 317) Ann 2–15 cm. **STS** very slender, openly branched above or throughout, soft-hairy (esp below), glandular puberulent above. **LVS** 1–2.5 cm, often clustered at nodes, sometimes ± toothed, finely appressed- to spreading-bristly. **INFL**: heads solitary or few, in dense cymes; peduncles 1–12 mm, ± thread-like, gen bracted below; involucre 2–4 mm, widely top-shaped or obovoid; phyllaries loosely appressed, tips not flat, backs minutely stalked-glandular (glands golden to black); chaff scales strongly fused. **RAY FLS** 3–5, inconspicuous; corolla tubes 0.5–1 mm, ligules 0.5–1 mm, pale yellow. **DISK FLS** 1–2, fertile; corolla 1–2.3 mm; anthers yellow. **FR**: ray achenes 1.8–2.8 mm, compressed front-to-back, strongly bowed out, not angled, hairy, black, beaked, pappus 0; disk achenes ± cylindric or club-shaped, pappus 0. $2n$=32. Open forest, shrubland; 550–2600 m. KR, NCoR, CaR, SN, se SnFrB, n WTR, SnBr, PR, MP, w DMoj; to B.C.

M. nutans (E. Greene) Keck (p. 317) NODDING MADIA Ann 5–25 cm. **STS** branched near base and above, flexible, bristly below, puberulent and minutely stalked-glandular above; glands ± black. **LVS** 1.5–3 cm, mostly alternate, ± bristly, strigose to long-soft-hairy. **INFL**: heads in open cymes or solitary, nodding in bud, esp in fr; involucre 4–7 mm, ± hemispheric; phyllaries linear, acuminate, soft-hairy, glandular or not, tips not flat; receptacle glabrous; chaff scales weakly fused ± 1/2 length. **RAY FLS** 4–8; corolla tubes ± 1 mm, ligules 3–7 mm, bright yellow, base often reddish. **DISK FLS** 7–30, fertile; corollas 3–4 mm; anthers yellow. **FR**: ray achenes 2.7–4.1 mm, club-shaped, slightly curved, beak 0, pappus 0; disk achenes 2.3–4.5 mm, slightly club-shaped, expanded at base, hairy, pappus scales ± 10, 2–3.7 mm, lanceolate, long-acuminate, fringed. $2n$=18. UNCOMMON. Rocky soils, chaparral, woodland; 150–950 m. s NCoR, n SnFrB. ❀DRN,IRR October to May, SUN:**7**,**14–17**,**19–24**.

M. radiata Kellogg (p. 317) SHOWY MADIA Ann 1–9 dm. **STS** simple below to branched throughout; branches ascending, glandular; glands yellow to brown. **LVS** 2–10 cm; upper often widest, sometimes ± clasping, entire or ± serrate, fine-glandular, often ± short-bristly. **INFL**: heads in open, cyme-like clusters; involucre 4.5–6.5 mm, depressed-hemispheric; phyllaries soft-hairy or ± bristly, acuminate, glandular, tips flat; chaff scales free, falling from head in fr. **RAY FLS** 8–16; corolla tubes 1–3 mm, ligules 6–19 mm, golden yellow. **DISK FLS** 20–65, fertile; corollas 3.5–5.5 mm; anthers yellow to black. **FR**: ray achenes 3.3–4.2 mm, plump or ± compressed side-to-side, strongly curved, glabrous, black, beaked, pappus 0; disk achenes 3.2–4.4 mm, pappus 0. $2n$=16. UNCOMMON. Grassy slopes; < 900 m. w SnJV, e SnFrB, SCoRI. ❀TRY.

M. rammii E. Greene Ann 1–6 dm. **STS**: branches ascending, moderately leafy, soft-hairy below, densely stalked-glandular above; glands golden. **LVS** 1.5–5 cm, mostly scattered, strigose to bristly. **INFL**: heads gen in open cymes; peduncles 3–10 cm; involucre 3.8–4.8 mm, widely urn-shaped; phyllaries acute, soft-hairy to short-bristly, not glandular, tips not flat. **RAY FLS** 15–20; tubes 0.5–1 mm, ligules 4–9 mm, bright yellow, sometimes ± purple- or brown-lined. **DISK FLS** 20–65, staminate; corollas 3–4 mm; anthers black. **FR**: ray achenes 2.4–3.5 mm, strongly bowed out, compressed side-to-side, beaked, pappus 0 or scales minute, fringed; disk ovaries linear, pappus scales 5–7, 3–4 mm, linear, bristle-like, ± bent, ciliate, purplish. $2n$=16. Grassy slopes; 400–1600(2700) m. n SN.

M. sativa Molina (p. 317) COAST TARWEED Ann 2–24 dm, ± strongly ill-scented. **STS** simple or branches stiff, ascending, very leafy, soft-glandular-hairy throughout, gen very sticky; glands yellow to black. **LVS**: lower 2–15 cm, linear to lanceolate, entire or toothed, finely appressed- to spreading-hairy, densely glandular to ± glandless; upper reduced or well developed, hairy, densely glandular. **INFL**: heads sessile to short-peduncled, in dense cymes or panicle-like clusters; involucre 7–15 mm, ovoid or urn-shaped; phyllaries soft-hairy or bristly, gen densely glandular, tips long, flat; chaff scales strongly fused. **RAY FLS** 6–9, gen 8; corolla tubes 1.5–4 mm, ligules 1.5–4 mm, ± = tube, greenish yellow, sometimes red-tinged. **DISK FLS** 11–14, fertile; corollas 1.7–5 mm; anthers black. **FRS** alike, 2.5–5 mm, strongly compressed side-to-side, weakly bowed out, glabrous, black or dark brown, sometimes mottled; beak 0; pappus 0. $2n$=32. Coastal grassland; < 950 m. NW, CW, SW; to AK, n Baja CA; also sw S.Am. [*M. capitata* Nutt.] Closely related to, partly interfertile with *M. gracilis*.

M. stebbinsii T.W. Nelson & J.P. Nelson (p. 317) STEBBINS' MADIA Ann 5–25 cm. **STS** branched near base and above, sparsely bristly; upper branches minutely stalked-glandular; glands black. **LVS** 1–2.2 cm, mostly crowded at st bases and branches, bristly; margins inrolled. **INFL**: heads in open cymes; involucre 4.5–5.5 mm, ovoid; phyllaries linear, acuminate, long-soft-hairy, black-glandular, tips purple; chaff scales ± free. **RAY FLS** 4–6; corolla tubes ± 1.5 mm, ligules 4–5 mm, bright yellow. **DISK FLS** 8–20; outer fertile; inner sterile; corollas ± 3.5 mm; anthers yellow. **FR**: ray achenes 2–3.5 mm, club-shaped, slightly curved, beak 0, pappus 0.5–1 mm, scales plumose; disk achenes 2–4 mm, slightly club-shaped, base expanded, glabrous, pappus scales 8–10, 1.2–3.5 mm, slender, plumose. $2n$=18. RARE. Serpentine soils; 1100–1400 m. NCoRI (Trinity & Tehama cos.).

M. subspicata Keck (p. 317) Ann 0.5–6 dm. **STS** sometimes sparingly branched above; branches gen short, ascending, soft-hairy below, sticky-glandular puberulent, stalked-glandular above; glands yellow. **LVS** 2–7 cm, narrowed to base, finely soft-appressed to spreading-hairy. **INFL**: heads subsessile or short-peduncled in spike-like clusters; involucre 6–7 mm, ovoid; phyllaries densely stalked-glandular, tips flat; chaff scales free to fused 1/2+ length. **RAY FLS** 5–8; corolla tubes 2.5–3.5 mm, ligules 1–2.5 mm, pale yellow. **DISK FLS** 5–15, fertile; corollas 3–3.5 mm; anthers black or yellow. **FRS** alike, 3–3.5 mm, widely club-shaped, compressed side-to-side, moderately curved, glabrous, dark brown to black, sometimes purple-spotted; beak 0; pappus 0. $2n$=16. Grassy slopes, oak woodland; 50–800 m. n&c SNF.

M. yosemitana A. Gray YOSEMITE TARWEED Ann 5–25 cm. **STS** very slender, simple below to branched throughout; branches gen ± ascending, moderately leafy, soft-hairy below, densely stalked-glandular above; glands golden to dark brown. **LVS** 1–3 cm, mostly scattered, strigose to bristly. **INFL**: heads gen in open cymes; peduncles gen 1–4.5 cm, thread-like; involucre 3–4.5 mm, widely top-shaped; phyllaries obtuse to acute, soft-hairy, not glandular, tips not flat; chaff scales ± strongly fused. **RAY FLS** 2–8; tube ± 1 mm; ligules ± 2 mm, pale yellow. **DISK FLS** 1–7, staminate; corollas 2–3 mm; anthers black. **FR**: ray achenes 2.7–3.8 mm, strongly bowed out, compressed side-to-side, beaked, pappus scales minute, fringed; disk ovaries linear, pappus scales 5–7, 2.5–3 mm, linear, bristle-like, ciliate, not purplish. 2*n*=16. RARE. Grassy slopes, meadows; 1200–2300 m. SN (exc Teh).

MALACOTHRIX

W.S. Davis

Ann or per < 60 dm, gen ± branched, gen erect; sap milky. **LVS** gen basal and cauline, alternate, sessile, gen reduced upward. **INFL**: heads ligulate, 2–10 mm diam; involucre 5–20 mm, gen bell-shaped; phyllaries in 3–6 series; receptacle naked or with fragile bristles < 5 mm. **FLS**: corollas readily withering; ligules yellow or white, gen ± purple-striped below. **FR** gen fusiform, straw-colored to purple-brown, truncate; veins 15 (5 gen prominent, 10 sometimes obscure); outer pappus ± 0 or of 0–6 smooth, persistent bristles, inner 12–32 bristles fused at base, readily deciduous, minutely barbed below. 21 ssp.: w N.Am, s S.Am. (Greek: soft hair) [Williams 1957 Am Midland Naturalist 58:494–517]

1. Per
 2. Ligules yellow; fr smooth, outer pappus 0 .... ***M. incana***
 2′ Ligules white; fr minutely spiny, outer pappus of irregular teeth .... ***M. saxatilis***
  3. Upper lvs gen 1–2-pinnately lobed
   4. Sts densely leafy; upper lvs (ob)ovate, 2-pinnately divided into linear segments .... var. ***implicata***
   4′ Sts ± sparsely leafy; upper lvs linear to lanceolate, sharply and narrowly lobed .... var. ***tenuifolia***
  3′ Upper lvs gen entire
   5. Upper lvs gen linear to ovate, sometimes obtuse — coastal bluffs .... var. ***saxatilis***
   5′ Upper lvs lanceolate to elliptic, gen acute
    6. Herbage tomentose .... var. ***arachnoidea***
    6′ Herbage glabrous to ± hairy .... var. ***commutata***
1′ Ann
 7. Cauline lvs with ear-like basal lobes; involucre gen spheric; outer phyllaries round .... ***M. coulteri***
 7′ Cauline lvs gen without ear-like basal lobes; involucre bell-shaped; outer phyllaries lanceolate to ovate
  8. Outer ligules gen exserted 1–4 mm
   9. St gen leafy; outer pappus 0
    10. Pls gen 10–30 cm, not mat-like; cauline lvs gen not fleshy, toothed or lobes narrow, sharp .... ***M. foliosa***
    10′ Pl gen 2–10 cm, mat-like; cauline lvs gen fleshy, lobes 1–3 pairs, short, wide, obtuse .... ***M. indecora***
   9′ St lvs gen much reduced upward; outer pappus present
    11. Basal lvs gen fleshy, lobes 3–8 pairs, wide, short, ± equal, gen toothed; outer pappus a scalloped crown, bristles 0 .... ***M. phaeocarpa***
    11′ Basal lvs gen not fleshy, toothed or narrowly lobed; outer pappus of needle-like teeth, bristle 1
     12. Basal lvs gen (ob)lanceolate; corolla gen 4.5–6 mm, (pale) yellow; fr gen 1.2–1.8 mm, veined to tip; pollen 3-pored .... ***M. clevelandii***
     12′ Basal lvs gen obovate; corolla gen 6–8 mm, white or pale yellow; fr gen 1.8–2.3 mm, very tip veinless; pollen 4-pored .... ***M. stebbinsii***
  8′ Outer ligules exserted > 5 mm
   13. St scapose; lvs gen long-hairy at base .... ***M. californica***
   13′ Sts leafy, ± branched; basal lvs gen not long-hairy at base
    14. Cauline lvs gen narrowly triangular, with 5–10 pairs of narrow, sharp teeth or lobes; scarious margins of outer phyllaries > 0.6 mm .... ***M. squalida***
    14′ Cauline lvs gen lanceolate or (ob)ovate, entire or with short teeth or lobes; scarious margins of outer phyllaries < 0.5 mm
     15. Basal lvs gen not fleshy, narrowly toothed or lobed
      16. Corolla gen < 14 mm; outer ligules exserted < 8 mm; ChI .... ***M. foliosa***
      16′ Corolla gen > 15 mm; outer ligules exserted > 9 mm; SnJV, SNE, D .... ***M. glabrata***
     15′ Basal lvs ± fleshy, with 3–8 pairs of ± equal, short, wide, gen toothed lobes
      17. Ligules gen white (yellow); fr gen 1–2 mm; outer pappus 0; not of desert habitats .... ***M. floccifera***
      17′ Ligules medium yellow; fr gen 2–4 mm; outer pappus present; desert habitats
       18. Cauline lvs widest at base; outer pappus a scalloped crown, bristles 0 .... ***M. sonchoides***
       18′ Cauline lvs gen narrowed to strap-like base; outer pappus of irregular teeth, bristles 0–6 .... ***M. torreyi***

M. californica DC. (p. 317) Ann 5–45 cm, scapose. **ST** gen glabrous. **LVS** basal, gen linear to oblanceolate; bases conspicuously long-hairy; teeth or narrow lobes well spaced. **INFL**: heads 5.2–6 mm diam; involucre gen 10–15 mm; outer phyllaries ± 1/2 inner, lanceolate, tangled-hairy. **FLS**: corollas 17–22 mm, gen yellow (white), outer corollas exserted 11–13 mm. **FR** 2–3.4 mm, ± light brown; outer pappus teeth irregular, bristles 2. 2*n*=14. Open, sandy soil in grassland, oak woodland, chaparral, desert margins; < 1700 m. SnJV, CW, SW, DMoj; Mex. ❀TRY.

M. clevelandii A. Gray (p. 317) Ann 4–36 cm, glabrous. **LVS**: basal not fleshy, gen (ob)lanceolate, teeth or short lobes ± equal, well spaced; cauline much reduced, upper 2–4-toothed at base. **INFL**: heads 2.5–3.5 mm diam; involucre 5–8.7 mm; outer phyllaries ± 1/2 inner, gen lanceolate. **FLS**: corollas 4.6–7.7 mm, pale yellow, outer corollas exserted 1–3 mm. **FR** 1.2–1.8 mm, ± light brown; veins ± equal, extending to very tip; outer pappus teeth needle-like, bristle 1. 2*n*=14. Cleared areas (burns, slides), gen chaparral; < 2100 m. NW, SNF, SnJV, CW, n ChI (Santa Rosa Island); to UT, Mex. ❀TRY.

M. coulteri A. Gray (p. 317) SNAKE'S-HEAD Ann 5–50 cm, ± branched, ascending or erect, ± glabrous, glaucous. **LVS**: basal oblanceolate to obovate, entire, toothed or short-lobed; cauline ear-like at base. **INFL**: heads 6–10 mm diam; involucre gen spheric, 10–20 mm; outer phyllaries widely overlapping, ± = inner, widely ovate to round. **FLS**: corollas 8–12 mm, gen pale yellow (white), outer corollas exserted 2–5 mm. **FR** 1.6–3.2 mm, ± light brown; 5 veins ± winged; outer pappus teeth uneven, bristles 2–6. 2*n*=14. Sandy, open areas, in coastal sage, grassland, deserts; < 1500 m. SnJV, CW, (formerly n ChI), WTR, PR, DMoj. ❀SUN,DRN:**7–11**,**14–16**,17,18,**19–24**.

M. floccifera (DC.) S.F. Blake (p. 317) Ann 5–42 cm. **ST** glabrous. **LVS** gen white hair-patched at lobe bases; basal oblanceolate to obovate, ± fleshy, ± equally and widely 6–16-lobed; upper cauline 2–4-toothed at base. **INFL**: heads 2.5–4.5 mm diam; involucre 5–9 mm; outer phyllaries ± 1/2 inner, lanceolate to ovate, glabrous. **FLS**: corollas 7–15 mm, gen white (± yellow), outer corollas exserted 5–9 mm. **FR** 1.2–2 mm, red- or purple-brown or gray; outer pappus 0. 2*n*=14. Open burns, slides, roadcuts, gen in loose soil in chaparral, pinyon/juniper woodland, yellow-pine forest; < 2000 m. NW, CaR, n&c SNH, ScV, CW, WTR. ❀SUN,DRN:2,3,**7**,8,9,**14**,15,16,**18**,19–24.

M. foliosa A. Gray (p. 317) LEAFY MALACOTHRIX Ann 4–45 cm, gen glabrous. **ST** gen leafy. **LVS**: basal obovate, sharply and narrowly 4–6-lobed. **INFL**: heads 3–7 mm diam; involucre 7–11 mm; outer phyllaries < inner, lanceolate to ovate. **FLS**: corollas 10–17 mm, yellow, outer corollas exserted 4–8 mm. **FR** 1–1.5 mm, ± brown; veins ± equal; outer pappus gen 0. 2*n*=14, 28. ± sandy open areas or among shrubs; < 150 m. UNCOMMON. n ChI (Anacapa Island), s ChI. Three closely related races incl here.

M. glabrata A. Gray (p. 317) DESERT DANDELION Ann 6–40 cm, ± glabrous. **LVS**: basal oblanceolate to obovate, with well spaced teeth or long narrow lobes, base sometimes ± hairy; cauline gen long-lobed. **INFL**: heads gen 4–7 mm diam; involucre 9–17 mm; outer phyllaries gen 1/2 inner, lanceolate, sometimes short-white-hairy. **FLS**: corollas 15–23 mm, gen pale yellow (white), outer corollas gen exserted 9–13 mm. **FR** 2–3.3 mm; outer pappus of irregular teeth, bristles 1–5. 2*n*=14. Coarse soils in open areas or among shrubs; < 2000 m. SnJV, SNE, D; to OR, ID, UT, Mex. ❀SUN,DRN:1,2,**3**,**7–10**,11,**14–16**,**18–21**,22,23.

M. incana (Nutt.) Torrey & A. Gray (p. 317) DUNEDELION Per 12–70 cm, mounded, glabrous to densely hairy. **LVS**: basal and cauline similar, obovate, entire or (gen blunt-) lobed. **INFL**: heads 4–8 mm diam; involucre 10–14 mm; outer phyllaries < inner, gen ± ovate, glabrous. **FLS**: corollas 11–20 mm, medium yellow, outer corollas exserted 5–10 mm. **FR** 1.5–2.2 mm, smooth, ± dark brown; veins ± equal; outer pappus 0. 2*n*=14. UNCOMMON. Dunes; < 300 m. CCo, SCo, n ChI, s ChI (San Nicolas Island).

M. indecora E. Greene (p. 317) SANTA CRUZ ISLAND MALACOTHRIX Ann gen 2–10 cm, mat-like, gen glabrous. **LVS**: basal and cauline similar, obovate, gen fleshy, obtusely 4–8-lobed. **INFL**: heads 2–6 mm diam, base rounded; involucre 6–8 mm; outer phyllaries ± 1/2 inner, gen ovate. **FLS**: corollas 4–8 mm, medium yellow, outer corollas exserted 1–4 mm. **FR** 1.2–1.5 mm, gen dark brown; veins sometimes ± equal; outer pappus gen 0. 2*n*=14. RARE. Shallow soils of ocean bluffs or open rocky areas; < 30 m. n ChI (1 extant population on Santa Cruz Island).

M. phaeocarpa W. Davis (p. 317) Ann 10–44 cm. **STS** ± glaucous. **LVS** gen with white hair-patch at base of each lobe; basal obovate, ± fleshy, ± equally 6–16-lobed; upper cauline 2–4-toothed at base. **INFL**: heads 2–4.5 mm diam; involucre 5–8 mm; outer phyllaries gen 1/2 inner, gen ovate, glabrous. **FLS**: corollas 5–8 mm, white, outer corollas gen exserted 1–3 mm. **FR** 1.2–2 mm, (red-)brown; outer pappus a scarious, rounded-toothed crown, bristles 0. 2*n*=14. Open chaparral, burns, Bishop-pine woodland; 300–1300 m. SnFrB, SCoR, WTR.

M. saxatilis (Nutt.) Torrey & A. Gray Per 3–60 dm, from rhizome or caudex, glabrous to tomentose. **LVS** linear to obovate, entire to irregularly lobed; basal withering early. **INFL**: heads 6–8 mm diam, base gen tapered; involucre 10–12 mm; outer phyllaries gen 1/2 inner, gen lanceolate. **FLS**: corollas 13–20 mm, white, outer corollas exserted 8–14 mm. **FR** 1.3–2.5 mm, minutely spiny, straw-colored to purple-brown; outer pappus of irregular teeth, bristles 0. 2*n*=18. Open coastal sage, chaparral, sand dunes; < 2000 m. SCoR, SCo, ChI, WTR. Vars. intergrade.

var. ***arachnoidea*** (McGregor) E. Williams (p. 321) CARMEL VALLEY MALACOTHRIX Pl rhizomed, leafy, tomentose. **LVS**: cauline lanceolate to elliptic, gen entire, tip acute. **INFL** dense. RARE. Rocky, open banks; < 200 m. c SCoRO (Carmel Valley, Monterey Co.).

var. ***commutata*** Ferris (p. 321) Pl rhizomed, leafy, glabrous to medium-hairy. **LVS**: cauline lanceolate to elliptic, gen entire, tip acute. **INFL** dense. Crumbling shale along road-cuts, in canyons, chaparral, foothill woodland; 200–1600 m. SCoR. ❀TRY.

var. ***implicata*** (Eastw.) H.M. Hall (p. 321) Pl rhizomed or caudexed, leafy, glabrous to lightly hairy. **LVS**: cauline 2-pinnately divided, lobes linear, tip obtuse. **INFL** dense. ± clay flats or canyon slopes; < 330 m. n ChI, s ChI (San Nicolas Island). ❀SUN:5,7,14,**15–17**,**19–24**.

var. ***saxatilis*** (p. 321) Pl rhizomed, leafy, glabrous to lightly hairy. **LVS**: cauline linear to lanceolate, entire or obscurely toothed, tip obtuse or acute. **INFL** dense. Coastal bluffs, on flats or in crevices; < 200 m. n SCo, sw WTR (Santa Barbara Co.).

var. ***tenuifolia*** (Nutt.) A. Gray (p. 321) Pl gen caudexed, not very leafy near heads, gen glabrous. **LVS**: cauline narrow, gen toothed or lobes sharp, tip acute. **INFL** open. Canyons, coastal-sage scrub; < 2000 m. SCo, s ChI (Santa Catalina Island; introduced on San Clemente and San Nicolas islands), TR. [var. *altissima* (E. Greene) Ferris]

M. sonchoides (Nutt.) Torrey & A. Gray (p. 321) Ann 6–40 cm, ± glabrous. **ST** glaucous. **LVS**: basal obovate, ± fleshy, ± equally and widely 6–16-lobed; cauline widely lobed at base. **INFL**: heads 4–6 mm diam; involucre 7–13 mm; outer phyllaries ± 1/2 inner, ovate, margins sometimes with tack-shaped hairs. **FLS**: corollas 10–14 mm, lemon yellow, outer corollas exserted 6–10 mm. **FR** 1.8–3 mm, dark or red-brown; outer pappus a scarious, rounded-toothed crown, bristles 0. 2*n*=14. Gen deep, sandy soils; 400–1400 m. SNE, DMoj; to WY, NM. ❀TRY.

M. squalida E. Greene (p. 321) ISLAND MALACOTHRIX Ann 3–21 cm, glabrous. **ST** ± glaucous. **LVS**: basal obovate, sharply 2–6-lobed; cauline 6–10-lobed, esp near base. **INFL**: heads 4–10 mm diam; involucre 9–12 mm; outer phyllaries ± 2/3 inner, widely ovate. **FLS**: corollas 12–19 mm, light yellow, outer corollas exserted 6–11 mm. **FR** 1.3–2.1 mm, ± dark brown; outer pappus of irregular teeth, bristles 0(1). 2*n*=28. UNCOMMON. Shallow soils, canyon flats or slopes; < 200 m. n ChI (Anacapa, Santa Cruz islands).

M. stebbinsii W. Davis & Raven (p. 321) Ann 5–60 cm, gen glabrous. **LVS**: basal gen obovate, ± narrowly lobed; upper cauline ± 2–4-lobed near base. **INFL**: heads ± 3–4 mm diam; involucre 7–10 mm; outer phyllaries ± 1/2 inner, lanceolate to ovate. **FLS**: corollas 6–8 mm, white or pale yellow, outer corollas exserted 1–2 mm. **FR** 1.7–2.3 mm, ± light brown; veins ± equal, very tip veinless; outer pappus of narrow teeth, bristles 1(2). 2*n*=28. Uncommon. Gravelly soils beneath shrubs; 300–1300 m. SNE, D; OR, NV, AZ. *M. similis* W. Davis & Raven (fr veined to very tip), mostly of Mex, collected twice in CA (n SCo, n ChI).

M. torreyi A. Gray (p. 321) Ann 4–40 cm, often with some tack-shaped hairs. **ST** ± glaucous. **LVS**: basal obovate, ± fleshy, ± equally and widely 6–16-lobed; cauline sometimes entire, gen narrowed to strap-like base. **INFL**: heads 4–5 mm diam; involucre 8–14 mm; outer phyllaries ± 1/2 inner, gen lanceolate, with tack-shaped hairs. **FLS**: corollas 14–20 mm, medium yellow, outer corollas exserted 7–10 mm. **FR** ± 2.5–4 mm, brown; outer pappus of irregular teeth, bristles 0–6. 2*n*=14. Coarse soils in desert scrub; 1500–1800 m. GB; to OR, ID, WY, Colorado, AZ. ❀TRY.

MALPERIA

A. Michael Powell

1 sp. (Anagram of Palmeri) [King & Robinson 1987 Monogr Syst Bot Missouri Bot Garden 22:235–237]

M. tenuis S. Watson (p. 321) BROWN TURBANS Ann < 40 cm, glabrous to puberulent. **LVS** opposite below, alternate above, gen sessile, < 5 cm, linear, gen entire. **INFL** open; heads discoid, many, 8–14 mm, 5–6 mm wide; involucre cylindric to narrowly bell-shaped; phyllaries 20–25, unequal, in several series, lanceolate; receptacle flat, naked. **FLS** ± 30; corollas 5–6 mm, cylindric, ± white, pink-tinged; anther bases rounded, tips oblong; style branches long, club-shaped, rounded. **FR** 3 mm, slender, 5-ribbed; pappus of 3 bristles 5 mm and 3 scales 0.5 mm. 2*n*=20. RARE in CA. Sandy creosote-bush scrub. DSon; n Mex.

MICROPUS

James D. Morefield

Ann, gen grayish, cobwebby to tomentose. **STS** 1–several, erect; central axis dominant. **LVS** simple, alternate, ± sessile, narrowly oblanceolate to elliptic, entire. **INFL**: heads disciform, ± sessile, solitary or in groups of 2–5; bracts like lvs; phyllaries 4–6, equal, obovate, rounded, scarious, persistent; receptacle chaffy; chaff scales subtending the pistillate fls (pistillate chaff) > phyllaries, each enclosing a fl, falling with a fr, bulged out and up, ± compressed side-to-side, ± thickened, hard (at least in part), ± woolly, tip ± offset to inner edge, scarious-winged; chaff scales subtending disk fls (disk chaff) 0 or reduced, open or folded, glabrous, scarious. **PISTILLATE FLS** 4–8(12) in 1(2) series; corollas tubular. **DISK FLS** staminate, 2–5; ovary vestigial. **FR** obovate, bulged out and up, compressed side-to-side, smooth; style ± offset to inner edge; pappus 0; disk pappus 0 or of 1–5 smooth, deciduous bristles, incl. 5 spp.: CA-FP, Medit. (Greek: small foot) [Morefield 1992 Madroño 39:114–130] Ongoing research may show CA spp. fit best in Eur *Bombycilaena*, a conclusion compatible with this generic description.

1. Pistillate chaff scales thick and hard only near midvein, tip ± flat to concave, projected inward from near top of scale, wing prominent, ovate; receptacle 1.2–1.8 × > wide; disk fls 4-lobed, pappus bristles 1–5; disk chaff present .... ***M. amphibolus***
1′ Pistillate chaff scales thick and hard throughout, tip ± cylindric, projected upward from inner edge of scale, wing obscure, ± linear; receptacle not > wide; disk fls 5-lobed, pappus bristles 0 or 1; disk chaff 0 .. ***M. californicus***
  2. Longest chaff scale 3–4 mm, wool dense, loose .... var. ***californicus***
  2′ Longest chaff scale 2–3 mm, wool thin, ± appressed .... var. ***subvestitus***

M. amphibolus A. Gray (p. 321) MOUNT DIABLO COTTONWEED **INFL**: heads 3.5–5 mm, ± spheric; receptacle 1.2–1.8 × > wide; pistillate chaff scales 2–3 mm, thick and hard near midvein, tip projected inward from near top of scale, ± flat to concave, wing prominent, ovate, not withering; disk chaff present. **PISTILLATE FLS** 8–12 in 2 series. **DISK FLS** 1.2–1.9 mm, 4-lobed. **FR** 1–1.5 mm; pappus bristles of disk fr 1–5, 1.7–2 mm. UNCOMMON. Bare, grassy, or rocky slopes; 50–800 m. s NCoR, SnFrB, s SCoRO (Santa Barbara Co.), rare in n SNF. [*Stylocline a.* (A. Gray) J. Howell] Distinct but ± intermediate between *M. californicus* and *Stylocline* spp.; perhaps of hybrid origin.

M. californicus Fischer & C. Meyer SLENDER COTTONWEED **INFL**: heads 2–4 mm, depressed; receptacle not > wide; pistillate chaff scales 2–4 mm, thickened and hard throughout, tip projected upward from inner edge of scale, ± cylindric, beak-like, wing obscure, ± linear, withering; disk chaff 0. **PISTILLATE FLS** 4–6 in 1 series. **DISK FLS** 1–2 mm, 5-lobed. **FR** 1.4–2.6 mm; pappus bristles of disk fr 0 or 1, 0.9–1.5 mm. Dry or moist, bare or grassy places; < 1700 m. CA-FP; w OR, nw Baja CA.

var. ***californicus*** (p. 321) **INFL**: heads 3–4 mm, 4.5–6 mm wide; pistillate chaff scales 3–4 mm, wool dense, loose. **DISK FLS** 1.3–2 mm. **FR** 1.8–2.6 mm; pappus of disk fr (0)1.3–1.5 mm. Common. Habitats and range of sp.

var. ***subvestitus*** A. Gray (p. 321) **INFL**: heads 2–3 mm, 3–4.5 mm wide; pistillate chaff scales 2–3 mm, wool thin, ± appressed. **DISK FLS** 1–1.5 mm. **FR** 1.4–2.2 mm; pappus of disk fr (0)0.9–1.3 mm. Dry places; 50–1000 m. c SNF (very uncommon), CW (exc SCoRI).

MICROSERIS

Kenton L. Chambers

Ann or per, ± mealy (hairs drying as minute white scales); sap milky. **LVS** mostly basal, gen linear to (ob)lanceolate, gen variably entire to pinnately lobed. **INFL**: heads solitary, ligulate, ± nodding in bud; involucre gen fusiform to spheric; phyllaries in 2–several series, outer overlapping, inner often ± black-hairy; receptacle naked. **FLS** 5–many; corollas white to orange, often reddish below, ligules = to >> involucre. **FR** cylindric to fusiform, gen square-topped, not beaked; ribs ± 10, ± scabrous; pappus of gen 5–many ± lanceolate, bristle-tipped scales. (Greek: small chicory) [Chambers Contr Dudley Herb Stanford U: 1955 4:207–312 & 1957 5:57–68] Hybridization common. Self-pollinating (ann) or self-incompatible and ± complex (per).

1. Per; fls >> involucre; pappus parts often > 5 (subg. *Scorzonella*)
  2. Pappus bristles many, brownish, barbed; basal scales 0; lvs basal; sphagnum bogs, NCoRO .... ***M. borealis***
  2′ Pappus bristles 5–many, each subtended by 1 small to large scale; lvs gen basal to cauline; various habitats
    3. Pappus parts 10–30; scales silvery; bristles barbed to plumose
      4. Pappus parts 10–24; bristles barbed, not plumose; KR .... ***M. laciniata*** ssp. ***siskiyouensis***
      4′ Pappus parts 15–30; bristles plumose; widespread .... ***M. nutans***
    3′ Pappus parts 5–10, scales often dull white or brownish, bristles smooth or barbed
      5. Pappus scales 5–10 mm; bristles brownish, barbed — outer phyllaries recurved; GV, adjacent foothills .... ***M. sylvatica***
      5′ Pappus scales gen < 5 mm; bristles white to brownish, often barely barbed
        6. Outer phyllaries narrowly ovate or wider, gen glabrous .... ***M. laciniata*** ssp. ***laciniata***

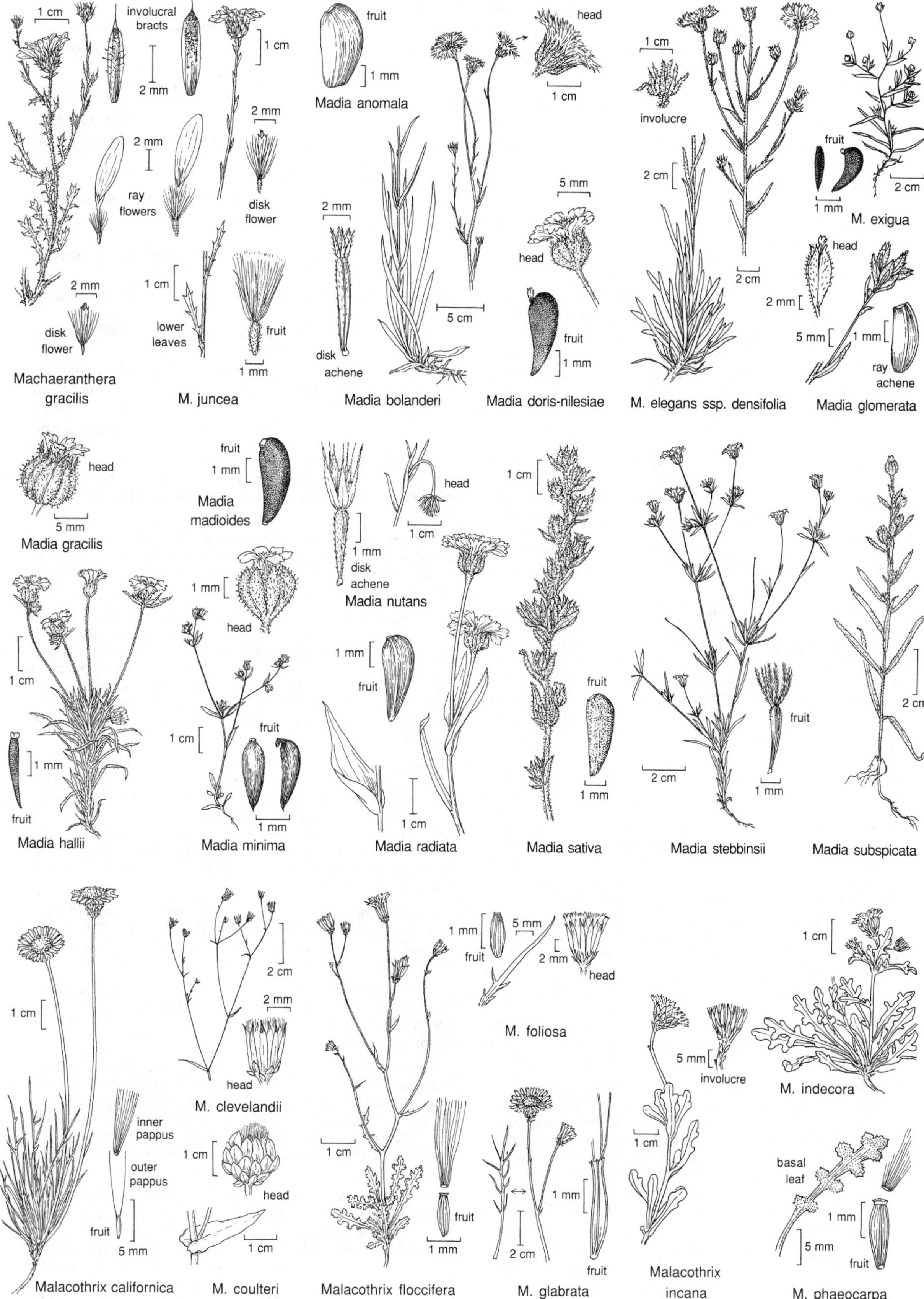
1 cm
involucral bracts
2 mm
1 cm
2 mm
2 mm
ray flowers
disk flower
2 mm
disk flower
1 cm
lower leaves
fruit
1 mm
Machaeranthera gracilis
M. juncea
fruit
1 mm
Madia anomala
head
1 cm
2 mm
disk achene
5 cm
Madia bolanderi
5 mm
head
fruit
1 mm
Madia doris-nilesiae
1 cm
involucre
2 cm
2 cm
M. elegans ssp. densifolia
fruit
1 mm
2 cm
M. exigua
head
2 mm
5 mm
1 mm
ray achene
Madia glomerata
head
5 mm
Madia gracilis
fruit
1 mm
Madia madioides
1 mm
head
1 cm
1 mm
fruit
Madia hallii
1 cm
fruit
1 mm
Madia minima
head
1 cm
1 mm
disk achene
Madia nutans
1 mm
fruit
1 cm
Madia radiata
1 cm
fruit
1 mm
Madia sativa
2 cm
fruit
1 mm
Madia stebbinsii
2 cm
Madia subspicata
1 cm
inner pappus
outer pappus
fruit
5 mm
Malacothrix californica
2 cm
2 mm
head
M. clevelandii
1 cm
head
1 cm
M. coulteri
1 cm
fruit
1 mm
Malacothrix floccifera
1 mm
fruit
5 mm
2 mm
head
M. foliosa
2 cm
1 mm
fruit
M. glabrata
5 mm
involucre
1 cm
Malacothrix incana
1 cm
M. indecora
basal leaf
1 mm
5 mm
fruit
M. phaeocarpa

6′ Outer phyllaries linear to lanceolate, often mealy and blackish hairy
7. Pappus scales 0.5–1.5 mm; bristles white or brownish; NW .... ***M. laciniata*** ssp. ***leptosepala***
7′ Pappus scales 2–4 mm; bristles brownish; n CW .... ***M. paludosa***
1′ Ann; fls ± = involucre; pappus parts < 6 (subg. *Microseris*)
8. Fr gen < 3 mm
9. Fr widest near middle, tapered to base and (slightly) to tip
10. Pappus scales 5, 1+ mm; NCo, CCo .... [2]***M. bigelovii***
10′ Pappus scales < or = 5, gen < 1 mm; widespread .... [2]***M. douglasii*** ssp. ***tenella***
9′ Fr widest at tip, tapered to base
11. Pappus scales > 2 mm, strongly inrolled; pappus bristles stout, barbed; SCo, s ChI .... [2]***M. douglasii*** ssp. ***platycarpha***
11′ Pappus scales < or = 2 mm, nearly flat; pappus bristles fine, barely barbed; widespread .... ***M. elegans***
8′ Fr gen > 3 mm
12. Pappus scales < 1 mm .... [2]***M. douglasii*** ssp. ***tenella***
12′ Pappus scales averaging 1 mm or longer
13. Pappus scales 3.5–11 mm, linear-lanceolate, barely inrolled, midrib > 1/5 scale width .... ***M. acuminata***
13′ Pappus scales < 7 mm (if linear-lanceolate midrib < 1/5 scale width)
14. Pappus scales curved only at base, midrib linear, thicker at very base — NCo, CCo .... [2]***M. bigelovii***
14′ Pappus scales finally ± curved throughout, midrib evenly tapered from thick base
15. Pappus scales 5, barely or not pigmented, finally slightly inrolled — s SNF, SnJV, e CW .... ***M. campestris***
15′ Pappus scales 5 or fewer, darkly to not pigmented, finally strongly inrolled .... ***M. douglasii***
16. Pappus scales gen < fr (if > fr, then fr > 4.5 mm); widespread .... ssp. ***douglasii***
16′ Pappus scales = or > fr; fr < or = 4.5 mm; SCo, s ChI .... [2]ssp. ***platycarpha***

M. acuminata E. Greene (p. 321) Ann 5–35 cm, scapose. **LF** 3–20 cm, gen linear-lobed. **INFL**: involucre 10–22 mm; outer phyllaries deltate, << inner. **FLS** 5–50; ligules yellow. **FR**: body 4.5–7 mm, flared at tip, evenly brown, glabrous; pappus scales 5, 3.5–11 mm, barely inrolled, white to brownish, hairy or not, midrib > 1/5 scale width, bristle-tips 4–7 mm, barbed. $2n$=36. Grassy, open rocky or clay soil; < 500 m. NCoR, n&c SNF, ScV, n SnJV, e SnFrB; OR.

M. bigelovii (A. Gray) Schultz-Bip. (p. 321) Ann 6–60 cm, scapose. **LF** 2–25 cm, gen lobed; tip often wide, blunt. **INFL**: involucre 6–14 mm; outer phyllaries deltate, << inner. **FLS** 5–100; ligules yellow or orange. **FR**: body 2.5–5 mm, widest near middle, ± brown; outermost frs gen hairy; pappus silvery to blackish, scales 5, 1–4 mm, curved at base, glabrous, midrib abruptly wider at base, bristles 3–8 mm, slightly barbed. $2n$=18. Open sandy soil, or soil pockets on rocky coastal headlands; < 100 m. NCo, CCo; to B.C. ❀TRY.

M. borealis (Bong.) Schultz-Bop. (p. 321) NORTHERN MICROSERIS Per 15–60 cm; caudex rhizome-like; roots several, fleshy. **LVS** basal, 6–30 cm, entire to few-toothed. **INFL**: involucre 10–18 mm; outer phyllaries narrow, << inner. **FLS** 20–50; ligules yellow. **FR**: body 4–8 mm, brown; ribs smooth; tip not flared; pappus scales 0; bristles 30–60, 5–10 mm, brown, barbed. $2n$=18. RARE in CA. Wet meadows, sphagnum bogs; 1000–2000 m. NCoRO (Bald Mtn., Humboldt Co.); to AK.

M. campestris E. Greene (p. 321) Ann 5–50 cm, scapose. **LF** 3–20 cm, gen lobed. **INFL**: involucre 5–10 mm; outer phyllaries deltate, << inner. **FLS** 5–100+; ligules yellow or white. **FR** 3–5+ mm, widest at tip, gray or pale brown, often dark-spotted; outermost frs gen glabrous; pappus scales 5, 1–4.5 mm, becoming curved throughout, barely inrolled, not or barely pigmented, glabrous, midrib tapered from base, bristles 3.5–5.5 mm, barbed. $2n$=36. Open clay grasslands, often near vernal pools; < 500 m. c&s SNF, SnJV, e CW. Derivative of *M. douglasii* × *M. elegans*.

M. douglasii (DC.) Schultz-Bip. Ann 5–60 cm, scapose. **LF** 3–25 cm. **INFL**: involucre 7–16 mm; outer phyllaries deltate, << inner. **FLS** 5–100+; ligules yellow or white. **FR** 3–10 mm, gray to brown or blackish, often dark-spotted; outermost frs gen hairy; pappus scales < or = 5, 0.5–6.5 mm, silvery to blackish, often hairy, bristles 1–8.5 mm, ± strongly barbed. $2n$=18. Inland clay soils, grassland, often near vernal pools or serpentine outcrops; < 1000 m. NCoR, SNF, Teh, GV, CW, SCo, ChI; OR, Baja CA. Highly variable; vars. intergrade.

ssp. ***douglasii*** (p. 321) **FR** 4–10 mm, widest at tip; pappus scales 1–6 mm (gen < fr), becoming curved throughout and strongly inrolled, bristles 3–8 mm. Habitat and range of sp. (exc ChI, Baja CA).

ssp. ***platycarpha*** (A. Gray) Chambers (p. 321) **FR** 3–4.5 mm, widest at tip; pappus scales 2–7 mm (gen > fr), becoming curved and ± inrolled, bristles 1–4 mm. Habitats of sp. c&s SCo, s ChI; Baja CA.

ssp. ***tenella*** (A. Gray) Chambers (p. 321) **FR** 3–6.5 mm, gen widest near middle; pappus scales < 1 mm, bristles 3–8.5 mm. Habitats of sp. w-c GV, CW, SCo, n ChI. Hybridizes with *M. bigelovii*; often confused with *M. elegans*.

M. elegans A. Gray (p. 321) Ann 5–35 cm, scapose. **LF** 2–20 cm, gen lobed. **INFL**: involucre 4–8(10) mm; outer phyllaries deltate, << inner. **FLS** 5–100; ligules yellow or orange. **FR** 1.5–3 mm, widest at tip, evenly brown to blackish; outermost frs sometimes hairy; pappus scales 5, 0.2–2 mm, ± flat, white, glabrous, midrib dark, bristles 3–5 mm, barely barbed. $2n$=18. Gen inland clay grassland, often near vernal pools; < 700 m. NCoR, SNF, GV, CW, SCo, ChI; Baja CA.

M. laciniata (Hook.) Schultz-Bip. Per 15–100 cm, ± branched and leafy below. **LF** 10–50 cm. **INFL**: involucre 14–30 mm; outer phyllaries linear to deltate, < to << inner. **FLS** 13–100+; ligules yellow, >> involucre. **FR** 3.5–8 mm, not or barely wider at tip, gray to brown, smooth or outermost frs scabrous on ribs; pappus scales 5–24, < 8 mm, white, bristles 4–12 mm, smooth to barbed. $2n$=18. Open grassland, meadows, rocky slopes, forest edge; < 2000 m. NW, CaRH, MP; OR.

ssp. ***laciniata*** (p. 321) **INFL**: outer phyllaries narrowly ovate or wider, thin, gen glabrous, keel 0. **FLS** 25–100+. **FR**: pappus scales 5–10, < 8 mm, bristles smooth below. Habitats and range of sp. Intergrades with ssp. *leptosepala* in NCoR. ❀TRY.

ssp. ***leptosepala*** (Nutt.) Chambers **INFL**: outer phyllaries linear to lanceolate, mealy, ± black-hairy, keel fleshy. **FLS** 13–70. **FR**: pappus scales 5–10, < 1.5 mm; bristles smooth below. Habitats of sp. NW. ❀TRY.

ssp. ***siskiyouensis*** Chambers **INFL**: outer phyllaries deltate to lanceolate, mealy, ± black-hairy, keel fleshy. **FLS** 13–50. **FR**: pappus scales 10–24, < 1.5 mm, bristles barbed throughout. Habitats of sp. KR. [*M. howellii* A. Gray misapplied] Intergrades with ssp. *leptosepala*. Ssp. *detlingii* Chambers of s OR (pappus scales 4–8 mm, outer phyllaries ovate) should be sought near Siskiyou Pass.

M. nutans (Hook.) Schultz-Bip. (p. 321) Per 10–70 cm, branched and leafy below. **LF** 5–30 cm. **INFL**: involucre 8–22 mm, gen ± mealy and black-hairy; outer phyllaries linear to deltate, << inner. **FLS** 13–50; ligules yellow. **FR** 3.5–8 mm, not widest at tip, pale brown to reddish, smooth or outermost frs scabrous on ribs; pappus scales 15–30, 1–3 mm in CA, silvery, bristles 4–7 mm, plumose. $2n$=18. Common. Moist, rocky meadows, open coniferous woodlands, sagebrush scrub; 1000–3000 m. KR, NCoRH, CaRH, SNH, MP; to B.C., SD, Colorado. Highly variable. ❀TRY.

M. paludosa (E. Greene) J. Howell (p. 321) Per 15–70 cm, gen branched and leafy only near base. **LF** 5–35 cm. **INFL**: involucre 10–20 mm, gen mealy and ± black-hairy; outer phyllaries narrowly ovate, tapered, not recurved, < inner. **FLS** 25–70; ligules yellow. **FRS** 4–7 mm, not wider at tip, straw-colored or dull white, smooth or outermost frs scabrous on ribs; pappus scales 2–4 mm, brownish, bristles 6–9 mm, barbed. 2*n*=18. Uncommon. Moist grassland or open woods; < 300 m. CCo, SnFrB. Like *M. laciniata* ssp. *leptosepala* exc pappus.

M. sylvatica (Benth.) Schultz-Bip. (p. 321) Per 15–75 cm, gen few-branched and leafy below. **LF** 10–35 cm, wavy-margined to lobed, long-tapered. **INFL**: involucre 12–25 mm, glabrous or mealy; outer phyllaries ovate, < inner, recurved. **FLS** 25–70; ligules yellow. **FR** 5–11 mm, not wider at tip, straw-colored or dull white, smooth or outermost scabrous on ribs; pappus scales 5–10 mm, bristles 6–9 mm, barbed. 2*n*=18. Uncommon. Grassland, open woods; < 1500 m. NCoRI, ScV, SNF, Teh, e SnFrB, SCoR.

MONOLOPIA

Dale E. Johnson

Ann 1–8 dm, sometimes decumbent, ± woolly. **LVS** opposite below, alternate above, linear to (ob)lanceolate, entire to wavy-dentate. **INFL**: head solitary, radiate; involucre hemispheric; phyllaries in 1 series, free or fused into lobed cup, tips black-hairy; receptacle conic, naked. **RAY FLS** 1 per phyllary; corolla gen yellow, with small lobe opposite ligule. **DISK FLS** yellow; lobe hairs large, nonglandular. **FR**: ray frs 3-angled (disk frs 4-angled), ± obconic; pappus 0. 4 spp.: CA. (Greek: single husk, from phyllaries) [Johnson 1991 Novon 1:119–124]

1. Ligules ± equally 3-lobed; disk achenes compressed front-to-back
 2. Fr ± uniformly gray-strigose; phyllaries free to ± 1/2 fused .... ***M. lanceolata***
 2′ Fr glabrous or hairs concentrated at top; phyllaries fused into cup with triangular lobes .... ***M. major***
1′ Ligules entire to slightly toothed (or middle lobe < outer); disk achenes not compressed
 3. Fr glabrous or sparsely hairy, ± 2 mm; branches ± spreading, gen near top of pl .... ***M. gracilens***
 3′ Fr ± uniformly gray-strigose, 2.5–3 mm; branches ± erect .... ***M. stricta***

M. gracilens A. Gray (p. 321) **STS**: branches mostly near top, ± spreading. **INFL**: peduncles 2–12 cm; involucre 5–7 mm; phyllaries 7–11, elliptic-oblanceolate, free. **RAY FLS**: ligules 5–10 mm, entire or slightly lobed. **FR** ± 2 mm, not compressed, ± glabrous. 2*n*=24?,26. Serpentine grassland, open chaparral, oak woodland; 100–1200 m. SnFrB, SCoR.

M. lanceolata Nutt. (p. 321) **STS** simple or branched above and below. **INFL**: peduncles 1–13 cm; involucre 6–10 mm; phyllaries ± 8, elliptic-oblanceolate, free or fused to ± 1/2. **RAY FLS**: ligules 10–20 mm, 3-lobed. **FR** 2–4 mm, compressed front-to-back, strigose. 2*n*=20. Grassland, bare clay, open chaparral, woodland; 50–1600 m. s SNF, Teh, c SCo, WTR, SnGb, nw PR. ❀SUN,DRN, IRR October to May :**7,14–24**.

M. major DC. (p. 321) **STS** simple or branched above and below. **INFL**: peduncles 1–13 cm; involucre 8–13 mm; phyllaries gen 8, fused into cup with triangular lobes. **RAY FLS**: ligules 8–20 mm, 3-lobed, sometimes cream-colored. **FR** 2.5–4 mm, compressed front-to-back, ± glabrous. 2*n*=24. Grassland, bare clay; < 1100 m. NCoRI, sw ScV, n SnJV, SnFrB, n SCoR. ❀TRY.

M. stricta Crum (p. 321) **STS** simple or branches ± erect. **INFL**: peduncles 3–5 cm; involucre 5–7 mm; phyllaries ± 8, oblanceolate. **RAY FLS**: ligules 2–7 mm (w SnJV) or 9–17 mm (s&e SnJV), entire or slightly lobed. **FR** 2.5–3 mm, not compressed, strigose. 2*n*=26. Grassland, bare clay, open chaparral, woodland; 50–800 m. s SNF, Teh, w&s SnJV, SCoRI, n WTR. ❀TRY.

MONOPTILON DESERT STAR

Ann, gen prostrate, often minute, gen ± short-bristly. **LVS** alternate, often tufted below heads, linear or oblanceolate, entire. **INFL**: heads radiate, ± subsessile; involucre bell-shaped or hemispheric; phyllaries many in 1 series, equal, ± folded, gen acuminate, purplish; receptacle convex, naked. **RAY FLS** many; ligule white to purple, often dark-veined. **DISK FLS** many; corolla yellow; style tips short-triangular. **FR** compressed, oblong to obovate, finely appressed-hairy, light brown; pappus of scales and gen 1 or more slender bristles. 2 spp.: sw N.Am. (Greek: 1 feather, from pappus)

1. Pappus a minute crown of fused scales 0.1–0.2 mm and 1 plume-tipped bristle .... ***M. bellidiforme***
1′ Pappus of 1–several non-plumose bristles and several narrow scales ± dissected into bristles .... ***M. bellioides***

M. bellidiforme A. Gray (p. 321) **STS** < 6 cm. **LVS** 4–10 mm, gen oblanceolate, obtuse. **INFL**: phyllaries 4–4.5 mm. **RAY FLS**: corolla 5–7 mm; ligule 3–5 mm. **DISK FLS**: corolla 3–4.5 mm. **FR** 1.5–2 mm; pappus a minute crown of fused scales 0.1–0.2 mm and 1 plume-tipped bristle 3–4 mm. 2*n*=16. Sandy deserts, washes; 660–1000 m. DMoj; to sw UT, w AZ.

M. bellioides (A. Gray) H.M. Hall (p. 321) **STS** < 25 cm. **LVS** 5–10 mm, gen linear, obtuse to subacute. **INFL**: phyllaries 4–6 mm. **RAY FLS**: corolla 6–11.5 mm; ligule 5–8.5 mm. **FR** 1.5–2 mm; pappus of 0–12 bristles 1–2 mm and several slender scales 0.5–1 mm, ± dissected into bristles. 2*n*=16. Sandy deserts, washes; 200–1200 m. D; to sw UT, w AZ, nw Mex.

NICOLLETIA

Ann, per; herbage ill-scented, glabrous, often glaucous. **LVS** gen alternate, simple, pinnately lobed; each lobe with an embedded oil gland, bristle-tipped. **INFL**: heads radiate, solitary or in cymes; peduncles short, stout; involucre hemispheric to bell-shaped; phyllaries gen in 2 series, free, outer 0–6, short, lanceolate to triangular, inner 8–12, oblong, membrane-margined, gland-dotted; receptacle convex, naked. **RAY FLS** gen 8; ligules showy, white to pink. **DISK FLS** many; corollas yellow, sometimes purple-tipped; style-tips long, thread-like. **FR** narrowly club-shaped, short-hairy; pappus of 5 bundles of bristles alternating with 5 lanceolate scales. 5 spp.: sw US, n Mex. (J.N. Nicollet, French astronomer, explorer, 1786–1843) [Strother 1978 Sida 369–374]

N. occidentalis A. Gray (p. 325) Per from deep taproot, glaucous. **STS** erect, 12–30 cm. **LVS** ± fleshy, 2–7 cm, divided into 5–11 pairs of short lobes. **INFL**: heads solitary; peduncles gen 2–10 mm; involucre ± cylindric; outer phyllaries 4–8 mm; inner phyllaries 14–18 mm, linear to ovate, acute to acuminate, gland-dotted. **RAY FLS** 8–12; ligules 4.5–8.5 mm. **DISK FLS**: corollas 8–9.5 mm. **FR** 7–9 mm; pappus bristles 3–7 mm, scales 6–8 mm. 2*n*=20. Sandy desert soils, often dunes, washes; 600–1400 m. s SNH, DMoj; also n Baja CA.

NOTHOCALAIS FALSE-AGOSERIS

Kenton L. Chambers

Per from ± thick caudex; sap milky. **STS** ± scapose. **LVS** basal, entire to pinnately lobed, ± glabrous. **INFL**: heads ligulate, solitary, erect; outer phyllaries ± = inner, glabrous or margins and midrib soft-white-hairy; receptacle convex, naked. **FLS** 13–many; corollas yellow, soon withering, ligules >> involucre, often reddish below. **FR** cylindric, ± 10-ribbed, tapered, not beaked, brown; pappus of many soft bristles or tapered, bristle-tipped scales. 4 spp.: N.Am. (Greek: false *Calais*) [Chambers 1957 Contr Dudley Herb 5:57–68]

1. Outer phyllaries lanceolate to ovate, evenly purple-dotted; lvs narrow to wide, often lobed, gen glabrous; pappus of soft bristles .... ***N. alpestris***
1′ Outer phyllaries linear-lanceolate, midrib purple-lined; lvs narrow, entire to wavy-margined, often ciliate; pappus of narrow, bristle-tipped scales .... ***N. troximoides***

N. alpestris (A. Gray) Chambers (p. 325) **STS** 3–25 cm, glabrous below head. **LVS** linear to widely oblanceolate, entire to lobed, gen glabrous (puberulent). **INFL**: involucre glabrous, 10–20 mm; outer phyllaries = inner but wider, green, evenly and finely purple-dotted. **FR** 5–10 mm, ± narrowed in upper 1–2 mm; pappus 6–10 mm, of 30–50 white, hair-like, minutely barbed bristles. $2n=18$. Uncommon. Rocky slopes; 1800–2500 m. KR, n SNH (El Dorado Co.); to WA. Often confused with *Agoseris glauca* var. *monticola*.

N. troximoides (A. Gray) E. Greene (p. 325) **STS** 5–40 cm, gen tomentose below head. **LVS** widely linear, entire to wavy-margined, often finely ciliate. **INFL**: involucre 12–25 mm; phyllaries ± linear, outer < or = inner, glabrous or soft-white-hairy, esp on margins and midrib, midrib gen purple-lined. **FR** 8–13 mm, cylindric-fusiform, pale brown; pappus 10–20 mm, silvery, of 10–30 narrow, tapered, bristle-tipped scales. $2n=18$. Open, rocky places; 700–2000 m. KR, MP; to WA, ID, MT.

ONOPORDUM

David J. Keil & Charles E. Turner

Bien, coarse. **ST** spiny-winged. **LVS** basal and cauline, alternate, dentate to deeply pinnately lobed, very spiny. **INFL**: heads discoid, 1–7, in cymes, in CA; involucre 3–7 cm diam, (hemi)spheric; phyllaries many, in several series, spine-tipped, outer gen spreading or reflexed; receptacle naked, deeply pitted. **FLS** many; corollas (nearly) radial, white to purple, tube slender, lobes linear; anther base acutely tailed, tips awl-shaped; style tip with minutely hairy node, terminal appendage subentire, long, cylindric, minutely papillate. **FR** ± cylindric, 4–5-angled, glabrous, gen ± cross-roughened, brown to gray-black, mottled; pappus bristles many, barbed or plumose, fused at base. ± 40 spp.: Eurasia, Medit. (Greek: name for cotton thistle) All CA spp. are NOXIOUS WEEDS.

1. Herbage green, ± sticky-glandular, short-hairy .... ***O. tauricum***
1′ Herbage ± canescent-tomentose
 2. Lvs dentate to shallowly lobed; phyllaries linear; corolla lobes not glandular ...... ***O. acanthium*** ssp. ***acanthium***
 2′ Lvs ± deeply 1–2 × lobed; phyllaries lanceolate to ovate; corolla lobes glandular-puberulent .... ***O. illyricum***

O. acanthium L. ssp. *acanthium* (p. 325) SCOTCH THISTLE Pl < 30 dm, canescent-tomentose. **LVS** 1–5 dm, dentate to shallowly lobed (lobes 8–10 pairs, widely triangular). **INFL**: phyllaries linear, puberulent, ± cobwebby-tomentose, spines < 5 mm. **FLS**: corollas 20–25 mm, purple or white, glabrous. **FR** 4–5 mm; pappus 7–9 mm, pink to reddish. $2n=34$. Disturbed sites; < 1600 m. NW, CaR, SN, SnJV, CW, SW, MP; native to Eur.

O. illyricum L. (p. 325) ILLYRIAN THISTLE Pl < 25 dm, canescent-tomentose. **LVS** 1–5 dm, (deeply) pinnately lobed (lobes 8–10 pairs, triangular, entire or again lobed). **INFL**: phyllaries lanceolate to ovate, glabrous or ± cobwebby, spines < 5 mm. **FLS**: corollas 25–35 mm, purple, lobes glandular-puberulent. **FR** 4–5 mm; pappus 10–12 mm, ± white. $2n=34$. Disturbed sites; < 500 m. s SnFrB; native to se Eur.

O. tauricum Willd. TAURIAN THISTLE Pl < 2 m, ± sticky-glandular. **LVS** 1–2.5 dm, pinnately lobed (lobes 6–8 pairs, acutely triangular). **INFL**: phyllaries lanceolate, glabrous, spines < 4 mm. **FLS**: corollas 25–30 mm, purplish pink, gen glabrous. **FR** 5–6 mm; pappus 8–10 mm. Disturbed sites; < 1400 m. SNF, n SCoRI, MP; native to Medit.

OROCHAENACTIS

1 sp. (Greek: mountain *Chaenactis*)

O. thysanocarpha (A. Gray) Cov. (p. 325) Ann 1–25 cm, short-hairy to ± tomentose and glandular. **STS** spreading to erect, simple or ± branched, very slender. **LVS** opposite below, alternate above, simple, 1–4 cm, linear to narrowly oblong-oblanceolate, entire. **INFL**: heads discoid, very small, sessile in small, leafy-bracted clusters at st tips; involucre cylindric; phyllaries 4–7 in 1 series, 3–6 mm, free, oblong, ± purple; receptacle naked. **FLS** 4–9; corollas 3–5 mm, yellow, lobes short-triangular; anther tips short-triangular; style tips tapered. **FR** 2.5–4 mm, club-shaped, ribbed, minutely puberulent; pappus of 11–17 scales, 1.5–2.5 mm, oblanceolate, obtuse, fringed, deciduous together in a ring. $2n=18$. Open forests, dry meadows, gravelly slopes; 2200–3800 m. s SNH.

OSMADENIA

Robert L. Carr & Gerald D. Carr

1 sp. (Greek: odor gland)

O. tenella Nutt. (p. 325) OSMADENIA Ann < 4 dm, ± densely glandular, ± scabrous and long-hairy, strongly scented; tack- or saucer-shaped glands 0. **ST** erect; branchlets many, thread-like, ± spreading. **LVS** alternate, sessile, reduced upward; lower 2–5 cm, linear, entire. **INFL**: heads many, radiate, in cymes; peduncles spreading, bracts few, 5–8 mm, narrowly lanceolate; phyllaries 4–7 mm, partly enfolding ray fr; chaff scales between ray and disk fls, 4–7 mm, fused. **RAY FLS** 3–5; ligules deeply 3-lobed, white to reddish, sometimes red-spotted near base. **DISK FLS** 3–10; corolla colored like ligule; anthers yellow. **FR**: ray fr 1.8–2.7 mm, back rounded to angled, rough, glabrous, inner surface ± flat, beak short, off-center, pappus 0; disk fr ± angled, tapered at base, ± thinly hairy, pappus of ± 8 scales, long and short. $2n=18$. UNCOMMON. Barren, often rocky, exposed hillsides, canyons; < 600 m. SW (exc ChI); Baja CA. [*Calycadenia t.* (Nutt.) Torrey & A. Gray]

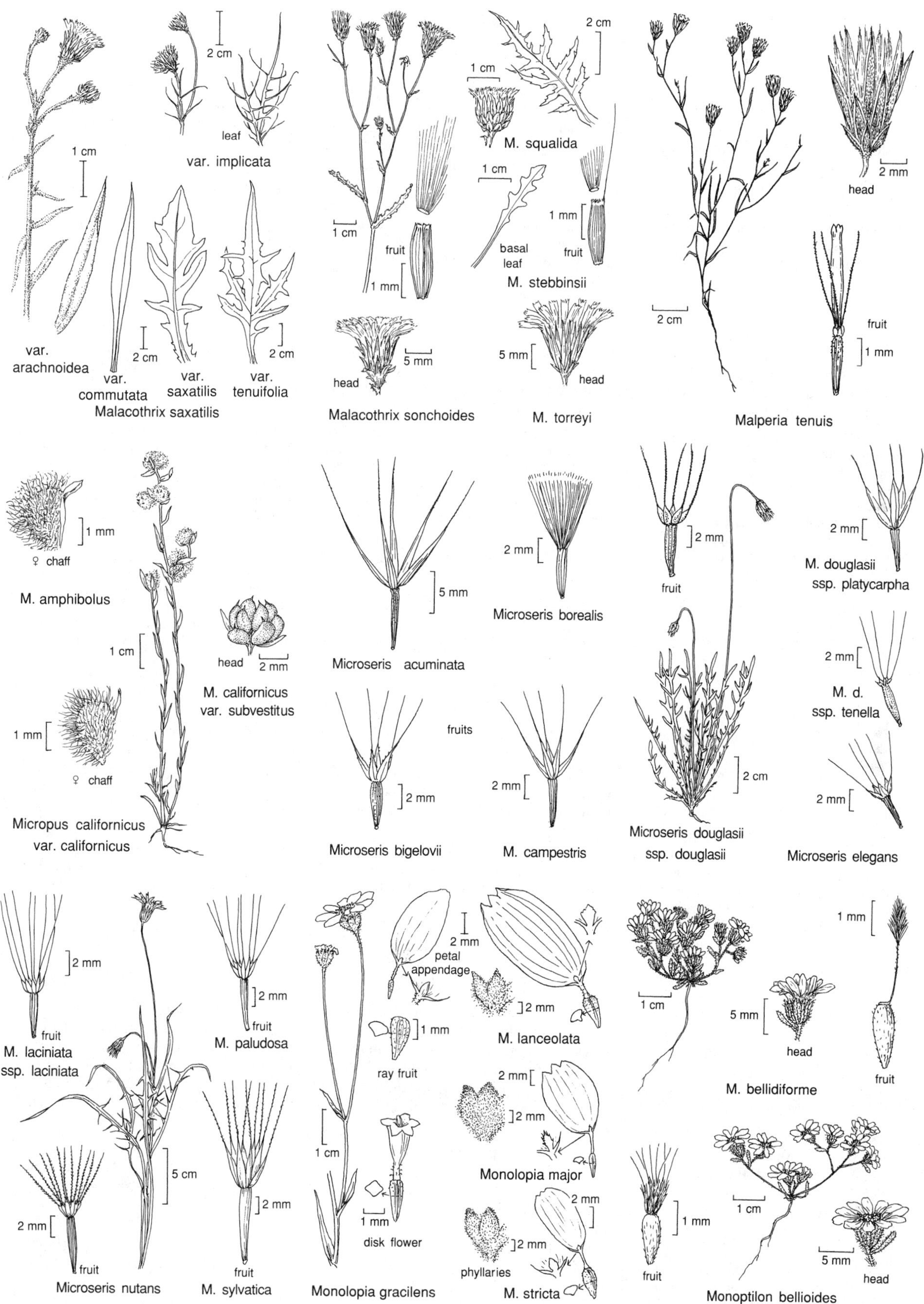

var. arachnoidea
var. commutata
var. saxatilis
var. tenuifolia
var. implicata
leaf
Malacothrix saxatilis
Malacothrix sonchoides
fruit
head
M. squalida
basal leaf
M. stebbinsii
M. torreyi
Malperia tenuis
head
fruit
♀ chaff
M. amphibolus
M. californicus var. subvestitus
Micropus californicus var. californicus
Microseris acuminata
Microseris borealis
fruits
Microseris bigelovii
M. campestris
Microseris douglasii ssp. douglasii
M. douglasii ssp. platycarpha
M. d. ssp. tenella
Microseris elegans
M. laciniata ssp. laciniata
M. paludosa
Microseris nutans
M. sylvatica
petal appendage
ray fruit
disk flower
Monolopia gracilens
M. lanceolata
Monolopia major
phyllaries
M. stricta
M. bellidiforme
Monoptilon bellioides

OSTEOSPERMUM

Elizabeth McClintock

Ann to shrubs. **LVS** gen alternate, entire to pinnately lobed. **INFL**: heads radiate, gen solitary, long-peduncled; involucre hemispheric; phyllaries in 1–3 series, margins narrowly scarious; receptacle flat or convex, naked. **RAY FLS** < 20, yellow or white to violet; ligule 3-toothed. **DISK FLS** many, staminate, blue; anther tip ovate or triangular-ovate, base sagittate, short-tailed; style branches short, not separating. **FR** < 10 mm, round or angled, smooth or tubercled; pappus 0. ± 70 spp.: esp s Afr. (Greek: bone seed, for hard achenes) [Norlindh 1943 Studies Calenduleae]

O. ecklonis (DC.) Norlindh AFRICAN DAISY Subshrub. **STS** branched from base, decumbent or ascending, sometimes > 1 m. **LVS** alternate, narrowed to a petiole, < 10 cm, oblong to oblanceolate, coarsely serrate, glandular-hairy. **INFL**: heads 1–few, 5–8 cm, closing at night. **RAY FLS** ± 15; ligule < 4 cm, white above, blue-purple beneath, or blue-purple throughout. **DISK FLS** blue-purple. **FR** ± 7 mm, narrowly triangular-ovoid, lightly net-sculptured. 2*n*= 20. Uncommon. Escape from cult in disturbed, urban, coastal areas; < 200 m. SCoRO, WTR, expected more widely; native to s Afr. [*Dimorphotheca e.* DC.]

PALAFOXIA

Ann, per, subshrubs. **STS** ascending or erect. **LVS** simple, opposite below, alternate above, gen petioled; blades entire. **INFL**: heads radiate or discoid, in cymes; involucre ± cylindric; phyllaries in 2–3 ± equal series; receptacle flat, naked. **RAY FLS** 0–few; corolla ± purple. **DISK FLS** few–many; corollas radial or ± bilateral, white to ± purple, lobes narrowly triangular; anther tips triangular; style branches linear. **FR** 4-angled, narrowly obpyramidal; pappus of 4–10 scales. 12 spp.: sw US, Mex. (J. Palafox, Spanish general, 1776–1847) [Turner & Morris 1976 Rhodora 78:567–628]

P. arida B. Turner & M. Morris (p. 325) Ann. **STS** gen erect, much-branched. **LVS** 2–12 cm; petioles 5–15 mm; blades linear to linear-lanceolate. **INFL**: heads discoid; cymes ± flat-topped; involucre cylindric or ± obconic; phyllaries linear, scabrous to densely glandular. **FLS** 9–40; corollas white to pink; anthers pink to purple. **FR** 10–15 mm, ± strigose; pappus of outer achenes 0 or of 3–8 scales of varying length; pappus of inner achenes of 4 scales 8–12 mm and 4 shorter scales. Sandy places; < 1000 m. D; NV, AZ, n Mex. [*P. linearis* (Cav.) Lagasca misapplied]

1. Pl gen 1–7 dm; main st < or = 5 mm diam; heads gen 20–25 mm .... var. ***arida***
1′ Pl gen 9–20 dm; main st 5–10 mm diam; heads 28–35 mm .... var. ***gigantea***

var. ***arida*** **STS** gen 1–7 dm; axis and branches < or = 5 mm diam, ± rough-hairy, glandular above. **LVS** 2–10 cm, ± rough-canescent. **INFL**: heads gen 5–40, gen 20–25 mm (incl fls); involucr, 5–10 mm diam, cylindric; phyllaries 10–20 mm. **FLS** 9–20; corollas 9–11 mm, outer ± bilateral. **FR** 10–15 mm. 2*n*=24. Common. Habitats and range of sp. [*P. linearis* var. *l.*] ❀STBL.

var. ***gigantea*** (M.E. Jones) B. Turner & M. Morris GIANT SPANISH-NEEDLE **STS** gen 9–20 dm, ± glabrous; axis and branches gen 5–10 mm diam, ± glabrous. **LVS** 6–12 cm, gen ± glabrous. **INFL**: heads gen 10–20, 28–35 mm (incl fls); involucre 10–20 mm diam, obconic; phyllaries 16–25 mm. **FLS** 18–40; corollas 10–13 mm, radial. **FR** 12–16 mm. 2*n*=24. RARE. Desert sand dunes; < 100 m. DSon (se Imperial Co.); sw AZ. [*P. linearis* var. *g.* M. E. Jones]

PECTIS

Ann, per, often scented. **STS** prostrate to erect. **LVS** simple, opposite, sessile, narrow, bristly-ciliate, dotted with embedded oil glands. **INFL**: heads radiate, peduncled, solitary or in leafy cymes; involucre cylindric to bell-shaped; phyllaries in 1 series, free, gland-dotted; receptacle naked. **RAY FLS** as many as and subtended by phyllaries; corollas yellow. **DISK FLS** few–many; corollas yellow, 4–5-lobed, gen 2-lipped. **FR** cylindric, gen puberulent; pappus of bristles, scales, or awns. ± 85 spp.: w US, Caribbean, S.Am. (Greek: comb, from ciliate lvs) [Keil 1977 Rhodora 79:32–78]

P. papposa Harvey & A. Gray var. ***papposa*** (p. 325) CHINCH-WEED Ann 1–20 cm, mound-shaped; herbage spicy-scented. **STS** 1–several from base, simple or much-branched. **LVS** narrowly linear, gland-dotted on margins. **INFL**: heads in dense cymes, 6–10 mm diam; peduncles 3–10 mm; phyllaries 8, 3–5 mm, linear, each with subterminal gland and several smaller submarginal glands. **RAY FLS** 8; corollas 3–6 mm. **DISK FLS** 6–14; corollas 2–3.5 mm. **FR** 2–4.5 mm; pappus of rays a low crown; pappus of disk ± 20 subplumose bristles. Arid plains, rocky slopes; < 1500 m. D; to sw UT, sw NM, nw Mex. ❀STBL.

PENTACHAETA

Meredith A. Lane

Ann from slender taproot, ± hairy (but appearing glabrous). **STS** gen simple or branching in lower half of pl, erect, gen flexible, glabrous to hairy, green to reddish. **LVS** gen narrowly linear, ciliate, green. **INFL**: heads radiate, disciform or discoid, nodding in bud, solitary, terminal; peduncles slender; involucres 3–7 mm, gen bell-shaped; phyllaries in 2–3 gen equal series, lanceolate to (ob)ovate, green, margins widely scarious; receptacle naked. **RAY FLS** present or 0, or corolla reduced to tube; corollas white, yellow, or reddish. **DISK FLS** few–many; corollas yellow to reddish or maroon; style tips linear, acute. **FR** 1.5–3 mm, oblong-fusiform, gen compressed, gen hairy; pappus of 0–20 fragile, slender bristles. 6 spp.: CA. (Greek: five bristles, from pappus) [Van Horn 1973 Univ Calif Publ Bot 65:1–41]

1. Well developed ray fls 0; ligules of pistillate fls 0 or pistillate fls 0
 2. Disk fls 4; corollas 3-lobed .... ***P. alsinoides***
 2′ Disk fls 6–34; corollas 5-lobed .... ***P. exilis*** ssp. ***exilis***
1′ Well developed ray fls present; ligules evident
 3. Ray fls white; pappus bristles present or 0
  4. Ray fls 7–16; peduncles glabrous to short-hairy .... ***P. bellidiflora***
  4′ Ray fls 0–3; peduncles soft-shaggy-hairy .... ***P. exilis*** ssp. ***aeolica***
 3′ Ray fls gen yellow; pappus bristles present
  5. Ray fls 7–12, disk fls 10–23; largest lvs < 2.5 cm .... ***P. fragilis***

5′ Ray and disk fls > 12; largest lvs 2.5–5.5 cm
6. Phyllaries ovate-elliptic, glabrous or sparsely short-hairy; pappus bristles 5–8 .... *P. aurea*
6′ Phyllaries lanceolate, soft-shaggy-hairy; pappus bristles 8–12 .... *P. lyonii*

P. alsinoides E. Greene (p. 325) **STS** 3–14 cm, hairy below heads. **LVS** < 3.5 cm, 1 mm wide; upper surface hairy; lower surface gen glabrous. **INFL**: heads < or = 200 per pl; peduncles hairy; involucres narrowly obconic; phyllaries elliptic, glabrous or sparsely hairy. **RAY FLS** gen 5; ligule < 1 mm, gen reduced to tube, light yellow, tips reddish. **DISK FLS** gen 4; corollas 3-lobed, colored as rays. **FR**: pappus bristles 3, slightly expanded at base. 2*n*=18. Grassy areas; < 320 m. NCoRO, SNF, SnFrB, SCoR, w WTR. [*Chaetopappa a.* (E. Greene) Keck]

P. aurea Nutt. (p. 325) GOLDEN-RAYED PENTACHAETA **STS** 5–36 cm, short-hairy. **LVS** < 5.5 cm, 3 mm wide; upper surface glabrous to sparsely long-hairy; lower surface glabrous to very hairy. **INFL**: heads < or = 22 per pl; peduncles short-hairy; involucres widely obconic; phyllaries ovate-elliptic, glabrous or sparsely short-hairy. **RAY FLS** 14–52; ligules 3–12 mm, yellow to brownish orange, rarely white. **DISK FLS** 30–90; corollas 5-lobed, yellow. **FR**: pappus bristles 5–8, expanded at base. 2*n*=18. Grassy areas; < 1850 m. SnGB, SnBr, PR. [*Chaetopappa a.* (Nutt.) Keck] ❀TRY.

P. bellidiflora E. Greene (p. 325) WHITE-RAYED PENTACHAETA **STS** 6–17 cm, sparsely hairy. **LVS** < 4.5 cm, 1 mm wide; upper surface glabrous; lower surface glabrous. **INFL**: heads < or = 4 per pl; peduncles glabrous to short-hairy; involucres widely bell-shaped; phyllaries elliptic to obovate, glabrous. **RAY FLS** 7–16; ligules 3–6 mm, white, lower surface sometimes reddish. **DISK FLS** 16–38; corollas 5-lobed, yellow. **FR**: pappus bristles 0 or 5, slightly expanded at base. 2*n*=18. RARE. Grassy or rocky areas; < 620 m. SnFRB. [*Chaetopappa b.* (E. Greene) Keck]

P. exilis (A. Gray) A. Gray **STS** 2–6 cm, hairy at base and below heads. **LVS** < 3.2 cm, 1 mm wide; upper surface sparsely hairy; lower surface glabrous. **INFL**: heads < or = 23 per pl; peduncles soft-shaggy-hairy; involucre ± bell-shaped; phyllaries elliptic to obovate, glabrous. **RAY FLS** 0–3; ligules 0–2 mm, white or corolla reduced to a yellow tube tinged with red. **DISK FLS** 6–34; corollas 5-lobed, yellow or reddish. **FR**: pappus bristles 0, 3, or 5, not expanded at base. 2*n*=18. Grasslands, woodlands; < 900 m. NCoR, SNF, GV, SnFrB, SCoR. [*Chaetopappa e.* (A. Gray) Keck]

ssp. ***aeolica*** Van Horn & Ornd. (p. 325) SLENDER PENTACHAETA **RAY FLS**: ligules well developed, white; corollas not reduced. **DISK FLS** yellow. RARE. Habitats of sp.; < 670 m. SCoR.

ssp. ***exilis*** **RAY FLS**: ligules 0; corollas reduced to yellow tubes tinged with red. **DISK FLS** reddish. Grasslands; < 900 m. NCoR, SNF, GV, SnFrB.

P. fragilis Brandegee (p. 325) **STS** 4–16 cm, glabrous to short-hairy (or hairy at base and below heads). **LVS** < 2.5 cm, 3 mm wide; upper surface hairy; lower surface glabrous. **INFL**: heads < or = 52 per pl; peduncles ± hairy; involucres widely bell-shaped; phyllaries elliptic to obovate, glabrous or hairy. **RAY FLS** 7–12; ligules < or = 5 mm, yellow. **DISK FLS** 10–23; corollas 5-lobed, yellow. **FR**: pappus bristles 8–20, not expanded at base. 2*n*=18. Grassy areas; 50–2100 m. s SNF, se SCoRO. [*Chaetopappa f.* (Brandegee) Keck]

P. lyonii A. Gray (p. 325) LYON'S PENTACHAETA **STS** 6–48 cm, hairy, esp below heads and at nodes. **LVS** < or = 5 cm, < or = 6 mm wide; upper surface hairy; lower surface glabrous. **INFL**: heads < or = 36 per pl; peduncles hairy; involucre widely bell-shaped; phyllaries lanceolate, soft-shaggy-hairy. **RAY FLS** 17–42; ligules 3–8 mm, yellow. **DISK FLS** 21–91; corollas 5-lobed, yellow. **FR**: pappus bristles 8–12, slightly expanded at base. 2*n*=18. **ENDANGERED** CA. Coastal habitats; < ± 150 m. c SCo (Los Angeles Co.), s ChI (Santa Catalina Island). [*Chaetopappa l.* (A. Gray) Keck]

PERICOME

Per, subshrub. **STS** many from base. **LVS** simple, gen opposite (or uppermost alternate), petioled, deltate-ovate, puberulent, gland-dotted; tip long-acuminate. **INFL**: heads discoid, small, few–many in ± flat-topped cymes, these often arrayed in leafy-bracted, compound clusters; peduncles slender; involucre cylindric to bell-shaped; phyllaries in 1 series, linear, ± fused; receptacle rounded, naked. **FLS** many; corollas 4-lobed, creamy yellow; anther tips triangular; style tips linear, tapered. **FR** oblanceolate, flat; surfaces black, puberulent, margins ± thickened, densely ciliate; pappus a low crown of fringed scales, sometimes with 1–2 bristles. 2 spp.: sw US, n Mex. (Greek: hairs around, from ciliate fr margin) [Powell 1973 Southw Naturalist 18:335–339]

P. caudata A. Gray (p. 327) **STS** < or = 2 m, much-branched, ± puberulent, resin-dotted. **LVS** many; petioles 1.5–5 cm; blades 3–12 cm, base rounded to cordate or hastate, margin entire or basal half toothed or shallowly few-lobed. **INFL**: peduncles 5–30 mm; involucre 4.5–6 mm diam; phyllaries 4.5–7 mm, fused in lower half, margins transparent, tips soft-hairy. **FLS**: corollas 3–5 mm. **FR** 3.5–5 mm; pappus scales ± 1 mm, bristles 0–2, 1–4.5 mm, gen unequal. 2*n*=36. Dry, rocky slopes; 1200–2400 m. s SN, SNE; to Colorado, TX, n Mex.

PERITYLE

Ann, per, subshrubs, shrubs. **LVS** opposite or alternate, simple to deeply divided or compound, sessile or petioled. **INFL**: heads radiate, discoid or disciform, solitary or in cymes; peduncles short or long; involucre cylindric, hemispheric, or bell-shaped; phyllaries in 2–3 ± equal series, linear to ovate; receptacle flat to conic, naked. **RAY FLS**: ligules yellow or white. **DISK FLS** many; corollas yellow or white, 4-lobed; anther tips triangular; style tips tapered. **FR** linear to oblanceolate, very flat, sometimes weakly 3–4-angled; surface dark brown or black, glabrous or puberulent; margins gen ± thick, puberulent to strongly ciliate; pappus 0 or a crown of fringed scales and 0–2 slender bristles. ± 75 spp.: sw N.Am. (Greek: around the margin, from thick fr margin) [Powell 1974 Rhodora 76:229–306] ❀STBL.

1. Ann; heads gen radiate .... *P. emoryi*
1′ Subshrubs; heads discoid
2. Herbage gen long-hairy
3. Lf margins serrate to lobed; lvs opposite or alternate .... *P. inyoensis*
3′ Lf margins entire or lobes 1–3 per margin; lvs alternate .... *P. villosa*
2′ Herbage short-rough-hairy .... *P. megalocephala*
4. Lvs many, narrowly ovate to ± round, 4–9 mm wide .... var. *megalocephala*
4′ Lvs few, linear to lanceolate, 1–4 mm wide .... var. *oligophylla*

P. emoryi Torrey (p. 327) Ann 2–60 cm, puberulent to rough-hairy and glandular. **STS** simple to much-branched. **LVS** gen alternate, petioled; blades 2–10 cm, ovate, round, or triangular, coarsely toothed to palmately lobed, teeth and lobes gen again toothed or lobed. **INFL**: heads radiate (rarely disciform), 1–many; peduncles 0.1–7 cm; involucre hemispheric to bell-shaped; phyllaries many, 5–6 mm, lanceolate or oblanceolate to ovate. **RAY FLS** gen 8–12; ligules 1.5–4 mm, white, rarely vestigial. **DISK FLS**: corollas 2–2.5 mm, yellow. **FR** gen 2–3 mm; margins thin, ciliate; surfaces of ray achenes gen ± puberulent; surfaces of disk achenes gen glabrous; pappus scales well developed or vestigial, bristle 0 or 1, 1–2.5 mm. $2n$=64–72, 100–116. Common. Desert plains, slopes, and washes; < 1000 m. D (uncommon SCo, ChI); NV, AZ, n Mex; also in w S.Am.

P. inyoensis (Ferris) A. Powell (p. 327) INYO LAPHAMIA Subshrub 12–25 cm. **STS** many; hairs soft, spreading. **LVS** opposite or alternate; petioles 0.5–2 mm; blade 1–2 cm, ovate to triangular or round, tip ± acute, margin serrate to lobed, surface soft-hairy and glandular. **INFL**: heads discoid, 1–3; peduncles 1–4 cm; involucre bell-shaped; phyllaries 14–21, 5.6–6.5 mm, linear-lanceolate. **RAY FLS** 0. **DISK FLS**: corollas 4–5 mm. **FR** 3–3.5 mm, 1 or both surfaces rounded or angled, puberulent; margins hairy; pappus 0 or a crown of vestigial scales. $2n$=±36. RARE. Dry, rocky slopes; 1800–2600 m. W&I (s Inyo Mtns). [*Laphamia i.* Ferris]

P. megalocephala (S. Watson) J.F. Macbr. Subshrub 15–55 cm, short-rough-hairy. **STS** many, much-branched. **LVS** gen alternate; petiole 1–6 mm; blade 7–15 mm, 1–10 mm wide, linear to widely ovate, tip acute to obtuse, margin gen entire, rarely serrate or lobed. **INFL**: heads discoid, 1–few, loosely clustered; peduncles gen 1–5 cm, bracts lf-like; involucre 3.5–7.5 mm diam, bell-shaped; phyllaries 14–19, 5–6 mm, lanceolate or oblong. **RAY FLS** 0. **DISK FLS**: corollas 3.5–4.2 mm. **FR** 2.5–3 mm; 1 or both surfaces rounded or ± angled, puberulent; margin hairy; pappus 0 or a crown of vestigial scales, rarely with 1 bristle. Rocky places; 1300–2800 m. W&I, DMtns; NV.

var. ***megalocephala*** Pl gen 30–60 cm. **LVS** many, 7–15 mm, 4–9 mm wide, lance-ovate to round. **INFL**: heads gen 5–6 mm diam. **FR**: pappus bristle 0. $2n$=34. Rocky slopes; 1500–2800 m. W&I, DMtns; NV. [*Laphamia megalocephala* S. Watson var. *m*]

var. ***oligophylla*** A. Powell Pl gen 15–35 cm. **LVS** few, 7–17 mm, gen 1–4 mm wide, linear to lanceolate. **INFL**: heads gen 3.5–7.5 mm diam. **FR**: pappus bristle 0 or rarely 1. $2n$=68. Rocky slopes; 1300–2600 m. W&I, DMtns. [*Laphamia m.* S. Watson var. *intricata* (Brandegee) Keck misapplied]

P. villosa (S.F. Blake) Shinn. HANAUPAH LAPHAMIA Subshrub 13–20 cm, soft-hairy. **LVS** alternate; petiole 3–6 mm; blade 12–22 mm, ovate to widely wedge-shaped, tip acute, margins entire or with 1–3 short, pointed lobes, surface soft-hairy. **INFL**: heads discoid, 1–3; peduncle gen 1–2 cm; involucre 5–7 mm diam, bell-shaped; phyllaries 13–23, 6 mm, linear-lanceolate to oblong. **RAY FLS** 0. **DISK FLS**: corollas 4–5 mm, yellow. **FR** 3–3.5 mm, puberulent; surfaces rounded; pappus 0 or a vestigial crown, sometimes with 1–2 bristles 1–2 mm. $2n(=3x)$=51. RARE. Dry, rocky slopes; 1700–2600 m. W&I (Inyo Mtns), n DMtns (Panamint, Grapevine mtns). [*Laphamia v.* S.F. Blake]

PETASITES

Per from stout rhizome, ± dioecious. **STS** erect, appearing before lvs. **LVS**: basal, large, long-petioled, blades wide; cauline gen sheathing, scale-like. **INFL**: heads disciform or weakly radiate, in raceme-like to ± flat-topped clusters; involucre ± cylindric to bell-shaped; main phyllaries in 1 series, equal; receptacle flat, naked. **PISTILLATE FLS** 0–few in staminate heads, many in pistillate heads; corollas of 2 kinds, white to pale yellow, sometimes purplish, outer often with short ligules, inner cylindric. **DISK FLS** staminate, 0–few at center of pistillate heads, many in staminate heads; corollas white to pale yellow, sometimes purplish; anther bases entire or short-sagittate, tips acute; style tips slightly thickened, entire or slightly lobed; pappus reduced. **FR** cylindric, 5–10-ribbed; pappus bristles many. 15 spp.: N.Am, Eurasia. (Greek: broad-brimmed hat, from large lvs) [Cronquist 1978 N.Am Fl II 10:174–179]

P. frigidus (L.) Fries var. ***palmatus*** (Aiton) Cronq. (p. 327) COLTSFOOT **ST** 2–6 dm. **LVS**: basal blade 10–40 dm wide, ± round to reniform, lobes palmate, coarsely toothed or again pinnately lobed, (sub)glabrous above, sometimes loosely tomentose below. **INFL**: bracts 2– 6 cm, entire to serrate, parallel-veined; heads gen 10–20; involucre 5–9 mm, bell-shaped, often purplish; phyllaries linear. **PISTILLATE FLS**: outer corollas with tube 4–6 mm, ligules 2–7 mm, white to pale pink; inner corollas 3–5 mm, cylindric. **DISK FLS**: corollas 3.5–5 mm, white to pale pink. **FR** 3–4.5 mm; pappus 6–13 mm. $2n$=60,61,62. Forests, gen wet soil; < 400 m. NW, nw&nc CW; to AK, ne US. [*P. p.* (Aiton) A. Gray] ❀IRR or WET:**4–6,15–17**&SHD:7,8,9,**14,19–24**;GRCVR (deciduous in winter); may be INV; also STBL.

PETRADORIA ROCK GOLDENROD

Meredith A. Lane

1 sp.: w N.Am. (Greek: rock goldenrod)

P. pumila (Nutt.) E. Greene ssp. ***pumila*** (p. 327) Per < 3 dm, light green, glabrous, taprooted. **STS** several, erect, striate, gummy. **LVS** alternate, 2–12 cm, linear to (ob)lanceolate, leathery, 3-veined, resin-dotted, entire; margins scabrous. **INFL**: heads radiate, many, in dense flat-topped clusters; involucres < 10 mm, < 3 mm diam, cylindric; phyllaries 10–21 in 3–6 series, oblong to ovate, light yellow, tips green; receptacle naked. **RAY FLS** 2–3; ligules yellow. **DISK FLS** 2–4, staminate; corollas < 6.2 mm, yellow. **FR** 4–5 mm, 6–9-veined, compressed, glabrous; pappus bristles thread-like, ± twisted, brownish. $2n$=18. Rocky soils, pine forest to juniper scrub, often on limestone; 2300–3400 m. DMtns; to ID, WY, NM. [Anderson 1963 Trans KS Acad Sci 66:632–684] ❀TRY.

PEUCEPHYLLUM

1 sp.: sw N.Am. (Greek: fir-leaf) [Strother 1978 N.Am Fl II 10:160–173]

P. schottii A. Gray (p. 327) PYGMY-CEDAR Shrub or small tree < 3 m, rounded. **ST** densely leafy, green. **LVS** alternate, gen 1–2 cm, narrowly linear, thick, gen entire, glabrous, gland-dotted, resin-varnished. **INFL**: heads solitary, discoid; peduncle 8–25 mm, leafy-bracted; involucre ± obconic; phyllaries 9–18 in 1 series, 8–12 mm, linear to lanceolate, thick, acuminate, gland-dotted near tip, margins often scarious; receptacle flat, naked. **DISK FLS** 12–21; corollas 6.5–8.5 mm, pale yellow, tube << throat; anther bases weakly tailed, tips lanceolate to ovate; style branches minutely papillate, rounded-truncate. **FR** 3–4 mm, narrowly obconic, weakly angled, blackish, bristly; pappus of many fine bristles, 2–5 mm (sometimes also 15–20 slender scales 4–6 mm), straw-colored to red-brown. Rocky slopes, often among boulders; < 1400 m. D; AZ, nw Mex. ❀TRY;DFCLT.

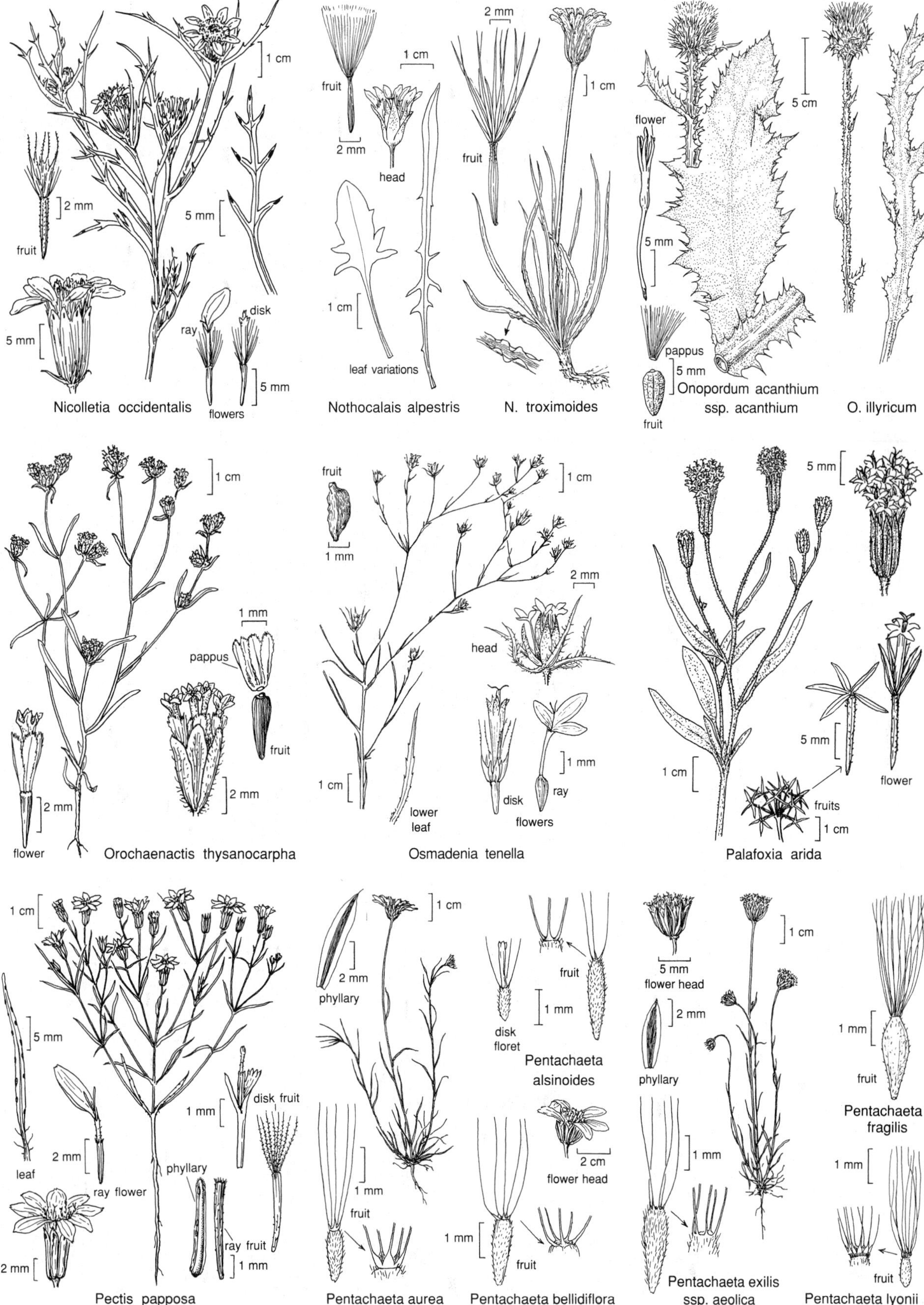
1 cm
2 mm
fruit
5 mm
5 mm
ray
disk
5 mm
flowers
Nicolletia occidentalis
fruit
2 mm
1 cm
head
1 cm
leaf variations
Nothocalais alpestris
2 mm
1 cm
fruit
N. troximoides
flower
5 cm
5 mm
pappus
5 mm
fruit
Onopordum acanthium
ssp. acanthium
O. illyricum
1 cm
1 mm
pappus
fruit
2 mm
2 mm
flower
Orochaenactis thysanocarpha
fruit
1 mm
1 cm
2 mm
head
1 cm
lower
leaf
1 mm
disk
ray
flowers
Osmadenia tenella
5 mm
5 mm
flower
1 cm
fruits
1 cm
Palafoxia arida
1 cm
5 mm
leaf
2 mm
ray flower
1 mm
disk fruit
phyllary
ray fruit
1 mm
2 mm
Pectis papposa
1 cm
2 mm
phyllary
1 mm
fruit
Pentachaeta aurea
fruit
disk
floret
1 mm
Pentachaeta
alsinoides
2 cm
flower head
1 mm
fruit
Pentachaeta bellidiflora
5 mm
flower head
2 mm
phyllary
1 cm
1 mm
Pentachaeta exilis
ssp. aeolica
1 mm
fruit
Pentachaeta
fragilis
1 mm
fruit
Pentachaeta lyonii

PHALACROSERIS

Kenton L. Chambers

1 sp. (Greek: bald chicory)

P. bolanderi A. Gray (p. 327) Per from ± woody caudex, glabrous; sap milky. **STS** 10–35 cm, scapose. **LVS** basal, 6–20 cm, linear to oblanceolate, entire. **INFL**: heads ligulate; involucre 7–12 mm, green, glabrous; phyllaries ± equal or outermost shorter, inner ± fused near base in fr. **FLS** 13–many; corollas yellow, soon withering, ligules >> involucre. **FR** 3 mm, oblong, ± 4-angled, flat-topped, smooth, pale brown, dark-spotted; pappus 0 or a low crown. $2n$=18. Wet meadows, sphagnum bogs; 1800–2500 m. n&c SNH. Pls with crown-like pappus have been called var. *coronata* H.M. Hall.

PICRIS OX-TONGUE

G. Ledyard Stebbins

Ann, bien; sap milky; herbage bearing rigid hairs, most of which have 2–4-hooked branches at tip, often with scattered prickles. **STS** solitary, branched, leafy. **LVS** basal and cauline, alternate, entire to pinnately lobed. **INFL**: heads ligulate, solitary, cymes, or panicles; phyllaries in 2–3 overlapping series; receptacle naked. **FLS** many; ligules yellow, readily withering. **FR** oblong, 5–10-ribbed, covered with minute spines, beaked; pappus 2 rows of plumose or simple bristles. ±45 spp.: Eurasia, Afr. (Greek: bitter)

P. echioides L. (p. 327) BRISTLY OX-TONGUE **STS** coarse, stout, 3–8 dm. **LVS** 5–20 cm, oblong, entire, coarsely toothed, or shallowly lobed; lower tapered to winged petioles; upper sessile, sometimes clasping. **INFL**: heads 2–4 cm diam, terminal and axillary, short- to long-peduncled in leafy-bracted cymes; involucre 15–20 mm; outer phyllaries spreading to ascending, lf-like, widely ovate, cordate, >> and often concealing the erect, lanceolate, tapered, inner phyllaries. **FR** 5–7.5 mm (incl beak); body ± = beak, brownish, transversely roughened; beak slender; pappus, 4–7 mm, plumose, white. $2n$=10. Common. Weed of waste places; < 450 m. CA-FP; native to Eur.

PLEUROCORONIS ARROW-LEAF

A. Michael Powell

Per or subshrub. **LVS** opposite below, alternate above. **INFL** open; heads discoid, few; phyllaries in several series, outer short, ovate, inner lanceolate; receptacle ± flat, naked. **FLS** 25–30; anther bases ± rounded, tips ovate or oblong; style branches, ± 1.5 mm, club-shaped. **FR** ellipsoid to obconic, 4–5-ribbed; sides densely hairy; pappus of bristles and scales. 3 spp.: sw US, nw Mex. (Greek & Latin: side crown, from pappus) [King & Robinson 1987 Monogr Syst Bot Missouri Bot Garden 22:237–239]

P. pluriseta (A. Gray) R. King & H. Robinson (p. 327) Subshrub < 60 cm. **LVS** thin; blades 3–10 mm (<< glandular petioles), lanceolate to ± diamond-shaped, gen few-toothed, glabrous or ± glandular. **INFL**: heads 6–11 mm; phyllaries ± 25, 2.5–6 mm, glandular, tips darker, gen recurved. **FLS**: corollas 4–5 mm. **FR** 3–4 mm; pappus bristles 10–16, 2.5–5 mm, scales 10–12, 1–2 mm. $2n$=18. Common. Rocky creosote-bush scrub; < 1300 m. D; to UT, AZ, nw Mex. [*Hofmeisteria p.* A. Gray]

PLUCHEA

G. Ledyard Stebbins

Ann to shrub, stiff. **LVS** alternate. **INFL**: heads disciform, many; involucre ± hemispheric; phyllaries in 3–5 unequal series, outer leathery above, inner narrower, ± membranous; receptacle naked. **PISTILLATE FLS** many; corolla very slender, 4–5-lobed. **DISK FLS** few, fruiting or staminate; corolla 5-lobed, pink or purple in CA; anther bases short-tailed, tips ± ovate; style branches 0 to short, obtuse. **FR** ± cylindric, grooved; pappus 1 series of slender bristles. ± 40 spp.: trop, warm temp. (N.A. Pluche, 18th century French naturalist)

1. Ann or per, glandular; lvs 4–12 cm, ovate, toothed, not crowded .... ***P. odorata***
1′ Shrub, nonglandular, silky-hairy; lvs 1–4 cm, linear to lanceolate, entire, crowded .... ***P. sericea***

P. odorata (L.) Cass. (p. 327) SALT MARSH FLEABANE Ann or per 5–12 dm, coarse, glandular, ill-scented. **LVS** 4–12 cm, ovate, not crowded, toothed. **INFL**: involucre 4.5–5.5 mm. **PISTILLATE FLS**: corollas 3.5–4 mm, purple. **DISK FLS**: corollas 4–5 mm, purple. **FR** ± 1 mm, minutely rough-hairy; pappus bristles slender to tip. $2n$=20. Moist, often saline valley bottoms; 0–300 m. s SNF, GV, SnFrB, SCo, ChI, D; to e US, Caribbean, n S.Am. [*P. purpurascens* (Sw.) DC. and *P. camphorata* (L.) DC. misapplied] ❀TRY.

P. sericea (Nutt.) Cov. (p. 327) ARROW WEED Shrub 1–5 m, finely silky, not scented, nonglandular. **LVS** 1–4 cm, lanceolate, crowded, entire. **INFL**: involucre 3.5–4.5 mm. **PISTILLATE FLS**: corollas 5–6.5 mm, pink to deep rose. **DISK FLS**: corollas 5–6 mm, pink to deep rose. **FR** 0.5–1 mm, smooth. $2n$=20. Forming thickets in stream bottoms, washes, canyons, around springs, sometimes in saline areas; < 600 m. SnJV, SCoRI, SCo, s ChI, TR, PR, D; to TX, nw Mex. ❀STBL;INV.

POROPHYLLUM

Ann, per, subshrubs; odor strong. **STS** erect. **LVS** simple, alternate or opposite; blade dotted with embedded oil glands. **INFL**: heads discoid, peduncled, solitary or in ± leafy cymes; involucre ± cylindric; phyllaries in 1 series, equal, free or fused at base, gland-dotted; receptacle naked. **FLS**: corollas white to greenish yellow or purple; style branches long, slender. **FR** cylindric; pappus of many bristles. ± 30 spp.: N.Am., S.Am. (Greek: pore-leaf, from gland dotted lvs) [Johnson 1969 Univ Kansas Sci Bull 48:225–267]

P. gracile Benth. (p. 327) ODORA Subshrub; herbage ill-scented. **STS** 1–many, 3–7 dm, glaucous, glabrous; branches slender, ascending. **LVS** 1–5 cm, narrowly linear, entire, glaucous. **INFL**: heads 1–few; involucre 4–8 mm diam; phyllaries 5, free, 10–16 mm, oblong, glaucous, dotted or streaked with glands. **FLS** 20–30; corolla 7–9 mm, purplish or whitish. **FR** 8–10 mm; pappus 6–7 mm, dull white to brownish. $2n$=48. Rocky slopes; < 1500 m. D; to NV, TX, n Mex. ❀STBL.

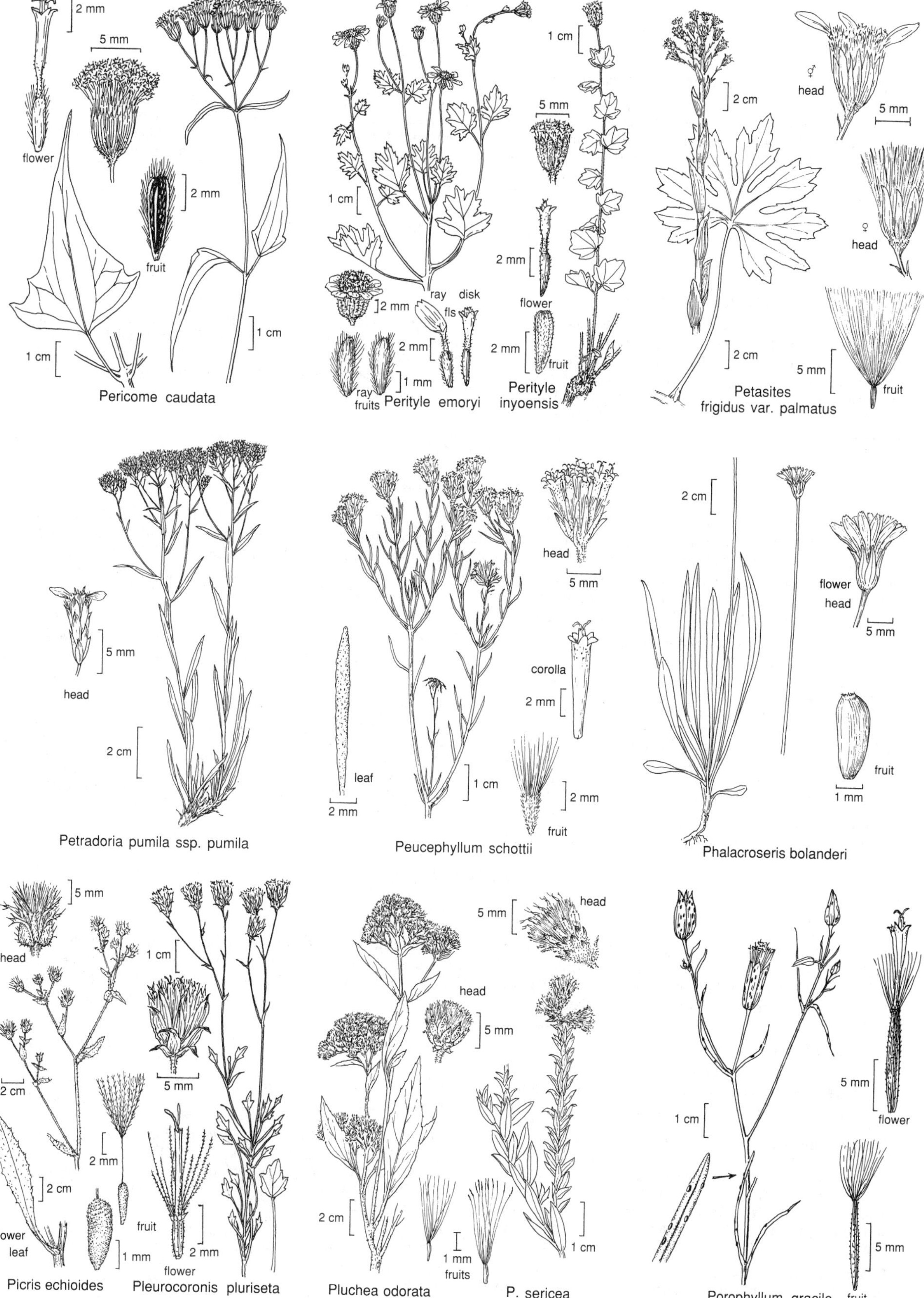
2 mm
5 mm
flower
2 mm
fruit
1 cm
1 cm
Pericome caudata
1 cm
1 cm
ray disk
2 mm
fls
2 mm
1 mm
ray fruits
Perityle emoryi
5 mm
2 mm
flower
2 mm
fruit
Perityle inyoensis
2 cm
head
5 mm
head
2 cm
5 mm
fruit
Petasites frigidus var. palmatus
head
5 mm
2 cm
head
Petradoria pumila ssp. pumila
head
5 mm
corolla
2 mm
leaf
2 mm
1 cm
2 mm
fruit
Peucephyllum schottii
2 cm
flower head
5 mm
fruit
1 mm
Phalacroseris bolanderi
5 mm
head
1 cm
2 cm
5 mm
2 mm
2 cm
lower leaf
1 mm
fruit
2 mm
flower
Picris echioides
Pleurocoronis pluriseta
5 mm
head
head
5 mm
2 cm
1 mm
fruits
1 cm
Pluchea odorata
P. sericea
5 mm
flower
1 cm
5 mm
fruit
Porophyllum gracile

PRENANTHELLA

G. Ledyard Stebbins

1 sp.: w N.Am. (Latin: diminutive of *Prenanthes*)

P. exigua (A. Gray) Tomb (p. 333) Ann, 1–4 dm; sap milky. **STS** slender, gen openly much-branched, sparsely glandular-puberulent. **LVS**: basal, 2–4 cm, oblanceolate, entire to pinnately lobed, ciliate, otherwise glabrous, often early-deciduous; cauline few, reduced, upper scale-like. **INFL**: heads ligulate, many, in open panicles; peduncles very slender; involucre cylindric, 4–5 mm; phyllaries in 2 series, outer 2–3, < 1 mm, inner 3–4, 4–5 mm, lanceolate; receptacle naked. **FLS** 3–4; ligules light pink or white, readily withering. **FR** 3–4 mm, cylindric, 5-ribbed, white; pappus of fine bristles, 2–4 mm, white. $2n$=14. Desert canyons and valleys, juniper woodlands; < 1850 m. SNE, D; to UT, Colorado, w TX. [*Lygodesmia e.* A. Gray] [Tomb 1972 Brittonia 24:223–228]

PRIONOPSIS

Gregory K. Brown

1 sp. (Greek: saw-like, from lf margin)

P. ciliata (Nutt.) Nutt. Ann or bien 5–15 dm, erect, glabrous, branched above. **LVS** alternate, sessile, 3–8 cm, oblong to ± obovate, spiny-dentate, obtuse. **INFL** open; heads radiate, few, 15–25 mm wide; involucre hemispheric; phyllaries in 3–5 series, linear-lanceolate, outer ± recurved, ± green; receptacle flat, naked. **RAY FLS** ± 25–30; corollas 12–18 mm, yellow. **DISK FLS** many; outer fruiting; corollas 7 mm, yellow. **FR** 2–4 mm, ± oblong, tan, glabrous, attached at hard, whitish ring; pappus of many bristles, ± 7 mm, yellow to red-brown, falling ± as a unit. Uncommon. Along railroads, disturbed sites; < 500 m. SnFrB, SCo; native to c US. [*Haplopappus c.* (Nutt.) DC.]

PSATHYROTES

Theodore M. Barkley

Ann to low, dense subshrub (< 40 cm diam in CA), much-branched, ± hairy and scaly; odor turpentine-like. **LVS** alternate, hairy; blade ± ovate to reniform. **INFL**: heads discoid, peduncled in axils; involucre ± obconic; phyllaries in 2 series; receptacle flat, naked. **DISK FLS** 9–50; corollas cylindric, in CA light yellow and often fading reddish, glandular, ± soft-hairy; anther bases ± sagittate, tips acute to blunt; style branches ± shaggy-papillate, ± truncate or with tapered appendage. **FR** cylindric to obconic, weakly 10-ribbed, densely hairy in CA; pappus of many bristles in 1–4 series. 5 spp.: sw N.Am. (Greek: brittleness, from sts) [Strother 1978 N.Am Fl II 10:142–146]

1. Lvs sparsely tomentose; outer phyllaries erect, like inner; fls gen 3–16; pappus bristles < 50, in 1 series .. ***P. annua***
1′ Lvs velvety; outer phyllaries recurved, wider than inner; fls gen 21–26; pappus bristles > 120, in 3–4 series .... ***P. ramosissima***

P. annua (Nutt.) A. Gray (p. 333) **STS** often purplish. **LVS** short-petioled; blade 8–16 mm, gen ± toothed upward, (gray-) green. **INFL**: outer phyllaries 4–5 mm, erect; inner phyllaries gen 8 or 13. **FLS** 13–16; corollas 4–4.5 mm. **FR** ± 2.5 mm; pappus bristles 35–50 in 1 series, coarse, red-brown. $2n$=34. Dry, sandy shadscale scrub, alkali flats; gen 800–2000 m. SNE, DMoj; to ID, UT, nw AZ.

P. ramosissima (Torrey) A. Gray (p. 333) TURTLEBACK **STS** woolly, becoming glabrous and shiny. **LVS** ± long-petioled; blade 8–20 mm, prominently few-toothed, brown- to gray-green, velvety and woolly-scaly. **INFL**: outer phyllaries 5–6 mm, recurved, wider than inner; inner phyllaries 12–15, deciduous. **FLS** 16–32; corollas 4.5–5 mm. **FR** 2–3+ mm; pappus bristles 120–140 in 3–4 series, fine, brownish. $2n$=34. Sandy creosote-bush scrub; gen < 1000 m. s SNE, D; s NV, AZ, nw Mex. ❀DRN,SUN:**10**,11,**12**,**13**&DRY: 1–3,8,9,19–21;DFCLT.

PSEUDOBAHIA

Dale E. Johnson

Ann, ± woolly. **LVS** alternate, entire to pinnately lobed. **INFL**: heads solitary, radiate; involucre bell-shaped or hemispheric; phyllaries in 1 series, equal, fused at base, margins sometimes translucent; receptacle conic or hemispheric, naked. **RAY FLS** 1 per phyllary; ligules 5–10 mm, ± ovate, yellow, tip entire or slightly toothed. **DISK FLS** yellow; bases long-hairy. **FR** oblanceolate, 3–4-angled, ± compressed front-to-back; pappus 0. 3 spp.: CA. (Greek: false *Bahia*) [Johnson 1991 Novon 1:119–124]

1. Largest lvs entire or 3-lobed .... ***P. bahiifolia***
1′ Largest lvs 1–2-pinnately lobed
 2. Largest lvs 1-pinnately lobed; phyllaries fused to 1/2 length .... ***P. heermannii***
 2′ Largest lvs 2-pinnately lobed (exc small pls); phyllaries fused only at base .... ***P. peirsonii***

P. bahiifolia (Benth.) Rydb. (p. 333) HARTWEG'S PSEUDOBAHIA Pls 5–20 cm. **LVS** 8–25 mm, linear-oblanceolate, entire or 3-lobed. **INFL**: peduncles 2–5 cm; involucre 5–6 mm; phyllaries 3–8, elliptic-lanceolate. **DISK FLS** ± 2.5 mm; lobes glabrous. **FR** 1.5–2.5 mm. $2n$=8. **ENDANGERED** CA. Grassland, open woodland, in clay soil; ± 150 m. c SNF (Madera Co.), e SnJV. [*Monolopia b.* Benth.] Threatened by agriculture.

P. heermannii (Durand) Rydb. (p. 333) Pls 1–3 dm. **LVS** 1–3 cm; lower 1-pinnately lobed; segments 0.5–1.5 mm wide. **INFL**: peduncles 2–5 cm, often reddish; involucre 5–6 mm; phyllaries ± 8, elliptic-lanceolate, fused ± 1/2, enclosing frs, hard, crested below sinus. **DISK FLS** ± 2.5 mm; lobes glabrous. **FR** 2–2.5 mm. $2n$=6+, 8+,10+. Sandy or rocky grassland, open chaparral, woodland, forest; 160–1600 m. CaRF (Butte Co.), SNF, SCoRO. [*Eriophyllum h.* (Durand) E. Greene] ❀TRY.

P. peirsonii Munz (p. 333) TULARE PSEUDOBAHIA Pls 2–7 dm. **LVS** 2–6 cm, triangular-ovate, gen 2-pinnately lobed; segments 1–5 mm wide. **INFL**: peduncles 2–8 cm; involucre 6–9 mm; phyllaries ± 8, oblanceolate, fused at base; receptacle rounded-conic. **DISK FLS** ± 3 mm; lobes sparsely glandular. **FR** ± 3 mm. $2n$=16.

ENDANGERED CA. Grassland, bare dark clay; 100–800 m. s SNF (Kern Co.), se SnJV (Fresno, Tulare cos.). Some ray fl tips reflect ultraviolet light. Threatened by agriculture, flood control.

PSILOCARPHUS WOOLLY-HEADS, WOOLLY MARBLES

James D. Morefield

Ann, gray to green, cobwebby to tomentose. **STS** gen several from base, gen spreading, < 20 cm, ± forked; central axis gen 0 or not dominant. **LVS** simple, gen opposite, ± sessile, < 3 cm, ± obtuse, entire (uppermost ± alternate or seeming whorled, gen ± appressed to heads). **INFL**: heads disciform, sessile, solitary or in small groups, gen spheric; bracts like lvs; phyllaries 0; receptacle 1–2 × > wide, ± obovoid, gen entire, chaffy; chaff scales each enclosing a pistillate fl, falling with a fr, length gen < 3 × width, gen ± obovoid, bulged upward, woolly, net-veined, tip scarious-winged, offset to inner edge, gen ± 2/3 the distance from base to top of scale, projected inward, beak-like, hidden in head. **PISTILLATE FLS** in many series, each subtended by a chaff scale; corollas tubular. **DISK FLS** staminate, 2–10, not subtended by chaff scales; ovary vestigial. **FR** smooth, shiny; pappus 0. 3–5 ill-defined spp.: w N.Am, s S.Am. (Greek: slender chaff) [Cronquist 1950 Res Stud State Coll WA 18:71–89] Spp. appear to intergrade, need detailed study.

1. Largest head 6–14 mm; longest chaff scale gen 2.8–4 mm, or if shorter then hidden by dense shaggy hairs
 2. Largest head 9–14 mm, ± ovoid, receptacle deeply lobed; chaff scales narrowly cylindric, length > 3 × width, wing ± 1/2 distance from base to tip of scale . . . . ***P. brevissimus*** var. ***multiflorus***
 2′ Largest head 6–9 mm, spheric, receptacle entire or shallowly lobed; chaff scales obovoid, < 3 × width, wing ± 2/3 distance from base to tip of scale
  3. Sts gen spreading, branched from base, central st 0 or not dominant; uppermost lvs gen lanceolate to ovate, longest gen 8–15 mm, gen 1.5–4 × > wide; wool dense, loose; fr ± obovoid . . . . . ***P. brevissimus*** var. ***brevissimus***
  3′ Sts ± erect, gen branched above, central st dominant; uppermost lvs gen linear to oblanceolate, longest gen 17–35 mm, gen 4.5–9 × > wide; wool thin, ± appressed; fr ± cylindric . . . . ***P. elatior***

1′ Largest head gen < 6 mm; longest chaff scale gen < 2.8 mm, visible through thin, ± appressed, cobwebby or silky hairs
 4. Lvs linear to narrowly oblanceolate, gen > 5 × width, gen > 3 × heads; fr ± cylindric . . . . ***P. oregonus***
 4′ Lvs gen ovate to obovate, gen < 5 × width, < 3 × heads; fr ± obovoid . . . . ***P. tenellus***
  5. Uppermost lvs ovate to widely elliptic, gen < 2 × width, ± appressed to heads; disk fls gen 4-lobed . . . . var. ***globiferus***
  5′ Uppermost lvs oblanceolate to obovate, gen > 2 × width, spreading below heads; disk fls gen 5-lobed . . . . var. ***tenellus***

P. brevissimus Nutt. Pls variously tomentose. **STS** several from base or branched above, spreading to erect; central axis various. **LVS**: uppermost gen lanceolate to ovate, longest 8–25 mm, 1.5–6 × > wide. **INFL**: largest head 6–14 mm; longest chaff scale gen 2.8–4 mm. **DISK FLS** 0.8–1.6 mm, (4)5-lobed. **FR** 0.8–1.9 mm, ± obovoid. Vernal pools and flats; < 2500 m. CA-FP, MP; to WA, MT, UT, nw Baja CA; s S.Am.

var. ***brevissimus*** DWARF WOOLLY-HEADS Pls gen densely shaggy-tomentose. **STS** gen several from base, spreading; central st 0 or not dominant. **LVS**: longest of uppermost gen 8–15 mm, gen 1.5–4 × > wide. **INFL**: largest head 6–9 mm, spheric; receptacle entire to shallowly lobed; longest chaff scale < 3 × > wide, ± obovoid, wing ± 2/3 distance from base to top of scale. **DISK FLS** (4) 5-lobed. $2n=28$. Habitats and range of sp. Possible intermediates to *P. tenellus* var. *t.* have been called *P. globiferus* Nutt.

var. ***multiflorus*** Cronq. (p. 333) DELTA WOOLLY MARBLES Pls gen thinly silky-tomentose. **STS** gen branched above, ± erect; central st dominant. **LVS**: longest of uppermost lvs gen 14–25 mm, gen 3–6 × > wide. **INFL**: largest head 9–14 mm, ± ovoid; receptacle deeply lobed; longest chaff scale > 3 × > wide, narrowly cylindric, wing ± 0.5 distance from base to top of scale. **DISK FLS** 5-lobed. UNCOMMON. Vernal pools, flats; 10–500 m. Deltaic GV, SnFrB. Like *P. elatior*.

P. elatior (A. Gray) A. Gray TALL WOOLLY-HEADS Pls gen thinly cobwebby-tomentose. **ST** gen branched above, ± erect; central axis dominant. **LVS**: uppermost gen linear to oblanceolate, longest 17–35 mm, gen 4.5–9 × > wide. **INFL**: largest head 6–8 mm; longest chaff scale gen 2.8–3.8 mm. **DISK FLS** 1.3–1.9 mm, 5-lobed. **FR** 0.9–1.7 mm, ± cylindric. UNCOMMON. Vernally moist places; < 1000 m. KR, CaR, MP; sw B.C., ID. A few specimens from within the range of *P. brevissimus* var. *b.* appear to be this sp., which may be a mere variant.

P. oregonus Nutt. (p. 333) OREGON WOOLLY-HEADS Pls gen silky-tomentose. **STS** gen several from base, spreading; central axis 0 or not dominant. **LVS**: uppermost linear to narrowly oblanceolate, longest 12–20 mm, gen 6–12 × > wide. **INFL**: largest head 4–6 mm; longest chaff scale 1.5–2.7 mm. **DISK FLS** 0.7–1.4 mm, gen 4-lobed. **FR** 0.6–1.2 mm, ± cylindric. Vernal pools, rarely moist slopes; 10–1500 m. n CA-FP (rare and doubtful in s CA-FP), MP; to WA, ID, Baja CA?. [misspelled *P. oreganus*] Highly variable; apparently intergrades with *P. tenellus* var. *t.*, esp in s CA-FP, Baja CA.

P. tenellus Nutt. Pls variously tomentose. **STS** gen several from base, spreading; central axis 0 or not dominant. **LVS**: uppermost oblanceolate to obovate, longest < 16 mm, gen 1.5–5 × > wide. **INFL**: largest head < 6 mm; longest chaff scale gen 1.8–2.7 mm. **DISK FLS** 0.8–1.5 mm, 4–5-lobed. **FR** 0.6–1.2 mm, ± obovoid. Vernal pools, coastal dunes, dry disturbed soil; < 2000 m. CA-FP; to B.C., ID, nw Baja CA; c Chile.

var. ***globiferus*** (DC.) Morefield (p. 333) ROUND WOOLLY MARBLES Pls variously tomentose. **LVS**: uppermost ± appressed to and concealing heads, ovate to widely elliptic, gen 1.5–2 × > wide. **DISK FLS** gen 4-lobed. UNCOMMON. Vernal pools, coastal dunes; 0–700 m. c&s SNF, SnJV, CCo, SnFrB; c Chile. [var. *tenuis* (Eastw.) Cronq.; *P. globiferus* Nutt. misapplied] A similar SnBr form seems best assigned to *P. brevissimus* var. *b.*

var. ***tenellus*** (p. 333) Pls gen thinly cobwebby-tomentose. **LVS**: uppermost ± spreading below heads, oblanceolate to obovate, gen 1.5–5 × > wide. **DISK FLS** gen 5-lobed. Dry slopes on gen disturbed soil, rarely vernal pools; 10–2000 m. CA-FP; to B.C., ID, nw Baja CA.

PSILOSTROPHE PAPER-DAISY

Per, subshrubs. **STS** 1–many, erect, ± hairy. **LVS** simple, alternate, entire or pinnately lobed, ± soft-hairy; upper often ± glandular. **INFL**: heads radiate, solitary or in cymes; peduncles short to long; involucre cylindric to bell-shaped; phyllaries 5–13 in 1–2 subequal series; receptacle flat, naked. **RAY FLS** 2–6; corollas yellow, fading to cream-colored, persistent on fr; ligules ovate, 3-lobed, reflexed when dry. **DISK FLS** few–many; corollas yellow, densely hairy; anther tips triangular; style tips truncate. **FR** cylindric (or ray achenes slightly flat), ribbed, glabrous, glandular or soft-hairy; pappus of 4–6 unequal transparent scales. 6 spp.: sw US, n Mex. (Greek: naked turn) [Brown 1973 PhD dissertation AZ State Univ]

P. cooperi (A. Gray) E. Greene (p. 333) Per, subshrub 2–6 dm. **STS** densely white-tomentose, openly branched. **LVS** 1–8 cm, linear, entire, tomentose or becoming glabrous. **INFL**: heads 1–few; peduncles 3–8 cm; involucre 3–5 mm diam, cylindric; phyllaries in 2 series, outer 5–8, 6–8 mm, lanceolate, ± soft hairy, inner 4–5, shorter, membranous. **RAY FLS** 3–6; ligules 8–18 mm. **DISK FLS** 10–25; corollas 4–5 mm. **FR** 2–3 mm, glabrous or sparsely glandular; pappus scales 1–2 mm. 2*n*=32. Dry plains, hillsides, washes; 150–1500 m. e DMoj, DSon; to UT, AZ, n Mex. ❀DRN:10,11, **12,13**,19–21&DRY,SUN:7–9,14,18.

PULICARIA

G. Ledyard Stebbins

Ann to per. **LVS** basal or some cauline, alternate. **INFL**: heads radiate, 1–many; involucre ± hemispheric; phyllaries in 2–several series, narrowly membranous-margined; receptacle naked. **RAY FLS** many; ligules linear, yellow. **DISK FLS** many; corolla yellow; anther base appendages bristle-like; style branches ± club-like, spreading. **FR** subcylindric, 5-ribbed, short-hairy; outer pappus a of short crown of ± fused scales, inner of barbed bristles. (Latin: flea-like) [Raven 1963 Aliso 5:251–253]

P. paludosa Link Pl 6–12 dm, from short rhizome, stiff, ± soft-hairy. **LVS** 1–3(8) cm, linear to oblong, entire, often rolled under, ± clasping. **INFL**: heads many; peduncles 1–3 cm; involucre 4–5 mm, 6–10 mm diam; phyllaries 2.7–4.8 mm, subequal in 2–3 series, narrowly linear. **RAY FLS**: corollas 3.5–4 mm, tube 1.5–2 mm, ligule 1.5–2 mm. **DISK FLS**: corollas 2.2–3 mm. **FR** ± 1 mm; outer pappus < 0.4 mm; inner pappus bristles 8–16, 2–3 mm. 2*n*= 18. Uncommon. Damp sand; < 700 m. w SnJV, SW, e DSon; native to Eur. [*P. hispanica* (Boiss.) Boiss.]

PYRROCOMA

Gregory K. Brown

Per from woody taproot. **STS** 1–many, decumbent to erect, gen red-tinged. **LVS** alternate, simple, glabrous to tomentose or glandular; basal petioled; cauline gen clasping, reduced. **INFL**: heads gen radiate, 1–many; involucre hemispheric to bell-shaped; phyllaries in 2–6 ± graduated series, herbaceous. **RAY FLS** 10–80; corollas 2–35 mm, yellow. **DISK FLS** 20–100+; corollas 5–15 mm, cylindric to funnel-shaped, yellow. **FR** 3–4-angled, gen hairy; pappus bristles 15–60, gen rigid, unequal. ± 10 spp.: w N.Am. (Greek: reddish pappus) [Mayes 1976 PhD Univ TX] Formerly incl in *Haplopappus*.

1. Ray fls obscure, often << disk fls (heads seeming discoid); involucre subtended by leafy bracts .... ***P. carthamoides*** var. ***cusickii***
1′ Ray fls conspicuous, >> disk fls; involucre not subtended by leafy bracts
 2. Pl glandular
  3. Glands sessile; pl glabrous below; infl crowded, spike-like .... ***P. lucida***
  3′ Glands stalked; pl gen hairy and red-tinged below; infl raceme- or panicle-like
   4. Infl panicle-like; pl sometimes becoming glabrous below .... ***P. lanceolata*** var. ***subviscosa***
   4′ Infl raceme-like; pl tomentose below .... ***P. hirta***
    5. Pl densely glandular; phyllaries equal, barely overlapping, herbaceous nearly to base .... var. ***hirta***
    5′ Pl sparsely glandular; phyllaries unequal, much overlapping, herbaceous only at tip .... var. ***lanulosa***
 2′ Pl glandless
  6. Head gen 1(–4)
   7. Phyllaries oblanceolate to oblong, in 3–4 graduated series, glabrous; fr glabrous .... ***P. apargioides***
   7′ Phyllaries ± linear, in 2 series, not overlapping, long-soft-hairy; fr silky .... ***P. uniflora***
    8. Herbage tufted-woolly; involucres 10–13 mm; phyllaries unequal .... var. ***gossypina***
    8′ Herbage tomentose, becoming glabrous; involucres 6–9 mm; phyllaries equal .... var. ***uniflora***
  6′ Heads gen 5+
   9. Infl panicle-like .... ***P. lanceolata*** var. ***lanceolata***
   9′ Infl raceme- or ± spike-like .... ***P. racemosa***
    10. Involucre 10–15 mm
     11. St cobwebby, hairs tangled; basal lvs lanceolate; phyllaries densely long-soft-hairy .... var. ***pinetorum***
     11′ St gen glabrous; basal lvs oblanceolate to elliptic; phyllaries ciliate .... var. ***racemosa***
    10′ Involucre 5–9 mm
     12. Infl a narrow raceme- to panicle-like cluster, not crowded .... var. ***paniculata***
     12′ Infl spike-like, gen crowded
      13. Involucre 12–16 mm diam, phyllaries yellow-green throughout, sparsely tomentose near base or throughout, tip glandular-ciliate .... var. ***congesta***
      13′ Involucre 5–7 mm diam, phyllaries green-tipped, leathery, ciliate (exc recurved tip), margin translucent .... var. ***sessiliflora***

P. apargioides (A. Gray) E. Greene (p. 333) **STS** 5–18 cm, 0–few-lvd, glabrous to sparsely tomentose. **LVS**: basal ± 4–10 cm, ± (ob)lanceolate, leathery, gen ± coarsely dentate or cut, glabrous; cauline reduced, gen entire. **INFL**: head gen 1; involucre 13–20 mm wide, hemispheric; phyllaries in 3–4 graded series, 6–10 mm, ± oblong to oblanceolate, green toward tip, glabrous. **RAY FLS** 11–40; ligules 7–16 mm. **DISK FLS** 45–90; corollas 5–7 mm. **FR** 2.5–6 mm, 3-angled, glabrous; pappus 5–7.5 mm, tan. 2*n*=12. Rocky slopes, meadows, forest openings; 2200–3700 m. SN, W&I; w NV. [*Haplopappus a.* A. Gray] ❀TRY.

P. carthamoides Hook. var. ***cusickii*** (A. Gray) J. Kartesz & K. Gandhi (p. 333) **STS** 6–43 cm. **LVS** ± 5–21 cm; blade ± narrowly (ob)lanceolate, ± entire to spiny-serrate, puberulent. **INFL**: heads obscurely radiate, 1–4, in raceme-like cluster, subtended by leafy bracts; involucre 11–20 mm, narrowly bell-shaped; phyllaries ± lanceolate, ± not overlapping, entire to serrate, herbaceous with pale, thin, dry margin. **RAY FLS** < 25; ligules 2–7 mm, gen < or = pappus. **DISK FLS** many; corollas 9–13 mm. **FR** 3–5.5 mm, 4-angled, glabrous; pappus ± 6–9 mm, tan to red-brown. 2*n*=12. Barren rocky areas, lava fields; 900–2800 m. CaRH, n SNH, MP; to WA, ID, n NV. [*Haplopappus ca.* (Hook.) A. Gray ssp. *cu.* (A. Gray) H.M. Hall]

P. hirta (A. Gray) E. Greene **STS** 10–45 cm, ± tomentose to woolly, glandular above. **LVS** ± 1–19 cm, elliptic to (ob)lanceolate, gen toothed, tomentose. **INFL**: heads gen 5–13, in raceme-like cluster; involucre 9–20 mm wide, hemispheric to bell-shaped; phyllaries linear-lanceolate. **RAY FLS** 10–30; ligules ± 8–15 mm. **DISK FLS** many; corollas 5–8 mm. **FR** 3-angled, ± 3–5 mm, silky; pappus 5–8 mm, tan. Meadows, open forests, sometimes in alkaline soils; 1200–2200 m. MP; to WA, ID, n NV. [*Haplopappus h.* A. Gray]

var. ***hirta*** (p. 333) Pl densely glandular. **LF** densely tomentose. **INFL**: phyllaries equal, barely overlapping, herbaceous ± to base. 2*n*=24. Habitats and ± range (exc OR) of sp.; 1200–1600 m. [*Haplopappus h.* A. Gray ssp. *h.*] ❀TRY.

var. ***lanulosa*** (E. Greene) G. Brown & Keil (p. 333) Pl sparsely glandular. **LF** woolly. **INFL**: phyllaries unequal, obviously overlapping, herbaceous only at tip. Habitats of sp.; 1700–2200 m. MP; to c OR. [*Haplopappus h.* A. Gray ssp. *l.* (E. Greene) H.M. Hall] ❀TRY.

P. lanceolata (Hook.) E. Greene Pl gen ± (sub)glabrous (densely glandular upward). **STS** 15–50 cm. **LVS** ± (ob)lanceolate, sometimes glandular; basal petioled, 6–30 cm, dentate; cauline sessile, reduced. **INFL**: heads 4–25, in flat-topped clusters; involucre 9–22 mm wide, hemispheric; phyllaries overlapping in 3–4 series, 6–11 mm, unequal, ± lanceolate. **RAY FLS** 18–40; ligules 6–12 mm. **DISK FLS** many; corollas 5–7 mm. **FR** ± 4 mm, 4-angled, silky; pappus ± 5–7 mm, tan. Alkaline meadows, marsh edges, open places; 1300–2800 m. MP, n SNE; to OR, ID, s-c Can, c US. [*Haplopappus l.* (Hook.) Torrey & A. Gray] Variable; needs study.

var. ***lanceolata*** Pl (sub)glabrous. 2*n*=12,24,36. Habitats and range of sp. [*H. l.* (Hook.) Torrey & A. Gray incl ssp. *tenuicaulis* (D. Eaton) H.M. Hall] ❀TRY.

var. ***subviscosa*** (E. Greene) G. Brown & Keil (p. 333) Pl copiously stalked-glandular, esp toward infl. 2*n*=12. Dry slopes and alkaline meadows; 1500–2000 m. MP; nw NV.

P. lucida (Keck) J. Kartesz & K. Gandhi (p. 333) Pl glabrous and ± shiny from gland-dots. **STS** 25–75 cm. **LVS** 1–25 cm, (ob)lanceolate, entire or serrate. **INFL**: heads 12–30+, in crowded spike-like cluster; involucre 6–13 mm wide, bell-shaped; phyllaries 7–13 mm, overlapping, linear or lanceolate, herbaceous. **RAY FLS** 12–20; ligules 7–14 mm. **DISK FLS** 25–40; corollas 6–10 mm. **FR** 2.5–4 mm, 4-angled, silky; pappus 5–7 mm, tan to brownish. 2*n*=24. Alkaline clay flats, open forest; 700–1600 m. n SNH. [*Haplopappus l.* (Keck) Keck]

P. racemosa (Nutt.) Torrey & A. Gray **STS** 15–90 cm, gen glabrous. **LVS**: basal 5–36 cm, (ob)lanceolate to widely elliptic, entire to serrate, petioles tomentose; cauline clasping, reduced, gen serrate, glabrous. **INFL**: heads 3–15+, in ± narrow clusters; involucre 5–18 mm diam, hemispheric or bell-shaped; phyllaries overlapping in 4–5 series, 6–13 mm, (ob)lanceolate to oblong, (sub)glabrous. **RAY FLS** 7–28; ligules 5–10 mm. **DISK FLS** 20–65; corollas 5–8 mm. **FR** 2.5–5.5 mm, 4-angled, glabrous to densely tomentose; pappus 6–9 mm, tan to brownish. Many habitats; < 2500 m. NW, CaR, SN, s ScV, CW, s Co, GB, DMoj; to OR, UT. [*Haplopappus r.* (Nutt.) Torrey]

var. ***congesta*** (E. Greene) G. Brown & Keil (p. 333) DEL NORTE HAPLOPAPPUS **INFL** gen crowded, spike-like; involucre 5–8.5 mm, 12–16 mm diam; phyllaries herbaceous and yellow-green throughout, acute, sparsely tomentose (at least near base), tip glandular-ciliate, not recurved. UNCOMMON. Chaparral, coniferous forests; 200–1000 m. nw KR; sw OR. [*H. r.* ssp. *c.* (E. Greene) H.M. Hall]

var. ***paniculata*** (Nutt.) J. Kartesz & K. Gandhi **STS** glaucous. **INFL** narrow; involucre 5–8.5 mm, 8–12 mm diam. 2*n*=12, 24. Alkaline flats, saline meadows; 150–2500 m. c&s SNH, Teh, GB, DMoj; to OR, ID, UT. Highly variable. [*H. r.* ssp. *glomeratus* (Nutt.) H.M. Hall]

var. ***pinetorum*** (Keck) J. Kartesz & K. Gandhi (p. 333) **STS** cobwebby. **LVS**: basal lanceolate. **INFL** raceme-like; involucre 10–15 mm, 12–15 mm diam; phyllaries densely tomentose. Meadows, open forest; 600–1700 m. KR. [*H. r.* ssp. *p.* Keck]

var. ***racemosa*** (p. 333) **LVS**: basal oblanceolate to elliptic. **INFL** raceme-like; involucre 10–15 mm, 11–18 mm diam; phyllaries ciliate. Coastal valleys, marshes, sometimes in saline soils; < 300 m. NCoR, s ScV, SnFrB, SCoRI, SCo; OR.

var. ***sessiliflora*** (E. Greene) G. Brown & Keil (p. 333) **STS** glaucous. **LVS**: basal linear-oblanceolate. **INFL** gen crowded, spike-like; involucre 5–8.5 mm, 5–7 mm diam; phyllary base pale, leathery, margin translucent, tip green, recurved. 2*n*=12. Dry alkaline flats or saline meadows; 300–2200 m. se SNE, DMoj; w NV. [*H. r.* ssp. *s.* (E. Greene) H.M. Hall]

P. uniflora (Hook.) E. Greene **STS** 7–38 cm. **LVS** ± tomentose or woolly; basal 3–12 cm, (ob)lanceolate, sharply dentate to cut; cauline few, clasping, reduced. **INFL**: heads 1(–4), in raceme-like cluster; involucre 6–13 mm, 11–20 mm diam, hemispheric; phyllaries barely or not overlapping in ± 2 series, 6–12 mm, ± linear, herbaceous, gen tomentose to woolly. **RAY FLS** 25–45; ligules 7– 11 mm. **DISK FLS** 35–60; corollas 5–8 mm. **FR** ± 3–4 mm, 3–4-angled, silky; pappus ± 5–8 mm, tan. Alkaline soils of mtn meadows, open forest, near hot springs; 1400–2900 m. SnBr, GB; to OR, ID, MT, WY, Colorado. [*Haplopappus u.* (Hook.) Torrey & A. Gray]

var. ***gossypina*** (E. Greene) J. Kartesz & K. Gandhi (p. 333) BEAR VALLEY HAPLOPAPPUS Herbage woolly-tufted. **INFL**: involucre 10–13 cm; phyllaries unequal. RARE. Habitats of sp.; 1600–2300 m. SnBr (near Baldwin Lake). [*H. u.* ssp. *g.* (E. Greene) H.M. Hall]

var. ***uniflora*** (p. 333) Herbage tomentose or becoming glabrous. **INFL**: involucre 6–9 mm; phyllaries equal. 2*n*=12. Habitats and range (exc SnBr) of sp. ❀TRY.

RAFINESQUIA

G. Ledyard Stebbins

Ann, glabrous; sap milky. **STS** erect, gen branched above. **LVS** basal and cauline, alternate, oblong to widely elliptic or oblanceolate, dentate or pinnately lobed; lower ± petioled or sessile, clasping; upper bract-like. **INFL**: heads ligulate, solitary or in ± flat-topped or panicle-like clusters; peduncles with scale-like or reduced lf-like bracts; involucre cylindric or obconic; phyllaries in 3–4 series, outer 2–3 series unequal, lanceolate to ovate, acute to acuminate, tips spreading, innermost series ± equal, linear-acuminate, >> outer, erect, gen ± membrane-margined; receptacle flat or convex, naked. **FLS** many; ligules white or cream-colored, often rose-tinged, esp beneath, readily withering. **FR** narrowly elliptic; body smooth or tubercled, weakly ribbed, tapered to a beak; pappus a ring of plumose bristles. 2 spp.: sw US, n Mex. (C.S. Rafinesque, eccentric US naturalist, 1783–1840)

1. Ligules 5–8 mm, 3–5 mm > phyllaries; fr beak very slender, ± = and clearly different from body; side hairs of pappus bristles stiff, straight; CA-FP, D .... ***R. californica***
1′ Ligules 15–20 mm, 10+ mm > phyllaries; fr beak stout, < and unclearly different from body; side hairs of pappus bristles soft, ± matted; D .... ***R. neomexicana***

R. californica Nutt. CALIFORNIA CHICORY **ST** gen 1 from base, erect, branched chiefly in upper half, 2–15+ dm, in larger pls 10+ mm diam near base. **LVS** 3–15 cm, oblong, dentate or coarsely and ± widely lobed. **INFL**: heads gen several–many; peduncles 1–8 cm; involucres 14–20 mm. **FLS**: ligules 5–8 mm, slightly > phyllaries. **FR** 9–11 mm (incl beak); body 4–5 mm, ± glabrous to short-rough-hairy; beak slender, ± = body; pappus bristles 6–10 mm, very fine, plumose to tip with straight hairs, dull white to brownish; bristles bearing straight hairs. 2*n*=16. Shrubby slopes, open woods, deserts, often common after fires; < 1500 m. NCoR, SNF, SnFrB, SCoR, SW, D; to sw UT, AZ, Baja CA. ❀TRY.

R. neomexicana A. Gray (p. 333) DESERT CHICORY **STS** 1–several from base, 1–4 dm, in larger pls gen < 5 mm diam near base. **LVS** 3–15 cm, dentate or ± narrowly lobed. **INFL**: heads gen 1–5 per st; peduncles 2–8(15) cm; involucres gen 17–30 mm. **FLS**: ligules 15–20 mm, >> phyllaries. **FR** 12–14 mm (incl beak); body 8–10 mm, puberulent or appressed-scaly; beak stout, gen < body; pappus bristles 12–17 mm, stiff, wider at base, plumose with tangled hairs, often merely barbed at tip, white. 2*n*=16. Gravelly and sandy desert soils, often partially supported in branches of shrubs; gen < 1400 m. D; to sw UT, w TX, n Mex. ❀TRY.

RAILLARDELLA

Bruce G. Baldwin

Per, ± scapose from gen branched rhizome; rosettes often clumped. **LVS** simple, sessile, ± basal. **INFL** glandular, sometimes hairy; heads radiate or discoid, gen solitary; phyllaries gen as many as ray fls, ± folded around ray ovaries; chaff scales in 1 involucre-like series, ± equal, ciliate-hairy. **RAY FLS** 0–13; corollas yellow to red; ligules often deeply lobed. **DISK FLS** 7–84; corollas yellow to red; style branches long, tips bristly. **FR** ± cylindric, ± straight, ascending-hairy, black; pappus of flat, ciliate-plumose bristles. 3 spp.: montane CA, w NV, OR. (Latin: small *Raillardia* or *Railliardia*) [Baldwin et al. 1991 Proc Natl Acad USA:88:1840–1843]

1. Lvs glabrous (or cauline glandular); corollas orange to reddish; rays gen > 5 .... ***R. pringlei***
1′ Lvs glandular or hairy; corollas yellowish; rays gen 0–5
 2. Lvs silky-hairy, glands 0 or inconspicuous; rays 0; pl 1–14.5 cm .... ***R. argentea***
 2′ Lvs glabrous to sparsely hairy, clearly glandular; rays 0–7; pl 6–53 cm .... ***R. scaposa***

R. argentea (A. Gray) A. Gray (p. 339) SILKY RAILLARDELLA Pl 1–14.5 cm. **LVS** basal, entire, 0.7–8 cm, gen oblanceolate, silky-hairy, minutely glandular or nonglandular. **INFL**: heads discoid, solitary, cylindric to bell-shaped; peduncles 0.1–12.5 cm, subtending bracts 0–1; phyllaries 0; chaff scales 5–15, 6–16 mm, ± fused. **RAY FLS** 0. **DISK FLS** 7–26; corollas 6–11 mm, yellow. **FR** 5–9.5 mm, ± linear; pappus bristles 16–30, 6–11 mm. 2*n*=34,36. Dry, open, gen gravelly sites; 2200–3900 m. CaRH, SNH, SnBr, Wrn, SNE; w NV, OR. Interfertile with *R. pringlei*. ❀DRN,IRR,SUN: 1–3,6,7,15,16,18;DFCLT.

R. pringlei E. Greene (p. 339) SHOWY RAILLARDELLA Pl ± 25–50 cm. **LVS** basal or mostly opposite, 2.5–15 cm, linear to lanceolate or oblanceolate, entire or minutely dentate, glabrous or upper glandular. **INFL**: heads radiate, 1–3, bell-shaped to hemispheric; peduncles 10–28 cm, often subtended by 1–few gen alternate bracts; phyllaries 6–13; chaff scales 16–24, 10–14 mm, free or ± fused, sometimes overlapping. **RAY FLS** 6–13; corollas 11–20 mm, orange to reddish. **DISK FLS** ± 45–84; corollas 8–11.5 mm, orange to reddish. **FR** 5–8 mm, oblanceolate; pappus bristles ± 8–17, 7–11 mm. 2*n*=34. RARE. Wet meadows, streambanks; 1200–2200 m. KR (Trinity Alps, Scott Mtns).

R. scaposa (A. Gray) A. Gray (p. 339) Pl 6–53 cm. **LVS** ± basal, 1–16 cm, linear to lanceolate or oblanceolate, entire, glandular, sometimes sparsely hairy. **INFL**: heads radiate or discoid, ± cylindric to hemispheric; peduncles 1–39 cm, gen subtended by 1–few bracts; phyllaries 0–7; chaff scales 8–20, 8–18 mm, free or fused, sometimes overlapping. **RAY FLS** 0–7; corollas < or = 26 mm, yellow to orange-yellow. **DISK FLS** 9–44; corollas 7.5–12 mm, yellow to orange-yellow. **FR** 4.5–10 mm, linear; pappus bristles 8–30 (0 in some ray achenes), 7.5–11.5 mm. 2*n*=68,70. Dry, exposed sites to shady, wet sites; 2000–3500 m. SNH; w NV, OR. ❀DRN, IRR:1–3,6,7,15,16,18;DFCLT.

RAILLARDIOPSIS

Bruce G. Baldwin

Per from caudex or woody rhizome, hairy to glandular-hairy. **STS** ascending or erect, gen < or = 6 dm, often branched. **LVS** opposite below, alternate above, simple, linear to lanceolate, entire, sessile. **INFL** glandular; heads radiate or discoid, solitary or in leafy cymes, cylindric to bell-shaped; peduncles 0–8 cm; phyllaries 0–3, folded around ray ovaries; chaff scales in 1 involucre-like series, ± equal, ± fused. **RAY FLS** 0–3; corollas yellow; ligules deeply lobed; style branches long, ± glabrous. **DISK FLS** 5–29; corollas yellow; style branches long, bristly to base. **FR** ± cylindric, black; ray achenes curved, glabrous to partly hairy; disk achenes ± straight, ascending-hairy; pappus of flat, ciliate-plumose bristles. 2 spp.: montane CA-FP. (Greek: like *Raillardia* or *Railliardia*) [Baldwin & Kyhos 1990 Madroño 37:43–54] Close to *Madia*; ancestral to Hawaiian silversword alliance (incl *Railliardia*).

1. Ray fls and phyllaries 0; chaff scales glandular-hairy on outer surface; pappus = or > mature achene ..... ***R. muirii***
1′ Ray fls and phyllaries 0–3; chaff scales and phyllaries with non-glandular hairs gen restricted to margins; pappus < or = mature achene .... ***R. scabrida***

R. muirii (A. Gray) Rydb. (p. 339) MUIR'S RAILLARDELLA Pl often matted; rhizome loosely branched. **LVS** < or = 42 mm, greenish, not clasping; tip acute. **INFL**: chaff scales 5–16, 8–13 mm, glandular-hairy on outer surface. **RAY FLS** 0. **DISK FLS** 7–29; corollas 6.5–10 mm. **FR** 3.5–7.5 mm, 0.5–1.2 mm wide, linear; pappus bristles 9–17, = or > mature achene. 2*n*=16. RARE. Dry, open sites on granitic soil; 1100–2500 m. s SNH, SCoRO (Ventana Double Cone, Monterey Co.). [*Raillardella m.* A. Gray]

R. scabrida (Eastw.) Rydb. (p. 339) SCABRID RAILLARDELLA Pl from branched caudex, not matted. **LVS** < or = 30 mm, blue-greenish, sometimes clasping; tip often obtuse. **INFL**: chaff scales 5–12, 6–12 mm, glandular and ciliate-hairy. **RAY FLS** 0–3; corollas 3.5–11.5 mm. **DISK FLS** 5–23; corollas 4.5–9.5 mm. **FR**: ray achenes 6–8 mm; pappus bristles 0(–7); disk achenes 5–9 mm, 1.3–2.4 mm wide, ± club-shaped to linear; pappus bristles 11–21, < or = mature achene. 2*n*=14. UNCOMMON. Dry, open ridges on metamorphic soil; 1650–2300 m. NCoRH, CaRH (Shasta Co.). [*Raillardella s.* Eastw.]

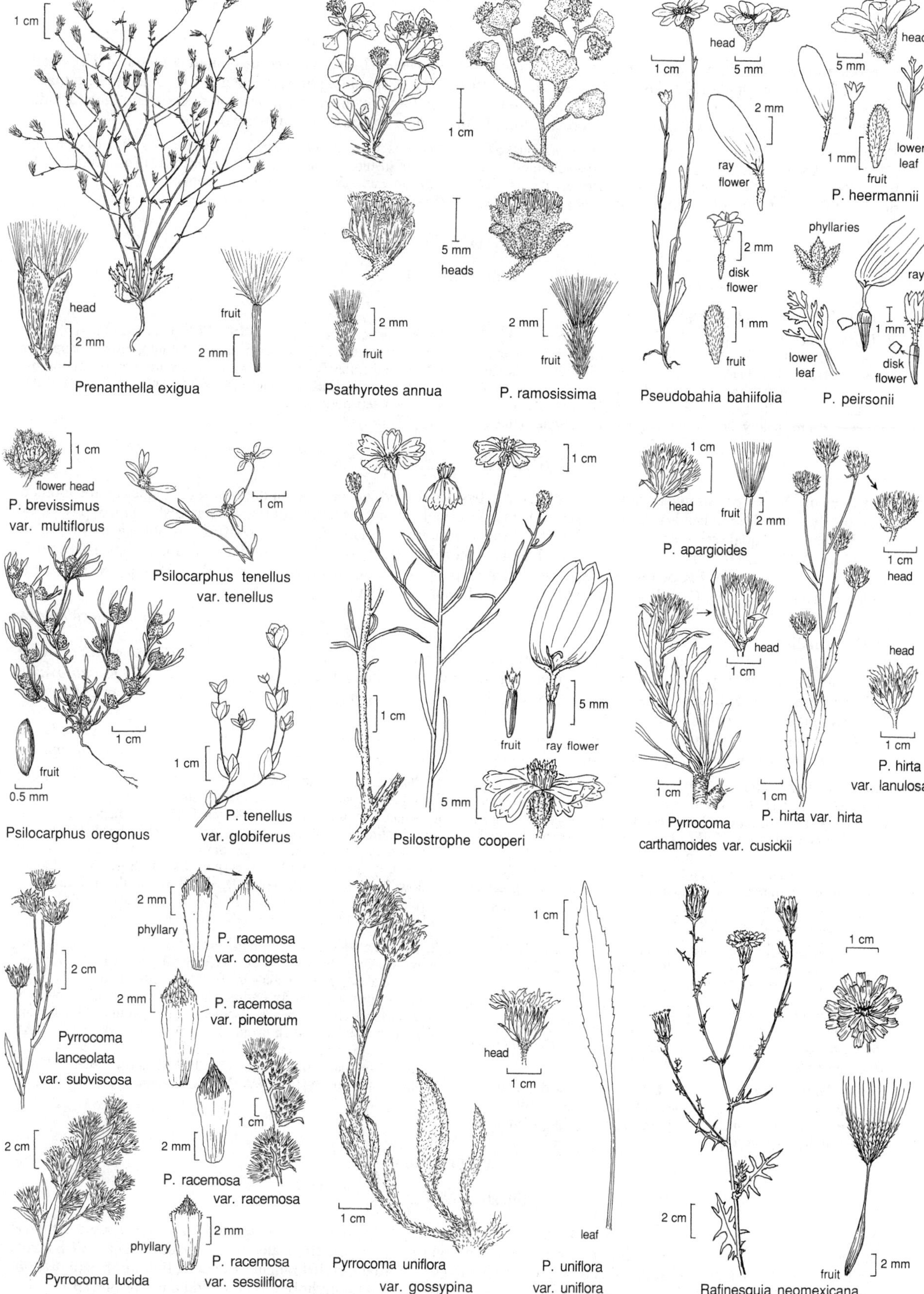
1 cm
head
2 mm
fruit
2 mm
Prenanthella exigua
1 cm
5 mm
heads
2 mm
fruit
2 mm
fruit
Psathyrotes annua
P. ramosissima
head
1 cm
5 mm
2 mm
ray flower
2 mm
disk flower
1 mm
fruit
Pseudobahia bahiifolia
head
5 mm
1 mm
fruit
lower leaf
P. heermannii
phyllaries
ray
1 mm
lower leaf
disk flower
P. peirsonii
1 cm
flower head
P. brevissimus var. multiflorus
1 cm
Psilocarphus tenellus var. tenellus
1 cm
fruit
0.5 mm
Psilocarphus oregonus
1 cm
P. tenellus var. globiferus
1 cm
1 cm
5 mm
fruit
ray flower
5 mm
Psilostrophe cooperi
1 cm
head
fruit
2 mm
P. apargioides
1 cm
head
head
1 cm
1 cm
Pyrrocoma carthamoides var. cusickii
1 cm
P. hirta var. hirta
head
1 cm
P. hirta var. lanulosa
2 mm
phyllary
P. racemosa var. congesta
2 cm
2 mm
P. racemosa var. pinetorum
Pyrrocoma lanceolata var. subviscosa
2 cm
2 mm
1 cm
P. racemosa var. racemosa
2 mm
phyllary
P. racemosa var. sessiliflora
Pyrrocoma lucida
1 cm
head
1 cm
1 cm
leaf
Pyrrocoma uniflora var. gossypina
P. uniflora var. uniflora
1 cm
2 cm
fruit
2 mm
Rafinesquia neomexicana

RHAGADIOLUS

G. Ledyard Stebbins

Ann; sap milky. **ST** solitary, branched. **LVS** basal and cauline, alternate, simple, toothed or pinnately lobed. **INFL**: heads ligulate, few–many; phyllaries in 2 rows, outer short, inner fused at base, spreading in fr; receptacle ± flat, naked. **FLS** many; ligules yellow, readily withering. **FR** cylindric, of 2 kinds; outer enclosed by inner phyllaries, long-persistent; inner deciduous; pappus 0. 7 spp.; Eur, w Asia. (Latin: crevice-like, from strongly folded inner phyllaries)

R. stellatus (L.) Gaertner Pls sparsely hairy throughout. **STS** 0.7–4 dm. **LVS** 2.5–14 cm, oblong to obovate, sparsely toothed or lobed. **INFL**: heads 5–8 mm in fl; outer phyllaries 5; inner phyllaries 5–8. **FR**: outer frs 10–15 mm, curved, spreading with attached phyllaries to form star-shaped structure. 2*n*=10. Uncommon. Weed of disturbed places; < 200 m. s NCoRO (Napa Co.), SnFrB; native to Eur. [*R. edulis* Willd. misapplied]

RIGIOPAPPUS

Robert Ornduff

1 sp.: w US. (Greek: stiff pappus)

R. leptocladus A. Gray (p. 339) Ann < 30 cm, ± hairy. **ST** simple or side branches > main axis. **LVS** alternate, < 3 cm, ± erect, linear, entire. **INFL**: heads radiate, terminal; involucre obconic; phyllaries in 2 series, < 8 mm, tapered, partly enclosing outer fr; receptacle flat, chaffy between ray and disk fls. **RAY FLS** < 15; corolla < 2 mm, pale yellow, often purple-tinged; style branches flat; stigmatic lines marginal. **DISK FLS** bisexual, < 60; corolla < pappus, ± cylindric, lobes gen 3–4. **FR** < 4 mm, brown; pappus of 0–5 tapered scales, < 5 mm. 2*n*=18. Grassy areas or among shrubs; < 2300 m. NW, CaR, SN, GV, CW, WTR, SnGb, MP; to WA, ID, NV. Gen self-pollinated.

RUDBECKIA CONE-FLOWER

Ann, bien, per. **STS** erect, simple or branched. **LVS** simple, basal and alternate, petioled or upper sessile; blades entire, toothed or pinnately lobed. **INFL**: heads radiate or discoid, 1–few, terminal; peduncles long, stout; involucre disk-shaped; phyllaries in 1–3 equal or unequal series, spreading or reflexed; receptacle conic to columnar, chaffy; chaff scales oblong, entire folded around frs. **RAY FLS** 0 or many, sterile. **DISK FLS** many; corolla yellow to dark brown, diam of tube and throat ± equal; style tips blunt-triangular. **FR** oblong, 4-angled, ± flattened; pappus a crown of fused scales or 0. ± 15 spp.: N.Am. (O.J. Rudbeck, 1630–1702, and O.O. Rudbeck, 1660–1740, professors of botany at Uppsala, Sweden)

1. Heads discoid .... ***R. occidentalis***
1' Heads radiate
  2. Pappus 0; disk fls dark purple; disk spheric or ovoid .... ***R. hirta*** var. ***pulcherrima***
  2' Pappus a crown of fused scales; disk fls greenish yellow; disk ± columnar .... ***R. californica***
    3. Lvs green; lower lf surface hairy .... var. ***californica***
    3' Lvs glaucous; lower lf surface glabrous
      4. Lvs entire or few-toothed; disk gen > 3 cm .... var. ***glauca***
      4' Lvs coarsely toothed; disk 1.5–3 cm .... var. ***intermedia***

R. californica A. Gray (p. 339) CALIFORNIA CONE-FLOWER Per; rhizome stout. **STS** 6–18 dm, erect, gen unbranched, glabrous. **LVS**: blade gen 10–30 cm, lanceolate to ovate or elliptic, entire, toothed or coarsely lobed. **INFL**: head, radiate, solitary; peduncles 2–6 dm; outer phyllaries 1–2 cm, linear-oblong, > inner; disk 1.5–6 cm, ovoid to columnar; chaff greenish. **RAY FLS** 8–21; ligules yellow, 2–6 cm, often reflexed. **DISK FLS**: corollas 4–5 mm, greenish yellow. **FR** 4–5 mm, glabrous; pappus scales 1–1.5 mm. Meadows, seeps, mires; < 2600 m. KR, NCoR, SNH; s OR.

var. ***californica*** **LF** 5–10 cm wide, entire to coarsely toothed or lobed; lower surface hairy. 2*n*=36. Meadows, seeps; 1650–2600 m. KR, SNH. ❀WET,SUN:1–5,**6**,7–10,**14–18**,19;DFCLT.

var. ***glauca*** S.F. Blake **LF** gen 3–6 cm wide, entire or shallowly few-toothed; both surfaces glabrous. Meadows, seeps, mires, stream-banks, often on serpentine; 60–1250 m. n NCo, w KR; sw OR. ❀WET or IRR,SUN:5,6,**7**,8–10,**14–18**,19.

var. ***intermedia*** R. Perdue **LF** 5–10 cm wide, entire to coarsely toothed or lobed, glabrous. Meadows, seeps; 1000–1600 m. KR. Differing from var. *californica* mainly by glabrous lvs. ❀WET,SUN:1–10,**14–18**,19.

R. hirta L. var. *pulcherrima* Farw. (p. 339) BLACK-EYED SUSAN Ann, bien, or short-lived per, stiffly hairy. **STS** 3–8 dm, simple or few-branched above. **LVS**: blade gen 5–10 cm, oblanceolate to elliptic, entire or shallowly crenate-serrate. **INFL**: heads radiate, 1–few; peduncle 5–20 cm; outer phyllaries 1–1.5 cm, linear, ± = inner; disk nearly spheric; chaff dark purple. **RAY FLS** gen 8–13; ligules yellow, 2–3 cm, gen spreading. **DISK FLS**: corollas ± 4 mm, dark purple. **FR** 1.5–2 mm; pappus 0. 2*n*=38. Meadows, fields, roadsides; 100–1200 m. c SNF, GV; native to midwestern US.

R. occidentalis Nutt. var. ***occidentalis*** (p. 339) WESTERN CONE-FLOWER Per; rhizome stout. **STS** 6–20 dm, simple or few-branched above. **LVS**: blade gen 10–30 cm, elliptic to widely ovate, coarsely serrate to nearly entire, glabrous or short-hairy on 1 or both surfaces. **INFL**: heads discoid, 1–few; peduncle 15–45 cm; phyllaries 1–1.5 cm; disk ovoid, 2–6 cm; chaff dark purple. **RAY FLS** 0. **DISK FLS**: corollas 3–4 mm, dark purple. **FR** 3–5 mm; pappus 0 or scales 0.5–1 mm. 2*n*=38. Meadows, seeps; 1200–1850 m. KR, n SNH; to WA, WY, Colorado.

SANTOLINA

Elizabeth McClintock

Per, shrubs, aromatic. **STS** erect, branched. **LVS** alternate, pinnately lobed or finely divided. **INFL**: heads discoid, solitary; phyllaries in several series, margins dry, scarious; receptacle slightly convex, chaffy, scales partly surrounding frs. **FLS** many, gen bisexual; corollas yellow; anther tips ovate, bases rounded or ± cordate; style branches truncate, tips shrub-like. **FR** oblong, glabrous, gen 3–4-angled; pappus 0. ± 5 spp: w Medit; 2–3 cult. (Latin: holy flax, the old name for one sp.)

S. chamaecyparisus L. LAVENDER-COTTON Shrub, evergreen, much-branched, densely tomentose, silvery-gray to white. **STS** 20–60 cm. **LVS** 2.5–4 cm, 4–5 mm wide, pinnately dissected and 3-dimensional. **INFL**: heads ± spheric, 1.5–2 cm diam; phyllaries in 3 series, strongly ribbed; chaff scales firm, flexible. **FR** ± 3 mm, 5-angled, glabrous. $2n$=18,36,±45. Uncommon. Escape from cult in disturbed coastal areas; < 500 m. NCoRO, SnFrB, SCoRI, SCo; native to Medit.

SANVITALIA

Ann, per. **STS** simple to much-branched, prostrate to erect. **LVS** simple, opposite. **INFL**: heads radiate, solitary or in few-headed cymes; peduncle slender; involucre disk-like to hemispheric; phyllaries in 1–2 series; receptacle conic, chaffy; chaff scales lanceolate, ± awn-tipped. **RAY FLS** 5–13; tube 0; ligules 2–3-lobed, cream-colored to orange, persistent on fr. **DISK FLS** many; corollas cream-colored to yellow or brown; lobes very small; style tips triangular. **FR** glabrous; ray achenes thick, pappus of short, stout awns; disk achenes short, pappus 0 or of 2 awns. 5 spp.: sw US, Mex, n C.Am, S.Am. (either Sanvital, a Spanish botanist, or the Italian Sanvital family) [Strother 1979 Madroño 26:173–179]

S. abertii A. Gray (p. 339) ABERT'S SANVITALIA Ann 2–29 cm. **STS** spreading or erect, simple to much-branched, strigose. **LVS** sessile or short-petioled, 2–5 cm, linear to lanceolate or narrowly elliptic, acute, scabrous. **INFL**: heads gen in cymes; peduncle 0–30 mm; phyllaries 5–11, prominently veined, acute, ± glabrous; awn-tips of chaff scales > disk fls. **RAY FLS** 1 per phyllary; corollas yellow, drying cream-colored; ligules thick, 2–3 mm, ± leathery, gen 2- lobed. **DISK FLS**: corollas 1–2 mm, cylindric, yellow, drying cream-colored. **FR**: ray achenes 3–4 mm, straw-colored, pappus awns 3, < or = 1 mm, stout; disk achenes 2.5–3.5 mm, brown, ± 4-angled, warty, pappus 0. $2n$=22. RARE. Dry slopes; 1800 m. DMtns (Clark, New York mtns); to TX, n Mex.

SAUSSUREA

David J. Keil & Charles E. Turner

Per. **LVS** basal and cauline, alternate. **INFL**: heads discoid, 1–many; involucre ovoid to ± obconic; phyllaries many, in several series, unequal, acute; receptacle flat or rounded, naked to bristly. **FLS** few–many; corollas blue to purple, tube slender, throat abruptly wider, lobes linear; anther bases short-tailed, tips linear, acute; style branches oblong, short-papillate, tips minutely hairy. **FR** oblong, ± angled, attached at base; pappus of 1–2 series, outer of short bristles or scales, inner of gen longer plumose bristles fused at base. ± 300 spp.: Eurasia, N.Am. (Theodore & Horace Saussure, Swiss naturalists, 18th century)

S. americana D. Eaton (p. 339) AMERICAN SAWWORT Pls 3–12 dm, from stout caudex, erect, leafy, ± loosely tomentose, ± glandular. **LVS**: lower blades 5–15 cm, lanceolate to triangular, sharply dentate, acute. **INFL**: peduncles 0–5 cm; involucre 10–15 mm; phyllaries pale green, ± loosely tomentose, margins ± membranous, tips gen dark. **FLS** 8–21; corollas ± 10 mm, white to dark purple. **FR** 4–6 mm, glabrous or gland-dotted; pappus bristles brownish, outer < 7 mm, inner ± 10 mm. RARE in CA. Meadows, slopes; ± 1750 m. KR; to AK, MT.

SCHKUHRIA

Ann, per. **STS** often much-branched. **LVS** simple or pinnately divided, gen gland-dotted; lower opposite; upper alternate. **INFL**: heads radiate or discoid, in open cymes; peduncles short to long; involucre obconic to bell-shaped; phyllaries 4–18, oblanceolate to obovate, obtuse, margins scarious; receptacle rounded, naked. **RAY FLS** 0–3; corollas yellow or white; ligules short, inconspicuous. **DISK FLS** few–many; corollas yellow; style tips acute, triangular. **FR** narrowly obpyramidal, gen 4-angled; angles hairy; pappus gen of 8 scales. 6 spp.: s N.Am, n S.Am (C. Schkuhr, 1741–1811, German botanist) [Heiser 1945 Ann Missouri Bot Gard 32:265–278]

S. multiflora Hook. & Arn. var. ***multiflora*** (p. 339) Ann. **STS** decumbent or erect, 5–25 cm, glandular, strigose, or becoming glabrous. **LVS** 2–4 cm, dissected into thread-like lobes 0.5–1 mm wide. **INFL**: heads discoid; cymes few-headed; peduncles 5–30 mm, glandular; involucre obconic; phyllaries 7–9, 5–6 mm, oblanceolate, green-centered, margins often red or yellow. **RAY FLS** 0. **DISK FLS** 15–30; corollas 1.5–2 mm. **FR** 3–4 mm; pappus scales 1–2 mm, obtuse to acute. $2n$=22. Dry, sandy soils; 1500–1700 m. e DMoj; to TX, n Mex; also in S.Am. [*Bahia neomexicana* (A. Gray) A. Gray]

SCOLYMUS GOLDEN THISTLE

G. Ledyard Stebbins

Bien, per, thistle-like, spiny; sap milky. **STS** erect, ± branched above. **LVS** basal and cauline, alternate, pinnately lobed; lobes spine-tipped, decurrent on st as spiny wings. **INFL**: heads ligulate, several–many, in spike-like to panicle-like clusters; phyllaries in several rows; receptacle chaffy, scales scarious, slightly winged, tightly enclosing frs. **FLS** many; ligules yellow, readily withering. **FR** flattened parallel to phyllaries, falling with chaff scales; pappus 0 or of a few rigid bristles. 3 spp.: Medit. (Greek: ancient name)

S. hispanicus L. **STS** 2–8 dm, stout. **LVS** 4–20 cm; basal soft, weakly spiny; cauline rigid, oblong to ovate, shallowly lobed, spiny and ± hairy. **INFL**: heads sessile or short-peduncled, solitary in upper lf axils or clustered in narrow panicles; involucre 15–20 mm. **FR** 3–5 mm, club-shaped; pappus bristles present. $2n$=20. Uncommon. Disturbed grassy areas; < 100 m. se ScV, SnFrB; native to Eur. NOXIOUS WEED; gen quickly eradicated.

SCORZONERA

G. Ledyard Stebbins

Per; sap milky. **LVS** basal and cauline, entire to pinnately lobed. **INFL**: heads ligulate, large, 1–many; phyllaries in many overlapping rows; receptacle ± flat, naked. **FLS** many; corollas yellow to white or purple; ligules readily withering. **FR** cylindric, not beaked; pappus of several rows of plumose bristles. ± 150 spp.: Medit, Eurasia. (Old French: viper)

S. hispanica L. SPANISH SALSIFY, VIPER'S GRASS **ST** 2.5–10 dm from vertical rootstock, sparsely woolly at base, few-branched, leafy. **LVS** 15–40 cm, linear to ovate-elliptic; margins wavy with a few teeth. **INFL**: heads 2–3 cm, < 4 cm in fr. **FLS** many; ligules yellow, sometimes purple-tinged below. **FR** 10–15 mm, ribbed; outer strongly ribbed, rough; inner weakly ribbed; pappus bristles ± = fr, dirty white. $2n$=14. Open fields; < 200 m. NCoRO; native to Eur.

SENECIO GROUNDSEL, RAGWORT, BUTTERWEED

Theodore M. Barkley

Ann to tree-like, gen ± loosely hairy, becoming ± subglabrous; roots gen fibrous, branched. **STS** gen 1–few per rosette. **LVS** alternate; lower gen petioled; middle gen weakly clasping; uppermost gen bract-like. **INFL** often ± flat-topped; heads radiate or discoid, 1–many; involucre cylindric to hemispheric; main phyllaries in 1 equal series (often subtended by few, much-reduced outer phyllaries). **RAY FLS** 0–13; ligules gen yellow to orange. **DISK FLS** gen < 40; corolla gen ± yellow; style tips truncate to obtuse, gen hair-tufted. **FR** cylindric; ribs shallow, often stiff-hairy; pappus of thin, minutely-barbed deciduous bristles ± = fr body. ± 1500 spp.: worldwide; some cult, some of unusual form. (Latin: old man, from white pappus) [Barkley 1978 N.Am Flora II 10:50–139] Among largest genera of fl pls.

1. Ligules gen pink to purple or blue (rarely white) — aliens of coast
 2. Herbage sticky; main lvs deeply toothed or lobed .... ***S. elegans***
 2′ Herbage not sticky; main lvs merely toothed .... ***S. hybridus***
1′ Ligules gen yellow (to brick-colored, rarely 0 or white)
 3. Vine; lvs ivy-like; weedy .... ***S. mikanioides***
 3′ Pls ± erect
  4. Subshrub or shrub (if weakly woody, branching aspect shrub-like); lvs or segments linear
   5. Outermost phyllaries well developed, some 1/2 × main phyllaries .... ***S. flaccidus***
    6. Herbage ± gray-tomentose .... var. ***douglasii***
    6′ Herbage ± glabrous .... var. ***monoensis***
   5′ Outermost phyllaries 0 or inconspicuous — herbage ± glabrous (exc sometimes lf lower surface or heads)
    7. Lvs pinnately lobed, ± hairy below — ChI .... ***S. lyonii***
    7′ Lvs gen linear, ± entire
     8. Phyllaries ± 13; lvs ± thread-like, < 2 mm wide; coastal .... ***S. blochmaniae***
     8′ Phyllaries ± 8; some lvs > 2 mm wide; e of SNH .... ***S. spartioides***
  4′ Ann or per; lvs or segments wider than linear
   9. Lvs ± equal, evenly distributed or crowded upward at fl
    10. Lvs gen 1–2-pinnately lobed; heads 20–60; sts 1–few, erect .... ***S. jacobaea***
    10′ Lvs entire to lobed (if 2-pinnately lobed, heads < 20, sts ascending)
     11. Ann from slender taproot
      12. Ligules conspicuous, spreading .... ***S. californicus***
      12′ Ligules 0 or small, often recurved or inrolled
       13. Main lvs cordate-clasping, bases 1–2 cm wide; herbage glabrous .... ***S. mohavensis***
       13′ Lvs petioled, little or not clasping; herbage glabrous or tomentose
        14. Outermost phyllaries conspicuous, black-tipped; main phyllaries ± 21; herbage ± glabrous — abundant weed .... ***S. vulgaris***
        14′ Outermost phyllaries 0 or inconspicuous, green-tipped; herbage glabrous to tomentose
         15. Main phyllaries gen ± 8, sometimes 13; pls gen < 2 dm; herbage gen glabrous; native ... ***S. aphanactis***
         15′ Main phyllaries ± 13 or 21; pls gen 3–8 dm; herbage sparsely rough-tomentose; naturalized weed .... ***S. sylvaticus***
     11′ Per from taproot, caudex, or rhizome
      16. Pls gen < 2 dm, often spreading; alpine or subalpine
       17. Lvs widely linear; pls rhizomed .... ***S. pattersonensis***
       17′ Lvs ± ovate; pls taprooted .... ***S. fremontii***
        18. Phyllaries (6)8–10 mm; upper lvs well developed; CaRH, Wrn .... var. ***fremontii***
        18′ Phyllaries 5–7(8) mm; upper lvs bract-like; SN, SnBr, SNE .... var. ***occidentalis***
      16′ Pls 5+ dm, erect; montane
       19. Lvs pinnately lobed or dissected; herbage tomentose with jointed hairs or becoming ± glabrous .... ***S. clarkianus***
       19′ Lvs dentate; herbage glabrous or sparsely tomentose below
        20. Lvs ± linear-lanceolate, tapered .... ***S. serra***
        20′ Lvs narrowly triangular, truncate or hastate .... ***S. triangularis***
   9′ Lvs reduced upward; lower lvs prominent, gen persistent
    21. Pl from button-like caudex; roots many, fleshy-fibrous, unbranched
     22. Pls glabrous or minutely hairy among heads
      23. Herbage glaucous; lvs fleshy; pls sometimes 10–20 dm, tolerant of standing saltwater .... ***S. hydrophilus***
      23′ Herbage not glaucous; lvs firm; pls < 10 dm, of damp soil but not standing water ...... ***S. hydrophiloides***
     22′ Pls hairy, sometimes becoming subglabrous
      24. Ray fls 0(–2); disk fls < 20; main phyllaries gen ± 8, sometimes 13; basal lf margins wavy or shallow-toothed .... ***S. aronicoides***
      24′ Ray fls gen present; disk fls 20–50; main phyllaries gen ± 21, sometimes 13; basal lvs gen subentire .... ***S. integerrimus***
       25. Ligules off-white to cream-colored .... var. ***ochroleucus***

25′ Ligules yellow, rarely 0
26. Main phyllaries 5–10 mm, lanceolate, often strongly black-tipped; ± hairy .... var. ***exaltatus***
26′ Main phyllaries 8–12 mm, ± linear, ± not black-tipped; densely hairy .... var. ***major***
21′ Pls from taproot, rhizome, or non-button-like caudex
27. Sts from thick, creeping or erect rhizome; roots fleshy, unbranched; lvs tapered to winged petiole, marginal teeth hard, translucent
28. Ray fls 0; main phyllaries ± 21; disk corollas orange-yellow .... ***S. astephanus***
28′ Ray fls present; main phyllaries ± 13; disk corollas yellow .... ***S. scorzonella***
27′ Sts from thin taproot, caudex, or rhizome; roots fibrous, branched; lvs subentire to pinnately divided, marginal teeth not hard, not translucent
29. Herbage (sub)glabrous; basal lvs entire to toothed
30. Ray fls 0 (or infl distinctly umbel-like)
31. Heads 8–20+; corollas yellow; lvs thin .... ***S. indecorus***
31′ Heads gen < 6; corollas deep yellow to brick-colored; lvs thick .... ***S. pauciflorus***
30′ Ray fls present (or infl cyme-like)
32. Head 1(–3) — pl from slender rhizome .... ***S. cymbalarioides***
32′ Heads gen 4–many
33. Pls sparsely short-hairy at maturity .... ***S. layneae***
33′ Pls ± completely glabrous at maturity
34. Lower lf bases cordate to truncate
35. Corollas orange-yellow to orange; basal lvs ± long as wide, rounded at tip .... ***S. ganderi***
35′ Corollas yellow; basal lvs widely lanceolate, blunt or pointed at tip .... ***S. pseudaureus***
34′ Lower lf bases obtuse to tapered
36. Herbage glaucous; lvs ± entire; pls < 10 dm .... ***S. clevelandii***
36′ Herbage not glaucous; lvs subentire to toothed; pls gen < 4 dm .... ***S. streptanthifolius***
29′ Young herbage hairy (sometimes becoming subglabrous) or basal lvs pinnately lobed or divided
37. Basal lvs pinnately divided or lobed; herbage ± glabrous
38. Terminal lobe of lower lvs = or > lateral; lateral lobes 1–3 pairs .... ***S. bolanderi***
38′ Terminal lobe of lower lvs < lateral (or lateral lobes > 6 pairs)
39. Heads small, main phyllaries gen 4–8 mm .... ***S. multilobatus***
39′ Heads large, main phyllaries gen 8–10 mm
40. Pls 4–10 dm; sts gen 1 from a short caudex; mid-cauline lvs ± prominent .... ***S. breweri***
40′ Pls 2–5(7) dm; sts often several from branched caudex; mid-cauline lvs reduced ***S. eurycephalus***
41. Lvs irregularly lobed or dissected, lobes toothed; sts several but aspect not shrub-like .... var. ***eurycephalus***
41′ Lvs deeply and finely 1–2-dissected; aspect ± shrub-like .... var. ***lewisrosei***
37′ Basal lvs entire to dentate (or if pinnate, loosely tomentose)
42. Heads gen > 6
43. Herbage permanently ± hairy; lf margins flat; blade length 1.5–2 × width .... ***S. canus***
43′ Herbage ± glabrous at fl; lf margins often weakly rolled under; blade length > 2 × width .... ***S. macounii***
42′ Heads 1–few (gen < 6)
44. Lvs lobed .... ***S. ionophyllus***
44′ Lvs subentire or dentate
45. Disk 18+ mm diam, ligules orange to brick-colored .... ***S. greenei***
45′ Disk < 15 mm diam, ligules yellow
46. Cauline lvs gradually reduced, conspicuous .... ***S. bernardinus***
46′ Cauline lvs bract-like (pls ± scapose) .... ***S. werneriifolius***

S. aphanactis E. Greene Ann 1–2 dm, taprooted, slender, branched upward, ± glabrous (exc heads). **LVS** 2–4 cm, linear to oblanceolate, subentire to lobed. **INFL**: heads barely radiate, 4–10+, ± urn-shaped; main phyllaries ± 8, sometimes 13, 5–6 mm, tips green. **RAY FLS** few; ligules barely > phyllaries. **DISK FLS** < 40. **FR** densely gray-hairy. $2n$=40. Drying alkaline flats; < 400 m. CW, SCo, ChI; Baja CA.

S. aronicoides DC. (p. 339) Bien or per 3–9 dm, from button-like caudex; roots fleshy, unbranched. **LVS**: lower blades 7–20 cm, oblanceolate to ovate, subentire, wavy, or shallowly toothed. **INFL**: heads gen discoid, 10–30; main phyllaries ± 8, sometimes 13, 4–8 mm, tips often black. **RAY FLS** 0(–2); ligules < 5 mm. **DISK FLS** < 20. **FR** glabrous; pappus sometimes well exserted. $2n$=40. Dry open foothill woodland, montane forest; < 2500 m. NW, CaR, SN (exc Teh), n CW; s OR. ❀TRY.

S. astephanus E. Greene (p. 339) Per 4–10 dm, from erect rhizome. **LVS** evenly distributed; lower tapered to winged petiole, 10–25 cm, oblanceolate to ovate, teeth minute, sharp, hard. **INFL**: heads discoid, 10–20; main phyllaries ± 21, 7–8 mm, tips green. **RAY FLS** 0. **DISK FLS** < 40; corollas orange-yellow. **FR** glabrous. $2n$=40. Steep rocky slopes; 400–1500 m. s SCoRO, TR.

S. bernardinus E. Greene (p. 339) SAN BERNARDINO RAGWORT Per 1–5 dm, from ± creeping caudex. **LVS**: lower blades 1–2.5 cm (<< petioles), (ob)ovate, toothed. **INFL**: heads radiate, 3–5 (8); main phyllaries ± 13, sometimes 21, 6–8 mm, obtuse, tips green. **RAY FLS** 8(13); ligules 8–10 mm. **DISK FLS** < 40. **FR** glabrous or minutely rough-hairy, esp on angles. $2n$=46. RARE. Pine forests; 1800–2300 m. e SnBr.

S. blochmaniae E. Greene Subshrub 5–12 dm, from thick taproot, glabrous. **LVS** crowded, < 12 cm, ± linear, ± thick, entire; lower drying, pendent. **INFL**: heads radiate, 5–20+; main phyllaries ± 13, 7–10 mm, tips green; outer phyllaries 0 or < 1/4 main phyllaries. **RAY FLS** 8; ligules 8–10 mm. **DISK FLS** < 40. **FR** soft-hairy. $2n$=40. Sand dunes, coastal floodplains; < 100 m. s CCo. ❀DRN,IRR:**14**,19–21,**22,23**,&SUN:**15–17,24**.

S. bolanderi A. Gray var. ***bolanderi*** (p. 339) Per 1–5 dm, from rhizome, unevenly jointed-hairy on phyllaries and lf lower surfaces. **LVS** thin (or fleshy near ocean); lower petioled, gen 5–15 cm, often 3–7-lobed, main lobe ± cordate, deeply crenate; cauline more lobed, midrib with few lobes. **INFL**: heads radiate, 8–15; main phyllaries ± 13, sometimes 21, 6–8 mm, tips green. **RAY FLS** 8 (13); ligules 10–15+ mm, showy. **DISK FLS** < 40. **FR** glabrous. $2n$=46. Wet cliffs, open forest; < 200 m. NCo; to WA. ❀TRY.

S. breweri Burtt Davy (p. 341) Bien or short-lived per 4–10 dm, from short caudex, glabrous or minutely hair-tufted in lower axils. **LVS**: lower 10–30 cm; blade gen > petiole, narrowly obovate, deeply dissected, terminal lobe ± ovate-cordate, lateral lobes 4–12+, margins dentate. **INFL**: heads radiate, (4)15–30+; main phyllaries ± 13, sometimes 21, 8–10 mm, tips green. **RAY FLS** 8(13); ligules 10–20 mm. **DISK FLS** < 40. **FR** glabrous. $2n$=46. Seasonally damp, protected woodlands; 200–1700. s SNF, Teh, e SnJV, c&s CW, n WTR. ❀TRY.

S. californicus DC. Ann 1–4 dm, from taproot, slender, ± glabrous. **ST**: branches arched upward. **LVS** thin (to fleshy near ocean); lower weakly petioled, 2–7 cm, ± linear to ovate, subentire to deeply dentate. **INFL**: heads radiate, (1)3–10+; main phyllaries ± 21, 5–7 mm, tips black. **RAY FLS** 13; ligules < 10 mm. **DISK FLS** < 40. **FR** soft-hairy. $2n$=40. Coastal strand to shrubland; < 1200 m. s SN, Teh, CW, SW, w DSon; Baja CA. Fleshy coastal pls have been called *S. ammophilus* E. Greene. ❀SUN,DRN:7,12,13,**14–24**.

S. canus Hook. Per 1–4 dm, from caudex or rhizome, ± felty-gray-tomentose. **LVS**: basal 2.5–5 cm (< petiole), ± lanceolate to ovate, entire to weakly dentate. **INFL**: heads radiate, 6–12; main phyllaries ± 13, sometimes 21, 5–6+ mm, tips green. **RAY FLS** 8 (13); ligules (5)8–10 mm. **DISK FLS** < 40. **FR** glabrous. $2n$=46, 92,+ polyploids. High rocky plains, sagebrush scrub; 1300–3600 m. e KR, CaR, SN, GB; to Can, c US. Locally abundant. ❀SUN, DRN:**17**&IRR:1–3,**7**,10,11,**14–16**,18,24.

S. clarkianus A. Gray (p. 341) Per 6–12 dm, from short caudex; hairs jointed. **LVS** evenly distributed; lower petioled, < 25 cm, blade lanceolate to oblong, lobed to dissected. **INFL**: heads radiate, 8–20; main phyllaries ± 21, sometimes 13, 5–8 mm, tips green; outer phyllaries well developed. **RAY FLS** (8)13; ligules 8–12 mm. **DISK FLS** < 40. **FR** glabrous. Damp meadows; 1400–2700 m. c&s SN. ❀TRY.

S. clevelandii E. Greene (p. 341) CLEVELAND'S RAGWORT Per 3–10 dm, from ± erect caudex. **LVS** fleshy; lower blades 3–10 cm (<< petiole), oblanceolate to obovate, ± entire; upper sometimes lobed. **INFL**: heads radiate, 12–24; main phyllaries ± 21, sometimes 13, 5–6 mm, tips green. **RAY FLS** 8(13); ligules 5–7 mm, yellow. **DISK FLS** < 40. **FR** glabrous. $2n$=46. UNCOMMON. Drying serpentine soils, esp among shrubs; 400–900 m. s NCoRI (Napa, Lake cos.), c SNF (Tuolumne Co.). [var. *heterophylla* Hoover] ❀IRR:5,15–17&SHD:7,14,19–24;DFCLT.

S. cymbalarioides J.N. Buek Per 1–3+ dm, from slender rhizome, appearing ± scapose, ± subglabrous. **LVS** ± fleshy; basal blades 1–3+ cm (< or = ± tapered petiole), ± ovate, subentire to shallowly crenate. **INFL**: heads radiate, 1(–3); main phyllaries (13) 21, 6–8 mm, often red-tinged, tips green. **RAY FLS** ± 13; ligules 5–10+ mm. **DISK FLS** < 40. **FR** glabrous. $2n$=46. Damp alpine meadows; 1700–3500 m. s CaRH, n&c SNH, MP; to WA, sw Can, WY. [*S. subnudus* DC. misapplied] ❀DRN,IRR:1–3,**7**,14,**18**& SUN:4–6,**15–17**,22–24.

S. elegans L. PURPLE RAGWORT Ann 3–6 dm, from taproot, branched upward, sticky. **LVS** 3–8 cm; lower, oblong, deeply toothed or lobed; petiole bases expanded. **INFL**: heads radiate, many; main phyllaries < 10 mm, tips black. **RAY FLS** ± 13; ligules 10–15 mm, (red-)purple. **DISK FLS** < 40; corolla yellowish. **FR** ribbed, glabrous. $2n$=20. Uncommon. Disturbed coastal habitats; < 100 m. CCo, SnFrB, SCo; native to s Afr; cult as orn.

S. eurycephalus A. Gray Per 2–7 dm, from creeping caudex, tomentose, becoming subglabrous. **LVS**: basal sometimes fallen by fl, blades 5–10+ cm, < or = petiole, ± ovate, dentate-lobed to dissected; cauline often more lobed. **INFL**: heads radiate, 5–20; main phyllaries ± 13, sometimes 21, 8–12 mm, tips green or yellow; outer phyllaries inconspicuous. **RAY FLS** 8(13); ligules 10–15 mm. **DISK FLS** < 40. **FR** glabrous. $2n$=46. Drying, disturbed sites, esp on serpentine; 300–1700 m. e NW, CaR, n SNH, MP; s OR.

var. ***eurycephalus*** (p. 341) Pl erect, becoming subglabrous. **LVS**: lower unevenly dissected, lobes toothed. $2n$=46. Woods, rocky streambeds, hillsides; 300–1700 m. Range of sp. (exc n SNH). ❀SUN,DRN:1,6,**7**,**14–16**,17,18.

var. ***lewisrosei*** (J. Howell) T. Barkley (p. 341) CUT-LEAVED RAGWORT Pl arched upward, silver-gray tomentose, becoming unevenly glabrous. **LVS** finely dissected (larger lobes 2-pinnately dissected). RARE. Serpentine slopes, canyons; 550–900 m. n SNH (Feather River drainage, e Butte, Plumas cos.). [*S. lewisrosei* J. Howell] In cult.

S. flaccidus Less. Subshrub (may fl 1st year) 3–15+ dm, ± taprooted, arching-branched above, glabrous to woolly. **LVS** thread-like, sometimes deeply divided into narrow segments, sometimes with axillary lf clusters. **INFL**: heads radiate, 3–10(20) per cluster; main phyllaries ± 13, sometimes 21, 5–10+ mm; outer phyllaries ± prominent. **RAY FLS** 8(13); ligules 10–20 mm, ± yellow. **DISK FLS** < 40. **FR** soft-hairy. $2n$=40. Dry, rocky or sandy sites; < 2000 m. CA-FP (exc NCo, KR), SNE, D; to Colorado, TX, w&c Mex.

var. ***douglasii*** (DC.) B. Turner & T. Barkley Pl gen appearing shrubby, persistently tomentose. **INFL**: main phyllaries ± 21, 7–10 mm. $2n$=40. Habitats of sp.; < 1600 m. CA-FP (exc NCo, KR); Baja CA. [*S. douglasii* DC. incl var. *tularensis* Munz] ❀DRN,SUN:5, **7–9**,10,**14–16**,17,18,**19–24**&IRR:11–13.

var. ***monoensis*** (E. Greene) B. Turner & T. Barkley Pl often not appearing shrubby, ± glabrous. **INFL**: main phyllaries ± 21, sometimes 13. $2n$=40. Exposed basins, foothills; 500–2000 m. SNE, D; to TX, nw Mex. [*S. douglasii* var. *monoensis* (E. Greene) Jepson] ❀TRY.

S. fremontii Torrey & A. Gray Per 1–2 dm, ± prostrate to arched upward from caudex, glabrous, often purplish below. **LVS** thickish, tapered to petiole or sessile and weakly clasping; larger 2–4.5 cm, oblanceolate to ovate, dentate. **INFL**: heads radiate, 1–5; main phyllaries ± 13, sometimes 8, 5–10 mm, tips green. **RAY FLS** 8; ligules 8–12 mm. **DISK FLS** < 40. **FR** glabrous or angles hairy. $2n$=40,40+,80. Talus, other rocky places; 2600–3600 m. CaRH, SN, SnBr, GB; to sw Can, WY, Colorado.

var. ***fremontii*** (p. 341) **LVS** ± even, 2.4–4.5 cm. **INFL**: main phyllaries gen 8–10 mm. Uncommon. Habitats of sp. CaRH, Wrn; to sw Can, Colorado. ❀TRY.

var. ***occidentalis*** A. Gray (p. 341) **LVS** few and much reduced above; lower gen < 3.5 cm. **INFL**: main phyllaries 5–8 mm. Habitats of sp. SN, SnBr, SNE; w NV. ❀TRY.

S. ganderi T. Barkley & Beauch. (p. 341) GANDER'S RAGWORT Per 3–5+ dm, from caudex, ± glabrous, purple-tinged below. **LVS** thickish; basal blades 4–8 cm, < petiole (gen ± perpendicular to it), ± round-cordate, often 2–6-lobed, toothed to shallowly lobed; cauline few, deeply sharply toothed. **INFL**: heads radiate, 4–8; main phyllaries ± 21, sometimes 13, 8–11 mm, tips green; outer phyllaries 0–few. **RAY FLS** 13+; ligules 10–15 mm, ± orange. **DISK FLS** < 40. **FR** glabrous. RARE. Chaparral, burns; 400–1200 m. sw PR (sw San Diego Co.). Uncommonly collected.

S. greenei A. Gray (p. 341) Per 2–3+ dm, from creeping rhizome. **LVS**: basal blades gen 2.5–6 cm, < or = petiole, ± ovate, subentire to toothed, narrowed at base, purplish below. **INFL**: heads radiate, 1–3(6), large, disk 18+ mm; main phyllaries ± 21, acuminate, 7–12 mm, tips green. **RAY FLS** 13; ligules < 20 mm, red-orange to brick-colored. **DISK FLS** < 40. **FR** glabrous. $2n$=40–46,92. Dry, open serpentine in scrub or; 400–1500 m. s KR, NCoR. ❀DRN,DRY:7,14,18–23&SUN:5,15–17,24;DFCLT.

S. hybridus Regel FLORISTS' CINERARIA Ann or per < 8 dm, from caudex, often red-purple below. **LVS**: lower blades 6–12 cm, = or > clasping petiole, ovate to round-cordate, wavy to dentate. **INFL**: heads radiate, often > 80; main phyllaries ± 13, sometimes 21, 4–6 mm, tips green. **RAY FLS** 13(21); ligules white to purple or 2-colored. **DISK FLS** < 40. **FR** glabrous or soft-hairy between ridges. $2n$=60. Damp, protected, disturbed sites; < 100 m. SnFrB; derived from Canary Island *S. cruentus* L'Her., etc. Cult as orn.

S. hydrophiloides Rydb. (p. 341) SWEET MARSH RAGWORT Bien or per 3–10+ dm, from button-like caudex, ± glabrous, sometimes reddish; roots fleshy, unbranched. **LVS** firm; lower blades 5–20 cm, < petiole, elliptic to lanceolate, subentire to shallowly dentate; middle and upper few. **INFL** dense; heads radiate or discoid, 10–25; main phyllaries ± 13, sometimes 21, 4–9 mm, tips black. **RAY FLS** (0)8(13); ligules 5–10 mm. **DISK FLS** < 40. **FR** glabrous.

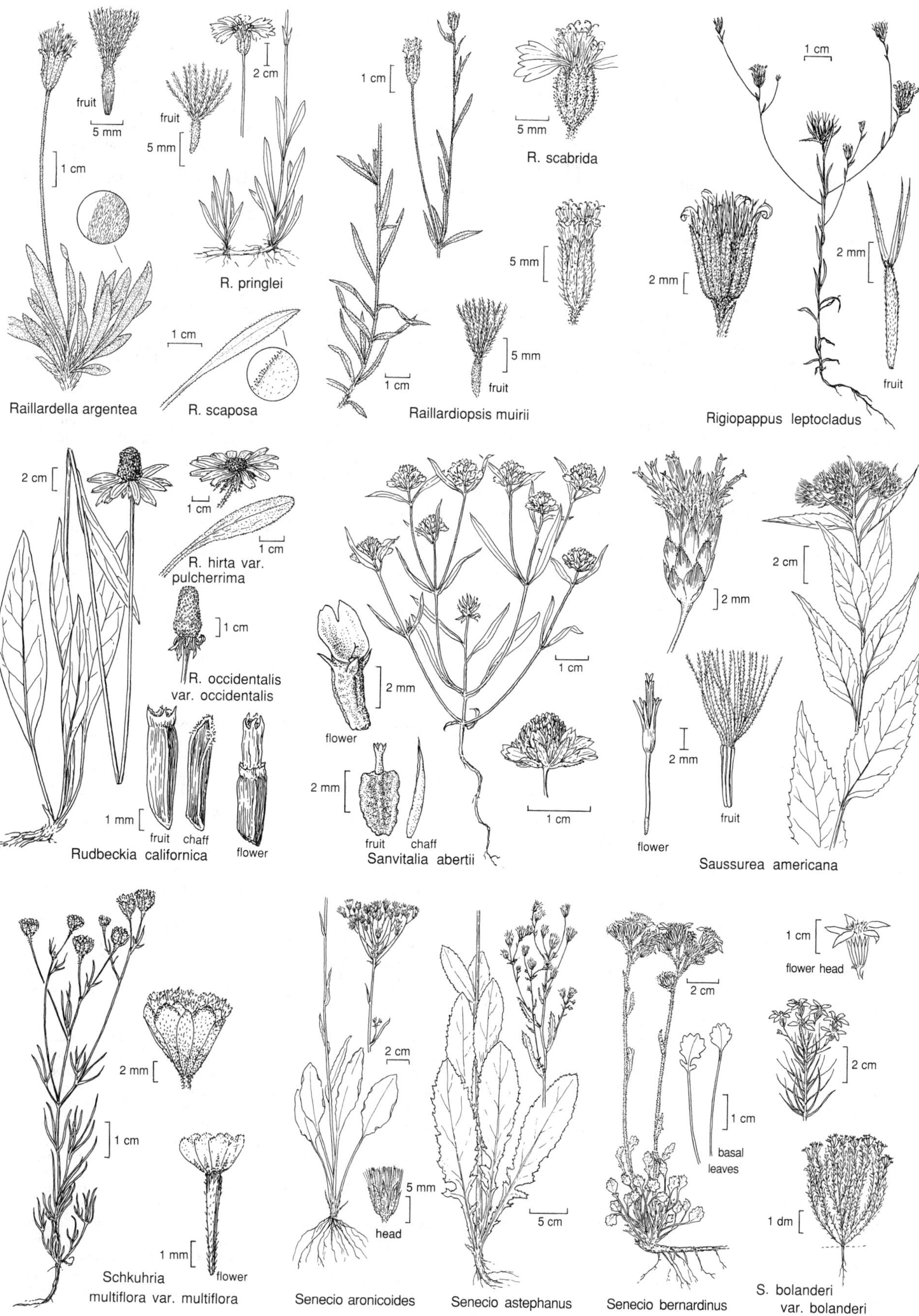
fruit
5 mm
1 cm
fruit
5 mm
2 cm
R. pringlei
1 cm
Raillardella argentea
R. scaposa
1 cm
5 mm
R. scabrida
5 mm
5 mm
1 cm
fruit
Raillardiopsis muirii
1 cm
2 mm
2 mm
fruit
Rigiopappus leptocladus
2 cm
1 cm
1 cm
R. hirta var. pulcherrima
1 cm
R. occidentalis var. occidentalis
1 mm
fruit
chaff
flower
Rudbeckia californica
2 mm
flower
2 mm
fruit
chaff
1 cm
1 cm
Sanvitalia abertii
2 mm
2 cm
2 mm
flower
fruit
Saussurea americana
2 mm
1 cm
1 mm
flower
Schkuhria multiflora var. multiflora
2 cm
5 mm
head
Senecio aronicoides
Senecio astephanus
2 cm
1 cm
basal leaves
5 cm
Senecio bernardinus
1 cm
flower head
2 cm
1 dm
S. bolanderi var. bolanderi

$2n$=40. UNCOMMON. Damp meadows, hillsides; 1500–2800 m. e CaRH, n SN, MP; to B.C., MT. [*S. foetidus* Howell incl var. *h.* (Rydb.) Cronq.]

S. hydrophilus Nutt. Bien or per 5–20 dm, from button-like caudex, glabrous, glaucous, light green; roots fleshy, unbranched. **ST** hollow. **LVS** thick; lower blades 5–20+ cm, > petiole, elliptic to oblanceolate, entire or shallowly toothed. **INFL**: heads radiate or discoid, 20–80+; main phyllaries 8(13), 5–8 mm, tips often black. **RAY FLS** (0)5; ligules 3–8 mm. **DISK FLS** < 40. **FR** glabrous. $2n$=40. Swamps, muddy sites, tolerant of standing saltwater; < 2300 m. s NCoR, CaR, SN, deltaic GV, n CW, GB; to B.C., Colorado. Reduced from wetland development. ❀TRY;STBL.

S. indecorus E. Greene Per 2–8 dm, from stout caudex, ± glabrous. **LVS** thin; basal blades 2–5 cm, < petiole, oblong to ± reniform, subentire to toothed. **INFL** umbel-like; heads gen discoid, 8–20+; main phyllaries reddish toward tip, ± 13, sometimes 21, 4–8 mm, tips green. **RAY FLS** gen 0 (or ligules gen < 5 mm). **DISK FLS** < 40. **FR** glabrous. $2n$=172–200, etc. Meadows, streambanks in open woods; 1600–2000 m. s MP (Pine Creek, Lassen Co.); also WA to AK, WY, e Can. [*S. pauciflorus* var. *fallax* Jepson (var. *jucundulus* Jepson misapplied)]

S. integerrimus Nutt. Bien or per 2–7 dm, from button-like caudex; roots fleshy, unbranched. **LVS**: lower petioled or tapered to base, 6–25 cm; blade gen lanceolate, subentire. **INFL**: heads radiate, 6–20(30), central head often largest, peduncle shorter; main phyllaries ± 21, sometimes 13. **RAY FLS** (8)13; ligules 6–15 mm, yellow or off-white. **DISK FLS** 20–50. **FR** glabrous. $2n$=40,80. Grassland, open forest; 150–3600 m. KR, NCoRI, CaR, SN (exc Teh), GB; to sw Can, n-c US, Colorado.

var. ***exaltatus*** (Nutt.) Cronq. (p. 341) **LVS**: petiole often indistinct. **INFL**: main phyllaries gen < 10 mm, ± strongly black-tipped. **RAY FLS** rarely 0; ligules yellow. Habitats of sp.; 1400–3200 m. CaR, SN (exc Teh), GB; to sw Can, WY, Colorado. ❀TRY.

var. ***major*** (A. Gray) Cronq. (p. 341) Pl persistently soft-hairy. **LVS**: petiole often indistinct. **INFL**: main phyllaries 8–12 mm, greenish or minutely black-tipped. **RAY FLS** rarely 0; ligules yellow. Forest; 150–3600 m. KR, NCoRI, CaRH, SN; s OR. ❀TRY.

var. ***ochroleucus*** (A. Gray) Cronq. **LVS**: petiole often distinct. **INFL**: main phyllaries < 10 mm, greenish or minutely black-tipped. **RAY FLS** present; ligules off-white to pale cream-colored. Coniferous forest; 600–1700 m. KR; to B.C., MT. ❀TRY.

S. ionophyllus E. Greene (p. 341) TEHACHAPI RAGWORT Per 1–5 dm, from taprooted caudex or rhizomed. **LVS** thick; lower blades 1–2.5 cm, < petiole, ovate to round, often 2–6-lobed; upper more deeply lobed. **INFL**: heads radiate, 1–6; main phyllaries ± 21, sometimes 8 or 13, 6–10 mm, tips green. **RAY FLS** (0)8(13); ligules 8–10 mm. **DISK FLS** < 40. **FR** glabrous. $2n$=46. UNCOMMON. Dry rocky coniferous forest, granite crevices; 1500–2700 m. s SN, SnGb, SnBr.

S. jacobaea L. TANSY RAGWORT Per 3–12 dm, from taprooted caudex, branched above. **LVS** evenly spaced, reduced upward; 5–20+ cm, ovate, deeply 1–2-pinnately dissected; lower soon deciduous. **INFL**: heads radiate, 20–60; main phyllaries ± 13, 3–5 mm, tips green or black. **RAY FLS** 13; ligules 8–12 mm. **DISK FLS** < 40. **FR** glabrous or angles hairy. $2n$=40. Pastures, roadsides, disturbed places; < 1500 m. NCo, w KR, sw CaR, n SN, n ScV, SnFrB; to B.C., e N.Am; native to Eurasia. NOXIOUS WEED. TOXIC to livestock.

S. layneae E. Greene (p. 341) LAYNE'S RAGWORT Per 4–7 dm, from caudex, subglabrous. **LVS**: basal thick, firm; blades 4–7 cm, < petiole, ± (ob)lanceolate, entire or unevenly shallow-toothed. **INFL**: heads radiate, 5–20; main phyllaries ± 21, sometimes 13, 7–10 mm, tips green. **RAY FLS** ± (5)8; ligules 12–16 mm. **DISK FLS** < 40. **FR** glabrous. $2n$=46. **RARE** CA. Dry pine/oak woodland, on serpentine; 200–1000 m. n SNF (El Dorado, Tuolumne cos.).

S. lyonii A. Gray (p. 341) Subshrub < 15 dm, from woody taproot, subglabrous (exc lf lower surfaces, tufted axils). **LVS** evenly spaced; lower soon deciduous, 4–12 cm, deeply 1–2-pinnately lobed, segments ± linear. **INFL**: heads radiate, in several 4–12-headed clusters; main phyllaries ± 21, sometimes 13, 5–8 mm, tips black. **RAY FLS** 13; ligules 8–10 mm. **DISK FLS** < 40. **FR** soft-hairy. $2n$=40. Hillsides; < 500 m. s ChI (Santa Catalina, San Clemente islands); Guadalupe Island, Mex. ❀TRY.

S. macounii E. Greene (p. 341) SISKIYOU MOUNTAINS RAGWORT Per 3–5 dm, from caudex. **LVS**: basal blades 3–7 cm, < petiole, ± lanceolate, entire or shallowly few-toothed, often weakly rolled under. **INFL**: heads radiate, 6–20; main phyllaries ± 13, sometimes 21, 5–7 mm, tips green. **RAY FLS** 8; ligules 8–10 mm. **DISK FLS** < 40. **FR** glabrous. $2n$=46,±92. RARE in CA. Rocky, disturbed streamsides, roadsides, clearings in coniferous forest; 400–900 m. nw KR (Del Norte Co.); to B.C. Intergrades with *S. canus* in OR. [*S. ligulifolius* E. Greene]

S. mikanioides Walp. GERMAN-IVY Per vine, luxuriant, glabrous. **LVS** evenly spaced; blades 3–8 cm, ± = petiole (often with 2 stipule-like lobes), ± round, sharply palmately 5–9-lobed. **INFL**: heads discoid, 20–40; main phyllaries ± 8, 3–4 mm, tips green. **RAY FLS** 0. **DISK FLS** < 40. **FR** ± glabrous. $2n$=40. Shady, ± disturbed places; < 200 m. NCo, CCo, SnFrB; native to s Afr. Highly invasive.

S. mohavensis A. Gray (p. 341) Ann 1–4 dm, from short, often twisted taproot, much-branched upward, glabrous, often purplish. **LVS**: lower blades 2–6 cm (petiole poorly defined), (ob)ovate, unevenly dentate to lobed; upper blades clearly clasping. **INFL**: heads radiate or disciform, 3–10; main phyllaries ± 8, sometimes 13, 6–7 mm, tips green. **RAY FLS** few, inconspicuous; ligules short, ± = phyllaries. **DISK FLS** < 40. **FR** soft-hairy. $2n$=40. Sandy washes, flats; < 1000 m. D; s NV, w AZ, nw Mex. Like *S. flavus* (Decne.) Schultz Bip. of Afr. [Liston et al. 1989 Amer J Bot 76:383–388].

S. multilobatus A. Gray Ann to per 2–5 dm, from taprooted caudex. **LVS**: lower blade 3–9+ cm, ± (ob)ovate, deeply pinnately dissected. **INFL**: heads gen radiate, 8–25; main phyllaries ± 13, sometimes 21, 4–8+ mm, tips green or yellow. **RAY FLS** (0)8(13); ligules < 10 mm. **DISK FLS** < 40. **FR** gen glabrous (or angles hairy). $2n$=46,92. Rocky or sandy soils, sagebrush or open woodland; 1400–3200 m. SNH (e slope), SNE, DMtns; to ID, WY, NM. [*S. stygius* E. Greene] ❀TRY.

S. pattersonensis Hoover (p. 341) MONO RAGWORT Per gen < 1 dm, from rhizome, arched upward, glabrous, sometimes reddish. **LVS** thick to ± fleshy, evenly spaced, 2–4 cm, ± narrowly lanceolate, sometimes with 1–2 lateral lobes, margins entire to wavy, often rolled under, sometimes decurrent. **INFL**: heads radiate, 1(–4); main phyllaries ± 13, 5–8 mm, tips green. **RAY FLS** 8; ligules 5–10 mm. **DISK FLS** < 40. **FR** glabrous. UNCOMMON. Talus slopes; 2900–3700 m. c SNH, n SNE (Sweetwater Mtns).

S. pauciflorus Pursh Per 1–5 dm, from stout caudex, becoming glabrous. **LVS**: basal thick, firm; blades 1.5–4 cm, < petiole ovate to ± reniform, ± toothed, bases truncate to tapered; cauline few, lower dissected. **INFL** umbel-like; heads gen radiate, 1–6+; main phyllaries ± 13, sometimes 21, 7–10 mm, reddish to purple-tinged, tips green. **RAY FLS** (0)8(13); ligules 5–7 mm, deep yellow to brick-colored. **DISK FLS** < 40. **FR** glabrous. $2n$=174–194,±260. Subalpine or alpine meadows; 2400–3500 m. SNH; also n Can to WA, WY.

S. pseudaureus Rydb. var. ***pseudaureus*** Per 2–7 dm, from weakly branched caudex. **LVS**: lower blades 2–10 cm, widely lanceolate, obtuse to cordate, toothed. **INFL** flat-topped; heads gen radiate, 5–20; main phyllaries ± 13, sometimes 21, 5–8 mm, tips green. **RAY FLS** (0)8(13); ligules 6–10+ mm. **DISK FLS** < 40. **FR** glabrous. $2n$=40?. Uncommon. Streambanks, meadows; 2400–3300 m. CaRH, SN, Wrn; to sw Can, MT, NV. [*S. pauciflorus* var. *jucundulus* Jepson; *S. pa.* var. *fallax* Jepson misapplied] ❀TRY.

S. scorzonella E. Greene (p. 347) Per 2–5 dm, from stout rhizome or caudex, gen evenly short-woolly; roots fleshy, unbranched. **LVS**: basal blades 10–24 cm, (ob)lanceolate, teeth fine, hard, dark, base tapered. **INFL**: heads gen radiate, 10–24; main phyllaries ± 13, 3–5 mm, tips often black. **RAY FLS** (0)5+; ligules 5–8+ mm. **DISK FLS** < 40. **FR** glabrous. Open forest, meadow edges; 1600–3500 m. CaRH, SNH, n W&I. [*S. covillei* E. Greene] ❀TRY.

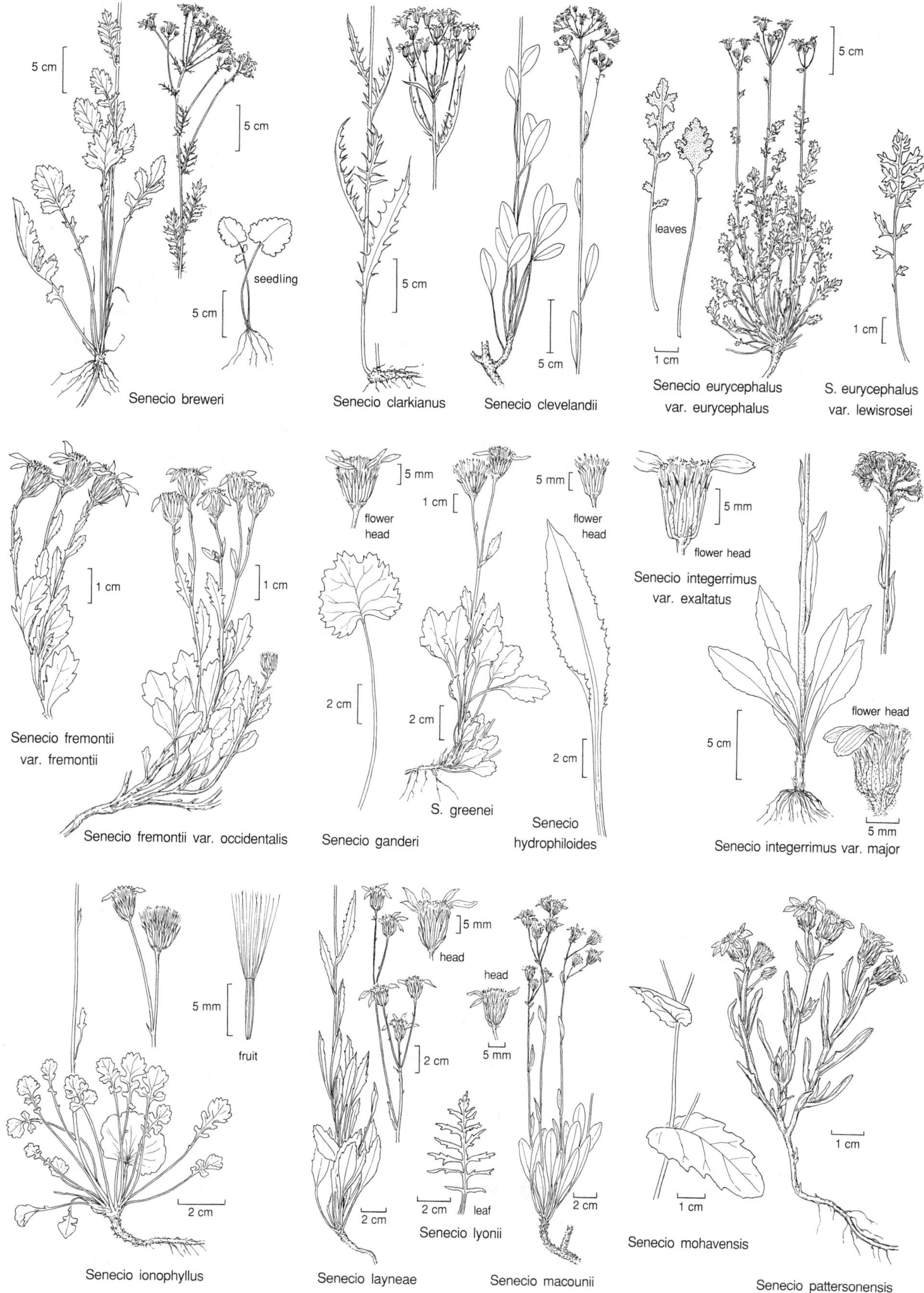

5 cm
5 cm
seedling
5 cm
Senecio breweri
5 cm
Senecio clarkianus
5 cm
Senecio clevelandii
5 cm
leaves
1 cm
Senecio eurycephalus
var. eurycephalus
1 cm
S. eurycephalus
var. lewisrosei
1 cm
1 cm
Senecio fremontii
var. fremontii
Senecio fremontii var. occidentalis
5 mm
flower
head
1 cm
2 cm
2 cm
Senecio ganderi
S. greenei
5 mm
flower
head
2 cm
Senecio
hydrophiloides
5 mm
flower head
Senecio integerrimus
var. exaltatus
flower head
5 cm
5 mm
Senecio integerrimus var. major
5 mm
fruit
2 cm
Senecio ionophyllus
5 mm
head
head
5 mm
2 cm
2 cm
2 cm
leaf
Senecio lyonii
2 cm
Senecio layneae
Senecio macounii
1 cm
Senecio mohavensis
1 cm
Senecio pattersonensis

S. serra Hook. var. ***serra*** Per 5–20 dm, from branched woody caudex, sometimes red-tinged below. **LVS** evenly spaced, ± equal; blades 5–20 cm, > petiole ± lanceolate, subentire to dentate; lowermost soon deciduous. **INFL**: heads radiate, 30–60+; main phyllaries ± 13, sometimes 8, 4–6 mm, tips often black. **RAY FLS** 8; ligules 5–7 mm. **DISK FLS** < 40. **FR** (sub)glabrous. 2*n*=40. Damp, open coniferous forest or sagebrush scrub; 1300–3200 m. s SNF, SNH (e slope); to WA, MT, WY. ❀TRY.

S. spartioides Torrey & A. Gray Subshrub 2–8 dm, arched upward from taprooted crown. **LVS** 5–10 cm, linear, entire, ± toothed, or 2–4-lobed. **INFL**: heads radiate, 10–20+; main phyllaries ± 8, sometimes 13, 6–9 mm, tips green. **RAY FLS** 5(8); ligules 10+ mm, ± yellow. **DISK FLS** < 40. **FR** soft-hairy. 2*n*=40. Dry, open rocky places; 1800–3200 m. SNH (e slope), SnBr, SNE, n DMtns; to SD, NM, nw Mex. May hybridize with *S. fremontii* var. *occidentalis* in n SNE (Sweetwater Mtns).

S. streptanthifolius E. Greene Per 1–6 dm, from caudex. **LVS**: basal blades gen 2–4 cm, < petiole, obovate to round, firm, subentire to shallowly dentate, base tapered to truncate. **INFL**: heads radiate, 4–20; main phyllaries ± 13, sometimes 8 or 21, 4–7+ mm, tips green. **RAY FLS** 8(13); ligules 5–10 mm. **DISK FLS** < 40. **FR** glabrous. 2*n*=46. Woodlands, rocky areas, to alpine barrens; 900–3100 m. CaR, SNH, MP; to nw Can, NM. [*S. cymbalarioides* DC. misapplied] ❀TRY.

S. sylvaticus L. Ann 3–8 dm, taprooted, mildly aromatic; hairs gen short, curled. **LVS** evenly spaced, 3–12 cm, ± linear to obovate, deeply dentate to lobed. **INFL**: heads disciform or inconspicuously radiate, 12–24; main phyllaries ± 13, sometimes 8 or 21, 4–7 mm, rarely minutely black-tipped; outer phyllaries inconspicuous. **RAY FLS** < 8; ligules barely > phyllaries. **DISK FLS** < 40. **FR** hairy, esp angles. 2*n*=40. Open, disturbed woodland, rocky sites, ± coastal; < 200 m. NCo, n CCo, SnFrB; to B.C., e US; native of Eur.

S. triangularis Hook. (p. 347) Per 5–12+ dm, from branched woody caudex. **LVS** evenly spaced, ± equal; blades 4–14 cm, < petiole ± triangular, subentire to dentate, base truncate to hastate. **INFL**: heads radiate, 10–30+; main phyllaries ± 13, sometimes 8 or 21, 6–10 mm, tips barely black. **RAY FLS** 8; ligules < 15 mm. **DISK FLS** < 40. **FR** glabrous. 2*n*=40,80. Wet meadows, streambanks in open, coniferous forest; 1000–3500 m. KR, CaR, SN, SnGb, SnBr, PR, MP; to AK, Colorado. ❀TRY.

S. vulgaris L. Ann 1–6 dm, arched upward, ± glabrous. **LVS**: blades 2–10 cm (petiole weakly defined), oblanceolate to obovate, deeply and unevenly dentate to lobed. **INFL**: heads discoid, 8–10; main phyllaries ± 21, 4–6 mm, tips often black; outer phyllaries 2–6, short, strongly black tipped. **RAY FLS** 0. **DISK FLS** < 40. **FR** subglabrous. 2*n*=40,80. Abundant. Gardens, farmlands, other disturbed sites; < 1500 m. CA (exc D); native to Eurasia.

S. werneriaefolius A. Gray Per gen 1–2 dm, ± tufted from branched, rhizome-like caudex, ± scapose. **LVS** thick, ± basal; blade < 4 cm, < (sometimes indistinct) petiole, linear-oblong, rolled under. **INFL**: heads gen radiate, 1–6; main phyllaries ± 13(21), 4–10 mm, tips green. **RAY FLS** (0)8(13); ligules 5–10 mm. **DISK FLS** < 40. **FR** glabrous. 2*n*=44,46. Talus, open sites, among trees near timberline, in loose soil; 3000–4000 m. n&c SNH, n W&I; to ID, MT, Colorado, n AZ. CA pls have narrower, more rolled lvs than pls from Rocky Mtns but intergrade completely. [*S. muirii* Greenman] ❀TRY;DFCLT.

SILYBUM MILK THISTLE

David J. Keil & Charles E. Turner

Ann or bien. **LVS** alternate, spiny-dentate, often coarsely lobed, dark green blotched white, ± glabrous; cauline reduced. **INFL**: heads discoid, large; peduncles bracted; involucre ovoid to spheric; phyllaries overlapping in several series, tips of outer and middle spreading, lanceolate to ovate, spiny-fringed and -tipped; receptacle flat, white-bristly. **FLS** many; corollas pink to purple, tube long, slender, throat abruptly wider, lobes linear; anther bases sharply short-sagittate, tips oblong; style tip with slightly swollen node, appendage long, cylindric, branches very short. **FR** ovoid, slightly compressed, glabrous, attachment slightly angled; pappus of many flat, minutely barbed bristles, falling in a ring. 2 spp.: Medit. (Greek: name for thistle-like pls)

S. marianum (L.) Gaertner (p. 347) **STS** 2–30 dm, glabrous or slightly woolly. **LVS**: basal 15–60+ cm; petiole winged; cauline bases clasping, coiled, spiny. **INFL**: body of heads 2–6 cm diam. **FR** 6–8 mm, brown and black-spotted; pappus 15–20 mm. 2*n*=34. Roadsides, pastures, waste areas; < 500 m. NCo, NCoR, GV, SnFrB, SCoR, SCo, ChI; native to Medit. Invasive. Sometimes cult as orn; seeds used as coffee substitute.

SOLIDAGO GOLDENROD

John C. Semple

Per from woody caudex or rhizome, branched above. **LVS** alternate, resinous, often sessile. **INFL**: heads radiate, few–many, in ± flat-topped to panicle-like, often ± 1-sided clusters; involucre cylindric to bell-shaped (wider when dry); phyllaries in 3–5 graduated, overlapping series, midrib gen ± swollen, translucent. **RAY FLS** few–many; ligules yellow. **DISK FLS** few–many; corollas yellow, gen glabrous; style branches finely papillate, appendage triangular. **FR** obconic, compressed; pappus of 25–45 long-barbed bristles in 1 series. ± 150 spp.: esp N.Am (S.Am, Eurasia). (Greek: make-well, from purported medicinal value) [Semple et al. 1990 Can J Bot 68:2070–2082]

1. Middle cauline lvs largest
  2. St glabrous below (infl subglabrous, ± waxy); infl pyramid-shaped, lower branches not crowded — rare, n SNH, MP .... ***S. gigantea***
  2' St ± hairy below and above, not waxy above; infl pyramid- or club-shaped
    3. Infl ± widely pyramid-shaped; involucre 3–5 mm; lvs weakly serrate; alien, disturbed sites .... ***S. altissima***
    3' Infl ± club-shaped; involucre 2.5–3.5 mm; lvs gen serrate; native, ± undisturbed sites .... ***S. canadensis*** ssp. ***elongata***

1' Lower cauline (or basal) lvs largest
  4. Infl flat- to round-topped; phyllaries often long-acuminate, not strongly graded; lf gen long-ciliate; alpine .... ***S. multiradiata***
  4' Infl panicle- or raceme-like or pyramid-shaped; phyllaries round to barely acuminate, ± strongly graded; lf not long-ciliate; up to subalpine
    5. Lvs and heads very resinous — NCo, n&c CCo .... ***S. spathulata***
    5' Lvs and heads not very resinous

6. Herbage ± short-hairy; upper nodes without axillary lf-clusters
7. Herbage ± densely hairy; lvs gen not 3-veined; phyllaries strigose; disk corolla throat obscure; CA-FP, MP .... ***S. californica***
7′ Herbage ± sparsely hairy; lvs ± 3-veined; phyllaries subglabrous; disk corolla throat obvious; GB .... ***S. sparsiflora***
6′ Herbage ± glabrous; upper nodes often with axillary lf-clusters
8. Phyllaries lanceolate to ± ovate, not inrolled, obtuse to acuminate; rays gen 8–12; esp SN, GB ... ***S. spectabilis***
8′ Phyllaries very narrowly triangular, inrolled near tip, sharply acute; rays gen 5–10; CW, SW, n DMtns
9. Lvs gen < 10 × longer than wide; infl gen club-shaped, panicle-like .... ***S. confinis***
9′ Lf gen 10+ × longer than wide; infl gen narrow, sometimes spike-like — s SCoRI .... ***S. guiradonis***

S. altissima L. var. *altissima* LATE GOLDENROD **STS** from rhizome, 2–15 dm, hairy to base. **LVS** 4–15 cm; mid-st lvs largest, (ob)lanceolate, strongly 3-veined, ± densely scabrous to short-hairy. **INFL** gen large, dense, pyramid-shaped; heads many; involucre 3–5 mm; phyllaries lanceolate, outer 1/4–1/3 length inner, margins ± strigose. **RAY FLS** 10–15; ligules 1–2 mm. **DISK FLS** 3–7; corollas 2.5–4.5 mm. **FR** 1–1.5 mm, ± strigose. $2n$=36,54. Uncommon. Disturbed sites; < 2800 m. NW, CaR, SN, CW, GB; to B.C.; native to e US.

S. californica Nutt. (p. 347) CALIFORNIA GOLDENROD Herbage gen ± densely short-soft-hairy. **STS** 2–15 dm, short-rhizomed. **LVS**: lower < 14 cm, oblanceolate to obovate, serrate, sometimes 3-veined, base tapered; upper much reduced. **INFL** long, 1-sided, pyramid- or wand-shaped, many-headed (or short, raceme-like, few-headed); involucre 3–5 mm; phyllaries gen narrow, acute, strigose, outer 1/4–1/3 length inner. **RAY FLS** 6–11; ligules 3–5 mm. **DISK FLS** 6–17; corollas 3–5 mm. **FR** 0.7–1.5 mm, ± densely strigose. $2n$=18,36. Woodland margins, grassland, disturbed soils; < 2300 m. CA-FP, MP; OR, Baja CA. ❀SUN:**4–6**&IRR:1–3,**15–23**&SHD:**7**,8,9,**14**.

S. canadensis L. ssp. ***elongata*** (Nutt.) Keck (p. 347) CANADA GOLDENROD Herbage ± sparsely strigose (more so above). **STS** 25–150 cm, rhizomed. **LVS**: midle cauline largest, 5–15 cm, ± lanceolate, gen 3-veined, toothed. **INFL** panicle-like; heads many; involucre 2.5–3.5 mm; phyllaries lanceolate, outer 1/4–1/3 length inner. **RAY FLS** 8–15; ligules 1.5–2 mm. **DISK FLS** 5–12; corollas 2.5–3.5 mm. **FR** 1–1.5 mm, ± strigose. $2n$=18,36. Meadows, thickets; < 2800 m. NW, CaR, SN, CW, GB; to B.C. Pls in CW have thicker, veinier lvs. *S. c.* incls many vague races across N.Am; one of the most difficult taxonomic problems in N.Am. ❀SUN:**4–6**&IRR:1–3,**7**,8–10,**14–24**;INV.

S. confinis Nutt. (p. 347) SOUTHERN GOLDENROD Pl ± glabrous. **STS** < 21 dm, often stout, from short, branched caudex. **LVS**: lowest largest, 5–25 cm, gen < 10 × longer than wide, entire, often ± fleshy, base nearly sheathing; uppermost sometimes scale-like or with axillary lf-clusters. **INFL** panicle-like; heads gen many; involucre 2.5–4 mm; phyllaries very narrowly triangular, inrolled near tip, sharply acute, outer 1/3–2/3 length inner; midrib gen enlarged, translucent. **RAY FLS** 3–13; ligules 1–2.5 mm. **DISK FLS** 10–20; corollas ± 3–4 mm. **FR** 1–1.5 mm, ± strigose. $2n$=18. Wet streambanks, springs, marshes; gen < 2500 m. Teh, CCo (formerly SnFrB), SCoR, SW, n DMtns. Involucres of n DMtns pls like those of *S. spectabilis*.

S. gigantea Aiton (p. 347) SMOOTH GOLDENROD Herbage glabrous to ± strigose. **STS** < 15 dm, rhizomed, ± waxy. **LVS**: middle cauline largest, 5–15 cm, narrowly elliptic to ± lanceolate, 3-veined, toothed. **INFL** ± open, pyramid-shaped; heads many; involucre 3–4 mm; phyllaries gen lanceolate, outer 1/4–1/3 length inner. **RAY FLS** 8–15; ligules 2–2.5 mm. **DISK FLS** 7–12; corollas 3–4 mm. **FR** 1–1.5 mm, sparsely strigose. $2n$=18,36,54. RARE in CA. Moist streambanks, lakesides; 1000–1500 m. n SNH, MP; to e N.Am. Like *S. canadensis*.

S. guiradonis A. Gray (p. 347) GUIRADO'S GOLDENROD Pl gen ± glabrous **STS** < 13 dm, from short caudex, often slender. **LVS**: lowest largest, 5–20 cm, (very) narrowly oblanceolate, entire, often ± fleshy, base nearly sheathing; uppermost reduced to linear scales, often with axillary lf-clusters. **INFL** narrow; heads few–many; involucre 2.5–4 mm; phyllaries very narrowly triangular, inrolled near tip, sharply acute, outer 1/3–1/2 length inner, midrib gen enlarged, translucent. **RAY FLS** 8–10; ligules 1–2.5 mm. **DISK FLS** 10–20; corollas ± 3–4 mm. **FR** 1–1.5 mm, ± strigose. $2n$=18. RARE. Near streams in asbestos-laden soils; 600–900 m. s SCoRI (San Benito, Fresno cos.). In cult.

S. multiradiata Aiton (p. 347) NORTHERN GOLDENROD Pl ± (sub)glabrous, hairier above. **STS** < 5 dm, from woody caudex. **LVS**: lower cauline < 12 cm, ± linear to spoon-shaped; upper reduced, ± linear to ovate, subclasping, gen long-ciliate. **INFL** flat- to round-topped; heads few–many; involucre 4–7 mm; phyllaries ± lanceolate, acute to long-acuminate, outer gen 2/3 to = length inner. **RAY FLS** 12–18; ligules 2–4 mm. **DISK FLS** 12–35; corollas 3–5 mm. **FR** 2–3 mm, ± sparsely strigose. $2n$=18,36. Alpine slopes, meadows; 2600–3700 m. KR, CaRH, SN, W&I; to AK, arctic Can, NM. ❀IRR,DRN,SUN:**1**,2–6,**7**,14,**15–18**,19–21.

S. sparsiflora A. Gray Pl ± sparsely short-hairy. **STS** 2–15 dm, short-rhizomed. **LVS**: lower largest, < 14 cm, oblanceolate to obovate, tapered at base, serrate, 3-veined, base tapered; upper much reduced, entire. **INFL** pyramid-shaped, ± sparsely branched, 1-sided at tip; heads gen many; involucre 3–6 mm; phyllaries lanceolate, rounded or obtuse, ± resinous. **RAY FLS** 6–9; ligules 4–6 mm. **DISK FLS** 5–12; corollas 3.5–5 mm, throat ± obvious. **FR** ± 1–2.5 mm, ± strigose. $2n$=18,36,54. Margins of dry woodlands, grasslands, disturbed soils; 500–2200 m. GB; to NM, Mexico.

S. spathulata DC. ssp. ***spathulata*** (p. 347) COAST GOLDENROD Pl ± glabrous (exc infl). **STS** 1–5 dm, decumbent to erect, from short, woody caudex. **LVS** glabrous, gen resinous-sticky; lower largest, 5–10 cm, spoon-shaped, crenate, obtuse; upper reduced, entire, acute. **INFL** wand- to club-like, strigose; heads few–many; involucre 4–7 mm; phyllaries very resinous-sticky, outer ovate, inner narrowly lanceolate. **RAY FLS** 4–10; ligules 2.3–4 mm. **DISK FLS** 10–18; corollas 4.3–6 mm. **FR** 2–3 mm, strigose. $2n$=18. Dunes, headlands; < 100 m. NCo, CCo. ❀IRR,DRN,SUN:**5,14–17**,19–24.

S. spectabilis (D. Eaton) A. Gray (p. 347) SHOWY GOLDENROD Pl ± glabrous. **STS** < 18 dm, from short caudex. **LVS**: basal largest, < 25 cm, oblanceolate, tapered, entire or serrate toward tip, ± fleshy; cauline reduced, entire, axils often with lf-clusters. **INFL** ± panicle-like, ascending to arching, tip sometimes 1-sided; heads gen many; involucre 3–4 mm; phyllaries lanceolate to narrowly ovate, obtuse to acuminate. **RAY FLS** 6–21; ligules 1.5–3.5 mm. **DISK FLS** 8–22; corollas 2.5–4.5 mm. **FR** 1–2 mm, sparsely strigose. $2n$=18. Bogs, alkaline meadow; 300–2300 m. CaRH, n SNH (e slope), GB; to OR, NV. [*S. missouriensis* Nutt., Missouri goldenrod, misapplied] ❀DFCLT.

SOLIVA

Martin F. Ray

Ann, ± depressed, glabrous to densely hairy. **LVS** pinnately lobed or compound; petiole bases wide, ± clasping. **INFL**: heads disciform, sessile in basal and cauline axils; phyllaries in 1–2 series, green to scarious; receptacle flat, naked. **PISTILLATE FLS**: corollas 0; style 2-branched. **DISK FLS** staminate; corollas 4-lobed; ovary sterile. **FR** flattened, often winged, tipped by a sharp, persistent stylar spine; pappus 0. ± 5 spp.: S.Am. (Dr. Salvador Soliva, 18th century physician to Spanish court) [Cabrera 1949 Not Mus La Plata 14:123–139; Ray 1984 MS thesis Stanford Univ]

S. sessilis Ruiz Lopez & Pavon (p. 347) Pl prostrate (< 25 cm diam) to ascending (2–7 cm), softly short-hairy. **STS** 1–many, often purple-spotted or dark-colored. **LVS** pinnate; lflets ± palmate, lobes 2–8, narrowly lanceolate. **INFL**: phyllaries 5–12, 2–3 mm, widely ovate to lanceolate, acute. **PISTILLATE FLS** 10–12. **DISK FLS** 4–6; corollas 2–3 mm, green-translucent. **FR** 3.5–5.2 mm (incl spine), ovate-lanceolate, with 0 or 2 teeth above, ± keeled, green to brown, often with tiny purple spots, glabrous to densely hairy, wingless and toothless to widely winged and long-toothed; wings sometimes notched in lower 1/3. $2n$=110–120. Disturbed areas, esp hard-packed paths, roadsides, lawns. NW, SNF, CW, SW, expected elsewhere; native to S.Am. [*S. daucifolia* Nutt.; *S. pterosperma* (A.L. Juss.) Less.]

SONCHUS SOW THISTLE

G. Ledyard Stebbins

Ann to shrubs; sap milky. **STS** erect, smooth, leafy. **LVS** basal and cauline, alternate, ± entire to toothed and coarsely pinnate-lobed; cauline gen sessile, clasping. **INFL**: heads ligulate, in cymes; involucre swollen at base; phyllaries gen in 3 series, outer many, short-triangular, those of inner 2 series linear, tapered; receptacle ± flat, naked. **FLS** many; ligules yellow, readily withering. **FR** gen ± flat, beakless; pappus of many fine, white bristles. 54 spp.: Eurasia, Afr. (Ancient Greek name) [Boulos 1972–74 Bot Not 125:287–319, 126:155–196; 127:7–37,402–451]

1. Per from long rhizomes; fr ± compressed, 3–4-angled, 2-ribbed between angles . . . . ***S. arvensis***
1′ Ann from taproots; fr weakly to strongly flattened
 2. Sts slender; lvs deeply lobed, lobes toothed and often with smaller secondary lobes; fr only slightly flattened, transversely roughened; ligule > corolla tube; uncommon . . . . ***S. tenerrimus***
 2′ Sts stout; lvs toothed or with wide lobes; fr flat, with thin edges, smooth or cross-roughened; ligule < or = corolla tube; widespread and common
  3. Basal lobes of lvs rounded; fr 3-ribbed on each side, otherwise smooth; ligule < corolla tube . ***S. asper*** ssp. ***asper***
  3′ Basal lobes of lvs acute; fr 2–4-ribbed and cross-wrinkled; ligule ± = corolla tube . . . . ***S. oleraceus***

S. arvensis L. PERENNIAL SOW THISTLE Per 4–15 dm, from long rhizomes that produce both fl and non-fl sts. **LVS**: basal 5–15 cm, entire to deeply lobed, short-petioled; cauline lvs sessile, coarsely lobed and toothed, basal clasping lobes curled. **INFL**: peduncles 2–8 cm, glandular-hairy; involucre 1.4–1.8 cm. **FLS**: ligule ± = tube. **FR** 2.5–3 mm, compressed, 2-ribbed on each face, rough on ribs; pappus bristles ± 4 × fr. $2n$=54. Uncommon. Weed in damp soil; < 1800 m. KR, CaRH, ScV, SCo; native to Eur. NOXIOUS WEED.

S. asper (L.) Hill ssp. *asper* (p. 347) PRICKLY SOW THISTLE Ann 1–12 dm. **STS** mostly unbranched below infl. **LVS**: basal tapered to gen sessile bases; cauline sessile, clasping, upper often widest near base, basal clasping lobes rounded, strongly curved to coiled; blades dentate, sometimes ± lobed, teeth and lobes tipped with soft spines. **INFL**: peduncles 0.5–5 cm, gen ± bristly-glandular; involucre 10–13 mm. **FLS**: ligule ± 1/3 < tube. **FR** 2–3 mm, very flat, 3-ribbed on each face, otherwise smooth; pappus bristles ± 3 × fr. $2n$= 18. Common. Weed in slightly moist waste places, gardens, along streams; < 1900 m. CA; to e US; native to Eur. Much like *S. oleraceus*.

S. oleraceus L. (p. 347) COMMON SOW THISTLE Ann 1–14 dm. **STS** often branched from below middle. **LVS**: basal gen < cauline, gen tapered or abruptly wing-petioled; cauline 5–35 cm, upper sessile, clasping, often widest at base, basal clasping lobes acute, not curled or coiled; blades nearly all lobed exc in dwarfed pls, lobes variable in width, terminal lobe often widely arrowhead-shaped. **INFL**: peduncles 0.5–7 cm, glabrous to bristly-glandular, sometimes cottony-tomentose just below heads; involucre 10–13 mm. **FLS**: ligule ± = tube. **FR** 2.5–3.8 mm, flat, 2–4-ribbed, cross-wrinkled on each face; pappus bristles ± 2 × fr. $2n$=32. Abundant. Weed in waste places, gardens, etc; < 1500 m. CA; to e US; native to Eur. Much like *S. asper*.

S. tenerrimus L. Ann to per 1–8 dm. **STS** slender, gen much-branched. **LVS**: basal gen < cauline, tapered to clasping bases; cauline 3–20 cm, sessile, clasping, basal clasping lobes acute; blades variously lobed, lobes linear to triangular, often secondarily lobed. **INFL**: peduncles 0.5–8 cm, tomentose below heads; involucre 10–12 mm. **FLS**: ligule ± 1/3 > tube. **FR** 2.5–3.3 mm, 1–3-ribbed on each face; ribs roughened; pappus bristles 6–8 mm. $2n$=14. Uncommon. Weed of disturbed sites; gen < 500 m. SnJV, ChI, PR; native to s Eur.

SPHAEROMERIA

Elizabeth McClintock

Per, subshrub, often from long, thick caudex, silky; hairs basally forked, attached off-center; herbage dotted with resin glands. **LVS** alternate, often crowded at base; simple and entire or 3-lobed to 1–3-pinnately dissected. **INFL**: heads disciform, few–many; involucre hemispheric; phyllaries in 2–3 overlapping series, scarious-margined and -tipped; receptacle hemispheric, naked or hairy. **PISTILLATE FLS** few, marginal; corolla cylindric to lance-ovoid. **DISK FLS** more numerous; corolla tubular to bell-shaped; anther tips ovate, bases rounded or ± cordate; style branches truncate, tips shrub-like. **FR** cylindric, gen 5–10-ribbed; pappus gen 0 or a narrow crown. 9 spp.: w. N.Am. (Greek: spherical division) [Holmgren et al 1976 Brittonia 28:255–272] Segregated from *Tanacetum*.

1. Subshrubs, gen > 8 cm; lvs gen simple or 3–4-lobed; heads gen in dense clusters; receptacle naked . . . . ***S. cana***
1′ Per with thick caudex; sts gen < 12 cm; lvs gen 1–3-pinnately divided; heads in loose clusters; receptacle hairy . . . . ***S. potentilloides***
 2. Sts nearly lfless; lvs gen basal, 1-pinnately divided; heads gen < 4 . . . . var. ***nitrophila***
 2′ Sts leafy; lvs gen scattered on sts; heads gen > 4 . . . . var. ***potentilloides***

S. cana (D.C. Eaton) A.A. Heller (p. 347) Subshrub. **STS** 15–30 cm, leafy, gen branched from base. **LVS** 5–12 mm, basal and cauline, sessile; basal and lower cauline gen 3–4-lobed, upper entire. **INFL**: heads gen 3–8, in dense clusters (rarely solitary); receptacle glabrous. **FR** < 2 mm, 10-ribbed; pappus 0. $2n$=18. Uncommon. Rocky places, ledges, trails, talus near or above timberline; 3000–4000 m. c&s SNH, SNE. [*Tanacetum c.* D.C. Eaton] Sage-like smell sometimes reported. ❀DRN,DRY,SUN:1–3,15,16;DFCLT.

S. potentilloides (A. Gray) A.A. Heller (p. 347) Per from elongate, thick woody caudex. **STS** 4–30 cm, branched from base. **LVS** < 10 cm, 1–3-pinnately divided; basal and lower cauline petioled, upper sessile. **INFL**: heads gen 3–5, in open clusters (rarely solitary); receptacle hairy. **FR** < 2 mm, 5-ribbed, swelling when wet; pappus 0. Mtn meadows, hot springs, alkaline areas; 1800–2400 m. n&c SNH, MP; to OR, ID, NV. [*Tanacetum p.* A Gray]

var. ***nitrophila*** (Cronq.) A. Holmgren, L. Schultz & T. Lowrey **STS** 4–15 cm, gen lfless. **LVS**: basal gen 1-pinnately divided. **INFL**: heads gen < 4 in an open cluster. $2n$=18. Uncommon. Gen alkaline areas; 2300–2400 m. c SNH; to ID, NV. ❀IRR,SUN, DRN:1–3,7,15,16;DFCLT.

var. ***potentilloides*** **STS** 10–30 cm, leafy. **LVS**: basal gen 2–3-pinnately divided. **INFL**: heads gen > 4, in a loose cluster. $2n$=18. Uncommon. Mtn meadows, hot springs, gen not alkaline areas; 1800–2400 m. n&c SNH, MP; OR.

STEBBINSOSERIS

Kenton L. Chambers

Ann, scapose; sap milky. **LVS** ± basal, entire to pinnately lobed, glabrous or mealy (hairs drying as minute, white scales). **INFL**: heads, ligulate, solitary, ± drooping in bud; involucre glabrous; phyllaries ± lanceolate, outer < 1/3 × inner. **FLS** 8–many; corollas soon withering, ± = involucre; ligules white or yellow, often reddish below. **FR** 10-ribbed, gray to brown or blackish; pappus scales 5, ± cut at tip, dull-white (silvery) to brownish, bristle-tip finely barbed. (Greek: Stebbins' chicory, for G.L. Stebbins, Jr., Am geneticist, evolutionist, 1906–). [Jansen et al. 1991 Amer J Bot 78:1015–1027]. 2 spp.: w N.Am. Derived by hybridization: *Uropappus lindleyi* × ann *Microseris.*

1. Fr widest at middle, ± dark brown to blackish, > pappus scales; n&c CCo, rare .... ***S. decipiens***
1′ Fr widest at tip, tapered to base, gen light-colored (or blackish in SCo), < or = pappus scales; widespread .... ***S. heterocarpa***

S. decipiens (Chambers) Chambers (p. 347) SANTA CRUZ MICROSERIS Pl 10–40 cm. **LF** 5–15 cm. **INFL**: involucre 10–18 mm. **FLS**: ligules yellow. **FR** 6–8 mm, widest at middle, dark-colored; pappus scales < fr, glabrous, midrib linear from near base, bristles 4–5 mm, hair-like, from irregularly cut scale tip. $2n$=36. RARE. Open, sandy, shaly, or serpentine sites; 10–500 m. n&c CCo. [*Microseris d.* Chambers] Derived from *Microseris bigelovii* × *Uropappus l.*

S. heterocarpa (Nutt.) Chambers (p. 347) Pl 8–60 cm. **LF** 5–35 cm. **INFL**: involucre 9–30 mm. **FLS**: ligules yellow or white. **FR** 4.5–12 mm, widest at tip, gen light-colored (or blackish in SCo); pappus scales < or = fr, sometimes hairy, midrib thicker at base, tapered, bristle 3.5–8 mm, arising from notched scale tip. $2n$=36. Open, sometimes disturbed, gen ± inland, rocky to clay soils; < 1000 m. NCoR, SNF, SnJV, CW, SCo, ChI, PR; n Mex, se AZ. [*Microseris h.* (Nutt.) Chambers; *M. lindleyi* (DC.) A. Gray misapplied; *Uropappus l.* (DC.) Nutt. misapplied] Derived from *Microseris douglasii* × *Uropappus l.* ❀TRY.

STENOTUS

Gregory K. Brown & David J. Keil

Per < 5 m diam, ± mat-forming. **LVS** gen crowded at branch tips, persistent for 2–3 years, < 10 cm, linear to oblanceolate, gen rigid, entire, scabrous; petiole indefinite. **INFL**: head, radiate solitary; peduncles < 15 cm, naked; involucre 5–10 mm, hemispheric; phyllaries in 2–3 series, linear to ± ovate, acute to acuminate. **RAY FLS** 6–15; ligules 7–12 mm, yellow. **DISK FLS** 25–50; corollas 6–7.5 mm, funnel-shaped, yellow. **FR** 3.5–7 mm, gen densely silky; pappus of soft bristles. ± 5 spp.: w N.Am. (Greek: narrow ear)

1. Herbage short-tomentose .... ***S. lanuginosus***
1′ Herbage glabrous to scabrous or short-bristly
 2. Lvs 2–10 cm, oblanceolate .... ***S. acaulis***
 2′ Lvs < 2 cm, linear .... ***S. stenophyllus***

S. acaulis Nutt. (p. 347) **LVS** 2–10 cm, oblanceolate, gen 3-veined, glabrous to short-bristly, sometimes sticky. **INFL**: involucre 6–10 mm. **RAY FLS** 6–15, 8–12 mm. **DISK FLS**: corollas 6–7.5 mm, tube glabrous. **FR** glabrous or densely silky; pappus white or tan. Dry, rocky open shrubland; 1900–3200 m. CaR, SN, GB, DMtns; to WA, MT, WY, Colorado, UT. [*Haplopappus a.* (Nutt.) A. Gray] ❀TRY;DFCLT.

S. lanuginosus A. Gray WOOLLY STENOTUS **LVS** 2–10 cm, linear-oblanceolate, 1–3-veined, loosely short-tomentose, ± glandular. **INFL**: involucre 7–12 mm. **RAY FLS** 10–20, 8–12 mm. **DISK FLS**: corollas 6–7.5 mm, tube puberulent. **FR** short-hairy; pappus white. RARE in CA. Shallow, rocky soils; ± 1500 m. s MP (Lassen Co.); to WA, ID, MT.

S. stenophyllus (A. Gray) E. Greene (p. 347) **LVS** < 2 cm, linear, 1-veined, scabrous, sometimes glandular. **INFL**: involucre 5–9 mm. **RAY FLS** 8–12, 7–11 mm. **DISK FLS**: corollas 6–7 mm, tube puberulent. **FR** densely silky; pappus white. Rocky, sagebrush scrub; 1200–1600 m. MP; n NV, n to OR, WA, ID. [*Haplopappus s.* A. Gray] ❀TRY;DFCLT.

STEPHANOMERIA

G. Ledyard Stebbins

Ann to shrubs, glabrous or hairy; sap milky. **STS** branched, 0.5–3 m. **LVS** cauline or some basal, alternate. **INFL**: heads ligulate; involucre cylindric; phyllaries in 2–several series; receptacle naked. **FLS** few–many; ligules lavender, pink, or whitish, readily withering. **FR** linear, club-shaped, or oblong; pappus gen of 9–30 stiff, plumose bristles. 24 spp.: w N.Am. (Greek: wreath division) [Gottlieb 1972 Madroño 21:463–481] Ann, tap-rooted, rosette-forming spp. highly variable and complexly interrelated.

1. Branching shrub; s ChI (San Clemente Island) .... ***S. blairii***
1′ Ann to subshrubs; mainland or islands
 2. Phyllaries graded; heads 12–26 mm; rosette-forming per; young herbage woolly .... ***S. cichoriacea***

2′ Phyllaries in 2 series, inner elongate, outer short; heads 6–15 mm; herbage glabrous or slightly hairy (rarely densely canescent)
3. Cauline lvs well developed at fl time, linear-lanceolate or wider; heads 12–15 mm; fls 7–14; per
4. Lvs entire to shallowly lobed; sts from a long rhizome; dry forests, n CA .... ***S. lactucina***
4′ Lvs deeply lobed; pls from a woody caudex; D .... ***S. parryi***
3′ Cauline lvs small, linear (exc *S. virgata*), often 0 at fl time; heads 6–10 mm; fls 3–9 (8–16 in *S. elata*); ann or per
5. Sts much-branched, prominent central axis 0; per from caudex
6. Sts forming sharp thorns; pappus of many non-plumose bristles .... ***S. spinosa***
6′ Sts not forming thorns; pappus of 20–30 plumose bristles
7. Branches slender, flexible, in fl gen bearing thin, linear, flexible lvs; dry montane forests, open subalpine slopes .... ***S. tenuifolia***
7′ Branches stout, rigid, gen lfless exc for a few short scales; low desert-like habitats .... ***S. pauciflora***
8. Herbage canescent .... var. ***parishii***
8′ Herbage glabrous .... var. ***pauciflora***
5′ Sts branched from a prominent central axis; ann, taprooted
9. Heads in an open panicle; peduncles gen 10–40 mm .... ***S. exigua***
10. Outer phyllaries reflexed; frs 3.2–4.3 mm .... ssp. ***carotifera***
10′ Outer phyllaries erect, appressed; frs 2.1–3.2 mm
11. Peduncles and involucres strongly glandular; frs 2.1–2.4 mm .... ssp. ***deanei***
11′ Peduncles and involucres glabrous or sparsely glandular; frs 2.6–3.2 mm .... ssp. ***exigua***
9′ Heads solitary or clustered on short, stiffly spreading branches; peduncles 1–10 mm
12. Sides of fr not grooved .... ***S. virgata***
13. Outer phyllaries erect, appressed; fls 5–6 .... ssp. ***pleurocarpa***
13′ Outer phyllaries reflexed; fls 8–9 .... ssp. ***virgata***
12′ Sides of fr each with a narrow, longitudinal groove
14. Frs 5–6.8 mm .... ***S. exigua*** ssp. ***macrocarpa***
14′ Frs 1.8–5 mm
15. Outer phyllaries reflexed
16. Pappus bristles plumose on upper 80–85% of their length; frs 1.8–2.4 mm .... ***S. diegensis***
16′ Pappus bristles plumose to the base; frs 2.5–5 mm .... ***S. elata***
15′ Outer phyllaries erect, appressed
17. Pappus bristles plumose on upper 60–80%; frs 2.3–3.1 mm .... ***S. exigua*** ssp. ***coronaria***
17′ Pappus bristles plumose throughout; frs 3.8–4.2 mm .... ***S. paniculata***

S. blairii Munz & I.M. Johnston (p. 351) BLAIR'S MUNZOTHAMNUS Shrub 1–1.5 m. **STS** ± fleshy, brownish tomentose. **LVS** 5–15 cm, tufted at branch tips, oblong-obovate, shallowly lobed, thinly tomentose, becoming glabrous. **INFL** densely stalked-glandular; heads many, in terminal panicles; involucre (6)8–9 mm, narrow; outer phyllaries << inner, erect; inner phyllaries 8–9. **FLS** 9–12; corollas light purple, ligules 6–8 mm. **FR** 3–3.5 mm, 5-angled, 1–2-grooved on each face; pappus of 30–40 non-plumose bristles, white. 2*n*=16. RARE. Rocky canyon walls; ± 200 m. s ChI (San Clemente Island). [*Malacothrix b.* (Munz & I.M. Johnston) Munz; *Munzothamnus b.* (Munz & I.M. Johnston) Raven] In cult.

S. cichoriacea A. Gray Per from large root crown, 5–15 dm. **STS** ± tomentose. **LVS**: basal in a well developed rosette, 5–18 cm, oblong-lanceolate, entire or with a few teeth, densely tomentose when young; cauline gradually reduced upward. **INFL**: heads short-peduncled from upper axils or in long terminal and axillary racemes or panicles; involucre 12–15 mm; phyllaries many, graduated in several series. **FLS** 10–15; ligules 15–20 mm, pink. **FR** ± 3 mm, weakly 5-angled, smooth; pappus plumose throughout, brownish. 2*n*=16. Dry rocky slopes; 0–2100 m. SCoR, SCo, ChI, TR. ❀SUN,DRN:14–24.

S. diegensis Gottlieb (p. 351) Ann 10–20 dm, puberulent. **STS** much-branched. **LVS**: basal rosette gen withered at fl; cauline lvs small, linear, often 0 at fl. **INFL**: heads solitary or clustered on short, stiffly spreading branches; involucre 6–9 mm; outer phyllaries reflexed, << inner. **FLS** 6–8; ligules 9–11 mm, pale pink to white. **FR** 1.8–2.4 mm; sides 5, each narrowly grooved; pappus bristles plumose 80–85% of length, white. 2*n*=16. Open clearings, near sand dunes; 0–200 m. SCo, s ChI, WTR, PR; Baja CA. Like *S. exigua*, *S. virgata*.

S. elata Nutt. Ann 5–12 dm, glabrous or puberulent. **STS**: branches long, slender. **LVS**: basal rosette gen withered at fl; cauline small, linear, often 0 at fl. **INFL**: heads solitary or clustered on short, stiffly spreading branches; involucre 4–6 mm; outer phyllaries << inner, gen reflexed; inner phyllaries 5–8. **FLS** 8–16; ligules 6–7 mm, pink. **FR** 2.5–5 mm; sides 5, each narrowly grooved; pappus plumose throughout, white. 2*n*=32. Dry, open slopes, roadsides; 150–1400 m. NW, CaRF, SN, CW; sw OR.

S. exigua Nutt. Ann 2–6 dm, glabrous to minutely glandular. **STS** much-branched. **LVS**: basal rosette gen withered at fl; cauline small, linear, often 0 at fl. **INFL** various; involucre 6–7 mm; outer phyllaries << inner, reflexed or erect and appressed; inner phyllaries 4–6. **FLS** 5–6; ligules 6–8 mm, white to pink. **FR** 2–6.8 mm, 5-angled; pappus bristles variously plumose, white. Desert scrub, dry disturbed ground; < 2000 m. Common. CA; to WA, ID, Colorado, Tex, Baja CA. Highly variable.

ssp. ***carotifera*** (Hoover) Gottlieb **INFL** puberulent; peduncles 1–3 mm; outer phyllaries reflexed. **FR** 3–4.5 mm; pappus bristles plumose throughout. 2*n*=16. Dry open slopes; < 1400 m. CCo, SCoR, WTR. [*S. c.* Hoover]

ssp. ***coronaria*** (E. Greene) Gottlieb **INFL** glabrous or sparsely hairy; peduncles 3–5 mm; outer phyllaries appressed. **FR** 2.3–3.1 mm; pappus bristles plumose on upper 60–85%. 2*n*=16. Dry steppes, shrubby slopes; 200–800 m. KR, SNH, Teh, SnJV, CCo, ChI, TR, MP; to OR, sw ID, w NV. [var. *c.* (E. Greene) Jepson]

ssp. ***deanei*** (J.F. Macbr.) Gottlieb **INFL** glandular; peduncles 10–40 mm; phyllaries appressed. **FR** 2–2.5 mm; pappus bristles plumose on upper 55–60%. 2*n*=16. Dry slopes; < 1700 m. SCo, PR; n Baja CA. [var. *d.* J.F. Macbr.]

ssp. ***exigua*** (p. 351) **INFL** glabrous or sparsely glandular; peduncles 10–40 mm; outer phyllaries erect, appressed. **FR** 2.5–3.5 mm; pappus bristles plumose on upper 45–55%. 2*n*=16. Deserts, dry slopes; 100–2000 m. GB, D; to Colorado, TX, Baja CA [var. *pentachaeta* (D.C. Eaton) H.M. Hall]

ssp. ***macrocarpa*** Gottlieb **INFL** glabrous to minutely glandular; peduncles 5–10 mm; outer phyllaries reflexed. **FR** 5–6.8 mm; pappus bristles plumose on upper 50–75%. 2*n*=16. Dry, open places; 300–1200 m. c&s SNF.

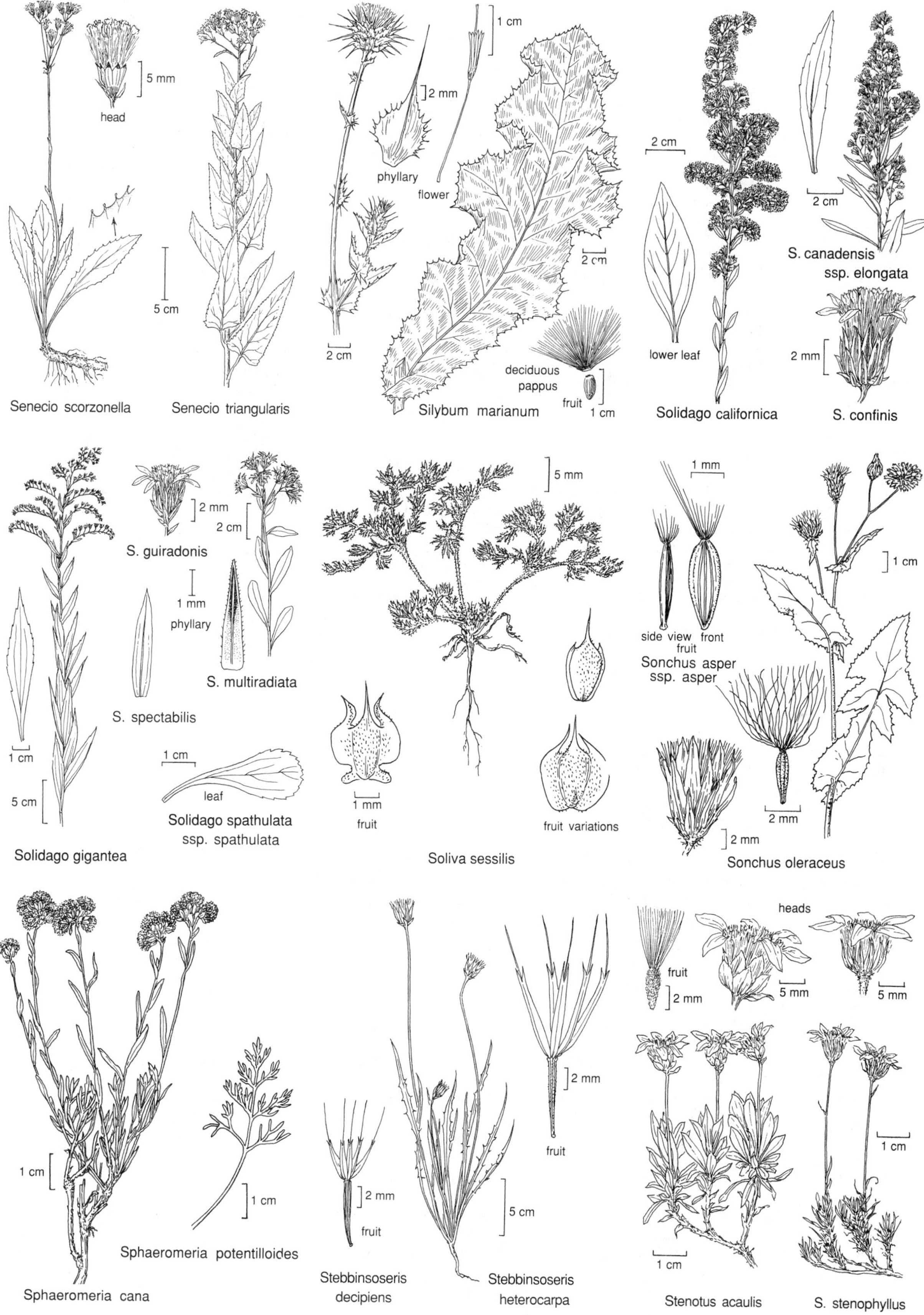
head
5 mm
5 cm
Senecio scorzonella
Senecio triangularis
1 cm
2 mm
phyllary
flower
2 cm
2 cm
deciduous
pappus
fruit
1 cm
Silybum marianum
2 cm
lower leaf
Solidago californica
2 cm
S. canadensis
ssp. elongata
2 mm
S. confinis
2 mm
2 cm
S. guiradonis
1 mm
phyllary
S. multiradiata
S. spectabilis
1 cm
5 cm
Solidago gigantea
1 cm
leaf
Solidago spathulata
ssp. spathulata
5 mm
1 mm
fruit
fruit variations
Soliva sessilis
1 mm
side view
front
fruit
Sonchus asper
ssp. asper
1 cm
2 mm
2 mm
Sonchus oleraceus
1 cm
1 cm
Sphaeromeria potentilloides
Sphaeromeria cana
2 mm
fruit
Stebbinsoseris
decipiens
2 mm
fruit
5 cm
Stebbinsoseris
heterocarpa
heads
fruit
2 mm
5 mm
5 mm
1 cm
1 cm
Stenotus acaulis
S. stenophyllus

S. lactucina A. Gray (p. 351) Per from creeping rhizome, 1–3 dm; herbage glabrous. **STS** solitary. **LVS** 2–8 cm, linear to lanceolate, few-toothed. **INFL**: heads few on separate peduncles; involucre 12–15 mm; outer phyllaries erect, << inner; inner phyllaries 8–11. **FLS** 7–10; ligules 8–11 mm, pink. **FR** 5–6 mm, ± 4–5-angled, grooved between angles; pappus bristles plumose, thicker and fused at base, white. 2*n*=16. Dry forests; 900–2400 m. KR, NCoRO, CaRH, n&c SNH, MP; OR. ❀TRY;DFCLT.

S. paniculata Nutt. Ann 2–9 dm; herbage glabrous. **ST** solitary, short-branched above. **LVS**: cauline small, linear, often 0 at fl. **INFL**: heads solitary or clustered on short, stiffly spreading branches; peduncles 5–7 mm; involucre 6–9 mm; outer phyllaries erect, appressed, << inner; inner phyllaries 4–6. **FLS** 4–6; ligules 6–14 mm, tinged light lavender-bluish. **FR** 3.8–4.2 mm; sides 5, each narrowly grooved; pappus bristles plumose throughout, white. 2*n*=16. Open scrub, deserts, gen sagebrush; 900–1400 m. CaRH, n SNH; to WA, ID.

S. parryi A. Gray Per from strong taproot, 2–4 dm, glabrous. **STS** 1–few. **LVS** 2–8 cm, lobed. **INFL**: heads in open cyme; peduncles 4–8 mm; involucre 12–16 mm; inner phyllaries 7–10; outer phyllaries erect to reflexed, << inner. **FLS** 10–14; ligules 11–15 mm, whitish. **FR** 3–4 mm; pappus bristles plumose above, minutely barbed at base, brownish or pale yellowish. 2*n*=32. Deserts; 680–2000 m. SNE, DMoj.

S. pauciflora (Nutt.) Nelson WIRE-LETTUCE Per or subshrub, 3–6 dm, glabrous or densely short-tomentose. **STS**: branches many, ± stout, rigid. **LVS**: cauline often reduced to scales. **INFL**: heads at tips of side branches; peduncles 1–13 mm (or 0 above); involucre 8–10 mm; outer phyllaries << inner, appressed or spreading; inner phyllaries 5. **FLS** 3–5; ligules lavender-pink. **FR** 3–4 mm; pappus plumose but not to base, pale brownish. 2*n*=16. Dry flats, deserts; < 2400 m. s SNH, Teh, SnJV, SCoRI, TR, SNE, D; to UT, TX, nw Mex; also in KS.

var. ***pauciflora*** Herbage glabrous, glaucous. Habitat and range of sp. [*S. myrioclada* D.C. Eaton] ❀TRY.

var. ***parishii*** (Jepson) Munz Herbage densely short-tomentose. Gravelly soil; < 1000 m. DMoj; w NV. ❀TRY.

S. spinosa (Nutt.) Tomb Subshrub 1–4 dm, from thick caudex, woolly at base and in lower axils, otherwise glabrous. **STS** few–several; branches intricate, becoming sharp thorns. **LVS** linear to scale-like. **INFL**: heads terminal on branchlets; peduncles < 7 mm; involucre 8–9 mm; outer phyllaries unequal, ± triangular, ascending; inner phyllaries 4–6. **FLS** 3–5; ligules pink. **FR** 4 mm; pappus of many non-plumose bristles, white to tawny. 2*n*=16. Deserts, scrub, dry mtn slopes; 1200–3300 m. SNH, SnGb, SnBr, MP, W&I, DMtns; NV, s OR. [*Lygodesmia s.* Nutt.]

S. tenuifolia (Torrey) H.M. Hall WIRE LETTUCE Per 1–6 dm, from woody caudex, glabrous. **STS** several; branches many, slender, ascending. **LVS** 1.5–5 cm, linear or thread-like, gen present at fl. **INFL**: heads solitary, nearly sessile at branch tips; involucre 8–10 mm; outer phyllaries << inner, erect or ascending; inner phyllaries gen 5. **FLS** 5. **FR** 2–3 mm; pappus bristles plumose throughout, white to tawny. 2*n*=16. Scrub, dry mtn slopes; 1100–3360 m. CaRH, SNH, MP; to WA, ID, UT, NM, nw Mex.

S. virgata Benth. Ann 0.5–3 m, glabrous or ± tomentose. **STS**: branches many, long. **LVS**: basal rosette gen withered at fl; cauline oblong or wider above middle, margin wavy or shallowly lobed. **INFL**: heads solitary or clustered on short, stiffly spreading branches; peduncles 3–7 mm; involucre 6–8 mm; outer phyllaries << inner, appressed or reflexed; inner phyllaries 6–9. **FLS** 5–9; ligules white above, purplish pink below. **FR** 2.2–3.6 mm; pappus bristles plumose throughout, white. Common. Dry, open slopes; < 2100 m. CA–FP; sw OR, NV.

ssp. ***pleurocarpa*** (E. Greene) Gottlieb (p. 351) **INFL**: outer phyllaries appressed. **FLS** 5. 2*n*=16. Habitats and range of sp. (exc SCoRO, SCo, TR).

ssp. ***virgata*** **INFL**: outer phyllaries strongly reflexed. **FLS** 8–9. 2*n*=16. Habitats of sp.; 0–1800 m. SCoRO, SW. [var. *tomentosa* (E. Greene) Munz] ❀SUN,DRN:**7,14–17**,18,**19–24**.

STYLOCLINE NEST STRAW

James D. Morefield

Ann, gen grayish, cobwebby to tomentose. **STS** 1–several from base, ± spreading to ascending, ± forked, gen ± evenly leafy below, ± lfless between upper forks. **LVS** simple, alternate, ± sessile, gen elliptic to oblanceolate, entire. **INFL**: heads disciform, ± sessile in groups of 2–10; bracts like lvs; phyllaries 0 or vestigial (or 1–4 reduced, unequal scales); receptacle (2.8)4–8 × > wide, ± cylindric to club-shaped; chaff scales subtending pistillate fls phyllary-like, each gen enclosing a fl (or outermost open), falling with a fr, gen woolly, ± scarious-winged, wing gen elliptic to obovate, base gen acute; chaff scales subtending disk fls reduced, open or folded, glabrous to cobwebby, scarious. **PISTILLATE FLS** in 3–several series; corollas tubular. **DISK FLS** 2–6, staminate; ovary vestigial. **FR** obovoid, smooth, shiny; style ± at tip; pappus 0; disk pappus gen of 1–12 bristles. 6 spp.: sw N.Am. (Greek: female column, from long receptacle) Close to *Filago* subg. *Oglifa*, esp *F. depressa*. See also *Ancistrocarphus, Micropus*.

1. Longest chaff scales scarious-winged throughout, wing widest near or below middle of scale; phyllaries 2–4, scarious throughout, ± persistent
  2. Wing of each chaff scale elliptic to ± obovate, base acute; disk ovary 0.2–0.6 mm, pappus bristles gen 6–12; receptacle ± club-shaped .... ***S. citroleum***
  2′ Wing of each chaff scale widely ovate, base round or cordate; disk ovary 0–0.2 mm, pappus bristles gen 1–5; receptacle cylindric .... ***S. gnaphaloides***

1′ Longest chaff scales scarious-winged only toward tip, wing widest well above middle of scale; phyllaries 0 or vestigial, or 1–3, deciduous, not scarious throughout
  3. Receptacle club-shaped, length 2.8–3.5 × width; disk ovary 0.3–0.6 mm; outer fr 0.6–0.8 mm; heads ± spheric, 3–4 mm wide; longest chaff scale < 3.4 mm; lowest lvs obtuse .... ***S. sonorensis***
  3′ Receptacle ± cylindric, length > 3.5 × width; disk ovary 0–0.3(0.4) mm; outer fr, heads, chaff, and lowest lvs various
    4. Heads spheric, largest 5–9 mm wide; longest chaff scales 3.4–4.5 mm, body various; outermost chaff scales each enclosing a fl, woolly; fr various
      5. Body of longest chaff scale hard, splitting lengthwise if forced; fr compressed front-to-back; uppermost lvs gen 4–11 mm, ± elliptic to oblanceolate or obovate .... ***S. intertexta***
      5′ Body of longest chaff scale (exc midvein) membranous, easily tearing as wool is pulled; fr compressed side-to-side; uppermost lvs gen 11–18 mm, awl-like to lanceolate .... ***S. micropoides***
    4′ Heads ± ovoid, largest 1.5–4 mm wide; longest chaff scales < 3.4 mm, body hard; outermost chaff scales open, nearly glabrous; fr compressed front-to-back

6. Heads 1.5–2.5 mm wide; longest chaff scale 2–2.7 mm; fr 0.7–1 mm; disk fls 0.8–1.1 mm, gen 4-lobed; lvs narrowly obtuse . . . . *S. masonii*
6′ Heads 2.5–4 mm wide; longest chaff scale 2.8–3.3 mm; fr 1.1–1.6 mm; disk fls 1.2–2 mm, 5-lobed; lvs gen acute . . . . *S. psilocarphoides*

S. citroleum Morefield (p. 351) OIL NEST STRAW **STS** < 13 cm. **LVS** < 14 mm, gen obtuse; uppermost < 13 mm, ± elliptic to obovate. **INFL**: heads 4–5.5 mm, 3.5–5 mm wide, ± spheric; phyllaries 2–4, 1.5–2.5 mm, elliptic to obovate, scarious throughout, ± persistent; receptacle ± club-like, length 4–6 × width; longest chaff scale 2.5–3.5 mm, body (exc midvein) membranous, winged throughout, wing widest near middle, base acute; outermost chaff scales ± closed, woolly. **DISK FLS** 1.2–2.1 mm. **FR** 0.8–1 mm, compressed side-to-side; disk ovary 0.2–0.6 mm, pappus bristles gen 6–12, 1.5–1.7 mm. RARE. Flats, clay soils in oil-producing areas; 50–400 m. s SnJV (Kern Co.), s SCo (San Diego Co.). Last collected in 1935. Some features suggest *S. gnaphaloides* × *Filago californica*; collections gen mixed with 1 or both.

S. gnaphaloides Nutt. (p. 351) EVERLASTING NEST STRAW **STS** < 24 cm. **LVS** < 14 mm, gen obtuse; uppermost < 13 mm, ± elliptic to obovate. **INFL**: heads 3–6 mm, spheric; phyllaries 2–4, 1–3.5 mm, ± ovate, scarious throughout, ± persistent; receptacle cylindric, length 5–8 × width; longest chaff scale 1.8–4.5 mm, widely ovate, body (exc midvein)membranous, winged throughout, wing widely ovate, widest below middle, rounded to cordate at base; outermost chaff scales ± closed, cobwebby. **DISK FLS** 1–2 mm. **FR** 0.8–1 mm, compressed side-to-side;disk ovary 0–0.2 mm, pappus bristles gen 1–5, 1.3–1.9 mm. *n*=14.Common. Open, gen sandy soil; < 1200 (1700 m). c AZ, nw Mex.[Misspelled *S. gnaphalioides*]

S. intertexta Morefield (p. 351) **STS** < 11 cm. **LVS** < 15 mm, gen acute; uppermost gen 4–11 mm, ± elliptic to obovate. **INFL**: largest head 5–6 mm, spheric; phyllaries 0 or vestigial; receptacle ± cylindric, length 4–7 × width; longest chaff scale 3.4–4.2 mm, winged toward tip, body hard; outermost chaff scales ± closed, woolly.**DISK FLS** 1.1–2.3 mm. **FR** 1–1.4 mm, compressed front-to-back; disk ovary 0–0.3 mm, pappus bristles 0 or 1–4(8), 1.1–2 mm. Rocky or sandy, often calcareous soils; 50–1400 m. ne DMoj, nw DSon; to sw UT,w-most AZ. Perhaps derived from *S. micropoides* × *S. psilocarphoides*.

S. masonii Morefield MASON NEST STRAW **STS** < 11 cm. **LVS** < 11 mm, narrowly obtuse; uppermost < 6 mm, linear-oblong to narrowly elliptic. **INFL**: heads 2–5 mm, 1.5–2.5 mm wide, ± cylindric to ovoid; phyllaries 0 or vestigial (or 1–3, 1–2 mm, obovate, deciduous), body hard, not scarious; receptacle ± cylindric, length 5–8 × width; longest chaff scale 2–2.7 mm, winged toward tip, body hard;outermost chaff scales open, nearly glabrous. **DISK FLS** 0.8–1.1 mm,gen 4-lobed. **FR** 0.7–1 mm, compressed front-to-back; disk ovary 0–0.1 mm, pappus bristles 0 or 1, < 1 mm. RARE. Sandy washes; 100–400(1200) m. nw Teh, s SnJV, c&s SCoRO (Kern, Monterey, San Luis Obispo cos.). Perhaps derived from *S. psilocarphoides*. Last collected in 1971.

S. micropoides A. Gray (p. 351) DESERT NEST STRAW **STS** < 21 cm. **LVS** < 20 mm, acute; uppermost gen 11–18 mm, awl-like to lanceolate. **INFL**: largest head 5–9 mm, spheric; phyllaries 0 or vestigial; receptacle ± cylindric, length 5–8 × width; longest chaff scale 3.4–4.5 mm, winged toward tip, body (exc midvein) membranous; outermost chaff scales ± closed, woolly. **DISK FLS** 1.4–2 mm. **FR** 1–1.4 mm, compressed side-to-side; disk ovary 0–0.3 mm, pappus bristles gen 2–5(10), 1.1–2 mm. *n*=14. Rocky or sandy, often calcareous soils; 50–1600 m. s SNE, ne DMoj, nw DSon; to w TX, n-most Mex.

S. psilocarphoides M. Peck (p. 351) PECK NEST STRAW **STS** < 18 cm, ± lfless between lower forks. **LVS** gen acute, < 18 mm; uppermost < 10 mm, ± elliptic to obovate. **INFL**: heads 3.5–5 mm, 2.5–4 mm wide, ± ovoid; phyllaries 0 or vestigial (or 1–3, 1.5–2.5 mm, obovate, deciduous), body hard, not scarious; receptacle ± cylindric, length 5–8 × width; longest chaff scale 2.8–3.3 mm, winged toward tip, body hard; outermost chaff scales open, nearly glabrous. **DISK FLS** 1.2–2 mm. **FR** 1.1–1.6 mm, compressed front-to-back; disk ovary 0.1–0.4 mm, pappus bristles 0 or 1–3, 1.1–1.5 mm. Sandy or rocky soils; (150)600–1800 m. SNE, DMoj, w edge DSon; to se OR, sw ID, sw UT. CA pls long mistaken for other *Stylocline* or *Filago* spp. See *S. masonii*. ❀TRY.

S. sonorensis Wiggins (p. 351) MESQUITE NEST STRAW **STS** < 15 cm. **LVS** < 13 mm, gen obtuse; uppermost < 10 mm, elliptic to ± lanceolate, acute. **INFL**: heads 3.5–4.5 mm, 3–4 mm wide, ± spheric; phyllaries 0 or vestigial; receptacle club-shaped, length 2.8–3.5 × width; longest chaff scale 1.9–3.1 mm, winged toward tip, body (exc midvein) gen membranous; outermost chaff scales ± closed, woolly. **DISK FLS** 1.2–1.9 mm. **FR** 0.6–0.8 mm, ± compressed side-to-side; disk ovary 0.3–0.6 mm, pappus bristles 3–8, 0.9–1.3 mm. RARE in CA. Open sandy drainages; ± 400 m. DSon (Hayfields, Riverside Co., in 1930); se AZ, ne Sonora. Like *Filago depressa*, which may have been involved in its origin.

SYNTRICHOPAPPUS

Dale E. Johnson

Ann 2–10 cm, gen loosely woolly. **LVS** alternate (or lowest opposite), simple. **INFL**: heads solitary, radiate; involucre subcylindric; phyllaries in 1 series, spreading in fr, oblanceolate, acute, alternate phyllaries scarious-margined; receptacle convex, naked. **RAY FLS** 1 per phyllary; ligules 3-lobed, yellow (or white and red-veined). **DISK FLS** narrowly funnel-shaped, yellow, glabrous; stamen tips narrowly triangular. **FR** narrowly obconic; pappus 0, or of many bristles fused at base. 2 spp.: sw US. (Greek: fused bristly pappus) [Johnson 1991 Novon 1:119–124] ❀TRY.

1. Ligules yellow . . . . *S. fremontii*
1′ Ligules white above with red veins . . . . *S. lemmonii*

S. fremontii A. Gray (p. 351) Pl ± decumbent. **LVS** 5–20 mm, wedge-shaped below, spoon-shaped above, 3-lobed (margins rolled under) or entire. **INFL**: peduncles 3–25 mm; involucre 5–7 mm; phyllaries gen 5. **RAY FLS**: ligules 3–5 mm, yellow. **FR** 2–3.5 mm, strigose; pappus of 30–40 bristles, ± 2 mm. 2*n*=12. Open, sandy to gravelly areas; 600–2500 m. DMoj; to sw UT, nw AZ.

S. lemmonii (A. Gray) A. Gray (p. 351) LEMMON'S SYNTRICHOPAPPUS Pl ± erect, sometimes becoming glabrous. **LVS** 3–8 mm, linear; tips obtuse. **INFL**: peduncles 3–15 mm, reddish; involucre 4–5 mm; phyllaries 5–8. **RAY FLS**: ligules 2–3 mm, white above, pinkish purplish below, red-veined. **FR** 2–2.5 mm, (sub)glabrous; pappus 0 or of < 30 bristles, ± 1 mm. 2*n*=14. UNCOMMON. Open, sandy to gravelly areas; 900–1500 m. SCoRI (Monterey Co.), e WTR, SnGb, SnBr, SnJt.

TAGETES MARIGOLD

Ann, per, shrubs, gen scented. **STS** erect. **LVS** pinnately divided or compound in CA, opposite below, opposite or alternate above, sessile or petioled, dotted with embedded oil glands. **INFL**: heads gen radiate, small to large and showy, peduncled, solitary or in terminal, leafy cymes; involucre cylindric to bell-shaped; phyllaries in 1 series, all equal, fused, gland-dotted;

receptacle naked. **RAY FLS** (0)1–many; corollas white to yellow, orange, or brown; style tips long, tapered. **DISK FLS** 3–many; corollas yellow to orange. **FR** cylindric; pappus of scales, sometimes with 1 or more awns. ± 50 spp.: N.Am., S.Am. ± TOXIC to root-parasite nematodes.

1. Heads few, gen large; involucre widely cylindric or bell-shaped, gen >> 10 mm; ray fls 8–many .... ***T. erecta***
1′ Heads many, small; involucre narrowly cylindric, 7–10 mm; ray fls 1–3 .... ***T. minuta***

T. erecta L. AFRICAN or FRENCH MARIGOLD Ann 1–10 dm, erect, glabrous. **LVS**: lflets serrate or dentate. **INFL**: heads 1–few; peduncles 3–10 cm, stout; involucre 13–19 mm, widely cylindric or bell-shaped; phyllaries 7–11, sometimes splitting apart in age. **RAY FLS** 8–many; corollas yellow to orange or brown; ligules 10–30 mm, showy. **DISK FLS** many; corollas 10–15 mm, yellow to orange or brown, radial to strongly bilateral. **FR** 6.5–10 mm; pappus of 1–2 acuminate scales (each 8–11 mm) and 2–3 blunt scales (each 3–6 mm). $2n$=24,48. Uncommon. Waif in disturbed places; < 1000 m. CW, SW; native to Mex. [*T. patula* L.] Many cult forms.

T. minuta L. Ann 2–10 dm, erect, glabrous. **LVS**: lflets serrate or dentate. **INFL**: heads many in terminal cymes; peduncles 5–5.5 mm, slender; involucre 7–10 mm, narrowly cylindric; phyllaries 3–5, not splitting apart. **RAY FLS** 1–3; corollas pale yellow; ligules 1–2 mm, inconspicuous. **DISK FLS** 3–5; corollas 3–4 mm, yellow. **FR** 4.5–7 mm; pappus of 1–2 acuminate scales (each 2–3 mm) and 3–5 ovate to lanceolate scales (each 0.5–1 mm). $2n$=24. Disturbed places; < 1000 m. s SNF, sSnJV (esp Tulare Co.), SnFrB, s SCoR, w WTR; native to w S.Am. NOXIOUS WEED. Sometimes cult. May cause contact dermatitis in susceptible individuals.

TANACETUM TANSY

Elizabeth McClintock

Ann, per, sometimes woody at base, often aromatic, glabrous or hairy. **STS** decumbent or erect, branched. **LVS** alternate, dissected. **INFL**: heads disciform, radiate, or discoid, few–many; phyllaries in 2–3 overlapping series; receptacle convex to conic, gen dotted with tubercles. **PISTILLATE** or **RAY FLS** many; corollas tubular or with inconspicuous to well developed ligules, yellow or white. **DISK FLS** many; corollas tubular, yellow; anther tips ovate, bases rounded or ± cordate; style branches truncate with shrub-like tips. **FR** gen 3–10-ribbed; pappus a crown of short scales. 70 spp.: Eur, Asia, N.Am. (Latin: immortality) See also *Sphaeromeria*.

1. Sts decumbent, not ribbed .... ***T. camphoratum***
1′ Sts erect, ribbed
2. Heads gen < 20, radiate; ligules short, white; 1° lflets gen < 5, ovate, with irregular rounded divisions .... ***T. parthenium***
2′ Heads gen > 20, disciform; 1° lflets gen > 5, lanceolate, sharply toothed .... ***T. vulgare***

T. camphoratum Less. (p. 351) DUNE TANSY Per, stout, aromatic, hairy, esp on young growth; rhizome creeping. **STS** 10–25 cm, branched, decumbent. **LVS** 7–25 cm, 3–5 cm wide, thick, ovate, sessile, clasping at base, 2–3-pinnate, each 1° lflet with 1–5 divisions, flat or edges curled under, glandular. **INFL**: heads disciform or radiate, 3–15, in flat-topped clusters, 10–15 mm diam; phyllaries firm, margins scarious. **PISTILLATE FLS**: corolla tubular or with ligule < 3 mm. **DISK FLS**: corollas, 3–4 mm, tubular, 5-lobed. **FR** 3–4 mm, indistinctly 5-ribbed, glandular; pappus a short, irregularly toothed crown. $2n$=54. Uncommon. Coastal dunes; < 30 m. NCo, n CCo; OR to s B.C. [*T. douglasii* DC.] ❀DRN,SUN:4,**5,17**&IRR:8,9,14,**15,16**,19–21,**22–24**.

T. parthenium (L.) Schultz-Bip. FEVERFEW Per, strongly aromatic, ± glabrous to finely hairy; caudex woody. **STS** < 1 m, ridged, branched. **LVS** 4–10 cm, 1.5–4 cm wide, ovate in outline, petioled, 2-pinnate; uppermost gen 1-pinnate, 1° lflets ovate, gen 3–5, coarsely rounded or acute. **INFL**: heads radiate, 10–30 in rounded or flat-topped clusters, < 2 cm diam; phyllaries in 2–3 series, leathery, sparingly hairy, margins transparent. **RAY FLS**: ligules < 8 mm, white. **DISK FLS**: corolla 2 mm, yellow. **FR** 2 mm, cylindric, 5–10-ribbed; pappus a crown, < 0.5 mm or 0. $2n$=18. Common. Disturbed urban areas, roadsides, fields; gen < 250 m. KR, NCoR, n SNF, GV, SnFrB, SCoRO, SCo; native to Eur. Long cult in Eur and US as orn and for medicine. [*Chrysanthemum p.* (L.) Bernardi]

T. vulgare L. TANSY Per, coarse, aromatic, glabrous to sparsely hairy; rhizome stout, creeping. **STS** < 1 m, ridged, leafy, branched. **LVS** 4–10 cm, ovate in outline, sessile or short-petioled, pinnately divided; 1° lf divisions gen 4–10 pairs, lanceolate, regularly toothed or lobed, gland-dotted. **INFL**: heads disciform, < 70 in a much divided, gen flat-topped cluster; phyllaries in 3 series, outermost lanceolate, inner oblong. **PISTILLATE FLS**: corollas 3–4-lobed, yellow. **DISK FLS**: corollas 2–3 mm, yellow. **FR** 1 mm, 5-angled, gland-dotted; pappus a narrow-toothed crown. $2n$=18. Uncommon. Escape from cult in disturbed, gen urban areas. NCo, NCoRO, CaRH, SCoRO; native to Eur. TOXIC: dried lvs and fls have been used medicinally, esp in home remedies; overdoses may be very toxic; also causes contact dermatitis.

TARAXACUM

G. Ledyard Stebbins

Per from taproot; sap milky. **STS** naked, hollow. **LVS** all basal, toothed or lobed; lobes acute. **INFL**: heads ligulate, solitary, scapose; phyllaries many, outer ovate to lanceolate, gen reflexed, inner erect, linear; receptacle convex, naked. **FLS** many; ligules yellow, readily withering. **FR** fusiform; ribs rough; beak slender, >> body; pappus of many, white, slender bristles, not plumose. Many named spp., mostly reproducing clonally by asexual seeds: Eurasia, CA. (Greek: ancient name) [Taylor 1987 Bull Torrey Bot Club 114:109–120]

1. Lvs toothed or shallowly lobed; outer phyllaries erect, lanceolate to ovate; rare, SnBr .... ***T. californicum***
1′ Lvs of fl pls gen lobed, lobes often toothed; outer phyllaries reflexed; widespread weed .... ***T. officinale***

T. californicum Munz & I.M. Johnston (p. 351) CALIFORNIA DANDELION **LVS** toothed or shallowly lobed, oblanceolate. **INFL**: heads 2.2–3 cm diam; outer phyllaries erect, 5–7 mm; inner phyllaries rounded to truncate. **FR** pale brown; body tubercled near tip. RARE. Moist meadows; 1950–2400 m. SnBr.

T. officinale Wigg. (p. 351) **LVS** variously toothed or lobed. **INFL**: heads 2–3.3 cm diam; outer phyllaries reflexed, 5–8 mm; tips of inner phyllaries acute to minutely truncate. **FR** grayish or olive-brown; body tubercled near tip. $2n$=16,24,48. Abundant. Weed, esp of lawns, meadows; 0–3300 m. CA-FP, GB; native to Eur. [*T. laevigatum* (Willd.) DC.]

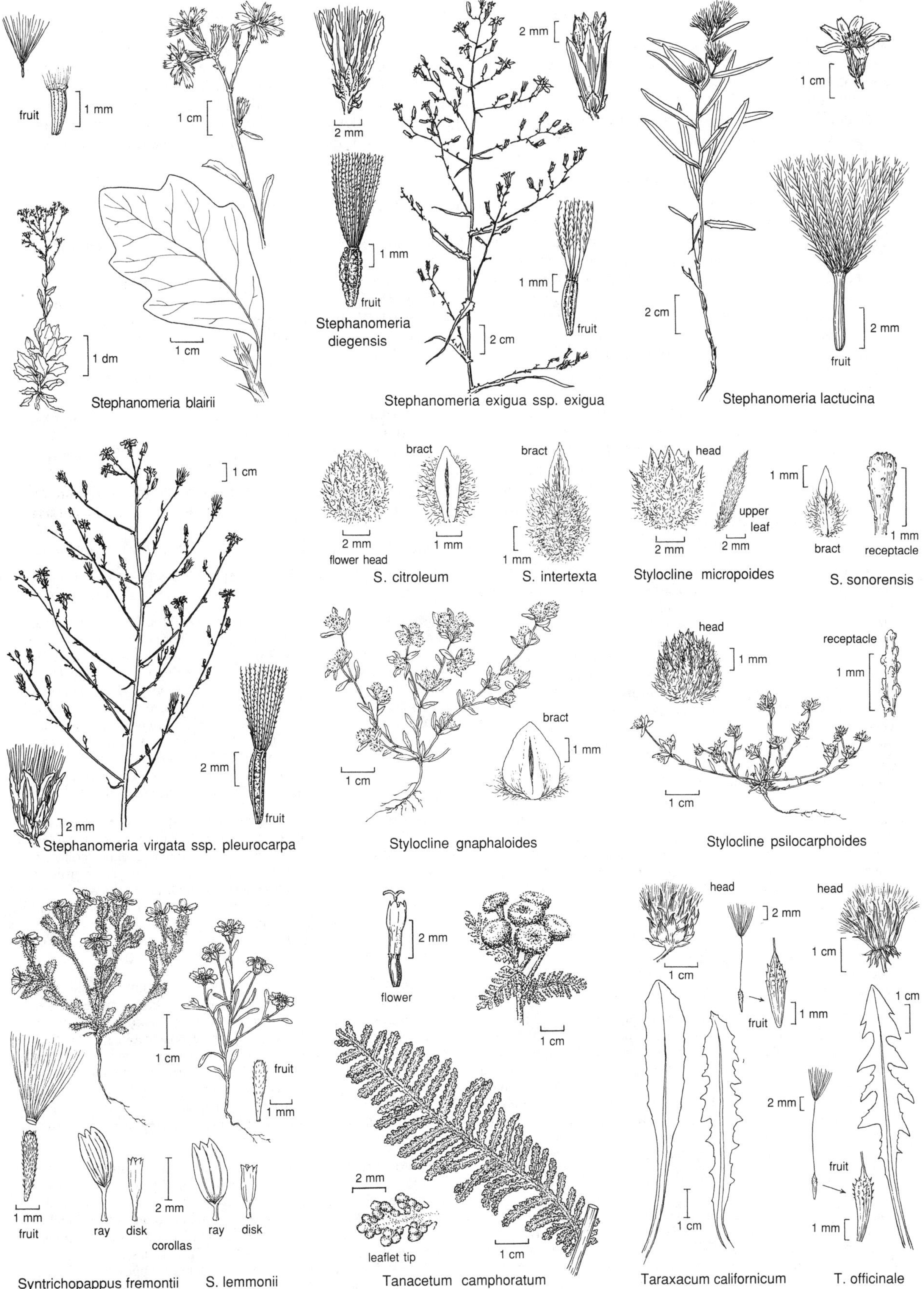
fruit
1 mm
1 cm
1 dm
1 cm
Stephanomeria blairii
2 mm
1 mm
fruit
Stephanomeria diegensis
2 mm
2 cm
1 mm
fruit
Stephanomeria exigua ssp. exigua
1 cm
2 cm
2 mm
fruit
Stephanomeria lactucina
1 cm
2 mm
2 mm
fruit
Stephanomeria virgata ssp. pleurocarpa
bract
2 mm
flower head
1 mm
S. citroleum
bract
1 mm
S. intertexta
head
upper leaf
2 mm
2 mm
Stylocline micropoides
1 mm
bract
1 mm
receptacle
S. sonorensis
1 cm
bract
1 mm
Stylocline gnaphaloides
head
1 mm
receptacle
1 mm
1 cm
Stylocline psilocarphoides
1 cm
fruit
1 mm
1 mm
fruit
ray
disk
2 mm
ray
disk
corollas
Syntrichopappus fremontii
S. lemmonii
2 mm
flower
1 cm
2 mm
leaflet tip
1 cm
Tanacetum camphoratum
head
2 mm
1 cm
fruit
1 mm
1 cm
Taraxacum californicum
head
1 cm
1 cm
2 mm
fruit
1 mm
T. officinale

TETRADYMIA COTTON-THORN, HORSEBRUSH

Shrubs. **STS** ± tomentose. **LVS** alternate and gen clustered in axils, linear to (ob)lanceolate, sometimes persisting as stiff spines, glabrous to tomentose. **INFL**: heads discoid, axillary or in ± rounded, terminal clusters; involucre cylindric to hemispheric; phyllaries in 1–2 ± equal series, often keeled; receptacle naked. **DISK FLS** gen 4–8; corollas cream-colored to yellow, lobes long, spreading; anther bases ± sagittate, tips obtuse or acute; style branches papillate to short-bristly, tips truncate to conic. **FR** obconic or fusiform, often angled; pappus 0 or of gen many bristles or slender scales. 10 spp.: w N.Am. (Greek: 4 together, from 4-fld heads of some) [Strother 1974 Brittonia 26:177–202] Esp fl buds TOXIC to sheep (toxicity poorly understood).

1. Sts unarmed (lvs not forming spines)
 2. Fls gen 6–7; pappus 0 (but fr covered with long, pappus-like hairs); axillary lf clusters gen 0 .... [2]*T. comosa*
 2′ Fls 4(5); pappus present; axillary lf clusters gen present
  3. Main lvs densely tomentose or silvery; clustered lvs ± (ob)lanceolate .... *T. canescens*
  3′ Main lvs glabrous or sparsely tomentose; clustered lvs narrowly linear
   4. Main lvs narrowly awl-shaped, stiffly ascending or appressed, gen 5–10 mm; clustered lvs gen 3–10 mm, glabrous; fr gen 3–4 mm, short-stiff-hairy; pappus of ± 100 bristles .... *T. glabrata*
   4′ Main lvs ± thread-like, soft, gen 10–30 mm; clustered lvs gen 10–20 mm, loosely tomentose; fr 5–6 mm, long-soft-hairy; pappus of ± 20 stiff bristles or slender scales (± hidden by fr hairs) .... *T. tetrameres*
1′ Sts armed with spines derived from main lvs
 5. Fr glabrous, 2.5–3.5 mm .... *T. argyraea*
 5′ Fr ± long-soft-hairy, 4–8 mm
  6. Main lvs narrowly (ob)lanceolate, finally deciduous, base not hardened; pappus 0 .... [2]*T. comosa*
  6′ Main lvs tapered to tip, persistent as stout spines, base expanded, hardened; pappus present
   7. Clustered lvs silvery-hairy; fr ± short-hairy; pappus of many fine bristles 9–12 mm .... *T. stenolepis*
   7′ Clustered lvs ± glabrous; fr densely long-hairy; pappus of ± 25 slender scales 6–9 mm
    8. Spines sharply recurved, < 2 cm, tomentose; involucres gen 8–12 mm, bell-shaped; fr gen 6–8 mm . *T. spinosa*
    8′ Spines straight, gen 2–4 cm, becoming glabrous; involucres gen 7–9 mm, obconic; fr gen 4–5 mm . *T. axillaris*
     9. Phyllaries and peduncles glabrous; fr hairs gen 6–8 mm .... var. *axillaris*
     9′ Phyllaries and peduncles tomentose; fr hairs gen 9–11 mm .... var. *longispina*

T. argyraea Munz & Roos STRIPED HORSEBRUSH Pl < 20 dm, spiny. **STS** becoming ± glabrous in stripes below spines. **LVS**: main lvs 1–2(3) cm, canescent or becoming glabrous, linear, forming straight or ± upturned spines; clustered lvs 3–8(20) mm, thread- to club-like, ± glabrous. **INFL**: heads gen 2–5; peduncles gen 1–4 mm, tomentose, bracts 0; involucre ± 7 mm, obconic; phyllaries 5, narrowly elliptic, tomentose. **FLS** 5; corollas ± 9 mm, pale yellow. **FR** 2.5–3.5 mm, glabrous; pappus of many fine bristles, ± 8 mm. 2*n*=60. UNCOMMON. Pinyon/juniper woodland; 1400–2100 m. DMtns.

T. axillaris Nelson Pl < 15 dm, spiny. **STS** evenly tomentose. **LVS**: main lvs 1–5 cm, forming straight spines, tomentose, becoming glabrous; clustered lvs 2–12(20) mm, thread- to club-like, ± glabrous. **INFL**: heads 1–3 in axils of previous year's growth; peduncles gen 4–15 mm, gen bracted; involucre 7–9 mm, obconic; phyllaries 5, ovate. **FLS** 5–7; corollas 7.5–9 mm, pale yellow. **FR** gen 4–5 mm, densely long-white-hairy; pappus of ± 25 slender scales, 6–7.5 mm. Sagebrush or saltbush scrub; 1200–2300 m. s SNE, DMoj; to sw UT.

var. ***axillaris*** **INFL**: peduncles and phyllaries glabrous. **FR**: hairs 6–8(10) mm. 2*n*=60. Habitats of sp. ne DMoj; s NV.

var. ***longispina*** (M.E. Jones) Strother (p. 357) **INFL**: peduncles and phyllaries tomentose. **FR**: hairs 9–11(14) mm. 2*n*=60,62. Habitats of sp. s SNE, w DMoj; to sw UT.

T. canescens DC. (p. 357) Pl 1–8 dm, unarmed. **STS** unevenly tomentose, becoming ± glabrous in stripes below nodes. **LVS**: main lvs < 4 cm, ± (ob)lanceolate, sparsely tomentose to silvery; clustered lvs like (gen <) main lvs. **INFL**: heads gen 3–6 in flat-topped clusters; peduncles 5–15(25) mm, bracts 0; involucre 6–8(12) mm, cylindric to obconic; phyllaries 4, oblong to ovate. **FLS** 4; corollas 7–15 mm, creamy to bright yellow. **FR** 2.5–5 mm, glabrous or short-stiff-hairy; pappus of many fine bristles, 6–11 mm. 2*n*=60, 90,120. Sagebrush scrub, pinyon/juniper woodland, forest; (400) 1600–3300 m. TR, s PR, GB, DMoj; to B.C., MT, WY, NM. ❀TRY.

T. comosa A. Gray Pl 3–12 dm, ± spiny or not. **STS** tomentose. **LVS**: main lvs 2–6 cm, gen narrowly lanceolate, becoming rigid, spine-tipped but finally deciduous, tomentose; clustered lvs gen 0 (8–15 mm, ± oblanceolate). **INFL**: heads gen 3–6; peduncles gen 3–8 mm, tomentose, bract gen 1, narrow; involucre ± 8 mm, ± widely bell-shaped; phyllaries 5(6), elliptic to (ob)ovate, often unequal. **FLS** 5–9; corollas 8–9 mm, (brownish) yellow. **FR** ± 4 mm; hairs long, white, pappus-like; pappus 0. 2*n*=60. Coastal scrub, chaparral, sagebrush scrub; 300–1500 m. TR, PR; n Baja CA.

T. glabrata Torrey & A. Gray (p. 357) Pl < 12 dm, unarmed. **STS** unevenly tomentose, becoming ± glabrous in stripes below nodes. **LVS**: main lvs 5–10(20) mm, narrowly awl-shaped, ascending to appressed, sparsely woolly or becoming glabrous; clustered lvs 3–10(15) mm, thread- to ± club-like, glabrous. **INFL**: heads gen 3–7; peduncles gen 5–10 mm, bracts 0; involucre 7–10 mm, obconic; phyllaries 4, lanceolate to obovate. **FLS** 4; corollas 9–10 mm, cream-colored to golden yellow. **FR** 3–5 mm, short-stiff-hairy; ribs glandular; pappus of many fine bristles, 6–8 mm. 2*n*=60,62,120,180. Sagebrush scrub, pinyon/juniper or Joshua-tree woodland; 800– 2400 m. GB, DMoj; to OR, ID, UT. ❀TRY.

T. spinosa Hook. & Arn. Pl < 10 dm, spiny. **STS** tomentose. **LVS**: main lvs 5–25 mm, tomentose, forming rigid, recurved spines; clustered lvs 3–15(25) mm, thread-like to ± oblanceolate, ± glabrous. **INFL**: heads gen 1–2 in axils of previous year's growth; peduncles gen 5–30 mm, tomentose, bracted; involucre 8–12 mm, ± bell-shaped; phyllaries 4–6, oblong to ovate. **FLS** 5–8; corollas 6–10 mm, ± yellow. **FR** 6–8 mm, densely long-white-hairy; pappus of ± 25 slender scales, 6–9 mm. 2*n*=60. Gen saltbush scrub; 800–2400 m. s MP, n SNE; to OR, ID, MT, WY, n NM.

T. stenolepis E. Greene (p. 357) Pl < 12 dm, spiny. **STS** unevenly tomentose, becoming ± glabrous in stripes below spines. **LVS**: main lvs 2–3 cm, tomentose or becoming glabrous, forming ± straight spines; clustered lvs 10–30 mm, ± oblanceolate, tomentose or silvery-hairy. **INFL**: heads gen 4–7; peduncles gen 5–12 mm, tomentose, bracts 0; involucre 8–10 mm, narrowly obconic; phyllaries (4)5, narrowly ovate. **FLS** (4)5; corollas 10–12 mm, pale yellow. **FR** 5–8 mm, ± short-soft-hairy; pappus of many fine bristles, 9–12 mm. 2*n*=60. Joshua-tree woodland, creosote-bush scrub; 600–1500 m. SNE, DMoj; s NV

T. tetrameres (S.F. Blake) Strother Pl < 20 dm, unarmed. **STS** tomentose. **LVS** sparsely tomentose; main lvs 1–4 cm, linear, thread-like, soft; clustered lvs 10–20 mm, thread-like to linear-oblanceolate. **INFL**: heads gen 4–6 on short side-branches; peduncles gen 1–3 mm, tomentose, bracts 0–2; involucre 8–9 mm, obconic; phyllaries 4(5), widely elliptic. **FLS** 4(5); corollas ± 8 mm, pale yellow. **FR** 5–6 mm, densely long-soft-hairy; pappus of ± 20 stiff bristles or slender scales, 3–5 mm, ± hidden by fr hairs. 2*n*=60. Dunes, deep sand, sagebrush scrub; 1200–2100 m. n SNE; w NV.

THELESPERMA

Ann to per. **LVS** opposite (or upper alternate). **INFL**: heads 1–few, gen discoid; involucre bell-shaped to hemispheric; phyllaries in 2 series, outer spreading, thick, free, inner erect, thin, ± fused; receptacle flat; chaff scales curved around disk fls. **RAY FLS** 0–several, sterile; ligules yellow to red. **DISK FLS** gen many; corollas yellow to red, ± bilateral (inner lip 4-lobed), tube slender, throat widely cylindric; anther tips triangular; style tips triangular, minutely bristly. **FR** ± linear, ± compressed front-to-back, rounded, ± rough; pappus of 2 awns with reflexed barbs.14 spp.: w N.Am, s S.Am. (Greek: nipple seed, from rough frs)

T. megapotamicum (Sprengel) Kuntze Per 3–6 dm, few-branched, glabrous, glaucous. **LVS** 3–12 cm; lower 1–2-pinnately dissected, segments linear. **INFL**: involucre 5–8 mm diam; inner phyllaries purplish, white-margined, ± = chaff scales. **RAY FLS** gen 0 (or ligules yellow, 5–6 mm). **DISK FLS**: corollas 7.5–8.5 mm, yellow or orange. **FR** 4–8 mm, dark; pappus awns 1–2 mm, stout. $2n$=11,22. Disturbed ground; < 500 m. SCo, s ChI (Santa Catalina Island); native WY to TX, S.Am.

THYMOPHYLLA

Ann, per, subshrubs. **LVS** simple or pinnately divided, opposite or alternate, dotted with embedded oil glands. **INFL**: heads radiate or discoid, sessile or peduncled, solitary or in few-headed cymes; involucre hemispheric to bell-shaped; phyllaries in 2–3 series (outer, if present, few, free, inner fused, gland-dotted); receptacle flat to rounded, naked. **RAY FLS** 0–few; corollas yellow or white. **DISK FLS** many; corollas yellow; style tips truncate or conic. **FR** obpyramidal, obconic, or cylindric; pappus of scales, each awn-tipped or dissected into bristles. ± 17 spp.: s N.Am, Caribbean. (Greek: thyme-leaved) [Strother 1986 Sida 11:371–378]

T. pentachaeta (DC.) Small var. ***belenidium*** (DC.) Strother (p. 357) Subshrub 1–3 dm. **STS** gen many, slender, very leafy. **LVS** opposite, 1–2 cm, pinnately divided into 3–5 stiff, linear lobes, dotted with tiny glands, puberulent. **INFL**: heads radiate, solitary; peduncles 2.5–4.5 cm; involucre 3.5–6 mm diam; outer phyllaries 3–5, short-triangular; inner phyllaries 13 in 2 series, 5–6 mm, gland-dotted, outer linear, ciliate, inner wider. **RAY FLS** gen 13; ligules 2–3.5 mm, yellow. **DISK FLS**: corollas 2.5–3 mm. **FR** 2–3 mm, puberulent; pappus of 10 scales, each 2.5–3 mm, dissected into 3 awns (or 5 of the 10 truncate, ± 1 mm). $2n$=26. Dry roadsides, gravelly slopes; 900–1800 m. SnJt, e DMoj; to TX, n Mex; also in s S.Am. [*Dyssodia p.* (DC.) Robinson var. *b.* (DC.) Strother; *D. thurberi* (A. Gray) Robinson] ❀TRY.

TOLPIS

G. Ledyard Stebbins

Ann, per; sap milky. **STS** 1–many, gen branched. **LVS** basal and cauline, entire to lobed. **INFL**: heads ligulate, few–many, long-peduncled; phyllaries in 2–3 series; receptacle naked. **FLS** few–many; ligules gen yellow (greenish when dry), readily withering. **FR** ± cylindric, ribbed, not beaked; pappus of 4 bristle-like scales. 20 spp.: Medit, Afr, Atlantic islands. (Greek: ball, from involucre)

T. barbata (L.) Gaertner (p. 357) **ST** < 7 dm. **LVS** 2–10 cm, oblong, entire to lobed; cauline smaller. **INFL**: phyllaries 6–15 mm, linear, spreading; pappus of outer frs << those of inner frs. $2n$=18. Uncommon. Weed in waste places; < 200 m. NCoRO, n CCo; native to s Eur. [*T. umbellata* Bertol.]

TONESTUS

Gregory K. Brown

Per from taproot or branched caudex, glandular. **LVS**: basal persistent; cauline alternate, reduced above, oblanceolate, 1–3-veined. **INFL** loose; heads, radiate, 1–5; involucre bell-shaped; phyllaries in 3–4 ± equal series, lanceolate, 1-veined, outer graduated into upper lvs; receptacle convex, naked. **RAY FLS** 10–25; ligules < 12 mm, yellow. **DISK FLS** many; corolla flaring slightly, yellow. **FR** narrowly oblong, ± cylindric, fusiform, or compressed, hairy; pappus of many white bristles in 1 series. ± 4 spp.: w N.Am. (Anagram of *Stenotus*) [Nesom & Morgan 1990 Phytologia 68:174–180]

1. Lvs entire .... ***T. lyallii***
1′ Lvs toothed
  2. Involucre 7.5–15 mm; disk corollas 6–7 mm, = pappus .... ***T. eximius***
  2′ Involucre 15–20 mm; disk corollas 8–10 mm, > pappus .... ***T. peirsonii***

T. eximius (H.M. Hall) Nelson & J.F. Macbr. (p. 357) **STS** < 13 cm. **LVS** 5 cm, toothed. **INFL**: involucre 7.5–15 mm; phyllaries green. **DISK FLS**: corollas 6–7 mm. **FR** 3.5–4 mm; pappus = disk corolla. $2n$=18. UNCOMMON. Alpine and subalpine sites; 2500–3300 m. n SNH (Alpine, El Dorado, Inyo cos.); NV. [*Haplopappus e.* H.M. Hall]

T. lyallii (A. Gray) Nelson (p. 357) LYALL'S HAPLOPAPPUS **STS** 5–15 cm. **LVS** < 6 cm, entire. **INFL**: involucre 8–11 mm; phyllaries green to reddish. **DISK FLS**: corollas 6.5–7.5 mm. **FR** 3–4 mm; pappus = disk corolla. $2n$=18. RARE in CA. Alpine talus, barrens; 2500–2700 m. c KR (Trinity Alps); to B.C., MT, WY, Colorado. [*Haplopappus l.* A. Gray]

T. peirsonii (Keck) G. Nesom & R. Morgan (p. 357) **STS** < 20 cm. **LVS** < 8 cm, toothed. **INFL**: involucre 15–20 mm; phyllaries green. **DISK FLS**: corollas 8–10 mm. **FR** 5 mm; pappus < disk corolla. $2n$=90. Rocky alpine and subalpine slopes; 2900–3700 m. c SNH (Inyo, Fresno cos.). [*Haplopappus p.* (Keck) J. Howell] ❀TRY:DFCLT.

TOWNSENDIA

Geraldine A. Allen

Ann to per. **STS** 0 to erect, < 30 cm. **LVS** alternate, entire, petioled. **INFL**: head, radiate gen solitary; involucre conic to hemispheric; phyllaries in 2–6 series, outer gen < inner, free, margins scarious to ciliate; receptacle flat, naked. **RAY FLS** many; ligules white, pink, blue, or yellow. **DISK FLS** many; corollas ± yellow; style branches flat, tip hairy. **FR** ± compressed, brown, glabrous to hairy; pappus of minutely barbed, gen flat bristles (ray frs sometimes also with small outer series). 21 spp.: w N.Am. (D. Townsend, amateur US botanist, 1787–1858) [Reveal 1970 GB Naturalist 30:23–52] Some spp. reproduce by asexual seed.

1. Lvs long-soft-woolly; pappus readily dehiscent .... ***T. condensata***
1' Lvs ± finely strigose; pappus ± firmly attached to fr
  2. Fr glabrous; lvs gen linear to oblanceolate, acute .... ***T. leptotes***
  2' Fr ± spreading-hairy; lvs oblanceolate to obovate, ± obtuse
    3. Sts (2)5–30 cm, leafy to middle or above .... ***T. parryi***
    3' Sts 2–8 cm, leafy only at base .... ***T. scapigera***

T. condensata Eaton (p. 357) Bien or per < 5 cm; caudex taprooted. **LVS** basal, gen 1–1.5 cm, ± narrowly obovate, rounded, entire, long-soft-woolly. **INFL**: phyllaries ± subequal, lanceolate, ± acuminate, scarious-margined, ± long-hairy. **RAY FLS**: ligules white to pink or violet. **FR** short-hairy; pappus readily dehiscent. $2n=18$. Gravel slopes; 3200–3500 m. n SNE (Mono Co); to sw Can, WY, Colorado.

T. leptotes (A. Gray) G.E. Osterh. (p. 357) Per < 3 cm; caudex taprooted. **LVS** basal, gen 1–2 cm, ± linear to oblanceolate, gen acute, entire, gen strigose. **INFL**: phyllaries lanceolate, acuminate, margins scarious-ciliate, glabrous to ± hairy. **RAY FLS**: ligules white to pink or blue. **FR** glabrous; pappus ± persistent. $2n=18$. Rocky slopes; 3500–3700 m. W&I; to MT, NM.

T. parryi Eaton PARRY'S TOWNSENDIA Per (2)5–30 cm; caudex taprooted. **LVS** basal and cauline, gen 2–6 cm, oblanceolate to obovate, obtuse to rounded, entire, gen strigose. **INFL**: phyllaries lanceolate, acuminate, margins scarious-ciliate, glabrous to ± hairy. **RAY FLS**: ligules violet to blue. **FR** ± spreading-hairy; pappus ± persistent. $2n=18,36$. RARE in CA. Open places; 3000–3500 m. n SNE (Sweetwater Mtns, Mono Co.); to sw Can, Colorado.

T. scapigera Eaton (p. 357) Per 2–8 cm; caudex taprooted. **LVS** basal and cauline, 1–5 cm (< petiole), obovate, ± obtuse to rounded, entire, strigose. **INFL**: phyllaries narrowly ovate to oblong, ± acute, scarious-margined, ± hairy. **RAY FLS**: ligules white to pink or violet. **FR** ± spreading-hairy; pappus ± persistent. Rocky slopes; 1400–3500 m. SNE; NV. ❀DRN,SUN:1–3,7,18;DFCLT.

TRACYINA

Robert Ornduff

1 sp.: CA. (Joseph P. Tracy, nw California botanist, 1879–1953) [Blake 1937 Madroño 4:73–77]

T. rostrata S.F. Blake (p. 357) BEAKED TRACYINA Ann < 35 cm. **ST** gen branched above, erect, glabrous. **LVS** alternate, entire, < 2.5 cm, narrowly lanceolate; margins hairy. **INFL**: heads radiate, terminal; involucre, < 7 mm, cylindric to widely obovoid; phyllaries in 2–4 series, ± linear, falling separately. **RAY FLS** < 22 in 1 series; corolla < 5.5 mm, thread-like, inconspicuous, pale or greenish yellow, red-tinged. **DISK FLS** < 26, < ray fls, pale or greenish yellow. **FR** < 5.5 mm, cylindric or fusiform, brown; beak short, hairy; pappus of < 40 slender bristles, < 4 mm, many > disk fls. RARE. Grassy slopes; < 500 m. NCoR (Humboldt, Lake cos.). Gen self-pollinated.

TRAGOPOGON GOAT'S BEARD

G. Ledyard Stebbins

Bien or per from strong taproot; herbage ± glabrous; sap milky. **STS**: branches few, strongly ascending. **LVS** basal and cauline, alternate, entire, grass-like, parallel-veined. **INFL**: heads ligulate, solitary at branch tips; peduncles long, naked; involucre cylindric or narrowly conic; phyllaries in 1 series; receptacle naked. **FLS** yellow to bronze or purple; ligules readily withering. **FR** 2.5–3 cm; beak stout, > body; pappus of stout plumose bristles, 2° bristles tangled, tips of a few bristles > the rest, unbranched; frs spreading, forming a spheric head 4–5 cm diam. ± 45 spp.: Eurasia. (Greek: goat's beard)

1. Fls purple; phyllaries > fls .... ***T. porrifolius***
1' Fls pale to bright yellow; phyllaries < to >> fls
  2. Peduncles much wider toward tip; phyllaries >> pale yellow fls .... ***T. dubius***
  2' Peduncles not or little wider toward tip; phyllaries < bright yellow fls .... ***T. pratensis***

T. dubius Scop. Ann, bien 3–10 dm. **LVS** 2–5 dm. **INFL**: peduncle much wider towared tip; phyllaries 8–13, 2.5–4 cm in fl heads, >> fls, 4–7 cm in fr heads. **FLS**: ligules pale lemon-yellow. **FR** 25–35 mm; pappus ± white. $2n=12$. Uncommon. Weed in waste places; 0–2700 m. n SN, SnJV, SnFrB, SCoRO, SnBr, GB; native to Eur.

T. porrifolius L. (p. 357) SALSIFY, OYSTER PLANT Bien 4–10 dm. **LVS** 2–4 dm. **INFL**: peduncle much wider toward tip; phyllaries 5–11, 2.5–4 cm in fl heads, > fls, 4–7 cm in fr heads. **FLS**: ligules purple. **FR** 2.5–4 cm; pappus ± brown. $2n=12$. Common. Weed in waste places; 0–1700 m. Most of CA-FP; native to Eur.

T. pratensis L. Bien 1.5–8 dm, slightly hairy when young, soon glabrous. **LVS** 2–4 dm. **INFL**: peduncle gen not becoming wider upward; phyllaries gen 8, 12–30 mm in fl heads, < fls, 18–45 mm in fr heads. **FLS**: ligules bright yellow. **FR** 15–25 mm; pappus ± white. $2n=12$. Gen ± moist, disturbed places; 900–1600 m. KR, CaRH, SNH, MP; native to Eur.

TRICHOCORONIS

A. Michael Powell

Ann or per < 30 cm, ascending, hairy. **LVS** sessile; lower opposite; blades < 2.5 cm, oblong, serrate. **INFL** open; heads discoid, small, 1–few; involucre hemispheric to bell-shaped; phyllaries ± 30, subequal in 2–3 series; receptacle convex to conic, naked. **FLS** 75–125; corolla tube narrow; style branches slender. **FR** 4–5-ribbed; pappus of 2–6 short bristles, ± 1 mm. 2 spp.: sw US, Mex. (Greek: hair crown, from pappus) [King & Robinson 1987 Monogr Syst Bot Missouri Bot Garden 22:188–190]

T. wrightii (A. Gray) A. Gray var. *wrightii* (p. 357) **INFL** few-branched; heads ± 5 mm diam. **FLS**: corolla ± 1 mm, throat white below, maroon above, lobes white. **FR** ± 1.5 mm. $2n$=30. Uncommon. Moist places. GV, SCo; native to s TX, ne Mex. Another var. in Baja CA.

TRICHOPTILIUM

1 sp. (Greek: feathery bristle, from pappus)

T. incisum A. Gray (p. 357) Ann, short-lived per 5–25 cm, gen ± tomentose. **STS** 1–several from base. **LVS** simple, basal and cauline, alternate or subopposite, mostly clustered in lower half of pl, 1–3 cm, sessile or tapered to a short, winged petiole; blade acute, sharply dentate or shallowly lobed, densely tomentose, resin-dotted. **INFL**: heads discoid, solitary; peduncles 3–11 cm, slender, bractless, glandular-puberulent; involucre 6–12 mm diam, hemispheric or bell-shaped, often appearing ± disk-like when pressed; phyllaries in 2 ± equal series, free, 5–7 mm, linear-elliptic, acute; receptacle rounded, naked. **FLS** many; corolla 4.5–7 mm, yellow, outermost sometimes enlarged and ± bilateral; style-tips truncate. **FR** 2–3 mm, obpyramidal, glabrous to densely strigose; pappus of 5 scales, each 5–7 mm, dissected into many bristles. $2n$=26. Dry slopes, plains; < 1000 m. s DMoj, DSon; s NV, sw AZ, n Baja CA. ❀TRY.

TRIMORPHA

Guy L. Nesom

Ann, bien, per, fibrous-rooted, gen from short rhizome. **LVS** oblanceolate to obovate or spoon-shaped, entire. **INFL**: heads disciform or inconspicuously radiate, solitary or in few-headed clusters; buds erect; involucre hemispheric; phyllaries ± graduated, narrowly lanceolate outer and middle with 3 orange-resinous veins, at least near the base; receptacle naked. **PISTILLATE FLS** in ± 2 series; outer gen with very narrow, sometimes coiled ligules; inner gen without ligules (exc *T. lonchophyllus*). **DISK FLS**: corollas yellow, narrowly tubular, not hard or inflated; veins orange-resinous; style tips deltate. **FR** narrowly oblong, 2–4-ribbed, ± flat; pappus of barbed bristles in 1–2 series, outer 0 or bristles few, short, inner at maturity > involucre. ± 45 ssp.: N.Am, Eurasia. (Latin & Greek: 3 forms, from fl types) [Nesom 1989 Phytologia 67:61–66] Formerly treated as *Erigeron* sect. *Trimorpha*.

1. Inner pistillate fls without ligules; sts and phyllaries stalked-glandular; infl gen wider near tip .. ***T. acris*** var. ***debilis***
1′ Pistillate fls with ligules; sts and phyllaries not glandular; infl gen raceme-like .... ***T. lonchophylla***

T. acris (L.) S.F. Gray var. ***debilis*** (A. Gray) G. Nesom (p. 361) NORTHERN DAISY Bien, per. **STS** 10–35 cm, sparsely hairy, stalked-glandular. **LVS**: basal 3–8 cm, gradually reduced upwards. **INFL** wider near top or of only 1 head; phyllaries sparsely soft-hairy, stalked-glandular, greenish. **PISTILLATE FLS** in ± 2 series, outer with ligules 3–4.5 mm, inner without ligules. **DISK FLS**: corollas 4–5 mm. **FRS** 2–2.4 mm. RARE in CA. Volcanic rocks, meadows; 2700–2900 m. CaR; to AK, Colorado. [*Erigeron d.* (A. Gray) Rydb.]

T. lonchophylla (Hook.) G. Nesom (p. 361) Ann, bien. **STS** 4–20 cm, sparsely to densely bristly or soft-hairy; glands 0. **LVS**: basal 2–8 cm, gradually reduced and becoming linear upward. **INFL** gen raceme-like or of 1 head; phyllaries ± rough-hairy, without glands, gen purple-tipped. **PISTILLATE FLS** in 2–3 series, all with very narrow ligules 2–3 mm, barely extending beyond involucre. **DISK FLS**: corollas 3–4.5 mm. **FR** 1.3–1.5 mm. $2n$=18. Meadows, creek and ditch banks; 1800–3550 m. SN, SnBr; to AK, e Can, NM. [*Erigeron l.* Hook.]

TRIXIS

Shrubs. **LVS** simple, alternate, sessile or petioled; blade long, entire or toothed, gen ± glandular. **INFL**: heads discoid (but often appearing radiate), sessile or short-peduncled, in flat-topped or panicle-like cymes; involucre cylindric; phyllaries gen in 1 series, linear, keeled; receptacle flat, short-hairy. **FLS** few–many; corolla yellow, 2-lipped, outer lip spreading, ligule-like, 3-lobed, sometimes recurved, inner lip 2-lobed, recurved or coiled; anthers exserted, base with tail-like appendages, tip oblong, acute; style tips truncate. **FR** ± cylindric, 5-ribbed, black or brown, short-hairy and glandular; pappus of many bristles. ± 65 spp. N.Am, S.Am. (Greek: 3-fold, from outer corolla lip) [Anderson 1972 Mem NY Bot Gard 22(3):1–68]

T. californica Kellogg var. ***californica*** (p. 361) Pl glandular and short-hairy. **STS** 2–20 dm, erect, stiff, much-branched. **LVS** many; petiole 1.5–5 mm, winged; blade 2–11 cm, linear-lanceolate to lanceolate, acute at both ends, entire or serrate, margin often rolled under. **INFL**: peduncles gen < or = 5 mm, bearing 5–7 bracts; phyllaries 8–10, 8–14 mm, green, tips acute. **FLS** 11–25; corolla glandular, tube 6–9 mm, outer lip 5–8 mm, spreading, inner lip 4–5.5 mm, gen coiled. **FR** 6–10.5 mm; pappus 7.5–12 mm, dull white. $2n$=54. Dry, rocky slopes, washes, desert flats; gen < 1000 m. D; to TX, n Mex. ❀DRN,DRY,SUN:10,11,**12,13,19–21**,22–24.

UROPAPPUS SILVER PUFFS

Kenton L. Chambers

1 sp. (Latin: tailed pappus, from awn-tipped scales)

U. lindleyi (DC.) Nutt. (p. 361) Ann 5–70 cm, ± scapose; sap milky. **LF** 5–30 cm, ± basal, ± linear, long-tapered, entire to narrowly lobed, ± soft-hairy (esp petiole base). **INFL**: head, ligulate solitary, erect; involucre 10–40 mm, glabrous; phyllaries narrowly lanceolate, outer progressively shorter; receptacle naked. **FLS** 5–many; corollas yellow (often reddish below), soon withering, < or = involucre. **FR** 7–17 mm, slender, tapered to tip in CA, obscurely 10-ribbed, gen blackish; outermost frs scabrous; pappus scales 5, 5–15 mm, deciduous, smooth, silvery, bristle-tip 4–6 mm, slender, smooth, from notched scale tip. $2n$=18. Common. Open grassland, woods, chaparral, deserts, gen in loose soils; < 1800 m. CA (exc NCo); to WA, ID, UT, w TX, n Mex. [*Microseris l.* (DC.) A. Gray; *M. linearifolia* (DC.) Schultz-Bip.] With ann *Microseris* spp. a parent of *Stebbinsoseris*. ❀TRY.

UROSPERMUM

G. Ledyard Stebbins

Ann to per; sap milky. **ST** gen 1 from base. **LVS** mostly cauline, entire to pinnately lobed. **INFL**: heads ligulate, solitary, long-peduncled; phyllaries in 1 series, fused at base; receptacle naked. **FLS** many; ligules yellow, readily withering. **FR** ± oblong, beaked; pappus of 2 rows of plumose bristles. 2 spp.: Medit. (Latin: tailed seed, from beak)

U. picroides (L.) Schmidt (p. 361) **ST** 3–4 dm, long-hairy and with slender spine-like bristles. **LVS**: lower tapered to base; upper sessile, clasping; blades toothed or lobed (or upper entire). **INFL**: peduncles stout; involucre 1.3–2.2 cm; phyllaries 7–8, widely lanceolate, long-acuminate, bristly. **FR** 11–14 mm; beak ± > body. $2n$=8,10. Uncommon. Disturbed places; < 100 m. SnFrB, SCo; native to Eur.

VENEGASIA CANYON-SUNFLOWER

1 sp. (M. Venegas, Mexican writer, missionary to CA, 1680–1764)

V. carpesioides DC. (p. 361) Shrub 5–15 dm. **STS** gen several, ± puberulent. **LVS** simple, alternate; petiole 1–5 cm; blade 3–15 cm, triangular-ovate, 3-veined from base, acute to acuminate, entire to coarsely toothed, minutely puberulent, resin-dotted below. **INFL**: heads radiate, solitary or in ± leafy cymes; peduncles gen 2–3 cm, slender, bracts 0; involucre 1–3 cm diam, hemispheric; phyllaries unequal, in several series, 5–14 mm, oblong to widely obovate, obtuse, ± veiny, outer spreading, middle > inner; receptacle flat or rounded, naked. **RAY FLS** 12–21; ligule 15–30 mm, yellow. **DISK FLS** many; corolla 5–6 mm, yellow, tube base glandular; stamen tips triangular; style tips ± obtuse-triangular. **FR** 2–3 mm, ± cylindric, ribbed, ± curved, dark brown, glabrous; pappus 0. $2n$=38. Canyons, moist, wooded slopes; < 900 m. SCoRO, SW; Baja CA. ❀15–17,**24**SHD:14,**20–23**&IRR:19;also STBL.

VENIDIUM

Elizabeth McClintock

Ann, per, tomentose. **STS** erect, branched from base. **LVS** alternate. **INFL**: heads radiate, solitary; peduncles long; phyllaries in several overlapping series; receptacle flat, pitted, gen naked. **RAY FLS**: ligules gen yellow. **DISK FLS** many; corollas yellow; anthers with ovate tips, weakly sagittate at base; style slender below, thickened above a swollen node, branches short. **FR** 3–4-winged or -ribbed, glabrous; pappus 0 or of minute scales. 20–30 spp.: s Afr. (Latin: vein, from the ribs) [Stapf 1928 Curtis's Bot Mag Tab 9127]

V. fastuosum (Jacq.) Stapf MONARCH-OF-THE-VELD, CAPE DAISY, NAMAQU-LAND DAISY Ann. **STS** 15–35(90) cm, finely ribbed, cobwebby-hairy when young. **LVS** < 9 cm; basal gen in loose rosettes; lower and middle cauline alternate, irregularly 4–5-pinnately lobed, petioles flattened, esp wide at base; uppermost lvs with reduced lobes to ± entire, sessile. **INFL**: heads 4–10 cm diam; phyllaries cobwebby-hairy, outer spreading to reflexed, awl-shaped, inner appressed, obovate. **RAY FLS** 35–40; ligule 3–5.5 cm, bright orange-yellow, purple at base. **DISK FLS** many; corolla yellow-brown to brown-purple, bisexual or inner staminate. **FR** < 2 mm, ovoid, flattened, glabrous; outer side smooth; inner slit longitudinally, with 2 rows of teeth; pappus 0 or a minute, scaly ring. $2n$=18. Uncommon. Escape from cult in disturbed urban areas, roadsides; < 500 m. SCo; native to s Afr.

VERBESINA CROWNBEARD

Ann, per, shrubs. **LVS** simple, opposite or alternate. **INFL**: heads radiate or discoid, solitary or in few–many-headed cymes; phyllaries in 2–6 equal to very unequal series, free; receptacle flat to conic, chaffy; chaff scales entire, folded around ovaries. **RAY FLS** fertile or sterile; corollas yellow to white. **DISK FLS** many; corollas yellow to orange; style branches acute to acuminate. **FR**: ray achenes triangular or 0; disk achenes flattened, wing-margined, obovate; pappus 0 or of 2 awns. (Derived from *Verbena*) [Coleman 1966 Madroño 18:129–137; Coleman 1966 Amer Midl Naturalist 76:475–481]

1. Subshrub; lvs sessile, bright green; ray fls sterile .... ***V. dissita***
1′ Ann; lvs petioled, dull green to canescent; ray fls fertile .... ***V. encelioides*** ssp. ***exauriculata***

V. dissita A. Gray (p. 361) CROWNBEARD Subshrub gen < 1 m. **STS** glabrous or short-hairy near infl. **LVS** sessile; blade ovate, 1-veined, entire or with few, short teeth, bright green, glabrous to densely scabrous. **INFL**: cyme, ± flat-topped; peduncles subtended by scale-like entire bracts; heads 3–16; phyllaries in 3–4 unequal series, 5–13 mm, oblong, obtuse to acute, short-hairy; chaff scales 10–13 mm, acute. **RAY FLS** sterile; ligule orange-yellow, ± entire; style 0, ovary 2.5–3 mm. **DISK FLS**: corolla 7–9 mm; anthers dark brown. **FR** 8–9 mm, dark brown, glabrous; wing thin, brown; pappus of 2 awns, 3–4 mm. RARE. Shrubby coastal slopes; < 100 m. s SCo (Orange Co.), naturalized in SnBr; Baja CA.

V. encelioides (Cav.) A. Gray ssp. *exauriculata* (Robinson & Greenman) J. Coleman (p. 361) GOLDEN CROWNBEARD Ann 15–130 cm; odor unpleasant. **STS** densely short-hairy. **LVS** petioled; blade lanceolate to triangular-ovate, 3-veined from base, coarsely dentate, dull green, ± strigose-canescent. **INFL**: heads 1–many; peduncles subtended by lf-like bracts; phyllaries in 2 or 3 ± equal series, 8–10 mm, linear-lanceolate, acute, ± strigose; chaff scales 6–8 mm, abruptly acuminate. **RAY FLS** fertile; ligule orange-yellow, ± 3-lobed; style present. **DISK FLS**: corolla 5–6 mm; anthers yellow to light brown. **FR**: ray achenes 3–4 mm, obovoid, triangular, wing 0, pappus 0; disk achene 4–6.5 mm, obovate, flattened, brown or black, soft-hairy, wing wide, ± white, corky, pappus awns 2, 1–2.5 mm. $2n$=34. Disturbed areas, roadsides, fields; gen < 200 m. SnJV, SCoR (Salinas Valley), SCo, DSon; w N.Am; native AZ to Great Plains, Mex. [*V. en.* var. *ex.* Robinson & Greenman] TOXIC to livestock but unpalatable.

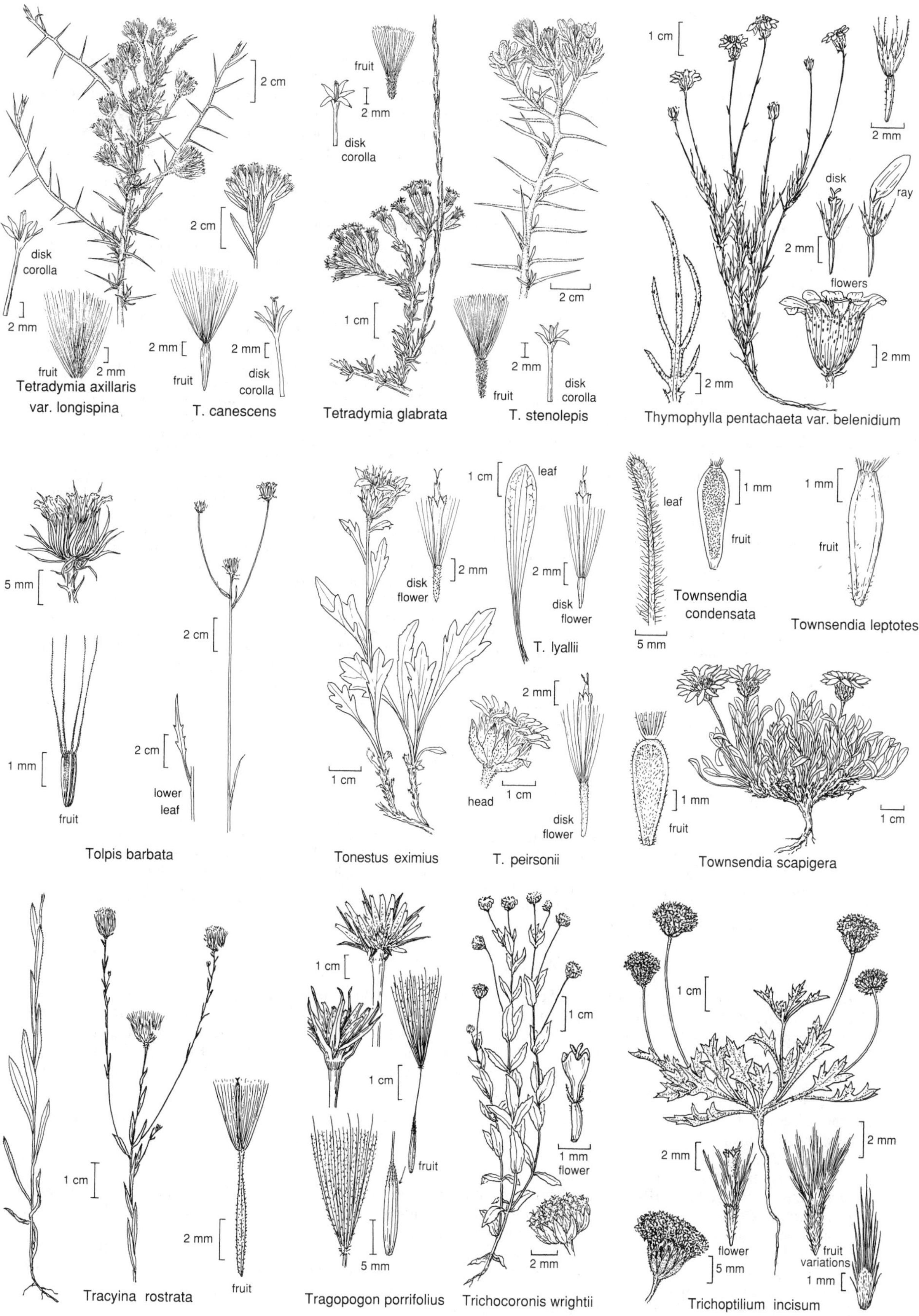
2 cm
disk corolla
2 mm
fruit
2 mm
Tetradymia axillaris var. longispina
2 cm
2 mm
2 mm
fruit
disk corolla
T. canescens
fruit
2 mm
disk corolla
1 cm
Tetradymia glabrata
2 cm
2 mm
fruit
disk corolla
T. stenolepis
1 cm
2 mm
disk
ray
2 mm
flowers
2 mm
2 mm
Thymophylla pentachaeta var. belenidium
5 mm
1 mm
fruit
2 cm
2 cm
lower leaf
Tolpis barbata
2 mm
disk flower
1 cm
Tonestus eximius
1 cm
leaf
2 mm
disk flower
T. lyallii
2 mm
head
1 cm
disk flower
T. peirsonii
leaf
5 mm
1 mm
fruit
Townsendia condensata
1 mm
fruit
Townsendia leptotes
1 mm
fruit
1 cm
Townsendia scapigera
1 cm
2 mm
fruit
Tracyina rostrata
1 cm
1 cm
fruit
5 mm
Tragopogon porrifolius
1 cm
1 mm
flower
2 mm
Trichocoronis wrightii
1 cm
2 mm
2 mm
flower
5 mm
fruit variations
1 mm
Trichoptilium incisum

VIGUIERA

Ann, per, shrubs. **STS** gen several from base. **LVS** simple, alternate or opposite, sessile or petioled. **INFL**: heads radiate or discoid, solitary or in few-headed, terminal cymes; peduncles long or short; involucre hemispheric or bell-shaped; phyllaries in several series, equal to very unequal; receptacle rounded to conic, chaffy; chaff scales entire or 3-lobed, folded around frs. **RAY FLS** 0–many, sterile; corolla yellow; ligules entire to 3-lobed. **DISK FLS** many; corollas yellow or orange; anther tips triangular; style tips triangular. **FR** ± flattened, obovate, glabrous or ± hairy; pappus of scales, gen 1 or more lanceolate. ± 150 spp.: New World. (L.G.A. Viguier, 1790–1867, French physician, botanist) [Shilling 1990 Madroño 37:149–170]

1. Lvs ± lanceolate, shiny green, resinous . . . . *V. laciniata*
1′ Lvs ovate, dull green to canescent-tomentose
  2. Lvs short rough-hairy on both surfaces, 1–2.5 cm, triangular-ovate, often toothed; veins not strongly raised beneath . . . . *V. parishii*
  2′ Lvs densely canescent-tomentose above, dull green and loosely tomentose beneath, 3–8 cm, not triangular, entire; veins prominently raised beneath . . . . *V. reticulata*

V. laciniata A. Gray (p. 361) SAN DIEGO COUNTY VIGUIERA Shrub < 2 m diam, short-rough hairy, covered with a varnish like resin throughout. **STS** 6–13 dm, slender. **LVS** gen alternate; petiole 0–10 mm; blade 1–5 cm, oblong to narrowly lanceolate, gen 3-veined, base rounded to truncate or hastate, tip obtuse to acute, margin entire to coarsely round-toothed or shallowly lobed, often ± rolled under, veins ± prominently raised on undersurface. **INFL**: heads solitary or in ± flat-topped cymes; peduncles 0.5–5 cm with 1–several, scattered, leafy or scale-like bracts; heads radiate; involucre 8–10 mm diam, hemispheric; phyllaries in 2–3 unequal series, 3–7 mm, ovate to lanceolate; chaff scales 5.5–7.3 mm. **RAY FLS** 5–13; ligules 6–12 mm. **DISK FLS**: corolla 3.5–4.5 mm. **FR** 2–4 mm, obovate; pappus of 2 fringed scales, 0.4–1 mm and 1–2 readily deciduous lanceolate scales, 1.7–3 mm. $2n=36$. RARE in CA. Shrubby slopes; 90–750 m. s SCo, sw PR (sw San Diego Co.); Baja CA, w Sonora. In cult.

V. parishii E. Greene (p. 361) Shrub < 2 m diam, short-rough-hairy throughout. **ST** 6–13 dm, much-branched. **LVS** opposite below, alternate above; petiole 2–8 mm; blade gen 1–3.5 cm, triangular-ovate, 3-veined from obtuse to truncate or subcordate base, tip obtuse to acute, entire or teeth few, short, surfaces green to lightly canescent. **INFL**: heads solitary or in open, few-headed cymes; peduncles 3–15 cm, slender, bracts 0 or few and lf-like; heads radiate; involucre 10–13 mm diam, hemispheric or appearing disk-like when pressed; phyllaries in 2–3 equal or unequal series, 3–9 mm, lance-oblong, tips abruptly narrowed, surface green to canescent; chaff scales 5–6 mm. **RAY FLS** 8–15; ligules 1–1.5 cm. **DISK FLS**: corollas 3.5–5 mm. **FR** 2.7–3.8 mm; pappus of 2 fringed scales (each 0.5–1 mm) and 2–3 lanceolate scales (each 2–3 mm). $2n=36$. Common. Washes, dry, rocky slopes; < 1500 m. D; AZ, NV, nw Mex. [*V. deltoidea* A. Gray var. *p.* (E. Greene) Vasey & Rose] ❀DRN:10,11,**12–14**,18,**19–24**&DRY,SUN:15–17; continued fls & lvs with IRR;GRNCVR;also STBL.

V. reticulata S. Watson (p. 361) Shrub < 1.5 m diam. **STS** many, 5–15 dm, soft-hairy; bark peeling in age. **LVS** opposite below, alternate above; petioles 3.5–30 mm; blades 2–9 cm, ovate, 3-veined from truncate to cordate base, tip acute, margin entire, upper surface densely canescent-tomentose, undersurface with veins prominently raised, gray-green, loosely tomentose. **INFL**: cyme few-headed, on long, ± naked branch; bracts reduced to scales 3–10 mm; peduncles 5–50 mm; heads radiate; involucre hemispheric or appearing disk-like when pressed; phyllaries in 2–3 unequal series, 2–5 mm, oblong to ovate, obtuse, short-white-hairy; chaff scales 4–5.5 mm. **RAY FLS** 10–15; ligules 7–15 mm. **DISK FLS**: corollas 3–4 mm. **FR** 2.5–4 mm, obovate; pappus of 2 scales (each 0.5–1 mm) and 1–2 lanceolate scales (each 1–2.8 mm). Arid slopes; < 1500 m. DMoj; w NV. ❀TRY.

WHITNEYA

1 sp. (J.D. Whitney, CA geologist, 1819–1896)

W. dealbata A. Gray (p. 361) Per 15–35 cm, densely covered with short, curly hairs and sessile glands. **STS** erect from long, branched rhizome. **LVS** basal and opposite; lower petioled, 5–10 cm, elliptic to oblanceolate or obovate; upper sessile, smaller, narrower, all entire. **INFL**: heads radiate, solitary or in few-headed, terminal cymes; peduncles slender; involucre 1–2 cm diam, hemispheric or bell-shaped; phyllaries 5–12 in 1 series, 8–11 mm, oblong or elliptic; receptacle rounded, naked. **RAY FLS** 5–12 (1 per phyllary); ligules 10–25 mm, yellow, when dry papery and persistent on fr. **DISK FLS** many, functionally staminate; corolla 6–7 mm, yellow; style incl, ovary not maturing. **FR** 4–5 mm, oblong, ± flattened, brown or black, short-hairy; pappus 0. Open forests, meadows, slopes; 1200–2400 m. CaR, n&c SN. ❀TRY.

WYETHIA MULES EARS

Per from stout, taprooted caudex, gen ± unbranched. **LVS** alternate, gen reduced upward; petiole gen << blade, often winged. **INFL**: heads gen radiate, 1–few, ± large; involucre hemispheric or bell-shaped; phyllaries ± many in 2–3 series, outer often ± lf-like; receptacle ± convex; chaff scales ± linear, entire, gen ± hairy. **RAY FLS** (0)5–20; ligules yellow. **DISK FLS** many; corollas yellow, lobes short; anthers brown or yellow, tips triangular; style tips linear, tapered. **FR**: pappus 0 or crown of scales; ray achenes 3-angled; disk achenes 4-angled, sometimes ± compressed. (Nathaniel J. Wyeth, US explorer, 1802–1856) [Weber 1946 Amer Midl Naturalist 35:400–452]

1. Heads discoid (or rays rarely 1–3) . . . . *W. invenusta*
1′ Heads radiate, rays 5–21+
  2. Basal lvs 0 at fl (or gen < cauline lvs)
    3. Outer phyllaries many, gen < 6 mm wide; involucre hemispheric; pls gen > 50 cm
      4. Pl densely soft-hairy; fr 10–12 mm, glabrous or short-hairy above; pappus 1–3 mm . . . . *W. elata*
      4′ Pl scabrous and sparsely bristly; fr 5–6 mm; pappus 0 or minute . . . . *W. reticulata*
    3′ Outer phyllaries few, lf-like, gen 6–17 mm wide; involucre bell-shaped; pls 10–30 cm
      5. Head 1; lvs glabrous; pappus 0 . . . . *W. bolanderi*
      5′ Heads gen several; young lvs tomentose; pappus a crown of low triangular scales . . . . *W. ovata*
  2′ Basal lvs present at fl, gen > cauline lvs
    6. Outer phyllaries wide, lf-like, gen >> inner; head gen 1, large

7. Pl shiny green, glabrous to sparsely short-hairy, glandular; phyllaries glabrous or glandular; fr 10–12 mm .... ***W. glabra***
7′ Pl densely tomentose, becoming glabrous; phyllaries persistently tomentose; fr 12–15 mm ..... ***W. helenioides***
6′ Outer phyllaries narrow, gen barely > inner; heads 1–few
8. Pls ± tomentose, becoming glabrous; phyllaries gen few .... ***W. mollis***
8′ Pls short-hairy; phyllaries gen many
9. Phyllaries ± soft-hairy, ciliate; head gen 1 .... ***W. angustifolia***
9′ Phyllaries puberulent, not ciliate; heads 1–4 .... ***W. longicaulis***

W. angustifolia (DC.) Nutt. Pl 3–9 dm, ± rough-hairy. **LVS**: basal blades 10–50 cm, oblanceolate. **INFL**: peduncle 20–30 cm, soft-hairy; involucre 3–4 cm diam, hemispheric; phyllaries 20–30 mm, linear to ± lanceolate, gen not lf-like, soft-hairy, ciliate; chaff scales 13–15 mm. **RAY FLS** 8–21; ligules 15–45 mm. **DISK FLS**: corollas 10–11 mm. **FR** 7–8 mm, ± hairy; pappus of 1–4 awn-tipped scales, 7–9 mm (and sometimes crown of smaller scales). $2n=38$. Grassland, < 2000 m. NW, SN, SCoR; to WA. ❀SUN, DRN:1,**4–7,14–18**&IRR:8,9,19–24.

W. bolanderi (A. Gray) W.A. Weber Pl 1–3 dm, ± sticky. **LVS** ± equal; blades 4–12 cm, widely ovate-cordate, obtuse, veiny below. **INFL**: peduncle 0–2 cm; involucre bell-shaped; outer phyllaries 4–6, 18–30 mm, obovate, veiny, lf-like, > inner; chaff scales ± 15 mm, tip tomentose. **RAY FLS** ± 13; ligules 20–25 mm. **DISK FLS**: corollas 10 mm. **FR** 7 mm; pappus 0. $2n=38$. Grassland, chaparral; 300–1000 m. n&c SNF. ❀TRY.

W. elata H.M. Hall HALL'S WYETHIA Pl 6–10 dm, densely soft-hairy. **LVS** cauline; blades 8–20 cm, triangular-ovate, 1–3-veined, obtuse-cordate, acute. **INFL** sometimes flat-topped, leafy-bracted; peduncles 4–12 cm; involucre hemispheric; phyllaries 25–35 mm, lanceolate to ovate, acuminate, outer sometimes reflexed, gen not lf-like, inner erect, short-rough-hairy. **RAY FLS** 13–20; ligules 50–60 mm. **DISK FLS**: corollas 12 mm. **FR** 10–12 mm, glabrous or short-hairy above; pappus an uneven crown of scales, 1–3 mm. $2n=38$. UNCOMMON. Open woodland, forest; 1000–1400 m. c SNF.

W. glabra A. Gray (p. 361) Pls 1–4 dm, glabrous or sparsely short-hairy, glandular. **LVS**: basal blades 25–45 cm, oblanceolate to obovate. **INFL**: peduncle 1–6 cm; outer phyllaries 40–70 mm, lanceolate, obtuse to acute, lf-like, spreading, gen > rays, inner shorter, ± erect; chaff scales 15–20 mm. **RAY FLS** 12–21; ligules 25–50 mm. **DISK FLS**: corollas 10–11 mm. **FR** 10–12 mm, sparsely hairy; pappus scales 2–several, 1–5 mm, lanceolate to triangular, glabrous. Gen shady sites; < 800 m. s NCoR, CW. ❀TRY.

W. helenioides (DC.) Nutt. Pls 3–7 dm, densely tomentose, often becoming ± glabrous. **LVS**: basal blades 25–45 cm, elliptic to obovate. **INFL**: peduncle 1–6 cm; outer phyllaries 40–70 mm, narrowly ovate, lf-like, spreading, often > rays, inner shorter, ± erect; chaff scales 15–20 mm. **RAY FLS** 13–21; ligule 25–50 mm. **DISK FLS**: corollas 10–11 mm. **FR** 12–15 mm, ± hairy; pappus scales 2–several, 1–5 mm, lanceolate to triangular, ± hairy. $2n=38$. Open grassland, shrubland; < 2000 m. s NCoR, c SNF, CW. ± glabrous pls approach *W. glabra*. ❀DRN,DRY:**7**,14–17,22–24&SHD:8,9, 19–21;DFCLT.

W. invenusta (E. Greene) W.A. Weber (p. 361) Pls 6–10 dm, short-rough-hairy, ± sticky. **LVS** cauline, 7–15 cm, ovate to triangular; base obtuse-cordate. **INFL**: heads gen discoid; involucre 2–3 cm diam; phyllaries 20–35 mm, lanceolate to ovate, acuminate, striate, not lf-like, ciliate; chaff scales 12 mm. **RAY FLS** 0(–3). **DISK FLS**: corollas 10 mm. **FR** 7–8 mm, glabrous; pappus 0. Open forest, clearings; 1100–1900 m. s SN.

W. longicaulis A. Gray HUMBOLDT COUNTY WYETHIA Pl 3–5 dm, glabrous or puberulent, ± gland-dotted. **LVS**: basal blades 10–20 cm, oblong-oblanceolate. **INFL**: peduncle 3–8 cm; involucre 15–30 mm diam, bell-shaped; phyllaries ± equal, 15–30 mm, lanceolate to ovate, barely lf-like, margins puberulent; chaff scales 15–17 mm. **RAY FLS** 5–10; ligule 20–30 mm. **DISK FLS**: corollas 10–11 mm. **FR** 8–10 mm, glabrous or tip puberulent; pappus scales < 1 mm, obtuse. UNCOMMON. Grassland, open forest; 750–1500 m. KR, n NCoR.

W. mollis A. Gray (p. 361) Pls 3–5 dm, ± tomentose, often becoming glabrous. **LVS**: basal > cauline; blade 20–40 cm, oblanceolate to widely obovate, base acuminate to obtuse. **INFL**: peduncle 1–10 cm, leafy-bracted; involucre ± 2 cm diam, ± bell-shaped; phyllaries gen few, ± equal, 15–35 mm, oblong-lanceolate, acute to obtuse; chaff scales 15–16 mm. **RAY FLS** 5–11; ligules 30–45 mm. **DISK FLS**: corollas 10 mm. **FR** 9–11 mm; upper half puberulent to tomentose; pappus of short scales, < 1 mm, sometimes also 1–few lanceolate scales < 8 mm. $2n=38$. Open forest, dry rocky slopes; 1200–3400 m. KR, CaR, n&c SN; se OR, w NV. ❀TRY: DFCLT.

W. ovata Torrey & A. Gray Pl < 5 dm, often branched, soft-hairy or becoming glabrous, ± sticky-glandular. **LVS** ± equal; blades 8–20 cm, elliptic to ovate, gland-dotted, base obtuse to acute. **INFL** flat-topped, ± leafy-bracted; peduncles 1–20 cm; involucre 1.5–2 cm diam, bell-shaped; outer phyllaries 4–6, erect, 30–50 mm, ± linear to ovate, ± lf-like, inner shorter, narrower; chaff scales 13–15 mm. **RAY FLS** 5–8; ligules ± 15 mm. **DISK FLS**: corollas 8 mm. **FR** 10–11 mm, puberulent near tip; pappus scales ± 1 mm. $2n=38$. Grassland, open woodland and forest; 300–2300 m. s SN, SW; n Baja CA. ❀TRY.

W. reticulata E. Greene (p. 361) EL DORADO COUNTY MULE EARS Pl 4–7 dm, soft-hairy, ± sticky-glandular. **LVS** cauline; lower > upper; blades 5–15 cm, lanceolate to triangular, scabrous above, base obtuse to ± cordate, veiny below. **INFL**: involucre 2–3 cm diam, hemispheric to bell-shaped; phyllaries ± equal, 20–30 mm, oblong or lanceolate, gen not lf-like, outer often spreading or reflexed; chaff scales 12–13 mm. **RAY FLS** 10–21; ligules 20–25 mm. **DISK FLS**: corollas 10 mm. **FR** 5–6 mm, glabrous; pappus 0 or minute. $2n=38$. **ENDANGERED** CA. Wooded slopes, chaparral; 300–500 m. n SNF (El Dorado Co.). Threatened by development. In cult.

XANTHIUM COCKLEBUR

Willard W. Payne

Ann. **LVS** alternate, petioled, gen ± lobed, ± hairy. **INFL**: staminate heads in clusters, involucre 0, receptacle chaffy; pistillate heads 2-fld, 2-beaked, spiny, bur-like; phyllaries 0 or minute; chaff scales many, spirally arrayed, fused below, free tips spiny, hooked. **STAMINATE FLS** many; corolla translucent; filaments fused, attached to corolla tube base, anthers free; ovary slender. **PISTILLATE FLS** gen 2 per head; corolla 0; style branches long. **FR** 2, enclosed in bur, germinating in successive years; pappus 0. 2 spp.: worldwide. (Greek: yellow, from fr-extract dye)

1. Lvs white-hairy below; sts spiny .... ***X. spinosum***
1′ Lvs green; sts unarmed .... ***X. strumarium***

X. spinosum L. (p. 365) SPINY COCKLEBUR **STS** < 10 dm, slender, tough; axillary spines < 3 cm, 3-branched, stout, golden. **LVS** short-petioled; blades gen 3–10 cm, linear to elliptic, subentire to pinnately lobed, white-hairy below, gray-green above; base gen 1-veined. **INFL**: pistillate heads gen 1 (few in upper axils). **FR**: bur cylindric; spines slender. $2n=36$. Common. Disturbed areas; < 900 m. CA-FP; world-wide.

X. strumarium L. (p. 365) COCKLEBUR **STS** < 15 dm, thick, ± fleshy, gen ± red- or black-spotted, unarmed. **LVS** long-petioled; blades < 15 cm, widely triangular, gen ± 3-lobed, coarsely toothed, green below and above, ± glandular, scabrous; base gen 3-veined. **INFL**: pistillate heads clustered below staminate heads. **FR**: bur cylindric to barrel-shaped; spines gen stout, ± glandular. 2*n*=36. Common. Disturbed areas; gen < 500 m. CA-FP, D; worldwide. [vars. *canadense* (Miller) Torrey & A. Gray & *glabratum* (DC.) Cronq.] Highly variable; populations show founder effects.

XYLORHIZA DESERT-ASTER

Per, subshrubs, shrubs. **STS** gen white, glabrous or hairy. **LVS** simple, alternate, entire or toothed; midrib white. **INFL**: heads radiate, solitary, peduncled; involucre bell-shaped or hemispheric; phyllaries graduated in several series; receptacle convex, naked. **RAY FLS** gen many; corollas white to blue or purple. **DISK FLS** many; corolla yellow; style tips linear, acute. **FR** linear to club-shaped, weakly compressed, covered with long, appressed hairs; pappus of many, unequal bristles. (Greek: woody root) [Watson 1977 Brittonia 29:199–216]

1. Herbs or subshrubs; sts gen branched only in the basal half; peduncles 8–22 cm .... ***X. tortifolia*** var. ***tortifolia***
1′ Shrubs; sts gen branched throughout; peduncles 0–11 cm
 2. Younger sts and peduncles glandular; phyllaries loosely appressed, outermost glandular, innermost < immediately preceding series .... ***X. cognata***
 2′ Younger sts and peduncles glabrous; phyllaries tightly appressed, essentially glabrous; innermost phyllaries = or > immediately preceding series .... ***X. orcuttii***

X. cognata (H.M. Hall) T.J. Watson (p. 365) MECCA-ASTER Shrub < 1.5 m. **STS** gen short-glandular, glabrous in age. **LVS** 1–5 cm, oblanceolate to ovate, glabrous to glandular, obtuse or acute, ± spiny-dentate or entire, not reduced upward. **INFL**: peduncles 0–11 cm; phyllaries 8–19 mm, 0.8–2.2 mm wide, subglabrous to densely glandular, innermost < immediately preceding series. **RAY FLS** 20–30; tube 5–8 mm; ligule 1.8–2.5 cm, light blue. **DISK FLS** 40–80; corolla 7–9 mm. **FR** 3–4.5 mm; pappus bristles < or = 9.5 mm. 2*n*=12. RARE. Arid canyons; 20–240 m. n DSon (Riverside Co.). [*Machaeranthera c.* (H.M. Hall) Cronquist & Keck]

X. orcuttii (Vasey & Rose) E. Greene (p. 365) ORCUTT'S WOODY-ASTER Shrub < 1.5 m. **STS** glabrous (rarely puberulent below heads). **LVS** 2–6 cm, oblanceolate to oblong or lanceolate, glabrous or sparsely hairy on margins, obtuse or acute, ± spiny-dentate or entire, not much reduced upward. **INFL**: heads sessile (or peduncle < or = 11 cm); phyllaries 5–12 mm, 1.5–3.5 mm wide, glabrous, innermost = or > immediately preceding series. **RAY FLS** 25–40; tube 4–6 mm; ligule 1.2–3.2 cm, light blue. **DISK FLS** 55–140; corolla 8–10.5 mm. **FR** 3–4 mm; pappus bristles < or = 12 mm. 2*n*=12. RARE. Arid canyons; 20–300 m. s DSon (Imperial, San Diego cos.). [*Machaeranthera o.* (Vasey & Rose) Cronq. & Keck]

X. tortifolia (Torrey & A. Gray) E. Greene var. ***tortifolia*** (p. 365) MOJAVE-ASTER Per, subshrubs 2–6 dm from much-branched caudex. **STS** with long, non-glandular hairs and shorter, stalked glands. **LVS** 2.5–10 cm, linear to lanceolate, oblanceolate or elliptic, gen soft-hairy and glandular, acute to spine-tipped, ± spiny-dentate, reduced upward. **INFL**: peduncles 8–22 cm; phyllaries 5–25 mm, 0.7–2.5 mm wide, soft-hairy and glandular, innermost > or < immediately preceding series. **RAY FLS** 25–60; tube 4–6 mm; ligule 1–3.3 cm, light blue or white. **DISK FLS** 70–110; corolla 5.5–8.5 mm. **FR** 3–6 mm; pappus bristles < or = 9 mm. 2*n*=12, 24. Desert slopes, canyons; 240–2000 m. DMoj, n DSon; to sw UT, w AZ. [*Machaeranthera t.* (Torrey & A. Gray) Cronq. & Keck] ❀DRN, SUN:2,7–10,11–13,18–24.

BASELLACEAE BASELLA FAMILY

Elizabeth McClintock

Per vines, ± fleshy, glabrous. **LVS** simple, alternate, petioled, entire; stipule 0. **INFL**: raceme, panicle, or spike, terminal or axillary; pedicel subtended by bract, with 2 fused bractlets immediately subtending fl. **FL** bisexual, radial; sepals 2; petals 5, ± fused at base to form a shallow cup; stamens 5, from petal cup; ovary superior, chamber 1, styles 3, sometimes ± fused, stigmas ± spheric or elongate. **FR** indehiscent, fleshy or papery, enclosed in perianth. **SEED** 1. 4 genera, ± 20 spp.: trop, subtrop, mostly Am; some cult (*Anredera*, orn; *Basella*, edible lvs; *Ullucus*, edible tubers). [Bogle 1969 J Arnold Arbor 50: 590–598]

ANREDERA MADEIRA VINE

INFL: panicle of spike-like racemes. **FL**: calyx, corolla spreading, ± similar. ± 14 spp.: trop, subtrop Am. (Derivation unknown) [Van Steenis 1957 Fl Malesiana I 5(3):302–304]

A. cordifolia (Ten.) Steenis MIGNONETTE VINE Root tuber-like. **ST** with small axillary tubers. **LF** 2–8 cm; blade ± cordate. **INFL** ± 30 cm; fls many; pedicel < 2 mm. **FL** ± 6 mm diam; perianth white, fragrant, black in fr. **FR** not seen. Uncommon. Disturbed urban areas; < 500 m. CCo, SCo; native to S.Am. [*Boussingaultia gracilis* Miers var. *pseudobaselloides* L. Bailey] Spreads rapidly by st tubers; fr 0 in CA.

BATACEAE SALTWORT FAMILY

William J. Stone

Shrub, low, bushy, maritime, dioecious or monoecious. **LVS** opposite, simple, narrow, fleshy; stipules minute. **INFL**: spike, cone-like, small, axillary, ± sessile; stipules minute. **FLS** unisexual. **STAMINATE FLS** 8–12; fl initially enclosed in sac-like organ which may be sepals or pair of bractlets, splitting with age; perianth parts 4, alternate stamens; stamens 4, anthers opening by longitudinal slits. **PISTILLATE FLS** 4–12, not enclosed; perianth 0; ovary superior, fleshy, chambers 4, ovule 1

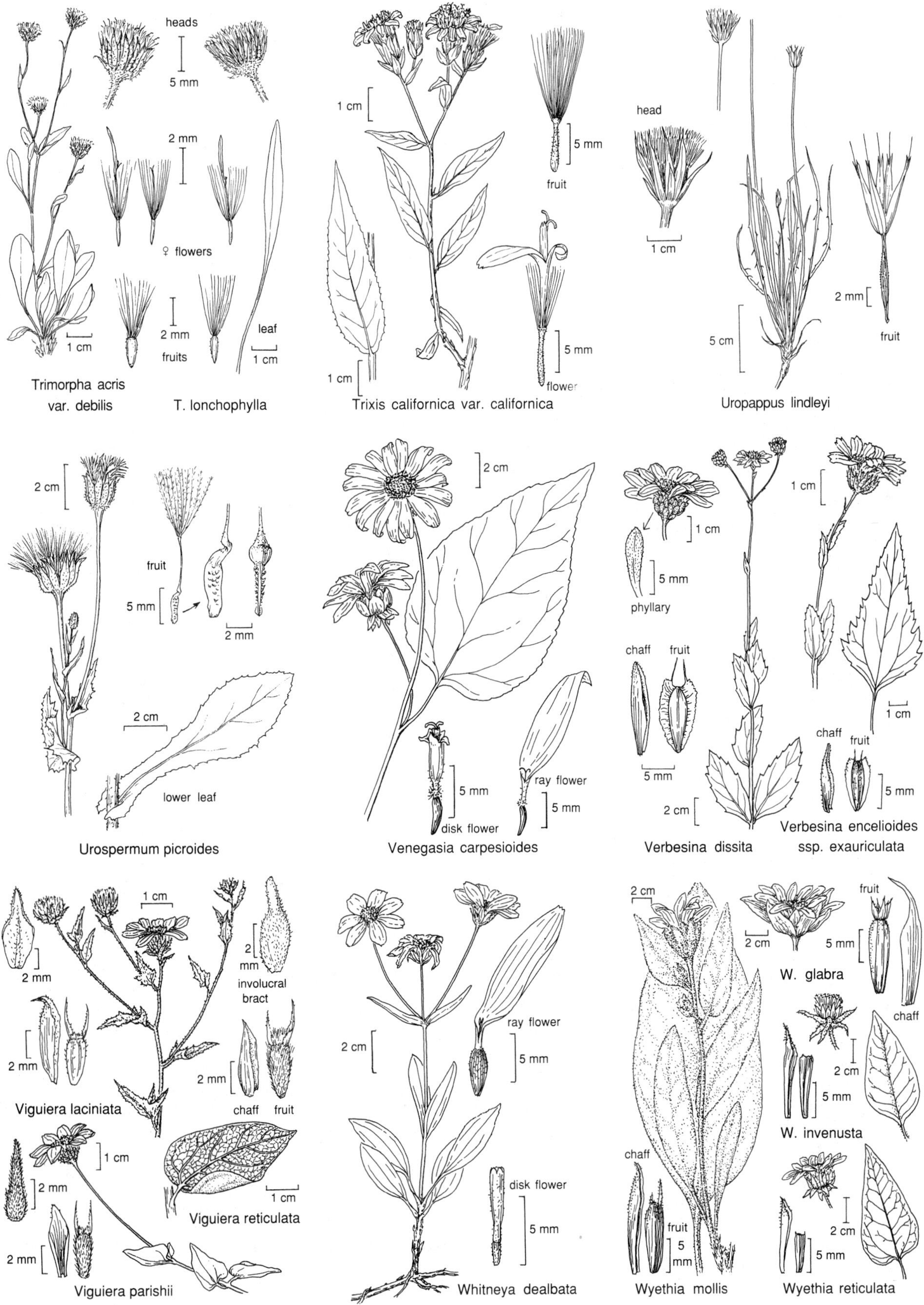
heads
5 mm
2 mm
♀ flowers
2 mm
fruits
1 cm
leaf
1 cm
Trimorpha acris var. debilis
T. lonchophylla
1 cm
5 mm
fruit
5 mm
flower
1 cm
Trixis californica var. californica
head
1 cm
2 mm
fruit
5 cm
Uropappus lindleyi
2 cm
fruit
5 mm
2 mm
2 cm
lower leaf
Urospermum picroides
2 cm
5 mm
disk flower
ray flower
5 mm
Venegasia carpesioides
1 cm
5 mm
phyllary
chaff
fruit
5 mm
2 cm
Verbesina dissita
1 cm
1 cm
chaff
fruit
5 mm
Verbesina encelioides ssp. exauriculata
2 mm
1 cm
2 mm
involucral bract
2 mm
2 mm
chaff
fruit
Viguiera laciniata
1 cm
2 mm
1 cm
Viguiera reticulata
2 mm
Viguiera parishii
2 cm
ray flower
5 mm
disk flower
5 mm
Whitneya dealbata
2 cm
2 cm
fruit
5 mm
W. glabra
chaff
2 cm
5 mm
W. invenusta
chaff
fruit
5 mm
Wyethia mollis
2 cm
5 mm
Wyethia reticulata

per chamber, stigmas 2, head-like, sessile. **FR** drupe-like, water dispersed. **SEEDS** 1–4, hard-walled. 2*n*=18,22. 1 genus, 2 spp.: trop, subtrop Am., Pacific, Australia. (Greek: name of some seashore plant) Perianth parts in staminate fls also interpreted as staminodes.

BATIS

The only genus; 2 spp. (*B. argillicola* P. Royen of New Guinea, ne Australia is monoecious)

B. maritima L. (p. 365) SALTWORT, BEACHWORT Dioecious. **STS** prostrate to ascending, < 1.5 m; base woody. **LF** 1–2 cm, ± cylindric, linear-oblanceolate. **INFL**: staminate 5–10 mm, ovoid-cylindric, bractlets rounded; pistillate < 1 cm in fr, short peduncled. **STAMINATE FL**: perianth parts white, triangular; stamens exserted. Salt marshes along coast; < 10 m. SCo; to se US, Caribbean, n S.Am, Hawaii. ❀WET:**24**;salt marsh STBL.

BERBERIDACEAE BARBERRY FAMILY

Michael P. Williams

Per, shrub, gen from rhizomes; caudex sometimes present, glabrous, glaucous, or hairy. **STS** spreading to erect, branched or not. **LVS** simple, 1–3-ternate, or pinnately compound, basal and cauline, gen alternate, deciduous or evergreen, petioled. **INFL**: gen raceme, spike, or panicle, scapose, terminal, or axillary. **FL**: sepals 6–18 or 0, gen in whorls of 3; petals gen 6, in 2 whorls of 3, or 0; stamens 6–12, free or fused at base, 2-whorled or not, anthers dehiscent by flap-like valves or longitudinal slits; ovary superior, chamber 1, ovules gen 1–10, style 1 or 0, stigma flat or spheric. **FR**: berry, capsule, or achene. 16 genera, ± 670 spp.: temp, trop worldwide; some cult (*Berberis, Epimedium, Nandina* (Heavenly bamboo), *Vancouveria*). [Ernst 1964 J Arnold Arbor 45:1–35]

1. Shrub; lvs pinnately compound, lflets leathery, gen spine-toothed; fr a berry .... **BERBERIS**
1′ Per, gen scapose, caudex short, erect; lvs 1–3-ternate, lflets herbaceous, not spine-toothed; fr a capsule or achene
  2. Infl a spike; lflets triangular to fan-shaped; perianth 0; fr an achene .... **ACHLYS**
  2′ Infl a raceme or panicle; lflets ± ovate; perianth parts 12+; fr a 2-valved capsule .... **VANCOUVERIA**

ACHLYS VANILLA LEAF, DEER FOOT

Per from scaly rhizomes; caudex short, erect. **LVS** 1–few, basal, long-petioled, ternate; lflets triangular to fan-shaped, base tapered. **INFL**: spike, ± scapose, dense, long-peduncled; lateral fls gen unisexual, terminal fls bisexual. **FL**: perianth 0; stamens gen 9, anther valves flap-like, curled inward; ovule 1, style 0, stigma ± flat, furrowed. **FR**: achene, curved, furrowed, brown to red-purple. 2 spp.: w N.Am, Japan. (Greek: thin mist or obscurity, from inconspicuous fls) [Terabayashi 1981 Bot Mag Tokyo 94:141–157]

1. Central lflet gen 6–8-lobed, 7–16 cm, 8–17 cm wide; stamens 4–6 mm; ovary 1.5–2 mm; fr brown ***A. californica***
1′ Central lflet gen 3-lobed, 4–11 cm, 4–8 cm wide; stamens 3–4 mm; ovary 1–1.5 mm; fr red-purple .... ***A. triphylla*** ssp. ***triphylla***

A. californica I. Fukuda & H.G. Baker (p. 365) Pl 3–5 dm; rhizome internodes (7.5)9–10 cm. **LF**: terminal lflet gen 7–16 cm, 8–17 cm wide, (3)6–8(12)-lobed. **INFL** (exc peduncle) < 4 cm. **FL**: stamens 4–6 mm; ovary 1.5–2 mm. **FR** 3.5–5 mm, brown. 2*n*=24. Moist, shaded sites, coniferous forest; < 1200 m. NCo, w KR, w NCoRO; to B.C. ❀TRY.

A. triphylla (Smith) DC. ssp. ***triphylla*** (p. 365) Pl 2–4 dm; rhizome internodes 2.5–5 cm. **LF**: terminal lflet gen 4–11 cm, 4–8 cm wide, 3(–5)-lobed. **INFL** (exc peduncle) 2.5–5 cm. **FL**: stamens 3–4 mm; ovary 1–1.5 mm. **FR** 3–4.5 mm, red-purple. 2*n*=12. Moist shaded sites, coniferous forest; < 1500 m. NCo, w KR, w NCoRO; to B.C. Other ssp. in Japan. ❀SHD,IRR,DRN:**4,5**,6,15,**16,17**; GRNCVR;DFCLT.

BERBERIS OREGON-GRAPE, BARBERRY

Shrub, gen from rhizomes. **STS** spreading to erect, branching, spiny or not, sometimes vine-like; inner bark, wood gen bright yellow; bud bracts deciduous or persistent. **LVS** simple or pinnately compound, cauline, alternate, deciduous or evergreen; lflets gen 3–11, ± round to lanceolate, gen spine-toothed. **INFL**: raceme, axillary or terminal. **FL**: sepals 9 in 3 whorls of 3; petals 6 in 2 whorls of 3, base gen glandular; stamens 6, anther valves pointed down to ± spreading; ovules 2–9, stigma ± spheric. **FR**: berry, spheric to elliptic, gen purple-black. ± 600 spp.: temp worldwide. (Latin: ancient Arabic name for barberry) [see Moran 1982 Phytologia 52:221–226 for relationship between *Berberis* and *Mahonia*] Roots often TOXIC; spines may inject fungal spores into skin.

1. Bud bracts persistent among upper lvs, 15–45 mm, thick, lanceolate; lflets ± palmately veined .... ***B. nervosa***
1′ Bud bracts gen deciduous, < 5 mm, thin, ovate to deltate; lflets gen pinnately veined
  2. Infl open, fls gen < 10; largest lflets < 4 cm, < 2 cm wide; petiole < 2 cm; terminal lflets gen lanceolate, oblong, or narrowly elliptic — SW, D
    3. Lflets flat to ± wavy, not folded along midrib, margin serrate, teeth gen > 8; spine tips ± 1 mm .... ***B. nevinii***
    3′ Lflets wavy, gen folded along midrib, margin ± lobed, teeth gen 3–8; spine tips 1–3 mm
      4. Terminal lflet gen ovate-lanceolate, length gen < 3 × width; fls 8–12; fr yellowish red to reddish purple .... ***B. fremontii***

4′ Terminal lflet gen narrowly lanceolate, length gen > 3 × width; fls 3–5; fr reddish brown to dark purple . . . . **B. haematocarpa**

2′ Infl dense, fls > 10; largest lflets gen > 4 cm, > 2 cm wide; petiole gen > 2 cm (0–3 cm in *B. pinnata*); terminal lflets gen ovate to widely elliptic

5. Lflet margin strongly wavy, thick, gen hard, teeth gen < 10 per margin; spine tips 2–5 mm . . . . **B. aquifolium** var. ***dictyota***

5′ Lflet margin wavy to flat, ± thin, not hard, teeth gen > 10 per margin; spine tips 1–2 mm

6. Petioles gen > 3 cm; lflets gen 5–9

7. Sts ascending to erect, gen > 1 m; lflets gen shiny above . . . . **B. aquifolium** var. ***aquifolium***

7′ Sts spreading, sometimes erect, gen < 1 m; lflets gen dull above . . . . **B. aquifolium** var. ***repens***

6′ Petioles 0 or < 3 cm; lflets gen 7–11 . . . . **B. pinnata**

8. Upper sts reclining or vine-like; pl 2–8 m . . . . ssp. ***insularis***

8′ Upper sts erect; pl gen < 2 m . . . . ssp. ***pinnata***

B. aquifolium Pursh **STS** spreading to erect, 0.1–2 m; bud scales gen deciduous. **LVS** cauline, not crowded, 8–24 cm; petiole 1–6 cm; lflets 5–9, 2–7.5 cm, 1.5–4.5 cm wide, ± round to elliptic, ± flat to strongly wavy, base slightly lobed to wedge-shaped, tip acute to obtuse (exc tooth), margin serrate, spine-tipped teeth 6–24 (40), 2–5 mm. **INFL** 3–6 cm, dense; axis internodes 2–4 mm in fl, fr. **FR** 4–7 mm diam, ovoid to obovoid, glaucous, dark blue to purple. **SEEDS** 4–5 mm. Slopes, canyons, coniferous forest, oak woodland, chaparral; < 2200 m. CA-FP, MP; to Can, w Great Plains, n Mex. Vars. intergrade; variation needs study.

var. ***aquifolium*** **STS** ascending to erect, < 2 m. **LF** 10–24 cm; petiole 1–2.5(5) cm; lflets gen 5–9, 3–7.5 cm, 2–4.5 cm wide, ovate to elliptic, ± flat, base oblique to obtuse; marginal teeth 12–24, 1–2 mm. **FR** ovoid to obovoid, dark blue. $2n=28$. Coniferous forest; 400–2100 m. KR, NCoR, CaR, SN; to B.C. [*B. piperiana* (Abrams) McMinn] ❀**4–6**&SHD:**7,15–17**&IRR:**8,9**,10,**14**,18,**19–24**;CVS incl. GRNCVR.

var. ***dictyota*** (Jepson) Jepson (p. 365) **STS** erect, < 1 m. **LF** 9–15 cm; petiole 1–5 cm; lflets gen 7–9, 2–6 cm, 2–3.5 cm wide, ± round to ovate, strongly wavy, base slightly lobed to wedge-shaped; marginal teeth 6–12, 2–5 mm. **FR** ± ovoid, blue-purple. $2n=28$. Habitat of sp.; 150–1900 m. SN, ScV (Sutter Buttes), SCoR, TR, PR. [*B. dictyota* Jepson] ❀DRN:15,**16**,17,24&SHD:**7**,8,9,**14**,18,**19–23**;CV.

var. ***repens*** (Lindley) H. Scoggan **STS** spreading to erect, gen < 0.8 m. **LVS** 8–18 cm; petiole 1–6 cm; lflets gen 5–7, 3–7 cm, 2.5–4.5 cm wide, round, widely elliptic, or ovate, ± flat, base oblique to slightly rounded; marginal teeth 11–15, 0.5–1 mm. **FR** obovoid to elliptic, blue to dark blue. $2n=28$. Habitats of sp.; 300–2200 m. KR, NCoR, CaR, SN, PR, DMoj; to B.C., Great Plains. [*B. repens* Lindley; *B. amplectens* (Eastw.) Wheeler; *B. pumila* E. Greene; *B. sonnei* (Abrams) McMinn; *Mahonia s.* Abrams Truckee barberry **ENDANGERED** CA, US] In cult.

B. fremontii Torrey (p. 365) **STS** erect, 0.1–4(5) m; bud bracts < 5 mm, gen deciduous. **LVS** 3–6 cm, crowded on short lateral sts; petiole < 1 cm; lflets 3–7(9); terminal lflet 1.5–2.5 cm, 1–1.5 cm wide, ovate, oblong, or lanceolate, wavy, gen folded along midrib, base truncate to wedge-shaped, tip gen acute, margin ± lobed, spine-tipped teeth 3–8, 2–3 mm. **INFL** 4–5.5 cm, open; axis internodes 2–10 mm, 5–10 mm in fr; fls 8–12. **FR** 6–15 mm diam, ± spheric, glaucous, yellowish or purplish red to dark purple. **SEEDS** 3–4 mm. Rocky slopes, pinyon/juniper woodland, chaparral; 900–1850 m. e&s DMoj, PR; to Colorado, NM, Mex. [*B. higginsiae* Munz; *Mahonia higginsiae* (Munz) Ahrendt, Higgin's barberry] Intergrades with *B. haematocarpa*, esp in e DMoj. ❀4–6,**7,14–16**,17,**23,24**&IRR:2,3,**8–10**,11–13,**18–22**.

B. haematocarpa Wooton **STS** erect, 0.5–4 m; bud bracts < 5 mm, gen deciduous. **LVS** 3–6 cm, crowded on short lateral sts; petiole < 1 cm; lflets 3–5; terminal lflet 3–3.5 cm, 0.8–1.2 cm wide, narrowly lanceolate, wavy, gen folded along midrib, base truncate to wedge-shaped, tip gen acuminate, margin ± lobed, spine-tipped teeth 3–8, 2–3 mm. **INFL** 2–3.5 cm, open; axis internodes 2–10 mm, 5–10 mm in fr; fls 3–5. **FR** 8–10 mm diam, ± spheric, reddish brown to dark red. **SEEDS** 3–4 mm. Rocky slopes, pinyon/juniper woodland, chaparral; 1000–1700 m. e&s DMoj; to TX, Mex. Perhaps best treated as part of *B. fremontii*. ❀4–6,**7,14–16,23,24**&IRR:2,3,**8–10**,11–13,**18–22**.

B. nervosa Pursh **STS** erect, 0.1–0.6(2) m; bud bracts 15–45 mm, lanceolate, ± leathery, persistent among upper lf bases. **LVS** ± crowded distally, 12–45 cm; petiole gen 2–7 cm; lflets 7–23, 2.5–8 cm, 1.5–3 cm wide, lanceolate to ovate, flat, base ± oblique to rounded, tip acute, serrate, bristle-tipped teeth 10–24, 1–2 mm. **INFL** 4–15 cm, ± open; axis internodes 2–8 mm, 4–10 mm in fr; fls > 20. **FR** 8–12 mm diam, ovoid to obovoid, subglaucous, blue-purple. **SEEDS** 4–6 mm. $2n=56$. Coniferous forest; < 2000 m. NW, n SNH (Sierra Co.), SnFrB, n SCoR; to B.C., ID. [*Mahonia n.* (Pursh) Nutt. var. *mendocinensis* Roof, Hardy Creek barberry] ❀SHD:**4–6**&IRR:2,3,**7**,8–10,**14–17**,18–24.

B. nevinii A. Gray (p. 365) NEVIN'S BARBERRY **STS** erect, 1–4 m; bud bracts gen deciduous. **LVS** cauline or crowded on short, lateral sts, 3.5–7(12) cm; petiole gen 0.5–2 cm; lflets gen 3–5, 2.5–4 cm, 1.2–2 cm wide, narrowly elliptic to lanceolate, flat to wavy, base ± obtuse, tip acute to acuminate, serrate, spine-tipped teeth 8–10, tips ± 1 mm. **INFL** 3.5–6.5 cm, open; axis internodes 5–10 mm in fl; fls 3–5. **FR** 5–8 mm diam, spheric, reddish. **SEEDS** 3.5–4 mm. **ENDANGERED** CA. Sandy to gravelly soils, washes, chaparral; < 650 m. SW. [*Mahonia n.* (A. Gray) Fedde] In cult.

B. pinnata Lagasca **STS** erect, reclining on tree branches, or vine-like, 0.4–8 dm; bud bracts gen deciduous. **LVS** cauline, not crowded, 9–20 cm; petiole 1–3 cm; lflets gen 7–11, 3–7 cm, 2–4.5 cm wide, ovate to widely elliptic, wavy, base slightly lobed to truncate, tip acute to obtuse exc tooth, dentate to serrate, sometimes entire, spine-tipped teeth 15–23, 1–2 mm. **INFL** 3–4 cm, dense; axis internodes 2–4 mm in fl. **FR** 6–8 mm diam, ovoid to obovoid, glaucous, blue-purple. **SEEDS** 3–4 mm. Rocky slopes, coniferous forest, oak woodland; < 1200 m. NW, CW, n ChI, WTR, SnGb, PR; to B.C., Baja CA. Relationship to *B. aquifolium* needs study.

ssp. ***insularis*** Munz (p. 365) ISLAND BARBERRY **STS** reclining or vine-like, 2–8 m. **LF**: lflets dentate, serrate, or subentire. **INFL** 3–7.5 cm. **ENDANGERED** CA. Closed-cone-pine forest; < 400 m. n ChI. [*Mahonia* p. (Lagasca) Fedde ssp. *i.* (Munz) Roof] In cult.

ssp. ***pinnata*** **STS** erect, 0.4–2 m. **LF**: lflets serrate, sometimes entire. **INFL** 3.5–7.5 cm. Habitat of sp.; < 1200 m. NW, CW, WTR, SnGb, PR; to B.C., Baja CA. Intergrades with *B. aquifolium* var. *dictyota*. ❀4–6,**17**&IRR:**7,15,16**&SHD:**8,9,14,18–24**;CVS.

VANCOUVERIA

Per from scaly rhizomes. **ST** gen erect, short, underground. **LVS** basal, long-petioled, 1–2-ternate; 1° divisions sometimes pinnately compound; lflets ovate to ± cordate, base ± lobed. **INFL**: raceme or panicle, ± scapose, open, long-peduncled; fls spreading to pendulent. **FL**: sepals gen 12–15, 8–9 mm, outer 6–9 << inner 6, bract-like, deciduous, inner petal-like, persistent, becoming reflexed; petals 6, < inner sepals, reflexed, distally glandular; stamens gen 6, anther valves flap-like, pointed tipward; ovules 2–10, style 1, < ovary, persistent, beak-like in fr, stigma cup-like. **FR**: capsule, 2-valved, gen elliptic. 3 spp.: temp w N.Am. (Captain George Vancouver, British explorer, 1757–1798) [Stearn 1938 J Linn Soc Bot 51:409–535]

1. Pedicel glabrous throughout; lflet margin thin, membranous; petiole becoming straw-colored; lvs deciduous in fr .... ***V. hexandra***
1′ Pedicel base short-glandular-hairy; lflet margin thick, whitish; petiole becoming reddish brown; lvs persistent in fr
 2. Infl gen raceme, sometimes branched below; corolla yellow; petal tip reflexed, hood-like; ovary, pedicel short glandular-hairy throughout .... ***V. chrysantha***
 2′ Infl panicle; corolla white, drying yellow; petal tip flat, notched; ovary glabrous, pedicel glabrous distally .... ***V. planipetala***

V. chrysantha E. Greene (p. 365) SISKIYOU INSIDE-OUT-FLOWER **LF** 10–18 cm; upper surface glabrous to sparsely short-hairy; lower surface sparsely hairy; petiole sparsely hairy, becoming reddish brown. **INFL**: raceme, sometimes branched below; upper axis, pedicels short-glandular-hairy. **FL**: outer sepals 2–4 mm, inner sepals 6–10 mm; petals 4–6 mm, yellow, tip strongly reflexed, hood-like; filaments glandular-hairy. **FR**: body 9–12 mm, glandular puberulent. $2n=12$. UNCOMMON. Dry sites, chaparral, coniferous forest; < 1500 m. n NCo, w KR; sw OR. ❀SHD,DRN,IRR:**16,17**; DFCLT.

V. hexandra (Hook.) Morren & Decne. (p. 365) **LF** 8–25 cm; upper surface glabrous; lower surface sparsely hairy; petiole gen glabrous, becoming straw-colored. **INFL**: raceme, sometimes branched below; axis, pedicels glabrous. **FL**: outer sepals 2–4 mm; inner sepals 5–7 mm; petals 4–6 mm, white, tip strongly reflexed, ± hood-like; filaments red-glandular; ovary glandular. **FR**: body 8–10 (15) mm, ± red-glandular-puberulent. $2n=12$. Coniferous forest; < 1700 m. NCo, w KR, n NCoRO; to w WA. ❀SHD:**4–6**&IRR:7, 14,**15–17**,19–24;slow,GRNCVR.

V. planipetala Calloni (p. 365) REDWOOD IVY **LF** 10–30 cm; upper surface glabrous; lower surface glabrous to sparsely hairy; petiole sparsely hairy, becoming glabrous, reddish brown. **INFL**: panicle; upper axis short-glandular-hairy: pedicel short-glandular-hairy in lower 1/3. **FL**: outer sepals 2–4 mm; inner sepals 4–5 mm; petals 3–4 mm, white or tinged lavender, tip flat, notched; filaments glabrous. **FR**: body 5–7 mm, glabrous. $2n=24$. Coastal coniferous forest; < 1300 m. NW (exc NCoRI), SnFrB, SCoRO (Santa Lucia Mtns); sw OR. ❀SHD:4,**5**,6&IRR:7,14,**15–17**;slow,GRNCVR.

BETULACEAE BIRCH FAMILY

John O. Sawyer, Jr.

Tree, shrub, monoecious. **ST**: trunk < 35 m; bark ± smooth; lenticels present. **LVS** simple, alternate, petioled, deciduous; stipules deciduous; blade ovate to elliptic, gen serrate, ± doubly so. **INFL**: catkin, gen appearing before lvs, often clustered; bracts each subtending 2–3 fls and 3–6 bractlets. **STAMINATE INFL** pendent, ± elongate. **PISTILLATE INFL** pendent or erect, developing variously in fr (see key to genera). **STAMINATE FL**: sepals 0–4, minute; petals 0; stamens 1–10; pistil vestigial or 0. **PISTILLATE FL**: sepals 0–4; petals 0; stamens 0; pistil 1, ovary inferior, chambers 2, each 1-ovuled, stigmas 2. **FR**: nut or nutlet, sometimes winged, subtended or enclosed by 1–2 bracts. 6 genera, 105 spp.: gen n hemisphere; some cult.

1. Lf base cordate; frs 1–2 per catkin, not winged, each enclosed in a papery involucre of 2 fused bracts **CORYLUS**
1′ Lf base ± truncate, rounded, subcordate, or tapered; frs many per catkin, winged, each subtended but not enclosed by 1 bract
 2. Pistillate catkin cone-like, with woody bracts remaining attached after fr release .... **ALNUS**
 2′ Pistillate catkin not cone-like, with papery, lobed bracts released with but not attached to fr .... **BETULA**

ALNUS ALDER

Tree, shrub. **ST**: trunk < 35 m; bark smooth, gray to brown; twigs glabrous to finely hairy, reddish gray; lenticels small; winter buds stalked, 2-scaled. **LF** glabrous to finely hairy; blade 3–15 cm, elliptic to ovate, base ± truncate to tapered, sometimes subcordate. **STAMINATE INFL** 5–20 cm; bracts each subtending 3 fls and 4 bractlets. **PISTILLATE INFL** 5–20 mm; bracts each subtending 2 fls and 4 fused bractlets. **STAMINATE FL**: sepals 4; stamens 1–4. **PISTILLATE FL**: sepals 0. **FRS** many, in cone-like catkin, bracts 3 mm, woody, winged. 30 spp.: n hemisphere, S.Am. (Latin: alder) [Furlow 1979 Rhodora 81:1–121, 151–248] Root nodules contain nitrogen-fixing bacteria; wood used for interior finishing, to smoke fish, meats.

1. Fresh lf thin, ± translucent, bright green on both surfaces, hairs on lower restricted to or denser in major vein axils; montane shrub .... ***A. viridis*** ssp. ***sinuata***
1′ Fresh lf thick, opaque, dull green on both surfaces, hairs on lower, if present, not restricted to or denser in major vein axils; coastal or lowland tree or montane shrub
 2. Lf margin tightly rolled under; coastal tree .... ***A. rubra***
 2′ Lf margin gen ± flat; lowland tree or montane shrub
  3. Lf midrib, major veins on upper surface gen indented; montane shrub, 1200–2400 m ***A. incana*** ssp. ***tenuifolia***
  3′ Lf midrib, major veins on upper surface gen not indented; gen lowland tree, 100–2400 m .... ***A. rhombifolia***

A. incana (L.) Moench ssp. ***tenuifolia*** (Nutt.) Breitung MOUNTAIN ALDER Shrub. **ST**: trunk < 10 m. **LF**: blade thick, base rounded to subcordate, tip round to acute, margin gen ± flat, upper surface dark green, dull, midrib and major veins gen indented, lower surface yellow-green. Wet places; 1200–2400 m. KR, NCoRH, CaRH, SNH; to AK, w Can, WY, NM. [*A. t.* Nutt.] 2 other sspp. in Can, AK, Eur. ❀WET&SUN:**1–3**,**7**,10,14–24;STBL.

A. rhombifolia Nutt. (p. 365) WHITE ALDER Tree. **ST**: trunk < 35 m. **LF**: blade thick, base tapered to round, tip round to acute, margin gen ± flat, upper surface green, midrib and major veins not indented, lower surface yellow-green. Along permanent streams; 100–2400 m. CA-FP (except GV), MP (uncommon); to WA, ID. ❀SUN:**1–3**,4–6,**7**,8,**9**,10,**14–18**,19,**22–24**;STBL.

A. rubra Bong. (p. 365) RED ALDER Tree. **ST**: trunk < 25 m. **LF**: blade thick, base tapered to round, tip acute, margin tightly rolled under, upper surface gray-green, midrib and major veins indented, lower surface ± gray-green, rusty-hairy, or with rusty, sessile glands. Wet places, esp after logging; 0–1000 m. NCo, NCoRO, CCo, SnFrB; to s AK, ID. [*A. oregona* Nutt.] ❀WET:**4–7**,**16**, **17**,24&SHD:2,3,14,**15**,20–23; STBL;timber.

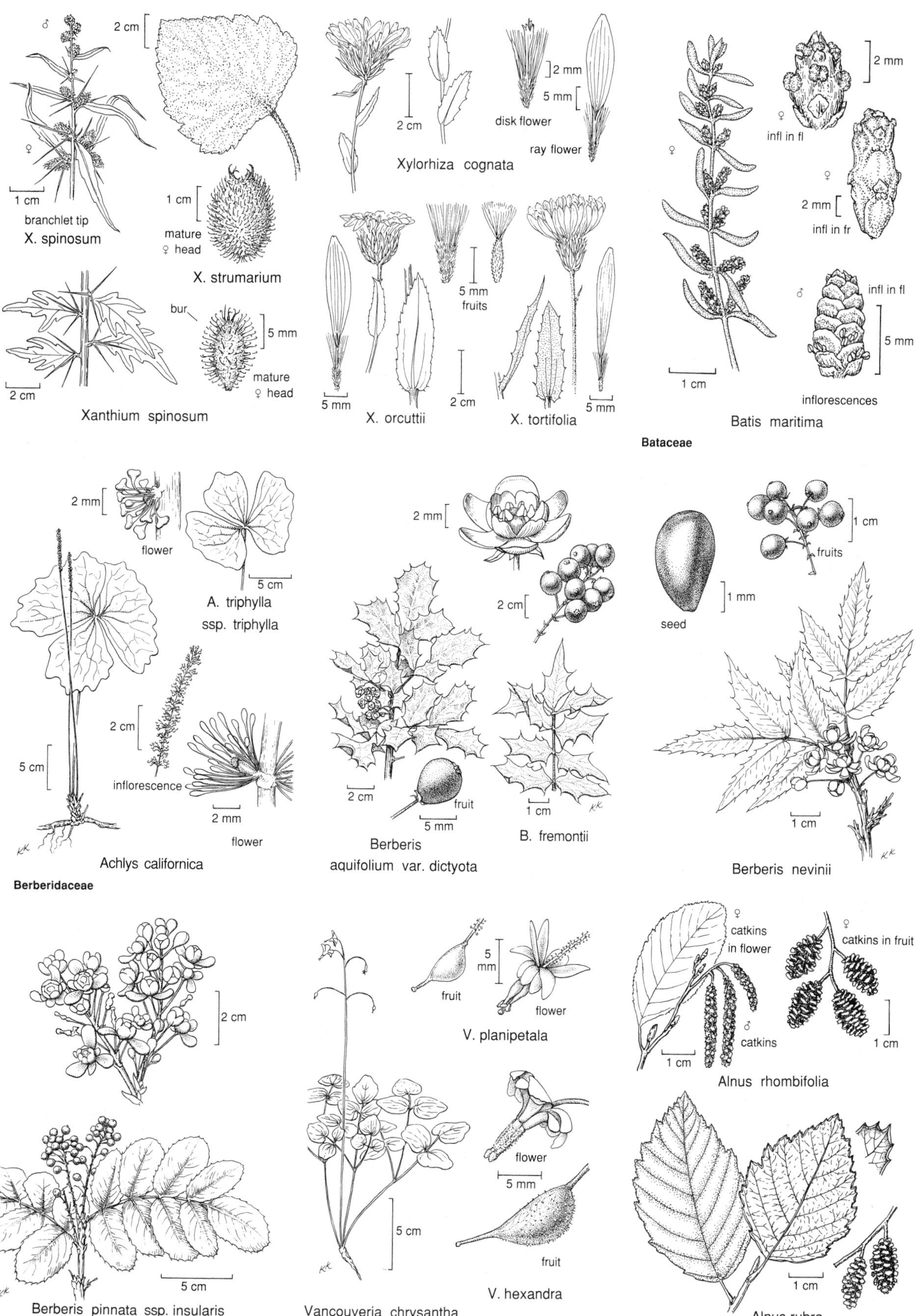
2 cm
1 cm
branchlet tip
X. spinosum
1 cm
mature
♀ head
X. strumarium
bur
5 mm
mature
♀ head
2 cm
Xanthium spinosum
2 cm
2 mm
5 mm
disk flower
ray flower
Xylorhiza cognata
5 mm
fruits
5 mm
2 cm
5 mm
X. orcuttii
X. tortifolia
2 mm
infl in fl
2 mm
infl in fr
infl in fl
5 mm
1 cm
inflorescences
Batis maritima
Bataceae
2 mm
flower
5 cm
A. triphylla
ssp. triphylla
2 cm
5 cm
inflorescence
2 mm
flower
Achlys californica
Berberidaceae
2 mm
2 cm
2 cm
fruit
5 mm
Berberis
aquifolium var. dictyota
1 cm
B. fremontii
seed
1 mm
1 cm
fruits
1 cm
Berberis nevinii
2 cm
5 cm
Berberis pinnata ssp. insularis
5
mm
fruit
flower
V. planipetala
flower
5 mm
fruit
5 cm
V. hexandra
Vancouveria chrysantha
catkins
in flower
catkins
1 cm
catkins in fruit
1 cm
Alnus rhombifolia
1 cm
Alnus rubra
Betulaceae

A. viridis (Chaix) DC. ssp. ***sinuata*** (Regel) A. Löve & D. Löve THINLEAF or SITKA ALDER Shrub. **ST**: trunks < 8 m. **LF**: blade thin, sticky when young, base tapered to subcordate, tip acute to tapered, margin ± flat, upper surface yellow-green, shiny, lower surface green, hairs restricted to or denser in major vein axils. **INFL** appearing with or before lvs. Along creeks, seeps, meadow margins; 1000–2700 m. KR, NCoRO, NCoRH; to AK, w Can [*A. s.* (Regel) Rydb.] 2 other sspp. in Can, Eurasia. ❀WET&SUN:**1–3,7**,14–15;STBL.

BETULA BIRCH

Tree, shrub. **ST**: trunk < 30 m; bark smooth or scaly, aromatic, often peeling in thin layers; twigs puberulent, glandular, or both; lenticels prominent; winter buds sessile, 3-scaled. **LF** glandular-hairy; blade 2–5 cm, widely elliptic, base ± truncate to tapered. **STAMINATE INFL** 2–7 cm; bracts each subtending 3 fls and 3 bractlets. **PISTILLATE INFL** 2–3 cm; bracts each subtending 3 fls and 3 bractlets. **STAMINATE FL**: sepals 4; stamens 2. **PISTILLATE FL**: sepals 0. **FRS** many, in a non-cone-like catkin, winged; bracts lobed, papery, released with but not attached to fr. 50 spp.: circumboreal. (Latin: birch) Important wildlife food; wood used for interior finishing; many spp. cult.

1. Pl < 3 m; lf thick, tip round, margin crenate . . . . ***B. glandulosa***
1′ Pl > 3 m; lf thin, tip acute, margin serrate . . . . ***B. occidentalis***

B. glandulosa Michaux (p. 375) RESIN BIRCH Shrub. **ST**: trunks < 2 m; bark brown to gray, not peeling; twigs waxy-gray, with resin glands. **LF**: petiole < 6 mm, hairy; blade 1–2 cm, elliptic to widely ovate, leathery, glands on both surfaces, base tapered, tip round, margin crenate. **PISTILLATE INFL** 1–3 cm; bracts resin dotted. $2n$=28. Uncommon. Streams, meadow edges; ± 2100 m. CaRH, Wrn; to AK, e N.Am. ❀WET&SUN:**1–3,7**,14–16;STBL.

B. occidentalis Hook. (p. 375) WATER BIRCH Tree, shrub. **ST**: trunks < 10 m; bark black, red-brown, not peeling; twigs with large resin glands, hairy. **LF**: petiole < 15 mm, hairy; blade 2–5 cm, widely ovate, thin, glands esp on upper surface, base ± truncate to tapered, tip acute, margin doubly serrate exc at base. **PISTILLATE INFL** 3–5 cm; bract fringed with hairs. Streamsides, springs; 600–2500 m. KR, CaRH, SNH, GB, DMtns; scattered in w N.Am. ❀WET:**1–3**,4–6,**7,9,10**,14–17,&SHD:**18,19**,20–24;STBL.

CORYLUS HAZELNUT, FILBERT

Shrub, small tree. **ST**: trunk < 6 m; bark smooth or scaly, dark brown; twigs glandular-hairy, becoming glabrous, brown; lenticels small; winter buds ciliate. **LF** hairy; blade 4–10 cm, oblong to ovate, base obliquely cordate. **STAMINATE INFL** 4–7 cm; bracts each subtending 3 fls and 3 bractlets. **PISTILLATE INFL** < 1 cm, appearing as terminal bud; bracts each subtending 2 fls and 6 bractlets. **STAMINATE FL**: sepals 0; stamens 4. **PISTILLATE FL**: sepals 4; stigmas showy, red. **FR** 2–3 cm, not winged, 1–2 per catkin, each enclosed in a papery involucre of 2 fused bracts. 15 spp.: n hemisphere. (Latin: hazelnut, filbert) Flexible sts used in basket-making; some cult as food crop.

C. cornuta Marsh var. ***californica*** (A. DC.) W. Sharp (p. 375) **ST**: trunks < 4 m. **LF**: petiole 5–10 mm; blade ± velvety-hairy, base cordate, tip acute to acuminate. **FR**: involucre vase-shaped. Common. Many habitats, esp moist, shady places; < 2100 m. NW, CaR, SN, SnFrB; to B.C. ❀4–6,17&IRR:**7,15,16,23,24**&SHD:1–3,8,**14**,18–20;STBL;edible nuts.

BIGNONIACEAE BIGNONIA FAMILY

Lawrence R. Heckard

Per to tree (many woody vines). **LVS** gen 1–3-pinnately or -palmately compound, rarely simple, gen opposite or whorled. **FL** bisexual, showy; calyx gen 5-lobed, sometimes 2-lipped; corolla funnel- or bell-shaped, 5-lobed, gen 2-lipped; stamens epipetalous, gen 4, paired, a 5th vestigial or 0; ovary superior, chambers gen 2, placentas 4, axile (or chamber 1, placentas 2–4, parietal), style long, stigma 2-lobed. **FR**: gen capsule, long, cylindric, 2-valved. **SEEDS** many, flat, gen winged. 110 genera, 800 spp.: gen trop, esp S.Am.; many orn (*Campsis*, trumpet creeper; *Catalpa*; *Jacaranda*).

CHILOPSIS DESERT-WILLOW

1 sp. (Greek: resembling lips, from fl shape) [Henrickson 1985 Aliso 11:179–197]

C. linearis (Cav.) Sweet ssp. ***arcuata*** (Fosb.) Henrickson (p. 375) Shrub or tree 1.5–7 m, willow-like. **LVS** deciduous, gen alternate (often some opposite to whorled on same pl); blade 10–26 cm, ± linear, curved. **INFL**: panicle or raceme, terminal. **FL**: calyx 8–14 mm, inflated, 2-lipped, gen soft-hairy, purplish; corolla 2–5 cm, sweetly fragrant, gen light pink to lavender with yellow ridges and purple lines on throat and lower lobes, lobes spreading, margins jagged, wavy; stamens incl; stigma lobes closing when touched. **FR** < 35 cm, linear, round in X-section. **SEED** 6–12 mm, oblong; both ends long-hairy. $2n$=40. Common. Sandy washes; < 1500 m. D, adjacent TR, PR; to UT, NM, n Mex. ❀SUN,DRN:7,**14**,15–17,24&IRR:**8,9**,10,11,**12,13,18–23**.

BORAGINACEAE BORAGE FAMILY

Ann, per, shrubs, gen bristly or sharply hairy. **ST** prostrate to erect. **LVS** cauline, often with basal rosette, gen simple, alternate; lower sometimes opposite, entire. **INFL**: cyme, gen elongate, panicle-, raceme- or spike-like, coiled in fl, gen uncoiled in fr or fls 1–2 per axil. **FLS** gen bisexual, gen radial; sepals 5, free or fused in lower half; corolla 5-lobed, gen salverform, top of tube gen appendaged, appendages 5, alternating with stamens, sometimes arching over tube; stamens 5, epipetalous;

ovary superior, gen 4-lobed, style gen entire. **FR**: nutlets 1–4, smooth to variously roughened, sometimes prickly or bristled. ± 100 genera, ± 2000 spp.: trop, temp, esp w N.Am, Medit; some cult (*Borago, Echium, Myosotis, Symphytum*). Almost all genera may be TOXIC from alkaloids or accumulated nitrates. Family description, key to genera by Timothy C. Messick.

1. Style deeply divided; stigmas 2 .... **TIQUILIA**
1′ Style simple or 0; stigma 1
 2. Style attached atop ovary, deciduous in fr .... **HELIOTROPIUM**
 2′ Style attached to receptacle, gen persistent
  3. Mature nutlets spreading widely
   4. Corolla blue to reddish purple; nutlet subspheric, prickles barbed; gen per .... **CYNOGLOSSUM**
   4′ Corolla white; nutlet gen compressed, prickles hooked at tip, not barbed; ann
    5. Sepals in fr very unequal, upper 2 >> others, partly fused, arched over 1 nutlet, ± bur-like, with 5–10 stout spines each with hooked bristles, lower 3 sepals distinct; nutlets 2 .... **HARPAGONELLA**
    5′ Sepals in fr ± equal or , if ± unequal, upper 2 > others, sepals without spines but with hooked or straight bristles, distinct, not arched over 1 nutlet; nutlets gen 4 .... **PECTOCARYA**
  3′ Mature nutlets ± erect
   6. Corolla rotate; anthers adherent to each other — stamens together conic in outline .... **BORAGO**
   6′ Corolla salverform; anthers separate
    7. Infl of 1–2 reflexed fls in each axil; calyx lobes irregularly toothed in fr .... **ASPERUGO**
    7′ Infl a panicle, raceme, or spike, coiled in fl; calyx lobes without teeth
     8. Nutlet scar flat to strongly convex, surrounded by a thick ring- or collar-like rim
      9. Corolla appendages oblong to ovate, gen above anthers; corolla throat slightly expanded above tube or corolla salverform .... **ANCHUSA**
      9′ Corolla appendages linear-lanceolate to lanceolate, at same height as anthers; corolla throat abruptly expanded above tube .... **SYMPHYTUM**
     8′ Nutlet scar flat, curved, or grooved, not surrounded by a thick, circular rim
      10. Receptacle ± flat; nutlet scar gen basal
       11. Shrubs, 1–3 m .... [2]**ECHIUM**
       11′ Ann, per
        12. Corolla white to yellow
         13. Corolla tube > lobes; fls subtended by bracts in fr .... **LITHOSPERMUM**
         13′ Corolla tube < lobes; fls not bracted .... ***Myosotis laxa***
        12′ Corolla blue to purple
         14. Corolla ± bilateral .... [2]**ECHIUM**
         14′ Corolla radial
          15. Corolla lobes erect, tips erect to slightly spreading .... **MERTENSIA**
          15′ Corolla lobes abruptly and widely spreading .... **MYOSOTIS**
      10′ Receptacle ± conic or elongate; nutlet scar gen lateral
       16. Nutlet margins prickled, prickles barbed
        17. Gen per; pedicel recurved in fr, receptacle wide, pyramidal, ± 1/2 nutlet length .... **HACKELIA**
        17′ Ann; pedicel erect in fr, receptacle, narrow, tapered, ± = nutlet .... **LAPPULA**
       16′ Nutlet margins prickled or not, prickles if present not barbed
        18. Corolla limb and tube bright yellow or orange
         19. Ann .... **AMSINCKIA**
         19′ Per .... ***Cryptantha confertiflora***
        18′ Corolla limb and tubes white to creamy yellow
         20. Nutlet wall grooved above scar, groove gen flared open to 2-forked at base; scar gen recessed or depressed .... **CRYPTANTHA**
         20′ Nutlet wall keeled above scar; scar gen elevated .... **PLAGIOBOTHRYS**

AMSINCKIA FIDDLENECK

Fred R. Ganders

Ann; hairs gen bristly, often with bulbous bases. **ST** gen erect, 2–12 dm, gen green. **LVS** basal and cauline, alternate, sessile or lower short-petioled, gen linear to narrowly lanceolate or oblong, gen ± entire. **INFL** spike-like, gen ± terminal; tip coiled. **FL** gen radial; calyx lobes 5, sometimes appearing to be 2–4 from fusion; corolla orange or yellow, limb gen with 5 red-orange marks. **FR**: nutlets erect, ± triangular, gen with oval lateral scar, gen with round or sharp tubercles. 10 spp.: w N.Am, sw S.Am, widely alien elsewhere. (W. Amsinck, patron of Hamburg Botanic Garden, early 19th century) [Ray & Chisaki 1957 Amer J Bot 44:529–554] Self-compatible; often heterostylous; large-fld taxa gen cross-pollinated, small-fld self-pollinated. Seeds and herbage TOXIC to livestock (esp cattle) from alkaloids and high nitrate concentrations. Sharp plant hairs irritate human skin.

1. Calyx lobes 2–4; corolla tube 20-veined near base; nutlet surface smooth, cobblestone-like, or round-tubercled
 2. St (exc infl) gen ± glabrous, glaucous, ± pink below; lf glaucous; nutlet surface smooth, with a longitudinal groove, scar not obvious .... ***A. vernicosa***
  3. Corolla 12–18 mm, 8–14 mm wide at top, orange; gen heterostylous; nutlet groove forked at base ... var. ***furcata***
  3′ Corolla 8–12 mm, 2–8 mm wide at top, yellow; not heterostylous; nutlet groove unforked .... var. ***vernicosa***

2′ St hairy, not glaucous, green; lf not glaucous; nutlet surface smooth, cobblestone-like, or round-tubercled, ungrooved, with lateral scar
4. Corolla 14–22 mm, 10–16 mm wide at top; heterostylous
5. Nutlet surface cobblestone-like; corolla yellow-orange; SCoR, WTR .... ***A. douglasiana***
5′ Nutlet surface smooth; corolla red-orange; w-c GV .... ***A. grandiflora***
4′ Corolla 8–16 mm, 2–10 mm wide at top; not heterostylous .... ***A. tessellata***
6. Corolla 12–16 mm, 6–10 mm wide at top; anthers not appressed to, often below stigma .... var. ***gloriosa***
6′ Corolla 8–12 mm, 2–6 mm wide at top; anthers appressed to stigma .... var. ***tessellata***
1′ Calyx lobes 5; corolla tube 10-veined near base; nutlet surface gen sharp-tubercled, often ridged
7. Nutlet 1–2 mm; lf margin finely toothed; coastal or on inland dunes .... ***A. spectabilis***
8. St erect; all calyx lobes free; nutlet 1–1.5 mm; heterostylous .... var. ***microcarpa***
8′ St gen decumbent; 2–3 calyx lobes ± half fused; nutlet 1.5–2 mm; heterostylous or not .... var. ***spectabilis***
7′ Nutlet 2–4 mm; lf margin gen entire; inland, incl dunes
9. Corolla bilateral, tube bent, limb with 2 red-orange marks; heterostylous or anthers in upper, lower groups .... ***A. lunaris***
9′ Corolla radial, tube straight, limb with 0 or 5 red-orange marks; not heterostylous
10. Corolla ± closed by hairy bulges at top of tube; stamens, style incl in tube .... ***A. lycopsoides***
10′ Corolla open; stamens, style exserted from tube, incl in throat
11. Corolla 10–20 mm, 8–14 mm wide at top; anthers not appressed to, gen below stigma ...... ***A. eastwoodiae***
11′ Corolla 4–11 mm, 2–10 mm wide at top; anthers gen appressed to stigma .... ***A. menziesii***
12. Corolla 7–11 mm, 4–10 mm wide at top, ± orange .... var. ***intermedia***
12′ Corolla 4–7 mm, 2–3 mm wide at top, pale yellow .... var. ***menziesii***

A. douglasiana A. DC. (p. 375) **LF** sometimes ± spoon-shaped. **FL**: calyx lobes 2–4; corolla 15–22 mm, 10–16 mm wide at top, yellow-orange, tube 20-veined near base. **FR** 3–5 mm; surface cobblestone-like. 2*n*=12. Uncommon. Loose, shaly slopes; 100–600 m. SCoR, WTR. Heterostylous. ❀SUN:7–9,**14**,15–17,**18–23**,24.

A. eastwoodiae J.F. Macbr. (p. 375) **FL**: calyx lobes 5; corolla 10–20 mm, 8–14 mm wide at top, orange, tube 10-veined near base; anthers not appressed to, gen below stigma, barely exserted from tube. **FR** 2.5–4 mm, tubercled, sometimes ridged. 2*n*=24. Open valleys, hills; 50–500 m. NCoRI, SNF, Teh, GV, SCoR, SW. [*A. intermedia* var. *e.* (J.F. Macbr.) Jepson & Hoover] Not heterostylous. Like large-fld pls of *A. menziesii* var. *intermedia*. ❀SUN: **7–9**,10,11,**14**,15–17,**18–23**,24.

A. grandiflora A. Gray (p. 375) LARGE-FLOWERED FIDDLENECK **LF** linear to narrowly ovate. **FL**: calyx lobes 2–4; corolla 14–20 mm, 10–15 mm wide at top, red-orange, tube 20-veined near base. **FR** 3–4 mm; surface smooth, shiny. 2*n*=12. **ENDANGERED** CA, US. Grassy slopes; < 300 m. c-w GV (near Corral Hollow, San Joaquin Co.; presumed extinct near Antioch, Contra Costa Co.). Heterostylous. Being re-established (apparently successfully) in several sites. In cult.

A. lunaris J.F. Macbr. (p. 375) BENT-FLOWERED FIDDLENECK **FL**: calyx lobes 5; corolla bilateral, 7–10 mm, 5–7 mm wide at top, orange, tube bent, 10-veined near base, limb with 2 red-orange marks. **FR** 2.5–4 mm; surface tubercled, sometimes ridged. 2*n*=8. UNCOMMON. Open woods; 50–500 m. NCoRI, w-c GV, SnFrB. Heterostylous or anthers in upper and lower group. Fl size variable.

A. lycopsoides Lehm. (p. 375) **FL**: calyx lobes 5; corolla 7–11 mm, 5–10 mm wide at top, yellow, tube 10-veined near base, ± closed by hairy bulges at top; stamens and style incl in tube. **FR** 2.5–3 mm; surface tubercled, gen not ridged. 2*n*=30. Common. Open, gen disturbed places; < 400 m. CA-FP; to B.C., ID. [*A. glomerata* Suksd.] Not heterostylous. Hybridizes with *A. menziesii* var. *intermedia*. ❀STBL;SUN:2,3,**7**,**14**,**18**.

A. menziesii (Lehm.) Nelson & J.F. Macbr. RANCHER'S FIRE-WEED **FL**: calyx lobes 5; corolla 4–11 mm, 2–10 mm wide at top, tube 10-veined near base, limb with red-orange marks or not; anthers appressed to stigma in corolla throat. **FR** 2–3.5 mm; surface tubercled, sometimes ridged. Abundant. Open, gen disturbed places; < 1700 m. CA; to B.C., ID, UT; also in S.Am; alien in e US, e hemisphere. 100+ named, mostly indistinct variants; self-pollinated; different variants may grow together and remain distinct but intergrade over their ranges. Not heterostylous.

var. ***intermedia*** (Fischer & C. Meyer) Ganders **FL**: corolla 7–11 mm, 4–10 mm wide at top, ± orange, limb gen with 5 red-orange marks. 2*n*=30,34,38. Habitat and elevation of sp. CA; to B.C., ID, Baja CA. [*A. i.* Fischer & C. Meyer, incl var. *echinata* (A. Gray) Wiggins] Hybridizes with *A. lycopsoides*; large-fld pls like *A. eastwoodiae*. ❀STBL;SUN:2,3,**7–10**,**14**,15–17,**18–23**,24.

var. ***menziesii*** (p. 375) **FL**: corolla 4–7 mm, 2–3 mm wide at top, pale yellow, limb gen without red-orange marks. 2*n*=16,26,34. Habitat and elevation of sp. CA; to B.C., ID, AZ; naturalized in c&e US; also in S.Am. [*A. helleri* Brand; *A. micrantha* Suksd.; *A. retrorsa* Suksd.]

A. spectabilis Fischer & C. Meyer **LF**: margin finely toothed. **FL**: calyx lobes 5, sometimes 2–3 ± half fused; corolla 7–19 mm, 5–14 mm wide at top, yellow, tube 10-veined near base. **FR** 1–2 mm; surface tubercled, sometimes ridged. Sandy places, incl coastal dunes; < 100 m. NCo, CCo, SCo, ChI; to B.C., Baja CA. Vars. intergrade in s CCo (near Morro Bay).

var. ***microcarpa*** (E. Greene) Jepson & Hoover (p. 375) **ST** erect. **FL**: all calyx lobes free; corolla 12–19 mm, 8–14 mm wide at top. **FR** 1–1.5 mm, not ridged; tubercles sharp. Stabilized dunes; < 100 m. s CCo (< 20 km inland from coast, San Luis Obispo, Santa Barbara cos.). Heterostylous. ❀SUN:14,15,**16**,**17**,**24**.

var. ***spectabilis*** (p. 375) **ST** gen decumbent. **FL**: 2–3 calyx lobes ± half fused; corolla 7–15 mm, 5–12 mm wide at top. **FR** 1.5–2 mm, sometimes ridged; tubercles sharp or obscure. 2*n*=10. Coastal bluffs, dunes; < 100 m. Range of sp. [*A. s.* var. *nicolai* (Jepson) Munz; *A. scouleri* I.M. Johnston] Heterostylous or not. ❀SUN:14,15,**16**,**17**,**24**.

A. tessellata A. Gray DEVIL'S LETTUCE **FL**: calyx lobes 2–4; corolla 8–16 mm, 2–10 mm wide at top, yellow or orange, tube 20-veined near base. **FR** 2.5–4 mm; surface cobblestone-like or round-tubercled, ridged or not. 2*n*=24. Common. Often disturbed places; 50–2200 m. NCoRI (Colusa Co.), GV, SnFrB, SCoR, WTR, GB, D; to WA, ID, AZ, S.Am. Not heterostylous.

var. ***gloriosa*** (Suksd.) Hoover (p. 375) **FL**: corolla 12–16 mm, 6–10 mm wide at top, orange; anthers not appressed to, gen below stigma. 2*n*=24. Sandy or shaly soils; 50–1700 m. NCoRI, ScV (Colusa Co.), SnFrB, SCoR, WTR. [*A. g.* Suksd.] ❀SUN: TRY.

var. ***tessellata*** (p. 375) **FL**: corolla 8–12 mm, 2–6 mm wide at top, yellow; anthers appressed to stigma. 2*n*=24. Rocky or sandy soils; 50–2200 m. SnJV, SnFrB, SCoR, GB, D; to WA, ID, AZ, S.Am. [*A. t.* var. *elegans* (Suksd.) Hoover]

A. vernicosa Hook. & Arn. **ST** (exc infl) gen ± glabrous, glaucous, ± pink below. **LF** glaucous. **FL**: calyx lobes 2–4; corolla 8–22 mm, 2–14 mm wide at top, yellow or orange, tube 20-veined near base. **FR** 3.5–6 mm; surface smooth, shiny, with a longitudinal groove, scar not obvious. $2n=14$. Uncommon. Loose, shaly slopes; 50–1400 m. w SnJV, SnFrB, SCoR, DMoj; to e OR. Vars. sometimes intergrade.

var. ***furcata*** (Suksd.) Hoover (p. 375) FORKED FIDDLENECK **FL**: corolla 12–22 mm, 8–14 mm wide at top, orange. **FR**: groove forked at base. $2n=14$. RARE. Habitat of sp.; 50–1000 m. w SnJV, SCoRI (San Benito Co. to Kern Co.) [*A.f.* Suksd.] Heterostylous exc at s end of range.

var. ***vernicosa*** (p. 375) **FL**: corolla 8–12 mm, 2–8 mm wide at top, yellow. **FR**: groove unforked. $2n=14$. Habitat of sp.; 50–1400 m. w SnJV, SnFrB, SCoR, DMoj; to e OR. [*A. carinata* Nelson & J.F. Macbr.] Not heterostylous.

ANCHUSA ALKANET, BUGLOSS

Dieter H. Wilken

Ann, per; hairs bristly to strigose, bases sometimes bulbous. **ST** ± erect. **LVS** basal and cauline, petioled to clasping; blades oblong to oblanceolate. **INFL** axillary or terminal, gen spike-like; tip coiled in fl. **FL**: corolla gen salverform, white to red or blue, throat appendages 5, ovate to oblong, ± puberulent; stamens incl to exserted. **FR**: nutlets 1–4, erect, ± ovoid, irregularly angled or wrinkled, scar basal or oblique, rim of scar thickened, ring-like. ± 35 spp.: Eurasia, Afr. Orn, cult for drugs, dyes. (Greek: ancient name for alkanet) [Greuter 1965 Candollea 20:192–210]

1. Corolla tube slightly curved, lobes unequal .... ***A. arvensis***
1′ Corolla tube straight, lobes equal
 2. Calyx lobes in fl > tube, linear; nutlet 5–10 mm .... ***A. azurea***
 2′ Calyx lobes in fl 0.5 × tube, lanceolate; nutlet 2–3 mm .... ***A. officinalis***

A. arvensis (L.) M. Bieb. (p. 375) Ann, bristly; hair base bulbous. **ST** 5–9 dm. **LVS**: upper 10–20 mm wide, oblong to ovate. **FL**: calyx 4–5 mm in fl, 8–10 mm in fr, lobes acuminate; corolla tube 4–7 mm, limb 4–6 mm wide, lobes unequal, blue; stamens inserted below mid tube, incl. **FR**: nutlets 2–3 mm, base 3–4 mm wide, tip erect, scar ± oblique. $2n=48$. Uncommon as waif. Disturbed areas; < 500 m. n SNF, c SCo, SnGb; widespread US; native to Eur. [*Lycopsis a.* L.]

A. azurea Miller Per (may fl 1st year), bristly. **ST** 5–8 dm. **LVS**: upper 5–15 mm wide, linear-lanceolate. **FL**: calyx 6–9 mm in fl, 8–12 mm in fr, lobes linear; corolla tube 6–10 mm, limb 10–15 mm wide, lobes equal, blue; stamens inserted on upper tube, partly exserted. **FR**: nutlet 5–10 mm, base 2–4 mm wide, angled, tip erect, scar basal. Uncommon as waif. Open sites, grassland, shrubland; < 500 m. CCo, SnFrB; widespread US; native to c&s Eur.

A. officinalis L. (p. 375) Per (may fl 1st year), bristly. **ST** 4–7 dm. **LVS**: upper 5–10 mm wide, linear-lanceolate. **FL**: calyx 5–7 mm in fl, 8–12 mm in fr, lobes lanceolate; corolla tube 5–10 mm, limb 6–12 mm wide, lobes equal, blue; stamens inserted on upper tube, partly exserted. **FR**: nutlet 2–3 mm, base 2–3.5 mm wide, angled, tip incurved, scar basal. $2n=16$. Uncommon as waif. Open sites, fields; < 1000 m. KR, CaRH; widespread US; native to Eur. [*A. procera* Link]

ASPERUGO MADWORT

Dieter H. Wilken

1 sp. (Latin: rough, from hairs)

A. procumbens L. (p. 375) Ann, bristly; hairs curved to reflexed, base bulbous. **ST** prostrate to weakly decumbent, 3–8 dm, much branched, angled. **LVS** simple, ± cauline, opposite at base, alternate above, petioled to sessile above; blades 2–7 cm, oblanceolate to elliptic. **INFL** axillary; fls 1–2; pedicels reflexed in fr. **FL**: calyx 5-lobed, 2–3 mm in fl, compressed in fr, 5–8 mm, 10–15 mm wide, papery, with 1–2 teeth per lobe, enclosing nutlets; corolla = calyx, purplish; stamens incl, anthers sessile. **FR**: nutlets 2–4, 2–3 mm, compressed, ovoid in outline, shiny, minutely tubercled, scar lateral. $2n=48$. Disturbed areas, roadsides, ditches; 1000–1500 m. MP; widespread US; native to e&c Eur.

BORAGO

Dieter H. Wilken

Ann, per, bristly to rough-hairy. **STS** erect. **LVS** cauline, ± petioled; blade ovate to oblanceolate, entire. **INFL**: cymes, terminal, 2–3-fld; pedicels ± spreading to pendent. **FL**: calyx deeply lobed; corolla rotate to bell- shaped, lobes widely spreading, throat appendages erect, glabrous; stamens strongly exserted, together conic in outline, filament base dilated, anthers adherent, enclosing style. **FR**: nutlets erect, stout, obovoid, irregulary tubercled, rim of scar thickened. 3 spp.: Eur. (Latin: ancient name of unknown origin)

B. officinalis L. (p. 375) Ann, gen branching above. **ST** ascending to erect, 2–7 dm. **LVS**: lower blades 8–20 cm, 3–8 cm wide. **FL**: calyx lobes 8–10 mm in fl, linear; corolla rotate, bright blue, lobes 8–12 mm; anthers 5–8 mm, dark brown. **FR**: nutlets 5–7 mm, ± oblong. $2n=16$. Open, often disturbed sites; < 300 m. NCo, n&c SNF, CCo, SnFrB, SCo; widespread US; native to s Eur. Orn, cult for bees (nectar source), potherb.

CRYPTANTHA

Walter A. Kelley and Dieter H. Wilken

Ann, bien, per. **ST** simple or branched; branches gen ascending to erect, hairy. **LVS** strigose, rough-hairy, or bristly, largest bristles (esp lower surface) bulbous-based; basal whorled; cauline gen opposite below, alternate above. **INFL** gen terminal, gen elongated in fr, open (fls in fr not overlapping or touching side to side) or dense (fls in fr overlapping or touching side to side). **FL**: sepals ± free; corolla gen white, tube gen 1–13 mm, appendages 5, white to yellow, limb 1–5 mm wide in ann, 6–12 mm wide in per; anthers incl; ovary gen 4-lobed, **FR**: nutlets 1–4, back gen grayish brown, smooth and shiny or granular, tubercled, or rough at 10×, margin rounded, sometimes sharp-edged, groove on inside surface narrow, open to closed, sometimes raised, edges inrolled to sharp-angled, gen forked or flared open at base. ± 160 spp.: w Am. (Greek: hidden fls, from cleistogamous fls of some spp.) [Higgins 1971 Brigham Young Sci Bull Biol Ser 13(4):1–63] Ann spp. gen self-pollinating; per spp. homostylous or heterostylous. Many ann spp. difficult to separate; observation of nutlets and hairs requires magnification at 20×.

Key to Groups

1. Bien or per, gen with persistent basal rosette or tuft of lvs; caudex present in per, gen woody (exc *C. virginensis*) ..... **Group 1**
1′ Ann, basal lvs gen not persistent, gen not green in fr; caudex 0
 2. Most fls with 1 nutlet (sometimes 2) ..... **Group 2**
 2′ Most fls with 4 nutlets (sometimes 2–3) ..... **Group 3**

Group 1: Bien or Per

1. Sepals 1–2 mm, 2–3 mm in fr; infl gen open in fr, fls pedicelled to subsessile; basal lvs not persistent, not green in fr; woody caudex 0
 2. Infl dense in fr; fls subsessile, ascending to erect in fr; nutlets gen equal ..... [4]***C. holoptera***
 2′ Infl open in fr; fls pedicelled, spreading in fr; nutlets gen of 2 kinds, 1 > other 2–3 ..... [3]***C. racemosa***
1′ Sepals 2–10 mm, 3.5–15 mm in fr; infl ± dense in fr, fls subsessile; basal lvs persistent, gen green in fr; caudex present, gen woody (not so in *C. virginensis*)
 3. Corolla tube > sepals
  4. Corolla and appendages yellow; nutlet back smooth, groove closed, groove edges ± overlapping ..... ***C. confertiflora***
  4′ Corolla white, appendages light yellow; nutlet back rough, groove open, narrowed near middle, groove edges elevated ..... ***C. flavoculata***
 3′ Corolla tube = or < sepals
  5. Sts prostrate to ascending; nutlet back clearly curved, surface smooth; groove closed, edges gen overlapping ..... ***C. cinerea*** var. ***abortiva***
  5′ Sts ± erect (gen cespitose in *C. humilis*); nutlet back straight, surface smooth or rough; groove closed or open, edges not overlapping
   6. Sts 1–2 cm, < or = basal lvs; basal lvs gen < 1 cm ..... ***C. roosiorum***
   6′ Sts > 2 cm, > basal lvs; basal lvs gen >> 1 cm
    7. Nutlet 5–6 mm; sepals in fr 9–15 mm — volcanic soils, c SNH ..... ***C. crymophila***
    7′ Nutlet < 5 mm; sepals in fr 4–11 mm
     8. Nutlet lanceolate to narrowly ovate, inner surface (between groove and margin) gen smooth, shiny
      9. Corolla tube 2–3 mm; nutlet 2–3 mm; east slope c&s SNH, W&I, n DMtns ..... ***C. nubigena***
      9′ Corolla tube 3–4 mm; nutlet 3–4 mm; KR, CaRH, MP, ne SNE (Sweetwater Mtns) ..... ***C. sobolifera***
     8′ Nutlet ovate, inner surface gen rough or wrinkled
      10. Bien or short-lived per; caudex not woody ..... [3]***C. virginensis***
      10′ Long-lived per; caudex woody
       11. Edge of nutlet groove not elevated; GB ..... ***C. humilis***
       11′ Edge of nutlet groove elevated; DMtns ..... ***C. tumulosa***

Group 2: Ann; nutlet gen 1(2)

1. Basal lvs tufted, some green and persistent in fr; st hairs soft-downy under densely spreading bristles [3]***C. virginensis***
1′ Basal lvs not tufted, gen withered in fr; st hairs strigose, rough-hairy, or bristly, not soft-downy
 2. Nutlet back granular, tubercled, or papillate, shiny or dull, not smooth
  3. Sepals 2–6 mm, 4–10 mm in fr
   4. Sepals densely white-silky-hairy, hairs appressed to ascending; nutlet back gen fine-granular to papillate, tip ± beaked; groove raised — n CaRF (Shasta, Tehama cos.) ..... ***C. crinita***
   4′ Sepals strigose to ascending-rough-hairy and spreading-bristly; nutlet back granular to tubercled, tip not beaked; groove not raised
    5. Sepals 4–6 mm, 5–10 mm in fr; corolla limb gen < 3 mm wide; appendages white to light yellow ***C. barbigera***
    5′ Sepals 2–3 mm, 3.5–5 mm in fr; corolla limb 3–6 mm wide; appendages gen bright yellow ..... ***C. intermedia***
  3′ Sepals 1.5–2.5 mm, 2–4 mm in fr
   6. Fl spreading to recurved in fr; sepal tips recurved in fr; nutlet curved ..... ***C. recurvata***
   6′ Fl ascending to ± erect in fr; sepal tips ± straight, erect in fr; nutlet ± straight
    7. Infl with some bracts lf-like, = or > fl; nutlets gen 2, 1 nutlet tubercled, other smooth or shiny ..... [2]***C. maritima***
    7′ Infl bracts 0 or << fl; nutlet gen 1
     8. Sepals ovate to elliptic in fr; nutlet margin sharp-angled to narrowly winged near tip; throat appendages gen bright yellow; style gen = nutlet ..... ***C. utahensis***
     8′ Sepals lanceolate to linear-oblong in fr; nutlet margin rounded; throat appendages white to light yellow; style < nutlet
      9. Nutlet axis ± parallel to sepals; nutlet groove expanded at base into narrow, ± flat triangle with bottom side curved inward ..... ***C. decipiens***
      9′ Nutlet axis ± perpendicular to sepals; nutlet groove expanded at base into wide, excavated triangle with 3 straight sides ..... ***C. excavata***
 2′ Nutlet back smooth throughout, shiny
  10. Sepal hairs curved to ± hooked ..... ***C. flaccida***

10′ Sepal hairs straight
11. Sepals gen soft-hairy, not bristly, hairs appressed to stiffly ascending — GB, D .... *C. gracilis*
11′ Sepals strigose and bristly, bristles reflexed, spreading, or ascending
12. Sepals 6–10 mm in fr; longest sepal bristles 3–4 mm in fr — w DSon (Borrego Valley) .... *C. ganderi*
12′ Sepals 1–6 mm in fr; longest sepal bristles < 3 mm in fr
13. Nutlet groove clearly off-center (nearer one margin than center), closed (edges overlapping), abruptly flared open at base; infls axillary and terminal, ± spheric in fl .... *C. glomeriflora*
13′ Nutlet groove ± central, open to closed, gen forked at base; infl terminal, gen elongate in fl
14. Sepals gen 3–6 mm in fr
15. Sepal tips with bristles reflexed .... [2]*C. nemaclada*
15′ Sepal tips with bristles mostly spreading or ascending
16. Sepals strigose and clearly bristly, bristles spreading to reflexed; nutlet groove flat (edges not sharp-angled); st hairs ± soft, ascending to spreading .... [3]*C. clevelandii*
16′ Sepals strigose and rough-hairy, hairs gen ascending; nutlet groove raised, edges sharp-angled; st strigose to spreading-rough-hairy .... [2]*C. milobakeri*
14′ Sepals gen 1–3 mm in fr
17. Sepals 1–2 mm in fr .... *C. microstachys*
17′ Sepals 2–3 mm in fr
18. Infl with evident bracts, some lf-like, = or > fl .... [2]*C. maritima*
18′ Infl bracts 0
19. Nutlet tip beaked; serpentine soils .... *C. hispidula*
19′ Nutlet tip acute; nonserpentine soils .... [2]*C. milobakeri*

Group 3: Ann; nutlets (2–3)4

1. Pls gen wider than tall, rounded to cushion-like — branches many, dense, spreading to ascending
2. Calyx circumscissile below middle in fr, limb and upper tube falling as 1 unit, lower part of tube scarious, persistent, cup-like in fr .... *C. circumscissa*
2′ Calyx intact in fr, not circumscissile .... [2]*C. micrantha*
1′ Pls gen taller than wide, not rounded or cushion-like
3. Basal lvs tufted, some green, persistent in fr; st hairs soft-downy under densely spreading bristles [3]*C. virginensis*
3′ Basal lvs not tufted, gen withered in fr; st hairs strigose, rough-hairy, or bristly, not soft-downy
4. Nutlet margin clearly sharp-angled (in part or throughout; acute in X-section) or winged
5. Nutlet back smooth or minutely rippled to weakly tubercled, shiny
6. Nutlet back smooth
7. Corolla limb 3–5 mm wide .... [2]*C. mohavensis*
7′ Corolla limb 1–2 mm wide .... [2]*C. watsonii*
6′ Nutlet back minutely rippled to weakly tubercled
8. Nutlets gen of 2 kinds, 1 gen > other 2–3 .... [3]*C. racemosa*
8′ Nutlets gen equal
9. Nutlet inner surface smooth, flat; groove edges not elevated .... *C. costata*
9′ Nutlet inner surface weakly tubercled, not flat; groove edges gen elevated .... [4]*C. holoptera*
5′ Nutlet back clearly tubercled to granular, dull or shiny
10. Nutlet margin sharp-angled, not clearly winged (exc sometimes at tip in *C. racemosa*)
11. Nutlets gen equal; infl dense in fr; fls subsessile in fr .... [4]*C. holoptera*
11′ Nutlets gen of 2 kinds, 1 gen > other 2–3; infl open in fr; fls pedicelled in fr .... [3]*C. racemosa*
10′ Nutlet margin clearly winged (1 nutlet gen not winged in *C. pterocarya*)
12. Nutlets gen of 2 kinds, 3–4 winged, 1 sometimes not winged; nutlet wing gen lobed at tip, 0.5–1 mm wide; sepals ovate to lanceolate and 4–6 mm in fr, densely strigose, bristles few or 0 .... *C. pterocarya*
12′ Nutlets all winged; nutlet wing not lobed, < 0.5 mm wide; sepals oblong to linear and 2.5–4 mm in fr, bristly to rough-hairy, hairs spreading to ascending
13. Corolla limb 1–2 mm wide; nutlet margin evenly and narrowly winged, back sparsely tubercled, groove opened widely below middle; e DMoj, DSon .... [4]*C. holoptera*
13′ Corolla limb 4–6 mm wide; nutlet margin narrowly winged near base, widely winged above middle, back fine-tubercled to white-granular; groove narrowly open at base; s CA-FP, w DMoj ...... *C. oxygona*
4′ Nutlet margin gen rounded throughout, not sharp-angled or winged
14. Nutlets clearly of 2 kinds, 1 gen > other 2–3 (back surface also different from other 3 in *C. micromeres*)
15. Calyx ± spheric in fr, sepals 1–1.5 mm in fr; nutlets deltate to ovate, shiny, 3 sparsely tubercled, 1 smooth; s CA-FP .... *C. micromeres*
15′ Calyx ± oblong to lanceolate in fr, sepals 2–4 mm in fr; nutlets lanceolate to narrowly ovate, ± dull, all white-granular; SNE, D
16. Distal st part ascending to erect, hairs appressed and spreading; infl gen dense, fls spreading to ascending in fr .... *C. angustifolia*
16′ Distal st part spreading, prostrate, or sprawling over rocks, other pls, hairs appressed to ascending; infl open, fls ± appressed in fr .... *C. dumetorum*
14′ Nutlets gen equal, back surface similar
17. Infls axillary and terminal, fls ± evenly distributed along st
18. Style > nutlet; sepals 1–3 mm in fr; infl clearly bracted — bracts gen = sepals .... [2]*C. micrantha*
18′ Style = or < nutlet; sepals 4–6 mm in fr; infl bracts 0 or sparse

19. Infl bracts 0; nutlet densely papillate ..... [2]*C. echinella*
19′ Infl bracts sparse; nutlet tubercled to sparsely papillate ..... [4]*C. hooveri*
17′ Infl mostly terminal; lower fls sometimes axillary, subtended by reduced lvs
20. Nutlet back smooth, shiny
21. Nutlet groove clearly off-center, ± curved parallel to nearest margin ..... *C. affinis*
21′ Nutlet groove central, straight
22. Nutlet margin ± angled (obtuse in X-section), esp near tip
23. Corolla limb 3–5 mm wide ..... [2]*C. mohavensis*
23′ Corolla limb 1–2 mm wide ..... [2]*C. watsonii*
22′ Nutlet margin gen rounded throughout, sometimes angled below middle
24. Infl gen dense, most fls touching in fr
25. Style = or > nutlets; main sts gen prostrate to decumbent in fr; fls gen subtended by bracts ..... *C. leiocarpa*
25′ Style < nutlets (gen 2/3 nutlet length); main st, branches ascending to erect; fl bracts gen 0
26. Nutlets oblong-lanceolate, nutlet back rounded ..... [3]*C. clevelandii*
26′ Nutlets oblong-ovate, nutlet back ± flat ..... [2]*C. torreyana*
24′ Infl gen open in fr, most fls not touching in fr
27. Sepals ± 2 mm, 2.5–4 mm in fr, gen rough-hairy, hairs ascending, spreading bristles gen 1–4 per sepal; nutlet grayish — s SN (Tulare Co.) ..... *C. incana*
27′ Sepals 1.5–3 mm, 3–6 mm in fr, strigose and bristly, spreading bristles > 5 per sepal; nutlet brownish or mottled
28. Nutlet base ± truncate, groove flared open at base, sometimes forked; — serpentine soils, n&c SN ..... [3]*C. mariposae*
28′ Nutlet base ± rounded, groove ± closed throughout, ± forked at base
29. Nutlet back flat, ± triangular in X-section ..... [2]*C. torreyana*
29′ Nutlet back flat to rounded, not triangular in X-section
30. Sepal tip bristles spreading to ascending ..... [3]*C. clevelandii*
30′ Sepal tip bristles reflexed ..... [2]*C. nemaclada*
20′ Nutlet back papillate, granular, or tubercled, shiny or dull
31. Nutlet ± 1.5 mm
32. Nutlet back granular to tuberculate throughout; SnJV ..... [4]*C. hooveri*
32′ Nutlet back fine-granular to smooth below middle, fine-granular to minutely tubercled above middle; s ChI (San Clemente, San Nicolas islands) ..... *C. traskiae*
31′ Nutlet gen > 1.5 mm
33. Corolla limb 3–6 mm wide, appendages bright yellow (white to light yellow in *C. mariposae*)
34. Nutlet beaked, base ± truncate; corolla limb ± 3 mm wide ..... [3]*C. mariposae*
34′ Nutlet tapered (tip acute or acuminate), base ± rounded; corolla limb gen > 3 mm wide
35. Style = or < nutlet ..... [2]*C. intermedia*
35′ Style > nutlet ..... [2]*C. muricata*
33′ Corolla limb gen < 3 mm wide, appendages white to light yellow
36. Most st hairs ascending or appressed, spreading hairs 0 or few
37. Nutlets lanceolate, tip tapered, back densely tubercled; sepals 6–10 mm in fr ..... *C. nevadensis*
37′ Nutlets ovate to deltate, tip acuminate, back tubercled (also fine-granular in *C. simulans*); sepals 3–7 mm in fr
38. Infl dense in fr, fls gen clustered near axis tip
39. Nutlet back strongly tubercled; SnJV, < 150 m ..... [4]*C. hooveri*
39′ Nutlet back sparsely low-tubercled and densely fine-granular; mtns > 650 m ..... [2]*C. simulans*
38′ Infl open in fr, fls well spaced on axis
40. Nutlet back densely high-tubercled, tubercle tips translucent; style ± > nutlet; c DMoj (near Barstow) ..... [2]*C. clokeyi*
40′ Nutlet back sparsely low-tubercled, densely fine-granular; style = or < nutlet; mtns > 650 m ..... [2]*C. simulans*
36′ Most st hairs spreading, sometimes also strigose
41. Nutlet back densely-fine-papillate ..... [2]*C. echinella*
41′ Nutlet back tubercled (sometimes weakly so in *C. ambigua, C. mariposae*)
42. Nutlet beaked, base ± truncate, back gen brown — serpentine soils, n&c SN ..... [3]*C. mariposae*
42′ Nutlet tapered to tip, acute, or acuminate, base ± rounded, back grayish brown to light gray
43. Sepals 1–2 mm, 2–4 mm in fr ..... [2]*C. muricata*
43′ Sepals 2–4 mm, 4–7 mm in fr
44. Style > nutlet; nutlet ± 3 mm; c DMoj (near Barstow) ..... [2]*C. clokeyi*
44′ Style = or < nutlet; nutlet 1–2.5 mm; CaR, n&c SN, SnJV, GB
45. Nutlet back weakly tubercled to sparsely papillate; CaR, n&c SN, GB ..... *C. ambigua*
45′ Nutlet back tubercled; SnJV ..... [4]*C. hooveri*

C. affinis (A. Gray) E. Greene Ann 5–40 cm. **ST** simple or few branched throughout, gen strigose; some hairs curved upward. **LVS** 1–4 cm, oblanceolate to oblong, short-bristly; upper oblong-lanceolate. **INFL** open in fr. **FL**: sepals 1.5–2 mm, 2.5–4 mm and ± lanceolate in fr, bristles spreading; corolla limb 1–2 mm wide. **FR**: nutlets 4, ± 2 mm, ovate, back smooth, shiny, groove ± closed, off-center near margin, sometimes short-forked at base. Open areas, gen coniferous forest; 750–2900 m. KR, NCoR, CaR, SN, SnBr, PR; to WA, WY.

C. ambigua (A. Gray) E. Greene Ann 10–35 cm. **ST** branched throughout, strigose and spreading-hairy. **LF** 1–4 cm, linear to oblong, strigose to bristly; bristles ascending to spreading. **INFL** dense to open below in fr. **FL**: sepals 2–3 mm, 4–7 mm and linear in fr, densely strigose and spreading-bristly; corolla limb 3–4 mm wide. **FR**: nutlets 4, 1.5–2.5 mm, ± ovate, back tubercled, groove ± closed, flared open or forked at base. Open, sandy or gravelly areas, sagebrush, scrub, coniferous forest; 1300–2400 m. CaR, n&c SN, GB; to B.C., MT, Colorado.

C. angustifolia (Torrey) E. Greene Ann 5–60 cm. **ST** simple to branched throughout, strigose and rough-hairy to bristly; hairs spreading. **LF** 1–4 cm, linear to oblong; bristles spreading. **INFL** dense to open below in fr. **FL**: sepals 1–2 mm, 2.5–4 mm and ± linear in fr, densely spreading-bristly. **FR**: nutlets 4 (1 slightly > other 3), ± 1 mm, lanceolate to narrowly ovate, back brown, white-granular, dull, groove ± open throughout to flared open at base. Sandy to rocky soils, creosote scrub, desert woodland; < 1400 m. SNE, D; to TX, n Mex. [*C. inaequata* I.M. Johnston]

C. barbigera (A. Gray) E. Greene Ann 10–50 cm. **ST** simple to branched gen below middle, strigose and densely spreading-bristly. **LF** 1–5(8) cm, linear-oblong to narrowly lanceolate; bristles spreading. **INFL** gen dense to open in fr. **FL**: sepals 4–6 mm, 5–10 mm and narrowly oblanceolate to oblong in fr, gen spreading-bristly esp at base; corolla tube 3–5 mm, limb 1–2 mm wide. **FR**: nutlets 1–4, 1.5–2 mm, lanceolate to ovate, back white-tubercled, groove ± open, esp at forked base, to slightly flared open at base. Open, sandy to rocky soils; 300–2250 m. s SN, Teh, SNE, D; to UT, NM, Baja CA.

C. cinerea (E. Greene) Cronq. var. ***abortiva*** (E. Greene) Cronq. (p. 375) BOWNUT CRYPTANTHA Per 3–20 cm; caudex woody. **ST** branched from base, prostrate to ascending, ± strigose. **LF** 1.5–9 cm, linear to linear-oblanceolate; upper surface ± strigose; lower surface also bulbous-based bristly. **INFL** gen dense, cylindric; branches 3–many, elongate and straight in fr. **FL**: sepals 2–4 mm, 4–7 mm in fr, hairs gen soft, bristles few; corolla tube 2–4 mm, limb 5–9 mm wide, throat gen yellow. **FR**: nutlets 1–4, 1.5–2.5 mm, ± ovate, slightly curved inward, back smooth, groove closed, edges overlapping. Common. Sandy soils; 1800–3300 m. SNE, DMtns; s NV. [*C. jamesii* (Torrey) Payson var. *a.* (E. Greene) Payson] Other vars. in Great Basin, Rocky Mtns. ❀TRY,DFCLT.

C. circumscissa (Hook. & Arn.) I.M. Johnston (p. 379) Ann < 10 cm, cushion-like; taproot red, purple when dry. **ST** much-branched throughout, strigose and bristly or rough-hairy; hairs gen ascending. **LF** 0.3–1.5 cm, linear to narrowly oblanceolate, bristly to rough-hairy; hairs ± ascending. **INFL** axillary or in branch forks; fls 1–5 per cluster, dense in fr. **FL**: sepals 1.5–2 mm, 2.5–4 mm and circumscissile below middle in fr, hairs ± like lvs; corolla limb 1–6 mm wide, appendages gen yellow. **FR**: nutlets 3–4, ± 1.5–2 mm, ovate, back gen smooth, shiny, gen mottled gray and brown, groove ± open below middle, forked at base. $2n$=24,36. Sandy soils; 300–3700 m. SN, s SnJV, SCoRI (se Monterey Co.), e SCo, TR, e PR, GB, D; to WA, Colorado; also in s S.Am. [vars. *hispida* (J.F. Macbr.) I.M. Johnston, *rosulata* J. Howell] Pls from n SnGb, w SnBr, sw DMoj with corolla limb 3.5–6 mm wide, throat yellow, $2n$=12, have been called *C. similis* K. Mathew & Raven. Corolla limb width intergrades in some places.

C. clevelandii E. Greene Ann 10–60 cm. **ST** simple to branched throughout, strigose and ± spreading-soft-hairy. **LF** 1–5 cm, ± linear, gen appressed-hairy, sometimes also fine-bristly. **INFL** gen dense; lowest fls sometimes not touching in fr. **FL**: sepals 1.5–2.5 mm, 3–6 mm and linear in fr, strigose and bristly, distal bristles spreading, lower bristles reflexed; corolla limb 1–6 mm wide. **FR**: nutlet 1(2–4), ± 1 mm, lanceolate to ovate, back flat, smooth, shiny, gen mottled gray-brown, groove ± closed, sometimes open only at forked base. Sandy or rocky (sometimes serpentine) soils, chaparral; < 1100 m. s NCo, s NCoRI, CW, SCo, ChI; Baja CA. [var. *florosa* I.M. Johnston] Pls on serpentine from s NCoRI with corolla limb 4–6 mm wide have been called var. *dissita* (I.M. Johnston) Jepson & Hoover.

C. clokeyi I.M. Johnston (p. 379) CLOKEY'S CRYPTANTHA Ann 8–15 cm. **ST** gen branched throughout, mostly strigose. **LF** 0.5–3 cm, linear-lanceolate to oblong, stiffly strigose to spreading-rough-hairy; hair bases gen bulbous. **INFL** gen dense above middle; lower fls not touching in fr. **FL**: sepals 2–3 mm, 5–7(10) mm and ± linear in fr, densely strigose and spreading-bristly; corolla limb 1–2 mm wide. **FR**: nutlets 4, 2–2.5 mm, ± ovate, back, inside surface tubercled, tubercle tips translucent, groove open, flared at base, edge slightly elevated. RARE. Sandy or gravelly soils; 800–900 m. c DMoj (near Barstow). [*C. muricata* (Hook. & Arn.) Nelson & J.F. MacBr. var. *c.* (I.M. Johnston) Jepson]

C. confertiflora (E. Greene) Payson YELLOW CRYPTANTHA Per 13–440 cm; caudex woody. **ST** simple, erect, tomentose below, strigose and sparsely bristly above. **LF** 3–12 cm, ± oblanceolate; upper surface strigose; lower surface also with bulbous-based bristles. **INFL** gen dense, ± head-like at tip; axillary clusters below. **FL**: sepals 6–10 mm, 9–14 in fr, strigose and spreading-bristly; corolla yellow, tube 9–13 mm, limb 8–12 mm wide. **FR**: nutlets gen 4, 3–4 mm, ovate to deltate, back smooth, groove closed, edges ± overlapping. Common. Dry, rocky soils; 1200–2700 m. SNE, DMtns; to sw UT, ne AZ. Heterostylous. ❀TRY;DFCLT.

C. costata Brandegee RIBBED CRYPTANTHA Ann 10–20 cm. **ST** branched throughout, densely strigose and spreading-bristly. **LF** 1–3(4) cm, ± linear, bristly; longest bristles spreading to ascending. **INFL** dense in fr. **FL**: sepals 3–4 mm, 4–6 mm and ± oblong in fr, hairs like lvs; corolla limb 1–2 mm wide. **FR**: nutlets gen 4, 1.5–2 mm, lanceolate, back smooth to minutely rippled, inner surface smooth, flat, margin sharp-angled, groove gen open, flared at base, edge flat. UNCOMMON. Sandy soils, creosote scrub; < 500 m. e DMoj, DSon; AZ, Baja CA.

C. crinita E. Greene (p. 379) SILKY CRYPTANTHA Ann 10–30 (40) cm. **ST** branched gen throughout, strigose and rough-hairy; hairs appressed to ascending. **LF** 1–3 cm, narrowly oblanceolate to oblong, rough-hairy to bristly; hairs appressed to ascending. **INFL** dense in fr. **FL**: sepals 2–4 mm, 4–6 mm and linear-oblong in fr, densely soft-hairy, hairs appressed to ascending; corolla tube ± 3 mm, limb 2–3 mm wide. **FR**: nutlet gen 1, ± 3 mm, lanceolate to ovoid, back finely granular, ± shiny, tip ± beaked, groove raised, edges sharp-angled, flared open at base. RARE. Sandy stream-banks, gravel bars; 150–300 m. n CaRF (Shasta, Tehama cos.).

C. crymophila I.M. Johnston SUBALPINE CRYPTANTHA Per 13–30 cm; caudex woody. **ST** simple, erect, stiffly bristly. **LF** 0.4–1 cm, oblanceolate to spoon-shaped, appressed-stiff-bristly; lower surface bristles bulbous-based. **INFL** dense, ± head-like at tip and axillary clusters below, gen not elongate in fr. **FL**: sepals 4–5 mm, 9–15 mm in fr, spreading-bristly; corolla tube 3–5 mm, limb 4–8 mm wide, appendages yellow. **FR**: nutlets 4, 5–6 mm, ± ovoid, inner surface gen smooth, back low-ridged to smooth, groove open. UNCOMMON. Rocky volcanic soils; 2600–3200 m. n&c SNH (Alpine, Tuolumne cos.).

C. decipiens (M.E. Jones) A.A. Heller (p. 379) Ann 10–40 cm. **ST** simple to branched throughout, strigose to rough-hairy or bristly; hairs appressed to spreading. **LF** 0.5–5 cm, linear to narrowly lanceolate; bristles appressed to spreading. **INFL** dense to open in fr; lower fls not touching, appressed to ascending in fr. **FL**: sepals 1.5–2.5 mm, 2–4 mm and linear-oblong in fr, densely bristly to rough-hairy, hairs ascending to spreading; corolla limb 1–5 mm wide. **FR**: nutlets 1–2, 1.5–2 mm, lanceolate, gen white-granular throughout, sometimes smooth, shiny, white-granular near tip, groove closed to ± open, sometimes flared open at base, raised, edges sharp-angled. Open, sandy areas, grassland, shrubland; < 1500 m. Teh, s SnJV, SCoR, WTR, DMoj, n DSon; to sw UT, AZ. Pls from w of D, with corolla limb 2–3 mm wide and st gen hairs gen appressed, have been called *C. corollata* (I.M. Johnston) I.M. Johnston. Pls from n SCoR, e of Monterey Bay, with corolla limb 3–4 mm wide and st hairs gen spreading have been called *C. rattanii* E. Greene, Rattan's cryptantha.

C. dumetorum (E. Greene) E. Greene (p. 379) Ann 10–60 cm. **ST** branched throughout; branches prostrate to ascending, gen sprawling over rocks, other pls, ± fragile, strigose to appressed-stiff-hairy. **LF** 0.5–3 cm, linear to lanceolate; bristles ascending to ± appressed. **INFL** axillary and terminal, open in fr; fls ± appressed. **FL**: sepals ± 1 mm, 2–3 mm and oblong in fr, gen bristly, some bristles > sepal; corolla limb 0.5–1 mm wide. **FR**: nutlets 4 (1 gen > other 3), 2–2.5 mm, lanceolate to narrowly ovate, back gen white-granular, groove ± open throughout, not gen flared or forked at base. Sandy or gravelly soils, gen under other pls; 200–1500 m. SNH (e slope), TR (n slope), SNE, DMoj, n DSon; to sw UT.

C. echinella E. Greene Ann 5–35 cm. **ST** simple or branched throughout, strigose and spreading-bristly. **LF** 0.5–5 cm, oblong to oblanceolate, strigose and bristly; bristles ascending. **INFL** dense to open. **FL**: sepals 2–3 mm, 5–6 mm and ± linear in fr, strigose and spreading-bristly; corolla limb 1–1.5 mm wide. **FR**: nutlets gen 4, 1.5–2 mm, ovate, back densely fine-papillate, groove closed to ± open, gen forked at base. Open, gravelly or rocky soils, woodland, coniferous forest; 700–2800 m. SN, TR, GB, n DMtns; to OR, ID.

C. excavata Brandegee (p. 379) DEEP-SCARRED CRYPTANTHA Ann 5–30 cm. **ST** gen branched throughout; branches gen few, strigose and rough-hairy, hairs appressed to spreading. **LF** 0.6–2 cm, linear to narrowly oblong, rough-hairy to bristly; lower surface bristles bulbous-based, hairs appressed to spreading. **INFL** open in fr. **FL**: sepals < 2 mm, 2–2.5 mm and lanceolate in fr, densely soft-hairy, spreading bristles 1–2 or 0. **FR**: nutlet 1, perpendicular to sepal axis, tip gen exserted between sepals, ± 2 mm, ovate, back sparsely low-tubercled, papillate to granular between tubercles, groove open below middle, widely triangular, edges ± inrolled. UNCOMMON. Sandy, gravelly soils, dry streambanks, woodland; 100–500 m. s NCoRI (Colusa, Lake, Yolo cos.).

C. flaccida (Lehm.) E. Greene Ann 5–50 cm. **ST** branched above to throughout; branches ascending, strigose. **LF** 0.5–4(5) cm, linear, oblong, or narrowly oblanceolate; bristles ascending to spreading; lower surface bristles sometimes bulbous-based. **INFL** gen open in fr; upper fls sometimes touching. **FL**: sepals 1.5–3 mm, 3–4 mm in fr, linear to narrowly oblong, densely spreading-bristly, bristles gen curved to ± hooked (esp at sepal tips); corolla limb 1–4 mm wide. **FR**: nutlet 1, 1.5–2.5 mm, lanceolate, back smooth, shiny, groove closed, edges overlapping to ± open, esp at base. Open, dry sites; < 2000 m. CA-FP; to WA, ID. Pls from SnJV foothills, with ± compressed nutlets, groove ± closed, have been called *C. sparsiflora* (E. Greene) E. Greene. Pls from n CA-FP with nutlets ± plump, groove ± closed have been called *C. rostellata* (E. Greene) E. Greene incl var. *spithamea* (I.M. Johnston) Jepson.

C. flavoculata (Nelson) Payson (p. 379) Per 5–35 cm; caudex woody. **ST** simple, erect; bristles ± soft, spreading. **LF** 3–11 cm, linear-oblanceolate to spoon-shaped; upper surface densely strigose and with bulbous-based bristles or only silky-strigose; lower surface strigose and with bulbous-based bristles or only silky-strigose. **INFL** dense, cylindric or ± head-like, elongate in fr. **FL**: sepals 5–7 mm, 8–10 mm in fr, spreading-bristly; corolla tube 7–12 mm, limb 7–12 mm wide, appendages yellow. **FR**: nutlets 4, 2.5–4 mm, lanceolate to ± ovate, back rough, groove open, narrowed near middle, edges elevated. Common. Loose soils; 1300–3200 m. SNE, DMtns; to ID, Colorado. Heterostylous. ❀TRY;DFCLT.

C. ganderi I.M. Johnston (p. 379) GANDER'S CRYPTANTHA Ann 10–40 cm. **ST** simple to branched throughout, strigose and rough-hairy to bristly; hairs spreading. **LF** 1–3(4) cm, linear to narrowly lanceolate; bristles spreading, some bulbous-based. **INFL** open in fr. **FL**: sepals 3–4 mm, 6–10 mm and linear-oblong in fr, densely strigose to rough-hairy, midvein densely spreading-long-bristly; corolla limb 1–2 mm wide. **FR**: nutlets 1–2, 2.5–3 mm, lanceolate, back smooth, shiny, with a faint longitudinal ridge, gen mottled gray-brown, groove ± open, clearly forked at base. RARE. Open, sandy soils, creosote scrub; < 400 m. w DSon (Borrego Valley); nw Mex.

C. glomeriflora E. Greene Ann < 15 cm. **ST** gen branched throughout, gen strigose. **LF** 0.5–2 cm, linear to oblong-lanceolate, strigose to appressed-bristly; some bristles bulbous-based. **INFL** axillary and terminal, dense, spheric in fl, open, elongate in fr. **FL**: sepals 1.5–2 mm, 2–2.5 mm and linear-lanceolate in fr, strigose at base, ascending-bristly near tip; corolla limb gen < 1 mm wide. **FR**: nutlet gen 1, 1.5–2 mm, ± ovate, back smooth, shiny, groove closed, abruptly flared open at base, edges strongly overlapping esp near base. Open slopes, meadows; 1800–3400 m. SNH (e slope), W&I.

C. gracilis Osterh. Ann 10–35 cm. **ST** simple or branched throughout; branches decumbent to erect, strigose and rough-hairy to short-bristly, hairs spreading. **LF** 1–3.5 cm, linear to narrowly oblanceolate, densely bristly. **INFL** gen dense in fr; lowest fls sometimes not touching. **FL**: sepals < 2 mm, 2–3 mm and lanceolate to narrowly ovate in fr, gen soft-hairy, hairs appressed to stiffly ascending; corolla limb 1–2 mm wide. **FR**: nutlet gen 1, ± 2 mm, lanceolate to narrowly ovate, back smooth, shiny, groove ± closed, ± flared open at base, edges ± overlapping. Sandy to rocky soils, slopes; 900–2150 m. GB, e DMoj; to OR, ID, Colorado.

C. hispidula Brand Ann 10–30(50) cm. **ST** simple or few-branched above middle; branches ascending, strigose to spreading rough-hairy. **LF** 0.5–2 cm, linear to narrowly oblong, appressed-hairy to sparsely fine-bristly. **INFL** gen open in fr. **FL**: sepals 1.5–2 mm, 2–3 mm and ± linear in fr, rough-hairy to bristly, hairs spreading to ascending; corolla limb 2–3 mm wide. **FR**: nutlet gen 1, 1.5–2 mm, lanceolate to ovate, back smooth, shiny, gen mottled gray-brown, groove ± open, short-forked at base. Uncommon. Serpentine soils; 200–1000 m. KR (near Peanut, Trinity Co.), s NCoRI (Colusa, Lake, Napa cos.). Like *C. microstachys*.

C. holoptera (A. Gray) J.F. Macbr. (p. 379) WINGED CRYPTANTHA Ann (per), 10–50 cm. **ST** gen branched throughout, strigose and bristly; bristles spreading to ascending, some bulbous-based. **LF** 1–3.5(5) cm, linear, narrowly lanceolate, or oblong, short-bristly; bristles spreading. **INFL** gen dense in fr; lowest fls not touching. **FL**: sepals < or = 2 mm, 2.5–3(4) mm and linear-oblong in fr, densely bristly to rough-hairy, hairs spreading; corolla limb 1–2 mm wide. **FR**: nutlets 4, ± 1–2 mm, subequal, ovate to triangular, back and gen inner face sparsely tubercled, margin gen narrowly winged, groove opened widely below middle, edge gen elevated. UNCOMMON. Sandy to rocky soils; 100–1200 m. e DMoj, DSon; w NV, AZ. [*C. inaequata* I.M. Johnston]

C. hooveri I.M. Johnston (p. 379) HOOVER'S CRYPTANTHA Ann 5–20 cm. **ST** simple or branched at base; branches gen erect, strigose. **LF** 1–2.5 cm, linear, gen folded to inrolled, strigose to ascending-hairy. **INFL** dense in fr, gen head-like; fls gen clustered near axis tip. **FL**: sepals 2.5–4 mm, 4–6 mm and linear in fr, strigose and spreading-bristly; corolla limb gen < 1 mm wide. **FR**: nutlets 4, 1–1.5 mm, ovate to deltate, tip acuminate, back tubercled, shiny, groove closed distally, gen flared open at base. UNCOMMON. Sandy soils; < 150 m. n&c SnJV.

C. humilis (E. Greene) Payson (p. 379) Per 5–30 cm, gen cespitose; caudex woody. **ST** simple, erect, spreading bristly. **LF** 1.5–5 cm, oblanceolate to spoon-shaped, strigose to ± tomentose and bristly; bristles appressed, bulbous-based. **INFL** dense, cylindric. **FL**: sepals 2.5–5 mm, 5–10 mm in fr; corolla limb 4–10 mm wide, appendages white to yellow. **FR**: nutlets 1–4, 2.5–4.5 mm, ovate, back rough, groove open, narrow to triangular, edges not elevated. Common. Gravelly soils; 1700–3600 m. GB; to ID, Colorado. [*C. alpicola* Cronq.] Variable; further study needed. ❀TRY;DFCLT.

C. incana E. Greene Ann 15–50 cm. **ST** branched throughout, thinly strigose to appressed-rough-hairy; some hairs ascending to spreading. **LF** 1–3.5 cm, lanceolate to oblong; upper surface ± appressed-hairy; lower surface bristly, some bristles bulbous-based. **INFL** open in fr. **FL**: sepals ± 2 mm, 2.5–4 mm and oblong in fr, rough-hairy, hairs ascending, bristles 2–4 per sepal, spreading; corolla limb 3–4 mm wide. **FR**: nutlets gen 4, sometimes 2–3, 1–1.5 mm, lanceolate to ovate, back smooth, shiny, grayish, groove ± closed, flared open or forked at base, scar open. Open, gravelly or rocky areas; 1800–2000 m. s SN (Tulare Co.).

C. intermedia (A. Gray) E. Greene (p. 379) Ann 10–60 cm. **ST** branched throughout; branches gen ascending, strigose and rough-hairy to bristly, hairs spreading. **LF** 1.5–5 cm, linear to lanceolate, spreading-bristly. **INFL** open in fr. **FL**: sepals 2–3 mm, 3.5–5 mm and linear-oblong in fr, rough-hairy and bristly, bristles ascending to spreading; corolla limb 3–6 mm wide, throat appendages bright yellow. **FR**: nutlets 1–4, 1.5–2 mm, widely lanceolate to ovate, back gen tubercled to sparsely rough-granular, shiny, groove ± open, esp at forked base. Sandy to rocky soils, oak woodland, coniferous forest; 300–2800 m. $2n$=24. CA (exc coast, GV, D); to B.C., ID, Baja CA. [*C. hendersonii* (Nelson) Piper]

C. leiocarpa (Fischer & C. Meyer) E. Greene Ann 5–30 cm. **ST** branched from base; branches erect, gen becoming prostrate or decumbent, gen strigose, sometimes sparsely rough-hairy, hairs spreading. **LF** 1–3.5 cm, linear to narrowly oblanceolate, strigose to rough-hairy or bristly; hairs ascending; some bristles bulbous-based. **INFL** ± dense in fr, gen bracted; axillary fls 1–3. **FL**: sepals 1.5–2 mm, 2–3 mm and linear-lanceolate in fr, densely strigose and rough-hairy to bristly, hairs ascending; corolla limb 1–2.5 mm

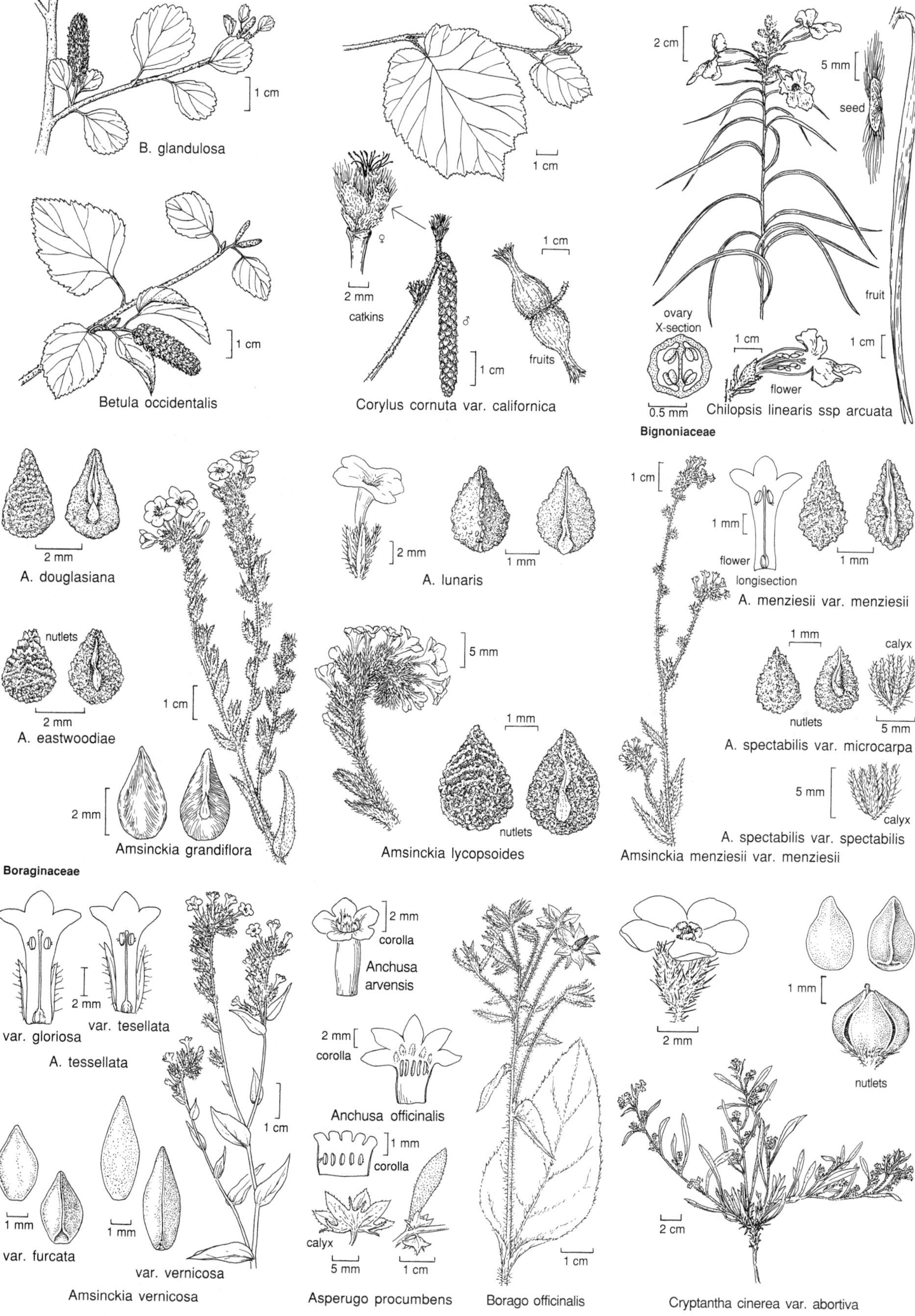

1 cm
B. glandulosa
1 cm
Betula occidentalis
1 cm
♀
2 mm
catkins
♂
1 cm
1 cm
fruits
Corylus cornuta var. californica
2 cm
5 mm
seed
fruit
ovary
X-section
1 cm
1 cm
flower
0.5 mm
Chilopsis linearis ssp arcuata
Bignoniaceae
2 mm
A. douglasiana
nutlets
2 mm
A. eastwoodiae
1 cm
2 mm
Amsinckia grandiflora
Boraginaceae
2 mm
1 mm
A. lunaris
5 mm
1 mm
nutlets
Amsinckia lycopsoides
1 cm
1 mm
flower
longisection
1 mm
A. menziesii var. menziesii
1 mm
calyx
nutlets
5 mm
A. spectabilis var. microcarpa
5 mm
calyx
A. spectabilis var. spectabilis
Amsinckia menziesii var. menziesii
2 mm
var. gloriosa
var. tesellata
A. tessellata
1 cm
1 mm
var. furcata
1 mm
var. vernicosa
Amsinckia vernicosa
2 mm
corolla
Anchusa
arvensis
2 mm
corolla
Anchusa officinalis
1 mm
corolla
calyx
5 mm
1 cm
Asperugo procumbens
1 cm
Borago officinalis
2 mm
1 mm
nutlets
2 cm
Cryptantha cinerea var. abortiva

wide. **FR**: nutlets 3–4, 1.5–2 mm, lanceolate to narrowly ovate, back smooth, shiny, gen mottled, groove ± closed, short-forked at base. Sandy soils, dunes; < 200 m. NCo, CCo, n SCo; s OR.

C. mariposae I.M. Johnston (p. 379) MARIPOSA CRYPTANTHA Ann 10–25 cm. **ST** simple or branched gen below middle; branches few, ascending, strigose to rough-hairy; some hairs spreading. **LF** 1–2 cm, ± oblong, short-bristly; bristles gen ascending. **INFL** dense to ± open below middle in fr; lowest fls sometimes not touching. **FL**: sepals 2–3 mm, 4–6 mm and linear in fr, strigose, densely spreading-bristly; corolla limb 2–3 mm wide. **FR**: nutlets gen 4, sometimes 2–3, ± 2 mm, ± ovate, back smooth to weakly tubercled, shiny, tip beaked, base ± truncate, groove closed above, ± flared open or forked at base. UNCOMMON. Serpentine soils; 200–650 m. n&c SN (Calaveras, Mariposa, Tuolumne cos.).

C. maritima (E. Greene) E. Greene (p. 379) Ann 10–30(40) cm. **ST** simple to branched throughout; branches ascending to erect, gen strigose and spreading-rough-hairy. **LF** 1–4 cm, linear to narrowly oblanceolate, appressed bristly to rough-hairy, some or all hairs bulbous-based. **INFL** gen dense in fr; lowest fls sometimes not touching, bracted. **FL**: sepals 1.5–2 mm, 2–3 mm and linear in fr, densely rough- to long-soft-hairy, hairs ascending, sometimes spreading-bristly. **FR**: nutlets 1–2, 1.5–2 mm, lanceolate, smooth, shiny (if 2, 1 with back smooth, shiny, other fine-granular), groove ± closed to near base, edges overlapping. Sandy to gravelly soils; < 1500 m. SW, D; AZ, Baja CA. [var. *pilosa* I.M. Johnston]

C. micrantha (Torrey) I.M. Johnston (p. 379) Ann < 15 cm, rounded to cushion-like or taller than wide; taproot red, gen purple when dry. **ST** branched throughout; branches spreading to erect, strigose. **LF** < 1 cm, linear to narrowly oblanceolate, short-bristly; hairs ± ascending. **INFLS** axillary and terminal, gen dense in fr, bracted. **FL**: sepals ± 1–1.5 mm, ± 2 mm and linear-oblong in fr, hairs like lvs; corolla limb 0.5–3.5 mm wide. **FR**: nutlets gen 4, ± 1 mm, ± lanceolate, back white-granular to tubercled, sometimes smooth, shiny, groove gen flared open only at base. $2n=24$. Sandy soils; 200–2300 m. c&s SN (e slope), SW (exc ChI), SNE, D; to UT, TX, n Mex. [var. *lepida* (A. Gray) I.M. Johnston]

C. micromeres (A. Gray) E. Greene (p. 379) Ann 10–50 cm. **ST** gen branched throughout, short-rough-hairy; hairs gen only spreading. **LF** 1–4 cm, linear to oblong, short-bristly. **INFL** dense to open in fr; lower fls gen not touching. **FL**: sepals < 1 mm, 1–1.5 mm and narrowly lanceolate in fr, bristly, bristles spreading to ascending; corolla limb 0.5–1 mm wide. **FR**: nutlets 4 (1 slightly > other 3), < 1 mm, back gen shiny, brown, odd nutlet back smooth, other 3 fine-tubercled, groove flared open at base, edges raised. Open places, chaparral, woodland; < 950 m. SNF, SnJV, CW, SW; Baja CA.

C. microstachys (A. Gray) E. Greene Ann 10–50 cm. **ST** simple to few-branched above middle, strigose and rough-hairy to bristly; hairs spreading to ascending. **LF** 0.5–4 cm, linear to oblong, gen ascending bristly. **INFL** open in fr. **FL**: sepals < 1 mm, 1–2 mm and ± linear in fr, rough-hairy to bristly, hairs spreading to ascending; corolla limb 0.5–1 mm wide. **FR**: nutlet 1, lanceolate, back smooth, shiny, groove ± closed, gen short-forked and open at base. Open sites, chaparral, woodland; 200–1700 m. NCoRI, s SN, SCoR, SW (exc ChI); Baja CA.

C. milobakeri I.M. Johnston Ann 15–40 cm. **ST** gen branching throughout, strigose and spreading-rough-hairy. **LF** 0.5–3 cm, densely rough-hairy to bristly; hairs ascending. **INFL** gen dense in fr; lowest fls sometimes not touching. **FL**: sepals ± 2 mm, 3–4 mm and linear-oblong in fr, densely rough-hairy and bristly, hairs gen ascending; corolla limb 2–4 mm wide. **FR**: nutlet 1, 1.5–2 mm, ovate, back smooth, shiny, groove ± closed exc forked base, edges of groove raised, sharp-angled. Rocky or gravelly soils, gen coniferous forest; < 300–1500 m. NW, CaR, n SNH; s OR.

C. mohavensis (E. Greene) E. Greene Ann 10–40 cm. **ST** simple to branched throughout, rough-hairy; hairs ascending to spreading. **LF** 0.5–4 cm, linear to oblong, densely bulbous-based bristly. **INFL** gen open in fr. **FL**: sepals ± 2 mm, 3–5 mm and lanceolate in fr, densely rough-hairy and sparsely spreading-bristly; corolla limb 3–5 mm wide. **FR**: nutlets 4, ± 2 mm, widely lanceolate to ovate, back smooth, shiny, margin sharp-angled (esp above middle), groove ± closed, 2-forked at base, scar open, ± triangular. Open areas; 600–2800 m. c&s SNH (e slope), Teh, w DMoj.

C. muricata (Hook. & Arn.) Nelson & J.F. Macbr. (p. 379) Ann 10–100 cm. **ST** branched throughout or above middle, thinly strigose and spreading-soft- to rough-hairy. **LF** 0.5–4 cm, ± linear, rough-hairy to bristly; some bristles bulbous-based. **INFL** gen open in fr; upper fls sometimes touching or overlapping. **FL**: sepals 1–2 mm, 2–4 mm and ± lanceolate in fr, strigose and spreading-bristly; corolla limb 2–6 mm wide. **FR**: nutlets 4, 1–2 mm, ovate to deltate, back tubercled, groove ± closed, forked to flared open at base. Sandy or gravelly, open areas; 150–2700 m. CA-FP (exc NCo, KR); Baja CA. [vars. *denticulata* (E. Greene) I.M. Johnston, *jonesii* (A. Gray) I.M. Johnston]

C. nemaclada E. Greene (p. 379) Ann 10–35 cm. **ST** simple or branched above middle; branches gen few, ascending, strigose to ascending-rough-hairy. **LF** 0.5–2.5(3) cm, linear to oblong, rough-hairy to bristly; hairs ascending. **INFL** open in fr. **FL**: sepals 2–3 mm, 3–6 mm and linear-oblanceolate in fr, bristles below middle spreading to reflexed, bristles near tip fine, reflexed. **FR**: nutlets 1–2(4), 2–2.5 mm, lanceolate, back smooth, shiny, groove ± closed throughout or slightly flared open at base. Slopes, shaded sites, chaparral, woodland; 200–1000 m. s NCoRI, n Teh, SnFrB, SCoRI, n WTR (ne Ventura Co.).

C. nevadensis Nelson & Kenn. Ann 10–60 cm. **ST** simple to branched throughout, strigose; some hairs ascending. **LF** 1–4(5) cm, linear to oblong, gen bristly; bristles ± ascending. **INFL** gen dense in fr; lowest fls sometimes not touching. **FL**: sepals 3–3.5 mm, (4)6–10 mm and linear in fr, densely rough-hairy and bristly, hairs ascending, bristles spreading; corolla limb 1–2 mm wide. **FR**: nutlets 4, 2–2.5 mm, lanceolate, back densely tubercled, groove ± closed, forked or flared open at base. Sandy, gravelly soils; < 2100 m. s SNF, Teh, SnJV, SCoR, D; to s UT, Baja CA. [var. *rigida* I.M. Johnston] Pls from Ne DMoj, with sepals ± 4–6 mm in fr and nutlet back ± fine-tubercled have been called *C. scoparia* A. Nelson, desert cryptantha.

C. nubigena (E. Greene) Payson Per 2–30 cm; caudex gen woody. **ST** simple, erect, spreading-bristly. **LF** 2–5 cm, linear-oblanceolate to spoon-shaped; upper surface spreading-bristly; lower surface bristles bulbous-based. **INFL** dense, ± head-like to cylindric, not elongated in fr. **FL**: sepals 2–3 mm, 4–7 mm in fr, strigose and spreading-bristly; corolla tube 2–3 mm, 4–7 mm in fr, limb 3.5–4 mm wide, appendages gen yellow. **FR**: nutlets gen 4, 2–3 mm, lanceolate to narrowly ovate, back rough, inner surface gen smooth, groove ± closed. Common. Volcanic gravel, talus; 2400–3800 m. c&s SNH (e slope), W&I, n DMtns. ❀TRY;DFCLT.

C. oxygona (A. Gray) E. Greene Ann 10–40 cm. **ST** branched throughout, strigose and sparsely rough-hairy; hairs spreading. **LF** 0.5–3(4) cm, narrowly oblanceolate to oblong, bristly; bristles ascending. **INFL** gen dense in fr; lowest fls not touching. **FL**: calyx ± 2 mm, 3–4 mm and narrowly lanceolate in fr, rough-hairy to few-bristled, hairs gen ascending; corolla limb 4–6 mm wide. **FR**: nutlets 4, ± 2 mm, ± ovate, back finely tubercled to white-granular, sometimes mottled gray-brown, margin winged, groove narrowly open at base. Open sites, slopes; < 1950 m. c&s SNF, SNH (e slope), Teh, SnJV, SCoRI, SW (exc ChI), w DMoj.

C. pterocarya (Torrey) E. Greene (p. 379) Ann 10–40(50) cm. **ST** gen branched throughout; branches few, strigose and rough-hairy, hairs gen ascending. **LF** 0.5–5 cm, linear to oblong, bristly; bristles ascending. **INFL** gen open in fr; lowest fls not touching. **FL**: sepals 2–2.5 mm, 4–6 mm and widely lanceolate to ovate in fr, densely strigose to rough-hairy, hairs ascending, bristles few or 0. **FR**: nutlets 3–4, ± 2 mm, winged or 3 clearly winged, 1 not, back tubercled to white-granular, groove flared open to 2-forked at base; winged nutlet widely ovate to ± round, wing gen minutely white-lobed; wingless nutlet narrowly ovate. Sandy to gravelly soils; < 2500 m. s SN, s SnJV, e PR, GB, D; to WA, ID, Colorado, TX. Pls with all 4 nutlets winged are called var. *cycloptera* (E. Greene) J.F. Macbr. [var. *purpusii* Jepson]

C. racemosa (S. Watson) E. Greene (p. 379) Per 20–100 cm, sometimes rounded, sometimes shrubby, gen fl first year. **ST** branched intricately throughout; branches spreading to ascending;

distal branches ± wiry, canescent and sparsely spreading-bristly. **LF** 0.5–3.5 cm, narrowly oblanceolate to linear, bristly. **INFL** open in fr; fls in fr pedicelled. **FL**: sepals 1–2 mm, 2–3 mm and narrowly lanceolate in fr, bristly in lower 1/2, bristles spreading to reflexed; corolla limb 0.5–1 mm wide. **FR**: nutlets gen 4 (1 gen slightly > other 2–3), ± 1–1.5 mm, lanceolate-ovate, back sparsely tubercled to white-granular, margin sharp-angled, minutely winged at tip, groove opening widely, scar triangular. Rocky slopes, washes, canyons; < 1500 m. SnBr (n slope), PR, s SNE, D; s NV, AZ, Baja CA.

C. recurvata Cov. (p. 379) Ann 5–35 cm. **ST** branched from base or throughout, strigose. **LF** 1–2(3) cm, linear to narrowly oblanceolate, with many bulbous-based bristles. **INFL** open in fr; fls spreading to recurved. **FL**: sepals 1.5–2 mm, 2–3.5 mm and curved in fr, densely spreading-bristly, hairs collectively light yellow-brown; corolla limb 1–2 mm wide. **FR**: nutlet 1–2 (if 2, 1 > other), 1.5–2 mm, lanceolate, curved, back fine-granular, brown, groove ± closed, flared open at base. Sandy to rocky soils; 700–2500 m. SNE, DMoj; to OR, ID, Colorado.

C. roosiorum Munz BRISTLECONE CRYPTANTHA Per 1–2 cm, cespitose; caudex woody. **ST** simple, erect, barely > basal lvs, spreading-soft-bristly. **LF** 0.5–1.2 cm, oblanceolate to spoon-shaped, densely strigose to tomentose; bristles ± appressed, bulbous-based. **INFL** dense, ± head-like. **FL**: sepals 2.5–3 mm, 3.5–4.5 mm in fr, densely strigose and spreading-bristly; corolla tube ± 2.5–3 mm, limb 4.5–5.5 mm wide, appendages yellow. **FR**: nutlets gen 4, 2.5–3 mm, lanceolate-ovate, back rough, groove ± open, ± triangular. **RARE** CA. Rocky soils, high ridges; > 3000 m. W&I (n Inyo Mtns).

C. simulans E. Greene Ann 5–40 cm. **ST** branched throughout, strigose and rough-hairy; hairs gen ascending. **LF** 0.5–3.5 cm, linear to narrowly oblanceolate, rough-hairy to bristly; hairs ascending to ± appressed. **INFL** gen open; fls ascending to erect. **FL**: sepals 2.5–3 mm, 3–6 mm and linear-lanceolate in fr; corolla limb 1–2 mm wide. **FR**: nutlets gen 4, ± 2 mm, ovate, tip acuminate, back densely granular, sparsely tubercled, groove ± closed, forked at base. Open sites, coniferous forest; 650–2300 m. KR, NCoRH, CaR, SN, TR, PR, GB; to WA, ID.

C. sobolifera Payson (p. 379) Per 5–20 cm; caudex woody. **ST** sparsely to densely bristly. **LF** 1–4 cm, oblanceolate to spoon-shaped, strigose to tomentose and bristly; lower surface bristles bulbous-based. **INFL** dense, head-like to cylindric. **FL**: sepals 3–4 mm, 6–9 mm in fr, densely bristly; corolla tube 3–4 mm, limb 4–9 mm wide, throat yellow. **FR**: nutlets 2–4, 3–4 mm, lanceolate, back rough to smooth, inner surface gen smooth, groove closed to ± open. Common. Volcanic soils, pumice; 1300–3100 m. KR, CaRH, MP, ne SNE (Sweetwater Mtns); to e OR, MT, nw NV. [*C. schoolcraftii* Tiehm, *C. subretusa* I.M. Johnston, Mount Eddy cryptantha] Variable; pls in ne SNE like *C. nubigena*; needs study.

C. torreyana (A. Gray) E. Greene (p. 379) Ann 10–40 cm. **ST** simple or branched throughout, weakly strigose to rough-hairy or bristly; hairs ascending to spreading. **LF** 0.3–3(4) cm, linear to oblanceolate, strigose to rough-hairy, sometimes bristly; hairs ascending. **INFL** open in fr. **FL**: sepals 2–2.5 mm, 3.5–6 mm and linear-lanceolate in fr, strigose and spreading-bristly; corolla limb 1–2 mm wide. **FR**: nutlets 4, 1.5–2 mm, ± ovate, back flat, smooth, shiny, gen mottled, groove closed, slightly raised, forked and closed at base. Open areas, slopes, gen coniferous forest; 350–2000 m. KR, NCoR, CaRF, SN, GB; to B.C., MT, Colorado. [var. *pumila* (A.A. Heller) I.M. Johnston]

C. traskiae I.M. Johnston TRASK'S CRYPTANTHA Ann 5–20 cm. **ST** gen branched throughout, gen only strigose; few hairs ascending. **LF** 0.5–2 cm, linear to narrowly oblong, strigose to rough-hairy; hairs ascending to curving outward. **INFL** gen dense above middle in fr; lower fls not touching. **FL**: sepals ± 2 mm, 3–4 mm in fr, densely bristly, bristles spreading to ascending; corolla limb 1–2 mm wide. **FR**: nutlets 4, ± 1.5 mm, ovate, back dull below middle, finely tubercled to granular above middle, mottled gray-brown, groove ± closed, flared open to forked at base. Rocky, open sites; < 400 m. RARE. s ChI (San Clemente, San Nicolas islands).

C. tumulosa (Payson) Payson NEW YORK MOUNTAINS CRYPTANTHA Per 7–25 cm; caudex woody. **ST** simple, erect, densely stiff- to soft-hairy. **LF** 3–6 cm, oblanceolate to ± spoon-shaped, tomentose to bulbous-based bristly. **INFL** cylindric, not gen elongated in fr. **FL**: sepals 3.5–5 mm in fl, 7–10 mm in fr, linear-lanceolate, densely bristly; corolla tube 3.5–4.5 mm, limb 6–8 mm wide, appendages yellow. **FR**: nutlets 1–3, 3–4.5 mm, ovate, both surfaces rough, back ridged down middle, groove open-triangular, edges elevated. UNCOMMON. Gravel or clay, granitic or limestone soils; 1400–2100 m. n&e DMtns; sw NV (Spring Mtns). Like per form of *C. virginensis*.

C. utahensis (A. Gray) E. Greene Ann 10–30 cm. **ST** branched throughout; branches gen ascending, strigose. **LF** 0.3–3(5) cm, linear to oblong, ± appressed-bristly; some bristles on lower surface bulbous-based. **INFL** gen dense in fr; lower fls not touching. **FL**: sepals 2–2.5 mm, 2.5–3 mm and elliptic to ovate in fr, densely strigose to rough-hairy near margin, hairs ascending; corolla tube 2–2.5 mm, limb 2–4 mm wide, throat appendages gen bright yellow. **FR**: nutlets 1(2), ± 2 mm, lanceolate, ± 3-sided, margin sharp-angled to narrowly winged distally, back finely tubercled to white-granular, groove flared open at base. Sandy to gravelly soils, creosote scrub, pinyon/juniper woodland; < 2000 m. s SNE, D; to sw UT, AZ.

C. virginensis (M.E. Jones) Payson (p. 379) Ann to short-lived Per 10–40 cm; caudex not woody. **ST** simple, erect, downy-hairy under densely spreading, ± stiff hairs. **LF** 2–12 cm, oblanceolate to spoon-shaped, strigose to tomentose and bulbous-based bristly. **INFL** dense, gen cylindric, gen elongated in fr. **FL**: sepals 3–5 mm, 5–11 mm in fr, densely bristly; corolla tube 3–5 mm, limb 5–10 mm wide, appendages yellow. **FR**: nutlets 1–4, 2.5–4.5 mm, ovate, back rough, gen ridged down middle, groove open, edges elevated. Common. Loose soils; 1900–3100 m. W&I, DMtns; to sw UT, nw AZ. Ann or bien pls from high W&I, with 1 st and infl not elongated in fr, have been called *C. hoffmannii* I.M. Johnston. ❀TRY;DFCLT.

C. watsonii (A. Gray) E. Greene Ann 10–40 cm. **ST** branched throughout, thinly strigose, spreading-rough-hairy. **LF** 0.5–3.5 cm, ± linear, densely strigose to appressed-rough-hairy above; lower surface bristly. **INFL** dense to open in fr. **FL**: sepals ± 2 mm, 2.5–3.5 mm and lanceolate in fr, densely strigose, spreading bristly; corolla limb 1–2 mm wide. **FR**: nutlets 4, 1.5–2 mm, lanceolate to narrowly ovate, back smooth, shiny, margin sharp-angled, groove ± closed, short-forked at base. Rocky areas, pinyon/juniper woodland, coniferous forest; 1500–3200 m. e SNH, GB; to WA, ID, Colorado.

CYNOGLOSSUM HOUND'S TONGUE

Ronald B. Kelley

Per, bien, ± hairy, taprooted. **ST** erect. **LVS** entire; basal petioled; cauline petioled or not. **INFL**: panicle, ± terminal, bracted or not. **FL** radial; calyx ± deeply 5-lobed, enlarging in fr; corolla 5-lobed, funnel-shaped or salverform, appendages large; style entire. **FR**: nutlets gen 4, subspheric, covered with short, barbed prickles. 80 spp.: worldwide. (Greek: dog tongue)

1. Infl ± overlapping lvs, bracts ± lf-like; corolla dull reddish purple; nutlet outer surface flattened, margin raised; bien; local, alien . . . . ***C. officinale***

1′ Infl held above lvs, bracts scale-like or 0; corolla bluish; nutlet outer surface rounded, margin not raised; per; common, native

2. St hairs 0; basal lf blade abruptly narrowed to petiole; cauline lvs few, petioled; corolla gen 10–15 mm wide . . . . ***C. grande***

2′ St hairs spreading; basal lf blade tapered to petiole; cauline lvs ± many, sessile; corolla 4–9 mm wide ***C. occidentale***

C. grande Lehm. (p. 379) Per. **ST** gen 1, 3–9 dm; hairs 0. **LVS**: lower surfaces hairy, upper surfaces ± glabrous; basal petioles 8–15 cm, ± unwinged; blades 8–15 cm, 3–10 cm wide, ± ovate to elliptic, bases truncate or cordate; cauline few, petioled. **INFL** held above lvs; bracts scale-like or 0; pedicels 10–25 mm. **FL**: corolla 8–12 mm, gen 10–15 mm wide, ± salverform, blue, tube often violet, appendages white. **FR**: nutlets ascending-spreading, outer surfaces rounded, margins not raised. $2n$=24. Shaded or open areas, woodland or chaparral; < 1500 m. NW (exc NCoRH), CaR, SNF (uncommon), n SNH, SnFrB, SCoR; to B.C. ❀DRN:**4,5**&SHD:**6, 15–17**&IRR:**7**,14.

C. occidentale A. Gray (p. 379) Per. **STS** 1–several, clustered, 1.5–5 dm; hairs spreading. **LVS** harshly hairy; basal petioles 4–10 cm, winged; blades 5–15 cm, 1–4 cm wide, ± oblanceolate, bases tapered; cauline ± many, sessile. **INFL** held above lvs; bracts scale-like or 0; pedicels 4–8 mm. **FL**: corolla 6–9 mm, 4–9 mm wide, funnel-shaped, bluish, tinged rose to brown, appendages whitish. **FR**: nutlets ascending-spreading, outer surfaces rounded, margins not raised. Open coniferous forests, often with ponderosa pine; 900–2100 m. KR, NCoRH, NCoRI, CaRH, SNH, MP; w OR. ❀TRY.

C. officinale L. Bien. **ST** 1, 3–12 dm, ± soft-hairy. **LVS** soft-hairy; basal petioles 4–10 cm, unwinged; blades 8–20 cm, oblanceolate or narrowly elliptic, bases tapered; cauline many, sessile. **INFL** ± overlapping lvs; bracts ± lf-like; pedicels 5–12 mm. **FL**: corolla 3–5 mm, 4–9 mm wide, ± salverform, dull reddish purple, sometimes drying bluish, appendages purple. **FR**: nutlets descending-spreading, outer surfaces flat, margins raised. $2n$=24. Disturbed places; 850–1000 m. CaRH (se Siskiyou, adjacent Shasta cos.); native to Eurasia. TOXIC to cattle (no reports from CA).

ECHIUM

Dieter H. Wilken and Ronald B. Kelley

Ann, per, shrubs, bristly to strigose. **LVS** basal and cauline, linear to lanceolate, entire. **INFL**: panicle, terminal; branches 3–many, ± spike-like. **FL** radial to ± bilateral; calyx deeply lobed; corolla throat straight or slightly curved, lobes equal or unequal; stamens inserted below mid-tube, incl or exserted; style exserted. **FR**: nutlet erect, short, ovoid, 3-angled, scar flat. 40 spp.: s Eurasia, Afr. Cult for orn. (Greek: viper, from nutlet shape, which resembles viper's head) [Bramwell 1972 Lagascalia 2:37–115]

1. Shrub; corolla ± radial, tube ± = calyx . . . . ***E. candicans***
1′ Ann; corolla ± bilateral, tube ± 2 × calyx . . . . ***E. plantagineum***

E. candicans L.f. Shrub 1–3 m. **LF** persistent, 6–20 cm, narrowly elliptic, densely strigose. **INFL** 3–4 dm; branches many, spreading. **FL** ± radial; calyx 3–5 mm; corolla 5–9 mm, blue to violet; stamens all exserted. **FR**: nutlets rough, fine-tubercled. Open, dry slopes and bluffs; < 300 m. CCo, SnFrB, SCo; native to Madeira, Canary Islands. Several spp. cult on CA coast, > 1 probably naturalized, some may be hybrids. Pls with pink to pale blue corollas and nutlets sharply tubercled are called *E. strictum* L.f. Pls 2–3 m with basal lf rosette and ± cylindric infl 1+ m are called *E. pininana* Webb & Berth. [*E. fastuosum* Aiton misapplied]

E. plantagineum L. Ann or bien 4–8 dm. **LF** 2–10 cm, narrowly elliptic to oblong, bristly; hair bases bulbous. **INFL** 5–15 cm; branches 2–8, ascending. **FL** ± bilateral; calyx 5–9 mm; corolla 15–20 mm, blue; 2–3 stamens exserted. **FR**: nutlets rough, wrinkled to fine-tubercled. $2n$=16. Disturbed areas, fields; < 300 m. c CCo, SCo; native to s Eur.

HACKELIA STICKSEED

Robert L. Carr

Per (rarely bien); hairs appressed to spreading; caudex gen branched in age, often ± woody, taprooted. **ST** ascending or erect. **LVS**: lowest gen with petioles ± = blades, ± winged; other lvs gen sessile, becoming bract-like toward infl. **INFL**: cymes, gen ± terminal, gen > 3, gen arrayed in panicles, coiled at tips. **FL** radial; corolla rotate-salverform, with appendage near base of each lobe. **FR**: nutlets erect, > style, gen with lateral-medial scar, gen with barb-tipped prickles on margin and exposed face. 40 spp.: gen w N.Am, se Asia. (J. Hackel, Czech botanist, born 1783) [Gentry & Carr 1976 Mem New York Bot Gard 26:121–227] Difficult genus needing much work, esp in n CA, se Asia; sometimes merged with *Lappula*. ❀TRY;DFCLT.

1. Corolla white or pink and gen turning blue in age
 2. Corolla pink, gen turning blue in age . . . . ***H. mundula***
 2′ Corolla white
  3. Nutlet facial prickles << marginal; calyx lobes ± 4–5 mm; corolla limb ± 12–19 mm wide, appendages gen longer than wide . . . . ***H. bella***
  3′ Nutlet facial prickles ± as long as marginal; calyx lobes ± 2–3 mm; corolla limb ± 5–14 mm wide, appendages gen wider than long . . . . ***H. californica***
1′ Corolla light to dark or purplish blue
 4. Nutlet facial prickles ± as long as marginal; corolla tube >> calyx
  5. Corolla limb 5–11(14) mm wide; st hairs gen sparse below middle, very sparse to 0 above . . . . ***H. nervosa***
  5′ Corolla limb 12–20 mm wide; st hairs ± dense below middle, ± dense to ± sparse above . . . . ***H. velutina***
 4′ Nutlet facial prickles < marginal or 0; corolla tube ± = calyx
  6. Cymes 1–3 per infl; > 3000 m — s SNH (near Mount Whitney; Fresno, Inyo, Tulare cos.) . . . . ***H. sharsmithii***
  6′ Cymes gen > 3 per infl; gen < 3000 m
   7. Hairs at mid-st gen ± strongly appressed — MP, W&I
    8. Nutlets 2–3 mm, facial prickles 0–7; hairs at mid-st appressed downward; corolla limb gen 5–7.5 mm; calyx lobes 1–2 mm; W&I . . . . ***H. brevicula***
    8′ Nutlets 3–4.5 mm, facial prickles 8–15; hairs at mid-st appressed upward; corolla limb gen 7–10 mm; calyx lobes 4–5 mm; MP . . . . ***H. cusickii***

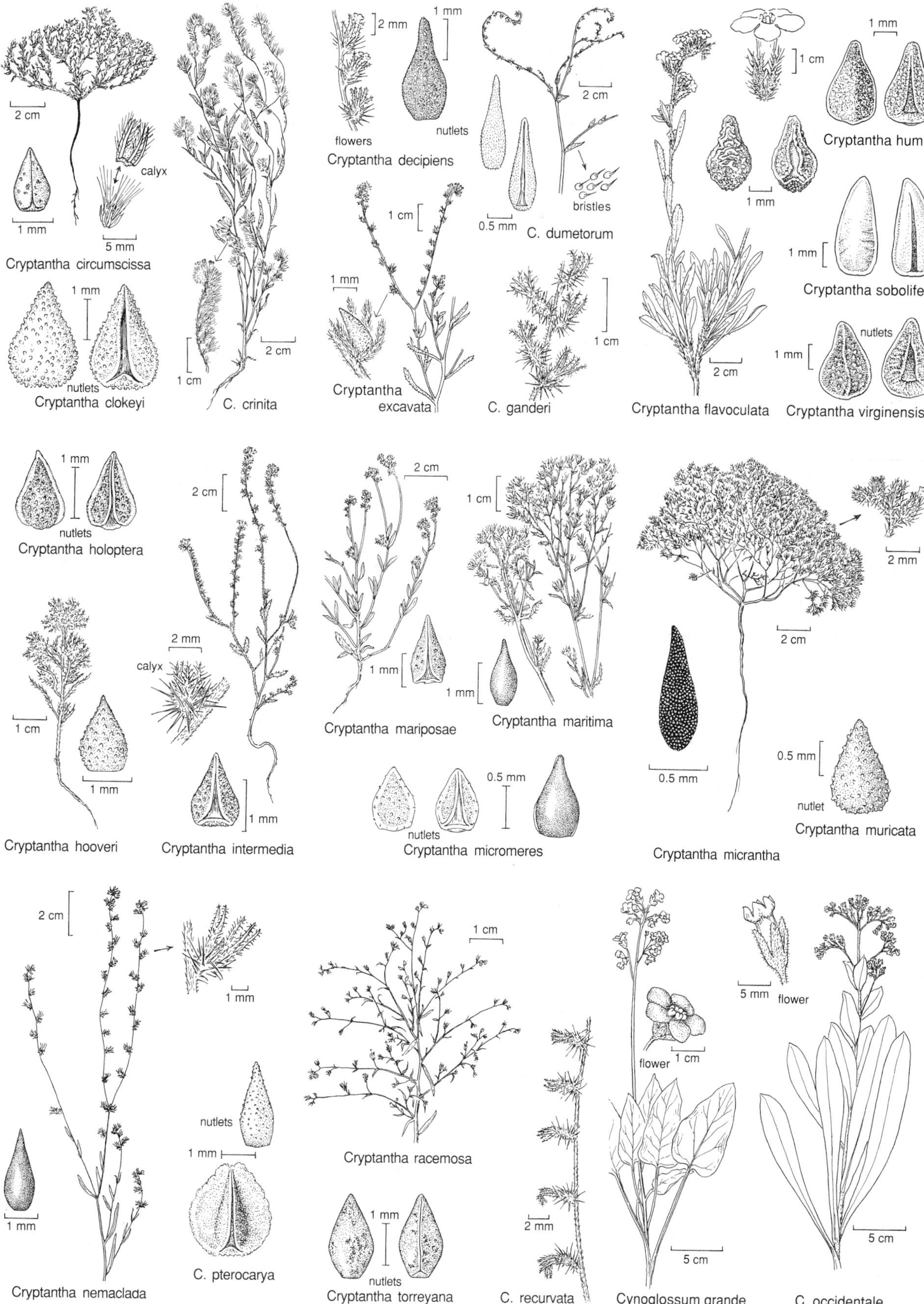
2 cm
1 mm
calyx
5 mm
Cryptantha circumscissa
1 mm
nutlets
Cryptantha clokeyi
1 cm
C. crinita
2 mm
1 mm
flowers
nutlets
Cryptantha decipiens
1 cm
1 mm
2 cm
Cryptantha excavata
2 cm
bristles
0.5 mm
C. dumetorum
1 cm
C. ganderi
1 cm
2 cm
1 mm
Cryptantha flavoculata
1 mm
Cryptantha humilis
1 mm
Cryptantha sobolifera
nutlets
1 mm
Cryptantha virginensis
1 mm
nutlets
Cryptantha holoptera
1 cm
1 mm
Cryptantha hooveri
2 cm
2 mm
calyx
1 mm
Cryptantha intermedia
2 cm
1 mm
Cryptantha mariposae
1 cm
1 mm
Cryptantha maritima
0.5 mm
nutlets
Cryptantha micromeres
2 mm
2 cm
0.5 mm
Cryptantha micrantha
0.5 mm
nutlet
Cryptantha muricata
2 cm
1 mm
1 mm
Cryptantha nemaclada
nutlets
1 mm
C. pterocarya
1 cm
Cryptantha racemosa
1 mm
nutlets
Cryptantha torreyana
2 mm
C. recurvata
flower
1 cm
5 cm
Cynoglossum grande
5 mm
flower
5 cm
C. occidentale

7′ Hairs at mid-st gen ± spreading
9. Nutlet facial prickles gen 0, rarely to 3; infl often elongate, narrow; robust bien to short-lived per **H. floribunda**
9′ Nutlet facial prickles gen > 3; infl gen open, wide; per
10. Individual hairs of lower lvs clearly visible to naked eye, gen >> 1 mm, esp on margins; lf bases above mid-st tapered, not clasping; largest caudex lvs gen < 20 cm, gen < 20 mm wide .... **H. setosa**
10′ Individual hairs of lower lvs not clearly visible to naked eye, gen < 1 mm, exc sometimes on margins; lf bases above mid-st tapered to obtuse, truncate, or cordate and subclasping; largest caudex lvs often > 20 cm, often > 20 mm wide
11. Lf bases near mid-st (just below infl) truncate to cordate, subclasping, blade often lanceolate-ovate; st hairs gen ± dense, gen < 1 mm; nutlet facial prickles gen 10–17 .... **H. amethystina**
11′ Lf bases near mid-st (just below infl) tapered to obtuse, not subclasping, blade gen elliptic; st hairs gen ± 0 to ± sparse, often > 1 mm; nutlet facial prickles gen 4–10 .... **H. micrantha**

H. amethystina Eastw. (p. 385) AMETHYST STICKSEED **ST** 4–8 dm; hairs gen < 1 mm, gen ± dense. **LVS**: caudex 10–30 cm, 13–40 mm wide, narrow-elliptic; lower cauline smaller, gen more linear, ephemeral; mid-cauline ovate-lanceolate, subclasping. **FL**: corolla blue to pinkish, limb 7–12 mm wide. **FR**: nutlets 4–5 mm, facial prickles gen 10–17, < marginal. 2*n*=24. UNCOMMON. Meadows, forest clearings, along stream banks, roadsides; 1500–2100 m. NCoRH, n SNH (Plumas Co.).

H. bella (J.F. Macbr.) I.M. Johnston (p. 385) **ST** 5–7 dm; hairs gen spreading, strigose in infl. **LVS**: caudex often many, 15–26 cm, 22–45 mm wide, narrow-elliptic, obtuse; lower cauline 8–13 cm, 6–17 mm wide, linear to narrow-elliptic, ephemeral; mid to upper cauline 4–10 cm, 14–42 mm wide, lanceolate to ovate, bases cordate, subclasping. **FL**: corolla white, appendages gen longer than wide, limb 12–19 mm wide. **FR**: nutlets 5–6 mm, facial prickles ± 10, << marginal. 2*n*=24. Stream banks, roadsides, forest openings; 900–2000 m. KR, NCoR.

H. brevicula (Jepson) J. Gentry (p. 385) POISON CANYON STICKSEED **ST** 2–6 dm; hairs ± strongly appressed downward, ± stiff, coarse. **LVS**: caudex 6–18 cm, 5–18 mm wide, narrow-elliptic; cauline ± similar but gen smaller, esp above. **FL**: corolla pale blue, limb 5–8 mm wide. **FR**: nutlets 2–3 mm, facial prickles 0–7, < marginal. RARE. Open hillsides, dry stream beds, open aspen stands; 2600–3200 m. W&I (Mono Co.). [*H. patens* (Nutt.) I.M. Johnston misapplied]

H. californica (A. Gray) I.M. Johnston (p. 385) **ST** 3–10 dm; hairs strigose to ± long, soft. **LVS**: caudex, lowest cauline 3–17 cm, 6–30 mm wide, lanceolate to oblanceolate; mid to upper cauline lanceolate to ovate, often clasping. **FL**: corolla white or rarely pinkish, limb 5–14 mm wide, appendages gen wider than long. **FR**: nutlets 4–4.5 mm, facial prickles 10–26, ± as long as marginal. 2*n*=24. Meadows, forest openings; 900–1800 m. NCoRH, CaRH, n SNH.

H. cusickii (Piper) Brand (p. 385) CUSICK'S STICKSEED **ST** 1–5 dm, slender; hairs gen strongly appressed, strigose in infl. **LVS**: caudex 5–18 cm, 4–28 mm wide, narrow-elliptic; lower cauline 5–10 cm, 4–11 mm wide, narrowly elliptic to lanceolate. **FL**: corolla blue, limb 5–13 mm wide. **FR**: nutlets 3–4.5 mm, facial prickles 8–15, < marginal. 2*n*=48. UNCOMMON. In shelter of low junipers; 1200–2000 m. MP.

H. floribunda (Lehm.) I.M. Johnston (p. 385) Bien to short-lived per. **ST** gen 4–12 dm; hairs gen spreading, ± coarse. **LVS**: caudex gen < cauline, ephemeral; lower cauline gen 5–24 cm, 5–35 mm wide, oblanceolate to narrow-elliptic. **INFL** elongate, narrow. **FL**: corolla blue, limb gen 4–8 mm wide, sometimes smaller. **FR**: nutlets 2–4 mm, marginal prickles sometimes fused basally, ± forming a wing, facial prickles gen 0–3, < marginal. *n*=±12. Meadows, stream banks, other vernally wet areas, less often open slopes, forests; 700–3100 m. SN, SNE.

H. micrantha (Eastw.) J. Gentry (p. 385) **ST** 3–11 dm; hairs gen ± 0 to ± sparse, gen ± spreading, often > 1 mm, ± strigose in infl. **LVS**: caudex 6–33 cm, 7–37 mm wide, narrowly elliptic to oblanceolate; lower cauline gen 5–23 cm, 6–24 mm wide. **FL**: corolla blue, limb gen 5–11 mm wide, sometimes smaller. **FR**: nutlets 3–5 mm, facial prickles gen 4–10, < marginal. *n*=12. Meadows, along streams, open slopes, forests; 700–3400 m. KR, NCoR, CaR, SN, MP; w N.Am. [*H. jessicae* (MacGregor) Brand]

H. mundula (Jepson) Ferris (p. 385) **ST** 4–8 dm; hairs ± dense, gen spreading, long, soft, sometimes strigose. **LVS**: caudex gen 6–22 cm, 5–28 mm wide, oblanceolate; lower cauline variable, often smaller, becoming ovate to lanceolate, gen clasping to subclasping above. **FL**: corolla pink, gen turning blue in age, limb 10–16 mm wide. **FR**: nutlets 4.5–6.5 mm, facial prickles many, ± as long as marginal. Open slopes, forest clearings, roadsides; 1600–2800 m. SN.

H. nervosa (Kellogg) I.M. Johnston (p. 385) **ST** 4–7 dm; hairs gen sparse below middle, very sparse to 0 above, often ± spreading. **LVS**: caudex 3–12 cm, 6–25 mm wide, ± oblong to oblanceolate; lower cauline gen smaller, often sessile, becoming ovate to lanceolate upward; uppermost subclasping. **FL**: corolla dark to purplish blue, tube gen >> calyx, limb 5–11(14) mm wide. **FR**: nutlets 4–6 mm, facial prickles many, ± as long as marginal. Open slopes, forests, forest clearings; 1700–3000 m. CaRH, SN.

H. setosa (Piper) I.M. Johnston (p. 385) **ST** 3–6 dm; hairs gen ± spreading, strigose in infl. **LVS**: caudex many, 9–22 cm, 10–25 mm wide, narrow-oblong-ovate; cauline sessile, 5–9 cm, 4–10 mm wide, linear to lanceolate or oblanceolate. **FL**: corolla blue, throat whitish, limb 8–13 mm wide. **FR**: nutlets 3.2–3.7 mm, facial prickles 9–13, < marginal. 2*n*=48. Open, wooded ridges; 300–1800 m. KR, n SNH (Sierra Valley).

H. sharsmithii I.M. Johnston (p. 385) SHARSMITH'S STICKSEED **ST** 1–3 dm, strigose, sometimes slightly so; hairs appressed downward below, upward above. **LVS**: caudex 3–14 cm, 6–23 mm wide, elliptic to lanceolate; cauline sessile, 1–4 cm, 5–16 mm wide, oblanceolate to ovate. **INFL**: cymes 1–3. **FL**: corolla pale blue, limb 5.5–8 mm wide. **FR**: nutlets 2–3 mm, facial prickles 0–5, < marginal. RARE in CA. Cracks, crevices in granite cliffs, large boulder talus; 3000–3700 m. s SNH (near Mount Whitney; Fresno, Inyo, Tulare cos.).

H. velutina (Piper) I.M. Johnston (p. 385) **ST** 4–8 dm; hairs gen spreading, ± dense below middle, ± dense to ± sparse above. **LVS**: caudex gen 5–17 cm, 5–20 mm wide, narrowly elliptic to oblanceolate; lower cauline ± similar, often smaller, becoming ovate to narrow-lanceolate above, ± clasping above middle. **FL**: corolla dark to purplish blue, tube gen >> calyx, limb 12–20 mm wide. **FR**: nutlets 4.5–6.5 mm, facial prickles many, ± as long as marginal. Dry, open slopes, forest clearings, roadsides; 1300–2500 m. CaRH, SN. [*H. longituba* I.M. Johnston]

HARPAGONELLA

Timothy C. Messick

1 sp. (Latin: small grappling hook, from calyx spines)

H. palmeri A. Gray (p. 385) PALMER'S GRAPPLING HOOK **ST** ascending to erect, 3–30 cm. **INFL**: pedicels in fr 0.5–1 mm, twisted. **FL**: sepals in fr > nutlets, upper 2 >> others, partly fused, arched over 1 nutlet, ± bur-like, with 5–10 stout spines, each with hooked bristles, lower 3 sepals distinct. **FR**: nutlets 2, 1–4 mm, dissimilar, ± oblanceolate, margins entire. *n*=12. RARE in CA. Dry sites in chaparral, coastal scrub, grassland; < 450 m. SCo, PR; AZ, nw Mex.

HELIOTROPIUM HELIOTROPE

Dieter H. Wilken

Ann, per, shrub, glabrous to bristly or strigose. **STS** prostrate to erect, branched. **LVS** gen cauline, petioled to sessile, gen entire. **INFL**: fl 1 and axillary or terminal spikes with many fls, coiled in fl. **FL**: corolla rotate to bell-shaped, white to purple; stamens inserted on upper tube, incl, anthers ± sessile; style attached atop ovary, stigma linear to disk-like. **FR**: nutlets 2 or 4, erect, gen ovoid to spheric, smooth, roughened or hairy, scar gen lateral. ± 250 spp.: temp, trop. Orn, cult for medicinal drugs. (Greek: sun turning, because some spp. fl at summer solstice)

1. Per
  2. Sts, lvs not fleshy, soft-hairy; corolla purple .... ***H. amplexicaule***
  2′ Sts, lvs fleshy, glabrous; corolla white to bluish .... ***H. curassavicum***
1′ Ann
  3. Corolla limb 8–12 mm wide; fls axillary, solitary .... ***H. convolvulaceum*** var. ***californicum***
  3′ Corolla limb 3–5 mm wide; infl with many fls in terminal coiled spikes .... ***H. europaeum***

H. amplexicaule M. Vahl Per. **ST** decumbent to ascending, 2–6 dm, short-hairy. **LF** 4–9 cm, oblong to oblanceolate, short-petioled to subsessile, acute, short-hairy. **INFL**: spikes 3–5, terminal, coiled in fl. **FL**: calyx lobes ± linear-lanceolate, bristly; corolla 4–6 mm, bell-shaped, limb 3–4 mm wide, purple. **FR**: nutlets 2, irregularly roughened, faintly tubercled. Open sites, fields; < 300 m. n&c SCo; native to Argentina.

H. convolvulaceum (Nutt.) A. Gray var. ***californicum*** (E. Greene) I.M. Johnston (p. 385) Ann, taprooted. **ST** ascending to erect, 7–18 cm, canescent. **LF** 1–4 cm, elliptic to ovate, gen petioled, acute, densely strigose. **INFL**: fls 1, axillary. **FL**: calyx lobes lanceolate, long-tapered, densely bristly; corolla 7–10 mm, ± rotate, limb 8–12 mm wide, papery, white. **FR**: nutlets 4, long-soft-hairy. 2*n*=42. Sandy soils; < 700 m. D; to Great Plains, n Mex. ❀TRY.

H. curassavicum L. (p. 385) Per, sometimes from rhizome-like root. **ST** prostrate to weakly ascending, 1–6 dm, fleshy, glabrous. **LF** 1–6 cm, gen oblanceolate, short-petioled to subsessile, acute to obtuse, fleshy, glabrous. **INFL**: spikes 2–4, terminal, coiled in fl. **FL**: calyx lobes oblong to narrowly ovate, glabrous; corolla 3–5 mm, bell-shaped, limb 3–4 mm wide, white to bluish. **FR**: nutlets 4, smooth. Moist to dry, saline soils; < 2100 m. CA; NV, AZ, subtrop, trop Am. [*H. c.* var. *oculatum* (A.A. Heller) I.M. Johnston] ❀6,**7**, **14–17**,**24**IRR:2,3,**8**,**9**,10–13,**18–23**;STBL;INV.

H. europaeum L. Ann, taprooted. **ST** ascending to erect, 5–40 cm, puberulent to short-soft-hairy. **LF** 1.5–5 cm, elliptic to ovate, petioled, obtuse, appressed-short-hairy. **INFL**: spikes 2–4, terminal, coiled in fl. **FL**: calyx lobes linear to lanceolate, bristly; corolla 2–4 mm, salverform, limb 3–5 mm wide, white. **FR**: nutlets 4, irregularly roughened, faintly tubercled. Open, often disturbed sites, fields; < 1400 m. n&c SNF, GV, CCo, SnFrB, MP; e US; native to s&e Eur, n Afr.

LAPPULA STICKSEED

Ronald B. Kelley

Ann, hairy, taprooted. **ST** ± erect; branches 0–many. **LVS** basal and cauline, sessile, entire. **INFL**: raceme, long, ± terminal; bracts lf-like. **FL**: calyx ± deeply 5-lobed, enlarging in fr; corolla 5-lobed, funnel-shaped, appendages present; style entire. **FR**: nutlets gen 4, ovate, covered with ± long, barbed prickles. 12–14 spp.: n hemisphere. (Latin: little bur)

1. Nutlet marginal prickles in 2 or 3 rows, bases not or slightly widened, not in contact, not fused; uncommon alien .... ***L. squarrosa***
1′ Nutlet marginal prickles in 1 row, bases slightly or very widened, gen in contact, gen fused; common native .... ***L. redowskii***
  2. Nutlet marginal prickles much wider at base, fused to form a crown .... var. ***cupulata***
  2′ Nutlet marginal prickles slightly wider at base, slightly fused, not forming a crown .... var. ***redowskii***

L. redowskii (Hornem.) E. Greene **ST** 0.5–3.5 dm. **LF** 1–4 cm, linear to lanceolate. **INFL**: pedicel 1–2 mm. **FL**: calyx 3–3.5 mm in fr, lobes lanceolate, ± erect in fr; corolla 1.5–2.5 mm wide, blue to white. **FR**: nutlets 2–3 mm. Dry, open, rocky, often disturbed sites; 600–3300 m. SNH, SnBr, SnJt, GB, DMoj; w N.Am, Eurasia. Var. *c.* may occupy warmer, drier habitats than var. *r.*

var. ***cupulata*** (A. Gray) M.E. Jones (p. 385) **FR**: nutlet margin prickles much wider at base, fused to form a crown. Habitat of sp.; 600–2100 m. GB, DMoj; w N.Am. [var. *desertorum* (E. Greene) I.M. Johnston]

var. ***redowskii*** (p. 385) **FR**: nutlet marginal prickles slightly wider at base, slightly fused, not forming a crown. 2*n*=48. Habitats of sp.; 1300–3300 m. SNH, SnBr, SnJt, GB, DMtns; w N.Am, Eurasia.

L. squarrosa (Retz.) Dumort. (p. 385) **ST** 1.5–8 dm. **LF** 2–5 cm, linear to oblong. **INFL**: pedicel 1–3 mm. **FL**: calyx 2.5–3 mm in fr, lobes widely linear, spreading in fr; corolla 2–4 mm wide, blue. **FR**: nutlets 3–4 mm. 2*n*=48. Moist, disturbed, cult soils, weedy, urban areas; < 200 m. SCo (Santa Monica, Upland); native to Eurasia. [*L. echinata* Gilib.]

LITHOSPERMUM STONESEED

Ronald B. Kelley

Per, ann, hairy, taprooted. **ST** erect. **LVS** gen cauline, ± sessile, entire. **INFL**: bracted cymes, open panicles, or fls solitary in upper lf axils. **FL**: calyx deeply 5-lobed, enlarging in fr; corolla 5-lobed, funnel-shaped or salverform, gen ± yellow, appendages present or not; style entire. **FR**: nutlets 1–4, large, smooth to pitted or wrinkled. 75 spp.: worldwide, gen temp or mtn. (Greek: stone seed) [Baker 1961 Rhodora 63:229–235]

1. Ann; caudex 0; corolla ± white to bluish white; fr brownish, wrinkled, ± tubercled, ± dull; alien, weedy .... ***L. arvense***
1′ Per; caudex ± woody; corolla ± yellow; fr white to gray, smooth to pitted, shiny; native, not weedy
 2. Corolla 15–35 mm, 10–20 mm wide, appendaged, lobe margins jagged; cleistogamous fls present; DMtns .... ***L. incisum***
 2′ Corolla 9–18 mm, 7–13 mm wide, appendages 0, lobes entire; cleistogamous fls 0; n CA
  3. Lf blades lanceolate to linear, many; fls in many, dense cymes; pedicels 1–3 mm, not recurved in fr; corolla 9–12 mm, 1–1.4 × calyx; not heterostylous .... ***L. ruderale***
  3′ Lf blades oblong to lance-ovate, not many; fls in open panicles or solitary in upper lf axils; pedicels 4–7 mm, recurved in fr; corolla 12–18 mm, 1.5–2 × calyx; heterostylous .... ***L. californicum***

L. arvense L. GROMWELL Ann, strigose; caudex 0. **STS** 1–few, 1–7 dm, ± not clustered, branched at base. **LVS** ± few; blade 2–6 cm, ± linear to lanceolate or oblong. **INFL**: fls solitary in upper lf axils; pedicels 1 mm, not recurved in fr. **FL**: corolla 5–8 mm, ± = calyx, 2–4 mm wide, funnel-shaped, ± white to bluish white, appendages 0. **FR**: nutlets wrinkled, ± tubercled, ± dull, brownish. $2n$=28,42. Disturbed areas; < 1200 m. SnFrB, MP(?); to OR, e N.Am; native to Eurasia. [*Buglossoides a.* I.M. Johnston] Not heterostylous; cleistogamous fls 0.

L. californicum A. Gray (p. 385) Per; hairs spreading, ± coarse; caudex ± woody. **STS** 1–several, 1.5–4 dm, clustered, ± branched. **LVS** ± few; blade 2.5–5 cm, oblong to lance-ovate. **INFL**: open panicles ± few (or fls solitary in upper lf axils); pedicels 4–7 mm, recurved in fr. **FL**: corolla 12–18 mm, 1.5–2 × calyx, 7–9 mm wide, salverform to funnel-shaped, yellow, unappendaged, lobes entire. **FR**: nutlets smooth, shiny, white. $2n$=28. Open, dry slopes, yellow-pine forests, pine/oak woodland, chaparral; 250–1800 m. KR, NCoRH, NCoRI, CaRH, n SNH, Wrn(?); sw OR. Heterostylous; cleistogamous fls 0. ❀DRN:6,17&IRR:7,15,16;DFCLT.

L. incisum Lehm. (p. 385) Per, strigose; caudex woody. **STS** few–several, 1–3 dm, clustered, ± unbranched. **LVS** many; blade 1.5–6 cm, linear to linear-oblong. **INFL**: cymes many, in upper axils; pedicels 2–5 mm, ± recurved in fr. **FL**: corolla 15–35 mm, 2–3.5 × calyx, 10–20 mm wide, salverform, yellow, appendaged. **FR**: nutlets ± pitted, shiny, gray. $2n$=24,36. Sandy, rocky slopes, piny-on-juniper woodland; 1650–1700 m. DMtns (Keystone Canyon, New York Mtns, San Bernardino Co.); to B.C., MT, Great Plains, s NV. Not heterostylous; cleistogamous fls present. ❀DRN:1–3, 16,17;DFCLT.

L. ruderale Lehm. (p. 385) Per; hairs ± spreading; caudex ± woody. **STS** 1–several, 2–5 dm, clustered, ± unbranched. **LVS** many; blade 3–8 cm, lanceolate to linear. **INFL**: cymes many, dense, in upper lf axils; pedicels 1–3 mm, not recurved in fr. **FL**: corolla 9–12 mm, 1–1.4 × calyx, 7–13 mm wide, ± salverform, pale yellowish, unappendaged. **FR**: nutlets smooth, shiny, whitish. $2n$= 24. Open, dry slopes, plains, sagebrush steppe, coniferous woodland, chaparral; (750)1200–1800 m. CaRH, n SNH, MP; nw N.Am. Not heterostylous; cleistogamous fls 0.

MERTENSIA BLUEBELLS, LUNGWORT

Elaine Joyal

Per from branched caudex, glabrous to coarsely hairy. **ST** ± erect. **LVS** cauline and gen basal, alternate, gen petioled (upper gen sessile). **INFL**: cyme, gen panicle- or raceme-like; bracts 0. **FL**: calyx gen deeply lobed; corolla blue, gen abruptly expanded at throat, limb often ± cylindric or flared; filaments often ± flat, gen attached ± below obvious corolla appendages, anthers incl. **FR**: nutlets gen wrinkled, each attached near or below middle to convex receptacle. ± 50 spp.: N.Am, temp Eurasia. (F.C. Mertens, Germany, 1764–1831) [Milek 1988 PhD U Northern Colorado; Strachan 1988 PhD U Montana] Hybrids common; identification sometimes difficult, esp in MP. ❀TRY;DFCLT.

1. Corolla not sharply divided into tube and limb, 6–10 mm; nw KR (Sect. *Neuranthia*) .... ***M. bella***
1′ Corolla ± sharply divided into tube and limb, often > 10 mm; more widely distributed (Sect. *Mertensia*)
 2. Herbage with spreading hairs .... ***M. cusickii***
 2′ Herbage glabrous to ± strigose
  3. Pl 4–15 dm; lateral veins of cauline lvs conspicuous; wet places in mtns, fl late spring, summer .... ***M. ciliata***
  3′ Pl gen < 4 dm; lateral veins of cauline lvs obscure; gen spring-moist, drying places of plains, foothills, fl spring
   4. Sts gen 1–2, easily detached from shallow, tuber-like root; basal lvs rare on fl pls; cauline lf length 1.5–4 × width; corolla tube 2–3 × limb .... ***M. longiflora***
   4′ Sts many, firmly attached to deep, sometimes fleshy caudex; basal lvs gen well developed; cauline lf length gen 2.5–7 × width; corolla tube gen 1.3–2 × limb .... ***M. oblongifolia***
    5. Lf ± glabrous; st gen < 20 cm .... var. ***nevadensis***
    5′ Lf hairy on 1 or both surfaces; st gen > 20 cm
     6. Lf hairy on both surfaces .... var. ***amoena***
     6′ Lf glabrous on lower surface .... var. ***oblongifolia***

M. bella Piper (p. 385) OREGON LUNGWORT Pl 2–5 dm, glabrous or sparsely hairy. **ST** 1, slender, branched or not. **LVS**: lateral veins conspicuous; basal not persisting; cauline petioled, strigose on upper surface. **INFL** raceme-like, open. **FL**: calyx strigose; corolla 6–10 mm, bright blue, not sharply divided into tube and limb, appendages ± 0; filaments slender, attached ± 1 mm above corolla base, ± > anthers; style incl. UNCOMMON. Wet meadows, under taller herbs; 1500–2000 m. nw KR; to OR, ID, MT.

M. ciliata (Torrey) G. Don (p. 385) STREAMSIDE BLUEBELLS Pl 4–15 dm, glabrous, sometimes glaucous. **STS** clustered on thick, branched caudex, leafy. **LVS**: basal gen > cauline; cauline with conspicuous lateral veins, lower petioled; blades lanceolate to ovate, acute. **INFL** panicle-like, open. **FL**: corolla 10–17 mm, tube ± = or > limb, often with ring of hairs below middle inside; filaments wide, gen > anthers; style exserted 2–5 mm. $2n$=24,48. Streamsides, wet meadows, damp thickets, wet cliffs; 1700–3600 m. n SNH, MP, W&I; to s OR, NV, UT. Pls with style ± incl have been called var. *stomatechoides* (Kellogg) Jepson. ❀DRN,IRR:1,2,17 &WET:14–16;DFCLT.

M. cusickii Piper TOIYABE BLUEBELLS Pl 3–5 dm; hairs spreading. **STS** ± clustered on branched caudex. **LVS**: basal gen few; cauline ± veiny, lower large, petioled, upper gradually smaller, sessile. **INFL** panicle-like, open or ± dense. **FL**: corolla 10–16 mm, tube 5–8 mm, with ring of hairs inside near base, limb 5–8 mm; filaments wide, flat, ± = anthers; style ± incl. Uncommon. Streamsides, dry drainage-bottoms, wooded slopes, dry meadows; < 2500

m. Wrn; to e OR, ID, nw NV. Morphologically and ecologically intermediate between "short" and "tall" mertensias. [*M. toiyabensis* J.F. Macbr.]

M. longiflora E. Greene (p. 385) LONG BLUEBELLS Pl gen < 4 dm, glabrous to ± strigose. **STS** gen 1–2, easily detached from short, shallow, tuber-like root. **LVS**: basal rarely developed on fl pls; cauline gen sessile, few, gen 1.5–4 × longer than wide, lateral veins obscure. **INFL** ± panicle-like, dense. **FL**: corolla 15–25 mm, tube gen 2–3 × longer than limb, glabrous inside; filaments wide, > anthers; style ± incl. Uncommon. Open, gen spring-moist, drying places of plains, foothills, esp with sagebrush or ponderosa pine; 1600–2200 m. MP; to B.C., MT.

M. oblongifolia (Nutt.) G. Don SAGEBRUSH BLUEBELLS Pl gen < 4 dm, glabrous to strigose. **STS** many, firmly attached to stout, deep, sometimes fleshy caudex. **LVS**: basal gen well developed; cauline gen 2.5–7 × longer than wide, lateral veins obscure, lower lvs gen petioled. **INFL** ± panicle-like, gen dense. **FL**: corolla 10–20 mm, tube gen 1.3–2 × longer than limb, sometimes with ring of hairs inside; filaments wide, ± = anthers; style ± incl. Open slopes, drier meadows, gen spring-moist places, esp with sagebrush; 1000–3000 m. CaRH, n&c SNH, MP; to WA, WY, Colorado. Polyploid complex (2*n*=24,48); CA vars. sometimes found together in Wrn.

var. ***amoena*** (Nelson) L.O. Williams **ST** gen > 20 cm. **LF** hairy on both surfaces. **FL**: corolla tube hairy inside. 2*n*=48. Habitats and elev. gen of sp. Wrn; to WA, WY, UT.

var. ***nevadensis*** (Nelson) L.O. Williams (p. 385) **ST** gen < 20 cm. **LF** glabrous, sometimes bumpy. **FL**: corolla tube glabrous inside. 2*n*=24. Habitats and elev. gen of sp. CaRH, n&c SNH, Wrn; to e OR, ID, Colorado. Most common var. in CA.

var. ***oblongifolia*** **ST** gen > 20 cm. **LF** hairy on upper surface, glabrous on lower. **FL**: corolla tube hairy inside. 2*n*=48. Habitats and elev. gen of sp. MP (esp Wrn); to e WA, WY, UT.

MYOSOTIS FORGET-ME-NOT

Elaine Joyal

Ann or per, glabrous to rough-hairy; roots gen diffuse. **ST** decumbent to erect. **LVS**: basal gen oblong or oblanceolate; cauline gen linear to elliptic. **INFL** coiled, gen raceme-like, eventually ± open; bracts 0 (or lf-like in lower half); fls gen in upper half of pl. **FL**: calyx lobes 5, gen ± equal; corolla salverform or widely funnel-shaped, white, yellow, or gen blue, tube 5-appendaged at top, limb abruptly spreading, 5-lobed; stamens 5, incl; style gen incl. **FR**: nutlets smooth, shiny, each with raised outer margin, scar lateral, at base, small. 50 spp.: temp, boreal. (Greek: mouse ear, from lf) [Grau 1964 Osterr Bot Zeitschr 111:561–617] Gen fls in spring.

1. Calyx tube hairs appressed, not hooked
 2. Corolla limb 2–5 mm diam; calyx tube ± = or < lobes; style < nutlets; st often decumbent but base not stolon-like; ann, bien, short-lived per ..... ***M. laxa***
 2′ Corolla limb 5–10 mm diam; calyx tube >> lobes; style ± = or > nutlets; st base often creeping or stolon-like; per ..... ***M. scorpioides***
1′ Calyx tube hairs spreading, some or all hooked
 3. Calyx lobes ± unequal; corolla white, limb 1–2 mm diam ..... ***M. verna***
 3′ Calyx lobes ± equal; corolla gen blue (sometimes initially yellowish) or pink; limb 1–10 mm diam
  4. Corolla limb 5–10 mm diam; pedicel in fr > or = calyx; per ..... ***M. latifolia***
  4′ Corolla limb 1–5 mm diam; pedicel in fr < to << calyx; ann, winter ann, or bien
   5. Fls ± restricted to upper half of pl; hairs on lower lf surface straight; style = or > nutlets ..... ***M. discolor***
   5′ Fls ± from base of pl; hairs on lower lf surface hooked; style << nutlets ..... ***M. micrantha***

M. discolor Pers. Ann, bien, puberulent to rough-hairy; roots diffuse. **ST** 1–5 dm, slender, branched or not. **LVS** sparse, gen 1–4 cm, 2–8 mm wide; hairs straight; basal oblanceolate; cauline ± linear to oblong. **INFL**: bracts 0 or 1–2 near base; pedicel in fr < calyx; fls ± restricted to upper 1/2 of pl. **FL**: calyx 3–5 mm, tube hairs spreading, hooked, lobes strigose or puberulent-strigose; corolla yellowish, turning blue, limb gen 1–2 mm diam, not completely flat. **FR** gen = or < style, dark brown or blackish. 2*n*=64. Roadsides, moist ground, wet meadows; 0–1300 m. CA-FP (esp NCoR, SN); to e N.Am; native to Eur. [*M. versicolor* (Pers.) Smith]

M. latifolia Poiret (p. 391) Per. **ST** < 70 cm; base woody. **LVS**: basal large, ovate; cauline oblong. **INFL** bracted at base; pedicel in fr ascending to spreading, at least 5 mm, > or = calyx. **FL**: calyx 3–6 mm in fr, tube hairs gen hooked; corolla blue or pink, tube 2 × calyx, limb 5–10 mm diam. **FR** 2–3 mm, > style, widely ovate, dark brown. 2*n*=18. Moist, disturbed, shady places; 0–300 m. Scattered in CA-FP; native to nw Afr.

M. laxa Lehm. (p. 391) Ann, bien, short-lived per. **ST** slender, weak, often decumbent but base not creeping or stolon-like, 1–4 dm, ± strigose. **LVS** 1.5–8 cm, 3–15 mm wide; basal oblanceolate; cauline oblong to lanceolate. **INFL** bracted at base; pedicel in fr spreading, > calyx. **FL**: calyx 3–8 mm, closely but not densely strigose, tube ± = or < lobes, lobes equal or not, narrowly triangular; corolla blue, limb 2–5 mm diam, ± flat. **FR** < 2 mm, > style, brown to black. Moist soil, shallow water; 0–2000 m. NW (Del Norte Co.), SN (El Dorado, Kern cos.); ± circumboreal. Highly localized in CA. ❀WET:1–3,**4–6**,7,14–16,**17**.

M. micrantha Lehm. Ann or winter ann. **ST** < 2 dm, simple or branched from near base, ± puberulent or rough-hairy. **LVS** gen < 2 cm, 7 mm wide; hairs on lower surface hooked, esp along midrib; basal oblanceolate or wider; cauline oblong or elliptic. **INFL** bracted in lower half; pedicel in fr ascending or ± spreading, << calyx; fls from near pl base, where 1 per axil. **FL**: calyx 3–5 mm, strigose at least above, tube hairs spreading, hooked; corolla blue, limb 1–2 mm diam, not completely flat. **FR** >> style, brown, sometimes paler. 2*n*=±36–40. Uncommon. Roadsides, streambanks, disturbed open places; 500–1000 m. KR (1 locality in Siskiyou Co.); n US, s Can, native to Eurasia. Other taxa often misidentified as this in CA.

M. scorpioides L. (p. 391) Per. **ST** 2–6 dm, gen unbranched, ± strigose; base often creeping or stolon-like. **LVS** 2.5–8 cm, 7–20 mm wide; basal oblanceolate; cauline oblong or elliptic to lance-elliptic. **INFL**: bracts 0; pedicel in fr spreading, ± = or > calyx. **FL**: calyx 3–5 mm, sparsely strigose, lobes << tube, sometimes unequal, widely triangular, in fr < 6 mm; corolla blue, limb 5–10 mm diam, flat. **FR** > 1.5 mm, ± = or < style, ovoid, ± black, marginal rim faint. 2*n*=64. Shallow water, wet soil, drainage ditches; 0–2000 m. NW, SN (esp Plumas Co.); native to Eur.

M. verna Nutt. (p. 391) Ann, bien. **ST** erect, 1–3 dm; hairs rough, coarse. **LVS** 1–4 cm, 1–10 mm wide; basal oblanceolate; cauline linear to oblong or oblanceolate, sessile or lower short-petioled. **INFL**: bracts at base or to middle; pedicel in fr suberect, < calyx. **FL**: calyx 4–6 mm, hairs spreading, hooked, lobes unequal; corolla white, limb 1–2 mm diam. **FR** convex on outer side, keeled on inner. Waif in moist grain fields; 0–200 m. s KR; to B.C., Atlantic Coast, native to e N.Am, Pacific NW. [*M. virginica* (L.) Britton, Sterns & Pogg.]

PECTOCARYA

Timothy C. Messick and Barbara Veno

Ann. **ST** 2–40 cm, strigose, breaking apart at nodes or not. **LVS** gen alternate, gen 0.5–4 cm, ± linear, strigose to sharp-bristled. **INFL**: pedicel in fr gen free from nutlets, gen recurved. **FL**: sepals gen < fr, upper 2 in fr gen > others; corolla 0.8–3 mm, white; style attached to receptacle, unbranched, gen persistent, stigma 1, head-like. **FR**: nutlets gen 4, spreading, 1–4.5 mm, gen paired, gen compressed, marginal prickles hooked at tip, not barbed; nutlet pairs or all 4 often dissimilar in shape, ornamentation, margin width. 15 spp.: CA to B.C., WY, TX, n Mex; also S.Am. (Greek: comb nut, from dentate nutlet margins in some spp.) [Veno 1979 PhD dissertation UCLA] ❀STBL.

1. Calyx radial, sepals ± equal, > nutlets; nutlet margins ± entire, not dentate; lower cauline lvs opposite, fused at base (sect. *Gruvelia*)
  2. Sepal tip with hooked bristles; sts and sepals with appressed, short hairs; nutlet margins not membranous-winged; pedicel in fr 1–2.3 mm; nutlets 4-sided, not paired ..... ***P. pusilla***
  2′ Sepal tip with straight bristles; sts and esp sepals with appressed, short and several spreading, long, stiff hairs (incl bristles); nutlet margin gen widely membranous-winged; pedicel in fr ± 0.5 mm; nutlets ± obovate to ± round, paired ..... ***P. setosa***

1′ Calyx bilateral, upper 2 sepals > others, all < nutlets; nutlet margins entire to gen dentate; lower cauline lvs mostly alternate, free (sect. *Pectocarya*)
  3. Basal fls cleistogamous; basal nutlet margins less ornamented than cauline
    4. Pedicel in fr partly fused to 1 nutlet; cauline nutlets gen curved; lower 3 sepals unequal ..... ***P. heterocarpa***
    4′ Pedicel in fr free from nutlets; cauline nutlets gen straight; lower 3 sepals ± equal ..... ***P. peninsularis***
  3′ Basal fls not cleistogamous; basal, cauline nutlet margins similarly ornamented
    5. Nutlet margins ± entire, not or barely dentate ..... ***P. penicillata***
    5′ Nutlet margins dentate
      6. Nutlet strongly recurved to coiled, linear; nutlet margin teeth distinct ± to base ..... ***P. recurvata***
      6′ Nutlet straight to moderately recurved, ± oblanceolate; nutlet margin teeth distinct ± to or fused at base
        7. Sts prostrate to decumbent; nutlet margin teeth gen narrower at base than length, distinct ± to base; nutlets straight or slightly recurved at tip ..... ***P. linearis*** ssp. ***ferocula***
        7′ Sts ascending to erect; nutlet margin teeth ± as wide at base as length, fused at base; nutlets slightly to moderately recurved ± throughout ..... ***P. platycarpa***

P. heterocarpa (I.M. Johnston) I.M. Johnston (p. 391) **ST** prostrate to ascending, 2–25 cm. **INFL**: pedicel in fr 1.3–2.8 mm, partly fused to 1 nutlet. **FLS**: basal cleistogamous; in fr lower 3 sepals unequal, < others. **FR**: nutlet 1.2–3 mm, oblong to oblanceolate; basal nutlets 2–4 per fl, not paired, reflexed, 1 unmargined, others narrowly entire- to ± dentate-margined; cauline nutlets paired, curved, margins ± entire (and narrowly to widely membranous) to dentate. *n*=12. Washes, roadsides, openings in creosote bush scrub, Joshua tree woodland; < 1400 m. SW, D; to UT, TX, nw Mex.

P. linearis (Ruiz Lopez & Pavon) DC. ssp. ***ferocula*** (I.M. Johnston) Thorne (p. 391) **ST** prostrate to decumbent, 6–26 cm. **INFL**: pedicel in fr 1.5–3 mm. **FR**: nutlet body 2–3.8 mm, straight or slightly recurved near tip, linear-oblanceolate; margin teeth distinct ± to base, width at base gen < length. *n*=24. Roadsides, grassy slopes, clearings; 5–2000 m. GV, CW, SW, DMoj; Baja CA; also s S.Am.

P. penicillata (Hook. & Arn.) A. DC. (p. 391) **ST** prostrate to decumbent, 2–25 cm. **INFL**: pedicel in fr ascending, 1.3–2.5 mm. **FR**: nutlets straight, 1.1–3.3 mm, oblanceolate; margins erect to incurved, ± entire, all in a fl ± equal in width, bristled only above ± middle. *n*=12. Disturbed sites, roadsides in many communities; 90–2100 m. CA-FP, D; to B.C., WY, AZ, n Baja CA. Pls with margins of 1–2 nutlets narrower than others in same fl, bristles from nutlet base to tip, *n*=24, may be distinct sp.

P. peninsularis I.M. Johnston (p. 391) **ST** prostrate to ascending, 2–24 cm. **INFL**: pedicel in fr 1–1.5 mm. FLS: basal cleistogamous; in fr lower 3 sepals ± equal. **FR**: nutlets 1.1–2 mm, elliptic to ovate; basal nutlets 2–4 per fl, not paired, reflexed, 1–3 unmargined, others narrowly ± dentate-margined; cauline nutlets paired, gen straight, margins narrowly to widely membranous, ± dentate. *n*=12. Washes, roadsides, clearings; 30–300 m. DSon; Baja CA.

P. platycarpa (Munz & I.M. Johnston) Munz & I.M. Johnston (p. 391) **ST** ascending to erect, 4–25 cm. **INFL**: pedicels in fr 2.5–4 mm. **FR**: nutlet slightly to moderately recurved ± throughout, 2.5–4.5 mm, oblanceolate to narrowly obovate; margin teeth fused at base, width at base ± = length. *n*=24. Washes, roadsides in creosote-bush scrub, Joshua-tree woodland; 150–1500 m. SW, D; s NV, AZ, nw Mex.

P. pusilla (A. DC.) A. Gray (p. 391) **ST** ascending to erect, 3–38 cm. **LVS**: lower opposite, fused at base; upper alternate. **INFL**: pedicel in fr 1–2.3 mm. **FL**: sepals in fr ± equal, > nutlets, with appressed, short hairs, tips with hooked bristles. **FR**: nutlets not paired, 1.5–3 mm, ± 4-sided; margins entire, not membranous-winged. *n*=12. Dry sites in grassland, woodland, roadsides; 100–1800 m. NW, SNF, CW; to WA; waif in Chile.

P. recurvata I.M. Johnston (p. 391) **ST** ascending to erect, 3.5–21 cm. **FR**: pedicels in fr 2–3 mm. **FR**: nutlets strongly recurved to coiled, 2.5–4 mm, linear; margin teeth distinct ± to base, linear (or width at base < length). *n*=12. Shelter of rocks, bases of shrubs, sometimes roadsides, creosote bush scrub, Joshua tree woodland; 10–1600 m. D; to s NV, sw NM, nw Mex.

P. setosa A. Gray (p. 391) **ST** ascending to erect, 2–23 cm. **LVS**: lower opposite, fused at base; upper alternate. **INFL**: pedicels in fr ascending or reflexed, ± 0.5 mm. **FL**: sepals in fr ± equal, > nutlets, with appressed, short, and several spreading, long, stiff, hairs (incl bristles), tips with straight bristles. **FR**: nutlets 1.5–4 mm, ± obovate to ± round; margins ± entire, membranous-winged, wide on 3 nutlets, narrow on 1. *n*=12. Clearings in sagebrush scrub, creosote-bush scrub, pinyon/juniper woodland, grassland; 150–2300. s SN, CW, SW, D; to WA, ID, UT, AZ, n Baja CA.

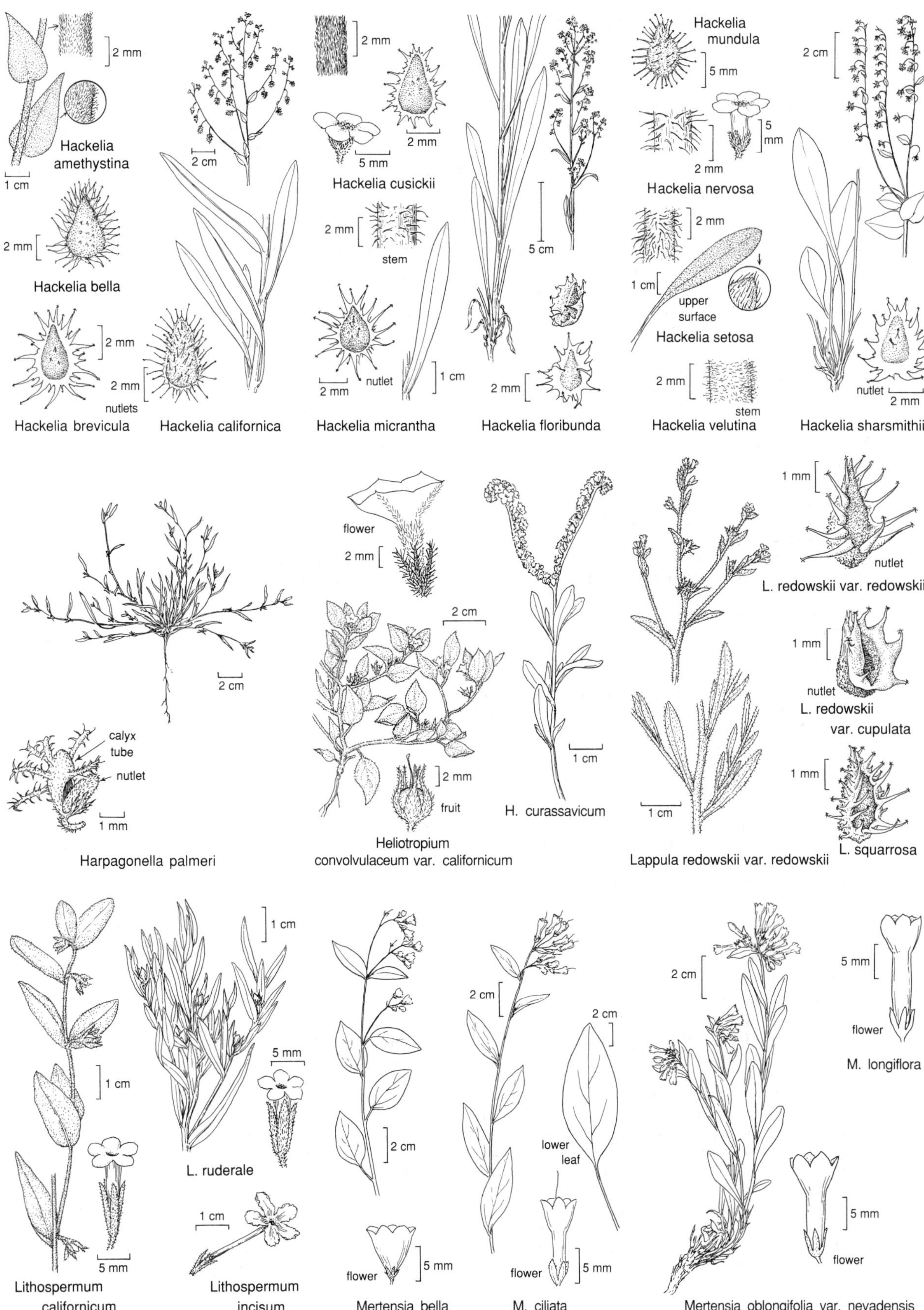
2 mm
Hackelia amethystina
1 cm
2 mm
Hackelia bella
2 mm
2 mm
nutlets
Hackelia brevicula
2 cm
Hackelia californica
2 mm
2 mm
5 mm
Hackelia cusickii
2 mm
stem
2 mm
nutlet
1 cm
Hackelia micrantha
5 cm
2 mm
Hackelia floribunda
Hackelia mundula
5 mm
5 mm
2 mm
Hackelia nervosa
2 mm
1 cm
upper surface
Hackelia setosa
2 mm
stem
Hackelia velutina
2 cm
nutlet
2 mm
Hackelia sharsmithii
2 cm
calyx tube
nutlet
1 mm
Harpagonella palmeri
flower
2 mm
2 cm
2 mm
fruit
Heliotropium convolvulaceum var. californicum
1 cm
H. curassavicum
1 cm
Lappula redowskii var. redowskii
1 mm
nutlet
L. redowskii var. redowskii
1 mm
nutlet
L. redowskii var. cupulata
1 mm
L. squarrosa
1 cm
5 mm
Lithospermum californicum
1 cm
5 mm
L. ruderale
1 cm
Lithospermum incisum
2 cm
flower
5 mm
Mertensia bella
2 cm
2 cm
lower leaf
flower
5 mm
M. ciliata
2 cm
5 mm
flower
5 mm
flower
M. longiflora
Mertensia oblongifolia var. nevadensis

PLAGIOBOTHRYS POPCORNFLOWER

Timothy C. Messick

Ann, per, gen strigose. **ST** prostrate to erect, branched at base or above, < 5 dm. **LVS** simple, 0.5–10 cm, gen smaller upward; all cauline (lower opposite, linear to oblong, upper gen alternate) or both basal (often in rosettes) and cauline (alternate, linear to oblanceolate). **INFL**: raceme or spike, coiled in bud, gen elongate in fr; bracts 0–many; pedicels gen 0–1 mm. **FL** bisexual; sepals fused below middle, 2–10 mm in fr; corolla 1–12 mm wide, all white or yellow inside tube. **FR**: nutlets gen 4, 1–3.5 mm; back gen with midrib, lateral ribs, cross-ribs, interspaces, gen tubercled, sometimes prickled or bristled; scar gen lateral (on side) near middle or base, sometimes basal (on bottom) or oblique (between side and bottom), sometimes on a stalk or short peg, gen ovate to triangular. ± 65 spp.: temp w N.Am, w S.Am. (Greek: sideways pit, from position of nutlet attachment scar) [Higgins 1974 Great Basin Natur 34(2):161–166; Johnson 1932 Contr Arnold Abroretum 3:1–102] Fully mature nutlets critical for identification; intergradation common in some spp. groups; sect. *Allocarya* often treated as a separate genus; many spp. need further study.

1. Nutlet scar at tip of oblique stalk (sect. *Echidiocarya*) .... ***P. collinus***
 2. Pl cespitose; infl short, < lvs .... var. ***ursinus***
 2′ Pl not cespitose; infl elongate, > lvs
  3. Corolla 4–7 mm wide; st hairs gen ± fine, appressed .... var. ***californicus***
  3′ Corolla 1–3 mm wide; st hairs gen coarse, spreading
   4. Lf oblanceolate, 3–5 mm wide; gen > 600 m .... var. ***fulvescens***
   4′ Lf linear, 2–2.5 mm wide; gen < 200 m .... var. ***gracilis***

1′ Nutlet scar sessile, lateral to oblique, or on short, basal to rarely ± oblique peg (gen < stalks in 1.), not on oblique stalk
 5. Lower cauline lvs alternate, sometimes ± paired, upper alternate; basal rosette present or 0; corolla tube white; nutlet scar lateral near middle; habitat gen dry, upland
  6. Scar gen above middle of nutlet; infl short, coiled in fr (sect. *Sonnea*) .... ***P. hispidus***
  6′ Scar at or slightly below middle of nutlet; infl gen elongate in fr
   7. Nutlet scar long, narrow, along crest of keel; basal rosette 0 (sect. *Amsinckiopsis*)
    8. Corolla 1–3 mm wide; nutlet covered with cobblestone-like bulges, tubercles 0 .... ***P. jonesii***
    8′ Corolla 4–7 mm wide; nutlet irregularly ribbed, tubercled .... ***P. kingii***
     9. Infl ± short, coiled in fr; st 0.5–1.5 dm .... var. ***harknessii***
     9′ Infl ± elongate in fr; st 1–4 dm .... var. ***kingii***
   7′ Nutlet scar short, ± round, below keel; basal rosette gen conspicuous (sect. *Plagiobothrys*)
    10. Nutlet scar deeply concave
     11. Bracts 0–few, near base of infl; basal rosette persistent .... ***P. fulvus***
     11′ Bracts many, throughout infl; basal rosette short-lived .... ***P. infectivus***
    10′ Nutlet scar convex to shallowly concave
     12. Interspaces of nutlet back gen wide, flat, between narrow, sometimes ± toothed cross-ribs
      13. Bracts few, near base of infl; corolla 3–9 mm wide; nutlet slightly arched in profile .... ***P. nothofulvus***
      13′ Bracts many, throughout infl; corolla 2–3 mm wide; nutlet gen strongly arched in profile
       14. Calyx circumscissile in fr; nutlets gen 2, strongly attached .... ***P. arizonicus***
       14′ Calyx not circumscissile in fr; nutlets gen 4, weakly attached .... ***P. canescens***
     12′ Interspaces of nutlet back gen narrow, groove-like, sometimes obscure, between wide, sometimes ± tubercled cross-ribs
      15. Nutlet ± cross-shaped, abruptly, ± equally narrowed near tip and base; sap gen not purple
       16. Infl not or only slightly coiled toward tip, fls gen < 10 per branch; calyx 5–7 mm; sts 1–few; nutlet 2–3 mm .... ***P. shastensis***
       16′ Infl ± strongly coiled toward tip, fls gen > 10 per branch; calyx 3–5 mm; sts several; nutlet 1–2 mm .... ***P. tenellus***
      15′ Nutlet ± ovoid, ± narrowed near tip but not base; sap gen purple
       17. Sepal hairs minutely hooked at tip .... ***P. uncinatus***
       17′ Sepal hairs not hooked at tip
        18. Nutlet cross-ribs rounded, tubercled to papillate, interspaces ± wider than crack-like, their bottoms narrow but visible .... ***P. myosotoides***
        18′ Nutlet cross-ribs flat, smooth, interspaces crack-like, their bottoms not visible .... ***P. torreyi***
         19. Sts few–many, decumbent to ascending; cauline lvs crowded; widespread .... var. ***diffusus***
         19′ Sts few, erect; cauline lvs scattered; uncommon .... var. ***torreyi***
 5′ Lower cauline lvs opposite, upper gen alternate; basal rosette 0; corolla tube often yellow inside; nutlet scar lateral near base or basal, sometimes oblique; habitat gen seasonally wet (sect. *Allocarya*)
  20. Per; st stout, coarse, rooting at nodes; hairs of st, lf long, soft, spreading .... ***P. mollis***
   21. Nutlet gray, ribs on back irregular; n SNH .... var. ***mollis***
   21′ Nutlet brown, ribs on back netted; SnFrB (s Sonoma Co.) .... var. ***vestitus***
  20′ Ann; st slender; hairs of st, lf not simultaneously long, soft, and spreading
   22. Fls gen present near base of st; pedicels near base of st stout, recurved; sepal midrib thickened; st prostrate
    23. Nutlet widely ovoid; scar 20–35% nutlet length, gen surrounded by funnel-like collar; lower lvs 1–2 cm .... ***P. scriptus***
    23′ Nutlet narrowly ovoid; scar < 20% nutlet length, surrounded by low ridge but not funnel-like collar; lower lvs 3–8 cm .... ***P. humistratus***

22′ Fls gen not present near base of st; pedicels gen not stout, not recurved; sepal midrib sometimes thickened (gen when nutlet scar basal); st prostrate to erect
24. Scar 25–60% nutlet length, gen concave
25. Nutlet sometimes bristled, but without large prickles
26. Nutlet scar gen linear, sometimes triangular near base .... ***P. strictus***
26′ Nutlet scar ± pear-shaped or ± triangular
27. Nutlet scar ± pear-shaped; nutlet ± 1.5 mm, cross-ribs wide; st prostrate to ascending; lower lvs 0.5–2.5 cm .... ***P. distantiflorus***
27′ Nutlet scar ± triangular; nutlet ± 2 mm, cross-ribs narrow; st ascending to erect; lower lvs 4–8 cm .... ***P. glyptocarpus***
28. Corolla 5–9 mm wide .... var. ***glyptocarpus***
28′ Corolla 2–3 mm wide .... var. ***modestus***
25′ Nutlet with large prickles (or rarely without them)
29. Nutlet cross-ribs 0 or obscure
30. Nutlet prickles stout, on midrib and lateral ribs only .... ***P. austinae***
30′ Nutlet prickles slender, all over nutlet back .... ***P. greenei***
29′ Nutlet cross-ribs present, gen prominent
31. Prickles of nutlet back without bristles .... [2]***P. trachycarpus***
31′ Prickles of nutlet back with minute bristles
32. Nutlet prickles long, slender, rarely ± 0; bristles ± sparse on nutlet prickles, 0 or sparse on surfaces between .... ***P. acanthocarpus***
32′ Nutlet prickles short, stout; bristles dense on nutlet prickles and surfaces between .... ***P. hystriculus***
24′ Scar < 20% nutlet length, gen not or only slightly concave
33. Nutlet scar basal or rarely ± oblique, often on short peg; sepal midrib ± fleshy
34. St decumbent; calyx gen strongly bent .... ***P. leptocladus***
34′ St ascending to erect; calyx straight or slightly bent
35. Calyx base fleshy; pedicel thick, hollow; uncommon .... ***P. glaber***
35′ Calyx base not fleshy; pedicel slender, gen solid; common .... ***P. stipitatus***
36. Corolla 2–3 mm wide .... var. ***micranthus***
36′ Corolla 5–12 mm wide .... var. ***stipitatus***
33′ Nutlet scar lateral near base or sometimes oblique, sessile; sepal midrib gen slender
37. Nutlet scar ± linear, slit- or keel-like
38. Nutlet smooth, shiny, without cross-ribs, 2.5–3 mm; corolla 2–4 mm wide — NCoRI .... ***P. lithocaryus***
38′ Nutlet tubercled, dull, with cross-ribs, 1–2 mm; corolla gen < 2 mm or > 5 mm wide
39. Corolla 1.5–2 mm wide; pedicel inconspicuous, gen 0–1 mm; nutlet back gen ± flat, interspaces narrower than or as wide as cross-ribs; NW to PR .... ***P. undulatus***
39′ Corolla 5–10 mm wide; pedicel conspicuous, gen 2-10 mm; nutlet back convex, interspaces wider than cross-ribs; CCo, SnFrB .... ***P. chorisianus***
40. Bases of lower lf pairs gen fused, loosely sheathing st; sts branched from upper axils; corolla 6–10 mm wide; pedicels gen > calyx .... var. ***chorisianus***
40′ Bases of lower lf pairs gen ± separate, not or only slightly sheathing st; sts branched from lower axils; corolla 5–6 mm wide; pedicels gen < calyx .... var. ***hickmanii***
37′ Nutlet scar gen elongate to triangular, not linear, not slit- or keel-like
41. St hairs spreading
42. Nutlets < 2 mm, cross-ribs prominent, scar lateral near base; bracts few, near base of infl .... ***P. parishii***
42′ Nutlets gen > 2 mm, cross-ribs low, scar gen oblique; bracts many, throughout infl .... ***P. salsus***
41′ St hairs appressed
43. Nutlet scar gen oblique
44. Nutlet scar and keel above not in a trough
45. Nutlet tubercled, not scabrous-bristled, not papillate .... ***P. bracteatus***
45′ Nutlet scabrous-bristled to ± papillate .... ***P. cognatus***
44′ Nutlet scar and keel above gen in a trough .... ***P. reticulatus***
46. Nutlet interspace tubercles gen sparse or small, cross-ribs gen irregularly net-like; trough around scar gen wide, distinct .... var. ***reticulatus***
46′ Nutlet interspace tubercles gen dense or large, cross-ribs ± linear at middle of nutlet, trough around scar gen narrow, obscure .... var. ***rossianorum***
43′ Nutlet scar gen lateral
47. Nutlet scar narrowly elliptic to obovate, ribs not toothed
48. Nutlet not bristled, not scabrous, scar elongate with thin, incurved margins .... ***P. cusickii***
48′ Nutlet minutely bristled or scabrous, scar narrowly elliptic .... ***P. hispidulus***
47′ Nutlet scar ovate to triangular, ribs ± toothed
49. Bracts 0–few, near base of infl; corolla 3–7 mm wide .... ***P. tener***
49′ Bracts many, throughout infl; corolla 1–2 mm wide .... [2]***P. trachycarpus***

P. acanthocarpus (Piper) I.M. Johnston (p. 391) ADOBE ALLOCARYA Ann, strigose. **ST** spreading to erect, 1–4 dm. **LVS** cauline; lower 2–6 cm. **INFL** bracted throughout; pedicels 1–2 mm. **FL**: calyx 3–6 mm; corolla 1–2.5 mm wide. **FR**: nutlet 1.5–2.5 mm, ovoid; cross-ribs gen narrow; prickles gen all over back, slender, long, rarely ± 0; bristles ± sparse, gen on prickles only; scar lateral near base, ovate to triangular, concave. Vernal pools, moist clay soil; < 700 m. s ScV, SnJV, SCo (mesas near San Diego); Mex. May intergrade with *P. greenei*.

P. arizonicus (A. Gray) A. Gray (p. 391) Ann; sap purple. **ST** ascending to erect, 1–4 dm; hairs rough, sharp, some spreading.

LVS: basal in rosette, 1.5–5 cm; cauline alternate. **INFL** bracted throughout. **FL**: calyx ± 3 mm, gen circumscissile in fr; corolla 2–2.5 mm wide. **FR**: nutlets gen 2, ± 2 mm, ovoid, strongly arched in profile, strongly attached; midrib, lateral ribs, cross-ribs narrow; interspaces wide; scar lateral near middle, round. Common. Dry, coarse soils in scrub or woodlands; < 2100 m. Teh, e SnFrB, SCoRI, TR, D; to NM, n Mex. Intergrades with *P. canescens*, *P. nothofulvus*. ❀TRY.

P. austinae (E. Greene) I.M. Johnston (p. 391) Ann, strigose. **ST** decumbent to erect, 1–4 dm. **LVS** cauline; lower 1–5 cm. **INFL** bracted near base. **FL**: calyx 5–8 mm; corolla 1–2.5 mm wide. **FR**: nutlet ± 2.5 mm, widely ovoid, angled, narrowed near tip; cross-ribs 0 or obscure; prickles on midrib and lateral ribs only, stout, bristled; scar lateral near base, ovate or triangular, concave. Vernal pools, wet sites; < 500 m. CaRF, SNF, e ScV, ne SnJV.

P. bracteatus (J. Howell) I.M. Johnston (p. 391) Ann, sparsely strigose. **ST** gen ascending, gen 1–4 dm. **LVS** cauline; lower 3–10 cm. **INFL** bracted below middle. **FL**: calyx 2–4 mm in fr; corolla 1–3 mm wide. **FR**: nutlet 1–2 mm, lanceolate-ovate; midrib short, near tip; lateral ribs below middle; cross-ribs irregular, ± continuous above middle, broken into tubercles below middle; scar oblique, ± ovate to deltate. Vernal pools, wet places in grassland, coastal-sage scrub, chaparral; < 1550 m. CA-FP (esp SW); sw OR, nw Mex. [var. *aculeolatus* (Piper) I.M. Johnston]

P. canescens Benth. (p. 391) Ann; sap purple. **ST** prostrate to erect, 1–6 dm, purple-dyed; hairs long, rough or bristled. **LVS**: basal in rosette, 1.5–5 cm; cauline alternate. **INFL** bracted throughout. **FL**: calyx 4–6 mm, gen not circumscissile in fr; corolla 2–3 mm wide. **FR**: nutlets ± 2 mm, round-ovoid, gen strongly arched in profile, gen weakly attached; midrib, lateral ribs, cross-ribs narrow; interspaces wide; scar lateral near middle, round. Common. Grasslands, woodlands, coastal scrub; < 1400 m. CaRF, SNF, GV, SW, w DMoj. Intergrades with *P. arizonicus*, *P. nothofulvus*; a variant on s ChI has been called var. *catalinensis* Jepson. ❀TRY.

P. chorisianus (Cham.) I.M. Johnston Ann, sparsely short-strigose. **ST** prostrate to erect, 1–4 dm. **LVS** cauline; lower 3–7 cm. **INFL** bracted near base; pedicels conspicuous, gen 2–10 mm. **FL**: calyx ± 4 mm; corolla 5–10 mm wide. **FR**: nutlet 1–2 mm, ovoid, dull; cross-ribs irregular to nearly 0; interspaces wider than cross-ribs; scar lateral near base, linear, set in deep groove. Grassy and moist places, coastal scrub, chaparral; < 100 m. CCo, SnFrB. May intergrade with *P. undulatus* in SnFrB. Vars. intergrade; if differences are environmentally induced, recognition is not warranted.

var. ***chorisianus*** (p. 391) CHORIS'S POPCORNFLOWER **ST** decumbent to erect, branched from upper axils. **LVS**: lower pairs gen fused at bases, loosely sheathing st. **INFL**: pedicel gen > calyx. **FL**: corolla 6–10 mm wide. UNCOMMON. Habitat, elevation of sp. n CCo, sw SnFrB.

var. ***hickmanii*** (E. Greene) I.M. Johnston **ST** prostrate, branched from lower axils. **LVS**: lower pairs gen ± separate at bases, not or only slightly sheathing st. **INFL**: pedicel gen < calyx. **FL**: corolla 5–6 mm wide. Habitat, elevation of sp. c CCo.

P. cognatus (E. Greene) I.M. Johnston (p. 391) Ann, strigose. **ST** decumbent to erect, 0.5–3 dm. **LVS** cauline; lower 2–7 cm. **INFL** bracted below middle. **FL**: calyx 2–4 mm; corolla 1–1.5 mm wide. **FR**: nutlet 1–2 mm, lanceolate-ovate, scabrous-bristled to ± papillate; midrib, lateral ribs short, near tip; cross-ribs irregular, rounded; scar oblique, ± triangular. Moist places in meadows, forests; < 2100 m. c SN, NCoRI; to WA, Rocky Mtns, AZ. [*P. scouleri* (Hook. & Arn.) I.M. Johnston var. *penicillatus* (E. Greene) Cronq., in part] With *P. cusickii*, *P. hispidulus*, *P. reticulatus*, *P. scouleri* (OR, WA to Rocky Mtns, Can), and possibly others, forming a widespread, highly variable complex of poorly defined taxa in need of further study.

P. collinus (Philbr.) I.M. Johnston (p. 391) Ann; hairs appressed to spreading, fine or coarse. **ST** prostrate to ascending, gen 1–4 dm. **LVS** cauline; lower gen opposite, 1–4 cm; upper alternate. **INFL** gen elongate, bracted throughout. **FL**: calyx ± 3 mm; corolla 1–7 mm wide. **FR**: nutlet 1–2 mm, ovoid; midrib, lateral ribs, cross-ribs sharp; scar at tip of oblique stalk. Dry places, many habitats; < 1200 m. s CA-FP, D; Mex, Chile. Vars. *californicus*, *fulvescens*, *gracilis* ± intergrading.

var. ***californicus*** (A. Gray) Higgins Pl not cespitose. **ST**: hairs gen ± fine, appressed. **LF** 1–3 cm, 2–5 mm wide, oblancelate. **INFL** > lvs. **FL**: corolla 4–7 mm wide. Grassland, coastal scrub; < 400 m. CW, SW; Mex. [*P. californicus* (A. Gray) E. Greene]

var. ***fulvescens*** (I.M. Johnston) Higgins Pl not cespitose. **ST**: hairs gen coarse, spreading. **LF** 1–3 cm, 3–5 mm wide, oblanceolate. **INFL** > lvs. **FL**: corolla ± 2 mm wide. Dry places, chaparral, coniferous forest; gen 600–2000 m. SCoRO, SW; AZ, Mex, Chile. [*P. californicus* var. *f.* I.M. Johnston]

var. ***gracilis*** (I.M. Johnston) Higgins Pl not cespitose. **ST**: hairs gen coarse, spreading. **LF** 1–4 cm, 2–2.5 mm wide, linear. **INFL** > lvs. **FL**: corolla 1.5–2 mm wide. Dry places; gen < 200 m. s SCo (near San Diego), ChI. [*P. californicus* var. *g.* I.M. Johnston]

var. ***ursinus*** (A. Gray) Higgins Pl cespitose. **ST**: hairs gen coarse, spreading. **LF** 1–2.5 cm, 3–5 mm wide, oblanceolate. **INFL** < lvs, dense. **FL**: corolla 1–2 mm wide. Sandy or gravelly soils, coniferous forest; 1400–2100 m. SnBr, SnJt; Mex. [*P. californicus* var. *u.* (A. Gray) I.M. Johnston]

P. cusickii (E. Greene) I.M. Johnston (p. 391) Ann, strigose. **ST** prostrate to ascending, 0.5–2 dm. **LVS** cauline; lower 3–10 cm. **INFL** bracted below middle. **FL**: calyx 1.5–4 mm; corolla 1–1.5 mm wide. **FR**: nutlet 1–2 mm, lanceolate-ovate; midrib, lateral ribs short, near tip; cross-ribs irregular, rounded; scar lateral near base, elongate, with thin, incurved margins. Montane flats, meadows; 1200–2100 m. GB; to WA, NV. [*P. scouleri* (Hook. & Arn.) I.M. Johnston var. *penicillatus* (E. Greene) Cronq., in part] See note under *P. cognatus*.

P. distantiflorus (Piper) M. Peck (p. 391) Ann, strigose. **ST** prostrate to ascending, 1.5–3 dm. **LVS** cauline; lower 0.5–2.5 cm. **INFL** bracted ± throughout. **FL**: calyx 2.5–3 mm; corolla 1–1.5 mm wide. **FR**: nutlet ± 1.5 mm, narrow near tip, ± arched in profile; cross-ribs wide, tubercled; interspaces shallow; scar lateral near base, ± pear-shaped, concave. Moist places in grasslands, woodlands; < 400 m. n&c SNF. May intergrade with *P. glyptocarpus*.

P. fulvus (Hook. & Arn.) I.M. Johnston (p. 391) Ann, tomentose; some hairs spreading, long, rough or sparse; sap purple. **ST** erect, 3–6 dm. **LVS**: basal in loose, persistent rosette, 2–8 cm; cauline alternate. **INFL**: bracts 0–few, near base. **FL**: calyx 5–6 mm; corolla 3–4 mm wide. **FR**: nutlets 2–4, 2.5–3 mm, triangular-ovoid, ± narrowed below tip; cross-ribs narrow, straight; interspaces wide; scar lateral near middle, ± round, deeply concave. Common. Woodlands, grasslands, sandy or gravelly soil; < 500 m. NCoR, SNF, SnFrB; OR, Chile. N.Am pls have been called var. *campestris* (E. Greene) I.M. Johnston. ❀TRY.

P. glaber (A. Gray) I.M. Johnston (p. 391) HAIRLESS POPCORNFLOWER Ann. **ST** ascending to erect, thick, hollow. **LVS** cauline; lower 2–11 mm. **INFL**: pedicels 1–2 mm, thick, hollow. **FL**: calyx 8–10 mm in fr, bottom 2–3 mm fused into fleshy cylinder (sepal midribs also fleshy); corolla ± 3 mm wide. **FR**: nutlet 1.5–2.5 mm, lanceolate to lance-ovate; midrib, lateral ribs low, rounded, below middle obscure; cross-ribs above middle, rounded, oblique or arched, sometimes bristled; scar basal, gen on short peg. **PRESUMED EXTINCT**. Wet, alkaline soils in valleys, coastal marshes; < 100 m. CCo, s SnFrB (esp near Hollister). Perhaps a var. of *P. stipitatus*.

P. glyptocarpus (Piper) I.M. Johnston (p. 391) Ann, strigose. **ST** ascending to erect, 1–5 dm. **LVS** cauline; lower 4–8 cm. **INFL** bracted near base. **FL**: calyx 3–5 mm; corolla 2–9 mm wide. **FR**: nutlet ± 2 mm, ovoid; lateral ribs, cross-ribs narrow, high, ± toothed, irregular; interspaces deep; scar lateral near base, ± triangular, concave. Moist places, grasslands, woodlands; < 600 m. NW, CaR, n SNF; OR. May intergrade with *P. distantiflorus*.

var. ***glyptocarpus*** **FL**: corolla 5–9 mm wide. NW, CaR; OR.

var. ***modestus*** I.M. Johnston CEDAR CREST POPCORNFLOWER **FL**: corolla 2–3 mm wide. n SNF (near Grass Valley, Nevada Co.). UNCOMMON. May be a minor variant or hybrid.

P. greenei (A. Gray) I.M. Johnston (p. 391) Ann, strigose. **ST** decumbent to erect, 1–4 dm. **LVS** cauline; lower 1–5 cm. **INFL** bracted near base. **FL**: calyx 5–8 mm; corolla 1–2.5 mm wide. **FR**: nutlet ± 2.5 mm, widely ovoid, narrowed near tip; cross-ribs 0 or obscure; prickles all over back, long, slender, bristled; scar lateral near base, ovate or triangular, concave. Wet sites, grassland to woodland; < 900 m. NCoR, CaRF, SNF, ScV, n SnJV; OR. May intergrade with *P. acanthocarpus*.

P. hispidulus (E. Greene) I.M. Johnston (p. 391) Ann, strigose. **ST** prostrate or ± ascending, 0.5–4 dm. **LVS** cauline; lower 1–5 cm. **INFL** bracted below middle. **FL**: calyx ± 3 mm; corolla 1–2 mm wide. **FR** 1.5–2 mm, ± ovoid, minutely bristled or scabrous; midrib, cross-ribs narrow, low; interspaces wide; scar lateral near base, ± narrowly elliptic. Moist or drying sites; 1200–3400 m. KR, SN, TR, PR; to WA, WY, NV. [*P. scouleri* (Hook. & Arn.) I.M. Johnston var. *penicillatus* (E. Greene) Cronq., in part] See note under *P. cognatus*.

P. hispidus A. Gray (p. 391) Ann, sharply hairy and sparsely short-tomentose. **ST** erect, 0.5–2 dm. **LVS** mostly cauline, alternate, 1.5–4 cm, ± blistered. **INFL** short, coiled in fr, bracted below middle. **FL**: calyx 2 mm; corolla 1 mm wide. **FR**: nutlet gen 1, 1–1.5 mm, ovoid; midrib, lateral ribs low, wide; cross-ribs 0; scar lateral gen above middle, ± round. Dry places, gen in sandy soil; 1200–2800 m. CaR, GB; OR, w NV.

P. humistratus (E. Greene) I.M. Johnston (p. 391) Ann, sparsely strigose. **ST** prostrate, 1–4 dm. **LVS** cauline; lower 3–8 cm. **INFL** bracted ± throughout; fls gen present near base of st; lower pedicels < 2 mm, recurved. **FL**: calyx 6–10 mm; corolla 1–2 mm wide. **FR**: nutlet 2–2.5 mm, lance-ovoid; cross-ribs irregular, often bristled; interspaces wide; scar lateral near base, < 20% nutlet length, triangular to ovate, gen concave, surrounded by low ridge. Vernal pools, wet places, grassland; < 200 m. GV (foothills). May intergrade with the similar *P. scriptus*.

P. hystriculus (Piper) I.M. Johnston (p. 391) BEARDED POPCORNFLOWER Ann, strigose. **ST** spreading to erect, 1–4 dm. **LVS** cauline; lower 2–6 cm. **INFL** bracted throughout; pedicels 1–2 mm. **FL**: calyx 3–6 mm; corolla 1–2.5 mm wide. **FR**: nutlet 1.5–2.5 mm, ovoid; cross-ribs gen narrow; prickles short, stout; bristles dense, minute, all over back; scar lateral near base, ovate to triangular, concave. **PRESUMED EXTINCT**. Grassland, probably vernal pools, wet sites; < 50 m. sw ScV (Solano Co.).

P. infectivus I.M. Johnston Ann, tomentose, some hairs spreading, long, rough or sparse; sap purple. **ST** erect, 1–5 dm. **LVS**: basal in short-lived rosette, 2–8 cm; cauline alternate. **INFL** bracted throughout. **FL**: calyx 5–7 mm; corolla 3–4 mm wide. **FR**: nutlets 2–4, 3–3.5 mm, triangular-ovoid, ± narrowed below tip; cross-ribs few, irregular; scar lateral near middle, gen oval, deeply concave. Clay soils; < 300 m. SnFrB, SCoRI. Perhaps only a var. of the similar *P. fulvus*.

P. jonesii A. Gray (p. 391) Ann; hairs sharp, bristled, spreading. **ST** ascending to erect, < 4 dm. **LVS** cauline, alternate; lower 2–10 cm. **INFL** bracted near base. **FL**: calyx 4–8 mm; corolla 1–3 mm wide. **FR**: nutlets 3–4, 2–3 mm, triangular-ovate, covered with cobblestone-like bulges; tubercles 0; midrib, lateral ribs weak; cross-ribs 0; scar lateral near middle, along crest of keel, long, narrow. Sandy, gravelly, or rocky slopes, creosote bush scrub to pinyon/juniper woodland; < 1800 m. SNE, DMoj; to UT, w AZ, Mex.

P. kingii (S. Watson) A. Gray (p. 391) GREAT BASIN POPCORNFLOWER Ann; hairs coarse, stiff, spreading, unequal. **ST** ascending to erect, 0.5–4 dm. **LVS** cauline, alternate; lower crowded, < 6 cm; upper scattered. **INFL** bracted near base. **FL**: calyx 3–4 mm; corolla 4–7 mm wide. **FR**: nutlet 2–3 mm, ovoid, arched in profile; midrib, lateral ribs, cross-ribs irregular, coarse; scar lateral near middle, along crest of keel, long, narrow. Dry, open slopes, sagebrush scrub, saltbush scrub, juniper woodland; 1200–2300 m. GB, DMoj. Vars. intergrade.

var. ***harknessii*** (E. Greene) Jepson **STS** all ascending, 0.5–1.5 dm. **LF** < 1 cm wide. **INFL** ± short, coiled in fr. Habitat and elevation as in sp. GB; se OR, NV.

var. ***kingii*** **STS** ascending at pl sides to erect at pl center, 1–4 dm. **LF** < 2 cm wide. **INFL** ± elongate in fr. Habitat and elevation as in sp. SNE, DMoj; to se OR, w UT.

P. leptocladus (E. Greene) I.M. Johnston (p. 391) ALKALI PLAGIOBOTHRYS Ann, sparsely strigose to ± glabrous. **ST** decumbent, 1–3 dm. **LVS** cauline; lower 3–10 cm. **INFL** bracted below middle. **FL**: calyx 4–8 mm, gen strongly bent, turning corolla skyward, sepal midribs ± fleshy; corolla 1–2 mm wide. **FR**: nutlet 1.5–2.5 mm, ± lanceolate; midrib above middle; cross-ribs 0 or few, weak, above middle; back often minutely bristled; scar basal, often on short peg. Gen alkaline clay soils in vernal pools, wet places; < 2500 m. SW, w DMoj; to AK, c Can, Mex.

P. lithocaryus (E. Greene) I.M. Johnston (p. 391) MAYACAMAS POPCORNFLOWER Ann, short-strigose. **ST** erect, 1–3 dm. **LVS** cauline; lower 3–6 cm. **INFL** bracted throughout; pedicels 1–4 mm. **FL**: calyx 4 mm; corolla 2–4 mm wide. **FR**: nutlet 2.5–3 mm, shiny; cross-ribs 0; tubercles 0; scar lateral near base, at base of long, narrow slit, linear. **PRESUMED EXTINCT**. Moist sites; 300–450 m. s NCoRI (valleys near Mayacamas Range).

P. mollis (A. Gray) I.M. Johnston Per; hairs dense, long, soft, spreading. **ST** decumbent, 1–3 dm, rooting at nodes. **LVS** cauline; lower 4–8 cm; upper opposite or alternate. **INFL**: bracts 0. **FL**: calyx 4–5 mm; corolla 5–10 mm wide. **FR**: nutlet ± 1.5 mm, ovoid; ribs on back irregular or netted; scar lateral near middle, ovate or triangular. Sagebrush scrub, grassland; < 50 or 1200–1700 m. n SNH, SnFrB; OR, NV. Relationships to per spp. of S.Am in sect. *Allocarya* should be studied.

var. ***mollis*** (p. 401) **FR**: nutlet gray; ribs on back irregular, ± as wide as interspaces. Uncommon. Moist, alkaline places in sagebrush scrub; 1200–1700 m. n SNH; OR, NV.

var. ***vestitus*** (E. Greene) I.M. Johnston (p. 401) PETALUMA POPCORNFLOWER **FR**: nutlet brown; ribs on back netted, gen narrower than interspaces. **PRESUMED EXTINCT**. Wet sites in grassland, possibly coastal marsh margins; < 50 m. SnFrB (near Petaluma, s Sonoma Co).

P. myosotoides (Lehm.) Brand FORGET-ME-NOT POPCORNFLOWER Ann; hairs stiff, spreading; sap purple. **ST** erect, 0.5–2 dm. **LVS**: basal in loose rosette, 1–2 cm; cauline alternate. **INFL** bracted throughout. **FL**: calyx 2.5 mm; corolla 1.5–2 mm wide. **FR**: nutlet ± 1.5 mm; cross-ribs rounded, tubercled to papillate; interspaces ± wider than crack-like, with narrow bottoms; scar lateral near middle, ± round. UNCOMMON. Chaparral; 500–2000 m. s SNF (Fresno, Tulare cos.), SnFrB (Santa Clara Co.); Peru, Chile.

P. nothofulvus (A. Gray) A. Gray (p. 401) POPCORNFLOWER Per; hairs spreading, rough, sharp; sap purple. **ST** ± erect, 2–7 dm. **LVS**: basal in rosette, 3–10 cm; cauline few, alternate. **INFL** bracted near base. **FL**: calyx 2–3 mm, often circumscissile in fr; corolla 3–9 mm wide. **FR**: nutlets gen 1–3, ± 2 mm, round-ovoid, abruptly narrowed below acute tip, slightly arched in profile, sometimes strongly attached; midrib, lateral ribs, cross-ribs narrow; interspaces wide; scar lateral near middle, round. $2n=24$. Abundant. Grasslands, woodlands; gen < 800 m. CA-FP, rarely edge of D; to WA, Mex. Intergrades with *P. arizonicus*, *P. canescens*. ❀TRY.

P. parishii I.M. Johnston (p. 401) PARISH'S POPCORNFLOWER Ann; hairs short, spreading. **ST** prostrate, 0.5–3 dm. **LVS** cauline; lower 1–5 cm, lower surface blistered. **INFL** bracted near base. **FL**: calyx 2–3 mm; corolla 3–5 mm wide. **FR**: axial nutlet 1–1.8 mm, > others, ovoid or lance-ovoid; midrib, lateral ribs, cross-ribs narrow, irregular; scar lateral near base, gen linear to narrowly triangular. Uncommon. Wet, alkaline soil around desert springs; 750–1400 m. SNE, DMoj.

P. reticulatus (Piper) I.M. Johnston (p. 401) Ann, ± strigose. **ST** decumbent to erect, 1–4 dm. **LVS** cauline; lower 3–8 cm. **INFL** bracted below middle. **FL**: calyx 2–4 mm; corolla 1.5–3.5 mm wide. **FR**: nutlet 0.7–1.7 mm, narrowly ovoid; midrib, lateral ribs, cross-ribs irregular; scar oblique, in a trough, narrow-elliptic to triangular. Common. Moist places in forests, grasslands; gen < 300 m. NW, SnFrB; OR. See note under *P. cognatus*; vars. intergrade.

var. ***reticulatus*** **FR**: cross-ribs gen irregularly net-like; interspace tubercles gen sparse or small; trough around scar gen wide, distinct. Habitat, elevation, range of sp.

var. ***rossianorum*** I.M. Johnston **FR**: cross-ribs gen ± linear at middle of back; interspace tubercles gen dense or large; trough around scar gen narrow, obscure. Habitat, elevation of sp. NW, SnFrB. [*P. scouleri* (Hook. & Arn.) I.M. Johnston var. *penicillatus* (E. Greene) Cronq., in part] Pls from Presidio of San Francisco with nutlet back densely tubercled and nutlet walls thick have been called *P. diffusus* (E. Greene) I.M. Johnston, San Francisco popcornflower, **ENDANGERED** CA, but appear to intergrade with other minor variants not recognized taxonomically.

P. salsus (Brandegee) I.M. Johnston Ann; hairs stiff, spreading. **ST** decumbent to erect, 0.6–1.6 dm. **LVS** cauline; lower 3–6 cm. **INFL** bracted throughout. **FL**: calyx 2–5 mm; corolla 2–4 mm wide. **FR**: nutlet 1.5–2.5 mm, lanceolate; cross-ribs few, low, above middle, arched or irregular; scar gen oblique, narrowly ovate. Moist, alkaline mud flats; ± 700 m. ne DMoj (e Inyo Co); se OR, NV.

P. scriptus (E. Greene) I.M. Johnston (p. 401) Ann, sparsely strigose. **ST** prostrate, 1–2 dm. **LVS** cauline; lower 1–2 cm; upper 0.5–2 cm. **INFL** bracted ± throughout; fls present near base of st; lower pedicels < 2 mm, stout, recurved. **FL**: calyx 4–8 mm; corolla ± 2 mm wide. **FR**: nutlet 2 mm, widely ovoid; cross-ribs narrow, irregular, minutely bristled; scar lateral near base, 20–35% nutlet length, triangular, concave, surrounded by funnel-like collar. Moist sites in grassland; gen < 150 m. n SNF, c SNF (Stanislaus Co.). Pls with obscure collar around nutlet scar may intergrade with *P. humistratus*.

P. shastensis A. Gray (p. 401) Ann; hairs straight. **STS** erect, 1–few, 0.5–3 dm. **LVS**: basal in rosette, 1–3 cm; cauline few, alternate. **INFL** not or only slightly coiled toward tip, bracted throughout; fls gen < 10 per branch, gen paired. **FL**: calyx 5–7 mm; corolla 1–3 mm wide. **FR**: nutlet 2–3 mm, thickly cross-shaped; cross-ribs wide, gen tubercled; interspaces narrow, groove-like; scar lateral near middle. Uncommon. Dry slopes, flats, grassland, woodland; < 800 m. NW, n&c SNF, SnFrB, SCoRI; sw OR.

P. stipitatus (E. Greene) I.M. Johnston Ann, short-strigose. **ST** ascending to erect, 1–5 dm, often fleshy, hollow. **LVS** cauline; lower 2–11 cm. **INFL** bracted below middle; pedicels gen solid. **FL**: calyx 5–8 mm, sepal midribs ± fleshy; corolla 2–12 mm wide. **FR**: nutlet 1.5–2.5 mm, lanceolate to lance-ovate; midrib, lateral ribs low, rounded, below middle obscure; cross-ribs above middle, rounded, oblique or arched, sometimes bristled; scar basal or rarely ± oblique, gen on short peg. Vernal pools, wet sites; < 1500 m. CA-FP, GB; OR. Vars. intergrade in SnJV.

var. ***micranthus*** (Piper) I.M. Johnston (p. 401) **FL**: corolla 2–3 mm wide. Common. Vernal pools, wet sites in grasslands to conifer forests; < 1500 m. CA-FP, GB; se OR.

var. ***stipitatus*** **FL**: corolla 5–12 mm wide. Vernal pools, wet sites in grasslands; < 450 m. CA-FP (esp GV, NW); OR.

P. strictus (E. Greene) I.M. Johnston (p. 401) CALISTOGA POPCORNFLOWER Ann, ± glabrous, exc calyx strigose. **ST** erect, 1–4 dm. **LVS** cauline; lower 4–9 cm. **INFL**: branches paired; bracts 0 or few near base. **FL**: calyx 3 mm; corolla 4–6 mm wide. **FR**: nutlet ± 1.5 mm, ovoid; midrib above middle; cross-ribs narrow, irregular; interspaces irregular; scar lateral near base, gen linear, sometimes triangular near base, flanked by oblique ribs. RARE. Moist sites near hot springs; 100–150 m. s NCoRI (Calistoga, Napa Co.).

P. tenellus (Nutt.) A. Gray (p. 401) Ann; hairs spreading, shaggy. **STS** several, erect, 0.5–3 dm. **LVS**: basal in rosette, 1–5 cm; cauline few, alternate. **INFL** ± strongly coiled toward tip, bracted near base; fls gen > 10 per branch. **FL**: calyx 3–5 mm; corolla 1–3 mm wide. **FR**: nutlet 1–2 mm, thickly cross-shaped; cross-ribs wide, gen tubercled; interspaces narrow, groove-like, often obscure; scar lateral near middle. Common. Dry slopes in grassland, scrub, woodland, or forest; < 1700 m. CA-FP, GB (uncommon), D; to B.C., ID, UT, AZ, Mex.

P. tener (E. Greene) I.M. Johnston Ann, sparsely strigose. **ST** gen erect, 1–3 dm. **LVS** cauline; lower 2–6 cm. **INFL** bracted near base. **FL**: calyx lobes 2–3.5 mm; corolla 3–7 mm wide. **FR**: nutlet 1.5–2.2 mm, lance-ovoid, slightly arched in profile; midrib, lateral ribs gen obscure; cross-ribs mostly above middle, low, irregular, ± toothed; interspaces wide; scar lateral near base, narrowly ovate to triangular. Moist meadows, stream banks in chaparral, oak woodlands, conifer forests; < 1700 m. NCoR, CaR, n SN, MP.

P. torreyi (A. Gray) A. Gray (p. 401) Ann; hairs stiff, spreading; sap purple. **ST** decumbent to erect, 0.5–2 dm. **LVS**: basal in loose rosette, 1–2 cm; cauline alternate. **INFL** bracted throughout. **FL**: calyx 2.5 mm; corolla 1.5–2 mm wide. **FR**: nutlet 1.5–2.2 mm, round-ovoid, narrowed near tip; midrib, lateral ribs low, sparsely tubercled; cross-ribs wide, flat; interspaces crack-like, without visible bottoms; scar lateral near middle, ± round. Moist meadows, flats, forest edges; 1200–3400 m. SNH, SnBr.

var. ***diffusus*** I.M. Johnston **STS** few–many, decumbent to ascending. **LVS**: cauline crowded. Common. Moist meadows, flats, forest edges; 1200–3400 m. SNH, SnBr.

var. ***torreyi*** **STS** few, erect. **LVS**: cauline scattered. Uncommon. Moist meadows, flats, forest edges; ± 1200 m. SNH (Yosemite Valley area).

P. trachycarpus (A. Gray) I.M. Johnston (p. 401) Ann, strigose. **ST** prostrate to ascending, 0.5–4 dm. **LVS** cauline; lower 5–10 cm. **INFL** bracted throughout. **FL**: calyx 4–6 mm; corolla 1–2 mm wide. **FR**: nutlet ± 2 mm, ovoid; midrib, lateral ribs, cross-ribs gen narrow, ± toothed, rarely ± prickled; interspaces wide; scar lateral near base, ovate or triangular, often concave. Vernal pools, wet places in grassland, scrub, chaparral, woodland; < 1000 m. SnJV, SnFrB, SCo, WTR.

P. uncinatus J. Howell (p. 401) HOOKED POPCORNFLOWER Ann; hairs stiff, spreading; sap purple. **ST** decumbent to erect, 0.5–2 dm. **LVS**: basal in loose rosette, 1–2 cm; cauline alternate. **INFL** bracted throughout. **FL**: calyx 2–2.5 mm, hairs minutely hooked at tip; corolla 1.5–2 mm wide. **FR**: nutlet 1–1.3 mm, widely ovoid; midrib, lateral ribs narrow; cross-ribs wide, irregular; interspaces narrow; scar lateral near middle, ± round. UNCOMMON. Canyon sides, chaparral; 300–600 m. n SCoR (Gabilan Range, Santa Lucia Mtns).

P. undulatus (Piper) I.M. Johnston (p. 401) Ann, sparsely short-strigose. **ST** spreading to erect, 1–3 dm. **LVS** cauline; lower 2–6 cm. **INFL**: bracts few; pedicels inconspicuous, gen 0–1 mm. **FL**: calyx ± 2 mm; corolla 1.5–2 mm wide. **FR**: nutlet 1–1.6 mm, ovoid to lance-ovoid; back gen ± flat; midrib, lateral ribs 0 to obscure; cross-ribs low, ± curved; interspaces narrower than or as wide as cross-ribs; scar lateral near base, linear, in groove below keel. Vernal pools, wet places; < 400 m. NCo, NCoRO, s ScV, CCo, SnFrB, SCoRO, SCo, WTR, PR. May intergrade with *P. chorisianus* in SnFrB.

SYMPHYTUM COMFREY

Dieter H. Wilken and Ronald B. Kelley

Per; root thick, carrot-like. **STS** ascending to erect. **LVS** gen cauline; lower petioled; upper short-petioled to sessile; blade lanceolate to ovate. **INFL** terminal or axillary, gen peduncled, coiled in fl. **FL**: calyx deeply lobed, bristly; corolla bell-shaped to ± urn-shaped, throat appendages 5, ± = stamens, papillate, alternating with stamens; stamens inserted on upper tube; style exserted. **FR**: nutlets 1–4, ovoid; tip ± incurved, scar at base ± flat with thick, ring-like, minutely toothed rim. 35 spp.: Eurasia. (Greek: growing together, from putative healing properties) [Gadella 1984 Ann Missouri Bot Gard 71:1061–1067] Orn, folk medicine, cult for forage.

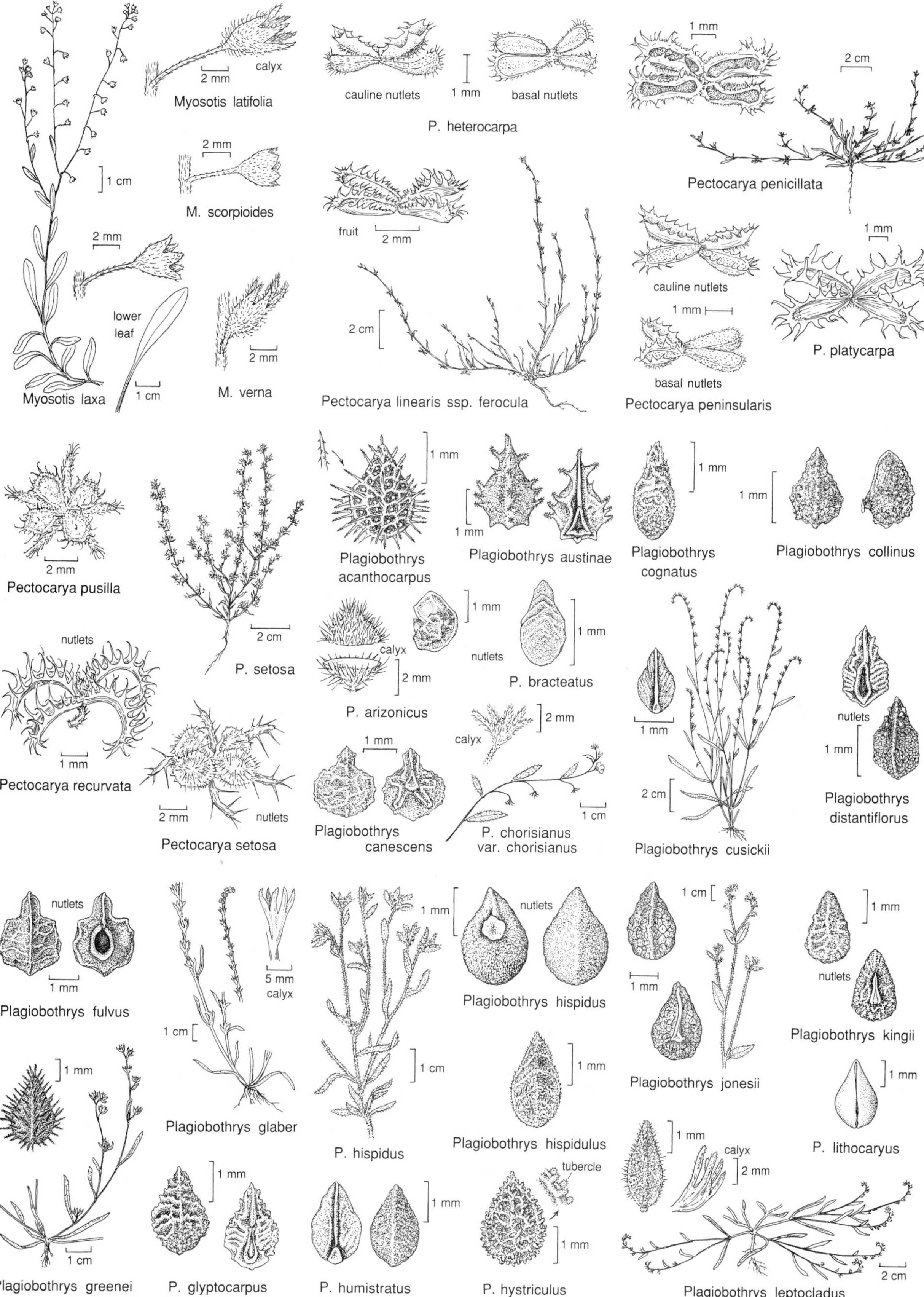
calyx
2 mm
Myosotis latifolia
1 cm
2 mm
M. scorpioides
2 mm
lower leaf
1 cm
2 mm
M. verna
Myosotis laxa
cauline nutlets
1 mm
basal nutlets
P. heterocarpa
fruit
2 mm
2 cm
Pectocarya linearis ssp. ferocula
1 mm
2 cm
Pectocarya penicillata
cauline nutlets
1 mm
basal nutlets
Pectocarya peninsularis
1 mm
P. platycarpa
2 mm
Pectocarya pusilla
2 cm
P. setosa
nutlets
1 mm
Pectocarya recurvata
2 mm
nutlets
Pectocarya setosa
1 mm
Plagiobothrys acanthocarpus
1 mm
Plagiobothrys austinae
1 mm
Plagiobothrys cognatus
1 mm
Plagiobothrys collinus
1 mm
calyx
2 mm
P. arizonicus
1 mm
nutlets
P. bracteatus
1 mm
Plagiobothrys canescens
calyx
2 mm
1 cm
P. chorisianus var. chorisianus
1 mm
2 cm
Plagiobothrys cusickii
nutlets
1 mm
Plagiobothrys distantiflorus
nutlets
1 mm
Plagiobothrys fulvus
5 mm
calyx
1 cm
Plagiobothrys glaber
1 mm
1 cm
P. hispidus
nutlets
Plagiobothrys hispidus
1 mm
Plagiobothrys hispidulus
1 cm
1 mm
Plagiobothrys jonesii
1 mm
nutlets
Plagiobothrys kingii
1 mm
P. lithocaryus
1 mm
1 cm
Plagiobothrys greenei
1 mm
P. glyptocarpus
1 mm
P. humistratus
tubercle
1 mm
P. hystriculus
1 mm
calyx
2 mm
2 cm
Plagiobothrys leptocladus

1. Internodes not winged; lf base not decurrent, sometimes decurrent < 1 cm; nutlets brown, fine-granular **S. asperum**
1′ Internodes clearly winged; lf base decurrent to wing; nutlets black, shiny .... ***S. officinale***

S. asperum Lepechin **ST** 6–10 dm, branched, sparsely sharp-bristly. **LF** 5–15 cm, sparsely bristly. **FL**: calyx 2–4 mm in fl, lobes gen linear-oblong in fl, triangular in fr; corolla 10–15 mm, pink, turning deep blue or purple, throat appendages linear-lanceolate. **FR**: nutlets 3–4 mm, 2–2.5 mm wide, brown, fine-granular. $2n=32$. Wet, open sites, ditches; < 100 m. NCo, n NCoRO, waif elsewhere (ScV, Yolo Co.); to B.C., ne US, Eur; native to sw Asia. NOXIOUS WEED.

S. officinale L. **ST** 5–10 dm, simple to branched, sharp-bristly. **LF** 5–30 cm, bristly. **FL**: calyx 3–6 mm in fl, lobes lanceolate in fl, ± triangular in fr; corolla gen red to purple, throat appendages lanceolate. **FR**: nutlets 4–5 mm, ± 3 mm wide, black, shiny. $2n=40$. Waste places, fields; < 100 m. SnFrB, expected elsewhere; to Can, e US; native to Eur.

TIQUILIA

Ronald B. Kelley

Ann, per, subshrub, variously hairy, ± taprooted; rhizome gen 0. **ST** spreading to prostrate. **LVS** cauline, alternate, gen clustered, evergreen, petioled; margin rolled under, entire or ± crenate. **INFL** ± axillary; fls solitary or clustered, sessile. **FL**: calyx ± deeply 5-lobed, not enlarging in fr; corolla 5-lobed, gen ± funnel-shaped, tube yellow when young, appendages 0; style branches 2. **FR**: nutlets 1–4, sometimes ± tubercled. 27 spp.: w hemisphere deserts. (native S.Am. name for flower) [Richardson 1977 Rhodora 79:467–572] Separated from *Coldenia* of e hemisphere.

1. Branches alternate; per; lf veins obscure, blade ovate to narrowly elliptic; fls gen ± solitary; style branched < 1/3 from tip; fr 4-grooved (sect. *Stegnocarpus*) .... ***T. canescens***
2. Corolla 4–7.5 mm, at top 2.5–4.5 mm wide .... var. ***canescens***
2′ Corolla 8–12 mm, at top 5–8 mm wide .... var. ***pulchella***
1′ Branches opposite; ann or per; lf veins obvious, ± sunken, blade ovate, round, or obovate; fls clustered; style branched 1/2–4/5 from tip; fr 4-lobed (sect. *Tiquiliopsis*)
3. Rhizome present; st ± glandular; lf veins deeply sunken, 4–7 pairs; hairs within calyx long .... ***T. plicata***
3′ Rhizome 0; st ± non-glandular; lf veins shallowly sunken, 2–3 pairs; hairs within calyx short or 0
4. Ann; style < calyx; lf margin entire, lateral veins ± 30° from midvein; corolla pink to white; seed oblong-ovoid .... ***T. nuttallii***
4′ Per; style > calyx; lf margin ± crenate, lateral veins ± 45° from midvein; corolla blue, purple, or lavender; seed spheric .... ***T. palmeri***

T. canescens (DC.) A. Richardson Per, subshrub. **ST**: branches alternate; hairs ± spreading. **LVS** sometimes clustered, white-tomentose; blade 5–13 mm, ovate to narrowly elliptic, veins obscure, margin entire, spiny-ciliate. **INFL**: fls ± solitary; bracts 0. **FL**: calyx 3–5 mm, free 2/3–3/4 length; style branched < 1/3 from tip, shortly exserted from calyx. **FR** spheric, 4-grooved, not lobed. **SEED** 2–2.5 mm, ovoid, minutely tubercled, hairy or not. Slopes, ridges of broken granite, limestone, gneiss; 500–1500 m. DMtns, DSon; sw N.Am., n Mex. [*Coldenia c.* DC.]

var. ***canescens*** (p. 401) **FL**: corolla 4–7.5 mm, 2.5–4.5 mm wide at top, lavender, pink, or white. $n=9$. Habitats and range of sp. ❀DFCLT;TRY.

var. ***pulchella*** (I.M. Johnston) A. Richardson **FL**: corolla 8–12 mm, 5–8 mm wide at top, blue or lavender. Habitats of sp. DSon (Imperial, Riverside cos.); sw AZ. ❀TRY.

T. nuttallii (Hook.) A. Richardson (p. 401) Ann. **ST**: branches opposite; hairs ± appressed. **LVS** clustered; hairs ± spreading; blade 3.5–9 mm, ovate to round, margin entire, veins 2–3 pairs, shallowly sunken, ± 30° from midvein. **INFL**: fls clustered, bracted. **FL**: calyx 3–5 mm, free 2/3–3/4 length, hairs within short; corolla 3–4 mm, 2–2.5 mm wide, pink to white; style < calyx, branched 1/3–1/2 from tip. **FR** deeply 4-lobed. **SEED** oblong-ovoid, smooth, shiny. $n=8$. Sandy plains, washes, slopes, saline flats; < 2400 m. Teh, e MP, SNE, DMoj; to OR, UT; also in Argentina. [*Coldenia n.* Hook.]

T. palmeri (A. Gray) A. Richardson (p. 401) Per, ± woody; bark white. **ST**: branches opposite; hairs ± shaggy. **LVS** clustered, grayish strigose; blade 3.5–11 mm, ovate to round, margin ± crenate, veins 2–3 pairs, shallowly sunken, ± 45° from midvein. **INFL** bracted; fls clustered. **FL**: calyx 2–3.5 mm, free ± 1/2 length, hairs within short or 0; corolla 5–9 mm, 4–5 mm wide, blue, purple, or lavender; style > calyx, branched 1/2 from tip. **FR** deeply 4-lobed. **SEED** spheric, smooth, shiny. $n=8,9$. Sandy gravel soils; < 900 m. D (esp w edge DSon and near Colorado River); sw NV, w AZ, n Mex. [*Coldenia p.* A. Gray]

T. plicata (Torrey) A. Richardson (p. 401) Per, ± woody; rhizome present. **ST**: branches opposite, ± glandular. **LVS** clustered, white-canescent; blade 3–12 mm, obovate to widely ovate, margin entire, veins 4–7 pairs, deeply sunken. **INFL**: fls clustered; bracts 0. **FL**: calyx 2–3 mm, free ± total length, hairs within long; corolla 4–6 mm, 2–3 mm wide, blue to lavender; style > calyx, branched 1/2–4/5 from tip. **FR** deeply 4-lobed. **SEED** ovoid, smooth, shiny. $n=8$. Dune sand, sandy gravel flats; < 900 m. D; w AZ, s NV, n Mex. [*Coldenia p.* (Torrey) Cov.]

BRASSICACEAE [Cruciferae] MUSTARD FAMILY

Ann to subshrub. **LVS** gen basal and cauline, alternate, gen simple; stipules 0. **INFL**: gen raceme. **FL** bisexual; sepals 4, free; petals (0)4, free, gen white or yellow, often clawed; stamens gen (2,4) 6, gen 4 long, 2 short; ovary 1, superior, chambers gen 2, septum membranous, connecting 2 parietal placentas, style 1, stigma simple or 2-lobed. **FR**: gen capsule ("silique") with 2 deciduous valves, sometimes breaking transversely or indehiscent. **SEEDS** 1–many per chamber. 300+ genera, 3000+ spp.: worldwide, esp cool regions; some cult for food (esp *Brassica, Raphanus*) and orn. [Al-Shehbaz 1984 J Arnold Arbor 65:343–373] Family description, key to genera by Robert A. Price.

Reed C. Rollins, except as specified

Key to Groups

1. Fr gen at least 4 × longer than wide, gen linear .... **Group 1**
1′ Fr < 4 × longer than wide, not linear
2. Fr round to ± flat in X-section and sometimes inflated .... **Group 2**

2′ Fr flat, not inflated
3. Fr 1-seeded, 1-chambered, indehiscent .... **Group 3**
3′ Fr gen 2+-seeded, 2-chambered, gen dehiscent
4. Fr flat perpendicular to septum .... **Group 4**
4′ Fr flat parallel to septum .... **Group 5**

Group 1

1. Lf hairs branched (sometimes also simple)
2. Lf deeply lobed or compound; fr < 3 cm
3. Lf segments rigid, often sharp-pointed; fr often flat perpendicular to septum .... [2]**POLYCTENIUM**
3′ Lf segments ± flexible, not sharp-pointed; fr ± cylindric or ± flat parallel to septum
4. Fr ± cylindric, valves midribbed ± to tip; pedicel gen thread-like; petals yellow or whitish ... [2]**DESCURAINIA**
4′ Fr gen flat, valves midribbed only at base; pedicel linear, width nearly = fr width; petals white to purplish .... [2]**SIBARA**
2′ Lf entire to gen shallowly lobed; fr often > 3 cm
5. Fr flat
6. Hairs appressed, forked to multibranched; petals gen > 15 mm, blade much wider than claw ... [3]**ERYSIMUM**
6′ Hairs appressed and not, simple to multibranched; petals < 15 mm, blade little wider than claw
7. Seeds 1 row per chamber
8. Fr not long-tapered; taproot often ± slender; basal lvs entire or dentate, gen < 10 cm; seed often winged .... [4]**ARABIS**
8′ Fr long-tapered to tip; taproot very thick; basal lvs entire, 3–15 cm; seed unwinged ...... **PHOENICAULIS**
7′ Seeds 2 rows per chamber
9. Fr 5–7 mm wide, tapered, sharp-pointed; seed coat silvery .... [2]**ANELSONIA**
9′ Fr < 5 mm wide (or tip not tapered and sharp-pointed); seed coat not silvery
10. Fr gen > 3 cm, > 10 × longer than wide; petals not yellow .... [4]**ARABIS**
10′ Fr gen < 2 cm, < 10 × longer than wide; petals often yellow .... [3]**DRABA**
5′ Fr ± flat to ± 4-sided or round in X-section
11. Stigma lobes pointed, decurrent .... **MALCOLMIA**
11′ Stigma lobes not pointed, decurrent or not
12. Fl showy (petals gen > 15 mm, blade much wider than claw); fr gen > 4 cm (inflated if shorter)
13. Lf and st hairs appressed, forked to multibranched; petals often yellow or orange .... [3]**ERYSIMUM**
13′ Lf and st hairs appressed and not, simple to multibranched; petals not yellow or orange
14. Fr < 3 cm, inflated; style 4–8 mm; stigma lobes not decurrent .... **AUBRIETA**
14′ Fr > 3 cm, not inflated; style < 4 mm; stigma lobes ± decurrent
15. Lf dentate; lf hairs ± sparse, simple and forked, fr hairs gen 0 .... **HESPERIS**
15′ Lf ± entire; lf and fr hairs dense, multibranched, rarely also simple .... **MATTHIOLA**
12′ Fl not showy (petals gen < 10 mm, blade little wider than claw); fr gen < 4 cm (exc *Arabis glabra*), not inflated
16. Hairs appressed, forked to multibranched; petals yellow .... [3]**ERYSIMUM**
16′ Hairs appressed or not, simple to multibranched; petals white or cream-colored, rarely lilac
17. Fr 4–10 cm; upper st lvs clasping .... ***Arabis glabra***
17′ Fr < 4 cm; upper st lvs not clasping
18. Ann; fr glabrous; seeds ± 1 row per chamber .... **ARABIDOPSIS**
18′ Bien or per; fr sometimes hairy; seeds 2 rows per chamber .... **HALIMOLOBOS**
1′ Lf hairs 0 or simple (rarely branched in *Caulanthus*)
19. Fr stalk above pedicel 1–3 cm; stamens ± equal .... **STANLEYA**
19′ Fr stalk above pedicel 0(–1) cm; stamens gen 4 long, 2 short
20. Calyx ± urn- or flask-shaped, nearly closed at full fl, often brightly colored; petals gen strap-shaped (exc *Streptanthella*)
21. Fr gen ± cylindric or ± flat parallel to septum; seed gen unwinged
22. Sepals 3–18 mm, gen not green; petals 5–23 mm, blade base as wide as or narrower than claw; stigma lobes gen 2; fr 1.5–18 cm, ± cylindric or ± flat; seed 1–4 mm, gen ± brown, wing sometimes narrow at 1 end or 0 .... **CAULANTHUS**
22′ Sepals 1–11 mm, greenish or not; petals 3–15 mm, blade base sometimes much wider than claw; stigma lobes 0 or 2; fr 1–9 cm, ± cylindric; seed < 2 mm, brownish or yellowish, wing 0 .... [3]**GUILLENIA**
21′ Fr gen flat parallel to septum; seed gen winged
23. Fr pendent, valves persistent to beak .... **STREPTANTHELLA**
23′ Fr ascending or pendent, valves dehiscent, falling, beak 0 .... **STREPTANTHUS**
20′ Calyx not urn- or flask-shaped, fully opening, gen greenish; petals gen clawed or spoon-shaped, sometimes linear to obovate
24. Bracts subtending at least several fls per infl
25. Fr ± 4-sided .... **ERUCASTRUM**
25′ Fr flat .... **TROPIDOCARPUM**
24′ Bracts gen 0 (may subtend a few lower fls or main branches)
26. Fr not opening by valves
27. Fr flat, winged, oblong to oblanceolate, indehiscent, not breaking into segments .... [2]**ISATIS**
27′ Fr ± flat or ± round in X-section, unwinged, not oblong to oblanceolate, gen breaking into transverse segments
28. Fr beak narrow, > 5 mm; fr breaking into 2–several 0–5-seeded segments

29. Pl glandular; fr often upcurved, not grooved .... **CHORISPORA**
29′ Pl not glandular; fr straight, grooved below .... **RAPHANUS**
28′ Fr beak either wide or < 5 mm; fr breaking into 2 unlike segments, each gen 1-seeded
30. Pl gen glabrous; lf fleshy; ocean dunes; petals white to lavender; fr 1–3 cm .... [2]**CAKILE**
30′ Pl coarsely stiff-hairy; lf not fleshy; disturbed areas (not dunes); petals yellow; fr < 1 cm **RAPISTRUM**
26′ Fr opening by valves
31. Fr beak prominent, gen nearly as wide as valves
32. Seeds 2 rows per chamber; petals gen prominently dark-veined .... **ERUCA**
32′ Seeds 1 row per chamber; petals not dark-veined
33. Fr valves with 1 midvein .... [3]**BRASSICA**
33′ Fr valves with 3–7 veins, esp when young
34. Fr appressed to st, beak base abruptly swollen .... **HIRSCHFELDIA**
34′ Fr not appressed to st, beak base gen not abruptly swollen .... **SINAPIS**
31′ Fr beak gen 0 or obscure (exc *Brassica*), narrower than valves
35. Fr gen ± flat
36. Fr clearly stalked above receptacle .... [2]**THELYPODIUM**
36′ Fr ± not stalked above receptacle
37. Fr valves without prominent midrib
38. Fr valves gen opening elastically, sometimes coiling from base .... **CARDAMINE**
38′ Fr valves not opening elastically .... [2]**SIBARA**
37′ Fr valves with prominent midrib
39. Lvs entire to toothed; petals white or straw to deep purple; cotyledons not doubly folded ... [4]**ARABIS**
39′ Some lvs lobed; petals pale to lemon yellow; cotyledons doubly folded .... **DIPLOTAXIS**
35′ Fr ± 4-sided to round in X-section
40. Seeds 2 rows per chamber .... [3]**RORIPPA**
40′ Seeds 1 row per chamber
41. Petals white to purple
42. Cauline lvs compound; wet places .... ***Rorippa gambellii***
42′ Cauline lvs ± entire to dissected but not compound; drier places
43. Pl woody at base; fr not clearly stalked above receptacle — n W&I .... **CAULOSTRAMINA**
43′ Pl gen not woody at base; fr ± stalked above receptacle
44. Fr stalk < 1 mm, ± as wide as body; fr often reflexed, not densely clustered; cauline lvs gen not clasping .... [3]**GUILLENIA**
44′ Fr stalk < to > 1 mm, narrower than body; fr not reflexed, often densely clustered; cauline lvs often clasping .... [2]**THELYPODIUM**
41′ Petals ± yellow (sometimes very pale or pinkish in *Guillenia*)
45. Cauline lvs clasping st
46. Fr beak > 5 mm; seed spheric .... [3]**BRASSICA**
46′ Fr beak 0 or obscure, any remaining style < 3 mm; seed not spheric
47. Lf pinnately lobed to ± compound; pl not glaucous; fr < 8 cm .... **BARBAREA**
47′ Lf ± entire; pl glaucous; fr 8–13 cm .... **CONRINGIA**
45′ Cauline lvs not clasping st
48. Fr beak > 2 mm; seed spheric .... [3]**BRASSICA**
48′ Fr beak gen 0, any remaining style < 3 mm; seed not spheric
49. Fr ascending to reflexed, 1–7 cm; petals 3–6 mm, white, pale yellow, pinkish .... ***Guillenia lasiophylla***
49′ Fr erect to ascending, 0.8–10 cm; petals 1.5–10 mm, (pale) yellow .... **SISYMBRIUM**

Group 2

1. Lf gen lobed to compound, sometimes entire
2. Fr transversely jointed into 2 segments; lf very fleshy; ocean dunes .... [2]**CAKILE**
2′ Fr not transversely jointed; lf not very fleshy; inland
3. Petals white to purplish; fr ± flat parallel to septum to ± plump; alpine .... **SMELOWSKIA**
3′ Petals gen yellow (whitish); fr gen ± cylindric or plump; not alpine
4. Cauline lf 1–many-pinnately lobed to compound, 1° segments at least again lobed; hairs gen multibranched .... [2]**DESCURAINIA**
4′ Cauline lf entire to deeply lobed (rarely compound), 1° segments ± entire to toothed; hairs simple or 0 .... [3]**RORIPPA**
1′ Lf entire to shallowly lobed
5. Hairs 0 or simple
6. Lower lvs 3–5 dm; st 6–20 dm; frs and seeds gen abortive .... **ARMORACIA**
6′ Lower lvs < 3 dm; st gen < 5 dm; frs and seeds gen fertile
7. Pl aquatic; lf narrowly awl-like .... **SUBULARIA**
7′ Pl terrestrial; lf wider than awl-like
8. Lower petioles >> blades; coastal (NCo) .... **COCHLEARIA**
8′ Lower petioles 0 or not > blades; gen not near coast
9. Lf entire, < 5 mm wide; seed gen 1 per fr .... [2]**CUSICKIELLA**

9′ Lf gen toothed or lobed, > 5 mm wide; seeds 2–many per fr
10. Fr ± indehiscent, ± flattened or plump; seeds 2–4 .... [2]**CARDARIA**
10′ Fr dehiscent, plump; seeds several–many .... [3]**RORIPPA**
5′ Some hairs gen forked, stellate, or multibranched, esp near st base (simple in *Cardaria*)
11. Upper cauline lvs clasping st
12. Fr dehiscent, gen > 4-seeded; infl gen not flat-topped, longer than wide .... **CAMELINA**
12′ Fr ± indehiscent, 2–4-seeded; infl ± flat-topped, wider than long .... [2]**CARDARIA**
11′ Upper cauline lvs 0 or not clasping st
13. Herbage hairs mostly stellate; seeds gen 2+ per chamber
14. Fr not notched .... [2]**LESQUERELLA**
14′ Fr tip (and often base) ± deeply notched .... **PHYSARIA**
13′ Herbage hairs mostly simple or multibranched; seed gen 1 per chamber
15. Lvs gen ± basal, < 5 mm wide; fr not beaked .... [2]**CUSICKIELLA**
15′ Lvs cauline, often > 10 mm wide; fr beak stout .... **EUCLIDIUM**

Group 3

1. Fr > 2 × longer than wide, oblong to oblanceolate .... [2]**ISATIS**
1′ Fr ± as long as wide, gen round to elliptic
2. Fr 2–2.5 mm wide, hooked-hairy, wing 0 .... **ATHYSANUS**
2. Fr gen > 3 mm wide, hooked hairs 0, wing obvious, often perforated, radiating rays 0 to very distinct .... **THYSANOCARPUS**

Group 4

1. Infl axillary; fr of 2 nutlet-like halves, wrinkled or ridged, spiny .... **CORONOPUS**
1′ Infl often terminal; fr not of nutlet-like halves, not wrinkled, ridged, or spiny
2. Seed 1 per chamber
3. Fr deeply lobed at tip and base between ± round halves; herbage hairs multibranched, stellate .... **DITHYREA**
3′ Fr sometimes notched at tip but not deeply lobed; herbage hairs 0 or simple .... **LEPIDIUM**
2′ Seeds 2+ per chamber
4. Upper cauline lvs ± clasping st
5. Fr obtriangular-obcordate, corners acute; basal lvs gen lobed to deeply dissected .... **CAPSELLA**
5′ Fr ± obovate to round, corners rounded; basal lvs entire to dentate .... **THLASPI**
4′ Upper cauline lvs 0 or not clasping st
6. Lf deeply lobed to compound, segments linear, often sharp-pointed .... [2]**POLYCTENIUM**
6′ Lf entire to deeply lobed, lobes gen wider than linear, not sharp-pointed
7. Lf hairs dense, multibranched; petals 15–25 mm; fr ± 10–20 mm wide .... **LYROCARPA**
7′ Lf hairs 0 or sparse, simple or branched; petals < 5 mm; fr < 5 mm wide
8. Cauline lvs several; filaments scaleless; fr tip not notched; lf hairs 0 or branched .... **HUTCHINSIA**
8′ Cauline lvs 0–few; filaments with a white basal scale; fr tip shallowly notched; lf hairs 0 or simple .... **TEESDALIA**

Group 5

1. Fr 2 × longer than wide
2. Fr 5–7 mm wide, tip ± tapered, ± sharp-pointed; seed 2.5–3 mm, coat silvery .... [2]**ANELSONIA**
2′ Fr gen < 5 mm wide, tip not tapered or sharp-pointed; seed gen < 2 mm, coat not silvery .... [3]**DRABA**
1′ Fr < 2 × longer than wide (often ± round)
3. Lvs basal
4. Fls in racemes; seed wing 0 .... [3]**DRABA**
4′ Fls solitary; seed wing ± 1 mm wide .... **IDAHOA**
3′ Lvs basal and cauline
5. Fl and fr > 10 mm wide; fr stalk above receptacle 7–12 mm .... **LUNARIA**
5′ Fl and fr < 5 mm wide; fr stalk above receptacle 0
6. Infl 1-sided; fr often ± twisted .... **HETERODRABA**
6′ Infl not 1-sided; fr not twisted
7. Lf hairs forked at base, appressed; fr valves with prominent midvein .... **LOBULARIA**
7′ Lf hairs often multibranched or stellate, often spreading; fr valves without prominent midvein
8. Style < 1.6 mm .... **ALYSSUM**
8′ Style > 2 mm .... [2]**LESQUERELLA**

ALYSSUM

Ann, per; hairs simple or branched, often stellate. **STS** 1–several. **LVS** simple, entire. **FL**: sepals erect to spreading, bases not sac-like; petals yellow to cream-colored, often fading to ± white; long filaments paired, gen winged, short ones single, gen appendaged. **FR** round to widely oblong, very flat parallel to septum. **SEEDS** 1–2(6) per chamber; embryonic root at edges of both cotyledons. ± 180 spp.: n N.Am, Eurasia. (Greek: without rabies, from supposed cure for hydrophobia)

1. Fr glabrous; lf linear to narrowly oblanceolate; pedicel ascending-spreading .... ***A. desertorum***
1′ Fr hairs dense; lf narrowly oblanceolate to obovate; pedicel spreading (sometimes ascending in *A. alyssoides*)
2. Sepals persistent in fr; filaments not winged, not appendaged .... ***A. alyssoides***
2′ Sepals shed early; filaments winged, appendaged
3. Fr hairs of ± 1 kind (coarse, stellate, not appressed) .... ***A. minus*** ssp. ***micranthum***
3′ Fr hairs of 2 kinds (stiff, spreading, once-forked and smaller, stellate, appressed) .... ***A. strigosum***

A. alyssoides (L.) L. (p. 401) Ann, bien; hairs appressed, stellate. **STS** decumbent to ascending, 1–2(4) dm. **FL**: sepals persistent in fr; petals 3–4 mm, linear, cream-colored fading to ± white; filaments not winged, not appendaged. **FR** 2–5 mm, round, bulged over seed, hairy; pedicel spreading to ascending; style 0.3–0.6(1) mm, glabrous. **SEED**: wing narrow. $2n$=32. Local, uncommon. Roadsides, disturbed areas, waste places; 300–1800 m. e KR, n SNH, expected elsewhere; native to Eurasia.

A. desertorum Stapf Ann; hairs stellate, appressed. **STS** ascending to erect, 1–2 dm. **FL**: sepals shed early; petals 2–3 mm, light yellow. **FR** 3–4 mm, round, shallowly notched, glabrous; pedicel ascending-spreading; style 0.5–1 mm, glabrous. **SEED**: wing narrow. $2n$=32. Disturbed areas, rocky sagebrush flats, waste places; 1000–1500 m. MP; to WA, NE; native to Eur.

A. minus (L.) Roth ssp. *micranthum* (C. Meyer) Dudley Ann; hairs stellate, coarse, appressed or not. **STS** 1–3 dm. **FL**: sepals shed early; petals 2–3.5 mm, light yellow. **FR** 3.5–6 mm, ± round; hairs of ± 1 kind (coarse, stellate, not appressed); pedicel spreading; style 0.7–1.6 mm, hairy. **SEED** winged. $2n$=16. Roadcuts, waste places, foothills, open rangeland; 1000–1500 m. MP; to MT, Colorado; native to Eurasia.

A. strigosum Banks & Sol. Ann; hairs stiff, once-forked, and stellate. **STS** ascending, 2–4 dm. **FL**: sepals shed early; petals 2–3.5 mm, yellowish. **FR** 4–5 mm, round; hairs of 2 kinds (stiff, spreading, once-forked and smaller, stellate, appressed); pedicel spreading; style < 1 mm, gen glabrous. **SEED** winged. $2n$=16. Disturbed areas, waste places; < 300 m. SnFrB; native to se Eur.

ANELSONIA

1 sp. (A. Nelson, Rocky Mountain botanist, 1859–1952)

A. eurycarpa (A. Gray) J.F. Macbr. & Payson (p. 401) Per; roots deep; caudex loosely branched; hairs branched, dense on lvs, infl axes, sepals. **LVS** basal, dense, overlapped, entire, canescent; base tapered. **INFL** scapose, umbel-like. **FL**: sepals early deciduous, bases not sac-like; petals small, white; stigma 2-lobed. **FR** 2–3 cm, 5–7 mm wide, lanceolate-elliptic, flat parallel to septum, glabrous, gen purplish; base ± obtuse or rounded; tip ± tapered, sharp-pointed; valves leathery; pedicel erect to ascending; style 1–2 mm. **SEEDS** several–many, 2 rows per chamber, 2.5–3 mm, ± flat, oblong; coat silvery; wing 0; hairs dense, minute, club-shaped; embryonic root at edges of both cotyledons. Broken rock, talus, slopes, ridges; > 3000 m. SN, SNE; to ID, NV. [*Phoenicaulis e.* (A. Gray) Abrams]

ARABIDOPSIS

Ann, per. **STS** 1–few, branched or not, slender. **LVS**: basal in rosette, petioled; cauline sessile, base tapered or lobed, clasping st or not. **INFL** terminal, open. **FL** small; sepal bases not sac-like; petals white, pale purple, or yellowish. **FR** linear, glabrous; pedicel ascending; style short or 0. **SEEDS** many, ± 1 row per chamber, plump; wing 0; embryonic root at back of 1 cotyledon. Few but controversial number of spp.: Eur, Asia, 1 native to N.Am. (Greek: resembling *Arabis*)

A. thaliana (L.) Heynh. (p. 401) MOUSE-EAR CRESS, THALE CRESS Ann, freely branched. **ST** 1–4 dm; hairs below simple, spreading, above 0. **LVS**: basal oblanceolate, entire or minutely dentate, hairs coarse, branched; cauline few, base tapered. **FL**: petals white. **FR** 1–1.5 cm, < 1 mm wide, linear, ± cylindric to ± flat; pedicel erect to ascending, 5–10 mm, slender. **SEED** < 1 mm, oblong. $2n$=10. Uncommon. Disturbed, open ground; < 1000 m. NCo, NCoRI, SnFrB, expected elsewhere; to e US (where abundant); native to Eur. Often used in experimental genetics, physiology, biochemistry.

ARABIS ROCK CRESS

Bien, per; base woody or not; hairs 0 to dense, simple, forked, stellate, or multibranched; caudex branched or not. **ST** branched or not, cylindric, leafy. **LVS**: basal petioled, entire or dentate; cauline gen sessile, entire or dentate, base often lobed, often clasping st. **INFL**: bracts 0. **FL** erect to reflexed; sepals erect; petals spoon-shaped to oblong and narrowed at base or narrowly obovate, white to deep purple, rarely straw-colored. **FR** erect to reflexed, linear, straight to curved, flat parallel to septum, rarely ± cylindric. **SEEDS** ± many, gen 1 row per chamber, flat or plump, winged or not; embryonic root at edges of both cotyledons. ± 120 spp.: temp N.Am, Eurasia, Afr. (Latin: of Arabia)

1. Seed 2.5–5 mm wide incl 1–3 mm wing; fr gen > 3 m wide
 2. Pedicel in fr erect to ± spreading
  3. Lower cauline petioles winged; basal lvs oblanceolate to widely spoon-shaped, 10–30 mm wide; petals = or ± > sepals ..... ***A. repanda***
   4. St 1–3.5 dm; caudex gen branched; basal lvs ± entire ..... var. ***greenei***
   4′ St 4–7 dm; caudex simple; basal lvs wavy-dentate ..... var. ***repanda***
  3′ Lower cauline petioles 0; basal lvs linear to oblanceolate, < 8 mm wide; petals > sepals
   5. Lf, lower st whitish hairy; pedicel hairy ..... [2]***A. dispar***
   5′ Lf, lower st glabrous or hairy but not whitish; pedicel glabrous
    6. Basal lvs linear, ± 1 mm wide, fleshy, lower surface grooved on both sides of midvein; hairs coarse, gen simple or forked ..... ***A. pygmaea***
    6′ Basal lvs narrowly lanceolate to oblanceolate, > 2 mm wide, ± not fleshy, lower surface not grooved; hairs 0 or multibranched, not coarse
     7. Hairs multibranched, dense on basal lvs, ± dense on upper cauline; basal lvs < 3 mm wide ..... ***A. pinzlae***
     7′ Hairs gen 3–5 rayed, ± dense or 0 on basal lvs, 0 on upper cauline; basal lvs > 3 mm wide ... ***A. platysperma***
      8. Basal lvs glabrous; pl 0.5–2(3) dm; 3000–3600 m ..... var. ***howellii***
      8′ Basal lvs hairy; pl 1–4 dm; 1300–3350 m ..... var. ***platysperma***
 2′ Pedicel in fr reflexed or recurved (spreading in *A. suffrutescens* var. *horizontalis*)
  9. Lf, st hairs ± dense throughout; fr tip obtuse; seeds 2 rows per chamber ..... ***A. glaucovalvula***
  9′ Lf, st hairs 0 to dense below, 0 to sparse above; fr tip acute to acuminate; seeds 1 row per chamber
   10. Anthers exserted; style 2.5–4.5 mm ..... ***A. constancei***
   10′ Anthers incl; style < 1 mm or 0
    11. Petals 9–11 mm, cream-white; fr 2–3 mm wide; seeds oblong, wing on ends, on sides narrow or 0; embryonic roots < 1 mm ..... ***A. rollei***

11′ Petals 5–7 mm, rose to purplish; fr 3–6 mm wide; seeds round, wing all around; embryonic roots ± 2 mm .... ***A. suffrutescens***
12. Fr spreading to ± recurved; pl 1–3 dm; hairs of basal lvs, lower sts ± dense .... var. ***horizontalis***
12′ Fr pendent to reflexed; pl 2–5 dm; hairs of basal lvs, lower sts sparse to ± 0 .... var. ***suffrutescens***
1′ Seed gen < 2 mm wide incl 0–1 mm wing; fr gen < 3 mm wide
13. Basal lvs obovate to widely oblanceolate (rarely oblong), rosetted, gen thin, tips gen obtuse to rounded, fr erect to ± spreading; outer sepals sac-like at base (exc *A. glabra*)
14. Seeds 2 rows per chamber; fr ± cylindric; petals cream-colored, rarely lilac; cauline lvs lanceolate to ovate, gen glaucous .... ***A. glabra***
15. St base hairs spreading to gen ± appressed, gen branched, small .... var. ***furcatipilis***
15′ St base hairs spreading, simple, large .... var. ***glabra***
14′ Seeds 1 row per chamber; fr flat; petals white to purple; cauline lvs lanceolate to obovate, rarely or not glaucous
16. Hairs 0 or on teeth of basal lvs .... ***A. macdonaldiana***
16′ Hairs at least on lower sts, basal lvs
17. Petals white to cream-white or pinkish, < 1 cm; pedicel hairs 0, rarely sparse .... ***A. hirsuta***
18. Petals 5–9 mm, white to pinkish; cauline lvs gen spaced, lower ± glabrous; fr ± ascending to ± spreading; outer sepals strongly sac-like at base .... var. ***glabrata***
18′ Petals < 5 mm, white to cream-white; cauline lvs gen overlapped, lower ± hairy; fr erect; outer sepals weakly sac-like at base .... var. ***pycnocarpa***
17′ Petals purple or rose- or pink-purple, 1–2 cm; pedicel hairy
19. Pl 0.5–2 dm; fr 2–4 cm, tip obtuse, style gen ± stout — NCoRO, SnFrB .... ***A. blepharophylla***
19′ Pl gen > 2 dm; fr 3.5–6.5 cm, tip acute, style slender or 0
20. Lower st hairs appressed; basal lvs not ciliate, hairs small, multibranched .... ***A. modesta***
20′ Lower st hairs spreading; basal lvs ciliate, hairs large, simple, forked, or multibranched
21. Basal lvs 1–4 cm, ciliate with simple (rarely forked) hairs; cauline lvs 5–15 mm; caudex branched .... ***A. aculeolata***
21′ Basal lvs 3–8 cm, ciliate with forked or multibranched hairs; cauline lvs 15–50 mm; caudex branched or not .... ***A. oregana***
13′ Basal lvs linear to oblanceolate, if wider then hairy or fr reflexed or both, sometimes clustered but not rosetted, ± thick, tips gen acute; outer sepals gen not sac-like at base
22. Sts, pedicels, lvs grayish white with hairs dense, minute, multibranched; often of deserts or arid mtns nearby
23. Pedicel recurved to pendent
24. Seeds 1 row per chamber; cauline lvs oblong to widely lanceolate, ± crowded, often ± toothed or ± lobed
25. Fr tip gen obtuse; style 0; petals 7–10 mm .... ***A. puberula***
25′ Fr tip gen acute; style ± 1 mm or ± 0; petals 10–14 mm .... ***A. subpinnatifida***
24′ Seeds 2 rows per chamber; cauline lvs linear, not crowded, entire .... [2]***A. pulchra***
26. Fr reflexed-appressed, hairs dense, pedicel reflexed .... var. ***pulchra***
26′ Fr pendent to spreading, hairs dense or ± 0, pedicel spreading-recurved
27. Fr, upper st, pedicel ± glabrous .... var. ***gracilis***
27′ Fr, upper st, pedicel densely hairy .... var. ***munciensis***
23′ Pedicel erect to spreading
28. Style 1–8 mm; basal lvs linear to linear-oblanceolate
29. Style 1.5–3 mm; seed wing wide; fr 3–5 cm .... ***A. johnstonii***
29′ Style 4–8 mm; seed wing narrow; fr 1–2 cm .... ***A. parishii***
28′ Style 0–1 mm; basal lvs linear-oblanceolate to spoon-shaped
30. Seeds 2 rows per chamber, ± unwinged, plump, ± 1 mm wide; cauline lvs overlapped, base ± lobed or not .... ***A. shockleyi***
30′ Seeds 1 row per chamber, winged, flat, 1–2.5 mm wide; cauline lvs gen spaced, base lobed or not
31. Seed wing > 0.5 mm wide; fr 2.5–3.5 mm wide, spreading-ascending .... [2]***A. dispar***
31′ Seed wing < 0.5 mm wide; fr ± 2 mm wide, gen spreading .... [3]***A. inyoensis***
22′ Sts, pedicels, at least cauline lvs greenish with hairs 0 to dense, minute to coarse, simple to multibranched; not gen of deserts or arid mtns nearby
32. Pedicel in fr spreading to reflexed-appressed
33. Basal lvs ciliate with coarse, simple or forked hairs; pedicel glabrous, reflexed-appressed .... ***A. rectissima*** var. ***rectissima***
33′ Basal lvs not ciliate, hairs fine to coarse, multibranched; pedicel glabrous or not, spreading to reflexed
34. Seeds 2 rows per chamber; cauline lvs linear; petals 8–20 mm, gen showy, limbs spreading; fr hairs dense .... [2]***A. pulchra*** (see 24′ for sspp.)
34′ Seeds 1 row per chamber; cauline lvs gen lanceolate or ovate to oblong, rarely linear or narrowly lanceolate; petals < 12 mm, limbs ± erect; fr hairs 0 or very sparse
35. Basal lvs linear to linear-oblanceolate; petals white, ± 4 mm; pedicel widely arched with ± pendent fr — SNE, DMtns .... ***A. cobrensis***
35′ Basal lvs widely spoon-shaped to oblanceolate (linear to narrowly oblanceolate in *A. microphylla* var. *microphylla*); petals gen purplish to pinkish, > 4 mm; pedicel gen not widely arched with pendent fr (exc in *A. holboellii* var. *pinetorum*)
36. Pedicel 2–4(6) mm; cauline lvs ± ovate, hairy or not .... [2]***A. lemmonii*** (see 57. for sspp.)
36′ Pedicel (4)6–20 mm; cauline lvs linear-lanceolate to oblong, gen hairy
37. Pedicel in fr recurved to reflexed, ± straight; fr gen ± straight, reflexed and gen appressed to pendent (exc var. *pinetorum*) .... ***A. holboellii***

38. Cauline lf base tapered, not clasping st; pl 1–2 dm; basal lvs < 3 mm wide .... var. ***pendulocarpa***
38′ Cauline lf base lobed, often ± clasping st; pl gen 2–9 dm; basal lvs > 3 mm wide
39. Pedicel in fr arched to gently recurved; fr ± curved; lf hairs not minute .... var. ***pinetorum***
39′ Pedicel in fr reflexed, gen straight; fr gen straight; lf hairs minute .... var. ***retrofracta***
37′ Pedicel in fr spreading, straight or spreading-recurved; fr straight, spreading to widely recurved
40. Sts 1–2(3) dm, many, slender; cauline lvs gen few, spaced .... [3]***A. microphylla*** var. ***microphylla***
40′ Sts 2–9 dm, 1–several, ± stout; cauline lvs gen many, overlapped below (exc sometimes *A. perennans*)
41. Basal lvs entire, grayish, hairs fine, dense; st hairs dense at least below, appressed; pedicel hairs 0 to sparse .... [3]***A. inyoensis***
41′ Basal lvs, or at least outer, dentate, rarely entire, greenish, hairs coarse, sparse to dense; st hairy gen below, spreading, rarely appressed; pedicel hairs 0 or not dense
42. Outer basal lvs oblanceolate to widely oblanceolate, tip ± obtuse; pedicel slender, 10–20 mm, in fr glabrous; petals 6–9 mm, 1.5–2.5 mm wide .... ***A. perennans***
42′ Outer basal lvs linear-oblanceolate, tip acute; pedicel stout, 5–15 mm, in fr hairy; petals 9–12 mm, 2–4 mm wide .... [2]***A. sparsiflora*** (see 61′ for vars.)
32′ Pedicel in fr erect to spreading
43. Lower pedicels in fr 2–4 cm, glabrous; fr ± straight or ± curved, seeds 2 rows per chamber .... ***A. hoffmannii***
43′ Lower pedicels in fr < 2 cm, glabrous or not; fr straight or if ± curved then seeds 1 row per chamber
44. Basal, lower cauline lvs greenish, hairs gen 0 or sparse, simple to multibranched; sts gen 1–several, from gen simple caudex
45. Fr < 1.5 mm wide, gen curved; seeds round, ± 1 mm wide; st hairy below [3]***A. microphylla*** var. ***microphylla***
45′ Fr 1.5–3.5 mm wide, gen straight; seeds round to oblong, 1.2 mm wide; st hairy below or not
46. Seeds 2 rows per chamber, gen oblong, wing gen on 1 side and 1 end, narrower or 0 elsewhere; fr erect, tip obtuse, rarely subacute; basal lf hairs narrowly boat shaped, centrally attached, rarely 0; petals white, rarely pinkish .... ***A. drummondii***
46′ Seeds 1 row per chamber, oblong to round, wing on 1 end or gen all around; fr spreading to erect, tip gen acute; basal lf hairs 0 or simple to multibranched, not boat-shaped; petals pink to rose or purplish, rarely white
47. Sts 3–9 dm, 1–several from simple or branched caudex; fr ascending to spreading, rarely ± reflexed
48. Fr 1.5–2.5 mm wide, margins not wavy; seed 1–1.5 mm wide .... ***A. ×divaricarpa***
48′ Fr 2.5–3.5 mm wide, margins wavy; seed 2–2.5 mm wide .... ***A. rigidissima*** var. ***rigidissima***
47′ Sts < 3 dm, several–many from gen branched caudex; fr erect to ± spreading
49. Caudex not thick, without old lf bases; fr straight; basal lvs not abruptly pointed at tip, not long-petioled .... ***A. lyallii***
50. Basal lvs oblanceolate, 3–6 mm wide, tip obtuse to acute .... var. ***lyallii***
50′ Basal lvs linear to linear-oblanceolate, 1–2.4 mm wide, tip acute to acuminate .... var. ***nubigena***
49′ Caudex thick, old lf bases present; fr ± straight to ± curved; basal lvs abruptly pointed at tip or long-petioled
51. Basal lvs 3–8 cm, long-petioled, tip obtuse, without a hair; seed round to widely oblong, wing all around .... ***A. davidsonii***
51′ Basal lvs 1.5–2.5 cm, gen short-petioled, tip acute, often with a hair; seed oblong, wing on 1 end .... ***A. tiehmii***
44′ Basal, lower cauline lvs grayish or greenish, hairs dense, stellate or multibranched; sts gen many, from branched caudex
52. Basal lvs oblanceolate, 5–15 mm wide; fr ± curved .... ***A. breweri***
53. Pedicel 3–4 mm, glabrous; fr 2–3 cm .... var. ***pecuniaria***
53′ Pedicel 5–15 mm, gen hairy; fr gen 3–7 cm
54. Petals 10–13 mm; cauline lvs 2–4 cm; pedicel 10–15 mm .... var. ***austinae***
54′ Petals 6–9 mm; cauline lvs gen < 2 cm; pedicel 5–9 mm .... var. ***breweri***
52′ Basal lvs gen linear to linear-oblanceolate, gen < 5 mm wide; fr gen ± straight (exc ± curved sometimes in *A. bodiensis, A. koehleri, A. microphylla, A. sparsiflora*)
55. Petals < 7 mm; style gen ± 0, rarely < 1 mm
56. Basal lvs, lower sts greenish, hairs not matted; cauline lvs linear to narrowly lanceolate, ± spaced; pedicel ± glabrous .... [3]***A. microphylla*** var. ***microphylla***
56′ Basal lvs, lower sts grayish, hairs ± matted; cauline lvs oblong to narrowly ovate, crowded at base (exc in *A. fernaldiana* var. *stylosa*); pedicel gen hairy
57. Fr tip obtuse; style ± 0; cauline lvs with basal lobes .... [2]***A. lemmonii***
58. Fr ascending to ± spreading; basal lvs narrowly oblanceolate to lanceolate; infl in fr not 1-sided .... var. ***depauperata***
58′ Fr spreading to ± reflexed; basal lvs obovate to widely oblanceolate; infl in fr gen 1-sided .... var. ***lemmonii***
57′ Fr tip acuminate; style < 1 mm; cauline lvs with 0 or inconspicuous basal lobes
59. Frs ± curved to ± straight, many per st, spreading to ± ascending .... ***A. bodiensis***
59′ Fr ± straight, few per st, ± erect .... ***A. fernaldiana*** var. ***stylosa***
55′ Petals 7–14 mm; style 0 or < 1 mm
60. Basal lvs linear-oblanceolate, minutely hairy, grayish, tip obtuse; fr ± straight; caudex branches not elongate .... [3]***A. inyoensis***

60′ Basal lvs linear to linear-oblanceolate, ± coarsely hairy, greenish, tip acute; fr gen curved; caudex branches elongate

61. Pl < 3(4) dm; pedicel glabrous; lower st hairs 0, rarely sparse; caudex with many lf base remnants .... ***A. koehleri*** var. ***stipitata***

61′ Pl 3–9 dm; pedicel glabrous or not; lower st gen hairy; caudex without lf base remnants [2]***A. sparsiflora***

62. Pedicel hairs ± appressed, dense; fr pedicel spreading-recurved; st hairs ± appressed var. ***californica***

62′ Pedicel hairs gen spreading, dense to ± 0; fr pedicel ascending to spreading; st hairs spreading, at least below

63. Hairs of upper lvs, upper sts ± dense .... var. ***arcuata***

63′ Hairs of upper lvs, upper sts 0 to very sparse

64. Basal lvs entire, linear-oblanceolate; pedicel ascending, hairs ± 0; st gen branched above .... var. ***sparsiflora***

64′ Basal lvs gen ± dentate, oblanceolate to wider; pedicel spreading, hairs gen spreading; st rarely branched above .... var. ***subvillosa***

A. aculeolata E. Greene (p. 401) WALDO ROCK CRESS Per; caudex branched, often covered with old lf bases; hairs large, simple or rarely forked. **STS** few–several, simple, gen 2–3.5 dm; hairs sparse to dense, spreading. **LVS**: basal rosetted, 1–4 cm, obovate, entire to ± wavy, ciliate with simple (rarely forked) hairs, hairs sparse to dense, tip obtuse; cauline well spaced, sessile, 5–15 mm, entire to sparsely dentate. **FL**: petals spoon-shaped, purple. **FR** erect, 3.5–6.5 cm, glabrous; pedicel erect to ascending, 1–1.5 cm; style 1–2 mm, slender. **SEED** 1.5–2 mm wide, round to ± oblong; wing narrow all around. $2n$=32. RARE in CA. Semi-isolated serpentine areas; 900–1800 m. KR (n Del Norte Co.); s OR.

A. blepharophylla Hook. & Arn. (p. 401) COAST ROCK CRESS Per; caudex branches 0–few; hairs coarse, forked or multibranched. **STS** 1–few, simple, 0.5–2 dm, hairy esp above. **LVS**: basal rosetted, many, petioled, 2–8 cm, oblanceolate to obovate, entire to dentate, hairy on surfaces and margins or surfaces glabrous, tip obtuse; cauline few, sessile. **FL**: petals widely spoon-shaped, rose-purple. **FR** erect, 2–4 cm, glabrous; pedicel erect, 5–20 mm, stout, hairy; style 1–2 mm. **SEED** ± round; wing narrow. $2n$=16. UNCOMMON. Rocky outcrops, grassy slopes; < 500 m. NCoRO, SnFrB.

A. bodiensis Rollins BODIE HILLS ROCK CRESS Per; caudex branched; hairs minute, multibranched, some large, simple. **STS** several–many, erect, 1.5–3.5 dm; hairs dense below, 0 above. **LVS** acute at tip; basal petioled, 1–3 cm, linear to linear-oblanceolate, grayish with dense hairs, entire; cauline sessile, 1–2.5 cm, oblong, base lobes 0 or inconspicuous. **FR** spreading to ± ascending, 3–5 cm, ± curved to ± straight; pedicel 3–6 cm, ascending to ± spreading. **SEED** widely oblong to ± round; wing narrow. RARE. Rock crevices, open slopes; 2500–3100 m. SNE (n Mono Co.); w NV.

A. breweri S. Watson Per, cespitose; caudex woody, many-branched. **STS** several–many, simple, 0.5–2 dm; hairs dense below, often 0 above. **LVS**: basal short-petioled, 1–3 cm, oblanceolate, entire to few-toothed, hairs on both surfaces gen 3-forked, tip acute; cauline sessile, 1–2(4) cm, oblong-lanceolate to oblong, base lobed, clasping st. **FL**: petals spoon-shaped, reddish purple to pink. **FR** spreading-ascending, ± curved to ± straight; pedicel erect to ascending, 3–15 mm, gen hairy. **SEED** round; wing narrow. Rocky places; 450–3200 m. KR, NCoRI, CaR, n SNF, ScV, SnFrB, SCoRI, SnBr; s OR.

var. ***austiniae*** (E. Greene) Rollins **LVS**: cauline 2–4 cm, ± glabrous to hairy. **FL**: sepals purple, hairs sparse; petals 10–13 mm. **FR** 5–7 cm; pedicel 1–1.5 cm, gen hairy. Rock outcrops, canyon walls; 600–2300 m. CaRF, n SNF, ScV. ❀DRN,DRY:7,8,**9**,**14**,**18–24**&SUN:6,**15**,**16**,17.

var. ***breweri*** **LVS**: cauline gen < 2 cm, hairy. **FL**: sepals grayish green, hairs dense; petals 6–9 mm. **FR** gen 3–7 cm; pedicel 5–9 mm, gen hairy. $2n$=14. Rock crevices, slopes; 450–2300 m. KR, NCoRI, CaR, SnFrB, SCoRI; s OR. [var. *figularis* Jepson] ❀DRN,DRY:7,8,9,**14**,18–21,**22–24**&SUN:**4–6**,**15–17**.

var. ***pecuniaria*** Rollins SAN BERNARDINO ROCK CRESS **LVS**: cauline < 2 cm, hairs dense. **FL**: sepals grayish, hairs sparse; petals 6–8 mm. **FR** 2–3 cm; pedicel 3–4 mm, glabrous. RARE. Ledges, talus; 2700–3200 m. SnBr (Dollar Lake).

A. cobrensis M.E. Jones (p. 401) MASONIC ROCK CRESS Per; caudex branched; hairs multibranched, minute. **STS** several–many, simple or gen branched above, 2–5 dm, slender. **LVS**: basal many, 2–5 dm, gen linear, entire, gray, hairs dense, fine, tip acute; cauline few, sessile, base ± lobed, ± clasping st. **FL**: petals ± 4 mm, white. **FR** ± pendent, 2–4 cm, ± straight; tip obtuse; hairs 0 or very sparse; pedicel hairs sparse; style very short or 0. **SEED** ± oblong to ± round; wing ± wide. $2n$=14. RARE in CA. Sandy soils, sagebrush; 1375–2800 m. n SNE (Mono Co.), n DMtns (Panamint Mtns); to WY.

A. constancei Rollins (p. 401) CONSTANCE'S ROCK CRESS Per; caudex branched, ± woody; hairs simple or forked. **STS** erect, 1–several, 1.5–3 dm, simple, glabrous. **LVS**: basal densely rosetted, 1.5–3 cm, linear-oblanceolate, entire, stiff, thick, bluish green, hairs 0 to sparse, some gen marginal, tip acute; cauline sessile, base tapered. **INFL** 5–10-fld. **FL**: petals erect, ± white; anthers exerted. **FR** pendent to reflexed, 4–5.5 cm, ± straight; margins uneven; tip acute to acuminate; pedicel reflexed or recurved, 6–10 mm; style 2.5–4.5 mm. **SEED** ± round, winged. $2n$=28. RARE. Rocky places, serpentine; 1100–1300 m. n SN (Plumas Co.).

A. davidsonii E. Greene Per; caudex thick, simple or branched, old lf bases present. **STS** several, simple, 5–15 cm, slender, glabrous. **LVS**: basal 3–8 cm, long-petioled, oblanceolate, entire or rarely few-toothed near tip, glabrous, thickish, tip obtuse; cauline few, sessile, 6–15 mm, oblong, base tapered. **FL**: petals spoon-shaped, white to pale purple. **FR** ascending, 3–5 cm, ± straight to ± curved; pedicel ascending, 1–1.5 cm; style very short. **SEED** round; wing narrow. Uncommon. Rock crevices, large granite outcrops; 1500–3500 m. SN; e OR, n NV. ❀DRN,SHD:6,17&IRR: 1–3,**7**,**14–16**.

A. dispar M.E. Jones Per; caudex branched; hairs minute, multibranched. **STS** several, simple or branched above base, 1–2.5 dm; hairs below, dense. **LVS** whitish hairy; basal many, erect, 1.5–2.5 cm, slender-petioled, linear-oblanceolate to oblanceolate, entire; cauline 1–2 cm, sessile, widely linear. **FL**: petals > sepals, obovate, purplish. **FR** ascending, 5–7 cm, 2.5–3.5 mm wide, glabrous; tip acute; pedicel ± erect to ascending, 1–2 cm, hairy. **SEED** ± 2.5 mm, round; wing wide. Loose gravelly slopes, compact talus; 1200–2400 m. s SNH, n SnBr, SNE, DMtns; sw NV.

A. ×divaricarpa Nelson (p. 405) Bien, rarely per; caudex branched or not; hairs branched, appressed, gen 3–5-rayed. **STS** 1–3, simple or branched above, 3–9 dm; hairs 0 or below. **LVS**: basal 2–6 cm, widely oblanceolate to narrowly spoon-shaped, dentate to subentire, hairs sparse, tip ± acute; cauline sessile, narrowly oblong to lanceolate, often sagittate, clasping st, entire or lower ± dentate, hairs 0 or sparse. **FL**: petals spoon-shaped, pink to purplish. **FR** ascending to spreading, rarely ± reflexed, 2–8 cm, straight; pedicel ascending to ± reflexed, 6–12 mm; style very short or 0. **SEED** ± round; wing narrow. $2n$=14,13+2,21,20+2,28. Gravelly, calcareous soils; 2100–3300 m. CaR, SN, GB; to Rocky Mtns, e Can. [var. *interposita* (E. Greene) Rollins] (see Rollins 1983 Amer J Bot 70:625–634 for discussion of hybridity)

A. drummondii A. Gray (p. 405) Bien, per; caudex gen simple; hairs sessile, narrowly boat-shaped, centrally attached. **STS** 1–few, simple or branched above, 3–9 dm; hairs 0 or below, sparse. **LVS**: basal petioled, 2–8 cm, narrowly oblanceolate, entire to sparsely dentate, hairs rarely 0, tip gen acute; cauline sessile, clasping st, 2–7 cm, oblong to oblong-lanceolate, glabrous, tip acute. **FL**: petals white, rarely pinkish. **FR** erect, 4–10 cm, gen crowded, straight, glabrous; tip gen obtuse, rarely subacute; pedicel erect, 1–2 cm, glabrous; style short or 0. **SEEDS** 2 rows per chamber, gen oblong; wing gen on 1 end and 1 side, narrow or 0 elsewhere. 2*n*=14,20,28. Calcareous gravels, talus, open disturbed areas; 1800–3200 m. SN, GB; to AK, e N.Am. [*A. confinis* S. Watson]

A. fernaldiana Rollins var. ***stylosa*** (S. Watson) Rollins STYLOSE ROCK CRESS Per; caudex branched; hairs gen minute, many-branched. **STS** several–many, gen simple, 1–3 dm. **LVS**: basal many, often in sterile clusters, 1–2 cm, spoon-shaped to linear-oblanceolate, entire, grayish, hairs dense, petioles short, ± ciliate; cauline sessile, lobes at base 0 or inconspicuous, hairs dense on lower, gen 0 on upper. **FL**: petals 5–7 mm, pink to lavender. **FR** ± erect, 4–6 cm, ± straight, glabrous, sessile above receptacle; pedicel spreading-ascending, 5–10 mm, hairy; style < 1 mm. **SEED** oblong; wing narrow or 0. RARE. Ridges, rocky, limestone soils, sagebrush; 2300–2800 m. GB; NV. var. *f.* in NV, UT, Colorado.

A. glabra (L.) Benth. Bien, rarely per; hairs forked or multi-branched, spreading, rarely appressed. **STS** 1 or few, erect, simple, rarely branched above, 4–12 dm, stout. **LVS**: basal petioled, 6–15 cm, widely oblanceolate to spoon-shaped or oblong, dentate to ± lobed, hairs coarse; cauline sessile, 4–15 cm, lanceolate to ovate, sagittate, clasping st, entire or lower toothed, hairs 0 to sparse. **FL**: petals linear to narrowly spoon-shaped, cream-colored, rarely lilac. **FR** erect, 4–10 cm, ± cylindric, glabrous; style short, stout. **SEEDS** 2 rows per chamber, oblong to ± round; wing 0 to very narrow. Open fields, meadows, slopes; < 2300 m. CA-FP; temp N.Am, Eur.

var. ***furcatipilis*** M. Hopk. **ST**: hairs below, spreading to gen ± appressed, gen branched, small. Uncommon. Meadows, dry slopes; 500–1000 m. s SnFrB (Santa Cruz Co.), SCoRO (Santa Lucia Mtns); UT.

var. ***glabra*** (p. 405) **ST**: hairs, at least at base, spreading, simple, large. 2*n*=12,16,32. Open fields, woodlands, meadows, shaded mtn slopes; < 2300 m. CA-FP; temp N.Am, Eur. [*A. perfoliata* Lam.]

A. glaucovalvula M.E. Jones Per; caudex branched; hairs minute, multibranched. **STS** 1–several, simple or branched above, 1.5–4 dm, densely canescent. **LVS**: basal 2–5 cm, linear to ± wider, entire, tip obtuse to ± acute, hairs dense; cauline 1–4 cm, sessile, lanceolate to linear-lanceolate, tapered below, gray. **FL**: petals white to pinkish. **FR** reflexed, 2–4.5 cm, 5–8 mm wide, oblong, glabrous, glaucous; base, tip obtuse; pedicel stout, recurved; hairs dense; style < 1 mm. **SEEDS** 2 rows per chamber, round; wing very wide. 2*n*=14. Rocky, limestone soils, open sites, summits, hill slopes, desert shrubland; 1000–1300 m. e SNE, DMoj; NV. ❀TRY.

A. hirsuta (L.) Scop. Bien, weak per; caudex branched or not; hairs simple or forked, coarse, spreading. **STS** 1–several, simple or branched above, erect, 2–7 dm. **LVS**: basal short-petioled, 2–8 cm, oblong to obovate, entire to dentate, tip obtuse; cauline sessile, 1–5 (7) cm, lanceolate to obovate, sagittate, clasping st. **FR** 3–6 cm, flat, glabrous; pedicel 0.5–1.5 cm; hairs 0, rarely sparse; style 0.5–1 mm. **SEED** ± round to ± rectangular; wing prominent or narrow on 1 end to 0. Gravelly soils, swales, disturbed sites; 1000–2600 m. KR, NCoRO, CaR, SN, SnBr; to AK, e Can, Colorado. Var. *eschscholtziana* (Andrz.) Rollins OR to AK, e Asia; var. *hirsuta* native of Eur, Asia, not in N.Am.

var. ***glabrata*** Torrey & A. Gray (p. 405) **ST**: hairs sparse below, 0 above. **LVS**: basal with hairs sparse to 0; cauline gen well spaced, lower ± glabrous. **FL**: outer sepals strongly sac-like at base; petals 5–9 mm, white to pinkish. **FR** ± ascending to ± spreading. Gravelly soils, swales, open woods; 1200–2600 m. KR, NCoRO, CaR, SN, SnBr; to B.C., Colorado.

var. ***pycnocarpa*** (M. Hopk.) Rollins **ST**: hairs ± throughout. **LVS**: basal hairy; cauline gen overlapped, lower ± hairy. **FL**: outer sepals weakly sac-like at base; petals < 5(–8) mm, white to cream-white. **FR** erect. 2*n*=16,32. Meadows, gravel bars, shady slopes, moist, disturbed sites; 1000–1900 m. CaR, SN, SnBr; to AK, e Can. [*A. p.* M. Hopk.]

A. hoffmannii (Munz) Rollins HOFFMAN'S ROCK CRESS Per, often coarse; caudex scaly with old lf bases; hairs multibranched. **STS** 1–several, branched above, 5–7 dm. **LVS**: upper surface ± glabrous; lower surface hairy; tip obtuse; basal many, crowded, 5–10 cm, linear-lanceolate, leathery; cauline crowded, ± clasping st, 3–6 cm, linear-oblong. **FL**: petals linear-oblong, white. **FR** ascending to ± spreading, 6–10 cm, ± straight or curved, thick, leathery, glabrous; pedicel ascending, 1–4 cm, glabrous; style short, stout or ± 0. **SEEDS** 2 rows per chamber, round; wing narrow. RARE. Cliff ledges; < 200 m. n ChI (Santa Cruz Island). [*A. maxima* E. Greene var. *h.* Munz]

A. holboellii Hornem. Bien, per; caudex branched or not; hairs gen multibranched. **STS** 1–several, erect. **LVS**: basal 1–5 cm, linear-oblanceolate to widely spoon-shaped, tip acute, hairs dense; cauline sessile, 2–4 cm, oblong to lanceolate, entire, lower surface densely hairy, upper surface ± hairy to not. **FL**: petals purplish pink to whitish. **FR** 3–7 cm, gen ± straight, glabrous; pedicel arched or recurved to ± reflexed, (4)6–16 mm, straight to ± curved, slender, hairy or not. **SEED** round; wing narrow all around. Rocky slopes, open sites; 500–3500 m. NW, CaR, SN, SnFrB, GB; to AK, e Can, Colorado. 5 vars. in N.Am, difficult to distinguish, esp in fl.

var. ***pendulocarpa*** (Nelson) Rollins **ST** gen simple, 1–2 dm, slender; hairs present below, ± 0 above. **LVS**: basal < 3 mm wide, entire; cauline base tapered. **FR** pendent. 2*n*=14. Rocky slopes; 1800–3000 m. c SNH, GB; to B.C., Colorado.

var. ***pinetorum*** (Tidestrom) Rollins **ST** gen branched above, 3–9 dm, hairy below. **LVS** minutely hairy; basal oblanceolate. **FR** 4–7 cm, ± curved, glabrous; pedicel arched to gently recurved. 2*n*=21. Rocky slopes; 2000–3500 m. CaR, SN; to B.C., NE.

var. ***retrofracta*** (Graham) Rydb. (p. 405) **ST** branched above, 2–9 dm; hairs throughout or ± 0 above. **LVS** minutly hairy; basal > 3 mm wide, entire to ± dentate; cauline base lobed, often ± clasping st. **FR** gen straight; pedicel reflexed, gen straight. 2*n*=14, 21+1. Common. Rocky slopes, open areas; 500–2400 m. NW, CaR, SN, SnFrB, GB; to AK, e Can.

A. inyoensis Rollins Per; caudex branched; hairs minute, multibranched. **STS** several, erect, 2–5 dm, rigid; hairs dense, esp below. **LVS**: basal many, 2–3 cm, linear-oblanceolate to spoon-shaped, entire, grayish, hairs dense, tip acute or obtuse; cauline sessile, 1–2.5 cm, gray, base lobed, clasping st, or tapered. **FL**: petals spoon- to tongue-shaped, pink to purplish. **FR** gen spreading, ± 2 mm wide, ± straight, glabrous; pedicel gen spreading, 6–12 mm. **SEED** round; wing < 0.5 mm wide. 2*n*=21,23. Rocky ridges, slopes; 1500–3500 m. SNH, SNE; NV. ❀TRY.

A. johnstonii Munz JOHNSTON'S ROCK CRESS Per; caudex branched, woody; hairs fine, multibranched. **STS** several, branched or not, erect or ascending, 1–2 dm, hairy. **LVS** densely gray-hairy; basal petioled, 1–2 cm, linear-oblanceolate, entire; cauline sessile, 1–1.5 cm, linear-oblong to lanceolate, entire, base tapered. **FL**: petals purple. **FR** erect, 3–5 cm, glabrous; pedicel ascending, 6–10 mm, hairy; style 1–3 mm, slender, persistent. **SEED** ± round; wing wide. RARE. Dry, rocky slopes; 1350–1500 m. SnJt.

A. koehleri Howell var. ***stipitata*** Rollins KOEHLER'S STIPITATE ROCK CRESS Per; caudex woody, with many lf base remnants; branches elongate. **STS** many, simple, 5–30(40) cm, slender; hairs 0 or rarely sparse below. **LVS**: basal many, 1–2 cm, linear to narrowly oblanceolate, tip acute, hairs stellate; cauline sessile, 1–2 cm, overlapped, lanceolate, ± glabrous, base lobed, ± clasping st. **FL**: petals scarlet to deep purple. **FR** ascending to ± spreading, 5–8 cm, curved, glabrous; tip tapered; pedicel ascending, 1–2 cm, glabrous; style short or 0. **SEED** round; wing narrow. UNCOMMON. Dry, rocky serpentine slopes, ridges; 500–800 m. nw KR (n Del Norte Co.); OR. Var. *k.* in s OR. ❀TRY:DFCLT.

A. lemmonii S. Watson Per; caudex branched; hairs multibranched, minute. **STS** several–many, simple, gen prostrate or decumbent, 6–20 cm, slender, hairy or gen glabrous above. **LVS**: basal 1–2 cm, entire or few-toothed, hairs dense, tip gen acute; cauline sessile, 4–10 mm, oblong-lanceolate to ± ovate, hairy or not, base lobed, gen clasping st. **FL**: petals pink to purplish. **FR** 2–4 cm, glabrous; pedicel 2–4(6) mm, glabrous or not; style ± 0. **SEED** round; wing narrow. Rocky to gravelly soils; 2400–4300 m. CaR, SNH, W&I; to Yukon, Colorado. 4 vars. total.

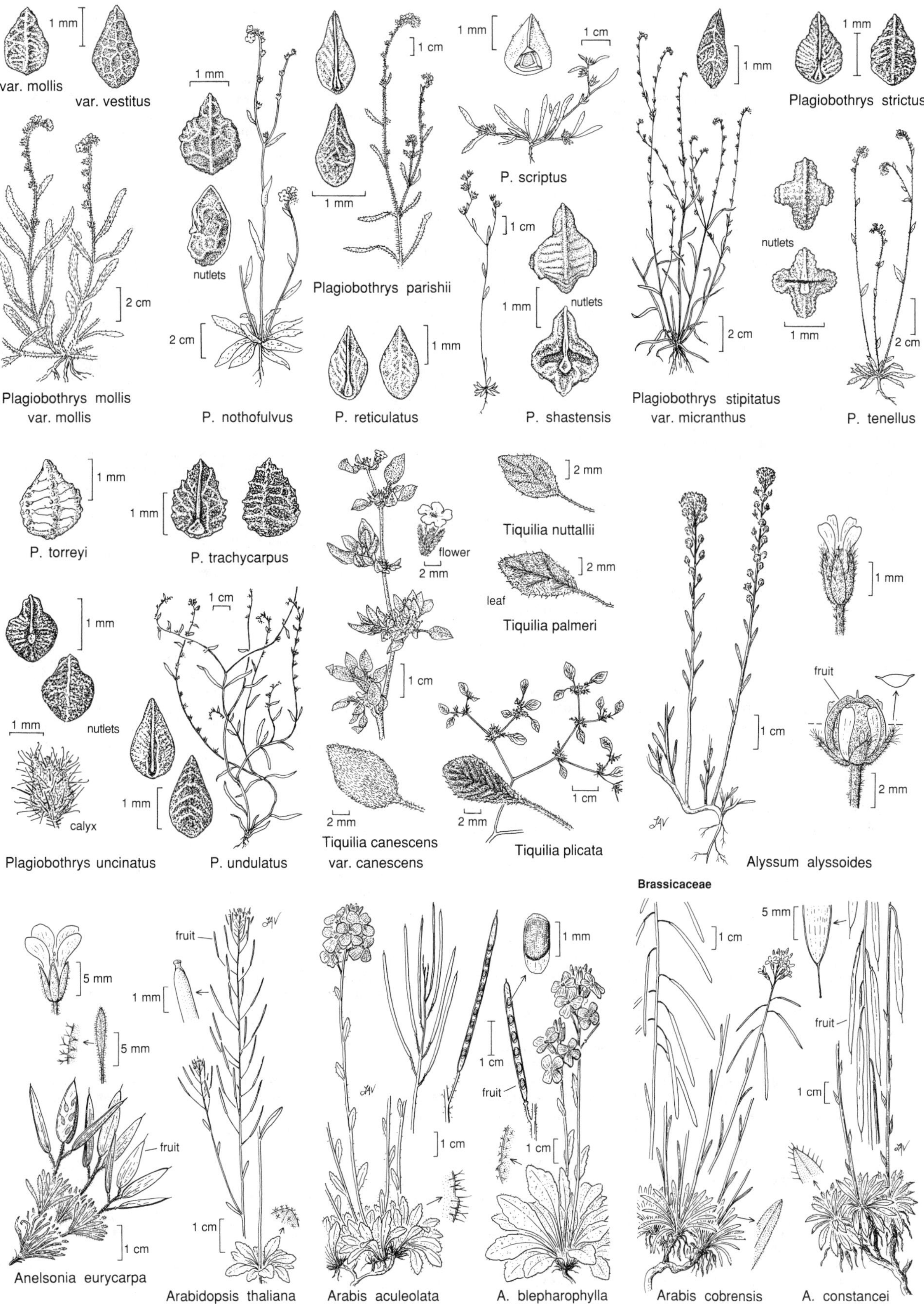
1 mm
var. mollis
var. vestitus
2 cm
Plagiobothrys mollis
var. mollis
1 mm
nutlets
2 cm
P. nothofulvus
1 cm
1 mm
Plagiobothrys parishii
1 mm
P. reticulatus
1 mm
1 cm
P. scriptus
1 cm
1 mm
nutlets
P. shastensis
1 mm
2 cm
Plagiobothrys stipitatus
var. micranthus
1 mm
Plagiobothrys strictus
nutlets
1 mm
2 cm
P. tenellus
1 mm
P. torreyi
1 mm
P. trachycarpus
1 mm
1 mm
nutlets
calyx
Plagiobothrys uncinatus
1 cm
1 mm
P. undulatus
flower
2 mm
1 cm
2 mm
Tiquilia canescens
var. canescens
2 mm
Tiquilia nuttallii
2 mm
leaf
Tiquilia palmeri
1 cm
2 mm
Tiquilia plicata
1 cm
1 mm
fruit
2 mm
Alyssum alyssoides
Brassicaceae
5 mm
5 mm
fruit
1 cm
Anelsonia eurycarpa
fruit
1 mm
1 cm
Arabidopsis thaliana
1 cm
Arabis aculeolata
1 mm
1 cm
fruit
1 cm
A. blepharophylla
1 cm
Arabis cobrensis
5 mm
fruit
1 cm
A. constancei

var. ***depauperata*** (Nelson & Kenn.) Rollins **LVS**: base narrowly oblanceolate to lanceolate. **INFL** in fr not 1-sided. **FR** ascending to ± spreading. Talus, rocky slopes, gravelly soils; 2400–4300 m. SNH, W&I; NV. [*A. d.* Nelson & Kenn.]

var. ***lemmonii*** **LVS**: base obovate to widely oblanceolate. **INFL** in fr gen 1-sided. **FR** spreading to reflexed. 2*n*=14. Talus, rock fields, ridges, outwash gravels; 2400–3700 m. CaR, SNH, W&I; to Yukon, Colorado.

A. lyallii S. Watson Per; caudex branched; hairs multibranched, ± minute. **STS** few–many, < 15 cm, glabrous. **LVS**: basal narrowly petioled, 1–3 cm, entire; cauline few, well spaced, sessile, 1–2 cm, lanceolate to oblong, base gen tapered, tip acute. **FL**: petals spoon-shaped, rose to purplish. **FR** erect to ± spreading, 3–5 cm, straight, glabrous; tip tapered; style short or 0. **SEED** round, winged. Rock crevices, slopes, ridges; 2400–3800 m. CaRH, SNH, Wrn, W&I; to B.C., WY. 2 vars. total.

var. ***lyallii*** **ST** erect, stout. **LVS**: basal 3–6 mm wide, oblanceolate, tip obtuse to acute. Rock crevices, talus, meadows; 2400–3700 m. CaRH, SNH, Wrn; to B.C., WY. [*A. drummondii* A. Gray var. *alpina* S. Watson] ❀TRY.

var. ***nubigena*** (J.F. Macbr. & Payson) Rollins **ST** ± prostrate to ascending, slender. **LVS**: basal 1–2.4 mm wide, linear to linear-oblanceolate, tip acute to acuminate. Rock slides, open ridges; 2900–3800 m. SNH, W&I; to ID, WY.

A. macdonaldiana Eastw. MCDONALD'S ROCK CRESS Per; caudex branched. **STS** few–many, simple, 1–3 dm, slender, glabrous. **LVS**: basal rosetted, 1–2 cm, wavy-margined to few-toothed, glabrous or teeth bristle-tipped; cauline well spaced, sessile, 4–7 mm, oblong, entire, base tapered. **FL**: petals narrowly spoon-shaped, rose-purple, tip truncate or ± rounded. **FR** erect to ascending, 2.5–3.5 cm, ± straight, flat, glabrous; pedicel ascending, 7–8 mm, glabrous. **SEED** widely oblong; wing 0. 2*n*=16. **ENDANGERED** CA, US. Deep reddish soils, steep slopes, dry ridges, serpentine areas; ± 1200 m. NCoRO. [*A. blepharophylla* var. *m.* Jepson; *A. serpentinicola* Rollins] In cult.

A. microphylla Nutt. var. ***microphylla*** SMALL-LEAVED ROCK CRESS Per; caudex branched, underground; hairs gen multibranched, some simple. **STS** several–many, branched or gen not, 1–2(3) dm in CA, slender; hairs spreading below, 0 above. **LVS**: basal 5–29 mm, linear to narrowly oblanceolate, entire, greenish, tip acute, hairs minute; cauline few, sessile, 1–2 cm, narrowly lanceolate, base lobed, ± clasping st. **FL**: petals spoon-shaped, pale rose to purplish. **FR** ± ascending to spreading, 2–6 cm, ± 1 mm wide, straight to downcurved, glabrous; pedicel spreading to ascending, 5–15 mm, slender, ± glabrous; style < 1 mm or ± 0. **SEED** ± 1 mm, round; wing narrow. 2*n*=14. UNCOMMON. Rock crevices, basaltic or granitic outcrops; 1700–2700 m. n SNH, GB; to WA, MT, WY.

A. modesta Rollins MODEST ROCK CRESS Per; caudex simple or closely branched; hairs forked to multibranched, small, often appressed. **STS** 1 or few, simple or gen branched above, 2.5–6 dm, hairy. **LVS**: hairs dense; basal petioled, 2–6 dm, obovate, margin wavy or entire, lower surface often purplish, tip obtuse; cauline few, well spaced, sessile, 1–2.5 cm, oblong, shallowly dentate to entire. **FL**: petals spoon-shaped, purple to pink-purple. **FR** ascending to ± spreading, 3.5–6 cm, glabrous; tip acute; pedicel ascending to ± spreading, hairy; style 1.5–2 mm. **SEED** oblong; wing on 1 end. 2*n*=16. UNCOMMON. Deep soil on steep slopes, cliffs, shaded canyon ledges; < 800 m. KR, NCoR; s OR. Isolated pls near Yolo-Napa Co. line need study. ❀DRN,SHD:7,14–16; DFCLT.

A. oregana Rollins OREGON ROCK CRESS Per; caudex branched or not; hairs simple, forked, or multibranched, large. **STS** 1 or few, branched above or simple, 3–5 dm; hairs below simple, spreading, above forked. **LVS**: basal petioled, 3–8 cm, obovate, base abruptly narrowed, tip obtuse, entire or wavy-margined to coarsely toothed, hairs on surfaces and margins forked or multibranched; cauline sessile, 1.5–5 cm, oblong to obovate, ± hairy, base unlobed. **FL**: petals purple. **FR** erect to ascending, 4–5 cm, straight, glabrous; pedicel ascending-spreading, 1–2 cm, hairy; style < 2 mm. **SEED** oblong; sides narrowly winged. 2*n*=16. UNCOMMON. Rocky hillsides, steep banks, n-facing slopes; 600–1400 m. KR, NCoRO; s OR.

A. parishii S. Watson PARISH'S ROCK CRESS Per; caudex branched underground; hairs multibranched, fine. **STS** several–many, tufted, simple, 3–14 cm, slender; hairs dense below, ± sparse above. **LVS** gray; hairs dense; basal many, 5–15 mm, linear to linear-oblanceolate, entire, tip acute; cauline few, sessile, 5–10 mm, linear. **FL**: sepals dark purple; petals purple to deep lavender, rarely ± white. **FR** ± erect to ascending, 1–2 cm, glabrous; pedicel ± erect to ascending, 3–7 mm, stout, hairy; style 4–8 mm, very slender. **SEED** elliptic to ± round; wing narrow. 2*n*=14. RARE. Rocky limestone or marble, steep hillsides, ridges; 1950–2900 m. SnBr.

A. perennans S. Watson (p. 405) Per; caudex branched or not, gen above ground; hairs coarse, multibranched. **STS** several, simple or branched above, 2–6 dm, hairy below, glabrous above. **LVS**: basal many, petioled, 2–6 cm, oblanceolate to widely oblanceolate, dentate, rarely entire, hairs dense; cauline sessile, 1–3 cm, lanceolate, base lobed, ± clasping st. **FL**: petals 6–9 mm, spoon-shaped, purple to pinkish. **FR** spreading or recurved, 4–6 cm; pedicel spreading-recurved, 1–2 cm, slender, gen glabrous. **SEED** round; wing all around. 2*n*=14. Calcareous rocks, canyon walls, gravelly slopes, pinyon/juniper woodland; 300–2200 m. SnGb, SnBr, e PR, SNE, DMtns; to TX. ❀TRY.

A. pinzlae Rollins PINZL'S ROCK CRESS Per; caudex branched or not; hairs multibranched, minute. **ST** simple, 3–8 cm, slender; hairs below, sparse to 0 above. **LVS**: basal tufted, erect, 6–10 mm, linear-lanceolate, densely gray-hairy; cauline 3–5, sessile, 4–6 mm, narrowly oblong, hairy. **FL**: petals erect, ± 5 mm, spoon-shaped, purple at and whitish below tip. **FR** erect, 2–4 cm, 2–3 mm wide, linear, glabrous; margins uneven; tip acuminate; pedicel erect to spreading-ascending. **SEED** ± round; wing wide all around. RARE. Rocky slopes; 3000–3350 m. W&I (White Mtns); NV.

A. platysperma A. Gray Per; caudex branched; hairs gen 3–5-rayed. **STS** several–many, simple or often branched above, ± decumbent to erect, hairy to glabrous. **LVS**: basal many, 2–6 cm, oblanceolate, entire, tip acute to obtuse; cauline few, well spaced, sessile, 1–1.5 cm, oblong to linear-lanceolate, upper glabrous. **FL**: petals spoon-shaped, pink, rarely whitish. **FR** erect to ascending, 3–7 cm, 3–5 mm wide, straight, flat; tip acuminate; pedicel ascending to ± spreading, 5–15 mm, glabrous; style < 1 mm to 0. **SEED** round; wing wide. Rocky slopes, ridges; 1300–3600 m. NCoRO, CaR, SNH, GB; OR, NV.

var. ***howellii*** (S. Watson) Jepson **ST** simple, 0.5–2(3) dm. **LVS**: basal glabrous; cauline bases clasping st or not. Rocky slopes, ridges; 3000–3600 m. NCoRO, CaR, SNH, GB; OR, NV.

var. ***platysperma*** (p. 405) **ST** often branched above, 1–4 dm. **LVS**: basal hairy; cauline base tapered, not clasping st. Slopes, granitic outcrops, limestone ridges; 1300–3350 m. s SNH, W&I; OR, NV. ❀TRY.

A. puberula Torrey & A. Gray Bien, per; caudex simple; hairs dense, multibranched, fine. **STS** 1 or few, simple or branched above, 1.5–5 dm, often stout, gray. **LVS**: basal 1–3 cm, oblanceolate to linear-oblanceolate, gray, entire or few-toothed, tip acute; cauline many, ± crowded, sessile, 1–3 cm, oblong to lanceolate, often ± toothed or ± lobed, base ± lobed, ± clasping st. **FL**: petals 7– 10 mm, spoon-shaped to narrower, rose to purplish or ± white. **FR** pendent to reflexed, 3–6 cm; tip gen obtuse; hairs ± dense; pedicel recurved or reflexed, 4–8 mm, hairs dense; style 0. **SEED** round, plump; wing narrow. Rocky sites, sagebrush, juniper woodland; 1200–3200 m. GB; to WA, ID, UT.

A. pulchra M.E. Jones Per; base subshrubby; caudex branched or not, elevated; hairs multibranched, often appressed, minute. **STS** 1–several, branched or not, 2–6 dm; hairs dense or 0 above. **LVS**: hairs dense; basal 4–8 cm, linear, entire; cauline sessile, 2–6 cm, linear, entire, base lobed or not, clasping st or not. **FL**: petals 8–20 mm, widely spoon-shaped, purple to reddish purple. **FR** 4–7 cm; pedicel 8–20 mm, hairs dense; style very short or 0. **SEED** ± round; wing prominent. Canyons, slopes, washes; 600–2200 m. SN, GB, D; to UT, AZ, Mex. 5 vars. total.

var. ***gracilis*** M.E. Jones **ST**: upper ± glabrous. **FR** pendent to ± spreading, ± glabrous; pedicel spreading-recurved, ± glabrous. $2n$=14. Limestone soils; 850–1850 m. SN, W&I, D; AZ, n Mex. [vars. *glabrescens* Wiggins & *viridis* Jepson]. ❀TRY.

var. ***munciensis*** M.E. Jones DARWIN ROCK CRESS **ST**: upper densely hairy. **FR** pendent to spreading; hairs dense; pedicel spreading-recurved, hairs dense. $2n$=21. RARE in CA. Canyon slopes, ledges, rock outcrops, desert shrublands; 1100–2000 m. GB, DMoj; to UT.

var. ***pulchra*** (p. 405) **ST**: upper densely hairy. **FR** reflexed-appressed; hairs dense; pedicel reflexed; hairs dense. $2n$=14. Canyons, slopes, rocky washes; 600–2200 m. GB, DMoj; NV, Mex. ❀TRY.

A. pygmaea Rollins (p. 405) TULARE COUNTY ROCK CRESS Per; caudex branched or not; hairs gen simple or forked, coarse, spreading. **STS** several, simple, erect, 5–10 cm, slender; hairs dense below, ± dense to 0 above. **LVS**: basal tufted, persistent, 1–2 cm, ± 1 mm wide, linear, entire, hairs on margins gen simple, lower surface grooved both sides of midvein; cauline few, well spaced, sessile, 1–2 cm, linear. **FL**: petals > sepals, white. **FR** erect, 2–4 cm, 4–5 mm wide, straight, glabrous; pedicel ascending, 5–8 mm, hairs sparse to 0. **SEED** round; wing wide. UNCOMMON. Volcanic sand, gravel, barren flats; 2600–3400 m. s SNH (Tulare Co.).

A. rectissima E. Greene var. ***rectissima*** Bien, ± per; caudex rarely branched; hairs simple or forked, coarse. **STS** 1–few, simple or branched above, 2–8 dm, ± stout or not, often purplish; hairs 0 to below, sparse. **LVS**: basal many, short-petioled, spoon-shaped to oblanceolate, entire, hairy; cauline crowded below, well spaced above, sessile, 1–2 cm, oblong to ± lanceolate, base lobed or not, clasping st or not. **FL**: petals spoon- to narrowly tongue-shaped, white, rarely pinkish. **FRS** crowded, reflexed-appressed, 5–8 cm, straight; pedicel reflexed-appressed, 4–12 cm, glabrous. **SEED** round; wing all around. Open pine forests, rocky, open knolls, granitic sand; 1200–2800 m. SN, SnBr, MP; to OR, NV. 2 vars. total.

A. repanda S. Watson Per; hairs forked or multibranched. **STS** 1 or few, branched above; hairs dense below, sparse to 0 above. **LVS**: basal rosetted, petioled, 3–7 cm, 1–3 cm wide, oblanceolate to widely spoon-shaped, tip obtuse, hairs dense; cauline 1–6 cm, widely oblanceolate to ± linear, lower petioles winged, upper petioles 0. **FL**: petals = or ± > sepals. **FR** ascending, 4–10 cm, straight or curved, leathery, glabrous to hairy; pedicel ascending to erect, 3–6(10) mm, stout; style ± 1 mm, slender. **SEED** round to ± elliptic; wing wide. Gravel, rocks, talus, slopes, pine forests; 1400–3600 m. SNH, e WTR, SnGb, SnBr, SnJt; NV.

var. ***greenei*** Jepson Caudex gen branched. **ST** 1–3.5 dm. **LVS**: basal ± entire. Granitic gravel, crevices, talus of metamorphic rocks; 2700–3600 m. c SNH.

var. ***repanda*** Caudex simple. **ST** 4–7 dm. **LVS**: basal wavy-dentate. n=7. Moist gravel, loose soil on steep slopes, volcanic rocky areas, open pine forests; 1400–2800 m. Range of sp.

A. rigidissima Rollins var. ***rigidissima*** TRINITY MOUNTAINS ROCK CRESS Per; caudex branched or not, woody; hairs multibranched or forked, spreading, fine. **STS** 1–several, 2–4 dm; hairs 0 or below, sparse. **LVS**: basal narrowly petioled, 1.5–3 cm, spoon-shaped, hairy; cauline well spaced, sessile, 1–2 cm, ovate to oblong, entire, glabrous, base lobed and clasping st. **FL**: petals spoon-shaped, pink. **FR** ascending, 4–7 cm, 2.5–3.5 mm wide, straight, glabrous; margin wavy; pedicel ascending, 5–10 mm, glabrous; style < 1 mm or 0. **SEED** oblong to round, wing narrow. UNCOMMON. Open, gravelly or rocky soil; 1600–2200 m. KR. 2 vars. total.

A. rollei Rollins (p. 405) Per; caudex elevated, branched, woody. **ST** simple, 1.5–2.5 dm, glabrous. **LVS**: basal ± overlapped, 2–4 cm, entire, hairs 0 to sparse, gen forked or 3–4-rayed; cauline overlapped below, well spaced above, 1.2 cm, base of lower tapered, of upper lobed, clasping st. **FL**: petals 9–11 mm, spoon-shaped, cream-white. **FR** pendent to reflexed, 3.5–6 cm, 2–3 mm wide, glabrous; margin irregular; pedicel reflexed to recurved, lower 4–6 mm; style 0.5–1 mm. **SEEDS** 1 row per chamber, oblong; wing on ends ± 0.5 mm, on sides narrow or 0. Open, forested slopes among rocks; 1500–1800 m. n KR (n Siskiyou Co.).

A. shockleyi Munz SHOCKLEY'S ROCK CRESS Per; caudex simple; hairs multibranched, fine, dense. **STS** 1–few, simple or branched above, 1.5–4 dm, stout, grayish white. **LVS** grayish white; basal crowded, 1–2 cm, spoon-shaped to obovate, tip acute; cauline many, overlapped, 1–2 cm, widely lanceolate, base ± lobed or not, clasping st or not. **FL**: petals 8–11 mm, ± linear, pink. **FRS** crowded, ascending, 5–8 cm, straight to outcurved; hairs sparse to 0; pedicel ascending, 8–12 mm, straight, hairs dense; style ± 0. **SEED** oblong, plump; wing ± 0. RARE in CA. Limestone and quartzite ridges, gravel; 1000–2000 m. SnBr, DMoj; UT.

A. sparsiflora Torrey & A. Gray Per; caudex branched or not; hairs multibranched to simple, spreading or appressed, coarse. **STS** 1–several, simple or branched above, 3–9 dm, gen stout; hairs below, sometimes above. **LVS**: basal many, 3–10 cm, linear-oblanceolate to wider, entire or ± dentate, hairy, tip acute; cauline many, sessile, 2–8 cm, linear-lanceolate or wider, base lobed, gen sagittate, clasping st. **FL**: petals 9–12 mm, 2–4 mm wide, spoon-shaped, pink to purple. **FR** ascending to recurved, 6–12 cm, straight or curved, glabrous; pedicel ascending to spreading-recurved, 5–15 mm, often stout, glabrous to hairy. **SEED** round; wing narrow. Rocky slopes, valleys; < 2800 m. CaR, SNF, SW, GB; to ID, UT.

var. ***arcuata*** (Nutt.) Rollins (p. 405) **ST**: hairs ± dense, spreading, branched or not. **LVS**: basal linear-oblanceolate, entire or toothed. **FR**: pedicel ascending; hairs ± dense, spreading. Shrubland, woodland; 700–1900 m. CaR, SNF, SW. [*A. a.* A. Gray; incl var. *rubicundula* Jepson]

var. ***californica*** Rollins (p. 405) **ST**: hairs ± appressed, multibranched. **LVS**: basal oblanceolate, coarsely toothed. **FR**: pedicel spreading-recurved, hairs ± appressed. $2n$=22,23. Rocky canyons, gravelly outwash, chaparral; < 1600 m. SCo, SnGb, PR; Mex.

var. ***sparsiflora*** (p. 405) **ST** gen branched above; hairs spreading, simple below, ± 0 to very sparse above. **LVS**: basal linear-oblanceolate, entire. **FR**: pedicel ascending; hairs ± 0 (to sparse, spreading). Steep, gravelly slopes, grassy sagebrush, basaltic talus; 1300–2800 m. GB; to ID, UT.

var. ***subvillosa*** (S. Watson) Rollins (p. 405) **ST** rarely branched above; hairs spreading, simple or forked, ± 0 above. **LVS**: basal oblanceolate to wider, gen ± dentate. **FR**: pedicel spreading; hairs gen spreading. $2n$=21,21+1,22. Canyon slopes, talus, among granitic boulders, sagebrush; 500–1100 m. CaR, n SNF; to B.C., MT, WY.

A. subpinnatifida S. Watson Per; caudex gen branched; hairs multibranched, minute. **STS** 1–few, simple or branched above, 1.5–4 dm; hairs dense or 0 above. **LVS**: basal 1–3 cm, linear to linear-oblanceolate, whitish, hairs dense, dentate to pinnately lobed, tip acute; cauline margins often inrolled, ± dentate to pinnately lobed. **FL**: petals spoon- to tongue-shaped, purple to light lavender. **FR** pendent, 5–7 cm; hairs 0 to sparse; pedicel spreading-recurved, 6–12 mm, ± stout, hairy; style ± 1 mm or ± 0. **SEED** round to ± oblong; wing on sides or all around. $2n$=14. Basaltic knolls, serpentine cliffs, slopes; 800–2300 m. KR, NCoR, Wrn; s OR. ❀DRN, SUN:**6**&IRR:1–3,**7**,**14–16**,18.

A. suffrutescens S. Watson Per; base woody; caudex widely branched. **STS** several–many, simple, rarely branched above; hairs ± 0 or below, multibranched or stellate. **LVS**: hairs 0 to dense; basal 1–4 cm, linear to ± spoon-shaped; cauline few, sessile, 1–3 cm, lanceolate, base lobed, clasping st. **FL**: petals 5–7 mm, spoon-shaped, rose to purplish. **FR** 3–6 mm wide, glabrous; tip acute; pedicel 4–10 mm, slender, glabrous; style < 1 mm or 0. **SEED** round; wing all around, wide. Gravelly to rocky slopes, basalt, pumice, serpentine; ± 1500 or 1700–2800 m. e KR, n NCoRH, CaR, SN; to OR, ID.

var. ***horizontalis*** (E. Greene) Rollins **ST** 1–3 dm, hairy below. **LVS**: basal hairy. **FR** spreading to ± recurved, 2–3 cm. Gravelly or stony slopes, serpentine, dry pumice; ± 1500 m. n NCoRH (Grouse Mtn), n CaR; s OR.

var. ***suffrutescens*** (p. 405) **ST** 2–5 dm; hairs sparse to ± 0. **LVS**: basal hairs sparse to ± 0. **FR** pendent to reflexed, 4–7 cm. Dry, rocky slopes, basaltic, serpentine outcrops; 1700–2800 m. e KR, CaR, SN; to OR, ID. [var. *perystylosa* Rollins]

A. tiehmii Rollins (p. 405) TIEHM'S ROCK CRESS Per; caudex simple or densely branched, old lf bases present; hairs simple or forked, large. **ST** gen ascending, 8–17 cm, slender, wavy or straight, glabrous. **LVS**: basal tufted, erect, gen short-petioled, 1.5–2.5 cm; tip acute, often with a hair; cauline 3–5, sessile, 8–12 mm, oblong, hairs 0 or few, on margin, bases tapered or upper sometimes minutely lobed, tip acute. **FL**: petals spoon-shaped, white. **FR** erect to ± ascending, 1.5–2 cm, glabrous; tip acute; pedicel erect to ± ascending, ± 3.5 mm, straight, slender; style < 0.5 mm. **SEED** oblong, plump; wing on 1 end, narrow. RARE. Rock outcrops, decomposed granite slopes; 3200–3500 m. n SNH; NV.

ARMORACIA

Per, glabrous. **ST** gen simple, < 2 m, ± coarse. **LVS** simple, entire to pinnately lobed, sometimes dissected underwater. **FL**: sepal bases not sac-like; petals short-clawed, white; filaments wider toward base. **FR**: septum gen incomplete; stigma bilobed. **SEEDS** 2 rows per chamber; wing 0; embryonic root at edges of both cotyledons. 4 spp.: 3 Eurasia, 1 N.Am. (Latin: horseradish)

A. rusticana P. Gaertner, Meyer, & Scherb. (p. 405) HORSERADISH Per, gen erect; roots deep, swollen, spongy. **ST** often branched, 6–10(20) dm. **LVS**: basal long-petioled, 3–5 dm, widely ovate to ovate-oblong, crenate to ± toothed; cauline petioled or not, lowest like basal, middle and upper often lobed. **FL**: petals 5–7 mm. **FR** ± round, ± flat perpendicular to septum, gen abortive; style ± 0.3 mm, stigma prominent, persistent. **SEEDS** 4–6 per chamber, gen abortive. $2n$=32. Ditches, roadsides, moist places; < 200 m. CCo, GV, SnFrB, expected elsewhere; temp N.Am; native to Eurasia.

ATHYSANUS

1 sp. (Greek: without wings, from fr, in contrast to fr of *Thysanocarpus*)

A. pusillus (Hook.) E. Greene (p. 405) Ann, small, leafy toward base. **ST** 5–35 cm, often prostrate; branches gen from near base, slender; hairs simple or 2–5-branched. **LVS**: basal 6–30 mm, 2–10 mm wide, entire or obscurely dentate, short-petioled, hairs coarse; cauline like basal, sessile, bases not lobed. **INFL** terminal, slender, 1-sided, open; bracts 0. **FL** inconspicuous; sepals ± 1 mm, early deciduous; petals 1–2 mm, spoon-shaped, white, sometimes 0. **FR** 2–2.5 mm wide, ± circular, flat parallel to septum, indehiscent; larger hairs flat, hooked, smaller hairs branched or 0; pedicel recurved, 1–4 mm; style < 0.3 mm, chamber 1. **SEED** 1, flat; embryonic root at edges of both cotyledons. Grassy, open slopes, rocky outcrops; gen < 1525 m. CA-FP (exc SN); to B.C., MT, Mex.

AUBRIETA

Per, cespitose to prostrate; caudex branched; hairs simple or branched, stellate or not. **INFL** terminal. **FL**: filaments of single, outer stamen with a dentate appendage at base. **FR** 2–many × longer than wide, flat to ± cylindric or inflated; valves with a median vein. **SEEDS** 2 rows per chamber. ± 15 spp.: mtns, Italy to Iran. (C. Aubriet, French painter of fls, animals, 1651–1743)

A. deltoidea (L.) DC. (p. 409) Low; hairs dense. **LVS**: cauline spoon-shaped to obovate, entire to coarsely dentate. **FL** showy; sepals 6–10 mm; petals 12–28 mm, reddish purple to violet, rarely white. **FR** 6–22 mm, inflated; hairs dense, gen stellate; style 4–8 mm. Open, disturbed areas; 1500–2000 m. c NCoR (Lake, Mendocino cos.); native to Eur.

BARBAREA

Bien, per, erect; hairs simple, sparse, stiff, or 0. **LVS**: basal, lower cauline pinnately lobed to ± compound, terminal lobe >> lateral; cauline smaller upward, upper sessile, base ± lobed, clasping st. **INFL** terminal; bracts 0. **FL**: petals yellow. **FR** linear, ± cylindric to ± 4-sided; valves strongly 1-veined base to tip; style 0.5–3 mm, beak-like. **SEEDS** many, 1 row per chamber; wing 0; embryonic root at edges of both cotyledons. ± 20 spp.: e Eur, sw Asia. (Saint Barbara)

1. Basal lvs with 4–10 pairs of lateral lobes; pedicel > 1 mm thick; fr > 4 cm .... ***B. verna***
1' Basal lvs gen with < 4 pairs of lateral lobes; pedicel < 1 mm thick; fr gen < 3 cm
  2. Style in fr stout, < 1 mm; petals pale yellow; upper cauline lvs pinnately lobed; native .... ***B. orthoceras***
  2' Style in fr slender, 1.5–3 mm; petals bright yellow; upper cauline lvs entire to coarsely dentate; alien ***B. vulgaris***

B. orthoceras Ledeb. (p. 409) Bien, per; caudex simple. **ST** 1–6 dm, branched, stiff, angled. **LVS** pinnately lobed; basal < 12 cm, with 2–3 pairs of lateral lobes, terminal lobe ovate, entire or irregularly toothed; cauline less lobed upward, often clasping st. **FL**: petals pale yellow. **FR** erect to spreading-ascending, (1.5)2.5–5 cm, straight; pedicel 2–3 mm, < 1 mm thick; style 0.5–1 mm. **SEED** ± 1.5 mm. $2n$=16. Damp meadows, wet rocks, streambanks, moist woods; 700–3350 m. CA-FP (exc SN, GV); to AK, e N.Am, Mex, also in temp Asia. [var. *dolichocarpa* Fern.]

B. verna (Miller) Asch. (p. 409) EARLY WINTER CRESS, SCURVY GRASS Bien, rarely per. **ST** 3–8 dm, branched. **LVS** pinnately lobed; basal ± 2 dm, with (4)6–10 pairs of lateral lobes; cauline with 6–16 pairs of lateral lobes. **FL**: petals 5–7, bright yellow. **FR** ascending, 4.5–8 cm, straight; pedicel stout, > 1 mm thick, nearly as thick as fr; style 1–2.5 mm. **SEED** 1–1.5 mm wide, grayish brown. $2n$=16. Uncommon. Damp soils, fields, roadsides; < 500 m. CCo (Monterey Co.), expected elsewhere; widespread e N.Am; native to Eurasia.

B. vulgaris R.Br. (p. 409) COMMON WINTER CRESS, YELLOW ROCKET Bien. **ST** 2–8 dm, branched above, coarse. **LVS**: basal with 1–4 pairs of lateral lobes; upper cauline entire to coarsely dentate, glabrous. **FL**: petals 6–8 mm, bright yellow. **FR** erect to ascending, 1–3 cm, straight; pedicel 3–5 mm, < 1 mm thick; style gen 2–3 mm. **SEED** 1–1.5 mm, oblong. $2n$=16. Uncommon. Disturbed sites; < 1000 m. GB, expected elsewhere; native to Eurasia.

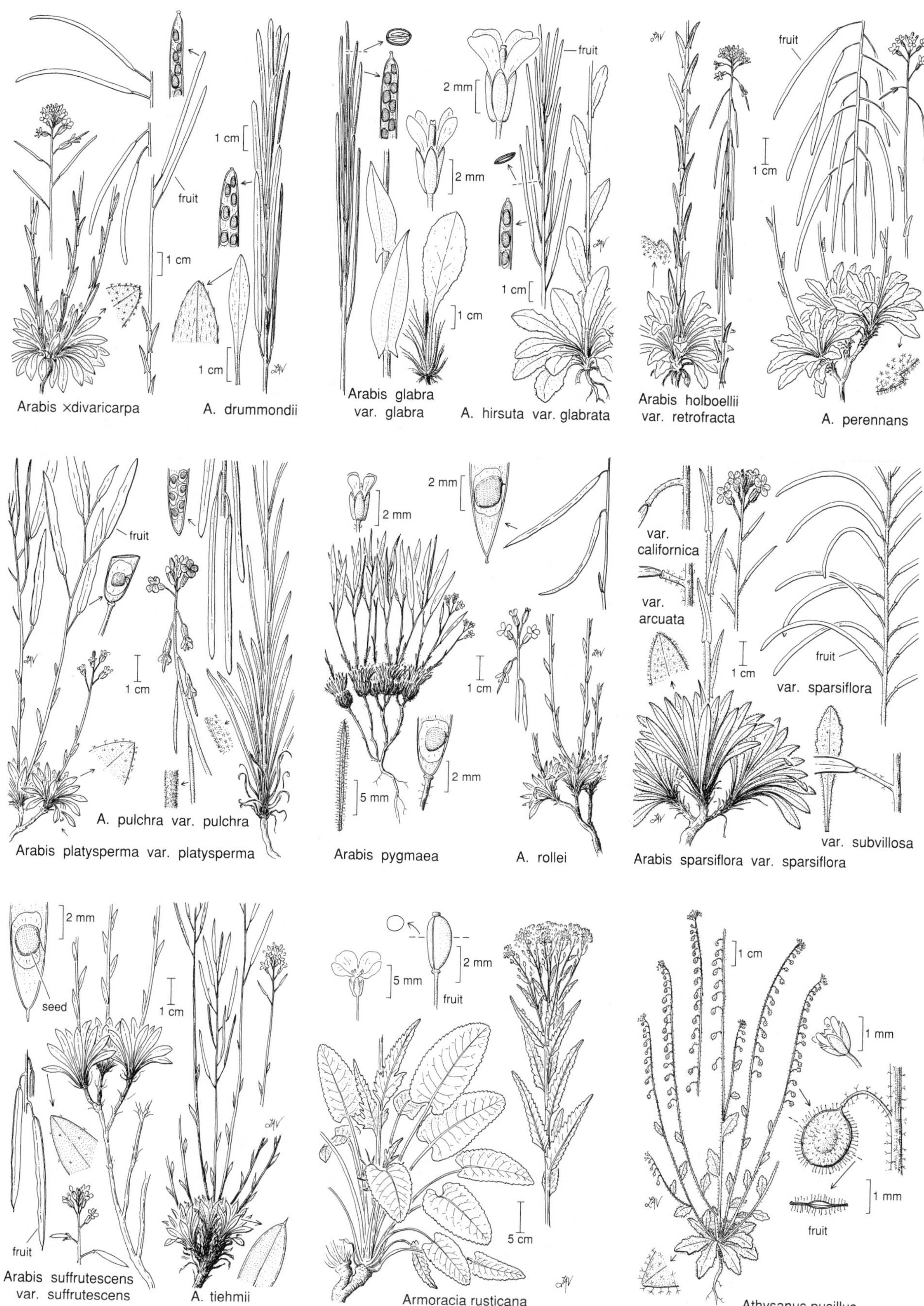
fruit
1 cm
1 cm
1 cm
Arabis ×divaricarpa
A. drummondii
2 mm
2 mm
1 cm
1 cm
Arabis glabra var. glabra
A. hirsuta var. glabrata
Arabis holboellii var. retrofracta
A. perennans
A. pulchra var. pulchra
Arabis platysperma var. platysperma
5 mm
Arabis pygmaea
A. rollei
var. californica
var. arcuata
var. sparsiflora
var. subvillosa
Arabis sparsiflora var. sparsiflora
seed
Arabis suffrutescens var. suffrutescens
A. tiehmii
5 cm
Armoracia rusticana
1 mm
Athysanus pusillus

BRASSICA MUSTARD, TURNIP

Ann, per; hairs simple. **ST** erect, branched, glabrous above. **LVS**: basal and lower cauline petioled, dentate to pinnately lobed, lateral lobes < terminal. **INFL** terminal; bracts ± 0. **FL**: sepals erect; petals gen yellow. **FR** linear; valves 1–veined; beak conic or cylindric, with seeds 0 or rarely 1–2. **SEEDS** many, 1 row per chamber, spheric, finely to coarsely netted. ± 35 spp.: Medit, Eurasia, some naturalized ± worldwide. (Latin: cabbage) Naturalizing CVS soon lose desirable food properties. *B. oleracea* L., cabbage, with thick, glaucous lvs and open infl, is established on se-facing seacliffs, n CCo, c&s NCo.

1. Upper cauline lvs sessile, base lobed, partly or fully clasping st
 2. Petals 6–11 mm; fr beak 10–15 mm; lvs greenish .... ***B. rapa***
 2′ Petals 10–14 mm; fr beak 7–11 mm; lvs glaucous .... ***B. napus***
1′ Upper cauline lvs short-petioled or sessile, base tapered
 3. Pedicel in fr erect, ± appressed; fr 1–2 cm .... ***B. nigra***
 3′ Pedicel in fr spreading to ascending, not appressed; fr gen > 3 cm
  4. Basal lvs, lower st glabrous; basal lvs early deciduous, lobes in 2–3 pairs, ± pinnately arranged, or 0; petals > 6 mm .... ***B. juncea***
  4′ Basal lvs, lower st with stiff hairs; basal lvs persistent, lobes in > 3 pairs, pinnately arranged; petals < 5 mm .... ***B. tournefortii***

B. juncea (L.) Czernov INDIAN MUSTARD Ann 4–10 dm, glabrous, ± glaucous. **LVS**: basal early deciduous, < 2.5 dm, lobes in 2–3 pairs or 0; cauline dentate to lobed, upper reduced, base tapered, petiole short or 0. **FL**: petals > 6 mm, pale yellow. **FR** ± spreading to ascending, 2–4 cm, subcylindric; pedicel spreading to ascending, 1–1.5 cm, slender; beak 5–10 mm, slender. **SEED** finely netted. $2n$=16. Uncommon. Fields, disturbed areas; < 300 m. GV; native to Eurasia. Cult for seed.

B. napus L. SWEDE RAPE, RAPESEED Ann, bien < 15 dm, leafy. **LVS** glaucous; basal and lower cauline ± pinnately lobed, hairs 0 or sparse; middle and upper cauline sessile, base lobed, often clasping st. **FL**: petals 10–14 mm, golden to dull yellow. **FR** ascending, 5–10 cm; pedicel spreading to ascending, slender; beak 7–11 mm, seedless. **SEED** ± 1.8 mm wide, finely netted. $2n$=36. Roadsides, fields, disturbed areas; < 500 m. SnJV, SCo; native to Eur.

B. nigra (L.) Koch (p. 409) BLACK MUSTARD Ann; hairs sparse to dense, stiff, esp below. **ST** 4–20 dm, gen branched above. **LVS**: basal pinnately lobed, serrate-dentate; cauline similar to basal but upper smaller, sessile, base tapered. **FL**: petals 7–11 mm. **FR** erect, 1–2 cm; pedicel erect, ± appressed. **SEED** ± 2 mm wide, coarsely netted. $2n$=16. Abundant. Fields, disturbed areas; < 1500 m. CA-FP; native to Eur.

B. rapa L. (p. 409) TURNIP, FIELD MUSTARD Ann, erect; hairs 0 or very sparse, not stiff. **ST** simple to freely branched, 2–10 dm. **LVS**: lower cauline ± pinnately lobed, lateral lobes 2–4, terminal lobe obovate, wavy-dentate; middle, upper lvs sessile, base lobed, ± clasping st. **FL**: petals 6–11 mm, yellow. **FR** ascending to ± spreading, 3–7 cm; pedicel ± ascending, 7–25 mm; beak (8)10–15 mm, narrowed to a slender style. **SEED** ± 1.5 mm wide, very finely netted. $2n$=20. Grainfields, orchards, disturbed areas; < 1500 m. CA-FP, SNE; widespread US, native to Eur. [*B. campestris* L.]

B. tournefortii Gouan (p. 409) Ann, branched ± from base, widely above. **ST** < 7 dm; hairs on lower st ± dense, stiff, white. **LVS**: basal rosetted, persistent, petioled, pinnately lobed, serrate-dentate; cauline few, base tapered, uppermost bract-like. **FR** 3–7 cm, cylindric, narrowed between seeds; pedicels spreading, lower 8–15 mm; beak 1–1.5 cm, stout. **SEED** ± 1 mm wide, finely netted. $2n$=20. Roadsides, washes, open areas; < 800 m. SW, D; to s NV, TX; native to Medit. Locally abundant.

CAKILE SEA ROCKET

Ann, per, fleshy, many-branched, gen glabrous. **ST** erect or decumbent. **LF** fleshy. **INFL** terminal, elongating; bracts gen 0. **FL**: petals lavender to white. **FR** fleshy, green, in age dry, corky, with 2 transversely jointed, indehiscent segments, each falsely 1-chambered, gen 1-seeded. **SEEDS**: upper > lower. ± 4 spp.: shores, N.Am, Eur, Afr. (Arabic name) [Rodman 1974 Contr Gray Herb 205:3–146]

1. Lower fr segment without lateral lobes or horns; early lvs wavy-margined or deeply dentate or irregularly shallowly lobed but not pinnately lobed; petals 0 or < 3 mm wide, gen white, less often pale lavender ***C. edentula***
1′ Lower fr segment with 2 lateral lobes or horns; early lvs deeply pinnately lobed; petals 3–6 mm wide, lavender, rarely white .... ***C. maritima***

C. edentula (Bigelow) Hook. (p. 409) Ann, gen ascending to erect. **LF** widely ovate to obovate or spoon-shaped, petioled or not. **FR** 1–3 cm; upper segment with ± flat beak, tip acute to notched. $2n$=18. Beach dunes; < 50 m. NCo, CCo, SCo; native to e N.Am. [ssp. *californica* (A.A. Heller) Hultén] Formerly more common in CA, now being replaced by *C. maritima*.

C. maritima Scop. (p. 409) Ann, prostrate or mound-forming to erect. **LF** obovate to spoon-shaped, petioled. **FR** 1.5–3 cm; upper segment conic to cylindric, tip acute to blunt. $2n$=18. Beach dunes; < 50 m. NCo, CCo, SCo; to B.C., e N.Am, Mex.; native to Eur.

CAMELINA FALSE FLAX

Ann, bien, ± erect. **LVS** gen cauline, sessile, entire to dentate or shallowly lobed; basal lobes clasping st. **FL**: petals white to yellow. **FR** obovate to pear-shaped, plump, ± stalked above receptacle; septum wide; valves sturdy, ± winged. **SEEDS** 2 rows per chamber. (Greek: low flax, from inhibition of flax pls)

C. microcarpa Andrz. (p. 409) Ann 3–10 dm. **ST** simple to ± branched above; hairs simple, spreading, below also forked, stellate. **LVS** oblong to lanceolate, ± entire; tip acute to acuminate; lower lf hairs as on sts. **INFL** elongate. **FL**: petals 4–5 mm, pale yellow. **FR** 5–8 mm, obovate; pedicel spreading to ascending, 1–1.5 cm, slender; style 2–2.5 mm, beak-like. **SEED** ± longer than wide, plump. $2n$=40. Grainfields, roadsides, hillsides; < 1000 m. CA-FP; N.Am; native to Eurasia. [*C. sativa* (L.) Crantz misapplied]

CAPSELLA SHEPHERD'S PURSE

Ann, bien; hairs 0, simple, or branched. **LVS**: basal clustered or rosetted, entire to dissected or lobed; cauline sessile, sagittate, ± clasping st. **FL**: sepals green or reddish, bases not sac-like; petals ± 2 mm, obovate to spoon-shaped, white or pinkish. **FR** obtriangular-obcordate, flat perpendicular to septum; valves keeled. **SEEDS** many; wing 0; embryonic root at back of 1 cotyledon. ± 4 spp.: Eur. (Latin: little box)

C. bursa-pastoris (L.) Medikus (p. 409) Ann; hairs simple and stellate. **ST** erect, 1–5 dm, branched or not. **LVS**: basal rosetted, petioled, 3–6 cm, oblanceolate, subentire to pinnately lobed or dissected. **INFL** many-fld. **FL**: petals white, distinctly clawed. **FR** 4–8 mm; pedicel spreading to ascending. $2n$=32. Disturbed sites, gardens; < 2300 m. CA; N.Am; native to Eur.

CARDAMINE BITTER-CRESS, TOOTHWORT

Ann, bien, per, from taproots, fibrous roots, or tuber-like rhizomes; hairs 0 or simple. **LVS** entire or palmately or pinnately lobed to compound; rhizome lvs often present, separate from others. **INFL** bracted or not. **FL**: sepals equal at base; petals white to pink or rose. **FR** linear, gen flat; valves gen opening elastically, sometimes by coiling from base; septum margins intruding on valves. **SEEDS** many, 1 row per chamber, wingless (± margined in *C. oligosperma*); embryonic root at edges of both cotyledons. ± 170 spp.: most temp parts of world. (Greek: for a cress with medicinal uses)

1. Lvs simple, shallowly lobed to gen entire, rarely ± dissected
 2. Pl ± scapose, rarely with 1–2 cauline lvs, 2–12 cm; basal lvs many, blade base ± tapered; infl in fr ± umbel-like ..... ***C. bellidifolia*** var. ***pachyphylla***
 2′ Pl with cauline lvs, 15–50 cm; basal lvs few, blade base abruptly narrowed; infl in fr elongate
  3. Cauline lvs from base to infl; rhizome lvs 0; sts amid gen dense rhizomes ..... ***C. cordifolia*** var. ***lyallii***
  3′ Cauline lvs near infl; rhizome lvs present; sts from ends of elongate rhizomes ..... ***C. pachystigma***
   4. Cauline lvs 1-palmately or -pinnately ± dissected ..... var. ***dissectifolia***
   4′ Cauline lvs entire or shallowly lobed ..... var. ***pachystigma***
1′ Lvs or at least lower cauline palmately or pinnately lobed to compound
 5. St widened to strong attachment with rhizome or rhizome 0; rhizome lvs gen 0; cauline lvs pinnately lobed to compound
  6. Ann or bien from fibrous roots or weak taproot
   7. Basal lvs 0–few, not rosetted, gen deciduous; lower st hairs few or gen 0 ..... ***C. pensylvanica***
   7′ Basal lvs many, gen ± rosetted, persistent; lower st hairs 0–many
    8. Sts 1–many, erect, 5–30 cm, outer 0 or decumbent; fr < 1 mm wide; alien ..... ***C. hirsuta***
    8′ Sts 1–several, erect to ascending, gen > 20 cm; fr 1–2 mm wide; native ..... ***C. oligosperma***
  6′ Per from rhizomes or strong taproot
   9. Petals 8–14 mm; style 1.5–6 mm
    10. Rhizome elongate, > 3 cm, not tuber-like, ± spreading; style 1.5–3(4) mm; lflets widely dentate to lobed ..... ***C. angulata***
    10′ Rhizome short, < 2 cm, tuber-like, ± spreading to ± erect; style 3–6 mm; lflets or lobes entire to wavy or dentate ..... [2]***C. californica*** (see 17. for vars.)
   9′ Petals 4–6 mm; style ± 1–1.2 mm
    11. Basal lvs pinnately lobed; rhizome ± tuber-like at st base ..... ***C. occidentalis***
    11′ Basal lvs gen entire to shallowly palmately lobed; rhizome not tuber-like at st base ..... ***C. breweri***
     12. Lflets of lower cauline lvs wedge-shaped to truncate at base; rhizome ± elongate, spreading ..... var. ***breweri***
     12′ Lflets of lower cauline lvs cordate at base; rhizome gen short, ± erect ..... var. ***orbicularis***
 5′ St narrowed to weak attachment with rhizome; rhizome lvs present; cauline lvs pinnately or gen palmately lobed to compound
  13. St slender, < 2 dm; rhizome ovoid to elongate, gen with ovoid, tuber-like swelling from which lf arises, 0.5–3 cm; infl in fr ± umbel-like ..... ***C. nuttallii***
   14. Rhizome lf simple, ± entire to shallowly lobed ..... var. ***nuttallii***
   14′ Rhizome lf subpalmately compound, lflets 3–5, ± entire to ± lobed
    15. Rhizome lflets (3)5, oblong, ± entire, 6–10 mm wide ..... var. ***covilleana***
    15′ Rhizome lflets 3–5, ovate to ± elliptic to obovate, toothed or lobed, 10–20 mm wide
     16. Rhizome lflets ± sessile, toothed or lobed; rhizome elongate, gen whitish ..... var. ***dissecta***
     16′ Rhizome lflets stalked, toothed; rhizome ovoid to slightly elongate, gen yellowish ..... var. ***gemmata***
  13′ St ± stout, 2–7 dm; rhizome ovoid to subspheric, tuber-like, < 2 cm; infl in fr elongate ..... [2]***C. californica***
   17. Rhizome lvs 1–2-pinnate, lflets 5–9, often lobed ..... var. ***cuneata***
   17′ Rhizome lvs simple, entire, or lflets 3, rarely 5 and ± pinnately arranged
    18. Rhizome lvs gen with 3 lflets, rarely with 5 or simple, entire
     19. Lflets or lobes of cauline lvs gen ovate or wider, thin, gen sharp-toothed ..... var. ***californica***
     19′ Lflets or lobes of cauline lvs gen oblong to linear, thickish, gen entire or teeth small, few or only 1 at tip ..... var. ***integrifolia***
    18′ Rhizome lvs simple
     20. Cauline lvs simple; rhizome lf blade margin shallowly wavy ..... var. ***cardiophylla***
     20′ Cauline lvs gen compound; rhizome lf blade margin wavy to deeply dentate ..... var. ***sinuata***

C. angulata Hook. Per; rhizome elongate, > 3 cm, ± spreading. **ST** simple, 40–100 cm; hairs gen 0 or sparse. **LVS**: rhizome lvs with 3 lflets, long-petioled; cauline lflets > 3, ovate-lanceolate to ovate, widely dentate to lobed. **FL**: petals 8–14 mm, white to pinkish. **FR** erect to ascending, 15–30 mm; pedicel ascending, 10–18 mm; style 1.5–3(4) mm. Shady thickets, streambanks, deep woods; < 400 m. NW; to AK. ❀TRY.

C. bellidifolia L. var. ***pachyphylla*** Cov. & Leiberg (p. 409) Per, ± cespitose, ± scapose, glabrous; taproot 1–4 mm thick near top; caudex branches 0–many, < 10 cm, rough, lf bases persistent. **STS** several, erect, 2–12 cm. **LVS** simple; basal many, ± rosetted, 1–3 cm, thick, fleshy, entire or shallowly 1–3-lobed, blades 5–30 mm, diamond-shaped-elliptic to ovate, base ± tapered; cauline 0, rarely 1–2. **INFL** ± umbel-like, few-fld. **FL**: petals 5–6 mm, white. **FR** erect, 15–35 mm; pedicel ascending, 4–10(14) mm; style 2–3 mm, stout. Near large rocks, ledges, rocky, moist soils, alpine slopes, volcanic peaks; 1900–2800 m. CaR; OR. ❀TRY.

C. breweri S. Watson Per. **ST** 15–50(70) cm; hairs gen 0 or below. **LF** simple or lflets 3 or 5; simple lvs and terminal lflets >> lateral lflets, ovate to cordate, wavy-margined to shallowly lobed. **FL**: petals ± 5 mm, white. **FR** erect, 15–30 mm; pedicel ascending, 5–20 mm; style 1.2 mm. **SEED** 1.5–2 mm. Wet sites, coniferous forest; < 3200 m. n NW, CaRH, SNH, WTR, SnBr, SNE; to B.C., MT, Colorado.

var. ***breweri*** (p. 409) Rhizome ± elongate, spreading, 1–4 mm thick. **STS** 1 or few, decumbent or erect, rarely prostrate and rooting at nodes. **LVS**: lower cauline lflets gen 3, wedge-shaped to truncate at base. $2n$=42–48. Margins of streams, seeps, lakes, swamps; 1200–3200 m. CaRH, SNH, WTR, SnBr, SNE; to B.C., MT, Colorado. ❀IRR,DRN:1,4,5,6,15–17&SHD:**2,3**,7,14,**18**.

var. ***orbicularis*** (E. Greene) Detl. Rhizome gen short, ± erect, 0.5–2 mm thick. **ST** gen 1, erect, 1.5–4 dm. **LVS**: lower cauline lflets cordate at base. Muddy streambanks, swamps, wet woods; < 500 m. n NW (Humboldt Co.); to WA.

C. californica (Torrey & A. Gray) E. Greene MILK MAIDS, TOOTH WORT Per; rhizome < 2 cm, tuber-like; hairs 0, rarely minute, simple. **ST** 20–70 cm. **LVS**: lflets or lobes of cauline lvs entire to wavy or dentate; lflets or lobes of rhizome lvs gen 3, 2–6 cm wide, ovate to ± cordate; cauline lvs alternate, lower long-petioled, upper short-petioled to ± sessile, lflets or lobes 3–5, widely ovate to oblong. **INFL** elongate in fr. **FL**: petals 9–14 mm, white to pale rose. **FR** erect, 20–50 mm, 1.5–2 mm wide; pedicel ascending, 10–30 mm; style 3–6 mm. Gen shaded sites, canyons, woods; < 1200 m. CA-FP; OR, Baja CA. [*Dentaria c.* Torrey & A. Gray]

var. ***californica*** (p. 409) **ST**: hairs 0, rarely simple. **LVS** gen thin; rhizome gen with 3 lflets, rarely with 5 or simple, entire, lflet stalks < blades; cauline lflets or lobes ± 3, gen sharply toothed. $2n$=16,32. Common. Shady slopes, wooded ravines, cliffs, shaded rock crevices; < 1200 m. CA-FP; Baja CA. ❀SHD,DRN,IRR:**5**,6,7,9,14,**15–17**,18–24.

var. ***cardiophylla*** (E. Greene) Rollins **ST**: hairs sparse or 0. **LVS** simple; rhizome lvs cordate or ± round, blade margin shallowly wavy; cauline lvs cordate to widely ovate, entire; petioles 1.5–3 cm exc above. Forest floor, streambanks, moist slopes; < 600 m. NW. [*D. c.* var. *cardiophylla* (E. Greene) Detl.] ❀SHD,DRN,IRR:**5**,7,14,**15–17**,22–24.

var. ***cuneata*** (E. Greene) Rollins **ST**: hairs sparse or 0. **LVS**: rhizome lvs 1–2-pinnate, lflets 5–9, often lobed; cauline lvs gen 1-pinnate, lflets ± linear to obovate, ± tapered to base. Open woods, moist hillsides; < 900 m. SCoR. [*D. c.* var. *cuneata* (E. Greene) Detl.] ❀DRN,SHD:**15–17**&IRR:**7,14**,18–21,**22–24**.

var. ***integrifolia*** (Torrey & A. Gray) Rollins **ST**: hairs gen 0, rarely simple. **LVS**: rhizome lvs gen with 3 lflets, rarely with 5 or simple, entire, blade elliptic to ± round, ± tapered to base; cauline lf lobes or lflets gen 3, rarely 5, lflets short-petioled, gen linear to oblong, gen entire or teeth small, few or only 1 at tip; upper cauline lvs linear to oblong, entire, sessile. $2n$=32. Common. Open meadows, hill slopes, canyons; < 500 m. CA-FP; OR. [*D. c.* var. *i.* (Torrey & A. Gray) Detl.] ❀DRN:**4,5**,6&SHD,IRR:7,14,**15–17**.

var. ***sinuata*** (E. Greene) O. Schulz **LVS**: rhizome lvs simple, round or cordate, margin wavy to deeply dentate, lower surface gen purplish; cauline lvs deeply lobed or gen with 3 lflets, lower rarely simple. Forest floor, hillsides, stream valleys; < 300 m. NW; OR. [*D. c.* var. *s.* (E. Greene) Detl.] ❀SHD,DRN:**5**&IRR:14,**15–17**.

C. cordifolia A. Gray var. ***lyallii*** (S. Watson) Nelson & J.F. Macbr. Per; rhizome extensive, 2–25 mm thick; hairs 0, rarely on st bases, lf margins, minute, simple. **ST** amid gen dense rhizomes, simple, 20–50 cm. **LVS** simple, fleshy, cordate or reniform, petioled; basal and lower cauline 2–10 cm wide, gen wider than long, gen crenate; cauline arrayed from base to infl, upper reduced. **INFL**: bracts 0. **FL**: petals 7–12 mm, obovate or narrower, long-clawed, white. **FR** ascending to erect, 20–34 mm, straight or curved; pedicel ascending, straight, 10–20 mm; style 0.5–2(6) mm. **SEED** flat, not winged, smooth. Streambanks, moist aspen groves, meadows; 1500–2300 m. KR, CaR, n SNH; to B.C. [*C. l.* S. Watson] ❀STBL.

C. hirsuta L. Ann. **ST** gen 1, erect, 5–30 cm, stiff; hairs toward base, gen 0. **LVS** compound, crowded below; basal often rosetted; petioles ciliate; lflets 2–3 pairs, 0.5–4 mm wide, ovate to ± round, wavy-margined to shallowly few-lobed, terminal > lateral. **FL**: petals 1.5–2 mm, white; stamens gen 4. **FR** erect; pedicel ascending; style 0.3–1 mm, stout; valves at dehiscence rolling into tight rings. **SEEDS** 22–36. $2n$=16. Lawns, roadsides, ditches; 600–800 m. KR, expected elsewhere; e N.Am, C.Am; native to Eur.

C. nuttallii E. Greene Per, glabrous; rhizome 0.5–3 cm, 2–5 mm thick, ovoid to elongate, whitish. **ST** simple, < 2 dm, slender. **LVS**: rhizome 1, simple, ± entire to shallowly lobed, or with 3–5 ± palmately arranged lflets; cauline above, gen 2, petioled, lflets 2–5, 1–4 cm, ± equal, oblong to oblanceolate, ± sessile, entire. **INFL** ± umbel-like. **FL** conspicuous; petals 1–1.5 cm, pale pink to purple, rarely ± white. **FR** 2–4 cm, 2–2.5 mm wide; pedicels ascending, lower 1–2 cm; style 3–5 mm. Gen moist sites, canyons, forest; < 2200 m. NW; to B.C. 4 vars. total.

var. ***covilleana*** (O. Schulz) Rollins (p. 409) **LVS**: rhizome lflets (3–)5, ± subpalmately arranged, rarely 2 below and 3 above, 6–10 mm wide, oblong, ± entire, with 1 prominent vein ending in a sharp tip, round at base. **FL**: petals purple or white with purple veins. Moist, shaded hillsides, wet, open pine forests; 1200–2200 m. KR, NCoR; OR. [*Dentaria tenella* Pursh var. *palmata* Detl.] ❀SHD,IRR,DRN:1,**4–6**,7,14–16,**17**.

var. ***dissecta*** (O. Schulz) Rollins **LVS**: rhizome lflets 3(–5), ± subpalmately arranged, 1.5–3 cm, 10–20 mm wide, ovate to obovate, ± sessile, toothed or lobed; cauline 3(–5), lflets 3(–5), 2–4 cm, 3–6 mm wide, linear to lanceolate, ± entire. Mossy slopes, pine-covered hillsides, damp woods; < 1900 m NW; to WA. [*C. tenella* (Pursh) O. Schulz var. *d.* O. Schulz] ❀SHD,DRN:**4–6**&IRR:7,14,**15–17**.

var. ***gemmata*** (E. Greene) Rollins (p. 409) **LVS**: rhizome lflets 3–5, subpalmately arranged, widely ovate to ± elliptic, stalked, thickish, toothed; cauline lflets 3–5, ± linear. Redwood forest; < 100 m. NCo (Del Norte Co.); sw OR. [*C. g.* E. Greene, yellow-tubered toothwort; *Dentaria g.* (E. Greene) Howell] ❀SHD, DRN:**5**&IRR:14,**15–17**.

var. ***nuttallii*** **LVS**: rhizome lvs simple, ± round or cordate, ± entire to shallowly lobed, palmate veins ending in sharp tips. Moist humus, shaded bottomlands, damp forests; < 1500 m. NW; to B.C. [*C. pulcherrima* (Robinson) E. Greene var. *tenella* (Pursh) Hitchc.]

C. occidentalis (Robinson) Howell Per; rhizomes short, 0.5–1 mm thick, ± tuber-like at st base; hairs gen sparse or 0. **ST** erect or decumbent, rooting at nodes, ± branched, 20–40 cm. **LVS**: basal few, petioled, pinnately lobed, lateral lobes (2)4–6, 3–8 mm, ovate to cordate-ovate, gen entire, terminal lobe 1–2 cm, cordate-ovate, wavy-margined; cauline several, upper ± sessile. **FL**: petals 4–6 mm, white. **FR** erect, 20–30 mm, 1–1.5 mm wide; pedicels ascending, lowest 10–15 mm; style ± 1 mm. $2n$=32,64. Wet soils, margins of lakes, creeks; < 1000 m. NW; to AK.

C. oligosperma Torrey & A. Gray (p. 415) Ann, bien, taprooted. **STS** erect to ascending, 1–several, gen > 20 cm, branched; hairs short or 0. **LVS** many; basal ± rosetted, 1-pinnate, lateral lflets 5–9, 3–20 mm, short-stalked, widely obovate to ± round, sparsely, shallowly dentate, < terminal. **INFL** 3–10 cm, elongate. **FL**: petals 2–4 mm, white. **FR** ascending to erect, 15–25 mm, 1–2 mm wide; hairs 0 or sparse; pedicel ascending to erect, 5–15 mm; style < 0.5 mm. **SEED** oblong-ovate, narrowly margined. $2n$=16. Wet meadows, shady banks, creek bottoms; < 1100 m. CA-FP; B.C., MT, Colorado.

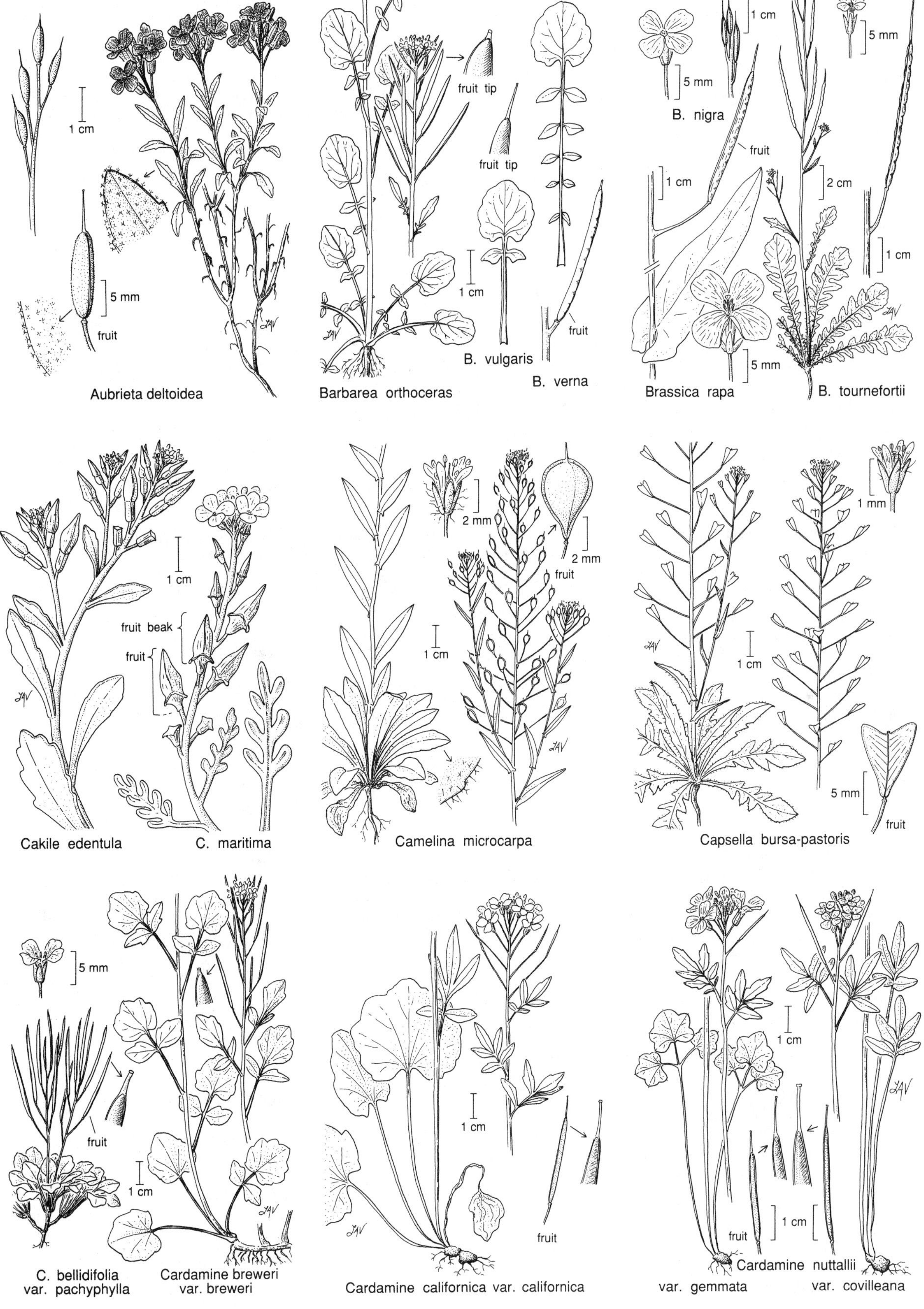
1 cm
5 mm
fruit
Aubrieta deltoidea
fruit tip
fruit tip
1 cm
fruit
B. vulgaris
B. verna
Barbarea orthoceras
1 cm
5 mm
B. nigra
5 mm
fruit
1 cm
2 cm
1 cm
5 mm
Brassica rapa
B. tournefortii
1 cm
fruit beak
fruit
Cakile edentula
C. maritima
2 mm
2 mm
fruit
1 cm
Camelina microcarpa
1 mm
1 cm
5 mm
fruit
Capsella bursa-pastoris
5 mm
fruit
1 cm
C. bellidifolia var. pachyphylla
Cardamine breweri var. breweri
1 cm
fruit
Cardamine californica var. californica
1 cm
1 cm
fruit
Cardamine nuttallii
var. gemmata
var. covilleana

C. pachystigma (S. Watson) Rollins Per, glabrous; rhizome ± 3 mm wide. **STS** from ends of elongate rhizomes, erect, 15–30 cm, simple, often stoutish. **LVS**: rhizome lvs simple, rarely with 1–2 small lflets, 4–5 cm, widely ovate to ± round, thick, fleshy, margin entire or wavy, base cordate; cauline lvs near infl, 2–3, petioled. **INFL** short. **FL**: petals 4–11 mm, pink to purple, rarely white. **FR** ± erect, 30–60 mm, 2–4 mm wide, narrowed to tip; pedicels ascending to erect, lower 10–20 mm; style 3–5 mm, ± thick. Rocky outcrops, slopes; < 2900 m. NCo, NCoR, CaR, SN, ScV. [*Dentaria p.* S. Watson]

var. ***dissectifolia*** (Detl.) Rollins DISSECTED-LEAF TOOTHWORT **LVS**: cauline 1-palmately or -pinnately ± dissected, segments 3–5, 1.5–4 cm, narrowly lanceolate. UNCOMMON. Serpentine outcrops; < 900 m. NCo (Mendocino Co.), ScV (Butte Co.). [*Dentaria p.* var. *d.* Detl.] ❀TRY:DFCLT.

var. ***pachystigma*** **LVS**: cauline entire or shallowly lobed, base gen cordate or truncate, teeth wide, abruptly pointed. Lava slides, cliffs, rocky slopes; 1500–2900 m. NCoR, CaR, SN. ❀SHD,DRN:1,6,7,14–16;DFCLT.

C. pensylvanica Willd. (p. 415) Bien; roots fibrous. **STS** erect or decumbent, 1–few, gen branched, 10–70 cm; hairs few or gen 0 toward base. **LVS** 1-pinnate, gen glabrous; basal 0–few, not rosetted; basal, lower cauline lflets 5–11, widely elliptic to obovate, lateral < terminal. **FL**: petals 2–4 mm, white. **FR** ± erect, 20–30 mm, ± 1 mm wide; pedicel ascending, 5–10 mm, slender, glabrous; style ± 1 mm. **SEED** ± 1 mm; wing 0. $2n=32$. Muddy streambanks, seepages; 750–1900 m. CaR, n SNF; to Can, se US.

CARDARIA WHITE-TOP, HOARY CRESS

Per, gen strongly rhizomed; hairs 0 to dense, simple, fine. **STS** 1–several, gen erect. **LVS** simple, toothed; base often lobed, clasping st. **INFL** ± flat-topped, wider than long, gen dense. **FL**: petals white. **FR** cordate or ovate to ± round to obovate, indehiscent or tardily dehiscent, partly inflated or flat perpendicular to septum; style 1–2 mm, slender. **SEEDS** 1–2 per chamber, ovate, ± flat; wing 0; embryonic root at back of 1 cotyledon. 5 spp.: Eurasia. (Greek: heart-shaped, from fr of *C. draba*) [Mulligan & Findley 1974 Canad J Pl Sci 54:149–160]

1. Sepals, fr hairy .... ***C. pubescens***
1′ Sepals, fr glabrous
 2. Fr widely ovate or widely obovate to ± round, not narrowed at septum .... ***C. chalepensis***
 2′ Fr widely cordate or widely ovate, gen narrowed at septum .... ***C. draba***

C. chalepensis (L.) Hand.-Mazz. (p. 415) LENS-PODDED HOARY CRESS **ST** 2–4 dm; hairs below, sparse to ± 0 above. **LVS** widely oblanceolate to obovate; basal short-petioled, ± toothed to entire; middle and upper cauline sessile, base lobed, clasping st. **FL**: sepals glabrous; petals 3–4 mm. **FR** 2.5–6(8) mm, ± round to widely ovate or widely obovate, inflated, not narrowed at septum, glabrous. $2n=80$. Disturbed, gen saline soils, fields; < 1500 m. CA (exc D); to Can, c US; native to e Asia. NOXIOUS WEED.

C. draba (L.) Desv. (p. 415) HEART-PODDED HOARY CRESS **ST** 2–5 dm; hairs below, sparse or ± 0 above. **LVS** widely oblanceolate to obovate; basal lvs short-petioled, ± toothed to entire; middle and upper cauline lvs sessile, base lobed, clasping st. **FL**: sepals glabrous, margin white; petals 3–4 mm. **FR** 3–5 mm wide, widely cordate or ovate, gen narrowed at septum, glabrous. $2n=62,64$. Disturbed, gen saline soils, fields, roadsides; < 1200 m. CA (exc D); to Can, c US; native to Eurasia. NOXIOUS WEED. Pls called var. *repens* (Shrenk) O. Schulz are apparent hybrids with *C. chalepensis*.

C. pubescens (C. Meyer) Jarmol. (p. 415) WHITE-TOP **ST** 1–4 dm, hairy. **LVS** ± dentate; basal short-petioled; cauline sessile, oblong-linear to -lanceolate, base lobed, clasping st. **FL**: sepals hairy, margin white; petals 2–3.5 mm. **FR** 3–4.5 mm, ovate, round or obovate, inflated, not narrowed at septum, hairy. $2n=16$. Saline soils, fields, ditchbanks; < 2000 m. CA; to e N.Am; native to Asia. NOXIOUS WEED. [var. *elongata* Rollins]

CAULANTHUS JEWELFLOWER

Roy E. Buck

Ann to per, gen tapered-hairy on lvs and lower st. **ST** gen ascending to erect, ± glaucous. **LVS** ± entire to deeply cut; basal gen rosetted, withering, gen oblanceolate to obovate; cauline gen linear to obovate, clasping, reduced. **INFL** becoming more open; bracts gen 0. **FL** biradial to ± bilateral; calyx ± urn-shaped, sepals often ± pouched below, gen not green, gen not darker in bud, gen erect after fl; petal (and sepal) margins often scarious, wavy or not; filaments gen in 3 pairs, gen free (or longest 1–2 pairs ± fused below); style < 4 mm, stigma gen 2-lobed. **FR** ascending to reflexed, gen ± cylindric. **SEED** gen ± oblong, gen compressed, gen ± brown. ± 14 spp.: ± sw N.Am. (Greek: st fl, from use of some as cauliflower-like vegetable)

1. Middle and upper cauline lvs not clasping, petioled or not
 2. Sepal hairs ± dense; ann .... ***C. hallii***
 2′ Sepals gen glabrous or hairs sparse (if dense, per from woody caudex)
  3. Ann to weak per, caudex gen 0 or weak; lvs and lower st conspicuously hairy .... ***C. pilosus***
  3′ Per from woody caudex; lvs and lower st ± glabrous
   4. Basal and lower cauline lvs gen < 3 × longer than wide, narrowed abruptly to petiole, basal ± not rosetted; seed yellowish or light brown .... ***C. glaucus***
   4′ Basal and lower cauline lvs gen > 3 × longer than wide, tapered gradually to petiole (unless base deeply lobed), basal rosetted; seed dark brown
    5. St gen not inflated; sepals glabrous or tips sparsely hairy; stigma lobes < 0.5 mm .... ***C. major***
     6. Sepals greenish white or cream-colored .... var. ***major***
     6′ Sepals purple .... var. ***nevadensis***
    5′ St ± inflated; sepals glabrous or densely hairy; stigma lobes > 0.5 mm .... ***C. crassicaulis***
     7. Sepals ± densely hairy .... var. ***crassicaulis***
     7′ Sepals glabrous .... var. ***glaber***
1′ Middle and upper cauline lvs clasping, sessile

8. Sepals gen obviously darker in bud (exc sometimes *C. coulteri*); longest filament pair fused at least at base (exc sometimes *C. californicus*)
9. St ± inflated; pl glabrous or sparsely hairy below .... ***C. inflatus***
9′ St not inflated; pl gen obviously hairy below (exc sometimes *C. coulteri* var. *lemmonii*)
10. Sepals 4–8(10) mm; petals 6–11 mm; fr 1–6 cm, compressed perpendicular to septum; seed ± spheric .... ***C. californicus***
10′ Sepals (5)7–18 mm; petals (8)10–31 mm; fr 4–13 cm, cylindric or ± compressed parallel to septum; seed compressed .... ***C. coulteri***
11. Lvs and lower sts gen densely hairy, gen some lf hairs branched (exc gen SW); lower and middle cauline lvs lanceolate to oblong, gen coarsely dentate to deeply cut; fr gen reflexed .... var. ***coulteri***
11′ Lvs and lower sts ± glabrous to sparsely hairy, lf hairs unbranched; lower and middle cauline lvs lanceolate-oblong to ovate, entire to finely dentate; fr gen erect or ascending .... var. ***lemmonii***
8′ Sepals not or barely darker in bud; longest filament pair free
12. Basal and lower cauline lf length gen > 10 × width, sides ± parallel .... ***C. heterophyllus***
13. Sepals purple; petals whitish with purple veins or purplish .... var. ***heterophyllus***
13′ Sepals yellowish or cream-colored, reddish blushed or not; petals yellowish green, purple-veined or not .... var. ***pseudo-simulans***
12′ Basal and lower cauline lf length gen << 10 × width, sides gen not parallel
14. Pl ± conspicuously spreading-hairy below .... ***C. simulans***
14′ Pl glabrous (or st inconspicuously appressed-hairy)
15. St gen inconspicuously appressed-hairy; upper lvs ± narrowly oblong, hastate, tips acute .... ***C. cooperi***
15′ Pl glabrous; upper lvs widely (ob)ovate, not hastate, tips ± obtuse .... ***C. amplexicaulis***
16. Sepals purple .... var. ***amplexicaulis***
16′ Sepals yellowish or cream-colored .... var. ***barbarae***

C. amplexicaulis S. Watson Ann, glabrous. **ST** gen branched below. **LVS** < 10 cm, lighter on main veins; basal coarsely dentate, tapered to short, winged petiole or ± sessile; cauline widely (ob) ovate, entire to coarsely dentate, sessile, clasping, tips obtuse. **INFL** open. **FL**: sepals 4–9 mm, ± pouched below; petals 7–18 mm, tips gen reflexed, upper pair longer, purplish, lower gen straw-colored or paler purple, margins wavy; style < 0.5 mm, stigma lobes 0 or shallow. **FR** ascending to spreading, < 15 cm, often curved. Open, sandy or rocky areas; 1000–2800 m. s SCoRO, TR.

var. ***amplexicaulis*** (p. 415) **FL**: sepals purple. **FR** 5–12 cm. Habitats and elevations of sp. TR.

var. ***barbarae*** (J. Howell) Munz SANTA BARBARA JEWELFLOWER **FL**: sepals yellowish or cream-colored. **FR** < 15 cm. RARE. Serpentine; ± 1000 m. s SCoRO (San Rafael Mtns).

C. californicus (S. Watson) Payson (p. 415) CALIFORNIA JEWELFLOWER Ann, ± glabrous or sparsely bristly below. **ST** decumbent to erect, often branched ± throughout. **LVS** < 11 cm; basal coarsely wavy-dentate to shallowly cut, tapered to short, winged petiole; cauline ovate to ± round, entire to coarsely dentate, clasping. **INFL** often 1-sided. **FL**: sepals erect or spreading, 4–10 mm, pouched below, keeled, darker in bud; petals 6–11 mm, whitish, veins purple, margins wavy; filaments free or longest pair ± fused; style < 3 mm. **FR** ascending to reflexed, 1–6 cm, compressed perpendicular to septum. **SEED** ± spheric. **ENDANGERED** CA, US. Flats, gentle slopes, gen in non-alkaline grassland, open juniper woodland; 70–1000 m. s SnJV. Formerly more widespread in s SnJV. Sometimes treated as *Stanfordia c.* S. Watson because seed spheric, fr compressed perpendicular to septum.

C. cooperi (S. Watson) Payson Ann. **ST** often irregularly twisted or weakly twining, gen branched, glabrous or inconspicuously branched-appressed-hairy. **LVS** < 6 cm, glabrous; basal entire or shallowly cut, tapered to short, winged petiole; cauline ± lanceolate-oblong, entire to dentate, sessile, clasping, hastate, tips acute. **FL**: sepals 5–7 mm, not pouched below; petals 7–10 mm, yellowish or purplish, margins scarious, not wavy; style 1–2 mm. **FR** spreading to reflexed, 2–4.5 cm, often curved. Common. Open, sandy or gravelly soil, gen among shrubs; 300–2500 m. n TR, e PR, s SNE, D; to sw UT, w AZ, Baja CA. Pls with st and fr greenish or straw-colored often mixed with pls with st and fr dark purple.

C. coulteri S. Watson Ann, ± glabrous or ± bristly below. **ST** gen branched above. **LVS** < 13 cm; hairs branched or not; basal ± entire to deeply cut, tapered to short, winged petiole; cauline oblong to ovate, entire to cut, sessile, clasping. **FL**: sepals erect or spreading, (5)7–18 mm, ± pouched below, keeled, glabrous or ± bristly, gen ± darker in bud; petals 8–31 mm, whitish, cream-colored and purple-veined, purplish, or brownish, margins wavy, gen scarious; longest 1–2 pairs of filaments ± fused; style < 1 mm. **FR** erect to reflexed, 4–13 cm. Dry, exposed slopes; 80–2000 m. s SNF, Teh, sw SnJV, se SnFrB, e SCoRO, SCoRI, nw WTR, sw edge DMoj (Kern Co.). Vars. intergrade in se SCoRO, n WTR.

var. ***coulteri*** (p. 415) Gen densely hairy; gen some lf and sepal hairs branched. **LVS**: basal sometimes deeply cut; lower cauline gen coarsely dentate to deeply cut. **FL**: gen only longest filament pair ± fused; stigma lobes 0.5–2 mm. **FR** gen reflexed. Habitats of sp. s SNF, Teh, se SCoRO, nw WTR, sw DMoj (Kern Co.). Variable; some distinct local races. ❀SUN,DRN:**7**,8,9,11,**14–16**,17, **18**,19–24.

var. ***lemmonii*** (S. Watson) Munz Gen glabrous or sparsely hairy; lf and sepal hairs unbranched. **LVS**: basal gen wavy-dentate; lower cauline gen entire to ± finely dentate. **FL**: gen longest 2 filament pairs ± fused; stigma lobes 1–2.5 mm. **FR** gen erect or ascending. Habitats of sp.; 80–800 m. sw SnJV, se SnFrB, e SCoRO, SCoRI. Apparently extirpated from many places.

C. crassicaulis (Torrey) S. Watson THICK-STEM WILD CABBAGE Per from woody caudex; herbage ± glabrous. **ST** ± inflated, branched or not. **LVS** < 10 cm, petioled; rosette persistent; basal and lower cauline oblanceolate, entire to deeply cut; upper much-reduced, ± linear, entire. **FL**: sepals 8–15 mm, ± pouched below; petals 10–20 mm, purplish or brownish, margins scarious, not wavy; style < 0.5 mm, stigma lobes > 0.5 mm. **FR** erect to ascending, 6–13 cm. Dry sagebrush scrub, pinyon/juniper woodland; 900–2900 m. W&I, n&c DMtns; to ID, WY, Colorado, AZ.

var. ***crassicaulis*** (p. 415) **FL**: sepals greenish, gen purplish near tips, ± densely hairy. Uncommon. Habitats and range of sp.

var. ***glaber*** M.E. Jones **FL**: sepals cream-colored in CA, glabrous. Uncommon. Habitats of sp.; 900–2200 m. e DMtns (Providence Mtns); to UT. CA pls with cream-colored sepals may be distinct from more e pls with greenish or purplish sepals. [*C. g.* (M.E. Jones) Rydb.]

C. glaucus S. Watson Per from woody caudex, glabrous. **ST** gen branched above. **LVS**: basal ± not rosetted, blades 2–10 cm (± > petioles), widely oblong to widely obovate, entire or base few-lobed, abruptly narrowed to petiole; lower cauline sometimes ± reduced; upper linear to lanceolate, petioled. **FL**: sepals 6–10 mm, not pouched below; petals 9–18 mm, yellowish green or purplish, margins scarious, not wavy; style < 1.3 mm. **FR** spreading, 4–15 cm, gen curved. **SEED** yellowish or light brown. Uncommon. Open, rocky slopes, often in crevices; 1400–2500 m. W&I, n DMtns (Grapevine, Last Chance Mtns); w NV. Only CA sp. with stigma lobes over placentas rather than valves.

C. hallii Payson Ann. **ST** gen not inflated, hollow, glabrous, sometimes branched above. **LVS** ± short-petioled; basal blades 3–18 cm, oblanceolate or oblong, ± cut, ± glabrous or bristly; cauline blades oblong, ± entire to cut, ± glabrous. **FL**: sepals 6–8 mm, ± pouched below, ± densely white-hairy; petals 8–9 mm, whitish or pale yellowish, purple-veined or not, margins not or barely wavy; style < 2 mm. **FR** ascending to spreading, 6–11 cm. Uncommon. Dry, open areas; 150–1800 m. e PR, s DMtns (Little San Bernardino Mtns), w edge DSon; n Baja CA.

C. heterophyllus (Nutt.) Payson Ann, bristly below. **ST** gen branched above. **LVS**: basal ± not rosetted, 3–12 cm, ± narrowly oblong, coarsely dentate to lobed, tapered, petiole 0 or short; upper cauline ± entire, clasping. **FL**: sepals 8–9 mm, barely pouched below; petals 12–14 mm, margins scarious, not wavy; style 1–2 mm, stigma lobes ± 0 or shallow. **FR** reflexed, 5–8 cm, compressed or 4-angled. **SEED** gen narrowly winged. Dry, open scrub, chaparral, often after fire or disturbance; 0–1300 m. SCoRO (La Panza Range), SW (exc Ventura Co.); Baja CA. [*Streptanthus h.* Nutt.]

var. ***heterophyllus*** (p. 415) **FL**: sepals purple; petals whitish with purple veins or purplish. Habitats and elevations of sp. e SCoRO (La Panza Range), SCo and s PR (common), TR (uncommon); Baja CA. [*C. stenocarpus* Payson, slender-pod jewelflower, **RARE** CA.]

var. ***pseudosimulans*** R. Buck **FL**: sepals yellowish or cream-colored, reddish blushed or not; petals yellowish green, purple-veined or not. Habitats and elevations of sp. TR, n PR. Often confused with *C. simulans*.

C. inflatus S. Watson (p. 415) DESERT CANDLE Ann, glabrous or sparsely hairy below. **ST** ± inflated, sometimes branched above. **LVS**: basal 2–7 cm, ± entire to finely dentate esp toward tips, tapered to short, winged petiole; cauline widely oblong to ovate, entire or ± dentate, sessile, clasping. **FL**: sepals erect or spreading, 8–10 mm, ± pouched below, darker in bud; petals 8–14 mm, purplish esp on veins, margins scarious, wavy; longest 1–2 filament pairs ± fused; style < 0.5 mm. **FR** ascending, 5–11 cm. Open, sandy plains to rocky slopes; 150–1500 m. s SNF (Kern Co.), c&s SnJV (w edge), SCoRI, n WTR (± uncommon in preceding), sw DMoj (common). Only sp. with ± purplish stigma. ❀TRY;DFCLT.

C. major (M.E. Jones) Payson Per from woody caudex, ± glabrous. **ST** erect, gen not inflated, hollow, sometimes branched. **LVS**: rosette persistent; basal and lower cauline blades 1–9 cm, oblanceolate or elliptic, entire to deeply cut, tapered to long petiole; upper lvs much reduced, ± linear, entire. **FL**: sepals sometimes sparsely hairy at tips; petals 9–16 mm, darker on veins, margins scarious, not wavy; style < 0.5 mm, stigma lobes < 0.5 mm. **FR** erect to ascending, 4–13 cm. Dryish, often rocky slopes, sometimes in shade; 1500–2500 m. n SNH (Alpine Co.), SnGb, SnBr, s MP, e DMtns (Providence, New York mtns); se OR, n&w NV; also e UT. Possibly not distinct from *C. crassicaulis*.

var. ***major*** (p. 415) **ST** gen ± stout. **INFL**: pedicel gen glabrous (rarely bristly). **FL**: sepals 8–10 mm, pouched below, greenish white or cream-colored; petals brownish. Habitats and elevations of sp. SnGb, SnBr, e DMtns (Providence, New York mtns); also e UT.

var. ***nevadensis*** Rollins **ST** gen slender. **INFL**: pedicel gen bristly (rarely glabrous). **FL**: sepals 6–9 mm, barely pouched below, purple; petals purplish. Uncommon. Habitats and elevations of sp. n SNH (Alpine Co.), s MP; se OR, n&w NV.

C. pilosus S. Watson (p. 415) CHOCOLATE DROPS, HAIRY WILD CABBAGE Ann? to weak per; hairs slender, barely tapered. **ST**: branches 0–many. **LVS**: lower blades 1–25 cm (± > petioles), oblanceolate to oblong, ± cut; upper blades linear to oblanceolate, entire or few-lobed, petioled. **INFL**: lowest 1–4 fls often bracted. **FL**: sepals 3–10 mm, 2 or 4 pouched at base, greenish to purple, darker in bud or not; petals 5–13 mm, whitish or purplish, margins wavy, scarious or not; style gen < 0.5 mm. **FR** ascending to spreading, 2–18 cm, often curved. **SEED** 1–3 mm. Uncommon. Open, dry areas; 600–2800 m. c&s SNH (e slope), s MP (Honey Lake), SNE, n DMtns; to se OR, s ID, w UT. Pls from c&s SNH with dark purple sepals and seeds > 2 mm are an undescribed sp.

C. simulans Payson PAYSON'S JEWELFLOWER Ann, ± conspicuously spreading-bristly below. **ST** gen branched above. **LVS** < 10 cm; basal coarsely dentate to shallowly cut, petiole 0 or short, winged; cauline oblong to ovate, entire to coarsely dentate, sessile, clasping. **FL**: sepals 3–7 mm, barely pouched at base; petals 5–10 mm, cream-yellow, margins not scarious, not wavy; style 0 or < 4 mm. **FR** ± reflexed, 2–8 cm, often curved. RARE. Open, dry areas; 400–2200 m. e SCo (w Riverside Co.), e PR.

CAULOSTRAMINA

1 sp. (Latin: straw-colored stem)

C. jaegeri (Rollins) Rollins (p. 415) JAEGER'S CAULOSTRAMINA Per, deeply rooted; base woody. **STS** few–many, 1–3 dm, branched at base, wavy, glabrous, gray-green. **LVS** all similar, glabrous, grayish; basal deciduous; petioles 1–2 cm, slender; blades 1.5–4 cm, elliptic to ovate, wavy to few-lobed, base tapered to truncate, reniform, or cordate, veins conspicuous. **INFL** open; bracts 0 or rarely below. **FL**: petals 8–14 mm, spoon-shaped, white to pale lavender, purple-veined. **FR** spreading to ± erect, 3–5 cm, linear, ± cylindric, glabrous; pedicel spreading to ascending, 8–10 mm; style < ± 1 mm. **SEEDS** 1 row per chamber, < ± 1 mm; wing 0. RARE. Rock crevices, cliffs; 1800–2400 m. s W&I (Inyo Mtns). [*Thelypodium j.* Rollins]

CHORISPORA

Ann to per; hairs simple, gen glandular and not. **LVS** entire to pinnately lobed; basal not rosetted. **INFL**: bracts 0. **FL**: inner sepals sac-like at base; petals purplish blue to white or yellow; anthers ± exserted. **FR** cylindric, sessile above receptacle, tapered to sharp beak, indehiscent but breaking into 1-seeded segments. **SEEDS** 1 row per chamber, ± flat, embedded in cavities of septum; embryonic root at edges of both cotyledons. ± 12 spp.: Eurasia. (Latin: from breaking between seeds)

C. tenella (Pallas) DC. (p. 415) Ann. **ST** 1–5 dm, branched from near base. **LVS** elliptic-oblong to lanceolate or oblanceolate; basal and lower cauline petioled, blades 3–8 cm, wavy-dentate to pinnately lobed; upper cauline sessile, ± entire to dentate. **FL**: sepals erect, 6–8 mm, free but forming a tube; petals narrowly clawed, magenta. **FR** spreading to ascending, 35–45 mm, lanceolate, often upcurved; beak 7–20 mm; pedicel ascending, 2–4 mm, stout; style 0, stigma minute, entire. $2n=14$. Grainfields, roadsides, waste places; < 1300 m. CaR, GV, SCo, GB; widespread to c US; native to Eur.

COCHLEARIA SCURVEYGRASS

Ann, bien, per, low, ± fleshy, glabrous, taprooted. **LVS** simple, thick. **FL**: petals obovate to wedge-shaped, white, yellowish, or purplish. **FR** ± inflated, flat perpendicular to septum, sessile above receptacle, opening from base; valves 1-veined; stigma head-like. **SEEDS** 2 rows per chamber; embryonic root at edges of both cotyledons. ± 20 spp.: gen n coasts. (Latin: spoon, from basal lvs of some spp.)

C. officinalis L. var. ***arctica*** (DC.) Gelert (p. 415) GREENLAND'S COCHLEARIA Bien. **ST** spreading to erect, (0.5)1–3 dm. **LVS**: basal rosetted, blades << petioles, bases truncate or tapered; upper cauline sessile, entire to obscurely dentate. **FR** 3–7 mm, longer than wide, ovate to obovate; pedicel erect or gen ascending, 5–15 mm; style < 0.5 mm. $2n$=14. RARE in CA. Seabird nesting areas on offshore rocks; < 50 m. n NCo (Del Norte Co.); to AK. Local. [*C. groenlandica* L. misapplied]

CONRINGIA HARE'S EAR

Ann, per, glabrous, gen glaucous. **ST** erect. **LVS**: basal gen entire, ± fleshy; cauline lanceolate or oblong to ± round, base cordate, clasping st. **INFL** ± flat-topped; bracts 0. **FL**: outer 2 sepals linear to narrowly oblong, inner 2 wider; petals yellow or white. **FR** linear; stigma head-like, entire or 2-lobed. **SEEDS** 1 row per chamber, oblong, plump; wing 0; embryonic root at back of 1 cotyledon or twisted. 6 spp.: Eurasia. (H. Conring, professor at Helmstedt, Germany, 1606–1681)

C. orientalis (L.) Dumort. (p. 415) Ann. **ST** 3–7 dm, gen simple. **LVS**: basal 5–9 cm, oblanceolate to obovate, narrowed to base, ± entire; cauline sessile, oblong-lanceolate. **FL**: sepals erect, 6–8 mm, acute; petals 7–12 mm, lemon to creamy yellow, claws slender. **FR** 8–13 cm, ± beaded, 4-angled to ± cylindric; pedicel ascending, 10–15 mm; style ± 1 mm, thick. $2n$=14. Fields, roadsides, open, disturbed places; < 900 m. SCo, SNE, DSon, expected elsewhere; c US; native to Eurasia.

CORONOPUS WART or SWINE CRESS

Ann, bien; odor often foul; hairs 0 or simple. **STS** several–many. **LVS** entire to very deeply pinnately lobed; basal and lower cauline petioled; upper subsessile. **INFL** from lf axils, dense. **FL** minute: sepals spreading; petals small or 0, white or purplish; stamens 2 or 4. **FR** of 2 nutlet-like halves, inflated exc at septum, flat perpendicular to septum, sessile above receptacle, indehiscent; valves hardened, each ± surrounding 1 seed. **SEED** oblong to ± reniform; embryonic root at back of 1 cotyledon. ± 10 spp.: gen Medit, few n S.Am. (Greek: crow foot, from lf shape)

1. Style 0 or incl in fr tip notch; fr base cordate, tip notched; valve surface net-wrinkled .... ***C. didymus***
1′ Style exceeding fr tip; fr base cordate to ± truncate, tip acute; valve surface with coarse ridges, spine-like projections .... ***C. squamatus***

C. didymus (L.) Smith (p. 419) Ann. **ST** prostrate to decumbent, much-branched, 20–50 cm. **LVS** many, ovate-oblong, very deeply pinnately lobed; lobes narrow, entire to dentate or dissected. **FL** ± 0.5 mm; petals linear, white. **FR**: pedicel 1.5–2.5 mm. $2n$=32. Common. Disturbed areas, gardens, fields; < 2000 m. CA; N.Am; native to Eurasia.

C. squamatus (Forsskal) Asch. (p. 419) SWINE CRESS Ann, bien. **ST** prostrate to decumbent, gen much-branched, 5–30 cm. **LVS** many, oblanceolate to obovate, very deeply pinnately lobed; lobes narrow, gen coarsely toothed or ± dissected. **FL** 1–1.5 mm; petals obovate, white. **FR**: pedicel 1–2 mm. $2n$=32. Disturbed areas, fields; < 300 m. SnFrB, ScV, DSon; native to Eur. NOXIOUS WEED.

CUSICKIELLA

Per, low-cespitose, taprooted. **LVS** gen basal, clustered at tips of caudex branches, sessile, entire. **FL**: sepals hairy or not, bases not sac-like; petals erect, white or yellowish, distinct claws 0; ovules 2 per chamber. **FR** sessile above receptacle, ovate to ± oblong or fusiform, not or ± flat toward tip; valves rounded or keeled; pedicel ascending. **SEED** gen 1, 2–3 mm, brown, plump, ± round or ovate-oblong; wing 0; embryonic root at back of 1 cotyledon. 2 spp.: w US. (W.C. Cusick, Oregon pl collector, 1842–1922) [Rollins 1988 J Jap Bot 63:65–69]

1. Bracts subtending pedicels 0; fr valves rounded; lf 5–12 mm; petals white .... ***C. douglasii***
1′ Bracts subtending lower pedicels, lf-like; fr valves keeled; lf 2–4 mm; petals yellowish .... ***C. quadricostata***

C. douglasii (A. Gray) Rollins (p. 419) Caudex branches underground. **ST** 3–5 cm, hairy or not. **LF** oblanceolate; hairs gen on margins, some not, gen simple. **FR** 3–7 mm, ± flat; hairs gen simple or 0; pedicel hairy or not; style 0.5–1.5 mm. Rocky ridges, slopes; 1500–2450 m. KR, NCoR, SN, SnBr, GB; to WA, ID, UT. [*Draba d.* A. Gray var. *crockeri* (Lemmon) C. Hitchc.]

C. quadricostata (Rollins) Rollins (p. 419) Caudex branches gen underground. **ST** 2–5 cm, hairy. **LF** linear to oblong; hairs on margins, surfaces, simple or multibranched. **FR** 3–4 mm, 4-sided; hairs simple or forked; pedicel hairy; style 0.5–1 mm. Rocky flats, sagebrush, slopes; 2400–2800 m. SNE (Mono Co.); w NV. [*Draba q.* Rollins]

DESCURAINIA TANSY MUSTARD

Ann, bien, rarely per; hairs gen minute, gen multibranched, fewer simple, some glandular. **ST** branched. **LVS** 1–many-pinnately lobed to compound; basal often rosetted but withering. **INFL** elongating; bracts gen 0. **FL**: petals < 3 mm, yellow or whitish, blades obovate, obtuse. **FR** linear to ± obovate, straight or uneven-margined, gen ± cylindric; style < 0.8 mm or 0, stigma gen head-like, entire. **SEEDS** 1–2 rows per chamber, gen < 1 mm, elliptic, plump, gelatinous when wet; wing 0; embryonic root at back of 1 cotyledon. ± 40 spp.: worldwide temp. (F. Descourain, French botanist, 1658–1740) [Detling 1939 Amer Midl Naturalist 22: 481–520] Relationships, characters of spp. difficult. May be TOXIC to livestock.

1. Fr fusiform, base as acute as tip; style gen 0.5–0.8 mm .... ***D. californica***
1′ Fr linear, oblong, or ± widely elliptic to ± club-shaped or ± obovate, base as acute as to more acute than tip; style < 0.8 mm or 0
 2. Fr oblong or ± widely elliptic to ± club-shaped to ± obovate, tip obtuse or rounded; style < 0.5 mm or 0; seeds gen 2 rows per chamber
  3. Fr 2.5–4 mm, ± widely elliptic to ± obovate; basal lvs not rosetted, not shed early .... ***D. paradisa***

3′ Fr 4–20 mm, oblong to ± club-shaped; basal lvs rosetted, shed early .... ***D. pinnata***
4. Pedicel in fr ± 45° from infl axis; herbage hairs ± not dense .... ssp. ***intermedia***
4′ Pedicel in fr gen ± 60–110° from infl axis; herbage hairs dense
5. Lf lobes or lflets obtuse at tip, oblanceolate to obovate .... ssp. ***glabra***
5′ Lf lobes or lflets gen acute at tip, linear to obovate
6. Petals 1–2 mm, pale yellow to whitish; branches 0, below, or above; basal and lower cauline lvs canescent .... ssp. ***halictorum***
6′ Petals 2–3.5 mm, bright yellow; branches 0 or above; lvs ± greenish .... ssp. ***menziesii***
2′ Fr linear to narrowly oblong, tip acute to acuminate; style gen 0.5–0.8 mm; seeds gen 1 row per chamber
7. Pedicel in fr erect to ascending .... ***D. incana***
7′ Pedicel in fr ascending to spreading
8. Lf very deeply 2–3-pinnately lobed to compound; fr 10–35 mm, septum with (2)3 veins; alien, weedy .... ***D. sophia***
8′ Lf 1–2-pinnately lobed to ± 1-compound, 1° lobes or 1° lflets dentate to deeply lobed; fr 5–16 mm, septum with 0 or 1 indistinct vein; native, often weedy
9. Pl whitish; fr gen glabrous (exc when young), straight .... ***D. obtusa*** ssp. ***adenophora***
9′ Pl greenish; fr glabrous, gen ± curved .... ***D. incisa***
10. Pedicels ± ascending, (10)12–20 mm, lower > upper; fr ascending to gen erect .... ssp. ***filipes***
10′ Pedicels ± spreading, 5–10 mm, lower ± = upper; fr ascending to ± spreading .... ssp. ***incisa***

D. californica (A. Gray) O. Schulz (p. 419) Winter ann to bien. **ST** 1, 3–8 dm; glands 0; branches above, slender. **LVS** oblanceolate to obovate, greenish, 1-pinnately lobed; lobes 2–4 pairs, 1.5–6 cm, lanceolate to oblanceolate, tip acute or obtuse, margins serrate to dissected; lower cauline lvs ± hairy, upper ± glabrous. **INFL**: bracts subtending branches, lf-like. **FL**: petals spreading, < 2 mm, barely > sepals, bright yellow. **FR** erect to spreading, 2–5 mm, fusiform, ± straight or curved; pedicel ascending to spreading, 3–7 mm, slender; style gen 0.5–0.8 mm. **SEEDS** 1–3, 1 row per chamber. $2n$=14. Open sites, sagebrush, shrubland, aspen groves, open woodlands; 2100–3400 m. SNH, GB, DMtns; to WY, NM.

D. incana (Fischer & C. Meyer) Dorn (p. 419) Bien, canescent to greenish. **ST** 3–12 dm, slender, branched, with glandular hairs or not. **LVS** 1.5–10 cm, widely lanceolate or oblanceolate to ovate; basal and lower cauline lvs 1–2-pinnately lobed or -compound, upper simple to 1-pinnate, entire. **FL**: petals ± 2 mm, yellow. **FR** 5–25 mm, 0.5–1.5 mm wide, linear, gen straight; both ends acute; pedicel erect to ascending, 2–4(7) mm; style < 1 mm. **SEEDS** 4–8, 1 row per chamber. $2n$=14. Open sites, meadows, sagebrush, open aspen groves; 1500–3400 m. KR, SN, SnBr, GB; to AK, e N.Am. [*D. richardsonii* (Sweet) O. Schulz; *D. r.* ssp. *viscosa* (Rydb.) Detl.]

D. incisa (A. Gray) Britton Ann; hairs gen nonglandular. **ST** 1, gen branched above, rarely below. **LVS**: basal ± 1–2-pinnately lobed or –compound, 5–10 cm, obovate, soon withering; cauline reduced, gen simpler upward. **INFL** elongate; hairs on axes minute, glandular and not. **FL**: petals ± 2 mm, spoon-shaped, yellow. **FR** < 1 mm wide, linear to narrowly oblong, gen ± curved, cylindric, glabrous; both ends acute to acuminate; pedicel slender. **SEEDS** 1 row per chamber. Open sites, shrubland; 1000–3100 m. CaR, TR, PR, GB; to B.C., NM. 4 sspp. total.

ssp. ***filipes*** (A. Gray) Rollins Pl hairs sparse to moderate, minute, multibranched. **FR** ascending to gen erect, 10–16 mm; pedicels ± ascending, (10)12–20 mm, lower > upper. Shaly soils at cliff bases, dry washes, often sagebrush, pinyon/juniper woodland; 1200–2500 m. CaR, GB; to WA, WY, NM. [*D. pinnata* ssp. *f.* (A. Gray) Detl.]

ssp. ***incisa*** Pl hairs ± dense to ± 0, multibranched, minute. **FR** ± spreading to ascending, 6–12 mm; pedicels ± spreading, 5–10 mm, lower ± = upper. Hillsides, dry water courses, granitic sands, often with sagebrush, juniper woodland; 1000–3100 m. CaR, TR, PR, GB; to B.C., NM. [*D. richardsonii* (Sweet) O. Schulz ssp. *i.* (A. Gray) Detl.]

D. obtusa (E. Greene) O. Schulz ssp. ***adenophora*** (Wooton & Standley) Detl. Bien, erect, coarse, whitish. **ST** gen 1, simple or branched above. **LF** 1–6 cm, 1-pinnately lobed or -compound, canescent; hairs minute, multibranched, rarely glandular; lobes or lflets 2–5 pairs, linear to oblanceolate, tip gen obtuse. **FL**: sepals 2–2.5 mm, ± < petals, sparsely glandular; petals yellowish. **FR** 5–15 mm, linear, straight, gen glabrous; both ends tapered abruptly; pedicel ascending to spreading, 12–25 mm, sparsely glandular. **SEEDS** 24–32 per chamber, 1–2 rows per chamber, closely packed, 0.7–1 mm. Gravelly flats, open woods, lake margins; 900–2200 m. e PR, DMoj; to NM, Baja CA.

D. paradisa (Nelson & Kenn.) O. Schulz Ann, greenish to gray-green; hairs gen dense, minute, multibranched. **ST** 1, 1.5–2.5 (3.5) dm, branched from base. **LF** 2–3 cm, canescent to greenish, deeply 1-pinnately lobed to ± compound; lobes or lflets linear to oblong. **INFL**: axis hairs glandular and not. **FL**: petals ± 1 mm, pale yellow. **FR** 2.5–4 mm, ± widely elliptic to ± obovate, glabrous to sparsely hairy; tip rounded; pedicel ± ascending, 3–5 mm. **SEEDS** 2–4 per chamber, 2 rows per chamber. Sandy washes, dunes, sagebrush; 900–2300 m. GB; to OR, NV. [*D. pinnata* ssp. *p.* (Nelson & Kenn.) Detl.]

D. pinnata (Walter) Britton Ann; hairs sparse to dense, multibranched. **ST** 1–7 dm; branches 0, below, or above. **LVS** lanceolate or oblanceolate to ovate; lower cauline lvs 2-pinnately lobed; upper lvs 1–10 cm, ± 1–2-pinnately lobed or 1-compound; lobes or lflets linear to widely obovate, hairs glandular or not. **FL**: petals 1–3.5 mm, bright yellow to cream-colored. **FR** 4–20 mm, oblong to ± club-shaped; pedicel ascending to spreading, 4–20 mm. **SEEDS** 5–20 per chamber, 2 rows per chamber, 0.5–1 mm. Washes, slopes, slopes, often saline soils; < 2500 m. CaR, s SNH, GV, CCo, SnFrB, SCoR, SW, GB, DMoj; widespread N.Am. 8 difficult sspp.

ssp. ***glabra*** (Wooton & Standley) Detl. **ST** 1–4 dm, gen branched from base, ± canescent below, ± glabrous above. **LF** ± 1–2-pinnately lobed; lobes oblanceolate to obovate, obtuse at tip, greenish; hairs dense. **FL**: sepals 0.5–1.5 mm, ± < petals, yellow or rose; petals yellow. **FR** 5–8 mm, oblong to ± club-shaped; pedicel ± 60–110° from infl axis, 4–12 mm. **SEEDS** 8–12 per chamber, ± 0.5 mm. $2n$=28. Sandy or saline soils, dry washes, desert shrubland; < 1500 m. s SnJV, SCoR, DMoj; to NM, n Mex.

ssp. ***halictorum*** (Cockerell) Detl. (p. 419) **ST** 1.5–5 dm, often glandular above; branches 0, below, or above. **LVS** lanceolate to oblanceolate; basal and lower cauline deeply 2-pinnately lobed, rarely 1-compound, canescent, lobes or lflets gen acute at tip, linear to narrowly oblong, upper simpler, hairs gen sparser. **FL**: sepals 1–2 mm, ± < petals, yellowish or rose; petals < 2 mm, pale yellow to whitish. **FR** ± ascending, 5–20 mm, ± club-shaped; pedicel ± 60–110° from infl axis, 8–12 mm. **SEEDS** 8–20 per chamber, 0.7–1 mm. $2n$=14,28,42. Dry streambeds, stable dunes, juniper woodland; < 1800 m. CaR, SW, GB, DMoj; to TX, Baja CA.

ssp. ***intermedia*** (Rydb.) Detl. **ST** 2–6 dm, gen branched above, gen glabrous above. **LVS**: basal, lower cauline 1–2-pinnately lobed, linear to obovate, hairy; upper lvs 1-pinnate to ± entire, hairs gen 0, rarely glandular. **FL**: sepals 1.5–2.5 mm, yellow; petals erect, 2–3 mm, exceeding sepals by ± 0.5 mm, yellow. **FR** 8–12 mm, ± club-shaped, rarely oblong; pedicel 6–12 mm, ± 45° from infl axis. **SEEDS** 5–13 per chamber, crowded or not, appearing as 1 row per chamber, ± 1 mm. $2n$=28. Dry streambeds, meadows, slopes, sagebrush; 900–2200 m. GB; to B.C., s Colorado.

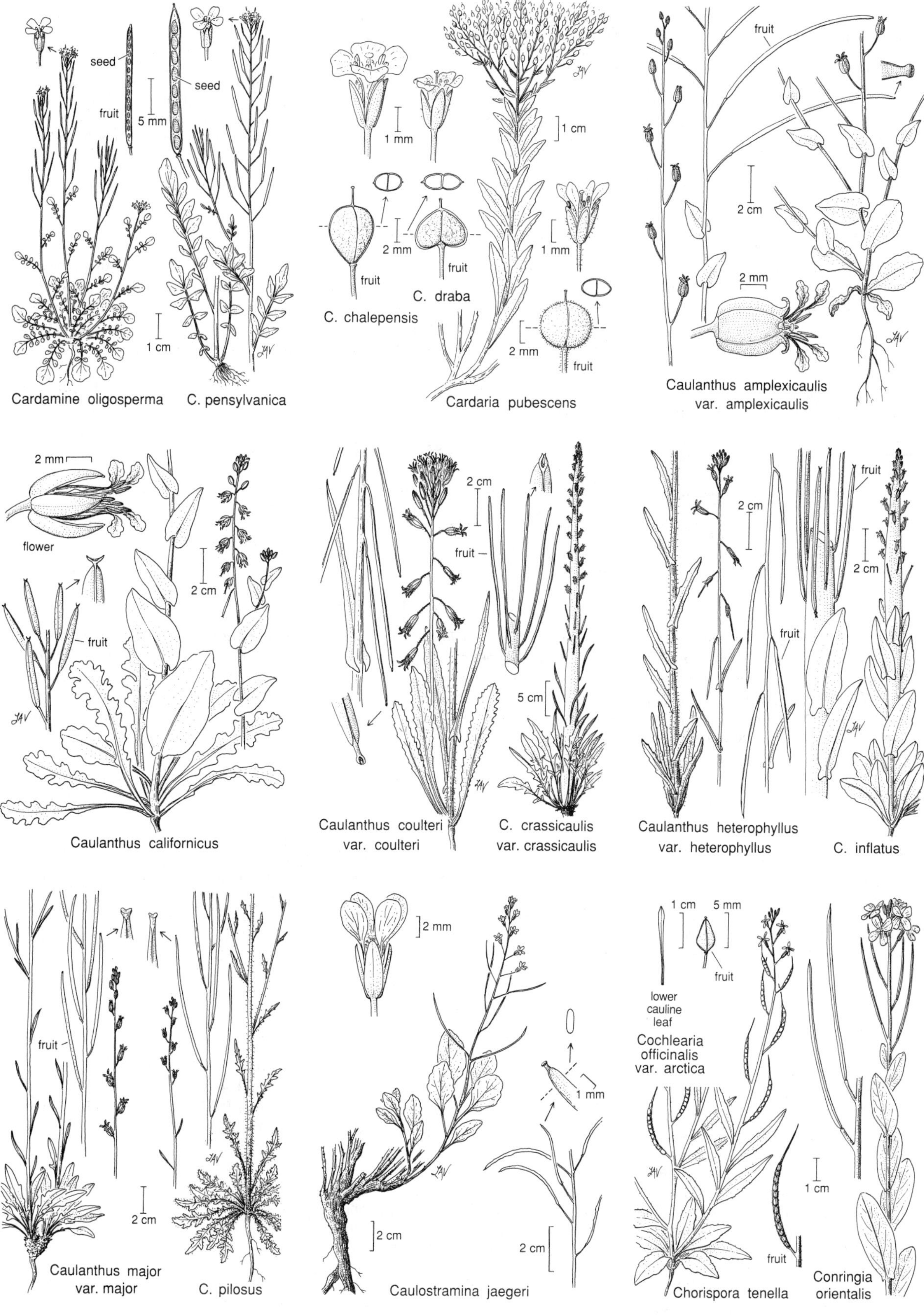
seed
fruit
5 mm
seed
1 cm
Cardamine oligosperma
C. pensylvanica
1 mm
fruit
2 mm
fruit
C. draba
C. chalepensis
1 cm
1 mm
2 mm
fruit
Cardaria pubescens
fruit
2 cm
2 mm
Caulanthus amplexicaulis
var. amplexicaulis
2 mm
flower
2 cm
fruit
Caulanthus californicus
2 cm
fruit
5 cm
Caulanthus coulteri
var. coulteri
C. crassicaulis
var. crassicaulis
fruit
2 cm
2 cm
fruit
Caulanthus heterophyllus
var. heterophyllus
C. inflatus
fruit
2 cm
Caulanthus major
var. major
C. pilosus
2 mm
1 mm
2 cm
2 cm
Caulostramina jaegeri
1 cm
5 mm
fruit
lower
cauline
leaf
Cochlearia
officinalis
var. arctica
fruit
1 cm
Chorispora tenella
Conringia
orientalis

ssp. ***menziesii*** (DC.) Detl. **ST** 1–6 dm; branches 0 or above; hairs glandular and not. **LVS** ± greenish; hairs dense; basal and lower cauline 2(3)-pinnately lobed, lobes gen obovate; upper lvs 1 (2)-pinnately lobed, lobes acute at tip, linear to oblanceolate. **FL**: sepals 1.5–2.5 mm, ± < petals, bright yellow; petals erect, 2–3.5 mm, bright yellow. **FR** 5–12(15) mm, ± club-shaped; pedicel 5–10 mm, ± 60–110° from infl axis. **SEEDS** 6–8, 0.7–1 mm. 2*n*=28. Rocky flats, disturbed areas, washes, chaparral; < 2500 m. s SNH, GV, CCo, SnFrb, SW, DMtns.

D. sophia (L.) Webb (p. 419) Ann, bien. **ST** 2.5–7.5 dm, short-branched above; hairs sparse to dense, minute, branched, nonglandular, sometimes also larger, simple. **LF** 1–9 cm, oblanceolate to widely ovate, very deeply 2–3-pinnately lobed to ± 2–3-compound; ultimate lobes of lflets gen linear (rarely obovate). **FL**: sepals 2–2.5 mm, yellow; petals erect, ± = sepals, yellow. **FR** 10–35 mm, ± 1 mm wide, linear, straight to ± curved, ± cylindric; pedicel ± ascending, 8–15 mm; septum (2)3-veined. **SEEDS** 10–20, 1 row per chamber, 0.7–1.5 mm, oblong-elliptic. 2*n*=20,28,38. Common. Disturbed areas, fields, roadsides, canyon bottoms, desert; < 2600 m. CA; native to Eurasia.

DIPLOTAXIS WALL ROCKET

Ann, bien, per; base woody or not. **ST** ascending to erect, gen several, gen branched. **LVS**: basal 0 or rosetted or not, pinnately lobed or toothed, petioled; cauline 0 or entire to lobed, petiole short or 0. **INFL** dense. **FL**: inner sepals sac-like at base; petals ± obovate, yellow, rarely violet to white. **FR** linear, often narrowed between seeds, cylindric to ± flat parallel to septum, ± stalked above receptacle; valves cleary 1-veined, glabrous; beak 3-veined, style short or 0. **SEEDS** many, 2 rows per chamber, 0.4–0.7(1) mm; cotyledons doubly folded. ± 25 spp.: mid latitudes, Old World. (Greek: 2-rowed, from seeds)

1. Ann, bien; st not woody; lvs gen basal; fr ± sessile above receptacle .... ***D. muralis***
1′ Per; st woody below; lvs gen cauline; fr stalk above receptacle 1–2 mm .... ***D. tenuifolia***

D. muralis (L.) DC. (p. 419) **ST** < 4 dm; hairs 0 or scattered, simple, bristle-like. **LF** oblanceolate to oblong. **FL**: petals 5–7 mm, pale yellow. **FR** 25–45 mm (incl 1.5–3 mm beak), straight or curved, narrowed between seeds, ± flat; pedicel ascending to ± erect, 5–20 mm. 2*n*=22,42. Uncommon. Waste places, dry streambeds; < 200 m. SCo, expected elsewhere; native to Eur.

D. tenuifolia (L.) DC. (p. 419) **ST** 3–7 dm; branches 0–many; hairs 0 or scattered, simple, ± not bristle-like. **LVS**: lower linear-oblong to spoon-shaped, entire to coarsely lobed, strong-smelling when crushed; upper lvs entire or lobes linear. **FL**: petals ± 11 mm, widely obovate, lemon-yellow. **FR** 20–45 mm (incl 1–2.5 mm beak), straight, gen not narrowed between seeds, ± cylindric; pedicel ± erect to ascending, 20–45 mm. 2*n*=22. Uncommon. Disturbed sites; < 300 m. ScV, SCo, expected elsewhere; native to Eur.

DITHYREA SPECTACLE-POD

Ann, per; herbage hairs dense, stellate, multibranched. **LVS** oblanceolate to widely obovate, entire to and shallowly dentate to lobed; basal and lower cauline petioled; middle and upper cauline petioles 0 or short. **INFL**: bracts 0. **FL** fragrant; sepals overlapped; petals erect, narrowly tongue-shaped, white to lavender. **FR** indehiscent, flat perpendicular to septum; 1-seeded halves ± round, each bordered by a raised rim; hairs simple, forked, club-like or not, esp dense on rim; pedicel < 3 mm, stout. **SEED** widely oblong, flat; wing 0; embryonic root at edges of both cotyledons. 2 spp.: CA, Mex. (Greek: with 2 shields, from fr) [Rollins 1979 Publ Bussey Inst Harvard 3–32]

1. Ann, taprooted; fr half 3–7 mm wide; infl open .... ***D. californica***
1′ Per, rhizomed; fr half 8–11 mm wide; infl dense .... ***D. maritima***

D. californica Harvey (p. 419) Ann canescent; bases of cauline lvs, pedicels with paired glands. **STS** arising below lf clusters, decumbent to erect, 1–7 dm; branches 0 below, sparse or 0 above. **LVS** not fleshy, canescent; basal 3–15 cm, oblanceolate to widely obovate, dentate to shallowly lobed, tip obtuse; cauline 1–4 cm, widely oblong to ovate. **FL**: petals 12–15 mm, prominently 3- veined, whitish to light lavender. **FR**: hairs simple, club-like or not; pedicel spreading. **SEED** 3–4 mm. 2*n*=20. Abundant. Sandy places, washes, shrubland; < 1400 m. D; w AZ, nw Mex. ❀TRY.

D. maritima (Davidson) Davidson (p. 419) BEACH SPECTACLE-POD Per, densely hairy; bases of cauline lvs, pedicels with paired glands or not. **STS** decumbent, < 2 dm; lower often under sand, lfless, branches 0 or few. **LVS** widely obovate to ± round, fleshy; basal 2–4 cm, entire to shallowly dentate or wavy-margined, tip ± rounded; cauline < 2 cm. **FL**: petals 7–9 mm, obscurely veined, white to cream-colored or purplish. **FR**: hairs simple; pedicel ascending to spreading. **SEED** 3–3.5 mm. 2*n*=60,±80. RARE. Sandy soils, dunes; < 50 m. s CCo, SCo; Baja CA.

DRABA

Robert A. Price

Ann to per, often cushion- or mat-forming; hairs often branched. **LVS** basal and sometimes cauline, entire or shallowly toothed. **FL**: sepal bases equal; petals < 10 mm, yellow or white, claw and limb gen distinct. **FR** < 30 mm, gen lanceolate to ovate, gen flat parallel to septum, less often partially inflated, sometimes twisted or wavy. **SEEDS** 2 rows per chamber; wing gen 0. 350+ spp.: n hemisphere, mtns of S.Am. (Greek: acrid) [Rollins & Price 1988 Aliso 12:17–27]

1. Ann (may germinate in winter); < 2500 m; styles gen << 0.2 mm
  2. Petals divided > 1/4 to base, white; lvs basal .... ***D. verna***
  2′ Petals entire or divided < 1/8 to base, white or light yellow; lvs basal and gen also cauline
    3. Lowest pedicel > 2 × fr; petals light yellow, not markedly smaller in lateral infls .... ***D. nemorosa***
    3′ Lowest pedicel < 2 × fr; petals white, often smaller or 0 in lateral infls
      4. Lf gen dentate; frs lanceolate to oblong, often loosely clustered; infl axes gen hairy .... ***D. cuneifolia***
      4′ Lf gen entire (to slightly dentate); frs ± linear, densely clustered; infl axes glabrous .... ***D. reptans***
1′ Bien to per; gen > 2000 m; styles gen conspicuous, gen > 0.2 mm (if shorter, go to 22. and 22′)

5. Longest styles > 1 mm; corolla yellow
6. St hairs coarse; basal lvs gen > 10 mm; cauline lvs gen several; frs gen > 30 per infl, sometimes densely clustered
7. Fr (3)4–6 mm wide, not twisted, tip obtuse; KR, CaRH ..... ***D. aureola***
7′ Fr (1.5)2–3(4) mm wide, gen twisted, tip acute; TR, PR ..... ***D. corrugata***
8. Cauline lvs 3–15, often crowded; frs densely clustered ..... var. ***corrugata***
8′ Cauline lvs 0–3, not crowded; frs sparsely clustered ..... var. ***saxosa***
6′ St hairs gen 0 above or fine (exc some *D. howellii*); basal lvs gen < 10 mm (exc some *D. howellii*); cauline lvs gen 0 (or 1–3, bract-like); frs gen < 30 per infl, not densely clustered
9. Lf narrowly ovate, tip gen acute; fr lanceolate, strongly twisted; seed wing 0; s SNH ..... ***D. sharsmithii***
9′ Lf oblanceolate to obovate, tip gen rounded; fr elliptic to ovate, little twisted; seed winged (exc *D. howellii*); KR, n SN
10. Lf with tangled, gen irregularly branched hairs on lower surface, hairs on margin long ..... ***D. pterosperma***
10′ Lf with non-tangled, gen 2–7-branched, gen stellate hairs (or glabrous) on lower surface, hairs on margin 0 or seldom long
11. Seed wing 0; cauline lvs gen 1–3, bract-like; fr gen hairy ..... ***D. howellii***
11′ Seed wing wide; cauline lvs (or bracts) gen 0; fr gen glabrous
12. Lf surface hairs gen 4–7-branched-stellate, margin glabrous; sts few–many from below; n SNH ..... ***D. asterophora*** var. ***macrocarpa***
12′ Lf surfaces glabrous, margin hairs 2–4-branched; sts 1–few from base; KR ..... ***D. carnosula***
5′ Longest styles < 1 mm; corolla white to yellow
13. Corolla yellow; lvs basal; sts ± cushion- or mat-forming; seeds gen > 1 mm
14. Lvs entire to minutely dentate, not persistent, not densely overlapped; fr narrowly lanceolate, hairs gen 0 or sparse, short ..... ***D. cruciata***
14′ Lvs entire, persistent, densely overlapped; fr gen ± ovate, hairs 0 or often some long
15. Lf obovate, thick, ± fleshy, gen > 3 mm wide
16. Lf hairs gen 4-branched-stellate, margin glabrous; seed winged ..... ***D. asterophora*** var. ***asterophora***
16′ Lf hairs simple or forked, gen prominent on margin; seed wing 0
17. Herbage glabrous or sparsely hairy; fr gen not wavy, not twisted ..... ***D. incrassata***
17′ Herbage hairy; fr often wavy or twisted ..... ***D. lemmonii*** var. ***lemmonii***
15′ Lf linear to oblanceolate, ± thin, 0.5–3 mm wide
18. Lf margin prominently stiff-hairy
19. Lf lower surface glabrous or sparsely short-branched-hairy ..... ***D. densifolia***
19′ Lf lower surface with long, branched hairs ..... ***D. paysonii*** var. ***treleasei***
18′ Lf margin glabrous (or hairs not very prominent)
20. St hairs 0 or stellate; lf hairs not bush-like, main axis obvious ..... ***D. oligosperma*** var. ***oligosperma***
20′ St hairs bush-like; lf hairs ± bush-like or stellate, main axis not obvious
21. Fr strongly compressed, gen twisted ..... ***D. sierrae***
21′ Fr swollen at base, not twisted ..... ***D. subumbellata***
13′ Corolla gen white to cream-colored (exc *D. albertina*); lvs basal and gen also cauline; sts sometimes ± tufted (not mat-forming); seeds gen < 1 mm
22. Stellate hairs with 4 or fewer branches
23. Lf lower surface with forked and cross-shaped hairs; corolla yellow; st 1–40 cm; frs 4–20 mm, not in umbel-like clusters; seed 0.7–1 mm ..... ***D. albertina***
23′ Lf lower surface with simple and forked hairs; corolla white; st < 10 cm; frs 3–5 mm, in umbel-like clusters; seed 0.6–0.8 mm ..... ***D. monoensis***
22′ Some stellate hairs with > 4 branches
24. Fr linear, gen > 10 mm, 1–2 mm wide ..... ***D. lonchocarpa*** var. ***lonchocarpa***
24′ Fr narrowly lanceolate to oblong or ± ovate, gen < 10 mm, gen > 2 mm wide
25. Lf lower surface with 4–8-branched, stellate hairs, gen gray-green; lower pedicels > 3 mm, ascending; fr not twisted, glabrous or hairs simple and forked
26. St < 10(15) cm, with fr ± throughout; cauline lvs overlapped ..... ***D. californica***
26′ St 10–35 cm, with fr only above middle; cauline lvs not overlapped ..... ***D. praealta***
25′ Lf lower surface with dense, many-branched, stellate hairs, gen whitish; lower pedicels gen < 3 mm, appressed; fr twisted or not, with gen > 3-branched, stellate hairs
27′ Cauline lvs gen entire; lower fls not bracted ..... ***D. breweri***
27. Cauline lvs toothed; lower fls gen bracted ..... ***D. cana***

D. albertina E. Greene Bien to short-lived per. **STS** 1–several from base, 1–40 cm, often branched from lf axils; hairs near base, coarse, simple and forked. **LVS**: basal 3–40 mm, oblanceolate to obovate, entire to minutely dentate; lower surface hairs forked and cross-shaped, upper surface hairs 0 or simple and forked; cauline 0–7. **INFL** < 35-fld, gen glabrous. **FL**: petals 2–3 mm, yellow. **FR** 4–20 mm, ± linear to narrowly ovate; hairs 0, rarely short; style < 0.2 mm. **SEEDS** < 40, 0.7–1 mm; wings 0. 2*n*=24. Moist meadows, streambanks; > 1500 m. CaRH, SNH, SnGb, MP, W&I; w US, w Can. [*D. stenoloba* Ledeb. vars. *nana* (O. Schulz) C. Hitchc. and *ramosa* C. Hitchc. (branched draba)] Small, lfless, high-elevation pls often confused with *D. crassifolia* Graham (esp var. *nevadensis* C. Hitchc., dolomite draba) of n&w N.Am outside CA, sparsely hairy, lf hairs not cross-shaped.

D. asterophora Payson (p. 419) Per. **STS** few–many from below, ± mat-forming, < 10 cm, glabrous above base. **LVS** basal, 5–15 mm, obovate, entire, with gen 4–7-branched, stellate hairs on surfaces; margins glabrous. **INFL** < 30-fld. **FL**: petals 4–7 mm, yellow. **FR** 5–15 mm, elliptic-ovate, ± wavy, gen glabrous; style 0.2–2 mm. **SEEDS** < 15, 1.5–3 mm; wings wide. Rock crevices, alpine barrens; > 2500 m. n&c SNH; w NV.

var. ***asterophora*** TAHOE DRABA **FR** gen 5–10 mm; style < 1 mm. RARE. Habitats and range of sp.

var. ***macrocarpa*** C. Hitchc. CUP LAKE DRABA **FR** < 15 mm; style 1–2 mm. RARE. Rock crevices; > 2500 m. n SNH (near Cup Lake, El Dorado Co.).

D. aureola S. Watson (p. 419) GOLDEN DRABA Bien or gen per. **STS** 1–few, 5–15 cm, branched or not; hairs coarse, gen simple and forked. **LVS**: lower surface hairs gen 4-branched, stellate; upper surface hairs gen simple and forked; basal lvs dense, 10–30 mm, oblanceolate, entire; cauline several, < basal. **INFL** 20–80-fld, dense. **FL**: petals ± 5 mm, yellow. **FR** 5–15 mm, (3)4–6 mm wide, ± oblong, not twisted, ± wavy, obtuse, hairs forked and stellate; style 1–2 mm. **SEEDS** ± 20–30, 1.5–2 mm; wings 0. $2n=20$. RARE. Scree, talus, gen volcanic substrates; > 2000 m. KR, s CaRH (Lassen Peak); to WA. In cult.

D. breweri S. Watson (p. 419) Per. **STS** gen several from base, < 15 cm; hairs dense, stellate. **LVS**: basal 5–15 mm, oblanceolate to obovate, gen whitened, hairs dense, stellate, main axis obscure (branches again branched); cauline 1–10, gen entire. **INFL** < 30-fld, elongate; lower pedicels gen < 3 mm, appressed, not bracted. **FL**: petals 2–4 mm, white. **FR** 4–10 mm, ± lanceolate, gen twisted; hairs > 3-branched, stellate; style < 0.5 mm. **SEEDS** < 35, 0.6–1 mm; wings 0. Open, rocky areas, gen above timberline; > 2500 m. CaRH, SNH, W&I; NV.

D. californica (Jepson) Rollins & R.A. Price (p. 419) CALIFORNIA DRABA Per. **STS** 1–several, < 10(15) cm; hairs dense, gen forked and stellate. **LVS**: basal 10–50 mm, ± lanceolate, gen entire, lower surface gen grayish green, hairs 4–7-branched, stellate; cauline 0–3, overlapped. **INFL** < 35-fld; lower pedicels ascending, > 3 mm. **FL**: petals 2–4 mm, white. **FR** ± throughout st, 5–12 mm, narrowly elliptic to lanceolate, not twisted, not wavy; hairs simple and forked or 0; style < 0.5 mm. **SEEDS** < 25, 0.8–1.2 mm, unwinged (tip gen with small extension). UNCOMMON. Open, rocky areas; > 3000 m. n W&I (White Mtns, Mono Co.). [*D. cuneifolia* Torrey & A. Gray var. *c.* Jepson] Locally common.

D. cana Rydb. HOARY DRABA Per. **STS** 1–several from base, < 25 cm, often branched from lf axils; hairs dense, simple and stellate. **LVS**: basal 10–40 mm, oblanceolate, gen dentate, gen whitened; hairs dense, simple and branched, gen stellate with branches branched; cauline 2–12, toothed. **INFL** < 60-fld, elongate in fr; lower pedicels gen bracted, appressed, gen < 3 mm. **FL**: petals 3–5 mm, white. **FR** 4–12 mm, lanceolate to oblong, ± not wavy, ± not twisted; hairs gen > 3-branched, stellate; style < 1 mm. **SEEDS** < 60, 0.6–1 mm; wings 0. $2n=32$. RARE in CA. Subalpine to alpine meadows, rock crevices; > 3000 m. s SNH (e slope, Inyo Co.); n&w US, Can. [*D. lanceolata* Royle misapplied]

D. carnosula O. Schulz (p. 423) MOUNT EDDY DRABA Per. **STS** 1–few from base, < 12 cm, simple, glabrous. **LVS** basal, 4–8 mm, obovate, entire; hairs 0 on surfaces, 2–4-branched on margins. **INFL** < 10-fld. **FL**: petals 5–6 mm, yellow. **FR** 12–22 mm, ± lanceolate, ± not wavy, ± not twisted, glabrous; style 1.5–3 mm. **SEEDS** < 13, ± 3 mm diam; wings wide. RARE. Rocky slopes; 2000–3000 m. KR (Mount Eddy area, Trinity and Siskiyou cos.). [*D. howellii* S. Watson var. *c.* (O. Schulz) C. Hitchc.]

D. corrugata S. Watson Bien to per. **STS** 1–several from base, < 25 cm; hairs coarse, simple and forked. **LVS**: hairs coarse, simple and forked; basal in dense cushion-like rosettes, 10–30 mm, oblanceolate, entire, grayish; cauline 0–15. **INFL** 10–100+-fld. **FL**: petals 3–5 mm, yellow. **FR** 5–20 mm, (1.5)2–3(4) mm wide, elliptic, gen twisted; tip acute; hairs 0 or simple, forked, and stellate; style 1.5–3.5 mm. **SEEDS** < 35, 1.2–1.6 mm; wings 0. Slopes among rocks; 2000–3800 m. TR, PR; Baja CA.

var. ***corrugata*** **ST** branched. **LVS**: cauline 3–15, often crowded. **FRS** densely clustered. Habitats and elevations of sp. SnGb, SnBr. Pls with frs more finely, densely hairy have been called forma *vestita* (Davidson) C. Hitchc.

var. ***saxosa*** (Davidson) Munz & I.M. Johnston **ST** gen simple. **LVS**: cauline 0–3, not crowded. **FRS** sparsely clustered. Habitats of sp.; 2500–3600 m. e PR (San Jacinto, Santa Rosa mtns).

D. cruciata Payson (p. 423) MINERAL KING DRABA Per. **STS** ± mat-forming, < 15 cm, glabrous above base. **LVS** basal, 5–20 mm, oblanceolate to ovate, entire to minutely dentate; surface hairs gen 4-branched, stellate; margins glabrous. **INFL** 5–20-fld. **FL**: petals 4–6 mm, yellow. **FR** 4–12 mm, narrowly lanceolate, not (or slightly) twisted; hairs gen 0 or sparse, short; style 0.1–0.8 mm. **SEEDS** 8–16, 1–1.4 mm; wings 0. UNCOMMON. Gravelly slopes; > 2500 m. s SNH (near Mineral King, Tulare Co.; doubtfully near Lake Tahoe).

D. cuneifolia Torrey & A. Gray (p. 423) Ann. **STS** 1–few from base, < 40 cm; hairs short, simple and branched, often stellate, rarely 0. **LVS** gen dentate; basal 5–70 mm, oblanceolate to obovate, surface hairs stellate; cauline 1–4. **INFL** < 75-fld; axes gen hairy; lowest pedicel < 2 × fr. **FL**: petals < 5 mm, often smaller or 0 in lateral infls, divided < 1/8 to base, white. **FRS** often loosely clustered, 3–12 mm, lanceolate to oblong; hairs 0 or simple, forked, and cross-shaped; style < 0.2 mm. **SEEDS** < 100, 0.5–0.7 mm; wings 0. $2n=16,32$. Open or disturbed places; < 2000 m. s SN, SnJV, SW, D; w US, n Mex. Vars. *cuneifolia, integrifolia* S. Watson, *sonorae* (E. Greene) Parish recognized by Hartman et al. [1976 Brittonia 27: 317–327] but ± intergrade in CA.

D. densifolia Nutt. (p. 423) Per. **STS** cushion-forming, < 15 cm, glabrous to hairy. **LVS** basal, 2–15 mm, linear to ± oblanceolate, entire; midrib prominent below; surface hairs 0 or sparse, short, branched; margin hairs prominent, stiff, simple. **INFL** < 20-fld. **FL**: petals 2–6 mm, yellow. **FR** 2–7 mm, ± ovate; hairs 0 or short, simple to stellate; style 0.5–1 mm. **SEEDS** < 15, 1.2–1.8 mm; wings 0. $2n=36$. Alpine barrens, rocky slopes; > 2000 m. SNH, W&I; nw US, w Can. Reproduces by asexual seed.

D. howellii S. Watson HOWELL'S DRABA Per. **STS** ± tufted, < 15 cm; hairs 0 or simple and forked, coarse or not. **LVS**: basal 5–25 mm, obovate, entire or finely toothed; tip gen rounded; surface hairs 0 or gen 3–4-branched, gen stellate; margin hairs gen 0; cauline gen 1–3, bract-like. **INFL** 10–30-fld. **FL**: petals 4–8 mm, yellow. **FR** 4–10 mm, ± ovate; hairs 0 or gen simple, forked, and cross-shaped; style 1.2–3.2 mm. **SEEDS** ± 10–20, 1.2–1.6 mm; wings 0. UNCOMMON. Rock crevices; 2000–3000 m. KR; sw OR. ❀DRN,SUN:**4–6**,17&IRR:1–3,7,15,16;DFCLT.

D. incrassata (Rollins) Rollins & R.A. Price SWEETWATER MOUNTAINS DRABA Per. **STS** ± mat-forming, < 15 cm; hairs 0 or sparse, simple. **LVS** basal, < 15 mm, obovate, thick, entire; surface hairs 0 or sparse, simple and forked; margin hairs gen prominent. **INFL** < 30-fld. **FL**: petals ± 3–5 mm, yellow. **FR** 3–12 mm, ± ovate, gen not wavy, not twisted; hairs 0 or sparse, marginal; style 0.2–0.8 mm. **SEEDS** < 20, 1.2–1.8 mm; wings 0. UNCOMMON. Alpine barrens, rocky slopes; > 2500 m. n SNE (Sweetwater Mtns, Mono Co.). [*D. lemmonii* S. Watson var. *i.* Rollins]

D. lemmonii S. Watson var. ***lemmonii*** (p. 423) Per. **STS** ± mat-forming, < 15 cm; hairs coarse, simple and forked. **LVS** basal, 5–30 mm, ± obovate, thick, entire; surface hairs simple and forked; margin hairs gen prominent. **INFL** < 30-fld. **FL**: petals 4–7 mm, yellow. **FR** 4–12 mm, ± ovate, often wavy or twisted; hairs 0 or gen simple and forked; style < 1 mm. **SEEDS** < 20, 1–1.5 mm; wings 0. Common. Talus, rock crevices, rocky meadows; > 2500 m. SNH. Another var. in e OR. ❀TRY;DFCLT.

D. lonchocarpa Rydb. var. ***lonchocarpa*** (p. 423) SPEAR-FRUITED DRABA Per. **STS** ± tufted, < 10 cm; hairs stellate or 0. **LVS**: basal 3–15 mm, ± oblanceolate; hairs dense, irregularly branched and stellate; cauline 0–2. **INFL** 5–20-fld. **FL**: petals 2–5 mm, white. **FR** gen > 10 mm, 1–2 mm wide, linear; hairs stellate or 0; style < 0.5 mm. **SEEDS** 10–30, ± 1 mm; wings 0. $2n=16$. Very uncommon. Calcareous scree; > 3000 m. c SNH (Convict Creek Basin, Mono Co.); to AK, Colorado. [*D. nivalis* Lilj. var. *elongata* S. Watson]

D. monoensis Rollins & R.A. Price (p. 423) WHITE MOUNTAINS DRABA Per. **STS** 1–few from base, < 10 cm; hairs gen dense, simple and forked. **LVS**: basal 5–20 mm, oblanceolate, entire or teeth few; surface hairs simple and forked; cauline 0–2. **INFL** 10–20-fld, umbel-like in fr. **FL**: petals 2–3 mm, white. **FR** 3–5 mm, ovate; hairs 0 or sparse, simple; style < 0.3 mm. **SEEDS** 10–20, 0.6–0.8 mm; wings 0. RARE. Moist gravel, rock crevices; > 3000 m. n W&I (White Mtns, Mono Co.), probably also in SNH. [*D. fladnizensis* Wulfen misapplied]

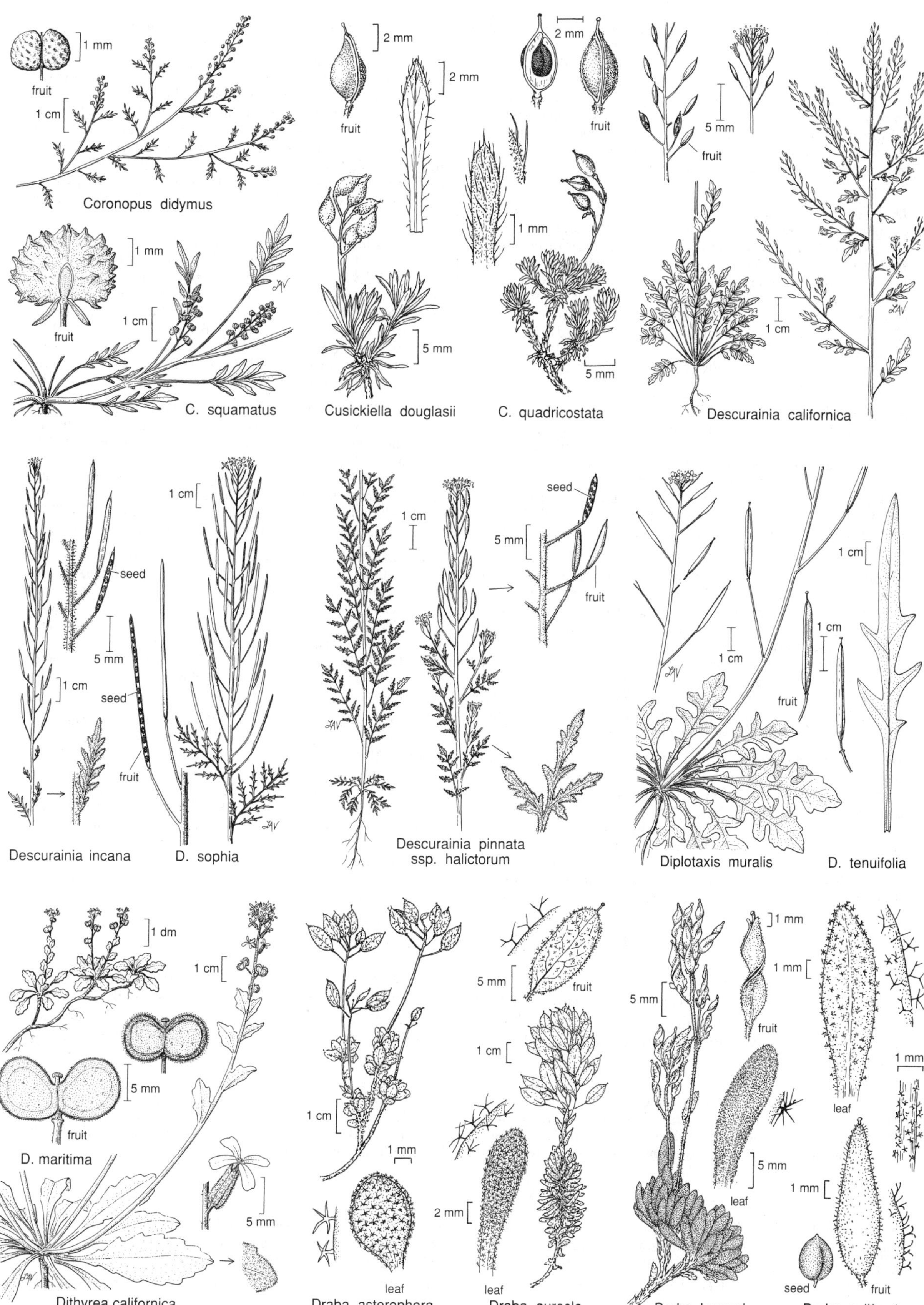
1 mm
fruit
1 cm
Coronopus didymus
1 mm
fruit
1 cm
C. squamatus
2 mm
2 mm
fruit
1 mm
5 mm
Cusickiella douglasii
2 mm
fruit
5 mm
C. quadricostata
5 mm
fruit
1 cm
Descurainia californica
seed
5 mm
1 cm
seed
fruit
1 cm
Descurainia incana
D. sophia
1 cm
seed
5 mm
fruit
Descurainia pinnata ssp. halictorum
1 cm
1 cm
fruit
1 cm
Diplotaxis muralis
D. tenuifolia
1 dm
1 cm
5 mm
fruit
D. maritima
5 mm
Dithyrea californica
1 cm
1 mm
leaf
Draba asterophora
5 mm
fruit
1 cm
2 mm
leaf
Draba aureola
1 mm
fruit
5 mm
5 mm
leaf
Draba breweri
1 mm
1 mm
leaf
1 mm
seed
fruit
Draba californica

D. nemorosa L. Ann (may germinate in winter); herbage hairs stiff, simple, forked, and stellate. **STS** 1–few from base, 3–30 cm, branched or not. **LVS**: basal 10–30 mm, widely lanceolate to obovate; cauline 0–7, gen dentate. **INFL** 5–40-fld; lowest pedicel > 2 × fr. **FL**: petals ± 2–4 mm, light yellow, gen entire. **FR** 3–11 mm, narrowly oblong to obovate; hairs 0 or short, stiff, simple; style < 0.1 mm. **FRS** becoming widely separated on elongate axis. **SEEDS** 20–50, < 1 mm; wings 0. 2*n*=16. Uncommon. Open, disturbed area; ± 2000 m. e KR (Siskiyou Co.); circumboreal, probably waif in CA.

D. oligosperma Hook. var. ***oligosperma*** (p. 423) Per. **STS** cushion-forming, < 12 cm; hairs 0 or stellate. **LVS** basal, 3–10 mm, linear-oblanceolate, entire; midrib prominent on lower surface; surface hairs sparse to dense, gen appressed; main axis obvious, branches shorter; margin hairs 0 or not very prominent. **INFL** < 20-fld. **FL**: petals 2.5–4.5 mm, yellow. **FR** 2.5–7 mm, ± ovate; hairs 0 or simple, hooked, or branched like lf hairs; style < 1 mm. **SEEDS** < 12, 1.3–1.8 mm. 2*n*=64. Common. Alpine barrens, dry slopes; > 2000 m. SNH, W&I; w US, w Can. Dwarfed pls of high elevations have been called var. *subsessilis* (S. Watson) O. Schulz; gen reproduces by asexual seed.

D. paysonii J.F. Macbr. var. ***treleasei*** (O. Schulz) C. Hitchc. (p. 423) Per. **STS** cushion-forming, < 6 cm; hairs tangled, simple and forked. **LVS** basal, 4–8 mm, ± linear, entire; midrib prominent on lower surface; hairs on lower surface long, forked and cross-shaped; margin hairs prominent, stiff, simple and forked. **INFL** < 10-fld. **FL**: petals 3–5 mm, yellow. **FR** 3–8 mm, ± ovate; hairs simple, forked, and stellate; style 0.4–1 mm. **SEEDS** < 10, 1.3–1.8 mm. 2*n*=42. Uncommon. Open, rocky slopes; > 2000 m. n SNH; nw US, w Can.

D. praealta E. Greene Bien to short-lived per. **STS** 1–few from base, 10–35 cm; hairs dense, simple and stellate. **LVS** gray-green; hairs simple, forked, and 4–8-branched, stellate; basal 10–30 mm, oblanceolate, entire to dentate; cauline 1–6, not overlapped. **INFL** < 35-fld; lower pedicels ascending, > 3 mm. **FL**: petals 2–4 mm, white to yellowish cream-colored. **FR** above middle of st, 5–14 mm, narrowly lanceolate, ± not twisted, ± not wavy; hairs simple and forked; style 0.1–0.4 mm. **SEEDS** < 75, ± 0.8 mm. 2*n*=56. Uncommon in CA. Montane or subalpine moist meadows, streambanks; > 2500 m. c SNH (e slope, Mono and Inyo cos.); w US, Can.

D. pterosperma Payson WINGED-SEED DRABA Per. **STS** cushion- or mat-forming, < 15 cm; hairs gen irregularly branched, fine. **LVS** basal, 3–10 mm, oblanceolate to obovate, entire; tip gen rounded; lower surface hairs tangled, gen irregularly branched; margin hairs long, simple and forked. **INFL** < 15-fld. **FL**: petals 4–8 mm, yellow. **FR** 4–12 mm, ovate to ± elliptic; hairs forked and stellate; style 1.5–3.5 mm. **SEEDS** < 15, ± 2–3 mm; wings wide. UNCOMMON. Marble or limestone crevices, talus, scree; 1800–2500 m. c KR (Marble Mtns, Siskiyou Co.). ❀TRY; DFCLT.

D. reptans (Lam.) Fern. Ann. **STS** 1–several from base, < 15 cm; hairs near base simple, forked, and stellate, near infl 0. **LVS**: basal 5–20 mm, oblanceolate to obovate, gen entire; lower surface hairs forked and stellate; upper surface hairs simple and forked; cauline 1–5. **INFL** < 30-fld; axes glabrous; lowest pedicel < 2 × fr. **FL**: petals 0.5–5 mm, white, often smaller or 0 in lateral infls. **FRS** densely clustered in upper 1/3 of st, 5–20 mm, ± linear; hairs 0 or short, stiff, simple; style < 0.2 mm. **SEEDS** < 75, 0.5–0.7 mm; wings 0. 2*n*=16,30,32. Very uncommon. Open or disturbed areas; < 2500 m. n SNH (Emigrant Gap, n Placer Co.), n DMtns (w Panamint Mtns); US, s-c Can. [*D. micrantha* Nutt.] Probably a waif in CA.

D. sharsmithii Rollins & R.A. Price (p. 423) MOUNT WHITNEY DRABA Per. **STS** densely mat-forming, < 10 cm; hairs 0 or near base, forked. **LVS** basal, 2–10 mm, narrowly ovate, entire; tip gen acute; surface hairs 2–4-branched, often stellate; margins hairy. **INFL** < 15-fld. **FL**: petals 4–6 mm, yellow. **FR** 6–20 mm, lanceolate, strongly twisted, glabrous; style 0.6–2 mm. **SEEDS** < 15, 1.2–1.7 mm; wings 0. UNCOMMON. Protected rock crevices; > 3500 m. s SNH (Fresno, Inyo cos.). [*D. cruciata* Payson var. *integrifolia* C. Hitchc. & Sharsm.]

D. sierrae Sharsm. (p. 423) SIERRA DRABA Per. **STS** cushion-forming, < 6 cm; hairs dense, bush-like. **LVS** basal, 2–7 mm, linear-oblanceolate, entire, grayish; surface hairs ± bush-like or stellate with branches branched; margin hairs 0 or not very prominent. **INFL** < 10-fld. **FL**: petals ± 4–5 mm, yellow. **FR** 3–8 mm, ± ovate, strongly compressed, gen twisted; hairs small, forked and stellate; style 0.2–0.9 mm. **SEEDS** < 10, 1–1.2 mm; wings 0. UNCOMMON. Rock crevices; > 3500 m. c SNH (Fresno, Inyo cos.).

D. subumbellata Rollins & R.A. Price (p. 423) WHITE MOUNTAINS CUSHION DRABA Per. **STS** cushion- or mat-forming, < 5 cm; hairs dense, bush-like. **LVS** basal, 2–6 mm, oblong to obovate, entire; surface hairs dense, bush-like; margin hairs 0 or not very prominent. **INFL** 2–10-fld. **FL**: petals 3–5 mm, yellow. **FR** 2–7 mm, ovate, inflated at base, not twisted; hairs branched; style 0.2–0.7 mm. **SEEDS** < 10, ± 1.5 mm; wings 0. RARE. Scree, talus, among rocks; > 3000 m. s SNH (e slope, nw Inyo Co.), n W&I (White Mtns, Mono Co.).

D. verna L. (p. 423) Ann. **ST** < 25 cm; hairs 0 or simple and forked. **LVS** basal, < 30 mm, ± oblanceolate; hairs simple, forked, or stellate. **INFL** < 35-fld. **FL**: petals 2–3 mm, divided > 1/4 to base, white. **FR** 3–10 mm, elliptic to obovate, glabrous; style < 0.15 mm. **SEEDS** < 70, 0.4–0.7 mm; wings 0. 2*n*=14,±24–64,94. Open or disturbed areas; < 2000 m. CA-FP, MP; circumboreal. Complex with many minor variants; largely self-pollinating. Pls with fr ± 4–5 mm, obovate, have been called var. *aestivalis* Lej., which intergrades with var. *v.* in CA.

ERUCA GARDEN-ROCKET, SALAD-ROCKET

Ann, per. **LVS**: basal petioled, deeply pinnately lobed to ± compound, rarely simple, entire; cauline short-petioled to sessile, shallowly lobed to entire. **INFL** gen dense, many-fld; bracts 0. **FL**: sepals erect, linear to oblong; petals widely obovate to oblanceolate, veins prominent, dark. **FR** linear to oblong or elliptic, beaked; pedicel ascending to erect, subappressed. **SEEDS** 2 rows per chamber, orange or brown; cotyledons doubly folded. 3 spp.: Eurasia. (Latin: perhaps burn, from pl taste)

E. vesicaria (L.) Cav. ssp. *sativa* (Miller) Thell. (p. 423) **ST** gen branched from base, 2–10 dm; hairs toward base, simple, often reflexed. **LF** 5–15 cm, gen widely oblanceolate; hairs on upper surface 0, on lower along midrib, few, simple. **FL**: petals erect to ascending, 1.5–2 cm, ± 2 × sepals, white to yellowish. **FR** 1–2.5 cm, ± cylindric, 4-ribbed; pedicel 2–5 mm; beak ± flat, 1/2 to = valve. Disturbed areas, fields, roadsides; < 400 m. CaR, GV, SCoR, DSon; N.Am.; native to Eur. [*E. s.* Miller]

ERUCASTRUM

Ann to rarely subshrub; hairs simple, rarely 0. **LVS**: basal petioled, entire or gen deeply pinnately lobed to compound; cauline ± as basal, gen simpler, upper gen sessile, ± clasping st or not. **FL**: sepals erect, linear to oblong, bases not sac-like; petals gen obovate, yellowish or white. **FR** linear, ± 4-sided, ± narrowed between seeds, sessile or short-stalked above receptacle, beaked, hairy or not. **SEEDS** 1 row per chamber, oblong to ovate, plump; wing 0; cotyledons doubly folded. ± 18 spp.: esp Afr. (Latin: ± resembling *Eruca*)

E. gallicum (Willd.) O. Schulz Ann, bien; hairs stiff, recurved. **ST** simple to much-branched, 1.5–8 dm. **LF** deeply pinnately lobed, 3–20 cm; margins often wavy. **INFL** open; bracts subtending each fl, lower lf-like. **FL**: petals 4–7 mm, pale yellow. **FR** 1–4.5 cm; pedicel ascending, < 1 cm, slender; valves strongly 1-veined; style 1.5–3 mm. 2*n*=30. Disturbed areas, roadsides; < 200 m. s SnFrB; e US; native to Eurasia.

ERYSIMUM WALLFLOWER

Robert A. Price

Ann to subshrub; hairs appressed, forked to many-branched. **LVS** in basal rosettes and cauline, simple, entire to lobed. **FL**: petals clawed, gen cream-colored to orange; stigma 2-lobed. **FR** narrow, round, 4-sided, or ± flattened (gen parallel to septum); style 0.2–5 mm. **SEEDS** 1–2 rows per chamber, 1–4 mm, often ± winged. 160+ spp.: temp n hemisphere, esp Eurasia. (Greek: to help, from medicinal uses) [Rossbach 1958 Madroño 14:261–267] Incl *Cheiranthus*; native CA taxa all related to *E. capitatum*.

1. Ann (rarely bien); petal < 10 mm, < 3 mm wide
2. Pedicel < fr in width, 1/3–1/2 as long; fr ascending, gen 1–4 cm; basal lvs entire to shallowly toothed .... ***E. cheiranthoides***
2′ Pedicel ± = fr in width, < 1/10 as long; fr stiffly spreading, 4–10 cm; basal lvs deeply toothed ...... ***E. repandum***
1′ Bien to subshrub; petal > 10 mm, > 3 mm wide
3. Stigma lobes longer than wide; sepal purplish, base green; lf 6–20 mm wide, gen entire with acute tip; alien .... ***E. cheiri***
3′ Stigma lobes ± as long as wide; sepal greenish or purplish (not purplish with green base); lf < 6 mm wide if entire with acute tip; native
4. Lf very narrowly linear, < 3 mm wide, finely toothed; herbage dull purplish; inland sand, sw SnFrB ***E. teretifolium***
4′ Lf linear-lanceolate or wider, gen > 3 mm wide, entire to coarsely toothed; herbage gen not purplish; widespread
5. Pl gen much-branched below infl; caudex elongate, woody; elongate non-fl sts present; coastal
6. Lf gen coarsely toothed; fr gen flattened parallel to septum; seed winged at tip and along 1 or both sides; NCo, nw CW .... [2]***E. franciscanum***
6′ Lf ± entire; fr gen 4-sided or flattened perpendicular to septum; seed wing 0 or at tip; s CA ...... ***E. insulare***
7. Fr flattened perpendicular to septum; n ChI .... ssp. ***insulare***
7′ Fr gen ± 4-sided (slightly flattened); gen dunes, s CCo, n SCo .... ssp. ***suffrutescens***
5′ Pl gen not much-branched below infl; caudex gen not elongate, gen not woody; elongate non-fl sts 0; coastal or inland
8. Immature fr slightly fleshy, mature fr ± 4-sided or barely flattened; seed wing 0 or at tip; gen inland .... ***E. capitatum***
9. Basal lvs ± spoon-shaped, tip obtuse; fr narrowed between seeds; style gen > 2 mm; petals yellow — KR, CaRH, SNH .... ssp. ***perenne***
9′ Basal lvs gen (ob)lanceolate, tip gen acute; fr gen not narrowed between seeds; style often < 2 mm; petals orange to yellow
10. Caudex gen short; lower lvs often < 10 cm, (ob)lanceolate .... ssp. ***capitatum***
10′ Caudex gen elongate; lower lvs gen > 10 cm, narrowly lanceolate
11. Fr stiffly ascending, ± 4-sided, ± straight; ne CCo .... ssp. ***angustatum***
11′ Fr spreading, ± flattened, gen irregularly curved; s SCoRO .... ssp. ***lompocense***
8′ Immature fr sometimes very fleshy, mature fr flattened; seed wing at tip and along 1 or both sides; coastal
12. Basal lvs narrowly (ob)lanceolate, gen > 10 × longer than wide
13. Lf entire or slightly toothed; fr spreading to ascending; style gen < 1 mm in fr; petals rich yellow; dunes .... ***E. ammophilum***
13′ Lf gen coarsely toothed; fr gen strongly ascending (spreading); style gen > 1 mm in fr; petals cream-colored to yellow; gen on rocks .... [2]***E. franciscanum***
12′ Basal lvs ± spoon-shaped, ± 5 × longer than wide .... ***E. menziesii***
14. Lower frs gen erect, pedicels gen > 10 mm; petals cream-colored to yellow; gen headlands, cliffs .... ssp. ***concinnum***
14′ Lower frs spreading, pedicels gen < 10 mm; petals yellow; foredunes
15. Often per (caudex branched), fls summer — c CCo .... ssp. ***yadonii***
15′ Bien, fls gen winter or spring
16. Lf not very fleshy; pedicel (5)9–15 mm; longest frs gen > 8 cm; n NCo .... ssp. ***eurekense***
16′ Lf fleshy; pedicel 3–9(13) mm; longest frs gen < 8 cm; c NCo, c CCo .... ssp. ***menziesii***

E. ammophilum A.A. Heller COAST WALLFLOWER Bien or short-lived per. **STS** 1–several, 5–60+ cm. **LVS**: basal 4–15 cm, 2–10 mm wide, narrowly (ob)lanceolate, entire or slightly toothed; cauline wider, esp near infl; hairs 2–several-branched. **FL**: petals 14–25 mm, rich yellow. **FR** spreading to ascending, 2–12 cm, 1.5–3.5 mm wide, fleshy when immature, flattened when dry; style gen < 1 mm. **SEED** 1.6–3.2 mm, 1–1.8 mm wide; wing at tip and gen along sides. *n*=18. UNCOMMON. Coastal dunes; < 50 m. c CCo (Monterey Bay), n ChI (Santa Rosa Island). Threatened by development. Pls intermediate to *E. capitatum* formerly in s SCo. ❀DRN:5,**15–17**&IRR:14,22–24&SHD:8,19–21.

E. capitatum (Douglas) E. Greene WESTERN WALLFLOWER Bien or short-lived per. **STS** 1–few, 0.5–100+ cm. **LVS** ± linear to spoon-shaped, entire to toothed; tip gen acute; lower ± 2–25 cm; hairs 2–several-branched. **FL**: petals 12–30 mm, gen orange to yellow (cream-colored to reddish). **FR** gen ascending, 3–15 cm, 1–4 mm wide, slightly fleshy when immature, ± 4-sided or slightly flattened when mature; style 0.2–5 mm. **SEED** 1–4 mm, 0.7–2 mm wide; wing gen 0 or at tip, rarely along sides. *n*=18. Common. Many habitats, gen inland; 0–4000 m. CA (exc GV); to e-c US. Highly variable, with many intergrading local variants.

ssp. ***angustatum*** (E. Greene) R.A. Price CONTRA COSTA WALLFLOWER Caudex often elongate. **LVS** narrowly lanceolate, dentate; hairs gen 2-branched; lower (6)10–20 cm. **FL**: petals yellow or yellow-orange. **FR** stiffly ascending, ± 4-sided, ± straight. **ENDANGERED** CA, US. Dunes; < 20 m. ne CCo (Antioch Dunes, Contra Costa Co.). [var. *a*. (E. Greene) Rossbach]. In cult.

ssp. ***capitatum*** (p. 423) Caudex gen short. **LVS** (ob)lanceolate, entire or dentate; tip sometimes obtuse; hairs 2–several-branched; lower lvs 2–10(25) cm. **FL**: petals orange to yellow. **FR** ascending, gen 4-sided, straight or slightly curved. Common. Many habitats; 0–4000 m. CA (exc GV, SNH); w N.Am. [*E. asperum* (Nutt.) DC. var. *stellatum* J. Howell; *E. moniliforme* Eastw.] In CA gen bien, orange-fld, hairs 2–4-branched; intergrades with ssp. *perenne* in high KR, SNH; serpentine NCo pls may resemble *E. franciscanum*. Pls from DMoj with 2–3-branched hairs, ± entire lvs, ± flattened frs have been called var. *bealianum* (Jepson) Rossbach; yellow-fld GB pls with gen 2-branched hairs have been called *E. argillosum* (E. Greene) Rydb. ❀SUN,DRN:**4–6,15–17**&IRR:1–3,**7,14,22–24**&SHD:8–10,12,**18–21**.

ssp. ***lompocense*** (Rossbach) R.A. Price SAN LUIS OBISPO WALLFLOWER Caudex elongate. **LVS** gen narrowly lanceolate, dentate; hairs gen 2-branched; lower lvs (4)10–25 cm. **FL**: petals yellow (-orange). **FR** spreading, ± flattened, gen irregularly curved. UNCOMMON. Sandy hillsides, mesas; < 500 m. s SCoRO (San Luis Obispo, Santa Barbara cos.). [*E. suffrutescens* (Abrams) Rossbach var. *l*. Rossbach] Like *E. teretifolium*, but lvs wider; intergrades locally with ssp. *c*. and *E. insulare* ssp. *suffrutescens*. ❀SUN,DRN:**15–17**&IRR:7,**14,22–24**&SHD:8,9,**19–21**.

ssp. ***perenne*** (Cov.) R.A. Price (p. 427) Caudex short. **LVS** gen oblanceolate, subentire to dentate; tip obtuse; hairs gen 2–3-branched; lower lvs 2–10 cm, ± spoon-shaped. **FL**: petals yellow. **FR** ascending, partly flattened, narrowed between seeds. Montane slopes to alpine barrens; 2000–4000 m. KR, CaRH, SNH; OR, NV. [*E. p*. (Cov.) Abrams] See ssp. *c*. ❀IRR,DRN,SUN:1–3,**4–7**,8,9,**14–24**.

E. cheiranthoides L. (p. 427) WORMSEED MUSTARD Ann (bien). **ST** 5–100 cm. **LVS**: basal ± 2–10 cm, oblong-lanceolate, entire to shallowly toothed; hairs gen 3(2–4)-branched. **FL**: petals 2–6 mm, light yellow. **FR** ascending, gen 1–4 cm, 1–1.5 mm wide, ± round; pedicel 1/3–1/2 fr length, < fr width; style 0.3–1 mm. **SEED** 0.8–1.3 mm, 0.5–0.7 mm wide; wing 0. *n*=8. Uncommon. Disturbed areas; 0–2500 m. SN, GV, expected elsewhere; to e N.Am, circumboreal; native to Eurasia, possibly w N.Am.

E. cheiri (L.) Crantz (p. 427) WALLFLOWER Bien to ± subshrub. **STS** 1–few, gen much-branched, 15–80 cm. **LF** ± 4–20 cm, lanceolate to obovate, gen entire; tip acute; hairs gen 2-branched. **FL**: sepals purplish, base green; petals gen 20–35 mm, yellow to red or brown. **FR** ascending, 3–10 cm, > 3 mm wide, densely white-hairy; style 0.5–4 mm; stigma lobes longer than wide. **SEED** 2–4 mm, 1.5–3 mm wide, flattened; wing along sides or at tip only. *n*=6. Uncommon. Disturbed areas; 0–1500 m. s ChI (Santa Catalina Island), e WTR; native to s Eur.

E. franciscanum Rossbach SAN FRANCISCO WALLFLOWER Bien to subshrub. **STS** 1–several, sometimes much-branched, 5–50 cm. **LVS** ± linear to oblanceolate, gen coarsely toothed; hairs 2–several-branched; basal 3–20 cm. **FL**: petals 14–26 mm, cream-colored to yellow. **FR** gen strongly ascending (spreading), 4–13 cm, 2–4 mm wide, gen flattened parallel to septum, often tapered to tip; style 0.5–4 mm. **SEED** 2–4 mm, 1.2–2.5 mm wide; wing at tip and narrow along 1 or both sides. *n*=18. UNCOMMON. Serpentine outcrops, granitic cliffs, sometimes coastal dunes; 0–500 m. NCo, n&c CCo, SnFrB. Fleshy, coastal pls have been called var. *crassifolium* Rossbach; inland pls approach *E. capitatum*. ❀DRN,IRR,SUN:5,7,14, **15–17,19–24**.

E. insulare E. Greene Per or subshrub. **STS** 5–60 cm, much-branched; some elongate, non-fl. **LVS** linear to oblanceolate, ± entire; hairs 2–several-branched; lower 3–15 cm. **FL**: petals 13–20 mm, yellow (-orange). **FR** ascending or spreading, 2–10 cm, 1.4–4 mm wide, gen 4-sided to flattened perpendicular to septum, ± fleshy when immature; style 0.5–3 mm. **SEED** 1.5–3 mm, 0.9–1.8 mm wide, < 2 mm thick; wing 0 or at tip. *n*=18. Coastal dunes, cliffs; 0–300 m. s CCo, n SCo, n ChI.

ssp. ***insulare*** (p. 427) ISLAND WALLFLOWER **ST** 5–30 cm. **LF** 2–9 cm; hairs gen 2-branched. **FR** flattened perpendicular to septum. **SEED** 1–2 mm thick. UNCOMMON. Mesas, cliffs; 0–300 m. n ChI. Variable on Santa Rosa, Anacapa islands. ❀DRY,DRN,SUN:14–16,**17**,22,23,**24**

ssp. ***suffrutescens*** (Abrams) R.A. Price SUFFRUTESCENT WALLFLOWER **ST** 15–60 cm. **LF** 4–15 cm; hairs 2–4-branched. **FR** ± 4-sided to slightly flattened. **SEED** 0.4–1 mm thick. *n*=18. UNCOMMON. Coastal dunes (bluffs at Morro Rock); 0–150 m. s CCo, n SCo. [*E. s*. (Abrams) G. Rossbach] Hybridizes locally with *E. capitatum*. Sporadic wide-lvd pls have been called var. *grandifolium* Rossbach, large-lvd wallflower. ❀DRN,SUN:14,**15–17**,22,23,**24**.

E. menziesii (Hook.) Wettst. Bien or short-lived per. **STS** 1–several, 1–15(30) cm, gen little-branched. **LVS** gen in dense rosette, gen fleshy; hairs 2–several-branched; lower lvs 2–11 cm, ± spoon-shaped, entire to lobed. **FL**: petals 14–32 mm, gen yellow. **FR** (esp lower) gen stiffly spreading, 3–14 cm, 1.5–5 mm wide, slightly to very fleshy and round when immature, flattened when dry; style 0.5–3 mm. **SEED** 1.7–4 mm, 1–3 mm wide. *n*=18. Coastal foredunes, headlands, cliffs; 0–300 m. NCo, CCo.

ssp. ***concinnum*** (Eastw.) R.A. Price (p. 427) Fls gen winter, spring. **LF** subentire to sharply toothed, gen fleshy. **FL**: petals cream-colored to yellow. **FR** gen erect, 3–13 cm, 2–5 mm wide; pedicel gen > 10 mm. Gen headlands, cliffs; 0–300 m. NCo, n CCo (Point Reyes); s OR. [*E. c*. Eastw.] Pls at Fort Bragg, Mendocino Co., intermediate to ssp *m*. ❀DRN:**5,15–17**&IRR:7,**14,19–24**;fragrant.

ssp. ***eurekense*** R.A. Price Fls gen winter, spring. **LF** toothed, not obviously fleshy. **FL**: petals light yellow. **FR** (5)8–14 cm, 2–3 mm wide; pedicel (5)9–15 mm. Foredunes; 0–300 m. n NCo (Humboldt Bay, Humboldt Co.). Threatened by development. In cult.

ssp. ***menziesii*** (p. 427) MENZIES' WALLFLOWER Fls gen winter, spring. **LF** gen lobed or irregularly toothed, fleshy. **FL**: petals rich yellow. **FR** gen 3–8 cm, 2–4 mm wide; pedicel 3–9(13) mm. **ENDANGERED** CA. Foredunes; 0–300 m. c NCo (near Fort Bragg, Mendocino Co.), c CCo (Monterey Bay Area, s of Point Pinos). Endangered by habitat loss. In cult.

ssp. ***yadonii*** R.A. Price (p. 427) Often per (caudex branched), fls summer. **LF** subentire to lobed, fleshy. **FL**: petals rich yellow. **FR** 3–8 cm, 2–4 mm wide; pedicel 3–9 mm. Foredunes; 0–300 m. c CCo (± c Monterey Bay, Monterey Co.). Threatened by development, sand mining. In cult.

E. repandum L. (p. 427) Ann. **STS** 1–few, 5–50 cm. **LVS** linear to narrowly oblanceolate; most hairs 2-branched; basal 5–15 cm, gen deeply toothed. **FL**: petals 5–10 mm, light yellow. **FR** stiffly spreading, 4–10 cm, 1–1.5 mm wide, ± round, narrowed between seeds; pedicel 2–7 mm, ± = fr width; style 1–5 mm. **SEED** 1.2–1.7 mm, 0.5–0.8 mm wide; wing 0. *n*=8. Fields, disturbed areas; 0–2000 m. n&c SnJV, MP, expected elsewhere; native to Eur.

E. teretifolium Eastw. (p. 427) SANTA CRUZ WALLFLOWER Bien or short-lived per, gen dull purplish. **STS** 1–several, gen unbranched, 15–100 cm. **LF** very narrowly linear (< 3 mm wide), finely toothed; most hairs 2-branched; basal 3–17 cm. **FL**: petals 15–25 mm, yellow (-orange). **FR** ascending to spreading, 4–15 cm, 1.2–2.5 mm wide, partly flattened, often irregularly curved; style 0.5–2.5 mm. **SEED** 1.5–2.7 mm, 0.9–1.5 mm wide; wing gen at tip. **ENDANGERED** CA. Inland sand; 200–400 m. sw SnFrB (Santa Cruz Co.). Endangered by development, sand mining. In cult.

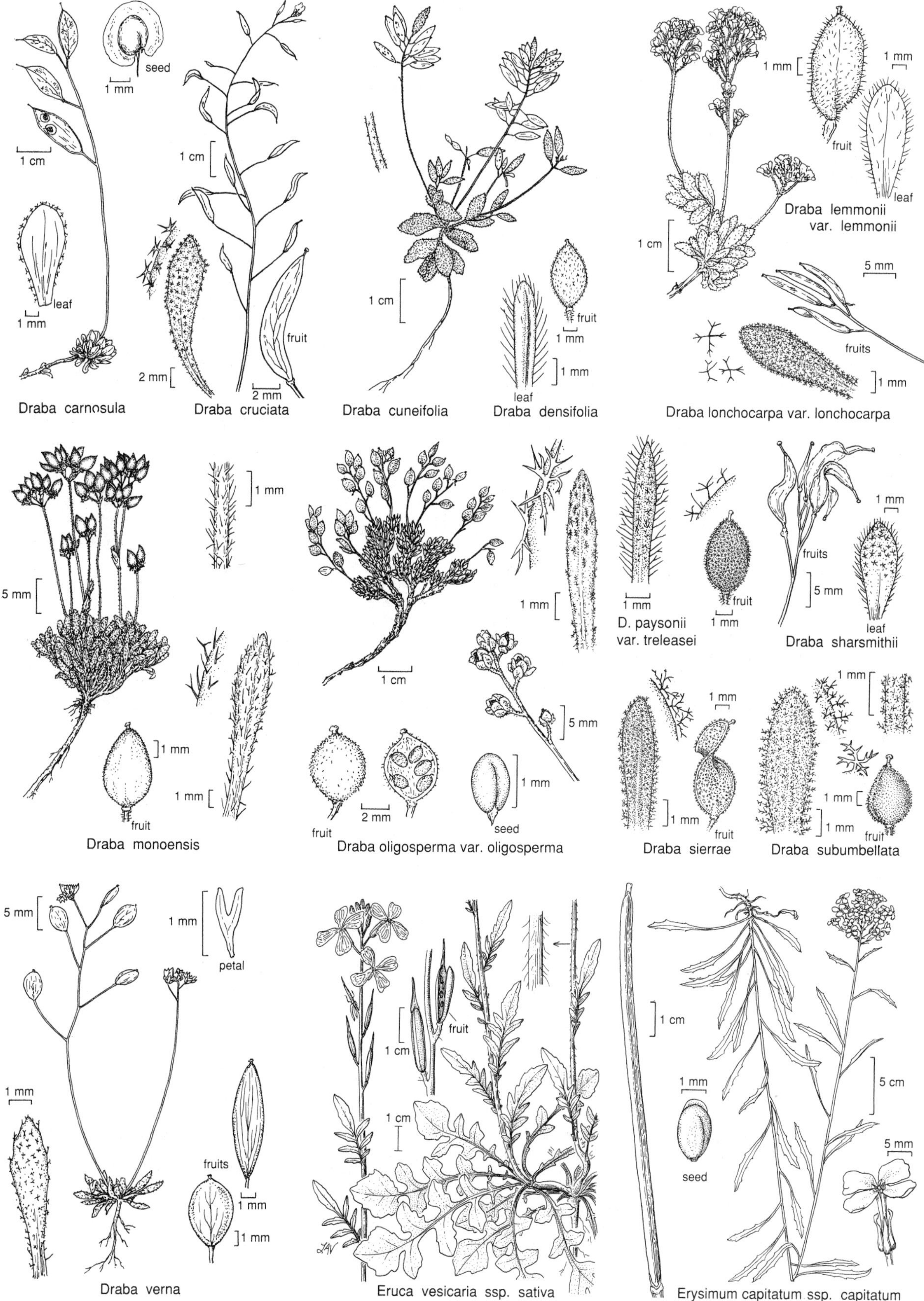
seed
1 mm
1 cm
leaf
1 mm
Draba carnosula
1 cm
2 mm
fruit
2 mm
Draba cruciata
1 cm
Draba cuneifolia
fruit
1 mm
1 mm
leaf
Draba densifolia
1 mm
fruit
1 mm
leaf
Draba lemmonii var. lemmonii
1 cm
5 mm
fruits
1 mm
Draba lonchocarpa var. lonchocarpa
1 mm
5 mm
1 mm
1 mm
fruit
Draba monoensis
1 mm
1 cm
5 mm
1 mm
fruit
2 mm
seed
Draba oligosperma var. oligosperma
1 mm
fruit
1 mm
D. paysonii var. treleasei
1 mm
fruits
5 mm
leaf
Draba sharsmithii
1 mm
1 mm
fruit
Draba sierrae
1 mm
1 mm
1 mm
fruit
Draba subumbellata
5 mm
1 mm
petal
1 mm
fruits
1 mm
1 mm
Draba verna
fruit
1 cm
1 cm
Eruca vesicaria ssp. sativa
1 cm
1 mm
seed
5 cm
5 mm
Erysimum capitatum ssp. capitatum

EUCLIDIUM

Ann; hairs simple or not. **LVS** simple; basal lobes 0. **INFL** spike-like, open. **FL** inconspicuous, ± sessile; sepal bases not sac-like; petals spoon-shaped, white. **FR** sessile above receptacle, 2-chambered, tardily dehiscent; beak short-tapered. **SEEDS** 2; embryonic root at edges of both cotyledons. 2 spp.: Eurasia. (Greek: tightly closed, from fr)

E. syriacum (L.) R.Br. (p. 427) Pl branched from near base. **ST** 1–4 dm, rigid. **LVS**: basal rosette 0; cauline with dense, gen simple and forked hairs, lower 1–5 cm, lanceolate, entire to dentate, petioled, upper oblong to oblanceolate, entire to obscurely dentate, sessile. **FL**: petals 1–1.3 mm, white. **FR** 2–3 mm, ovate to ± round, plump; hairs dense, simple and forked, whitish; pedicel erect, appressed, stout, 0.7–1 mm; beak curved away from axis. 2*n*=14. Roadsides, disturbed pastures; 1200–1500 m. s MP (Lassen Co.); to WA, Colorado; native to Eurasia.

GUILLENIA

Roy E. Buck

Ann, glabrous to ± hairy below. **ST** often hollow, ± glaucous. **LVS**: basal rosetted, often withering in fl, gen ± oblanceolate, entire to deeply cut, petioles < blades; upper lvs reduced. **INFL** longer in fr; bracts gen 0. **FL**: sepals pouched at base or not, greenish or not; petals ± linear to obovate; anthers coiled or ± curved when open; style gen ± tapered, stigma small, entire or shallowly 2-lobed. **FR** ascending to reflexed, ± cylindric; stalk-like base < 1 mm, = body width. **SEED** ± oblong, brownish or yellowish; wing 0. 3 spp.: w N.Am. (Father C. Guillen, Jesuit missionary, Mexico, born 1677)

1. Petal blade channeled, ± wavy-margined, claw width = or > blade width at junction, blade flared to tip or not ..... ***G. flavescens***
1′ Petal blade not channeled, not wavy-margined, claw width << blade width at junction
  2. Fl parts ± erect; sepals gen greenish (rarely pinkish); petals gen white or pale yellow (rarely pinkish); pedicel in fr 0.5–3(4) mm ..... ***G. lasiophylla***
  2′ Fl parts spreading to ascending; sepals pink or purplish; petals pink or whitish with pink veins; pedicel in fr (2)3–9 mm ..... ***G. lemmonii***

G. flavescens (Hook.) E. Greene (p. 427) **LVS**: basal blades 5–22 cm, (ob)lanceolate, wavy-dentate to deeply cut; cauline gen not clasping, ± entire to cut. **FL**: sepals spreading to erect, 5–11 mm, ± pouched at base, gen cream-colored or pale yellow (to purple); petals erect, 8–15 mm, whitish, purple-veined or not, blade channeled, ± wavy-margined, flared to tip or not, claw often > blade, width = or > blade width at junction; style 2–3.5 mm. **FR** ascending to reflexed, 3–9 cm, ± straight. **SEED** ± 1.5 mm, brownish. *n*=14. Dry, exposed slopes, often on serpentine; 80–750 m. s NCoRI, ScV (Montezuma Hills), se SnFrB, SCoRI. [*Thelypodium f.* (Hook.) S. Watson] ❀TRY.

G. lasiophylla (Hook. & Arn.) E. Greene CALIFORNIA MUSTARD **LVS**: lower blades < 22 cm, lanceolate to ± oblong, entire to cut; lower cauline gen > basal; uppermost lvs ± subsessile. **INFL**: pedicel in fr 0.5–4 mm. **FL**: parts ± erect; sepals 1.5–4 mm, not pouched at base, greenish (pinkish), narrowly scarious-margined; petals 3–6 mm, ± oblanceolate, gen white or pale yellow (pinkish), blade not channeled, not wavy-margined, narrowed to claw; style 0.1–2.4 mm. **FR** gen reflexed, 1–7 cm, straight or outcurved. **SEED** ± 1 mm, yellowish or brownish. *n*=14. Common. Dry, open, sometimes disturbed areas; < 2500 m. CA (exc MP); to B.C., UT, nw Mex. [*Thelypodium l.* (Hook. & Arn.) E. Greene incl vars. *inalienum* Robinson, *rigidum* (E. Greene) Robinson, and *utahense* (Rydb.) Jepson] Highly variable; needs study.

G. lemmonii (E. Greene) R. Buck **LVS**: lower blades 2–16 cm, (ob)lanceolate, entire to shallowly cut, often 1–2-lobed at base; uppermost lvs gen ± subsessile. **INFL**: pedicels in fr (2)3–9 mm. **FL** sweetly fragrant; parts ascending to spreading; sepals 2.5–6 mm, not pouched at base, pink or purplish; petals 3–9 mm, oblanceolate to obovate, pink or whitish with pink veins, blade not channeled, not wavy-margined, narrowed to short claw; stamens nearly equal; style > 1 mm. **FR** ascending to reflexed, 2–7 cm, ± straight. **SEED** ± 1 mm, yellowish or brownish. *n*=14. Open slopes, plains, often alkaline soil; 300–1600 m. s SNF, sw SnJV, se SCoRO, SCoRI, n WTR. [*Thelypodium l.* E. Greene] ❀SUN,DRN,IRR:7,**8**,**9**,**14–24**.

HALIMOLOBOS

Bien, per, hairy or not. **ST** erect, herbaceous. **LVS** simple, entire to deeply lobed; basal often shed early. **FL**: sepals erect, bases not sac-like. **FR** 0.5–3 cm, linear, cylindric, hairy or not. **SEEDS** many, (1)2 rows per chamber, 0.4–1.5 mm, elliptic; wing 0; embryonic root at back of 1 cotyledon. 19 spp.: gen Mex. (Greek: sea pod, from resemblance to *Alyssum halimifolium*) [Rollins 1943 Contr Dudley Herb 3:241–265]

1. Cauline lf base tapered, not eared; fr densely hairy; basal lvs gen 0 ..... ***H. jaegeri***
1′ Cauline lf base eared, not tapered; fr glabrous; basal lvs gen present ..... ***H. virgata***

H. jaegeri (Munz) Rollins (p. 427) Per; hairs multibranched; base woody. **STS** several–many, many-branched, 2–6 dm. **LVS**: basal gen 0; cauline 1–6 cm, oblanceolate to ovate, coarsely toothed to deeply lobed, lower short-petioled or sessile, upper sessile. **INFL** dense. **FL**: petals 4–5.5 mm, linear-oblanceolate to spoon-shaped, white; stamens exserted. **FR** spreading, 1.5–2.5 cm, < 1 mm wide, gen slightly upcurved; pedicels spreading, 2–7 mm, densely hairy; style 0.5–2.0 mm. **SEED** ± 1 mm, embedded in septum. Limestone cliffs, steep rock outcrops, sagebrush/juniper areas; 1500–2500 m. W&I, DMoj; NV. [*H. diffusa* (A. Gray) O. Schulz var. *j.* (Munz) Rollins]

H. virgata (Nutt.) Schulz VIRGATE HALIMOLOBOS Bien, per; hairs simple or multibranched; base not woody. **ST** gen 1, simple below, branched above, 1–4 dm. **LVS**: basal gen present, 2–6 cm, oblanceolate, entire or toothed; cauline 1.5–4 cm, lanceolate to oblong, entire to toothed, sessile. **INFL** open to dense. **FL**: petals 3–4 mm, spoon-shaped, white; stamens not exserted. **FR** erect, 2–4 cm, ± 1 mm wide, straight; pedicels ascending, 5–12 mm, ± hairy; style ± 0.5 mm. **SEED** ± 1 mm, not embedded in septum. RARE in CA. Meadows, near aspen groves, pinyon/ juniper woodland; 2000–3000 m. SNE; to w Can, Colorado.

HESPERIS ROCKET

Bien, per. **LVS** entire to lobed. **INFL**: bracts 0; pedicel base with a large gland on each side. **FL** showy, fragrant; sepals erect, inner pair sac-like at base; petals clawed, white to purple. **FR** linear, cylindric to 4-sided, sessile above receptacle, tardily dehiscent; valves 1–3-veined; stigma deeply 2-lobed. **SEEDS** many, 1 row per chamber; wing 0; embryonic root at back of 1 cotyledon. ± 30 spp.: Old World, esp Eurasia. (Greek: evening, from time some fls are most fragrant)

H. matronalis L. (p. 427) DAME'S ROCKET **STS** 1–few, 5–13 dm; branches 0–few; hairs simple and forked, some glandular. **LVS** many, 5–20 cm, ± lanceolate, dentate, hairy; lower cauline petioled, upper ± sessile. **INFL**: panicle of racemes. **FL**: sepals hairy; petals 2–2.5 cm, purple, rose, or white; style < 4 mm, stigma lobes decurrent. **FR** 4–10 cm, ± cylindric, gen ± narrowed between seeds. **SEED** 3–4 mm. $2n$=24,26,28. Roadsides, slopes; 1000–1500 m. KR (Trinity Co.), n SNH (Plumas Co.), expected elsewhere; to Can, e US; native to Eur.

HETERODRABA

1 sp. (Greek: different *Draba*)

H. unilateralis (M.E. Jones) (p. 427) E. Greene Ann. **ST** hairy; branches few, near base, prostrate to ascending, 1–4 dm. **LVS** basal and cauline, 1–2 cm, widely oblanceolate to obovate, entire to sparsely dentate; hairs dense, minute, multibranched; cauline gen near base. **INFL** terminal, 1-sided, open; bracts 0. **FL** inconspicuous; sepals purplish; petals white; stamens 6. **FR** widely separated, 3–5 mm, flat parallel to septum, widely ovate, often twisted; hairs dense, 0 above or not; pedicel recurved, 1–2 mm; style < 0.2 mm. **SEEDS** 6–12, 1.2–1.5 mm, plump, oblong; wing 0; embryonic root at edges of both cotyledons. Uncommon. Grassy, open slopes, flats, clay soils; < 800 m. CaRF, s SNF, Teh, GV, SCoR; OR, Mex.

HIRSCHFELDIA

Ann, bien, per; hairs simple, dense. **ST** erect. **LVS** basal, cauline, lobed. **INFL** narrow, greatly elongating in fr; bracts 0. **FL**: sepals spreading, bases not or barely sac-like; petals obovate, clawed, pale yellow to white; stigma expanded. **FR** cylindric; beak abruptly swollen at base. 2 spp.: s Eur. (C. Hirschfeldt, horticulturist, 1742–1792)

H. incana (L.) Lagr.-Fossat (p. 427) Bien, per, canescent. **ST** branched from base and above, 2–10 dm. **LVS**: basal rosetted, flat on ground, pinnately lobed, terminal lobe > lateral, ± crenate-dentate; cauline ± sessile, ± simple, not clasping st. **FR** erect, appressed, 1–1.5 cm, glabrous; valves 3–7-veined; pedicel erect, 3–4 mm, stout, club-like, beak 3–6 mm, flat. **SEEDS** 1 row per chamber, spheric, reddish brown. $2n$=14. Roadsides, creek bottoms, disturbed areas; < 1600 m. NCo, SNF, GV, CW, SCo; native to Medit. [*Brassica geniculata* (Desf.) Ball; *H. adpressa* Moench]

HUTCHINSIA

Ann; hairs 0 or sparse, minute, branched. **INFL** open; bracts 0. **FL**: petals white. **FR** ± elliptic, ± flat perpendicular to septum; wing 0; notch 0. **SEEDS** several per chamber; wing 0; embryonic root at back of 1 cotyledon. 3 spp.: Eur, 1 apparently native in N.Am as well. (E. Hutchins, Irish botanist, 1785–1815)

H. procumbens (L.) Desv. (p. 427) **ST** ± decumbent to ± erect, (3)5–10(15) cm, many-branched. **LVS**: basal and lower cauline petioled, 5–20 mm, obovate, entire to lobed; upper lvs fewer, sessile, reduced. **FL**: sepals ± 1 mm; petals ± 1 mm, obovate or narrower. **FR** 2.5–3.5 mm, elliptic to ± obovate; valve veins prominent, netted; style < 0.2 mm. $2n$=12. Alkali flats, saline seeps, shaded sites, sagebrush, juniper woodland; < 2600 m. CA (exc KR, SNH); to B.C., Labrador, Baja CA, also Eur.

IDAHOA FLAT-POD

1 sp. (State of Idaho)

I. scapigera (Hook.) Nelson & J.F. Macbr. (p. 427) Ann, scapose, glabrous. **LVS** rosetted, several–many, petioled; blades 1–3 cm, ovate, entire to deeply lobed, lobes few, oblanceolate, terminal > others. **INFL**: peduncles many, (2)3–13 cm, 1-fld, slender, bractless. **FL**: sepals 1.4–2 mm, gen reddish or purplish; petals = or ± > sepals, white. **FR** 6–12 mm, round to widely ovate or elliptic, flat parallel to septum; beak < 1 mm; style ± 0. **SEED** 3.5–5 mm wide incl ± 1 mm wide wing, ± round, flat; embryonic root at edges of both cotyledons. $2n$=16. Moist ledges, slopes, meadows, foothills; 600–1900 m. CaR, SnFrB (Mount Hamilton), NCoRI (Mount Saint Helena), n SN, s SNF, SCoRI, GB; to B.C., MT. ❀TRY.

ISATIS WOAD

Ann to per, erect. **LVS** simple. **FL** small; sepals < petals. **FR** pendent, oblong to round, flat perpendicular to septum, indehiscent, ribbed, 1-chambered; wing ± inflated or not; style 0. **SEED** 1, pendent, in place of septum. ± 60 spp.: gen Eurasia, esp Iraq, Iran. (Greek: pl with dark dye)

I. tinctoria L. (p. 427) Bien, per, glaucous. **ST** 1, 4–12 dm, simple below, glabrous, ± grayish. **LVS**: basal several, clustered, long-petioled, < 2 dm, ± oblanceolate, ± toothed or wavy-margined, hairs sparse, simple; cauline sessile, sagittate, clasping st, gen entire, glabrous. **INFL**: panicle of racemes; bracts subtending main branches, lf-like. **FL**: petals ± 3.5 mm, spoon-shaped. **FR** 8–18 mm, 5–7 mm wide, oblong to oblanceolate, black; tip obtuse to rounded; pedicel ascending to reflexed, < fr, slender. **SEED** yellowish; embryonic root at back of 1 cotyledon. $2n$=28. Roadsides, fields, disturbed sites; < 1000 m. KR, CaR, n SNH, MP; to ID, UT, also e US; native to Eur. Once cult as source of blue dye.

LEPIDIUM PEPPERGRASS, PEPPERWORT

Ann to shrub; hairs 0 or simple. **LVS**: basal not rosetted, gen petioled, entire to pinnately lobed; cauline short-petioled to sessile, sometimes clasping or surrounding st. **FL** small; sepals erect or spreading, oblong to ovate, shed early or persistent; petals linear to obovate, gen white, rarely yellowish, sometimes bristle-like or 0; stamens 6, 4, or 2. **FR** dehiscent, oblong to elliptic or obcordate, flat perpendicular to septum; pedicel cylindric or flat, winged or not. **SEEDS** 1 per chamber, gelatinous when wetted; wing narrow or 0; embryonic root at back of 1 cotyledon, rarely at edges of both. ± 175 spp.: ± worldwide. (Greek: little scale, from fr) [Hitchcock 1936 Madroño 3:265–300]

1. Upper cauline lvs clasping or surrounding st
 2. Upper cauline lvs surrounding st, ovate to round; basal lvs 2–3-pinnately lobed or -divided; pedicel in fr glabrous; petals yellowish; seed winged ..... ***L. perfoliatum***
 2′ Upper cauline lvs clasping st, linear to oblong; basal lvs rarely lobed; pedicel in fr hairy; petals white; seed wing 0
  3. Ann, bien; st gen 1; fr with scale- or sac-like hairs; style 0.2–0.6 mm, < to ± > notch; anthers yellow ..... ***L. campestre***
  3′ Per; sts several; fr with 0 or few scale- or sac-like hairs; style ± 1 mm, gen > notch; anthers purple ..... ***L. heterophyllum***
1′ Upper cauline lvs not clasping, not surrounding st
 4. Stigma gen exceeding fr (exc in *L. pinnatifidum*); fr tip notched to not
  5. St prostrate to ascending; fr valves with prominent, ascending-spreading tips; style < to > 1/2 fr; petals yellow ..... ***L. flavum***
   6. Fr 3–4.5 mm, 3–3.5 mm wide; style < 1/2 fr ..... var. ***felipense***
   6′ Fr 2–3 mm, 1.5–2.2 mm wide; style > 1/2 fr ..... var. ***flavum***
  5′ St gen erect, outer decumbent or not; fr valves without prominent tips; style < 1/2 fr; petals white, rarely pale cream-colored (yellow fading whitish in *L. jaredii*)
   7. Pls rhizomed, individuals often in contact, in colonies; basal lvs < 3 dm, toothed ..... ***L. latifolium***
   7′ Pls not rhizomed, individuals not in contact, not in colonies; basal lvs gen < 1 dm, gen lobed or compound
    8. Style ± = fr tip notch, < 0.2 mm — fr widely elliptic to round, ± 2 mm, hairs sparse, weak, simple; alien ..... ***L. pinnatifidum***
    8′ Style > fr tip notch (or notch 0), > 0.3 mm
     9. Ann; petals yellow, fading whitish; lower pedicels 4–5 × fr ..... ***L. jaredii***
     9′ Per, rarely bien or ann; petals white, rarely pale cream-colored; lower pedicels gen 3 × fr
      10. Pl glabrous, grayish, ± woody; fr obovate to round ..... ***L. fremontii***
       11. Fr widely ovate to ± round or obovate, stalk above receptacle ± 0 ..... var. ***fremontii***
       11′ Fr widely obovate to ± round, stalk above receptacle ± 1 mm ..... var. ***stipitatum***
      10′ Pl with at least some hairs, not grayish, base rarely woody; fr elliptic to round
       12. St hairs a mixture of long, flat and short, club-shaped; upper cauline lvs gen pinnately lobed; winter ann, bien ..... ***L. thurberi***
       12′ St hairs all long, flat or short, club-shaped or scale-like, rarely ± 0; upper cauline lvs gen entire; bien to per ..... ***L. montanum***
        13. Hairs of sts, pedicels linear, long (several–many × longer than wide) ..... var. ***canescens***
        13′ Hairs of sts, pedicels oblong, club-shaped, or scale-like, short (gen not > 2 × longer than wide), or ± 0
         14. Hairs of sts, pedicels ± dense, oblong; per, rarely bien ..... var. ***cinereum***
         14′ Hairs of sts, pedicels sparse to ± dense, club-shaped or scale-like, or ± 0; per, ± woody at base or not ..... var. ***montanum***
 4′ Stigma exceeded by fr; fr tip notched
  15. Pedicel in fr very flat, > (2)3 × wider than thick, gen = or < fr
   16. Fr tip notch deep, gen very narrow, > 1/3 seed pouch; fr valves persistent; raceme in fr 1–1.5 cm wide, < or > basal lvs, or basal lvs 0 or shed early ..... ***L. latipes***
    17. St branched above base, nodes spaced; raceme in fr > basal lvs; basal lvs 0 or shed early ..... var. ***heckardii***
    17′ St branched at or near base, nodes not spaced; raceme in fr gen < basal lvs; basal lvs gen present ..... var. ***latipes***
   16′ Fr tip notch shallow, often wide, gen < 1/5 × seed pouch; fr valves readily shed; raceme in fr gen < 1 cm wide, gen > basal lvs
    18. Lf oblanceolate to oblong, toothed or with obovate to oblong lobes; fr margin not upturned ..... ***L. lasiocarpum*** var. ***lasiocarpum***
    18′ Lf linear or with linear lobes; fr margin upturned or not
     19. Fr hairs ± throughout, rarely 0; pedicel suberect to ascending, rarely ± reflexed ..... ***L. dictyotum***
      20. Fr valve tips > 1 mm, acuminate, ± ascending, winged; fr tip notch U- or V-shaped ..... var. ***acutidens***
      20′ Fr valve tips < 1 mm, rounded to acute, erect, unwinged; fr tip notch ± closed or narrowly U- or V-shaped ..... var. ***dictyotum***
     19′ Fr hairs 0, rarely few, marginal; pedicel ± recurved to ± ascending ..... ***L. nitidum***
      21. Fr valve tips ascending, ± beak-like ..... var. ***oreganum***
      21′ Fr valves tips erect or ± ascending, not beak-like
       22. Fr gen fringed with hairs; st hairs dense ..... var. ***howellii***
       22′ Fr glabrous; st hairs 0 to ± dense ..... var. ***nitidum***
  15′ Pedicel in fr cylindric to flat, gen < or ± 2 × wider than < thick, = or > fr (< fr in *L. strictum*)
   23. Lf linear or with linear segments; fr valve tips gen acute, notch gen ± 90–160° — ephemeral ann ..... ***L. oxycarpum***

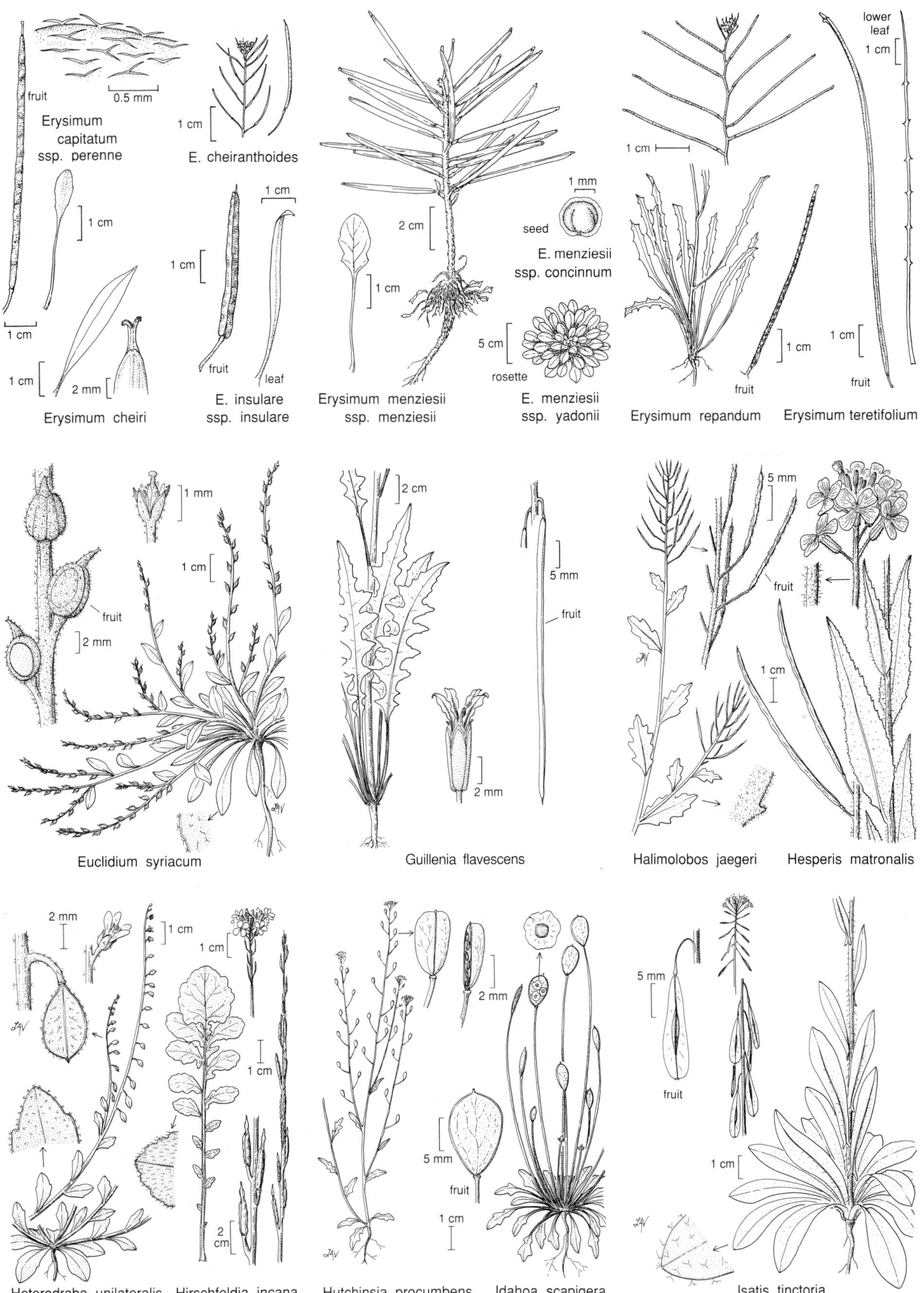
fruit
0.5 mm
Erysimum capitatum ssp. perenne
1 cm
E. cheiranthoides
1 cm
1 cm
1 cm
1 cm
2 mm
Erysimum cheiri
1 cm
fruit
leaf
E. insulare ssp. insulare
2 cm
1 cm
Erysimum menziesii ssp. menziesii
1 mm
seed
E. menziesii ssp. concinnum
5 cm
rosette
E. menziesii ssp. yadonii
1 cm
1 cm
fruit
Erysimum repandum
lower leaf
1 cm
1 cm
fruit
Erysimum teretifolium
1 mm
1 cm
fruit
2 mm
Euclidium syriacum
2 cm
5 mm
fruit
2 mm
Guillenia flavescens
5 mm
fruit
1 cm
Halimolobos jaegeri
Hesperis matronalis
2 mm
1 cm
1 cm
1 cm
2 cm
Heterodraba unilateralis
Hirschfeldia incana
2 mm
5 mm
fruit
1 cm
Hutchinsia procumbens
Idahoa scapigera
5 mm
fruit
1 cm
Isatis tinctoria

23′ Lf sometimes narrow but not linear, gen without linear segments; fr valve tips gen rounded, notch < 45° (± 90° in *L. strictum*)
24. Lvs lobed or divided below, gen above; st gen prostrate to ascending, branched, 0.5–2(3) dm
25. Fr ± ovate, veins prominent, valve tips ascending, gen winged .... ***L. strictum***
25′ Fr ± elliptic, round, or obovate, veins 0 or obscure, valve tips erect, unwinged .... ***L. oblongum***
26. Fr round to obovate, hairy on margin; pedicel hairs gen on both sides .... var. ***insulare***
26′ Fr ± elliptic to obovate-elliptic, hairs 0 on margin; pedicel hairs 0 or on upper surface ...... var. ***oblongum***
24′ Lvs entire to lobed below, entire to toothed, rarely ± lobed or dissected above; st erect, branched or not, gen > 2 dm
27. Petals = to 2(3) × sepals; fr hairs 0 .... ***L. virginicum***
28. Pedicel ± cylindric, hairs sparse to ± dense; embryonic root at edges of both cotyledons .... var. ***virginicum***
28′ Pedicel ± flat, gen ± winged, hairs 0 to ± dense; embryonic root at back of 1 cotyledon
29. Infl and upper st glabrous .... var. ***medium***
29′ Infl and upper st hairy
30. Cauline lvs not divided, not lobed, entire to shallowly dissected; st 2–7 dm .... var. ***pubescens***
30′ Cauline lvs divided or lobed; st gen 1–2 dm .... var. ***robinsonii***
27′ Petals < sepals, vestigial, or 0; fr hairs 0 or minute
31. Seed not margined; fr valve tips ± acute; axillary racemes in fr many .... ***L. ramosissimum*** var. ***bourgeauanum***
31′ Seed gen margined or winged; fr valve tips rounded; axillary racemes in fr 0 .... ***L. densiflorum***
32. Fr ± 2.5 mm, glabrous; pedicels very slightly flat, > 9 per cm .... var. ***densiflorum***
32′ Fr ± = or > 3 mm, hairy at least on margin or glabrous; pedicels flat at least on 1 side, < 9 per cm
33. Pedicel flat on upper, lower sides, ± 2 × wider than thick; fr glabrous .... var. ***ramosum***
33′ Pedicel flat gen on lower side, < ± 2 × wider than thick; fr hairy at least on margin or glabrous
34. Fr glabrous .... var. ***macrocarpum***
34′ Fr hairy at least on margin
35. Fr hairy on margin .... var. ***elongatum***
35′ Fr hairy ± throughout .... var. ***pubicarpum***

L. campestre (L.) R.Br. (p. 433) Ann, bien; hairs dense, spreading. **ST** gen 1, erect, branched above, 2–5 dm, leafy. **LVS**: basal petioled, 5–7 cm, oblanceolate, entire to lobed; cauline sagittate, upper clasping st. **FL**: sepals ± 1.5 mm, hairy; petals ± 2 mm, narrowly spoon-shaped, white; anthers yellow. **FR** 5–6 mm, ± 4 mm wide, widely oblong to ovate; hairs scale- or sac-like; margins, tips winged; upper surface concave; pedicel spreading, 4–8 mm, slender, ± flat, hairy; style 0.2–0.6 mm, incl in to exserted from notch. $2n=16$. Common. Waste places, fields, roadsides; 900–1900 m. KR, CaR, n SNF, expected elsewhere; widespread US; native to Eur.

L. densiflorum Schrader Ann, bien. **ST** gen 1, erect, branched above, 1–3(4) dm; hairs minute, flat, obtuse. **LVS**: basal rosetted, 3–10 cm, toothed to pinnately lobed; cauline entire to lobed, upper sessile, not clasping st. **FL**: sepals ± 1 mm; petals 0, rarely vestigial, white; stamens 2(4). **FR** ± 2.5–3.5 mm, oblong-obovate; hairs 0 or minute; valve tips rounded; pedicel ± ascending, slender, gen < ± 2 × wider than thick; style ± 0. **SEED** gen margined or winged. Sandy soils, plains, slopes, gen disturbed sites; < 1800 m. KR, CaR, GV, GB, DMoj; to AK, Can, e US. Vars. difficult.

var. ***densiflorum*** (p. 433) Ann. **FR** ± 2.5 mm, 2 mm wide, widest at middle or just above, glabrous; pedicel very slightly flat. $2n=32$. Grassy slopes, flood plains, disturbed areas; < 1200 m. CaR, GV, DMoj, expected elsewhere; to AK, se Can, e US.

var. ***elongatum*** (Rydb.) Thell. Bien. **FR** 3–3.5 mm, 2.5–3 mm wide, hairy on margins; pedicel ± flat. $2n=32$. Sandy banks, alkaline soils, open areas; 700–1800 m. SCo, GB; to AK, ID.

var. ***macrocarpum*** G. Mulligan Bien. **FR** ± 3 mm, 2.5–3 mm wide, glabrous; pedicel flat. $2n=32$. Roadsides, rocky knolls, sagebrush; < 2200 m. KR, n SNH, ScV, SNE, expected elsewhere; to B.C., n-c US.

var. ***pubicarpum*** (Nelson) Thell. Ann, winter ann. **FR** 3–3.5 mm, 2.5–3 mm wide, hairy ± throughout; pedicel ± flat. $2n=32$. Roadbanks, meadows; < 1200 m. KR, CaR, GB, DMtns, expected elsewhere; to s-c Can, UT.

var. ***ramosum*** (Nelson) Thell. Ann. **FR** ± 3.5 mm, glabrous; pedicel flat on upper, lower sides, ± 2 × wider than thick. Sandy soils, slopes, flats, sagebrush; < 800 m. SNE, DMoj; to WY, AZ.

L. dictyotum A. Gray Ann, low; hairs dense, spreading. **ST**: branched from base; outer branches ± decumbent, 2–20 cm. **LVS**: basal pinnately lobed, lobes linear; cauline 1–2.5 mm wide, linear, gen entire. **FL**: sepals 0.7–1 mm; petals ± > sepals, white, gen 0; stamens 4 or 6. **FR** 3–4.5 mm, ± 2.3 mm wide, notch < 1/5 × seed pouch; hairs ± throughout, rarely 0; pedicel suberect to ascending, rarely ± reflexed, 1.5–3.5 mm, very flat; style 0. **SEED** ± 2 mm. Saline soils, dry streambeds, fields; < 1000 m. CA (exc NW, SN); to WA, UT, Baja CA.

var. ***acutidens*** A. Gray **FR** 3.5–4.5 mm; valve tips ± ascending, > 1 mm, acuminate, ± spreading, winged, notch U- or V-shaped. Alkaline flats, streambeds; < 1000 m. CaR, GV, SW, MP; to WA, Baja CA. [*L. oxycarpum* var. *a.* Jepson]

var. ***dictyotum*** (p. 433) **FR** ± 3 mm; valve tips erect, < 1 mm, rounded to acute, unwinged, notch ± closed or narrowly U- or V-shaped. Saline areas, playas, alkaline soils; < 1000 m. CaR, GV, SnFrB, SCoR, SW, MP, DMoj; to WA, UT, Baja CA.

L. flavum Torrey Ann, glabrous. **ST** prostrate to ascending, 1–4 dm, branched from base. **LVS**: basal rosetted, 2–5 cm, oblong-lanceolate to spoon-shaped, margin irregularly to pinnately lobed, lobes ± rounded; cauline often toothed or entire toward tip. **FL**: petals 2–3 mm, yellow; stamens 6. **FR** ovate or ± round, ± winged; tip notched; pedicel cylindric or ± flat; style 1–1.5 mm, stigma exceeding fr. Alkaline or sandy soils; < 1400 m. D; n NV, Baja CA.

var. ***felipense*** C. Hitchc. BORREGO VALLEY PEPPERGRASS **FR** 3–4.5 mm, 3–3.5 mm wide; hairs ± 0; style < 1/2 fr. RARE. Sandy soils, creosote-bush shrub; ± 100 m. DSon (Borrego Valley).

var. ***flavum*** (p. 433) **FR** 2–3 mm, 1.5–2.2 mm wide; hairs sac- or scale-like or 0; style > 1/2 fr. $2n=32$. Common. Alkaline soils, flats; < 1400 m. D; NV, Baja CA. ❀TRY;DFCLT.

L. fremontii S. Watson Per, gen shrubby, 4–10 dm, glabrous, gray. **ST** many-branched. **LVS** 3–10 cm, linear, gen pinnately lobed; lobes 3–9, 1–3 mm wide; upper 2–3 mm wide, entire or not. **INFL** many-branched, ± leafy. **FL**: petals ± 3 mm, obovate to spoon-shaped, white, claw slender. **FR** 5–8 mm; valves thin; veins faint. Sandy washes, barren knolls, gravelly soils, rocky slopes, ridges; < 1600 m. s SNE, D; to s UT, w AZ.

var. ***fremontii*** (p. 433) **FR** widely ovate to ± round or obovate; stalk above receptacle ± 0; style 0.5–0.8 mm. $2n=64$. Sandy washes, barren knolls, gravelly soils; < 1600 m. Range of sp. ❀TRY.

var. ***stipitatum*** Rollins **FR** widely obovate to ± round; stalk above receptacle ± 1 mm; style ± 1 mm. Uncommon. Rocky slopes, ridges; 1200–1500 m. W&I, ne DMtns (Last Chance Mtns), DSon.

L. heterophyllum Benth. Per. **STS** several, often decumbent, simple or branched above, 1.5–4.5 dm; hairs spreading. **LVS**: basal petioled, oblanceolate to elliptic, entire to sparsely dentate; cauline sessile, sagittate, clasping st, oblong to narrowly lanceolate, dentate, glabrous to hairy. **FL**: petals 2.5–3 mm, spoon-shaped, white, claw narrow; anthers purple. **FR** 4–7 mm, 3–6 mm wide, ovate-oblong; hairs 0 or few, scale- or sac-like; pedicel spreading, 3–6 mm, ± flat, hairs dense; style ± 1 mm, gen > notch. Uncommon. Fields, roadsides, open slopes; 1000–1500 m. CaR; to B.C., e N.Am; native to Eur.

L. jaredii Brandegee (p. 433) JARED'S PEPPERGRASS Ann 1–7 dm. **ST** erect; branches 0 to few, slender, hairy or not. **LF** 3–10 cm, 0.5–1 mm wide, lanceolate, entire to sparsely dentate. **FL**: sepals ± 2.5 mm, yellow, hairs few; petals ± 3 mm, yellow, fading whitish; stamens 6. **FR** 3–4 mm, widely ovate, glabrous to densely papillate; pedicels ascending to spreading, cylindric, lower 4–5 × fr; style 0.5–1.2 mm, stigma exceeding fr. $2n=16$. RARE. Alkali bottoms, slopes, washes; < 500 m. SCoRI, SnJV.

L. lasiocarpum Torrey & A. Gray var. ***lasiocarpum*** (p. 433) Ann; hairs spreading, rigid. **ST** prostrate to erect, 1–2(3) dm, gen branched from base. **LVS** 2–15 cm, oblanceolate to oblong, toothed or with obovate to oblong lobes; upper sometimes subentire. **FL**: sepals 1–1.5 mm, widely oblong, purplish, margin thin, white, lower side gen hairy; petals minute or 0; stamens 2, 4 or 6. **FR** 2.5–4 mm, (ob)ovate to oblong-elliptic, gen hairy on surfaces or margins; notch < 1/5 × seed pouch; valve tips gen winged; pedicel gen < fr, > 2 × wider than thick, hairs on gen both sides, or 0 on lower; style 0. Dry flats, washes, roadsides, sagebrush; < 600 m. CCo, SW, D; to Colorado, AZ, Baja CA. [var. *georginum* (Rydb.) C. Hitchc. misapplied] 3 other vars. outside CA.

L. latifolium L. Per 4–10(20) dm, ± glabrous, grayish, rhizomed. **LVS**: basal < 3 dm, 6–8 cm wide, toothed, long-petioled; cauline reduced but many 1–4 cm wide, lower petioled, upper sessile. **INFL**: panicle; hairs sparse or 0. **FL**: sepals < 1 mm, margins wide, white; petals white; stamens 6. **FR** ± 2 mm, ± round; notch 0; hairs sparse; pedicel >> fr, slender, cylindric, hairs sparse or 0; style 0. $2n=24$. Beaches, tidal shores, saline soils, roadsides; < 1900 m. CA (exc KR, D); widespread US; native to Eurasia.

L. latipes Hook. Ann. **ST** 3–10(25) cm, stout; hairs dense. **LF** 5–10 cm, linear or with linear segments. **INFL** in fr very dense, 2–5 (12) cm, 1–1.5 cm wide. **FL**: sepals ± 1.3 mm, oblong-ovate, densely covered to fringed with spreading hairs; petals 2–4 mm, oblong, greenish, ciliate, hairs sparse on lower side; stamens 4. **FR** 5–7 mm, oblong-ovate; hairs gen large, flat and minute, cylindric; valve tip wings acuminate; notch > 1/3 seed pouch, gen very narrow; pedicel ascending, 2–3 mm, very flat. Alkaline soils; < 800 m. NCo, NCoR, GV, CCo, SnFrB, SCoRI, SCo; Baja CA.

var. ***heckardii*** Rollins HECKARD'S PEPPERGRASS **ST** erect, branched above base; nodes spaced. **LVS**: basal 0 or shed early; cauline entire. **INFL** in fr > basal lvs. RARE. Alkaline flats; < 200 m. s ScV (Yolo Co.).

var. ***latipes*** (p. 433) DWARF PEPPERGRASS **ST** prostrate or decumbent, branched at or near base; nodes not spaced. **LVS**: basal gen present, entire to pinnately dissected, segments dissected or not; cauline entire. **INFL** in fr gen < basal lvs. UNCOMMON. Alkaline soils, fields, vernal pools, grasslands; < 800 m. Range of sp.

L. montanum Nutt. Bien to shrub-like, rounded; root crown simple or branched. **STS** gen ± erect, 1–several, (1)2–4 dm; branches 0 or many. **LVS**: basal 3–15 cm, pinnately lobed, lobes gen dentate or dissected; cauline reduced, upper gen entire, petioled to not. **INFL** 2–4 cm, many-fld. **FL**: sepals 1.2 mm, hairs 0 to sparse; petals ± 2 mm, white to pale cream-colored; stamens (2)6. **FR** 2.5–4 mm, ± ovate, glabrous; pedicel ± cylindric, slender; style > notch, 0.3–1 mm. Sandy, gravelly, often saline soils; 800–2100 m. CaR, GB, DMoj; to OR, MT, AZ, n Mex. 11 often indistinct vars.

var. ***canescens*** (Thell.) C. Hitchc. Bien, sometimes fl 1st year; hairs of sts, pedicels sparse to dense, long, linear, wavy. **LVS**: basal pinnately divided, segments entire, ± 1.5 mm. **FR** ± 2.5 mm, ovate, glabrous, narrowly wing-margined above; notch minute; pedicel 5–8 mm, slender. Saline flats, sagebrush, greasewood; 1300–2100 m. CaR, MP; to OR, UT.

var. ***cinereum*** (C. Hitchc.) Rollins Per, rarely bien; hairs of sts, pedicels ± dense, minute, short, oblong. **LVS**: basal pinnately lobed, lobes often dentate; upper gen linear, entire. **FR** hairy or not. Sandy areas, saline soils, ravines; 800–1900 m. DMoj; NV. [ssp. *c.* C. Hitchc.]

var. ***montanum*** Per (± bien outside CA), ± woody at base or not; hairs of sts, pedicels sparse to dense, short, club-shaped, scale-like, or ± 0. **LVS**: basal pinnately divided, segments lobed or toothed. **FR** 2.5–3 mm, elliptic to ovate, glabrous. Clay or gravelly soils, washes, greasewood, saltbush; 1400–1850 m. GB; to AZ, MT.

L. nitidum Torrey & A. Gray Ann. **ST** erect to spreading, 1–4 dm, slender, puberulent near infl, ± glabrous below; branches 0–many. **LVS**: basal 3–10 cm, deeply pinnately divided, segments 6–14, linear, entire to coarsely toothed or lobed; cauline less divided to simple, entire. **FL**: sepals ± 1 mm, ovate, hairy or not, not persistent in fr; petals < 1.5 mm, spoon-shaped, white; stamens gen 6. **FR** 2.5–4(6) mm, ovate-elliptic to ± round, smooth, shiny; notch narrow, 0.2–0.5 mm deep; hairs 0, rarely few, marginal, minute; pedicel ± recurved to ascending, very flat, ± densely puberulent; style 0. Alkaline soils, flats, slopes; < 1500 m. CA (exc e D); to c US.

var. ***howellii*** C. Hitchc. **ST**: hairs dense. **INFLS** many, dense. **FR** gen fringed with hairs; valves 3.5–5 mm, tips erect or ± ascending, not beak-like. Slopes, flats; 750–1500 m. w DMoj.

var. ***nitidum*** (p. 433) **ST**: hairs 0 to ± dense. **FR** gen glabrous; valves 3.5–6 mm, tips erect or ± ascending, not beak-like. Meadows, alkaline flats, vernal pools; < 1500 m. Range of sp.

var. ***oreganum*** (E. Greene) C. Hitchc. **ST**: hairs 0 to ± dense. **FR** glabrous; valves ± 3.5 mm, tips ascending, ± beak-like. Dry lake beds, alkaline soils, rain pools, flats, valleys; < 400 m. NCo, GV, CCo, SCoRO; OR. Possibly a hybrid involving *L. dictyotum.*

L. oblongum Small Ann. **ST** prostrate to ascending, 0.5–2(3) dm, gen branched from base; hairs long, weak. **LF** oblanceolate to obovate, pinnately lobed to 2-pinnate; segments gen 1–3 mm. **FL**: sepals ± 1 mm, hairy on lower surface; petals 0 or vestigial, linear; stamens 2. **FR** 2–3.3 mm; notch gen open, 1/8–1/6 fr; veins 0 or obscure, netted; valve tips erect, unwinged; pedicel gen flat, wing-margined; style 0. Bluffs, slopes, disturbed areas, roadsides; < 500 m. GV, SW; to c US, C.Am.

var. ***insulare*** C. Hitchc. **FR** gen 2–2.5 mm, obovate to round; hairs on margin; pedicel hairs gen on both surfaces. Bluffs, low hills, slopes; < 200 m. s CCo, SCo, ChI; Baja CA.

var. ***oblongum*** **FR** 2.5–3.2 mm, ± elliptic to obovate-elliptic; hairs 0 or on margin; pedicel hairs 0 or on upper surface. Disturbed areas, roadsides, slopes; < 500 m. Range of sp.

L. oxycarpum Torrey & A. Gray (p. 433) Ann, ephemeral. **ST** erect to spreading, 1–1.5 dm, branched gen from base, slender; hairs spreading or 0. **LVS** linear or with linear segments; basal 2–6 cm, simple, entire to pinnately lobed or divided, hairs sparse or 0; cauline gen linear, entire. **FL**: sepals ± 0.5–1 mm; petals 0 or ± 0.5–1 mm, white; stamens gen 4. **FR** 2.5–3.5 mm wide, ovate, glabrous; valve tips gen acute, ascending, notch V-shaped; pedicel ascending to ± recurved, ± 2.5 mm, = or > fr, flat, narrowly wing-margined; style 0. Fields, vernal pool margins, alkaline flats; < 600 m. GV, CW; probably alien in B.C.

L. perfoliatum L. (p. 433) Ann, bien. **ST** (1)2–6 dm, ± puberulent below, glabrous above; branches 0–many. **LVS**: basal 2–3-pinnately lobed or divided, segments linear; upper cauline ovate to round, entire or minutely dentate, surrounding st. **FL**: petals ± 1.5 mm, narrowly spoon-shaped, yellowish; stamens 6. **FR** ± 4 mm, diamond-shaped-ovate; hairs 0, rarely few; pedicel gen > fr, slender, cylindric, glabrous; style ± 0.2 mm, ± = notch. 2*n*=16. Roadsides, fields; 600–1500 m. CaR, GV, SCoRI, GB, D; widespread N.Am; native to Eurasia.

L. pinnatifidum Ledeb. Ann, bien, per. **ST** gen erect, 1, 2–5 dm; branches above, rarely near base; hairs 0 or above, minute, simple. **LVS**: basal pinnately divided; cauline entire or lower pinnately lobed to dentate. **FL**: sepals ± 1 mm; petals vestigial, white, or 0; stamens 2 or 4. **FR** 1.8–2 mm, widely elliptic to ± round; hairs sparse, weak; pedicel ± ascending, slender, cylindric, hairy; style < 0.2 mm, ± = notch. Disturbed areas; < 200 m. CCo, SCo (expected elsewhere); native to Eurasia.

L. ramosissimum Nelson var. ***bourgeauanum*** (Thell.) Rollins Bien. **ST** erect, 1.5–6 dm, leafy; branches many, suberect; hairs sparse to dense, fine, ± reflexed. **LVS**: basal pinnately lobed, lobes ± dentate; upper cauline linear, entire, rarely minutely dentate. **INFL**: upper bractless, overtopping axillary ones. **FL**: sepals ± 1 mm; petals 0 or ± vestigial, white; stamens 2. **FR** 2.5–3 mm, 1.5–2 mm wide, ovate to obovate, glabrous; valve tips ± acute; notch ± 0.2 mm; pedicel spreading to ascending, = or > fr, ± flat, densely puberulent; style 0. **SEED** not margined. 2*n*=32. Roadsides, alkaline soils; < 1000 m. CaR, n ScV, SW; widespread N.Am. [*L. densiflorum* var. *b.* (Thell.) C. Hitchc.] Possibly alien in CA; 3 vars. total.

L. strictum (S. Watson) Rattan (p. 433) Ann, prostrate; hairs spreading, simple, pointed. **ST** branched from base, 5–20 cm. **LVS**: basal, lower cauline 1–2-pinnately lobed or divided, 3–7 cm, 1–2 cm wide, segments gen oblanceolate or oblong; upper cauline gen lobed to ± entire. **INFL** many, crowded. **FL**: sepals 1–1.5 mm, persistent in fr; petals 0 or vestigial; stamens 2. **FR** 2.5–3.5 mm, 2–3 mm wide, ovate to oblong-ovate, glabrous or sparsely ciliate; veins prominent, netted; valve tips ascending, gen winged, gen rounded; notch open, ± 0.4 mm; pedicel < fr, ± 2 × wider than thick; style 0. 2*n*=±32. Uncommon. Disturbed sites, gen urban; < 300 m. CA (exc D); to OR, Colorado.

L. thurberi Wooton Ann, bien. **ST** erect, 1–6 dm, branched; hairs long, flat, and short, club-shaped. **LVS**: basal 3–6 cm, petioled, pinnately divided, segments lobed or toothed; upper cauline gen pinnately lobed. **FL**: sepals 1–1.5 mm, hairs soft; petals 2–3 mm, white; stamens 6. **FR** 2–3 mm, 2–2.5 mm wide, elliptic to round, glabrous; pedicel gen < 3 × fr; style 0.4–0.7 mm, > notch. Saline flats, clay soils, grassland; < 1000 m. DMoj; to NM, Mex.

L. virginicum L. Ann; hairs 0 to dense. **ST** erect; branches above, gen 0 below. **LVS**: basal 5–15 cm, obovate, ± pinnately lobed to dissected; upper cauline reduced, entire, rarely dentate. **FL**: sepals ± 1 mm, hairs 0 or few, on lower surface; petals 1–2(3) mm, obovate, white; stamens 2(4). **FR** 2.5–4 mm, ± round, glabrous; notch shallow, gen > style; pedicel = or > fr, slender, cylindric to flat, puberulent or hairs large, gen 0 on lower surface. **SEED**: embryonic root gen at back of 1 cotyledon. Dry, gen open areas; < 2400 m. CA (exc KR, SNH); widespread N.Am. 9 vars. total, difficult.

var. ***medium*** (E. Greene) C. Hitchc. **ST** 1.5–4 dm, glabrous. **INFL**: pedicel ± flat, gen ± winged, glabrous. **FR** ± 2.5 mm. 2*n*= 32. Dry, disturbed areas, roadsides; < 2000 m. NW, CaR, MP; to WA, se US.

var. ***pubescens*** (E. Greene) Thell. **ST** 2–7 dm; hairs rigid. **LVS**: cauline entire to shallowly dissected. **INFL**: pedicel ± flat, gen ± winged, hairy. **FR** 3–4 mm. Common. Disturbed areas, abandoned fields, meadows, roadsides; < 2400 m. Range of sp.

var. ***robinsonii*** (Thell.) C. Hitchc. **ST** gen 1–2 dm; hairs dense, pointed. **LVS**: cauline divided or lobed, segments narrow, 1–2 mm wide. **INFL**: pedicel ± flat, gen ± winged, hairy. Uncommon. Dry soils, shrubland; < 500 m. SW; Baja CA.

var. *virginicum* **ST** 2–6 dm; hairs sparse to ± dense, minute, rounded. **LVS**: cauline dissected to ± entire. **INFL**: pedicel ± cylindric, hairs sparse. **SEEDS**: embryonic root at edges of both cotyledons. Old fields, roadsides; < 2400 m. Range of sp.; probably alien in CA.

LESQUERELLA BLADDERPOD

Bien, per; hairs ± dense, stellate, rarely simple, often silvery. **ST** gen arising laterally from basal lf cluster. **LVS** simple; basal petioled, linear to round, entire to pinnately lobed; cauline sessile or lower short-petioled, base tapered. **FL**: sepals erect or spreading, oblong to elliptic; petals widely obovate, entire, gen yellow; stamens 6. **FR** elliptic to ± round, plump or ± flat parallel or perpendicular to septum; hairs stellate or 0; pedicel slender, straight or not. **SEEDS** 2–10(14) per chamber, ± round, ± plump or flat; margin gen 0; embryonic root at edges of both cotyledons. ± 95 spp.: gen N.Am, ± 12 S.Am. (L. Lesquereux, Am botanist, 1805–1889) [Rollins & Shaw 1973 Harvard Univ Press: 1–228]

1. Fr ± flat parallel to septum, elliptic to obovate ..... ***L. occidentalis*** ssp. ***occidentalis***

1′ Fr plump or ± flat parallel or perpendicular to septum, elliptic to obcordate or round
- 2. Ann; basal rosette 0; caudex 0 ..... ***L. tenella***
- 2′ Per; basal rosette present; caudex present ..... ***L. kingii***
  - 3. Fr obovate to obcordate, often ± flat perpendicular to septum, sessile or short-stalked above receptacle, top truncate or shallowly notched; valves gen hairy inside; seeds 2–4 per chamber; petals 5.5–9.5 mm . ssp. ***kingii***
  - 3′ Fr ± round or elliptic to obovate, plump or ± flat parallel to septum, often short-stalked above receptacle, top obtuse; valves gen glabrous inside; seeds 2–8 per chamber; petals 9–13 mm
    - 4. Style 6–9 mm, slender; seeds 2–4 per chamber; pl erect; fr ± round; SnBr ..... ssp. ***bernardina***
    - 4′ Style 2–6 mm, stout; seeds 4–8 per chamber; pl prostrate to erect; fr ± round or elliptic to obovate; DMtns ..... ssp. ***latifolia***

L. kingii (S. Watson) S. Watson Per, caudexed; hairs dense, 5–7-rayed. **STS** prostrate, decumbent or erect, few–many, 0.5–1.5(4) dm. **LVS**: basal blades 2–6 cm, widely elliptic or diamond-shaped to round, entire to ± lobed; cauline 0.5–2 cm, elliptic to obovate, lower short-petioled. **FL**: petals 5.5–13 mm. **FR** 3.5–9 mm; valve hairs dense outside, 0 to dense inside; pedicel 5–10(15) mm, gen S-shaped, rarely straight or 1-curved; style (2)4–9 mm. **SEEDS** 2–8 per chamber, flat, not margined. Dry soils, rocky sites; 1500–2750 m. SnBr, SNE, DMoj; to UT, Mex. 4 sspp. total.

ssp. ***bernardina*** (Munz) Munz (p. 433) SAN BERNARDINO MOUNTAINS BLADDERPOD Pl erect, often purplish. **LVS** wavy-margined to shallowly dentate; outer basal lvs diamond-shaped to round, inner elliptic. **FR** ± round, short-stalked above receptacle; valve hairs spreading outside, gen 0 inside; septum entire or not; style 6–9 mm, slender. **SEEDS** 2–4 per chamber. RARE. Dry flats, pine forest; 1850–2150 m. SnBr (Bear Valley).

ssp. ***kingii*** Pl prostrate to erect. **LVS**: basal entire to ± lobed, petioles 2–3 × blades. **FR** obovate to obcordate, ± wider than long, often ± flat perpendicular to septum, often short-stalked above receptacle; top truncate or shallowly notched; valve hairs appressed outside, gen present inside; septum ± not entire. **SEEDS** 2–4 per chamber. Dry, rocky soils, pinyon/juniper woodland; 1500–2750 m. SNE, DMtns; NV. [var. *cordiformis* (Rollins) Maguire & Holmgren misapplied] ❀TRY.

ssp. ***latifolia*** (Nelson) Rollins & E. Shaw Pl prostrate to erect. **LVS**: basal widely elliptic or ± round, petioles = to 2 × blades. **FR** elliptic to round, ± flat parallel to septum, often short-stalked above receptacle; top obtuse; valve hairs dense outside, 0 inside; septum entire; style 2–6 mm, stout. **SEEDS** 4–8 per chamber. 2*n*=10. Gravelly soil, limestone outcrops, ridges; 1500–2300 m. DMtns; to UT, Mex. ❀TRY.

L. occidentalis (S. Watson) S. Watson ssp. ***occidentalis*** (p. 433) Per; hairs dense, 4–7-rayed. **ST** prostrate to erect, many, 0.3–1.5(3) dm, gen unbranched. **LVS**: basal 1–8 cm, wavy-margined or dentate to entire, elliptic to round, narrowed to slender petioles; cauline 0.5–1.5(2.5) cm, oblanceolate, entire or sparsely dentate, lower short-petioled. **FL**: petals 7–9(14) mm. **FR** elliptic to obovate, ± flat parallel to septum, esp near margins, often near tip, often ± beaked; valve hairs dense, appressed or ± spreading outside, 0 or sparse inside; pedicel 5–10(15) mm, C- or S-shaped; style 2– 6.5 mm, often hairy. **SEEDS** 2 per chamber, marginless. 2*n*=10. Gravelly soils, talus, ridges; 1600–3000 m. KR, NCoRH, CaRH, n SNH, MP; to OR, ID. [*L. cusickii* M.E. Jones] 2 sspp. total. ❀DRN,SUN,IRR:1–3,**7,14–17**,18.

L. tenella Nelson (p. 433) Ann; hairs ± dense, 4–7-rayed, some often simple. **STS** several, decumbent to erect, 1.5–6 dm, often stout, much-branched. **LVS**: basal not rosetted, blades (1.5)3–6.5 cm, elliptic, entire to dentate; cauline linear to obovate, lower short-petioled. **FL**: petals 6.5–9(11) mm, yellow to orange. **FR** obovate to round, plump, ± sessile above receptacle; valve hairs sparse, stellate outside, dense, simple or stellate inside; pedicel 5–15 mm, spreading to recurved-S-shaped; style 2–4.5 mm. **SEEDS** 2–6 per chamber, flat, margined. 2*n*=10. Sandy soils, washes, slopes; < 1200 m. D; to UT, Mex. [*L. palmeri* S. Watson misapplied]

LOBULARIA SWEET ALYSSUM

Ann, per. **LVS** simple, entire, narrow; hairs appressed, forked at base. **FL**: petals widely obovate, entire, white to purplish tinged; stamens 6, anthers ovate, filaments flat, not winged. **FR** very flat parallel to septum, ± sessile above receptacle, tardily dehiscent; valves with prominent midvein, lesser netted veins. **SEEDS** 1–5 per chamber, winged or not; embryonic root at edges of both cotyledons. 6 spp.: Medit, Eur. (Latin: small lobe, possibly from hairs)

L. maritima (L.) Desv. (p. 433) Per, woody near base or not, grayish. **ST** often prostrate or decumbent, branched from base, 1–4 dm. **LF** 1–5 cm, 1–4 mm wide, linear to linear-lanceolate; tip acute. **FL** fragrant; sepals 1.5–2 mm; petals 3–4 mm, widely obovate, white; glands at base of short stamens ± 1 mm. **FR** 2–3.5 mm, widely (ob) ovate to round, greenish to brown or purplish; valve hairs gen few; pedicel spreading to ± ascending, 5–10 mm, slender. **SEEDS** 1 per chamber; wing 0; style ± 0.3–0.5 mm. 2*n*=24. Waste places; < 200 m. NCo, CCo, SnFrB, SCo; native to Medit, Eur.

LUNARIA MOONWORT

Ann, bien, per; hairs simple. **LVS** simple, petioled, ± ovate, dentate. **INFL**: bracts 0 or subtending lower fls, lf-like. **FL**: sepals erect, inner pair sac-like at base; petals long-clawed, bluish or lilac to purplish, rarely white; stamens 6. **FR** widely elliptic to round, very flat parallel to septum, long-stalked above receptacle; valves falling; style ± beak-like. **SEEDS** 2 rows per chamber, very flat, winged; stalks fused to septum; embryonic root at edges of both cotyledons. 3 spp.: Eur. (Latin: moon, from shiny septum)

L. annua L. (p. 433) Ann, bien. **ST** 5–10 dm, branched. **LVS**: basal clustered; lower cauline petioled, upper sessile, blades 4–10 cm, bases cordate or reniform to rounded, upper clasping st. **FL**: petals ± 2 cm, bluish purple. **FR** 3.5–4.5 cm, oblong-ovate to ± round; stalk above receptacle 7–12 mm; style 6–8 mm, beak-like. **SEEDS** 3–5 per chamber, (5)7–9 mm wide. Disturbed areas, roadsides, urban areas; < 200 m. SnFrB, SCo, expected elsewhere; native to Eur.

LYROCARPA

Ann, per, woody at base or not; hairs dense ± throughout, multibranched. **ST** branched. **LVS** petioled, deeply pinnately lobed to wavy-margined. **INFL** open, elongate. **FL** fragrant; calyx cylindric, sepals erect, converging above, << petals, outer pair ± sac-like at base. **FR** obcordate to lyre-shaped, flat perpendicular to septum. **SEEDS** 3–10 per chamber; embryonic root at edges of both cotyledons. 3 spp.: sw US, Mex. (Greek: lyre fr, from fr shape)

L. coulteri Hook. & Harvey var. ***palmeri*** (S. Watson) Rollins (p. 433) COULTER'S LYREPOD Per; caudex ± woody, irregularly branched. **STS** several–many, 3–8 dm, straw-colored below, grayish above. **LF** 1–15 cm, gen < 1 cm wide, linear to ± ovate; lobes linear to oblong. **FL**: sepals 8–11 mm, ± 1.5 mm wide; petals 1.5–2.5 cm, linear to oblanceolate, brownish to dull purple, blades 1–3 mm wide, gen twisted, tip acute to acuminate. **FR** 1–2 cm, ± as wide as long; pedicel spreading or ascending, 3–7 mm; style ± 0, stigma with 2 large spreading lobes. **SEEDS** 3–5 per chamber, 2–3 mm, round; wing 0. UNCOMMON. Dry slopes, gravelly flats, washes; < 600 m. DSon; Baja CA.

MALCOLMIA

Ann, per; hairs simple or gen branched. **ST** prostrate to erect, gen branched. **LF** simple to 2-pinnate. **INFL** bracted or not. **FL**: sepals erect, inner pair gen sac-like at base; petals prominently clawed, reddish or purple, rarely white. **FR** linear, sessile above receptacle, dehiscent; style tapered or 0; stigma lobes decurrent, gen fused. **SEEDS** many, 1 row per chamber; margins 0. ± 35 spp.: s Eurasia, n Afr. (W. Malcolm, London nurseryman, 1769–1820)

M. africana R.Br. Ann, often ± prostrate, stiff; branches many. **ST** 1.5–5 dm; hairs dense, small. **LVS** simple; lower 3–6 cm, oblanceolate, sparsely dentate, petioled; upper reduced, graduated to bracts. **FL**: petals 6–9 mm, rose-violet to pink. **FR** ascending, 4–6 cm, 1–1.5 mm wide, cylindric to ± 4-sided, ± narrowed between seeds; pedicel 1–2 mm; style 0, stigma lobes pointed. 2*n*=28. Disturbed areas, desert shrubland; 1250–2000 m. SNE; to Rocky Mtns; native to Medit.

MATTHIOLA STOCK

Ann, per, gen grayish; hairs dense, multibranched, rarely also simple or glandular. **STS** 1–few, branched. **LVS** simple, entire to dentate. **FL**: sepals erect, inner pair sac-like at base; petals long-clawed; filaments often expanded at base, appendages 0; stigma with a deep, decurrent lobe above each valve. **FR** linear, cylindric or ± flat. **SEEDS** 1 row per chamber, winged. ± 50 spp.: Eur, Afr, Asia. (P.A. Matthioli, Italian physician, botanist, 1500–1577)

M. incana (L.) R.Br. Per; base often woody. **ST** erect, < 8 dm. **LF** linear-oblong to oblanceolate, thickish, ± entire; tip rounded. **FL** fragrant; sepals 9–15 mm; petals 2–2.6 cm, reddish to violet, rarely white. **FR** < 15 cm, ± flat, ± narrowed between seeds; style < 4 mm, stigma with short horns or bumps or not. $2n=14$. Sandy areas, beaches, ocean bluffs; < 100 m. s NCo, CCo, SCo; Mex; native to Eur.

PHOENICAULIS

1 sp. (Greek: visible stem)

P. cheiranthoides Torrey & A. Gray (p. 437) Per, ± cespitose; taproot thick, deep; caudex branches covered with old lf bases. **ST** ± scapose, 0.5–2 dm. **LVS** simple, entire, gray or whitish; hairs dense, multibranched; basal densely clustered, 3–15 cm, petioled. **INFL** many-fld. **FL**: sepals erect, 3–6 mm, often pink or purplish, esp on margin; petals 6–15 mm, obovate-oblanceolate, blade narrow, clawed, pinkish to reddish purple; anthers 1.5–2 mm. **FR** ± spreading, 1.5–8 cm, 2–5 mm wide, ± lanceolate, flat parallel to septum, glabrous; pedicel ± spreading, 1–3 cm; valve midrib prominent; style ± 1 mm, stigma ± unlobed. **SEEDS** 3–6, 2 rows per chamber, 3–4 mm, smooth; wing 0; embryonic root at edges of both cotyledons, rarely at edge of one. Common. Basalt outcrops, clay soils, slopes; 1500–3200 m. NW, CaR, GB; to WA, ID. [ssp. *glabra* (Jepson) Abrams] ❀TRY;DFCLT.

PHYSARIA DOUBLE BLADDERPOD

Per, silvery; taproot enlarged; hairs stellate, gen ± sessile, appressed. **STS** several–many, gen from below rosette; outer prostrate to decumbent. **LVS**: basal obovate to ± round, < petioles; cauline entire, upper gen sessile. **FL**: petals yellow. **FR** widely obovate to ± round, inflated; tip notched to ± middle, gen also notched from base; valves shed with seeds enclosed; style persistent. **SEEDS** 2–6 per chamber, ± flat; margins 0; embryonic root at edges of both cotyledons. 22 spp.: w N.Am, gen Rocky Mtns, Great Basin. (Greek: bellows, from inflated fr)

P. chambersii Rollins (p. 437) Pl cespitose. **ST** unbranched, 5–15 cm. **LVS**: basal 3–6 cm, 1–2 cm wide, entire to dentate, tip obtuse; cauline 1–2 cm, spoon-shaped, tip gen acute. **INFL** 2–10 cm, dense, ± umbel-like. **FR** 1–1.5 cm, hairy; notch in base shallow or ± 0; septum 4–6 mm; pedicel ascending to spreading, 8–15 mm; style 6–8 mm, exserted from notch. **SEEDS** gen 4 per chamber. $2n=8$, 10,16,24. Limestone soils; 1500–2500 m. n DMtns (Clark, Grapevine mtns); to OR, UT, AZ. ❀TRY:DFCLT.

POLYCTENIUM

1 sp. (Greek: many combs, from lvs)

P. fremontii (S. Watson) E. Greene Per, cespitose, often glaucous. **STS** erect, few–several, gen simple, 5–15 cm; caudex without persistent lf bases, branched. **LF** 1–2 cm, ± sessile, deeply pinnately lobed to compound, rigid; lobes or lflets ± 1–5 mm, linear, often sharp-pointed with terminal hair; hairs stiff, gen branched. **INFL** ± dense, ± umbel-like; bracts 0. **FL**: sepals erect, 2–3 mm, oblong; petals 5–6 mm, white to pale purple. **FR** ascending to erect, often flat perpendicular to septum, glabrous; pedicel ascending, 4–6 mm, straight. **SEEDS** 12–28, 1 row per chamber, ± 1 mm, oblong; wing 0; embryonic root at back of 1 cotyledon. Saline soils, playas, lake margins, wet meadows; 1000–2000. GB; to ID, NV.

1. Fr oblong, 4–6 mm, 1.5–2.3 mm wide, tip rounded; style << 0.5 mm .... var. ***confertum***
1′ Fr linear, 6–13(20) mm, 1–1.5 mm wide, tip tapered; style 0.5–1 mm .... var. ***fremontii***

var. ***confertum*** Rollins **FR** flat perpendicular to septum. Uncommon. Playas, lake margins; ± 2000 m. SNE; NV. ❀TRY.

var. ***fremontii*** (p. 437) **FR** ± flat perpendicular to septum to ± cylindric. Uncommon. Playas, wet meadows; 1000–1700 m. MP; to ID. ❀TRY.

RAPHANUS WILD RADISH

Ann, bien, erect, gen scabrous, taprooted; hairs simple, rigid. **ST** gen branched, esp above. **LVS**: basal and lower cauline petioled, gen pinnately lobed to compound, terminal lobe or lflet widely ovate to round, >> lateral, toothed; upper cauline short-petioled to sessile, ± dentate or few-lobed. **INFL** many-fld; bracts 0. **FL**: sepals erect, inner pair sac-like at base; petals long-clawed. **FR** indehiscent, longitudinally grooved, esp below, transversely jointed; lower part seedless, very short, or 0; upper part seeded, linear to ovate or dagger-shaped, beaked. **SEEDS** 1 row per chamber, ± spheric; wing 0; cotyledons doubly folded. 3 spp.: Medit. (Greek: appearing rapidly, from seed germination)

1. Fr very narrowed, often breaking between seeds, 2–12-seeded, upper part ± uniform in width, grooved; petals yellowish fading white or whitish, often dark-veined or -tinged .... ***R. raphanistrum***
1′ Fr not or ± narrowed, not breaking between seeds, 1–3(5)-seeded, upper part widest below middle, not to ± grooved; petals gen ± purple, rarely white, gen dark-veined .... ***R. sativus***

R. raphanistrum L. (p. 437) JOINTED CHARLOCK Pl hairs sparse. **ST** gen 1, 3–8 dm. **LVS**: basal 6–20 cm. **FL**: petals 15–20 mm. **FR** 4–8 cm (incl beak), 3–6 mm wide; pedicel ascending, 10–25 mm. $2n=18$. Disturbed areas, fields, roadsides; < 800 m. CA (exc D, mtns); N.Am; native to Medit Eur.

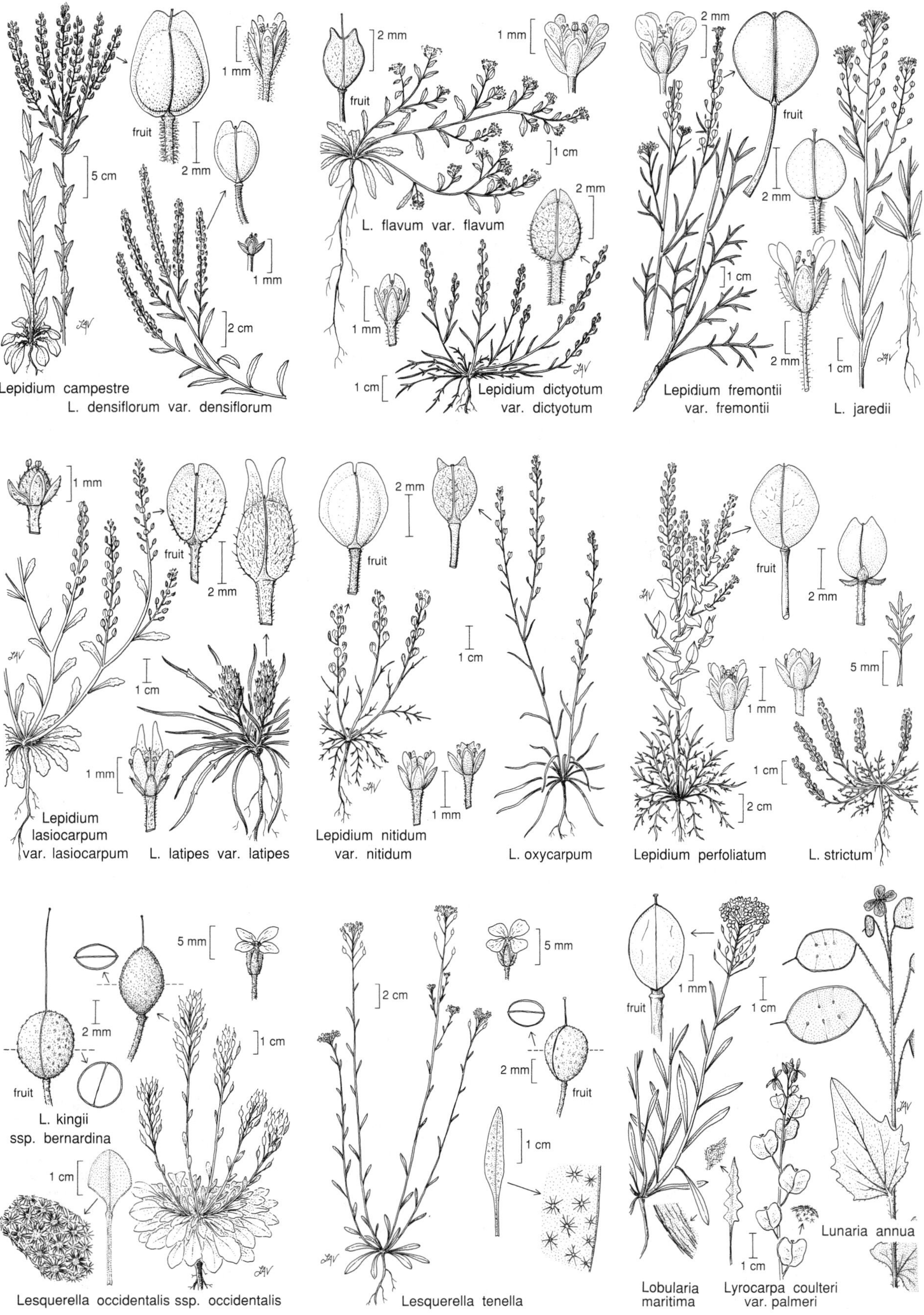
fruit
5 cm
2 mm
1 mm
2 cm
Lepidium campestre
L. densiflorum var. densiflorum
L. flavum var. flavum
1 cm
Lepidium dictyotum
var. dictyotum
Lepidium fremontii
var. fremontii
L. jaredii
Lepidium
lasiocarpum
var. lasiocarpum
L. latipes var. latipes
Lepidium nitidum
var. nitidum
L. oxycarpum
5 mm
Lepidium perfoliatum
L. strictum
L. kingii
ssp. bernardina
Lesquerella occidentalis ssp. occidentalis
Lesquerella tenella
Lobularia
maritima
Lyrocarpa coulteri
var. palmeri
Lunaria annua

R. sativus L. (p. 437) RADISH Pl hairs sparse, rarely 0. **ST** 1, 4–12 dm. **LVS**: basal 10–20 cm. **FL**: petals 15–25 mm. **FR** 3–6 cm (incl beak), 5–10 mm wide; pedicel spreading to ascending, 10–25 mm. 2*n*=18. Disturbed places, fields, roadsides; < 1000 m. CA-FP (exc mtns); temp worldwide; native to Medit Eur.

RAPISTRUM WILD TURNIP

Ann, per; hairs simple. **LVS**: lower petioled, pinnately lobed to compound; upper sessile or short-petioled, simple, subentire to dentate. **INFL** dense; bracts 0. **FL**: petals obovate, clawed. **FR** elliptic to widely oblong, breaking transversely into 2 indehiscent segments, sessile above receptacle; pedicel slender to thick; valves rigid, gen wrinkled. **SEEDS** gen 1–3, oblong or ovate, ± flat; cotyledons folded lengthwise. 2 spp.: Eur. (Latin: turnip-like)

R. rugosum (L.) All. (p. 437) Ann, erect; hairs stiff. **ST** 3–10 dm, branched. **LF** oblanceolate to (ob)ovate, scabrous. **FL**: sepals 2.5–3 mm; petals 5–9 mm, yellow, dark-veined. **FR** hairy or not; lower part 2.5–3.5 mm, upper 2.5–4 mm, ± round, ribbed; pedicel erect, appressed, 2.5–3 mm, tip thick; style 1.5–3.5 mm, stigma 2-lobed. 2*n*=16. Disturbed areas, fields; < 200 m. SnFrB; to OR, e US, Mex; native to s Eur.

RORIPPA YELLOW or WATER CRESS

Ann, bien, per; hairs simple, rarely sac-like. **STS** 1–many, prostrate to erect, branched or not, often from center of basal rosette, rooting at nodes or not. **LVS**: basal, lower cauline sessile or short-petioled, entire to pinnately compound; cauline reduced upward. **INFL** terminal and lateral. **FL** gen small; petals 0 or obovate to narrowly spoon-shaped, pale to bright yellow or white. **FR** linear to round, plump; valves 2 or 4(6); pedicel ascending to recurved, slender, gen with 2 minute glands at base; style 0 or prominent, persistent. **SEEDS** 10–200, (1)2 rows per chamber, dense, gen plump; wing 0; embryonic root at edges of both cotyledons. ± 75 spp.: worldwide. (Old Saxon: for these, perhaps other crucifers) [Stucky 1972 Sida 4:277–340]

1. Petals white; cauline lvs 1-pinnate; pl in aquatic to very wet places, rooting at nodes or not
  2. Lateral lflets prominently dentate; terminal lflet of lvs on fertile st ± = lateral; lower pedicels often bracted; lflets 7–13; pedicel junctions flat .... ***R. gambellii***
  2′ Lateral lflets ± entire or wavy-margined; terminal lflet of lvs on fertile st > lateral; pedicels not bracted; lflets gen 3–7; pedicel junctions not flat .... ***R. nasturtium-aquaticum***

1′ Petals yellow; cauline lvs simple, entire to deeply pinnately lobed, rarely 1-pinnate; pl often in wet but gen not aquatic places, rarely, if ever, rooting at nodes
  3. Per, often from creeping underground roots, rhizomes, or coarse lateral root system
    4. Cauline lvs entire to dentate, not lobed, sessile, clasping st; seeds gen 0 .... ***R. austriaca***
    4′ Cauline lvs deeply dentate to deeply pinnately lobed, rarely 1-pinnate, sessile to short-petioled; seeds present
      5. Fr narrowly oblong; sepals not persistent in fr; pedicel ascending to recurved, 5–11(15) mm .... ***R. sinuata***
      5′ Fr ± round to widely oblong; sepals persistent in fr; pedicel erect to ascending, 3–8 mm
        6. Fr hairs dense; fr stigma ± expanded; axillary infls elongate; sepals, style hairy .... ***R. columbiae***
        6′ Fr hairs ± 0; fr stigma ± unexpanded; axillary infls ± umbel-like to ± elongate; sepals, style glabrous .... ***R. subumbellata***
  3′ Ann, bien, rarely ± short-lived per, gen from taproot
    7. Fr ± as long as wide, 1–2.5 mm .... ***R. sphaerocarpa***
    7′ Fr gen > 2 × longer than wide, 1.5–15 mm
      8. Fr linear to oblong, not narrowed near middle, < 1.5 mm wide .... ***R. curvisiliqua***
      8′ Fr oblong to ± round, often ± narrowed near middle, gen > 1.5 mm wide
        9. Pl 4–10(14) dm, ± erect, with 1 dominant st from base .... ***R. palustris***
          10. Fr ovate to round, 2–6 mm; lf hairy both surfaces; st hairs ± dense .... var. ***hispida***
          10′ Fr oblong, 7–15 mm; lf hairs 0; st hairs 0 or below, sparse .... var. ***occidentalis***
        9′ Pl 1–5 dm, prostrate to erect, often with several–many sts from base
          11. Fr rough, minutely papillate, acute at tip, not narrowed near middle .... ***R. tenerrima***
          11′ Fr smooth, glabrous, acute to ± obtuse at tip, gen narrowed near middle .... ***R. curvipes***
            12. Fr ovate to lanceolate, gen 2 × longer than wide, tip acute to ± obtuse .... var. ***curvipes***
            12′ Fr oblong, 2–3.5 × longer than wide, tip ± truncate to widely obtuse .... var. ***truncata***

R. austriaca (Crantz) Besser AUSTRIAN FIELD-CRESS Per. **ST** ascending to ± erect, 4–10 dm, branched above; hairs below, minute. **LVS** glabrous, dull blue-green; cauline sessile, clasping st, 3–10 cm, linear-oblong to oblanceolate, entire to dentate. **FL**: sepals 2–2.5 mm; petals 3–5 mm, yellow. **FR** ± 3 mm, 2.4–3 mm wide, round-ovate; pedicels slender, spreading to ascending, lower 7–15 mm; style 1–2 mm. **SEED** gen 0, 0.7–0.9 mm, reddish brown, finely papillate. 2*n*=16. Disturbed or cult areas, roadsides, mud flats; 1200–2000 m. MP; to Can, e US; native to Eur.

R. columbiae (Robinson) Howell (p. 437) COLUMBIA YELLOW CRESS Per; hairs spreading, ± dense, short, stiff or longer, pliable. **ST** prostrate to erect, gen branched, 1–4 dm. **LVS**: basal and lower cauline gen petioled; middle and upper cauline sessile, ± clasping st, 2.5–5 cm, oblanceolate to oblong, deeply pinnately lobed. **INFL** axillary elongate. **FL**: sepals 2–3.5 mm, ovate to oblong, hairy, persistent in fr; petals 2.5–4 mm, ± tongue-shaped, yellow. **FR** 3–6 mm, widely oblong; hairs dense; pedicels erect to ascending, often appressed, lower 4–8 mm; style 1–3 mm, hairy, stigma ± expanded. **SEED** ± < 1 mm, ± reniform, finely papillate. RARE. Meadows, playas; 1200–1800 m. MP; to WA. [*R. calycina* (Engelm.) Rydb. var. *columbiae* (Robinson) Rollins]

R. curvipes E. Greene Ann, rarely ± short-lived per, gen ± glabrous. **ST** 1–5 dm, branched, often from near base; hairs often below, ± sparse. **LVS**: basal ± entire to deeply pinnately lobed, rarely partly ± 1-pinnate, gen petioled; cauline 0 or 4–8 cm, 5–15 mm wide, oblong, obovate, or oblanceolate, clasping st, entire to ± pinnately lobed. **FL**: petals ascending, oblong to ± tongue-shaped, yellow. **FR** straight to upcurved, gen narrowed near middle, smooth, glabrous; pedicel 2–5 mm; style 0.5–1 mm. **SEED** ± reniform, finely papillate. Wet sites; < 2300 m. CA-FP, SNE; N.Am.

var. ***curvipes*** Pl prostrate to decumbent, rarely ascending. **FL**: petals 0.5–1 mm, gen < sepals. **FR** 1.5–5 mm, gen 2 × longer than wide, lanceolate to ovate; tip acute to ± obtuse; pedicel gently recurved or spreading. Mud flats, stream beds, hillside seeps; 800–2300 m. SN, SNE; to c US.

var. ***truncata*** (Jepson) Rollins Pl ± decumbent to erect. **FL**: petals 1–1.5 mm, ± = sepals. **FR** 3–6(8) mm, 2–3.5 × longer than wide, oblong; tip ± truncate to widely obtuse; pedicel spreading to ascending. Wet streambanks, pond margins, mud flats; 2300 m. CA-FP; to Can, TX, Mex.

R. curvisiliqua (Hook.) Britton (p. 437) Ann, bien. **ST** prostrate to erect, 1–4(5) dm; branches many; hairs 0 or sparse. **LF** 2–7 cm, gen oblong-oblanceolate, ± entire to deeply pinnately lobed; lobes linear to ovate, entire or dentate. **FL** 1–2 mm; sepals spreading; petals yellow. **FR** 6–15 mm, < 1.5 mm wide, linear to oblong, ± cylindric, upcurved or straight; pedicel ascending to ± recurved, 2–4 mm; style 0.3–1 mm. **SEED** finely papillate. Uncommon. Streambanks, seepage areas, lake shores; < 900 m. CA-FP; to AK, Rocky Mtns.

R. gambellii (S. Watson) Rollins & Al-Shehbaz (p. 437) GAMBEL'S WATER CRESS Per. **ST** erect to decumbent, branched, 5–20 dm; hairs 0 to dense, flat. **LF** 1-pinnate; petiole with small lobes ± at base; lflets 7–13, sessile, 5–25 mm, ± linear to ± round, prominently dentate, tip acute. **FL**: petals 6–8 mm, narrowly spoon-shaped, white. **FR** erect to ascending, 15–25 mm, ± 1 mm wide, linear, ± cylindric, ± incurved, ± narrowed between seeds; pedicels ± spreading to ± reflexed, slender, lower often bracted; style 1.5–2 mm. **SEEDS** 1 row per chamber, ± round, plump but ± flat. RARE. Marshes, streambanks, lake margins; < 1250 m. s CCo, SCo; Mex. [*Cardamine g.* S. Watson]

R. nasturtium-aquaticum (L.) Hayek (p. 437) WATER CRESS Per, ± glabrous. **ST** submersed, ± floating, or prostrate on mud, 1–6 dm, rooting at nodes. **LVS** many, 1-pinnate; lflets gen 3–7, widely oblong to ovate, ± entire or wavy-margined. **FL**: petals 3–4 mm, white. **FR** spreading to ± erect, 10–15 mm, narrowly oblong, straight to upcurved; pedicels ± spreading, ± straight, not bracted, junctions not flat, lower 8–15 mm; style < 0.5 mm or 0. **SEEDS** 2 rows per chamber, ± 1 mm wide, ± round. $2n=32$. Streams, springs, marshes, lake margins; < 2700 m. CA-FP; temp worldwide. [*Nasturtium officinale* R.Br.] ❀WET-fresh water:1,**2–9**, 10–13,**14–24**; widely cult for edible greens.

R. palustris (L.) Besser (p. 437) Ann, bien, short-lived per, ± erect, with 1 dominant st from base. **ST** > 3 mm diam; branches 0 or ± from base. **LVS**: basal and lower cauline short-petioled to sessile, ± clasping st or not, 5–20(30) cm, oblong to oblanceolate, irregularly dentate to deeply pinnately lobed. **FL**: petals 1–2.5(3.5) mm, widely spoon-shaped, yellow. **FR** gen > 2 mm wide, straight to ± upcurved; pedicel spreading to ascending, 2–14 mm; style 0.2–1 mm, abruptly attached to fr tip. **SEED** 0.5–1 mm, widely reniform. Gen wet areas; < 2000 m. CA; temp N.Am, Eurasia. 10 vars. total.

var. ***hispida*** (Desv.) Rydb. **ST** 4–10 dm; hairs ± dense. **LF** hairy on both surfaces. **FR** 2–6 mm, widely ovate to round. $2n=32$. Wet woods, ditches, moist depressions; 1200–1500 m. MP; to AK, NM. [*R. islandica* (Oeder) Borbás var. *h.* (Desv.) F.K. Butters & Abbe]

var. ***occidentalis*** (S. Watson) Rollins **ST** 4–10(14) dm; hairs 0 or below, sparse. **LF**: hairs 0. **FR** 7–15 mm, oblong. Stream beds, sand bars, wet depressions; < 2000 m. CA; to NM, Mex, AK. [*R. islandica* (Oeder) Borbás vars. *o.* (S. Watson) F.K. Butters & Abbe, *fernaldiana* F.K. Butters & Abbe]

R. sinuata (Torrey & A. Gray) A. Hitchc. (p. 437) Per. **ST** prostrate to ascending, 1–4(5) dm; hairs sparse to dense, sac-like. **LVS**: basal and lower cauline short-petioled, oblanceolate, deeply dentate to deeply pinnately lobed, rarely 1-pinnate; middle, upper sessile, ± clasping st, 3–6(8) cm, with sac-like hairs on midribs. **FL**: sepals not persistent in fr; petals 3.5–5.5 mm, oblanceolate to spoon-shaped, yellow. **FR** 5–13 mm, narrowly oblong, upcurved; pedicel ascending to recurved, 5–11(15) mm, S-shaped or not; style 1–2 mm. **SEED** 0.7–1 mm, ± widely reniform, angled. Lake shores, playas, wet depressions; 900–1900 m. GB; to Can, c US, TX.

R. sphaerocarpa (A. Gray) Britton Ann. **ST** decumbent to erect, 1–4 dm; branches 0 or from base; hairs below, sparse. **LF** petioled to sessile, 4–10 cm, oblanceolate, deeply ± pinnately lobed; lobes dentate. **FL**: petals 0.5–1 mm, oblong to narrowly spoon-shaped, yellow. **FR** 1–2.5 mm, ± round; pedicel spreading to recurved, 1.5–4 mm, slender; style 0.2–1 mm. **SEED** 0.5–0.7 mm, widely reniform. Uncommon. Lake margins, muddy streambanks; 1200–2450 m. SW; to Rocky Mtns. [*R. obtusa* (Nutt.) Britton misapplied]

R. subumbellata Rollins (p. 437) TAHOE WATER CRESS Per. **STS** several, decumbent, 5–15(20) cm, branched; hairs gen crinkled. **LF** sessile to short-petioled, clasping st or not, 1–3 cm, oblong to widely oblanceolate, wavy-margined to deeply pinnately lobed; hairs 0 or sparse. **INFL** ± umbel-like to ± elongate. **FL**: sepals 2–3 mm, yellowish, glabrous, persistent in fr; petals 2.5–3.5 mm, oblong-oblanceolate to spoon-shaped, yellow. **FR** 3–5 mm, widely oblong to ± round, ± glabrous; pedicel erect to ascending, 3– 6 mm, straight; style 1–1.5 mm, glabrous, stigma ± unexpanded. **SEED** ± 1 mm, plump, ± angled. **ENDANGERED** CA. Sandy lake margins; 1800–2500 m. n SNH (Lake Tahoe Basin); w NV.

R. tenerrima E. Greene Ann, ± glabrous. **ST** prostrate to decumbent, 1–2 dm, branched from base. **LF** short-petioled, 2–5(8) cm, oblong to oblanceolate, gen deeply pinnately lobed; lobe margins entire. **FL**: sepals 0.7–1.2 mm, not persistent in fr; petals 0.6–0.8 mm, ± oblong, yellow. **FR** 3.5–8.5 mm, oblong, gen ± incurved, rough, minutely papillate; tip acute; pedicel ascending, 2–4 mm, sparsely to densely papillate; style 0.5–1 mm. **SEED** 0.5–0.7 mm, widely reniform. Marshes, wet meadows, streambanks; < 300 m. CW, SW; to B.C., c US, Baja CA.

SIBARA

Ann, bien. **STS** 1–several from base, gen erect; hairs 0 or simple or branched. **LVS**: basal and lower cauline pinnately lobed to 1-pinnate; upper often simpler, grayish green. **INFL** elongate, open. **FL** small; sepals ± oblong to ovate, bases gen not sac-like; petals spoon-shaped to ± oblong, white to purplish. **FR** linear, flat parallel to septum or ± cylindric; valve veined or not. **SEED** oblong or ± round, winged or not; embryonic root at back of 1 cotyledon or at edges of 1 or both. 10 spp.: N.Am. (Anagram of Arabis)

1. Seed ± round, 1–1.5 mm wide, wing narrow, conspicuous; fr 1.7–2 mm wide, hairs 0 .... ***S. virginica***
1′ Seed oblong, < 1 mm wide, wing 0 or very narrow, inconspicuous; fr < 1.5 mm wide, hairs sparse or 0
 2. Fr spreading to ± pendent, 1–1.5 cm, hairs sparse .... ***S. deserti***
 2′ Fr ascending, 1.5–4 cm, hairs 0
  3. Pedicel 6–12 mm, straight; basal lvs 0 or shed early; style not expanded toward tip .... ***S. filifolia***
  3′ Pedicel 2–3 mm, upcurved; basal lvs persistent; style expanded toward tip .... ***S. rosulata***

S. deserti (M.E. Jones) Rollins (p. 437) Ann; hairs minute, branched. **ST** 1, 1–3 dm; branches 0 or above. **LVS**: basal shed early; cauline 2–4 cm, hairs sparse; lower cauline deeply pinnately divided, segments 4–8 mm. **FL**: sepals 1.5–2 mm, oblong; petals 2–3 mm, spoon-shaped, white. **FR** spreading to ± pendent, 10–15 mm, < 1.5 mm wide, flat, ± curved; hairs sparse; pedicel spreading to reflexed, 3–4 mm, hairy or not; style 1–1.5 mm. **SEED** < 1 mm wide, oblong; wing 0; embryonic root at edges of both cotyledons. 2*n*=26. Washes, steep hillsides, dry flats; 350–1300 m. n DMoj; NV.

S. filifolia (E. Greene) E. Greene (p. 437) SANTA CRUZ ISLAND ROCK CRESS Ann; hairs 0. **ST** gen 1, 1.5–3 dm; branches above. **LVS**: basal 0 or shed early; cauline 2–4 cm, pinnately lobed, lobes 5–10 mm, linear. **FL**: sepals 2–3 mm, oblong; petals 3–5 mm, spoon-shaped, purplish. **FR** ascending, 25–40 mm, < 1 mm wide, glabrous; pedicel ascending, 6–12 mm, straight; style ± 1 mm, not expanded toward tip. **SEED** ± 1 mm, < 0.5 mm wide, oblong; wing 0 or very narrow, inconspicuous; embryonic root at back or edge of 1 cotyledon. RARE. Coastal scrub; < 100 m. ChI (Santa Catalina, Santa Cruz islands).

S. rosulata Rollins (p. 441) Ann. **STS** 1–few, 1–3 dm; branches above. **LVS**: basal rosetted, 3–5 cm, persistent, deeply pinnately lobed, lobes 4–8 mm, 1–2 mm wide, hairs 0 or sparse, simple or branched; lower cauline ± lobed, upper entire. **FL**: sepals 1.5–2 mm, oblong; petals 2.5–3 mm, narrowly spoon-shaped, white. **FR** ascending, 15–30 mm, < 1.5 mm wide, flat, glabrous; pedicel ascending, 2–3 mm, upcurved, glabrous; style 2–3 mm, expanded toward tip. **SEED** < 1 mm wide, oblong; wing 0; embryonic root at edges of both cotyledons. Sandy or gravelly washes, scree, calcareous rubble; 250–950 m. e DMoj.

S. virginica (L.) Rollins (p. 441) Ann, bien. **STS** 1–several, decumbent to ascending, 1–3 dm, branched; hairs below, simple, rarely forked. **LVS**: basal loosely rosetted, 2–6 cm, pinnately lobed, linear to oblong, hairs esp on petioles, sparse; cauline like basal. **FL**: sepals ± 1.5 mm, gen purplish; petals ± > sepals, spoon-shaped, white. **FR** ascending to erect, 2–2.5 cm, 1.7–2 mm wide, flat, straight, glabrous; pedicel ascending, 2–4 mm; style ± 0. **SEED** 1–1.5 mm wide, ± round; wing narrow, conspicuous; embryonic root at edges of both cotyledons. 2*n*=16. Borders of vernal pools, streambanks, open ground; < 300 m. GV; to e US, Baja CA.

SINAPIS

Reed C. Rollins

Ann, rarely per; hairs 0 or simple, reflexed or spreading. **ST** erect, leafy, ± branched. **LVS**: basal not rosetted, petioled, pinnately lobed or not, terminal lobe > lateral; upper cauline reduced, sessile, base tapered. **INFL** terminal, not flat-topped, elongate in fr; bracts 0. **FL**: sepals spreading, yellowish, bases not sac-like; petals yellow to pale yellow. **FR** cylindric to angled or ± flat, hairy or not; beak prominent. **SEEDS** 1 row per chamber, spheric; wing 0. 6 spp.: Medit, Eurasia. (Latin: mustard, from flavor of seeds)

1. Fr beak ± linear, flat, 1.5–3 cm, hairs of body dense, long, stiff, white; pedicel slender, ± spreading, 5–12 mm ..... ***S. alba***
1′ Fr beak conic, not flat, 0.5–1 cm cm, hairs of body ± 0; pedicel thick, erect or ascending, 3–5 mm ...... ***S. arvensis***

S. alba L. (p. 441) WHITE MUSTARD Ann 2–6 dm; hairs ± sharp, bristle-like, rarely 0. **LVS**: lower pinnately lobed, 1–2 dm, terminal lobe > lateral, coarsely dentate. **FL**: petals 8–11 mm, pale yellow to whitish. **FR** 20–30 mm, narrowed between seeds or not, few-seeded. **SEED** pale yellow. 2*n*=24. Fields, orchards, disturbed areas; < 1000 m. CCo, SCo, expected elsewhere; native to Eurasia. [*Brassica hirta* Moench]

S. arvensis L. (p. 441) CHARLOCK Ann 2–6(–10) dm; hairs at least below, awl-bristle-like. **LVS** coarsely toothed; lower 5–15 cm, obovate, irregularly dissected or lobed; upper simple. **FL**: petals ± 10 mm, ± yellow. **FR** 20–35 mm, often narrowed between seeds, 5–12-seeded. **SEED** reddish brown. 2*n*=16. Grainfields, orchards, disturbed areas; < 500 m. CA-FP (exc SNH); widespread US, native to Eur. [*Brassica kaber* (DC.) Wheeler, incl vars. *pinnatifida* (Stokes) Wheeler, and *schkuhriana* (Rchb.) Wheeler]

SISYMBRIUM

Ann, bien; hairs simple or 0. **ST** gen erect, branched. **LVS** petioled or sessile, variously lobed or dissected, green or glaucous. **INFL** many-fld; bracts 0. **FL**: sepals erect to ± spreading; petals yellow, clawed. **FR** ascending to erect, linear to ± awl-shaped, gen cylindric, straight to ± curved, hairy or not; valves prominently veined; style conic or 0, stigma 2-lobed. **SEEDS** many, gen 1 row per chamber; margin 0; not or ± gelatinous when wet; embryonic root at back of 1 cotyledon, sometimes obliquely so. ± 90 spp.: most continents. (Greek: for various mustards)

1. Fr narrowly awl-like, beaked; pedicel erect to ascending, ± appressed ..... ***S. officinale***
1′ Fr linear, beak 0; pedicel spreading to ascending, not appressed
  2. Upper cauline lvs with lobes or lflets thread-like to linear, terminal ± = lateral; pedicel in width ± = or > fr; outer 2 sepals with erect horns at tip ..... ***S. altissimum***
  2′ Upper cauline lvs ± entire or with lobes or lflets lanceolate to triangular, lanceolate, terminal > lateral; pedicel in width ± = to << fr; sepal horns 0
    3. Pedicel width ± = fr width
      4. Petals 1.5–3 mm; fr 2–4(5) cm, upper cauline lf margins dentate ..... ***S. erysimoides***
      4′ Petals gen 8–10 mm; fr 3–10 cm; upper cauline lf margins entire to few-toothed ..... ***S. orientale***
    3′ Pedicel in width < fr width
      5. St branched from near base; frs gen overtopping fls, gen 3–4 cm; hairs 0 or above, few, ± short ..... ***S. irio***
      5′ St branched esp above; frs gen not overtopping fls, 2–3.5 cm; hairs at least on lower sts, long ..... ***S. loeselii***

S. altissimum L. (p. 441) TUMBLE or JIM HILL MUSTARD Ann. **ST** 30–150 cm; branches many, esp above. **LVS** petioled, < 15 cm, widely lanceolate; basal, lower cauline ± pinnately lobed to 1-pinnate, lobes or lflets ± lanceolate, dentate; upper with thread-like to linear lobes or lflets, terminal ± = lateral. **FL**: sepals ± 4 mm, outer 2 with erect horns at tip; petals 6–8 mm, pale yellow. **FR** 5–10 cm, ± 1 mm wide, linear, rigid, branch-like; beak 0; pedicel gen spreading, 4–10 mm, width ± = or > fr width. **SEEDS** many, ± 1 mm; embryonic root obliquely at back of 1 cotyledon. 2*n*=14. Disturbed areas, fields, roadsides; < 2500 m. CA; N.Am; native to Eur.

S. erysimoides Desf. (p. 441) Ann; hairs sparse or 0. **ST** 10–60 (80) cm. **LF** petioled, pinnately lobed; lobes dentate, terminal > lateral. **FL**: sepals ± 3 mm; petals 1.5–3 mm, slightly < stamens, pale yellow. **FR** 2–4(5) cm, ± 1 mm wide, straight, linear, rigid; pedicel spreading, ± as wide as fr; style ± 1 mm, ± as wide as fr. **SEED** ± 1 mm, ± 0.5 mm wide, oblong, plump; embryonic root at back of 1 cotyledon. Disturbed areas, fields; < 300 m. SCo; native to w Medit.

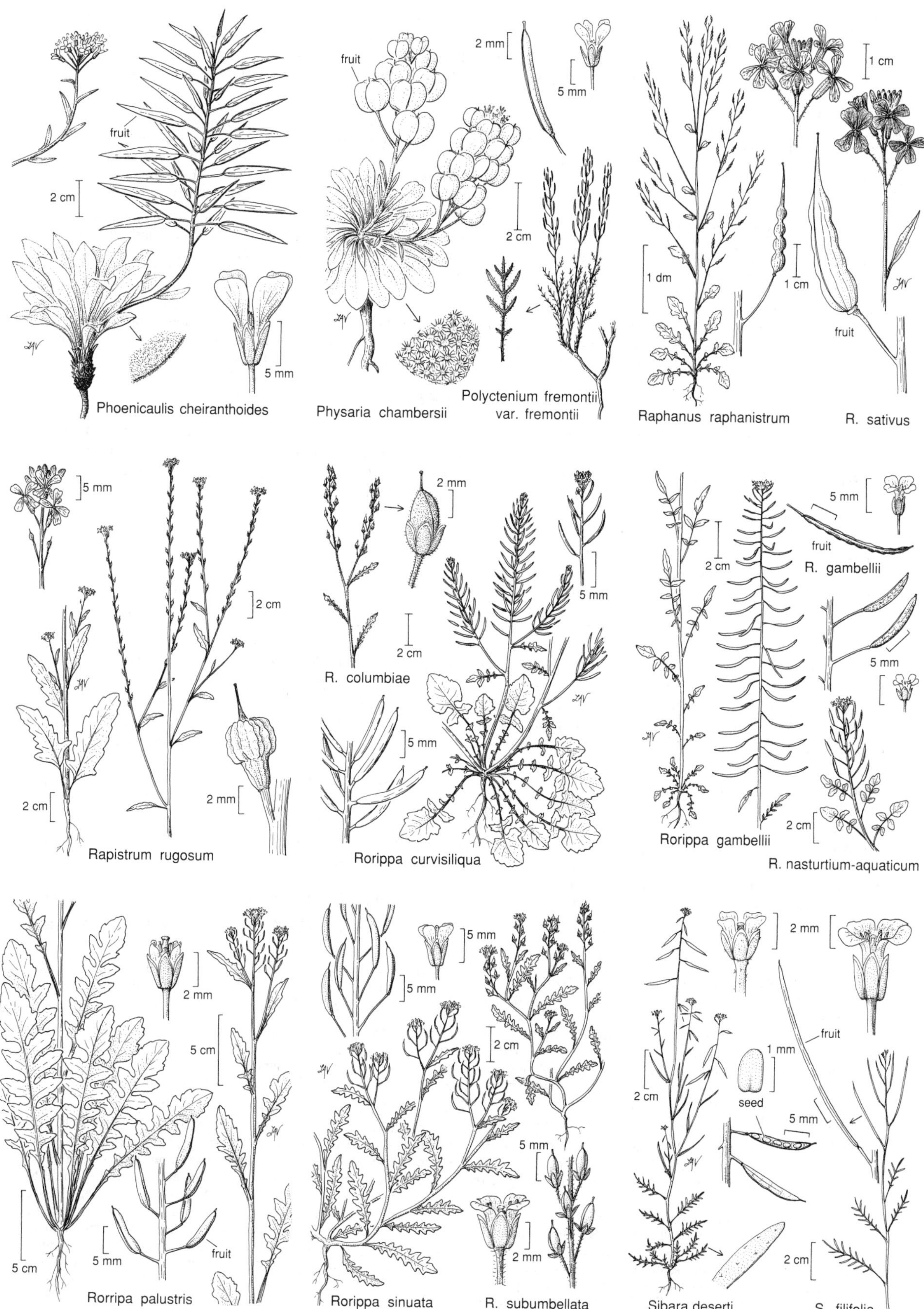
fruit
2 cm
5 mm
Phoenicaulis cheiranthoides
fruit
2 mm
5 mm
2 cm
Physaria chambersii
Polyctenium fremontii
var. fremontii
1 cm
1 dm
1 cm
fruit
Raphanus raphanistrum
R. sativus
5 mm
2 cm
2 cm
2 mm
Rapistrum rugosum
2 mm
2 cm
R. columbiae
5 mm
5 mm
Rorippa curvisiliqua
2 cm
5 mm
fruit
R. gambellii
5 mm
Rorippa gambellii
2 cm
R. nasturtium-aquaticum
2 mm
5 cm
5 cm
5 mm
fruit
Rorripa palustris
5 mm
5 mm
2 cm
5 mm
2 mm
Rorippa sinuata
R. subumbellata
2 mm
fruit
1 mm
seed
5 mm
2 cm
2 cm
Sibara deserti
S. filifolia

S. irio L. (p. 441) LONDON ROCKET Ann. **ST** 15–50 cm, branched from near base; hairs 0 or above, few, ± short, thin. **LVS**: basal not clustered, petioled, pinnately lobed, terminal lobe > lateral, often hastate; upper cauline pinnately lobed to ± entire. **FL**: petals 2.5–4 mm, barely > sepals, narrowly oblong, pale yellow, claws long. **FR** gen overtopping fls, gen 3–4 cm, ± 1 mm wide; pedicel ascending, 5–11 mm, width < fr width; style ± 0.5 mm. **SEED** < 1 mm, oblong, ± papillate; embryonic root obliquely at back of 1 cotyledon. 2*n*=14. Disturbed areas, orchards, roadsides; < 800 m. GV, SW; to TX, Baja CA; native to Eur.

S. loeselii L. (p. 441) Ann. **ST** 40–120 cm; branches esp above; hairs at least below, sparse to dense, spreading or reflexed, long. **LF** petioled, < 1.5 dm, widely deltate-lanceolate, pinnately to irregularly lobed; lobes ± linear to lanceolate, few-toothed. **FL**: sepals 2.5–3.5 mm, lanceolate; petals 6–8 mm, widely obovate, yellow, claws long, narrow. **FR** gen not overtopping fls, 2–3.5 cm, linear, straight or ± incurved; pedicel spreading to ascending, 1–2 cm, width < fr width; style ± 0.5 mm. **SEED** ± 0.7 mm; embryonic root at back of 1 cotyledon, sometimes obliquely so. 2*n*=14. Disturbed areas, fields, roadsides; 1200–2000 m. GB; to Can, n US; native to Eur.

S. officinale L. (p. 441) HEDGE MUSTARD Ann. **ST** 30–80(100) cm, stiff; branches 0 or more often above; hairs spreading or reflexed, sharp. **LVS**: basal < 2 dm, oblanceolate, deeply pinnately or irregularly lobed, terminal lobe ± ovate; cauline reduced, gen sessile, lobes narrow. **FL**: sepals ± 2 mm, oblong-oblanceolate; petals 3–4 mm, pale yellow. **FR** ascending to erect, appressed or not, 8–15 mm, narrowly awl-like, beaked, tardily dehiscent; pedicel erect to ascending, ± appressed, 2–3 mm, stout, club-shaped; style 1–2 mm. **SEED** ± 1.3 mm, plump, variable in shape; embryonic root obliquely at back of 1 cotyledon. 2*n*=14. Disturbed areas, gardens, roadsides; < 2200 m. CA-FP; N.Am; native to Eur.

S. orientale L. (p. 441) Ann; hairs ± soft, of different sizes. **ST** 3 dm, branched. **LVS**: basal clustered, deeply pinnately lobed or compound; cauline lanceolate, with 2 basal, lanceolate, spreading lobes, margins gen entire or few-toothed. **FL**: petals 8–10 mm, pale yellow. **FR** 3–10 cm, linear, straight; beak 0; hairs 0 or sparse; pedicel ascending, 3–6 mm, width ± = fr width; style 1–3 mm, club-shaped; embryonic root at back of 1 cotyledon, sometimes obliquely so. 2*n*=14. Disturbed areas, fields; < 1000 m. CA (exc NW, CaR, SN); to TX, Baja CA; native to Eur.

SMELOWSKIA

Per, ± cespitose; caudex matted with lvs, lf bases. **LVS** soft, pliable, gen 1-pinnate; hairs dense, branched or not; basal petioled; cauline short-petioled to sessile. **INFL** ± umbel-like, dense. **FL** showy; sepal bases not sac-like; petals spoon-shaped, white to purplish. **FR** dehiscent, linear to ovate, ± flat parallel to septum to ± plump, hairy or not; stigma disk-like. **SEEDS** several, 1 row per chamber; embryonic root at back of 1 cotyledon, sometimes obliquely so. ± 10 spp.: w N.Am, n&c Asia. (T. Smielowski, Russian botanist, 1769–1815)

S. ovalis M.E. Jones var. ***congesta*** Rollins (p. 441) LASSEN PEAK SMELOWSKIA Pl deeply rooted; caudex branched underground. **LF**: hairs simple. **FR** 3–6 mm; style 0.5–1 mm. RARE. Loose talus; 2900–3100 m. CaRH (Lassen Peak). Var. *ovalis* in OR, WA. In cult.

STANLEYA PRINCE'S PLUME

Ann, per, shrub, often glaucous; hairs 0 or simple. **ST** 2–15 dm, branched or not. **LVS**: basal clustered or not; cauline petioled or not, entire to deeply lobed. **INFL** dense, gen > 1 dm; buds club-shaped. **FL**: sepals spreading to reflexed, linear-oblong; petals gen conspicuous, yellow to white; filaments ± equal, >> petals. **FR** linear, flat parallel to septum or ± cylindric; stalk above receptacle 1–3 cm; style ± 0 or short. **SEED** oblong; margin 0; embryonic root at edges of 1 or both cotyledons. 6 spp.: w US. (E. Stanley, English ornithologist, 19th century) Concentrates selenium to TOXIC levels, but rarely eaten.

1. Middle and upper cauline lvs sessile, clasping st; lvs entire to few-toothed; petal hairs 0 .... ***S. viridiflora***
1′ Middle and upper cauline lvs petioled; lvs entire to deeply pinnately lobed; petal hairs 0 or on inner side of claws, dense
  2. Lower cauline lvs entire to sharp-toothed; petal hairs 0, blades yellow to whitish .... ***S. elata***
  2′ Lower cauline lvs deeply pinnately lobed; petal hairs on inner side of claws, dense, blades yellow ..... ***S. pinnata***
    3. Shrub; trunk short, 4–8 cm wide; lvs yellow-green .... var. ***inyoensis***
    3′ Per to subshrub; trunk 0, sts several–many from base; lvs gray-green, gen glaucous .... var. ***pinnata***

S. elata M.E. Jones Per, erect, coarse. **STS** 1–several, 6–15 dm, glabrous; branches 0 or above. **LF** petioled, thick; hairs 0 or sparse, minute; blade 8–15 cm, 1.5–2.5 cm wide, widely oblong to lance-ovate, entire to sharply toothed, tip obtuse to acute. **INFL** 6–20 cm, dense. **FL**: sepals reflexed, 8–12 mm, linear-oblong, yellowish, glabrous; petals 8–10 mm, ± 1 mm wide, yellow to whitish, glabrous, claw bases wide, blades reduced, ± 1 mm wide; filament bases enlarged, papillate. **FR** spreading to ± recurved, 5–10 cm, ± cylindric; stalk above receptacle ± 20 mm; pedicel spreading, 5–10 mm, glabrous; style ± 1 mm. **SEED** ± 2 mm, ± 1 mm wide, brown; embryonic root obliquely at back of 1 cotyledon. 2*n*=28. Among boulders, canyons, shrubland; 1300–2000 m. DMtns, SNE; NV.

S. pinnata (Pursh) Britton (p. 441) **STS** several–many, 4–15 dm, glaucous; hairs 0 or sparse; base branched, woody. **LVS** petioled; basal and lower cauline 5–15 cm, 2–5 cm wide, widely lanceolate, deeply pinnately lobed, hairs 0 or short; upper cauline linear-lanceolate to ovate, entire or few-lobed. **INFL** 1–3 dm, dense; buds yellowish. **FL**: sepals spreading or reflexed, 10–15 mm; petals 10–18 mm, yellow, claws with dense, long, wavy hairs on inner side; stamens >> petals, ± equal. **FR** spreading to ± downcurved, 3–8 cm, ± cylindric; stalk above receptacle 10–25 mm; pedicel spreading, 6–12 mm, hairs 0 or few. **SEED** ± 2 mm, oblong, plump; embryonic root at back of 1 cotyledon. Gen open sites, slopes, canyons; < 1850 m. SCo (Conejo Valley), WTR, GB, DMoj; to c US.

var. ***inyoensis*** (Munz & Roos) Rev. Shrub. **ST**: trunk short, 4–8 cm wide. **LF** yellow-green. Sandy areas, creosote bush scrub; 850–1000 m. W&I, n DMoj; NV. ❀TRY;DFCLT.

var. ***pinnata*** Per to subshrub. **STS** several–many from base; trunk 0. **LF** gray-green, gen glaucous. 2*n*=24. Open areas, chaparral, desert shrubland, woodland, seashore dunes; < 1850 m. Range of sp. ❀SUN,DRN,DRY:7–12,14–16,**18**,19–24;DFCLT.

S. viridiflora Torrey & A. Gray (p. 441) Per, glabrous throughout; caudex simple, covered with old lf bases. **ST** 1, erect, 3–12 dm; branches 0 or above. **LVS**: basal many, clustered, petioled, 1–3 dm, oblanceolate to obovate, entire or dentate, rarely dissected; middle and upper cauline sessile, clasping st, lanceolate to ovate, entire to few-toothed. **INFL** 1–5 dm. **FL**: sepals 12–16 mm, linear-oblong; petals 15–20 mm, 2–3 mm wide, lemon-yellow to ± white; stamens >> petals, ± equal. **FR** ± spreading, 4–7 cm, ± cylindric, arched; stalk above receptacle 15–25 mm; pedicel spreading, 4–7 mm, stout. **SEED** 2–3 mm, ± 1.5 mm wide, oblong; wing 0; embryonic root at edges of both cotyledons. 2*n*=24,28. Cliffs, shales, clay knolls, white ash deposits; ± 1300 m. s MP (Lassen Co.); to Rocky Mtns.

STREPTANTHELLA

1 sp. (Latin: small *Streptanthus*)

S. longirostris (S. Watson) Rydb. (p. 441) Ann, gen glaucous; hairs 0 or below, sparse, minute, simple. **ST** gen 1; branches many, rarely 0. **LVS**: basal, lower cauline (2)3–6 cm, narrowly oblanceolate, wavy-margined, entire to dentate, rarely lobed; upper cauline ± linear, clasping st or not, gen entire. **FL**: sepals 2–3 mm, upper surface greenish to purple, lower surface white; petals 3–4 mm, narrowly spoon-shaped, white to yellowish with purple veins, blades wavy; anthers < 1 mm, gen ± exserted. **FR** pendent, 35–45 mm, ± 1.5 mm wide, linear, flat parallel to septum; pedicel down-curved to reflexed, 1.5–3 mm; valves ± 3.5 mm, narrowed to a beak-like, indehiscent tip; style ± 0. **SEED** flat; wing narrow; embryonic root at back of 1 cotyledon. $2n=28$. Common. Sandy soils, desert shrubland, woodland; < 2000 m. s SnJV, SCoRI, SW, D; to WA, Colorado, NM, Baja CA. [var. *derelicta* Howell]

STREPTANTHUS JEWELFLOWER

Roy E. Buck, Dean W. Taylor, and Arthur R. Kruckeberg

Ann to per, glabrous to bristly, gen ± glaucous. **LVS** ± entire to pinnately compound; basal gen rosetted, gen ± petioled; cauline linear to (ob)ovate, often clasping. **INFL** gen ± open; bracts gen 0. **FL** biradial or bilateral; calyx gen ± urn-shaped, sepals erect, gen not green, bases ± pouch-like, gen keeled; petals gen exserted, blade gen narrower than claw, ± channeled, margins ± wavy, gen ± scarious; stamens gen in 3 free pairs; style 0 or short, stigma gen ± entire, blunt. **FR** long, gen strongly compressed parallel to septum. **SEEDS** gen compressed, gen ± winged. ± 40 spp.: sw US, n Mex. (Greek: twisted fl, from wavy-margined petals) [Dolan & LaPré 1989 Madroño 36:33–40; Kruckeberg & Morrison 1983 Madroño 30:230–244] *Caulanthus* sometimes incl here. Calluses on lf margins of some mimic pierid butterfly eggs, reducing larval herbivory. Variable, complex; needs study.

1. Filaments free to base; anthers of longest (upper) stamen pair > 1/2 others (subg. *Pleiocardia*)
 2. At least 1 fl subtended by a bract; cauline lvs gen barely reduced upward (exc *S. gracilis*)
  3. Basal and lowest 3 cauline lvs dissected to deeply lobed; ann — c&s SN
   4. Lf divisions ± thread-like or narrowly linear, gen < 1.5 mm wide; petals yellow, rarely tinged light purple; fr reflexed, 1–1.3 mm wide .... ***S. diversifolius***
   4′ Lf divisions oblong-lanceolate, often > 3 mm wide; petals violet or purplish
    5. Fr ± 3 mm wide; petals white with purple veins; bracts purple; calyx strongly narrowed at tip .... ***S. farnsworthianus***
    5′ Fr < 2 mm wide; petals rose-purple; bracts gen green exc at tip; calyx somewhat narrowed at tip ***S. fenestratus***
  3′ Basal and lowest 3 cauline lvs entire to deeply dentate; ann to per
   6. Fr ± 1 mm wide, straight, ascending; upper cauline lvs reduced .... ***S. gracilis***
   6′ Fr > 2 mm wide, curved, spreading; upper cauline lvs ± not reduced .... ***S. tortuosus***
    7. Branches gen many; lvs conspicuously darker than bracts; ann — NW, SN .... var. ***orbiculatus***
    7′ Branches gen 0–few; if lvs conspicuously darker than bracts, woody base gen strongly developed; ann to per
     8. Sepals bright yellow; petals lacking purple; ann; s SNH .... var. ***flavescens***
     8′ Sepals gen purple; petals ± purple, at least on veins near tip; gen bien or per; widely distributed
      9. Subshrub; sepals 8–10 mm .... var. ***suffrutescens***
      9′ Gen bien (ann to weak per); woody base 0 or weak; sepals < 8 mm .... var. ***tortuosus***
 2′ Bracts 0; upper cauline lvs gen ± strongly reduced upwards (exc *S. barbatus*)
  10. Buds and sepals glabrous; middle cauline lvs petioled or wedge-shaped at base, not clasping .... ***S. howellii***
  10′ Buds and sepals gen bristly at tip; middle cauline lvs clasping
   11. Petals light greenish yellow .... ***S. bernardinus***
   11′ Petals rose to light purple
    12. Basal petioles < 3 mm; lower and middle cauline lvs overlapping, barely reduced upwards .... ***S. barbatus***
    12′ Basal petioles > 3 mm; lower and middle cauline lvs not overlapping, gen strongly reduced upwards
     13. Fr ± strongly curved, spreading .... ***S. campestris***
     13′ Fr gen ± straight, erect or ascending
      14. Stigma lobes gen 0; pl rhizomed .... ***S. oliganthus***
      14′ Stigma ± 2-lobed; rhizomes 0 .... ***S. cordatus***
       15. Fr gen > 5 mm wide .... var. ***cordatus***
       15′ Fr < 3 mm wide
        16. Upper cauline lvs widely ovate; e&s SNH, W&I .... var. ***duranii***
        16′ Upper cauline lvs lanceolate-oblong; s-most SNH .... var. ***piutensis***

1′ Longest (upper) filament pair ± fused or closely adherent, their anthers << others (subg. *Euclisia*)
 17. Upper sepal much wider than others, expanded into a banner-like hood; SNF (sect. *Polygaloides*) ***S. polygaloides***
 17′ Upper sepal not much wider than others, not forming a hood; gen NCoR, SnFrB, SCoR, rarely n SNF
  18. Stigma ± 2-lobed; infl tip gen with a conspicuous tuft of elongated sepals of sterile fls; herbage ± bristly; calyx ± biradial (sect. *Insignes*)
   19. Lower cauline lvs wedge-shaped at base, not clasping, gen ± petioled; pls densely bristly; e SnFrB (Mount Diablo) .... ***S. hispidus***
   19′ Lower cauline lvs slightly narrowed below or not, but not wedge-shaped at base, base often ± clasping, petiole 0 or rudimentary; pls sparsely to moderately bristly; se SnFrB (Mount Hamilton Range), SCoRI
    20. Petals strongly flared at tip; fr < 2 cm, cylindric, curved upward; seeds rounded, wing 0; st < 1 dm; se SnFrB (Mount Hamilton Range) .... ***S. callistus***

20′ Petals not to somewhat flared at tip; fr > 2 cm, ± flat, gen ± straight; seeds compressed, winged; st gen > 1 dm; SCoRI . . . . **S. insignis**
21. Sterile sepals dark purple, fertile sepals lighter; widely distributed in SCoRI . . . . ssp. **insignis**
21′ Sterile and fertile sepals greenish yellow; w Merced Co. . . . . ssp. **lyonii**
18′ Stigma ± entire; infl lacking well developed tuft of terminal sterile fls (late-season fls may abort); if herbage bristly, calyx bilateral
22. Bien (non-fl rosettes present within fl population); cauline lvs gen gradually reduced upwards; calyx ± biradial; herbage glabrous — serpentine, gen NCoRI (sect. *Biennes*)
23. Pl gen 5–12 dm, branched only above; petals whitish or brownish, lower not or faintly darker than upper, lacking purple exc sometimes on veins of upper . . . . **S. morrisonii**
23′ Pl gen 2.5–5 dm, gen branched throughout; lower petals gen ± purple, upper whitish or with less purple . . . . **S. brachiatus**
24. Sepals glabrous, rose-purple; upper fused filaments orange . . . . ssp. **brachiatus**
24′ Sepals gen hairy, yellow or dark purple; upper fused filaments yellow . . . . ssp. **hoffmanii**
22′ Ann; cauline lvs gen abruptly reduced upwards (or calyx bilateral and herbage gen ± bristly below)
25. Calyx ± biradial; lower petals gen purple, upper lighter; herbage glabrous; fr ± narrowed between seeds (sect. *Hesperides*)
26. Middle cauline lvs ± linear, barely or not clasping; basal lvs ± not rosetted — st gen > 1.5 dm . . **S. barbiger**
26′ Middle cauline lvs gen narrowly lanceolate or wider, deeply clasping; basal lvs gen ± rosetted
27. Middle cauline lvs ± narrowly lanceolate; basal lvs often coarsely and sharply dentate
28. Longest lvs gen > 3 cm; st gen > 1.5 dm; lower lvs entire to dentate; s NCoR, NCoRH . . . . . . **S. barbiger**
28′ Longest lvs < 3 cm; st gen < 1.5 dm; lower lvs coarsely and sharply dentate; nw SnFrB, s NCoRO . . . . **S. batrachopus**
27′ Middle cauline lvs widely lanceolate or wider; basal lvs gen not coarsely and sharply dentate (exc some *S. breweri* var. *hesperidis*) — longest lvs gen > 3 cm
29. Lower cauline lvs strongly 2-ranked, gen ± round; upper cauline lvs ± ovate; infl crowded, internodes < sepals . . . . **S. drepanoides**
29′ Lower cauline lvs not strongly 2-ranked, gen widely (ob)ovate; upper cauline lvs lanceolate; infl not crowded, internodes gen > sepals . . . . **S. breweri**
30. Herbage blue-green; infl ± straight; calyx somewhat narrowed at tip, sepals greenish white to purple; NW, SnFrB, SCoRI . . . . var. **breweri**
30′ Herbage yellow-green; infl ± zig-zag; calyx strongly narrowed at tip, sepals always greenish white; s NCoRI . . . . var. **hesperidis**
25′ Calyx bilateral; all 4 petals uniformly colored or upper darker; herbage gen ± bristly below; fr not narrowed between seeds (sect. *Pulchelli*)
31. Pedicels gen > sepals; infl ± zig-zag; hairs 0–few — n CCo (Tiburon Peninsula) . . . . **S. niger**
31′ Pedicels gen < sepals; infl ± straight; gen strongly (sparsely) bristly below
32. Infl ± 1-sided; sepals whitish or greenish yellow to reddish purple; gen n SnFrB, NCoRO . . **S. glandulosus**
33. Sepals reddish purple; infl short, crowded; st gen < 3 dm; nw SnFrB . . . . ssp. **pulchellus**
33′ Sepals white or greenish yellow to rose; infl not crowded; st often > 3 dm; gen n SnFrB, NCoRO . . . . ssp. **secundus**
32′ Infl not 1-sided; sepals gen lilac-lavender to dark purple (or greenish white); gen NCoRH, NCoRI, CW
34. Sepals dark purple; herbage gen ± densely bristly below . . . . **S. glandulosus** ssp. **glandulosus**
34′ Sepals greenish white to lilac-lavender; herbage gen ± sparsely bristly below . . . . **S. albidus**
35. Sepals greenish white; serpentine, s SnFrB (Santa Clara Co.) . . . . ssp. **albidus**
35′ Sepals lilac-lavender; serpentine or not, s SnFrB, c SCoRO . . . . ssp. **peramoenus**

S. albidus E. Greene Ann 2–12 dm, simple to much-branched, gen sparsely (rarely ± densely) bristly below. **LVS** < 13(22) cm; basal narrowly oblanceolate, coarsely dentate, petioles short, winged; cauline linear-lanceolate, lower ± coarsely dentate, upper ± entire. **FL**: calyx ± bilateral, sepals 5–10 mm, glabrous; petals 8–14 mm, upper > lower; upper filament pair exserted, ± fused or adherent, with reduced anthers. **FR** spreading to ascending, 4–12 cm, straight. **SEED**: wing at 1 end. 2*n*=28. Open, grassy or ± barren slopes, often serpentine; ± 150–800 m. e SnFrB, c SCoRO.

ssp. ***albidus*** (p. 441) METCALF CANYON JEWELFLOWER Pl 5–12 dm. **FL**: sepals white or greenish white; petals whitish, veins brownish or purplish. RARE. Serpentine; 150–800 m. se SnFrB (Santa Clara Co.).

ssp. ***peramoenus*** (E. Greene) Kruckeb. UNCOMMON JEWEL-FLOWER Pl 2–8 dm. **FL**: sepals lavender to rose-purple; petals ± purplish. RARE. Habitats and range of sp. Pls from c SCoRO (previously treated as *S. glandulosus* ssp. *g.*) are indistinct from lavender-fld pls in SnFrB. In cult.

S. barbatus S. Watson (p. 441) Per < 7 dm, simple to few-branched, decumbent to erect, glabrous. **LVS** < 3 cm; basal obovate, short-petioled, with blunt teeth tipped with short stiff hairs; cauline overlapping (exc upper), ± 2-ranked, widely ovate or ± round, gen ± entire, glabrous or short-hairy at tip, obtuse, scarcely reduced upward. **FL**: sepals 4–7 mm, yellowish green in bud, purple in fl, tips short-bristly; petals 5–9 mm, barely exserted, ± purple; filaments free. **FR** ± recurved to spreading, 2–7 cm, gen curved. 2*n*=28,56. Serpentine, rocky, open Jeffrey-pine woods; ± 800– 2200 m. KR. ❀TRY.

S. barbiger E. Greene Ann 1–8 dm, gen branched throughout. **LVS**: basal ± not rosetted, lower blades < 7 cm, oblanceolate to widely obovate, gen coarsely dentate, petioled; middle and upper cauline < 11 cm, ± linear, gen ± entire, gen sessile, sometimes ± clasping. **FL**: sepals 3–6 mm, gen greenish yellow (purplish), gen ± glabrous; petals 5–10 mm, upper ± whitish, lower gen purple (rarely ± whitish); upper filament pair fused, anthers reduced. **FR** spreading, 2–7 cm, ± curved, ± narrowed between seeds. **SEED**: wing 0 or at 1 end. 2*n*=28. Serpentine barrens, chaparral; 200–500 m. s NCoR, NCoRH. Variable; intergrades with *S. batrachopus* in Sonoma Co., *S. breweri* in Napa, Lake cos. Pls from w-c Tehama Co. with hairy sepals and pls from Lake Co. with deeply clasping cauline lvs are apparently undescribed taxa.

S. batrachopus J. Morrison (p. 445) TAMALPAIS JEWELFLOWER Ann 0.5–2 dm, simple to branched throughout. **LVS** < 3 cm; lower ± obovate, coarsely dentate, short-petioled; middle and upper cauline ± lanceolate, entire or tip few-toothed. **FL**: sepals 3–4 mm, greenish or purple; petals 5–7 mm, upper whitish or purple-veined,

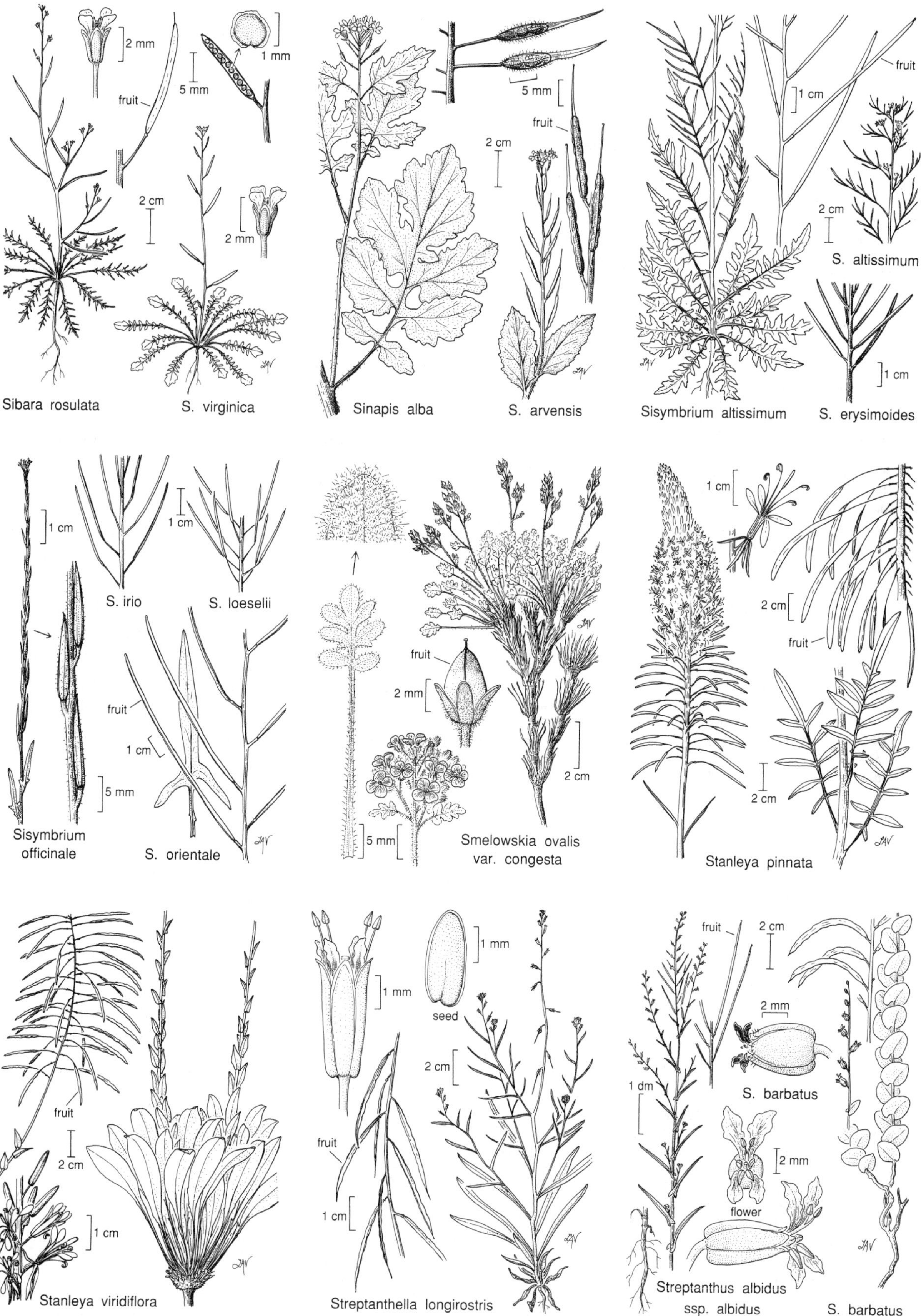
2 mm
1 mm
5 mm
fruit
2 cm
2 mm
Sibara rosulata
S. virginica
5 mm
fruit
2 cm
Sinapis alba
S. arvensis
fruit
1 cm
2 cm
S. altissimum
1 cm
Sisymbrium altissimum
S. erysimoides
1 cm
1 cm
S. irio
S. loeselii
fruit
1 cm
5 mm
Sisymbrium officinale
S. orientale
fruit
2 mm
5 mm
2 cm
Smelowskia ovalis var. congesta
1 cm
2 cm
fruit
2 cm
Stanleya pinnata
fruit
2 cm
1 cm
Stanleya viridiflora
1 mm
1 mm
seed
2 cm
fruit
1 cm
Streptanthella longirostris
fruit
2 cm
2 mm
S. barbatus
1 dm
2 mm
flower
Streptanthus albidus ssp. albidus
S. barbatus

lower purple; upper filament pair fused, with reduced anthers, lower fused in basal ± 1/2, anthers unreduced. **FR** ascending or spreading, 2–3 cm, ± straight or curved, ± narrowed between seeds. 2*n*=28. RARE. Serpentine barrens, chaparral or woodland; < 650 m. s NCoRO, nw SnFrB (Mount Tamalpais). NCoRO pls, some with hairy sepals, may be undescribed. Relationship to *S. barbiger* needs further study.

S. bernardinus (E. Greene) Parish LAGUNA MOUNTAINS JEWELFLOWER Per from woody root crown, 3–6 dm, simple to few-branched. **LVS**: basal 3–8 cm, 10–25 mm wide, oblanceolate, gen dentate above middle, rarely ciliate on petioles; cauline ± lanceolate, acuminate, margins often purple. **FL**: calyx biradial, sepals 5–9 mm, light green in bud becoming light yellow to white in fl, widely oblong, margins scarious, bristles at tip 0–few; petals 7–11 mm, white, slightly exserted; filaments free, anthers equal; stigma barely 2-lobed. **FR** ascending or spreading, 5–8 cm, 1.5–2 mm wide, gen straight. **SEED** oblong, winged. 2*n*=14. RARE. Montane conifer forest; 1200–2500 m. e TR, PR; Baja CA?

S. brachiatus F.W. Hoffm. Bien 2–6 dm, gen branched below. **LVS**: basal oblanceolate, toothed above middle, fleshy, blades 1.5–4 cm, purple below, petiole short, winged; lower cauline narrowly ovate, serrate, short-petioled, upper lanceolate. **FL**: calyx biradial, sepals 6–9 mm, yellow to light purple; petals 7–9 mm, upper white or purple-veined, lower ± light purple; longest filament pair fused, anthers reduced, 2nd longest fused at very base. **FR** ascending, 5–7 cm, ± narrowed between seeds. **SEED**: wing weak, at tip. 2*n*=28. RARE. Serpentine barrens, open chaparral or woodland; 600–900 m. s NCoRI.

ssp. ***brachiatus*** (p. 445) SOCRATES MINE JEWELFLOWER Pl 2–4 dm. **FL**: calyx glabrous, light purple. RARE. Habitats of sp. sw NCoRI (Mayacamas Mtns, Sonoma Co.)

ssp. ***hoffmanii*** R. Dolan & L. LaPré FREED'S JEWELFLOWER Pl 2–6 dm. **FL**: calyx hairy, yellow to light purple. RARE. Habitats of sp. s NCoRI (Sonoma, Lake cos.)

S. breweri A. Gray Ann < 10 dm, branched above or throughout, glabrous. **LVS**: basal gen widely (ob)ovate, entire to coarsely dentate, gen petioled; lower cauline clasping; middle and upper cauline gen ± lanceolate, entire. **FL**: sepals purplish or greenish white; upper petals whitish or purple-veined, lower ± purple; upper filament pair fused, with reduced anthers. **FR** ascending or spreading, 2–11 cm, gen ± curved, ± narrowed between seeds. **SEED**: wing 0 or at 1 end. Serpentine barrens in chaparral or woodland; 250–1700 m. s-most KR, NCoRH, c&s NCoRI, e SnFrB, SCoRI. Vars. intergrade.

var. ***breweri*** (p. 445) Pl < 10 dm, ± blue-green. **LVS** < 12 cm; lower entire to wavy- or blunt-dentate. **INFL** ± straight. **FL**: calyx somewhat narrowed at tip, sepals 3–9 mm, greenish white or purplish, gen glabrous; petals 5–13 mm. 2*n*=28. Habitats and range of sp. Variable. ❀DRN,SUN:7;DFCLT.

var. ***hesperidis*** (Jepson) Jepson Pl < 4 dm, ± yellow-green. **LVS** < 5 cm; lower entire to coarsely and sharply dentate. **INFL** ± 1-sided, zig-zag. **FL**: calyx strongly narrowed at tip, sepals 4–7 mm, greenish white, glabrous; petals 6–8 mm. **SEED**: wing 0. 2*n*=28. Uncommon. Habitats of sp.; 250–600 m. s NCoRI (Napa, s Lake cos.).

S. callistus J. Morrison (p. 445) MOUNT HAMILTON JEWELFLOWER Ann 3–8 cm, simple or branched throughout, sometimes decumbent, ± bristly, esp below. **LVS** < 2 cm, coarsely dentate; basal obovate, short-petioled; cauline > basal, widely (ob)ovate. **INFL** sterile at tip. **FL**: sepals ± bristly, fertile 3–8 mm, gen green (purplish), sterile ± lilac, narrow, elongated; petals 8–11 mm, strongly flaring at tip, purple, scarious margin weak; filaments purple, longest 2 pairs fused, anthers of longest pair shortest, those of shortest pair longest; stigma shallowly ± 2-lobed. **FR** spreading to erect, 1.5–2 cm, ± cylindric, gen ± curved, sparsely bristly. **SEED** not compressed; wing 0. 2*n*=28. RARE. Open chaparral; ± 600 m. se SnFrB (Arroyo Bayo, Mount Hamilton Range). Most endangered taxon in genus. In cult.

S. campestris S. Watson (p. 445) SOUTHERN JEWELFLOWER Bien? or short-lived per 6–15 dm, stout, simple to few-branched, glabrous. **LVS**: basal < 10 cm, oblanceolate to obovate, fleshy, obtuse, gen dentate, teeth bristly, petiole margins ± ciliate; cauline narrowly ovate, acute. **FL**: sepals 7–10 mm, purple, tips bristly, margins scarious; petals 9–12 mm, bases light yellow, tips light purple; filaments free, anthers exserted, equal; stigma 2-lobed. **FR** spreading to ascending, 6–14 cm, 2.5–3.5 mm wide, curved. **SEED** oblong, winged. RARE. Open, rocky coniferous forest, chaparral, woodland; 900–2300 m. TR, PR; n Baja CA. Pls from s CaR and n SNF, with mid-st lvs ± oblanceolate and yellow (purple-tipped) petals, are an undescribed taxon.

S. cordatus Nutt. Per 2–10 dm, gen simple, glabrous. **LVS**: basal widely obovate, toothed above middle, teeth often bristly, petioles = lvs, often ciliate; cauline few-toothed to entire, gen acute. **FL**: calyx biradial, sepals 8–13 mm, yellowish green in bud becoming purple in fl, tips gen bristly; petals exserted, 10–14 mm, linear, purple; stamens free, equal; stigma 2-lobed. **FR** ascending to ± erect, 5–10 cm, 2.5–6 mm wide, straight. *n*=12. Rocky or sandy sagebrush scrub, pinyon/juniper woodland, ponderosa-pine forest; 1200–3100 m. e CaRH, s SNH, GB, DMtns; to se OR, WY, n NM.

var. ***cordatus*** (p. 445) Pl 2–6 dm. **LVS**: upper cauline widely ovate, acute to obtuse, clasping. **FR** 5–8 cm, 4–6 mm wide. Common. Rocky or sandy sagebrush scrub or pinyon/juniper woodland; 1400–2800 m. e CaRH, MP, n SNE, e DMtns; to se OR, WY, n NM. ❀TRY;DFCLT.

var. ***duranii*** Jepson Pl 2–7 dm. **LVS**: upper cauline widely ovate, acute to obtuse, clasping. **FR** 5–9 cm, 2.5–4 mm wide. Uncommon. Talus, calcareous outcrops; 1800–3100 m. s SNE, W&I.

var. ***piutensis*** J. Howell PIUTE MOUNTAINS JEWELFLOWER Pl 5–10 dm. **LVS**: upper cauline lanceolate-oblong, acuminate, deeply clasping. **FL**: sepal tips short-hairy. RARE. Open chaparral, Piute-cypress stands; 1200–1650 m. s SNH (Piute Mtns).

S. diversifolius S. Watson (p. 445) VARIED-LEAVED JEWELFLOWER Ann 4–6 dm, simple or branched above. **LVS**: basal 7–12 cm, simple or 2-pinnate, segments thread-like, axis sometimes winged; upper cauline lvs entire, widely lanceolate to round. **INFL**: bracts 1–2, like upper cauline lvs. **FL**: calyx somewhat narrowed at tip, sepals 5–6 mm, yellow or light purple, margins whitish; petals 8–9 mm, white or yellowish, veins purple; filaments free, upper pair exserted, recurved, with ± reduced fertile anthers. **FR** reflexed, 4–8 cm, straight or ± curved, wide, flat. **SEED** narrowly winged at ends. 2*n*=28. Open woodlands, rocky slopes; < 1600 m. c&s SNF. ❀DRN,SUN,DRY:7,14–16;DFCLT.

S. drepanoides Kruckeb. & J. Morrison SICKLE-FRUIT JEWELFLOWER Ann < 4 dm, simple to spreading-branched, glabrous. **LVS** < 7 cm; lower gen ± round, entire or short-dentate, esp toward tip; basal ± sessile or short-petioled; lower cauline overlapping, 2-ranked, deeply clasping; middle and upper cauline gen ± ovate. **FL**: sepals 4–8 mm, gen greenish yellow (greenish white or purplish), gen glabrous; petals 5–10 mm, upper whitish or purple-veined, lower ± purple; upper filament pair exserted, fused, with reduced anthers, lower fused in basal 2/3 or less. **FR** spreading, 3–9 cm, ± straight to strongly curved, ± narrowed between seeds. **SEED**: wing 0 or at 1 end. 2*n*=28. UNCOMMON. Open chaparral or Jeffrey-pine woodland, on serpentine; 400–1600 m. s-most KR, NCoRH, n NCoRI, n SNF (Butte Co.).

S. farnsworthianus J. Howell FARNSWORTH'S JEWELFLOWER Ann 2–5 dm, simple or branched, ± purple. **LVS**: lower 7–15 cm, deeply lobed to 1–2-pinnate, petioled, segments linear to oblanceolate; middle and upper cauline entire. **INFL**: bracts 1–2, lf-like, purple. **FL**: calyx narrowed at tip, sepals 7–9 mm, violet-purple, tips recurved, acute; petals equal, 8–11 mm, white with purple veins, margins wavy, tips reflexed; filaments free, upper pair exserted, recurved, with reduced fertile anthers. **FR** ascending, 7–12 cm, 2.5–3.5 mm wide, straight or ± curved, flat. 2*n*=28. UNCOMMON. Foothill woodland; ± 400–1300 m. c SNF (Madera, Fresno cos.), s SNF. ❀TRY;DFCLT.

S. fenestratus (E. Greene) J. Howell TEHIPITE VALLEY JEWELFLOWER Ann 1–4 dm, simple or few-branched below. **LVS**: lower 2–5 cm, deeply 1–2-pinnately lobed, segments linear to oblanceolate, petiole long; uppermost ovate, entire to coarsely toothed. **INFL**: bracts 1–2, lf-like, green or with purple tips. **FL**: calyx somewhat narrowed at tip, sepals 7–8 mm, purple, tips recurved; petals purple, blades 9–13 mm, wide, flat; filaments free, upper pair exserted, recurved, with reduced anthers. **FR** reflexed to erect, 1.5–5 cm, 1.5–1.8 mm wide, flat. **SEED**: wing at 1 end, sometimes minute. 2*n*=28. RARE. Granite ledges, sand, open mixed-conifer/oak woods; 1200–1750 m. c SNH (Kings River Canyon, Fresno Co.).

S. glandulosus Hook. JEWELFLOWER Ann 1–10 dm, simple to branched throughout, gen ± densely bristly, esp below. **LVS** < 13 cm; basal ± oblanceolate, coarsely dentate to ± lobed, narrowed to ± winged petiole; cauline ± linear to (ob)lanceolate, entire to ± lobed. **INFL** ± straight; pedicels gen < sepals. **FL**: calyx bilateral, sepals 4–9 mm, glabrous or sparsely hairy; petals 6–15 mm, upper >, gen darker than lower; upper filament pair exserted, fused at least in basal 2/3, recurved, with reduced ± sterile anthers. **FR** spreading to ± erect (less commonly deflexed), 3–11 cm, straight to ± curved. 2*n*=28. Dry, open grasslands, chaparral, open conifer/oak woodland, sometimes on serpentine, mostly away from coast; 15–1300 m. s NCoR, SnFrB, n&c SCoR; sw OR?.

ssp. ***glandulosus*** (p. 445) **INFL** not 1-sided. **FL**: sepals dark purple; petals 6–13 mm, purple. Habitats ± of sp.; 200–1300 m. s NCoRO (uncommon), s-most NCoRH, s NCoRI, SnFrB, n SCoRO (uncommon), n&c SCoRI. Variable. ❀DRN,SUN,DRY:7,14–16; DFCLT

ssp. ***pulchellus*** (E. Greene) Kruckeb. MOUNT TAMALPAIS JEWELFLOWER **ST** gen 1–3 dm. **INFL** dense, ± 1-sided. **LVS** < 6 cm. **FL**: sepals 4–7 mm, reddish purple; petals 8–13 mm, ± purple. RARE. Habitats ± of sp.; 150–800 m. nw SnFrB (Marin Co.). [var. *p.* (E. Greene) Jepson]

ssp. ***secundus*** (E. Greene) Kruckeb. **LVS** < 9 cm. **INFL** open, ± 1-sided. **FL**: sepals white to greenish yellow, purplish, or rose; petals 7–15 mm, whitish to purplish. **FR** gen curved. Habitats ± of sp.; 15–800 m. s NCoRO, sw NCoRI, nw SnFrB, sw OR?. Variable; the following forms could be elevated to sspp.: pls from Austin Creek area, Sonoma Co. with sepals rose and petals ± purplish have been called var. *hoffmannii* Kruckeb. (secund jewelflower); pls from Sonoma Co. with sepals white to greenish yellow without purple and petals whitish have been called var. *sonomensis* Kruckeb. ❀DRN,SUN,DRY:7,14–16;DFCLT

S. gracilis Eastw. ALPINE JEWELFLOWER Ann 5–30 cm, few-branched. **LVS**: basal oblong to obovate, coarsely toothed to lobed, petiole = blade; cauline narrowly oblong to obovate, entire to lobed, lower petioled. **INFL**: bracts 1–2, entire, sessile. **FL**: sepals 4–5 mm, purple, tips spreading; petals exserted, spreading, 7–10 mm, pink; filaments free. **FR** ascending, 3–7 cm, straight to slightly curved, slender. **SEED**: wing 0 or small. UNCOMMON. Rocky slopes; 2800–3500 m. se SNH (Kings-Kern Divide region).

S. hispidus A. Gray MOUNT DIABLO JEWELFLOWER Ann < 3 dm, simple to few-branched, densely bristly. **LVS** < 5 cm; lower obovate, wedge-based, coarsely dentate, petioled to ± sessile; middle and upper cauline lanceolate to obovate, coarsely dentate, sessile, not or barely clasping. **INFL**: terminal sterile fls gen present. **FL**: sepals ± bristly, fertile 3–7 mm, brownish green, sterile purplish, narrow, elongated; petals 6–10 mm, ± purplish; upper filament pair exserted, fused, with sterile anthers; stigma shallowly ± 2-lobed. **FR** erect or ascending, 3–8 cm, ± straight, bristly. 2*n*=28. RARE. Rocky chaparral or grassland; 600–1200 m. ne SnFrB (Mount Diablo). In cult.

S. howellii S. Watson (p. 445) HOWELL'S JEWELFLOWER Per 3–7 dm. **STS** 1–few. **LVS**: basal 2–10 cm, obovate, entire to coarsely dentate, fleshy, petioled; cauline petioled to ± sessile, not clasping, lower narrowly obovate, upper narrower. **FL**: sepals 5–7 mm, purple, glabrous; petals 8–10 mm, purple at tip, yellow below, margins ± curled; filaments free. **FR** 5–9 cm, ascending to spreading, curved. 2*n*=28. RARE. Rocky serpentine in open conifer/hardwood forest; 600–1500 m. n KR (Del Norte, Siskiyou cos.); sw OR.

S. insignis Jepson SAN BENITO JEWELFLOWER Ann < 6 dm, gen branched above, bristly (esp below). **LVS** < 12 cm; basal ± oblanceolate, coarsely dentate to lobed, short-petioled; lower cauline sometimes ± clasping; middle and upper cauline ± lanceolate, dentate to lobed, ± clasping. **INFL**: terminal sterile fls present exc in early fl. **FL**: sepals gen ± bristly, fertile 3–7 mm, sterile narrow, elongated; petals 6–10 mm; longest filament pair fused, recurved, with anthers ± reduced and sterile, fertile filaments free; stigma ± 2-lobed. **FR** reflexed to erect, 3–11 cm, gen ± straight. 2*n*=28. Dry grassland or open chaparral, gen on serpentine; 200–1400 m. SCoRI.

ssp. ***insignis*** (p. 445) **FL**: sterile sepals dark purple, fertile gen paler; petals gen ± purple. Habitats and range of sp. Pls from w Merced Co. with greenish yellow fertile sepals and petals ± without purple (non-serpentine?) intermediate to ssp. *lyonii*. ❀TRY.

ssp. ***lyonii*** Kruckeb. & J. Morrison ARBURUA RANCH JEWELFLOWER **FL**: sepals greenish yellow; petals ± yellowish. RARE. Serpentine, gen grassland; 500–600 m. SCoRI (near Ortigalita Peak, w Merced Co.).

S. morrisonii F.W. Hoffm. (p. 445) Bien 5–12 dm, simple or branched above. **LVS**: basal 3–5 cm, oblanceolate, fleshy, petioled, toothed above middle, ± purple-mottled to uniformly purple below; cauline lanceolate to ovate or spoon-shaped, gradually reduced upwards. **FL**: calyx biradial; sepals 5–7 mm, pale yellow to pale purple; petals 6–10 mm, ± white to light yellow or ± brownish, lower sometimes ± darker, upper purple-veined or not; longest filament pair fused, their anthers sterile, 2nd longest fused basally, recurved. **FR** reflexed to erect, 4–8 cm, straight or ± curved, flat, ± narrowed between seeds. **SEED**: wing at 1 end. 2*n*=28. RARE. Serpentine barrens, chaparral, cypress/knobcone-pine woodlands, often among shrubs; 150–1000 m. c&s NCoR, c SCoRO?. Variable, infraspecific taxa need detailed study: pls from sw NCoRI with lvs purple-mottled above, sepals yellowish and nearly glabrous, upper cauline lvs 2–4 × longer than wide have been called ssp. *elatus* F.W. Hoffm. (Three peaks jewelflower); pls from s NCoRO with lvs purple-mottled above, sepals purple, densely hairy have been called ssp. *hirtiflorus* F.W. Hoffm. (Dorr's cabin jewelflower); pls from c NCoRI with lvs purple-mottled above, sepals yellowish and nearly glabrous, upper cauline lvs 1–2 × longer than wide have been called ssp. *kruckebergii* R. Dolan & L. LaPré (Kruckeberg's jewelflower); pls from s NCoRO with lvs unmottled above, sepals yellowish and nearly glabrous have been called ssp. *morrisonii* (Morrison's jewelflower); pls from c SCoRO (Burro Creek, Monterey Co.) apparently represent a related, undescribed taxon. In cult.

S. niger E. Greene (p. 445) TIBURON JEWELFLOWER Ann 2–7 dm, simple or branched above, glabrous or sparsely hairy below. **LVS** < 9 cm; basal ± oblanceolate, coarsely dentate or short-lobed, petioles short, winged; cauline ± lanceolate, gen ± entire. **INFL** ± zig-zag; pedicels gen > calyx. **FL**: calyx bilateral, sepals 5–7 mm, purplish black; petals 8–12 mm (upper > lower), dark purple; upper filament pair fused, anthers reduced. **FR** ± ascending, 2–7 cm, straight. 2*n*=28. RARE. Serpentine outcrops in grassland; < 150 m. n CCo (Tiburon Peninsula, Marin Co.). [*S. glandulosus* var. *n.* (E. Greene) Munz] Fls nearly closed at throat, self-pollinated.

S. oliganthus Rollins MASONIC MOUNTAIN JEWELFLOWER Per 2–5 dm, rhizomed, gen simple, glabrous, glaucous. **LVS** 2–8 cm; basal (ob)lanceolate, entire, finely ciliate, petiole > blade; cauline lanceolate to oblong. **FL**: calyx biradial, sepals 7–10 mm, yellow in bud, purple in fl; petals 10–13 mm, tips purple, margins curled; upper stamens free; stigma flat. **FR** spreading to ascending, 5–8 cm, 2.5–3.5 mm wide, straight or slightly curved, wide, flat. 2*n*=28. RARE. Rocky sites or talus; 2100–2800 m. n&c SNE (Sweetwater, Masonic mtns), n W&I (n White Mtns); w-c NV. [*S. cordatus* var. *exiguus* Jepson]

S. polygaloides A. Gray (p. 445) MILKWORT JEWELFLOWER Ann 2–10 dm, simple or branched above, glabrous, not glaucous. **LVS** < 8 cm, entire, ± narrowly (ob)lanceolate, narrower upward; lower ± petioled. **FL**: calyx strongly bilateral, sepals 3–8 mm, gen yellow-green (purple), uppermost larger, ± round, forming banner-like hood; petals 4–8 mm, barely exserted, purplish or brownish, esp on veins; upper filament pair barely exserted, fused, with reduced, ± sterile anthers. **FR** reflexed, 2–5 cm, ± straight, ± compressed to ± 4-angled. **SEED** often unwinged. 2*n*=28. Serpentine barrens, sparse pine/cypress woodland; 200–1100 m. SNF. Accumulates much nickel. The genus *Microsemia* has been based on this sp.

S. tortuosus Kellogg MOUNTAIN JEWELFLOWER Ann to subshrub 1–10 dm, simple or branched. **LVS**: basal gen oblong to widely ovate, gen toothed above middle, coarse, petiole winged; cauline gen round to oblong, entire to dentate, upper often larger, gen narrower. **INFL**: bract gen 1, ovate, entire. **FL**: sepals purple, gray-green, or yellowish; petals widely linear, purple to yellow, tips reflexed; filaments exserted, free, upper pair with ± unreduced, ± fertile anthers; stigma ± sessile, lobes weak. **FR** gen reflexed, 8–12 cm, curved. 2*n*=28. Gen rocky to sandy soils, in open coniferous forest; 200–3500 m. NW, CaRH, SN, n SnFrB, SCoRO; sw OR. Variable, needs study; vars. intergrade. Pls with lanceolate lvs from KR (Burnt Ranch, Trinity Co.) undescribed.

var. ***flavescens*** Jepson Ann 2–7 dm, gen few-branched. **LVS** and bracts equally pale green. **FL**: sepals 4–6 mm, yellowish; petals 6–8 mm, yellowish. Uncommon. Forests; 2600–3200 m. s SNH (Kings Canyon, Fresno Co.; Sawtooth Range, Tulare Co.).

var. ***orbiculatus*** (E. Greene) H.M. Hall Ann 1–3 dm, gen much-branched. **INFL**: bracts paler than lvs. **FL**: sepals 5–6 mm, purple; petals 7–9 mm, base yellow, tip white, purple-veined. Common. Dry slopes, forests; 1900–3500 m. KR, NCoRH, CaRH, SN; sw OR. ❀TRY.

var. ***suffrutescens*** (E. Greene) Jepson Subshrub 4–9 dm. **STS** 1–several, gen few-branched. **INFL**: bracts paler than lvs. **FL**: sepals 8–10 mm, purple; petals 11–13 mm, base yellow, tip white, purple-veined. Uncommon. Serpentine or volcanic cliffs; 200–850 m. s NCoR (ne Sonoma Co.), n&c SNF, SCoRO (Santa Lucia Mtns). [var. *optatus* Jepson, in part]

var. ***tortuosus*** (p. 445) Gen bien (ann to weak per) 2–10 dm, simple or few-branched. **LVS**: uppermost longest. **FL**: sepals 5–8 mm, gen purple; petals gen 6–8 mm. Dry slopes; 300–2200 m. NCoR, CaRF, SNF, n SnFrB. [var. *pallidus* Jepson] ❀SUN,DRN: 1,**15–17**&IRR:2,3,**7**,**14**,**18–21**,22–24.

SUBULARIA AWLWORT

Ann, tufted, ± scapose, glabrous. **LF** erect, linear to narrowly awl-like, entire. **FL**: sepals erect; petals 0 or white; stamens unappendaged; style 0, stigma entire. **FR** inflated, dehiscent; valves 1-ribbed. **SEEDS** 2–6, 2 rows per chamber; cotyledons linear; embryonic root at back of 1 cotyledon. 2 spp.: n hemisphere, aquatic or of sea shores. (Latin: awl, from shape of some lvs)

S. aquatica L. var. ***americana*** (G. Mulligan & Calder) Boivin (p. 445) **LF** 1–5(7) cm, narrowly awl-like. **INFL**: fl st 2–12(18) cm, lfless. **FL** minute; sepals erect; petals ± > sepals, white. **FR** 2–3 mm, narrowly elliptic to widely obovate; pedicels ascending to erect, 2–5 mm, straight. **SEED** ± 1 mm, oblong; wing 0. Uncommon. Shallow lake margins, streambanks; 2000–3100 m. SNH; to AK, ne US. Var. *aquatica* native to Eurasia.

TEESDALIA

Ann, gen scapose; hairs 0 or simple. **LVS** rosetted, petioled, gen dissected; cauline 0 or few, sessile. **FL**: sepals glabrous; petals not clawed, white; stamens 4 or 6, filaments with a white basal scale. **FR** ± round, flat perpendicular to septum, winged above, dehiscent; pedicel flared to cup-like tip; styles short or 0. **SEEDS** 2 per chamber; wing 0. 2 spp.: Eur. (R. Teesdale, English botanist, horticulturist, 1740–1804)

T. coronopifolia (Bergeret) Thell. (p. 445) Pl slender, glabrous. **LF** 1–5 cm, narrowly oblanceolate, pinnately few-lobed to entire. **INFL** dense; peduncles erect or ascending, 1–several, 6–16 cm. **FL**: sepals widely ovate; petals ± equal, narrowly obovate or tongue-shaped; stamens 4. **FR** 3–4 mm, widely obovate to obcordate; tip shallowly notched; pedicel ± ascending to ± reflexed; style 0. 2*n*=36. Uncommon. Wet, disturbed areas; < 100 m. s NCoRO (Sonoma Co.); native to Eur.

THELYPODIUM

Winter ann to bien, per; hairs 0 or simple. **ST** gen erect, branched or not. **LVS**: basal gen rosetted, petioled, gen shed early, entire to pinnately lobed; cauline petioled or sessile, often clasping. **FL**: sepals erect to reflexed, greenish, white, lavender, or purplish, bases sac-like or not; petals linear to oblanceolate, white, lavender, or purple; stamens equal or 4 long, 2 short, paired filaments rarely ± fused. **FR** erect ot spreading, narrowly linear, ± narrowed between seeds, cylindric or ± flat parallel to septum, gen stalked above receptacle; pedicel ± flat at base or not, gen expanded at tip; stigma in width < style tip, gen entire. **SEEDS** 1 row per chamber, ± flat; wing gen 0; embryonic root at edge or toward back of 1 cotyledon. 20 spp.: w N.Am. (Greek: female foot, from fr stalk above receptacle) [Al-Shehbaz 1973 Contr Gray Herb 204:1–148]

1. Cauline lvs petioled, lower and middle pinnately lobed, rarely just dentate; fr ± flat parallel to septum to cylindric, rarely ± 4-sided; style ± club-shaped or obconic
  2. Fr pedicel ± straight, spreading, rarely ascending; st solid, not inflated; fr spreading; petal blades linear ..... *T. laciniatum*
  2' Fr pedicel upcurved, gen erect at tip; st gen ± hollow to infl, inflated or not; fr ± erect, often appressed; petal blades oblanceolate to spoon-shaped ..... *T. milleflorum*
1' Cauline lvs sessile, lower and middle gen entire, rarely dentate; fr cylindric; style cylindric or tapered to tip
  3. Per, caudex thick, covered with many old petiole bases; st subdecumbent to erect, gen wavy; basal lvs persistent; petioles glabrous ..... *T. flexuosum*
  3' Winter ann, bien, with few old petiole bases at base or not, if per then rhizome slender, without old petiole bases; st gen erect, straight; basal lvs withering early; petioles often hairy
    4. Cauline lvs narrowed to base, ± without basal lobes, ascending ..... *T. integrifolium*
      5. Petals white; fr pedicel whitish, 6–13 mm, stalk above receptacle 1–3 mm ..... ssp. *affine*
      5' Petals lavender to purple, rarely whitish; fr pedicel not whitish, 2–5(6) mm, stalk above receptacle 0.5–1 mm ..... ssp. *complanatum*
    4' Cauline lvs expanded to base, with basal lobes, erect-appressed to ascending
      6. Infl dense, spike-like; petals gen white, rarely light lavender, blades often crinkled; fr pedicel erect, if ± ascending to spreading then < 2.5 mm
        7. Fr pedicel 1–2 mm, spreading, rarely ± ascending, stout, base flat; seed plump ..... *T. brachycarpum*

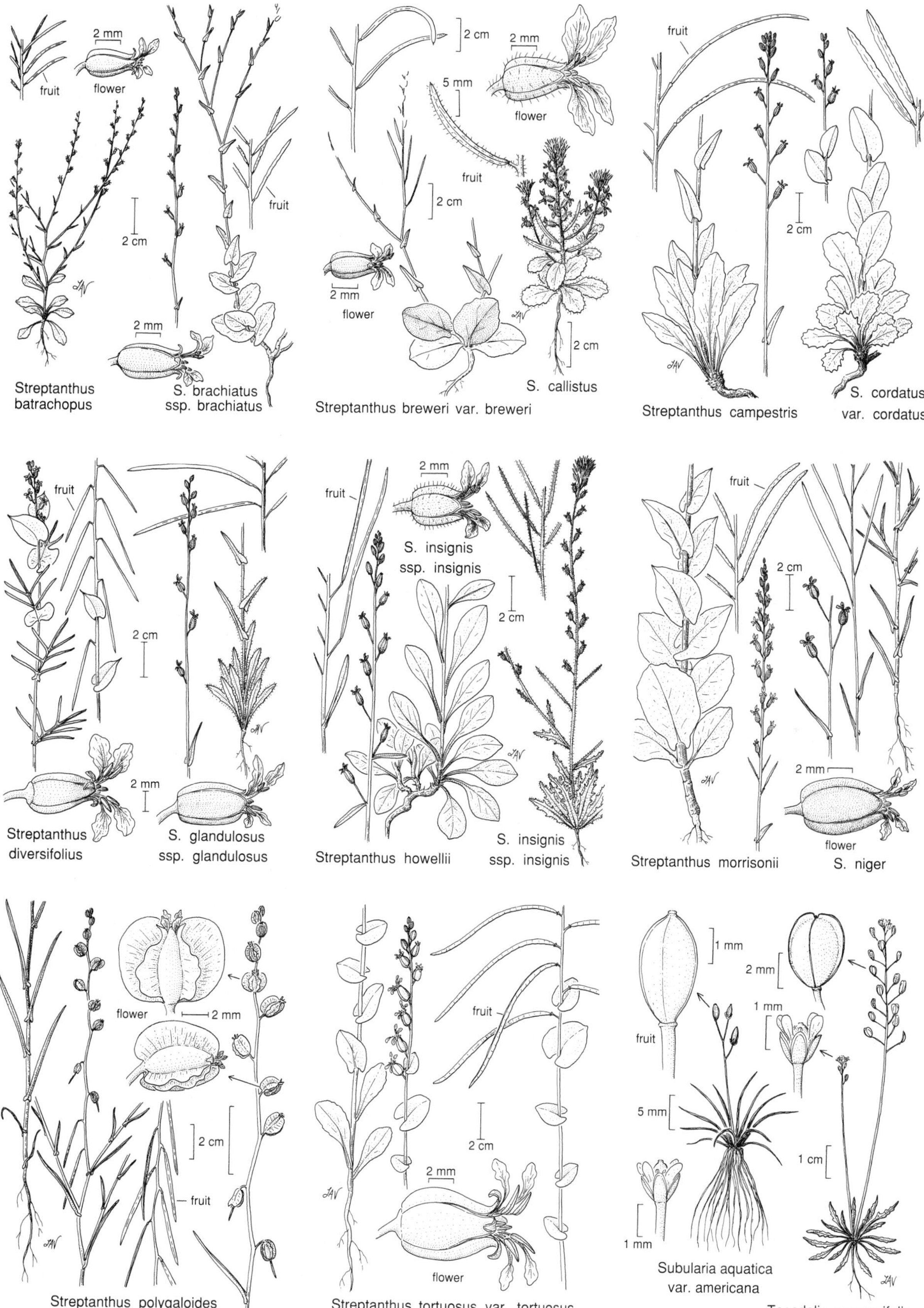

fruit
flower
2 mm
2 cm
Streptanthus batrachopus
S. brachiatus ssp. brachiatus
5 mm
Streptanthus breweri var. breweri
S. callistus
Streptanthus campestris
S. cordatus var. cordatus
Streptanthus diversifolius
S. glandulosus ssp. glandulosus
S. insignis ssp. insignis
Streptanthus howellii
Streptanthus morrisonii
S. niger
Streptanthus polygaloides
Streptanthus tortuosus var. tortuosus
1 mm
1 cm
Subularia aquatica var. americana
Teesdalia coronopifolia

7′ Fr pedicel 2–6 mm, erect to erect-ascending, partly or fully appressed, slender, base not flat; seed ± flat .... ***T. crispum***
6′ Infl open, not spike-like; petals lavender or purple, rarely white, blades not crinkled or only so at base; fr pedicel ascending, rarely ± spreading, 3–8 mm
8. Paired filaments partly to completely fused; petals spoon-shaped .... ***T. howellii*** ssp. ***howellii***
8′ Paired filaments free; petals linear .... ***T. stenopetalum***

T. brachycarpum Torrey (p. 449) SHORT-PODDED THELYPODIUM Bien, taprooted. **ST** 3–8(12) dm, branched or not; hairs ± 0 or near base, sparse. **LVS**: basal 4–14(20) cm, ± entire to pinnately lobed, thickish, glaucous, hairy or not; cauline sessile, 2–6 cm, sagittate, clasping st, entire, glaucous. **INFL** spike-like, dense. **FL**: petals white, crinkled. **FR** 1–3 cm, cylindric, narrowed between seeds; stalk above receptacle 1–2(5) mm, slender; pedicel spreading, rarely ± ascending, 1–2 mm, stout, base flat; style 0.5–1 mm, slender. **SEED** plump; embryonic root at back or toward edge of 1 cotyledon. UNCOMMON. Alkaline soils, adobe flats, pond margins; 800–2320 m. KR, NCoR, CaR; s OR. ❀TRY.

T. crispum Payson (p. 449) Winter ann, bien, per. **ST** 1–7 dm, branched; hairs 0 or toward base, sparse. **LVS**: basal 2–15 cm, ± oblanceolate to spoon-shaped, ± pinnately lobed, rarely entire, thickish, glaucous, hairy or not; cauline sessile, 1–5 cm, sagittate, clasping st, entire, glaucous. **INFL** spike-like, dense. **FL**: petals gen white, rarely light lavender with purplish veins, often crinkled. **FR** 1–2.5 cm, cylindric, incurved or straight, narrowed between seeds; stalk above receptacle 0.5–1.5(3.5) mm; pedicel erect to ascending-erect, partly or fully appressed, 2–6 mm, slender, base not flat; style 0.5–1(2.5) mm, slender. **SEED** ± flat; embryonic root at edge or back of 1 cotyledon. $2n=26$. Alkaline or sandy soils, lake margins, shrubland; 1300–3000 m. SNH, GB; NV.

T. flexuosum Robinson (p. 449) Per; taproots long; caudex thick, woody, covered with many old petiole bases. **STS** subdecumbent to erect, few–several, 2–6 dm, branched above, weak, slender, gen wavy, glabrous. **LVS** (ob)lanceolate, gen entire, glabrous; basal 4–16 cm, glaucous, persistent; cauline sessile, clasping st. **FL**: petals 5–8 mm, lavender to white. **FR** erect to ascending, 1–2.5(4) cm, cylindric, narrowed between seeds; stalk above receptacle 0.2–0.5 mm; pedicel spreading to ascending, 4–9 cm, slender, straight to ± curved; style 1–2 mm. **SEED**: embryonic root at back or toward edge of 1 cotyledon. $2n=26$. Among shrubs, canyons, slopes; 900–1800 m. MP; to ID, NV.

T. howellii S. Watson ssp. ***howellii*** Bien. **ST** 1, 1–9 dm, branched above; hairs sparse to dense below, 0 above. **LVS**: basal 2–10 cm, oblanceolate, few-lobed, rarely entire or dentate, hairs 0 or sparse; cauline sessile, 1–10 cm, linear to lanceolate, sagittate, ± clasping st, entire, glabrous, glaucous. **FL**: petals spoon-shaped, lavender to purple, blades 0.5–1.2 mm wide, ± crinkled at base; paired filaments partly to completely fused. **FR** 1.5–4.5 cm, cylindric, ± incurved, narrowed between seeds; stalk above receptacle 0.5–1(3.5) mm; pedicel ascending, 3–8 mm, stout, straight or ± curved; style 1–3 mm. **SEED** plump; embryonic root near back edge of 1 cotyledon. Alkaline meadows, sagebrush scrub; 1200–1550 m. MP; to WA.

T. integrifolium (Torrey & A. Gray) Endl. Bien. **ST** 4.5–17 dm, straight, glabrous, glaucous; branches 0 or above. **LVS**: basal 5–31 cm, oblong to obovate, ± entire, thickish, ± glaucous; cauline sessile, 2–8 cm, narrowed to base, ± without basal lobes, entire to ± dentate, thickish, glabrous. **FL**: petals 6–9 mm, not crinkled. **FR** spreading to ascending, cylindric to ± flat, ± narrowed between seeds; pedicel straight, rarely ± curved; style 0.5–1.5 mm, slender. **SEED**: embryonic roots ± near edge on back of 1 cotyledon. Gen sandy, silty, or alkaline soils; 700–2500 m. GB, DMoj; to WA, Rocky Mtns.

ssp. ***affine*** (E. Greene) Al-Shehbaz **FL**: petals white. **FR** 2–4 cm, upcurved; stalk above receptacle 1–3 mm; pedicel gen spreading, 6–13 mm, straight to ± curved, whitish, base ± flat. Among shrubs, low dunes, meadows; 700–1100 m. SNE, DMoj; to UT.

ssp. ***complanatum*** Al-Shehbaz (p. 449) **FL**: petals lavender to purple, rarely whitish. **FR** 1.5–3 cm, straight or curved; stalk above receptacle 0.5–1 mm; pedicel ± spreading, 2–5(6) mm, ± stout, not whitish, base flat. Alkaline or silty soils, woodland; 1100–2500 m. GB; to OR, UT.

T. laciniatum (Hook.) Endl. (p. 449) Bien. **ST** 2.5–10(14) dm, branched, glabrous, solid, not inflated. **LVS**: basal and lower cauline petioled, lanceolate to deltate-lanceolate, gen pinnately lobed, thickish; middle, upper cauline entire to dentate, base tapered, tip often acuminate. **INFL** dense. **FL**: petals linear, white to purplish, ± crinkled. **FR** spreading, 3.5–10 cm, flat parallel to septum to cylindric, rarely ± 4-sided, ± narrowed between seeds; stalk above receptacle 1–5 mm; pedicel spreading, rarely ascending, 3–7(15) mm, ± straight, stout; base flat; style 0.7–2.5 mm, stout. **SEED**: embryonic root at edge of 1 cotyledon. Rocky hillsides, basaltic cliffs; 600–1900 m. CaR, GB; to B.C., ID.

T. milleflorum Nelson Bien. **ST** 1, 4.5–13 dm, gen branched above, gen ± hollow to infl, inflated or not, glabrous, glaucous. **LVS**: basal and lower cauline petioled, 6–23 cm, ± narrowly oblong to lanceolate or ovate, pinnately lobed, rarely dentate or dissected, tip acute. **INFL** dense. **FL**: petals white, blades oblanceolate to spoon-shaped. **FR** erect to erect-ascending, often appressed, 3.5–8.5 cm, linear, ± flat parallel to septum, rarely cylindric, ± narrowed between seeds; stalk above receptacle 1–4 mm, stoutish; pedicel upcurved, gen erect at tip, 2.5–5 mm, gen stout, base ± flat, tip erect; style 0.5–0.8 mm, stout. **SEED**: embryonic root at edge of 1, rarely both cotyledons. $2n=26$. Sandy soils, shrubland; 1300–2500 m. GB; to WA, UT. [*T. laciniatum* (Hook.) Endl. var. *m.* (Nelson) Payson]

T. stenopetalum S. Watson SLENDER-PETALED THELYPODIUM Bien, glabrous, glaucous. **ST** ± decumbent, 3–8 dm, branched from base. **LVS**: basal 4–15 cm, oblanceolate to oblong, entire or wavy-margined, thickish; cauline ascending, sessile, 2–5 cm, ± oblong-lanceolate, sagittate, clasping st, ± entire to few-toothed or shallowly lobed. **FL**: petals linear, lavender, rarely white, blades crinkled at base; paired filaments free. **FR** 3–5 cm, straight to ± curved, cylindric or rarely 4-angled, ± narrowed between seeds; stalk above receptacle 0.5–3.5 mm, stout; pedicel ascending, rarely ± spreading, 4–8 mm, stout, base not flat; style 1–2 mm, slender. **SEED**: embryonic root at edge or back of 1 cotyledon. **ENDANGERED** CA,US. Alkaline flats, lake shores; 1900–2200 m. SnBr (Bear Valley).

THLASPI PENNY-CRESS

Ann, bien, per; hairs simple, gen 0. **ST** branched or not. **LVS** simple, entire to dentate; basal ± petioled; cauline sessile, clasping st. **FL**: sepal green or purple-tinged, bases not sac-like; petals 1–2 × sepals, white to purplish. **FR** obcordate or obovate to round, flat perpendicular to septum, tip rounded or notched; valves keeled, often winged. **SEEDS** 2–8 per chamber, ± striate; wing 0; embryonic root at edges of both cotyledons. ± 75 spp.: temp, gen n hemisphere. (Greek: to crush shield, from flat fr or perhaps use of crushed seeds as mustard) [Holmgen 1971 Mem NY Bot Gard 21:1–106]

1. Ann, sterile lf-clusters 0, caudices 0; basal lvs 0 or few, shed before fr complete; fr 10–15 mm wide .... ***T. arvense***
1′ Per, gen with sterile lf-clusters from branched caudices; basal lvs gen many; fr < 6 mm wide
2. Fr 2–3 × longer than wide, tip acute; pedicel ascending; basal lvs sparsely dentate .... ***T. californicum***
2′ Fr 1–2 × longer than wide, tip gen truncate or notched, rarely acute; pedicel spreading to ± ascending; basal lvs ± entire .... ***T. montanum*** var. ***montanum***

T. arvense L. (p. 449) FAN-WEED Ann; caudex 0. **ST** 1–5 dm; branches 0–many. **LVS**: basal few, 2–6 cm, short-petioled, oblanceolate, shed early; upper cauline sessile, wavy-margined to dentate, base lobed, clasping st. **FL**: sepals 1.5–2 mm; petals 3–4 mm, white. **FR** 1–1.5 cm wide, widely oblong to round, winged; tip notch 1.5–2.5 mm; pedicel spreading to upcurved, 7–15 mm, slender. **SEED** ± 2 mm, striate; wing 0. $2n$=14. Disturbed areas, fields, roadsides; < 500 m. CA; N.Am; native to Eur.

T. californicum S. Watson KNEELAND PRAIRIE PENNY-CRESS Per, glaucous or not; caudex branched or not. **ST** 1–12(20) cm, simple, slender. **LVS**: basal many, 2–4 cm, sparsely dentate; cauline sessile, 1–1.5 cm, entire to dentate, lobed at base, clasping st, green or purplish. **INFL** 5–10 cm, open. **FL**: sepals 2.5–4 mm, green to purplish; petals spoon-shaped, white. **FR** 7–10 mm, 2–3 × longer than wide, elliptic to obovate; tip acute; wing 0; pedicel ascending; style 1.5–2 mm. **SEEDS** 2–6, dark brown. RARE. Serpentine outcrops; 500–700 m. NCoRO (Kneeland Prairie). [*T. alpestre* L. var. *c.* (S. Watson) Jepson; *T. montanum* L. var. *c.* (S. Watson) P. Holmgren] In cult.

T. montanum L. var. ***montanum*** (p. 449) Per, ± glaucous; caudex branched or not. **STS** 1–many, 1–4 dm, rarely branched, slender. **LVS**: basal many, 1–3 cm, oblong to (ob)ovate, ± entire, green to purplish, petioles 1–2 × blades; cauline sessile, 0.5–2 cm, entire or shallowly dentate, lobed at base, ± clasping st. **INFL** 2–8(25) cm, ± open. **FL**: sepals 2–3.5 mm, green to purplish, margin membranous; petals spoon-shaped, white to purplish pink. **FR** 5–8(12) mm, 1–2 × longer than wide, obovate to obcordate; tip gen truncate or notched, rarely ± acute; wing 0; pedicel spreading to ± ascending; style 1–2 mm. **SEEDS** 4–8, faintly striate. $2n$=28. Talus, alluvial slopes, meadows; < 2000 m. KR, CaR, MP; to WA, Rocky Mtns, Mex. [*T. glaucum* Nelson var. *hesperium* Payson; *T. fendleri* A. Gray var. *h.* (Payson) C. Hitchc.] ❀DRN,IRR,SUN:1–3,7,10, 14–16,18–21;DFCLT.

THYSANOCARPUS LACEPOD, FRINGEPOD

Ann; hairs 0 or simple. **ST** erect to ascending, slender; branches 0–many. **LVS** simple, sessile, entire, dentate, or pinnately lobed. **INFL** elongate; bracts 0. **FL**: sepals ascending, bases not sac-like; petals white or purple; stamens 6, ± equal, ± exserted. **FR** gen elliptic to round, very flat, 1-chambered (septum 0), indehiscent; wing entire to lobed, often perforated; rays 0 to very distinct; pedicel very slender. **SEED** 1, smooth; embryonic root at edges of both cotyledons. 5 spp.: N.Am. (Greek: fringe fr)

1. Fr pedicel sharply bent near tip; fr ± round, (6)8–10 mm, wing wide, rays very distinct, vein-like .... ***T. radians***
1′ Fr pedicel curved to straight; fr elliptic to round, < 8 mm, wing narrow to ± wide, rays 0 or ± indistinct, not vein-like
  2. Fr bowl-shaped, wing strongly curved to flat or concave side; pedicel curved to straight .... ***T. conchuliferus***
  2′ Fr ± flat, wing flat to ± curved to convex side; pedicel curved
    3. Cauline lvs lobed at base, clasping st, often sagittate, (ob)lanceolate, lower dentate to shallowly lobed; style incl in or exserted from fr sinus; fr elliptic to round; wing often perforated .... ***T. curvipes***
    3′ Cauline lvs gen tapered at base, gen not clasping st, linear, lower ± entire to ± deeply lobed; style gen exserted from fr sinus; fr elliptic to obovate; wing perforated or not .... ***T. laciniatus***

T. conchuliferus E. Greene SANTA CRUZ ISLAND FRINGEPOD Pl delicate, glabrous, glaucous. **ST** (4)6–15(20) cm, gen branched. **LVS** sessile, clasping st, 1–4 cm, linear to lanceolate, dentate to pinnately lobed. **INFL** dense, many-fld. **FL**: sepals purplish to whitish; petals 2–2.5 mm, spoon-shaped. **FR** ± elliptic, hairy or not; wing strongly curved to flat or concave side, entire to lobed, perforated or not; rays 0 or ± indistinct; pedicel ± ascending to ± reflexed, 4–8 mm, curved to straight, hairs long-pointed or short-club-shaped; style ± 1 mm, incl in deep sinus. RARE. Open, dry slopes; < 500 m. n ChI (Santa Cruz Island). [*T. laciniatus* var. *c.* (E. Greene) Jepson]

T. curvipes Hook. (p. 449) **ST** 1.5–8 dm, branched or not; hairs gen below. **LVS** (ob)lanceolate; basal, lower cauline ± petioled or not, 1.5–5(7) cm, dentate to shallowly lobed; middle and upper cauline sessile, entire to dentate, base lobed, clasping st. **FL**: sepals ± 1 mm, often purplish, margin white; petals ± = sepals, narrow, white or purple-tinged. **FR** 5–8 mm, elliptic to round, hairy or not; wing entire, wavy-margined, or crenate, often perforated; rays 0 or ± indistinct; pedicel recurved, 4–7 mm; style 0.5–1.5(2) mm, incl in or exserted from sinus. $2n$=28. Common. Slopes, washes, moist meadows; < 1800 m. CA-FP; to B.C., Rocky Mtns, Baja CA. Vars. *elegans* (Fischer & C. Meyer) Robinson, *eradiatus* Jepson, and *longistylus* Jepson indistinct. ❀SUN,DRN:6,**7–9**,**14–24**&IRR:1–3, 10–12.

T. laciniatus Torrey & A. Gray (p. 449) Pl gen glaucous; hairs ± 0. **ST** 1–6 dm, slender, branched or not. **LF** 1–4 cm, linear, gen tapered at base, gen not clasping st, ± entire to deeply linear lobed. **FR** 3–6 mm, elliptic, rarely obovate or round; hairs 0 or ± club-shaped; wing entire, wavy-margined, or crenate, perforated or not; rays 0 or ± indistinct; pedicel recurved, 3–6 mm; style 0.4–0.6 mm, gen exserted from sinus. Common. Slopes, rocky ridges, shaded sites; < 2400 m. s SNF, Teh, SnFrB, SCoR, SW, SNE, D; Baja CA. Vars. *crenatus* (Nutt.) Brewer, *hitchcockii* Munz, *ramosus* (E. Greene) Munz and *rigidus* Munz indistinct. ❀SUN,DRN: **15–17**&SHD:1–3,7,8–12,**14**,**18–24**.

T. radians Benth. (p. 449) Pl glabrous, sometimes exc fr. **ST** 1.5–5 dm; branches few. **LVS**: basal and lower cauline 1.5–5 cm, wavy-margined to pinnately lobed, shed early; cauline ± ovate, sub-entire, base lobed, clasping st. **INFL** elongate, open. **FL**: sepals ± 2 mm, gen purplish; petals ± = sepals, white or purple-tinged. **FR** pendent, (6)8–10 mm, ± round; hairs 0 to dense; wing wide, entire, membranous, not perforated; rays very distinct, vein-like; pedicel ascending, reflexed near tip; style < 0.2 mm, gen not exserted from sinus. Moist slopes, pastures, open meadows; < 1200 m. NW, ScV, SnFrB; OR. ❀SUN,DRN:4,5,**6**,**15–17**&IRR:7,8,9,**14**,18–24.

TROPIDOCARPUM

Ann; hairs gen simple, spreading or reflexed. **ST** prostrate to ± erect, branched. **LVS** deeply pinnately lobed; segments linear to oblong, entire to dentate; basal ± present, not rosetted. **INFL** open; bract lf-like. **FL** < 6 mm; sepals gen spreading, ± 3 mm, ovate-oblong, bases not sac-like; petals obovate to spoon-shaped, yellow or yellowish, purplish-tinged or not; stamens 4 long, 2 short; style slender, stigma obscurely lobed. **FR** linear to oblong, flat perpendicular to narrow septum (or septum ± 0); valves 2 or 4; hairs reflexed. **SEEDS** 1 or 4 rows per chamber, oblong, compressed, brown; wing 0; embryonic root at back of 1 cotyledon. 2 spp.: CA, Baja CA. (Greek: keeled fr)

1. Fr oblong, 1.5–2.5 cm, 3–4 mm wide, septum ± 0, valves 4, seeds 4 rows per chamber .... ***T. capparideum***
1′ Fr narrowly linear, 2.5–5 cm, < 2 mm wide, septum present, valves 2, seeds 1 row per chamber .... ***T. gracile***

T. capparideum E. Greene (p. 449) CAPER-FRUITED TROPIDOCARPUM **ST** ± erect, 1, 2–5 dm, few-branched above. **LF** 2–5 cm. **FL**: sepals yellowish, hairs sparse; petals 4–5 mm, obovate to spoon-shaped, yellow, purplish tinged or not. **FR**: pedicels spreading to ascending, straight to slightly upcurved, lower 1.5–2 cm; valves not or ± keeled; hairs ± 0 to few; style 1–2.6 mm. **PRESUMED EXTINCT**. Alkaline soils, low hills, valleys; < 200 m. nw SnJV (near Mount Diablo). Last seen 1957.

T. gracile Hook. (p. 449) **STS** prostrate to decumbent, several, 1–5 dm, branched below, above. **LF** 1–5 cm. **FL**: sepals greenish yellow or purplish; petals ± 4 mm, obovate, yellow, purplish tinged or not. **FR**: pedicels ascending to ± erect, gen straight, lower 1–1.5 cm; valves keeled; hairs ± 0–few; style 0.5–3 mm. $2n=16$. Common. Grassy banks, open fields, roadsides, pastures; < 1150 m. NCoRI, CaRF, SNF, Teh, GV, SnFrB, SCoRI, SW, w DMoj; Baja CA. [var. *dubium* (Davidson) Jepson]

BUDDLEJACEAE BUDDLEJA FAMILY

Elizabeth McClintock

Shrub, tree, rarely herb; hairs gen stellate, branched, scale-like or glandular. **LVS** simple, gen opposite, entire, toothed or lobed; stipules at least partially on st, often ridge-like. **INFL**: cyme but appearing to be a panicle, head, raceme, or spike, terminal or axillary, gen dense. **FL** bisexual, sometimes functionally unisexual, ± radial; calyx lobes 4–5; corolla lobes 4–5; stamens 4–5, attached to corolla tube; ovary superior or half inferior, chambers 2 or 4, style 1, stigma elongate or ± spheric. **FR**: capsule, rarely berry. **SEEDS** small, often winged. ± 10 genera, 150 spp.: trop, subtrop; some cult for orn. [Rogers 1986 J Arnold Arb 67:143–185] Sometimes incl in Loganiaceae.

BUDDLEJA BUTTERFLY BUSH

Shrub, tree, deciduous or evergreen. **LVS** rarely alternate or ± whorled, lanceolate, oblong, or linear, short-petioled or sessile. **FL** gen fragrant; calyx ± bell-shaped, lobes gen 4, ± = or < tube; corolla bell-shaped, funnel-shaped, or salverform, lobes gen 4, < to << tube, abruptly spreading; stamens gen 4, anthers ± sessile. **FR** 2-parted; calyx persistent. **SEEDS** many, often winged. ± 100 spp.: Am, Afr, Asia. (Rev. Adam Buddle, England, 1660–1715) [Norman 1967 Gentes Herb 10:47–114] Often spelled *Buddleia*, perhaps incorrectly.

1. Pl densely hairy ± throughout; infl spheric, head-like .... ***B. utahensis***
1′ Pl densely hairy only on lower lf surfaces; infl cylindric or widely spreading, panicle-like
 2. Lf lanceolate, gen > 10 cm; corolla lilac to purple; stamens not exserted from corolla tube .... ***B. davidii***
 2′ Lf linear to oblong, gen < 10 cm; corolla white or cream-colored; stamens exserted from corolla tube ***B. saligna***

B. davidii Franchet BUTTERFLY BUSH, SUMMER LILAC Shrub < 5 m, deciduous or semi-evergreen. **LF** 5–30 cm, lanceolate, serrate to ± entire; upper surface dark green, hairs sparse, gen branched; lower surface white, hairs dense, gen branched. **INFL** panicle-like, 15–25 cm, slender, cylindric. **FL**: calyx ± 3 mm; corolla salverform, lilac to purple with central orange spot, tube ± 10 mm, lobes 2–3 mm; stamens incl. **FR** 5–6 mm. **SEEDS** winged at both ends. Disturbed, gen urban, often ± damp; < 200 m. NCo, CCo, SnFrB; native to China. Commonly cult in CA.

B. saligna Willd. Shrub, small tree < 7 m, deciduous. **ST**: twig ± 4-angled. **LF** 1.5–10 cm, linear to oblong, entire, ± rolled under; upper surface glabrous, shiny, olive-green; lower surface white, hairs dense, stellate. **INFL** panicle-like, ± 12 cm, widely spreading. **FL**: corolla ± salverform, white or cream-colored, sometimes with reddish spot in center, tube ± 2.5 mm, lobes ± 1 mm; stamens exserted. **FR** ± 2 mm, ovoid, exserted from persistent calyx. **SEEDS** not winged. Disturbed areas, chaparral; 200–700 m. SCo (Santa Monica Mtns.); native to S.Afr. [*Chilianthus oleaceus* Burchell] Rarely cult in CA.

B. utahensis Cov. (p. 449) PANAMINT BUTTERFLY BUSH Shrub < 5 dm, densely branched, deciduous, dioecious; hairs dense, stellate or branched ± throughout. **LF** 1.5–3 cm, linear-oblong, thickish; margin entire to wavy, rolled under. **INFL** paired at upper nodes into single, head-like, spheric, dense clusters ± 10–15 mm wide. **FL** unisexual; calyx 3–4 mm; corolla 4–5 mm, salverform, creamy yellow, becoming purplish or brown-purple, lobes ± 1 mm; stamens incl. **FR** spheric to oblong. **SEEDS** not winged. Uncommon. Slopes, often on dolomite, volcanic rocks, limestone; 900–1700 m. DMoj; to UT.

BURSERACEAE TORCHWOOD FAMILY

Niall F. McCarten

Trees, shrubs, dioecious or monoecious. **ST** gen erect, < 15 m. **LVS** simple or compound; cauline, gen alternate, deciduous, petioled; lf axis often winged, glabrous to densely hairy. **INFL**: panicle or fl solitary. **FL** radial; disk ring- or cup-shaped; sepals 3–5; petals 0–5; stamens gen 1–2 × number of petals; ovary superior, chambers 2–5, style 0 or 1. **FR**: drupe or capsule; stones 1–5, each 1-seeded. 17 genera; 500 spp.: worldwide esp trop; some cult (*Boswellia* frankincense, *Commiphora* myrrh, *Bursera*).

BURSERA ELEPHANT TREE, TOROTE

Trees, aromatic. **ST** < 10 m; bark smooth, shedding. **INFL**: panicle. **FL**: sepals 5; petals 5, arising from disk; stamens 10; ovary chambers 3. **FR**: valves 2–3. 60 spp.: trop Am. (J. Burser, born 1500's)

B. microphylla A. Gray (p. 449) ELEPHANT TREE **ST** < 4 m; branches spreading, gen red; mature bark white. **FL**: sepals ± 5 mm; petals ± 4 mm, white to cream-colored. **LF** pinnately compound, 2–8 cm, glabrous; lflets 7–33, 5–10 mm. **FR**: valves 3; stone yellow. RARE in CA. Rocky slopes; < 700 m. w edge DSon (San Diego Co.); to AZ, Mex. Population highly localized. *B. hindsiana* (Benth.) Engler (some lvs simple) has been reported but not confirmed from PR (s San Diego Co.).

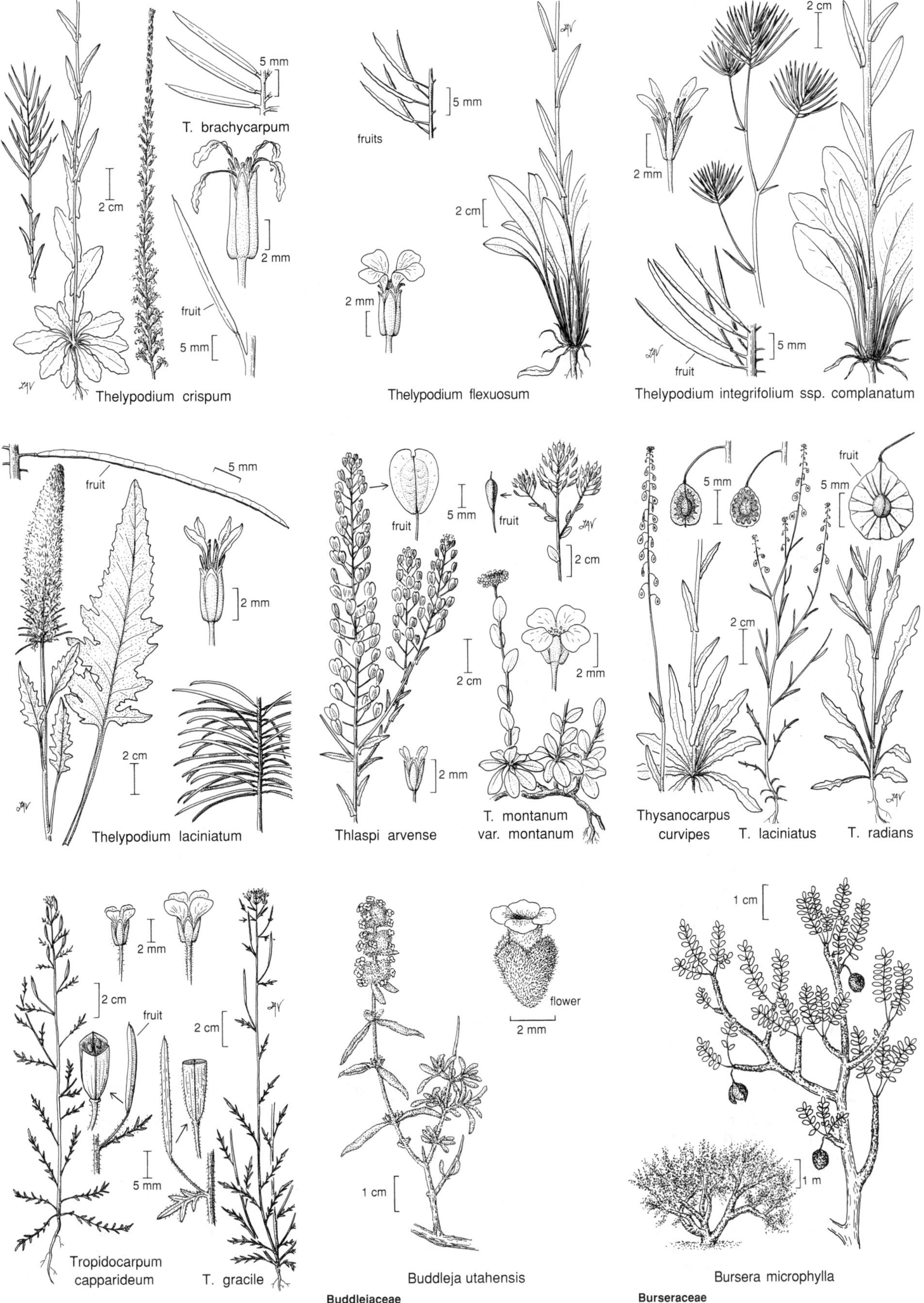

Buddlejaceae

Burseraceae

CABOMBACEAE WATERSHIELD FAMILY

William J. Stone

Aquatic per in freshwater; rhizomes in mud. **STS** elongate, leafy; ends floating. **LVS** of 2 types; alternate, long-petioled with floating, peltate blades; opposite to whorled, short-petioled, submersed blades deeply dissected. **FL** solitary, above surface of water, bisexual; sepals 2–4; stamens 3–6 (*Cabomba*), 12–18 (*Brasenia*), filaments slightly flattened, anthers opening lengthwise; ovary superior, pistils simple, 2–many, ovules 1–3, styles terminal or decurrent. **SEEDS** 1–3. 2 genera, 8 spp.: temp & trop Am, Afr, e Asia, Australia. Some *Cabomba* spp. cult for aquaria, may occur as waifs in CA.

BRASENIA WATERSHIELD

1 sp. (origin unknown). Sometimes placed in Nymphaeaceae, but differs in having simple pistils.

B. schreberi J. Gmelin (p. 453) **ST** slender, 3–20 dm, branching, reddish brown; submersed parts covered with thick gelatinous coating. **LVS** alternate, long-petioled, floating; blades 4–12 cm, 3–8 cm wide, oval, centrally peltate. **INFL**: fl solitary in axils. **FL** reddish purple; sepals, petals 3, 10–15 mm, linear-oblong; stamens 12–18, filaments thread-like; pistils 4–18, separate, styles decurrent. **FR** 6–8 mm, oblong, leathery, indehiscent. Ponds, ditches, slow streams; < 2200 m. NW, SNF, SNH, ScV, MP; temp N.Am. ❀shallow, fresh water:1–3,**4–7**,8,9,**14–18**,19–24.

CACTACEAE CACTUS FAMILY

Edward F. Anderson (except *Opuntia*)

Per, shrub, tree, gen fleshy. **ST** cylindric, spheric, or flat; surface smooth, tubercled, or ribbed (fluted); nodal areoles bear fls, gen bear spines from center ("central spines") and margin ("radial spines") (*Opuntia* areoles bear small, barbed, deciduous bristles sometimes called glochids, gen also bear spines). **LF** gen 0. **FL** gen solitary, bisexual, sessile, ± radial; perianth parts gen many, grading from scale-like to petal-like; stamens many; ovary appearing inferior, ± submerged in st, so gen with areoles on surface, style 1, stigma lobes gen many. **FR** gen fleshy, gen indehiscent, spiny, scaly, or smooth. **SEEDS** many. 93 genera, ± 2000 spp.: esp Am deserts; many cult. (Greek: thorny pl) [Benson 1982 Cacti of US & Can; Hunt & Taylor eds 1990 Bradleya 8:85–107]

1. St clearly jointed; small barbed bristles present in areoles; seed white, bone-like .... **OPUNTIA**
1′ St not clearly jointed; barbed bristles 0; seed black or brown
 2. St ribs 0 or inconspicuous, tubercles prominent
  3. Tubercle longitudinally grooved on top (indented in X-section); central spine not hooked .... **ESCOBARIA**
  3′ Tubercle round in X-section (not grooved); some central spine of areole hooked .... **MAMMILLARIA**
 2′ St ribs prominent, tubercles 0 to prominent
  4. Pl > 3 m; st > 30 cm diam, gen branching above 1.5 m; fl creamy white .... **CARNEGIEA**
  4′ Pl < 3 m; st < 30 cm diam, branching near ground or unbranched; fl yellow to red or magenta
   5. Sts length gen > 8 × width; empty fr long-persistent .... **BERGEROCACTUS**
   5′ Sts length gen < 8 × width; fr not long-persistent
    6. Ovary and young fr spiny, glabrous; st soft-fleshy; branches gen few–many .... **ECHINOCEREUS**
    6′ Ovary and young fr either spineless or woolly; st firm-fleshy; branches gen 0 (if present, then larger spines with ring-like ridges)
     7. Fr and st tip densely woolly; bracts sharp-tapered .... **ECHINOCACTUS**
     7′ Fr and st tip not woolly; bracts wide, obtuse to acute
      8. St > 15 cm diam; seed pitted .... **FEROCACTUS**
      8′ St < 15 cm diam; seed smooth or weakly tubercled .... **SCLEROCACTUS**

BERGEROCACTUS GOLDEN CEREUS

1 sp. (A. Berger, German succulent specialist, 1871–1931)

B. emoryi (Engelm.) Britton & Rose (p. 453) GOLDEN-SPINED CEREUS Pls ± spreading. **STS** many, gen ascending, densely clumped, each < 3 m, 3–6 cm diam, cylindric, ± obscured by dense spines; ribs gen 12–16, prominent; tubercles indistinct; spines 20–30, yellow, becoming darker; central spines 1–3, the longest curved down; radial spines straight. **FL** 2.5–4 cm diam; outer perianth parts yellow with green midveins, inner yellow. **FR** at first fleshy, becoming dry, spiny, 2.5–3 cm diam, spheric, dehiscing at tip. **SEED** 3 mm, ± flat-obovoid, shiny black. $2n=44$. RARE in CA. Sandy soils, dry hills along coast; < 100 m. SCo (San Diego Co.), s ChI; Baja CA. [*Cereus e.* Engelm.] Threatened by development, collecting, feral goats. In cult.

CARNEGIEA SAGUARO

Tree. **ST** erect, 3–16 m, 30–75 cm diam, cylindric, gen few-branched; ribs 12–30, prominent; tubercles indistinct; spines 15–30, ± dense, spreading, gray. **FL** 5–6 cm diam, nocturnal; perianth creamy white. **FR** scaly, 25–45 mm diam, obovoid, dehiscing vertically, red, edible. **SEED** 2 mm, obovoid, black. 12 spp.: CA, AZ, Mex. (Andrew Carnegie, Am industrialist, philanthropist, 1835–1919)

C. gigantea (Engelm.) Britton & Rose (p. 453) **ST** columnar, massive, fleshy. **FL** 8.5–12.5 cm; outer parts green, margins lighter; inner parts petal-like, white; ovary green, scaly, style 10–15 mm. $2n=22$. RARE in CA. Rocky hills, plains; < 1500 m. e DSon; AZ, Mex. [*Cereus g.* Engelm.] In cult.

ECHINOCACTUS CLUSTERED BARREL CACTUS

STS 1–30, often in ± 1 m diam clumps, each 3–6 dm, 10–20 cm diam, spheric to columnar; ribs 13–21, prominent; tubercles indistinct; spines dense. **FL** 4–5 cm diam; perianth yellow tinged with pink. **FR** dry, densely woolly. **SEED** 2–2.5 mm, 2–3.5 mm diam, black. 6 spp.: sw US, Mex. (Greek: hedgehog cactus)

E. polycephalus Engelm. & J. Bigelow var. ***polycephalus*** (p. 453) **ST** < 1 m; central spines 4, 6–7.5 cm, red or yellow, canescent at first, spreading, lowermost curved slightly; radial spines 6–8, 3–4.5 cm, spreading, slightly curved, red or yellow. **FL**: ovary densely woolly, with sharp-tapered bracts, style ± 20 mm. $2n=22$. Rocky hills, silty valleys; < 1000 m. e SnBr, DMoj, n DSon; to AZ, Mex. ❀DRN,DRY:2,3,**10–12**,13,19&SUN:20,21.

ECHINOCEREUS HEDGEHOG CACTUS

STS 1–many, often densely clumped, each < 1 m, 2–10 cm diam, cylindric; ribs 5–13, prominent; tubercles ± indistinct; spines straight or curved. **FLS** on old growth, often near upper margin of spine-bearing areoles; ovary spiny. **FR** spheric to ovoid, glabrous; spines deciduous. **SEED** ovoid to ± spheric, tubercled, gen black. 47 spp.: sw US, Mex. (Greek: hedgehog candle) [Taylor 1985 Genus Echinocereus]

1. Perianth purplish to lavender, closing at night, inner parts gen acuminate or bristle-tipped; anthers yellow; spines glabrous, at least 1 central spine gen angled to flat ..... ***E. engelmannii***
1′ Perianth orange to red, remaining open at night, inner parts round to notched; anthers pink to lavender; spines < 1 year old puberulent near tip, all ± angled ..... ***E. triglochidiatus***

E. engelmannii (Engelm.) Lemaire (p. 453) Pl branched, forming clumps or mounds < 1 m diam. **STS** < 60, ± erect, 5–60 cm, 4–9 cm diam, cylindric, green; ribs 10–13; tubercles indistinct; areole wool present in first year only; spines variable in color and shape, always present, glabrous; central spines 2–7, < 8 cm, spreading, straight to twisted; radial spines 6–14, 2–20 cm. **FL** 5–7.5 cm diam, short-funnel-shaped; perianth purplish to magenta or lavender, inner parts gen acuminate or bristle-tipped (sometimes round); anthers yellow. **FR** fleshy, 20–30 mm, spheric, red, edible; spine clusters deciduous. $2n=44$. Many dry habitats; < 2400 m. SnBr, PR, D; to UT, AZ, Mex. [*E. munzii* (Parish) L. Benson] Highly variable; sometimes divided into varieties, but not satisfactorily. More study needed. ❀DRN,DRY:2,3,7,10–24.

E. triglochidiatus Engelm. (p. 453) Pl gen forming dense mounds. **STS** 1–500, 5–40 cm, 5–15 cm diam, ± spheric to cylindric, light-to bluish green; ribs 5–12; tubercles ± obvious; areole wool persistent; spines highly variable, ± angled, gen gray, those < 1 year old puberulent near tip; central spines 1–6, difficult to distinguish from radial spines. **FL** < 9 cm, funnel-shaped; perianth orange to red, inner parts round or evenly notched. **FR** 20–25 mm, 10–15 mm diam, pink to red; spines deciduous. $2n=22$. Many habitats; 150–3000 m. D; to Rocky Mtns, TX, Mex. [*E. mojavensis* (Engelm. & J. Bigelow) Ruempler] Highly variable; sometimes divided into varieties, but not satisfactorily. More study needed. ❀DRN,DRY:2,3,10,11,14,18–23.

ESCOBARIA BEEHIVE CACTUS

STS 1–200, gen in ± 50 cm clumps, each 2–15 cm, 2–15 cm diam, ± depressed to cylindric; ribs inconspicuous; tubercles grooved on upper surface from areole to base; central spines straight, ± following tubercle axis; radial spines wide-spreading ± in 1 plane from tubercle tip. **FL** 1–3(6) cm diam; outer perianth parts ciliate. **FR** becoming dry, spheric to club-shaped, red or green; perianth persistent. **SEED** reniform, black or brown, pitted. 16 spp.: w US, Mex. (R. & N. Escobar, Mexico) [Taylor 1986 Cact Succ J Gr Brit 4:36–44]

E. vivipara (Nutt.) F. Buxb. (p. 453) **ST** 2–15 cm, 2–15 cm diam; tubercles 6–9 mm; spines dense, central spines 3–12, white, tip darker; radial spines 12–40, 9–25 mm, straight, white. **FL** 2.5–5 cm diam, pink, red, lavender, or yellow-green. **FR** 12–25 mm, elliptic in outline, green, sometimes with a few scales. Sandy to rocky soils; 75–2700 m. D; to UT, AZ, Mex. [*Coryphantha v.* (Nutt.) Britton & Rose; *Mammillaria v.* (Nutt.) Haw.]

1. Central spines < 8; perianth straw-yellow, or pink ..... var. ***deserti***
1′ Central spines > 8; perianth magenta, purplish, or pink
 2. St 6–8 cm diam, cylindric; fl 3 cm diam, magenta to pink; < 1500 m, s DMoj, DSon ..... var. ***alversonii***
 2′ St 7–15 cm diam, ovoid-spheric; fl 3–5 cm diam, magenta to purplish; > 1500 m, e DMoj ..... var. ***rosea***

var. ***alversonii*** (J. Coulter) D. Hunt (p. 453) FOXTAIL CACTUS **ST** gen 1, 10–15 cm, 6–8 cm diam, cylindric; central spines 8–10; radial spines 12–18. **FL** ± 3 cm diam, magenta to pink. RARE. Sandy or rocky areas, creosote bush scrub; 75–600 m. s DMoj, DSon. [*Coryphantha v.* var. *a.* (J. Coulter) L. Benson; *Mammillaria a.* (J. Coulter) H. Zeissold] Threatened by collecting. In cult.

var. ***deserti*** (Engelm.) D. Hunt **STS** 1–few, 7–15 cm, 7–9 cm diam, cylindric to ovoid; central spines 4–6; radial spines 12–20. **FL** 2–3 cm diam, straw-yellow, yellow-green, or pink. Limestone soils; 1000–2400 m. DMtns (e San Bernardino Co.); to sw UT, nw AZ. [*Coryphantha d.* (Engelm.) Britton & Rose; *Mammillaria d.* Engelm.] ❀DRN,DRY;DFCLT.

var. ***rosea*** (Clokey) D. Hunt VIVIPAROUS FOXTAIL CACTUS **STS** 1–several, 7–18 cm, 7–15 cm diam, ovoid-spheric; central spines 10–12; radial spines 12–18. **FL** 3–5 cm diam, magenta to purplish. $2n=22$. RARE. Limestone slopes, hills; 1500–2700 m. DMtns (ne San Bernardino Co.); s NV, nw AZ. [*Coryphantha v.* var. *r.* (Clokey) L. Benson] Threatened by collecting. In cult.

FEROCACTUS BARREL CACTUS, VISNAGA

ST gen 1, < 3 m, 20–35 cm diam, depressed-spheric to columnar; ribs 15–27; tubercles inconspicuous; spines dense. **FL** 3–6 cm diam; tube poorly developed; ring of hairs separating perianth and stamens; ovary densely scaly. **FR** 1–2 cm diam, scaly, gen yellow, opening by a pore near base. **SEED** 1.5–3 mm, black, pitted; scar basal. 23 spp.: sw US, Mex. (Latin: fierce cactus) [Taylor 1984 Bradleya 2:19–38]

1. Pl < 0.3 m; ribs 13–20; spines < 5 cm; coastal ..... ***F. viridescens***
1′ Pl > 1 m; ribs 18–27; spines gen > 5 cm; interior ..... ***F. cylindraceus***
 2. Central spines 7.5–17 cm; seed 2–3 mm; < 600 m ..... var. ***cylindraceus***
 2′ Central spines 5–7 cm; seed 1–2 mm; > 700 m ..... var. ***lecontei***

F. cylindraceus (Engelm.) Orc. CALIFORNIA BARREL CACTUS **ST** 10–30 dm, gen taller than wide, spheric or columnar; ribs 18–27; spines 10–18, erect and spreading, longest recurved, gen with some red, becoming gray. **FL**: perianth parts yellow with red base; ovary 9–12 mm, scales fringed, style 12–20 mm. **FR** yellow. **SEED** 1–3 mm. 2*n*=22. UNCOMMON. Gravelly, rocky, or sandy areas; 60–1500 m. D (esp e DMoj, w DSon); to UT, Mex. [*Echinocactus c.* Engelm.; *F. acanthodes* (Lemaire) Britton & Rose misapplied (rejected name)] Threatened by collecting; monitoring needed.

var. ***cylindraceus*** (p.453) **ST**: central spines 7.5–17 cm. **SEED** 2–3 mm. Gravelly or rocky places; < 600 m. D; sw AZ, Baja CA. ❀DRN,DRY:**12**,13,18–21&SUN:7,14,16,17,22–24.

var. ***lecontei*** (Engelm.) H. Brav.-Holl. **ST**: central spines 5–7 cm. **SEED** 1–2 mm. Gravelly, rocky, or sandy places; > 700 m. D; to sw UT, AZ, n Mex. ❀DRN,DRY:**10-12**,13,18–21&SUN:7,14, 16,17,22–24.

F. viridescens (Torrey & A. Gray) Britton & Rose (p. 453) COAST BARREL CACTUS **ST** 1.5–3 dm, gen wider than tall, depressed-spheric to cylindric; ribs 13–20; spines 14–24, gen spreading, longest slightly curved, red, becoming ± gray or yellow. **FL**: perianth parts pink, green, or yellow; ovary 5–7 mm, scales fringed-dentate, style 6–9 mm. **FR** yellow to reddish. **SEED** 1–2 mm. 2*n*=22. UNCOMMON. Sandy to rocky areas; 10–150 m. SCo (San Diego Co); Baja CA. Threatened by urbanization, off road vehicles, collecting. In cult.

MAMMILLARIA NIPPLE CACTUS, FISH-HOOK CACTUS

STS 1–many, 2–30 cm, 2–20 cm diam, spheric to cylindric; ribs inconspicuous; tubercles round in X-section; central spines gen hooked. **FL** 1–5 cm diam; ovary glabrous. **FR** ovoid to cylindric, red, glabrous. **SEED** black, pitted. 150 spp.: N.Am. (Latin: nipple) [Hunt 1984–87 Bradleya 2:65–96; 3:53–66; 5:17–48]

1. Radial spines > 30; seed with corky aril ..... ***M. tetrancistra***
1′ Radial spines < 30; seed without corky aril
 2. Radial spines white; perianth yellow to white; bristles present in axils of tubercles ..... ***M. dioica***
 2′ Radial spines light brown to red; perianth lavender to reddish purple; axillary bristles 0 ..... ***M. milleri***

M. dioica M.K. Brandegee (p. 457) Pl with either all bisexual or all pistillate fls. **ST** gen 1 (–many), 5–30 cm, 3–7 cm diam; bristles present in axils of tubercles; central spines 1–4, 8–15 mm; radial spines 11–22 in 1 rank, 4–10 mm, white. **FL** 10–22 mm, 20–40 mm diam; perianth yellow to white. **FR** 10–25 mm, club-shaped to ovoid. **SEED**: aril 0. 2*n*=44,66. Hillsides, washes, coastal scrub to creosote-bush scrub; 10–1500 m. SCo, w edge DSon; Baja CA. [var. *incerta* (Parish) Munz] ❀DRN,DRY:17,19–24.

M. milleri (Britton & Rose) Boed. (p. 457) **STS** gen several (1–many), 7–15 cm, 4–7 cm diam; axillary bristles 0; central spines 1–2, 12–15 mm, longer hooked; radial spines 18–28 in 1 rank, 6–12 mm, light brown to red. **FL** 15–25 mm, 20–30 mm diam; perianth lavender to reddish purple. **FR** 12–25 mm, cylindric, green, becoming red. **SEED**: aril 0. 2*n*=22. Uncommon. Sandy or rocky canyons, washes, plains, creosote-bush scrub; 300–900 m. ne DSon (se San Bernardino Co.); AZ, n Mex. [*M. microcarpa* Engelm. misapplied (invalid name)] ❀DRN,DRY;DFCLT.

M. tetrancistra Engelm. (p. 457) **ST** gen 1, 7–25 cm, 3.5–7.5 cm diam, cylindric; axillary bristles 0; central spines 3–4, 18–25 mm, gen hooked, tips dark; radial spines 30–60 in 2–3 ranks, 10–25 mm, white or tips dark. **FL** 30–40 mm, 25–40 mm diam; perianth deep pink to lavender. **FR** 15–32 mm, cylindric. **SEED**: aril corky. 2*n*=22. Uncommon. Sandy hills, valleys, plains, creosote bush scrub; 130–1400 m. D; to UT, AZ, n Mex. ❀DRN,DRY; DFCLT.

OPUNTIA PRICKLY-PEAR, CHOLLA

Bruce D. Parfitt & Marc A. Baker

Shrubs, trees; roots fibrous. **ST** gen erect, < 12 m; segments flat to cylindric, gen firmly attached; tubercles gen elongate along st; ribs sometimes present; spines 0–many, sometimes flat, tip smooth or barbed, epidermis persistent or separating as a papery sheath; small, barbed deciduous bristles gen many. **LF** small, conic, fleshy, deciduous, obvious on young sts and ovaries. **FR** juicy, fleshy or dry; wall thick, bearing areoles. **SEED** dark brown, encased in a bony, whitish aril. 200 spp.: Am; *O. ficus-indica* cult for food, others for orn. (Possibly from Papago Indian name ("opun") for this food pl; or named for a spiny pl of Opus, Greece) Spines smaller, fewer in shade forms; when yellow, blacken with age. Hybridization common within subgenera.

1. St segments cylindric to club-like, tubercled; spine epidermis sheath-like, gen deciduous
 2. Major spines distinctly flat, epidermal sheath separating only from tip (subg. *Corynopuntia*)
  3. Perianth yellow; largest spine with cross-rows of rough papillae; ovary bristles stiff, with reflexed barbs ..... ***O. parishii***
  3′ Perianth pink-purple; largest spine smooth; ovary bristles hair-like, with ascending barbs ..... ***O. pulchella***
 2′ Spines not flat, epidermal sheath separating from entire spine (subg. *Cylindropuntia*)
  4. Fr densely spiny
   5. St < 1 cm diam, tubercle < 2 mm high ..... ***O. ramosissima***

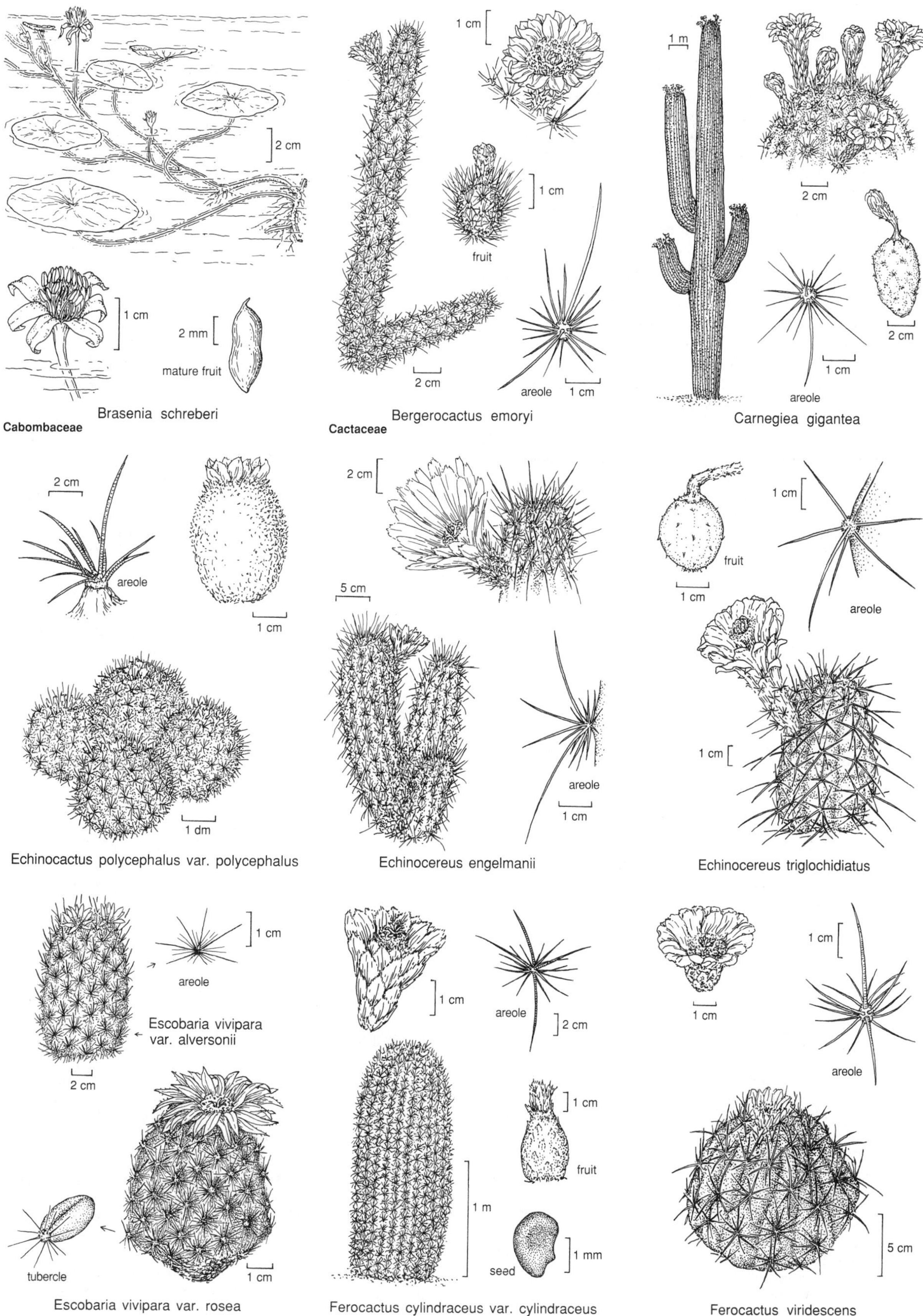
2 cm
1 cm
2 mm
mature fruit
Brasenia schreberi
Cabombaceae
1 cm
1 cm
fruit
2 cm
areole
1 cm
Bergerocactus emoryi
Cactaceae
1 m
2 cm
2 cm
1 cm
areole
Carnegiea gigantea
2 cm
areole
1 cm
1 dm
Echinocactus polycephalus var. polycephalus
2 cm
5 cm
areole
1 cm
Echinocereus engelmanii
1 cm
fruit
1 cm
areole
1 cm
Echinocereus triglochidiatus
1 cm
areole
Escobaria vivipara
var. alversonii
2 cm
tubercle
1 cm
Escobaria vivipara var. rosea
1 cm
areole
2 cm
1 cm
fruit
1 m
seed
1 mm
Ferocactus cylindraceus var. cylindraceus
1 cm
1 cm
areole
5 cm
Ferocactus viridescens

5′ St > 1.5 cm diam, tubercle > 3 mm high
6. Terminal st segment gen < 2 dm, tubercle length < 2 × width .... ***O. echinocarpa***
6′ Terminal st segment gen > 2 dm, tubercle length > 3 × width
7. St gen branched only above; trunk 1 .... ***O. acanthocarpa*** var. ***coloradensis***
7′ St gen branched near base; trunks several–many
8. St gen < 2 cm diam, filaments green .... ***O. parryi***
8′ St gen > 2.5 cm diam, filaments red- to green-purple .... ***O. wolfii***
4′ Fr spineless (sometimes with a few deciduous bristles)
9. Fr gen bearing fls, perianth purple-red .... ***O. prolifera***
9′ Fr not bearing fls, perianth yellow, greenish, or red-brown
10. Terminal st segment > 3 × longer than wide .... ***O. parryi***
10′ Terminal st segment gen < 3 × longer than wide
11. Tubercle length ± = width .... ***O. bigelovii***
11′ Tubercle length ± 2 × width
12. Spines 7–10 per tubercle; fr < 2.5 cm; seeds irregularly shaped, < 3 mm .... ***O. ×fosbergii***
12′ Spines 9–16 per tubercle; fr 2.5–3.5 cm; seeds ± spheric, ± 3 mm .... ***O. ×munzii***
1′ St segments flat, not tubercled; spine epidermis not separable (subg. *Opuntia*)
13. St gen minutely papillate-hairy, spines 0 (2–8 per areole in c Kern Co.); perianth pink to magenta; filaments red; fresh stigma white .... ***O. basilaris***
14. Spines 2–8, rigid, < 2.6 cm; rare, c Kern Co. .... var. ***treleasei***
14′ Spines 0 (barbed bristles present)
15. St segment length < 2 × width, width > 5 × thickness .... var. ***basilaris***
15′ St segment length > 2 × width, width < 2 × thickness .... var. ***brachyclada***
13′ St glabrous, gen bearing spines; perianth gen yellow to dull red (rarely pink); filaments white, yellow, red, or pink; fresh stigma green; not in Kern Co.
16. Terminal st segment easily detached, < 3.5 cm, width < 3 × thickness; w CaRH (Siskiyou Co.) .... ***O. fragilis***
16′ Terminal st segments firmly attached when fresh, > 4 cm, width > 4 × thickness; mostly SW, D
17. Fr becoming dry, tan, gen spiny; perianth yellow to pink-magenta; style slender, white, base slightly thicker .... ***O. erinacea***
18. Ovary spiny; spines gen in all st areoles; smaller spines at lower edge of st areole 3 or more, strongly reflexed .... var. ***erinacea***
18′ Ovary spineless; spines in < 50% of st areoles; smaller spines at lower edge of st areole 0–4, ± reflexed .... var. ***utahensis***
17′ Fr juicy, gen with some deep purple, spines 0; perianth yellow, orange, or red; style thick, white to red, base much thicker
19. Trunk 1, erect; areoles gen > 38 per ovary
20. Mature pl > 3 m; st bristles < 1 mm, inconspicuous; ovary clearly tubercled when fresh .... ***O. ficus-indica***
20′ Mature pl < 2.5 m; st bristles 2–12 mm, conspicuous; ovary smooth
21. Longest spines gen 32–47 mm; style white when fresh; seed 3 mm; se CA-FP, D .... ***O. chlorotica***
21′ Longest spines 19–25 mm; style red or pink when fresh; seed 3.5–4 mm; w CA-FP .... ***O. oricola***
19′ Basal branches several, ± decumbent to ascending; areoles < 36 per ovary
22. Style and filaments white
23. St segment > 15 cm wide; perianth base yellow; fr red inside .... ***O. engelmannii***
23′ St segment < 14 cm wide; perianth base red; fr green inside .... ***O. phaeacantha***
22′ Style pink or filaments yellow (or both)
24. St segment > 15 cm wide .... ***O. ×occidentalis***
24′ St segment < 15 cm wide
25. St segment oblong-elliptic or narrowly obovate; major spines gen round, 4–11 per areole .... ***O. littoralis***
25′ St segment obovate; major spines gen flat, 1–4 per areole .... ***O. ×vaseyi***

O. acanthocarpa Engelm. & J. Bigelow var. ***coloradensis*** L. Benson (p. 457) BUCKHORN CHOLLA **ST** ± tree-like, < 4 m; segments cylindric, terminal < 40 cm, 2–2.5 cm diam; tubercle 20–40 mm, 5–7 mm high; spines 12–21, < 5 cm, pale yellow- to red-brown, sheath pale yellow-brown. **FL**: inner perianth < 3.5 cm, yellow with purple to brown-red tint; filaments purple-red. **FR** dry, tubercled; base acute; lower tubercles >> upper. **SEED** < 6 mm. $2n=22$. With creosote bush, Joshua tree; < 1300 m. D; NV, AZ. *O. deserta* Griffiths is probably *O. a.* × *O. echinocarpa*. ❀DRN, DRY,SUN:**10**,11,**12**,14,18,**19–24**.

O. basilaris Engelm. & J. Bigelow BEAVERTAIL CACTUS **ST** clumped, ascending to erect, 7–40 cm; segments flat, gen erect, 5–21 cm, gen puberulent; spines gen 0(–8); bristles many. **FL**: inner perianth pink-magenta; filaments deep magenta-red; style white or pink, stigma white. **FR** 2–4 cm, becoming dry, green and purple becoming tan, gen puberulent; areoles 24–76. **SEED** 6.5–9 mm, ± spheric. Desert, chaparral, pinyon-juniper woodland; 120–2200 m. s SN, Teh, se SnJV, SnGb, SnBr (and adjacent SCo), e PR, s SNE, D; to UT, AZ, Mex.

var. ***basilaris*** (p. 457) BEAVERTAIL **ST**: segment 8–21 cm, 5–13 cm wide, flat, ± obovate; spines 0. $2n=22$. Desert to pinyon-juniper woodland; 150–2200 m (higher n). Range of sp. exc SnJV; < 900 m in SnGb, SnBr. [var. *ramosa* Parish, *O. whitneyana* E. Baxter] ❀DRN,DRY,SUN:2,**3**,**7–14**,16,17,**18–24**.

var. ***brachyclada*** (Griffiths) Munz (p. 457) SHORT-JOINT BEAVERTAIL **ST**: segment 5–13 cm, 1.5–5 cm wide, slightly flat, ± cylindric to ± club-shaped; spines 0. $2n=22$. RARE. Chaparral; 1200–1800 m. SnGb, SnBr. [*O. humistrata* Griffiths] Weakly differentiated from var. *basilaris*. Threatened by collecting.

var. ***treleasei*** (J. Coulter) Toumey (p. 457) BAKERSFIELD CACTUS **ST**: segment 9–20 cm, 5–7.5 cm wide, flat, gen obovate; spines 2–8, 7–26 mm, spreading, gen straight, yellow. **FL**: spines 0–6, < 11 mm, pale yellow. $2n=22,33$. **ENDANGERED** CA, US. Arid plains; 120–150 m. Teh, se SnJV (Kern Co.). [*O. treleasei* J. Coulter] Threatened by habitat loss.

O. bigelovii Engelm. (p. 457) TEDDY-BEAR CHOLLA **ST** 1–2 m; trunk 1; main branches few, short, spreading, becoming black; segments cylindric, terminal gen < 10 cm, 4–6 cm diam, easily detached; tubercle 4–11 mm, 3 mm high; spines 4–10, < 2.5 cm, very pale yellow-brown, sheath translucent to pale brown. **FL**: inner perianth 1.5–4 cm, yellow; filaments green. **FR** 1–2 cm, leathery, tubercled, yellow; spines 0 or bristles few. **SEED** gen sterile. 2*n*=33 (rarely 22). Rocky fans, benches, with creosote bush; < 1000 m. DSon; s NV, AZ, n Mex. Reproduces by rooting of detached segments. ❀DRN,DRY,SUN:10,11,**12,13,19–21**,22–24.

O. chlorotica Engelm. & J. Bigelow PANCAKE PRICKLY-PEAR **ST** ± tree-like, 1.5–2 m; segments flat, 13–20 cm, gen round; spines 3–8 in at least upper areoles, longest 2.5–5 cm, gen straight, flat, ± reflexed, translucent yellow. **FL**: inner perianth yellow; filaments white; style white, stigma yellow-green. **FR** ± 4 cm, juicy; exterior purple-red; middle layer white, seed-pulp pink; areoles 40–70. **SEED** 3 mm. 2*n*=22. Uncommon. Pinyon-juniper woodland, desert-scrub edge, chaparral; 600–1300 m. e PR, DMtns; to NV, NM, n Mex. Distinctive stabilized hybrids with *O. phaeacantha* in DMtns (New York Mtns, e San Bernardino Co.), s NV, w AZ may be called *O. curvospina* Griffiths (2*n*=44). ❀DRN,DRY,SUN:3,**7–12**,13,**14,16,18–21**,22–24.

O. echinocarpa Engelm. & J. Bigelow (p. 457) SILVER or GOLDEN CHOLLA **ST** ± tree-like, < 3 m; segments cylindric, terminal gen < 10 cm, 2–3 cm diam; tubercle 6–15 mm, 3–8 mm high; spines 9– 20, < 4 cm, pale gray to translucent yellow, sheath gen same color. **FL**: inner perianth < 2.5 cm, green-yellow, rarely red-brown; filaments pale green. **FR** dry, odor of rancid butter, tubercled; base obtuse; spines dense above. **SEED** 4–6 mm. 2*n*=22. Dry habitats; 300–1400 m. D; to NV, UT, AZ. Hybridizes with most co-occurring chollas. ❀DRN,DRY,SUN:**10**,11,**12**,14,16,18,**19–21**,22–24.

O. engelmannii Engelm. var. ***engelmannii*** ENGELMANN PRICKLY-PEAR Shrub, mound-shaped. **ST** gen < 1 m; lower branches gen decumbent, upper spreading to ascending; segments flat, 15–25 cm, gen obovate; spines 3–12 in all areoles, longest 4–5 cm, straight, spreading from areole, ± appressed to st, ± flat, yellow, coated chalky white, base often red-brown. **FL**: inner perianth yellow; filaments white; style white, stigma yellow-green to green. **FR** 4–6.5 cm, juicy, red-purple throughout; areoles 20–32. **SEED** 4–6 mm. 2*n*=66. Uncommon in CA. Desert scrub, dry oak woodland, etc.; 900–1500 m. SnJt, DMtns; to NV, TX, Mex. [*O. phaeacantha* var. *discata* (Griffiths) L. Benson & Walkington] Hybridizes with *O. phaeacantha*. ❀DRN,DRY,SUN:2,3,7–9,**10**,11,14,**18**,19–21.

O. erinacea Engelm. & J. Bigelow **ST** clumped; branches gen ascending to erect, gen < 0.5 m tall; segments, 5.5–18 cm, flat, elliptic to obovate; spines 1–24, longest 1.7–13 cm, flat to round, straight to wavy, gen whitish, base yellow-brown, surrounded by shorter, gen reflexed, whiter spines. **FL**: inner perianth yellow to pink-magenta; filaments gen white (magenta); style white, stigma green. **FR** 2.5–4 cm, gen spiny, green, tinted red, becoming dry, tan; areoles 14–68. **SEED** 5–6.5 mm. Creosote-bush to pine scrub; 900–3300 m. se SNH, SnBr, SnJt, SNE, DMoj (esp DMtns); to WA, Rocky Mtns, NM.

var. ***erinacea*** MOJAVE PRICKLY-PEAR **ST**: spines 4–24 per areole (0–3 in SnJt), gen in all areoles, longest 1.7–13 cm, spines on lower edge of areole 3 or more, straight to slender and wavy, strongly reflexed. **FR** spiny exc sometimes in SnJt; areoles 24–68. 2*n*= 44. Desert-scrub, joshua tree woodland, pinyon woodland; 900–2200 m. Range of sp. in CA; to UT, AZ. [*O. ursina*; not *O. rufispina* Engelm. & J. Bigelow] ❀DRN,DRY,SUN:2,**3**,7–9,**10**,11,14, **18**,19–21.

var. ***utahensis*** (Engelm.) L. Benson **ST**: spines 1–7 per areole, gen in < 50% of areoles, longest 1.7–5.5 cm, spines on lower edge of areoles 0–4, straight, ± reflexed. **FR**: spines 0(–4); areoles 14–26. 2*n*=88. Sage scrub to Jeffrey-pine woodland; 2000–2800 m. GB (Mono, nw Inyo cos.); to ID, UT, NM. [*O. rhodantha* Schumann, *O. xanthostemma* Schumann] ❀DRN,DRY,SUN:1,**2,3,7**, 10,18.

O. ficus-indica (L.) Miller (p. 457) INDIAN-FIG **ST** 4–5 m, tree-like; segments, 25–43 cm, flat, gen elliptic-obovate; spines gen 0 (–6), if present, longest ± 1–3 cm, flat, white-tan or brown-gray. **FL**: inner perianth yellow or orange; filaments pale green to pale pink; ovary tubercled, style white (pale pink), stigma green. **FR** 6–9 cm, juicy, yellow-orange or purple; areoles 43–71. **SEED** 4.5–5 mm. 2*n*=88. Uncommon. Dry coastal habitats; 6–450 m. s CCo, s SCoRO, SCo, w WTR, w PR, s ChI; cult in warm regions worldwide; native range unknown. [not *O. megacantha* Salm-Dyck]

O. ×fosbergii C. Wolf (p. 457) PINK TEDDY-BEAR CHOLLA **ST** 1.5–2.5 m; trunk 1; branches gen several, long; segments cylindric, terminal gen < 10 cm, 4–6 cm diam, easily detached; tubercle 10–20 mm, 3–5 mm high; spines 7–10, < 2.5 cm, pale red-brown, sheath pale yellow-brown. **FL**: inner perianth < 2.5 cm, pale red-brown; filaments green. **FR** < 2.5 cm, dry to leathery, tubercled; spines 0 or bristles few. **SEED** sterile, < 3 mm. 2*n*=33. Uncommon. Valley floors, alluvial fans; 300–450 m. DSon (e San Diego Co.). 2 sets of genes probably come from *O. bigelovii*. If other parent sp. is *O. echinocarpa*, *O.* × *munzii* should be considered a diploid form. [*O. bigelovii* Engelm. var. *hoffmannii* Fosb.] ❀DRN,DRY,SUN:8-11,**12,13**,14,16,**19-21**,22–24.

O. fragilis (Nutt.) Haw. (p. 457) BRITTLE or LITTLE PRICKLY-PEAR **ST** decumbent-sprawling, 6.5 cm; segments somewhat flat to ± round in X-section, 2–2.5 cm, 2–3 cm wide, gen elliptic-obovate, thickness ± = width, ± tubercled, terminal gen easily detached; spines 3–7 in all areoles, round, longest 3.5 cm, ± rigid, straight, spreading, gray, brown at tip. **FL**: inner perianth yellow, base sometimes red; filaments white or red; style white, stigma green. **FR** ± 1.3 cm, dry, tan; areoles ± 19–21, spines 1–5, 5–7 mm. 2*n*=66. RARE in CA. Juniper woodland; 880 m. w CaRH (Shasta Valley, Siskiyou Co.); to n&e Can, e-c US, TX. Most northern range of any cactus.

O. littoralis (Engelm.) Cockerell **ST** spreading to sprawling in clumps < 9 m diam, 1 m; segments 15–22 cm, flat, elliptic to narrowly obovate; spines 4–11 in all areoles, gen round, longest 2–4 cm, gen straight, upper spreading, lower ± reflexed, yellow, gen coated whitish, base yellow (brown). **FL**: inner perianth yellow to dull red; filaments orange-yellow; style pink or red, stigma yellow-green to green. **FR** 3.5–5 cm, juicy, dark red-purple throughout; areoles 22–36. **SEED** 3–4.5 mm. 2*n*=66. Coastal sage, chaparral; 8–400 m. SCo, s PR, ChI; Mex. Highly variable; hybridizes with other spp. of same chromosome number. [*O. semispinosa* Griffiths, not *O. occidentalis* Engelm.] ❀DRN,DRY:**14**,19-21,**22–24**& SUN:16,17.

O. ×munzii C. Wolf (p. 457) MUNZ'S CHOLLA **ST** < 2.4 m; trunk 1; branches several; segments cylindric, terminal easily detached, gen < 10 cm, 3–5 cm diam; tubercle 10–16 mm, 3–4 mm high; spines 9–16, < 4 cm, pale red-brown, sheath yellow-brown. **FL**: inner perianth 1.5–2 cm, yellow-green to red-brown; filaments pale green. **FR** 2.5–3.5 cm, dry, low tubercles, sometimes with deciduous spines. **SEED** 3 mm. 2*n*=22. RARE. Gravelly or sandy soils of washes, canyon walls; 150–600 m. DSon (Chocolate & Chuckwalla Mtns, Imperial & Riverside cos.) Probable hybrid of *O. bigelovii* and *O. echinocarpa* (see *O. ×fosbergii*). Most localities ± inaccessible. In cult.

O. ×occidentalis Engelm. **ST** sprawling to ± erect, < 1 m; segments flat, 19–35 cm, gen obovate; spines 3–6, in 90–100% of areoles, gen flat, gen straight, longest 2.5–5 cm, ± spreading, upper 1–2 yellow, whitish coated, base brown, lower 2–4 shorter, ± reflexed. **FL**: inner perianth yellow to deep pink; filaments gen yellow; style pink to white, stigma green. **FR** 4.5–5 cm, juicy, red-purple throughout; areoles 24–30. **SEED** 4–5.5 mm. 2*n*=66. Chaparral; 7–400 m. SCo, w edge PR. [not *O. ficus-indica*] *O. littoralis* × (*O. engelmannii* × *O. phaeacantha*). ❀DRN,DRY:**14**,19–21,**22–24**&SUN:16,17.

O. oricola Philbr. (p. 457) **ST** gen tree-like, 2–2.5 m; segments flat, 16–25 cm, gen round; spines 5–13 in nearly all areoles, flat, longest 2–2.5(5) cm, rigid, gen curved, reflexed, translucent yellow. **FL**: inner perianth yellow; filaments orange-yellow; style red, stigma green. **FR** 3.7–6 cm, juicy, spheric; exterior purple-red; interior white-yellow; seed-pulp gen red; areoles 23–63. **SEED** 3.5–4 mm. 2*n*=66. Coastal-sage scrub, chaparral; 3–450 m. SCo, ChI, WTR, w PR; Baja CA. [not *O. littoralis* Engelm.] *O. demissa* Griffiths could be a hybrid with this sp. as one parent. ❀DRN,DRY:14,19–21,**22,23**&SUN:16,17,**24**.

O. parishii Orc. CLUB or MAT CHOLLA **ST** decumbent; branches ascending, in clumps, < 2 m diam, 10–20 cm; segments ± obovoid, terminal 5–7.5 cm, 2–3 cm diam; tubercles 12–25 mm, 3–8 mm high; spines < 21, < 5 cm, largest distinctly flat, with cross-rows of rough papillae exc at tip, gray to brown, margin white, thick, sheath separating only from tip. **FL**: inner perianth 1.5–2.5 cm, yellow; filaments green. **FR** 4.5–8 cm, fleshy, smooth, yellow; spines 0 or easily detached; bristles many. **SEED** 3–4.5 mm. $2n=22$. Uncommon. Sandy flats; 900–1200 m. e&s DMoj; s NV, w AZ. [*O. stanlyi* var. *parishii* (Orc.) L. Benson] ❀DRN,DRY,SUN:**10**,11, **12**,18,19.

O. parryi Engelm. (p. 457) CANE CHOLLA, SNAKE CHOLLA **ST** decumbent to erect, < 3 m; segments cylindric, terminal < 30 cm, 1.5–2.5 cm diam; tubercle 12–35 mm, 4–6 mm high; spines 7–20, < 3 cm, gray to red-brown, sheath brown. **FL**: inner perianth < 3 cm, yellow to green-yellow, gen with purple; filaments pale green. **FR** leathery to dry; spine number variable. **SEED** 6 mm. $2n=22$. Common. Chaparral, coastal-sage scrub, beach; < 1600 m. s SCoRI, SW (exc ChI); n Baja CA. [var. *serpentina* (Engelm.) L. Benson; *O. echinocarpa* var. *parkeri* J. Coulter] Variable. Hybrids with *O. echinocarpa* have been called *O. ganderi* C. Wolf. ❀DRN,DRY: 14,18,**19–21**&SUN:16,**22–24**.

O. phaeacantha Engelm. (p. 457) Shrub. **ST** decumbent to ± spreading, 0.3–1 m; segments flat, 11–30 cm, gen obovate; spines 1–4(6) per areole on upper 30–70% of segment, largest 3–8 cm, gen flat, spreading, upper 1–2 red-brown near base, white or straw-colored exc at base, smaller 1–3 ± reflexed, gen white or gray. **FL**: inner perianth yellow, base red; filaments white; style white, stigma yellow-green to green. **FR** 2.5–6.5 cm, juicy, red-purple; interior gen green; areoles 15–32. **SEED** 3–6 mm. $2n=66$. Many habitats; 45–2220 m. SCoRO, SnBr, e PR, DMtns, DSon; to SD, KS, OK, TX, Mex. [var. *major* Engelm.; *O. littoralis* var. *piercei* (Fosb.) L. Benson; probably *O. mojavensis* Engelm. & J. Bigelow] ❀DRN, DRY:2,7–9,**10**,11,12,**14**,16,**18–24**.

O. prolifera Engelm. CHOLLA **ST** tree-like, < 3 m; segments cylindric, terminal < 15 cm, 3.5–5 cm diam; tubercle 1.5–2.5 cm, 4–8 mm high; spines 6–12, < 2 cm, pale red-brown to dark brown, sheath shining, pale yellow-brown. **FL**: inner perianth < 2 cm, purple-red; filaments yellow-green, gen tinted purple. **FR** fleshy, chained (fls produced from areoles of older fr), green; spines 0. **SEED** gen sterile, < 2 mm. $2n=22,33$. Ocean bluffs, inland coastal scrub; < 300 m. SCo, ChI; Baja CA, Guadalupe Island. ❀DRN, DRY:16,**22–24** &SHD:14,19–21.

O. pulchella Engelm. (p. 457) SAND CHOLLA **ST** clumped, 10–20 cm, arising from a bristle-covered tuber; segments narrowly club-shaped to cylindric, terminal < 10 cm, 0.5–2.5 cm diam; tubercle 6–9 mm, < 1.5 mm high; spines 8–15, densest near st tip, < 6 cm, bulbous at base, largest flat, sharply angled, without cross-rows of rough papillae; sheath separating only near tip; barbed bristles of tuber gen 1–1.5 cm. **FL**: inner perianth 1.5–2.5 cm, pink-magenta; filaments green to yellow. **FR** 2–3 cm, fleshy, smooth, red, with soft, upwardly barbed bristles. **SEED** 3–6 mm. RARE in CA. Dry-lake borders, sandy flats; 1500–1700 m. SNE; to UT. Highly variable; juvenile forms sometimes fl. In cult.

O. ramosissima Engelm. (p. 457) PENCIL CACTUS, DIAMOND CHOLLA **ST** decumbent to ± tree-like, < 1.5 m; segments cylindric, terminal < 10 cm, 4–6 mm diam; tubercle 4.5–7.5 mm, 0–1 mm high; spines 0–3 (gen 1, spreading, long, dark, straight), < 4 cm, pink-gray to dark brown, sheath ± white to pale yellow, translucent, tip darker. **FL**: inner perianth < 6 mm, orange-pink to red-brown; filaments pale green. **FR** < 2 cm, dry; spines dense (rarely 0). **SEED** 2–4 mm. $2n=22,44$. Desert flats; < 1100 m. D; NV, AZ. Variable. *O. wigginsii* L. Benson (sw AZ, reported from DSon) is possibly *O. ramosissima* ×*O. echinocarpa*. ❀DRN,DRY,SUN:8–11,**12**,**13**,14,19–21.

O. ×vaseyi (J. Coulter) Britton & Rose **ST** sprawling to spreading, < 1 m; segments flat, 9–22 cm, gen obovate; spines 0–4, in 0–70% of areoles, gen flat, longest gen 3–4.8 cm, gen straight, spreading to reflexed, yellow, gen ± whitish coated, base brown or yellow, smaller spines 0–4, ± 5 mm. **FL**: inner perianth yellow, orange, or dull red; filaments orange-yellow; style gen pink, stigma green to yellow-green. **FR** juicy, red-purple throughout; areoles 20–36. **SEED** 4.5–6 mm. $2n=66$. Chaparral, disturbed areas; 275–500 m. SCo (edges of adjacent zones). *O. littoralis* × *O. phaeacantha*. [*O. l.* var. *v.* (J. Coulter) L. Benson, *O. l.* var. *austrocalifornica* L. Benson, *O. occidentalis* var. *covillei* (Britton & Rose) Parish] ❀DRN, DRY:**14**,18,**19–24**.

O. wolfii (L. Benson) M. Baker WOLF'S CHOLLA **ST** gen ± erect, < 2 m, gen branched from base; segments cylindric, terminal < 40 cm, 2.5–4 cm diam; tubercle 15–25 mm, 9 mm high; spines 12–22, < 3 cm, pale to dark brown, sheath translucent to pale brown. **FL**: inner perianth 2–3.5 cm, pale purple-brown; filaments red-purple. **FR** 2.5–3 cm, dry, strongly tubercled; spines dense. **SEED** gen sterile. $2n=66$. Dry places above valley floors; 300–1200 m. w edge DSon; Baja CA. [*O. echinocarpa* var. *w.* L. Benson] Locally common. ❀DRN,DRY,SUN:**10**,11,**12**,14,18,**19–21**,22–24.

SCLEROCACTUS PINEAPPLE CACTUS, DEVIL-CLAW

ST gen 1, 5–15 cm, 2–12 cm diam, ovoid to cylindric; ribs 8–21, prominent; tubercles distinct; central spines 1–11, straight or hooked; radial spines gen 3–30, 6–30 mm, straight. **FL** 25–75 mm diam; perianth greenish yellow to magenta; ovary scaly. **FR** becoming dry, 6–25 mm, scaly. **SEED** reniform, tubercled, black. 19 spp.: sw US, Mex. (Greek: hard cactus)

1. Central spines 4–8, straight to curved but not hooked .... ***S. johnsonii***
1′ Central spines 9–11, some hooked at tip .... ***S. polyancistrus***

S. johnsonii (Engelm.) N.P. Taylor (p. 457) **STS** gen 1, 10–25 cm, 5–10 cm diam, ovoid to cylindric; ribs 17–21; spines yellow or pink to reddish, central 4–8, radial 9–10. **FL** 5–8 cm diam, greenish yellow, pink, or magenta. **FR** 7–15 mm, 3–5 mm diam; scales widely cordate, ciliate. $2n=22$. Granitic areas, creosote-bush scrub; 500–1200 m. n DMoj (Inyo Co.); to sw UT, nw AZ. [*Echinocactus j.* Engelm.; *Neolloydia j.* (Engelm.) L. Benson] ❀DRN,DRY, SUN;DFCLT.

S. polyancistrus (Engelm. & J. Bigelow) Britton & Rose (p. 457) MOJAVE FISH-HOOK CACTUS **ST** 1, 10–25 cm, 5–8 cm diam, cylindric; ribs 13–17; central spines 9–11, all but 1–2 hooked, red-brown or white; radial spines 10–15, white. **FL** 4–6 cm diam, rose-purple to magenta. **FR** 20–30 mm, 15–20 mm diam; scales narrow, ciliate near tip. UNCOMMON. Limestone areas, hills and canyons, creosote-bush scrub, Joshua tree woodland; 750–2100 m. DMoj; NV. [*Echinocactus p.* Engelm. & J. Bigelow] ❀DRN,DRY,SUN; DFCLT.

CALLITRICHACEAE WATER-STARWORT FAMILY

C. Thomas Philbrick

Ann, in water or on wet ground, monoecious. **ST** slender, gen ascending under water, floating on surface, or prostrate on ground, gen much-branched. **LVS** simple, gen opposite, 4-ranked, linear-lanceolate to spoon-shaped, entire; stipules 0. **INFL**: fls 1–3 per lf axil, the group subtended by gen 2 whitish, inflated bracts. **FLS** minute, unisexual; perianth 0.

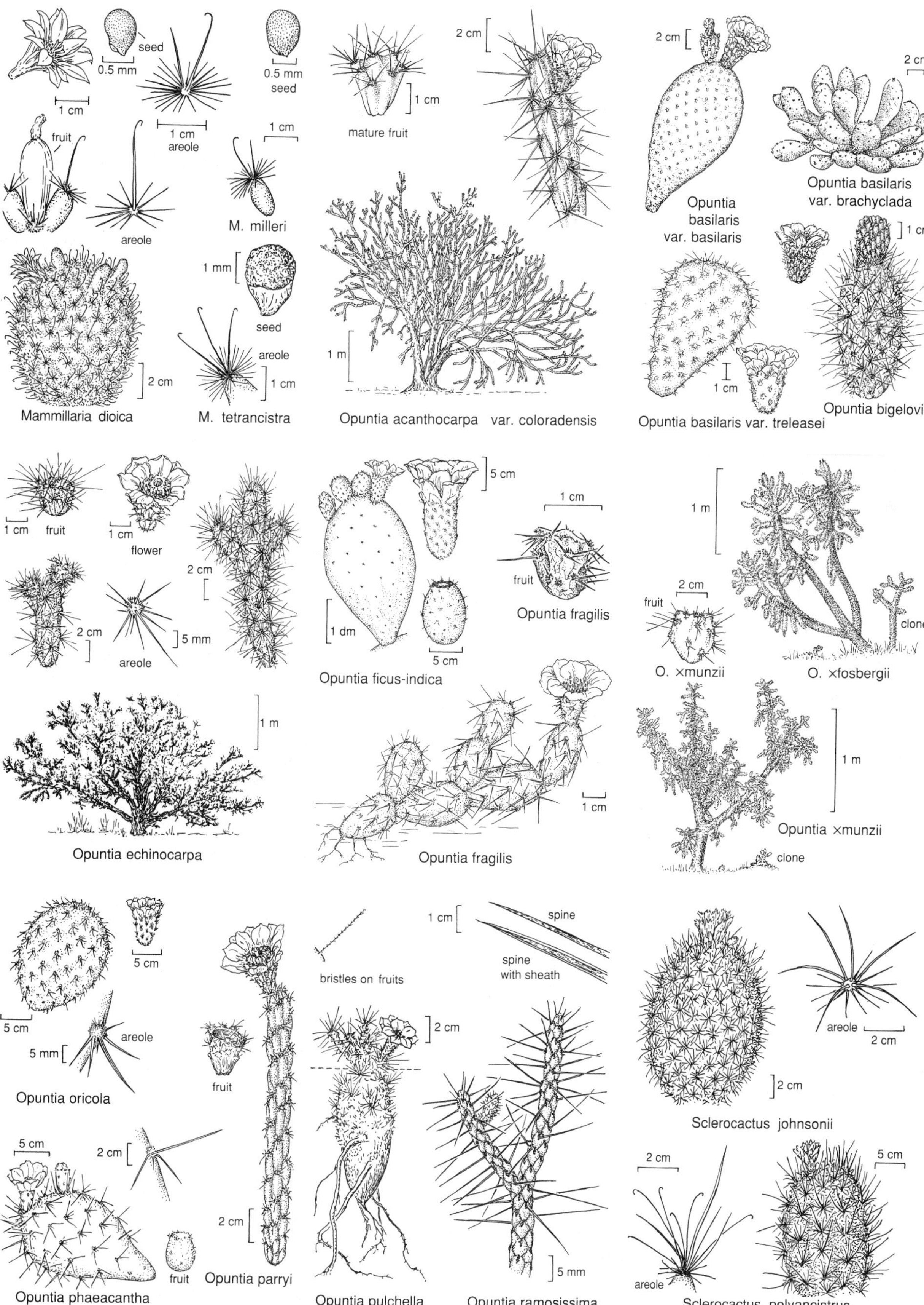
seed
0.5 mm
1 cm
0.5 mm
seed
fruit
1 cm
areole
1 cm
areole
M. milleri
1 mm
seed
areole
2 cm
1 cm
Mammillaria dioica
M. tetrancistra
1 cm
mature fruit
2 cm
1 m
Opuntia acanthocarpa var. coloradensis
2 cm
2 cm
Opuntia basilaris var. basilaris
Opuntia basilaris var. brachyclada
1 cm
1 cm
Opuntia basilaris var. treleasei
Opuntia bigelovii
1 cm
fruit
1 cm
flower
2 cm
2 cm
areole
5 mm
Opuntia echinocarpa
1 m
5 cm
1 dm
5 cm
Opuntia ficus-indica
1 cm
fruit
Opuntia fragilis
1 cm
Opuntia fragilis
1 m
2 cm
fruit
O. ×munzii
clone
O. ×fosbergii
1 m
Opuntia ×munzii
clone
5 cm
5 cm
areole
5 mm
Opuntia oricola
fruit
5 cm
2 cm
fruit
2 cm
Opuntia parryi
Opuntia phaeacantha
bristles on fruits
1 cm
spine
spine with sheath
2 cm
Opuntia pulchella
5 mm
Opuntia ramosissima
areole
2 cm
2 cm
Sclerocactus johnsonii
2 cm
areole
5 cm
Sclerocactus polyancistrus

STAMINATE FL: stamen gen 1, filament elongate. **PISTILLATE FL**: ovary superior, ± obcordate, chambers 4, styles 2, thread-like. **FR** 0.6–1.6 mm (width gen ± = length), ± dry, ± grooved longitudinally, splitting into 4 achene-like units. ❀shallow fresh water: TRY.

CALLITRICHE WATER-STARWORT

The only genus, ± 40 spp.: trop, temp. (Greek: beautiful hair, from slender sts) [Fassett 1951 Rhodora 53:137–155,161–182, 185–194,209–222; Philbrick & Jansen 1991 Syst Bot 16:478–491] Taxonomically difficult; mature fr and 10× magnification needed for identification.

1. Some lvs floating, emergent, or terrestrial and gen spoon-shaped; submersed lvs linear-lanceolate or spoon-shaped
  2. Pedicel in fr 1–25 mm ..... ***C. marginata***
  2′ Pedicel in fr gen 0–0.5 mm
    3. Fr margin wing gen uniformly wide from base to tip
      4. Transition from fr wall to wing gradual; fr 1.2–1.7 mm wide, light brown to tan, grooves deep ..... ***C. stagnalis***
      4′ Transition from fr wall to wing abrupt; fr 1–1.5 mm wide, dark brown to black, grooves shallow [2]***C. trochlearis***
    3′ Fr margin wing 0, only above middle, or wider above middle
      5. Fr longer than wide; fr margin wing wider above middle ..... [2]***C. verna***
      5′ Fr as wide as long; fr margin winged only above middle or 0 ..... [2]***C. heterophylla***
        6. Fr 0.8–1.4 mm diam ..... [2]var. ***bolanderi***
        6′ Fr 0.6–0.8 mm diam ..... [2]var. ***heterophylla***

1′ All lvs submersed and linear-lanceolate
  7. Fr margin wing 0 or only above middle
    8. Pedicel in fr > 1 mm ..... [2]***C. hermaphroditica***
    8′ Pedicel in fr 0–0.5 mm ..... [2]***C. heterophylla***
      9. Fr 0.8–1.4 mm diam ..... [2]var. ***bolanderi***
      9. Fr 0.6–0.8 mm diam ..... [2]var. ***heterophylla***
  7′ Fr margin winged from base to tip
    10. Fl and young fr not subtended by bracts ..... [2]***C. hermaphroditica***
    10′ Fl and young fr subtended by 2 whitish, inflated bracts
      11. Fr wing uniformly wide from base to tip; fr of equal width above and below middle ..... [2]***C. trochlearis***
      11′ Fr wing and fr gen wider above middle ..... [2]***C. verna***

C. hermaphroditica L. (p. 463) **LVS** submersed, linear-lanceolate. **INFL**: bracts subtending fls 0. **FR** 1.1–1.6 mm, 1.2–1.8 mm wide; wing variable, transition to fr wall abrupt; pedicel 1–3 mm. Submersed in quiet streams, ponds; < ± 2000 m. SN, GV, CW, MP; to Can, e US.

C. heterophylla Pursh **LVS** often with floating rosette; submersed lvs linear-lanceolate; floating, emergent, or terrestrial lvs sometimes present and spoon-shaped. **INFL**: bracts subtending fls 2, whitish, inflated, persistent in fr. **FR**: wing 0 or only above middle; pedicel 0–0.5 mm. Becoming stranded at edge of pools or streams or submersed < ± 15 dm; < 2000 m. CA-FP; MP; to e N.Am. Vars. intergrade.

var. ***bolanderi*** (Hegelm.) Fassett (p. 463) **FR** 0.8–1.4 mm diam. Habitats and elevations of sp. CA-FP, MP; to B.C. [ssp. *b.* (Hegelm.) Calder & Roy Taylor] More common in CA than var. *h.*

var. ***heterophylla*** (p. 463) **FR** 0.6–0.8 mm diam. Habitats and elevations of sp. NW, GV, CW, SW; to e N.Am.

C. marginata Torrey (p. 463) Pl mat-forming at water edge or with floating rosettes. **LF** narrowly to widely spoon-shaped. **INFL**: bracts subtending fls 2, whitish, inflated, persistent in fr or not, 0 on terrestrial pls. **FR** 0.7–1.4 mm, 0.9–1.4 mm wide, maturing underground on terrestrial pls; wing 0 to wide; pedicel 1–25 mm. Becoming stranded (often in vernal pools) or submersed < ± 6 dm; < ± 1500 m. CA-FP; to B.C., Baja CA. [*C. longipedunculata* Morong]

C. stagnalis Scop. Pl mat-forming at water edge or with floating rosettes. **LVS** extremely variable; floating, emergent, or terrestrial lvs gen spoon-shaped. **INFL**: bracts subtending fls 2–4, whitish, inflated, persistent in fr. **FR** 1.2–1.8 mm, 1.2–1.7 mm wide, light brown to tan; grooves deep; wing gen uniformly wide from base to tip, transition to fr wall gradual; pedicel 0–0.5 mm. Becoming stranded by streams, ponds, or ditches, or submersed < ± 9 dm; < 800 m. NCo, NCoR, n SNF; native to Eur. Scattered but spreading in nw US.

C. trochlearis Fassett (p. 463) Pl mat-forming at water edge or with floating rosettes. **LVS**: submersed lvs linear-lanceolate; floating, emergent, or terrestrial lvs spoon-shaped. **INFL**: bracts subtending fls 2, whitish, inflated, persistent in young fr. **FR** 1–1.5 mm long, wide, dark brown to black; groove shallow; wing uniformly wide from base to tip, transition to fr wall abrupt; pedicel gen 0–0.5 mm. Becoming stranded or submersed < ± 6 dm; < ± 500 m. NW, CW; w OR.

C. verna L. (p. 463) **LVS**: submersed lvs linear-lanceolate; floating, emergent, or terrestrial lvs sometimes present and spoon-shaped. **INFL**: bracts subtending fls 2, whitish, inflated, persistent in fr. **FR** 0.9–1.4 mm, 0.8–1.3 mm wide (wider above middle); wing from base to tip, gen wider above middle, transition to fr wall abrupt; pedicel 0–0.5 mm. Becoming stranded at water edge or submersed < ± 15 dm; < 4000 m. CA-FP, MP; to e N.Am, Eur.

CALYCANTHACEAE SWEET-SHRUB or CALYCANTHUS FAMILY

Fosiée Tahbaz

Shrubs, deciduous or evergreen, aromatic. **LVS** simple, entire, opposite, short-petioled. **INFL**: fls solitary, terminal on bracted branches. **FL** bisexual, radial, large; bracts grading into perianth; perianth parts many, spirally arrayed; stamens many, attached by short filaments to receptacle, becoming smaller and sterile inward; pistils many, simple, attached to inner face of hollow receptacle, ovules 1–2 per pistil, style thread-like, exserted from receptacle. **FR**: aggregate of achenes inside ± leathery receptacle. 3 genera, ± 6 spp.: N.Am, China; *Calycanthus, Chimonanthus* cult as orn and for fragrance.

CALYCANTHUS

Shrub, deciduous. **FL**: perianth reddish brown, inner parts shorter; stamens in several whorls. 2 spp.: CA, e US. (Greek: calyx flower, from hollow receptacle)

C. occidentalis Hook. & Arn. (p. 463) SWEET-SHRUB, SPICEBUSH Shrub erect, 1–3+ m, ± rounded. **LF**: petiole 4–6 mm; blade 3–15 cm, widely lanceolate to elliptic or oblong, firm, base gen rounded (or cordate), tip acute, lower surface hairy, upper surface scabrous. **FL** ± 5 cm diam; perianth ± fleshy, hairy; inner sterile stamens shaggy-hairy. **FR**: hollow receptacle ovoid; outer surface with distinctive scars of perianth parts; included achenes many, ± velvety, margins granular. *n*=11. Moist, shady places, canyons, streamsides; gen < 1500 m. s NCoRO, NCoRI, s CaR, SNF, c&s SNH. ❀**4–6,17**&IRR:1–3,**7,14–16,22–24** & part SHD:**8,9**,10,11, **18–21**.

CAMPANULACEAE BELLFLOWER FAMILY

Nancy Morin, except as specified

Ann to tree. **LVS** gen cauline, gen simple, gen alternate, petioled or not; stipules 0. **INFL**: panicle, raceme, spike, or fls solitary in axils, gen open; bracts lf-like or not. **FL**: bisexual, radial or bilateral, sometimes inverted (pedicel twisted 180°; hypanthium gen present, ± fused to ovary; sepals gen 5; corolla radial to 2-lipped, gen fused (tube sometimes split down back), lobes gen 5; stamens 5, free or ± fused (anthers and filaments fused into tube or filaments fused above middle); ovary inferior, sometimes half inferior, chambers 1–3, placentas axile or parietal, ovules many, style gen 1, 2–5-branched. **FR**: gen capsule, dehiscing on sides or at tip by pores or short valves. **SEEDS** many. ± 70 genera, ± 2000 spp.: worldwide. Some cult for orn (*Campanula, Jasione, Lobelia*). Subfamilies sometimes treated as different families.

1. Fl radial; filaments not fused (subfamily Campanuloideae)
 2. Per .... [2]**CAMPANULA**
 2′ Ann
  3. Fls in axils of lf-like bracts, sessile
   4. Corolla 3–5 mm, cylindric; sepals widely triangular, lf-like .... **HETEROCODON**
   4′ Corolla 5–10 mm, rotate; sepals narrowly triangular, not lf-like .... **TRIODANIS**
  3′ Fls terminal, subtended, but gen not immediately so, by 1–several, gen non-lf-like bracts, sessile or not
   5. Ovary oblong; fr dehiscing by lateral pores .... [2]**CAMPANULA**
   5′ Ovary obconic; fr dehiscing at top, within persistent sepals .... **GITHOPSIS**
1′ Fl bilateral; anthers and filaments fused into tube (or anthers free and filaments fused into tube above middle (subfamily Lobelioideae)
 6. Anthers free, filaments free at base, fused above middle; pls of dry areas
  7. Sepals linear to triangular; fr fusiform, obconic, or hemispheric, dehiscing by valves .... **NEMACLADUS**
  7′ Sepals oblanceolate; fr hemispheric, top dome-like, circumscissile .... **PARISHELLA**
 6′ Anthers and filaments fused; pls of wet areas
  8. Fls sessile; ovary long, narrow, pedicel-like; fr dehiscing on sides .... **DOWNINGIA**
  8′ Fls pedicelled; ovary slender to spheric, clearly wider than pedicel; fr dehiscing at tip
   9. Fr spheric; per .... **LOBELIA**
   9′ Fr obovate or cylindric; ann
    10. Corolla white or 0; fr cylindric .... **LEGENERE**
    10′ Corolla blue, base of lobes yellow or white; fr obovate .... **PORTERELLA**

CAMPANULA HAREBELL

Ann, per, from taproot, fibrous roots, or rhizome, glabrous to densely hairy. **ST** reclining or erect, branched, 5–150 cm, 4-angled. **LVS** cauline, sometimes also basal, gen lanceolate to ovate, thin, fleshy, or leathery, entire to toothed, sessile or petioled. **INFL**: raceme, panicle, or fls solitary, terminal or axillary. **FL**: corolla cylindric to funnel- or bell-shaped, white to deep blue, lobes linear to triangular; ovary inferior, hemispheric to obconic. **FR** dehiscing by 2–3 lateral pores. **SEEDS** 2 mm, oblong. ± 300 spp.: n hemisphere; many cult, some medicinal. (Latin: little bell, from corolla shape) [Morin 1980 Madroño 27:149–163]

1. Ann, in dry places
 2. Lf oblong-ovate; ovary obovoid to spheric .... ***C. angustiflora***
 2′ Lf ± linear or lanceolate; ovary oblong
  3. Corolla < 4 mm .... ***C. griffinii***
  3′ Corolla > 7 mm
   4. Upper lvs, bracts ± linear; ovary (hypanthium surface) hairy but not papillate .... ***C. exigua***
   4′ Upper lvs, bracts widely lanceolate; ovary papillate .... ***C. sharsmithiae***
1′ Per, in various habitats, incl dry places
 5. Style > corolla, well exserted from it
  6. Petiole gen < 5 mm; fr hemispheric, cordate at base .... ***C. prenanthoides***
  6′ Petiole 10–20 mm; fr obconic, tapered at base .... ***C. scouleri***
 5′ Style = or < corolla, not exserted

7. Pl hairy throughout
8. Lf ± 30 mm, entire .... *C. scabrella*
8′ Lf 6–7 mm, with few large teeth .... *C. shetleri*
7′ Pl glabrous or hairs only on lf margins, st angles, sepals
9. Lf margin entire; cauline lf 30–60 mm .... *C. rotundifolia*
9′ Lf margin ± sharply toothed or crenate; cauline lf < 20 mm
10. Lf ovate, margin crenate, with recurved, stiff hairs; corolla bell-shaped; coastal .... *C. californica*
10′ Lf narrowly oblong, margin with few sharp teeth, glabrous; corolla funnel-shaped; high elevations .... *C. wilkinsiana*

C. angustiflora Eastw. (p. 463) Ann, stiffly hairy; roots fibrous. **ST** erect, 5–20 cm. **LF** 4.5–9 mm, oblong-ovate, leathery, few-toothed, sessile. **FL**: pedicel 3–20 mm; sepals erect, in fr converging; corolla 2.5–6 mm, cylindric, pale blue to white, lobes spreading; stamens 2–2.5 mm, bases sparsely ciliate; ovary 2–3 mm, obovoid, style 2–3 mm, white, upper 50% papillate. **FR** spheric, strongly ribbed; pores near middle. **SEED** ± 0.7 mm, fusiform. *n*=15. Chaparral, burns, serpentine soil; 30–600 m. NCoRI, SnFrB.

C. californica (Kellogg) A.A. Heller (p. 463) SWAMP HAREBELL Per, stiffly recurved-hairy on st angles, lf margins, ovary; rhizome slender. **ST** clambering, 10–30 cm. **LF** 10–20 mm, ovate, thin, crenate; petiole 0 or short. **FL**: pedicel 1–20 mm; sepals spreading; corolla 8–15 mm, bell-shaped, pale blue, lobes reflexed; stamens 5 mm, bases sparsely ciliate; ovary 2–3 mm, hemispheric, style ± 9 mm, white, upper 95% papillate. **FR** spheric, weakly ribbed; pores basal. **SEED** 2 mm, oblong. RARE. Marshy areas; ± 5–10 m. s NCo, n CCo. In cult.

C. exigua Rattan (p. 463) CHAPARRAL HAREBELL Ann, glabrous or stiffly hairy; roots fibrous. **ST** erect, 5–20 cm. **LF** 5–11 mm, ± linear, leathery, few-toothed, sessile. **FL**: pedicel 3–20 mm; sepals erect; corolla 7–18 mm, funnel-shaped, lobes spreading; stamens 4–6 mm, bases ciliate; ovary 2–3 mm, oblong, base tapered, style 6–8 mm, blue, upper 66% papillate. **FR** oblong, strongly ribbed; pores near middle. **SEED** ± 0.7 mm, oblong. *n*=17. UNCOMMON. Talus slopes, gen serpentine soil; 300–1250 m. e SnFrB, n SCoRI. ❀DRN,SUN:**7,14–17,20–24**&IRR:8,**9**,18,**19**;DFCLT.

C. griffinii N. Morin (p. 463) Ann, glabrous or stiffly hairy; roots fibrous. **ST** erect, 2–20 cm. **LF** 2–9 mm, ± linear, leathery, serrate, sessile. **FL**: pedicel 1–6 mm; sepals erect; corolla 2.2–3.7 mm, cylindric, pale blue to white, lobes spreading; stamens 2 mm, bases glabrous; ovary 1.7–3.6 mm, oblong, base tapered, style 2.5 mm, white, upper 33% papillate. **FR** oblong, strongly ribbed; pores near middle. **SEED** ± 0.6 mm, oblong. *n*=17. Chaparral, serpentine soil; 30–600 m. NCoRI, SnFrB, n SCoR. [*C. angustiflora* var. *exilis* J. Howell]

C. prenanthoides Durand (p. 463) Per, ± sturdy, taprooted, glabrous to short-hairy. **ST** reclining to erect, 20–150 cm. **LF** 10–60 mm, lanceolate to ovate, thin to leathery, serrate; petiole gen < 5 mm. **FL**: pedicel 2–6(20) mm; sepals spreading; corolla 7–14 mm, cylindric to bell-shaped, bright blue, lobes erect in lower half, reflexed in upper; stamens 6 mm, bases ciliate; ovary 2.5–5 mm, hemispheric, style 15–18 mm, sometimes curved, blue, upper 55% papillate. **FR** hemispheric; base cordate, strongly ribbed; pores at or below middle. **SEED** 2 mm, oblong. *n*=16,17. Montane, redwood forests; 50–2000 m. NW, CaRH, w-n&c SN, w SnFrB, n SCoRO; OR. Pls with large lvs scattered in NW & esp n SN. ❀DRN:**4,5**&SHD:**6**&IRR:7,**14–17**,22–24.

C. rotundifolia L. (p. 463) HAREBELL Per, taprooted, glabrous. **ST** erect, 10–60 cm. **LVS** on st 30–60 mm, linear to lanceolate, thin, entire, sessile. **FL**: pedicel 10–20 mm; sepals spreading; corolla 12–20 mm, bell-shaped, deep blue, lobes reflexed; stamens 6 mm, bases ciliate; ovary 4–6 mm, oblong, base rounded, style 11–12 mm, white, upper 60% papillate. **FR** hemispheric, weakly ribbed; pores near base. **SEED** 2 mm, oblong. *n*=34. Moist slopes; 1400–2500 m. e KR (Siskiyou Co.); circumboreal. ❀IRR:1,2,**6**,**7**,**14**,19–21&SUN:**4,5,15–17**.

C. scabrella Engelm. (p. 463) ROUGH HAREBELL Per, cespitose, densely short-appressed-hairy; rhizome woody. **ST** < 6 cm. **LF** 30 mm, ± linear to ovate, stiff, entire; petiole wide-winged. **FL**: pedicel 10 mm; sepals erect; corolla 7.5–10 mm, funnel-shaped, powder-blue, lobes erect; stamens ± 8 mm, bases ciliate; ovary 2.5–4 mm, obconic, style 9.5 mm, blue, upper 50% papillate. **FR** elongated, weakly ribbed; pores between middle and top. **SEED** ± 1.5– 3.5 mm, oblong. UNCOMMON. Bare talus slopes; 2400–2800 m. KR, CaRH, Wrn. ❀IRR,DRN:1,2,**6,7**,14,17,19–21&SUN:**4,5**,15,16;DFCLT.

C. scouleri Engelm. (p. 463) SCOULER'S HAREBELL Per, glabrous to short-hairy; rhizome slender. **ST** reclining to erect, 20–30 cm. **LF** 10–60 mm, widely lanceolate to ± round, thin to leathery, serrate; petiole 1–2 cm, winged. **FL**: pedicel 5–20 mm; sepals spreading; corolla 8–15 mm, widely bell-shaped, pale blue, lobes reflexed; stamens 4–6 mm, bases ciliate; ovary 3 mm, hemispheric to obconic, style 12–15 mm, straight, blue, upper 20–40% papillate. **FR** obconic, weakly ribbed; pores near middle. **SEED** 2 mm, oblong. Shaded woods, streamsides; 400–1500 m. KR, n NCoRO, CaRH (Mount Shasta, Lassen Peak); to AK. Old collections also from n SN (n Butte, w Sierra cos.). ❀**4,5**IRR:**6,15–17**&SHD:1,7, 14,19–24;DFCLT.

C. sharsmithiae N. Morin (p. 463) SHARSMITH'S HAREBELL Ann, stiffly hairy; roots fibrous. **ST** erect, 5–25 cm. **LF** 5–11 mm, widely lanceolate, fleshy, serrate, sessile. **FL**: pedicel 1–3 mm; sepals erect; corolla 7–16 mm, funnel- to bell-shaped, deep purple, lobes reflexed; stamens 4–6 mm, bases ciliate; ovary 2–4.5 mm, oblong, papillate (protuberances conic), style ± 4.5–5.5 mm, blue, upper 75% papillate. **FR** oblong, papillate, strongly ribbed; pores near middle. *n*=17. RARE. Talus slopes; ± 400 m. s SnFrB, n SCoRI (Mount Hamilton Range).

C. shetleri Heckard (p. 463) CASTLE CRAG'S HAREBELL Per, cespitose, densely short-appressed-hairy; rhizome woody. **ST** < 5 cm. **LF** 6–7 mm, ± 6-sided, leathery; teeth few, large; petiole 1–2 mm. **FL**: pedicel 1–4 mm; sepals spreading (erect in fr); corolla 7–10.5 mm, pale blue to white, lobes reflexed; stamens 4–6.5 mm, bases ciliate; ovary 2.5–4 mm, oblong, style 7–8 mm, blue, upper 33% papillate. **FR** cup-shaped; base 3-lobed, weakly ribbed; pores between middle and top. **SEED** 1 mm, ovoid. *n*=17. RARE. Rock crevices; 1300–1500 m. CaRH (Shasta Co.). In cult.

C. wilkinsiana E. Greene (p. 463) WILKIN'S HAREBELL Per, forming dense colonies, glabrous; rhizome slender. **ST** reclining to erect, 5–30 cm. **LF** 12–20 mm, narrowly oblong, thin; teeth few, sharp; petiole 0. **FL**: pedicel 10–30 mm; sepals spreading (erect in fr); corolla 12–15 mm, funnel-shaped, deep blue, lobes erect; stamens 6 mm, bases ciliate; ovary 4–5 mm, narrowly obconic, style ± 12 mm, blue, upper 50% papillate. **FR** narrowly obconic, weakly ribbed; pores near top. **SEED** ± 0.7 mm, oblong. RARE. Wet meadows, streamsides; 1800–2600 m. KR (Trinity Mtns), CaRH (Mount Shasta), n SNH (reported from Deer Creek Meadows, Plumas Co.).

DOWNINGIA

Tina Ayers

Ann, glabrous. **ST** decumbent to erect, (10)20–40 cm. **LVS** cauline, often deciduous before fl, << fl bracts, 0.5–2(4) mm wide, lanceolate to awl-like (uppermost sometimes wider), sessile, gen entire. **INFL**: spike; terminal fls often aborted, overtopped by fertile. **FL** sessile, gen inverted at full bloom by twisted ovary; corolla gen >> calyx, blue to pink or white, gen with a symmetric white or yellow spot on lower lip, tube entire, limb strongly 2-lipped, gen 2 lobes of upper lip < 3 of lower; stamens fused (filaments, anthers in tubes), gen 2 smaller anthers each with terminal tuft of bristles, 1 triangular or horn-like, gen 0.2–0.5 mm, others linear, shorter; ovary pedicel-like, chambers 1–2, placentas parietal or axile. **FR** dehiscent by 3–5 lateral slits. 13 spp.: w N.Am, Chile. (A.J. Downing, American horticulturist, born 1815) [Weiler 1962 PhD Univ of CA Berkeley] Fl part positions (upper = next to st; lower = away from st) given at full bloom.

1. Ovary 1-chambered; placentas parietal
 2. Anthers ± 90° to filaments
  3. Lower corolla lobes obtuse, abruptly toothed, blue with 2 oblong, orange-yellow spots in central white field ..... ***D. bacigalupii***
  3′ Lower corolla lobes acute, blue with a central white field ..... ***D. elegans***
 2′ Anthers < 45° to filaments
  4. Larger 3 anthers hairy at tips; calyx lobes unequal ..... ***D. montana***
  4′ Larger 3 anthers glabrous at tips; calyx lobes subequal ..... ***D. yina***
1′ Ovary 2-chambered; placentas axile
 5. Anthers ± 90° to filaments ..... ***D. insignis***
 5′ Anthers < 45° to filaments
  6. 2 larger bristles at tips of 2 smaller anthers 0.6–2.7 mm, gen tightly twisted together ..... ***D. bicornuta***
   7. 2 larger bristles at tips of 2 smaller anthers 0.6–1.5 mm, < anthers; corolla tube dark blue-purple on upper side, without prominent veins ..... var. ***bicornuta***
   7′ 2 larger bristles at tips of 2 smaller anthers 1.6–2.7 mm, gen > anthers; corolla tube light blue or pale yellow on upper side, prominently net-veined ..... var. ***picta***
  6′ 2 larger bristles at tips of 2 smaller anthers < 0.5 mm, not twisted
   8. Sinus between 2 upper corolla lobes with a backward-projecting, short horn ..... ***D. ornatissima***
    9. Upper corolla lobes gen abundantly hairy within near tips ..... var. ***eximia***
    9′ Upper corolla lobes glabrous or sparsely hairy within ..... var. ***ornatissima***
   8′ Sinus between 2 upper corolla lobes plane, without a horn
    10. Corolla 2.5–7 mm, lobes narrowly triangular; fl not inverting, upper corolla lip 3-lobed
     11. Seed longitudinally striate, not appearing twisted; MP ..... ***D. laeta***
     11′ Seed spirally striate, appearing twisted; NCoRI, GV ..... ***D. pusilla***
    10′ Corolla > 7 mm, upper lobes narrowly triangular or elliptic, lower obovate; fl inverting, upper corolla lip 2-lobed
     12. Anthers 2.5–3.5 mm, exserted from corolla tube ..... ***D. pulchella***
     12′ Anthers < 2.5 mm, incl in corolla tube
      13. Seeds spirally striate, appearing twisted; corolla without purple spots at throat ..... ***D. cuspidata***
      13′ Seeds longitudinally striate, not appearing twisted; corolla with 1–3 purple spots at throat
       14. Upper corolla lobes glabrous ..... ***D. bella***
       14′ Upper corolla lobes ciliate-scabrous ..... ***D. concolor***
        15. Fr gen 12–25 mm ..... var. ***brevior***
        15′ Fr 30–50 mm ..... var. ***concolor***

D. bacigalupii Weiler (p. 463) **FL**: corolla 5–18 mm, lateral sinuses >> upper, lower lobes ovate, obtuse, abruptly toothed, blue with 2 oblong, orange-yellow spots in central white field; anthers ± 90° to filaments; ovary 1-chambered, placentas parietal. **FR** (15) 25–45(55) mm; lateral walls papery, easily ruptured when dry, dehiscent along translucent lines. **SEED** longitudinally striate. *n*=12. Vernal pools, wet ditches, grassy meadows; 100–2300 m. CaR, n SNH, MP; to s OR, sw ID. ❀TRY.

D. bella Hoover (p. 463) **FL**: corolla 10–12 mm, glabrous, lateral sinuses ± = upper, lower lip blue with a central white field incl 2 yellow spots, 2 raised nipples, gen alternate 3 small purple spots at throat, lower lobes obovate, obtuse, abruptly toothed; anthers incl in corolla tube, < 45° to filaments; ovary 2-chambered, placentas axile. **FR** 18–60 mm; lateral walls tough, tardily dehiscent with no evident translucent lines. **SEED** longitudinally striate. *n*=11. Uncommon. Vernal pools; < 200 m (GV), 1400–1600 m (SCo, WTR). GV (Colusa to Tulare cos), SCo (Santa Rosa Plateau, Riverside Co.), WTR (extreme n Ventura, s Kern cos.). Some pls in n Sutter Co. have a minute corolla horn similar to that of *D. ornatissima*; pls in SCo, WTR (fr < 30 mm, nectar guides ± unusual), possibly distinct. ❀TRY.

D. bicornuta A. Gray **FL**: corolla 7–19 mm, densely white-hairy in tube, lateral sinuses >> upper, lower lip blue-purple, with a central white field incl yellow-green spots, base purple with 2 nipple-like projections, lower lobes obovate, obtuse, abruptly toothed; anthers < 45° to filaments, 2 larger bristles at tips of 2 smaller ones 0.6–2.7 mm, gen tightly twisted together, rarely 0 or not twisted; ovary 2-chambered, placentas axile. **FR** 35–65(90) mm, tardily dehiscent with no evident translucent lines. **SEED** longitudinally striate. *n*=11. Vernal pools, roadside ditches, lake margins; < 1700 m. NCoRI (sporadic), CaRF, SN, MP; to s OR, sw ID, w NV.

var. ***bicornuta*** (p. 463) **FL**: corolla 9–19 mm, tube (2.5)3–4 (4.5) mm, 2 upper lobes erect or recurved, not crossed at tips, lower lip gen ± flat; 2 larger bristles of 2 smaller anthers 0.6–1.5 mm, < anthers. Habitats, range of sp. ❀WET in winter & spring, SUN: 1–3,6,**7**,8,9,**14–16**,17.

var. ***picta*** Hoover (p. 463) **FL**: corolla 7–10 mm, tube (1.5)2–2.6 mm, light blue or pale yellow on upper surface, prominently net-veined, upper lobes reflexed, appressed to tube, crossed at tips, lower lip strongly concave; 2 larger bristles at tips of 2 smaller anthers 1.6–2.7 mm, gen > anthers. Habitats of sp.; < 300 m. CaRF, n&c SNF.

D. concolor E. Greene **FL**: corolla 7–13 mm, glabrous exc upper lobes ciliate-scabrous, lateral sinuses ± = or slightly > upper, lower lip blue with central white field incl 1 purple, 4-sided spot at base, lower lobes obovate, obtuse, abruptly toothed; anthers incl in corolla tube, < 45° to filaments; ovary 2-chambered, placentas axile. **SEED** longitudinally striate. Vernal pools, mud flats, pond margins; < 1400 m. s NCoR, w ScV, s SnFrB, c PR.

var. ***brevior*** McVaugh (p. 463) CUYAMACA LAKE DOWNINGIA **FR** gen 12–25 mm, soon dehiscent, with translucent lines. *n*=9. **ENDANGERED** CA. Shores; 1400 m. c PR (Cuyamaca Lake). Threatened by development, grazing, recreation.

var. ***concolor*** **FR** 30–50 mm, tardily dehiscent, without translucent lines. *n*=8. Vernal pools, mud flats, pond margins; < 700 m. s NCoR, w ScV (Lake to Solano cos.) SnFrB (Monterey Co.). [*D. tricolor* E. Greene] ❀WET or IRR in winter & spring; SUN:7,8,9,**14–16**,17,22–24.

D. cuspidata Jepson (p. 463) **FL**: corolla 7–15 mm, glabrous, lateral sinuses slightly > upper, lower lip pale to bright blue or lavender with central white field incl 2 yellow spots near angle of throat, lower lobes obovate, obtuse, abruptly toothed; anthers incl in corolla tube, < 45° to filaments; ovary 2-chambered, placentas axile. **FR** 30–72 mm; lateral walls tough or papery with prominent translucent lines or not. **SEED** spirally striate, appearing twisted. *n*=11. Vernal pools, lake margins, meadows; < 450 m. NW, c SNF, SnJV, SCoRO, PR; Mex. Variable, needs field study. ❀TRY.

D. elegans (Lindley) Torrey (p. 463) **FL**: corolla 5–18 mm, lateral sinuses >> upper, lower lobes acute, blue with a central white field; anthers ± 90° to filaments; ovary 1-chambered, placentas parietal. **FR** (15)25–45(55) mm; lateral walls papery, easily ruptured when dry, dehiscent gen with no evident translucent lines. **SEED** longitudinally striate. *n*=10. Vernal pools, wet ditches, grassy meadows; < 2000 m. NW; to WA, ID. [var. *brachypetala* (Gandg.) McVaugh] Pls in Humboldt Co. with filaments 4.5–10.5 mm (vs < 4.5 mm) have been called var. *corymbosa* (A. DC.) Nelson & J.F. Macbr. ❀WET or IRR in winter & spring;SUN:1–3,**6**,**7**,8,9,**14–17**,18–24.

D. insignis E. Greene (p. 467) **FL**: corolla 9–15 mm, glabrous, lateral sinuses > upper, lower lip bright blue with prominent dark blue veins, central white field incl near throat 2 yellow, ovate spots and a dark purple band or 3 distinct spots, lower lobes ovate, obtuse, abruptly toothed; anthers ± 90° to filaments; ovary 2-chambered, placentas axile. **FR** (25)45–80 mm; lateral walls tough, tardily dehiscent with no evident translucent lines. **SEED** longitudinally striate. *n*=11. Vernal pools, roadside ditches, lake margins; < 1500 m. n SNH, c GV (Glenn to Stanislaus cos.), MP; w NV. ❀TRY.

D. laeta (E. Greene) E. Greene (p. 467) **FL** not inverting; corolla 4–7 mm, < or = calyx, glabrous, upper lip 3-lobed, pale blue or lavender with central white field incl 2 yellow spots alternate with 3 small purple spots or a purple band at angle of throat, lower lip 2-lobed, lateral sinuses ± = lower, lobes narrowly triangular, acute; anthers < 45° to filaments; ovary 2-chambered, placentas axile. **FR** 21–43 mm; lateral walls tough, tardily dehiscent with no evident translucent lines. **SEED** longitudinally striate. *n*=11. Ditches, ponds, streams, vernal pools; 1230–2200 m. MP; to s-c Can, MT, WY, UT. [*D. brachyantha* (Rydb.) Nelson & J.F. Macbr.] ❀TRY.

D. montana E. Greene (p. 467) **FL**: calyx lobes unequal; corolla 9–12 mm, > calyx, glabrous, lateral sinuses > upper, lower lip pale lavender-blue with central white field incl 2 small yellow spots and purple markings near throat, these gen minute, nipple-like projections, lower lobes obovate, obtuse, abruptly toothed; anthers < 45° to filaments, larger 3 hairy at tips; ovary 1-chambered, placentas parietal. **FR** 15–35(45) mm; lateral walls firm with 3 translucent lines. **SEED** longitudinally striate. *n*=11. Grassy meadows, roadside ditches in pine forest; 500–1700 m. CaRH, n&c SN. ❀TRY.

D. ornatissima E. Greene **FL**: corolla (7)8–13 mm, lateral sinuses >> upper, the one between 2 upper lobes with a backward-projecting, short horn, lower lip lavender-blue with central white field incl 2 small yellow and 2 minute purple spots at angle of throat, lower lobes obovate, obtuse, abruptly toothed; anthers < 45° to filaments; ovary 2-chambered, placentas axile. **FR** 25–65 mm; lateral walls tough, not easily ruptured even when dry, with no evident translucent lines. **SEED** longitudinally striate. *n*=12. Vernal pools, roadside ditches; < ± 150 m. GV.

var. ***eximia*** (Hoover) McVaugh (p. 467) **FL**: upper corolla lobes gen abundantly hairy within near tips, tips sometimes reflexed but not curved into a ring. Habitats and elevations of sp. s ScV, SnJV. ❀TRY.

var. ***ornatissima*** (p. 467) **FL**: upper corolla lobes glabrous or sparse-hairy within, tips curled outward, backward into a ring or strongly recurved. Habitats and elevations of sp. ScV, n SnJV. ❀TRY.

D. pulchella (Lindley) Torrey (p. 467) **FL**: corolla 7–19 mm, lateral sinuses > upper, lower lip deep blue with central white field incl 2 ovate yellow spots alternate 3 purple ones or with a purple band at angle of throat, lower lobes widely obovate, obtuse to ± truncate, abruptly toothed; anthers 2.5–3.5 mm, exserted < 45° to filaments, slightly long-tapered, pointed at tip; ovary 2-chambered, placentas axile. **FR** 35–75 mm; lateral walls tough, tardily dehiscent, without translucent lines. **SEED** longitudinally striate. *n*=11. Vernal pools, roadside ditches; < 300 m. s Teh, c&s ScV, n SnJV, s SnFrB. ❀WET or IRR in winter & spring, SUN:7,**8**,**9**,**14–17**,18, **19–24**.

D. pusilla (Don) Torrey (p. 467) **FL** not inverting; corolla 2.5–4 mm, < or = calyx, glabrous, upper lip 3-lobed, with 2 minute yellow spots near throat, lobes white or blue, narrowly triangular, acute; anthers < 45° to filaments; ovary 2-chambered, placentas axile. **FR** 20–27 mm; lateral walls tough, dehiscent along translucent lines. **SEED** spirally striate, appearing twisted. *n*=11. Vernal pools, roadside ditches; < ± 150 m. NCoRI, s ScV, n&c SnJV; Chile. [*D. humilis* E. Greene, dwarf downingia]

D. yina Appleg. (p. 467) **FL**: calyx lobes subequal; corolla 8–10 mm, slightly < calyx, glabrous, lateral sinuses gen > upper, lower lip blue with 2 round, yellow spots in a central white field alternate 3 small purple spots at angle of throat, lower lobes obovate, obtuse, abruptly toothed; anthers < 45° to filaments, 1.6–2 mm, 3 larger glabrous at tips; ovary 1-chambered, placentas parietal. **FR** 20–25 mm; lateral walls thin, easily fractured, 3 valves separated by thin translucent lines. **SEED** longitudinally striate. *n*=6,8,10,12. Boggy places near lakes, ponds, vernal pools, mtn meadows; < 1650 m. KR, CaR; to WA. [var. *major* McVaugh]

GITHOPSIS BLUECUP

Ann, glabrous to hairy; roots fibrous. **ST** erect, simple or branched, 2–40 cm, 4-angled. **LVS** cauline, widely linear to ovate, serrate, sessile. **INFL**: fls gen solitary, terminal. **FL** pedicelled or not; sepals 0.5–3 × ovary, linear to narrowly triangular; corolla cylindric, funnel-, or bell-shaped, throat white, lobes linear to widely ovate, white to deep purple; ovary inferior, obconic to cylindric and narrowed near middle. **FR** dehiscing at top, through opening formed when style base falls off. **SEEDS** ± 1 mm, angular-fusiform. 4 spp.: w N.Am. (Greek: *Githago*-like) [Morin 1983 Syst Bot 8:436–468]

1. Corolla cylindric, << 3 mm; fls gen cleistogamous; bracts ovate .... ***G. tenella***
1′ Corolla funnel- to bell-shaped, gen > 3.5 mm; fls not cleistogamous; bracts narrower than ovate
 2. Corolla showy, 1–1.5 × longer than wide; filament base wide, ciliate .... ***G. pulchella***
  3. Ovary 3–5.5 mm; corolla gen < 12 mm .... ssp. ***serpentinicola***
  3′ Ovary > 5.5 mm; corolla gen > 12 mm
   4. Ovary obconic, with dense, recurved, stiff hairs; upper branches 15–30 mm .... ssp. ***campestris***
   4′ Ovary cylindric to narrowly obconic, ± glabrous to finely hairy; upper branches > 32 mm .... ssp. ***pulchella***
 2′ Corolla inconspicuous, 2–4.5 × longer than wide at level of sinuses; filament base narrow, glabrous or sparsely ciliate
  5. Ovary obconic, top slightly narrowed, base long-tapered; pedicel rarely 0; corolla 4.5–14 mm ***G. specularioides***
  5′ Ovary cylindric to obconic, narrowed near middle, base swollen; pedicel 0; corolla 1.5–7.5 mm .... ***G. diffusa***
   6. Corolla lobes white, pale blue, or pink, becoming darker in drying
    7. Upper st 0.4–1 mm wide; corolla 5–7.5 mm .... ssp. ***candida***

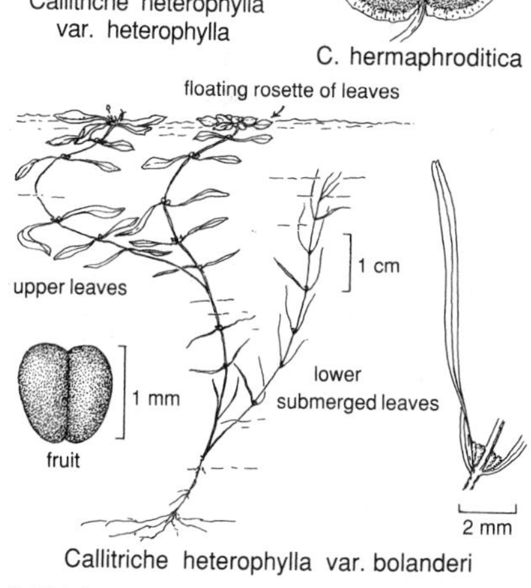

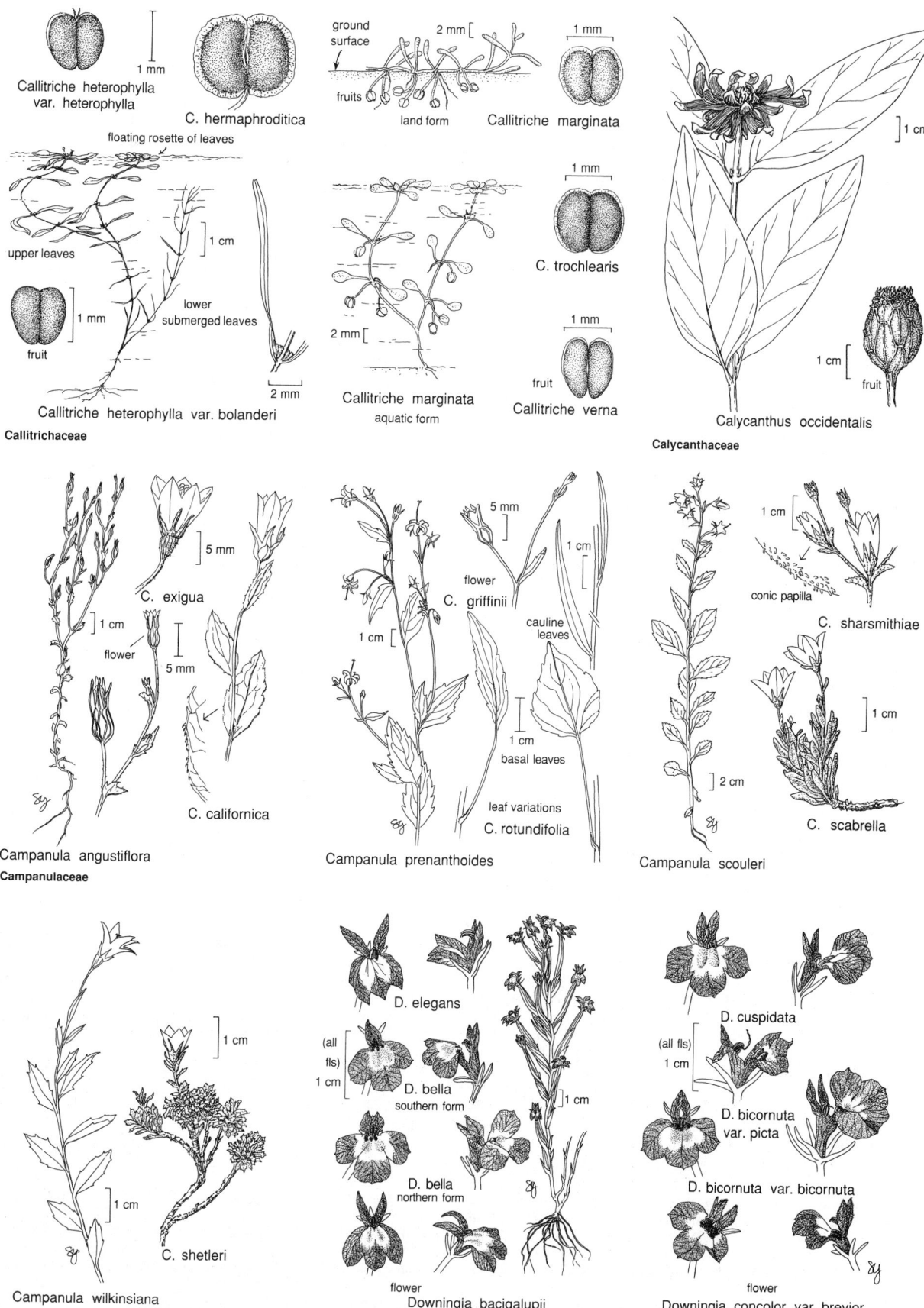

ground surface
2 mm
1 mm
fruits
land form
Callitriche marginata
1 mm
C. trochlearis
1 mm
2 mm
fruit
Callitriche marginata
aquatic form
Callitriche verna

Callitrichaceae

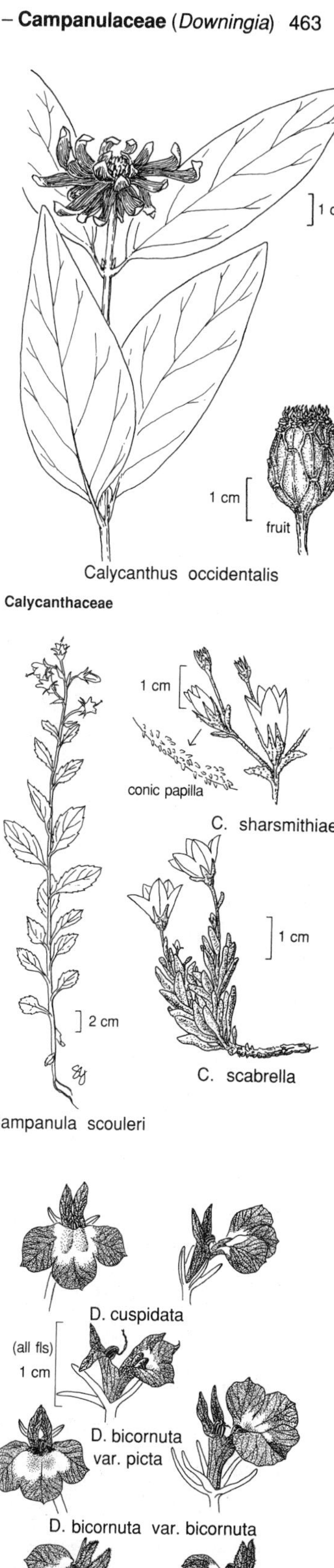

Calycanthaceae

5 mm
C. exigua
1 cm
flower
5 mm
C. californica
Campanula angustiflora

Campanulaceae

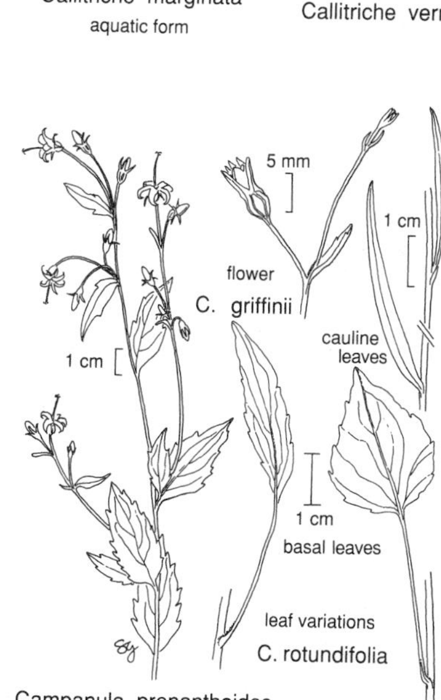

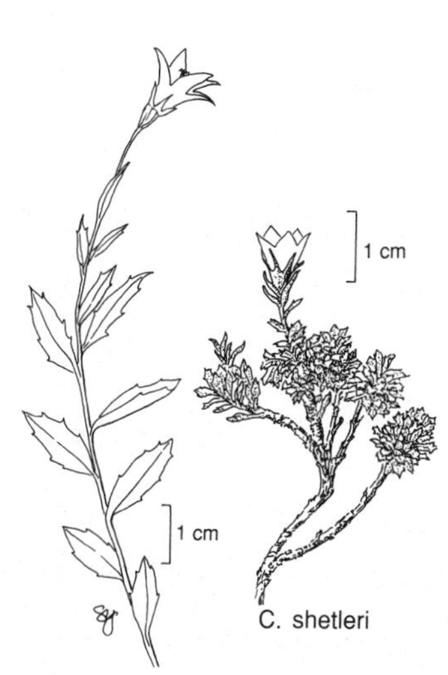

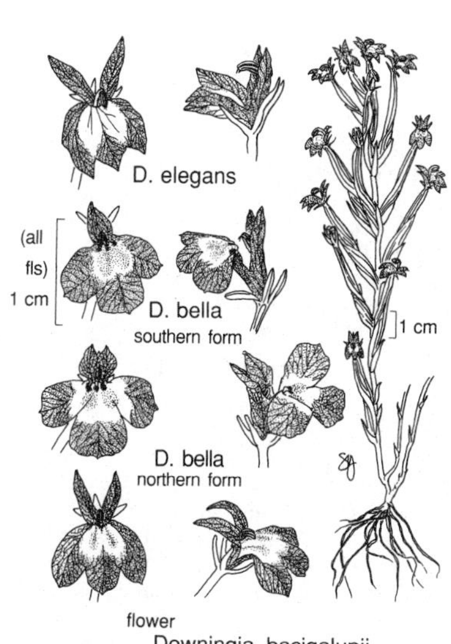

7′ Upper st 0.2–0.4 mm wide; corolla 1.5–5 mm .... ssp. ***filicaulis***
6′ Corolla lobes light violet, violet-blue, or deep blue
8. Ovary 5–6.5 × longer than wide at top .... ssp. ***diffusa***
8′ Ovary 2–5 × longer than wide at top .... ssp. ***robusta***

G. diffusa A. Gray Pl glabrous to hairy. **ST** clambering to erect, 2–30 cm. **LF** 3–15 mm; bracts 2.5–10 mm, linear or oblanceolate, < 5 mm apart. **FL**: pedicel 0; sepals 1.5–2 × hypanthium; corolla 1.5–7.5 mm, 2.5–4.5 × longer than wide at level of sinuses, narrowly funnel-shaped, lobes = or < ovary, white to deep blue; filament bases narrow, glabrous; ovary 4–9 mm, cylindric to obconic, 2–6.5 × longer than wide at top, narrowed near middle, base swollen, ribs 10, those ending at sinuses narrower, style 2–4.5 mm, upper 35–75% papillate. *n*=10,20. Shade or open areas; 50–2000 m. KR, NCoR, CaRF,SNF, SnFrB, SCoR, n ChI, TR, PR; Baja CA.

ssp. ***candida*** (Ewan) N. Morin **ST** erect, 4–20 cm; upper 0.4–1 mm wide. **LF** 6–13 mm; bract 4–10 mm. **FL**: corolla 5–7.5 mm, widely funnel-shaped, lobes 2–5 mm, white; ovary 5 × longer than wide at top. *n*=10. Chaparral, burned areas, gen on serpentine or similar soils; 600–1600 m. s PR; Baja CA. [*G. specularioides* ssp. *c.* Ewan]

ssp. ***diffusa*** (p. 467) **ST** often decumbent, 3–30 cm; upper 0.4–0.8 mm wide. **LF** 4–10 mm; bract 4–7 mm. **FL**: corolla 3–5 mm, base narrow, throat flared, lobes ± 1.5 mm, deep blue; ovary 5–6.5 × longer than wide at top. *n*=10. Moist, disturbed areas; 450–1100 m. s SNF, SCoR (Monterey Co.), n ChI, TR, PR; Baja CA. [*G. gilioides* Ewan]

ssp. ***filicaulis*** (Ewan) N. Morin MISSION CANYON BLUE-CUP **ST** decumbent, 8–25 cm; upper 0.2–0.4 mm wide. **LF** 3–10 mm; bract 1–5 mm. **FL**: corolla 1.5–5 mm, funnel-shaped, lobes 0.7–2.5 mm, white or pale blue; ovary 3–4 × longer than wide at top. RARE. *n*=10. Moist or disturbed areas; 450–700 m. PR (San Diego, Riverside cos.). [*G. f.* Ewan]

ssp. ***robusta*** N. Morin **ST** erect, 2–25 cm; upper 0.4–1 mm wide. **LF** 5.5–15 mm; bract 4–10 mm. **FL**: corolla 3–7 mm, funnel-shaped, lobes 1–3 mm, light violet or violet-blue; ovary 2–5 × longer than wide at top. *n*=20. Shaded or disturbed areas, burns; 50–2000 m. KR, NCoR, CaRF, n&c SNF, SnFrB, SCoR. Intergrades with ssp. *diffusa* in Monterey Co., with *G. specularioides* in Butte Co.

G. pulchella Vatke Pl glabrous to hairy. **ST** erect, 4–40 cm. **LF** 3–20 mm; bracts 2–15 mm, 2–30 mm apart, lanceolate. **FL**: pedicel 0–10 mm; sepals 2 × ovary; corolla 7–24 mm, 1–1.5 × longer than wide at level of sinuses, narrowly to widely bell-shaped, lobes < to > tube, light violet to deep blue; filament bases wide, ciliate; ovary 2.5–10.5 mm, cylindric to obconic, 2–5 × longer than wide at top, base long-tapered, top round, ribs 10, equal, style 4–12 mm, upper 66–90% papillate. *n*=10. Oak woodland; 30–1300 m. CaRH, n&c SN.

ssp. ***campestris*** N. Morin **ST** 8–13 cm; upper branches 1.5–3 cm. **LF** 8–13 mm; bracts < 5 mm apart. **FL**: pedicel 0; corolla 11–15 mm, 7–10 mm wide at sinuses, widely bell-shaped, lobes 7–10 mm, bright blue; ovary 5.5–7 mm, obconic, deeply ribbed in fr, hairs dense, recurved, stiff, style 4–9 mm, 0.3–0.6 mm wide, upper 66% papillate. Uncommon. Open places, volcanic soil; 200–1300 m. CaRH, n&c SNH.

ssp. ***pulchella*** (p. 467) **ST** 10–40 cm; upper branches 3.2–6 cm or longer. **LF** 7–18 mm; bracts 5–13 mm apart. **FL**: pedicel 0–7 mm; corolla 10–24 mm, 7–14 mm wide at sinuses, widely bell-shaped, lobes 5–19 mm, deep blue; ovary 5.5–10.5 mm, cylindric to narrowly obconic, glabrous to finely hairy, deeply ribbed in fr, style 5–12 mm, 0.5–1 mm wide, upper 75–90% papillate. Grassy openings; 30–1300 m. n&c SN. Glabrous pls on serpentine outcrops are var. *glabra* (Jepson) N. Morin. ❀TRY.

ssp. ***serpentinicola*** N. Morin **ST** 4–25 cm; upper branches 2–6 cm; hairs fine or 0. **LF** 3.5–9 mm; bracts 5–15 mm apart. **FL**: pedicel 0–10 mm; corolla 7–13 mm, 4–8 mm wide at sinus, narrowly bell-shaped, lobes 3–6 mm, light blue-violet; ovary 2.5–5.5 mm, obconic, shallowly ribbed in fr, style 5–9 mm, 0.2–0.6 mm wide, upper 70–80% papillate. Serpentine, similar outcrops, and Ione Formation; 450–540 m. c SNF (esp Amador Co.). ❀TRY.

G. specularioides Nutt. (p. 467) Pl glabrous to hairy. **ST** erect, 2–40 cm. **LF** 4–20 mm; bracts 3–12 mm, > 5 mm apart, lanceolate to oblong. **FL**: pedicel rarely 0; sepals 1–3 × ovary; corolla 4.5–14 mm, 2 × longer than wide at level of sinuses, funnel-shaped, lobes < to ± = tube, deep blue; filament bases narrow, sparsely ciliate; ovary 4–14 mm, obconic, ± 3 × longer than wide at top, top ± narrowed, base long-tapered, ribs 10, equal, style 2.5–6.5 mm, upper 50% papillate. *n*=19,20. Chaparral, oak woodlands; 70–1400 m. KR, NCoR, CaR, SN, ScV, SnFrB, s SCoRI (San Luis Obispo, Kern cos.), e SCo (San Bernardino Co.); to B.C. [*G. calycina* Benth.]

G. tenella N. Morin (p. 467) Pl hairy. **ST** clambering to erect, 3–9 cm. **LF** 4–6.5 mm; bracts 2–5.6 mm, < 5 mm apart, ovate. **FL** gen cleistogamous; pedicel 0–2 mm; sepals = ovary; corolla 2.5–3 mm, 2 × longer than wide at level of sinuses, cylindric, lobes < tube, deep blue; filament bases linear, glabrous; ovary 3–5 mm, obconic, ± 2–4.5 × longer than wide at top, middle ± narrowed, base long-tapered, ribs 10, those ending at sinuses ± raised, style ± 2–3 mm, upper 25% papillate. Uncommon. *n*=9. Moist places in oak woodlands; 1100–1900 m. s SNH (Kern, Tulare cos.). Possibly also in s SCoRI (Cholame Hills, Monterey Co.).

HETEROCODON

1 sp. (Greek: different bell, from cleistogamous and opening fls)

H. rariflorum Nutt. (p. 467) Ann; roots fibrous. **ST** erect, 5–30 cm, simple or branched from base, 4-angled, very thin, sparsely hairy. **LVS** cauline, 2–10 mm, round-cordate, thin, serrate, sessile. **INFL**: fls solitary in axils of lf-like bracts, sessile. **FLS**: lower cleistogamous; sepals 2–4 mm, widely triangular, lf-like, toothed; corolla 3–5 mm, cylindric, tube white to pale blue, lobes ± 1.5 mm, triangular, erect, deep blue; stamens ± 1.5 mm, filaments linear; ovary inferior, 2–3 mm, short-oblong, papery, gen with many long, stiff hairs, style 3 mm, upper 25% papillate. **FR** dehiscing by lateral pores near base. **SEED** 0.5 mm, elliptic. Vernally wet places; < 2300 m. CA-FP; to B.C., MT, NV.

LEGENERE

1 sp. (Anagram of E. L. Greene, American botanist, 1843–1915)

L. limosa (E. Greene) McVaugh (p. 467) LEGENERE Ann, emergent or terrestrial, glabrous. **ST** reclining, 10–30 cm; lateral branches erect, slender, stiff, sometimes fleshy. **LVS** cauline, narrowly triangular, entire, sessile, early deciduous. **INFL**: raceme, terminal; axis ± zig-zag; bract 1 per fl, 6–12 mm, ovate, lf-like, spreading; pedicels 6–20 mm in fr. **FL** not inverted; sepals 1/3–1/2 × ovary, triangular; corolla white, sometimes 0 in lower fls, tube ± 1.5 mm, linear, split down back ± to base, 2-lipped, upper lip 2-lobed, lobes ± 2 mm, narrow, erect, lower lip 3-lobed, lobes ± 2 mm, obovate; stamens fused into tube, anthers 0.5–1 mm, sometimes free with age, 2 shorter anthers minutely appendaged; ovary inferior, ± 3.5 mm, narrowly obconic, stigma head-like, smooth. **FR** 6–10 mm, 1–2 mm diam, cylindric, 1-chambered, slightly > hypanthium, tip rounded, dehiscing at tip. **SEEDS** 1 mm, elliptic, shiny, chestnut-brown. RARE. Wet areas, vernal pools; < 150 m. s NCoR, s ScV, n SnJV, SnFrB (Santa Cruz Mtns).

LOBELIA

Tina Ayers

Per, glabrous or hairy. **LVS** mostly basal or all cauline, 0.5–1.5 cm wide, linear-lanceolate to elliptic, sessile; margin with small, gland-tipped teeth. **INFL**: raceme. **FL** inverted in full bloom by twisted pedicel; corolla red or blue, rarely white, tube entire or with an upper sinus, limb strongly 2-lipped, 2 lobes of upper lip < 3 of lower; stamens fused, gen 2 smaller anthers each with terminal tuft of bristles, 1 sometimes triangular or horn-like, others linear, shorter; ovary ± spheric, chambers 2, placentas 2, axile. **FR** dehiscent by 2 valves at tip. ± 350 spp.: ± worldwide. (Matthias de l'Obel, Flemish botanist, 1538–1616) Fl part positions (upper, next to st; lower, away from st) given at full bloom.

1. Peduncle long; corolla red, rarely white, upper sinus >> lateral .... ***L. cardinalis*** var. ***pseudosplendens***
1′ Peduncle 0; corolla blue, upper sinus = lateral .... ***L. dunnii*** var. ***serrata***

L. cardinalis L. var. ***pseudosplendens*** McVaugh (p. 467) CARDINAL FLOWER **ST** erect, 4–20 dm, < 1.5 cm diam, purple-red. **FL**: corolla red, rarely white, glabrous, tube 15–20 mm, from upper sinus to base; anther tube 3.5–4.5 mm, triangular bristle at tips of 2 shorter anthers 0. *n*=7. Stream bottoms; 450–1600 m. SnGb, SnBr, PR, DMtns (Panamint Mtns); to w TX, Mex. Incl by McVaugh in ssp. *graminea* (Lam.) McVaugh, with 3 other vars. incl var. *multiflora* (Paxton) McVaugh (pls with dense, short hairs throughout; lvs lanceolate to ovate, probably not in CA). Seriously TOXIC, esp when used as a home remedy. ❀TRY.

L. dunnii E. Greene var. ***serrata*** (A. Gray) McVaugh (p. 467) **ST** decumbent to erect, 2–8.5 dm, < 4 mm diam, light green. **FL**: corolla blue, hairy, tube 12(14)–19 mm, entire, in age splitting incompletely from base to near middle; anther tube 2.3–3 mm, triangular bristle at tips of 2 shorter anthers present. *n*=7. Falls, seeps of cliffs; 30–1820 m. SCoRO, TR, SnJt; n Baja CA. ❀WET or IRR; SHD:7–9,**14–17**,18,**19–24**; or SUN:**14–17**;INV.

NEMACLADUS

Nancy R. Morin & Jennifer Milburn

Ann; roots fibrous. **STS** erect or spreading, simple or branched at base or below middle. **LVS** basal; petiole short or 0. **INFL** ± raceme-like; bract 1 per fl, small; pedicel gen thread-like. **FL** inverted; sepals linear to triangular; corolla nearly radial and 5- lobed or 2-lipped (upper lip 3-lobed, lower lip 2-lobed); filaments free at base, fused into tube above, sometimes appendaged at tube base, anthers free, all alike; ovary gen half-inferior in fr, sometimes 0, hemispheric to obconic, sometimes glandular, stigma 2-lobed, papillate. **FR** gen > hypanthium, hemispheric to fusiform; tip pointed or rounded, dehiscing at tip by 2 valves; chambers 2. **SEED** elliptic to oblong. 13 spp.: sw US, nw Mex. (Greek: thread-like branch) [McVaugh 1942 N Amer Flora 32A: 1–134]

1. Ovary superior; fr 2–3 × sepals .... ***N. longiflorus***
2. Corolla 3–3.5 mm, tube 2–2.5 mm; filament tube ± 2–3 mm .... var. ***breviflorus***
2′ Corolla 5–8 mm, tube 2.5–5 mm; filament tube 3.5–7.5 mm .... var. ***longiflorus***
1′ Ovary partly to completely inferior; fr < or slightly > sepals
3. Sts < 1 cm, cushion-like; infl gen head-like, dense .... ***N. twisselmanii***
3′ Sts gen > 1 cm, spreading, decumbent, or erect; infl an open raceme
4. St below branches silver-gray, shiny, sometimes dark; corolla yellow with brown marks .... ***N. rubescens***
4′ St below branches reddish, brown, or dark purple; corolla white, pink, lavender, or purplish
5. Sts spreading to decumbent; pedicels ± thread-like, 0.2–0.3 mm diam .... ***N. rigidus***
5′ Sts gen erect to stiffly ascending (spreading in *N. gracilis, N. secundiflorus, N. sigmoideus*); pedicels thread-like, < 0.2 mm diam
6. Infl axis straight or weakly zigzag
7. Bracts 2–9 mm, flat, not enfolding pedicel base; corolla white; fr bell-shaped .... ***N. ramosissimus***
7′ Bracts 2–4 mm, enfolding pedicel base; corolla pink or lavender; fr hemispheric
8. Corolla 1.5–2 mm, white with lavender lines; fr ± 1.5 mm; seeds elliptic .... ***N. gracilis***
8′ Corolla 3–5 mm, lavender or pink, yellow-blotched at base; fr ± 2–2.5 mm; seeds subspheric .... ***N. secundiflorus***
6′ Infl axis strongly zigzag
9. Corolla divided 1/2 or less to base .... ***N. pinnatifidus***
9′ Corolla divided nearly to base
10. Pedicels clearly S-curved; bracts ovate; fls spreading .... ***N. sigmoideus***
10′ Pedicels straight, slightly curved, or reflexed; bracts linear, elliptic, or lanceolate (sometimes ovate in *N. glanduliferus*); fls spreading or reflexed
11. Fr ± hemispheric (base and tip rounded); sepals spreading or reflexed .... ***N. glanduliferus***
12. Pedicels spreading, tip curved; sepals 1.5–2.5 mm .... var. ***glanduliferus***
12′ Pedicels stiffly ascending, tip straight; sepals ± 1–1.5 mm .... var. ***orientalis***
11′ Fr fusiform or obconic (base acute, tip acute, ± pointed, or rounded); sepals erect
13. Pedicels straight, base to tip; seed surface narrowly ridged .... ***N. capillaris***
13′ Pedicels straight below, tip curved; seed surface with clearly pitted rows or deeply impressed lines
14. Seed surface with clearly pitted rows; corolla 2.5–5 mm; lf blades irregularly serrate ...... ***N. interior***
14′ Seed surface with deeply impressed, vertical lines; corolla 1.5–2 mm; lf blades entire or obscurely toothed .... ***N. montanus***

N. capillaris E. Greene (p. 467) **STS** stiffly ascending, 7–18 cm; base brownish or purplish. **LF** 3–15 mm, ovate, narrowed abruptly to short petiole, entire, glabrous or hairy. **INFL**: axis zigzag, esp in fr; bracts 1–3 mm, lanceolate to elliptic; pedicels 8–12 mm, 1–1.5 mm diam, spreading, straight or slightly curved, tip not curved. **FL**: hypanthium ± 1 mm; sepals ± 0.5 mm, elliptic to lanceolate, erect; corolla 0.7–1.3 mm, divided ± to base, white, glabrous; filament tube ± 1 mm, tip slightly curved, glabrous, anthers 0.1–0.2 mm. **FR** 1.5–2.5 mm, obconic (base acute, tip rounded). **SEED** 0.5–0.7 mm, widely elliptic, narrowly ridged. *n*=9. Dry slopes, burned areas; 400–2100 m. NCoR, CaRH, n SNF, SNH, MP, DMoj; OR.

N. glanduliferus Jepson **STS** stiffly ascending, 5–25 cm; base brownish or purplish. **LF** 3–16 mm, oblanceolate to elliptic, narrowed gradually to petiole, toothed or pinnately lobed, hairy. **INFL**: axis strongly zigzag; bracts 1–3 mm, spreading, linear to ovate; pedicels 6–16 mm, 0.1–0.2 mm diam, spreading, straight or slightly curved, tip curved. **FL**: hypanthium ± 1 mm; sepals 1.5–2.5 mm, linear-elliptic to ± deltate, spreading; corolla 2–2.5 mm, divided ± to base, white, upper lobes erect, lower lobes ciliate; filament tube 1–2.3 mm, tip slightly curved, glabrous, anthers 0.2–0.4 mm. **FR** ± 2–4 mm, ± hemispheric (base and tip rounded). **SEED** ± 0.5 mm, cylindric; surface with impressed, vertical lines crossed by fine transverse lines. Dry slopes, sandy soils, washes; < 2400 m. SNE, D; to UT, NM, Baja CA.

var. ***glanduliferus*** (p. 467) **FL**: pedicels spreading, tip curved; sepals 1.5–2.5 mm; filament tube 1.6–2.3 mm. Sandy or gravelly soils, canyons; 150–1900 m. s SNE, D; NV, AZ, Baja CA.

var. ***orientalis*** McVaugh **FL**: pedicels stiffly ascending, tip straight; sepals ± 1–1.5 mm; filament tube 1–2 mm. Sandy soils, rocky slopes, washes; < 2400 m. SNE, D; to UT, NM, Baja CA.

N. gracilis Eastw. (p. 467) SLENDER NEMACLADUS **STS** spreading to ascending, 2.5–10 cm; base dull reddish brown. **LF** 2.5–8 mm, oblanceolate to oblong, narrowed to wide petiole, irregularly dentate to ± pinnately lobed, hairy. **INFL**: axis straight or weakly zigzag; bracts 2–4 mm, linear-oblong, tip recurved; pedicels 5–12 mm, 0.1 mm diam, reflexed, slightly S-curved, tip erect. **FL**: hypanthium 0.5 mm; sepals ± 0.5 mm, linear, spreading; corolla 1.5–2 mm, ± 1/2 divided, white with lavender veins, sparsely hairy, lobes erect; filament tube ± 1 mm, tip curved, fine-hairy; anthers 0.5 mm. **FR** ± 1.5 mm, hemispheric (base acute, tip rounded). **SEED** ± 0.5 mm, widely elliptic; surface with vertical zigzag ridges alternating with clearly pitted rows. UNCOMMON. Rocky slopes, sandy washes; < 1900 m. Teh, SnJV (sw Merced Co.), s SCoRO, WTR, w DMoj (Los Angeles Co.).

N. interior (Munz) G. Robb. **STS** stiffly ascending, 7–25 cm; base brownish or purplish. **LF** 10–20 mm, oblanceolate to elliptic, sessile or abruptly narrowed to petiole, irregularly serrate, glabrous. **INFL**: axis strongly zigzag; bracts 1–3 mm, linear to lanceolate, appressed, sometimes enfolding pedicel base; pedicels 7–13 mm, 0.1–0.2 mm diam, ascending, straight, tip curved. **FL**: hypanthium 0.5 mm; sepals ± 1 mm, narrowly triangular, erect; corolla 2.5–5 mm, deeply divided, pale lilac, white, or pink, wine-red band at base, yellow-tinted below band, glabrous, upper lobes erect, lower lobes spreading; filament tube ± 2 mm, tip slightly curved, glabrous; anthers 0.4–0.7 mm. **FR** 2–2.5 mm, fusiform (base and tip acute). **SEED** ± 0.5 mm, elliptic; surface with clearly pitted rows. Dry, gravelly slopes, yellow-pine forest; 150–2700 m. c SNF, SNH, Teh; OR.

N. longiflorus A. Gray **STS** erect, 5–21 cm; base brownish or purplish. **LF** 3–12 mm, oblanceolate to ovate or oblong, narrowed to a winged petiole, entire to finely crenate, hairy. **INFL**: axis ± zigzag; bracts 2–4 mm, elliptic to ovate; pedicels 6–23 mm, 0.1 mm diam, ascending, S-curved, tip abruptly curved. **FL**: hypanthium ± 0; sepals ± 1 mm, elliptic, erect; corolla 3–8 mm, divided < 1/2 to base, white, tube cylindric, lobes erect, upper lip puberulent, lower lip glabrous; filament tube 2–7.5 mm, tip abruptly curved, fine-hairy, anthers 0.2–0.6 mm diam; ovary superior in fr. **FR** ± 2.5–5 mm, fusiform (base and tip acute). **SEED** 0.2–0.5 mm, widely elliptic to ± round; surface with obscure, lengthwise, wavy ridges. Sandy or gravelly slopes and washes; 300–2400 m. s SNH, SCoR, SCo, SnGb, SnBr, PR; Baja CA.

var. ***breviflorus*** McVaugh **FL**: corolla 3–3.5 mm, tube 2–2.5 mm; filament tube ± 2–3 mm. **SEED** < 0.5 mm, ± round. Habitat of sp.; 800–1200 m. s SNH, SnGb, SnBr, PR; Baja CA.

var. ***longiflorus*** (p. 467) **FL**: corolla 5–8 mm, tube 2.5–5.5 mm; filament tube 3.5–7.5 mm. **SEED** ± 0.5 mm, widely elliptic. Habitats and elevations of sp. SCoR, SCo, SnGb, PR; Baja CA.

N. montanus E. Greene **STS** erect, 8–18 cm; base brownish or purplish. **LF** 6–18 mm, oblanceolate to elliptic, narrowed to short, wide petiole, entire or obscurely toothed, hairy. **INFL**: axis strongly zigzag; bracts 1–3 mm, linear to lanceolate, flat; pedicels 10–15 mm, 0.1 mm diam, ascending, gen straight, tip upturned. **FL**: hypanthium ± 1 mm; sepals ± 1 mm, triangular, erect; corolla 1.5–2 mm, divided ± to base, white or purplish, glabrous, lobes spreading; filament tube ± 1 mm, tip slightly curved, glabrous, anthers ± 0.5 mm. **FR** 2.5–3 mm, obconic (base acute, tip obscurely pointed). **SEED** ± 1 mm, elliptic, with deeply impressed, vertical lines. *n*=9. Serpentine soils; 300–1200 m. s NCoRI, n ScV, se SnFrB (Mount Hamilton).

N. pinnatifidus E. Greene (p. 467) **STS** stiffly erect, 6–20 cm, brownish or purplish. **LF** 5–20 mm, oblanceolate, narrowed to a long petiole, deeply pinnately lobed; lobes toothed or entire, glabrous. **INFL**: axis zigzag; bracts 2–5 mm, linear to elliptic; pedicels 5–15 mm, < 0.1 mm diam, spreading, S-curved, tip abruptly curved. **FL**: hypanthium 0.5–0.7 mm; sepals ± 1 mm, narrowly triangular, erect; corolla 1.5–2 mm, divided ± 1/3 to base, white or rose-purple, glabrous, lobes erect; filament tube ± 0.7 mm, tip slightly curved, glabrous, anthers 0.1–0.2 mm. **FR** 3–4 mm, fusiform (base and tip acute). **SEED** 0.5 mm, elliptic, with vertical, clearly pitted rows. Dry washes, burned areas, chaparral; 300–1300 m. SCo, SnGb, PR; Baja CA.

N. ramosissimus Nutt. **STS** erect, 5–32 cm; base brownish or purplish. **LF** 3–18 mm, oblanceolate, narrowed to slender or wide petiole, irregularly toothed or ± pinnately lobed, margins and base hairy. **INFL**: axis straight; bracts 2–9 mm, linear, flat; pedicels 6–22 mm, 0.7 mm diam, ± spreading, slightly S-curved, tip ± erect. **FL**: hypanthium 0.5 mm; sepals lobes ± 0.5 mm, deltate, erect; corolla 1.5–2 mm, divided > 1/2 length, white, glabrous, lobes erect; filament tube 1–2 mm, tip curved, anthers 0.2–0.3 mm. **FR** ± 1.5–2.5 mm, bell-shaped (base narrowed, tip rounded). **SEED** ± 0.5 mm, subspheric; surface with clearly pitted rows. Dry, sandy or gravelly soils; 150–1600 m. SnFrB (Mount Hamilton), SCoRO, SCo, WTR, SnGb, PR; to UT, NM, Baja CA.

N. rigidus Curran (p. 467) **STS** spreading to decumbent, 4–9 cm; base shiny, purple. **LF** 5–10 mm, elliptic to oblanceolate, narrowed to wide petiole, entire to scalloped, hairy. **INFL**: axis strongly zigzag; bracts 2–3 mm, widely elliptic; pedicels 5–12 mm, 0.2 mm diam, spreading, straight, becoming curved in age. **FL**: hypanthium 1 mm; sepals ± 1.5 mm, triangular, erect; corolla 1–1.5 mm, deeply divided, white or purplish, veins and margins red, lobes ovate, erect, glabrous; filament tube 1.2–1.6 mm, tip slightly curved, glabrous, anthers 0.2–0.3 mm. **FR** 3–4 mm, hemispheric (base oblique, tip pointed). **SEED** 0.6–0.7 mm, elliptic; surface with wide ridges alternating with pitted rows. Bare soil or sand; 200–2500 m. GB; e OR, ID, NV.

N. rubescens E. Greene **STS** erect, 5–20 cm; base shiny, silver-gray. **LF** 5–20 mm, elliptic to oblanceolate, narrowed abruptly to winged petiole, entire, toothed, or ± pinnately lobed, glabrous or coarsely hairy. **INFL**: axis weakly zigzag; bracts 1–2.5 mm, widely lanceolate; pedicels 8–15 mm, 0.1–0.2 mm diam, horizontal to ascending, slightly S-curved, tip slightly curved or not. **FL**: hypanthium 0.2–0.3 mm; sepals ± 1 mm, elliptic to deltate, erect; corolla 1.5–2 mm, divided > 1/2 length, yellow with purple or brown marks, upper lobes reflexed, slightly ciliate, lower lobes erect, purple-tipped, densely ciliate; filament tube 2–3 mm, straight or tip slightly curved, anthers 0.6 mm; ovary 1/4–1/2 inferior. **FR** 2–2.5 mm, ± bell-shaped (base narrowed, tip rounded). **SEED** 0.4 mm, widely elliptic; surface with wavy ridges alternating with weakly pitted rows. Dry, sandy or gravelly soils; < 1600 m. PR (e slope), SNE, D; NV, AZ, Baja CA. Pls from PR (e slope) & D, with narrow, deeply toothed lvs, filament appendage stalks >> processes have been called var. *tenuis* McVaugh.

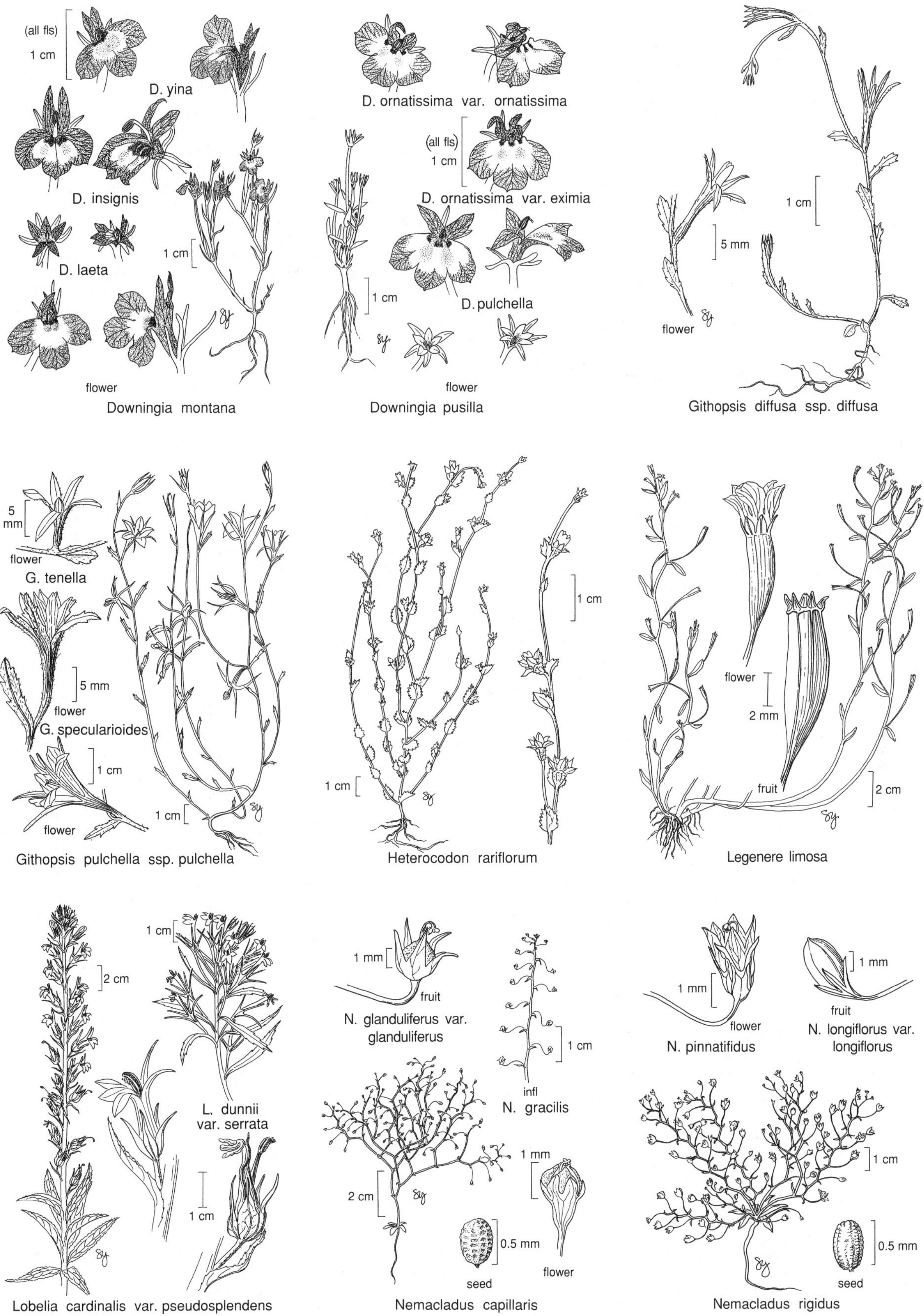
(all fls)
1 cm
D. yina
D. insignis
D. laeta
1 cm
flower
Downingia montana
D. ornatissima var. ornatissima
(all fls)
1 cm
D. ornatissima var. eximia
1 cm
D. pulchella
flower
Downingia pusilla
1 cm
5 mm
flower
Githopsis diffusa ssp. diffusa
5 mm
flower
G. tenella
5 mm
flower
G. specularioides
1 cm
flower
1 cm
Githopsis pulchella ssp. pulchella
1 cm
1 cm
Heterocodon rariflorum
flower
2 mm
fruit
2 cm
Legenere limosa
1 cm
2 cm
L. dunnii
var. serrata
1 cm
Lobelia cardinalis var. pseudosplendens
1 mm
fruit
N. glanduliferus var.
glanduliferus
1 cm
infl
N. gracilis
1 mm
2 cm
0.5 mm
seed
flower
Nemacladus capillaris
1 mm
flower
N. pinnatifidus
1 mm
fruit
N. longiflorus var.
longiflorus
1 cm
0.5 mm
seed
Nemacladus rigidus

N. secundiflorus G. Robb. **STS** spreading to ascending, 2.5–12.5 cm; base dull, reddish brown. **LF** 3–6 mm, narrowly oblanceolate to ± spoon-shaped, narrowed to wide petiole, irregularly dentate, hairy. **INFL** ± 1-sided; axis straight; bracts 2–3 mm, widely lanceolate to oblong; pedicels 9–12 mm, ± 0.1 mm diam, horizontal, ± S-curved, tip erect. **FL**: hypanthium 0.5 mm; sepals ± 0.8 mm, linear, erect; corolla 3–5 mm, divided < 1/2 length, lavender or pink with yellow blotch at lobe base, glabrous to hairy, upper lobes spreading, lower lobes reflexed; filament tube ± 1.5–2 mm, tip slightly curved, fine-hairy, anthers 0.5 mm. **FR** ± 2–2.5 mm, ± hemispheric (base and tip rounded). **SEED** 0.3 mm, subspheric; surface with zigzag ridges alternating with clearly pitted rows. Dry, gravelly slopes; 200–2000 m. s SNH, SCoR. Pls from s SNH have ± superior ovary and open (not 1-sided) infl. Pls in San Luis Obispo Co. have smaller fls.

N. sigmoideus G. Robb. (p. 473) **STS** widely spreading, 4–12 cm; base purplish brown. **LF** 1.5–10 mm, ovate to elliptic, sessile, entire or irregularly dentate, short-hairy. **INFL**: axis strongly zigzag; bracts 0.8–1.5 mm, ovate; pedicels 10–18 mm, < 0.1 mm diam, spreading, S-curved, tip erect. **FL**: hypanthium 0.5 mm; sepals ± 1.5 mm, lanceolate-deltate, erect, spreading in fr; corolla 2.5–3.5 mm, deeply divided, white, yellow at tips, hairy, upper lobes erect, lower lobes spreading; filament tube 1.5 mm, tip curved, fine-hairy, anthers ± 0.3 mm; ovary nearly superior. **FR** ± 2 mm, widely fusiform (base and tip acute). **SEED** ± 0.5 mm, widely elliptic; surface with zigzag ridges alternating with clearly pitted rows. Sandy or gravelly soils, Joshua-tree woodland; 50–2300 m. s SNH (e slope), Teh, e PR, SNE, DMoj, nw DSon; NV, AZ, Baja CA.

N. twisselmannii J. Howell (p. 473) TWISSELMANN'S NEMACLADUS **STS** forming small cushions, 0.5–1 cm; base dull, reddish brown. **LF** 2–3 mm, ± spoon-shaped, narrowed to wide petiole, entire, hairy. **INFL** ± head-like; axis obscure; bracts 1–1.5 mm, oblong, flat; pedicels 2–4 mm, 0.2 mm diam, erect, straight. **FL**: hypanthium 0.5 mm; sepals ± 1 mm, linear, erect; corolla 1–3 mm, divided 1/2 length, white, hairy, lobes erect; filament tube 0.5 mm, tip slightly curved, glabrous, anthers 0.3 mm; ovary 2/3 inferior in fr. **FR** ± 3.5 mm, hemispheric (base rounded, tip ± pointed). **SEED** 0.7–0.8 mm, elliptic to elliptic-oblong; surface with deeply impressed, vertical lines crossed by fine transverse lines. **RARE** CA. Granitic sands, rocks, yellow-pine forest; 2450 m. s SNH (n Kern Co.)

PARISHELLA

1 sp. (brothers Samuel B. (1838–1928) & William F. Parish, botanical collectors)

P. californica A. Gray (p. 473) Ann, glabrous to hairy. **STS** branched from base, reclining, 1–10 cm. **LVS**: basal rosette, persistent; petiole < 7 mm; blade 6–15 mm, 2–5 mm wide, oblanceolate, leathery, tip rounded, margin narrowly translucent. **INFL**: clusters head-like, terminal; bracts oblanceolate, spreading to erect; pedicels in fr 2–5 mm. **FL** not inverted; hypanthium ± 3.5 mm; sepals 2–6 mm, oblanceolate; corolla white, tube 2–3 mm, flaring, lobes 5, 1–1.5 mm, subequal, oblong; filaments 2.5–3.5 mm, free at base, fused above middle, anthers equal, ± round, lower 4 appendaged at junction with tube, upper not; appendages rod-like, minute, translucent; ovary ± 2-chambered, stigma head-like, smooth. **FR** 3–3.5 mm, 2–3 mm wide; top dome-like, circumscissile. **SEED** ± 0.7 mm, elliptic; surface with 8–10 vertical, pitted rows, pits 10–12 per row. Sandy or gravelly soils; 650–1500 m. sw SnJV (Caliente Mtns), se SCoRO, DMoj (San Bernardino Co.).

PORTERELLA

1 sp. (Thomas C. Porter, botanist, 1822–1901)

P. carnosula (Hook. & Arn.) Torrey (p. 473) Ann 2–30 cm, emergent or terrestrial, glabrous. **ST** erect, branched from base or not. **LVS** cauline, sessile; blade (5)12–15 mm, (1)3–5 mm wide, narrowly ovate (aerial) to narrowly triangular (submersed), entire, sometimes few-toothed. **INFL**: raceme; bract 1 per fl, lf-like, ascending; pedicels in fr 1–3 cm. **FL** inverted, fragrant; hypanthium 1–2.5 mm; sepals = or > fr, narrowly triangular; corolla tube 4–5 mm, linear, blue, throat with 2 folds, yellow, limb 2-lipped, upper lip 2-lobed, lobes 1–2 mm, narrowly triangular, erect, deep blue, lower lip 3-lobed, lobes 3–8 mm, ± round, spreading, blue, base yellow or white; stamens fused, anthers ± 2 mm, 2 short anthers appendaged, horn-like appendage ± 0.5 mm, comb-like appendage 4-pronged; ovary 2-chambered, stigma cup- or plate-like, papillate. **FR** 5–10 mm, hemispheric to cylindric; tip conic, dehiscing by valves. **SEED** 1 mm, smooth, fine striate. *n*=12. Moist, grassy roadsides, lake and pond edges; 1500–3100 m. CaRH, n SNH, GB; to WY, AZ. ❀TRY.

TRIODANIS VENUS LOOKING-GLASS

Ann; roots fibrous; hairs sparse, stiff, backward-pointing. **ST** erect, simple or branched from base, gen 5–40 cm, 4-angled. **LVS** cauline, thin, serrate, sessile. **INFL**: fls solitary to several in axils of lf-like bracts, sessile. **FLS**: lower cleistogamous; sepals spreading, narrowly triangular, not lf-like; corolla rotate, lobes deep blue to blue-violet; stamen bases wide, ciliate; ovary inferior, elliptic to obovoid, upper 50% papillate. **FR** dehiscing by lateral pores. **SEED** ± 0.5 mm, widely elliptic in outline. 7–8 spp.: N.Am, 1 Medit. (Greek: 3 teeth) [McVaugh 1945 Wrightia 1:13–52]

1\. Ovary elliptic to ovoid, narrowed at top; lf widely lanceolate to ovate, base not clasping st .... *T. biflora*
1′ Ovary oblong to obovoid, not narrowed at top; lf round-cordate, base gen clasping st .... *T. perfoliata*

T. biflora (Ruiz Lopez & Pavon) E. Greene (p. 473) **LF** 5–15 mm, widely lanceolate to ovate; veins on lower surface inconspicuous; base not clasping st; tip acute. **FLS** opening only in upper 1–3 bract axils (others cleistogamous); corolla 5–9 mm, lobes 4–7 mm; stamens ± 2.5 mm; ovary 4.5–7 mm, elliptic to ovoid, style 3–3.5 mm. **FR**: pores near top. Disturbed areas; < 2000 m. NW, SN GV, CW, SW; to c&s US, S.Am.

T. perfoliata (L.) Nieuwl. (p. 473) **LF** gen 8–11 mm, round-cordate; veins on lower surface conspicuous; base gen clasping st; tip blunt. **FLS** opening in most upper bract axils (lower fls cleistogamous); ovary 5–10 mm, oblong to obovoid; corolla 8–10 mm, lobes 6–8 mm; stamens 3.5 mm; style 4.5 mm. **FR**: pores near or below middle. Disturbed areas; < ± 900 m. NW, n SNH (Plumas Co.), SW (uncommon); to e US, S.Am.

CANNABACEAE HEMP FAMILY

Elizabeth McClintock

Ann, per, gen dioecious, wind-pollinated; epidermis with hardened hairs, glands, etc.; sap watery. **ST** erect or twining. **LVS** palmately compound or lobed, petioled; lower gen opposite (upper often less lobed, alternate). **STAMINATE INFL**: panicle or spike-like cluster, ± open. **PISTILLATE INFL**: spike-, cone-, or head-like cluster of often paired fls, dense. **STAMINATE FL**: perianth parts 5, free; stamens 5; pistil 0. **PISTILLATE FL**: perianth parts fused into a short, unlobed tube or ring; stamens 0; ovary superior, chamber 1, ovule 1, style 1, very short, stigmas 2, thread-like, plumose. **FR**: achene, gen ± enclosed in persistent perianth. 2 genera, 3 spp.: n hemisphere; cult, economically important. [Mitchell 1988 NY State Mus Bull 464:17–23]

1. St erect; lf gen palmately compound; pistillate infl a ± spike-like cluster, erect to spreading .... **CANNABIS**
1′ St twining; lf simple, often palmately lobed; pistillate infl a head- or cone-like cluster, gen pendent ... **HUMULUS**

CANNABIS HEMP

Ann. **ST** erect; hairs unbranched. **LF** gen palmately compound. **STAMINATE INFL**: panicle- or spike-like cluster, ± open. **PISTILLATE INFL**: spike-like cluster, erect to spreading, dense. **FR** sometimes ± enclosed in persistent perianth. 1–3 spp.: native range uncertain. (Greek, Latin: hemp) [Small & Cronquist 1976 Taxon 25:405–435]

C. sativa L. (p. 473) GRASS, HEMP, MARIJUANA, POT, ETC. **ST** branched, < 4 m; inner bark fibrous. **LF**: lflets gen 3–7, < 15 cm, narrowly lanceolate, coarsely serrate. **STAMINATE INFL** > 15 cm. **PISTILLATE INFL** > 2 cm. **FR** 1–2 mm. Disturbed or waste ground; probably < 600 m. CA-FP; possibly native to c Asia, but cult since pre-history. [*C. indica* Lam.] Highly variable complex. Possession gen illegal due to psychoactive resin, tetrahydrocannabinol (THC) that is concentrated in pistillate infls; used legally in medicine; st fibers used for rope, fabric, paper, etc.

HUMULUS HOP

Ann, per. **ST** twining; hairs sometimes forked. **LF** unlobed to palmately lobed. **STAMINATE INFL**: panicle, ± open. **PISTILLATE INFL**: head- or cone‹like cluster, gen pendent (rarely erect), dense. **FR** enclosed in persistent, enlarged perianth subtended by papery bract. 3 spp.: n temp. (Probably latinized from Low German: hop)

H. lupulus L. EUROPEAN HOP Per. **LF** gen 3–5-lobed; blade ± cordate, coarsely serrate. **PISTILLATE INFL** 2.5–5 cm, ± oblong in fr. Uncommon. Disturbed places, persisting from cult; < 3000 m. NCoRO, n&c SNF, SCo, WTR; native to Eurasia. Orn vine and cult as major source of aroma and flavor of beer.

CAPPARACEAE CAPER FAMILY

Staria S. Vanderpool

Ann, shrub, tree, ill-smelling. **LVS** gen 1-palmate, gen alternate, gen petioled; stipules gen minute, often bristle-like or hairy; lflets 3–7. **INFL**: raceme, head, or fls solitary, gen longer in fr; bracts gen 3-parted below, simple above, or 0. **FL** gen bisexual, radial to ± bilateral; sepals gen 4, free or fused, gen persistent; petals gen 4, free, ± clawed; stamens gen 6, free, exserted, anthers gen coiling at dehiscence; ovary superior, gen on stalk-like receptacle, chamber gen 1, placentas gen 2, parietal, style 1, persistent, stigma gen minute, ± head-like. **FR**: gen capsule, septicidal; valves gen 2, deciduous, leaving septum (frame-like placentas) behind; pedicel gen ± reflexed to spreading. 45 genera, 800 spp.: widespread trop to arid temp; some cult (*Capparis spinosa*, caper bush). [Ernst 1963 J Arnold Arbor 44:81–93] CA members placed in subfamily Cleomoideae. Alternate family name: Capparidaceae.

1. Shrub; fr an inflated capsule; petals > 4 mm wide .... **ISOMERIS**
1′ Ann; fr a slightly inflated capsule or pair of nutlets; petals < 4 mm wide
 2. Fls in axillary heads; style in fr stout, spine-like .... **OXYSTYLIS**
 2′ Fls in terminal racemes or solitary in lf axils; style in fr sometimes elongate but not spine-like
  3. Stamens 8–32; ovary ± sessile; fr valves not deciduous .... **POLANISIA**
  3′ Stamens 6; ovary on stalk-like receptacle; fr valves deciduous
   4. Fr a pair of 1–3-seeded nutlets; septum < 2 mm wide .... **WISLIZENIA**
   4′ Fr a 2–many-seeded capsule; septum > 2 mm wide
    5. Fr gen 12–55 mm, longer than wide; septum linear to oblong .... **CLEOME**
    5′ Fr 2–6 mm, often wider than long; septum elliptic to round .... **CLEOMELLA**

CLEOME

Ann, gen ± glabrous. **ST** gen branched from upper nodes. **LF**: petiole 5–45 mm; lflets gen 3. **INFL**: raceme, terminal, gen 1–4 cm in fl, gen 5–40 cm in fr; pedicels 4–20 mm. **FL** often ± unisexual (stamens or pistils vestigial), ± bilateral, most parts gen yellow; sepals free or fused; petals sessile to short-clawed. **FR**: capsule, longer than wide; septum linear to oblong; receptacle stalk-like, reflexed to ascending. **SEEDS** 10–40. 150–170 spp.: esp trop, subtrop Am, Afr; some trop weeds. (Early Eur name for a mustard-like pl)

1. Pl densely glandular-hairy; fr oblong, ± flat; sepals awl-shaped, gen 4–6 mm .... ***C. platycarpa***
1′ Pl glabrous to slightly hairy; fr ± linear, ± round in transverse section; sepals ovate to lanceolate, gen 1.7–3 mm
  2. Infl open, few-fld; sepals free, deciduous; anthers 3–6 mm .... ***C. sparsifolia***
  2′ Infl dense, many-fld; sepals fused in basal half, persistent; anthers 1–2.6 mm
    3. Lflets gen 5; petals yellow .... ***C. lutea***
    3′ Lflets gen 3; petals purple .... ***C. serrulata***

C. lutea Hook. (p. 473) YELLOW BEE PLANT Pl 2.5–13 dm, ± glabrous. **LF**: lflets gen 5, 1.5–6 cm, linear to elliptic. **FL**: sepals fused in basal half, persistent, 1.6–2.6 mm, lanceolate, minutely dentate, yellow; petals 5–8 mm, oblong to ovate, yellow; stamens 10–20 mm, yellow, anthers 1.9–2.6 mm. **FR** 15–40 mm, 2–5 mm wide, ± round in transverse section, striate; receptacle 5–17 mm. 2*n*=34. Dry, sandy flats, desert scrub, weedy roadsides; 1100–2400 m. SNE; to n WA, e Colorado. ❀SUN,DRN,IRR:TRY.

C. platycarpa Torrey (p. 473) GOLDEN BEE PLANT Pl densely branched from base, 1–6 dm, green with purple, densely glandular-hairy. **LF**: lflets 1–3.5 cm, ovate to obovate. **FL**: sepals free, deciduous, gen 4–6 mm, awl-shaped, entire, yellow; petals 6–12 mm, oblong, golden yellow; stamens 10–17 mm, yellow, anthers 1.8–2 mm; style 1–3 mm. **FR** 12–25 mm, 8–12 mm wide, ± flat, hairy; receptacle 10–18 mm. 2*n*=40. Alkaline, clay soils, volcanic tuff, dry foothills, often sagebrush scrub; 1200–1500 m. KR, CaR, n SN, MP; OR, w NV.

C. serrulata Pursh ROCKY MOUNTAIN BEE PLANT Pl open, 3–8 dm, ± glabrous. **LF**: lflets 2–6 cm, elliptic. **FL**: sepals fused in basal half, persistent, gen 1.7–3 mm, ovate, acuminate, minutely dentate, purple to green; petals 7–12 mm, oblong to ovate, purple, rarely white; stamens 18–24 mm, purple, anthers 2–2.3 mm, green; style 0.1–0.5 mm. **FR** 30–55 mm, 3–6 mm wide, ± round in transverse section, smooth; receptacle 10–15 mm. 2*n*=34,60. Sagebrush scrub, pinyon pine-juniper woodland; 1200–1700 m. KR, occasional waif in s CA; to Great Plains, s B.C. ❀TRY.

C. sparsifolia S. Watson Pl 1–9 dm, densely branched, glabrous, glaucous. **LVS** ± sparse; lflets 3 below, 1 (lvs simple) above, 0.4–1.5 cm, obovate. **INFL** open, few-fld, 2–10 cm, not much expanding in fr. **FL**: sepals free, deciduous, 1.4–3 mm, ovate, acuminate, minutely serrate, brown-green; petals 9–13 mm, strap-shaped, recurved, yellow with brown central streak; stamens 9–15 mm, yellow, anthers 3–6 mm, brown; style 0.1–0.4 mm. **FR** 15–45 mm, 1–3 mm wide, ± round in transverse section, smooth; receptacle 2–5 mm. 2*n*=32. Sand dunes, beaches; 900–2000 m. SNE, DMoj; w NV.

CLEOMELLA

Ann, gen glabrous. **ST** gen ascending to erect, gen branched from base, often red-tinged. **LVS** gen many; petiole gen 7–20 mm; lflets gen 3. **INFL**: raceme, ± terminal; fls solitary in lf axils, or both; pedicel gen 4–25 mm. **FL** radial to bilateral; parts gen yellow; sepals fused in basal third, gen entire; petals ± sessile, upper 2 often recurved. **FR**: capsule, often wider than long; septum elliptic to round; receptacle stalk-like. **SEEDS** < 10. ± 10 spp.: arid w N.Am. (Diminutive of *Cleome*) [Payson 1922 Univ Wyoming Publ Sci Bot 1:29–46] *C. hillmanii* Nelson, known from near Reno, NV; may be found in adjacent CA.

1. Pl hairy; older sts prostrate; receptacle in fr 6–8 mm, reflexed .... ***C. obtusifolia***
1′ Pl glabrous; older sts ascending to erect; receptacle in fr < 6 mm or 6–10 mm and spreading to ascending
  2. Fls solitary in lf axils; receptacle reflexed in fr .... ***C. brevipes***
  2′ Fls in terminal racemes, sometimes also solitary in lf axils; receptacle spreading to ascending in fr
    3. St branched gen from base; petals 1.8–2.2 mm; receptacle in fr 0.3–0.8 mm .... ***C. parviflora***
    3′ St branched gen from upper nodes; petals 3.5–7 mm; receptacle in fr 6–10 mm .... ***C. plocasperma***

C. brevipes S. Watson Pl glabrous, glaucous. **ST** 5–45 cm, rough. **LF**: petiole 0.5–3 mm; lflets 5–15 mm, obovate-linear, fleshy. **INFL**: fls solitary in lf axils, incl those near base; pedicels 1.5–3 mm. **FL**: sepals 0.8–1.2 mm, ovate, acuminate; petals 1.5–2 mm, pale yellow; stamens 1.5–2.2 mm, anthers 0.3–0.5 mm; style 0.1–0.3 mm. **FR** 2–3 mm, 2–3.2 mm wide, round; valves slightly conic; receptacle 0.5–3 mm, reflexed. Alkaline marsh, wet, salt-encrusted soil around thermal springs; 400–1400 m. SNE, DMoj; w NV.

C. obtusifolia Torrey & Frémont (p. 473) MOJAVE STINKWEED Pl hairy. **ST** 1–9 dm, rough; younger ascending to erect; older prostrate, forming circular mat < 9 dm wide. **LF**: lflets 5–15 mm, obovate. **INFL**: raceme on older sts, 1–10 cm, terminal, dense, on younger sts fls solitary in lf axils. **FL**: sepals 1–1.5 mm, ovate, green, hairy, margin with long hairs; petals 4–6 (9) mm, dark yellow, lower surface hairy; stamens 8–14 mm, anthers 1.5–2.3 mm; style 1.5–5 mm. **FR** 3–4 mm, hairy, striate; valves conic to horn-shaped; receptacle 6–8 mm, reflexed. Desert scrub, sandy, rocky alkaline flats; 300–1200(2000) m. D; w NV. [*C. o.* var. *pubescens* Nelson] Variable; deserves additional study.

C. parviflora A. Gray Pl glabrous. **ST** 3–45 cm, smooth. **LVS** few; lflets 5–35 mm, linear-elliptic, ± fleshy. **INFL**: raceme, 0.5–30 cm, terminal; fls sometimes also solitary in lf axils. **FL**: sepals 0.5–1 mm, lanceolate; petals 1.8–2.2 mm, pale yellow; stamens 1.9–2.5 mm, anthers 0.4–0.5 mm; style < 0.2 mm, stigma 2-lobed, 0.3 mm, purple. **FR** 3–4 mm; valves slightly conic; receptacle 0.3–0.8 mm, spreading to ascending. Wet, alkaline meadows about thermal springs in sagebrush desert; 1200–2000 m. GB, DMoj; w NV. Often occurs with *C. brevipes*, *C. plocasperma*.

C. plocasperma S. Watson (p. 473) Pl glabrous. **ST** branched gen from upper nodes, 10–55 (80) cm, smooth. **LF**: lflets 15–45 mm, linear-elliptic. **INFL**: raceme, 1–20 cm, terminal. **FL**: sepals 0.9–2.2 mm, lanceolate; petals 3.5–7 mm; stamens 8–12 mm, anthers 1.5–1.9 mm; style 0.8–1.2 mm. **FR** 4–5 mm; valves ± hemispheric to horn-shaped; receptacle 6–10 mm, spreading to ascending. Wet, alkaline meadows, greasewood flats, around thermal springs; 800–1400 m. GB, DMoj; to OR, ID, UT. Vars. *mojavensis* (Payson) Crum, *stricta* Crum of uncertain taxonomic status.

ISOMERIS

1 sp. (Greek: equal part)

I. arborea Nutt. (p. 473) BLADDERPOD Shrub, profusely branched, gen 5–20 dm, minutely hairy. **LF**: petiole 1–3 cm; lflets gen 3, 15–45 mm, oblong-elliptic. **INFL**: raceme, 1–30 cm, terminal; pedicels 8–15 mm, thicker in fr. **FL**: sepals fused in basal half, 4–7 mm, ± entire, green; petals 8–14 mm, 4–5 mm wide, sessile, yellow; stamens 15–25 mm, yellow, anthers 2–2.5 mm; style 0.9–1.2 mm or pistil aborting in bud. **FR**: capsule, tardily dehiscent, 3–4 cm, inflated, oblong to ± spheric, smooth, leathery, light brown;

valves 2–3; receptacle stalk-like, 1–2 cm, stout, reflexed. $2n$= 34,40. Common. Coastal bluffs, hills, desert washes, flats; 0–1300 m. s SNF, Teh, SnJV, CCo, SCo, ChI, D; to Baja CA. [*Cleome isomeris* E. Greene] Vars. *globosa* Cov., *insularis* Jepson, *angusta-ta* Parish have been recognized based on fr variation that may be loosely correlated with geography. ❀DRN,SUN&DRY:8,9,**14**, 15–17,**19–24**;IRR:**12,13**;STBL.

OXYSTYLIS

1 sp. (Greek: sharp style)

O. lutea Torrey & Frémont (p. 473) Ann, ± glabrous. **ST** branched from base, 5–15 dm. **LF**: petiole 2.5–7 cm; lflets 3, 2–6 cm, ± elliptic, thick, firm. **INFL**: head, 0.5–3 cm, axillary, ± spheric, dense, ± sessile, not elongating in fr; pedicels 1–3 mm, thicker and reflexed in fr. **FL** radial; sepals free, 1–2 mm, lanceolate, green; petals 2–3 mm, elliptic, straw-yellow, ± sessile; stamens 3–5 mm, yellow, anthers 1–1.2 mm; ovary 1–1.5 mm, ± sessile, lobes 2, nearly separate, each 1-ovuled; style 2–3 mm, wide, fleshy at base. **FR**: nutlets 2, 2.5 mm, ± spheric, smooth, white to deep purple; receptacle < 2 mm, ± stalk-like; style 4–11 mm, stiff, spine-like. **SEED** 1 per nutlet. $2n$=20,40. Rocky or sandy alkaline flats; > 600 m. DMoj; w NV (Amargosa Desert).

POLANISIA CLAMMY WEED

Ann, sticky, glandular-hairy. **LF**: lflets 3–5, oblanceolate to ovate. **INFL**: raceme, terminal, dense. **FL** bilateral; sepals ± free, deciduous; petals gen unequal, obovate, notched at tip, clawed; stamens 8–32, unequal, anthers not coiling; ovary ± sessile, subtended on 1 side by a prominent nectary. **FR**: capsule, dehiscent in upper half; valves persistent; receptacle not stalk-like. **SEEDS** many. 4 spp.: N.Am. (Greek: many, unequal, from stamens) [Iltis 1958 Brittonia 10:33–58] Taxonomic position problematic.

P. dodecandra (L.) DC. ssp. ***trachysperma*** (Torrey & A. Gray) Iltis (p. 473) **LVS** many; petiole 1.5–4.5 cm; lflets 1.6–4.5 cm. **INFL** appearing after lvs, 5–30 cm; pedicels 1–2.5 cm, green to purple. **FL**: sepals 4–7 mm, purple; petals 6–13 mm, white, claw 4–5 mm; stamens 10–20, 5–40 mm, filaments purple, anthers 1.0–1.3 mm; nectary bright orange, spheric, concave at top, 2–2.3 mm; style 5–17 mm, deciduous, stigma purple. **FR** 4–7 cm, 5–9 mm wide, green. $2n$=20. Waste areas, dry, sandy soil; 1400–1500 m. ScV, MP; to e Can, se US. [*P. trachysperma* Torrey & A. Gray]

WISLIZENIA

1 sp. (A. Wislizenius, pl collector in sw US, born 1810)

W. refracta Engelm. JACKASS CLOVER Ann or per, glabrous to puberulent. **ST** profusely branched from base, 0.5–24 dm. **LF**: petiole 3–25 mm; lflets gen 3. **INFL**: raceme, 1–3 cm, dense, terminal, in fr 4–20 cm; pedicels 5–10 mm. **FL** radial; sepals free, ± 2 mm, ± entire, green; petals 2.5–6.3 mm, elliptic, yellow, ± sessile but tapered to base; stamens 8–14 mm, yellow; ovary 0.3–0.6 mm, gen exserted, lobes 2, nearly separate, each gen 1-ovuled, style 2–5.5 mm. **FR**: nutlets 2; valves deciduous; receptacle stalk-like, reflexed; style elongate but not spine-like. **SEEDS** gen 1 per nutlet. $2n$=40. Desert washes and flats, fields, roadsides, esp alkaline soils; 0–800 m. c&s SNF, SnJV, D; to TX, nw Mex. [Keller 1979 Brittonia 31:333–351] Valuable honey plant. TOXIC but seldom eaten.

1. Gen per; lflets 3 below, 1 (lvs simple) above, linear-elliptic, 3–8 ± longer than wide; sepals ovate, < 1.75 × longer than wide .... ssp. ***palmeri***

1′ Ann; lflets 3, ovate to obovate, 2–3 × longer than wide; sepals lanceolate, > 1.75 × longer than wide

2. Lflets 2 × longer than wide; anthers 1.6–2.1 mm; c&s SNF, SnJV .... ssp. ***californica***

2′ Lflets 2–3 × longer than wide; anthers 0.9–1.2 mm; D .... ssp. ***refracta***

ssp. ***californica*** (E. Greene) C.S. Keller (p. 473) Ann. **ST** green, tan. **LF**: lflets 3, 4–20 mm, obovate. **FL**: receptacle 6–10 mm; sepals 1–2 mm, lanceolate; anthers 1.6–2.1 mm; style 3–5 mm. Sandy washes, weedy roadsides, cult and fallow fields; 0–100 m. c&s SNF, SnJV.

ssp. ***palmeri*** (A. Gray) C.S. Keller Per, short-lived. **ST** brown-gray. **LF**: lflets 3 below, 1 (lf simple) above, 17–35 mm, linear-elliptic. **FL**: receptacle 5–14 mm; sepals 1–1.7 mm, ovate; anthers 1.5–2.3 mm; style 2.5–5.5 mm. Sandy washes, beach dunes, desert scrub; 0–200 m. DSon; nw Mex.

ssp. ***refracta*** JACKASS CLOVER Ann. **ST** green, tan. **LF**: lflets 3, 7–30 mm, ovate. **FL**: receptacle 3–6 mm; sepals 1–1.5 mm, lanceolate; anthers 0.9–1.2 mm; style 2–5 mm. UNCOMMON. Sandy washes, roadsides, alkaline flats; 600–800 m. DMoj, n DSon; to w TX.

CAPRIFOLIACEAE HONEYSUCKLE FAMILY

Lauramay T. Dempster

Subshrub, shrub, vine, or small tree. **LVS** opposite, simple or compound; stipules gen 0. **FL**: calyx tube fused to ovary, limb gen 5-lobed; corolla radial or bilateral, rotate to cylindric, gen 5-lobed; stamens gen 5, epipetalous, alternate corolla lobes; ovary inferior, 1–5-chambered, style l. **FR**: berry, drupe, or capsule. ± 12 genera, 450 spp.: esp n temp.

1. Corolla rotate to deeply saucer-shaped, radial; style short; fls many, in rounded, dome-shaped or flat-topped infl; lf simple or compound

2. Lf compound; fr berry-like; seeds 3–5 .... **SAMBUCUS**

2′ Lf simple; fr a drupe; seed 1 .... **VIBURNUM**

1′ Corolla cylindric, funnel-shaped, or 2-lipped; style gen long; fls paired or several in elongated, raceme-like infls; lf simple

3. Creeping vine .... **LINNAEA**

3′ Erect or twining shrub

4. Corolla ± bilateral, cylindric, 2-lipped, swollen on 1 side at base; fr red or black; seeds gen > 2 .... **LONICERA**
4′ Corolla ± radial, openly bell-shaped to narrowly funnel-shaped, not or barely swollen at base; fr white; seeds 1–2 .... **SYMPHORICARPOS**

LINNAEA TWIN FLOWER

Per, shrub, creeping. **INFL**: fls gen 2 at forked tip of slender erect peduncle, nodding. **FL**: calyx lobes slender, tapered; corolla radial, slender, bell- to funnel-shaped, pink; stamens 4, 2 shorter and inserted near base of corolla tube; ovary chambers 3, 2 with abortive ovules, other with 1 developing ovule. **FR**: capsule, 3-valved. 1 sp., ± 3 vars.: circumboreal. (Named for Linnaeus)

L. borealis L. var. ***longiflora*** Torrey (p. 477) **ST** trailing, 15–20 cm, slender. **LF**: petiole ± 2 mm; blade 12–18 mm, ovoid, serrate above middle. **INFL**: peduncle ± 6 cm; inner bracts 2, small, with straight hairs; outer bracts 2, round, surrounding ovary, densely glandular-hairy. **FL**: corolla 10–13 mm, bell-shaped above slender tube. Moist shady places in coniferous forest; 200–2600 m. NW, CaR, n SNH, MP; to n AK. ❀SHD,DRN:1,2,4–6&IRR:14–17; DFCLT.

LONICERA HONEYSUCKLE

Shrub, erect or twining. **LVS** simple, entire, short-petioled; 1–2 pairs beneath infl often fused around st. **INFL**: spikes, interrupted, at ends of branches, or fls paired on axillary peduncles and subtended by 0–2 sets of bracts. **FL**: calyx-limb 0 or gen 5-toothed, gen persistent; corolla 5-lobed, ± radial or strongly 2-lipped (4 upper lobes, 1 lower), tube pouched at base; ovary chambers 2–3. **FR**: berry, gen round. ± 200 spp.: temp, subtrop N.Am, Eur, Asia, n Afr. (Adam Lonitzer, German herbalist, 16th century) [Rehder 1903 Rep Missouri Bot Gard 14:27–231] 2 collections (Del Norte Co., Eldorado Co.) have purplish, apparently sterile, variously distorted fls with long, slender ovary/hypanthium; probably alien (key 7.).

1. Infl a peduncled pair of fls
 2. Corolla > 25 mm; climbing vine .... ***L. japonica***
 2′ Corolla < 20 mm; erect shrub
  3. Ovaries of fl pair fused or appearing so
   4. Corolla yellowish, weakly 2-lipped; bracts gen 1–3, narrowly lanceolate; ovaries and berries tightly enclosed by fused inner bracts, appearing fully fused .... ***L. cauriana***
   4′ Corolla dark red, strongly 2-lipped; bracts 0 or minute; ovaries and berries fused ± 2/3 .... ***L. conjugialis***
  3′ Ovaries of fl pair obviously free
   5. Bracts not leafy, not obscuring ovaries .... ***L. tatarica***
   5′ Bracts leafy, forming a conspicuous involucre, ± enveloping ovaries .... ***L. involucrata***
    6. Pl gen < 9 dm; corolla yellow, tube ± wider upward; style and stigma well exserted; montane .... var. ***involucrata***
    6′ Pl gen > 15 dm; corolla yellow, strongly tinged orange or red, tube cylindric; stigma rarely exserted; coastal .... var. ***ledebourii***
1′ Infl a spike, gen ± interrupted — twining or trailing shrubs
 7. Ovary/hypanthium long, slender, gen sterile — see note after generic description
 7′ Ovary/hypanthium short, round, fertile
  8. Lf 5–10 cm; corolla 15–40 mm; infl a dense, short spike; upper lf pair fused around st; NW, CaR
   9. Corolla orange, weakly 2-lipped; stamens and style little exserted .... ***L. ciliosa***
   9′ Corolla ± yellowish white, strongly 2-lipped; stamens and style well exserted .... ***L. etrusca***
  8′ Lf 1–8 cm; corolla gen < 15 mm; infl a ± long, interrupted spike; upper lf pair fused around st or not; widespread
   10. Upper lf pairs fused around st; corolla hairy or not
    11. Lvs gen with ± obvious stipules, at least toward infl; corolla glandular-hairy .... ***L. hispidula*** var. ***vacillans***
    11′ Lvs without stipules; corolla glabrous .... ***L. interrupta***
   10′ Upper lf pairs not fused around st; corolla often hairy .... ***L. subspicata***
    12. Lf < 2 × longer than wide; widespread (incl Santa Barbara Co.) .... var. ***denudata***
    12′ Lf 3–4 × longer than wide; WTR (Santa Barbara Co.) .... var. ***subspicata***

L. cauriana Fern. (p. 477) Shrub, erect, 3–9 dm; herbage puberulent (lvs ciliate). **LF** 2–5 cm; blade oblong-ovate, ciliate, base tapered to petiole, tip round or obtuse. **INFL**: fls paired; peduncles ± 2 mm, axillary; bracts 1–3, narrowly lanceolate, inner fused, tightly enclosing ovaries. **FL**: calyx-limb exserted from sheathing bracts; corolla 6–9 mm, yellowish, bell-shaped, weakly 2-lipped, divided halfway; anthers exserted; ovaries appearing fused because of sheathing bracts, style ± = corolla, glabrous. **FR** ± 8 mm, red; 2 calyces apparent in fr. Uncommon. Bogs, wet meadows; 2200–3200 m. c&s SNH; also OR to AK, ID. ❀WET:1,2&SHD:15–16.

L. ciliosa (Pursh) Poiret (p. 477) ORANGE HONEYSUCKLE Pl trailing to high-climbing; herbage glabrous or soft-puberulent (lvs ciliate). **ST** 3–30 dm. **LF** deciduous, 6–10 cm; blade oval or ovate, ciliate, base tapered to petiole, tip round to sharp; upper 1–2 pairs fused around st. **INFL**: spike, short, dense; fls ± 20 in 2–4 whorls. **FL**: corolla 16–40 mm, ± cylindric, weakly 2-lipped, divided 1/6–1/4, orange; stamens, style, stigma slightly exserted. **FR** ± 8 mm, red, slightly glaucous. Forests, thickets; 700–1700 m. NW, CaR; to B.C., MT. Pollinated by hummingbirds. ❀IRR or WET:**4**,**5**&SHD:1,**6**,7,15–17.

L. conjugialis Kellogg (p. 477) Shrub, erect, slender, 6–18 dm; herbage puberulent. **LF** 2–8 cm; blade elliptic to round, base ± tapered to petiole, tip round to acute. **INFL**: fls paired; peduncles 14–30 mm, slender, axillary; bracts 0 or minute. **FL**: calyx limb 0 or inconspicuous; corolla 4–7 mm, strongly 2-lipped, dark red, glabrous or sparsely hairy, upper lip erect, shallowly 4-lobed, lower lip down-turned; anthers 3 in upper lip, 2 exserted from deepest corolla sinuses; ovaries fused 1/5–4/5, style hairy well below stigma, stigma exserted. **FRS** paired, 6–8 mm, bright red, translucent; the pair transversely oblong. Streambanks, moist places in coniferous forest, open rocky slopes, talus; 1400–3300 m. NW, SN, MP; to WA, w NV. ❀WET:1,2,4,5&SHD:6,7,15,16;DFCLT.

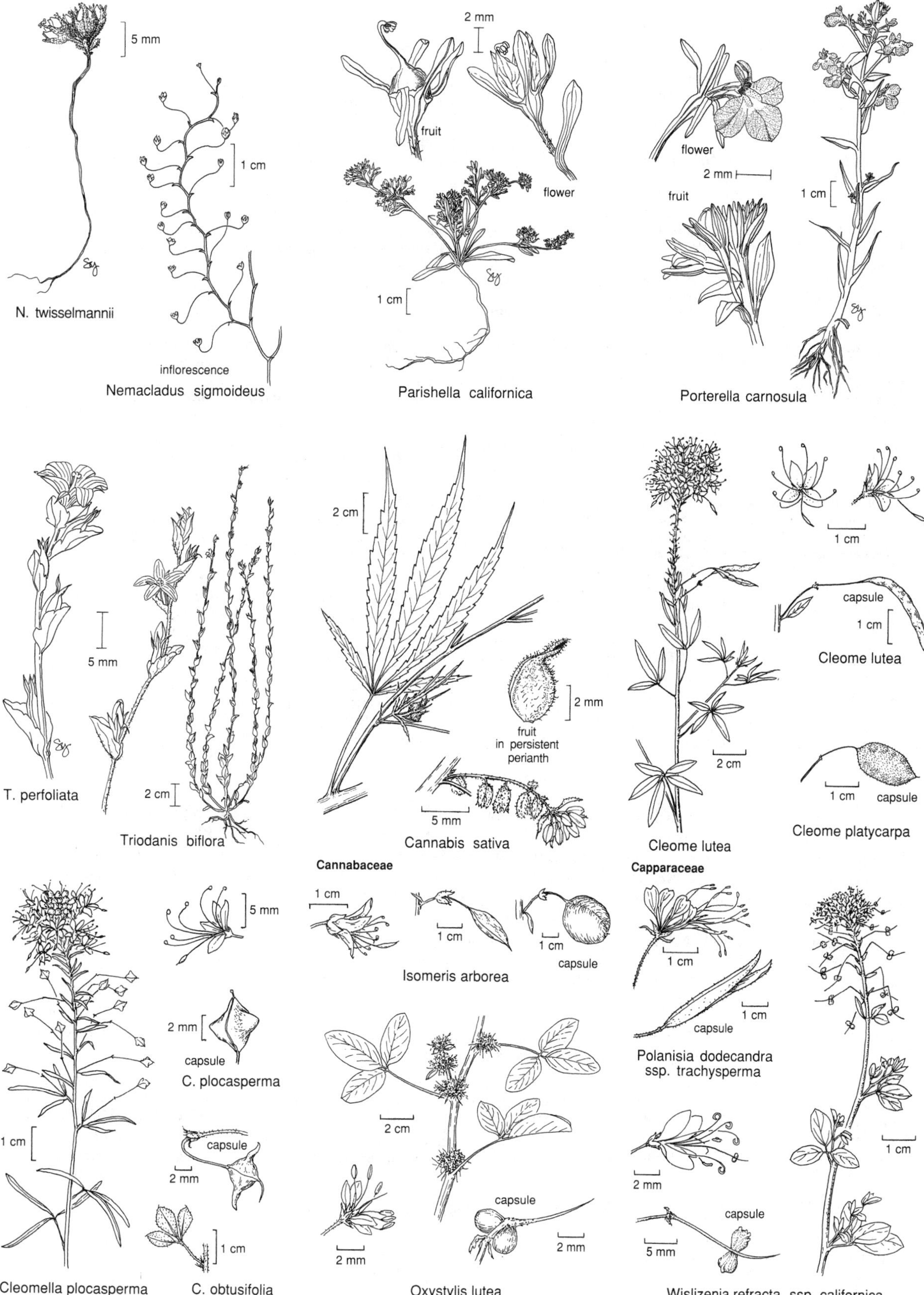

5 mm
N. twisselmannii
1 cm
inflorescence
Nemacladus sigmoideus
2 mm
fruit
flower
1 cm
Parishella californica
flower
2 mm
fruit
1 cm
Porterella carnosula
5 mm
T. perfoliata
2 cm
Triodanis biflora
2 cm
2 mm
fruit
in persistent
perianth
5 mm
Cannabis sativa
Cannabaceae
2 cm
Cleome lutea
1 cm
capsule
1 cm
Cleome lutea
1 cm
capsule
Cleome platycarpa
Capparaceae
5 mm
2 mm
capsule
C. plocasperma
1 cm
capsule
2 mm
1 cm
Cleomella plocasperma
C. obtusifolia
1 cm
1 cm
1 cm
capsule
Isomeris arborea
2 cm
capsule
2 mm
2 mm
Oxystylis lutea
1 cm
1 cm
capsule
Polanisia dodecandra
ssp. trachysperma
2 mm
capsule
5 mm
1 cm
Wislizenia refracta ssp. californica

L. etrusca Santi Vine; herbage glabrous. **LF** gen 6–9 cm; blade ovate, sessile to short-petioled, tip obtuse to rounded; upper pair fused around st. **INFL**: spike, short, dense, sometimes glandular; bracts round. **FL**: corolla 3–5 cm, strongly 2-lipped, divided 1/4–1/3, yellowish white tinged purplish red, tube very slender; stamens, style and stigma well exserted. Uncommon. Disturbed places; < 300 m. NCo (Del Norte, Humboldt cos.); OR; native to Eur.

L. hispidula Douglas var. ***vacillans*** A. Gray (p. 477) Pl sprawling or climbing, 18–60 dm; herbage puberulent. **LF** 4–8 cm; blade oblong to ovate, base truncate or subcordate, tip gen obtuse; upper pairs fused around st, others gen with green or merely scale-like stipules. **INFL**: spike, in upper axils, long, interrupted, very glandular, esp in fr; fls in pairs, sessile. **FL**: corolla 12–16 mm, strongly 2-lipped, pink, glandular-hairy, upper lip shallowly 4-lobed; stamens, style and stigma exserted. **FR** ± 8 mm, red. Canyons, streamsides, woodlands; < 1100 m. NW, SN, CW, SW; s OR. [var. *californica* Jepson] ❀15–17,IRR:**7**,24&SHD:8,9,**14**,18–23;STBL.

L. interrupta Benth. Shrub. **ST**: trunk rigid, woody, ± 3 dm; branches climbing or sprawling; herbage glabrous or puberulent. **LF** 2–2.5 cm; stipules 0; blade elliptic to round, base tapering or round, tip gen ± round; upper 1–3 pairs fused around st. **INFL**: spike, long, interrupted. **FL**: corolla 8–10 mm, strongly 2-lipped, deeply divided, cream-yellow, glabrous. **FR** ± 10 mm, red. Dry slopes, ridges, mixed forest; 500–1400 m. NW, w CaR, SN, CW, SW; AZ. ❀**7,14**,IRR:2,8–10,18–24;SUN:**15–17**;STBL.

L. involucrata (Richardson) Banks TWINBERRY Shrub, erect, 6–30 dm; herbage with stalked glands, sparsely hairy. **LF** 3–12 cm; petiole 2–10 mm; blade elliptic to ovate, base rounded or tapered, tip acute to acuminate. **INFL**: fls paired; peduncles 12–32 mm, axillary, subtended by conspicuous involucre of 2 pairs of bracts; outer bracts opposite fls, 7–12 mm, widely ovate, gen acute; inner bracts 4–8 mm, alternate fls, deeply 2-lobed, truncate, densely glandular. **FL**: calyx limb 0; corolla 12–18 mm, ± cylindric to narrowly bell-shaped, ± yellow, hairy, lobes ± 2 mm, subequal, slightly spreading; ovaries of a fl pair free. **FR** 6–10 mm, black, in enlarged, purple or red, spreading or reflexed involucre. Moist places; < 2900 m. NCo, SN, CCo, MP; to AK, e N.Am, Mex.

var. ***involucrata*** (p. 477) Pl 6–9 dm. **LF** thin. **FL**: corolla clear yellow, tube a little wider upward; stamens gen incl; style and stigma exserted. Moist places in mtns; 600–2900 m. SN, MP; to AK, e N.Am, Mex. ❀SHD&IRR or WET:1,2,4–7,10,15–19.

var. ***ledebourii*** (Eschsch.) Jepson (p. 477) Pl 15–36 dm. **LF** leathery; midribs often somewhat arched. **FL**: corolla heavily tinged orange or red outside, tube cylindric; stamens gen incl; stigma incl or slightly exserted. Moist coastal places; < 1500 m. NCo, CCo; s OR. ❀IRR or WET:**5**,6,**15–17**,24&SHD:7,**14**,19–23.

L. japonica Thunb. JAPANESE HONEYSUCKLE Vine, climbing; herbage glabrous or soft-hairy. **LF** gen 3–8 cm; blade oblong to ovate, base rounded, tip ± acute. **INFL**: fls paired, each pair subtended by 2 lf-like bracts and 4 ± round bractlets that are ± 1/2 ovary length; peduncles short, axillary. **FL**: corolla 25–40 mm, strongly 2-lipped, white turning yellow, often tinged purplish, tube hairy; stamens, style and stigma exserted. **FR** black. $2n$=18. Disturbed places; gen < 1000 m. CA; abundant in se US; native to Asia. Sporadic escape from cult.

L. subspicata Hook. & Arn. Pl gen climbing or reclining on shrubs, 9–24 dm; base woody; herbage glabrous to puberulent. **LF** gen 1–4 cm; blade oblong or ovate, base round or ± tapered, tip round or obtuse; upper pairs not fused around st. **INFL**: spike, long, interrupted, often ± glandular-hairy. **FL**: corolla 8–12 mm, pale yellow, strongly 2-lipped, often hairy; stamens, style, stigma exserted. **FR** ± 8 mm, red or yellow. Chaparral slopes; < 1800 m. n SNH (Butte Co.), Teh, CW, SW.

var. ***denudata*** Rehder (p. 477) **LF**: blade widely elliptic to ± round, < 2 × longer than wide. Habitat and range of sp. [var. *johnstonii* Keck] ❀**7,8,9,14–24**.

var. ***subspicata*** (p. 477) **LF**: blade narrowly elliptic, 3–4 × longer than wide. Chaparral; < 1000 m. WTR (Santa Ynez Mtns, Santa Barbara Co.) ❀**14–24**.

L. tatarica L. Shrub, erect, < 3 m; herbage glabrous or soft-hairy. **LF** gen 3–6 cm; blade ovate, round or ± cordate, tip obtuse; upper pairs not fused around st. **INFL**: fls in pairs, sessile; peduncles ± 15 mm, slender, axillary; bracts not forming an involucre, lower lanceolate, spreading, upper round-ovate, erect. **FL**: calyx limb ± deeply lobed; corolla ± 15 mm, white, fading yellowish, glabrous, weakly 2-lipped, lobes ± = tube, subequal, obovate; stamens well exserted; ovaries of a fl pair free, stigma barely exserted. **FR** ± 1 cm, red. Disturbed places; 700–1100 m. w CaR; native to Siberia. Berries possibly TOXIC, attractive to children.

SAMBUCUS ELDERBERRY

Large shrub (or small tree gen lacking main trunk), deciduous. **ST**: pith conspicuous, spongy. **LVS** pinnately (rarely bipinnately) compound, with terminal lflet; lflets serrate. **INFL**: panicle made up of cymes, terminal. **FL** small; calyx 5-toothed; corolla radial, rotate, 5-lobed, white or cream; ovary chambers 3–5, ovules 1 per chamber and suspended from its top, style short, stigmas 3–5. **FR**: drupe. **SEEDS** 3–5. (Greek, the name of a musical instrument made from wood of this genus) 20 spp.: temp, subtrop; some cult as orn. TOXIC in quantity (exc cooked frs).

1. Fr appearing blue (black and densely glaucous); infl ± flat-topped, central axis gen abruptly shorter and weaker than branches ..... ***S. mexicana***
1' Fr red or purplish black, not glaucous; infl ± dome-shaped, central axis dominant
  2. Fr purplish black — infl gen 4–7 cm diam ..... ***S. melanocarpa***
  2' Fr red ..... ***S. racemosa***
    3. Lf gen glabrous beneath; infl gen 3–6 cm diam; montane ..... var. ***microbotrys***
    3' Lf ± hairy beneath; infl gen 6–10 cm diam; ± coastal ..... var. ***racemosa***

S. melanocarpa A. Gray Shrub 1–2 m. **LF**: lflets 5–7, 4–12 cm, oblong to narrowly ovate, ± abruptly acuminate, glabrous beneath or ± stiff-hairy along veins. **INFL** gen 4–7 cm diam, dome-shaped. **FR** purplish black, not glaucous. Streamsides, edges of meadow or coniferous forest; 1800–3600 m. CaR, SNH; to w Can, UT, NM. ❀**4,5**,WET:1,2,6,7.

S. mexicana C. Presl (p. 477) BLUE ELDERBERRY Shrub 2–8 m, gen as wide as tall, lacking main trunk. **LF**: lflets 3–9, 3–20 cm, elliptic to ovate, glabrous to hairy, axis often curved or bowed, base often asymmetric, tip acute to acuminate. **INFL** 4–33 cm diam, ± flat-topped; central axis gen abruptly shorter and weaker than branches. **FR** nearly black and densely white glaucous, thus appearing bluish. $2n$=36. Common. Streambanks, open places in forest; < 3000 m. CA-FP, GB; to B.C., UT, NM. [*S. caerulea* Raf.] Variable, currently impossible to split into unified subgroups; detailed study warranted. ❀4,5,**6,7,14–17,24**,IRR:1–3,**8,9**,10,**18–23**.

S. racemosa L. RED ELDERBERRY Shrub 2–6 m. **LF**: lflets 5–7, 6–16 cm, lanceolate to oblong-ovate, base gen asymmetric, tip gradually acuminate. **INFL** 6–10 cm diam, ± dome-shaped; central axis remaining dominant throughout development. **FR** gen bright red, not glaucous. Moist places; < 3300 m. CA-FP; to AK, Colorado, AZ.

var. ***microbotrys*** (Rydb.) Kearney & Peebles (p. 477) **LF**: lflets gen glabrous. Common. Moist places, ± montane; 1800–3300 m. NW, CaR, SN, SnBr; to Colorado, AZ. [*S. microbotrys* Rydb.] ❀4,5,WET:1,2,6,7,15–17;DFCLT.

var. ***racemosa*** **LF**: lflets stiffly hairy beneath, esp along veins. Moist woods, open flats in ± coastal places; < 1800 m. NCo, w KR, NCoRO, SnFrB; to AK. ❀**4,5**,IRR:**6,15–17**,24&SHD:**7**, 14,19–23.

SYMPHORICARPOS WAXBERRY, SNOWBERRY

Shrub. **ST** decumbent to erect, slender. **LF** simple, deciduous, small, short-petioled; blade gen elliptic to round, some often ± lobed. **INFL**: gen raceme, gen ± terminal, gen few-fld; fl subtended by 2 fused bractlets. **FL** ± radial; hypanthium ± spheric; calyx with 5-toothed, persistent limb; corolla bell-shaped to ± salverform, gen 5-lobed, white or pink, often ± hairy inside; nectaries 1–5, ± basal; stamens gen incl; ovary chambers 4, styles gen incl, stigma head-shaped. **FR** gen berry-like, gen white. **SEEDS** 2 (1 per lateral ovary chamber), ± oblong, planoconvex. ± 10 spp.: N.Am, 1 in China. (Greek: to bear fr together, the berries borne in clusters) [Jones 1940 J Arnold Arbor 21:201–252]

1. Corolla lobes hairy inside, ± = throat (subg. *Symphoricarpos*)
 2. Pl 6–18 dm, erect; corolla swollen on 1 side, glandular within swelling; infl gen 8–16-fld ***S. albus*** var. ***laevigatus***
 2′ Pl 1.5–6 dm, sprawling; corolla scarcely or not swollen, nectary glands 5, below all corolla lobes; infl gen 2–8-fld .... ***S. mollis***
1′ Corolla lobes glabrous inside, << throat or tube (subg. *Anisanthus*)
 3. Corolla ± salverform, 8–15 mm, tube slender, glabrous inside, 3–4 × length of spreading lobes; nectary 1; lf ± similar above and below, blade gen 4–12 mm .... ***S. longiflorus***
 3′ Corolla bell-shaped, 6–10 mm, tube wide, ± hairy inside, 2–3 × length of erect lobes; nectaries 5; lf more obviously veined below, blade gen 8–20 mm .... ***S. rotundifolius***
  4. Pl trailing; upper 2/3 corolla throat sparsely hairy inside; s SNH, SW, SNE, DMtns .... var. ***parishii***
  4′ Pl erect, spreading; upper 1/3 of corolla tube glabrous inside, middle 1/3 hairy; CaR, SN, GB var. ***rotundifolius***

S. albus (L.) S.F. Blake var. ***laevigatus*** (Fern.) S.F. Blake (p. 477) SNOWBERRY Pl erect, 6–18 dm, glabrous or puberulent. **ST**: branches stiff, spreading; new shoots erect, unbranched, often with infl, their lvs larger, more variable. **LF**: blade gen 1–3 cm (those of new shoots –6 cm). **INFL**: fls 8–16. **FL**: calyx limb spreading, divided halfway; corolla 4–6 mm, bell-shaped, pink, swollen on lower side, glandular within swelling, lobes ± 1/2 corolla length, lobes and upper throat ± densely hairy inside, lobes ± erect. **FR** 8–12 mm, round. **SEED** 4–5 mm. Shady woods, streambanks, n slopes; < 1200 m. NW, w edge CaR, n SNF, CW, SW; to AK, MT. [*S. rivularis* Suksd.] Naturalized in e US. Fr may be TOXIC to humans. ❀**4,6,15–17**,SHD:**7,14,18–24**&IRR:**8,9**,10,11;STBL.

S. longiflorus A. Gray (p. 477) FRAGRANT SNOWBERRY Pl 9–12 dm, stiff, glabrous or puberulent, often dotted with minute glands. **ST**: branches often spiny; young bark red or brown, old whitish, shredding. **LF**: blade 0.5–2 cm, entire, ± thick, sometimes lanceolate, bluish, veins below not prominent. **INFL**: fls sometimes solitary in axils. **FL** very fragrant; calyx limb ± erect, unevenly and often shallowly lobed, sinuses often round; corolla 8–15 mm, ± salverform, pink or cream-colored, tube slender, often red or purple outside, glabrous inside, lobes 1/5–1/4 corolla length, spreading, ovate; nectary 1, long, slender; style gen hairy above middle. **FR** ± 7 mm, narrowly elliptic, dry. **SEED** ± 5 mm. Among rocks; 1350–1600 m. GB, DMtns; to Colorado, TX. ❀4,6,IRR:1,15& SHD:2,3,7,16;DFCLT.

S. mollis Nutt. (p. 483) CREEPING SNOWBERRY, TRIP VINE Pl trailing or creeping, 1.5–6 dm, ± glabrous to soft-hairy. **ST**: branches often rooting; root-crowns and old nodes often becoming very swollen. **LF**: blade 0.5–3 cm. **INFL**: fls 2–8. **FL**: calyx limb spreading, divided halfway; corolla ± 4 mm, bell-shaped, pink (often red outside), lobes ± erect, 1/2 corolla length, hairy inside; nectary glands below all 5 lobes. **FR** ± 8 mm, round. **SEED** 2–4 mm. Ridges, slopes, open places in woods; 9–3000 m. NW, CaR, SN, CW, SW, MP; to B.C., ID, NM. Two ill-defined forms may be recognized: *S. acutus* (A. Gray) Dieck (twig hairs spreading, lvs densely hairy below; chiefly SN); *S. hesperius* G. Jones (twigs puberulent, lvs thin, sparsely hairy below; chiefly KR). ❀**4,6,17**, SHD:**15,16**&IRR:1,2,7,9,**14,18**,19–24;STBL.

S. rotundifolius A. Gray Pl 6–12 dm, stiff, puberulent. **ST**: old bark shredding. **LF**: blade 8–20 mm, paler and more prominently veined below. **INFL**: fls 1–2 in axils. **FL**: calyx limb flaring, lobes deep, irregular, margin gen transparent; corolla 6–10 mm, narrowly bell-shaped, pink or white, middle 1/3 of tube lightly hairy, upper 1/3 glabrous, lobes ± erect, ± 1/5–1/3 corolla length; nectaries below all 5 lobes; style glabrous. **FR** 8–12 mm, ovoid. **SEEDS** 4–6 mm. Slopes, ridges, open places in forest; 1100–3300 m. CaR, SN, SW, GB, DMtns; to WA, WY, Colorado, w TX. [*S. vaccinioides* Rydb.]

var. ***parishii*** (Rydb.) Dempster Pl trailing, 3–6 dm. **ST**: branches often arched, tips rooting; twigs gen ± straight-hairy. **FL**: corolla 6–9 mm, inside of tube sparsely hairy throughout. Slopes, ridges; 1100–3300 m. s SNH, SW, SNE, DMtns; NM. [*S. parishii* Rydb.] Hard to distinguish from var. *rotundifolius* in Mono Co. ❀TRY.

var. ***rotundifolius*** (p. 483) Pl ± erect, 6–12 dm. **ST**: branches not arched; twigs finely puberulent. **FL**: corolla 7–10 mm, inside of tube hairy in middle 1/3. Rocky or sandy slopes, open places in coniferous forest; 1200–3200 m. CaR, SN, GB; to WA, WY, Colorado, w TX. ❀DRN:4–6,15,16&IRR:1–3,7,14,18;DFCLT.

VIBURNUM

Shrub. **LVS** simple. **INFL**: cyme, terminal, many-fld, ± flat-topped. **FL**: corolla deeply saucer-shaped, white; stamens 5; ovary 1-chambered, 1-ovuled. **FR**: drupe. **SEED** flat. ± 200 spp.: n temp & subtrop. (Classical Latin name)

V. ellipticum Hook. (p. 483) Pl slender, gen hairy and glandular. **LF** deciduous; petiole 6–12 mm; blade 2–6 cm, elliptic to round or cordate, coarsely dentate exc at base, gen 3-veined from base. **INFL** 1–3 cm, rounded or ± flat-topped, gen with oblanceolate bracts; peduncles 1.5–4 cm; rays gen 7. **FL** 6–8 mm diam. **FR** 10–12 mm, elliptic in outline. **SEED** 5-grooved. Chaparral, yellow-pine forest, gen on n-facing slopes; 300–1400 m. NW, n&c SNF, SnFrB; to WA. ❀**4–6,17**&SHD:**15,16**&IRR:1,2,**17**,14.

CARYOPHYLLACEAE PINK FAMILY

Ronald L. Hartman (except *Silene*)

Ann, bien, per, rarely dioecious, taprooted or rhizome gen slender. **LVS** simple, gen opposite; stipules gen 0; petiole gen 0; blade entire, sheath gen 0. **INFL**: cyme, gen open; fls few–many or fl solitary and axillary; involucre gen 0. **FL** gen bisexual, radial; hypanthium sometimes present; sepals gen 5, ± free or fused into a tube, tube gen herbaceous between lobes or teeth; awns gen 0; petals gen 5 or 0, gen tapered to base (or with claw long, blade expanded), entire to 2–several-lobed, blade gen without scale-like appendages (inner surface), gen without ear-like lobes at base; stamens gen 10, gen fertile, gen free, gen from ovary base; nectaries gen 0; ovary superior, gen 1-chambered, placentas basal or free-central, styles 2–5 or 1 and 2–3-branched. **FR**: capsule or utricle (rarely modified, dehiscent), gen sessile. **SEEDS**: appendage gen 0. 85 genera, 2400 spp.: widespread, esp arctic, alpine, temp, n hemisphere; some cult (*Agrostemma, Arenaria, Cerastium, Dianthus, Gypsophila, Lychnis, Saponaria, Silene, Vaccaria*).

1. Fr a utricle (± modified in *Achyronychia, Scopulophila*); petals 0 or < 0.6 mm, scale-like; stamens arising from hypanthium rim; stipules present (exc *Scleranthus*), scarious (subfamily Paronychioideae)
 2. Stipules 0; hypanthium in fr very hard .... **SCLERANTHUS**
 2′ Stipules present; hypanthium in fr herbaceous to ± hard
  3. Sepals widely ovate to reniform, margin ± wide, white, scarious, awns 0; sterile stamens 14–19 and thread-like or 5 and petal-like
   4. Ann; base ± glabrous; st prostrate to ascending; sterile stamens 14–19, ± 0.5 mm, thread-like, nectary 0; style 2-branched .... **ACHYRONYCHIA**
   4′ Per; base densely woolly; st erect; sterile stamens 5, 1–1.5 mm, oblong, petal-like, nectary wide; style 3-branched .... **SCOPULOPHILA**
  3′ Sepals lanceolate to oblong, if ovate or obovate then awned, margin green, membranous, scarious, or narrow; sterile stamens 4–5, thread-like or 0
   5. Stipules ± 0.5–1 mm; sepals awnless, hairs ± long, stiff, often hooked, margin herbaceous ...... **HERNIARIA**
   5′ Stipules 1–8 mm; sepals with awn 0.5–4 mm, hairs not as above, margin white, scarious
    6. Awn very stout, spine-tipped, 1.5–4 mm; fl densely woolly .... **CARDIONEMA**
    6′ Awn thread-like to ± stout, not esp spiny, 0.5–1.5 mm; fl with hairs ± straight or tips coiled .... **PARONYCHIA**

1′ Fr a capsule; petals present, sometimes 0; stamens from ovary base or on disk around ovary; stipules 0 (exc in couplet 16.)
 7. Sepals fused, tube prominent, lobes or teeth < tube (exc *Agrostemma*); petals long-clawed, gen appendaged near junction with claw (subfamily Silenoideae)
  8. Styles 3–5; fr valves or teeth 3–6 or 10
   9. Calyx lobes > tube, gen 15–50 mm, widely linear, tube strongly 10-ribbed, with long, ascending, appressed hairs .... **AGROSTEMMA**
   9′ Calyx lobes < tube, < 13 mm, triangular to awl-like, tube not as above (exc *Lychnis* with lobes 4–7 mm)
    10. Styles (4)5, fls bisexual; fr teeth (4)5; pl densely silky-hairy to tomentose .... **LYCHNIS**
    10′ Styles 3(4), if 5 then dioecious; fr teeth 3, 6, or 10; pl glabrous or variously hairy but not as above **SILENE**
  8′ Styles 2; fr valves gen 4
   11. Fl subtended by 2–6 bracts
    12. Calyx veins or ribs 20–40, tube green, scarious between lobes; involucral bracts herbaceous, linear to lanceolate (ovate, 1.5–2.5 mm wide in *D. deltoides*) .... **DIANTHUS**
    12′ Calyx veins 5–15, tube scarious between teeth; involucral bracts reddish to brown-scarious, widely ovate, gen 5–12 mm wide .... **PETRORHAGIA**
   11′ Fl with 0 involucre bracts
    13. Stamens 5; calyx narrowly cylindric, 0.8–1 mm diam; fl gen axillary, solitary .... **VELEZIA**
    13′ Stamens 10; calyx cup-shaped to widely tubular, 1.5–9 mm diam; fl few–many, in open to head-like, terminal cyme
     14. Calyx 1.5–5 mm, cup- to bell-shaped, tube white, scarious between teeth .... **GYPSOPHILA**
     14′ Calyx 9–25 mm, ovoid to widely tubular, tube green, herbaceous between teeth
      15. Infl very dense, pedicels 0–3 mm; calyx in fl 20–25 mm, rounded, keels 0; petal appendages 2 .... **SAPONARIA**
      15′ Infl open, pedicels 5–40+ mm; calyx in fl 10–17 mm, 5-angled or 5-keeled; petal appendages 0 .... **VACCARIA**
 7′ Sepals ± free; petals not long-clawed, not appendaged at junction with claw (subfamily Alsinoideae)
  16. Stipules 0.5–11 mm, ovate to bristle-like, scarious
   17. Lf blade oblanceolate to obovate, petiole ± present; pl glabrous; styles 0.1–0.3 mm, fused below, 3-branched .... **POLYCARPON**
   17′ Lf-blade awl-like to linear, petiole 0; pl glandular-hairy, esp infl (exc some *Spergularia*); styles gen 0.3–3 mm, free, 3–5 (if shorter, then stipules bristle-like)
    18. Lvs ± whorled, 10–30 per node; styles, fr valves 5 .... **SPERGULA**
    18′ Lvs opposite, often with axillary lf clusters; styles, fr valves 3
     19. Fls axillary, 1–2; sepals awn-tipped; petals 0; stipules bristle-like .... **LOEFLINGIA**
     19′ Fls in cyme, few–many; sepals awnless; petals present; stipules ovate to lanceolate ...... **SPERGULARIA**
  16′ Stipules 0
   20. Petals 2-lobed, often ± to base (sometimes 0 in *Stellaria, Cerastium*)
    21. Fr cylindric, often curved, tip 10-toothed .... **CERASTIUM**
    21′ Fr spheric to ± ovoid, tip not toothed, valves 6 (8 or 10 in *Stellaria calycantha*)
     22. Petals 2-lobed, lobes < 1/3 petal length; rhizomes with tuber-like thickenings 3–12 mm diam; pl glandular-hairy, esp below; fr spheric, valves rolled under; seeds 1–2 .... **PSEUDOSTELLARIA**

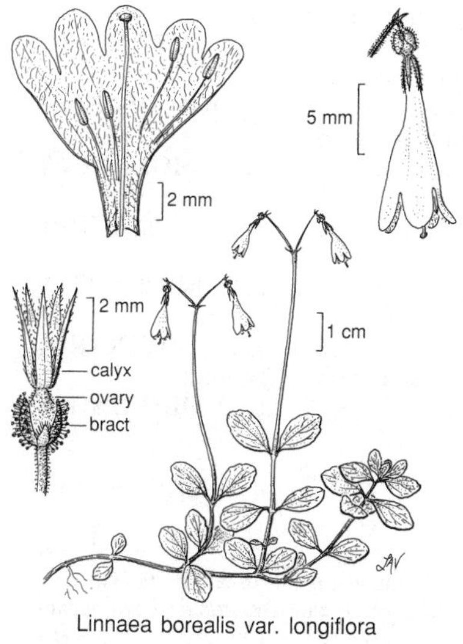

Linnaea borealis var. longiflora

Caprifoliaceae

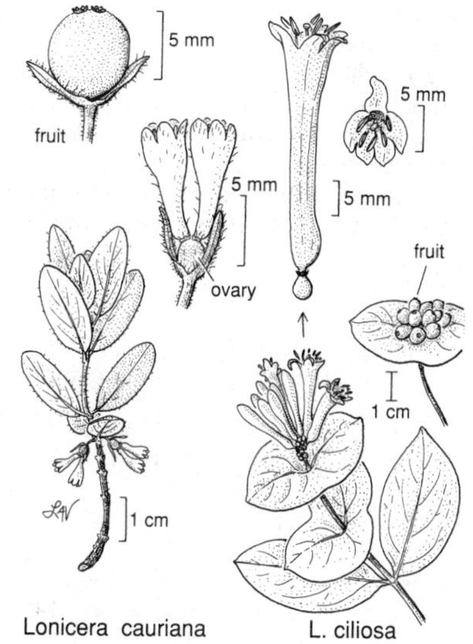

Lonicera cauriana

L. ciliosa

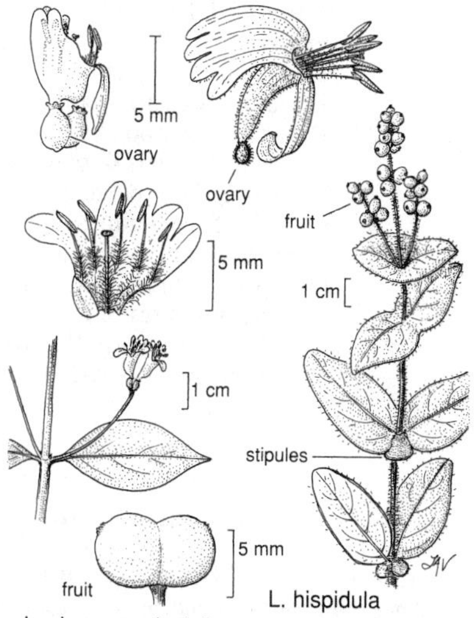

Lonicera conjugialis

L. hispidula var. vacillans

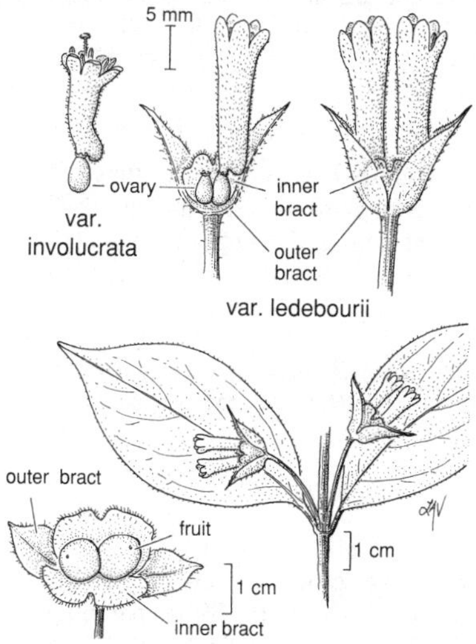

var. involucrata

var. ledebourii

Lonicera involucrata var. ledebourii

var. subspicata

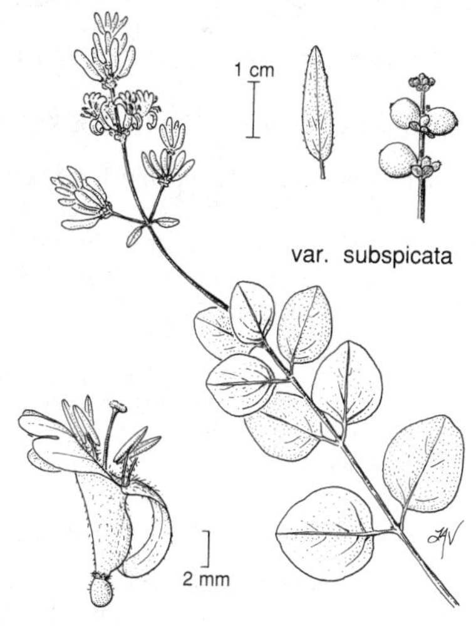

Lonicera subspicata var. denudata

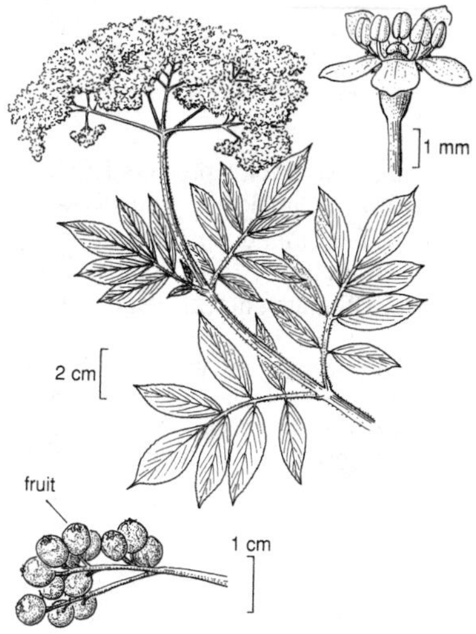

Sambucus mexicana

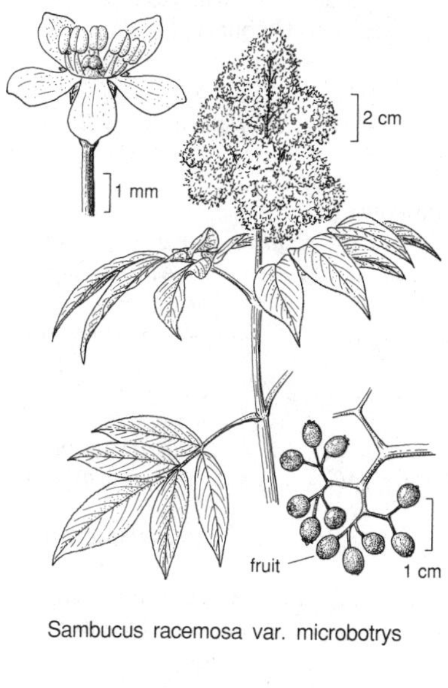

Sambucus racemosa var. microbotrys

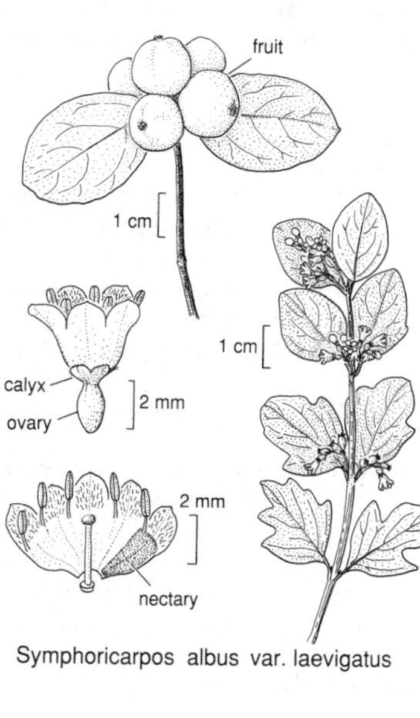

Symphoricarpos albus var. laevigatus

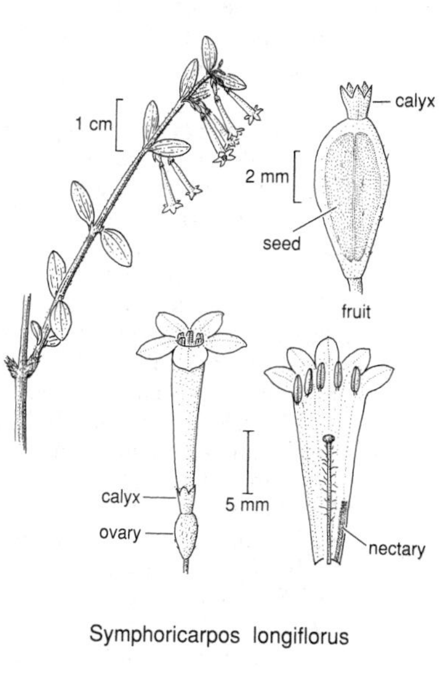

Symphoricarpos longiflorus

22′ Petals 2-lobed > 1/2 to base, sometimes 0; rhizomes not enlarged; pl not glandular; fr ± ovoid, valves not rolled under; seeds several–many .... **STELLARIA**
20′ Petals entire, irregularly toothed, or ± notched (sometimes 0 in *Minuartia, Moenchia, Sagina*)
23. Petals ± irregularly toothed; infl umbel-like; fr teeth 6, rolled under .... **HOLOSTEUM**
23′ Petals entire or notched; infl not umbel-like (exc *Arenaria congesta*); fr teeth 8 or valves 3–6
24. Fr teeth 8; petals 4; pl erect; lf blade ± linear, rigid, acute .... **MOENCHIA**
24′ Fr valves 3–6; petals 5 (if 4 then pl not erect, lf blade not linear, rigid, and acute)
25. Styles 5, sometimes 4, alternate with sepals; fr valves 5, sometimes 4 .... **SAGINA**
25′ Styles 3, opposite sepals; fr valves 3 or 6
26. Ovary sutures 3; fr valves 3 .... **MINUARTIA**
26′ Ovary sutures 6; fr valves 6
27. Seed appendage 0; rhizomes 0; styles 0.5–2 mm .... **ARENARIA**
27′ Seed appendage ± 0.7 mm, ± elliptic, spongy; pls from long, branched rhizome; styles 2–3 mm .... **MOEHRINGIA**

ACHYRONYCHIA

1 sp. (Greek: chaff fingernail, from silvery, chaffy sepals)

A. cooperi Torrey & A. Gray (p. 483) ONYX FLOWER, FROST-MAT Ann, prostrate to ascending, glabrous to ± hairy, taprooted. **STS** many, 3–17 cm. **LF**: stipules 0.1–0.4 mm, ± ovate, scarious, ± fringed, white; blade 3–20 mm, oblanceolate; vein 1. **INFL**: cyme, axillary; fls 20–60+; pedicels 0.5–2.5 mm. **FL** 2.5–3 mm; hypanthium in fr ± cylindric; calyx abruptly expanded above; sepals 5, free, ± 1.2–1.5 mm, ovate to reniform, green, fleshy, margin wide, scarious, ± jagged, white, deciduous; petals 0; fertile stamens 1–2, sterile stamens 14–19, ± 0.5 mm, thread-like, arising from hypanthium rim; ovary superior, style 2-branched in upper 1/2, 0.3–0.4 mm. **FR**: utricle, ovoid; teeth 8–10, minute. **SEED** 1, ± 1 mm, ovoid, ± compressed, tan, red dot near narrow end. Sandy slopes, flats, washes; 50–700 m. D; AZ, Mex. Closely related to and possibly same as *Scopulophila*.

AGROSTEMMA CORN-COCKLE

Ann, bien, erect, taprooted. **LF**: blade linear to narrowly lanceolate; vein 1 or lateral pair obscure. **INFL**: cyme, terminal; fls 1–few; peduncle and pedicels 4–20+ cm. **FL**: sepals 5, fused, hairs long, ascending and appressed, tube prominent, 9–15 mm, 7–12 mm diam, ovoid to widely cylindric, round in X-section, strongly 10-ribbed, lobes gen 15–50 mm, > tube, widely linear; petals 5, 24–36 mm, claw long, blade entire or notched; styles 5, 12–15 mm. **FR**: capsule, ovoid; teeth 5, ascending. **SEEDS** many, black. 2 spp.: Medit Eur. (Greek: field garland)

A. githago L. Pl 30–90+ cm; hairs dense, long, silky, ± appressed. **ST** simple or sparingly branched above. **LF** 5–15 cm. **INFL** leafy. **FL**: calyx lobes green; petals exserted 10–20 mm, obovate, rounded to truncate, purplish red; stamens exserted 8–10 mm. **SEED** 3–3.5 mm, widely ovate; tubercles thin, triangular. $2n=48$. Disturbed areas; < 1000 m. NCoRI, n SN, ScV, SCo; native to s Eur.

ARENARIA SANDWORT

Ann, per, erect to mat-forming, taprooted. **ST** gen round in X-section. **LF**: blades thread-like to ovate; veins 1–5. **INFL**: cyme, terminal or axillary, open to head- or umbel-like; fls 1–many; peduncles and pedicels 0–50+ mm. **FL**: hypanthium barely present; sepals 5, ± free, 1.5–8 mm, ± lanceolate to widely ovate, glabrous to glandular-hairy; petals 0 or 5, 1.5–10 mm, entire or notched; stamens inserted on obscure to prominent disk; ovary ± superior, styles 3, 0.5–2 mm. **FR**: capsule, ovoid to urn-shaped; teeth 6, ascending to recurved. **SEEDS** 1–15+, grayish, dark brown, reddish brown, yellowish tan, blackish purple, or blackish. 150 spp.: n temp, esp mtns, arctic Am, Eurasia. (Latin: sand, a common habitat) [McNeill 1980 Rhodora 82:495–502]

1. Ann; lf blade ± ovate, veins gen 3–5, palmate .... ***A. serpyllifolia*** ssp. ***serpyllifolia***
1′ Per; lf blade lanceolate to linear or needle-shaped, vein ± 1
2. Lf blade linear-lanceolate to oblanceolate, 2–5.5 mm wide, not spine-tipped
3. Infl cyme, terminal; petals 1.5–3.5 mm; st rounded, dull, hairs minute, curved downward .... ***A. lanuginosa*** ssp. ***saxosa***
3′ Fl 1, axillary; petals 5–6 mm; st angled or grooved, shiny, glabrous .... ***A. paludicola***
2′ Lf blade ± needle-like, 0.5–2 mm wide, often spine-tipped
4. Infl gen head-like cyme or umbel, dense to ± open; pedicels gen < 7 mm .... ***A. congesta***
5. Pedicels ± 0; infl tightly dense
6. Lf blade 3–8 cm, ± thread-like — Wrn, CaR, n SNH .... var. ***congesta***
6′ Lf-blade 1–3 cm, needle-like
7. Sepals 4.5–6 mm, acute; lf blade 1–2 cm, < 1 mm wide, not fleshy; DMtns .... var. ***charlestonensis***
7′ Sepals 3.5–4 mm, obtuse; lf blade 2–3 cm, 1–2 mm wide, ± fleshy; KR, NCoRH, MP .... var. ***crassula***
5′ Pedicels obvious; infl ± open
8. Infl umbel, bracts gen at base; sepals 3–4 mm, obtuse .... var. ***suffrutescens***
8′ Infl cyme, ± open, bracts scattered among fls; sepals 4–6.5 mm, acute
9. Sepals 5.5–6.5 mm .... var. ***simulans***
9′ Sepals 4–5 mm .... var. ***subcongesta***
4′ Infl ± open cyme or fls 1–2; pedicels some or all > 7 mm

10. Sepals obtuse to rounded, sometimes abruptly pointed
11. Lf blade 1–3.5 cm; sepals in fl 3–4.5 mm; pl glaucous, mat-forming; n SNH, MP .... ***A. aculeata***
11′ Lf blade 0.5–1 cm; sepals in fl 1.8–3 mm; pl green, tufted; SnBr .... ***A. ursina***
10′ Sepals acute to acuminate, sometimes ± spine-tipped
12. St gen 2–20 cm; lvs 0.5–2 cm; nectaries rounded, < 0.5 mm .... ***A. kingii*** var. ***glabrescens***
12′ St gen 20–40 cm; lvs 2–6 cm; nectaries 2-lobed, 0.7–1.5 mm .... ***A. macradenia***
13. Sepals 3–4.3 mm, in fr < 5.5 mm; infl open, branches spreading .... ssp. ***ferrisiae***
13′ Sepals 4.5–7.2 mm, in fr < 8 mm; infl ± compact, branches erect to ascending
14. Cauline lvs ± ascending, 0.8–1.2 mm wide .... var. ***macradenia***
14′ Cauline lvs curved downward, larger ones ± 1.2–2 mm wide
15. Infl and sepals glabrous .... var. ***arcuifolia***
15′ Infl and sepals densely glandular-hairy .... var. ***kuschei***

A. aculeata S. Watson (p. 483) Per, mat-forming, glaucous. **ST** 7–20 cm, ± dull, glandular-hairy. **LF** 10–35 mm, 0.5–1.5 mm wide, needle-like, herbaceous, sharp-pointed; vein 1. **INFL**: cyme, terminal, few–several-fld; pedicels 3–25 mm. **FL**: sepals 3–4.5 mm, in fr < 6 mm, gen obtuse to rounded; petals 4.5–10 mm; nectaries < 0.5 mm, rounded. **SEEDS** 6–8, 2–2.5 mm, elliptic-oblong, compressed, yellow-tan; tubercles low, rounded, elongate. Rocky slopes, alluvium, and volcanic areas; 2100–2600 m. n SNH, MP; to OR, MT, UT. Robust forms from n CA do not deserve recognition, have been called *A. pumicola* Cov. & Leiberg var. *californica* Maguire. ❀DRN,SUN,IRR:1–7,14–18.

A. congesta Nutt. Per, tufted, ± green. **ST** 8–40 cm, ± dull, often glandular-hairy. **LF** 10–80 mm, 0.5–2 mm wide, thread- to needle-like, herbaceous or ± fleshy, sharply acute to spine-tipped; vein 1. **INFL**: cyme, terminal, head- or umbel-like; fls few–many, dense to ± open; pedicels < 7 mm. **FL**: sepals 3–6 mm, in fr < 6.5 mm, rounded to acute; petals 5–8 mm; nectaries < 0.2 mm, rounded. **SEEDS** 4–8, 1.4–3 mm, widely elliptic to ovate, compressed, reddish brown; tubercles low, rounded, often elongate. Dry, rocky or sandy slopes, ridges, rock crevices; 1200–3300 m. KR, NCoRH, CaRH, SNH, GB, DMtns; to WA, MT, Colorado. Vars. often intergrade.

var. ***charlestonensis*** Maguire **LF** 10–20 mm, < 1 mm wide, ± herbaceous, needle-like. **INFL**: cyme, dense; bracts closely enveloping sepals; pedicels ± 0. **FL**: sepals 4.5–5.5 mm, acute. Uncommon. Sandy ridges; 2200 m. DMtns (New York Mtns); sw NV.

var. ***congesta*** (p. 483) **LF** 30–80 mm, 0.5–1 mm wide, herbaceous, ± thread-like. **INFL**: cyme, tightly dense; bracts closely enveloping sepals; pedicels ± 0. **FL**: sepals 3.5–4.2 mm, obtuse. $2n=22$. Gravelly or sandy soil, in open; 1300–2000 m. CaRH, n SNH, Wrn; to WA, MT, Colorado. ❀DRN,SUN,IRR:1–3,6,7,14–16,18; DFCLT.

var. ***crassula*** Maguire **LF** 20–30 mm, 1–2 mm wide, ± fleshy, needle-like. **INFL**: cyme, tightly dense; bracts closely enveloping sepals; pedicels ± 0. **FL**: sepals 3.5–4 mm, obtuse. Uncommon. Dry ridges, rock crevices; 2200–2500 m. KR, NCoRH, MP; sw OR.

var. ***simulans*** Maguire **LF** 20–35 mm, ± 0.5 mm wide, herbaceous, needle-like. **INFL**: cyme, ± open; bracts scattered among fls; pedicels obvious. **FL**: sepals 5.5–6.5 mm, acute. Uncommon. Open, rocky slopes; 1300–1700 m. n SNH, MP; nw NV.

var. ***subcongesta*** (S. Watson) S. Watson (p. 483) **LF** 10–30 mm, 0.5–1 mm wide, herbaceous, needle-like. **INFL**: cyme, ± open; bracts scattered among fls; pedicels obvious. **FL**: sepals 4–5 mm, acute. Uncommon. Open rocky slopes, flats, often on volcanics; 1350–2750 m. CaRH, n SNH, GB; to UT. ❀DRN,SUN, IRR:1–3,6,7,14–16,18;DFCLT.

var. ***suffrutescens*** (A. Gray) Robinson **LF** 15–80 cm, 0.5–1.5 mm wide, herbaceous, needle-like. **INFL** umbel-like, ± open; bracts gen confined to base; pedicels obvious. **FL**: sepals 3–4 mm, obtuse. Rocky slopes and outcrops; 1200–3300 m. KR, SNH; sw OR. ❀DRN,SUN,IRR:1–3,6,7,14–16,18;DFCLT.

A. kingii (S. Watson) M.E. Jones var. ***glabrescens*** (S. Watson) Maguire (p. 483) Per, tufted, green. **ST** 1–20 cm, ± dull, glandular-hairy. **LF** 3–20 mm, 0.5–1.2 mm wide, needle-like, herbaceous, gen sharp-pointed; vein 1. **INFL**: cyme, terminal; fls few–many (fl 1–2 in alpine pls); pedicels 2–15 mm. **FL**: sepals 2.5–4 mm, in fr < 4.5 mm, acute to acuminate; petals 4–7 mm; nectaries < 0.5 mm, rounded. **SEEDS** 2–5, 1.2–1.8 mm, elliptic-oblong to ovate, compressed, reddish brown to dark purple; tubercles low, rounded, often elongate. Rocky slopes, summits, canyon floors; 2100–4050 m. CaRH, SNH, SNE; to OR, ID, UT. High elevation pls 2–6 cm, called ssp. *compacta* (Cov.) Maguire, with lvs 3–6 mm and sepals 2.5–3.5 mm, intergrade completely, do not deserve recognition. ❀DRN,IRR,SUN:1–3,7,14–16.

A. lanuginosa (Michaux) Rohrb. ssp. ***saxosa*** (A. Gray) Maguire (p. 483) Per, tufted or sts trailing, green. **ST** 10–40 cm, rounded, dull; hairs minute, down-curved, in lines. **LF** 8–22 mm, 2–6 mm wide, gen narrowly lanceolate to oblanceolate, herbaceous, obtuse to acute; vein 1. **INFL**: cyme, terminal or axillary; fls few–many; pedicels 3–25 mm. **FL**: sepals 1.5–2.8 mm, in fr < 3.5 mm, acute to acuminate; petals 1.5–3.5 mm; nectaries not apparent. **SEEDS** 8–12, 0.7–0.8 mm, ± circular, compressed, smooth, dark brown. $2n=44$. Uncommon. Moist, sandy soil along streams; 1800–2600 m. SnBr, PR; to UT, NM, Mex. [*A. confusa* Rydb.; *Stellaria lagunensis* M.E. Jones]

A. macradenia S. Watson DESERT SANDWORT Per, tufted, green. **ST** 20–40 cm, rounded, ± dull, sometimes glandular-hairy. **LF** 20–60 mm, 0.5–2 mm wide, needle-like, herbaceous, blunt to sharp-pointed; vein 1. **INFL**: cyme, terminal; fls several–many, compact to open; pedicels 3–55 mm. **FL**: sepals 3–7.2 mm, in fr < 8 mm, acute to acuminate; petals 6–11 mm; nectaries 2-lobed, 0.7–1.5 mm. **SEEDS** 4–9, 1.8–2.7 mm, ± spheric to ovate, compressed, reddish brown to blackish; tubercles low, rounded to conic. Open woodlands, sagebrush flats, dry rocky slopes; 1100–2500 m. c&s SNH, SnJV, SnGb, SnBr, SNE, DMoj; to UT, AZ.

var. ***arcuifolia*** Maguire (p. 483) **LVS** curved downward, 0.8–2 mm wide. **INFL** ± compact; branches erect to ascending, ± glabrous. **FL**: sepals 4.5–6.5 mm, in fr < 7 mm, ± glabrous. Uncommon. Dry, often gravelly canyon slopes, dry yellow-pine and oak forest, ridges, summits; 650–2400 m. s SN, SnGb. Intergrades with var. *macradenia* and might be considered the same as the latter.

ssp. ***ferrisiae*** Abrams **LVS** ± ascending, 0.5–1 mm wide. **INFL** open; branches spreading, glabrous to moderately glandular-hairy. **FL**: sepals 3–4.3 mm, in fr < 5.5 mm, glabrous to sparsely glandular-hairy. Pine and oak woodlands, granitic alluvium; 1450–2500 m. c&s SNH, SNE; to UT.

var. ***kuschei*** (Eastw.) Maguire FOREST CAMP SANDWORT **LVS** curved downward, ± 1.2–2 mm wide. **INFL** ± compact; branches erect to suberect, densely glandular-hairy. **FL**: sepals in fl 5–7 mm, densely glandular-hairy. **FR** unknown. UNCOMMON. Habitat, elevation unknown. DMoj (Forest Camp, Inyo Co.).

var. ***macradenia*** (p. 483) **LVS** ± ascending, 0.8–1.2 mm wide. **INFL** ± compact; branches erect to ascending, ± glabrous. **FL**: sepals 4.5–7.2 mm, in fr < 8 mm, ± glabrous. Open woodlands, sagebrush flats, dry rocky slopes, alluvial deposits, often on carbonates; 1100–2200 m. SnJV, SnBr, SnGb, SNE, DMoj; to UT, AZ. [Pls called var. *parishiorum* Robinson, with petals ± = sepals and < 5 cauline lf pairs, intergrade completely and should be considered a synonym]

A. paludicola Robinson (p. 483) MARSH SANDWORT Per, erect or not, often supported by surrounding vegetation, green. **ST** 25–90 cm, angled or grooved, shiny, glabrous exc at nodes. **LVS** 20–55 mm, some 2–7 mm wide, ± lanceolate, herbaceous, narrowly acute;

vein 1. **INFL**: fl solitary, axillary; pedicels 20–50 mm. **FL**: sepals 2.8–3.5 mm, in fr < 4 mm, obtuse to rounded; petals 5–6 mm; nectaries not apparent. **SEEDS** 15–20, 0.8–0.9 mm, widely reniform, ± compressed, smooth, dark brown. RARE. Boggy meadows and marshes; < 300 m. s CCo (Nipomo Mesa, San Luis Obispo Co.), SCo (Santa Ana River), probably undercollected; to WA. Threatened by development.

A. serpyllifolia L. ssp. *serpyllifolia* (p. 483) Ann, tufted or sts sometimes trailing, green. **ST** 3–25 cm, dull; hairs minute, downcurved. **LF** 2–7 mm, 1–4 mm wide, ± ovate, ± herbaceous, acute to acuminate; veins gen 3–5, palmate. **INFL**: cyme, terminal; fls few–many; pedicels 1–12 mm. **FL**: sepals 2.5–3 mm, in fr < 3.8 mm, narrowly acute to acuminate; petals 0.6–2.7 mm; nectaries not apparent. **SEEDS** 10–15, 0.5–0.6 mm, widely reniform, plump, grayish; tubercles low, elongate. 2*n*=40. Disturbed areas, sand, gravel bars, dry woods; 150–1800 m. NW, c SNF, SCo; native to Eur.

A. ursina Robinson (p. 483) BEAR VALLEY SANDWORT Per, tufted, green. **ST** 10–18 cm, dull to ± shiny, often glandular-hairy. **LF** 5–10 mm, 0.5–1 mm wide, needle-like, herbaceous, sharp-pointed; vein 1. **INFL**: cyme, terminal; fls few–many, ± open; pedicels ± 0.5–1.5 mm. **FL**: sepals 1.8–3 mm, in fr < 4.2 mm, obtuse or rounded; petals 2–4.5 mm; nectaries < 0.5 mm, rounded. **SEEDS** 1–2, 2.2–2.5 mm, ± spheric to widely elliptic, compressed, dark purple; tubercles low, rounded, often elongate. RARE. Rocky soil, pinyon/juniper woodland; 1950–2100 m. e SnBr (Bear Valley, s San Berbardino Co.). Threatened by development, grazing, vehicles.

CARDIONEMA

Per, ± prostrate, taprooted. **LVS**: stipules 4–8 mm, lanceolate to ovate, scarious, ± entire, white; blade needle-like; vein 1. **INFL**: fls axillary, 1–5, ± sessile. **FL**: hypanthium cup-shaped, not abruptly expanded above; sepals 5, free, 1.2–2.8 mm (exc awn), oblong to obovate, densely woolly, margin scarious below, tip awned, awn 1.5–4 mm, very stout, spine-tipped; petals 5, 0.3–0.5 mm, scale-like; stamens 3–5, arising from hypanthium rim; ovary superior, styles 2, 0.2 mm. **FR**: utricle, elliptic. **SEED** 1, tan. 6 spp.: w N.Am, Chile. (Greek: heart thread, from stamen shape)

C. ramosissimum (J.A. Weinm.) Nelson & J.F. Macbr. (p. 483) **ST** 5–30+ cm, often concealed by stipules. **LF** 5–13 mm, glabrous, finely spine-tipped. **FL**: sepal margin incurved near tip. **SEED** 1.4–1.6 mm, narrowly ovate, not compressed. Sandy beaches and hills, dunes, bluffs; < 150 m. NCo, CCo, SCo; to WA, Mex, also in Chile.

CERASTIUM MOUSE-EAR CHICKWEED

Ann, per, erect to mat-forming; taproot or rhizomes present. **LF**: blade linear to ovate; vein 1. **INFL**: cyme, terminal or axillary; fls few-many, open to tightly dense; pedicels 1–36+ mm. **FL**: sepals 5, 3.5–12 mm, free, lanceolate to ovate, hairy to glandular-hairy; petals 0 or 5, 2.5–15 mm, ± 2-lobed; stamens (5)10; styles 5, 0.5–3.3 mm. **FR**: capsule, cylindric, ± curved in upper 1/2; teeth 10, spreading to recurved. **SEEDS** several–many, pale brown to reddish brown. 60 spp.: worldwide. (Greek: horn, from fr shape)

1. Ann; sts gen all fl, ascending to erect
  2. Calyx 8.5–12 mm; fr 14–22 mm; seeds ± 1 mm . . . . ***C. dichotomum***
  2′ Calyx 3.5–7 mm; fr 3.5–11 mm; seeds ± 0.4–1 mm
    3. Sepals with hairs < tip, scarious margin of most outer sepals 0.2–0.6 mm wide; upper bracts scarious-margined; pedicels in fr gen > sepals . . . . ***C. fontanum*** ssp. ***vulgare***
    3′ Sepals with some long hairs 0.2–0.8 mm > tip; scarious margin of most outer sepals < 0.1 mm wide; all bracts herbaceous; pedicels in fr (exc lowest) < sepals . . . . ***C. glomeratum***

1′ Per; sts both vegetative (mat-forming) and fl (± erect)
  4. Petals gen < to 0.5 mm > sepals . . . . ***C. fontanum*** spp. ***vulgare***
  4′ Petals ± 1.2–6 mm > sepals
    5. Lvs clustered in fl st axils (esp lower); petals 3–6 mm > sepals; fr 9–16 mm; seeds ± 1–1.3 mm; < 2500 m . . . . ***C. arvense***
    5′ Lvs not clustered in fl st axils; petals 1.2–2.8 mm > sepals; fr 5.5–8.5 mm; seeds 0.7–0.8 mm; 2900–4300 m . . . . ***C. beeringianum*** var. ***capillare***

C. arvense L. (p. 483) FIELD CHICKWEED Per, gen not fl 1st year, 8–45 cm, glandular-hairy above, ± with long hairs below. **STS** both vegetative (mat-forming) and fl (± erect). **LVS**: those on fl st gen 8–45 mm, linear or lanceolate, ± widely so, sometimes ± glabrous; lf clusters in axils (esp lower). **INFL**: bracts gen scarious in upper 1/4; pedicels in fr 1–4 ± sepals. **FL**: calyx 4.2–9 mm, glandular-hairy, rarely with long hairs > tip, scarious margin of outer sepals < 0.2 mm wide; petals 3–6 mm > sepals. **FR** 9–16 mm. **SEED** 1–1.3 mm. 2*n*=36,72,90. Moist seeps, shaded areas, grassy, gen rocky or sandy slopes; < 2500 m. NCo, KR, NCoRO, c SNF, n&c SNH, CCo; to AK, e N.Am, Greenland, Eurasia. ❀SUN:**4–6,17**& IRR or SHD:1–3,**7**,8,9,**14–16,18**,19–21,**22–24**;also STBL.

C. beeringianum Cham. & Schldl. var. ***capillare*** Fern. & Wieg. (p. 483) Per, gen not fl first year, 1.5–10 cm, glandular-hairy. **STS** both vegetative (mat-forming) and fl (± erect). **LVS**: those on fl st 5–15 mm, lanceolate to elliptic; axillary lf clusters 0. **INFL**: bracts ± completely herbaceous; pedicels in fr 0.3–2.2 ± sepals. **FL**: calyx 4.5–6 mm, ± glandular-hairy, with 0 long hairs > tip, scarious margin of outer sepals < 0.2 mm wide; petals 1.2–2.8 mm > sepals. **FR** 5.5–8.5 mm. **SEED** 0.7–0.8 mm. 2*n*=72. Moist, rocky areas, grassy meadows, open slopes; 2900–4300 m. c SNH, W&I; to WA, MT, Colorado. [ssp. *earlei* (Rydb.) Hultén]

C. dichotomum L. Ann 7–18 cm, glandular-hairy. **ST**: fl st ascending to erect. **LF** 10–35 mm, lanceolate; axillary lf clusters 0. **INFL**: bracts herbaceous; pedicels in fr 0.3–0.9 ± sepals. **FL**: calyx 8.5–12 mm, glandular-hairy, tip ± glabrous, scarious margin of outer sepals < 0.2 mm wide; petals 1.5–3 mm < sepals. **FR** 14–22 mm. **SEED** 0.9–1.1 mm. 2*n*=38. Fields, roadsides, disturbed areas; 750–850 m. e KR (Siskiyou Co.); native to sw Eur.

C. fontanum Baumg. ssp. *vulgare* (Hartman) Greuter & Burdet Per, often fl first year and appearing ann, 6–35 cm; hairs non-glandular. **ST** gen both vegetative (mat-forming) and fl (± erect). **LVS**: those on fl st 8–25 mm, lanceolate to oblong, ± widely so; axillary lf clusters 0. **INFL**: bract margins gen scarious; pedicels in fr 1–4 × sepals. **FL**: calyx 4.5–7 mm, non-glandular, hairs < tip, ascending, scarious margin of most outer sepals 0.2–0.6 mm wide; petals 1.5 mm, < to 0.5 mm > sepals. **FR** 6.5–11 mm. **SEED** 0.6–0.7 mm. 2*n*=72,126,136,144,160,180. Disturbed areas, grassy slopes, damp woods, marshy ground; < 2200 m. NW, CaRH, c SNF, n&c SNH; native to Eur. [ssp. *triviale* (Spenner) Jalas; *C. vulgatum* L. misapplied]

C. glomeratum Thuill. (p. 483) MOUSE-EAR CHICKWEED Ann 3–40 cm, hairy (± glandular). **ST**: fl sts ascending to erect. **LVS**: those on fl st 5–35 mm, lanceolate or oblanceolate to ovate; axillary lf clusters 0. **INFL**: bracts herbaceous; pedicels gen 0.5–0.9 × sepals, (exc lowest). **FL**: calyx 3.5–5 mm, glandular-hairy, also with some long, non-glandular hairs 0.2–0.8 mm > tip, scarious margin of most outer sepals < 0.1 mm wide; petals 1.5 mm < to 1 mm > sepals, often 0 on lateral branches. **FR** 3.5–8 mm. **SEED** 0.4–0.6 mm. $2n$=72. Dry hillsides, grasslands, chaparral, disturbed areas; < 1600 m. CA-FP (exc s SNH); native to Eur. [*C. viscosum* L. misapplied]

DIANTHUS PINK, CARNATION

Ann, bien, per, erect, taprooted or rhizomed. **LF**: blade linear to oblanceolate; vein 1 or lateral pairs less prominent. **INFL**: cyme, terminal; fls few–many, tightly dense, or fls 1–few, loosely arranged; involucral bracts 2–6, linear to ovate; pedicels 0–3 mm or 10+ mm. **FL**: sepals 5, fused, glabrous to hairy, tube prominent, 1.3–2 cm, 1.8–3.3 mm diam, ± cylindric, weakly 10–40-ribbed, lobes 3–8 mm, < tube, triangular to lanceolate; petals 5, 13–24 mm, claw long, blade irregularly toothed; stamens fused with petals to stalk; styles 2, 5–12 mm. **FR**: capsule, ± tubular; stalk 1–4 mm; teeth 4, ascending. **SEEDS** many, black. 300 spp.: Eurasia to s Afr. (Greek: divine fl, from beauty or fragrance of fl)

1. Fls 1–few; pedicels mostly 5–10+ mm; bracts ovate, 1/3–1/2 × calyx tube, tip acute to short-tapered ..... ***D. deltoides*** ssp. ***deltoides***
1′ Fls few–many; pedicels 0–3 mm; bracts linear to lanceolate, mostly = or > calyx tube, tip long-tapered
 2. Calyx moderately hairy; lf blade linear; fr stalk ± 1 mm ..... ***D. armeria*** ssp. ***armeria***
 2′ Calyx glabrous; lf blade ± lanceolate; fr stalk 3–4 mm ..... ***D. barbatus*** ssp. ***barbatus***

D. armeria L. ssp. *armeria* GRASS PINK, DEPTFORD PINK Ann, bien 15–60 cm; taproot slender. **LVS**: basal blades lanceolate to oblanceolate; cauline blades ± linear. **INFL** few–several-fld, ± open; bracts mostly = or > calyx tube, linear to lanceolate, long-tapered; pedicels 0–3 mm. **FL**: calyx 1.5–2 cm, moderately hairy, hairs long, ± appressed, ribs 20–25, lobes long-tapered; petal blade 4–5 mm, pink or rose with white dots. **FR**: stalk ± 1 mm. $2n$=30. Disturbed areas; 400–1200 m. KR, NCoRO, CaR, c SNH; native to s Eur.

D. barbatus L. ssp. *barbatus* SWEET WILLIAM Per 30–60 cm; rhizome ± stout. **LVS**: basal blades lanceolate to oblanceolate; cauline blades ± lanceolate. **INFL**: fls many, tightly dense; bracts mostly = or > calyx tube, ± linear, long-tapered; pedicels 0–3 mm. **FL**: calyx 1.5–1.8 cm, glabrous, ribs 40, lobes acute to short-tapered; petal blade 6–10 mm, white to pink, purple, violet, or 2-colored. **FR**: stalk 3–4 mm. $2n$=30. Disturbed areas; < 1500 m. NW, CaRF, CCo; native to s Eur.

D. deltoides L. ssp. *deltoides* MAIDEN or MEADOW PINK Per 18–40 cm; rhizomes slender. **LVS**: basal blades oblanceolate; cauline blades linear to linear-lanceolate. **INFL**: fls 1–few, not dense; bracts 1/3–1/2 × calyx tube, ovate, short-tapered; pedicels mostly 5–10+ mm. **FL**: calyx 1.3–1.7 cm, glabrous to minutely hairy above, ribs 25–30, lobes linear to ± triangular; petal blade 4–9 mm, deep pink with darker zigzag band near base. **FR**: stalk ± 3 mm. $2n$=30. Wet meadows, disturbed areas; < 2500 m. CaRH, n&s SNH, SnJV, MP; native to Eur.

GYPSOPHILA BABY'S-BREATH

Ann, bien, per, erect, taprooted or rhizomed. **LVS**: blade ± lanceolate; veins 1–3, often obscure. **INFL**: cyme, gen panicle-like, terminal; fls ± few–many; pedicels 1–30+ mm. **FL**: sepals 5, fused, glabrous or glandular-hairy, tube ± prominent, ± 1.3–4 mm, 0.8–2 mm diam, cup- to bell-shaped, round to angled in X-section, white-scarious between teeth, veins ± 5, teeth 0.2–1 mm, < tube, lanceolate to triangular; petals 5, ± 2.2–9 mm, claw long, blade entire to notched; styles 2, 2–5 mm. **FR**: capsule, oblong to spheric; teeth 4, ascending to recurved. **SEEDS** 2–several, black. 125 spp.: temp Eurasia, n Afr. (Greek: gypsum lover, from habitat of 1 sp.) [Barkoudah 1962 Wentia 9:1–203]

1. Sepals, pedicels glandular-hairy; largest lvs gen 10–35 mm wide ..... ***G. scorzonerifolia***
1′ Sepals, pedicels glabrous; largest lvs gen 2–9 mm wide
 2. Calyx 3–5 mm; ann, bien; taproot slender ..... ***G. elegans*** var. ***elegans***
 2′ Calyx 1.5–2.1 mm; per; rhizome stout ..... ***G. paniculata*** var. ***paniculata***

G. elegans M. Bieb. var. *elegans* Ann, bien 15–50 cm; taproot slender. **LF**: blade 2–5 mm wide, ± linear-lanceolate. **INFL** ± open; fls ± few; pedicels glabrous. **FL**: calyx 3–5 mm, glabrous; petals 2–5 × calyx, white to pink, veins purple. $2n$=26,34. Open pine-fir forest; ± 2000 m. n SNH (Placer Co.); native to sw Eur, se Asia.

G. paniculata L. var. *paniculata* BABY'S BREATH Per 50–90 cm; rhizome stout. **LF**: blade 2–9 mm wide, ± lanceolate. **INFL** openly branched; fls many; pedicels glabrous. **FL**: calyx 1.5–2.1 mm, glabrous; petals 1.5–2 × calyx, white. $2n$=26,28,34. Disturbed areas; 1200–2000 m. NCoRI, CaRH, n SNH, SnJV, SCoRO, SCo, MP, DMoj; native to e&c Eur, adjacent Asia. NOXIOUS WEED; infestations widely scattered.

G. scorzonerifolia Ser. Per 30–90 cm; rhizome stout. **LF**: blade 3–35 mm wide, linear-oblong to lanceolate. **INFL** ± open; fls ± few; pedicels glandular-hairy. **FL**: calyx 2–2.5 mm, glandular-hairy; petals 1.5–2 × calyx, white to light pink. $2n$=34,68. Disturbed areas; ± 1200 m. SNE (Inyo Co.); native to e Eur.

HERNIARIA

Ann, ± prostrate, taprooted. **LVS** opposite below, alternate above; stipules 0.4–1 mm, ovate to deltate, scarious, ciliate, white; blade oblanceolate to obovate; vein 0–1. **INFL**: cyme, axillary; fls 3–10, dense, ± sessile. **FL**: hypanthium cup-like, not abruptly expanded above; sepals 5, 0.6–1.2 mm, free, lanceolate to oblong, hairy, margin entire, herbaceous; petals 0; fertile stamens 2–5, sterile stamens 4–5, ± 0.5 mm, ± thread-like, arising from hypanthium rim; styles 2 or 2-branched in upper 2/3, 0.1–0.4 mm. **FR**: utricle, obovoid. **SEED** 1, dark reddish brown. 20 spp.: Eur, s Asia, Afr. (Latin: rupture, 1 sp. being a supposed cure) [Chaudhri 1968 Meded Bot Mus Herb Rijks Univ Utrecht 285:297–398]

H. hirsuta L. **ST** gen 4–20 cm. **LF**: stipules 0.4–1 mm; blade 1–13 mm. **INFL**: fls 3–8. **FL**: sepals ± equal to unequal; stamens 2–5; styles 2 or 2-branched. **FR** minutely papillate. **SEED** ± compressed, smooth; margin with prominent rim. Disturbed, sandy or clay soils; < 1750 m. KR, c&s SNF, c SNH, SnJV, SnFrB, SCo; native to s Eur, n Afr, sw Asia.

1. Sepals in fr ± unequal; fl hairs of 2 sizes, long hairs 1/2–2/3 × sepals, tips of some or all hooked or tightly coiled, short hairs 1/4–1/3 × sepals, gen on hypanthium, tips recurved; stamens 2–3 .... ssp. ***cinerea***

1′ Sepals ± equal; fl hairs of ± 1 size, 1/5–1/3 × sepals, ± shorter on hypanthium, tips ± straight; stamens gen 5 .... ssp. ***hirsuta***

ssp. *cinerea* (DC.) Cout. **ST** gen 5–20 cm. **INFL**: fls 3–8. **FL** ± 1.2–1.8 mm; hairs of 2 sizes, long hairs 1/2–2/3 ± sepals, tips of some or all hooked or tightly coiled, short hairs 1/4–1/3 × sepals, gen on hypanthium, tips recurved; sepals in fr ± unequal; stamens 2–3; styles 2, 0.2–0.4 mm. **SEED** 0.5–0.6 mm. $2n$=36. Disturbed areas, alkaline hills, clay flats; < 800 m. s SNF, SnJV, SnFrB, SCo; native to s Eur, n Afr, sw Asia. [*H. c.* DC.; *Paronychia pusilla* E. Greene] Possibly best treated as separate sp.

ssp. *hirsuta* **ST** gen 4–15 cm. **INFL**: fls 3–6. **FL** 0.9–1.1 mm; hairs of ± 1 size, 1/5–1/3 × sepals, ± shorter on hypanthium, tips ± straight; sepals ± equal; stamens gen 5; style 2-branched, < 0.1 mm. **SEED** 0.6–0.7 mm. $2n$=18,36. Sandy flats, roadsides, woodlands; 200–1750 m. KR, c SNF; native to s Eur, n Afr, sw Asia.

HOLOSTEUM JAGGED CHICKWEED

Ann, ascending to erect, taprooted. **LVS**: basal blades linear to oblanceolate, short-petioled; cauline blades narrowly oblanceolate to oblong-ovate, sessile; vein 1. **INFL**: cyme, umbel-like, terminal; fls 3–12; involucral bracts 5–15, ± 0.3–1.8 mm; pedicels 0.5–2.5 cm. **FL**: sepals 5, 3–4.5 cm, lanceolate to ovate, glabrous; petals 5, 3.3–5 mm, gen minutely toothed; stamens 5–10; styles 3, 0.5–1.4 mm. **FR**: capsule, ovoid or ovoid-cylindric, straight; teeth 6, recurved. **SEEDS** many, reddish brown. 6 spp.: temp Eurasia. (Greek: all bone, humorous reference to frailty of pl) [Shinners 1965 Sida 2:119–128]

H. umbellatum L. ssp. *umbellatum* Pl 5–25 cm. **STS** 1–many, unbranched, glandular-hairy near middle. **LVS**: cauline 1–4 pairs in lower 1/2; blades 2–25 mm, glabrous or margin glandular-hairy. **INFL**: bracts whorled, 1–2 mm, partly scarious; pedicels ascending to erect, reflexed in fr, slender. **FL**: petals narrowly elliptic, white. **SEEDS** 0.7–0.8 mm, compressed, with marginal ridge opposite an elongate depression; tubercles low, rounded. $2n$=20. Uncommon. Disturbed areas; < 850 m. KR, CaR, CCo; native to c, e, & s Eur.

LOEFLINGIA

Ann, erect to prostrate, taprooted. **LF**: stipules 0.4–1.2 mm, bristle-like, scarious, entire, ± white; blade awl-like to oblong; vein 0–1. **INFL** axillary, sessile; fls 1–2. **FL**: sepals 5, 2.7–6 mm, ± free, ± lanceolate, glandular-hairy; petals 0 or rudimentary; stamens 3–5; styles 3, < 0.1 mm. **FR**: capsule, lanceolate to ovoid; valves 3, ± recurved at tip. **SEEDS** many, tan with reddish brown band on curved edge. 7 spp.: N.Am., Medit. (P. Loefling, Swedish botanist & explorer, 1729–1756) [Barneby & Twisselmann 1970 Madroño 20:398–408]

L. squarrosa Nutt. Pl 1–7 cm, much-branched at base, glandular-hairy, ± fleshy. **ST** stiff. **LF**: blade 2–7 mm, erect to ± recurved, bristly; base gen fused into a short scarious sheath; tip blunt to spine-tipped. **FL** cleistogamous; sepals spine-tipped, becoming hardened, margin often scarious. **FR** 0.5–0.8 × sepals, 3-angled. **SEED** 0.4–0.6 mm, minutely papillate on flat edge. Sandy, gravelly areas; < 1200 m. Teh, SnJV, SnFrB, SCo, PR, GB, DMoj; to OR, NE, TX.

1. Sepals ± 2.7–3 mm, in fr ± equal and tip ± straight, lateral spurs 0; fr gen 1.5–2.5 mm; stipules 0.4–0.6 mm .... var. ***artemisiarum***

1′ Sepals gen 3.5–6 mm, in fr unequal and tip strongly recurved, lateral spurs often present, bristly; fr ± 2.6–3.7 mm; stipules 0.5–1.2 mm .... var. ***squarrosa***

var. ***artemisiarum*** (Barneby & Twisselm.) R. Dorn (p. 483) SAGE-LIKE LOEFLINGIA **LF**: stipules 0.4-0.6 mm; blade 2–4 mm, oblong, erect to ± spreading; tip blunt or short-spined. **FL**: sepals 2.7–3 mm, in fr ± equal and tip ± straight, lateral spurs 0. **FR** gen 1.5–2.5 mm, 2–2.7 × longer than wide. UNCOMMON. Sand dunes, sandy flats; 700–1200 m. GB, DMoj (se Kern, ne Los Angeles cos.); to OR, WY. [ssp. *a.* Barneby & Twisselm.]

var. ***squarrosa*** (p. 483) **LF**: stipules 0.5–1.2 mm; blade gen 3–7 mm, needle- to awl-like, gen ± recurved; tip with spine. **FL**: sepals gen ± 3.5–6 mm, in fr unequal and tip strongly recurved, lateral spurs often present, bristly. **FR** 2.6–3.7 mm, 3–4 × longer than wide. Sandy and gravelly soil of hills, mesas, dunes, sometimes disturbed areas; < 1200 m. Teh, SnJV, SnFrB (Santa Cruz Co.), SCo, PR; n Baja CA. [*L. pusilla* Curran]

LYCHNIS CAMPION

Bien, per, erect, taprooted or roots fibrous. **LF**: petiole present or 0; blade oblanceolate to narrowly elliptic; vein 1, prominent. **INFL**: cyme, terminal; fls few; pedicels 10–55+ mm. **FL**: sepals 5, fused, densely silky-hairy to tomentose, tube prominent, 12-14 mm, 7–10 mm diam, elliptic to ovoid, rounded, strongly 10-ribbed (obscured by hairs), lobes 4–7 mm, < tube, linear to ± lanceolate; petals 5, 22–30 mm, claw long, blade widely notched, appendages 2; styles (4)5, 16–18 mm. **FR**: capsule, ovoid; stalk 0–0.5 mm; teeth (4)5, ascending. **SEEDS** many, gray to blackish purple. 35 spp.: n temp, arctic. (Greek: lamp, from flame-colored fl of some spp.)

L. coronaria (L.) Desr. ROSE CAMPION, MULLEIN PINK Pl densely silky-hairy to tomentose. **ST** sparingly branched above. **LVS**: basal many, 9–15 cm, oblanceolate; cauline fewer, 5–15 cm, gen narrowly elliptic. **INFL** leafy; branches, pedicels spreading, often arching upward. **FL**: calyx lobes twisted; petals obovate, reddish purple, appendages 2–3 mm, awl-like, thickened. **SEED** 1–1.3 mm, ± spheric; tubercles rounded, elongate. $2n$=24. Disturbed areas, open slopes, redwood/douglas fir forests; < 500 m. NCoR, SnFrB, WTR; native to se Eur.

MINUARTIA SANDWORT

Ann, per, erect to mat-forming, taprooted or rhizomed. **LF**: blade thread-like to awl-shaped or narrowly oblong; veins or ribs 1–3. **INFL**: cyme, terminal or axillary; fls 2–many, open to ± dense, or fl solitary; peduncles and pedicels 0.5–35+ mm. **FL**: hypanthium short, obscure; sepals 5, ± free, 1.9–7 mm, ± lanceolate to ovate, glabrous to glandular-hairy; petals 5 or 0,

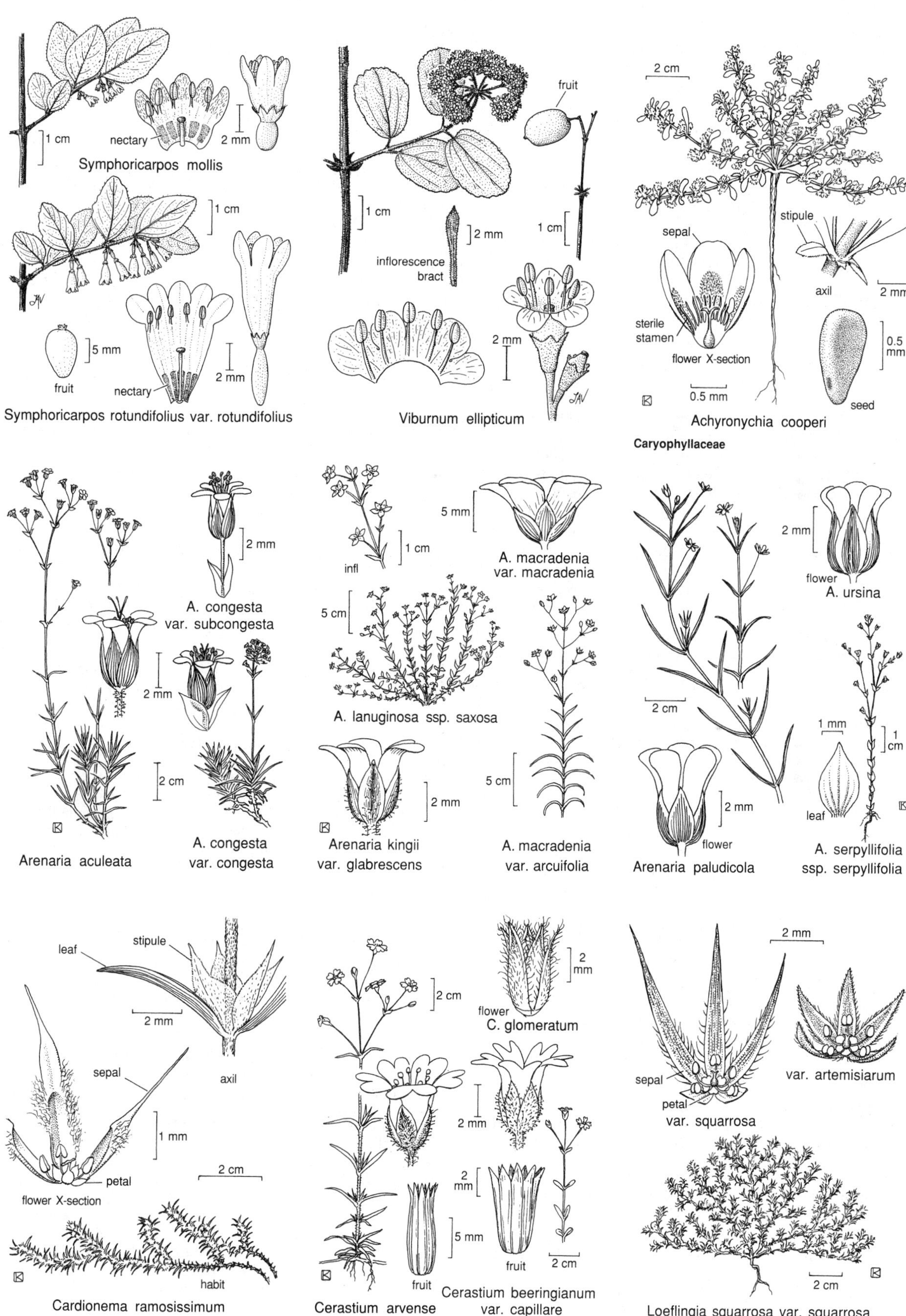
1 cm
nectary
2 mm
Symphoricarpos mollis
1 cm
5 mm
fruit
nectary
2 mm
Symphoricarpos rotundifolius var. rotundifolius
fruit
1 cm
2 mm
inflorescence bract
1 cm
2 mm
Viburnum ellipticum
2 cm
stipule
sepal
axil
2 mm
sterile stamen
flower X-section
0.5 mm
0.5 mm
seed
Achyronychia cooperi
Caryophyllaceae
2 mm
A. congesta var. subcongesta
2 mm
2 cm
Arenaria aculeata
A. congesta var. congesta
infl
1 cm
5 mm
A. macradenia var. macradenia
5 cm
A. lanuginosa ssp. saxosa
2 mm
Arenaria kingii var. glabrescens
5 cm
A. macradenia var. arcuifolia
2 mm
flower
A. ursina
2 cm
1 mm
1 cm
leaf
2 mm
flower
Arenaria paludicola
A. serpyllifolia ssp. serpyllifolia
leaf
stipule
2 mm
axil
sepal
1 mm
petal
flower X-section
2 cm
habit
Cardionema ramosissimum
2 cm
2 mm
flower
C. glomeratum
2 mm
2 mm
5 mm
fruit
Cerastium arvense
fruit
2 cm
Cerastium beeringianum var. capillare
2 mm
sepal
petal
var. squarrosa
var. artemisiarum
2 cm
Loeflingia squarrosa var. squarrosa

0.7–10 mm, entire or notched; stamens arising from an obscure to prominent disk; styles 3, 0.3–2 mm. **FR**: capsule, narrowly ovoid to widely elliptic; teeth 3, ascending to recurved. **SEEDS** 1–many, reddish tan to reddish, purplish, or blackish brown. 120 spp.: arctic to Mex, n Afr, s Asia. (J. Minuart, Spanish botanist & pharmacist, 1693–1768) [McNeill 1980 Rhodora 82: 495–502]

1. Ann, not cespitose, not mat-forming
 2. Petals 0.5–1 × sepals .... ***M. pusilla***
 2′ Petals 1.4–2.3 × sepals
  3. Lf blade 2–5 mm, linear-lanceolate to -oblong; infl glabrous; seeds 0.4–0.5 mm .... ***M. californica***
  3′ Lf blade 5–40 mm, thread-like to narrowly linear; infl glandular-hairy; seeds 1.3–2 mm
   4. Lvs ± evenly distributed, blade thread-like, gen 0.3 mm wide, flexible, becoming curled; seeds reddish brown, margin thin, wing-like .... ***M. douglasii***
   4′ Lvs mostly near base, blade linear-lanceolate, gen 1–1.5 mm wide, rigid, recurved; seeds brown to blackish brown, margin thick .... ***M. howellii***
1′ Per, cespitose or mat-forming
 5. Sepal tip narrowly rounded, margin incurved, hood-like .... ***M. obtusiloba***
 5′ Sepal tip acute to acuminate, margin not incurved
  6. Pl ± densely cespitose, st decumbent to erect; taproot < 1.5 mm diam
   7. Petals nearly 0.8–1 × sepals; pl glandular-hairy .... ***M. rubella***
   7′ Petals 0; pl glabrous .... ***M. stricta***
  6′ Pl mat-forming, rhizomes elongate or st trailing; taproot > 3 mm diam
   8. Pl densely glandular-hairy .... ***M. nuttallii***
    9. Lvs prominently recurved; sepals 3-ribbed .... ssp. ***fragilis***
    9′ Lvs straight or slightly recurved; sepals 1- or 3-ribbed, lateral pair less prominent
     10. Petals 0.7–0.9 × sepals; SNH, SNE .... ssp. ***gracilis***
     10′ Petals 1.1–1.6 × sepals; NW, CaRH, Wrn .... ssp. ***gregaria***
   8′ Pl glabrous or ± sparsely glandular-hairy esp in infl; st, lvs ± glabrous
    11. Sepals 5–6 mm, in fl gen 3-ribbed; petals 0.7–0.9 × sepals .... ***M. decumbens***
    11′ Sepals 2.5–4.8 mm, in fl obscurely veined; petals 1.4–2.2 × sepals
     12. Pl glaucous; fl st from branching rhizomes; lvs > internodes; axillary lvs well developed .... ***M. rosei***
     12′ Pl gray-green; fl st from stolons; lvs mostly < internodes; axillary lvs weakly developed .... ***M. stolonifera***

M. californica (A. Gray) Mattf. (p. 489) Ann, simple or much-branched, (1)2–12 cm, green, glabrous; taproot thread-like. **STS** spreading to erect. **LVS** 2–5 mm, 0.2–1.5 mm wide, >> internode, linear to awl-shaped or narrowly oblong, ± straight, flexible, ± evenly spaced; axillary lvs 0. **FL**: sepals 1.9–4 mm, rounded to widely acute or acuminate, margin not incurved, veins 1 or 3, often prominent; petals 1.4–1.8 × sepals. **SEED** 0.4–0.5 mm; margin thick, reddish brown. 2*n*=26. Gravelly or sandy slopes, grassy ridges, chaparral, sometimes on serpentine; < 1500 m. NW, CaRF, SNF, c SNH, ScV, CW, TR; s OR. [*Arenaria c.* (A. Gray) Brewer; *A. pusilla* S. Watson var. *diffusa* Maguire]

M. decumbens T.W. Nelson & J.P. Nelson (p. 489) THE LASSICS SANDWORT Per, mat-forming, 4–15 cm, green, gen ± glabrous; taproot > 3 mm diam. **STS** in fl ascending to erect, trailing st < 30+ cm. **LVS** 3–6(9) mm, 0.7–2 mm wide, < to > internodes, needle-like to awl-shaped, ascending, ± rigid, ± evenly spaced; axillary lvs well developed. **INFL** ± sparsely glandular-hairy. **FL**: sepals 5–6 mm, acute to acuminate, margin not incurved, ribs gen 3; petals 0.7–0.9 × sepals. **SEED** 1.8–2.2 mm; margin thick, purplish brown. RARE. Serpentine soils in Jeffrey-pine forest; 1500–1600 m. n NCoRH (Mule Ridge, Trinity Co.).

M. douglasii (Torrey & A. Gray) Mattf. (p. 489) Ann, simple or often branched from base, 4–30 cm, green or st purple, finely glandular-hairy at least above; taproot thread-like. **STS** erect to spreading. **LVS** 5–40 mm, gen 0.3 mm wide, ± thread-like, becoming curled, flexible, ± evenly spaced; axillary lvs 0. **FL**: sepals 2.5–3.7 mm, obtuse to acute, margin not incurved, ribs gen 3; petals 1.7–2.1 × sepals. **SEED** 1.3–2 mm; margin thin, wing-like, reddish brown. Rocky and sandy slopes and flats in chaparral, oak, or pine woodlands, often on serpentine; 100–1800 m. NW, CaR, SNF, c SNH, GV, CW, SW, MP; s OR, AZ. [*Arenaria d.* Torrey & A. Gray; variation in presence or absence of petal notch precludes recognition of *A. d.* var. *emarginata* H. Sharsm.] ❀DRN,SUN:1,2,6,**7**,8, **9**,10,11,**14–24**.

M. howellii (S. Watson) Mattf. (p. 489) HOWELL'S SANDWORT Ann, simple or often branched from base, 12–30 cm, green, in fr becoming purple, finely glandular-hairy; taproot < 2 mm. **STS** erect to spreading. **LVS** 5–15 mm, 1–1.5 mm wide, linear-lanceolate, recurved, rigid, mostly restricted near base; axillary lvs 0. **FL**: sepals 1.9–3 mm, ± acute, margin not incurved, ribs ± 3 near base; petals 1.8–2.3 × sepals. **SEED** 1.4–1.7 mm; margin thick, blackish brown. UNCOMMON. Chaparral, Jeffrey-pine/oak woodland, serpentine; 550–1000 m. KR; s OR. [*Arenaria h.* S. Watson]

M. nuttallii (Pax) Briq. Per, mat-forming, 2–20 cm, ± green, densely glandular-hairy; taproot > 3 mm diam; rhizomes and trailing st < 60+ cm. **STS** in fl ascending to erect. **LVS** 4–12(15) mm, ± 0.3–1.1 mm wide, > internodes, needle-like to awl-shaped, straight to recurved, ± rigid, ± evenly spaced; axillary lvs well developed. **FL**: sepals 3.5–7 mm, acute to acuminate, margin not incurved, ribs 1 or 3; petals 0.7–1.6 × sepals. **SEED** 1.5–2.2 mm; margin thick, reddish brown to dark brown. Sandy and rocky slopes and ridges, barren rock, chaparral, open pine woodland, often on serpentine; 650–3800 m. NW, CaRH, SNH, GB; OR, NV. [*Arenaria n.* Pax]

ssp. ***fragilis*** (Maguire & A. Holmgren) McNeill (p. 489) **LF** prominently recurved. **FL**: sepals 3-ribbed; petals 0.8–1.1 × sepals. Uncommon. Basins, limestone talus; 1650–2400 m. GB; NV, OR. [*Arenaria n.* ssp. *f.* Maguire & A. Holmgren]

ssp. ***gracilis*** (Robinson) McNeill (p. 489) **LF** straight or slightly recurved. **FL**: sepals 1-ribbed, sometimes obscurely 3-ribbed; petals 0.7–0.9 × sepals. Loose talus, sandy flats, gravelly areas, barren rock; 1500–3800 m. SNH, SNE; w NV. [*Arenaria n.* ssp. *g.* (Robinson) Maguire] ❀TRY.

ssp. ***gregaria*** (A.A. Heller) McNeill (p. 489) **LF** straight or slightly recurved. **FL**: sepals 1- or 3-ribbed, lateral pair less obvious; petals 1.1–1.6 × sepals. Sandy, rocky slopes and ridges, scree, barren rock, serpentine, chaparral, open Jeffrey-pine woodland; 650–3200 m. NW, CaRH, Wrn; s OR. [*Arenaria n.* ssp. *g.* (A.A. Heller) Maguire] ❀TRY.

M. obtusiloba (Rydb.) House (p. 489) ALPINE SANDWORT Per, cespitose to mat-forming, 1–12 cm, green, glandular-hairy; taproot > 3 mm diam. **ST** in fl ± erect; trailing st 2–20+ cm. **LVS** 1.5–8 mm, 0.4–1 mm wide, > internodes, needle- to awl-shaped, ± straight, flexible, ± evenly spaced; axillary lvs well developed. **FL**: sepals 2.9–4 mm, narrowly rounded; margin incurved, hood-like, ribs gen 3; petals 1.2–1.5 × sepals. **SEED** 0.6–0.7 mm; margin thick, reddish tan. 2*n*=26,52, ±78. UNCOMMON. Among dwarf willows, unglaciated granitics and metamorphics, alpine; 3150–3700 m. c&sSNH; to Ak, Colorado. [*Arenaria o.* (Rydb.) Fern.]

M. pusilla (S. Watson) Mattf. (p. 489) Ann, simple or branched, 1–5 cm, slender, green, glabrous; taproot thread-like. **ST** spreading to erect. **LVS** 1.5–5 mm, 0.2–1.5 mm wide, << internode, awl-shaped to lanceolate, ± straight, flexible, ± evenly spaced; axillary lvs 0. **FL**: sepals 1.5–3.5 mm, acute to acuminate; margin not incurved, ribs 1, sometimes 3; petals 0.5–1 × sepals or 0. **SEED** 0.4–0.7 mm; margin thick, purplish brown. Plains, open pine forest, chaparral slopes; 800–2400 m. KR, n SNH, SnFrB; to WA. [*Arenaria p.* S. Watson]

M. rosei (Maguire & Barneby) McNeill (p. 489) PEANUT SANDWORT Per, mat-forming, 5–20 cm, glaucous, gen glabrous; taproot > 3 mm diam; rhizomes and trailing sts 5–20+ cm. **ST** in fl ascending to erect. **LVS** 4–15 mm, ± 0.5–1.2 mm wide, > internodes, needle-like, straight or curved, ± flexible, ± evenly spaced; axillary lvs well developed. **INFL** often glandular-hairy. **FL**: sepals 2.5–4 mm, acute to acuminate, margin not incurved, vein obscure; petals 1.4–2.2 × sepals. **SEED** 2.3–2.8 mm; margin thick, reddish brown to brown. UNCOMMON. Open serpentine slopes with scattered oak and Jeffrey pine; 750–1350 m. KR, NCoRH. [*Arenaria r.* Maguire & Barneby] ❀TRY;DFCLT.

M. rubella (Wahlenb.) Hiern (p. 489) Per, cespitose, 2–8(10) cm, green, densely glandular-hairy; taproot < 1.5 mm diam; rhizomes and trailing sts 0. **ST** ascending to erect. **LVS** 1.5–10 mm, 0.3–0.8 mm wide, < to > internodes, needle-like, ± straight, flexible, mostly near base; axillary lvs well developed. **FL**: sepals 2.5–3.2 mm, acute to acuminate, margin not incurved, ribs 3; petals 0.8–1 × sepals. **SEED** 0.3–0.5 mm; margin thick, reddish brown. $2n=24$. Rocky ridges and slopes, unglaciated metamorphics and granitics; 2400–3800 m. KR, c&s SNH; to AZ, Colorado, circumboreal. [*Arenaria r.* (Wahlenb.) Smith]

M. stolonifera T.W. Nelson & J.P. Nelson (p. 489) SCOTT MOUNTAIN SANDWORT Per, mat-forming, 10–20 cm, gray-green, gen ± glabrous; taproot > 3 mm diam; trailing st or stolon 6–20+ cm. **ST** in fl ± erect. **LVS** 5–9 mm, 0.5–0.9 mm wide, mostly < internodes, needle-like, ± straight, rigid, ± evenly spaced; axillary lvs weakly developed. **INFL** glandular-hairy. **FL**: sepals 3.5–4.8 mm, narrowly acute to acuminate, margin not incurved, vein obscure, in fr 1 or 3; petals 1.6–1.8 × sepals. **SEED** 2–2.4 mm; margin thick, reddish brown to brown. RARE. Serpentine soils, Jeffrey-pine forest; 1250–1400 m. s KR (Scott Mtn, Siskiyou Co.).

M. stricta (Sw.) Hiern (p. 489) Per, cespitose, 0.8–2.5 cm, green, glabrous; taproot < 1.5 mm diam; rhizomes and stolons 0. **ST** decumbent to erect. **LVS** 2–9 mm, 0.3–0.6 mm wide, > internodes, needle-like, ± straight, flexible, mostly near base; axillary lvs well developed. **FL**: sepals 2–3.2 mm, acute to acuminate, margin not incurved, vein 1 or 3, often prominent; petals 0. **SEED** 0.5–0.6 mm; margin thick, reddish brown. $2n=26,30$. Uncommon. Granitic gravels, sandy wet spots, sedge meadows, alpine; 3500–3900 m. c&s SNH, W&I; circumboreal; to Colorado. [*Arenaria rossii* Richardson misapplied] ❀TRY.

MOEHRINGIA

Per, ascending to erect, rhizomed. **LF**: petiole very short or 0; blade ± lanceolate or elliptic; vein 1. **INFL**: cyme, terminal or axillary; fls 2–5; pedicels 2–15+ mm. **FL**: hypanthium short, obscure; sepals 5, ± free, 2.8–5.5 mm, ± ovate, minutely ciliate; petals 5, 2–8 mm, entire; stamens arising from a disk; styles 3, 2–3 mm. **FR**: capsule, widely ovoid; teeth 6, ± recurved. **SEEDS** few, reddish brown to blackish; appendage ± 0.7 mm, ± elliptic, spongy. 25 spp.: n temp. (P.H.G. Moehring, Danzig naturalist, 1710–1791) [McNeill 1980 Rhodora 82:495–502]

M. macrophylla (Hook.) Fenzl (p. 489) **ST** simple to branched, 2–18 mm, ± angled or grooved; hairs minute, peg-like; rhizome slender, branching. **LVS** 1.5–5 cm, ± evenly spaced, ± thin; margin smooth to minutely granular, gen ciliate in lower half. **INFL**: bracts 1–4 mm, margin widely scarious, minutely ciliate; pedicels ascending to erect, sometimes ± spreading in fr. **FL**: sepals acute to acuminate, margin scarious, midrib ± keeled; petals ± round. **FR** black. **SEED** 1.8–2.2 mm, widely elliptic; tubercles minute, low, rounded. $2n=48$. Moist, shaded slopes, rocky ridges, summits, pine and oak forests, serpentine; 300–1800 m. NW, n&s SNH, SnFrB, SCoRO, PR; to B.C., ne N.Am, NM, Asia. [*Arenaria m.* Hook.]

MOENCHIA

Ann, erect, taprooted. **LF**: petiole present or 0; blade linear to oblanceolate; vein 1. **INFL**: cyme, terminal; fls few, ± open, or fl solitary; peduncles and pedicels 1–7+ cm. **FL**: sepals gen 4, free, 3.8–7 mm, lanceolate, glabrous; petals (0)4, 2.5–6 mm, entire; stamens 4 or 8; styles gen 4, 0.2–0.4 mm. **FR**: capsule, cylindric, straight in upper half; teeth 8, rolled under. **SEEDS** many, reddish brown. 6 spp.: w&c Eur, Medit. (C. Moench, German naturalist, 1744–1805)

M. erecta (L.) P. Gaertner, Meyer, & Scherb. ssp. *erecta* Pl glabrous, glaucous. **ST** 2.5–15 cm. **LVS**: basal ± petioled, 5–18 mm, ± oblanceolate; cauline sessile, 3–12 mm, linear to linear-lanceolate, ascending, rigid. **FL**: sepals 3-veined, lateral veins obscure; petals lanceolate, white; styles opposite sepals. $2n=38$. Disturbed areas; ± 200 m. n SNF (Butte Co.); native to sw Eur.

PARONYCHIA WHITLOW-WORT, NAILWORT

Ann, per, erect or ± prostrate, taprooted. **LF**: stipules 1–6 mm, lanceolate to ovate, scarious, ± entire, white; blade elliptic to oblanceolate; vein ± 1. **INFL**: clusters axillary; fls 1–12 per cluster, dense; pedicels 0–2 mm. **FL**: hypanthium cup-shaped; calyx sometimes abruptly expanded above; sepals 5, free, 0.7–4.4 mm (exc awn), lanceolate to ovate, ± hairy, margin narrow, white, scarious; awn thread-like to stiff, wavy to stiff; petals 0; fertile stamens 5, sterile stamens 0 or 5, 0.5–1 mm, thread-like, arising from hypanthium rim; styles 2 or 2-branched in upper 1/2, 0.2–0.5 mm. **FR**: utricle, ovoid to spheric. **SEED** 1, brown. 50 spp.: worldwide. (Greek: inflammation of finger, esp beneath nail [whitlow], from ailment the pl was believed to cure) [Chaudhri 1968 Meded Bot Mus Herb Rijks Univ Utrecht 285:64-297]

1. Per; st prostrate, mat-forming; taproot > 3 mm diam; fl sparsely hairy near tip, hairs ± straight ...... ***P. franciscana***
1′ Ann; st ± erect; taproot < 1 mm diam; fl ± moderately hairy on lower 1/2, hairs hooked or tightly coiled
  2. St 0.5–2 cm, ± concealed by lvs; fl 1, axillary, 4.2–5 mm; sepals awned from back (tip above awn 1–1.5 mm), awn 1–1.5 mm; sepal margin scarious, 0.5–1 mm wide .... ***P. ahartii***
  2′ St 2–15 cm, readily visible between lvs; fls 3–12, axillary, 2–2.5 mm; sepals awned at tip, awn 0.6–0.9 mm; sepal margin scarious, 0.1–0.2 mm wide .... ***P. echinulata*** var. ***echinulata***

P. ahartii B. Ertter (p. 489) AHART'S PARONYCHIA Ann, inconspicuous, ± spheric; taproot < 1 mm diam. **ST** ± erect, 0.5–2 cm, ± concealed by lvs. **LF**: stipules 3–6 mm; blade 2–7.5 mm, ± narrowly oblanceolate, smooth, green; tip a bristle; margin and midrib scabrous. **FL** 1, axillary, 4.2–5 mm, ± moderately hairy in lower 1/2; hairs 0.3–0.6 mm, tightly coiled; sepals 3.8–4.4 mm (exc awn), ± lanceolate, tip erect, 1–1.5 mm, margin 0.5–1 mm wide, awned from back, awn 1–1.5 mm, thread-like, wavy, ± spreading; sterile stamens ± 1 mm. **SEED** ± 1 mm, lenticular. RARE. Well-drained, rocky outcrops, often vernal pool edges, volcanic uplands; < 500 m. CaRF, ScV.

P. echinulata Chater var. *echinulata* Ann, slender; taproot < 1 mm diam. **ST** ± erect, 2–15 cm, readily visible between lvs. **LF**: stipules 1–3 mm; petiole ± present; blade 3–7 mm, elliptic to oblong, granular, brown; tip abruptly pointed; margin green, scabrous. **FLS** 3–12, axillary, 2–2.5 mm, ± moderately hairy in lower 1/2; hairs 0.1–0.2 mm, hooked; sepals 0.7–1 mm (exc awn), obovate to oblong, margin 0.1–0.2 mm wide, awned from tip, awn 0.6–0.9 mm, straight, stout, spreading; sterile stamens 0.5–0.6 mm. **SEED** ± 0.7 mm, ± spheric. $2n$=14,24,28. Disturbed clay soil; ± 25 m. n SnJV (La Grange, Stanislaus Co.); native to s Eur, nw Afr, w Asia.

P. franciscana Eastw. (p. 489) Per, mat-forming; taproot > 3 mm diam. **ST** prostrate, 5–50 cm, ± concealed by lvs. **LF**: stipules 3–6 mm; petiole 5–10 mm, ± elliptic to oblanceolate, ± smooth, green, gen moderately hairy; tip a bristle; margin green. **FLS** 2–6, axillary, 2–2.3 mm, sparsely hairy near tip; hairs 0.3–0.6 mm, ± straight; sepals 1.2–1.3 mm (exc awns), oblong to ovate, margin ± 1 mm wide, awned from tip, awn 0.5–0.7 mm, ± straight, slender, erect; sterile stamens 0. **SEED** 1–1.2 mm, ± spheric. Grassy hills; < 250 m. n CCo (around San Francisco Bay), SnFrB; native to Chile.

PETRORHAGIA

Ann, erect, taprooted. **LF**: base sheathing, 1–9+ mm; blade linear to linear-lanceolate; veins 3. **INFL**: cyme, terminal, head-like; fls few–several; involucral bracts 2–6 per fl, widely ovate; pedicels 0–3 mm, concealed. **FL**: sepals 5, fused, glabrous to sparsely, minutely hairy, tube prominent, 8–13 mm, 1–2 mm diam, cylindric, scarious between teeth, veins 5–15, teeth 0.5–1.8 mm, < tube, rounded; petals 5, 10–14 mm, claw long, blade entire or 2-lobed; styles 2, 9–12 mm. **FR**: capsule, ovoid; stalk 0.2–0.7 mm; valves 4, ascending to recurved. **SEEDS** many, blackish brown to black. 20 spp.: Medit to s Asia. (Greek: rock fissure, from habitat of some spp.) [Rabeler 1985 Sida 11:6–44]

1. Lf sheath length ± = width, gen 1–2 mm; petals truncate or shallowly notched .... ***P. prolifera***
1′ Lf sheath length 1.5–3 × width, gen 3–9 mm; petals obcordate to 2-lobed
 2. Seeds 1–1.4 mm, conic-papillate; lf sheath gen 4–9 mm; inner infl bracts like outer, all abruptly pointed .. ***P. dubia***
 2′ Seeds (1.3)1.5–1.8 mm, tubercled; lf sheath (2)3–4 mm; inner infl bracts obtuse or abruptly pointed .. ***P. nanteuilii***

P. dubia (Raf.) G. Lopez & Romo Ann, erect, 9.5–60 cm. **ST**: middle internodes often densely glandular-tomentose. **LF**: sheath 2–3 × as long as wide, (3)4–9 mm; blade 10–60 mm, linear to oblong, lowermost oblanceolate. **INFL** head-like; bracts abruptly pointed. **FL**: petals 2-lobed, sometimes obcordate, pink or pinkish, veins 5–6, dark. **SEED** 1–1.4 mm; surface conic-papillate. $2n$=30. Disturbed areas, woodland savanna; < 1800 m. NCoR, CaRF, n SNF, n&c SNH, ScV, SCoRO; native to s Eur. [*Kohlrauschia velutina* (Guss.) Reichb.]

P. nanteuilii (Burnat) P. Ball & Heyw. Ann, erect, 21–52 cm. **ST**: middle and lower internodes minutely hairy. **LF**: sheath 1.5–2 × as long as wide, (2)3–4 mm; blade 10–25 mm, gen linear. **INFL** head-like; outer bracts abruptly pointed; inner bracts often obtuse. **FL**: petals obcordate or ± 2-lobed, pink or slightly purplish, veins 1–3, center vein gen darker. **SEED** (1.3)1.5–1.8 mm; surface tubercled. $2n$=12,36,60. Disturbed areas; ± 200 m. s NCoRO (Sonoma Co.); native to sw Eur.

P. prolifera (L.) P. Ball & Heyw. Ann, erect, 6–60 cm. **ST**: internodes glabrous or middle ones slightly scabrous. **LF**: sheath ± as long as wide, gen 1–2 mm; blade 12–30 mm, linear to linear-lanceolate. **INFL** head-like; bracts obtuse or outer abruptly pointed. **FL**: petals truncate or notched, pink to slightly purplish, vein 1, dark. **SEED** 1.1–1.8 mm; surface sculpture worm-like. $2n$=30. Disturbed areas; ± 400 m. c SNH (Mariposa Co.); native to sw Eur. [*Tunica p.* (L.) Scop.]

POLYCARPON POLYCARP

Ann, matted or tufted, taprooted. **LVS** opposite, sometimes appearing whorled; stipules 0.4–2.8 mm, lanceolate to triangular, scarious, entire to irregularly toothed or cut, white; petiole short or 0; blade oblanceolate to obovate; vein 1. **INFL**: cyme, axillary; fls few–many, open to dense; pedicels 0.2–2 mm. **FL**: sepals 5, ± free, 1–2.2 mm, lanceolate to ovate, glabrous, awn conic to widely triangular; petals 5, 0.5–1.1 mm, entire or notched; stamens 3–5, ± fused at base; style 1, 3-branched, 0.1–0.3 mm. **FR**: capsule, ovoid to spheric; valves 3, margin rolled inward. **SEEDS** several, brown. 16 spp.: worldwide. (Greek: many fr, from capsule number)

1. Stipules 0.4–1.2 mm; lvs clearly opposite; sepals 1–1.5 mm, obscurely keeled, awn ± 0.1 mm .... ***P. depressum***
1′ Stipules 1.8–2.8 mm; lvs often appearing to be in whorls of 4; sepals 1.8–2.2 mm, prominently keeled, awn 0.3–0.7 mm .... ***P. tetraphyllum***

P. depressum Nutt. Pl glabrous. **ST** prostrate, repeatedly branched, 1–6 cm. **LVS** clearly opposite; stipules 0.4–1.2 mm, ovate to widely triangular; petiole slender; blade 3–7(9) mm, oblanceolate to obovate. **FL**: sepals 1–1.5 mm, elliptic to ovate, margin scarious, white, obscurely keeled, awn ± 0.1 mm, conic; petals linear to oblong. **SEED** 0.4–0.5 mm, obliquely triangular, minutely granular. Bluffs, gravelly or sandy soil, chaparral, fields, disturbed areas; < 500 m. SCoRO, SCo; Mex.

P. tetraphyllum (L.) L. FOUR-LEAVED ALLSEED Pl glabrous. **ST** prostrate to erect, often much-branched, esp above, 3–17 cm. **LVS** opposite but often appearing to be in whorls of 4; stipules 1.8–2.8 mm, lanceolate to widely triangular; petiole 0 or tapered into blade; blade 4–12 mm, obovate. **FL**: sepals 1.8–2.2 mm, lanceolate to ovate, margin scarious, white, prominently keeled, awn 0.3–0.7 mm, widely triangular; petals linear to elliptic. **SEED** 0.4–0.5 mm, obliquely triangular, granular. $2n$=32,48,64. Disturbed areas, roadsides, shaded waste areas; < 450 m. NCoRO, n SNF, ScV, CCo, SCo, SnGb; native to s Eur.

PSEUDOSTELLARIA

Per, sprawling to erect, rhizomed, with spheric to elongate, tuber-like thickenings 3–12 mm diam. **LF**: blade ± lanceolate; vein 1. **INFL**: cyme, terminal or axillary; fls few–many; pedicels 5–30+ mm. **FL**: hypanthium short, obscure; sepals 5, ± free, 3–6 mm, lanceolate to oblong-ovate, glandular-hairy; petals 5, 6–12 mm, shallowly 2-lobed; stamens arising from a narrow disk; styles gen 3, 4–4.5 mm. **FR**: capsule, spheric; valves recurved to rolled under. **SEEDS** 1–2, reddish brown. 16 spp.: N.Am, Eur, c&e Asia. (Latin: false *Stellaria*, from incorrect placement of sp.) [Weber & Hartman 1979 Phytologia 44:313–314]

P. jamesiana (Torrey) W.A. Weber & R.L. Hartman (p. 489) Pl 12–45 cm, glandular-hairy, esp below. **ST** simple to much-branched, ascending to erect, 4-angled. **LVS** 15–100 mm, ± smaller above, thick; margin ± smooth to roughened. **INFL**: bracts leafy; pedicels ± straight, ascending to spreading. **FL** often cleistogamous in lower lf axils. **SEED** 2–3.4 mm; tubercles ± prominent, elongate. $2n$=±96. Meadows, dry understory of coniferous forest; 1400–2700 m. KR, NCoRH, CaRH, c&s SNF, SNH, Teh, WTR, MP; to WA, MT, NM. [*Stellaria j.* Torrey]

SAGINA PEARLWORT

Ann, per, tufted to matted, taprooted. **LF**: blade linear to awl-shaped; vein 0–1. **INFL**: fl solitary, terminal or axillary; pedicels 2–30 mm. **FL**: sepals 4–5, free, 1.3–3.5 mm, lanceolate to ovate, glabrous to glandular-hairy; petals 0 or 4–5, 1–3 mm, entire or sometimes notched; stamens 4, 5, 8, or 10; styles 4–5, 0.1–0.6 mm. **FR**: capsule, ovoid; valves 4–5, spreading to recurved. **SEEDS** few–many, brown or reddish brown. 25 spp.: n temp, trop mtns. (Latin: fattening, once applied to *Spergula*, used as early forage) [Crow 1978 Rhodora 80:1–91]

1. Ann, often glandular-hairy on pedicels or sepals or both; sterile basal rosettes 0; st thread-like; pedicels ± straight
 2. Upper lvs minutely ciliate near base; sepals 4(5); petals 0 or rarely 4, minute, < sepals .... ***S. apetala***
 2′ Upper lvs glabrous; sepals gen 5; petals gen 5, ± = sepals .... ***S. decumbens*** ssp. ***occidentalis***

1′ Per, glabrous; sterile basal rosettes often present; st stouter, not thread-like; pedicels often recurved below fl
 3. Sepals 4(5), spreading to ascending after fr dehiscence; petals 1/4–1/2 × sepals or sometimes 0 ... ***S. procumbens***
 3′ Sepals 5, ± appressed after fr dehiscence; petals 3/4–1 × sepals
  4. Lvs fleshy, blade widely linear; sepals gen 2.5–3.5 mm; seeds ± reniform, plump, 0.4–0.5 mm, without groove; coastal .... ***S. maxima*** ssp. ***crassicaulis***
  4′ Lvs not fleshy, blade narrowly linear; sepals 1.5–2.5 mm; seeds obliquely triangular, ± compressed, 0.2–0.4 mm, back grooved; montane .... ***S. saginoides***

S. apetala Ard. (p. 489) DWARF PEARLWORT Ann (1.5)3–8 cm, ± glandular-hairy above; sterile basal rosettes 0. **ST** thread-like, erect to decumbent. **LVS** not fleshy; upper minutely ciliate near base; blade 3–9(12) mm, narrowly linear. **INFL**: pedicels 2–8(12) mm, thread-like, gen straight, glandular-hairy. **FL**: sepals 4(5), ± appressed in fr, 1.5–2 mm, ± glandular-hairy; petals 0, rarely 4, minute; stamens 4. **FR** 1–1.2 × sepals. **SEED** 0.2–0.3 mm, obliquely triangular, ± compressed, roughened or papillate, brown; back grooved. $2n$=12. Sandy waste areas, river bars, streamsides; < 700 m. NW, CaR, n&c SN, GV, CCo, SCo, n ChI; to WA; scattered in e N.Am. [var. *barbata* Fenzl]

S. decumbens (Elliott) Torrey & A. Gray ssp. ***occidentalis*** (S. Watson) G. Crow WESTERN PEARLWORT Ann (2)4–16 cm, glabrous or ± glandular-hairy above; sterile basal rosettes 0. **ST** thread-like, gen erect or ascending. **LF** not fleshy, glabrous; blade 4–20 mm, narrowly linear. **INFL**: pedicels 2–14(20) mm, thread-like, gen straight, glabrous or ± glandular-hairy. **FL**: sepals gen 5, ± appressed in fr, 1.7–2.1(2.5) mm, ± glandular-hairy; petals gen 5, ± = sepals; stamens 5 or 10. **FR** 1.2–1.7 × sepals. **SEED** ± 0.4 mm, obliquely triangular, ± compressed, smooth to slightly roughened, brown; back grooved. Dry streams, chaparral, grassy areas, rock outcrops, vernal pools; < 2000 m. NW, n&c SN, GV, CCo, SCo, ChI, PR; to BC. [*S. occidentalis* S. Watson]

S. maxima A. Gray ssp. ***crassicaulis*** (S. Watson) G. Crow (p. 489) Per 3–18 cm, glabrous; sterile basal rosettes present. **ST** ± stout, spreading to decumbent. **LF** fleshy; blade 7–22 mm, widely linear. **INFL**: pedicels 5–25 mm, slender to stout, gen straight. **FL**: sepals 5, ± appressed in fr, gen 2.5–3.5 mm; petals 5, slightly < sepals; stamens 10. **FR** 1.3–1.6 × sepals. **SEED** 0.4–0.5 mm, ± reniform, plump, smooth or slightly roughened, reddish brown; back not grooved. $2n$=44,46,66. Sandy bluffs, rock crevices; < 30 m. NCo, CCo; to AK. [*S. crassicaulis* S. Watson] ❀**4,5**,IRR:14,**15–17**,24.

S. procumbens L. ARCTIC PEARLWORT Per 2–18 cm, glabrous; sterile basal rosettes often present. **ST** slender, gen decumbent to ascending. **LF** not fleshy; blade 3–10(20) mm, linear. **INFL**: pedicels 5–20(30) mm, thread-like, recurved in fr. **FL**: sepals 4(5), spreading to ascending after fr dehiscence, 1.3–2 mm; petals 4(5), 1/4–1/2 × sepals, sometimes 0; stamens gen 4. **FR** 1.2–1.4 × sepals. **SEED** 0.3–0.4 mm, obliquely triangular, ± compressed, smooth or slightly roughened, brown; back grooved. $2n$=22. Wet, gravelly or sandy soil, roadsides, sidewalk cracks, waste areas; < 15 m. NCo, CCo; to AK; also in ne N.Am, Mex. ❀TRY;INV.

S. saginoides (L.) Karsten (p. 489) Per (1)2–12 cm, glabrous; sterile basal rosettes often present. **ST** slender, ascending or sometimes decumbent. **LF** not fleshy; blade (3)5–15 mm, narrowly linear. **INFL**: pedicels 10–30 mm, thread-like, recurved in fr. **FL**: sepals 5, ± appressed in fr, 1.5–2.5 mm; petals 5, 3/4–1 × sepals; stamens gen 10. **FR** 1.5–2 × sepals. **SEED** 0.2–0.4 mm, obliquely triangular, ± compressed, smooth or slightly roughened, brown; back grooved. $2n$=22. Moist banks, streamsides, dry creeks; (100) 1000–3800 m. KR, NCoRO, CaRH, SN, TR, PR, MP; to MT, WY, NM, Mex; circumboreal. [var. *hesperia* Fern.] ❀TRY.

SAPONARIA BOUNCING BET, SOAPWORT

Per, erect, rhizomed. **LF** petioled or not; blade oblanceolate to ovate; veins 3. **INFL**: cyme, terminal or axillary; fls 20–40+, dense; pedicels gen 0–3 mm. **FL**: sepals 5, fused, tube prominent, 15–20 mm, 4–8 mm diam, lanceolate to oblong, rounded, veins 20, obscure, teeth 5, 1.5–5 mm, < tube, triangular, tapered; petals 5, 25–40 mm, claw long, blade entire to ± obcordate, appendages 2; stamens fused with petals to ovary stalk; styles 2, 2–2.5 cm. **FR**: capsule, ± ovoid; stalk 2–3 mm; valves 4, ascending to recurved. **SEEDS** many, purplish to black. 30 spp.: Eurasia. (Latin: soap, because juice lathers with water)

S. officinalis L. Pl 3–9+ dm, ± glabrous. **ST** simple or branched above. **LF** 3–10 cm. **INFL**: peduncles ascending or erect, bracted. **FL**: calyx tube base depressed, indented, lobes unequal, thinly scarious, triangular-acuminate; petals 8–12 mm wide, obovate, pink, appendages 1–2 mm, linear-lanceolate. **SEED** 1.5–2 mm, ± spheric, notched, ± compressed, tubercled. $2n$=28. Roadsides, oak woodlands, streambeds, disturbed areas; < 1500 m. NW, CaRH, n SNF, SnFrB, SCoRO, SCo, PR, MP; native to s Eur.

SCLERANTHUS KNAWEL

Ann, prostrate to erect, taprooted. **LF**: blade needle-like; vein 1. **INFL**: cyme, axillary; fls 2–5, dense or fl solitary, ± sessile. **FL**: hypanthium widely obovate to urn-shaped, abruptly expanded above; sepals 5, 1.5–2.2 mm, narrowly triangular to awl-shaped, glabrous, margin thinly scarious; petals 0; stamens 1–10, arising from hypanthium rim; style 2-branched. **FR**: utricle, ovoid. **SEED** 1. 10 spp.: Eurasia, Afr, Australia. (Greek: hard fl, from the extremely hard hypanthium)

S. annuus L. ssp. *annuus* **ST** prostrate to erect, much-branched, gen 4–15 cm, rigid; hairs in lines, fine, recurved. **LF** 4–13 mm; sheath scarious, ciliate; tip sharp-pointed. **INFL** 3–15 mm diam. **FL** 3–4.2 mm; hypanthium 10-ribbed, in fr very hard; sepals narrowly triangular to awl-shaped, erect to spreading; style 2-branched > 1/2 length, ± 0.5 mm. **SEED** 1.4–1.6 mm, widely ovoid, tan exc red crescent near acute tip. $2n$=22,44. Meadows, stream margins, serpentine areas, disturbed areas; 300–1200 m. KR, NCoRH, NCoRO, CaR, n SN, PR; native to Eur.

SCOPULOPHILA

Per, erect, dioecious, taprooted. **LF**: stipules 0.8–3.5 mm, triangular, scarious, jagged to ciliate, white; blade linear to lanceolate; vein ± 1. **INFL** axillary; fls 1–4, sessile. **FL** unisexual (appears bisexual); hypanthium in fr conic to urn-shaped, abruptly expanded above; sepals 5, free, 1.1–2.1 mm, elliptic to round, ± glabrous, margin wide, scarious, white; petals 0; stamens 5, sterile in pistillate fl, 1–1.5 mm, oblong, petal-like, arising from hypanthium rim; nectaries wide; ovary sterile in stamiante fl, style 3-branched in upper 1/3, ± 1.5 mm. **FR**: utricle, modified, ovoid; teeth 3, minute. **SEED** 1, tan. 2 spp.: sw US, Mex. Closely related to and possibly same as *Achyronychia*. (Greek: fond of high places, from habitat)

S. rixfordii (Brandegee) Munz & I.M. Johnston (p. 495) RIXFORD ROCKWORT Pl glabrous exc base; lf axils densely woolly. **STS** many, branched above, 10–30 cm. **LF**: blade 8–25 mm, ± fleshy. **INFLS** eventually many. **FL** 2.2–4.2 mm; hypanthium green, becoming brown, thickened, ± hard, ± angled; sepals erect to spreading, ± concave, often unequal, central portion linear to oblong, fleshy, green, margin much wider, entire to irregular, white, possibly deciduous. **SEED** 0.9–1.1 mm, ovoid, ± compressed, red dot near narrow end. Uncommon. Limestone outcrops; 1200–1550 m. SNE, n DMoj; w NV.

SILENE CATCHFLY, CAMPION

Dieter H. Wilken

Ann, bien, per, ± erect, rarely dioecious, taprooted or rhizomed. **LVS** petioled or not; blade linear to oblanceolate; vein 1. **INFL**: cyme, gen terminal, sometimes axillary, open to dense; fls few–many, gen erect, gen with pedicels 5–40+ mm. **FL** gen bisexual; sepals 5, fused, tube prominent, 4–25 mm, 2–13 mm diam, cylindric to bell-shaped, rounded, hairs various or 0, veins gen 10+, lobes or teeth 1–13 mm, < tube, triangular to linear; petals 5, 6–48 mm, claw long, blade entire or 2–6-lobed, appendages 0–6 at junction of claw and blade; basal lobes present or 0; stamens gen fertile, fused with petals to stalk; ovary chamber 1 or ± incompletely 3–5, styles 3–5, 1–35 mm. **FR**: capsule, cylindric to ovoid; stalk 0–7 mm, gen glabrous; teeth 3, 6, or 10, ascending to recurved. **SEEDS** many, gray to red, brown, or black. 500 spp.: n hemisphere. (Greek: Probably from mythological Silenus, intoxicated foster-father of Bacchus, who was covered with foam, from sticky secretions of many spp.) [Hitchcock & Maguire 1947 Univ Wash Publ Biol 13:1–73; Showers 1987 Madroño 29–40]

1. Ann or bien, taproot gen slender
 2. Calyx tube clearly 16–30-veined
  3. Styles 5; corolla blade clearly 2-lobed; fr valves 10; fl pistillate .... [3]***S. latifolia*** ssp. ***alba***
  3′ Styles gen 3; corolla blade entire to notched; fr valves 3; fl bisexual
   4. Calyx 18–26 mm; petal appendages 2–5 mm .... ***S. conoidea***
   4′ Calyx 8–12 mm; petal appendages 0 .... ***S. multinervia***
 2′ Calyx tube clearly 10-veined
  5. Pedicel and calyx glabrous; upper lvs linear to oblanceolate; most upper internodes sticky .... ***S. antirrhina***
  5′ Pedicel and calyx hairy; upper lvs lanceolate to narrowly oblong; internodes not sticky
   6. Pedicels 0–5 mm; fls gen 1 per node; infl ± 1-sided
    7. Calyx densely short-hairy, not glandular; petal blade 5–12 mm, lobed ± 1/2 length; style exserted; infl bracts thin, ± translucent .... ***S. dichotoma***
    7′ Calyx glandular-hairy esp on veins; petal blade 2–5 mm, entire to notched; style incl; infl bracts thick, opaque .... ***S. gallica***
   6′ Some pedicels > 5 mm; some nodes with 2–3 fls; infl not 1-sided
    8. Styles 5; fr not developing; calyx tube veins not clearly net-like, lobes gen 3–6 mm, acute to acuminate; fls staminate .... [3]***S. latifolia*** ssp. ***alba***
    8′ Styles gen 3; fr developing, valves 3; calyx tube veins net-like above middle, lobes 6–13 mm, gen long-tapered; fls bisexual .... ***S. noctiflora***
1′ Per, from short rhizome or simple to branched, woody caudex
 9. Corolla bright red; sts ± weak, prostrate to decumbent
  10. Cauline lvs widely lanceolate to ovate; fr ovoid .... ***S. californica***
  10′ Cauline lvs linear to narrowly lanceolate; fr ± oblong to ovoid .... ***S. laciniata***
 9′ Corolla white, reddish, purple, or yellowish white to green; sts stiff, decumbent to gen erect
  11. Calyx clearly ± inflated (papery in *S. vulgaris*)
   12. Calyx 12–20 mm, short-hairy, 10- or 20-veined, lobes 3–6 mm; fr ovoid .... [3]***S. latifolia*** ssp. ***alba***
   12′ Calyx 7–10 mm, gen glabrous, faintly 10–15-veined, lobes 1–3 mm; fr subspheric .... ***S. vulgaris***

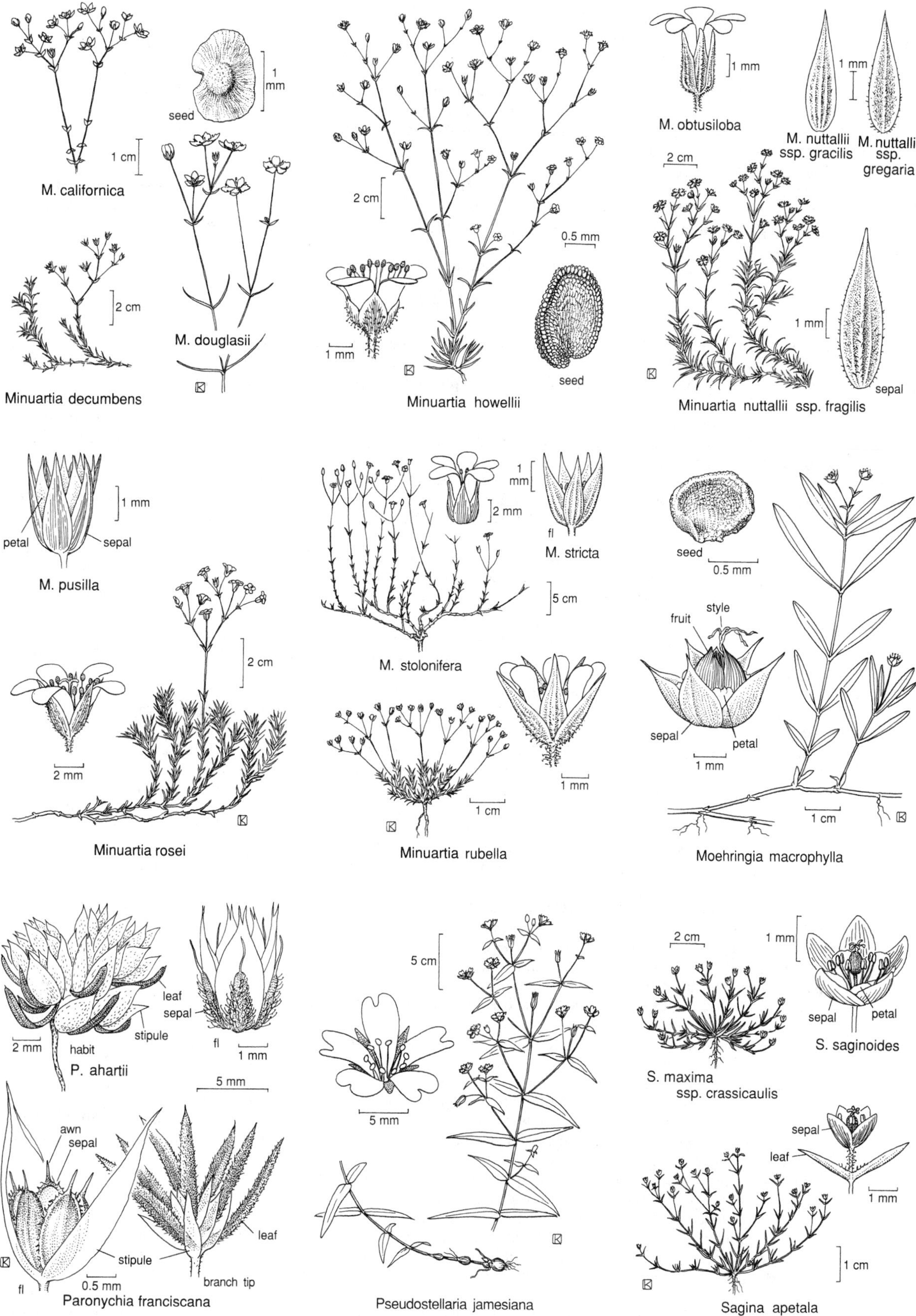

seed
1 mm
1 cm
M. californica
2 cm
M. douglasii
Minuartia decumbens
2 cm
1 mm
0.5 mm
seed
Minuartia howellii
1 mm
M. obtusiloba
1 mm
M. nuttallii ssp. gracilis
M. nuttallii ssp. gregaria
2 cm
1 mm
sepal
Minuartia nuttallii ssp. fragilis
1 mm
petal
sepal
M. pusilla
2 cm
2 mm
Minuartia rosei
1 mm
2 mm
fl
M. stricta
5 cm
M. stolonifera
1 mm
1 cm
Minuartia rubella
seed
0.5 mm
style
fruit
sepal
petal
1 mm
1 cm
Moehringia macrophylla
leaf
sepal
stipule
2 mm
habit
fl
1 mm
P. ahartii
5 mm
awn
sepal
leaf
stipule
fl
0.5 mm
branch tip
Paronychia franciscana
5 cm
5 mm
Pseudostellaria jamesiana
2 cm
1 mm
sepal
petal
S. saginoides
S. maxima ssp. crassicaulis
sepal
leaf
1 mm
1 cm
Sagina apetala

11′ Calyx not clearly inflated or papery in fr
13. Petal claw tapered to blade; appendages 0
14. Calyx lobes 4–5 mm, slightly < tube; petal 4-lobed, claw ± fusiform, short-woolly below middle; s SNH .... ***S. aperta***
14′ Calyx lobes 1–2 mm, << tube; petals entire to slightly notched, claw gradually tapered to blade, glabrous to sparsely short-hairy below middle; s CaRH, n SNH .... ***S. invisa***
13′ Petal claw and blade junction gen abrupt, often narrowed; appendages 2–6, 0.5–4 mm
15. Petal blade gen 4–8-lobed
16. Calyx 15–40 mm; corolla 20–30 mm
17. Petals pink to rose-red, blades 14–20 mm; s CaRH .... ***S. occidentalis*** ssp. ***longistipitata***
17′ Petals yellowish white, blades 7–10 mm; TR, PR .... ***S. parishii***
16′ Calyx 5–20(25) mm; corolla gen < 20 mm (< 30 mm in *S. occidentalis* ssp. *occidentalis*)
18. Calyx tube gen < or = 10 mm
19. Calyx tube cylindric, length > width; stamens and styles clearly exserted; petal claw short-woolly .... ***S. lemmonii***
19′ Calyx tube bell-shaped, length ± = width; stamens and styles = corolla or slightly exserted; petal claw glabrous or ciliate at base
20. Petals 5–10 mm, blade 4-lobed; fr stalk glabrous; fls gen ascending to erect; calyx lobes gen < 2 mm .... [2]***S. menziesii***
20′ Petals 12–20 mm, blade 4–6(8)-lobed; fr stalk puberulent; fls nodding to ± spreading; calyx lobes 2–5 mm .... ***S. campanulata***
21. Lvs 2–10 mm wide, linear to lanceolate .... ssp. ***campanulata***
21′ Lvs 10–30 mm wide, lanceolate to ± round .... ssp. ***glandulosa***
18′ Calyx tube gen > 10 mm
22. Upper cauline lvs gen overlapping or exceeding pedicels; calyx canescent to soft-white-hairy; petal blade ± = claw length, lobes > 10 mm .... ***S. hookeri***
22′ Fls located well above cauline lvs; calyx short-glandular-hairy; petal blade 1/3–1/2 × claw length, lobes < 8 mm
23. Calyx ± elliptic; caudex much-branched; sts < 30 cm; basal lvs ± fleshy, densely tufted ...... [2]***S. grayi***
23′ Calyx cylindric; caudex 1 or few-branched; sts gen > 30 cm; basal lvs gen thin, not fleshy, not tufted
24. Fr stalk 4–7 mm; petal blade 7–17 mm; calyx > 15 mm .... ***S. occidentalis*** ssp. ***occidentalis***
24′ Fr stalk 2–5 mm; petal blade 3–7 mm; calyx gen < 15(18) mm
25. Petal appendages 2, deeply cut or fringed; petal claw and stamen filament ciliate at base .... ***S. bernardina***
25′ Petal appendages 4–6, entire; petal claw and stamen filament glabrous .... ***S. oregana***
15′ Petal blade 2-lobed, sometimes short-toothed near base
26. Basal and lower cauline lvs densely tufted, fleshy, linear to narrowly oblanceolate, gen < 40 mm, < 5 mm wide; fl sts gen < 30 cm
27. Seeds 2–3 mm; calyx 8–10 mm; petal blades 3–5 mm; basal lvs 1–5 mm wide .... [2]***S. grayi***
27′ Seeds 1–2 mm; calyx 9–15 mm; petal blades 2.5–3.5 mm; basal lvs 1.5–3.5 mm wide
28. Walls between calyx hair cells clear; seed surface honeycombed throughout; petal lobes gen entire; SNH, SNE .... ***S. sargentii***
28′ Walls between calyx hair cells purple; seed surface honeycombed, margins papillate; petal lobes short-toothed near base; CaRH (Mount Shasta, Lassen Peak) .... ***S. suksdorfii***
26′ Basal lvs not densely tufted, not fleshy, gen lanceolate to oblanceolate, some > 30 mm, > 5 mm wide; some fl sts > 30 cm
29. Petals < 10 mm, calyx 5–7 mm; petal appendages < 0.5 mm .... [2]***S. menziesii***
29′ Petals > 10 mm, calyx 7–20 mm; petal appendages > 1 mm
30. Fls nodding; stamens and style much exserted .... ***S. bridgesii***
30′ Fls spreading to gen erect; stamens and style incl to = corolla
31. Pedicel and calyx puberulent, not glandular
32. Calyx base truncate to rounded in fr; basal lvs and lower internodes glabrous to sparsely puberulent .... ***S. douglasii***
32′ Calyx base narrowed in fr; basal lvs and lower internodes moderately to densely puberulent, sometimes ± finely scabrous .... ***S. verecunda***
33. St scabrous; petal claws ciliate throughout; lvs ± thick, stiff .... ssp. ***andersonii***
33′ St short-soft-hairy or puberulent; petal claws ciliate at base; lvs ± thin, flexible
34. Pedicel and calyx puberulent; lower lvs 5–9 cm; mtns > 500 m .... ssp. ***platyota***
34′ Pedicel and calyx ± short-soft-hairy; lower lvs 3–6 cm; coastal < 400 m .... ssp. ***verecunda***
31′ Pedicel and calyx glandular-hairy
35. Most basal lvs > 10 mm wide, oblanceolate, lanceolate, or elliptic
36. Cauline lvs << basal lvs; cauline lf pairs < 3; fr stalk 1–2 mm, glabrous or sparsely puberulent; calyx lobe veins wider than tube veins; > 1200 m .... ***S. nuda***
36′ Cauline lvs gradually reduced upward; cauline lf pairs > 4; fr stalk 3–6 mm, puberulent to woolly; calyx lobe and tube vein width ± equal, ± puberulent; < 300 m .... ***S. scouleri*** ssp. ***grandis***
35′ Basal and lower cauline lvs 0 or withered (if present, most < 10 mm wide, linear to narrowly oblanceolate)
37. Middle cauline lvs 5–10 mm wide; calyx lobe margins membranous .... ***S. marmorensis***
37′ Middle cauline lvs < 6 mm wide; calyx lobe margin gen translucent ... ***S. verecunda*** (see 33 for sspp.)

S. antirrhina L. (p. 495) Ann 12–80 cm. **ST** erect, glabrous or puberulent; upper internodes gen sticky. **LVS** gradually reduced upward, 1–3(6) cm, 3–5 mm wide; lower ± oblanceolate; upper linear to oblanceolate. **FL**: calyx 4–9 mm, gen glabrous, 10-veined, lobes 1–2 mm; petal claw glabrous, appendages 0, blade 2-lobed, white to pink; stamens incl; styles 3, ± incl. **FR** ± ovoid; stalk ± 1 mm. **SEED** < 1 mm, black. 2*n*=24. Open areas, burns; < 1800 m. CA-FP, D (uncommon); to B.C., e US.

S. aperta E. Greene (p. 495) Per 15–60 cm; caudex much-branched. **ST** erect, puberulent. **LVS** ± abruptly reduced gen above middle; basal and middle 5–12 cm, 1–4 mm wide, linear to oblanceolate; upper few, < 8 cm, 1–2 mm wide, linear. **FL**: calyx 6–10 mm, puberulent, 10-veined, lobes 4–5 mm; petal claw short-woolly below middle, appendages 0, blade 4-lobed, white to yellowish green; stamens ± = petals; styles 3, ± = petals. **FR** ovoid; stalk 1–2 mm, puberulent. **SEED** < 1 mm, brown. 2*n*=48. Uncommon. Open areas, coniferous forest; 1800–2800 m. s SNH (Tulare Co.).

S. bernardina S. Watson (p. 495) Per 15–55 cm; caudex gen few-branched. **ST** erect, puberulent to short-hairy, glandular above or throughout. **LVS** gradually reduced upward; lower 2–8 cm, 2–6 mm wide, linear to oblanceolate; upper 1–6 cm, 1–4 mm wide, ± linear. **INFL** axillary and terminal. **FL**: calyx 12–15 mm, glandular-puberulent, 10-veined, lobes 2–3.5 mm; petal claw ciliate at base, appendages 2, blade gen 4-lobed, white, pink, or purple; stamens slightly > petals; styles 3–4, ± = stamens. **FR** slightly elliptic; stalk 2–5 mm, puberulent. **SEED** 1.5–2 mm, brown. 2*n*=48. Rocky slopes, shrubland, coniferous forest, alpine; 1350–3600 m. KR, NCoR, CaR, SN, GB, n DMtns; OR, NV. [*S. montana* S. Watson ssp. *bernardina* (S. Watson) C. Hitchc. & Maguire, ssp. *maguirei* Bocq., ssp. *montana* vars. *sierrae* C. Hitchc. & Maguire & *rigidula* (Robinson) Tiehm] ❀TRY.

S. bridgesii Rohrb. Per 16–50 cm; caudex 1 or few-branched. **ST** decumbent to erect, puberulent, sticky or glandular above. **LVS** gradually reduced above middle; lower gen withering; middle 2–6 (8) cm, 6–15 mm wide, oblanceolate to elliptic; upper 1–5 cm, 3–12 mm wide, lanceolate, elliptic, or oblong. **INFL** axillary and terminal; fls nodding. **FL**: calyx 7–14 mm, glandular-puberulent, 10-veined, lobes 2–3 mm; petal claw gen glabrous, appendages gen 2, blade 2-lobed, white; stamens > petals; styles 3, much exserted. **FR** ovoid; stalk 2–3 mm, puberulent. **SEED** 1–1.5 mm, brown. 2*n*=48. Open areas, coniferous forest; 600–2400 m. s CaR, SN. ❀TRY.

S. californica Durand (p. 495) Per 20–50 cm; caudex 1 or few-branched. **ST** ± prostrate or reclining, gen sparsely puberulent. **LVS** gradually reduced upward; lower withering; middle 2–8 cm, 5–25 mm wide, oblanceolate to ovate; upper 1–6 cm, 5–12 mm wide, widely lanceolate to ovate. **FL**: calyx 15–25 mm, glandular-puberulent, faintly 10-veined, lobes 3–6 mm; petal claw glabrous to ciliate, appendages 2, blade 4–6-lobed, bright red; stamens > petals; styles 3, gen exserted. **FR** ovoid; stalk 2–3 mm, glabrous to puberulent. **SEED** 2–2.5 mm, reddish brown. 2*n*=48,72. Chaparral, oak woodland, coniferous forest; < 2200 m. NW, CaR, SN, CCo, SnFrB, n SCoRO, WTR (n slope), w SnGb. Pls from Teh, WTR, w SnGb intermediate to (sometimes difficult to separate from) *S. laciniata*. ❀DRN:5,6,15–17&SHD:**7,14,18**&IRR:1–3,8,9,19–24; DFCLT.

S. campanulata S. Watson Per 5–40 cm; caudex gen much-branched. **ST** erect, glabrous to puberulent, glandular or not. **LVS** slightly reduced upward, 1–5 cm, 2–30 mm wide; lower lanceolate to ± round; upper linear to ovate. **INFL** axillary and terminal; fls gen nodding, subsessile to short-pedicelled. **FL**: calyx 6–10 mm, puberulent, glandular or not, faintly 10-veined, lobes 2–5 mm; petal claw glabrous to ciliate, appendages 2, blade 4–6(8)-lobed, white to greenish or pinkish; stamens > petals; styles 3, exserted. **FR** ovoid; stalk 1–2.5 mm, puberulent. **SEED** 2–2.5 mm, brown. Chaparral, coniferous forest; 300–1900 m. KR, NCoR, n CaR; OR.

ssp. ***campanulata*** (p. 495) RED MOUNTAIN CATCHFLY Pl 5–20 cm. **ST** puberulent. **LF** 1.5–4 cm, 2–10 mm wide, linear to lanceolate. **ENDANGERED** CA. Serpentine, chaparral, coniferous forest; 500–1000 m. c NCoR (Red Mtn, Mendocino Co.; Cook's Springs, Colusa Co.). Some pls intermediate to ssp. *glandulosa*; relationship needs careful study.

ssp. ***glandulosa*** C. Hitchc. & Maguire (p. 495) Pl 15–40 cm. **ST** glabrous to puberulent, glandular or not. **LF** 1–3(5) cm, 10–30 mm wide, lanceolate to ± round. 2*n*=48. Open or shaded areas, coniferous forest; 300–1900 m. KR, NCoR, n CaR; OR. [ssp. *greenei* (S. Watson) C. Hitchc. & Maguire] ❀DRN,SHD:1–3,**6**,**7**, 14–18.

S. conoidea L. Ann 50–80 cm. **ST** erect, puberulent below, glandular-puberulent above. **LVS** gradually reduced upward; lower 5–12 cm, 5–10 mm wide, lanceolate to oblanceolate; upper 1–8 cm, 3–8 mm wide, lanceolate. **FL**: calyx 18–26 mm, puberulent, sometimes glandular, ± 30–veined, lobes 5–10 mm; petal claw glabrous, appendages 2, blade entire, fine-toothed, or notched, white, pink, or purplish; stamens exserted; styles 3, slightly > petals. **FR** ovoid to conic; stalk 1–2.5 mm. **SEED** 1–1.5 mm, grayish brown. Uncommon. Disturbed, open areas; < 500 m. n SN, SCo, DSon; to WA, e US; native to Eur.

S. dichotoma Ehrh. Ann 20–80 cm. **ST** erect, gen short-hairy or puberulent above. **LVS** gradually reduced upward; lower 6–8 cm, 15–30 mm wide, lanceolate to oblanceolate; upper 2–5 cm, 3–20 mm wide, gen lanceolate. **INFL**: fls gen spreading, sessile to subsessile. **FL**: calyx 9–14 mm, densely short-hairy, 10-veined, lobes 2–4 mm; petal claws glabrous, appendages 2, minute, blade deeply 2-lobed, white to red; stamens > petals; styles 3, exserted. **FR** ovoid; stalk 1–2 mm. **SEED** 1–1.5 mm, dark brown. Fields, roadsides; < 1000 m. CaR, SnFrB, SCo; to e US; native to Eur.

S. douglasii Hook. (p. 495) Per 10–40(70) cm; caudex few- to many-branched. **ST** decumbent to erect, puberulent. **LVS** ± gradually reduced upward; lower 2–6 cm, 2–8 mm wide, lanceolate to oblanceolate; upper 1–4 cm, 1–5 mm wide, ± linear. **FL**: calyx 10–14 mm, glabrous to puberulent, 10-veined, lobes 1.5–2.5 mm; petal claw ciliate at base, appendages 2, blade 2-lobed, white to pink or purplish; stamens ± = corolla; styles 3–4, incl to = corolla. **FR** oblong to ovoid; stalk 3–4 mm, puberulent. **SEED** 1–1.5 mm, brown. 2*n*=48. Open areas, shrubland, oak woodland, coniferous forest; 1400–2900 m. KR, n NCoR, CaR, n&c SN, MP; to B.C., MT. [var. *monantha* (S. Watson) Robinson] Difficult to separate from *S. verecunda*; variation, esp in SN, needs study. ❀TRY.

S. gallica L. Ann 10–40 cm. **ST** decumbent to erect, short-rough-hairy to minutely bristly. **LVS** gradually reduced upward; lower 1–3.5 cm, 3–5 mm wide, oblanceolate; upper 0.8–2.5 cm, 2–4 mm wide, oblanceolate to oblong. **INFL** axillary and terminal; fls ascending to erect, sessile to short-pedicelled. **FL**: calyx 6–10 mm, glandular-hairy (esp on veins), 10-veined, lobes ± 1 mm; petal claw glabrous, appendages 2, blade entire to notched, white, pink, or lavender; stamens incl; styles 3, incl. **FR** ovoid; stalk < 1 mm, puberulent. **SEED** ± 1 mm, gray. 2*n*=24. Fields, disturbed areas; < 1000 m. CA-FP; to B.C., e US; native to Eur.

S. grayi S. Watson Per 10–20(30) cm; caudex much-branched. **ST** decumbent to erect, puberulent below, glandular-puberulent above. **LVS** abruptly reduced upward, fleshy; basal tufted, 1.5–4 cm, 1–5 mm wide, narrowly oblanceolate; cauline 0.4–1 cm, 1–3 mm wide, linear to oblanceolate. **INFL**: fls sometimes nodding. **FL**: calyx 8–10 mm, short-glandular-hairy, 10-veined, lobes 1–2 mm; petal claw ciliate at base, appendages 2, blade (2)4-lobed, pink to purplish; stamens ± = petals; styles 3, ± = petals. **FR** ± ovoid; stalk 1.5–3 mm, puberulent. **SEED** 2–3 mm, brown. 2*n*=48. Chaparral, coniferous forest, alpine; 1200–2900 m. KR, n CaR; s OR. ❀TRY.

S. hookeri Nutt. (p. 495) Per 5–20 cm; caudex much-branched. **ST** decumbent to erect, appressed hairy below, sometimes glandular-puberulent above. **LVS** slightly reduced upward, 2–8(9) cm, 8–25 mm wide, oblanceolate. **INFL**: fls ascending to erect. **FL**: calyx 12–20 mm, canescent to soft-white-hairy, faintly 10-veined, lobes 4–7 mm; petal claw ± ciliate at base, appendages 2, blade 4-lobed, lobes equal or outer < inner, white, pink, or purple; stamens ± = corolla; styles 3, slightly > petals. **FR** oblong to ovoid; stalk 2–5 mm, glabrous to puberulent. **SEED** ± 2 mm, black. 2*n*=72. Rocky slopes, open areas, oak woodland, coniferous forest; < 1400 m. n NW; OR. [ssp. *bolanderi* (A. Gray) Abrams; ssp. *pulverulenta* (M.E. Jones) C. Hitchc. & Maguire, Hooker's glandular campion] ❀DRN:1,4,5&IRR,SUN:2,**6**,**7**,14–17;DFCLT;CVS.

S. invisa C. Hitchc. & Maguire (p. 495) SHORT-PETALED CAMPION Per 10–40 cm; caudex simple to much-branched. **ST** erect, puberulent, ± glandular above. **LVS** slightly reduced upward, 1.5–5 cm, 2–6 mm wide, linear to narrowly oblanceolate. **FL**: calyx 7–9 mm, glandular-puberulent, 10-veined, lobes 1–2 mm; petal claw gen glabrous, appendages 0, blade entire to slightly notched, pink to lavender; stamens incl; styles 3, incl. **FR** ± ovoid to cylindric; stalk < 1 mm, sparsely puberulent. **SEED** ± 1 mm, brown. $2n=48$. UNCOMMON. Open areas, coniferous forest; 900–2800 m. s CaRH, n SNH.

S. laciniata Cav. ssp. ***major*** C. Hitchc. & Maguire (p. 495) Per 30–70 cm; caudex simple or few-branched. **ST** reclining or decumbent, ± glabrous below, glandular-puberulent above. **LVS** slightly reduced upward, 1.5–10 cm, 2–10(20) mm wide; lower oblanceolate to lanceolate; upper ± linear. **INFL**: fls ascending to erect. **FL**: calyx 12–26 mm, glandular-puberulent, faintly 10-veined, lobes 3–5 mm; petal claw glabrous to ciliate, appendages 2, blade 4–6-lobed, bright red; stamens slightly > petals; styles 3, exserted. **FR** oblong to ovoid; stalk 2–4 mm. **SEED** 1–1.5 mm, reddish brown. $2n=96$. Chaparral, oak woodland; < 1200 m. s SCoRO, SW; Baja CA. [vars. *angustifolia* C. Hitchc. & Maguire, *latifolia* C. Hitchc. & Maguire] Other ssp. in sw US, Mex; closely related to *C. californica*. ❀DRN:**15–17**&IRR:**7,14,19–24**.

S. latifolia Poiret ssp. *alba* (Miller) Greuter & Burdet Bien, per 30–100 cm, dioecious (fls appear bisexual); caudex few-branched. **ST** erect, gen rough-hairy, ± glandular above. **LVS** gradually reduced upward; lower 5–10 cm, 6–25 mm wide, oblanceolate; upper 1.5–8 cm, 3–15 mm wide, lanceolate. **INFL**: fls ascending to erect, subsessile to short-pedicelled. **FL**: calyx 12–20 mm, short-hairy, lobes 3–6 mm; staminate calyx 10-veined; pistillate calyx ± 20-veined, much inflated in fr; petal claw glabrous, appendages 2, blade 2-lobed, white; stamens = or > petals; styles 5, exserted. **FR** ovoid; stalk 1–2 mm. **SEED** 1–2 mm, brown. $2n=24$. Fields, roadsides; < 1000 m. NCo, NCoRO, s SNF, GV, SnFrB, GV, s SNF; to e US; native to Eur. [*Lychnis a.* Miller] Often mistaken for *S. noctiflora* L.

S. lemmonii S. Watson (p. 495) Per 15–45 cm; caudex gen few-branched. **ST** decumbent to erect, short-hairy below, glandular above; some short branches at base spreading to decumbent, only vegetative. **LVS** ± abruptly reduced upward, elliptic to oblanceolate; basal 2–3.5 cm, 8–10 mm wide; cauline few, 0.5–2.5 cm, 3–5 mm wide. **INFL**: fls nodding. **FL**: calyx 6–10 mm, ± short-glandular-hairy, 10-veined, lobes 1–2 mm; petal claw short-woolly, appendages 2, blade 4-lobed, yellowish white to pink; stamens exserted; styles 3, much exserted. **FR** oblong to ovoid; stalk 2–3 mm, puberulent. **SEED** 1–1.5 mm, straw-colored. $2n=48$. Oak woodland, coniferous forest; 850–2800 m. NW, CaR, SN, w SnFrB (Santa Cruz Mtns), SCoR, TR, PR, MP; OR. ❀TRY.

S. marmorensis Kruckeb. (p. 495) MARBLE MOUNTAIN CAMPION Per 25–40 cm; caudex 1 or few-branched. **ST** erect, puberulent below, glandular above. **LVS** gradually reduced above, lanceolate; lower withering; middle 2–4.5 cm, (3)5–10 mm wide; upper 1.5–4 cm, 2–8(10) mm wide. **INFL** axillary and terminal; fls spreading to erect. **FL**: calyx 12–14 mm, glandular-puberulent, faintly 10-veined, lobes 3–5 mm; petal claw glabrous, appendages 2, blade 2-lobed, pink to yellowish; stamens ± = petals; styles 3, ± = petals. **FR** ovoid; stalk 3–4 mm, puberulent. **SEED** 2–3 mm, black. $2n=48$. RARE. Oak woodland, coniferous forest; 850–1000 m. KR (Humboldt, Siskiyou cos.). Closely related to *S. bridgesii*. [Kruckeberg 1960 Madroño 15:172–177]

S. menziesii Hook. (p. 495) Per 5–20(50) cm, ± mat-like, gen dioecious (fls appear bisexual); caudex much-branched. **ST** decumbent to erect, puberulent to short-hairy, gen glandular above. **LVS** slightly reduced upward, 2–6 cm, 3–20 mm wide, lanceolate to elliptic. **INFL**: fls ascending to erect. **FL**: calyx 5–7 mm, puberulent, ± glandular, faintly 10-veined, lobes 1–2 mm; petal claw glabrous, appendages 2, blade 2–4-lobed, white; stamens deeply incl (pistillate fl) or ± = corolla (staminate fl); styles 3(4), << petals (staminate fl) or > petals (pistillate fl). **FR** ovoid; stalk 1–2 mm. **SEED** 0.5–1 mm, reddish brown. $2n=24,48$. Coniferous forest, pinyon/juniper woodland; 900–2900 m. KR, NCoR, CaR, SN, SnBr, GB; to AK, Rocky Mtns. [ssp. *dorrii* (Kellogg) C. Hitchc. & Maguire; ssp. *menziesii*, var. *viscosa* (E. Greene) C. Hitchc. & Maguire] ❀TRY.

S. multinervia S. Watson (p. 495) Ann 20–65 cm. **ST** erect, gen glandular-short-hairy. **LVS** gradually reduced upward; lower 3–8 (10) cm, 5–13 mm wide; upper 0.8–3 cm, 2–8 mm wide, lanceolate. **INFL** axillary and terminal. **FL**: calyx 8–12 cm, glandular-short-hairy, 16–30-veined, lobes 1–3 mm; petal claw glabrous, appendages 0, blade ± notched, white to pink; stamens incl; styles 3, incl. **FR** ovoid; stalk 1–1.5 mm. **SEED** 0.5–1 mm, black. Open areas, burns; < 2000 m. s NCoRO, CCo, SnFrB, SCoRO, SW; Baja CA.

S. noctiflora L. Ann 20–60(80) cm. **ST** gen erect, rough-hairy, glandular above. **LVS** gradually reduced upward; lower 6–12(14) cm, 20–45 mm wide, elliptic to oblanceolate; upper 1–7 cm, 3–12 mm wide, lanceolate. **INFL**: fls ascending to erect, pedicelled. **FL**: calyx 14–22 mm, glandular-hairy, 10-veined, veins branching above middle, net-like, lobes 6–13 mm; petal claw glabrous, appendages 2, blade 2-lobed, white to pinkish, stamens gen incl; styles 3, incl. **FR** ovoid; stalk 1–3 mm. **SEED** ± 1 mm, reddish brown. $2n=24$. Open, disturbed areas, fields; < 1900 m. CaR, SNF, SCo (expected elsewhere); to e US; native to Eur. Often mistaken for *S. latifolia* ssp. *alba*.

S. nuda (S. Watson) C. Hitchc. & Maguire Per 15–50 cm; caudex simple or few-branched. **ST** erect, glandular-hairy. **LVS** abruptly reduced above base; basal 6–15 cm, 10–30 mm wide, oblanceolate to elliptic; cauline few, 0.8–4 cm, 3–8 mm wide, linear to lanceolate. **INFL**: fls pedicelled or subsessile. **FL**: calyx 12–16 mm, glandular-puberulent, 10-veined, lobe veins wider than tube veins, lobes 2–5 mm; petal claw glabrous to ciliate at base, appendages 2, blade 2-lobed, pink; stamens ± incl; styles 3, incl. **FR** conic to elliptic; stalk 1–2 mm. **SEED** 1–1.5 mm, brown. $2n=48$. Shrubland, juniper woodland, coniferous forest; 1200–1900 m. n&c SNH, MP; s OR, w NV. [ssp. *insectivora* (L. Henderson) C. Hitchc. & Maguire]

S. occidentalis S. Watson Per 30–60 cm; caudex simple or branched. **ST** erect, short-soft-hairy, glandular above. **LVS** gradually reduced upward; lower 5–12 cm, 7–20 mm wide, gen oblanceolate; upper 2–8.5 cm, 3–18 mm wide, oblanceolate. **INFL** axillary and terminal; fls ascending to erect. **FL**: calyx 15–38 mm, short-glandular-hairy, 10-veined, lobes 2–4 mm; petal claw ciliate at base, appendages 2, blades 4-lobed, pink to rose-red; stamens slightly > petals; styles 3, ± = petals. **FR** oblong to ovate; stalk 4–18 mm, short-hairy. **SEED** 1–1.5 mm, grayish brown. Chaparral, coniferous forest; 700–2300 m. s CaRH, n SNH, MP.

ssp. ***longistipitata*** C. Hitchc. & Maguire (p. 495) WESTERN CAMPION **FL**: calyx 27–38 mm; petal claws 30–40 mm, blades 14–20 mm. **FR**: stalk 10–18 mm. UNCOMMON. Habitat of sp.; 1000–2000 m. s CaRH (Butte, Tehama cos.). Relationship to ssp. *occidentalis* needs study.

ssp. ***occidentalis*** (p. 495) **FL**: calyx 15–25 mm; petal claws 12–28 mm, blades 7–15 mm. **FR**: stalk 4–7 mm. $2n=48$. Habitat and range of sp.

S. oregana S. Watson Per 30–50(70) cm; caudex simple or few-branched. **ST** erect, puberulent throughout, glandular above. **LVS** gradually reduced upward; lower 5–8 cm, 7–12 mm wide, oblanceolate; upper 1–6(8) cm, 2–6 mm wide, lanceolate. **INFL** axillary and terminal. **FL**: calyx 9–15 mm, glandular-puberulent, 10-veined, lobes 2–3 mm; petal claw glabrous, appendages 4–6, blade 4–6-lobed, white to pink; stamens incl to slightly > petals; styles 3 (4), gen incl. **FR** elliptic to ovoid; stalk 3–4 mm, puberulent. **SEED** 1–2 mm, brown. $2n=48$. Sagebrush, subalpine coniferous forest; 1500–2500 m. Wrn; to WA, MT, WY.

S. parishii S. Watson (p. 495) Per 10–40 cm; caudex gen much-branched. **ST** ascending to erect, strigose to short-hairy, glandular below or throughout. **LVS** slightly reduced upward, 1.5–6 cm, 5–15 mm wide, lanceolate to ± ovate. **INFL**: fls subsessile to short-pedicelled. **FL**: calyx 24–29 mm, short-glandular-hairy, 10-veined, lobes 4–8 mm; petal claw ciliate to puberulent, appendages 2, blade ± 6-lobed, yellowish white; stamens ± = petals; styles 3, slightly > petals. **FR** ovoid to elliptic; stalk 2–3 mm. **SEED** 1.5–2 mm, brown. $2n=48$. Open, rocky to gravelly slopes, coniferous forest, alpine; 1800–3350 m. SnGb, SnBr, e PR (SnJt, Santa Rosa Mtns, Hot Springs Mtn). [vars. *viscida* C. Hitchc. & Maguire, & *latifolia* C. Hitchc. & Maguire] ❀TRY;DFCLT.

S. sargentii S. Watson Per 10–15(20) cm; caudex much-branched. **ST** decumbent to erect, puberulent above, gen glandular above. **LVS** ± abruptly reduced upward; basal tufted, fleshy, 1.5–3 cm, 1–3 mm wide, oblanceolate; cauline 1–2.5 cm, 1–2 mm wide, ± linear. **FL**: calyx 9–15 mm, glandular-puberulent, 10-veined, lobes 2–3 mm; petal claw ciliate at base, appendages 2, blade 2-lobed, white to red-purple; stamens ± = petals; styles 3(4), < or = petals. **FR** ± ovoid; stalk 1.5–3 mm, woolly-puberulent. **SEED** 1–2 mm, brown. 2*n*=48. Subalpine forest, alpine; 2400–3800 m. SNH, SNE (Sweetwater & White Mtns); NV.

S. scouleri Hook. ssp. ***grandis*** C. Hitchc. & Maguire Per 15–70 cm; caudex gen simple. **ST** gen erect, densely puberulent, glandular above. **LVS** gradually reduced upward; lower 5–10 cm, 12–20 mm wide, oblanceolate to elliptic; upper 2–6 cm, 3–15 mm wide, lanceolate to ovate. **INFL** axillary and terminal; fls ascending to erect, subsessile to pedicelled. **FL**: calyx 12–16 mm, short-glandular-hairy, 10-veined, lobes 2–4 mm; petal claw ± ciliate at base, appendages 2, blade 2(4)-lobed, white to rose; stamens < to ± = petals; styles 3(4), ± = corolla. **FR** elliptic to ovoid; stalk 3–6 mm, puberulent to woolly. **SEED** 1–1.5 mm, grayish brown. 2*n*=48. Rocky slopes, coastal bluffs; < 300 m. NCo, CCo, SnFrB; OR. Other ssp. in nw US, sw B.C. ❀DRN,SUN4,**5**,**17**&IRR:14,**15**,**16**, 24.

S. suksdorfii Robinson (p. 495) CASCADE ALPINE CAMPION Per 3–10 cm; caudex gen much-branched. **ST** decumbent to erect, puberulent, glandular above. **LVS** ± abruptly reduced upward; basal tufted, fleshy, 0.5–4.5 cm, 1.5–3.5 mm wide; cauline 0.5–1.5 cm, 1–2 mm wide, ± linear. **FL** calyx 9–12 mm, short-glandular-hairy, 10-veined, lobes 1–2 mm; petal claws ciliate at base, appendages 2, blades 2-lobed, white to purplish; stamens ± = petals; styles 3(4), ± = petals. **FR** ± ovoid; stalk 2–3.5 mm, puberulent. **SEED** 1–2 mm, brown. RARE in CA. Rocky slopes, alpine; 2400–3100 m. CaRH (Mount Shasta, Lassen Peak); to WA.

S. verecunda S. Watson Per 10–55 cm; caudex branched. **ST** erect, densely puberulent, sometimes glandular above. **LVS** ± gradually reduced upward; lower 3–9 cm, 2–9 mm wide, gen lanceolate; upper 1–4.5 cm, 2–6 mm wide, linear to lanceolate. **FL**: calyx 10–15 mm, gen glandular-puberulent, 10-veined, lobes 2–5 mm; petal claw puberulent at base, appendages 2, blade 2-lobed, white to rose; stamens ± = petals; styles 3(4), ± = petals. **FR** oblong to ovoid; stalk 2–5 mm, puberulent. **SEED** 1–1.5 mm, dark brown. Open areas, chaparral, sagebrush, oak woodland, pinyon/juniper woodland, coniferous forest; < 3400 m. c&s NCoR, SN, CW, TR, PR, W&I, DMtns; to UT, Baja CA.

ssp. ***andersonii*** (Clokey) C. Hitchc. & Maguire **ST** ± scabrous. **LVS** ± thick, stiff; lower gen 4–8 cm; middle ascending to erect. **FL**: calyx short-hairy; petal claw ciliate throughout. 2*n*=48. Open areas, sagebrush, pinyon/juniper woodland, coniferous forest; 1700–2700 m. SNH (e slope), W&I, DMtns; to UT. Sometimes difficult to separate from *S. douglasii*.

ssp. ***platyota*** (S. Watson) C. Hitchc. & Maguire (p. 495) **ST** ± short-soft-hairy. **LVS** thin, flexible; lower gen 5–9 cm; middle ± spreading to ascending. **FL**: calyx ± densely puberulent; petal claw ciliate at base. 2*n*=48. Open areas, oak woodland, coniferous forest; 300–3400 m. c&s NCoR, c&s SN, SnFrB, SCoR, TR, PR; Baja CA. ❀DRN:**17**&IRR:1–3,**7**,**14–16**,**18–24**.

ssp. ***verecunda*** (p. 495) SAN FRANCISCO CAMPION **ST** puberulent. **LVS** ± thin, flexible; lower gen 3–6 cm; middle ± spreading to ascending. **FL**: calyx ± sparsely short-hairy; petal claw ciliate at base. 2*n*=48. RARE. Sandy soils, coastal bluffs, chaparral; < 400 m. n CCo, SnFrB. Intergrades with and difficult to separate from ssp. *platyota*; probably an ecological form. In cult.

S. vulgaris (Moench) Garcke Per 20–80 cm; caudex few-branched; rhizome short. **ST** decumbent to erect, glabrous to glaucous. **LVS** gradually reduced upward; lower 4–8 cm, 5–20 mm wide, lanceolate to oblanceolate; upper 3–4.5 cm, 5–15 mm wide, lanceolate to ovate. **INFL** gen flat-topped. **FL**: calyx 7–10 mm, gen glabrous, inflated and papery in fr, faintly 10–15-veined; petal claw glabrous, appendages 2 and minute or 0, blade 2-lobed, white; stamens exserted; styles 3, exserted. **FR** subspheric; stalk 2–3 mm; **SEED** 1–1.5 mm, brown. 2*n*=24,48. Open areas, fields; < 1200 m. CaR, GV, SnFrB, SCo (expected elsewhere); to e US; native to Eur. [*S. cucubalus* Wibel]

SPERGULA SPURREY

Ann, ascending to erect, taprooted. **LVS** appearing whorled (axillary clusters of 8–15, ± unequal); stipules 1–2 mm, ovate to triangular, scarious, entire, white; blade ± linear; vein 1. **INFL**: cyme, terminal, several–many-fld; pedicels 40+ mm. **FL**: sepals 5, free, 2.5–6 mm, elliptic to ± ovate, glandular-hairy; petals 5, 2.5–4 mm, entire; stamens 5 or 10; styles 5, 0.3–0.7 mm. **FR**: capsule, ovoid; valves 5, spreading to ± recurved. **SEEDS** several, blackish. 5 spp.: temp Eurasia. (Latin: to scatter, from sowing seeds for early forage in Eur)

S. arvensis L. ssp. *arvensis* (p. 495) STICKWORT, STARWORT Pl glabrous or gen glandular-hairy. **ST** 10–40 cm or >; base ± branched. **LF** 1–5 cm, ± linear; tip blunt to abruptly pointed; margin often rolled under, giving cylindric appearance. **INFL**: bracts like stipules, often purplish; pedicels erect to ascending, in fr spreading to reflexed. **FL**: sepals ± acute to rounded, margin widely scarious, ribs often 3, weak; petals ovate, persistent in fr. **SEED** 1–1.5 mm diam, ± spheric, with whitish, club-shaped papillae or minutely roughened. 2*n*=18,36. Open slopes, pine woods, sand dunes, fields, disturbed areas; < 200 m. NCo, NCoRO, SnJV, CCo, SnFrB, SCoRO, SCo; native to Eur.

SPERGULARIA SAND-SPURREY

Ann, per, erect to sprawling; taprooted. **LF**: stipules 1–11 mm, lanceolate and acuminate to widely triangular, scarious, ± entire or splitting ± at tip, white to tan; blade thread-like to linear; vein 1. **INFL**: gen cyme, terminal, few–many-fld, open to dense; pedicels 0.5–28+ mm. **FL**: sepals 5, ± free, 1.5–10 mm, lanceolate to ovate, glabrous to glandular-hairy; petals 5, 0.6–9 mm, entire; stamens 2–10; styles 3, 0.3–1.9 mm. **FR**: capsule, ovoid; valves 3, spreading with tip recurved. **SEEDS** few–many, dark brown, reddish brown, or black. 40 spp.: worldwide. (Latin: derivative of *Spergula*) [Rossbach 1940 Rhodora 42:57–83,105–143,158–193,203–213]

1. Pl strongly per; stamens 7–10
 2. Seeds 0.4–0.5 mm; calyx lobes 2.5–4 mm, in fr < 5 mm; styles 0.4–0.6 mm ..... ***S. villosa***
 2′ Seeds 0.6–0.9 mm; calyx lobes 4.5–7 mm, in fr < 8 mm; styles 0.6–3 mm ..... ***S. macrotheca***
  3. Petals pink or rosy; styles 0.5–1.2 mm ..... var. ***macrotheca***
  3′ Petals white; styles 1.2–3 mm
   4. Fr 1.2–1.4 × calyx; styles 1.2–1.8 mm ..... var. ***leucantha***
   4′ Fr 0.8–1 × calyx; styles 2–3 mm ..... var. ***longistyla***
1′ Ann (*S. rubra, S. media* short-lived per); stamens 2–10
 5. Seeds black ..... ***S. atrosperma***

5′ Seeds light to dark brown or reddish brown
6. Stamens 6–10
7. Seeds 0.8–1.1 mm, smooth, not papillate, gen winged; fr gen 5.5–8 mm; pl stout, lower main st gen 1–3 mm diam, lvs markedly fleshy . . . . ***S. media***
7′ Seeds 0.4–0.6 mm, roughened or variously sculptured, papillate, not winged; fr 2.7–5.4 mm; pl ± delicate, lower main st gen 0.3–1 mm diam, lvs scarsely to ± fleshy
8. Lvs gen 0–2 per cluster, ± fleshy; seeds light brown; stipules mostly deltate, gen 1.5–4.5 mm, tip acute to short-acuminate, dull white, gen inconspicuous . . . . ***S. bocconii***
8′ Lvs 2–4+ per axillary cluster, scarcely fleshy; seeds reddish brown to dark brown; stipules lanceolate, gen 3.5–5 mm, tip ± long-acuminate, shiny white, conspicuous . . . . ***S. rubra***
6. Stamens 2–5
9. Seeds 0.9–1.1 mm . . . . ***S. canadensis*** var. ***occidentalis***
9′ Seeds 0.3–0.7(0.8) mm
10. Infl simple or 1–3+ × compound, glandular-hairy; fr 2.8–6.4 mm; calyx lobes 1.8–4.5 mm, < 4.8 mm in fr; seeds 0.5–0.8 mm . . . . ***S. marina***
10′ Infl gen 4–7+ × compound, glabrous; fr 1.4–2.6 mm; calyx lobes 0.9–1.6 mm, < 2 mm in fr; seeds ± 0.3–0.4 mm . . . . ***S. platensis***

S. atrosperma R. Rossbach (p. 499) Ann, delicate. **ST**: lower main 0.3–1 mm diam. **LF** fleshy; axillary clusters 0; stipules 1–2.5 mm, widely triangular, dull white to tan, inconspicuous, tip acute to short-acuminate. **INFL** simple or 1–3+ × compound, gen glandular-hairy. **FL**: sepals fused 0.2–0.5 mm, lobes 1.8–2.5 mm, < 2.9 mm in fr; petals white to rosy; stamens 4–8; styles 0.4–0.8 mm. **FR** 3.3–5 mm, 1–1.3 × calyx. **SEED** 0.6–0.8 mm, ± shiny black, wingless or rarely partly winged; surface sculpture worm-like, not papillate. Uncommon. Alkaline areas, mud flats, streambeds, sandy areas; < 1500 m. GV, PR, MP; NV.

S. bocconii (Scheele) Merino Ann, ± delicate. **ST**: lower main gen 0.5–1 mm diam. **LF** ± fleshy; axillary clusters 0 or 1–2-leaved; stipules 1.5–4.5 mm, mostly deltate, dull white to tan, gen inconspicuous, tip acute to short-acuminate. **INFL** gen 1–6+ × compound; fls of upper branches often ± on 1 side, glandular-hairy. **FL**: sepals fused 0.4–0.6 mm, lobes 1.9–2.5 mm, in fr < 3(4) mm; petals white or pink to rosy; stamens 8–10; styles 0.4–0.6 mm. **FR** 2.7–5.3 mm, 1–1.2 × calyx. **SEED** 0.4–0.6 mm, light brown, wingless; surface ± sculptured, minutely papillate. $2n=36$. Salt marshes, alkaline areas, sandy soils; < 400 m. c SNF, GV, CCo, SCo, s ChI; OR; native to sw Eur, Medit.

S. canadensis (Pers.) G. Don var. ***occidentalis*** R. Rossbach (p. 499) Ann, delicate to ± stout. **ST**: lower main gen 0.4–1.8 mm diam. **LF** fleshy; axillary clusters gen 0; stipules 1–2.7 mm, widely triangular, dull white, inconspicuous, tip obtuse to acute. **INFL** 1–2+ × compound or fl gen axillary, solitary, glabrous to glandular-hairy. **FL**: sepals fused 0.5–0.6 mm, lobes 2.5–3.5 mm, in fr < 4.5 mm; petals white or pink; stamens 2–4; styles 0.3–0.7 mm. **FR** 3.5–5.3 mm, 1.2–1.3 × calyx. **SEED** 0.9–1.1 mm, reddish brown, often winged; surface ± smooth, minutely glandular-hairy. $2n=36$. Uncommon. Salt marshes; < 3 m. n NCo (Humboldt Bay); to B.C.

S. macrotheca (Hornem.) Heynh. Pl strongly per, stout. **ST**: lower main 0.8–3 mm diam. **LF** fleshy; axillary cluster 0 or 1–2+-leaved; stipules 4.5–11 mm, narrowly triangular, dull white to tan, ± conspicuous, tip long-acuminate. **INFL** simple or 1–3+ × compound or fl axillary, solitary, glandular-hairy. **FL**: sepals fused 0.5–1.8 mm, lobes 4.5–7 mm, < 8 mm in fr; petals white or pink to rosy; stamens 9–10; styles 0.5–3 mm. **FR** 4.6–10 mm, 0.8–1.4 × calyx. **SEED** 0.6–0.9 mm, ± reddish brown, gen winged; surface smooth, tubercled, or sculpture worm-like or of low rounded mounds, not papillate. $2n=36,72$. Alkaline marshes, meadows, fields, salt flats, mud flats, coastal bluffs, outcrops; < 800 m. NCo, NCoRI, GV, CCo, SnFrB, SCoRO, SCo, ChI, DMoj; to B.C., Baja CA.

var. ***leucantha*** (E. Greene) Robinson (p. 499) Pl 10–40 cm. **FL**: calyx lobes 4.5–5.5 mm, < 6.5 mm in fr; petals white; styles 1.2–1.8 mm. **FR** 1.2–1.4 × calyx. **SEED** 0.7–0.8 mm. Alkaline soils, floodplains, vernal pools and meadows, marshy ground; < 800 m. GV, SnFrB, SCoRO, SCo, DMoj.

var. ***longistyla*** R. Rossbach (p. 499) Pl 10–30 cm. **FL**: calyx lobes (4.5)5–5.5 mm, < 7 mm in fr; petals white; styles 2–3 mm. **FR** 0.8–1 × calyx. **SEED** 0.8–0.9 mm. Alkaline marshes, mud flats, meadows, hot springs; < 200 m. NCo, NCoRI, GV.

var. ***macrotheca*** Pl 5–35 cm. **FL**: calyx lobes (4.5)5–7 mm, < 8 mm in fr; petals pink to rosy; styles 0.5–1.2 mm. **FR** 0.8–1.2 × calyx. **SEED** 0.6–0.8 mm. Salt flats and marshes, dunes, rocky outcrops, sandy or rocky coastal bluffs, gravelly ridges, alkaline fields; < 250 m. NCo, CCo, SCo, ChI; to B.C., Baja CA.

S. marina (L.) Griseb. (p. 499) Ann, delicate. **ST**: lower main 0.6–2 mm. **LF** fleshy; axillary clusters 0; stipules 1.2–3 mm, widely triangular, dull white, inconspicuous, tip ± acute. **INFL** 1–3+ × compound or fl axillary, solitary, glandular-hairy. **FL**: sepals fused 0.5–1 mm, lobes 1.8–4.5 mm, < 4.8 mm in fr; petals white or pink to rosy; stamens 2–5; styles 0.4–0.6 mm. **FR** 2.8–6.4 mm, 1–1.5 × calyx. **SEED** 0.5–0.8 mm, light brown to reddish brown, gen wingless; surface smooth or slightly roughened, papillate or not. $2n=36$. Mud flats, alkaline fields, sandy river bottoms, sandy coasts, salt marshes; < 700 m. NCo, NCoRO, c SNF, GV, CCo, SnFrB, SCo, ChI, PR, D; to WA, e US, S.Am; Eurasia. [var. *tenuis* (E. Greene) R. Rossbach] *S. salina* J.S. Presl & C. Presl may prove to be the correct name for this sp.

S. media (L.) Griseb. Ann or short-lived per, stout. **ST**: lower main gen 1–3 mm diam. **LF** distinctly fleshy; axillary clusters gen 0; stipules 2.6–6 mm, deltate, dull white, inconspicuous, tip obtuse to ± acute. **INFL** simple or 1–2+ × compound, glabrous or sparsely glandular-hairy. **FL**: sepals fused 0.3–1 mm, lobes 2.5–4 mm, < 5 mm in fr; petals white; stamens 9–10; styles 0.5–1 mm. **FR** (4.5) 5.5–8 mm, 1.2–1.4 × calyx. **SEED** 0.8–1.1 mm, dark brown, gen winged; surface ± smooth, not papillate. $2n=18,36$. Salt flats, salt marshes, sandy beaches; < 2 m. n CCo (Marin, Contra Costa cos.); OR, e N.Am, S.Am; native to coastal Eur. *S. maritima* (All.) Chiov. may prove to be the correct name for this sp.

S. platensis (Cambess.) Fenzl Ann, delicate. **ST**: lower main gen 0.3–1 mm diam. **LF** ± not fleshy; axillary clusters gen 0; stipules 1.5–3.5 mm, deltate, dull white, inconspicuous, tip ± acute. **INFL** gen 4–7+ × compound, glabrous. **FL**: sepals fused 0.1–0.3 mm, lobes 0.9–1.6 mm, < 2 mm in fr; petals white; stamens 5; styles 0.3–0.4 mm. **FR** 1.4–2.6 mm, 1.3–1.5 × calyx. **SEED** 0.3–0.4 mm, light to dark or reddish brown, wingless; surface sculpture worm-like, papillae minute, cup-shaped. Uncommon. Dried or brackish mud flats and adobe mesas; < 400 m. SCo, PR; to TX, Baja CA; probably native to s S.Am (Argentina).

S. rubra (L.) J.S. Presl & C. Presl (p. 499) Ann or short-lived per, delicate. **ST**: lower main 0.3–0.5 mm diam. **LF** scarcely fleshy, 2–4+ per axillary cluster; stipules gen 3.5–5 mm, lanceolate, shiny white, conspicuous, tip ± long-acuminate. **INFL** 1–3+ × compound or fl axillary, solitary, glandular-hairy. **FL**: sepals fused 0.5–0.7 mm, lobes 2–3.2 mm, < 4 mm in fr; petals pink; stamens 6–10; styles 0.6–0.8 mm. **FR** 3.5–5 mm, 1–1.2 × calyx. **SEED** 0.4–0.6 mm, reddish to dark brown, wingless; surface sculpture worm-like, minutely papillate. $2n=18,36,54$. Open forests, gravelly glades, meadows, mud flats, disturbed areas; < 2400 m. NW, CaR, c SNF, n&c SNH, ScV, CW, SCo, SnGb, PR; to B.C., e N.Am, S.Am; native to Eur.

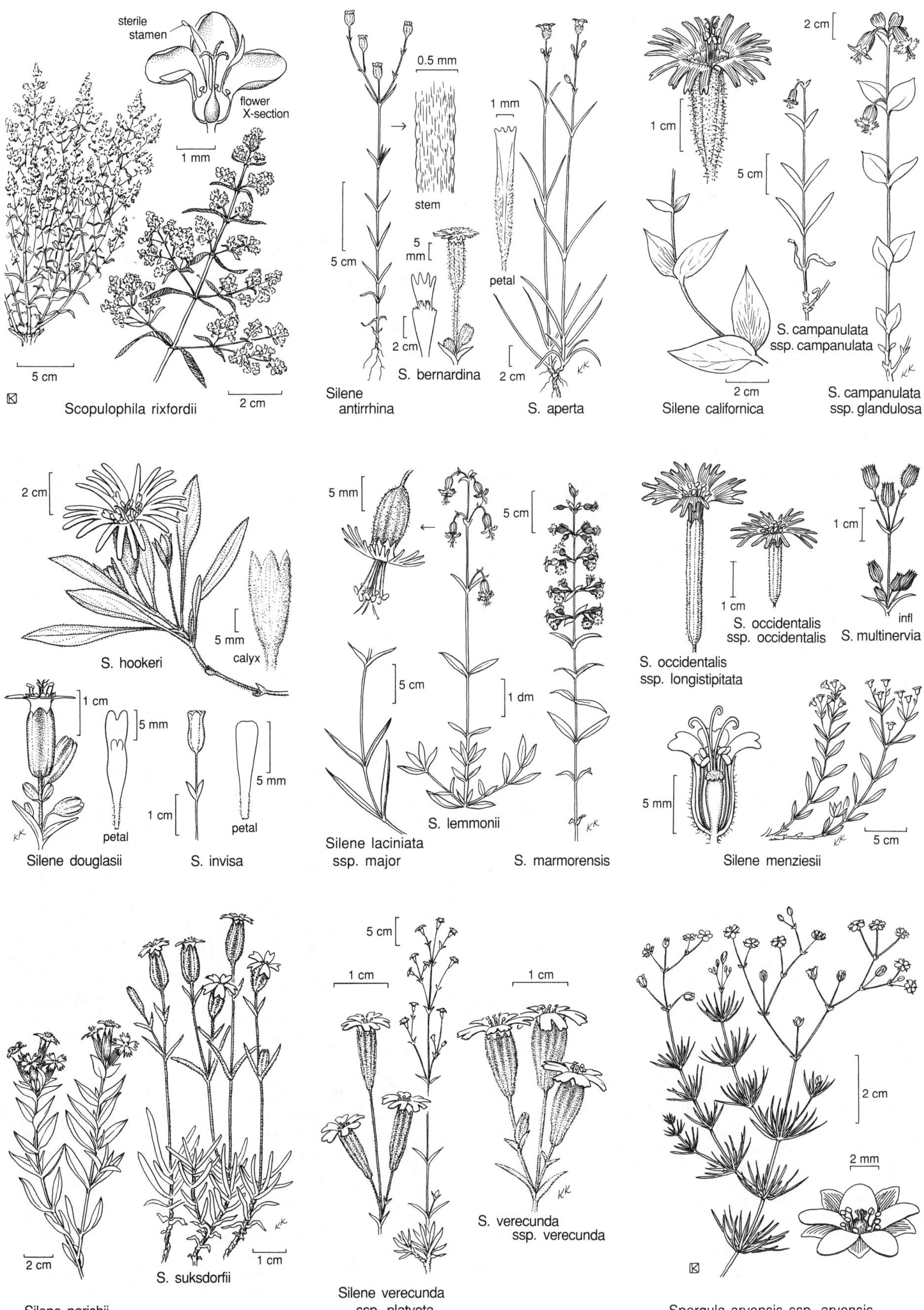

sterile stamen
flower X-section
1 mm
5 cm
2 cm
Scopulophila rixfordii
0.5 mm
stem
5 cm
5 mm
2 cm
S. bernardina
Silene antirrhina
1 mm
petal
2 cm
S. aperta
2 cm
1 cm
5 cm
S. campanulata ssp. campanulata
2 cm
Silene californica
S. campanulata ssp. glandulosa
2 cm
5 mm
calyx
S. hookeri
1 cm
5 mm
petal
Silene douglasii
1 cm
5 mm
petal
S. invisa
5 mm
5 cm
5 cm
1 dm
S. lemmonii
Silene laciniata ssp. major
S. marmorensis
1 cm
S. occidentalis ssp. occidentalis
1 cm
infl
S. multinervia
S. occidentalis ssp. longistipitata
5 mm
5 cm
Silene menziesii
2 cm
Silene parishii
1 cm
S. suksdorfii
5 cm
1 cm
1 cm
Silene verecunda ssp. platyota
S. verecunda ssp. verecunda
2 cm
2 mm
Spergula arvensis ssp. arvensis

S. villosa (Pers.) Cambess. Pl strongly per, stout. **ST**: lower main 0.5–2 mm diam. **LF** ± not fleshy, 2–4+ per axillary cluster; stipules 3–8 mm, lanceolate, dull white, ± conspicuous, tip acuminate. **INFL** 1–3+ × compound, glandular-hairy. **FL**: sepals fused 0.5–0.7 mm, lobes 2.5–4 mm, < 5 mm in fr; petals white; stamens 7–10 mm; styles 0.4–0.6 mm. **FR** (4)5–6.5 mm, 1.1–1.3 × calyx. **SEED** 0.4–0.5 mm, reddish to dark brown, often winged; surface smooth or minutely roughened, sometimes glandular-hairy. Sandy slopes and bluffs, clay ridges and plains, disturbed areas; < 450 m. NCoRO, n SNF, ScV, CCo, SCoRI, s ChI; Baja CA; native to s S.Am.

STELLARIA CHICKWEED, STARWORT

Ann, per, erect to prostrate; taproot and rhizomes present. **LF**: petiole present or 0; blade linear to ovate; vein 1. **INFL**: cyme, terminal or axillary, few–many-fld, open to dense or umbel-like or fl axillary, solitary; peduncles, pedicels 0.8–50+ mm. **FL**: sepals 4 or gen 5, free, 1.5–5.5 mm, lanceolate to ovate, glabrous to glandular-hairy; petals 0 or 5, 0.8–7 mm, gen 2-lobed ± to base; stamens 10, sometimes fewer; styles 3(4–5 in *S. calycantha*), 0.2–2.8 mm. **FR**: capsule, ± ovoid to cylindric-oblong; teeth 6(8,10), ascending to recurved. **SEEDS** several–many, brown to yellowish, reddish, or purplish brown. 120 spp.: worldwide. (Latin: star, from fl shape) [Chinnappa & Morton 1991 Rhodora 93:129–135; Morton & Rabeler 1989 Canad J Bot 67:121–127]

1. Ann, from slender taproot
  2. Lvs crowded near base, blade gen linear-lanceolate (to ovate); internodes glabrous or hairs scattered; bracts scarious ..... ***S. nitens***
  2′ Lvs ± evenly spaced, blade ± ovate; internodes with line of wavy hairs; bracts leafy
    3. Sepals 3–4.5 mm, < 6 mm in fr; petals gen present; seeds 0.9–1.3 mm, dark reddish or purplish brown ***S. media***
    3′ Sepals 2–3 mm, < 4 mm in fr; petals 0; seeds 0.7–0.8 mm, light reddish brown ..... ***S. pallida***
1′ Per, from slender, white rhizomes
  4. Pl covered (exc ± lvs) with long wavy hairs — coastal marshes and bluffs, NCo, CCo ..... ***S. littoralis***
  4′ Pl glabrous or hairs restricted to st or lf margin
    5. Petals > sepals
      6. Infl ± narrow; pedicels ascending to erect; fl gen 1–7 ..... ***S. longipes*** var. ***longipes***
      6′ Infl open, branches widely spreading; pedicels spreading or reflexed; fl gen many
        7. Sepals 4–5.5 mm, margin of some densely ciliate; infl terminal; seeds with prominent, elongate tubercles; lf margin ± smooth, shiny ..... ***S. graminea***
        7′ Sepals 3–4 mm, margin ± glabrous; infl lateral; seeds minutely roughened; lf margin papillate (at 20× magnification), dull ..... ***S. longifolia***
    5′ Petals < 0.8 × sepals or 0
      8. Fls many (exc in alpine), in umbel-like cyme; bracts scarious; pedicels in fr widely spreading; petals 0 ..... ***S. umbellata***
      8′ Fls solitary in axils or few–many in terminal cyme; bracts leafy; pedicels in fr often recurved to reflexed; petals present or 0
        9. Fls solitary in axils; lf blade ± ovate
          10. Sepals gen 5, acute or acuminate, ribs in fr 3; lf margin ± wavy, glabrous; seeds 0.8–1 mm ..... ***S. crispa***
          10′ Sepals gen 4, ± obtuse, ribs obscure; lf margin ± flat, gen ciliate near base; seeds 0.6–0.7 mm ..... ***S. obtusa***
        9′ Fls gen few–many in terminal or axillary cymes; lf blade lanceolate to ovate
          11. Lf blade lanceolate to widely so, primary blades mostly > 2 cm, margin papillate (at 20× magnification), dull; st gen finely papillate; sepals 3–3.5 mm, < 4.5 mm in fr, ribs in fr gen 3 ..... ***S. borealis*** ssp. ***sitchana***
          11′ Lf blade elliptic to ovate, primary blades mostly < 2 cm, margin ± smooth, shiny; st not finely papillate; sepals 1.5–3 mm, < 2.5 mm in fr, veins gen obscure ..... ***S. calycantha***

S. borealis Bigelow ssp. ***sitchana*** (Steudel) Piper (p. 499) Per, sprawling to erect, 15–50 cm, ± glabrous; rhizome white. **ST**: internodes gen finely papillate. **LVS** ± evenly spaced; blade 15–45 mm, ± lanceolate; margin roughened at 20× magnification, flat, dull sometimes ciliate near base. **INFL** terminal or axillary, several–many-fld; bracts leafy; pedicels erect to ascending, in fr curved to reflexed. **FL**: sepals 5, 3–4.5 mm, lanceolate, acute, glabrous, margin ± widely scarious, ribs in fr 3, prominent; petals (0)5, 0.5–0.8 × sepals. **SEED** 1–1.2 mm, dark reddish brown; tubercles low, elongate. $2n$=52. Sedge stands, meadows, streambank, swamps, moist woods; < 2100 m. NCo, NCoRO, CaRH, c&s SNF, SNH; to AK, SD. [*S. s.* Steudel var. *bongardiana* (Fern.) Hultén]

S. calycantha (Ledeb.) Bong. (p. 499) Per, prostrate to erect, 5–25 cm, often glabrous; rhizome white. **ST**: internodes glabrous or with wavy scattered hairs. **LVS** ± evenly spaced; blade 3–25 mm, elliptic to ovate; margin ± smooth, flat to wavy, shiny, sometimes ciliate. **INFL** terminal or axillary, 1–few-fld; bracts leafy; pedicels ascending to erect, in fr often recurved. **FL**: sepals 5, 1.5–3 mm, ovate to elliptic, ± acute, glabrous, margin scarious, veins gen obscure; petals (0)5, 0.3–0.5 × sepals. **SEED** 0.7–0.9 mm, reddish brown; surface ± smooth or minutely roughened. $2n$=26. Mossy banks, bogs, dry creeks, wet meadows, shaded areas; 1700–3800 m. KR, NCoRO, CaRF, n&c SN, Teh, SnGb, MP; to AK, NM; ne Asia. [ssp. *interior* Hultén; var. *simcoei* (J. Howell) Fern.]

S. crispa Cham. & Schldl. (p. 499) Per, prostrate to trailing, 10–40 cm, gen glabrous; rhizome white. **ST**: internodes rarely with scattered, wavy hairs. **LVS** ± evenly spaced; blade 8–20 mm, ± ovate; margin ± smooth, ± wavy, shiny, glabrous. **INFL**: fl axillary, solitary; pedicels ascending, in fr spreading to recurved. **FL**: sepals gen 5, 2.5–4 mm, lanceolate, acute to acuminate, glabrous, margin scarious, ribs in fr 3; petals gen 0. **SEED** 0.8–1 mm, reddish brown; tubercles low, elongate. $2n$=26,52. Shaded, damp areas; < 1700 m. NCo, KR, CaRH, n&c SNH; to AK, MT.

S. graminea L. Per, sprawling to erect, 10–60 cm, gen glabrous; rhizome white. **ST**: internodes glabrous. **LVS** ± evenly spaced; blade 10–35 mm, linear to lanceolate; margin ± smooth, flat, shiny, ± ciliate near base. **INFL** terminal; fls many; bracts scarious; pedicels spreading to erect, in fr gen widely branched. **FL**: sepals 5, 4–5.5 mm, ± lanceolate, acute, margin widely scarious, some densely ciliate, ribs in fr 3, prominent; petals 5, 1–1.4 × sepals. **SEED** 0.9–1.1 mm, dark brown; tubercles prominent, elongate. $2n$=26,39,52. Lawns, gardens, disturbed areas; < 40 m. SnJV, SnFrB, SCo; native to Eur.

S. littoralis Torrey (p. 499) Per, sprawling, 10–60 cm, covered with long wavy hairs; rhizome white. **ST** ± uniformly hairy. **LVS** ± evenly spaced; blade 10–45 mm, ± ovate; margin flat to wavy, ± shiny, densely ciliate. **INFL** terminal; fls few–many; bracts leafy;

pedicels ascending to erect, in fr spreading to reflexed. **FL**: sepals 5, 2.8–5 mm, lanceolate, sharply acute, hairy, margin widely scarious, ciliate near base, ribs in fr 1 or 3; petals 5, 1–1.2 × sepals. **SEED** ± 1 mm, reddish brown; surface minutely roughened. Uncommon. Marshy fields, marshes, coastal bluffs; < 40 m. NCo, CCo. ❀IRR,SUN:5,14,**15–17,22–24**.

S. longifolia Willd. (p. 499) LONG-LEAVED STARWORT Per, sprawling to ascending, 15–40 cm, gen glabrous; rhizome white. **ST**: internodes ± scabrous. **LVS** ± evenly spaced; blade 15–35 cm, linear to linear-lanceolate; margin papillate at 20× magnification, flat, dull, sometimes ciliate near base. **INFL** lateral, several- to gen many-fld; bracts scarious; pedicels spreading to erect, in fr widely branched. **FL**: sepals 5, 3–4 mm, lanceolate to narrowly elliptic, ± acute, glabrous, margin widely scarious, ribs in fr ± 3 near base; petals 5, 1–1.2 × sepals. **SEED** 0.8–0.9 mm, reddish brown; surface minutely roughened. 2*n*=26. RARE in CA. Moist areas; ± 900 m. CaRH (Goose Valley, Shasta Co.); to WA, n-c US, NM; circumboreal.

S. longipes Goldie var. ***longipes*** (p. 499) Per, ascending to erect, 5–35 cm, gen glabrous; rhizomes white. **ST**: internodes glabrous or hairs scattered, wavy. **LVS** ± evenly spaced; blade 10–40 cm, linear to linear-lanceolate; margin smooth, flat, shiny, sometimes ciliate near base. **INFL** terminal or axillary, gen 1–7-fld, ± narrow; bracts leafy or ± scarious; pedicels ascending to erect, in fr ± straight. **FL**: sepals 5, 3–5.5 mm, lanceolate to ± ovate, ± acute, glabrous, margin widely scarious, in fr ± 3-ribbed; petals 5, 1–1.2 × sepals, 2-lobed > 1/2 to base. **SEED** 0.7–1 mm, reddish brown; surface minutely roughened. 2*n*=52,65,72,78,84,91,104. Streambanks, moist to boggy meadows, seeps; 1250–3500 m. KR, NCoRI, CaRH, SN, SnBr, Wrn; to MT, circumboreal. [var. *laeta* S. Watson]

S. media (L.) Villars COMMON CHICKWEED Ann but often over-wintering, prostrate to erect, 7–50 cm; taproot slender. **ST**: internodes hairy in line. **LVS** ± evenly spaced; blade 8–45 mm, ± ovate; margin ± smooth, ± flat, shiny, often ciliate near base. **INFL** terminal or axillary, few-fld, ± dense; bracts leafy; pedicels spreading to erect, in fr curved to reflexed. **FL**: sepals 5, 3–4.5 mm, > 6 mm in fr, lanceolate to ovate, acute to obtuse, glabrous or ± hairy and glandular, margin ± widely scarious, ribs often 1 or 3 near base; petals 5, 0.7–0.9 × sepal. **SEED** 0.9–1.3 mm, reddish or purplish brown; surface papillate. 2*n*=40,42,44. Oak woodlands, meadows, disturbed areas; < 1300 m. NW, CaRH, c SNF, GV, CCo, SnFrB, SCo, ChI, DSon; native to sw Eur. Often a pernicious urban weed.

S. nitens Nutt. (p. 499) SHINING CHICKWEED Ann, ascending to erect, 3–25 cm, glabrous or sparsely hairy; taproot thread-like. **ST**: internodes glabrous or hairs scattered. **LVS** crowded near base; blade 5–15 mm, margin smooth, shiny, often ciliate; lower oblanceolate to obovate; upper linear-lanceolate. **INFL** terminal, few–several-fld; bracts scarious; pedicels ascending to erect, in fr ± straight. **FL**: sepals 5, 2.8–4.2 mm, ± lanceolate, acuminate, glabrous, margin widely scarious, ribs in fr ± 3; petals 5, 0.2–0.5 × sepals or 0. **SEED** 0.5–0.7 mm, brown; surface minutely tubercled. 2*n*=40. Sand dunes, streambanks, open woodlands, beneath boulders, disturbed areas; < 1500 m. NCo, KR, NCoR, n&c SNF, SNH, SnJV, CW, SW; to B.C., MT, Mex.

S. obtusa Engelm. (p. 499) OBTUSE STELLARIA Per, gen prostrate, 4–20 cm, gen glabrous; rhizome white. **ST**: internodes glabrous. **LVS** ± evenly spaced; blade 5–12 mm, ± ovate; margin ± smooth, ± flat, shiny, gen ciliate near base. **INFL** axillary; fl solitary; pedicels ascending, in fr spreading to reflexed. **FL**: sepals gen 4, 1.5–3.5 mm, ± ovate, ± obtuse, glabrous, margin ± thinly scarious, ribs obscure; petals 0. **SEED** 0.6–0.7 mm, dark brown; tubercles low, ± elongate. 2*n*=26,52,±65,±78. UNCOMMON. Moist areas in woods, shaded edges of creek; 1600–2000 m. NCoRO, CaR, c SNF, n&c SNH, MP; to B.C., MT, Colorado.

S. pallida (Dumort.) Piré Ann but often over-wintering, prostrate to erect, 7–50 cm; taproot slender. **ST**: internodes hairy in line. **LVS** ± evenly spaced; blade 8–45 mm, ± ovate, margin ± smooth, ± flat, shiny, often ciliate near base. **INFL** terminal or axillary, few-fld, ± dense; bracts leafy; pedicels spreading to erect, in fr curved to reflexed. **FL**: sepals 5, 2–3 mm, < 4 mm in fr, lanceolate to ovate, obtuse to acute, glabrous or ± hairy and glandular, margin ± thinly scarious, ribs often 1 or 3 near base; petals 0. **SEED** 0.7–0.8 mm, light reddish brown; surface papillate. 2*n*=22. Oak woodlands, streambanks, grassy hills and flats, disturbed areas; < 450 m. c SNF, ScV, CCo, s ChI, PR; to e N.Am; native to sw Eur. [*S. media* ssp. *p.* (Dumort.) Asch. & Graebn.]

S. umbellata Karelin & Kir. (p. 499) Per, prostrate to erect, 2–20 cm, glabrous; rhizome white. **ST**: internodes glabrous. **LVS** ± evenly spaced; blade 5–20 mm, elliptic to ovate, margin smooth, gen flat, shiny, glabrous. **INFL** terminal, gen umbel-like, few to gen many-fld; bracts scarious; pedicels erect to spreading, in fr spreading to reflexed. **FL**: sepals 5, 1.5–2.5 mm, lanceolate to ovate, ± acute, margin widely scarious, ribs in fr 3, weak; petals minute or 0. **SEED** 0.6–0.7 mm, brown; surface ± smooth. 2*n*=26. Moist meadows, rocky summits, stream sides; 1800–2300 m. CaRH, c&s SNH; to Siberia, MT, NM.

VACCARIA COW-HERB, COCKLE

1 sp. (Latin: cow, from use as fodder or prevalence in pastures)

V. hispanica (Miller) Rauschert Ann (8)20–100 cm, glabrous, glaucous, taprooted. **LF** 2–12 cm; petiole present or 0; blade lanceolate to ovate, base rounded to cordate-clasping; veins 3–7. **INFL**: cyme, terminal, 10–70-fld +, ± flat-topped; bracts leafy; pedicels 5–40+ mm. **FL**: sepals 5, fused, glabrous, tube prominent, 7.5–17 mm, 1.5–9 mm diam, cylindric to urn-shaped, 5-angled or 5-keeled, veins 10, teeth 1.5–3 mm, < tube, ovate to triangular; petals 5, 15–25 mm, claw long, blade oblanceolate to obovate, entire or obcordate, pink to reddish; styles 2, 9–21 mm. **FR**: capsule, ovoid; stalk 0.5–1 mm; teeth 4, ascending to recurved. **SEEDS** many, 1.6–1.8 mm, ± spheric; tubercles fine, low, reddish brown to black. 2*n*=24,30,60. Fields, disturbed areas; < 2800 m. KR, NCoR, CaRH, c SNF, n SNH, Teh, ScV, CW, SCo, PR, GB; native to Eurasia, Medit. [*V. segetalis* (Necker) Asch.]

VELEZIA

Ann, erect, taprooted. **LF**: blade linear to awl-shaped; veins 3. **INFL** axillary; fl gen solitary; pedicels 1.5–3.5 mm. **FL**: sepals 5, fused, tube prominent, 10–14 mm, 0.8–1 mm diam, narrowly cylindric, rounded, ribs 15, teeth 1–1.2 mm, < tube, lanceolate-acuminate; petals 5, 11–16 mm, claw long, blade entire or notched, appendages 6–8, linear to lanceolate; stamens 5; styles 2, 7–8 mm. **FR**: capsule, narrowly cylindric; stalk 0.2–0.7 mm; teeth 4, ascending. **SEEDS** 6–8, black. 6 spp.: Medit, s Asia. (C. Velez, friend of botanist Loefling, 18th century)

V. rigida L. Pl 7–40 cm, glandular-hairy at least above. **ST**: branches widely spreading or repeatedly 2-forked, rigid, green to purplish. **LF** 5–20 mm; margin ciliate, scarious near and fused at base. **FL**: calyx base swollen, hardened; petals, narrowly obovate, pink to purple, appendages 0.4–0.6 mm. **FR**: tip rounded. **SEEDS** in a single row, 1.3–1.8 mm, ovate-oblong, with abrupt, rounded point; papillae fine, low. 2*n*=28. Oak woodlands, open ridges, gravelly streambeds, serpentine; 100–800 m. KR, n&c SNF; native to s Eur.

CELASTRACEAE STAFF-TREE FAMILY

Barry A. Prigge

Shrub (sometimes climbing), tree, sometimes thorny, gen glabrous. **LVS** simple, opposite or alternate, ephemeral to persistent, subsessile or petioled; veins pinnate. **INFL**: cluster, cyme, raceme, panicle, or fl solitary, axillary or terminal, bracted. **FL** gen bisexual, radial, small; hypanthium ± cup-shaped; sepals 4–5; petals (0)4–5, free; stamens 4–5, alternate petals, attached below or to rim of disk; ovary superior or ± embedded in disk, 2–5-chambered, placentas axile or basal, style gen 1, short, stigma ± head-like, 2–5-lobed. **FR**: capsule, winged achene, berry, drupe, or nutlet, often 1-chambered. **SEED** gen 1 per chamber, arilled. 50 genera, 800 spp.: worldwide, esp se Asia; some orn (*Celastrus, Euonymus, Maytenus,Paxistima*). [Brizicky 1964 J Arnold Arbor 45:206–234]

1. Lvs alternate, entire; infl terminal, many-fld; petals white .... **MORTONIA**
1′ Lvs opposite, ± toothed; infl axillary, 1–5-fld; petals ± reddish
 2. Deciduous shrub or small tree, 2–6 m; lf 3–14 cm; fl parts in 5's .... **EUONYMUS**
 2′ Evergreen shrub, < 1 m; lf 1–3 cm; fl parts in 4's .... **PAXISTIMA**

EUONYMUS BURNING BUSH

Shrub, small tree, erect, glabrous. **ST**: twig gen 4-angled, with corky ridges. **LVS** opposite, deciduous, gen scalloped or finely toothed. **INFL** axillary, few-fld; pedicel jointed to peduncle. **FL**: parts in 5's; petals ± green or purple; disk fused to hypanthium, flat, ± 5-lobed; stamens short, attached to disk margin; ovary embedded in disk, sometimes bumpy or warty, style 0 or short, stigma lobes 3–5, obscure. **FR**: capsule, loculicidal; valves 3–5. **SEED** white, red, or black, enclosed by orange or red aril. 180 spp.: esp trop s hemisphere. (Greek: good name) [Blakelock 1961 Kew Bull 210–290]

E. occidentalis Torrey WESTERN BURNING BUSH Pl 2–6 m. **ST**: branches slender, often climbing. **LF**: petiole 3–15 mm; blade 3–14 cm, ovate to obovate, thin, base truncate to tapered. **INFL** 1–5-fld; peduncle 2–7 cm, slender; pedicel 5–15 mm. **FL**: sepals 1–1.5 mm, 1.5–2.5 mm wide; petals 4–6.5 mm, brown-purple, finely dotted, margin transparent; disk ± 3 mm wide. **FR** depressed, deeply 3-lobed, smooth. **SEED** 4–6 mm, brownish; aril ± red. Shaded streambanks, canyons; < 2000 m. NW, CW, PR; to WA

1. Lf tip abruptly acuminate; twig ± green; infl 1–5-fld .... var. ***occidentalis***
1′ Lf tip ± obtuse or round; twig ± white; infl 3–5-fld .... var. ***parishii***

var. ***occidentalis*** (p. 499) Shaded streambanks, canyons; 20–1600 m. NW, CW; to WA. ❀SHD,IRR or WET:1,2,7,14&SUN:**4,6,15–17**.

var. ***parishii*** (Trel.) Jepson Shaded canyons; 1300–2000 m. PR. ❀SHD,IRR or WET:1–3,7,14&SUN:**4–6,15–17**.

MORTONIA

Shrub, erect, scabrous. **LVS** alternate, persistent, ascending, leathery, entire; margin gen thicker. **INFL**: panicle, terminal; fls many. **FL**: parts in 5's; hypanthium obconic; petals white; disk fused to hypanthium exc at top, fleshy, ± white, becoming red-purple; ovary superior, narrowly ovoid, stigma lobes 5, slender, spreading. **FR**: nutlets 5, oblong-cylindric, light brown. **SEED** 1 per nutlet, straw-colored, very difficult to separate from nutlet; aril 0. 5 spp.: sw US, Mex. (S.G. Morton, 19th century N.Am naturalist)

M. utahensis (Trel.) Nelson (p. 499) Pl 3–12 dm, coarsely scabrous. **ST**: twigs creamy white, turning gray. **LF**: petiole ± 0–1 mm; blade 6–16 mm, ovate to round, from above transversely concave and longitudinally convex, base rounded to tapered, tip round to acute, sometimes with small point. **INFL** 8–65 mm, 6–23 mm wide. **FL**: hypanthium 1.5–2 mm; sepals 1–2.3 mm, tips often acute, keeled; petals 2.2–3 mm, ovate. **FR** 5–7 mm, glabrous. Limestone slopes, canyon bottoms; 900–2100 m. DMtns (Inyo, ne San Bernardino cos.); to sw UT. ❀DRN,DRY:1–3,10;DFCLT.

PAXISTIMA

Shrub, prostrate to ascending. **ST**: twig 4-angled, with corky ridges. **LVS** opposite, persistent, leathery, finely toothed. **INFL** axillary, 1–3-fld. **FL**: parts in 4's; petals red-brown; disk fused to hypanthium, ± square, fleshy; stamens short, attached to disk margin; ovary embedded in disk, stigma lobes 2, obscure. **FR**: capsule, loculicidal; valves 2. **SEED** shiny, dark brown to black; aril surrounding base and 1 side of seed, white, thin, fringed. 2 spp.: N.Am. (Greek: thick stigma)

P. myrsinites (Pursh) Raf. (p. 499) OREGON BOXWOOD Pl 3–10 dm. **ST** prostrate to spreading, ± stiff, densely branched. **LF**: petiole ± 1 mm; blade 8–34 mm, ovate to oblanceolate, base tapered, tip rounded to acute. **INFL**: peduncle 2–3 mm. **FL**: petals ± 1 mm, ovate. **FR** 4–7 mm, obovoid. Shaded places; 600–2000 m. NW, CaR, n&c SN, n SnFrB (Marin Co.), MP; to B.C., Rocky Mtns, n Mex. ❀SHD,DRN,IRR:1–7,15,16;DFCLT.

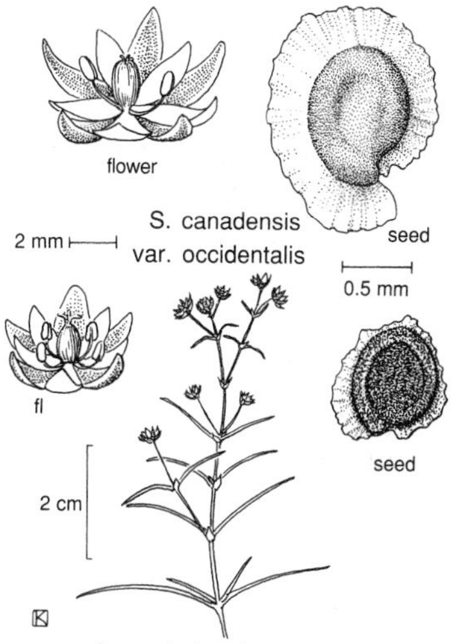

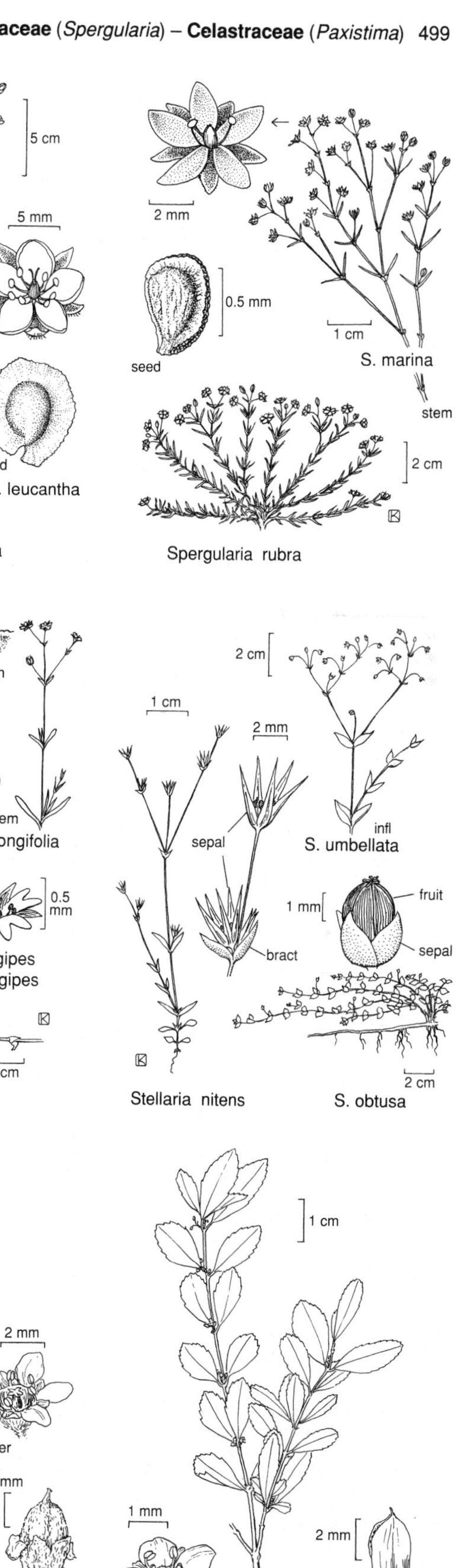

Spergularia atrosperma

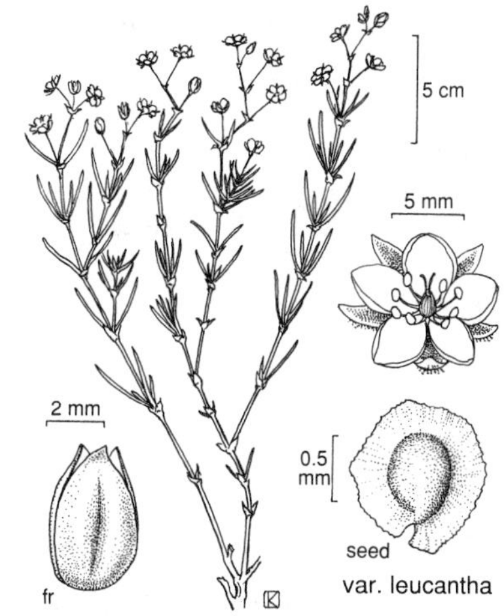

Spergularia macrotheca

Spergularia rubra

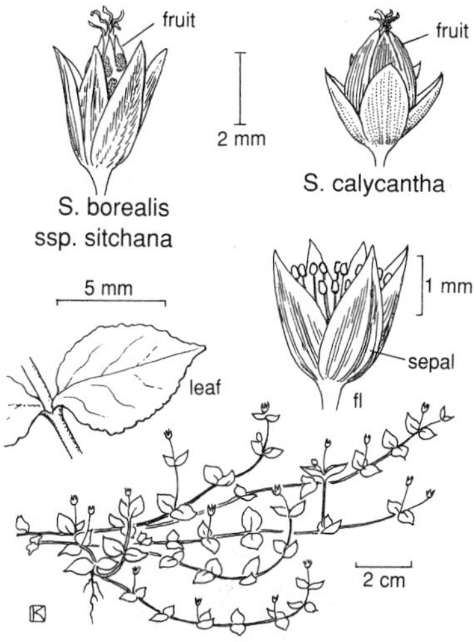

Stellaria crispa

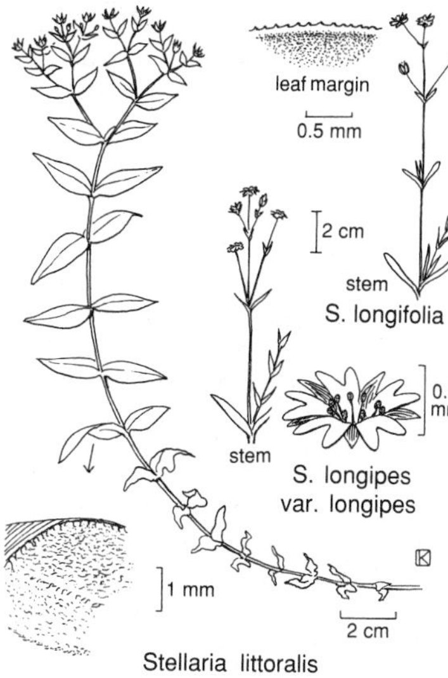

Stellaria littoralis

Stellaria nitens

S. obtusa

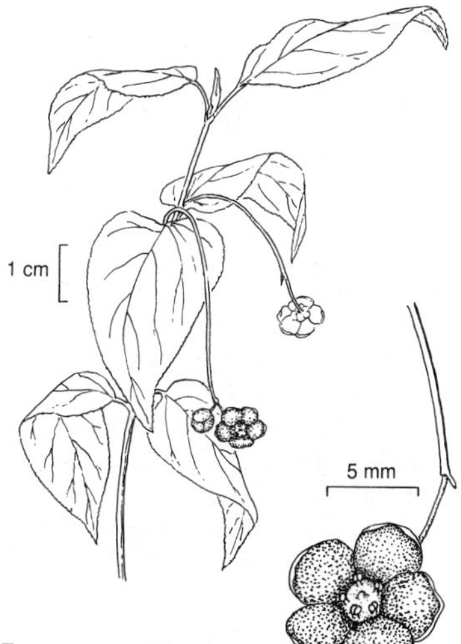

Euonymus occidentalis
var. occidentalis

Celastraceae

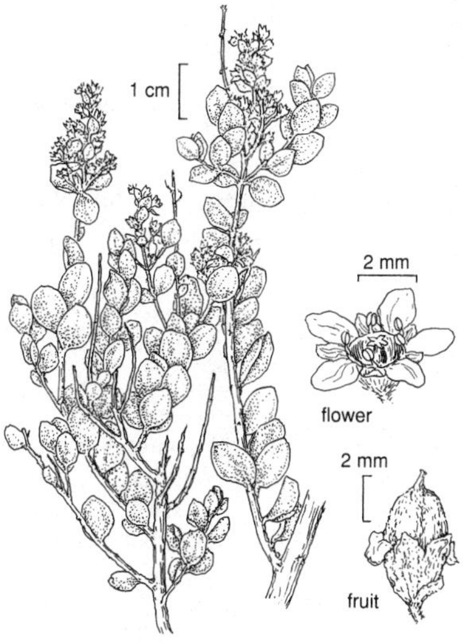

Mortonia utahensis

Paxistima myrsinites

CERATOPHYLLACEAE HORNWORT FAMILY

Fosiée Tahbaz

Ann, per, submerged aquatic, monoecious. **ST** slender, well branched. **LVS** whorled, repeatedly forked; divisions thread-like or narrow, ± stiff, minutely serrate. **INFL**: fls solitary in axils, sessile. **FL** minute; perianth thin, many-parted; stamens 10–20, free; pistil gen 1, ovary sessile, chamber 1, ovule 1, style persistent. **FR**: achene, smooth to spiny. 1 genus.

CERATOPHYLLUM HORNWORT

The only genus; ± 3 spp.: worldwide. (Greek: horn leaf, from stiff lf divisions) Cult in pools, aquaria.

C. demersum L. (p. 507) **ST**: branches 5–25 dm. **LVS** 5–12 per whorl, 1–2.5 cm, ± prickly-serrate on 1 surface. **FR** ± 5 mm, widely elliptic, smooth to spiny-tubercled. $2n$=24. Ponds, ditches, slow streams; gen < 2000 m. CA; worldwide.

CHENOPODIACEAE GOOSEFOOT FAMILY

Dieter H. Wilken, except as specified

Ann to tree, sometimes monoecious or dioecious, glandular or with bead-like hairs that collapse with age, becoming scaly or powdery. **ST** often fleshy. **LVS** gen alternate, entire to lobed; veins gen pinnate. **INFL**: raceme, spike, catkin-like, or spheric cluster, or fl 1; bracts 0–few. **FL**: sepals 1–5, often 0 in pistillate fls, free or fused, gen persistent in fr; petals 0; stamens 0–5; ovary gen superior, chamber 1, ovule 1, styles 1–3. **FR**: gen utricle. **SEED** 1, vertical (fr compressed side-to-side) or horizontal (fr compressed top-to-bottom). 100 genera, 1300 spp.: worldwide, esp deserts, saline or alkaline soils; some cult for food (*Beta*, beets, chard; *Chenopodium*, quinoa).

1. Upper sts clearly jointed, ± fleshy; lvs scale-like
  2. Lvs alternate; shrub ..... **ALLENROLFEA**
  2′ Lvs opposite; ann or per ..... **SALICORNIA**
1′ Upper sts not jointed, not fleshy; lvs gen with blades
  3. Lvs opposite, clasping; sepals strongly overlapping side-to-side; pls from rhizomes ..... **NITROPHILA**
  3′ Lvs mostly alternate, not clasping; sepals not overlapping; rhizomes 0
    4. Shrubs or subshrubs
      5. Herbage stellate-hairy, gen becoming rust-colored; lf margins inrolled ..... **KRASCHENINNIKOVIA**
      5′ Herbage glabrous, puberulent, minutely scaly, or powdery, green or grayish; lf margins gen flat
        6. Pistillate fls gen ± enclosed by 2 tightly appressed or ± fused bractlets; pistillate calyx 0; lf blades gen flat
          7. Bractlets compressed, gen triangular, tip obtuse to acute; lvs minutely scaly ..... [2]**ATRIPLEX**
          7′ Bractlets thick, ± round, tip rounded; lvs glabrous or with soft, branched hairs ..... **GRAYIA**
        6′ Fls not ± enclosed by 2 bractlets (sometimes bisexual); lf blade gen ± cylindric (exc *Kochia, Suaeda*)
          8. St branches becoming spine-like; staminate fls in spikes with peltate bracts, calyx 0; pistillate fls 1–3, axillary, perianth cup-like, winged in fr ..... **SARCOBATUS**
          8′ St branches not spine-like; fls gen bisexual, axillary; calyx lobes rounded to winged
            9. Calyx lobes rounded, hooded, or keeled in fr; ovary with neck-like extension; fl-st gen glabrous [2]**SUAEDA**
            9′ Calyx lobes winged in fr; ovary top rounded to ± depressed, not neck-like; fl-st gen densely hairy
              10. Lf blade linear to lanceolate, glabrous to silky-hairy ..... [3]**KOCHIA**
              10′ Lf blade oblong to ovate, puberulent ..... ***Salsola vermiculata***
    4′ Ann or per (sometimes woody below ground)
      11. Pistillate fls enclosed by 2 bracts; pistillate calyx 0; staminate calyx segments 3–5 ..... [2]**ATRIPLEX**
      11′ Fls sometimes bisexual, bracts 0–1 (if 2, not enclosing fl); calyx segments 1–5
        12. Lvs or infl bracts spine- or bristle-tipped
          13. Lvs and infl bracts cylindric, abruptly bristle-tipped; sepal tips winged in fr ..... **HALOGETON**
          13′ Lvs and infl bracts thread- to awl-like, acute; infl bract widely lanceolate, spine-tipped; sepal back winged in fr ..... [2]**SALSOLA**
        12′ Lvs or infl bracts not spine- or bristle-tipped
          14. Calyx bilateral, 1 lobe > others ..... [2]**SUAEDA**
          14′ Calyx radial, lobes gen equal
            15. Calyx winged in fr
              16. Calyx lobes ± free, wing-like in fr; fr obovoid to ± spheric, gen longer than wide ..... [2]**SALSOLA**
              16′ Calyx lobes ± = tube, top of tube winged in fr; fr compressed-spheric, wider than long
                17. Lf ± wavy-toothed; st finely tomentose; wing continuous around calyx, irregularly toothed ..... **CYCLOLOMA**
                17′ Lf entire; st glabrous to straight-hairy; wings 5, opposite calyx lobes ..... [3]**KOCHIA**
            15′ Calyx not winged in fr (fr winged or not)
              18. Calyx lobes tubercled or curved- to hooked-spiny
                19. Calyx densely and finely hairy, lobes curved- to hooked-spiny ..... **BASSIA**
                19′ Calyx glabrous to ciliate, lobe tips tubercled ..... [3]**KOCHIA**
              18′ Calyx lobes not appendaged

20. Ovary half-inferior, embedded in receptacle; sepals thick, becoming hard in fr; frs adherent in clusters .... **BETA**
20′ Ovary superior; sepals papery to membranous in fr; frs not adherent
21. Calyx tube > lobes .... [2]**CHENOPODIUM**
21′ Calyx tube 0 or << lobes
22. Calyx lobes 4–5; stamens gen 5 .... [2]**CHENOPODIUM**
22′ Calyx lobes 1–3(5); stamens 1–3
23. Lvs mostly ovate, upper ± clasping; style branches 3; fr depressed-spheric; seed horizontal .... **APHANISMA**
23′ Lvs mostly linear to oblanceolate, not clasping; style branches 2; fr ovoid to spheric; seed vertical
24. Fr ± elliptic, ± flat, winged; lvs subtending fls membranous, scarious-margined **CORISPERMUM**
24′ Fr ovoid to ± spheric, not winged; lvs subtending fls herbaceous, often fleshy ..... **MONOLEPIS**

ALLENROLFEA IODINE BUSH

1 sp. (Robert Allen Rolfe, English botanist, 1855–1921)

A. occidentalis (S. Watson) Kuntze (p. 507) Shrub 50–200 cm, glabrous. **ST** much-branched, jointed; internodes 5–20 mm, green to ± glaucous, fleshy. **LVS** sessile, ± decurrent, scale-like, triangular. **INFL**: spike, 5–25 mm, cylindric, sessile; fls spirally arranged; bracts peltate. **FL** bisexual; calyx 1–1.5 mm, 4–5-lobed, enclosing and falling with fr; stamens 1–2, exserted; stigmas 2. **FR** ± 1 mm, ovoid. **SEED** red-brown. Saline soils, flats, bluffs; < 1300 m. SnJV, e SnFrB, SCoRI, n WTR, e PR, s SNE, D; to OR, ID, TX, n Mex. ❀TRY.

APHANISMA

1 sp. (Greek: inconspicuous)

A. blitoides Moq. (p. 507) APHANISMA Ann, fleshy, glabrous. **STS** 1–many from base, 10–55 cm, decumbent to erect. **LVS** gradually reduced; lower sessile; upper ± clasping; blade 8–40 mm, elliptic to ovate. **INFL** axillary, sessile; fls 1–3(5). **FL** bisexual; calyx lobes 3(–5), ± 1 mm, subequal, concave; stamen 1; style branches 3. **FR** ± 1 mm diam, depressed-spheric; equatorial margin ± thick, ring-like. **SEED** horizontal, black. RARE. Coastal shrubland, bluffs, sand; < 100 m. s CCo, SCo, ChI; Baja CA.

ATRIPLEX SALTBUSH

Dean Taylor & Dieter H. Wilken

Ann (gen monoecious) to shrub (gen dioecious), often scaly. **LVS** gen alternate, gen entire; lower gen ± short-petioled; upper gen sessile, ± reduced. **STAMINATE INFL**: spike or spheric cluster; bracts 0. **PISTILLATE INFL**: clusters to spike- or panicle-like; bracts 2 per fr, free to fused, gen compressed, gen sessile. **STAMINATE FL**: calyx lobes 3–5; stamens 3–5. **PISTILLATE FL**: calyx ± 0; ovary ovoid to spheric, style branches 2. **SEED** gen erect. ± 250 spp.: temp to subtrop worldwide. (Latin: ancient name) Gen in alkaline or saline soils; some weedy; some accumulate selenium.

1. Per, subshrub, or shrub
2. Per; central st herbaceous to ± woody at base
3. Sts gen erect
4. Fr bracts 2–3 mm, obovate; coastal .... [2]***A. coulteri***
4′ Fr bracts 3–5 mm, obovate to subspheric; inland .... [2]***A. fruticulosa***
3′ Sts gen prostrate to decumbent
5. Cauline lvs gen opposite; pls mat-like, monoecious or dioecious
6. Lvs thin, gray-scaly, sometimes thinly so above; fr bracts sharply dentate; gen monoecious .... [2]***A. coulteri***
6′ Lvs thick to ± fleshy, densely white-scaly; pistillate fr bracts (sub)entire; gen dioecious .... ***A. watsonii***
5′ Cauline lvs gen alternate; pls not mat-like; gen dioecious
7. Fr bracts ± free, ± compressed, ± thin; staminate and pistillate fls gen in same clusters .... ***A. californica***
7′ Fr bracts fused to middle or above, thick, spongy to fleshy; pistillate infl gen below staminate infl
8. Fr bracts fleshy, reddish, 3–6 mm .... ***A. semibaccata***
8′ Fr bracts spongy, gray to white, 4–12 mm
9. Lf blade elliptic to widely ovate; fr bracts fused ± to middle, body ± subspheric; seeds alike, 2–3 mm; coastal .... ***A. leucophylla***
9′ Lf blade oblanceolate to diamond-shaped; fr bracts fused to near top, body ± widely obconic; seeds of 2 forms, ± 1 mm or ± 1.5 mm; inland .... [2]***A. lindleyi***
2′ Subshrub or shrub, central st aboveground clearly woody
10. Fr bracts together 4-winged base to top .... ***A. canescens***
11. Lf blade 3–8 mm wide; fr bracts stalked, wings gen 3–6 mm wide .... ssp. ***canescens***
11′ Lf blade 1–3 mm wide; fr bracts sessile, wings gen < 3.5 mm wide .... ssp. ***linearis***
10′ Fr bracts together not clearly 4-winged
12. Lvs gen ± wavy to toothed
13. Fr bracts free, short-stalked; lf blade silver-scaly; SNE, D (incl edges) .... ***A. hymenelytra***
13′ Fr bracts fused ± to middle, sessile; lf blade gray-scaly; SCo .... ***A. nummularia***
12′ Lvs gen entire

14. Sts several–many from base, gen unbranched; fr bracts fusiform to ovoid — tips long-tapered ... ***A. gardneri*** var. ***falcata***
14′ St gen 1 from base, much-branched, stiffly spreading to erect; fr bracts ovate to ± round or ± spheric (exc tips)
15. Fr bracts 5–20 mm
16. Fr bracts ± free, widely elliptic to ± round, few-toothed below middle, entire above middle . ***A. confertifolia***
16′ Fr bracts fused ± to middle, entire and spheric below middle, compressed and irregularly toothed above middle ... ***A. spinifera***
15′ Fr bracts 2–6 mm
17. Upper lvs gen sessile to clasping, base truncate to cordate ... ***A. parryi***
17′ Upper lvs gen subsessile to short-petioled, base tapered to round
18. Lf blade oblong to narrowly oblanceolate, gen 2–4 mm wide ... ***A. polycarpa***
18′ Lf blade ovate to deltate, gen 10–50 mm wide ... ***A. lentiformis***
19. Twigs gen smooth, not striate ... ssp. ***lentiformis***
19′ Twigs gen sharp-angled, striate ... ssp. ***torreyi***
1′ Ann
20. Lvs gen green on both surfaces, glabrous to sparsely powdery or fine-scaly
21. Fr bracts ovate to ± round; seeds on same pl of 2 sizes and colors (± 1–2 mm and black or 2–4 mm and brown)
22. Fr bracts clearly net-veined, 8–18 mm; black seeds horizontal ... ***A. hortensis***
22′ Fr bracts not net-veined, 2–7 mm; black seeds vertical
23. Fr bracts elliptic to ± round, surfaces smooth, margins entire ... ***A. heterosperma***
23′ Fr bracts ovate to widely triangular, surfaces 2-tubercled, margins few-toothed ... ***A. subspicata***
21′ Fr bracts ± triangular; seeds on same pl ± all alike
24. Fr bracts fused to near top ... [2]***A. serenana***
24′ Fr bracts gen ± free above middle
25. Fr bract base cordate to hastate; lvs fleshy, brittle, base hastate; sts faintly striate ... ***A. phyllostegia***
25′ Fr bract base truncate to obtuse; lvs not fleshy, not brittle, base tapered to truncate; sts gen green-yellow-striate
26. Fr bract margin narrow, minutely dentate (esp below middle)
27. Lower lvs oblong to lanceolate; fr bract base tapered to rounded, surfaces gen smooth ***A. patula*** var. ***patula***
27′ Lower lvs lanceolate to triangular; fr bract base truncate, surfaces gen tubercled ... ***A. triangularis***
26′ Fr bract margin wide, entire
28. Fr bract 3–4 mm, surfaces ribbed; lvs ovate to triangular, irregularly wavy-toothed ... [2]***A. joaquiniana***
28′ Fr bract 6–12 mm, surfaces smooth or tubercled; lvs linear to lanceolate, entire to finely dentate ... ***A. patula*** var. ***obtusa***
20′ Lvs white to gray, densely and finely scaly (esp below)
29. Fr bracts ± free — sts striate ... [2]***A. joaquiniana***
29′ Fr bracts fused from middle, above middle, or to near top
30. Fr bracts gen widest below middle, ovate, deltate, or diamond-shaped (sometimes ± round in *A. cordulata*)
31. Lvs coarsely wavy-toothed; fr bracts hard ... ***A. rosea***
31′ Lvs gen entire; fr bracts flexible
32. Sts 0–few-branched, erect, 1–8 dm; fr bracts 2.5–5 mm
33. Lower lvs with cordate base; fr bracts 3.5–5 mm; s ScV, SnJV ... ***A. cordulata***
33′ Lower lvs with rounded base; fr bracts 2.5–3.5 mm; s SnJV (Kern Lake) ... ***A. tularensis***
32′ Sts much-branched, prostrate to erect, < 4 dm; fr bracts 1–3.5 mm
34. Sts prostrate to decumbent
35. Sts glabrous to densely scaly near tips; lvs alternate or opposite ... ***A. depressa***
35′ Sts woolly near tips; lvs gen opposite ... ***A. parishii***
34′ Sts erect (branches widely spreading)
36. Fr bracts ± 2–3 mm, ± 2–2.5 mm wide, surfaces densely tubercled; sts brittle; SnJV ... ***A. minuscula***
36′ Fr bracts 1–2 mm, ± 1 mm wide, surfaces smooth; sts rigid, not brittle; GB ... ***A. pusilla***
30′ Fr bracts (incl stalk) gen widest at or above middle, wedge-shaped to ± round
37. Staminate fls in terminal spikes, pistillate fls in axillary clusters
38. Sts prostrate to decumbent
39. Lf blade elliptic to oblanceolate; fr bracts 1–1.5 mm ... ***A. pacifica***
39′ Lf blade linear-lanceolate; fr bracts > 3 mm ... [2]***A. fruticulosa***
38′ Sts gen ascending to erect ... [2]***A. serenana***
40. Lf blades 10–20 mm; staminate infl spheric; fr bracts faintly 3-veined; s SCo ... var. ***davidsonii***
40′ Lf blades 15–40 mm; staminate infl spike, elongate; fr bracts ± 1-veined; widespread ... var. ***serenana***
37′ Staminate fls in axillary clusters (or staminate and pistillate fls mixed, gen in axillary clusters)
41. Fr bracts spongy, 5–12 mm, top-shaped to ± spheric — SnJV, s SCo ... [2]***A. lindleyi***
41′ Fr bracts not spongy, 2–6 mm (–8 mm in *A. argentea*), wedge-shaped to round, gen compressed (if spheric then densely tubercled)
42. Fr bracts round, ± dentate to cut ... ***A. elegans***
43. Fr bract margin 0.5–1 mm wide, dentate to cut ... var. ***elegans***
43′ Fr bract margin 0.3–0.5 mm wide, minutely crenate to finely toothed ... var. ***fasciculata***
42′ Fr bracts gen wedge-shaped to obovate or ± spheric, entire below middle, toothed above
44. Fr bract tip gen truncate, 4–5-toothed, margin entire — gen SNE, DMoj (incl edges) ... ***A. truncata***

44′ Fr bract tip acute to rounded in outline, margin toothed
45. Lf margin coarsely serrate .... ***A. suberecta***
45′ Lf margin entire, wavy or flat
46. Lf blade base gen hastate; fr bracts 4–8 mm, fused to near top .... ***A. argentea***
47. Sts decumbent, 15–30 cm; fr bracts sessile .... var. ***hillmanii***
47′ Sts erect, 30–80 cm; fr bracts short-stalked
48. Upper lvs short-petioled; lowest lvs alternate; GB .... var. ***argentea***
48′ Upper lvs sessile; lowest lvs opposite; GV, e SnFrB, e SCo, D .... var. ***mohavensis***
46′ Lf blade base tapered to ± obtuse; fr bracts 2.5–6 mm, fused ± to middle
49. Fr bracts 2.5–4 mm, ± 2–3 mm wide .... ***A. vallicola***
49′ Fr bracts 4.5–6 mm, 3–4 mm wide .... ***A. coronata***
50. Sts decumbent to ascending; fr bracts ± compressed, tubercles gen few, < marginal teeth var. ***coronata***
50′ Sts ± erect; fr bracts together ± spheric, tubercles dense, ± = marginal teeth .... var. ***notatior***

A. argentea Nutt. SILVERSCALE Ann 1.5–8 dm. **STS** decumbent to erect, densely branched, finely gray-scaly, peeling. **LF**: blade 7–40 mm, elliptic to deltate, gray-scaly, wavy-margined, base gen ± subhastate. **PISTILLATE INFL**: bracts in fr 4–8 mm, fused to near top, widely deltate to ± round, gen tubercled, margins green, toothed. **SEED** 1.5–2 mm, brown. Saline soils; < 1700 m. n SNH, GV, e SnFrB, e SCo, GB, D; to se OR, c U.S., TX, n Mex.

var. ***argentea*** (p. 507) **STS** 3–8 dm, erect; branches ascending. **LVS**: blade 7–20 mm, elliptic to deltate; upper short-petioled. Habitats of sp.; < 1500 m. GB; to c US, n Mex. ❀TRY.

var. ***hillmanii*** M.E. Jones **STS** 1.5–3 dm, decumbent; branches spreading. **LF**: blade 8–20 mm, widely lanceolate to ovate. Habitats of sp.; 1200–1700 m. n SNH, e&s MP; se OR, w NV. ❀TRY.

var. ***mohavensis*** M.E. Jones (p. 507) **STS** 3–8 dm, erect; branches ascending. **LVS**: blade 7–40 mm, lanceolate to ovate, often curled toward st; lowest opposite. Habitats of sp.; < 1000 m. GV, e SnFrB, e SCo, D; to TX, n Mex. [ssp. *expansa* (S. Watson) H.M. Hall & Clements]. ❀TRY.

A. californica Moq. Per < 3 dm, < 8 dm wide, monoecious; caudex thick, ± fleshy. **STS** ± many from base, spreading to decumbent; branches decumbent to ascending. **LVS**: blade 5–24 mm, lanceolate to elliptic, gray-scaly; lower opposite. **PISTILLATE INFL**: bracts in fr 3–4.5 mm, ± free, ovate to ± round, ± thin, smooth, entire. **SEED** ± 2 mm. Sandy soils, coastal dunes, shrubland, salt marshes; < 50 m. s NCo, CCo, SCo, ChI; Baja CA. ❀TRY.

A. canescens (Pursh) Nutt. FOURWING SALTBUSH, SHAD-SCALE Shrub < 20 dm, erect; branches many, spreading to ascending. **LF**: blade 8–50 mm, linear to oblanceolate, densely white-scaly. **PISTILLATE INFL** terminal; bracts in fr 4–25 mm, gen fused to near top, ovoid to spheric, hard, wings 4, 3–6 mm wide, wavy to deeply sharp-dentate. **SEED** 1.5–2.5 mm. $2n$=18,36. Clay to gravelly flats, slopes, shrubland; < 2400 m. SNH (e slope), Teh, SCoRI, SCo, n TR, PR, GB, D; to w Can, SD, n Mex. [var. *laciniata* Parish a form of both CA sspp.] Sspp. intergrade.

ssp. ***canescens*** (p. 507) **LF**: blade linear to oblanceolate. **PISTILLATE INFL**: bracts in fr 6–25 mm, stalked, wings gen 3–6 mm wide. Habitats and range of sp. [var. *macilenta* Jepson]. ❀DRN,DRY,SUN:1,**2**,**3**,**7–13**,14–17,**18–23**,24;CVS.

ssp. ***linearis*** (S. Watson) H.M. Hall & Clements **LF**: blade linear. **PISTILLATE INFL**: bracts in fr sessile, 4–8 mm, wings gen < 3.5 mm wide. Sandy soils, dunes, flats; < 100 m. n WTR?, DSon; n Mex. ❀TRY.

A. confertifolia (Torrey & Frémont) S. Watson (p. 507) Shrub < 10 dm, gen rounded, ± erect, much-branched; twigs gen spreading, stiff, becoming spine-like. **LF** short-petioled; blade 8–24 mm, elliptic to widely ovate, firm, densely gray-scaly. **PISTILLATE INFL** sometimes terminal; bracts in fr 5–12(20) mm, free, widely elliptic to ± round, smooth, entire to few-toothed. **SEED** 1.5–2 mm. $2n$= 18,36,54. Alkaline flats, gravelly slopes in shrubland, pinyon/juniper woodland; < 2400 m. SNH (e slope), GB, DMoj; to OR, n-c US, n Mex. May hybridize with *A. canescens*. ❀TRY.

A. cordulata Jepson (p. 507) HEARTSCALE Ann 1–5 dm. **STS** 1–few from base, erect, rigid; branches ascending to erect, gray-scaly, tips tomentose. **LVS**: blade 6–15 mm, ovate, gray-scaly, lower bases cordate, upper bases rounded. **PISTILLATE INFL**: bracts in fr 3.5–5 mm, fused to middle, ovate to ± round, smooth to barely tubercled, margin thin, soft, deeply toothed. **SEED** ± 1.5–2 mm, red-brown. RARE. Saline or alkaline soils; < 200 m. s ScV, SnJV.

A. coronata S. Watson Ann 1–3 dm. **STS** 1–few from base, decumbent to erect; branches ascending to erect, stiff, gen gray-scaly, becoming glabrous, straw-colored. **LVS**: blade 8–20 mm, elliptic to ovate; upper gray-scaly, base tapered to ± obtuse. **PISTILLATE INFL**: bracts in fr 4.5–6 mm, fused ± to middle, widely deltate to ± round, ± compressed to spheric, smooth to tubercled, margin greenish, toothed. **SEED** 1–1.5 mm, dark brown. Alkaline soils; < 500 m. s ScV, SnJV, e SCoRI, SCo.

var. ***coronata*** (p. 507) CROWNSCALE **STS** decumbent to ascending. **PISTILLATE INFL**: bracts in fr 4.5–6 mm, 3–3.5 mm wide, ± compressed, tubercles few, gen < marginal teeth. n=18. UNCOMMON. Fine, alkaline soils; < 200 m. s ScV, SnJV, e SCoRI.

var. ***notatior*** Jepson (p. 507) SAN JACINTO VALLEY CROWNSCALE **STS** ± erect. **PISTILLATE INFL**: bracts in fr 4.5–5 mm, 3–5 mm wide, ± spheric, tubercles dense, gen ± = marginal teeth. RARE. Alkaline flats; 400–500 m. e SCo (San Jacinto Valley, Riverside Co.).

A. coulteri (Moq.) D. Dietr. COULTER'S SALTBUSH Per < 5 dm, monoecious. **STS** 1–several from base, decumbent to erect; branches many, spreading to ascending. **LVS** sometimes opposite; blade 7–20 mm, narrowly elliptic to ovate, gray-scaly (esp below). **PISTILLATE INFL**: bracts in fr 2–3 mm, fused ± to middle, obovate, smooth or few-tubercled, sharply dentate above middle. **SEED** 1–1.5 mm. RARE. Alkaline or clay soils, open sites, coastal shrubland; < 50 m. SCo, ChI; Baja CA. Perhaps native only to s SCo.

A. depressa Jepson BRITTLESCALE Ann < 2 dm. **STS** prostrate to decumbent, glabrous to scaly, white, gen brittle. **LVS** sometimes opposite; blade 4–8 mm, ovate to cordate, entire, gen densely white-scaly, tip acute; lower gen sessile. **PISTILLATE INFL**: bracts in fr 2–3.5 mm, fused to near top, ovate or diamond-shaped, gen white-scaly, tubercled, entire to few-toothed. **SEED** ± 1–1.5 mm, reddish. RARE. Alkaline or clay soils; < 200 m. s ScV, SnJV. Perhaps best considered part of *A. parishii*.

A. elegans (Moq.) D. Dietr. WHEELSCALE Ann ± 1–5 dm. **STS** decumbent to ascending, ± branched, finely scaly, becoming glabrous. **LF**: blade 5–25 mm, elliptic to oblanceolate, densely white-scaly below, entire to dentate, tapered to base. **PISTILLATE INFL**: bracts in fr 2–3.5 mm, fused to near top, round, smooth or with 1 low tubercle, toothed. **SEED** 1–1.5 mm, brown. Saline or alkaline soils, dry lakes; < 1000 m. s SNE, D; to TX, n Mex. Vars. intergrade.

var. ***elegans*** (p. 507) **LF**: blade entire to dentate. **PISTILLATE INFL**: bract in fr deeply dentate, margin 0.5–1 mm wide. ± saline creosote-bush scrub; < 800 m. e D; to TX, n Mex.

var. ***fasciculata*** (S. Watson) M.E. Jones (p. 507) **LF**: blade gen entire. **PISTILLATE INFL**: bract in fr minutely toothed, margin 0.3–0.5 mm wide. n=9. Habitats of sp. s SNE, D; AZ, n Mex.

A. fruticulosa Jepson (p. 507) Per < 5 dm, monoecious. **ST** gen simple below; branches many, ± decumbent to erect, scaly, becoming glabrous. **LF**: blade 5–12(20) mm, narrowly lanceolate to ellictic, densely gray-scaly. **PISTILLATE INFL**: bracts in fr 3–5 mm, fused to middle or above, widely obovate to subspheric, ± hard, smooth to few-tubercled below middle, margin irregularly dentate to sharply tubercled. **SEED** ± 1.5 mm. Clay or alkaline soils, open sites, shrubland; < 700 m. Teh, s ScV, SnJV, SnFrB, SCoRI, w DMoj. ❀TRY.

A. gardneri (Moq.) D. Dietr. var. ***falcata*** (M.E. Jones) Welsh (p. 507) Subshrub or shrub 1–6 dm. **STS** several–many from base, gen unbranched, decumbent to erect, gray-scaly. **LF**: blade 15–45 mm, oblong to narrowly oblanceolate, gray-green, densely scaly. **PISTILLATE INFL**: bracts (sub)sessile, in fr 4–8 mm, fused to near top, fusiform to ovoid, smooth to few-tubercled, tip long-tapered, margin entire, dentate or tubercled. **SEED** 1.5–2 mm. Open, often alkaline soils, sagebrush scrub; 1200–1700 m. MP; to WA, MT, WY. Other 5+ vars. in e GB, Rocky Mtns. [*A. nuttallii* S. Watson var. *f.* M.E. Jones]. ❀TRY.

A. heterosperma Bunge Ann 7–15 dm. **ST** erect, stiff; branches gen ascending, striate, green, sparsely scaly. **LVS**: blade 10–65 mm, triangular, green, sparsely fine-scaly, base hastate; lower opposite; upper abruptly reduced, petioled. **INFL** panicle-like; branches spike-like, terminal. **PISTILLATE INFL**: bracts in fr either 2–2.5 mm or 5–7 mm, free, ± round, smooth, net-veined, entire. **SEEDS** of 2 forms (1–1.5 mm and black or 2–3 mm and yellow-brown). Open, gen disturbed places; < 1200 m. NCo, NCoRO, CaRF, GV (expected elsewhere); to e US; native to Eurasia.

A. hortensis L. GARDEN ORACLE Ann 5–18 dm. **ST** gen 1, erect; branches ascending to erect, striate, sparsely scaly. **LVS**: blade 15–120 mm, ovate to triangular, entire to finely toothed, green, glabrous to sparsely scaly; lower opposite; upper abruptly reduced, petioled. **PISTILLATE INFL** terminal; bracts in fr 0 or 2, free, 8–18 mm, widely elliptic to ± round, smooth, clearly net-veined, entire. **PISTILLATE FL**: calyx 3–5-lobed (if bracts 0) or 0 (if bracts 2). **SEEDS** of 2 forms (1–1.5 mm and black if bracts 2 or 2–4 mm and brown if bracts 0). *n*=9. Open, disturbed places; < 1300 m. NCo, SnFrB, SCo, MP (expected elsewhere); to B.C., e US; native to Eurasia.

A. hymenelytra (Torrey) S. Watson (p. 507) Shrub 4–10 dm, rounded, silver-scaly. **ST** simple below, erect; branches many, spreading to ascending. **LVS** petioled; blade 12–45 mm, widely ovate to round, thick, irregularly and sharply dentate. **PISTILLATE INFLS** sometimes terminal; bracts in fr short-stalked, 6–20 mm, free, round to ± reniform, entire to ± crenate. **SEED** ± 2 mm. Slopes, washes, shrubland; < 1500 m. se PR, SNE, D; to sw UT, AZ, n Mex. ❀DFCLT.

A. joaquiniana Nelson (p. 507) SAN JOAQUIN SALTBUSH Ann 1–10 dm. **ST** erect, gen striate; branches ascending, sparsely scaly, becoming glabrous. **LVS**: blades 10–70 mm, ovate to triangular, finely gray-scaly or green above, gen irregularly wavy-toothed, base truncate to tapered; upper abruptly reduced. **STAMINATE INFL** spike- or panicle-like, terminal, dense. **PISTILLATE INFL**: bracts in fr 3–4 mm, fused below middle, ± round-deltate, ribbed, entire. **SEED** ± 1–1.5 mm, dark brown. *n*=9. RARE. Alkaline soils; < 300 m. s ScV, SnJV, SCoRI (e slope). [*A. patula* L. ssp. *spicata* (S. Watson) H.M. Hall & Clements]

A. lentiformis (Torrey) S. Watson BIG SALTBUSH Shrub 8–30 dm, gen wider than tall. **ST** erect; branches many, spreading to ascending; twigs densely fine-scaly, becoming glabrous. **LVS**: blade 7–50 mm, ovate to deltate or hastate, densely gray-scaly. **PISTILLATE INFL** ± panicle-like, terminal; bracts in fr 2–6 mm, fused ± to middle, widely ovate to round, entire to minutely crenate. **SEED** ± 1.5 mm. Alkaline or saline washes, dry lakes, shrubland; < 1500 m. s SN, deltaic GV, SnJV, SCoRI, SCo, n WTR, SNE, D; to sw UT, n Mex.

ssp. ***lentiformis*** (p. 507) Monoecious or dioecious. **ST**: twigs smooth. **LF**: blade 15–50 mm. **PISTILLATE INFL**: bracts in fr 2.5–6 mm. *n*=9. Habitats and range of sp. Coastal and ChI pls with large lvs and frs have been called ssp. *breweri* (S. Watson) H.M. Hall & Clements. ❀SUN,DRN:**2,3,7,14-24**&IRR:**8-13**;STBL; CVS.

ssp. ***torreyi*** (S. Watson) H.M. Hall & Clements (p. 507) Gen dioecious. **ST**: twigs sharply angled, striate. **LF**: blade 7–30 mm, base sometimes hastate. **PISTILLATE INFL**: bracts in fr 2–4.5 mm. Alkaline clay soils, dry lakes, washes; 300–1500 m. SNE, n DMoj; to sw UT. [*A. torreyi* (S. Watson) S. Watson]. ❀TRY.

A. leucophylla (Moq.) D. Dietr. (p. 507) Per < 3 dm, densely white-scaly, gen monoecious. **STS** 3–10 dm, prostrate to decumbent; branches decumbent to erect. **LF**: blade 9–40 mm, elliptic to widely ovate. **PISTILLATE INFL**: bracts in fr 4–6 mm, fused ± to middle, ± subspheric, ± spongy, low-tubercled, entire to dentate above middle. **SEED** 2–3 mm. Sandy soils, dunes; < 50 m. NCo, CCo, SnFrB, SCo, ChI; Baja CA. ❀TRY.

A. lindleyi DC. Ann or short-lived per 1–4 dm. **STS** spreading to decumbent, brittle, sparsely white-scaly, becoming glabrous. **LVS**: blade 8–30 mm, oblanceolate to diamond-shaped, green to finely white-scaly, entire to coarsely wavy-dentate, base narrowed; upper short-petioled. **PISTILLATE INFL**: bracts in fr 5–12 mm, fused to near top, widely top-shaped, spongy. **SEEDS** of 2 forms (± 1 mm diam and black or 1.5 mm diam and red-brown). Open, disturbed places, fields; < 200 m. SnJV, s SCo (expected elsewhere); native to Australia.

A. minuscula Standley LESSER SALTBUSH Ann < 4 dm. **STS** many from base, ascending to erect; branches spreading, brittle, reddish, peeling. **LVS** opposite; blade 1–4 mm, ovate to cordate, white-scaly below, green above, base oblique; lower sessile. **PISTILLATE INFL**: bracts in fr ± 2–3 mm, fused to near top, ovate to diamond-shaped, densely tubercled, entire. **SEED** 0.8–0.9 mm, reddish. RARE. Sandy, alkaline soils; < 200 m. s SnJV. Related to *A. depressa* or *A. parishii*.

A. nummularia Lindley Shrub 15–30+ dm, gray-scaly, monoecious or dioecious. **STS** 1–few from base; branches many, ascending to erect. **LVS**: blade 15–65 mm, widely ovate to ± deltate, thick, ± wavy to irregularly toothed; upper petioled. **PISTILLATE INFL** terminal; bracts in fr 4.5–9(12) mm, fused ± to middle, round to ± reniform, entire to irregularly toothed. **SEED** ± 2 mm. Sandy soils, open, disturbed places, coastal bluffs; < 100 m. SCo; native to Australia.

A. pacifica Nelson (p. 509) SOUTH COAST SALTBUSH Ann gen 1–3 dm, mat-like. **STS** prostrate to decumbent; branches ascending, scaly, becoming ± glabrous. **LF**: blade 4–18 mm, elliptic to oblanceolate, gray- to white-scaly below, greenish above, base tapered. **STAMINATE INFL**: spike, terminal. **PISTILLATE INFL**: bracts in fr 1–1.5 mm, fused ± to middle, obovate to ± round, smooth or few-tubercled, entire, tip ± truncate, minutely 3–5-toothed. **SEED** ± 0.8 mm, brown. *n*=9. RARE. Bluffs, shrubland; < 100 m. SCo, ChI; Baja CA.

A. parishii S. Watson (p. 509) PARISH'S BRITTLESCALE Ann < 2 dm. **STS** prostrate to decumbent, gen flexible, white, scaly to densely woolly near tips. **LVS** gen opposite; blade 4–8 mm, ovate to cordate, gen densely white-scaly, tip acute; lower gen sessile. **PISTILLATE INFL**: bracts in fr 2.5–3 mm, fused to near top, ovate or diamond-shaped, ± densely tubercled, entire to few-toothed. **SEED** ± 1–1.5 mm, reddish. **PRESUMED EXTINCT**. Alkaline or clay soils; < 1900 m. SW (exc ChI), w DMoj; Baja CA.

A. parryi S. Watson (p. 509) Shrub 2–5 dm, rounded, gen dioecious. **ST** erect; branches many, spreading to erect; twigs slender, stiff, becoming spine-like, densely scaly. **LF**: blade 5–20 mm, elliptic to widely ovate, densely white-scaly. **PISTILLATE INFL**: bracts in fr 2.5–4 mm, fused ± at middle or above, round to reniform, smooth, margin entire to low-tubercled above middle. **SEED** 1–1.5 mm. Alkaline soils, flats, dry lakes; < 1500 m. s SNE, DMoj, n DSon; NV. ❀TRY.

A. patula L. SPEAR ORACLE Ann 4–15 dm. **ST** 1+ from base, erect; branches ascending to spreading, stiff, glabrous to sparsely fine-scaly. **LF**: blade 15–70 mm, linear to lanceolate, green, glabrous to sparsely scaly, entire to minutely toothed, base tapered or rounded. **STAMINATE INFL**: spike or panicle, gen terminal. **PISTILLATE INFL**: bracts in fr 3–12 mm, 3–7 mm wide, fused at base, ovate to deltate, smooth to tubercled, entire to minutely dentate. **SEEDS** of 2 forms (± 1–2 mm and black or ± 2.5–4 mm and brown). Saline soils; < 1400 m. NCo, CaRF, CCo, SnFrB; to AK, e N.Am, Eurasia. Vars. intergrade.

var. ***obtusa*** (Cham.) C. Hitchc. (p. 509) **LVS**: lower with 2–4 ascending lobes, tip obtuse. **PISTILLATE INFL**: bracts in fr 6–12 mm, smooth or tubercled, margin wide, entire. Habitats of sp. NCo, CCo, SnFrB; to AK, e N.Am, Eurasia.

var. ***patula*** (p. 509) **LVS**: lower with base tapered to ± rounded, tip acute. **PISTILLATE INFL**: bracts in fr 3–8 mm, smooth, margin narrow, minutely dentate. Habitats and range (exc n Can, AK) of sp.

A. phyllostegia (Torrey) S. Watson (p. 509) ARROWSCALE Ann 1–4 dm. **STS** much-branched from base, faintly striate, green, glabrous to sparsely fine-scaly. **LF**: blades 10–40 mm, lanceolate to deltate, fleshy, brittle, base tapered to hastate. **PISTILLATE INFL**: bracts in fr 5–20 mm, fused at base, lanceolate to triangular, smooth or tubercled, base often ± hastate. **SEED** ± 1 mm, brown. *n*=9. Saline soils, meadows, flats; < 1500 m. SnJV, SNE, DMoj; to OR, UT.

A. polycarpa (Torrey) S. Watson Shrub 5–20 dm, densely gray-scaly. **ST** erect; branches many, spreading to ascending; twigs slender, becoming ± spine-like. **LVS** gen subsessile; blade 3–25 mm, oblong to narrowly oblanceolate, ± thick. **PISTILLATE INFL** sometimes terminal; bracts in fr 2–3 mm, fused ± to middle, ± subspheric, smooth to tubercled, minutely crenate to toothed. **SEED** 1–1.5 mm. Alkaline flats, dry lakes; < 1500 m. SnJV and margins, n TR, e PR, s SNE, D; to UT, n Mex. ❀STBL;may be INV.

A. pusilla (Torrey) S. Watson (p. 509) Ann < 3 dm. **STS** 1–several from base, rigid, not brittle, gen scaly; tips reddish. **LVS**: blade 3–15 mm, elliptic to ovate, thick, ± fleshy, ± rounded, gen sparsely scaly; lower sessile. **STAMINATE INFL** terminal; fls 1–2. **PISTILLATE INFL**: fls 1–2; bracts in fr 1–2 mm, fused to near top, ovate, smooth, entire, tip acute. **SEED** 0.8 mm, brown. Alkaline soils, hot springs; 1300–2000 m. MP, n SNE; OR, NV.

A. rosea L. TUMBLING ORACLE Ann 4–15 dm. **ST** erect; branches ascending, gen ± glabrous. **LVS** firm, persistent; blades 10–60 mm, 4–30 mm wide, densely fine-scaly below, greenish above, becoming red, coarsely wavy-toothed, base tapered. **PISTILLATE INFL**: bracts in fr 4–8 mm, fused ± to middle, diamond-shaped to deltate, hard, tubercled, dentate. **SEED** 2–2.5 mm, brown. *n*=9. Common. Open, disturbed places, fields; < 2400 m. CA-FP, D (uncommon); to e N.Am; native to Eurasia.

A. semibaccata R.Br. AUSTRALIAN SALTBUSH Per or subshrub < 3.5 dm, monoecious. **STS** ± several, 3–10 dm, ± spreading; branches spreading to ascending, ± white-scaly or becoming glabrous. **LVS**: blade 8–30 mm, oblong to narrowly elliptic, entire to wavy-toothed, ± scaly (esp below); upper short-petioled. **PISTILLATE INFL**: fls few; bracts in fr 3–6 mm, fused to middle or above, ovate to ± diamond-shaped, fleshy, reddish, net-veined, (sub)entire. **SEED** 1.5–2 mm. 2*n*=18. Waste places, shrubland, woodland; < 1000 m. ± CA-FP (exc CaR, n&c SN), D; to UT, TX, n Mex; native to Australia.

A. serenana Nelson BRACTSCALE Ann 3–10 dm, mat-like. **STS** decumbent to ascending, sparsely scaly; tips flexible. **LVS** subsessile; blade 10–40 mm, elliptic to lanceolate, ± greenish, sparsely fine-scaly above, dentate. **STAMINATE INFL**: spikes 1–many in panicles (or clusters spheric, terminal). **PISTILLATE INFL**: bracts in fr 2–3.5 mm, fused to middle, ± round to wedge-shaped, smooth or tubercled, toothed above middle. Alkaline flats, coastal bluffs; < 2000 m. s SN, SnJV, SCoRO, SW, SNE (naturalized), w DMoj, DSon; Baja CA.

var. ***davidsonii*** (Standley) Munz DAVIDSON'S SALTBUSH **LF**: blade 10–20 mm. **STAMINATE INFL**: clusters terminal, spheric. **PISTILLATE INFL**: fr bract faintly 3-veined. RARE. Bluffs; < 200 m. s SCo.

var. ***serenana*** (p. 509) **LF**: blade 15–40 mm. **STAMINATE INFL**: panicle, elongate. **PISTILLATE INFL**: fr bract ± 1-veined. *n*=9. ± habitats and range of sp.

A. spinifera J.F. Macbr. (p. 509) Shrub 3–20 dm, densely white- to gray-scaly. **ST** erect; branches many, ascending to erect; twigs spreading, ± stiff, becoming spine-like. **LVS**: blade 6–28 mm, elliptic to widely ovate, entire to 2-toothed below middle; upper short-petioled. **PISTILLATE INFL**: bracts in fr 5–20 mm, 3–11 mm wide, fused ± to middle, smooth, below middle spheric and densely scaly, above middle toothed and sparsely scaly. **SEED** 2–3 mm. Saline soils, flats, dry lakes; < 800 m. s SnJV, SCoRI, n WTR, DMoj.

A. suberecta I. Verd. Ann 2–6 dm, sprawling. **STS** decumbent to ascending, densely scaly below. **LVS**: blades 12–30 mm, ovate to ± diamond-shaped, ± fine-scaly below, ± glabrous above, coarsely serrate, base ± obtuse; upper short-petioled. **PISTILLATE INFL**: bracts in fr 2–5 mm, fused to middle, diamond-shaped, entire below middle, widest and 2–4-toothed above middle. *n*=9. Disturbed places, fields; < 300 m. SnJV, n SCo (Ventura Co.), DSon; native to Australia.

A. subspicata (Nutt.) Rydb. Ann 5–15 dm. **ST** branched at base, gen erect; branches ascending, reddish to straw-colored at base, green-striate above. **LVS**: blade 20–80(190) mm, lanceolate, green to reddish, sparsely fine-scaly, becoming glabrous, entire to toothed, base hastate; upper petioled. **PISTILLATE INFL**: bracts in fr 3–7 mm, fused at base, ovate to widely triangular, surfaces with 2 spongy tubercles, base truncate to obtuse, margins few-toothed. **SEEDS** of 2 forms (2–3 mm wide and brown, 1.5–2 mm wide and black). Moist, saline or alkaline soils; < 200 m. s ScV, SnJV, CCo, SnFrB; to e N.Am.

A. triangularis Willd. (p. 509) SPEARSCALE Ann 3–9 dm. **STS** branched from base, ascending to ± erect, finely white-scaly, becoming glabrous. **LVS**: blades 10–70 mm, lanceolate to triangular-hastate, green, glabrous to sparsely fine-scaly, entire to wavy-dentate, base truncate to rounded; upper petioled. **INFL** gen panicle-like; branches spike-like, dense. **PISTILLATE INFL**: bracts in fr 3–7 mm, fused at base, ovate to deltate, base truncate, short-tubercled, finely dentate. **SEED** 1.5–2.5 mm, black. Wet places, marshes; < 300 m. NCo, s NCoRO, GV, CCo, SCo, ChI; to e N.Am, Eurasia. [*A. patula* L. ssp. *patula* misapplied]

A. truncata (Torrey) A. Gray WEDGESCALE Ann < 7 dm. **STS** 1–few, erect; branches ascending, gray-scaly. **LVS**: blade 10–40 mm, ovate to deltate, glabrous to sparsely gray-scaly, base truncate to ± lobed. **PISTILLATE INFL**: bracts in fr 2–4 mm, fused to near top, widely wedge-shaped, smooth to barely tubercled, entire, tip ± truncate to notched, 4–5-toothed. **SEED** 1–1.5 mm, brown. *n*=9. Alkaline soils, flats; 600–2500 m. SNH (e slope), TR (desert slopes), SNE, DMoj; to B.C., Colorado, NM.

A. tularensis Cov. (p. 509) BAKERSFIELD SMALLSCALE Ann 1–8 dm. **ST** 0–few-branched, erect; branches rigid, brittle, white-scaly, uppermost overlapping, tips reddish. **LVS**: blade 6–29 mm, lanceolate to ovate, base rounded; lower opposite, sessile. **PISTILLATE INFL**: bracts in fr 2.5–3.5 mm, fused to middle or above, ovate to diamond-shaped, tip acute, margins thin, toothed. **SEED** ± 1 mm, red-brown. **ENDANGERED** CA. Alkaline soils, lake shore; < 200 m. s SnJV (Kern Lake bed, sw Kern Co.).

A. vallicola Hoover (p. 509) LOST HILLS CROWNSCALE Ann < 2 dm. **STS** 1–few from base, gen erect; branches ascending to erect, stiff, gray-scaly, becoming glabrous. **LVS**: blades 8–11 mm, elliptic to ovate, green to gray-scaly, entire, base tapered to obtuse. **PISTILLATE INFL**: bracts in fr 2.5–4 mm, widely deltate, fused to ± middle, flat to spheric, smooth to tubercled, toothed, greenish. **SEED** 1–1.5 mm, dark brown. Dried ponds, alkaline soils; < 200 m. SnJV. Intergrades with *A. coronata*, perhaps best considered a ssp. of it. Pls from sw SnJV (Carrizo Plain) are undescribed: green-yellow-scaly; lvs petioled, blades ovate; fr bracts ± 3–4 mm, smooth.

A. watsonii Nelson (p. 509) Per < 10 dm, 5–25 dm diam, often mat-like, densely white-scaly. **STS** several–many, prostrate to decumbent; twigs slender. **LVS** opposite; blade 8–25 mm, widely elliptic to ovate, thick to ± fleshy. **PISTILLATE INFL**: bracts in fr 4–8 mm, fused to middle or above, ovate to diamond-shaped, thick, (sub)entire. **SEED** ± 1 mm. Sand dunes, salt marshes, shrubland; < 50 m. s CCo, SCo, ChI; Baja CA. ❀TRY.

BASSIA

Ann, gen hairy. **ST**: axis gen erect; branches ascending to erect. **LVS** linear to lanceolate, reduced upward. **INFL**: spike; bracts lf-like; fls 1–few per axil. **FLS** gen bisexual; calyx lobes 5, incurved, hooked-spiny in fr; stamens gen 5; stigmas gen 2. **FR** ± depressed-spheric. **SEED** horizontal. ± 5 spp.: warm temp Eurasia. (Ferdinando Bassi, Italian botanist, 1710–1774) Perhaps best incl in *Kochia*.

B. hyssopifolia (Pallas) Kuntze (p. 507) Pl < 1 m. **LVS**: lower 5–60 mm, 1–3.5 mm wide, flat, often withered in fr. **INFL** 5–50 mm; bracts 2–5 mm, ± oblong. **FL**: calyx densely tan-woolly, base in fr leathery, spines ± 1 mm. **FR** 1–1.5 mm diam. **SEED** dark brown. Disturbed sites, fields, roadsides; < 1200 m. CA (exc NW, SNH); widespread N.Am; native to Eurasia. Sometimes confused with *Kochia scoparia*.

BETA BEET

Ann or per, gen glabrous. **ST** decumbent to erect, simple to branched. **LVS** basal and cauline, petioled, ovate to diamond-shaped. **INFL**: spike, axillary or in terminal, panicle-like clusters, gen bracted. **FL** bisexual; sepals 5, persistent and thickened in fr; stamens 5; ovary half-inferior, sunken in receptacle, stigmas 2–3. **FRS** achene-like, ± circumscissile, hard, adherent in clusters. **SEED** subspheric, dark brown. ± 5 spp.: Medit, w Asia. (Greek: probably from Celtic name for red root)

B. vulgaris L. Pl < 1 m, from thick taproot. **LF** 5–14 cm; petiole ± = blade; blade base tapered to truncate. **INFL** 11–45 cm; bracts (0)2–8 mm. **FL**: sepals 2–2.5 mm, incurved, back keeled in fr, margin scarious; stigmas 2. **FRS** 3–5 mm diam, 5–11 per cluster. $2n=18$. Disturbed places, fields, gen as waif; < 200 m. CCo, SnFrB, SCo, ChI; to e US; native to s Eur. Cult for food, source of sugar.

CHENOPODIUM PIGWEED, GOOSEFOOT

Ann or per, glabrous, glandular, or powdery. **ST**: branches 0 to gen ± spreading. **LVS** gen petioled, linear to deltate, entire to lobed, reduced upward; base gen tapered. **INFL**: spheric clusters, spikes, or panicle-like, gen dense; bracts gen 0; fls gen sessile. **FL**: calyx segments gen 5, fused or not, persistent, flat to keeled; stamens gen 5; ovary lenticular to spheric, stigmas 2–5. **SEED** vertical or horizontal, red-brown to black; wall very thin. ± 150 spp.: temp; some cult for food or grain. (Greek: goose foot, from lf shape of some) [Wahl 1954 Bartonia 27:1–46; Crawford 1975 Brittonia 27:279–288] Fr gen required for identification.

1. Herbage ± glandular, sometimes sticky or strong-smelling
 2. Pl ± sprawling; calyx minutely 3–5-toothed, obovoid, netted, tube enclosing fr .... ***C. multifidum***
 2′ Pl gen erect; calyx deeply 5-lobed, ± spheric, not netted, lobes gen folded over fr
  3. Infl ± spheric, 3–6 mm diam, axillary; sepals yellow-glandular; seed gen vertical; longest lvs < 30 mm ***C. pumilio***
  3′ Infl spike or panicle-like, 10+ mm, sometimes terminal; sepals glabrous to short-stalked-glandular; seed gen horizontal; longest lvs 30–120 mm
   4. Infl branches straight; fls sessile; sepals glabrous to sparsely sessile-glandular; fr wall free from seed .... ***C. ambrosioides***
   4′ Infl branches curved; fls short-pedicelled; sepals densely short-stalked-glandular; fr wall adherent to seed .... ***C. botrys***
1′ Herbage glabrous to minutely scaly or powdery, gen not strong-smelling
 5. Per from stout, fleshy caudex; calyx tube gen > lobes in fr .... ***C. californicum***
 5′ Ann; calyx tube gen < lobes
  6. Most seeds vertical; calyx lobes gen 3(–4) — horizontal seeds few, with 4–5 calyx lobes
   7. Upper infls axillary, ± spheric, < 10 mm diam
    8. Lower lf blades lanceolate to deltate, gen glabrous below; sepals thickened or ± fleshy in fr ...... ***C. foliosum***
    8′ Lower lf blades oblong to ovate, densely powdery below; sepals thin or membranous in fr — vertical and horizontal seeds gen ± equal in number .... [2]***C. glaucum***
   7′ Upper infls axillary or terminal, spike-like, 15+ mm
    9. Vertical seeds enclosed by calyx tube; style lobes ascending
     10. Young lvs glabrous or sparsely powdery; calyx lobes abruptly pointed .... ***C. chenopodioides***
     10′ Young lvs densely white-powdery; calyx lobes rounded to acute .... ***C. macrospermum*** var. ***halophilum***
    9′ Vertical seeds subtended by calyx tube; style lobes spreading
     11. Lf densely powdery below; vertical & horizontal seeds often ± equal in number .... [2]***C. glaucum***
     11′ Lf gen glabrous below; vertical seeds many more than horizontal .... ***C. rubrum***
  6′ Most or all seeds horizontal; calyx lobes gen 5
   12. Lf bright green, deeply 3–5-lobed, base truncate to cordate, tip acuminate to long-tapered; fr 1.5–3 mm diam .... ***C. simplex***
   12′ Lf dull green to grayish, entire to toothed, base tapered to acute, tip rounded to acute; fr gen < 1.5 mm diam
    13. Lower lf blades 1-veined from base
     14. Lower lf blades > 4 mm wide; calyx lobes strongly keeled in fr; fr wall free from seed, easily detached with dissecting needle .... [2]***C. desiccatum***
     14′ Lower lf blades < 3.5 mm wide; calyx lobes weakly keeled in fr; fr wall adherent to seed, firm or fracturing when teased with dissecting needle .... ***C. leptophyllum***
    13′ Lower lf blades 3-veined from base
     15. Fr wall free from seed, easily detached with dissecting needle
      16. Lf length < 1.5 × width

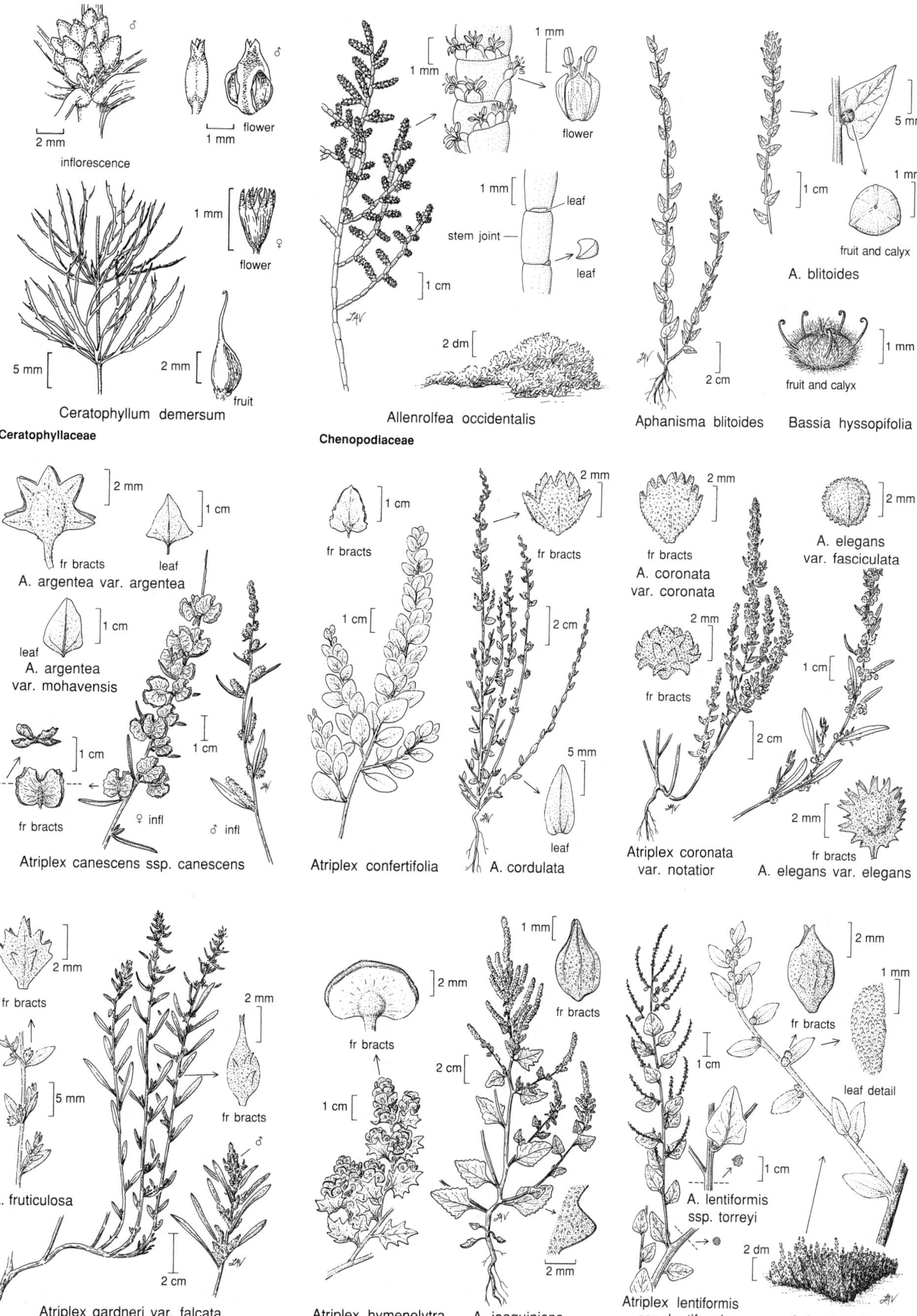
♂
inflorescence
2 mm
♂
flower
1 mm
1 mm
flower
♀
5 mm
2 mm
fruit
Ceratophyllum demersum
Ceratophyllaceae
1 mm
1 mm
flower
1 mm
leaf
stem joint
leaf
1 cm
2 dm
Allenrolfea occidentalis
Chenopodiaceae
2 cm
Aphanisma blitoides
5 mm
1 cm
1 mm
fruit and calyx
A. blitoides
1 mm
fruit and calyx
Bassia hyssopifolia
2 mm
fr bracts
1 cm
leaf
A. argentea var. argentea
leaf
1 cm
A. argentea
var. mohavensis
1 cm
1 cm
fr bracts
♀ infl
♂ infl
Atriplex canescens ssp. canescens
1 cm
fr bracts
1 cm
Atriplex confertifolia
2 mm
fr bracts
2 cm
5 mm
leaf
A. cordulata
2 mm
fr bracts
A. coronata
var. coronata
2 mm
fr bracts
2 cm
Atriplex coronata
var. notatior
2 mm
A. elegans
var. fasciculata
1 cm
2 mm
fr bracts
A. elegans var. elegans
2 mm
fr bracts
5 mm
A. fruticulosa
2 mm
fr bracts
♂
2 cm
Atriplex gardneri var. falcata
2 mm
fr bracts
1 cm
Atriplex hymenelytra
1 mm
fr bracts
2 cm
2 mm
A. joaquiniana
1 cm
1 cm
A. lentiformis
ssp. torreyi
Atriplex lentiformis
ssp. lentiformis
2 mm
1 mm
fr bracts
leaf detail
2 dm
A. leucophylla

17. Fr top visible between calyx lobes; fr 1–1.5 mm diam; lf blade thin .... ***C. fremontii***
17′ Fr top hidden by calyx lobes; fr ± 1 mm diam; lf blade thick, becoming ± leathery .... ***C. incanum***
16′ Lf length gen 1.5+ × width
18. Sts several from base, branches ± dense; fr ± 1 mm diam, top hidden by calyx lobes ..... [2]***C. desiccatum***
18′ Sts 1–few from base, branches well spaced; fr 1–1.5 mm diam, top visible between calyx lobes
19. Lower lvs entire, length 1.5–3 × width .... ***C. atrovirens***
19′ Lower lvs ± 1–2-lobed (-toothed), length 3–5 × width .... ***C. pratericola***
15′ Fr wall adherent to seed, firm or fracturing when teased with dissecting needle
20. Sts mostly prostrate to ascending
21. Lf blade oblong to lanceolate, length gen 2+ × width, base tapered; petiole < blade .... ***C. carnosulum*** var. ***patagonicum***
21′ Lf blade widely elliptic to ovate, length < 1.5 × width, base obtuse to rounded; petiole ± = blade .... ***C. vulvaria***
20′ Sts erect
22. Lower lf blade gen entire or 1–2-toothed at base
23. Pl gen much-branched, ± wide; infl often rounded in fr; fr < 1 mm diam .... ***C. nevadense***
23′ Pl gen few-branched, slender; infl mostly spikes, ascending to erect; fr 1–1.5 mm diam
24. Lower lvs narrowly lanceolate to elliptic; fr ± 1 mm diam .... ***C. hians***
24′ Lower lvs ovate to ± deltate; fr 1–1.5 mm diam .... ***C. incognitum***
22′ Lower lf blade few–many-toothed above base
25. Fr wall surface minutely honeycombed at 20×; calyx lobes widely keeled in fr .... ***C. berlandieri***
25′ Fr wall surface smooth or irregularly rough at 20×; calyx lobes smooth to narrowly keeled
26. Lf gen shiny dark green above, glabrous to sparsely powdery below; seed sharply angled at equator .... ***C. murale***
26′ Lf dull green to grayish above, powdery below; seed obtusely angled or rounded at equator
27. Fr top gen visible between calyx lobes; terminal infl branches often stiffly ascending or erect; lower st branches often decumbent .... ***C. strictum*** var. ***glaucophyllum***
27′ Fr top gen hidden by calyx lobes; terminal infl branches weak, sometimes pendent; lower st branches 0 or gen ascending
28. Fr 1–1.5 mm diam; lf blade length < 1.5 × width; infl branches gen straight .... ***C. album***
28′ Fr ± 1 mm diam; lf blade length gen 1.5–2 × width; infl branches gen curved or pendent .... ***C. missouriense***

C. album L. (p. 509) PIGWEED, LAMB'S QUARTERS Ann 18–100+ cm. **LF**: blade 15–70 mm, lanceolate to ± deltate, entire to irregularly wavy-toothed, dull green above, powdery below. **FL**: sepals gen enclosing fr, gen keeled, powdery. **FR** 1–1.5 mm diam; wall ± rough at 20×, adherent to seed. **SEED** horizontal. 2*n*=54. Common. Disturbed places, fields, roadsides; < 1800 m. CA; ± temp worldwide; probably native to Eur. Often confused with *C. berlandieri*, *C. missouriense*, and *C. strictum*. Pls with lf blade as wide as long and calyx lobes strongly keeled have been called *C. opulifolium* Schrader.

C. ambrosioides L. MEXICAN TEA Ann to per 25–130 cm, strong-smelling. **LF**: blade 15–100 mm, gen densely glandular, lanceolate to ovate below, serrate to pinnately lobed, oblong and entire above. **INFL**: branches straight. **FL**: sepals enclosing fr, smooth, glabrous to sparsely glandular. **FR** ± 0.5 mm diam; wall free from seed. **SEED** gen horizontal. 2*n*=16,32,64. Disturbed places; < 1400 m. CA-FP; to Can, e US; native to trop Am. [vars. *anthelminticum* (L.) A. Gray and *vagans* (Standley) J. Howell] Per pls have been called var. *suffruticosum* (Willd.) Aellen.

C. atrovirens Rydb. (p. 509) Ann 7–60 cm. **LF**: blade 9–35 mm, oblong to widely ovate, entire or 1–2-toothed to -lobed at base, ± glabrous to sparsely powdery. **FL**: sepals not enclosing fr, barely keeled, sparsely powdery. **FR** 1–1.5 mm diam; wall free from seed. **SEED** horizontal. 2*n*=18. Open places, shrubland, woodland, coniferous forest; 300–3500 m. NW, CaR, SN, SnBr, PR, GB; to WY, c US, NM. Much like *C. pratericola*.

C. berlandieri Moq. (p. 509) PITSEED GOOSEFOOT Ann 18–85+ cm. **LF**: blade 15–50 mm, lanceolate to ± deltate, gen irregularly wavy or toothed, powdery. **FL**: sepals enclosing fr, back strongly keeled, powdery. **FR** 1–1.5 mm diam; wall surface honeycombed at 20×, adherent to seed. **SEED** horizontal. 2*n*=36. Common. Open, often disturbed places; < 2700 m. CA; to e N.Am, n Mex. [vars. *sinuatum* (Murray) Murray and *zschackei* (Murray) Murray] Often confused with *C. album*.

C. botrys L. (p. 509) JERUSALEM OAK Ann 14–65 cm, strong-smelling. **LF**: blade 3–65 mm, short-stalked-glandular, ovate to elliptic and wavy to pinnately lobed below, oblong and gen entire above. **INFL**: branches arched or curved; fls short-pedicelled. **FL**: sepals weakly enclosing fr, ± flat, densely short-stalked-glandular. **FR** ± 0.5 mm diam; wall adherent to seed. **SEED** horizontal. 2*n*=18. Disturbed places; < 2100 m. Most of CA; to Can, e US; native to Eur.

C. californicum (S. Watson) S. Watson (p. 509) Per 20–90+ cm; caudex stout, fleshy. **STS** several from base, decumbent to ascending. **LF**: blade 15–90+ mm, ± deltate, coarsely dentate to wavy-toothed, green to glaucous, base truncate to hastate, tip acute. **INFL**: axillary, ± spheric, < 10 mm diam; terminal spikes, 8–20 cm, interrupted. **FL**: calyx tube enclosing fr, lobes ± erect, flat, ± glabrous. **FR** 1.5–2 mm diam; wall adherent to seed. **SEED** vertical. Gen open sites, sandy to clay soils; < 2000 m. s NCo, NCoRO, c&s SNF, Teh, GV, CW, SW, s SNE, w DMoj; Baja CA.

C. carnosulum Moq. var. *patagonicum* (Philippi) Wahl Ann 5–20 cm. **STS** several from base, prostrate to ascending. **LF**: blade 12–20 mm, oblong to lanceolate, entire, gen sparsely powdery below, ± glabrous above. **INFL**: spike, axillary, 5–15 mm. **FL**: sepals ± enclosing fr, ± flat, powdery. **FR** ± 1 mm diam; wall gen adherent to seed. **SEED** horizontal. Gen sandy soils, dunes; < 200 m. s CCo, n SCo; native to S.Am. Other vars. in Mex, S.Am.

C. chenopodioides (L.) Aellen (p. 509) Ann 10–45 cm. **LF**: blade 8–32 mm, ovate to deltate, entire to irregularly few-toothed, ± glabrous. **FL**: calyx tube enclosing fr, > lobes, ± glabrous. **FR** 0.5–1 mm diam; wall free from seed. **SEEDS** mostly vertical; calyx of vertically seeded fls 3-lobed; calyx of horizontally seeded fls 4–5-lobed. Saline soils, drying ponds, mudflats; < 2500 m. NCoRO, SN, SnBr, PR, MP, expected elsewhere; to WA; native to S.Am.

C. desiccatum Nelson (p. 509) Ann 12–35 cm. **STS** gen several from base. **LF**: blade 5–25 mm, oblong to elliptic, ± fleshy, margin entire, lower surface densely powdery, ± glabrous above. **FL**: sepals enclosing fr, keeled, powdery. **FR** ± 1 mm diam; wall free from seed. **SEED** horizontal. 2*n*=18. Uncommon. Open places, shrubland, coniferous forest; < 2900 m. SN, SnJV, GB, n DMoj; to WY, Colorado.

C. foliosum (Moench) Asch. (p. 509) Ann 5–60 cm. **LVS**: blades 7–38 mm, entire to irregularly toothed, ± glabrous; lower lanceolate to deltate, base 2-lobed to hastate; upper oblong to lanceolate, base

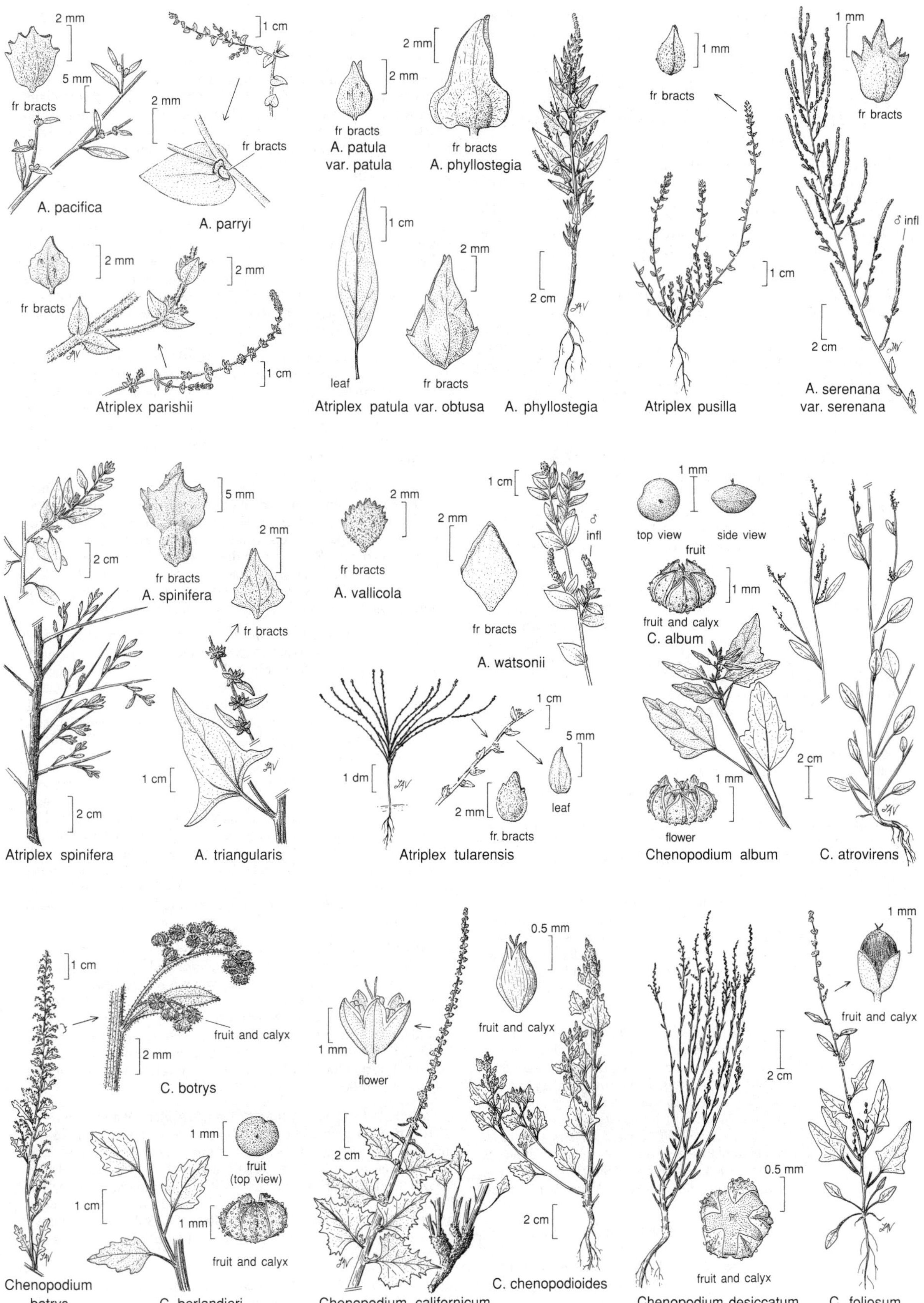
2 mm
5 mm
fr bracts
A. pacifica
1 cm
2 mm
fr bracts
A. parryi
2 mm
2 mm
fr bracts
1 cm
Atriplex parishii
2 mm
2 mm
fr bracts
A. patula
var. patula
fr bracts
A. phyllostegia
1 cm
2 mm
leaf
fr bracts
Atriplex patula var. obtusa
2 cm
A. phyllostegia
1 mm
fr bracts
1 cm
Atriplex pusilla
1 mm
fr bracts
♂ infl
2 cm
A. serenana
var. serenana
5 mm
2 cm
fr bracts
A. spinifera
2 mm
fr bracts
1 cm
2 cm
Atriplex spinifera
A. triangularis
2 mm
fr bracts
A. vallicola
1 cm
2 mm
♂
infl
fr bracts
A. watsonii
1 cm
5 mm
1 dm
leaf
2 mm
fr. bracts
Atriplex tularensis
1 mm
top view
side view
fruit
1 mm
fruit and calyx
C. album
2 cm
1 mm
flower
Chenopodium album
C. atrovirens
1 cm
2 mm
fruit and calyx
C. botrys
Chenopodium
botrys
1 mm
fruit
(top view)
1 cm
1 mm
fruit and calyx
C. berlandieri
1 mm
flower
2 cm
Chenopodium californicum
0.5 mm
fruit and calyx
2 cm
C. chenopodioides
2 cm
0.5 mm
fruit and calyx
Chenopodium desiccatum
1 mm
fruit and calyx
C. foliosum

tapered. **INFL** 3–5 mm diam, axillary, ± spheric, leafy-bracted. **FL**: sepals gen 3, ± enclosing fr, smooth, gen glabrous, becoming ± reddish; stamens 3–4. **FR** 1–1.5 mm diam; wall ± adherent to seed. **SEED** gen vertical. 2*n*=16,18. Open, gravelly or sandy soils; < 3700 m. NCoR, CaRF, c SNH, TR, GB, w DMoj; to w Can, c US; native to Eur. [*C. overi* Aellen] Pls with infl clusters 10+ mm diam, reddish, fleshy, leafy bracts 0, have been called *C. capitatum* (L.) Asch., reported from nw CA.

C. fremontii S. Watson (p. 513) Ann 22–55 cm. **LF**: blade 8–40 mm, widely ovate to deltate, gen 1–3-lobed at base, sometimes entire, ± powdery below, glabrous above. **FL**: sepals not enclosing fr, keeled, ± powdery. **FR** 1–1.5 mm diam; wall free from seed. **SEED** horizontal. 2*n*=18. Gen shaded places, shrubland, coniferous forest; 700–3100 m. SN, TR, PR, GB, DMtns; to e N.Am, n Mex. Much like *C. incanum*.

C. glaucum L. Ann 8–20 cm. **STS**: lower often prostrate; upper ascending. **LF**: blade 5–35 mm, oblong to ovate, margin entire to toothed, densely powdery below, glabrous to sparsely powdery above. **FL**: sepals gen 3, not enclosing fr, back flat, glabrous; stamens gen 3. **FR** ± 1 mm diam; wall free from seed. **SEEDS** vertical and horizontal. 2*n*=18. Open places, often saline soils, drying ponds, streambanks; < 2200 m. CaR, SNH (e slope), SnBr, GB, expected elsewhere; to Can, e US; native to Eurasia. [ssp. *salinum* (Standley) Aellen]

C. hians Standley (p. 513) Ann 30–80 cm. **LF**: blade 8–30 mm, narrowly lanceolate to elliptic, entire, ± densely powdery below, glabrous to sparsely powdery above. **FL**: sepals not enclosing fr, smooth, powdery. **FR** ± 1 mm diam; wall adherent to seed. **SEED** horizontal. 2*n*=18. Open places, shrubland, woodland; 300–2700 m. SN, SCoRI, TR, PR, GB; to WY, NM. Much like *C. leptophyllum*.

C. incanum (S. Watson) A.A. Heller var. ***occidentale*** D.J. Crawford (p. 513) Ann 8–30 cm. **LF**: blade 6–22 mm, narrowly deltate to diamond-shaped, entire to weakly 2-lobed at base, powdery below, glabrous to sparsely powdery above. **FL**: sepals gen enclosing fr, strongly keeled, powdery. **FR** ± 1 mm diam; wall free from seed. **SEED** horizontal. Open places, sandy or gravelly soils; 700–2300 m. SNH (e slope), GB, DMoj; to ID, sw UT. [*C. fremontii* S. Watson var. *i.* S. Watson] Other vars. in Rocky Mtns, w Great Plains. Much like *C. fremontii*.

C. incognitum Wahl Ann 13–80+ cm. **LVS**: blades 7–30 mm, entire, densely powdery below, glabrous to sparsely powdery above; lower blades ovate to ± deltate; upper blades gen narrowly elliptic. **FL**: sepals not enclosing fr, ± keeled, powdery. **FR** ± 1–1.5 mm diam; wall adherent to seed. **SEED** horizontal. 2*n*=18. Open places, shrubland, coniferous forest; < 800–2700 m. NW, CaR, SN, SCoR, TR, PR, MP; to WY, NM. Much like *C. atrovirens* & *C. hians*.

C. leptophyllum Moq. Ann 20–60 cm. **LF**: blade 8–25 mm, linear to narrowly lanceolate, ± thin, entire, densely powdery below, less so above. **FL**: sepals enclosing fr or tips reflexed, ± keeled, gen densely powdery. **FR** ± 1 mm diam; wall adherent to seed. **SEED** horizontal. 2*n*=18. Open, gravelly soils, shrubland; 300–3200 m. s NCoRH, c&s SN, W&I, n DMoj; to e US, n Mex. Much like *C. hians*.

C. macrospermum Hook. f. var. *halophilum* (Philippi) Standley Ann 8–75 cm. **ST**: lower branches decumbent; upper branches ascending. **LF**: blade 8–50 mm, ovate to ± deltate, thick to fleshy, base rounded to tapered, gen toothed, ± densely powdery (but becoming glabrous) below, glabrous to sparsely powdery above. **FL**: calyx tube enclosing fr, > lobes, ± glabrous; stamens 3–5. **FR** ± 1 mm diam; wall adherent to seed. **SEEDS**: calyx of vertically seeded fls 3-lobed; calyx of horizontally seeded fls 4–5-lobed. Wet places, marshes; < 100 m. NCo, deltaic GV, CCo, SnFrB, SCo; to WA; probably native to S.Am. [var. *farinosum* (S. Watson) J. Howell] Other vars. in S.Am.

C. missouriense Aellen Ann 40–200+ cm. **LF**: blade 15–95 mm, ovate to deltate, irregularly wavy-dentate to deeply toothed, powdery, base wedge-shaped. **INFL**: branches ± curved or pendent in fr. **FL**: sepals enclosing fr, back ± flat, sparsely powdery. **FR** ± 1 mm diam; wall adherent to seed. **SEED** horizontal. 2*n*=54. Open, often disturbed places; < 1000 m. n SNF, s SN, SnJV, SCo, PR; native to c&e US. Similar to *C. album*.

C. multifidum L. (p. 513) Ann 8–30 cm. **ST** 15–60 cm; branches spreading to decumbent, glandular. **LF**: blade 3–20 mm, oblong, pinnately lobed, minutely glandular. **INFL**: axillary < 5 mm diam, ± spheric; terminal spike- or panicle-like, 10–25 mm, leafy-bracted. **FL**: calyx minutely 3–5-lobed, obovoid, tube enclosing fr, surface netted; stigmas 3. **FR** ± 1 mm diam; wall ± free from seed. **SEED** vertical. 2*n*=32. Disturbed places; < 1400 m. NW, c SNF, CW, SCo, ChI (San Nicolas, Santa Rosa islands), s WTR; to e N.Am; native to S.Am.

C. murale L. Ann 15–70+ cm. **LF**: blade 15–60 mm, widely ovate to deltate, toothed, shiny dark green above, sparsely powdery below, base truncate to wedge-shaped. **FL**: sepals ± enclosing fr, keeled, powdery. **FR** 1–1.5 mm diam; equatorial margin sharply angled; wall adherent to seed. **SEED** horizontal. 2*n*=18. Common. Disturbed places, fields; < 2900 m. CA-FP, D (uncommon); to Can, e US, n Mex; native to Eur.

C. nevadense Standley (p. 513) Ann 10–48 cm. **LVS**: blades 6–18 mm, entire, ± powdery (esp below), sometimes becoming ± glabrous; lower blades ovate to diamond-shaped; upper blades elliptic. **INFL** panicle-like, often rounded in fr. **FL**: sepals ± enclosing fr, back flat, densely powdery. **FR** < 1 mm diam; wall adherent to seed. **SEED** horizontal. Washes, shrubland; 1400–2000 m. c SNH (e slope), SNE; se OR, NV.

C. pratericola Rydb. Ann 16–65 cm. **LVS**: blades 10–30 mm, thin, narrowly elliptic to lanceolate, densely powdery below, glabrous to sparsely powdery above; lower blades gen 1–2-lobed or -toothed. **FL**: sepals not enclosing fr, back keeled, moderately to densely powdery. **FR** 1–1.5 mm diam; wall free from seed. **SEED** horizontal. 2*n*=18. Open, dry places; < 2500 m. CaR, SN, GV, e SnFrB, SCoR, WTR, SNE, D; to c US. [*C. desiccatum* Nelson var. *leptophylloides* (Murray) Wahl] Much like *C. atrovirens*.

C. pumilio R. Br. (p. 513) Ann 12–25 cm, ± strong-smelling. **LF**: blade 4–20 mm, lanceolate to ovate, pinnately toothed to lobed, gray-glandular-puberulent (esp below); base gen wedge-shaped. **INFL** 3–6 mm diam, axillary, ± spheric, dense. **FL**: sepals weakly enclosing fr, gen keeled, yellow-glandular; stamens 1–3. **FR** ± 1 mm diam; wall free from seed. **SEED** vertical. Disturbed places; < 3000 m. KR, NCoR, CaR, SN, GV, SnFrB, SCo, TR, w PR, MP; to e US, n Mex; native to Australia.

C. rubrum L. Ann 20–58+ cm. **LF**: blade 15–90+ mm, lanceolate to widely ovate, toothed, glabrous to sparsely powdery below, ± glabrous above; base gen wedge-shaped. **INFL**: spike, axillary, 10–30 mm. **FL**: sepals gen 3, weakly enclosing fr, back flat, glabrous to sparsely powdery. **FR** 0.5–1 mm diam; wall ± adherent to seed. **SEEDS** vertical and horizontal. 2*n*=36. Open, saline places, drying mudflats; < 1100 m. NCoRO, CaR, Teh, GV, CCo, SnFrB, SCo, GB; to Can, e US, Mex, Eurasia. [*C. humile* Hook.] Some forms may be introduced.

C. simplex (Torrey) Raf. (p. 513) LARGE-SEEDED GOOSEFOOT Ann 35–150 cm, gen few-branched above middle. **LF**: blade 25–150+ mm, widely ovate to deltate, wavy-lobed (lobes 3–5), glabrous to sparsely powdery, base truncate to ± cordate. **FL**: sepals not enclosing fr, back flat, glabrous to sparsely powdery. **FR** 1.5–3 mm diam; wall free or adherent to seed, netted. **SEED** horizontal. 2*n*=36. UNCOMMON. Disturbed or open places, shrubland, coniferous forest; 1400–2400 m. n SN, MP, W&I; to e US. [*C. gigantospermum* Aellen]

C. strictum Roth var. *glaucophyllum* (Aellen) Wahl Ann 45–120+ cm. **LVS** glabrous or sparsely powdery above; basal 25–40 mm, gen ovate, irregularly serrate; cauline 12–20 mm, lanceolate to narrowly ovate, entire. **FL**: sepals not enclosing fr, back flat or weakly keeled, sparsely powdery. **FR** ± 1 mm diam; wall gen adherent to seed. **SEED** horizontal. 2*n*=36. Open, disturbed places; < 1700 m. CaRF, SN, GV, SnFrB, SCoRO, SCo, SnBr; to Can; native to e US. Perhaps best considered part of *C. album*. Var. *strictum* native to Eurasia.

C. vulvaria L. Ann 5–20 cm. **STS** 10–75 cm, prostrate to decumbent. **LF**: blade 3–20 mm, ovate to widely elliptic, entire, densely powdery below, glabrous to sparsely powdery above, base obtuse to rounded. **FL**: calyx tube >> lobes, ± enclosing fr, back flat, powdery. **FR** 1–1.5 mm diam; wall adherent to seed. **SEED** horizontal. $2n$=18. Uncommon. Open, disturbed places; < 1400 m. CaRF, s SN, GV, CW, WTR; to e US; native to Eurasia.

CORISPERMUM

Ann, gen erect. **ST**: branches 0–few, spreading to ascending. **LF** gen linear. **INFL**: spike; bracts lf-like. **FL** bisexual; sepals 0–5, scarious; stamens 1–5; ovary superior, chamber 1, stigmas 2. **FR** elliptic to obovate, ± flat. **SEED** vertical, gen black. 60 spp: n temp. (Latin & Greek: leathery seed) [Maihle & Blackwell 1978 Sida 7:382–391]

C. hyssopifolium L. (p. 513) Pl 5–30 cm. **LF** 9–25 mm, 1–2.5 mm wide. **INFL** 1–4 cm, dense; bracts in fr 3–10 mm, margin scarious. **FR** 3–4 mm diam, round, dark green to black; wing white to translucent. Uncommon. Sandy soils, dunes; 100–1100 m. n SNE, n DSon; to AK, e N.Am, n Mex; native to Eurasia.

CYCLOLOMA WINGED PIGWEED

1 sp. (Greek: circular wing, from calyx in fr)

C. atriplicifolium (Sprengel) J. Coulter (p. 513) Ann 12–75 cm, rounded. **ST** with many spreading branches, slender, striate, finely woolly, becoming glabrous. **LVS** gradually reduced upward; petiole 0–12 mm; blade 5–65 mm, lanceolate to ovate, ± wavy-toothed. **INFL** panicle-like, terminal, open in fr; bracts 0; fls sessile. **FL**: gen bisexual (pistillate); calyx enclosing fr, 2–3 mm diam in fr, winged, lobes 5, ± keeled; stamens 5; ovary densely and finely tomentose, style deeply 2–3-lobed. **FR** ± 2 mm diam. **SEED** 1.5–2 mm, horizontal, black. Fields, disturbed sites; < 800 m. GV, s SCo, w PR, e DMoj; native to c N.Am.

GRAYIA HOP-SAGE

1 sp. (Asa Gray, eminent Am botanist, Harvard University, 1810–1888)

G. spinosa (Hook.) Moq. (p. 513) Shrub gen < 1 m, gen dioecious; hairs simple or scaly. **ST**: branches many, stiff; bark straw-colored to gray; twigs becoming spine-like. **LF** 5–35 mm, gen elliptic to oblanceolate, flat, entire, tapered to short-petioled. **STAMINATE INFL** spike-like, terminal, 7–18 mm; bract ± lf-like; fls < 8 per axil. **PISTILLATE INFL** ± spike-like, axillary or terminal, 6–18 cm in fr; bracts 3–10 mm, ± lf-like; bractlets 2, 7–15 mm, fused, together sac-like, ± round, flat, winged, white to red-tinged. **STAMINATE FL**: calyx lobes 4; stamens 4. **PISTILLATE FL**: calyx 0; stigmas 2. **FR** gen 2–3 mm. $2n$=36. Sandy to gravelly soils, shrubland, pinyon/juniper woodland; 500–2800 m. SNH (e slope), Teh, s SCoRI, WTR (n slope), GB, DMoj, nw DSon; to WA, MT, NM. Sometimes incl in *Atriplex*. ❀TRY;alsoSTBL.

HALOGETON

Ann, glabrous to papillate. **ST** gen branched at rase, prostrate to erect. **LVS** ± cylindric, fleshy, abruptly pointed to bristle- or spine-tipped. **INFL** axillary; fls densely clustered; bractlets 0–2. **FL** bisexual or pistillate; calyx lobes 5, gen enclosing fr, tip winged in fr; stamens 2–5; stigmas 2. **FR**: wall adherent to seed. **SEED** vertical or horizontal. 3 spp.: Eurasia. (Greek: salty neighbor, from habitat)

H. glomeratus (M. Bieb.) C. Meyer (p. 513) **STS** 6–25 cm, curved, leafy throughout. **LF** 4–22 mm, 1–1.5 mm wide, sessile, withered or deciduous in fr; bristle 1–2 mm, stiff. **INFL**: bracts 1.5–2 mm, ± glaucous; fls many, throughout st. **FL**: calyx lobes 1–2 mm, wings 2–3.5 mm, fan-like, membranous, veiny. **FR** 1–2 mm. Alkaline soils, open flats, shrubland; 800–1800 m. CaR, GB, DMoj; to ID, Colorado, NV; native to Eurasia. NOXIOUS WEED; TOXIC to livestock from concentrated oxalates.

KOCHIA

Ann to subshrub, gen erect, gen short-hairy. **LVS** alternate or opposite, linear to lanceolate, flat to cylindric, fleshy or not. **INFL**: spike or clusters; bracts lf-like; fls 1–few per axil. **FL** gen bisexual; calyx lobes 5, incurved, keeled and tubercled to winged in fr; stamens 5; stigmas 2–3. **FR** ± compressed-spheric. **SEED** horizontal. ± 20 spp.: w N.Am, Eurasia. (Wilhelm D. Koch, German physician & botanist, 1771–1849)

1. Ann; lower cauline lvs short-petioled, gen 3–5-veined below middle .... ***K. scoparia***
1′ Per or subshrub; lvs sessile, vein 1 or obscure
  2. Sts many from base, gen simple, glabrous to finely white-tomentose; lvs gen overlapping .... ***K. americana***
  2′ Sts 1–few from base, gen branched throughout, gray- to brown-puberulent; lvs not overlapping .... ***K. californica***

K. americana S. Watson (p. 513) Subshrub 8–40 cm. **STS** many from base, gen simple, ascending to erect, gen finely white-tomentose, sometimes becoming glabrous. **LF** 5–20 mm, 1–2 mm wide, gen overlapping, ± cylindric, ± fleshy, glabrous to sparsely silky-hairy; vein obscure or 1. **INFL**: fls 1–3 per axil. **FL**: calyx lobes gen white-tomentose, wings in fr < 2 mm, < 4 mm wide. Alkaline soils, flats, dry lake margins; 600–2100 m. GB, DMoj; to OR, ID, NM.

K. californica S. Watson (p. 513) Per or subshrub 20–60 cm. **STS** 1–few from base, erect, branched throughout, densely gray- to brown-puberulent. **LF** 3–12 mm, 1.5–3 mm wide, gen well-spaced, gen flat, ± fleshy, silky-hairy; vein obscure or 1. **INFL**: fls 1–5 per axil. **FL**: calyx lobes densely short-hairy, wings in fr ± 1–2 mm, < 3 mm wide. Alkaline soils, flats; < 1000 m. s SnJV, DMoj; s NV.

K. scoparia (L.) Schrader Ann 30–120 cm. **ST** simple to much-branched, glabrous to silky-hairy. **LF** 8–50 mm, 1–3.5 mm wide, flat, glabrous to short-soft-hairy, 3–5-veined below middle. **INFL**: fls gen 3–7 per axil. **FL**: calyx lobes short-hairy, with tubercles or wings < 1 mm, < 1 mm wide in fr. Disturbed places, fields, roadsides; < 1500 m. GV, n SnFrB, SCo, GB, DMoj, expected elsewhere; to e US, Eur; native to Asia. [vars. *culta* Farw. & *subvillosa* Moq.]

KRASCHENINNIKOVIA

Per or shrub, monoecious or dioecious, gen erect, stellate or long-hairy. **LF** linear to lanceolate, flat, entire. **INFL** spike-like, terminal; staminate fls above pistillate fls, bracts ± lf-like; pistillate fls few, clustered, subtended by 2 bractlets ± fused at base. **STAMINATE FL**: calyx lobes 4; stamens 4. **PISTILLATE FL**: calyx lobes 0; stigmas 2. **FR**: wall free or adherent to seed. **SEED** vertical. ± 8 spp.: n Medit, temp Asia, w N.Am. (Stephan P. Krascheninnikov, Russian botanist, 1713–1755)

K. lanata (Pursh) A.D.J. Meeuse & Smit (p. 513) WINTER FAT Shrub gen 5–10 dm, gen monoecious; hairs white, becoming ± rust-colored. **LF** 6–30 mm, 1.5–5 mm wide; margins inrolled. **INFL** 3–19 cm; staminate fls many; pistillate fls 1–4 in lower axils; bractlets densely hairy, 4–6 mm in fr. **STAMINATE FL**: calyx lobes 1–2 mm, densely hairy; stamens exserted. **PISTILLATE FL**: stigmas exserted. **FR** ± 2 mm, white-hairy. Rocky to clay soils, flats, gentle slopes; 100–2700 m. SNH (e slope), Teh, s SnJV, s SCoRI, WTR (n slope), GB, DMoj; to WA, n-c US, NM, n Mex. [*Eurotia l.* (Pursh) Moq.; *Ceratoides l.* (Pursh) J. Howell — both invalid] ❀DRN, SUN:1,**2,3,7–12**,13,**14,18–24**.

MONOLEPIS POVERTY WEED

Ann, gen glabrous. **LVS** alternate, gen reduced upward. **INFL**: clusters gen axillary, 1–15+-fld; bracts lf-like. **FL** bisexual or pistillate; sepals 1–3; stamens 0–1; style branches 2. **FR**: wall pitted to tubercled, sometimes adherent to seed. **SEED** gen vertical. 3 spp.: w N.Am. (Greek: 1 scale, from sepal number in most spp.)

1. Sts gen much-branched; fls 1–3 per cluster, axillary and terminal; lf blade elliptic to ± oblong .... ***M. pusilla***
1′ Sts 2–many from base, not branched above; fls 4–15+ per cluster, axillary; lf blade (ob)lanceolate
 2. Lf blade 2-toothed to hastate; fr 1.5–2 mm, wall minutely pitted, adherent to seed .... ***M. nuttalliana***
 2′ Lf blades entire; fr ± 0.5 mm, wall minutely papillate, free from seed .... ***M. spathulata***

M. nuttalliana (Schultes) E. Greene (p. 513) Pl 4–40 cm. **STS** 2–many from base, ascending to erect, fleshy. **LF** 10–45 mm, lanceolate, fleshy, 2-toothed to hastate. **INFL**: fls gen 5–15+ per cluster. **FL**: sepal oblanceolate to obovate. **FR** 1.5–2 mm; wall minutely pitted, adherent to seed. **SEED** dark brown. Open, disturbed, often wet places; < 3500 m. CA (exc NW); to c N.Am, n Mex. Often confused with *Chenopodium*.

M. pusilla S. Watson (p. 513) Pl 4–14 cm. **STS** 1–5 from base, simple to ± intricately branched. **LF** 3–6 mm, elliptic to ± oblong. **INFL**: fls 1–3 per cluster. **FL**: sepals 1–3, oblanceolate. **FR** 0.5–1 mm; wall minutely tubercled, ± free from seed. **SEED** dark brown. Open places, shrubland; 1500–2400 m. n SNH (e slope), MP, n SNE; to ID, Colorado.

M. spathulata A. Gray (p. 513) Pl 2–25 cm. **STS** 3–11 from base, decumbent to erect, fleshy. **LF** 3–120 mm, narrowly oblanceolate, fleshy, entire. **INFL**: fls 4–15+ per cluster. **FL**: sepal oblanceolate. **FR** ± 0.5 mm; wall minutely papillate, ± free from seed. **SEED** red-brown. Open places, shrubland, coniferous forest; 1400–2600 m. SNH, SnBr, n SNE (Sweetwater Mtns); to ID, NV, Baja CA.

NITROPHILA

Per, glabrous, rhizomed. **ST** decumbent to erect; branches paired. **LVS** opposite, linear to ovate, fleshy, sessile to clasping. **INFL** axillary; bracts gen 2, unequal; fls 1–3 per cluster. **FL**: sepals 5(–7), enclosing fr, papery; sides overlapping; back ribbed; stamens 5, incl; stigmas 2, persistent in fr. **FR** ± 2 mm. **SEED** vertical, black. ± 8 spp.: temp Am. (Greek: soda loving)

1. Lf 3–4.5 mm, ± ovate; pl 3–10 cm .... ***N. mohavensis***
1′ Lf 5–16 mm, linear to oblong; pl 7–30 cm .... ***N. occidentalis***

N. mohavensis Munz & Roos (p. 519) AMARGOSA NITROPHILA Pl 3–10 cm. **ST** gen erect; internodes < lvs. **LF** 3–4.5 mm, ± ovate. **FL**: sepals 1.5–2 mm, pink, becoming white. RARE. Alkaline flats; 600–750 m. n DMoj (Amargosa Desert); w NV.

N. occidentalis (Nutt.) Moq. (p. 519) Pl 7–30 cm. **ST** decumbent to erect; internodes = or > lvs. **LF** 5–16 mm, linear to oblong. **FL**: sepals 1–2 mm, white or pink, becoming white. Moist, alkaline soils; < 1500 m. CaRF, GV, SCo, w PR, SNE, DMoj; to e OR, UT, n Mex.

SALICORNIA PICKLEWEED

Ann to subshrub, glabrous. **ST** gen many-branched, jointed; internodes green to glaucous, fleshy. **LVS** opposite, sessile, ± decurrent; lf pairs gen fused at base, clasping or together ring-like. **INFL**: spike, terminal, cylindric, dense; bracts scale-like; fls gen 3 per axil, sessile to sunken in axis. **FL**: calyx bladder-like, slitted, ± deciduous in fr; stamens gen 2; stigmas 2–3. **FR**: wall free from seed. **SEED** vertical. ± 13 spp.: ± worldwide. (Greek: salt horn) Needs further study.

1. Ann; central fl of a node gen higher than lateral fls
 2. Pl ± slender; st branched above middle; spike 4–6 mm wide .... ***S. bigelovii***
 2′ Pl gen rounded or ovoid; st branched throughout; spike 2–4 mm wide .... ***S. europaea***
1′ Per from rhizomes; fls of a node gen at same level
 3. Spike 2–3 mm wide, distal internodes lacking fls .... ***S. subterminalis***
 3′ Spike 3–5 mm wide, fld to tip
  4. St axis gen erect, branches gen ascending; ne DMoj .... ***S. utahensis***
  4′ St axis gen creeping, rooting at nodes, branches decumbent to erect; coastal, s SnJV .... ***S. virginica***

S. bigelovii Torrey (p. 519) Ann 9–45 cm, slender. **ST** erect, branching above middle; internodes 10–30 mm, 2.5–4 mm wide. **INFL** 15–90 mm, 4–6 mm wide; lower bracts gen abruptly pointed; central fl of a node gen above lateral fls. **SEED** 1–1.5 mm, puberulent. Salt marshes; < 20 m. SCo; to e US, Caribbean, Mex. ❀SUN,IRR,saline:22,23,**24**;oil-seed crop plant.

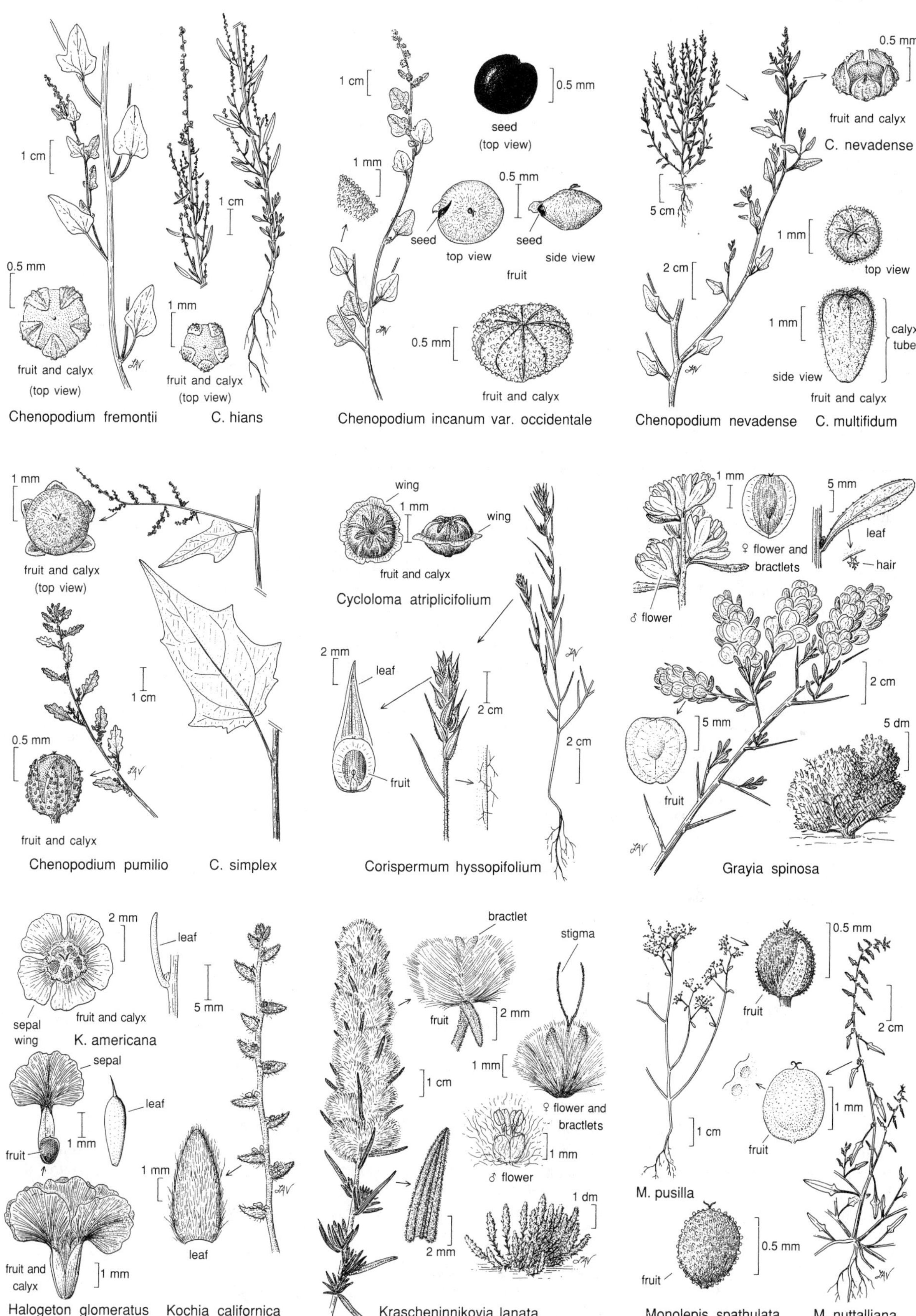
1 cm
0.5 mm
fruit and calyx
(top view)
Chenopodium fremontii
1 cm
1 mm
fruit and calyx
(top view)
C. hians
1 cm
0.5 mm
seed
(top view)
1 mm
0.5 mm
seed
seed
top view
side view
fruit
0.5 mm
fruit and calyx
Chenopodium incanum var. occidentale
0.5 mm
fruit and calyx
C. nevadense
5 cm
1 mm
top view
2 cm
1 mm
calyx
tube
side view
fruit and calyx
Chenopodium nevadense
C. multifidum
1 mm
fruit and calyx
(top view)
1 cm
0.5 mm
fruit and calyx
Chenopodium pumilio
C. simplex
wing
1 mm
wing
fruit and calyx
Cycloloma atriplicifolium
2 mm
leaf
2 cm
2 cm
fruit
Corispermum hyssopifolium
1 mm
5 mm
leaf
♀ flower and
bractlets
hair
♂ flower
2 cm
5 mm
5 dm
fruit
Grayia spinosa
2 mm
leaf
5 mm
sepal
wing
fruit and calyx
K. americana
sepal
leaf
1 mm
fruit
1 mm
fruit and
calyx
1 mm
leaf
Halogeton glomeratus
Kochia californica
bractlet
stigma
fruit
2 mm
1 cm
1 mm
♀ flower and
bractlets
1 mm
♂ flower
1 dm
2 mm
Krascheninnikovia lanata
0.5 mm
fruit
2 cm
1 mm
fruit
1 cm
M. pusilla
0.5 mm
fruit
Monolepis spathulata
M. nuttalliana

S. europaea L. (p. 519) Ann 8–25 cm, rounded to ovoid. **ST** many-branched throughout; internodes 10–18 mm, 1–3 mm wide. **INFL** 25–70 mm, 2–4 mm wide; lower bracts gen obtuse to rounded; central fl of a node above lateral fls. **SEED** ± 1–2 mm, puberulent. $2n$=18,36. Salt marshes, alkaline flats; < 1200 m. Teh, CCo, SCo, MP; to AK, e US, Baja CA, Eur. Pls with infl internodes as long as wide have been called *S. rubra* Nelson. ❀STBL.

S. subterminalis Parish (p. 519) Per 7–30 cm. **ST** spreading to erect; branches ascending; internodes 6–18 mm, 2–3 mm wide. **INFL** 10–40 mm, 2–3 mm wide; distal 5–14 nodes lacking fls; lower bracts gen obtuse to rounded; fls at same level. **SEED** ± 1 mm, glabrous. Salt marshes, alkaline flats; < 800 m. SnJV, CCo, SnFrB, SCo, ChI, w DMoj, DSon; n Mex. ❀STBL.

S. utahensis Tidestrom Per or subshrub 14–28 cm. **ST**: axis ascending to erect; branches ascending; internodes 10–40 mm, 3–6 mm wide. **INFL** 10–38 mm, 3–5 mm wide; lower bracts acute to obtuse; fls at same level. **SEED** puberulent. Alkaline soils; < 100 m. ne DMoj (Death Valley); also UT. Differences between CA and UT pls need study. ❀STBL.

S. virginica L. (p. 519) Per or subshrub 20–70 cm. **ST**: axis gen creeping, rooting at nodes; branches decumbent to erect; internodes 7–30 mm, 2.5–7 mm wide. **INFL** 15–60 mm, 3–4 mm wide; lower bracts gen obtuse to rounded; fls ± at same level. **SEED** 0.5–1 mm, puberulent. Salt marshes, alkaline flats; < 100 m. NCo, SnJV, CCo, SCo, ChI; to AK, Baja CA, e US, Caribbean, Medit. ❀STBL.

SALSOLA

Ann to subshrub. **ST** simple to much-branched. **LVS** gen reduced upward, thread-like to subcylindric, gen becoming thick, ridged, spine-tipped. **INFL** axillary; bracts 1–2; fls gen 1 per axil. **FL** bisexual; sepals 4–5, in fr thickened, persistent, gen tubercled to winged; stamens gen 5, exserted, style branches gen 2, exserted. **FR** spheric to obovoid; top ± depressed. **SEED** gen horizontal. ± 100 spp.: ± worldwide. (Latin: salty, from habitats)

1. Shrub; lvs 3–9 mm, not reduced upward, oblong to ovate, puberulent .... ***S. vermiculata***
1′ Ann; lvs 5–55 mm, gen reduced upward, thread-like to lanceolate, base gen wide in fr
  2. Pls fleshy in fr; calyx 3.5–5 mm, outer sepals winged in fr, inner (facing st) tubercled in fr .... ***S. soda***
  2′ Pls not fleshy in fr; calyx 2.5–3.5 mm, all sepals winged in fr
    3. Sepal wings 2.5–4.5 mm in fr; lvs yellow-green; branchlets often minutely papillate .... ***S. paulsenii***
    3′ Sepal wings 0.5–2.5 mm in fr; lvs ± green; branchlets often short-stiff-hairy .... ***S. tragus***

S. paulsenii Litv. (p. 519) Ann gen < 50 cm, glabrous to minutely papillate. **ST** much-branched. **LF** 5–32 mm, thread-like, becoming yellow-green, thick; base becoming wide, leathery; tip spiny; margin at base white to translucent. **INFL**: bract subcylindric, spiny. **FL**: calyx 2.5–3.5 mm, wings in fr gen 2.5–4.5 mm, entire to irregularly, minutely toothed. Common. Disturbed places; 700–1800 m. n WTR, SNE, DMoj; to UT; native to se Eur, c Asia. May hybridize with *S. tragus*; needs further study.

S. soda L. Ann 15–45 cm, slender to rounded, fleshy, gen glabrous. **LF** 6–55 mm, narrowly oblong; base wing-margined; tip rigid, ± acute; margin becoming ± white-translucent. **INFL**: bract lanceolate to ovate; base wing-margined, tip rigid, ± acute, margin white-translucent. **FL**: calyx 3.5–5 mm, fleshy, outer sepals with wings < 1.5 mm, inner short-tubercled. Mudflats, open areas in salt marshes; < 50 m. n CCo (San Francisco Bay); native to s Eur.

S. tragus L. (p. 519) RUSSIAN THISTLE, TUMBLEWEED Ann < 1 m, gen rounded, glabrous to short-stiff-hairy. **ST** gen much-branched, in fr deciduous at base. **LF** 8–52 mm, thread-like, becoming rigid; base becoming wide, leathery, tip sharp-pointed to spiny, margin at base translucent. **INFL**: bract subcylindric, spiny. **FL**: calyx 2.5–3 mm, wings in fr 0.5–2.5 mm, gen minutely toothed to crenate. Common. Disturbed places; < 2700 m. CA; to e N.Am, Mex; native to Eurasia. [*S. australis* R. Br.; *S. iberica* Sennen & Pau; *S. kali* L. var. *tenuifolia* Tausch. — all misapplied] NOXIOUS WEED. *S. pestifer* Nelson may be the valid name.

S. vermiculata L. Shrub 60–100+ cm, gen puberulent. **ST** branched ± throughout; branches ascending to erect. **LVS** 3–9 mm, lower oblong; upper ovate, puberulent; tip obtuse to acute. **INFL**: bract lf-like, wing-margined. **FL**: calyx 2–3 mm, wing ± 2 mm. Clay soils, flats; ± 1000 m. se SCoRI (Temblor Range); native to nw Medit.

SARCOBATUS GREASEWOOD

1 sp. (Greek: fleshy bramble)

S. vermiculatus (Hook.) Torrey (p. 519) Shrub 5–30 dm, often rounded, gen monoecious. **STS** yellowish to light gray; twigs short, spreading, rigid, often spine-tipped. **LVS** opposite below, alternate above, deciduous; blades 5–28 mm, ± linear, subcylindric, entire, fleshy, glabrous to hairy. **INFL**: spike, 5–19 mm, cylindric, dense, catkin-like; staminate fls gen many, spirally arranged, bracts peltate; pistillate fls 1–3, below staminate fls, bract lf-like. **STAMINATE FL**: perianth 0; stamens 2–3. **PISTILLATE FL**: perianth cup-like, in fr winged, 4–8 mm wide, persistent; ovary half-inferior. **FR** 4–5 mm, conic above wing, ± glabrous. Alkaline soils, dry lakes, washes, shrubland; 100–2100 m. GB, DMoj; to w Can, Great Plain, n Mex. [var. *baileyi* (Cov.) Jepson]. ❀STBL.

SUAEDA SEA-BLITE, SEEPWEED

Wayne R. Ferren, Jr.

Ann, per, shrubs, glabrous to hairy. **LVS** gen alternate; blade entire, sometimes cylindric or upper surface flat, fleshy, gen glaucous, tip acute or pointed. **INFL**: cyme; clusters sessile, gen arrayed in compound spikes; bracts lf-like or reduced; bractlets subtending fls 1–3, minute, membranous; fls 1–12. **FL** gen bisexual; calyx radial or bilateral, lobes 5, rounded, hooded, keeled, horned, or wing-margined; ovary ± lenticular, rounded, conic or with a neck-like extension, stigmas 2–4. **FR**: utricle, enclosed in calyx. **SEED** horizontal or vertical, lenticular or flat. 115 spp.: worldwide, saline and alkaline soils. (ancient Arabic name) [Ferren & Whitmore 1983 Madroño 30:181–190] ❀STBL.

1. Ann, per, or subshrub; sts 1–several, prostrate to erect; calyx gen bilateral (1 lobe gen larger), lobes gen hooded and keeled, horned, or wing-margined; stigmas glabrous, from top of rounded to lenticular ovary; seeds horizontal, lenticular and shiny, or flat and dull (sect. *Heterosperma*)
  2. Ann; calyx lobes with horns and lateral wings; CA .... ***S. calceoliformis***

2′ Per or subshrub; calyx lobes with hoods and keels, horns and wings 0; SCo .... ***S. esteroa***
1′ Subshrub or shrub; sts gen several from woody base; calyx ± radial, lobes gen rounded, hooded or keeled, not horned or wing-margined; stigmas hairy-papillate, from pit in top of neck-like extension of ovary; seeds horizontal or vertical, lenticular and shiny (sect. *Limbogermen*)
3. Bracts gen < lvs, gen not overlapping or covering internodes at st tips; old lf scars on sts and branches ± smooth; seeds < 1 mm; fls ± 1–2.5 mm, gen on upper branches; infl branches slender, 0.7–2 mm diam; gen interior .... ***S. moquinii***
3′ Bracts gen = lvs, overlapping, covering internodes at st tips; old lf bases on sts knobby; seeds > 1 mm; fls 1–4 mm, gen throughout pl; infl branches coarse, 1–4 mm diam; coastal
4. Pl glabrous or sparsely hairy, gen green; ovary ± conic, without obvious neck; salt marshes, CCo ***S. californica***
4′ Pl gen densely hairy, glaucous; ovary pear-shaped with obvious neck; coastal bluffs, margins of salt marshes, SCo, ChI .... ***S. taxifolia***

S. calceoliformis (Hook.) Moquin (p. 519) HORNED SEA-BLITE Ann < 8 dm, glabrous, glaucous. **STS** prostrate to erect, 1–several; branches ascending or spreading, gen striped. **LVS** often tightly ascending, < 40 mm, linear, sessile; upper surface flat, green or reddish. **INFL** gen dense, branched; fls 3–5 per cluster; bracts subtending branches = lvs; bracts subtending fls < lvs, wider at base. **FL** bilateral, 1–4 mm incl horns; calyx lobes horned and ± keeled, ± wing-margined; ovary rounded to lenticular, stigmas gen 2, glabrous. **SEED** horizontal; lenticular form 0.8–1.7 mm, shiny, gen black; flat form 1–1.5 mm, dull, brown. Dry, saline or alkaline wetland soils; < 2200 m. CA; to AK, e Can, TX. [*S. depressa* (Pursh) S. Watson var. *erecta* S. Watson; *S. d.* var. *d.* misapplied] Yellow-green pls (gen GB) with lf base ± tapered, infl gen open and slender, branches spreading, calyx lobes wing-margined have been called *S. occidentalis* S. Watson.

S. californica S. Watson (p. 519) CALIFORNIA SEA-BLITE Shrub 3–8 dm, mound-like, glabrous or sparsely hairy, gen green. **STS** decumbent, several, dull gray-brown; branches spreading, many, pale green or reddish, herbaceous. **LVS** overlapping, subsessile; petioles ± 1 mm, forming prominent knobs on lower sts and branches; blades 5–35 mm, ± lanceolate, subcylindric to flat. **INFL**: clusters scattered throughout pl; branches thick, 2–4 mm diam; fls 1–5 per cluster; bracts gen = lvs, densely overlapping at branch tips. **FL** bisexual or lateral pistillate, radial, 2–3 mm; calyx lobes rounded or hooded, glabrous; ovary ± conic, stigmas 3, hairy-papillate. **SEED** horizontal or vertical, 1.5–2 mm, biconvex, shiny, black. RARE. Margins of coastal salt marshes; < 5 m. CCo.

S. esteroa W. Ferren & S. Whitmore (p. 519) ESTUARY SEA-BLITE Per or subshrub 1–6 dm, glabrous, gen glaucous. **ST** decumbent to erect; branches gen ascending. **LVS** ascending, sessile; upper < 60 mm, linear-lanceolate, upper surface flat, overlapping, green or reddish, gen glaucous; lower = mid-st but gen withered, straw-colored, persistent or breaking apart into fibers. **INFL**: clusters confined to upper sts; fls gen 3–5 per cluster; bracts < lvs. **FL** bilateral, 1.5–3 mm; calyx lobes hooded and keeled; ovary rounded to lenticular, stigmas 2–3, linear, glabrous. **SEED** horizontal; biconvex form 1–1.2 mm, shiny, black or reddish; flat form 1.2–2 mm, dull, brown. UNCOMMON. Coastal salt marshes; < 5 m. SCo; n Mex.

S. moquinii (Torrey) E. Greene (p. 519) BUSH SEEPWEED Subshrub, shrub 2–15 dm, glabrous or hairy, glaucous. **STS** spreading or erect, several, base gen woody; ann sts shiny, yellow-brown; branches spreading. **LVS** ascending to widely spreading, gen not overlapping; petiole 1 mm; blade 10–30 mm, subcylindric to flat, linear to narrowly lanceolate, base narrow, yellow-green or red. **INFL** gen open; clusters confined to upper sts; branches thin, 0.7–2 mm diam; fls 1–12 per cluster; bracts gen < lvs. **FL** bisexual or lateral pistillate, radial, 0.7–2 mm; calyx lobes rounded; ovary ± pear-shaped, stigmas 3, hairy-papillate. **SEED** horizontal or vertical, 0.5–1 mm, biconvex, shiny, black. Interior and desert (rarely coastal), alkaline and saline places; < 1600 m. GV, SnFrB, SW, GB, D; to w Can, TX, Mex. [*S. torreyana* S. Watson incl var. *ramosissima* (Standley) Munz; *S. fruticosa* (L.) Forsskal misapplied]

S. taxifolia (Standley) Standley (p. 519) WOOLLY SEA-BLITE Subshrub or shrub < 15 dm, glabrous to gen densely hairy, glaucous. **STS** spreading or erect, several, dull, gray-brown; branches spreading. **LVS** ascending to widely spreading, subsessile; base of lower deciduous lvs forming small knob; blades < 30 mm, lanceolate to short-elliptic, subcylindric or upper surface flat. **INFL**: clusters gen throughout pl; branches gen thick, 2–4 mm diam; fls gen 1–3 per cluster; bracts gen = lvs, gen overlapping. **FL** bisexual or lateral pistillate, radial, 1–3 mm; calyx lobes rounded or hooded, gen hairy; ovary ± pear-shaped, stigmas 3–4, hairy-papillate. **SEED** horizontal or vertical, 1–2 mm, lenticular, shiny, black or brown. Coastal bluffs, margins of salt marshes; < 15 m. SCo, ChI; Baja CA. [*S. californica* vars. *t.* (Standley) Munz & *pubescens* Jepson]

CISTACEAE ROCK-ROSE FAMILY

Elizabeth McClintock

Per, shrub; hairs short and in stellate clusters, or short or long and simple, peltate, or glandular. **ST** erect, gen branched. **LVS** simple, gen opposite, entire, petioled or not; stipules present or 0. **INFL**: cyme (raceme- or panicle-like), or fls solitary. **FL** gen bisexual, radial; sepals 5, outer 2 often narrower, bract-like, or 3, often persistent in fr; petals gen 5, gen ephemeral; stamens gen many, free, ± persistent in fr or not; ovary superior, chamber 1 (or appearing as 3–10 from intruded parietal placentas), style 1 or 0, stigma gen 1, entire or 3–10-lobed. **FR**: capsule, loculicidal, 3–10-valved. **SEEDS** 3–many. 8 genera, 165 spp.: warm temp, esp Medit; some cult (*Cistus*; *Helianthemum*; *Tuberaria*). [Brizicky 1964 J Arnold Arbor 45:346–357] Fls open in sunshine for < 1 day.

1. Petals white or pink to purple, gen with dark or yellow blotch at base; fl > 2.5 cm wide; lvs ovate to oblong, gen opposite .... **CISTUS**
1′ Petals yellow; fl < 2.5 cm wide; lvs gen linear to lanceolate or oblanceolate, alternate or opposite
2. Per or subshrub; cauline lvs gen alternate, basal 0; style < 2 mm .... **HELIANTHEMUM**
2′ Ann; cauline lvs gen opposite, basal in a rosette, often withering early; style 0 .... **TUBERARIA**

CISTUS ROCK-ROSE

Shrub, evergreen, hairy. **ST** < 2.5 m. **LVS** gen opposite, ovate to oblong, petioled or not; stipules 0. **INFL** panicle-like or fls solitary. **FL**: sepals 3 or 5; petals white or pink to purple, gen with dark or yellow blotch at base; ovary 1-chambered, placentas 5, style 1 or 0, stigma large, hemispheric, 5–10-lobed. **FR** 5–10-valved. (Ancient Greek name) [Bean 1970 Trees & shrubs Brit Isles I:615–629] Some hybrids and cultivars.

1. Sepals 3; fls solitary .... *C. ladanifer*
1′ Sepals 5; fls gen in panicle-like cymes
 2. Style thread-like, ± = stamens; petals rose to purple .... *C. creticus*
 2′ Style 0; petals white
  3. Lf sessile, linear to linear-lanceolate, gen with 3 main veins from base; pedicel < 1 cm .... *C. monspeliensis*
  3′ Lf petioled, blade ovate to elliptic, gen with 1 main vein from base; pedicel gen > 1 cm .... *C. salvifolius*

C. creticus L. **ST** < 1.3 m; hairs long, in short, stellate clumps, or 0. **LF**: petiole 3–15 mm; blade elliptic, ± wrinkled on upper surface, with gen 1 main vein from base, margin ± wavy. **INFL** gen panicle-like, 1–7-fld; pedicel < 2 cm. **FL**: sepals 5, ovate-lanceolate, acuminate; petals 2–3 cm, rose to purple; style thread-like, ± = stamens. Uncommon. Disturbed places; < 500 m. n&s CCo, n SCo; native to s Eur. [*C. villosus* L., incl vars.]

C. ladanifer L. GUM CISTUS **ST** 1–2.5 m, shiny, resinous, sticky, with sessile glands. **LF** sessile, 4–8 cm, linear-lanceolate to narrow-oblong, gen with 3 main veins from base. **INFL**: fls solitary at ends of lateral branchlets. **FL**: sepals 3; petals 3–5 cm, white, gen with red or yellow blotch at base; style short. Uncommon. Disturbed places; < 300 m. s CCo, SnGb; native to sw Eur. [sometimes mispelled *ladaniferus*]

C. monspeliensis L. **ST** < 1 m, glandular-hairy, ± sticky. **LF** sessile, 15–50 mm, linear to linear-lanceolate, wrinkled on upper surface, gen with 3 main veins from base; margin rolled under. **INFL** gen panicle-like, 2–8-fld; pedicel < 1 cm. **FL**: sepals 5; petals 1–1.5 cm, white; styles 0. Uncommon. Disturbed places; < 300 m. NCo; native to s Eur, n Afr.

C. salvifolius L. **ST** ± spreading, < 1 m; hairs long, in short, stellate clumps, or glandular. **LF** petioled; blade 1–4 cm, ovate to elliptic, wrinkled on upper side, gen with 3 main veins from base. **INFL** gen panicle-like, 1–4-fld; pedicel gen > 1 cm. **FL**: sepals 5; petals 1.5–2.5 cm, white, with yellow blotch at base; style 0. Uncommon. Disturbed places; < 300 m. s CCo; native to s Eur.

HELIANTHEMUM SUN-ROSE, RUSH-ROSE

Per, shrub, evergreen; hairs gen in stellate clusters, rarely glandular (exc infl), very sparse to dense. **ST** gen erect, ± broom-like. **LVS** cauline, gen alternate in CA, gen linear to lanceolate or oblanceolate, ± sessile; stipules present or 0. **INFL** raceme- or panicle-like. **FL**: sepals 5, outer 2 gen narrower; petals yellow; stamens 10–many; style < 2 mm, stigma ± hemispheric. **FR** gen 3-valved. **SEEDS** 3–many. ± 120 spp.: ± range of family. (Greek: sun fl) [Daoud & Wilbur 1965 Rhodora 67:63–82, 201–216, 255–312] CA spp. esp abundant after fire.

1. Infl hairs dense, gen 0.5–0.7 mm, mostly glandular, dark red, some nonglandular, white; outer 2 sepals lanceolate .... ***H. greenei***
1′ Infl hairs sparse to dense, gen < 0.5 mm, nonglandular, white; outer 2 sepals linear .... ***H. scoparium***

H. greenei Robinson ISLAND RUSH-ROSE **ST** 15–30 cm. **LF** 7–30 mm, 1–4 mm wide, linear-lanceolate to oblanceolate. **FL**: outer sepals 2.5–4 mm, 0.5–1 mm wide, lanceolate; inner sepals 4.5–7 mm, 3–4 mm wide, ovate, acuminate; petals 5–8 mm, obovate; stamens 20–25. **FR** 4–6 mm, ovoid. **SEEDS** ± 15. RARE. Dry, rocky slopes, ridges, chaparral; < 300 m. ChI. Threatened by grazing.

H. scoparium Nutt. (p. 519) PEAK RUSH-ROSE **ST** 12–45 cm. **LF** 5–40 mm, 0.5–6 mm wide, gen linear to narrow-oblanceolate. **FL**: outer sepals 0.5–4.5 mm , < 0.5 mm wide, linear; inner sepals 2.5–7 mm, 2–3.5 mm wide, ovate, acuminate; petals 3–11 mm, obovate; stamens 12–45. **FR** 2.5–4 mm, ovoid. **SEEDS** 4–10. Dry sandy or rocky soil of hills, slopes, ridges; < 1500 m. NCoR, n&c SNF, n SNH, SnJV, CW, SCo, ChI, SnBr, PR. [*H. suffrutescens* B. Schreiber, Bisbee Peak rush-rose] Threatened by mining.

TUBERARIA

Ann, per; hairs spreading, long, white or short, red. **ST** erect. **LVS** in basal rosettes, without stipules, often withering early; cauline, gen opposite, stipuled or not. **INFL** raceme-like. **FL**: sepals 5, outer 2 gen narrower, bract-like; petals yellow; stamens 10–many; style 0, stigma large, hemispheric. **FR** 3-valved. 12 spp.: Eur. (Latin: from tuber-like swellings on roots of some)

T. guttata (L.) Fourr. **ST** < 30 cm. **LF** 1–5 cm, linear; margin ± rolled under. **FL**: pedicel 7–15 mm, slender; petals 7–10 mm, gen with a dark spot at base. Uncommon. Disturbed places; 80–150 m. c SNF; native to Eur. [*Helianthemum g.* (L.) Miller]

CONVOLVULACEAE MORNING-GLORY FAMILY

Lauramay T. Dempster (except *Calystegia*)

Per (ann), gen twining or trailing. **LVS** alternate. **INFL**: cyme or fls solitary in axils; pedicels often with 2 bracts. **FL** bisexual, radial; sepals 5, ± free, overlapping, persistent, often unequal; corolla gen showy, gen bell-shaped, ± shallowly 5-lobed, gen pleated and twisted in bud; stamens 5, epipetalous; pistil 1, ovary superior, chambers gen 2, ovules gen 2 per chamber, styles 1–2. **FR**: gen capsule. **SEEDS** 1–4 (6). 50 genera, 1,000 spp.: warm temp to trop; some cult as orn. (Family description, key to genera by L.T. Dempster)

1. Lf < 1 cm, ± sessile, ± elliptic; st not twining .... CRESSA
1′ Lf gen > 1 cm, gen petioled, blade gen reniform, hastate, or deeply lobed (sometimes oblanceolate); st often twining
 2. Stigma 1, head-like .... IPOMOEA
 2′ Stigmas 2, ± linear
  3. Densely matted herb; ovary gen ± 2-lobed; styles 2 .... DICHONDRA

3′ Tufted or twining herbs; ovary not lobed; style 1
4. Calyx > 7 mm; corolla gen > 3 cm; stigma lobes cylindric or oblong, ± flattened .... **CALYSTEGIA**
4′ Calyx < 5 mm; corolla < 3 cm; stigma lobes cylindric or thread-like, not flattened .... **CONVOLVULUS**

CALYSTEGIA MORNING-GLORY

Richard K. Brummitt

Per, subshrub from caudex or rhizome, glabrous to tomentose. **ST** very short to high-climbing, gen twisting and twining. **LF** gen > 1 cm, linear to reniform, often sagittate to hastate, rarely deeply divided. **INFL**: peduncle gen 1-fld; bractlets small and remote from calyx to large and concealing calyx, sometimes lobed. **FL** gen showy; corolla glabrous, white or yellow to pink or purple; ovary chamber 1 (septa gen incomplete), stigma lobes 2, gen swollen, cylindric or oblong, ± flattened. **FR** ± spheric, ± inflated. **SEEDS** gen ± 4. ± 150 spp.: temp, worldwide. (Greek: concealing calyx, from bractlets of some) [Brummitt 1980 Kew Bull 35(2):327–328] Intergradation common; intermediate forms often difficult to identify. Appears similar to *Convolvulus*, but anatomy suggests that the 2 genera are not very closely related.

1. Bractlets attached > 1 mm below calyx and not concealing it, < 4 mm wide or variously toothed or lobed or lf-shaped
2. Pl glabrous throughout
3. Bractlets elliptic to widely elliptic-oblong, entire .... ***C. peirsonii***
3′ Bractlets linear, entire or with basal lobes
4. St ± stiffly erect or intertwining; subshrub; lf linear to narrowly triangular, entire or lobes ± linear with 1 tip, sinus rounded .... ***C. longipes***
4′ St strongly climbing; vine; lf triangular to reniform, lobes wide, truncate or with 2 tips, sinus V-shaped or nearly closed .... ***C. purpurata***
5. St climbing < 7 m; lf triangular, tip acute, sinus V-shaped, bractlets rarely lobed .... ssp. ***purpurata***
5′ St trailing or weakly climbing; lf ovate-triangular to reniform, tip gen rounded to notched, sinus almost closed; bractlets gen with small lobes .... ssp. ***saxicola***
2′ Pl puberulent to tomentose at least around lf sinus or at top of peduncle
6. Bractlets entire, not lf-shaped, ± sessile .... ***C. occidentalis*** ssp. ***occidentalis***
6′ Bractlets toothed or lobed or lf-shaped, petioled
7. Lf lobes 7–9, linear, finger-like .... ***C. stebbinsii***
7′ Lf ± entire or lobes 2, not linear nor finger-like
8. Lf blade widely triangular to ± reniform, tip acute to acuminate, lobes 2-tipped; hairs ± shaggy .... ***C. malacophylla*** ssp. ***malacophylla***
8′ Lf blade gen narrowly triangular, tip ± acute, lobes 1-tipped; puberulent or hairs fine .... ***C. occidentalis*** ssp. ***fulcrata***
1′ Bractlets attached just below calyx and ± concealing it, often > 4 mm wide, not lobed or lf-shaped
9. Bractlets < 4 mm wide
10. Pl densely hairy .... ***C. collina*** ssp. ***tridactylosa***
10′ Pl glabrous or minutely appressed-hairy
11. Lf blade narrowly triangular, lobes 1/2 blade length, ± abruptly spreading and curved backward ... ***C. peirsonii***
11′ Lf blade triangular-hastate to reniform
12. Bractlets ± obtuse; pl glabrous .... [2]***C. atriplicifolia*** ssp. ***buttensis***
12′ Bractlets acute; pl gen minutely appressed-hairy .... [2]***C. subacaulis*** ssp. ***episcopalis***
9′ Bractlets > 4 mm wide
13. Pl strongly climbing, st > 1 m
14. Pl from rhizomes, st not woody; peduncle 1-fld; marshes, river banks .... ***C. sepium***
15. Bractlets < sepals; corolla ± 40 mm .... ssp. ***binghamiae***
15′ Bractlets gen > sepals; corolla 35–70 mm .... ssp. ***limnophila***
14′ Pl from woody caudex, lower st often woody; peduncle often > 1-fld; gen dry places .... ***C. macrostegia***
16. Bractlets 16–30 mm wide, strongly keeled to ± sac-like at base
17. Bractlets 19–37 mm; calyx 16–25 mm; corolla 47–68 mm; stamens 23–35 mm; s ChI ..... ssp. ***amplissima***
17′ Bractlets 13–26 mm; calyx 10–22 mm; corolla 36–60 mm; stamens 17–26 mm; n ChI .... ssp. ***macrostegia***
16′ Bractlets 4–16 mm wide, flat or slightly keeled
18. Bractlets obtuse .... ssp. ***cyclostegia***
18′ Bractlets acute
19. Lf densely short-hairy .... ssp. ***arida***
19′ Lf glabrous to puberulent
20. Lf blade triangular, > 7 mm wide .... ssp. ***intermedia***
20′ Lf blade ± linear, < 7 mm wide .... ssp. ***tenuifolia***
13′ Pl not or only weakly climbing, st < 1 m
21. Pl glabrous or sparsely appressed-puberulent; lf triangular-hastate to reniform
22. Corolla deep pink or purplish; lf reniform, glabrous, slightly fleshy; ocean beaches, dunes ..... ***C. soldanella***
22′ Corolla white or yellowish, rarely pink-tinged; lf various; inland
23. Bractlets acute; pl gen minutely appressed-hairy .... [2]***C. subacaulis*** ssp. ***episcopalis***
23′ Bractlets obtuse; pl glabrous .... [2]***C. atriplicifolia*** ssp. ***buttensis***
21′ Pl with dense, spreading hairs or lf not triangular-hastate to reniform
24. St well developed, < 100 cm, without basal rosette of lvs

25. Pl finely tomentose; lf margin wavy .... [2]*C. collina* ssp. ***venusta***
25′ Pl tomentose to shaggy-hairy; lf margin not wavy .... ***C. malacophylla***
26. Lf widely triangular to ± reniform, lobes gen each 2-tipped, lf tip gen obtuse to notched; hairs gen brown or golden-brown; CaR, SN .... ssp. ***malacophylla***
26′ Lf ± narrowly triangular, lobes gen 1-tipped (2nd tip 0 or poorly developed), lf tip gen acute; hairs gen grayish; SnFrB, SCoR, WTR .... ssp. ***pedicellata***
24′ St poorly developed, < 15 cm, < or = basal rosette of lvs
27. Hairs sparse to moderately dense, short, irregular, spreading; lf margin not or barely wavy .... ***C. subacaulis*** ssp. ***subacaulis***
27′ Hairs dense, short to long, matted; lf margin wavy .... ***C. collina***
28. Sepals finely tomentose .... [2]ssp. ***venusta***
28′ Sepals glabrous or strigose along center
29. Lf tip obtuse, lobes gen indistinct, blade triangular to reniform .... ssp. ***collina***
29′ Lf tip acute, lobes clearly defined, blade triangular .... ssp. ***oxyphylla***

C. atriplicifolia Hallier f. ssp. ***buttensis*** Brummitt (p. 519) BUTTE COUNTY MORNING-GLORY Per from rhizome, gen glabrous. **ST** decumbent to ± erect, 10–50 cm. **LF**: blade gen 2–4 cm at midrib, gen ± equilaterally triangular-hastate. **INFL**: peduncle < 5 cm, gen < subtending lf; bractlets 8–12 mm, 4–8 mm wide, ± cordate-elliptic, entire, obtuse, attached just below calyx, flat, not fully concealing calyx. **FL**: sepals 10–15 mm; corolla 30–45 mm, white or pink-tinged. UNCOMMON. Dry, rocky places in open forest or chaparral; 600–1200 m. KR (Del Norte Co.), CaRF (Butte Co.), SnFrB (Mt. Diablo). [*Convolvulus nyctagineus* E. Greene, in part] Scattered, uncommon, and variable in CA. Ssp. *atriplicifolia* of OR & WA differs by lvs gen 4–6 cm at midrib, more rounded, less hastate; bractlets gen 15–22 mm; corolla gen 50–70 mm. KR pls are intermediate to ssp. *a*. SnFrB pls are ± intermediate to *C. subacaulis*. Also intergrades with *C. occidentalis* ssp. *o*. ❀TRY.

C. collina (E. Greene) Brummitt Per from rhizome, gen densely tomentose; hairs gen long, brownish. **ST** decumbent, 8–30 cm, not or weakly climbing. **LF** gen < 3 cm at midrib, reniform (basal lobes indistinct) to distinctly lobed; margin gen wavy. **INFL**: peduncle < 6 cm, sometimes > subtending lf; bractlets 7–17 mm, 2–14 mm wide, ± linear to widely ovate, entire, attached just below calyx, partly to entirely concealing it. **FL**: sepals 8–13 mm; corolla 25–55 mm, white. Open grassy or rocky places or in open oak-pine woods, often on serpentine; < 600 m. NCoR, SnFrB, SCoR. [*Convolvulus c.* E. Greene; *C. malacophyllus* ssp. *c.* (E. Greene) Abrams]

ssp. ***collina*** (p. 519) Pl hairs sometimes grayish. **ST** 8–15 cm. **LF** reniform to triangular; lobes gen not clearly defined. **INFL**: bractlets 8–15 mm, 6–14 mm wide, widely ovate, ± concealing calyx. **FL**: sepals glabrous or strigose at center; corolla 30–54 mm. Habitat of sp. NCoR, SnFrB. ❀7,8,9,**14**;SUN;**15–17**;GRCVR; INV.

ssp. ***oxyphylla*** Brummitt (p. 519) MOUNT SAINT HELENA MORNING-GLORY **ST** 8–15 (20) cm. **LF**: basal lobes 2, distinct. **INFL**: bractlets 8–17 mm, 5–11 mm wide, lanceolate to widely ovate, ± concealing calyx. **FL**: sepals glabrous or strigose at center; corolla 27–53 mm. UNCOMMON. Habitat of sp. NCoR (Lake, Sonoma, Napa cos.). Intermediate between sspp. *collina* and *tridactylosa*. ❀TRY.

ssp. ***tridactylosa*** (Eastw.) Brummitt (p. 519) **ST** gen 15–30 cm. **LF**: basal lobes 2, distinct; margin barely wavy. **INFL**: bractlets 7–10 mm, 2–3.5 mm wide, linear-elliptic to lanceolate, not fully concealing calyx. **FL**: outer sepals tomentose; corolla 27–33 mm. Habitat of sp. NCoR (Mendocino, Lake cos.) [*Convolvulus t.* Eastw.] May intergrade with *C. occidentalis* ssp. *o.*, *C. malacophylla* ssp. *m.*

ssp. ***venusta*** Brummitt (p. 519) SOUTH COAST RANGE MORNING-GLORY Pl finely tomentose or ± silky; hairs short, brownish to whitish. **ST** 8–30 cm. **LF** ± triangular to ± reniform; lobes distinct or not. **INFL**: bractlets 8–16 mm, 4–10 mm wide, lanceolate to widely ovate, ± flat, spreading, so not concealing calyx. **FL**: sepals finely tomentose; corolla 25–44 mm. UNCOMMON. Habitat of sp. SCoR. Probably intergrades with *C. subacaulis* ssp. *s.* and *C. malacophylla* ssp. *pedicellata*. ❀TRY.

C. longipes (S. Watson) Brummitt (p. 523) Subshrub from woody caudex, ± hemispheric, glabrous. **ST** stiffly erect or intertwining, 3–10 dm. **LF** < 6 cm, linear to narrowly triangular; lobes linear; sinus rounded; upper lvs less lobed. **INFL**: peduncle gen 1-fld, < 20 cm, >> subtending lf; bractlets 3–17 mm, 0.2–3 mm wide, linear, often with basal lobes, gen alternate, attached 4–50 mm below calyx. **FL**: sepals 8–11 mm; corolla 28–36 mm, white or cream-colored to pale pink or lavender. Dry, rocky places, desert scrub; 600–1300 m. s SnJV, w edge D; NV, AZ. [*Convolvulus l.* S. Watson, *C. linearilobus* Eastw.] Intergrades with *C. macrostegia* ssp. *tenuifolia* in San Diego Co., *C. peirsonii* in Los Angeles Co., *C. malacophylla* ssp. *pedicellata* in San Luis Obispo Co. ❀DRY:7,8, **9**,10–12,**14,18–23**&SUN:**15–17,24**.

C. macrostegia (E. Greene) Brummitt Per or subshrub from woody caudex, glabrous to densely short-hairy. **ST** slender and weakly climbing to woody and high-climbing, 1–9 m. **LF** < 13 cm, gen widely triangular and lobed. **INFL**: peduncle 1–several-fld, gen > subtending lf; bractlets 6–37 mm, 4–30 mm wide, lanceolate to ± round, entire, flat to sac-like, attached just below calyx and ± concealing it. **FL**: sepals 7–25 mm; corolla 22–68 mm, white or fading pink. Gen, dry, rocky ± coastal places; < 1000 m. CCo, SCoR, SCo, ChI; to Mex. [*Convolvulus macrostegius* E. Greene, *C. occidentalis* var. *macrostegius* (E. Greene) Munz]

ssp. ***amplissima*** Brummitt (p. 523) ISLAND MORNING-GLORY Pl becoming ± glabrous. **LF**: lobes 2–3-tipped; sinus widely rounded. **INFL**: bractlets 19–37 mm, 17–30 mm wide, widely ovate, strongly keeled to sac-like, tip round to acuminate. **FL**: sepals 16–25 mm; corolla 47–68 mm. UNCOMMON. Rocky slopes, canyon walls; gen < 100 m. s ChI. Most like, but larger than, ssp. *macrostegia*. ❀7,9,**14–17**,18,19,**20–24**;GRCVR.

ssp. ***arida*** (E. Greene) Brummitt (p. 523) Pl short-hairy. **ST** slender, trailing or low-climbing. **LF**: lobes 1-tipped; sinus rounded to square. **INFL**: bractlets 10–21 mm, 6–10 mm wide, lanceolate, not or barely keeled, tip acute. **FL**: sepals 9–12 mm; corolla 24–34 mm. Coastal scrub, chaparral; < 1000 m. SnGb, SnBr, n PR. [*Convolvulus a.* E. Greene] Intergrades with ssp. *intermedia*, ssp. *tenuifolia*, *C. occidentalis* ssp. *fulcrata*, probably *C. peirsonii*. ❀TRY.

ssp. ***cyclostegia*** (House) Brummitt (p. 523) Pl glabrous to puberulent. **ST** trailing or climbing, < 2.5 m. **LF** triangular; lobes gen ± 2-tipped; sinus acute, rounded or squarish. **INFL**: bractlets 6–20 mm, 7–16 mm wide, widely ovate to ± round, flat to barely sac-like, tip notched to acuminate. **FL**: sepals 9–15 mm; corolla 28–52 mm. Coastal scrub; < 350 m. CCo, n SCo. [*Convolvulus c.* House] Intergrades with ssp. *macrostegia*, ssp. *intermedia*, *C. purpurata* ssp. *p.* ❀7–9,**14–17**,18,19,**20–24**.

ssp. ***intermedia*** (Abrams) Brummitt (p. 523) Pl glabrous to puberulent. **ST** trailing or climbing. **LF**: lobes rounded to 2-angled; sinus ± acute to rounded or square. **INFL**: bractlets 10–20 mm, 6–12 mm wide, lanceolate, not or slightly keeled. **FL**: sepals 9–16 mm; corolla 2–40 mm. Coastal or inland hills; gen < 200 m. SCo, s ChI, s WTR, SnGb, n PR. [*Convolvulus aridus* ssp. *i.* Abrams; *C. a.* ssp. *longilobus* Abrams] Intergrades with other sspp., probably with *C. peirsonii*. ❀TRY.

ssp. ***macrostegia*** Pl becoming ± glabrous. **ST** climbing, often > 4 m. **LF**: lobes gen 2-tipped; sinus widely rounded. **INFL**: bractlets 13–26 mm, 16–27 mm wide, widely ovate to ± round, strongly keeled to sac-like, tip notched to acuminate. **FL**: sepals 10–22 mm; corolla 36–60 mm. Coastal scrub; gen < 500 m. n ChI; Guadalupe & San Martin Islands, Baja CA. Intergrades with sspp. *amplissima*, *cyclostegia*, *intermedia*. ❀7–9,**14–17**,18,19,**20–24**.

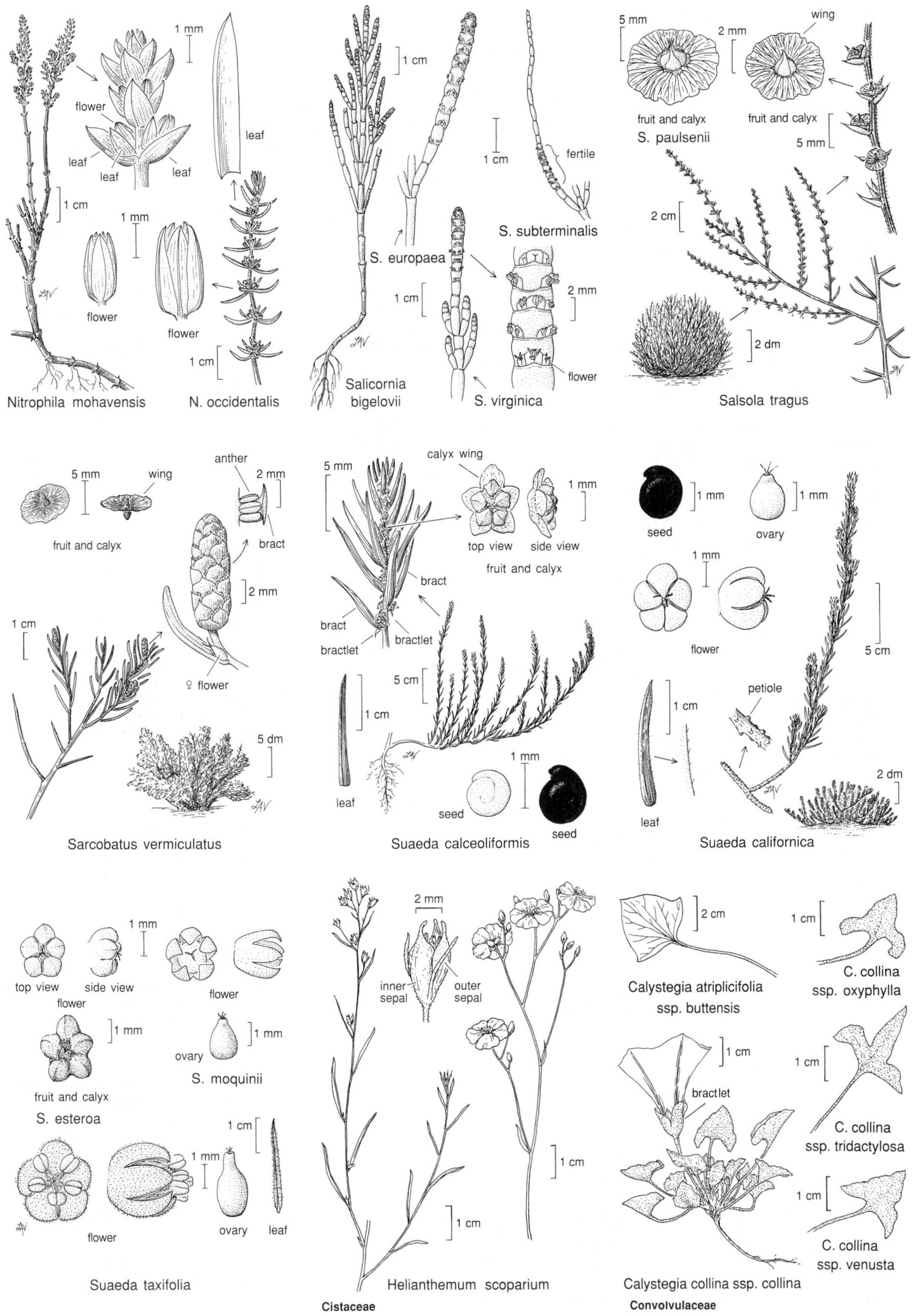

Cistaceae

Convolvulaceae

ssp. ***tenuifolia*** (Abrams) Brummitt (p. 523) Pl glabrous to puberulent. **ST** trailing or low-climbing. **LF**: blade linear to very narrowly triangular; lobes linear; sinus rounded or square. **INFL**: bractlets 8–14 mm, 4–8 mm wide, narrowly lanceolate, flat, tip acute. **FL**: sepals 7–10 mm; corolla 22–40 mm. Dry, ± inland places; gen < 500 m. c&s SCoR, SCo, w PR; Baja CA. [*Convolvulus aridus* ssp. *t.* Abrams] Intergrades with ssp. *intermedia*, ssp. *arida, C. longipes*. ❀TRY.

C. malacophylla (E. Greene) Munz Per from rhizome, densely shaggy-tomentose. **ST** decumbent to ascending, 10–100 cm, not or barely climbing. **LF** < 6 cm, < 9 cm wide (gen much smaller), narrowly triangular to ± reniform; lobes 1–2-tipped; margin not wavy; tip notched to acute. **INFL**: peduncle 6–9 cm, < or > subtending lf; bractlets 7–20 mm, 5–15 mm wide, entire, lanceolate to widely ovate, gen acute, gen attached just below calyx and ± concealing it. **FL**: sepals 9–15 mm, ± densely hairy; corolla 20–45 mm, white. Dry slopes, chaparral; 300–3500 m. CaR, SN, SnFrB, SCoR, WTR. [*Convolvulus m.* E. Greene]

ssp. ***malacophylla*** (p. 523) Pl hairs gen brown or golden-brown (grayish). **LF** 3–6 cm, 4–9 cm wide, widely triangular to ± reniform; lobes gen 2-tipped; tip notched to acute. **INFL**: peduncle 3–5 cm, < subtending lf. Habitat of sp.; gen 1000–2400 m. CaR, SN. Intergrades with *C. occidentalis* and perhaps *C. collina* ssp. *tridactylosa*. Pls with bractlets attached well below calyx from s SN may be called var. *berryi* (Eastw.) Brummitt [Berry's morning-glory; UNCOMMON]. ❀DRN:1,2,7,14–17.

ssp. ***pedicellata*** (Jepson) Munz (p. 523) Pl hairs gray (brownish). **LF** 3–4.5 cm, 3–4 cm wide, ± narrowly triangular; lobes gen 1-tipped; tip ± obtuse to acute. **INFL**: peduncle 6–9 cm, often > subtending lf. Habitat of sp.; 300–1900 m. SnFrB, SCoR, WTR. [*Convolvulus m.* ssp. *p.* (Jepson) Abrams] Intergrades with ssp. *malacophylla, C. occidentalis* ssp. *fulcrata, C. collina* ssp. *venusta, C. purpurata* ssp. *p., C. macrostegia* ssp. *cyclostegia, C. longipes, C. subacaulis*. ❀DRN:7,14–17.

C. occidentalis (A. Gray) Brummitt Per from woody caudex; puberulent to finely tomentose. **ST** decumbent to strongly climbing. **LF**: blade gen 1.5–4 cm at midrib; lobes gen ± distinct, rounded to 2-tipped; sinus rounded to ± square or tapered. **INFL**: peduncle 1–4-fld, < to ± > subtending lf; bractlets 5–12 mm, linear to nearly round, sessile or stalked, entire or lobed like lvs, attached 1–15 mm below calyx, gen ± overlapping but not concealing it. **FL**: sepals 9–15 mm; corolla 20–48 mm, white to creamy-yellow. Dry slopes, chaparral, pine woods; 300–2700 m. CA-FP, GB; OR. [*Convolvulus o.* A. Gray].

ssp. ***fulcrata*** (A. Gray) Brummitt (p. 523) **ST** trailing or weakly climbing, gen < 1 m. **LF**: lobes distinct, gen 1- or 2-tipped. **INFL**: peduncle 1-fld; bractlets attached 2–15 mm below calyx, 5–30 mm, 2–7 mm wide, narrowly lanceolate to widely triangular, lobed like lvs. Habitat of sp. SNF, TR, PR. [*Convolvulus fulcratus* (A Gray) E. Greene; *C. deltoideus* E. Greene] Intergrades with ssp. *occidentalis, C. malacophylla*, and *C. macrostegia* ssp. *arida*. ❀TRY.

ssp. ***occidentalis*** (p. 523) **ST** decumbent to strongly climbing, gen > 1 m. **LF**: lobes gen indistinct, gen 2-tipped. **INFL**: peduncle often > 1-fld; bractlets attached 1–7 mm below calyx, 4–18 mm, 1–5 mm wide, gen ± linear-oblong, entire. Habitat of sp.; < 1200 m. KR, NCoR, CaRH, SNH, SnFrB, MP; OR. [*Convolvulus fruticetorum* E. Greene, *C. polymorphus* E. Greene] Intergrades with ssp. *fulcrata, C. atriplicifolia* ssp. *buttensis, C. collina* ssp. *tridactylosa, C. malacophylla, C. peirsonii, C. subacaulis, C. purpurata* ssp. *p.* Pls from CaR, s SN (Greenhorn Mtns), and Teh with lvs ± equilateral, bractlets attached ± 1 mm below calyx, may be called var. *tomentella* (E. Greene) Brummitt [*Convolvulus t.* E. Greene]. ❀DRN:1–3,7,14&SUN:15–17.

C. peirsonii (Abrams) Brummitt (p. 523) PEIRSON'S MORNING-GLORY Per from rhizome, glabrous, ± glaucous. **ST** decumbent to weakly, climbing, < 0.4 m. **LF**: blade < 2 cm at midrib, narrowly triangular; lobes sometimes = blade, broadest at tip, gen distinctly 2-tipped; sinus rounded to ± square. **INFL**: peduncle 2–8 cm, gen < subtending lf; bractlets attached 0–3 mm below calyx and overlapping it, 3–7 mm, 2.5–4 mm wide, elliptic to widely oblong, entire, flat. **FL**: calyx 9–13 mm; corolla 25–40 mm, white. UNCOMMON. Rocky slopes; 1000–1500 m. n SnGb, adjacent DMoj (Antelope Valley). [*Convolvulus p.* Abrams] Intergrades with *C. longipes*, mainland coastal *C. macrostegia, C. occidentalis* ssp. *o.*, perhaps *C. sepium*. ❀TRY.

C. purpurata (E. Greene) Brummitt Per or subshrub from woody caudex, glabrous, often glaucous. **ST** decumbent to strongly climbing, < 7 m. **LF**: blade gen 1.5–5 cm at midrib, triangular; lobes spreading, gen 2–3-tipped; sinus V-shaped or ± closed. **INFL**: peduncle 1–5-fld, gen >> subtending lf; bractlets attached 3–16 mm below calyx, gen not overlapping it, 2–16 mm, 0.4–1.5 mm wide, gen linear. **FL**: sepals 7–14 mm; corolla 23–52 mm, white or cream-colored to purple (often ± purple-striped). Chaparral, coastal scrub; < 300 m. NCo, NCoRO, GV (Sutter Buttes), CCo, SnFrB, SCoRO, n SCo, w WTR. Sspp. intergrade.

ssp. ***purpurata*** (p. 523) **ST** gen robust, strongly climbing, > 1 m. **LF**: blade ± triangular; sinus V-shaped; tip acute; lobes gen strongly angled; margin not wavy. **INFL**: bractlets opposite, gen entire. Habitat and range of sp. [*Convolvulus occidentalis* var. *p.* (E. Greene) J. Howell; *C. o.* var. *solanensis* (Jepson) J. Howell] Intergrades with *C. macrostegia* ssp. *cyclostegia, C. occidentalis* ssp. *o.*, perhaps *C. malacophylla* ssp. *pedicellata*. ❀9,14,**15–17**.

ssp. ***saxicola*** (Eastw.) Brummitt (p. 523) **ST** trailing or weakly climbing, < 1 m. **LF** ovate-triangular to reniform; sinus gen ± closed; tip gen notched to rounded; lobes rounded; margin ± wavy. **INFL**: bractlets often ± alternate, gen lobed ± like lvs. Rocky coastal scrub; < 100 m. s&c NCo, n SnFrB. [*Convolvulus occidentalis* var. *s.* (Eastw.) J. Howell] ❀17.

C. sepium (L.) R.Br. HEDGE BINDWEED Per from rhizome, glabrous to hairy. **ST** climbing, < 4 m. **LF**: blade gen 4–8 cm at midrib, lobed. **INFL**: peduncle < subtending lf; bractlets attached just below calyx and ± concealing it, entire, flat or keeled. **FL**: sepals gen 10–18 mm; corolla 30–70 mm, white or pink. $2n$=20,22,24. Salt and freshwater marshes; < 500 m. SnFrB, SCoR, SCo, TR; temp regions worldwide. Highly variable, many geographic sspp.

ssp. ***binghamiae*** (E. Greene) Brummitt Pl glabrous to sparsely hairy. **LF**: lobes not or barely spreading, ± entire to ± obscurely 2-tipped; sinus widely rounded; tip acute to ± obtuse. **INFL**: bractlets 8–12 mm, < sepals, 4–8 mm wide, elliptic, ± flat. **FL**: corolla gen ± 40 mm, white. Coastal marshes; < 20 m. n&c SCo. [*Convolvulus b.* E. Greene; *C. s.* var. *dumetorum* Posp.] ❀TRY.

ssp. ***limnophila*** (E. Greene) Brummitt (p. 523) Pl gen glabrous (densely hairy). **LF**: lobes ± abruptly spreading; sinus rounded to square; tip acute. **INFL**: bractlets 13–28 mm (= or > sepals), 8–18 mm wide, narrowly ovate, flat or keeled. **FL**: corolla 35–70 mm, white or pink-tinged. Marshes; < 500 m. Deltaic GV, SnFrB, SCoR, TR, e DMoj (Amargosa River, 500 m); to e US, n Mex. [*Convolvulus l.* E. Greene; *C. s.* var. *repens* (L.) A. Gray] Intergrades with *C. peirsonii*. ❀TRY.

C. soldanella (L.) R.Br. (p. 523) Per from rhizome, glabrous. **ST** decumbent, < 0.6 m. **LF**: blade 1–3 cm at midrib, 1.5–2 × wider than long, reniform, slightly fleshy. **INFL**: peduncle 3–6 cm, gen > subtending lf; bractlets attached just below calyx and ± concealing it, 7–16 mm, 5–10 mm wide, ovate to ± round, flat or enfolding calyx; tip notched to obtuse. **FL**: sepals 10–16 mm; corolla 32–52 mm, pink. $2n$=22. Sandy seashores, coastal strand; < 50 m. NCo, CCo, n&c SCo; worldwide, esp temp. [*Convolvulus s.* L.] ❀DRN,IRR:**17,24**;DFCLT.

C. stebbinsii Brummitt (p. 523) STEBBINS' MORNING-GLORY Per from rhizome or barely woody caudex; hairs appressed to spreading, whitish. **ST** trailing to climbing, < 1 m. **LF**: lobes 7–9, 7–55 mm, deep, palmate, linear, middle longest; margin gen rolled under. **INFL**: peduncle 3–13 cm, gen >> subtending lf; bractlets attached well below calyx and not concealing it, < 1.8 cm, 3–9-lobed like lvs. **FL**: sepals 7–11 mm, gen glabrous; corolla 30–35 mm, creamy-yellow, ± pink-tinged. Chaparral; 300 m. **ENDANGERED** CA. n SNF (El Dorado Co.). 2 populations known.

C. subacaulis Hook. & Arn. Per from rhizome or woody caudex; ± hairy. **ST** decumbent or ascending, 2–20 cm. **LF**: blade 3–4 cm at midrib, ± triangular-hastate, sometimes with small backwardly directed lobes; tip acute to rounded; base tapered. **INFL**: peduncle < 3.3 cm, < subtending lf; bractlets attached just below calyx and ± concealing it, 7–17 mm, 4–9 mm wide, widely ovate to lanceolate or oblong. **FL**: sepals 10–13 mm; corolla 33–62 mm, white or cream-colored, often purple-tinged along midveins. Dry, open scrub or woodland; < 500 m. s NCoR, SnFrB, SCoR. [*Convolvulus s.* (Hook. & Arn.) E. Greene] Sspp. intergrade.

ssp. ***episcopalis*** Brummitt SAN LUIS OBISPO COUNTY MORNING-GLORY Pl hairs sparse or appressed, inconspicuous. **ST** decumbent to ascending, ± 20 cm, **LF**: basal rosette 0; blade ± narrowly triangular-hastate, acute. **INFL**: bractlets gen ± 12 mm, gen ± 4 mm wide, lanceolate, acute. UNCOMMON. Habitat of sp. c SCoRO (San Luis Obispo Co.). ❀TRY.

ssp. ***subacaulis*** (p. 523) Pl hairs ± dense, spreading or deflexed, obvious. **ST** gen ± 2 cm. **LF**: basal rosette > fls; blade ± widely rounded to acuminate. **INFL**: bractlets 7–17 mm, 4–9 mm wide, widely ovate to oblong, rounded. Habitat of sp. s NCoR, SnFrB. Intergrades with *C. collina* ssp. *venusta*, *C. occidentalis* ssp. *o.*, *C. malacophylla* ssp. *pedicellata*. ❀TRY.

CONVOLVULUS MORNING-GLORY, BINDWEED

Ann, per from caudex or rhizome, gen ± glabrous. **ST** trailing to high-climbing, gen twisting and twining. **LF** gen > 1 cm, gen petioled; blade gen cordate or hastate. **INFL**: bracts gen 2, below calyx. **FL** gen showy; corolla gen funnel-shaped, pleated, 5-angled or lobed; stamens incl; ovary chambers 2, septa complete, stigma lobes 2, linear or thread-like, not flattened. **FR** spheric, ± inflated. **SEEDS** gen 4. (Latin, entwine) 250 spp.: gen temp. Not easily distinguished from *Calystegia*.

1. Lf oblanceolate, entire, narrowed to base; corolla ± 0.6 cm, deeply lobed; ann .... ***C. simulans***
1′ Lf hastate or lobed, petioled; corolla 2–6 cm, not cleft; per
 2. Lf, at least upper, deeply lobed; corolla purple to deep pink; n SNF, SW .... ***C. althaeoides***
 2′ Lf hastate; corolla white or pinkish; widespread weed .... ***C. arvensis***

C. althaeoides L. Per. **ST** climbing. **LVS** variable, 3–4 cm; at least some blades deeply 3–many-lobed. **INFL**: peduncle 3–4 cm, 1-fld, reflexed below fl. **FL**: calyx ± 10 mm, lobes unequal, widely ovate or spoon-shaped; corolla 3–4 cm, funnel-shaped, deep pink. $2n$=40. Disturbed places; < 1000 m. n SNF (Nevada Co.), SW; native to Medit. Localized weed.

C. arvensis L. (p. 523) BINDWEED, ORCHARD MORNING-GLORY Per from deep persistent root, ± glabrous to minutely hairy. **ST** prostrate or twining. **LF** gen 2–3 cm; blade round, oblong or ovate, base hastate, tip gen rounded. **INFL**: peduncles gen 2.5–6 cm, 1–several-fld, in fr gen reflexed; bracts ± linear. **FL**: calyx ± 5 mm, lobes oblong; corolla 2–2.5 cm, open funnel-shaped, white, purplish outside, particularly at folds, margin entire. **FR** ± 8 mm. n=24. Abundant. Orchards, gardens; gen < 1500 m. CA; native to Eur. NOXIOUS WEED.

C. simulans Perry (p. 523) Ann, strigose. **ST** diffusely branched, 10–40 cm. **LF** entire, gen < 6 cm, oblanceolate; blade narrowed gradually to petiole. **INFL**: peduncles 1-fld, short, ± sharply nodding in fr; bracts 3–7 mm, linear or narrowly oblanceolate, well below calyx. **FL**: calyx lobes 3–4 mm, oblong-obovate; corolla ± 0.6 cm, bell-shaped, pinkish or bluish, lobes ± 1/2 tube length, ± ascending. **FR** ± 7 mm. Wet clay, serpentine ridges; 30–700 m. SnJV, CW, SW, s ChI; Baja CA. [not *C. pentapetaloides* L.] Probably native.

CRESSA ALKALI WEED

Per, subshrub, canescent. **ST** prostrate to erect, not twining. **LF** entire. **INFL**: fls solitary in upper axils, appearing like a 1-sided, leafy raceme. **FL**: calyx ± erect, concealing corolla tube; corolla lobes ± = tube, exserted; ovary chambers 2, ovules 4, styles 2, stigmas head-like. 5 spp.: trop & subtrop. (Greek: a Cretan woman)

C. truxillensis Kunth (p. 523) ALKALI WEED **ST** 7–25 cm, much branched from base, densely silky-canescent. **LF** gen < 1 cm, ± sessile, ± elliptic. **INFL**: peduncle short, bracted. **FL** 5–8 mm; sepals elliptic; corolla white, persistent, lobes ovate, acute; stamens and styles exserted. **SEED** often 1, by abortion of others. $2n$=28. Saline and alkaline soils; < 1200 m. CA-FP, DMoj; to OR, TX, Mex. [var. *vallicola* (A.A. Heller) Munz; not *C. cretica* L.] ❀DRY,SUN,ALKALINE,SALINE:**7–11**,13,**14**,15–17,**18–24**;INV.

DICHONDRA

Per, matted from creeping stolons; hairs equally forked. **LF**: petiole long; blade ± reniform. **INFL**: fls solitary in axils; bracts 0; upper peduncle recurved in fr. **FL** inconspicuous; calyx lobes 5, deep, ± equal, ovate to obovate; corolla barely > calyx, lobes > tube; ovary 2-lobed, each lobe with 2 ovules, styles 2, separate or fused at base, stigmas head-shaped. **FR** spheric to ± 2-lobed. 9 spp.: temp & trop. (Greek: double grain, from deeply lobed fr of some) [Tharp & Johnston 1961 Brittonia 13: 346–360] Cult as ground cover.

1. Fr gen deeply 2-lobed, gen separating into 2, gen 1-seeded nutlets; n CA .... ***D. donelliana***
1′ Fr a spheric to ± notched capsule, each valve often 2-seeded; s CA .... ***D. occidentalis***

D. donelliana Tharp & M. Johnston (p. 523) Pl densely brownish silky. **LF** gen < 65 mm. Uncommon. Open slopes, moist fields; 50–300 m. NCo, n SN, CW, probably introduced in SCo, n ChI. Much like *D. micrantha* Urban of TX, Mex. ❀IRR:2,3,7–9,**14**, 18,19,**23**&SUN:**15–17**;GRCVR.

D. occidentalis House (p. 523) WESTERN DICHONDRA **ST** densely brownish silky. **LF** nearly hairless, longest gen > 60 mm. UNCOMMON. Slopes, headlands, gen under shrubs; 50–500 m. SCo, s ChI; Baja CA. ❀DRN,SHD:**17**,18,19,22,23,**24**.

IPOMOEA MORNING-GLORY

Ann, per from rhizome or caudex. **ST** trailing to high-climbing. **LF** petioled; blade cordate, sometimes lobed. **INFL**: bracts 0. **FL**: corolla gen ± funnel-shaped, not or barely lobed; style 1, stigma head-like or of 2–3 spheric lobes. **FR** spheric; valves 2–4. (Greek: worm-like) 500 spp.: trop and warm temp; some cult as orn or for food (*I. batatas*, sweet-potato).

1. Lf palmately compound .... ***I. cairica***
1′ Lf simple, lobed or entire

2. Sepal 16–30 mm, tips slender; ann or per
3. Corolla 5–7 cm; lf strigose to canescent; sepals < 30 mm, gradually tapered, inconspicuously hairy (hairs ± 1 mm); per .... ***I. mutabilis***
3′ Corolla 2–4 cm; lf sparsely strigose; sepals < 16 mm, abruptly narrowed, conspicuously hairy (hairs > 2 mm); ann .... ***I. nil***
2′ Sepals 8–16 mm, tips acute to acuminate; ann
4. Corolla > 5 cm; lf not lobed, ± hairy .... ***I. purpurea***
4′ Corolla < 2 cm; lf often lobed, glabrous .... ***I. triloba***

I. cairica (L.) Sweet Per. **LF** palmately compound, ± glabrous; central lflet gen < 5 cm. **INFL**: peduncles gen 1-fld. **FL**: sepals < 1 cm, ± ovate, tip rounded; corolla ± 9 cm, white to purple. Disturbed places; ± 300 m. SCoRO (Santa Margarita Lake, San Luis Obispo Co.); native to S.Am; cult as orn. [*I. palmata* Forsskal]

I. mutabilis Lindley Per. **LF** simple; blade gen 10–15 cm, ± deeply 3-lobed, strigose to canescent. **INFL**: peduncles 1–10-fld. **FL**: sepals ± 3 cm, gradually tapered to long, slender tip, hairs near base ± 1 mm; corolla 5–7 cm, rose, becoming blue. Disturbed places; ± 50 m. CW, SW; native to Mex; cult as orn. [*I. acuminata* (M. Vahl) Roemer & Schultes]

I. nil (L.) Roth IVY MORNING-GLORY Ann. **LF** simple; blade 3–8 cm, ± deeply 3-lobed, sparsely strigose. **INFL**: peduncles 1–3-fld. **FL**: sepals gen 16–20 mm, abruptly narrowed to long, linear tip, hairs near base gen > 2 mm; corolla 2–4 cm, light blue, purple, or multi-colored, margin sometimes fringed or 5-lobed. 2*n*=30. Fields, orchards, disturbed places; 250 m. s CA-FP; native to se US; cult as orn. [not *I. hederacea* Jacq.]

I. purpurea (L.) Roth (p. 529) COMMON MORNING-GLORY Ann. **LF** simple; blade 7–12 cm, cordate, short-acuminate, entire, hairy. **INFL**: peduncles 1–5-fld. **FL**: sepals 12–16 mm, lanceolate-oblong, acute, hairy; corolla 5–6 cm, purple, blue, pink or white. 2*n*=30. Disturbed or waste places, fields, orchards; 15–100 m. SNF, GV, CW, SW; native to trop Am; cult as orn.

I. triloba L. (p. 529) Ann. **LF** simple; blade 3–6 cm, cordate, ± acuminate, entire or 3-lobed, glabrous. **INFL**: peduncles 1–5-fld. **FL**: sepals ± 8 mm, ± narrowly ovate, acuminate; corolla 1.2–2 cm, narrowly bell-shaped, shallowly 5-lobed. 2*n*=30. Fields, orchards, other disturbed places; –34–50 m. DSon; native to trop Am. NOXIOUS WEED.

CORNACEAE DOGWOOD FAMILY

James R. Shevock

Per, shrubs, trees, sometimes dioecious. **LVS** gen opposite, simple, gen entire, gen deciduous; stipules 0. **INFL**: cyme or racemes, gen umbel- or head-like, sometimes subtended by showy, petal-like bracts. **FL** gen small, gen bisexual; calyx gen 4-lobed; petals 0 or 4(5), free; stamens gen as many as and alternate petals; ovary inferior, chambers 1–4, each 1-ovuled, style simple, stigma 1–4-lobed. **FR**: gen drupe or berry. **SEEDS** gen 1–2. ± 12 genera, ± 100 spp.: esp n temp (also s trop, subtrop). Cult as orn (*Cornus, Aucuba*); some timber spp. Genera diverse; many have been treated as constituting families, but trend is to treat Cornaceae broadly. [Eyde 1987 Syst Bot 12:505–518]

CORNUS DOGWOOD

Per, shrubs, trees. **LVS** gen opposite or whorled, simple, gen deciduous; both ends gen tapered. **INFL** small, head- or umbel-like, and surrounded by showy bracts (or cyme, large, open, lacking showy bracts). **FL** gen minute; sepals 4, fused at base; petals 4; stamens 4, attached to receptacle; style 1, thread-like, stigma simple. **FR**: drupe; stone 1–2-chambered. ± 50 spp.: n temp (rare in s hemisphere); many cult as orn, some for autumn color. Some fr used for jam, syrup. (Latin: horn, from the hard wood) Divided by some into at least 6 genera.

1. Infl a cyme, gen appearing with or after lvs; petal-like bracts 0
2. Lf gen 3–5 cm, veins 3–4 pairs .... ***C. glabrata***
2′ Lf gen 5–10 cm, veins 4–7 pairs .... ***C. sericea***
3. Lf gen densely rough-hairy beneath; petals 3–4.5 mm; stone gen with 3 wide ridges on each face .... ssp. ***occidentalis***
3′ Lf gen ± glabrous or strigose beneath; petals 2–3 mm; stone gen smooth .... ssp. ***sericea***
1′ Infl a head or umbel, gen appearing before lvs; ± petal-like bracts present
4. Infl subtended by 4 ephemeral, ± brownish, not very petal-like bracts .... ***C. sessilis***
4′ Infl subtended by 4–7 persistent, whitish, showy, petal-like bracts
5. Herb or subshrub from rhizomes; bracts 4, ± 1.5 cm; fr gen ± 8 mm, spheric .... ***C. canadensis***
5′ Shrub or tree; bracts 4–7, ± 5 cm; fr gen 10–15 mm, elliptic in outline .... ***C. nuttallii***

C. canadensis L. (p. 529) BUNCHBERRY Per from rhizomes. **ST** < 2 dm, gen with 4–6 whorled lvs below infl, pair of lvs near middle of st. **LVS** 2.5–7 cm, elliptic to obovate, glabrous to strigose. **INFL**: head solitary, 2–4 cm, slender; bracts 4, 0.8–1.6 cm, ovate, whitish, petal-like. **FL**: sepals 0.4 mm; petals 1.5 mm, yellowish or ± purplish; style 1.5–2 mm. **FR** gen ± 8 mm, spheric, red; stone smooth. Moist forest, bogs; 1100 m. NW; to AK, ne Asia, e N.Am. ❀SHD,DRN,WET to moist:1–3,**4–6**,14–17;GRCVR, DFCLT:best on rotted organic matter.

C. glabrata Benth. (p. 529) BROWN DOGWOOD Shrub, small tree 1.5–6 m, gen ± glabrous, gen forming clonal thickets. **ST** slender, brownish to reddish purple. **LVS** deciduous; blade gen 2–5 cm, lanceolate to elliptic, gray-green on both surfaces or paler beneath, veins 3–4 pairs; petiole 3–8 mm. **INFL**: cymes, 2.5–4.5 cm wide; pedicels 2–3 mm. **FL**: sepals gen < 1 mm; petals 4.5–5 mm, dull white; styles 3.5 mm, slightly hairy. **FR** 8–9 mm, white to bluish; stone 5–6 mm wide, nearly smooth. Many, gen moist communities; < 1550 m. CA-FP (uncommon in s CA); OR. ❀WET or IRR:**4–9**, **14–17**,**19–24**.

C. nuttallii Audubon (p. 529) MOUNTAIN DOGWOOD Tree < 25 m. **ST**: twigs green; bark becoming dark red to almost black, hairy. **LVS** deciduous; blade 6–12 cm, narrowly elliptic to obovate, appressed-puberulent above, paler and hairier beneath; petioles 5–10 mm. **INFL**: head; subtending bracts 4–7, 4–6 cm, 3–6 cm wide, showy, petal-like, white to pinkish; receptacle convex. **FL**: sepals 2.5 mm; petals 4 mm, greenish to white; style 2 mm. **FR** 1–1.5 cm,

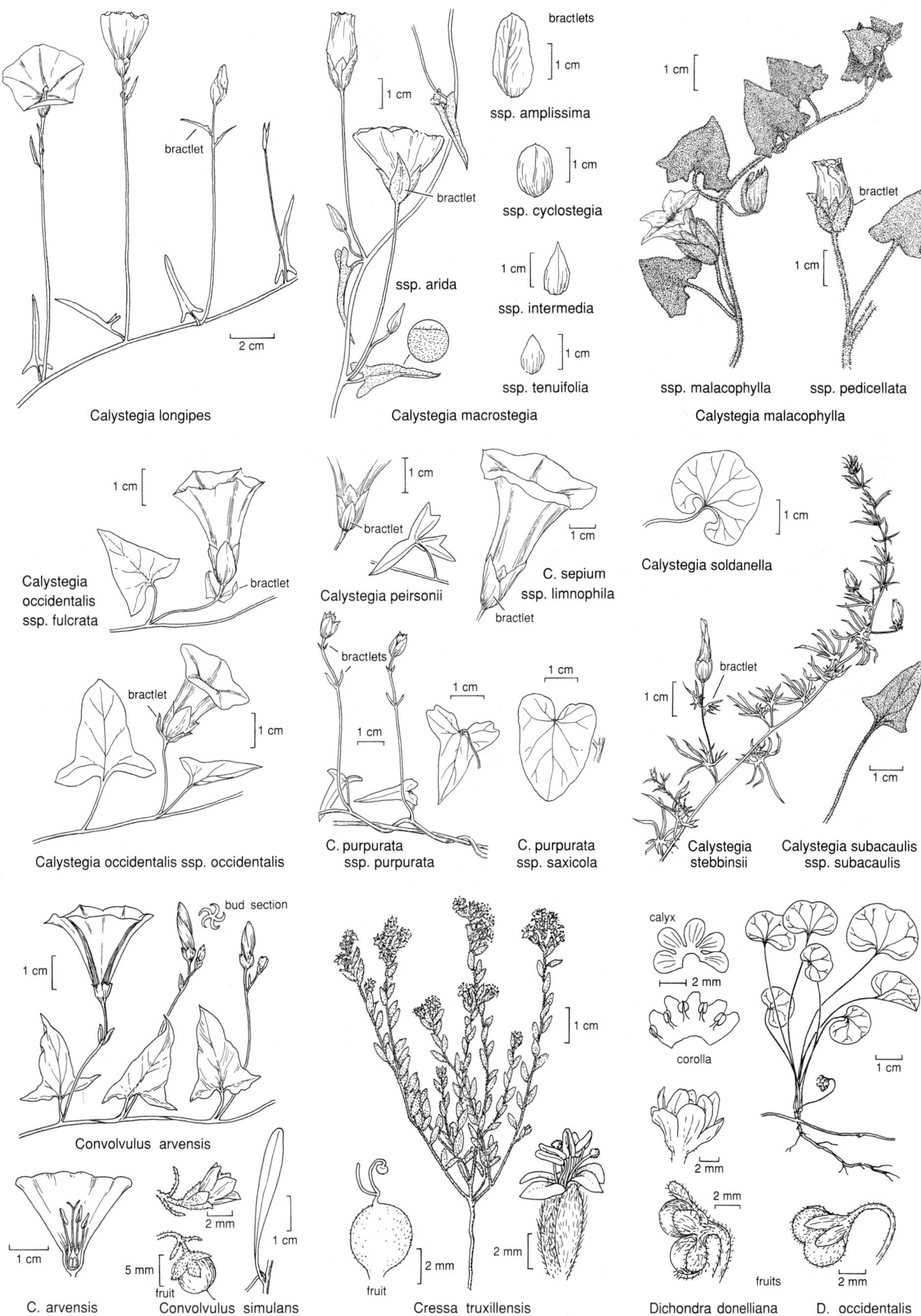
bractlet
2 cm
Calystegia longipes
1 cm
bractlet
ssp. arida
bractlets
1 cm
ssp. amplissima
1 cm
ssp. cyclostegia
1 cm
ssp. intermedia
1 cm
ssp. tenuifolia
Calystegia macrostegia
1 cm
bractlet
1 cm
ssp. malacophylla
ssp. pedicellata
Calystegia malacophylla
1 cm
Calystegia occidentalis ssp. fulcrata
bractlet
bractlet
1 cm
Calystegia occidentalis ssp. occidentalis
1 cm
bractlet
Calystegia peirsonii
1 cm
C. sepium ssp. limnophila
bractlet
bractlets
1 cm
1 cm
C. purpurata ssp. purpurata
1 cm
C. purpurata ssp. saxicola
1 cm
Calystegia soldanella
bractlet
1 cm
Calystegia stebbinsii
1 cm
Calystegia subacaulis ssp. subacaulis
bud section
1 cm
Convolvulus arvensis
1 cm
C. arvensis
2 mm
5 mm
fruit
1 cm
Convolvulus simulans
1 cm
2 mm
fruit
2 mm
Cressa truxillensis
calyx
2 mm
corolla
1 cm
2 mm
2 mm
fruits
2 mm
Dichondra donelliana
D. occidentalis

elliptic in outline, gen angled from crowding, red; stone smooth. Forests; < 2000 m. CA-FP (less common in s CA); to B.C., ID. ❀DRN:**4,5**&IRR:**6,**&SHD:1,2,7,14–17,22,23;DFCLT.

C. sericea L. AMERICAN DOGWOOD Shrub gen 1.5–4 m. **ST**: branches reddish to purple, ± glabrous to minutely strigose; older sts grayish green, gen glabrous. **LVS** deciduous; blade gen 5–10 cm, lanceolate to ovate or elliptic, paler beneath, sparsely strigose, veins 4–7 pairs. **INFL**: cyme, strigose. **FL**: petals 2–4.5 mm; style 1–3 mm. **FR** 7–9 mm, white to cream-colored; stone smooth to grooved on face, furrowed on sides. Many, gen moist habitats; < 2800 m. CA-FP; to AK, e N.Am, Mex. Highly variable complex with many local forms, treated broadly here. Sspp. intergrade widely. ❀IRR: 1–3,**4–9**,10,**14–23**,24;rather INV.

ssp. ***occidentalis*** (Torrey & A. Gray) Fosb. Pl ± rough hairy. **FL**: petals 3–4.5 mm; style 2.5–3 mm. **FR**: stone gen 3-ridged on faces, furrowed on sides. Habitats of sp.; < 2500 m. CA-FP; to AK, MT. [*C. o.* (Torrey & A. Gray) Cov.]

ssp. ***sericea*** (p. 529) Pl glabrous to ± strigose, not rough-hairy. **FL**: petals gen 2–3 mm; style 1–2 mm. **FR**: stone smooth on faces, furrowed on sides. Habitats and range of sp. (uncommon in s CA). [*C.* ×*californica* C.A. Meyer, incl var. *nevadensis* Jepson; *C. stolonifera* Michaux]

C. sessilis Durand Shrub, small trees < 5 m; herbage subglabrous. **ST** gray or yellowish brown. **LVS** deciduous; blade 4.5–9 cm, gen obovate to elliptic, strigose below (± tomentose at vein axils); petiole 5–10 mm. **INFL**: cluster, umbel-like, sessile; subtending bracts 4, ± 1 cm, ephemeral, brownish, gen with yellow margins; pedicel ± 1 cm, soft-white-hairy; fls few–several. **FL** yellowish; sepals 0.5 mm; petals 3 mm; style 1 mm. **FR** 1–1.5 cm, elliptic in outline, greenish white becoming yellow, then red, then shiny purple-black. Streambanks; < 1550 m. NW, CaR, n SN. ❀IRR:**4–7**,8–10,**14–17**,19–24.

CRASSULACEAE STONECROP FAMILY

Ann to subshrubs, fleshy. **LVS** gen simple, gen basal and cauline, alternate or opposite, gen reduced upward. **INFL**: gen cyme, gen bracted. **FL**: sepals gen 3–5, gen ± free; petals gen 3–5, ± free or fused; stamens = to >> sepals, free or epipetalous; pistils gen 3–5, simple (sometimes fused at base), ovary 1-chambered, placenta 1, parietal, ovules 1–many, style 1. **FR**: follicles gen 3–5. **SEEDS** 1–many, small. ± 30 genera, ± 1500 spp.: ± worldwide, esp dry temp; many cult for orn. Family description and generic key by Melinda F. Denton and Reid Moran.

1. Ann; lvs << 1 cm; fl 1–4 mm
 2. Lvs opposite, pairs fused around st; fls axillary; petals < 2 mm .... **CRASSULA**
 2′ Lvs alternate above, free; fls in terminal cyme; petals 1.5–4.5 mm .... **PARVISEDUM**
1′ Gen per to shrub; lvs > 1 cm; fl gen > 10 mm (if ann, fl > 4 mm)
 3. Shrub or subshrub
  4. Lvs alternate, many in rosette, ciliate; sepals 6–16; petals ± free .... **AEONIUM**
  4′ Lvs opposite, few, not ciliate; sepals 5; petals fused, tube > sepals .... **COTYLEDON**
 3′ Per (ann or bien in *Sedum radiatum*)
  5. Infls axillary; cauline lvs different from rosette lvs .... **DUDLEYA**
  5′ Infls terminal; cauline lvs ± like rosette lvs, or basal lvs brown and scale-like .... **SEDUM**

AEONIUM

Reid Moran

Per, shrub. **LVS** crowded at branch tips, alternate, gen obovate to oblanceolate, gen ciliate. **INFL**: cyme, terminal; branches many, 1-sided. **FL**: sepals 6–16, ± free; petals ± free, > sepals; stamens 2 ± sepal number. **SEEDS** many, small, striate, brown. 31 spp.: Afr, Yemen, Madeira, Cape Verde, and esp Canary islands. (Latin: name given to *A. arboreum* by Dioscorides, Greek botanist) [Liu 1989 Natl Mus Nat Sci Taiwan Spec Publ 3]

1. Lf rosette 1–2 dm wide, lvs 50–75; fl open, petals 9–11, yellow; calyx puberulent .... ***A. arboreum***
1′ Lf rosette < 1 dm wide, lvs 15–25; fl bell-shaped, petals 7–9, cream-colored; calyx glabrous .... ***A. haworthii***

A. arboreum (L.) Webb. & Berth. var. *arboreum* Subshrub 1 (–2) m, ± open; adventitious prop roots 0. **ST** 10–40 mm diam, fleshy, ± smooth; branches few, above. **LVS** 50–75, 5–9(15) cm, 1.5–3 mm thick, oblong-oblanceolate, green or dark purple. **INFL**: pedicel puberulent. **FL** ± 2 cm wide, open; petals 9–11, yellow. $2n$=36,72. Bluffs, dunes; < 100 m. SCo; native to Canary Islands.

A. haworthii Salm-Dyck Shrub ± 7 dm, closely dome-shaped; adventitious prop roots many. **ST** 3–10 mm diam, woody, rough-netted; branches many, throughout. **LVS** 15–25, 3–6 cm, 2–5 mm thick, obovate, gray-green; margin gen red. **INFL**: pedicel glabrous. **FL** ± 1 cm wide, bell-shaped; petals 7–9, cream-colored. $2n$=72. Sea cliffs, dunes; < 100 m. c&s SCo; native to Canary Islands.

COTYLEDON

Reid Moran

Shrub. **LVS** simple, opposite, sessile; margin entire. **INFL**: cyme, terminal. **FL**: sepals 5, ± free; corolla tube >> sepals, often > lobes; stamens 10, hairy at base. **SEEDS** many, elliptic in outline, ridged. 9 spp.: esp s Afr; some cult for orn. (Greek: cavity) [Tolken 1985 Fl S Afr 14:3–17]

C. orbiculata L. var. *oblonga* (Haw.) DC. Pl < 1 m, erect, branched below, glabrous in CA. **LF** 5–12 cm, obovate to wedge-shaped, often red-margined, glaucous; tip rounded to obtuse. **INFL**: peduncle 20–40 cm, thick, nearly lfless; fls pendent. **FL**: sepals 2–5 mm; corolla orange, tube 20–25 mm, cylindric, lobes 10–15 mm, recurved. $2n$=18. Bluffs; < 100 m. c SCo (Newport Beach, Orange Co.); native to s Afr. Highly TOXIC to sheep and goats but rarely eaten.

CRASSULA

Reid Moran

Ann in CA, terrestrial, sometimes submersed and later stranded in dry ponds, glabrous in CA. **ST** decumbent to erect, branched or not. **LVS** basal and cauline, opposite; lf bases fused, ± sheathing. **INFL**: cyme, axillary; fls 1–many. **FL**: sepals 3–5, ± fused at base; petals free or ± fused at base; stamen number = sepal number; pistils 3–5. **FR**: follicles 3–5, gen erect. **SEEDS** 1–many, in CA gen < 0.5 mm, elliptic to elliptic-oblong, red-brown. ± 300 spp.: esp Afr (ann spp. ± worldwide). (Latin: diminutive of thick) Some subshrubs (e.g. *C. tetragona* L.) cult, sometimes persisting.

1. Sepals acute to acuminate, gen > petals; fls gen 2 per lf pair; follicles 1–2-seeded; terrestrial
  2. Sepals gen 4, erect; fls sessile or pedicelled .... ***C. connata***
  2′ Sepals gen 3, tip often outcurved; fls subsessile .... ***C. tillaea***

1′ Sepals rounded to obtuse, << petals; fl 1 per lf pair; follicles 6–17-seeded; aquatic, later gen stranded in dry ponds
  3. Follicles oblong, subtruncate; suture erect, straight, abruptly outcurved in upper 1/4 .... ***C. aquatica***
  3′ Follicles lanceolate, obliquely acute; suture outcurved from near middle .... ***C. solieri***

C. aquatica (L.) Schönl. (p. 529) **STS** decumbent, later ± erect if stranded, gen branched at base, rooting at lower nodes. **LF** 2–6 mm, oblanceolate to linear; tip acute. **INFL**: fl 1 per lf pair; pedicel < 19 mm. **FL** 1–2 mm; sepals 4, ± 0.5–1.5 mm, ovate to oblong, tip rounded to obtuse; petals 1–2 mm, lanceolate; style minute, spreading in fr. **FR**: follicles erect, oblong, subtruncate. **SEEDS** 6–17, elliptic-oblong. $2n=42$. Salt marshes, vernal pools, ponds; < 3000 m. NCo, SNF, GV, CW, SW (exc n ChI); to AK, ne US, Mex, n Eurasia. [*Tillaea a.* L.] Pls from interior habitats with pedicels elongated in fr have been called *C. saginoides* (Maxim.) Bywater & Wick.

C. connata (Ruiz Lopez & Pavon) A. Berger (p. 529) PYGMY-WEED **STS** erect, 2–6(10) cm, branched or not, in age red. **LF** 1–3 (6) mm, ovate to oblong; tip obtuse, acute, or abruptly fine-pointed. **INFL**: fls (1)2 per lf pair, gen crowded; pedicel < 6 mm. **FL** 0.5–2 mm; sepals (3)4, 0.5–2 mm, lanceolate, tip acute to acuminate; petals gen < sepals, narrow-triangular. **FR**: follicles ascending, ovoid, tapered to styles. **SEEDS** 1–2, elliptic. $2n=16$. Open areas; < 1500 m. NW, SNF, GV, CW, SW, DSon; to OR, TX, n C.Am; also in w S.Am. [*Tillaea erecta* Hook. & Arn.] Locally abundant.

C. solieri (C. Gay) F. Meigen (p. 529) **STS** erect when stranded, 2–7 cm, ± branched. **LF** 1–5 mm, oblong to linear; tip ± obtuse. **INFL**: fl 1 per lf pair; pedicel 0.5–6(10) mm. **FL** ± 1–2 mm; sepals 4, ± 0.5–1 mm, ovate to lanceolate, tip rounded to obtuse; petals ± 1–2 mm, lanceolate. **FR**: follicles ascending, lanceolate, obliquely acute, bases oblique. **SEEDS** 9–14, elliptic-oblong. Vernal pools; < 500 m. NCoRI, CaRF, n SNF, GV, SCo; to OR, WY, TX, Baja CA; also in Chile.

C. tillaea Lester-Garl. **STS** gen erect, 1–6 cm, branched or not, in age red. **LF** 1–3 mm, oblong; tip ± acute. **INFL**: fls (1)2 per lf pair, crowded; pedicel gen < 0.5 mm. **FL** ± 1–1.5 mm; sepals gen 3, ± 1–1.5 mm, lanceolate, tip acuminate; petals 0.5–1 mm, narrowly lanceolate. **FR**: follicles ascending, ovoid, tapered to styles. **SEEDS** (1)2, elliptic. Open, gravelly sites; < 500 m. NCoRI, n&c SNF, GV, CCo, SW; native to Medit. [*Tillaea muscosa* L.]

DUDLEYA

Jim A. Bartel

Per, fleshy, glabrous. **ST** gen caudex or corm-like, gen vertical, branched or not, gen covered with dried lvs or lf bases. **LVS** simple, in basal rosettes, evergreen vernal and ephemeral. **INFL**: cyme, axillary, 1-sided; bracts lf-like, alternate. **FL**: sepals 5, fused below; petals 5, fused at base, erect to spreading above; stamens 10, epipetalous; pistils 5, ± fused below. **FR**: follicles 5, erect to spreading, many-seeded. **SEED** < 1 mm, narrowly ovoid, brown, striate. ± 45 spp.: sw N.Am. (W.R. Dudley, western US botanist, 1849–1911) [Moran in Jacobsen 1960 Handb Succ Pl:344–359] Unopened follicles gen most reliable for posture (erect, spreading, etc).

1. St corm-like, below surface, simple; lvs vernal, base gen < 1 cm wide — petals, follicles gen spreading (subg. *Hasseanthus*)
  2. Fl odorless; petals yellow; follicles spreading
    3. Lf linear, slightly narrowed above base, sharply acute at tip, 4–15 cm, base > 4 mm wide; petals fused 1–2 mm .... ***D. multicaulis***
    3′ Lf oblanceolate to spoon-shaped, strongly narrowed above base, acute to obtuse at tip, 1–7 cm, base gen < 3 mm wide; petals fused 0.5–1 mm .... ***D. variegata***
  2′ Fl odor musky-sweet; petals white; follicles ascending to spreading
    4. Petals erect to ascending, 7–14 mm, 3.5–5.5 mm wide, fused 1–2 mm; follicles ascending; lf base 4–12 mm wide .... ***D. nesiotica***
    4′ Petals spreading, 5–10 mm, 2–4 mm wide, fused gen < 1 mm; follicles spreading; lf base 1–4 mm wide .... ***D. blochmaniae***
      5. Caudex oblong, 4–8 × longer than wide; lower bracts < 1.5 × longer than wide; lf 7–15 mm, spoon-shaped .... ssp. ***brevifolia***
      5′ Caudex spheric to fusiform, gen < 3 × longer than wide; lower bracts > 2 × longer than wide; lf 10–60 mm, ± oblanceolate
        6. Lvs gen < 12 per rosette; peduncle > 4 cm; mainland .... ssp. ***blochmaniae***
        6′ Lvs > 15 per rosette; peduncle 3–7 cm; n ChI .... ssp. ***insularis***

1′ St a caudex, above surface, often elongate, often branched; lvs gen evergreen, base gen 1–5 cm wide
  7. Petals erect below, variously spreading from near middle; follicles ascending to spreading (subg. *Stylophyllum*)
    8. Lf ± round in X-section, exc at base
      9. Infl 1° branches gen 2–3, simple; caudex gen < 1 cm wide; follicles 30–45° < erect, slightly swollen near base .... ***D. attenuata*** ssp. ***orcuttii***

9′ Infl 1° branches 3–several, forked 1–2 ×; caudex gen > 1 cm wide; follicles 45–80° < erect, strongly swollen near base
10. Lf covered with mealy powder; styles 2–3 mm; pedicel > 2 mm; SnGb . . . . **D. densiflora**
10′ Lf not covered with mealy powder; styles 1–2 mm; pedicel < 2 mm; SCo, PR . . . . **D. edulis**
8′ Lf ± flat to ± elliptic in X-section
11. Lf sticky, appearing oily, odor resinous . . . . **D. viscida**
11′ Lf nonsticky, not appearing oily, odorless
12. Petals bright yellow, tips erect or outcurved; lf 4–6 ± wider than thick . . . . **D. traskiae**
12' Petals white to reddish, tips ascending to spreading; lf 2–5 ± wider than thick
13. Lvs 15–30 per rosette, 3–10 cm, 5–15 mm wide, ± covered with mealy powder; peduncle 1–3 dm . . . . **D. hassei**
13′ Lvs 20–45 per rosette, 8–20 cm, 15–30 mm wide, green or glaucous; peduncle 2–6 dm . . . . **D. virens**
7′ Petals erect below, slightly spreading at tips; follicles gen erect (subg. *Dudleya*)
14. Petals fused ± 1/2 length; pedicel in fr often sharply bent . . . . **D. pulverulenta**
15. Caudex 4–9 cm wide; lvs 40–60 per rosette, gen 8–25 cm; pedicel 5–30(35) mm . . . . ssp. **pulverulenta**
15′ Caudex 1–4 cm wide; lvs 15–25 per rosette, gen 5–15 cm; pedicel 5–15(20) mm . . . . ssp. **arizonica**
14′ Petals fused < 1/3 length; pedicel in fr not sharply bent
16. Pl branched by stolons; follicles ascending; SCo (Orange Co.) . . . . **D. stolonifera**
16′ Pl unbranched or branched by forked caudex; follicles ± erect
17. Rosettes 1–5 dm wide, gen 1 per pl; lvs gen 20–45 per rosette; n ChI . . . . **D. candelabrum**
17′ Rosettes gen < 1.5 dm wide, 1–many per pl; lvs < 30 per rosette; gen mainland
18. Caudex gen 0.5–10 cm, erect to prostrate, gen many-branched; lf 2–8 mm thick
19. Lf ± 2 × longer than wide, oblong-ovate, 2.5–6 cm; petal margins gen not touching . . . . **D. farinosa**
19′ Lf gen > 2 × longer than wide, not oblong-ovate, 2–20 cm; petal margins gen touching
20. Petals pale yellow or whitish, purple-tinged near tip or not, 8–13 mm
21. Caudex < 2 cm wide; lf < 10 mm wide; SCoRO (San Luis Obispo Co.) . . . . . . [2]**D. abramsii** ssp. **bettinae**
21′ Caudex 2–5 cm wide; lf > 10 mm wide; ChI . . . . **D. greenei**
20′ Petals lemon-yellow or bright yellow to yellow with red, 8–16 mm
22. Rosette gen < 5 cm wide; peduncle 0.5–1.5 dm; petals lemon-yellow, tips spreading . . . . **D. verityi**
22′ Rosette 5–20 cm wide; peduncle 1.5–6.5 dm; petals yellow to yellow with red, tips erect
23. Petals yellow; lf 10–20 mm wide; infl terminal branches 3–15 cm . . . . **D. caespitosa**
23′ Petals yellow with red; lf 15–50 mm wide; infl terminal branches 5–8 cm . . . . **D. palmeri**
18′ Caudex gen < 5 cm, erect, simple to many-branched; lf 1–6 mm thick
24. Pedicel gen 0.5–7 mm
25. Petals bright yellow to red, keels similarly colored, margins entire
26. Peduncle 5–15(20) cm; lf (1)2–8 cm, < 8 mm wide; s SNH (Tulare Co.) . . . . . [2]**D. cymosa** ssp. **costafolia**
26′ Peduncle 15–75 cm; lf 5–30 cm, 10–40 mm wide; SCoR, TR, PR, DMtns . . . . [2]**D. lanceolata**
25′ Petals pale yellow, keel often reddish, margin often jagged
27. Infl 1° branches 2–4, gen spreading, branched 0–3 ×; pedicel 2–12 mm; petals darker yellow to red-tinged on keel; SN . . . . **D. calcicola**
27′ Infl 1° branches 0 or 2–3, ascending, gen simple; pedicel gen < 5 mm; petals often with fine, purple to red lines on keel; CW, SW
28. Lower bracts 10–40 mm; peduncle 5–25 cm, 2–6 mm wide; serpentine; SnFrB, SCoRO (San Luis Obispo Co. n)
29. Petals not purple-tinged, without red lines; pl not exuding purple dye when crushed; SnFrB (Santa Clara Co.) . . . . **D. setchellii**
29′ Petals purple-tinged or red-lined; pl exuding purple dye when crushed; SCoRO (San Luis Obispo Co.) . . . . **D. abramsii**
30. Rosettes gen > 10 per pl; lf < 7 mm wide, ± cylindric; infl 1° branches gen < 4; petals often purple-tinged near tips, with few purple flecks . . . . [2]ssp. **bettinae**
30′ Rosettes < 10 per pl; lf > 7 mm wide, upper side ± flat; infl 1° branches gen > 4; petals with purple keel, purple flecks throughout . . . . ssp. **murina**
28′ Lower bracts gen 4–15 mm; peduncle 2–18 cm, 1–3 mm wide; nonserpentine; SW . . . . **D. abramsii**
31. Lvs vernal; roots narrowed at irregular intervals; petals fused 1–2 mm . . . . ssp. **parva**
31′ Lvs gen evergreen; roots gen not narrowed at intervals; petals fused 1.5–4.5 mm
32. Lf oblong to oblong-lanceolate or tapered base to tip; caudex few–many-branched; petals fused 2–4.5 mm . . . . ssp. **abramsii**
32′ Lf elliptic to oblanceolate; caudex gen simple; petals fused 1.5–2.5 mm . . . . ssp. **affinis**
24′ Pedicel 5–20 mm
33. Lf lanceolate to oblong-lanceolate to oblong-triangular, tip acute to subacuminate; terminal infl branches gen 3–12 cm
34. Lf 5–30 cm, 1–4 cm wide; peduncle 1.5–7.5 dm; petals fused 1–2 mm . . . . [2]**D. lanceolata**
34′ Lf 1–15 cm, 0.3–2.5 cm wide; peduncle 0.3–3.5(4) dm; petals fused 1–4 mm
35. Petals greenish yellow to bright yellow, often red-tinged; pedicel 5–20 mm . . . . **D. saxosa**
36. Caudex 1–1.5 cm wide; lf 3–9 cm; peduncle 0.5–2 dm; petals gen red-tinged . . . . ssp. **saxosa**
36′ Caudex 1–3 cm wide; lf 4–15 cm; peduncle 1–3.5 dm; petals rarely red-tinged . . . . ssp. **aloides**
35′ Petals pale yellow; pedicel 2–14 mm
37. Infl 1° branches 2–4, gen spreading, branched 0–3 ×; pl glaucous; petals sometimes red-tinged on keel . . . . [2]**D. calcicola**

37′ Infl 1° branches 2–3, ascending, gen simple; pl ± glaucous; petals not red-tinged on keel .... ***D. setchellii***

33′ Lf elliptic to oblanceolate to spoon-shaped (rarely linear), tip acute, or acuminate to abruptly sharp; terminal infl branches 1–5(17) cm .... ***D. cymosa***

38. Lvs vernal, 1.5–4 cm, 5–12 mm wide; caudex 2–10 mm wide; peduncle < 1 dm ...... ssp. ***marcescens***

38′ Lvs evergreen, 2–17 cm, gen 10–60 mm wide; caudex gen > 10 mm wide; peduncle 0.5–4.5 dm

39. Peduncle gen 1–4.5 dm; lf 3–17 cm

40. Peduncle bracts 20–50; lf elliptic to spoon-shaped; SnGb .... ssp. ***crebrifolia***

40′ Peduncle bracts < 20; lf not elliptic, rarely spoon-shaped; NW, CaR, SN, CW

41. Petals bright yellow to red; lf 1–6 cm wide .... ssp. ***cymosa***

41′ Petals pale yellow; lf 0.5–2 cm wide .... ssp. ***paniculata***

39′ Peduncle gen < 1.5 dm; lf gen < 7 cm

42. Lf linear to linear-oblanceolate, < 8 mm wide; rosettes 5–40 per pl; SNH .... [2]ssp. ***costafolia***

42′ Lf spoon-shaped to oblanceolate to ovate, > 10 mm wide; rosettes < 10 per pl; CW, SW

43. Lvs oblong to elliptic to ovate, acute to acuminate at tip, gen 6–10 per rosette; petals bright yellow .... ssp. ***ovatifolia***

43′ Lvs diamond-shaped-oblanceolate to spoon-shaped, gen short-acuminate to abruptly sharp at tip, gen 10–25 per rosette; petals bright yellow to red .... ssp. ***pumila***

D. abramsii Rose **ST**: caudex simple to gen branched, cespitose. **LF** 1.5–11 cm, 5–20 mm wide, oblong-lanceolate or tapered base to tip, rarely elliptic to oblanceolate, gen glaucous, upper surface gen ± flat; rosette 2–6 cm wide. **INFL**: 1° branches 0 or gen 2–3, gen simple, ascending; terminal branches 3–15 cm, 2–15-fld; lower bracts 4–40 mm; pedicel 0.5–5(11) mm. **FL**: sepals 2–5 mm, deltate; petals 8–13 mm, 1.5–3.5 mm wide, fused 1–4.5 mm, elliptic, acute, pale yellow, keel gen with fine, purple to red lines, margin often jagged. *n*=17. Rocky outcrops; 50–2600 m. s SCoRO, SW; n Baja CA. Sspp. *bettinae, murina, parva* exude purple dye when crushed.

ssp. ***abramsii*** **ST**: caudex 10–15 mm wide; branches few–many. **LF** < 6 cm, glaucous, oblong to oblong-lanceolate or tapered base to tip. **INFL**: peduncle 2–15 cm, 1–3 mm wide; lower bracts 4–15 mm. **FL**: petals fused 2–4.5 mm, keel gen red-lined. Outcrops, gen granitic; 750–1750 m. PR; n Baja CA. ❀DRN,DRY:15–17,22–24&SHD:14,18–21;DFCLT.

ssp. ***affinis*** K. Nakai (p. 529) SAN BERNARDINO MOUNTAINS DUDLEYA **ST**: caudex 10–15 mm wide, gen simple. **LF** 2–4 cm, 7–15 mm wide, glaucous, oblanceolate to elliptic. **INFL**: peduncle 5–11 cm, 1–3 mm wide; lower bracts 5–6 mm. **FL**: petals fused 1.5–2.5 mm, keel gen red-lined. UNCOMMON. Outcrops, granitic or quartzite, rarely limestone; 1800–2600 m. SnBr.

ssp. ***bettinae*** (Hoover) J. Bartel SAN LUIS OBISPO SERPENTINE DUDLEYA, BETTY'S LIVEFOREVER **ST**: caudex 3–20 mm wide, many-branched, cespitose. **LF** 2–7 cm, 2–7 mm wide, ± glaucous, ± cylindric; tip acute; rosettes 10–40 per pl. **INFL**: peduncle 5–25 cm, 2–5 mm wide; lower bracts 10–20 mm. **FL**: petals fused 1–2 mm, often purple-tinged near tip, with few purple flecks. RARE. Rocky outcrops in serpentine grassland; 50–180 m. s SCoRO (San Luis Obispo Co.). [*D. bettinae* Hoover] In cult.

ssp. ***murina*** (Eastw.) Moran SAN LUIS OBISPO or MOUSE-LEAVED DUDLEYA **ST**: caudex 10–30 mm wide; branches few–many. **LF** 3–11 cm, 7–20 mm wide, ± glaucous, oblong-lanceolate; upper surface ± flat; rosettes < 10 per pl. **INFL**: peduncle 5–25 cm, 2–6 mm wide; lower bracts 10–40 mm. **FL**: petals fused 1.5–3 mm, purple-flecked throughout, keel purple. UNCOMMON. Serpentine outcrops; 120–300 m. s SCoRO (San Luis Obispo Co.). ❀DRN, DRY:14,16,17,19–24;DFCLT.

ssp. ***parva*** (Rose & Davidson) J. Bartel CONEJO DUDLEYA Roots narrowed at irregular intervals. **ST**: caudex 2–7 mm wide, often branched, often laterally. **LF** vernal, 1.5–4 cm, 3–6 mm wide, glaucous esp when young, oblanceolate. **INFL**: peduncle 5–18 cm, 1–2 mm wide; lower bracts 5–15 mm. **FL**: petals fused 1–2 mm, keel often red-flecked. RARE. Clay grassland; 60–450 m. s WTR (w Santa Monica Mtns, Ventura Co.). [*D. p.* Rose & Davidson] In cult.

D. attenuata (S. Watson) Moran ssp. ***orcuttii*** (Rose) Moran (p. 529) ORCUTT'S DUDLEYA **ST**: caudex 3–15 (gen < 10) mm wide, branched. **LF** 2–10 cm, 2–5 mm wide, cylindric exc at base, glaucous, linear to linear-oblanceolate; base 5–15 mm wide; rosette 2–5 cm wide. **INFL**: 1° branches (1)2–3, simple, ascending; peduncle 5–25 cm, 1–3 mm wide; terminal branches 2–11 cm, 3–15-fld; pedicel 0.5–3 mm. **FL**: sepals 1.5–4 mm, deltate-ovate; petals 6–10 mm, fused 0.5–3 mm, elliptic, acute, spreading from middle, white, often rose-flushed, red-lined. **FR**: follicles 30–45° < erect, slightly swollen near base. *n*=17,34. RARE in CA. Coastal bluffs; < 50 m. s SCo (San Diego Co.); n Baja CA. In cult.

D. blochmaniae (Eastw.) Moran **ST** corm-like. **LF** 1–6 cm, ± oblanceolate to spoon-shaped, base 1–4 mm wide; tip acute to rounded. **INFL**: 1° branches 2–3, simple or forked 1 ×; peduncle 2–22 cm, 0.5–2 mm wide; terminal branches 1–6 cm, 3–10-fld; pedicel < 1 mm. **FL**: sepals 1.5–4 mm, deltate-ovate; petals spreading from base, 5–10 mm, elliptic, acute, white, keel often red-lined. **FR**: follicles spreading. Open, coastal; < 450 m. CCo, SCo, n ChI; n Baja CA.

ssp. ***blochmaniae*** BLOCHMAN'S DUDLEYA **ST**: caudex 7–25 mm, 4–15 mm wide, ± spheric to fusiform. **LVS** 3–12 per rosette, 1–6 cm, 3–8 mm wide, 2–4 mm thick, ± oblanceolate. **INFL**: peduncle > 4 cm; bract deltate-lanceolate to -ovate. *n*=17,34,51. RARE. Open, rocky slopes, often serpentine or clay-dominated; < 450 m. s CCo, SCo; n Baja CA. In cult.

ssp. ***brevifolia*** Moran (p. 529) SHORT-LEAVED DUDLEYA **ST**: caudex 1.5–3.5 cm, 1–6 mm wide, oblong. **LVS** 5–15 per rosette, 0.7–1.5 cm, 2–7 mm wide, 2–4 mm thick, spoon-shaped. **INFL**: bract deltate-ovate to subcircular. *n*=17. **ENDANGERED** CA. Bare sandstone terraces; < 250 m. s SCo (San Diego Co.). [*D. brevifolia* (Moran) Moran] In cult.

ssp. ***insularis*** Moran (p. 529) SANTA ROSA ISLAND DUDLEYA **ST**: caudex 1–2 cm, 5–20 mm wide, ± spheric to oblong. **LVS** 15–50 per rosette, 1–3.5 cm, 2–7 mm wide, 1–3 mm thick, ± oblanceolate. **INFL**: peduncle 3–7 cm; bract deltate-lanceolate to -ovate. *n*=17. RARE. Coastal bluffs; ± 3 m. n ChI (Santa Rosa Island). In cult.

D. caespitosa (Haw.) Britton & Rose (p. 529) **ST**: caudex 1.5–4 cm wide, < 2 dm, branched. **LVS** 15–30 per rosette, 5–20 cm, 1–2 cm wide, 3–8 mm thick, oblong-lanceolate; tip acute; rosette 5–20 cm wide. **INFL**: 1° branches 3–5, forked 0–2 ×; peduncle 1.5–6 dm, 4–8 mm wide; terminal branches 3–15 cm, 4–15-fld; pedicel 1–6 mm. **FL**: sepals 2–5 mm, deltate-ovate, acute; petals 8–16 mm, 3–5 mm wide, fused 1.5–2.5 mm, elliptic, yellow, tips erect, acute. *n*=17,51,68. Coastal; gen < 100 m. CCo, SCo, n ChI. Poorly defined, difficult complex of several named and unnamed entities; evidently intergrades with *D. cymosa, D. farinosa, D. lanceolata, D. palmeri*. ❀DRN,SUN:**15–17**&IRR:**24**&SHD:9,**14,19–23**;GRCVR.

D. calcicola J. Bartel & J.R. Shevock (p. 529) LIMESTONE DUDLEYA **ST**: caudex 1–2 cm wide, cespitose; branches 0 to gen many. **LF** 1–10 cm, 3–13 mm wide, oblong-lanceolate or tapered base to tip, glaucous; tip acute to subacuminate; rosette 1–9 cm wide. **INFL**: 1° branches 2–4, gen spreading, branched 0–3 ×; peduncle 3–18 cm, 1.5–4 mm wide; terminal branches 1–6 cm, gen spreading, 2–8-fld; pedicel 2–14 mm. **FL**: sepals 2.5–6 mm, triangular-ovate to -lanceolate; petals 9–15 mm, 3.5–5 mm wide, fused 1–3 mm, lanceolate, narrowly acute, pale yellow, keel darker yellow to red-tinged. *n*=17. UNCOMMON. Open, rocky outcrops; 500–2600 m. s SNH, Teh. [*D. abramsii* spp. *c.* (J. Bartel & J.R. Shevock) K. Nakai] Intermediate between *D. abramsii*. ssp. *a.* and *D. cymosa* ssp. *c.* ❀DRN,DRY:7,14–24;DFCLT.

D. candelabrum Rose (p. 529) CANDLEHOLDER DUDLEYA **ST**: caudex 2–8 cm wide, simple. **LVS** 20–45 per rosette, 6–17 cm, 3–7 cm wide, obovate to oblong-oblanceolate, nonglaucous; base 2–4.5 cm wide; tip acuminate; rosette 1–5 dm wide. **INFL**: 1° branches gen 3, forked 1–2 ×; peduncle 1.5–3.5 dm; terminal branches 2.5–13 cm, spreading, 5–25-fld; pedicel 2–6 mm. **FL**: sepals 5–8 mm, deltate-ovate, acute; petals 8–12 mm, 2.5–3.5 mm wide, fused 1.5–3.5 mm, oblong, acute, pale yellow. *n*=17. RARE. Rocky places, n-facing slopes; < 380 m. n ChI (Santa Cruz, Santa Rosa islands). In cult.

D. cymosa (Lemaire) Britton & Rose **ST**: caudex 1–3.5 cm wide; branches 0–many. **LF** 1.5–17 cm, 2.5–50 mm wide, gen oblanceolate to spoon-shaped, rarely ovate or linear-oblanceolate, ± glaucous; tip acute, often acuminate to abruptly sharp; rosette 1–20 cm wide. **INFL**: 1° branches gen 2–4, gen spreading, branched 0–3 ×; peduncle 4–45 cm; terminal branches 1–5(17) cm, 2–10(20)-fld; pedicel gen 5–15 mm. **FL**: sepals 1.5–5 mm, deltate-ovate, acute; petals 7–14 mm, 1.5–3.5 mm wide, fused 1–3 mm, elliptic to lanceolate, narrowly acute, yellow to red. *n*=17. Rocky outcrops, slopes, talus; 50–2700 m. NW, SN, CW, SW; sw OR.

ssp. ***costafolia*** J. Bartel & J.R. Shevock PIERPOINT SPRINGS LIVEFOREVER **ST**: caudex 1.5–2 cm wide, cespitose, many-branched. **LF** (1)2–8 cm, 2.5–8 mm wide, linear to linear-oblanceolate, glaucous; rosettes 5–40 per pl, densely packed, 1–5 cm wide. **INFL**: peduncle 5–15(20) cm; terminal branches 1–4 cm, 2–7-fld; pedicel 2.5–9 mm. **FL**: petals bright yellow. RARE. Limestone outcrops; 1450–1600 m. s SNH (Tule River, Tulare Co.).

ssp. ***crebrifolia*** K. Nakai & Verity SAN GABRIEL RIVER DUDLEYA **ST**: caudex 1–2 cm wide; branches 0 to rarely few. **LF** 4–10 cm, 20–50 mm wide, elliptic to spoon-shaped, rarely glaucous. **INFL**: peduncle gen 10–30 cm; bracts 20–50, closely-spaced; terminal branches 3–15 cm, 2–10(20)-fld; pedicel 3–8 mm. **FL**: petals fused 1–1.5 mm, mustard-yellow. UNCOMMON. Granitic slopes; 400 m. SnGb (Fish Canyon, Los Angeles Co.). ❀DRN,DRY:14, 16–24.

ssp. ***cymosa*** (p. 529) **ST**: caudex 1–3.5 cm wide; branches few–many. **LF** 3–17 cm, 10–60 mm wide. **INFL**: peduncle 5–45 cm; terminal branches 1–17 cm, 4–20-fld. **FL**: petals bright yellow to red. Rocky outcrops, talus slopes, less often shaded canyon slopes; 100–2700 m. NCoR, CaR, SN, SnFrB; sw OR. [ssp. *gigantea* (Rose) Moran] ❀DRN,DRY:7,8,9,**14–24**.

ssp. ***marcescens*** Moran (p. 529) SANTA MONICA MOUNTAINS or MARCESCENT DUDLEYA **ST**: caudex 2–10 mm wide; branches 0–few. **LF** vernal, 1.5–4 cm, 5–12 mm wide. **INFL**: 1° branches 1–2, simple; peduncle 4–10 cm; terminal branches 1–3 cm, 3–5-fld. **FL**: petals 2.5–3.5 mm wide, bright yellow, rarely orange or red-marked. **RARE** CA. Shaded, rocky slopes; 150–500 m. s WTR (Santa Monica Mtns). In cult.

ssp. ***ovatifolia*** (Britton) Moran SANTA MONICA MOUNTAINS DUDLEYA **ST**: caudex 1–1.5 cm wide; branches 0–few. **LVS** gen 6–10 per rosette, 2–5 cm, 15–25 mm wide, oblong to elliptic to ovate. **INFL**: peduncle 4–15 cm; terminal branches 1–3 cm, 3–5-fld. **FL**: petals bright yellow, rarely orange or red-marked. RARE. Shaded, rocky slopes; 150–500 m. s WTR (Santa Monica Mtns). [ssp. *agourensis* K. Nakai]

ssp. ***paniculata*** (Jepson) K. Nakai **ST**: caudex 1–2 cm wide; branches gen few. **LF** 3–10 cm, 5–20 mm wide, oblong-lanceolate, glaucous. **INFL**: peduncle 0.5–2 dm; terminal branches 1–5 cm, 4–10-fld; pedicel 3–12 mm. **FL**: petals 1.5–2.5 mm wide, pale yellow. Rocky outcrops, slopes; 30–1200 m. SnFrB, SCoRI. [*D. caespitosa* var. *p.* Jepson] ❀DRN:7,9,**14,16,17,19–24**.

ssp. ***pumila*** (Rose) K. Nakai **ST**: caudex 1–2(3.5) cm wide; branches 0–few. **LF** 1.5–5 cm, 10–30 mm wide, diamond-shaped-oblanceolate to spoon-shaped; tip gen short-acuminate to abruptly sharp. **INFL**: peduncle 0.5–2.5 dm; terminal branches 1–3 cm, 3–6-fld. **FL**: petals bright yellow to red. Rocky cliffs, slopes; 50–2600 m. SCoRO, TR. ❀DRN,DRY:**15–17,22–24**&SHD:**14, 19–21**.

D. densiflora (Rose) Moran (p. 529) SAN GABRIEL MOUNTAINS DUDLEYA **ST**: caudex < 2 dm, 1–2.5 cm wide, branched. **LF** 6–15 cm, 6–12 mm wide, linear, ± cylindric exc at base, covered with mealy powder; base 1–2 mm wide; tip abruptly pointed. **INFL**: 1° branches 3–several, forked 1–2 ×; peduncle 1–3 dm, 2–4 mm wide; terminal branches 2–4 cm, 2–8-fld; pedicel 2–5 mm. **FL**: sepals 1.5–2.5 mm, deltate-ovate; petals 5–10 mm, 2–3 mm wide, fused 0.5–2 mm, narrowly ovate, acute, white or pink; styles 2–3 mm. **FR**: follicles 45–80° < erect, strongly swollen near base. *n*=17. RARE. Steep, granitic canyon walls; 300–520 m. SnGb (Los Angeles Co.). In cult.

D. edulis (Nutt.) Moran **ST**: caudex 1.5–4.5 cm wide, branched. **LF** 8–20 cm, 4–10 mm wide, ± cylindric exc at base, pale green, ± to nonglaucous; base 1.5–3 cm wide; tip acute to abruptly pointed. **INFL** open; 1° branches several, forked 1–2 ×; peduncle 2–5 dm, 3–10 mm wide; terminal branches 4–10 cm, 3–11-fld; pedicel 1–2 mm. **FL**: sepals 2.5–4.5 mm, oblong-ovate; petals 7–10 mm, 2–3 mm wide, fused 1–2 mm, elliptic-oblong, acute, spreading or ± reflexed from middle, white; styles 1–2 mm. **FR**: follicles 45–60° < erect, strongly swollen at base. *n*=17. Rocky slopes, ledges; < 1300 m. SCo, PR; n Baja CA. ❀DRN,DRY:**15–17,24**&IRR:14, 19,**20–23**;GRCVR.

D. farinosa (Lindley) Britton & Rose (p. 533) **ST**: caudex 1–3 cm wide, gen many-branched, often elongate. **LF** 2.5–6 cm, 1–2.5 cm wide, 3–6 mm thick, oblong-ovate, glaucous or not; base 1–2.5 cm wide; tip acute. **INFL**: 1° branches 3–5, close-set, simple or forked 1 ×; peduncle 1–3.5 cm, 3–10 mm wide; terminal branches 1–3.5 cm, ascending in age, 3–11-fld; pedicel 1–3 mm. **FL**: sepals 3–7 mm, deltate-ovate; petals 10–14 mm, 3–5 mm wide, fused 1–2 mm, oblanceolate, acute to obtuse, pale yellow. *n*=17. Coastal; < 150 m. NCo, n&c CCo; sw OR. ❀DRN:5,**15–17**&SHD,IRR:7, **14,19–24**.

D. greenei Rose (p. 533) GREENE'S DUDLEYA **ST**: caudex 2–5 cm wide, branched, often elongate. **LF** 3–11 cm, 1–3.5 cm wide, 4–8 mm thick, oblong-oblanceolate to -obovate, nonglaucous to ± covered mealy powder; base 1–3 cm wide; tip acute to abruptly pointed. **INFL**: 1° branches 3–6, branched 0–2 ×; peduncle 1.5–4 dm, 3–13 mm wide; terminal branches 1–9 cm, 2–15-fld; pedicel 1– 5 mm. **FL**: sepals 1.5–5 mm, deltate; petals 8–12 mm, 3–5 mm wide, fused 1.5–2.5 mm, elliptic, acute, pale yellow or whitish. *n*= 34,51. UNCOMMON. Coastal cliffs; < 150 m. ChI (San Miguel, Santa Rosa, Santa Cruz islands.) A variable complex barely distinct from *D. caespitosa*. ❀DRN,DRY:5,15,**16,17,22–24**&SHD: 7,**14,19–21**.

D. hassei (Rose) Moran **ST**: caudex 1–3 cm wide, many-branched. **LF** 3–10 cm, 5–15 mm wide, 2–4 mm thick, linear-lanceolate, ± covered with mealy powder; base 8–15 mm wide; tip obtuse. **INFL**: 1° branches 2–4, simple or forked 1 ×; peduncle 1–3 dm, 1.5–4 mm wide; terminal branches 2–8 cm, 4–14-fld; pedicel 1–5 mm. **FL**: sepals 2–3 mm, deltate-ovate; petals 8–10 mm, fused 1.5–2 mm, spreading from middle, white. *n*=34. Rocks, cliffs; < 400 m. s ChI (Santa Catalina Island). ❀15,16,**17,24**&SHD:**22,23** &IRR:**14,20,21**;GRCVR.

D. lanceolata (Nutt.) Britton & Rose (p. 533) **ST**: caudex erect, 1–3 cm wide, branches 0 or few. **LF** 5–30 cm, 1–4 cm wide, 1.5–6 mm thick, oblong-lanceolate, glaucous or not; base 1–3 cm wide; tip acute to abruptly pointed. **INFL**: 1° branches 2–3, simple or forked 1 ×; peduncle 15–75 cm, 3–12 mm wide; terminal branches 2–25 cm, 2–20-fld; pedicel spreading, 2–12 mm, becoming erect. **FL**: sepals 3–6 mm, deltate-ovate; petals 10–16 mm, 3.5–5 mm wide, fused 1–2 mm, elliptic to oblanceolate, acute, yellow to gen red. *n*=34. Rocky slopes; 30–1250 m. SCoR, TR, PR, DMtns; n Baja CA. [*D. cymosa* ssp. *minor* (Rose) Moran] ❀DRN,DRY:14, 15,**16,17,22–24**&IRR:7,9,**18–21**.

D. multicaulis (Rose) Moran (p. 533) MANY-STEMMED DUDLEYA **ST** corm-like, 1.5–5 cm, 3–18 mm wide, simple, oblong. **LF** 4–15 cm, 2–6 mm wide, linear, cylindric exc at base; base 4–10 mm wide; tip narrowly acute. **INFL**: 1° branches 2–many, simple or forked 1 ×; peduncle 5–35 cm, 2–4 mm wide; terminal branches 2–10 cm, 3–15-fld; pedicel 0–3 mm. **FL**: sepals 2–3 mm, deltate-acute; petals 5–9 mm, 2–3 mm wide, fused 1–2 mm, elliptic-lanceolate, acute, yellow. **FR**: follicles spreading. *n*=17. RARE. Heavy soils, often clayey, coastal plain; < 600 m. SCo (Los Angeles, Orange, Riverside, San Diego, San Bernardino cos.). In cult.

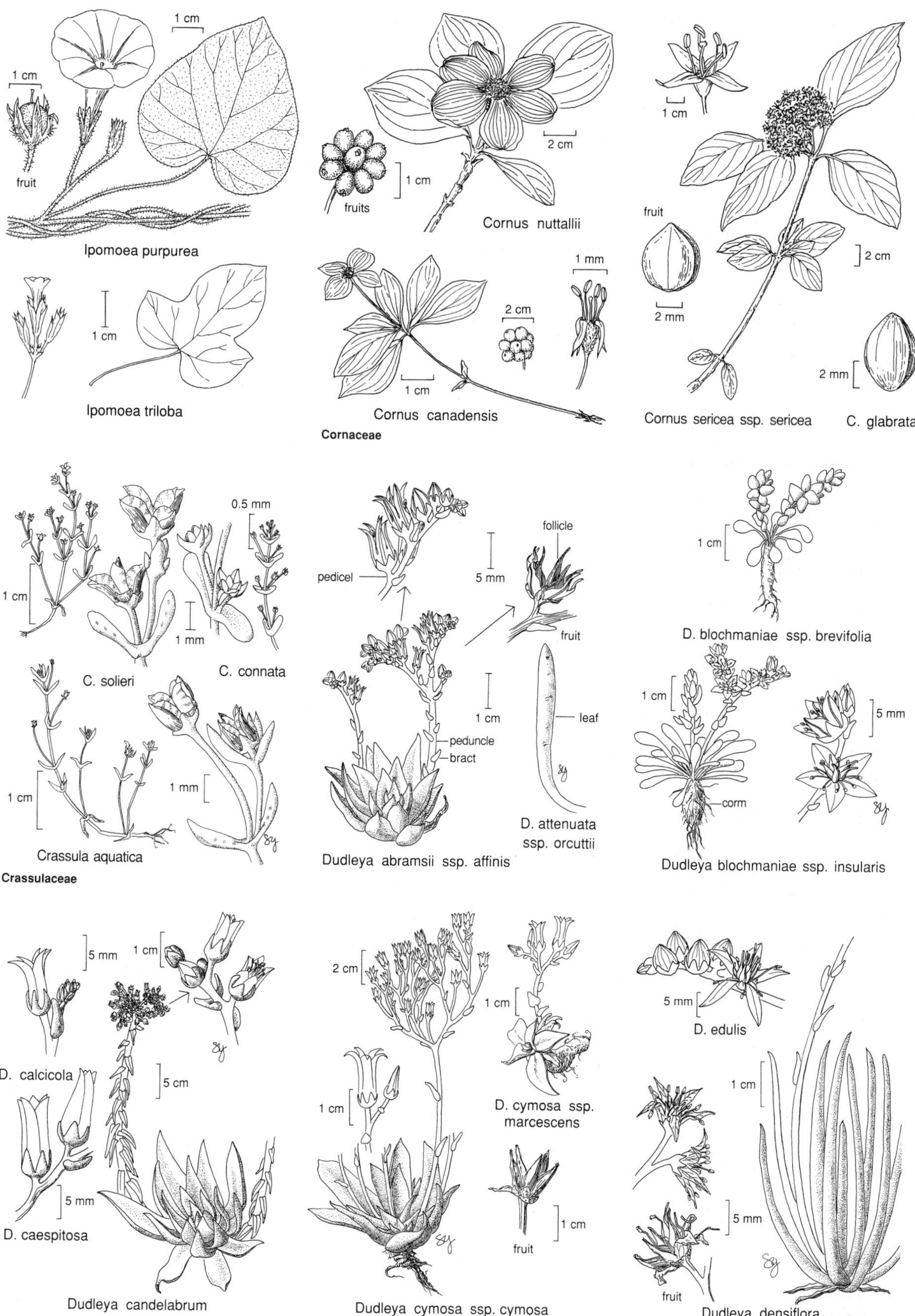
1 cm
1 cm
fruit
Ipomoea purpurea
1 cm
Ipomoea triloba
2 cm
1 cm
fruits
Cornus nuttallii
1 mm
2 cm
1 cm
Cornus canadensis
Cornaceae
1 cm
fruit
2 cm
2 mm
2 mm
Cornus sericea ssp. sericea
C. glabrata
0.5 mm
1 cm
1 mm
C. solieri
C. connata
1 cm
1 mm
Crassula aquatica
Crassulaceae
pedicel
5 mm
follicle
fruit
1 cm
peduncle
bract
leaf
D. attenuata ssp. orcuttii
Dudleya abramsii ssp. affinis
1 cm
D. blochmaniae ssp. brevifolia
1 cm
5 mm
corm
Dudleya blochmaniae ssp. insularis
5 mm
1 cm
D. calcicola
5 cm
5 mm
D. caespitosa
Dudleya candelabrum
2 cm
1 cm
1 cm
D. cymosa ssp. marcescens
1 cm
fruit
Dudleya cymosa ssp. cymosa
5 mm
D. edulis
1 cm
5 mm
fruit
Dudleya densiflora

D. nesiotica Moran (p. 533) SANTA CRUZ ISLAND DUDLEYA **ST** corm-like, 1–3 cm, 7–20 mm wide, ± spheric. **LF** 2.5–5 cm, 5–25 mm wide, 2–5 mm thick, oblanceolate to spoon-shaped; base 4–12 mm wide; tip acute to obtuse. **INFL**: 1° branches gen 2, simple; peduncle 3–10 cm, 1–3 mm wide; terminal branches 3–8-fld; pedicel 1–2 mm. **FL**: sepals 3–4 mm, deltate-ovate; petals 7–14 mm, 3.5–5.5 mm wide, fused 1–2 mm, elliptic, acute, white. **FR**: follicles ascending. *n*=34. **RARE** CA. Coastal bluffs; < 50 m. n ChI (Fraser Point, Santa Cruz Island). In cult.

D. palmeri (S. Watson) Britton & Rose (p. 533) **ST**: caudex < 2 dm, 2–4 cm wide, loosely branched. **LVS** 15–25 per rosette, 5–20 cm, 1.5–5 cm wide, 3–8 mm thick, oblong-lanceolate; tip acute to acuminate; rosette 5–20 cm wide. **INFL**: 1° branches ± 3, forked 0–2 ×; peduncle 2–6.5 dm, 5–11 mm wide; terminal branches 5–8 cm, 5–14-fld; pedicel 2–10 mm. **FL**: sepals 3–5 mm, deltate-ovate, acute; petals 11–16 mm, 3–5 mm wide, fused 1.5–2 mm, elliptic, yellow with red, tips erect, acute. *n*=68,85,119. Coastal areas; gen < 100 m. CCo, SCo (San Luis Obispo, Santa Barbara cos.).

D. pulverulenta (Nutt.) Britton & Rose Pl covered with dense, mealy powder to chalky wax. **ST**: caudex simple. **LF** oblong to oblong-obovate. **INFL**: 1° branches 3–10, simple or forked 1 ×; terminal branches twisted at base. **FL**: sepals deltate-ovate; petals acute to obtuse. *n*=17. Rocky places; < 1500 m. c&s CCo, SCoRO, SCo, TR, PR, DMtns; s NV, w AZ, nw Mex.

ssp. ***arizonica*** (Rose) Moran **ST**: caudex 1–4 cm wide. **LVS** 15–25 per rosette, (3)5–15(17) cm, 1–5 cm wide, 2–4 mm thick, oblong to oblong-obovate; base 1–3.5 cm wide; tip gen long-acuminate. **INFL**: peduncle 1.5–6 dm, 2–6 mm wide; terminal branches ascending, 3–6, 4–27 cm, 3–6-fld; pedicel 5–15(20) mm, gen erect to ascending. **FL**: petals 9–15 mm, fused 4–8 mm, red or yellow, lobes 1.5–2 mm wide. Rocky slopes; 600–1500 m. DMtns; s NV, w AZ, nw Mex. [*D. a.* Rose] ❀DRN:2,3,7,10,11,14,18,**19–23**&DRY:15–17,**24**.

ssp. ***pulverulenta*** (p. 533) **ST**: caudex 4–9 cm wide. **LVS** 40–60 per rosette, 8–25(27) cm, 3–10 cm wide, 3–10 mm thick, oblong; base 3–8 cm wide; tip acuminate to abruptly sharp. **INFL**: peduncle 3–10 dm, 5–15 mm wide; terminal branches erect to spreading, 3–many, 10–40 cm, 10–30-fld; pedicel 5–30(35) mm, in bud pendent, in fr often sharply bent, erect. **FL**: petals 11–19 mm, fused 6–10 mm, red, lobes 2–4 mm wide. Rocky cliffs, canyons; gen < 1000 m. c&s CCo, SCoRO, SCo, TR, PR; n Baja CA. ❀DRN,DRY:15,16,**17,22–24**&SHD or IRR:14,18,**19–21**.

D. saxosa (M.E. Jones) Britton & Rose **ST**: caudex short; branches 0–few. **LF** oblong-lanceolate, glaucous in youth; tip acute. **INFL** often reddish; 1° branches 2–3, simple or forked 1 ×; pedicel 5–20 mm. **FL**: sepals 4–8 mm, deltate, acute; petals 2.5–4 mm wide, oblong-lanceolate, acute. Rocky, gen n-facing slopes; 240–2200 m. PR, DMtns; c AZ, n Baja CA.

ssp. ***aloides*** (Rose) Moran (p. 533) **ST**: caudex 1–3 cm wide. **LF** 4–15 cm, 6–20 mm wide, 2–5 mm thick; base 10–25 mm wide. **INFL**: peduncle 1–3.5(4) dm, 4–9 mm wide; terminal branches wavy, 1–12 cm, 2–20-fld. **FL**: petals 8–15(20) mm, fused 1.5–3 mm, greenish yellow to yellow, rarely red-tinged. *n*=17. Rocky, shaded slopes; 240–1700 m. PR, DMtns; n Baja CA. [*D. alainae* Reiser, Banner dudleya] ❀DRN,DRY:7,10–24;DFCLT.

ssp. ***saxosa*** PANAMINT DUDLEYA **ST**: caudex 1–1.5 cm wide. **LF** 3–9 cm, 5–15 mm wide, 1.5–3 mm thick; base 5–15 mm wide. **INFL**: peduncle 0.5–2 dm, 2–4 mm wide; terminal branches not wavy, 1–4 cm, 2–9-fld. **FL**: petals 9–12 mm, fused 1–4 mm, bright yellow, gen red-tinged. *n*=68,85. UNCOMMON. N-facing, granitic or limestone slopes; 1100–2200 m. n DMtns (w Panamint Mtns). ❀DRN,DRY:2,3,7,10,11,14–24;DFCLT.

D. setchellii (Jepson) Britton & Rose (p. 533) SANTA CLARA VALLEY DUDLEYA **ST**: caudex 10–20 mm wide; branches few. **LF** 3–8 cm, 7–15 mm wide, ± glaucous, oblong-triangular. **INFL**: 1° branches 2–3, gen simple, ascending; peduncle 5–20 cm, 2–3.5 mm wide; lower bracts 10–25 mm; pedicel 2–7 mm. **FL**: sepals 2–5 mm, deltate; petals 8–13 mm, 2.5–3.5 mm wide, fused 1–2.5 mm, elliptic, pale yellow, keel not red or purple, margin jagged or not, tip acute. *n*=17. RARE. Rocky outcrops in serpentine grassland; 120–300 m. se SnFrB (Santa Clara Co.). [*D. cymosa* ssp. *s.* (Jepson) Moran in part] Intermediate between *D. abramsii* ssp. *murina* and *D. cymosa* ssp. *paniculata*. In cult.

D. stolonifera Moran (p. 533) LAGUNA BEACH DUDLEYA Pl branched by stolons. **ST**: caudex 1.5–3 cm wide, simple. **LF** 3–7 cm, 1.5–3 cm wide, 3–4 mm thick, oblong-obovate, nonglaucous; base 1–2 cm wide; tip short-acuminate. **INFL**: 1° branches gen 2, rarely more, simple to rarely forked 1 ×, ascending; peduncle 8–25 cm, 2–4 mm wide; terminal branches 1–6 cm, 3–9-fld; pedicel 5–8 mm. **FL**: sepals 2–3 mm, deltate, wider than long; petals 10–11 mm, 3–3.5 mm wide, fused 1–2 mm, elliptic, yellow, tips outcurved, acute. **FR**: follicles ascending. *n*=17. **THREATENED** CA. N-facing cliffs, outcrops; < 250 m. c SCo (San Joaquin Hills, Orange Co.). In cult.

D. traskiae (Rose) Moran (p. 533) SANTA BARBARA ISLAND DUDLEYA **ST**: caudex 1–3 cm wide, branched. **LF** 4–15 cm, 1–4 cm wide, 4–6 mm thick, oblong-oblanceolate, glaucous; base 1–4 cm wide; tip acute to acuminate. **INFL**: 1° branches ± 3, forked 0–2 ×; peduncle 2–3 dm, 3–8 mm wide; terminal branches 4–10 cm, 7–15-fld; pedicel 1–4 mm. **FL**: sepals 2.5–4 mm, deltate, acute; petals 8–10.5 mm, 3–4 mm wide, fused 1–2 mm, narrowly ovate, yellow, tips gen outcurved, acute. **FR**: follicles ascending. *n*=34. **ENDANGERED** US, CA. Steep slopes; < 110 m. s ChI (Santa Barbara Island). In cult.

D. variegata (S. Watson) Moran (p. 533) VARIEGATED DUDLEYA **ST** corm-like, 1–3 cm, 3–15 mm wide, ± spheric to oblong. **LF** 1–7 cm, 3–11 mm wide, oblanceolate to spoon-shaped; base gen 0.5–3 mm wide; tip acute to obtuse. **INFL**: 1° branches 2–3, ascending; peduncle 5–20 cm, 0.5–3 mm wide; terminal branches 2–15 cm, 3–11-fld; pedicel 0.5–3 mm. **FL**: sepals 2–3.5 mm, deltate-ovate; petals 5–8 mm, 2–3.5 mm wide, fused 0.5–1 mm, elliptic, acute, yellow. **FR**: follicles spreading. *n*=17. UNCOMMON. Dry hillsides, mesas; < 300 m. s SCo, s PR (San Diego Co.); n Baja CA. ❀DRN,DRY:14,16,17,19–24;DFCLT.

D. verityi K. Nakai VERITY'S DUDLEYA **ST**: caudex 2–10 cm, 2–10 mm wide, branched. **LF** 2–5 cm, 4–8 mm wide, oblong-lanceolate, glaucous; base 5–8 mm wide; tip acute to acuminate. **INFL**: 1° branches 2–3, simple to gen forked 1 ×, ascending; peduncle 5–15 cm, 3–6 mm wide; terminal branches 2–5 cm, ascending in age, 2–10-fld; pedicel ascending to erect, 3–5 mm. **FL**: sepals 4–5 mm, triangular, acute; petals 10–14 mm, 2.5–4 mm wide, fused 1–2 mm, lemon-yellow, tips spreading, acute. *n*=17. RARE. N-facing volcanic outcrops; 60–120 m. s WTR (w Santa Monica Mtns, Ventura Co.). In cult.

D. virens (Rose) Moran BRIGHT GREEN DUDLEYA, GREEN LIVEFOREVER **ST**: caudex 1–4 cm wide, branched. **LF** 8–20 cm, 15–30 mm wide, oblong, green or glaucous; tip acute. **INFL**: 1° branches many, forked 1–2 ×; peduncle 2–6 dm, 4–15 mm wide; terminal branches 1–4 cm, 3–12-fld; pedicel 2–4 mm. **FL**: sepals 2–4 mm, deltate-ovate, acute; petals 7–10 mm, 2–3 mm wide, fused 1–2 mm, elliptic, acute, white. *n*=17. UNCOMMON. Coastal bluffs; < 400 m. c SCo (Los Angeles Co.), s ChI (San Clemente, San Nicolas, Santa Catalina islands); Guadalupe Island, Mex. ❀DRN:15,**16, 17,24**&IRR:9,14,18,19,**20–23**;GRCVR; fls frangrant.

D. viscida (S. Watson) Moran (p. 533) STICKY DUDLEYA, STICKY-LEAVED LIVEFOREVER **ST**: caudex short, 1–4 cm wide, branched. **LF** 6–15 cm, 5–15 mm wide, 3–5 mm thick, ± elliptic in X-section, linear-deltate, sticky, appearing oily; odor resinous; base 1–2 cm wide; tip abruptly pointed. **INFL**: 1° branches 3–many, forked 1–2 ×; peduncle 2–4 dm, 2–8 mm wide; terminal branches 2–6 cm, 3–10-fld; pedicel 1–4 mm. **FL**: sepals 1.5–4 mm, ovate, acute; petals 6–9 mm, 2.5–3.5 mm wide, elliptic, acute, spreading from middle, white, red-lined. **FR**: follicles ascending. *n*=17. RARE. Bluffs, rocky cliffs; < 450 m. s SCo (Orange, San Diego cos.). In cult.

PARVISEDUM

Melinda F. Denton

Ann, small, gen branched near base, glabrous. **LVS** cauline, alternate, early deciduous, oblong-elliptic to ovoid; tip rounded to obtuse. **INFL**: fls solitary or in 0–3-branched cyme. **FL**: sepals 5; petals 5, linear to narrowly ovate, slightly fused at base, yellow, midrib often reddish; stamens 5 or 10; pistils 5, oblong, stigmas ± 0.1 mm diam. **FRS** erect to curved out, glabrous or glandular. **SEED** 1 per follicle, 0.7–2 mm, elliptic, brown. 4 spp.: CA. (Latin: small Sedum)

1. Stamens 10
 2. Petals 2–3 mm; fr 1.2–2 mm, (widely) spreading; seeds ± 1 mm; anthers ± 0.2 mm .... ***P. congdonii***
 2′ Petals 3–4.5 mm; fr 2–2.5 mm, ± erect; seeds ± 1.5 mm; anthers 0.3–0.4 mm .... ***P. pumilum***
1′ Stamens 5
 3. Pl < 3.5 cm; petals ± 3–4 mm; fr 1.5–2 mm, (sub)glabrous; seeds 1.2–1.5 mm .... ***P. leiocarpum***
 3′ Pl < 8 cm; petals ± 1.5–2 mm; fr ± 1 mm, glandular-hairy; seeds ± 0.7 mm .... ***P. pentandrum***

P. congdonii (Eastw.) R.T. Clausen Pl 1–9 cm. **LF** 2–4.5 mm, ± 1–2 mm wide. **FL**: sepals 0.5–0.6 mm, ± 0.3 mm wide; petals 2–2.8 mm, 0.4–0.6 mm wide; stamens 10, anthers ± 0.2 mm, yellow or reddish brown. **FR** 1.2–2 mm, ± glandular on inner surface, (widely) spreading. **SEED** 0.8–1.2 mm. *n*=9. Open, rocky, often wet sites; 50–1500 m. SN. ❀SUN,winterWET,summerDRY:7,9,14–24.

P. leiocarpum (H. Sharsm.) R.T. Clausen (p. 533) LAKE COUNTY STONECROP Pl 1–3.5 cm, gen simple. **LVS** 2–4.5 mm, ± 1–2 mm wide. **FL**: sepals 0.6–1 mm, ± 0.5 mm wide; petals 2.8–3.8 mm, 0.5 mm wide; stamens 5, anthers ± 0.2 mm, yellow. **FR** 1.5–2 mm, erect, slightly papillate. **SEED** 1.2–1.5 mm. RARE. Dry vernal pools, rocky depressions; 500–600 m. s NCoRI (Lake Co.).

P. pentandrum (H. Sharsm.) R.T. Clausen Pl 2–8 cm, often simple. **LF** 4–5 mm, 2–3 mm wide. **FL**: sepals 0.5–0.8 mm, 0.3–0.5 mm wide; petals 1.3–2 mm, 0.5–0.8 mm wide; stamens 5, anthers ± 0.2 mm, yellow. **FR** 1–1.2 mm, erect, densely glandular. **SEED** ± 0.7 mm. *n*=9. Compacted soil, sandstone or serpentine outcrops; 300–700 m. NCoRI, e ScV, ne SnJV, CW. ❀SUN,DRY:7,8,**9**,14–24.

P. pumilum (Benth.) R.T. Clausen (p. 533) Pl 2–17 cm, sometimes simple. **LF** 2.5–5 mm, 1–2.5 mm wide. **FL**: sepals 0.5–0.8 mm, 0.4 mm wide; petals 3–4.5 mm, 0.7–1 mm wide; stamens 10, anthers 0.3–0.4 mm, yellow or reddish brown. **FR** 2–2.5 mm, erect to slightly spreading, glabrous or sparsely glandular on inner surface. **SEED** ± 1.5 mm. *n*=9. Rock outcrops, clay soils, vernal pools; 30–1200 m. s NCoRO, NCoRI, n&c SNF, GV. ❀SUN, winterWET,summerDRY:**7–9**,**14**,15–24.

SEDUM

Melinda F. Denton

Gen per from rhizomes or stout, scaly caudex, gen glabrous. **LVS** sessile, gen alternate, gen obovate to spoon-shaped. **INFL** gen raceme- to panicle-like. **FL**: sepals 4–5, free to fused below, < petals, obtuse to long-tapered; petals 4–5, free or fused below, erect to spreading; stamens 8 or 10, in 2 whorls, barely epipetalous; pistils 4–5, free or fused below, **FR** erect or spreading. **SEEDS** many, elliptic, often winged at both ends. (Latin: to assuage, from healing properties of houseleek, to which *Sedum* was applied by some authors) [Denton 1982 Brittonia 34:48–77]

1. Fls gen unisexual; sepals and petals 4(5), gen deep purple; lvs cauline, upper often dentate; caudex short, thick, fleshy, rhizome 0 .... ***S. roseum*** ssp. ***integrifolium***
1′ Fls bisexual; sepals & petals gen 5, white, pink, or yellow; lvs often basal and cauline, upper entire, gen deciduous; rhizomes slender, often much-branched
 2. Lvs gen widest ± below middle, tapered to tip; rosette lvs ± = cauline lvs
  3. Petals white, midvein pink, fused > 1 mm; fls 1–9; fr erect .... ***S. niveum***
  3′ Petals yellow or white, ± free; fls 3–44; fr erect to spreading
   4. Lvs becoming clearly scarious
    5. Lf obtuse to acute; fr 3–5 mm, spreading to strongly reflexed; bulblets 0; ann or bien .... ***S. radiatum***
    5′ Lf long-tapered; fr 5–8 mm, ± erect; sts often with bulblets in place of upper lvs and fls; per ***S. stenopetalum***
   4′ Lvs remaining green, fleshy
    6. Cauline lvs ± elliptic to round, surfaces convex; fr strongly curved out .... ***S. divergens***
    6′ Cauline lvs lanceolate, surfaces ± flat; fr ± erect, tips curved out .... ***S. lanceolatum***
 2′ Lvs gen widest above middle, tapered to base; rosette lvs > cauline lvs
  7. Infl gen ± 3-branched; petals long-acuminate and frs erect (or petals acute and frs gen curved out)
   8. Petals long-acuminate, 8–13 mm, fused < 3 mm; fr erect; outer rosette lvs ± like inner .... ***S. oreganum***
   8′ Petals acute, 7–10 mm, ± free; fr gen outcurved; outer rosette lvs >> inner .... ***S. spathulifolium***
  7′ Infl panicle-like or flat-topped; petals obtuse to abruptly pointed; fr erect
   9. Rosette loose (internodes visible, some 3–5 mm)
    10. Petals cream-colored or light yellow, rounded to obtuse; anthers gen yellow (red-brown); infl < 7.5 cm, gen 13–40-fld .... ***S. obtusatum*** ssp. ***retusum***
    10′ Petals gen yellow (cream-colored), acute; anthers yellow or red-brown; infl < 14 cm, gen 20–70-fld .... ***S. oregonense***
   9′ Rosette dense (internodes not visible, < 3 mm)
    11. Infl strongly white-glaucous; rosette lvs < 6 cm, gen > 3.5 × longer than wide; anthers gen yellow (some turning red-brown)
     12. Cauline lvs 14–30 mm; rosettes 3–9 cm diam, lvs < 60 mm, 7–16 mm wide, widest well below tip; petals ± pale yellow; anthers yellow or red-brown .... ***S. albomarginatum***

12′ Cauline lvs 7–13 mm; rosettes 1–6 cm diam, lvs < 35 mm, 3–9 mm wide, widest just below tip; petals ± cream-colored; anthers yellow .... ***S. oblanceolatum***
11′ Infl not or barely glaucous; rosette lvs gen < 4 cm, gen < 3 × longer than wide (or anthers red-purple); anthers yellow to reddish black
13. Petals dark pink to dark red; cauline lf surfaces convex, base rounded or truncate .... ***S. eastwoodiae***
13′ Petals white to yellow or pink; cauline lf surfaces flat, base truncate or ear-like and decurrent
14. Petals white, pink, or pale yellow .... ***S. laxum***
15. Cauline lf length >> width; pl 7–40 cm; rosette lvs 10–50 mm .... ssp. ***laxum***
15′ Cauline lf length < to > width; pl 9–24 cm; rosette lvs 9–32 mm
16. Petals pale yellow; cauline lf bases often truncate; sepals < 1/3 petals .... ssp. ***flavidum***
16′ Petals white to pink; cauline lf base ± clasping; sepals 1/3–2/3 petals .... ssp. ***heckneri***
14′ Petals cream-colored or deep yellow
17. Petals cream-colored; sepals 3.5–6 mm, acute to long-tapered .... ***S. paradisum***
17′ Petals yellow (sometimes pinkish or red-veined); sepals 1.8–5 mm, obtuse to acute .... ***S. obtusatum***
18. Cauline lvs 8–15 mm; petals pale yellow or pinkish white; infl 15–38-fld .... ssp. ***boreale***
18′ Cauline lvs gen 4–9(13) mm; petals yellow (or red-veined); infl gen 8–26(46)-fld .... ssp. ***obtusatum***

S. albomarginatum R.T. Clausen (p. 533) FEATHER RIVER STONECROP Pl 14–25 cm, glabrous, glaucous; rosettes 3–9 cm diam, internodes between lvs < 3 mm. **LVS**: rosette lvs 14–60 mm (widest 6–9 mm from tip), ± 1–4 mm thick, rounded or barely notched; cauline lvs 14–30 mm, base truncate or obtuse. **INFL** 4–10 cm, 20–55-fld. **FL**: petals ± 7–10 mm, obovate, light yellow, obtuse or minutely and abruptly pointed; anthers yellow or red-brown. **FR** 7–9 mm, erect. **SEED** 1–1.2 mm. *n*=15. RARE. Steep serpentine slopes; 300–900 m. n SNF (Plumas Co.). In cult.

S. divergens S. Watson (p. 533) CASCADE STONECROP Pl 5–12 cm, matted, glabrous; ascending sterile shoots many. **LVS** opposite, 3–8 mm, obovate to nearly spheric, glabrous or finely papillate; tip gen rounded. **INFL** gen 1–5 cm, 3–17-fld. **FL**: petals 5–7 mm, oblong-lanceolate, long-tapered and abruptly pointed, yellow; anthers yellow. **FR** 5–7 mm, widely spreading. **SEED** ± 0.5 mm. *n*=8. RARE in CA. Gravelly flats or slopes; 1600–2000 m. KR; to B.C. In cult.

S. eastwoodiae (Britton) A. Berger (p. 533) RED MOUNTAIN STONECROP Pl 7–19 cm, glabrous; rosettes 1–6 cm diam, internodes between lvs < 3 mm. **LVS**: rosette lvs 10–29 mm, widest ± 6 mm below tip, 2–5 mm thick, rounded to barely notched; cauline lvs 4–17 mm, base truncate to rounded. **INFL** ± 2–7 cm, 10–26-fld, sometimes flat-topped. **FL**: petals 6–9 mm, obovate, acute, dark pink to dark red; anthers light red or ± red-purple. **FR** ± 8 mm, erect. **SEED** 1.1–1.8 mm. *n*=30. RARE. Serpentine soils among rocks; 600–1200 m. c NCoRO (Red Mtn, Mendocino Co.). [*S. laxum* ssp. *e.* (Britton) R.T. Clausen] In cult.

S. lanceolatum Torrey (p. 533) Pl 3–20 cm, glabrous; rosettes 3–10 mm diam, internodes between lvs ± 1 mm. **LVS** linear to ovate; rosette lvs 5–30 mm, 1.5–2 mm thick, acute; cauline lvs 3–10 mm, mostly fallen by fl. **INFL** 1–4 cm, 3–24-fld. **FL**: petals 5–8 mm, lanceolate, acute to acuminate, yellow (midrib often reddish); anthers yellow. **FR** 4–9 mm, erect to widely spreading. **SEED** ± 1 mm. *n*=8,16,24. Granite outcrops, rocky soils; 1800–2800 m. KR, CaR, SN; to AK, SD, NM. ❀DRN,SUN:1,4,5,**6**,17&IRR:2,3,**7**, 14–16,18–24;DFCLT.

S. laxum (Britton) A. Berger Pl 7–40 cm, glabrous, sometimes glaucous; rosettes 1–6 cm diam, internodes between lvs < 3 mm. **LVS** ± (ob)ovate; rosette lvs 9–50 mm, 1–5 mm thick, gen widest 3–8 mm below tip, rounded or barely notched; cauline lvs 5–20 mm. **INFL** 2–11 cm, 12–80+-fld, sometimes flat-topped. **FL**: sepals gen 1/3–2/3 petals, gen acute; petals 7–13 mm, obovate, obtuse; anthers gen ± red-brown to purplish black. **FR** 6–12 mm, erect. **SEED** 1–2 mm. Outcrops; 50–2000 m. NCo, n NCoR, KR; sw OR.

ssp. ***flavidum*** Denton (p. 533) PALE YELLOW STONECROP Pl 9–24 cm. **LVS**: rosette lvs 9–25 mm; cauline lf length < or > width, base truncate or barely sheathing and ear-like. **FL**: sepals < 1/3 petals, obtuse; petals pale yellow; anthers yellow to ± red-brown. *n*=30. UNCOMMON. Serpentine or basalt outcrops; 800–2000 m. n NCoRH, KR. ❀DRN,SHD:4–6,**17**&IRR:7,15,16.

ssp. ***heckneri*** (M. Peck) R.T. Clausen (p. 533) HECKNER'S STONECROP Pl 9–24 cm. **LVS**: rosette lvs 11–32 mm; cauline lf length < or > width, base sheathing and ear-like. **FL**: petals white to light pink. *n*=15. UNCOMMON. Gen steep serpentine or gabbro outcrops; 100–1800 m. n NCoRO, KR; sw OR. ❀DRN,IRR, SHD:5,6,15,16,**17**.

ssp. ***laxum*** (p. 533) Pl 7–40 cm. **LVS**: rosette lvs 10–50 mm; cauline lf length >> width, base obtuse, often slightly decurrent. **FL**: petals gen white to light pink (light yellow). *n*=15. On serpentine, sometimes on basalt; 50–2000 m. NCo, KR; sw OR. [sspp. *latifolium* R.T. Clausen and *perplexum* R.T. Clausen] ❀DRN, SHD,IRR:4–7,**15–17**.

S. niveum Davidson (p. 537) DAVIDSON'S STONECROP Pl 3–9 cm, matted, glabrous. **LVS** 5–9 mm, 1–3 mm thick, oblong to spoon-shaped, rounded to widely acute. **INFL** 1–2 cm, 1–9-fld. **FL**: petals spreading to reflexed, 5–8 mm, lanceolate, acute, white or pink-tinged or -veined; anthers red to black. **FR** 5–7 mm, erect. **SEED** ± 0.5 mm. *n*=16. UNCOMMON. Rocky ledges, crevices; 2200–3000 m. SnBr, e PR (Santa Rosa Mtns); Baja CA.

S. oblanceolatum R.T. Clausen (p. 537) APPLEGATE STONECROP Pl 7–14 cm, glabrous, strongly glaucous; rosettes 1–6 cm diam, internodes between lvs < 3 mm. **LVS** obtuse to notched; rosette lvs 11–35 mm, widest 2–3 mm from tip, 2–3 mm thick; cauline lvs 7–13 mm. **INFL** 1–5 cm, 7–50-fld, gen flat-topped. **FL**: petals 7–11 mm, obovate, slightly abruptly pointed, very pale yellow or cream-colored; anthers yellow. **FR** 6–10 mm, erect. **SEED** 0.9–1.5 mm. *n*=15. RARE. Rocky slopes; 400–2000 m. n KR (nw Siskiyou Co.); sw OR. In cult.

S. obtusatum A. Gray Pl 3–22 cm, glabrous, glaucous; rosettes 1–6 cm diam, internodes between lvs < 3 mm. **LVS**: rosette lvs 6–30 mm, 1–4 mm thick, widest 2–8 mm below tip, rounded to slightly notched; cauline lvs 4–17 mm, obtuse or truncate. **INFL** 2–12 cm, 8–60-fld, often flat-topped. **FL**: sepals gen acute; petals 4–11 mm, obovate, rounded to obtuse, gen abruptly pointed; anthers tawny or yellow to dark red-brown. **FR** 5–10 mm, erect. **SEEDS** 1–1.6 mm. Outcrops; 400–3700 m. NCoRH, KR, CaRH, SN; sw OR.

ssp. ***boreale*** R.T. Clausen (p. 537) **LVS**: cauline lvs 8–15 mm. **INFL** 2–12 cm. **FL**: petals ± 6–9 mm, pale yellow or white, often pink-tinged or -veined. *n*=15. Uncommon. Outcrops; 1400–2000 m. CaRH, SN. ❀DRN:15–17&IRR:1,2,7,14;DFCLT.

ssp. ***obtusatum*** (p. 537) **LVS**: cauline lvs gen 4–9(13) mm. **INFL** 2–7(12) cm. **FL**: petals 4–10 mm, yellow (sometimes drying reddish or midvein red). *n*=15. Outcrops; 1200–3700 m. KR, SN. ❀DRN:4–6,**17**&IRR:**1**,2,7,15,16,18.

ssp. ***retusum*** (Rose) R.T. Clausen (p. 537) **LVS**: cauline lvs (5)8–17 mm. **INFL** 2–8 cm. **FL**: sepals obtuse; petals ± pale yellow or cream-colored. *n*=30. Outcrops; 400–2300 m. NCoRH, KR; sw OR. [*S. laxum* ssp. *r.* (Rose) R.T. Clausen] NCoRH. ❀DRN:4–6,**17**&IRR:7,14–16.

S. oreganum Nutt. (p. 537) Pl 6–15 cm, glabrous, strongly rhizomed; rosettes 12–16 mm diam, internodes between lvs gen 1–2 mm. **LVS** obovate to spoon-shaped; rosette lvs 5–25 mm, ± 2 mm thick, widest 1–3 mm below tip, rounded to barely notched; cauline lvs 4–12 mm, base truncate. **INFL** dense, 3-parted, 4–12 cm, 3–16-fld. **FL**: petals 8–13 mm, narrowly lanceolate, long-acuminate, yellow (or midvein red); anthers light yellow-brown to red-brown. **FR** 5–8 mm, erect. **SEED** ± 1 mm. *n*=12. Rocky ledges, gravelly ridges; 1100–1800 m. n KR; to AK. ❀DRN,IRR,SHD:**4,5,17**.

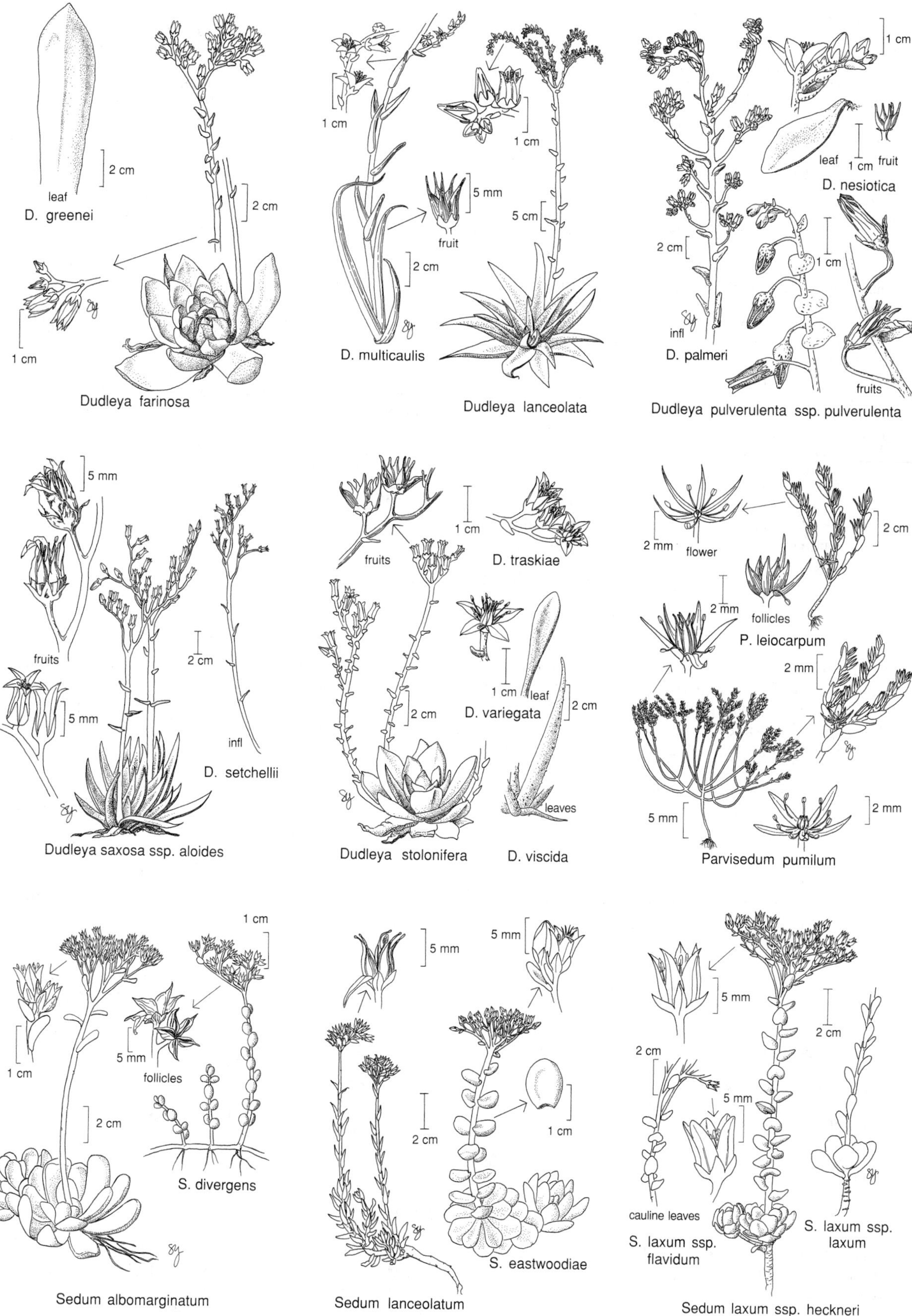
leaf
2 cm
D. greenei
2 cm
1 cm
Dudleya farinosa
1 cm
5 mm
fruit
2 cm
D. multicaulis
1 cm
5 cm
Dudleya lanceolata
1 cm
leaf
1 cm
fruit
D. nesiotica
2 cm
1 cm
infl
D. palmeri
fruits
Dudleya pulverulenta ssp. pulverulenta
5 mm
fruits
2 cm
5 mm
infl
D. setchellii
Dudleya saxosa ssp. aloides
1 cm
fruits
D. traskiae
2 cm
1 cm
leaf
D. variegata
2 cm
leaves
Dudleya stolonifera
D. viscida
2 mm
flower
2 cm
2 mm
follicles
P. leiocarpum
2 mm
5 mm
2 mm
Parvisedum pumilum
1 cm
1 cm
5 mm
follicles
2 cm
S. divergens
Sedum albomarginatum
5 mm
5 mm
2 cm
1 cm
S. eastwoodiae
Sedum lanceolatum
5 mm
2 cm
2 cm
5 mm
cauline leaves
S. laxum ssp. flavidum
S. laxum ssp. laxum
Sedum laxum ssp. heckneri

S. oregonense (S. Watson) M. Peck (p. 537) Pl 4–21 cm, glabrous; rosettes 1–8 cm diam, internodes between lvs gen 3–5 mm. **LVS**: rosette lvs 12–41 mm, widest 3–9 mm below tip, 1–4 mm thick, rounded or barely notched; cauline lvs 5–16 mm, base obtuse or truncate. **INFL** gen panicle-like, 1–14 cm, 12–120-fld. **FL**: petals ± 5–11 mm, obovate, obtuse and gen slightly abruptly pointed, pale yellow above, gen red-marked or whitish below; anthers yellow or red-brown. **FR** 6–9 mm, erect. **SEED** 1–1.8 mm. *n*=45. Rock outcrops; 900–2200 m. KR; OR. ❀DRN,IRR:2,**7**,14,**15,16**&SUN:1,4,**5,6,17**.

S. paradisum (Denton) Denton (p. 537) CANYON CREEK STONECROP Pl 10–20 cm, glabrous, glaucous; rosettes 2–5 cm diam, internodes between lvs < 3 mm. **LVS**: rosette lvs 10–33 mm, widest 3–4 mm below tip, 2–4 mm thick, obtuse to notched; cauline lvs 9–19 mm, base obtuse or truncate. **INFL** gen 3–9 cm, 10–58-fld, sometimes flat-topped. **FL**: petals 7–10 mm, obovate, rounded to obtuse, gen abruptly short-pointed, cream-colored; anthers tawny or yellow to ± red-brown. **FR** 9–10 mm, erect. **SEED** 1–1.6 mm. *n*=15. RARE. Granite outcrops; 300–1400 m. se KR (Trinity Co.). [*S. obtusatum* ssp. *p.* Denton] In cult.

S. radiatum S. Watson Ann or bien 4–20 cm, glabrous. **LF** 5–11 mm, oblong to narrowly ovate, obtuse to acute, flat; veins red-scarious. **INFL** 2–4 cm, 1–44-fld. **FL**: petals 5–10 mm, lanceolate, acuminate, yellow to white, reflexed; anthers yellow to yellow-brown. **FR** 3–5 mm, spreading to strongly reflexed. **SEED** ± 1 mm. *n*=8. Rocky ledges, gravelly serpentine slopes; 60–2100 m. KR, NCoR, SNH, SnFrB, SCoRO; sw OR. [*S. stenopetalum* ssp. *r.* (S. Watson) R.T. Clausen] ❀DRN,IRR,SHD:1,2,**5–7**,14–16,**17**.

S. roseum (L.) Scop. ssp. ***integrifolium*** (Raf.) Hultén (p. 537) Pl 2–30 cm, glabrous; caudex short, thick, fleshy. **LF** 7–25 mm, oblanceolate to obovate, widest 3.5–5 mm below tip, flat, acute, sometimes dentate above middle. **INFL** ± 1–3 cm, 7–50-fld, dense. **FLS** 4(5)-parted, some or all unisexual; petals free, slightly spreading, 2–3.5 mm, oblong, fleshy, obtuse to acute, deep purple; anthers light brown to red-purple. **FR** 3–6 mm, erect. **SEED** 1.5–2 mm. *n*= 18. Cliffs, talus, alpine ridges; 1800–4000 m. KR, c SNF, SNH, Wrn, W&I; to AK, ne N.Am, Colorado, Eurasia. ❀DRN,IRR:1,2,4–7,15–17;DFCLT.

S. spathulifolium Hook. (p. 537) Pl 5–22 cm, glabrous, often glaucous; rosettes 1–6 cm diam, internodes between lvs gen 1–2 mm. **LVS**: rosette lvs 11–22 mm, outer >> inner, widest 2–5 mm below tip, 1–2 mm thick, rounded to obtuse; cauline lvs 6–11 mm, elliptic, base truncate. **INFL** 3–8 cm, 5–48-fls. **FLS** petals 5–8 mm, ± erect to widely spreading, lanceolate, acute, yellow; anthers yellow or red-brown. **FR** 4–8 mm, erect until mature, then strongly spreading. **SEED** ± 1 mm. *n*=15. Outcrops, often in shade; 50–2500 m. NW, CaR, SN, CW, TR; to B.C. Variable and intergrading complex. [sspp. *anomalum* (Britton) R.T. Clausen & C.Uhl, *pruinosum* (Britton) R.T. Clausen & C. Uhl; *S. purdyi Jepson*] ❀DRN:**4–6**&IRR:1,**17**,24&SHD:2,**7**,8,9,**14–16**,18–23;CVS.

S. stenopetalum Pursh (p. 537) Pl 5–27 cm, glabrous, often with bulblets in place of upper lvs and fls. **LF** 6–16 mm, gen ± lanceolate, long-tapered, base scarious, often ± 3-lobed. **INFL** raceme-like, gen ± 1.5 cm, 1–44-fld; petals 4–10 mm, lanceolate, long-tapered, yellow (or midrib ± red); anthers yellow or light brown. **FR** 5–8 mm, ± erect. **SEED** ± 1 mm. *n*=8,16,24. Well drained, rocky or gravelly soil; 1400–2600 m. KR, CaR, n SNH, MP; to B.C., MT. [*ssp. monanthum (Suksd.)* R.T. Clausen] ❀DRN:**4–6,17**&IRR:1,2,7,14,**15,16**.

CROSSOSOMATACEAE CROSSOSOMA FAMILY

James R. Shevock

Shrubs. **ST** gen glabrous; twigs or branchlets gen thorny. **LVS** gen deciduous, simple, small, gen alternate, entire; stipules minute or 0. **INFL**: fls solitary. **FL** gen bisexual, radial; hypanthium short, mostly forming thick nectary-disk; sepals gen 5 (3–6), free; petals ephemeral, gen 5(3–6), free, gen white; stamens 4–50, attached to nectary-disk; pistils 1–9, simple, styles short, stigma head-like, ovules gen 2–many. **FR**: follicles 1–9. **SEEDS** brown to black, with an aril. 3 genera, 8 spp.: w US, Mex (*Apacheria* of AZ, NM has 1 sp.). [Thorne & Scogin 1978 Aliso 9:171–178]

1. Lf gen elliptic; stamens 15–50; fr 1–9, 8–20 mm, stalked .... **CROSSOSOMA**
1′ Lf gen oblanceolate; stamens 4–10, fr 1–3, 1–5 mm, sessile .... **GLOSSOPETALON**

CROSSOSOMA

Shrubs. **LVS** deciduous (or dry during dormant periods), gen ± narrowly elliptic; petiole 0–short. **INFL**: peduncle sometimes bracted. **FL**: petals 9–15 mm, rounded or oblong, gen white; stamens 15–50 in several series. **FR**: follicles 8–20 mm, cylindric. **SEEDS** gen > 2 per follicle, black, shiny, round to flat; aril conspicuous, fringed, yellowish. 2 spp.: s CA, e to AZ, s to Mex. (Greek: fringe body, from aril)

1. Lvs 5–15 mm, clustered; sepals 4–5 mm .... ***C. bigelovii***
1′ Lvs 25–90 mm, clearly separated; sepals > 10 mm .... ***C. californicum***

C. bigelovii S. Watson Shrub 1–2 m. **ST** much branched; branchlets thorny. **LVS** clustered, 5–15 mm, gray-green. **FL**: sepals 4–5 mm, rounded; petals 9–12 mm, oblong, white to purplish, distinctly clawed. **FR**: follicles 1–3, 8–10 mm, ± straight. **SEEDS** gen 2–5 per follicle, ± 2 mm diam. Dry, rocky slopes, canyons; < 1250 m. e SnBr, D; NV, AZ, Baja CA. ❀DRN:14,18–21&SHD:13; DFCLT.

C. californicum Nutt. (p. 537) CATALINA CROSSOSOMA Shrub small tree, 1–5 m. **ST**: branchlets ± thorny. **LVS** gen clearly separated, 25–90 mm, sometimes nearly round, pale green. **FL**: sepals ± 10 mm, round-ovate; petals 12–15 mm, rounded, white, scarcely clawed. **FR**: follicles 2–9, 15–20 mm; tip ± recurved. **SEEDS** gen > 20 per follicle, ± 2.5 mm diam. UNCOMMON. Dry, rocky slopes, canyons; < 500 m. c SCo (Palos Verdes Peninsula), s ChI (San Clemente, Santa Catalina islands); Guadalupe Island, Mex. ❀DRN:14–17,**19–24**.

GLOSSOPETALON

Small shrubs, gen densely branched. **ST** greenish, glabrous to sparsely hairy, angled; branchlets thorny. **LVS** small, gen deciduous, gen oblanceolate to obovate, ± entire. **FL**: petals narrowly oblanceolate, white; stamens 4–10. **FR**: follicles 1–3, ovoid, sessile, gen striate, gen beaked. **SEEDS** gen 1–2 per follicle, gen brown; aril ± inconspicuous, gen whitish. ± 5 spp.: w US, n Mex, esp limestone in desert mtns. (Greek: tongue petal, from petal shape) [Holmgren 1988 Brittonia 40:269–274] Formerly called *Forsellesia* and assigned to Celastraceae.

1. Lf spine-tipped; stipules 0; fls terminal; fr < 1 mm .... ***G. pungens***
1′ Lf tip rounded or short-pointed; stipules minute; fls axillary; fr 3–5.5 mm .... ***G. spinescens***

G. pungens Brandegee PUNGENT FORSELLESIA Shrub 5–20 cm, low, matted. **ST** not thorny. **LF** 6–10 mm, oblanceolate or narrowly elliptic, sharply spine-tipped; veins prominent below; stipules 0. **INFL**: fls terminal. **FL**: sepals acuminate, 2–3 spine-tipped; petals 6–8 mm; stamens 10 in 2 series, longer opposite sepals. **FR**: follicle 1, < 1 mm. **SEED** 1 per follicle, < 1 mm, light brown. RARE. Limestone cliffs; 1700–2000 m. e DMtns (Clark Mtns); s NV (Sheep Mtns). [*Forsellesia p.* (Brandegee) A.A. Heller, incl var. *glabra* Ensign]

G. spinescens A. Gray (p. 537) NEVADA GREASEWOOD Shrub < 2 m, ± erect. **ST**: tips ± thorny. **LF** 5–17 mm, oblong to obovate; veins ± inconspicuous below; stipules appearing as 2 minute bristles near petiole base. **INFL**: fls axillary. **FL**: sepals rounded, none spine-tipped; petals 3–7 mm; stamens 6–10. **FR**: follicles 1–2, 3–5.5 mm. **SEEDS** 1–2 per follicle, 2–3 mm, gen shiny brown. Limestone; 850–2200 m. s KR, s SNH (Piute Mtns, Kern Co.), SnBr (n base), W&I, DMtns; to WA, WY, TX, n Mex. Highly variable; if the 4 weak vars. are recognized, CA pls are var. *aridum* M.E. Jones. [*Forsellesia arida* (M.E. Jones) A.A. Heller; *F. nevadensis* (A. Gray) E. Greene; *F. stipulifera* (St. John) Ensign, stipule-bearing forsellesia] ❀TRY.

CUCURBITACEAE GOURD FAMILY

Robert L. Schlising

Ann, per, gen monoecious; hairs often hardened by calcium deposits. **STS** trailing or climbing, 1–many; tendril gen 1 per node, often branched. **LVS** gen simple, alternate, gen palmately lobed, veined, petioled; stipule 0. **INFLS** at nodes; staminate fls in racemes, panicles, small clusters, rarely solitary; pistillate fls gen solitary. **FL** unisexual in CA, radial; hypanthium > ovary; calyx (apparently 0 or) gen 5-lobed; corolla rotate or cup-shaped, gen 5-lobed; stamens 3–5 (or appearing 1–3 from fusion), anthers often > filaments, twisted together; ovary ± inferior, chambers gen 5, placentas parietal, ± growing into chambers, styles 1–3, stigmas gen lobed, large. **FR**: berry (sometimes drying) or capsule (irregularly dehiscent), gen gourd- or melon-like. **SEEDS** 1–many. 100 genera, 700 spp.: esp trop; some cult (*Citrullus*; *Cucumis*; *Cucurbita*; *Sechium*, chayote).

1. Corolla white, greenish, or cream-colored, < 2 cm wide; staminate fls gen in racemes, panicles, or small clusters, pistillate fls solitary or in small clusters, gen at same nodes as staminate; fr gen prickly (exc *Bryonia*), sometimes irregularly dehiscent; seeds gen not flat
  2. Staminate fl < 3 mm wide; fr asymmetric, beak ± = body; seed 1; D .... **BRANDEGEA**
  2′ Staminate fl > 3 mm wide; fr symmetric, beak < body or 0; seeds gen 2–many; CA-FP
    3. Staminate infl axis ± glandular; tendril unbranched; fr a berry, < 1 cm wide; dioecious .... **BRYONIA**
    3′ Staminate infl axis nonglandular; tendril branched; fr a capsule, > 2 cm wide; monoecious .... **MARAH**
1′ Corolla yellow, gen 2–12 cm wide; staminate and pistillate fls 1–few, gen at different nodes; fr unarmed (weakly prickly in some *Cucumis*), gourd- or melon-like, indehiscent; seeds ± flat
  4. Corolla 3–12 cm wide, deeply cup-shaped, fused portion > 3 cm; per from large, tuber-like root .... **CUCURBITA**
  4′ Corolla gen < 3 cm wide, shallowly cup-shaped or rotate, fused portion < 1 cm; ann
    5. Tendril branched; lf ± palmately lobed, 1° lobes ± pinnately lobed .... **CITRULLUS**
    5′ Tendril unbranched; lf angled to ± palmately lobed, 1° lobes ± entire to irregularly lobed .... **CUCUMIS**

BRANDEGEA

1 sp. (T.S. Brandegee, CA botanist & engineer, 1843–1925)

B. bigelovii (S. Watson) Cogn. (p. 537) Per from taproot. **ST** ± glabrous; tendril unbranched. **LF** round, cordate to ± square, gen deeply lobed; central lobe gen longest; upper surfaces dotted with white glands. **INFLS**: staminate fls in small axillary clusters; pistillate fls 1 per node. **FL**: corolla 1.5–3 mm wide, rotate or shallowly cup-shaped, cream-colored. **FR** dry, indehiscent, asymmetric, ± prickly; body 5–6 mm, ± = beak. **SEED** 1. Canyons, washes; < 900 m. D; sw AR, Mex.

BRYONIA BRYONY, WILD HOP

Per from tuber, gen dioecious. **ST** gen smooth, glabrous; tendril unbranched in CA. **LF** wide, lobed or angled. **INFLS**: racemes, panicles, or clusters, small; staminate infl > pistillate infl. **FL**: corolla > 1 cm wide, rotate or shallowly cup-shaped; stigmas 3, each simple or 2-lobed. **FR**: berry, round, unarmed. **SEEDS** several, not flat. 12 spp.: Eur, n Afr, w Asia. (Greek: swelling, from sprouting of tuber each year)

B. dioica Jacq. **LF** rough. **FL**: corolla 1–2 cm wide, yellow-green; stigmas 2-lobed. **FR** < 1 cm wide, smooth, red or orange. **SEEDS** 3–6, ± 3 mm. $2n=20$. Disturbed places; < 100 m. n CCo (San Francisco); native to Eur, w Asia. Similar to *Marah*.

CITRULLUS WATERMELON

Ann, per. **ST** ± hairy; tendril branched. **LF** deeply, ± palmately lobed; 1° lobes ± pinnately lobed. **INFLS**: staminate fls and pistillate fls solitary at different nodes. **FL**: corolla 3–4 cm wide, rotate or shallowly cup-shaped, yellow, deeply 5-lobed, fused portion < 1 cm; anthers 3, free; stigmas 3, reniform. **FR** melon-like, indehiscent, round to oblong; rind hard, smooth. **SEEDS** many, ± 1 cm. 4 spp.: trop Afr, Asia. (Latin: diminutive of citrus)

C. colocynthis (L.) Schrader var. *lanatus* (Thunb.) Matsum. & Nakai **FR** large, green, gen striped or mottled with darker green. **SEEDS** white to black. 2*n*=22. Disturbed areas; < 300 m. CA-FP; native to Afr, widely cult and naturalized. [*C. l.* (Thunb.) Mansf.; *C. vulgaris* Schrader] Pls with small, gen round frs, white fr flesh, green or tan seeds assignable to var. *citroides* (L. Bailey) Mansf., citron, an uncommon weed.

CUCUMIS CUCUMBER, MELON

Ann, per. **ST** gen scabrous; tendril unbranched. **LF** angled to ± palmately lobed; 1° lobes ± entire to irregularly lobed. **INFLS**: staminate fls 1–several per node; pistillate fls gen 1 at different nodes. **FL**: corolla 2–3 cm wide, rotate or shallowly cup-shaped, yellow, deeply 5-lobed, fused portion < 1 cm; anthers 3, free; styles 3–5, ± reniform. **FR** gourd- or melon-like, indehiscent, cylindric to round; rind firm, net-veined or hairy, prickles 0 or weak. **SEEDS** many, < 1 cm, ± flat; margin plain. ± 40 spp.: Afr, s Asia. (Greek: cucumber)

1. Lf angled or shallowly lobed; fr 3–6 cm wide, prickles 0 . . . . **C. melo** var. ***dudaim***
1′ Lf deeply lobed; fr 2–3 cm wide, prickles weak . . . . **C. myriocarpus**

C. melo L. var. *dudaim* (L.) Naudin DUDAIM MELON **FR** cylindric to ± round, orange, sometimes irregularly blotched or striped. Fields, roadsides; < 200 m. n SCo (Santa Barbara Co.), se DSon (Imperial Co.); native to Afr. NOXIOUS WEED.

C. myriocarpus Naudin PADDY MELON **FR** ± round, yellow-green, green-striped. 2*n*=24. Fields, disturbed areas; < 300 m. c&s SnJV, s SCoRO (Santa Barbara Co.); native to s Afr. NOXIOUS WEED.

CUCURBITA GOURD, SQUASH

Ann, per (from large, fleshy, tuber-like root). **ST** smooth to scabrous; tendril gen branched. **LF** lanceolate to round, entire to deeply lobed. **INFL**: fls 1 per node, staminate and pistillate at different nodes. **FLS**: corolla > 3 cm wide (staminate gen < pistillate), deeply cup- or bell-shaped, yellow, fused portion 4–12 cm, lobes gen recurved; stigmas 3, each 2-lobed. **FR** gourd-like, indehiscent, ± round to ± flat; rind firm, smooth, rough, or grooved. **SEEDS** many, < ± 2 cm, ± ovate, ± flat; margin thick or raised. ± 30 spp.: warm Am. (Latin: gourd) [Rhodes et al. 1968 Brittonia 20:251–266]

1. Lf blade angled, finely toothed, or weakly lobed at base; tendril branched > 1 cm from base . . . . **C. foetidissima**
1′ Lf blade deeply lobed; tendril branched ± from base
 2. Lf blade lobes ± linear-lanceolate, distinct nearly to petiole . . . . **C. digitata**
 2′ Lf blade lobes triangular or widely lanceolate, distinct ± 1/2 to petiole . . . . **C. palmata**

C. digitata A. Gray (p. 537) FINGER-LEAVED GOURD Herbage ± scabrous, hairy; tendril branched ± from base. **LF** 3–9 cm; blade ± cordate in outline, green, main veins lighter, lobes gen 5, distinct nearly to petiole, ± linear-lanceolate, lateral 2 lobes gen coarsely toothed. **FL**: corolla 3–5 cm. **FR** 7–8 cm wide, round to oblong, dark green, ± mottled, with several, narrow, well defined, whitish stripes. **SEED** 10–11 mm, white. 2*n*=40. Uncommon. Sandy, open or shrubby places; < 1200 m. PR, DSon; to NM, Mex. ❀TRY.

C. foetidissima Kunth (p. 537) CALABAZILLA Herbage coarsely scabrous, hairy; tendril branched gen > 1 cm above base. **LF** 15–30 cm; blade gen triangular-ovate, ± cordate or truncate at base, gray-green, ill-smelling, angled, finely toothed or weakly lobed at base. **FL**: corolla 9–12 cm. **FR** gen 7–8 cm wide, ± round, green, mottled, with coarse, white stripes. **SEED** 12–14 mm, white. 2*n*=40. Sandy, gravelly places; < 1300 m. GV, CW, SW, D; to NE, TX, n Mex. Other localities possibly due to human transport. ❀SUN, DRN:**7,14–16**,17,**18–24**&IRR:**8–13**;deciduous,GRNCVR.

C. palmata S. Watson (p. 537) COYOTE MELON Herbage scabrous, hairy; tendril branched ± from base. **LF** 8–15 cm; blade ± cordate, gray-green, main veins whitish, lobes gen 5, distinct ± half way to petiole, middle 3 widely triangular, gen without prominent teeth. **FL**: corolla 6–8 cm. **FR** 8–9 cm wide, round, dull green, mottled, with poorly-defined, whitish stripes. **SEED** 10–14 mm, white. 2*n*=40. Sandy places; < 1300 m. SnJV, CW, SW, D; AZ, Baja CA. ❀DRN,SUN:7,14,**19–23**,24&IRR:**8,9**,10,11,**12,13**;deciduous, GRNCVR.

MARAH MAN-ROOT, WILD CUCUMBER

Per, sometimes temporarily dioecious; tuber large. **ST** ± scabrous or hairy, becoming glabrous; tendril branched. **LF** ± round, cordate, ± 5–7-lobed. **INFLS**: staminate fls in racemes or panicles with nonglandular axes (or 1 fl per axil early in season); pistillate fl 1 per axil (gen same axil as staminate). **FL**: sepals 0; corolla 3–15 mm wide (wider in pistillate), cup-shaped to rotate, white or cream-colored to yellowish green; stamens fused, anthers twisted together; stigma 1, ± hemispheric. **FR**: capsule, irregularly dehiscent, ± symmetric, 3–20 cm, round, ovate, or oblong, sometimes tapered to a beak, ± prickly. **SEED** gen > 1 cm. 7 spp.: w N.Am. (Latin: bitter, from taste of all parts) [Schlising 1969 Amer J Bot 56:552–561] Extremely variable in habit, lvs, sexual expression; presumed hybrids occur where spp. overlap. Sometimes incl in *Echinocystis*.

1. Corolla rotate, yellowish green, cream-colored, or white; ovary and fr ± round, prickles sparse to ± dense, ± stiff, unhooked . . . . **M. fabaceus**
1′ Corolla ± cup-shaped, white; ovary and fr sometimes not round, prickles sparse to dense, stiff or flexible, hooked or unhooked
 2. Staminate fls gen < 8 mm wide; herbage glaucous; ovary and fr ± round, prickles sparse to dense, flexible, gen hooked; seeds 1–4, round . . . . **M. watsonii**
 2′ Staminate fls gen > 8 mm wide; herbage not glaucous; ovary and fr ovate, oblong, or tapered to a beak, prickles gen dense, stiff or flexible, unhooked; seeds gen 3–24, disc-shaped, ovate, or oblong
  3. Mature fr (and ovary) tapered to a beak, often striped dark green, prickles sparse to dense, flexible; seeds gen 3–6, disc-shaped, ± flat; NW, SnFrB . . . . **M. oreganus**

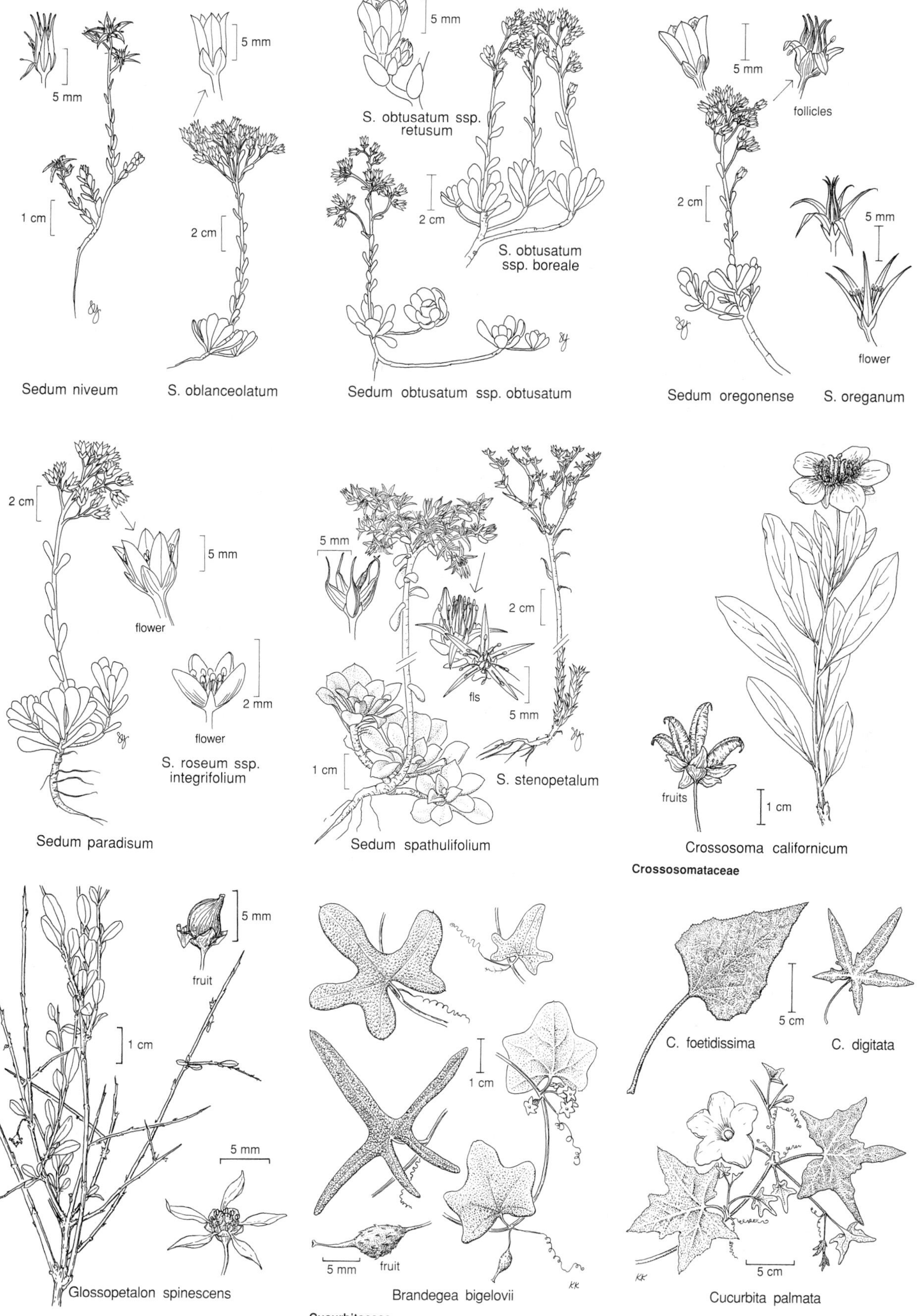
5 mm
5 mm
1 cm
2 cm
Sedum niveum
S. oblanceolatum
5 mm
S. obtusatum ssp. retusum
2 cm
S. obtusatum ssp. boreale
Sedum obtusatum ssp. obtusatum
5 mm
follicles
2 cm
5 mm
flower
Sedum oregonense
S. oreganum
2 cm
5 mm
flower
2 mm
flower
S. roseum ssp. integrifolium
Sedum paradisum
5 mm
2 cm
fls
5 mm
1 cm
S. stenopetalum
Sedum spathulifolium
fruits
1 cm
Crossosoma californicum
Crossosomataceae
5 mm
fruit
1 cm
5 mm
Glossopetalon spinescens
1 cm
5 mm
fruit
Brandegea bigelovii
Cucurbitaceae
5 cm
C. foetidissima
C. digitata
5 cm
Cucurbita palmata

3′ Mature fr and ovary oblong or ovate, gen not tapered to a beak, not striped green, prickles ± dense, stiff; seeds gen > 6, ovate or oblong, sometimes flat at 1 end; SNF, SW, D
4. Corolla deeply cup-shaped; seeds very flat at 1 end; c&s SNF, Teh ..... ***M. horridus***
4′ Corolla shallowly cup-shaped; seeds not very flat at either end; SW, DSon ..... ***M. macrocarpus***
5. Staminate fls 8–13 mm wide; seeds 15–20 mm; mainland ..... var. ***macrocarpus***
5′ Staminate fls 14–30 mm wide; seeds 20–33 mm; ChI ..... var. ***major***

M. fabaceus (Naudin) E. Greene (p. 543) CALIFORNIA MAN-ROOT Herbage gen not glaucous. **FL**: corolla rotate, yellowish green, cream-colored or (esp inland) white. **FR** 4–5 cm, ± round; prickles sparse to dense, < 12 mm, ± stiff, unhooked. **SEEDS** 2–4, 18–24 mm, ovate to oblong, ± flat on sides or not. $2n$=32. Streamsides, washes, shrubby and open areas; < 1600 m. CA-FP (exc n NW, n CaR). Highly variable; pls outside SnFrB and CCo are assignable to var. *agrestis* (E. Greene) K.M. Stocking (fr prickles flexible, < 5 mm; seeds gen 2–3, not flat on sides), which intergrades ± completely with the pls considered var. *f.*; more study needed. ❀DRN: 7,8–10,**14–24**;INV.

M. horridus (Congdon) Dunn (p. 543) Herbage not glaucous. **FL**: corolla deeply cup-shaped, white. **FR** 9–20 cm, oblong, rounded at both ends; prickles dense, stiff. **SEEDS** gen 6–16(24), 26–32 mm, oblong or ovate, very flat at 1 end. Shrubby and open areas; < 1000 m. c&s SNF, Teh.

M. macrocarpus (E. Greene) E. Greene (p. 543) Herbage not glaucous. **FL**: corolla shallowly cup-shaped, white. **FR** 5–12 cm, oblong, gen rounded at both ends (sometimes with sharp beak); prickles ± dense, stiff. **SEEDS** gen 4–12(24), 13–33 mm, ± round, oblong, or ovate, angled at tip or not. $2n$=32,64. Washes, shrubby or open areas; < 900 m, SW, DSon; Baja CA.

var. ***macrocarpus*** **FL**: staminate 8–13 mm wide. **SEED** 13–20 mm. Habitats of sp. SW mainland, DSon; Baja CA.

var. ***major*** (Dunn) K.M. Stocking **FL**: staminate 14–30 mm wide. **SEED** 20–33 mm. Habitats of sp. ChI. Larger, but much like var. *macrocarpus*; more study warranted.

M. oreganus (Torrey & A. Gray) Howell (p. 543) COAST MAN-ROOT Herbage not glaucous. **FL**: corolla deeply cup-shaped, white. **FR** 4–8 cm, ovate, tapered to a beak, gen striped dark green; prickles sparse to dense (gen 0 at tip), flexible. **SEEDS** gen 3–6, 16–22 mm, disc-shaped, ± flat. $2n$=32. Shrubby or open areas, forest edges; < 1800 m. NW, SnFrB; to B.C.

M. watsonii (Cogn.) E. Greene (p. 543) Herbage glaucous. **FL**: corolla deeply cup-shaped, white. **FR** 2–3.5 cm, ± round, often striped dark green; prickles ± 0 to dense, flexible, often hooked. **SEEDS** 1–4, 11–14 mm, ± round. Shrubby areas, forest edges; < 1200 m. NCoRI, CaRF, n SNF, ScV.

CUSCUTACEAE DODDER FAMILY

Tania Beliz

Ann, parasitic vine. **ST** twining, ± thread-like, yellow-green to bright orange, gen glabrous. **LVS** 0 or scale-like, ± 2 mm, gen triangular to lanceolate. **INFL**: cyme or cluster (rarely fls solitary), gen head- or spike-like, axillary, sometimes bracted. **FL** bisexual, radial; calyx gen persistent, lobes gen 4–5, gen overlapped; corolla gen deciduous, < 6 mm, mostly white, tube gen appendaged opposite stamens, lobes 4–5; stamens 4–5, alternate corolla lobes; ovary superior, chambers 2(3), 2-ovuled, styles gen 2, stigma gen 1 per style, gen ± head-like. **FR**: capsule (circumscissile or irregularly dehiscent) or berry-like. 1 genus, ± 150 spp.: esp Am trop; some crop pests. Sometimes incl in Convolvulaceae.

CUSCUTA DODDER

The only genus (Arabic: ancient name)

1. Corolla appendages 0–0.1 mm ..... ***C. californica***
2. Ovary and fr conic, top acute ..... var. ***apiculata***
2′ Ovary and fr obovoid, top depressed
3. Perianth papillate ..... var. ***papillosa***
3′ Perianth not papillate
4. Corolla bulged out between stamens ..... var. ***breviflora***
4′ Corolla not bulged out between stamens ..... var. ***californica***
1′ Corolla appendages 0.7–2.5 mm
5. Stigma cylindric ..... ***C. approximata***
5′ Stigma ± head-like
6. Corolla shallowly bell- to urn-shaped, tube ± shorter than wide
7. Corolla appendage divisions gen 0–few, knob-like ..... ***C. denticulata***
7′ Corolla appendage divisions few–many, finger-like
8. Ovary and fr top unthickened ..... ***C. pentagona***
8′ Ovary and fr top thickened ..... ***C. indecora***
9. Pedicel and calyx not papillate ..... var. ***indecora***
9′ Pedicel and calyx ± papillate ..... var. ***neuropetala***
6. Corolla funnel- or bell-shaped, tube longer than wide
10. Anthers sessile; corolla lobes < tube ..... ***C. subinclusa***
10′ Anthers on filaments; corolla lobes < to > tube
11. Perianth parts obtuse, gland-dotted, not papillate; corolla appendage divisions few, scattered, finger-like; near streams, rivers, lakes ..... ***C. cephalanthi***

11′ Perianth parts acute, sometimes papillate and gland-dotted; corolla appendage divisions many, finger- or knob-like; salty marshes, flats, or ponds, vernal pools

12. Corolla appendages 0.7–1.2 mm, divisions finger-like; vernal pools .... ***C. howelliana***

12′ Corolla appendages 1.2–1.5 mm, divisions knob-like; salty marshes, flats, ponds .... ***C. salina***

13. Perianth papillate; fl ± 3 mm .... var. ***papillata***

13′ Perianth not papillate; fl 2–4.5 mm

14. Fl 3–4.5 mm; coastal salt marshes .... var. ***major***

14′ Fl 2–3 mm; inland salt flats .... var. ***salina***

C. approximata Bab. (p. 543) **INFL** head-like; pedicels 0–0.5 mm. **FL**: calyx gen 2–2.5 mm, lobes gen 4, 1–1.5 mm; corolla 2–4 mm, shallowly bell-shaped (urn-like in fr), lobes 5, spreading to ascending, each with a fleshy keel, appendages 1.5–2 mm, divisions gen knob-like; filaments 0.3–0.7 mm, anthers ± 0.5 mm; ovary 1–1.5 mm, obovoid, top depressed, style ± 0.4 mm, stigma cylindric. *n*=14. Uncommon. On alfalfa, other crops; gen < 1500 m. NCoR, GV, MP; to UT; native to Old World. [*C. epithymum* Murray; var. *urceolata* (Kunze) Yuncker misapplied]

C. californica Hook. & Arn. **INFL** spike-like. **FL** 2.6– 4 mm; calyx persistent, lobes 5, spreading to recurved, lanceolate, acute to acuminate, 0.5–1 × corolla tube; corolla persistent, shallowly bell-shaped, gland-dotted, lobes 5, ± 3–6 mm, reflexed to spreading, lanceolate, acute, appendages 0–0.1 mm; filaments 0.7– 1.4 mm, anthers 0.2–1.1 mm; ovary 1–2 mm, gen obovoid, gland-dotted, top gen depressed< styles 0.7–3 mm. **FR** 1.5–2 mm, enveloped by perianth, gen obovoid; top gen depressed. On herbs and shrubs on roadsides, chaparral, grassland, yellow-pine forest; gen < 2500 m. CA-FP, DSon; to WA, Colorado, Mex.

var. ***apiculata*** Engelm. **FL**: corolla not papillate; ovary (and fr) conic, top acute. On herbs; probably < 500 m. e DSon (near Colorado River).

var. ***breviflora*** Engelm. **FL**: corolla bulged out between stamens, not papillate; ovary (and fr) obovoid, depressed. On herbs, habitats of sp. CA-FP; to WA, Colorado, Mex. [*C. brachycalyx* (Yuncker) Yuncker incl var. *apodanthera* (Yuncker) Yuncker; *C. occidentalis* Millsp.; *C. suksdorfii* Yuncker incl var. *subpedicellata* Yuncker]

var. ***californica*** (p. 543) **FL**: corolla not bulged between stamens, not papillate; ovary (and fr) obovoid, depressed. Habitats of sp. CA-FP; to WA, NV, Baja CA.

var. ***papillosa*** Yuncker **FL**: perianth (and pedicel, receptacle) papillate; ovary (and fr) obovoid, depressed. *n*=14. On herbs and shrubs in chaparral; < 1500 m. SNF, GV, SCoRO, SCo.

C. cephalanthi Engelm. (p. 543) **INFL** head- or spike-like; pedicels 0–1.2 mm. **FL** 3–4 mm; perianth parts obtuse, gland-dotted, not papillate; calyx appressed, lobes 3–5, gen free, 1–2 mm, unequal, margins irregular; corolla sometimes persistent, 2–4 mm, funnel-shaped, tube longer than wide, lobes 3–4, spreading to erect, 0.6–1.3 mm, unequal, appendages 1.3–2 mm, divisions few, scattered, finger-like; filaments 0.2–0.6 mm, anthers 0.4–0.7 mm; ovary 0.9–2 mm, ± spheric to ovoid, top depressed or not, thickened or not. **FR** sometimes asymmetric; top as ovary. 2*n*=60. Uncommon. On herbs, near streams, rivers, lakes; probably < 3000 m. KR, CaR, Wrn; to WA, e US, Mex.

C. denticulata Engelm. **INFL** loose, spike-like, few-fld; pedicels 0–3.3 mm. **FL** 2–4 mm; calyx persistent, lobes 5, 0.3–1.5 mm, finely toothed, acute to obtuse; corolla gen persistent, 2–3 mm, shallowly bell- to urn-shaped, tube gen shorter than wide, lobes 5, becoming reflexed, tube, widely ovate, obtuse, appendages 0.7–1.4 mm, divisions 0–few, knob-like; ovary ± 1–2 mm, conic, top acute. On herbs or esp shrubs in creosote-bush scrub, Joshua-tree woodland; gen < 1300 m. D; to UT, AZ, Baja CA. Pls with acute calyx lobes have been called *C. veatchii* Brandegee.

C. howelliana Rubtzoff BOGGS LAKE DODDER **INFL** spike- or head-like; pedicels gen 0. **FL**: perianth parts papillate, acute; calyx 1.5–3 mm, lobes 4–5, not overlapped, loose around corolla tube, 1–1.4 mm, partly divided, triangular, tips spreading, acute to acuminate; corolla gen persistent, funnel-shaped, tube longer than wide, lobes 4–5, sometimes spreading, 1–1.5 mm, triangular, acute to acuminate, appendages many, 0.7–1.2 mm, divisions finger-like; filaments 0.1–0.5 mm, anthers 0.6–1.1 mm; ovary 0.5–1 mm, ± obovoid, top depressed, thickening inconspicuous. UNCOMMON. Esp on *Eryngium* in vernal pools; < 100 m. NCoRI, CaRF, n SNF, GV.

C. indecora Choisy (p. 543) **INFL** spike- or panicle-like; pedicels 0–4.4 mm. **FL** 3–5 mm; calyx 1–2 mm, lobes 5, not overlapped, ± 1/2 corolla tube, acute; corolla gen persistent, 3–4.5 mm, shallowly bell-shaped, tube gen shorter than wide, lobes 5, < tube, triangular, erect, tips incurved, appendages 1.5–2.5 mm, divisions many, finger-like; ovary 2– 3 mm, ovoid-spheric, top with thickening, styles 3–4 mm. **FR** sometimes gland-dot-lined; top as ovary. Common. On herbs, often in moist fields, roadsides; probably < 1500 m. NCo, NCoR, SN, GV, D; to c&se US, Mex; also Caribbean, S.Am.

var. ***indecora*** **INFL**: pedicel and calyx not papillate. *n*=15. Habitats and range of sp. [*C. jepsonii* Yuncker; *C. suaveolens* Ser. misapplied]

var. ***neuropetala*** (Engelm.) Hitchc. **INFL**: pedicel and calyx ± papillate. Common. Habitats of sp. SNF, GV, D.

C. pentagona Engelm. (p. 543) **INFL**: pedicels 0.8–1 mm. **FL**: calyx ± 1–2 mm, lobes 4–5, 0.6–1.3 mm, ovate to round, gland-dotted; corolla ± 2–3 mm, shallowly bell-shaped, tube gen wider than long, lobes 4–5, erect, 0.9–1.3 mm, tips incurved to reflexed, acute, often papillate outside, appendages ± 1–2 mm, curved over ovary, divisions few–many, finger-like; filaments 0.3–0.8 mm, anthers 0.3–0.6 mm; ovary 1–1.6 mm, ± spheric, unthickened, top depressed, styles 0.4–0.5 mm. **FR** 1–3.6 mm, ± spheric; top as ovary. Common. On herbs and shrubs on roadsides; < 500 m. NCo, SNF, GV, CCo, SCo; to se US, Caribbean, n Mex. [*C. campestris* Yuncker]

C. salina Engelm. (p. 543) **INFL** spike-like; pedicels 0–2.5 mm. **FL** 2–4.5 mm; perianth parts ± acute, gland-dotted; calyx 1.3–2.7 mm, lobes 5, erect to spreading, triangular-lanceolate; corolla 3–5 mm, bell-shaped, tube longer than wide, lobes 5, erect to spreading, lanceolate, appendages 1.2–1.5 mm, divisions many, knob-like; filaments 0.1–0.7 mm, anthers 0.4–0.7 mm; ovary 1–3 mm, gland-dotted, thickening at top more conspicuous in fr, styles 0.4–1 mm. Common. On herbs in salty marshes, flats, ponds; gen < ± 100 m. NCo, KR, GV, CCo, SnFrB, SCo; to B.C., UT, AZ, Baja CA.

var. ***major*** Yuncker **FL** 3–4.5 mm; perianth not papillate. Gen on *Salicornia* in salt marshes; gen < ± 100 m. NCo, CCo, SCo; to B.C.

var. ***papillata*** Yuncker **FL** ± 3 mm; perianth (and sometimes ovary) papillate. Uncommon. Salt flats, saline ponds; gen < 100 m. NCo; to UT, AZ.

var. ***salina*** **FL** 2–3 mm; perianth not papillate. Inland salt flats; gen < ± 100 m. Range of sp. (exc coast).

C. subinclusa Durand & Hilg. **INFL** spike- or head-like; pedicels 0–1 mm. **FL**: calyx persistent, 2–3.4 mm, lobes (4)5, ± 1.5–2.5 mm, lanceolate, acute to acuminate; corolla ± 4.5–5.5 mm, funnel-shaped, tube longer than wide, lobes spreading, 1–1.5 mm, < tube, triangular, tip often papillate outside, appendages 1.6–2.1 mm, spoon-shaped, divisions short, knob- to finger-like; anthers sessile, ± 1–2 mm; ovary 1–1.5 mm, obovoid to elliptic, top thickened, styles 1.1–1.5 mm. Common. Gen on shrubs, in forests near streams, rivers; < 1600 m. NCoR, SN, GV, SnFrB, SCoR, SNE.

DATISCACEAE DATISCA FAMILY

William J. Stone

Per, tree, gen dioecious. **LVS** simple, alternate, gen pinnate. **INFL**: axillary spike, raceme, or cluster. **STAMINATE FL**: sepals 3–9; petals 6–8 or 0; stamens < 25. **PISTILLATE OR BISEXUAL FL**: sepals 3–8; petals 0; stamens ± functional; ovary inferior, chamber 1, placentas 3, parietal, styles 3. **FR**: capsule, opening at top between styles. **SEEDS** many, minute. 3 genera, 4 spp.: Asia, w N.Am.

DATISCA

Per, glabrous. **LVS** pinnately divided. **INFL**: small axillary clusters. **STAMINATE FL**: calyx very short, lobes 4–9, unequal; corolla 0; stamens 8–12, filaments short. **PISTILLATE FL**: calyx 3-toothed; stamens, if present, 2–4; ovary ovoid, 3-angled, styles 3, thread-like, 2-forked. 2 spp.: 1 Asia, 1 w N.Am. (Derivation unknown)

D. glomerata (C. Presl) Baillon (p. 543) DURANGO ROOT **STS** gen clustered, erect, branched above, 1–2 m. **LVS** alternate above but appearing opposite or somewhat whorled below, ± 15 cm, ovate to lanceolate, acuminate, unequally pinnate; petioles 2–3(4) cm. **STAMINATE FL**: calyx 2 mm; anthers 4 mm, ± sessile, yellow. **PISTILLATE FL**: calyx 5–8 mm; styles ± 6 mm. **FR** ± 8 mm. **SEEDS** ± 1 mm, light brown, pitted in rows. Dry stream beds or washes in many communities; < 2000 m. CA-FP (exc GV); w NV, Baja CA. All parts toxic.

DIPSACACEAE TEASEL FAMILY

Elizabeth McClintock

Ann to per, rarely shrub, sometimes armed with prickles. **ST** gen branched. **LVS** simple, gen in basal rosettes and cauline, gen opposite, ± fused at base around st, entire, toothed, or pinnately lobed or dissected, petioled or sessile; stipules 0. **INFL**: head, terminal, on long peduncle, many-fld, dense, ± spheric or cylindric, subtended by involucre; each fl gen enclosed ± at base by an involucel of 1–2 gen fused bractlets, this gen expanded above or in fr, gen subtended by a receptacle bract. **FL** bisexual, ± bilateral, esp outermost; calyx limb cup-shaped or divided into 4–5(10) linear or bristle-like segments; corolla ± funnel-shaped, lobes 4–5, < tube, gen unequal; stamens gen 4, attached to corolla tube, alternate lobes; ovary inferior, 1-chambered, style slightly exserted from corolla, stigma 2-lobed. **FR**: achene, enclosed by sometimes enlarged involucel, gen topped by persistent calyx. 10–11 genera, 270–350 spp.: Eur to e Asia, c&s Afr; several cult for orn. [Moore 1976 Flora Europaea 4:56–72]

1. Pl armed with prickles or spines; involucre bracts > fls, stiff; calyx, corolla 4-parted .... **DIPSACUS**
1′ Pl unarmed; involucre bracts < fls, flexible; calyx, corolla 5-parted .... **SCABIOSA**

DIPSACUS TEASEL

Bien, armed with prickles, sometimes ± throughout. **ST** erect, gen < 2 m, stout, few-branched, rough-hairy. **LF** gen < 5 dm, entire or toothed. **INFL** gen ± cylindric; involucre bracts unequal, > fls, linear, stiff; receptacle bract ending in a spine. **FL**: calyx limb cup-shaped, 4-lobed, persistent; corolla gen lavender, rarely white, tube long, lobes 4, unequal; stamens 4. **FR** 4-angled, hairy. ± 15 spp.: Eur, w Asia, Afr. (Greek: thirst, from lf bases that in some spp. hold water) [Ferguson & Brizicky 1965 J Arnold Arb 46:362–365]

1. Involucre bract gen erect or curved upward; receptacle bract ± flexible, ending in a straight spine .... ***D. fullonum***
1′ Involucre bract ± spreading, sometimes reflexed; receptacle bract very stiff, ending in a recurved spine .. ***D. sativus***

D. fullonum L. (p. 543) WILD TEASEL **LVS**: pairs gen only shallowly fused around st. **INFL** 5–10 cm, ovoid-cylindric. **FR** 6–8 mm. $2n$=16,18. Roadsides, pastures, old fields, sometimes moist sites; < 1700 m. NCo, KR, c&s SNF, SnFrB; native to Eur. [*D. sylvestris* Hudson]

D. sativus (L.) Honck. (p. 543) FULLER'S TEASEL **LVS**: pairs fused around st, sometimes deeply so. **INFL** 5–10 cm, ovoid to ovoid-cylindric. **FR** ± 6–8 mm. $2n$=16,18. Disturbed areas, fields, vacant lots, old pastures; < 800 m. NCo, NCoRO, SnFrB, SCoR, PR; native to Eur. [*D. fullonum* L., *D. sylvestris* Hudson misapplied] Fr head long used to raise nap on woolen cloth.

SCABIOSA PINCUSHION FLOWER

Ann to per, unarmed. **ST** erect, < 6 dm, branched, glabrous or hairy. **LF** gen < 1 dm, pinnately lobed or dissected. **INFL** ± spheric; outermost fls gen larger; peduncle often long; involucre bracts ± equal, gen < fls, lanceolate, flexible; involucel tubular, sometimes with expanded, fan-like, many-veined limb. **FL**: calyx limb divided into 5 bristles; corolla gen blue, purple, pink, or white, lobes 5, upper 2 smaller; stamens 4, rarely 2. **FR** ± angled, gen hairy. ± 80 spp.: temp Eurasia, Afr. (Latin: itch, from medicinal use)

1. St ± glabrous; corolla purple, white, or pink; pl gen bien .... ***S. atropurpurea***
1′ St shaggy-hairy; corolla blue; pl gen ann .... ***S. stellata***

S. atropurpurea L. **ST** < 6 dm. **LVS** on st pinnately dissected. **INFL** < 30 mm wide, elongating at maturity; limb of involucel < 1 mm, ± urn-shaped. **FR**: calyx bristles >> limb of involucel. Disturbed, gen urban areas; < 600 m. CCo, SnFrB, SCoRO, SnBr, PR; native to s Eur. Cult as orn.

S. stellata L. **ST** < 4 dm. **LVS** on st dentate to pinnately dissected. **INFL** 20–50 mm wide, not elongating at maturity; limb of involucel 3–4 mm, funnel-shaped. **FR**: calyx bristles scarcely > expanded, fan-like, scarious, many-veined limb of involucel. Disturbed urban areas; < 250 m. SCo; native to sw Eur. Cult as orn, rarely naturalized.

DROSERACEAE SUNDEW FAMILY

William J. Stone

Ann, per, subshrub, carnivorous; roots weak. **LVS** gen basal rosette, often coiled in bud; blade with insect-catching hairs on upper surface, hairs either gland-tipped and sticky or sensitive bristles. **INFL**: cyme or raceme-like; fls 1–few, on long peduncle. **FL** bisexual, radial; calyx lobes gen 5; petals gen 5, free or slightly fused; stamens 4–20; pistil 1, ovary superior, chamber 1, placentas gen 3(5), parietal, style branches gen 3(5), each gen 2-lobed. **FR**: capsule, loculicidal; valves gen 3(5). **SEEDS** gen many, spindle-shaped. 4 genera (3 with 1 sp. each), 100 spp.: temp, trop, esp Australia, New Zealand, esp in bogs, swamps; some cult as novelties (*Dionaea*, Venus' fly-trap, of se US). (Greek: dewy)

DROSERA SUNDEW

Ann, per, often brownish or reddish. **LF**: petiole long; blade hairs on upper surface stalked, gland-tipped (glands secrete sticky, insect-trapping fluid, lf folds around prey, which is digested by enzymes, ribonucleases, and bacteria). **INFL**: raceme-like cyme. **FL**: sepals, petals, stamens gen 5; main style branches, placentas, valves gen 3. 100 spp.: range of family. *D. filiformis* Raf. and *D. linearis* Goldie were planted in NCo (Mendocino Co.) and may persist.

1. Lf ascending to erect, blade ± oblanceolate . . . . ***D. anglica***
1′ Lf gen spreading, blade ± round . . . . ***D. rotundifolia***

D. anglica Hudson (p. 543) ENGLISH or LONG-LEAVED SUNDEW **LF**: blade 15–35 mm, 2–7 mm wide. **INFL**: peduncle gen 1, 6–18 cm. **FL**: calyx 4–6 mm, fused ± 1/3 length; petals 8–12 mm, white; stamens < petals; style branches 2-lobed for 2/3 length. **SEED** 1–1.5 mm, ± ovoid, black. $2n$=40. RARE in CA. Swamps, bogs; 1300–2000 m. Scattered in KR, CaRH, n SNH (n of Lake Tahoe), s Wrn; circumboreal. [*D. longifolia* L.] Sterile hybrids with *D. rotundifolia* may be called *D. ×obovata* Mert. & Koch. In cult.

D. rotundifolia L. (p. 543) ROUND-LEAVED SUNDEW **LF**: blade 4–12 mm, 4–12 mm wide, base sometimes ± cordate. **INFL**: peduncles 1–several, 5–25 cm. **FL**: calyx 4–6 mm, fused only at base; petals 4–6 mm, white to pink; style branches 2-lobed nearly to base. **SEED** 1–1.5 mm, spindle-shaped, light brown. $2n$=20. Uncommon. Sphagnum bogs; 0–2000 m. Scattered in NW (esp near coast), CaR, SNH; to e US, circumboreal. ❀WET,bog & pure water: 4–7,15–17; DFCLT.

ELAEAGNACEAE OLEASTER FAMILY

Elizabeth McClintock

Shrub, tree, sometimes ± dioecious, often thorny, gen densely silvery-hairy throughout; hairs often scale-like. **LVS** simple, alternate or opposite, gen deciduous, entire; petiole gen short; stipules 0. **INFL** gen umbel-like, axillary; fls 1–few. **FL** radial; hypanthium rotate to salverform, lower part gen receptacle-like, persistent, with a disk, becoming fleshy; sepals (hypanthium lobes) gen 4, ± petal-like; petals 0; anthers 4 or 8, ± sessile; ovary superior (appearing inferior), chamber 1, style 1. **FR**: achene enclosed in fleshy hypanthium, the whole drupe- or berry-like. 3 genera, ± 45 spp.: N.Am, Eur, Asia, e Australia; esp temp, subtrop. [Graham 1964 J Arnold Arbor 45:274–278]

1. Lvs alternate; fls bisexual, stamens 4 . . . . **ELAEAGNUS**
1′ Lvs opposite; fls unisexual (spp. dioecious), stamens 8 . . . . **SHEPHERDIA**

ELAEAGNUS

Shrub, tree. **LVS** alternate. **FL** bisexual; hypanthium bell-shaped to salverform, lobes 4; stamens 4, barely exserted; disk flask-shaped, enclosing base of style; stigma ± elongate, on 1 side of style. ± 40 spp.: N.Am, s Eur, Asia. (Greek: olive, chaste-tree)

E. angustifolius L. OLEASTER, RUSSIAN OLIVE Pl < 7 m, sometimes thorny. **LF** 4–8 cm, lanceolate to oblong, more silvery on lower surface. **FL** 5–10 mm, ± as wide at top, fragrant; hypanthium ± dark yellow inside, tube ± = lobes. **FR** 10–20 mm, elliptic in outline, yellow. Uncommon. Disturbed, sometimes moist places; gen < 1500 m. SnJV, SnFrB, SNE, DMoj; native to temp Asia. Cult as orn.

SHEPHERDIA

Shrub, dioecious. **LVS** opposite or some alternate. **STAMINATE FL**: hypanthium rotate; stamens 8; disk lobes alternate stamens. **PISTILLATE FL**: hypanthium urn-shaped, nearly closed at top; disk lobes 8, nearly meeting above ovary; stigma cap-like. 3 spp.: N.Am. (John Shepherd, 1764–1836, curator of Liverpool Botanic Garden)

S. argentea Nutt. (p. 543) BUFFALO BERRY Pl 2–6 m, much branched, thorny. **LF** 2–6 cm, gen oblong, more silvery on lower surface. **FL**: hypanthium greenish yellow inside. **STAMINATE FL** < 2 mm, ± 4 mm wide at top; tube < lobes. **PISTILLATE FL** ± 2 mm, 1 mm wide; tube > lobes. **FR** 5 mm, elliptic in outline, red. Along streams, river bottoms, slopes; 1000–2000 m. c SNH, SCoR, WTR, SnBr; to Can, c US. Fr sour, made into sauce eaten with buffalo meat along Overland Trail; sometimes cult as orn. ❀SUN&DRN4–6,**10**&IRR:**1–3,7–10,14–22**,23,24;STBL.

ELATINACEAE WATERWORT FAMILY

Gordon C. Tucker

Ann, per, in or near water; roots fibrous, from a taproot or not, gen from lower lf axils as well. **ST** gen soft. **LVS** simple, opposite, ± 4-ranked; stipules scarious. **INFL**: fls axillary or terminal, solitary or clustered. **FLS** small, inconspicuous, radial, bisexual; sepals and petals gen free, 3–5, equal in number; ovary spheric, styles 3–5, very short. **FR**: capsule, septicidal, ± spheric, ovoid, or depressed-ovoid, walls thin; chambers 2–5, each several–many-seeded. **SEED** very small; surface net-like or glossy. 2 genera, 50 spp.: ± worldwide. [Tucker 1986 J Arnold Arbor 67:471–483]

1. Pl glandular-hairy; sepals and petals 5; sepals acute, midvein rib-like .... **BERGIA**
1' Pl glabrous; sepals and petals 2–4; sepals obtuse, midvein not apparent .... **ELATINE**

BERGIA

Ann, per, glandular-hairy. **ST** prostrate to ascending, simple to much-branched; base sometimes ± woody. **LVS** 4-ranked; petiole < blade; blade elliptic, serrate, base wedge-shaped, tip acute. **INFL**: fls axillary or terminal, solitary or clustered. **FL**: sepals 5, deltate; petals 5, oblong, membranous, glabrous; stamens 5–10, filaments ± = petals, anthers elliptic; styles 5. **FR** ovoid; chambers 5, each many-seeded. **SEEDS** not visible through fr wall, oblong, ± curved, brown; surface obscurely net-like. ± 25 spp.: gen Old World trop, 1 N.Am. (P.J. Bergius, student of Linnaeus, 1730–1790)

B. texana (Hook.) Seub. (p. 553) Ann or short-lived per, glandular-hairy exc petals. **FL**: sepals acute, midvein rib-like, green, margins scarious; petals whitish. Moist, disturbed soils, sand bars along rivers, margins of pools; < 200 m. GV, SCo; to e WA, c&s US, ne Mex.

ELATINE WATERWORT

Ann, short-lived per, glabrous. **ST** erect underwater, ± prostrate on wet ground, branched or not; base not woody. **LVS** opposite, ± 4-ranked; petiole < 1/3 blade, flat, ± blade-like; blades narrowly elliptic to ± round, ± entire, bases wedge-shaped to ± rounded, tips rounded. **INFL**: fls 1(2) per node, 0–1 per upper lf axil. **FL**: sepals 3–4, widely elliptic, membranous, very pale green; petals gen as many as and ± = sepals, widely elliptic, membranous, pale greenish white; stamens 3, 6, 8, rarely 1, filaments ± 1/2 × to ± = petals, anthers widely ovoid; styles 3–4. **FR** ± spheric or depressed-ovoid; chambers 3–4, each 3–15-seeded; pedicel gen ± 0. **SEEDS** ± visible through fr wall, elliptic, straight or curved, brown to yellowish brown; surface net-like. ± 25 spp.: worldwide. (Greek: fir tree, from a Eur sp. that suggests such a pl in miniature) At least 20× magnification needed for pits on seeds.

1. Sepals and petals 4; stamens 8; pedicels elongating in fr, becoming 2–3 × fr; seeds curved 90–180° .... ***E. californica***
1' Sepals 2–3, petals 3; stamens 3, 6, rarely 1; pedicels gen ± 0, less often slightly elongating in fr, becoming at most 1/2 fr; seeds straight or curved < 30°
  2. Fl subsessile to stalked, ± erect; pedicel of fr turned to 1 side; sepals 3, ± equal .... ***E. ambigua***
  2' Fl sessile, erect; pedicel of fr not turned to 1 side; sepals 2 or, if 3, 1 < others
    3. Stamens 6, rarely 1 (if 3, opposite petals) .... ***E. heterandra***
    3' Stamens 3, opposite sepals
      4. Seed pits 10–15 per row .... ***E. brachysperma***
      4' Seeds pits 16–35 per row
        5. Seed pits wider than long .... ***E. chilensis***
        5' Seed pits ± as wide as long .... ***E. rubella***

E. ambigua Wight **ST** ± erect, 1–8 cm. **LF** lanceolate to oblong-elliptic; petiole 1/4 blade. **INFL**: fls 1 per node. **FL**: sepals 3, ± equal; petals 3, equal, ± > sepals. **FR**: chambers 3; pedicel turned to 1 side, 1/4–1/2 fr length. **SEED** narrowly oblong, curved 15–30°; pits ± 20 per row, wider than long. Rice fields, irrigation canals; 10–100 m. GV; native to e&s Asia.

E. brachysperma A. Gray (p. 553) **ST** decumbent to erect, 1–5 (12) cm. **LF** ovate, narrowly oblong; petiole 1/4–1/3 blade or indistinct. **INFL**: fls 1 per node. **FL**: sepals 2 or, if 3, 1 < others; petals 3, equal, = and ± wider than sepals; stamens 3, opposite sepals. **FR**: chambers 3. **SEED** widely oblong, curved ± 15°; pits 10–15 per row, wider than long. Muddy shores, shallow pools; 50–500 m. CA; c&s US. [*E. obovata* (Fassett) H. Mason]

E. californica A. Gray (p. 553) **ST** decumbent to erect, 1–5 cm. **LF** oblong, tapered to base; petiole ± 1/2 × blade. **INFL**: fls 1 per node. **FL**: sepals 4, fused in basal ± 1/3, separating in fr; petals 4, > and wider than sepals; stamens 8. **FR**: chambers 4; pedicel recurved, 2–3 × fr. **SEED** oblong to elliptic, curved 90–180°; pits ± 15 per row, as wide as or wider than long. Pools, ponds, rice fields, stream banks; 50–1900 m. s NCoRH (Snow Mtn), GV; to WA.

E. chilensis C. Gay (p. 553) **ST** decumbent to erect, 0.5–10 cm. **LF** narrowly oblong to widely elliptic, tapered to base; petiole < 1/4 blade. **INFL**: fls 1(2) per node. **FL**: sepals 2 or, if 3, 1 < others; petals 3, ovate; stamens 3, opposite sepals. **FR**: chambers 3. **SEEDS** 20–40 per chamber, elliptic, ± straight; pits 25–35 per row, wider than long. Muddy shores of ponds; ± 600–1700 m. n SNH, MP (Madeline Plains, Lassen Co.; Sierra Valley, Plumas Co.), PR; temp S.Am. *E. gracilis* H. Mason, vernal pools, 1200 m, SN (Little Truckee River, Nevada Co.) differs only in fewer seeds per chamber, pits per row, and probably should be incl here.

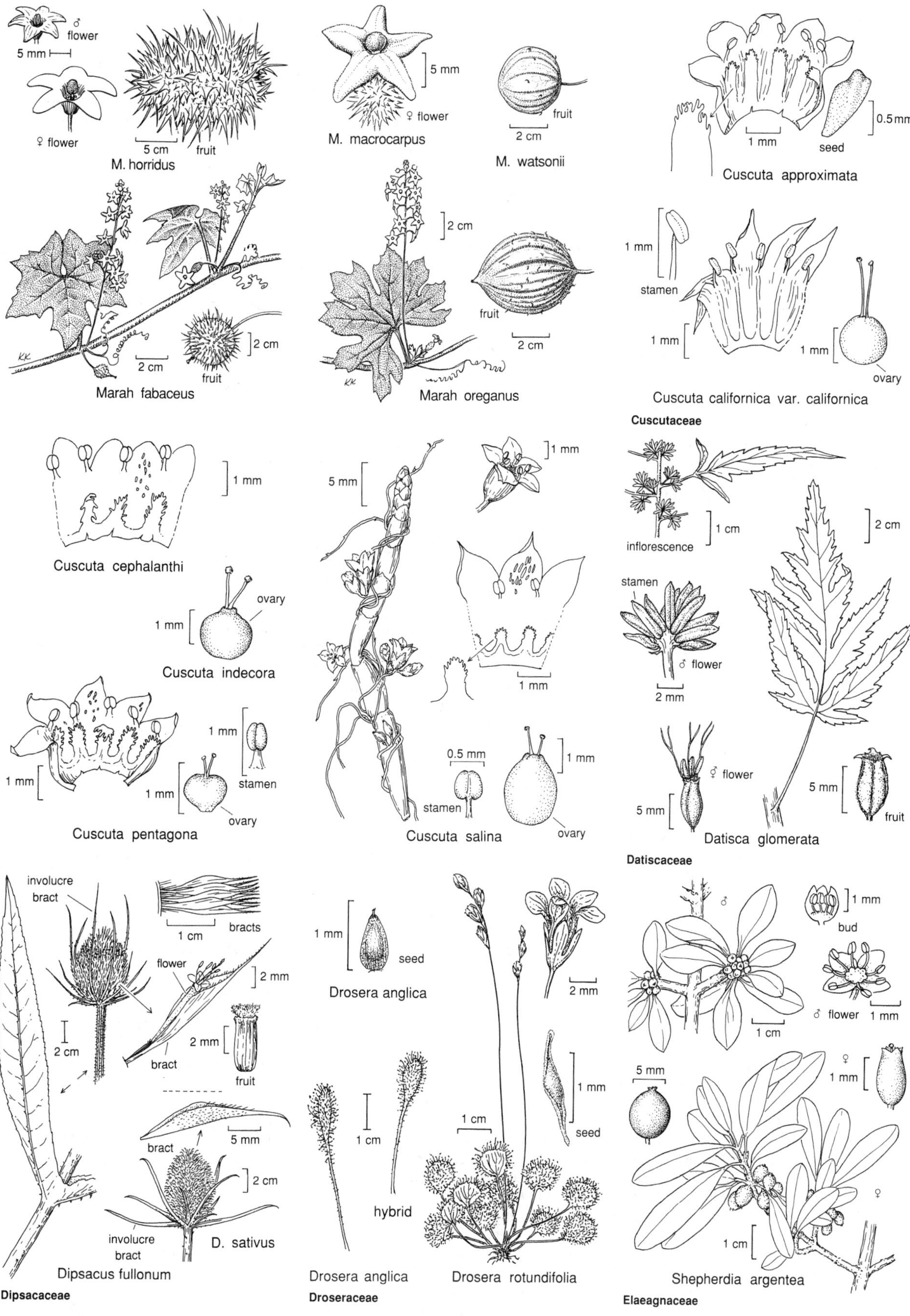
♂ flower
5 mm
♀ flower
5 cm
fruit
M. horridus
5 mm
♀ flower
M. macrocarpus
fruit
2 cm
M. watsonii
0.5 mm
1 mm
seed
Cuscuta approximata
2 cm
2 cm
fruit
Marah fabaceus
2 cm
fruit
2 cm
Marah oreganus
1 mm
stamen
1 mm
1 mm
ovary
Cuscuta californica var. californica
Cuscutaceae
1 mm
Cuscuta cephalanthi
1 mm
ovary
Cuscuta indecora
1 mm
1 mm
stamen
1 mm
ovary
Cuscuta pentagona
5 mm
1 mm
1 mm
0.5 mm
stamen
1 mm
ovary
Cuscuta salina
1 cm
inflorescence
2 cm
stamen
♂ flower
2 mm
♀ flower
5 mm
5 mm
fruit
Datisca glomerata
Datiscaceae
involucre bract
1 cm
bracts
flower
2 mm
2 cm
2 mm
bract
fruit
bract
5 mm
2 cm
involucre bract
D. sativus
Dipsacus fullonum
Dipsacaceae
1 mm
seed
Drosera anglica
2 mm
1 mm
seed
1 cm
1 cm
hybrid
Drosera anglica
Drosera rotundifolia
Droseraceae
♂
1 mm
bud
♂ flower
1 mm
1 cm
5 mm
♀
1 mm
♀
1 cm
Shepherdia argentea
Elaeagnaceae

E. heterandra H. Mason **ST** decumbent, 2–5 cm. **LF** obovate to widely oblong-elliptic; petiole ± 1/4 blade. **INFL**: fls 1 per node. **FL**: sepals 2; petals 3, ovate; stamens 6, rarely 1 (if 3, then opposite petals). **FR**: chambers 3. **SEED** widely elliptic, curved 15–30°; pits 12–16 per row, wider than long. Pond edges; 1000–1500 m. NCoR, SN.

E. rubella Rydb. **ST** prostrate to erect, 3–6(15) cm, often tinted reddish. **LF** oblong-lanceolate; petiole < 1/4 blade; tip blunt to notched. **INFL**: fls 1–2 per node. **FL**: sepals 2 or, if 3, 1 < others; petals 3, equal, widely elliptic; stamens 3, opposite sepals. **FR**: chambers 3. **SEED** narrowly oblong, straight or curved < 15°; pits 16–35 per row, ± as wide as long. Muddy shores, shallow vernal pools, ditches, rice fields; < 500 m. CA; widespread in w N.Am.

EMPETRACEAE CROWBERRY FAMILY

Fosiée Tahbaz

Subshrub or shrub, low, evergreen, gen dioecious. **LVS** alternate, linear to oblong, stiff, deeply grooved beneath (lf seeming rolled under). **INFL**: fls solitary or few, axillary or terminal, ± sessile. **FL**: perianth segments 3–6, sometimes differentiated into sepals and petals. **STAMINATE FL**: stamens 2–4. **PISTILLATE FL**: ovary superior, chambers 2–9, ovule 1 per chamber, style deeply 2–9-lobed. **FR**: drupe, dry or juicy; stones 2–9. 3 genera, ± 5 spp.: cold temperate; most cult as orn.

EMPETRUM CROWBERRY

FL small, inconspicuous; sepals 3; petals 3; stamens 3 per staminate fl, alternate petals. **FR** berry-like, black or red. ± 2 spp.: n&s cold temperate zones. (Greek: on rocks, from habitat)

E. nigrum L. (p. 553) BLACK CROWBERRY **ST** ± decumbent; branches many, 15–40 cm. **LVS** crowded, 3–6 mm, glabrous exc along groove. **FL**: perianth dark purplish red; filaments < 4 mm. **FR** 4–6 mm diam, black. $2n=26$. RARE in CA. Rocky sea cliffs, in coastal scrub; < 200 m. n NCo (Del Norte, Humboldt cos.); to AK; circumboreal. CA populations scattered, small. Pls with bisexual fls have been called ssp. *hermaphroditum* (Lange) Böcher; most CA pls seem dioecious. In cult.

ERICACEAE HEATH FAMILY

Gary D. Wallace, except as specified

Per, shrub, tree. **ST**: bark often peeling distinctively. **LVS** simple, gen cauline, alternate, opposite, rarely whorled, evergreen or deciduous, often leathery, petioled or not; stipules 0. **INFL**: raceme, panicle, cyme, or fls solitary, gen bracted; pedicels often with 2 bractlets. **FL** gen bisexual, gen radial; sepals gen 4–5, gen free; petals gen 4–5, free or fused; stamens 8–10, free, filaments rarely appendaged, anthers awned or not, dehiscent by pores or slits; nectary gen at ovary base, disk-like; ovary superior or inferior, chambers gen 1–5, placentas axile or parietal, ovules 1–many per chamber, style 1, stigma head- to funnel-like or lobed. **FR**: capsule, drupe, berry. **SEEDS** gen many, sometimes winged. ± 100 genera, 3000 spp.: gen worldwide exc deserts; some cult, esp *Arbutus, Arctostaphylos, Rhododendron, Vaccinium.* [Wallace 1975 Wasmann J Biol 33:1–88; 1975 Bot Not 128:286–298] Subfamilies Monotropoideae, Pyroloideae, Vaccinioideae sometimes treated as families. Nongreen pls obtain nutrition from green pls through fungal intermediates.

1. Pl nongreen; lvs scale-like (Monotropoideae)
 2. Ovary chamber 1 (or appearing > 1 by intrusion of parietal placentas); fr a berry; fr axis fleshy, not long-persisting
  3. Petals fused; anther elongate, dehiscent by separate slits ..... **HEMITOMES**
  3′ Petals free; anther elongate or not, dehiscent by separate or fused slits
   4. Fl densely hairy inside, ± glabrous outside; anther horseshoe-shaped, not elongate; stigma < 5 mm wide, ± funnel-shaped, subtended by ring of hairs ..... **PITYOPUS**
   4′ Fl glabrous inside and out; anther elongate; stigma < 2.5 mm wide, crown-like, not subtended by hairs ..... **PLEURICOSPORA**
 2′ Ovary chambers 4–5; fr a capsule; fr axis fibrous, gen persisting long after seed dispersal
  5. Petals free; fr dehiscent tip to base; seed fusiform
   6. Infl emerging from soil erect, with red or maroon stripes; pedicels not recurved when anthers open; sepals gen 0, petals present; stamens exserted ..... **ALLOTROPA**
   6′ Infl emerging from soil nodding, without red or maroon stripes; pedicels recurved to spreading when anthers open; sepals, petals gen present; stamens incl ..... **MONOTROPA**
  5′ Petals fused; fr dehiscent base to tip or indehiscent; seed ovoid, winged or not
   7. Pl pinkish, becoming brown, tall and slender (infl 1.5–17 dm, axis < 1.5 cm wide below lowest fl); corolla 6–9 mm, cream-colored to yellowish; anther awned; fr dehiscent; seed winged ..... **PTEROSPORA**
   7′ Pl red, short and stout (infl gen < 4 dm, axis gen > 1 cm wide below lowest fl); corolla 12–18 mm, red; anther unawned; fr indehiscent; seed unwinged ..... **SARCODES**
1′ Pl green; lvs gen normally expanded (exc some *Pyrola*)
 8. Pl not or barely woody; lvs basal or on short sts, alternate to whorled; petals free (Pyroloideae)
  9. Fls 1–few in ± head- or umbel-like raceme; fr dehiscent tip to base
   10. Fls in raceme; style within depressed center of ovary, stout; stigma not crown-like; nectary present; lvs cauline, leathery, elliptic to oblanceolate ..... **CHIMAPHILA**
   10′ Fls solitary; style projecting beyond ovary, slender; stigma crown-like; nectary 0; lvs ± basal, thin, ovate ..... **MONESES**

9′ Fls few–many in elongate raceme; fr dehiscent from base to tip
11. Infl 1-sided, arching, becoming ± erect in fr; petal tubercles 2, basal, inside; nectary present ..... **ORTHILIA**
11′ Infl symmetric, ± erect; petal tubercles 0; nectary 0 ..... **PYROLA**
8′ Pl woody; lvs cauline, alternate, rarely opposite or whorled; petals gen fused
12. Ovary inferior; fr a berry ..... **VACCINIUM**
12′ Ovary superior; fr a berry, drupe, or capsule
13. Fr ± enclosed by fleshy calyx; lvs evergreen ..... **GAULTHERIA**
13′ Fr not enclosed by calyx; lvs evergreen or deciduous
14. Corolla persistent in fr; lf needle-like, gen < 1 mm wide ..... **ERICA**
14′ Corolla not persistent in fr; lf gen not needle-like, gen > 1 mm wide
15. Fr a drupe or berry; pls gen of lower-elevation, dry habitats (exc *Arbutus*, some *Arctostaphylos*)
16. Lvs opposite or whorled, ± linear; fr a drupe, stones fused, each 2-seeded — panicles many-fld ..... **ORNITHOSTAPHYLOS**
16′ Lvs (exc some *Xylococcus*) alternate, wider than linear; fr a berry or drupe, stones free or fused, each 1–few-seeded
17. Ovary and fr smooth, not papillate; fr not juicy; infl gen dense
18. Lf margin not rolled under; filament swollen at base; fr flesh gen mealy, stones gen free (or some or all fused) ..... **ARCTOSTAPHYLOS**
18′ Lf margin rolled under; filament not swollen at base; fr flesh dry, stones fused into a unit ..... **XYLOCOCCUS**
17′ Ovary and fr papillate; fr ± juicy; infl gen open
19. Fr a berry; seeds few per chamber ..... **ARBUTUS**
19′ Fr a drupe; seeds 1 per stone ..... **COMAROSTAPHYLIS**
15′ Fr a capsule; pls gen of higher-elevation or moister habitats (*Rhododendron, Menziesia* from low-elevation, moister habitats)
20. Anther awned or not; fr loculicidal
21. Lvs boat-shaped, opposite, < 5 mm, appressed; fls solitary in upper lf axils; anther awns elongate ..... **CASSIOPE**
21′ Lvs not boat-shaped, alternate, > 10 mm, spreading; fls in raceme; anther awns 0 or rudimentary ..... **LEUCOTHOE**
20′ Anther unawned; fr septicidal
22. Fr dehiscent base to tip; petals free exc sometimes at base ..... **LEDUM**
22′ Fr dehiscent tip to base; petals ± fused
23. Corolla gen cup-shaped to rotate, with pockets holding anthers until anthers open ..... **KALMIA**
23′ Corolla gen not cup-shaped to rotate, without pockets
24. Lf < 5 mm wide, margin strongly rolled under, lower surface < 1/3 visible; infl without bud scales ..... **PHYLLODOCE**
24′ Lf > 5 mm wide, margin not strongly rolled under, lower surface > 1/3 visible (exc sometimes in bud); infl with deciduous or ± persistent bud scales
25. Corolla cylindric to urn-shaped, < 1 cm; lvs deciduous ..... **MENZIESIA**
25′ Corolla shallowly bell- to funnel-shaped, 1–5 cm; lvs evergreen or deciduous ... **RHODODENDRON**

ALLOTROPA SUGAR STICK

1 sp. (Greek: different turned, from erect infl)

A. virgata A. Gray (p. 553) Per, nongreen, rhizomed, glabrous; roots brittle. **ST** 0. **LF** scale-like. **INFL** raceme-like, < 5 dm, white with red or maroon stripes, emerging from soil erect, persistent after seed dispersal; bracts < 3 cm; bractlets 0; pedicels not recurved. **FL**: sepals gen 0, rarely 2–4; corolla cup-shaped, white, petals 5, free, concave; stamens 10, exserted, anthers dehiscent by short separate slits, maroon; nectary disk-like, lobes 10, short; ovary superior, chambers 5, placentas axile, style < 2 mm, stigma disk-like. **FR**: capsule, loculicidal. **SEEDS** many per chamber, fusiform. $2n=26$. Oak, mixed, or coniferous forests; 75–3000 m. NW, CaRH, SNH; to B.C., ID, MT. [Wallace 1975 Wasmann J Biol 33:1–88]. Petals incorrectly considered sepals by some.

ARBUTUS MADRONE

Shrub, tree, glabrous to hairy, burled or not. **ST** erect; bark smooth at first, then shredding or fissured. **LVS** alternate, evergreen, leathery. **INFL**: panicle, bracted; bractlets 2. **FL**: sepals 5, fused at base; petals 5, fused, urn-shaped; stamens 10, anthers dehiscent by short separate gaping slits, awns elongate; ovary superior, papillate, chambers 5. **FR**: berry. **SEEDS** few per chamber, large. 20 spp.: N.Am, C.Am, w Eur, Medit, w Asia. (Latin: name for *A. unedo*, strawberry tree)

A. menziesii Pursh (p. 553) PACIFIC MADRONE **ST** < 40 m; bark reddish; twigs stout. **LF**: blade < 12 cm, ovate to oblong, glabrous, rounded to pointed at tip, entire to minutely serrate, upper surface bright green, lower surface whitish. **FL** < 8 mm; corolla yellowish white or pinkish. **FR** < 12 mm wide, spheric, orange-red, papillate. $2n=26$. Coniferous, oak forests; 100–1500 m. NW, CaRH, n&c SNH, CW, n ChI (Santa Cruz Island), WTR, SnGb, PR; to B.C., Baja CA. ❀DRN,DRY&SUN:4,**5**,6,15–17;DRN,IRR:3,7,15,24&SHD:9,14,18,23;DFCLT.

ARCTOSTAPHYLOS MANZANITA

Philip V. Wells

Shrubs, small trees. **ST** prostrate to erect; fire-resistant burl sometimes present at base; bark gen reddish, smooth or gray, rough, and shredded; hairs gen alike on twig, infl axis, bract. **LVS** alternate, spreading to ascending, evergreen; blade surfaces gen alike, sometimes convex, differing in color (stomata restricted to lower surface) or hairiness; margin flat to rolled.

INFL: raceme or panicle-like, terminal; branches raceme-like; fls bracted; bracts lf-like, gen flat or scale-like, gen folded, keeled; immature infl present late summer through winter. **FL** radial; sepals gen 5, free, persistent; corolla gen 5-lobed, urn-shaped to ± spheric, white to pink; stamens gen 10, incl, filament base glabrous or hairy, anther 2-pored, awns 2, recurved; ovary superior, base surrounded by nectary disk, chambers 2–10, ovule 1 per chamber, style 1, stigma head-like. **FR**: drupe, berry-like, gen ± spheric; pulp gen thick, mealy; stones 2–10, free, separable, or strongly fused. ± 60 spp.: N.Am (esp CA) to C.Am, Eurasia. (Greek: bear berries) [Wells 1988 Madroño 35:330–341] Observation of hairs requires 10× magnification. Distribution of many spp. local; hybridization occurs in areas of overlap. ❀Beautiful but mostly DFCLT due to fungus and often salinity and alkali. Avoid overhead watering in hot weather. CVS are the easier garden subjects.

1. Fr ± cylindric, 3–4 mm, 2–3 mm wide; skin ribbed vertically; pulp 0 or not mealy; stones 2–5; blade gen 0.5–2.5 cm; petiole 1–3 mm (subg. *Micrococcus*) ..... **Group 1**
1′ Fr spheric or ovoid, 4–15 mm wide; skin smooth, not ribbed; pulp gen mealy; stones gen 8–10, separable or strongly fused into 1 unit; blade gen 2–4+ cm; petiole gen > 3 mm (subg. *Arctostaphylos*)
2. Most or all infl bracts scale-like, folded, back keeled, deltate to awl-like, all exc lowermost bract 2–5 mm; sepals ± appressed (sect. *Arctostaphylos*) ..... **Group 2**
2′ Bracts lf-like, lanceolate to ovate, gen 5–15 mm, gen green, sometimes pink; sepals ± reflexed in fr (sects. *Foliobracteata, Pictobracteata*) ..... **Group 3**

Group 1 (subg. Micrococcus)

1. Sepals 4; corolla lobes 4; fr 2–4-ribbed; blade surfaces not alike, upper surface ± convex, darker than lower, glabrous, margin ± rolled; lower surface sparsely fine-bristly; NCo, NCoRO, SnFrB
2. Blade elliptic to oblong-elliptic, 5–12 mm, 3–7 mm wide; infl axis hairs not glandular; pedicel straight; corolla 4–5 mm, narrow ..... ***A. mendocinoensis***
2′ Blade ± round to round-ovate, 10–22 mm, 8–18 mm wide; infl axis hairs ± gland-tipped; pedicel recurved in fr; corolla 5–6 mm, spheric ..... ***A. nummularia***
1′ Sepals 5; corolla lobes 5; fr 5-ribbed; blade surfaces alike, bright green or glaucous, flat; n&c SNF
3. Infl bracts 1–2 mm, scale-like, folded, keeled, red; pedicel 1–3 mm, ± recurved in fr; blade bright green, shiny; twigs, infl axis fine-glandular-bristly; bark red, smooth, glaucous ..... ***A. myrtifolia***
3′ Infl bracts 3–9 mm, lf-like, flat, gen gray; pedicel 5–7 mm, straight; blade glaucous, dull; twigs, infl axis short-hairy, not glandular; bark gray, rough ..... ***A. nissenana***

Group 2 (sect. Arctostaphylos)

1. Infl ± a raceme, sometimes with 1–2 (2–5 in *A. edmundsii*) short (5–15 mm) branches at base; blades mostly 1–2 cm; pls gen low, sts decumbent (erect in *A. pungens, A. rudis*)
2. Lf surfaces not alike, upper surface darker than lower, slightly convex, margin cupped or ± rolled; lvs spreading to ± ascending
3. Infl branches 2–5; lf blade ± round to round-ovate, base truncate to slightly lobed; filament base hairy ..... ***A. edmundsii***
3′ Infl raceme, sometimes with 1 short branch; lf blade obovate to oblanceolate, base ± wedge-shaped; filament base ± glabrous
4. Pedicel, ovary ± white-hairy; lower lf blade surface gray-tomentose; fr dull reddish brown ..... ***A. pumila***
4′ Pedicel, ovary ± glabrous; lower lf blade surface ± glabrous, ± shiny, not gray-tomentose; fr bright red ..... ***A. uva-ursi***
2′ Lf surfaces alike, flat, margin flat; lvs ascending to erect
5. Infl axis stout, club-like near tip; fls, bracts crowded distally; infl ± umbel-like; pedicel 5–10 mm; fr depressed-spheric; pl 1–3 m, sts erect ..... ***A. pungens***
5′ Infl axis slender, not club-like; fls, bracts evenly spaced; infl not umbel-like; pedicel 2–6 mm; fr gen spheric (exc *A. rudis*) pl gen < 1 m, sts spreading to decumbent (erect, 1–2 m in *A. rudis*)
6. Bark gray, rough, persistent; basal burl present; pls 1–2 m, sts erect; fr 8–14 mm wide, depressed-spheric ..... ***A. rudis***
6′ Bark red, smooth; basal burl 0; pls < 1 m, mat- to mound-like; sts gen prostrate to decumbent; fr 3–8 mm wide, spheric to slightly depressed
7. Lower 1–2 infl bracts ± lf-like, lanceolate-linear, 5–10 mm; lf ± obovate; lf tip obtuse to rounded; NCoR, KR, CaRH, SNH (conifer forest)
8. Lf glaucous, dull, midvein sparsely fine-glandular; pedicel, ovary finely glandular-bristly; stones fused into a ribbed, pitted unit ..... ***A. klamathensis***
8′ Lf bright green, shiny, glabrous; pedicel, ovary glabrous; stones separable or sticking together in groups ..... ***A. nevadensis***
7′ Lower infl bracts scale-like, ± awl-like, 2–5 mm; lf ± elliptic or round; lf tip acute to abruptly pointed; CCo, SnFrB ..... ***A. hookeri***
9. Fr 6–8 mm wide; corolla 5–7 mm; immature infl axis 5–10 mm — serpentine soils
10. Pls gen mat-like, sts prostrate; lf blade 0.5–1 cm wide, gen oblanceolate; twigs, infl axis puberulent; (San Francisco — presumed extinct) ..... ssp. ***franciscana***
10′ Pls mound-like, sts decumbent; lf blade 1–1.5 cm wide, gen elliptic; twigs, infl axis ± densely tomentose; (n CCo, Mount Tamalpais) ..... ssp. ***montana***
9′ Fr 3–6 mm wide; corolla 3–6 mm; immature infl axis 3–5 mm
11. Pls gen 0.5–1 m; blade 2–3 cm; petiole 4–8 mm ..... ssp. ***hookeri***
11′ Pls gen < 0.5 m; blade 0.8–2 cm; petiole 1–4 mm

12. Blade elliptic to lanceolate-elliptic, 0.8–1.2 cm, 4–7 mm wide; infl 3–6-fld; sandy terraces, CCo (Arroyo de la Cruz) . . . . ssp. ***hearstiorum***

12′ Blade ± round to round-elliptic, 1–2 cm, 10–15 mm wide; infl 6–10-fld; serpentine, SnFrB (San Francisco Presidio) . . . . ssp. ***ravenii***

1′ Infl a panicle, branches 3–10, 15–30 mm; blades mostly 2–5 cm; pls gen shrub, tree-like, sts gen erect

13. Infl bracts, fls evenly spaced on slender, long axes; immature infl gen ascending, axis < 1 mm wide, ± thread-like; buds gen exposed, not concealed by bracts

14. Blade surfaces not alike, upper surface ± convex, darker above; pedicel 6–9 mm, glandular-hairy; n ChI (Santa Cruz Island) . . . . ***A. insularis***

14′ Blade surfaces alike, both surfaces dark green to glaucous; pedicel gen 2–6 mm, glabrous; mainland

15. Blade gen 3–5 cm, 1.5–2.5 cm wide; infl bracts 1–1.5 mm, deltate; corolla gen deep pink . . . . . ***A. stanfordiana***

16. Lf slightly glaucous, ± papillate, ± rough; twigs, infl axis, bracts finely glandular-bristly . . . . ssp. ***raichei***

16′ Lf bright green, shiny, smooth; twigs, infl axis glabrous or puberulent, not glandular

17. Pl 0.5–1 m; sts decumbent; petiole 4–8 mm; twigs, infl axis puberulent, dull . . . . ssp. ***decumbens***

17′ Pl 1–2 m; sts ± erect; petiole 8–12 mm; twigs, infl axis glabrous, shiny . . . . ssp. ***stanfordiana***

15′ Blade gen 1–3 cm, 0.5–2 cm wide; infl bracts 1–4 mm, deltate to awl-like, lowest 5–10 mm, ± lf-like; corolla white to pink-tinged

18. Lf blade widely ovate to elliptic, base ± obtuse, rounded, or truncate; pedicel in fr 6–8 mm; fr 8–10 mm wide, strongly depressed, reddish brown; stones 3–4 mm wide, side facing fr wall deeply wrinkled, ribbed — NCoRO (Sonoma Co.) . . . . ***A. bakeri***

19. Twigs, infl axis, petiole densely finely glandular-bristly; lf blade glandular, rough; petiole 3–6 mm; immature infl axis 10–15 mm; NCoRO (near Camp Meeker, Vine Hill) . . . . ssp. ***bakeri***

19′ Twigs, infl axis, petiole gen puberulent to fine tomentose and sparsely glandular; lf blade not glandular, sparsely puberulent, smooth or slightly rough; petiole 4–8 mm; immature infl axis 15–20 mm; serpentine, NCoRO (The Cedars) . . . . ssp. ***sublaevis***

18′ Lf blade lanceolate-elliptic to oblanceolate, base wedge-shaped; pedicel in fr 4–6 mm; fr 5–7 mm wide, subspheric, whitish to tan, becoming brown; stones 2 mm wide, side facing fr wall faintly wrinkled

20. Twigs, infl axis, petiole finely tomentose or puberulent, not glandular; lf blade bright green, shiny, not glandular, smooth; filaments glabrous — NCoRO (Vine Hill) . . . . ***A. densiflora***

20′ Twigs, infl axis, petioles gen glandular-bristly; lf blade dark green, ± dull, ± glandular, papillate, rough; filaments hairy . . . . ***A. hispidula***

13′ Infl bracts, fls ± crowded on distal part of axis; immature infl gen nodding or abruptly turned downward, axis > 1 mm wide; bracts overlapping, concealing buds (exc *A. mewukka*)

21. Lf gray-green to white-glaucous, dull; blade base gen truncate to slightly lobed (± rounded in *A. mewukka*)

22. Stones fused into 1 spheric unit (10–15 mm wide); pulp scanty, not mealy (fr wall leathery, sticky); SnFrB, SCoR, SW, DMtns . . . . ***A. glauca***

22′ Stones 3–8, not fused; pulp mealy, thick; NW, CaR, SN

23. Pedicel glabrous; twigs, infl axis, fl buds, bracts not sticky; fr 10–16 mm wide, dark mahogany-brown, glabrous, smooth . . . . ***A. mewukka***

24. Burl present; pl 1–2 m; blade lanceolate-ovate to oblong-ovate, grayish, 2–4 cm wide; petiole 6–10 mm; immature infl bracts spreading . . . . ssp. ***mewukka***

24′ Burl 0; pl 2–4 m; blade ± round to round-ovate, white-glaucous, 4–7 cm wide; petiole 10–15 mm; immature infl bracts appressed, overlapping . . . . ssp. ***truei***

23′ Pedicel finely glandular-bristly; twigs, infl axis, fl buds, bracts glandular, gen sticky; fr 6–8 mm wide, reddish brown, gen sticky . . . . ***A. viscida***

25. Lf blade finely glandular-hairy, papillate, rough; twigs, infl axis densely glandular-bristly . . . . ssp. ***mariposa***

25′ Lf blade ± glabrous, smooth; twigs, infl axis gen not glandular-bristly

26. Ovary and fr finely glandular-bristly, very sticky; KR, NCoR . . . . ssp. ***pulchella***

26′ Ovary and fr ± glabrous, not glandular, not sticky; KR, CaR, SN . . . . ssp. ***viscida***

21′ Lf bright green, shiny (subglaucous in *A. manzanita* ssp. *glaucescens*); blade base rounded, obtuse, or wedge-shaped (truncate to slightly lobed in *A. patula*)

27. Fr ± spheric or ovoid; stones mostly fused into 1 ribbed unit; TR

28. Burl present; sts erect, not forming circular clones; pedicels, ovary puberulent; corolla pink . . ***A. gabrielensis***

28′ Burl 0; sts decumbent, rooting when buried, forming circular clones; pedicels, ovary glabrous; corolla white . . . . ***A. parryana***

27′ Fr depressed-spheric; stones free, separable; NW, CaR, SN, CW, SnGb, SnBr, SnJT

29. Twigs, infl axis, petiole with golden, glistening, gland-tipped hairs; sts ± decumbent; pls forming circular clones; burl ± 0 or present (SN); fr dark chestnut-brown . . . . ***A. patula***

29′ Twigs, infl axis, petiole mostly finely tomentose, glands not golden, glistening; sts erect; pls not forming circular clones; burl gen 0 (exc ssp. *roofii*); fr reddish . . . . ***A. manzanita***

30. Lf slightly glaucous, dull — NCoR . . . . ssp. ***glaucescens***

30′ Lf bright green, shiny

31. Twigs, infl axis, bracts finely glandular-bristly; blade ± fine papillate, finely scabrous, veins finely bristly — CaRF . . . . ssp. ***wieslanderi***

31′ Twigs, infl axis, bracts ± glabrous or puberulent, not glandular; blade smooth, glabrous

32. Ovary, fr hairs gland-tipped; fr ± sticky — NCoRH . . . . ssp. ***elegans***

32′ Ovary, fr glabrous or short-white-hairy, not glandular; fr not sticky

33. Infl branches gen 5–7; immature infl axis slender, 15–25 mm . . . . ssp. ***manzanita***

33′ Infl branches gen 3–5; immature infl axis stout, < 15 mm

34. Burl 0; ovary, fr glabrous; SnFrB (Mount Diablo), s NCoRI (Vaca Mtns) .... ssp. ***laevigata***
34′ Burl present; ovary, fr ± white-hairy, fr becoming glabrous; NCoRI (nw Glenn Co.), s CaRF (n Butte Co.) .... ssp. ***roofii***

Group 3 (sects. Foliobracteata, Pictobracteata)

1. Infl bracts thin, deep pink in fl, becoming tan, withering; sepals ± oblong, margin flat, densely glandular-ciliate; pedicels in fr 15 + mm — SnBr, PR (sect. *Pictobracteata*) .... ***A. pringlei*** ssp. ***drupacea***
1′ Infl bracts ± thick, green, sometimes reddish, not withering; sepals widely ovate, margin inrolled, not densely glandular-ciliate; pedicels gen < 10 mm (sect. *Foliobracteata*)
2. Burl present at st base or root crown, resprouting after fire; immature infl nodding, branches 3–8, evident, not framed by long bracts
3. Blade surfaces alike, margin flat; lvs ± ascending to erect; petiole 5–10 mm .... ***A. glandulosa***
4. Twigs, petioles, infl axis, bracts short-soft-hairy, with long, white bristles; glandular hairs gen 0
5. Lf bright green, ± shiny; WTR (Santa Monica, Santa Ynez Mtns) .... ssp. ***mollis***
5′ Lf gray-green, dull; SnBr, SnGb, SCo
6. Middle, lower infl bracts lf-like; bristles 0 in 50% of pls; SCo (San Diego Co.) .... ssp. ***crassifolia***
6′ Middle, lower infl bracts scale-like; bristles present in all pls; SnGb, SnBr .... ssp. ***glaucomollis***
4′ Twigs, petioles, infl axis, bracts puberulent, finely tomentose, or finely glandular-bristly; bristles or long white hairs 0
7. Infl bracts mostly scale-like, deltate to awl-like, lowest bracts lf-like; pedicel in fr 5–10 mm, > bracts; lf white-glaucous — e PR .... ssp. ***adamsii***
7′ Infl bracts mostly lf-like, oblong-lanceolate; pedicel in fr 2–5 mm; lf deep green to glaucous
8. Lf deep to grayish green; blade gen ovate, base rounded, obtuse, or wedge-shaped; KR, NCoR, SnFrB .... ssp. ***glandulosa***
8′ Lf glaucous, dull gray; blade round-ovate to ± round, base truncate or ± lobed to obtuse; SCoRO, SCo, w PR .... ssp. ***zacaensis***
3′ Blade surfaces not alike, upper surface convex, darker than lower; margin cupped or ± rolled; lvs ± spreading; petiole 2–5 mm .... ***A. tomentosa***
9. Trunk, lower st bark gray, rough, persistent as flat shreds — CCo
10. Lower surface lf veins glandular-bristly or glabrous; twigs, infl axis, bracts glandular-bristly or finely tomentose — Jack's Peak, near Monterey .... ssp. ***bracteosa***
10′ Lower surface lf blade densely tomentose; twigs, infl axis, bracts tomentose, not glandular
11. Twigs, infl axis, bracts with long, white hairs; Los Osos Valley, near San Luis Obispo .... ssp. ***daciticola***
11′ Twigs, infl axis, bracts tomentose; Monterey, San Luis Obispo cos. .... ssp. ***tomentosa***
9′ Trunk, lower st bark red, gen smooth (± rough in ssp. *rosei*)
12. Hairs (twigs, petioles, infl axis, bracts, pedicels) finely glandular; lvs ± papillate, scabrous; n ChI (Santa Cruz, Santa Rosa islands) .... ssp. ***subcordata***
12′ Hairs tomentose to long-white-bristly, not glandular; lvs smooth; mainland
13. Hairs (twigs, petioles, infl axis, bracts) long-white-bristly
14. Lower lf surface densely tomentose; twigs densely tomentose; SnFrB (Santa Cruz Mtns) ..... ssp. ***crinita***
14′ Lower lf surface glabrous to sparsely tomentose; twigs ± puberulent; CW, WTR .... ssp. ***crustacea***
13′ Hairs not long-white-bristly
15. Lower lf surface tomentose; pedicel, ovary tomentose; n ChI (Santa Cruz, Santa Rosa islands) .... ssp. ***insulicola***
15′ Lower lf surface ± glabrous; pedicel, ovary short-white-hairy or becoming glabrous; mainland
16. Pedicel, ovary becoming glabrous; lower st bark ± smooth; CCo (La Purisima Hills) .... ssp. ***eastwoodiana***
16′ Pedicel, ovary white-hairy; lower st bark rough; CCo (Mount Tamalpais to s Monterey Co.) .... ssp. ***rosei***
2′ Burl 0, not resprouting after fire; immature infl gen ascending or reflexed, gen framed by long bracts
17. Lvs ± well spaced, not overlapping, not clasping; petioles gen 5–10 mm
18. Immature infl bell-shaped (± enclosed by spreading or recurved outer bracts); twigs, infl axis, lvs gen canescent, finely tomentose, or white-downy, not long-bristly; pedicels gen recurved in fr
19. Immature infl gen ascending to erect, peduncle curved or abruptly bent; infl a panicle with 2–5 branches, framed by appressed or stiffly spreading bracts; lf base truncate to slightly lobed
20. Pedicel, ovary, fr glabrous; twigs, infl axis finely tomentose, not glandular; pedicel strongly recurved in fr; immature infl bracts 7–14 mm, linear-lanceolate, acuminate; SCoRO (serpentine, Santa Lucia Mtns) .... ***A. obispoensis***
20′ Pedicel finely glandular-bristly; ovary densely white-hairy; fr hairy; twigs, infl axis finely tomentose, glandular; pedicel straight in fr; immature infl bracts 3–6 mm, lanceolate, acute; NCoRI .... ***A. malloryi***
19′ Immature infl gen nodding, peduncle recurved; infl a raceme or with 1–2 branches, framed by long, recurved bracts; lf base gen ± wedge-shaped, obtuse, or rounded
21. Pedicel, ovary, fr glabrous; filament glabrous; infl branches 0–1; blade oblong-elliptic, tip gen rounded, base ± wedge-shaped; SnFrB (Santa Cruz Mtns) .... ***A. silvicola***
21′ Pedicel, ovary, fr white-hairy; filament base hairy; infl branches 0–3; lf blade ovate to round-ovate, tip acute to abruptly soft-pointed, base obtuse to rounded; NCoR, SnFrB (Mt. Tamalpais, Loma Prieta) .... ***A. canescens***
22. Twigs, infl axis, pedicel, ovary, fr not glandular .... ssp. ***canescens***
22′ Twigs, infl axis, pedicel, ovary, fr finely glandular .... ssp. ***sonomensis***

18′ Immature infl not bell-shaped (exc *A. nortensis*); twigs, infl axis long-bristly or densely glandular-bristly; lvs gen becoming glabrous; pedicels gen straight in fr

23. Infl a raceme, sometimes with 1 short branch; pedicel, ovary, fr glabrous (see also *A. nortensis*); glands 0

24. Corolla 8–9 mm, conic, throat tapered; filament base hairy; blade 1–3 cm, glaucous, dull, base truncate to slightly lobed; s SCoRO (near Santa Margarita) .... ***A. pilosula***

24′ Corolla 6–7 mm, subspheric, throat not tapered; filament base glabrous; blade 2–4 cm, grayish green, ± shiny, base rounded to wedge-shaped; s CCo (hills se of San Luis Obispo) .... ***A. wellsii***

23′ Infl branches 3–10 (1–2 in *A. nortensis*); pedicel, ovary, fr white-hairy or finely glandular-bristly

25. Twigs, petiole, infl axis gen with long, white bristles; ovary densely white-tomentose; immature infl branches ± concealed, framed by long, outer bracts

26. Infl branches 3–8; pedicel finely glandular-bristly; blade 4–6 cm, 2–3 cm wide; petiole 4–10 mm; immature infl bracts 10–18 mm, straight, acuminate .... ***A. columbiana***

26′ Infl branches only 1–2, short; pedicel puberulent; blade 2–4 cm, 1–2 cm wide; petiole 4–5 mm; immature infl bracts 5–12 mm, recurved, obtuse .... ***A. nortensis***

25′ Twigs, petiole, infl axis ± finely glandular-bristly, long, white bristles 0; ovary finely glandular-bristly; immature infl branches clearly evident, not concealed by long bracts

27. Fr depressed-spheric, 8–12 mm wide, stones ± separable; blade dark green, slightly glaucous, ± finely glandular-bristly, papillate, scabrous; immature infl branches nodding; c CCo (Fort Ord), SCoRO (Toro Mtn, nw Monterey Co.) .... ***A. montereyensis***

27′ Fr ± spheric, 6–8 mm wide, stones ± fused into 1 unit; blade glaucous, finely tomentose, becoming glabrous, smooth; immature infl branches stiffly ascending; sw PR .... ***A. otayensis***

17′ Lvs crowded, strongly overlapping, clasping or not, sessile or petiole gen < 6 mm and gen < sinus between basal lobes

28. Bark rough, persistent as flat shreds, gray or ± reddish (upper st bark ± smooth, reddish in *A. cruzensis*)

29. Blade surfaces not alike, upper surface gen convex, darker than lower, margin curved down; infl axis ± densely white-bristly

30. Petioles 2–5 mm; blade oblong-ovate to oblong-elliptic, upper surface dark green, shiny, lower surface white-tomentose; pedicel ± glabrous; s CCo (near Morro Bay) .... ***A. morroensis***

30′ Petioles < 2 mm; blade ovate to triangular-ovate, upper surface slightly glaucous, ± dull, lower surface glabrous; pedicel white-hairy; n-c CCo, s SnFrB (Pajaro Hills) ..... ***A. pajaroensis***

29′ Blade surfaces alike, flat, not different in color or hairiness; twigs, infl axis tomentose, bristles 0 or sparse

31. Upper st bark ± smooth, reddish; twigs, infl axes (2–3) ± sparsely bristly; blade oblong-ovate; pedicel, ovary, fr white-tomentose .... ***A. cruzensis***

31′ Upper st bark rough, grayish; twigs, infl axis not white-bristly; blade ovate to round-ovate; pedicel, ovary, fr glabrous .... ***A. osoensis***

28′ Bark smooth, red throughout

32. Blade base ± truncate to slightly lobed, lvs not clasping; petiole mostly 3–6 mm

33. Infl a raceme, sometimes with 1 short branch; blade narrowly lanceolate-oblong, scabrous; twigs, petioles, infl axes finely short-glandular-hairy, not long-white-hairy; fr 6–8 mm wide .... ***A. virgata***

33′ Infl a panicle, branches 3–10; blade widely lanceolate to round-ovate, ± smooth; twigs, petioles, infl axes long-white-hairy and glandular-bristly; fr 8–15 mm wide

34. Pl < 2 m, sts decumbent to erect; infl axes 3–5 (short, crowded); fr 8–11 mm wide — n ChI (Santa Rosa Island) .... ***A. confertiflora***

34′ Pl 2–8m+, tree-like; infl axes 4–10; fr 10–15 mm wide

35. Lvs lanceolate-ovate to elliptic, flat; pedicels 2–5 mm; pedicel, ovary densely white-hairy, sparsely glandular; fr spheric, sparsely hairy; s ChI (Santa Catalina Island) .... ***A. catalinae***

35′ Lvs ± oblong to ovate, ± swaybacked; pedicels 5–15 mm; pedicel, ovary finely glandular-bristly; fr depressed-spheric, glandular-hairy, sticky; SCoRO (Santa Lucia Mtns) .... [2]***A. hooveri***

32′ Blade base deeply lobed, lvs clasping; petiole gen < 4 mm (3–6 mm in *A. hooveri*)

36. Pedicel in fr strongly recurved; immature infl bell-shaped, ± enclosed by large, outer bracts; lvs, twigs, infl axis gen finely tomentose, canescent, or short-soft-hairy; bristles present or 0

37. Pedicel, ovary, fr glabrous; infl axes 1–3; twigs, petioles, infl axis finely tomentose; bristles, glands 0; lower infl bracts ± linear-lanceolate, narrowly acuminate; SCoRO (Santa Lucia Mtns) .... ***A. luciana***

37′ Pedicel, ovary, fr densely white-hairy to finely glandular-bristly; infl axes 3–5; twigs, petioles, infl axis, bracts tomentose and long-white-bristly; lower infl bracts ovate to widely oblong-lanceolate, acute; SnFrB

38. Pedicel, ovary, fr densely white-hairy, glands 0; long, white hairs gen not frayed at tip, not glandular; Mount Diablo .... ***A. auriculata***

38′ Pedicel, ovary, fr finely glandular-bristly; long, white hairs frayed at tip, sometimes plumose, glandular hairs present; Santa Cruz Mtns .... ***A. glutinosa***

36′ Pedicel in fr gen straight; twigs, infl axis mostly bristly, gland-tipped or not

39. Infl ± a raceme, branches < 3, compact; pedicel, ovary, fr glabrous or hairy, not glandular, not sticky; twigs, infl axis, bracts densely bristly, not glandular

40. Pedicel, ovary densely white-hairy; twigs, infl axis densely tomentose; lf base slightly lobed; n ChI (Santa Cruz Island) .... ***A. viridissima***

40′ Pedicel, ovary, fr glabrous; twigs, infl axis finely short-hairy; lf base deeply lobed; mainland

41. Blade 2–4 cm, ± oblong-ovate, slightly glaucous; infl bracts linear-lanceolate, narrowly acuminate; corolla 6–8 mm, conic; filament base hairy; fr 8–12 mm wide; s CCo (Pecho Hills, sw San Luis Obispo Co.) .... ***A. pechoensis***

41′ Blade 1–2.5 cm, round to round-ovate, bright green; infl bracts ovate to lanceolate-ovate, acute; corolla 4–6 mm, ± spheric; filament glabrous; fr 5–8 mm wide; s SCoRO (w Santa Barbara Co.) .... ***A. purissima***

39′ Infl a panicle, branches 3–10; pedicel finely glandular-bristly; ovary, fr finely glandular-bristly, sticky (glabrous in *A. refugioensis*); twigs, infl axis, bracts gen ± glandular-bristly

42. Fr ovoid to spheric, glabrous, smooth, tip abruptly soft-pointed; stones fused into a spheric unit; infl bracts elliptic, obtuse; lvs glabrous, glaucous .... ***A. refugioensis***

42′ Fr depressed-spheric, glandular-hairy, ± sticky, tip flat or depressed; stones separable; infl bracts lanceolate, acute or acuminate; lvs gen hairy

43. Blade oblong, 4–7 cm, surfaces not alike, upper surface ± convex; immature infl axes spreading — w SnFrB (Santa Cruz Mtns) .... ***A. andersonii***

43′ Blade gen ovate, 3–5 cm, surfaces alike, ± flat; immature infl axes crowded, ± clustered

44. Twigs, infl axis, petiole bristly to fine-bristly, not glandular; infl bracts sparsely ciliate or becoming glabrous; blade glabrous, smooth, glaucous — e SnFrB (Berkeley and Oakland Hills) .... ***A. pallida***

44′ Twigs, infl axis, petiole glandular-bristly; infl bracts finely glandular-ciliate; blade ± glandular-hairy, rough, papillate

45. Blade gray-hairy, ± sticky, 4–6 cm, oblong-ovate, ± sway-backed, tip curved up; pedicel in fr 6–15 mm

46. Petiole 3–6 mm; blade base lobed, ± not clasping; pedicel in fr 10–15 mm; fr 10–15 mm wide, dark brown; SCoRO (Santa Lucia Mtns) .... [2]***A. hooveri***

46′ Petiole < 3 mm; blade base deeply lobed, strongly clasping; pedicel in fr 6–10 mm; fr 6–8 mm wide, light orange- to red-brown; w SnFrB (Kings, Montara Mtns) .... ***A. regismontana***

45′ Blade green, sparsely hairy, not sticky, 2–4 cm, ± round to ovate, flat; pedicel in fr 3–6 mm — w SnFrB

47. Sts prostrate to decumbent, pls < 1 m; lvs spreading, blade round-ovate; infl bract ovate to widely lanceolate; pedicel 3–5 mm; corolla 3–5 mm, ± spheric, throat short, abrupt; San Bruno Mtn .... ***A. imbricata***

47′ Sts erect, pls 1–5 m; lvs ascending, blade oblong-ovate; infl bract lanceolate; pedicel 5–6 mm; corolla 6–9 mm, ± conic, throat tapered; Montara and San Bruno Mtns .... ***A. montaraensis***

A. andersonii A. Gray SANTA CRUZ MANZANITA Shrub, ± tree-like, 2–5+ m; burl 0. **ST**: twigs densely short-bristly and with long, white, ± gland-tipped bristles. **LVS** overlapping, clasping; petiole < 4 mm; blade 4–7 cm, 1.5–2.5 cm wide, ± oblong, subglaucous, base deeply lobed, margin ± cupped, serrate, surfaces not alike, upper glabrous, convex, lower midrib ± bristly. **INFL** open; branches 4–6; bracts 8–15 mm, lf-like, lanceolate; pedicel 6–8 mm, finely glandular-bristly; immature axes 2–3 cm, ± spreading, not framed by long bracts. **FL**: ovary finely glandular-bristly. **FR** 6–8 mm wide, finely glandular-bristly, sticky. *n*=13. RARE. Open sites, redwood forest; < 700 m. w SnFrB (Santa Cruz Mtns). In cult.

A. auriculata Eastw. (p. 553) MOUNT DIABLO MANZANITA Shrub 1–4.5 m; burl 0. **STS** erect; twigs densely white-tomentose and with long, white bristles; glands 0. **LVS** overlapping, clasping; petiole < 2 mm; blade 1.5–4.5 cm, 1.5–3 cm wide, oblong-ovate to round-ovate, base deeply lobed, margin entire, surfaces alike, canescent, becoming ± glabrous, glaucous. **INFL** dense; branches 3–5; bracts 5–15 mm, lf-like, ovate to lanceolate-ovate; pedicel 4–10 mm, tomentose, recurved in fr; immature axes 10–15 mm, crowded, ± obscured by large, lower bracts, peduncle curved. **FL**: ovary densely white-tomentose. **FR** 5–10 mm wide, short-hairy. *n*=13. RARE. Sandstone, chaparral; 150–650 m. e SnFrB (Mount Diablo). In cult.

A. bakeri Eastw. Shrub 1–3 m; burl 0. **ST**: twigs finely glandular-bristly, puberulent, or finely tomentose. **LVS** ± spreading or erect; petiole 3–7 mm; blade 1–3 cm, 1–2 cm wide, gen elliptic, some oblong-ovate or widely ovate, base gen wedge-shaped, sometimes truncate to rounded, margin entire, surfaces alike, dark green, dull or ± shiny, finely glandular-bristly, papillate, scabrous, or appressed puberulent, ± smooth. **INFL**: branches 2–5, crowded; pedicel 3–5 mm, glabrous; bracts gen 2–4 mm, > buds, scale-like, deltate to awl-like; lowest bract 6–10 mm, ± lf-like, narrowly lanceolate; immature axes 1–2 cm, ± slender, spreading outward. **FL**: ovary glabrous. **FR** 8–10 mm wide, glabrous. Open sites, gen serpentine, chaparral, woodland; 200–600 m. s NCoRO (Sonoma Co.).

ssp. ***bakeri*** (p. 553) BAKER'S MANZANITA **STS** erect; twigs glandular-bristly. **LF**: blade finely glandular-bristly, papillate, scabrous; petiole 3–6 mm, hairs like twig hairs. **INFL**: immature axes 10–15 mm, ± stout. *n*=26. **RARE** CA. Serpentine outcrops; 200–300 m. s NCoRO (between Camp Meeker and Occidental, Sonoma Co.). [*A. stanfordiana* ssp. *b.* (Eastw.) J. Adams] In cult.

ssp. ***sublaevis*** P. Wells (p. 553) THE CEDARS MANZANITA **STS** ± spreading; twigs puberulent to finely tomentose, ± finely glandular. **LF**: blade appressed-puberulent, ± sparsely papillate; petiole 4–8 mm, hairs like twig hairs. **INFL**: immature axes 15–20 mm, slender. RARE. Serpentine outcrops, ridges; 300–600 m. s NCoRO (between The Cedars and Healdsburg, Sonoma Co.). In cult.

A. canescens Eastw. Shrub 0.3–2 m; burl 0. **ST**: twigs densely short-soft-hairy to white-tomentose. **LVS** erect; petioles 3–10 mm; blade 2–5 cm, 1–3 cm wide, round-ovate, ovate, or elliptic, base rounded to wedge-shaped, tip acute to abruptly soft-pointed, margin entire, surfaces alike, glaucous, canescent, becoming strigose. **INFL**: raceme or densely 1–3-branched; pedicel 5–9 mm, recurved in fr, tomentose, sometimes glandular; bracts 6–20 mm, lf-like, wide-lanceolate; immature infl bell-shaped, axes 1–2 cm, crowded, ± concealed by large, recurved bracts, peduncle ascending, erect, or recurved. **FL**: ovary densely white-tomentose, sometimes glandular. **FR** 5–10 mm wide, hairy. Ridges, slopes, chaparral, forest; 200–1500 m. w KR, NCoRO, w SnFrB. Hybrids with *A. viscida* are called *A. ×cinerea* Howell and *A. ×oblongifolia* Howell.

ssp. ***canescens*** **ST**: twigs not glandular. **INFL**: axes, pedicel not glandular; bracts 10–20 mm. *n*=13. Ridges, slopes, chaparral, forest; 400–1500 m. NCoR, w SnFrB (Mount Tamalpais, Loma Prieta); sw OR. [var. *candidissima* (Eastw.) Munz] ❀DRN,DRY: 7,14&SUN:5,15–17&IRR,afternoon SHD:19–24;DFCLT.

ssp. ***sonomensis*** (Eastw.) P. Wells SONOMA MANZANITA **ST**: twigs finely glandular. **INFL**: axes, pedicel finely glandular; bracts 6–12 mm. RARE. Ridges, slopes, chaparral, forest; 200–1500 m. w KR, NCoRO. [var. *s.* (Eastw.) J. Adams] In cult.

A. catalinae P. Wells (p. 553) SANTA CATALINA ISLAND MANZANITA Shrub, tree-like, ?–5+ m; burl 0. **ST**: twigs densely finely glandular-bristly and with long, white bristles. **LVS** overlapping; petioles 2–6 mm; blade 2–5 cm, 1.5–3 cm wide, lanceolate-ovate or elliptic, base truncate to ± slightly lobed, margin entire or ± serrate, surfaces alike, light green, dull, finely bristly, sparsely glandular-bristly, ± papillate, finely scabrous. **INFL** open; branches 4–10; bracts 6–10(15) mm, lf-like, lanceolate to narrowly ovate, acute, overlapping on distal half; pedicel 2–5 mm, densely white-hairy, sparsely glandular; immature axes 2–3 cm, ± spreading, bracts gen overlapping on distal half. **FL**: ovary hairs like pedicel. **FR** 8–15 mm wide, ± spheric, sparsely hairy. RARE. Volcanic outcrops, ridges; 300–600 m. s ChI (Santa Catalina Island). In cult.

A. columbiana Piper (p. 553) Shrub, tree-like, 2–10 m; burl 0. **ST**: twigs densely tomentose, gen with long, white bristles, gland-tipped or not. **LVS** spreading; petiole 4–10 mm; blade 4–6 cm, 2–3 cm wide, gen ovate to elliptic, base wedge-shaped to ± rounded, margin entire, surfaces alike, dark green, dull, finely tomentose, ± sparsely glandular-bristly, papillate, becoming glabrous. **INFL** ± open; branches 3–8; bracts 10–18 mm, lf-like, oblong-lanceolate, acuminate; pedicel 2–4 mm, finely glandular-bristly; immature axes 15–25 mm, spreading, framed by large, stiff bracts. **FL**: ovary densely white-hairy, sparsely glandular or not. **FR** 8–11 mm wide, sparsely hairy. *n*=13. Rocky slopes, coniferous forest; < 800 m. NCo, w KR, NCoRO; to B.C. Pls without long, white, gland-tipped hairs are called f. *tracyi* (Eastw.) P. Wells [var. *t.* (Eastw.) J. Adams]. Hybrids with *A. uva-ursi* are called *A.* ×*media* E. Greene. ❀DRN, SUN:4,5,17&IRR:7,14–16;DFCLT.

A. confertiflora Eastw. (p. 553) SANTA ROSA ISLAND MANZANITA Shrub 0.1–2 m; burl 0. **STS** decumbent or erect; twigs densely tomentose and with long, white, gland-tipped bristles. **LVS** overlapping; petiole 3–6 mm; blade 2–4.5 cm, 1.5–3.5 cm wide, ± round to widely ovate, base truncate to slightly lobed, margin entire or short-ciliate, surfaces alike, gen light green, dull, glandular-puberulent, ± papillate, finely scabrous, ± glandular-bristly near base, or upper surface convex, becoming glabrous, margins cupped. **INFL** dense; branches 3–5; bracts 8–14(18) mm, lf-like, ovate to oblanceolate, acute; pedicel 3–5 mm, finely glandular-bristly; immature axes 15–20 mm, crowded, bracts crowded, ± densely overlapping near axis tip. **FL**: ovary densely white-hairy, sparsely glandular. **FR** 8–11 mm wide, sparsely hairy. *n*=13. RARE. Sandstone outcrops; < 500 m. n ChI (Santa Rosa Island). [*A. subcordata* Eastw. var. *c.* (Eastw.) Munz] In cult.

A. cruzensis Roof (p. 553) LA CRUZ MANZANITA Shrub 0.1–1 (3) m; burl 0. **ST**: trunk, lower branch bark reddish, rough, ± peeling; twigs tomentose, sparsely bristly. **LVS** strongly overlapping, clasping; petiole < 2 mm; blade 1.5–3 cm, 1–2.5 cm wide, oblong-ovate, base deeply lobed, margin entire or toothed, bristly-ciliate near base, surfaces alike, bright green, finely tomentose, becoming glabrous, smooth. **INFL** dense; branches 1–3; bracts 5–15 mm, lf-like, lanceolate to lanceolate-ovate, ± acuminate; pedicel 4–5 mm, hairy; immature infl ± bell-shaped, axes 5–15 mm, crowded, ± concealed by large, lower bracts. **FL**: corolla 7–8 mm, throat tapered; ovary white-tomentose. **FR** 8–10 mm wide, hairy; stones separable, angular. *n*=13. RARE. Sandy bluffs; < 150 m. c CCo (s Monterey, nw San Luis Obispo cos.). In cult.

A. densiflora Baker (p. 553) VINE HILL MANZANITA Shrub < 1 m; burl 0. **STS** ± decumbent; twigs finely tomentose. **LVS** erect; petiole 4–5 mm; blade 1–2.5 cm, 0.5–1.5 cm wide, elliptic to narrowly lanceolate-elliptic, base wedge-shaped to obtuse, margin entire, surfaces alike, bright green, shiny, sparsely puberulent, smooth. **INFL**: branches 3–5, crowded; fls well spaced; bracts gen 1.5–3 mm, scale-like, deltate, sharp-pointed; lowest bract 5–10 mm, ± lf-like, linear to elliptic; pedicel 4–5 mm, glabrous; immature axes 10–15 mm, slender, ± spreading. **FL**: ovary glabrous. **FR** 5–6 mm wide, subspheric, glabrous, whitish tan. *n*=13. **ENDANGERED** CA. Shale outcrops; ± 100 m. s NCoRO (Vine Hill, near Forestville, Sonoma Co.). In cult.

A. edmundsii J. Howell (p. 553) LITTLE SUR MANZANITA Shrub mat- or mound-like, < 1.2 m; burl 0. **STS** decumbent; twigs finely tomentose. **LVS** spreading; petiole 2–3 mm; blade 1–2.5 cm, 1–1.5 cm wide, ± round to round-ovate, base truncate to ± lobed, margin entire, ± cupped, surfaces not alike, upper convex, dark green, ± shiny, lower sparsely puberulent, becoming glabrous. **INFL** dense; branches 2–5; bracts 3–5 mm, gen scale-like, deltate to awl-like, green; lowest bract 5–10 mm, ± lf-like, gen oblanceolate; pedicel 3–4 mm, sparsely puberulent or glabrous; immature axes 5–10 mm, crowded. **FL**: ovary sparsely puberulent or glabrous. **FR** 6–8 mm wide, subspheric to subovoid, glabrous, reddish brown. *n*=13. **RARE** CA. Sandy terraces, bluffs; < 100 m. c CCo (nw Monterey Co.). In cult.

A. gabrielensis P. Wells SAN GABRIEL MANZANITA Shrub 1–2 m; burl large, spheric. **STS** erect; twigs finely tomentose. **LVS** erect; petiole 5–8 mm; blade 2–4 cm, 1.5–2.5 cm wide, elliptic, ovate, or oblong-ovate, base ± wedge-shaped to rounded, margin entire, surfaces alike, bright green, shiny, sparsely puberulent, becoming glabrous, smooth. **INFL**: branches 4–7; axes ± crowded; bracts gen 3–5 mm, scale-like, deltate, acuminate, sharp-pointed; lowest bract 10–15 mm, lf-like, ± lanceolate; pedicel 5–10 mm, white-hairy; immature axes 5–20 mm, crowded. **FL**: ovary white-hairy. **FR** 8–14 mm wide, subspheric, sparsely puberulent; stones ± separable or fused into 1 ± depressed-spheric unit, surface pitted between ribs. RARE. Rocky outcrops, chaparral; ± 1500 m. SnGb (Mill Creek Summit). In cult.

A. glandulosa Eastw. (p. 553) Shrub 1–2.5 m; burl large, gen wide, ± flat-topped. **STS** erect; twigs tomentose or short-bristly, glandular or not, sometimes with long, soft or stiff hairs. **LVS** ascending or erect; petiole 5–10 mm; blade 2–4.5 cm, 1–2.5 cm wide, gen elliptic to ovate, base wedge-shaped to rounded, sometimes ± lobed or truncate, margin entire or toothed, surfaces alike, bright green to strongly glaucous, shiny or dull, glandular-puberulent to -bristly, papillate, scabrous, or puberulent to finely tomentose, or becoming glabrous, smooth. **INFL**: branches 3–6, ± crowded; bracts gen 5–10 mm, lf-like, upper bracts sometimes 3–5 mm, ± deltate or awl-like; lower bracts 8–15 mm, lf-like, lanceolate to ± ovate; pedicel 3–10 mm, finely tomentose to short-glandular-bristly; immature axes 1–3 cm, ± crowded, not concealed by long bracts. **FL**: ovary hairs gen like pedicel hairs. **FR** 6–10 mm wide, ± hairy, smooth or finely glandular-bristly, sticky. Rocky outcrops, slopes, ridges; < 2200 m. KR, NCoR, SnFrB, SCoRO, SW; sw OR, Baja CA.

ssp. ***adamsii*** (Munz) Munz **ST**: twigs fine-tomentose or finely glandular-bristly. **LF**: blade white-glaucous, smooth, base rounded, sometimes ± lobed to truncate. **INFL**: bracts gen 3–5 mm, deltate to ± awl-like, acuminate; lower bracts 5–12 mm, lf-like; pedicel in fr 5–10 mm. **FL**: corolla pink. Rocky outcrops, ridges, chaparral, coniferous forest; 200–2200 m. e PR; Baja CA. [var. *adamsii* Munz] ❀DRN,DRY,afternoon SHD:**16**.

ssp. ***crassifolia*** (Jepson) P. Wells (p. 557) COSTA BAJA MANZANITA **ST**: twigs finely tomentose, gen with long, white hairs, not glandular. **LF**: blade slightly glaucous, gray, smooth. **INFL**: bracts 5–15 mm, lf-like; pedicel in fr 3–5 mm. **FL**: corolla white. RARE. Sandstone outcrops, bluffs; < 100 m. s SCo (San Diego Co.); Baja CA. [var. *c.* Jepson] In cult.

ssp. ***glandulosa*** (p. 557) **ST**: twigs gen finely glandular-bristly or finely tomentose. **LF**: blade bright green, sometimes slightly glaucous, glandular-puberulent, papillate, scabrous or sparsely puberulent, smooth. **INFL**: bracts 5–15 mm, lf-like; pedicel in fr 3–5 mm. **FL**: corolla white. *n*=26. Slopes, ridges, chaparral, forest; 300–1900 m. KR, NCoR, SnFrB; sw OR. Pls with fine-tomentose twigs are called f. *cushingiana* (Eastw.) P. Wells. [var. *c.* (Eastw.) McMinn; *A. intricata* Howell] Possible hybrids with *A. nevadensis* from n Del Norte Co. are called *A.* ×*parvifolia* Howell. ❀DRN, DRY,SUN:5,7,14,**15–17**,18–24.

ssp. ***glaucomollis*** P. Wells **ST**: twigs bristly, bristles long, white, not glandular. **LF**: blade ± glaucous or grayish puberulent, smooth. **INFL**: bracts gen 3–5 mm, ± scale-like; lowest bract 6–15 mm, lf-like; pedicel in fr 4–8 mm. **FL**: corolla white. Rocky slopes, ridges, coniferous forest; 600–1800 m. SnGb, w SnBr.

ssp. ***mollis*** (J. Adams) P. Wells **ST**: twigs bristly, bristles long, white, not glandular. **LF**: blade bright green, shiny, smooth. **INFL**: bracts 5–15 mm, lf-like; pedicel in fr 3–6 mm. **FL**: corolla white. Rocky slopes, ridges, woodland; 750–2000 m. WTR (Santa Ynez, Santa Monica mtns). ❀DRN,DRY,SUN:5,7,14,**15–17**,18–24.

ssp. ***zacaensis*** (Eastw.) P. Wells (p. 557) **ST**: twigs puberulent, tomentose or finely glandular-bristly. **LF**: blade glaucous, dull, finely glandular-bristly, scabrous or sparsely puberulent, papillate, base rounded to wedge-shaped, sometimes ± lobed to truncate. **INFL**: bracts gen 5–15 mm, lf-like; pedicel in fr 3–7 mm. **FL**: corolla gen white. *n*=26. Rocky slopes, ridges, chaparral, woodland; 300–1800 m. SCoRO (Santa Lucia, San Rafael Mtns), w PR; Baja CA. [var. *howellii* (Eastw.) McMinn; var. *z.* (Eastw.) McMinn] ❀DRN,DRY,SUN:7,14,**15–23**,24.

A. glauca Lindley (p. 557) Shrub, tree-like, 2–8+ m; burl 0. **ST**: twigs glabrous, ± glaucous, sometimes finely bristly. **LVS** erect; petiole 7–15 mm; blade 2.5–5 cm, 2–4 cm wide, oblong-ovate to ± round, base rounded, truncate, or slightly lobed, margin entire to toothed, surfaces alike, white-glaucous, dull, smooth. **INFL** open; branches 4–8; bracts 3–6 mm, scale-like, deltate to ± awl-like, weakly keeled; lowest bract 10–15 mm, lf-like, widely lanceolate; pedicel finely glandular-bristly, 8–10 mm in fr; immature axes 2–3 cm, spreading; bracts ± spreading, buds exposed. **FL**: ovary glandular. **FR** 12–15 mm wide, spheric or ovoid, sticky; pulp thick, leathery, not mealy; stones fused into 1 subspheric unit (± 10 mm wide), surface with vertical seams. *n*=13. Rocky slopes, chaparral, woodland; < 1400 m. ne SnFrB (Mount Diablo), SCoRI, TR, PR, sw DMtns (Little San Bernardino Mtns); Baja CA. [var. *puberula* J. Howell] ❀DRN,SUN,DRY:7,11,14–24.

A. glutinosa B. Schreiber (p. 553) SCHREIBER'S MANZANITA Shrub 1–2 m; burl 0. **STS** erect; twigs short-soft-hairy to tomentose, ± glandular-bristly. **LVS** overlapping, clasping; petiole < 4 mm; blade 2–5 cm, 1–3 cm wide, oblong to oblong-ovate, base deeply lobed, margin entire, toothed near base, surfaces alike, glaucous, dull, canescent, becoming weakly strigose. **INFL** dense; branches 2–4; bracts 5–15 mm, lf-like, oblong-lanceolate; pedicel 5–8 mm, finely glandular-bristly, recurved in fr; immature axes 15–25 mm, ± spreading, obscured by subtending bracts, peduncle curved. **FL**: ovary densely fine-glandular-bristly, glands red. **FR** 7–14 mm wide, glandular-hairy, ± sticky. *n*=13. RARE. Diatomaceous shale outcrops, chaparral; 400–500 m. sw SnFrB (The Chalks, nw Santa Cruz Co.). In cult.

A. hispidula Howell HOWELL'S MANZANITA Shrub 1–2 m; burl 0. **STS** spreading or erect; twigs finely glandular-bristly. **LVS** erect; petiole 3–6 mm; blade 1.5–3 cm, 0.5–1.5 cm wide, ± elliptic to oblanceolate, base gen wedge-shaped, margin entire or teeth few, surfaces ± alike, dark green, dull or ± shiny, gen finely glandular-bristly, papillate, scabrous. **INFL**: branches 3–6, crowded; fls well spaced; bracts 2–4 mm, scale-like, awl-like; lowest bract 8–10 mm, lf-like, linear-lanceolate; pedicel 3–5 mm, glabrous; immature axes 1–2 cm, slender, ascending to erect. **FL**: ovary glabrous. **FR** 5–7 mm wide, subspheric, glabrous, whitish tan, becoming ± glaucous. UNCOMMON. Rocky serpentine soils or sandstone, open sites, forest; 300–600 m. KR, NCoRO; sw OR. [*A. stanfordiana* ssp. *h.* (Howell) J. Adams]

A. hookeri G. Don Shrub, mat- to mound-like, gen < 1 m, sometimes 2–3 m; burl 0. **ST**: twigs finely tomentose. **LVS** erect; petiole 1–8 mm; blade 0.8–3 cm, 0.4–1.5 cm wide, ± round, widely elliptic, or oblanceolate, base rounded to wedge-shaped, margin entire, surfaces alike, bright green, shiny, ± puberulent, becoming glabrous, smooth. **INFL** ± spheric, dense; raceme, sometimes 1-branched; fls dense; bracts 2–5 mm, scale-like, deltate, sharp-pointed; pedicel 2–6 mm, glabrous; immature axes 3–10 mm, ± slender. **FL**: ovary glabrous. **FR** 3–8 mm wide, ± spheric, glabrous. *n*=13,26. Coastal scrub, woodland; < 600 m. n&c CCo, w SnFrB.

ssp. ***franciscana*** (Eastw.) Munz (p. 557) FRANCISCAN MANZANITA Pl mat-like. **STS** prostrate. **LF**: petiole 3–5 mm; blade 1.5–2 cm, 0.5–1 cm wide, oblanceolate. **INFL**: axis puberulent; fls gen 10. **FL**: corolla 5–7 mm, urn-shaped. **FR** 6–8 mm wide. *n*=13. **PRESUMED EXTINCT** (in CA natural habitat; now cult). Serpentine outcrops; < 300 m. n CCo (San Francisco Peninsula). In cult.

ssp. ***hearstiorum*** (Hoover & Roof) P. Wells (p. 557) THE HEARSTS' MANZANITA Pl mat-like. **STS** prostrate, rooting at lower nodes. **LF**: petiole 1–3 mm, blade 0.8–1.2 cm, 4–7 mm wide, elliptic to lanceolate-elliptic. **INFL**: fls gen 4–6. **FL**: corolla 3–5 mm, narrowly urn-shaped. **FR** 3–4 mm wide. *n*=13. **ENDANGERED** CA. Sandy terraces, coastal grassland; < 200 m. c CCo (Arroyo de la Cruz, nw San Luis Obispo Co.). [*A. hearstiorum* Hoover & Roof] In cult.

ssp. ***hookeri*** HOOKER'S MANZANITA Pl gen mound-like. **STS** < 1 m, decumbent, or 1–2(3) m, erect. **LF**: petiole 4–8 mm; blade 2–3 cm, 1–1.5 cm wide, narrowly to widely elliptic. **INFL**: fls gen 10. **FL**: corolla 4–6 mm, ± spheric. **FR** 4–6 mm wide. *n*=13. RARE. Sandy soils, sandy shales, sandstone outcrops; < 300 m. n&c CCo, w SnFrB (Santa Cruz Mtns s to Carmel). In cult.

ssp. ***montana*** (Eastw.) P. Wells (p. 557) TAMALPAIS MANZANITA Pl gen mound-like. **STS** decumbent or ± erect, < 2 m. **LF**: petiole 3–6 mm; blade 1–2.5 cm, 1–1.5 cm wide, round-elliptic to elliptic. **INFL**: axis densely tomentose. **FL**: corolla 5–7 mm, urn-shaped. **FR** 6–8 mm wide. *n*=26. RARE. Gen serpentine outcrops; 200–600 m. n CCo, nw SnFrB (Mount Tamalpais, Marin Co.). [*A. pungens* var. *m.* (Eastw.) Munz] In cult.

ssp. ***ravenii*** P. Wells (p. 557) PRESIDIO MANZANITA Pl mat-like. **STS** prostrate, < 1 m. **LF**: petiole 2–4 mm; blade 1–2 cm, 1–1.5 cm wide, ± round to round-elliptic. **INFL**: axis puberulent. **FL**: corolla 4–5 mm, ± spheric. **FR** 4–5 mm wide. *n*=26. **ENDANGERED** CA,US. Serpentine outcrop; < 200 m. n CCo (San Francisco Presidio). Pls apparently belong to a single clone. In cult.

A. hooveri P. Wells HOOVER'S MANZANITA Shrub, tree-like, 2–8+ m; burl 0. **ST**: twigs densely fine-bristly and with long, white, gland-tipped hairs. **LVS** overlapping; petiole 3–6 mm; blade 4–6 cm, 2–3 cm wide, ± sway-backed, ± oblong to ovate, base lobed, margin entire or coarsely toothed, surfaces alike, glaucous, dull, glandular-hairy, sticky, becoming ± glabrous, papillate, scabrous. **INFL** open; branches 4–6; bracts 8–15(20) mm, lf-like, lanceolate, acuminate; pedicel finely glandular-bristly, 10–15 mm in fr; immature axes 15–25 mm, curved, ± concealed by large bracts. **FL**: ovary finely glandular-bristly. **FR** 10–15 mm wide, glandular-hairy, sticky. *n*=13. UNCOMMON. Rocky slopes, woodland; 700–1000 m. SCoRO (Santa Lucia Mtns). In cult.

A. imbricata Eastw. (p. 557) SAN BRUNO MOUNTAIN MANZANITA Pl mat- to mound-like, < 1 m; burl 0. **STS** ± creeping; twigs densely fine-bristly with long, gland-tipped bristles. **LVS** strongly overlapping, clasping; petiole < 2 mm; blade 2.5–4 cm, 2–3 cm wide, ± round to round-ovate, base deeply lobed, margin entire or toothed, surfaces alike, light green, dull, sparsely glandular-bristly, papillate, ± scabrous. **INFL** dense; branches 3–5; bracts 5–10 mm, lf-like, ± ovate, acute; pedicel 3–5 mm, finely glandular-bristly; immature axes 5–10 mm, densely clustered, ± sessile. **FL**: corolla 3–5 mm, ± spheric; ovary finely glandular-bristly. **FR** 6–7 mm wide, glandular-hairy, ± sticky. *n*=13. **ENDANGERED** CA. Chaparral; 200–400 m. w SnFrB (San Bruno Mtn). [*A. andersonii* A. Gray var. *i.* (Eastw.) McMinn] In cult.

A. insularis E. Greene (p. 557) Shrub, tree-like, 2–5+ m; burl 0. **ST**: bark reddish, ± glaucous; twigs glabrous, finely tomentose, or finely glandular-bristly. **LVS** spreading or erect; petiole 4–8 mm; blade 2.5–4.5 cm, 1–3 cm wide, oblong-elliptic, base ± rounded, margin entire or toothed, ± cupped, bright green, shiny, ± glabrous, smooth, surfaces not alike, upper darker than lower. **INFL** open; branches 5–10; fls well spaced; bracts gen 1.5–3 mm, scale-like, deltate, reddish; lowest bract 7–10 mm, elliptic or spoon-shaped; pedicel 6–9 mm, finely glandular-bristly; immature axes 2–3 cm, slender, ascending or erect. **FL**: ovary hairs like twig hairs. **FR** 8–15 mm wide, becoming glabrous, bright orange-brown. *n*=13. Uncommon. Rocky slopes, chaparral; woodland; < 400 m. n ChI (Santa Cruz Island). [var. *pubescens* Eastw.] ❀DRN,SUN:15–24;CVS.

A. klamathensis S. Edwards, T. Keeler-Wolf & W. Knight (p. 557) KLAMATH MANZANITA Pl mat- or mound-like < 0.5 m; burl 0; **ST**: twigs finely glandular-bristly. **LVS** erect; petiole 4–7 mm; blade 1–3.5 cm, 0.5–2.5 cm wide, obovate to oblanceolate or widely elliptic, base wedge-shaped or obtuse, margin entire or teeth few, surfaces alike, ± glaucous, dull, papillate, ± scabrous, midvein sparsely finely glandular-bristly. **INFL** spheric; raceme or 1-branched, dense; bracts 2–3(5) mm, scale- to awl-like; lowest bract sometimes ± lf-like, lanceolate; pedicel 3–6 mm, finely glandular-bristly; immature axis 5–10 mm. **FL**: ovary finely glandular-bristly. **FR** 6–7 mm wide, ± spheric; stones fused into 1 unit, pitted, vertically ribbed. RARE. Rocky outcrops, slopes, subalpine forest; 1600–2000 m. e KR (Scott Mtn Divide, Slate Mtn).

A. luciana P. Wells SANTA LUCIA MANZANITA Shrubs, tree-like, 2–5+ m; burl 0. **ST**: twigs finely tomentose. **LVS** strongly overlapping, clasping; petiole < 2 mm; blade 2–4 cm, 1.5–2.5 cm wide, ovate to ± round, base deeply lobed, margin entire, surfaces alike, glaucous, dull, appressed-canescent to finely tomentose, becoming

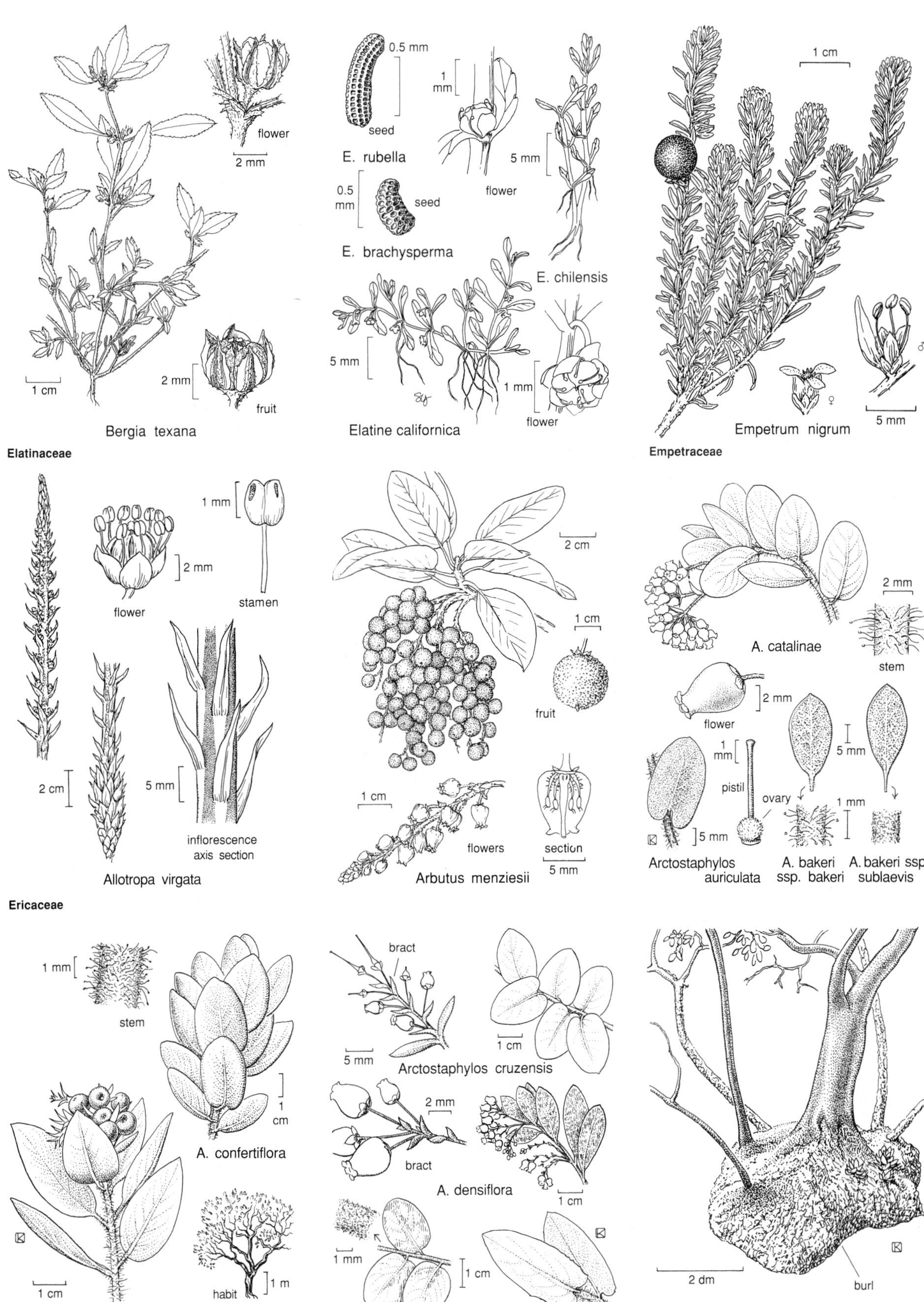
flower
2 mm
1 cm
2 mm
fruit
Bergia texana
0.5 mm
seed
1 mm
E. rubella
5 mm
flower
0.5 mm
seed
E. brachysperma
E. chilensis
5 mm
1 mm
flower
Elatine californica
1 cm
♂
♀
5 mm
Empetrum nigrum
Elatinaceae
Empetraceae
1 mm
2 mm
flower
stamen
2 cm
5 mm
inflorescence axis section
Allotropa virgata
2 cm
1 cm
fruit
1 cm
flowers
section
5 mm
Arbutus menziesii
2 mm
A. catalinae
stem
2 mm
flower
1 mm
pistil
5 mm
ovary
1 mm
5 mm
Arctostaphylos auriculata
A. bakeri ssp. bakeri
A. bakeri ssp. sublaevis
Ericaceae
1 mm
stem
1 cm
A. confertiflora
1 cm
habit
1 m
Arctostaphylos columbiana
bract
5 mm
1 cm
Arctostaphylos cruzensis
2 mm
bract
A. densiflora
1 cm
1 mm
1 cm
A. edmundsii
A. glutinosa
2 dm
burl
Arctostaphylos glandulosa

glabrous, smooth. **INFL**: raceme or 1–2-branched, dense; bracts 5–10 mm, lf-like, linear-lanceolate, acuminate; pedicel 5–10 mm, glabrous, recurved in fr; immature axes 5–10 mm, crowded, concealed by long bracts, peduncle curved. **FL**: ovary glabrous. **FR** 6–12 mm wide, glabrous, becoming glaucous. *n*=13. RARE. Shale outcrops, slopes, chaparral; 500–700 m. SCoRO (se of Cuesta Pass, Santa Lucia Mtns, near San Luis Obispo). In cult.

A. malloryi (W. Knight & R. Gankin) P. Wells MALLORY'S MANZANITA Shrub < 1.5 m; burl 0. **STS** erect; twigs tomentose, finely glandular-bristly. **LVS** erect; petiole 5–10 mm; blade 2–3 cm, 1.5–2.5 cm wide, ± round to ovate, base rounded, truncate, or ± lobed, margin entire, surfaces alike, canescent or densely white-tomentose, becoming glabrous, glaucous. **INFL**: branches 2–5, spreading; bracts 3–5 mm, ± lf-like, linear-lanceolate; lowest bract 6–10 mm, spreading to recurved; pedicel 6–9 mm, straight, finely glandular-bristly; immature axes 1–2 cm, spreading to ascending, bracts appressed. **FL**: ovary densely white-hairy. **FR** 7–9 mm wide, hairy, becoming glaucous. UNCOMMON. Volcanic soils, chaparral; 800–1200 m. n&c NCoRI (Colusa, Shasta, Trinity cos.). ❀TRY.

A. manzanita C. Parry Shrub, tree 2–8+ m; burl 0 (exc ssp. *roofii*). **ST**: twigs smooth to finely tomentose. **LVS** erect; petiole 6–12 mm; blade 2–5 cm, 1–3.5 cm wide, gen widely ovate or oblong-ovate to obovate, base ± rounded to ± wedge-shaped, margin entire, surfaces alike, bright green, shiny, (± glaucous, dull in ssp. *glaucescens*), glabrous, smooth (scabrous, veins finely bristly in ssp. *wieslanderi*). **INFL** open; branches 3–7; axes, bracts finely tomentose (finely glandular-bristly in ssp. *wieslanderi*); bracts 2–4 mm, scale-like, deltate, sharp-pointed to acuminate; pedicel 3–8 mm, glabrous; immature axes 15–45 mm, stout or slender, ± abruptly turned downward, bracts appressed. **FL**: ovary glabrous or glandular. **FR** 8–12 mm wide, glabrous to hairy; stones free, separable. Slopes, rocky soils, chaparral, woodland, forest; < 1500 m. s NCoRO, NCoRI, s CaRF, n&c SNF, SnFrB.

ssp. ***elegans*** (Jepson) P. Wells (p. 557) **INFL**: immature axes > 2 cm, slender. **FL**: ovary hairs gland-tipped. **FR**: hairs gland-tipped, becoming glabrous, ± sticky. *n*=26. Slopes, chaparral, woodland; 600–1400 m. s NCoRO. [*A. e.* Jepson] ❀DRN,DRY,SUN:7,14–19;CVS.

ssp. ***glaucescens*** P. Wells **ST**: twigs smooth or ± finely glandular-bristly. **LVS** ± glaucous, ± dull, smooth. **INFL**: axis hairs like twig hairs; immature axes > 2 cm, slender. Slopes, chaparral, woodland; 200–600 m. s NCoRO. ❀DRN,DRY,SUN:7,14–17.

ssp. ***laevigata*** (Eastw.) Munz **INFL**: immature axes < 15 mm, stout, crowded. Rocky slopes, chaparral; 500–1100 m. e SnFrB (Mount Diablo), s NCoRI (Vaca Mtns). ❀DRN,DRY,SUN:7,14–19.

ssp. ***manzanita*** (p. 557) **INFL**: immature axes 15–25 mm, slender. *n*=26. Chaparral, woodland, coniferous forest; < 1200 m. NCoRI, s CaRF, n&c SNF. ❀DRN,SUN,DRY:7,8,9,**14–17**,18–24;some forms are DFCLT;CVS.

ssp. ***roofii*** (R. Gankin) P. Wells Burl present. **INFL**: immature axes < 15 mm, stout, crowded. **FL**: ovary glabrous or white-hairy. **FR** ± hairy, becoming glabrous. Uncommon. Chaparral, woodland; 800–1300 m. NCoRI (nw Glenn Co.), s CaRF (n Butte Co.). ❀TRY.

ssp. ***wieslanderi*** P. Wells (p. 557) **ST**: twigs finely glandular-bristly. **LF**: blade finely papillate, scabrous, veins finely bristly. **INFL**: axis, bract hairs like twigs; immature axes > 2 cm, slender. Chaparral, coniferous forest; < 1500 m. CaRF (Shasta, Tehama cos.). ❀DRN,DRY,SUN:7,14–19.

A. mendocinoensis P. Wells PYGMY MANZANITA Shrub, mat- to mound-like, < 0.5 m; burl 0. **ST**: bark reddish, ± peeling; twigs sparsely fine bristly. **LF**: petiole < 1 mm; blade 5–12 mm, 3–7 mm wide, oblong-elliptic, base obtuse to wedge-shaped, margin entire, surfaces not alike, upper surface ± convex, dark green, shiny, glabrous. **INFL** dense; branches 1–4; bracts 0.5–1 mm, scale-like, deltate, acuminate; pedicel 2–3 mm, glabrous; immature axes 3–6 mm, thread-like. **FL**: sepals, corolla lobes 4; corolla narrowly urn-shaped; ovary white-hairy. **FR** ± 3 mm, 2 mm wide, ± cylindric; pulp ± 0; stones 4. *n*=13. RARE. Acidic, sandy-clay soils, dwarf coniferous forest; < 200 m. c NCo (Mendocino Co.). In cult.

A. mewukka Merriam Shrub 1–4 m; burl present or 0. **ST**: twigs glabrous or puberulent. **LVS** erect; petiole 6–15 mm; blade 3–7 cm, 2–7 cm wide, ± round, round-, oblong-, or lanceolate-ovate, base ± rounded, margin entire, surfaces alike, glabrous, subglaucous or white-glaucous, dull, smooth. **INFL** open; branches 2–7; axis, bract hairs like twig hairs or ± glandular; bracts 3–6 mm, gen scale-like, deltate to linear-lanceolate, ± keeled, acute or acuminate; lowest bract 8–12 mm, lf-like, lanceolate; pedicel 4–8 mm, glabrous; immature axes 2–3 cm, stout, spreading. **FL**: ovary glabrous. **FR** 10–16 mm wide, ± spheric, glabrous, dark mahogany-brown. *n*=26. Uncommon. Chaparral, coniferous forest; 500–1800 m. SN.

ssp. ***mewukka*** (p. 557) Shrub 1–2 m; burl present. **STS** ± spreading. **LF**: petiole 6–10 mm; blade 2–4 cm·wide, oblong- to lanceolate-ovate, ± acute, gray-glaucous. **INFL**: immature axes 3–5, bracts spreading, exposing buds, ± flat, tip acute. *n*=26. Chaparral, coniferous forest; 800–1800 m. SN. ❀TRY.

ssp. ***truei*** (W. Knight) P. Wells (p. 557) TRUE'S MANZANITA Shrub 2–4 m; burl 0. **STS** erect. **LF**: petiole 1–1.5 cm; blade 4–7 cm wide, ± round to round-ovate, white-glaucous. **INFL**: immature axes 4–7, bracts appressed, concealing buds, keeled, tip acuminate. *n*=26. UNCOMMON. Chaparral, coniferous forest; 500–850 m. n SNF.

A. montaraensis Roof (p. 557) MONTARA MANZANITA Shrub, tree-like, 1–5 m; burl 0. **ST**: twigs densely fine-glandular-bristly, and with long, gland-tipped bristles. **LVS** strongly overlapping, clasping; petiole < 2 mm; blade 2.5–4.5 cm, 1.5–2.5 cm wide, ovate, base deeply lobed, margin entire or toothed, surfaces alike, light green, dull, sparsely glandular-puberulent, ± scabrous. **INFL** dense; branches 4–6; bracts 6–9 mm, lf-like, lanceolate, acuminate, ± spreading; pedicel 5–6 mm, finely glandular-bristly; immature infl sessile, axes 10–15 mm, densely clustered. **FL**: corolla 6–9 mm, ± conic, throat ± slender; ovary finely glandular-bristly. **FR** 6–7 mm wide, glandular-hairy, ± sticky. RARE. Slopes, ridges, chaparral; 200–500 m. w SnFrB (San Bruno, Montara mtns). In cult.

A. montereyensis Hoover MONTEREY MANZANITA Shrub 1–2 m; burl 0. **STS** erect; twigs finely glandular-bristly. **LVS** erect; petiole 4–6 mm; blade 2–3 cm, 1.5–3 cm wide, ± round to oblong-ovate, base rounded, truncate or slightly lobed, margin entire, surfaces alike, dark green or slightly glaucous, dull, finely glandular-bristly, papillate, scabrous. **INFL** dense; branches 4–10, densely glandular; bracts 3–12 mm, lf-like, lanceolate, acuminate; pedicel 5–6 mm, finely glandular-bristly; immature axes 1–2 cm. **FL**: ovary finely glandular-bristly. **FR** 8–12 mm wide, finely glandular-bristly; stones separable or fused into a lobed unit. *n*=13. RARE. Sandy soils, chaparral; < 350 m. c CCo (Fort Ord), n SCoRO (Toro Mtn, nw Monterey Co.). In cult.

A. morroensis Wiesl. & B. Schreiber (p. 557) MORRO MANZANITA Shrub 1–4 m; burl 0. **STS** ± spreading; bark gray, persistent as flat shreads; twigs tomentose with long, white bristles. **LVS** ± overlapping; petiole 2–5 mm; blade 1.5–3 cm, 1–2 cm wide, oblong-ovate to oblong-elliptic, base lobed to ± truncate, margin cupped, entire or 1–2-toothed near base, surfaces not alike, upper convex, dark green, ± shiny, ± glabrous, lower gray-tomentose. **INFL** dense; branches 2–5; bracts 5–8 mm, lf-like, linear-lanceolate to narrowly oblong, acuminate; pedicel 4–6 mm, puberulent or becoming ± glabrous; immature axes 5–8 mm, partly framed by lower bracts. **FL**: ovary densely white-tomentose. **FR** 7–10 mm wide, ± puberulent. *n*=13. RARE. Sand dunes; < 200 m. s CCo (Morro Bay, San Luis Obispo Co.). In cult.

A. myrtifolia C. Parry (p. 557) IONE MANZANITA Shrub < 1.2 m; burl 0. **ST**: bark reddish, glaucous, smooth; twigs finely glandular-bristly. **LVS**: erect; petiole 1–3 mm; blade 6–15 mm, 3–8 mm wide, narrowly elliptic, base ± wedge-shaped, margin entire, finely glandular-ciliate, surfaces alike, bright green, shiny, sparsely finely glandular-bristly, papillate, rough. **INFL**: raceme, slender; bracts 1–2 mm, scale-like, deltate, acuminate; pedicel 1–3 mm, glabrous, recurved in fr; immature axis 5–10 mm, thread-like, bright red. **FL**: sepals, corolla lobes 5; ovary white-hairy. **FR** 3–4 mm, 2–2.5 mm wide, ± cylindric; ribs 5; pulp ± 0; stones 5. *n*=13. RARE. Acidic sandy or clay soils, chaparral, woodland; 100–300 m. c SNF (Amador, Calaveras cos.). Hybrids with *A. viscida* are called *A.* ×*helleri* Eastw. In cult.

A. nevadensis A. Gray (p. 557) Shrub < 0.6 m, mat- to mound-like; burl 0. **STS** spreading to decumbent, twigs finely tomentose. **LVS** ± erect; petiole 3–7 mm; blade 1–3 cm, 1–1.5 cm wide, obovate or oblanceolate, base ± wedge-shaped, tip ± obtuse, margin entire, fine ciliate, surfaces alike, bright green, shiny, ± puberulent, becoming glabrous, smooth. **INFL** spheric, dense; raceme, sometimes weakly 1-branched; axis, bract hairs like twig hairs or sometimes finely glandular-bristly; bracts 2–3 mm, scale-like, linear or linear-lanceolate, acuminate, sharp-pointed, lowest bract gen 5–10 mm, lf-like, linear-lanceolate; pedicel 3–5 mm, glabrous; immature axis 5–10 mm. **FL**: ovary glabrous. **FR** 6–8 mm wide, ± spheric, glabrous; stones ± separable. *n*=26. Rocky soils, coniferous forest; 900–3000 m. KR, NCoRH, CaRH, SNH; to WA. Possible hybrids with *A. glandulosa* in NCoRH are called *A. ×knightii* R. Gankin & W. Hildreth and *A. ×parvifolia* Howell. ❀DRN:1,2,4–6,15–17&IRR,SHD:3,7,14,18;DFCLT.

A. nissenana Merriam (p. 557) NISSENAN MANZANITA Shrub 0.6–1.5 m; burl 0. **ST**: bark gray, rough, persistent; twigs short-soft-hairy. **LVS** erect; petiole 1–3 mm; blade gen 1–2 cm, 0.8–1.5 cm wide, elliptic to oblong-elliptic, base wedge-shaped to rounded, margin entire, surfaces alike, ± glaucous, gray, dull, sparsely appressed-puberulent, becoming glabrous, smooth. **INFL**: raceme, slender; upper bracts 3–5 mm, gen awl-like; lower bracts 5–9 mm, lf-like, lanceolate, acute; pedicel 5–7 mm, glabrous or sparsely hairy; immature axis 2–5 mm, slender, framed by long bracts. **FL**: sepals, corolla lobes 5; ovary white-hairy. **FR** 3–4 mm, 2–3 mm wide, cylindric; ribs 5; pulp 0; stones 5. *n*=13. RARE. Open, rocky ridges, chaparral, woodland; 450–1100 m. n&c SNF (El Dorado Co. to Tuolumne Co.). In cult.

A. nortensis (P. Wells) P. Wells DEL NORTE MANZANITA Shrub 1–2 m; burl 0. **ST**: twigs tomentose, bristly or not. **LVS** ascending or erect; petiole 4–5 mm; blade 2–4 cm, 1–2 cm wide, narrowly elliptic-ovate, base truncate to rounded, margin entire, slightly cupped, surfaces ± alike, finely white-tomentose, becoming dark green, dull, glabrous, smooth. **INFL** dense; raceme, sometimes 1-branched; bracts 5–10 mm, lf-like, ± ovate, ± obtuse; pedicel 2–4 mm, puberulent; immature infl bell-shaped, axis 5–12 mm, ± concealed by recurved to reflexed bracts, peduncle recurved. **FL**: ovary densely white-hairy. **FR** 6–8 mm wide, sparsely hairy; stones thin. *n*=13. UNCOMMON. Rocky slopes, sometimes on serpentine, chaparral, coniferous forest; 500–800 m. nw KR (n Del Norte Co.). ❀IRR,DRN:**16**.

A. nummularia A. Gray (p. 557) Shrub 1–2 m; burl 0. **ST**: bark reddish, smooth; twigs densely fine-bristly, and with white, ± gland-tipped bristles. **LF**: petiole 1–3 mm; blade 1–2.2 cm, 0.8–1.8 cm wide, ± round to round-ovate, base truncate to ± lobed, margin ± fine toothed, cupped or ± rolled, finely ciliate, surfaces not alike, upper convex, dark green, shiny, lower light green, midvein bristly. **INFL** dense; branches 4–12; bracts 0.5–1 mm, scale-like, deltate, sharp-pointed; pedicel 3–5 mm, glabrous, recurved in fr; immature axes 5–10 mm, thread-like. **FL**: sepals, corolla lobes 4; corolla spheric; ovary white-hairy. **FR** 3–4 mm, 2–3 mm wide, cylindric; ribs 2–4; pulp ± 0; stones 2–4. *n*=13. Rocky sites, woodland, coniferous forest; < 600 m. NCo, NCoRO, w SnFrB (Mount Tamalpais, Santa Cruz Mtns). [var. *sensitiva* (Jepson) Munz] ❀DRN:5,**17**&IRR:**15,16**&SHD:14;CV.

A. obispoensis Eastw. BISHOP MANZANITA Shrub, tree-like, 1–4+ m; burl 0. **ST**: twigs finely tomentose. **LVS** erect; petiole 5–7 mm; blade 2–4.5 cm, 1–2.5 cm wide, oblong- to lanceolate-ovate, base rounded to truncate or ± lobed, margin entire, surfaces alike, glaucous, appressed-canescent, becoming glabrous, smooth. **INFL** dense; branches 2–4; bracts 7–14 mm, lf-like, linear-lanceolate, narrowly acuminate, rigid; pedicel 8–10 mm, glabrous, recurved in fr; immature infl bell-shaped, axes 10–25 mm, peduncle erect or ascending, curved. **FL**: ovary glabrous. **FR** 9–14 mm wide, glabrous, becoming glaucous. *n*=13. UNCOMMON. Rocky, gen serpentine soils, woodland, coniferous forest; 150–900 m. s SCoRO (s Santa Lucia Mtns). In cult.

A. osoensis P. Wells OSO MANZANITA Shrub 1–4+ m; burl 0. **STS** spreading; trunk, branch bark gray, rough, persistent as flat shreds; twigs short-tomentose. **LVS** strongly overlapping, clasping; petiole < 2 mm; blade 1.5–3 cm, 1.5–2.5 cm wide, ovate to round-ovate, base deeply lobed, ± obtuse, margin entire or sharply toothed, surfaces alike, dark green, ± shiny, sparsely short-hairy or glabrous, smooth. **INFL**: raceme or 1-branched; bracts 4–8 mm, lf-like, lanceolate to ± ovate, acute; pedicel 8–9 mm, glabrous, ± recurved in fr; immature axis 5–10 mm. **FL**: corolla ± 6 mm, throat abrupt; ovary glabrous. **FR** 5–8 mm wide, glabrous; stones separable, rounded. RARE. Chaparral, woodland; 300–500 m. s CCo (w Los Osos Valley, San Luis Obispo Co.). In cult.

A. otayensis Wiesl. & B. Schreiber (p. 557) OTAY MANZANITA Shrub 1–2.5 m; burl 0. **STS** erect; twigs finely tomentose, finely glandular-bristly. **LVS** erect; petiole 5–8 mm; blade gen 2–3 cm, 1–2 cm wide, narrowly elliptic or oblong-elliptic to ovate, base rounded or truncate, margin entire, surfaces alike, glaucous, finely appressed-tomentose, becoming ± glabrous, smooth. **INFL** ± open; branches 4–7; bracts 6–12 mm, lf-like, linear-lanceolate, acuminate; pedicel 3–8 mm, finely glandular-bristly; immature axes 15–25 mm, ± parallel, stiffly ascending, bracts slightly spreading, ± rigid. **FL**: ovary finely glandular-bristly. **FR** 6–8 mm wide, ± spheric, becoming glabrous; stones ± separable or fused into a lobed, ribbed unit. *n*=13. RARE. Volcanic rock outcrops, chaparral, woodland; 500–1700 m. sw PR (mtns near San Diego). In cult.

A. pajaroensis J. Adams PAJARO MANZANITA Shrub 1–4+ m; burl 0. **STS** erect; bark gray or reddish, rough, persistent; twigs tomentose, bristles, long, white. **LVS** strongly overlapping, clasping; petiole < 2 mm; blade 2–4 cm, 1–2 cm wide, ovate to triangular-ovate, base deeply lobed, margin entire to ± toothed, cupped or ± rolled, gen reddish, smooth, surfaces not alike, upper convex, dark green, ± glaucous, dull, lower ± light green, midvein bristly. **INFL** open; branches 2–5; axes, bracts finely bristly; bracts 5–10 mm, lf-like, ± linear lanceolate, acuminate; pedicel 5–8 mm, fine bristly; immature axes 10–15 mm, spreading. **FL**: ovary densely white-tomentose. **FR** 6–8 mm wide, short-hairy. *n*=13. RARE. Chaparral; < 200 m. n-c CCo, s SnFrB (Pajaro Hills). In cult.

A. pallida Eastw. (p. 557) PALLID MANZANITA Shrub 2–4+ m; burl 0. **STS** erect; twigs bristly. **LVS** strongly overlapping, clasping; petiole < 2 mm, bristly; blade 2.5–4.5 cm, 2–3 cm wide, ovate or oblong-ovate, base deeply lobed, margin entire or toothed, ± red, surfaces alike, glaucous, dull, glabrous, smooth. **INFL** dense; branches 3–5; bracts 5–9 mm, lf-like, widely lanceolate, acute; pedicel 8–12 mm, finely glandular-bristly; immature infl reclining on subtending lf, axes 5–10 mm, concealed by bracts. **FL**: ovary finely glandular-bristly. **FR** 8–10 mm wide, sticky. *n*=13. **ENDANGERED** CA. Siliceous shales, slopes, ridges; 200–350 m. e SnFrB (Sobrante, Huckleberry ridges, Berkeley-Oakland hills; Alameda and Contra Costa cos.). [*A. andersonii* var. *p.* (Eastw.) McMinn] In cult.

A. parryana Lemmon Shrub, forming circular clones, 1–2 m; burl 0. **STS** erect or lower spreading, rooting on ground; twigs finely tomentose or puberulent. **LVS** erect; petiole 5–10 mm; blade 1.5–5 cm, 1.5–2.5 cm wide, oblong-ovate to wide-elliptic, base obtuse to truncate, margin entire, surfaces alike, bright green or becoming ± glaucous, shiny, glabrous or sparsely puberulent, smooth. **INFL** dense; branches 3–5; bracts 2–4 mm, gen scale-like, deltate to wide-ovate, acuminate; lowest bract 5–10 mm, lf-like, ± linear; pedicel 4–9 mm, glabrous; immature axes 5–15 mm, stout, bracts ± appressed. **FL**: ovary glabrous. **FR** 8–10 mm wide, spheric to ovoid, ± abruptly soft-pointed, glabrous; stones fused into 1 ± spheric, vertically ribbed unit. *n*=26. Chaparral, coniferous forest; 1200–2350 m. n WTR, SnGb. ❀DRN:2,3,15,16,18.

A. patula E. Greene (p. 557) Shrub, forming circular clones, 1–2 m; burl gen 0 (exc c&s SNH). **STS** ± decumbent; lower branches rooting on ground; twigs glandular to finely glandular-bristly; glands golden, glistening. **LVS** erect; petiole 7–15 mm; blade 2.5–6 cm, 1.5–4 cm wide, widely-ovate to round, base rounded, truncate, or slightly lobed, tip ± rounded, margin entire, surfaces alike, bright green, shiny, glabrous, smooth. **INFL** open; branches 4–8; bracts 4–6 mm, gen scale-like, ± deltate, acuminate; lowest bract 1–2 cm, lf-like, linear-lanceolate; pedicel 2–7 mm, glabrous; immature axes 15–30 mm, spreading, bracts appressed or incurved. **FL**: ovary glabrous. **FR** 7–10 mm wide, glabrous, dark chestnut-brown. *n*=13. Open, coniferous forest; 750–3350 m. NCoRH, CaRH, SNH, SnGb, SnBr, SnJt; to WA, MT, Colorado, Baja CA. [*A. acutifolia* Eastw.; *A. parryana* Lemmon var. *pinetorum* (Rollins) Wiesl. & B. Schrieber; *A. pungens* Kunth var. *platyphylla* A. Gray] ❀DRN:1–3,7,15,16&IRR:14,18;DFCLT.

A. pechoensis Abrams PECHO MANZANITA Shrub, tree-like, 2–5+ m; burl 0. **ST**: twigs finely tomentose with long, white bristles. **LVS** strongly overlapping, clasping; petiole 1–4 mm; blade 2–4 cm, 1–2.5 cm wide, ± oblong-ovate, base deeply lobed, margin entire or toothed, ciliate near base, surfaces alike, ± glaucous, dull to ± shiny, puberulent, becoming glabrous, smooth. **INFL** dense; branches 1–4; bracts 8–15 mm, lf-like, ± linear-lanceolate, tip acuminate, rigid, margin ciliate; pedicel 5–10 mm, ± recurved in fr, glabrous; immature axes 10–15 mm, dense, framed by lower bracts. **FL**: corolla 6–8 mm, ± conic in outline; ovary glabrous. **FR** 8–12 mm wide, glabrous, becoming glaucous. *n*=13. RARE. Shale outcrops, chaparral, coniferous forest; < 850 m. s CCo (Pecho Hills, sw of San Luis Obispo). In cult.

A. pilosula Jepson & Wiesl. (p. 557) SANTA MARGARITA MANZANITA Shrub 1–2 m; burl 0. **ST**: twigs finely tomentose, sparsely bristly. **LVS** erect; petiole 5–8 mm; blade 1–3 cm, 1–2 cm wide, round to round-ovate, base truncate to slightly lobed, tip obtuse, margin entire, surfaces alike, glaucous, dull, smooth. **INFL** ± spheric; raceme, sometimes 1-branched; bracts 10–15 mm, lf-like, ± lanceolate, bristles < 5 per margin; pedicel 2–3 mm, glabrous; immature axis 1–2 cm, bracts stiffly ascending. **FL**: corolla 8–9 mm, conic in outline; ovary glabrous. **FR** 8–10 mm wide, red-brown. RARE. Shale outcrops, slopes, chaparral; 300–1100 m. s SCoRO (near Santa Margarita, San Luis Obispo Co.). In cult.

A. pringlei C. Parry ssp. ***drupacea*** (C. Parry) P. Wells Shrub, tree-like, 2–5+ m; burl 0. **ST**: twigs densely finely glandular-bristly, sticky. **LVS** erect; petiole 5–10 mm; blade 2–5 cm, 1–4 cm wide, elliptic, ovate or ± round, base rounded, truncate, or ± lobed, margin entire, papillate, ciliate, surfaces alike, glaucous, finely glandular-bristly, papillate, scabrous. **INFL**: raceme, sometimes 1- branched, large; bracts 6–10 mm, lf-like, lanceolate, acute to acuminate, deep pink; pedicel 5–10(15) mm, finely glandular-bristly; immature axis 10–15 mm, concealed by crowded bracts. **FL**: ovary finely glandular-bristly. **FR** 6–12 mm wide, spheric to ovoid, ± short-pointed, finely glandular-bristly, sticky; stones fused into a subspheric, vertically ribbed, pitted unit. Uncommon. Rocky slopes, open coniferous forest; 1200–2400 m. SnBr, PR. Ssp. *pringlei* in n AZ, Baja CA.

A. pumila Nutt. (p. 557) SANDMAT MANZANITA Shrub, mat- to mound-like, < 1.5 m; burl 0. **ST**: bark reddish, ± rough, persistent; twigs fine-gray-tomentose. **LVS** spreading; petiole 2–3 mm; blade 1–2 cm, 0.5–1.5 cm wide, narrow-obovate to oblanceolate, base wedge-shaped, tip ± obtuse, margin entire, cupped or slightly rolled, surfaces not alike, upper convex, dark green, ± shiny, sparsely puberulent, becoming glabrous, lower fine gray-tomentose. **INFL** dense; raceme or 1-branched; bracts 2–3 mm, scale-like, acute to ± sharp-pointed; lowest bract 5–10 mm, lf-like, lanceolate or oblanceolate; pedicel 3–4 mm, white-hairy, rarely glandular, sometimes becoming glabrous; immature axis 3–5 mm, slender. **FL**: ovary hairs like pedicel hairs. **FR** 5–6 mm wide, ± spheric, brownish, becoming glabrous. *n*=13. RARE. Sandy soils, hills, woodland, coniferous forest; < 200 m. c CCo (around Monterey Bay). In cult.

A. pungens Kunth Shrub, erect, 1–3 m; burl 0. **ST**: twigs tomentose. **LVS** erect; petiole 4–8 mm; blade 1.5–4 cm, 1–1.8 cm wide, elliptic to lanceolate-elliptic, base obtuse to wedge-shaped, sometimes rounded, margin entire, surfaces alike, bright or dark green, shiny, ± finely tomentose, becoming glabrous, smooth. **INFL** spheric; raceme, sometimes 1-branched; bracts 2–4 mm, scale-like, ovate-deltate, acuminate, tip sharp-pointed; pedicel 5–10 mm, glabrous; immature axis 5–15 mm, stout, club-like, bracts recurved, crowded near tip. **FL**: ovary glabrous. **FR** 5–8 mm wide, glabrous. *n*=13. Rocky slopes, ridges, chaparral, coniferous forest; 900–2250 m. SCoRI (Monterey, San Benito cos.), SnBr, PR, e DMtns (New York, Providence mtns); to UT, TX, Mex. ❀DRN, SUN,DRY:2,3,**7**,9,10,14–17,**18**,19–23.

A. purissima P. Wells LA PURISIMA MANZANITA Shrub 1–4+ m; burl 0. **STS** prostrate to erect; twigs densely fine-bristly and with long, white bristles. **LVS** strongly overlapping, clasping; petiole < 2 mm; blade 1–2.5 cm, 1–2 cm wide, round-ovate to ± round, base deeply lobed, margin entire, surfaces alike, bright green, ± shiny, glabrous, smooth. **INFL** spheric, dense; raceme, sometimes 1–2-branched; bracts 5–8 mm, lf-like, ovate to lanceolate-ovate; pedicel 3–5 mm, straight, glabrous; immature axes 5–10 mm, concealed by densely overlapping bracts. **FL**: corolla 4–6 mm, ± spheric; ovary glabrous. **FR** 5–8 mm wide, glabrous. *n*=13. RARE. Sandstone outcrops, sandy soils, chaparral; < 300 m. s SCoRO (w Santa Barbara Co.). In cult.

A. refugioensis R. Gankin REFUGIO MANZANITA Shrub 2–4+ m; burl 0. **STS** erect; twigs densely fine-bristly, and with long, ± gland-tipped bristles. **LVS** strongly overlapping, clasping; petiole < 2 mm; blade 3–4.5 cm, 2–3 cm wide, wide- to oblong-ovate, base deeply lobed, margin entire or toothed, red, surfaces alike, ± glaucous, dull, glabrous, smooth, or midrib ± fine bristly. **INFL** open; branches 5–10; bracts 4–10 mm, lf-like, ovate to wide-elliptic, tip obtuse; pedicel 7–9 mm, straight, finely glandular-bristly; immature axes 2–3 cm, spreading, bracts strongly overlapping. **FL**: corolla 8–10 mm, conic in outline; ovary glabrous. **FR** 10–15 mm wide, ovoid or spheric; top ± abruptly pointed, glabrous, reddish, shiny; stones fused into 1 unit. *n*=13. RARE. Sandstone outcrops, chaparral; 300–700 m. s SCoRO, w WTR (Santa Ynez Mtns). In cult.

A. regismontana Eastw. Shrub 2–4+ m; burl 0. **STS** erect; twigs densely glandular-bristly, sticky. **LVS** strongly overlapping, clasping; petiole < 3 mm; blade 3–6 cm, 2–3 cm wide, oblong-ovate, sway-backed, tip curved up, base deeply lobed, margin entire or ± sharply toothed, surfaces alike, pale green, dull, glandular-hairy, sticky, becoming glabrous, smooth or ± finely scabrous. **INFL** open; branches 4–6; bracts 5–12 mm, lf-like, narrowly oblong-lanceolate, long-tapered; pedicel 6–10 mm, finely glandular-bristly; immature axes 15–20 mm, straight, erect, clustered, dense. **FL**: corolla 6–9 mm, ± conic; ovary finely glandular-bristly. **FR** 6–8 mm wide, glandular-hairy, sticky. *n*=13. Uncommon. Granite, sandstone outcrops; ± 500 m. w SnFrB (n Santa Cruz Mtns).

A. rudis Jepson & Wiesl. (p. 557) SAND MESA MANZANITA Shrub 1–2 m; burl gen present. **ST**: bark grayish, rough, persistent as flat shreds; twigs tomentose. **LVS** erect; petiole 3–8 mm; blade 1–3 cm, 1–2 cm wide, ± elliptic, base ± wedge-shaped to rounded, margin entire, fine ciliate, surfaces alike, bright green, shiny, ± puberulent, becoming glabrous, smooth. **INFL** spheric, dense; raceme or 1-branched; bracts 2–6 mm, scale-like, deltate, ± keeled, acuminate, abruptly pointed, lowest bract ± lf-like, ± lanceolate; pedicel 3–6 mm, glabrous; immature axis 5–10 mm. **FL**: ovary glabrous. **FR** 8–14 mm wide, glabrous. *n*=13. RARE. Sandy soils, chaparral; < 100 m. s CCo (Nipomo, Burton mesas, Point Sal, sw San Luis Obispo, nw Santa Barbara cos.). In cult.

A. silvicola Jepson & Wiesl. (p. 563) BONNIE DOON MANZANITA Shrub, tree-like, 1–6 m; burl 0. **ST**: twigs finely tomentose or finely appressed-canescent. **LVS** erect; petiole 3–8 mm; blade 1.5–3.5 cm, 1–1.5 cm wide, narrowly obovate or oblong-elliptic, base ± wedge-shaped, tip gen obtuse, margin entire, surfaces alike, gray, dull, appressed-canescent, becoming glabrous, glaucous, smooth. **INFL**: raceme or 1-branched, dense; bracts 5–12 mm, lf-like, lanceolate, tip acute to obtuse; pedicel 5–7 mm, recurved in fr, hairy or glabrous; immature infl bell-shaped, axis 5–10 mm, framed by recurved bracts, peduncle ascending to recurved. **FL**: ovary glabrous. **FR** 6–12 mm wide, glabrous. RARE. Sandy soils, chaparral, coniferous forest; < 600 m. sw SnFrB (s Santa Cruz Mtns). In cult.

A. stanfordiana C. Parry Shrub 0.5–2 m; burl 0. **ST**: twigs glabrous or finely glandular-bristly, rarely puberulent. **LVS** erect; petiole 4–12 mm; blade 3–5 cm, 1.5–2.5 cm wide, gen elliptic, ± oblong, or oblanceolate, base wedge-shaped or ± obtuse, margin entire, surfaces alike, bright green to slightly glaucous, ± shiny, glabrous, smooth (finely glandular-puberulent, finely scabrous in ssp. *raichei*). **INFL** open; branches 3–5; bracts 1–1.5 mm, scale-like, deltate, acuminate; pedicel glabrous; immature axes 20–25 mm, slender, ascending or erect. **FL**: corolla ± deep pink; ovary glabrous. **FR** 6–8 mm wide, ± asymmetric, glabrous. Slopes, ridges, chaparral; 100–1300 m. c&s NCoRO.

ssp. ***decumbens*** (P. Wells) P. Wells RINCON MANZANITA **STS** decumbent; twigs, infl axes puberulent. **LF**: petiole 4–8 mm; blade bright green, shiny, glabrous. UNCOMMON. Open areas, chaparral; 100 m. s NCoRO (Rincon Ridge, near Santa Rosa, Sonoma Co.). [var. *repens* Roof]

ssp. ***raichei*** W. Knight (p. 563) RAICHE'S MANZANITA **STS** ± spreading or erect; twigs, infl axes, bracts finely glandular-bristly.

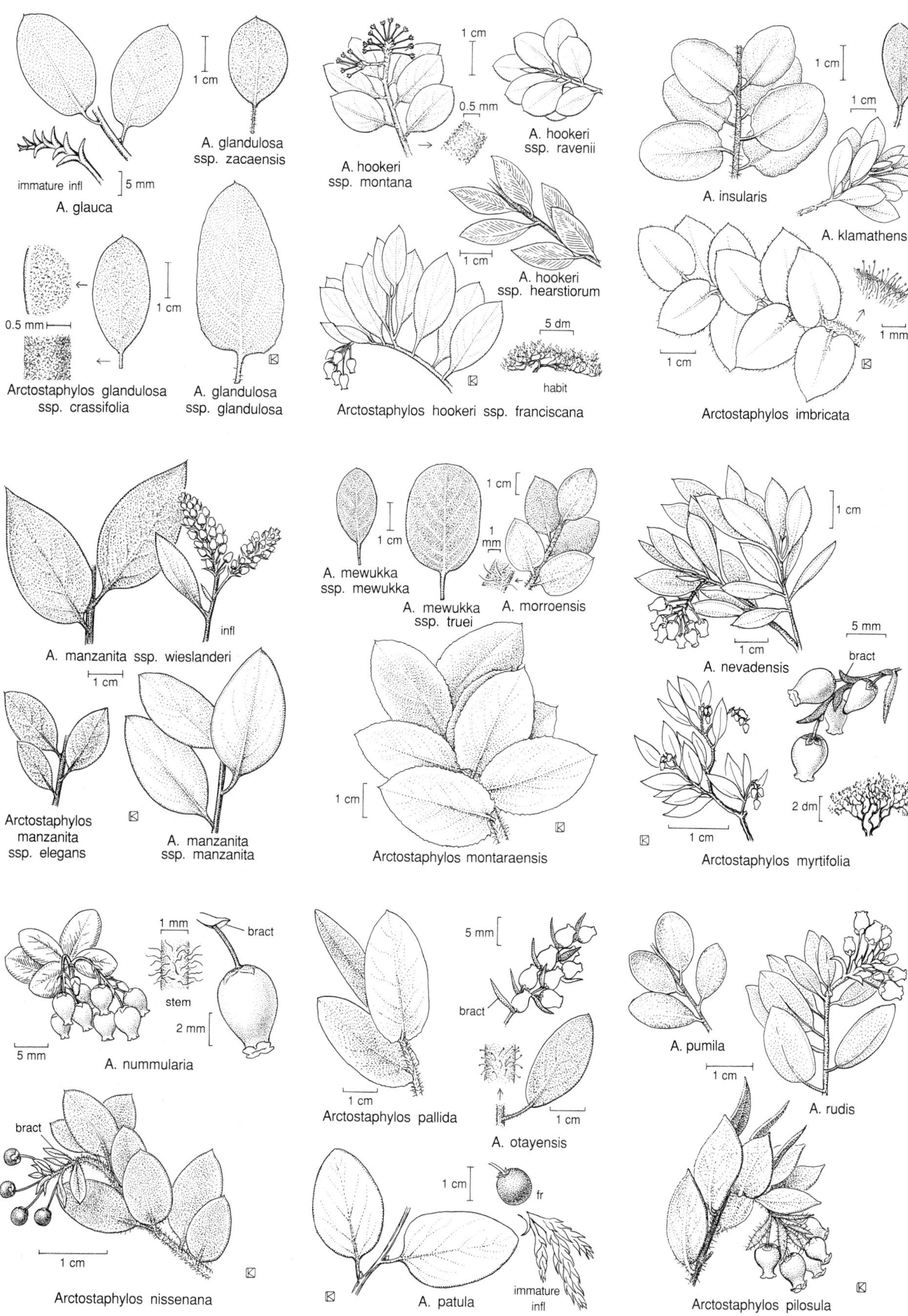
1 cm
immature infl
5 mm
A. glauca
A. glandulosa ssp. zacaensis
0.5 mm
Arctostaphylos glandulosa ssp. crassifolia
A. glandulosa ssp. glandulosa
1 cm
0.5 mm
A. hookeri ssp. montana
A. hookeri ssp. ravenii
A. hookeri ssp. hearstiorum
5 dm
habit
Arctostaphylos hookeri ssp. franciscana
A. insularis
A. klamathensis
1 mm
Arctostaphylos imbricata
infl
A. manzanita ssp. wieslanderi
Arctostaphylos manzanita ssp. elegans
A. manzanita ssp. manzanita
A. mewukka ssp. mewukka
A. mewukka ssp. truei
A. morroensis
Arctostaphylos montaraensis
A. nevadensis
5 mm
bract
2 dm
Arctostaphylos myrtifolia
stem
2 mm
A. nummularia
Arctostaphylos nissenana
Arctostaphylos pallida
A. otayensis
fr
immature infl
A. patula
A. pumila
A. rudis
Arctostaphylos pilosula

LF: petiole 4–8 mm; blade slightly glaucous, ± shiny, papillate, ± rough, margin, midvein sparsely fine-glandular-bristly. RARE. Rocky, ± serpentine soils, chaparral; 450–1000 m. s NCoRO (s Mendocino, w Lake cos.). In cult.

ssp. ***stanfordiana*** (p. 563) **STS** erect; twigs, infl axes, bracts glabrous, shiny, rarely puberulent. **LF**: petiole 8–12 mm; blade bright green, shiny, glabrous, smooth. *n*=13. Slopes, ridges, chaparral; 300–1300 m. c&s NCoRO. ❀DRN:**15**,16,17&IRR:7,14, 21–24 & afternoon SHD:18–20;DFCLT.

A. tomentosa (Pursh) Lindley Shrub 1–2.5 m; burl present, often wide, ± flat-topped. **STS** erect; bark smooth, reddish, or sometimes rough, gray; twigs gen tomentose, sometimes glandular or long-bristly. **LVS** ± spreading; petiole 2–5 mm; blade 2–5 cm, 1.5–2.5 cm wide, oblong-ovate to oblong-lanceolate, base truncate to ± lobed, margin entire, sometimes toothed, cupped or ± rolled, blade surfaces not alike, upper convex, dark to bright green, ± shiny, lower ± tomentose, sometimes finely glandular-bristly, papillate, scabrous or ± glabrous. **INFL**: branches 2–8, crowded; bracts 8–15 mm, lf-like, lanceolate; pedicel 2–5 mm, finely tomentose, sometimes finely glandular-bristly or glabrous; immature axes 10–25 mm, ± crowded. **FL**: ovary gen densely white-tomentose, sometimes glandular or glabrous. **FR** 6–10 mm wide, ± hairy. *n*=26. Rocky or sandy soils, slopes, chaparral, coniferous forest; < 1100 m. CCo, SnFrB, SCoRO, n ChI, WTR.

ssp. ***bracteosa*** (DC.) J. Adams (p. 563) **ST**: bark gray, rough, persistent as flat shreds; twigs densely glandular-bristly or tomentose. **LF**: lower blade surface sparsely glandular-bristly, papillate, scabrous or ± glabrous. **FL**: pedicel hairs like twig hairs; ovary white-tomentose, ± sparsely glandular. Uncommon. Open, shale outcrops; 200–300 m. c CCo (Jack's Peak, near Monterey). [var. *trichoclada* (DC.) Munz; var. *hebeclada* (DC.) McMinn] Pls with twigs tomentose, lvs ± glabrous are called f. *hebeclada* (DC.) P. Wells. Pls with twigs, lvs ± glandular-bristly are called f. *bracteosa*. Both forms, some intermediates found together. In cult.

ssp. ***crinita*** (McMinn) R. Gankin (p. 563) **ST**: bark red, smooth; twigs tomentose and with long, white bristles. **LF**: lower blade surface tomentose, smooth. **FL**: pedicel, ovary tomentose. *n*=26. Chaparral, forest; < 500 m. sw SnFrB (Santa Cruz Mtns). [var. *tomentosiformis* (J. Adams) Munz] ❀DRN,DRY,SUN:15–17;DFCLT.

ssp. ***crustacea*** (Eastw.) P. Wells **ST**: bark red, smooth; twigs sparsely puberulent, with long, white bristles. **LF**: lower blade surface ± glabrous, smooth. **FL**: pedicel, ovary tomentose. *n*=26. Chaparral; < 1100 m. CCo, SnFrB, SCoRO, WTR. [*A. c.* Eastw.; *A. glandulosa* var. *campbellae* (Eastw.) McMinn] ❀DRN,DRY,SUN: 3,7,14–24.

ssp. ***daciticola*** P. Wells DACITE MANZANITA **ST**: bark gray, rough, persistent as flat shreads; twigs short-tomentose and with long, white bristles. **LF**: lower blade surface tomentose, smooth. **FL**: pedicel, ovary tomentose. *n*=26. RARE. Chaparral; < 300 m. s CCo (w Los Osos Valley, San Luis Obispo Co.). In cult.

ssp. ***eastwoodiana*** P. Wells EASTWOOD'S MANZANITA **ST**: bark red, smooth; twigs tomentose. **LF**: blade surface ± glabrous, smooth. **FL**: pedicel, ovary ± glabrous. RARE. Sandy soils, mesas, chaparral; < 200 m. s CCo (nw Santa Barbara Co.). In cult.

ssp. ***insulicola*** P. Wells ISLAND LOVING MANZANITA **ST**: bark red, smooth; twigs tomentose, sometimes sparsely bristly. **LF**: lower blade surface tomentose, smooth. **FL**: pedicel, ovary tomentose. *n*=26. UNCOMMON. Chaparral, coniferous forest; 100–500 m. n ChI (Santa Cruz, Santa Rosa islands). ❀TRY.

ssp. ***rosei*** (Eastw.) P. Wells **ST**: lower bark reddish, ± rough, ± peeling; twigs ± tomentose. **LF**: lower blade surface ± glabrous, smooth. **FL**: pedicel, ovary tomentose. *n*=26. Uncommon. Gen sandy soils, chaparral; < 1100 m. n CCo (Lake Merced, San Francisco Co.), c CCo (Monterey Co.), nw SnFrB (Mount Tamalpais). [*A. crustacea* var. *r.* McMinn]

ssp. ***subcordata*** (Eastw.) P. Wells (p. 563) SUBCORDATE MANZANITA **ST**: bark red, smooth; twigs finely tomentose, densely glandular-bristly. **LF**: lower blade surface ± finely tomentose, sparsely fine-glandular-bristly, ± papillate, scabrous. **FL**: pedicel finely glandular-bristly; ovary white-tomentose, ± sparsely glandular. *n*=26. UNCOMMON. Chaparral; 100–600 m. n ChI (Santa Cruz, Santa Rosa islands). [*A. s.* Eastw.] ❀DRN,DRY,SUN:**15–17,22–24**&SHD:**19–21**.

ssp. ***tomentosa*** (p. 563) **ST**: bark gray, rough, persistent as flat shreds; twigs tomentose. **LF**: lower blade surface tomentose to finely so, smooth. **FL**: pedicel, ovary tomentose. *n*=26. Uncommon. Sandy soils, chaparral, coniferous forest; < 200 m. c&s CCo (Monterey, San Luis Obispo cos.).

A. uva-ursi (L.) Sprengel (p. 563) BEARBERRY Shrub, mat-like, rarely mound-like, < 0.5 m; burl gen 0. **ST** prostrate to decumbent; bark ± rough; twigs finely tomentose, becoming glabrous, or finely glandular-bristly. **LVS** spreading; petiole 2–4 mm; blade 1–2.5 cm, 0.5–1.5 cm wide, gen oblanceolate to obovate, base wedge-shaped, tip gen rounded or obtuse, margin cupped or ± rolled, entire, surfaces not alike, upper convex, dark green, shiny, lower lighter green, sparsely puberulent, becoming glabrous, smooth. **INFL** spheric, dense; raceme or 1-branched; bracts ± appressed, 2–6 mm, scale-like, narrowly deltate, narrowly acuminate; pedicel 2–4 mm, glabrous, rarely glandular; immature axis 3–10 mm, slender. **FL**: ovary glabrous. **FR** 6–12 mm wide, glabrous, bright red. 2*n*=26,39,52,65,78. Rocky outcrops, slopes, sandy soils, chaparral, coniferous forest; gen < 100 m, (2400–3200 m in c SNH). NCo, c SNH (above Convict Lake, Mono Co.), CCo, SnFrB; to AK, ne US, Rocky Mtns, Eurasia. Many taxa recognized at and below ssp. level incl: f. *adenotricha* (Fern. & J.F. Macbr.) P. Wells (twigs, infl axis finely glandular-bristly; c SNH); f. *coactilis* (Fern. & Macbr.) P. Wells (twigs, infl axis finely tomentose; NCo, CCo, SnFrB). Some hybrids with *A. glandulosa* from SnFrB gen have small burls; some have been called *A.* ×*pacifica* Roof. ❀DRN,SUN:**4,5**,6&IRR:1, **15–17**,24 & afternoon SHD:7,14,19–23;CVS.

A virgata Eastw. MARIN MANZANITA Shrub, tree-like, 2–5+ m; burl 0. **ST**: twigs densely fine-glandular-bristly, ± sticky. **LVS** ascending, overlapping; petiole 2–4 mm; blade 3–5 cm, 1–2.5 cm wide, narrowly oblong-ovate to oblong-lanceolate, base truncate to slightly lobed, margin entire or ± toothed at base, surfaces alike, bright green, ± shiny, sparsely glandular-hairy, papillate, scabrous. **INFL** dense; raceme or 1-branched; bracts 8–20 mm, lf-like, linear-lanceolate, acuminate; pedicel 3–8 mm, finely glandular-bristly; immature infl reflexed, axis 15–20 mm, ± concealed by long bracts. **FL**: ovary densely fine-glandular-bristly. **FR** 6–8 mm wide, fine-glandular-bristly, ± sticky. *n*=13. RARE. Sandstone, granite outcrops; < 500 m. n CCo, nw SnFrB (Marin Co.). In cult.

A. viridissima (Eastw.) McMinn (p. 563) MCMINN'S MANZANITA Shrub 1–4+ m; burl 0. **STS** erect; twigs finely tomentose, and with long, white bristles. **LVS** strongly overlapping, ± clasping; petiole 1–4 mm; blade 2–3.5 cm, 1.5–2.5 cm wide, narrowly ovate to oblong-ovate, base ± truncate to lobed, margin entire or toothed at base, surfaces alike, bright green, shiny, puberulent, becoming glabrous, smooth. **INFL** dense; raceme or 1-branched; bracts 6–10 mm, lf-like, ± lanceolate, acute; pedicel 2–3 mm, finely tomentose; immature axis 10–15 mm, ± concealed by crowded bracts. **FL**: ovary densely white-tomentose. **FR** 10–15 mm wide, sparsely hairy. *n*=13. UNCOMMON. Shale outcrops, chaparral, coniferous forest; 100–500 m. n ChI (e Santa Cruz Island). [*A. pechoensis* Dudley var. *v.* Eastw.] In cult.;CVS.

A. viscida C. Parry Shrub, tree-like, 1–5+ m; burl 0. **ST**: twigs glabrous or densely fine-glandular-bristly. **LVS** erect; petiole 5–12 mm; blade 2–5 cm, 2–4 cm wide, ± ovate to ± round, base rounded, truncate or ± lobed, margin entire or ± toothed, sometimes ciliate, surfaces alike, white-glaucous, dull, glabrous, smooth, or ± sparsely finely glandular-bristly, papillate, scabrous. **INFL** ± open; branches 4–7; bracts 3–4 mm, gen scale-like, deltate, acute to acuminate, lowest bract 8–10 mm, lf-like, linear-lanceolate; pedicel 6–10 mm, finely glandular-bristly; immature axes 1–3 cm, spreading, bracts overlapping, appressed. **FL**: ovary glabrous or finely glandular-bristly. **FR** 6–8 mm wide, ± spheric, red-brown, glabrous or finely glandular-bristly and sticky; stones separable. Rocky slopes, woodland, chaparral, coniferous forest; 150–1850 m. NW, CaRF, SN; sw OR.

ssp. ***mariposa*** (Dudley) P. Wells (p. 563) **ST**: twigs densely glandular-bristly, ± sticky. **LF**: blade sparsely fine-glandular-bristly, ciliate, papillate, scabrous. **INFL**: axis, bract hairs like twig hairs. **FL**: ovary finely glandular-bristly. **FR** finely glandular-bristly, sticky. n=13. Rocky slopes, woodland, chaparral, coniferous forest; 600–1850 m. SN. [*A. mariposa* Dudley] Hybrids with *A. patula* are called *A.* ×*jepsonii* Eastw. ❀DRN,DRY,SUN:1,**7**,**14**, 15–17.

ssp. ***pulchella*** (Howell) P. Wells **ST**: twigs gen glabrous, sometimes finely glandular near infl. **LF**: blade glabrous, entire, smooth. **INFL**: axes, bracts fine-glandular, sticky. **FL**: ovary finely glandular-bristly. **FR** finely glandular-bristly, sticky. Chaparral, gen serpentine; 200–1000 m. KR, NCoR; sw OR. ❀DRN,DRY, SUN:15,16.

ssp. ***viscida*** **ST**: twigs glabrous, smooth. **LF**: blade, petiole glabrous, entire, smooth. **INFL**: axes, bracts sticky. **FL**: ovary glabrous. **FR** glabrous, smooth. n=13. Chaparral, coniferous forest; 150–1500 m. KR, CaRF, SN. ❀DRN,DRY,SUN:1,6,**7**,**8**,**14**, 15–23.

A. wellsii W. Knight WELLS' MANZANITA Shrub, tree-like, 2–5+ m; burl 0. **ST**: twigs finely tomentose, densely long-bristly. **LVS** erect; petiole 4–7 mm; blade 2–4 cm, 1–2.5 cm wide, narrowly elliptic to ovate, base obtuse to wedge-shaped, tip acute, margin entire, red-tinged, surfaces alike, bright green or slightly glaucous, ± shiny, smooth. **INFL**: raceme or 1-branched; bracts 10–17 mm, lf-like, linear, acuminate, sharp-pointed, bristles 5–10 per margin; pedicel 3–5 mm, glabrous; immature axis 5–10 mm, bracts recurved, densely bristly. **FL**: corolla 6–8 mm, ± spheric; ovary glabrous. **FR** 6–8 mm wide, becoming light tan, obscurely striped vertically. n=13. UNCOMMON. Sandstone outcrops, chaparral; < 400 m. s CCo (hills se of San Luis Obispo). In cult.

CASSIOPE MOSS HEATHER

Shrub, small, glabrous to hairy. **ST** decumbent or prostrate, often rooting. **LVS** opposite, appressed, evergreen, leathery or thin. **INFL**: fls solitary in upper lf axils; bracts 0; bractlets 4–6; pedicels jointed to fl. **FL**: sepals 4–5, free; petals 4–5, ± 2/3 fused, gen white; stamens 10, anthers dehiscent by gaping pores, awns elongate; ovary superior, chambers 5, placentas near top. **FR**: capsule, loculicidal. **SEEDS** several per chamber. ± 14 spp.: s&e Asia, N.Am. (Greek: mother of Andromeda)

C. mertensiana (Bong.) Don (p. 563) WHITE HEATHER Pl low, densely branched. **ST** < 3 dm, glabrous or finely hairy. **LF** ± peltate, 2–5 mm, boat-shaped, elliptic, concave, leathery, glabrous; lower surface not grooved; margin entire, ciliate or minutely glandular, not rolled under. **INFL**: pedicels glabrous or hairy. **FL**: corolla widely bell-shaped, white, lobes 5; filaments glabrous. Moist, subalpine slopes, around rocks and areas of late snow; 1800–3505 m. KR, CaRH, SNH; to AK, w Can, MT. Pls have been assigned to ssp. *californica* Piper [lf 3–5 mm, margin minutely glandular-ciliate; s CaRH (Lassen Peak), SNH] or ssp. *ciliolata* Piper [lf 2–3 mm, margin with white, ephemeral hairs; KR, n CaRH (Mount Shasta)], but study needed. ❀IRR:1,2&SHD:6,7,14–17;SUN:4,5;DFCLT.

CHIMAPHILA PRINCE'S PINE, PIPSISSEWA

Erich Haber

Per, ± shrubby, evergreen, rhizomed. **LVS** gen cauline, ± whorled, lanceolate to oblanceolate, leathery, gen prominently toothed, petioled. **INFL**: ± head- or umbel-like raceme; fls 1–10; peduncle gen papillate to glandular-hairy; bracts narrowly lanceolate to widely ovate. **FL** radial, nodding, parts in 5's, free; petals spreading; stamens 10, filaments widened at base, ± hairy, anther pores on tubes; disk present; ovary superior, style in depressed center of ovary, stigma wide, peltate, lobes 5, flattened, spreading. **FR**: capsule, erect; valves opening from tip to base, margins not fibrous. 4–5 spp.: circumboreal, N.Am, C.Am, Eurasia. (Greek: winter loving, from evergreen habit)

1. Lf lanceolate to elliptic; infl 1–3-fld; sepals ± 5 mm, ovate; bracts widely ovate to obovate, persistent .... ***C. menziesii***

1′ Lf oblanceolate; infl 3–10-fld; sepals < 2 mm, widely ovate; bracts narrowly lanceolate, deciduous ... ***C. umbellata***

C. menziesii (D. Don) Sprengel (p. 563) LITTLE PRINCE'S PINE **ST** < 15 cm, slender. **LVS** 1–several per node, gen 1–3 (5) cm, toothed or entire; main veins ± white-bordered. **INFL** ± glabrous to minutely papillate. **FL**: petals white, turning pink; filament base entirely hairy. Uncommon. Montane conifer forest; 1000–2500 m. KR, NCoR, CaRH, SNH, SCoRO, SnBr, SnGb, PR; to B.C., MT.

C. umbellata (L.) Bartram (p. 563) PRINCE'S PINE **ST** < 30 cm, stout. **LVS** many per node, gen 3–7 cm, toothed, esp toward tip; veins not bordered. **INFL** ± densely glandular-hairy. **FL**: petals pink to red; filament base hairy on margins. $2n$=26. Common. Dry conifer forest; 300–2900 m. NCoR, SN, SnBr, SnJt; to e N.Am, C.Am, Eurasia. [var. *occidentalis* (Rydb.) S.F. Blake]

COMAROSTAPHYLIS

Shrub, small tree, gen hairy to glandular, densely, rigidly branched, burled. **ST**: bark often shredding. **LVS** alternate, evergreen, leathery, entire or serrate. **INFL**: raceme or panicle, bracted; bractlets 2. **FL**: sepals (4)5, fused, lobes > tube; petals (4) 5, fused, urn-shaped; stamens (8)10, anthers dehiscent by short separate slits, awned; ovary superior, chambers 4–6, placentas pendent, axile. **FR**: drupe, juicy, papillate, red or black; stones 4–6, fused into a unit. **SEEDS** 1 per stone. 10 spp.: subtrop, trop Am. (Greek: arbutus cluster, from strawberry-tree-like frs)

C. diversifolia (C. Parry) E. Greene **ST** erect, < 5 m; twigs gray-tomentose; bark shredding. **LF** obovate, entire or serrate. **INFL**: raceme, gen gray-tomentose; bracts < 7–10 mm, lance-linear to oblong-ovate. **FL**: sepals lanceolate or narrowly triangular. **FR** red. Chaparral; 100–550 m. SCo, ChI, WTR, PR; n Baja CA.

1. Lf margin clearly rolled under; bracts < 3 mm, lance-linear .... ssp. ***diversifolia***

1′ Lf margin not or inconspicuously rolled under; bracts gen 3–10 mm, oblong-ovate .... ssp. ***planifolia***

ssp. ***diversifolia*** (p. 563) SUMMER HOLLY **INFL** 3.5–8 cm. RARE. Chaparral; 100–550 m. SCo, PR; to n Baja CA.

ssp. ***planifolia*** (Jepson) G.D. Wallace (p. 563) **INFL** 6–14 cm. Uncommon. Chaparral; 100–600 m. n ChI (Santa Rosa, Santa Cruz islands), s ChI (Santa Catalina Island), WTR. [var. *planifolia* Jepson] ❀DRN:**14–17**,**24** & afternoon SHD:7,18–21,**22**,**23**.

ERICA HEATH

Shrub, glabrous to hairy. **ST** gen erect. **LVS** alternate, opposite, or gen whorled, evergreen, leathery; margin entire or minutely dentate, gen rolled under, hiding lower surface. **INFL**: gen raceme, panicle, or umbel, gen bracted; bractlets gen 3; pedicels not jointed to fl. **FL**: sepals gen 4, gen free; petals gen 4, fused, spheric to tubular, gen persistent in fr; stamens gen 8, anthers gen 2-lobed, dehiscent gen by pores or separate slits, gen awned; nectary gen disk-like; ovary superior, chambers 4 (8). **FR**: capsule, loculicidal, dehiscent tip to base. **SEEDS** gen several per chamber. ± 630 spp.: s&e Afr, Eur. (Latin: heath)

E. lusitanica Rudolphi **ST** < 2 m; branchlets short, hairs dense. **LF** < 1 cm, gen < 1 mm wide, needle-like, bright green. **INFL**: fls solitary in lf axils. **FL**: sepals fused in basal 1/3, petal-like; petals white to pinkish. **FR** > 4 mm. Disturbed, open, sandy areas; < 50 m. NCo (Humboldt Co.); native to sw Eur.

GAULTHERIA

Shrub, glabrous to hairy, gen rhizomed. **ST** erect to prostrate, rooting at nodes or not. **LVS** gen alternate, evergreen, leathery. **INFL**: raceme or fls solitary, bracted; bractlets 2 or more per pedicel; pedicels jointed to fl. **FL**: sepals gen 5, fused; petals 5, fused, urn- or rarely bell-shaped, gen white; stamens (5,8)10, anthers dehiscent by short separate slits, awned or not; ovary superior, chambers 5, placentas gen axile, at top of chamber. **FR**: capsule, loculicidal, ± enclosed by fleshy calyx. **SEEDS** few–many per chamber. ± 200 spp.: w Asia to Australia, temp Am. (J.-F. Gaulthier, botanist, physician, Quebec, 1708–1756)

1. Lf (3)5–10 cm, margin minutely serrate, glabrous; pl gen > 20 cm; st coarse; fls in racemes; corolla urn-shaped, glandular-hairy; filaments hairy; anther awns 4 .... ***G. shallon***

1′ Lf 1–3 cm, margin minutely serrate or ± entire, hairs few; pl gen < 20 cm, st slender; fls solitary in lf axils; corolla bell-shaped, glabrous; filaments glabrous, anther awns 0

2. St glabrous to puberulent, not glandular; lf minutely serrate or, esp toward base, ± entire; sepals glabrous .... ***G. humifusa***

2′ St hairs spreading, brownish, glandular and not; lf minutely serrate; sepals glandular-hairy .... ***G. ovatifolia***

G. humifusa (Graham) Rydb. (p. 563) ALPINE WINTERGREEN **ST** low, < 2 dm, glabrous to puberulent, not glandular. **LF** 1–2 cm; base rounded to truncate; tip rounded; margin minutely serrate or, esp near base, ± entire, hairs few. **INFL**: fls solitary in lf axils. **FL**: sepals glabrous; corolla bell-shaped, glabrous; filaments glabrous, anther awns 0. **FR** red. Wet subalpine forests; 1600–4000 m. KR, SNH; to B.C., Colorado. ❀IRR,DRN:4,5&SHD:1,2,6,7,14–17; acidic soil.

G. ovatifolia A. Gray **ST** low, < 3.5 dm; hairs spreading, brownish, glandular and not. **LF** 2–3 cm; base truncate to rounded or reniform; tip acute; margin minutely serrate, hairs few. **INFL**: fls solitary in lf axils. **FL**: sepals glandular-hairy; corolla bell-shaped, glabrous; filaments glabrous, anther awns 0. **FR** red. Wet fir forests; 400–1900 m. KR, n NCoRH, CaRH, n SNH; to B.C., ID. ❀IRR: **4,5** or WET,DRN:1,2&SHD:6,14–17;acidic soil.

G. shallon Pursh (p. 563) SALAL **STS** clumped, erect, < 2 m, glabrous to ± glandular. **LF** (3)5–10 cm; base truncate to reniform; tip acute, with a sharp point or not; margin minutely serrate, glabrous. **INFL**: raceme, glandular-hairy. **FL**: perianth glandular-hairy; corolla urn-shaped; filaments hairy, anther awns 4. **FR** dark purple. Moist forest margins; < 800 m. NCo, KR, NCoRO, CCo, SnFrB, s SCoRO; to AK. ❀**4,5**&IRR:**6**&SHD:**15–17**:&WET:7, 14,19–24; acidic soil;GRNCVR.

HEMITOMES GNOME PLANT

1 sp. (Greek: half eunuch, from 1 anther sac thought sterile)

H. congestum A. Gray (p. 563) Per, nongreen, fleshy, mostly glabrous, rhizomed; roots brittle. **ST** 0. **LF** scale-like. **INFL**: gen dense cyme, raceme, or fls solitary, 2–10 cm, gen pink, cream-colored, emerging from soil erect, not persisting after seed dispersal; bracts < 2 cm, margins ciliate; bractlets gen 0(1–2). **FL**: sepals (2)4, free, lateral 2 often folded, clasping corolla, other 2 flat if present; petals 4(5), ± 2/3 fused, cylindric to flask-shaped, cream-colored or gen pink, inside densely hairy; stamens gen 8, filaments densely hairy, anthers 1–2 mm, elongate, dehiscent by longitudinal, separate slits, unawned; nectary 8–10-lobed; ovary superior, chamber 1 or appearing > 1 by intrusion of parietal placentas, style < 5 mm, hairy, stigma 1.5–2.5 mm wide, disk-like, yellow, subtended by dense hairs. **FR**: berry, < 1 cm. **SEEDS** many, ovoid. Uncommon. Mixed or coniferous forests; 30–2700 m. NCo, KR, NCoRO, CaRH, n&s SNH, SnFrB, SCoRO; to B.C.

KALMIA LAUREL

Shrub, small tree, glabrous to hairy, rhizomed or not. **ST** prostrate to erect. **LVS** alternate, opposite, or whorled, gen evergreen, leathery; margins entire, rolled under or not. **INFL**: panicle, raceme, or fls solitary; bracts lf-like; bractlets 2; pedicels jointed to fl. **FL**: sepals 5, fused near base; petals 5, ± 1/2 fused, gen cup-shaped to rotate; stamens 10, filaments reflexed, anthers held in corolla pockets until dehiscing by separate slits, unawned; ovary superior, chambers 5, placentas axile. **FR**: capsule, septicidal, dehiscent tip to base. **SEEDS** many per chamber, ovoid or not, winged or not. 10 spp.: N.Am. (P. Kalm, student of Linnaeus, traveler in e N.Am., 1716–1779) [Southall & Hardin 1974 J Elisha Mitchell Sci Soc 90:1–23]

K. polifolia Wangenh. Pl gen rhizomed. **ST** ascending, 1–5 dm, glabrous or puberulent. **LF**: blade 4–60 mm, 3–25 mm wide, ovate to oblong or obovate, upper surface glabrous, lower surface densely canescent. **INFL**: fls solitary in upper lf axils. **FL**: petals 7–11 mm, pink to rose-purple. **FR** 4–5 mm wide. $2n=24,48$. Bogs, moist meadows, rock crevices; gen 1000–3500 m. KR, CaRH, SNH, Wrn; to AK, Colorado, e N.Am. Sspp. overlap in range and intergrade; more study needed to clarify status of ssp. *microphylla*, sometimes considered a sp. ❀DRN:4,5&IRR or WET:1,2,17& SHD:6,14–16; acidic soil;DFCLT.

1. St gen 1–3 dm; lf 4–20 mm, 3–6 mm wide; margin not or gen loosely rolled under; pedicels gen < 15 mm .... ssp. ***microphylla***

1' St gen 3–5 dm; lf 20–60 mm, 15–25 mm wide; margin gen tightly rolled under; pedicels gen > 20 mm .... ssp. ***polifolia***

ssp. ***microphylla*** (Hook.) Calder & Roy Taylor (p. 563) Pl gen rhizomed. $2n$=24. Habitats of sp.; 2000–3500 m. e KR, CaRH, SNH; to AK, Colorado.

ssp. ***polifolia*** (p. 563) Pl rhizomed or not. $2n$=48. Bogs, moist meadows; gen 1000–2000 m. KR, n SNH, Wrn; to AK, WY, e N.Am.

LEDUM LABRADOR TEA

Shrub, gen hairy. **ST** decumbent to erect, rooting. **LVS** alternate, reflexed in age, evergreen, leathery; margin entire, rolled under or not. **INFL**: raceme, ± flat-topped, terminal, bracted; bractlets 1–2, deciduous; pedicels not jointed to fl. **FL**: sepals 5, fused; petals 5, free exc sometimes at base, when corolla ± rotate; stamens 8–10, anthers dehiscent by pores, unawned; ovary superior, chambers 5, placentas axile. **FR**: capsule, septicidal, dehiscent base to tip. **SEEDS** many per chamber, fusiform, unwinged. 2–3 spp.: n hemisphere. (Greek: for pl now known as *Cistus*) [Kron & Judd 1990 Syst Bot 15:57–68 (where incl in *Rhododendron*)]

L. glandulosum Nutt. (p. 565) WESTERN LABRADOR TEA **ST** gen erect, < 1.5 m; bark smooth; twigs puberulent to glandular. **LF** 1–3.5 cm, oblong to elliptic; margin not or ± rolled under; lower surface finely hairy, with sessile, flat glands. **FL** cream-yellow to whitish. Common. Boggy areas; < 3600 m. NCo, KR, s NCoRO, CaRH, SNH, CCo, SnFrB, n SCoRI; to w Can, Colorado. [var. *californicum* (Kellogg) C. Hitchc.; ssp. *columbianum* (Piper) C. Hitchc. var. *australe* C. Hitchc.; ssp. *olivaceum* C. Hitchc.; *Rhododendron neoglandulosum* Harmaja] ❀WET or IRR,DRN:1,2,**4–6,17**&SHD:7,14–16; acidic soil;CV.

LEUCOTHOE

Shrub, small tree, coarsely branched, hairy to glabrous. **ST** erect. **LVS** alternate, evergreen or deciduous, serrate to entire. **INFL**: raceme, bracted; bractlets 2; pedicels jointed to fl. **FL**: sepals 5, fused at base; petals 5, fused, cylindric to urn-shaped; stamens 10, epipetalous, anthers dehiscent by pores, awns 0–4; ovary superior, chambers 5, placentas axile. **FR**: capsule, loculicidal, persistent after seed dispersal, dehiscent tip to base. **SEEDS** many per chamber, small, winged or not. ± 8 spp.: US, n C.Am, Japan, China. (Greek: named for a princess of Babylon)

L. davisiae Torrey (p. 565) SIERRA LAUREL **ST** < 1.5 m. **LF** 1–6 cm, oblong to elliptic, evergreen, leathery, glabrous. **INFL** in upper lf axils, < 15 cm, many-fld, bracted; bractlets basal; pedicels recurved. **FL**: corolla 6–8 mm, urn-shaped, >> calyx, white; anther awns 0 or rudimentary. **FR** < 6 mm wide, thin-walled. **SEEDS** winged. Uncommon. Bogs, wet areas; 1300–2600 m. KR, SNH, Wrn; OR. ❀WET or IRR:1,2,4,5&SHD:6,14–17; acidic soil; DFCLT.

MENZIESIA MOCK AZALEA

Shrub, open, glabrous to hairy or glandular. **ST** gen erect; bark finely shredding. **LVS** alternate, deciduous, elliptic, thin, gen hairy; margin rolled under in bud. **INFL** ± umbel-like; bud scales ± persistent; bracts scarious; bractlets gen 0. **FL** ± bilateral; sepals 4–5, ± 3/4 fused; petals 4–5, ± 2/3 fused, bell-shaped to cylindric or urn-shaped; stamens 5, 8, or 10, anthers dehiscent by short separate slits, unawned; ovary superior, chambers 4–5, placentas axile. **FR**: capsule, septicidal, dehiscent tip to base. **SEEDS** many per chamber, fusiform. 7 spp.: temp Asia, Am. (A. Menzies, naturalist on Vancouver expedition, 1754–1842)

M. ferruginea Smith (p. 565) **ST** sprawling or erect, < 4 m; twigs, pedicels glandular-hairy, less often short-hairy. **LF** 1.5–4 cm; tip with small point; margin, surfaces, esp upper, often glandular. **FL**: corolla 6–10 mm, cylindric to urn-shaped, yellowish green to pinkish or orange; stamens 8, glabrous. **FR** < 8 mm wide; valves thick. Occasional in coniferous woods; < 300 m. NCo, KR; to AK, MT. ❀WET,SHD:1,**4,5**,6,15–17; acidic soil;DFCLT.

MONESES WOODNYMPH

Erich Haber

1 sp. (Greek: single delight, from single fl)

M. uniflora (L.) A. Gray (p. 565) WOODNYMPH Per < 10 cm, evergreen, not rhizomed (but root rhizome-like). **LVS** ± basal, < 3.5 cm, ovate to obovate, not leathery, finely round- or sharp-toothed, petioled. **INFL**: fl solitary; peduncle minutely papillate above; bracts 1–2, ovate. **FL**: radial, nodding, parts in 5's, free; sepals fringed; petals < 1 cm, spreading, entire or minutely fringed, waxy-white to pinkish; stamens 10, filaments ± widened at base, glabrous, anther pores on tubes; disk 0; ovary superior, style straight, projecting beyond ovary, stigma ± peltate, lobes 5, prominent, erect, marginal. **FR**: capsule, erect; valves opening from tip to base, margins not fibrous. $2n$=26. Uncommon. Moist, mossy conifer forests; 100–1000 m. n NCo, w KR, n NCoRO, CaRH, c SNH (Fresno Co.); circumboreal, N.Am, Eurasia. [var. *reticulata* (Nutt.) S.F. Blake]

MONOTROPA

Per, nongreen, glabrous to glandular-hairy; roots brittle, main often elongate. **ST** 0. **LF** scale-like. **INFL**: raceme or fls solitary; emerging from soil nodding, erect in fr, persistent after seed dispersal, bracted; bractlets 1–2; pedicels gen recurved to spreading when anthers open, erect in fr, jointed to fl. **FL**: sepals gen 5, free; petals gen 5, free, oblong-cup-shaped, each ± bulged at base; stamens gen 10, anthers dehiscent by 1 or 2 slits, unawned; nectary lobes (8)10, ± clasping stamen bases; ovary superior, lines of dehiscence evident, chambers (4)5, placentas axile. **FR**: capsule, loculicidal, erect, dehiscent tip to base. **SEEDS** many per chamber, fusiform. 2 spp.: n hemisphere. (Greek: 1 direction, from 1-sided infl)

1. Infl raceme-like, fls rarely solitary; pl yellowish to pink or red; stigma shallowly depressed, often subtended by ring of bristly hairs; nectary lobes short, stout; fr segments thin-walled, often irregularly deciduous . . . ***M. hypopitys***
1′ Fls solitary; pl white, rarely pink or reddish orange; stigma widely funnel-like, not subtended by hairs; nectary lobes elongate, slender; fr segments thick-walled, persistent . . . ***M. uniflora***

M. hypopitys L. (p. 565) PINESAP **FL**: sepals ± unlike petals; style 1–2 mm wide. Uncommon. Mixed or coniferous forests; 120–2200 m. NCo, KR, NCoRO, NCoRH; to AK, e US and adjacent Can; also Mex, Eurasia. [*Hypopitys monotropa* Crantz; *H. lanuginosa* Michaux; *H. fimbriata* (A. Gray) Howell]

M. uniflora L. INDIAN PIPE **FL**: sepals ± like petals; style 2–5 mm wide. RARE in CA. Low mixed or coniferous forests; < 200 m. NCo, KR; to B.C., e N. Am; also C.Am, n S.Am, e Asia.

ORNITHOSTAPHYLOS

1 sp. (Greek: bird cluster, for obscure reasons)

O. oppositifolia (C. Parry) Small (p. 565) BAJA CALIFORNIA BIRDBUSH Shrub, glabrous to hairy, rigidly branched, burled. **ST** erect, < 2 m; bark thin. **LVS** opposite to whorled, 2.5–8 cm, 3–6 mm wide, ± linear, evergreen, leathery; margins rolled under. **INFL**: panicle, bracted; bractlets 2; pedicels not jointed to fl. **FL**: sepals (4)5, ± 1/2 fused; petals (4)5, ± 2/3 fused, ± spheric to urn-shaped, white; stamens gen 10, anthers dehiscent by separate gaping slits, awned; ovary superior, chambers 5, placentas axile. **FR**: drupe; stones 5, fused into a unit. **SEEDS** gen 10. RARE in CA. Chaparral; 100–800 m. SCo (San Diego Co.); n Baja CA. In cult.

ORTHILIA

Erich Haber

1 sp. (Greek: straight spiral, from 1-sided raceme) [Haber & Cruise 1974 Can J Bot 52:877–883] Traditionally placed in *Pyrola*.

O. secunda (L.) House (p. 565) ONE-SIDED WINTERGREEN Per, sometimes ± shrubby, < 20 cm, evergreen; rhizomed. **LVS** ± cauline, gen near base, 1.5–6 cm, ovate-elliptic, leathery or not, entire to finely round-toothed, petioled. **INFL**: raceme, 1-sided, arching, becoming erect in fr; peduncle densely papillate; bracts several, gen lanceolate. **FL** radial, ± closed, parts in 5's, free; petals with 2 basal tubercles on upper surface, greenish to cream-white; stamens 10, filaments ± narrow throughout, glabrous, anther pores not on tubes; disk present; ovary superior, style straight, exserted, stigma peltate, lobes 5, shallow, domed. **FR**: capsule, pendent; valves opening from base to tip, margins fibrous. $2n=38$. Dry, shady, conifer forests; 1000–3200 m. KR, NCoRH, s CaRH, SNH, SnBr, SnJt, Wrn; circumboreal, subarctic, N.Am, C.Am, Eurasia. [*Pyrola s.* L.]

PHYLLODOCE MOUNTAIN HEATHER

Shrub, gen matted, glabrous to glandular, gen rhizomed. **ST** decumbent, rooting, rough from persistent, decurrent lf bases. **LVS** alternate, crowded, < 5 mm wide, linear, needle-like, evergreen, leathery; margin strongly rolled under; lower surface channeled. **INFL**: fls solitary in axils of lf-like bracts, ± clustered near st tip; bractlets 2; pedicels jointed to fl. **FL**: sepals 5, fused at base; petals 5, ± 4/5 fused, urn- to cup-shaped; stamens gen 10, anthers elongate, dehiscent by short separate slits, unawned; ovary superior, chambers 5, placentas axile. **FR**: capsule, septicidal, dehiscent tip to base. **SEEDS** many per chamber, narrowly winged. 4–7 spp.: circumboreal. (Greek: a sea nymph; possibly "lf similar," from resemblance to *Erica*)

1. Stamens exserted; sepals, corolla lobes ± = corolla tube . . . ***P. breweri***
1′ Stamens incl; sepals, corolla lobes < corolla tube . . . ***P. empetriformis***

P. breweri (A. Gray) Maxim. (p. 565) **FL**: sepals < 4.5 mm; corolla cup-shaped, pink to rose-purple, tube < 4.5 mm, lobes < 3.5 mm; filaments > 2 × anthers. Moist rocky slopes, meadows, subalpine; 1200–3500 m. s CaRH (Magee Peak, nnw of Lassen Peak), SNH, SnBr. ❀IRR,DRN:1,2,**14–17**;acidic soils DFCLT.

P. empetriformis (Smith) D. Don (p. 565) **FL**: sepals < 2.5 mm; corolla cup-shaped, pink to rose-purple, tube < 5 mm, lobes < 2 mm; filaments < 2 × anthers. $2n=24$. Moist slopes, meadows, subalpine; 1450–2650 m. KR, n CaRH (Mount Shasta); to AK, MT, WY. ❀IRR:1,2,4,17&SHD:6,14–16; acidic soil;DFCLT.

PITYOPUS

1 sp. (Greek: pine foot, from habitat)

P. californicus (Eastw.) H. Copel. (p. 565) CALIFORNIA PINEFOOT Per, nongreen, fleshy; roots brittle. **ST** 0. **LF** scale-like. **INFL**: raceme or fls solitary, 1–10 cm, cream-colored to yellowish, emerging from soil; erect, not persistent after seed dispersal, bracted; bractlets 0. **FL**: sepals 4(5), free, lateral 2 often folded, clasping corolla, others flat against corolla; petals 4(5), free, cylindric, cream-colored to yellowish, outside ± glabrous, inside densely hairy; stamens gen 8, anthers erect, horseshoe-shaped, dehiscent by 1 unified slit, unawned; nectary lobes 8–10, among stamen bases; ovary superior, chamber 1 or appearing > 1 by intrusion of parietal placentas, style < 5 mm, stigma < 5 mm wide, ± funnel-shaped, yellowish, subtended by ring of hairs. **FR**: berry, < 1 cm. **SEEDS** many, ovoid. UNCOMMON. Mixed or coniferous forests; < 1800 m. NCo, KR, NCoRO, s SNH, CCo, SnFrB; OR.

PLEURICOSPORA

1 sp. (Greek: seeds at side, from parietal placentas)

P. fimbriolata A. Gray (p. 565) Per, nongreen, fleshy, glabrous; roots brittle. **ST** 0. **LF** 0. **INFL**: raceme, gen 6–10 cm, yellowish cream-colored, emerging from soil erect, not persistent after seed dispersal, bracted, drying brown or black at tips; bractlets 0. **FL**: sepals 4, free, 5–8 mm; petals 4(5), free, 8–10 mm, cylindric, glabrous, yellowish cream-colored, margins jagged; stamens gen 8, anthers 3–4 mm, elongate, dehiscent by long, separate lateral slits, unawned; ovary superior, chamber 1, placentas parietal, style <

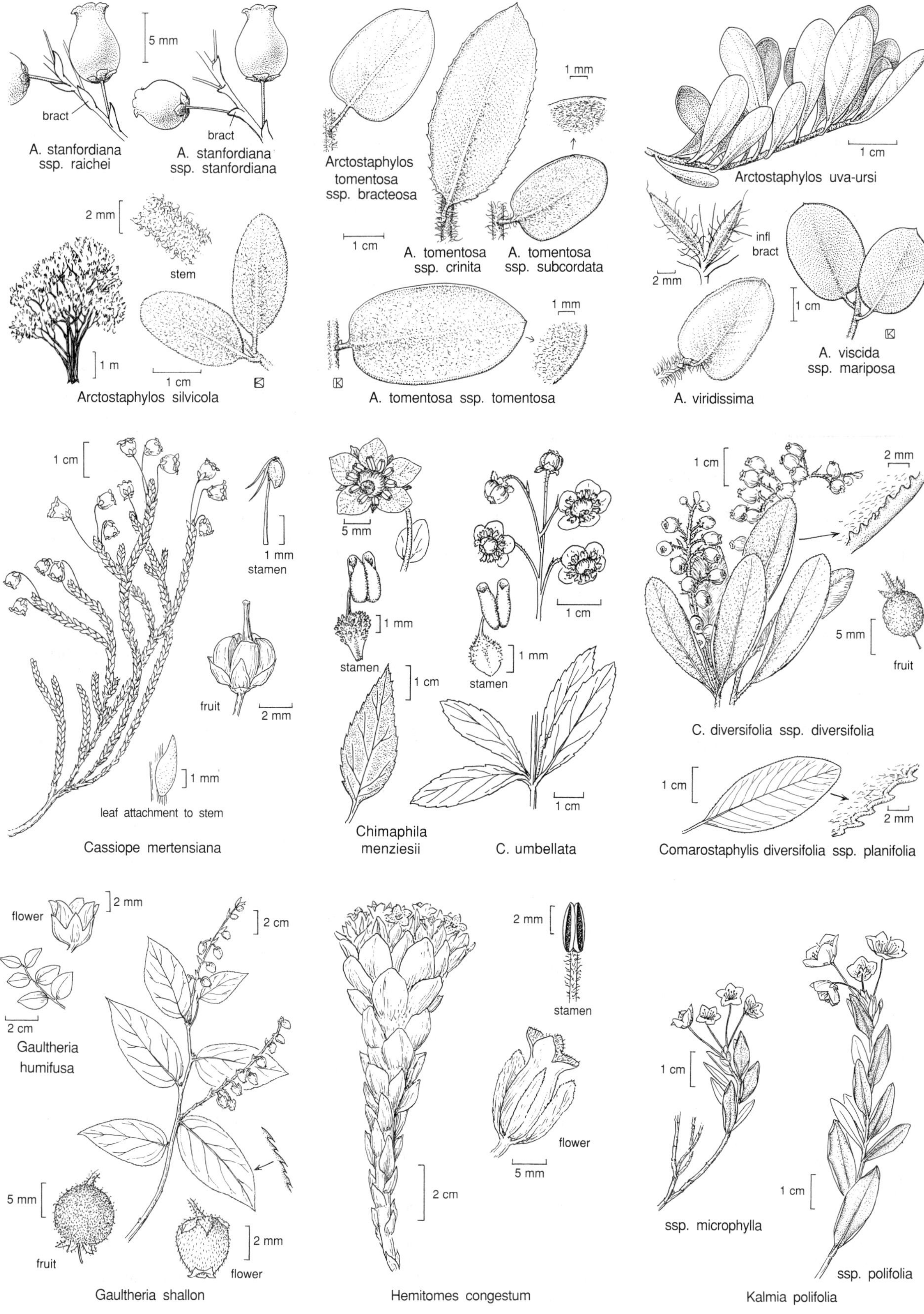

5 mm
bract
bract
A. stanfordiana ssp. raichei
A. stanfordiana ssp. stanfordiana
Arctostaphylos tomentosa ssp. bracteosa
1 mm
1 cm
A. tomentosa ssp. crinita
A. tomentosa ssp. subcordata
1 cm
Arctostaphylos uva-ursi
2 mm
stem
1 m
1 cm
Arctostaphylos silvicola
1 mm
A. tomentosa ssp. tomentosa
infl bract
2 mm
1 cm
A. viscida ssp. mariposa
A. viridissima
1 cm
1 mm
stamen
fruit
2 mm
1 mm
leaf attachment to stem
Cassiope mertensiana
5 mm
1 mm
stamen
1 cm
1 cm
1 mm
stamen
1 cm
Chimaphila menziesii
C. umbellata
1 cm
2 mm
5 mm
fruit
C. diversifolia ssp. diversifolia
1 cm
2 mm
Comarostaphylis diversifolia ssp. planifolia
flower
2 mm
2 cm
2 cm
Gaultheria humifusa
5 mm
fruit
2 mm
flower
Gaultheria shallon
2 mm
stamen
flower
5 mm
2 cm
Hemitomes congestum
1 cm
1 cm
ssp. microphylla
ssp. polifolia
Kalmia polifolia

3 mm, stigma 1.5– 2.5 mm wide, crown-like, not subtended by hairs. **FR**: berry, < 1 cm wide, cream-colored to whitish. **SEEDS** many, ovoid. 2*n*=52. Mixed or coniferous forests; 150–2800 m. NCo, KR, NCoRO, CaRH, SNH, SnFrB; to B.C..

PTEROSPORA PINEDROPS

1 sp. (Greek: winged seed)

P. andromedea Nutt. (p. 569) Per, nongreen, densely sticky-glandular; roots brittle. **ST** 0. **LF** scale-like. **INFL**: raceme, 1.5–17 dm, pink to reddish, emerging from soil erect, persistent after seed dispersal, bracted; axis < 1.5 cm wide below lowest fl; bractlets 0. **FL** pendent; sepals 5, free; petals 5, ± 4/5 fused, 6–9 mm, urn-shaped, cream-colored to yellowish, lobes recurved; stamens 10, anthers dehiscent by separate slits, awned; ovary superior, chambers 5, placentas axile, style < 3 mm, jointed to ovary, stigma gen 1.5 mm wide, disk-like. **FR**: capsule, loculicidal, dehiscent base to tip, pendent, < 1.3 cm wide. **SEEDS** many per chamber, < 0.2 mm wide, ovoid; wing terminal, < 1 mm wide, membranous. Mixed or coniferous forests; 60–3700 m. KR, NCoRH, CaRH, SNH, Teh, TR, SnJt, Wrn; to B.C., Mex; also e N.Am.

PYROLA WINTERGREEN, SHINLEAF

Erich Haber

Per, evergreen, rhizomed. **LVS** ± basal, reniform, ovate, ± round, elliptic, or obovate, ± entire to round- or sharp-toothed, petioled. **INFL**: raceme, symmetric; peduncle smooth, glabrous; bracts gen 1–several, ovate or lanceolate. **FL**: radial, ± closed or bilateral, ± open, parts in 5's, free; petals without tubercles, upper 2 gen forming hood over upturned stamens; stamens 10, filaments gen widened at base, smooth, glabrous, anther pores gen on tubes; disk 0; ovary superior, style straight, ± incl, or downwardly curved, exserted, stigma peltate, with 5 spreading lobes above a prominent, reflexed collar or not peltate, with 5 ± erect lobes projecting beyond a delicate, reflexed collar. **FR**: capsule, pendent; valves opening from base to tip, margins fibrous. ± 15–20 spp.: gen circumboreal, high mtns of C.Am, Sumatra. (Latin: little pear, ± from lf shape)

1. Style straight, ± incl; fl radial, ± closed; anthers < 1 mm, pores not on tubes; stigma peltate, lobes spreading (subg. *Amelia*) .... ***P. minor***
1′ Style downwardly curved, exserted; fl bilateral, ± open; anthers >> 2 mm, pores on tubes; stigma not peltate, lobes ± erect (subg. *Pyrola*)
  2. Fl bract gen >> pedicel, ovate; sepal ovate to gen lanceolate or oblong-lanceolate; pore tube < ± 1/5 × anther .... ***P. asarifolia***
    3. Fl bract < ± 1.5 × pedicel; sepal < 3.5 mm, ovate to gen lanceolate; lf ovate, round, or obovate, entire to ± round-toothed; anther 2–3 mm .... ssp. ***asarifolia***
    3′ Fl bract gen >> 2 × pedicel; sepal gen >> 3.5 mm, oblong-lanceolate; lf ovate to elliptic, gen sharp-toothed; anther 2.5–3.5 mm .... ssp. ***bracteata***
  2′ Fl bract < pedicel, lanceolate; sepal deltate to ovate; pore tube ± 1/3 × anther
    4. Lf gen < 4 cm, ovate-elliptic, veins not white-bordered; petals pale green; sepal < 1.8 mm, deltate-ovate, gen blunt; n SNH (Sierra Co.) .... ***P. chlorantha***
    4′ Lf gen >> 4 cm, ovate to elliptic, veins white-bordered (sometimes oblanceolate, veins not white-bordered); petals greenish, cream-white, or pink; sepal gen >> 2 mm, ovate, acute; n CA .... ***P. picta***

P. asarifolia Michaux **LF** < 10 cm; lower surface often purple. **INFL** < 6 dm incl peduncle; bracts gen >> pedicels, ovate. **FL** bilateral, ± open; sepals ovate to gen lanceolate or oblong-lanceolate, acute to acuminate; petals pink to deep red; anthers 2–3.5 mm, pore tubes < ± 1/5 as long; style downwardly curved, exserted, stigma lobes ± erect. Moist to dry forests, wetlands; 100–3000 m. NCo, KR, NCoRO, SNH, SnBr; to AK, e N.Am, e Asia. [Haber 1983 Syst Bot 8:277–298]

ssp. ***asarifolia*** (p. 569) BOG WINTERGREEN **LF** ovate, round, or obovate, entire to obscurely round-toothed. **INFL**: bracts < 1.5 × pedicels. **FL**: sepals 2–3.5 mm, ovate to gen lanceolate; anthers 2–3 mm. 2*n*=46. Common. Moist forests, swamps, bogs, stream banks; 1000–3000 m. KR, CaRH, SNH, SnBr, MP; to AK, e N.Am, e Asia. [var. *purpurea* (Bunge) Fern.; *P. californica* Krísa] ❀TRY.

ssp. ***bracteata*** (Hook.) E. Haber (p. 569) LONG-BRACTED WINTERGREEN **LF** ovate to elliptic, gen sharp-toothed. **INFL**: bracts gen >> 2 × pedicels. **FL**: sepals 3–5.8 (gen >> 3.5) mm, oblong-lanceolate; anthers 2.5–3.5 mm. Uncommon. Moist to dry forests; 100–2000 m. NCo, KR, NCoRO, n SNH; to s AK, MT. [*P. b.* Hook.] ❀TRY.

P. chlorantha Sw. GREEN-FLOWERED WINTERGREEN **LF** gen < 4 cm, ovate-elliptic, obovate, or reduced, bract-like; veins not white-bordered. **INFL** < 3 dm (incl peduncle); bracts < pedicels, narrowly lanceolate. **FL** bilateral, ± open; sepals gen < 1.8 mm, deltate-ovate, ± blunt; petals pale green; anthers 2–4 mm, pore tubes ± 1/3 as long; style downwardly curved, exserted, stigma lobes ± erect. 2*n*=46. Gen coniferous or mixed forest (no data from 1 CA record); ± 900 m. n SNH (near Downieville, Sierra Co.); circumboreal, N.Am, Eurasia. [*P. virens* Schreber] Possibly extirpated in CA.

P. minor L. (p. 569) LESSER WINTERGREEN **LF** < 5 cm, ovate to oblong-obovate, entire to obscurely round-toothed. **INFL** < 2 dm (incl peduncle); bracts >> pedicels, ovate-lanceolate, gen larger, wider at base of peduncle. **FL** ± radial, ± closed; sepals 1.5 mm, deltate, acute; petals white to pinkish; anthers < 1 mm, pores not on tubes; style straight, incl, stigma peltate, lobes spreading. 2*n*=46. Uncommon. Moist, mossy sites, high montane conifer forest; 2400– 3000 m. KR, c&s SNH, SnBr, SnJt, Wrn; circumboreal, to AK, e N.Am, Eurasia. [Haber 1984 Can J Bot 62:1054–1061]

P. picta Smith (p. 569) WHITE-VEINED WINTERGREEN Pl sometimes ± lfless. **LF** gen < 10 cm; blade ovate to elliptic, ± entire; upper surface dark green, lower surface often purple; veins white-bordered (lf blade sometimes oblanceolate, entire to prominently toothed; upper surface dull green, lower surface often bluish waxy when young, veins not bordered). **INFL** < 4 dm incl peduncle; bracts < pedicels, lanceolate. **FL** bilateral, ± open; sepals gen >> 2 mm, ovate, acute; petals greenish, cream-white, or pink; anthers 2.0–5.5 mm, pore tubes ± 1/3 as long; style downwardly curved, exserted, stigma lobes ± erect. 2*n*=46. Common. Dry ponderosa-pine forest; 400–2400 m. NW, CaR, SN, SnFrB, SCoRO, TR, PR, Wrn; to sw Can, NM. [f. *aphylla* (Smith) Camp; ssp. *dentata* (Smith) Piper] [Haber 1987 Syst Bot 12:324–335] Variable, may hybridize with other spp. (pollen, seeds sometimes abortive).

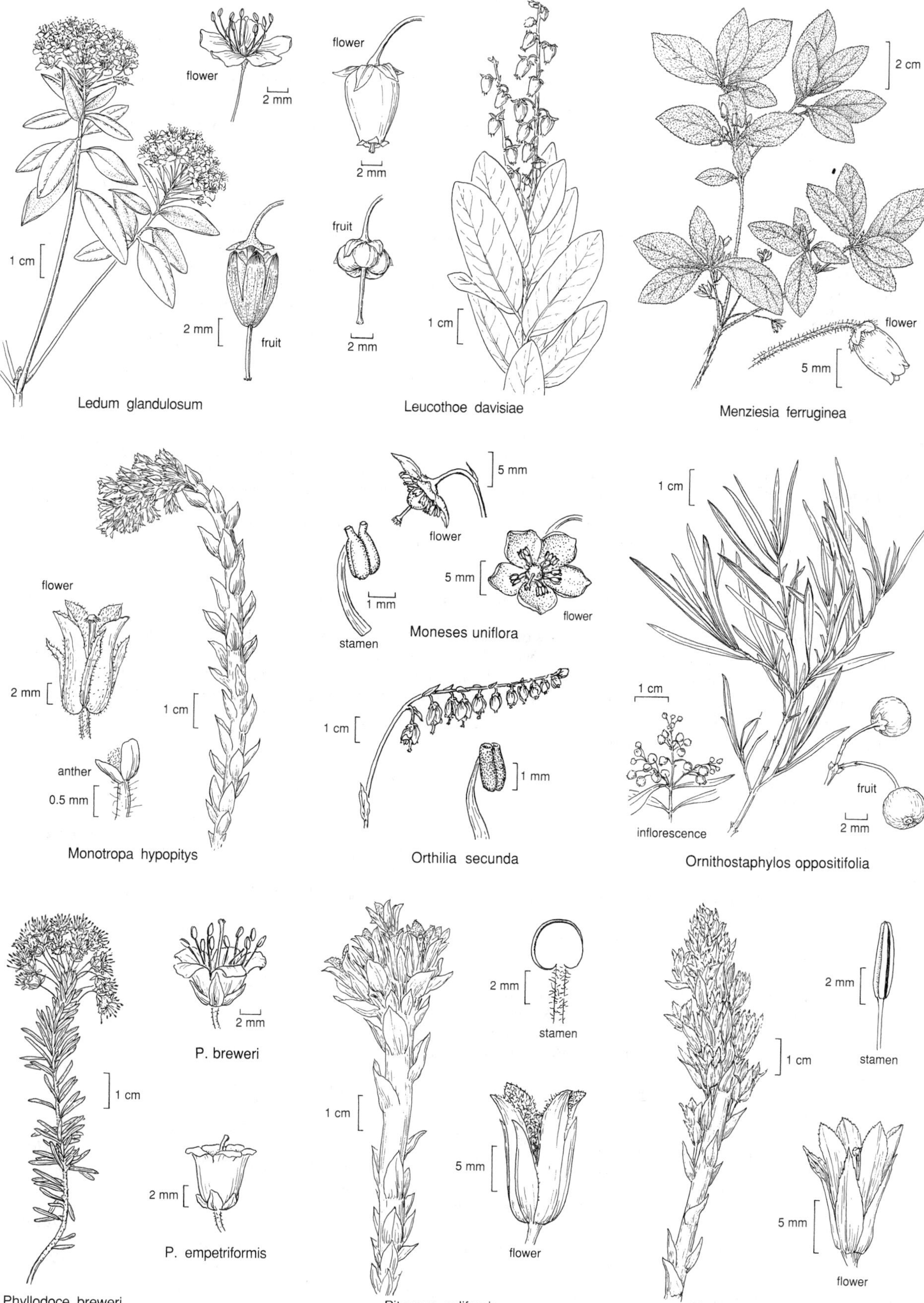
flower
2 mm
1 cm
2 mm
fruit
Ledum glandulosum
flower
2 mm
fruit
2 mm
1 cm
Leucothoe davisiae
2 cm
flower
5 mm
Menziesia ferruginea
flower
2 mm
1 cm
anther
0.5 mm
Monotropa hypopitys
5 mm
flower
stamen
1 mm
5 mm
flower
Moneses uniflora
1 cm
1 mm
Orthilia secunda
1 cm
1 cm
inflorescence
fruit
2 mm
Ornithostaphylos oppositifolia
2 mm
P. breweri
1 cm
2 mm
P. empetriformis
Phyllodoce breweri
2 mm
stamen
1 cm
5 mm
flower
Pityopus californicus
2 mm
1 cm
stamen
5 mm
flower
Pleuricospora fimbriolata

RHODODENDRON

Shrub, tree, glabrous to hairy. **ST** prostrate to erect; bark gen thin. **LVS** alternate, evergreen or deciduous; margin entire to serrate. **INFL** umbel-like; bud scales deciduous; bracts, bractlets scarious; pedicels not jointed to fl. **FL** radial to bilateral, gen showy; sepals gen 5, fused at base; petals gen 5, ± 3/4 fused, shallowly bell- to funnel-shaped; stamens 5–10(16), gen exserted, anthers elongate, each chamber dehiscent by terminal pore, unawned; ovary superior, chambers gen 5 (4–12), placentas axile. **FR**: capsule, septicidal, dehiscent tip to base. **SEEDS** many per chamber, fusiform or not; wing 0 or wide. ± 1000 spp.: temp n hemisphere, Australia. (Greek: rose tree)

1. Lvs evergreen, leathery; corolla rose to purple; stamens 10 .... ***R. macrophyllum***
1′ Lvs deciduous, thin; corolla white to pinkish, upper petal with yellow blotch at base; stamens 5 ..... ***R. occidentale***

R. macrophyllum D. Don (p. 569) CALIFORNIA ROSE-BAY **ST** coarsely branched, < 4 m; twigs stout. **LF** 7–15 cm, 3–6 cm wide, ovate to oblong-obovate to elliptic; upper surface dark green, midvein sunken; base gen wedge-shaped; margin not ciliate. **FL** < 4 cm. Moist coniferous forests; < 1100 m. NCo, KR, CCo, SnFrB; to B.C. ❀IRR,DRN:**4,5**,17&SHD:**6**,7,14–16; acidic soil.

R. occidentale (Torrey & A. Gray) A. Gray (p. 569) WESTERN AZALEA **ST** densely branched, < 5 m; twigs slender. **LF** 3–9 cm, 1–3 cm wide, elliptic to obovate; upper surface light green, midvein not sunken; base wedge-shaped; margin ciliate. **FL** < 5 cm. 2*n*=26. Streambanks, seeps, coniferous forests; < 2200 m. NCo, KR, NCoRO, NCoRI, CaRH, SNH, CCo, SnFrB, n SCoRI, PR, SnJt; OR. ❀IRR or WET,DRN:**4–6,15–17**&SHD:3,**7,14**,19–24;CVS.

SARCODES SNOW PLANT

1 sp. (Greek: flesh-like, from red infl)

S. sanguinea Torrey (p. 569) Per, nongreen, fleshy, glandular-hairy; roots thick, brittle. **ST** 0. **LF** scale-like. **INFL**: raceme, gen 1.5–3 dm, stout, bright red to orange-red, emerging from soil erect, persistent after seed dispersal; axis gen > 1 cm below lowest fl; bracts < 8 cm, margins long-ciliate; bractlets 0. **FL**: sepals 5, free; petals 5, ± 3/4 fused, 12–18 mm, urn-shaped, red; stamens 10, anthers < 4 mm, dehiscent by short separate slits, unawned; nectaries barely visible at ovary base; ovary superior, chambers 5, placentas axile, style < 8 mm, stigma 2–3 mm wide, head-like. **FR**: capsule, < 2.5 cm wide, indehiscent, brittle. **SEEDS** many per chamber, < 1 mm wide, ovoid, unwinged. 2*n*=64. Coniferous or mixed forests; 1000–3100 m. KR, NCoRO, NCoRH, CaRH, SNH, TR, SnJt; sw OR, n Baja CA.

VACCINIUM BLUEBERRY, HUCKLEBERRY, CRANBERRY

Shrub, tree, glabrous to hairy, rhizomed or not; burls gen 0. **ST** trailing to erect. **LVS** alternate. **INFL**: raceme or fls solitary, bracted; bud scales present; bractlets gen 2. **FL**: sepals 4–5, 2/3 to fully fused (lobes then 0); petals gen 4–5, ± 2/3 fused, cylindric to urn- or cup-shaped, gen white; stamens 8 or 10, filaments gen glabrous, anthers elongate, dehiscent by pores on small tubes, awned or not; ovary inferior, chambers 4–5 (or appearing 10 by intrusion of ovary wall), placentas axile, stigma head-like. **FR**: berry. **SEEDS** gen many. 400+ spp.: temp n hemisphere, trop mtns, Afr. (Latin: for *V. myrtillus* L.) [Vander Kloet 1988 The genus *Vaccinium* in N.Am]

1. Lvs evergreen (but see *V. parvifolium*), leathery, veins not prominent on lower side; pedicels jointed to fl; filaments ± hairy
  2. Pl 5–30 dm, branches stout; lf > 20 mm, serrate; corolla lobes < tube, erect to spreading when anthers open .... ***V. ovatum***
  2′ Pl < 1.5 dm, branches slender; lf gen 7–17 mm, ± entire; corolla lobes >> tube, reflexed when anthers open .... ***V. macrocarpon***
1′ Lvs deciduous, thin or ± thickened, veins gen prominent on lower side; pedicels not jointed to fl; filaments glabrous
  3. Lf entire; calyx lobes gen > tube, persistent; twigs not angled, not green; fls solitary or few on lfless older shoots .... ***V. uliginosum*** ssp. ***occidentale***
  3′ Lf serrate or minutely so (but see *V. parvifolium*); calyx lobes ± 0 or gen < tube, deciduous; twigs angled or not, green or not; fls often solitary in axils of lowest lvs of youngest shoots
    4. Twigs strongly angled, green; fr red (or dark purple when dry)
      5. Shrubby shrub, < 5 dm, rhizomed; lf serrate, lower surface glabrous .... ***V. scoparium***
      5′ Erect shrub, > 5 dm, not rhizomed; lf entire to serrate or with only a deciduous, sharp point at tip, lower surface, esp midvein, puberulent .... ***V. parvifolium***
    4′ Twigs not or weakly angled, not green but sometimes greenish or yellowish green; fr not red (but see *V. membranaceum*)
      6. Pl gen 5–15 dm; twigs weakly angled; lf gen ovate to elliptic or obovate, 2–5 cm, very thin, membranous, often rounded to truncate at base, acute at tip .... ***V. membranaceum***
      6′ Pl gen < 5 dm; twigs not or weakly angled; lf gen oblong or obovate to oblanceolate, rarely elliptic, gen 1–3.5 cm, gen thin but not membranous, tapered at base, seldom acute at tip
        7. Youngest twigs gen puberulent or glandular; lf not glaucous, gen oblong or obovate to elliptic; corolla narrowly urn-shaped; fr < 9 mm wide .... ***V. caespitosum***
        7′ Youngest twigs gen glabrous, glaucous; lf glaucous, obovate or oblanceolate, rarely elliptic; corolla ± spheric; fr > 9 mm wide .... ***V. deliciosum***

V. caespitosum Michaux (p. 569) DWARF BILBERRY Shrub, gen hairy, rhizomed. **ST** prostrate to erect, < 5 dm, gen rooting; twigs not or weakly angled, greenish, youngest gen puberulent or glandular. **LVS** deciduous, gen 1–3 cm; blade gen oblong or obovate to elliptic, gen thin, not membranous, minutely serrate, base tapered, tip seldom acute, lower surface gen glandular. **INFL**: fls solitary in axils of lowest lvs of youngest shoots; bractlets 0; pedicels not jointed to fl. **FL**: calyx lobes ± 0; corolla < 6 mm, whitish to pinkish, narrowly urn-shaped; anthers awned. **FR** < 9 mm wide, gen blue-glaucous. 2*n*=24. Margins of wet meadows, mtn slopes; < 3400 m. NCo, KR, NCoRO, NCoRH, CaRH, SNH, CCo, SnFrB, Wrn; to AK, MT, also ne US and adjacent Can. [*V. arbuscula* (A. Gray) Merriam; *V. nivictum* Camp] ❀WET or IRR:1,2,**4**,**5**,17& SHD:**6**,14–16; acidic soil;DFCLT;GRNCVR.

V. deliciosum Piper (p. 569) CASCADE BILBERRY Shrub, glabrous, rhizomed. **ST** matted, < 4 dm, gen rooting; twigs weakly or gen not angled, youngest gen glabrous, glaucous. **LVS** deciduous, gen 1.5–3.5 cm; blade obovate or oblanceolate, rarely elliptic, glaucous, gen thin, not membranous, gen serrate in upper 2/3, base tapered, tip seldom acute, lower surface not glandular. **INFL**: fls solitary in lf axils; pedicels not jointed to fl. **FL**: calyx lobes ± 0; corolla < 6 mm, ± spheric, pinkish; anthers awned. **FR** > 9 mm, gen blue-glaucous. 2*n*=48. Alpine meadows, subalpine coniferous woods, near coast; 600–2000 m. KR, n SNH, Wrn; to B.C.

V. macrocarpon Aiton CRANBERRY Shrub, glabrous to hairy, rhizomed. **STS** prostrate and erect, < 1.5 dm, slender, rooting; twigs not angled, brownish. **LVS** evergreen, gen 7–17 mm; blade narrowly elliptic to oblong, leathery, entire, lower surface glaucous. **INFL**: fls solitary in axils of reduced lowest lvs of youngest shoots; bractlets 2, lf-like; pedicels recurved, jointed to fl. **FL**: sepals 4, fused; petals 4, fused at base, lobes >> tube, reflexed when anthers open; filaments ± hairy, anthers unawned. **FR** 9–14 mm diam, red. 2*n*= 24. Boggy soil at abandoned placer mine; ± 900 m. n SNH (near North Columbia, n Nevada Co.); native to e N.Am.

V. membranaceum Hook. (p. 569) THINLEAF HUCKLEBERRY Shrub, ± glabrous; rhizome gen 0. **ST** erect, gen 5–15 dm, not rooting; twigs weakly angled, yellowish green. **LVS** deciduous, 2–5 cm; blade gen ovate to elliptic or obovate, very thin, membranous, serrate, teeth gen with a gland-tipped hair, base often rounded to truncate, tip acute, lower surface with prominent veins. **INFL**: fls solitary in axils of lowest lvs of youngest shoots; bractlets 2, scale-like; pedicels not jointed to fl. **FL**: calyx lobes ± 0; corolla < 6 mm, cylindric to urn-shaped, pinkish; anthers awned. **FR** 9–11 mm diam, gen black (dark red). Wet meadows, mtn slopes; 1100–2200 m. KR, NCoRH, n SNH, Wrn; to AK, SD. Study needed of *V. coccineum* Piper (Siskiyou Mountains huckleberry), a sporadic red-fruited form not treated by Vander Kloet. ❀WET:1,2,4–6; acidic soil;DFCLT.

V. ovatum Pursh (p. 569) CALIFORNIA HUCKLEBERRY Shrub, hairy; rhizome 0. **ST** erect, 0.5–3 m, stout, gen not rooting; twigs not angled, grayish. **LVS** evergreen, 2–5 cm; blade elliptic to lanceolate, leathery, serrate, lower surface with sparse, dark glandular hairs, veins not prominent. **INFL** umbel-like, axillary, dense; bractlets 2; pedicels jointed to fl. **FL**: sepals 5, fused at base, lobes deltate; corolla < 8 mm, urn-shaped, lobes < tube, erect to spreading when anthers open; filaments ± hairy, anthers often short-awned. **FR** 6–9 mm, black, sometimes glaucous. 2*n*=24. Edges, clearings in coniferous woods; 3–800 m. NCo, KR, NCoRO, CCo, SnFrB, n ChI (Santa Cruz, Santa Rosa islands), WTR, PR (uncommon); to B.C. [var. *saporosum* Jepson] ❀DRN:**4**,**5**&IRR:**6**,**17**&SHD:7, 14,**15**,**16**,22–24; acidic soil;CVS.

V. parvifolium Smith (p. 569) RED HUCKLEBERRY Shrub, glabrous to puberulent; rhizome 0. **ST** erect 1–4 m, gen not rooting; twigs strongly angled, green. **LVS** deciduous, rarely evergreen in young pls, 10–25 mm; blade elliptic to ovate, thin, entire to serrate or with only a deciduous, sharp point at tip, lower surface (esp midvein) puberulent, veins prominent. **INFL**: fls solitary in axils of lowest lvs of youngest shoots; bractlets 0; pedicels not jointed to fl. **FL**: calyx lobes ± 0 or < tube, rounded, deciduous; corolla < 5 mm, greenish or pinkish; anthers awned. **FR** 6–10 mm diam, bright red. 2*n*=24. Moist, shaded woods; 3–1400 m. NCo, KR, NCoRO, CaRH, n&c SNH, SnFrB; to se AK. ❀SHD,IRR:**4**–**6**,14–17; acidic soil.

V. scoparium Cov. LITTLELEAF HUCKLEBERRY Shrub, shrubby, glabrous, rhizomed. **ST** gen erect, < 5 dm, rooting; twigs strongly angled, green. **LVS** deciduous, 8–15 mm, ovate to elliptic, serrate; lower surface glabrous. **INFL**: fls solitary in axils of lowest lvs of youngest shoots; bractlets 0; pedicels not jointed to fl. **FL**: calyx lobes ± 0 or < tube, rounded, deciduous; corolla < 4 mm, urn-shaped, pink; anthers awned. **FR** 3–6 mm diam, red. RARE in CA. Rocky subalpine woods; 1800–2200 m. KR; to w Can, SD, UT.

V. uliginosum L. ssp. ***occidentale*** (A. Gray) Hultén (p. 569) WESTERN BLUEBERRY Shrub, glabrous, rhizomed or not. **ST** erect, < 6 dm, or prostrate, gen rooting; twigs not angled, not green. **LVS** deciduous, 1–2 cm; blade elliptic to ovate, ± thick, entire, glaucous or not, lower surface visibly but not prominently veined. **INFL**: gen raceme; fls 1–4; bractlets 0; pedicels not jointed to fl. **FL**: calyx lobes 4–5, > tube, triangular, persistent; anthers awned. **FR** < 6 mm diam, blue-black, glaucous. Bogs, wet meadows; < 3400 m. NCo, KR, CaRH, SNH; to B.C., MT, NV. [*V. o.* A. Gray] Circumboreal complex; needs study. Some NCo pls with prominently veined lvs and fr > 6 mm diam may be ssp. *u.*, which is characteristic of lowland n Eur. ❀WET,DRN:14–17.

XYLOCOCCUS

1 sp. (Greek: wood berry, from stone of fr)

X. bicolor Nutt. (p. 569) Shrub, burled. **ST** erect, < 2.5 m; bark shredding; twigs canescent. **LVS** gen alternate, evergreen, 2.5–4.5 cm, elliptic to oblong, leathery; margin entire, rolled under; upper surface glabrous, dark green; lower surface densely white- to gray-hairy. **INFL**: panicle, ± reflexed, dense, bracted; bractlets 2; pedicels not jointed to fl. **FL**: sepals gen 5; petals 5, fused, < 9 mm, urn-shaped, white or pinkish, hairy; stamens gen 10, filaments woolly in lower 1/2, anthers dehiscent by separate slits, awned; ovary superior, hairy, chambers gen 5, placentas axile. **FR**: drupe, < 9 mm wide, smooth; stones gen 5, fused into a smooth, 3–5-seeded unit. 2*n*=26. Chaparral; < 650 m. SCo, s ChI (Santa Catalina Island); n Baja CA. ❀SUN,DRN:14–17,19–24.

EUPHORBIACEAE SPURGE FAMILY

Grady L. Webster, except as specified

Ann, per, shrub, tree, vine, monoecious or dioecious. **ST** gen branched, sometimes fleshy or spiny. **LVS** gen simple, alternate or opposite, gen stipuled, petioled; blade entire, toothed, or palmately lobed. **INFL**: cyme, panicle, raceme, spike; fls sometimes in clusters (dense, enclosed by involucre, fl-like in *Chamaesyce, Euphorbia*), terminal or axillary. **FL** unisexual, ± radial; sepals gen 3–5, free or fused; petals gen 0; stamens 1–many, free or filaments fused; ovary superior, chambers 1–4, styles free or fused, simple or lobed. **FR**: gen capsule. **SEEDS** 1–2 per chamber; seed scar appendage sometimes present, pad- to dome-like. 300 genera, 7500 spp.: ± worldwide esp trop; some cult (*Aleurites*, tung oil; *Euphorbia* ssp.; *Hevea*, rubber; *Ricinus*). [Webster 1967 J Arnold Arbor 48:303–430] Many spp. ± highly TOXIC.

1. Lvs opposite or 3-whorled
 2. Sts prostrate or erect, sap milky; lf base oblique; infl dense, enclosed by involucre, ± fl-like, bisexual .... CHAMAESYCE

2′ Sts erect, sap clear; lf base not oblique; infl ± open, not enclosed by involucre, unisexual, not fl-like
3. Ann; staminate infl spike-like, axillary; ovary 2-chambered, bristly .... **MERCURIALIS**
3′ Shrub; staminate infl raceme or panicle, axillary or clustered at tips of short, lateral twigs; ovary 4–5-chambered, smooth or tomentose .... [2]**TETRACOCCUS**
1′ Lvs alternate (sometimes opposite in *Euphorbia*)
4. Infl dense, enclosed by involucre, ± fl-like, bisexual .... **EUPHORBIA**
4′ Infl ± open, not enclosed by involucre, not fl-like, unisexual
5. Lvs palmately lobed, 1–5 dm wide; infl panicle, terminal .... **RICINUS**
5′ Lvs pinnately veined, entire or toothed, << 1 dm wide; infl various
6. St and lf hairs 2-branched or stellate
7. Hairs 0 or 2-branched; petals present; filaments fused; seeds pitted .... **DITAXIS**
7′ Hairs stellate; petals 0; filaments free; seeds smooth
8. Ovary 1-chambered; lvs prominently 3-veined, veins raised .... **EREMOCARPUS**
8′ Ovary 3-chambered; lvs pinnately veined, not raised
9. Lf margin bluntly toothed; stipules persistent; staminate infl axillary .... **BERNARDIA**
9′ Lf margin entire; stipules inconspicuous; staminate infl terminal .... **CROTON**
6′ St and lf hairs simple
10. Pistillate bract toothed, > fr .... **ACALYPHA**
10′ Pistillate bract entire, < fr
11. Shrub; staminate infl cyme; ovules 2 per chamber .... [2]**TETRACOCCUS**
11′ Per; staminate infl spike or raceme; ovule 1 per chamber
12. Lvs glabrous; stamens 2; base of pistillate fl bracts glandular; st sap milky .... **STILLINGIA**
12′ Lvs with stinging, nettle-like hairs; stamens 3–6; base of pistillate bracts not glandular; st sap clear .... **TRAGIA**

ACALYPHA

Ann, per, shrub, < 2 m, mostly monoecious; sap clear. **STS**: central erect, gen much-branched; lateral spreading to ascending. **LVS** simple, cauline, alternate, stipuled; hairs simple, sometimes glandular. **INFL**: spike, terminal or axillary; staminate bracts minute; pistillate bracts lf-like, toothed. **STAMINATE FL**: sepals 4; petals 0; stamens 4–8, filaments free or fused at base; nectary disk 0. **PISTILLATE FL**: sepals 3(–5); petals 0; nectary 0; ovary 3-chambered, styles 3, deeply cut. **FR** ± spheric, smooth or ± lobed. **SEEDS** 1 per chamber, smooth to pitted; scar appendage minute. ± 400 spp.: trop, warm temp worldwide. (Greek: ancient name for a kind of nettle)

A. californica Benth. (p. 575) Shrub < 1.5 m, hairy, ± glandular. **LF**: stipules 2–5 mm, linear; petiole < 1.5 cm; blade 1–2 cm, ovate to ± deltate, base truncate to ± lobed, margin crenate. **STAMINATE INFL** 1.5–4 cm, slender. **PISTILLATE INFL** < 2 cm; bracts together cup-like, hairy, margin glandular. **STAMINATE FL**: sepals ± 0.5 mm, puberulent; stamens >> sepals. **PISTILLATE FL**: sepals ± 1 mm, puberulent; ovary ± 1 mm diam, puberulent, styles reddish. **FR** 1–3 mm diam, puberulent. Rocky slopes, chaparral, oak woodland; 200–1300 m. PR (San Diego Co.), w DSon; Baja CA. ❀DRN:15,16,18–21,**22–24**.

BERNARDIA

Shrub, monoecious or dioecious; sap clear. **STS** erect, gen much-branched. **LVS** simple, cauline, alternate, stipuled; hairs simple or stellate. **STAMINATE INFL**: spike or raceme, axillary. **PISTILLATE INFL** terminal; fl sometimes solitary. **STAMINATE FL** sessile or short-pedicelled; calyx splitting into 3–4 segments; petals 0; stamens 3–25, filaments free; nectar disk minute or 0. **PISTILLATE FL** sessile; sepals 4–6; petals 0; nectar disk 0; ovary 3-chambered, styles 3, free, 2-lobed or -toothed. **FR** 3-lobed. **SEED** 1 per chamber; scar appendaged. 30–40 spp.: trop, subtrop Am. (Latin: Bernard de Jussieu, French taxonomist, 1699–1776)

B. myricifolia (Scheele) S. Watson (p. 575) Shrub < 2.5 m, dioecious, hairy. **LF**: stipules ± 1 mm, deciduous; petiole 1–5 mm; blade 0.5–3 cm, elliptic, tip obtuse or rounded, margin crenate. **STAMINATE INFL**: raceme; pedicel 3–4 mm. **PISTILLATE INFL**: fl 1, sessile. **STAMINATE FL**: stamens 12–15; nectar disk of small glands. **PISTILLATE FL**: sepals 5, ± 2 mm, unequal; ovary tomentose, styles jagged. **FR** 8–10 mm diam, tomentose. **SEEDS** 5 mm, smooth; back ribbed. Washes, rocky canyons; < 1200 m. s DMoj, DSon; to TX, Mex. [*B. incana* C. Morton] ❀TRY.

CHAMAESYCE PROSTRATE SPURGE

Daryl L. Koutnik

Ann, per, gen monoecious, glabrous to hairy; sap milky. **ST** prostrate to erect, < 5 dm; branches alternate. **LVS** cauline, opposite, short-petioled; stipules present; blade base gen asymmetric, veins dark green. **INFL** fl-like, gen 1 per node; involucre ± bell-shaped, bracts 5, fused; glands 4, distal appendages gen colorful, petal-like; fls central. **STAMINATE FLS** 3–many, gen in 5 clusters around pistillate fl, each fl a stamen. **PISTILLATE FL** 1, central, stalked; ovary chambers 3, ovule 1 per chamber, styles 3, separate or fused at base, divided to entire. **FR**: capsule, round to 3-angled or -lobed in X-section. **SEED** gen 4-angled, smooth or sculptured. ± 250 spp.: dry temp, subtrop worldwide, esp Am. Often treated as subg. of *Euphorbia*. (Greek: ancient name for kind of prostrate plant) [Wheeler 1941 Rhodora 43:97–154, 168–286] ❀STBL.

1. Infls in dense axillary cyme-like clusters; st erect — sparsely hairy .... ***C. nutans***
1′ Infls 1 per node, sometimes crowded on lateral branches; st prostrate to ascending (sometimes erect)

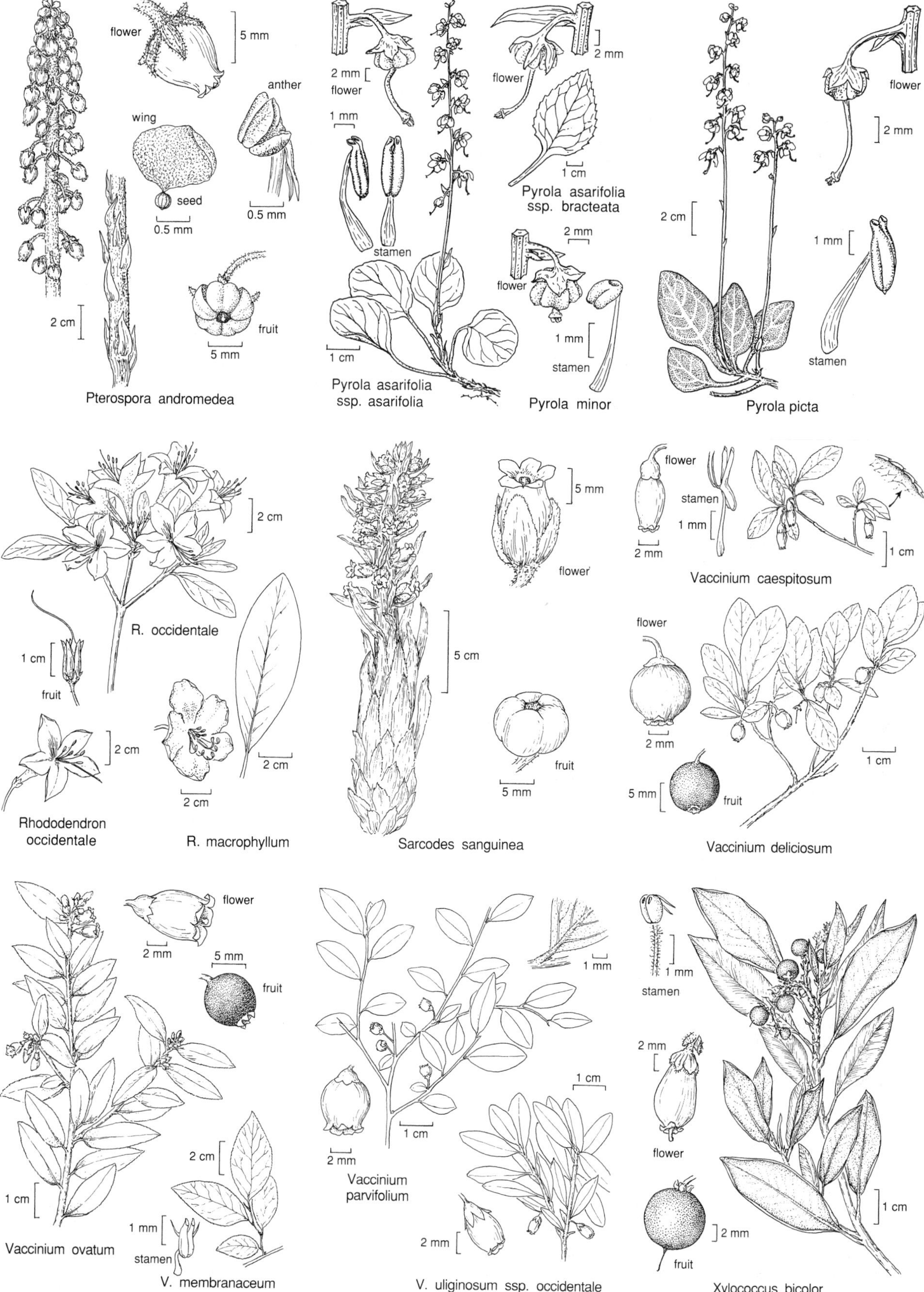
flower
5 mm
anther
wing
seed
0.5 mm
0.5 mm
2 cm
fruit
5 mm
Pterospora andromedea
2 mm
flower
1 mm
stamen
1 cm
Pyrola asarifolia
ssp. asarifolia
2 mm
flower
1 cm
Pyrola asarifolia
ssp. bracteata
2 mm
flower
1 mm
stamen
Pyrola minor
2 cm
flower
2 mm
1 mm
stamen
Pyrola picta
2 cm
R. occidentale
1 cm
fruit
2 cm
Rhododendron
occidentale
2 cm
2 cm
R. macrophyllum
5 mm
flower
5 cm
fruit
5 mm
Sarcodes sanguinea
flower
stamen
1 mm
2 mm
1 cm
Vaccinium caespitosum
flower
2 mm
1 cm
5 mm
fruit
Vaccinium deliciosum
flower
2 mm
5 mm
fruit
1 cm
Vaccinium ovatum
2 cm
1 mm
stamen
V. membranaceum
1 mm
1 cm
2 mm
Vaccinium
parvifolium
1 cm
2 mm
V. uliginosum ssp. occidentale
1 mm
stamen
2 mm
flower
1 cm
2 mm
fruit
Xylococcus bicolor

2. Involucre, fr hairy; st, lvs gen hairy or becoming glabrous
3. Per
4. Involucre urn-shaped; staminate fls < 12 — DSon .... ***C. arizonica***
4′ Involucre ± bell-shaped; staminate fls > 15
5. Seed 3-angled, transversely 4–5-ridged, ridges rounded .... ***C. pediculifera***
5′ Seed 4-angled, smooth to ± wrinkled
6. St and lf hairs short, straight .... [2]***C. polycarpa***
6′ St and lf hairs tomentose or long, appressed, dense
7. Involucre glands red .... ***C. melanadenia***
7′ Involucre glands green to yellow .... ***C. vallis-mortae***
3′ Ann
8. Staminate fls > 30; fr > 2 mm; n ScV .... ***C. ocellata*** ssp. ***rattanii***
8′ Staminate fls < 20; fr < 2 mm; s CA
9. Gland appendage deeply 3–5-lobed; involucre urn-shaped .... ***C. setiloba***
9′ Gland appendage entire, shallowly scalloped, or 0; involucre bell-shaped or obconic
10. Gland appendage 0; seed smooth to slightly wrinkled .... [2]***C. micromera***
10′ Gland appendage present; seed wrinkled or ridged
11. Fr > 1.5 mm; seed > 1 mm, wrinkled; gland appendage narrower than gland ... ***C. serpyllifolia*** ssp. ***hirtula***
11′ Fr < 1.5 mm; seed gen < 1 mm, transversely ridged or wrinkled; gland appendage = to or wider than gland
12. Seeds transversely wrinkled; fr evenly strigose; appendage = to gland width .... ***C. maculata***
12′ Seeds transversely ridged; fr hairy on edges only; appendage wider than gland .... ***C. prostrata***
2′ Involucre, fr, st, lvs glabrous (sometimes hairy)
13. Stipules fused into wide, membranous scale
14. Per; staminate fls > 12 .... ***C. albomarginata***
14′ Ann; staminate fls < 10 .... ***C. serpens***
13′ Stipules separate or fused below
15. Per
16. Glands round, appendage 0; lvs < 5 mm .... ***C. parishii***
16′ Glands elliptic or oblong, appendage present; lvs > 3 mm
17. Fr > 2 mm; stipules separate; seed > 2 mm .... ***C. fendleri***
17′ Fr < 2 mm; lower stipules fused; seed < 2 mm .... [2]***C. polycarpa***
15′ Ann
18. Lvs linear, base symmetric; st ascending to erect
19. Glands 1–4, oval, appendage narrower than gland; fr and seed > 1.5 mm; seed smooth .... ***C. parryi***
19′ Glands ± 4, round, appendage wider than gland or 0; fr and seed < 1.5 mm; seed transversely ridged .... ***C. revoluta***
18′ Lvs lanceolate to round, base asymmetric; st prostrate to decumbent
20. Lf margin toothed, at least toward tip
21. Style entire; gland appendages deeply 3–5-lobed — vernal pools .... ***C. hooveri***
21′ Style divided; gland appendages entire or shallowly lobed
22. Pl hairy; seed transversely ridged .... [2]***C. abramsiana***
22′ Pl glabrous; seed smooth, wrinkled, or transversely ridged
23. Seed transversely ridged; appendage wider than gland .... ***C. glyptosperma***
23′ Seed smooth to wrinkled; appendage narrower than gland .... ***C. serpyllifolia*** ssp. ***serpyllifolia***
20. Lf margin entire
24. Seed flattened top to bottom, surface smooth; gland ovate .... ***C. platysperma***
24′ Seed 3–4-angled, surface smooth, wrinkled, or transversely ridged; glands disc-like or elliptic
25. Gland appendage present; seed transversely ridged .... [2]***C. abramsiana***
25′ Gland appendage 0; seed smooth or wrinkled
26. Fr < 2 mm; seed < 1 mm; staminate fls < 10 .... [2]***C. micromera***
26′ Fr > 2 mm; seed > 1 mm; staminate fls > 30 .... ***C. ocellata***
27. Lf > 10 mm, lanceolate to ovate; DMoj .... ssp. ***arenicola***
27′ Lf < 10 mm, ovate, sickle-shaped; CA-FP .... ssp. ***ocellata***

C. abramsiana (Wheeler) Koutnik Ann. **ST** prostate, hairy to subglabrous. **LF** 2–12 mm; stipules separate, 2–5-parted; blade ovate to elliptic-oblong, hairy to glabrous, tip obtuse, margin entire to finely toothed. **INFL** dense on short lateral branches; involucre < 1 mm, obconic, glabrous; gland < 0.5 mm, round to elliptic; appendage wider than gland, entire or shallowly 2-lobed, white. **STAMINATE FLS** 3–5. **PISTILLATE FL**: style divided 1/2 length. **FR** 1.5–2 mm, oblong, round, glabrous. **SEED** 1–1.5 mm, ovoid, transversely 4–6-ridged, white. Sandy flats; < 200 m. DSon; to AZ, Mex. [*Euphorbia a.* Wheeler]

C. albomarginata (Torrey & A. Gray) Small (p. 575) RATTLESNAKE WEED Per. **ST** prostrate, glabrous. **LF** 3–8 mm; stipules fused, triangular, ciliate; blade round to oblong, glabrous, tip obtuse, margin entire. **INFL**: involucre < 2.5 mm, bell-shaped to obconic, glabrous; gland < 1 mm, oblong; appendage wider than gland, entire to slightly scalloped, white. **STAMINATE FLS** 15–30. **PISTILLATE FL**: style divided 1/2 length. **FR** 2–2.5 mm, ovoid, 3-angled, glabrous. **SEED** 1–2 mm, oblong, smooth, white. Common. Dry slopes; < 2300 m. s SnJv, SW, D; to UT, TX, Mex. [*Euphorbia a.* Torrey & A. Gray]

C. arizonica (Engelm.) J.C. Arthur ARIZONA SPURGE Per. **ST** prostrate to erect, hairy. **LF** 2–10 mm; stipules gen separate, minute; blade ovate, hairy, tip acute, margin entire. **INFL**: involucre < 2 mm, urn-shaped, hairy; gland < 0.5 mm, ovate, appendage wider than gland, entire, white to pink. **STAMINATE FLS** 5–10. **PISTILLATE FL**: style divided 1/2 length. **FR** < 2 mm, spheric, lobed, hairy. **SEED** ± 1 mm, ovoid, transversely ridged, white to brown. RARE in CA. Sandy flats; < 300 m. DSon; to TX, Mex. [*Euphorbia a.* Engelm.]

C. fendleri (Torrey & A. Gray) Small Per. **ST** decumbent, glabrous. **LF** 3–11 mm; stipules separate, linear, entire; blade ovate, glabrous, tip acute, margin entire. **INFL**: involucre < 2 mm, bell-shaped to obconic, glabrous; gland < 1 mm, elliptic; appendage narrower than gland, scalloped, white. **STAMINATE FLS** 25–35. **PISTILLATE FL**: style divided 1/2 length. **FR** 2–2.5 mm, ovoid, lobed, glabrous. **SEED** 2–2.5 mm, ovoid, smooth to slightly wrinkled, white. Uncommon. Dry slopes, woodland; 1500–2300 m. W&I, DMtns; to NE, TX, Mex. [*Euphorbia f.* Torrey & A. Gray]

C. glyptosperma (Engelm.) Small Ann. **ST** prostrate, glabrous. **LF** 3–15 mm; stipules separate, thread-like; blade ovate to ovate-oblong, glabrous, tip rounded, margin finely toothed. **INFL**: involucre < 1 mm, obconic, glabrous; gland < 0.5 mm, elliptic; appendage wider than gland, scalloped, white. **STAMINATE FLS** 1–5. **PISTILLATE FL**: style divided < 1/2 length. **FR** 1.5–2 mm, ovoid, lobed, glabrous. **SEED** 1–1.5 mm, ovoid, transversely 3–4- ridged, white to light brown. Uncommon. Dry ground; 200–300 m. n CaRF (near Redding); to B.C., e N.Am. [*Euphorbia g.* Engelm.]

C. hooveri (Wheeler) Koutnik (p. 575) HOOVER'S SPURGE Ann. **ST** prostrate to decumbent, glabrous. **LF** 2–5 mm; stipules fused, fringed; blade round, glabrous, papillate, tip rounded, margin coarsely toothed. **INFL**: involucre ± 2 mm, bell-shaped, glabrous; gland < 1 mm, round; appendage wider than gland, deeply 3–5-lobed, white. **STAMINATE FLS** 30–35. **PISTILLATE FL**: style undivided. **FR** 1.5–2 mm, spheric, lobed, glabrous. **SEED** ± 1.5 mm, ovoid, widely 4-angled, shallowly ridged, white. RARE. Vernal pools; < 250 m. GV (Butte, Tehama, Tulare cos.) [*Euphorbia h.* Wheeler] Threatened by habitat loss.

C. maculata (L.) Small (p. 575) SPOTTED SPURGE Ann. **ST** prostrate, hairy. **LF** 4–17 mm; stipules separate, fringed; blade ovate to oblong, hairy or becoming glabrous, tip acute to obtuse, margin finely toothed. **INFLS** dense on short, lateral branches; involucre < 1 mm, obconic, hairy; gland < 0.5 mm, elliptic; appendage width = gland width, scalloped, white to pink. **STAMINATE FLS** 4–5. **PISTILLATE FL**: style divided < 1/2 length. **FR** < 1.5 mm, ovoid, lobed, hairy. **SEED** < 1.5 mm, ovoid, transversely wrinkled, light brown. Waste places, gardens; < 200 m. CA-FP; native to e US. [*Euphorbia m.* L., *E. supina* Raf.]

C. melanadenia (Torrey) Millsp. Per. **ST** decumbent to ascending, tomentose or becoming glabrous. **LF** 2–9 mm; stipules separate, linear; blade ovate, tomentose, tip acute, margin entire. **INFL**: involucre 1–1.5 mm, bell-shaped, tomentose; gland < 1 mm, oblong; appendage width = gland width, scalloped, white. **STAMINATE FLS** 15–20. **PISTILLATE FL**: style divided > 1/2 length. **FR** 1.5–2 mm, ovoid, lobed, tomentose. **SEED** 1–1.5 mm, ovoid, slightly wrinkled, white. Dry, stony slopes or flats; < 1300 m. SW, DSon; AZ, Baja CA. [*Euphorbia m.* Torrey]

C. micromera (Engelm.) Wooton & Standley Ann. **ST** prostrate, glabrous to hairy. **LF** 2–7 mm; stipules fused below, separate above, triangular, ciliate; blade ovate to oblong, glabrous to hairy, tip acute to obtuse, margin entire. **INFL**: involucre < 1 mm, bell-shaped, glabrous to hairy; gland << 0.5 mm, round, red or pink; appendage 0. **STAMINATE FLS** 2–5. **PISTILLATE FL**: style divided 1/2 length. **FR** < 1.5 mm, spheric, angled, glabrous to hairy. **SEED** 1–1.5 mm, ovoid, smooth to slightly wrinkled, white to brown. Sandy places; < 1000 m. D; to UT, TX, n Mex. [*Euphorbia m.* Engelm.]

C. nutans (Lagasca) Small Ann. **ST** erect, sparingly hairy or becoming glabrous. **LF** 8–35 mm; stipules fused, triangular; blade oblong, glabrous to hairy, tip obtuse, margin finely toothed, glabrous to hairy. **INFLS** 1 per node or dense in axillary cyme-like clusters; involucre 1–2 mm, obconic, glabrous; gland < 0.5 mm, oblong; appendage wider than gland, entire, white to red. **STAMINATE FLS** 5–11. **PISTILLATE FL**: style divided 1/2 length. **FR** 2–2.5 mm, ovoid, lobed, glabrous. **SEED** 1–1.5 mm, ovoid, shallowly wrinkled, black to brown. Waste areas; < 300 m. CA-FP; native to se US, S.Am. [*Euphorbia n.* Lagasca]

C. ocellata (Durand & Hilg.) Millsp. Ann. **ST** prostrate, glabrous to hairy. **LF** 4–15 mm; stipules separate, thread-like; blade ovate to lanceolate, glabrous to hairy, tip acute to obtuse, margin entire, rolled down. **INFL**: involucre 1.5–2 mm, obconic to bell-shaped, glabrous to hairy; gland < 1 mm, round, appendage wider than gland or 0. **STAMINATE FLS** 40–60. **PISTILLATE FL**: style divided 1/2 length. **FR** 2–2.5 mm, spheric, lobed, glabrous to hairy. **SEED** 1.5–2 mm, ovoid, widely 3-angled, smooth to shallowly wrinkled, white. Sandy soils; < 800 m. CA-FP, DMoj; to UT, AZ. [*Euphorbia o.* Durand & Hilg.]

ssp. ***arenicola*** (Parish) Thorne **ST** glabrous. **LF** < 16 mm; blade ovate to lanceolate, tip acute, glabrous. **INFL**: involucre glabrous; gland appendage 0. **FR** glabrous. **SEED** smooth. Sandy places; < 800 m. DMoj; to UT, AZ. [*Euphorbia o.* var. *a.* (Parish) Jepson]

ssp. ***ocellata*** **ST** glabrous. **LF** < 10 mm; blade ovate, sickle-shaped, glabrous. **INFL**: involucre glabrous; gland appendage 0. **FR** glabrous. Dry, sandy places; < 500 m. CA-FP (exc PR).

ssp. ***rattanii*** (S. Watson) Koutnik STONY CREEK SPURGE **ST** hairy. **LF** < 10 mm; blade ovate, acute, hairy. **INFL**: involucre hairy; gland appendage wider than gland, white. UNCOMMON. Sandy or stony ground; < 100 m. n ScV (Glenn, Tehama cos.). [*Euphorbia o.* var. *r.* (S. Watson) Wheeler]

C. parishii (E. Greene) Millsp. Per. **ST** prostrate, glabrous. **LF** 2–4 mm; stipules separate, linear, ciliate; blade ovate, glabrous, tip abruptly pointed, margin entire, glabrous. **INFL**: involucre ± 1 mm, bell-shaped, glabrous; gland < 0.5 mm, round, yellow to red; appendage 0. **STAMINATE FLS** 40–50. **PISTILLATE FL**: style divided 1/2 length. **FR** < 2 mm, spheric, lobed, glabrous. **SEED** < 1.5 mm, ovoid, 4-angled, slightly wrinkled, white. Uncommon. Sandy washes; < 1000 m. D; NV. [*Euphorbia p.* E. Greene]

C. parryi (Engelm.) Rydb. Ann. **ST** ascending to prostrate, glabrous. **LF** 5–28 mm; stipules separate, linear; blade linear, glabrous, tip acute to obtuse, margin entire. **INFL**: involucre 1.5–2 mm, bell-shaped, glabrous; glands 1–4, < 0.5 mm, oval; appendage narrower than gland, entire, white. **STAMINATE FLS** 40–55. **PISTILLATE FL**: style divided 1/2 length. **FR** 2 mm, spheric, lobed, glabrous. **SEED** ± 2 mm, ovoid, 3-angled, smooth, brown to white. Sand dunes; < 700 m. DMoj; to Rocky Mtns, TX, Mex. [*Euphorbia p.* Engelm.]

C. pediculifera (Engelm.) Rose & Standley Per. **ST** prostrate to erect, hairy or becoming glabrous. **LF** 2–20 mm; stipules separate, thread-like; blade ovate to spoon-shaped, hairy to glabrous, tip acute, margin entire. **INFL**: involucre 1.5–2 mm, bell-shaped, hairy to glabrous; gland 0.5 mm, oblong, appendages unequal, entire to almost 0 on some glands, largest wider than gland. **STAMINATE FLS** 22–25. **PISTILLATE FL**: style divided to base. **FR** 2 mm, ovoid, lobed, hairy. **SEED** 1–1.5 mm, ovoid, white, 3-angled, transversely 4–5-ridged, ridges rounded. Uncommon. Dry slopes; < 500 m. DSon; AZ, Mex. [*Euphorbia p.* Engelm.]

C. platysperma (S. Watson) Shinn. FLAT SEEDED SPURGE Ann. **ST** prostrate, glabrous. **LF** 5–10 mm; stipules separate, 2–3-lobed; blade oblong to obovate, glabrous, tip obtuse to rounded, margin entire. **INFL**: involucre 1.5–2 mm, bell-shaped, glabrous; gland 1 mm, ovate, glabrous; appendage 0. **STAMINATE FLS** ± 50. **PISTILLATE FL**: style divided to base. **FR** < 4.5 mm, widely ovoid, slightly lobed, glabrous. **SEED** 2.5–3 mm, ovoid, flattened top to bottom, smooth, white. Sandy soil; < 100 m. DSon (Coachella Valley); to sw AZ, n Sonora. [*Euphorbia p.* S. Watson] Not seen in CA since 1914.

C. polycarpa (Benth.) Millsp. (p. 575) Per. **ST** prostrate to ascending, glabrous to hairy. **LF** 1–10 mm; stipules separate, triangular; blade round to ovate, glabrous to hairy, tip acute to obtuse, margin entire. **INFL**: involucre 1–1.5 mm, bell-shaped, glabrous to hairy; gland < 1 mm, oblong, appendage wider to narrower than gland, entire to scalloped, white to red. **STAMINATE FLS** 15–32. **PISTILLATE FL**: style divided > 1/2 length. **FR** 1–1.5 mm, spheric, lobed, glabrous to hairy. **SEED** 1–1.5 mm, ovoid, smooth, white to light brown. Common. Dry, sandy slopes and flats; < 1000 m. SW, D; NV, Mex. [*Euphorbia p.* Benth.] DMoj pls with hairy sts, lvs, involucre have been called var. *hirtella* (Boiss.) Parish; relationship to glabrous pls needs further study.

C. prostrata (Aiton) Small (p. 575) Ann. **ST** prostrate, hairy or becoming glabrous. **LF** 3–11 mm; stipules separate, linear; blade ovate to elliptic, glabrous to hairy, tip obtuse, margin finely toothed. **INFLS** gen on short, lateral branches; involucre < 1.5 mm, obconic, hairy to glabrous; gland < 0.5 mm, oval, appendage = to or wider than gland, entire to scalloped, white. **STAMINATE FLS** 4. **PISTILLATE FL**: style divided 1/2 length. **FR** 1–1.5 mm, spheric, lobed, lobes hairy. **SEED** ± 1 mm, ovoid, transversely ridged, white to gray. Waste areas; < 250 m. s CA-FP; to e US; native to W. Indies, S.Am. [*Euphorbia p.* Aiton]

C. revoluta (Engelm.) Small Ann. **ST** erect, glabrous. **LF** 3–26 mm; stipules separate, linear; blade linear, glabrous, tip acute to obtuse, margin entire, rolled down. **INFL**: involucre < 1.5 mm, obconic, glabrous; gland < 0.5 mm, round, appendage wider than gland or 0, entire, white. **STAMINATE FLS** 5–10. **PISTILLATE FL**: style divided 1/2 length. **FR** slightly < 1.5 mm, spheric, lobed, glabrous. **SEED** 1–1.5 mm, ovoid, 3-angled, transversely 2–3-ridged, white to gray. Uncommon. Rocky slopes; < 3100 m. DMtns; to Rocky Mtns, Mex. [*Euphorbia r.* Engelm.]

C. serpens (Kunth) Small Ann. **ST** prostrate, rooting at nodes; glabrous. **LF** 2–7 mm; stipules fused, triangular; blade ovate to oblong, glabrous, tip obtuse, margin entire. **INFL**: involucre < 1.5 mm, obconic, glabrous; gland < 0.5 mm, oblong; appendages wider than gland, scalloped, white. **STAMINATE FLS** 5–10. **PISTILLATE FL**: style divided 1/2 length. **FR** < 1.5 mm, spheric, lobed, glabrous. **SEED** < 1.5 mm, ovoid, smooth, white to brown. Waste areas; < 200 m. SnJV, s SCoR, SW (expected elsewhere); to e US; native to S.Am. [*Euphorbia s.* Kunth]

C. serpyllifolia (Pers.) Small THYME-LEAFED SPURGE Ann. **ST** prostrate to ascending, glabrous to hairy. **LF** 3–14 mm; stipules separate, linear; blade ovate to oblong, glabrous to hairy, tip rounded, margin finely toothed. **INFL**: involucre ± 1 mm, bell-shaped, glabrous to hairy; gland < 0.5 mm, oblong; appendage narrower than gland, entire to scalloped, white. **STAMINATE FLS** 5–18. **PISTILLATE FL**: style divided 1/2 length. **FR** 1.5–2 mm, ovoid, lobed, glabrous to hairy. **SEED** 1–1.5 mm, ovoid, smooth to wrinkled, white to brown. Common. Dry habitats; < 2500 m. CA; to B.C., e N.Am., Mex. [*Euphorbia s.* Pers.]

ssp. ***hirtula*** (S. Watson) Koutnik **ST** hairy. **LF** 3–7 mm, hairy. **INFL**: involucre hairy. **FR** hairy. Dry places, woodland; < 2500 m. c SN, SW; Baja CA. [*Euphorbia s.* Pers. var. *h.* (S. Watson) Wheeler]

ssp. ***serpyllifolia*** **ST** glabrous. **LF** glabrous. **INFL**: involucre glabrous. **FR** glabrous. Common. Habitat and range of sp.

C. setiloba (Torrey) Millsp. (p. 575) Ann. **ST** prostrate, hairy. **LF** 2–7 mm; stipules separate, thread-like; blade oblong to ovate, hairy, tip acute, margin entire. **INFL**: involucre < 1.5 mm, urn-shaped, hairy; gland < 0.5 mm, oblong, appendage wider than gland, 3–5-lobed, white. **STAMINATE FLS** 3–7. **PISTILLATE FL**: style divided to base. **FR** < 1.5 mm, spheric, lobed, hairy. **SEED** ± 1 mm, ovoid, wrinkled, white to brown. Uncommon. Sandy places; < 1500 m. D; to TX, Mex. [*Euphorbia s.* Torrey]

C. vallis-mortae Millsp. Per. **ST** prostrate to decumbent, tomentose or becoming glabrous. **LF** 4–6 mm; lower stipules fused; upper stipules separate, thread-like, tomentose; blade oblong to ovate, tomentose, tip obtuse, margin entire. **INFLS** gen clustered at branch tips; involucre < 2.5 mm, bell-shaped, tomentose; gland < 1 mm, oblong; appendage = to or wider than gland, entire to scalloped, white. **STAMINATE FLS** 17–22. **PISTILLATE FL**: style divided 1/2 length. **FR** 2 mm, ovoid, 3-lobed, tomentose. **SEED** < 2 mm, ovoid, smooth, white. Uncommon. Dry, sandy places; < 1300 m. SNE, DMoj. [*Euphorbia v-m.* (Millsp.) J. Howell]

CROTON

Ann, per, shrub, tree, monoecious or dioecious; sap clear or colored. **STS** gen erect. **LVS** gen simple, cauline, alternate; hairs gen stellate. **INFL**: spike or raceme, gen terminal. **STAMINATE FL** gen pedicelled; sepals gen 5; petals 5 or 0; stamens 8–50(300), filaments free, bent inward in bud; nectar disk gen divided. **PISTILLATE FL**: pedicel short or 0, becoming longer in fr; sepals gen 5, entire to lobed; petals gen 0; nectar disk entire; ovary 3-chambered, styles 2-lobed or toothed. **FR** spheric or 3-lobed, smooth or tubercled. **SEEDS** 1 per chamber, smooth to ribbed or pitted; scar appendaged. 900–1000 spp.: trop, warm temp, worldwide. (Greek: from resemblance of seed to a tick)

1. Seeds 3.5–5.5 mm; pedicels in fr < 2 mm; staminate sepals ± 2–2.5 mm .... ***C. californicus***
1′ Seeds 6.5–7 mm; pedicels in fr 4–7 mm; staminate sepals 2.5–3 mm — sand dunes (se Imperial Co.) ***C. wigginisii***

C. californicus Muell. Arg. (p. 575) Per or subshrub < 1 m, dioecious; hairs stellate, scale-like. **LF**: petiole 1–4 cm; blade 2–5.5 cm, elliptic to narrowly oblong, tip rounded to obtuse, margin entire. **INFL**: raceme. **STAMINATE FL**: pedicels 1–5.5(7) mm; petals 0; stamens 10–15, filaments hairy. **PISTILLATE FL**: pedicel < or = 1 mm, 1–1.5(3) mm in fr; sepals ± 2 mm, entire; styles 2-lobed, lobes 2-forked. **SEED** 3.5–5.5 mm, smooth. Sandy soils, dunes, washes; < 900 m. CCo, SCo, D; AZ, Baja CA. [vars. *mohavensis* A. ferg and *tenuis* (S. Watson) A. Ferg.] ❀DRN:15–24; DFCLT.

C. wigginsii Wheeler (p. 575) WIGGINS' CROTON Shrub or subshrub, < 1 m, dioecious; hairs stellate, scale-like. **LF**: petiole 1–4 cm; blade 2–8.5 cm, narrowly elliptic to linear-oblong, tip rounded to obtuse, margin entire. **INFL**: raceme. **STAMINATE FL**: pedicels 1–5.5(7) mm; petals 0; stamens 10–15, filaments hairy. **PISTILLATE FL**: pedicel < 2 mm, 4–7 mm in fr; sepals ± 2 mm, entire; styles 2-lobed, lobes 2-forked. **SEED** 6.5–7 mm, smooth. RARE in CA. Sand dunes; < 100 m. se DSon (se Imperial Co.); AZ, nw Mex.

DITAXIS

Ann, per, subshrub, gen monoecious; sap clear; hairs 0 or gen 2-forked, gen appressed. **STS** spreading to erect, 1–10 dm. **LVS** simple, alternate, stipuled. **INFL**: raceme, axillary; staminate fls gen above pistillate fls; axis gen densely appressed-hairy; bracts entire. **STAMINATE FL**: sepals 5, edges abutting in bud; petals 5; stamens 5–15, gen in 2 sets, some > others, filaments fused into a column, staminodes 0–3 at column tip. **PISTILLATE FL**: sepals 5, overlapping in bud; petals 5; nectar disk ± dissected; ovary 3-chambered, styles 3, 2-lobed. **FR** smooth. **SEEDS**: surface net-like to finely pitted; scar not appendaged. ± 50 spp.: trop, warm temp Am. (Greek: 2-ranked, from 2 sets of anthers)

1. Pl glabrous .... ***D. californica***
1′ Pl hairy, hairs 2-forked, appressed, sometimes simple or glandular
  2. Margin of stipules, bracts, and pistillate sepals with stalked glands; infl axis minutely spreading-hairy .... ***D. clariana***
  2′ Margin of stipules, bracts, and pistillate sepals glabrous or faintly glandular, glands not stalked; infl axis appressed-hairy
    3. Subshrub, sts brittle; pistillate sepals ± = petals; fr appressed-hairy; style lobes expanded .... ***D. lanceolata***
    3′ Ann or per (sometimes woody at base), sts not brittle; pistillate sepals clearly > petals; fr ± spreading-hairy; style branches entire to barely toothed

4. Seeds angled in X-section, clearly pitted; lvs gen lanceolate, not densely hairy, entire to faintly toothed . . . . *D. neomexicana*

4′ Seeds round in X-section, ± striate, not clearly pitted; lvs gen widely elliptic, densely hairy, clearly toothed distally . . . . *D. serrata*

D. californica (Brandegee) Pax & K. Hoffm. (p. 575) CALIFORNIA DITAXIS Ann or per. **ST** 1.5–5 dm, glabrous. **LF** 1–5 cm; stipules ± 1 mm, faintly gland-toothed; blade lanceolate to elliptic, glabrous, margin finely toothed. **STAMINATE FL**: sepals 1.5–2.5 mm; petals 2–2.5 mm; stamen column 1 mm. **PISTILLATE FL**: sepals 2.5–4.5 mm, faintly gland-toothed; petals ± 1.5 mm, glabrous; ovary glabrous, styles gen free, lobe tips not expanded. **FR** ± 3.5 mm. **SEED** ± 2 mm, ± angled, ± pitted. RARE. Washes, canyons; 50–1000 m. DMoj (Eagle Mtn), nw DSon (Coachella Valley).

D. clariana (Jepson) Webster (p. 575) GLANDULAR DITAXIS Ann or per. **ST** 1–5 dm; some hairs simple and spreading, others 2-forked and appressed. **LF** 1–4 cm; stipules 1.5–3 mm, gland-toothed; blade lanceolate, lower surface hairy, margin finely gland-toothed. **STAMINATE FL**: sepals 3.5–5 mm, hairy; petals ± = sepals, glabrous or hairy; stamen column 1.5–2 mm. **PISTILLATE FL**: sepals 3.5–5.5 mm, unequal, glabrous or hairy; petals ± = sepals; ovary sparsely hairy, styles fused below, lobe tips expanded. **FR** ± 4.5 mm. **SEED** ± 2 mm, angled, faintly pitted. RARE in CA. Sandy soils, creosote-bush scrub; < 100 m. DSon (Coachella Valley). [*D. adenophora* (A. Gray) Pax & K. Hoffm. misapplied]

D. lanceolata (Benth.) Pax & K. Hoffm. (p. 575) Subshrub. **STS** gen erect, 1–5 dm, brittle, appressed-hairy. **LF** 2–6 cm; stipules ± 1 mm, entire; blade lanceolate, densely hairy, margin entire. **STAMINATE FL**: sepals 2.5–3 mm, hairy; petals 3–3.5 mm, back hairy; stamen column ± 1.5 mm. **PISTILLATE FL**: sepals 3–4 mm, entire; petals ± = sepals, lanceolate to ovate, back hairy; ovary densely appressed-hairy, styles gen free, lobe tips expanded. **FR** 3–5 mm. **SEED** 2–2.5 mm, angled, pitted. Rocky soils, slopes, canyons; < 600 m. DMoj (Eagle Mtn), DSon; AZ, Mex.

D. neomexicana (Muell. Arg.) A.A. Heller (p. 575) Ann or per. **ST** 1–3.5 dm, densely appressed-hairy. **LF** 1–3.5 cm; stipules 1–1.5 mm, entire; blade lanceolate, ± hairy, margin ± entire. **STAMINATE FL**: sepals 2–2.5 mm; petals ± 2 mm, glabrous; stamen column ± 1 mm. **PISTILLATE FL**: sepals 3–4 mm, back hairy, margin ± entire, ± white; petals ± 2.5 mm, lanceolate, glabrous or appressed-hairy, hairs not exceeding petal tip; ovary stiff-hairy; styles free, lobe tips not expanded. **FR** 3–4 mm. **SEED** ± 2 mm, angled, pitted. Slopes, creosote-bush scrub; < 300 m. DMoj (s edge), DSon; to TX, Mex.

D. serrata (Torrey) A.A. Heller (p. 575) Ann or per. **ST** 1–3.5 dm, densely appressed-hairy. **LF** 1–3 cm; stipules 1–1.5 mm, entire; blade widely elliptic to ovate, densely hairy, margin clearly toothed distally. **STAMINATE FL**: sepals 2–2.5 mm; petals ± 2 mm, glabrous; stamen column ± 1 mm. **PISTILLATE FL**: sepals 3–4 mm, back hairy; petals ± 2.5 mm, obovate, back clearly appressed-hairy, hairs exceeding petal tip; ovary stiff-hairy, styles free, lobe tips not expanded. **FR** ± 2 mm. **SEED** ± 2 mm, rounded, ± striate, not clearly pitted. Sandy or rocky soils, creosote-bush scrub; < 200 m. DSon; AZ, Mex.

EREMOCARPUS TURKEY MULLEIN, DOVE WEED

1 sp. (Greek: solitary fr)

E. setigerus (Hook.) Benth. (p. 575) Ann < 2 dm, < 8 dm wide, mound-like, monoecious; sap clear. **ST** much-branched from base, spreading to ascending. **LVS** simple, cauline, alternate; stipules vestigial; petiole 1–5 cm; blade 1–6 cm, ovate, base obtuse to wedge-shaped, margin entire, densely soft stellate-hairy, 3-veined, veins raised. **STAMINATE INFL**: cyme, terminal; pedicel 2–3 mm. **PISTILLATE INFL** axillary, below staminate infl; fls 1–3. **STAMINATE FL**: receptacle finely bristly; sepals 5–6; petals 0; stamens 6–10, free, exserted, filaments 1.5–2 mm; nectary 0. **PISTILLATE FL**: sepals and petals 0; glands below ovary 4–5; ovary 1-chambered, puberulent, style slender. **FR** ± 4 mm diam. **SEED** 1, 3–4 mm, smooth or ± ridged; scar not appendaged. 2n=20. Dry, open, often disturbed areas; < 1000 m. CA-FP, w D; to WA. Seeds eaten by birds; herbage TOXIC to livestock, esp in hay. ❀SUN, DRN:**7–9**,**14**,15–17,**18–24**;rather INV.

EUPHORBIA SPURGE

Daryl L. Koutnik

Ann, per, gen monoecious, glabrous or hairy. **ST** ascending to erect, < 1 m; branches forked, forks equal. **LVS** cauline, gen alternate; stipules 0 or gland-like; petiole present or 0; lf base symmetrical. **INFL** fl-like or not, gen clustered; clusters gen umbel-like or cyme-like; involucre ± bell-shaped; bracts 5, fused; glands gen 4, distal appendages gen 0; fls central. **STAMINATE FLS** 5–many, gen in 5 clusters around pistillate fl. **PISTILLATE FL** 1, central, stalked; ovary chambers 3, ovule 1 per chamber, styles 3, separate or fused at base, divided or entire. **FR**: capsule, round to 3-angled or -lobed in X-section. **SEED** round or angled in X-section; surface smooth or sculptured, gen with a knob-like structure at attachment scar. ± 1500 spp.: warm temp to trop, worldwide. See *Chamaesyce*. (Latin: Euphorbus, Physician to the King of Mauritania, 1st century) [Wheeler 1936 Bull S Calif Acad Sci 35:127–147] ❀STBL.

1. Involucre with petal-like appendages on glands; stipules present, gland-like, minute
  2. Lvs opposite, margin finely toothed; ann; seed with a knob at tip . . . . *E. exstipulata*
  2′ Lvs alternate, margin entire; woody; seed without a knob . . . . *E. misera*

1′ Involucre without petal-like appendages on glands; stipules 0
  3. Infls few–many in cyme-like clusters at branch tips; involucre glands 1–3, cup-shaped
    4. Lvs alternate below, opposite above; involucre, fr glabrous; seed ovoid . . . . *E. dentata*
    4′ Lvs alternate throughout; involucre, fr hairy; seed flattened . . . . *E. eriantha*
  3′ Infl clusters umbel-like below, cyme-like above; involucre glands 4, ± flat
    5. Lvs opposite, 4-ranked; fr > 8 mm . . . . *E. lathyris*
    5′ Lvs alternate; fr < 6 mm
      6. Lf margin finely toothed; gland margin entire or truncate
        7. Per; seed surface smooth or dotted
          8. St hairy; fr tubercled . . . . *E. oblongata*
          8′ St glabrous; fr smooth . . . . *E. serrata*

7′ Ann; seed surface sculptured, net-like
9. Fr smooth — disturbed, waste places ..... *E. helioscopia*
9′ Fr tubercled ..... *E. spathulata*
6′ Lf margin entire; gland horned or scalloped
10. Lvs linear, sometimes oblanceolate; seed surface smooth ..... *E. esula*
10′ Lvs obovate to ovate or elliptic; seed surface dotted or sculpture net-like
11. Seed dotted; fr 2-keeled on lobes — waste places, gardens ..... *E. peplus*
11′ Seed sculpture low, net-like; fr lobes smooth
12. Ann (rarely bien); gland horns thin, > gland width ..... *E. crenulata*
12′ Per; gland horns thick, < gland width or 0
13. Gland hornless, margin scalloped ..... *E. incisa*
13′ Gland 2-horned, margin entire to shallowly scalloped ..... *E. palmeri*

E. crenulata Engelm. (p. 579) CHINESE CAPS Ann or bien. **ST** erect, 1.2–6 dm, glabrous. **LF** 1.5–3.5 cm, petioled to sessile; blade obovate to spoon-shaped, glabrous, tip obtuse to abruptly pointed, margin entire. **INFL**: involucre 2 mm, bell-shaped, glabrous; gland 1–2 mm, crescent-shaped, 2-horned. **STAMINATE FLS** 11–18. **PISTILLATE FL**: style divided < 1/2 length. **FR** 3–3.5 mm, oblong, lobed, glabrous. **SEED** 2–2.5 mm, oblong-ovoid, round; surface net-like to almost smooth. Common. Dry places; < 1600 m. CA-FP; OR.

E. dentata Michaux TOOTHED SPURGE Ann. **ST** erect or ascending, 2–5 dm, hairy. **LVS** gen opposite, 1–6 cm, petioled; blade linear to lanceolate to ovate, hairy, tip acute, margin coarsely toothed to lobed. **INFL**: clusters cyme-like at branch tips; involucre 2–3 mm, bell-shaped, glabrous; glands 1–2, < 1 mm, cupped. **STAMINATE FLS** < 50. **PISTILLATE FL**: style divided > 1/2 length. **FR** 4–5 mm, spheric, lobed, glabrous. **SEED** 2–3 mm, ovoid, 3- angled, tubercled, gray to brown; knob reduced. Waste places, disturbed sites; < 300 m. SnJV, CCo, SnFrB; native to e US, Mex.

E. eriantha Benth. (p. 579) BEETLE SPURGE Ann. **ST** erect, 1.5–5 dm, glabrous or hairy and becoming glabrous. **LF** 2–7 cm, petioled; stipules minute, obscure; blade linear, hairy, becoming glabrous, tip acute to obtuse and abruptly pointed, margin entire. **INFL** 1 or few clustered at branch tips; involucre 1.5–2 mm, obconic, hairy; glands 1–3, ± 1.5 mm, round, cupped, lobes 5–7, curved over gland. **STAMINATE FLS** 23–36. **PISTILLATE FL**: style undivided. **FR** 4–5 mm, oblong, lobed, hairy. **SEED** 3.5–4 mm, flattened top to bottom, 4-angled, tubercled, white to gray. Canyons, rocky slopes; < 100 m. DSon; to TX, Mex.

E. esula L. LEAFY SPURGE Per. **ST** erect, 3–8 dm, glabrous to hairy. **LF** 2–6 cm, sessile; blade linear to oblanceolate, glabrous, tip acute, margin entire. **INFL**: involucre 1.5–2.5 mm, bell-shaped, glabrous; gland 1.5–2 mm, crescent-shaped, 2-horned. **STAMINATE FLS** 11–21. **PISTILLATE FL**: style divided 1/2 length. **FR** 3–5 mm, spheric, lobed, granular to smooth. **SEED** 2–2.5 mm, oblong, round, smooth, yellow-brown. Fields, pastures; < 1400 m. s NCo (Sonoma Co.), e KR, CaR, MP; to e US; native to Eur. NOXIOUS WEED.

E. exstipulata Engelm. var. ***exstipulata*** CLARK MOUNTAIN SPURGE Ann. **ST** erect, < 2.5 dm, hairy. **LVS** opposite, 2–5 cm, petioled; stipules gland-like; blade linear to lanceolate, hairy, tip acute, margin finely toothed. **INFL** 1 per node; involucre 1–2 mm, bell-shaped, hairy; gland < 0.5 mm, oblong, appendage = to or wider than gland, 2–4-lobed. **STAMINATE FLS** 8–14. **PISTILLATE FL**: style divided > 1/2 length. **FR** 2.5–3.4 mm, spheric, lobed, hairy. **SEED** 2–3 mm, oblong, 4-angled, transversely 2–3-ridged, tubercled, gray to brown or white. RARE in CA. Rocky slopes; 1800–2000 m. e DMtns (Clark Mtns); to TX, Mex. Other vars. in sw US, Mex.

E. helioscopia L. WARTWEED Ann. **ST** ascending to erect, 1–5 dm, sparsely hairy. **LF** 1–3 cm, petioled to sessile; blade obovate, glabrous, tip obtuse to notched, margin finely toothed. **INFL**: involucre ± 2 mm, obconic, glabrous; gland < 1 mm, oblong, entire. **STAMINATE FLS** 8–12. **PISTILLATE FL**: style divided < 1/2 length. **FR** 2.5–3 mm, spheric, lobed, glabrous. **SEED** 2–2.5 mm, spheric; surface coarsely net-like, dark gray to brown. Waste places; < 200 m. n&c NCo, n&c NCoRO, SnFrB, SCo; native to Eur.

E. incisa Engelm. MOJAVE SPURGE Per. **ST** ascending to erect, 1–4 dm, glabrous to slightly hairy. **LF** 0.6–2 cm, sessile; blade obovate to elliptic, glabrous, tip acute to abruptly pointed, margin entire. **INFL**: involucre 2–3 mm, bell-shaped, glabrous; gland 1–2 mm, crescent-shaped, margin scalloped. **STAMINATE FLS** < 20. **PISTILLATE FL**: style divided < 1/2 length. **FR** 4–5 mm, oblong, lobed, glabrous. **SEED** 2–3 mm, oblong, round, white to gray; surface low net-like to almost smooth. Rocky or sandy slopes; 1000–2300 m. DMoj; NV, AZ. Intergrades with *E. palmeri*, esp in gland shape.

E. lathyris L. (p. 579) CAPER SPURGE, GOPHER PLANT Bien. **ST** erect, 5–10 dm, glabrous. **LVS** opposite, 5–15 cm, sessile; blade lanceolate to linear, glabrous, tip acute, margin entire. **INFL**: involucre 2.5–4 mm, bell-shaped, glabrous; gland ± 2 mm, crescent-shaped, shortly 2-horned. **STAMINATE FLS** 15–40. **PISTILLATE FL**: style divided 1/2 length. **FR** 8–15 mm, spheric, glabrous. **SEED** 4–5 mm, oblong, round; surface shallowly net-like, brown. Waste areas; < 200 m. NCo, CCo, GV, SCo (expected elsewhere); to e US; native to Eur.

E. misera Benth. CLIFF SPURGE Shrub. **ST** erect, 5–10 dm, hairy, becoming glabrous. **LF** 0.4–1.5 cm, petioled; stipule thread-like; blade ovate to round, hairy, tip round, margin entire. **INFL** 1 at branch tip; involucre 2–3 mm, bell-shaped, hairy; glands 5, 1.5–2 mm, oblong, appendage = to gland width, scalloped, white. **STAMINATE FLS** 30–40. **PISTILLATE FL**: style divided 1/2 length. **FR** 4–5 mm, spheric, lobed, becoming glabrous. **SEED** 2.5–3 mm, ovoid, round, wrinkled, white to gray; knob 0. RARE in CA. Rocky slopes, coastal bluffs; < 500 m. SCo, s ChI, w DSon; Baja CA. In cult.

E. oblongata Griseb. Per. **ST** ascending to erect, 5–8 dm, hairy. **LF** 4–6.5 cm, sessile; blade oblong to lanceolate, glabrous, tip obtuse, margin finely toothed. **INFL**: involucre 1.5–2.5 mm, bell-shaped, glabrous; gland ± 1 mm, elliptic. **STAMINATE FLS** 20–40. **PISTILLATE FL**: style divided < 1/2 length. **FR** 3–4.5 mm, spheric, tubercled. **SEED** 2.5 mm, ovoid, round, smooth, brown. Waste places; < 200 m. GV, SnFrB (expected elsewhere); native to Eur. NOXIOUS WEED.

E. palmeri S. Watson Per. **ST** ascending to erect, 1–3.5 dm, glabrous. **LF** 0.5–2 cm, petioled to sessile; blade obovate to ovate, glabrous, tip obtuse to rounded, margin entire. **INFL**: involucre < 3 mm, obconic, glabrous; gland < 1.5 mm, oblong, margin entire to scalloped, shortly 2-horned. **STAMINATE FLS** < 20. **PISTILLATE FL**: style divided < 1/2 length. **FR** ± 4 mm, ovoid, round, glabrous. **SEED** ± 2.5 mm, oblong, round, gray; surface net-like. Common. Dry slopes, flats; 1300–2800 m. SW; to UT, AZ. Intergrades with *E. incisa*, esp in deserts.

E. peplus L. (p. 579) PETTY SPURGE Ann. **ST** ascending to erect, 1–4.5 dm, glabrous. **LF** 1–3.5 cm, petioled; blade obovate to ovate, glabrous, tip obtuse to notched, margin entire. **INFL**: involucre 1–1.5 mm, bell-shaped, glabrous; gland < 0.5 mm, crescent-shaped, 2-horned. **STAMINATE FLS** 10–15. **PISTILLATE FL**: style divided 1/2 length. **FR** ± 2 mm, spheric, glabrous, lobed; lobes 2-keeled. **SEED** 1–1.5 mm, oblong, 4-angled, dotted, white to gray. Common. Waste places, gardens; < 300 m. CA-FP; to Can, e US; native to Eur.

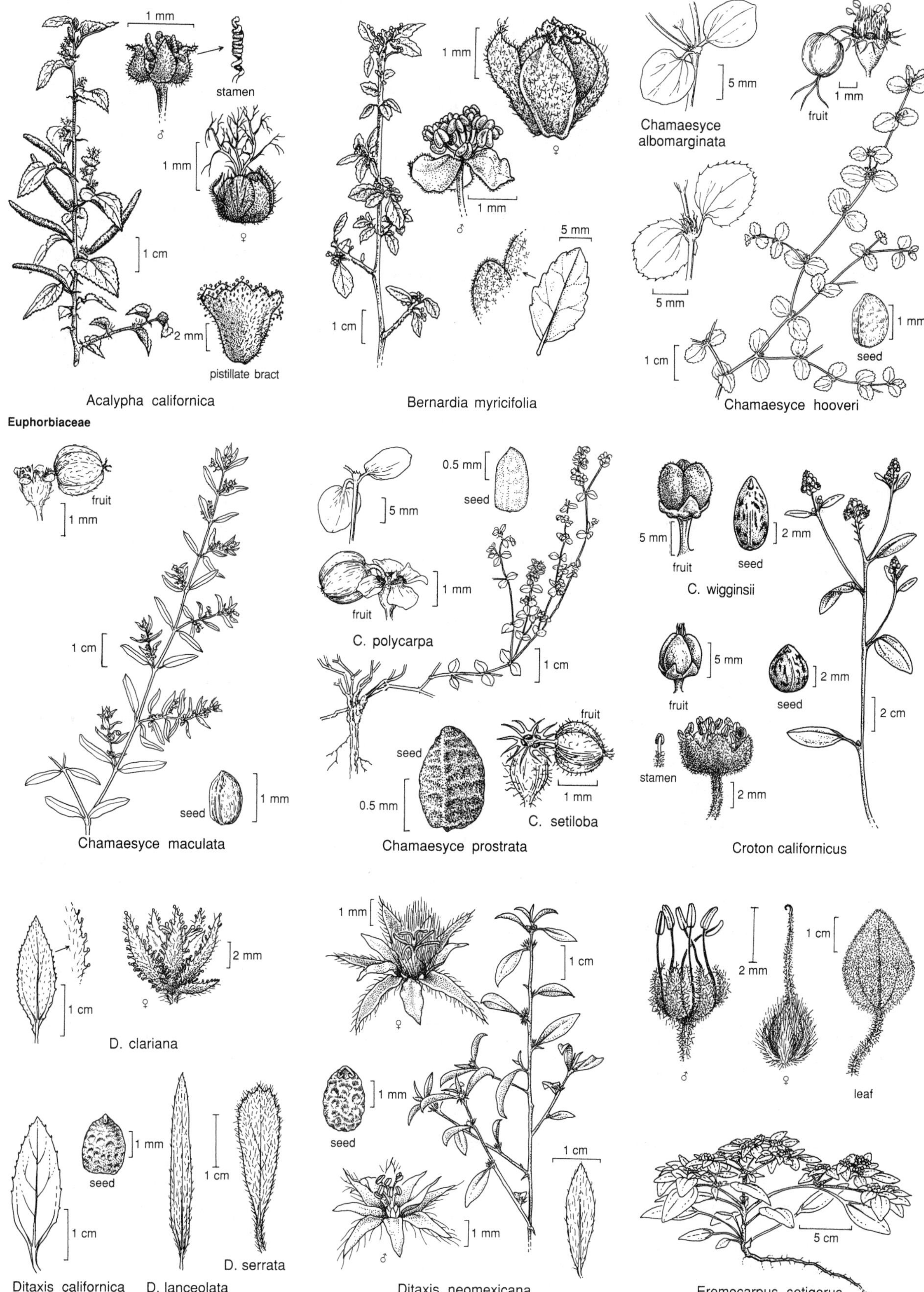
1 mm
stamen
♂
1 mm
♀
1 cm
2 mm
pistillate bract
Acalypha californica
1 mm
♀
1 mm
♂
5 mm
1 cm
Bernardia myricifolia
5 mm
Chamaesyce albomarginata
1 mm
fruit
5 mm
1 mm
seed
1 cm
Chamaesyce hooveri
Euphorbiaceae
fruit
1 mm
1 cm
seed
1 mm
Chamaesyce maculata
0.5 mm
seed
5 mm
1 mm
fruit
C. polycarpa
1 cm
seed
0.5 mm
fruit
1 mm
C. setiloba
Chamaesyce prostrata
5 mm
fruit
seed
2 mm
C. wigginsii
5 mm
fruit
seed
2 mm
2 cm
stamen
2 mm
Croton californicus
1 cm
2 mm
♀
D. clariana
1 cm
seed
1 mm
1 cm
1 cm
Ditaxis californica
D. lanceolata
D. serrata
1 mm
♀
1 cm
seed
1 mm
1 cm
1 mm
♂
Ditaxis neomexicana
2 mm
1 cm
♂
♀
leaf
5 cm
Eremocarpus setigerus

E. serrata L. SAW-TOOTHED SPURGE Per. **ST** erect, 2–6 dm, glabrous. **LF** 2–5 cm, sessile; blade linear to ovate-lanceolate, glabrous, tip acute to obtuse, margin finely toothed. **INFL**: involucre 2–4 mm, bell-shaped, glabrous; gland ± 1 mm, oval; margin smooth to truncate or 2-horned. **STAMINATE FLS** 5–20. **PISTILLATE FL**: style ± divided < 1/2 length. **FR** 4–6 mm, spheric, smooth, glabrous. **SEED** 2.5–3 mm, smooth or dotted, gray. Waste places; < 200 m. SnFrB; native to Eur. NOXIOUS WEED.

E. spathulata Lam. (p. 579) Ann. **ST** erect, 1.5–4.5 dm, glabrous. **LF** 1–3 cm, petioled to sessile; blade obovate to spoon-shaped, glabrous, tip obtuse to notched, margin finely toothed. **INFL**: involucre < 1.5 mm, bell-shaped, glabrous; gland < 1 mm, oblong. **STAMINATE FLS** 5–8. **PISTILLATE FL**: style divided ± 1/2 length. **FR** 2–3 mm, spheric, lobed, tubercled. **SEED** 1.5–2 mm, elliptic, round, yellow-brown; surface net-like. Open, gen disturbed places; < 1300 m. CA-FP; to WA, e N.Am, S. Am.

MERCURIALIS MERCURY

Ann, per, monoecious or dioecious, gen glabrous; sap clear. **STS**: central erect, branched; lateral < central, spreading to ascending. **LVS** simple, cauline, opposite; stipules persistent; hairs simple. **STAMINATE INFL** spike-like, axillary; fls clustered. **PISTILLATE INFL**: fls clustered, axillary, below staminate fls. **STAMINATE FL**: calyx 3-lobed; petals 0; stamens 8–15(20), free; nectary 0. **PISTILLATE FL**: sepals 3; petals 0; staminodes 2, elongated; ovary 2-chambered, bristly, styles free, simple. **FR** bristly. **SEEDS** 1 per chamber, gen pitted; scar appendaged. 8 spp.: Eurasia, n Afr. (Greek: Greek God Mercury)

M. annua L. Ann, 1–3 dm, dioecious, ± glabrous. **LF**: stipules 1–1.5 mm, lanceolate-deltate; petiole 0.5–2(2.5) cm; blade 2–5 cm, ± ovate, margin serrate, ciliate. **STAMINATE INFL** < 2 cm, short-peduncled. **PISTILLATE INFL**: fls 2–3; pedicel < 5 mm. **STAMINATE FL**: calyx ± 1 mm; stamens exserted. **PISTILLATE FL**: calyx ± 1 mm; ovary puberulent, styles ± 1 mm. **FR** 2–3 mm diam, finely bristly. **SEEDS** 1.5–2 mm, shiny, pitted. 2n=16, 32. Open, disturbed areas, fields, roadsides; < 200 m. SnFrB; native to Eur.

RICINUS CASTOR BEAN

1 sp. (Latin: tick, from seed shape)

R. communis L. (p. 579) Shrub, sometimes tree-like, 1–3 m, monoecious, ± glabrous; sap clear. **ST**: trunk ascending to erect, branched above. **LVS** simple, cauline, alternate, peltate; stipules fused, 1–1.5 cm, sheath-like, deciduous; petiole 1–3 dm, glandular distally; blade 1–5 dm, ± round, palmately 7–11-lobed, sharply toothed. **INFL**: panicle, terminal, 1–3 dm; staminate fls below pistillate fls. **STAMINATE FL**: sepals 3–5; petals 0; stamens many, clustered; nectary 0. **PISTILLATE FL**: sepals 3–5; petals 0; ovary 3-chambered, bristly, styles 2-lobed, plumose, reddish. **FR** 1.2–2 cm diam, ± spiny. **SEED** 9–22 mm, smooth, shiny, mottled; scar appendaged. 2n=20. Disturbed areas, fields, roadsides; < 300 m. GV, CCo, SCo, expected elsewhere; e US; native to Eur. Highly TOXIC: seeds attractive to children, fatal when ingested.

STILLINGIA

Ann, per, < 2 m, monoecious; sap clear or milky. **STS** erect. **LVS** alternate, simple, entire or toothed; stipules minute, petioled; blade base gen with 2 glands. **INFL**: spike, axillary or terminal; bracts glandular. **STAMINATE FL**: calyx 2-lobed; petals 0, stamens 2; nectary disk 0. **PISTILLATE FL**: sepals 3 (overlapping in bud), reduced, or 0; petals 0; ovary 3-chambered, styles free, fused below, lobes 0. **FR** gen 3-lobed, separating into 3 1-seeded segments; central axis persistent. **SEEDS** pointed; scar not appendaged. 30 spp.: trop, warm temp. (Latin: Benjamin Stllingfleet, British botanist, 1702–1771) [Johnston & Warnock 1963 Southw Naturalist 8:100–106]

1. Lvs elliptic to ovate, sharply toothed, 3-veined; spikes axillary; seeds striate .... ***S. spinulosa***
1′ Lvs linear, entire or sparsely toothed below middle, 1-veined; spikes terminal; seeds smooth
 2. Spikes not projecting above lvs, open in lower 1/3, pistillate fls well separated; base of lf blade entire; seeds ± 2 mm .... ***S. linearifolia***
 2′ Spikes projecting above lvs, dense in lower 1/3, pistillate fls crowded; base of lf blade few-toothed; seeds 2.5–3 mm .... ***S. paucidentata***

S. linearifolia S. Watson (p. 579) Per < 7 dm. **LF**: blade 1–4 cm, < 2 mm wide, linear, margin entire. **INFL** 2–7 cm; glands of pistillate bracts stalked, ± 1 mm. **PISTILLATE FLS** 3–6 per infl, well separated; styles ± 1 mm. **FR** ± 3.5 mm. **SEED** ± 2 mm, smooth. Dry slopes, washes; < 1500 m. SW, D; AZ, Mex.

S. paucidentata S. Watson (p. 579) Per < 5 dm. **LF**: blade 3–8 cm, 1.5–5 mm wide, linear, margin few-toothed below middle, teeth ± spiny. **INFL** 2–7 cm; glands of pistillate bracts subsessile, ± 1–1.5 mm. **PISTILLATE FLS** 2–3 per infl, crowded; styles ± 2.5 mm. **FR** 4–4.5 mm. **SEED** 2.5–3 mm, smooth. Slopes, flats, creosote-bush scrub; < 1500 m. DMoj, n DSon; AZ.

S. spinulosa Torrey (p. 579) Ann or per < 10 dm. **LF**: blade gen 2–4 cm, 5–12 mm wide, elliptic to ovate, margin sharply toothed. **INFL** 1–2 cm; glands of pistillate bracts stalked, ± 2 mm. **PISTILLATE FLS** 1–2 per infl, ± open; styles 3–3.5 mm. **FR** 4–5 mm. **SEED** 3–3.5 mm, striate, minutely roughened. Sandy soils, dunes, creosote-bush scrub; < 900 m. D; s NV, AZ.

TETRACOCCUS

Shrub, gen 0.5–2 m, dioecious; sap clear. **ST**: axis erect; branches gen many, spreading to erect; twigs gen reddish, becoming gray, gen hairy, becoming glabrous; young lateral twigs short, sometimes becoming spine-like. **LVS** simple, cauline, alternate, opposite, or whorled in 3's, gen clustered at short, lateral branch tips; stipules 0; petiole < 2 mm; blade leathery, entire or toothed, base obtuse to acute. **STAMINATE INFL**: cyme, raceme, or panicle, axillary, sometimes clustered on short, lateral twigs, minutely bracted. **PISTILLATE INFL** axillary; fl 1. **STAMINATE FL**: sepals 4–10, 0.5–2 mm; petals 0; stamens 5–10, filaments glabrous or hairy; nectary disk ± minutely lobed. **PISTILLATE FL**: sepals 4–13, 2–5 mm; petals 0; nectary disk minutely lobed; ovary (2)3–5-chambered, styles = chambers, free, ± flattened, gen spreading. **FR** ± spheric, gen lobed, glabrous or short hairy, gen brown. **SEEDS** 1–2 per chamber, smooth, shiny; scar gen appendaged. 5 spp.: CA, AZ, Mex. (Latin: 4 seeds, from 4-lobed ovary in *T. dioicus*) [Dressler 1954 Rhodora 56:45–61]

1. Pistillate pedicel gen < 3 mm; ovary gen 3-lobed, 3-chambered; styles gen 3; stamen filaments glabrous; lvs gen alternate — se DMoj, DSon .... ***T. hallii***
1′ Pistillate pedicel 6–15 mm; ovary gen 4-lobed, 4-chambered; styles gen 4; stamen filaments soft-hairy at base; lvs opposite or whorled in 3's
 2. Lf blade linear to narrowly oblanceolate, entire to sparsely fine-toothed; s SCo, PR .... ***T. dioicus***
 2′ Lf blade ovate, lanceolate to ovate, sharp-toothed; n DMtns (Grapevine, Panamint mtns) .... ***T. ilicifolius***

T. dioicus C. Parry (p. 579) PARRY'S TETRACOCCUS **ST**: twigs sparsely fine-tomentose near axils, becoming glabrous. **LVS** gen opposite or 3-whorled; blade 10–30 mm, narrowly oblanceolate, tip rounded to acute, margin entire or sparsely fine-toothed, sometimes inrolled. **STAMINATE INFL**: gen raceme; pedicel 3–10 mm. **STAMINATE FL**: sepals 6–10, ovate to lanceolate; stamens 5–10, filaments 2.5–4 mm, base soft-hairy. **PISTILLATE FL**: pedicel 6–15 mm; sepals 7–13, 3–5 mm, widely lanceolate to ovate; ovary finely tomentose, chambers 4(5), style 3–3.5 mm. **FR** ± 6 mm, 7–9 mm wide, sparsely fine-tomentose. RARE. Dry slopes, chaparral; < 1000 m. s SCo (San Diego Co.), w PR; Baja CA. In cult.

T. hallii Brandegee (p. 579) **ST**: twigs sparsely short-strigose, becoming glabrous. **LVS** gen alternate, gen clustered on short, lateral twigs; blade 2–12 mm, oblanceolate to obovate, tip obtuse to rounded, margin entire. **STAMINATE INFL**: cyme; fls gen 1–5, gen clustered on short, lateral twigs; pedicel 3–5.5 mm. **STAMINATE FL**: sepals 4–6, ± round; stamens 4–8, filaments 1.5–2.5 mm, glabrous. **PISTILLATE FL**: pedicel 0.5–1(3) mm; sepals gen 5, 2–5 mm, ovate to deltate; ovary densely and finely gray-tomentose, chambers gen 3, sometimes 2 or 4, style 1.5–2 mm. **FR** 8–12 mm, 6–10 mm wide, finely tomentose. Rocky slopes, washes; < 1200 m. se DMoj, DSon; w AZ. ❀TRY.

T. ilicifolius Cov. & Gilman (p. 579) HOLLY-LEAVED TETRACOCCUS **ST**: twigs sparsely and finely brown-tomentose, becoming glabrous. **LVS** gen opposite or 3-whorled; blade 15–30, ovate to widely elliptic, sometimes narrowly ovate, tip obtuse to acute, margin toothed, teeth 8–20 per lf. **STAMINATE INFL**: panicle, ± dense; fls ± sessile; pedicel << 0.5 mm. **STAMINATE FL**: sepals 7–9, ± linear to lanceolate; stamens 7–9, filaments 2–3 mm, base soft-hairy. **PISTILLATE FL**: pedicel 8–15 mm; sepals 5–8, 2–4 mm, widely lanceolate to ovate; ovary densely tomentose, chambers 4, style ± 3 mm. **FR** 8–9 mm, 6–8 mm wide, brown tomentose. RARE. Dry, rocky slopes; 600–1700 m. n DMtns (Grapevine, Panamint mtns). In cult.

TRAGIA NOSEBURN

Per < 0.5 m, monoecious; hairs with stinging, nettle-like. **STS** spreading to erect, branched, sometimes twining. **LVS** gen simple, cauline, alternate; stipules persistent. **INFL**: raceme, terminal or opposite lf; staminate fls above pistillate fls. **STAMINATE FL**: sepals 3(–5); petals 0; stamens 3–6(50); nectary 0. **PISTILLATE FL**: sepals 4–8; petals 0; ovary 3-chambered, styles simple, ± fused at base. **FR** ± spheric. **SEEDS** 1 per chamber, smooth of ± rough; scar not appendaged. ± 100 sp.: trop, warm temp worldwide. (Latin: Tragus, name for Hieronymus Bock, German herbalist, 1498–1554) [Miller & Webster 1967 Rhodora 69:241–305]

T. ramosa Torrey (p. 579) Pl rough-hairy. **ST** 1–3 dm. **LF**: stipules 1–4.5 mm, lanceolate to ovate; petiole 2–20 mm; blade 1–2 cm, lanceolate to ovate, base truncate to ± lobed, margin coarsely, sharply toothed. **INFL** 0.5–1 cm, ± spreading; pedicels 1–2 mm; staminate fls 2–4; pistillate fl 1. **STAMINATE FL**: sepals 4–5, ± 1 mm, recurved; stamens 3–6, = sepals, filaments ± flattened. **PISTILLATE FL**: sepals 5, 1.5–2 mm; ovary < 2 mm diam, puberulent to finely bristly, styles fused in lower 1/3. **FR** 3–4 mm, 6–8 mm wide, depressed-spheric, sparsely and finely bristly. **SEED** 2.5–3.5 mm, ± spheric. Dry, rocky slopes, shrubland, pinyon/juniper woodland; 900–1700 m. DMtns (Clark, New York, Providence mtns); to c US, TX, Mex. [*T. stylaris* Muell. Arg.]

FABACEAE [Leguminosae] LEGUME FAMILY

Ann to tree. **LVS** gen compound, alternate, stipuled; lflets gen entire. **INFL**: gen raceme, spike, umbel or head; fls sometime 1–2 in axils. **FLS** gen bisexual, gen bilateral; hypanthium gen flat or cup-like; sepals gen 5, fused; petals gen 5, free, or the 2 lower ± fused; stamens 1–many, often 10 with 9 filaments at least partly fused, 1 (uppermost) free; pistil 1, ovary superior, gen 1-chambered, ovules 1–many, style, stigma 1. **FR**: legume, sometimes incl a stalk-like base above receptacle, dehiscent, or indehiscent and breaking into 1-seeded segments, or indehiscent, 1-seeded, and achene-like. **SEEDS** 1–several, often ± reniform, gen hard, smooth. ± 650 genera, 18,000 spp.: worldwide; with grasses, requisite in agriculture and most natural ecosystems. Many cult, most importantly *Arachis*, peanut; *Glycine*, soybean; *Phaseolus*, beans; *Medicago*; *Trifolium*; and many orns. [Polhill & Raven (eds) 1981 Advances in legume systematics; Allen & Allen 1981 Leguminosae] Family description and key to genera by Duane Isely.

Key to Groups

1. Fl radial; calyx, corolla gen inconspicuous; petals gen fused, lobes not overlapping in bud; stamens 5–many, often long-exserted; lf 2-pinnate (simple in alien *Acacia*) (Mimosoideae) .... **Group 1**
1′ Fl gen bilateral, sometimes ± radial; calyx, corolla gen conspicuous; petals overlapping in bud, free or 2 lowermost ± fused; stamens 1–10, gen ± incl; lf 1- or 2-pinnate (sometimes simple)
 2. Fl slightly bilateral (evidently bilateral in *Cercis*, upper petal inside lateral ones in bud); sepals ± free (fused in *Cercis)*; stamens exposed, visible without manipulation, free; lf 1- or 2-pinnate (simple in *Cercis*) (Caesalpinoideae) .... **Group 2**
 2′ Fl evidently bilateral, upper petal (banner) outside lateral ones (wings) in bud (only banner present in *Amorpha*, petal position not evident in some *Dalea*); sepals fused basally; stamens gen hidden within lower petals (keel) (exposed in *Amorpha*, some *Dalea*), gen all or 9 filaments fused (free in *Thermopsis*, *Pickeringia*); lf 1-pinnate (lflets often 3), some palmately compound (esp *Lupinus*) or simple (Papilionoideae) .... **Group 3**

Group 1: Mimosoideae

1. Stamens 10; 1° lflets gen 1–2 pairs .... **PROSOPIS**
1′ Stamens > 10; 1° lflets gen several pairs, or lf simple

2. Filaments free; lf 2-pinnate or simple .... **ACACIA**
2' Filaments fused below, free above; lf 2-pinnate
3. Infl a spike; petiole with a gland; fr indehiscent or slowly dehiscent but then the 2 valves not recurving; urban areas .... **ALBIZIA**
3' Infl a head; petiole without a gland; fr dehiscent, the 2 valves recurving; D .... **CALLIANDRA**

Group 2: Caesalpinioideae

1. Lf simple; corolla pink .... **CERCIS**
1' Lf compound; corolla gen yellowish, less often reddish or with orange or reddish marks
2. Lf 1-compound, pl unarmed or main lf axis gen > 2 cm, tip with 1 weak spine .... **SENNA**
2' Lf 2-compound, main axis gen < 2 cm, a strong spine, or, if lf interpreted as 1-compound (see *Parkinsonia*), subtended by a strong spine gen < 2 cm, main lf axis not a spine at tip
3. Pl unarmed
4. Shrub, > 2 m .... [2]**CAESALPINIA**
4' Per, < 30 cm .... **HOFFMANNSEGGIA**
3' Pl armed
5. Pl with scattered prickles; sepals not all alike; fr indehiscent .... [2]**CAESALPINIA**
5' Pl with spines at nodes or thorns in lf axils (see lf scars); sepals all alike; fr dehiscent
6. Pl with thorns in lf axils (see lf scars); main lf axis, main 1° lflet axes falling with 2° lflets .... **CERCIDIUM**
6' Pl with spines at nodes; main lf axis persistent as spine, main 1° lflet axis persistent as ribbon-like streamer after 2° lflets fall (or, if lf interpreted as 1-compound, subtended by a spine, main lf axis persistent as ribbon-like streamer after 1° lflets fall) .... **PARKINSONIA**

Group 3: Papilionoideae

1. Shrub, tree
2. Pl armed with prickles, spines, or thorns
3. Lvs simple, sometimes small or falling early
4. Pl gland-dotted; corolla blue to pink-purple; seed 1 .... [4]**PSOROTHAMNUS**
4' Pl not gland-dotted; corolla yellow to reddish; seeds few–several
5. Corolla reddish; fr narrowed between seeds, glabrous .... **ALHAGI**
5' Corolla yellow; fr not narrowed between seeds, hairy .... **ULEX**
3' Lvs all or mostly compound, gen persistent
6. Lf palmately compound; all filaments free .... **PICKERINGIA**
6' Lf pinnately compound; all or 9 filaments fused
7. Pl gland-dotted; fr indehiscent; seed 1 .... [4]**PSOROTHAMNUS**
7' Pl not gland-dotted; fr dehiscent (sometimes slowly so); seeds gen several
8. Lf odd-pinnate; corolla white or pink — fr flat .... [2]**ROBINIA**
8' Lf even-pinnate; corolla not white
9. Corolla yellow; fr flat; lflets 8–12 .... [2]**CARAGANA**
9' Corolla reddish to pink, or 2-colored; fr plump; lflets as follows
10. Lflets 4; fr not narrowed between seeds; deciduous shrub .... **HALIMODENDRON**
10' Lflets 8–19; fr gen narrowed between seeds; evergreen shrub or tree .... **OLNEYA**
2' Pl unarmed
11. Petal 1 (the banner) — infl spike-like; fr indehiscent; seed 1 .... **AMORPHA**
11' Petals 5
12. Filaments all fused; lvs ternately palmately compound to simple
13. Lvs tiny, simple; sts gen ± lfless .... **SPARTIUM**
13' Lvs mostly not tiny, simple or gen compound; sts clearly leafy
14. Lf palmately compound, lflets 5–11; corolla blue, white, or yellow, banner glabrous, keel beaked .... [2]**LUPINUS**
14' Lf sometimes appearing simple, lflets 1–3; corolla yellow (or if white then banner hairy), keel obtuse
15. Calyx cylindric, 8–9 mm; banner white, purple-lined, or spotted, gen hairy .... **CHAMAECYTISUS**
15' Calyx bell-shaped, < ± 7 mm; banner yellow, gen not hairy
16. Style gen abruptly curved ± at middle; upper lip of calyx barely 2-lobed .... **CYTISUS**
16' Style ± abruptly bent at tip; upper lip of calyx strongly 2-lobed .... **GENISTA**
12' Filaments all free or 9 fused, 1 (uppermost) free or 0; lvs gen pinnately compound
17. Lf even-pinnate; corolla 1.6–2.2 cm, yellow .... [2]**CARAGANA**
17' Lf odd-pinnate, if clearly neither even nor odd-pinnate (as sometimes in *Lotus*), corolla < 1.6 cm, yellow or not
18. Infl an umbel, sometimes a cluster of 1–3 fls; corolla gen yellow .... [4]**LOTUS**
18' Infl a raceme; corolla yellow or not
19. Corolla gen yellow; fr large, papery, inflated .... **COLUTEA**
19' Corolla not yellow; fr small or large, not papery, not inflated
20. Fl 15–25 mm; pl not gland-dotted; fr dehiscent; seeds several .... [2]**ROBINIA**
20' Fl 6–10 mm; pl gland-dotted; fr indehiscent; seed 1 .... [4]**PSOROTHAMNUS**
1' Ann, per, subshrub
21. Lflets 0, but stipules lflet-like — lf axis ending as a tendril .... [3]**LATHYRUS**
21' Lflets 2–many
22. Lf palmately compound; lflets gen 3–9

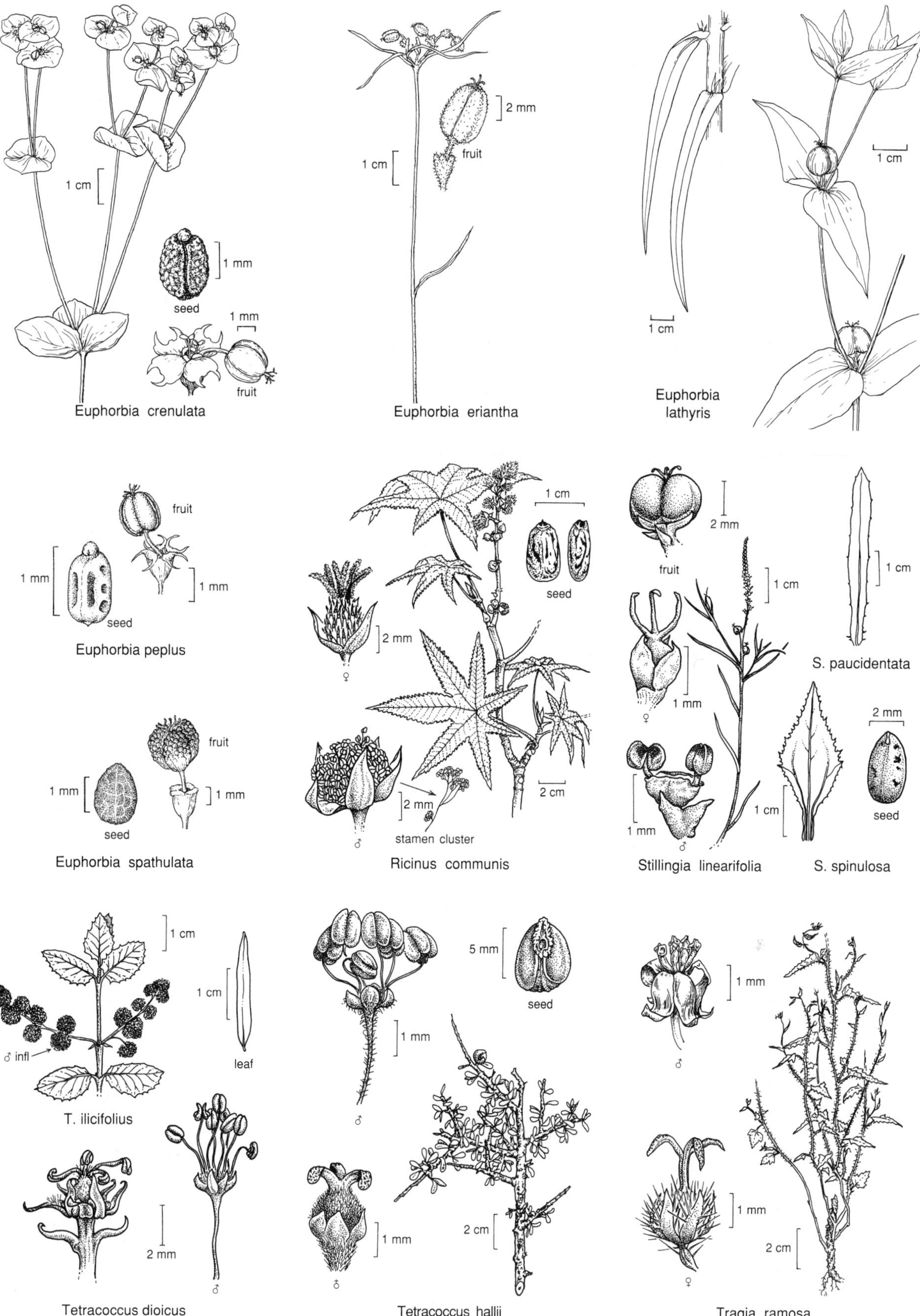
1 cm
1 mm
seed
1 mm
fruit
Euphorbia crenulata
2 mm
fruit
1 cm
Euphorbia eriantha
1 cm
1 cm
Euphorbia lathyris
fruit
1 mm
1 mm
seed
Euphorbia peplus
fruit
1 mm
1 mm
seed
Euphorbia spathulata
1 cm
seed
2 mm
♀
2 mm
stamen cluster
2 cm
♂
Ricinus communis
2 mm
fruit
1 cm
1 cm
S. paucidentata
♀
1 mm
2 mm
1 cm
seed
1 mm
♂
Stillingia linearifolia
S. spinulosa
1 cm
1 cm
leaf
♂ infl
T. ilicifolius
2 mm
♂
Tetracoccus dioicus
5 mm
seed
1 mm
♂
1 mm
2 cm
♀
Tetracoccus hallii
1 mm
♂
1 mm
♀
2 cm
Tragia ramosa

23. All filaments free; corolla yellow, 15–25 mm .... **THERMOPSIS**

23′ All or 9 filaments fused; corolla not yellow or, if so, gen < 15 mm

24. Lflet gland-dotted; fr indehiscent or breaking transversely; seed 1

25. Lvs mostly basal or clustered; lflets widely obovate to oblanceolate; fr incl in calyx exc for conspicuous beak .... **PEDIOMELUM**

25′ Lvs cauline; lflets obovate to linear; fr exserted from calyx .... **PSORALIDIUM**

24′ Lflets not gland-dotted; fr indehiscent or dehiscent through longitudinal sutures; seeds gen several

26. Filaments of all stamens fused at least in basal 1/2; lflets gen 5–9, entire .... [2]**LUPINUS**

26′ Filaments of 9 stamens fused, the 10th (uppermost) free; lflets 3–5, entire, or margin toothed or wavy

27. Lflets 5, lower 2 in stipular position, others palmate, stipules reduced to bumps or not apparent; infl an umbel, corolla yellow .... [4]**LOTUS**

27′ Lflets gen 3, lower 2 not in stipular position, stipules gen papery or membranous, not reduced to bumps, rarely lflet-like, if so then infl not an umbel and corolla not yellow

28. Lflet entire; fr not enclosed in corolla .... [3]**ASTRAGALUS**

28′ Lflet margin ± toothed or wavy; fr enclosed in persistent corolla .... [2]**TRIFOLIUM**

22′ Lf pinnately to subpalmately compound (axis apparent beyond lowermost lflets); lflets 2–many

29. Lflets 2, main axis ending as a tendril or bristle .... [3]**LATHYRUS**

29′ Lflets 3 or more, main axis ending as a lflet, tendril, or bristle

30. Lflets 3; fr (exc *Phaseolus*) gen indehiscent

31. Keel petals spirally coiled; lflet gen lobed — trailing or twining vine .... **PHASEOLUS**

31′ Keel petals not spirally coiled; lflet margin entire, toothed, or wavy

32. Lflet margin toothed or wavy — fr ovoid or reniform, gen 1-seeded, or ± coiled, several-seeded

33. Corolla persistent, enclosing fr, yellow, 3.5–5 mm .... [2]**TRIFOLIUM**

33′ Corolla deciduous, various in color and size, incl above

34. Fr spirally coiled, rarely only sickle-shaped, gen prickly; seeds > 2 .... [2]**MEDICAGO**

34′ Fr not spirally coiled, not prickly; seeds 1–2

35. Fr reniform, with curved ridges; corolla 2–3 mm .... [2]**MEDICAGO**

35′ Fr ovoid, with transverse ridges or a ± distinct network of lines; corolla 2.5–7 mm **MELILOTUS**

32′ Lflet margins not toothed, not wavy

36. Pl ± gland-dotted; seed 1; fr, exc beak, gen incl in calyx

37. Corolla at least partly blue to purple; calyx not swollen in fr .... **HOITA**

37′ Corolla white to yellow; calyx conspicuously swollen in fr .... **RUPERTIA**

36′ Pl not gland-dotted; seeds gen 2–several; fr exserted from calyx or incl exc for beak

38. Infl a raceme; fr exserted from calyx .... [3]**ASTRAGALUS**

38′ Infl an umbel; fr exserted from calyx or incl exc for beak .... [4]**LOTUS**

30′ Lflets > 3 on all or most lvs; fr dehiscent or indehiscent

39. Lf even-pinnate, main axis ending as a bristle or tendril

40. Calyx lobes > 2 × tube; pl hairy — fls 1–3 at peduncle tip .... **LENS**

40′ Calyx lobes, some or all, < 2 × tube; pl glabrous or hairy

41. Lflets 30–60; fr either 15–20 cm or wavy-margined and breaking into 1-seeded segments

42. Fr wavy-margined and breaking into 1-seeded segments; fls 8–11 mm .... **AESCHYNOMENE**

42′ Fr 15–20 cm, neither wavy-margined nor breaking into 1-seeded segments; fls 10–18 mm **SESBANIA**

41′ Lflets < 30; fr < 8 cm, not wavy-margined or 1-seeded

43. Stipules lflet-like, often > lflets; style longitudinally folded, opening on convex side; lflets 4–6, glabrous — waif .... **PISUM**

43′ Stipules gen not lflet-like but sometimes ± = lflets; style not folded; lflets 4–many, hairy or glabrous

44. Style ± flat, puberulent on concave side ± 1/3–1/2 length; lflets ± rolled in bud .... [3]**LATHYRUS**

44′ Style gen round in X-section, puberulent all around at tip or mostly on convex side; lflets folded in bud .... **VICIA**

39′ Lf odd-pinnate, main axis ending as a lflet

45. Corolla wings tiny, << keel; lflet upper sides finely red-dotted; fr 1-seeded, leathery, strongly net-ridged .... **ONOBRYCHIS**

45′ Corolla wings not tiny, ± = keel; lflets not red-dotted (may be dark gland-dotted); fr without above combination of traits

46. Pl gland-dotted on sts, lflets, or both; fr indehiscent

47. Fr several-seeded, long-exserted from calyx, gen prickly; lflets 6–10 mm wide .... **GLYCYRRHIZA**

47′ Fr 1-seeded, incl in calyx, glandular; lflets gen < 6 mm wide

48. Petals all arising from receptacle; stamens 10; infl sometimes head-like .... [4]**PSOROTHAMNUS**

48′ Petals, exc banner, arising from side or top of column of fused filaments; stamens 5 or 9–10; infl not head-like

49. Pl prostrate or decumbent

50. St glandular .... [2]**DALEA**

50′ St not glandular .... [3]**MARINA**

49′ Pl erect or ascending

51. Infl a dense spike; stamens 5 .... [2]**DALEA**

51′ Infl an open raceme; stamens 9–10 .... [3]**MARINA**

46′ Pl not obviously gland-dotted; fr dehiscent or indehiscent

52. Infl an umbel; fls many, sometimes 2–3 .... [4]**LOTUS**

52′ Infl a spike or raceme, sometimes head-like; fls many, rarely 1–2
53. Seed 1; lflet with a single, large gland at tip; infl a dense spike, ± 1 cm — fl 5–5.5 mm .... [3]**MARINA**
53′ Seeds several; lflet without a gland at tip; infl a raceme or spike
54. Fr indehiscent, breaking into 1-seeded segments; corolla 6–9 mm; lflet gen < 6 mm **ORNITHOPUS**
54′ Fr gen dehiscent, sometimes slowly so, not breaking into 1-seeded segments; corolla often > 9 mm; lflet often > 6 mm
55. Stigma or style tip finely hairy — fr base stalk-like
56. Stipules spiny; fr not inflated; fls white to pink .... **PETERIA**
56′ Stipules 0 or spiny; fr inflated; fls red .... **SPHAEROPHYSA**
55′ Stigma and style glabrous
57. Keel tip rounded to acute, rarely short-beaked; fr 1-chambered or, if 2-chambered, septum arising from lower suture (rarely a narrow flange from upper suture also present) .... [3]**ASTRAGALUS**
57′ Keel tip distinctly beaked; fr gen 2-chambered, the partial or complete septum arising from upper suture .... **OXYTROPIS**

ACACIA

Elizabeth McClintock

Tree, shrub, armed or unarmed. **LVS** even-2-pinnate or, if simple, true blades 0, petioles and midribs blade-like (comprising phyllodia), gen alternate, gen evergreen; axes with prominent raised glands or not. **INFL**: heads, spheric, gen axillary, these solitary or in racemes or panicles, or fls in spikes. **FL** radial; sepals, petals inconspicuous; stamens many, conspicuous, exserted, free. **FR** gen dehiscent, sometimes tardily so, flat or ± cylindric. ± 1200 spp.: trop, subtrop, esp Australia. (Greek: sharp point) [Whibley 1980 Acacias of South Australia; Clarke et al. 1989 Systematic Botany 14:549–564] Australian spp. cult, sometimes naturalized and spreading in CA (seed arilled, stalk often elongated, encircing seed or not).

1. Lvs 2-pinnate, with blades and axes
2. Shrub, armed; fls in spikes or heads
3. Spines (modified stipules), straight, paired, only at nodes; infls heads, gen < lvs .. ***A. farnesiana*** var. ***farnesiana***
3′ Prickles, curved, not paired, irregularly scattered on st; infls spikes, gen > lvs .... ***A. greggii***
2′ Tree, unarmed; fls in heads
4. Lf with 3–6 pairs of 1° lflets
5. Lf silvery blue; 2° lflets linear, 5–7 mm, touching .... ***A. baileyana***
5′ Lf dark green; 2° lflets lanceolate, 2–8 cm, gen not touching .... ***A. elata***
4′ Lf with > 6 pairs of 1° lflets
6. Twig prominently winged, ± glabrous; 1° lflets 5–15 pairs, 3–7 cm, not overlapped; 2° lflets 15–35 pairs, 5–15 mm, ± overlapped .... ***A. decurrens***
6′ Twig ± angled, without wing-like ridges, hairy; 1° lflets 8–25 pairs, 3–5 cm, overlapped; 2° lflets 20–70 pairs, 2–5 mm, overlapped
7. Lf main axis with glands at junction of each pair of 1° lflets; twig, lf gen silvery hairy .... ***A. dealbata***
7′ Lf main axis with glands at junction of and gen between each pair of 1° lflets; twig, lf ± green .... ***A. mearnsii***
1′ Lvs simple, with no blades but petioles and midribs blade-like
8. Fls in spikes
9. Lf blade-like, with 2–3 veins more prominent than others, abruptly narrowed to a short, sharp tip ... ***A. longifolia***
9′ Lf needle-like or slightly flat, with 1 vein (midrib) more prominent than others, tapered to a spine-like tip .... ***A. verticillata***
8′ Fls in heads, these solitary or in racemes or panicles
10. Shrub with stipular spines; lf < 2 cm; heads solitary .... ***A. paradoxa***
10′ Tree, shrub without stipular spines; lf > 2 cm; heads in racemes or panicles
11. Lf with 2 or more prominent, longitudinal veins
12. Shrub, densely branched, spreading from base; lf narrowly oblong to obovate, gen 6–12 mm wide, simple throughout .... ***A. cyclops***
12′ Tree, gen with a single trunk; lf lanceolate or oblanceolate, gen 1–3 cm wide, even-2-pinnate at tip or not .... ***A. melanoxylon***
11′ Lf with 1 prominent, longitudinal midvein
13. Twig not ridged; lf widely lancolate to oblanceolate, curved to sickle-shaped; seed stalk ± 1/2 length of seed .... ***A. pycnantha***
13′ Twig ridged; lf linear-lanceolate to narrow-oblanceolate, straight or slightly curved, not sickle-shaped; seed stalk encircling seed in double fold .... ***A. retinodes***

A. baileyana F. Muell. COOTAMUNDRA WATTLE Tree < 6 m, unarmed. **ST**: twig ridged, ± hairy. **LF** gen 2-pinnate, silvery blue; main axis with glands at junctions of upper pairs of 1° lflets; 1° lflets 3–6 pairs, overlapped; 2° lflets 12–20 pairs, touching, 5–7 mm, linear. **INFL**: raceme of 8–10 heads, axillary, > lf. **FL** bright yellow. **FR** 4–10 cm, ± straight, slightly narrowed between seeds, brown, glaucous when young. **SEED** arilled; stalk short. Uncommon. Roadsides, disturbed areas; 50–300 m. SnFrB, SCoRO, WTR; native to se Australia.

A. cyclops G. Don Shrub gen 1–3 m, unarmed. **ST**: twig ± angled. **LF** simple, 4–9 cm, narrowly oblong or obovate, ± straight; tip ± oblique with blunt point; veins 3–5, longitudinal, not prominent. **INFL**: raceme of 2–3 heads, axillary, < 1/2 lf. **FL** bright yellow. **FR** 4–9 cm, curved or twisted, not narrowed between seeds, leathery to ± woody, gen gray-brown, often persistent after seed release. **SEED**: stalk thick, encircling seed in double fold. Uncommon. Disturbed areas, coastal dunes; 50 m. SCo; native to sw Australia. Often cult along highways.

A. dealbata Link SILVER WATTLE Tree < 12 m, unarmed. **ST**: twig angled, minutely, silvery hairy. **LF** 2-pinnate, silvery hairy; petiole not decurrent on winged st; main axis with glands at junction of each pair of 1° lflets; 1° lflets 10–25 pairs, overlapped; 2° lflets 20–50 pairs, overlapped, 2–5 mm, linear. **INFL**: raceme or panicle of 25–30 heads, axillary, gen > lf. **FL** bright yellow. **FR** 5–8 cm, straight or ± curved, slightly narrowed between seeds, glaucous or light purplish brown. **SEED** arilled; stalk short. Locally common. Disturbed areas, often roadsides; < 500 m. NCoRO, SnFrB, SCoRO, SCo; native to se Australia. [*A. decurrens* (Wendl.) Willd. var. *dealbata* (Link) F. Muell.] Often reported as *A. decurrens*.

A. decurrens (Wendl.) Willd. GREEN WATTLE Tree < 15 m, unarmed; twig prominently winged, ± glabrous. **LF** 2-pinnate, olive-green; petiole ± angled, decurrent on twig; main axis with glands at junction of each pair of 1° lflets; 1° lflets > 6 pairs, not overlapped, 3–7 cm; 2° lflets 15–35 pairs, ± overlapped, 5–15 mm, linear, gen glabrous. **INFL**: raceme or panicle of 20–25 heads, axillary, > lf or not. **FL** bright yellow. **FR** 4–10 cm, straight or ± curved, slightly narrowed between seeds, glabrous, dark brown. **SEED** arilled; stalk short. Disturbed, coastal, urban areas?. Native to e Australia. Sometimes confused with *A. dealbata*. Reports of naturalization unconfirmed.

A. elata Benth. CEDAR WATTLE Tree gen < 15 m, unarmed. **ST**: twig round in X-section, with short, gold hairs when young. **LF** 2-pinnate, dark green; petiole with 1 gland at tip, just below lowest pair of 1° lflets; 1° lflets 3–6 pairs, not touching; 2° lflets gen 8–14 pairs, gen not touching, 2–8 cm, lanceolate. **INFL**: panicle of < 20 heads, axillary, < 25 cm. **FL** pale yellow. **FR** 10–15 cm, ± straight, barely narrowed between seeds. **SEED** arilled; stalk short. Uncommon. Disturbed urban areas; < 100 m. SCoRO (Santa Barbara Co.); native to e Australia.

A. farnesiana (L.) Willd. var. *farnesiana* SWEET ACACIA Shrub < 3 m; stipular spines straight, becoming white in age. **ST**: twig ± ridged, hairy. **LVS** gen 2-pinnate, deciduous; 1° lflets 2–5 pairs, 0.5–2.5 cm; 2° lflets 8–25 pairs, ± overlapped, 2–5 mm, oblong. **INFL**: heads gen 1–3 per axil, gen < lf. **FL** golden yellow to dull orange. **FR** indehiscent, 7–15 cm, ± curved, ± plump, ± narrowed between seeds, very hard when dry, brown. **SEEDS** embedded in pulp; stalk 0. Dry slopes in chaparral; < 300 m. SCo (s San Diego Co.); native from w Florida to s TX, s AZ, Mex, trop Am, alien worldwide, often a troublesome weed. [*A. minuta* (M.E. Jones) Beauch.ssp. *minuata*, coastal scrub acacia, misapplied] Cult in s Eur for fl oils used in perfumes.

A. greggii A. Gray (p. 593) CATCLAW Shrub, small tree 2.5–7 m, with curved prickles on st. **ST**: twig ± angled, hairy. **LVS** 2-pinnate, alternate, clustered on short-shoots or not, deciduous, gray-green; 1°lflets 2–3 pairs, separated, 1–1.5 cm; 2° lflets < 10 pairs, ± overlapped, < 6 mm, oblong. **INFL**: spikes 1 or more, clustered with lvs on short-shoots, gen > lf. **FL** light yellow. **FR** 5–15 cm, recurved or twisted, flat, narrowed between seeds, glaucous, brown. **SEED**: stalk 0. Uncommon. Flats, washes; 100–1400 m. D; to TX, Mex. ❀SUN,DRN:**7–10,12,14–16**,17,18,**19–24**&IRR:**11,13**.

A. longifolia (Andrews) Willd. SYDNEY GOLDEN or GOLDEN WATTLE Shrub, small tree 3–7 m, unarmed. **ST**: twig angled, glabrous or minutely hairy when young. **LF** simple, 5–15 cm, linear-lanceolate or obovate; 2–3 longitudinal veins more prominent than others. **INFL**: spike, 2–4 cm, axillary, < lf. **FL** bright yellow. **FR** 5–10 cm, ± straight, ± cylindric, narrowed between seeds, ending in a curved beak, brown. **SEED**: stalk short, aril thick, cup-shaped. Uncommon. Disturbed places, esp sandy, coastal areas; < 150 m. SnFrB, SCoRO, SCo; native to e Australia.

A. mearnsii De Wild. BLACK WATTLE Tree 7–10 m, unarmed, all parts softly hairy. **ST**: bark often exuding gum; twig angled, green. **LF** 2-pinnate, olive-green; main axis with glands at junction of and sometimes between each pair of 1° lflets; 1° lflets 8–25 pairs, ± overlapped, 3–5 cm; 2° lflets 30–70 pairs, overlapped, < 5 mm, linear. **INFL**: raceme or panicle of 30–35 heads, axillary, ± = lf. **FL** pale yellow, ± fragrant. **FR** 5–10 cm, straight, flat, ± narrowed between seeds, almost black; hairs fine, white. **SEED** arilled; stalk short. Disturbed, urban areas?. Native to e Australia. Confused with *A. dealbata*. Reports of naturalization unconfirmed.

A. melanoxylon R.Br. BLACKWOOD ACACIA Tree 8–15 m, unarmed. **ST**: twig ± angled. **LVS**: juvenile (on seedlings, young branches) 2-pinnate; adult simple (sometimes 2-pinnate at tips), 6–15 cm, lanceolate, veins 2–5, longitudinal. **INFL**: raceme of 2–8 heads, axillary, < and ± obscured by lvs. **FL** pale yellow. **FR** 4–12 cm, curved or twisted, flat, not or barely narrowed between seeds, brown. **SEED**: stalk stout, wavy, encircling seed in an irregular, double fold. Uncommon. Disturbed areas; < 200 m. SnFrB, SCoRO, SCo, s ChI; native to e Australia. Often cult.

A. paradoxa DC. KANGAROO THORN Shrub < 3 m; spines straight, stipular. **ST**: twig angled. **LF** simple, 1–2 cm, oblong or lanceolate; margin ± wavy; midvein off-center. **INFL**: solitary heads, stalked, axillary, ± = lf. **FL** bright yellow. **FR** 4–7 cm, ± straight, cylindric, little narrowed between seeds, brown, gen softly hairy. **SEED**: stalk in 2–3 thick folds at tip of seed. Abundant. Plantings, spreading to other disturbed areas; ± 200 m. SnFrB, SCoRO, SCo; native to s Australia. [*A. armata* R.Br.] NOXIOUS WEED.

A. pycnantha Benth. GOLDEN WATTLE Tree < 8 m, unarmed. **ST**: twig not ridged. **LF** simple, 5–20 cm, widely lanceolate to oblanceolate, sickled-shaped, gen widest above middle; 1 vein more prominent than others. **INFL**: raceme or panicle of 20–25 heads, axillary, gen ± = lf. **FL** golden yellow. **FR** 5–12 cm, gen straight, flat, slightly narrowed between seeds, brown. **SEED** arilled; stalk ± 1/2 seed. Uncommon. Disturbed areas; < 200 m. SCoRO, SCo; native to se Australia. Cult.

A. retinodes Schldl. EVERBLOOMING ACACIA Tree < 8 m, unarmed. **ST**: twig gen ridged. **LF** simple, 3–20 cm, linear-lanceolate to narrow-oblanceolate, gen widest at and above middle; 1 longitudinal vein more prominent than others. **INFL**: raceme of 6–8 heads, axillary, < lf. **FL** pale yellow. **FR** 3–14 cm, 6–8 mm wide, flat, gen straight, slightly narrowed between seeds, brown. **SEED** arilled; stalk encircling seed in a double fold. Uncommon. Disturbed areas; < 900 m. SnFrB, SCoRO, SCo, WTR; native to se Australia.

A. verticillata (L'Her.) Willd. STAR ACACIA, PRICKLY MOSES Shrub 1–4 m; lf tips spine-like. **ST**: twig ridged, ± hairy. **LF** simple, gen whorled, < 15 mm, needle-like; 1 vein more prominent than others. **INFL**: group of 1–3 stalked spikes, axillary, ± = lf. **FL** pale yellow. **FR** 3–8 cm, 3–4 mm wide, straight, flat, barely narrowed between seeds. **SEED** arilled; stalk short, curved. Uncommon. Disturbed areas; < 150 m. SnFrB, SCoRO, SCo; native to se Australia.

AESCHYNOMENE

James C. Hickman

Per or shrub, unarmed. **ST** erect. **LVS** even-1-pinnate; main axis ending in a tiny bristle; lflets alternate or opposite, many, small, sensitive to touch. **FL**: corolla white to red or purple; filaments partly fused, gen splitting into 2 groups of 5. **FR** ± narrowed between seeds, breaking into 1-seeded segments. ± 150 spp.: esp standing water, trop and warm regions.

A. rudis Benth. Pl somewhat bristly-hairy, finely gland-dotted. **ST** emergent. **LF**: stipule ± peltate, deciduous, lower lobe rounded to fringed; lflets gen 9–10 mm, entire or minutely serrate. **FL**: corolla 8–15 mm, whitish, purple-tinged. **FR** 4–7 mm wide; lower margin indented between seeds; stalk-like base 4–10 mm. Rice fields; < 20 m. ScV (ne Colusa Co.), expected more widely; se US; native to Asia. Range as rice-field weed apparently expanding.

ALBIZIA

Elizabeth McClintock

Shrub, tree, unarmed. **LVS** even-2-pinnate, alternate. **INFL**: spike or spheric head, axillary or terminal. **FL** radial, ± yellow, white, or pink; sepals, petals inconspicuous; stamens many, conspicuously exerted, filaments fused below into tube ± = petals, free above. ± 150 spp.: esp trop. (F. del Albizzi, Florentine nobleman, 18th century)

A. lophantha (Willd.) Benth. PLUME ACACIA Shrub, small tree. **LVS** evergreen; petiole with a gland; 1° lflets 7–15 pairs; 2° lflets ± 25 pairs, < 1 cm. **FL** greenish yellow. **FR** indehiscent or slowly dehiscent, 5–10 cm, oblong, flat exc over seeds, barely narrowed between seeds, glabrous, brown. Gen disturbed, coastal, urban areas; < 300 m. CCo, SnFrB, SCoRO, SCo; native to sw Australia. [*A. distachya* (Vent.) Macbr.]

ALHAGI

Duane Isely

Per, shrub; thorns axillary; rhizome spreading. **LVS** simple, small. **INFL**: raceme, axillary; main axis a thorn. **FL**: 9 filaments fused, 1 free. **FR** indehiscent, oblong, round in X-section, narrowed between seeds. ± 3 spp.: Medit, w Asia. (Arabic: pilgrim)

A. pseudalhagi (M. Bieb.) Desv. (p. 593) CAMEL THORN **ST** much branched, 3–10 dm, greenish. **LF** 7–20 mm, elliptic or obovate. **FL**: corolla 8–9 mm, red-purple. **FR** 1–3 cm, narrowed between seeds, becoming glabrous; stalk-like base short. $2n$=16. Uncommon. Arid agricultural areas; esp < 500 m. GV, s SNE, D; sporadic to w TX; native to w Asia. [*A. camelorum* Fischer; *A. maurorum* Medikus] Desert forage, source of manna. NOXIOUS WEED. Most infestations have been eradicated.

AMORPHA FALSE INDIGO

Duane Isely

Shrub, unarmed, gland-dotted. **LVS** odd-1-pinnate; stipules bristle-like, ephemeral; lflets with tiny stipule-like appendages. **INFL**: raceme, terminal, spike-like. **FL**: petal 1 (banner); stamens 10, exserted, filaments fused near bases; style puberulent. **FR** indehiscent. **SEED** 1. ± 15 spp.: N.Am. (Greek: deformed, from single petal) [Wilbur 1975 Rhodora 77:337–409]

1. Main lf axis without prickle-like glands — esp PR .... ***A. fruticosa***
1′ Main lf axis with prickle-like glands .... ***A. californica***
 2. Pl hairy; widespread .... var. ***californica***
 2′ Pl glabrous or nearly so; s NCoR, n SnFrB .... var. ***napensis***

A. californica Nutt. **LF**: main axis with prickle-like glands; lflet midrib gen ending as a sessile gland. **INFLS** gen scattered. $2n$=20. Wooded, shrubby, open slopes or chaparral; < 2300 m. s NCoR, CaRF, n SNF, n SnFrB, SCoR, TR, n&e PR; AZ, n Baja CA.

var. ***californica*** (p. 593) Pl hairy. Wooded, shrubby, or open slopes; < 2300 m. Range of sp. ❀STBL.

var. ***napensis*** Jepson Pl glabrous or nearly so. Chaparral; < 2000 m. s NCoR (Napa, Sonoma cos.), n SnFrB (Marin Co.). ❀DRN:**15–17**&SHD:**14**;STBL.

A. fruticosa L. (p. 593) **LF**: main axis without prickle-like glands; lflet midrib often ending as a short bristle, gland-tipped or not. **INFLS** gen clustered. $2n$=20,40. Along streams, canyons; < 1200 m. c&s SCo, SnBr, PR; to s Can (Manitoba), e US, n Mex. [var. *occidentalis* (Abrams) Kearney & Peebles] Highly variable, esp outside CA. ❀DRN:**6,15–17**&IRR:1,**2,3,7,14,18–24**;STBL.

ASTRAGALUS

Richard Spellenberg

Ann or per from crown, glabrous to hairy; hairs sometimes forked at base, branches parallel with lf surface, sometimes very unequal. **ST** 0 or prostrate to erect. **LVS** odd-1-pinnate; lflets gen jointed to midrib; stipules membranous, sometimes fused around st at st base. **INFL**: raceme, axillary, sometimes head- or umbel-like; fls 2–many. **FL** bilateral; calyx 5-lobed; banner outside wings in bud, keel blades with small protrusion at base locking into pit on adjacent wing; 9 filaments fused, 1 free; ovary (and fr) gen sessile, style slender, stigma minute. **FR** gen 1- or ± 2-chambered, often mottled, gen becoming ± dry; placenta on upper suture. **SEEDS** 2–many, smooth, compressed, ± notched at attachment scar. > 2000 spp.: ± worldwide (380 in N.Am, 94 in CA incl many rare taxa). (Greek: ankle-bone or dice, perhaps from rattling of seeds within fr) [Barneby 1964 Mem NY Bot Gard 20:1–1188; Isely 1986 Iowa State J Res 61:157–289] Very difficult; both fl and fr needed for identification; many good spp. appear similar; some spp. complexes need study. Taxa near province boundaries may appear in > 1 key. Vars. keyed under spp. for simplicity; spp. with vars. so identified in key. Fr length incls beak and any stalk-like base unless fr body specified.

Key to Groups

1. Corolla bright red, banner 35–41 mm; calyx 18–24 mm; fr 25–40 mm, densely woolly .... ***A. coccineus***
1′ Corolla not red, banner < 35 mm; calyx < 18 mm; fr often < 25 mm, often not woolly
 2. Lflet ± 1 mm wide, spine-tipped; raceme 1–3-fld, ± incl in axil; peduncle inconspicuous or 0; fl and fr 4–9 mm .... ***A. kentrophyta*** vars.

2′ Lflet gen > 1 mm wide, not spine-tipped; raceme often > 3-fld, well exserted from axil; peduncle conspicuous; fl 3–30 mm; fr 2–60 mm
3. Pls of CA-FP ..... **Group 1**
3′ Pls of GB or D
4. Pls of D ..... **Group 2**
4′ Pls of GB ..... **Group 3**

Group 1: plants of California Floristic Province

1. Ann
2. Banner ± 3 mm; fls early dense, ascending, then well separated, reflexed; fr reflexed, 2.8–4.2 mm, width > depth — frs overlapping, ovate or ± round in top view — common ..... ***A. gambelianus***
2′ Banner gen > 3 mm; fls remaining dense, erect if banner ± 3 mm, otherwise erect, spreading, or reflexed; fr gen > 4 mm, if not then width ± = or < depth
3. Fl, fr strongly ascending, in dense, head-like racemes; fr 2–4 mm, 2-lobed in X-section, 2-chambered ..... ***A. didymocarpus***
3′ Fl, fr often ascending early, then spreading or reflexed, often in loose racemes; fr > 4 mm, gen unlobed in X-section, 1–2-chambered
4. Infl 10–40-fld; fr inflated, papery, 1-chambered, 4–17 mm wide, sides gen strongly convex
5. Corolla gen white; fls, frs in dense, oblong or subspheric, head-like racemes ..... ***A. hornii*** var. ***hornii***
5′ Corolla gen pink-purple; fls early dense, then ± well separated, frs loosely clustered in elongate racemes ..... ***A. palmeri***
4′ Infl 2–14-fld; fr narrow, ± 2-chambered, < 4 mm wide, sides parallel or slightly convex
6. Pl silvery-canescent; local, Cushenbury Canyon (ne SnBr, adjacent DMoj) ..... ***A. albens***
6′ Pl greenish, ± hairy; widespread, not of Cushenbury Canyon
7. Fr body 5–10 mm, tapered to spine-like or hooked beak ± = body; immature seeds 2–6; frs in head- or umbel-like racemes ..... ***A. breweri***
7′ Fr body > 10 mm or beak < 1/4 × body; immature seeds 5–20; frs in sometimes head- or umbel-like racemes
8. Keel > wings; fr base stalk-like, receptacle with peg-like extension (evident in calyx after fr drop) ..... ***A. clarianus***
8′ Keel < wings; fr base sometimes ± stalk-like, receptacle without peg-like extension
9. Infl 2–7-fld, open, axis in fr 7–20 mm; fr blushed or ± purple-mottled ..... ***A. pauperculus***
9′ Infl 2–12-fld, ± dense, axis in fr 1–8 mm; fr green, maturing straw-colored or black
10. Fr on erect peduncles; seeds pitted; NCoR ..... ***A. rattanii*** vars.
10′ Fr on spreading or reflexed peduncles; seeds smooth; GV, CCo, SnFrB, SCo ..... ***A. tener*** vars.

1′ Per
11. Hairs forked at base; corolla cream-colored; fr erect; MP, n SNE (possibly adjacent CA-FP) ..... ***A. canadensis*** var. ***brevidens***
11′ Hairs simple; corolla often pink-purple, white, or yellow; fr erect to reflexed; often not of MP, n SNE, and adjacent CA-FP
12. Calyx 10–12 mm, base with pouch on upper side; fls reflexed — petals dull yellow; fr hanging from stalk-like base; n SNH ..... ***A. gibbsii***
12′ Calyx gen < 10 mm, base gen ± symmetric, without pouch; fls ascending or spreading if calyx > 10 mm
13. Terminal lflet not or only obscurely jointed to midrib, joint unlike that of lateral lflet; lateral lflets often reduced or 0, spaces between gen >> 1.5 × lflet width; lvs few, sparse
14. Calyx 8–10 mm; petals whitish; fr ± 4-sided ..... ***A. bicristatus***
14′ Calyx 4–5 mm; petals reddish pink and yellow; fr 2-sided ..... ***A. inversus***
13′ Terminal lflet clearly jointed to midrib, joint like that of lateral lflet; lateral lflets present, spaces between rarely > 1.5 × lflet width; lvs gen many, not sparse
15. Pl tufted or matted; st < 15 cm (gen much shorter), from crown at surface; hairs dense, pl grayish or silvery; fls strongly ascending
16. Banner, wings hairy on outside; fr 5–7 mm, ± incl in calyx ..... ***A. austiniae***
16′ Banner, wings glabrous on outside; fr > 7 mm, well exserted from calyx
17. Calyx 4–6 mm; fr 3-sided, < 4 mm wide; hairs of st, lf sparse, ± straight, gen < 0.5 mm ..... ***A. obscurus***
17′ Calyx 6–14 mm; fr swollen, ± round or 2-lobed in X-section, > 4 mm wide; hairs of st, lf dense, curly, often > 0.5 mm
18. Fr hairs gen > 1.5 mm, completely hiding surface ..... ***A. purshii*** vars.
18′ Fr hairs rarely > (gen <<) 1 mm, sometimes not hiding surface
19. Petals whitish, with some lilac, banner 11–13.3 mm; fr 8–15 mm; lflets 7–13 ..... ***A. subvestitus***
19′ Petals pink-purple, banner 16–18.5 mm; fr 13–25 mm; lflets 7–19 ..... ***A. leucolobus***
15′ Pl gen open; st gen > 15 cm, gen from crown at or sometimes below surface; if pl low and tufted then fls ultimately spreading or reflexed, or hairs sparse, or both
20. Lower stipules fused around st into sheath, (sometimes very low and ± inconspicuous or ruptured early and seemingly not fused)
21. Lflets of at least some lvs > 22; petals ± white or cream-colored
22. Banner 7–10 mm; fr 6–11 mm, beak long, slender, hooked ..... ***A. pycnostachyus*** vars.
22′ Banner 10–19 mm (slightly shorter when dry); fr often > 11 mm, beak not long, not hooked
23. Pl grayish white-woolly; stipules finely tomentose; ChI ..... ***A. miguelensis***

23′ Pl greenish even if hairy; stipules gen glabrous or sparsely hairy, sometimes densely strigose; ChI or mainland
24. Fr base not stalk-like; sts not more densely hairy than lvs; gen coastal .... ***A. nuttallii*** vars.
24′ Fr base stalk-like; sts sometimes more densely hairy than lvs; often non-coastal
25. Fls early spreading, then quickly reflexed; stalk-like fr base 5–40 mm, ± strigose
26. Stalk-like fr base 14–40 mm .... ***A. asymmetricus***
26′ Stalk-like fr base 5–17 mm .... ***A. trichopodus*** vars.
25′ Fls early ascending or spreading, then often reflexed; stalk-like fr base 2–12 mm, glabrous, strigose, or shaggy
27. Stalk of ovary, fr with ± wavy, ± ascending hairs; lower stipules sometimes densely shaggy .... ***A. oxyphysus***
27′ Stalk of ovary, fr glabrous or sparsely strigose; lower stipules ± glabrous or sparsely shaggy
28. Stalk-like fr base 5–12 mm, jointed at base; fr ± 2-lobed in X-section, 2-chambered; SNH .... ***A. bolanderi***
28′ Stalk-like fr base 2–6 mm, jointed at top; fr ± round in X-section, 1-chambered; s CCo, w SCoR, n ChI .... ***A. curtipes***
21′ Lflets gen < 22; petals white, cream-colored, pinkish, or purplish
29. Stalk-like fr base at least 2 mm
30. Fr ± narrowly oblong in side view, strongly compressed side-to-side, hairy
31. Calyx tube 5.2–7 mm, hairs ± spreading, wavy; fr body 27–43 mm .... ***A. californicus***
31′ Calyx tube 3–3.5 mm, hairs ± appressed; fr body 17–30 mm .... ***A. filipes***
30′ Fr widely oblong, ovoid, or roundish in side view, swollen, often glabrous
32. Fr bladdery, thinly papery .... ***A. whitneyi*** vars.
32′ Fr swollen, often ± inflated, but not bladdery, not thinly papery
33. Fr body stiffly leathery or ± woody when dry, pendent on a down-curved stalk-like base, 1-chambered .... ***A. bicristatus***
33′ Fr body stiffly papery when dry, erect or ± spreading, at angle on ascending to ± reflexed stalk-like base, 2-chambered .... ***A. bolanderi***
29′ Stalk-like fr base 0 or < 2 mm
34. Calyx 8–13 mm; banner 14–19 mm
35. Fls spreading, 10–30 per infl .... ***A. sepultipes***
35′ Fls ascending to ± spreading, 6–14 per infl .... ***A. webberi***
34′ Calyx < 9 mm; banner 5.2–14 (gen < 12) mm
36. St 1–10 cm, from crown below surface; fr swollen but not bladdery, walls becoming stiffly papery or leathery
37. Infl 7–17-fld; peduncle ± stout, hairs spreading; s SNH, n Teh, 1750–1900 m .... ***A. ertterae***
37′ Infl 2–8-fld; peduncle ± thread-like, hairs appressed; s SNH, 3400–3450 m .... ***A. ravenii***
36′ St > 10 cm, gen from crown at surface (1–35 cm if from crown below surface); fr bladdery with thinly papery walls or swollen with stiffly papery or leathery walls
38. Fr 1-chambered; crown below surface
39. St ± prostrate; fr bladdery, blunt at both ends, 10–20 mm .... ***A. pulsiferae*** vars.
39′ St decumbent to ascending; fr narrow, ± 3-sided near base, tapered at both ends, 13–31 mm .... ***A. shevockii***
38′ Fr 2-chambered; crown at surface
40. Infl 5–10-fld; fr 5–9 mm .... ***A. lentiformis***
40′ Infl 12–26-fld; fr 10–18 mm .... ***A. andersonii***
20′ Lower stipules not fused into sheath or sometimes fused into a sheath around st but ruptured early and seemingly not fused (see 20.) (underground stipules, if present, sometimes fused)
41. Stalk-like fr base > 3 mm; corolla white to yellow
42. ChI (San Nicolas, Santa Barbara, San Clemente islands), fr ± 2-chambered, not bladdery — lower side of fr grooved; lvs softly hairy
43. Banner 10.6–12.7 mm; ovary, fr glabrous .... ***A. nevinii***
43′ Banner 14.2–17.5 mm; ovary, fr hairy .... ***A. traskiae***
42′ Mainland, if insular then fr 1-chambered, bladdery
44. Fls, frs strongly ascending; lflets linear or linear-oblong, upper side silvery-strigose, lower less densely strigose, greenish or grayish; lvs few, sparse — pl bushy, sts wiry ...... ***A. pachypus*** vars.
44′ Fls, frs spreading to reflexed; lflets linear-lanceolate to widely obovate, both sides greenish or ± gray-strigose; lvs many, not sparse
45. Infl 7–15-fld; fr fleshy; KR, CaRF .... ***A. accidens*** var. ***hendersonii***
45′ Infl 10–65-fld; fr papery; not of KR, CaRF
46. Stalk-like fr base spreading-, wavy-hairy .... ***A. oxyphysus***
46′ Stalk-like fr base strigose or ± glabrous
47. Lflets 9–19; fr strongly compressed side-to-side .... ***A. filipes***
47′ Lflets 15–39; fr strongly compressed side-to-side to swollen and bladdery .... ***A. trichopodus*** vars.
41′ Stalk-like fr base 0 or < 3 mm; corolla white to yellow, pinkish, or purplish
48. Lflets 5–9; pl silvery; Cushenbury Canyon (ne SnBr, adjacent DMoj) .... ***A. albens***
48′ Lflets on most lvs > 9; pls often green; widespread, incl Cushenbury Canyon
49. Fr 5–7 mm, 2-chambered; calyx 3–4 mm; banner 4–7 mm

50. St 3–10 dm; fls reflexed, 20–95 per infl .... ***A. clevelandii***
50′ St 1–4 dm; fls strongly ascending, 2–13 per infl .... ***A. lemmonii***
49′ Fr 9–60 mm, 2- or often 1-chambered; calyx 4–13 mm; banner > 7 mm
51. Fr bladdery, walls ± thinly papery
52. Middle of fr 2-chambered, beak strongly compressed side-to-side, triangular in side view, 1-chambered; banner recurved 30–50° .... ***A. lentiginosus*** vars.
52′ Middle of fr 1-chambered, beak often not strongly compressed, not triangular in side view, sometimes 2-chambered; banner recurved 40–90°, often > 60°
53. Fls, frs crowded in head-like clusters 2.5–3.5 cm wide in fr; fr 12–18 mm, ascending, hairs ± spreading .... ***A. hornii*** var. ***hornii***
53′ Fls, frs, if crowded, in clusters > 3.5 cm wide in fr; fr gen > 20 mm, often spreading, hairs ± appressed or becoming 0
54. Banner gen > 11 mm, recurved ± 40–45°; lflets often > 25 (see also *A. macrodon*, lead 56.)
55. Lflets 19–29, not clearly smaller from lf base to tip; midrib of lflet prominently raised on lower side; immature seeds 29–40; 250–300 m, PR (sw San Diego Co.) ..... ***A. deanei***
55′ Lflets 25–41, clearly, gradually smaller from lf base to tip; midrib of lflet not prominently raised on lower surface; immature seeds 35–55; 50–700 m, s SCoRO, SCo, PR, nw Baja CA .... ***A. pomonensis***
54′ Banner gen < 11 mm, recurved 60–90°; lflets gen < 25 (see also *A. deanii*, lead 55.)
56. Lf, fr minutely strigose or becoming glabrous; lflet margin, midrib green; calyx lobes 0.7–2.6 mm; immature seeds 51–71; s SN, Teh, GV, CW, SW .... ***A. douglasii*** vars.
56′ Lf, fr ± wavy- or incurled-hairy; lflet margin, midrib gen purple; calyx lobes 2.5-4.3 mm; immature seeds 29–52; c&s CW .... ***A. macrodon***
51′ Fr narrow or swollen but not bladdery, walls stiffly papery, leathery, or woody (fr inflated but not bladdery, walls stiffly papery in *A. oocarpus, A. palmeri*)
57. St 3.5–10 cm, widely spreading-hairy, from crown below surface; s SNH, n Teh .... ***A. ertterae***
57′ St gen > 10 cm, ascending- or appressed-strigose or ± glabrous, gen from crown at surface; widespread, incl s SNH, n Teh
58. St erect to widely ascending, 60–150 cm; sw SW
59. Fl reflexed; corolla dull lilac; fr reflexed, deciduous, < 10 mm, densely wavy-hairy .... ***A. brauntonii***
59′ Fl widely ascending; corolla cream-colored; fr erect, persistent, 15–25 mm, ± glabrous ***A. oocarpus***
58′ St decumbent to erect, < 60 cm (> 60 cm in n CA-FP); widespread
60. Fr 1-chambered throughout, thickly papery, ± bladdery; fls spreading; corolla often pink-purple (or cream-colored); w edge DSon, adjacent foothills of SnBr, PR .... ***A. palmeri***
60′ Fr ± 2-chambered, at least in middle, often throughout, thinly papery to stiffly leathery or woody, linear to very swollen and bladdery; fls ascending to reflexed; corolla cream-colored or pink-purple; widespread
61. Beak of fr triangular in side view, compressed strongly side-to-side, 1-chambered; fr > 5 mm wide, swollen, both sutures ± sunken; corolla purplish or whitish ... ***A. lentiginosus*** vars.
61′ Beak of fr acuminate to ± triangular in side view, gen not compressed strongly, 2-chambered; fr gen < 5 mm wide, ± 3-sided, lower suture gen sunken, upper raised; corolla pinkish or greenish white or whitish
62. Fr, ovary glabrous
63. Pl ± glabrous; fr body 14–24 mm, spreading or reflexed; KR, NCoRO .... ***A. umbraticus***
63′ Pl minutely strigose, esp above; fr body 20–42 mm, ascending; SnBr, DMtns
64. Lflets 7–19; banner 7–10.2 mm; calyx tube 2.7–4.1 mm .... ***A. bernardianus***
64′ Lflets 17–27; banner 12.6–15.7 mm; calyx tube 4.1–5 mm .... ***A. tricarinatus***
62′ Fr, ovary hairy
65. Fls 6–14 per infl, strongly ascending; fr erect; st 5–15 cm; se KR, MP .... ***A. obscurus***
65′ Fls gen 10–40 per infl, spreading or reflexed; fr loosely ascending or reflexed; st > 15 cm; SNE, SNF, NCoRO
66. Herbage silvery-silky, hairs wavy; banner 13–18 mm, pink-white — SNE, perhaps e edge SNH .... ***A. sepultipes***
66′ Herbage green, ± glabrous (grayish hairy in c&s SNF); banner often < 14 mm, whitish
67. Calyx lobes 3–5 mm; fr 11–15 mm; NCoRO .... ***A. agnicidus***
67′ Calyx lobes 1–3 mm; fr 15–35 mm; c&s SNF .... ***A. congdonii***

Group 2: plants of Desert Floristic Province

1. Terminal lflet not or only obscurely jointed to midrib, joint unlike that of lateral lflet; lateral lflets often reduced or 0, spaces between gen >> 1.5 × lflet width; lvs few, sparse
2. Banner 10–14.2 mm; lflets 3–13; fr half-ovoid to ± spheric, bladdery .... ***A. magdalenae*** var. ***peirsonii***
2′ Banner 17–26 mm; lflets 5–11; fr plumply oblong, not bladdery .... ***A. serenoi*** var. ***shockleyi***
1′ Terminal lflet clearly jointed to midrib, joint like that of lateral lflet; lateral lflets present, spaces between rarely > 1.5 × lflet width; lvs gen many, not sparse
3. Hairs dense, forked at base; lflets 1–7, crowded near lf tip .... ***A. calycosus*** var. ***calycosus***
3′ Hairs 0 to dense, not forked; lflets often > 7, gen not crowded near lf tip
4. Ann (or fl 1st year and appearing so)

5. Fl, fr strongly ascending in dense, head-like racemes; fr ± spheric, 2–4 mm .... ***A. didymocarpus*** vars.
5′ Fl, fr ± ascending to spreading in often open racemes; fr linear to ± ovate-inflated, > 4 mm
6. Keel 10–21 mm
7. Pl densely shaggy-hairy; st gen << 7 cm, ± prostrate, often nearly 0; fr 1-chambered .... ***A. tidestromii***
7′ Pl ± glabrous to strigose; st > 9 cm, prostrate to erect; fr 1- or 2-chambered
8. Keel 10–13 mm; fr 2-chambered at middle .... ***A. lentiginosus*** vars.
8′ Keel 17–21 mm; fr 1-chambered .... ***A. crotalariae***
6′ Keel 2–10.5 mm
9. Fr bladdery, walls thinly papery, translucent
10. Infl gen 10–35-fld
11. Lflets 11–33; fr 1-chambered throughout .... ***A. hornii*** var. ***hornii***
11′ Lflets 11–19; fr 2-chambered below beak .... ***A. lentiginosus*** vars.
10′ Infl gen 2–10-fld
12. Corolla whitish; lflets 11–19; upper suture of fr convex; immature seeds 13–21 .... ***A. allochrous*** var. ***playanus***
12′ Corolla purplish; lflets 7–19; upper suture of fr straight or convex; immature seeds 7–24
13. Keel 4.8–6 mm; lflets gen 11–19; upper suture of fr much less convex than lower; immature seeds 7–14 .... ***A. insularis*** var. ***hardwoodii***
13′ Keel 5.9–6.6 mm; lflets gen 7–13; upper and lower sutures of fr ± equally convex; immature seeds 19–24 .... ***A. nutans***
9′ Fr linear or swollen, not bladdery, walls papery (but not esp translucent) or leathery
14. Infl 20–40-fld .... ***A. palmeri***
14′ Infl < 20-fld
15. Fr 1-chambered, swollen, half-ovate or -elliptic in side view, ± ovate or round in X-section
16. Pl, fr hairs appressed, ± straight, ± obscuring surface; immature seeds 3–7 .... ***A. aridus***
16′ Pl, fr hairs ascending or spreading, ± wavy, not obscuring surface; immature seeds 10–19 .... ***A. sabulonum***
15′ Fr 2-, rarely 1-chambered, then often ± linear, curved, ± 2- or 3-sided in X-section
17. Upper surface of lflets sparsely hairy; infl 1–6-fld; keel < 6 mm; fr often < 3 mm wide
18. Lflets of upper lvs gen blunt, notched at tip; fr quickly deciduous, maturing pale brown . ***A. acutirostris***
18′ Lflets of upper lvs gen acute at tip; fr persistent, maturing dark brown or blackish .... ***A. nuttallianus*** vars.
17′ Upper surface of lflets silvery-canescent; infls 3–16-fld; keel > 6 mm; fr often > 3 mm wide
19. Fr 10–18 mm, 2.8–3.5 mm wide, incurved, ± 3-sided; immature seeds 8–11; Cushenbury Canyon (of ne SnBr, adjacent DMoj) .... ***A. albens***
19′ Fr 13–32 mm, 3.5–8.5 mm wide, straight, ± 2-sided, or incurved, ± 3-sided; immature seeds 20–30; widespread in DMoj .... ***A. mojavensis*** vars.
4′ Per
20. Fr, gen ovary, glabrous (*A. douglasii, A. nutans* also sometimes glabrous, under 20′)
21. Lflet hairs 0 or few, restricted to margins, midrib on 1 or both surfaces; fr base stalk-like
22. Fls early ascending, then sometimes reflexed; keel 9.5–10.6 mm; fr ± ascending, ± 2-chambered .... ***A. cimae*** vars.
22′ Fls strongly ascending; keel 11–19 mm; fr erect, 1-chambered .... ***A. preusii*** vars.
21′ Lflet hairs ± throughout 1 or both surfaces; fr base stalk-like or not
23. Keel 6–9.4 mm; fr 2-chambered, at least at middle
24. Pl often coarse, leafy, gen in open; st not wiry; fr sessile — fr swollen, beak strongly compressed side-to-side, triangular in side view, 1-chambered .... ***A. lentiginosus*** vars.
24′ Pl slender, sparsely leafy, often in shelter of shrubs; st ± wiry; stalk-like fr base 1–5 mm
25. Fls 10–25 per infl, ascending; fr ascending, 3-sided .... ***A. bernardianus***
25′ Fls 5–15 per infl, early ascending, then reflexed; fr pendent, ± 2-sided .... ***A. jaegerianus***
23′ Keel 9.7–19 mm; fr 1- or 2-chambered
26. Calyx glabrous; banner recurved ± 85°; fr bladdery, 1-chambered, base stalk-like .... ***A. oophorous*** var. ***oophorous***
26′ Calyx strigose; banner recurved < 50°; fr ovoid or narrow, not bladdery, ± 2-chambered base, stalk-like or not
27. Corolla pink-purple; lflets very narrow, upper surface often paler than lower (silvery vs greenish); fr erect, sessile; n Inyo Co. .... ***A. serenoi*** var. ***shockleyi***
27′ Corolla white or cream-colored; lflets narrow or wide, upper surface not paler than lower; fr ascending to pendent, stalk-like base 1–10 mm; w or sw edge of D
28. Banner 15–22 mm; fr 15–27 mm, swollen, ± compressed side-to-side; lflets sometimes > 17 .... ***A. pachypus*** vars.
28′ Banner 12.6–15.7 mm; fr 24–42 mm, ± linear, thin-walled, 3-sided; lflets 17–27 .... ***A. tricarinatus***
20′ Fr, ovary with at least some hairs (*A. douglasii, A. nutans* sometimes glabrous)
29. Pl conspicuously woolly- or wavy-hairy; sts < 9 cm, gen densely tufted
30. Lflets > 21, at least on some lvs; infl 10–35-fld; se DMoj .... ***A. tephrodes*** var. ***brachylobus***
30′ Lflets not > 21 on any lf (rarely to 23 in *A. layneae*); infl often < 10-fld (10–45-fld in *A. layneae*); widespread
31. Keel 18–28 mm; fr 1-chambered, densely white-hairy

32. Calyx with more black than white hairs; fr 25–50 mm; lf bases not notably persistent on crown; ne DMtns (e of Death Valley) .... ***A. funereus***
32′ Calyx gen with more white than black hairs; fr 13–28 mm; lf bases persistent on crown; widespread, incl DMtns .... ***A. newberryi*** var. ***newberryi***
31′ Keel gen < 20 mm; fr 2-chambered if densely white-hairy
33. Infl 10–45-fld; corolla 2-colored, whitish with keel tip, wing tips, and sometimes banner tip purplish; fr hairs spreading — pls from deep rhizomes (± stless pls of *A. minthorniae* will key out here; see 37.) .... ***A. layneae***
33′ Infl 3–16-fld; corolla pink-purple or ± 2-colored (± as at 33.); fr hairs appressed or densely white-cottony
34. Keel 11.5–20.8 mm; petals pink-purple; fr often ± 2-chambered, hairs spreading or curly, densely white-cottony .... ***A. purshii*** var. ***tinctus***
34′ Keel 10–12 mm; petals whitish, tinged dull purple; fr 1-chambered, hairs gen appressed .. ***A. tidestromii***
29′ Pl hairy to ± glabrous; sts gen > 9 cm (if < 9 cm, pl not woolly- or wavy-hairy), tufted or not
35. St < 16 cm, from deep rhizome; corolla 2-colored, whitish with keel tip, wing tips, and sometimes banner tip purplish .... ***A. layneae***
35′ St often > 15 cm, from crown at or slightly below surface; corolla 1- or 2-colored
36. Lflet hairs 0 or few on margins, midrib on 1 or both sides; stalk-like fr base 0 or 2–7 mm .... ***A. preussii*** vars.
36′ Lflet hairs at least few, ± throughout 1 or both sides; stalk-like fr base 0 or < 3 mm
37. St, lf hairs spreading, sometimes curly, 0.8–1.5 mm; fr shaggy-hairy, ± erect .... ***A. minthorniae*** var. ***villosus***
37′ St, lf hairs appressed, if spreading then < 0.8 mm; fr often strigose or silky-canescent, erect to reflexed or pendent
38. St slender, wiry, incurved, < 15 cm; petioles persistent, wiry; infl 1–4-fld; — pls of limestone, resembling an unkempt nest .... ***A. panamintensis***
38′ St slender or coarse, if wiry then not incurved, often > 15 cm; petioles deciduous, not wiry; infl often > 4-fld
39. Petals dingy white at least at base, dull lilac on margins or tips; wings ± twisted, tips shortly fringed or notched; fr pendent, ± compressed side-to-side, 2-chambered ..... ***A. atratus*** var. ***mensanus***
39′ Petals white to ± purple; wings plane or curved, tips not fringed, not notched; fr erect to reflexed, ± 3-sided or round in X-section, sometimes plump and widely grooved on upper side, 1- or 2-chambered
40. Fr 2-chambered in middle, beak often 1-chambered; banner recurved 30–50°
41. Lflets gen > 11; fr X-section ± round, upper suture gen ± sunken in a wide channel; pl gen robust — fr beak widely triangular, 1-chambered .... ***A. lentiginosus*** vars.
41′ Lflets 3–11; fr ± 2- or 3-sided, upper suture raised; pl robust or delicate
42. Fr 10–18 mm, 2.8–3.5 mm wide, incurved, ± 3-sided; immature seeds 8–11; Cushenbury Canyon (ne SnBr, adjacent DMoj) .... ***A. albens***
42′ Fr 13–32 mm, 3.5–8.5 mm wide, straight, ± 2-sided, or incurved, ± 3-sided; immature seeds 20–30; widespread in DMoj .... ***A. mojavensis*** vars.
40′ Fr 1-chambered throughout; banner recurved 45–90°
43. Of the following, all true: most lvs with > 19 lflets; fr bladdery; infl > 14-fld ..... ***A. douglasii*** vars.
43′ Of the following, 1 true: most lvs with < 19 lflets; fr swollen but not bladdery; infl < 12-fld
44. Banner > 11 mm; calyx gen > 7 mm; fr stiffly papery or leathery
45. Keel < 14 mm; fr reflexed, 5–10 mm wide; SNE, DMtns (Inyo Co.) .... ***A. casei***
45′ Keel > 17 mm; fr ascending or spreading, 10–14 mm wide; DSon .... ***A. crotalariae***
44′ Banner < 11 mm; calyx 4–7 mm; fr bladdery, ± papery
46. Infl gen 10–40-fld; petals gen bright pink-purple; w edge DSon, adjacent foothills of SnBr, PR .... ***A. palmeri***
46′ Infl often < 10-fld; petals whitish or pink-purple; c&e DSon, e DMoj
47. Petals whitish; lflets 11–19; fr beak short, obscure .... ***A. allochrous*** var. ***playanus***
47′ Petals pink-purple; lflets 7–13; fr beak deltate, strongly compressed side-to-side
48. Banner 6.1–8 mm; immature seeds 8–12; n DMoj (Panamint Mtns), > 2000 m ..... ***A. gilmanii***
48′ Banner 7.8–10.4 mm; immature seeds 19-24; e DMoj, < 2000 m .... ***A. nutans***

Group 3: plants of Great Basin Floristic Province

1. Terminal lflet not or only obscurely jointed to midrib, joint unlike that of lateral lflet; lateral lflets often reduced or 0, spaces between gen >> 1.5 × lflet width; lvs few, sparse
2. Fr strongly compressed side-to-side, base stalk-like; banner 9.4–12.2 mm .... ***A. inversus***
2′ Fr cylindric or ± compressed side-to-side, sessile; banner 17–26 mm .... ***A. serenoi*** var. ***shockleyi***
1′ Terminal lflet clearly jointed to midrib, joint like that of lateral lflet; lateral lflets present, spaces between rarely > 1.5 × lflet width; lvs gen many, not sparse
3. Ann, slender, rarely persisting into second season
4. Banner 3–5 mm; fr < 4 mm .... ***A. didymocarpus*** var. ***dispermus***
4′ Banner 4.7–7.6 mm; fr > 12 mm
5. Fr not bladdery, not inflated, 2–3 mm wide; lflet upper surface ± appressed-hairy .... ***A. acutirostris***
5′ Fr inflated, 6–10 mm wide; lflet upper surface ± glabrous .... ***A. geyeri*** var. ***geyeri***
3′ Per, often coarse — fl, fr > 5 mm

6. Lf hairs forked at base
7. Lflets 1–7 .... ***A. calycosus*** var. ***calycosus***
7′ Lflets 7–25 .... ***A. canadensis*** var. ***brevidens***
6′ Lf hairs simple
8. Fr densely white-hairy, resembling a small ball of cotton; st 0–14 cm; pl hairs dense, of 2 kinds or not
9. Sts forming a thickened crown covered by persistent lf bases; longer hairs of lf gen straight, some spreading; some fr hairs curly and short, some long and ± straight (see also *A. platytropis*) .... ***A. newberryi*** var. ***newberryi***
9′ Sts tufted or matted, lf bases not persistent; longer hairs of lf wavy, tangled; fr hairs all ± wavy or all straight .... ***A. purshii*** vars.
8′ Fr sometimes silvery-hairy, but not resembling a small ball of cotton; st often > 10 cm; pl hairs sparse or dense, gen not of 2 kinds
10. Calyx base strongly asymmetric, pedicel attached at lower side, upper ± pouched; petals white to yellowish; fr curved 1/4 to ± full circle, base stalk-like
11. St hairs subappressed or incurved; fr 2.5–4.5 mm wide; lowest stipules not fused .... ***A. curvicarpus*** var. ***curvicarpus***
11′ St hairs spreading or reflexed; fr 4–8 mm wide; lowest stipules fused around st into low sheath .... ***A. gibbsii***
10′ Calyx base ± symmetric, pedicel attached ± at middle, upper side not conspicuously pouched; petals white, yellowish, or purplish pink; fr straight to curved ± full circle, base stalk-like or not
12. Fr pendent, early ± flat, then bulged around seeds, base slender, stalk-like; corolla white or cream-colored
13. Calyx, lf hairs wavy, ± ascending or spreading; calyx > 6 mm .... ***A. californicus***
13′ Calyx, lf hairs ± straight, appressed; calyx gen < 6 mm .... ***A. filipes***
12′ Fr erect to pendent, early ± swollen then ± round, triangular, or 2-lobed in X-section, base stalk-like or not; corolla whitish, yellowish, or purplish
14. Lower stipules fused around st into short sheath (often ruptured by st expansion or missing on specimens without base)
15. Stalk-like fr base > 0.5 mm
16. Banner < 6 mm; fr body narrow, 7–11 mm, 3-sided .... ***A. johannis-howellii***
16′ Banner > 8 mm; fr body bladdery, > 15 mm, X-section ± round .... ***A. whitneyi*** vars.
15′ Stalk-like fr base 0 (see also *A. johannis-howellii*, 16.)
17. Banner often 4.8–9 mm; fr 1-chambered (± 2-chambered in *A. ravenii*); st delicate, 1–20 cm
18. Fr gen compressed side-to-side, ± 3-sided, 3.5–4.5 mm, hairs subappressed, ± straight; banner 4.8–5.2 mm, recurved ± 60–80° .... ***A. anxius***
18′ Fr bladdery, ± round in X-section, 10–20 mm, hairs spreading, wavy; banner 5.2–8.5 mm, recurved 90–100° .... ***A. pulsiferae*** vars.
17′ Banner often > 9 mm; fr ± 2-chambered; st often coarse, 0–35 cm
19. Herbage very sparsely hairy or ± glabrous; fls, frs strongly ascending, crowded in ovoid, head-like racemes .... ***A. agrestis***
19′ Herbage densely hairy; fls ultimately spreading or reflexed, frs ± ascending to reflexed, crowded or not
20. Pls cespitose; st < 2 cm; fr bladdery .... ***A. platytropis***
20′ Pls tufted or loosely matted; st > 5 cm; fr not bladdery
21. Crown at surface; lowest few internodes above surface, densely white-tomentose; fr X-section bluntly triangular .... ***A. andersonii***
21′ Crown below surface; lowest few internodes below surface, ± glabrous; fr X-section oblong or bluntly triangular
22. Keel 6.7–8 mm; fr oblong in X-section, not fully 2-chambered, ± narrowed at sutures .... ***A. monoensis***
22′ Keel 10–12.2 mm; fr bluntly 3-sided in X-section, fully 2-chambered, not narrowed at sutures .... ***A. sepultipes***
14′ Lower stipules not fused around st into sheath
23. Fls 1–4 per infl, strongly ascending; corolla pink-purple; sts low, tufted or matted, softly gray-hairy .... ***A. argophyllus*** var. ***argophyllus***
23′ Fls gen > 4 per infl, strongly ascending to reflexed; corolla ± purple to white; sts low or not, tufted or not, rarely matted, ± glabrous to strigose or spreading-hairy
24. Fr 1-chambered at mid-section, stalk-like base 0 to > 3 mm
25. Ovary, fr glabrous; fr bladdery, widely ovoid, base stalk-like .... ***A. oophorous*** vars.
25′ Ovary, fr ± hairy; fr bladdery or only swollen, ovoid or incurved, base stalk-like or not
26. Stalk-like fr base 2–5 mm; corolla pink-purple .... ***A. inyoensis***
26′ Fr sessile; corolla white to purple
27. Keel 3–5 mm; fr 10–20 mm, bladdery .... ***A. pulsiferae*** vars.
27′ Keel 8–14 mm; fr 12–55 mm, swollen but not bladdery
28. St ascending or erect, wiry; lflets > 4 × longer than wide, spaces between >> width; keel > 10 mm .... ***A. casei***
28′ St ± prostrate to decumbent, ± mat-forming; lflets < 2.5 × longer than wide, spaces between ± = or < width; keel < 10 mm
29. Banner 10–15.5 mm; herbage, fr hairs stiff, straight, appressed, < 0.7 mm .... ***A. iodanthus*** var. ***iodanthus***
29′ Banner 9–10 mm; herbage, fr hairs soft, wavy, spreading, ± 1 mm .... ***A. pseudiodanthus***

24′ Fr ± 2-chambered at mid-section, stalk-like base 0 or < 3 mm
30. Banner 4.8–6.1 mm; peduncles often in 2's or 3's in upper axils .... ***A. lemmonii***
30′ Banner > 7 mm; peduncles gen in 1's in upper axils
31. Stipules 7–17 mm, whitish or straw-colored; fr, herbage hairs ± 1.5–2.5 mm, spreading ... ***A. malacus***
31′ Stipules < 6 mm, often greenish; fr, herbage hairs 0 or gen < 1.5 mm, ± appressed
32. Fls, frs in ovoid, head-like racemes, strongly ascending; herbage sparsely hairy or ± glabrous .... ***A. agrestis***
32′ Fls, frs not in head-like racemes, gen spreading or reflexed; herbage ± glabrous to densely hairy
33. Fr erect, < 4 mm wide, straight, ± bluntly 3-sided, beak slender, ± tooth-like .... ***A. obscurus***
33′ Fr spreading or reflexed, gen > 4 mm wide, straight to curved, sides rounded, beak ± triangular
34. Crown below surface; fr curved at least 1/2 circle, wider than deep .... ***A. pseudiodanthus***
34′ Crown at surface; fr at most slightly curved, gen ± as deep as wide .... ***A. lentiginosus*** vars.

A. accidens S. Watson var. ***hendersonii*** (S. Watson) M.E. Jones (p. 593) Per, ± bushy; hairs sparse, minute, curved. **ST** ± ascending, 3–6 dm. **LF** 3–12 cm; lflets 15–29, 6–22 mm, ± oblong, tips gen notched. **INFL**: fls 7–15, ± spreading. **FL**: petals ± white, banner 13.8–19.3 mm, recurved ± 45°, keel 9.5–12.3 mm. **FR** pendent; body 16–25 mm, 8–12 mm wide, plump, fleshy, glabrous, drying leathery or woody; stalk-like base 6–12 mm; beak sharp; upper and lower sutures thick, raised; chambers 2. 2*n*=26. Shrubland, open woods, roadbanks; < 1250 m. KR, CaRF; sw OR.

A. acutirostris S. Watson (p. 593) Ann; hairs ± appressed, ± curved. **ST** prostrate or ascending, 2–30 cm. **LF** 1–4 cm; lflets 7–15, 2–8 mm, ± oblong, tips gen notched. **INFL**: fls 1–6, ± spreading. **FL**: petals ± white, banner 4.7–7 mm, recurved ± 45°, keel 4.3–5.8 mm. **FR** 12–30 mm, 2–3 mm wide, gently curved, ± 3-sided, pale brown, thin-walled but not bladdery, quickly deciduous; chambers 2. Sandy or gravelly areas; 600–1500 m. SNE, DMoj. Like and often growing with *A. nuttallianus* (which has fr persistent, ultimately dark brown or blackish, curved most strongly near base; at least upper lflets acute).

A. agnicidus Barneby (p. 593) HUMBOLDT MILKVETCH Per, coarse, leafy; base ± glabrous; hairs above sparse, fine, spreading, wavy. **ST** erect, 3–9 dm. **LF** 4–16 cm; lflets 13–27, 3–22 mm, ± oblong to ovate. **INFL**: fls 10–40, dense, soon reflexed. **FL**: calyx lobes 3–5 mm; petals white, banner 9.1–11 mm, recurved 45°, keel 7–7.4 mm. **FR** reflexed, 11–15 mm, ± 3 mm wide, ± 3-sided (lower narrow, grooved), becoming papery; hairs sparse, ascending, curved; chambers 2. **ENDANGERED** CA. Open soil in woodland; ± 750 m. NCoRO (s Humboldt Co.). Like the more n *A. umbraticus* (which has fr glabrous; lvs ± glabrous).

A. agrestis G. Don (p. 593) PURPLE LOCO Per, ± glabrous to sparsely strigose. **ST** decumbent or ascending, 4–30 cm, underground for 0–9 cm, forming small patches, often supported by grasses. **LF** 2–10 cm; lower stipules fused around st into sheath; lflets 9–23, 5–20 mm, ± lanceolate to ovate, tips acute to notched. **INFL** head-like, 0.5–2.5 cm, ovoid; fls 5–15, ascending, dense. **FL**: petals pink-purple to ± white, banner 15–22 mm, recurved ± 25°, keel 11.4–14 mm. **FR** erect, 7–10 mm, 3–4 mm wide, ± ovoid, pointed at tip, ± 3-sided (lower deeply grooved), stiffly papery, turning black; hairs white; chambers 2. 2*n*=16. Vernally-moist soil in sagebrush; ± 1650 m. MP; to Rocky Mtns, also Asia.

A. albens E. Greene (p. 593) CUSHENBURY or SILVERY-WHITE MILKVETCH Ann, sometimes ± per, delicate; hairs dense, appressed, flat, silvery. **ST** ± prostrate, 2–30 cm, loosely matted. **LF** 1–5.5 cm; lflets 5–9, 2–10 mm, ovate to obovate, tips blunt, ± notched. **INFL**: fls 5–14, widely spreading or reflexed. **FL**: petals pink-purple, banner 7.3–9.5 mm, recurved ± 40°, keel ± = banner, > wings. **FR** 10–18 mm, 2.8–3.5 mm wide, crescent-shaped, ± 3-sided (lower grooved), stiffly papery, densely strigose; chambers 2. RARE. Rocky areas; 1200–1800 m. Cushenbury Canyon (ne SnBr, adjacent DMoj). Threatened by limestone mining.

A. allochrous A. Gray var. ***playanus*** (M.E. Jones) Isely (p. 593) Ann or per, leafy, thinly, minutely strigose. **ST** ± prostrate to erect, 1–5 dm. **LF** 2–12 cm; lflets 11–19, 5–20 mm, ± oblanceolate or oblong, tips shallowly notched or obtuse. **INFL**: fls 4–10, early ascending, then spreading or reflexed. **FL**: petals whitish, banner 5–7 mm, recurved ± 45°, keel 4–6 mm. **FR** spreading, 15–30 mm, 12–20 mm wide, bladdery, papery, minutely strigose; beak short, obscure; chamber 1. 2*n*=22. Sandy flats; ± 800 m. e DMoj; to TX, Mex, where fl color, fr size, vary. [*A. wootoni* E. Sheldon var. *w.*] In CA, var. *playanus* seems ± distinct from other members of *Inflatai*, which are geog isolated; in AZ and NM it intergrades thoroughly with other vars. of *A. allochrous*; entire complex needs study.

A. andersonii A. Gray (p. 593) Per, loosely tufted; hairs spreading or ascending, wavy, grayish, on lower internodes dense, interwoven, whitish. **ST** decumbent to ascending, 7–20 cm. **LF** 2–10 cm; lower stipules fused around st into scarious sheath; lflets 9–21, 3–14 mm, elliptic to ± oblanceolate. **INFL**: fls 12–26, early ascending, then reflexed. **FL**: petals white, sometimes tinged dull purple, banner 9.5–14.5 mm, recurved ± 45°, keel 6.6–9 mm. **FR** spreading or pendent, 10–18 mm, 3–5 mm wide, widely oblong, ± 3-sided (lower narrow, shallowly grooved), stiffly papery; hairs wavy, long; upper suture thick, raised; chambers 2. 2*n*=24. Gen disturbed flats, slopes; 1300–2200 m. n&e-c SNE, MP; w NV.

A. anxius R. Meinke & T. Kaye (p. 593) Per, delicate, ± matted; hairs sparse, finely wavy. **ST** prostrate, 3–20 cm, slender. **LF** 1–3.5 cm; lower stipules barely fused around st; lflets 9–15, 2.5–12 mm, obovate, tips blunt or notched. **INFL**: fls 7–15, crowded or congested, spreading to reflexed. **FL**: petals purple and white, with pale lilac veins, banner 6.5–10 mm, recurved 60–80°, keel 3.5–3.7 mm. **FR** spreading or reflexed, 3.5–4.5 mm, 3.2–4.2 mm wide, ovoid, weakly compressed side-to-side, ± 3-sided (lower side bluntly angled), thinly papery, sparsely strigose; chamber 1. RARE. Gravelly volcanic soil among pines, sagebrush; 1550 m. MP (Lassen Co.).

A. argophyllus Torrey & A. Gray var. ***argophyllus*** (p. 593) SILVERLEAF MILKVETCH Per, cespitose, from heavy crown; hairs on lvs dense, appressed or ascending, silvery. **ST** ± prostrate, < 15 cm. **LF** 2–15 cm; lflets 9–21, 4–15 mm, ± elliptic or ovate, tips acute or obtuse. **INFL**: fls 1–4, ascending. **FL**: petals bright pink-purple, banner 22–24 mm, keel 17–20 mm. **FR** 15–25 mm, 7–12 mm wide, ± widely lanceolate, straight or curved, densely, loosely strigose, early fleshy, then stiffly leathery; chamber 1. RARE in CA. Heavy alkaline or saline soil; 1400–2350 m. MP (e Lassen Co.), SNE (nw Inyo Co.); to ID, WY, UT.

A. aridus A. Gray (p. 593) Ann, silvery silky-canescent. **ST** decumbent or ascending, 3–30 cm. **LF** 2–9 cm; lflets 7–17, 4–16 mm, ± oblong or oblanceolate, tips acute to notched. **INFL**: fls 3–9, ascending. **FL**: petals whitish, pink-tinged, or ± tannish pink, banner 3.3–6.5 mm, recurved ± 40°, keel 3.5–5 mm. **FR** ascending, 10–17 mm, 4–7 mm wide, ± half-elliptic in side view, round in X-section, grayish, thickly papery; hairs appressed; beak triangular, strongly compressed side-to-side; chamber 1. Sandy places; < 350 m. DSon; AZ, nw Mex. Highly variable in stature: low, slender (drier spring seasons) or robust, coarse (moist seasons).

A. asymmetricus E. Sheldon (p. 593) Per, clumped, finely silky-strigose; some hairs near top longer, more spreading. **ST** erect, 5–12 dm, stout, often hollow. **LF** 5–20 cm; midrib stiff, often curved; lower stipules fused around st into sheath; lflets 17–35, 6–25 mm, linear to ± elliptic, tips obtuse or shallowly notched. **INFL**: fls 15–45, early spreading, quickly reflexed. **FL**: calyx tube 5–7 mm; petals ± cream-colored, banner 12.6–17.6 mm, ± recurved, margins abruptly folded back, keel 11.5–14.7 mm. **FR** pendent; body 25–40 mm, 13–18 mm wide, bladdery; upper suture ± concave to ± convex, lower strongly convex; stalk-like base 1.4–4 cm, arched, sparsely strigose, not jointed at base; beak triangular; chamber 1; immature seeds 16–30. 2*n*=22. Grassy areas, open woodlands, disturbed sites; 50–900 m. SnJV, SnFrB, SCoRI. Like *A. curtipes, A. oxyphysus, A. trichopodus*, but note calyx length; stalk length, hairs, position of joint; seed number.

A. atratus S. Watson var. ***mensanus*** M.E. Jones (p. 593) DARWIN MESA MILKVETCH Per, wiry, loosely matted or scrambling through shrubland, minutely strigose. **ST** 3–25 cm. **LF** 1.5–15 cm; lflets 7–15, well separated, 3–16 mm, linear to ± ovate, tips acute to shallowly notched. **INFL**: fls 4–18, reflexed. **FL**: petals dingy white at least at base, dull lilac on margins or tips, banner 9.8–13.4 mm, recurved ± 90°, blade pinched to appear fiddle-shaped, wings irregularly toothed, ± twisted, keel 8.2–10 mm. **FR** pendent, 16–22 mm, ± 4 mm wide, ± linear-oblong, ± compressed side-to-side, stiffly papery or leathery, minutely strigose; upper suture more convex than lower; tip slightly reflexed; chambers 2. RARE. With pinyon pine; 1700–1850 m. DMtns (n and w of Panamint Valley, Inyo Co.); other vars. in OR, ID, NV.

A. austiniae Brewer & S. Watson (p. 593) Per, dwarfed, cespitose; hairs dense, wavy, silvery. **ST** < 11 cm. **LF** 1–5 cm; lower stipules fused around st into often overlapping sheaths; lflets 5–13, 1–7 mm, ± elliptic to oblanceolate, keeled on lower surface. **INFL** ± head-like; fls 4–14, erect to ascending. **FL**: petals whitish, dull lilac-tinged, banner 8.4–11.3 mm, recurved ± 35°, it and wings finely hairy on outside, keel 6.2–8.1 mm. **FR** ascending or spreading, ± incl in calyx, 5–7 mm, 3–4 mm wide, oblong-ovoid, finely tomentose; chambers ± 2 in lower 2/3. Exposed ridges, slopes near, above timberline; 2700–3200 m. n SNH (near Lake Tahoe).

A. bernardianus M.E. Jones (p. 593) Per, often twining among sagebrush, wiry, sparsely leafy; minutely strigose, esp above. **ST** 1–5 dm, slender. **LF** 3–14 cm; lflets 7–19, ± well separated, 4–20 mm, ± lanceolate, tips acute to notched. **INFL** ± among lvs; fls 10–25, well separated, ascending. **FL**: calyx tube 2.7–4.1 mm; petals pale to dark lilac, banner 7–10.2 mm, recurved 45–90°, keel 6.8–9.4 mm, < 0.5 mm < banner. **FR** ascending, 20–30 mm, 4–5 mm wide, narrowly oblanceolate in side view, straight or ± curved, 3-sided (lower shallowly channeled or ± flat), pale, papery, glabrous; chambers 2. Stony areas among desert shrubs, junipers; 900–2000 m. SnBr, DMtns (New York, Ivanpah mtns).

A. bicristatus A. Gray (p. 593) CRESTED MILKVETCH Per, sparsely leafy, often minutely grayish strigose. **ST** ascending or sprawling, < 5 dm. **LF** 3–14 cm; lower stipules fused around st into low sheath; lflets 11–23, often ± well separated, 5–20 mm, linear to narrowly oblong, tips gen obtuse or notched, terminal not or only obscurely jointed to midrib. **INFL**: fls 5–20, ascending. **FL**: petals ± whitish, banner 15–19 mm, recurved ± 50°, keel 12–13 mm. **FR** ± pendent; body 20–43 mm, 6–9 mm wide, incurved, glabrous, early ± round in X-section, fleshy, then ± 4-sided (prominently flanged on upper, lower), stiffly leathery or ± woody; stalk-like base downcurved, 6–10 mm, stout; chamber 1. UNCOMMON. Open, rocky areas in pine forest; 1700–2500 m. SCo (w Riverside and San Bernardino cos.), SnGb, SnBr.

A. bolanderi A. Gray (p. 593) Per, stiff, sparsely leafy; hairs fine, wavy or curly. **ST** ± decumbent to erect, 1.5–4 dm, basally lfless. **LF** 3–16 cm; lower stipules fused around st into ± loose sheath; lflets 13–27, 5–20 mm, ± oblanceolate to ± oblong, midrib prominent on lower surface, extending to tip as small, hard point. **INFL**: fls 7–18, ± crowded, spreading. **FL**: petals ± white, ± faintly lilac-tinged, banner 13–17.6 mm, recurved ± 40°, keel 9.8–12.4 mm. **FR**: body erect or ± spreading, at angle to stalk, 10–30 mm, 7–12 mm wide, swollen, incurved, stiffly papery, glabrous; stalk ascending to ± reflexed, 5–12 mm, jointed at base; chambers 2 below beak, a wing-like outgrowth protruding into each from central septum. $2n=22$. Dry meadows, rocky areas; 1500–3100 m. n&c SNH; w-c NV.

A. brauntonii Parish (p. 595) BRAUNTON'S MILKVETCH Per, coarse; hairs whitish, dense, entangled, some longer, spreading. **ST** ± erect, 7–15 dm, stout, hollow. **LF** 3–16 cm; lflets 25–33, 3–20 mm, ± obovate, tips acute to obtuse. **INFL** spike-like; fls 35–60, overlapping, reflexed. **FL**: petals dull lilac, persisting withered, banner 9.1–11.7 mm, recurved ± 40°, keel 6.4–8.5 mm. **FR** reflexed, ± 1/2 incl in calyx, 6.5–9 mm, 3–4 mm wide, oblong in side view, bluntly 3-angled, not bladdery, hairs dense, wavy; chambers 2 in lower 1/2. RARE. Disturbed areas in chaparral; < 450 m. c SCo, n PR (Los Angeles Basin). Threatened by development.

A. breweri A. Gray (p. 595) BREWER'S MILKVETCH Ann, sparsely leafy, thinly, minutely strigose. **STS** 4–30 cm, few, slender. **LF** 1.5–7.5 cm; lflets 7–13, well separated, 3–12 mm, narrowly to widely ± obovate, tips notched. **INFL** head- or umbel-like; fls 4–10, ascending and spreading. **FL**: petals pale yellow to white, sometimes ± streaked with lavender, banner 7.8–11.4 mm, recurved 35–45°, keel 5–7.8 mm. **FR** ascending and spreading; body 5–10 mm, 2.5–4 mm wide, ovoid; beak spine-like or hooked, ± = body; chambers 2; immature seeds 2–6. UNCOMMON. Open slopes, grassy areas, sometimes on serpentine; < 650 m. c&s NCoR, n SnFrB. In fl, like *A. rattanii* (which has 8–20 immature seeds).

A. californicus (A. Gray) E. Greene (p. 595) Per, robust, open and widely branched; hairs wavy, grayish. **STS** clumped, ± ascending, 2–5 dm. **LF** 3–8.5 cm; lower stipules fused around st into fragile, papery sheath; lflets 13–23, 5–20 mm, ± narrowly oblanceolate, tips obtuse or blunt, shallowly notched. **INFL**: fls 10–30, not crowded, ultimately reflexed. **FL**: petals cream, banner 11.5–17.4 mm, recurved ± 45°, keel 9.5–12.4 mm. **FR** pendent; body 27–43 mm, 3.5–5 mm wide, linear, strongly compressed side-to-side, stiffly papery, sparsely to densely minutely strigose; stalk-like base 7–15 mm; chamber 1. Dry, open areas in shrubland, woodland; 300–1300 m. e KR, n CaRH; s OR.

A. calycosus S. Watson var. ***calycosus*** (p. 595) Per, ± stless, tufted, silvery-strigose; hairs forked at base. **LF** 1–7 cm; lflets 1–7, gen 3, crowded near lf tip, 5–19 mm, elliptic to obovate, tips gen obtuse or acute. **INFL**: fls 1–8, ascending or spreading. **FL**: petals ± white to bright purple, wing tips white, banner 10–13 mm, keel 7.4–9.4. **FR** ascending, 10–25 mm, 3–4 mm wide, oblong, ± 3-sided (lower narrow, grooved), strigose; chambers ± 2. $2n=22$. Rocky areas, sagebrush shrublands to pine forests; 1500–3550 m. SNE, n DMtns; to ID, WY.

A. canadensis L. var. ***brevidens*** (Gand.) Barneby (p. 595) Per from rhizome, leafy, ± strigose; hairs forked at base. **ST** 1.5–5.5 dm. **LF** 5–23 cm, stipules fused around st into sheath; lflets 7–25, 5–40 mm, widely lanceolate or elliptic, ± glabrous, tips obtuse or shallowly notched, small-pointed. **INFL** spike-like; fls 20–many, reflexed, overlapping. **FL**: petals cream, rarely tinged dull purple, banner 11.7–17.5 mm, recurved 40–90°, keel 8.9–13.6 mm. **FR** erect; body 10–15 mm, 3–4 mm wide, ± cylindric, stiffly papery, gen minutely strigose; beak recurved, ± 3 mm, stiff; chambers 2. Heavy soil where moist at least in spring; 1500–2450 m. MP, n SNE, adjacent CA-FP; to WA, WY, Colorado.

A. casei A. Gray (p. 595) Per, slender, open and widely branched, wiry, sparsely leafy, minutely strigose. **ST** ascending or erect, 1–4 dm, often zig-zag. **LF** 3–10 cm, midrib rigid, tapered; lflets 5–15, well separated, 3–25 mm, linear to oblanceolate, tips gen obtuse or notched. **INFL**: fls 8–25, well separated, ultimately reflexed. **FL**: petals pink-purple, wing, keel tips white, rarely petals all white, banner 12–18 mm, recurved ± 45°, keel 10.6–13.3 mm. **FR** pendent; 20–55 mm, 5–10 mm wide, slightly incurved, ± half-lanceolate, wider than deep, minutely strigose, early pulpy, then tough, wrinkled; beak sharp, rigid, compressed side-to-side; sutures keel-like; chamber 1. Dry, gravelly soils or on dunes, with sagebrush or pinyon; 1200–2000 m. SNE, DMtns (Inyo Co.); w NV.

A. cimae M.E. Jones Per, ± coarse, ± fleshy; hairs ± 0. **ST** ± spreading, 2–25 cm. **LF** 4.5–11 cm; lflets 11–23, 5–20 mm, obovate to ± round, tips obtuse or notched. **INFL**: fls 10–25, early ascending, then sometimes reflexed. **FL**: petals reddish purple, white- or pale-tipped, banner 12–15 mm, keel 9.5–10.6 mm. **FR** ± ascending or spreading; body ± oblong in side view, bladdery or not, glabrous; stalk-like base 5–12 mm, tapered to narrow base; chambers ± 2. Calcareous soils, gen among sagebrush, sometimes pinyon pine; 1400–1850 m. n&e DMtns.

1 . Fr 8–12 mm wide, sides drying stiffly leathery or woody .... var. ***cimae***

1′ Fr 13–21 mm wide, sides drying thinly papery var. ***sufflatus***

var. ***cimae*** (p. 595) CIMA MILKVETCH **FR**: body 15–25 mm, gen incurved > 90° (so beak is perpendicular to infl axis), fleshy when green; stalk-like base 6–8 mm. RARE. Gen among sagebrush; 1400–1850 m. e DMtns (e San Bernardino Co.).

var. ***sufflatus*** Barneby **FR**: body 20–37 mm, straight to incurved 90° (so beak is ± parallel to infl axis), bladdery, becoming papery; stalk-like base 5–12 mm. Habitats of sp.; 1500–1850 m. n DMtns (e slope Inyo Mtns).

A. clarianus Jepson (p. 595) CLARA HUNT'S MILKVETCH Ann, very slender, sparsely leafy, thinly and minutely strigose. **ST** ascending, 3–12 cm. **LF** 1.5–5 cm; lflets 5–9, 2–10 mm, obovate, tips deeply notched. **INFL** head-like, black-hairy; fls 2–7, early spreading, then reflexed. **FL**: petals whitish, banner and keel bright purple-tipped, banner 8.9–12 mm, recurved ± 20°, keel 7.4–9.1 mm, > wings. **FR** spreading or reflexed, ephemeral; body 17–25 mm, 1.6–3.1 mm wide, crescent-shaped, tapered to both ends, stiffly papery, finely strigose, lower part slender, stalk-like, attached to peg-like, 1.5–2.5 mm extension of receptacle (evident in calyx after fr drop); beak prominent, ± curved; chambers 2. RARE. Open grassy areas and thin clay soil; 100–150 m. s NCoR (Sonoma, Napa cos.). Threatened by development.

A. clevelandii E. Greene (p. 595) CLEVELAND'S MILKVETCH Per, ± robust, bushy, leafy. **ST** erect, 3–10 dm, glabrous. **LF** 2–14 cm; lflets 13–27, 3–23 mm, elliptic to oblanceolate, ± glabrous on upper surface, sparsely, finely strigose on lower, tips ± obtuse. **INFL** spike-like, long; fls 20–100, reflexed. **FL**: petals white or cream-colored, banner 4.8–6 mm, recurved ± 45°, keel 3.8–4.6 mm. **FR** reflexed, 4.5–7 mm, ± 2 mm wide, incurved, half-ovoid in side view, 3-sided, stiffly papery, glabrous; upper suture raised, lower in a groove; chambers 2. $2n$=26. UNCOMMON. Moist serpentine; 200–1500 m. s NCoRI, e SCoRI.

A. coccineus Brandegee (p. 595) SCARLET MILKVETCH Per, tufted, ± stless; hairs dense, whitish. **LF** 3–10 cm; lflets 7–15, 3–14 mm, ± oblanceolate, tips ± acute. **INFL**: fls 3–10, ascending. **FL**: calyx 18–24 mm; petals scarlet, banner 35–41 mm, recurved 20–30°, keel 35–40 mm. **FR** 25–40 mm, 10–12 mm wide, plump, ± narrowly ovoid in side view, often curved, ± strongly compressed top-to-bottom, early fleshy, then leathery, hairs dense, shaggy, white or tawny; chamber 1. $2n$=22. Gravelly places, gen sagebrush or pinyon woodland; 750–2450 m. SNE, w edge D; s NV, w AZ, n Baja CA. Longer fr, remnants of fl ± distinguish this in fr from *A. newberryi*. ❀DFCLT;TRY.

A. congdonii S. Watson (p. 595) Per, often coarse; hairs abundant, ± spreading. **ST** ± ascending, 2–7 dm. **LF** 3–14 cm; lflets 11–37, 3–15 mm, ± elliptic to round, tips blunt or notched. **INFL**: fls 8–35, well separated, spreading to reflexed. **FL**: petals ± cream-colored, banner 10.4–16.6 mm, recurved ± 45°, keel 7.4–12.7 mm. **FR** reflexed, 15–35 mm, ± 3 mm wide, linear, straight or curved, ± 3-sided (lower grooved), drying stiffly papery; hairs sparse to dense, appressed and spreading; chambers 2. $2n$=26. Open, disturbed sites in shrubland, woods; 150–750 m. c&s SNF.

A. crotalariae (Benth.) A. Gray (p. 595) SALTON MILKVETCH Per (often fl 1st season and seemingly ann), bushy-clumped, coarse, ± ill-scented, ± strigose. **ST** ± erect, 1.5–6 dm, often hollow. **LF** 5–16.5 cm; lflets 9–19, 5–35 mm, ± obovate to round, flat, thick, tips notched. **INFL**: fls 10–25, ascending or spreading. **FL**: calyx 7.6–12.3 mm; petals bright reddish purple, sometimes white, banner 21–28 mm, recurved ± 40°, keel 17–21 mm. **FR**: body 20–30 mm, 10–14 mm wide, inflated, ovate in side view, sparsely to densely strigose, with fine, net-like pattern, drying thickly papery; stalk-like base 1–1.5 mm, stout; beak erect or incurved, tipped by persistent style-base; sutures ± raised; chamber 1. $2n$=24. UNCOMMON. Sandy, gravelly areas; –60–250 m. DSon; w AZ, n Baja CA.

A. curtipes A. Gray (p. 595) Per, minutely strigose and ± tomentose, often grayish. **STS** ascending, clumped, 2–4 dm. **LF** 4–16 cm; lower stipules fused around st into sparsely hairy sheath; lflets 25–39, 2–25 mm, linear-oblong to narrowly obovate, tips blunt or shallowly notched. **INFL**: fls 15–35, ± dense, early spreading, then reflexed. **FL**: calyx tube 4–5 mm; petals cream-colored or keel faintly lilac-tipped, banner 13–16 mm, recurved ± 50°, keel 10.7–12.7 mm. **FR** ascending or ± spreading; body 23–36 mm, 12–20 mm wide, bladdery, ± half-ovate in side view, thinly papery, minutely, sparsely strigose; stalk-like base 2.3–6 mm, stiff, sparsely strigose, jointed at top; chamber 1; immature seeds 26–37. Shrubland, grassy or disturbed areas near coast; 0–150 m. s CCo, w SCoR, n ChI. Like *A. asymmetricus, A. oxyphysus, A. trichopodus*, but note calyx length; stalk length, hairs, position of joint; ovule number. Phase with fr body 4–7 cm very uncommon in San Benito Co.

A. curvicarpus (A.A. Heller) Macbr. var. ***curvicarpus*** (p. 595) Per, leafy; hairs ± appressed, spreading, or curly, often grayish. **ST** decumbent to ascending, 1.5–4 dm, often stout. **LF** 2.5–9 cm; lflets 7–21, 3–23 mm, obovate, tips obtuse or notched. **INFL**: fls 5–35, not crowded, reflexed. **FL**: petals early white, then pale yellow, banner 15–21 mm, recurved ± 45°, keel 11–15.2 mm. **FR** pendent; body 20–35 mm, 2.5–4.5 mm wide, narrowly oblong, curved 1/4 to ± full circle, ± strongly compressed side-to-side, sparsely wavy-hairy or rarely glabrous, drying stiffly papery; stalk-like base 9–20 mm, jointed to receptacle; chamber 1. Loose soil, often with sagebrush; 1300–2900 m. GB; to OR, ID, NV.

A. deanei (Rydb.) Barneby (p. 595) DEAN'S MILKVETCH Per, coarse; hairs ± 0. **ST** ± erect, 3–6 dm. **LF** 8–18 cm; lflets 19–29, 4–21 mm, lanceolate to oblong, midrib prominently raised on lower surface, esp in basal 1/2, tips ± obtuse. **INFL**: fls 15–25, ± spreading, not crowded; pedicel at first slender becoming thick in fr. **FL**: petals whitish, banner 9.5–15.2 mm, recurved ± 45°, keel 7.8–10.5 mm. **FR** ± ascending, 15–30 mm, 10–20 mm wide, bladdery, thinly, minutely strigose, drying papery; upper suture nearly straight, lower suture strongly convex; chamber 1; immature seeds 29–40. RARE. Open shrubby slopes; 250–300 m. sw PR (sw San Diego Co.). Threatened by development.

A. didymocarpus Hook. & Arn. TWO-SEEDED MILKVETCH Ann, gen slender, ± minutely grayish strigose. **ST** prostrate to ± erect, 3–30 cm. **LF** 0.8–7.5 cm; lflets 9–17, 2–14 mm, linear to oblanceolate, tips notched. **INFL** head-like; fls 5–30, < 9 mm, erect or ascending. **FL**: calyx hairs ± mixed black, white; petals whitish, purple-tinged. **FR** ascending, ± incl in calyx, 2–4 mm, ± 2 mm wide, ± spheric, 2-lobed in X-section, ± minutely strigose, rarely glabrous, coarsely wrinkled, drying stiffly papery; chambers 2. Open grassy, gravelly, or sandy areas; 0–1550 m. c&s SNF, Teh, GV, CW, SW, D; s NV, w AZ, nw Mex. Vars. intergrade

1. Keel 2.4–4.5 mm; s ± 1/2 CA
  2. Calyx hairs mostly black, lobes gen 0.8–1.5 mm, < (rarely =) tube; lflets ± glabrous on upper side; st ± erect .... var. ***didymocarpus***
  2′ Calyx hairs mostly white, lobes gen 1.5–2.4 mm, > tube; lflets ± canescent on upper sides; st ± prostrate .... var. ***dispermus***
1′ Keel 4.7–7.2 mm; coast, mtns of s CW, SW
  3. Calyx hairs mostly black; keel abruptly curved, tip short, blunt .... var. ***milesianus***
  3′ Calyx hairs mostly white; keel crescent-shaped, tip narrowly triangular, acute .... var. ***obispoensis***

var. ***didymocarpus*** (p. 595) **ST** ± erect, 25–45 cm; herbage ± green, sparsely ± strigose. **INFL**: fls 5–25. **FL**: calyx hairs mostly black, lobes gen < 1.5 mm; banner 2.8–6.1 mm, keel 2.4–4.5 mm, abruptly curved, tip bluntly pointed. $2n$=24. Grassy areas; < 700 m. c&s SNF, Teh, GV, CW, SW exc ChI, DMoj; s NV. [var. *daleoides* Barneby]

var. ***dispermus*** (A. Gray) Jepson **ST** ± prostrate, 15–27 cm; herbage ± grayish hairy. **INFL**: fls 7–20. **FL**: calyx hairs mostly white, lobes gen > 1.5 mm; banner 3.4–5.4 mm, keel 3.4–4.5 mm, abruptly incurved, tip bluntly or sharply triangular. $2n$=26. Sandy, gravelly areas; 30–1200 m. D.

var. ***milesianus*** (Rydb.) Jepson **ST** ± erect, 25–45 cm; herbage ± green, sparsely ± strigose. **INFL**: fls 5–25. **FL**: calyx hairs mostly black, lobes 1–2 mm; banner 7.5–10 mm, keel 4.7–6.9 mm, abruptly incurved, tip short, blunt. Grassy areas near coast; < 60 m. s SCo.

var. ***obispoensis*** (Rydb.) Jepson **ST** ascending, 2.5–25 cm; herbage ± grayish hairy. **INFL**: fls 5–30. **FL**: calyx hairs mostly white, lobes 1–2.5 mm; banner 6.2–8.6 mm, keel 5.2–7.2 mm, crescent-shaped, tip narrowly triangular, acute. Grassy hills, openings in chaparral; 400–950 m. SW exc WTR; Baja CA.

A. douglasii (Torrey & A. Gray) A. Gray Per, leafy, ± minutely strigose. **STS** ± decumbent or erect, gen many, 2–10 dm. **LF** 5–18 cm; lflets 7–25, 5–25 mm, elliptic, ovate, or obovate, midrib often raised on lower surface, margins greenish, tips obtuse or shallowly notched. **INFL**: fls 10–30, well separated, spreading or ascending. **FL**: calyx green, lobes 0.7–2.6 mm; petals whitish to pale yellow, banner 8–13 mm, recurved 60–90°, keel 6.2–10.8 mm, keel and wing claws 2.2–4.9 mm, keel blade 4.3–6.4 mm, > keel claw.

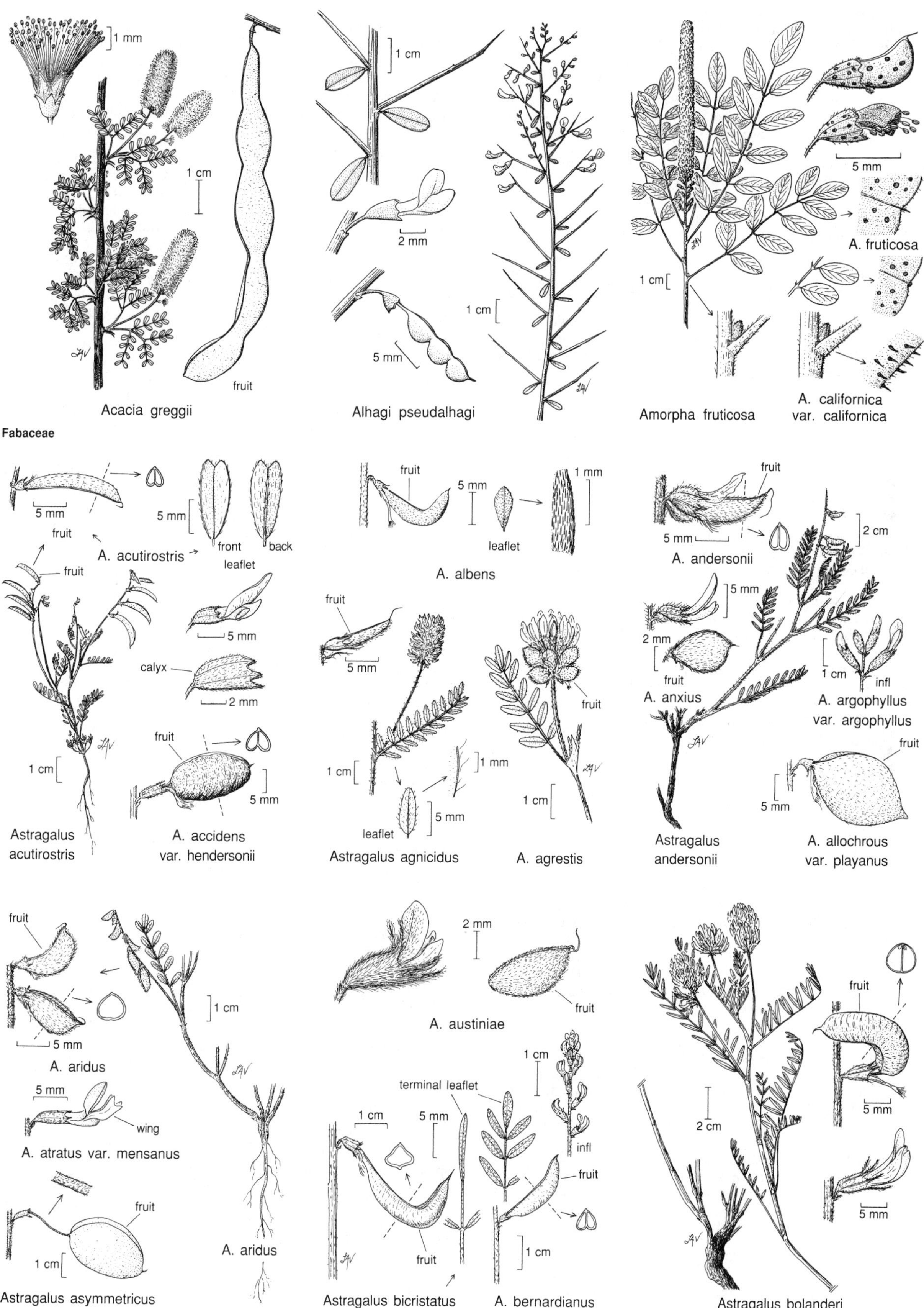
1 mm
1 cm
fruit
Acacia greggii
1 cm
2 mm
5 mm
1 cm
Alhagi pseudalhagi
5 mm
A. fruticosa
1 cm
Amorpha fruticosa
A. californica var. californica
Fabaceae
5 mm
fruit
A. acutirostris
5 mm
front
back
leaflet
fruit
5 mm
calyx
2 mm
fruit
1 cm
5 mm
Astragalus acutirostris
A. accidens var. hendersonii
fruit
5 mm
leaflet
1 mm
A. albens
fruit
5 mm
1 cm
1 mm
5 mm
leaflet
Astragalus agnicidus
fruit
1 cm
A. agrestis
fruit
5 mm
2 cm
A. andersonii
5 mm
2 mm
fruit
A. anxius
1 cm
infl
A. argophyllus var. argophyllus
fruit
5 mm
Astragalus andersonii
A. allochrous var. playanus
fruit
5 mm
A. aridus
1 cm
5 mm
wing
A. atratus var. mensanus
fruit
1 cm
Astragalus asymmetricus
A. aridus
2 mm
fruit
A. austiniae
1 cm
terminal leaflet
1 cm
5 mm
infl
fruit
fruit
1 cm
Astragalus bicristatus
A. bernardianus
fruit
2 cm
5 mm
5 mm
Astragalus bolanderi

FR spreading or ± ascending, 25–60 mm, 12–32 mm wide, bladdery, very thinly hairy, drying thinly papery; beak erect, pointed, strongly compressed side-to-side; chamber 1; deciduous. Open areas; 150–2350 m. c&s SN, Teh, GV, CW, SW; Baja CA. Much like *A. macrodon* (with hairs ± spreading, ± wavy; lflet margins red), which may have nomenclatural priority; like *A. pomonensis* (lflets 25–41; keel blade = or < keel claw); vars. ± intergrade.

1. Calyx tube ± evenly hairy throughout, lobes narrow, pointed, gen 1.4–2.5 mm (calyces intergrade in SnBr with those of var. *parishii*) .... var. ***douglasii***
1′ Calyx tube most densely hairy between lobes; lobes triangular, gen ± wide as long, gen 0.7–2.2 mm
 2. St prostrate to ± ascending; peduncle incurved-ascending, in fr often prostrate .... var. ***parishii***
 2′ St ± stiffly erect; peduncle erect, even in fr var. ***perstrictus***

var. ***douglasii*** (p. 595) **ST** prostrate to ± ascending, 2–7 dm. **LF**: lflets 11–25. **FL**: calyx tube ± evenly hairy throughout, lobes gen 1.4–2.5 mm, narrow, pointed. **FR** 25–60 mm. $2n$=22. Open areas; 50–1500 m. c&s SN, Teh, s ScV, SnJV, SCoR.

var. ***parishii*** (A. Gray) M.E. Jones **ST** prostrate to ± ascending, 2–6 dm. **LF**: lflets 15–25. **FL**: calyx tube most densely hairy between lobes, lobes gen 0.7–2.2 mm, triangular, ± wide as long. **FR** 25–50 mm. $2n$=22. Open areas; 1200–2350 m. SCo, SnBr, PR.

var. ***perstrictus*** (Rydb.) Munz JACUMBA MILKVETCH **ST** ± stiffly erect, 4–10 dm. **LF**: lflets 13–19. **FL**: calyx tube most densely hairy between lobes, lobes gen 0.7–2.2 mm, triangular, ± wide as long. **FR** 35–60 mm. $2n$=22. **RARE** CA. Rocky areas in open oak woodland; 900–1200 m. e PR; n Baja CA.

A. ertterae Barneby & J.R. Shevock (p. 595) ERTTER'S or WALKER PASS MILKVETCH Per; hairs spreading, long, soft, shaggy. **ST** 3.5–10 cm, buried for ± 1/2 length, ± prostrate above ground. **LVS** 4–5, 3–6.5 cm, crowded on upper part of st; underground, bladeless stipules fused around st into sheath; lflets 9–13, 6–13 mm, ± oblanceolate, tips blunt or shallowly notched. **INFL**: fls 7–17, crowded, ascending. **FL**: petals cream-colored, banner 10–12 mm, recurved ± 45°, keel 8.5–9 mm. **FR** spreading, 16–22 mm, 7–9 mm wide, oblong, gently incurved, swollen, abruptly narrowed at each end, bluntly 3- sided in X-section (lower depressed), glabrous, becoming stiffly leathery; upper suture prominently raised, lower a low, wavy ridge; chamber 1. RARE. Open, sandy, granitic soil among pines, live oaks; 1750–1900 m. s SNH (Kern Co.).

A. filipes A. Gray (p. 595) Per, sparsely leafy, ± glabrous to densely strigose. **STS** clumped, 3–9 dm. **LF** 2.5–12 cm; lower stipules fused around st into low, papery sheath; lflets 5–23, ± well separated, 3–25 mm, linear or narrowly oblong, tips ± obtuse or notched. **INFL**: fls 4–30, not crowded, gen reflexed. **FL**: petals ± dull white or pale yellow, banner 10–15.5 mm, recurved 50–85°, keel 6.7–12 mm. **FR** spreading or pendent; body 17–30 mm, 3–6.5 mm wide, flat, ± oblong, tapered at each end, glabrous or minutely strigose, becoming papery, dehiscing from base; stalk-like base 6–16 mm, minutely hairy; chamber 1. $2n$=24. Dry, open areas in sagebrush or pine (MP) or chaparral (SW); 1000-1700 m. WTR, SnBr, SnJt, MP; to B.C. ID, NV.

A. funereus M.E. Jones (p. 599) BLACK MILKVETCH Per; hairs grayish, dense, stiff, some short, ± wavy. **STS** prostrate, loosely tufted, 2–8 cm. **LF** 2.5–7 cm; lflets 7–17, ± crowded, 3–12 mm, obovate, tips blunt or notched. **INFL**: fls 3–10, ascending. **FL**: calyx hairs black or mixed black, white; petals pink-purple, banner 22–29 mm, recurved ± 40°, keel 21–28 mm. **FR** ascending, 25–50 mm, 10–15 mm wide, ± lanceolate in side view, straight, ± strongly compressed top-to-bottom at base, gently incurved near tip, leathery; hairs dense, 1.5–2.5 mm, wavy, white; chamber 1. RARE. Gravelly, clayey, or rocky areas; 1300–1500 m. ne DMtns (Funeral Mtns e of Death Valley). [*A. purshii* Hook. var. *f.* Jepson]

A. gambelianus E. Sheldon (p. 599) Ann, slender; hairs minute, ± incurved. **ST** erect to ± decumbent, 2–30 cm, slender. **LF** 1–4 cm; lflets 7–15, 1–9 mm, ± oblanceolate, tips blunt or notched. **INFL** black-hairy; fls 4–15, intially dense, ascending, then well separated, reflexed. **FL**: petals whitish, tinged with purple, banner 2.5–3.3 (6.5) mm, recurved ± 25°, keel 2–2.5(4.2) mm. **FR** reflexed, 2.8–4.2 mm, 2.4–3.6 mm wide, ovate or ± round in top view, ± wrinkled on edges, minutely hairy; chambers ± 2. $2n$=22. Open, grassy areas, shrubland; 50–900 m. CA-FP; sw OR, n Baja CA. Fr hairs of pls near coast straight, elsewhere incurved; pls from Amador, Marin cos. with banners 5.5–6.5 mm (gen growing with small-bannered pls), have been called var. *elmeri* (E. Greene) J. Howell.

A. geyeri A. Gray var. ***geyeri*** (p. 599) GEYER'S MILKVETCH Ann, rarely persistent into 2nd season, often very slender, minutely strigose. **ST** prostrate to ascending, 1–20 cm. **LF** 1.5–10 cm; lflets 3–13, ± well separated, 5–15 mm, ± linear to oblong, terminal often notably longest. **INFL** among lvs; fls 3–8, well separated, early ascending, then reflexed. **FL**: petals whitish, sometimes blushed with lilac, keel tip purple, banner 5.2–7.6 mm, recurved ± 45(–80)°, keel 3.8–4.8 mm. **FR** 15–25 mm, 6–10 mm wide, inflated, half-ovate in side view, thinly papery, minutely strigose; beak triangular; upper suture straight or ± concave; chamber 1. RARE in CA. Sandy areas; ± 1500 m. e SNE; to WA, WY, UT.

A. gibbsii Kellogg (p. 599) Per, leafy; hairs gen ± spreading, minute, wavy, ± grayish. **STS** ± prostrate, many, 1.5–3.5 dm. **LF** 1.5–9.5 cm; lower stipules small, fused around st into sheath; lflets 7–19, 4–20 mm, obovate or oblong. **INFL**: fls 10–30, reflexed. **FL**: calyx base with pouch on upper side; petals yellowish, wings and banner ± equal, banner 14–18 mm, shallowly S-shaped, keel 12–15 mm. **FR** pendent; body 22–30 mm, 4–8 mm wide, oblong, incurved 1/4–1/2 circle, swollen but ± compressed side-to-side, densely wavy-hairy, early fleshy, then leathery; stalk-like base 7–22 mm; sutures prominent, raised, keel-like; chamber 1. Among sagebrush or pine; 1300–1650 m. n SNH, GB; w NV.

A. gilmanii Tidestrom (p. 599) GILMAN'S MILKVETCH Per or ± ann; hairs ± spreading or curved, minute. **ST** ± ascending, 5–25 cm. **LF** 1.5–7.5 cm; lflets 7–13, 2–12 mm, ± oblanceolate, margins often purple, tips notched or not. **INFL** ± among lvs; fls 4–9, well separated, early ascending, then reflexed. **FL**: petals pink-purple, banner 6.1–8 mm, recurved ± 45°, keel 4.7–6.1 mm. **FR** 14–25 mm, 8–16 mm wide, ± ovoid, bladdery, papery; hairs minute, curved; chamber 1; immature seeds 8–12. UNCOMMON. Gravelly areas; 2000–3050 m. n DMtns (Panamint Mtns); NV.

A. hornii A. Gray var. ***hornii*** (p. 599) Ann, open and widely branched; hairs ± appressed or ascending. **ST** 3–12 dm, slender-solid or stout-hollow. **LF** 1.5–13 cm, often reflexed; lflets 11–33, 5–20 mm, ± elliptic. **INFL** dense, head-like; fls 10–35, spreading and ascending. **FL**: petals white to pale lilac, banner 7.8–10.2 mm, recurved ± 40°, keel 5.9–8.4 mm. **FRS** crowded, spreading, in cylindric or subspheric heads (2.5–3.5 cm wide), each fr 12–18 mm, 7–9 mm wide, ± ovoid, inflated, bladdery, papery; hairs spreading, coarse; beak prominent, pointed; chamber 1. Salty flats, lake shores; 60–150 (850 in w DMoj) m. s SnJV, WTR, w edge DMoj; w-c NV.

A. insularis Kellogg var. ***harwoodii*** Munz (p. 599) HARWOOD'S RATTLEWEED or MILKVETCH Ann, ± grayish strigose. **ST** decumbent to ascending, 5–40 cm, slender. **LF** 2–12 cm; lflets 9–21, ± well separated, 4–20 mm, ± narrowly elliptic or oblong, tips gen notched. **INFL** among lvs; fls 4–9, well separated, early spreading, then reflexed. **FL**: petals pink-violet, banner 5.5–7.4 mm, recurved ± 50–60°, keel 4.8–6 mm. **FR** spreading or ± reflexed, 15–24 mm, 5–15 mm wide, half-ovoid, bladdery, papery, sparsely strigose; beak conspicuous; upper suture straight or ± convex; lower suture strongly convex; chamber 1; immature seeds 7–14. RARE in CA. Sandy or gravelly areas; 0–300 m. DSon; AZ, nw Mex.

A. inversus M.E. Jones (p. 599) SUSANVILLE MILKVETCH Per, wiry, sparsely leafy, thinly, minutely strigose. **ST** prostrate to spreading, 2–5 dm, slender. **LF** 3–12 cm; lflets 5–11, well separated, 4–20 mm, narrow, arched, terminal not or only obscurely jointed to midrib. **INFL**: fls 5–12, well separated, reflexed. **FL**: petals reddish pink, banner white-tipped, wings and keel with buff-yellow, banner 9.4–12.2 mm, recurved 35–45°, keel 8.2–10 mm. **FR** pendent; body 20–35 mm, 4–5 mm wide, straight or down-curved, strongly compressed side-to-side, papery, minutely strigose or glabrous; stalk-like base 6–14 mm, minutely hairy; chamber 1. UNCOMMON. Dry soils, sagebrush or pine forest; 1200–1850 m. e CaRH, MP.

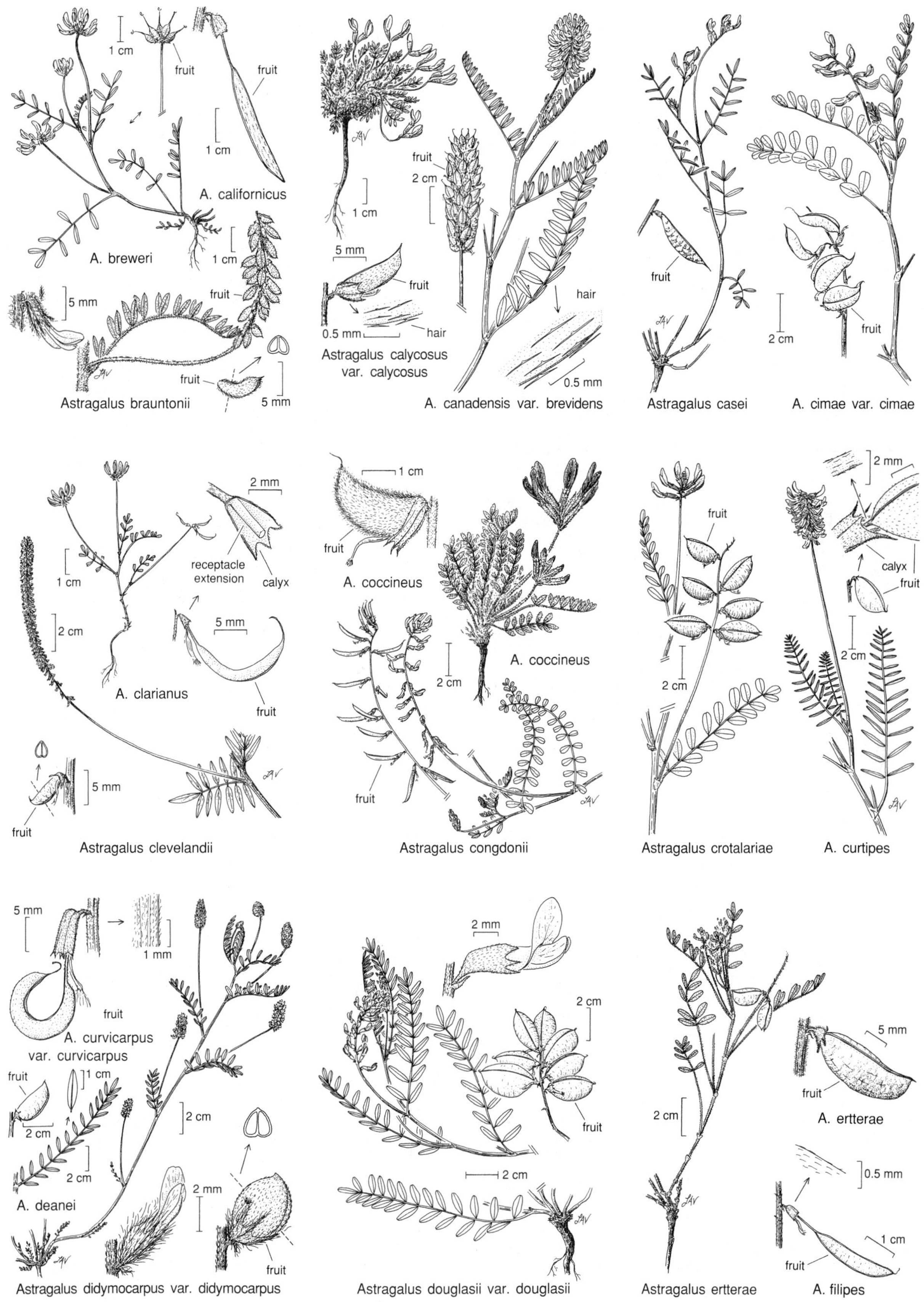
1 cm
fruit
fruit
1 cm
A. californicus
A. breweri
1 cm
5 mm
fruit
fruit
5 mm
Astragalus brauntonii
1 cm
fruit
2 cm
5 mm
fruit
0.5 mm
hair
Astragalus calycosus
var. calycosus
hair
0.5 mm
A. canadensis var. brevidens
fruit
2 cm
fruit
Astragalus casei
A. cimae var. cimae
2 mm
1 cm
receptacle
extension
calyx
2 cm
5 mm
A. clarianus
fruit
5 mm
fruit
Astragalus clevelandii
1 cm
fruit
A. coccineus
A. coccineus
2 cm
fruit
Astragalus congdonii
fruit
2 cm
Astragalus crotalariae
2 mm
calyx
fruit
2 cm
A. curtipes
5 mm
1 mm
fruit
A. curvicarpus
var. curvicarpus
fruit
1 cm
2 cm
2 cm
2 cm
A. deanei
2 mm
fruit
Astragalus didymocarpus var. didymocarpus
2 mm
2 cm
fruit
2 cm
Astragalus douglasii var. douglasii
2 cm
5 mm
fruit
A. ertterae
0.5 mm
1 cm
fruit
Astragalus ertterae
A. filipes

A. inyoensis Cov. (p. 599) Per, sparsely leafy, minutely strigose, grayish green. **STS** prostrate, loosely matted, 1–6 dm, slender, zig-zag. **LF** 1.5–4.5, widely spreading; lflets 9–21, crowded, 3–10 mm, narrowly obovate, tips blunt or notched. **INFL**: fls 6–15, well separated, spreading to reflexed. **FL**: petals pink-purple, banner 8.6–10.8 mm, recurved ± 45°, keel 8.2–9.6 mm. **FR** pendent; body 12–15 mm, 6–8 mm wide, lanceolate but strongly incurved 1/8–1/2 circle, wider than deep, ± 3-sided (lower side deeply grooved), stiffly leathery, minutely strigose; stalk-like base 2–5 mm, stout; chamber ± 1. 2*n*=22. Gravelly areas; 1500–2300 m. W&I.

A. iodanthus S. Watson var. ***iodanthus*** (p. 599) Per, sparsely strigose. **ST** ± prostrate, 5–40 cm. **LF** 2–7 cm; lflets 11–21, 3–18 mm, obovate or ± round, tips blunt or notched. **INFL**: fls 7–25, early crowded-ascending, then well separated and reflexed. **FL**: petals purple, whitish with purple keel tip, or cream, banner 10–15.5 mm, recurved ± 45°, ± = or > keel. **FR** pendent, 20–40 mm, 5–8.5 mm wide, incurved 1/4-full circle, wider than deep or ± 3-sided; chambers ± 2 in lower half, 1 in upper. Dry areas, sagebrush shrubland; 1200–2100 m. se MP, SNE; to OR, ID, UT.

A. jaegerianus Munz (p. 599) LANE MOUNTAIN MILKVETCH Per, sparsely leafy; hairs minute, ± flat, scale-like. **ST** weak, often scrambling through shrubland, 3–7 dm. **LF** 2–5 cm, stiffly spreading or reflexed; lflets 7–15, ± well separated, 3–15 mm, narrow, hairier on upper surface. **INFL**: fls 5–15, well separated, early ascending, then reflexed, twisted to 1 side. **FL**: petals dull pale purplish with darker veins, fading cream, often dingy, banner 6.5–10 mm, recurved 50–75°, keel 6.4–8.5 mm. **FR** pendent; body 18–25 mm, 3–5 mm wide, plump, stiffly papery or leathery, glabrous; stalk-like base 3–5 mm; sutures thick, raised, ± wavy; chambers 2. RARE. Among desert shrubs, in sand or gravel; 900–1200 m. c DMoj (near Barstow). Threatened by military activity, grazing, vehicles.

A. johannis-howellii Barneby (p. 599) LONG VALLEY MILKVETCH Per, open and widely branched; hairs ± appressed to ascending. **ST** ± prostrate or decumbent, 3–20 cm, slender. **LF** 4–6 cm; lower stipules fused around st into sheath; lflets 13–23, 2–6 mm, narrowly obovate, upper surface glabrous. **INFL** ± among lvs; fls 6–12, very well separated, reflexed in age. **FL**: petals whitish, banner 5–5.5 mm, recurved 90°, keel 3.3–3.9 mm. **FR** pendent; body 7–11 mm, ± 3 mm wide, half-ellipsoid, 3-sided (lower side deeply, openly grooved), thinly papery, minutely strigose; stalk-like base 0.5–2.5 mm; chambers 2. 2*n*=22. **RARE** CA. Sandy areas, sagebrush shrubland; 2150 m. SNE (sw Mono Co.). Threatened by grazing, mining.

A. kentrophyta A. Gray Per, tufted or matted, gen spiny, strigose; hairs attached by base or on side. **ST** < 3 dm. **LF** 2–26 mm; stipules sometimes like lflets, gen minutely spine-tipped, at least lower fused around st into sheath; lflets 3–9, 1–17 mm, linear or narrowly lanceolate, ± spine-tipped. **INFL**: fls 1–3. **FL**: petals white to pink-purple, banner 3.9–9.2 mm, recurved ± 45°, keel 2.9–6.3 mm. **FR** 4–9 mm, 1.5–4 mm wide, compressed side-to-side, finely hairy; sutures prominent; chamber 1. Open areas, clay, gravel, talus, rock; 2280–3660 m. SNH, SNE; to s Can, WY, ND, NM.

1. Hairs forked at base (1 branch gen short, sometimes ± 0); immature seeds 2–4; pl not a cushion or mat .... var. ***elatus***
1′ Hairs simple; immature seeds 5–8; pl a cushion or mat
 2. Lflets 3 (sometimes 5 on lower lvs), rigid, spine-tipped; pl a dense, rounded cushion .... var. ***danaus***
 2′ Lflets 5–9, ± soft, minutely spine-tipped; pl a ± flat mat .... var. ***tegetarius***

var. ***danaus*** (Barneby) Barneby (p. 599) SWEETWATER MOUNTAINS MILKVETCH Pl a dense, rounded, very spiny cushion, strigose; hairs simple. **ST** gen < 5 cm. **LF** 4–20 mm; lflets 3 (sometimes 5 on lower lvs), 3–7 mm, stiff, spine-tipped. **FL**: petals pale purple or whitish with purple keel-tip, banner 4–5.6 mm, keel 3.3–4.1 mm. **FR**: immature seeds 5–8. UNCOMMON. Rocky places at and above timberline; 3000–3600 m. c&s SNH.

var. ***elatus*** S. Watson (p. 599) SPINY-LEAVED MILKVETCH Pl not a cushion or mat, spiny, strigose; hairs forked at base (1 branch gen short, sometimes ± 0). **ST** ± erect, ± 1 dm, open and widely branched. **LF** 10–26 mm; lflets 3–7, stiff, spine-tipped. **FL**: petals gen white or faintly purple-veined, banner 4.8–6.2 mm, keel 3.7–4.1 mm. **FR**: immature seeds 2–4. RARE in CA. Open rocky areas; 3000 m. W&I; to WY, Colorado, NM. CA pls differ in aspects of habit, fr from e pls and may be taxonomically distinct.

var. ***tegetarius*** (S. Watson) Dorn (p. 599) Pl a ± flat mat, ± spiny, ± strigose; hairs simple. **LF** 2–20 mm; lflets 3–9, ± soft, minutely spine-tipped. **FL**: petals gen purple, sometimes whitish, banner 3.9–6.2 mm, keel 2.9–4 mm. **FR**: immature seeds 5–8. Open, rocky areas; 2700–3350 m. W&I; to OR, MT, NM. [var. *implexus* (Canby) Barneby] ❀DRN,DRY:1–3,14–16,18;DFCLT.

A. layneae E. Greene (p. 599) Per from deep rhizome (gen overlooked); hairs coarse, gen grayish. **ST** erect, 2.5–16 cm. **LF** 4–16 cm; lflets 11–23, 5–23 mm, ovate to ± round. **INFL**: fls 10–45, early ascending, then spreading. **FL**: calyx hairs mostly black, some white; petals whitish with keel tip, wing tips purple, banner often lilac-blushed, banner 12.5–18 mm, recurved ± 50°, keel 10.4–16.5 mm. **FR** 20–65 mm, 3.5–8 mm wide, incurved 1/4-full circle, leathery; hairs spreading, wavy; chambers ± 2 in lower half. 2*n*=44. Sandy flats, washes; 450–1550 m. DMoj; s NV, nw AZ.

A. lemmonii A. Gray (p. 599) Per, open and widely branched, sparsely strigose. **ST** ± prostrate, 1–5 dm, slender. **LF** 1–4.5 cm; lflets 7–15, 2–11 mm, narrowly elliptic, tips acute. **INFLS** in 2's or 3's in upper axils; fls 2–13, clustered at ends of peduncles, ascending. **FL**: petals whitish or dull lilac, banner 4.8–6.1 mm, recurved 45–85°, keel 3.4–4 mm. **FR** ± spreading, 4–7 mm, ± 2 mm wide, elliptic, 3-sided (lower narrow), papery; chambers ± 2. Meadows, lake shores; 1300–2200 m. GB (lower in MP), adjacent edge c SNH; s-c OR.

A. lentiformis Brewer & S. Watson (p. 599) LENS-POD MILKVETCH Per, open; hairs fine, spreading, grayish. **ST** spreading, 1–2 dm, slender. **LF** 1.2–3.5 cm; lower stipules fused around st into sheath; lflets 7–15, 2–10 mm, narrowly ± obovate. **INFL**: fls 5–10, early ascending, then reflexed. **FL**: petals pale yellow, drying darker, banner 6.2–7 mm, recurved ± 50°, keel 4–4.9 mm. **FR** spreading or reflexed, 5–9 mm, ± 3 mm wide, ± lens-shaped in side view, 3-sided (lower side very narrow), papery; hairs fine, wavy; chambers 2. RARE. Dry sandy soil in sagebrush or pine; ± 1500 m. n SNH (se Plumas Co.).

A. lentiginosus Hook. FRECKLED MILKVETCH Per (sometimes fl 1st year or ann), moderately leafy, ± glabrous to silvery-strigose. **LF** 1–15 cm; lflets linear to widely ± ovate. **INFL**: fls 3– ± 50, ascending or spreading. **FL**: petals purplish, cream, whitish, or mixed purplish and whitish, keel 0.65–0.8 × banner, banner recurved 30–50°. **FR** ovoid or spheric, widely grooved above and below, gen ± bladdery, ± papery, deciduous; beak gen triangular, flat; chambers 2 below beak. 2*n*=22. Gen dry, open places; –30–3600 m. CA; to OR, WY, n Mex. Highly variable; vars. often very distinct, yet intermediates common; fl, fr both needed for identification.

1. Lflets 3–5, linear-oblanceolate .... var. ***piscinensis***
1′ Lflets gen 7 or more, often wider than linear-oblanceolate
 2. Keel 8.4–15 mm; calyx often > 6.5 mm
  3. Pls of CA-FP
   4. Infl 7–20-fld, 1–4 cm in fr; petals pink-purple; herbage hairs subappressed .... var. ***idriensis***
   4′ Infl 10–32-fld, 4–15 cm in fr; petals purple, ± yellow, or white; herbage hairs spreading or subappressed
    5. Petals cream to ± yellow; hairs spreading .... var. ***nigricalycis***
    5′ Petals purple, rarely white; hairs subappressed .... var. ***variabilis***
  3′ Pls of GB or D
   6. Petals whitish, tinged with pink; MP var. ***platyphyllidius***
   6′ Petals gen purple or pinkish; SNE (Inyo Co.), D
    7. Herbage sparsely hairy, greenish, or if densely hairy then gen either fr sparsely hairy or calyx lobes < 1.4 mm
     8. St prostrate, > 60 cm; moist alkaline flats, extreme n DMoj .... var. ***sesquimetralis***
     8′ St ascending or spreading, < 50 cm; gen sand, widespread .... var. ***variabilis***

7′ Herbage hairs dense, silvery, grayish, or whitish; fr densely hairy; calyx lobes > 1.4 mm
9. Fr slightly swollen, < 7 mm wide, upper suture concave in side view; e DMoj, s DSon .... var. ***borreganus***
9′ Fr very swollen, often bladdery, > 8 mm wide, upper suture convex in side view; n DMoj, nw DSon
10. Pl base not woody; DSon (Coachella Valley) .... var. ***coachellae***
10′ Pl base ± woody; n DMoj (Eureka Valley) .... var. ***micans***
2′ Keel 5.5–9 mm (±10 mm in vars. *ineptus, salinus, sierrae*); calyx rarely > 6.5 mm
11. Lflet densely hairy, ashy-gray or silvery, at least on 1 side (see also 11′)
12. Fr beak curved down, away from st; clayey, alkaline flats, seeps; SNE, w DMoj ..... var. ***albifolius***
12′ Fr beak straight or curved upward, toward st; various habitats, gen exc clayey, alkaline flats, seeps; DMoj, CA-FP
13. Fr cluster < 4 cm; pine forest; SnGb .... var. ***antonius***
13′ Fr cluster > 4 cm; gravel or sand; DMoj, s CA-FP
14. Keel 5.6–8.5 mm; fr glabrous to sparsely hairy; gen e DMoj .... var. ***fremontii***
14′ Keel > (8)8.5 mm; fr sparsely to thinly to densely hairy; gen w DMoj .... var. ***variabilis***
11′ Lflet sparsely hairy or ± glabrous, green (sometimes densely hairy, grayish in var. *ineptus*)
15. Pls of MP and n SNE (from c Mono Co.), barely in ne CA-FP
16. Fr stiffly papery, ± opaque, minutely hairy at least when young, often little inflated, strongly incurved .... var. ***lentiginosus***
16′ Fr thinly papery, translucent, ± glabrous when young (and afterward), bladdery, ± straight
17. Fr cluster dense, cylindric or spheric; st gen with several–many branches in lower 1/2, gen > 25 cm .... var. ***floribundus***
17′ Fr cluster ± open; st gen with 0–few branches in lower 1/2, gen < 30 cm .... var. ***salinus***
15′ Pls of s SN, SnBr, SNE, D
18. Petals gen bright pink-purple; SNE, D
19. Keel 6–8.5 mm; gen e DMoj .... var. ***fremontii***
19′ Keel > 8 mm; gen w DMoj .... var. ***variabilis***
18′ Petals gen ± white, sometimes blushed with lilac, rarely purple; SN, SNE, W&I, SnBr
20. Fr beak ± cylindric, slender; calyx lobes gen < 1.2 mm, acute, slender — s SNH ... var. ***kernensis***
20′ Fr beak flat, ± triangular; calyx lobes often > 1.2 mm, acute, slender (if < 1.2 mm, then blunt, triangular)
21. Calyx lobes < 1.2 mm, blunt, triangular; SnBr .... var. ***sierrae***
21′ Calyx lobes > 1.2 mm, acute, slender; SNH, SNE, W&I
22. Lf 1.5–5.5 cm, lflets 9–21, crowded; SNH, SNE .... var. ***ineptus***
22′ Lf 4–9 cm, lflets 13–27, ± well separated; W&I .... var. ***semotus***

var. ***albifolius*** M.E. Jones Per. **ST** ± prostrate, 3–7 dm. **LF**: lflets 9–21, 3–18 mm, narrowly oblanceolate, densely strigose. **INFL** dense; fls 9–35; axis in fr 5–40 mm. **FL**: petals gen whitish with some purple, banner 8.2–11.5 mm, keel 6–8.5 mm. **FR** 10–17 mm, 8–14 mm wide, inflated, bladdery, thinly papery; beak curved down, 3–5 mm. Alkaline flats, seeps; 600–1500 m. SNE, w DMoj (Los Angeles Co.).

var. ***antonius*** Barneby SAN ANTONIO MILKVETCH Per. **ST** prostrate or spreading, 1–3 dm. **LF** 3–8 cm; lflets 11–21, 3–11 mm, ± obovate, densely strigose. **INFL**: fls 10–15; axis in fr < 5 cm. **FL**: petals purple, banner 9–10.5 mm, keel 7.2–8.2 mm. **FR** 14–30 mm, 10–18 mm wide, plumply ovoid to ± spheric, bladdery, papery, sparsely strigose, straw-colored, ± shiny; beak erect, 3–6 mm. RARE. Pine forest; 1500–2600 m. SnGb.

var. ***borreganus*** M.E. Jones (p. 599) BORREGO MILKVETCH Ann (sometimes per), ± densely silvery-hairy. **ST** ascending, 1–3 dm. **LF** 6–16 cm; lflets 7–19, 4–21 mm, ± obovate. **INFL**: fls 13–50; axis in fr 4.5–26 cm. **FL**: petals pink-purple, banner 12–14.8 mm, keel 10–13 mm. **FR** 15–23 mm, 4.5–6 mm wide, swollen, not bladdery, papery, silky-hairy, in side view lanceolate or narrowly ovate, gently incurved; upper suture concave, tapered to short, ± triangular, tooth-like beak. UNCOMMON. Sand; 30–250 m. e DMoj, s DSon; nw Mex. [var. *coulteri* (Benth.) M.E. Jones]

var. ***coachellae*** F. Shreve & Wiggins (p. 599) COACHELLA VALLEY MILKVETCH Ann or per, densely silvery-hairy. **STS** ascending, clumped, 1–3 dm. **LF** 5–11.5 cm; lflets 7–21, 5–17 mm, ± widely ovate. **INFL**: fls 11–25; axis in fr 3–10 cm. **FL**: petals pink-purple, banner 12.7–14.5 mm, keel 10.8–11.6 mm. **FR** 16–21 mm, 9–14 mm wide, greatly inflated, stiffly papery, grayish strigose; beak 3.5–6 mm. RARE. Sand; 0–350 m. DSon (Coachella Valley). Threatened by vehicles, development.

var. ***floribundus*** A. Gray (p. 599) Per, gen short-lived. **ST** ± ascending, 2–5 cm, gen with several–many branches in lower half. **LF** 3–11 cm; lflets 11–19, 5–15 mm, obovate, ± glabrous. **INFL** dense; fls 10–40; axis in fr 1–4(10) cm. **FL**: petals white, ± lilac-tinged, banner 8.8–11.6 mm, keel 6.6–8 mm. **FR** 8–21 mm, 6–12 mm wide, bladdery-inflated, thinly papery, glabrous or sparsely strigose, straw-colored; beak 3–7 mm, incurved or erect. Often among sagebrush; 1150–1600 m. n SNH s MP, SNE; s-c OR, w-c NV.

var. ***fremontii*** (A. Gray) S. Watson (p. 599) Ann or per, densely or sparsely hairy. **ST** decumbent to erect, 1–5 dm. **LF** 3–12 cm; lflets 9–19, 5–19 mm, obovate. **INFL**: fls 8–30; axis in fr 2.5–16 cm. **FL**: petals ± purple, banner 9.1–12.4 mm, keel 5.6–8.5 mm. **FR** 14–36 mm, 5–18 mm wide, gen bladdery, thinly papery, glabrous to thinly hairy; beak 2–10 mm, ± incurved. $2n=22$. Open sand, gravel; 900–2900 m. SNE, e DMoj; s NV. At lower elevations in DMoj. Esp hairy in SNE.

var. ***idriensis*** M.E. Jones Per; herbage hairs subappressed. **ST** open and widely branched, 1–4 dm. **LF** 2–11 cm; lflets 7–29, 3–18 mm, widely obovate. **INFL**: fls 7–20; axis in fr 1–4 cm. **FL**: petals pink-purple, banner 12–20 mm, keel 9–14 mm. **FR** 12–30 mm, 5–16 mm wide, ± half-ovoid, often incurved, greatly or ± inflated, ± fleshy, minutely strigose, drying thickly papery or ± leathery; beak 3–10 mm. Dry, grassy hillsides, often with oak; 650–1200 m. SnFrB, SCoRI, WTR. Highly variable.

var. ***ineptus*** (A. Gray) (p. 599) M.E. Jones Per; hairs ± ascending, grayish. **ST** decumbent, 1–3 dm. **LF** 1.5–5.5 cm; lflets 9–21, 2–10 mm, crowded, obovate. **INFL**: fls 10–21; axis in fr 10–25 mm. **FL**: calyx lobes > 1.2 mm; petals cream-colored, banner 8.8–12.2 mm, keel 7.2– 9.3 mm. **FR** 10–18 mm, 6–12 mm wide, bladdery, thinly papery, minutely strigose, rarely glabrous; beak erect or incurved. Open gravelly places; 2000–3600 m. SNH (barely in n SNH), SNE. Intergrades with var. *semotus*.

var. ***kernensis*** (Jepson) Barneby KERN PLATEAU MILKVETCH Per, minutely strigose. **ST** prostrate or decumbent, 2–12 cm, slender. **LF** 1–5 cm; lflets 7–19, 2–7 mm, ± ovate, upper surface ± glabrous. **INFL**: fls 2–9; axis in fr 3–15 mm. **FL**: calyx lobes gen < 1.2 mm; petals whitish, banner 9.3–11.3 mm, keel 7.3–8.7 mm. **FR** 6–10 mm, 6–9 mm wide, ± spheric, bladdery, thinly papery; beak ± cylindric, slender. RARE in CA. Sandy areas; 2350–2600 m. s SNH; s NV.

var. ***lentiginosus*** Per, sparsely strigose. **ST** prostrate or decumbent, 1–5 dm. **LF** 3–10 cm; lflets 5–19, 5–18 mm, widely obovate. **INFL**: fls 8–22; axis in fr 5–35 mm. **FL**: petals cream-colored, keel tip and wing bases ± lilac-tinged, banner 7.4–11 mm, keel 6.3–8.4 mm, very rarely > wings. **FR** 10–23 mm, 4–10 mm wide, lanceolate or ovate in side view, strongly incurved, often little inflated, stiffly papery, ± opaque, ± glabrous; beak 4–9 mm. $2n=22$. Dry open areas among sagebrush or pines; 900–1500 m. MP; to WA, ID, NV. [var. *carinatus* M.E. Jones]

var. ***micans*** Barneby SHINING MILKVETCH Per; hairs dense, silvery or white, silky. **STS** clumped, ascending or erect, 2–4 dm. **LF** 4.5–9.5 cm; lflets 11–17, 5–14 mm, widely ± ovate. **INFL**: fls 12–35; axis in fr 4–10 cm. **FL**: petals white, tips blushed with pink-lavender, banner 12.2–14.3 mm, keel 9.6–10 mm. **FR** 15–20 mm, 8–10 mm wide, very swollen, stiffly papery, densely silky-hairy; beak 2.5–4 mm. RARE. Dunes; 900 m. n DMoj (Eureka Valley). Threatened by vehicles. More strongly per, with longer hairs (1.1–2 mm) than similar pls of var. *variabilis*.

var. ***nigricalycis*** M.E. Jones Per; hairs ± spreading, curly or wavy, blackish among fls, frs. **ST** ± ascending, 2.5–5 dm, stout. **LF** 4–16 cm; lflets 19–25, 6–25 mm, oblong to obovate. **INFL**: fls 10–32; axis in fr 4–11 cm. **FL**: petals cream to ± yellow, yellow when dry, banner 12–18.5 mm, keel 10.8–13 mm. **FR** 20–35 mm, 10–20 mm wide, greatly inflated, bladdery, papery; hairs ± ascending; beak 4–8 mm. Dry, grassy areas, roadcuts; 100–750 m. s SNF, SnJV, SCoRI.

var. ***piscinensis*** Barneby (p. 599) FISH SLOUGH MILKVETCH Per, sparsely leafy, ± canescent. **ST** prostrate, < 1 m. **LF** 3–4 cm; lflets 3–5, linear-oblanceolate, lateral 7–20 mm, terminal 14–32 mm. **INFL**: fls 5–12; axis in fr 1.5–4 cm. **FL**: petals lavender, banner 13 mm, keel 9 mm. **FR** 20–24 mm, 8–12 mm wide, strongly inflated, stiffly papery, densely strigose; beak 4.5–7 mm. RARE. Wet soil; 1300 m. c SNE (Mono, Inyo cos., near Bishop).

var. ***platyphyllidius*** (Rydb.) M. Peck Per, ± glabrous. **ST** ± ascending, 1–3.5 dm, coarse. **LF** 4–11 cm; lflets 7–19, 5–20 mm, widely obovate. **INFL**: fls 5–15; axis in fr 1–3.5 cm. **FL**: petals whitish, keel, sometimes wing tips tinged with pink, banner 12.6–21.4 mm, keel 11–15 mm, very rarely > wings. **FR** 13–48 mm, 7–14 mm wide, greatly or ± inflated, straight or curved 3/4 circle, leathery or stiffly papery, ± glabrous; beak 5–15 mm. Arid plains, hillsides, often volcanic soil; 1050–1350 m. MP; to OR, ID, WY, Colorado. Intergrades with vars. *lentiginosus, salinus*.

var. ***salinus*** (Howell) Barneby (p. 599) Per, gen short-lived, glabrous to sparsely strigose. **ST** ascending, gen with 0–few branches in lower half, 1–3 dm. **LF** 4–10 cm; lflets 9–19, 5–20 mm, widely obovate. **INFL**: fls 10–25; axis in fr 1.5–9 cm. **FL**: petals whitish, keel and wing tips sometimes tinged lilac, banner 9.5–13.3 mm, keel 6–9.6 mm. **FR** 14–30 mm, 6–14 mm wide, bladdery, thinly papery, gen glabrous; beak 3–9 mm. $2n=22$. Gen among sagebrush; 1050–1350 m. MP; to OR, WY. Intergrades with vars. *floribundus, lentiginosus, platyphyllidius*.

var. ***semotus*** Jepson Per, ± strigose; hairs subappressed. **STS** loosely tufted, < 15 cm. **LF** 4–9 cm; lflets 13–27, 2–9 mm, ± well separated, ± narrowly elliptic, upper surface ± glabrous. **INFL**: fls 6–10; axis in fr 1–3 cm. **FL**: calyx lobes > 1.2 mm; petals whitish, banner 10.4–12 mm, keel 7.7–8.5 mm. **FR** 10–20 mm, 6–12 mm wide, bladdery, papery, sparsely strigose; beak 4–7 mm, incurved. $2n=22$. Sandy or gravelly flats, hillsides; 2450–3350 m. W&I. Intergrades with var. *ineptus*.

var. ***sesquimetralis*** (Rydb.) Barneby SODAVILLE MILKVETCH Per, sparsely strigose. **ST** prostrate, 6–8 dm. **LF** 2–5 cm; lflets 7–17, 6–18 mm, oblanceolate, upper surface ± glabrous. **INFL**: fls 5–12; axis in fr 1–2.5 cm. **FL**: petals purple, banner 12–14.5 mm, keel 9.3–9.5 mm. **FR** 12–26 mm, 5–12 mm wide, ± inflated, stiffly papery, sparsely strigose; beak 4–8 mm. **ENDANGERED** CA. Moist, alkaline flats; 950 m. n DMoj (n Death Valley, e slope Last Chance Mtns); NV.

var. ***sierrae*** M.E. Jones BEAR VALLEY MILKVETCH Per, thinly strigose. **STS** open and widely branched, ± matted, 1–3.5 dm. **LF** 2–5 cm; lflets 15–21, 3–8 mm, obovate, gen crowded, midribs gen ± arched. **INFL**: fls 5–15; axis in fr 1–3 cm. **FL**: calyx lobes < 1.2 mm; petals whitish, tips ± pink-tinged, banner 10.4–14.5 mm, keel 8.1–9.8 mm. **FR** 15–22 mm, 8–15 mm wide, plumply ovoid, bladdery, papery, sparsely strigose; beak 3–6 mm. Rocky meadows, pine woodlands; 1800–2600 m. SnBr.

var. ***variabilis*** Barneby Ann or per, gen robust, coarse; fl-st sometimes single, weak; hairs sparse, appressed, straight to ± dense, spreading, wavy. **ST** ascending, 1–4 dm. **LF** 2.5–13 cm; lflets 11–25, 4–17 mm, obovate. **INFL**: fls 10–30; axis in fr 4–17 cm. **FL**: petals purple, rarely white, banner 11.1–15 mm, keel 8.4–12.3 mm. **FR** 12–30 mm, 8–15 mm wide, bladdery, ± firmly papery, thinly to densely strigose or ± wavy-hairy; beak 3–9 mm, gently incurved. $2n=22$. Sand; 140–1600 m. s SNF, Teh, s SnJV, s-most SNE, w&s DMoj. Intergrades with vars. *fremontii, micans* to n; *coachellae, nigricalycis* to w.

A. leucolobus M.E. Jones (p. 599) BEAR VALLEY WOOLLYPOD Per, ± cespitose; herbage hairs gen 0.6–1.3 mm, straight and wavy, entangled, grayish. **ST** < 7 cm. **LF** 1.5–9 cm; lflets 7–19, 3–13 mm, ± widely obovate. **INFL**: fls 5–13, ascending. **FL**: petals pink-purple, wings barely < banner, banner 16–18.5 mm, recurved ± 40°, keel 14.3–16.8 mm. **FR** ascending, 13–25 mm, 5–8 mm wide, ± incurved, ± strongly compressed top-to-bottom, abruptly bent at tip, leathery; hairs dense, gen 0.7–1.5 mm, white; chambers ± 2. RARE. Dry, rocky areas among sagebrush or pines; 1750–2450 m. SnGb, SnBr, SnJt. Much like vars. of *A. purshii* (which have wings << banner, fr hairs > 1.5 mm, or banner 10–15 mm).

A. macrodon (Hook. & Arn.) A. Gray (p. 599) SALINAS MILKVETCH Per; hairs ± ascending or spreading, wavy or curved. **STS** clumped, ± ascending, 5–10 dm. **LF** 5.5–15 cm; lflets 11–29, 7–20 mm, ± elliptic, margins, midribs gen purple. **INFL**: fls 8–35, early spreading, then reflexed. **FL**: calyx lobes 2.5–4.3 mm, purple; petals ± cream, tinged with red-brown at tips, banner 8.3–11.4 mm, streaked with red-brown, recurved ± 90°, keel 7.5–9.1 mm. **FR** ± spreading or reflexed, 20–40 mm, 14–20 mm wide, widely ovoid, bladdery, thinly papery; hairs ± spreading or ascending; chamber 1. UNCOMMON. Eroded pale shales or sandstone, or serpentine alluvium; 300–950 m. c SCoR. See note under *A. douglasii*.

A. magdalenae E. Greene var. ***peirsonii*** (Munz & J. McBurney) Barneby (p. 599) PEIRSON'S MILKVETCH Ann or per, silvery-canescent. **ST** ascending or erect, 2–9 dm. **LF** 1–15 cm; midrib ± flattened; lflets 3–15, ± well separated, 2–8 mm, narrow, oblong, terminal not jointed to midrib. **INFL**: fls 5–20, ascending or spreading. **FL**: petals pink-purple, often white-tipped, banner 10–14.2 mm, recurved ± 40°, keel 8.5–10 mm. **FR** spreading, 20–35 mm, 10–20 mm wide, half-ovoid to ± spheric, bladdery, papery, finely strigose; chamber 1. **ENDANGERED** CA. Sand dunes; 50–250 m. DSon; reported from w AZ, n Baja CA. Seeds to 5 mm, largest of Am spp. of *Astragalus*.

A. malacus A. Gray (p. 603) Per; hairs 1.5–2.5 mm, spreading. **ST** ascending or erect, 1–4 dm. **LF** 4–15 cm; lflets 7–21, 5–20 mm, elliptic to obovate, tips ± notched. **INFL**: fls 9–35, reflexed in age. **FL**: petals reddish violet, banner 15–21 mm, recurved ± 45°, keel 12.3–16 mm. **FR** pendent; body 18–38 mm, 5–6 mm wide, incurved, ± 3-sided, stiffly papery; hairs long; stalk-like base 1–3 mm, stout; chambers 2. Dry rocky or stiff soils, with sagebrush, pinyon woodland; 1050–2350 m. e MP, SNE; to OR, ID, NV.

A. miguelensis E. Greene (p. 603) SAN MIGUEL MILKVETCH Per, open; hairs dense, short, woolly, ± gray. **ST** ± ascending, 1–4 dm. **LF** 2.5–12 cm; lower stipules fused around st into finely tomentose sheath; lflets 17–27, 6–22 mm, narrowly oblong or ± obovate, tips notched or blunt. **INFL**: fls 10–30, dense, ± spreading. **FL**: petals whitish to yellowish, banner 12.5–16 mm, recurved ± 45°, keel 9–12 mm. **FR** spreading, 16–26 mm, 13–23 mm wide, bladdery, papery, hairs fine; chamber 1. $2n=22$. UNCOMMON. Slopes, bluffs, coastal beaches; < 500 m. n ChI, s ChI (San Clemente Island).

A. minthorniae (Rydb.) Jepson var. ***villosus*** Barneby (p. 603) Per, robust or ± cespitose; hairs ± spreading, coarse. **ST** ascending or erect, 3–30 cm. **LF** 4–17 cm; lower stipules long-woolly; lflets 7–17, 8–25 mm, ± obovate. **INFL**: fls 7–35, ascending to reflexed. **FL**: petals cream (keel tip purple) or purplish (wing tips pale), banner 12–18 mm, recurved ± 45°, keel 9.5–13 mm. **FR** ± erect, 15–30 mm, 4–6 mm wide, ± compressed side-to-side but sides convex, leathery; hairs ± spreading; chambers 2. Rocky, calcareous hillsides, washes, gen with pinyon, juniper; 1350–2300 m. SnBr, W&I, DMtns; s NV.

A. mojavensis S. Watson Ann or ± per, open and widely branched or tufted, grayish or silvery-strigose or canescent. **ST** weakly ascending, 5–35 cm. **LF** 2–12.5 cm; lflets 3–11, 3–18 mm, ovate to ± round, tips gen blunt, sometimes (rarely on upper) shallowly notched. **INFL**: fls 3–15, early ascending, then reflexed. **FL**: petals pink-purple, banner 7–12.5 mm, recurved ± 45°, keel 6.4–10.5 mm. **FR** reflexed, 13–32 mm, 3.5–8.5 mm wide, stiffly leathery, densely, minutely strigose; chambers ± 2 below, 1 near tip; immature seeds 20–30. Dry, rocky areas; 750–2300 m. DMoj; s NV.

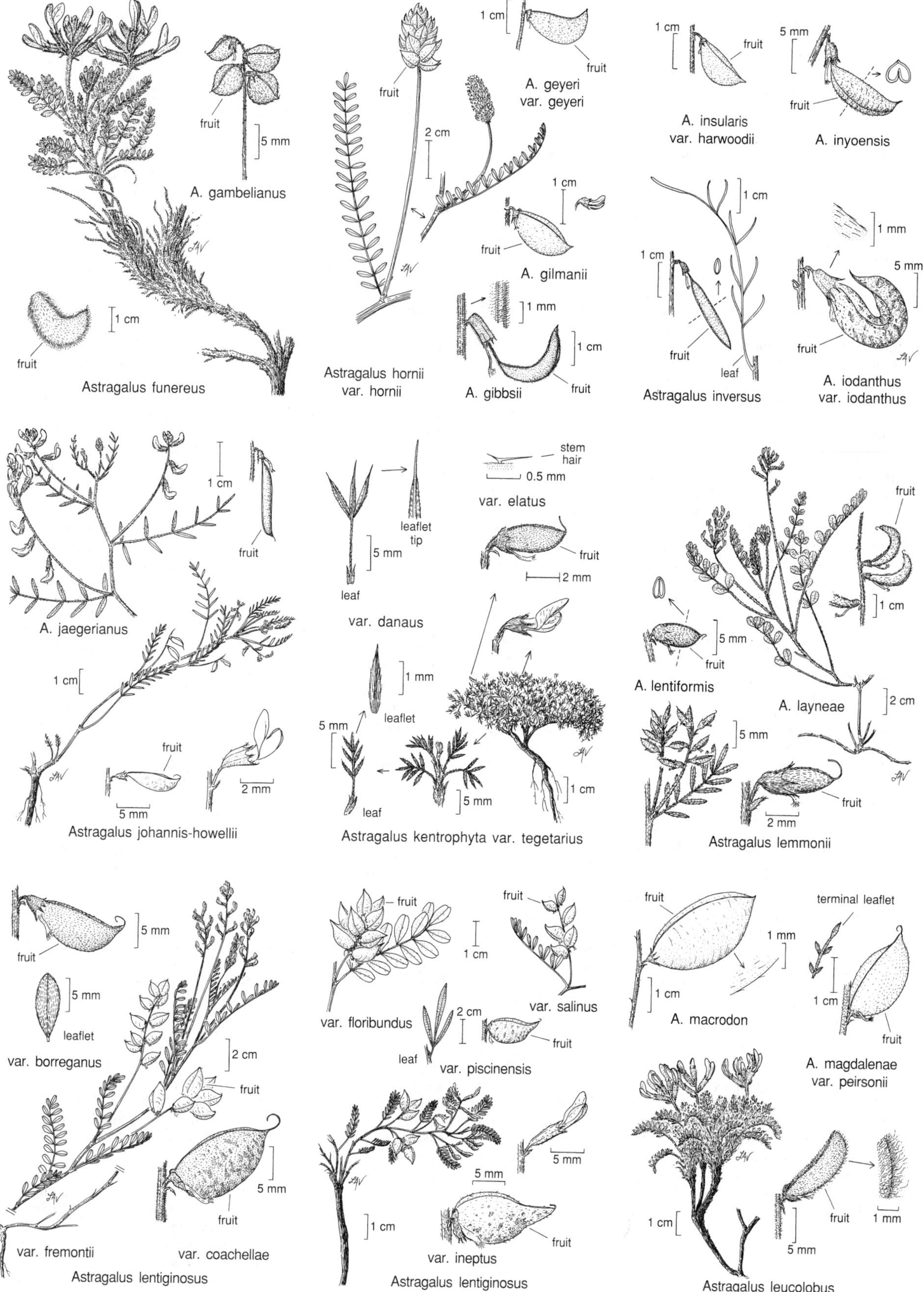
fruit
5 mm
A. gambelianus
1 cm
fruit
Astragalus funereus
fruit
2 cm
Astragalus hornii
var. hornii
1 cm
fruit
A. geyeri
var. geyeri
1 cm
fruit
A. gilmanii
1 mm
1 cm
A. gibbsii
fruit
1 cm
fruit
A. insularis
var. harwoodii
5 mm
fruit
A. inyoensis
1 cm
1 cm
fruit
leaf
Astragalus inversus
1 mm
5 mm
fruit
A. iodanthus
var. iodanthus
1 cm
fruit
A. jaegerianus
1 cm
fruit
5 mm
2 mm
Astragalus johannis-howellii
leaflet
tip
5 mm
leaf
var. danaus
1 mm
leaflet
5 mm
leaf
5 mm
stem
hair
0.5 mm
var. elatus
fruit
2 mm
1 cm
Astragalus kentrophyta var. tegetarius
fruit
5 mm
fruit
A. lentiformis
1 cm
A. layneae
2 cm
5 mm
fruit
2 mm
Astragalus lemmonii
5 mm
fruit
5 mm
leaflet
var. borreganus
2 cm
fruit
5 mm
fruit
var. fremontii
var. coachellae
Astragalus lentiginosus
fruit
1 cm
var. floribundus
fruit
var. salinus
2 cm
leaf
fruit
var. piscinensis
5 mm
5 mm
1 cm
fruit
var. ineptus
Astragalus lentiginosus
fruit
1 mm
1 cm
A. macrodon
terminal leaflet
1 cm
fruit
A. magdalenae
var. peirsonii
1 cm
fruit
1 mm
5 mm
Astragalus leucolobus

1 . Fr incurved 1/4–1/2 circle, ± 3-sided, upper suture raised, lower suture in wide, shallow groove; DMtns (immediately w of Death Valley) .... var. ***hemigyrus***
1′ Fr straight or incurved gen < 1/4 circle, ± compressed side-to-side or 3-sided, upper and lower sutures gen both raised; DMoj .... var. ***mojavensis***

var. ***hemigyrus*** (Clokey) Barneby (p. 603) HALF-RING-POD, CURVED-POD or MOJAVE MILKVETCH **PRESUMED EXTINCT** in CA; extant in s NV. Limestone; 1250–1600 m. DMtns (Darwin Mesa, w of Death Valley, Inyo Co., in 1941); s NV.

var. ***mojavensis*** Gen limestone; 750–2300 m. DMoj.

A. monoensis Barneby (p. 603) MONO MILKVETCH Per from rhizome; crown below soil surface; hairs dense, silky, wavy. **ST** 7–20 cm, 2–6 cm buried, ± glabrous; above ground part decumbent, hairy. **LF** 7–30 mm; lower stipules, esp underground stipules, fused around st into sheath; lflets 9–15, crowded, 2–8 mm, ovate. **INFL** head-like; fls 6–12, spreading. **FL**: petals ± pale pink-tinged but drying ± yellow, banner 10–13 mm, recurved ± 50°, keel 6.7–8 mm. **FR** spreading or ascending, 15–20 mm, 6–9 mm wide, widely incurved-lanceolate, papery; hairs short, wavy; chambers ± 2; immature seeds 18–20. 2*n*=22. **RARE** CA. Open pumice sand or gravel; 2250–2400 m. SNE (c Mono Co). Threatened by vehicles, road maintenance, grazing.

A. nevinii A. Gray (p. 603) SAN CLEMENTE ISLAND MILKVETCH Per; hairs fine, soft, kinky, woolly, intertangled. **ST** ascending, bushy-branched, 1–3 dm. **LF** 2–8 cm; lflets 11–25, 3–12 mm, ± oblong or obovate, tips blunt or notched. **INFL**: fls 15–30, dense, ultimately reflexed. **FL**: petals cream-colored, banner 10.6–12.7 mm, recurved 30–40°, keel 9–10 mm. **FR** pendent; body 14–20 mm, 3–5 mm wide, incurved, 3-sided (lower side openly grooved), stiffly papery, glabrous; stalk-like base 5–10 mm, slender; beak stout; chambers ± 2. RARE. Coastal sand, bluffs; probably < 50 m. ChI (San Clemente Island).

A. newberryi A. Gray var. ***newberryi*** (p. 603) Per, ± cespitose; old lf bases persistent; hairs silky, longer ± straight, partly spreading. **LF** 1.5–15 cm; lflets 3–15, 5–20 mm, obovate, tips acute or blunt, sometimes notched. **INFL**: fls 3–8, ascending. **FL**: calyx with more white than black hairs; petals pink-purple or whitish, tipped with pink-purple, banner 21.5–30 mm, recurved ± 40°, keel 18.5–26 mm. **FR** 13–28 mm, 7–13 mm wide, ovoid, gen incurved; hairs dense, longer ones 2–4.5 mm, white, woolly, some curly, some ± straight; chamber 1. 2*n*=22. Rocky areas; 1300–2350 m. SNE, ne DMoj; to OR, UT, NM. Like *A. purshii* (which has hairs extremely fine, entangled).

A. nutans M.E. Jones (p. 603) PROVIDENCE MOUNTAIN MILKVETCH Ann or ± per, minutely strigose. **ST** prostrate to erect, 6–15 cm. **LF** 2–8 cm; lflets 7–13, 5–15 mm, ± narrowly elliptic or obovate, tips acute or shallowly notched. **INFL**: fls 6–10, ascending to reflexed. **FL**: petals pink-purple, wing tips often paler, banner 7.8–10.4, recurved ± 90°, keel 5.9–6.6 mm. **FR** spreading, 15–25 mm, 11–15 mm wide, ovoid, bladdery, thinly papery, sparsely strigose, rarely ± glabrous; beak strongly compressed side-to-side, widely triangular; chamber 1; immature seeds 19–24; seeds suspended from flange 1–2.5 mm wide. 2*n*=22. UNCOMMON. Sandy, gravelly places; 450–1950 m. se DMtns, DSon.

A. nuttallianus A. DC. Ann, slender, minutely strigose. **ST** prostrate or weakly ascending, 4–45 cm. **LF** 1.5–6.5 cm; lflets 5–13, 2–10 mm, at least upper elliptic, acute at tips, lower sometimes blunt, notched at tip. **INFL**: fls 1–4, 4–7 mm. **FL**: corolla whitish, faintly lilac-tinged (rarely purplish). **FR** 10–20 mm, 2–3 mm wide, linear in side view, gently curved near base, ± straight toward tip, ± 3-sided (laterals ± convex, bottom grooved), maturing dark brown or blackish, glabrous or minutely strigose; chambers ± 2. Gravelly or sandy areas; 0–1950 m. D; to Colorado, n Mex. See note under *A. acutirostris*; vars. intergrade.

1 . Calyx lobes gen 1.8–2.8 mm, hairs 0.7–1.2 mm, spreading, very stiff .... var. ***austrinus***
1′ Calyx lobes gen 1–1.7 mm, hairs 0.5–0.8 mm, ascending, not very stiff
2 . Lflet tips notched on lower lvs, acute on upper; calyx tube 1.4–1.7 mm .... var. ***cedrosensis***
2′ Lflet tips acute; calyx tube 1.9–2.8 mm .. var. ***imperfectus***

var. ***austrinus*** (Small) F. Shreve & Wiggins **LF**: lflet tips acute. **FL**: calyx tube ± 2.5 mm, lobes gen 1.8–2.8 mm, hairs spreading, 0.7–1.2 mm, very stiff, silvery. 2*n*=22. Calcareous soils; gen 600–2150 m. Approaches DSon near AZ border, possibly a waif; to OK, TX, nw Mex.

var. ***cedrosensis*** M.E. Jones **LF**: lflet tips notched on lower lvs, acute on upper. **FL**: calyx tube 1.4–1.7 mm, ± as wide, lobes 0.7–1.6 mm, hairs ascending, short, not very stiff. Rocky flats, washes; < 300 m. DSon; AZ, nw Mex.

var. ***imperfectus*** (Rydb.) Barneby (p. 603) **LF**: lflets tips acute. **FL**: calyx tube 1.9–2.8 mm, ± 1/2 as wide, lobes 1–2 mm, hairs ascending, short, not very stiff. 2*n*=22. Sandy, gravelly flats, washes; 300–1950 m. e DMoj; to UT, AZ.

A. nuttallii (Torrey & A. Gray) J. Howell Per, gen robust, leafy; hairs fine, incurved-ascending or curly, grayish, or ± 0. **STS** prostrate to erect, often in dense tangles, 2–10 dm. **LF** 2.5–17 cm, gen spreading, arched; stipules fused around st into sheath, sometimes early ruptured, seemingly free; lflets 21–43, gen ± crowded, 3–25 mm, ± obovate or oblong, tips notched. **INFL**: fls 20–125, dense, early reflexed. **FL**: petals ± cream-colored, sometimes lavender-tinged, banner 10–15 mm, recurved ± 40°, wings 1 mm < to 1.5 mm > banner, keel 9.7–14 mm. **FR** 20–60 mm, 15–27 mm wide, bladdery, papery, glabrous to thinly hairy; chamber 1; immature seeds 14–38. Open bluffs, dunes, sandy areas; < 150 m. c&s NCo, CCo; Baja CA.

1 . Lflets gen ± grayish hairy (esp in s) on upper surface; c&s CCo .... var. ***nuttallii***
1′ Lflets gen glabrous on upper surface; c&s NCo, n CCo .... var. ***virgatus***

var. ***nuttallii*** (p. 603) **ST** gen prostrate or decumbent (ascending in sheltered places). **FR** ± hairy, at least when young; immature seeds 22–38. 2*n*=22. Rock, sandy areas, bluffs; < 70 m. c&s CCo. ❀DRN:**15,16**&SUN:**17**.

var. ***virgatus*** (A. Gray) Barneby **ST** ascending or erect, at coast sometimes prostrate. **FR** glabrous or sometimes sparsely hairy; immature seeds 14–21. Sandy soil, bluffs; < 150 m. c&s NCo, n CCo. ❀DRN:**15,16**&SUN:**17**.

A. obscurus S. Watson (p. 603) Per, minutely strigose. **ST** ± prostrate, open and widely branched or tufted, < 15 cm. **LF** 2.5–10 cm; lflets 5–15, ± well separated, 2–15 mm, narrowly to widely elliptic or oblong, ± thick. **INFL**: fls 5–15, ± dense, ascending. **FL**: petals barely graduated, cream-colored or dirty white, sometimes tinged dirty lilac, banner 7–10.5 mm, recurved ± 45°, keel 6.3–10 mm. **FR** erect, 10–25 mm, 2.4–3.3 mm wide, linear-oblong, ± bluntly 3- sided (lower side grooved), leathery, finely strigose; upper suture a thick ridge; chambers ± 2. Rocky, basaltic areas; 800–1950 m. se KR, CaR, MP; to OR, ID, NV.

A. oocarpus A. Gray (p. 603) SAN DIEGO MILKVETCH Per, stout, leafy; hairs ± glabrous. **ST** widely ascending to erect, 6–13 dm, hollow, glabrous. **LF** 4.5–17 mm; midrib grooved; lflets 17–35, 6–33 mm, widely ± lanceolate, glabrous (exc sparsely strigose margins, midveins), prominently veined, midvein raised on lower surface, in groove on upper surface. **INFL**: fls 20–75, widely ascending. **FL**: calyx 4.5–6 mm; petals cream-colored, banner 10.5–12.5 mm, recurved 70–90°, keel 9–10.8 mm. **FR** erect, 15–25 mm, 10–16 mm wide, greatly inflated, stiffly papery, glabrous or minutely strigose, persistent; chamber 1. 2*n*=22. RARE. Openings in chaparral, oak woodland; 600–1500 m. PR (c San Diego Co.). Like *A. douglasii* (which has all tissues thinner; fr much less persistent).

A. oophorous S. Watson Per, ± robust, ± glabrous. **ST** decumbent or ascending, 1–3 dm. **LF** 5–15 cm; lflets 7–21, 4–20 mm, ovate to ± round. **INFL**: fls 4–10, not crowded, spreading. **FL**: calyx glabrous; corolla red-purple with white wing tips, or cream-colored, or white drying cream-colored, banner 16–23 mm, recurved ± 85°, keel 10– 16.5 mm. **FR** spreading or pendent; body 25–55 mm, 10–20 mm wide, widely ± ovoid, bladdery, glabrous; stalk-like base 3.5–10 mm; chamber 1. 2*n*=24. Dry, open areas, often among sagebrush, pinyon; 1500–3100 m. SNE, n DMtns; w&c NV.

1. Petals cream-colored, or white drying cream-colored; lflets 7–11 . . . . var. ***lavinii***
1′ Petals red-purple, wing tips white; lflets > 11 on longer lvs . . . . var. ***oophorous***

var. ***lavinii*** Barneby LAVIN'S MILKVETCH UNCOMMON. Habitats of sp.; ± 2450–3050 m. SNE (Bodie Hills, Mono Co.); w&c NV.

var. ***oophorous*** (p. 603) Habitats of sp.; 1500–3100 m. SNE, n DMtns.

A. oxyphysus A. Gray (p. 603) Per, robust, bushy-clumped; hairs very fine, ± spreading, wavy. **ST** ± erect, 3–8 dm. **LF** 4.5–17 cm; lower stipules fused around st into ± hairy sheath gen ruptured by st expansion; lflets 11–31, 5–32 mm, ± lanceolate. **INFL**: fls 20–65, well separated, early ascending, then reflexed. **FL**: calyx tube 6–8.5 mm; petals cream-colored, banner 15.3–19 mm, recurved ± 50°, keel 13.1–14.7 mm. **FR** spreading or pendent; body 25–40 mm, 8– 14 mm wide, inflated but compressed side-to-side, ± half-elliptic in side view, thinly papery, translucent, sparsely strigose; stalk-like base 3–11 mm, minutely hairy, jointed at top; chamber 1; immature seeds 8–18. 2*n*=22. Arid grassland or shrubland; 100–1200 m. s SNF, SnJV, SCoRI. Like *A. asymmetricus, A. curtipes, A. trichopodus*, but note calyx length; stalk length, hairs, position of joint; seed number.

A. pachypus E. Greene Per, robust, rigid, bushy, sparsely leafy; hairs < 0.3 mm, appressed, ± scale-like, ± gray. **ST** ± erect, 2–8 dm, wiry. **LF** 2.5–16.5 cm; lflets 11–27, ± well separated, 3–34 mm, narrow. **INFL**: fls 4–28, well separated, ascending. **FL**: petals white or cream-colored, banner 15–22 mm, recurved ± 45°, keel 10.7–15.3 mm. **FR** ascending or spreading; body 12–28 mm, 4–8 mm wide, straight or ± curved, compressed side-to-side and stiffly leathery when mature, gen glabrous; stalk-like base 4–8 mm, stout; beak short, sharp, rigid, persistent; chambers 2. Open slopes in grassland or shrubland; 500–1900 m. Teh, s SnJV, SCoRI, WTR, PR, w edge D.

1. Petals yellow when fresh, or dry, banner 15–17 mm; calyx tube 3.7–4.3 mm . . . . var. ***jaegeri***
1′ Petals white when fresh, drying yellow, banner 15–22 mm; calyx tube 4–5.2 mm . . . . var. ***pachypus***

var. ***jaegeri*** Munz JAEGER'S MILKVETCH **LF**: lflets 15–25. RARE. Rocky or sandy areas; 500–750 m. n PR (incl SnJt), nw edge DSon.

var. ***pachypus*** (p. 603) **LF**: lflets 11–21. 2*n*=22. Open areas or in shrubland, often on gravelly clay, shale, or sandstone; 500–1900 m. Teh, s SnJV, SCoRI, WTR, w edge D. San Diego Co. pls have smaller fls but are otherwise indistinguishable.

A. palmeri A. Gray (p. 603) Ann or per, low, open and widely branched, sparsely to densely silvery-strigose. **ST** decumbent, 2–5 dm. **LF** 2–16 cm; lflets 9–21, 5–25 mm, widely ± elliptic. **INFL**: fls gen 20–40, early dense, then ± well separated, spreading or widely ascending. **FL**: petals gen pink-purple, sometimes cream-colored with purple veins and petal tips, banner 7–10.3 mm, recurved 90°, keel 6.2– 8.8 mm. **FR** 10–25 mm, 5–14 mm wide, moderately to very swollen (then ± bladdery), ± ovoid or half-ellipsoid, papery but not esp thin and translucent, thinly to densely strigose; beak ± erect, 1/5– 1/3 × body, triangular; chamber 1. Sandy or rocky places; 150– 1650 m. w edge DSon, adjacent foothills of SnBr, PR; Baja CA. [*A. vaseyi* S. Watson]

A. panamintensis E. Sheldon (p. 603) Per, resembling an unkempt nest due to wiry, incurved branches and persistent petioles, silvery-canescent. **ST** 1–15 cm, slender. **LF** 1.5–12 cm; lflets 5–11, well separated, 2–14 mm, linear-elliptic, late deciduous, tips acute. **INFL**: fls 1–4, ascending, well separated. **FL**: petals pink-purple, banner 8–14 mm, recurved ± 45°, keel 7–9 mm. **FR** spreading or ascending, 8–18 mm, 3–5 mm wide, ± oblong-elliptic in side view, bluntly triangular in X-section, densely strigose, becoming papery; chambers ± 2 in lower 1/2–2/3. In cracks, limestone ledges; 1200–2150 m. n DMtns (Inyo Co.).

A. pauperculus E. Greene (p. 603) DEPAUPERATE MILKVETCH Ann, small, delicate, finely strigose. **ST** gen incurved-ascending, < 1 dm. **LF** 1.5–5 cm; lflets 5–11, ± well separated, 2–8 mm, ± oblanceolate, tips notched or blunt. **INFL**: fls 2–7, well separated, gen reflexed in age. **FL**: petals purple, inner margins of wings pale or white, banner 5.4–10.5 mm, recurved ± 40°, keel 4.3–5.8 mm. **FR** ascending to reflexed, 12–20 mm, ± 3 mm wide, narrowly crescent-shaped, round at base, otherwise ± 3- or 4-sided, often ± purple-mottled, glabrous to sparsely strigose; beak short; chambers ± 2. UNCOMMON. Open, vernally-moist, volcanic clay; 150–600 m. CaRF, n ScV.

A. platytropis A. Gray (p. 603) BROAD-KEELED MILKVETCH Per, cespitose; hairs dense, silvery or gray. **ST** < 2 cm. **LF** 1–9 cm; lower stipules barely fused around st into sheath; lflets 5–15, 4–11 mm, elliptic to obovate, tips gen blunt. **INFL** head-like; fls 4–9. **FL**: petals whitish or pale purple, banner 7.2–9.5 mm, recurved 30–45°, keel 7.8–8.6 mm. **FR** ascending, 15–33 mm, 10–18 mm wide, bladdery, ± strongly compressed top-to-bottom, strigose; chambers 2 by inflexion of lower suture, which meets seed-bearing flange in middle of fr, seeds therefore along center of partition. RARE in CA. Rocky areas above timberline; 2800–3550 m. SNE; to ID, MT, NV.

A. pomonensis M.E. Jones (p. 603) Per, clumped, leafy. **ST** ± ascending, 2.5–8 dm, stout, hollow; hairs 0. **LF** 5–20 cm; lflets 25–41, 6–30 mm, ± elliptic, thinly strigose on lower surface, sometimes only on midrib. **INFL**: fls 10–45, spreading or reflexed. **FL**: petals cream-colored, banner 11.1–15.3 mm, recurved ± 40°, keel 9.4–13.2 mm, keel and wing claws 5.1–6.9 mm, keel blade 4.6–6.9 mm, = or < keel claw. **FR** spreading or ascending, deciduous, 18–45 mm, 10– 20 cm wide, bladdery, ± ovoid to half-ovoid, ± transparent, papery, sparsely strigose or ± glabrous; chamber 1; immature seeds 34–55. 2*n*=22. Shrubby, grassy, weedy areas; 50–700 m. s SCoRO, SCo, PR; nw Baja CA. Like *A. douglasii* (which has lflets fewer; keel blade > keel claw).

A. preussii A. Gray Per, robust, ill-scented, ± glabrous. **ST** ± erect, 1–3.5 dm. **LF** 3.5–18 cm; lflets 7–25, ± well separated, 2–27 mm, linear to ± round, hairs 0 or only on margins, midribs. **INFL**: fls 4–22, ascending. **FL**: petals pink-purple or ± pale, banner 14–24 mm, recurved ± 40°, keel 11–19 mm. **FR** erect or ascending; body 12–40 mm, 7–13 mm wide, inflated, oblong-ellipsoid, ± round in X-section, stiffly papery, glabrous or minutely hairy; stalk-like base 0 or 2–7 mm; chamber 1. Alkaline clay flats, gravelly washes; 700–750 m. DMoj; to UT, AZ.

1. Fr base not stalk-like; infl open, axis in fr 4–23 cm; banner ± 14 mm . . . . var. ***laxiflorus***
1′ Stalk-like fr base 2–7 mm; infl ± dense, axis in fr 1–9 cm; banner 17–24 mm . . . . var. ***preussii***

var. ***laxiflorus*** A. Gray (p. 603) LANCASTER MILKVETCH UNCOMMON (possibly extinct in CA). Alkaline flats; 700 m. sw DMoj (Antelope Valley); s NV, sw UT, w AZ.

var. ***preussii*** DESERT MILKVETCH UNCOMMON. Clay flats; ± 750 m. e DMoj (se Inyo, ne San Bernardino cos.); to UT, AZ.

A. pseudiodanthus Barneby (p. 603) TONOPAH MILKVETCH Per from underground crown; hairs soft, ± spreading, curved or curly. **STS** ± prostrate, loosely matted, 2–3 dm. **LF** 2.5–5 cm; lflets 7–19, ± crowded, 3–10 mm, ± obovate. **INFL**: fls 7–25, early crowded, spreading, then ± well separated, reflexed. **FL**: petals reddish lilac, banner 9–10 mm, recurved ± 40°, keel 8.5–9.3 mm. **FR** reflexed, 12–24 mm, 5–8 mm wide, curved > 1/2 circle, wider than deep, early fleshy, then leathery; hairs spreading; chambers 1 or ± 2. RARE. Sand; 2050 m. SNE (e Mono Co.); w NV. [*A. iodanthus* S. Watson var. *p.* (Barneby) Isely] Possibly a sand ecotype of *A. iodanthus*.

A. pulsiferae A. Gray Per, delicate; hairs fine, spreading or ± subappressed, grayish. **ST** ± prostrate, open and widely branched, 1–3 cm, fan-like, slender. **LF** 1–5.5 cm; lower stipules often fused around st into sheath; lflets 3–13, crowded, 2–12 mm, ± obovate, tips gen notched. **INFL**: fls 3–13, reflexed in age. **FL**: corolla whitish, lavender-veined, keel tipped with lilac, banner 5.2–8.5 mm, recurved 90–100°, keel 3.4–5.3 mm. **FR** 10–20 mm, 6–11 mm wide, ± spheric or half-ovoid, bladdery, thinly papery, translucent; hairs ± sparse, long, wavy; chamber 1. Dry, open, rocky or sandy flats, dunes; 1300–1650 m. s CaRH, n SNH, MP; NV.

1. Hairs of peduncle, upper st ± spreading, 1–1.6 mm; hairs of fr 0.7–1.7 mm .... var. ***pulsiferae***
1′ Hairs of peduncle, upper st ± subappressed, 0.5–0.75 mm; hairs of fr ± 0.5 mm .... var. ***suksdorfii***

var. ***pulsiferae*** (p. 603) AMES MILKVETCH RARE. $2n$=24. Sandy or rocky soil, often with pines or sagebrush; 1300–1650 m. n SNH, e MP; NV.

var. ***suksdorfii*** (Howell) Barneby SUKSDORF'S MILKVETCH RARE. Loose, often rocky soil, often with pines or sagebrush; 1300 m. s CaRH.

A. purshii Hook. Per, sparsely to densely cespitose; herbage hairs gen 1–2.3 mm, extremely fine, cottony, entangled, silvery or gray. **ST** 0–14 cm. **LF** 1–15 cm; lflets 3–17, 2–20 mm, narrowly elliptic to ± round, tips blunt to notched. **INFL** ± among lvs; fls 1–11, ascending. **FL**: petals white, cream-colored, pink-purple, or purple, banner 9–26 mm, recurved ± 40°, keel 8–21.2 mm. **FR** ascending, 7– 27 mm, 4–13 mm wide, ovoid or widely lanceolate in side view; hairs gen very dense, gen 1.5–5 mm, white (fr resembling a cotton boll), all ± wavy or all straight; chambers 1 or 2. Dry flats, slopes; 450–3350 m. NW, CaR, SNH, Teh, SCoRI, SnBr, GB, DMoj; to Can, ND, Colorado. Locally and regionally variable. Like *A. newberryi* var. *n.* (which has longer, ± straight, partly spreading hairs).

1. Calyx < 10 mm; fr 7–17 mm
 2. Fr incurved at least 1/2 circle; ne MP ...... var. ***lagopinus***
 2′ Fr incurved only near beak; SNH, SnBr, perhaps w edge SNE .... var. ***lectulus***
1′ Calyx > 10 mm; fr 13–27 mm
 3. Petals white or cream-colored, sometimes tipped pale lilac; infl 1–6-fld .... var. ***purshii***
 3′ Petals pink-purple or purple; infl 3–11-fld .... var. ***tinctus***

var. ***lagopinus*** (Rydb.) Barneby (p. 603) **ST** 0–8 cm. **LF** 1–7 cm; lflets 3–11, 5–15 mm. **INFL**: fls 2–7. **FL**: calyx 5.5–9 mm; petals pink-purple, sometimes pale, purple-tipped, banner 9–13.2 mm, keel 8–11.3 mm. **FR** 7–17 mm, 3.8–7 mm wide, incurved at least 1/2 circle; chamber 1; immature seeds 14–20. Dry plains, slopes, often in pumice, often with sagebrush; 900–1300 m. ne MP; OR.

var. ***lectulus*** (S. Watson) M.E. Jones **ST** 0–10 cm. **LF** 1–5 cm; lflets 3–11, 2–10 mm. **INFL**: fls 1–5. **FL**: calyx 5.6–8.8 mm; petals pink or pale purple, banner 10.3–15 mm, keel 9.4–11.7 mm. **FR** 7.5–15 mm, 4–8 mm wide, incurved only near beak; chamber 1; immature seeds 24–32. Dry, open flats, slopes, often with juniper, pines, to rocky slopes above timberline; 1800–3350 m. SNH, SnBr, perhaps w edge SNE.

var. ***purshii*** **ST** 0–1 cm. **LF** 1.5–15 cm; lflets 5–17, 2–10 mm. **INFL**: fls 1–6. **FL**: calyx 12–19 mm; petals white or cream-colored, sometimes pale lilac-tipped, banner 19–26 mm, keel 15–21.2 mm. **FR** 13–27 mm, 5–13 mm wide, straight or curved near tip; chamber 1; immature seeds 20–34. $2n$=22. Dry flats, often with sagebrush; 1050–2100 m. MP; to sw Can, ND, Colorado. ❀DRN:DRY:1, 16;DFCLT.

var. ***tinctus*** M.E. Jones (p. 603) **ST** 0–10 cm. **LF** 2–11 cm; lflets 3–17, 2–14 mm. **INFL**: fls 3–11. **FL**: calyx 12–19 mm; petals pink-purple or purple, banner 14.6–25 mm, keel 11.5–20.8 mm. **FR** 13–27 mm, 5–13 mm wide; chambers often ± 2 (esp in DMoj); immature seeds 18–46. Gravelly, sandy flats, slopes, often with pines or sagebrush; 450–2900 m. KR, NCoRH, CaRH, n&s SNH, Teh, SCoRI, GB, w DMoj; OR, NV. [var. *longilobus* M.E. Jones] ❀DRN,DRY:1,2,16,18;DFCLT.

A. pycnostachyus A. Gray Per, stout, clumped, leafy, canescent or white-woolly. **ST** ± erect, 4–9 dm, hollow. **LF** 3–15 cm; lower stipules fused around st into sheath; lflets 23–41, crowded, 5–30 mm, narrow. **INFL**: fls many, overlapping, reflexed. **FL**: petals greenish white or cream-colored, banner 7–10 mm, recurved ± 35°, keel 7.1–9.1 mm. **FR** 6–11 mm, 3.5–6 mm wide, ovoid, ± inflated, papery; hairs 0 or sparse; beak 5–8 mm, stiff, slender, hooked, persistent; chamber 1. Coastal salt marshes, coastal seeps; < 30 m. NCo, n CCo, c SCo.

1. Peduncle 2–4 cm; calyx tube 3–4 mm; c SCo .... var. ***lanosissimus***
1′ Peduncle 4–10 cm; calyx tube 3.5–5 mm; NCo, n CCo .... var. ***pycnostachyus***

var. ***lanosissimus*** (Rydb.) Munz (p. 607) VENTURA MARSH MILKVETCH Pl white-woolly. **FR** gen thinly strigose; immature seeds 8–12. **PRESUMED EXTINCT** (last seen 1967). Coastal marshes or seeps; < 30 m. c SCo. Habitat lost to urbanization.

var. ***pycnostachyus*** Pl canescent. **FR** glabrous; immature seeds 2–5. Coastal marshes or seeps; < 30 m. NCo, n CCo.

A. rattanii A. Gray Ann, minutely, thinly strigose; hairs often blackish above. **ST** decumbent to erect, 4–30 cm, gen slender. **LF** 1.5–5 cm; lflets 5–13, 2–12 mm, narrowly to widely ± obovate, tips gen blunt or notched. **INFL** head-like; fls 2–10, spreading. **FL**: petals pink-purple, wing tips paler, banner 7.2–12 mm, recurved ± 40°, keel 3.8–8.1 mm. **FR** ascending, 15–57 mm, 2–3 mm wide, narrowly linear, ± round in X-section, tapered to a narrow, sharp beak, papery, finely strigose; chambers ± 2; immature seeds 8–20. **SEEDS** pitted. Open grassy or gravelly areas; 50–750 m. c&s NCoR.

1. Calyx 2.5–3.4 mm, tube 2–2.5 mm; fr 1.5–3 cm .... var. ***jepsonianus***
1′ Calyx 3.7–5 mm, tube 2.6–3.5 mm; fr 2.1–5.7 cm .... var. ***rattanii***

var. ***jepsonianus*** Barneby (p. 607) JEPSON'S MILKVETCH **LF**: lflets gen 7–9. **INFL**: fls 4–9. **FL**: petals gen 2-colored (white exc purplish banner tip and keel). UNCOMMON. Often serpentine; 350–600 m. s NCoRI.

var. ***rattanii*** (p. 607) RATTAN'S MILKVETCH **LF**: lflets 5–13. **INFL**: fls 2–10. **FL**: petals pink-purple, claws, wing tips paler, rarely white. UNCOMMON. Often river banks or sand bars; 50–750 m. n&c NCoR.

A. ravenii Barneby (p. 607) RAVEN'S MILKVETCH Per, delicate, open and widely branched, minutely strigose. **ST** prostrate, 1–10 cm, weak, slender; buried parts 1–6 cm. **LF** 0.5–3 cm; lower stipules fused around st into sheath; lflets 7–13, 1–4 mm, ovate to ± round, tips notched. **INFL**: fls 2–8, spreading. **FL**: petals whitish, veins faint lilac, keel tip lilac, banner 5.5–8.4 mm, recurved ± 85°, keel 4.5–5.5 mm. **FR** ascending, 8–17 mm, 5–8 mm wide, ± incurved, swollen, papery, sparsely strigose; chamber 1. $2n$=22. RARE. Gravel; 3400–3450 m. s SNH (Sawmill Pass, Fresno and Inyo cos.); reported from SNE (Mono Craters, Mono Co.). [*A. monoensis* var. *r.* (Barneby) Isely] Closely related to *A. monoensis*.

A. sabulonum A. Gray (p. 607) Ann, coarse, leafy; hairs ± dense, ascending or spreading, ± wavy. **ST** erect or decumbent, 2–26 cm. **LF** 1.5–6.5 cm; lflets 5–15, 2–13 mm, oblanceolate, tips blunt, ± notched. **INFL**: fls 2–7, well separated, spreading or ascending. **FL**: petals dingy cream-colored, lilac-tinged, banner 5.2–7.2 mm, recurved 50–70°, keel 5–6.5 mm. **FR** 9–20 mm, 5–11 mm wide, ± ovoid, incurved (often abruptly so near tip), leathery; hairs ± dense, stiff, spreading, wavy; chamber 1. $2n$=24. Sandy or gravelly areas; -30–200 m. DSon; to UT, NM. As high as 2000 m outside CA.

A. sepultipes (Barneby) Barneby (p. 607) Per, open and widely branched, loosely clumped; hairs ± dense, ± appressed, silvery- or grayish silky. **ST** ascending, 1.5–3.5 dm; buried parts 1–5 cm, ± glabrous. **LF** 2–8 cm; lower stipules fused around st into sheath; lflets 7–17, 3–14 mm, obovate, tips notched or ± blunt. **INFL**: fls 10–30, spreading. **FL**: petals pale pink- or lilac-white, wing tips whitish, banner 12.7–17.5 mm, recurved ± 40°, keel 10–12.2 mm. **FR** spreading to reflexed, 15–20 mm, 3–6 mm wide, ± incurved, ± 3-sided, stiffly papery, hairs fine, shaggy; chambers 2; immature seeds 14–20. Dry granitic sand among sagebrush; 1800–1950 m. SNE, perhaps e edge c&s SNH.

A. serenoi (Kuntze) E. Sheldon var. ***shockleyi*** (M.E. Jones) Barneby (p. 607) NAKED MILKVETCH Per, bushy-clumped, minutely, thinly strigose, often grayish. **ST** 1.5–4.5 dm. **LF** 5–15 cm; lflets 5–11, well separated, 5–30 mm, very narrow, upper surface more densely strigose, paler than lower, terminal lflet not or only obscurely jointed to midrib. **INFL**: fls 3–25, well separated, ascending. **FL**: petals ± purple with pale wing tips, strongly graduated, banner 17–26 mm, recurved 35–40°, keel 12.4–18.5 mm. **FR** erect, 17–31 mm, 7–12 mm wide, plumply oblong, leathery, glabrous;

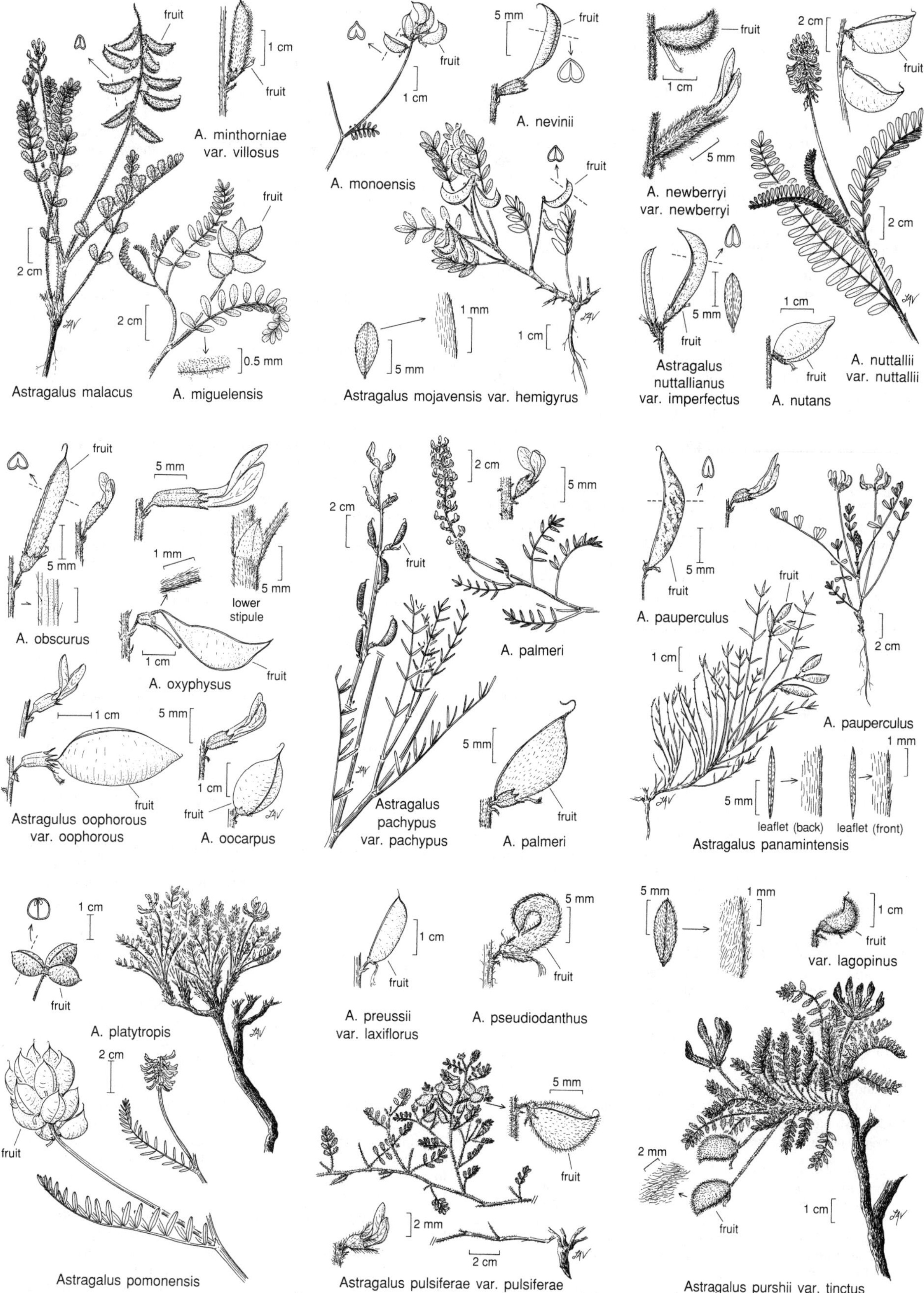
fruit
1 cm
fruit
2 cm
A. minthorniae
var. villosus
fruit
2 cm
0.5 mm
Astragalus malacus
A. miguelensis
fruit
1 cm
A. monoensis
5 mm
fruit
A. nevinii
fruit
1 mm
5 mm
1 cm
Astragalus mojavensis var. hemigyrus
fruit
1 cm
5 mm
A. newberryi
var. newberryi
5 mm
fruit
Astragalus
nuttallianus
var. imperfectus
2 cm
fruit
2 cm
1 cm
fruit
A. nutans
A. nuttallii
var. nuttallii
fruit
5 mm
A. obscurus
5 mm
1 mm
5 mm
lower
stipule
1 cm
fruit
A. oxyphysus
1 cm
5 mm
1 cm
fruit
fruit
Astragulus oophorous
var. oophorous
A. oocarpus
2 cm
fruit
Astragalus
pachypus
var. pachypus
2 cm
5 mm
A. palmeri
5 mm
fruit
A. palmeri
5 mm
fruit
A. pauperculus
fruit
1 cm
2 cm
A. pauperculus
1 mm
5 mm
leaflet (back)
leaflet (front)
Astragalus panamintensis
1 cm
fruit
A. platytropis
1 cm
fruit
A. preussii
var. laxiflorus
5 mm
fruit
A. pseudiodanthus
5 mm
1 mm
1 cm
fruit
var. lagopinus
2 cm
fruit
Astragalus pomonensis
5 mm
fruit
2 mm
2 cm
Astragalus pulsiferae var. pulsiferae
2 mm
fruit
1 cm
Astragalus purshii var. tinctus

beak stout, sharp; chambers ± 2. $2n=22(24?)$. UNCOMMON. Open, dry, alkaline gravelly-clay soil, gen with sagebrush or pinyon; 1500–2250 m. W&I, n DMtns (Inyo Co.); NV.

A. shevockii Barneby (p. 607) SHEVOCK'S MILKVETCH Per, wiry, open; hairs sparse. **ST** decumbent to ascending, 1–3.5 dm, slender; hairs at base spreading, those above appressed. **LF** 2.5–6.5 cm; lowest stipules fused around st into sheath; lflets 9–17, well separated, 2–10 mm, ± elliptic. **INFL**: fls 2–13, well separated, ascending. **FL**: petals ± cream-colored, banner 9–9.8 mm, recurved ± 75°, keel 9–9.8 mm. **FR** ascending, 13–31 mm, 3–4 mm wide, tapered at both ends, incurved, ± 3-sided near base, papery, hairy; upper suture a low keel, lower suture a keel near tip, depressed in a wide groove in lower half; chamber 1. RARE. Granitic sand in Jeffrey-pine forest; 1900 m. s SNH (se Tulare Co.).

A. subvestitus (Jepson) Barneby (p. 607) KERN COUNTY MILKVETCH Per, cespitose or matted, leafy; herbage hairs gen 0.7–1.2 mm, dense, woolly, ± curly, grayish. **ST** prostrate, 1–8 cm. **LF** 1.5–6.5 cm; lflets 7–13, 2–9 mm, ± elliptic or obovate. **INFL** among lvs, short; fls 3–8. **FL**: petals whitish, with some lilac, keel pink-tipped, banner 11–13.3 mm, recurved ± 40°, keel 9.5–10.7 mm. **FR** ascending, 8–15 mm, 4–7 mm wide, ± ovoid, compressed from top-to-bottom in lower half, stiffly papery; hairs dense, gen < 1 mm, curly; beak erect or incurved, triangular, compressed side-to-side; chamber 1; immature seeds 11–14. UNCOMMON. Gravelly, sandy areas, with sagebrush; 2450–2750 m. s SNH (Tulare, Kern cos.).

A. tener A. Gray Ann, delicate, ± sparsely strigose to ± glabrous. **ST** erect or ascending, 2–30 cm. **LF** 2–9 cm; lflets 7–17, ± well separated, 3–16 mm, lanceolate to obovate, glabrous on upper surface, tip notched or pointed. **INFL** dense; fls 2–12, spreading. **FL**: petals pink-purple, banner 5.2–11.8 mm, recurved 35–40°, keel 3.4–6.4 mm. **FR** reflexed, 6–50 mm, 1.7–3.5 mm wide, ± narrowly lanceolate, straight or curved, openly grooved on lower side, stiffly papery, glabrous, base ± stalk-like or not; chambers ± 2. **SEEDS** smooth. Open, moist areas, coastal bluffs; < 60 m. GV, CCo, SnFrB, SCo.

1. Banner 5.2–6 mm; immature seeds 5–11; mostly coastal .... var. ***titi***
1' Banner 7.8–11.8 mm; immature seeds 8–16; mostly inland
  2. Fr 2.7–5 cm, strongly incurved, base ± stalk-like, 3–5 mm .... var. ***ferrisiae***
  2' Fr 1–2.5 cm, ± incurved, base round, not stalk-like .... var. ***tener***

var. ***ferrisiae*** A. Liston FERRIS' MILKVETCH **ST** 6–26 cm. **LF** 2–6 cm; lflets 7–15. **INFL**: fls 3–12. **FL**: banner 7.8–9.6 mm; keel 4.2–5.1 mm. **FR** 2.7– 5 cm, strongly incurved; base ± stalk-like, 3–5 mm; immature seeds 10–16. **PRESUMED EXTINCT.** Alkaline flats, vernally-moist meadows; < 60 m. n ScV. Possibly extinct.

var. ***tener*** (p. 607) ALKALI MILKVETCH **ST** 4–30 cm. **LF** 2–9 cm; lflets 7–17. **INFL**: fls 3–12. **FL**: banner 8.2–11.8 mm; keel 4.7–6.4 mm. **FR** 10–25 mm, ± incurved; base round, not stalk-like; immature seeds 8–14. RARE. Alkaline flats, vernally-moist meadows; < 60 m. s ScV, n SnJV, e SnFrB (where mostly extirpated).

var. ***titi*** (Eastw.) Barneby COASTAL DUNES MILKVETCH **ST** 2–12 cm. **LF** 2–7 cm; lflets 7–13. **INFL**: fls 2–7. **FL**: banner 5.2–6 mm; keel 3.4–3.9 mm. **FR** 6–14 mm, straight or outcurved; base round, not stalk-like; immature seeds 5–11. **ENDANGERED** CA. Moist sandy depressions near coast, coastal bluffs, dunes; < 20 m. c CCo (Monterey Co.), SCo (where possibly extirpated). Threatened by urbanization, recreation, alien pls.

A. tephrodes A. Gray var. ***brachylobus*** (Gray) Barneby (p. 607) Per, tufted, coarse, greenish gray to silvery-silky; hairs appressed or ± ascending. **ST** prostrate, 0–8 cm. **LF** 4–16 cm; lflets 11–31, 4–17 mm, widely ± obovate. **INFL**: fls 10–35, ascending. **FL**: petals ± pink-purple, banner 14–24 mm, recurved ± 45°, keel 14.7–21 mm. **FR** ascending, 17–30 mm, 6–10 mm wide, plumply lanceolate, stiffly leathery; hairs appressed or ± spreading, rarely 0; chamber 1. Open, dry ground; 150 m. se DMoj (near Needles, San Bernardino Co.); to sw UT, AZ. Possibly a waif in CA.

A. tidestromii (Rydb.) Clokey (p. 607) Per (or fl 1st year and appearing ann), tufted; hairs dense, ± stiff, often curly, entangled, shaggy. **ST** 0 or ± prostrate, < 7 cm. **LF** 3–15 cm; lflets 7–19, 4–17 mm, widely obovate. **INFL**: fls 5–16, well separated, spreading-ascending. **FL**: petals whitish, tinged with dull purple, wings, keel tipped dark purple, banner 12–17.7 mm, recurved ± 40°, keel 10–12 mm. **FR** ascending, 15–55 mm, 6–16 mm wide, ± lanceolate, curved 1/4–5/4 circle, ± 4-sided, compressed side-to-side at both ends, from top-to-bottom in middle, stiffly leathery, minutely strigose; beak long, narrowly triangular; chamber 1. Open, calcareous gravel; 600–1500 m. DMoj (extreme e-c DMtns and mouth of Cushenbury Canyon near n edge SnBr); s NV.

A. traskiae Eastw. (p. 607) TRASK'S MILKVETCH Per, open and widely branched, leafy; hairs dense, grayish. **ST** ± ascending, 1–4 dm. **LF** 4–10 cm; lflets 21–29, 5–15 mm, ± obovate or ovate. **INFL**: fls 12–30, well separated, ascending or spreading. **FL**: petals cream-colored, banner 14.2–17.5 mm, recurved 30–40°, keel 10.2–12.9 mm. **FR** pendent or ascending; body 8–16 mm, 3–5.5 mm wide, ± half-elliptic in side view, ± 3-sided (lower grooved), thinly fleshy, then leathery; hairs minute, wavy; stalk-like base 4–9 mm; chambers ± 2. **RARE** CA. Sandy coastal bluffs, dunes; < 60 m. s ChI (San Nicolas, Santa Barbara islands). Threatened by military activity on San Nicolas Island.

A. tricarinatus A. Gray (p. 607) TRIPLE-RIBBED MILKVETCH Per, loosely tufted, finely strigose. **ST** ± erect, 5–25 cm, stiff. **LF** 7–20 cm; lflets 17–27, well separated, 3–12 mm, narrowly ± obovate, silvery-strigose on upper surface, greenish on lower. **INFL**: fls 5–15, well separated, spreading-ascending. **FL**: calyx 6.1–7.6 mm, tube 4.1–5 mm; petals cream-colored, banner 12.6–15.7 mm, recurved ± 45°, keel 9.7–11 mm. **FR** ascending; body 24–42 mm, 3.5–5.5 mm wide, ± linear, 3-sided, thinly papery, glabrous; stalk-like base 1– 2.5 mm, stout, jointed at top; upper suture a narrow ridge; chambers 2. RARE. Exposed rocky slopes, canyon walls; 450–550 (1250) m. e SnBr (Whitewater, Morongo Valley), adjacent edges D.

A. trichopodus (Nutt.) A. Gray Per, robust, bushy-branched, ± minutely strigose, gen also spreading-hairy. **ST** ± erect, 2–10 dm. **LF** 2.5–20 cm; lower stipules rarely fused around st into sheath; lflets 15–39, 2–25 mm, ± lanceolate. **INFL**: fls 10–50, spreading or reflexed. **FL**: calyx 5–8.7 mm, tube 3.6–5.4 mm; petals ± cream-colored, sometimes faintly lilac-blushed or -lined, banner 11.3–19 mm, recurved 40–45°, keel 8.6–13.7 mm. **FR** pendent; body 13–45 mm, 4.8–21 mm wide, linear-elliptic and compressed side-to-side to ovate and bladdery, persistent, thinly papery; hairs minute, sparse; dehiscence from tip, along top suture; stalk-like base 5–17 mm, slender, strigose or ± glabrous, joint between body base and stalk 0; chamber 1; immature seeds 10–30. Open, grassy areas, bluffs, rocky sites; 0–1220 m. SCoRO, s SCoRI, SCo, ChI, WTR, w PR. Like *A. asymmetricus, A. curtipes, A. oxyphysus*, but note calyx length; stalk length, hairs, position of joint; ovule number.

1. Fr compressed side-to-side, sides ± flat or low-convex .... var. ***phoxus***
1' Fr bladdery or ± compressed side-to-side but sides very convex
  2. Fr 8–21 mm wide, hairy n of San Diego, sometimes glabrous s, upper suture much less convex than lower .... var. ***lonchus***
  2' Fr 5–13 mm wide, glabrous, upper and lower sutures ± equally convex .... var. ***trichopodus***

var. ***lonchus*** (M.E. Jones) Barneby (p. 607) **INFL**: fls 12–36. **FL**: banner 11.3–19 mm; keel 8.6–13 mm. **FR**: body 17–45 mm, 8–21 mm wide, bladdery, with very convex sides, hairy n of San Diego, sometimes glabrous s; upper suture much less convex than lower. $2n=22$. Coastal bluffs, fields; < 300 m. SCo, n ChI; Baja CA. [*A. leucopsis* (Torrey) Torrey & A. Gray]

var. ***phoxus*** (M.E. Jones) Barneby **INFL**: fls 10–50. **FL**: banner 11.4–16.7 mm; keel 9.3–12.7 mm. **FR**: body 15–36 mm, 5–9 mm wide, compressed side-to-side, with ± flat or low-convex sides, glabrous or rarely minutely strigose. Gen inland, grassy or shrubby hillsides; 50–1220 m. s SCoR, n SCo (Santa Barbara, Ventura cos.), WTR, w edge DMoj. [*A. antiselli* A. Gray] var. ***trichopodus*** **INFL**: fls 10–50. **FL**: banner 11.5–15.4 mm; keel 9.5–11.5 mm. **FR**: body 13–35 mm, 5–13 mm wide, bladdery, with very convex sides, gen glabrous; upper and lower sutures ± equally convex. Coastal bluffs, hills; < 100 m. n half SCo.

A. umbraticus E. Sheldon (p. 607) BALD MOUNTAIN MILKVETCH Per, open and widely branched, ± glabrous. **ST** ascending, 2–5 dm. **LF** 4–12 cm; lflets 11–23, 4–20 mm, widely oblong to ± round, tips notched. **INFL**: fls 10–25, ± spreading. **FL**: petals greenish white, banner 10–14 mm, recurved 40–60°, keel 7.7–10 mm. **FR** spreading or reflexed; body 14–24 mm, 2.5–3.5 mm wide, linear-lanceolate, incurved 1/4–1/2 circle, firmly papery, glabrous, turning black; stalk-like base 0.8–1.9 mm; chambers 2. UNCOMMON. Dry woodlands; 200–1250 m. w KR, n NCoRO (Humboldt Co.); w OR.

A. webberi Brewer & S. Watson (p. 607) WEBBER'S MILKVETCH Per, open and widely branched, leafy, densely, minutely strigose (giving dry lflets a satiny sheen). **ST** ± decumbent, 1.5–5 dm; buried parts 0–6 cm. **LF** 2.5–15 cm; lower stipules fused around st into sheath; lflets 9–25, 5–35 mm, narrowly to widely obovate. **INFL**: fls 6–14, ± well separated, ascending to ± spreading. **FL**: petals cream-colored, banner 15.4–18.8 mm, recurved 50–75°, keel 11.4– 13.7 mm. **FR** ascending to ± spreading, 20–35 mm, 7–12 mm wide, ± round in X-section, slightly compressed top-to-bottom, stiffly leathery, glabrous; chamber 1. RARE. Open, shrubby slopes, dry woodland; 800–1050 (1500 ?) m. n SNH (Plumas, Sierra cos.).

A. whitneyi A. Gray Per, gen open and widely branched, silvery-hairy or not. **ST** ± ascending or erect, 4–40 cm. **LF** 1.5–10 cm; lower stipules fused around st into sheath; lflets 5–21, 2–21 mm, oblong to obovate. **INFL**: fls 3–16, ± well separated, spreading or reflexed. **FL**: petals cream-colored to pink-purple, banner 8.3–17.2 mm, recurved 50–80°, keel 7.3–13.8 mm. **FR** pendent; body 15–60 mm, 10–25 mm wide, bladdery, papery, glabrous to minutely strigose; stalk-like base 2–9 mm, slender; chamber 1. Open, sandy, gravelly, or rocky places; 800–3660 m. KR, NCoRH, SNH, w WTR, MP, W&I; to WA, ID, NV. Intergrading complex.

1. Ovary, young fr hairs minute, curved .... var. ***confusus***
1′ Ovary, young fr glabrous
 2. Herbage hairs spreading or spreading-ascending; lf 2–4 cm, with ± 0 spaces between lflets .... var. ***lenophyllus***
 2′ Herbage hairs ± appressed or ± ascending; lf often > 4 cm, with spaces between lflets
  3. Fr body 1.5–6 cm; KR, NCoRH ...... var. ***siskiyouensis***
  3′ Fr body 1.5–3 cm; s SNF, SNH, n WTR, W&I .... var. ***whitneyi***

var. ***confusus*** Barneby Pl gray or silvery; hairs short-curly, long-straight. **LF** 2–10 cm, with spaces between lflets. **INFL**: fls 5–16. **FL**: petals whitish, tinged with pink or lilac, banner 12.8–17.2 mm. **FR**: body 1.7–6 cm, hairs minute, curved; stalk-like base 4–9 mm. Open areas, often with sagebrush; 1350–2400 m. n SNH, MP; to ID, NV.

var. ***lenophyllus*** (Rydb.) Barneby Pl gray-green; hairs ± stiff, spreading. **LF** 2–4 cm, with ± 0 spaces between lflets. **INFL**: fls 5–9. **FL**: petals cream-colored, banner 8.3–16.5 mm. **FR**: body 1.5–4.2 cm, glabrous; stalk-like base 5–9 mm. Open, rocky places; 2650–3050 m. n SNH.

var. ***siskiyouensis*** (Rydb.) Barneby (p. 607) Pl greenish or grayish; hairs ± strigose. **LF** 3–9 cm, with spaces between lflets. **INFL**: fls 4–16. **FL**: petals cream-colored, banner 9.5–13.5 mm. **FR**: body 1.5–6 cm, glabrous; stalk-like base 4–16 mm. Open, gravelly or rocky areas, often on serpentine; 750–2650 m. KR, n NCoRH; s OR.

var. ***whitneyi*** (p. 607) Pl greenish to grayish; hairs strigose. **LF** 3–11 cm, with spaces between lflets. **INFL**: fls 3–15. **FL**: petals pink or lilac, wing tips pale, banner 8.3–16.5 mm. **FR**: body 1.5–3 cm, glabrous; stalk-like base 3–15 mm. $2n$=22. Open, rocky areas; 2050–3660 m. s SNF, SNH, n WTR, W&I; w NV.

CAESALPINIA

Elizabeth McClintock

Per, shrub, tree, armed or not, gland-dotted or not. **LVS** odd- or even-2-compound, alternate. **INFL**: gen raceme, axillary or terminal. **FL** ± bilateral; sepals ± free, overlapped above; stamens 10, ± exserted, free. **FR** dehiscent or not, inflated or flat. ± 200 spp.: trop, warm temp; some cult. (A. Caesalpini, Italy, 1519–1603) [Isely 1975 Mem New York Bot Garden 25(2):33–51]

1. St ± lfless; lf odd-2-ternate-pinnate, 1° lflets 3, 1 terminal, 1 pair lateral; raceme few-fld; fr sickle-shaped, 1.5–2.5 cm .... ***C. virgata***
1′ St leafy; lf gen even-2-pinnate, 1° lflets 3–12 pairs; racemes many-fld; fr ± straight, 6–12 cm
 2. Petals > 2 cm; stamens red, strongly exserted; sepals all alike .... ***C. gilliesii***
 2′ Petals < 1 cm; stamens yellow, barely exserted; sepals not all alike (lowest 1 boat-shaped, with many long, marginal teeth) .... ***C. spinosa***

C. gilliesii (Hook.) D. Dietr. BIRD-OF-PARADISE Shrub, small tree < 4 m. **ST** ± glandular. **LVS** evergreen; stipules small, persistent; 1° lflets 8–12 pairs, not always opposite, 1.5–3 cm; 2° lflets 7– 11 pairs, < 8 mm. **INFL** < 10 cm, wider below, many-fld, glandular. **FL**: sepals all alike, 1.5–2 cm, oblong-elliptic; petals 2–3.5 cm, yellow with orange marks; stamens 8–9 cm, strongly exserted, red. **FR** dehiscent, 6–12 cm, oblong, flat, ± curved to straight, gland-dotted. Uncommon. Disturbed areas, urban and rural areas; < 300 m. SnJV, expected elsewhere; native to Argentina. [*Poinciana g.* Hook.] Fr and seeds TOXIC.

C. spinosa (Molina) Kuntze TARA Shrub, small tree < 2 m; prickles scattered. **ST**: twigs brown-hairy. **LVS** evergreen; stipules 0; 1° lflets 3–10 pairs, < 8 cm; 2° lflets 5–7 pairs, 1.5–4 cm, elliptic. **INFL** 15–20 cm, not wider below, many-fld, puberulent. **FL**: sepals 5–7 mm, not all alike (lowest boat-shaped, with many, long, marginal teeth); petals 6–7 mm, yellow to reddish; stamens unequal, irregular in length, barely exserted, yellow. **FR** indehiscent, 6–10 cm, oblong, becoming more withered, dry, spongy to leathery. Uncommon. Dry, disturbed, often urban areas; < 50 m. SCo (Ventura Co.), expected elsewhere; native to S.Am.

C. virgata E.M. Fisher (p. 607) Shrub 0.5–2 m, hairy; branches slender, gen lfless, rush-like, green. **LVS** deciduous; 1° lflets 3, ternately arranged, lateral pair 0.5–1 cm, with 3–6 pairs of 2° lflets, terminal 1.5–4 cm, with 8–10 pairs of 2° lflets. **INFL** 5–15 cm, few-fld, puberulent. **FL**: sepals 5–6 mm; petals 6–8 mm, banner yellow with reddish marks, later entirely reddish; stamens ± 8 mm, slightly exserted, yellow. **FR** dehiscent, 1.5–2.5 cm, sickle-shaped, with sessile or short-stalked glands esp when young, margins ciliate. Uncommon. Gravelly or sandy desert gullies, washes, or canyon slopes; 100–500 m. DSon; to TX, Baja CA. [*Hoffmannseggia microphylla* Torrey] ❀DRN,DRY,SUN:8–14,19–21;DFCLT.

CALLIANDRA FAIRYDUSTER, MOCK MESQUITE

Elizabeth McClintock

Per, shrub, tree, unarmed. **LVS** even-2-pinnate, alternate. **INFL**: gen head, axillary, few-fld. **FL** radial, purplish red or white; sepals, petals inconspicuous; stamens many, strongly exserted, filaments fused below, free above. **FR** dehiscent; 2 valves recurving; margins thickened. ± 200 spp.: trop Am, Madagascar, India. (Greek: beautiful stamens)

C. eriophylla Benth. (p. 607) FAIRYDUSTER Shrub < 30 cm; branches many, spreading from base. **LVS** deciduous; 1° lflets gen 2–4 pairs; 2° lflets gen 7–9 pairs, ± 2–3.5 mm. **FL** reddish purple; stamens 18–22 mm. **FR** 5 cm, flat. RARE in CA. Sandy washes, slopes, mesas; ± 1500 m. DSon; AZ, n Mex. In cult.

CARAGANA

Duane Isely

Shrub, tree, spiny or not. **LVS** clustered on spurs (alternate on new growth), even-1-pinnate; stipules membranous, bristle- or spine-like. **INFL**: clusters of 1–4 fls on spurs. **FL**: calyx lobes << tube; corolla yellow; 9 filaments fused, 1 free. **FR** dehiscent, oblong, flat. **SEEDS** few. 50+ spp.: mostly c Asia. (Mongolian name)

C. arborescens Lam. SIBERIAN PEA TREE Pl glabrous or puberulent. **LF**: lflets 8–12, 1–2.5 cm, elliptic to oblong. **FL**: corolla 1.6–2.2 cm. **FR** 4–5.5 cm, oblong, glabrous. $2n$=16. PR (reportedly naturalized on Mount Palomar, 1200–1500 m); native to temp Asia. Cult in Eur, e&c US.

CERCIDIUM PALO VERDE

Elizabeth McClintock

Tree, shrub; branches with pointed tips; thorns in lf axils (see lf scars). **LVS** even-2-pinnate, alternate, falling early; 1° lflets gen 1 pair. **INFL**: raceme, axillary, < 7-fld. **FL** slightly bilateral; sepals ± free, all alike, reflexed; petals ± equal, clawed, yellow or cream-white; stamens 10, exserted, free. **FR** dehiscent or not, flat, narrowed between seeds or not. 4 spp.: deserts, se CA; AZ, nw Mex. (Greek: weaver's shuttle, from fr) [Carter 1974 Proc Calif Acad Sci 40(2):17–57]

1. Lf petioled; 2° lflets 1–3 pairs, 4–8 mm; fr not or very slightly narrowed between seeds ***C. floridum*** ssp. ***floridum***
1′ Lf sessile; 2° lflets 4–8 pairs, 1–5 mm; fr narrowed between seeds .... ***C. microphyllum***

C. floridum A. Gray ssp. ***floridum*** (p. 611) BLUE PALO VERDE Tree gen < 8 m; branches spreading, ± zig-zagged, ± glabrous. **LF** blue-green; 1° lflets 1 pair, < 1 cm. **FL**: banner 9–15 mm, widely ovate, orange-dotted or not. **FR** 3–11 cm; tip beak-like. Uncommon. Washes, flood plains; ± 1100 m. DSon; to AZ, nw Mex. Fls gen 2 weeks before *C. microphyllum*. ❀SUN,DRN:7,**8**,**9**,10,**14**,**19–23**,24&IRR:11,**12**,**13**; also STBL.

C. microphyllum (Torrey) Rose & I.M. Johnston (p. 611) Shrub, small tree 3–4(9) m; branches gen ascending or spreading, broom-like, hairy. **LF** yellow-green; 1° lflets 1 pair, 3–6.5 cm. **FL**: banner < 10 mm, widely ovate, gen cream-white. **FR** < 11 cm; tip beak-like, gen ending in a spine. Uncommon. Rock slopes; ± 600 m. DMoj; to AZ, nw Mex. Branches used as livestock feed; seeds edible. Hybrids with *C. floridum* reported. ❀SUN,DRN:**8**,**9**,10,**14**,15,16,**19–21**,22,23&IRR:11,**12**,**13**; also STBL.

CERCIS REDBUD

Elizabeth McClintock

Shrub, tree, unarmed. **LVS** simple, alternate, cordate to reniform, ± leathery, glabrous. **INFL** umbel-like, axillary on short spur or ± sessile on woody branches. **FLS** bilateral, appearing before lvs; sepals fused at base; petals pink, upper one inside lateral ones in bud, keel petals free; stamens 10, gen incl, free. **FR** dehiscent, oblong, flat. 5–7 spp.: n hemisphere. (Greek: applied perhaps to a poplar, but also to *C. siliquastrum*, Judas tree) [Isley 1975 Mem New York Bot Gard 25(2):134–150]

C. occidentalis Torrey (p. 611) WESTERN REDBUD Tree < 7 m, glabrous. **LF** deciduous, < 10 cm; petiole 15–20 mm. **INFL** 2–5-fld. **FL**: keel petals 12–13 mm, > wings and banner. **FR** 5–8 cm. Dry, shrubby slopes, canyons, ravines, stream banks, chaparral, foothill woodlands to yellow-pine forest; 100–1500 m. NW, CaR, SN, SnJV, PR. ❀DRN,SUN:4–6,**14–16**,17,**18**,**22–24**;&IRR or part SHD:1–3,**7–9**,12,**19–21**;also STBL;CVS.

CHAMAECYTISUS

Elizabeth McClintock

Shrub, small tree, unarmed. **ST**: twig round in X-section. **LVS** ternately 1-compound, alternate, petioled; stipules free or 0. **INFL**: cluster, peduncled, axillary; fls 1–4(7). **FL** bilateral; calyx cylindric, upper of 2 lips gen widely 2-lobed, gen ± 1/2 tube length, lower lip shallowly 3-lobed; petals 5, white or gen yellow, banner gen obovate, reflexed or not, keel oblong-sickle-shaped, curved on lower surface, claw gen < 1/2 keel length; stamens 10, filaments fused; style curved upward ± at middle. **FR** dehiscent, gen leathery, black or brown; pedicel short. **SEEDS** few–many, arilled. 30 spp.: Eur, Canary Islands. (Greek: dwarf *Cytisus*)

C. proliferus (L.) Link Shrub < 5 m. **ST**: twigs hairy. **LF**: petiole gen 5–12 mm, hairy; lflets 10–30 mm, lanceolate to ovate, upper surface gen glabrous or sparsely hairy, lower surface silky-hairy. **INFL**: pedicel hairy. **FL**: calyx 8–9 mm, silky-hairy, tube 5–6 mm; corolla white, banner ± 15 mm, ± reflexed, ± hairy outside, purple-lined or spotted inside. **FR** 3.5–5 cm, silky-hairy, black. Uncommon. Disturbed places; < 100 m. CCo, SnFrB; native to Canary Islands. [*Cytisus p.* L.]

COLUTEA BLADDER SENNA

Duane Isely & Elizabeth McClintock

Shrub, unarmed. **LVS** odd-1-pinnate, alternate. **INFL**: raceme, axillary, few-fld. **FL** bilateral; sepals 5; petals 5, gen yellow; 9 filaments fused, 1 free. **FR** indehiscent, inflated, bladder-like; walls papery, making a popping sound when squeezed. **SEEDS** many. ± 28 spp.: Medit to w China, w Afr. (Greek: ancient name) Lvs with cathartic properties.

C. arborescens L. (p. 611) Pl gen < 3 m, rounded; branches many. **LF** deciduous; lflets 9–13, 1.5–3 cm, elliptic, sparsely hairy. **FL** ± 3 cm. **FR** 2–3 cm, glabrous. Uncommon. Disturbed areas; esp < 500 m. n SNF, SnFrB; native to Eur. Cult as orn.

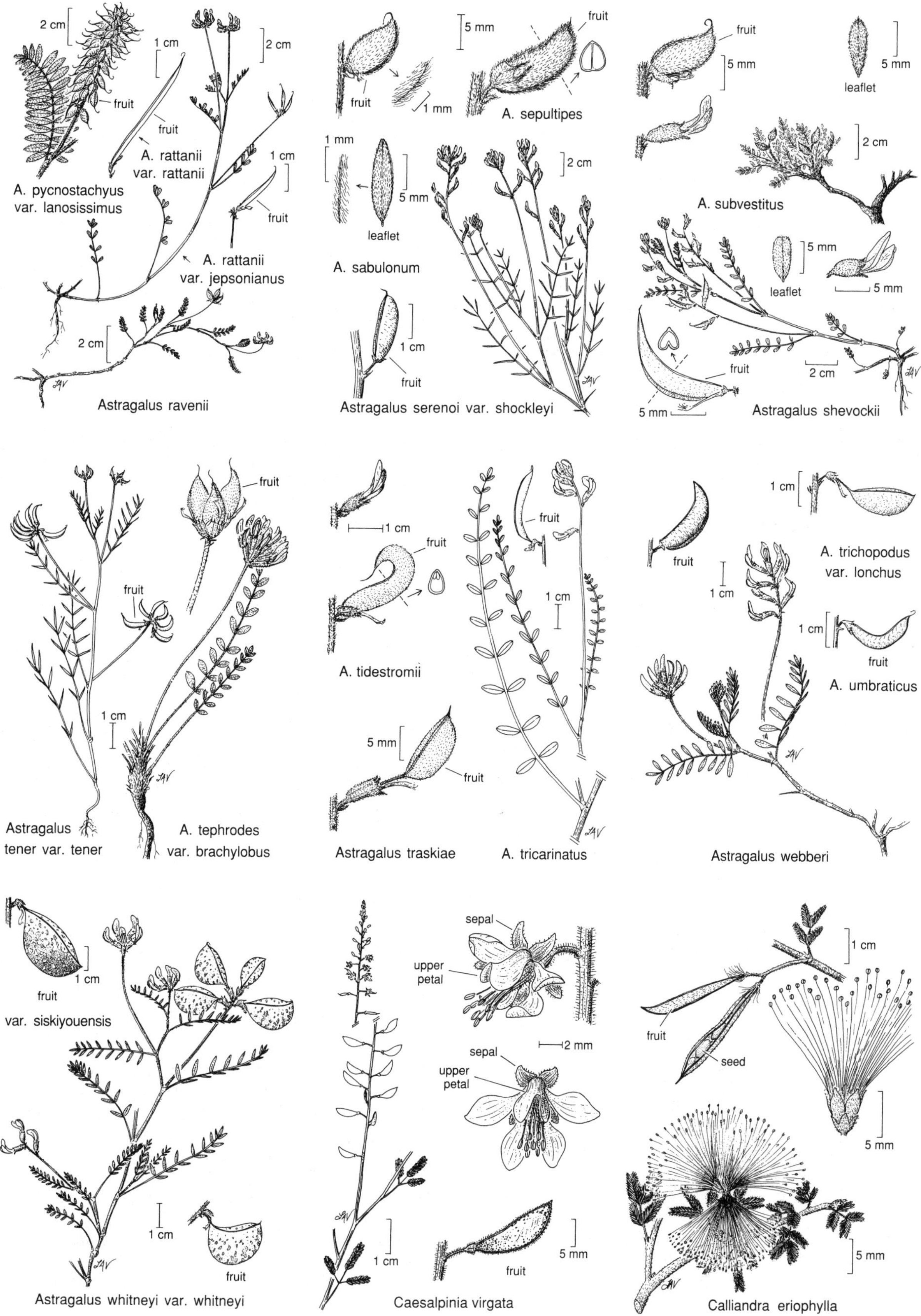
2 cm
fruit
1 cm
fruit
2 cm
A. rattanii
var. rattanii
A. pycnostachyus
var. lanosissimus
1 cm
fruit
A. rattanii
var. jepsonianus
2 cm
Astragalus ravenii
5 mm
fruit
fruit
1 mm
A. sepultipes
1 mm
5 mm
leaflet
A. sabulonum
2 cm
1 cm
fruit
Astragalus serenoi var. shockleyi
fruit
5 mm
5 mm
leaflet
2 cm
A. subvestitus
5 mm
leaflet
5 mm
fruit
2 cm
5 mm
Astragalus shevockii
fruit
fruit
1 cm
Astragalus
tener var. tener
A. tephrodes
var. brachylobus
1 cm
fruit
A. tidestromii
fruit
1 cm
5 mm
fruit
Astragalus traskiae
A. tricarinatus
1 cm
fruit
1 cm
A. trichopodus
var. lonchus
1 cm
fruit
A. umbraticus
Astragalus webberi
1 cm
fruit
var. siskiyouensis
1 cm
fruit
Astragalus whitneyi var. whitneyi
sepal
upper
petal
2 mm
sepal
upper
petal
1 cm
fruit
5 mm
Caesalpinia virgata
1 cm
fruit
seed
5 mm
5 mm
Calliandra eriophylla

CYTISUS

Elizabeth McClintock

Shrub, small tree, unarmed. **ST** often ribbed, persistently green. **LVS** 1-compound (gen ternate), gen alternate, petioled; stipules free or 0. **INFL**: gen racemes, terminal or fls 1–3 in axillary, peduncled clusters. **FL**: calyx bell-shaped, 2-lipped, upper lip barely 2-lobed, lower lip minutely 3-lobed; petals white or gen yellow, banner yellow, gen not hairy, keel oblong-sickle-shaped to ± 1/2 circular, curved on lower surface, claw ± 1/4 keel; stamens 10, filaments fused; style gen abruptly curved at ± middle or gently curved entire length. **FR** dehiscent, gen oblong, papery to ± leathery; pedicel short. **SEEDS** few–many, gen arilled. 33 spp.: Eur, w Asia, n Afr, Canary Islands; some cult. (Greek: name for several woody Fabaceae) [Viciosa 1955 Inst For de Investi y Exper no. 72]

1. Corolla white, banner < 10 mm . . . . *C. multiflorus*
1′ Corolla pale or golden yellow, banner > 10 mm
 2. Calyx glabrous; fr flat, glabrous exc margin; branches gen 5-angled . . . . *C. scoparius*
 2′ Calyx appressed-hairy; fr ± inflated, densely white-hairy; branches gen 8–10-angled . . . . *C. striatus*

C. multiflorus (L'Heritier) Sweet SPANISH or PORTUGUESE BROOM Shrub 3–4+ m; branches many, 5-angled, flexible, broom-like, gen lfless in fl, silvery-silky-hairy when young, then ± glabrous. **LVS**: on lower branches petioled, lflets 3; on upper branches 0 or sessile, lflets 1; lflets < 10 mm, linear-lanceolate or oblong, silvery-silky-hairy. **INFL**: cluster, axillary; fls 1–3; pedicels < 10 mm. **FL**: calyx ± 5 mm, silky-hairy; corolla white, banner < 10 mm, dark-striate in lower center, gen ± reflexed. **FR** 25–30 mm, linear-oblong, appressed-hairy. **SEEDS** gen 4–6. Uncommon. Disturbed roadsides; ± 100 m. SCoRO (Monterey Co.); native to Spain, Portugal.

C. scoparius (L.) Link (p. 611) SCOTCH BROOM Shrub 2–2.5 m. **ST**: branches gen 5-angled, green, hairy when young, then gen glabrous. **LVS**: lflets 5–20 mm, obovate to oblong, hairs appressed or 0; lflets 1, sessile on young branches, 3 on older. **INFL**: cluster, axillary; fls 1–2 per cluster; pedicels < 10 mm, glabrous. **FL**: calyx glabrous, < 5 mm; corolla golden yellow, banner gen 15–18 mm, reflexed or not. **FR** 25–40 mm, flat, brown or black, glabrous exc margins. **SEEDS** 5–12. Common. Disturbed places; < 1000 m. NW, CaRF, n&c SNF, GV, SnFrB, SCo; native to s Eur, n Afr. NOXIOUS WEED.

C. striatus (Hill) Rothm. Shrub 2–3 m. **ST**: branches many, slender, gen 8–10-angled, silky-hairy when young, then ± glabrous. **LVS**: lflets 5–15 mm, obovate, upper surface glabrous, glaucous, lower surface silky-hairy; on upper branches sessile, lflets 1 or 3, on lower branches petioled, lflets 3. **INFL**: cluster, axillary; fls gen 1–2 per cluster; pedicels 5–10 mm. **FL**: calyx ± 5 mm, hairs appressed; corolla pale yellow, banner 10–25 mm, not reflexed. **FR** 15–40 mm, ± inflated, densly white-hairy. **SEEDS** several. Uncommon. Disturbed places; < 300 m. SnFrB, SCo, PR, expected elsewhere; native to Spain, Portugal. May be confused with *C. scoparius*. Locally abundant.

DALEA

Duane Isely

Ann, per, unarmed, gland-dotted. **LVS** gen odd-1-pinnate; stipules inconspicuous, thread-like or glandular. **INFL**: spike in CA; bracts gen ± conspicuous. **FL**: calyx tube 10-ribbed; banner arising from receptacle, other petals from side or top of filament column; stamens 9–10 or 5, filaments fused; ovules 2. **FR** indehiscent, incl in or slightly exserted from calyx. **SEED** 1. ± 165 spp.: w US, Mex, s S.Am. (T. Dale, English botanist, 18th century) [Barneby 1977 Mem New York Bot Gard 27:135–582, 650–877] Incl spp. sometimes placed in *Petalostemon*; exc others found here in *Marina*, *Psorothamnus*.

1. Pl hairy, gen prostrate, sometimes tiny; stamens 9–10
 2. Calyx 3–7 mm, lobes ± = tube; infl 8–15 mm wide; corolla > or ± = calyx; lflet margin gen entire, flat . . . . *D. mollis*
 2′ Calyx 7–8 mm, lobes gen > tube; infl gen 14–16 mm wide; corolla < calyx; lflet margin gen shallowly lobed or wavy . . . . *D. mollissima*
1′ Pl (exc infl) glabrous, ascending; stamens 5
 3. Infl (minus corollas) ± 12–16 mm wide, axis not visible between frs; calyx tube thinly papery, with long, silky, appressed hairs . . . . *D. ornata*
 3′ Infl (minus corollas) 8–12 mm wide, axis visible between frs; calyx tube leathery, puberulent . . . . *D. searlsiae*

D. mollis Benth. (p. 611) Ann, diminutive or mat-forming, hairy. **LF**: lflets 8–12, 3–7 mm, ± round to obovate-oblong, flat or folded, entire, flat. **INFL** 8–15 mm wide, ovoid or short-cylindric; bracts < ± 1 mm wide. **FL**: calyx 3–7 mm, lobes needle-like, ± = tube, shaggy-hairy; petals often > sepals, whitish or lavender; stamens 9–10. $2n$=16. Common. Creosote bush flats, washes, roadsides; < 800 m. D; to AZ, Mex. ❀DRN,SUN:**10–13**.

D. mollissima (Rydb.) Munz (p. 611) Ann or per, diminutive or mat-forming, hairy. **LF**: lflets 8–12, 3–10 mm, ± round to obovate-oblong, often folded, shallowly lobed, wavy. **INFL** gen 14–16 mm wide, ovoid or short-cylindric; bracts gen 1–1.5 mm wide. **FL**: calyx ± 7–8 mm, lobes needle-like, gen > tube, shaggy-hairy; petals < sepals, whitish or lavender; stamens 9–10. $2n$=16. Common. Desert flats, washes; < 900 m. D; s NV, w AZ. ❀DRN,SUN:**10–13**.

D. ornata (Hook.) Eaton & J. Wright ORNATE DALEA Per, glabrous. **STS** clustered, ascending, 3–5 dm. **LF**: lflets 5–7, 10–20 mm, widely ovate to elliptic. **INFL** (minus corolla) ± 12–16 mm wide, ovoid to cylindric, compact, not exposing axis in fr. **FL**: calyx 4–6 mm, tube not slit, thinly papery, with long, ± appressed, silky hairs; petals 7–9 mm, lavender to purple; stamens 5. RARE in CA. Open rocky hillsides; 1700 m. MP (Shaffer Mtn, Lassen Co.); to WA, ID, NV. [*Petalostemon o.* Hook.]

D. searlsiae (A. Gray) Barneby (p. 611) Per, glabrous. **STS** clustered, ascending, 3–5 dm. **LF**: lflets 5–7, 10–16 mm, obovate to oblong. **INFL** (minus corollas) ± 8–12 mm wide, oblong or narrowly oblong, initially compact, exposing axis in fr. **FL**: calyx 3.5– 4.5 mm, tube recessed or slit on upper side, becoming leathery, puberulent; petals 5–7 cm, purple; stamens 5. Juniper/sagebrush, slopes, bluffs; 1200–2000 m. W&I (Inyo Mtns), DMtns; to UT, AZ. [*Petalostemon s.* A. Gray] ❀DRN,DRY:**1–3,10,11**.

GENISTA BROOM

Elizabeth McClintock

Shrub, small tree, unarmed. **STS** gen ribbed or angled, persistently green. **LVS** ternately 1-compound, alternate, petioled; stipules fused to lf bases, rarely 0. **INFL**: racemes or heads, terminal or axillary, or clusters on axillary short-shoots. **FL** bilateral; calyx gen < corolla, 2-lipped, upper lip strongly 2-lobed, lobes ± 1/3 tube, lower lip 3-lobed, lobes < upper lobes; petals 5, yellow, banner gen ovate or rounded, outside gen glabrous or variously hairy, keel narrow-oblong, ± straight on lower surface, often hairy; stamens 10, filaments fused; style ± abruptly bent at tip. **FR** dehiscent, narrow-oblong, slightly inflated or not; pedicel < 7 mm. **SEEDS** several, gen arilled. 87 spp.: Eur, w Asia, n Afr, Canary Islands. (Latin: from *planta genista*, from which English Plantagenet monarchs took their name) [Gibbs & Dingwall 1971 Bol Soc Brot 45:269–316] Most naturalized CA pls are hybrids involving *G. canariensis, G. monspessulana, G. stenopetala.*

1. Lflet narrowly oblanceolate to narrowly elliptic, length gen 5 × width, margins rolled under; banner silky-hairy . . . . ***G. linifolia***

1′ Lflet oblanceolate or obovate to elliptic or ± round, length gen < 5 × width, margins flat; banner gen glabrous or nearly so

 2. Lflet < 10 mm, length 1–2 × width; banner glabrous exc gen for ± V-shaped hairy area from base to tip . . . . ***G. canariensis***

 2′ Lflet gen > 10 mm, length 2–5 × width; banner glabrous or hairy only on midrib

  3. Stipules persistent, giving older twigs scaly appearance; twigs gen dull-yellow-brown-hairy . . . . . . ***G. maderensis***

  3′ Stipules deciduous; twigs silvery-hairy, at least when young

   4′ Lflet gen < 15 mm, length ± 2 × width; petiole < 5 mm; terminal or central fl of cluster gen opening last; banner glabrous . . . . ***G. monspessulana***

   4. Lflet gen > 15 mm, length ± 3 × width; petiole > 5 mm; terminal fl of raceme gen opening first; banner glabrous or ± hairy only on midrib . . . . ***G. stenopetala***

G. canariensis L. Shrub < 3 m. **ST**: twigs silky-hairy. **LF**: stipules < 2 mm; petiole < 6 mm; lflets 5–10 mm, obovate or ± round, length 1–2 × width, upper surface sparsely hairy, lower surface densely so. **INFL**: racemes 10–60 mm, terminal and axillary; fls 5–20; pedicels very short. **FL**: calyx 4–6 mm, densely silky-hairy; banner 10–12 mm, ovate, glabrous exc gen for ± V-shaped hairy area from base to tip. **FR** 15–25 mm, short-hairy. Uncommon. Disturbed places; < 300 m. WTR; native to Canary Islands. [*Cytisus c.* (L.) Kuntze]

G. linifolia L. Shrub < 2.5 m. **ST**: twigs densely silky-hairy. **LF**: stipules 1–6 mm; petiole < 5 mm or 0; lflets 10–60 mm, narrowly oblanceolate to narrowly elliptic, length gen 5 × width, upper surface sparsely white-hairy, lower surface densely so, margins rolled under. **INFL**: racemes < 30 mm, terminal, dense; fls 5–20; pedicels 2–7 mm. **FL**: calyx 5–15 mm, densely silky-hairy; banner 10–15 mm, silky-hairy. **FR** 15–35 mm, hairy. Uncommon. Disturbed places; < 200 m. SCoRO (Santa Barbara Co.), s ChI (Catalina Island); native to Canary Islands, w Medit, n Afr. [*Cytisus l.* (L.) Lam.]

G. maderensis (Webb & Berth.) Lowe Shrub < 2 m. **ST**: twigs gen dull-yellow-brown-hairy. **LF**: stipules 3–6 mm, persistent, giving older twigs a scaly appearance; petiole 5–10 mm; lflets 10–15 mm, obovate to oblanceolate, length ± 2 × width, upper surface sparsely hairy, lower surface densely so. **INFL**: racemes, 15–35 mm, axillary; fls 5–15; pedicels 2–4 mm. **FL**: calyx 5–7 mm, densely silky-hairy; banner 10–15 mm, widely ovate, glabrous. **FR** 20–35 mm, densely silky-hairy. Uncommon. Disturbed areas; < 300 m. Reported but unconfirmed in SnFrB (Santa Clara Co.), SCoRO (Monterey Co.); native to Madeira. [*Cytisus m.* (Webb & Berth.) Masferer]

G. monspessulana (L.) L. Johnson (p. 611) Shrub < 3 m. **ST**: twigs silvery-silky-hairy. **LF**: stipules < 2 mm, deciduous; petiole < 5 mm; lflets gen 10–15 mm, oblanceolate to obovate, length ± 2 × width, upper surface gen glabrous, lower surface appressed- or spreading-hairy. **INFL**: clusters on axillary short-shoots, 15–60 mm, dense; fls 4–10, terminal or central gen opening last; pedicels 1–3 mm. **FL**: calyx 5–7 mm, silky-hairy; banner 10–15 mm, ovate, glabrous. **FR** 15–25 mm, densely silky-hairy. Common. Disturbed places, < 500 m. NCoRO, NCoRI, SnFrB, SCoRO, s ChI, WTR, PR (San Diego Co.); native to Medit, the Azores, occurrence in Canary Islands questionable. [*Cytisus m.* L.] Most pls reported as this sp may be hybrids. Fls (perhaps all parts) TOXIC. NOXIOUS WEED.

G. stenopetala Webb & Berth. Shrub < 6 m. **ST**: twigs silvery-hairy when young, becoming glabrous. **LF**: stipules 1–3 mm, deciduous; petiole 5–20 mm; lflets gen 10–40 mm, oblanceolate to narrow-elliptic, length gen 3 × width, both surfaces ± glabrous or lower sparsely silky-hairy. **INFL**: racemes 5–10 cm, terminal or axillary; fls 10–30, terminal gen opening first; pedicels 2–4 mm. **FL**: calyx 4–7 mm, silky-hairy; banner 10–15 mm, ovate, ± notched at tip, glabrous or ± hairy only on midrib. **FR** 20–35 mm, silky-hairy. Uncommon. Disturbed places; < 200 m. SCoRO; native to Canary Islands. [*Cytisus s.* (Webb & Berth.) Christ]

GLYCYRRHIZA LICORICE

Duane Isely

Per, unarmed or ± prickly on axes, fr, sometimes glandular-hairy. **LVS** odd-1-pinnate; stipules deciduous; lflets gland-dotted. **INFL**: raceme, spike-like, axillary. **FL**: calyx lobes unequal, < or ± = tube; corolla white-yellow to blue; 9 filaments fused, 0 or 1 free. **FR** indehiscent, ellipsoid, prickly or rarely glabrous. ± 20 spp.: esp Eurasia. (Greek: sweet root) Several spp. cult.

1. Fr glabrous or covered with straight bristles; lflet length ± 2 × width . . . . ***G. glabra***

1′ Fr covered with hooked prickles; lflet length ± 3 × width . . . . ***G. lepidota***

G. glabra L. LICORICE Pl ± glabrous or finely hairy, ± glandular or not. **LF**: lflets 9–13, widely ovate or elliptic. **INFL** open. **FL**: corolla 9–11 mm, bluish to purple. **FR** 12–25 mm. $2n=16$. Sporadic. Waste areas; esp < 500 m. CA-FP; occasional US; native to Eurasia. Licorice of commerce obtained from roots.

G. lepidota Pursh (p. 611) WILD LICORICE Pl glabrous to very glandular-hairy. **LF**: lflets 9–19, lanceolate to narrowly ovate. **INFL** dense. **FL**: corolla 9–14 mm, yellowish or green-white. **FR** 12–15 mm. $2n=16$. Moist, gen open, disturbed sites, such as creek banks, roadsides; < 2400 m. CA; to c US, Can, Mex. ❀DRN:1–5, **6,7,14–24**&IRR:**8–10**,11,12;INF;DFCLT;STBL.

HALIMODENDRON

Duane Isely

1 sp. (Greek: salt tree)

H. halodendron (L.) Voss (p. 611) RUSSIAN SALT TREE Shrub, spiny. **ST** sprawling or erect, 1–3 m. **LVS** mostly clustered on spurs, even-1-pinnate; main axes and stipules of those subtending spurs becoming woody spines; lflets 4, obovate to oblanceolate, tips pointed. **INFL**: raceme, on spurs, 2–4-fld. **FL**: calyx lobes << tube; corolla 15–18 mm, red-purple to pink. **FR** slowly dehiscent, 1.5–2 cm, 0.8–1 cm wide, ellipsoid to obovoid, plump, woody, becoming black; stalk-like base 2–4 mm. **SEEDS** few. $2n$=16. Cult land; < 200 m. GV, c SCo (Los Angeles basin); to c US; native to sw Asia. Cult. NOXIOUS WEED. Known infestations have been eradicated.

HOFFMANNSEGGIA

Elizabeth McClintock

Subshrub from spreading roots, unarmed. **LVS** 2-pinnate, ± basal. **INFL**: raceme, scapose. **FL** slightly bilateral; sepals ± free, equal; petals ± equal, yellow to orange-red; stamens 10, exserted, free, filaments often glandular. **FR** tardily dehiscent. **SEEDS** several. 28 spp.: Am, S. Afr. (J. Centurius, Count of Hoffmannsegg, Germany, 1766–1849) [Isley 1975 Mem New York Bot Gar 25(2):162–168]

H. glauca (Ortega) Eifert (p.615) PIG-NUT, HOG POTATO Pl erect, < 30 cm; stalked glands throughout; roots deep, tubered. **ST**: branches from base, slender. **LF** 5–12 cm; 1°lflets 5–11, odd-pinnate, 5–20 mm; 2° lflets 10–20, even-pinnate, 4–6 mm. **INFL** 5–15 cm, often glandular. **FL**: petals spreading, ± 1 cm, orange-red. **FR** 1.5–4 cm, ± curved; glands scattered, short-stalked, deciduous or not. Uncommon. Dry, alkaline flats in deserts, disturbed areas; < 900 m. SnJV, SCoRO, SCo, WTR, D; to TX, Mex, S.Am. [*H. densiflora* A. Gray] Aggressive weed, spreading by edible tubers.

HOITA

James W. Grimes

Per, unarmed, gland-dotted; hairs or stalked glands, or both, at least above; caudex woody. **ST** prostrate (incl stolon) to erect; base green to gray-brown. **LVS** odd-1-pinnate, cauline; stipules free; lflets 3. **INFL**: raceme, axillary; bract 1, deciduous; fls 2–3 per node, pedicelled. **FL**: calyx lobes > tube, lowest ± keeled, > others, tube flaring only slightly in fr; corolla at least partly blue to purple; 9 filaments fused, 1 less so or free; ovary hairy, ovule 1, style tip curved or bent, stigma feathery. **FR** indehiscent, unevenly elliptic, brown to black, hairy; veins obvious. **SEED** ± reniform. 3 spp.: CA, Baja CA. (native American name) [Grimes 1990 Mem New York Bot Gard 61:1–114]

1. St prostrate (incl stolon) to decumbent, erect parts < 6.5 cm; lflet obovate to round .... ***H. orbicularis***
1' St erect, < 2 m; lflet lanceolate to round
  2. Fl 9–10 mm .... ***H. macrostachya***
  2' Fl 13–19 mm .... ***H. strobilina***

H. macrostachya (DC.) Rydb. **ST** erect, < 2 m, much-branched, gen striate; base hollow. **LF**: stipule 1.5–5 mm; petiole 7–62 mm; lflet 2–10 cm, lanceolate, both surfaces glandular, hairy or not. **INFL** < 30 cm; bract 3–11 mm. **FL** 9–10 mm; calyx 4.5–9 mm; banner 5.5–10 mm. **FR** 6–8 mm, brown; veins prominent. **SEED** 5–7 mm. Streamsides, marshes, spring-moist places; < 2500 m. CA-FP (exc GV); Baja CA. [*Psoralea m.* DC.] ❀WET:**6–9, 14,18–24**&SUN:5,**15–17**;STBL.

H. orbicularis (Lindley) Rydb. (p. 615) **ST** prostrate to decumbent, sometimes faintly striate; erect parts < 6.5 cm. **LF**: stipule 4.5–10 mm; petiole 12–75 mm; lflet 3–11 cm, obovate to round, both surfaces brown-glandular, hairy or not. **INFL** 6–35 cm; bract 7–27 mm. **FL** 12–23 mm; calyx 15–19 mm; banner 10–16 mm. **FR** 6–9 mm, brown; veins obvious. **SEED** 4–5 mm. Meadows, streamsides, moist hillsides; < 2250 m. CA-FP (exc GV). [*Psoralea o.* Lindley] ❀WET:**6–9,14,18–24**&SUN:5,**15–17**;GRNCVR.

H. strobilina (Hook. & Arn.) Rydb. **ST** erect, rarely to 1 m, smooth to striate. **LF**: stipule 7–16 mm, becoming reflexed; petiole 6–14 mm; lflet 4.5–8 cm, lanceolate to round. **INFL** 7–13 cm; bract 15–21 mm. **FL** 13–19 mm; calyx 13–17 mm; banner 12–13 mm. **FR** ± 10 mm, dark brown to black; veins obvious. **SEED** 6–7 mm. Uncommon. Chaparral, oak woodlands; < 600 m. SnFrB. [*Psoralea s.* Hook. & Arn.) ❀IRR,DRN:**14–16**&SUN:**17**;STBL.

LATHYRUS WILD PEA

Duane Isely

Ann or per, unarmed, glabrous or hairy, rarely glandular, gen rhizomed. **ST** sprawling, climbing, or erect; st angled, flanged, or winged. **LVS** even-1-pinnate; stipules persistent, upper lobe > lower; main axis ending as a tendril or short bristle; lflets 0–16, ± opposite or alternate, linear to widely ovate. **INFL**: raceme, gen axillary, 1–many-fld. **FL**: upper calyx lobes gen < and wider than lower; corolla 8–30 mm, pink-purple or pale, sometimes white or yellow; 9 filaments fused, 1 free; style flat, finely hairy on concave side. **FR** dehiscent, oblong, ± flat. ± 150 spp.: temp N.Am, Eurasia. (Ancient Greek name) [Broich 1987 Syst Bot 12:139–153] Some spp. variable, intergrading with others; some hybridization probable. Seeds of most alien spp. TOXIC to humans (esp young males) and livestock (esp horses).

1. Lflets 0; stipules lflet-like; corolla bright yellow .... ***L. aphaca***
1' Lflets 2–16; stipules gen not lflet-like; corolla purple to white (sometimes fading to dull yellow) or (in *L. sulphureus*) yellow to bronze- orange (drying tan or dark)

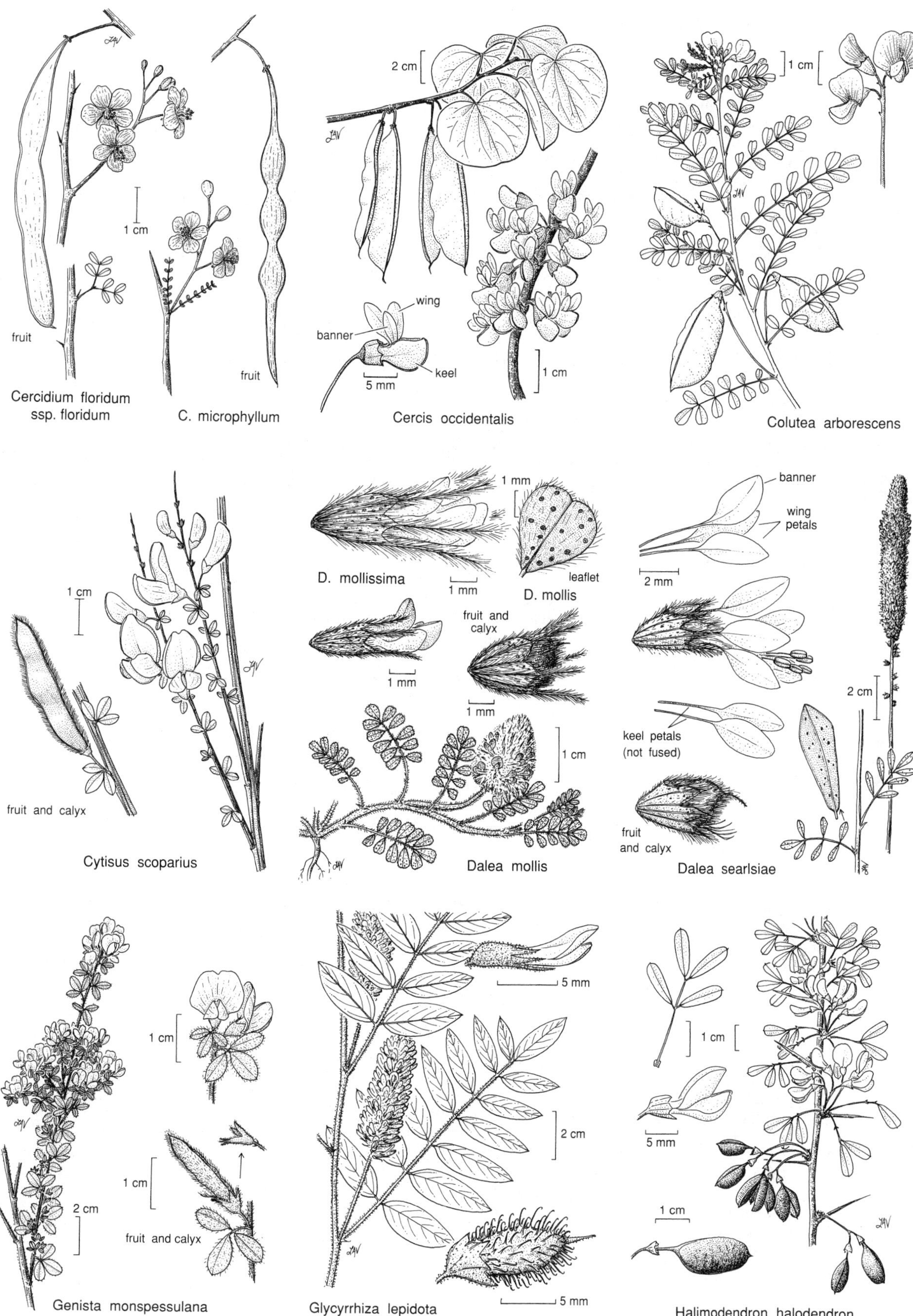
fruit
1 cm
fruit
2 cm
wing
banner
keel
5 mm
1 cm
1 cm
Cercidium floridum ssp. floridum
C. microphyllum
Cercis occidentalis
Colutea arborescens
1 mm
banner
wing petals
1 cm
D. mollissima
1 mm
leaflet
D. mollis
2 mm
fruit and calyx
1 mm
1 mm
2 cm
1 cm
keel petals (not fused)
fruit and calyx
fruit and calyx
Cytisus scoparius
Dalea mollis
Dalea searlsiae
5 mm
1 cm
1 cm
2 cm
5 mm
1 cm
1 cm
2 cm
fruit and calyx
5 mm
Genista monspessulana
Glycyrrhiza lepidota
Halimodendron halodendron

2. Lflets 2; aliens of disturbed habitats
3. Corolla 20–30 mm
4. Infl 4–15-fld; wings of st, petiole often > 2 mm wide . . . . *L. latifolius*
4′ Infl 1–4-fld; wings of st, petiole < 2 mm wide
5. Fr hairy; lflets ovate to elliptic, length 1.2–3 × width . . . . *L. odoratus*
5′ Fr glabrous; lflets ovate to lanceolate, length 1.2–6.5 × width . . . . *L. tingitanus*
3′ Corolla 8–15 mm
6. Fr hairy; infl 1–2-fld; calyx tube ± = lobes . . . . *L. hirsutus*
6′ Fr glabrous; infl 1-fld; calyx tube << lobes
7. Infl axis not extended beyond fl; calyx tube << lobes; corolla red-purple . . . . *L. cicera*
7′ Infl axis extended as a bristle beyond fl; calyx tube ± = or < lobes; corolla not red-purple
8. Peduncle 2–8 cm; corolla lavender or purple; fr veins indistinct, netted . . . . *L. angulatus*
8′ Peduncle 0.5–2 cm; corolla orange-red; fr veins distinct, longitudinal . . . . *L. sphaericus*
2′ Lflets 2–16 (> 2 at least on some lvs); natives of diverse natural habitats
9. Lflet, fr gland-dotted; lflets ± 14–16 — NCoR . . . . *L. glandulosus*
9′ Lflet, fr not gland-dotted; lflets often < 14
10. St winged, the wings ± 1 mm or more wide
11. Lflets gen 6, opposite; coastal marshes, NCo . . . . [2]*L. palustris*
11′ Lflets gen 8–16, irregularly arranged; not of immediate coast, exc in SnFrB
12. Corolla 10–15 mm, white (sometimes dark-striate), fading dull yellow; KR . . . . [2]*L. delnorticus*
12′ Corolla 15–20 mm, pink to pink-purple; widespread . . . . *L. jepsonii*
13. Pl gen puberulent; KR to SCoR, s SN, exc SnFrB . . . . var. *californicus*
13′ Pl glabrous; GV but gen coastal marshes in SnFrB . . . . var. *jepsonii*
10′ St angled or flanged, the angles or flanges < 1 mm wide
14. Tendrils not coiled or branched, often reduced to a short bristle
15. Stipule ± = or > lflet; gen dunes or beaches of immediate coast — NCo
16. Pl glabrous or puberulent; lflet 2.5–4.5 cm . . . . [2]*L. japonicus*
16′ Pl gray-hairy; lflet 1–2 cm . . . . *L. littoralis*
15′ Stipule < lflet; not of immediate coast
17. Infl 1–2-fld; lflet gen 0.5–2 cm; fr glabrous or puberulent
18. Lflets 10–15, ovate, 1–2 cm; fr puberulent — NCo, NCoR, SnFrB . . . . *L. torreyi*
18′ Lflets 4–8, gen oblong or narrowly lanceolate to linear; fr glabrous
19. Lflets 4, oblong-lanceolate, gen ± 0.5–1 cm; corolla greenish white, dark-striate; NCoRO (Humboldt Co.) . . . . *L. biflorus*
19′ Lflets of at least some lvs 4–8, narrowly lanceolate to linear, 1.5–8 cm; corolla pale lavender to purple; KR, SNH, MP . . . . [4]*L. lanszwertii* var. *aridus*
17′ Infl 2–16-fld; lflet 1–8 cm; fr glabrous
20. Sts erect, many, clustered; fls crowded; corolla white or pink-tinged, 17–23 mm; MP . . . . *L. rigidus*
20′ Sts erect or sprawling, solitary or a few clustered; fls crowded or not; corolla lavender to pink-purple or, if white, < 17 mm; n CA-FP
21. Lflets 10–16, subopposite to alternate; stipules often ± as wide as lflets; pl glabrous; KR, n&c NCoR . . . . [2]*L. polyphyllus*
21′ Lflets 4–10, irregularly arranged or opposite; stipules gen narrower than (to ± as wide as) lflets; pl often hairy; widespread
22. Corolla 13–25 mm; lflets gen widely ovate or elliptic, gen opposite . . . . . [2]*L. nevadensis* var. *nevadensis*
22′ Corolla 7–12 mm; lflets gen linear to lanceolate, irregularly arranged or opposite . . . . [4]*L. lanszwertii*
23. Corolla pale lavender to purple; KR, SNH, MP . . . . var. *aridus*
23′ Corolla white to cream-colored, often dark-striate; KR, NCoR . . . . var. *tracyi*
14′ Tendrils, at least on some lvs, coiled and gen branched
24. Lflets ± = stipules, 6–8; coastal beaches, dunes — NCo (Del Norte Co.) . . . . [2]*L. japonicus*
24′ Lflets gen > stipules, various in number, incl 6–8; inland (exc *L. palustris* of coastal marshes)
25. Corolla whitish yellow to bronze-orange (drying tan or dark), 11–14 mm; banner reflexed above middle; infl 10-15-fld, gen dense, 1-sided — fl appearing wide and blocky; NCoR, SNF . . . . *L. sulphureus*
25′ Corolla not whitish yellow to bronze- orange (sometimes fading tan-yellow), various in size, incl 11–14 mm; banner reflexed or bent from middle or below; infl 1–15-fld, gen ± open, rarely 1-sided
26. Corolla wine-red to crimson, 25–35 mm; banner reflexed 120–180° — PR (San Diego Co.) *L. splendens*
26′ Corolla white to purple, 8–25 mm; banner bent ± 90°
27. Lflets gen 6, opposite, glabrous or ± puberulent, and fls 3–6 per infl, 15–20 mm; coastal marshes, NCo . . . . [2]*L. palustris*
27′ Lflets, fls not entirely as above; immediate coast only in SnFrB
28. Stipules > 1 cm wide; lflets 10–16; KR, NCoR . . . . *L. polyphyllus*
28′ Stipules < 1 cm wide and lflets 12 or fewer (or not of KR, NCoR)
29. Infl gen 2–4-fld; DMoj . . . . *L. hitchcockianus*
29′ Infl sometimes 2–16-fld; not DMoj
30. Corolla white or cream-colored, fading yellow-tan — NCo, KR, NCoR
31. St flanged or narrowly winged; stipules widely triangular, 0.5–0.8 cm wide; KR . . . . [2]*L. delnorticus*
31′ St angled or ± flanged; stipules gen lanceolate, < 0.5 mm wide; NCo, KR, NCoR
32. Lflets gen 4–6, glabrous or puberulent; corolla 9–12 mm . . . . [4]*L. lanszwertii* var. *tracyi*

32′ Lflets gen 8–12, glabrous; corolla 14–16 mm .... [2]***L. vestitus*** var. ***ochropetalus***
30′ Corolla purple, lavender, or reddish
33. Stipules > 1 cm wide; lflets 10–16 — KR, NCoR .... [2]***L. polyphyllus***
33′ Stipules < 1 cm wide; lflets 4–12
34. Fls often crowded, gen 8–14, and corolla 14–20 mm; lflets 8–12, gen subopposite to alternate; w CA-FP .... [2]***L. vestitus***
35. Corolla dark purple to wine-red, 16–20 mm; banner gen reflexed > 90°; SCo, s ChI, PR .... var. ***alefeldii***
35′ Corolla pale lavender to purple, 14–18 mm; banner bent ± 90°; TR and n .... var. ***vestitus***
34′ Fls gen loosely spaced, either 2–10 and corollas 8–15 mm or 2–4 and corollas 13–25 mm; lflets 4–10, opposite or subopposite; esp KR, CaR, SN
36. Lflets gen opposite, gen widely ovate or elliptic; infl 2–4-fld; corolla 13–25 mm .... [2]***L. nevadensis*** var. ***nevadensis***
36′ Lflets gen subopposite or irregularly arranged, ovate to linear; infl 2–10-fld; corolla 8–15 mm
37. Stipules widely ovate-triangular, often toothed, conspicuous; calyx glabrous exc lobes often ciliate; corolla gen 10–14 mm, banner little upcurved above other petals .... ***L. brownii***
37′ Stipules lanceolate, not toothed, often inconspicuous; calyx gen hairy; corolla 10–16 mm, banner gen more reflexed than other petals .... [4]***L. lanszwertii*** var. ***lanszwertii***

L. angulatus L. Ann, glabrous. **ST** often flanged, scarcely winged. **LF**: stipules small, narrow; lflets 2, 2–6 cm, thread-like to narrowly lanceolate; tendril coiled. **INFL** 1-fld; peduncle 2–8 cm, axis extended beyond fl as a bristle. **FL**: calyx lobes subequal, ± = or > tube; corolla 9–12 mm, lavender or purple. **FR** glabrous, faintly net-veined. $2n$=14. Uncommon. Disturbed places; esp < 300 m. CA-FP; OR; native to Eur.

L. aphaca L. (p. 615) Ann, glabrous. **ST** angled or flanged, not winged. **LF**: stipules 1.5–5 cm, lflet-like; lflets 0; tendril coiled. **INFL** 1–2-fld. **FL**: calyx lobes > tube; corolla 10–13 mm, yellow. **FR** glabrous. $2n$=14. Uncommon. Disturbed places; esp < 500 m. NW; to WA, e US; native to Eur.

L. biflorus T.W. Nelson & J.P. Nelson (p. 615) TWO-FLOWERED PEA Per, very small, puberulent. **ST** wiry, angled or flanged, not winged. **LF**: stipules small; lflets 4, gen 0.5–1(2) cm, oblong-lanceolate; tendril bristle-like, not branched or coiled. **INFL** 2-fld. **FL**: calyx tube > lobes; corolla 8–10 mm, greenish white, dark-striate. **FR** glabrous. RARE. Serpentine; ± 1300 m. NCoRH (Lassics Range, Humboldt Co.).

L. brownii Eastw. (p. 615) Per, glabrous. **ST** angled or flanged, not winged. **LF**: stipules widely ovate-triangular, often toothed; lflets 6–10, 1–5 cm, gen ovate or elliptic, sometimes oblong-lanceolate; tendril variable, gen branched, coiled, at least on some lvs. **INFL** 2–6-fld. **FL**: calyx lobes < tube, often finely ciliate; corolla 10–14 mm, blue-purple to pinkish, all petals upcurved, the banner to ± 90°, keel ± = wings. **FR** glabrous. Dry, open woodlands, streambanks; 1000–1700 m. NCo, KR, CaR, n&c SN; s OR. [*L. pauciflorus* Fern. ssp. *b.* (Eastw.) Piper] Keel << wings in *L. pauciflorus* of e OR, etc. ❀TRY.

L. cicera L. (p. 615) Ann, glabrous. **ST** narrowly winged. **LF**: stipules lanceolate; lflets 2, 3–6 cm, linear to linear-lanceolate; tendril branched, coiled. **INFL** 1-fld; axis not extended beyond fl. **FL**: calyx tube << lobes; corolla 10–15 mm, red-purple. **FR** glabrous; upper suture furrowed. $2n$=14. Uncommon. Disturbed areas; esp < 500 m. CA-FP; native to Eur.

L. delnorticus C. Hitchc. (p. 615) DEL NORTE PEA Per, glabrous. **ST** flanged or winged. **LF**: stipules small but wide, toothed; lflets 10–16, gen subopposite, 3–5 cm, elliptic or lanceolate; tendril branched, coiled. **INFL** 8–10-fld, ± dense. **FL**: calyx tube gen < lower lobes; corolla 10–15 mm, white, gen lavender-striate, fading yellow-tan. **FR** glabrous. $2n$=14. UNCOMMON. Streambanks, serpentine; 250–700 m. nw KR (Del Norte, w Siskiyou cos.); OR.

L. glandulosus Broich (p. 615) STICKY PEA Per, glandular-puberulent. **ST** often strongly angled or flanged, but gen not winged. **LF**: stipules small, narrow; lflets ± 14–16, gen subopposite to alternate, 3–5 cm, ovate to lanceolate, conspicuously glandular on lower surface; tendril branched, coiled. **INFL** 5–12-fld. **FL**: calyx tube > upper lobes, ± = or < lower; corolla 10–14 mm, lavender to purple. **FR** glandular. $2n$=14. UNCOMMON. Roadsides, oak woodlands; < 800 m. NCoR (Humboldt, n Mendocino cos.).

L. hirsutus L. (p. 615) CALEY PEA Ann, ± glabrous, exc fr. **ST** winged. **LF**: stipules narrow; lflets 2, 3–6 cm, narrowly elliptic to oblong-lanceolate; tendril branched, coiled. **INFL** 1–2-fld; axis not extended beyond fls. **FL**: calyx tube ± = lobes; corolla 9–14 mm, pink to blue-purple or 2-colored. **FR**: hairs each with a bulbous base. $2n$=14. Uncommon. Disturbed places, wet meadows, creekbeds; < 1000 m. SnFrB, sporadic elsewhere in CA-FP; to se US; native to Eur. Cult.

L. hitchcockianus Barneby & Reveal Per, glabrous or puberulent. **ST** angled or flanged, not winged. **LF**: stipules small; lflets ± 4–6, 1–1.4 cm, linear to lanceolate; tendril coiled, sometimes branched. **INFL** gen 2-fld, open. **FL**: calyx tube >> lobes; corolla 8–12 mm, lilac to purple. **FR** glabrous. Washes, desert scrub; 1500 m. ne DMoj (Grapevine Mtns, Inyo Co.); w NV. Expected, but last collected in CA a century ago.

L. japonicus Willd. (p. 615) Per, gen glabrous. **ST** sprawling, angled or flanged, not winged. **LF**: stipules ± = lflets; lflets 6–8, 2.5–4.5 cm, ovate or elliptic, fleshy; tendril coiled, branched, or neither. **INFL** 3–8-fld. **FL**: calyx tube < lower lobes; corolla 1.8–2.2 mm, blue-purple or 2-colored. **FR** puberulent. $2n$=14. Uncommon. Coastal beaches, dunes; < 30 m. NCo (Del Norte Co.); to AK, circumboreal, also Chile. [var. *glaber* (Ser.) Fern.] ❀DRN, SUN:15,**16,17**;INV,STBL.

L. jepsonii E. Greene Per, glabrous to puberulent. **ST** climbing, winged. **LF**: stipules small, gen narrow; lflets 10–16, gen subopposite to alternate, 3.5–5.5 cm, lanceolate or oblong-lanceolate; tendril branched, coiled. **INFL** 6–15-fld. **FL**: calyx tube > upper lobes, ± = lower; corolla 15–20 mm, gen pink to pink-purple. **FR** glabrous. $2n$=14. Forests, open areas, coastal and estuarine marshes; < 1500 m. KR, NCoR, CaR, n SN, GV, CW, SCoRO.

var. ***californicus*** (S. Watson) Hoover Pl gen puberulent, sometimes glabrous. Forests, open areas; < 1500 m. Range of sp. Intermediates to *L. vestitus* may be hybrids. ❀IRR,DRN:**15–17**& SHD:**7–9,14**&STBL.

var. ***jepsonii*** (p. 615) DELTA TULE PEA Pl glabrous, often robust (st < 2.5 m). RARE. Coastal and estuarine marshes; < 30 m. GV, esp SnFrB.

L. lanszwertii Kellogg Per, glabrous or puberulent. **ST** angled or ± flanged, not winged. **LF**: stipules small, gen narrow; lflets 4–10, opposite or subopposite, 2–8 cm, linear to lanceolate; tendril branched, coiled, reduced to bristle, or 0. **INFL** 2–10-fld. **FL**: calyx tube gen > upper lobes; corolla 7–16 mm. **FR** glabrous. $2n$=14,28. Open, dry, gen coniferous woodlands, meadows; 200–2000 m. KR, NCoR, SNH, GB; to WA, WY, Colorado, NM.

var. ***aridus*** (Piper) Jepson (p. 615) Pl gen puberulent. **ST** gen ascending, not climbing. **LF**: lflets narrowly lanceolate to linear; tendril often a bristle or 0. **FL**: corolla 7–11 mm, pale lavender to purple. $2n$=14. Open, dry woodlands, meadows; 1200–2000 m. KR, SNH, MP; to WA. Intergrades with var. *lanszwertii* in CA; some KR pls uncertainly distinct from *L. bijugatus* T. White of e WA, ID. ❀TRY.

var. ***lanszwertii*** (p. 615) Pl gen puberulent. **ST** climbing or ascending. **LF**: lflets elliptic-lanceolate to lanceolate; tendril gen coiled. **FL**: corolla 10–16 mm, purple to lavender. $2n$=28. Open slopes, pine forests; 1200–2000 m. KR, NCoR, GB; to WA, ID, UT. ❀TRY.

var. ***tracyi*** (Bradshaw) Isely Pl glabrous or puberulent. **ST** climbing or erect. **LF**: lflets linear to elliptic; tendril well developed or reduced to a bristle. **FL**: corolla 9–12 mm, white to cream-colored, banner often dark-striate. $2n$=14. Conifer forests; 200–1100 m. KR, NCoR. [*L. tracyi* Bradshaw] Seemingly differs from var. *lanszwertii* only in fl color, range. ❀TRY.

L. latifolius L. (p. 615) PERENNIAL SWEET PEA Per, glabrous, often robust. **ST**: wings often 2 mm or more wide. **LF**: stipules small or large; lflets 2, 5–14 cm, lanceolate to ovate; tendril branched, coiled. **INFL** 4–15-fld. **FL**: calyx tube > lobes; corolla 20–25 mm, pink, pink-purple, or red. **FR** glabrous. $2n$=14. Disturbed areas, esp roadsides; gen < 2000 m. CA-FP; sporadic to e US; native to Eur. Cult as orn. Locally conspicuous.

L. littoralis (Nutt.) Endl. (p. 615) Per, densely gray-hairy. **ST** prostrate or ascending, angled or flanged, not winged. **LF**: stipules = or > lflets; lflets 4–8, 1–2 cm, obovate to short-oblong; tendril bristle-like. **INFL** 4–8-fld, dense. **FL**: calyx tube < to > lobes; corolla 15–18 mm, 2-colored (pink-purple and white). **FR** elliptic to oblong, hairy. Open, coastal dune areas; < ± 5 m. NCo, n&c CCo; to B.C. ❀Sand:4,**5**,16,**17**;DFCLT.

L. nevadensis S. Watson var. ***nevadensis*** (p. 615) Per; glabrous or puberulent. **ST** angled or flanged, not winged. **LF**: stipules small, narrow or wide; lflets 4–8, gen opposite, 1.5– ± 4 cm, gen elliptic or widely ovate; tendril branched and coiled, or bristle-like. **INFL** 2–4-fld. **FL**: calyx tube > lobes; corolla 13–25 mm, white to blue-purple. **FR** glabrous. $2n$=28. Coniferous, mixed forests; 450–2300 m. KR, NCoR, CaR, n&c SN. [ssp. *lanceolatus* (J. Howell) C. Hitchc.] Infraspecific taxa based on tendrils, lflet number, fl size are unsubstantial in CA. ❀TRY.

L. odoratus L. SWEET PEA Ann, hairy. **ST** sprawling or climbing, gen winged. **LF**: stipules small; lflets 2, 3–6 cm, narrowly elliptic to ovate, length 1.2–3 × width; tendril branched, coiled. **INFL** 2–4-fld, open. **FL**: calyx tube ± = lobes; corolla 20–30 mm, white, pink, or purple. **FR** < 6 cm, hairy. $2n$=14. Uncommon. Disturbed areas; esp < 1000 m. CA-FP; to e US; native to Eur. Widely cult.

L. palustris L. (p. 615) MARSH PEA Per, gen glabrous or ± puberulent. **ST** angled, flanged, or narrowly winged. **LF**: stipules lanceolate, lower lobe conspicuous; lflets gen 6, sometimes 4 or 8, opposite, 2.5–5.5 cm, elliptic to oblong-lanceolate; tendril branched, coiled. **INFL** 3- 6-fld. **FL**: calyx tube ± = lower lobes; corolla 15–20 mm, banner bent ± 90°, pink-purple or purple, rarely white. **FR** glabrous or initially glandular. $2n$=14. UNCOMMON Moist coastal areas; gen < 100 m. n NCo; to AK, ne US, circumboreal.

L. polyphyllus Nutt. (p. 615) Per, glabrous; often robust. **ST** angled or flanged, not winged. **LF**: stipules gen conspicuous, gen widely ovate, > 1 cm wide; lflets 10–16, subopposite to alternate, 3–6 cm, elliptic to ovate; tendril branched, coiled, or reduced to a bristle. **INFL** 6–12-fld, ± 1-sided. **FL**: lower calyx lobe narrow, ± = tube, lateral lobes triangular, not widened above base; corolla 16–20 mm, purple. **FR** glabrous. $2n$=14. Forests and margins; < 1500 m. KR, n&c NCoR; to WA. ❀**5**,IRR:**6**,**15–17**&SHD:**14**.

L. rigidus T. White (p. 615) Per, glabrous. **STS** erect, clump-forming, angled or flanged, not winged. **LF**: stipules gen narrow, ± 1/2 lflet length; lflets 6–10, 1–3 cm, lanceolate to elliptic; tendril bristle-like. **INFL** 2–5-fld, dense. **FL**: calyx tube gen ± = lower lobes, > upper; corolla 17–23 mm, white to pink. **FR** glabrous. $2n$=14. Sagebrush scrub, disturbed areas; 800–1200 m. MP (Modoc Co.); to OR, ID.

L. sphaericus Retz. Ann, glabrous. **LF**: stipules small, narrow; lflets 2, 2–6 cm, linear to narrowly lanceolate; tendril gen coiled. **INFL** 1-fld; peduncle 1–2 cm, axis extended beyond fl as short bristle. **FL**: calyx lobes ± = or > tube; corolla 8–14 mm, orange-red. **FR** glabrous, longitudinally striate. $2n$=14. Uncommon. Disturbed places; < 1000 m. CA-FP; OR; native to Eur.

L. splendens Kellogg PRIDE OF CALIFORNIA, CAMPO PEA Per, glabrous or puberulent. **ST** gen climbing, angled or flanged, not winged. **LF**: stipules gen small, sometimes wide, wavy-margined; lflets gen 6–8, 2–4 cm, linear to widely ovate; tendril coiling. **INFL** 4–6-fld, open. **FL**: calyx tube ± = lower lobe, > others; corolla 25–35 mm, wine-red to crimson, banner reflexed 120–180°; ovary glandular, puberulent. **FR** glabrous. $2n$=14. UNCOMMON. Chaparral; < 1050 m. s SCo, PR (San Diego Co.); Baja CA. ❀DRN,DRY,SUN:14–24;DFCLT.

L. sulphureus A. Gray Per, glabrous. **ST** angled or flanged, not winged. **LF**: stipules 1–2.5 cm, gen wide; lflets 6–12, often alternate, 1.5–4 cm, elliptic to ovate; tendril branched, coiled. **INFL** 10–15-fld, often 1-sided, dense. **FL**: calyx tube > lobes; corolla 11–14 mm, whitish yellow to bronze- orange, drying tan or dark, appearing wide and blocky; banner reflexed above middle. **FR** glabrous. $2n$=14. Foothill woodlands to fir forests; 600–2500 m. KR, NCoR, SN; sw OR. [var. *argillaceus* Jepson] ❀17,SHD:1,2,**6**& IRR:**7**,**15**,**16**.

L. tingitanus L. TANGIER PEA Ann, glabrous. **ST** winged. **LF**: stipules small or wide and conspicuous; lflets 2, 2–6 cm, lanceolate to widely ovate, length 1.2–6.5 × width; tendril branched, coiled. **INFL** 2–3-fld, open. **FL**: calyx tube > lobes; corolla 25–30 mm, maroon to crimson. **FR** < 10 cm, glabrous. $2n$=14. Gen disturbed areas, esp coastal; gen < 500 m. CA-FP; OR, sporadic in w US; native to Eur. Cult as orn.

L. torreyi A. Gray (p. 621) Per, puberulent. **ST** angled or flanged, not winged. **LF**: stipules small, narrow or wide; lflets 10–15, 1–2 cm, elliptic to ovate; tendril reduced to a bristle or 0. **INFL** 1–2-fld. **FL**: calyx tube gen < lower lobes; corolla 8–10 mm, lilac to pale purple-blue. **FR** puberulent. $2n$=14. Open woodland; < 800 m. NCo, NCoR, n SnFrB; to WA.

L. vestitus Nutt. Per, glabrous or puberulent. **ST** often sharply angled or flanged. **LF**: stipules small and entire to wide and wavy-margined; lflets 8–12, 2–4.5 cm, linear to elliptic; tendril branched, coiled. **INFL** 8–15-fld, often dense. **FL**: calyx tube < or > lower lobes, which in some phases may be slightly wider above base; corolla 14–20 mm, gen purple or lavender, sometimes white; banner bent or reflexed ± 90° or more. **FR** glabrous or initially puberulent. $2n$=14. Chaparral, oak woodland, coniferous or mixed forest; < 1500 m. NCo, KR, NCoR, CW, SW. Intergrading complex of taxa and local variants (see Broich).

var. ***alefeldii*** (T. White) Isely Pl glabrous. **FL**: corolla 16–20 mm, wine red to dark purple, banner gen reflexed > 90°. Chaparral; < 1200 m. SCo, s ChI, PR. [*L. laetiflorus* ssp. *a.* (T. White) Bradshaw] Intergrades with var. *vestitus*. ❀DFCLT.

var. ***ochropetalus*** (Piper) Isely Pl glabrous. **FL**: corolla 14–16 mm, lavender or white. Coniferous or mixed forests; gen < 150 (–400) m. NCo, KR, NCoRO. [ssp. *bolanderi* (S. Watson) C. Hitchc., in part] ❀TRY.

var. ***vestitus*** (p. 621) Pl glabrous or puberulent. **FL**: corolla 14–18 mm, pale lavender to purple, banner bent ± 90°. Coniferous forest in n to chaparral and oak woodland in s; < 1500 m. NCo, NCoR, CW, SW. [ssp. *bolanderi* (S. Watson) C. Hitchc., in part; ssp. *puberulus* (E. Greene) C. Hitchc.; *L. laetiflorus* E. Greene, incl ssp. *barbarae* (T. White) C. Hitchc.] Variable: see Broich for more detailed classification. May hybridize with *L. jepsonii* var. *californicus*. ❀**16**,**17**,SHD:**15**&IRR:**7**,**9**,**14**;INV,STBL.

LENS

Duane Isely

Ann, unarmed. **ST** erect or climbing. **LVS** even-1-pinnate; stipules entire or lobed, persistent; main axis ending in a tendril or bristle; lflets alternate or opposite. **INFL**: raceme, axillary, 1–3-fld near tip. **FL**: calyx lobes >> tube; corolla pink-white to blue; 9 filaments fused, 1 free; style puberulent on concave side near tip. **FR** dehiscent, elliptic to oblong. ± 5 spp.: Medit, adjacent regions. (Latin: lentil) Lentil seeds provided the concept of a biconvex "lens"-shape ("lenticular"), from which optical lenses derived their name.

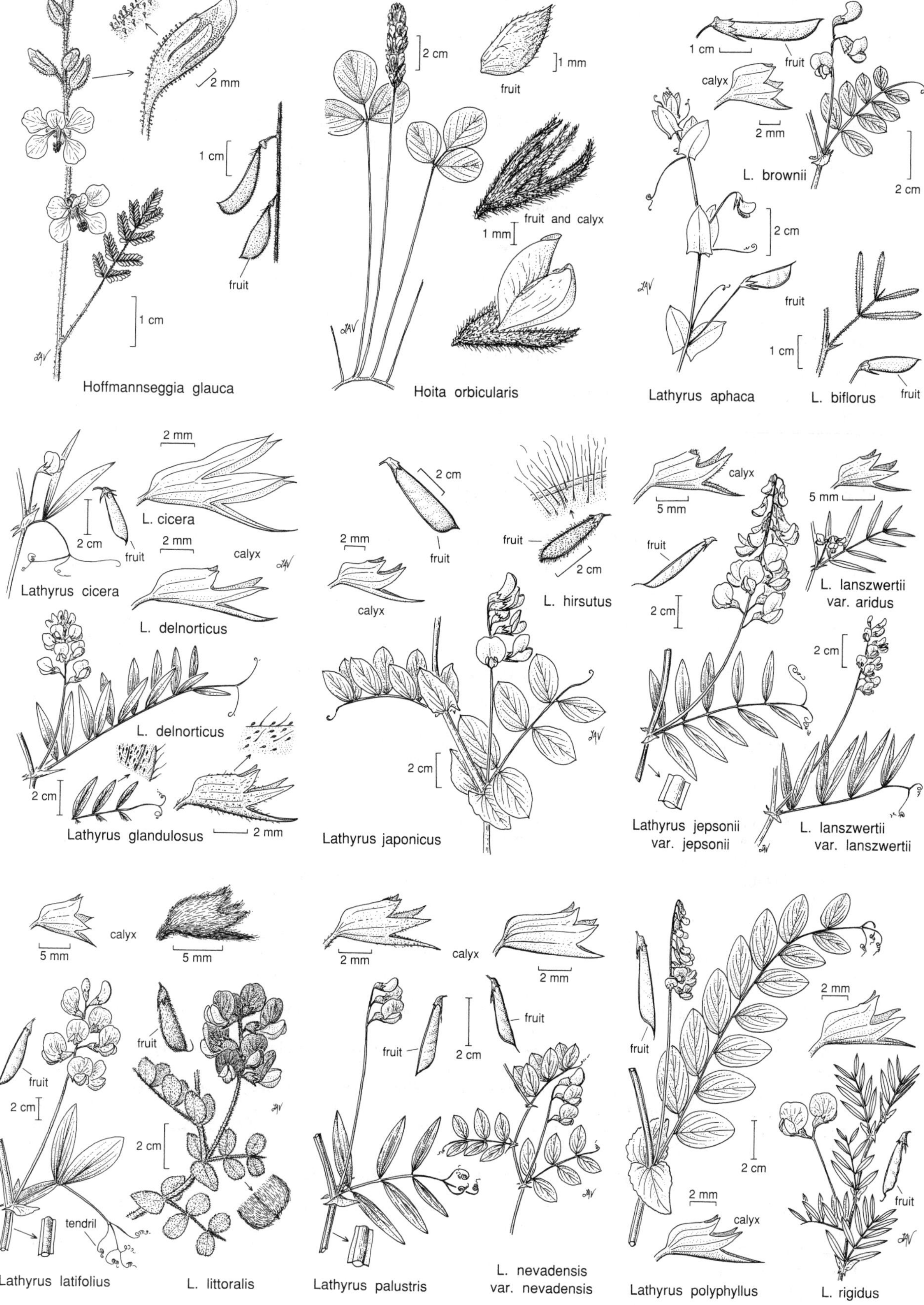
2 mm
1 cm
fruit
1 cm
Hoffmannseggia glauca
2 cm
1 mm
fruit
fruit and calyx
1 mm
Hoita orbicularis
1 cm
fruit
calyx
2 mm
L. brownii
2 cm
2 cm
fruit
1 cm
fruit
Lathyrus aphaca
L. biflorus
2 mm
L. cicera
2 cm
fruit
2 mm
calyx
Lathyrus cicera
L. delnorticus
L. delnorticus
2 cm
Lathyrus glandulosus
2 mm
2 cm
2 mm
fruit
calyx
fruit
2 cm
L. hirsutus
2 cm
Lathyrus japonicus
calyx
5 mm
fruit
2 cm
5 mm
L. lanszwertii
var. aridus
2 cm
Lathyrus jepsonii
var. jepsonii
L. lanszwertii
var. lanszwertii
5 mm
calyx
5 mm
fruit
fruit
2 cm
2 cm
tendril
Lathyrus latifolius
L. littoralis
2 mm
calyx
2 mm
fruit
fruit
2 cm
Lathyrus palustris
L. nevadensis
var. nevadensis
fruit
2 mm
2 cm
2 mm
calyx
fruit
Lathyrus polyphyllus
L. rigidus

L. culinaris Medikus (p. 621) LENTIL Ann, gen finely hairy. **LF**: lflets gen 10–14, ± 7–15 mm, narrowly oblong to elliptic. **FL**: corolla ± 6 mm. **FR** pendent, 1–1.4 cm, ovate or oblong, flat. **SEEDS** 1–2, lenticular. $2n$=14. Uncommon. Disturbed areas; < 2000 m. CW; sporadic in US; native to Old World. Cult, esp WA, ID, for edible seed, the lentil of commerce.

LOTUS

Duane Isely

Ann, per, shrub, unarmed. **LVS** gen odd-1-pinnate (sometimes ± palmately compound, rarely some or most simple); stipules conspicuous or not; lflets 3–many, often irregularly arranged. **INFL**: umbel or 1–2-fld, axillary, gen peduncled, often bracted. **FL**: corolla gen yellow (sometimes white or pink), fading darker; 9 filaments fused, 1 free. **FR** dehiscent or not, exserted from calyx or not, ovoid to oblong, ± beaked. **SEEDS** 1–several. (Greek: derivation unclear) [Isely 1981 Mem New York Bot Garden 25:128–206] Spp. gen variable; intermediates may be hybrids. Key below separates natural groups.

Key to Groups

1. Lflets 5, 3 near axis tip, lower pair large, stipular in position; stipules gland-like, often not apparent ...... **Group 1**
1′ Lflets 3–15, lowest not stipular in position; stipules as above or not
 2. Stipules scarious or lflet-like, sometimes fragile and quickly lost; lflets 3–15, pinnately arranged, often opposite or subopposite .... **Group 2**
 2′ Stipules gland-like, bump-like, or conic, often not apparent; lflets 3–9, pinnately or ± palmately arranged, gen irregularly arrayed on lf axis
  3. Fr indehiscent, gen not flat, gen incl or moderately exserted from calyx (sometimes conspicuously exserted and curved most of length), gen with curved, 2–3 mm beak; gen per .... **Group 3**
  3′ Fr dehiscent, gen flat, gen almost entirely exserted and straight, gen with straight or curved, 0.5–1.5 mm beak; ann or per
   4. Ann; fr gen straight; wings < or ± = keel; stigma glabrous .... **Group 4**
   4′ Per (exc *L. strigosus*); fr straight or ± curved; wings gen > keel; stigma nearly glabrous or finely hairy .... **Group 5**

Group 1: Leaflets 5, lower pair stipular in position

1. Ann; hairy; infl gen 1–3-fld, 5–8 mm — NCo (Sonoma Co.) .... ***L. angustissimus***
1′ Per; glabrous or hairy; infl 3–12-fld, 8–14 mm
 2. Infl 3–8-fld; calyx lobes not outcurved in bud; rhizome 0; st gen solid, glabrous or strigose ...... ***L. corniculatus***
 2′ Infl 8–12-fld; calyx lobes often outcurved in bud; rhizome conspicuous; st hollow, hairs often spreading .... ***L. uliginosus***

Group 2: Stipules scarious or leaflet-like

1. Lflets 3–5; fl 8–10 mm; pl very small, decumbent; NCoR .... ***L. yollabolliensis***
1′ Lflets (3)5–15; fl gen 8–15 mm; pl various, gen erect or ascending; if lflets < 5, not of NCoR
 2. Lf conspicuously hairy, gray or silvery, with soft, wavy hairs; fr glabrous .... ***L. incanus***
 2′ Lf glabrous or hairy, rarely gray or silvery; fr gen hairy if lf hairy
  3. Peduncle bract 0 or just below umbel, gen simple or 2–3-parted; fr 1.5–3 mm wide
   4. Petal claws incl to slightly exserted from calyx; pl often hairy .... ***L. oblongifolius***
    5. Lflets 3–7, glabrous; corolla 8–9 mm; local, Tulare Co. .... var. ***cupreus***
    5′ Lflets 7–11, glabrous or hairy; corolla 9–13 mm; more widespread, incl Tulare Co. .... var. ***oblongifolius***
   4′ Petal claws clearly exserted from calyx; pl ± glabrous
    6. Corolla wings pink or purple, fading white; bracts gen present and 3-parted .... ***L. formosissimus***
    6′ Corolla wings white; bracts 0 or gen simple .... ***L. pinnatus***
  3′ Peduncle bract gen well below umbel, gen 3–5-parted, sometimes fragile and lost; fr 3–5 mm wide
   7. Hairs gen persistent, spreading, sometimes glandular; stipules often wide, lflet-like, persistent and sometimes becoming scarious; corolla pink to red-purple; n&c CA .... ***L. stipularis***
    8. St wiry; stipules wide, clasping st .... var. ***ottleyi***
    8′ St often fleshy; stipules narrow or wide, not or slightly clasping st .... var. ***stipularis***
   7′ Hairs gen nearly 0 or appressed, rarely glandular; stipules scarious, fragile, not lflet-like, gen some lost; corolla initially white, pink, or yellow-green
    9. Fls gen 6–10; corolla 10–12 mm, white or pink, becoming striate, darker; pedicel 2–3 mm, slightly longer in fr; lflet length gen 2–3 × width; gen n NCoRO .... ***L. aboriginus***
    9′ Fls gen 12–20; corolla 12–17 mm, yellow-green, becoming dark-blotched; pedicel gen 3–6 mm (> 3 mm at least in fr); lflet length gen 1–2.5 × width; widespread, esp inland .... ***L. crassifolius***
     10. Pl glabrous or strigose; widespread, incl n PR (Riverside Co.) .... var. ***crassifolius***
     10′ Pl soft-hairy; s PR (San Diego Co.) .... var. ***otayensis***

Group 3: Stipules often gland-like; fruit indehiscent

1. Ann, prostrate or decumbent, sometimes ascending; young growth not conspicuously hairy; fl 3–7 mm; fr bodies exserted; s SCoR, SW, gen s of Los Angeles
 2. Infl ± sessile; s SCoR, SW .... ***L. hamatus***
 2′ Infl conspicuously peduncled; s SCo (San Diego Co.) .... ***L. nuttallianus***
1′ Per or shrub, prostrate to erect; young growth conspicuously hairy if pl blooming 1st year or fls < 7 mm; fr bodies incl or exserted; widespread, incl SW

3. Per or shrub; ChI
4. Calyx nearly glabrous or strigose; fr body soon exserted; lf green or gray, sparsely or densely strigose or silky .... ***L. dendroideus***
5. Lflets 3, densely strigose or silky, gen gray; n ChI (San Miguel Island) .... var. ***veatchii***
5′ Lflets 3–5, finely or sparsely strigose, green; ChI (exc San Miguel Island)
6. Fr 1–1.5 cm; peduncle bract 0; ChI (exc San Miguel, San Clemente islands) .... var. ***dendroideus***
6′ Fr 2.5–5 cm; peduncle bract gen present; s ChI (San Clemente Island) .... var. ***traskiae***
4′ Calyx shaggy-hairy; fr incl (exc slowly elongating beak); lf gen silvery-silky with fine, straight, ultimately wavy or tangled hairs .... [2]***L. argophyllus***
7. Infls peduncled, not crowded; calyx lobes 1.5–2.5 mm; st decumbent to ascending .... var. ***argenteus***
7′ Infls ± sessile, densely crowded at st tip; calyx lobes 2.5–5 mm; st ascending to erect
8. St erect; lvs overlapping; calyx lobes ± 2.5–3.5 mm; s ChI (San Clemente Island) .... var. ***adsurgens***
8′ St ascending to erect; lvs not overlapping; calyx lobes 2.5–5 mm; n ChI (Santa Cruz Island) .... var. ***niveus***
3′ Per, rarely shrub (exc *L. scoparius*); mainland
9. Lflets 3; infl 1–3-fld, ± sessile
10. Lflet 2–5 mm; fr 6–9 mm; uncommon, se PR, sw DSon .... ***L. haydonii***
10′ Lflet 4–12 mm; fr 10–15 mm; widespread, incl PR
11. Calyx glabrous (see couplet 15 for vars.) .... [2]***L. scoparius***
11′ Calyx strigose .... ***L. procumbens***
12. Corolla 9–12 mm; calyx 4–6 mm, lobes ± = tube; s SN .... var. ***jepsonii***
12′ Corolla 6–8 mm; calyx 2–3 mm, lobes << tube; not s SN .... var. ***procumbens***
9′ Lflets > 3 on some or most lvs; infl 2–15-fld, sessile or peduncled
13. Lflets glabrous or finely strigose, gen green; hairs straight or slightly wavy
14. St gen ascending or erect (some of immediate coast prostrate), bushy-branched, often ± 1 m; st slightly woody (but green), sparsely leafy; corolla yellow; calyx lobes gen 1–2 mm, not curved outward; coastal and inland .... [2]***L. scoparius***
15. Keel prominent, > wings; fl gen 8–9 mm; gen inland, SW, DSon .... var. ***brevialatus***
15′ Keel not prominent, ± = wings; fl 7–12 mm; coastal and inland, widespread .... var. ***scoparius***
14′ Pls prostrate or low-ascending, < 1 m; st not woody, leafy; corolla pink-white or yellow; calyx lobes gen < 1 mm or curved outward; esp coastal
16. Calyx lobes slender, some or all curved outward or hooked; corolla gen white to pinkish (to brick-red when dry) .... ***L. benthamii***
16′ Calyx lobes wide, not curved outward, not hooked; corolla yellow .... ***L. junceus***
17. Peduncle 8–25 mm; st wiry, gen decumbent; fr body well exserted .... var. ***biolettii***
17′ Peduncle 1–5 mm; st gen stout, prostrate to ascending; fr body slightly exserted .... var. ***junceus***
13′ Lflets conspicuously hairy, green, gray, or silvery; hairs straight or wavy
18. St, esp near tip, gen with conspicuous, spreading or obliquely directed, often straight and stiff hairs < 0.5 mm; corolla 4–6 mm; esp coastal CA, inland to D edge .... ***L. heermannii***
19. Corolla gen 4–5 mm; ovary, young fr gen finely strigose; SCo, PR, D edge .... var. ***heermannii***
19′ Corolla gen 5–6 mm; ovary, young fr gen with spreading hairs; NCo, NCoRO, CCo, SCoRO .... var. ***orbicularis***
18′ St gen strigose or with ± spreading, often wavy hairs gen 0.2–0.4 mm; corolla 5–12 mm; esp inland CA
20. Fr body exserted; lflet hairs gen wavy, not obscuring lflet surface; lvs gray or green .... ***L. nevadensis***
21. Fls gen 3–5 per infl, 6–7 mm, appearing blocky; banner abruptly upcurved 90°; TR, ne PR .. var. ***davidsonii***
21′ Fls gen 5–12 per infl, 5–10 mm, not appearing blocky; banner upcurved 30–90°; not of TR .... var. ***nevadensis***
20′ Fr body incl or shortly exserted; lflet hairs at first gen straight, in age tangled and gen obscuring lflet surface; lvs silky-silvery .... [2]***L. argophyllus***
22. Infl gen < 1 cm wide, 4–8-fld; corolla 6–10 mm; esp c SN and s .... var. ***argophyllus***
22′ Infl often 1.5 cm wide, 10–15-fld; corolla 8–12 mm; n SN .... var. ***fremontii***

Group 4: Stipules often gland-like; fruit dehiscent; annuals

1. Fls gen 1 per lf axil; peduncle << 1 cm, bracts 0
2. Corolla yellow, reddening with age; pl prostrate to low-ascending and calyx lobes 0.8–2 × tube
3. Calyx lobes 1–2 × tube; fr gen 3–4 mm wide; pl gen with soft, spreading hairs .... ***L. humistratus***
3′ Calyx lobes ± 0.8–1.2 × tube; fr gen 2.3–3 mm wide; pl inconspicuously strigose or with soft, spreading hairs .... ***L. wrangelianus***
2′ Corolla white, pale yellow, or pink, often darkening with age; pl gen ascending to erect and calyx lobes < 2 × tube, or rarely prostrate and calyx lobes > 2 × tube
4. Pl gen 1–3 dm; calyx lobes ± 1.5 × tube; common .... ***L. denticulatus***
4′ Pl < 1 dm; calyx lobes ± 2 × tube; uncommon .... ***L. rubriflorus***
1′ Fls 1–several per lf axil; peduncle gen > 1 cm, gen bracted
5. Infl gen 2–4-fld (1st-formed sometimes 1–2-fld); lflets 3–7, obovate or ± round, terminal gen largest .... ***L. salsuginosus***
6. Corolla 3.5–5 mm, keel > other petals; fr 1–1.5 cm, becoming narrowed between seeds .... var. ***brevivexillus***
6′ Corolla 6–10 mm, keel ± = other petals; fr 1.5–3 cm, not narrowed between seeds .... var. ***salsuginosus***
5′ Infl gen 1-fld; lflets 1–5, lanceolate, elliptic, or obovate, ± equal
7. Calyx lobes >> tube; lflets gen 3 (often 1 on upper lvs) .... ***L. purshianus***
7′ Calyx lobes ± = tube; lflets 3–9 throughout

8. St erect or ascending; lf axis not conspicuously flattened; corolla pink or salmon, quickly fading; fr straight, margin often wavy .... ***L. micranthus***

8′ St prostrate, often mat-forming; lf axis conspicuously flattened; corolla yellow; fr curved near tip, margin not wavy .... ***L. strigosus***

Group 5: Stipules often gland-like; fruit dehiscent; perennials

1. Ann, gen prostrate and mat-forming; infl gen 1–2-fld; corolla 5–10 mm; lf axis flat, ± blade-like; fr gen curved only at or near tip .... ***L. strigosus***

1′ Per or shrub-like, prostrate (mat-forming or not) to erect; infl 1-9-fld; corolla 8–25 mm; lf axis not flat or blade-like; fr gen straight (sometimes curved throughout, rarely only at or near tip)

2. Lf axis incl petiole 10–35 mm; lflets 7–9, gen 1–2 cm, length 1.5–3 × width .... ***L. grandiflorus***

3. Pl puberulent or soft-hairy; lflet gen 1–1.5 cm .... var. ***grandiflorus***

3′ Pl ± glabrous or puberulent; lflet gen 1.6–2.2 cm .... var. ***macranthus***

2′ Lf axis incl petiole 1–8 mm; lflets 3–5, < 1 cm or length > 3 × width, or both

4. Shrub-like, ascending, < 1.5 m, finely strigose; corolla 12–22 mm .... ***L. rigidus***

4′ Per, prostrate or low-ascending, < 0.3 m, silvery-silky or gray-puberulent; corolla 8–12 mm .... ***L. argyraeus***

5. St prostrate, mat-forming; calyx lobes ± 2 mm; SnBr, SnJt, DMtns .... var. ***argyraeus***

5′ St decumbent to low-ascending; calyx lobes 2–3 mm; DMtns

6. Lflet oblanceolate to obovate, length ± 3–4 × width; New York Mtns .... var. ***multicaulis***

6′ Lflet obovate, length ± 2 × width; Providence Mtns .... var. ***notitius***

L. aboriginus Jepson Per, glabrous or puberulent. **ST** spreading or erect. **LF**: stipules scarious, inconspicuous; lflets 9–15, 1.5–3 cm, elliptic, length 2–3 × width. **INFL** 6–10-fld; peduncle bract well below umbel. **FL**: calyx 4–6 mm, lobes < tube; corolla 10–12 mm, white or ± pink, claws often exceeding calyx tube. **FR** dehiscent, 3–5 cm, 3–4 mm wide, oblong, glabrous. **SEEDS** several. Banks, streamsides, burns, logged areas; 0–800 m. NCo, NCoRO; to WA. [*L. "aboriginum"* Jepson] ❀TRY.

L. angustissimus L. Ann, hairy. **ST** ± prostrate. **LF**: stipules inconspicuous; lflets 5, 8–12 mm, elliptic to narrowly obovate, 3 ± palmately arrayed near axis tip, lower 2 stipular in position. **INFL** 1–3-fld; peduncle bracted. **FL**: calyx 6–7 mm, lobes > tube; corolla 5–8 mm. **FR** dehiscent, 1.5–3 cm, linear, glabrous. **SEEDS** several. $2n$=12,24. Disturbed grasslands, roadsides; < 1000 m. NCo (Sonoma Co.); native to Eur. Established at 1 site 20 years or more without spreading.

L. argophyllus (A. Gray) E. Greene Per (sometimes ± woody), hairy, silvery or gray. **ST** prostrate to erect. **LF** irregularly pinnate to ± palmate; stipules not lflet-like; lflets 3–7, 6–12 mm, obovate to elliptic-lanceolate, densely silky, hairs straight, ultimately wavy or tangled. **INFL** head-like, 4–many-fld; peduncle sometimes ± 0, sometimes bracted. **FL**: calyx 4–8 mm, lobes < or ± = tube, densely hairy; corolla 6–12 mm. **FR** indehiscent, incl exc for curved beak. **SEED** gen 1. $2n$=14. Chaparral, canyons, openings in woodlands; < 1600 m. SN, SCoR, SCo, SnGb, SnBr, PR. Also dry, open slopes, bluffs; < 400 m. ChI. Major vars. geog distinct; mainland vars. intergrade with related spp.

var. ***adsurgens*** Dunkle SAN CLEMENTE ISLAND BIRD'S-FOOT TREFOIL Pl ± woody, silvery. **ST** ascending to erect, densely leafy; branches slender. **INFLS** congested at st tips, subsessile. **FL**: calyx lobes ± 2.5–3.5 mm. **ENDANGERED** CA. Hot, dry, rocky slopes, bluffs; < 300 m. s ChI (San Clemente Island). [ssp. *adsurgens* (Dunkle) Raven] Related to var. *niveus*. In cult.

var. ***argenteus*** Dunkle Pl ± woody. **ST** prostrate or ascending, not densely leafy. **INFL** 1.5–2 cm wide, peduncled. **FL**: calyx lobes 1.5–2.5 mm. Chaparral, bluffs; < 400 m. ChI (exc Santa Cruz Island). [var. *ornithopus* (E. Greene) Ottley, in part] Resembles var. *argophyllus*. ❀DRN,DRY.SUN:14–17,20–24;DFCLT.

var. ***argophyllus*** (p. 621) Pl not woody. **ST** prostrate or decumbent-ascending, not densely leafy. **INFL** < 1 cm wide, 4–8-fld, sessile or not. **FL**: calyx lobes 1.5–3 mm; corolla 6–10 mm. Dry slopes in chaparral, canyons; < 1600 m. c&s SN, SCoR, SCo, SnGb, SnBr, PR. [var. *decorus* (I.M. Johnston) Ottley] Variable, intergrading with other spp. in SnGb, SnBr. ❀DRN,DRY,SUN:7, 14–17,18–24;DFCLT.

var. ***fremontii*** (A. Gray) Ottley (p. 621) Pl not woody. **ST** prostrate or decumbent-ascending. **INFL** < 1.5 cm wide, spheric, 10–15-fld, gen sessile. **FL**: calyx lobes 3–5 mm; corolla 8–12 mm. Openings along rivers, trails, on canyon slopes; 600–1200 m. n SN. ❀TRY.

var. ***niveus*** (E. Greene) Ottley SANTA CRUZ ISLAND BIRD'S-FOOT TREFOIL Pl ± woody. **ST** gen ascending, not densely leafy. **INFLS** congested at st tips, ± sessile. **FL**: calyx lobes 2.5–5 mm. **ENDANGERED** CA. Rocky slopes, dry riverbeds; < 300 m. ChI (Santa Cruz Island). [ssp. *n.* (E. Greene) Munz] In cult;DFCLT.

L. argyraeus (E. Greene) E. Greene Per, hairy, silvery or gray. **ST** prostrate to ascending. **LF** irregularly pinnate or ± palmate; stipules gland-like; lflets 3–5, 4–12 mm, oblanceolate or obovate. **INFL** 1–3-fld; peduncle bract small or 0. **FL**: calyx lobes ± = or < tube; corolla 7–12 mm, wings unequal, > keel; stigma finely puberulent. **FR** dehiscent, 1–2.5 cm, oblong, straight or slightly curved. **SEEDS** several. Mtn slopes, sometimes in pine/juniper woodland; 1200–2400 m. SnBr, SnJt, DMtns; to Mex. Vars. similar but geog distinct.

var. ***argyraeus*** **ST** gen prostrate, mat-forming. **LF** canescent; lflets obovate. **FL**: calyx lobes ± 2 mm; corolla 7–10 mm. **FR** 1.5–2 cm, gen straight. Open granitic slopes or with pine; 1500–2400 m. SnBr, SnJt, DMtns; to Mex. [ssp. *a.*] ❀TRY.

var. ***multicaulis*** (Ottley) Isely **ST** decumbent to ascending. **LF** green or canescent; lflets oblanceolate to obovate, length ± 3–4 × width. **FL**: calyx lobes 2.5–3 mm; corolla 8–12 mm. **FR** 2–2.5 cm, curved at tip. Uncommon. Pinyon/juniper woodland; 1200–1500 m. DMtns (New York Mtns). [ssp. *m.* (Ottley) Munz] ❀TRY.

var. ***notitius*** Isely **ST** low-spreading or ascending. **LF** green or canescent; lflets obovate, length ± 2 × width. **FL**: calyx lobes 2–3 mm; corolla 8–12 mm. **FR** 1–2.5 cm, straight. Uncommon. Pinyon/juniper woodland; 1200–2000 m. DMtns (Providence Mtns). ❀TRY.

L. benthamii E. Greene Per, glabrous or strigose. **ST** mat-forming or ascending. **LF** irregularly pinnate or palmate; stipules gland-like; lflets 3–5, 5–12 mm, elliptic or obovate. **INFL** 3–10-fld; peduncle bracted. **FL**: calyx 3.5–6 mm, lobes < tube, slender, gen spreading or curved downward or hooked; corolla gen white or pink, often dark-striate. **FR** indehiscent, incl (exc beak). **SEEDS** 1–2. $2n$=14. Coastal dunes, slopes, bluffs; 0–200 m. CCo, SnFrB, SCoRO. Evidently hybridizes with other spp. ❀DRN,DRY,SUN: **17,24**.

L. corniculatus L. (p. 621) BIRDFOOT TREFOIL Per, ± glabrous or strigose. **ST** decumbent or ascending. **LF**: stipules gland-like; lflets 5, 5–20 mm, linear to obovate, 3 ± palmately arrayed at axis tip, lower 2 stipular in position. **INFL** 3–8-fld; peduncle bracted. **FL**: calyx 2–3.5 mm, lobes ± = tube, not outcurved in bud; corolla 8–14 mm, bright yellow. **FR** dehiscent, 1.5–3.5 cm, narrowly oblong. **SEEDS** few–several. $2n$=12,24. Probably naturalized in open, disturbed areas; < 1000 m. CA (exc D); to n US; native to Eurasia. In Eur, diploid *L. tenuis* Willd. is segregated; it seems indistinguishable in CA. Some pls TOXIC by production of cyanide-releasing compounds.

L. crassifolius (Benth.) E. Greene Per, glabrous to hairy, often robust. **ST** sprawling or erect, < 1.5 m. **LF**: stipules scarious, inconspicuous; lflets 9–15, 2–3 cm, elliptic or obovate, length gen 1–2.5 × width, lower surface pale. **INFL** gen 12–20-fld; peduncle bract gen well below umbel. **FL**: calyx 5–8 mm, lobes < tube; corolla 12–17 mm, yellow-green, becoming dark-blotched, claws exserted from calyx tube. **FR** dehiscent, 3.5–7 cm, 3–5 mm wide, oblong, glabrous. **SEEDS** several. $2n$=14. Chaparral, pine or mixed woodlands and margins, roadsides, other disturbed places; 300–2100 m. KR, NCoR, SN, CW, SW (exc ChI); to WA.

var. ***crassifolius*** (p. 621) Pl glabrous or strigose. **FL**: ovules ± 14–35 (> 24 in s). Common. Habitats and elevations of sp. KR, NCoR, SN, SCoR, SCo, TR, n PR (Riverside Co.); to WA. Conspicuous. ❀DRN,DRY,SUN:**7–9,14–24**;STBL.

var. ***otayensis*** Isely OTAY MOUNTAIN LOTUS Pl conspicuously hairy. **FL**: ovules 14–20. RARE. Disturbed areas; 900 m. s PR (Otay Mtn, San Diego Co.).

L. dendroideus (E. Greene) E. Greene Per or shrub, ± glabrous or strigose, green or gray. **ST** decumbent to ascending, < 2 m. **LF**: stipules gland-like; lflets 3–5, irregularly arranged, 7–15 mm, elliptic or obovate. **INFL** 3–10-fld; peduncle short or 0, sometimes bracted. **FL**: calyx 4–6 mm, thinly strigose to ± glabrous, lobes < tube; corolla 8–10 mm. **FR** indehiscent, oblong-tapering, exserted; beak slender. $2n$=14. Bluffs, inland canyons, open sites near ocean; < 350 m. ChI. Island-to-island variants have been recognized variously; the following are distinct morphologically and geog.

var. ***dendroideus*** (p. 621) ISLAND BROOM **LF**: lflets 3–5, finely strigose, greenish. **INFL**: peduncle bract 0. **FR** 1–1.5 cm. UNCOMMON. Habitats and elevations of sp. ChI (exc San Miguel, San Clemente islands). [*L. scoparius* var. *d.* (E. Greene) Ottley] ❀DRN,DRY,SUN:14–17,20–24;DFCLT.

var. ***traskiae*** (Noddin) Isely TRASK'S ISLAND LOTUS **LF**: lflets 3–5, finely strigose, greenish. **INFL**: peduncle bracted. **FR** 2.5–5 cm. **ENDANGERED** CA, US. Habitats and elevations of sp. ChI (San Clemente Island). [*L. scoparius* ssp. *t.* (Noddin) Raven]

var. ***veatchii*** (E. Greene) Isely SAN MIGUEL ISLAND DEERWEED **LF**: lflets 3, densely strigose, almost silky, gen gray. UNCOMMON. Habitats and elevations of sp. ChI (San Miguel Island). [*L. scoparius* var. *v.* (E. Greene) Ottley]

L. denticulatus (Drew) E. Greene (p. 621) Ann, ± glabrous or puberulent. **ST** decumbent to erect, 1–4 dm. **LF** subpinnate or palmate; stipules gland-like or 0; lflets 2–4, ± alternate, 1–2 cm, elliptic or obovate; axis flat, ± blade-like. **INFL** axillary, 1–2-fld, ± sessile. **FL**: calyx 3–5 mm, lobes ± = to 1.5 × tube; corolla 6–8 mm, cream-white or pale yellow, wings ± = keel. **FR** dehiscent, 0.5–1.5 cm, widely oblong, gen strigose. **SEEDS** 2–3. $2n$=12. Open meadows, slopes, streambanks; < 1400 m. NCo, KR, NCoRO, CaRF, n SNF, ScV, e SnFrB, MP; to B.C.

L. formosissimus E. Greene (p. 621) Per, ± glabrous, stoloned or rhizomed. **ST** sprawling to ascending, often spongy-based, 1–5 dm. **LF**: stipules large, triangular, ± scarious, fragile; lflets 3–7, ± opposite, 6–20 mm, elliptic or obovate. **INFL** 3–9-fld; peduncle bract just below umbel, gen 3-parted. **FL**: calyx 5–6 mm, lobes ± = or < tube; corolla 10–16 mm, banner yellow, wings pink-purple, fading white, claws exserted from calyx tube. **FR** dehiscent, 2–3 cm, 2–3 mm wide, oblong. **SEEDS** few. In water or springy areas, shores, meadows, roadside ditches; < 650 m. NCo, NCoRO, n CCo, SnFrB, n SCoRO. ❀IRR,SUN:**15–17**.

L. grandiflorus (Benth.) E. Greene Per, hairy or glabrous. **ST** decumbent to erect. **LF**: stipules gland-like, black, conic; lflets 7–9, gen 1–2 cm, elliptic or obovate, length 1.5–3 × width; axis (incl petiole) 10–35 mm. **INFL** 3–9-fld; peduncle gen 1–6 cm, bract near tip. **FL**: calyx 6–9 mm, lobes ± = or < tube; corolla 15–25 mm, greenish white or yellow, wings ± = or > keel; stigma puberulent. **FR** dehiscent, 2.5–6 cm, oblong, gen straight. **SEEDS** several. $2n$= 14. Gen dry, open, disturbed sites, chaparral to yellow-pine forest, sometimes moist river bottoms; 300–1800 m. NCo, NCoR, CaR, SN, CW, SW (exc ChI); Baja CA. Vars. are geog separated.

var. ***grandiflorus*** (p. 621) Pl finely hairy. **LF**: lflets gen 1– 1.5 cm. Gen dry, open, disturbed sites, chaparral to yellow-pine forest; 300–1500 m. Range of sp., (exc CaR, n&c SN). [var. *mutablis* Ottley] Hairiness variable. ❀DRN,DRY,SUN:7,8,14-16;DFCLT.

var. ***macranthus*** (E. Greene) Isely Pl ± glabrous, finely strigose, or puberulent. **LF**: lflets gen 1.6–2.2 cm. Yellow-pine forests, moist river bottoms; 900–1800 m. CaR, n&c SN.

L. hamatus E. Greene Ann, glabrous or puberulent. **ST** prostrate to ascending. **LF** subpinnate or palmate; stipules gland-like or 0; lflets 4–7, opposite to alternate, 4–10 mm, obovate or elliptic; axis sometimes flat, ± blade-like. **INFL** 2–5-fld; peduncle < 5 mm or 0; bract 0. **FL**: calyx 2–2.5 mm, lobes < tube; corolla 3–5 mm, keel > other petals. **FR** indehiscent, 1–1.5 cm, linear, bent, tapered, exserted; beak long, hooked. **SEEDS** 2. Coastal scrub, desert canyons, washes, disturbed areas; 0–600 m. s SCoR, SCo, s ChI, PR; Baja CA.

L. haydonii (Orc.) E. Greene HAYDON'S LOTUS Per, finely strigose. **ST** ascending or sprawling, bushy, sparsely leafy, rush-like. **LF** subpalmate, tiny, deciduous; stipules gland-like or 0; lflets 3, 2–5 mm, elliptic. **INFL** 1–2-fld; peduncle < 3 mm or 0, bract 0. **FL**: calyx 2.5–3 mm, lobes < tube; corolla 4–5 mm, keel > other petals. **FR** indehiscent, exserted, ascending or reflexed, 6–9 mm, oblong, curved, beaked. **SEEDS** 1–2. RARE. Creosote-bush scrub to pinyon/juniper woodland; 600–1200 m. se PR, sw DSon (esp San Diego Co.); Baja CA. Local, little known.

L. heermannii (Durand & Hilg.) E. Greene Per (or fl 1st year and appearing ann). **ST** prostrate, often mat-forming; hairs spreading or obliquely directed, often straight and stiff, < 0.5 mm, esp near tip. **LF** subpalmate; stipules gland-like; lflets 4–6, irregularly arrayed, 4–16 mm, ovate or obovate; axis sometimes flat, ± blade-like. **INFL** 3–8-fld; peduncle gen < 5 mm, bracted. **FL**: calyx 2–4 mm, lobes < tube; corolla 4–7 mm, dark-tipped, wings > other petals. **FR** indehiscent, narrowly oblong, gen curved, tapered to long beak. **SEEDS** 1–2. Washes, riverbanks, coastal scrub, chaparral; < 2000 m. NCo, NCoRO, CCo, SCoRO, SCo, SnBr, PR, DSon. Vars. based esp on habitat, geography.

var. ***heermannii*** **FL**: corolla gen 4–5 mm; ovary gen strigose. Washes, riverbanks, chaparral; < 2000 m. SCo, SnBr, PR, DSon.

var. ***orbicularis*** (A. Gray) Isely (p. 621) **FL**: corolla gen 5–6 mm; ovary hairs soft, spreading. Coastal scrub, chaparral; < 250 m. NCo, NCoRO, CCo, SCoRO. [var. *eriophorus* (E. Greene) Ottley] ❀DRN,SUN:**14–24**;GRCVR.

L. humistratus E. Greene (p. 621) Ann, often ± fleshy, hairy. **ST** mat-forming to ascending. **LF** subpinnate or palmate; stipules gland-like; lflets gen 4, ± alternate, 4–12 mm, elliptic or obovate; axis flat, ± blade-like. **INFL** axillary, 1-fld, ± sessile. **FL**: calyx 3–6 mm, lobes 1–2 × tube; corolla 5–9 mm, yellow, wings ± = keel. **FR** dehiscent, ascending, 6–12 mm, gen 3–4 mm wide, oblong. **SEEDS** few. $2n$=12. Abundant. Grassland, oak and pine woodland, desert flats and mtns, roadsides; < 1700 m. CA (exc GB); to sw UT, w NM, Mex. ❀DRN,DRY:**7–10**,11,12,**14–24**;STBL.

L. incanus (Torrey) E. Greene Per, silky or canescent, gray or silvery. **STS** erect, gen clustered. **LF**: stipules scarious; lflets 7–13, 0.8–1.6 cm, elliptic to obovate. **INFL** 3–8-fld; peduncle bract separated from umbel or not. **FL**: calyx 5.5–7.5 mm, lobes < tube; corolla 12–15 mm, 2-colored (red and white), claws exserted from calyx tube. **FR** dehiscent, 1.5–3.5 cm, linear, glabrous. Dry slopes, open pine forests; 800–1700 m. n&c SN. [*L. neo-incanus* Munz] ❀TRY.

L. junceus (Benth.) E. Greene Per (or fl 1st year and appearing ann), finely strigose. **ST** prostrate or ascending. **LF** small; stipules gland-like; lflets 3–5, 5–10 mm, oblanceolate to obovate; axis sometimes flat, blade-like. **INFL** 2–8-fld; peduncle 1–25 mm; bract present or 0. **FL**: calyx 3–5 mm, lobes < tube, triangular, straight; corolla 6–8 mm, yellow. **FR** indehiscent, oblong-tapered; beak recurved. **SEEDS** 1–2. Coastal sand, chaparral, disturbed areas, sometimes on serpentine; < 500 m. NCo, NCoRO, CCo, SCoRO.

var. ***biolettii*** (E. Greene) Ottley **ST** gen decumbent, often wiry. **INFL**: peduncle 8–25 mm. **FR**: body well exserted. Coastal sand, chaparral, disturbed areas; < 500 m. Range of sp. ❀DRN, SUN:14,**15–17**;STBL.

var. ***junceus*** **ST** decumbent or ascending, gen stout. **INFL**: peduncle 1–5 mm. **FR**: body slightly exserted. Chaparral, sometimes on serpentine; < 500 m. CCo, SCoRO.

L. micranthus Benth. (p. 621) Ann, glabrous or sparsely strigose. **ST** spreading to erect. **LF**: stipules gland-like; lflets 3–5, ± alternate, 4–12 mm, elliptic or obovate; axis sometimes flat, blade-like. **INFL** axillary, 1-fld; peduncle bracted. **FL**: calyx 1.5–2.5 mm, lobes ± = tube; corolla 5–6 mm, pink or salmon, not opening, quickly fading, wings ± = keel. **FR** dehiscent, 1.5–2.5 cm, oblong, wavy-margined, glabrous. **SEEDS** few–several. Abundant. Coastal bluffs to oak/pine or fir woodland, open or disturbed areas; < 1300 m. NCo, NCoR, n&c SN, ScV, CCo, SCoR, SCo, n ChI, PR; to B.C.

L. nevadensis (S. Watson) E. Greene Per, conspicuously hairy. **STS** mat-forming or ascending. **LF** subpinnate or palmate; stipules gland-like; lflets 3–5, irregularly arrayed, 4–12 mm, obovate to oblong, gen green. **INFL** 3–12-fld; peduncle gen bracted. **FL**: calyx 4–9 mm, lobes < tube; corolla 5–10 mm. **FR** indehiscent; body exserted, tapered-oblong, bent; beak slender, recurved. **SEEDS** 1–2. $2n$=14. Oak, yellow-pine, lodgepole, and fir forests, open bracken meadows; 850–2750 m. KR, NCoR, CaR, SN, CCo, SCo, TR, ne PR, DMtns; to B.C., ID, NV. Forms intermediates with *L. argophyllus, L. scoparius*. Many named variants.

var. ***davidsonii*** (E. Greene) Isely **LF**: axis (incl petiole) gen 2–5 mm. **INFL** gen 3–5-fld. **FL** 6–7 mm, blocky; banner upcurved ± 90°, gen drying dark. **FR** gen 2–2.2 mm wide. Open places, oak to pine forests; 1200–2750 m. TR, ne PR. [*L. davidsonii* E. Greene] Merges with var. *nevadensis* in adjacent areas.

var. ***nevadensis*** (p. 621) **LF**: axis (incl petiole) gen 5–10 mm. **INFL** gen 5–12-fld. **FL** 5–10 mm, not blocky; banner upcurved 30–90°, gen drying orange-yellow. **FR** gen 1.8–2 mm wide. Pine and fir forests, bracken meadows, dry slopes; 850–2750 m. KR, NCoR, CaR, SN, CCo, SCo, w PR, DMtns; to B.C., ID, NV. [*L. douglasii* E. Greene] NW pls gen have larger fls (corolla 8–10 mm). ❀**1–3,7**;STBL.

L. nuttallianus E. Greene (p. 621) NUTTALL'S LOTUS Ann, glabrous or strigose. **ST** prostrate or ascending. **LVS** pinnate or palmate, well spaced; stipules gland-like; lflets 3–6, 4–10 mm, oblanceolate to obovate. **INFL** 3–8-fld; peduncle 1–3 cm, bracted. **FL**: calyx 2–3 mm, lobes << tube; corolla 5–7 mm. **FR** indehiscent, widely spreading or reflexed, exserted 1–1.5 cm, linear, curved; beak tapered. **SEEDS** 2. RARE. Beaches, coastal scrub, urban weedy areas; < 30 m. s SCo (w San Diego Co.); Baja CA.

L. oblongifolius (Benth.) E. Greene Per, glabrous or hairy. **ST** sprawling or ascending. **LF**: stipules scarious, fragile; lflets 3–11, ± opposite, 1–2.5 cm, gen elliptic to oblong. **INFL** 2–6-fld; peduncle bract 0 or just below umbel, simple or divided. **FL**: calyx 4–6 mm, lobes < tube; corolla 8–13 mm, gen whitish yellow, claws incl to slightly exserted from calyx tube. **FR** dehiscent, 2.5–5 cm, 1.5–2 mm wide, oblong. **SEEDS** few. Open, moist forests, river bottoms, marshy meadows, open pine woodland; 200–2600 m. CA-FP, DMoj; s OR, w&s NV, Mex.

var. ***cupreus*** (E. Greene) Ottley COPPER-FLOWERED BIRD'S-FOOT TREFOIL **LF**: lflets 3–7, glabrous. **FL**: corolla 8–9 mm. UNCOMMON. Meadows, open pine woodland; 2400–2600 m. s SNH (Tulare Co.). [*L. c.* E. Greene]

var. ***oblongifolius*** **LF**: lflets 7–11, hairy or glabrous. **FL**: corolla 9–13 mm. Locally common. Open, moist forests, river bottoms, marshy meadows; 200–2400 m. Range of sp. [var. *nevadensis* (A. Gray) Munz] ❀IRR:1–3,**7**,8–11,**14–24**;STBL.

L. pinnatus Hook. (p. 621) Per, ± glabrous. **ST** solitary, decumbent or ascending; base often spongy-thickened. **LF**: stipules scarious; lflets 5–7, ± opposite, 1–2.5 cm, elliptic or obovate. **INFL** 4–10-fld; peduncle bract 0 or just below umbel, gen simple. **FL**: calyx 5–7 mm, lobes < tube; corolla 10–13 mm, banner bent or recurved 90–180°, yellow, other petals white, claws slightly exserted from calyx tube. **FR** dehiscent, 3–5 cm, 1.5–2.5 mm wide, linear. **SEEDS** few. Wet meadows, bogs, ditches; 600–1700 m. KR, NCoR, CaR, n SN, CCo; to WA, ID.

L. procumbens (E. Greene) E. Greene Per or stiff subshrub, puberulent or strigose, often gray. **ST** decumbent to erect, gen much-branched, < 1 m. **LF** subpalmate; stipules gland-like; lflets gen 3, 4–12 mm, oblanceolate to obovate. **INFL** 1–3-fld; peduncle 0 or < 3 mm, bract 0. **FL**: calyx 2–6 mm, strigose, lobes << or ± = tube; corolla 6–12 mm, yellow or with red, wings > keel. **FR** indehiscent, exserted, pendent, 1–1.5 cm, initially curved, then gen straight. **SEEDS** 2–3. Chaparral to pine forest, open slopes, ridges, flats, roadsides; < 2300 m. s SN, ScV, SCoR, TR, PR, DMoj.

var. ***jepsonii*** (Ottley) Ottley **FL**: calyx 4–6 mm, lobes ± = tube; corolla 9–12 mm. Local. Open slopes, ridges; 1800–2000 m. s SN (Tulare, Kern cos.).

var. ***procumbens*** (p. 629) **FL**: calyx 2–3 mm, lobes << tube; corolla 6–8 mm. Chaparral to Jeffrey-pine forest, sandy flats and slopes, roadsides; < 2300 m. ScV, SCoR, TR, PR, DMoj. Stiff subshrubs are found esp in Los Angeles Co.

L. purshianus (Benth.) Clements & E.G. Clements var. ***purshianus*** (p. 629) Ann, gen hairy. **ST** prostrate to erect, simple or openly branched, 0.5–6 dm. **LF** pinnate or ± simple; stipules gland-like; lflets gen 3, gen 1–2 cm, lanceolate to elliptic. **INFL** 1-fld; peduncle bracts simple. **FL**: calyx 3–6.5 mm, lobes >> tube; corolla 5–9 mm, yellow to pink, wings ± = keel. **FR** dehiscent, widely spreading or pendent, 1.5–3 cm, oblong. **SEEDS** few. $2n$=14. Coast, chaparral, mtn forests, water courses, roadsides, other weedy areas; 0–2400 m. CA (exc DSon); to Can, c US, Mex. [var. *glaber* (Nutt.) Munz] Many races and ecological forms. ❀**1–11,14–24**; STBL & forage.

L. rigidus (Benth.) E. Greene (p. 629) Per, shrub-like, finely strigose. **STS** ascending, clustered, branched, 0.5–1.5 m. **LVS** irregularly pinnate to ± palmate, well spaced; stipules gland-like or 0; lflets 3–4, 0.5–1.5 cm, oblanceolate to obovate; axis (incl petiole) 1–8 mm. **INFL** 1–3-fld; peduncle 3–6 cm; bract near top or 0. **FL**: calyx 5–8 mm, lobes < tube; corolla 12–22 mm, wings > keel; stigma ± glabrous. **FR** slowly dehiscent, spreading or erect, 2–4 cm, oblong, ± straight, much exserted, gen glabrous. **SEEDS** several–many. $2n$=14. Chaparral, desert flats, washes, foothills; < 1550 m. PR, D; to s NV, se UT, AZ, Baja CA. ❀TRY.

L. rubriflorus Sharsm. RED-FLOWERED BIRD'S-FOOT TREFOIL Ann, very small, gen densely hairy. **LF** irregularly pinnate; stipules gland-like; lflets 4, ± alternate, 3–10 mm, lanceolate; axis flat, ± blade-like. **INFL** 1-fld; peduncle ± 0. **FL**: calyx 5.5–6.5 mm, lobes ± 2 × tube; corolla ± = calyx, pink-red, wings ± = keel. **FR** dehiscent, 8–9 mm, widely oblong, hairy. UNCOMMON. Oak woodland, grassland; ± 200 m. NCoRI (Colusa Co.), SnFrB (Stanislaus Co.).

L. salsuginosus E. Greene Ann, often fleshy, very small to robust, glabrous or strigose. **STS** clustered, prostrate or ascending. **LF** irregularly pinnate; stipules gen not apparent; lflets 3–7, ± alternate, 0.5–1.5 cm, obovate or ± round, terminal largest; axis flat, ± blade-like. **INFL** 2–4-fld (fr often only 1); peduncle bract lf- or lflet-like or perhaps 0. **FL**: calyx 2–4 mm, lobes < tube; corolla 3.5–10 mm, wings ± = or < keel. **FR** dehiscent, gen 1.5–3 cm, narrowly oblong, often curved; beak short, hooked. **SEEDS** few–many. $2n$=14. Coastal scrub, foothill woodlands, washes, talus, deserts incl mtns; < 1850 m. CW, SCo, TR, PR, D; NV, AZ, Mex.

var. ***brevivexillus*** Ottley **FL**: calyx ± 2.5 mm; corolla 3.5–5 mm, keel > other petals. **FR**: 1–1.5 cm, becoming narrowed between seeds. Deserts, incl mtns; < 1850 m. D; NV, AZ, Mex. [ssp. *b.* (Ottley) Munz]

var. ***salsuginosus*** **FL**: calyx 3–4 mm; corolla 6–10 mm, keel ± = other petals. **FR**: 1.5–3 cm, not narrowed between seeds. Coastal scrub, foothill woodlands, washes, talus; < 1200 m. CW, SCo, TR, PR, DSon; Mex. [ssp. *s.*]

L. scoparius (Nutt.) Ottley CALIFORNIA BROOM Per, often shrubby, glabrous or finely strigose. **STS** clustered, gen ascending to erect (sometimes prostrate and mat-forming), bushy-branched, 0.5–2 m, greenish. **LVS** ± pinnate, well spaced, often deciduous;

2 mm
2 cm
2 cm
L. torreyi
fruit
2 cm
2 mm
calyx
Lathyrus vestitus var. vestitus
5 mm
fruit
1 cm
2 cm
2 mm
Lens culinaris
5 mm
L. a. var. fremontii
1 cm
5 mm
Lotus argophyllus var. argophyllus
fruit
2 cm
fruit
Lotus corniculatus
L. crassifolius var. crassifolius
1 cm
5 mm
L. denticulatus
5 mm
fruit
Lotus dendroideus var. dendroideus
1 cm
2 cm
1 cm
fruit
1 cm
Lotus formosissimus
L. grandiflorus var. grandiflorus
2 cm
2 mm
fruit
L. heermannii var. orbicularis
2 cm
fruit
5 mm
2 mm
Lotus humistratus
2 mm
5 mm
fruit
5 mm
fruit
L. micranthus
2 cm
Lotus nevadensis var. nevadensis
1 cm
fruit
5 mm
L. pinnatus
5 mm
fruit
Lotus nuttallianus

stipules gland-like or 0; lflets 3–6 (gen 3 on upper st), 6–15 mm, elliptic. **INFL** 2–7-fld; peduncle gen 0. **FL**: calyx 2.5–5 mm, glabrous, lobes < tube, not curved outward; corolla 7–11 mm. **FR** indehiscent, widely spreading or pendent, 1–1.5 cm, curved, long-beaked. **SEEDS** gen 2. 2*n*=14. Chaparral, roadsides, coastal sand, desert slopes, flats, washes; < 1500 m. NCo, NCoR, n SNF, CCo, SCo, TR, PR, DSon; AZ, Mex. May hybridize with *L. junceus, L. benthamii*.

var. ***brevialatus*** Ottley **FL** gen 8–9 mm; keel > wings. Desert slopes, flats, washes; < 1400 m. SCo, TR, PR, DSon; AZ, Mex. [ssp. *b.* (Ottley) Munz] Intergrades with var. *scoparius* in Los Angeles Co.

var. ***scoparius*** (p. 629) **FL** 7–12 mm; keel ± = wings. Chaparral, roadsides, coastal sands; < 1500 m. Common. NCo, NCoR, n SNF, CCo, SCo, WTR, PR; Baja CA. [ssp. *s.*] Gen erect, may be trailing in shade or mat-forming on beaches. Island forms here referred to *L. dendroideus*. ❀DRN,SUN:**7–9,14,18–24**&DRY:**15–17**;STBL.

L. stipularis (Benth.) E. Greene Per, gen soft-hairy, sometimes glandular. **ST** gen erect. **LF**: stipules conspicuous or reduced; lflets 9–15, 5–20 mm, oblong to ovate. **INFL** 4–9-fld; peduncle bract separated from umbel. **FL**: calyx 5–6.5 mm, lobes < tube; corolla 10–12 mm, pink to red-purple. **FR** dehiscent, 2–2.5 cm, 3–4 mm wide, widely oblong. **SEEDS** many. Open pine forests, stream beds, ditches, thickets, chaparral, logged areas; 200–1200 m. NW, CaR, n&c SN, CCo, SCoR. Vars. geographically distinct.

var. ***ottleyi*** Isely **ST** wiry. **LF**: stipules wide, clasping st. Open pine forests, stream beds, ditches; 600–1200 m. KR, CaR, n&c SN.

var. ***stipularis*** **ST** often fleshy. **LF**: stipules narrow or wide, often slightly clasping st. Thickets, chaparral, logged areas; 200–1000 m. NCo, NCoR, CCo, SCoR. Variable, sometimes glandular; some pls intermediate to *L. aboriginus*.

L. strigosus (Nutt.) E. Greene (p. 629) Ann, often fleshy, hairy or not. **ST** prostrate, gen branched from base. **LF**: stipules gland-like; lflets 4–9, gen alternate, 3–10 mm, oblanceolate to obovate; axis ± flat, ± blade-like. **INFL** 1–2-fld; peduncle gen bracted. **FL**: calyx 3–5.5 mm, lobes < tube; corolla opening or not, 5–10 mm, yellow, turning orange or reddish, wings gen > keel; stigma puberulent. **FR** dehiscent, 1–3.5 cm, gen ± curved only at or near tip. **SEEDS** several. 2*n*=14. Coastal scrub, chaparral, foothills, deserts, roadsides, other disturbed areas; < 2300 m. GV, CW, SW, D; AZ, Mex. [var. *hirtellus* (E. Green) Ottley; *L. tomentellus* E. Greene] Several variants often recognized (see Isely, pp. 193–198); pls in CA-FP gen ± strigose, with narrow lflets; pls in D fleshy, gen canescent, with wide lflets. Conspicuous in spring.

L. uliginosus Schk. (p. 629) Per, often fleshy, rhizomed; hairs spreading or 0. **ST** ascending, gen hollow. **LF**: stipules gland-like or not apparent; lflets 5, 3 ± palmately arranged near axis tip, lower 2 stipular in position. **INFL** 8–12-fld; peduncle gen 8–14 cm, with 3-parted bract at tip. **FL**: calyx 7–8 mm, lobes < tube, often outcurved in bud; corolla 11–12 mm, yellow. **FR** dehiscent, 2–3 cm, narrowly oblong. **SEEDS** few–several. 2*n*=24. Wet fields, roadsides, ditches; < 800 m. NCo, NCoRO, SnFrB, probably elsewhere; to WA, sporadic in e US; native to Eur.

L. wrangelianus Fischer & C. Meyer Ann, ± strigose or hairs soft, spreading. **ST** gen prostrate, branched at base. **LF**: stipules gland-like or ± 0; lflets gen 4, ± alternate, 4–15 mm, elliptic or obovate; axis flat, ± blade-like. **INFL** 1-fld; peduncle ± 0, slightly longer in fr. **FL** sometimes not opening; calyx 2.5–5 mm, lobes ± 0.8–1.2 × tube; corolla 5–9 mm, yellow, wings ± = keel. **FR** dehiscent, 10–18 mm, 2.2–3 mm wide, oblong. **SEEDS** few–several. 2*n*=12. Abundant. Coastal bluffs, chaparral, disturbed areas; 0–1500 m. CA-FP, probably naturalized in MP, DSon through agriculture. [sometimes mistaken for *L. subpinnatus* Lagasca]

L. yollabolliensis Munz (p. 629) YOLLA BOLLY MOUNTAINS BIRD'S-FOOT TREFOIL Per, very small, glabrous. **STS** decumbent, clustered. **LF** gen pinnate; stipules inconspicuous, scarious; lflets 3–5, 3–10 mm, oblanceolate to obovate. **INFL** 1–3-fld; peduncle bracted. **FL**: calyx 2.5–4 mm, lobes < tube; corolla 8–10 mm, white or yellowish, banner margin fringed. **FR** dehiscent, 1.5–2.5 cm, narrowly oblong, glabrous. **SEEDS** few. UNCOMMON. Open, dry slopes, fir forests; 1700–2100 m. n NCoRH (Yolla Bolly Mtns, S. Fork Mtn, Trinity and Humboldt cos.).

LUPINUS LUPINE

Rhonda Riggins (ann) & Teresa Sholars (per to shrubs)

Ann to shrubs; cotyledons gen petioled, withering early. **ST** gen erect. **LVS** palmately compound in CA, gen ± cauline; stipules fused to petiole; lflets 3–17, gen ± oblanceolate, entire. **INFL**: raceme; fls scattered or whorled; bracts persistent or not. **FL**: calyx 2-lipped, lobes entire or toothed, gen appendaged between lobes; banner centrally grooved, sides reflexed, wing tips slightly fused, keel gen pointed; stamens 10, filaments fused, 5 long with short anthers, 5 short with long anthers; style shrubby. **FR** dehiscent, gen oblong. **SEEDS** 2–12, gen smooth. ± 200 spp.: esp w N.Am, w S.Am to e US, also trop S.Am, Medit. (Latin: wolf, from mistaken idea that pls rob soil of nutrients) Some cult for fodder, green manure, edible seed, orn; some naturalized from CA in e N.Am, S.Am, Australia, s Afr; some (e.g. *L. arboreus, L. latifolius, L. leucophyllus*) have alkaloids (esp in seeds, frs, young herbage) TOXIC to livestock (esp sheep). [Barneby 1989 Intermountain Flora 3(B):237–267] Infl length does not incl peduncle. ❀Many *Lupine* taxa need seed pre-treatment (scarification, stratification, innoculation) for successful germination.

1. Ann, rarely living > 1 growing season
  2. Cotyledons sessile, persistent at pl base, disk- or cup-like, or leaving a circular scar; fr gen ovoid; seeds gen 2, gen tubercled or wrinkled (fr oblong; seeds 2–6 in *L. odoratus*)
    3. Fls distinctly whorled; infl bracts reflexed; upper keel margins ciliate near claw
      4. St not noticeably hollow; lower keel margins as densely ciliate as upper; upper surface of lflets gen hairy; seed dark brown, tubercled .... ***L. luteolus***
      4′ St hollow at least below; lower keel margins less ciliate than upper or glabrous; upper surface of lflets glabrous; seed gen mottled, wrinkled or smooth .... ***L. microcarpus***
        5. Wings widely elliptic, not withering, becoming translucent, upper (and gen lower) margins ciliate near claw; lower keel margins ciliate near claw; calyx appendages 1–2 mm .... var. ***horizontalis***
        5′ Wings linear to oblanceolate, withering, not becoming translucent, upper margins (and rarely lower) gen ciliate near claw; lower keel margins sometimes sparsely ciliate near claw; calyx appendages gen 0
          6. Calyx and bracts ± sparsely short-appressed- to long-spreading-hairy; fr ± spreading, gen on 1 side of infl axis .... var. ***densiflorus***
          6′ Calyx and bracts long-shaggy-hairy; fr gen ± erect, gen not on 1 side of infl axis .... var. ***microcarpus***
    3′ Fls not distinctly whorled; infl bracts not reflexed; upper keel margins glabrous
      7. Lvs basal; herbage sparsely hairy when young, gen glabrous at fl; pedicel 3–5 mm .... ***L. odoratus***

7′ Lvs cauline, often crowded near base; herbage remaining sparsely hairy to canescent; pedicel gen < 3 mm
8. Herbage canescent; upper calyx lip ± = lower; upper suture of fr gen wavy and densely stiff-ciliate; fr sides with short inflated hairs that become scale-like when dry ... ***L. shockleyi***
8′ Herbage sparsely to densely ± long-hairy; upper calyx lip < lower; upper suture of fr not obviously wavy and densely stiff-ciliate; fr sides with long, uninflated hairs
9. Peduncle 0–1 cm; infl < lvs; fr oblong, narrowed between seeds ... ***L. pusillus*** var. ***intermontanus***
9′ Peduncle 2–5 cm; infl > lvs; fr ovoid, not narrowed between seeds
10. Infl 1–2.5 cm; pedicel 0.5–1.5 mm; calyx appendages present; seed smooth ... ***L. brevicaulis***
10′ Infl 3–8 cm; pedicel 2–3 mm; calyx appendages 0; seed wrinkled ... ***L. flavoculatus***
2′ Cotyledons petioled, gen withering, gen deciduous; fr oblong; seeds gen > 2, smooth
11. Lower (and often upper) margins of keel ciliate near claw, both margins glabrous near tip
12. Fls distinctly whorled — petals gen blue-purple (white, lavender, or pink) ... [2]***L. succulentus***
12′ Fls not whorled (if appearing ± whorled, pedicel bases clearly scattered, not whorled)
13. Infl < peduncle — banner yellow, wings gen pink (whitish), keel white ... ***L. stiversii***
13′ Infl > peduncle
14. Petals gold-yellow or white; pedicel recurving; rare — c SNF ... ***L. citrinus***
15. Petals gold-yellow; Fresno, Madera cos. ... var. ***citrinus***
15′ Petals white; Mariposa Co. ... var. ***deflexus***
14′ Petals blue or dark pink to magenta; pedicel not recurving; common, at least locally
16. Herbage stiff-spreading-stinging-hairy; lflets 10–20 mm wide ... ***L. hirsutissimus***
16′ Herbage not stiff-spreading-stinging-hairy; lflets gen < 10 mm wide
17. Petiole flat, lflet-like; keel tip stout, blunt, upper margins densely ciliate near claw ... ***L. truncatus***
17′ Petiole not flat, not lflet-like; keel tip slender, pointed, upper margins gen glabrous
18. Pedicel 5–9 mm; fl bract 10–15 mm, >> fl bud ... ***L. benthamii***
18′ Pedicel < 5 mm; fl bract 3–8 mm, < to slightly > fl bud
19. Lflets 5–10 mm wide, upper surface glabrous; petals dark pink to magenta ... ***L. arizonicus***
19′ Lflets 2–4 mm wide, upper surface hairy at least near margins; petals gen blue (pinkish) ... ***L. sparsiflorus***
11′ Lower and upper keel margins glabrous throughout or upper ciliate only near tip
20. Fls not whorled; infl bracts gen persistent; upper keel margins glabrous
21. Peduncles decumbent; rare — dunes, s CCo ... ***L. nipomensis***
21′ Central peduncle erect, lateral often decumbent; common ... ***L. concinnus***
20′ Fls whorled at some or all nodes; infl bracts deciduous; upper keel margins gen ciliate near tip
22. Banner longer than wide; pedicel gen < 3 mm
23. Fr 6–9 mm wide; keel glabrous; uncommon ... ***L. pachylobus***
23′ Fr gen 3–6 mm wide; upper keel margins gen ciliate near tip; abundant ... ***L. bicolor***
22′ Banner as wide as or wider than long; pedicel gen > 3 mm
24. Upper keel margins with (sometimes inconspicuous) tooth near middle ... ***L. affinis***
24′ Upper edges of keel without tooth
25. Fr 4–7 mm wide; abundant ... ***L. nanus***
25′ Fr 8–10 mm wide; rare
26. Pedicel 4–5 mm; herbage sparsely hairy; ChI (San Clemente Island) ... ***L. guadalupensis***
26′ Pedicel 6–8 mm; herbage densely hairy; c SNF ... ***L. spectabilis***
1′ Per to shrub
27. Upper keel margin ciliate from claw to middle (glabrous near tip)
28. Pl fleshy (in fact ann); infl 9–15 cm; open or disturbed areas; fr 35–50 mm ... [2]***L. succulentus***
28′ Pl not fleshy; infl 16–60 cm; open to shady, moist areas; fr 20–45 mm ... ***L. latifolius***
29. Fl 13–18 mm
30. St densely hairy; SnFrB ... var. ***dudleyi***
30′ St subglabrous to sparsely strigose; s SN, SW ... var. ***parishii***
29′ Fl 8–14 mm
31. Fl 8–10 mm
32. St glabrous to strigose; MP ... var. ***barbatus***
32′ St strigose; KR, CaR, MP ... var. ***viridifolius***
31′ Fl 10–14 mm
33. Wings covering most of keel; pl 4–24 dm; SN ... var. ***columbianus***
33′ Wings not covering most of keel; pl 3–12 dm; NW, CW, SW ... var. ***latifolius***
27′ Upper keel margin glabrous or ciliate middle to tip
34. Calyx spur 1–3 mm
35. Wings with dense patch of hair on outside near tip; lf green, strigose ... ***L. arbustus***
35′ Wings glabrous; lf silver-silky ... ***L. argenteus*** var. ***heteranthus***
34′ Calyx bulge or spur 0–1 mm
36. Banner back gen hairy (best seen in buds)
37. Upper keel margin ± glabrous
38. Shrub — coastal strand, dunes ... ***L. chamissonis***
38′ Per
39. Lf green
40. Petals pale to orange-yellow ... ***L. angustiflorus***
40′ Petals white to purple — n SNH ... ***L. apertus***

39′ Lf silver-hairy
41. Petals gen yellow; n SNH (Plumas Co.); st long-hairy .... ***L. dalesiae***
41′ Petals cream-colored, pale yellow, lavender, or blue; s SNH, TR, SNE; st long-hairy or not
42. Petals lavender to blue; st not long-hairy; 1500–3000 m; s SNH, TR .... [2]***L. elatus***
42′ Petals cream-colored to pale yellow; st long-hairy; 2500–4000 m; s SNH, SNE (n Inyo, Tulare cos.) .... ***L. padre-crowleyi***
37′ Upper keel margin ciliate
43. Subshrub or shrub
44. Pl < 2 dm, matted; fl 6–11 mm; 2000–3500 m .... ***L. breweri*** var. ***grandiflorus***
44′ Pl gen > 2 dm, erect; fl gen > 10 mm; < 3000 m
45. Keel gen lobed near base; fl 9–18 mm; s SNH, Teh, SW, SNE, D .... ***L. excubitus***
46. Subshrub, gen < 7 dm
47. Infl 14–40 cm; Teh, SnGb, SnBr .... [2]var. ***austromontanus***
47′ Infl 3–14 cm; SnGb, SnBr, DSon
48. Petiole < 12 cm; SnGb, SnBr .... var. ***johnstonii***
48′ Petiole gen > 12 cm; DSon .... var. ***medius***
46′ Shrub, gen > 7 dm
49. Fl 9–13 mm; lf silvery-hairy; DMoj, adjacent s SNH, SNE .... var. ***excubitus***
49′ Fl 14–18 mm; lf greenish hairy; PR, SnGb, SnBr .... var. ***hallii***
45′ Keel gen unlobed near base; fl 9–16 mm; CA-FP .... [2]***L. albifrons***
50. Subshrub gen < 5 dm (base, some lower branches woody)
51. Lf woolly to shaggy; SCoRO .... [2]var. ***abramsii***
51′ Lf appressed-hairy, not woolly or shaggy; NW, SNF, CW .... [2]var. ***collinus***
50′ Shrub gen > 5 dm, woody more nearly throughout
52. Lf woolly to shaggy; SCoRO .... [2]var. ***abramsii***
52′ Lf hairy but not woolly or shaggy; widespread
53. Infl bracts 4–8 mm; NCoR, SNF, CW, SCo .... [2]var. ***albifrons***
53′ Infl bracts 10–24 mm; CCo, SnFrB, n ChI .... [2]var. ***douglasii***
43′ Per
54. Lflet 10–30 mm at widest point
55. Petals yellow — SnGb .... [2]***L. peirsonii***
55′ Petals ± purple or blue (rarely straw-colored)
56. Peduncle 13–20 cm; lf long-spreading-hairy; petals light blue, pink, or straw-colored; SCoRO .... [2]***L. cervinus***
56′ Peduncle 8–15 cm; lf short-appressed-hairy; petals ± purple; s NCoRI .... [2]***L. sericatus***
54′ Lflet gen < 10 mm at widest point
57. Pl < 1 dm — uncommon
58. Fls in many crowded whorls; banner patch cream-colored or white; SNE .... [3]***L. duranii***
58′ Fls in few separated whorls; banner patch yellow; KR .... [2]***L. lapidicola***
57′ Pl > 1 dm
59. Calyx gen bulged or spurred < 1 mm; GB, CaR, SNH .... ***L. argenteus*** (part)
60. Lvs basal and cauline; petioles (5)7–15 cm — c SNH, SNE .... var. ***montigenus***
60′ Lvs cauline; petioles 1–10 cm
61. St hairs appressed; lf gen appearing green; fl (8)10–12 mm .... var. ***argenteus***
61′ St hairs spreading; lf gray to silver; fl 8–10 mm .... var. ***palmeri***
59′ Calyx not bulged or spurred
62. Lf hairs appressed, sometimes sparse
63. Bracts deciduous; lflets 20–50 mm, silvery-silky; NW, CaR, n SNH .... ***L. obtusilobus***
63′ Bracts persistent; lflets 30–130 mm, green, hairy; SNE .... ***L. pratensis*** var. ***eriostachyus***
62′ Lf hairs ± spreading, dense, tomentose to woolly
64. Fl gen 12–15 mm; pl 2–3.5 dm; st prostrate to matted — common; SN .... [2]***L. grayii***
64′ Fl gen 10–13 mm; pl 3–9 dm; st erect
65. Petals lavender to purple (often turning brown); banner back densely hairy; n CA .... ***L. leucophyllus***
65′ Petals bluish to purple; banner back sometimes glabrous; s SCoRO .... [2]***L. ludovicianus***
36′ Banner back ± glabrous
66. Subshrub or shrub
67. Petiole gen < 3 cm; NCo, CCo, esp immediate coast
68. Shrub, gen 5–20 dm; woody sts erect; immediate coast and more inland .... ***L. arboreus***
68′ Subshrub, < 5 dm; woody sts prostrate to decumbent; immediate coast
69. Fl 10–13 mm; petals purple (exc banner patch whitish); n&c NCo .... [2]***L. littoralis***
69′ Fl 11–16 mm; petals white, yellow, rose, or purple (often on 1 pl); NCo, CCo .... [2]***L. variicolor***
67′ Petiole > 3 cm; gen inland
70. Low subshrub; st prostrate to decumbent
71. Keel unlobed near base; fl 10–15 mm; NW, SNF, CW .... [2]***L. albifrons*** var. ***collinus***
71′ Keel lobed near base; fl 14–18 mm; Teh, SnGb, SnBr .... [2]***L. excubitus*** var. ***austromontanus***
70′ Tall shrub; st erect
72. Lflets 30–60 mm; petiole 4–10 cm — SW .... ***L. longifolius***
72′ Lflets 10–45 mm; petiole 1–5 cm .... ***L. albifrons*** (part)
73. Bract 4–8 mm; lflets 10–30 mm; NCoR, SNF, CW, SCo .... [2]var. ***albifrons***

73′ Bracts 10–24 mm; lflets 25–45 mm; CCo, SnFrB, n ChI .... [2]var. ***douglasii***
66′ Per (rarely barely woody)
74. Lf upper surface glabrous to ± puberulent, green
75. Upper keel margin ± ciliate
76. Pl > 3.5 dm; lvs cauline — NCo .... ***L. rivularis***
76′ Pl < 3.5 dm; lvs gen basal or cauline, clustered near base
77. Fl 8–11 mm, bracts 3–4 mm; lf lower surface silky-hairy; KR, CaRH, n SNH .... ***L. onustus***
77′ Fl 12–18 mm, bracts 4–7 mm; lf lower surface long-stiff-hairy; MP .... ***L. saxosus***
75′ Upper keel margin glabrous
78. Petals ± yellow — NW, CaR
79. Infl bract 2–7 mm; pl 4–6 dm; petals bright yellow to orange-yellow; KR, CaR .... [2]***L. croceus***
79′ Infl bract 7–13 mm; pl 6–9 dm; petals pale yellow; NCoRH — uncommon .... ***L. elmeri***
78′ Petals not yellow
80. St slender; lvs cauline; lflets < 40 mm; dry areas; KR .... ***L. tracyi***
80′ St stout; lvs basal and cauline; lflets > 4 cm; wet areas .... ***L. polyphyllus***
81. Lflet upper surface puberulent to minutely rough-hairy (appearing ± glabrous); n CA-FP .... [2]var. ***pallidipes***
81′ Lflet upper surface glabrous; CA-FP, GB
82. Lflets 5–11; SNH, SnGb, SnBr, SnJt, GB .... var. ***burkei***
82′ Lflets 9–17; NW, w CW .... var. ***polyphyllus***
74′ Lf upper surface hairy, greenish gray to silver
83. Infl gen dense, bracts gen persistent (exc *L. constancei*)
84. Infl gen < lvs; pl matted — fls in 2–3 whorls; n W&I .... ***L. lepidus*** var. ***utahensis***
84′ Infl gen > lvs; pl matted or not
85. Cauline lvs 0 or clustered near base (appearing ± basal)
86. Infl bracts deciduous; petals pink — n NCoRH .... [3]***L. constancei***
86′ Infl bracts persistent; petals ± violet to blue .... ***L. lepidus*** (part)
87. Infl ± head-like, 2–8 cm, gen < some lvs; pl < 1 dm .... var. ***lobbii***
87′ Infl elongate, 4.5–11 cm, > lvs; pl 1.2–3.5 dm .... var. ***sellulus***
85′ Cauline lvs gen spread along st
88. Pl ± prostrate, matted; petals pink — n NCoRH .... [3]***L. constancei***
88′ Pl gen erect, gen not matted; petals blue or violet (sometimes pink)
89. Largest lflets > 40 mm
90. Infl open; petals light blue, banner patch yellow; infl bracts 7–15 mm, ± linear; pedicels 2–5 mm; hairs > 1 mm; c&s SNH .... [2]***L. covillei***
90′ Infl dense; petals violet to dark blue, banner patch orange to red; infl bracts 5–10 mm, lanceolate; pedicels 1–3 mm; hairs < 1 mm; c&s SNH, SNE .... ***L. pratensis*** var. ***pratensis***
89′ Largest lflets < 40 mm .... ***L. lepidus*** (part)
91. Infl dense, 5–30 cm, whorls > 7, ± crowded; lvs cauline — meadows, vernally moist areas .... var. ***confertus***
91′ Infl gen open, 2–12 cm, whorls 3–7, ± well spaced; at least some lvs basal
92. Lflets gen 10–30 mm; fl 9–11 mm; 2500–3000 m, s SNH .... var. ***culbertsonii***
92′ Lflets gen 5–15(23) mm; fl gen 7–9(12) mm; 3000–4000 m; c&s SNH, SNE .... var. ***ramosus***
83′ Infl gen ± open, bracts gen ± deciduous (exc *L. covillei*)
93. Upper keel margin gen glabrous
94. Pl < 2 dm; fl 4–11 mm; keel ± straight
95. Stipule 6–11 mm; pedicel 4–5 mm; lvs basal; petiole (2)3–8 cm; fl 8–11 mm; rare .... [3]***L. duranii***
95′ Stipule 2–5 mm; pedicel 1–3(4) mm; lvs clustered near base; petiole 1–5(6) cm; fl 4–11 mm; common .... ***L. breweri*** (part)
96. Fl 6–9 mm; lflet > 6 mm .... var. ***breweri***
96′ Fl 4–7 mm; lflet 3–5 mm .... var. ***bryoides***
94′ Pl > 2 dm; fl gen > 9 mm; keel gen upcurved
97. Stipules lf-like, green — SN .... ***L. fulcratus***
97′ Stipules not lf-like, green to silvery
98. Petals yellow to ± orange
99. Fl 9–12 mm; petals pale yellow; NCoR, SnFrB, SN .... [2]***L. adsurgens***
99′ Fl 12–15 mm; petals bright yellow to orange-yellow; KR, CaR .... [2]***L. croceus***
98′ Petals ± blue, lavender, violet, purple, pink, or ± yellowish white
100. Lvs basal and cauline; moist to wet places .... [2]***L. polyphyllus*** var. ***pallidipes***
100′ Lvs cauline; dry places
101. Banner narrow, wings narrow, not covering keel tip .... ***L. albicaulis***
101′ Banner ovate, wings wide, covering keel tip
102. Lf green, sparsely to densely hairy
103. Fl 9–12 mm; NW, SNH, SNE, WTR, SnBr .... ***L. andersonii***
103′ Fl 13–16 mm; SnGb, SnBr, PR .... ***L. hyacinthinus***
102′ Lf gray-hairy to silvery-silky
104. Pl resprouting below ground, rhizomed; gen valleys .... ***L. formosus***
105. Fl 10–14 mm; st 3–4 mm diam .... var. ***formosus***
105′ Fl 16–18 mm; st 5–7 mm diam .... var. ***robustus***

104′ Pl resprouting above ground, not rhizomed; mtns
106. Petals white (banner patch turning tawny); seeds 7–11 mm — NCoRH, uncommon .... ***L. antoninus***
106′ Petals not white; seeds 4–6 mm
107. Fl 9–14 mm; petals pale yellow to lavender or violet; lflets widest above middle; NCoR, SN, SnFrB .... [2]***L. adsurgens***
107′ Fl 10–14 mm; petals lavender to blue; lflets widest below middle; s SNH, TR ..... [2]***L. elatus***
93′ Upper keel margin ± ciliate
108. Pl ± prostrate to decumbent, gen < 3 dm
109. Petals pink; NCoRH .... [3]***L. constancei***
109′ Petals purple, ± violet, lavender, rose, light blue, yellow, or white
110. Lvs gen basal (if cauline, clustered near base); KR, SN, SNE
111. Infl 10–16 cm; fl 10–16 mm; SN .... [2]***L. grayii***
111′ Infl 2–7 cm; fl 7–12 mm
112. Stipule 6–11 mm; peduncle < 3.5 cm; SNE .... [3]***L. duranii***
112′ Stipule 4–5 mm; peduncle 5–10 cm; KR .... [2]***L. lapidicola***
110′ Lvs cauline; NCo, CCo
113. Lflets 3–5; st weak — rare; s NCo (Sonoma Co.), CCo (Marin, Monterey cos.) ...... ***L. tidestromii***
113′ Lflets 5–9; st not weak
114. Fl 10–13 mm; petals purple exc banner patch whitish; n&c NCo .... [2]***L. littoralis***
114′ Fl 11–16 mm; petals white, yellow, rose, or purple (often on 1 pl); NCo, CCo ...... [2]***L. variicolor***
108′ Pl gen erect, gen > 2 dm
115. Lf densely woolly
116. St hairs < 1 mm, not sharp or stiff; petiole 5–12 cm; fl 10–15 mm; s SCoRO ...... [2]***L. ludovicianus***
116′ Some st hairs 1–3 mm, sharp, stiff; petiole 6–30 cm; fl 10–18 mm; SNE, DMoj ...... ***L. magnificus***
115′ Lf sometimes densely hairy but not woolly
117. Lvs clustered near base
118. Petals yellow; SnGb .... [2]***L. peirsonii***
118′ Petals sometimes straw-colored or drying so, not yellow; NCoRI, SCoRO
119. Peduncle 13–20 cm; petals light blue, pink, or straw-colored (often drying straw-colored) SCoRO .... [2]***L. cervinus***
119′ Peduncle 8–15 cm; petals purple to violet; s NCoRI .... [2]***L. sericatus***
117′ Some lvs spread along st
120. Fl 5–7(10) mm; n&c SNH, SNE .... ***L. argenteus*** var. ***meionanthus***
120′ Fl 8–18 mm; c&s SNH, GB, DMtns
121. Lflets 30–100 mm
122. Infl bracts 7–15 mm, persistent; lf strigose to shaggy; c&s SNH .... [2]***L. covillei***
122′ Infl bracts 4–10 mm, ± deciduous; lf puberulent to ± hairy; c SNH (Yosemite) .... ***L. gracilentus***
121′ Lflets 15–50 mm
123. Pl 1–4 dm; fl 10–12 mm; GB, DMtns .... ***L. nevadensis***
123′ Pl 4–7 dm; fl 13–15 mm; n DMtns .... ***L. holmgrenanus***

L. adsurgens E. Drew (p. 629) Per 2–6 dm, hairy, silver to dull green. **ST** erect. **LVS** cauline; stipules 5–17 mm; petiole 2–6 cm; lflets 6–9, 20–50 mm. **INFL** 2–23 cm, ± open; peduncle 2–8 cm; pedicel 2–6 mm; fls not to subwhorled; bracts 2–8 mm, deciduous. **FL** 9–14 mm; calyx upper lip 4–6.5 mm, 2-toothed, lower lip 3–7 mm, entire to minutely 3-toothed; petals pale yellowish to lavender or violet, banner back glabrous, patch yellow to white, keel upcurved, glabrous. **FR** 2–4 cm, silky. **SEEDS** 3–6, 4–6 mm, mottled brown. $2n$=48. Dry slopes, montane forests; 500–3500 m. NCoR, SnFrB (Santa Clara Co.), SN; s OR. Pls with spreading hairs, fls 12–14 mm have been called var. *lilacinus* A.A. Heller (see *L. antoninus*); pls at 1000–3500 m in SN with wings barely > keel have been called var. *undulatus* C.P. Smith. ❀TRY.

L. affinis J. Agardh (p. 629) Ann, 2–5 dm, hairy. **LF**: petiole 3–10 cm; lflets 5–8, 20–50 mm, 4–11 mm wide. **INFL** 4–20 cm; fls whorled; peduncle 5–18 cm; bracts 5–7.5 mm, deciduous; pedicels 3–6 mm. **FL** 8–12 mm; calyx 5–7 mm, lips ± equal; petals blue, banner spot white, upper keel margins toothed (sometimes inconspicuously) near middle, ciliate from tooth to near tip. **FR** 3–5 cm, 5–9 mm wide, coarsely hairy. **SEEDS** 5–8. Uncommon. Open areas; < 800 m. NCoR, SnFrB. Intergrades with *L. nanus*. ❀TRY.

L. albicaulis Hook (p. 629) Per 3–12 dm; hairs puberulent to silky-appressed. **ST** erect. **LVS** cauline; stipules 5–18 mm; petiole 2–7 cm; lflets 5–10, 20–70 mm. **INFL** 10–44 cm, open; peduncle 2–12 cm; pedicel 2–7 mm; fls gen whorled; bracts 6–16 mm, deciduous. **FL** (8)12–16 mm; calyx upper lip 6-12 mm, 2-toothed, lower lip 7–13 mm, entire to 3-toothed; petals gen purple (yellowish white), banner back glabrous, patch indistinct, keel strongly upcurved, glabrous, tip not covered. **FR** 2–5 cm, silky. **SEEDS** 3–7, 4–7 mm, gray to tan, mottled tan. n=24. Dry slopes, openings, ± montane; 500–3000 m. KR, SNH, WTR; to WA. Doubtfully distinct from *L. andersonii*; pls with fls 8–11 mm have been called var. *shastensis* (A.A. Heller) C.P. Smith. ❀TRY.

L. albifrons Benth. Subshrub or shrub < 50 dm, hairy, gen silver (greenish). **ST** decumbent to erect. **LVS** cauline, clustered near base or not, both surfaces hairy; stipules 6–20 mm; petiole 1–8 cm; lflets 6–10, 10–45 mm. **INFL** 4–30 cm; peduncle 5–13 cm; fls gen not to loosely whorled; pedicel 3–10 mm; bracts 4–15 mm, deciduous. **FL** 9–16 mm; calyx upper lip 6–8 mm, deeply divided, lower lip 6–10 mm, entire to 3-toothed; petals violet to lavender, banner back gen hairy, patch gen yellow (to white) turning purple, keel gen unlobed near base, upper margins gen ciliate middle to tip, lower margins glabrous. **FR** 3–5 cm, hairy. **SEEDS** 4–9, 4–6 mm, mottled tan. Open sand or rocks; < 2000 m. CA-FP; s OR. Variable, ± indistinct from *L. excubitus*.

var. ***abramsii*** (C.P. Smith) (p. 629) Hoover ABRAM'S LUPINE Subshrub or shrub 2–10 dm, woolly to shaggy. **ST** decumbent to erect. **LF**: stipules 9–10 mm; petiole 3–5 cm; lflets 7–10, 10–30 mm. **INFL** 15–25 cm; peduncle 6–10 cm; pedicel 4–7 mm; bract 10–15 mm. **FL** 11–16 mm; banner back hairy to glabrous. UNCOMMON. Open woods; 600–2000 m. SCoRO (Santa Lucia Mtns). [*L. a.* C.P. Smith] ❀SUN,DRN,DRY:7,8,9,14,**15**,**16**,17–19.

var. ***albifrons*** (p. 629) Shrub 5–50 dm, gen with distinct trunk, green to silvery. **ST** erect. **LF**: stipules 7–10 mm; petiole 2–5 cm; lflets 6–10, 10–30 mm. **INFL** 8–30 cm; peduncle 5–13 cm; pedicel 4–9 mm; bract 4–8 mm. **FL** 9–14 mm. Common. Chaparral, foothill woodland; 0–1500 m. NCoR, SNF, CW, SCo. [var. *eminens* (E. Greene) C.P. Smith] ❀DRN,SUN:7,8,9,**14–17**,18,**19–24**.

var. ***collinus*** E. Greene (p. 629) Subshrub 2–4 dm, appressed-silvery. **ST** prostrate to decumbent. **LVS** cauline, clustered near base; stipules 6–8 mm; petiole 3–8 cm; lflets 6–9, 10–20 mm. **INFL** 4–14 cm; peduncle 6–10 cm; pedicel 3–5 mm; bract 7–9 mm. **FL** 10–15 mm; banner back gen hairy, keel upper margin glabrous or ciliate. Cliffs, openings in forests; 0–2000 m. NW, SNF, CW. ± indistinct from var. *flumineus* C.P. Smith. ❀SUN,DRN,DRY:7,8,9, **14–24**.

var. ***douglasii*** (J. Agardh) C.P. Smith (p. 629) Shrub 10–20 dm, gen silvery-silky. **ST** erect. **LF**: stipules 10–20 mm; petiole 1–4 cm; lflets 7–9, 25–45 mm. **INFL** 10–15 cm; peduncle 9–10 cm; pedicel 6–10 mm; bract 10–24 mm. **FL**: banner back sparsely hairy. Common. Coastal scrub, chaparral, open woodland; < 500 m. CCo, nw SnFrB, n ChI. ❀SUN,DRN:7–9,**14–17**,19–21,**22–24**.

L. andersonii S. Watson (p. 629) Per 2–9 dm, green, hairy. **ST** erect. **LVS** cauline; stipules 3–15 mm; petiole 2–6 cm; lflets 6–9, 20–60 mm. **INFL** 2–23 cm, open; peduncle 1–8.5 cm; pedicel 1.5–5 mm; fls ± whorled; bracts 2–10 mm, deciduous. **FL** 9–12 mm; calyx upper lip 5–7 mm, 2-toothed, lower lip 3–8 mm, 2–3-toothed; petals gen lavender to purple (yellowish), banner back glabrous, patch white turning purple, keel glabrous. **FR** 2–4.5 cm, silky. **SEEDS** 4–6, 4–6 mm, mottled tan, brown. Dry slopes; 1500–3000 m. NW, SNH, WTR, SnBr, SNE; s OR, w NV. ± indistinct from *L. albicaulis*. ❀TRY.

L. angustiflorus Eastw. (p. 629) Per 5–12 dm, green, glabrous to hairy. **ST** erect. **LVS** cauline; stipules 5–13 mm; petiole ± 1–5 cm; lflets 6–9, 20–60 mm. **INFL** 6–34 cm, open; peduncle ± 1–8 cm; pedicel 2–4 mm; bracts 3–7 mm, ± persistent. **FL** 8–10(12) mm; calyx upper lip 4–8 mm, 2-toothed, lower lip 4–9 mm, entire to 3-toothed; petals pale yellow to orange-yellow, banner back gen hairy, patch orange to yellow, keel glabrous, tip lavender. **FR** 2.5–4 cm, hairy. **SEED** 4.5–5.5 mm, speckled tan and brown. Gen volcanic soils; 1000–3500 m. CaRH, n&c SNH, GB. [*L. andersonii* var. *christinae* (A.A. Heller) Munz] ❀TRY.

L. antoninus Eastw. ANTHONY PEAK LUPINE Per 2–5 dm, gray- to silvery-hairy. **ST** erect. **LVS** cauline; stipules 10–12 mm; petiole 1–2 cm; lflets 6–7, 15–25 mm. **INFL** 4–20 cm, open; peduncle ± 1–4 cm; pedicel 3–4 mm; fls not whorled; bracts 7–8 mm, semideciduous. **FL** 12–14 mm; calyx upper lip 6–8 mm, 2-toothed, lower lip 6–8 mm, 3-toothed; petals white, banner back glabrous, patch turning tawny, keel glabrous. **FR** 2.5–3.5 cm, 14–17 mm wide, silky. **SEEDS** 4–5, 7–11 mm, mottled brown. UNCOMMON. Open fir forests; ± 2000 m. NCoRH (Anthony Peak, Mendocino Co.) [*L. adsurgens* var. *lilacinus* A.A. Heller in part] ❀TRY.

L. apertus A.A. Heller Per 2–6 dm, green, puberulent to sparsely appressed-hairy. **ST** erect. **LVS** cauline; stipules 5–10 mm; petiole 2–5 cm; lflets 7–9, 25–55 mm. **INFL** 8–11 cm; peduncle ± 1–8 cm; pedicel 3–6 mm; fls ± whorled to not; bracts 3.5–5 mm, deciduous. **FL** 10–12 mm; calyx upper lip 3.5–6 mm, 2-toothed, lower lip 4.5–7 mm, entire to 3-toothed; petals white to purple, banner back hairy, patch gen white, keel glabrous. **FR** 2–3 cm, hairy. **SEEDS** 3–4, 5–6 mm. Dry rocky soil; 1500–3000 m. n SNH. [*L. andersonii* var. *a.* (A.A. Heller) C.P. Smith] Reportedly toxic.

L. arboreus Sims (p. 629) YELLOW BUSH LUPINE Shrub < 20 dm, green-glabrous to silver-hairy. **ST** erect. **LVS** cauline; stipules 8–12 mm; petiole 2–3(6) cm; lflets 5–12, 20–60 mm. **INFL** 10–30 cm; peduncle 4–10 cm; pedicel 4–10 mm; fls whorled or not; bracts 8–10 mm, deciduous. **FL** 14–18 mm; calyx upper lip 5–9 mm, 2-toothed, lower lip 5–7 mm, entire; petals gen yellow (lilac to purple, esp n of c NCo), banner back glabrous, patch darker or not to white, upper keel margins ciliate from claw to tip, lower margins glabrous. **FR** 4–7 cm, brown to black, hairy. **SEEDS** 8–12, 4–5 mm, black to tan, often striped lighter. Coastal bluffs, dunes, or more inland; < 100 m. NCo, CCo (probably native Sonoma to Ventura cos., naturalized farther n). Grades ± into *L. rivularis* in NCo (see *L. r.*). Pls with yellow petals, sweet-smelling fls widely cult as sand binder. Hairier pls from w SnFrB (yellow banner, blue wings) have been called var. *eximius* (Burtt Davy) C.P. Smith; pls with glabrous lflets and purple petals have been called *L. propinquus* E. Greene; study needed ❀DRN,SUN:**4,5,17,24**&IRR:14,**15,16**,19,21,**22,23**;INV, STBL.

L. arbustus Lindley (p. 629) SPUR LUPINE Per 2–7 dm, green or gray-silky. **ST** erect. **LVS** cauline and sometimes basal; stipules 4–9 mm; petiole 2–16 cm; lflets 7–13, 20–70 mm. **INFL** 3–18 cm, open; peduncle 2–5 cm; pedicel 1–7 mm; fls whorled; bracts 3–6 mm, deciduous. **FL** 8–14 mm; calyx spur distinct, 1–3 mm, upper lip 2–4 mm, 2-toothed, lower lip 2.5–5 mm, 3-toothed; petals blue, purple, pink, white, or yellowish, banner back hairy, patch white, yellowish, or 0, wings with dense hair patch outside near tip, upper keel margins ciliate, lower keel margins glabrous. **FR** 2–3 cm, silky. **SEEDS** 3–6, 5–6 mm, tan. Open sagebrush, mixed-conifer forest; 1500–3000 m. SNH, SnGb, GB; to OR, ID, UT. [sspp. *silvicola* (A.A. Heller) Dunn and *calcaratus* (Kellogg) Dunn; var. *montanus* (Howell) Dunn] Like *L. argenteus* exc wing hairs. ❀TRY.

L. argenteus Pursh Per 1–15 dm, green-glabrous to silvery-hairy. **ST** erect. **LVS** basal to cauline; stipules 2–12 mm; petiole gen 1–15 cm; lflets 5–9, 10–60 mm, < 10 mm wide, glabrous to hairy above, hairy below. **INFL** 5–16(25) cm; peduncle 1–10 cm; pedicel 1–6 mm; fls whorled to not; bracts gen deciduous. **FL** 5–14 mm; calyx upper lip 4–8 mm, 2-toothed to entire, lower lip 4–8 mm, entire to 3-toothed, bulge or spur 0–3 mm (may be variable on 1 pl); petals blue, violet, or white, banner back gen hairy, patch yellowish to whitish to 0, upper keel margins ciliate, lower keel margins glabrous. **FR** 1–3 cm, hairy or silky. **SEEDS** 2–6, tan, brown, or red. Montane forest, sagebrush scrub; 1000–3500 m. CaR, SNH, GB, DMtns; to s Can, SD, NM. Highly variable; vars. intergrade. Var. *rubricaulis* (E. Greene) Welsh [*L. alpestris* Nelson] not in CA. ❀TRY.

var. ***argenteus*** (p. 629) Pls 2–15 dm. **LVS** cauline, green, appressed-hairy. **FL** (8)10–12 mm; calyx bulge < 1 mm; petals blue or purple to white. Esp dry sagebrush scrub; 1000–2000 m. CaR, n SNH, GB; to s Can, SD, NM. [var. *tenellus* (G. Don) D. Dunn; *L. sublanatus* Eastw.]

var. ***heteranthus*** (S. Watson) Barneby (p. 629) Pls 2–8 dm. **LVS** basal or some cauline, densely silver-silky. **FL** 8–14 mm; calyx spur 1–2 mm; petals violet or blue to white, banner back silky, wings glabrous. Dry open slopes; 1000–3000 m. GB; to OR, ID, UT. [*L. caudatus* Kellogg; *L. inyoensis* A.A. Heller]

var. ***meionanthus*** (A. Gray) Barneby (p. 629) Per or subshrub 2–9 dm. **LVS** cauline, appressed-silvery to gray-greenish. **FL** 5–7(10) mm; petals dull blue to lilac, banner back glabrous, patch yellow. Dry banks; 1500–3500 m. n&c SNH, SNE; NV. [*L. m.* A. Gray]

var. ***montigenus*** (A.A. Heller) Barneby Pl, < 4 dm, densely silvery-hairy. **LVS** basal and cauline; lower petioles gen 7–15 cm. **FL** 9–12(14) mm; calyx bulge < 1 mm; petals blue to violet, banner patch yellow to cream-colored. Dry, open montane forests, sagebrush scrub; 2500–3500 m. c SNH, SNE; NV. [*L. m.* A.A. Heller]

var. ***palmeri*** (S. Watson) Barneby (p. 629) Pls 3–6 dm. **LVS** cauline, densely gray-spreading-hairy and silvery-silky. **FL** 8–10 mm; calyx bulge or spur < 1 mm; petals blue, banner back hairy. Dry open montane forests; 2000–2500 m. SNH, SNE; to WA, UT, NM. Like var. *argenteus* exc hairs. [*L. p.* S. Watson]

L. arizonicus (S. Watson) S. Watson (p. 631) ARIZONA LUPINE Ann 1–5 dm, short-appressed- and long-spreading-hairy. **LF**: petiole 2.5–7 cm; lflets 6–10, 10–40 mm, 5–10 mm wide, upper surface glabrous. **INFL** 4–24 cm; fls ± whorled or not; peduncle 2–6 cm; bracts 4–8 mm, gen persistent; pedicels 2–4 mm. **FL** 7–10 mm; calyx 3–6 mm, lips ± equal, upper lip deeply lobed; banner, wings dark pink to magenta, drying blue-purple or whitish, banner spot yellowish, becoming darker magenta, upper keel margins glabrous, lower keel margins ciliate near claw. **FR** 1–2 cm, ± 5 mm wide, coarsely hairy, often on 1 side of infl axis. **SEEDS** 4–6. Sandy washes, open areas; < 1100 m. e DMoj, DSon; NV, AZ, Mex. [*L. concinnus* var. *brevior* (Jepson) D. Dunn] Robust pls have been called var. *barbatulus* (J. Thornber) I.M. Johnston. Locally common. ❀TRY.

L. benthamii A.A. Heller (p. 631) SPIDER LUPINE Ann 2–7 dm, short-appressed- and long-spreading-hairy. **LF**: petiole 3–12 cm;

lflets 7–10, 20–50 mm, 1.5–3.5 mm wide, linear, upper surface glabrous. **INFL** 6–30 cm; fls ± whorled or not; peduncle 4–7 cm; bracts 10–15 mm, >> buds, deciduous; pedicels 5–9 mm. **FL** 10–18 mm; calyx 5–6.5 mm, lips ± equal, upper lips deeply lobed; petals bright blue, banner spot whitish, becoming magenta, upper keel margins glabrous, lower keel margins ciliate near claw. **FR** ± 3 cm, 5 mm wide, coarsely hairy. **SEEDS** 5–8. Rocky slopes, open areas; < 1500 m. SNF, Teh, deltaic GV, SCoRO. [var. *opimus* C.P. Smith] Locally common. ❀SUN,DRN:3,**7–9**,10,11,**14–16**,17, **18–23**,24.

L. bicolor Lindley (p. 631) MINIATURE LUPINE Ann (may live 2 seasons in NCo), 1–4 dm, hairy. **LF**: petiole 1–7 cm; lflets 5–7, 10–40 mm, 1–5 mm wide, sometimes linear, upper surface gen (sub)glabrous. **INFL** 1–8 cm; whorls (0) gen 5; peduncle 3–10 cm; bracts 4–6 mm, deciduous; pedicels 1–3.5 mm. **FL** 4–10 mm; calyx upper lip 2–4 mm, deeply lobed, lower lip 4–6 mm; petals gen blue (rarely light blue, pink, or white), banner spot white, becoming magenta, keel whitish, gen pointed, upper keel margins gen ciliate near tip. **FR** 1–3 cm, 3–6 mm wide, hairy. **SEEDS** 5–8. $2n=48$. Abundant. Open or disturbed areas; < 1600 m. CA-FP; to B.C, Baja CA, naturalized in AZ. [sspp. *marginatus* D. Dunn, *microphyllus* (S. Watson) D. Dunn, *pipersmithii* (A.A. Heller) D. Dunn, *tridentatus* (C.P. Smith) D. Dunn, and *umbellatus* (E. Greene) D. Dunn; vars. *rostratus* (Eastw.) Jepson and *trifidus* (S. Watson) C.P. Smith; *L. congdonii* (C.P. Smith) D. Dunn; *L. polycarpus* E. Greene; *L. micranthus* Guss. misapplied] Highly variable, needs study; named sspp. and vars. ± indistinct in geography, morphology. Vigorous pls with larger fls may be confused with *L. nanus*. ❀SUN,DRN:**7–9**, 10–12,**14–17**,18,19,**20–24**;STBL.

L. brevicaulis S. Watson (p. 631) SAND LUPINE Ann < 1 dm, hairy; cotyledons disk-like, persistent. **LVS** cauline, crowded near base; petioles 2–5 cm; lflets 6–8, 10–15 mm, 3–6 mm wide, linear to oblanceolate, upper surface glabrous. **INFL** 1–2.5 cm, exceeding lvs; fls not whorled, dense; peduncle 2–5 cm; bracts 2–3 mm, not reflexed, persistent; pedicels 0.5–1.5 mm. **FL** 6–8 mm; calyx upper lip ± 3 mm, 2-toothed, lower lip ± 6 mm, appendages present; petals bright blue, banner spot white or yellow, keel glabrous. **FR** ± 1 cm, ± 5 mm wide, ovoid, hairy. **SEEDS** 1–2. Sandy washes, open areas; < 2300 m. MP (Lassen Co.), e DMoj; to OR, Colorado, Mex. Like small-fld *L. flavoculatus*. ❀TRY.

L. breweri A. Gray Per or subshrub < 2 dm, matted or tufted, silvery-silky. **ST** prostrate; base ± woody. **LVS** cauline, clustered near base: stipules 2–5 mm; petiole 1–5(6) cm; lflets 5–10, 3–20 mm. **INFL** 1–10 cm, ± dense; peduncle 1–8 cm; pedicel 1–3(4) mm; bracts 3–5 mm, deciduous. **FL** 4–11 mm; calyx upper lip 4–7 mm, 2-toothed, lower lip 4–6 mm, entire to 3-toothed; petals blue to violet, banner back glabrous to densely hairy, patch white or yellow, keel straight, upper keel margins glabrous or ciliate, lower keel margins glabrous. **FR** 1–2 cm, silky. **SEEDS** 3–4, 3–4 mm, mottled tan, brown. Common. Gen open montane forest; 1000–4000 m. CA-FP, SNE; s OR, NV.

var. ***breweri*** A. Gray Per. **LF**: lflets > 6 mm. **INFL** 1–2 cm; peduncle 1–3 cm. **FL** 6–9 mm; keel glabrous; banner back ± glabrous. Habitats of sp. CA-FP; s OR, w NV. ❀DRN,SUN:6,**7**, 17&IRR:14–16,18;DFCLT.

var. ***bryoides*** C.P. Smith (p. 631) Per. **LF**: lflets 3–5 mm. **INFL** < 2.5 cm; peduncle < 2 cm. **FL** 4–7 mm; keel and banner glabrous. Habitats of sp.; 2500–4000 m. s SNH, WTR, SNE; NV. ❀TRY.

var. ***grandiflorus*** C.P. Smith (p. 631) Subshrub, matted. **LF**: lflets 6–11 mm. **INFL** 2–10 cm; peduncle 2–8 cm. **FL** 6–11 mm; keel ciliate, banner back ± silky. Volcanic sand; 2000–3500 m. SN, SnBr, SNE. ❀TRY.

L. cervinus Kellogg (p. 631) SANTA LUCIA LUPINE Per 1.5–3 dm, gray-green, spreading-hairy. **ST** erect. **LVS** cauline, clustered near base; stipules 5–6 mm; petiole 13–15 cm; lflets 4–8, 40–80 mm, 10–30 mm wide. **INFL** < 20 cm, open; peduncle 13–20 cm; pedicel 3–6 mm; fls ± whorled to not; bracts 3–4 mm, deciduous. **FL** 14–16 mm; calyx upper lip 6–7 mm, entire to 2-toothed, lower lip 8–10 mm, entire to 2-toothed; petals light blue, pink, or straw-colored (esp when dry), banner back glabrous to ± hairy, patch yellow, upper keel margins ciliate throughout, lower keel margins ciliate near claw. **FR** 3–6 cm, silky. **SEEDS** 4–8, 2–4 mm, light brown with brown line or mottled tan. UNCOMMON. Dry places; 500–1000 m. SCoRO (Santa Lucia Mtns). ❀DRN,DRY:**7**,14–16.

L. chamissonis Eschsch (p. 631) Shrub 5–20 dm, silver, densely appressed-hairy. **ST** erect. **LVS** cauline; stipules 8–10 mm; petiole 1–3.5 cm; lflets 5–9, 10–25 mm. **INFL** 5–20 cm; peduncle 2–6 cm; pedicel 4–8 mm; fls ± whorled; bracts 7–10 mm, deciduous. **FL** 8–16 mm; calyx upper lip 5–7 mm, deeply lobed, lower lip 7–9 mm, entire; petals light violet to blue, banner back densely hairy, patch persistently yellow, upper keel margins glabrous, lower keel margins ± ciliate. **FR** 2.5–3.5 cm, hairy. **SEEDS** 4–8, 4–5 mm, mottled brown. Coastal strand, dunes; < ± 10 m. s NCo, CCo, SCo. ❀SUN,DRN:5,**15–17**&IRR:14,**22–24**.

L. citrinus Kellogg Ann 1–6 dm; hairs < 2 mm, soft, white, matted or not. **LF**: petiole 2–7 cm; lflets 6–9, 15–35 mm, 3–10 mm wide. **INFL** 4–15 cm; fls ± whorled or not; peduncle 1–9 cm; bracts 2.5–5 mm, deciduous; pedicels 2.5–5 mm, recurving. **FL** 8.5–12 mm; calyx 3–5 mm, lips ± equal; petals gold-yellow or white, upper keel margins glabrous, lower keel margins short-ciliate near claw. **FR** 1–2 cm, 3–5 mm wide, glabrous or becoming so. **SEEDS** 3–8, like bits of granite. Granitic soils, open yellow-pine forest; 400–1700 m. c SNF.

var. ***citrinus*** (p. 631) ORANGE LUPINE **FL**: petals gold-yellow, drying translucent, purplish. RARE. Habitats of sp.; 600–1700 m. c SNF (Fresno, Madera cos.)

var. ***deflexus*** (Congdon) Jepson MARIPOSA LUPINE **FL**: petals white, pink- or lavender-tinged or not, banner sometimes drying translucent, yellow. RARE. Granitic soils; 400–600 m. c SNF (Mariposa Co.). [*L. d.* Congdon]

L. concinnus J. Agardh (p. 631) BAJADA LUPINE Ann 1–3 dm, hairy. **ST** erect or decumbent. **LF**: petiole 2–7 cm; lflets 5–9, 10–30 mm, 1.5–8 mm wide, sometimes linear. **INFL** 1.5–9 cm, open or dense; whorls 0; some fls gen also in lower lf axils; peduncle 0–8 cm; bracts 2.5–4 mm, not reflexed, persistent; pedicels 0.7–2 mm. **FL** 5–12 mm; calyx 3–5 mm, lips ± equal, upper lip deeply lobed; petals pink to purple, rarely white, banner spot white or yellowish, keel gen glabrous. **FR** 1–1.5 cm, 3–5 mm wide, hairy. **SEEDS** 3–5. $2n=48$. Common. Open or disturbed areas, burns; < 1700 m. c&s CW, SW, D; to UT, TX, Mex. [vars. *agardhianus* (A.A. Heller) C.P. Smith, *desertorum* (A.A. Heller) C.P. Smith, *optatus* C.P. Smith, *orcuttii* (S. Watson) C.P. Smith, and *pallidus* (Brandegee) C.P. Smith] Highly variable, gen self-pollinated, needs study; named vars. ± indistinct; pls in D with linear, coarsely hairy lflets and barely ciliate lower keel margins may be confused with *L. sparsiflorus*. ❀DRN,DRY,SUN:**7–12**,13,**14–16**,17,**18–23**,24;STBL.

L. constancei T.W. Nelson & J.P. Nelson (p. 631) THE LASSICS LUPINE Per < 1.5 dm, matted, long-shaggy-hairy. **ST** ± prostrate. **LVS** cauline, clustered near base; stipules < 6 mm; petiole 6–8(14) cm; lflets 6–7, 10–20 mm, 8–10 mm wide. **INFL** 3–5 cm, dense; peduncle < 4 cm; pedicel 1–4 mm; bracts 2.5–3 mm, deciduous. **FL** 8–12 mm; calyx upper lip 4–5 mm, notched, lower lip 4–5 mm, entire; petals pink, banner back glabrous, strongly reflexed, patch light yellow, keel dark rose (white at claw), upper keel margins ciliate, lower keel margins glabrous. **FR** 1.5–2.5 cm, 0.5–1 cm wide, shaggy. **SEEDS** 3–5, tan. RARE. Serpentine barrens; 1500–2000 m. n NCoRH (Lassics range, se Humboldt and nw Trinity cos.).

L. covillei E. Greene (p. 631) Per 2–9 dm, strigose to shaggy. **ST** erect. **LVS** cauline; stipules 12–30 mm; petiole 1–10 cm; lflets 4–9, 30–100 mm. **INFL** 2–6 cm; peduncle 2–6 cm; pedicel 2–5 mm; fls whorled or not; bracts 7–15 mm, persistent. **FL** 10–14 mm; calyx upper lip 6–8 mm, 2-toothed, lower lip 6–11 mm, entire to 3-toothed; petals light blue, banner back glabrous, patch yellow, upper keel margins ± ciliate from ± middle to tip, lower keel margins glabrous. **FR** 2.5–4 cm, woolly. **SEEDS** 4–6, 3–4 mm, beige, mottled dark. Depressions, meadow edges, moist rocky slopes; 2500–3500 m. c&s SNH. ❀TRY.

L. croceus Eastw. (p. 631) Per 4–6 dm, green, hairy. **ST** erect. **LVS** cauline; stipules 4–10 mm; petiole 2–8 cm; lflets 5–9, 30–60 mm. **INFL** 6–28 cm; peduncle 2–6 cm; pedicel 3–6 mm; fls whorled or not; bracts 2–7 mm, finally deciduous. **FL** 12–15 mm; calyx upper lip 4–6 mm, 2-toothed, lower lip 6–7 mm, 2–3-toothed; petals bright yellow to orange-yellow, banner back gen glabrous (sparsely hairy on ridge), keel glabrous. **FR** 2–3.5 cm, hairy.

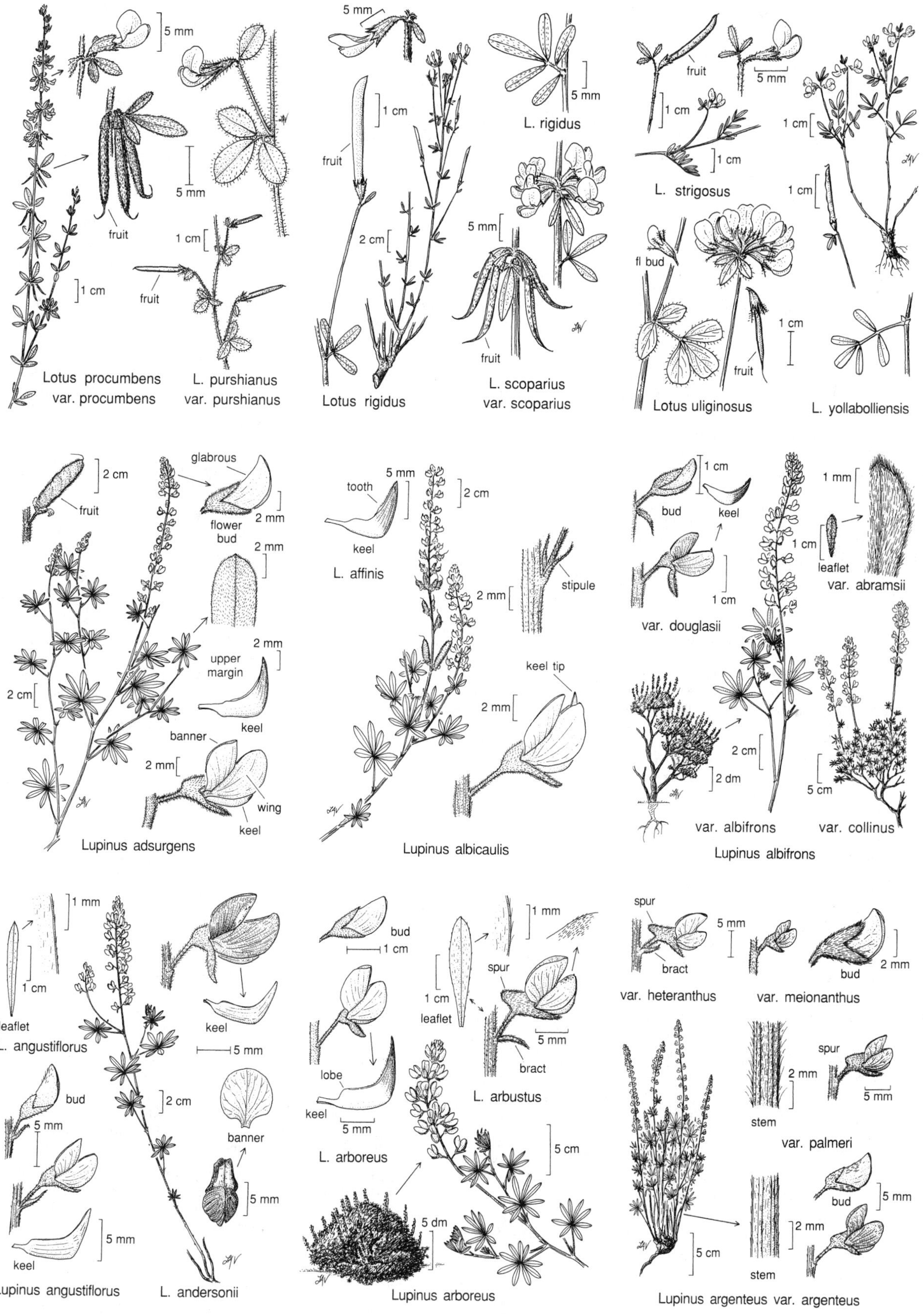
5 mm
fruit
5 mm
1 cm
fruit
1 cm
Lotus procumbens
var. procumbens
L. purshianus
var. purshianus
5 mm
1 cm
fruit
2 cm
5 mm
L. rigidus
5 mm
fruit
Lotus rigidus
L. scoparius
var. scoparius
fruit
5 mm
1 cm
1 cm
L. strigosus
1 cm
1 cm
fl bud
1 cm
fruit
Lotus uliginosus
L. yollabolliensis
2 cm
fruit
glabrous
2 mm
flower
bud
2 mm
2 mm
upper
margin
2 cm
keel
banner
2 mm
wing
keel
Lupinus adsurgens
5 mm
tooth
2 cm
keel
L. affinis
2 mm
stipule
keel tip
2 mm
Lupinus albicaulis
1 cm
bud
keel
1 mm
1 cm
leaflet
var. abramsii
1 cm
var. douglasii
2 cm
2 dm
5 cm
var. albifrons
var. collinus
Lupinus albifrons
1 mm
1 cm
leaflet
L. angustiflorus
keel
5 mm
bud
5 mm
2 cm
banner
5 mm
5 mm
keel
Lupinus angustiflorus
L. andersonii
bud
1 cm
1 mm
spur
1 cm
leaflet
5 mm
bract
lobe
keel
5 mm
L. arbustus
L. arboreus
5 cm
5 dm
Lupinus arboreus
spur
5 mm
bract
bud
2 mm
var. heteranthus
var. meionanthus
spur
2 mm
5 mm
stem
var. palmeri
bud
5 mm
2 mm
5 cm
stem
Lupinus argenteus var. argenteus

SEEDS 3–5, 6–8 mm, mottled tan. Rocky dry places; 900–2700 m. KR, CaR. Pls with spreading hairs from 900–1700 m have been called var. *pilosellus* (Eastw.) Munz, saffron-fld lupine. ❀TRY.

L. dalesiae Eastw. (p. 631) QUINCY LUPINE Per 2–5 dm, long-white-spreading-hairy. **ST** erect. **LVS** cauline, tomentose; stipules 6–16 mm; petiole 1–3 cm; lflets 6–9, 20–45 mm. **INFL** 5–16 cm; peduncle 2–5 cm; pedicel 2–5.5 mm; fls ± whorled; bracts 5–9 mm, deciduous. **FL** 9–12 mm; calyx upper lip 4–7 mm, 2-toothed, lower lip 3–7 mm, 3-toothed; petals gen yellow, banner back hairy, keel ± glabrous. **FR** 2–3 cm, strigose. **SEEDS** 3–5, 3–5 mm, tan. RARE. Dry pine forests; 1000–2500 m. n SNH (Plumas Co.). [*L. adsurgens* var. *undulatus* C.P. Smith in part]

L. duranii Eastw. (p. 631) MONO LAKE LUPINE Per 5–12 cm, robust, tufted, shaggy. **LVS** basal; stipules 6–11 mm; petiole 2–8 cm; lflets 5–8, 5–20 mm. **INFL** 2–6 cm; peduncle < 3.5 cm; pedicel 4–5 mm; whorls many, crowded; bracts 4–5 mm, ± deciduous. **FL** 8–11 mm; calyx upper lip 5–7 mm, deeply 2-toothed, lower lip 6–7 mm, ± entire; petals violet, banner back glabrous, patch cream-colored or white, upper keel margins glabrous to sparsely ciliate, lower keel margins glabrous. **FR** 1–2 cm, 8–20 mm. **SEEDS** 3–5, white. RARE. Dry volcanic pumice, gravel; 2000–2500 m. SNE (Mono Co.). [*L. tegeticulatus* var. *d.* (Eastw.) Barneby] Reports from Madera Co. questionable.

L. elatus I.M. Johnston (p. 631) SILKY LUPINE Per 5–9 dm, silvery-woolly to -silky. **ST** erect. **LVS** cauline; stipules 5–17 mm; petiole 2–5 cm; lflets 6–8, 20–80 mm. **INFL** 5–40 cm; peduncle 2–8 cm; pedicel 2–4 mm; fls ± whorled; bracts 6–11 mm, deciduous. **FL** 10–14 mm; calyx upper lip 5–7 mm, notched, lower lip 6–8 mm, 3-toothed; petals lavender to blue, banner back gen glabrous, patch pale yellowish, keel glabrous. **FR** 2–3 cm. **SEEDS** 4–6, 5–6 mm, mottled olive-brown. UNCOMMON. Dry forest; 1500–3000 m. s SNH, TR. Like *L. adsurgens, L. andersonii*; needs study. ❀TRY.

L. elmeri E. Greene (p. 631) Per 6–9 dm, green, hairy. **ST** erect, emerging from ground stout, red. **LVS** cauline; stipules 6–20 mm; petiole 1–7 cm; lflets 6–10, 15–60 mm, green, ± puberulent. **INFL** 15–20 cm; peduncle 3–9 cm; pedicel 2–6 mm; whorls 0; bracts 7–13 mm, ± persistent. **FL** 8–14 mm; calyx upper lip 7–9 mm, notched, lower lip 6–10 mm, 3-toothed; petals pale yellow, banner back glabrous, keel upcurved, glabrous. **FR** 2.5–5 cm, hairy. **SEEDS** 3–6. Uncommon. Gravel or clay soils; 1500–2000 m. nw NCoRH (South Fork Mtn, Humboldt and Trinity cos.). [*L. albicaulis* var. *sylvestris* Drew in part] ❀TRY.

L. excubitus M.E. Jones GRAPE SODA LUPINE Subshrub or shrub 2–20 dm, greenish to silver-hairy. **ST** prostrate to erect. **LVS** cauline, clustered at base or not, gen silver-hairy; stipules 5–20 mm; petiole 4–15 cm; lflets 7–10, 5–50 mm. **INFL** < 70 cm; peduncle < 30 cm; pedicel 2–7 mm; fls whorled or not; bracts 8–9 mm, deciduous. **FL** 9–18 mm, with distinctive sweet smell; calyx upper lip 6–8 mm, deeply notched, lower lip 6–8 mm, entire to 3-toothed; petals violet to lavender, banner back gen hairy, patch bright yellow (turning purple at fl), keel gen lobed near base, upper keel margins ciliate from middle to tip, lower keel margins glabrous. **FR** 3–5 cm, silky. **SEEDS** 5–8, mottled yellow-brown with lateral lines. Dry areas; < 3000 m. s SNH, Teh, SW, SNE, D.

var. ***austromontanus*** (A.A. Heller) C.P. Smith (p. 631) Subshrub 2–5 dm, gen silver-hairy. **FL** 14–18 mm; banner back glabrous or hairy. Dry slopes; 1000–3000 m. Teh, SnGb, SnBr. ❀TRY.

var. ***excubitus*** (p. 631) Shrub 10–15 dm, silver-hairy. **INFL**: axis gen persistent in winter. **FL** 9–13 mm. Desert slopes, washes; < 2500 m. DMoj, adjacent s SNH, SNE. ❀TRY.

var. ***hallii*** (Abrams) C.P. Smith Shrub 5–15 dm, greenish hairy. **FL** 14–18 mm. Sagebrush scrub; < 1500 m. SnGb, SnBr, PR. ❀SUN,DRN,DRY:7,8,9,**14–16**,17,**18–24**;CV.

var. ***johnstonii*** C.P. Smith INTERIOR BUSH LUPINE Subshrub 1–3 dm, silver-hairy. **FL** 14–18 mm. UNCOMMON. Dry slopes under pines; 1500–2500 m. SnGb, SnBr. Grades into var. *a.* ❀TRY.

var. ***medius*** (Jepson) Munz (p. 631) MOUNTAIN SPRINGS BUSH LUPINE Subshrub < 7 dm, silver-tomentose. **FL** 10–13 mm. RARE. Desert washes; < 1000 m. sw DSon (se San Diego, sw Imperial cos.).

L. flavoculatus A.A. Heller (p. 631) Ann 0.5–2 dm, hairy; cotyledons disk-like, persistent. **LVS** cauline, crowded near base; petioles 3–6 cm; lflets 7–9, 10–20 mm, 5–8 mm wide, upper surface glabrous. **INFL** 3–8 cm, > lvs, gen < peduncle, dense; whorls 0; peduncle 3–5 cm, 10 cm in fr; bracts 2–3 mm, not reflexed, persistent; pedicels 2–3 mm. **FL** 7–10 mm; calyx upper lip 1–3 mm, deeply lobed, lower lip 4–5 mm, appendages 0; petals bright blue, banner spot yellow, keel blunt, glabrous. **FR** 0.5–1 cm, ± 5 mm wide, ovoid, hairy, often on 1 side of infl axis. **SEEDS** 1–2, wrinkled. Sand or gravel; < 2200 m. e DMoj; NV. Like a hairy *L. odoratus*. ❀TRY.

L. formosus E. Greene Per 2–8 dm, densely hairy to tomentose, gray to silver; rhizomes 3–7 mm diam. **ST** spreading to erect. **LVS** cauline; stipules 4–15 mm; petiole 2–7 cm; lflets 7–9, 25–70 mm. **INFL** 10–30 cm; peduncle 3–7 cm; pedicel 3–7 mm; fls ± whorled; bracts 4–14 mm, deciduous. **FL** 10–18 mm; calyx upper lip 7–11 mm, 2-toothed, lower lip 8–12 mm, entire to 3-toothed; petals purple, banner back glabrous, patch white or not, keel glabrous, upcurved. **FR** 3–4.5 cm, hairy. **SEEDS** 5–7, 4–7 mm, mottled brown. Dry clay soils, grasslands, open areas under pines, gen in valleys; < 3000 m. CA-FP.

var. ***formosus*** (p. 631) **ST** 3–4 mm diam. **FL** 10–14 mm. Habitats and range of sp. ❀DRY:7,8,9,**14,18–21**&SUN:5,6,**15, 16**,17,**22–24**.

var. ***robustus*** C.P. Smith (p. 631) **ST** 5–7 mm diam. **FL** 16–18 mm. Valley grasslands; < 1500 m. SNF, c SNH, GV, SCoRO, SW. ❀TRY.

L. fulcratus E. Greene (p. 635) Per 3–8 dm, green, spreading-hairy. **ST** erect. **LVS** cauline; stipules 6–30 mm, 2–10 mm wide, lf-like, green; petiole 3–6 cm; lflets 6–9, 20–60 mm. **INFL** 3–20 cm; peduncle 1–11 cm; pedicel 2–7 mm; fls ± whorled; bracts 4–10 mm, deciduous. **FL** 10–14 mm; calyx upper lip 5–10 mm, 2-toothed, lower lip 5–12 mm, entire to 3-toothed; petals blue, banner back glabrous, patch white, keel upcurved, gen glabrous (sparsely hairy near middle of upper margins). **FR** 2–4 cm, silky. **SEEDS** 2–6, 4–5 mm, beige, mottled brown. Under pines in granitic soils; 1500–3000 m. SN. Stipules characteristic. May be part of *L. andersonii*. ❀TRY.

L. gracilentus E. Greene (p. 635) SLENDER LUPINE Per 2–8 dm, green, puberulent to hairy. **ST** suberect. **LVS** cauline; stipules 10–15 mm; petiole 3–14 cm; lflets 5–8, 40–80 mm. **INFL** 6–20 cm; peduncle 6–12 cm; pedicel 2–4 mm; whorls 4–6, distinct; bracts 4–10 mm, finally deciduous. **FL** 8–18 mm; calyx upper lip 7 mm, 2-toothed, lower lip 5–7 mm, 2–3-toothed to entire; petals blue, banner back glabrous, patch yellowish, upper keel margins gen ciliate, lower keel margins glabrous. **FR** 2–3 cm, densely hairy. **SEEDS** 6–8. UNCOMMON. Subalpine forest; 2500–3500 m. c SNH (Yosemite National Park).

L. grayii S. Watson (p. 635) Per 2–3.5 dm, spreading-tomentose to -woolly. **ST** prostrate to matted. **LVS** gen basal; stipules 4–10 mm; petiole 5–12 cm; lflets 5–11, 10–35 mm. **INFL** 10–16 cm; peduncle 3–15 cm; pedicel 2–4 mm; fls ± whorled; bracts 4–5(10) mm, deciduous. **FL** 10–16 mm; calyx upper lip 5–10 mm, deeply 2-toothed, lower lip 7–12 mm, entire to 3-toothed; petals deep purple to light blue, banner back glabrous to hairy, patch yellow turning reddish, upper keel margins densely hairy, lower keel margins gen ciliate near base. **FR** 2–3.5 cm, hairy. **SEEDS** 4–6, 3–4 mm, mottled gray-brown with dark lateral line. Common. Open forest slopes; 500–2500 m. SN. Sometimes fragrant. ❀TRY;DFCLT.

L. guadalupensis E. Greene (p. 635) GUADALUPE ISLAND LUPINE Ann 2–6 dm, sparsely hairy. **LF**: petiole 3–7 cm; lflets 7–9, 20–50 mm, 3–5 mm wide, sometimes linear. **INFL** 5–15 cm; fls gen whorled; peduncle 5–8 cm; bracts 8–10 mm, deciduous; pedicels 4–5 mm. **FL** 10–12 mm; calyx 6–10 mm, lips ± equal, upper lip deeply lobed; petals blue, banner spot white, upper keel margins

bud
5 mm
L. arizonicus
1 cm
keel
claw
keel base
2 cm
1 cm
leaflet
leaflet
Lupinus arizonicus
L. benthamii
5 mm
5 mm
1/2 width
length
bud
banner
2 mm
keel tip
fruits
1 cm
1 cm
cotyledon
Lupinus bicolor
5 mm
1 cm
fruits
1 dm
5 mm
L. breweri
var. grandiflorus
calyx appendage
1 cm
cotyledons
bud
2 mm
keel
L. brevicaulis
Lupinus breweri var. bryoides
bud
banner
5 mm
keel
L. chamissonis
1 mm
1 mm
2 cm
2 cm
1 mm
5 dm
Lupinus cervinus
L. chamissonis
5 mm
5 mm
keel
fruit
1 mm
2 cm
L. citrinus var. citrinus
banner patch
5 mm
L. citrinus
var. citrinus
1 cm
2 cm
axillary fruit
Lupinus concinnus
5 mm
2 cm
L. croceus
1 dm
L. covillei
5 mm
2 cm
5 mm
leaflet
Lupinus constancei
2 mm
5 mm
stem
keel
L. dalesiae
1 cm
bract
5 mm
L. elmeri
2 cm
5 mm
5 mm
1 cm
keel
2 mm
stem
leaflet
Lupinus duranii
L. elatus
1 cm
keel
5 mm
var. excubitus
5 mm
var. medius
5 cm
5 cm
1 dm
var. austromontanus
var. medius
Lupinus excubitus
2 mm
seed
L. flavoculatus
5 mm
5 mm
keel
5 cm
2 cm
L. formosus
var. robustus
2 mm
1 cm
leaflet
Lupinus
flavoculatus
L. formosus var. formosus

barely ciliate near tip. **FR** 3–6 cm, ± 10 mm wide, densely hairy. **SEEDS** 6–8. RARE. Sand or gravel; < 300 m. s ChI (San Clemente Island); Guadalupe Island, Mex. [*L. moranii* Dunkle] Intergrades with *L. nanus*.

L. hirsutissimus Benth (p. 635) STINGING LUPINE Ann 2–10 dm (often > 10 dm after fire), short-appressed- and stiffly spreading-blister-based-stinging-hairy. **LF**: petiole 4–9 cm; lflets 5–8, 20–50 mm, 10–20 mm wide. **INFL** 10–30 cm; whorls 0; peduncle 5–8 cm; bracts 4–5 mm, gen persistent; pedicels 2–5 mm. **FL** 12–18 mm; calyx 6–10 mm, lips ± equal, upper lip deeply lobed; petals dark pink to magenta, drying purplish, banner spot yellowish, becoming magenta, upper keel margins glabrous, lower keel margins densely ciliate from middle to near claw. **FR** 2–4 cm, ± 8 mm wide, coarsely hairy. **SEEDS** 3–6. Locally common. Dry, rocky areas, burns; < 1400 m. c&s CW, SW; Baja CA. ❀DRN,DRY,SUN:**7–9**,11,**14**,15–17,**18–23**,24;STBL.

L. holmgrenanus C.P. Smith (p. 635) HOLMGREN'S LUPINE Per 4–7 dm, long-hairy. **ST** erect. **LVS** basal and cauline; stipules 5–20 mm; lower petioles 2–17 cm, smaller upward; lflets 4–7, 15–50 mm. **INFL** 10–26 cm, open to ± dense; peduncle 3–10 cm; pedicel 6–10 mm; whorls 0; bracts 8–10 mm, gen deciduous. **FL** 13–15 mm; calyx bulge or spur < 1 mm, upper lip 6–7 mm, 2-toothed, lower lip 7–9 mm, entire; petals violet, banner back glabrous, patch yellow, upper keel margin slightly ciliate, lower keel margins glabrous. **FR** 4–5 cm, hairy. **SEEDS** 5–7. RARE in CA. Dry desert slopes; 1500–2500 m. ne DMtns (Last Chance, Grapevine mtns, Inyo Co.); w NV.

L. hyacinthinus E. Greene Per 4–10 dm, gray then green, ± hairy. **ST** erect. **LVS** cauline; stipules 5–16 mm; petiole 3–6 cm; lflets 7–12, 30–80 mm, 4–8 mm wide. **INFL** 4–22 cm; peduncle 3–12 cm; pedicel 2–6 mm; fls ± whorled; bracts 5–9 mm, deciduous. **FL** 13–16 mm; calyx upper lip 6–10 mm, 2-toothed, lower lip 7–11 mm, entire to 3-toothed; petals light blue to purple, banner back glabrous, patch yellowish to white, keel glabrous. **FR** 3–4 cm, silky. **SEEDS** 3–7, 4–6 mm, beige, speckled brown. Dry slopes under pines; 2000–3500 m. SnGb, SnBr, PR. [*L. formosus* var. *h.* (E. Greene) C.P. Smith] ❀TRY.

L. lapidicola A.A. Heller MOUNT EDDY LUPINE Per < 1 dm, silver-silky. **ST** ± prostrate, short. **LVS** cauline, clustered near base; stipules 4–5 mm; petiole 2–4.5 cm; lflets 6–8, 10–20 mm. **INFL** 2–7 cm; peduncle 5–10 cm; pedicel 2–4 mm; whorls few, separated; bracts 4–5 mm, gen deciduous. **FL** 9–12 mm; calyx upper lip 4–5 mm, notched, lower lip 5–6 mm, obscurely 3-toothed; petals ± violet, banner back gen hairy, patch yellow, upper keel margins ciliate, lower keel margins glabrous. **FR** not seen. UNCOMMON. Dry granite gravel; 1500–3000 m. e KR (Mount Eddy, Siskiyou Co.), NCoRO (King Range, sw Humboldt Co). ❀TRY.

L. latifolius J. Agardh Per 3–24 dm, green, glabrous to hairy. **ST** erect. **LVS** cauline; stipules 5–10 mm; petiole 4–20 cm; lflets 5–11, 40–100 mm, upper surface glabrous to hairy, lower surface ± hairy. **INFL** 16–60 cm, open; peduncle 8–20 cm; pedicel 2–12 mm; fls whorled or not; bracts 8–12 mm, deciduous. **FL** 8–18 mm; calyx upper lip 5–10 mm, entire to 2-toothed, lower lip 4–8 mm, entire or notched; petals blue or purple to white; banner back glabrous, patch gen white to yellowish turning purple, upper keel margins ciliate from claw to middle, lower keel margins gen ciliate. **FR** 2–4.5 cm, ± densely hairy. **SEEDS** 6–10, 3–4 mm, mottled dark brown. Common. Gen moist areas in woods, shady to open areas; < 3500 m. CA; to B.C., UT, AZ, Baja CA. [*L. rivularis* Douglas in part] TOXIC: causes birth defects in livestock.

var. ***barbatus*** (L. Henderson) Munz **ST** glabrous to strigose. **FL** 8–10 mm. Wet places; 1500–2500 m. MP. ❀TRY.

var. ***columbianus*** (A.A. Heller) C.P. Smith (p. 635) **ST** subglabrous to strigose. **FL** 10–14 mm; keel mostly covered by wings. Moist slopes, streamsides; 1000–3500 m. SN; to WA. ❀TRY.

var. ***dudleyi*** C.P. Smith (p. 635) **ST** densely hairy. **FL** 13–16 mm. Chaparral; < 1000 m. SnFrB. ❀TRY.

var. ***latifolius*** (p. 635) **ST** glabrous to puberulent. **FL** 10–14 mm; keel mostly not covered wings. Moist areas, open woods; < 2500 m. NW, CW, SW; to WA. ❀IRR:1,**4–6,17**&SHD:7,14,**15, 16,18**.

var. ***parishii*** C.P. Smith **ST** subglabrous to sparsely strigose. **FL** 14–18 mm. Moist areas; < 3500 m. s SN, SW. ❀TRY.

var. ***viridifolius*** (A.A. Heller) C.P. Smith (p. 635) **ST** strigose. **FL** 8–10 mm. Moist areas; 1000–2000 m. KR, CaR, MP. ❀TRY.

L. lepidus Douglas DWARF LUPINE Per < 6 dm, matted, hairy. **ST** 0 or prostrate to ± erect. **LVS** gen basal; stipules 3–25 mm; petiole 2–10 cm; lflets 5–8, 5–40 mm. **INFL** < 30 cm, gen dense; peduncle < 14 cm; pedicel 1–3 mm; bracts 4–15 mm, persistent. **FL** 6–11 mm; calyx upper lip 3–7 mm, entire to 2-toothed, lower lip 4–7 mm, entire to 3-toothed; petals pink, violet, or blue, banner back glabrous, upper keel margins ciliate, lower keel margins glabrous. **FR** 1–2 cm, hairy. **SEEDS** 2–4, 2–4 mm, ± mottled tan or green to brown. Montane to alpine open places; 1500–4000 m. CaRH, SNH, SW, GB, DMtns; to B.C., MT, Colorado. Variable complex best characterized by habit, infl, bracts, habitats. Other vars. throughout w N.Am.

var. ***confertus*** (Kellogg) C.P. Smith (p. 635) Pl, 25–60 cm, hairy. **ST** decumbent to erect. **LVS** cauline. **INFL** 5–30 cm, > lvs; whorls > 7, ± crowded; bracts 5–14 mm. **FL** 7–9 mm; banner patch yellowish fading brown to red. $2n$=48. Common. Meadows, vernally moist areas; 1500–3000 m. CaRH, SNH, WTR, SnBr, GB, DMtns; NV. [*L. c.* Kellogg] ❀TRY.

var. ***culbertsonii*** (E. Greene) C.P. Smith (p. 635) HOCKETT MEADOWS LUPINE Pl 15–40 cm, greenish. **ST** 0 to short. **LVS** gen basal; lflets gen 10–30 mm. **INFL** 4–12 cm, > lvs, gen open; whorls 3–7, well spaced; bracts 4–5 mm. **FL** 9–11 mm; banner patch white to light yellow. RARE. Rocky slopes; 2500–3000 m. s SNH (Kaweah River, Tulare and Fresno cos.) [*L. c.* E. Greene]

var. ***lobbii*** (S. Watson) C. Hitchc (p. 635) Pl, < 10 cm, hairy to shaggy. **ST** prostrate. **LVS** gen basal. **INFL** 2–8 cm, partly < lvs; bracts 5–6 mm. **FL** 6–10 mm; banner patch white. Dry rocks, meadows; 2000–3500 m. c SNH, SNE; to WA, NV. [*L. lobbii* S. Watson; *L. lyalli* A. Gray incl var. *danaus* (A. Gray) S. Watson] ❀TRY.

var. ***ramosus*** Jepson Pl, 13–30 cm, shaggy. **ST** decumbent to erect. **LVS** basal to cauline; lflets gen 5–15 mm. **INFL** 2–10 cm, open; whorls 3–7, ± well spaced; bracts 4–9 mm. **FL** 7–9(12) mm, fragrant; banner patch white to yellow. $2n$=48. Subalpine; 3000–4000 m. c&s SNH, SNE. [*L. hypoplasius* E. Greene]

var. ***sellulus*** (Kellogg) Barneby (p. 635) Pl, 12–35 cm. **ST** short, prostrate to ± erect. **LVS** subbasal; lflets 10–30 mm. **INFL** 4.5–11 cm, > lvs; bracts 4–8 mm. **FL** 8–9 mm; banner patch yellow to white turning red. $2n$=48. Dry rocks, open woods; 500–3500 m. NW, CaR, SN, GB; to OR, ID, NV. [*L. sellulus* Kellogg incl var. *artulus* (Jepson) Eastw. and ssp. *ursinus* (Eastw.) Munz] ❀TRY; DFCLT.

var. ***utahensis*** (S. Watson) C. Hitchc (p. 635) Pl, 10–25 cm, short-lived, matted, densely hairy. **ST** very short. **LVS** appearing basal. **INFL** 3–6 cm, < lvs; whorls 2–3; bracts 8–15 mm. **FL** 7–10 mm; banner patch white, upper keel margins ciliate near tip. $2n$=48. Sand or rocks with sagebrush, lodgepole pine; 1500–3500 m. n W&I (White Mtns); to OR, MT, Colorado. [*L. caespitosus* Nutt.] ❀TRY.

L. leucophyllus Lindley (p. 635) Per 4–9 dm, white-woolly and long-stiff-hairy. **ST** erect, branched from just above ground. **LVS** cauline, some clustered at base; stipule lobes 6–15 mm; petiole 6–20 cm; lflets 7–11, 30–70 mm. **INFL** 8–30 cm, dense; peduncle 2–8 cm; pedicel 1–2 mm, stout; bracts 3–12 mm, gen persistent. **FL** 10–13 mm; calyx upper lip 3–6 mm, 2-toothed, lower lip 3–8 mm, entire; petals lavender to purple, often turning brown, banner back densely hairy, patch yellow to brown, upper keel margins ciliate throughout, lower keel margins glabrous. **FR** 2–3 cm, hairy. **SEEDS** 3–6, mottled gray-tan. n=24,48. Grassy hillsides, sagebrush flats; 500–2000 m. NW, CaRH, MP; to WA, MT, UT. [var. *canescens* (Howell) C.P. Smith] ❀TRY.

L. littoralis Douglas (p. 635) Per or subshrub < 3 dm, greenish to silver, strigose to shaggy, long-hairy esp at nodes. **ST** prostrate to decumbent. **LVS** cauline; stipules 8–16 mm; petiole 2–6 cm; lflets 5–9, 10–35 mm. **INFL** 12–15 cm, ± open; peduncle 3–9 cm; pedicel 3–5 mm; fls whorled; bracts 5–9 mm, deciduous. **FL** 10–13 mm; calyx upper lip 5–6 mm, 2-toothed, lower lip 6–7 mm, 3-toothed to entire; petals purple, banner back glabrous, patch whitish, upper keel margins ciliate, lower keel margins glabrous. **FR** 3–4 cm, hairy. **SEEDS** 8–12. Coastal sand; < 100 m. n&c NCo; to B.C. Doubtfully distinct from *L. variicolor*. ❀SUN,DRN:**4,5**&IRR:**15–17**,24.

L. longifolius (S. Watson) Abrams (p. 635) Shrub 5–15 dm, silvery to green, soft-short-hairy. **ST** erect. **LVS** cauline; stipules 5–14 mm; petiole 4–10 cm; lflets 5–10, 30–60 mm. **INFL** 20–45 cm; peduncle 5–12 cm; pedicel 5–10 mm; fls ± whorled or not; bracts 4–11 mm, deciduous. **FL** 12–18 mm; calyx upper lip 8–10 mm, 2-toothed, lower lip 10–15 mm, entire; petals violet to blue, banner back glabrous, patch yellowish to white or 0, upper keel margins ciliate from middle to tip, lower keel margins glabrous. **FR** 4–6 cm, dark, hairy. **SEEDS** 6–8, 5–6 mm, brownish to gray. Scrub, canyons; < 500 m. SW. ❀DRN,DRY:**14**&SUN:**15–17**.

L. ludovicianus E. Greene (p. 635) SAN LUIS OBISPO COUNTY LUPINE Per 3–6 dm, woolly-tomentose. **ST** erect, branched from just above ground. **LVS** cauline, clustered at base; stipules 7–12 mm; petiole 5–12 cm; lflets 5–9, 15–40 mm, oblanceolate. **INFL** 10–40 cm; peduncle < 16 cm; pedicel 2–5 mm, stout; fls ± whorled or not; bracts 7–8 mm, deciduous. **FL** 10–15 mm; calyx upper lip 6–7 mm, deeply notched, lower lip 6–8 mm, 3-toothed; petals bluish to purple, banner back glabrous to ± hairy, patch yellow turning purple to white, upper keel margins ciliate from middle to tip, lower keel margins glabrous. **FR** 2–3 cm, hairy. **SEEDS** 3–4, 4–7 mm, mottled grayish. RARE. Open, grassy limestone in oak woodland; 50–500 m. s SCoRO (San Luis Obispo Co.).

L. luteolus Kellogg (p. 635) BUTTER LUPINE Ann 3–16(20) dm, sparsely hairy or becoming glabrous, appearing glaucous; cotyledons disk-like, persistent, or leaving a circular scar. **ST** hard, rigid. **LF**: petiole 2–5 cm; lflets 7–9, 10–30 mm, 4–9 mm wide, upper surface gen hairy (exc e KR). **INFL** 5–22 cm; whorls gen crowded; peduncle 4–15 cm; bracts 5–11 mm, reflexed, hairy, persistent; pedicels 1–3 mm. **FL** 10–16 mm; calyx upper lip 3–5 mm, lower lip 6–10 mm, appendages gen 0; petals gen pale yellow (pinkish or bright blue), wings gen ciliate on upper, rarely lower margins near claw, upper and lower keel margins equally (± densely) ciliate. **FR** 1–1.5 cm, ± 10 mm wide, ovoid, hairy, spreading. **SEEDS** 2, dark brown, tubercled. Clearings, open or disturbed areas; < 1900 m. NW, e SnFrB and n SCoRI (Diablo Range), w WTR; OR. Locally common. Unusually tall pls in c NCoRO (Round Valley, ne Mendocino Co.) with blue corollas have been called *L. milo-bakeri* C.P. Smith (Milo Baker lupine, **THREATENED** CA), but probably do not warrant taxonomic recognition.

L. magnificus M.E. Jones (p. 635) PANAMINT MOUNTAINS LUPINE Per 6–12 dm, white-woolly. **ST** erect, branched from base; long-sharp-stiff-hairy. **LVS** gen basal; stipules 10–24 mm; petiole 6–30 cm; lflets 5–9, 20–55 mm, 6–15 mm wide. **INFL** 20–50 cm; peduncle 10–50 cm; pedicel 2–8 mm; fls whorled or not; bracts 4–5 mm, deciduous. **FL** 10–18 mm, fragrant; calyx upper lip 5–9 mm, 2-toothed, lower lip 5–11 mm, entire; petals lavender to rose, banner back glabrous, patch yellow turning purple, upper keel margins ciliate from middle to tip, lower keel margins glabrous. **FR** 3–7 cm, densely hairy. **SEEDS** 5–8, 3–4 mm, tan. RARE. Desert slopes, washes; 1500–2500 m. SNE, n DMoj. Small-fld pls from Coso Mtns, Inyo Co., have been called var. *glarecola* M.E. Jones, Coso Mtn lupine; straight-keeled pls from SNE have been called var. *hesperius* (A.A. Heller) C.P. Smith, McGee Meadows lupine.

L. microcarpus Sims CHICK LUPINE Ann 1–8 dm, sparsely to densely hairy; cotyledons disk-like, persistent, or leaving a circular scar. **ST** hollow, at least below. **LF**: petiole 3–15 cm; lflets 5–11, gen 9, 10–50 mm, 2–12 mm wide, sometimes linear, upper surface glabrous. **INFL** 2–30 cm; whorls crowded or not; peduncle 2–30 cm; bracts 3.5–12 mm, reflexed, persistent; pedicels 0.5–5 mm. **FL** 8–18 mm; calyx upper lip 2–6 mm, lower lip 5–10 mm, appendages gen 0; petals white to dark yellow, pink to dark rose, or lavender to purple, wings gen ciliate on upper (less often lower) margins near claw, upper keel margins ciliate, lower keel margins less so near claw. **FR** 1–1.5 cm, ± 10 mm wide, ovoid, hairy, erect to spreading, on 1 side of infl axis or not. **SEEDS** 2, tan to brown, gen mottled, wrinkled or smooth. $2n$=48. Abundant. Open or disturbed areas, sometimes seeded on roadbanks; < 1600 m. CA–FP, MP, w DMoj; to B.C., Baja CA, S.Am. Highly variable; vars. intergrade.

var. ***densiflorus*** (Benth.) Jepson (p. 635) **INFL**: bracts short-appressed- to long-spreading-hairy. **FL**: calyx ± sparse, appressed- to spreading-hairy, appendages gen 0; petals gen white to yellow, pink- or lavender-tinged or not, rarely rose or purple, wings oblanceolate, withering, upper margins (and rarely lower margins near claw) gen ciliate, lower keel margins sometimes sparsely ciliate near claw. **FR** ± spreading, gen on 1 side of infl axis. Habitats of sp. NW (exc Siskiyou Co.), SNF, ScV, CW, e SCo, TR. [*L. d.* Benth., incl vars. *aureus* (Kellogg) Munz and *lacteus* (Kellogg) C.P. Smith (as used by Munz, in part)] ❀SUN:3–6,**7–9,14–17**,18,**19–24**&IRR:**10,11**,12,13;CV.

var. ***horizontalis*** (A.A. Heller) Jepson (p. 635) **INFL**: bracts short- to long-spreading-hairy. **FL**: calyx appressed- to spreading-hairy, appendages 1–2 mm, gen purplish; petals lavender to purple, becoming translucent, wings widely elliptic, not withering, ciliate on upper (and gen lower) margins near claw, lower keel margins ciliate near claw. Washes, sand or gravel; < 1500 m. s SnJV, e DMoj. [*L. h.* A.A. Heller incl var. *platypetalus* C.P. Smith; *L. arenicola* A.A. Heller; *L. densiflorus* Benth. var. *glareosus* (Elmer) C.P. Smith] ❀TRY.

var. ***microcarpus*** (p. 635) **INFL**: bracts (and calyx) long-shaggy-hairy. **FL**: calyx appendages gen 0; petals gen pink to purple (yellowish or white), wings linear to lanceolate, withering, upper (and rarely lower) margins gen ciliate near claw, lower keel margins sometimes sparsely ciliate near claw. **FR** ± erect, gen not on 1 side of infl axis. Habitats and range of sp. [*L. subvexus* C.P. Smith (incl vars.); *L. ruber* A.A. Heller; *L. densiflorus* Benth. vars. *austrocollium* C.P. Smith, *palustris* (Kellogg) C.P. Smith (as used by Munz), and *persecundus* C.P. Smith] ❀SUN:3–6,**7–9,14–17**,18,**19–24**.

L. nanus Benth. (p. 635) Ann 1–6 dm, hairy. **LF**: petiole 2–8.5 cm; lflets 5–9, gen 7, 10–40 mm, 1–12 mm wide, sometimes linear. **INFL** 2–20 cm; fls gen whorled; peduncle 2–15 cm; bracts 4–12 mm, deciduous; pedicels 2.5–7 mm. **FL** 6–15 mm; calyx 4–8 mm, lips ± equal, upper lip deeply lobed; petals blue, rarely light blue, lavender, pink, or white, banner spot white, upper keel margins ciliate near tip, lower keel margins glabrous. **FR** 2–4 cm, 4–7 mm wide, hairy. **SEEDS** 4–12. $2n$=48. Abundant. Open or disturbed areas; < 1300 m. CA-FP (exc s SW); to B.C. [sspp. *latifolius* (Benth.) D. Dunn and *menkerae* (C.P. Smith) D. Dunn; var. *maritimus* Hoover; *L. vallicola* A.A. Heller incl ssp. *apricus* (E. Greene) D. Dunn] Highly variable, needs study; n CA pls (*L. v.*) with smaller fls may be confused with *L. bicolor*. ❀SUN,DRN:3–5,**6–9,14–24**&IRR:10–12.

L. nevadensis A.A. Heller Per 1–4 dm, long-hairy. **ST** erect. **LVS** basal and cauline; stipules 8–10 mm; basal petioles < 14 cm, cauline < 4 cm; lflets 6–10, 20–50 mm. **INFL** 5–17 cm; peduncle 3–6 cm; pedicel 4–8 mm; whorls 0; bracts 4–5 mm, deciduous. **FL** 10–12 mm; calyx upper lip 3–4 mm, 2-toothed, lower lip 4–5 mm, 3-toothed; petals blue, banner back glabrous, patch white to yellowish, keel strongly upcurved, upper keel margins ciliate, lower keel margins glabrous. **FR** 2.5–4 cm, densely hairy. **SEEDS** 3–4, light. Hillsides, valleys, sagebrush; 1000–3000 m. GB, DMtns; OR, NV. ❀TRY.

L. nipomensis Eastw. (p. 635) NIPOMO MESA LUPINE Ann 1–2 dm, hairy. **ST** decumbent. **LF**: petiole 2–3 cm; lflets 5–7, 10–15 mm, 5–6 mm wide. **INFL** 1–5 cm; whorls 0; fls dense, not in lf axils; peduncle 2–3.5 cm; bracts 3–3.5 mm, gen persistent; pedicels 1–1.5 mm. **FL** 6–7 mm; calyx 4–5.5 mm, lips ± equal, upper lip deeply lobed; petals pink, banner spot white or yellowish, keel glabrous. **FR** 1.5–2 cm, 5 mm wide, hairy or becoming glabrous. **SEEDS** 3–4. **ENDANGERED** CA. Stabilized sand dunes; < 25 m. s CCo (Nipomo Dunes, sw San Luis Obispo Co.) Intergrades with *L. concinnus*. Threatened by coastal development.

L. obtusilobus A.A. Heller (p. 635) Per 1.5–3 dm, appressed-silvery-silky. **ST** decumbent to erect. **LVS** cauline; stipules 7–14 mm; petiole 2–5 mm; lflets 6–7, 20–50 mm. **INFL** 3–7 cm; peduncle 2–4 cm; pedicel 2–5 mm; fls ± whorled; bracts 3–4 mm, deciduous. **FL** 11–13 mm; calyx upper lip 6–7 mm, 2-toothed, lower lip 6–7 mm, 3-toothed; petals blue to lilac, banner back hairy, patch yellow, upper keel margins ciliate, lower margins glabrous. **FR** 2.5–4 cm, silky. **SEEDS** 4–5, 3–4 mm, mottled brown. Gravelly summits; 2500–3500 m. NW, CaR, n SNH. ❀TRY.

L. odoratus A.A. Heller (p. 635) MOJAVE LUPINE Ann 1–3 dm, sparsely short-hairy when young, becoming ± glabrous; cotyledons disk-like, persistent. **LVS** basal; petioles 2–12 cm; lflets 7–9, 10–20 mm, 3–8 mm wide, sometimes obovate, bright green. **INFL** 4–13 cm; whorls 0; peduncle 6–15 cm, hollow; bracts 2–4 mm, not reflexed, persistent, tips sparsely ciliate; pedicels 3–5 mm. **FL** with violet odor, 7–10 mm; calyx tips ciliate or not, upper lip 3–3.5 mm, rounded or 2-toothed, lower lip 4–5 mm; petals deep blue-purple, banner spot white or yellow becoming magenta, keel glabrous. **FR** 1.5–2.5 cm, ± 8 mm wide, oblong, upper suture wavy, densely long-ciliate, sides with a few short hairs that become scale-like when dry. **SEEDS** 2–6, wrinkled. Sandy flats, open areas; < 1600 m. GB, DMoj; NV, AZ. Long-soft-hairy pls have been called var. *pilosellus* C.P. Smith, may be confused with *L. flavoculatus*. ❀TRY.

L. onustus S. Watson (p. 635) Per 2–3 dm; green, silky. **ST** short, decumbent. **LVS** cauline, clustered near base, silky below, gen glabrous above; stipules 8–10 mm; petiole 5–13 cm; lflets 5–9, 15–50 mm. **INFL** 5–15 cm; peduncle 4–8 cm; pedicel 3–5 mm; whorls 0; bracts 3–4 mm, deciduous. **FL** 8–11 mm; calyx upper lip 2–5 mm, 2-toothed, lower lip 3.5–6 mm, entire; petals violet, banner back glabrous, upper keel margins ciliate, lower keel margins glabrous. **FR** 3–4.5 cm, hairy. **SEEDS** 5–6, 6–7 mm, brown. Dry banks, open forest, serpentine; 1000–2000 m. KR, CaRH, n SNH. ❀TRY.

L. pachylobus E. Greene (p. 639) BIG POD LUPINE Ann 1.5–4 dm, hairy. **LF**: petiole 4–8 cm; lflets gen 7, 20–25 mm, 2–5 mm wide. **INFL** 1–4 cm; fls gen whorled; peduncle 3–12 cm; bracts ± 6 mm, deciduous; pedicels 1–2.5 mm. **FL** 7–9 mm; calyx 4.5–6 mm, lips ± equal; petals blue, banner spot white, becoming dark magenta, keel blunt, glabrous. **FR** 3 cm, 6–9 mm wide, densely hairy, ± fleshy. **SEEDS** gen 5. Uncommon. Open or disturbed areas; < 600 m. NCoRO, CaRF, SNF, SCoRO. Occurs with and intergrades with *L. bicolor*. ❀STBL.

L. padre-crowleyi C.P. Smith (p. 639) FATHER CROWLEY'S LUPINE Per 5–7.5 dm, silver- to white-woolly. **ST** erect, from a wooly mat. **LVS** basal and cauline; stipules 5–11 mm; petiole 2–3 cm; lflets 6–9, 25–75 mm. **INFL** 7–21 cm; peduncle 2–5.5 cm; pedicel 2–3.5 mm; fls ± whorled; bracts 4–9 mm, deciduous or not. **FL** 10–14 mm; calyx upper lip 5–7 mm, 2-toothed, lower lip 5.5–8 mm, 3-toothed; petals cream-colored to pale yellow, banner back gen hairy, keel glabrous. **FR** 2–3 cm, silky. **SEEDS** 2–3, 4–5 mm, white, mottled black. **RARE** CA. Decomposed granite; 2500–4000 m. s SNH, SNE (n Inyo, Tulare cos.). [*L. dedeckerae* Munz & Dunn]

L. peirsonii H. Mason (p. 639) PEIRSON'S LUPINE Per 3–6 dm, silver-silky. **ST** erect, branched from just above ground. **LVS** cauline, clustered at base, ± fleshy; stipules l5–20 mm; petiole 2–15 cm; lflets 5–8, 25–70 mm, widely oblanceolate. **INFL** < 10 cm; peduncle < 24 cm; pedicel 1–2 mm; fls ± whorled; bracts 5–7 mm, deciduous. **FL** 10–12 mm; calyx upper lip 4–6 mm, obscurely 2-toothed, lower lip 5–7 mm, entire; petals yellow, banner back gen hairy, upper keel margins ciliate from middle to tip, lower keel margins glabrous. **FR** 3–4 cm, silky. **SEEDS** 3–5. UNCOMMON. Loose gravel; 1000–2000 m. SnGb.

L. polyphyllus Lindley Per 2–15 dm, green, glabrous or sparsely hairy. **ST** erect, stout. **LVS** basal and cauline; stipules 5–30 mm; petioles 3–45 cm, upper shorter; lflets 5–17, 40–150 mm. **INFL** 6–40 cm, open; peduncle 3–13 cm; pedicel 3–15 mm; fls ± whorled; bracts 7–11 mm, deciduous. **FL** 9–15 mm; calyx lips entire, upper lip 4–7 mm, lower lip 4–7 mm; petals violet to lavender to pink to white, banner back glabrous, patch yellow to white sometimes turning red-purple, keel upcurved, gen glabrous. **FR** 2.5–4 cm, hairy. **SEEDS** 3–9. $2n$=48. Moist areas to bogs; < 3000 m. CA-FP, GB; to B.C.

var. ***burkei*** (S. Watson) C. Hitchc. **LF**: lflets 5–11, glabrous above, sometimes sparsely hairy below. Wet places; 1500–3000 m. SNH, SnGb, SnBr, SnJt, GB; OR, ID, NV. [sspp. *superbus* (A.A. Heller) Munz and *bernardinus* (Abrams) Munz] ❀SUN,IRR, DRN:1–3,6,**7**,14,**15–16**,**18**.

var. ***pallidipes*** (A.A. Heller) C.P. Smith **LF**: lflets 9–17, gen minutely hairy (appearing ± glabrous). Moist to wet places; < 2500 m. KR, NCoRH, CaR; to B.C. ❀TRY.

var. ***polyphyllus*** (p. 639) **LF**: lflets 9–17, glabrous above, ± strigose below. Habitats of sp. NW, CCo, SnFrB, SCoRO; to B.C. ❀SUN,IRR:4,5,**6**,**7**,14,**15–17**.

L. pratensis A.A. Heller Per 3–7 dm, green, hairy. **ST** erect. **LVS** basal and cauline; stipules 5–20 mm; petiole 3–21(25) cm; lflets 5–10, 30–80 (130) mm. **INFL** 5–28 cm, dense; peduncle 4–17 cm; pedicel 1–3 mm; bracts 5–10 mm, persistent. **FL** 10–12 mm; calyx upper lip 4–7 mm, 2-toothed, lower lip 5–6 mm, entire; petals violet to dark blue, banner back glabrous or hairy, patch orange to red, upper keel margins densely ciliate, lower keel margins glabrous. **FR** 1.5–2 cm, hairy to woolly. **SEEDS** 4–6, 3–4 mm, mottled tan, brown. $2n$=48. Meadows, streambanks; 1000–3500 m. c&s SNH, SNE.

var. ***eriostachyus*** C.P. Smith **FL**: banner back hairy. Moist places; 1500–3500 m. SNE (Big Pine Creek, Inyo Co.) ❀TRY.

var. ***pratensis*** **FL**: banner back glabrous. Habitats and range of sp. ❀TRY.

L. pusillus Pursh var. ***intermontanus*** (A.A. Heller) (p. 639) C.P. Smith Ann < 1 dm, hairy; cotyledons disk-like, persistent. **LVS** cauline, crowded near base; petioles 3–6 cm; lflets 5–6, 10–20 mm, ± 5 mm wide, upper surface glabrous. **INFL** ± 3 cm, < lvs, dense; whorls 0; peduncle 0–1 cm; bracts ± 3 mm, not reflexed, persistent; pedicels < 1 mm. **FL** 6 mm; calyx upper lip ± 2.5 mm, lower lip ± 5 mm; petals pale blue, fading pinkish or whitish, banner spot white or yellow, keel glabrous. **FR** 1.5 cm, 6 mm wide, oblong, narrowed between seeds, hairy. **SEEDS** 2; surface wrinkled; margin ridged. $2n$=48. Uncommon. Open sandy areas; < 1600 m. GB (Modoc, Inyo cos.); to WA, c US, AZ.

L. rivularis Lindley (p. 639) Per (subshrub), 3.5–10 dm, green, ± glabrous. **ST** erect, dark brown to red, gen hollow. **LVS** cauline; stipules 7–15 mm; petiole 3–5 cm; lflets 5–9, 20–40 mm, glabrous above. **INFL** 15–50 cm, open; peduncle 3–15 cm; pedicel 5–10 mm; fls ± whorled or not; bracts 8–10 mm, deciduous. **FL** 12–16 mm; calyx upper lip 7–8 mm, 2-toothed, lower lip 7–9 mm, entire to ± 3-toothed; petals violet, banner back glabrous, upper keel margins ciliate from claw to tip, lower keel margins glabrous. **FR** 3–7 cm, dark, sparsely hairy. **SEEDS** 7–8, 3–4 mm, mottled brown with black line. Sand or gravel; < 500 m. NCo; to B.C. Grades ± into blue-fld *L. arboreus*, but fls earlier (late winter, spring), not sweet-smelling. ❀SUN,IRR:**4**,**5**,**15–17**,24.

L. saxosus Howell (p. 639) Per 2–3 dm, green. **ST** erect. **LVS** gen basal; stipules 10–20 mm; petiole 4–12(14) cm; lflets 7–12, 10–40 mm, long-stiff-hairy below, glabrous above. **INFL** 5–20 cm, dense; peduncle 2–15 cm; pedicel 4–8 mm; fls ± whorled or not; bracts 4–7 mm, ± deciduous. **FL** 12–18 mm; calyx upper lip 5–7 mm, 2-toothed, lower lip 5–6 mm, 3-toothed; petals blue, banner back glabrous, patch yellow turning violet, upper keel margins ± sparsely ciliate, lower keel margins glabrous. **FR** 2–4 cm, shaggy. **SEEDS** 4–6, brown. $n(2n?)$=48. Sagebrush scrub, open areas; 1000–2500 m. MP; OR, WA, NV. [*L. polyphyllus* var. *s.* (Howell) Barneby] ❀TRY.

L. sericatus Kellogg (p. 639) COBB MOUNTAIN LUPINE Per 1.5–5 dm, silver to gray-green, short-appressed-hairy. **ST** gen erect. **LVS** cauline, clustered near base; stipules 2–7 mm; petiole 5–15 cm; lflets 4–7, 30–50 mm, widely spoon-shaped. **INFL** 10–30 cm, open to dense; peduncle 8–15 cm; pedicel 4–6 mm; fls ± whorled; bracts 3–4 mm, deciduous. **FL** 12–16 mm; calyx upper lip 6–10 mm, 2-toothed, lower lip 7–10 mm, 3-toothed; petals purple to violet, banner back glabrous to ± hairy, upper keel margins ciliate from claw to tip, lower margins gen ± glabrous. **FR** 2–3 cm, hairy. **SEEDS** 3–7, 3–4 mm, light brown. RARE. Open wooded slopes; 500–1500 m. s NCoRI (Napa, Lake, Sonoma cos.). In cult.

L. shockleyi S. Watson (p. 639) DESERT LUPINE Ann 0.5–3 dm, canescent; cotyledons disk-like, persistent. **LVS** cauline, crowded near base; petioles 4–12 cm; lflets 8–10, 10–30 mm, 4–10 mm wide, upper surface glabrous. **INFL** 2–6 cm; whorls 0; peduncle 1–10 cm; bracts 2–4 mm, not reflexed, persistent; pedicels 1–3 mm.

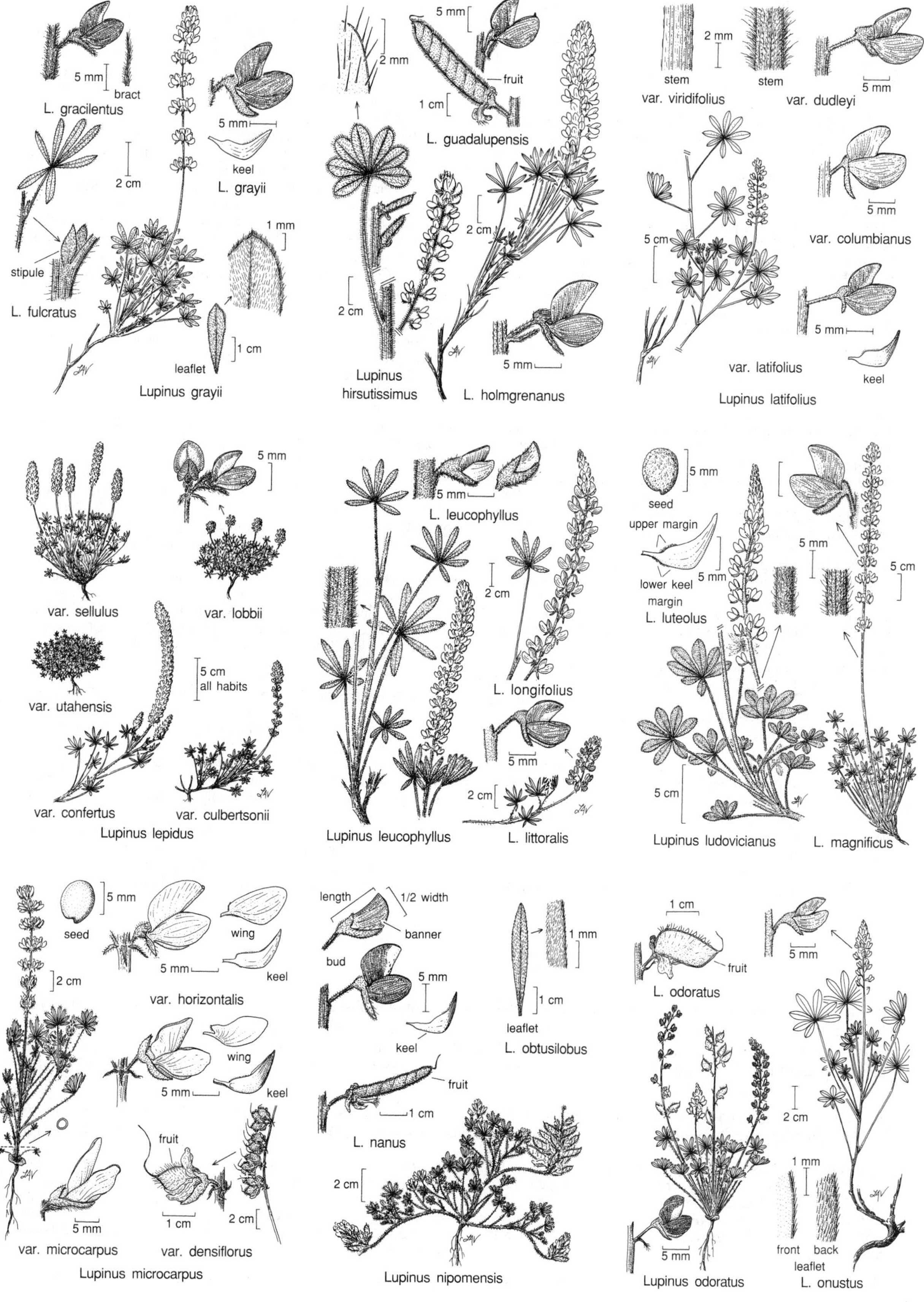
5 mm
bract
L. gracilentus
5 mm
keel
L. grayii
2 cm
1 mm
stipule
L. fulcratus
leaflet
1 cm
Lupinus grayii
2 mm
5 mm
fruit
1 cm
L. guadalupensis
2 cm
2 cm
5 mm
Lupinus hirsutissimus
L. holmgrenanus
stem
2 mm
stem
var. viridifolius
var. dudleyi
5 mm
5 mm
5 cm
var. columbianus
5 mm
var. latifolius
keel
Lupinus latifolius
5 mm
var. sellulus
var. lobbii
var. utahensis
5 cm all habits
var. confertus
var. culbertsonii
Lupinus lepidus
5 mm
L. leucophyllus
2 cm
L. longifolius
5 mm
2 cm
Lupinus leucophyllus
L. littoralis
5 mm
seed
upper margin
5 mm
lower keel margin
L. luteolus
5 mm
5 cm
5 cm
Lupinus ludovicianus
L. magnificus
5 mm
seed
wing
5 mm
keel
2 cm
var. horizontalis
wing
5 mm
keel
fruit
5 mm
1 cm
2 cm
var. microcarpus
var. densiflorus
Lupinus microcarpus
length
1/2 width
banner
bud
5 mm
keel
1 mm
1 cm
leaflet
L. obtusilobus
fruit
1 cm
L. nanus
2 cm
Lupinus nipomensis
1 cm
fruit
L. odoratus
5 mm
2 cm
1 mm
5 mm
front
back
leaflet
Lupinus odoratus
L. onustus

FL 4.5–6 mm; calyx 3–6 mm, lips ± equal; petals dark blue-purple, banner spot yellow, keel blunt, glabrous. **FR** 1.5–2 cm, 8–12 mm wide, ovoid, upper suture gen ± wavy, densely and stiffly long-ciliate, sides with short, inflated hairs that become scale-like when dry. **SEEDS** 2, wrinkled. Open, sandy areas; < 1200 m. D; NV, AZ. ❀TRY.

L. sparsiflorus Benth (p. 639) COULTER'S LUPINE Ann 2–4 dm, short-appressed- and long-spreading-hairy. **LF**: petiole 3–4 cm; lflets 7–11, 15–30 mm, 2–4 mm wide, linear to oblanceolate, upper surface hairy at least near margins. **INFL** 15–20 cm; fls ± whorled or not; peduncle 2–4 cm; bracts 3–5 mm, < buds, gen deciduous; pedicels 2–4 mm. **FL** gen 10–12 mm; calyx 3–6 mm, lips ± equal, upper lip deeply lobed; petals gen blue (pinkish), drying darker, banner spot whitish becoming magenta, lower (and often upper) keel margins ciliate near claw. **FR** 1–2 cm, ± 5 mm wide, coarsely hairy. **SEEDS** 4–5. Washes, sandy areas; < 1300 m. s SCoR, SW, DMoj; to UT, AZ, n Mex. Locally common. Pls in DMoj (often with smaller fls) have been called ssp. *mohavensis* Dziekanowski & D. Dunn; pls in sw SCo (with pinkish fls, truncate lflets) have been called var. *inopinatus* (C.P. Smith) Dziekanowski & D. Dunn. ❀STBL.

L. spectabilis Hoover (p. 639) SHAGGYHAIR LUPINE Ann 2–6 dm; densely short-appressed- and long-spreading-hairy. **LF**: petiole 4–9 cm; lflets gen 9, 10–40 mm, 4–9 mm wide. **INFL** 5–30 cm; fls whorled; peduncle 5–12 cm; bracts 8–9 mm, deciduous; pedicels 6–8 mm. **FL** 11–14 mm; calyx 4–7 mm, lips ± equal; petals blue, rarely white, banner spot white, upper keel margins ciliate near tip. **FR** 3–5 cm, 8–9 mm wide, densely hairy. **SEEDS** 5–10. RARE. Serpentine; < 800 m. c SNF (Mariposa, Tuolumne cos). Intergrades with *L. nanus*.

L. stiversii Kellogg (p. 639) HARLEQUIN LUPINE Ann 1–5 dm, sparsely hairy. **LF**: petiole 2–8 cm; lflets gen 7, 20–50 mm, 5–15 mm wide, upper surface bright green. **INFL** 3–8 cm (often longer in fr), dense; whorls 0; peduncle 5–10 cm; bracts 3–5 mm, finally deciduous; pedicels 1.5–4 mm. **FL** 13–18 mm; calyx upper lip 4–6 mm, deeply lobed, lower lip 5–7 mm; banner yellow, wings pink, rarely whitish, keel white, upper and lower margins ciliate from claw to ± middle. **FR** ± 2 cm, 5–7 mm wide, gen glabrous. **SEEDS** gen 5. Clearings, open areas; < 2100 m. SN, n SCoRO (Monterey Co.), SnGb, SnBr. Locally common. ❀DRN:**7**,14,**18**&SUN:15, 16;DFCLT.

L. succulentus Koch (p. 639) ARROYO LUPINE Ann (often appearing per), 2–10 dm, fleshy, sparsely hairy. **LF**: petiole 6–15 cm; lflets 7–9, 20–60 mm, 7–20 mm wide, upper surface glabrous. **INFL** 9–15 cm; whorls distinct; peduncle 5–9 cm; bracts 3–5 mm, deciduous; pedicels 3–7 mm. **FL** 12–18 mm; calyx 4–7 mm, lips ± equal, upper lip lobed; petals gen blue-purple (white, pink, or lavender), banner spot white, becoming magenta, wings sparsely ciliate on upper margins near claw, upper and lower keel margins ciliate near claw. **FR** 3.5–5 cm, 8–10 mm wide, coarsely hairy to tomentose. **SEEDS** 6–9. $2n$=48. Abundant. Open or disturbed areas, often seeded on roadbanks; < 800 m. c&s NW, GV, CW, SW; Baja CA. ❀SUN:**7–9,14–24**&IRR:10–12;also STBL.

L. tidestromii E. Greene (p. 639) TIDESTROM'S LUPINE Per 1–3 dm, white-shaggy-hairy. **ST** ± prostrate, weak. **LVS** cauline; stipules 8–12 mm; petiole 1–3 cm; lflets 3–5, 5–20 mm. **INFL** 2–10 cm, open; peduncle 4–8 cm; pedicel 3–5 mm; fls whorled; bracts 4–5 mm, deciduous. **FL** 11–13 mm; calyx upper lip 5–6 mm, deeply notched, lower lip 5–6 mm, entire to notched; petals light blue to lavender, banner back glabrous, patch white to yellow turning violet, upper keel margins ciliate from claw to tip, lower keel margins glabrous. **FR** 2–3 cm, shaggy. **SEEDS** 5–8, 3–4 mm, mottled brown, tan. **ENDANGERED** CA. Dunes, beaches; < 100 m. s NCo (Sonoma Co.), n&c CCo (Marin, Monterey cos.) Shaggier pls from n NCo have been called var. *layneae* (Eastw.) Munz, Point Reyes lupine.

L. tracyi Eastw. (p. 639) TRACY'S LUPINE Per 4–7 dm, glabrous, glaucous. **ST** erect, slender. **LVS** cauline; stipules 7–9 mm; petiole < 1 cm; lflets 6–7, 10–40 mm. **INFL** 4–16 cm; peduncle 2– 6 cm; pedicel 5–6 mm; fls ± whorled or not; bracts 8–10 mm, deciduous. **FL** 8–10(12) mm; calyx upper lip 3–8 mm, 2-toothed, lower lip 3–5 mm, 3-toothed; petals whitish to dull blue, banner back glabrous, keel glabrous. **FR** 1.5–2.5 cm, white-hairy (dark when dry). **SEEDS** 3–4, 4–5 mm. UNCOMMON. Dry, open montane forest; 1500–2000 m. KR; s OR.

L. truncatus Hook. & Arn. (p. 639) Ann 2–3(5) dm, finely hairy (gen appearing glabrous). **LF**: petiole 3–10 cm, flat, lflet-like; lflets 5–8, 20–40 mm, 2–5 mm wide, linear, upper surface glabrous, tip gen truncate. **INFL** 3–25 cm; fls sparse, not whorled; peduncle 3–10 cm; bracts 2–5 mm, persistent; pedicels 2–4 mm. **FL** 8–13 mm; calyx 3–4 mm, lips ± equal, upper lip deeply lobed; banner, wings magenta, banner spot yellowish, becoming dark magenta, keel stout, blunt, upper and lower keel margins ciliate from middle to claw. **FR** ± 3 cm, 5 mm wide, hairy. **SEEDS** 6–8. Openings in chaparral or woodlands, burns; < 1000 m. c&s CW, SW; Baja CA. Locally common. ❀SUN,DRN:**7**,8,9,14–17,**18**,19–24; DFCLT.

L. variicolor Steudel (p. 639) Per or subshrub 2–5 dm, gen densely appressed- or spreading-silver-hairy. **ST** prostrate to decumbent, not weak. **LVS** cauline, often appearing clustered near base 1st year; stipules 7–8 mm; petiole gen < 3 cm; lflets 6–9, 20–35 mm. **INFL** 6–15 cm; peduncle 4–12 cm; pedicel 4–12 mm; fls ± whorled or not; bracts 4–7 mm, deciduous. **FL** 11–16 mm; calyx upper lip 7–8 mm, 2-toothed, lower lip 8–9 mm, gen entire; petals white, yellow, rose, or purple (often on 1 pl), banner back glabrous, patch 0, upper keel margins ciliate, lower keel margins glabrous. **FR** 3–4 cm, dark, hairy. **SEEDS** 7–9, 3–4 mm, mottled dark. Coastal terraces, beaches; < 500 m. NCo, n&c CCo. Doubtfully distinct from *L. littoralis*. ❀SUN,DRN:4,**5**&IRR:**15–17**,24.

MARINA

Duane Isely

Per in CA, unarmed, hairy, gen gland-dotted. **LVS** odd-1-pinnate; stipules obscure; lflets opposite. **INFL**: raceme, terminal. **FL**: calyx lobes < tube; corolla gen 2-colored (blue-violet and white), petals (exc banner) arising from side of filament column; stamens gen 10, filaments fused; ovule 1. **FR** indehiscent, incl or slightly exserted. **SEED** 1. 38 spp.: esp Mex. (Marina, interpreter for Mexican conqueror Cortez, 16th century) [Barneby 1977 Mem New York Bot Gard 27:55–133; 608–649]

1. St decumbent, glands 0; infl dense, fls soon spreading or reflexed; e PR (Santa Rosa Mtns) ***M. orcuttii*** var. ***orcuttii***
1′ St ascending or erect, gland-dotted; infl open, fls ascending or spreading; D .... ***M. parryi***

M. orcuttii (S. Watson) Barneby var. ***orcuttii*** (p. 639) CALIFORNIA MARINA Per < 2 dm (often very small), strigose, gen gray. **LF**: lflets 11–15, crowded, 2–5 mm, oblanceolate to obovate, folded, with a large gland at tip. **INFL** 1–2 cm. **FL**: corolla 5–5.5 mm. RARE in CA. Rocky slopes; 1050–1150 m. e PR (Santa Rosa Mtns); Baja CA.

M. parryi (Torrey & A. Gray) Barneby (p. 639) Per 2–8 dm, strigose, gen gland-dotted. **LVS** well spaced; lflets 11–23, 0.5–6 mm, oblong obovate to orbicular to ± round, not folded, ± gland-dotted. **INFL** 2–10 cm. **FL**: corolla 5–7 mm. $2n$=20. Open desert washes, stony slopes, roadsides; < 800 m. D; AZ, Mex. [*Dalea p.* Torrey & A. Gray]

MEDICAGO

Duane Isely

Ann or per, unarmed. **ST** prostrate to erect. **LVS** gen odd-1-pinnate, sometimes subpalmately compound; stipules somewhat fused with petiole, entire or deeply cut; lflets 3, gen toothed near tip. **INFL**: raceme, axillary or terminal, few–many-fld. **FL**: calyx lobes ± equal or not; corolla yellow or purple; 9 filaments fused, 1 free. **FR** indehiscent, gen coiled 1.5–5 turns, gen prickly. **SEEDS** 1–several. ± 55 spp.: Eurasia (esp Medit); several cult and naturalized worldwide. (Medea, source of alfalfa, which then bore Greek name Medice) [Small & Jomphe 1989 Can J Bot 67:3260–3294]

1. Fr 1-seeded, reniform, not prickly, black; fl 2–3 mm ..... ***M. lupulina***
1′ Fr several-seeded, spirally coiled, rarely sickle-shaped, prickly or not, tan or dark; fl 3–11 mm
 2. Corolla 6–11 mm, violet or violet-green to greenish yellow, rarely yellow; per, gen erect; fr not prickly ***M. sativa***
 2′ Corolla 3–6 mm, yellow; ann, gen decumbent or low; fr prickly or not
  3. Fr flat, not prickly, ± 1–1.5 cm wide, papery; stipules gen divided nearly to base into a cluster of bristles ..... ***M. orbicularis***
  3′ Fr thick, gen prickly, < 1 cm wide, leathery or woody; stipules not divided nearly to base
   4. Stipules entire to slightly dentate; pl gen densely hairy ..... ***M. minima***
   4′ Stipules sharply cut; pl ± glabrous to sparsely hairy
    5. Lflet with a dark central spot, gen obcordate, length ± = width; lf sometimes subpalmate; stipule gen cut 1/4–1/2 width ..... ***M. arabica***
    5′ Lflet without a dark central spot, gen wedge-shaped or obovate, length 1–2 × width; lf pinnate; stipule gen cut ± 1/2 width or deeper
     6. Lflet 8–20 mm, gen glabrous; corolla 3.5–6 mm; CA-FP; common ..... ***M. polymorpha***
     6′ Lflet 3–10 mm, gen strigose; corolla 2.5–4 mm; SN, SnFrB, SCo, perhaps elsewhere; uncommon ..... ***M. praecox***

M. arabica (L.) Hudson BURCLOVER, SPOTTED BURCLOVER Ann glabrous to sparsely hairy. **ST** gen sprawling, 1–4 dm. **LF** pinnate or subpalmate; stipules gen cut 1/4–1/2 width; lflets gen 1–2.5 cm, widely obovate to obcordate, length ± = width, gen with a central dark spot. **INFL** 2–4-fld. **FL**: calyx 2–2.5 mm; corolla 4–5 mm, yellow. **FR** coiled 4–6 turns, spheric or short-cylindric; prickles curved, hooked. **SEEDS** few. $2n=16$. Disturbed and agricultural areas; < 2000 m. NCo, NCoR, n SN, CW, probably elsewhere; abundant in se US, elsewhere; native probably to Medit.

M. lupulina L. BLACK MEDICK, YELLOW TREFOIL Ann, hairy, sometimes finely strigose. **ST** decumbent to ascending, gen 1–4 dm. **LF**: stipules entire or toothed; lflets gen 1–1.5 cm, obovate. **INFL** ovoid, 10–20-fld, dense. **FL**: calyx 1–1.5 mm; corolla 2–3 mm, yellow. **FR** 2–2.5 mm, reniform, black; veins strong, concentric; prickles 0. **SEED** 1. $2n=16,32$. Disturbed areas, sometimes in forests, mtns; < 2500 m. CA-FP, GB; most of US, widespread elsewhere; native to Eur. [var. *cupaniana* (Guss.) Boiss.]

M. minima (L.) Bartal. BURCLOVER Ann, hairy. **ST** decumbent or ascending, 1–4 dm. **LF**: stipules gen entire, sometimes dentate; lflets 4–12 mm, elliptic or obovate, gen notched at tip. **INFL** 3–6-fld. **FL**: calyx 2–2.5 mm; corolla 3–5 mm, yellow. **FR** coiled 3–4 turns, spheric to short-cylindric; prickles hooked. **SEEDS** few. $2n=16$. Disturbed and waste areas; < 1000 m. CA-FP; to s-c US, sporadic in e US; native probably to Medit. Possibly naturalized in CA.

M. orbicularis (L.) Bartal. Ann, ± glabrous. **ST** spreading, 1–4 dm. **LF**: stipules gen divided nearly to base into a cluster of bristles; lflets 1–1.5 cm, obovate. **INFL** 1–3-fld, gen open. **FL**: calyx 2.5–3 mm; corolla 3–6 mm, yellow. **FR** coiled 3–5 loose turns, gen disk-like, papery; prickles 0. **SEEDS** few. $2n=16$. Uncommon. Disturbed areas; esp < 500 m. CA-FP; sporadic elsewhere in US; native probably to Medit. Probably not naturalized in CA.

M. polymorpha L. (p. 639) CALIFORNIA BURCLOVER Ann, gen glabrous. **ST** prostrate, mat-forming, or ascending, 1–4 dm. **LF**: stipules gen cut 1/2 width or deeper; lflets gen 1–2 cm, gen wedge-shaped or obovate, length 1–2 × width. **INFL** 2–6-fld. **FL**: calyx ± 3 mm; corolla 3.5–6 mm, yellow. **FR** coiled 2–6 turns, ovoid to short-cylindric; prickles ± straight, hooked, or 0. **SEEDS** few. $2n=14$. Abundant. Disturbed and agricultural areas; < 1500 m. CA-FP; widespread in s US, elsewhere; native to Medit. [*M. hispida* Gaertner, incl var. *confinus* (Koch) Burnat] Provides spring pasture in CA; bur-like frs lodge in animal fur.

M. praecox DC. Ann, often very small, strigose. **ST** decumbent, 1–3 dm. **LF**: stipules gen cut ± 1/2 width; lflets 3–10 mm, wedge-shaped or obovate, length 1–2 × width, tip often shallowly notched. **INFL** gen axillary, 1–3-fld. **FL**: calyx 1.5–2 mm; corolla 2.5–4 mm, yellow, ephemeral. **FR** coiled 3–5 turns, spheric or disk-like; prickles slender, hooked. **SEEDS** few. $2n=14$. Uncommon. Creek beds, oak parkland, disturbed areas; < 1500 m. SN, SnFrB, SCo, perhaps elsewhere; native to Medit.

M. sativa L. (p. 639) ALFALFA, LUCERNE Per, ± glabrous or puberulent. **ST** decumbent to gen erect, 2–8 dm. **LF**: stipules entire to sharply toothed; lflets 1–1.5 cm, narrowly lanceolate to obovate. **INFL** spike-like, 8–25-fld, longer in fr. **FL**: calyx 4–4.5 mm; corolla 8–10 mm, purple or multicolored (i.e., violet, violet-green, greenish yellow, rarely yellow). **FR** gen coiled 2–3 turns (rarely only sickle-shaped), leathery; prickles 0. $2n=16,32$. Disturbed and agricultural areas; < 1500 m. CA (exc D); most of US exc se; native to Eurasia. [Isely 1983 Iowa State J Res 57:207–220] Cult; variable polyploid complex in US, incl genetic components from several spp.; often divided into several spp. or sspp.

MELILOTUS SWEETCLOVER

Duane Isely

Ann or bien, unarmed. **ST** gen erect. **LVS** odd-1-pinnate; stipules gen narrow or bristle-like, bases fused to petiole; lflets 3, margin toothed or wavy. **INFL**: raceme, axillary or terminal, slender or short-cylindric, many-fld. **FL**: calyx lobes ± equal; corolla yellow or white; 9 filaments fused, 1 free. **FR** indehiscent, 2–4 mm, ovoid, compressed but thick, leathery, ridged or bumpy. **SEEDS** 1–2. 20 spp.: Eurasia, esp Medit; several spp. widely cult for soil improvement and naturalized. (Greek: honey-Lotus) [Isely 1954 Proc Iowa Acad Sci 61:119–131] TOXIC: inclusion in hay enhances production of mold toxins that may cause cattle death from hemorrhaging.

1. Corolla 2.5–3 mm; fr bumpy or with faint lines .... ***M. indica***
1′ Corolla 4–7 mm; fr irregularly cross-ridged or with network of lines
 2. Corolla white, 4–5 mm; fr with network of lines .... ***M. alba***
 2′ Corolla yellow, 4.5–7 mm; fr irregularly cross-ridged .... ***M. officinalis***

M. alba Medikus (p. 639) WHITE SWEETCLOVER Ann or bien, ± glabrous or strigose. **ST** erect, 0.5–2 m. **LF**: lflets 1–2.5 cm, elliptic-oblong to obovate, ± toothed. **INFL** slender; axis gen 3–8 cm when fls open. **FL**: calyx ± 2 mm; corolla 4–5 mm, white. **FR** 3–5 mm, ovoid, with network of lines. **SEED** 1. $2n$=16,24. Locally abundant. Open, disturbed sites; < 1500 m. CA (exc D); most of n US, adjacent Can; native to Eurasia. Indistinguishable from *M. officinalis* prior to fl.

M. indica (L.) All. (p. 639) SOURCLOVER Ann, ± glabrous. **ST** spreading or erect, 1–6 dm. **LF**: lflets 1–2.5 cm, oblanceolate to wedge-shaped-obovate, gen sharply dentate. **INFL** slender, compact; axis gen 1–2 cm when fls open. **FL**: calyx 1–1.5 mm; corolla 2.5–3 mm, yellow. **FR** 2–3 mm, ovoid, bumpy or with obscure lines. **SEED** 1. $2n$=16. Open, disturbed areas; < 1500 m. CA (exc D), most common s; to se US; native to Medit.

M. officinalis (L.) Pall. (p. 639) YELLOW SWEETCLOVER Bien, ± glabrous to strigose. **ST** erect, 0.5–2 m. **LF**: lflets 1–2.5 cm, elliptic-oblong to obovate, ± toothed. **INFL** slender; axis gen 3–8 cm when fls open. **FL**: calyx 2–2.5 mm; corolla 4.5–7 mm, yellow. **FR** 3–5 mm, ovoid, irregularly cross-ridged. **SEED** 1. $2n$=16. Open, disturbed sites; < 1500 m. CA (exc D); most of n US, adjacent Can; native to Eurasia.

OLNEYA

Duane Isely

1 sp. (S.T. Olney, Am botanist, 1812–1878)

O. tesota A. Gray (p. 639) IRONWOOD Shrub or tree with stipular spines, canescent. **LVS** even-1-pinnate, alternate or clustered; stipular spines sometimes 0 or breaking off, leaving scar, sometimes persisting above lvs and appearing internodal; lflets 8–19, opposite or subopposite, obovate or elliptic, thick; axis extending beyond lflets, pointed. **INFL**: raceme, axillary, 2–4-fld. **FL**: corolla 10–12 mm, wings purple, other petals yellow-white to pink; 9 filaments fused, 1 free. **FR** slowly dehiscent, oblong or elliptic, plump, irregularly narrowed between seeds, persistent. **SEEDS** 1–3. $2n$= 18. Often abundant in washes; < 600 m. PR, DSon; AZ, Mex. Fls erratically. ❀DRN,DRY:8,9,14,19,21,23,IRR:10,11,**12,13**;DFCLT.

ONOBRYCHIS

Duane Isely

LVS odd-1-pinnate; stipules papery, lower clasping st. **INFL**: raceme, spike-like. **FL**: calyx lobes > tube, narrowly lanceolate; corolla wings tiny, << keel; 9 filaments fused, 1 free. **FR** indehiscent, 1-seeded. ± 125 spp.: Eurasia. (Greek: ancient name)

O. viciifolia Scopoli SANFOIN Per, ascending, slightly hairy. **LF**: lflets 15–21, 1–2.5 cm, narrowly elliptic to obovate, upper surface finely red-dotted. **FL**: corolla ± 1 cm. **FR** ± 6–7 mm, obovate, leathery, strongly net-ridged, short-prickly near margin. Uncommon. Disturbed places; 900–1200 m. n SNF; to B.C., sporadic in US; native to Eurasia. Cult for forage, soil improvement.

ORNITHOPUS

Duane Isely

Ann, unarmed, ± glabrous or hairy. **ST** sprawling or ascending, 2–7 dm. **LF** odd-1-pinnate; stipules small or 0; lflets many, tips each with a small point. **INFL** head-like, axillary, 1–5-fld, subtended by a cluster of bracts or not. **FL**: calyx lobes ± equal, < tube; corolla 6–9 mm; 9 filaments fused, 1 free. **FR** indehiscent, linear or oblong, flat or round in X-section, separating into 1-seeded segments. 6–7 spp.: Eur. (Greek: birdfoot)

1. Fr round in X-section, narrowed slightly between seeds or not; infl not subtended by bracts; lflets 7–15 .... ***O. pinnatus***
1′ Fr flat, narrowed conspicuously between seeds; infl subtended by a cluster of bracts; lflets ± 19–31 ..... ***O. sativus***

O. pinnatus (Miller) Druce (p. 639) Pl glabrous or puberulent. **LF**: lflets linear to oblanceolate. **FL**: corolla yellow. **FR** 2–3.5 cm, linear, gen curved. $2n$=14. Disturbed areas; 150–200 m. CCo (Santa Cruz Co.); native to Eur. Collected in 1955, 1964, when reportedly common along roadsides; probably naturalized only at this site.

O. sativus Brot. (p. 639) Pl hairy. **LF**: lflets ovate to lanceolate. **FL**: corolla white or pink. **FR** 1–2.5 cm, oblong, straight. Disturbed areas; 150–200 m. CCo (Santa Cruz Co.); native to Eur. [*O. roseus* Dufour] Many collections since 1942; reportedly spreading.

OXYTROPIS

Duane Isely

Per, unarmed, hairy. **LVS** odd-1-pinnate, basal, sometimes also cauline; stipules gen partly fused to petiole, initially forming a sheath, or free. **INFL**: raceme, gen scapose, spike- or head-like, or 1–2-fld; bracts gen persistent. **FL**: calyx lobes < tube; corolla pink-purple, white, or yellowish, keel tip beaked; 9 filaments fused, 1 free; style glabrous. **FR** ascending or reflexed, gen persistent, lanceolate or inflated, ± 2-chambered, septum arising from upper suture, ± incomplete. ± 300 spp.: Eurasia, N.Am. (Greek: sharp keel) [Barneby 1952 Proc Calif Acad Sci Series IV 27:177–309] Seriously TOXIC: causes "staggers" in livestock, mostly outside CA.

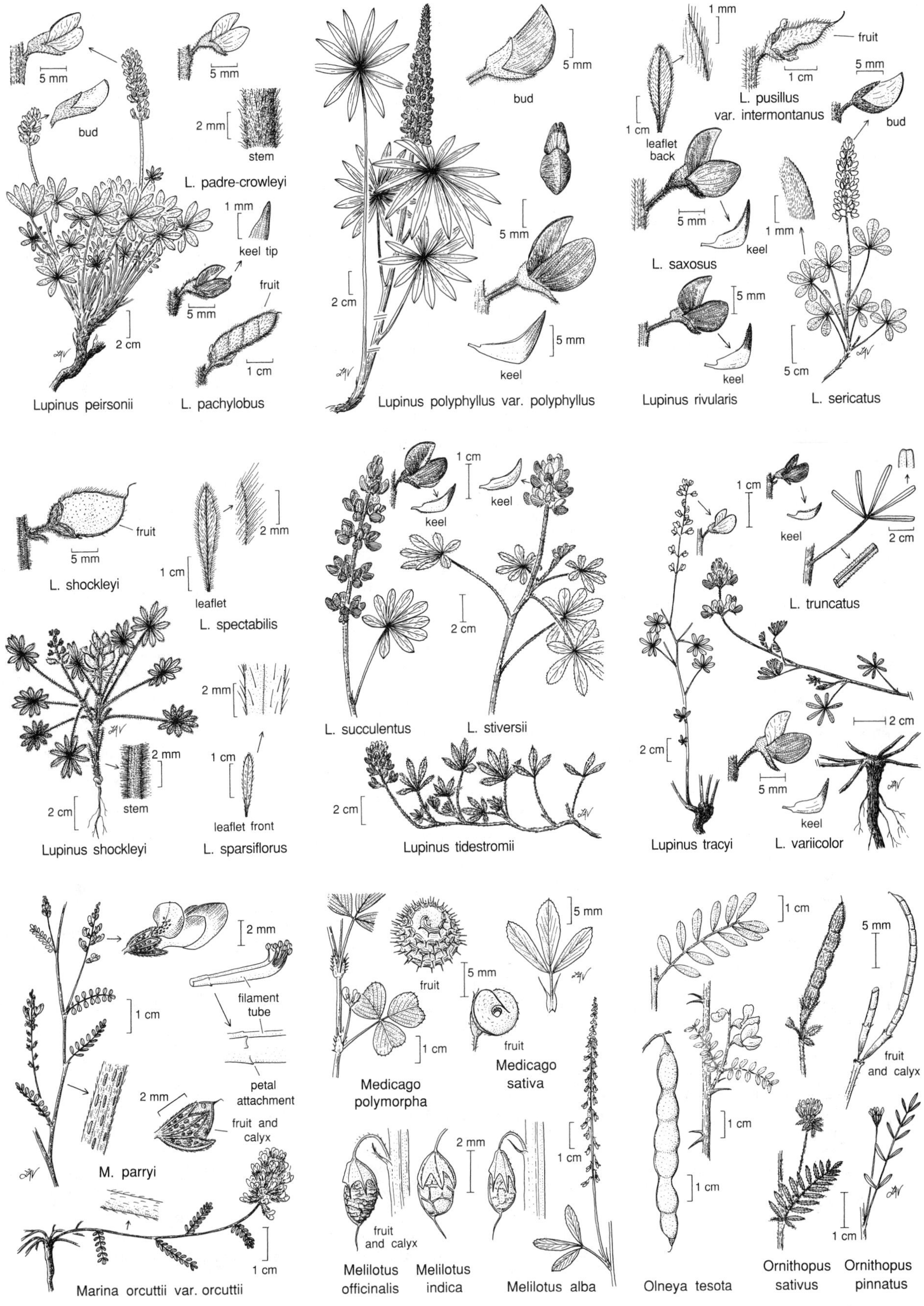
5 mm
bud
5 mm
2 mm
stem
L. padre-crowleyi
1 mm
keel tip
5 mm
fruit
2 cm
1 cm
Lupinus peirsonii
L. pachylobus
5 mm
bud
5 mm
2 cm
5 mm
keel
Lupinus polyphyllus var. polyphyllus
1 mm
fruit
1 cm
L. pusillus
var. intermontanus
5 mm
bud
1 cm
leaflet
back
5 mm
keel
1 mm
L. saxosus
5 mm
keel
5 cm
Lupinus rivularis
L. sericatus
fruit
5 mm
L. shockleyi
2 mm
1 cm
leaflet
L. spectabilis
2 mm
2 mm
stem
2 cm
1 cm
leaflet front
Lupinus shockleyi
L. sparsiflorus
1 cm
keel
keel
2 cm
L. succulentus
L. stiversii
2 cm
Lupinus tidestromii
1 cm
keel
2 cm
L. truncatus
2 cm
2 cm
5 mm
keel
Lupinus tracyi
L. variicolor
2 mm
filament
tube
1 cm
petal
attachment
2 mm
fruit and
calyx
M. parryi
1 cm
Marina orcuttii var. orcuttii
5 mm
fruit
5 mm
1 cm
fruit
Medicago
sativa
Medicago
polymorpha
2 mm
1 cm
fruit
and calyx
Melilotus
officinalis
Melilotus
indica
Melilotus alba
1 cm
1 cm
1 cm
Olneya tesota
5 mm
fruit
and calyx
1 cm
Ornithopus
sativus
Ornithopus
pinnatus

1. Fl quickly reflexed, fr reflexed; lvs basal and gen 1–3 cauline; infl a loose raceme .... ***O. deflexa*** var. ***sericea***
1′ Fl and fr ascending or erect; lvs all basal; infl densely spike- or head-like
  2. Lflets 23–39; pl green, puberulent, sticky-glandular; corolla 12–18 mm .... ***O. borealis*** var. ***viscida***
  2′ Lflets 7–17; pl silvery or gray, silky or tomentose, not glandular; corolla 7–10 mm
    3. Fls 2–12 per infl; fr ovoid-inflated, ± 7–9 mm wide, papery .... ***O. oreophila***
    3′ Fls 1–3 per infl; fr lanceolate, ± 5–6 mm wide, leathery .... ***O. parryi***

O. borealis DC. var. ***viscida*** (Nutt.) Welsh (p. 645) Pl green, puberulent, glandular-sticky, sometimes tiny and cespitose. **LVS** basal; lflets 23–39, 3–10 mm, oblong-lanceolate to ovate, flat or folded. **INFL** spike-like, exserted, sometimes longer in fr; fls 4–many, ascending or erect. **FL**: corolla 12–18 mm, red-purple or pale. **FR** ascending or erect, 8–14 mm, lanceolate to ovate in outline, somewhat inflated, papery, slightly 2-chambered; stalk-like base 0. Aspen meadows to alpine; 1200–3900 m. s SNH, W&I; circumboreal. [*O. viscida* Nutt.] ❀TRY.

O. deflexa (Pall.) DC. var. ***sericea*** Torrey & A. Gray (p. 645) BLUE PENDENT-POD OXYTROPE Pl green or gray. **LVS** basal, often also 1–3 cauline; stipules ± free; lflets 15–31, 3–20 mm, lanceolate to ovate, flat or folded. **INFL** spike-like; fls few–many, quickly reflexed; peduncle gen > 15 cm, often curved. **FL**: corolla 5–10 mm, dull white to pale lilac or blue, ± = or > calyx. **FR** reflexed, 3– 4.5 cm, elliptic or oblong, membranous or papery, slightly 2-chambered; stalk-like base gen short. 2*n*=16. RARE in CA. Moist meadows, forest openings; ± 2800 m. W&I (White Mtns, Mono Co.); to AK, NV, NM.

O. oreophila A. Gray (p. 645) Pl silvery or gray, silky, cespitose. **LVS** basal, congested; lflets 7–17, 2–10 mm, elliptic to oblong, folded. **INFL** head-like, incl or exserted; fls 2–12, ascending or erect. **FL**: corolla 7–10 mm, pink-purple, sometimes white. **FR** ascending or erect, 9–15 mm, 7–9 mm wide, ovoid-inflated, thinly papery, slightly 2-chambered; stalk-like base 0. 2*n*=16. Uncommon. Alpine; 3400–3800 m. SnBr, W&I; to UT, AZ. Variable; upper alpine pls only a few cm tall.

O. parryi A. Gray (p. 645) Pl gray, silky or tomentose, cespitose. **LVS** basal, clustered; lflets 11–15, 2–12 mm, ovate to oblong, folded. **INFL** head-like, gen exserted; fls 1–3, ascending or erect. **FL**: corolla 7–10 mm, purple. **FR** ascending or erect, 15–20 mm, 5–6 mm wide, lanceolate, leathery, nearly 2-chambered; stalk-like base 0. Near and above timberline; 3400–3800 m. SNE, W&I; to ID, Colorado, NM. ❀TRY.

PARKINSONIA

Elizabeth McClintock

Tree, armed with spines at nodes. **ST** ± zig-zag; bark smooth, green. **LVS** 2-pinnate, alternate; main axis a spine gen < 2 cm, persistent; 1° lflets 1–3 pairs, crowded, main axis flat, persistent; 2° lflets many. **INFL**: raceme. **FL** slightly bilateral; sepals ± free, all alike; petals yellow; stamens 10, free, exserted. **FR** dehiscent, oblong, ± inflated, gen narrowed between seeds. 2 spp.: 1 Am, 1 Afr; cult. (J. Parkinson, London, apothecary and author, 1567–1650) [Carter 1974 Proc Calif Acad Sci 40(2): 18–20; 50–54] Lvs sometimes considered 1-pinnate, alternate or in clusters of 1–6 in spine axils, with main axis flat, persistent, 1° lflets many.

P. aculeata L. MEXICAN PALO VERDE Tree < 12 m. **LF** 1–3 dm; main 1° lflet axis < 30 cm, conspicuous, persistent as ribbon-like streamer; 2° lflets 30–60, scattered, 3–5 mm, 1–1.5 mm wide, elliptic, ephemeral. **INFL** < lf. **FL** ± 2 cm wide; sepals reflexed; petals ± round, banner red-spotted at base, becoming entirely red in age. **FR** 5–10 cm, ± thickened, leathery. **SEEDS** several. Uncommon. Disturbed, dry places; < 400 m. SnJV, SCo, WTR, PR; native to deserts of AZ, Baja CA, S.Am.

PEDIOMELUM INDIAN BREADROOT

James W. Grimes

Per, unarmed, gland-dotted; hairs, stalked glands, or both; roots deep, woody, enlarged near ground surface. **ST**: main axis erect, nearly 0 to short; branches short, decumbent to ascending, sometimes underground. **LVS** ± palmately compound, ± basal (or cauline at branch tips); stipules at base of pl fused, those above free; lflets 5–7. **INFL**: basal, axillary, or terminal on branches, raceme with 1 sometimes tardily deciduous bract and 2–3 fls per node; pedicel sometimes very short. **FL**: calyx base swollen on top, tube enlarging in fr; corolla at least partly blue to purple; 9 filaments fused, 1 less so or free; ovary ± hairy, ovule 1, style tip curved to bent, stigma head-like. **FR** transversely dehiscent, beaked, hairy, rarely glandular. **SEED** elliptic, smooth or ridged. 22 spp.: N.Am. (Greek: plain apple) [Grimes 1990 Mem New York Bot Gard 61:1–114] *Bituminaria bituminosa* (L.) Stirton (*Psoralea b.* L.) possibly naturalized in SnBr; *Pediomelum mephiticum* (S. Watson) Rydb. incorrectly reported for s CA.

1. Lf palmately compound; pedicel 4–6 mm; seed smooth .... ***P. californicum***
1′ Lf palmately compound, but often with 1 lflet extended beyond others; pedicel 0–1.5 mm; seed ridged .... ***P. castoreum***

P. californicum (S. Watson) Rydb. (p. 645) **LF**: stipule 7–10 mm; petiole 8–11 cm; lflets 5–7, 15–28 mm, obovate or widest in middle. **INFL**: bract 6.5–8 mm. **FL** 8–12 mm; calyx 9–10.5 mm; banner 10–11 mm. **FR** ovate to round in outline; body 4–9 mm; beak 1–4 mm, straight, linear. **SEED** 5–5.5 mm, reniform, smooth, red-brown. 2*n*=22. Open chaparral, woodlands; 1000–2500 m. NCoRI, CW, TR, SnJt; Baja CA. [*Psoralea c.* S. Watson] ❀TRY.

P. castoreum (S. Watson) Rydb. **LF**: stipule 5–13.5 mm; petiole 6.8–15 cm; lflets 5–6, 25–42 mm, elliptic to oblanceolate. **INFL**: bract 3.5–8 mm. **FL** 9–13 mm; calyx 10–12 mm; banner 9–13 mm. **FR** ovate to elliptic in outline; body 6–8 mm; beak 8–11 mm, straight to curved, triangular. **SEED** 6 mm, reniform, ridged, gray. Open areas, roadcuts; < 1750 m. DMoj (San Bernardino Co.); NV, AZ. [*Psoralea c.* S. Watson] ❀TRY.

PETERIA

Duane Isely

Per; spines stipular. **ST** decumbent to erect. **LVS** odd-1-pinnate; stipules spiny or 0; lflets 9–many. **INFL**: raceme, spike-like, terminal. **FL**: calyx tube sometimes bulged on 1 side near base, lobes < or > tube, upper pair fused 1/2 or more; corolla pink or white; 9 filaments fused, 1 free; style tip finely hairy around stigma. **FR** dehiscent, oblong, ± flat but plump, leathery; base stalk-like. 4 spp.: s US, Mex. (R. Peter, 19th century Kentucky botanist) [Porter 1956 Rhodora 58:344–354]

P. thompsoniae S. Watson (p. 645) SPINE-NODED MILKVETCH Pl rhizomed; taproot often swollen. **ST** 2–6 dm. **LF**: lflets 13–21, elliptic or obovate. **INFL** often glandular-hairy. **FL**: calyx 11–15 mm, tube cylindric, gen darkly glandular-puberulent; corolla gen 15–20(25) mm; style hairs often hidden by pollen. **FR** 4–6 cm, glabrous. RARE in CA. Sandy alluvial fans; 800 m. DMtns (California Valley, se Inyo Co.); to ID, UT, sw Colorado, AZ.

PHASEOLUS WILD BEAN

Duane Isely

Ann, per, vine, unarmed; hairs gen incl minute, hooked ones. **LVS** odd-1-pinnate; main axis extended beyond basal lflets, with stipule-like appendages, at least at tip; stipules persistent; lflets 3, entire or lobed. **INFL**: raceme; bracts persistent. **FL**: calyx lobes << tube; corolla incurved, sickle-shaped in bud, keel coiled 2–3 turns; 9 filaments fused, 1 free. **FR** dehiscent, linear to oblong, flat or round in X-section. **SEEDS** few–several. ± 50 spp.: neotrop, warm regions; those of economic value on all continents. (Classical name, presumably for a bean) *P. vulgaris* L., *P. lunatus* L., and others may be found as waifs from cult.

P. filiformis Benth. (p. 645) WRIGHT'S PHASEOLUS, SLENDER-STEMMED BEAN **LF**: lflets ovate-triangular in outline, gen lobed. **INFL**: peduncle 2–10 cm. **FL**: corolla pink-purple. **FR** 2.5–3.5 cm, 5–7 mm wide, oblong, gen curved, ± glabrous. UNCOMMON. Washes; ± 125 m. DSon (Coachella Valley, Riverside Co.); to TX, n Mex. [*P. wrightii* A. Gray] ❀TRY.

PICKERINGIA

Duane Isely

1 sp. (C. Pickering, Am naturalist, 1805–1875)

P. montana Nutt. Shrub, rhizomed, leafy or with fls; thorns terminal or axillary. **ST** intricately branched, 1–3 m. **LF** palmate or sometimes simple, evergreen; stipules 0; lflets 1–3, 1–2 cm, elliptic or ovate; petiole ± 0. **INFL**: raceme, terminal or axillary. **FL**: calyx 5–7 mm, barely lobed; corolla 1.5–1.8 cm, gen purple, keel petals free; stamens 10, free. **FR** dehiscent, 3–6 cm, 4–5 mm wide, oblong; margins narrowed between seeds, often wavy. **SEEDS** 1–8. $2n$=28. Chaparral, open woodlands, washes; gen < 1700 m. NCoR, n SNF, SnFrB, SCoR, n ChI, TR, PR; Baja CA. Vars. intergrade but are geog separate.

1. Pl glabrous to inconspicuously strigose ...... var. ***montana***
1′ Pl strigose to tomentose .... var. ***tomentosa***

var. ***montana*** (p. 645) **FR** rarely abundant. Chaparral, open woodlands; gen < 660 m (to 1700 m in s). NCoR, n SNF, SnFrB, SCoR, n ChI, TR. ❀DRN,SUN:3,**7**,8,9,**14-16**,17,**18-23**,24;DFCLT.

var. ***tomentosa*** (Abrams) I.M. Johnston (p. 645) **FR** abundant. Chaparral, washes; < 1700 m. SnBr, PR; Baja CA. ❀TRY.

PISUM PEA

Duane Isely

Ann, bushy or climbing, unarmed, glabrous. **LVS** even-1-pinnate; stipules like and often > lflets; lflets 4–6, opposite; main axis ending in a tendril. **INFL**: raceme, axillary, 1–3-fld. **FL**: corolla white or 2-colored; 9 filaments fused, 1 free; style longitudinally folded, puberulent on concave side. **FR** dehiscent, oblong, flat. **SEEDS** few–several. 2 spp.: Medit. (Ancient name)

P. sativum L. (p. 645) GARDEN PEA, FIELD PEA, SUGAR PEA **LF**: stipule margins toothed or wavy; lflets 2–4 cm, ovate or elliptic. **FL** 10–25 mm, white, purple, or both. $2n$=14. Uncommon. Urban or agricultural areas; < 1000 m. CA-FP; to e US, elsewhere; native to Medit. [var. *arvense* (L.) Poir.] Cult worldwide.

PROSOPIS MESQUITE

Elizabeth McClintock

Shrub, tree; stipular spines gen 2 per node; roots long, spreading. **LVS** even-2-pinnate, alternate, deciduous; 1° lflets gen 1–2 pairs, opposite; 2° lflets gen many, opposite. **INFL**: raceme, axillary, spike-like or spheric head, many-fld. **FL** radial, small, greenish white or yellow; calyx shallowly bell-shaped, lobes very short; petals gen inconspicuous; stamens 10, exserted, free; style exserted, gen appearing before stamens. **FR** indehiscent, ± flat, ± narrowed between seeds or tightly coiled, pulpy when young, then woody. **SEEDS** several. ± 44 spp.: esp Am (also sw Asia, Afr). (Greek: burdock, for obscure reasons) [Burkhart 1976 J Arnold Arbor 57:220–524; Holland 1987 Madroño 34:324–333] Used for timber, firewood, shade, orn, bee, human, and livestock food.

1. Fr tightly coiled; petals fused
  2. Infl a spike-like raceme; lf hairy .... ***P. pubescens***
  2′ Infl a head; lf glabrous or short-hairy .... ***P. strombulifera***
1′ Fr not coiled; petals free
  3. Pl ± glabrous; 2° lflets gen > 15 mm, length 7–9 × width .... ***P. glandulosa*** var. ***torreyana***
  3′ Pl with short, dense hairs; 2° lflets gen < 15 mm, length 3–4 × width .... ***P. velutina***

P. glandulosa Torrey var. ***torreyana*** (L. Benson) M. Johnston (p. 645) MESQUITE, HONEY MESQUITE Shrub, tree < 7 m, ± glabrous; crown often wider than tall. **ST**: branches arched, crooked; spines 0.5–4 cm. **LF** glabrous; 1° lflets gen 1 pair, 6–17 cm; 2° lflets 7–17 pairs, 1–2.5 cm, oblong, length 7–9 × width. **INFL**: raceme, 6–10 cm, spike-like. **FL**: petals free, 2.5–3.5 mm. **FR** 5–20 cm, slightly narrowed between seeds, glabrous. **SEEDS** gen 5–18, 6–7 mm, oblong. Common. Grasslands, alkali flats, washes, bottomlands, sandy alluvial flats, mesas; 0–1700 m. SnJV, PR, D. ❀SUN, DRN:7,**9**,10,**12**,**14–16**,17,18,**19–24**&IRR:**8**,**11**,**13**;also STBL.

P. pubescens Benth. SCREW BEAN, TORNILLO Shrub, tree < 10 m; crown gen ± narrow. **ST**: branches ascending; spines 4–12 mm. **LF** hairy; 1° lflets 1 or 2 pairs, 3–5 cm; 2° lflets 5–8 pairs, 2–10 mm, oblong, length 2–3 × width. **INFL**: raceme, 4–8 cm, spike-like. **FL**: petals fused, 2–3 mm. **FR** 3–5 cm, tightly coiled. **SEEDS** gen 3 mm, ovoid. $2n$=28. Uncommon. Creek, river bottoms, sandy or gravelly washes or ravines; 100–1300 m. SnBr, D; sw US, n Mex. Fr used for food and as a coffee substitute. ❀SUN:7,**14**,15, 16,18,**19–21**,22,23&IRR:**8**,**9**,10,11,**12**,**13**;also STBL.

P. strombulifera (Lam.) Benth. Shrub < 3 m; crown not seen; roots long, spreading. **ST** ± zig-zag; spines 1–2 cm. **LF** glaucous, glabrous or short-hairy; 1° lflets 1 pair, 1–3 cm; 2° lflets 2–8 pairs, 2–10 mm. **INFL**: head, ± 15 mm wide. **FL**: petals fused, 3–4 mm. **FR** 1.5–5 cm, tightly coiled. $2n$=28. Uncommon. Disturbed places; ± 50 m. se DSon (Bard, Imperial Co.); native to Argentina.

P. velutina Wooton Shrub, tree < 15 m; crown ± spreading, rounded; hairs short, dense, on ± all parts, incl fr. **ST** often crooked; spines 1–2 cm. **LF**: 1° lflets 1–2 pairs, 2–9 cm; 2° lflets gen 15–30 pairs, 4–15 mm, oblong, length 3–4 × width. **INFL**: raceme, 5–15 cm, spike-like. **FL**: petals free, 2–3 mm. **FR** 8–15 cm, linear, ± flat, barely narrowed between seeds. **SEEDS** 5–7 mm, ovate. $2n$=28. Uncommon. Sandy, rocky soils in canyons, washes; < 1700 m. SnJV, CCo, SCo; native to sw US (exc CA), n Mex.

PSORALIDIUM SCURF-PEA

James W. Grimes

Per, unarmed, gland-dotted, glabrous to ± sparsely hairy; rhizomes or roots (or both) woody. **ST** erect, < 7.5 dm, green or yellow toward base. **LVS** palmately compound, cauline; stipules free; lflets 3–5. **INFL**: raceme, axillary, with 1 deciduous bract and 1–3 fls per node. **FL** pedicelled, 6–7 mm; calyx flaring back, tearing along 1 lateral sinus in fr; corolla white, yellow, or purple; 9 filaments fused, 1 less so or free; ovary glabrous to hairy, ovule 1, style tip bent, stigma head-like. **FR** indehiscent, ± spheric. **SEED** elliptic to round in outline. 4 spp.: N.Am. (Greek: diminutive of *Psoralea*) [Grimes 1990 Mem New York Bot Gard 61:1–114]

P. lanceolatum (Pursh) Rydb. (p. 645) **LF**: petiole 9–21 mm; lflet 17–33 mm, linear to oblanceolate. **FL** 4–7 mm; calyx 2–2.5 mm; petals white to purple-blue. **FR** 4–6 mm, papillate-glandular to glandular, ± hairy. **SEED** 4–5 mm, smooth, shiny. Alluvial plains, sand; < 2500 m. GB; to c Can, c US. [*Psoralea l.* Pursh ssp. *scabra* (Nutt.) Piper]

PSOROTHAMNUS

Duane Isely

Per, shrub, small tree, gen with thorns, gland-dotted. **ST** gen intricately branched. **LVS** odd-1-pinnate or simple. **INFL**: raceme, sometimes spike- or head-like, axillary or terminal; pedicels gen with bractlets. **FL**: calyx lobes gen unequal, upper pair often largest; petals all arising from receptacle, violet, blue, or 2-colored (purple and white); stamens 10, filaments partly fused; ovules gen 2. **FR** indehiscent, incl in or exserted from calyx, gen glandular. **SEED** 1. 9 spp.: sw US, Mex. (Greek: scabshrub) [Barneby 1977 Mem New York Bot Garden 27:21–54, 598–607]

1. Lf simple, linear or oblanceolate, rarely with 3 lflets
  2. Lvs persistent; infl axis not a thorn .... ***P. schottii***
  2′ Lvs deciduous by early summer; infl axis extended beyond fls as a thorn .... ***P. spinosus***
1′ Lvs all odd-1-pinnate or some simple
  3. Fr 2–4.5 mm, incl; infl a spike or raceme, dense, spheric or short-cylindric; pedicel lacking bractlets
    4. Terminal lflets of major lvs (or some of them) > lateral lflets; glands of twigs << 0.5 mm wide .... ***P. emoryi***
    4′ Terminal lflets of all lvs ± = lateral lflets; glands of twigs ± 0.5 mm wide .... ***P. polydenius***
  3′ Fr 7–10 mm, exserted; infl a ± open, sometimes spike-like raceme; pedicel with bractlets
    5. Fr glabrous, with many small glands forming longitudinal lines; D (e half San Bernardino Co.) ..... ***P. fremontii***
      6. Lflet linear, < ± 1 mm wide; DSon (Whipple Mtns, extreme se San Bernardino Co.) .... var. ***attenuatus***
      6′ Lflet narrowly elliptic to ovate or obovate, ± 1.5–3 mm wide; DMtns (e-c San Bernardino Co.) .. var. ***fremontii***
    5′ Fr glabrous or finely hairy, with few or several large, scattered glands; SNE to SnBr, DMoj (exc e half of San Bernardino Co.) .... ***P. arborescens***
      7. Lflets gen continuous with axis .... var. ***simplicifolius***
      7′ Lflets gen (or at least some) jointed to axis
        8. Calyx 7–9 mm, gen conspicuously hairy .... var. ***arborescens***
        8′ Calyx 5–7 mm, glabrous or sparsely hairy .... var. ***minutifolius***

P. arborescens (A. Gray) Barneby Shrub < 1 m, sometimes unarmed, glabrous to puberulent. **LF**: lflets gen 5–7, linear to ovate, 3–10 mm, terminal (sometimes all) often continuous with axis. **INFL**: raceme, open; pedicels with bractlets. **FL**: calyx 5–9 mm, lobes ± equal, gen < tube; corolla 6–10 mm, violet-purple. **FR** exserted, 7–10 mm, glabrous or finely hairy, with large, scattered glands. Desert mtns, slopes, canyons, flats, washes; 100–1900 m. SnBr, SNE, DMoj; s NV, Mex. Vars. similar morphologically, distinct geog. Close to *P. fremontii*, which might be considered to constitute two additional vars. ❀TRY.

var. ***arborescens*** (p. 645) MOJAVE INDIGO BUSH **LF**: lflets lanceolate to ovate, larger 2.5 mm or more wide, upper often continuous with axis. **FL**: calyx 7–9 mm, gen conspicuously hairy. UNCOMMON. Desert, incl mtns; 400–800 m. sw DMoj (Kern, San Bernardino cos.); Mex. [*Dalea a.* (A. Gray) A.A. Heller; *D. fremontii* var. *saundersii* (Parish) Munz]

var. ***minutifolius*** (Parish) Barneby (p. 645) **LF**: lflets lanceolate to ovate, larger 2.5 mm or more wide, gen jointed to axis. **FL**: calyx 5–7 mm, glabrous or sparsely hairy. Desert mtn slopes, canyons, talus; 150–1900 m. SNE (Mono Co.), n&c DMoj; s NV. [*Dalea fremontii* Torrey var. *m.* (Parish) Benson]

var. ***simplicifolius*** (Parish) Barneby (p. 645) **LF**: lflets oblanceolate, gen 1–2 mm wide, gen continuous with axis. **FL**: calyx 5–7 mm, puberulent. Lower desert mtn slopes, flats, washes; 100–1100 m. SnBr, adjacent DMoj (s San Bernardino, adjacent Riverside cos.). [*Dalea californica* S. Watson]

P. emoryi (A. Gray) Rydb. (p. 645) Subshrub < 1 m, < 2 m wide, gen canescent, (sometimes becoming glabrous); glands of twigs << 0.5 mm wide. **LF**: lflets of middle lvs 5–9, 2–10 mm, narrowly oblong to obovate, terminal lflet gen > others. **INFL** spike-like, ovoid or spheric, dense; bractlets 0. **FL**: calyx 4–7 mm, lobes ± = or > tube; corolla 4–6 mm, 2-colored (purple and white), puberulent. **FR** incl, ± 2.5–3 mm. $2n$=20. Desert flats, washes, dunes; < 700 m. s DMoj, DSon; sw AZ, Mex. [*Dalea e.* A. Gray] ❀DRN:11,**12**, **13**.

P. fremontii (A. Gray) Barneby Shrub < 1 m, gen silvery-strigose. **LF**: lflets 3–25 mm, linear to ovate, gen jointed to axis, exc uppermost. **INFL**: raceme, ± open; pedicels with bractlets. **FL**: calyx 5–9 mm, lobes unequal, < tube; corolla 7–9.5 mm, violet-purple. **FR** exserted, 7–10 mm, glabrous; small glands forming longitudinal lines. $2n$=20. Granite and volcanic slopes, flats, canyons; 250–1350 m. DMtns; to s UT, AZ. Like *P. arborescens* exc in fr, geography. ❀TRY.

var. ***attenuatus*** Barneby **LF**: lflets linear, < ± 1 mm wide. Uncommon. Habitats of sp.; 450–900 m. e DSon (se San Bernardino Co.); s NV, AZ.

var. ***fremontii*** (p. 645) **LF**: lflets narrowly elliptic to ovate or obovate, ± 1.5–3 mm wide. Habitats of sp.; 250–1350 m. e DMtns (ne San Bernardino Co.); to s UT, AZ. [*Dalea f.* A. Gray]

P. polydenius (S. Watson) Rydb. Shrub < 1.5 m, widely spreading, finely strigose; hairs pointing down, or puberulent; glands of twigs ± 0.5 mm wide. **LF**: lflets 7–13, 1–4.5 mm, obovate to ± round, terminal ± = lateral. **INFL**: raceme, spike-like, ovoid or short-cylindric, dense; pedicels < 1 mm, bractlets 0. **FL**: calyx 5–6 mm, lobes unequal, < tube; corolla 4.5–6 mm, pink-purple, persistent after fl. **FR** incl, < 4.5 mm. $2n$=20. Locally abundant. Desert flats, hills; 900–2250 m. SNE, DMoj; to UT. [*Dalea p.* S. Watson] ❀TRY.

P. schottii (Torrey) Barneby Shrub < 2 m, green or gray-strigose. **LF** simple, persistent, 1–3 cm, linear, gland-dotted. **INFL**: raceme, open, 5–15-fld; axis not extended beyond fls, not a thorn. **FL**: calyx 4–6 mm, lobes < tube; corolla 7–10 mm, bright blue. **FR** exserted; glands large, spaced. $2n$=20. Slopes, benches, washes; 0–600 m. DSon; AZ, Mex. [*Dalea s.* Torrey, incl var. *puberula* (Parish) Munz] ❀TRY.

P. spinosus (A. Gray) Barneby (p. 645) SMOKE TREE Shrub or tree 1.5–8 m, gray-canescent. **LF** simple, early deciduous, 0.5–2 cm, oblanceolate, thick, gland-dotted. **INFL**: raceme, gen dense, 5–15-fld; axis extended beyond fls, gen as a thorn; pedicels ± 1 mm. **FL**: calyx 4.5–5 mm, lobes < tube; corolla 6–8 mm, blue-purple. **FR** slightly exserted. $2n$=10. Common. Desert washes; < 400 m. D; AZ, nw Mex. [*Dalea s.* A. Gray] ❀DRN:**12,13**.

ROBINIA LOCUST

Duane Isely & Elizabeth McClintock

Shrub or tree, gen spreading from underground parts, gen with stipular spines not gland-dotted. **LVS** odd-1-pinnate, alternate, deciduous. **INFL**: raceme, axillary. **FL**: calyx bell-shaped, lobes 5; petals 5, white or pink, banner reflexed; 9 filaments fused, 1 free. **FR** flat or plump, dehiscent. 4 spp.: US. (J. & V. Robin, who introduced pls to Eur, 16–17th century) [Isely & Peabody 1984 Castanea 49:187–202]

1. Corolla pink; infl hairs coarse, glandular; fr glandular-hairy; probably native, DMoj .... ***R. neomexicana***
1′ Corolla white; infl hairs fine, nonglandular; fr glabrous; naturalized, CA-FP, GB .... ***R. pseudoacacia***

R. neomexicana A. Gray (p. 651) DESERT LOCUST Shrub or small tree. **INFL** pendent or not. **FL**: corolla 2–2.5 cm. **FR** thick, not winged. Uncommon. Canyons in pinyon/juniper woodland; < 1500 m. e DMoj (Mid Hills, ne San Bernardino Co.); to w TX, n Mex. ❀15–17,IRR:1,**2,3,7–10**,11,**14,18,19**,20–24.

R. pseudoacacia L. (p. 651) BLACK LOCUST Tree. **INFL** pendent. **FL**: corolla 1.5–2 cm. **FR** flat; upper margin narrowly winged. $2n$=20. Locally abundant. Near abandoned houses, along roadsides, canyon slopes, stream banks; 50–1900 m. CA-FP, GB; native to e US. TOXIC: ingested seeds, lvs, bark may be fatal to humans and livestock.

RUPERTIA RUPERT'S SCURF-PEA

James W. Grimes

Per, unarmed, gland-dotted (esp lvs), hairy or not, caudexed, rhizomed, or stoloned; roots deep, woody, extensive. **ST** ± decumbent at base or erect. **LVS** odd-1-pinnate, cauline; stipules reflexed, deciduous; lflets 3. **INFL**: raceme, gen axillary, with 1 deciduous bract and 2–3 fls per node. **FL** pedicelled; calyx base swollen on top, tube enlarging in fr; corolla yellowish white to yellow; 9 filaments fused, 1 less so or free; ovary ± hairy, ovule 1, style tip bent, stigma head-like. **FR** indehiscent, elliptic to depressed-obovate in outline, beaked or not. **SEED** reniform, smooth. 3 spp.: w N.Am, esp CA. (Rupert C. Barneby, botanist, 1911-) [Grimes 1990 Mem New York Bot Gard 61:1–114]

1. Largest stipules 13–15 mm, widely elliptic to obtriangular; largest bracts 9–13 mm .... ***R. hallii***
1′ Largest stipules 4–10 mm, ± elliptic, linear-lanceolate to -oblanceolate, or triangular; largest bracts 3–7 mm
  2. Calyx in fl 6–8 mm; banner 10–14 mm; fr 4–7 mm, with an abrupt, small point at tip .... ***R. physodes***
  2′ Calyx in fl 9–10 mm; banner 14–15 mm; fr 9–13 mm, with widely attached, < 3 mm beak .... ***R. rigida***

R. hallii (Rydb.) Grimes (p. 651) Pl hairy or not. **ST** erect, < 1 m. **LF**: stipule 13–15 mm, widely elliptic to obtriangular; petiole 10–30 mm; main axis 2–2.8 cm; lflets 4–9 cm, lanceolate to widely ovate, with glands but no hairs on both surfaces. **INFL**: bract 9–13 mm, tardily deciduous; pedicel 2 mm. **FL**: calyx 14–15 mm; banner 11–12 mm. **FR** 7–10 mm, elliptic, with hairs and sparse, minute, golden glands that fade with age; beak 1–3 mm, widely attached. **SEED** 6–7 mm, red-brown. Very uncommon. Woodland openings; < 2250 m. SNF (Butte, Tehama cos.) [*Hoita h.* Rydb.] Fr rare. ❀TRY.

R. physodes (Hook.) Grimes (p. 651) Pl sparsely hairy, sometimes stoloned. **ST** erect or decumbent, ± 0.5 m. **LF**: stipule 4–10 mm, linear-lanceolate to -oblanceolate or ± elliptic; petiole 11–65 mm; main axis 9–21 mm; lflet 3.5–7 mm, triangular to lanceolate, with glands, sparse hairs on upper surface, becoming ± glabrous on lower surface. **INFL**: bract 3–7 mm, deciduous; pedicel 1.5–2.5 mm. **FL**: calyx 6–8 mm; banner 10–14 mm. **FR** 4–7 mm, depressed-obovate, golden-red, faintly net-sculptured, with red-brown hairs; tip with an abrupt, small point. **SEED** 5–6.5 mm, dark red-brown. $2n$=22. Woodlands; < 2500 m. KR, NCoR, CW, SCo; to B.C., ID. [*Psoralea p.* Hook.; *P. p.* Douglas] ❀**6,17**,SHD:1–3, **7–10,14–16**,18,19,**20–24**;GRCVR.

R. rigida (Parish) Grimes PARISH'S PSORALEA Pl hairy; caudex woody. **ST** erect, < 0.75 m; base purple. **LF**: stipule 4–10 mm, linear-lanceolate or triangular; petiole 40–60 mm; lflet 35–65 mm, lanceolate, with glands and hairs on both surfaces, much denser on upper. **INFL**: bract 3–7 mm, deciduous; 1.5–2 mm. **FL**: calyx 9–10 mm; banner 14–15 mm. **FR** 9–13 mm, elliptic, golden-brown, smooth to faintly net-sculptured, with glands and red-brown hairs; beak < 3 mm, widely attached. **SEED** 6.5–7 mm, red-brown. UNCOMMON. Woodlands, chaparral, lower montane coniferous forests; < 2500 m. SnBR, PR; Baja CA. [*Psoralea r.* Parish; *Hoita r.* (Parish) Rydb.] ❀TRY.

SENNA

Elizabeth McClintock

Per, shrub, tree, unarmed or spines weak. **LVS** even-1-pinnate, alternate; stipules sometimes small or ephemeral; lflets 2–10 (18) pairs. **INFL**: raceme or panicle, axillary or terminal. **FL** gen slightly bilateral, gen showy; sepals ± free; petals free, gen yellow; stamens free, 7 fertile, 3 sterile, anthers gen > filaments, opening by terminal pores. **FR** dehiscent or not. **SEEDS** few–many. ±260 spp.: esp Am trop, also warm temp, sometimes deserts. (Arabic: Sana) [Irwin & Barneby 1982 Mem New York Bot Gard 35:1–918] Some cult as orns. Dried lvs of some spp. cathartic.

1. Lflets 2–4 pairs
  2. St, lvs ± glabrous; lflets not close, not overlapped, 4–6 mm, exceeded by main lf axis .... ***S. armata***
  2′ St, lvs densely white-hairy; lflets close, overlapped, 1–2.5 cm, not exceeded by main lf axis .... ***S. covesii***
1′ Lflets 6–10(18) pairs
  3. Stipules large, conspicuous, gen persistent; unopened fls enveloped in large, dark-colored bracts; fr flat, dehiscent .... ***S. didymobotrya***
  3′ Stipules small, inconspicuous, ephemeral; unopened fls not enveloped in bracts; fr not flat, indehiscent .... ***S. multiglandulosa***

S. armata (S. Watson) H. Irwin & Barneby (p. 651) SPINY SENNA Shrub, armed with weak spines, ± lfless most of year, ± glabrous. **ST** 0.5–1 m; branches from base, grooved, ending in a weak thorn or not, green. **LF**: stipules minute or 0; lflets 2–4 pairs, not overlapped, ± opposite, ephemeral, ± sessile, 4–6 mm, asymmetric, oblong; main axis elongating after lflets fall, weakly spine-tipped. **INFL** terminal, raceme-like (fls 1–2 per axil of upper lvs). **FL**: petals 8–12 mm, obovate, ± irregular, yellow to salmon-red. **FR** dehiscent, 2.5–4 cm, lanceolate, straight. **SEEDS** few. Uncommon. Sandy or gravelly washes; 200–1000 m. D; NV, AZ, Baja CA. [*Cassia a.* S. Watson] ❀TRY.

S. covesii (A. Gray) H. Irwin & Barneby COUES' CASSIA Subshrub, unarmed, leafy, densely white-hairy. **ST** 3–6 dm. **LF**: stipules bristle-like, some persistent; lflets 2–3 pairs, overlapped, opposite, short-stalked, 1–2.5 cm, elliptic. **INFL**: raceme, axillary, 5–15 mm, few-fld. **FL**: petals ± 12 mm, oblong-obovate, prominently veined. **FR** dehiscent, 2–5 cm, oblong, ± straight. **SEEDS** several. RARE in CA. Dry, sandy desert washes, slopes; 500–600 m. DSon; NV, AZ, Baja CA. [*Cassia c.* A. Gray]

S. didymobotrya (Fresen.) H. Irwin & Barneby Shrub, unarmed, leafy, hairy, gen strongly scented. **LF**: stipules gen persistent or tardily deciduous, 6–17 mm, ovate-cordate; lflets 7–10(18) pairs, ± overlapped, opposite, 2–6.5 cm, sessile, oblong, tips abrupt, slender. **INFL**: raceme, axillary toward st tips, many-fld; bracts in cone-shaped cluster covering uppermost, unopened fls, 1–2 cm, widely ovate, brownish to blackish, falling after fls open. **FL**: petals concave, longest 1.5–2.5 cm, obovate. **FR** dehiscent, 8–12 cm, oblong, flat; transverse partitions between seeds. **SEEDS** many. Uncommon. Disturbed, urban areas; < 100 m. s CCo, expected elsewhere; native to trop Afr.

S. multiglandulosa (Jacq.) H. Irwin & Barneby Shrub, small tree, unarmed, leafy, densely hairy. **LF**: stipules ephemeral, 3.5–7 mm, lanceolate; lflets 6–8 pairs, gen not overlapped, opposite, short-stalked, 2.5–4.5 cm, oblong to narrowly elliptic, lower surface more densely hairy, paler. **INFL**: raceme, axillary, 5–15-fld; bracts falling before fls open. **FL**: petals ± flat, longest 12–19 mm, obovate. **FR** indehiscent, 8–12 cm, oblong, ± inflated. **SEEDS** many. Disturbed areas; < 500 m. CCo, SnFrB; native to Mex, Guatemala, S.Am. [*Cassia tomentosa* L.] Often planted along highways.

SESBANIA

Duane Isely

Ann, unarmed. **LVS** even-1-pinnate; stipules gen deciduous; lflets gen many. **INFL**: raceme, axillary; bractlets sometimes appressed to calyx base. **FL**: calyx lobes subequal, < tube; corolla gen yellow, dark-spotted; 9 filaments fused, 1 free. **FR** gen dehiscent, linear or oblong, 4-sided. **SEEDS** few–many. ± 60 spp.: trop, warm regions. (Ancient Arabic name)

S. exaltata (Raf.) Cory (p. 651) COFFEE WEED, COLORADO RIVER HEMP **ST** erect, simple or branched, 1–3 m. **LF**: lflets 30–60, 1–2.5 cm, oblong. **INFL** 2–6-fld. **FL**: corolla 1–1.5 cm, yellowish or mottled. **FR** 15–20 cm, 2–3 mm wide, linear. **SEEDS** many. $2n$=12. Along streams, other moist sites, often in cult or old fields; < 500 m. DSon, probably elsewhere; to se US. ❀IRR:**8,9,11–14**; STBL.

SPARTIUM

Elizabeth McClintock

1 sp.: Medit, Canary, Madeira, Azores islands. (Greek: ancient name)

S. junceum L. (p. 651) SPANISH BROOM Shrub, unarmed. **ST** erect; branches few, < 3 m, rush-like, not angled, green, gen ± lfless. **LVS** simple, alternate to subopposite, ephemeral, < 2.5 cm, linear to lanceolate; upper surface glabrous; lower surface appressed-hairy. **INFL**: raceme, terminal, open; fls several. **FL** ± 2.5 cm, ± fragrant; calyx split above ± to base (rarely ± 2-lipped, 5-lobed); petals yellow; stamens 10, filaments fused. **FR** dehiscent, 5–10 cm, ± 5 mm wide. Common. Disturbed areas; < 600 m. NCoRO, ScV, SnFrB, SCoRO, SCo, s ChI, WTR. Sts used for fiber, fls for yellow dye.

2 cm
5 mm
O. oreophila
fruit and calyx
O. parryi
fruit
5 mm
O. borealis var. viscida
2 mm
keel
Oxytropis deflexa var. sericea
1 mm
calyx
1 cm
1 cm
Pediomelum californicum
5 mm
1 mm
1 mm
style
ovary
5 mm
1 cm
fruit
2 cm
2 mm
stipule
Peteria thompsoniae
2 cm
2 cm
fruit and calyx
Pisum sativum
1 cm
5 mm
fruit
keel
2 cm
Phaseolus filiformis
1 cm
fruit and calyx
var. tomentosa
1 cm
1 cm
fruit and calyx
Pickeringia montana var. montana
1 cm
fruit
1 mm
Prosopis glandulosa var. torreyana
1 cm
1 mm
1 cm
1 mm
fruit
Psoralidium lanceolatum
1 cm
leaf
var. simplicifolius
5 mm
1 cm
leaf
fruit and calyx
var. minutifolius
2 mm
fruit and calyx
Psorothamnus arborescens var. arborescens
2 mm
fruit and calyx
1 cm
1 cm
leaf
P. emoryi
fruit
2 mm
P. fremontii var. fremontii
fruit and calyx
1 cm
2 mm
Psorothamnus spinosus

SPHAEROPHYSA

Duane Isely

Per, unarmed. **LVS** odd-1-pinnate; stipules tiny or 0. **INFL**: raceme, axillary; bractlets at base of calyx. **FL**: calyx lobes subequal, < tube; corolla reddish; 9 filaments fused, 1 free; style hairy at tip. **FR** bladdery-inflated; base stalk-like; flange longitudinal, internal, on upper side. **SEEDS** many. 2 spp.: arid regions of Eurasia. (Greek: sphere bladder)

S. salsula (Pall.) DC. (p. 651) AUSTRIAN PEAWEED Pl spreading by root sprouting. **ST** ascending, 4–15 dm. **LF**: lflets 15–23, 0.6–2 cm, oblong to ovate. **INFL**: pedicels 3–7 mm. **FL**: corolla 12–14 mm, brick- or orange-red, keel > wings. **FR** spreading or reflexed, 1.4–2.4 cm, spheric to ovoid, grooved on upper side, papery or membranous, often mottled, glabrous; stalk-like base 4–7 mm. Cult fields, disturbed sites; < 500 m. s SnJV, possibly elsewhere; sporadic in w US; native to Asia. NOXIOUS WEED. Most known infestations have been eradicated.

THERMOPSIS GOLDEN PEA, FALSE-LUPINE

Duane Isely

Per, unarmed, gen hairy, rhizomed. **ST** erect; lower nodes lfless. **LVS** palmately compound; lower stipules clasping st, scarious, others not clasping, green; lflets 3. **INFL**: raceme, terminal; bracts like stipules, ± persistent. **FL**: calyx lobes 5, but upper 2 variously fused; corolla yellow; stamens 10, free. **FR** slowly dehiscent, short or oblong, flat; base stalk-like; margin often wavy. ± 10 spp.: N.Am, Asia. (Greek: like a lupine) [Isely 1951 Mem New York Bot Garden 25(3):104–122]

T. macrophylla Hook. & Arn. Pl green, gray, or silvery. **ST** 0.3–2 m. **LF**: lflets 2.5–6 cm, oblanceolate to ovate. **INFL** 1–5 dm, 6–many-fld; bracts conspicuous. **FL**: calyx 7–10 mm, lobes ± = or < tube; corolla 15–25 mm; ovary gen hairy. **FR** spreading to erect, oblong, often curved, often narrowed in places by poor seed set. **SEEDS** few–several. 2*n*=18,36. Grassland, chaparral, sandy scrub, woodlands, open ridges, disturbed areas; < 2100 m. NW, CaR, CW, SW, MP; OR. Regionally diverse complex that has been classified variously.

1. Pl canescent, 3–5 dm; s PR (San Diego Co.) . . . var. ***semota***
1′ Pl glabrous to silvery, gen 3–10+ dm; not PR
  2. Pl glabrous or puberulent, often without dominant axis; infl often open, 5–12-fld; KR, NCoR . . . . var. ***venosa***
  2′ Pl hairy, often with dominant axis; infl often dense, many-fld; gen not of KR, NCoR
    3. Pl conspicuously hairy, hairs appressed, straight; gen CaR, MP . . . . var. ***argentata***
    3′ Pl conspicuously hairy or not, hairs wavy or incurved; NW, CW, n SW . . . . var. ***macrophylla***

var. ***argentata*** (E. Greene) Jepson (p. 651) SILVERY FALSE-LUPINE Pl silvery-hairy. **ST** gen > 3 dm. **LF**: hairs straight, appressed. **FR** ascending to erect. 2*n*=18. UNCOMMON. Pine forests, open ridges; 900–1500 m. CaR (Siskiyou, Shasta cos.), MP (Modoc, Lassen cos.). [*T. argentata* E. Greene] ❀TRY.

var. ***macrophylla*** (p. 651) Pl green or gray-hairy. **ST** 5–20 dm. **LF** sometimes tomentose; hairs wavy or incurved. **FR** gen ascending, sometimes out-curved or erect. 2*n*=18. Grasslands, chaparral, open woodlands, disturbed areas; < 2100 m. NW, CW, n SW. Variable in habit, size, hairiness; robust pls (1–2 m) of Santa Ynez Mtns (Santa Barbara Co.) have been called var. *agnina* J. Howell, Santa Ynez false-lupine. ❀DRN,IRR:1–3,**6,7,15–17,22–24**& SHD:**14,18–21**;STBL.

var. ***semota*** Jepson (p. 651) VELVETY FALSE-LUPINE Pl canescent. **ST** 3–5 dm. **LF**: hairs matted, curved. **FR** divergent to ascending. RARE. Grasslands, sandy scrub; 1200–1500 m. s PR (San Diego Co.).

var. ***venosa*** (Eastw.) Isely Pl glabrous or puberulent. **ST** gen 4–8 dm, irregularly branched. **FR** spreading to ascending. 2*n*=36. Open, gen dry sites, disturbed woodlands; 300–1300 m. KR, NCoR; OR. [*T. gracilis* Howell] ❀TRY;STBL.

TRIFOLIUM CLOVER

Duane Isely

Ann or per, unarmed. **LVS** gen palmately compound; stipules conspicuous, partly fused to petiole; lflets gen 3, sometimes 5–9, ± serrate or dentate. **INFL**: raceme (often umbel-like), head, or spike, axillary or terminal, gen many-fld, often involucred, gen peduncled; fls bracted or not. **FL** gen spreading to erect, often becoming reflexed; corolla gen purple to pale lavender, sometimes yellow, persistent after fl; 9 filaments fused, 1 free. **FR** gen indehiscent, but often breaking, short, plump, gen incl in corolla; base often stalk-like. **SEEDS** 1–6. (Latin: 3 lvs) [Gillett 1980 Can J Bot 58: 1425–1558; Zohary & Heller 1984 Genus Trifolium]

Key to Groups

1. Involucral bracts gen fused, gen > 1 mm, often forming a cup or bowl about base of infl; infl not sessile above 1–2 lvs . . . . **Group 1**
1′ Involucral bracts 0 or sometimes abortive, then < 1 mm or forming a vestigial ring; infl sometimes ± sessile above 1–2 ± reduced lvs (stipules of which are sometimes involucre-like) . . . . **Group 2**

Group 1: Involucre obvious

1. Calyx or entire banner soon inflated in fr; involucre bracts gen ± free
  2. Calyx hairy, inflated in fr; infl becoming a fuzzy ball, ± 2 cm wide; banner not inflating in fr; per, stoloned . . . . [2]***T. fragiferum***
  2′ Calyx glabrous, not inflated in fr; infl not as above; banner inflating in fr; ann, decumbent to erect
    3. Fl 10–20 mm . . . . ***T. fucatum***
    3′ Fl 4–9 mm . . . . [2]***T. depauperatum***

4. Fr 3–4 mm, oblong, stalk-like base 0 or very short; involucre bracts wide, irregularly shaped, often ± half fused, margins wide, scarious .... var. ***amplectens***
4′ Fr 2–3 mm, ovoid or obovoid, stalk-like base present; involucre bracts not as above
5. Involucral bracts < 1 mm, fused at base; infl often 1–1.5 cm wide; corolla 6.5–9 mm; local, ScV, CW .... var. ***hydrophilum***
5′ Involucral bracts 2–2.5 mm, often ± free; infl gen 0.5–1 cm wide; corolla 4.5–7.5 mm; widespread .... var. ***truncatum***
1′ Calyx, corolla not inflated in fr (exc lower portion of banner in *T. barbigerum*); involucre bracts gen fused into a cup-, bowl-, or wheel-shaped, deeply toothed or cut ring (exc gen ± free in *T. monanthum*)
6. Involucre cup- or bowl-shaped, when pressed often partly hiding fls (exc in *T. barbigerum* var. *andrewsii*)
7. Calyx lobes < 1/2 tube, not bristle-tipped, not branched .... ***T. microdon***
7′ Calyx lobes > 1/2 tube, bristle-tipped, lower often branched or toothed
8. Calyx lobes, involucre lobes ± entire; hairs on pl gen fine, wavy .... ***T. microcephalum***
8′ Calyx lobes, involucre lobes, or both toothed, cut, or branched; hairs on pl various or 0
9. Bristle-tips of lower 3 calyx lobes irregularly, often secondarily forked; involucre margin wavy, gen finely toothed rather than lobed .... ***T. cyathiferum***
9′ Bristle-tips of lower 3 calyx lobes not secondarily divided; involucre margin coarsely, irregularly toothed, often cut 1/2 way to involucre base
10. Calyx lobes finely toothed, terminal bristle 1–1.5 mm, glabrous; involucre gen glabrous; banner not inflated .... ***T. buckwestiorum***
10′ Calyx lobes entire, terminal bristle 3–4 mm, plumose (or lower lobes divided near base into 2–3 bristles, each 3–4 mm, gen plumose); involucre gen puberulent; banner inflated in lower portion, persistent as twisted beak above .... ***T. barbigerum***
11. Corolla 8–16 mm, > calyx, stalk-like fr base 1–3 mm; infl not bristly .... var. ***andrewsii***
11′ Corolla 5–10 mm, ± = calyx (or, if longer, fr base stalk-like, < 1 mm); infl often bristly from calyx lobes .... var. ***barbigerum***
6′ Involucre wheel-shaped or bracts ± free, when pressed hiding only bases of fls
12. Involucre inconspicuous, bracts 2–5, gen ± free, 1–3 mm; infl 1–6-fld .... [3]***T. monanthum***
13. Lflet tip gen acute, length 4–6 × width; infl 2–6-fld; SW .... var. ***grantianum***
13′ Lflet tip gen rounded or truncate, length 2–4 × width; infl 1–3-fld; CaR, SN, SNE .... var. ***monanthum***
12′ Involucre gen conspicuous, bracts indefinite in number, fused, gen > 5 mm, if inconspicuous (e.g. 2–3 mm in *T. oliganthum*) bracts fused at least at base; infl gen 8–many-fld
14. Per; infl gen 2–3 cm wide, not lobed; seeds 2–4 .... ***T. wormskioldii***
14′ Ann; infl < 2 cm wide (or gen conspicuously lobed); seeds gen 1–2
15. Calyx lobes < or ± = tube, often toothed or shouldered below tapered tip; calyx tube splitting between upper lobes
16. Fl 5–8 mm; involucre 2–3 mm, gen cut > 1/2 way to base; infl 6–10 mm wide, 5–15-fld .... ***T. oliganthum***
16′ Fl 9–18 mm; involucre > 3 mm, cut < 1/2 way to base; infl gen 12–30 mm wide, often > 15-fld
17. Calyx glandular-hairy or bumpy (sometimes becoming glabrous), lobes gen entire; lflet sharply serrate, the longer teeth ± 1 mm .... ***T. obtusiflorum***
17′ Calyx glabrous, not bumpy, lobes entire or some often 3-toothed; lflet serrate, the teeth < 1 mm .... ***T. willdenovii***
15′ Calyx lobes > tube, gen entire below tapered tip; calyx tube not splitting between upper lobes
18. Calyx hairy; involucre small, cut from middle nearly to base; CCo (Monterey Co.) .... ***T. trichocalyx***
18′ Calyx glabrous; involucre gen well developed, cut to middle; widespread — diverse in habit, fl, infl size .... ***T. variegatum***

Group 2: Involucre absent or inconspicuous

1. Lflets 3–9 (at least some lvs with > 3 lflets)
2. Lflets gen 7–9; infl 3–5 cm wide .... ***T. macrocephalum***
2′ Lflets 3–7; infl 1–2.5 cm wide
3. Pl silvery or gray, densely hairy or tomentose; calyx lobe hairs ± 1 mm .... ***T. andersonii***
4. Longer petiole hairs 1.5–2 mm; peduncles gen < petioles, infl incl .... var. ***andersonii***
4′ Longer petiole hairs 0.4–1.2 mm; peduncles gen > petioles, infl exserted .... var. ***beatleyae***
3′ Pl green, strigose or puberulent; calyx lobe hairs < 1 mm
5. Lflet elliptic, serrate, the teeth 0.1–0.3 mm; infl gen incl, 1–1.5 cm wide .... ***T. gymnocarpon***
5′ Lflet obovate, coarsely serrate, the larger teeth 0.4–0.5 mm; infl gen exserted, 1.5–2.5 cm wide .... ***T. lemmonii***
1′ Lflets 3
6. Infl mostly of stalk-like, bristle-tipped, sterile fls without petals (2–8 fls with petals), forming a bur, delivered to ground by growing peduncle; pl prostrate .... ***T. subterraneum***
6′ Infl entirely of unmodified, fertile fls with petals, not forming a bur, not delivered to ground by peduncle; pls various in habit, incl prostrate
7. In fr, calyx inflated, infl becoming a red or brown fuzzy ball ± 2 cm wide; pl creeping and rooting or cespitose .... [2]***T. fragiferum***
7′ In fr, calyx rarely inflated, infl then << 2 cm wide; pls various in habit, incl creeping and rooting and cespitose
8. Lf pinnately compound; corolla bright yellow
9. Corolla clearly striate; infl 0.8–1.3 cm wide, gen > 20-fld; petioles, exc of uppermost lvs, > lflets .... ***T. campestre***

9′ Corolla weakly striate; infl 0.4–0.8 cm wide, 5–20-fld; petioles of middle lvs gen < lflets .... ***T. dubium***

8′ Lf palmately compound; fl not bright yellow (exc in *T. aureum*)

10. Lflets 4–10 cm, stipules 1–2.5 cm, green; calyx glabrous; corolla greenish white to pink .... ***T. howellii***

10′ Lflets or stipules (or both) < above, stipules gen not green; calyx hairy or glabrous; corollas of various colors incl above

11. Fls gen 1–5 per infl, gen with tiny bracts; corolla white to lavender-striate; pl gen very small (see Group 1, couplet 13 for vars.) .... [3]***T. monanthum***

11′ Fls > 5 per infl unless pl starved; corolla various in color, incl white to lavender-striate; pl various in stature, incl very small

12. Pedicels often initially short but quickly 1–4 mm, some or all fls becoming reflexed

13. Corolla bright yellow, 5–7 mm; calyx 1.5–2 mm; stalk-like fr base 2–3 mm .... ***T. aureum***

13′ Corolla not bright yellow, gen > 7 mm; calyx gen > 2 mm; stalk-like fr base < 2 mm or 0

14. Lflet lanceolate, lvs all basal or 2–3 cauline, pls of s SNH, SNE — RARE .... ***T. macilentum*** var. ***dedeckerae***

14′ Lflet wider than lanceolate (or lvs not gen basal or pls not of SNH and SNE)

15. Pl creeping and rooting, turf-forming; all petioles from ground level, ± equal; corolla white .... ***T. repens***

15′ Pl not creeping and rooting, not turf-forming; some or 0 petioles from ground level, lower > upper; corolla color various, incl white

16. Calyx tube hairy (sometimes slightly so)

17. Corolla 5–8 mm; pedicel 2–5 mm .... ***T. breweri***

17′ Corolla 10–18 mm; pedicel 1–2 mm .... [2]***T. longipes***

18. Sts gen well developed, not cespitose; petals gen white, short-tapered, acute, or obtuse; 1200–1800 m .... var. ***oreganum***

18′ Sts often poorly developed, cespitose; petals lavender to purple, longer ones long-tapered, acute; 1400–2700 m .... var. ***shastense***

16′ Calyx tube glabrous

19. Calyx lobes ciliate with short, flat bristles or bristle bases; bracts tiny, often in a ring at base of infl .... ***T. ciliolatum***

19′ Calyx lobes not ciliate but sometimes hairy; bracts subtending individual fls or 0

20. Lvs basal exc for 1 pair; corolla 12–17 mm .... [2]***T. beckwithii***

20′ Lvs cauline; corolla 5–11 mm

21. Infl 1.5–3 cm wide; bracts initially conspicuous, soon falling; calyx lobes sharply tapered but not bristle-tipped .... ***T. hybridum***

21′ Infl 0.5–1.5 cm wide; bracts 0; calyx lobes bristle-tipped

22. Calyx lobes ± 0.5 mm wide at base, glabrous, tube 1.5–2.5 mm .... ***T. gracilentum***

23. Lflet obovate to obcordate, length 1.5–2.5 × width; widespread, not of ChI ... var. ***gracilentum***

23′ Lflets narrowly elliptic or lanceolate, length 3.5–5 × width; ChI .... var. ***palmeri***

22′ Calyx lobes ± 0.2 mm wide at base, gen with a few hairs, tube 0.8–1.4 mm .... ***T. bifidum***

24. Lflet length 2.5–5.5 × width, notched at tip 1/10 to 4/10 lflet length, otherwise gen entire .... var. ***bifidum***

24′ Lflet length 1.5–2.5 × width, truncate or notched at tip < 2/10 length, otherwise dentate or entire .... var. ***decipiens***

12′ Pedicels remaining < 1 mm, fls remaining erect-spreading or becoming reflexed

25. Infls sessile in a succession of lf axils, 7–10 mm wide, < lvs; corolla 5–6 mm .... ***T. glomeratum***

25′ Infls peduncled (or immediately above 1–2 reduced lvs, gen > 10 mm wide, gen > lvs); corolla gen > 6 mm

26. Banner inflated in fr; involucre bracts often forming vestigial ring; corolla gen 6–9 mm .... [2]***T. depauperatum*** var. ***depauperatum***

26′ Banner not inflated in fr; involucre bracts 0; corolla various in size, incl above

27. Fls mostly or all reflexed soon after opening; infl often turned to side or downward

28. Calyx hairy, lobes twisted or bent, plumose .... ***T. eriocephalum*** var. ***eriocephalum***

28′ Calyx glabrous, lobes not twisted or bent, not plumose

29. Infl 1–2 cm wide; calyx 3.5–4.5 mm, gen dark or purple-black; corolla 10–12 mm ..... ***T. bolanderi***

29′ Infl 1.5–3 cm wide; calyx 3–7 mm, not dark or purple-black; corolla 10–17 mm

30. Calyx 4.5–7 mm; infl erect, axis not exserted .... [2]***T. beckwithii***

30′ Calyx 3–4 mm; infl gen turned to side or downward, axis exserted, gen forked .... ***T. kingii*** var. ***productum***

27′ Fls mostly remaining erect to spreading, not reflexed after opening; infl gen ± erect

31. Infl ± sessile above a pair of reduced lvs, stipules, or both (in *T. macraei*, sometimes 1 of a pair short-peduncled)

32. Infls paired, ± sessile, or sometimes 1 of a pair short-peduncled .... ***T. macraei***

32′ Infls single above subtending lvs

33. Corolla 2-colored (purple with white tips); local and uncommon, s NCoR (Sonoma Co.) .... [2]***T. amoenum***

33′ Corolla not 2-colored, red-purple or pink; sporadic or locally common, CA-FP, GB

34. Calyx lobes densely plumose, lower one ± 1–1.3 × others; corolla pink .... ***T. hirtum***

34′ Calyx lobes sparsely plumose, lower one ± 2 × others; corolla red-purple .... ***T. pratense***

31′ Infl peduncled

35. Corolla ± = or < calyx; calyx lobes (at least lower) > tube, plumose
36. Lflet oblanceolate to obovate, length 1.2-3 × width; corolla 2-colored (purple and white) . . . . [2]***T. albopurpureum***
37. Calyx 4–8 mm, lobes 3–6.5 mm . . . . var. ***albopurpureum***
37′ Calyx 8–14 mm, lobes 6.5–12 mm . . . . var. ***olivaceum***
36′ Lflet linear, narrowly oblong, or oblanceolate, length 3–8 × width; corolla not 2-colored, pale pink to white
38. Corolla 10–12 mm, ± = calyx . . . . ***T. angustifolium***
38′ Corolla 3–4 mm, < or << calyx . . . . ***T. arvense***
35′ Corolla > calyx; calyx lobes > tube or not, plumose or not
39. Ann; calyx lobes densely plumose, hairs gen straight
40. Corolla crimson, 10–14 mm; infl a ± cylindric, 2–6 cm spike . . . . ***T. incarnatum***
40′ Corolla purple, white-tipped, 7–16 mm; infl a spheric to ovoid, < 4 cm head
41. Corolla 7–12 mm; calyx 4–8 mm . . . . [2]***T. albopurpureum*** var. ***dichotomum***
41′ Corolla 12–16 mm; calyx 10–12 mm . . . . [2]***T. amoenum***
39′ Per; calyx lobes glabrous (or hairs few and curved or wavy)
42. Infl < 1 cm wide, fls gen 1–5, erect (see Group 1, lead 13, for vars.) . . . . [3]***T. monanthum***
42′ Infl 1.5–2.5 cm wide, fls many, ascending-spreading . . . . [2]***T. longipes***
43. Calyx lobes 3–5 × tube; lflet length often 6–10 × width; nw CA . . . . var. ***elmeri***
43′ Calyx lobes 1–3 × tube; lflet length 2–6 × width, at least in nw CA . . . . var. ***nevadense***

T. albopurpureum Torrey & A. Gray Ann, hairy. **ST** decumbent to erect. **LVS** cauline; stipules small; lflets 1–3 cm, oblanceolate to obovate. **INFL**: spike, 5–20 mm wide, ovoid to short-cylindric. **FL**: calyx 4–15 mm, lobes > or >> tube, tapered or bristle-like, plumose; corolla < or > calyx, 2-colored (purple and white). **SEEDS** 1–2. $2n=16,32$. Coastal dunes, grasslands, wet meadows, open slopes, oak chaparral, pine woodlands, roadsides, other disturbed places; < 2100 m. CA-FP; to B.C., Baja CA. Formidable complex here ± arbitrarily divided into 3 intergrading vars.; probably should incl *T. amoenum*.

var. ***albopurpureum*** (p. 651) **FL**: calyx 4–8 mm, lobes 3–6.5 mm; corolla 5–8 mm, ± = calyx. Abundant. Grassland, oak chaparral, pine forests, roadsides; gen < 1100 m (< 2100 m in s). CA-FP; to B.C., Baja CA.

var. ***dichotomum*** (Hook. & Arn.) Isely (p. 651) **FL**: calyx 4–8 mm, lobes 2.5–5.5 mm; corolla 7–12 mm, > calyx. Coastal dunes, open slopes, wet meadows, oak woodlands, disturbed areas; < 1300 m. NW, c SN, GV, SnFrB, SCoR; to WA. [*T. d.* Hook. & Arn.]

var. ***olivaceum*** (E. Greene) Isely (p. 651) **FL**: calyx 8–14 mm, lobes 6.5–12 mm; corolla 4–7 mm, << calyx. Uncommon. Grassy slopes, valley meadows, disturbed areas; < 500 m. NCoRI, CaR, n SN, GV, SnFrB. [*T. o.* E. Greene] Once abundant, esp GV.

T. amoenum E. Greene (p. 651) SHOWY INDIAN CLOVER Ann, often robust, hairy. **ST** erect. **LVS** cauline; stipules gen conspicuous; lflets widely obovate. **INFL**: head, ovoid to spheric. **FL**: calyx 10–12 mm, lobes > tube, slender, plumose; corolla 12–16 mm, purple, white-tipped. **SEEDS** 1–2. **PRESUMED EXTINCT** (last collected in 1969). Moist, heavy soils, disturbed areas; < 100 m. s NCoR, SnFrB. Probably belongs to *T. albopurpureum* complex. Last collected in Sonoma Co.

T. andersonii A. Gray Per, short-tufted or cushion-forming, soft-hairy or tomentose, silvery or gray. **ST** 0 (peduncle ascending or erect). **LVS** basal; stipules entire, persistent; lflets 3–7, 5–20 mm, oblanceolate to obovate, entire. **INFL** head-like, 1.5–2.5 cm wide; pedicel 0.5–1 mm. **FL**: calyx 8–10 mm, lobes slender, > tube, plumose, hairs ± 1 mm; corolla 10–15 mm, pink-purple or 2-colored. **SEEDS** 1–2. $2n=16$. Meadows, talus, washes, yellow-pine forest, rocky alpine slopes; 900–4000 m. n SNH, GB; se OR, NV.

var. ***andersonii*** (p. 651) **LF**: lflets gen 1–2 cm; longer petiole hairs 1.5–2 mm. **INFL** gen < (or =) lvs, gen incl; peduncle 1–4 cm. Rocky slopes, meadows, esp yellow-pine forest; 900–2000 m. n SNH, MP; NV.

var. ***beatleyae*** (J.M. Gillett) Isely BEATLEY'S FIVE-LEAVED CLOVER **LF**: lflets gen 0.5–1.4 cm; longer petiole hairs 0.4–1.2 mm. **INFL** > lvs, gen exserted; peduncle 2–10 cm. RARE in CA. Washes, talus, pine forest to alpine slopes; 1300–4000 m. SNE (Mono Co.); NV. [*T. monoense* E. Greene]

T. angustifolium L. (p. 651) Ann, hairy. **ST** gen erect. **LVS** cauline; stipules bristle-tipped; lflets 2–4.5 cm, linear to narrowly lanceolate. **INFL**: spike, 1–5 cm, cylindric. **FL**: calyx 10–12 mm, 10-ribbed, lobes spreading, unequal, all or lower > tube, needle-like, plumose, hardening in fr; corolla 10–12 mm, ± = calyx, pink. **SEED** 1. $2n=14$. Disturbed areas; esp < 200 m. CCo, SCoR; OR; native to Medit. Possibly naturalized in CA.

T. arvense L. (p. 651) RABBITFOOT CLOVER Ann, hairy. **ST** gen erect. **LVS** cauline; stipules lanceolate, sometimes bristle-tipped; lflets 0.5–2 cm, narrowly oblong or oblanceolate. **INFL**: spike, 1–3 cm, ovoid to short-cylindric, dull green to gray. **FL**: calyx 4.5–7.5 mm, lobes > tube, needle-like, plumose; corolla 3–4 mm, < or << calyx, pale pink to white. **SEED** 1. $2n=14,28$. Disturbed, waste areas; esp < 300 m. CA-FP; to B.C, se US; native to Eur. Possibly naturalized in CA.

T. aureum Pollich (p. 651) HOP CLOVER Ann or bien, glabrous or puberulent. **ST** ascending or erect. **LVS** cauline; stipules lanceolate to ovate; lflets 1–2.5 cm, oblanceolate or obovate. **INFL**: head, 1–1.5 cm wide, ovoid; fls quickly reflexed. **FL**: calyx 1.5–2 mm, 5-veined, glabrous, lobes unequal, ± = tube; corolla 5–7 mm, striate, initially bright yellow. **FR**: stalk-like base 2–3 mm; style 1–2 mm, persistent. **SEED** 1. $2n=14$. Disturbed areas; esp < 200 m. SnJV; to B.C., ne US; native to Eur. [not *T. agrarium* L.] Probably naturalized in CA.

T. barbigerum Torrey Ann, glabrous or puberulent. **ST** decumbent to erect. **LVS** ± cauline; stipules conspicuous, lanceolate to ovate, lower often overlapping, upper gen sharply cut; lflets 1.5–2.5 cm, oblanceolate to obovate, tips sometimes notched. **INFL**: head, 5–25 mm wide, 5–many-fld; involucre bowl-shaped, sharply toothed or deeply cut, gen puberulent. **FL**: calyx tube 5-veined, lobes simple, entire, terminal bristle 3–4 mm, plumose, or lower lobes divided near base into 2–3 bristles, each 3–4 mm, gen plumose; corolla pink-purple or 2-colored (white and purple), rarely all white, lower portion of banner inflating. **SEEDS** 1–2. Wet meadows, ditches, foothill slopes to pine woodland, open, disturbed places; gen < 700 m. NW, n&c SN, GV, CW; OR. Vars. intergrade.

var. ***andrewsii*** A. Gray (p. 651) GRAY'S CLOVER **INFL** not bristly. **FL**: calyx tube 3–5 mm, often ± = lobes; corolla 8–16 mm, > calyx. **FR**: stalk-like base 1–3 mm. UNCOMMON. Wet meadows, foothill slopes to pine woodland; < 600 m. NCoRO, n&c SN, ScV, SnFrB, SCoRO. [*T. grayi* Lojacono] Locally abundant. Variant with "gold" fls recently found in SNF (Butte Co.). ❀TRY.

var. ***barbigerum*** Torrey **INFL** bristly. **FL**: calyx tube gen 2–3 mm, < lobes; corolla 5–10 mm, gen ± = calyx. **FR**: stalk-like base < 1 mm. Wet meadows, ditches, open, disturbed areas; gen < 700 m. NW, n&c SN, GV, CW; OR.

T. beckwithii S. Watson (p. 651) Per, glabrous. **ST** ascending. **LVS** basal exc 1 pair; cauline stipules conspicuous, green; lflets

1–4 cm, oblong to ovate. **INFL** ± scapose, head-like, 2–3 cm wide; axis not exserted; pedicel 1–2 mm; fls soon reflexed. **FL**: calyx 4.5–7 mm, lobes unequal, ± = tube; corolla 10–17 mm, pink to purple or 2-colored. **SEEDS** 2–4. 2*n*=±48. Meadows, evergreen forests; 1200–2100 m. n SN, MP; to SD.

T. bifidum A. Gray Ann, glabrous or sparsely fine-hairy. **ST** sprawling to erect. **LVS** cauline; stipules short-awned; lflets 1–2 cm, oblanceolate to obovate, tip gen notched. **INFL** head- or umbel-like, 7–15 mm wide, 5–many-fld; axis tip often short-exserted; pedicel 1–2 mm; fls becoming reflexed. **FL**: calyx 3.5–5 mm, tube glabrous, 0.8–1.4 mm, lobes >> tube, ± 0.2 mm wide at base, awn-like, bristle-tipped, often with a few fine hairs; corolla 6–9 mm, dull yellow to pink-purple. **SEEDS** 1–2. 2*n*=16. Open grassy areas, oak chaparral, forests; < 1000 m. CA-FP (exc SNH, Teh); to WA.

var. ***bifidum*** (p. 651) **LF**: lflet length 2.5–5.5 × width, tip notched 1/10–4/10 length, otherwise gen entire. Habitats of sp.; < 600 m. NCoR, n SNF, GV, CW.

var. ***decipiens*** E. Greene **LF**: lflet length 1.5–2.5 × width, tip truncate or notched < 2/10 length, otherwise dentate or entire. Habitats of sp.; < 1000 m. Range of sp.

T. bolanderi A. Gray (p. 651) BOLANDER'S CLOVER Per, glabrous. **STS** clumped, short, ascending; elongated internodes 1–2. **LVS** gen basal; stipules ovate, papery or thick; lflets 7–15 mm, elliptic to obcordate, finely serrate. **INFL** head-like, often turned to side, 1–2 cm wide; axis not exserted; peduncle 6–15 cm; pedicel 0.5–1 mm; fls quickly reflexed. **FL**: calyx 3.5–4.5 mm, 5-veined, gen dark or purple-black, lobes unequal, ± = or < tube; corolla 10–12 mm, pale purple to rose. **SEEDS** 2. UNCOMMON. Moist montane meadows; 2100–2300 m. c SNH.

T. breweri S. Watson Per, puberulent. **ST** ascending or decumbent. **LVS** cauline; stipules sheathing to lanceolate; lflets 5–20 mm, gen obovate, ± entire to serrate. **INFL** umbel-like, often turned to side, 1–2 cm wide, 5–15-fld; pedicel 2–5 mm; fls or some becoming reflexed. **FL**: calyx 3.5–5 mm, ± 10-veined, lobes > tube; corolla 5–8 mm, yellowish white to pink-lavender. **FR** often exserted from corolla. **SEED** 1. 2*n*=16. Dry forests, open areas, roadsides; 200–1800 m. KR, CaR, n&c SN; OR.

T. buckwestiorum Isely (p. 651) Ann, glabrous. **ST** decumbent to ascending. **LVS** cauline; stipules with many bristle-tipped teeth; lflets 0.5–1.5 cm, elliptic to obovate, finely serrate. **INFL** head-like, 8–12 mm wide, 5–many-fld; involucre bowl-shaped, irregularly toothed and cut. **FL**: calyx tube 4–5 mm, 10-veined, lobes < tube, with 3–5 tiny, lateral teeth, ending in a 1–1.5 mm bristle; corolla 6–7 mm, pale pink or white, keel darker. **SEED** 1(2?). Uncommon. Grassy or waste areas; < 200 m. sw SnFrB (Santa Cruz Co.) First heads enclosed in stipules, reduced, seeming cleistogamous.

T. campestre Schreber (p. 651) HOP CLOVER Ann, puberulent. **ST** decumbent to erect. **LVS** 1-pinnate, cauline; stipules ovate; lflets 0.6–1.5 cm, obovate. **INFL** head-like, ovoid to spheric; fls quickly reflexed. **FL**: calyx 1.5–2 mm, upper lobes < lower; corolla 4–5.5 mm, bright yellow, becoming brown, striate. **FR** fragile; style persistent, < 1 mm. **SEED** 1. 2*n*=14. Disturbed areas, roadsides, lawns; esp < 300 m. Gen n CA-FP; most US exc Rocky Mtns, sw; native to Eur. [not *T. procumbens* L.] Probably naturalized in CA.

T. ciliolatum Benth. (p. 655) Ann, ± glabrous. **ST** erect; internodes long. **LVS** mostly cauline; upper stipules bristle-tipped; lflets 1–3 cm, oblanceolate to obovate, serrate. **INFL** head-like, 7–20 mm wide, ovoid or spheric, a ring of tiny bracts often at base; axis sometimes exserted; pedicel becoming 3–6 mm; fls soon reflexed. **FL**: calyx 5–6 mm, tube glabrous, lobes unequal, ciliate with flat bristles or, through breakage, bristle-bases; corolla ± = or > calyx, pink to purple. **SEED** 1. 2*n*=16. Locally common. Grassland, chaparral, disturbed areas; 150–1700 m. NCoR, CaR, n SN, ScV, CW, SW; to WA, Baja CA.

T. cyathiferum Lindley (p. 655) Ann, miniature to robust, ± glabrous. **ST** ascending to erect. **LVS** cauline; stipules entire or deeply toothed, tip acuminate; lflets 1–3 cm, oblanceolate to obovate. **INFL** head-like, 6–20 mm wide, 3–many-fld; involucre bowl-shaped, margin wavy, finely toothed. **FL**: calyx 7–11 mm, lower and lateral lobes ± = tube, with bristle tips irregularly and secondarily forked; corolla ± = or < calyx, white or yellowish, pink-tipped. **FR**: stalk-like base short. **SEEDS** 1–2. 2*n*=16. Gen spring-moist valleys, chaparral, roadcuts, ditches, forests; < 2500 m. NW, CaR, n&c SN, GV; to B.C., ID, NV.

T. depauperatum Desv. Ann, often very small, glabrous. **ST** decumbent to erect. **LVS** cauline; lower stipules oblong; upper stipules bristle-tipped; lflets 0.5–2 cm, narrowly oblong to obovate, entire or toothed, sometimes lobed, tip often truncate. **INFL** head-like, 0.5–1.5 mm wide, 3–many-fld; involucre present or vestigial. **FL**: calyx 2.5–5 mm, glabrous; corolla 4–9 mm, pink-purple, white-tipped, banner inflated in fr. **FR**: stalk-like base short to 0; style gen persistent. **SEEDS** 1–5. 2*n*=16. Common. Salt marshes, grasslands, coastal woodlands, openings, wet meadows, ditches, roadsides, other disturbed places, open alkaline or spring-moist, heavy soils; < 900 m. NCoR, CaRF, SNF, n&c SNH, Teh, GV, CW, SCo, ChI, PR; w N.Am, S.Am. Vars. *depauperatum, truncatum* predominant.

var. ***amplectens*** (Torrey & A. Gray) L.F. McDermott (p. 655) **INFL**: involucre bracts 4–5, free or basally fused, margins widely scarious, toothed near tip. **FL**: calyx 3–4 mm; corolla 5–6 mm. **FR** > style, oblong; stalk-like base ± 0. **SEEDS** 2–5. Grasslands, coastal woodlands; < 800 m. GV, SnFrB, SCoR. [*T. a.* Torrey and A. Gray] ❀**7–9,14**,SUN:**15–17**.

var. ***depauperatum*** (p. 655) **LF**: lflets sometimes notched at tip or lobed. **INFL**: involucre vestigial, ring-like. **FL**: calyx 2.5–3.5 mm; corolla gen 6–9 mm. **FR** > style, ovoid or oblong; stalk-like base 0.5–1.5 mm. **SEEDS** 2–4. Wet meadows, ditches, grasslands, roadsides, open spring-moist, heavy soils; < 900 m. NCoR, CaRF, n&c SN, GV, SnFrB; to B.C.

var. ***hydrophilum*** (E. Greene) Isely Pl often fleshy. **INFL** often 1–1.5 cm wide; involucre bracts basally fused, < 1 mm. **FL**: calyx 2.5–5 mm; corolla 6.5–9 mm, striate. **FR** ± = style, ovoid to oblong; stalk-like base 0.5–1 mm. **SEEDS** 1–2. Possibly extinct. Salt marshes, open areas in alkaline soils; < 300 m. ScV, CW. [*T. amplectens* Torrey & A. Gray var. *h.* (E. Greene) Jepson]

var. ***truncatum*** (E. Greene) Isely (p. 655) **LF**: lflet tip often truncate, toothed. **INFL**: involucre bracts ± separate, 2–2.5 mm, margins ± membranous. **FL**: calyx 2.5–3 mm; corolla 4.5–7.5 mm. **FR** ± = style; stalk-like base 0.5–1 mm. **SEEDS** 1–2. Grassy flats, disturbed slopes, openings in woodland; < 800 m. NCoR, CaRF, SNF, Teh, GV, CW, SCo, ChI, PR. [*T. amplectens* Torrey & A. Gray var. *truncatum* (E. Greene) Jepson]

T. dubium Sibth. (p. 655) LITTLE HOP CLOVER, SHAMROCK Ann, sparsely puberulent. **ST** decumbent to erect. **LVS** 1-pinnate, cauline; stipules ovate; petioles of middle lvs gen < lflets; lflets 0.6–1.2 cm, obovate. **INFL** head-to ± spike-like, < 1 cm wide, 5–10(20)-fld; fls quickly reflexed. **FL**: calyx 1.5–2 mm; corolla 3.5–4 mm, bright yellow, becoming brown, weakly striate. **FR**: style persistent, < 1 mm. **SEED** 1. 2*n*=14,16,28. Agricultural and disturbed areas, lawns; < 500 m. Esp CA-FP; much of US; native to Eur.

T. eriocephalum Nutt. var. ***eriocephalum*** (p. 655) Per, hairy. **ST** gen ascending or erect, gen unbranched. **LVS** basal and cauline; lower stipules sheathing st; upper stipules lanceolate; lflets 1–4 cm, elliptic-oblong to ovate. **INFL** head-like, 1.5–3 cm, often directed to side or downward; peduncle 5–15 cm; involucre 0; pedicel ± 0.5 mm; fls soon reflexed. **FL**: calyx 5–9 mm, hairy, lobes ± linear, often bent, > tube, plumose; corolla 8–14 mm, dull white or yellowish. **FR** hairy. **SEEDS** 1–3. 2*n*=16. Locally common. Moist meadows to dry, open slopes; 300–1500 m. NW, CaR, n SN; OR.

T. fragiferum L. (p. 655) STRAWBERRY CLOVER Per, stoloned or cespitose, glabrous. **LVS** ± basal; stipules gen overlapping; petiole >> blade; lflets 0.5–2 cm, elliptic to obovate. **INFL** head-like, 8–12 mm wide (in fr ± 2 cm wide, spheric, fuzzy, red or brown); involucre bracts tiny, ± free. **FL**: calyx in fr inflated, lobes linear, lower 2 bristle-like, > upper 3; corolla 5–6 cm, pink, quickly ± hidden by calyx. **SEEDS** 1–2. 2*n*=16. Lawns, roadsides etc., often in saline soil; < 1500 m. NCoR, ScV, SCoR, SW, MP, probably elsewhere; sporadic in much of US; native to Eur, Afr. Cult.

T. fucatum Lindley (p. 655) Ann, often robust, ± glabrous. **ST** decumbent to erect, often hollow. **LVS** cauline; stipules wide; lflets 1–2.5 cm, elliptic to ± round, ± entire to toothed. **INFL** head-like, 5–many-fld; involucre bracts ± free to 1/2 fused, margin scarious;

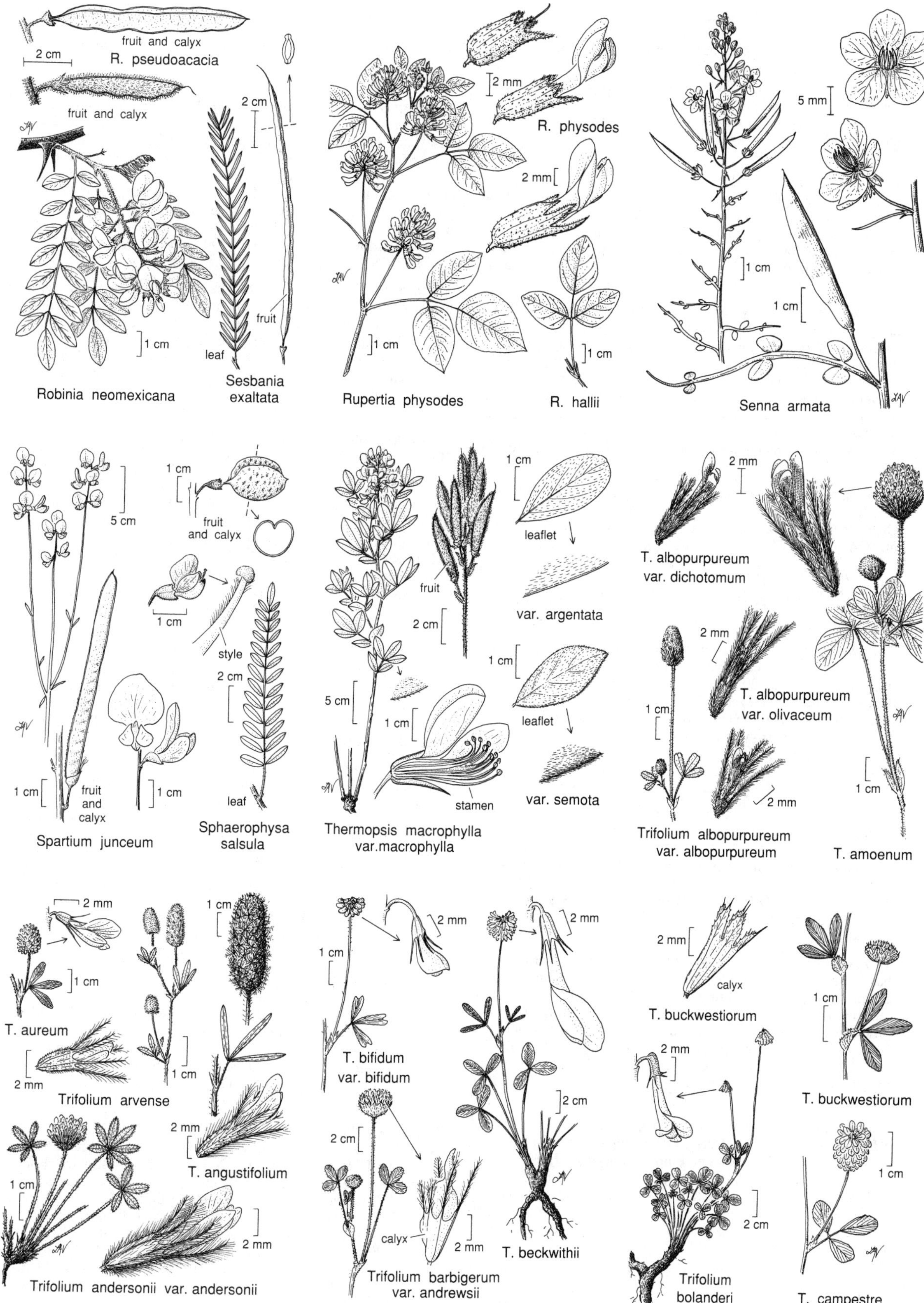
fruit and calyx
2 cm
R. pseudoacacia
fruit and calyx
2 cm
1 cm
leaf
fruit
Robinia neomexicana
Sesbania exaltata
2 mm
R. physodes
2 mm
1 cm
1 cm
Rupertia physodes
R. hallii
5 mm
1 cm
1 cm
Senna armata
5 cm
1 cm
fruit and calyx
1 cm
style
2 cm
1 cm
fruit and calyx
1 cm
leaf
Spartium junceum
Sphaerophysa salsula
fruit
2 cm
5 cm
1 cm
stamen
1 cm
leaflet
var. argentata
1 cm
leaflet
var. semota
Thermopsis macrophylla var.macrophylla
2 mm
T. albopurpureum var. dichotomum
2 mm
T. albopurpureum var. olivaceum
1 cm
2 mm
Trifolium albopurpureum var. albopurpureum
1 cm
T. amoenum
2 mm
1 cm
T. aureum
2 mm
1 cm
1 cm
Trifolium arvense
2 mm
T. angustifolium
1 cm
2 mm
Trifolium andersonii var. andersonii
2 mm
1 cm
T. bifidum var. bifidum
2 cm
calyx
2 mm
Trifolium barbigerum var. andrewsii
2 mm
2 cm
T. beckwithii
2 mm
calyx
T. buckwestiorum
2 mm
2 cm
Trifolium bolanderi
1 cm
T. buckwestiorum
1 cm
T. campestre

fls quickly spreading or reflexed. **FL**: calyx 4–7 mm, longer lobes bristle-like, > tube, often branched; corolla 10–20 mm, dull white or yellowish, purple-tipped, banner inflated in fr. **FR** 5–10 mm; stalk-like base 0.5–2 mm. **SEEDS** 2–4. 2*n*=16. Locally abundant. Moist, open grassland, ditches, marshes, roadsides, sometimes saline or serpentine soils; < 1100 m. NCo, NCoR, CaR, GV, SCo, ChI; OR. [var. *gambelii* (Nutt.) Jepson; var. *virescens* (E. Greene) Jepson] Many vars. based on insignificant variation in calyx. ❀SUN,WET:**5–9,14–24**.

T. glomeratum L. (p. 655) Ann, ± glabrous. **ST** decumbent to ascending. **LVS** cauline; stipules striate, shortly bristle-tipped; lflets 5–12 mm, obovate. **INFLS** in a succession of lf axils, head-like, 7–10 mm wide, many-fld; peduncle 0; pedicel < 1 mm. **FL**: calyx 3–5 mm, 10-ribbed, lobes triangular, recurved in fr; corolla pink, slightly > calyx. **SEEDS** 1–2. 2*n*=14,16. Uncommon. Disturbed areas; esp < 300 m. CA-FP; sporadic in US; native to Eur.

T. gracilentum Torrey & A. Gray Ann, glabrous or slightly hairy. **ST** prostrate to erect. **LVS** cauline; stipules ovate-tapered; lflets 5–15 mm. **INFL**: umbel, 1–1.5 cm wide, 3–many-fld, often turned to side; axis often ± exserted; involucre abortive; pedicels 1–2 mm; fls becoming recurved. **FL**: calyx 4–6 mm, glabrous, tube 10-veined, lobes ± 0.5 mm wide at base, > tube, shortly bristle-tipped; corolla 5–7 mm, pink to pink-purple. **FR** 4–6 mm, > petals. **SEEDS** 1–2. 2*n*=16. Open, disturbed, moist or dry places, grassy areas near ocean, sometimes serpentine; < 1800 m. CA-FP; to WA, AZ, Mex.

var. ***gracilentum*** (p. 655) **LF**: lflets obovate to obcordate, length 1.5–2.5 × width; lower stipules gen inconspicuous, < 1 cm. Open, disturbed, moist or dry places, sometimes serpentine; < 1800 m. CA-FP (exc ChI); to WA, AZ. [var. *inconspicuum* Fern.]

var. ***palmeri*** (S. Watson) L.F. McDermott PALMER'S CLOVER **LF**: lflets narrowly elliptic or lanceolate, length 3.5–5 × width; lower stipules often conspicuous, < 2 cm, fibrous. UNCOMMON. Grassy areas near ocean; < 15 m. ChI; Baja CA. [*T. p.* S. Watson]

T. gymnocarpon Nutt. var. ***plummerae*** (S. Watson) J.S. Martin (p. 655) Per, cespitose, mat- or cushion-forming, hairy. **LVS** ± basal; lflets 3–5, 0.8–20 mm, gen elliptic, thick, serrate, teeth 0.1–0.3 mm. **INFL**: umbel, gen incl, 10–15 mm wide, 4–many-fld; peduncle 1–4 cm; fls spreading-reflexed. **FL**: calyx 4–6.5 mm, tube 10-veined, lobes narrow, < to > tube; corolla 7–11 mm, pinkish to ± tan. **FR** 4–5 mm, exserted from corolla; stalk-like base ± 1 mm. **SEED** 1. With sagebrush, juniper; 1500–1800 m. MP; to OR, MT, NM. Merges clinally into var. *gymnocarpon* of Rocky Mtns.

T. hirtum All. (p. 655) ROSE CLOVER Ann, hairy. **ST** ascending or erect. **LVS** cauline; stipules bristle-tipped; lflets 1–2.5 cm, obovate. **INFL**: head, ± 1.5 cm wide, many-fld; sessile above 1–2 ± reduced lvs with involucre-like stipules; bristly in fr. **FL**: calyx 7–9 mm, lobes subequal, >> tube, bristle-like, plumose; corolla 11–15 mm, gen pink. **SEED** 1. 2*n*=10. Disturbed areas, roadsides; < 2060 m. CA-FP; sporadic to e US; native to Eurasia, n Afr. Cult; locally naturalized in CA.

T. howellii S. Watson (p. 655) HOWELL'S CLOVER Per, often robust, glabrous. **ST** erect. **LVS** gen cauline; stipules of middle lvs 1–2.5 cm, green; lflets 4–10 cm, elliptic to ovate. **INFL** head- or short-raceme-like, 1.5–3 cm; axis often exserted 2–3 mm; fls becoming reflexed. **FL**: calyx 3.5–4.5 mm, tube 10-veined, lobes slender, ± = or > tube; corolla 9–14 mm, greenish white to pink. **FR**: stalk-like base ± 1 mm. **SEEDS** 1–3. UNCOMMON. Wet or shady places, meadows with sedges, alder swamps; ± 800–1800 m. n KR, n NCoR, n CaR; OR.

T. hybridum L. (p. 655) ALSIKE CLOVER Ann or per, ± glabrous. **ST** sprawling to erect. **LVS** cauline; stipules ovate-lanceolate, strongly veined; lflets 1–4 cm, elliptic to obovate. **INFL**: umbel, 1.5–3 cm wide; fl bract < 1 mm; pedicel 2–6 mm; fls becoming reflexed. **FL**: calyx 3–5 mm, tube 10-veined, lobes lanceolate, ± = or > tube; corolla 6–11 mm, pink. **FR**: stalk-like base ± 0. **SEEDS** 2–4. 2*n*=16. Disturbed areas; < 1500 m. NW, sporadic elsewhere; US; native to Eur. Cult.

T. incarnatum L. (p. 655) CRIMSON CLOVER Ann, hairy. **ST** erect, often unbranched. **LVS** cauline; stipules ovate to oblong, strongly veined; lflets 1–2.5 cm, obovate or obcordate. **INFL**: spike, 2–6 cm, cylindric or ± conic. **FL**: calyx 7–10 mm, tube strongly 10-veined, lobes bristle-like, ± = or > tube; corolla 10–14 mm, crimson. **SEED** 1. 2*n*=14. Uncommon. Disturbed areas; esp < 300 m. CA-FP; naturalized in se US; native to s Eur. Possibly naturalized in CA.

T. kingii S. Watson var. ***productum*** (E. Greene) Jepson (p. 655) Per, glabrous. **ST** ascending; internodes 1–3 below peduncle. **LVS** basal and cauline; lower stipules papery; lflets 1–5.4 cm, lanceolate or elliptic, coarsely serrate. **INFL** head-like or short spike, turned to side or downward, 1.5–3 cm wide; axis exserted, gen forked; pedicel < 1 mm; fls quickly reflexed. **FL**: calyx 3–4 mm, glabrous, tube gen 10-veined, lobes slender, ± = tube; corolla 10–14 mm, reddish purple or pink. **FR**: stalk-like base 2–4 mm. 2*n*=16. Coniferous forests, meadows, to open ridges; 1100–2400 m. KR, CaRH, n&c SN, MP; OR. [*T. p.* E. Greene] ❀TRY.

T. lemmonii S. Watson (p. 655) LEMMON'S CLOVER Per, cespitose, strigose. **LVS** basal and cauline; basal stipules papery; lflets 3–7, 1–2 cm, obovate, thick, coarsely serrate, larger teeth 0.4–0.5 mm. **INFL**: umbel, 1.5–2.5 cm wide, ± spheric, gen many-fld; peduncle gen bent or curved; fls soon reflexed. **FL**: calyx 4–5 mm, lobes slender, ± bristle-like, > tube; corolla 10–13 mm, pink-white. **FR**: stalk-like base 1–1.5 mm. **SEEDS** 1–2. UNCOMMON. Pine forests, sagebrush flats; 1500–1800 m. n SNH; NV.

T. longipes Nutt. Per, rhizomed or not, cespitose or not, gen puberulent. **ST** decumbent to erect, sometimes ± 0. **LVS** basal and cauline; stipules oblong-lanceolate to ovate, < 2 cm; lflets 2–5 cm, linear to obovate. **INFL** head-like, incl or exserted, 1.5–3 cm wide; peduncle often bent or curved at tip. **FL**: calyx 5–10 mm, ± puberulent, lobes lanceolate to bristle-like, gen > tube; corolla 10–18 mm, dull white, purple, or 2-colored. **FR**: stalk-like base 0–1 mm. **SEED** gen 1. 2*n*=16,24,32,48. Meadows, streambanks, forests, open slopes, sometimes on serpentine; 600–3000 m. KR, NCoR, CaR, SN, SnBr, SnJt, GB; to WA, MT, Colorado. Intergrading complex [Gillett 1969 Can J Bot 47:93–113]; differentiating characters often indiscernible on herbarium specimens. Var. *longipes* not present in CA.

var. ***elmeri*** (E. Greene) L.F. McDermott (p. 655) **LF**: lflet length often 6–10 × width. **INFL**: pedicel 0.3–1 mm; fls ascending to spreading. **FL**: calyx lobes 3–5 × tube; corolla whitish. Moist, shaded areas, streambanks, meadows; 600–1700 m. KR, NCoR, CaR; OR. ❀TRY.

var. ***nevadense*** Jepson (p. 655) **LF**: lflet length 2.5–6 × width in CA-FP, to 10 × width in GB. **INFL**: pedicel 0.3–1 mm; fls spreading to ascending. **FL**: calyx lobes 1–3 × tube; corolla white to purple. Dry or boggy meadows, open slopes, woodlands, subalpine; 1100–3000 m. KR, NCoR, CaR, SN, SnBr, SnJt, GB; OR, NV. Intergrades in s SN with var. *atrorubens* (E. Greene) Jepson, which is isolated geog in SnBr, SnJt (hairier calyx, darker corolla). Pls very small at high elevation. ❀DRN,IRR:**1–3,6**,7,15–17.

var. ***oreganum*** (Howell) Isely **LF**: lflet length 2–5 × width. **INFL**: pedicel 1–2 mm; fls quickly reflexed. **FL**: calyx lobes 1–2 × tube; corolla probably white or yellowish, petals acute or obtuse. Forested slopes, gravelly meadows, serpentine; 1200–1800 m. KR, NCoR; OR. [*T. oreganum* Howell] ❀TRY.

var. ***shastense*** (House) Jepson Pl often cespitose. **LF**: lflet length 1.5–4 × width. **INFL**: pedicel 1–2 mm; some or most fls becoming reflexed. **FL**: calyx lobes 1–3 × tube; corolla lavender to purple, longer petals tapered to point. Coniferous forest to alpine slopes; 1400–2700 m. KR, CaR, GB; to WA. [var. *multipedunculatum* (Kenn.) Isely] ❀TRY.

T. macilentum E. Greene var. ***dedeckerae*** (J.M. Gillett) Barneby (p. 655) DEDECKER'S CLOVER Per, cespitose, glabrous. **ST** ± ascending. **LVS** basal, sometimes 2–3 cauline; basal stipules clasping st; lflets 0.5–4 cm, lanceolate, thick, ± serrate; petiole gen >> blades. **INFL** head-like, 1.5–3 cm wide; pedicel 1–2 mm; fls soon reflexed. **FL**: calyx 5–8 mm, glabrous, lobes unequal, gen > tube, narrow or bristle-like; corolla 12–15 mm, pink or pale violet. **FR**: stalk-like base < 2 mm. **SEEDS** probably 2. RARE. Pinyon woodland to alpine crests, rock crevices; 2100–3500 m. s SNH, SNE. [*T. dedeckerae* J.M. Gillett] Basal lvs sometimes ± deteriorated on herbarium specimens.

T. macraei Hook. & Arn. (p. 655) Ann. **ST**: decumbent to erect. **LVS** cauline; stipules ovate-acuminate or bristle-tipped; lflets 1–2 cm, narrowly elliptic to ovate. **INFLS**: heads gen 2, sessile (or 1 short-peduncled) above (1) 2 subtending lvs, each 5–15 mm wide, few–many-fld, ovoid or spheric; peduncle 0 or 1; pedicel 0. **FL**: calyx 4–7 mm, hairy, lobes tapered or bristle-like, >> tube, plumose; corolla 5–9 mm, purple or 2-colored. **SEED** 1. $2n$=16. Disturbed areas, dunes; < 600 m. NCo, NCoRO, CCo, SnFrB, ChI; S.Am.

T. macrocephalum (Pursh) Poiret (p. 655) Per, rhizomed, hairy. **ST** ascending. **LVS** basal and cauline; basal stipules brown-papery, others green, 1–3.5 cm; lflets 7–9, 1–2.5 cm, obovate, thick. **INFL** head- or raceme-like, 3–6 cm. **FL**: calyx 1.5–2 cm, tube 10-veined; lobes bristle-like, >> tube, plumose; corolla 2–2.7 cm, light pink-purple or 2-colored. **SEEDS** 1–3. $2n$=32, also inexact high number. Rocky flats, slopes with sagebrush, juniper, to mtn ridges; 650–2500 m. KR, CaR, n SNH, MP; to WA, ID, NV. ❀DRN, DRY,SUN:**1–3**,6,7,15–17;DFCLT.

T. microcephalum Pursh (p. 655) Ann; hairs gen soft, wavy. **ST** decumbent to erect. **LVS** cauline; stipules often ± bristle-tipped; lflets 0.8–2 cm, gen obovate, tip notched. **INFL**: head, 7–many-fld, often bur-like in fr; involucre bowl-shaped, lobes ± entire. **FL**: calyx 4–6 mm, tube 10-veined, lobes ± = or > tube, entire, unbranched, bristle-tipped; corolla 4–7 mm, pink to lavender. **FR** rupturing corolla. **SEEDS** 1–2. $2n$=16. Streambanks, moist, disturbed areas, roadsides, serpentine, conifer forest; 200–2700 m. NW, SN, GV, SCo, ChI, SnBr, PR, DMoj; to B.C., ID, AZ. ❀STBL.

T. microdon Hook. & Arn. (p. 655) Ann, puberulent. **ST** decumbent to erect. **LVS** cauline; stipules wide, upper deeply cut or not; lflets 0.5–1.5 cm, oblanceolate to obovate, tip truncate or notched. **INFL** head-like, 0.7–1.4 cm wide, 5–many-fld; involucre cup-shaped, divisions bristle-tipped. **FL**: calyx 2–3 mm, tube 10-veined, lobes triangular, << tube; corolla 4–6 mm, white to pink. **SEEDS** 1–2. $2n$=16. Common locally. Open, moist or dry, gen disturbed areas; 50–1100 m. NCo, NCoR, n&c SNF, GV, CW, ChI, PR; to B.C.; S.Am. [var. *pilosum* Eastw.]

T. monanthum A. Gray Per, very small, often cespitose, glabrous or puberulent. **ST** slender or reduced. **LVS** gen basal; stipules lanceolate to ovate; lflets 2–12 mm, elliptic-oblanceolate to widely obovate. **INFL**: reduced head, incl or exserted from lvs, 1–6-fld; involucre vestigial or bracts 2–5, inconspicuous, ± free, 1–3 mm. **FL**: calyx 4–5 mm, lobes ± = tube, bristle-tipped; corolla 7–12 mm, white or lavender-striate. **FR** sometimes rupturing corolla. **SEEDS** 1–2. $2n$=16. Pinyon pine belt upwards, mtn forests near streams, wet meadows with aspen or willows, coniferous woodlands; 1500–3800 m. CaR, SN, SCo, SnGb, SnBr, SnJt, SNE; NV.

var. ***grantianum*** (A.A. Heller) Parish **LF**: lflet length 4–6 × width, tip gen acute. **INFL** 2–6-fld. Mtn forests near streams; 1500–2700 m. SCo, SnGb, SnBr, SnJt.

var. ***monanthum*** (p. 655) **LF**: lflet length 2–4 × width, tip gen rounded or truncate. **INFL** 1–3-fld. Pinyon pine belt upwards, wet meadows with aspen or willows, coniferous woodland; 1500–3800 m. CaR, SN, SNE; NV. [var. *parvum* (Kellogg) L.F. McDermott; var. *eastwoodianum* J.S. Martin]

T. obtusiflorum Hook. & Arn. (p. 655) Ann, gen robust, glandular-hairy, often sticky. **ST** erect. **LVS** cauline; stipules deeply cut; lflets 2–4 cm, ± narrowly elliptic, sharply serrate, the longer teeth ± 1 mm. **INFL** head-like, 1.5–3 cm wide; involucre cut < 1/2 to base. **FL**: calyx 9–11 mm, tube slitting between upper lobes, lobes < or ± = tube, gen entire, bristle-tipped; corolla 14–18 mm, pale lavender to dull purple, tips white. **SEEDS** 1–2. $2n$=16. Moist disturbed areas, gravel bars, marshes; 200–1800 m. NW, n&c SN, SnJt, CW, SCo, SnGb, PR; sw OR.

T. oliganthum Steudel (p. 655) Ann, gen glabrous. **ST** ascending to erect. **LVS** cauline; stipules toothed or deeply cut; lflets 1–2 cm, linear to obovate. **INFL** head-like, 6–10 mm wide, 5–15-fld; involucre 2–3 mm, wheel-like, inconspicuous, gen cut > 1/2 to base into narrow lobes. **FL**: calyx 5–7 mm, tube slitting between upper lobes, lobes < tube, tapered to a sometimes slightly forked bristle; corolla 5–8 mm, ± = or slightly > calyx. **SEEDS** 1–2. Woody or shrubby slopes, roadsides; < 1000 m. KR, NCoR, SN, GV, SnFrB, SCoRO; to B.C.

T. pratense L. (p. 659) RED CLOVER Per, gen hairy. **ST** ascending. **LVS** cauline; stipules bristle-tipped; lflets 1.5–3.5 cm, elliptic to obovate. **INFL** terminal, above pair of reduced lvs, head-like, 2–3 cm wide; peduncle 0. **FL**: calyx 4.5–7.5 mm, tube 10-veined, lobes ± = or > tube, bristle-like, sparsely plumose; corolla 11–15 mm, red-purple. **FR** circumscissile. **SEEDS** 1–2. $2n$=14, 28. Disturbed areas; esp < 1000 m. CA-FP, GB, esp n; US, esp n, Can; native to Eur. Important forage crop.

T. repens L. (p. 659) WHITE CLOVER Per, gen ± glabrous. **ST** creeping, rooting. **LVS** cauline from ground level, clearly alternate or clustered (from lack of st elongation); stipules white-membranous; petioles >> blades; lflets 0.5–2.5 cm, obovate. **INFL** umbel-like, 1–2.5 cm wide; pedicel 1–3 mm; fls becoming reflexed. **FL**: calyx 3–6 mm, lobes tapered, all or lower > tube; corolla 7–11 mm, white. **SEEDS** 3–4. $2n$=16,28,48. Agricultural, disturbed, urban areas; esp < 1500 m. CA-FP, GB, esp n; n US; native to Eurasia. Important forage crop.

T. subterraneum L. (p. 659) SUBTERRANEAN CLOVER Ann, hairy. **ST** prostrate, creeping, rooting. **LVS** cauline from ground level, clearly alternate or clustered (from lack of st elongation); stipules wide, tapered; petioles > blades; lflets 10–15 mm, obovate or obcordate. **INFL** head-like, ± 1 cm wide, bur-like in fr; peduncle in fr curving, growing into ground. **FLS**: outer 2–8, fertile, corolla 8–14 mm, yellow and purple; inner many, sterile, calyx stalk-like, with 4–5 bristles at tip, elongating, recurving to surround frs and form bur, corolla 0. **SEED** 1. $2n$=16. Meadows, roadsides, urban waste areas; < 1000 m. NCoR, n SN, GV, SCoR, probably elsewhere; sporadic US; native to s Eur, n Afr. Fr of this and related spp. unique in family in US. TOXIC in excess: estrogen-like compounds may sterilize livestock.

T. trichocalyx A.A. Heller (p. 659) MONTEREY CLOVER Ann, initially sparsely hairy, becoming ± glabrous below infl. **ST** spreading. **LVS** cauline; stipules toothed or lobed; lflets 5–10 mm, oblanceolate to obovate. **INFL** head-like, incl or exserted, 4–14 mm wide, 1–10-fld; involucre small, cut beyond middle, glabrous or hairy. **FL**: calyx 6–7 mm, lobes gen > tube, bristle-tipped, sometimes slightly forked, hairy; corolla 6 mm, pink or lavender, ± = or < calyx. **FR** not seen. **ENDANGERED** CA. Pine woods, roadsides; < 100 m. CCo (Monterey Peninsula, Monterey Co.) Possibly hybrid of *T. variegatum* and a non-involucrate sp. Seriously threatened by urbanization.

T. variegatum Nutt. Ann or possibly short-lived per, gen ± glabrous. **ST** prostrate to erect, wiry to fleshy. **LVS** cauline; lower stipules gen entire; upper stipules deeply cut; lflets gen obovate or wedge-shaped, sometimes narrower. **INFL** head-like, incl or exserted from lvs, 0.5–2.5 cm wide, 1–many-fld; involucre wheel-like, gen well developed. **FL**: calyx 3–10 mm, tube 10–many-veined, all or some lobes gen > tube, bristle-tipped; corolla 3.5–16 mm, lavender to purple, tips gen white. **FR**: stalk-like base short or 0. **SEEDS** 1–2. $2n$=16. CA-FP, SNE; sporadic to B.C., MT, Colorado, AZ, Baja CA. Most variable of CA clovers; ± 30 names available; most conspicuous CA variants treated as phases below, with commonly used names indicated for each. Keel shape seems taxonomically insignificant (acute to beaked in phases 1 and 3); additional research needed.

1. Infl 1–5-fld; corolla 3.5–9 mm .... **phase 5**
1′ Infl 5–many-fld; corolla 6–16 mm
  2. Infl ± 10–15 mm wide, ± 8–10-fld; corolla ± 6–10 mm; CA-FP, SNE .... **phase 1**
  2′ Infl 15–25 mm wide, > 10-fld; corolla 10–16 mm; parts of CA-FP
    3. St 3–15 cm; pl tiny — SN, CW .... **phase 3**
    3′ St 1–5(10) dm; pl often large on coast
      4. Calyx lobes entire; widespread .... **phase 2**
      4′ Calyx lobes 3-divided or -toothed; CCo (Monterey Co.) .... **phase 4**

phase 1 (p. 659) Ann, very small to robust. **ST** decumbent to erect, sometimes mat- or tangle-forming, 0.5–5 dm. **INFL** ± 1–1.5 cm wide, ± 8–10-fld. **FL**: calyx 4–6 mm; corolla ± 6–10 mm. Open fields, wet forest meadows, roadsides; 50–2500 m. Range of sp. [*T. appendiculatum* Lojacono var. *rostratum* (E. Greene) Jepson] Gen considered "typical" *T. variegatum*; merging with phases 2 and 3. ❀WET:1–4,**5–9**,**14–24**.

phase 2 Ann or possibly short-lived per, gen robust. **ST** gen ascending, 1–5(10) dm. **INFL** 1.5–2.5 cm wide, many-fld. **FL**: calyx 6–10 mm; corolla 10–16 mm. Permanently wet or inundated sites, incl meadows, marshes, ditches; 200–2200 m. NCo, NCoR, SN, GV, CW, SW, SNE; sw OR. [var. *melananthum* (Hook. & Arn.) E. Greene; *T. appendiculatum* Lojacono var. *a.*] Commonly confused with *T. wormskioldii*. ❀WET:1–3,**4–9,14–24**.

phase 3 Ann, tiny. **ST** prostrate to ascending, 3–15 cm. **INFL** 1.5–2.5 cm wide, few–many-fld. **FL**: calyx 6–8 mm; corolla 10–16 mm. Gen moist soil, open areas; 400–1100 m. SN, CW. [var. *trilobatum* (Jepson) L.F. McDermott; *T. appendiculatum* Lojacono in part] Merging with phases 1 and 2; tiny pls with large infls probably = phase 2 reduced by environment. ❀WET:1–3,**6,7**,14–18.

phase 4 Ann. **FL**: calyx lobes ± = or < tube, 3-divided, -toothed. Closed-cone pine forest; < 300 m. CCo (Monterey Peninsula, Monterey Co.) [*T. polyodon* E. Greene, Pacific Grove clover, **RARE** CA] Otherwise as phase 2; taxonomic position questionable; possibly of hybrid origin.

phase 5 (p. 659) Ann, tiny or not. **ST** mat-forming or not, 1–30 cm. **LF**: lflets 3–8 mm, gen wedge-shaped. **INFL** < 1 cm wide, 1–5-fld; involucre sometimes reduced. **FL**: calyx 3–5 mm; corolla 3.5–9 mm. Open, disturbed, grassy slopes, woodlands, often in dry soil; 200–2500 m. NCo, NCoR, n&c SN, CW, PR; to WA, ID. [*T. geminiflorum* E. Greene; not *T. pauciflorum* Nutt.] Incl low-elevation, probably stunted phase 1 pls as well as mtn pls that may deserve separate recognition. ❀1–3,**5–7,14–17**.

T. willdenovii Sprengel (p. 659) Ann, tiny to robust, glabrous. **ST** sprawling to erect. **LVS** cauline; stipules bristle-tipped, upper gen sharply cut; lflets 1–5 cm, linear to obovate. **INFL** head-like, 1.5–3 cm wide, gen lobed, many-fld exc on tiny pls; involucre wheel-like, sharply lobed or dissected. **FL**: calyx 6–10 mm, shiny, tube slitting between upper lobes, lobes < tube, entire or some often 3-toothed, bristle-tipped; corolla 8–15 mm, lavender to purple, tip gen white. **SEEDS** 1–2. $2n$=16. Abundant. Disturbed, gen spring-moist, heavy soils, sometimes serpentine; < 1700 m. CA-FP, MP; sporadic to B.C., ID, NM, Baja CA, S.Am. [*T. tridentatum* Lindley, incl var. *aciculare* (Nutt.) L.F. McDermott] ❀WET:1–3,**5–7**,8,9, **14–24**.

T. wormskioldii Lehm. (p. 659) Per, cespitose or not, glabrous. **ST** decumbent or ascending. **LVS** gen basal; lower stipules bristle-tipped; upper stipules wide, toothed or sharply lobed; lflets 1–3 cm, narrowly elliptic to widely ovate. **INFL** head-like, incl or exserted from lvs, 2–3 cm wide; involucre wheel-like, segments or lobes many. **FL**: calyx 7–11 mm, lobes tapered, tips bristled; corolla 12–16 mm, pink-purple or magenta, tip white. **FR**: stalk-like base 0–1 mm. **SEEDS** 2–4. $2n$=16,32. Beaches to mtn meadows, ridges, gen open moist, or marshy places; < 3200 m. NW, CaR, SN, SnJV, CW, SCo, PR, SNE; to B.C., WY, NM, Mex. The only involucred per sp. in Pacific coast states; incl matted, rhizomed form (dry coastal sands); lush, long-stemmed form (lower to middle elevations); slender, often tiny form (middle to higher elevations). ❀IRR or WET,SUN:1–3,**4–9,14–17**.

ULEX GORSE, FURZE

Elizabeth McClintock

Shrub, heavily armed, not gland-dotted. **ST** much-branched from base, stiffly spreading, striate. **LVS** simple, alternate; juvenile (on seedlings, young shoots near ground) linear; adult awl-like, stiff, becoming spines. **INFL**: gen cluster, axillary near twig tips, few-fld. **FL**: calyx 2-lipped, membranous, yellow, persistent; petals ± equal, yellow, persistent. **FR** ± exserted from calyx, ovoid or oblong, explosively dehiscent. **SEED** with small basal outgrowth. (Latin: ancient name) ± 20 spp.: w Eur, n Afr.

U. europaea L. (p. 659) **ST** < 3 m; twigs hairy when young, becoming stiff, thorn-like, intricately intertwined. **FL**: calyx 15 mm; petals < 20 mm. **FR** 1–2 cm, densely hairy. Common. Disturbed places, esp old fields, pastures; 0–400 m. NCo, NCoRO, CaRF, n&c SNF, SnFrB, SCoRI, SCo, PR; native to w Eur. Old pls very flammable. NOXIOUS WEED.

VICIA VETCH

Duane Isely

Ann or per, unarmed. **ST** gen sprawling or climbing, ridged or angled. **LVS** even-1-pinnate; stipules with an upper (often toothed or lobed) and smaller lower segment; lflets 4–many, alternate to opposite (often on 1 pl), linear to ovate; main lf axis gen ending as a tendril. **INFL**: raceme or cluster, axillary; peduncle or pedicels present; bracts small or 0. **FL**: corolla gen lavender to purple, sometimes white or yellow; 9 filaments fused, 1 free; style gen round in X-section, hairs tufted at tip. **FR** dehiscent, gen ± oblong, gen flat; base stalk-like or not. **SEEDS** 2 or more. ± 130 spp.: N.Am, Eurasia. (Latin: vetch) [Herman 1960 USDA Handb 168] Best separated from *Lathyrus* by style characters.

1. Lflet 5–12 cm; pl erect; tendrils 0 .... ***V. faba***
1′ Lflet gen 1–4 cm; pl sprawling or climbing; tendrils gen present
  2. Infl a sessile or barely peduncled cluster of 1–4 pedicelled fls
    3. Fl 5–6 mm .... ***V. lathyroides***
    3′ Fl 10–30 mm
      4. Corolla pink-purple to whitish; banner glabrous, fr initially hairy, quickly glabrous .... ***V. sativa***
        5. Calyx 7–12 mm; corolla 1–1.8 cm, pink-purple to ± white; naturalized, not cult .... ssp. ***nigra***
        5′ Calyx 10–15 mm; corolla 1.8–3 cm, gen pink-purple; cult and naturalized .... ssp. ***sativa***
      4′ Corolla yellow or yellow-white, sometimes purple-blotched; banner puberulent on back, fr persistently hairy, or both
        6. Banner puberulent on back; calyx lobes ± equal; fr hairs not from conspicuous tubercles .... ***V. pannonica***
        6′ Banner glabrous; calyx lobes unequal; fr hairs from tubercles conspicuous on lower margin of fr .... ***V. lutea***
  2′ Infl a peduncled raceme
    7. Fl 3–9 mm
      8. Fr hairy, stalk-like base 0; lflets 10–16; fls gen 2–6 .... ***V. hirsuta***
      8′ Fr glabrous or ± strigose, stalk-like base ± 1–2 mm; lflets, fls various, incl as above
        9. Fr rounded at tip, with a short point on 1 side, 1–1.3 cm; lflets 4–10, linear to elliptic; peduncle thread-like, 1–3 cm, often curved, 1–3-fld .... ***V. tetrasperma***

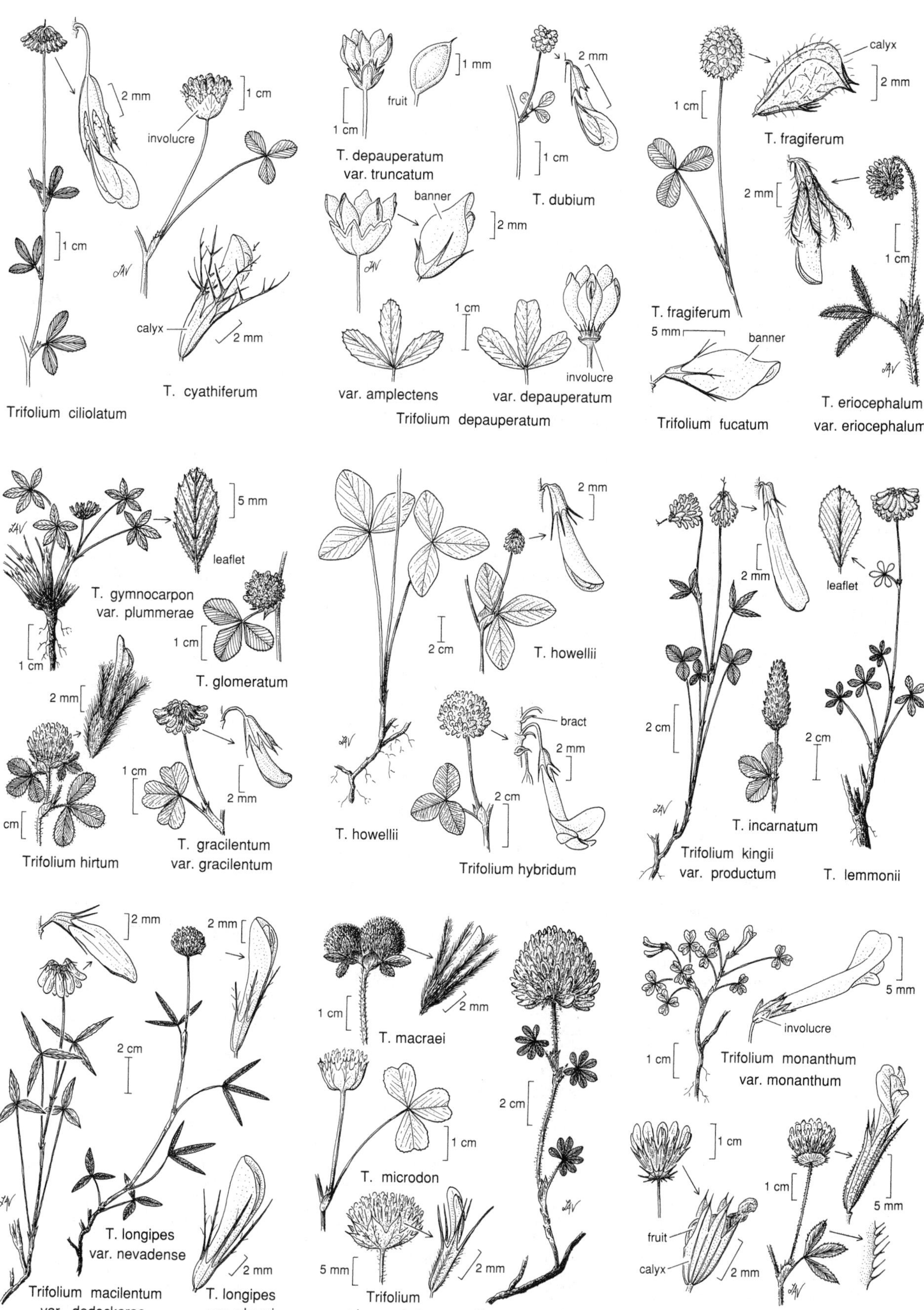
2 mm
1 cm
involucre
1 cm
calyx
2 mm
Trifolium ciliolatum
T. cyathiferum
1 mm
fruit
1 cm
T. depauperatum
var. truncatum
banner
2 mm
1 cm
var. amplectens
var. depauperatum
involucre
Trifolium depauperatum
2 mm
1 cm
T. dubium
calyx
2 mm
1 cm
T. fragiferum
T. fragiferum
5 mm
banner
Trifolium fucatum
2 mm
1 cm
T. eriocephalum
var. eriocephalum
5 mm
leaflet
T. gymnocarpon
var. plummerae
1 cm
1 cm
T. glomeratum
2 mm
1 cm
Trifolium hirtum
1 cm
2 mm
T. gracilentum
var. gracilentum
2 mm
2 cm
T. howellii
T. howellii
bract
2 mm
2 cm
Trifolium hybridum
2 mm
leaflet
2 cm
2 cm
T. incarnatum
Trifolium kingii
var. productum
T. lemmonii
2 mm
2 mm
2 cm
T. longipes
var. nevadense
2 mm
Trifolium macilentum
var. dedeckerae
T. longipes
var. elmeri
1 cm
2 mm
T. macraei
1 cm
T. microdon
5 mm
2 mm
Trifolium
microcephalum
2 cm
T. macrocephalum
5 mm
involucre
1 cm
Trifolium monanthum
var. monanthum
1 cm
fruit
calyx
2 mm
Trifolium oliganthum
1 cm
5 mm
T. obtusiflorum

9′ Fr tapered at tip, 1–3.5 cm; lflets or peduncle not as above

10. Lflets ± 12–18; fls 2–6; seeds 2 .... ***V. disperma***

10′ Lflets 4–10; fls 1–3; seeds gen > 2

11. Style hairs clumped at tip, gen on lower, convex side; lflet tip gen truncate; fls well separated or 1 .... ***V. hassei***

11′ Style hairs at tip and below, on all sides; lflet tip gen tapered; fls crowded or 1 .... ***V. ludoviciana*** var. ***ludoviciana***

7′ Fl > 9 mm

12. Lf gen 1–1.5 dm; lflets 16–24, larger ones 2.5–4 cm — fr often upcurved, ± 12–13 mm wide .... ***V. gigantea***

12′ Lf rarely > 1 dm; lflets either fewer or smaller than above

13. Fls not crowded, on > 1 side of axis (exc sometimes when pressed), 1.5–2.5 cm, length when pressed 2.5–3.5 × width; stalk-like fr base 2–5 mm .... ***V. americana*** var. ***americana***

13′ Fls crowded, gen on 1 side of axis, gen < 1.6 cm, length when pressed gen 4–6 × width; stalk-like fr base 1.5–3 mm

14. Infl ± = subtending lf, 3–12-fld; corolla dark or red-purple; fr puberulent .... ***V. benghalensis***

14′ Infl gen > subtending lf, gen 10–25-fld; corolla blue-purple or lavender to ± white; fr glabrous

15. Calyx often obliquely attached but only slightly lopsided-swollen at base; fl length when pressed ± 3-3.5 × width; fr 6–7 mm wide; per .... ***V. cracca***

15′ Calyx both obliquely attached and lopsided-swollen at base; fls when pressed 4–6 × width; fr 6–10 mm wide; ann .... ***V. villosa***

16. Hairs on sts, lvs 0 or inconspicuous, ± 1 mm; lower calyx lobe narrowly lanceolate, gen 1–2 mm .... ssp. ***varia***

16′ Hairs on upper sts, lvs conspicuous, 1–2 mm; lower calyx lobe needle- or thread-like, 2–4 mm .... ssp. ***villosa***

V. americana Willd. var. ***americana*** (p. 659) AMERICAN VETCH Per, hairy or glabrous. **ST** sprawling or short and erect. **LF**: stipules gen sharply lobed; lflets 8–16, 1–3.5 cm, widely elliptic, wedge-shaped, to narrowly oblong, tip acute, truncate, notched, or 1–5-toothed. **INFL** ± = subtending lf; fls 3–9, gen loosely spaced, on > 1 side of axis. **FL**: corolla 1.5–2.5 cm (length when pressed ± 2.5–3.5 × width), gen blue-purple to lavender. **FR** 2.5–3 cm, 5–7 mm wide, glabrous or hairy; stalk-like base 2–5 mm. 2*n*=14. Gen open, moist forest, along streams, disturbed areas; < 2400 m. CA-FP, GB; N.Am (exc se US). [ssp. *oregana* (Nutt.) Abrams; var. *linearis* S. Watson; var. *truncata* (Nutt.) Brewer; *V. californica* E. Greene] Attempts to use lflet form and hairs to define infraspecific taxa are untenable [see Gunn 1968 Iowa State Coll J Sci 42:171–214]. Often mistaken for a *Lathyrus*. ❀STBL.

V. benghalensis L. PURPLE VETCH Ann, hairy. **ST** sprawling or climbing. **LF**: lower stipules toothed or lobed; lflets 10–16, 1.5–3 cm, elliptic to oblong. **INFL** gen ± = subtending lf; fls 3–12, gen on 1 side of axis. **FL**: calyx lopsided-swollen at base or not, lobes plumose; corolla 12–16 mm (length when pressed 4–6 × width), dark or red-purple. **FR** 2.5–3.5 cm, 8–12 mm wide, puberulent; stalk-like base ± 1.5 mm. 2*n*=14. Disturbed areas; esp < 200 m. CA-FP; sporadic to e US; native to Eur. Cult.

V. cracca L. (p. 659) Per, puberulent. **ST** gen climbing. **LF**: stipules gen entire; lflets 14–22, 1–2.5 cm, linear to elliptic. **INFL** gen > subtending lf; fls 15–25, crowded, gen on 1 side of axis. **FL**: calyx often obliquely attached but gen not lopsided-swollen at base; corolla 1–1.5 cm (length when pressed ± 3–3.5 × width), gen blue-purple or lavender, sometimes ± white. **FR** 1.5–3 cm, 6–7 mm wide, glabrous; stalk-like base ± 1.5 mm. 2*n*=12,14,28. Disturbed areas, woodlands and borders; < 1500 m. NCoR, MP; to ne US, e Can; native to Eur. Perhaps not naturalized in CA.

V. disperma DC. Ann, glabrous or puberulent. **ST** sprawling. **LF**: stipules entire, lower segment divergent or ascending; lflets ± 12–18, 7–18 mm, oblong-elliptic, ± rounded, tip often with 1 minute tooth. **INFL** < subtending lf; fls 2–6. **FL**: corolla 3.5–4.5 mm, blue. **FR** 1–2 cm, 5–7 mm wide, glabrous; stalk-like base ± 1 mm. **SEEDS** 2. 2*n*=14. Disturbed areas; esp < 100 m. n CCo, SnFrB; native to Eur. Probably not naturalized in CA.

V. faba L. BROAD BEAN, HORSE BEAN Ann, ± glabrous. **ST** erect, 1–2 m. **LF**: stipules toothed or cut; lflets 4–6, ± 5–12 cm, elliptic to narrowly ovate; tendril 0. **INFL** << subtending lf; fls 2–6, crowded, subsessile. **FL**: corolla 2–3 cm, white with dark blotches. **FR** 5–25 cm, 1–2 cm wide, cylindric, often narrowed between seeds. 2*n*=12, 14. Disturbed sites; esp < 200 m. NW, CW, SW, probably elsewhere; sporadic to e N.Am; native to Eurasia. Food, forage crop; probably not naturalized in CA. Seeds and pollen TOXIC to certain enzyme-deficient humans (some fatalities in young males).

V. gigantea Hook. (p. 659) Per, robust, glabrous or puberulent. **ST** sprawling or climbing, 1–2 m. **LF**: stipules toothed or not; lflets 16–24, 1.5–4 cm, elliptic or oblong, tip rounded to acute. **INFL** gen < subtending lf; fls 6–15, crowded, gen on 1 side of axis. **FL**: corolla 12–14 mm, red-purple or variegated to pale yellow. **FR** 2–4.5 cm, 1.2–1.3 cm wide, often upcurved, glabrous, black; stalk-like base 1–2 mm. Coastal bluffs, beaches, woodlands, roadsides; < 200 m. NCo, CCo, SnFrB, SCo; to AK. ❀STBL.

V. hassei S. Watson (p. 659) Ann, ± glabrous. **LF**: lflets 4–10, 1–2.5 cm, narrowly oblong to elliptic, tip gen truncate, notched, or 1–5-toothed. **INFL** < subtending lf; fls at or near tip, 1 (or if 2, gen 4–8 mm apart). **FL**: corolla 7–8 mm, lavender to white; style hairs in a clump at tip, gen on lower, convex side (opposite banner). **FR** 2–3 cm, 5–6 mm wide, oblong to saber-shaped, initially with a few hairs; stalk-like base 1–2 mm. 2*n*=14. Grassy or shrubby slopes, canyons; 100–600 m. NCo, NCoR, GV, CW, SW, PR; to Baja CA, also (alien?) in s OR. [*V. exigua* Nutt. var. *h* (S. Watson) Jepson] [Lassetter 1975 Madroño 23:73–78] In CA, easily confused with *V. ludoviciana*.

V. hirsuta (L.) S.F. Gray (p. 659) Ann, ± glabrous or finely hairy. **ST** slender. **LF**: stipules linear, sometimes toothed; lflets 10–16, 0.5–1.5 cm, linear to narrowly elliptic, tip gen truncate, notched, or 1–5-toothed. **INFL** gen < subtending lf; fls 2–8, crowded near tip. **FL**: corolla 3–4.5 mm, dull white or pale blue, ephemeral. **FR** 6–10 mm, 3–5 mm wide, widely oblong or elliptic, hairy; stalk-like base 0. **SEEDS** gen 2. 2*n*=14. Disturbed sites, open or wooded slopes; esp < 200 m. NCo, NCoR, SnFrB, SW, probably elsewhere; to B.C., se US; native to Eur.

V. lathyroides L. Ann, tiny, glabrous or hairy. **LF**: stipules entire to slightly toothed; lflets 4–6, 3–15 mm, lanceolate to obovate; tendril simple or 0. **INFL** << subtending lf; fls 1(2) in lf axils on short pedicels. **FL**: calyx 3.5–4.5 mm; corolla 5–6 mm, white-lavender or dull blue. **FR** 1.5–2.5(3) cm, 3–4 mm wide, oblong, glabrous; stalk-like base 0. Disturbed areas; < 1000 m. c SNF, probably elsewhere; to WA, se US; native to Eur. Pls look like miniature *V. sativa* ssp. *nigra*; well established.

V. ludoviciana Nutt. var. ***ludoviciana*** (p. 659) Ann, glabrous or hairy. **ST** sprawling or low-climbing. **LF**: stipules small; lflets 4–10, 1–2.5 cm, narrowly oblong to elliptic, tip acute, 1-toothed, or sometimes truncate. **INFL** < subtending lf; fls 1–3, near tip, gen crowded. **FL**: corolla 4.5–7 mm, pale blue; style hairs at tip and below on all sides. **FR** 1.5–2.5 cm, 4–7 mm wide, oblong or saber-shaped, glabrous; stalk-like base ± 1–1.5 mm. 2*n*=14. Woodland margins, open shrubby areas, disturbed sites; < 1000 m. SCo, ChI, PR; to se US, Mex. [*V. exigua* Nutt. var. *e.*] [Lassetter 1984 Rhodora 86:475–505] CA pls of this variable sp. ± identical to a race in s TX and closely related to *V. hassei*.

V. lutea L. Ann, glabrous or sparsely hairy. **ST** ascending or climbing. **LF**: stipules lobed or not; lflets 8–16, 1–2 cm, oblong or those of uppermost lvs linear, tip rounded or truncate, 1-toothed. **INFL** << subtending lf; fls in sessile clusters of 1–3, pedicelled. **FL**: calyx lobes unequal; corolla 2–3 cm, yellow, often purple-tinged. **FR** 2.5–3.5 cm, 7–14 mm wide, elliptic-oblong; hairs from tubercles conspicuous on lower margin of fr; stalk-like base 1–2 mm. $2n=14$. Disturbed areas; < 200 m. SnFrB; sporadic to s US; native to Eur. Possibly naturalized in CA; similar to *V. grandiflora* (calyx lobes subequal; fr glabrous; widely naturalized in e US, sometimes referred to CA).

V. pannonica Crantz HUNGARIAN VETCH Ann, hairy. **ST** ascending or tangled. **LF**: lflets 10–20, 1–2.5 cm, elliptic-oblong to ovate, tip rounded to truncate, often with 1 slender tooth. **INFL** << subtending lf; fls in subsessile clusters of 2–4, pedicelled. **FL**: calyx lobes ± equal; corolla 1.5–2 cm, pale yellow or yellow with purple blotches, banner puberulent on back. **FR** 2–3 cm, 7–11 mm wide, elliptic to widely oblong, hairy; stalk-like base 1–2 mm. $2n=12$. Disturbed areas; < 200 m. NCoRO (Sonoma Co.), probably elsewhere; sporadic US; native to Eur. Cult, esp in OR.

V. sativa L. Ann, glabrous or hairy. **LF**: stipules often toothed; lflets 8–14, 1.5–3.5 cm, linear to wedge-shaped, tip acute, truncate, or notched, often with 1 slender tooth. **INFL** << subtending lf; fls in sessile or barely peduncled clusters of 1–3, pedicels short. **FL**: calyx 7–15 mm; corolla 1–3 cm, pink-purple to whitish. **FR** 2.5–6 cm, 2.5–8 mm wide, initially hairy, quickly glabrous; stalk-like base 0. $2n=12,14$. Disturbed areas, fields; < 1500 m. CA-FP; to e US; native to Eur. Following sspp. sometimes treated as spp. in US

ssp. *nigra* (L.) Erhart (p. 659) NARROW-LEAVED VETCH, COMMON VETCH **LF**: lflets 5–7 mm wide, length 4–10 × width, linear to gen oblong-lanceolate. **FL**: calyx 7–12 mm; corolla 1–1.8 cm, pink-purple to ± white. **FR** black. **SEED** 2.5–4 mm wide, spheric. Disturbed sites, woodland borders, clearings; < 1500 m. CA-FP; to se US. [*V. angustifolia* Reichard] Widespread spring weed in s US.

ssp. *sativa* COMMON VETCH, SPRING VETCH **LF**: lflets 4–10 mm wide, length 2–6 × width, wedge-shaped to oblong. **FL**: calyx 10–15 mm; corolla 1.8–3 cm, gen pink-purple. **FR** brown to black. **SEEDS** 6–8 mm wide, gen lens-shaped. Disturbed areas; esp < 500 m. CA-FP; to e US. Hay and winter cover-crop.

V. tetrasperma (L.) Schreber (p. 659) Ann, ± glabrous. **ST** slender. **LF**: lflets 4–10, 0.6–2 cm, linear to elliptic. **INFL** > subtending lf; fls gen 1–3, at tip of axis; peduncle thread-like. **FL**: corolla 4–6 mm, pale lavender to light purple. **FR** 1–1.3 cm, 3–4 mm wide, glabrous; tip rounded, abruptly short-pointed on 1 side, becoming membranous; stalk-like base ± 1 mm. $2n=14$. Disturbed areas, woodlands or borders; esp < 1000 m. NW, SnFrB, PR, probably elsewhere; US, s Can; native to Eur.

V. villosa Roth HAIRY VETCH, WINTER VETCH Ann, puberulent-strigose or hairy. **ST** sprawling, ascending, or climbing. **LF**: lflets 12–18, 1–2.5 cm, narrowly oblong to elliptic, tip rounded, 1-toothed. **INFL** gen > subtending lf; fls gen > 9, gen crowded, sometimes spaced, gen on 1 side of axis. **FL**: calyx obliquely attached, lopsided-swollen at base, lower lobes lanceolate to wispy; corolla 1–1.8 cm, length when pressed 4–6 × width, violet-purple or lavender to white. **FR** 1.5–4 cm, 6–10 mm wide, widely oblong or elliptic, glabrous; stalk-like base 2–3 mm. $2n=14$. Sporadic, locally abundant. Roadsides, fields, urban waste areas; esp < 1000 m. CA-FP; to US; native to Eur. Important green manure, cover crop.

ssp. *varia* (Host) Corbière Pl nearly glabrous to gen puberulent; hairs ± 1 mm. **INFL** gen 10–20-fld, often open. **FL**: lower calyx lobe 1–2.5 mm, narrowly lanceolate; corolla 1–1.6 cm. [*V. dasycarpa* Ten.] Habitats, elevation, range of sp.

ssp. *villosa* (p. 659) At least upper sts and lvs with conspicuous, 1–2 mm hairs. **INFL** gen > 19-fld, dense. **FL**: lower calyx lobes 2–4 mm, thread-like or wispy, often curved; corolla 1.4–1.8 cm. Habitats, elevation, range of sp.

FAGACEAE OAK FAMILY

John M. Tucker

Shrub or tree, monoecious, deciduous or evergreen. **LVS** simple, alternate, petioled; margin entire to lobed; stipules small, gen deciduous. **STAMINATE INFL**: catkin or stiff spike; fls many. **PISTILLATE INFL** 1–few-fld, gen above staminate infl; involucre in fr gen cup-like or lobed and bur-like, bracts many, gen overlapping, flat or cylindric. **STAMINATE FL**: sepals gen 5–6, minute; petals 0; stamens 4–12+. **PISTILLATE FL**: calyx gen 6-lobed, minute; petals 0; ovary inferior, style branches gen 3. **FR**: acorn (nut subtended by scaly, cup-like involucre) or 1–3 nuts subtended by spiny, bur-like involucre; nut maturing in 1–2 years. **SEED** gen 1. 7 genera, ± 900 spp.: gen n hemisphere. Wood of *Quercus* critical for pre-20th century ship-building, charcoal for metallurgy; some now supply wood (*Fagus, Quercus*), cork (*Q. suber*), food (*Castanea*, chestnut).

1. Fr bur-like, spiny, enclosing 1–3 ovoid, ± angled nuts .... **CHRYSOLEPIS**
1' Fr an acorn (scaly, cup-like involucre subtending 1 smooth nut)
 2. Acorn cup with slender, spreading scales; staminate infl a stiff spike, erect or spreading .... **LITHOCARPUS**
 2' Acorn cup with closely appressed scales; staminate infl a catkin, slender, pendent .... **QUERCUS**

CHRYSOLEPIS CHINQUAPIN

Evergreen. **LF** leathery; margin entire or slightly wavy above middle; upper surface ± glabrous, green; lower surface golden, with densely spaced, minute, appressed scales; stipules gen deciduous. **STAMINATE INFL** simple, sometimes branched, stiff, solitary or clustered, ascending to erect. **PISTILLATE INFL** clustered below staminate infl on same or separate stalk; fls 1–3. **STAMINATE FL**: sepals gen 6, minute; stamens gen 8–10+. **PISTILLATE FL**: sepals gen 6, minute. **FR**: nuts 1–3, enclosed by spiny, bur-like involucre, ovoid to subspheric, ± angled, maturing in 2 years. 2 spp.: w N.Am. (Greek: golden scale, from lower lf surface) [Hjelmqvist 1960 Bot Not 113:373–380] Asian *Castanopsis* has involucre with 7 free valves (5 outer, 2 inner separating 3 nuts).

1. Lf tip gen obtuse or rounded; blade ± elliptic; lvs 2–8 cm; shrub gen << 3 m, rounded; bark gen ± thin, smooth .... ***C. sempervirens***
1' Lf abruptly long-pointed; blade oblong to lanceolate; lvs 5–13 cm; shrub or tree, top conic; bark ± thick, rough, furrowed .... ***C. chrysophylla***
 2. Tree, 15+ m; lf blade ± flat .... var. ***chrysophylla***
 2' Shrub (or tree-like), < 5(10) m; lf blade folded, margins upturned .... var. ***minor***

C. chrysophylla (Hook.) Hjelmq. GIANT CHINQUAPIN Shrub or tree < 30(45) m; top ± conic. **ST**: trunk bark ± thick, rough, furrowed. **LF**: petiole 5–12 mm; blade 5–15 cm, lanceolate to oblong, base tapered, tip abruptly long-tapered; upper surface dark green; lower surface golden. **FR**: bur 3–5 cm diam; nut 6–15 mm. Coniferous forest, closed-cone-pine forest, chaparral; < 2000 m. NW, n CaR, n SNH (El Dorado Co.), CW (exc SCoRI); to WA. ❀DRN:**4–6**,15–17&IRR:1–3,7,14,18;DFCLT.

var. ***chrysophylla*** (p. 659) Tree < 30(45) m. **LF**: blade gen flat. Coniferous forest; < 2000 m. Range of sp. (exc SCoRO). [*Castanopsis c.* (Hook.) A. DC. var *c.*]

var. ***minor*** (Benth.) Munz (p. 659) Shrub or small tree < 5(10) m. **LF**: blade ± folded, margins turned upward lengthwise. Habitats of sp.; < 1800 m. NW, CW (exc SCoRI). [*Castanopsis c.* (Hook.) A. DC. var. *m.* (Benth.) A. DC.]

C. sempervirens (Kellogg) Hjelmq. (p. 659) BUSH CHINQUAPIN Shrub < 3(10) m; top rounded. **ST**: trunk bark gen thin, ± smooth, gen not furrowed. **LF**: petiole 4–15 mm; blade 2–8(12) cm, ± elliptic, base tapered to rounded, tip obtuse to rounded; upper surface dull green; lower surface golden to rusty. **FR**: bur 2–3.5 cm diam; nut 8–13 mm. Rocky slopes, coniferous forest, chaparral; 700–3300 m. KR, NCoRH, CaRH, SNH, SnGb, SnBr, SnJt, w MP; s OR. [*Castanopsis s.* (Kellogg) Dudley] ❀DRN,SUN:4–6&IRR:1–3,7,14–17;DFCLT.

LITHOCARPUS TAN OAK, TANBARK OAK

Evergreen. **LF**: stipules early deciduous. **STAMINATE INFL**: spike, elongate, simple, stiff, spreading or erect; fls many. **PISTILLATE INFL** below staminate infl on same or separate stalk; fl 1. **STAMINATE FL**: sepals 5–6, minute; stamens 10–12. **PISTILLATE FL**: calyx 6-lobed. **FR**: acorn, maturing in 2 years; nut enclosed by cup-like involucre. ± 100 spp.: w N.Am (1 sp.), esp se Asia. (Greek: rock fr, from hard fr wall)

L. densiflorus (Hook. & Arn.) Rehder TAN OAK, TANBARK OAK Shrub or tree < 30(45) m; trunk bark grayish brown. **LVS** evergreen; petioles 10–25 mm; blade 3–14 cm, oblong to ± ovate, base ± rounded, tip obtuse, margin entire to serrate, upper surface sparsely stellate-hairy, becoming ± glabrous, lower surface finely woolly, becoming ± glabrous. **STAMINATE INFL** stiff, spreading to erect, densely fld. **FR**: cup (1.5)2–3 cm diam, saucer-shaped, scales slender, ± tapered, reflexed to spreading; nut 20–35 mm, ovoid to subspheric. Redwood to red-fir forests; < 2000 m. NW, CaR, SN, CW, WTR; s OR.

1. Tree < 30(45) m; lf margin gen serrate, lower surface with main veins prominent, ending in teeth ..... var. ***densiflorus***

1′ Shrub < 3 m; lf margin entire to few-toothed, lower surface with main veins obscure ..... var. ***echinoides***

var. ***densiflorus*** (p. 663) Tree < 30(45) m. **LF** 4–14 cm, 12–40 mm wide; margin gen serrate; lower surface with main veins prominent, ending in teeth. Habitats of sp.; < 1500 m. NW, CaR, SN, CW, WTR; s OR. ❀DRN:4,**5**,6,**17**&IRR,part SHD:1–3,**7**,14,**15**,**16**,18–24;CV.

var. ***echinoides*** (R.Br. Campst.) Abrams (p. 663) Shrub 1–3 m. **LF** 3–5(8) cm, 10–30 mm wide; margin entire, slightly wavy, or few-toothed; lower surface with main veins obscure. Mixed-conifer and red-fir forests; 600–2000 m. KR, CaR, SN; s OR. ❀DRN:**4–6**,17&IRR:1–3,**7**,14–16,18–21.

QUERCUS OAK

Evergreen or deciduous. **LF**: stipules small, gen early deciduous. **STAMINATE INFLS**: catkins, 1–several, slender, on proximal part of twig. **PISTILLATE INFL** axillary among upper lvs, short-stalked; fl gen 1. **STAMINATE FL**: calyx 4–6-lobed, minute; stamens 4–10. **PISTILLATE FL**: calyx minute, gen 6-lobed; ovary enclosed by involucre. **FR**: acorn, maturing in 1–2 years; nut enclosed by cup-like involucre with thin or tubercled scales. $2n$=24 for all reports. ± 600 spp.: n hemisphere, to n S.Am, India. (Latin: ancient name for oak) Many more hybrids have been named but are not incl here. Reproduction of many spp. declining.

1. Cup scales thin, flat; bark dark gray or black; acorn shell woolly inside
 2. Lf moderately to deeply lobed, sinuses between lobes 1/2+ width to midrib, lobes 1–4-toothed, teeth gen bristle-tipped; pls deciduous ..... ***Q. kelloggii***
 2′ Lf entire to toothed, teeth abruptly pointed to spine-tipped, not bristle-tipped; pls evergreen
  3. Lf upper surface convex, margin sometimes inrolled, obscuring marginal teeth, lower vein axils often hair-tufted, blade gen widely elliptic to round; fr maturing in 1 year (on current twig) ..... ***Q. agrifolia***
   4. Lf lower surface glabrous to sparsely hairy (exc vein axils) ..... var. ***agrifolia***
   4′ Lf lower surface densely tomentose ..... var. ***oxyadenia***
  3′ Lf ± flat, margin not inrolled, lower surface glabrous, blade gen lanceolate to oblong; fr maturing in 2 years (on previous year's growth)
   5. Lvs 3–9 cm; lf lower surface gen ± dull, olive-green; nut oblong below middle, abruptly tapered to tip ..... ***Q. parvula***
    6. Shrub 1–3 m; n ChI, n SCo ..... var. ***parvula***
    6′ Tree < 17 m; w SnFrB, SCoRO ..... var. ***shrevei***
   5′ Lvs 2–5 cm; lf lower surface gen ± shiny, yellow-green; nut gradually tapered from below middle to tip ..... ***Q. wislizenii***
    7. Shrub 2–4(6) m ..... var. ***frutescens***
    7′ Tree 10–22 m ..... var. ***wislizenii***

1′ Cup scales gen thick, ± flat to gen clearly tubercled; bark light gray to whitish; acorn shell glabrous to woolly inside
 8. Acorn shell ± woolly inside; fr maturing in 2 years; pls evergreen
  9. Lvs mostly 3–8 cm; acorn cup thick; gen tree
   10. Lf lateral veins not clearly impressed on upper surface, lower surface golden-hairy, becoming glabrous ..... ***Q. chrysolepis***

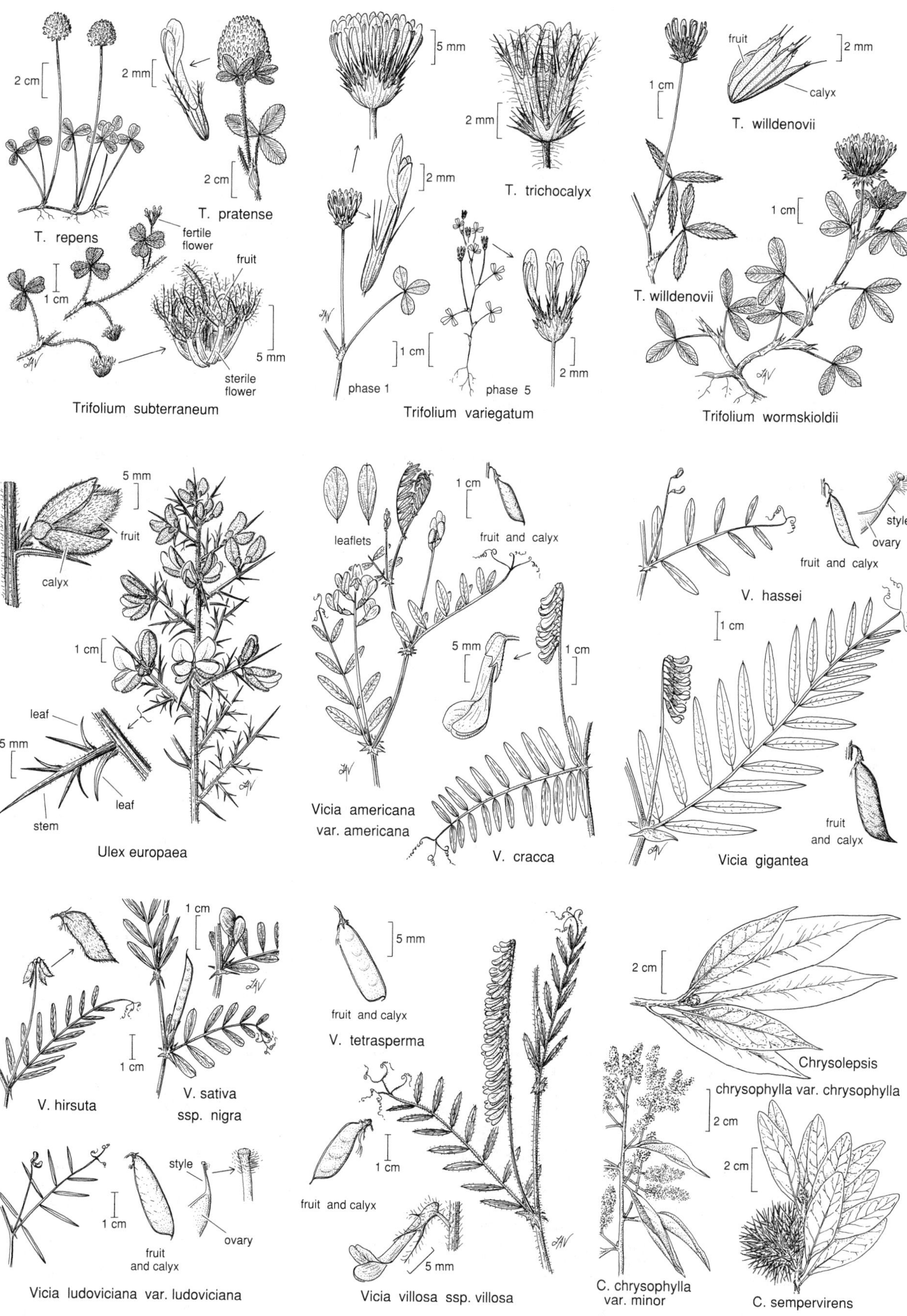
2 cm
2 mm
2 cm
T. repens
T. pratense
fertile flower
fruit
1 cm
5 mm
sterile flower
Trifolium subterraneum
5 mm
2 mm
2 mm
T. trichocalyx
1 cm
phase 1
phase 5
2 mm
Trifolium variegatum
fruit
2 mm
calyx
1 cm
T. willdenovii
1 cm
T. willdenovii
Trifolium wormskioldii
5 mm
fruit
calyx
1 cm
leaf
5 mm
leaf
stem
Ulex europaea
leaflets
1 cm
fruit and calyx
5 mm
1 cm
Vicia americana var. americana
V. cracca
style
ovary
fruit and calyx
V. hassei
1 cm
fruit and calyx
Vicia gigantea
1 cm
1 cm
V. hirsuta
V. sativa ssp. nigra
5 mm
fruit and calyx
V. tetrasperma
style
1 cm
fruit and calyx
ovary
Vicia ludoviciana var. ludoviciana
1 cm
fruit and calyx
5 mm
Vicia villosa ssp. villosa
2 cm
Chrysolepsis chrysophylla var. chrysophylla
2 cm
2 cm
C. chrysophylla var. minor
C. sempervirens
Fagaceae

10′ Lf lateral veins impressed on upper surface, lower surface densely tomentose, becoming sparsely tomentose — ChI . . . . . *Q. tomentella*

9′ Lvs mostly 1–3.5 cm; acorn cup thin; shrub (or tree < 7 m)

11. Twigs very stiff, rigid; lf margin ± wavy, clearly spine-toothed; blade elliptic to round-ovate; pls 2–6 m, erect . . . . . *Q. palmeri*

11′ Twigs pliable; lf margin gen entire, sometimes irregularly serrate, not spine-tipped; blade oblong to ovate; pls < 1.5 m, prostrate to spreading . . . . . *Q. vaccinifolia*

8′ Acorn shell glabrous inside; fr maturing in 1 year; pls evergreen or deciduous

12. Lf lobed, sinuses between lobes gen > 1/4 distance from midrib to lobe tip; pls deciduous

13. Lf moderately to deeply lobed, some sinuses between lobes gen > 1/2 distance to midrib; upper surface ± shiny, dark green

14. Acorn cup 10–30 mm deep, scales all clearly tubercled; nut 30–50 mm, gen long-conic in outline, gradually tapered to pointed tip (also see 17′) . . . . . *Q. lobata*

14′ Acorn cup 4–9 mm deep, upper scales ± flat, lower scales slightly tubercled; nut 20–30 mm, ovoid to subspheric, tip ± rounded . . . . . *Q. garryana*

15. Shrub 1–5 m; petiole 5–20 mm; lf blade 4–9 cm . . . . . var. *breweri*

15′ Tree 8–20 m; petiole 10–25 mm; lf blade 7–14 cm . . . . . var. *garryana*

13′ Lf weakly lobed, sinuses between lobes gen < 1/2 distance to midrib; upper surface gen dull, green to bluish green

16. Upper lf surface green; acorn cup scales clearly tubercled; ChI . . . . . *Q. ×macdonaldii*

16′ Upper lf surface bluish or grayish green; acorn cup scales slightly tubercled; mainland

17. Small tree (sometimes shrubby), semi-evergreen; lvs 1.5–5 cm, gen irregularly and coarsely toothed, upper surface bluish or grayish green . . . . . [2]*Q. ×alvordiana*

17′ Tree, deciduous; lvs gen 3–6 cm, slightly lobed or subentire, upper surface bluish green . . . . . *Q. douglasii*

12′ Lf entire to toothed, not lobed; pls evergreen

18. Lvs gen 7–11 cm, clearly serrate (teeth 20–32), lateral veins 20–28, prominent; stipules persistent, > 10 mm, silky-hairy . . . . . *Q. sadleriana*

18′ Lvs 1–6 cm, entire to dentate, lateral veins gen < 20, not prominent; stipules early deciduous, < 10 mm, not silky-hairy

19. Trees gen > 5 m; lvs 2–6 cm, oblong to obovate, entire or slightly toothed, upper surface often bluish green

20. Tree < 6(10) m; lvs gen irregularly toothed to subentire, upper surface green to bluish green, ± glossy to dull . . . . . [2]*Q.×acutidens*

20′ Tree 5–18 m; lvs gen entire or wavy-dentate, upper surface blue-green, dull . . . . . *Q. engelmannii*

19′ Shrub gen < 5 m; lvs 1–5 cm, elliptic to ovate or roundish, variously toothed, upper surface gen green

21. Lf 2-colored, upper surface yellow- or gray-green, lower surface whitish, densely finely tomentose, hairs obscuring lateral veins . . . . . *Q. cornelius-mulleri*

21′ Lf 1–2-colored, upper surface green to gray- or bluish green, lower surface gen green, not densely tomentose, hairs not obscuring lateral veins

22. Lf gen oblong to obovate, margin wavy to obtusely toothed — PR . . . . . [2]*Q. ×acutidens*

22′ Lf elliptic to ovate or roundish, margin clearly toothed, teeth abruptly pointed to spine-tipped

23. Acorn cup thin, scales ± flat to slightly tubercled; upper lf surface gray- or bluish green, dull

24. Acorn stalked (stalk < 15 mm), nut cylindric-ovoid to elliptic in outline, abruptly tapered at tip, 12–23 mm, gen yellow-brown; lvs oblong to elliptic, regularly spine-toothed — e DMtns *Q. turbinella*

24′ Acorn sessile, nut conic-ovoid, gradually tapered to tip, 20–40 mm, gen dark brown; lvs oblong to elliptic or obovate, irregularly toothed, teeth blunt to weakly spiny

25. Shrub or small tree; lvs 1.5–5 cm, marginal teeth not spiny, upper surface bluish to grayish green . . . . . [2]*Q. ×alvordiana*

25′ Shrub; lvs 1.3–2.8 cm, irregularly spine-toothed, upper surface grayish green . . . . . *Q. john-tuckeri*

23′ Acorn cup thick, scales tubercled; upper lf surface gen green, ± shiny (sometimes dull in *Q. durata*)

26. Upper lf surface convex, margin often inrolled, obscuring teeth, dull . . . . . *Q. durata*

27. Upper lf surface strongly convex; lower lf surface short-hairy; gen serpentine soils, NCoR, n SNF, SCoR . . . . . var. *durata*

27′ Upper lf surface slightly convex; lower lf surface densely short-hairy when young; nonserpentine soils, s SnGb . . . . . var. *gabrielensis*

26′ Upper lf surface flat to ± wavy, not convex, margin gen not inrolled, ± shiny

28. Fr 10–30 mm; acorn cup thick, 12–20 mm wide, scales strongly tubercled; nut ovoid; chaparral, 300–1500 m . . . . . *Q. berberidifolia*

28′ Fr < 20 mm; acorn cup thin, 8–15 mm wide, scales slightly to moderately tubercled; nut conic-ovoid; chaparral or coastal-sage scrub, < 200 m . . . . . *Q. dumosa*

Q. ×acutidens Torrey (p. 663) Shrub or small tree < 6(10) m, evergreen. **LF** 2–6 cm; petiole 3–7 mm; blade gen oblong to obovate, ± leathery, tip obtuse to short-toothed, margin wavy to irregularly toothed, teeth ± obtuse, upper surface shiny and green to dull and bluish green, lower surface ± densely puberulent, becoming glabrous, dull, pale green. **FR**: maturing in 1 year; cup 10–18 mm wide, 6–9 mm deep, bowl-shaped, scales ± tubercled; nut 20–25 mm, oblong to ovoid, tip ± obtuse, shell glabrous inside. Slopes, chaparral, woodland; < 1600 m. PR. Hybrids, involving *Q. cornelius-mulleri* and *Q. engelmannii*, considered by Torrey as sp.

Q. agrifolia Nee COAST LIVE OAK, ENCINA Tree (6)10–25 m, evergreen; top wide; trunk bark becoming furrowed, widely ridged, checkered, dark gray. **LF** 2.5–6(9) cm; petiole 4–15 mm; blade oblong, elliptic, or ± round, tip rounded to spine-toothed, margin weakly spine-toothed, upper surface strongly convex, ± dull green, lower surface glabrous to densely tomentose, dull, pale green. **FR** maturing in 1 year; cup 10–16 mm wide, 8–15 mm deep, obconic, scales thin, ± flat, ± glabrous, brownish; nut 25–35 mm, slender, ovoid, tip pointed, shell woolly inside. Valleys, slopes, mixed-evergreen forest, woodland; < 1500 m. NCoRO, CW, SW; Baja CA.

var. ***agrifolia*** (p. 663) **LF**: lower surface glabrous to sparsely hairy, esp at vein axils. Habitats of sp.; < 900 m. NCoRO, CW, SW; Baja CA. Hybridizes with *Q. kelloggii, Q. parvula* var. *shrevei, Q. wislizenii.* ❀DRN,SUN:5,7,**14–17,22–24**&IRR:8,**9**,10–13, **18–21**.

var. ***oxyadenia*** (Torrey) J. Howell (p. 663) **LF**: lower surface densely tomentose. Gen granitic soils; 600–1500 m. PR; Baja CA. Hybridizes with *Q. kelloggii.* ❀DRN,SUN:4–6,**7,14–17,18–24**& IRR:1–3,8,**9**,10–13.

Q. ×alvordiana Eastw. (p. 663) Shrub or tree < 3 m, some pls evergreen. **LF** 1.5–5 cm; petiole 2–5 mm; blade gen oblong to widely elliptic, tip obtuse to abruptly pointed, margin irregularly and coarsely toothed, upper surface dull, bluish to grayish green, lower surface finely hairy, dull, pale green. **FR** maturing in 1 year; cup 10–16 mm wide, 8–10 mm deep, cup- to bowl-shaped, scales slightly tubercled, light brown; nut 20–40 mm, gen narrowly ovoid, tapered to pointed tip, shell glabrous inside. Dry slopes, hills; 400–1300 m. Teh, SCoRI. Hybrids, involving *Q. douglasii* and *Q. john-tuckeri*, considered by Eastwood as sp.

Q. berberidifolia Liebm. (p. 663) SCRUB OAK Shrub 1–3 m, sometimes tree-like and taller, evergreen. **LF** 1.5–3 cm; petiole 2–4 mm; blade oblong, elliptic, or ± round, tip gen rounded, margin ± spiny to toothed, teeth abruptly pointed, upper surface ± flat to wavy, ± shiny, green, lower surface appressed-puberulent, dull, pale green. **FR** maturing in 1 year; cup 12–20 mm wide, 5–10 mm deep, hemispheric to bowl-shaped, thick, scales clearly tubercled; nut 10–30 mm, ovoid, tip obtuse to acute, shell glabrous inside. Dry slopes, chaparral; 300–1500 m. NCoR, SNF, Teh, CW, SW; Baja CA. [*Q. dumosa* Nutt. misapplied] Hybridizes with *Q. durata, Q. engelmannii, Q. garryana, Q. john-tuckeri, Q. lobata.* ❀DRN, SUN:5,7,8–10,**14–16**,17,**18–24**;STBL.

Q. chrysolepis Liebm. (p. 663) MAUL OAK, CANYON LIVE OAK Tree < 20 m, sometimes shrub-like, evergreen; trunk bark becoming narrowly furrowed, scaly, pale gray; twigs golden-tomentose, becoming ± glabrous. **LF** (1.5)3–6 cm, leathery; petiole 3–10 mm; blade oblong to oblong-ovate, sometimes round-ovate, tip acute to abruptly pointed, margin entire or spine-toothed, upper surface dark green, lower surface golden-puberulent, becoming glabrous, dull, grayish. **FR** maturing in 2 years; cup 17–30 mm wide, 5–10 mm deep, saucer- to bowl-shaped, scales thick, flat to slightly tubercled, golden-tomentose; nut 25–30 mm, 14–20 mm wide, ± ovoid, tip rounded to pointed, shell woolly inside. Canyons, shaded slopes, chaparral, mixed-evergreen forest, woodland; 200–2600 m. CA-FP (exc GV), e DMtns; OR, AZ, Baja CA. Highly variable. Hybridizes with *Q. palmeri, Q. tomentella, Q. vaccinifolia.* Shrubs with lvs 2–4 cm have been called var. *nana* (Jepson) Jepson. ❀DRN,SUN:**4–7,14–18,22,23**,24&IRR:1–3,8,**9**,10,11,**19–21**.

Q. cornelius-mulleri K. Nixon & K. Steele (p. 663) MULLER'S OAK Shrub 1–2.5 m, evergreen, densely branched; twigs finely tomentose. **LF** 2.5–3.5 cm, leathery; petiole 2–5 mm; blade oblong, ovate, or narrowly obovate, tip acute to rounded, margin entire or 4–6-toothed, upper surface sparsely puberulent, dull, yellow- to gray-green, lower surface densely and finely tomentose, whitish, midrib yellow. **FR** maturing in 1 year; cup 12–20 mm wide, 5–8 mm mm deep, hemispheric to cup-shaped, scales ± flat, gray-canescent; nut 20–30 mm, elliptic in outline to widely conic, tip obtuse, puberulent, shell glabrous inside. Slopes, gen granitic soils, chaparral, pinyon woodland; 1000–1800 m. SnBr (n slope), PR (e slope), s DMtns (Little San Bernardino Mtns); Baja CA. Hybridizes with *Q. engelmannii, Q. lobata.* ❀DRN,SUN:**7,9,14**,15,16, **18–21**,22,23&IRR:2,3,8,10,11;STBL.

Q. douglasii Hook. & Arn. (p. 663) BLUE OAK Tree 6–20 m, deciduous; trunk bark checkered into thin scales, grayish. **LF** 3–6 (8) cm; petiole 3–9 mm; blade oblong to obovate, tip gen rounded, margin ± entire, wavy, or slightly lobed, upper surface dull, bluish green, lower surface puberulent, pale bluish green. **FR** maturing in 1 year; cup 12–20 mm wide, 6–10 mm deep, cup- to bowl-shaped, scales slightly tubercled; nut 20–35 mm, ovoid, tip pointed, shell glabrous inside. Dry slopes, interior foothills, woodland; < 1200 m. NCoRI, CaRF, SNF, Teh, SCoRI, WTR (n slope). Hybridizes with *Q. garryana, Q. john-tuckeri, Q. lobata.* ❀DRN,SUN:1–3,**7**,8,**9**, 11,**14–16,18–21**,22–24&DRY:4–6,10,17.

Q. dumosa Nutt. (p. 663) NUTTALL'S SCRUB OAK Shrub 1–3 m, evergreen; twigs slender (1–1.5 mm diam), sparsely short-hairy, becoming glabrous, dark reddish brown. **LF** 1–2.5 cm; petiole < 5 mm; blade oblong, elliptic, or ± round, tip obtuse to abruptly pointed, margin ± spiny to toothed, teeth abruptly pointed, upper surface slightly shiny, green, lower surface finely tomentose, becoming glabrous, dull, pale green. **FR** maturing in 1 year; cup 8–15 mm wide, 5–8 mm deep, gen bowl-shaped, scales ± tubercled; nut 10–20 mm, ± slender, gen ovoid, tapered to tip, shell glabrous inside. RARE. Gen sandy soils near coast, sandstone, chaparral, coastal-sage scrub; < 200 m. SCo; Baja CA. Hybridizes with *Q. berberidifolia.* In cult.

Q. durata Jepson LEATHER OAK Shrub 1–3 m, evergreen; twigs tomentose, sometimes becoming glabrous. **LF** 1.5–3 cm; petiole < 5 mm; blade oblong to elliptic, convex above, tip spiny or abruptly pointed, margin rolled under, toothed, teeth spiny or abruptly pointed, upper surface puberulent, dull green, lower surface short-hairy, pale green. **FR** maturing in 1 year; cup 12–18 mm wide, 4–6 mm deep, bowl-shaped, scales tubercled; nut 15–25 mm, ovoid to cylindric, tip obtuse or rounded, shell glabrous inside. Chaparral; 150–1500 m. NCoR, n SNF, SCoR, SnGb. Hybridizes with *Q. berberidifolia, Q. garryana.*

var. ***durata*** (p. 663) **LF**: upper surface strongly convex; lower surface short-hairy. Gen serpentine soils; 150–1500 m. NCoR, n SNF, SCoR. ❀DRN,SUN:5,**7**,8–10,**14–16**,17,**18–24**.

var. ***gabrielensis*** K. Nixon & C.H. Muller **LF**: upper surface slightly convex; lower surface densely short-hairy. Granitic soils; 450–1000 m. SnGb (s slope). ❀TRY.

Q. engelmannii E. Greene (p. 663) ENGELMANN or MESA OAK Tree 5–18 m, evergreen; trunk bark becoming narrowly furrowed, scaly, grayish; young twigs finely tomentose, becoming glabrous. **LF** 2–6 cm; petiole 3–7 mm; blade oblong to obovate, tip obtuse to rounded, margin gen entire or wavy, sometimes toothed, upper surface dull, bluish green, lower surface soft-hairy when young, becoming glabrous, pale blue-green. **FR** maturing in 1 year; cup 10–15 mm wide, 6–8 mm deep, cup- to bowl-shaped, scales ± tubercled; nut 15–25 mm, oblong-cylindric to ovoid, tip rounded to obtuse, shell glabrous inside. UNCOMMON. Slopes, foothills, woodland; < 1300 m. SCo, s ChI (1 tree on Santa Catalina Island), SnGb, PR; Baja CA. Hybridizes with *Q. berberidifolia, Q. cornelius-mulleri.* ❀DRN,SUN:3,5,7,**14–16**,17,**18–21**,22–24&IRR:8,9.

Q. garryana Hook. OREGON OAK Tree 8–20 m or shrub 1–5 m, deciduous; trunk bark thin, becoming widely ridged, scaly, grayish; twigs short-hairy, becoming glabrous, reddish brown. **LF** 5–15 cm; petiole 5–25 mm; blade elliptic to obovate, tip obtuse to rounded, margin deeply 5–7-lobed, lobes entire or 2-toothed, upper surface shiny, dark green, lower surface short-hairy, dull, light green. **FR** maturing in 1 year; cup 12–25 mm wide, 4–9 mm deep, cup- to bowl-shaped, scales ± flat to slightly tubercled; nut 20–30 mm, ± ovoid to subspheric, tip rounded, shell glabrous inside. Slopes, woodland, mixed-evergreen or conifer forest; 300–1800 m. NW, CaR, SNF, Teh, SnFrB, ne WTR (Liebre Mtn); to B.C.

var. ***breweri*** (Engelm.) Jepson (p. 663) Shrub 1–5 m, spreading. **ST**: terminal buds 3–5 mm. **LF**: petiole 5–20 mm; blade 4–9 cm. Slopes, mixed-evergreen or conifer forest; 600–1800 m. KR, NCoRH, SNF, Teh, ne WTR (Liebre Mtn); to OR. [var. *semota* Jepson] Hybridizes with *Q. sadleriana.* ❀DRN:4,5,**6,14–16**,17 &IRR or part SHD:1–3,**7**,,8–11,**18**,19–23.

var. ***garryana*** (p. 663) Tree 8–20 m. **ST**: terminal buds 5–12 mm, fusiform, densely hairy. **LF**: petiole 10–25 mm; blade 7–14 cm. Slopes, woodland; 300–1800 m. NW, CaRF, SnFrB; to B.C. Hybridizes with *Q. berberidifolia, Q. douglasii, Q. durata, Q. lobata.* ❀DRN,SUN:**4–6,15–17**&IRR or part SHD:1–3,**7**,8,9,**14,18**, 19–23.

Q. john-tuckeri K. Nixon & C.H. Muller (p. 665) TUCKER'S OAK Shrub 2–5 m (sometimes tree-like, < 7 m), evergreen; young twigs finely tomentose. **LF** 1.3–2.8 cm; petiole 1–4 mm; blade oblong, elliptic, or obovate, base rounded to widely wedge-shaped, tip obtuse to rounded, margin irregularly spine-toothed, upper surface dull, gray-green, lower surface finely hairy, pale gray-green.

FR maturing in 1 year; cup 10–15 mm wide, 5–7 mm deep, thin, obconic to hemispheric, scales flat to slightly tubercled; nut 20–30 mm, ovoid to conic, tapered to tip, shell glabrous inside. Slopes on desert borders, chaparral, pinyon/juniper woodland; 900–2000 m. Teh (e slope), SCoRI, WTR (n slope), SnGb (n slope). [*Q. turbinella* E. Greene ssp. *californica* J. Tucker] Hybridizes with *Q. berberidifolia, Q. douglasii, Q. lobata.* ❀DRN,SUN:**7**,8,**9**,11,**14–16**,17, **18–23**,24.

Q. kelloggii Newb. (p. 665) CALIFORNIA BLACK OAK Tree < 25 m, deciduous; trunk bark becoming deeply furrowed, checkered, dark gray-brown to black. **LF** (6)9–20 cm; petiole (3)10–40 mm; blade widely elliptic, obovate, or ± round, tip gen acute, bristled, margin gen deeply 6-lobed, each lobe with 1–4 coarse, bristle-tipped teeth, upper surface glabrous, bright green, lower surface finely tomentose when young, becoming ± glabrous, pale green. **FR** maturing in 2 years; cup 16–25 mm wide, 15–25 mm deep, gen cup-shaped, scales thin, flat, glabrous to puberulent; nut 20–35 mm, oblong-ovoid, puberulent, tip gen obtuse, shell woolly inside. Slopes, valleys, woodland, coniferous forest; 200–2400 m. CA-FP (exc GV, SCo, ChI); OR, Baja CA. Hybridizes with *Q. agrifolia, Q. wislizenii.* ❀DRN:1,4,5,**6**,**15**,**16**,17&IRR or part SHD:2,3,**7**,8,9, **14**,**18**,19–21.

Q. lobata Nee (p. 665) VALLEY OAK, ROBLE Tree < 35 m, deciduous; trunk bark becoming deeply checkered into squarish sections, light grayish. **LF** 5–12 cm; petiole 5–12 mm; blade obovate, tip obtuse to rounded, margin deeply 6–10-lobed, lobes obtuse, gen coarsely 2–3-toothed at tip, upper surface often ± shiny, dark green, lower surface finely tomentose, dull to pale green. **FR** maturing in 1 year; cup 14–30 mm wide, 10–30 mm deep, hemispheric, scales clearly tubercled; nut 30–50 mm, 12–20 mm wide, gen long-conic, tapered to pointed tip, shell glabrous inside. UNCOMMON. Slopes, valleys, savannah; < 1700 m. NCoR, CaRF, SNF, Teh, GV, SnFrB, SCoR, nw SCo, ChI (Santa Cruz, Santa Catalina islands), WTR, w SnGb. Hybridizes with *Q. berberidifolia, Q. cornelius-mulleri, Q. douglasii, Q. engelmannii, Q. garryana, Q. john-tuckeri.* ❀SUN:**4–6**,**14–16**,17&IRR,DRN:1–3,**7–9**,**18–21**,22–24.

Q.* ×*macdonaldii E. Greene (p. 665) MACDONALD OAK Tree 5–15 m, deciduous; trunk bark scaly, grayish; twigs tomentose. **LF** 4–7 cm; petiole 3–10 mm; blade oblong to obovate, tip obtuse to rounded, margin 2–6(8)-lobed, lobes gen pointed, upper surface glabrous to sparsely hairy, green, ± shiny, lower surface densely appressed-stellate-hairy, ± pale green. **FR** maturing in 1 year; cup 10–20 mm wide, 6–10 mm deep, hemispheric, scales clearly tubercled, canescent; nut 20–35 mm, conic-oblong to ovoid, tip acute, shell glabrous inside. Slopes, canyons, woodland; < 600 m. ChI (exc San Clemente Island). Considered by Greene as sp. but derived from hybrids between *Q. berberidifolia* and *Q. lobata*; perhaps other spp. involved; needs further study. ❀DRN,SUN:5,**7**, 8–10,**14–16**,17,**18–24**.

Q. palmeri Engelm. (p. 665) PALMER'S OAK Shrub 2–6 m, evergreen; twigs spreading, rigid. **LF** 1–3 cm, very stiff; petiole 2–5 mm; blade elliptic to round-ovate, tip gen spiny, margin wavy and spine-toothed, upper surface glabrous to sparsely puberulent, ± shiny, olive-green, lower surface densely glandular-puberulent when young, pale gray-green. **FR** maturing in 2 years; cup 10–25 mm wide, 6–12 mm deep, gen bowl-shaped, rim ± spreading, scales flat, densely hairy; nut 20–30 mm, ± ovoid, tip gen obtuse, shell densely woolly inside. Uncommon. Rocky slopes, flats; 700–1300 m. e NCoRI (Colusa Co.), nw SnJV (Alameda, Contra Costa cos.), SCoR (San Luis Obispo, Santa Barbara cos.), SnGb (n slope), e PR, DMtns (Little San Bernardino Mtns); AZ, Baja CA. [*Q. dunnii* Kellogg] Hybridizes with *Q. chrysolepis.* ❀DRN,SUN:7,8,**9**,10, **14–16**,17,**18–23**,24&IRR:11;STBL.

Q. parvula E. Greene Shrub 1–3 m or tree < 17 m, evergreen. **LF** 3–9 cm; petiole 2–10 mm; blade oblong, lanceolate, or ovate, tip acute to acuminate, margin gen entire, sometimes toothed, upper surface glabrous, olive-green, lower surface glabrous, dull, light olive-green. **FR** maturing in 2 years; cup 12–15 mm wide, 6–10 mm deep, gen bowl-shaped, scales flat, ± thin; nut 30–45 mm, barrel-shaped to ovoid, tip abruptly tapered, puberulent, shell woolly inside. Canyons, slopes, chaparral, woodland; < 1000 m. SnFrB, SCoRO, n SCo, n ChI (Santa Cruz Island).

var. ***parvula*** (p. 665) SANTA CRUZ ISLAND OAK Shrub 1–3 m. UNCOMMON. Habitats of sp.; < 500 m. n SCo (Santa Barbara Co.), n ChI (Santa Cruz Island). ❀DRN,SUN:5,7,**14–17**,**22–24**& IRR or part SHD:8,9,**19–21**.

var. ***shrevei*** (C.H. Muller) K. Nixon (p. 665) Tree < 17 m. Rather moist woodland; < 1000 m. w SnFrB, SCoRO. Hybridizes with *Q. agrifolia, Q. kelloggii.* ❀DRN:5,**15–17**&IRR:**7**,8,**9**,**14**,18, **19–24**.

Q. sadleriana R.Br. Campst. (p. 665) DEER OAK Shrub 1–3 m, evergreen; twigs glabrous. **LF** 7–11(18) cm; petiole 10–20 mm; blade elliptic to oblong-obovate, tip ± acute, margin serrate, teeth 20–32, upper surface green, somewhat shiny, lower surface sparsely, finely appressed-hairy, pale green, lateral veins 20–28, prominent. **FR** maturing in 1 year; cup 10–18 mm wide, 7–9 mm deep, thin, cup-shaped to obconic, scales flat to slightly tubercled; nut 15–20 mm, elliptic in outline to ± spheric, tip rounded, shell glabrous inside. Open, rocky slopes, ridges, coniferous forest; 600–2200 m. KR; sw OR. Hybridizes with *Q. garryana* var. *breweri.* ❀DRN:**4–6**&IRR:1,17&SHD:2,3,**7**,9,14–16,18–21;DFCLT.

Q. tomentella Engelm. (p. 665) ISLAND OAK Tree < 20 m, evergreen; trunk bark becoming furrowed, scaly, gray, sometimes reddish brown; young twigs tomentose. **LF** 5–8 cm; petiole 5–18 mm; blade oblong to oblong-ovate, tip acute to obtuse or abruptly pointed, margin toothed, teeth abruptly pointed, upper surface ± finely tomentose, becoming glabrous, dark green, lower surface densely tomentose, becoming sparsely woolly, dull, grayish green. **FR** maturing in 2 years; cup 20–30 mm wide, 6–8 mm deep, saucer- to bowl-shaped, scales thick, tubercled; nut 20–35 mm, widely ovoid, tip rounded, shell ± woolly inside. UNCOMMON. Canyons, slopes, woodland; < 600 m. ChI; Baja CA (Guadalupe Island). Hybridizes with *Q. chrysolepis.* ❀DRN:5,**15–17**&IRR:7, **14**,**22–24**&SHD:8,9,**19– 21**.

Q. turbinella E. Greene (p. 665) SHRUB LIVE OAK Shrub 2–5 m (sometimes tree-like, < 7 m) evergreen; twigs densely and finely tomentose. **LF** 1.5–3 cm; petiole 1–3 mm; blade oblong to elliptic, base rounded to subcordate, tip acute to obtuse, sometimes spiny, margin spine-toothed, upper surface dull, gray-green, lower surface with both appressed-stellate and glandular, yellowish hairs. **FR** maturing in 1 year; stalk < 15 mm; cup 9–12 mm wide, 4–6 mm deep, ± hemispheric, scales flat to slightly tubercled, thin; nut 12–23 mm, cylindric-ovoid to elliptic in outline, tapered abruptly to tip, yellow-brown, shell glabrous inside. Pinyon/juniper woodland; 1200–2000 m. e DMtns (New York Mtns); to Colorado, TX, Baja CA. ❀DRN:**7**,**8**,**9**,10,11,**14**,**18–21**&SUN:5,15–17,22–24;STBL.

Q. vaccinifolia Kellogg (p. 665) HUCKLEBERRY OAK Shrub < 1.5 m, prostrate to spreading, evergreen; twigs slender, pliable, glabrous. **LF** 1.5–4 cm; petiole 3–6 mm; blade ± oblong, tip obtuse to acute, margin mostly entire, sometimes low-serrate, upper surface glabrous, green, lower surface glabrous, dull, pale green. **FR** maturing in 2 years; cup 10–12 mm wide, 4–6 mm deep, thin, gen cup-shaped, scales flat to slightly tubercled; nut 10–15 mm, ovoid to subspheric, tip rounded; shell thin, subglabrous to sparsely tomentose inside. Steep slopes, ridges, coniferous forest, subalpine; 900–2800 m. KR, NCoRH, SNH; OR. Hybridizes with *Q. chrysolepis.* ❀DRN,SUN:1,4–6&IRR:2,3,**7**,14,**15**,**16**,17,18.

Q. wislizenii A.DC. INTERIOR LIVE OAK Tree gen 10–22 m (or shrub, 2–6 m) evergreen; trunk bark becoming furrowed, ± checkered, grayish. **LF** 2–5 cm; petiole 3–15 mm; blade gen oblong to elliptic or lanceolate, tip gen acute, abruptly pointed, margin entire to spine-toothed, upper surface glabrous, shiny, gen dark green, lower surface glabrous, shiny, light or yellow-green. **FR** maturing in 2 years; cup 12–18 mm wide, 12–16 mm deep, cup-shaped to hemispheric, scales flat, ± thin; nut 20–40 mm, cylindric-ovoid, ovoid, or ± obconic, tapered to tip, shell woolly inside. Interior canyons, slopes, valleys, chaparral, pine/oak woodland; < 2000 m. NCoR, CaRF, SNF, Teh, SCoR, SW (exc ChI); Baja CA.

var. ***frutescens*** Engelm. (p. 665) Shrub 2–6 m. **LF**: blade 1.8–4 cm. Chaparral; 300–2000 m. NCoR, Teh, SCoR, SW (exc ChI); Baja CA. ❀DRN,SUN:1–6,7,9,**14–24**;STBL.

var. ***wislizenii*** (p. 665) Tree 10–22 m. **LF**: blade 2–5 cm. Canyons, slopes, pine/oak woodland; < 1600 m. NCoR, CaRF, SNF, SCoR, TR, PR; Baja CA. Hybridizes with *Q. agrifolia, Q. kelloggii.* ❀DRN,SUN:4–6,**7**,**14–17**,**22–24**&IRR:8,**9**,**18–21**.

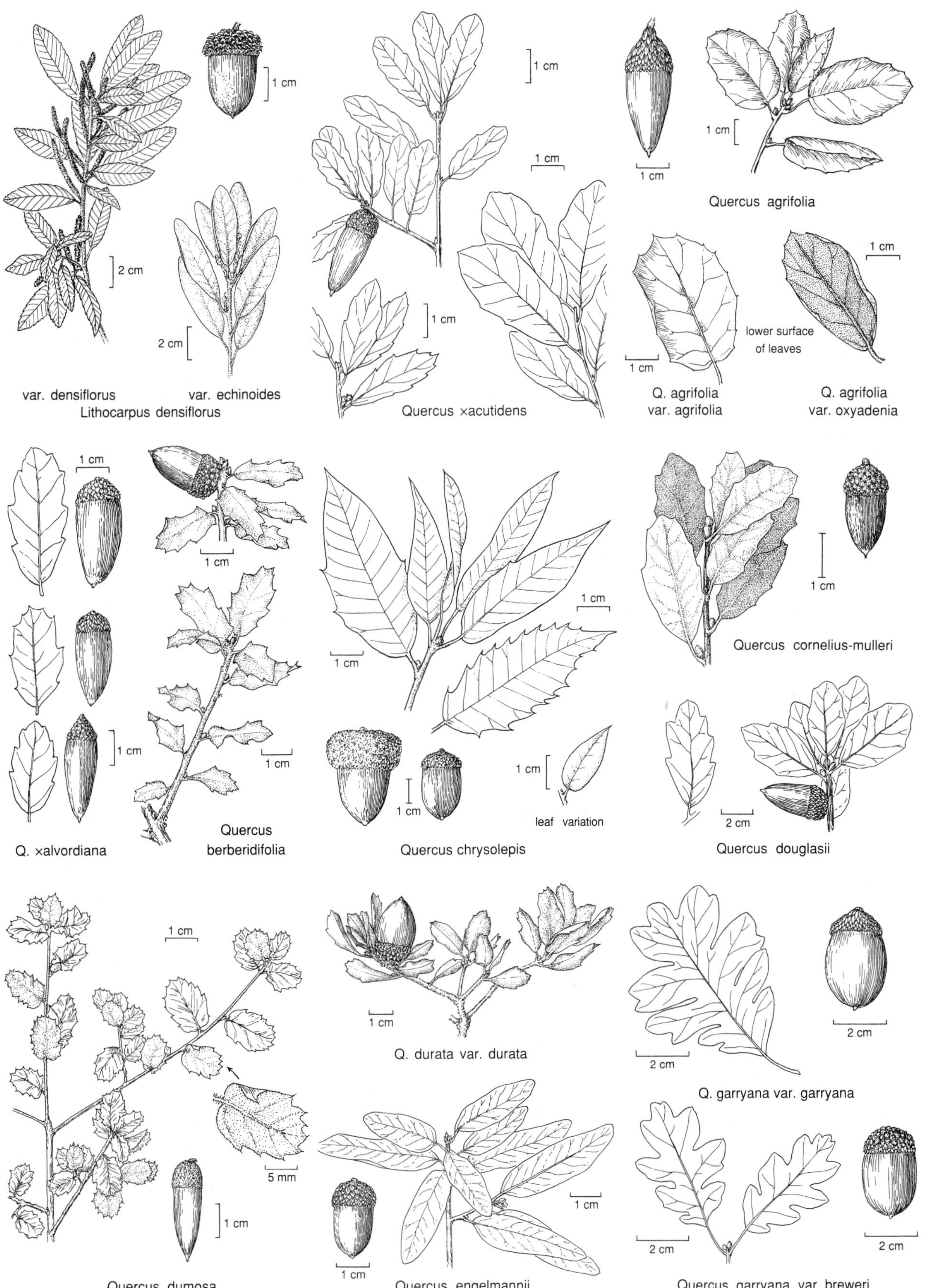
1 cm
2 cm
1 cm
1 cm
1 cm
1 cm
1 cm
1 cm
lower surface
of leaves
var. densiflorus
var. echinoides
Lithocarpus densiflorus
Quercus ×acutidens
Quercus agrifolia
Q. agrifolia
var. agrifolia
Q. agrifolia
var. oxyadenia
Q. ×alvordiana
Quercus
berberidifolia
leaf variation
Quercus chrysolepis
Quercus cornelius-mulleri
Quercus douglasii
5 mm
Quercus dumosa
Q. durata var. durata
Quercus engelmannii
Q. garryana var. garryana
Quercus garryana var. breweri

FOUQUIERIACEAE OCOTILLO FAMILY

William J. Stone

Shrubs, trees, spiny. **ST** branched near base, or trunk single, thick, fleshy. **LVS** simple, alternate, small, somewhat fleshy, glabrous, of 2 types: primary soon deciduous after rains, petiole long, it and midrib develop into persistent spine after blade drops; secondary lvs clustered in axil of developing spine. **INFL**: spike, raceme, or panicle, terminal; fls many. **FL** showy; sepals 5, unequal, overlapping, persistent; corolla tube cylindric, lobes 5, spreading, bright red or yellow; stamens 10–20, in 1–2 whorls, filaments free, epipetalous; pistil 1, ovary superior, incompletely 3-chambered, placenta axile at base, parietal above, ovules 3–6 per chamber. **FR**: capsule. **SEEDS** elliptic, angled. 1 genus (includes *Idria*), 11 spp.: sw US, Mex. (P.E. Fouquier, French professor of medicine) [Henrickson 1972 Aliso 7:439–537]

FOUQUIERIA OCOTILLO, CANDLEWOOD, BOOJUM

The only genus.

F. splendens Engelm. ssp. ***splendens*** (p. 665) OCOTILLO **STS** branched near base, erect to outwardly arching or ascending, 6–100, 2–10 m, gen < 6 cm diam, cane-like, lfless most of year; bark gray with darker furrows; spines 1–4 cm. **LVS**: primary 1–5 cm, petioles 1–2.5 cm; secondary 2–6 per cluster, 1–2 cm, 4–9 mm wide, petioles 2–8 mm, blade spoon-shaped to obovate, tip rounded to notched. **INFL**: panicle, gen 10–20 cm, widely to narrowly conic. **FL**: corolla 1.8–2.5 cm, bright red. **FR** ± 2 cm. $2n=24$. Dry, gen rocky soils; 0–700 m. DSon; to TX, c Mex, Baja CA. Sts used for fences, huts; bark for waxes, gums. ❀DRN,DRY:**10–13**,19–21.

FRANKENIACEAE FRANKENIA FAMILY

R. John Little

Subshrubs, shrubs, gen from rhizome; salt-secreting glands present. **ST** prostrate to erect, nodes swollen, often rooting; petioles or dead lvs persisting on older sts. **LVS** opposite, 4-ranked, ± clustered; blade entire, gen leathery or fleshy, glabrous to hairy, margins rolled under. **INFL**: cyme, axillary; fls 1–25. **FL** gen bisexual, radial; sepals 4–7, fused; petals 4–7, free, overlapping, clawed (together appearing salverform), white to blue-purple, petal blade with a scale-like appendage near base; stamens 3–12 in two whorls, outer shorter; ovary superior, chambers 1–4, style branches 1–4; ovules 1–many. **FR**: loculicidal capsule. **SEED** ivory to gold-brown. 1 genus, 90 spp.: temp saline and gypsum soils. (Whalen 1987 Syst Bot Monogr 17:1–93)

FRANKENIA FRANKENIA

The only genus. (Possibly named for J. Franke, Swedish botanist born 1590 or for Johann Frankenius, colleague of Linnaeus)

1. Lf blade 2–7 mm, gen < 1 mm wide, margins tightly rolled under, obscuring lower surface .... ***F. palmeri***
1′ Lf blade 4–15 mm, 1–6 mm wide, margins gently rolled under, exposing lower surface .... ***F. salina***

F. palmeri S. Watson PALMER'S FRANKENIA Shrub 1–10 dm diam. **ST** decumbent, 1–3 dm; twig gen with scattered hairs. **LF** sparsely to densely hairy. **INFLS** in few uppermost axils. **FL**: calyx tube 3–5 mm; petals 3–4 mm, white, lower half often pink; stamens gen 4, 4–9 mm; style branches gen 2. **FR** 2–2.5 mm. **SEED** 1, 1.4–2 mm, ovoid-conic. RARE in CA. Alkali flats, coastal marshes, dunes; < 450 m. SCo (sw San Diego Co.); n Mex.

F. salina (Molina) I.M. Johnston (p. 671) ALKALI HEATH Subshrub forming mats < 3 m diam. **ST** ± prostrate, 1–6 dm; twig glabrous to hairy. **LF** glabrous to densely hairy. **INFLS** in most upper axils. **FL**: calyx tube 4–9 mm; petals 5–14 mm, white to dark pink or blue-purple; stamens gen 6, 5–12 mm; style branches gen 3. **FR** 3–5 mm. **SEEDS** 1–20, 1–1.5 mm, ± ellipsoid. Salt marshes, alkali flats; < 750 m. GV, CCo, SCo, ChI, SNE, DMoj; to NV, Mex, S.Am. [*F. grandifolia* Cham. & Schldl.; *F. g.* var. *campestris* A. Gray] ❀SUN,IRR:**8**,10,**11**,12,13,**14**,15,**16–24**;GRCVR.

GARRYACEAE SILK TASSEL FAMILY

Thomas F. Daniel

Shrub, small tree, dioecious. **LVS** simple, opposite, evergreen, petioled; blade ± leathery, flat to concave-convex, margin entire, flat, rolled under, or strongly wavy. **INFL** catkin-like, pendent; fls small, in axils of opposite, 4-ranked, basally fused bracts. **STAMINATE FLS** (1–)3(4) per bract, pedicelled; perianth parts 4, gen fused at tips; stamens 4, alternate perianth parts, filaments free, anthers 2-chambered. **PISTILLATE FLS** (1–)3 per bract; pedicel ± 0 or short; perianth 0 or vestigial with 2 small appendages; ovary inferior, chamber 1, styles 2(3). **FR**: berry, spheric to ovoid, green, fleshy, becoming dark blue, black, or whitish gray, dry, brittle, not or irregularly dehiscent. **SEEDS** gen 2. 1 genus, 14 spp.: w US, C.Am, Caribbean; some cult. [Dahling 1978 Contr Gray Herb 209:1–104]

GARRYA SILK TASSEL BUSH

The only genus. (N. Garry, 1st secretary of Hudson Bay Co., friend of David Douglas, 1782?–1856) Intergradation among CA spp. suggests some may be unworthy of that status.

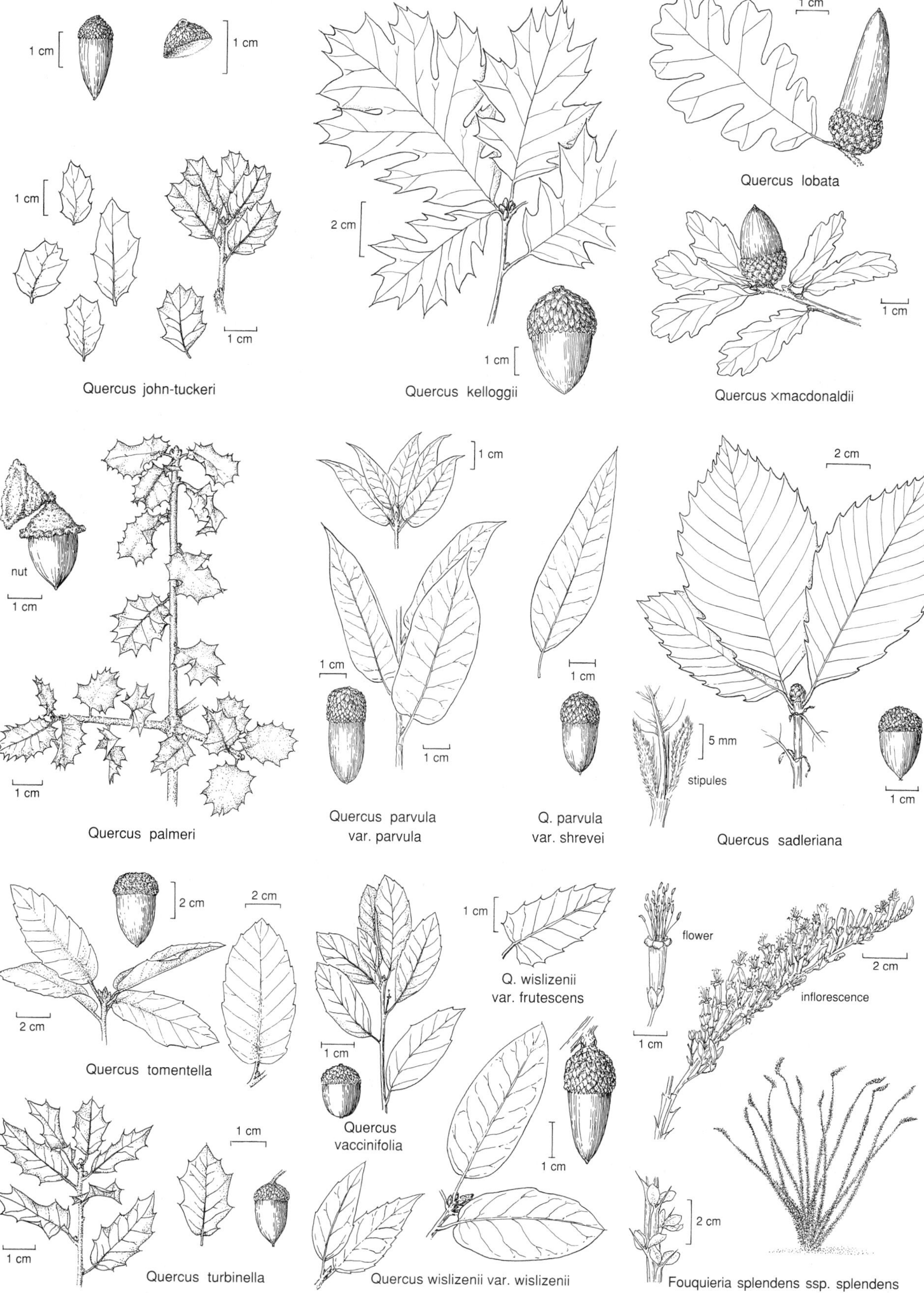

Quercus john-tuckeri

Quercus kelloggii

Quercus lobata

Quercus ×macdonaldii

Quercus palmeri

Quercus parvula var. parvula

Q. parvula var. shrevei

Quercus sadleriana

Quercus tomentella

Quercus vaccinifolia

Q. wislizenii var. frutescens

Quercus turbinella

Quercus wislizenii var. wislizenii

Fouquieria splendens ssp. splendens

Fouquieriaceae

1. Hairs of lower lf surface dense, curly-wavy, interwoven, not appressed toward lf tip (felt-like)
  2. Lf margin slightly to conspicuously rolled under and wavy; pistillate fls (1–)3 per bract; infl in fr 18–28 mm wide; n&c-w CA (to n SCoRO) .... ***G. elliptica***
  2′ Lf margin gen flat to slightly rolled under, sometimes slightly wavy; pistillate fls 1(–3) per bract; infl in fr 13–18 mm wide; ± s CA (to c SCoRO, San Luis Obispo Co.) .... ***G. veatchii***
1′ Hairs of lower lf surface 0 to dense, straight to wavy, appressed toward lf tip (not felt-like)
  3. Hairs of lower lf surface 0 (rarely sparse on young lvs); fr glabrous (sometimes sparsely hairy near tip) .... ***G. fremontii***
  3′ Hairs of lower lf surface sparse to dense; fr densely hairy throughout, glabrous throughout, or sparsely hairy near tip
    4. Hairs of lower lf surface dense, soft, wavy .... ***G. congdonii***
    4′ Hairs of lower lf surface ± 0 to dense, ± coarse, straight to slightly curved
      5. Fr glabrous or sparsely hairy near tip; hairs of lower lf surface gen dense; nw CA .... ***G. buxifolia***
      5′ Fr densely hairy, gen throughout; hairs of lower lf surface ± 0 to ± dense; esp c&s CA (sometimes nw CA) .... ***G. flavescens***

G. buxifolia A. Gray (p. 671) Shrub < 3 m. **LF** 14–66 mm, 9–33 mm wide, 1.3–2.3 × longer than wide, flat to ± concave-convex, ovate- to obovate-elliptic; margin flat; lower surface hairs gen dense, straight to slightly curved, appressed toward lf tip. **FR** glabrous or sparsely hairy near tip. Chaparral to yellow-pine forest; 150–2100 m. NW; w OR. ❀DRN,IRR:1–7,14–19;DFCLT.

G. congdonii Eastw. (p. 671) Shrub < 3 m. **LF** 15–70 mm, 7–35 mm wide, 1.3–2.6 × longer than wide, flat to ± concave-convex, ovate- to obovate-elliptic; margin flat to strongly wavy; lower surface hairs dense, soft, wavy, appressed toward lf tip. **FR**: hairs dense. Chaparral; 180–750 m. NW, n&c SNF, SnFrB. [*G. flavescens* ssp. *c.* (Eastw.) G.V. Dahling] Appears to intergrade equally with *G. elliptica*, *G. flavescens*. ❀DRN,DRY:7,9,14,19&SUN: 15–17.

G. elliptica Lindley (p. 671) Shrub, small tree, < 8 m. **LF** 18–107 mm, 14–72 mm wide, 1.2–2.6 × longer than wide, flat to ± concave-convex, elliptic; margin slightly to strongly wavy, often rolled under and appearing crenate or dentate; lower surface hairs dense, curly-wavy, interwoven, becoming sparse or not. **INFL** in fr 18–28 mm wide. **PISTILLATE FLS** (1–)3 per bract. **FR**: hairs dense. $2n=22$. Seacliffs, sand dunes, chaparral, foothill-pine woodland; < 800 m. NCo, NCoRO, n SNF, ScV, CCo, SnFrB, SCoRO; w OR. ❀DRN:4,**5**,6,**15–17**;IRR:**7,24**&SHD:8,9,**14**,18, 19,**20–23**;MALE CVS.

G. flavescens S. Watson Shrub < 3 m. **LF** 19–75 mm, 9–45 mm wide, 1.3–3.3 × longer than wide, flat to ± concave-convex, elliptic to obovate-elliptic; margin flat to wavy; lower surface hairs ± 0 to ± dense, ± coarse, straight to slightly curved, appressed toward lf tip. **FR**: hairs gen dense. Desert slopes, chaparral, pine/oak woodland; 650–2350 m. NW, SN, CW, SW, DMtns; to UT, AZ, n Baja CA. [var. *pallida* (Eastw.) Ewan] Variation in hairiness not related to geog in CA. ❀DRN,DRY,SUN:**7**,8–11,**14**,15–17,**18–23**,24.

G. fremontii Torrey Shrub < 3 m. **LF** 21–120 mm, 11–70 mm wide, 1.3–2.4 × longer than wide, gen flat, ovate- to obovate-elliptic; margin flat; lower surface hairs 0, rarely sparse, straight to ± wavy, appressed toward lf tip on young lvs. **FR**: hairs 0 (sometimes sparse near tip). Chaparral, foothill woodland, montane forest; 300–2300 m. NW, CaR, n SNF, n&c SNH, ScV, SnFrB, PR, Wrn; to WA, w NV. ❀DRN:4,5,**6**,15–17&IRR:1–3,**7,14**,22,23& SHD:9,18–21.

G. veatchii Kellogg (p. 671) Shrub < 2 m. **LF** 24–90 mm, 8–50 mm wide, 1.3–3.9 × longer than wide, flat to ± concave-convex, lance-ovate to obovate-elliptic; margin flat to slightly rolled under, sometimes slightly wavy; lower surface hairs dense, curly-wavy, interwoven. **INFL** in fr 13–18 mm wide. **PISTILLATE FLS** 1(–3) per bract. **FR**: hairs dense. Chaparral; 250–1750 m. SCoRO, SW; n Baja CA. ❀DRN,SUN:**7,14**,15–17,**18**,19–21.

GENTIANACEAE GENTIAN FAMILY

James S. Pringle (except as specified)

Ann, per. **ST** decumbent to erect, < 2 m. **LVS** simple, cauline (sometimes also basal), opposite or whorled, entire, sessile or basal ± petioled; stipules 0. **FL** bisexual, radial, parts in 4's or 5's, exc pistil 1; sepals fused, persistent; petals fused, persistent or deciduous, corolla gen without fringes or scales on inner surface, sinuses between lobes gen unappendaged; stamens epipetalous, alternate corolla lobes; ovary superior, chamber 1, placentas parietal, often intruding. **FR**: capsule, 2-valved. **SEEDS** many. ± 80 genera, 900 spp.: worldwide; some cult (*Eustoma*, *Exacum*, *Gentiana*). [Wood & Weaver 1982 J Arnold Arbor 63:441–487]

1. Sinuses between corolla lobes with a variously shaped, sometimes fringed appendage, or ± truncate (base of sinus gen > 1 mm wide) .... **GENTIANA**
1′ Sinuses between corolla lobes unappendaged and acutely tapered
  2. Corolla ± rotate, rarely bell-shaped, lobed ± to base, nectary pits prominent, margins fringed .... **SWERTIA**
  2′ Corolla not rotate, with distinct tube, nectary pits 0 (but corollas sometimes bearing fringes or scales)
    3. Corolla > 2 cm, widely bell-shaped; corolla lobes >> tube .... **EUSTOMA**
    3′ Corolla gen < 2 cm, not widely but sometimes narrowly ± bell-shaped; corolla lobes < or = tube
      4. Corolla < 1 cm, pink or yellow, lobes without fringes or scales on inner surface
        5. Calyx very deeply lobed, closely appressed to corolla tube .... **CENTAURIUM**
        5′ Calyx shallowly lobed, much wider than corolla tube .... **CICENDIA**
      4′ Corolla gen > 1 cm, blue to violet, rarely white, lobes with or without fringes or scales on inner surface near base
        6. Corolla < 2 cm, lobes ± entire, with fringes or scales on inner surface near base .... **GENTIANELLA**
        6′ Corolla > 2 cm, lobes ± entire to conspicuously serrate, jagged, or fringed, without fringes or scales on inner surface .... **GENTIANOPSIS**

CENTAURIUM CENTAURY

James C. Hickman

Ann, erect, glabrous. **LVS** opposite, often ± sheathing. **INFL**: cyme or panicle-like. **FL**: calyx lobes gen 5, ± linear-keeled, appressed to corolla; corolla salverform or funnel-shaped, gen pink, lobes ± entire, scales 0, sinus appendages 0; stamens gen ± exserted, dehisced anthers spirally twisted; style thread-like, stigma oblong to fan-shaped. **FR** ± cylindric-fusiform. **SEED** gen < 0.5 mm, ± rounded, ± brown, netted. ± 30 spp.: N.Am, Eurasia, Afr. (Latin: centaur, mythological discoverer of its medicinal properties) Variable and difficult; worldwide study needed.

1. Fls ± sessile (immediately subtended by 2 bracts); infl gen dense, units often ± flat-topped
 2. Bien; basal rosette conspicuous; corolla lobes 5–7 mm, gen puberulent outside; alien, NCo .... ***C. erythraea***
 2′ Ann; basal rosette 0 or weak; corolla lobes 2–5 mm, glabrous outside; native, n CA
  3. Corolla lobes 2–5 mm; stigma oblong .... ***C. muehlenbergii***
  3′ Corolla lobes 8–10 mm; stigma tapered, minute .... [2]***C. trichanthum***
1′ Most fls ± clearly pedicelled above subtending bracts; infl gen ± open
 4. Corolla not very showy (lobes gen 2–7 mm, gen < tube); undehisced anthers 1–1.5 mm
  5. Pedicels gen < 10 mm; corolla lobes elliptic, overlapping, gen flat; ± coastal .... ***C. davyi***
  5′ Pedicels 10–50 mm; corolla lobes appearing ± linear, not overlapping, often inrolled; inland ...... ***C. exaltatum***
 4′ Corolla ± showy (lobes gen 6–20 mm, often > tube); undehisced anthers 3–6 mm (exc stunted pls)
  6. Corolla lobes gen 5–9 mm; stigma minute, appearing ± unlobed — NCoRI, SnFrB .... [2]***C. trichanthum***
  6′ Corolla lobes gen 8–20 mm (exc stunted pls); stigma lobes separately stalked, 1–2 mm when fresh, fan-shaped
   7. Basal lvs gen rosetted, gen linear-oblanceolate; corolla lobes 8–12 mm; uncommon; e DSon .... ***C. calycosum***
   7′ Rosette 0; lvs narrowly oblong to ovate; corolla lobes (2)10–20 mm; CA-FP, DMoj .... ***C. venustum***

C. calycosum (Buckley) Fern. Gen bien?, gen 20–50 cm. **LVS**: basal often rosetted, gen 15–70 mm, gen oblanceolate. **INFL** open; pedicels 10–40 mm. **FL**: corolla lobes 8–12 mm, gen rose-purple; undehisced anthers 3–6 mm; stigma lobes 1–2 mm when fresh, separately stalked, fan-shaped, spreading. Uncommon. Damp places, esp riverbanks; 50–100 m. e DSon; to Colorado, TX, Mex. Intergrades with *C. exaltatum* & *C. venustum* in DMoj. ❀TRY.

C. davyi (Jepson) Abrams (p. 671) Ann 2–25 cm. **LF** 3–20 mm, gen ± elliptic. **INFL** open; pedicels gen < 10 mm (some ± 0). **FL**: corolla lobes 3–7 mm, gen widely elliptic, overlapping, gen flat; undehisced anthers ± 1 mm; stigma lobes fan-shaped. Moist coastal bluffs, dunes, open forest; < 1000 m. NCo, w CW, n ChI (Santa Cruz Island). Relatively uniform. ❀TRY.

C. erythraea Raf. Bien, 20–60 cm, ± coarse. **LVS**: basal rosette gen conspicuous at fl; basal lvs 20–40 mm, widely oblong, rounded; cauline acute. **INFL** dense; units often ± flat-topped; fls ± sessile (immediately subtended by 2 bracts). **FL**: corolla lobes 5–7 mm, rose, gen puberulent; undehisced anthers 2–3 mm; stigma oblong. Fields, roadsides; < 200 m. n&c NCo (esp near Crescent City, Eureka, Fort Bragg); to WA; native to Eur.

C. exaltatum (Griseb.) Piper Ann 10–35 cm. **LF** 10–30 mm, ± oblong, acute. **INFL** open; pedicels 10–50 mm. **FL**: corolla lobes 3–7 mm, gen to inrolled and appearing ± linear, not overlapping, white to rose; undehisced anthers ± 1 mm; stigma lobes fan-shaped. Moist, gen alkaline scrub; gen < 1500 m. SCoR, SnGb, PR, GB, DMoj, DSon (w edge); to e WA, UT. [*C. namophilum* Rev. et al. var. *nevadense* Broome, Nevada centaury] CA (not NV) pls that have been considered *C. n.* var. *namophilum*, spring-loving centaury, **THREATENED** US, with denser infl, from e DMoj (Ash Meadows), are apparently all *C. exaltatum*. [Holmgren 1984 Intermtn Flora 4:5–6] Much variation apparently environmental.

C. muehlenbergii (Griseb.) Piper (p. 671) Ann (or bien if damaged), 10–90 cm, ± slender. **LVS**: basal rosette 0 or weak; lvs 5–25 mm, ± oblong. **INFL** dense; units gen ± flat-topped; fls ± sessile (immediately subtended by 2 bracts). **FL**: corolla lobes 2–5 mm, pink, glabrous; undehisced anthers 1–2 mm; stigma oblong. Common. Moist, open forest; < 500 m. NW (esp Humboldt Co.), CaRF, n SNF, s ScV, n SnJV, n&c CW, n MP; to ID, w NV. [*C. curvistamineum* (Wittr.) Abrams; *C. floribundum* (Benth.) Robinson] Variable; may also = *C. pulchellum* (Sw.) Druce & *C. tenuiflorum* (Hoffsgg. & Link) Fritsch of Eur. Often confused with *C. erythraea*. ❀IRR,SUN:1–3,**4–7**,8,**9**,10,**14–17**,18,**19–24**.

C. trichanthum (Griseb.) Robinson (p. 671) Ann 5–45 cm. **LF** 10–30 mm, lanceolate to ovate. **INFL** dense to ± open; units gen ± flat-topped; pedicels 0 or gen short. **FL**: corolla lobes 8–10 mm, often inrolled and appearing linear, pink, glabrous; undehisced anthers ± 3.5 mm; stigma ± 0.5 mm, obscurely lobed. Alkaline or saline flats, moist chaparral or open forest; < 800 m. NCoRI, SnFrB. Likely a form of *C. venustum*; ± inseparable pls scattered in NCoRI, c SNF, DMoj. ❀SUN,IRR:6,**7–9**,10,11,**14–24**.

C. venustum (A. Gray) Robinson (p. 671) CANCHALAGUA Ann 3–50 cm. **LF** 5–25 mm, narrowly oblong to ovate. **INFL** dense or open; pedicels often short. **FL**: corolla throat ± white, lobes (2)10–20 mm, gen rose-purple (white); undehisced anthers 3–6 mm; stigma lobes 1–1.5 mm, separately stalked, fan-shaped, spreading. Common. Dry scrub, grassland, forest; < 1300 m. e KR, NCoRI, s NCoRO, CaRF, SNF, e SnJV, SW (exc n ChI, esp San Diego Co.), DMoj. [var. *abramsii* Munz] Highly variable; intergrades with *C. calycosum*, *C. exaltatum*, & *C. trichanthum*. Much variation apparently environmental. ❀SUN,IRR:2,3,**7–11**,**14–24**.

CICENDIA

Ann, glabrous. **ST** erect. **LVS** basal and cauline, opposite. **INFL**: cyme or fls solitary. **FL** parts in 4's; calyx midveins narrowly keeled; corolla salverform, yellow, lobes < tube, nectary pits 0 (or nectaries possibly indistinct, at base of corolla); ovary sessile, style distinct, stigma 2-lobed. 2 spp.: w N.Am, w S.Am, Eur. (Latin (probably): candelabrum)

C. quadrangularis (Lam.) Griseb. (p. 671) Pl < 9 cm. **ST** simple or branched from near base. **LVS** 4–9 mm, ovate, lanceolate, elliptic, or oblanceolate; cauline < internodes. **INFL**: pedicels < 5.5 cm. **FL**: calyx < 7 mm, tube hemispheric, much wider than corolla tube, lobes << tube, minutely triangular, acute, sinuses between lobes wide, truncate; corolla < 10 mm, lobes < 4 mm, ovate-elliptic, tip rounded. $2n$=26. Open places; 0–2700 m. NCo, NCoRO, c SNF, GV, CCo; to OR; S.Am. [*Microcala q.* (Lam.) Griseb.]

EUSTOMA CATCHFLY GENTIAN

Ann, short-lived per, glabrous, ± glaucous. **ST** erect, branched. **LVS** basal and cauline, opposite, clasping. **INFL**: cyme or fls solitary. **FL**: parts in 5's; calyx lobed ± to base, lobes linear, narrowly keeled, acuminate; corolla > 2 cm, bell-shaped, lobes >> tube, nectary pits 0; nectaries at base of ovary; ovary sessile, style distinct, slender, stigma 2-lobed. 2 weakly differentiated spp.: N.Am, C.Am, West Indies. (Greek: beautiful mouth, from corolla tube)

E. exaltatum (L.) Don (p. 671) Pl 1.5–10 dm. **LVS**: basal 2–10 cm, spoon-shaped-obovate to elliptic-oblong, obtuse; cauline 1.5–9 cm, elliptic to lanceolate, ± obtuse, upper acute. **INFL**: pedicels 2–14 cm. **FL**: calyx 10–21 mm; corolla 2–4.5 cm, pale to deep violet-blue, rarely rose-violet or white, throat whitish with dark purple blotches, lobes elliptic to obovate, rounded to subacute. $2n=\pm72$. Roadsides, alkaline marshes, other open, wet places; 100–600 m. SCo, PR, DSon; to se US, C.Am, West Indies. ❀IRR&SUN:7–9, 13,14,16,17,19–22,**23**,**24**.

GENTIANA GENTIAN

Ann, per, gen glabrous. **ST** gen simple below infl. **LVS** cauline, opposite (sometimes also basal). **INFL**: compact cyme or fls solitary. **FL** parts in 4's or 5's; calyx tube obvious; corolla tube narrowly bell-shaped, lobes spreading, < tube, sinuses between lobes ± truncate, unappendaged or gen with a variously shaped, sometimes fringed appendage, nectary pits 0 (nectaries on ovary stalk); ovary stalked, style short or indistinct, stigma 2-branched. ± 300 spp.: temp to subarctic and alpine Am, Eurasia. (Gentius, king of ancient Illyria, who might have found medicinal value in the pls)

1. Ann; st < 1 dm; lf < 15 mm; fl < 2 cm (sect. *Chondrophyllae*)
  2. Lvs dissimilar, upper cauline strongly ascending, much narrower than basal, margins of all conspicuously white; corolla lobes white to pale blue, exterior often dark blue ..... ***G. fremontii***
  2′ Lvs all ± similar, ± spreading, margins not or obscurely white; corolla medium to deep blue on both sides ..... ***G. prostrata***
1′ Per; st gen > 1 dm; lf > 15 mm; fl > 2 cm (sect. *Pneumonanthe*)
  3. Sts arising laterally from caudex, below a rosette
    4. Sts > 1 dm; lvs of middle to upper st < internodes; corolla ± uniformly blue ..... ***G. setigera***
    4′ Sts gen < 1 dm; lvs of st gen > internodes; corolla white to deep blue, exc exterior on and below lobes sometimes dark brownish purple ..... ***G. newberryi***
      5. Corolla medium to deep blue ..... var. ***newberryi***
      5′ Corolla white to pale blue exc exterior on and below lobes sometimes dark brownish purple ..... var. ***tiogana***
  3′ Sts arising terminally from caudex, rosettes 0
    6. Sinuses between corolla lobes ± truncate ..... ***G. sceptrum***
    6′ Sinuses between corolla lobes with a conspicuous, toothed or fringed appendage
      7. Corolla sinus appendages divided ± to base into several thread-like parts; lf base sheaths at mid-st gen > 5 mm ..... ***G. plurisetosa***
      7′ Corolla sinus appendages divided ± to base or < 1/2 way to base into often 2 parts, each thread-like only toward tip; lf base sheaths at mid-st < 5 mm
        8. Lf, calyx lobe margins minutely ciliate; calyx lobes gen <1.5 mm wide; corolla lobes gen < 7 mm; infls terminal and at 1–6 upper nodes, fls 1–several per infl ..... ***G. affinis*** var. ***ovata***
        8′ Lf, calyx lobe margins gen obscurely, minutely dentate, not ciliate; calyx lobes gen > 2 mm wide; corolla lobes gen > 7 mm; infls gen terminal (sometimes also at 1–3 upper nodes), fls 1–few ..... ***G. calycosa***

G. affinis Griseb. var. ***ovata*** A. Gray (p. 671) Per. **STS** from caudex, decumbent to erect, several–many, 5–70 cm, puberulent. **LVS**: basal ± 0; cauline many, 10–45 mm, 4–20 mm wide, minutely ciliate, lower < internodes, elliptic to ovate, upper > internodes, ovate to lanceolate, ± acute. **INFLS** terminal and gen also at 1–6 upper nodes; fls 1–several per infl. **FL**: calyx 5–23 mm, lobes linear to narrowly elliptic (sometimes some vestigial), gen < 1.5 mm wide, acute, minutely ciliate; corolla (12)25–45 mm, blue, rarely white, lobes gen < 7 mm, oblong-ovate, abruptly pointed, sinus appendages divided ± to base into 2 triangular parts, each thread-like only toward tip. **SEED** winged. $2n=26$. Meadows, shrubby places; 0–2300 m. NCo, KR, CaRH, CCo, SnFrB, Wrn; to OR. [*G. oregana* A. Gray, at least as applied to CA pls] Var. *parvidentata* Kusn. may be applied to pls with some or all calyx lobes small or vestigial; some pls in n CA approach var. *affinis* of Rocky Mtns. ❀IRR: **1**,2,4,**5**,6,16,17&SHD:14,15;DFCLT.

G. calycosa Griseb. (p. 671) Per. **STS** from caudex, ± decumbent, 2–many, 5–45 cm. **LVS**: basal ± 0; cauline many, ± evenly spaced, gen > 1 × internodes, 8–50 mm, 6–30 mm wide, ovate to nearly round, obtuse to acute, gen obscurely, minutely dentate, not ciliate. **INFLS** gen terminal (sometimes also at 1–3 upper nodes); fls 1–few per infl. **FL**: calyx 10–20 mm, lobes narrowly to widely ovate, gen > 2 mm wide, sometimes much reduced, obtuse to acute, gen obscurely, minutely dentate, not ciliate; corolla 25–50 mm, deep blue, rarely violet, lobes 5.5–13 mm, oblong-ovate to round, short-acuminate, sinus appendages divided < 1/2 way to base into gen 2 triangular parts, each thread-like only toward tip. **SEED** wingless. $2n=26$. Wet mtn meadows, slopes; 1300–3900 m. KR, CaRH, SNH; to B.C., MT. Hybridizes with *G. newberryi* var. *tiogana* in s SNH (Tulare Co.). ❀IRR&DRN&acidic:1,2,4–6,16,17& SHD:14,15;DFCLT.

G. fremontii Torrey (p. 671) MOSS GENTIAN Ann. **STS** decumbent to erect, 1–several, 2–10 cm. **LVS** conspicuously white-margined; basal 2–13 mm, 1.5–8 mm, widely spoon-shaped to round, abruptly pointed; lower cauline crowded, < 6 mm wide, spoon-shaped to oblanceolate, upper cauline strongly ascending, < internodes, < 2 mm wide, oblanceolate to linear, acute. **INFL**: fl 1. **FL**: calyx 4–12 mm, lobes narrowly triangular, acute; corolla 7–15 mm, lobes 2.2–4 mm, white to pale blue, exterior often dark blue, ovate, acuminate, sinus appendages white, triangular, ± entire or jagged-serrate, acute. **SEED** wingless. UNCOMMON. Wet mtn meadows; 2400–2700 m. SnBr; to w Can, Colorado. [*G. humilis* Steven; not *G. aquatica* L.] ❀TRY.

G. newberryi A. Gray (p. 671) Per. **STS** arising laterally, below rosette, decumbent, 1–several, 0.5–10(23) cm. **LVS**: basal, lower cauline 8–50(75) mm, 2–25 mm wide, widely spoon-shaped to oblanceolate, often ± petioled, rounded or abruptly pointed; upper cauline few, gen > internodes, 6–30(45) mm, 1–8(13) mm wide, oblanceolate to linear, acute. **INFL**: fls 1–5. **FL**: calyx (10)14–30 mm, lobes linear to narrowly ovate, acute; corolla 23–55 mm, lobes 7–17 mm, elliptic-obovate, short-acuminate, sinus appendages deeply divided into 2 triangular, ± serrate-jagged parts tapered to thread-like points. **SEED** winged. Wet mtn meadows; 1200–4000 m. KR, CaRH, SNH, W&I; OR, w NV. Size variation ± correlated with altitude, moisture; vars. intergrade in n SNH.

var. ***newberryi*** St, lf, fl gen in upper part of size ranges for sp. **FL**: corolla medium to deep blue. Wet mtn meadows; 1200– 2200 m. KR, CaRH, n SNH; OR. ❀WET&DRN:**1**,**2**,4–6,16,17& SHD:14,15.

var. ***tiogana*** (A.A. Heller) J. Pringle St, lf, fl size often in lower to middle part of ranges for sp. **FL**: corolla white to pale blue exc exterior dark brownish purple on and below lobes. $2n=26$. Wet mtn meadows; 1500–4000 m. SNH, W&I; w NV. [*G. tiogana* A.A. Heller] ❀WET&DRN:**1**,**2**,4–6,16,17&SHD:14,15.

G. plurisetosa C. Mason (p. 671) KLAMATH GENTIAN Per. **STS** from caudex, decumbent to erect, 2–several, 5–40 cm. **LVS**: basal ± 0; cauline many, ± evenly spaced, gen > 0.8 × internodes, 12–60 mm, 7–38 mm wide, elliptic to round; base sheaths at mid-st gen > 5 mm; tips obtuse to acute. **INFL** terminal; fls 1–5 (also sometimes solitary at 1–3 nodes). **FL**: calyx 17–25(35) mm, lobes lance-elliptic, rarely ± lf-like, not ciliate, obtuse to acute; corolla 35–50 mm, deep blue, lobes 7–14 mm, oblong-obovate to round, short-acuminate, sinus appendages divided ± to base into several thread-like parts. **SEED** winged. UNCOMMON. Wet mtn meadows; 1200–1900 m. KR, NCoRO; to sw OR. [not *G. setigera* A. Gray]

G. prostrata Haenke (p. 671) PIGMY GENTIAN Ann. **STS** ± prostrate to decumbent, 1–several, 0.5–8 cm. **LVS** basal and cauline similar, spreading, 2–6 mm, 1–4 mm wide, spoon-shaped-obovate to broadly oblanceolate, obscurely or not white-margined, abruptly pointed; cauline ± evenly spaced, gen 0.5–2 × internodes. **INFL**: fl 1. **FL**: calyx 3.5–10 mm, lobes triangular, abruptly pointed; corolla 7–18 mm, medium to deep blue, lobes 2.5–5 mm, ovate, acute to acuminate, sinus appendages triangular, ± entire or ± irregularly jagged toward tip, acute. **SEED** wingless. $2n$=36. UNCOMMON. Wet mtn meadows; 3500–3800 m. W&I (White Mtns); to AK, Colorado, Eurasia.

G. sceptrum Griseb. (p. 671) Per. **STS** from caudex, decumbent to erect, 1–several, 5–90 cm. **LVS**: basal ± 0; cauline many, 10–55 (70) mm, 5–18 mm wide, elliptic to narrowly ovate, lower ± crowded, upper < internodes, tip obtuse to acute. **INFL**: fls 1–few. **FL**: calyx 15–27 mm, lobes elliptic-ovate, rarely ± lf-like, not ciliate, acute; corolla 35–50 mm, blue, lobes 5–10 mm, ovate-oblong to ± round, abruptly pointed, sinuses ± truncate, ± unappendaged. **SEED** winged toward ends. $2n$=26. Bogs, wet meadows; 0–1200 m. NCo, NCoRO; to B.C. ❀WET&DRN:**4–6**,15–17; DFCLT.

G. setigera A. Gray MENDOCINO GENTIAN Per. **STS** arising laterally below rosette, decumbent, 1–few, 20–45 cm. **LVS**: basal 25–85 mm, 5–15 mm wide, spoon-shaped-obovate, tip obtuse; cauline many, 10–30 mm, 5–17 mm wide, elliptic, base sheathing, tip obtuse to acute, lower crowded, wider than upper, upper < internodes. **INFL**: fls 1–4. **FL**: calyx 14–23 mm, lobes ovate-oblong, subacute; corolla (25)35–50 mm, blue, lobes 10–16 mm, elliptic-obovate, acuminate, sinus appendages deeply divided into 2–3 thread-like parts. **SEED** winged. RARE. Wet mtn meadow; ± 1065 m. NCoRO (Red Mtn, Mendocino Co.); sw OR. [*G. bisetaea* Howell] [Chambers & Greenleaf 1989 Madroño 36:49–50] In cult.

GENTIANELLA

Ann, glabrous. **LVS** basal and cauline, opposite. **INFL**: cyme or fls solitary in axils or on branches. **FL**: parts in 4's or 5's; calyx tube < lobes; corolla < 2 cm, narrowly funnel-shaped, lobes spreading, < tube, subentire, with fringes or scales on inner surface near base, nectary pits 0 (nectaries on lower part of corolla tube); ovary sessile, style indistinct or 0, stigma 2-branched, persistent. ± 250 spp.: ± worldwide, esp temp to alpine, exc Afr. (Latin: little *Gentiana*) [Gillett 1957 Ann Missouri Bot Gard 44:195–269]

1. Fls in cymes and solitary in axils, parts gen in 5's; pedicels gen < internode below (subg. *Gentianella*) .... ***G. amarella*** ssp. ***acuta***
1′ Fls solitary, terminal on branches, parts gen in 4's; pedicels >> internode below (subg. *Comastoma*) .... ***G. tenella*** ssp. ***tenella***

G. amarella (L.) Boerner ssp. ***acuta*** (Michaux) J.M. Gillett (p. 671) Pl 5–80 cm. **ST** erect, 1, often with long branches near base, simple or with shorter branches above. **LVS**: basal withering early; lower cauline < 45 mm, spoon-shaped to elliptic-oblong, crowded, upper cauline < 60 mm, oblong-lanceolate to ovate, gen << internodes. **INFL**: pedicels gen < 2.5(–5) cm. **FL**: calyx 5–10(18) mm, lobes 1–6 × tube, linear to lanceolate; corolla blue to rose-violet or white, lobes ovate-triangular, a fringe of hairs across each. $2n$=18. Wet meadows, bogs; 1500–3500 m. KR, CaRH, SNH, SnBr, W&I; to AK, e N.Am, Baja CA, e Asia. [*Gentiana a.* L. var. *acuta* (Michaux) Herder] ❀TRY.

G. tenella (Rottb.) Boerner ssp. ***tenella*** (p. 671) Pl < 15(–25) cm. **STS** decumbent to erect, 1–25, simple or branched near base. **LVS** mostly basal, < 20 mm, elliptic-oblong to spoon-shaped; cauline 1–4 pairs, smaller. **INFL**: pedicels 2–10 cm. **FL**: calyx 4–11 mm, lobes >> tube, ovate to lanceolate; corolla pale violet-blue to white, lobes ovate-oblong, with 2 fringed scales at base of each. $2n$=10. Open, wet places; 3200–3900 m. c&s SNH, W&I; to AK, Colorado; circumpolar. [*Gentiana t.* Rottb.] ❀TRY.

GENTIANOPSIS FRINGED GENTIAN

Ann, per, glabrous. **LVS** basal and cauline, opposite. **INFL**: fl 1 per st or branch. **FL**: parts in 4's; calyx tube distinct, lobes lanceolate, acuminate; corolla > 2 cm, funnel-shaped, blue, rarely white, lobes < or = tube, oblong to elliptic-obovate, obtuse or rounded, ± entire to conspicuously serrate, jagged, or fringed, nectary pits 0 (nectaries on corolla tube near base); ovary stalked, style short or indistinct, persistent, stigma 2-branched. ± 15 spp.: temp N.Am, Eurasia. (Greek: resembling *Gentiana*)

1. St gen branched, << peduncle if simple; corolla lobes ± entire, sometimes shallowly irregular near tip .... ***G. holopetala***
1′ St simple, gen > peduncle; corolla lobes gen conspicuously serrate, jagged, or fringed in upper half or more .... ***G. simplex***

G. holopetala (A. Gray) Iltis (p. 675) Ann or per from root-sprouts or rooting sts, 3–45 cm. **ST** ± decumbent to erect. **LVS**: basal, lower cauline < 70 mm, < 15 mm wide, spoon-shaped, rounded to acute; upper cauline few, lance-elliptic to linear, acute. **INFL**: peduncle (0.4)2.5–21 cm, gen > 2.5 × subtending internode, gen > whole st. **FL**: calyx (10)14–36 mm; corolla 20–55 mm. **SEED** papillate, obtuse. $2n$=78. Wet meadows; 1800–4000 m. c&s SNH, W&I; w NV. [*Gentiana h.* (A. Gray) Holm] ❀TRY.

G. simplex (A. Gray) Iltis (p. 675) Ann, 4–40 cm. **ST** erect. **LVS**: basal, lower cauline < 20 mm, < 5 mm wide, spoon-shaped, withering early (esp basal); upper cauline several, lance-elliptic, obtuse to acute. **INFL**: peduncle 1.3–11(14) cm, gen < 2.5 (–4) × subtending internode, < whole st. **FL**: calyx 9–27 mm; corolla 17–45 mm. **SEED** striate-ridged, pointed. Wet meadows; 1200–3400 m. KR, NCoRH, CaRH, SNH, SnBr; to ID. [*Gentiana s.* A. Gray] ❀TRY.

SWERTIA

Per (dying after fl in *S. albomarginata*, *S. fastigiata*, *S. parryi*, *S. puberulenta*, *S. radiata*; non-fl rosettes preceding fl-sts in these spp., accompanying fl-sts in others). **LVS**: basal ± petioled; cauline < basal, gen whorled or opposite. **INFL**: cyme or panicle of dense clusters. **FL**: parts in 4's (gen 5's in *S. perennis*, many spp. outside CA); calyx fused only near base, lobes lanceolate; corolla rotate, rarely bell-shaped, sometimes with fringed ridges or scales between stamen bases, lobes >> tube, nectary pits prominent, 1–2 per lobe, pit margin fringed; ovary sessile, style short, persistent, stigma 2-branched. ± 120 spp.: temp N.Am, Eurasia, Afr. (E. Sweert, Dutch herbalist, born 1552) [St. John 1941 Amer Midl Naturalist 21:1–29] *Frasera* sometimes segregated.

1. Nectary pits 2 per corolla lobe (sometimes appearing as 1 due to overlapping fringes)
 2. St gen < 1 m, < 1 cm wide; cauline lvs opposite or some alternate; corolla gen bluish, 4- or 5-lobed . . . ***S. perennis***
 2′ St gen > 1 m, > 1.5 cm wide; cauline lvs whorled; corolla yellowish green, 4-lobed . . . . ***S. radiata***
1′ Nectary pit 1 per corolla lobe
 3. Lvs at low to mid st widely elliptic, not white-margined — infl dense, < 0.5 dm wide . . . . ***S. fastigiata***
 3′ Lvs at low to mid st narrower than widely elliptic, white-margined (sometimes narrowly so)
  4. Infl ± open, gen > 1 dm wide or lowest branches < 1 dm from st base and widely divergent, or both
   5. Cauline lvs whorled, uppermost sometimes opposite . . . . ***S. albomarginata***
   5′ Cauline lvs opposite
    6. Sts, lower lf surfaces glabrous; basal, lower cauline lvs gen > 1 cm wide; nectary pits U-shaped . . . . . ***S. parryi***
    6′ Sts, lower lf surfaces puberulent; basal, lower cauline lvs gen < 1 cm wide; nectary pits oblong-obovate . . . . ***S. puberulenta***
  4′ Infl dense, sometimes interrupted, < 1 dm wide, lowest branches > 1 dm from st base, short, strongly ascending — upper pedicels gen arising directly from main st
   7. Nectary-pit margin with 2 oblong, fringed lobes . . . . ***S. tubulosa***
   7′ Nectary-pit margin fringed but not lobed
    8. Nectary pits round to ± square; stamen bases with a low, fringed ridge between . . . . ***S. neglecta***
    8′ Nectary pits oblong; stamen bases with ± 2 mm scales between . . . . ***S. albicaulis***
     9. Sts, lf lower surfaces puberulent; larger lvs rounded at tip . . . . var. ***albicaulis***
     9′ Sts, lvs glabrous; larger lvs gen acute . . . . var. ***nitida***

S. albicaulis (Griseb.) Kuntze (p. 675) Pl 1–6.5 dm. **STS** 1–few; rosettes several. **LVS** narrowly white-margined; basal 4–23 cm, 3–12(20) mm wide, oblanceolate; cauline opposite, upper linear-oblong. **INFL** dense, interrupted below; pedicels 2–8(30) mm. **FL**: calyx 3–7(12) mm; corolla 6–12 mm, with ± 2 mm, triangular to ovate-oblong, ± coarsely jagged scales between stamen bases, greenish white to pale blue, sometimes dark blue-dotted, lobes elliptic-oblong, acute to short-acuminate; nectary pit 1 per corolla lobe, oblong. Dry, ± open places; 150–2200 m. KR, NCoR, CaR, n SN, MP; to WA, MT, w NV. [*Frasera a.* Griseb.] Vars. intergrade in CaRH, MP.

var. ***albicaulis*** Sts and lf lower surfaces puberulent; larger lvs not arching, lf tips obtuse to rounded. Dry, shrubby places; 900–1600 m. CaRH, MP; to WA, MT, nw NV. ❀TRY.

var. ***nitida*** (Benth.) Jepson Sts and lvs glabrous; larger lvs arching, lf tips acute. Dry, open woodlands, chaparral; 150–1900 m. KR, NCoR, CaR, n SN, MP; to OR. [*Frasera a.* var. *n.* (Benth.) C. Hitchc.] ❀TRY.

S. albomarginata (S. Watson) Kuntze (p. 675) Pl 3–6 dm, glabrous or sts puberulent. **STS** 1–few. **LVS** white-margined; basal 2–9 cm, 5–10 mm wide, oblanceolate, tips acute; cauline whorled, upper sometimes opposite, linear-lanceolate, tips acuminate. **INFL** open; pedicels 5–50 mm. **FL**: calyx 5–12 mm; corolla 8–14 mm, with a low, fringed ridge between stamen bases, greenish white, often purple-dotted, lobes lance-oblong, abruptly acuminate; nectary pit 1 per corolla lobe, oblong, wider, 2-lobed at tip. Dry, open woodlands; 1500–2200 m. DMtns; to Colorado. [*Frasera a.* S. Watson] ❀DRN&SHD:2,6,15–17.

S. fastigiata Pursh (p. 675) CLUSTERED GREEN-GENTIAN Pl 3–14 dm, glabrous. **ST** 1. **LVS**: basal 15–30 cm, 3–10 cm wide, spoon-shaped-obovate, tips acute; cauline widely elliptic, tips acuminate, lower whorled, upper often opposite, lower to middle not white-margined. **INFL** dense, sometimes interrupted below; pedicels 2–10 mm. **FL**: calyx 8–12 mm; corolla 8–12 mm, with a long fringe between stamen bases, pale yellowish green, sometimes blue-tinged, lobes elliptic-ovate, obtuse to acute; nectary pit 1 per corolla lobe, round. Uncommon. Mtn meadows; 1700–1900 m. KR; to OR, ID. [*Frasera f.* (Pursh) A.A. Heller; *F. umpquaensis* M. Peck & Appleg., Umpqua green-gentian]

S. neglecta (H.M. Hall) Jepson (p. 675) PINE GREEN-GENTIAN Pl 2–5.5 dm, glabrous. **STS** 1–several; rosettes several. **LVS** narrowly white-margined, linear to narrowly oblanceolate, tips acute; basal 2.5–20 cm, 3–9 mm wide; cauline opposite. **INFL** dense, interrupted; pedicels 2–25 mm. **FL**: calyx 5–8 mm; corolla 7–15 mm, with a low, fringed ridge between stamen bases, greenish white, purple-streaked, lobes oblong-obovate, abruptly acuminate; nectary pit 1 per corolla lobe, round to ± square. UNCOMMON. Dry, open woodlands; 1400–2500 m. SCoRO, TR. [*Frasera n.* H.M. Hall]

S. parryi (Torrey) Kuntze (p. 675) Pl 6–16 dm, glabrous. **STS** 1–2. **LVS** white-margined; basal 5–25 cm, 8–40 mm wide, strap-shaped to elliptic-oblanceolate, tips acute; cauline opposite, ovate to lance-oblong, tips acute to acuminate. **INFL** ± open; pedicels 6–32 mm. **FL**: calyx 8–17 mm; corolla 9–20 mm, with a low, fringed ridge between stamen bases, light green, purple-dotted, lobes obovate, short-acuminate; nectary pit 1 per corolla lobe, U-shaped, margin fringed. Dry, open woodlands, chaparral; 100–1900 m. SCo, SnGb, SnBr, PR; AZ, Baja CA. [*Frasera p.* Torrey]

S. perennis L. (p. 675) Pl 1–5 dm, glabrous. **ST** gen 1; rosettes 1–few. **LVS**: basal 3–16 cm, 12–30 mm wide, spoon-shaped-obovate, tips obtuse; cauline opposite or some alternate, elliptic-oblanceolate, tips subacute. **INFL** open; pedicels 2–30(60) mm. **FL** parts in 4's or 5's; calyx 4–8 mm; corolla 7–13 mm, with a low, ± fringed ridge between stamen bases, bluish white to violet-blue, veins darker, lobes lance-oblong, acute; nectary pits 2 per corolla lobe, round. $2n$=28. Wet meadows, bogs; 2300–3200 m. s SNH; Wallowa Mtns of OR, Rocky Mtns of Colorado to AK, Eurasia.

S. puberulenta (Davidson) Jepson (p. 675) Pl 1–3 dm; sts, lower lf surfaces puberulent. **STS** 1–several. **LVS** narrowly white-margined; basal 2–12 cm, 6–10(17) mm wide, narrowly obovate to elliptic-oblong, tips obtuse or abruptly pointed; cauline opposite, upper oblong to lanceolate, tips acute. **INFL** open; pedicels 5–35 mm. **FL**: calyx 6–11 mm; corolla 7–13 mm, with a low, ± fringed ridge between stamen bases, greenish white, purple-dotted, lobes oblong-obovate, abruptly acuminate; nectary pit 1 per corolla lobe, oblong-obovate. Dry, open coniferous woodlands; 1700–3400 m. c&s SNH, W&I. [*Frasera p.* Davidson; *S. albomarginata* var. *purpusii* Jepson]

S. radiata (Kellogg) Kuntze (p. 675) MONUMENT PLANT Pl 7–20 dm, glabrous or st and lvs puberulent. **ST** 1, > 1.5 cm wide. **LVS**: basal 7–50 cm, 1–15 cm wide, oblanceolate to elliptic-obovate, tips rounded to acute; cauline whorled, lance-oblong, tips acute. **INFL** elongate, ± dense above, open below; pedicels 1–10 cm. **FL**: calyx 12–25 mm; corolla 12–20 mm, with deeply, multi-divided scales between stamen bases, light yellowish green, purple-dotted, lobes elliptic-oblong, acute to short-acuminate; nectary pits 2 per corolla lobe, narrowly elliptic. $2n$=78. Mtn meadows, open woodlands; 1500–3000 m. KR, NCoRH, SNH, Wrn; to WA, SD, NM. [*Frasera speciosa* Griseb.]

S. tubulosa (Cov.) H. St. John (p. 675) Pl (0.6)2–11 dm, glabrous. **ST** gen 1; rosettes 0–few. **LVS** white-margined, curving

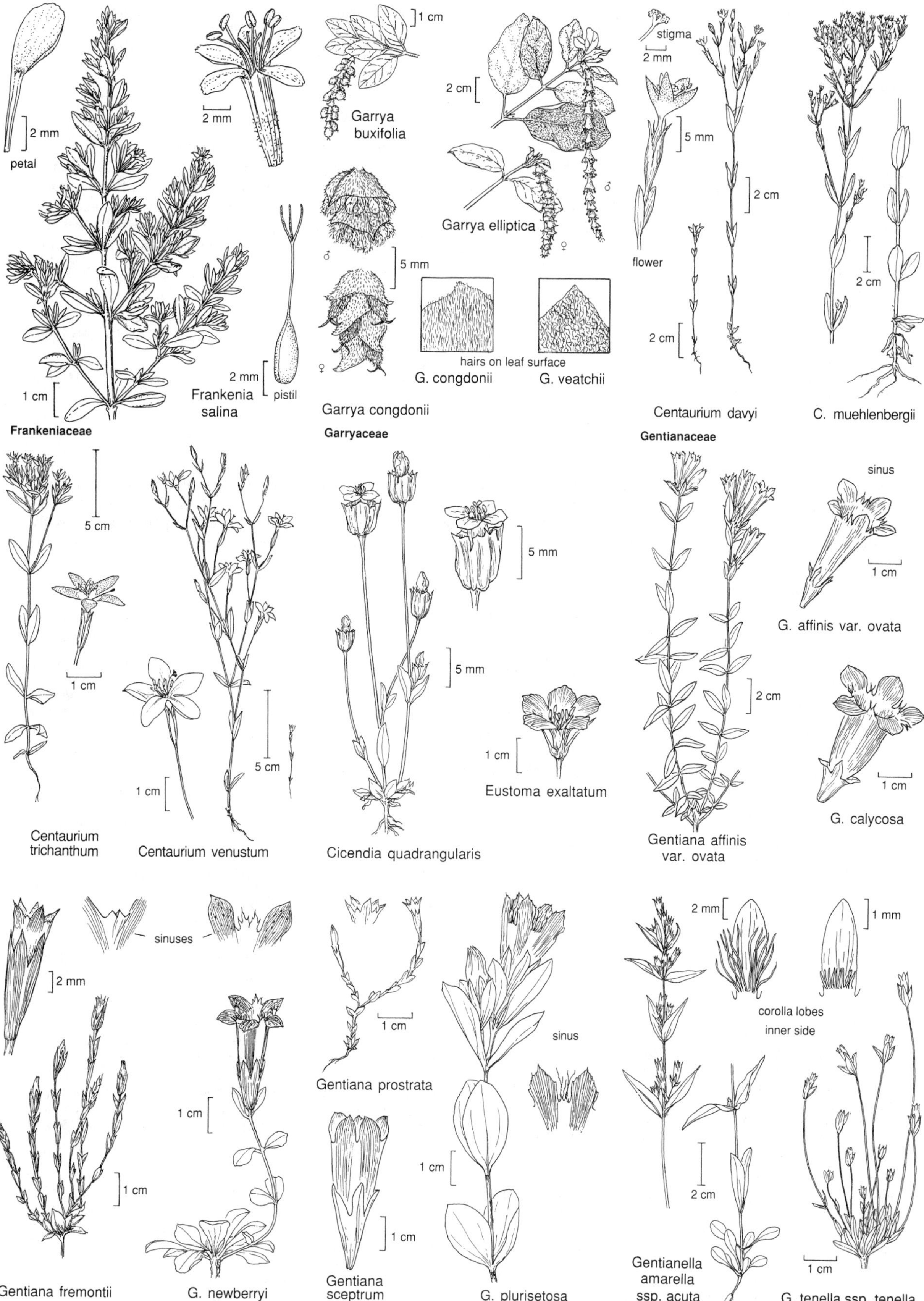
petal
2 mm
2 mm
1 cm
2 mm
pistil
Frankenia salina
Frankeniaceae
1 cm
Garrya buxifolia
2 cm
Garrya elliptica
5 mm
hairs on leaf surface
G. congdonii
G. veatchii
Garrya congdonii
Garryaceae
stigma
2 mm
5 mm
flower
2 cm
2 cm
Centaurium davyi
2 cm
C. muehlenbergii
Gentianaceae
5 cm
1 cm
Centaurium trichanthum
1 cm
5 cm
Centaurium venustum
5 mm
5 mm
Cicendia quadrangularis
1 cm
Eustoma exaltatum
2 cm
Gentiana affinis var. ovata
sinus
1 cm
G. affinis var. ovata
1 cm
G. calycosa
2 mm
sinuses
1 cm
Gentiana fremontii
1 cm
G. newberryi
1 cm
Gentiana prostrata
1 cm
Gentiana sceptrum
1 cm
sinus
G. plurisetosa
2 mm
1 mm
corolla lobes
inner side
2 cm
Gentianella amarella ssp. acuta
1 cm
G. tenella ssp. tenella

down toward tip; basal 2–9 cm, 3–10(15) mm wide, oblanceolate, tips abruptly pointed; cauline whorled, upper linear-oblong, tips acute. **INFL** elongate, ± continuous above; pedicels 2–25 mm. **FL**: calyx 6–11 mm; corolla 8–13 mm, bell-shaped, without ridges or scales between stamen bases, white to pale blue, veins darker, lobes elliptic-oblong to obovate, short-acuminate; nectary pit 1 per corolla lobe, ovate, margin extension tubular below, 2-lobed, fringed above. Open pine woodlands; 1800–2700 m. s SN. [*Frasera t.* Cov.]

GERANIACEAE GERANIUM FAMILY

Ann, per, or ± woody, gen hairy. **LVS** simple to compound, basal and cauline; cauline alternate or opposite, stipules present. **INFL**: cyme or umbel. **FL** bisexual, radial or ± bilateral; sepals 5, free, overlapping in bud; petals 5, free, with nectar glands at base; stamens gen 5 or 10; staminodes scale-like or 0; pistil 5-lobed, chambers 5, placentas axile, styles 5, fused to axis, columnar in fr, stigmas atop axis 5, free. **FR**: segments 5, dry, 1–2-seeded, separating from each other and then from column; fr body dehiscent on 1 side or not; part of style persistent atop ovary and separating with it, curved to tightly coiled when dry. 14 genera, ± 750 spp.: temp, ± trop. Some cult for orn, perfume oils. [Robertson 1972 J Arnold Arbor 53:182–201] Family description, key to genera by M.S. Taylor.

1. Petals unequal (sometimes upper 2 different from lower 3 in shape)
  2. Upper sepal not spurred — st canescent .... ***Erodium texanum***
  2′ Upper sepal spurred, spur fused to pedicel .... **PELARGONIUM**

1′ Petals equal
  3. Lf pinnate; fertile stamens 5 .... **ERODIUM**
  3′ Lf palmate; fertile stamens gen 10 .... **GERANIUM**

ERODIUM STORKSBILL, FILAREE

Mary Susan Taylor

Ann, per. **LVS** simple to pinnately compound; lower basal; upper opposite; blade lanceolate to reniform in outline, base cordate to truncate, short-hairy. **INFL**: umbel. **FL**: stamens 5, alternate 5 scale-like staminodes. **FR**: body indehiscent, fusiform, 1-seeded, base sharply pointed, top gen pitted, pits subtended by 1–2 furrows or not; style segment persistent to fr body stiffly hairy on side facing column. ± 75 spp.: temp Am, Eurasia, n Afr, Australia. (Greek: heron, from bill-like fr) [Guittonneau 1972 Boissiera 20:1–154] Some cult for forage, dyes.

1. Lower lvs pinnately compound, lflets 9–15
  2. Lower lflets deeply dissected; sepal tip bristly; petal base ciliate .... ***E. cicutarium***
  2′ Lower lflets ± toothed; sepal tip glabrous; petals glabrous .... ***E. moschatum***

1′ Lower lvs simple, entire to pinnately divided, lobes or divisions < 9
  3. Lf blade deeply 7–9-lobed or -divided
    4. St coarsely hairy; fr body 8–11 mm, pit atop body gen subtended by 2 deep furrows .... ***E. botrys***
    4′ St glandular-hairy; fr body 6–8 mm, pit atop body subtended by 1 shallow furrow .... ***E. brachycarpum***
  3′ Lf blade crenate to 3–5-divided
    5. Lf < 4 cm; sepals canescent; petals unequal .... ***E. texanum***
    5′ Lf > 4 cm; sepals ± hairy but not canescent; petals equal
      6. Lf blade deeply 3–5-divided, divisions wedge-shaped; petals blue .... ***E. cygnorum***
      6′ Lf blade entire to shallowly lobed; petals white to purple
        7. Lvs basal; blade reniform, ± entire .... ***E. macrophyllum***
        7′ Lvs basal and cauline; blade ovate, crenate .... ***E. malacoides***

E. botrys (Cav.) Bertol. (p. 675) Ann. **ST** prostrate to ascending, 1–9 dm, short-hairy. **LVS** lobed to dissected; lobes ± 8–10 mm wide; lower lvs 3–15 cm; blade ± = petiole, ovate to oblong in outline, glabrous to sparsely puberulent, veins short-appressed-hairy. **FL**: sepals 10–13 mm, tip bristly; petals slightly > sepals, lavender. **FR**: body 8–11 mm, pit ± round, deep, glabrous, gen subtended by 2 furrows; style column 5–12 cm. $2n=40$. Dry, open or disturbed sites; < 1000 m. CA-FP; native to s Eur.

E. brachycarpum (Godron) Thell. (p. 675) Ann. **ST** ascending, 1–6 dm, ± glandular-hairy. **LVS** lobed; lobes 7–11 mm wide; lower lvs 5–10 cm; blade ± = petiole, gen ovate in outline, glabrous to sparsely puberulent, veins short-appressed-hairy. **FL**: sepals 7–10 mm, tip bristly; petals slightly > sepals, lavender. **FR**: body 6–8 mm, pit transversely narrow, deep, puberulent, subtended by 1 shallow furrow; style column 5–8 cm. $2n=40$. Dry, open or disturbed sites; < 1000 m. CA-FP; OR; native to s Eur. [*E. obtusiplicatum* (Maire, Weiller, & Wilczek) J. Howell]

E. cicutarium (L.) L'Hér. (p. 675) Ann. **ST** decumbent to ascending, 1–5 dm, ± glandular-hairy. **LVS** compound; lower 3–10 cm, blade > petiole, ovate to oblanceolate in outline, sparsely hairy; lflets 9–13, deeply dissected, ultimate segments 1–2 mm wide. **FL**: sepals 3–5 mm, tip bristly; petals ± = sepals, red-lavender, base gen purple. **FR**: body 4–7 mm, pit ± round, glabrous, subtended by 1 shallow furrow or not; style column 2–5 cm. $2n=40$. Open, disturbed sites, grassland, shrubland; < 2000 m. CA; widespread US; native to Eurasia.

E. cygnorum Nees (p. 675) Ann. **ST** decumbent to ascending, 1–3 dm, ± soft-hairy. **LVS** deeply 3–5-divided; lower 4–8 cm; blade < petiole, ovate in outline, lower 2 divisions > 10 mm wide, wedge-shaped. **FL**: sepals 5–7 mm, tip glabrous to puberulent; petals slightly > sepals, blue. **FR**: body 5–6 mm, pit ± transversely elliptic, subtended by 1 furrow; style column 5–6 cm. Uncommon. Abandoned fields, disturbed sites; < 500 m. SCo; native to Australia.

E. macrophyllum Hook. & Arn. (p. 675) Ann, bien, gen scapose. **ST** < 5 cm, glandular-puberulent. **LF** simple, 10–15 cm; blade << petiole, reniform, crenate to shallowly lobed, puberulent. **FL**: sepals 8–10 mm, tip glabrous to puberulent; petals > sepals, gen white, tinged red to purple. **FR**: body 8–10 mm, tip truncate, pit ± round, puberulent, furrow 0; style column 3–5 cm. Open sites, grassland, shrubland; < 1200 m. ScV, n SnJV, CW, SCo, n ChI (Santa Cruz Island); to s UT, n Mex.

E. malacoides (L.) Willd. (p. 675) Ann, bien. **ST** decumbent to ascending, 1–6 dm, ± puberulent; nodes glandular. **LF** simple, 4–15 cm; blade gen < petiole, ovate to oblong, cordate, crenate to shallowly lobed, puberulent, veins short-appressed-hairy. **FL**: sepals 4–6 mm, tip bristly; petals ± = sepals, purple to lavender. **FR**: body 3–5 mm, pit ± round, glandular, gen subtended by 1 furrow; style column 2–3 cm. $2n=40$. Open sites, grassland, shrubland; < 500 m. n SnJV, SnFrB; native to s Eur.

E. moschatum (L.) L'Hér. (p. 675) Ann, bien. **ST** decumbent to ascending, 1–6 dm, short hairy. **LVS** compound; lower 5–15 cm; blade > petiole, oblong to obovate, sparsely hairy, veins short-appressed-hairy; lflets 11–15, lobed to shallowly divided, ultimate segments 1–4 mm wide. **FL**: sepals 6–9 mm, tip glabrous; petals 10–15 mm, red-lavender. **FR**: body 4–6 mm, pit ± round, glandular, subtended by 1 furrow; style column 2–4 cm. 2*n*=20. Open, disturbed sites; < 1500 m. CA-FP; native to Eur.

E. texanum A. Gray (p. 675) Ann, bien. **ST** prostrate to ascending, 1–5 dm, ± canescent. **LF** simple, 1.5–4 cm; blade gen < petiole, ovate, cordate, crenate to shallowly lobed, densely puberulent to strigose. **FL**: sepals 5–10 mm, tip strigose; petals unequal, 7–12 mm, lavender to purple. **FR**: body 5–8 mm, pit transversely elliptic, glabrous, furrow 0; style column 3–7 cm. 2*n*=20. Dry, open sites, shrubland; < 1500 m. s SnJV, s SCo, D; to TX, n Mex.

GERANIUM CRANESBILL, GERANIUM

Mary Susan Taylor

Ann, per. **LVS** palmately lobed or divided; upper alternate or opposite; blade gen round in outline, base gen cordate, ± hairy. **INFL**: cyme; fls (1)2. **FL**: sepals awned or not; stamens 10, outer 5 opposite petals, inner 5 alternate petals. **FR**: body dehiscent, gen ovoid, 1–2-seeded, base rounded; style column narrowed at top below free stigmas, forming a beak in fr; part of style persistent to fr body glabrous to puberulent on side facing column. 250–300 spp.: temp, trop mtns. (Greek: crane, from beak-like fr) [Jones & Jones 1943 Rhodora 45:5–26;32–53] Some orn, cult for oils. Native per (esp *G. californicum, G. richardsonii*) vary regionally, are often difficult to separate, need further study.

1. Ann (bien); root slender, short
 2. Lf blade ± compound — lflets 3, pinnately lobed; petals > 8 mm .... ***G. robertianum***
 2′ Lf blade deeply lobed to dissected; petals gen < 7 mm
  3. Sepals acute; sepals, petals gen < 4 mm
   4. Fr body transversely wrinkled, glabrous .... ***G. molle***
   4′ Fr body smooth, minutely strigose .... ***G. pusillum***
  3′ Sepals awned, awn 1–2 mm; sepals, petals gen > 4 mm
   5. Pedicel 10–30 mm, > fl; fr beak 3–5 mm .... ***G. bicknellii***
   5′ Pedicel 2–10 mm, ± = fl; fr beak 2–3 mm
    6. Pedicel hairs stiff, straight to curving downward; fr hairs ascending, ± 1 mm .... ***G. carolinianum***
    6′ Pedicel hairs glandular, spreading; fr hairs spreading, < 1 mm .... ***G. dissectum***
1′ Per; underground caudex thick, scaly, ± carrot-like
 7. Sepals 3–5 mm, petals 4–8 mm; fr beak < 1 mm
  8. St, petiole hairs ± spreading or curving downward, petals gen 6–8 mm .... ***G. potentilloides***
  8′ St, petiole hairs appressed, pointing downward; petals gen 4–6 mm .... ***G. retrorsum***
 7′ Sepals 6–14 mm, petals > 10 mm; fr beak > 2 mm
  9. Lf blade ± compound; sepal hairs coarse, purplish .... ***G. anemonifolium***
  9′ Lf blade deeply lobed to dissected; sepal hairs soft, glandular or not, white
   10. Petal surface glabrous, base ciliate; style branches < 3 mm .... ***G. oreganum***
   10′ Petal surface soft-hairy in lower half; style branches > 3 mm
    11. Pedicel hairs tipped with red to purple glands .... ***G. richardsonii***
    11′ Pedicel hairs tipped with white to yellow glands
     12. Petal surface soft-hairy over lower 1/2; style branches 6–9 mm; SN, SW .... ***G. californicum***
     12′ Petal surface soft-hairy over lower 1/3–1/5; style branches 4–5 mm; KR, MP .... ***G. viscosissimum***

G. anemonifolium L'Hér. (p. 675) Per. **ST** ascending to erect, 5–15 dm, glabrous to sparsely hairy, 5–10 mm wide in lower half. **LVS**: lower 15–30 cm; blades 8–20 cm wide; lflets gen 5, 6–10 cm, subsessile, pinnately lobed to dissected. **FL**: pedicel 5–10 mm; sepals 8–14 mm, soft-hairy, awned; petals 10–15 mm, rounded, pink to red-purple. **FR**: body 3–4 mm, with net-like ridges, glabrous; style column 16–20 mm, beak 7–9 mm. **SEED** finely pitted. Open to shaded sites; < 200 m. SnFrB; native to Madeira Island (off nw Afr).

G. bicknellii Britton (p. 675) Ann, bien. **ST** decumbent to erect, 1–6 dm, short-stiff-hairy, ± glandular above. **LVS**: lower 2–6 cm; blades 2–7 cm wide, divided into 3–5 wedge-shaped segments, upper half of segments toothed to lobed. **FL**: pedicel 1–3 cm; sepals 4–8 mm, short-awned; petals 5–7 mm, tip slightly notched, pale purple. **FR**: body 2–3 mm, minutely bristly; style column 1.5–2 cm, beak 3–5 mm. **SEED** finely pitted. Open, sunny to shaded sites, woodland, coniferous forest; 600–1500 m. NW, CCo, SnFrB; to B.C., ne N.Am. [var. *longipes* (S. Watson) Fern.]

G. californicum G. Jones & F. Jones (p. 675) Per. **ST** ascending to erect, 2–6 dm, soft-hairy, glandular above. **LVS**: lower 6–25 cm; blades 3–8 cm wide, divided into 4–6 wedge-shaped segments, upper half of segments lobed. **FL**: pedicel 1–12 cm; sepals 6–9 mm, short-awned; petals 10–15 mm, obtuse to slightly notched, white to rose, veins lavender to purple. **FR**: body 4–5 mm, sparsely glandular; style column 2–3 cm, beak 3–4 mm. **SEED** faintly pitted. Moist sites, streambanks, meadows, woodland; 1000–2800 m. c&s SN, TR, PR. Pls from s SNH with style branches > 5 mm, petals not purple-veined have been called *G. concinnum* G. Jones & F. Jones.

G. carolinianum L. (p. 675) Ann. **ST** gen erect, 1–7 dm; hairs dense, short, pointed downward. **LVS**: lower 10–15 cm; blades 2–6 cm wide, divided into ± 5 oblong to wedge-shaped segments, upper half of segments ± lobed. **INFL**: fls clustered atop st; pedicel 2–7 mm. **FL**: sepals 4–7 mm, short awned; petals ± = sepals, ± notched, white to rose-pink. **FR**: body 2–3 mm, puberulent and minutely bristly; style column 10–15 mm, beak 2–3 mm. **SEED** faintly pitted. 2*n*=52. Open to shaded sites, grassland, shrubland, forest; < 1700 m. CA-FP, n GB; to e N.Am.

G. dissectum L. (p. 675) Ann. **ST** ascending to erect, 2–8 dm; hairs rough, spreading to pointed downward, gen glandular above. **LVS**: lower 8–15 cm; blades 2–6 cm wide, deeply dissected into 5–7 linear to oblong segments, upper half of segments lobed. **FL**: pedicels 6–10 mm; sepals 4–7, short-awned; petals ± = sepals, ± notched, rose-purple. **FR**: body 2–3 mm, minutely bristly; style column 12–15 mm, beak 2–3 mm. **SEED** deeply pitted. 2*n*=22. Open, disturbed sites; < 1200 m. CA-FP; to e N.Am; native to Europe.

G. molle L. (p. 675) Ann, bien. **ST** decumbent to erect, 1–4 dm, ± soft-hairy. **LVS**: lower 4–12 cm; blades 1–5 cm wide, 5–7-lobed; lobes rounded, ± toothed. **FL**: pedicels 5–20 mm; sepals 3–4 mm, acute; petals ± = sepals, notched, red-purple. **FR**: body ± 2 mm, glabrous; style column 6–10 mm, beak 1–2 mm. **SEED** smooth. Open to shaded sites, disturbed ground; < 1000 m. CA-FP; widespread N.Am; native to Eur.

G. oreganum Howell (p. 675) Per. **ST** ascending to erect, 4–8 dm, glabrous to soft-hairy, glandular above. **LVS**: lower 1–4 dm; blade 6–15 cm wide, divided into 5–7 ± oblanceolate segments, upper 3/4 of segments lobed. **FL**: pedicels 2–5 cm; sepals 7–12 mm, short-awned; petals 15–20 mm, ± obtuse, red-purple. **FR**: body 6–8 mm, minutely bristly, sometimes glandular; style column 3–5 cm, beak 3–4 mm. **SEED** pitted. Moist, gen shaded sites, meadows, forest; 700–1800 m. NCo, KR, NCoRO; to WA.

G. potentilloides DC. (p. 681) Per. **ST** prostrate to decumbent, sometimes rooting at nodes, 3–5 dm; hairs short, stiff, spreading to pointing downward. **LVS**: lower 5–15 cm; blades 1–4 cm wide, divided into 3–7 wedge-shaped segments, segments lobed. **FL**: pedicel 1–2 cm; sepals 4–5 mm, awned; petals 6–8 mm, ± notched, pink to red. **FR**: body 2–3 mm, sparsely hairy; style column 12–15 mm, beak < 1 mm. **SEED** pitted. Moist, shaded sites, coniferous forest; < 500 m. NCo, SnFrB; native to Australia. [*G. pilosum* Forster f.; *G. microphyllum* Hook. f.]

G. pusillum Burm. f. Ann. **ST** prostrate to decumbent, 1–5 dm, often branched, puberulent. **LVS**: lower 10–25 cm; blades 2–5 cm wide, divided into 5–9 oblong to wedge-shaped segments, upper half of segments lobed or not. **FL**: pedicels 8–15 mm; sepals 3–4 mm, acute; petals ± = sepals, ± notched, pink to violet; fertile stamens 5–8. **FR**: body ± 2 mm, minutely strigose; style column 7–9 mm, beak < 1 mm. **SEED** smooth. Uncommon. Disturbed sites; < 500 m. NCo, s SNF, CCo, s MP, s SNE; to B.C., e N.AM; native to Eur.

G. retrorsum L'Hér. (p. 675) Per. **ST** ascending to erect, 2–5 dm; hairs sparse, stiff, appressed, pointed downward. **LVS**: lower 4–15 cm; blades 2–4 cm wide, divided into 3–5 broadly wedge-shaped segments, upper half of segments lobed. **FL**: pedicels 5–15 mm; sepals 3–4 mm, short-awned; petals 4–6 mm, ± obtuse, purple. **FR**: body 2–3 mm, minutely strigose; style column 10–12 mm, beak < 1 mm. Disturbed sites; < 500 m. NCo, SNF, CCo, n SCo, c PR; native to Australia.

G. richardsonii Fischer & Trautv. (p. 675) Per. **ST** ascending to erect, 2–8 dm, glabrous to sparsely hairy, glandular above. **LVS**: lower 5–30 cm; blades 5–15 cm wide, divided into ± 5 broadly wedge- to diamond-shaped segments, upper 3/4 of segments lobed. **FL**: pedicels 1–2 cm; sepals 6–12 mm, awned; petals 10–18 mm, rounded, white to lavender, gen purple-veined, basal half soft-hairy. **FR**: body 3–4 mm, sparsely hairy; style column 20–25 mm, beak 2–3 mm. **SEED** coarsely pitted. Moist sites, meadows, coniferous forest; 1200–2700 m. SNH, TR, SnJt, Wrn; to B.C., Rocky Mtns.

G. robertianum L. (p. 675) Ann, bien, branched at base. **ST** decumbent to ascending, 1–5 dm, ± hairy, 1–3 mm wide in lower half. **LVS**: lower 5–15 cm; blades 3–5 cm wide; lflets 3, < 5 cm, ± sessile to short-stalked, pinnately lobed to dissected. **FL**: pedicels 3–8 mm; sepals 6–9 mm, soft-hairy, awned; petals 8–13 mm, rounded, pink to red-purple. **FR**: body 2–3 mm, distally ridged, tip short hairy; style column 15–20 mm, beak 6–7 mm. **SEED** faintly pitted. Open to shaded sites; < 100 m. SnFrB; native to Eur.

G. viscosissimum Fischer & C. Meyer (p. 681) Per. **ST** ascending to erect, 3–8 dm, sparsely hairy to glandular. **LVS**: lower 10–30 cm; blades 3–10 cm wide, dissected into 5–7 broadly wedge- to diamond-shaped segments, upper half of segment lobed. **FL**: pedicel gen 2–4 cm; sepals 8–12 mm, awned; petals 12–20 cm, pink to red-purple, veins red to purple, basal 1/5–1/3 soft-hairy. **FR**: body 4–6 mm, ± glandular puberulent; style column 25–30 mm, beak 3–5 mm. **SEED** pitted. Meadows, open sites in sagebrush shrubland, coniferous forest; 1000–2500 m. e KR (Quartz Valley), MP; to B.C., Rocky Mtns. [*G. nervosum* Rydb., *G. attenuilobum* G. Jones & F. Jones]

PELARGONIUM GARDEN GERANIUM

Elizabeth McClintock

Ann, per, shrub, aromatic or strong-smelling. **ST** of shrubs sometimes soft-woody. **LVS** alternate to ± opposite above; blade lobed to dissected, margin gen crenate or serrate. **INFL**: umbel, dense to open; fls 3–many. **FL** bilateral; upper sepal with a spur fused to the pedicel; petals ± equal to strongly unequal, gen striped or splotched; upper 2 petals > lower 3, well separated, different in shape, position; fertile stamens 3–7. **FR**: body dehiscent, gen oblong, base acute, 1-seeded; part of style persistent to fr body stiff-hairy on side facing column. ± 250 spp.: s Afr, Australia. (Greek: stork, from beaked fruit) [Van der Walt 1985 Bothalia 15:345–385]

1. Lf peltate — gen 5-lobed, margin entire ..... *P. peltatum*
1' Lf not peltate
  2. Corolla gen > 3 cm wide; petals ± equal, overlapping
    3. Lf broadly ovate to ± triangular, base truncate to ± wedge-shaped; upper petals blotched or marked ..... ***P. ×domesticum***
    3' Lf round to reniform, base deeply notched; petals uniformly colored ..... ***P. ×hortorum***
  2' Corolla gen < 3 cm wide; petals dissimilar in size, shape, not overlapping
    4. Infl dense, pedicels < 6 mm
      5. Lf divided > 1/2 to midvein; divisions 5–7, narrow, each further divided ..... ***P. quercifolium***
      5' Lf divided < 1/2 to midvein; divisions gen 3, wide, entire ..... ***P. vitifolium***
    4' Infl open, pedicels gen > 6 mm
      6. Per; upper lvs opposite; st slender, angular ..... ***P. grossularioides***
      6' Shrub or subshrub; upper lvs alternate; st thick, round
        7. Lf blade pinnately divided ± or > 1/2 to midvein, ovate in outline ..... ***P. panduriforme***
        7' Lf blade irregularly divided < 1/2 to midvein, ± round in outline
          8. Petals gen scarlet, widely obovate, length ± 2 × width ..... ***P. inquinans***
          8' Petals gen pale rose-pink, red-striped, narrowly obovate, length ± 5 × width ..... ***P. zonale***

P. ×domesticum L. Bailey REGAL PELARGONIUM Subshrub. **ST** erect, < 1 m, soft-hairy. **LF**: blade ± 10 cm, 12 cm wide, obscurely angled or lobed, margin toothed. **INFL** dense to open; pedicels > 5 mm. **FL**: sepals 1–2 cm; petals gen red, purple, or pink, overlapping, upper 1.5–4.5 cm, 1.5–3.5 cm wide, dark-blotched and -veined. Disturbed sites; < 300 m. CCo, SCo. Hybrid origin complex, incompletely known.

P. grossularioides (L.) L'Hér. Per, sparsely short-hairy, glandular. **ST** ± erect, 2–5 dm. **LF**: blade < 6 cm wide, ± round to broadly ovate, shallowly lobed, margin coarsely toothed. **INFL** open; fls 3–50; pedicels 5–10 mm. **FL**: sepals < 5 mm; petals < 6 mm, pink to rose-purple. Disturbed sites: < 300 m. NCo, SnFrB, CCo, SCo; native to s Afr.

P. ×hortorum L. Bailey Subshrub, soft-woody. **ST** 3–6 dm, branched, glabrous or ± hairy. **LF**: blade 15–20 cm wide, round to reniform, margin crenate, gen with a color band or variegated. **INFL** open to dense; fls 20–40; pedicels < 20 mm. **FL**: petals ± equal, 1–2.5 cm, white to red. Disturbed sites; < 300 m. SCo, ChI. Hybrid between *P. inquinans, P. zonale.*

1 cm
5 mm
corolla lobe
1 cm
5 mm
corolla lobe
2 mm
2 cm
2 mm
corolla lobes with nectary pits
S. albicaulis
2 mm
S. fastigiata
2 mm
2 mm
corolla lobes with nectary pits
S. parryi
2 mm
S. perennis
2 cm
Gentianopsis holopetala
G. simplex
Swertia albomarginata
Swertia neglecta
2 mm
corolla lobes with nectary pits
2 mm
S. puberulenta
2 mm
S. tubulosa
2 cm
Swertia radiata
E. botrys
1 cm
fruit
1 cm
E. brachycarpum
Erodium botrys
E. cicutarium
1 cm
Erodium cygnorum
1 cm
E. macrophyllum
1 cm
1 cm
5 cm
E. malacoides
E. moschatum
E. texanum
Geraniaceae
5 mm
G. carolinianum
1 mm
carpel body
Geranium carolinianum
5 cm
5 mm
Geranium bicknellii
5 mm
G. californicum
Geranium carolinianum
1 mm
carpel body
5 mm
2 cm
Geranium dissectum
1 cm
1 mm
carpel body
Geranium molle
1 cm
Geranium oreganum
1 mm
carpel body
Geranium anemonifolium
1 cm
fruit
2 cm
Geranium richardsonii
1 mm
seed
Geranium retrorsum
2 cm
1 mm
carpel body
Geranium robertianum

P. inquinans (L.) L'Hér. Subshrub, soft-woody. **ST** 1–2 m, branched, soft-hairy. **LF**: blade gen 4–8 cm wide, round-cordate, gen irregularly 5–7-lobed, margins crenate, soft-hairy, glandular. **INFL** open; fls 5–30. **FL**: sepals 5–7 mm; petals ± equal, 15–20 mm, broadly obovate, (white to) gen scarlet. Disturbed sites; < 300 m. SnFrB; native to s Afr.

P. panduriforme Ecklon & Zeyher OAK-LEAFED GERANIUM Shrub. **ST** erect, > 1 m, branched. **LF**: blade < 12 cm, < 10 cm wide, gen pinnately divided to near midrib; divisions 3–7, unequal, entire; margins serrate. **INFL** open; fls 3–10; pedicels 10–20 mm. **FL**: sepals 8–14 mm; corolla bilateral, petals 20–35 mm, pink, upper with dark marks. Disturbed sites; < 300 m. CCo, SCo; native to s Afr. [*P. quercifolium* (L.) L'Hér. misapplied]

P. peltatum (L.) L'Hér. IVY GERANIUM Subshrub. **ST** prostrate or climbing, < 2 m. **LF** peltate, ± fleshy; blade 2–7 cm, glabrous. **INFL** open; fls 2–9; pedicels 3–20 mm. **FL**: sepals 10–15 mm; corolla strongly bilateral, petals 15–20 mm, pink or rose, upper dark-blotched or -striped. Disturbed sites; < 300 m. CCo, SCo; native to s Afr.

P. quercifolium (L.) L'Hér. Shrub. **ST** erect, < 1 m, much branched. **LF** ± pinnately divided > 1/2 to midvein; divisions 5–7, ± pinnately lobed; lobes coarsely serrate. **INFL** dense; fls 5–10; pedicels 1–2 mm. **FL**: sepals ± 10 mm; corolla bilateral, petals 20–25 mm, gen rose or pink, upper with dark markings. Disturbed sites; < 300 m. CCo, SCo; native to s Afr. [*P. graveolens* L'Hér. misapplied]

P. vitifolium (L.) L'Hér. Shrub. **ST** erect, < 1 m; distal branches soft-woody, soft-hairy. **LF**: blade ± 6 cm, ± 8 cm wide, 3–5-lobed, cordate, margin coarsely, irregularly toothed, stiff-glandular-hairy. **INFL** dense; fls < 15; pedicels < 3 mm. **FL**: sepals 10–12 mm; petals ± 12 mm, light to rose-pink, striped violet-purple. Disturbed sites; < 300 m. NCo, SnFrB; native to s Afr.

P. zonale (L.) L'Hér. Subshrub. **ST** spreading or erect, gen > 1 m; distal branches soft-woody. **LF**: blade 5–8 cm wide, 3–5-lobed, round, cordate, margin irregularly dentate, often with a dark horseshoe band on upper surface, ± glabrous. **INFL** open; fls 5–70; pedicels > 4 mm. **FL**: sepals < 10 mm; corolla strongly bilateral, petals < 20 mm, length gen 5 × width, narrowly obovate, pink to red, red-striped. Disturbed sites; < 300 m. CCo, SnFrB, SCo; native to s Afr.

GROSSULARIACEAE GOOSEBERRY FAMILY

Michael R. Mesler & John O. Sawyer, Jr.

Shrub gen < 2 m. **ST** gen erect; nodal spines 0–9; internodal bristles gen 0; twigs gen hairy, gen glandular. **LVS** simple, alternate, gen clustered on short, lateral branchlets, petioled, gen deciduous; blade gen palmately 3–5-lobed, gen thin, gen dentate or serrate, base gen cordate. **INFL**: raceme, axillary, gen pendent, 1–25-fld; pedicel gen not jointed to ovary, gen hairy or glandular; bract gen green. **FL** bisexual, radial; hypanthium tube exceeding ovary; sepals gen 5, gen spreading; petals gen 5, gen < sepals, gen flat; stamens gen 5, alternate petals, gen inserted at level of petals (hypanthium top), anthers gen free, gen glabrous, tips gen rounded; ovary inferior, chamber 1, ovules many, styles gen 2, gen fused exc at tip, gen glabrous. **FR**: berry. 1 genus, 120 spp.: n hemisphere, temp S.Am. Some cult as food, orn. Hypanthium data refer to part above ovary; statements about ovary hairs actually refer to the hypanthium around the ovary. Formerly incl in Saxifragaceae.

RIBES CURRANT, GOOSEBERRY

The only genus. (Arabic: for pls of this genus)

1. Nodal spines 0 (subg. *Coreosma*)
 2. Hypanthium disk- or saucer-shaped, barely exceeding ovary
  3. Lvs evergreen, lobes 0 or very shallow; PR, s ChI (Santa Catalina Island) .... ***R. viburnifolium***
  3' Lvs deciduous, lobes deep; mainland
   4. Ovary and lf blade lower side with stalked glands; st spreading or decumbent .... ***R. laxiflorum***
   4' Ovary and lf blade lower side with sessile glands; st ± erect
    5. Sepals green; NCo, w KR, NCoRO .... ***R. bracteosum***
    5' Sepals white; n MP .... ***R. hudsonianum* var. *petiolare***
 2' Hypanthium cup- to tube-shaped, clearly exceeding ovary
  6. Sepals yellow .... ***R. aureum***
   7. Fl odor spicy; petals yellow turning orange; sepals 5–8 mm .... var. ***aureum***
   7' Fl odor 0; petals yellow turning deep red; sepals 3–4 mm .... var. ***gracillimum***
  6' Sepals white or whitish green to pink, red, or purple
   8. Anther tip with cup-like depression
    9. Hypanthium < 2 × longer than wide; stamens inserted at level of petals; fr black to glaucous .... ***R. viscosissimum***
    9' Hypanthium > 2 × longer than wide; stamens inserted below level of petals; fr red .... ***R. cereum***
     10. Bract tip wide, with several prominent teeth; styles gen hairy .... var. ***cereum***
     10' Bract tip acute, with 1–3 shallow teeth on each side; styles glabrous .... var. ***inebrians***
   8' Anther tip rounded, blunt, or shallowly notched
    11. Styles glabrous at base
     12. Sepals erect; hypanthium ± as long as wide .... ***R. nevadense***
     12' Sepals spreading; hypanthium longer than wide .... ***R. sanguineum***
      13. Infl pendent; sepals pink to white .... var. ***glutinosum***
      13' Infl erect to spreading; sepals red .... var. ***sanguineum***
    11' Styles hairy, at least at base
     14. Hypanthium wider than long; styles free to base; infl dense, many-fld, spike- or head-like .... ***R. canthariforme***

14′ Hypanthium ± longer than wide; styles fused at least in basal 1/2; infl open
15. Hypanthium white, barely longer than wide .... ***R. indecorum***
15′ Hypanthium pink, ± 2 × longer than wide .... ***R. malvaceum***
16. Lf blade upper surface dull olive-green .... var. ***malvaceum***
16′ Lf blade upper surface bright green .... var. ***viridifolium***
1′ Nodal spines present (sometimes 0 on some shoots)
17. Infl gen > 5-fld; hypanthium disk- or saucer-shaped; pedicel jointed to ovary (subg. *Grossularioides*)
18. Lf hairs 0 or sparse, nonglandular; fr black .... ***R. lacustre***
18′ Lf hairy, glandular; fr orange-red .... ***R. montigenum***
17′ Infl < 5-fld; hypanthium cup- to tube-shaped; pedicel not jointed to ovary (subg. *Grossularia*)
19. Sepals 4, erect .... ***R. speciosum***
19′ Sepals 5, spreading or reflexed
20. Anthers not exserted from petals
21. St low, spreading; internode bristles glandular .... ***R. tularense***
21′ St erect or arched; internode bristles 0
22. Ovary hairs conspicuous, short and long, glandular and not .... ***R. velutinum***
22′ Ovary hairs gen 0 or inconspicuous, gen short, gen nonglandular
23. Hypanthium > 3 mm; 2100–3100 m .... ***R. lasianthum***
23′ Hypanthium < 3 mm; < 1350 m .... ***R. quercetorum***
20′ Anthers exserted from petals
24. Styles hairy at least at base
25. Calyx purple or purple-green; filaments exceeding petals by < 2 mm .... ***R. divaricatum***
26. Petals pink or red; styles 8–11 mm .... var. ***parishii***
26′ Petals white; styles 5–8 mm .... var. ***pubiflorum***
25′ Calyx greenish white, purple at base or not; filaments exceeding petals by > 3 mm .... ***R. inerme***
27. Lf hairs 0 to sparse, short, soft; sepal hairs 0 .... var. ***inerme***
27′ Lf hairs sparse to dense, long, soft to stiff; sepal hairs soft .... var. ***klamathense***
24′ Styles glabrous at base
28. Petals bright yellow, outer surface deeply concave, tip hooded .... ***R. marshallii***
28′ Petals white to pink, flat or outer surface shallowly concave or margins curled inward, sometimes nearly touching, tip not hooded
29. Petals flat or outer surface shallowly concave; st low, spreading .... ***R. binominatum***
29′ Petal margins curled inward, sometimes nearly touching; st ± erect
30. Anthers oblong, tips blunt or rounded; styles ± not exceeding anthers
31. Anthers slightly longer than wide when dehisced, with sessile glands on back; internodal bristles 0 .... ***R. lobbii***
31′ Anthers much longer than wide when dehisced, glabrous; internodal bristles present .... ***R. sericeum***
30′ Anthers lanceolate to ovate, tips acute or with a point; styles exceeding anthers
32. Lf blade lower surface glandular
33. Internodal bristles 0; hypanthium longer than wide .... ***R. amarum***
33′ Internodal bristles present; hypanthium gen ± as long as wide
34. Sepals purple; fr purple .... ***R. menziesii***
34′ Sepals greenish white; fr yellow .... ***R. victoris***
32′ Lf blade lower surface nonglandular
35. Hypanthium ± as long as wide .... ***R. californicum***
36. Lf hairs 0 or sparse; stamens > 2 × petals .... var. ***californicum***
36′ Lf hairy; stamens ± = to barely > petals .... var. ***hesperium***
35′ Hypanthium longer than wide
37. Sepals pinkish, spreading; n ChI (Santa Cruz Island) .... ***R. thacherianum***
37′ Sepals purple, reflexed; mainland .... ***R. roezlii***
38. Lf blade lower surface and hypanthium glabrous .... var. ***cruentum***
38′ Lf blade lower surface and hypanthium hairy or densely white-hairy
39. Hypanthium and lf blade lower surface densely white-hairy .... var. ***amictum***
39′ Hypanthium and lf blade lower surface hairy .... var. ***roezlii***

R. amarum McClatchie (p. 681) BITTER GOOSEBERRY **ST**: nodal spines 3. **LF**: blade 20–40 mm, hairy, glandular. **INFL** 1–3-fld. **FL**: hypanthium 5–6 mm, longer than wide; sepals reflexed, 2–4 mm, purple; petals 2 mm, white, margins curled inward; anthers exserted from petals, exceeded by styles, tips with a short, sharp, flexible point. **FR** 15–20 mm, purple; bristles stiff, glandular and not. Chaparral; < 1600 m. SNF, Teh, SnFrB, SCoRO, SW. Pls with hairy fr from Santa Barbara Co. have been called var. *hoffmannii* Munz. ❀DRN:4–6,**17**&SHD or IRR:7,8,9,**14–16,18–24**.

R. aureum Pursh GOLDEN CURRANT Shrub < 3 m. **ST**: nodal spines 0; internodes glabrous or puberulent. **LF**: blade firm, 15–50 mm, toothed or not, light green, gen glandular when young, gen glabrous when mature, base wedge-shaped to subcordate. **INFL** 5–15-fld. **FL**: hypanthium 6–10 mm, longer than wide; sepals 3–8 mm, yellow; petals 2–3 mm; styles fused base to tip. **FR** 6–8 mm, red, orange, or black, glabrous. 2*n*=16. Many habitats; < 3000 m. KR, NCoRI, CaR, SNH, SnJV, SnFrB, SCoR, SW, GB; to B.C., SD, NM.

var. ***aureum*** (p. 681) **FL**: odor spicy; hypanthium 1.5–2 × sepals; sepals 5–8 mm; petals yellow turning orange. Many habitats; < 3000 m. KR, CaR, SNH, GB; to B.C., SD, NM. ❀4,5,**6**,17; IRR:1–3,**7–10**,11,12,**14–16,18–24**.

var. ***gracillimum*** (Cov. & Britton) Jepson **FL**: odor 0; hypanthium 2–3 × sepals; sepals 3–4 mm; petals yellow turning deep red. Alluvial areas, forest edges; < 3000 m. NCoRI, SnFrB, SCoR, SW. ❀**6, 17**;IRR or SHD:**7–9**,10,**14–16,18–24**.

R. binominatum A.A. Heller (p. 681) TRAILING GOOSEBERRY **ST** low, spreading, rooting; nodal spines 3; internodes not bristly. **LF**: blade 20–50 mm, hairy, glandular. **INFL** 1–4-fld. **FL**: hypanthium 2–4 mm, ± as long as wide; sepals reflexed, 4–6 mm, greenish white to green with red margins; petals 2–3 mm, white to pink, flat or outer surface shallowly concave; anthers exserted from petals. **FR** 8–10 mm, yellow-green; prickles stout, nonglandular, yellow, hairs glandular or not. Montane, subalpine forests, meadows; 1000–2600 m. KR, NCoRH; OR. ❀4,5,**6**;IRR:1–3,**7**,14–17;DFCLT.

R. bracteosum Douglas (p. 681) STINK CURRANT Shrub < 4 m. **ST**: nodal spines 0; internodes sparsely hairy. **LF**: blade deeply 5–7-lobed, 4–20 cm, upper surface shiny, glabrous, lower surface dull, hairs sparse, glands sessile, yellow. **INFL** erect, 20–50-fld; lower bracts lf-like. **FL**: hypanthium 3–4 mm, saucer-shaped; sepals 3–5 mm, green; petals < 1 mm, white. **FR** 8–10 mm, black-glaucous; glands sessile. $2n=16$. Moist forests; < 1400 m. NCo, w KR, NCoRO; to AK. ❀IRR:**4–6**&SHD:1,2,**7**,14,**15–17**.

R. californicum Hook. & Arn. HILLSIDE GOOSEBERRY **ST**: nodal spines 3; internodes gen glabrous. **LF**: blade 10–30 mm, lower surface nonglandular. **INFL** 1–3-fld. **FL**: hypanthium 2 mm, ± as long as wide; sepals reflexed, 6–8 mm, green to red; petals 3 mm, white, margins curled inward; anthers exserted from petals, exceeded by styles, tips with a short, sharp, flexible point. **FR** 9–10 mm, red; shorter bristles glandular. Forest openings, chaparral, woodlands; < 1000 m. NCoR, SnFrB, SCoR, WTR, SnGb, w PR.

var. ***californicum*** **LF**: hairs 0 or sparse. **FL**: sepals green or red-tinged; stamens > 2 × petals. Forest openings, woodlands; < 300 m. NCoR, SnFrB, SCoR. ❀DRN:**17**&SHD:**7**,8,9,**14–16**, 19–21,**22–24**.

var. ***hesperium*** (McClatchie) Jepson **LF** hairy. **FL**: sepals red; stamens ± = to barely > petals. Chaparral, woodlands; < 1000 m. WTR, SnGb, w PR (Santa Ana Mtns). ❀DRN:6,**15–17**&SHD:**7**,**14**,**18**,19–24.

R. canthariforme Wiggins (p. 681) MORENO CURRANT **ST**: nodal spines 0. **LF**: blade 40–60 mm, thick, crenate, upper surface green, hairs long, soft, wavy, lower surface gray-green, hairs dense. **INFL** many-fld, dense, spike- or head-like. **FL**: hypanthium 1 mm, wider than long; sepals 2 mm, spoon-shaped, purple with darker veins; petals 1 mm, purple; styles free to sparsely hairy base. **FR** 5–6 mm, purple; hairs long, soft, wavy or glandular, 0 in age. RARE. Chaparral; 500–1200 m. sw PR (near Moreno Dam, sw San Diego Co.). In cult.

R. cereum Douglas WAX CURRANT **ST**: nodal spines 0. **LF**: odor spicy; blade 10–40 mm, round, shallowly lobed, finely toothed, upper surface glossy. **INFL** 3–7-fld. **FL**: hypanthium 6–8 mm, > 2 × longer than wide; sepals 1–2 mm, white to pink; petals < 1 mm, white to pink; stamens inserted below level of petals, anther tips with cup-like depression; styles fused ± to tip. **FR** 10–12 mm, red, glabrous to sparsely glandular. $2n=16$. Dry montane to alpine slopes, among rocks, forest edges; 1500–4000 m. KR, CaRH, SNH, Teh, TR, SnJt, GB, DMtns; to B.C., c US, AZ.

var. ***cereum*** (p. 681) **LF**: hairs 0 to dense, glandular. **INFL**: bract tip wide, with several prominent teeth. **FL**: styles gen hairy. Habitats of sp. KR, CaRH, SNH, Teh, TR, SnJt, GB, DMtns; to B.C. ❀DRN:4,5,**6**&IRR:1–3,**7**,14–18.

var. ***inebrians*** (Lindley) C. Hitchc. **LF**: hairs dense, nonglandular. **INFL**: bract tip acute, with 1–3 shallow teeth on each side. **FL**: styles glabrous. Open rocky areas; 2100–4000 m. s SNH, W&I; to ID, NE, NV, AZ. [*R. i.* Lindley] ❀DRN&IRR:1,2.

R. divaricatum Douglas Shrub < 3.5 m. **ST** arched; nodal spines 0–3; internodes bristly or not. **LF**: blade 20–60 mm, coarsely toothed. **INFL** < 5-fld. **FL**: hypanthium 1–4 mm; sepals reflexed, 5–7 mm, obovate, green, purple, or purple-green; petals 1–3 mm, white, pink, or red; filaments exceeding petals by < 2 mm; styles hairy at base. **FR** 6–10 mm, black, glabrous. $2n=16$. Coastal bluffs, forest edges, steamsides; < 650 m. NCo, KR, NCoRO, CaR, CW, SCo; to B.C. [Sinnott 1985 Rhodora 87:189–286] Var. *divaricatum* ranges from OR to B.C.

var. ***parishii*** (A.A. Heller) Jepson PARISH'S GOOSEBERRY **FL**: hypanthium 3–4 mm; petals 2–3 mm, pink or red; filaments 3–5 mm; styles 8–11 mm. RARE. Moist woodlands; 100 m. e SCo.

var. ***pubiflorum*** Koehne (p. 681) STRAGGLY GOOSEBERRY **FL**: hypanthium 1–2 mm; petals 1–2 mm, white; filaments > 5 mm; styles 5–8 mm. UNCOMMON. Coastal buffs, forest edges; < 650 m. NW (exc NCoRI), CW (exc SCoRI); sw OR. ❀IRR:**5**,**17**&SHD:14,**15**, **16**,24.

R. hudsonianum A. Richards var. ***petiolare*** (Douglas) Jancz. WESTERN BLACK CURRANT **ST** ± erect; nodal spines 0; internodes sparsely strigose. **LF**: blade 3–10 cm, coarsely double-dentate, upper surface glabrous, lower surface with few soft, shaggy hairs and sessile, yellow glands. **INFL** erect, 20–50-fld. **FL**: hypanthium 4–5 mm, saucer-shaped; sepals spreading, 4–7 mm, white; petals 2 mm, white. **FR** 9–10 mm, black; glands sessile, yellow. RARE in CA. Streamsides; 1500 m. n MP; to B.C., ID, NV. [*R. p.* Douglas]

R. indecorum Eastw. WHITE FLOWERING CURRANT Shrub < 3 m. **ST**: nodal spines 0. **LF** gen deciduous; blade 10–40 mm, thick, crenate, upper surface green, hairy, rough, lower surface white, tomentose. **INFL** 10–25-fld, open; bracts pink. **FL**: hypanthium 4–5 mm, barely longer than wide, white; sepals 1–2 mm, white; petals 1 mm, white; styles fused ± to tip, hairy at least at base. **FR** 6–7 mm, purple, hairy; glands stalked, sticky. Chaparral, coastal-sage scrub; < 2000 m. SCoRO, SCo, WTR, PR; n Baja CA. ❀DRN,DRY:**17**, **24**&SHD:**7**,8,9,11,**14–16**,**18–23**.

R. inerme Rydb. WHITE-STEMMED GOOSEBERRY **ST** scrambling; nodal spines 0–3; internodes gen glabrous. **LF**: blade 20–30 mm, coarsely toothed. **INFL** 1–5-fld. **FL**: hypanthium 2–3 mm, ± as long as wide; sepals reflexed, 3–4 mm, greenish white, purple at base or not; petals 1–2 mm, white; filaments exceeding petals by > 3 mm. **FR** 7–11 mm, purple, glabrous. $2n=16$. Forests, streamsides, meadow edges; 1200–3300 m. KR, NCoRO, CaRH, SNH, Wrn, SNE; to B.C., Rocky Mtns. [Sinnott 1985 Rhodora 87:189–286]

var. ***inerme*** **LF**: hairs 0 to sparse, short, soft. **FL**: hypanthium and sepal hairs 0. Forests, streamsides, meadow edges; 1200–3300 m. KR, CaRH, SNH, Wrn, SNE; to B.C., Rocky Mtns. [*R. divaricatum* var. *i.* (Rydb.) McMinn] ❀IRR:4,5,**6**,15–17&SHD:1–3,**7**, 14,18.

var. ***klamathense*** (Cov. & Britton) Jepson **LF**: hairs sparse to dense, long, soft to stiff. **FL**: hypanthium hairs ± 0 to soft; sepal hairs soft. Conifer forest edges; 1200–1500 m. KR, NCoRO, CaRH; sw OR. [*R. divaricatum* var. *k.* (Cov. & Britton) McMinn] ❀IRR:4,5,**6**,15–17&SHD:1–3,**7**,14,18.

R. lacustre (Pers.) Poiret (p. 681) SWAMP CURRANT **ST** prostrate to ascending; nodal spines 3–9; internode bristly. **LF**: blade 3–5 cm, deeply 3–7-lobed, toothed, hairs 0 or sparse, nonglandular, upper surface dark green, lower surface light green. **INFL** 5–15-fld; pedicel jointed to ovary. **FL**: hypanthium 1 mm, saucer-shaped; sepals 1.5 mm, green or purple; petals 1 mm, purple; styles free to base. **FR** 4–6 mm, black; hairs glandular. $2n=16$. Along creeks, seeps, meadow margins; 1200–1800 m. KR; to AK, e N.Am. ❀WET:**4–6**,15–17&SHD:1–3,**7**,14,18.

R. lasianthum E. Greene **ST**: nodal spines 1–3. **LF**: blade 10–20 mm, toothed, hairy, glandular. **INFL** erect, 2–4-fld. **FL**: hypanthium > 3 mm, longer than wide; sepals 2 mm, yellow; petals 1 mm, yellow; anthers not exserted from petals; styles fused ± to tip. **FR** 6–7 mm, red, ± glabrous. Open, rocky areas; 2100–3100 m. SNH, SnGb; OR, NV. ❀DRN,IRR:1–3.

R. laxiflorum Pursh (p. 681) TRAILING BLACK CURRANT **ST** spreading or decumbent; nodal spines 0. **LF**: blade 5–10 cm, deeply 5–7-lobed, ± round-toothed, upper surface dark green, glabrous, lower surface light green, sparsely hairy, glandular. **INFL** 6–12-fld. **FL**: hypanthium 6 mm, saucer-shaped; sepals 3–4 mm, purplish green; petals 1 mm, red. **FR** 4–6 mm black-glaucous; bristles glandular. UNCOMMON. Forests; < 300 m. n NCo; to AK, ID, also Siberia. ❀IRR:**4**,**5**,**17**&SHD:1–3,**6**,7,15,16.

R. lobbii A. Gray (p. 681) GUMMY GOOSEBERRY **ST**: nodal spines 3. **LF**: blade 2–3 mm, upper surface subglabrous, lower surface hairy, glandular. **INFL** 1–3-fld. **FL**: hypanthium 3–5 mm, longer than wide; sepals reflexed, 10–12 mm, red; petals 5–6 mm, white, margins curled inward; anthers exserted from petals, with sessile glands on back; styles ± not exceeding anthers. **FR** 10–15 mm, oblong, red; bristles glandular, dense. Montane, subalpine forests; 1500–2000 m. KR, NCoRH; to B.C. ❀DRN,IRR:4–6& SHD:1–3,7,15,16;DFCLT.

R. malvaceum Sm. (J.C.) CHAPARRAL CURRANT **ST**: nodal spines 0. **LF**: blade 20–50 mm, double-toothed, densely hairy, glandular. **INFL** 10–25-fld, open. **FL**: hypanthium 5–8 mm, ± 2 ×

longer than wide, pink; sepals 4–6 mm, pink to purple; petals 2–3 mm, pink to white; styles fused ± to tip, base hairy. **FR** 6–7 mm, purple-glaucous; hairs white, glandular. Chaparral, oak woodland; < 1500 m. NCoRI, SNF, SnFrB, SCoR, ChI, TR, PR; n Baja CA.

var. ***malvaceum*** (p. 681) **LF**: upper blade surface dull olive-green. Chaparral, oak woodland, < 800 m. SNF, NCoRI, SnFrB, SCoR, ChI, WTR. [var. *clementinum* Dunkle] ❀DRN, summer-DRY:6,**7,14–17,22–24**&winterIRR:8,9,**18–21**;CVS.

var. ***viridifolium*** Abrams **LF**: upper blade surface bright green. Chaparral; < 1500 m. SCoRO, TR, PR; n Baja CA. ❀TRY.

R. marshallii E. Greene (p. 681) MARSHALL'S GOOSEBERRY **ST** arched, rooting at tip; nodal spines 3. **LF**: blade 25–35 mm, toothed, subglabrous. **INFL** 1–3-fld. **FL**: hypanthium 3–4 mm, ± as long as wide; sepals reflexed, 10–15 mm, purple; petals 5–6 mm, bright yellow, outer surface deeply concave, tip hooded; anthers exserted from petals. **FR** 10–20 mm, oblong, dark red; prickles nonglandular; hairs sparse, appressed. UNCOMMON. Montane forests; 1200–2100 m. KR; sw OR. ❀SHD&IRR:1,2,5,6,**7**.

R. menziesii Pursh (p. 681) CANYON GOOSEBERRY Shrub < 3 m. **ST**: nodal spines 3; internode bristles dense, at least some glandular. **LF**: blade 1.5–4 cm, toothed, hairy, glandular. **INFL** 1–3-fld. **FL**: hypanthium 2–3 mm, gen ± as long as wide; sepals reflexed, 5–10 mm, purple; petals 2–5 mm, white, margins curled inward; anthers exserted from petals; styles exceeding anthers. **FR** 8–10 mm, purple; bristles dense, stiff, at least some glandular. Forest openings, chaparral; < 300 m. NCo, NCoRO, s SNF, CCo, SnFrB, SCoRO; sw OR. Variable; described vars. indistinct. Pls from SCoRO with petals = filaments and fr bristles glandular and nonglandular have been called var. *hystrix* (Eastw.) Jepson; pls from s SNF with aromatic lvs have been called var. *ixoderme* Quick, aromatic canyon gooseberry; pls from s NCoRO & SnFrB with glandular fr have been called var. *leptosmum* (Cov.) Jepson; pls from sw SnFrB with lvs barely glandular beneath have been called var. *senile* (Cov.) Jepson. ❀DRN&SHD or IRR:5,**7,14–17,22–24**.

R. montigenum McClatchie (p. 681) MOUNTAIN GOOSEBERRY **ST** spreading or decumbent; nodal spines 1–5; internodes ± bristly or not. **LF**: blade 1.5–2.5 cm, deep-lobed to -serrate, hairy, glandular. **INFL** gen > 5-fld; pedicel jointed to ovary. **FL**: hypanthium 1 mm, saucer-shaped; sepals 3–4 mm, green to greenish white; petals 1 mm, red. **FR** 4–5 mm, orange-red; bristles glandular. Many subalpine, alpine habitats; 2100–4800 m. KR, CaRH, TR, SnJt, Wrn, n DMtns; to B.C., ID, NV, AZ. ❀DRN&IRR:1–3.

R. nevadense Kellogg (p. 681) MOUNTAIN PINK CURRANT **ST**: nodal spines 0. **LF**: blade 3–8 cm. **INFL** erect to pendent, 8–20-fld, dense; bract pink. **FL**: hypanthium 2 mm, ± as long as wide; sepals erect, 4–5 mm, pink to red; petals 2–3 mm, white. **FR** 6–8 mm, blue-black-glaucous; hairs glandular. Forest margins; 900–1300 m. KR, NCoRH, CaR, SNH, Teh, TR, PR, Wrn; s OR, w NV. [vars. *glaucescens* (Eastw.) A. Berger & *jeageri* A. Berger] ❀DRN:4–6&IRR:**7**,15–17&SHD:1–3,14,18.

R. quercetorum E. Greene (p. 681) OAK GOOSEBERRY **ST** arched; nodal spines 1(3); internodes puberulent. **LF**: blade 10–20 mm, dentate, gen hairy, glandular. **INFL** 2–3-fld. **FL**: hypanthium < 3 mm, longer than wide; sepals reflexed, 3 mm, yellow; petals 1 mm, cream; anthers not exserted from petals; styles fused ± to tip. **FR** 7–8 mm, black, ± glabrous. Oak woodlands, chaparral; < 1350 m. c&s SNF, Teh, SnFrB, SCoR, WTR, PR, w edge D; AZ, n Baja CA. ❀DRN:**7**,8,9,11,**14,18–24**&SUN:5,**15–17**.

R. roezlii Regel SIERRA GOOSEBERRY **ST**: nodal spines 1–3. **LF**: blade 12–25 mm, toothed. **INFL** 1–3-fld. **FL**: hypanthium 5–7 mm, longer than wide; sepals reflexed, 7–9 mm, purple; petals 3–4 mm, white, margins curled inward; anthers exserted from petals, tips with a short, sharp, flexible point; styles exceeding anthers. **FR** 14–16 mm, red; prickles stout, nonglandular; hairs glandular. Forests, chaparral, woodlands; < 2800 m. KR, NCoRO, NCoRH, CaRH, SNH, Teh, TR, PR, MP; OR.

var. ***amictum*** (E. Greene) Jepson **LF**: blade lower surfaces densely white-hairy. **FL**: hypanthia, sepals densely white-hairy. Mixed-evergreen forest; < 2300 m. c NCoRO, NCoRH. ❀DRN: 5,**17**&SHD:6&IRR:7,14–16.

var. ***cruentum*** (E. Greene) Rehder **LF**: blade lower surface glabrous. **FL**: hypanthia, sepals glabrous. Forests, woodlands 150–2300 m. KR, NCoRO, NCoRH; OR. ❀DRN:5,17&SHD: **6**&IRR:**7**,14–16.

var. ***roezlii*** (p. 685) **LF**: blade lower surfaces hairy. **FL**: hypanthia, sepals hairy. Forests, chaparral, woodlands; 1000–2800 m. KR, CaRH, SNH, Teh, TR, PR, MP. ❀DRN:6&SHD:15,16&IRR:1–3,**7**,14,18–21.

R. sanguineum Pursh RED FLOWERING CURRANT Shrub < 4 m. **ST**: nodal spines 0. **LF** thin to moderately thick; blade 2–7 cm, irregularly toothed and finely serrate, upper surface puberulent, lower surface sparsely hairy to whitish tomentose. **INFL** 10–20-fld; bracts white to red. **FL**: hypanthium 2–7 mm, longer than wide; sepals 4–5 mm, white, pink, or red; petals 2–3 mm, white to red; styles fused ± to tip. **FR** 4–8 mm, blue-black-glaucous; hairs glandular. $2n=16$. Many habitats; < 2200 m. NW (exc NCoRI), CW (exc SCoRI); to B.C.

var. ***glutinosum*** (Benth.) Loudon (p. 685) **LF**: blade lower surface sparsely hairy. **INFL** pendent. **FL**: sepals pink to white. Many habitats; < 1000 m. NCo, NCoRO, CW (exc SCoRI); OR. [vars. *deductum* (E. Greene) Jepson and *melanocarpum* (E. Greene) Jepson] ❀DRN:4-6,**17**&IRR:**15,16**&SHD:**7**,8,9,**14**,18,**19-24**;CVS.

var. ***sanguineum*** **LF**: blade lower surface whitish hairy to finely tomentose. **INFL** erect to spreading. **FL**: sepals red. Montane forests; 300–2200 m. KR, NCoRH; to B.C. ❀DRN:**4,5**&IRR:**6**,17&SHD:1–3,**7**,14–16,18;CVS.

R. sericeum Eastw. SANTA LUCIA GOOSEBERRY **ST**: nodal spines 3; internodes densely hairy, bristles gen glandular. **LF**: blade 2–4 cm, toothed, upper surface subglabrous, lower surface hairy, glandular. **INFL** 1–3-fld. **FL**: hypanthium 3–4 mm, ± as long as wide; sepals reflexed, 6–8 mm, green or red; petals 3–4 mm, white, margins curled inward; anthers exserted from petals, much longer than wide when dehisced; styles ± not exceeding anthers. **FR** 10–20 mm, purple; bristles glandular. UNCOMMON. Forests, coastal scrub; < 800 m. SCoRO. ❀DRN:5,**17**&SHD:7,14–16.

R. speciosum Pursh (p. 685) FUCHSIA-FLOWERED GOOSEBERRY **ST**: nodal spines 3; internodes ± bristly. **LF** semi-deciduous; blade 1–3.5 cm, leathery, irregularly toothed, gen glabrous, upper surface shiny, dark green, lower surface light green. **INFL** 1–4-fld; pedicels bristly. **FL**: hypanthium 2–3 mm, wider than long; sepals 4, erect, 4–5 mm, red; petals 4, 4–5 mm, red, margins curled inward. **FR** 10–12 mm, bristles glandular, dense. Coastal-sage scrub, chaparral; < 500 m. CCo, SCoRO, SCo, WTR, PR; n Baja CA. ❀DRN:**17,24**&SHD or IRR:7–9,**14–16**,18,**19–23**.

R. thacherianum (Jepson) Munz SANTA CRUZ ISLAND GOOSEBERRY Shrub < 3 m. **ST**: nodal spines 0–3; internodes hairy, bristly. **LF**: blade 2–3 cm, shallowly lobed, toothed, upper surface sub-glabrous, lower surface hairy. **INFL** 1–2-fld; pedicels softly white-hairy. **FL**: hypanthium 4–5 mm, longer than wide; sepals spreading, 9–10 mm, pinkish; petals 6 mm, white, margins curled inward; anthers exserted from petals, tips with a short, sharp, flexible point; styles exceeding anthers. **FR** 7–10 mm, purple; bristles dense, hairs soft. RARE. Ravines; < 300 m. n ChI (Santa Cruz Island). In cult.

R. tularense (Cov.) Fedde SEQUOIA GOOSEBERRY **ST** low, spreading; nodal spines 3; internode hairs long, bristles glandular. **LF**: blade 2–5 cm, toothed, hairy, glandular. **INFL** 1–3-fld. **FL**: hypanthium 2–4 mm, ± as long as wide; sepals reflexed, 6 mm, green to white; petals 2–3 mm, white, outer surface shallowly concave; anthers not exserted from petals. **FR** 8–10 mm, light yellow, prickles nonglandular; bristles glandular. RARE. Montane forests; 1500–1800 m. s SNH (Tulare Co.). Closely related to *R. binominatum*.

R. velutinum E. Greene (p. 685) **ST** stout, arched; nodal spines 1(3). **LF**: blade 5–20 mm, crenate. **INFL** 1–4-fld. **FL**: hypanthium 2–3 mm, ± as long as wide; sepals 3 mm, white to yellow; petals 2 mm, white to yellow; anthers not exserted from petals; ovary hairs conspicuous, short and long, glandular and not. **FR** 6–7 mm, yellow becoming purple. Sagebrush steppe, juniper woodland, pine forest; 700–2500 m. KR, CaRH, SNH, Teh, TR, GB, DMtns; to UT, AZ. [var. *glanduliferum* (A.A. Heller) Jepson] ❀DRN,DRY: 1–3,**7**,10.

R. viburnifolium A. Gray (p. 685) SANTA CATALINA ISLAND or EVERGREEN CURRANT **ST** erect or arched; nodal spines 0; internode glands white, sessile. **LF** evergreen; blade 2–4 cm, (ob)ovate, leathery, lobes 0 or sparse, very shallow, teeth 0 or sparse, upper surface dark green, glabrous, lower surface with sessile, yellow glands, base acute, tip wide. **INFL** erect, branched at base, 6–15-fld. **FL**: hypanthium 4–5 mm, saucer-shaped; sepals 2–3 mm, brown; petals 2 mm, red. **FR** 5–6 mm, red, glabrous. $2n=16$. UNCOMMON. Chaparral; 30–300 m. PR (sw San Diego Co.), s ChI (Santa Catalina Island); Baja CA. ❀DRN:5,**17**&IRR:**24**&SHD: 7–9,**14–16**,**19–23**.

R. victoris E. Greene VICTOR'S GOOSEBERRY **ST**: nodal spines 1–3; internodes puberulent, sticky, sparsely bristly. **LF**: blade 1.5–5 cm, toothed, sparsely glandular. **INFL** 1–2-fld. **FL**: hypanthium 3 mm, ± as long as wide; sepals reflexed, 6–10 mm, greenish white; petals 3–5 mm, white, margins curled inward; anthers exserted from petals, tips with a short, sharp, flexible point; styles exceeding anthers. **FR** 8–10 mm, yellow; bristles ± glandular. UNCOMMON. Canyon forests, chaparral; < 300 m. NCoRO, SnFrB. [var. *minus* Jancz.] ❀DRN:5,**17**&SHD:14,**15**,**16**,22–24.

R. viscosissimum Pursh (p. 685) STICKY CURRANT **ST**: nodal spines 0. **LF** fragrant; blade 30–80 mm, round, thick, crenate, gray-green, glandular. **INFL** 4–15-fld. **FL**: hypanthium 5 mm, longer than wide by < 2 ×; sepals 6–7 mm, whitish green to pink; petals 3 mm; anther tip with cup-like depression; styles fused ± to tip. **FR** 10–12 mm, black to glaucous. Sagebrush, forests; 1200–3000 m. KR, NCoRH, CaRH, n&c SNH, MP; to B.C., MT, Colorado. CA pls with purplish sepals, glabrous fr have been called var. *hallii* Jancz. ❀DRN:4–6&SHD,IRR:1–3,**7**,15,16.

GUNNERACEAE GUNNERA FAMILY

Elizabeth McClintock

Per, often large, terrestrial or semi-aquatic, rhizomed, monoecious or dioecious. **ST** ± 0. **LVS** simple, gen alternate, from rhizome, often large; stipules 0 or scale-like; blades < to > petioles, ± round to reniform or ovate, gen toothed or lobed. **INFL**: spike or panicle of spikes, terminal or from upper axils, large; fls many, often pistillate below, staminate above, bisexual between. **FL** very small; sepals gen 2 or ± 0; petals 2 or 0; stamens 1–2; ovary inferior, chamber 1, styles 2. **FR**: drupe. 1 genus, ± 35 spp.: Mex, s hemisphere; several cult. Sometimes incl in Haloragaceae.

GUNNERA

(J.E. Gunner, Norwegian bishop, botanist, 1718–1773) [Osborne et al. 1991 Bioscience 41:224–234] Blue-green algae sometimes in live rhizome, or roots, fix nitrogen for host pl.

G. tinctoria (Molina) Mirbel **LF** covered with stiff, fleshy prickles; blade 1–2 m, ± round, thick, rough, palmately lobed, irregularly toothed; main veins palmate, prominent, esp on lower side; petiole 1–1.5 m. **INFL** 50–75 cm, < 10 cm wide; branches lateral, many, dense, 2–5 cm, stout. **FL**: petals 2. **FRS** many, red, conspicuous. Uncommon. Disturbed, shaded, damp areas; < 100 m. CCo (Marin, San Francisco cos.); native to Chile. [*G. chilensis* Lam.]

HALORAGACEAE WATER-MILFOIL FAMILY

Elizabeth McClintock

Ann, per, shrub, gen monoecious, gen aquatic. **LVS** cauline, opposite, alternate, or whorled; submersed blades with pinnate, thread-like divisions; aerial lvs simple, entire to divided. **INFL**: panicle, raceme, or spike; fls 1 or clustered, short-pedicelled to ± sessile. **FL** gen unisexual (bisexual in *Haloragis*), small, biradial; calyx tube short, fused to ovary, lobes 2–4; petals gen 2–4; stamens 4 or 8, filaments gen short; ovary inferior, chambers 1–4, styles 2–4, separate, stigmas gen plumose. **FR** fleshy or nut-like, dehiscent or not. **SEEDS** gen 1 per chamber. 6–8 genera, ± 100 spp.: esp s hemisphere, some cult.

1. Pl terrestrial; lvs opposite, simple, coarsely toothed .... **HALORAGIS**
1′ Pl aquatic; submersed lvs gen whorled, pinnately divided, divisions linear, thread-like; emergent lvs simple or divided .... **MYRIOPHYLLUM**

HALORAGIS

Ann, per, small shrub, terrestrial. **ST** erect. **LVS** simple, alternate or opposite; blade lanceolate to round, entire, toothed, or lobed. **INFL**: raceme or panicle, axillary; fl 1 or clustered. **FL** bisexual; calyx lobes 2–4, persistent in fr; petals 2–4; stamens 4 or 8, deciduous, filaments short; ovary 1–4-chambered, stigmas plumose. **FR** indehiscent, winged or ribbed. ± 70 spp.: s hemisphere. (Greek: wetland pl with clustered frs) [Forde 1964 New Zealand J Bot 2:425–453]

H. erecta (Murray) Eichler Per, small shrub. **STS** many, < 1 m, 4-angled. **LVS** opposite; lower petioled; upper subsessile; blade gen 2.5–5 cm, lanceolate, toothed. **INFL**: fls clustered, bracted; pedicel short. **FL**: calyx lobes 4; petals 4, deciduous; stamens 8, anthers > filaments; ovary 4-angled, 4-chambered. **FR** 4-winged. **SEED** 1 per chamber. Uncommon. Disturbed urban sites; < 100 m. SnFrB; native to New Zealand. Cult.

MYRIOPHYLLUM WATER-MILFOIL

Per from rhizomes, sometimes with overwintering bulblets, gen monoecious. **STS** simple or branched, gen open, gen green. **LVS**: submersed lvs whorled, 3–6 per node, pinnate divisions thread-like; emergent lvs opposite, lf- or bract-like, entire to pinnately divided. **INFL** spike-like, terminal, or fls clustered, axillary, gen emergent. **FL** unisexual; lower pistillate; middle

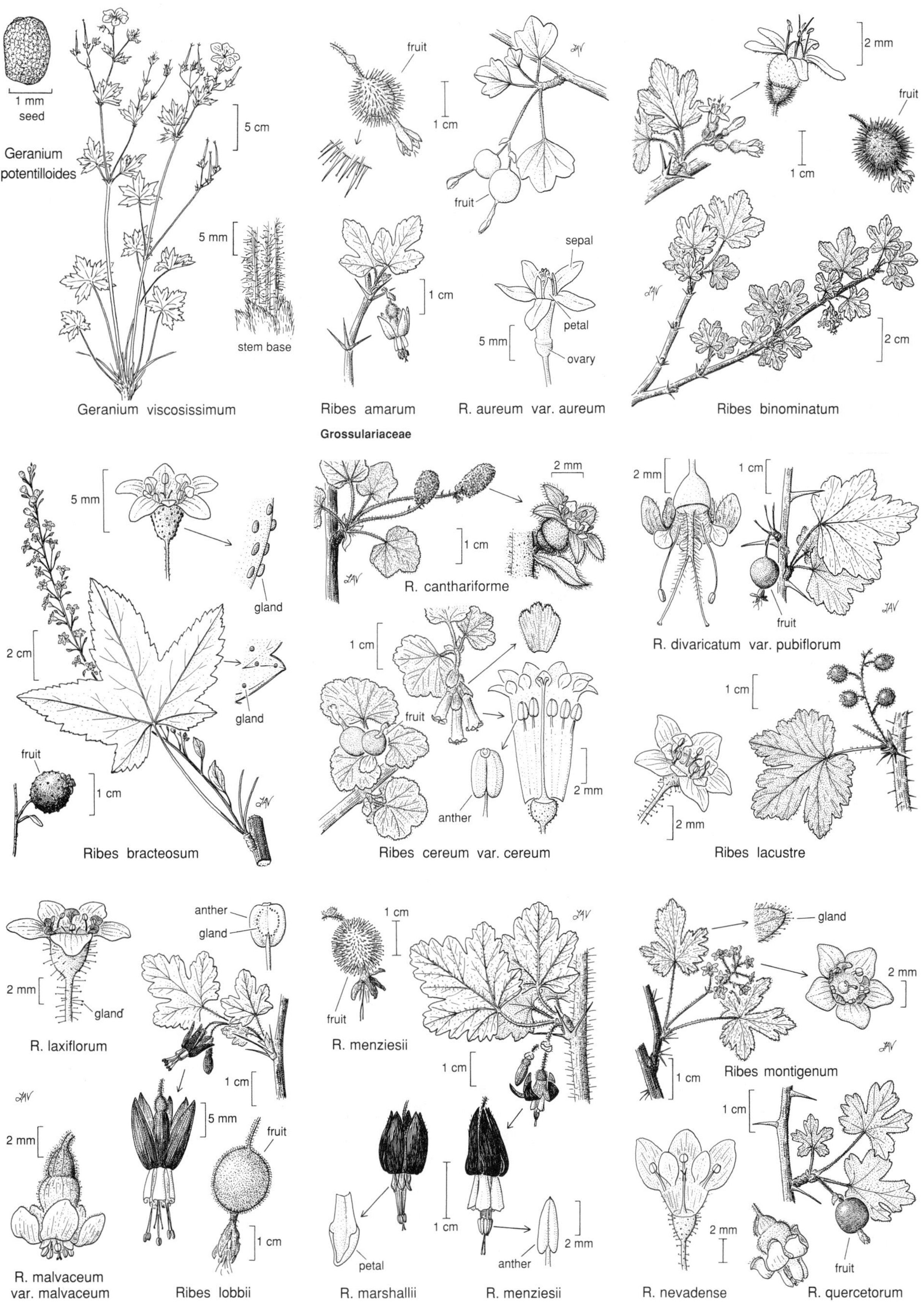
1 mm
seed
Geranium
potentilloides
5 cm
5 mm
stem base
Geranium viscosissimum
fruit
1 cm
Ribes amarum
fruit
sepal
petal
5 mm
ovary
R. aureum var. aureum
2 mm
fruit
1 cm
2 cm
Ribes binominatum
Grossulariaceae
5 mm
gland
2 cm
gland
fruit
1 cm
Ribes bracteosum
2 mm
1 cm
R. canthariforme
1 cm
fruit
anther
2 mm
Ribes cereum var. cereum
2 mm
1 cm
fruit
R. divaricatum var. pubiflorum
1 cm
2 mm
Ribes lacustre
2 mm
gland
R. laxiflorum
anther
gland
1 cm
5 mm
fruit
1 cm
2 mm
R. malvaceum
var. malvaceum
Ribes lobbii
1 cm
fruit
R. menziesii
1 cm
1 cm
petal
R. marshallii
anther
2 mm
R. menziesii
gland
2 mm
1 cm
Ribes montigenum
1 cm
2 mm
fruit
R. nevadense
R. quercetorum

sometimes bisexual; upper staminate; calyx lobes 4; petals gen 4, ephemeral on staminate fls, minute or 0 on pistillate fls; stamens gen 8; ovary 4-chambered, stigma plumose. **FR**: segments 4, nut-like. **SEED** 1 per chamber. ± 40 spp.: worldwide. (Greek: many lvs, from lf divisions) [Aiken & McNeill 1980 J Linn Soc Bot 80:213–222]

1. Lvs subtending fls pinnately dissected, like cauline lvs; spike 0; dioecious .... ***M. aquaticum***
1′ Lvs subtending fls gen simple, entire to lobed, < cauline lvs, lf- or bract-like; spike emergent; monoecious
 2. Emergent infl-lvs gen 4–15 mm, > 2 × fl, entire to lobed
  3. Emergent lvs gen ± entire to serrate .... ***M. hippuroides***
  3′ Emergent lvs gen pinnately lobed to divided .... ***M. verticillatum***
 2′ Emergent infl-lvs < 3 mm, < fl, bract-like, margin entire to finely toothed
  4. Submersed lvs with < 26 divisions; divisions thread-like, gen not paired .... ***M. sibiricum***
  4′ Submersed lvs with > 28 divisions; divisions linear, paired .... ***M. spicatum***

M. aquaticum (Vell. Conc.) Verdc. PARROT'S FEATHER Dioecious. **ST** < 2 m, glaucous or gray-green. **LVS**: submersed lvs 1.5–3.5 cm; midrib ± flat, wider than divisions, divisions < 7 mm, 20–30 per lf. **INFL** axillary, ± emergent. **STAMINATE FL**: stamens 8; ovary vestigial. **PISTILLATE FL**: stamens 0. **FR** gen 0 in CA. Common. Ponds, ditches, streams, lakes; < 500 m. NCo, CaRF, CW, SCo; warm temp, trop worldwide; native to S.Am. Cult and naturalized pls gen pistillate, clonal. [*M. brasiliense* Cambess.]

M. hippuroides Torrey & A. Gray (p. 685) WESTERN MILFOIL Monoecious. **ST** > 1 m. **LVS**: submersed lvs > 2 cm; midrib and divisions linear, divisions < 10 mm, 14–20(40) per lf. **INFL**: spike, 3–12 cm, emergent; lvs 5–15 mm, 1–2 mm wide, >> fls, gen entire to ± toothed. Ponds, ditches, small streams; < 1100 m. NCo, NCoRI, n SNH, SnJV; to WA, e US. ❀TRY.

M. sibiricum V. Komarov (p. 685) Monoecious; bulblets sometimes present. **ST** > 1 m, whitish when dry. **LVS**: submersed lvs 1–3 cm, midrib and divisions linear, divisions < 15 mm, < 26 per lf. **INFL**: spike, 3–8 cm, emergent; lvs 1–3 mm, < fls, bract-like, oblanceolate to ovate, entire to coarsely toothed. Ponds, ditches, streams, lakes; < 2600 m. NCo, KR, CaR, n SN, SnJV, CCo, SnFrB, GB, s DMoj (Mojave River); to B.C., e US, Eurasia. [*M. spicatum* L. ssp. *exalbescens* (Fern.) Hultén misapplied] ❀TRY.

M. spicatum L. EURASIAN MILFOIL Monoecious; bulblets 0. **ST** > 2 m, reddish or olive-green when dry. **LVS**: submersed lvs < 3 cm, midrib and divisions linear, divisions < 10 mm, > 28 per lf. **INFL**: spike, 4–8 cm, emergent; lvs 1–3 mm, < fls, bract-like, lanceolate, entire to toothed. Uncommon. Ditches, lake margins; < 150 m. SnFrB, SnJV; e US; native to Eurasia, n Afr.

M. verticillatum L. (p. 685) Monoecious; bulblets sometimes present. **ST** > 2.5 m. **LVS**: submersed lvs < 2.5 cm; midrib and divisions linear; divisions < 10 mm, 18–34 per lf. **INFL**: spike, 5–12 cm, emergent; lvs 3–8(10) mm, lower >> fls, upper 2–3 × > fls. Lakes, marshes; < 1500 m. KR, n SNH, MP; to B.C., e US, Eurasia. ❀TRY.

HIPPOCASTANACEAE BUCKEYE FAMILY

William J. Stone

Shrub or tree. **LVS** opposite, gen 4-ranked, palmately compound. **INFL**: panicle or raceme, terminal; fls many. **FLS** showy, ± bilateral, some staminate; sepals 5, free or fused into tube, lobes unequal; petals 4–5, clawed, unequal; stamens 5–8, filaments long, slender; ovary chambers 3, ovules gen 2 per chamber. **FR**: capsule, spheric or slightly 3-lobed, leathery, roughly spiny to shiny. **SEEDS** large, shiny. 3 genera, 18 spp.: n hemisphere. [Hardin 1957 Brittonia 9:145–170, 173–194]

AESCULUS BUCKEYE, HORSE CHESTNUT

Shrub or tree, 4–30 m, < 15 m diam. **LVS** deciduous. **INFL:** pedicels jointed; staminate fls, if any, at top of infl; seed-producing fls gen at base**. FL** ill-smelling; sepals fused into tube; style of seed-producing fls long, thick, of sterile fls short. ± 15 spp.: n hemisphere; some cult. (Latin: name of some oak)

A. californica (Spach) Nutt. (p. 685) CALIFORNIA BUCKEYE Large shrub or tree, 4–12 m, broad, rounded. **LF**: lflets 5–7, 6–17 cm, oblong-lanceolate, finely serrate, acute to acuminate; petiole 1–12 cm. **INFL** panicle-like, erect, 1–2 dm, finely hairy; pedicel 3–10 mm. **FL**: calyx 5–8 mm, 2-lobed; petals 12–18 mm, white to pale rose; stamens 5–7, 18–30 mm, exserted, anthers orange. **FR** gen 1 (sometimes 2–9), 5–8 cm diam, borne at infl tip. **SEED** gen 1, 2–5 cm, glossy brown. n=20. Dry slopes, canyons, borders of streams; < 1700 m. c&s NW, n&c CW, s CaR, SNF, Teh, sw DMoj, scattered in GV near foothills. All parts TOXIC. Native Americans used ground seed as fish poison; nectar and pollen toxic to honeybees. ❀4–6,**7**,**14–17**,**19–24**;IRR:3,**8**,**9**,10,**18**;CVS. Gen deciduous Jun–Feb.

HIPPURIDACEAE MARE'S-TAIL FAMILY

Elizabeth McClintock

Per from rhizome, emergent aquatic, glabrous. **ST** ± erect, unbranched, rooting at nodes, hollow. **LVS** simple, in whorls of gen 6–12, sessile, linear to elliptic, entire; stipules 0. **INFL**: fls solitary in upper axils, ± sessile. **FL** inconspicuous, gen bisexual (or staminate below pistillate); calyx a minute rim at ovary top; petals 0; stamen 1, off-center on top of ovary; ovary inferior, chamber 1, style 1, off-center, ± = stamen, slender, gen in groove between anther sacs, ± entirely stigmatic. **FR**: achene or thin-walled drupe. 1 genus, 1 variable sp.: temp and cool regions. Wind pollinated. Not closely related to Haloragaceae, to which the sp. is sometimes assigned.

HIPPURIS MARE'S-TAIL

1 sp. (Greek: horse tail)

H. vulgaris L. (p. 685) **ST** 3–6 dm; upper 1/4–1/2 emergent. **LF** 1–3.5 cm. **FR** 2–3 mm. 2n=32. Margins of shallow ponds, springs, marshy and swampy areas, roadsides, irrigation ditches; 0–2700 m. NW, CaRH, SN, SnFrB, SnBr, SnJt, MP; much of n hemisphere, Australia, s S.Am. ❀shallow water:**1–7**,**15–17**;INV.

HYDROPHYLLACEAE WATERLEAF FAMILY

Richard R. Halse, except as specified; Robert W. Patterson, Family Editor

Ann, per, shrub, gen hairy, gen taprooted. **ST** prostrate to erect. **LVS** simple to pinnately compound, basal or cauline, alternate or opposite; stipules 0. **INFL**: cyme (gen raceme-like and coiled) or fls solitary. **FL** bisexual, gen radial; calyx lobes gen 5, gen fused at base, gen persistent, enlarging in fr; corolla gen deciduous, rotate to cylindric, lobes gen 5, appendages in pairs on tube between filaments or 0; stamens gen 5, epipetalous, filament base sometimes appendaged, appendages scale-like; ovary gen superior, chamber 1, placentas 2, parietal, enlarged into chamber, sometimes meeting so ovary appears 2–5-chambered, styles 1–2, stigmas gen head-like. **FR**: capsule, gen loculicidal; valves gen 2. 20 genera, 300 spp.: esp w US; some cult (*Emmenanthe, Nemophila, Phacelia*).

1. Shrub; lvs evergreen ..... **ERIODICTYON**
1′ Ann, bien, or per; lvs deciduous
2. Calyx lobes strongly unequal, outer 3 wide, cordate, enlarged and veiny in fr, inner 2 linear ..... **TRICARDIA**
2′ Calyx lobes equal to subequal, if unequal, not wide and cordate
3. Pl scapose (root caudex-like); fls solitary on elongate peduncles ..... **HESPEROCHIRON**
3′ Pl gen not scapose, st clearly present; fls gen 2–many
4. Styles 2
5. Pl gen < 0.5 m; seeds gen cross-ridged, smooth, or with prominent depressions ..... [2]**NAMA**
5′ Pl 1–3 m; seeds striate ..... **TURRICULA**
4′ Style 1, simple or with 2 lobes
6. Per or bien, caudex gen well developed
7. Lvs 1–2-pinnate to deeply lobed, sinuses between lobes reaching midrib
8. Roots fibrous; ovary chamber 1 ..... **HYDROPHYLLUM**
8′ Tap-rooted; ovary chambers 2 ..... [3]**PHACELIA**
7′ Lvs entire, toothed, or lobed, sinuses between lobes not reaching midrib
9. Style entire or slightly 2-lobed; pl base bulb-like or from tubers; lvs reniform to round .. **ROMANZOFFIA**
9′ Style clearly, often deeply 2-lobed; pl base not bulb-like, tubers 0; lvs linear to ± round
10. Lvs opposite throughout ..... **DRAPERIA**
10′ Lvs alternate, lowermost sometimes opposite ..... [3]**PHACELIA**
6′ Ann, caudex 0
11. Herbage sticky, scented; ovules borne on 2 sides of placenta ..... **EUCRYPTA**
11′ Herbage gen not sticky, not strongly scented; ovules borne on 1 side of placenta
12. Ovary chamber 1; placenta lining inner capsule wall; reflexed appendages gen present between calyx lobes
13. Ovary and fr bristly-hairy ..... **PHOLISTOMA**
13′ Ovary and fr glabrous to hairy, not bristly ..... **NEMOPHILA**
12′ Ovary chambers 2; placentas enlarged into chamber, meeting in middle; reflexed appendages between calyx lobes 0
14. Stamens gen unequal, attached to corolla at different levels ..... [2]**NAMA**
14′ Stamens gen equal, attached ± at same level near corolla base
15. Fls pendent; corolla white, yellow, or pink, persistent, becoming paper-like ..... **EMMENANTHE**
15′ Fls spreading to erect; corolla blue to purple, gen deciduous (sometimes persistent and yellow) ..... [3]**PHACELIA**

DRAPERIA

1 sp. (J.W. Draper, Am historian and scientist, 1800's)

D. systyla (A. Gray) Torrey (p. 685) Per; caudex woody; hairs soft, long. **ST** decumbent to erect, sometimes rooting, 1–4 dm, slender. **LVS** simple, cauline, opposite; lower petioled; upper sessile; blade 1–7 cm, ovate, entire. **INFL**: pedicels 1–2 mm. **FL**: calyx lobes 4–6 mm in fl, 6–9 mm in fr, linear, hairy; corolla 7–14 mm, funnel-shaped, white to pink or lavender, hairy outside; stamens incl, unequal, unequally attached; ovary chambers 2, style 1, 3–4 mm, incl, lobes 2. **FR** 2–3 mm wide, spheric, long-hairy. **SEEDS** 1–4, ovoid, angled, dark brown; surface honeycombed. *n*= 9. Woodlands, talus, rock crevices; 200–3000 m. KR, CaRH, SN. ❀DRN,DRY:1,4–6&SHD:2,7,15–17;DFCLT.

EMMENANTHE

1 sp. (Greek: abiding fl, from persistent corolla)

E. penduliflora Benth. (p. 685) WHISPERING BELLS Ann, glandular, sticky, odorous. **ST** erect, simple to many-branched, 5–85 cm. **LVS** simple, basal, cauline, alternate; lower short-petioled; upper sessile, gen clasping, 1–12 cm, gen < 3 cm wide, toothed to deeply pinnately lobed. **INFL** terminal; pedicels 5–15 mm in fl, 12–25 mm in fr, thread-like, recurved. **FL**: calyx lobes 4–11 mm, 1–4 mm wide, lanceolate, glandular; corolla persistent, withering, enclosing fr, 6–15 mm, bell-shaped, white to pink, hairy, glandular; stamens incl; ovary chambers 2, style incl, lobes 2, 1–4 mm. **FR** 7–10 mm, 2–4 mm wide, glandular. **SEEDS** 6–15, flat, oval, brown; surface honeycombed. *n*=18. Dry, open slopes, common after burns, disturbances; < 2200 m. NCoRH, NCoRI, c&s SN, SnJV, CW, SW, SNE, D; to NV, UT, AZ.

1. Corolla yellow to cream ..... var. ***penduliflora***
1′ Corolla white to pink ..... var. ***rosea***

var. ***penduliflora*** **FL**: corolla yellow to cream. Chaparral to creosote-bush scrub, rocky, sandy, decomposed granite, serpentine soils; < 2200 m. Range of sp. ❀DRN,DRY:**7–10**,11–13,**14–16**, **18–23**&SUN:6,17,**24**.

var. ***rosea*** Brand **FL**: corolla white to pink. Talus slopes, rocky, sandy, or serpentine soils; 400–1800 m. SnFrB, SCoRI, n WTR (Mount Pinos). [*E. rosea* (Brand) Constance] ❀DRN, DRY:**7–9**,10–12,**14–16**,**18–23**&SUN:6,17,**24**.

ERIODICTYON YERBA SANTA

Shrub. **ST** erect; bark shredding. **LVS** simple, cauline, alternate, leathery; upper surface glabrous, shiny, sticky, or tomentose; lower surface tomentose. **INFL** gen open, terminal. **FL**: corolla funnel- to bell-shaped, white, lavender, or purple, hairy outside; stamens incl, filaments gen hairy; ovary chambers 2, styles 2, gen hairy. **FR** 1–3 mm wide; valves 4. **SEEDS** striate, dark brown or black. 9 spp.: sw US, Mex. (Greek: woolly net, from undersurface of some lvs) [Hannan 1988 Amer J Bot 75:579–588]

1. Lf upper surface densely hairy
 2. Calyx, corolla glandular; filaments glabrous
  3. Corolla 2–5 mm; calyx densely hairy, whitish ..... ***E. tomentosum***
  3′ Corolla 4–9 mm; calyx ciliate to slightly hairy, dark-colored ..... ***E. traskiae***
 2′ Calyx, corolla not glandular; filaments hairy ..... ***E. crassifolium***
  4. Lf tomentose, upper surface whitish ..... var. ***crassifolium***
  4′ Lf sparsely to densely hairy, upper surface greenish ..... [2]var. ***nigrescens***
1′ Lf upper surface glabrous to sparsely hairy
 5. Lf linear, 2–11 mm wide, gen entire (or teeth few and well separated)
  6. Infl head-like; s CCo, s SCoRO, n SCo, w WTR (Santa Barbara Co.) ..... ***E. capitatum***
  6′ Infl open, spreading; s SCoRO, DMoj (not in Santa Barbara Co.)
   7. Corolla 11–15 mm; s SCoRO (sw San Luis Obispo Co.) ..... ***E. altissimum***
   7′ Corolla 3–6 mm; DMoj ..... ***E. angustifolium***
 5′ Lf gen lanceolate to oblong, 5–60 mm wide, gen toothed
  8. Twigs gen densely hairy
   9. Lf gen tomentose below, with net-like pattern ..... [2]***E. crassifolium*** var. ***nigrescens***
   9′ Lf white-woolly below, lacking obvious net-like pattern — PR, w DSon ..... [2]***E. trichocalyx*** var. ***lanatum***
  8′ Twigs glabrous to sparsely hairy
   10. Corolla sparsely hairy; calyx ciliate to sparsely hairy; n CA ..... ***E. californicum***
   10′ Corolla densely hairy; calyx gen densely hairy; SW, D ..... ***E. trichocalyx***
    11. Lf densely white-woolly below, lacking obvious net-like pattern ..... [2] var. ***lanatum***
    11′ Lf tomentose below, with obvious net-like pattern ..... var. ***trichocalyx***

E. altissimum P. Wells (p. 685) INDIAN KNOB MOUNTAINBALM **ST** 2–4 m; twigs glabrous, sticky. **LF** 5–9 cm, 2–4 mm wide, linear, sessile, entire, glabrous, sticky above, white-tomentose below; margin strongly rolled under. **FL**: calyx lobes 2–3 mm, ciliate; corolla 11–15 mm, lavender, sparsely hairy; styles 5–7 mm. **SEEDS** many. **ENDANGERED** CA. Sandstone ridges, chaparral; ± 250 m. SCoRO (sw San Luis Obispo Co.).

E. angustifolium Nutt. NARROW-LEAVED YERBA SANTA **ST** < 2 m; twigs glabrous, sticky to sparsely hairy. **LF** 2–10 cm, 2–11 mm wide, linear to lance-linear, sessile to short-petioled, entire to coarsely toothed, glabrous or sticky to sparsely hairy above, hairy between veins forming net-like pattern below; margin rolled under. **FL**: calyx lobes 1–4 mm, glabrous or sparsely hairy, ciliate; corolla 3–6 mm, white, densely hairy; styles 2–3 mm. **SEEDS** 1–8. n=14. UNCOMMON. Washes, slopes; 1500–1900 m. e DMtns (New York Mtns); to NV, UT, AZ, Baja CA.

E. californicum (Hook. & Arn.) Torrey (p. 685) **ST** 1–3 m; twigs glabrous, sticky, rarely sparsely hairy. **LF** 4–15 cm, < 5 cm wide, lanceolate to oblong, short-petioled, entire to toothed, glabrous or sticky or sparsely hairy above, hairy between veins forming net-like pattern below; margin rolled under. **FL**: calyx lobes 2–4 mm, glabrous, ciliate or sparsely hairy; corolla 8–17 mm, white to purple, sparsely hairy; styles 3–8 mm. **SEEDS** 2–20. n=14. Slopes, fields, roadsides, woodland, chaparral; 60–1900 m. NW, CaR, SN, GV, CW; OR. ❀DRN,SUN,DRY or IRR:**7–9**,**14**, 15–17,**18–23**,24;INV,STBL.

E. capitatum Eastw. (p. 685) LOMPOC YERBA SANTA **ST** < 3 m; twigs glabrous, sticky. **LF** 2–9 cm, 2–5 mm wide, linear, sessile, entire, glabrous or sticky to sparsely hairy above, tomentose below; margin rolled under. **INFL** head-like. **FL**: calyx lobes 3–8 mm, densely long-hairy; corolla 6–15 mm, lavender, densely hairy; styles 3–6 mm. **SEEDS** 5. n=14. **RARE** CA. Ravines, chaparral; 40–900 m. s CCo, s SCoRO, n SCo, w WTR (all Santa Barbara Co.). In cult.

E. crassifolium Benth. (p. 689) **ST** 1–3 m; twigs densely hairy to tomentose. **LF** 3–17 cm, 1–6 cm wide, lance-ovate to oblong, short-petioled, entire to toothed, sparsely hairy to white-tomentose above, tomentose below. **FL**: calyx lobes 2–6 mm, densely hairy; corolla 5–16 mm, lavender, densely hairy; styles 3–8 mm. **SEEDS** 8–14. n=14. Slopes, roadsides, washes, river bottoms, mesas; 30–2500 m. Teh, SCoRO, SW. Lf hairiness variable; vars. not easily distinguished.

var. ***crassifolium*** **LF**: surfaces white tomentose. Habitats of sp.; 30–1600 m. SCo, WTR, SnGb, PR. ❀DRN,SUN,DRY or IRR:**7**,**8**,**9**,**14–18**,**19–24**;INV,STBL.

var. ***nigrescens*** Brand **LF**: upper surface sparsely to densely hairy, greenish; lower surface tomentose, greenish, with net-like pattern. Habitats of sp.; 300–2500 m. Teh, SCoRO, SCo, WTR, SnGb, SnJt. [var. *denudatum* Abrams] ❀TRY.

E. tomentosum Benth. (p. 689) **ST** 1–3 m; twigs densely white-tomentose. **LF** 3–10 cm, 1–5 cm wide, oblanceolate to oblong, short-petioled, entire to toothed, densely white-tomentose. **FL**: calyx lobes 1–4 mm, densely hairy, glandular; corolla 2–5 mm, white to lavender, sparsely hairy, glandular; filaments glabrous; styles 1–2 mm. **SEEDS** 10–12. n=14. Slopes, ridges, ravines, disturbed areas, chaparral; 150–1400 m. SCoR. ❀DRN,DRY,SUN:**7**,**8**,**9**, 14–24;DFCLT.

E. traskiae Eastw. TRASK'S YERBA SANTA **ST** < 2 m; twigs tomentose. **LF** 3–14 cm, 1–7 cm wide, oblanceolate, ovate-elliptic to oblong, short-petioled, entire to toothed, densely tomentose; margin rolled under. **FL**: calyx lobes 2–5 mm, slightly hairy to ciliate, glandular; corolla 4–9 mm, white to lavender, sparsely to densely hairy, glandular; filaments glabrous; styles 1–2 mm. **SEEDS** 2–4. n=14. UNCOMMON. Slopes, chaparral; 150–1300 m. SCoRO, s ChI (Santa Catalina Island), WTR. [ssp. *smithii* Munz] ❀DRN, SUN,DRYorIRR:7–9,14–21,**22–24**;STBL.

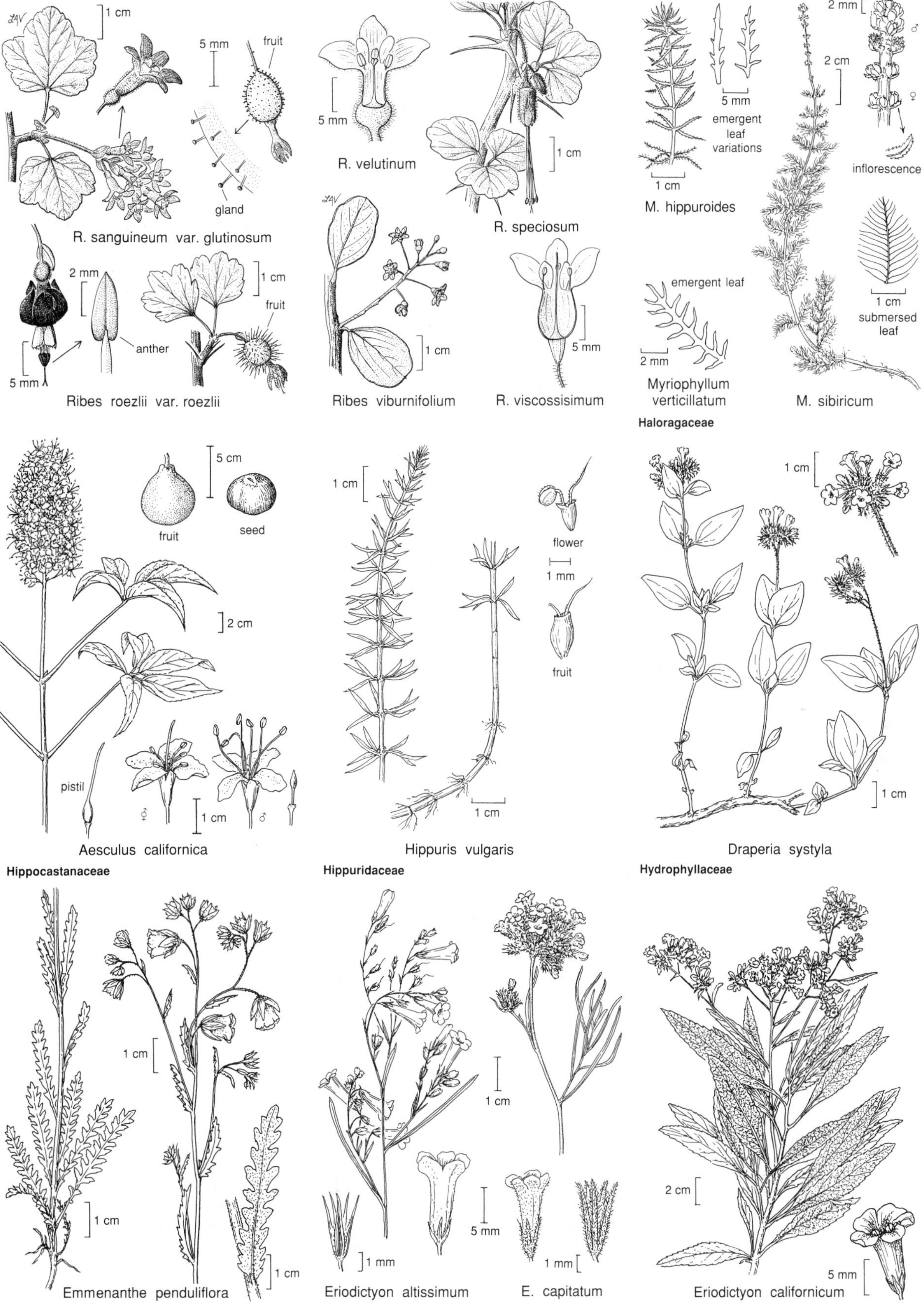
1 cm
5 mm
fruit
gland
R. sanguineum var. glutinosum
2 mm
1 cm
fruit
anther
5 mm
Ribes roezlii var. roezlii
5 mm
R. velutinum
1 cm
R. speciosum
1 cm
Ribes viburnifolium
5 mm
R. viscossisimum
5 mm
emergent leaf variations
1 cm
M. hippuroides
emergent leaf
2 mm
Myriophyllum verticillatum
2 mm
2 cm
♂
♀
inflorescence
1 cm
submersed leaf
M. sibiricum
Haloragaceae
5 cm
fruit
seed
2 cm
pistil
♀
1 cm
♂
Aesculus californica
Hippocastanaceae
1 cm
flower
1 mm
fruit
1 cm
Hippuris vulgaris
Hippuridaceae
1 cm
1 cm
Draperia systyla
Hydrophyllaceae
1 cm
1 cm
1 cm
Emmenanthe penduliflora
1 cm
1 mm
5 mm
1 mm
Eriodictyon altissimum
E. capitatum
2 cm
5 mm
Eriodictyon californicum

E. trichocalyx A.A. Heller **ST** < 2 m; twigs glabrous to hairy. **LF** 3–14 cm, 0.5–4 cm wide, linear-lanceolate to narrowly oblong, short-petioled, entire to toothed, glabrous, sticky to sparsely hairy above, sparsely to densely tomentose below; margin rolled under. **FL**: calyx lobes 1–5 mm, glabrous to hairy, ciliate; corolla 4–13 mm, white to lavender, gen densely hairy; styles 1–6 mm. **SEEDS** 4–8. *n*=14. Slopes, mesas, ravines, grasslands, chaparral; 100–2800 m. SCo, SnGb, SnBr, PR, D (w edge); Baja CA.

var. ***lanatum*** (Brand) Jepson **ST**: twigs sparsely to densely hairy. **LF** gen densely white-tomentose below, lacking net-like pattern. Habitats of sp.; 300–1300 m. PR, DSon (w edge); Baja CA. [*E. l.* (Brand) Abrams] ❀TRY.

var. ***trichocalyx*** **ST**: twig gen glabrous, sticky. **LF** gen tomentose below, with net-like pattern. Habitats and range of sp. ❀DRN,SUN,DRY or IRR:**7**,8,9,11,**14**,15–17,**18–24**;INV,STBL.

EUCRYPTA

Ann, glandular, sticky, scented. **ST** erect, much-branched. **LVS** simple, 1–3-pinnately toothed to dissected; lower cauline lvs opposite, petioled; upper lvs alternate, becoming smaller, sessile, clasping; petioles gen narrowly winged, ciliate. **INFL** terminal or axillary; pedicels thread-like, elongate in fr. **FL**: calyx < half-fused, bell-shaped, glandular, lobes oblong to spoon-shaped, ciliate; corolla bell-shaped, gen > calyx, with V-shaped transverse fold between each pair of filaments below throat; stamens incl, equal, equally attached; ovary chamber 1 (or appearing 5 from complex, enlarged placenta), ovules borne on both sides of placenta, style 1, stigmas 2. **FR** ovoid to spheric, bristly. **SEEDS** 5–15. 2 spp.: sw US. (Greek: well hidden, from seeds) [Constance 1938 Lloydia 1:143–152]

1. Calyx lobes erect, enclosing fr; seeds alike; lower lvs 1-pinnate ..... ***E. micrantha***
1′ Calyx lobes spreading below fr; seeds of 2 kinds; lower lvs 2–3-pinnate ..... ***E. chrysanthemifolia***
 2. Corolla = calyx ..... var. ***bipinnatifida***
 2′ Corolla > calyx ..... var. ***chrysanthemifolia***

E. chrysanthemifolia (Benth.) E. Greene **ST** erect to openly spreading, < 9 dm. **LVS**: lower 2–10 cm, 1–5 cm wide, petioles < 1/2 blade, widened, clasping, blade oblong to widely ovate, pinnate to deeply pinnately lobed, lobes 7–13, deeply 1–2-pinnately lobed, teeth obtuse; upper lvs smaller, narrower, less lobed, bases clasping. **INFL**: fls 4–15 per branch; pedicels gen recurved in fr. **FL**: calyx 2–4 mm; corolla 2–6 mm, lobes hairy on back; style < 3 mm. **FR** 2–4 mm wide, < spreading calyx. **SEEDS** 6–8, dark brown, of 2 kinds; some elliptic or round, disk-like, smooth; others oblong-ovoid, wrinkled. Canyons, chaparral, disturbed areas, slopes; 0–2300 m. s SNF, Teh, SnJV, CW, SW, SNE, D; NV, AZ, Baja CA.

var. ***bipinnatifida*** (Torrey) Constance **ST** openly spreading. **LVS**: lower 2–7 cm, 1–4 cm wide, lobes 7–9. **INFL**: fls 4–8 per branch. **FL**: corolla 2–3 mm, 2–3 mm wide, = calyx, white or bluish; style < 1 mm. *n*=10,20. Cliffs, rocky slopes, washes, crevices; 30–2300 m. s SNF, Teh, SnBr, e PR, SNE, D; NV, AZ, Baja CA.

var. ***chrysanthemifolia*** (p. 689) **ST** erect, stout. **LVS**: lower 2–10 cm, 1–5 cm wide, lobes 9–13. **INFL**: fls 8–15 per branch. **FL**: corolla 3–6 mm, 4–8 mm wide, > calyx, yellowish white; style 1–3 mm. *n*=10. Roadsides, burns, coastal bluffs, ravines; 0–1000 m. Teh, s SnJV, CW, SW; Baja CA.

E. micrantha (Torrey) A.A. Heller (p. 689) **ST** weak, < 3 dm, gen with stalked glands. **LVS**: lower 1–5 cm, < 2 cm wide, petiole short, widened to clasping base, blade oblong or ovate, deeply pinnately 7–9-lobed, lobes oblong or oblanceolate, straight or sickle-shaped, entire or few-toothed; upper lvs greatly reduced, lobed, toothed or entire. **INFL**: fls 4–12 per branch; pedicels gen erect in fr. **FL**: calyx 2–5 mm, gen black-glandular; corolla 2–4 mm, white or blue-purple, tube yellow; style 1–2 mm. **FR** 2–3 mm wide, < calyx. **SEEDS** 7–15, oblong, becoming incurved, worm-like, black or dark-brown, wrinkled. *n*=6,12. Canyons, hillsides, rocky crevices, washes, slopes; 60–2500 m. SnJt, SNE, D; to NV, UT, TX, Mex.

HESPEROCHIRON

Per, scapose; root caudex-like. **LVS** simple, in basal rosette, spreading or ascending; blade tapered to petiole, gen entire, margins gen ciliate. **INFL**: fl solitary; peduncle erect or spreading, 1–10 cm, slender. **FL**: calyx lobes gen unequal, 2–9 mm, glabrous to hairy, ciliate; corolla tube gen densely hairy inside, throat gen yellow, lobes glabrous to hairy, white or bluish, gen tinged or marked with lavender or purple; stamens incl, gen unequal, filament base widened; ovary hairy, chamber 1, style 1, stigmas 2, 2–5 mm. **FR** 5–11 mm, ovoid, hairy. **SEEDS** many, ovoid, angular, reddish brown; surface honeycombed or pitted. 2 spp.: w US, n Mex. (Greek: evening or western centaur)

1. Corolla bell- or funnel-shaped; lvs gen hairy on both surfaces ..... ***H. californicus***
1′ Corolla rotate; lvs gen glabrous at least on lower surface ..... ***H. pumilus***

H. californicus (Benth.) S. Watson (p. 689) **LVS** gen > 6, < 8 cm, < 3 cm wide, oblanceolate to elliptic or ovate; surfaces gen densely to sparsely spreading-hairy. **INFL**: fls gen > 5. **FL**: corolla < 3 cm, < 2 cm wide, bell- or funnel-shaped, lobes oblong, 3–10 mm. *n*=8. Wet meadows, flats, valleys; 1000–2900 m. KR, CaRH, SNH, Teh, WTR, SnBr, GB; to WA, MT, WY, UT, Baja CA. ❀IRR or WET,SUN,DRN:1–3,7,18;DFCLT.

H. pumilus (Griseb.) Porter (p. 689) **LVS** gen 2–10, 1–7 cm, < 2 cm wide, linear-oblong to oblanceolate or oblong; upper surface glabrous or hairy. **INFL**: fls gen 1–8. **FL**: corolla 5–15 mm, 7–30 mm wide, rotate, lobes 3–11 mm, rounded. *n*=8. Wet meadows, slopes, flats; 400–3000 m. KR, NCoRI, CaRH, SNH, Teh, WTR, GB; to WA, MT, UT, AZ. ❀DRN,DRYorIRR,SUN or part SHD: 1–3,7,15–17,**18**;DFCLT.

HYDROPHYLLUM

Per; roots fleshy-fibrous or tuber-like, attached to rhizomes. **ST** erect, fleshy. **LVS** simple, pinnately lobed, or compound, basal or cauline, alternate; petiole widened, clasping; lflets toothed or lobed, hairy, gen paler below. **INFL** gen branched, gen head-like; pedicels gen elongate, sometimes recurved in fr. **FL**: calyx bell-shaped, lobes linear to lanceolate, acute to obtuse, glabrous or hairy, gen ciliate; corolla lobed to middle, > calyx, bell-shaped, lobes hairy; stamens equal, exserted, filaments hairy; ovary chamber 1, style 1, exserted, stigmas 2, base persistent. **FR** 3–5 mm wide, spheric; tip gen bristly, loosely enclosed by calyx. **SEEDS** 1–4, oblong to spheric, brown; surface net-like. 8 spp.: N.Am. (Greek: water leaf) [Constance 1942 Amer Midl Nat 27:710–731]

1. Infl near ground, << subtending lvs; anthers << 1 mm ..... ***H. capitatum*** var. ***alpinum***
1′ Infl well above ground, = or > subtending lvs; anthers 1–2 mm
 2. Lf round to ovate; lflets gen 5 ..... ***H. tenuipes***
 2′ Lf ± oblong; lflets 7–15
  3. Lflets acuminate, teeth gen 4–8 per side, acuminate ..... ***H. fendleri*** var. ***albifrons***
  3′ Lflets obtuse to acute, teeth gen 2–4 per side, obtuse to acute ..... ***H. occidentale***

H. capitatum Douglas var. ***alpinum*** S. Watson (p. 689) WOOLEN-BREECHES Rhizome very short. **ST** very short, spreading-hairy. **LF** 4–12 cm; petiole 3–15 cm; blade ovate to oblong, deeply lobed; lflets 5–7, lower pair gen distinct, terminal lflets ± merged, lanceolate to obovate, obtuse or acute, short-pointed, entire. **INFL** near ground, < subtending lvs; peduncle 1–5 cm; pedicels 4–15 mm. **FL**: calyx lobes 3–4 mm in fl, < 8 mm in fr; corolla 4–10 mm, lobes 2–6 mm, white to purple or white with lavender marks; anthers < 1 mm; style 7–10 mm. **SEEDS** 1–3. *n*=9. Moist slopes, meadows, flats; 900–2500 m. CaRH, n SNH, MP; to OR, ID, UT. ❀DRN,IRR or DRY,SUN to SHD:1,2,7,14–18;DFCLT.

H. fendleri (A. Gray) var. ***albifrons*** (A.A. Heller) J.F. Macbr. (p. 689) Rhizome short. **ST** 25–90 cm, with reflexed bristles. **LF** 6–30 cm; petiole 3–18 cm; blade oblong to oval, deeply lobed; lflets 7–11, lower 2–3 pairs gen distinct, upper deeply lobed, lanceolate, acute to acuminate, serrate to deeply cut. **INFL** gen > subtending lvs; peduncle 3–18 cm; pedicels 2–10 mm. **FL**: calyx lobes 3–6 mm in fl, < 7 mm in fr; corolla 6–11 mm, lobes 4–5 mm, white, purple, or white with lavender marks; anthers 1–2 mm; style 9–14 mm. **SEEDS** gen 2. *n*=9. Moist, shady, wooded slopes; 1100–2000 m. KR; to B.C., ID. ❀IRR,SHD,DRN:1–4,7,18;DFCLT.

H. occidentale (S. Watson) A. Gray (p. 689) Rhizome short. **ST** 6–60 cm, short-hairy or with reflexed bristles. **LF** 5–40 cm; petiole 2–15 cm; blade oblong to oblong-ovate, deeply lobed to compound, lower pairs gen distinct, terminal widely merged, oblong, entire or deeply cut, obtuse or acute. **INFL** gen > subtending lvs; peduncle 5–30 cm; pedicels 2–5 mm. **FL**: calyx lobes 3–4 mm in fl, < 13 mm in fr; corolla 6–10 mm, lobes 4–6 mm, white to lavender or white with lavender marks; anthers 1–2 mm; style 7–19 mm. **SEEDS** gen 2. *n*=9. Moist, shaded slopes, woodlands, meadows, streambanks, chaparral; 600–3000 m. NW, CaRH, SN, ScV, SnFrB; to OR, UT, AZ. ❀SHD,DRN,DRYorIRR:1–7,18;DFCLT.

H. tenuipes A.A. Heller (p. 689) Rhizome long. **ST** 2–8 dm, with reflexed bristles. **LF** 8–20 cm wide; petiole 5–30 cm; blade widely ovate to round, lobed; lflets gen 3–5, lowest pair(s) gen smaller, distinct, terminal sometimes merged (or appearing 3-lobed), coarsely serrate to cut. **INFL** = or > subtending lvs; peduncle 2–14 cm; pedicels 4–12 mm. **FL**: calyx lobes 4–7 mm in fl, < 9 mm in fr; corolla 5–7 mm, lobes 3–4 mm, cream, greenish, purple or blue; anthers 1–2 mm; style 9–14 mm. **SEEDS** 1–3. *n*=9. Moist, shaded, wooded slopes, streambanks; < 1500 m. NCo, KR, NCoRO; to WA. ❀**4,5**;IRR,SHD:**6,7**,8,9,14,**15–17**,19–24;INV; GRCVR.

NAMA PURPLE MAT

John D. Bacon

Gen ann, hairy. **LVS** cauline, gen alternate, simple; margin entire, wavy, crenate, or rolled under. **INFL**: clusters (gen terminal, leafy) or fls solitary or paired in axils, not coiled. **FL**: corolla salverform to bell-shaped; stamens gen attached to corolla at different levels, gen unequal, portion fused to corolla gen narrowly winged; scales at filament base 0. **FR** gen loculicidal, ovoid to elliptic. **SEEDS** gen many, small, reddish brown, brown, black or yellow. ± 55 spp.: sw US, trop Am, Hawaii. (Greek: a stream) [Hitchcock 1933 Amer J Bot 20:415–430, 518–534]

1. Infls terminal heads; lf margins crenate-dentate ..... ***N. rothrockii***
1′ Infls not terminal heads; lf margins entire, flat, slightly wavy, or rolled under
 2. Per, rhizomed ..... ***N. lobbii***
 2′ Ann, taprooted
  3. Sepals fused to ovary in lower 1/3–1/2; lf margins wavy ..... ***N. stenocarpum***
  3′ Sepals free from ovary; lf margins flat or rolled under
   4. Corolla < 3 mm, bowl- or bell-shaped with a distinct basal tube ± 0.5 mm; styles 2; seeds < or = 4 ..... ***N. californicum***
   4′ Corolla > 2 mm, if bowl- or bell-shaped then without distinct basal tube; styles 1 or 2; seeds > 4
    5. Styles fused > 1/2 length
     6. Styles > 3 mm ..... ***N. aretioides*** var. ***multiflorum***
     6′ Styles < 2.5 mm
      7. Corolla limb > 4 mm wide; lobes = or > 2 mm wide ..... ***N. aretioides*** var. ***californicum***
      7′ Corolla limb < or = 4 mm wide; lobes < 2 mm wide ..... ***N. densum***
       8. Pl grayish, gen densely spreading-rough-hairy; style 0.3–1 mm ..... var. ***densum***
       8′ Pl greenish, bristly-strigose; style gen 1–2.5 mm ..... var. ***parviflorum***
    5′ Styles free
     9. Corolla limb < 4 mm wide
      10. Sepals glandular ..... ***N. dichotomum*** var. ***dichotomum***
      10′ Sepals not glandular
       11. Lf sessile; sepals gen pale greenish, not canescent; corolla slightly funnel-shaped or salverform; seeds gen oblong, surface gen cross-ridged ..... ***N. depressum***
       11′ Lf petioled; sepals white- or gray-canescent at least in basal 1/3–1/2; corolla ± cylindric; seeds irregular, surface slightly net-like ..... ***N. pusillum***
     9′ Corolla limb > 4 mm wide
      12. St ascending to erect; seeds fusiform, yellow to orange ..... ***N. hispidum*** var. ***spathulatum***
      12′ St prostrate; seeds spheric to ovoid, brown to black ..... ***N. demissum***
       13. Lf petioled; pl gen gray-green ..... var. ***covillei***
       13′ Lf sessile; pl gen green ..... var. ***demissum***

N. aretioides (Hook. & Arn.) Brand Hairs gen dense, coarse, appressed to spreading, gen swollen at base. **ST** prostrate, 3–12 cm, repeatedly forked. **LF** ± sessile, gen sickle-shaped. **INFL**: fls sessile. **FL**: sepals narrowly linear to lanceolate; corolla salverform;

fused parts of filaments unwinged; style 1, 2-lobed. **FR** 2–4 mm. **SEED** gen compressed, irregularly elliptic-ovoid, brown to black, smooth to minutely cross-ridged, with prominent depressions. Dry, sandy or loamy areas; 1200–2300 m. KR, NCoRO, CaR, n SNH, GB; to ID, NV.

var. ***californicum*** (Brand) Jepson **LF** 4–20 mm, narrowly linear to spoon-shaped. **FL**: corolla 5–9 mm, white or pink; stamens 1–4 mm, attached 1–3 mm above corolla base; style 1–2 mm. **SEED** < or = 1 mm, black, smooth or minutely roughened. 2*n*=14. Sandy-loamy flats, slopes; < 1700 m. MP; w NV.

var. ***multiflorum*** (A.A. Heller) Jepson (p. 689) **LF** 7–30 mm, lanceolate to spoon-shaped. **FL**: corolla 9–18 mm, gen pink or purple; stamens 3–8 mm, attached 2–5 mm above corolla base; style 3–7 mm. **SEED** 0.6–0.8 mm, brown to black, minutely cross-ridged. 2*n*=14. Dry, sandy areas; 1200–2300 m. KR, NCoRO, CaR, n SNH, GB; NV. ❀TRY;DFCLT.

N. californicum (A. Gray) J. Bacon (p. 689) Gen puberulent to finely strigose; hair bases gen swollen. **ST** prostrate, forked, 3–10 cm. **LF** gen sessile, 5–14 mm, 1–4 mm wide, long tapered, oblanceolate, spoon-shaped, or elliptic. **INFL**: fls ± sessile. **FL**: sepals 2–5 mm, linear-lanceolate, silky-hairy; corolla 1–3 mm, bell-shaped, white to pale pink, tube ± 0.5 mm, limb 1–2 mm diam, lobes 0.4–0.8 mm, 0.5–0.8 mm wide; stamens 0.8–1.3 mm, attached < or = 0.5 mm above corolla base, free filament abruptly expanded above attachment; styles 0.5–1 mm. **FR** 2–2.5 mm. **SEEDS** < or = 4, 0.8–1 mm, ovoid, minutely cross-ridged, with prominent depressions. 2*n*=14. Dry, sandy areas; 900–2400 m. NCoRI, Teh, GV, e SnFrB, SCoRI, SW, DMoj; w NV. [*Lemmonia c.* A. Gray]

N. demissum A. Gray Hairs gen dense, fine to coarse, gen mealy-glandular, bases swollen. **ST** prostrate, forked, 3–20 cm. **INFL**: pedicels 0–5 mm. **FL**: corolla funnel-shaped to salverform; stamens attached 2–4 mm above corolla base. **SEED** ± 0.5 mm, slightly cross-ridged, with depressions. Sandy or gravelly flats; < 1600 m. SNH, SW, SNE, D; to UT, AZ, Mex.

var. ***covillei*** Brand Hairs 0.5–1.2 mm, soft, shaggy. **LF**: blade 5–13 mm, elliptic or spoon-shaped to diamond-shaped; petiole 1.5–5 mm, winged. **FL**: sepals 4–8 mm, linear to oblanceolate; corolla 8–12 mm, blue-violet to pink, limb 8–9 mm diam, lobes 2–4 mm, 3–4 mm wide; stamens 3–6 mm; styles 2–4 mm. **FR** 2–5 mm. **SEED** ovoid to spheric, black. Dry, sandy flats and slopes; < 500 m. n DMoj (Death Valley region).

var. ***demissum*** (p. 689) Hairs < 1 mm, finely strigose. **LF** 1–4 cm, linear to spoon-shaped, long-tapered; petiole 0. **FL**: sepals 3–8 mm, linear-lanceolate; corolla 7–14 mm, blue-purple to rose-pink, limb 5–9 mm diam, lobes 2–3 mm, 2–4 mm wide; stamens 4–7 mm; styles 3–6 mm. **FR** 3–4 mm. **SEED** elliptic-ovoid, brown to black. 2*n*=14. Sandy or gravelly flats, slopes; 500–1600 m. s SNH, SW, SNE, D; NV, AZ, Mex. [var. *deserti* Brand]

N. densum Lemmon Hairs gen dense, stiff, gen appressed; bases gen swollen. **ST** prostrate, forked. **LF** gen long-tapered, lanceolate to oblanceolate; petiole 0. **INFL**: pedicels 0–1 mm. **FL**: sepals linear-lanceolate; corolla cylindric to funnel-shaped, white to pale purple; fused part of filament unwinged; style 2-lobed. **FR** 2–4 mm. **SEED** 0.6–0.9 mm, gen elliptic to ovoid, angled on underside, smooth to cross-ridged, brown to black. Sandy or gravelly flats, slopes; 1200–3400 m. NW, CaR, n&c SNH, GB; to WA, NV; also in WY, UT, Colorado.

var. ***densum*** Pl grayish canescent; hairs gen dense, spreading, rough. **FL**: corolla 2–5 mm, gen narrowed below limb; stamens attached ± 1 mm above corolla base; style 0.3–1 mm. **SEED** gen smooth, gen shiny. 2*n*=14. Habitats & elevations of sp. NW, CaR, n&c SNH, GB; to WA, NV.

var. ***parviflorum*** (Greenman) C. Hitchc. (p. 689) Pl green; hairs gen sparse, bristly-strigose. **FL**: corolla 3–8 mm, gen not narrowed below limb; stamens attached 1–2 mm above corolla base; style (0.6)1–2.5 mm. **SEED** minutely cross-ridged, with prominent depressions. 2*n*=14,28. Habitats of sp.; 1200–1800 m. Range of sp. (exc SNH).

N. depressum A. Gray Pl puberulent; hairs pointed, appressed to ascending. **ST** prostrate, forked, 2–10 cm; lower 1/2 of st gen lfless. **LF** 2–16 mm, long-tapered, oblanceolate to spoon-shaped; petiole 0. **INFL**: fls pediceled. **FL**: sepals 3–5 mm, linear to slightly spoon-shaped; corolla 3–6 mm, funnel-shaped-salverform, pink or white, limb 2–4 mm diam, lobes 0.6–1.2 mm, 0.8–1.2 mm wide; stamens 2–3 mm, attached 1–2 mm above corolla base; styles 1–2 mm. **FR** 2–4 mm. **SEED** 0.4–0.7 mm, oblong to ovoid, brown, slightly cross-ridged, gen with prominent depressions. 2*n*=14. Dry, sandy or gravelly flats, slopes; 600–1600 m. s SNH, SNE, DMoj; NV.

N. dichotomum (Ruiz Lopez & Pavon) Choisy var. ***dichotomum*** Pl short-glandular-strigose and short-nonglandular-hairy; nonglandular hairs sometimes spreading. **ST** gen erect, simple or forked, 5–20 cm. **LF** 6–20 mm, long-tapered, linear to spoon-shaped; petiole 0. **INFL**: pedicels 1–2 mm, slender. **FL**: sepals 4–10 mm, linear to spoon-shaped; corolla 3–8 mm, cylindric to bell-shaped, white to bluish, limb 2–4 mm diam, lobes 1–2 mm, 1–1.5 mm wide; stamens 2–4 mm, attached 0.5–1.2 mm above corolla base, filament wider just above attachment; styles 1–3 mm. **FR** 2–6 mm. **SEED** 0.5–0.7 mm, irregularly oblong to ovoid, brown, cross-ridged, with prominent depressions. 2*n*=28. Uncommon. Granite or limestone slopes, ridgetops; 1900–2200 m. e DMtns (New York Mtns); to TX, Mex.

N. hispidum A. Gray var. ***spathulatum*** (Torrey) C. Hitchc. Pl gen mealy-glandular; hairs gen dense, fine to bristly-strigose, base gen swollen. **ST** ascending to erect, simple or freely branched, 7–30 cm. **LF** 1–5 cm, linear to spoon-shaped, ± rolled under; base long-tapered; petiole 0. **INFL**: pedicels 0–4 mm. **FL**: sepals 4–10 mm, ± linear to oblanceolate; corolla 10–15 mm, salverform to narrowly bell-shaped, blue to purple-lavender, limb 5–14 mm diam, lobes 1–4 mm, 2–5 mm wide; stamens 2–6 mm, attached 1–4 mm from corolla base; styles 1–4 mm. **FR** 3–6 mm. **SEED** 0.3–0.7 mm, fusiform, yellow to orange; surface net-like. 2*n*=14. Dry, sandy or gravelly flats, slopes; < 600 m. SW, D; to TX, Mex. [var. *revolutum* Jepson]

N. lobbii A. Gray (p. 699) Per, rhizomed, gen forming low mats > 1 m diam, short-glandular-hairy and densely long-nonglandular-woolly-tomentose, with some stiff, spreading hairs. **ST** prostrate to ascending, freely branched, 16–50 cm. **LVS** 0.5–6 cm, 2–15 mm wide, oblanceolate to obovate, flat; smallest lvs clustered in lf axils, strongly rolled under, gen sticky above. **INFL**: pedicels 0–2 mm. **FL**: sepals 5–12 mm, linear to narrowly lanceolate; corolla 7–12 mm, widely funnel-shaped to narrowly bell-shaped, purple to pink, limb 7–9 mm diam, lobes 2–3 mm, 2–4 mm wide; stamens 4–8 mm, attached 2–4 mm above corolla base, filament gen with 1–4 short hairs; styles 4–6 mm. **FR** 2–4 mm, loculicidal and septicidal. **SEED** 1–2 mm, gen elliptic-ovoid, gen angled below, black, papillate. 2*n*=28. Dry, sandy or rocky alluvial slopes, ridges; 1300–2200 m. CaR, n&c SNH; s OR, w NV. ❀DRN,SUN:1–3,7, 14–16,18;DFCLT.

N. pusillum A. Gray Pl gen densely short-spreading-hairy; hairs slender, base gen swollen. **ST** prostrate, forked, 2–6 cm. **LF**: blade gen 2–11 mm, lanceolate to ovate, abruptly narrowed to winged petiole 1–6 mm. **INFL**: pedicels < or = 4 mm. **FL**: sepals 3–6 mm, linear to slightly spoon-shaped; corolla 3–5 mm, cylindric to slightly funnel-shaped, white to pale pink, limb < or = 2 mm diam, lobes < or = 0.5 mm, 0.5–1 mm wide; stamens 1–3 mm, attached 1–2 mm above corolla base; styles 1–2 mm. **FR** 2–4 mm. **SEED** ± 0.4 mm, irregular, brown-black; surface slightly net-like, with prominent depressions. 2*n*=14. Sandy to rocky flats; < 1700 m. SNE, DMoj; NV, AZ.

N. rothrockii A. Gray (p. 699) Per, rhizomed, forming colonies, gen densely short-glandular-hairy and long-bristly-nonglandular-hairy; longer hairs gen swollen at base. **ST** erect, simple to few-branched, 20–30 cm, sticky. **LF** 2–6 cm, lanceolate to elliptic, margin crenate-dentate. **INFL**: heads, terminal; pedicels ± 1 mm. **FL**: sepals 8–15 mm, linear-lanceolate; corolla 13–18 mm, ± funnel-shaped, pink, purple, or pale blue, limb 8–10 mm diam, lobes 4–6 mm, 3–6 mm wide; stamens 6–11 mm, attached 4–7 mm above corolla base; styles 7–11 mm. **FR** 3–6 mm. **SEED** 1–2 mm, ovoid, red-brown to brown, minutely pitted. 2*n*=34. Sandy alluvial flats, gravelly granitic slopes, meadows; 1700–4000 m. c&s SNH, SnBr; NV; nw AZ. ❀TRY;DFCLT.

N. stenocarpum A. Gray Pl short-soft-silky-hairy and short-glandular-hairy; some hairs stiff and swollen at base. **ST** prostrate

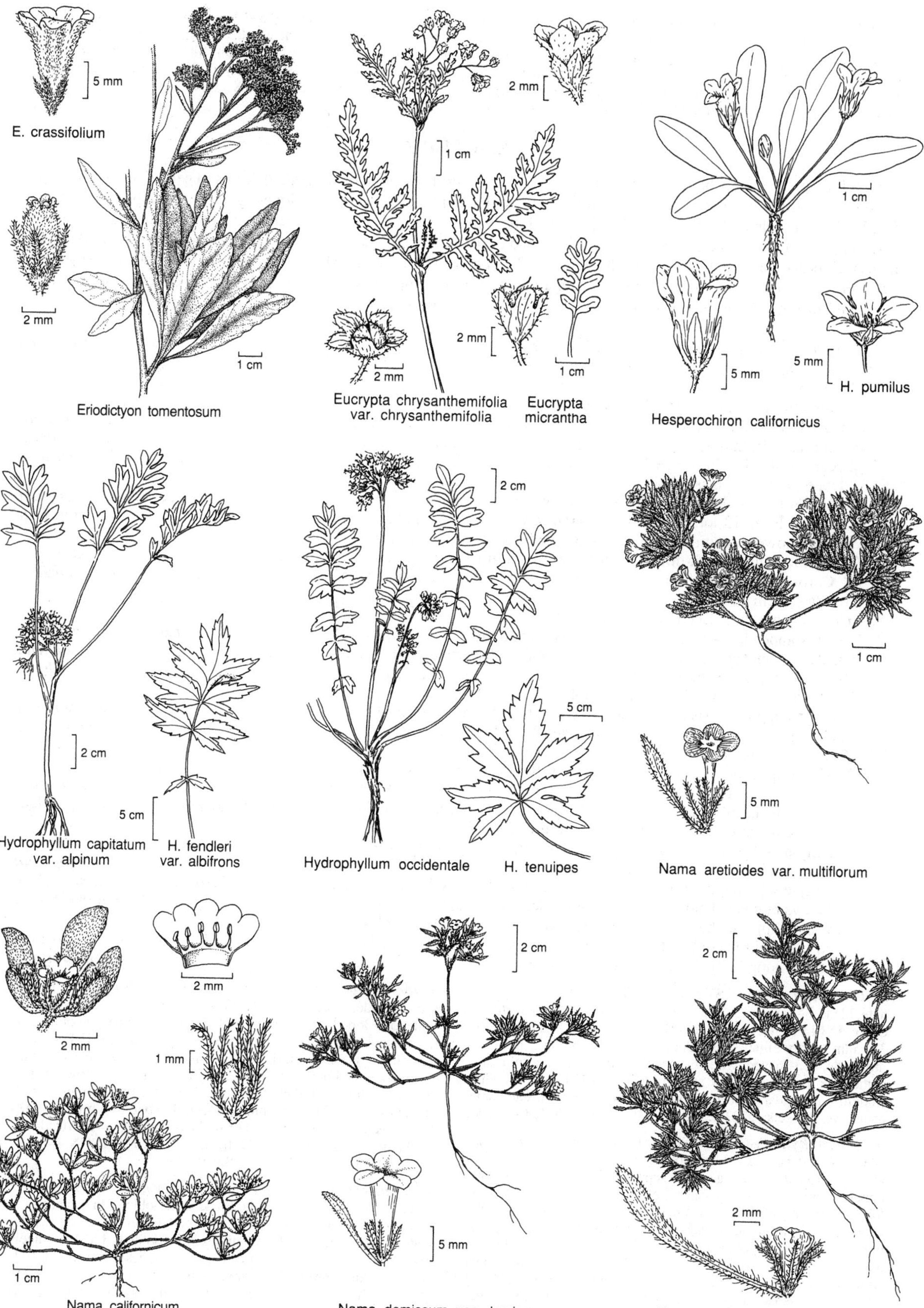
5 mm
E. crassifolium
2 mm
1 cm
Eriodictyon tomentosum
2 mm
1 cm
2 mm
2 mm
1 cm
Eucrypta chrysanthemifolia var. chrysanthemifolia
Eucrypta micrantha
1 cm
5 mm
5 mm
H. pumilus
Hesperochiron californicus
2 cm
5 cm
Hydrophyllum capitatum var. alpinum
H. fendleri var. albifrons
2 cm
5 cm
Hydrophyllum occidentale
H. tenuipes
1 cm
5 mm
Nama aretioides var. multiflorum
2 mm
2 mm
1 mm
1 cm
Nama californicum
2 cm
5 mm
Nama demissum var. demissum
2 cm
2 mm
Nama densum var. parviflorum

to ascending, freely branched, 8–40 cm. **LF** 5–30 mm, oblanceolate, oblong, or spoon-shaped, margins wavy, gen slightly rolled under; base gen slightly clasping st; petiole 0(–3) mm. **INFL**: pedicels < or = 5 mm. **FL**: sepals 3–6 mm (–9 mm in fr), oblanceolate to spoon-shaped, gen bristly; corolla 4–6 mm, ± funnel-shaped, white to cream, limb 3–5 mm diam, lobes 1–2 mm, 2–3 mm wide; stamens 2–4 mm, attached 0.5–1.5 mm above corolla base, filament gen toothed at attachment point; ovary partly inferior, styles 1–2 mm, gen fused at base. **FR** 3–9 mm. **SEED** ± 0.5 mm, irregularly angled, tan to brown; surface honeycombed. $2n$=14. Intermittently wet areas; < 500 m. SW; to TX, Mex.

NEMOPHILA

Ann. **ST** simple to openly branched, prostrate to erect, fleshy, brittle, angled or winged, glabrous to gen bristly. **LVS** simple, cauline, opposite or alternate; petiole gen bristly-ciliate; blade pinnately toothed or lobed, gen bristly, upper gen reduced. **INFL**: fls solitary in axils or opposite lvs; pedicels longer in fr, recurved. **FL**: calyx bell-shaped to rotate, sinuses gen with spreading or reflexed appendages; corolla bell-shaped to rotate, white, blue, or purple, sometimes spotted or marked; stamens incl; ovary chamber 1, style 1, gen lobed 1/3–1/2. **FR** gen 2–7 mm wide, spheric to ovoid, hairy, gen enclosed by calyx. **SEEDS** ovoid, smooth, wrinkled or pitted, with a conic, colorless appendage at 1 end. 11 spp.: se US, w N.Am. (Greek: woodland-loving) [Constance 1941 Univ CA Publ Bot 19:341–398]

1. St with minute reflexed prickles; lower lvs alternate .... *N. breviflora*
1′ St without reflexed prickles; lower lvs opposite
 2. Corolla > 1 cm wide
  3. Seeds smooth or pitted; purple spot gen at tip of corolla lobes
   4. Corolla 10–45 mm wide; style 3–6 mm .... *N. maculata*
   4′ Corolla 2–10 mm wide; style < 2 mm .... [2]*N. spatulata*
  3′ Seeds wrinkled and tubercled; purple spot 0 .... *N. menziesii*
   5. Upper, lower lvs dissimilar
    6. Lower lvs 5–7-lobed; corolla 6–15 mm wide .... [2]var. *integrifolia*
    6′ Lower lvs 6–13-lobed; corolla 10–40 mm wide .... [2] var. *menziesii*
   5′ Upper, lower lvs similar, deeply 5–13-lobed
    7. Corolla white, black-dotted from center to margin .... var. *atomaria*
    7′ Corolla blue with white center or blue-veined, gen black-dotted at center .... [2] var. *menziesii*
 2′ Corolla < 1 cm wide
  8. Corolla gen rotate; filaments > corolla tube
   9. Calyx appendages 1–4 mm; seeds wrinkled and tubercled .... [2]*N. menziesii* var. *integrifolia*
   9′ Calyx appendages 0–0.5 mm; seeds smooth or obscurely rough .... *N. pulchella*
    10. Corolla blue, > calyx; style 2–3 mm .... var. *pulchella*
    10′ Corolla white, = calyx; style < 2 mm
     11. Lvs opposite; basal in a rosette; style < to = 1 mm; seeds 2–4 .... var. *fremontii*
     11′ Lvs alternate; style 1–2 mm; seed gen 1 .... var. *gracilis*
  8′ Corolla bell- to bowl-shaped; filaments < or = corolla tube
   12. Calyx appendages > 1/3 lobes in fr; corolla gen marked black, blue or purple
    13. Lvs oblong to ovate, deeply 5–9-lobed, base truncate or weakly tapered .... *N. pedunculata*
    13′ Lvs oblanceolate or spoon-shaped, shallowly 3–5-lobed or toothed, base strongly tapered ...... [2]*N. spatulata*
   12′ Calyx appendages < 1/3 lobes in fr; corolla unmarked
    14. Style 2–4 mm; basal lvs deeply 5–7-lobed, lobes similar, distinct, round, stalked, sinuses wide .... *N. heterophylla*
    14′ Style < 2 mm; basal lvs cut or shallowly lobed, lobes dissimilar, merging or not stalked .... *N. parviflora*
     15. Lvs opposite below, alternate above, deeply lobed, lobes gen acute .... var. *parviflora*
     15′ Lvs opposite below and above, shallowly lobed, lobes gen obtuse
      16. Lf glabrous to bristly, base tapered .... var. *austinae*
      16′ Lf long-hairy, base truncate or cordate .... var. *quercifolia*

N. breviflora A. Gray **ST** prickles, minute, reflexed. **LVS** alternate; blade 7–30 mm, 15–40 mm wide, 3–6-lobed, lobes acute, gen entire. **INFL**: pedicels < 5 mm in fl, < 15 mm in fr. **FL**: calyx lobes 3–5 mm, stiffly ciliate, appendages 1–2 mm; corolla 2 mm, 1–4 mm wide, bell-shaped, white or purplish; filaments < corolla tube, anthers << 1 mm; style < 1 mm, tip lobed. **SEED** gen 1, reddish, smooth but regularly, deeply pitted in rows. n=9. Streambanks, meadows, thickets; 1500–2200 m. MP; to B.C., MT, Colorado.

N. heterophylla Fischer & C. Meyer (p. 699) **LVS**: lower opposite, 1–4 cm; petiole = blade, blade oblong to ovate, 5–7-lobed, lobes gen round, stalked, gen well separated, entire to 1–3-toothed; upper lvs alternate, nearly sessile, blade lanceolate to ovate, 3–5-lobed or entire. **INFL**: pedicels < 10 mm in fl, < 60 mm in fr. **FL**: calyx lobes 2–4 mm, appendages < 1 mm; corolla 3–10 mm, 4–12 mm wide, bowl-shaped, white or bluish; filaments = corolla tube, anthers < 1 mm; style 2–4 mm. **SEEDS** 2–5, yellow-brown, smooth or roughened. n=9. Forests, chaparral, roadsides, slopes, streambanks, talus; 30–1700 m. NW, CaR, SN, GV, CW; OR.

N. maculata Lindley (p. 699) FIVESPOT **LVS** opposite; petiole = or > blade; lower blades 8–30 mm, 3–15 mm wide, oblong to ovate, 5–9-lobed, lobes entire or 1–3-toothed; upper lvs oblanceolate to spoon-shaped, sessile, tip 3-toothed or entire. **INFL**: pedicels stout, 10–20 mm in fl, < 70 mm in fr. **FL**: calyx lobes 4–9 mm, appendages 1–4 mm; corolla 8–20 mm, 1–5 cm wide, bowl-shaped to rotate, white with dark veins and dots, lobe tips purple-spotted; filaments > corolla tube, anthers 1–3 mm; style 3–6 mm. **SEEDS** 2–12, greenish brown, smooth or shallowly pitted. n=9. Meadows, roadbanks, woodlands; 60–3100 m. SN, ScV. ❀SUN:**4–6,15–17**&SHD:**7,14,18,24**&IRR:1–3,8,9,**19–23**.

N. menziesii Hook. & Arn. BABY BLUE-EYES **LVS** opposite; lower 1–5 cm; petiole = blade, blade linear-oblong to ovate, 5–13-lobed, lobes obtuse, entire or 1–3-toothed; upper lvs nearly sessile, blade entire or less lobed (sometimes only toothed). **INFL**: pedicels 20–60 mm in fl, < 70 mm in fr. **FL**: calyx lobes 4–8 mm, appendages 1–4 mm; corolla 5–20 mm, 6–40 mm wide, bowl-shaped to rotate, bright blue with white center to white (and gen blue-veined and black-dotted); filaments = or > corolla tube, anthers 2–3 mm; style 2–7 mm. **FR** 5–15 mm wide. **SEEDS** 4–20, brown to black, wrinkled and tubercled. Meadows, fields, woodlands, roadsides, grasslands, canyons; 15–1900 m. CA-FP, SNE, DMoj; OR, Baja CA. Highly variable; vars. intergrade.

var. ***atomaria*** (Fischer & C. Meyer) Chandler (p. 699) **LVS**: upper and lower similar, deeply 5–13-lobed. **FL**: corolla 6–12 mm, 10–30 mm wide, white with black dots radiating from center almost to margins, sometimes faintly blue-tinted or -veined; filaments = corolla tube. **SEEDS** 8–12. *n*=9. Coastal bluffs, grassy slopes; 15–1500 m. NW, CCo, SnFrB; OR. ❀SUN:**4–5,15–17, 24**&SHD:7,14,20–23.

var. ***integrifolia*** Parish **LVS**: lower 5–7-lobed; upper entire to shallowly few-toothed, diamond-shaped, oblong or oblanceolate, sessile. **FL**: corolla 5–10 mm, 6–15 mm wide, blue, black-dotted at center or blue-veined; filaments > corolla tube. **SEEDS** 4–10. *n*=9. Grasslands, canyons, woodlands, burns, slopes; 100–1900 m. CCo, SCoR, SW, SNE, DMoj; Baja CA. ❀TRY.

var. ***menziesii*** (p. 699) **LVS**: lower 6–13-lobed; upper fewer, short-petioled, toothed or more narrowly lobed. **FL**: corolla 5–20 mm, 10–40 mm wide, bright blue with white center or blue-veined, center gen dotted; filaments = corolla tube. **SEEDS** 10–20. *n*=9. Meadows, grasslands, chaparral, woodlands, slopes, desert washes; 15–1600 m. CA-FP, DMoj. ❀SUN:**4–6,15–17,24**&SHD:1–3,**7–9**,10–12,**14,18–23**;CVS.

N. parviflora Benth. **ST** glabrous or soft- to bristly-hairy. **LVS** opposite or alternate; lower lvs cut or shallowly lobed, lobes dissimilar, merging or not stalked. **INFL**: pedicels 2–15 mm in fl, < 30 mm in fr. **FL**: calyx lobes 1–3 mm, appendages < 1 mm; corolla 2–4 mm, 1–5 mm wide, bell-shaped, white or blue; filaments < or = corolla tube; style < 2 mm. **SEEDS** 2–4, yellow to brick-red, smooth but shallowly pitted. Meadows, forests, woodlands, roadsides, slopes; < 2300 m. NW, SN, CW, MP; to B.C., ID, UT.

var. ***austinae*** (Eastw.) Brand (p. 699) **ST** ± glabrous to bristly. **LVS** opposite; lower 1–2 cm, base tapered, lobes 5–7, shallow, oblong or ovate, obtuse, entire; upper lvs short-stalked. **FL**: corolla 1–3 mm wide. *n*=9. Meadows, streambanks, roadsides, forests, ridges; 1100–2300 m. KR, NCoRH, n SNH, MP; to WA, ID, UT.

var. ***parviflora*** **ST** ± glabrous to bristly. **LVS**: lower opposite, 1–4 cm, base truncate or cordate, lobes gen 5, deep, gen acute, entire to toothed, upper 3 lobes sometimes ± merging; upper lvs alternate, gen sessile, asymmetric, shallowly lobed or toothed. **FL**: corolla 1–5 mm wide. *n*=9. Woodlands, forests, roadsides, slopes; < 1600 m. NW, CW; to B.C.

var. ***quercifolia*** (Eastw.) Chandler OAK-LEAVED NEMOPHILA **ST**: hairs soft, shaggy, or wavy. **LVS** opposite; base truncate or cordate; lobes 5–7, shallow, ± rounded; hairs dense, long, soft, wavy; upper lvs short-stalked. **FL**: corolla 3–5 mm wide. *n*=9. UNCOMMON. Forests, slopes, ravines; 700–2200 m. c SN, s SNF, Teh.

N. pedunculata Benth. (p. 699) **LVS** opposite; petiole < or = blade; blade 5–35 mm, oblong to ovate, lobes 5–9, deep, gen entire, obtuse or acute. **INFL**: pedicels 4–12 mm in fl, < 45 mm in fr. **FL**: calyx lobes 1–4 mm, appendages < 3 mm; corolla 2–5 mm, 2–8 mm wide, bowl- or bell-shaped, white or blue, gen veined or dotted with black, blue, or purple, lobe tips sometimes purple-spotted; filaments = corolla tube, anthers < 1 mm; style < 2 mm. **SEEDS** 2–8, black, brown or green, smooth, wrinkled or pitted. *n*=9. Common. Ocean bluffs, grasslands, slopes, meadows, sand bars, fields, woodlands, streambanks; 30–2400 m. CA-FP, MP; to B.C., ID, NV, Baja CA. ❀DRN:**4–6,15–17,24**&SHD:1–3,**7**,8,9,**14,18–23**.

N. pulchella Eastw. **LVS** opposite or alternate; lower lvs 2–5 cm, oblong to ovate, lobes 5, gen well separated, stalked, round, 1–5-toothed; upper lvs often sessile, oblong or lanceolate, entire to shallowly lobed. **INFL**: pedicels 10–30 mm in fl, < 50 mm in fr, slender. **FL**: calyx lobes 2–4 mm, appendages 0–0.5 mm; corolla 3–5 mm, 5–12 mm wide, rotate, white or blue; filaments > corolla tube, anthers < 1 mm; style < 3 mm. **SEEDS** 1–4, brown or greenish, smooth or rough. Meadows, forests, slopes, chaparral; 100–1900 m. s SN, c SNF, Teh, SnJV, SnFrB, SCoR, WTR, SnGb.

var. ***fremontii*** (Elmer) Constance **LVS** opposite; lower in rosette. **FL**: corolla = calyx, white; style < 1 mm, incl in calyx. **SEEDS** 2–4. *n*=9. Slopes, chaparral; 400–1400 m. s SN, Teh, SnJV, SnFrB, SCoR, WTR.

var. ***gracilis*** (Eastw.) Constance **LVS** alternate, shallowly lobed or toothed. **FL**: corolla slightly > calyx, white; style 1–2 mm, slightly > calyx. **SEED** gen 1. Woodlands, streambanks, slopes; 100–1000 m. c&s SNF.

var. ***pulchella*** (p. 699) **LVS** opposite (or uppermost alternate). **FL**: corolla >> calyx, blue with white center; style 2–3 mm, exserted. **SEEDS** 2–4. *n*=9. Slopes, forest, chaparral, woodland, meadows, streambanks; 100–1900 m. s SN, c SNF, SnJV, SnGb. ❀TRY.

N. spatulata Cov. (p. 699) **LVS** opposite, 5–30 mm; petiole winged, > blade; lower blades oblong to round, base tapered, obtuse, lobes 3–5, shallow, gen entire; upper blades oblanceolate to widely tapered, teeth 3–5, shallow, triangular, gen entire. **INFL**: pedicels 3–8 mm in fl, < 30 mm in fr. **FL**: calyx lobes 2–5 mm, appendages 1–2 mm; corolla 2–8 mm, 2–10 mm wide, bowl-shaped, white or blue, gen veined, dotted, lobes sometimes purple-spotted; filaments < corolla tube, anthers < 2 mm; style < 2 mm, tip lobed. **SEEDS** 5–7, brown, smooth but shallowly pitted. *n*=9. Meadows, roadsides, slopes; 1100–3200 m. CaRH, s SNF, SNH, Teh, WTR, SnBr, SnJt, Wrn; w NV. ❀TRY.

PHACELIA

Dieter H. Wilken, Richard R. Halse, & Robert W. Patterson

Ann, per, gen glandular-hairy, tap-rooted or from ± thick caudex. **LVS** gen alternate, simple to 2-pinnately compound, gen ± reduced upward. **INFL**: cyme, gen dense, coiled, gen 1-sided; pedicels gen short. **FL**: corolla rotate to bell-shaped, white to purple, tube base with scales free or fused to filaments; stamens gen attached at same level, equal; ovary chamber 1 (or 2 below middle), placentas parietal, enlarging and meeting in fr, style 2-lobed, gen hairy below lobes. **FR** oblong to spheric. **SEEDS** 1–many, oblong to spheric, gen brownish; back gen pitted or cross-furrowed. ± 175 spp.: Am; some cult for orn. (Greek: cluster, from the dense infl) [Halse 1981 Madroño 28:121–132; Heckard 1960 Univ Calif Publ Botany 32:1–126; Lee 1988 Syst Bot 13:16–20] Bristly hairs may cause severe dermatitis. CA pers often hybridize, difficult to separate. Bien and per spp. by Richard Halse.

1. Bien or per from ± woody taproot or branched caudex .... **Group 1**
1′ Ann from slender taproot (some key in both the following groups)
  2. Most cauline lf blades deeply lobed to compound (some sinuses reaching midrib) .... **Group 2**
  2′ Most cauline lf blades entire (lowest sometimes few-lobed at base, sinuses not reaching midrib) .... **Group 3**

Group 1: Bien or per

1. Ovules and seeds gen many (4+)
  2. Lf ± round, 5–25 mm; stamens incl; SNE, DMoj .... ***P. perityloides***

3. Corolla 12–15 mm; pedicels in fr 10–30 mm, reflexed .... var. ***jaegeri***
3′ Corolla 10–12 mm; pedicels in fr 5–15 mm, ascending to spreading .... var. ***perityloides***
2′ Lf longer than wide, gen > 20 mm; stamens gen exserted; NW, CaR, SN, MP
4. Corolla persistent in fr; stamens 10–15 mm .... ***P. sericea***
4′ Corolla deciduous; stamens 8–11 mm
5. Infl head-like, dense; filaments glabrous; pl 10–30 cm .... ***P. hydrophylloides***
5′ Infl ± elongate; filaments hairy; pls gen > 30 cm
6. Corolla 10–12 mm, blue to purplish .... ***P. bolanderi***
6′ Corolla 3–7 mm, cream to greenish or brownish .... ***P. procera***
1′ Ovules 4; seeds 1–4
7. Lf compound; lflets toothed or lobed .... ***P. ramosissima***
8. Herbage below infl glandular
9. St below infl with mostly long, coarse, stiff, bulb-based hairs (some hairs soft, spreading) .... var. ***latifolia***
9′ St below infl with long and short, soft, spreading hairs .... var. ***ramosissima***
8′ Herbage below infl not glandular
10. St below infl gen glabrous (or sparsely short-hairy) .... var. ***eremophila***
10′ St below infl hairy
11. St below infl with coarse, stiff, long, spreading, bulb-based hairs .... var. ***austrolitoralis***
11′ St below infl densely puberulent, hairs not bulb-based
12. St gen 100+ cm; NCo, CCo, n ChI .... var. ***montereyensis***
12′ St gen < 60 cm; SNH, Teh, SnGb, SNE .... var. ***subglabra***
7′ Lf gen simple (or basal lobed to compound); lflets entire (if any)
13. Fls few; pedicel 10–20 mm; pl 5–15 cm — KR .... ***P. dalesiana***
13′ Fls many; pedicel < 5 mm; pl (5)10–200 cm
14. Basal lvs simple, entire or 3–7-lobed, gen silvery to whitish
15. St densely glandular .... [2]***P. corymbosa***
15′ St gen not glandular
16. Lf gen ovate to ± round, veins deeply impressed; coastal dunes .... [2]***P. argentea***
16′ Lf gen lanceolate to widely elliptic, veins prominent but not deeply impressed; shrubland, open woodlands, 900–4000 m .... ***P. hastata***
17. St decumbent to ascending, 5–20(30) cm; hairs mostly spreading; calyx lobes gen glandular .... ssp. ***compacta***
17′ St ascending to ± erect, 20–50 cm; hairs mostly ± appressed; calyx lobes gen not glandular .... ssp. ***hastata***
14′ Basal lvs compound or dissected, gen ± green (grayish)
18. St densely glandular .... [2]***P. corymbosa***
18′ St gen not glandular
19. Lf gen entire (or with 2 basal lobes) — coastal dunes, n NCo .... [2]***P. argentea***
19′ Lvs gen dissected or compound
20. Basal lvs dissected, segments 3–15, lanceolate to triangular
21. Bien or weak per; st erect, gen 1 .... ***P. heterophylla*** ssp. ***virgata***
21′ Per; sts decumbent to ascending, gen > 1
22. Calyx lobes lanceolate to ovate, ± overlapping in fr .... ***P. imbricata***
23. Lf segments 7–15; outer calyx lobes narrowly ovate, often glandular .... ssp. ***imbricata***
23′ Lf segments 3–7; outer calyx lobes lanceolate, not glandular .... ssp. ***patula***
22′ Calyx lobes linear to narrowly ovate, not overlapping in fr
24. Corolla gen blue to lavender; stamens 7–10 mm; calyx lobes 6–8 mm in fr .... [2]***P. californica***
24′ Corolla white to cream; stamens 9–14 mm; calyx lobes 8–12 mm in fr .... ***P. egena***
20′ Basal lvs compound, lflets 3–7, lanceolate to ovate
25. St hairs finely stiff-hairy; pl 10–60 cm; inland mtns, 900–3100 m .... ***P. mutabilis***
25′ St hairs coarse, stiff, stinging; pls (15)50–200 cm; coastal foothills or mtns, < 1000 m
26. Style 8–12 mm; corolla gen blue or lavender .... [2]***P. californica***
26′ Style 6–9 mm; corolla greenish white to yellowish .... ***P. nemoralis***
27. Mid-st gen < 7 mm diam; corolla 3.5–5 mm .... ssp. ***nemoralis***
27′ Mid-st gen 7–10 mm diam; corolla 5–6 mm .... ssp. ***oregonensis***

Group 2: Ann; most cauline lvs deeply lobed to compound

1. Corolla persistent, ± enclosing fr, gen white to yellow (purplish to blue)
2. Anthers and style branches exserted; corolla rotate; lowest lvs opposite; SN
3. Corolla purple to violet; largest lvs 1–2-lobed at base .... ***P. marcescens***
3′ Corolla white to pale blue; largest lvs 2–6-lobed or -toothed at base .... ***P. stebbinsii***
2′ Anthers and style branches incl; corolla gen narrowly bell-shaped; lowest lvs alternate; GB
4. Corolla tube puberulent inside; filaments puberulent
5. Corolla 3.5–8 mm; style 1.5–3 mm; stamens 2.5–5 mm .... ***P. adenophora***
5′ Corolla 2–4 mm; style 0.5–1.5 mm; stamens 1.5–3 mm .... ***P. monoensis***
4′ Corolla tube glabrous inside; filaments glabrous
6. Seeds striate — seeds 5–30 per fr .... ***P. inundata***

6′ Seeds cross-furrowed, furrows 5–9
7. Sepals and petals gen 5; corolla 2–3 mm .... ***P. inyoensis***
7′ Sepals and petals gen 4; corolla 1.3–2 mm .... ***P. tetramera***
1′ Corolla gen deciduous, variously colored
8. Ovules 3–many per chamber; seeds gen 5+ per fr
9. Corolla 2–5 mm, narrowly bell-shaped; stamens and styles 1–3 mm
10. Upper st, infl axis, pedicels with dark, stalked glands .... ***P. glandulifera***
10′ Upper st, infl axis, pedicels short-hairy, glandular or not, glands not esp dark
11. Calyx lobes gen linear to oblong, 3–6 mm in fr, minutely ciliate, not glandular .... ***P. ivesiana***
11′ Calyx lobes oblanceolate to spoon-shaped, 4–10 mm in fr, short-stiff-hairy, ± glandular
12. Pedicels straight; seeds cross-furrowed; s SCoRI, TR, PR, W&I, D .... ***P. affinis***
12′ Pedicels curved; seeds pitted; SCo .... ***P. stellaris***
9′ Corolla 5–20 mm, rotate to bell-shaped; styles 2–8 mm; stamens 3–9 mm
13. Seeds cross-furrowed
14. Lower lvs 1–2-compound; filaments puberulent .... ***P. bicolor***
14′ Lower lvs deeply lobed to 1-compound; filaments glabrous
15. Corolla limb white to pink; corolla scales 0; seed < 1 mm .... ***P. brachyloba***
15′ Corolla limb blue to violet; corolla scales linear, mostly free from filament; seed 1–1.5 mm .... ***P. fremontii***
13′ Seeds pitted
16. St gen erect, 20–120 cm, 0–few-branched; lvs strongly glandular
17. Lf segments ± sharply toothed; infl densely dark-glandular-hairy; corolla 8–14 mm, white to rose; s SCo .... ***P. ixodes***
17′ Lf segments crenate; infl glandular-puberulent to short-stiff-hairy; corolla 5–7 mm, bluish; s ChI ... ***P. lyonii***
16′ St spreading to erect, 2–40 cm, gen branched at base; lvs short-spreading-hairy, not strongly glandular
18. Corolla scales free from filament base; fr 6–9 mm, obovoid, ± compressed below middle — NCo, n ChI .... ***P. insularis***
19. St gen decumbent to ascending; style 2.5–4 mm; seed 1.5–2 mm; NCo .... var. ***continentis***
19′ Sts gen erect; style 5–6 mm; seed 1–1.5 mm; n ChI .... var. ***insularis***
18′ Corolla scales fused along sides to filament base; fr 4–7 mm, ovoid, ± swollen below middle
20. Corolla tube and throat white, lobes ± violet, scales ± linear; ovules 9–17 per ovary; lf lobes ovate to triangular .... ***P. davidsonii***
20′ Corolla light blue to purple throughout, scales lanceolate to ovate; ovules 20–60 per ovary; lf lobes oblong to elliptic .... ***P. douglasii***
8′ Ovules 1–2 per chamber; seeds 1–4 per fr
21. Some lvs (esp upper cauline) entire or toothed
22. Lf blade ± deltate to ± round; corolla cream .... ***P. malvifolia***
22′ Lf blade linear to ovate; corolla bluish to lavender
23. Calyx lobes in fl 3–5 mm, becoming unequal in fr; stamens 2–4 mm; infl ± open in fr; mtns e of GV .... ***P. austromontana***
23′ Calyx lobes in fl 2–3 mm, subequal in fr; stamens 4–5 mm; infl dense in fr; mtns w of GV .... ***P. breweri***
21′ Lvs lobed to compound, segments gen toothed to lobed
24. Most calyx lobes pinnately lobed — s ChI .... ***P. floribunda***
24′ Calyx lobes entire to crenate
25. Calyx lobes puberulent, margin stiffly ciliate, veins raised
26. Calyx lobes 4–6 mm, 6–10 mm in fr; corolla 8–10 mm; stamens 8–12 mm; style 6–8 mm; w CA-FP .... ***P. ciliata***
26′ Calyx lobes 3–4 mm, 5–7 mm in fr; corolla 3–4 mm; stamens 1–3 mm; style ± 2 mm; e CaR, MP .... ***P. thermalis***
25′ Calyx lobes gen short- to long-hairy, margin ciliate but not stiffly so, veins not raised
27. 3 calyx lobes gen ± lanceolate, entire, other 2 ovate to ± round, sharply toothed to crenate; seed gen 1 .... ***P. platyloba***
27′ Calyx lobes ± alike, linear to oblanceolate, entire; seeds gen > 1
28. Stamens and style branches incl; stamens 3–5 mm; style 3–5 mm .... ***P. cryptantha***
28′ Some stamens (and gen style branches) exserted; stamens 4–15 mm; style 5–15 mm
29. Seed inner surface with central ridge separating 2 longitudinal grooves
30. Sepals 1–1.5 mm > fr; pedicels ± thread-like, densely long-hairy, not glandular .... ***P. pedicellata***
30′ Sepals ± = fr; pedicels stout, short-glandular-hairy
31. Corolla limb white, scales ovate; calyx lobes ± 1 mm wide; fr ovoid, sparsely stiff-hairy ... ***P. amabilis***
31′ Corolla limb mostly blue to purple, scales linear; calyx lobes ± 1.5 mm wide; fr ovoid to spheric, ± puberulent, glandular .... ***P. crenulata***
32. Corolla tube white, limb lavender to blue; calyx lobes 2–3 mm, 3–4.5 mm in fr ...... var. ***minutiflora***
32′ Corolla blue to purple ± throughout; calyx lobes 2.5–5.5 mm, 3.5–5.5 mm in fr
33. St glandular above middle; corolla 5–10 mm; stamens and style exserted 9+ mm ...... var. ***ambigua***
33′ St glandular throughout; corolla 4.5–7 mm; stamens and style exserted 5.5–11 mm ... var. ***crenulata***
29′ Seed cylindric or angled, inner surface not clearly ridged and grooved
34. Calyx lobes 4–8 mm in fr, ± straight, ± enclosing fr; corolla gen white to blue
35. Fr ± spheric, 2–3 mm; corolla deciduous; pedicels 1–3 mm in fr .... ***P. distans***
35′ Fr ovoid to elliptic, 2.5–4 mm; corolla ± persistent; fls ± sessile, pedicels < 1 mm

36. St stiffly erect; lflet tips acute; infl dense, gen 2–4-branched; fr glabrous below middle; seeds 1–2 .... *P. tanacetifolia*
36′ St decumbent to weakly erect; lflet tips rounded; infl ± open below, dense above, gen unbranched; fr ± puberulent below middle; seeds 2–4 .... *P. umbrosa*
34′ Calyx lobes 7–12 mm in fr, ± curved, not enclosing fr; corolla white to pink, lavender, or violet
37. Fr ± puberulent (stiff hairs few, gen < 1 mm, not bulb-based) .... *P. vallis-mortae*
37′ Fr hairs stiff or wavy, > 1 mm, bulb-based .... *P. cicutaria*
38. Infl axis wavy-hairy; infl dense, lower fls touching; calyx hairs shaggy .... var. *hubbyi*
38′ Infl axis stiff-hairy; infl open below, lower fls well separated; calyx hairs stiff
39. Corolla yellowish white; calyx lobes yellowish .... var. *cicutaria*
39′ Corolla lavender; calyx lobes grayish .... var. *hispida*

Group 3: Ann; most cauline lvs simple to weakly lobed

1. Corolla ± persistent, ± enclosing fr (exc *P. quickii*)
2. Seeds 1–4; stamens 3–7 mm, exserted 1–3 mm; corolla 3–5 mm; SN
3. Fr ± spheric; calyx lobes 4–7 mm in fr, strongly unequal; seeds 2–4; pollen yellow .... *P. quickii*
3′ Fr elliptic to ovoid, ± compressed; calyx lobes 3–4 mm in fr, ± equal; seeds 1–2; pollen white
4. Corolla widely bell-shaped, violet to purple; sepals 2–3 mm in fl; stamens 5–6 mm; style 5–6 mm; seed gen 2; fl deciduous from near pedicel tip .... *P. marcescens*
4′ Corolla rotate, white to pale blue; sepals 1–3 mm in fl; stamens 6–7 mm; style 4–5 mm; seed gen 1; fl and fr persistent .... *P. stebbinsii*
2′ Seeds gen 5–25; stamens 1–5 mm, gen incl; corolla gen < 3 mm (exc *P. adenophora*); SNE
5. Corolla puberulent inside; filaments puberulent .... *P. monoensis*
5′ Corolla tube glabrous inside; filaments glabrous
6. Sepals and petals gen 5; corolla 2–3 mm .... *P. inyoensis*
6′ Sepals and petals gen 4; corolla < 2 mm .... *P. tetramera*
1′ Corolla deciduous (exc *P. saxicola*)
7. Ovules 1–2(4) per chamber; seeds 1–4 per fr
8. Cauline lf blade gen as wide as long
9. Lvs, sts, and calyx lobes densely yellow-stiff-hairy; corolla 5–7 mm; stamens 5–10 mm .... [2]*P. malvifolia*
9′ Lvs and sts sparsely white- to clear-stiff-hairy; calyx lobes short-hairy; corolla 3–5 mm; stamens 2–3 mm .... *P. rattanii*
8′ Cauline lf blade gen longer than wide
10. Lf crenate to lobed (lobes again gen ± toothed); inner seed surface with central ridge separating 2 longitudinal grooves
11. Corolla 3–7 mm, stamens 2–5 mm; style 2–5 mm; stamens and style exserted 0–2 mm
12. Corolla 5–7 mm, widely bell-shaped; stamens 3–5 mm; style 3–5 mm; seeds 2.5–3.5 mm .... *P. anelsonii*
12′ Corolla 3–5 mm, bell-shaped; stamens 2–3 mm; style 2–3 mm; seeds 2–3 mm .... *P. coerulea*
11′ Corolla 6–10 mm; stamens 10–15 mm; style 12–15 mm; stamens and style gen exserted .... *P. crenulata*
13. Corolla tube white, limb lavender to blue; calyx lobes 2–3 mm, 3–4.5 mm in fr .... var. *minutiflora*
13′ Corolla blue to purple ± throughout; calyx lobes 2.5–5.5 mm, 3.5–5.5 mm in fr
14. St glandular above middle; corolla 5–10 mm; stamens and style exserted 9+ mm .... var. *ambigua*
14′ St glandular throughout; corolla 4.5–7 mm; stamens and style exserted 5.5–11 mm .... var. *crenulata*
10′ Lf blade entire (or lobes entire, obtuse to acute); inner seed surface not ridged and grooved
15. Upper st, infl axis, and pedicels not clearly glandular
16. Filaments clearly hairy; corolla scales deltate .... *P. humilis*
17. Calyx lobes 8–12 mm in fr; stamens 6–8 mm, clearly exserted .... var. *dudleyi*
17′ Calyx lobes 5–8 mm in fr; stamens 4–6 mm, barely exserted .... var. *humilis*
16′ Filaments glabrous; corolla scales linear to lanceolate
18. Calyx lobes 4–5 mm in fr, oblong; corolla 4–6 mm, light blue, lobes sparsely short-hairy outside; ne CW .... *P. breweri*
18′ Calyx lobes 8–10 mm in fr, linear; corolla 3–4 mm, lavender, lobes glabrous outside; s SNH .... [2]*P. novenmillensis*
15′ Upper st, infl axis, and pedicels glandular
19. Cauline lf blade elliptic to ovate, gen clearly petioled
20. Some calyx lobes in fr oblanceolate to obovate, 2–3 mm wide; corolla 6–8 mm; stamens 7–8 mm; style 6–9 mm .... [2]*P. purpusii*
20′ Calyx lobes in fr linear to narrowly oblanceolate, gen < 2 mm wide; corolla 2–6 mm; stamens 2–5 mm; style 1–5 mm
21. Calyx lobes in fr 8–10 mm, puberulent and stiffly ciliate, some hairs > 1 mm .... [2]*P. novenmillensis*
21′ Calyx lobes in fr 4–6 mm, short-stiff-hairy and ciliate, hairs < 1 mm
22. Lvs cauline; style gen divided 1/3–1/2 length .... *P. austromontana*
22′ Lvs clustered near st base; style divided ± to near base .... *P. eisenii*
19′ Cauline lf blade linear to narrowly oblanceolate, gen tapered to base
23. Mid- to lower cauline lvs opposite; CaR, n&c SN .... *P. racemosa*
23′ Mid- to lower cauline lvs alternate; KR, SnGb, SnBr
24. Calyx lobes 2–3 mm, 3–5 mm in fr, subequal; corolla rotate, limb violet to purple; serpentine soils, KR .... *P. greenei*

24′ Calyx lobes 3–5 mm, 5–15 mm in fr, strongly unequal; corolla bell-shaped, limb white to pale blue nonserpentine, SnGb, SnBr .... [2]***P. mohavensis***

7′ Ovules 3–many per chamber; seeds gen 5+ per fr

25. Lf blades, esp lower, gen as wide as long

26. Corolla 7–40 mm, scales 0 or fused to filament base and tooth-like, filament base slightly swollen; stamens (8)10–45 mm; style 8–45 mm

27. Filament base not toothed

28. Calyx lobes 5–8 mm, 10–12 mm in fr; corolla 12–25 mm, > 30 mm diam, violet to blue; style 15–25 mm .... ***P. grandiflora***

28′ Calyx lobes 3–4 mm, 5–10 mm in fr; corolla 8–18 mm, 15–25 mm diam, tube white or purple, limb blue; style 8–15 mm .... ***P. viscida***

27′ Filament base toothed

29. Filament teeth puberulent; corolla limb violet to deep purple

30. Corolla bell-shaped, 10–40 mm, uniformly purple; style 15–40 mm .... ***P. minor***

30′ Corolla rotate to widely bell-shaped, 10–20 mm, throat with white spots below lobes; style 10–20 mm .... ***P. parryi***

29′ Filament teeth glabrous; corolla limb white to bright blue

31. Corolla 7–12 mm, white to blue; seeds 10–20 per fr .... ***P. longipes***

31′ Corolla 10–40 mm, uniformly bright blue or throat white-spotted; seeds 40–80 per fr

32. Calyx lobes 3–4 mm, 5–8 mm in fr; stamens 12–20 mm; corolla 10–18 mm, tube white, limb blue, throat with white spots below sinuses; style 10–20 mm .... ***P. nashiana***

32′ Calyx lobes 6–8 mm, 9–11 mm in fr; stamens 20–45 mm; corolla 15–40 mm, bright blue; style 20–45 mm .... ***P. campanularia***

33. Corolla ± bell-shaped, 15–30 mm; petiole 1–10 cm .... ssp. ***campanularia***

33′ Corolla ± funnel-shaped, 25–40 mm; petiole 5–20 cm .... ssp. ***vasiformis***

26′ Corolla 3–12 mm, scales linear to ovate, free from filaments; base of filaments not swollen, gen not toothed, stamens 2–10 mm; style 1–12 mm

34. Lvs, sts, and calyx lobes densely yellow-stiff-hairy; fr 2–3 mm — NCo, CW, corolla cream ... [2]***P. malvifolia***

34′ Lvs, sts, and calyx lobes ± short-hairy, sometimes glandular; fr 3–7 mm

35. Seeds with 4–8 cross-furrows; calyx lobes 1.5–3 mm, 2–4 mm in fr; fr gen spheric

36. Corolla 8–12 mm; stamens 5–6 mm; style 5–6 mm .... ***P. calthifolia***

36′ Corolla 4–7 mm; stamens 2–4 mm; style 2–3 mm

37. Corolla cream; infl axis among lvs, finely glandular .... ***P. neglecta***

37′ Corolla violet to purple; infl axis gen > lvs, ± dark-glandular .... ***P. pachyphylla***

35′ Seeds not cross-furrowed; calyx lobes 2–5 mm, 4–9 mm in fr; fr ovoid to ± oblong

38. Corolla 6–12 mm; style 3–5 mm

39. Corolla narrowly bell-shaped, limb gen 3–5 mm wide; upper infl ± short-glandular-hairy (some hairs > 0.5 mm) .... ***P. mustelina***

39′ Corolla widely bell-shaped, limb gen > 5 mm wide; upper infl glandular-puberulent .... ***P. pulchella*** var. ***gooddingii***

38′ Corolla 3–6 mm; style 1–2 mm

40. Lf blade length gen > width, base tapered, margin entire to barely toothed .... [2]***P. parishii***

40′ Lf blade length < to = width, base truncate to slightly lobed, margin clearly toothed

41. Calyx lobes 6–8 mm in fr, some 1–2 mm wide; lf blade crenate to irregularly toothed .... ***P. peirsoniana***

41′ Calyx lobes 4–6 mm in fr, gen < 1 mm wide; lf blade dentate to weakly lobed, lobes obtuse .... ***P. rotundifolia***

25′ Lf blades gen longer than wide

42. St prostrate to ± ascending; fl 1–2 mm; infl axillary, partly hidden by lvs — n CaRH .... ***P. cookei***

42′ St gen ascending to erect; fl 3+ mm; infl gen terminal, not hidden by lvs

43. Ovules or seeds 20–80+

44. Corolla persistent in fr, 3–4 mm, tube white, limb blue to violet; stamens 1–2 mm .... ***P. saxicola***

44′ Corolla deciduous, 4–6 mm, tube yellow, limb white to lavender; stamens 2–4 mm

45. Lvs cauline, gradually reduced upward; blade slightly lobed to toothed; fr glandular-puberulent .... ***P. lemmonii***

45′ Lvs ± basal, ± abruptly reduced upward; blade entire to barely toothed; fr short-hairy .... [2]***P. parishii***

43′ Ovules or seeds < 20

46. Lower pedicels > upper (esp in fr), gen S-shaped or recurved in fr

47. Fr. obovoid, slightly flattened, 6–9 mm; corolla scales deltate to ovate; NCo, n ChI .... ***P. insularis***

48. St gen decumbent to ascending; style 2.5–4 mm; seed 1.5–2 mm; NCo .... var. ***continentis***

48′ St gen erect; style 5–6 mm; seed 1–1.5 mm; n ChI .... var. ***insularis***

47′ Fr ovoid, swollen at base, 4–7 mm; corolla scales linear; SN, TR, PR, DMoj, SNE

49. Corolla 4–8 mm; style 2–4 mm .... ***P. curvipes***

49′ Corolla 7–15 mm; style 4–8 mm .... ***P. davidsonii***

46′ Pedicels gen equal, gen straight in fr (exc *P. orogenes*)

50. Lf linear, oblong, or narrowly oblanceolate, tapered to indistinct petiole

51. Lower lvs opposite; calyx lobes 1–3 mm, 2–8 mm in fr; corolla 2–6 mm; stamens and style 2–5 mm

52. Corolla 2–3 mm; stamens and style ± 2 mm; style glabrous — KR .... ***P. leonis***

52′ Corolla 3–6 mm; stamens and style 3–5 mm; style short-hairy below branches

53. Lower pedicels 2–7 mm, S-shaped or recurved in fr, > upper pedicels; corolla tube white to yellowish, limb violet; s SNH .... ***P. orogenes***
53′ Lower pedicels 1–3 mm, straight in fr, gen = upper pedicels; corolla lavender; KR .... ***P. pringlei***
51′ Lower lvs alternate; calyx lobes 3–5 mm, 5–15 mm in fr; corolla 5–10 mm; stamens and style 5–14 mm
54. Pedicels < 2 mm in fr; calyx lobes subequal; fr 5–8 mm; ne CA .... ***P. linearis***
54′ Pedicels 2–5 mm, esp in fr; calyx lobes unequal (longest 1–2 mm > shortest); fr 3–5 mm; s CA
55. Stamens incl or barely exserted, subequal, 5–9 mm, filaments and pollen gen white; corolla tube base white to lavender, lobes ± spreading, lavender .... ***P. exilis***
55′ Stamens exserted, unequal, 5–14 mm, filaments violet, pollen brownish; corolla tube base yellow, lobes ± erect, white, aging pale blue .... [2]***P. mohavensis***
50′ Lf blade lanceolate to ovate, clearly petioled
56. Seeds with 7–9 cross-furrows; corolla scales 0 — n SNE .... ***P. gymnoclada***
56′ Seeds without transverse furrows; corolla scales linear to ovate
57. Corolla 6–15 mm
58. Lf blade slightly lobed to toothed; calyx lobes short-hairy, not clearly ciliate; corolla tube yellow, limb lavender to purple .... ***P. suaveolens***
59. Corolla 10–14 mm; seeds 8–10 per fr; nw PR .... var. ***keckii***
59′ Corolla 7–11 mm; seeds 10–16 per fr; NCoRI, c SNF, CW .... var. ***suaveolens***
58′ Lf blade entire or few-lobed near base; calyx lobes puberulent to short-hairy, margin clearly stiff-ciliate; corolla lavender to violet
60. Corolla 6–10 mm; stamens 3–5 mm; styles 3–5.5 mm; corolla scales 1–2.5 mm .... ***P. congdonii***
60′ Corolla 10–15 mm; stamens 8–10 mm; styles 6–10 mm; corolla scales 2.5–4 mm .... ***P. divaricata***
57′ Corolla 3–8 mm
61. Stamens 7–8 mm; style 5–9 mm; corolla widely bell-shaped
62. Sts stiff-hairy, many hairs > 1 mm; corolla gen white; SCoRO, WTR .... ***P. grisea***
62′ Sts puberulent to short-stiff-hairy, hairs < 1 mm; corolla lavender to violet; s CaRH, SN, MP .... [2]***P. purpusii***
61′ Stamens 2–4 mm; style 1–4 mm; corolla narrowly bell-shaped
63. Calyx lobes 4–6 mm in fr, short-glandular-hairy, stiff hairs on margin < 1 mm; infl open below; W&I, ne DMtns .... ***P. barnebyana***
63′ Calyx lobes 7–11 mm in fr, densely stiff-hairy, many hairs > 1 mm, weakly glandular; infl dense; s CaR, n&c SN, e SnFrB
64. Corolla white to light lavender, tube ± 1 mm wide; lf veins not impressed; corolla scales ovate, minutely toothed; fr obtuse; e SnFrB .... ***P. phacelioides***
64′ Corolla tube ± 2 mm wide, lavender, limb gen purple; lf veins impressed; corolla scales oblong; fr beaked; s CaR, n&c SN .... ***P. vallicola***

P. adenophora J. Howell Ann 10–40 cm. **ST** spreading to ascending, branched at base, short-hairy, glandular or not. **LF** 10–30 mm; blade = or > petiole, oblong to ± ovate, gen deeply lobed, lobes entire or toothed. **FL**: pedicel 1–3 mm; calyx lobes 2–5 mm, 4–7 mm in fr, narrowly oblanceolate, short-hairy; corolla 3.5–8 mm, narrowly bell-shaped, yellow or lobes ± purplish, persistent in fr, scales narrow; stamens 2.5–5 mm, puberulent; style 1.5–3 mm, puberulent. **FR** 2.5–6 mm, oblong, puberulent. **SEEDS** < 15, 1–1.6 mm; cross-furrows 8–12. n=12. Flats, slopes; 1200–2600 m. GB; OR, w NV.

P. affinis A. Gray (p. 699) Ann 6–30 cm. **ST** erect, simple to branched at base, short-hairy; hairs spreading to reflexed, glandular-puberulent above. **LF** 8–70 mm; blade > petiole, narrowly oblong, deeply lobed to compound, lobes entire, lflets ± lobed. **FL**: pedicel 1–2 mm, < 10 mm in fr; calyx lobes 4–5 mm, 6–10 mm in fr, oblanceolate, short-stiff-hairy, glandular; corolla 3–5 mm, narrowly bell-shaped, tube yellow, limb white to lavender, deciduous, scales ± 0; stamens 2–3 mm, glabrous; style 1–2 mm, gen glabrous. **FR** 4–6 mm, oblong to elliptic, puberulent below, sparsely short-glandular-hairy above. **SEEDS** 15–30, ± 1 mm; cross-furrows 5–8. n=11,12. Open, sandy or gravelly areas; < 3400 m. s SCoRI (Caliente Mtn), TR, PR, W&I, D; to UT, NM, Baja CA. Pls with spreading st hairs and glandular lower lvs have been called var. *patens* J. Howell.

P. amabilis Constance (p. 699) SALINE VALLEY PHACELIA Ann 10–60 cm. **ST** decumbent to erect, few-branched, sparsely soft-hairy, glandular-puberulent. **LVS** 20–120 mm; blades > petioles, lower gen oblong, deeply lobed, lobes crenate to toothed, upper ± ovate, coarsely toothed. **FL**: pedicel 1–2 mm; calyx lobes 3–4 mm, 4–5 mm in fr, oblong, short-hairy, glandular; corolla 6–8 mm, bell-shaped, white, deciduous, scales ovate; stamens 9–15 mm, glabrous; style 12–15 mm, puberulent below. **FR** 3–4 mm, ovoid, sparsely stiff-hairy. **SEEDS** 2–4, 3–4 mm; back finely pitted; inner surface with central ridge separating 2 longitudinal grooves. **PRESUMED EXTINCT**. Gravelly soils, canyons; 500–700 m. s W&I (Inyo Mtns).

P. anelsonii J.F. Macbr. AVEN NELSON'S PHACELIA Ann 10–50 cm. **ST** erect, gen simple, short-glandular-hairy; gland-tipped hairs dark. **LVS** 15–80 mm; blades > petioles, lower oblong to oblanceolate, deeply lobed, lobes gen crenate, upper ± ovate, toothed. **FL** gen sessile; calyx lobes 3–4 mm, 3.5–5.5 mm in fr, oblanceolate, sparsely short-hairy; corolla 5–7 mm, widely bell-shaped, white, pale blue, or lavender, deciduous, scales ± linear; stamens 3–5 mm, glabrous; style 3–5 mm, puberulent. **FR** 2–3.5 mm, ovoid, glandular-puberulent. **SEEDS** 2–4, 2.5–3.5 mm; back pitted; inner surface with central ridge separating 2 longitudinal grooves. RARE in CA. Sandy or gravelly soils, creosote-bush scrub, woodland; 1200–1500 m. e DMtns (New York Mtns); to sw UT.

P. argentea A. Nelson & J.F. Macbr (p. 699) SAND DUNE PHACELIA Per 10–45 cm, ± fleshy. **ST** prostrate to ascending, ± stiff-hairy, not glandular. **LF** thick; blade 20–120 mm, gen > petiole, elliptic to obovate, entire or 2-lobed at base, veins deeply impressed. **FL**: pedicel < 3 mm; calyx lobes 3–4 mm, 6–7 mm in fr, oblong to lanceolate; corolla 5–7 mm, bell-shaped, white to cream, scales oblong; stamens 6–9 mm, exserted, hairy; style 6–10 mm, exserted, hairy. **FR** 3–4 mm, ovoid, stiff-hairy. **SEEDS** 1–3, 1.5–3 mm, pitted in vertical rows. n=22. RARE. Sand dunes; < 20 m. n NCo (Del Norte Co.); OR. In cult.

P. austromontana J. Howell (p. 699) Ann 5–27 cm. **ST** decumbent to ascending, many-branched, puberulent, sparsely stiff-hairy, ± glandular. **LF** 10–30 mm; blade = or > petiole, narrowly elliptic to oblanceolate, entire or few-lobed. **FL**: pedicel 1–2 mm; calyx lobes 3–5 mm, 4–6 mm in fr, unequal, linear to narrowly oblanceolate, short-hairy, glandular; corolla 3–6 mm, bell-shaped, pale blue to lavender, deciduous, scales lanceolate; stamens 2–4 mm, glabrous to papillate; style 3–4 mm, puberulent. **FR** 3–5 mm, ovoid, beaked, puberulent. **SEEDS** 2–8, 1–2 mm, pitted. *n*=9. Open, sandy to rocky areas; 1800–3000 m. c&s SN, TR, SnJt, W&I, DMtns; to sw UT. ❀TRY.

P. barnebyana J. Howell Ann 5–30 cm. **ST** erect, simple to branched at base, glandular-puberulent. **LVS** 10–20 mm, little reduced upward; blade < petiole, gen ovate, entire to obscurely toothed. **FL**: pedicel 3–5 mm; calyx lobes 2–4 mm, 4–6 mm in fr, narrowly oblanceolate, short-hairy, glandular; corolla 3–5 mm, narrowly bell-shaped, tube white to yellow, limb pale violet, deciduous, scales lanceolate; stamens 2–3 mm, glabrous; style 1–2 mm, puberulent. **FR** 3.5–5.5 mm, elliptic, puberulent. **SEEDS** 15–20, ± 1 mm, pitted. Limestone scree; 1600–2700 m. W&I, ne DMtns (Clark Mtns); w NV. Closely related to *P. lemmonii*.

P. bicolor S. Watson var. ***bicolor*** (p. 699) Ann 6–40 cm. **ST** decumbent to erect, gen branched at base, short-hairy, glandular-puberulent. **LVS** 20–60 mm, little reduced upward, 1–2-compound; lflets toothed to irregularly lobed. **FL**: pedicel 1–4 mm; calyx lobes 3–5 mm, 4–6 mm in fr, ± linear, short-hairy, glandular; corolla 8–18 mm, funnel- to bell-shaped, tube yellow, limb lavender to purple, deciduous, scales linear; stamens 5–8 mm, puberulent; style 3–7 mm, puberulent. **FR** 3–4 mm, ovoid; tip short-stiff-hairy. **SEEDS** 12–20, 1–1.5 mm; cross-furrows 7–11. *n*=13. Sandy or alkaline soils, shrubland; 700–3400 m. SNH (e slope), GB, DMoj; OR, NV. Other var. in e OR. ❀TRY.

P. bolanderi A. Gray Per 12–100 cm. **ST** decumbent to ascending, glandular-hairy. **LF**: petiole 10–110 mm; blade 30–120 mm, oblong to widely ovate, coarsely toothed or 2-lobed at base. **FL**: pedicel 1–3 mm; calyx lobes 6–7 mm, 8–11 mm in fr, linear to oblong; corolla 10–12 mm, rotate to bell-shaped, pale blue to purplish, deciduous, scales narrow; stamens 9–11 mm, = corolla or slightly exserted, hairy; style 9–11 mm, = corolla or slightly exserted, base hairy. **FR** 6–8 mm, ovoid, rough-hairy. **SEEDS** 30–60, 1–1.5 mm, pitted. *n*=11. Bluffs, canyons, slopes; < 1400 m. NW; OR. ❀DRN,IRR:1–3,**7**,14,**15**,**16**&SUN:**4–6**,**17**.

P. brachyloba (Benth.) A. Gray Ann 8–60 cm. **ST** gen erect, simple or branched at base, short-hairy, ± glandular. **LF** 15–45 mm; blade > petiole, narrowly elliptic to oblanceolate, deeply lobed to compound, segments entire to toothed. **FL**: pedicel 1–4 mm; calyx lobes 3–4 mm, 4–5 mm in fr, narrowly oblanceolate, short-hairy, glandular; corolla 7–10 mm, funnel- to bell-shaped, tube yellow, limb white to pink, deciduous, scales 0; stamens 4–5 mm, glabrous; style 3–4 mm, puberulent. **FR** 4–5 mm, ovoid, short-beaked, sparsely puberulent below, short-hairy above. **SEEDS** 10–25, ± 0.5 mm; cross-furrows 5–7. *n*=12. Open or burned, sandy areas; < 2300 m. SCoR, SW (exc ChI); Baja CA. ❀TRY.

P. breweri A. Gray (p. 699) Ann 10–45 cm. **ST** spreading to ascending, gen many-branched, puberulent, short-stiff-hairy. **LVS** 10–40 mm; lower ± = upper; blades > petiole, lanceolate to ovate, lower lobed, upper lvs entire to few-lobed. **FL**: pedicel 1–2 mm; calyx lobes 2–3 mm, 4–5 mm in fr, subequal, ± oblong, hairy; corolla 4–6 mm, bell-shaped, light blue, deciduous, scales linear; stamens 4–5 mm, glabrous; style 2–3 mm, puberulent. **FR** 2–3 mm, ovoid, short-stiff-hairy. **SEEDS** 1–2, 1.5–2 mm, finely pitted. *n*=11. Rocky soils, slopes, chaparral, oak woodland; < 1400 m. e SnFrB, n SCoRI. ❀SUN,DRN:**7**,9,**14–16**,17–24.

P. californica Cham. Per 15–90 cm. **ST** decumbent to erect, gen densely stiff-hairy, not glandular. **LVS**: blade 50–200 mm, < or > petiole, ovate, gen compound, lflets 3–7; uppermost lvs often simple, entire. **FL**: pedicel 1–3 mm; calyx lobes 3–5 mm, 6–8 mm in fr, narrowly oblong to lanceolate-ovate; corolla 4–7 mm, bell-shaped, white, blue, or lavender, scales oblong; stamens 7–10 mm, exserted, hairy; style 8–12 mm, exserted. **FR** 3–4 mm, narrowly ovoid, stiff-hairy. **SEEDS** 1–2, 2–2.5 mm, pitted in vertical rows. *n*=22. Bluffs, open slopes, roadcuts, chaparral, woodland; < 1000 m. NCo, NCoRO, CCo, SnFrB. Intergrades with *P. imbricata*. ❀SUN,DRN,IRR:7,14,**15–17**.

P. calthifolia Brand Ann 10–30 cm, fleshy, ± brittle. **ST** spreading to erect, few-branched, short-glandular-hairy; gland-tipped hairs black. **LVS** 10–30 mm, ± basal; blade ± = petiole, ± round, entire to crenate, base lobed. **FL**: pedicel 1–3 mm; calyx lobes 2–3 mm, 3–4 mm in fr, ± linear, short-hairy, glandular; corolla 8–12 mm, funnel- to bell-shaped, violet to purple, deciduous, scales linear; stamens 5–6 mm, puberulent; style 5–6 mm. **FR** 4–5 mm, ± spheric, sparsely puberulent; tip sparsely dark-glandular. **SEEDS** 30–50, 1–1.5 mm; cross-furrows shallow, gen 5–8. *n*=11,12. Sandy soils, gen in creosote-bush scrub; < 1000 m. n DMoj; w NV. ❀TRY.

P. campanularia A. Gray Ann 18–55 cm. **ST** erect, gen simple, short-hairy, ± glandular. **LF**: petiole 20–40 mm; blade < or > petiole, ovate to ± round, clearly toothed. **FL**: pedicel 7–20 mm; calyx lobes 6–8 mm, 9–11 mm in fr, oblong, hairy, glandular; corolla 15–40 mm, funnel- to bell-shaped, bright blue, deciduous, scales fused to filament base, short, truncate to toothed; stamens 20–45 mm, puberulent; style 20–45 mm, short-hairy. **FR** 8–15 mm, ovoid, beaked, short-glandular-hairy. **SEEDS** 40–80, 1–1.5 mm, pitted. Open, sandy or gravelly areas; < 1600 m. DMoj, n&w DSon. Vars. intergrade in s DMtns.

ssp. ***campanularia*** **LF**: petiole 10–100 mm; blade 20–70 mm. **FL**: corolla 15–30 mm, ± bell-shaped; style parted 1/3–1/2 length. *n*=11. Habitats of sp. w DSon. ❀DRN,SUN:**7**,**14–24**&IRR:**8**,**9**,10,11,**12**,13.

ssp. ***vasiformis*** G. Gillett (p. 699) **LF**: petiole 50–200 mm; blade 30–100 mm. **FL**: corolla 25–40 mm, ± funnel-shaped; style parted 1/4–1/2 length. *n*=11. Habitats of sp. DMoj, n DSon. ❀TRY.

P. cicutaria E. Greene (p. 699) Ann 18–60 cm. **ST** ascending to erect, simple to branched, stiff-hairy, glandular. **LF** 20–150 mm; blade gen > petiole, ovate to ± oblong, deeply lobed to compound, segments toothed. **FL**: pedicel 0.5–3 mm; calyx lobes 6–8 mm, 9–12 mm in fr, ± linear, long-hairy, glandular; corolla 8–12 mm, bell-shaped, lavender to yellowish, deciduous, scales elliptic; stamens 8–12 mm, glabrous; style 8–12 mm, short-hairy. **FR** 3–4 mm, ± spheric, sparsely short-hairy. **SEEDS** 2–4, 2–3 mm, pitted. Gravelly or rocky slopes, chaparral, oak/pine woodland, grassland; < 1400 m. SNF, Teh, SCoR, SW, w D; Baja CA.

var. ***cicutaria*** **INFL**: axis long-stiff-hairy; lower fls ± well separated; upper fls dense. **FL**: calyx lobes yellowish, hairs stiff-hairy; corolla yellowish white. *n*=11. Rocky slopes, oak/pine woodland, grassland; < 1400 m. SNF, Teh.

var. ***hispida*** (A. Gray) J. Howell **INFL**: axis long-stiff-hairy; lower fls ± well separated; upper fls dense. **FL**: calyx lobes grayish, hairs stiff-hairy; corolla lavender. *n*=11. Rocky slopes, oak/pine woodland, grassland; < 1400 m. SCoR, SW, w D; Baja CA. ❀TRY.

var. ***hubbyi*** (J.F. Macbr.) J. Howell **INFL**: axis hairs long, wavy; fls dense. **FL**: calyx lobes grayish, hairs shaggy; corolla lavender. Gravelly or rocky slopes, chaparral, grassland; < 1000 m. n SCo, n ChI (Santa Cruz Island), WTR. ❀TRY.

P. ciliata Benth (p. 699) Ann 10–55 cm. **ST** gen erect, simple or branched at base, puberulent to short-hairy, ± glandular. **LF** 30–150 mm; blade gen > petiole, oblong to ovate, deeply lobed to compound, segments toothed to cut. **FL**: pedicel 0.5–2 mm; calyx lobes 4–6 mm, 6–10 mm in fr, ovate to lanceolate, opaque to ± translucent, puberulent, ciliate, veins raised; corolla 8–10 mm, funnel- to bell-shaped, tube pale blue, limb blue, deciduous, scales ± round; stamens 8–12 mm, ± glabrous; style 6–8 mm, short-hairy. **FR** 4–5 mm, ± spheric, short-hairy. **SEEDS** gen 4, 2.5–3.5 mm, pitted. *n*= 11. Clay or gravelly slopes in grassland, fields; < 1700 m. s NCoR, c&s SNF, Teh, s ScV (Sutter Buttes), SnJV, SnFrB, SCoRI, SW (exc ChI); Baja CA. Pls on clay soils from SnJV, with lanceolate, opaque calyx lobes, have been called var. *opaca* J. Howell, Merced phacelia. ❀SUN,DRN:**7–9**,**14–16**,17,18,**19–23**,24.

P. coerulea E. Greene Ann 12–40 cm. **ST** ascending to erect, simple to branched at base, puberulent to sparsely short-glandular-hairy. **LF** 15–70 mm; blade gen > petiole, oblong to oblanceolate, gen lobed toward base, crenate above middle. **FL**: pedicel 0.5–1 mm; calyx lobes 2–3 mm, 3–4 mm in fr, subequal, ± oblong, short-hairy; corolla 3–5 mm, bell-shaped, pale blue to pale purple, deciduous, scales narrow; stamens 2–3 mm, glabrous; style 2–3 mm, puberulent. **FR** 2.5–3.5 mm, spheric, sparsely puberulent. **SEEDS** gen 4, 2–3 mm; back pitted; inner surface with central ridge separating 2 longitudinal grooves. *n*=11. Open, sandy to rocky areas, gen in creosote-bush scrub; 1400–2000 m. e DMoj; to UT, TX, n Mex.

P. congdonii E. Greene (p. 699) Ann 5–35 cm. **ST** ascending to erect, simple or branching from below middle, sparsely short-hairy.

LVS ± basal, 10–50 mm; lower gen > upper; lower petioles > blades, upper < blades; blade elliptic to ovate, entire. **FL**: pedicel 1–2 mm; calyx lobes 4–6 mm, 5–12 mm in fr, linear to oblong, puberulent to short-hairy, margin ± stiffly ciliate; corolla 6–10 mm, funnel- to widely bell-shaped, lavender to violet, deciduous, scales deltate to ovate; stamens 3–5 mm, ± glabrous; style 3–5.5 mm, puberulent. **FR** 5–7 mm, ovoid, beaked, puberulent and short-hairy. **SEEDS** 8–15, 1–1.5 mm, pitted. *n*=10. Open areas, slopes, in chaparral, oak/pine woodland; 600–1700 m. c&s SNF, Teh, e WTR. [*P. divaricata* var. *c.* (E. Greene) Munz] ❀TRY.

P. cookei Constance & Heckard (p. 699) COOKE'S PHACELIA Ann 2–15 cm. **ST** prostrate to ascending, branched at base, glabrous to finely puberulent. **LF** 10–20 mm; lower ± = upper; blade ± = petiole, gen elliptic, entire. **FL** gen sessile; calyx lobes 1–1.5 mm, ± 2 mm in fr, linear, sparsely short-hairy; corolla 1–2 mm, narrowly bell-shaped, white, veins lavender, deciduous, scales 0; stamens < 1 mm, glabrous; style 0.5 mm, glabrous. **FR** 2–3 mm, ovoid, short-hairy. **SEEDS** 4–7, 1–1.5 mm, finely pitted. *n*=11. RARE. Open areas, volcanic, sandy soils, shrubland; 1400–1700 m. n CaRH (Mount Shasta).

P. corymbosa Jepson Per 15–40 cm. **ST** ascending to erect, ± stiff-hairy, densely glandular. **LVS**: basal, cauline few, reduced; blade 20–150 mm, = or > petiole, lanceolate to oblanceolate, entire to dissected, segments 3–7. **FL**: pedicel 0.5–4 mm; calyx lobes 3.5–6 mm, 6–10 mm in fr, linear to narrowly ovate; corolla 5–7 mm, ± cylindric, white, deciduous, scales oblong; stamens 10–12 mm, exserted, gen hairy; style 10–12 mm, exserted. **FR** 3–4 mm, narrowly ovoid, stiff-hairy. **SEEDS** 1–2, 1.5–3 mm, pitted in vertical rows. *n*=11,22. Gen serpentine soils, slopes, flats, ridges; 150–2600 m. KR, NCoR, CaRH; OR. Intergrades with *P. egena*. ❀DFCLT.

P. crenulata Torrey Ann 7–60 cm. **ST** gen erect, 0–few-branched at base, sparsely to densely short-hairy, glandular. **LVS** 20–80(120) mm, abruptly reduced upwards; blade > petiole, oblong to elliptic, crenate to deeply lobed. **FL**: pedicel 0.5–2 mm; calyx lobes 2–5 mm, 3–5.5 mm in fr, oblong, puberulent to short-hairy, glandular; corolla 4–10 mm, bell-shaped, blue to purple, throat ± white, deciduous, scales ± narrow; stamens 10–15 mm, glabrous; style 12–15 mm, glandular-puberulent. **FR** 2.5–4 mm, ovoid to spheric, puberulent. **SEEDS** gen 4, 2–3.5 mm; back finely pitted; inner surface with central ridge separating 2 longitudinal grooves. Sandy to gravelly washes, slopes; < 2200 m. se MP, SNE, D; to UT, AZ, nw Mex.

var. ***ambigua*** (M.E. Jones) J.F. Macbr. **ST** gen densely stiff-hairy, glandular above. **FL**: calyx lobes 3–5 mm, 4–5 mm in fr; corolla 5–10 mm, blue to purple; stamens and style exserted 9+ mm. **FR** 3.3–3.5 mm, spheric. *n*=11. Habitats of sp.; < 1600 m. D; to sw UT, AZ, nw Mex. ❀TRY.

var. ***crenulata*** **ST** gen glandular- and stiff-hairy throughout. **FL**: calyx lobes 2.5–4 mm, 3.5–5.5 mm in fr; corolla 4.5–7 mm, blue to violet; stamens and style exserted 5.5–11 mm. **FR** 2.5–4 mm, ovoid. *n*=11. Habitats of sp.; < 2200 m. se MP, SNE, e DMoj; to UT. [var. *funerea* J. Voss]

var. ***minutiflora*** (J. Voss) Jepson **ST** gen densely stiff-hairy, glandular above. **FL**: calyx lobes 2–3 mm, 3–4.5 mm in fr; corolla ± 4 mm, tube white, limb lavender to blue; stamens and style exserted < 2 mm. **FR** ± 3 mm, ± spheric. *n*=11. Habitats of sp.: < 500 m. DSon; AZ, Baja CA. [*P. minutiflora* J. Voss]

P. cryptantha E. Greene Ann 16–50 cm. **ST** gen erect, 0–few-branched, puberulent to sparsely stiff-hairy, glandular. **LVS** 20–150 mm; blade > petiole, elliptic to ovate, lower gen compound, upper lobed to compound, segments toothed. **FL**: pedicel 2–4 mm; calyx lobes 4–7 mm, 8–10 mm in fr, ± linear, ciliate; corolla 4–7 mm, narrowly bell-shaped, bluish to lavender, deciduous, scales linear; stamens 3–5 mm, glabrous; style 3–5 mm, short-hairy. **FR** 4–5 mm, ± spheric, short-stiff-hairy. **SEEDS** gen 4, 1.5–3 mm, pitted. *n*=11. Gravelly or rocky slopes, canyons; < 1900 m. s SN, SNE, n DMoj; to UT.

P. curvipes S. Watson Ann 4–15 cm. **ST** spreading to ascending, many-branched, short-hairy, glandular above. **LVS** 10–40 mm; lower ± = upper; blade gen > petiole, elliptic to oblanceolate, entire. **FL**: pedicels 1–6 mm, lower < 25 mm in fr; calyx lobes 3–6 mm, 7–10 mm in fr, unequal, narrowly oblong, puberulent, stiffly ciliate; corolla 4–8 mm, rotate to bell-shaped, tube and throat white, lobes blue to violet, deciduous, scales linear; stamens 2–6 mm, short-hairy; style 2–4 mm, short-hairy. **FR** 4–5 mm, ovoid, puberulent and appressed-short-hairy. **SEEDS** 6–16, 1–1.5 mm, pitted. *n*=11. Sandy to rocky slopes, chaparral, oak/pine woodland, coniferous forest; 500–2700 m. s SN, Teh, TR, SNE, DMoj; to sw UT, nw AZ.

P. dalesiana J. Howell (p. 699) SCOTT MOUNTAIN PHACELIA Per 5–15 cm. **STS** few, decumbent, densely glandular-hairy. **LVS**: basal in rosette; cauline few, reduced; blade 10–50 mm, ± = petiole, oblong to elliptic, entire. **FL**: pedicel 10–20 mm in fr; calyx lobes 3–5 mm, 4–7 mm in fr, oblanceolate; corolla 6–9 mm, rotate, white, throat purple-marked, deciduous, scales ± round; stamens 6–8 mm, incl to slightly exserted, glabrous; style 6–7 mm, incl to = corolla. **FR** 4 mm, subspheric, hairy. **SEEDS** 2–4, 2.5–4.5 mm; surface net-veined. *n*=8. RARE. Meadows, streambanks, coniferous forest; 1500–2000 m. KR.

P. davidsonii A. Gray Ann 5–20 cm. **ST** decumbent to erect, 0–few-branched at base, puberulent to sparsely stiff-hairy. **LVS** 8–70 mm; blade > petiole, elliptic to oblanceolate; lower deeply lobed to compound; upper entire to lobed, segments obtuse. **FL**: pedicels 3–8 mm, lower < 18 mm in fr; calyx lobes 4–5 mm, 5–8 mm in fr, oblanceolate, short-hairy; corolla 7–15 mm, rotate to bell-shaped, tube and throat white, lobes ± violet, deciduous, scales ± linear; stamens 3–6 mm, sparsely short-hairy; style 4–8 mm, short-hairy. **FR** 4–7 mm, ovoid, puberulent and short-hairy. **SEEDS** 7–15, 1–2 mm, pitted. *n*=10. Sandy to rocky soils, slopes, chaparral, coniferous forest; 700–2500 m. c&s SN, Teh, TR, PR. [*P. curvipes* var. *macrantha* (Parish) Munz] ❀TRY.

P. distans Benth (p. 699) Ann 15–80 cm. **ST** decumbent to erect, simple to branched at base, puberulent, sparsely stiff-hairy, finely glandular above. **LF** 20–100(150) mm; blade >> petiole, 1–2-compound, segments obtuse to toothed. **FL** gen subsessile; calyx lobes 3–4 mm, 4–5 mm in fr, gen unequal, oblong to oblanceolate, densely hairy, glandular; corolla 6–9 mm, funnel- to bell-shaped, ± blue, deciduous, scales ovate; stamens 8–12 mm, glabrous; style 7–12 mm, glabrous. **FR** 2–3 mm, ± spheric, puberulent. **SEEDS** 2–4, 2–3 mm, pitted. *n*=11. Common. Clay to rocky soils, slopes; < 2100 m. s NCoR, s ScV (Sutter Buttes), SnJV, CW, SW, SNE, D; s NV, n Mex. [*P. cinerea* Eastw.] ❀SUN,DRN:**7,14–24**&IRR:3,**8**,**9**,10,**11–13**.

P. divaricata (Benth.) A. Gray Ann 9–40 cm. **ST** decumbent to erect, simple to branched at base, puberulent to short-hairy. **LF** 10–80 mm; blade gen > petiole, elliptic to narrowly ovate, entire to irregularly 1–2-lobed at base. **FL**: pedicel 1–3 mm; calyx lobes 5–7 mm, 8–11 mm in fr, unequal, lanceolate to obovate, puberulent, margin stiffly ciliate; corolla 10–15 mm, funnel- to widely bell-shaped, lavender to violet, deciduous, scales ovate; stamens 8–10 mm, glandular-puberulent; style 6–10 mm, short-hairy. **FR** 5–10 mm, ovoid; tip short-hairy. **SEEDS** 8–16, 1–1.5 mm, pitted. *n*=10. Open areas, chaparral, woodland, grassland; < 1500 m. s NW, SnFrB, n SCoRI. ❀SUN,DRN,IRR:7,9,**14–17**,18–24.

P. douglasii (Benth.) Torrey Ann 6–40 cm. **ST** spreading to erect, branched at base, short-hairy, glandular. **LF** 5–80 mm, ± basal; blade ± = petiole, oblanceolate to ovate, deeply lobed to compound, segments rounded to obtuse. **FL**: pedicel 3–8 mm; calyx lobes 2–5 mm, 4–7 mm in fr, oblanceolate, short-hairy; corolla 6–12 mm, widely bell-shaped, light blue to purplish, deciduous, scales lanceolate to ovate; stamens 3–7 mm, glabrous to puberulent; style 2–7 mm, puberulent, glandular or not. **FR** 5–7 mm, ovoid, short-hairy. **SEEDS** 8–20, 0.5–1 mm, pitted. *n*=11. Open, gen sandy areas; < 1700 m. s SNF, Teh, s ScV, SnJV, CW, WTR, w DMoj. Pls from w SnJV, SCoRI, with glandular infl axis & glabrous style < 3.5 mm have been called var. *petrophila* Jepson. ❀TRY.

P. egena (Brand) J. Howell (p. 703) Per 15–60 cm. **ST** ascending to erect, stiff-hairy, not glandular. **LVS** mostly basal; blade 100–250 mm, > petiole, lanceolate to oblanceolate; basal dissected, segments 7–11(15); cauline gen entire. **FL**: pedicels 0.5–4 mm; calyx lobes 4–6 mm, 8–12 mm in fr, linear to oblanceolate; corolla 5–9 mm, bell-shaped, white to cream, scales oblong; stamens 9–14 mm, exserted, hairy; style 9–15 mm, exserted. **FR** ± 3 mm, ovoid, stiff-hairy. **SEEDS** 1–2, 2.5–3 mm, pitted in vertical rows. *n*=22.

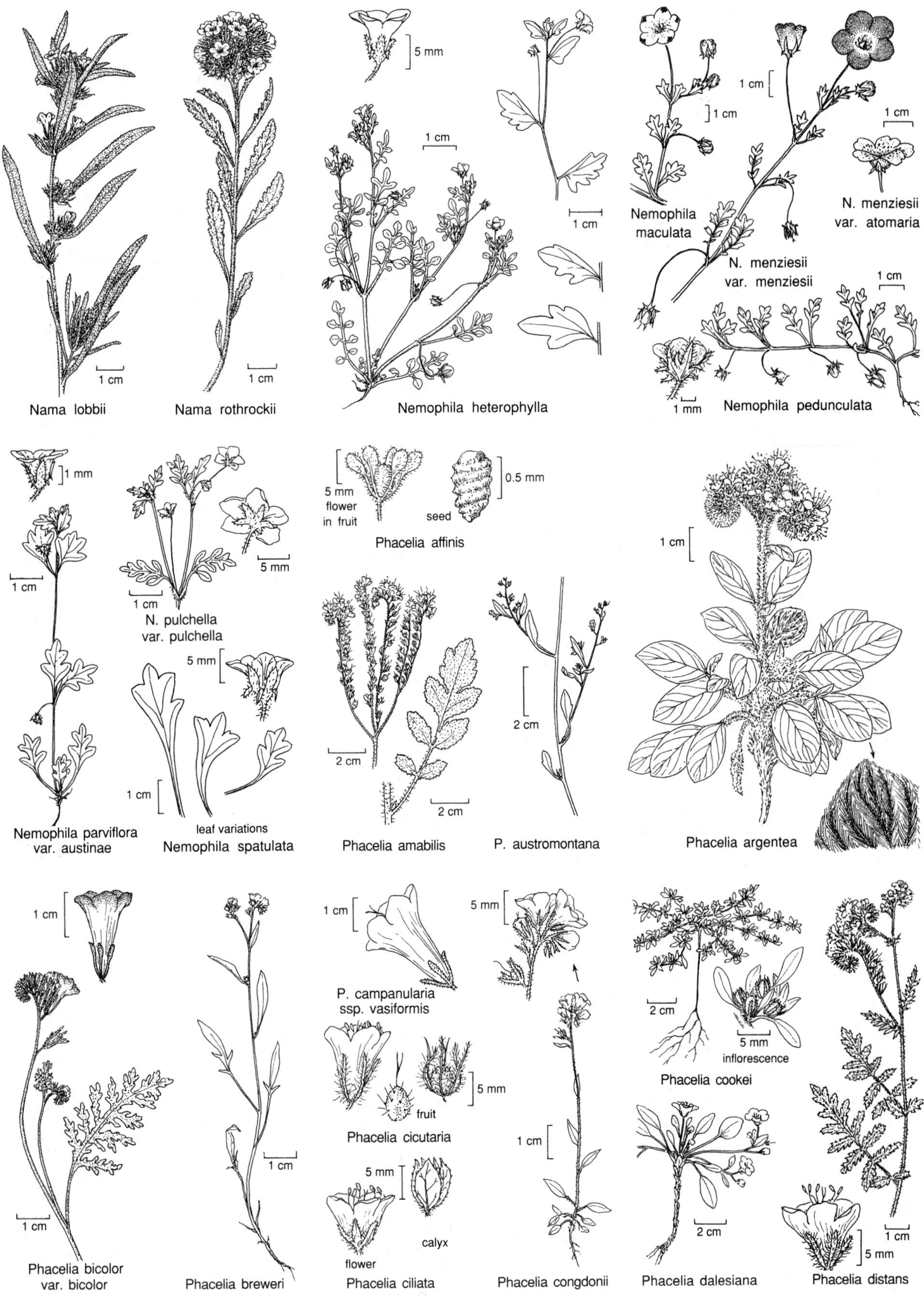

1 cm
Nama lobbii
1 cm
Nama rothrockii
5 mm
1 cm
1 cm
Nemophila heterophylla
1 cm
1 cm
1 cm
1 cm
Nemophila maculata
N. menziesii var. menziesii
N. menziesii var. atomaria
1 cm
1 mm
Nemophila pedunculata
1 mm
1 cm
Nemophila parviflora var. austinae
5 mm
1 cm
N. pulchella var. pulchella
5 mm
1 cm
leaf variations
Nemophila spatulata
5 mm flower in fruit
seed
0.5 mm
Phacelia affinis
2 cm
2 cm
Phacelia amabilis
2 cm
P. austromontana
1 cm
Phacelia argentea
1 cm
1 cm
Phacelia bicolor var. bicolor
1 cm
Phacelia breweri
1 cm
P. campanularia ssp. vasiformis
5 mm
fruit
Phacelia cicutaria
5 mm
calyx
flower
Phacelia ciliata
5 mm
1 cm
Phacelia congdonii
2 cm
5 mm
inflorescence
Phacelia cookei
2 cm
Phacelia dalesiana
1 cm
5 mm
Phacelia distans

Slopes, streambanks, flats, chaparral, woodland; 50–2500 m. NW (exc NCo), CaR, SN, GV, CW (exc CCo), WTR, SnGb. Intergrades with *P. imbricata* (esp in SnFrB), *P. corymbosa*, & *P. hastata*. Some pls in NCoR mat-forming. ❀TRY.

P. eisenii Brandegee (p. 703) Ann 2–10(15) cm. **ST** gen erect, gen simple, puberulent to short-stiff-hairy, glandular. **LVS** 10–25 mm; lowest opposite; blade = or > petiole, elliptic to oblanceolate, gen entire. **FL**: pedicel 0.5–2 mm; calyx lobes 1–3 mm, 4–5 mm in fr, unequal, linear to narrowly oblanceolate, short-hairy; corolla 2–4 mm, bell-shaped, white to lavender, deciduous, scales minute, fused to filaments; stamens 3–5 mm, glabrous; style 1–4 mm, glabrous to glandular-puberulent. **FR** 2–3 mm, ovoid to spheric, short-hairy. **SEEDS** 2–4, 1–1.5 mm, pitted. *n*=9. Sandy or gravelly soils, coniferous forest; 1300–3400 m. SN. Pls with some lower lvs lobed, corolla 3–4 mm and style 2–4 mm have been called var. *brandegeana* J. Howell.

P. exilis (A. Gray) G.J. Lee TRANSVERSE RANGE PHACELIA Ann 5–25 cm. **ST** decumbent to erect, simple to branched at base, short-hairy, ± glandular-puberulent. **LF** 10–35 mm; blade ± narrowly (ob)lanceolate, tapered to petiole, entire. **FL**: pedicel 2–4 mm; calyx lobes 3–5 mm, 5–15 mm in fr, unequal, linear to oblanceolate, puberulent, short-hairy; corolla 5–8 mm, bell-shaped, deciduous, lobes lavender, scales lanceolate; stamens 5–9 mm, subequal, puberulent; style 5–8 mm, puberulent. **FR** 3–5 mm, ovoid, short-hairy, gland-dotted. **SEEDS** 4–8, 1–2 mm, pitted. *n*=9. UNCOMMON. Sandy or rocky slopes, flats, meadows; 1100–2700 m. s SN, WTR (Lockwood Valley), SnGb, SnBr. [*P. mohavensis* var. *e.* A. Gray]

P. floribunda E. Greene (p. 703) SAN CLEMENTE ISLAND PHACELIA Ann 25–60 cm. **ST** erect, branched, soft-hairy, glandular-puberulent. **LF** (30)50–180 mm; blade >> petiole, ± ovate, ± compound, lflets scalloped to lobed. **FL** subsessile; calyx lobes 3.5–4.5 mm, 4–5.5 mm in fr, unequal, gen 3–5-lobed, 1–2 sometimes simple, oblong, short-hairy; corolla 5–7 mm, bell-shaped, pale blue to lavender, deciduous, scales oblanceolate; stamens 5–6 mm, glabrous; style 2–4 mm, glabrous. **FR** 2–3 mm, ovoid, short-hairy. **SEEDS** 1–4, 1.5–2 mm, pitted. *n*=11. RARE. Ravines, gen coastal-sage scrub; < 500 m. s ChI (San Clemente Island); Guadalupe Island, Mex.

P. fremontii Torrey (p. 703) Ann 7–30 cm. **ST** decumbent to erect, branched at base, puberulent, gen glandular above. **LVS** 15–50 mm, ± basal; blade > petiole, oblong to oblanceolate, deeply lobed to compound, segments gen rounded. **FL**: pedicel 1–3 mm; calyx lobes 3–5 mm, 4–6 mm in fr, subequal, linear to oblanceolate, short-hairy; corolla 7–15(20) mm, funnel- to bell-shaped, tube and lower throat yellow, upper limb blue to violet, deciduous, scales linear; stamens 3–8 mm, glabrous; style 3–5 mm, short-hairy. **FR** 4–6 mm, ovoid, puberulent below, short-stiff-hairy above. **SEEDS** 10–18, 1–1.5 mm; cross-furrows 6–9. *n*=13. Sandy or gravelly soils, shrubland, grassland; < 2300 m. s SN, SnJV, e SnFrB, SCoRI, TR, SNE, DMoj; to sw UT, AZ. ❀TRY.

P. glandulifera Piper (p. 703) Ann 5–25 cm. **ST** erect, 0–few-branched, short-hairy; gland-tipped hairs dark. **LVS** 10–50 mm; lower ± = upper; blade gen > petiole, oblong to oblanceolate, deeply lobed to compound, segments obtuse or toothed. **FL**: pedicel < 5 mm; calyx lobes 2–3 mm, 3–5 mm in fr, unequal, linear to oblanceolate, short-hairy; corolla 2–5 mm, narrowly bell-shaped, tube yellow, limb bluish, deciduous, scales 0; stamens 1–2.5 mm, glabrous; style 1–3 mm. **FR** 4–5.5 mm, elliptic, glabrous below, short-hairy above. **SEEDS** 7–14, 1–1.5 mm; cross-furrows 8–10. *n*=13. Gen sandy soils, shrubland, juniper woodland; 800–2500 m. CaR, GB; to e WA, WY.

P. grandiflora (Benth.) A. Gray Ann 50–100 cm. **ST** erect, simple or branched above, glandular-puberulent and sparsely stiff-hairy. **LF** 35–150 mm; blade < to > petiole, widely ovate to ± round, irregularly toothed. **FL**: pedicel 2–3 mm; calyx lobes 5–8 mm, 10–12 mm in fr, oblanceolate, hairy, glandular; corolla 12–25 mm, rotate to widely bell-shaped, violet to blue, deciduous, scales 0; stamens 10–20 mm, short-hairy; style 15–25 mm, short-hairy. **FR** 8–14 mm, ovoid, beaked, puberulent; tip short-glandular-hairy. **SEEDS** 50–100, 1–1.5 mm, pitted. *n*=11. Open areas or burns, gen sandy soils, chaparral or coastal-sage scrub; < 900 m. SCo, ChI, WTR (s slope), SnGb (s slope); Baja CA. ❀DRN,SUN:**14–24**.

P. greenei J. Howell (p. 703) SCOTT VALLEY PHACELIA Ann 2–15 cm. **ST** gen erect, simple to branched at base, ± short-glandular. **LVS** 8–30 mm; lowest opposite; blade ± narrowly (ob)lanceolate, tapered to petiole, entire. **FL**: pedicel 1–2 mm; calyx lobes 2–3 mm, 3–5 mm in fr, narrowly oblanceolate, short-hairy; corolla 5–6 mm, ± rotate, tube white to yellowish, limb violet to purple, deciduous, scales narrow; stamens 4–6 mm, glabrous; style 6–7 mm. **FR** 3–4 mm, ± spheric, ± beaked, short-hairy. **SEEDS** gen 2, ± 3 mm, pitted. *n*=10. RARE. Serpentine soils in coniferous forest; 800–1500 m. KR.

P. grisea A. Gray Ann (5)10–60 cm. **ST** gen erect, branched, stiff-hairy, finely glandular. **LF** 10–80 mm; blade > petiole, lanceolate to widely ovate, toothed to lobed. **FL**: pedicel 1–2 mm; calyx lobes 2.5–4 mm, 6–8 mm in fr, unequal, oblanceolate to obovate, densely short-hairy, glandular; corolla 5–7 mm, widely bell-shaped, white to pale lavender, deciduous, scales oblong; stamens 7–8 mm, papillate; style 5–9 mm, short-hairy. **FR** 4–5 mm, ovoid, short-glandular-bristly. **SEEDS** 5–10, 1–1.5 mm, pitted. *n*=9. Gravelly slopes, gen in chaparral; 300–1200 m. SCoRO, w WTR. Pls < 15 cm from w WTR with ± crenate lf blades and subequal calyx lobes have been called *P. hardhamiae* Munz.

P. gymnoclada S. Watson Ann 5–20 cm. **ST** spreading to ascending, branched at base, short-hairy, glandular-puberulent. **LVS** 10–40 mm, ± basal or reduced upward; blade < to > petiole, oblanceolate to ovate, wavy to obtusely lobed. **FL**: pedicel 1–7 mm; calyx lobes 3–5 mm, 5–8 mm in fr, ± linear, short-hairy; corolla 5–11 mm, funnel- to bell-shaped, tube yellow, limb lavender to purple, deciduous, scales 0; stamens 3–6 mm, short-hairy; style 2–5 mm, short-hairy. **FR** 3–4 mm, ± oblong, sparsely short-hairy. **SEEDS** 5–8, ovoid, 1–1.5 mm; cross-furrows 7–9. *n*=13. Clay to gravelly soils, gen in shrubland; 1700–2300 m. n SNE (Mono Co.); OR, NV. ❀TRY.

P. hastata Lehm. Per 5–50 cm. **ST** decumbent to ± erect, ± stiff-hairy, not glandular. **LVS** mostly basal; blade 15–120 mm, < to = petiole, lanceolate to widely elliptic, gen entire (or 2–4-lobed or compound with 3–5 lflets). **FL**: pedicels 0.5–3 mm; calyx lobes 3–7 mm, 5–9 mm in fr, linear to lanceolate; corolla 4–7 mm, urn- to bell-shaped, white to lavender, scales oblong; stamens 6–10 mm, exserted, glabrous to hairy; style 7–10 mm, exserted. **FR** 2–4 mm, ovoid, stiff-hairy. **SEEDS** 1–3, 1.5–2.5 mm, pitted in vertical rows. Flats, slopes, talus, shrubland, coniferous forest, alpine; 900–4000 m. KR, CaRH, SNH, SnBr, GB; to w Can, SD, Colorado. Ssp. intergrade.

ssp. ***compacta*** (Brand) Heckard (p. 703) **ST** decumbent to ascending, 5–20(30) cm; hairs mostly stiff, spreading. **FL**: calyx lobes gen glandular. *n*=22,33. Sandy to rocky slopes, flats, talus, coniferous forest, alpine; 1800–4000 m. CaRH, SNH, Wrn, SNE; to WA, NV. [*P. frigida* E. Greene sspp. *frigida* & *dasyphylla* (J.F. Macbr.) Heckard]

ssp. ***hastata*** (p. 703) **ST** ascending to ± erect, 20–50 cm; hairs ± appressed, some stiff, spreading. **FL**: calyx lobes gen not glandular. *n*=11,22. Sandy to rocky slopes, shrubland, coniferous forest; 900–2400 m. KR, CaRH, SNH, SnBr, GB; to w Can, SD, Colorado. [*P. oreopola* Heckard ssp. *simulans* Heckard]

P. heterophylla Pursh ssp. ***virgata*** (E. Greene) Heckard (p. 703) Bien, weak per, 20–120 cm. **STS**: central erect, > lateral; lateral ascending, stiff-hairy, not glandular. **LVS** mostly basal; blade 50–150 mm, gen > petiole, lanceolate to ovate, basal dissected, segments 5–9, upper simple to dissected. **FL**: pedicels 0.5–1 mm; calyx lobes 3–6 mm, 6–10 mm in fr, oblong to lanceolate; corolla 4–7 mm, bell-shaped, white to lavender, scales oblong; stamens 8–10 mm, exserted, hairy; style 8–10 mm, exserted. **FR** 2.5–3 mm, ovoid, stiff-hairy. **SEEDS** 1–3, 1.5–2.5 mm, pitted in vertical rows. *n*=11,22. Slopes, flats, roadsides; 100–2900 m. NW, CaR, n&c SN, GB; to OR, ID, WY. Sometimes intergrades with *P. mutabilis*. Ssp. *h.* in Rocky Mtns.. ❀SUN,DRN:1–3,6,**7**,**14–16**,17.

P. humilis Torrey & A. Gray (p. 703) Ann 5–20 cm. **ST** gen erect, simple to branched at base, short-stiff-hairy, sparsely glandular. **LVS** 10–40 mm; lowest opposite; blade gen >> petiole, elliptic to ovate, entire. **FL** gen subsessile; calyx lobes 2–3 mm, 5–12 mm in fr, ± linear, densely white-hairy; corolla 4–7 mm, bell-shaped, lavender to violet, deciduous, scales ± deltate; stamens 4–8 mm, short-hairy; style 4–7 mm, glabrous to short-hairy. **FR** 2–5 mm, ovoid, short-hairy. **SEEDS** 1–4, 1.5–3 mm, finely pitted. Flats, meadows; 800–2300 m. SNH (e slope), Teh, GB; se OR, nw NV, also c WA.

var. ***dudleyi*** J. Howell **FL**: calyx lobes 8–12 mm in fr; stamens 6–8 mm, clearly exserted. **SEEDS** 2.5–3 mm. n=11. Habitats of sp.; 800–1700 m. SNH (Mono Co.), Teh.

var. ***humilis*** **FL**: calyx lobes 5–8 mm in fr; stamens 4–6 mm, barely exserted. **SEEDS** 1.5–2.5 mm. n=11. Habitats of sp.; 1500–2300 m. SNH (e slope), GB; se OR, nw NV, also c WA.

P. hydrophylloides A. Gray (p. 703) Per 10–30 cm. **STS** decumbent to ascending, hairy, gen not glandular. **LF**: petiole 5–50 mm; blades 15–60 mm, oblong to ovate, gen coarsely toothed or lobed (1–2 lobes often deep). **INFL** head-like, dense. **FL**: pedicels 1–2 mm; calyx lobes 3–5 mm, 7–10 mm in fr, narrowly oblong; corolla 5–8 mm, bell-shaped, white to purplish blue, scales oblong to elliptic; stamens 8–11 mm, exserted, glabrous; style 8–10 mm, exserted. **FR** 5–7 mm, ovoid, finely strigose. **SEEDS** 3–16, 2–3 mm; surface net-like, pitted. n=11. Slopes, flats, meadows, coniferous forest; 1500–3100 m. CaRH, SNH; OR, NV.

P. imbricata E. Greene Per 20–120 cm. **STS** decumbent to erect, ± stiff-hairy, often glandular. **LVS** mostly basal; blade 50–150 mm, = or > petiole, narrowly lanceolate to ovate, dissected, segments 3–15; upper sometimes entire. **FL**: pedicels < 2 mm; calyx lobes overlapping, 3–6 mm, 5–10 mm in fr, ± unequal, lanceolate to ovate, often glandular; corolla 4–7 mm, cylindric to bell-shaped, white to lavender, scales oblong; stamens 9–13 mm, exserted, hairy; style 9–14 mm, exserted, hairy below. **FR** 3–4 mm, narrowly ovoid, stiff-hairy. **SEEDS** 1–3, 2–2.5 mm, pitted in vertical rows. Slopes, roadsides, flats, canyons, chaparral, woodland; 50–2600 m. NCoR, SN, GV, CW, SCo, TR, PR; Baja CA.

ssp. ***imbricata*** (p. 703) **ST** ascending to erect. **LF**: segments gen 7– 15. **FL**: outer calyx lobes narrowly ovate, often glandular. n=11,22. Habitats of sp.; 50–2300 m. NCoR, SN, GV, CW, SCo, TR. Intergrades with ssp. *patula* in TR; with *P. californica*; with *P. egena* esp in SnFrB. [ssp. *bernardina* (E. Greene) Heckard] ❀SUN,DRN:**7**, 8,9,**14–24**.

ssp. ***patula*** (Brand) Heckard **ST** decumbent to erect. **LF**: segments 3–7. **FL**: outer calyx lobes lanceolate, not glandular. n=11, 22. Habitats of sp.; 750–2600 m. SnGb, SnBr, PR; Baja CA. Pls from SnGb with decumbent sts and corolla pale lavender have been called *P. oreopola* Heckard ssp. *o.*

P. insularis Munz Ann 2–20 cm. **STS** spreading to erect, few-branched at base, short-stiff-hairy, glandular-puberulent. **LVS** 10–80 mm; blades < or = petiole, gen oblong to elliptic; lower entire to deeply lobed, lobes unequal; upper gen entire. **FL**: pedicels 2–15 mm, lower < 40 mm in fr; calyx lobes 5–6 mm, 6–9 mm in fr, oblanceolate, ciliate, veins ± raised; corolla 5–12 mm, rotate to bell-shaped, lavender to violet, deciduous, scales deltate to ovate; stamens 3–5 mm, glandular-puberulent; style 2.5–6 mm. **FR** 6–9 mm, obovoid, hairy. **SEEDS** 8–15, 1–2 mm, pitted. Sandy soils, dunes, bluffs; < 100 m. NCo, n ChI.

var. ***continentis*** J. Howell (p. 703) NORTHCOAST PHACELIA **STS** gen decumbent to ascending. **FL**: corolla 5–8 mm, ± bell-shaped; style 2.5–4 mm. **SEED** 1.5–2 mm. n=10. RARE. Sandy soils, bluffs; < 100 m. NCo. [*P. divaricata* var. *c.* (J. Howell) Munz]

var. ***insularis*** (p. 703) NORTH CHANNEL ISLANDS PHACELIA **STS** gen erect. **FL**: corolla 6–12 mm, ± rotate; style 5–6 mm. **SEED** 1–1.5 mm. RARE. Sand dunes; < 50 m. n ChI (Santa Rosa, San Miguel Islands). [*P. divaricata* var. *i.* (Munz) Munz]

P. inundata J. Howell (p. 703) PLAYA PHACELIA Ann 10–40 cm. **ST** spreading to erect, branched at base, short-stiff-hairy, glandular. **LVS** 10–30 mm; lower crowded; blade = or > petiole, oblong to narrowly ovate, deeply lobed to ± compound, segments rounded. **FL**: pedicel 1–4 mm; calyx lobes 3–4 mm, 5.5–8 mm in fr, narrowly oblong, short-hairy; corolla 3–5 mm, ± narrowly bell-shaped, yellow, persistent in fr, scales 0 or minute; stamens 1.5–3 mm, glabrous; style < 1.5 mm, puberulent. **FR** 4–7 mm, oblong, puberulent. **SEEDS** 5–30, 1–1.8 mm, cross-striate. n=12. RARE in CA. Alkaline flats, dry lake margins; 1500–2000 m. MP; OR, NV.

P. inyoensis (J.F. Macbr.) J. Howell (p. 703) INYO PHACELIA Ann 3–10 cm. **ST** decumbent to erect, branched at base, short-stiff-hairy, glandular-puberulent. **LVS** 5–20 mm, ± basal; blade > petiole, elliptic to obovate, entire to few-lobed. **FL**: pedicel 1–6 mm; calyx lobes 2–3 mm, 3.5–5 mm in fr, narrowly oblanceolate, short-hairy; corolla 2–3 mm, ± narrowly bell-shaped, pale yellow, persistent in fr, scales 0; stamens 1.5–3 mm, glabrous; style 1 mm, glabrous. **FR** 3–4 mm, oblong, puberulent. **SEEDS** 18–25, 0.5–1 mm; cross-furrows 5–8. n=12. UNCOMMON. Alkaline meadows; 1400–3200 m. SNE.

P. ivesiana Torrey Ann 5–25 cm. **ST** spreading to ascending, many-branched at base, finely short-hairy to glandular-puberulent. **LVS** 10–60 mm; lower ± = upper; blade = to > petiole, deeply lobed to ± compound, segments oblong to ± deltate. **FL**: pedicel 1–5 mm; calyx lobes 2.5–4 mm, 3–6 mm in fr, ± linear, minutely ciliate; corolla 2–4 mm, narrowly bell-shaped, tube yellow, lobes white, deciduous, scales 0 or minute; stamens 1–2 mm, glabrous; style ± 1 mm, ± glabrous. **FR** 3–5 mm, ovoid, puberulent; tip short-stiff-hairy. **SEEDS** 10–15, 1–1.5 mm; cross-furrows 5–12. n=11. Open, sandy areas; < 1000 m. D; to WY, Colorado, AZ. Pls with ± deltate lf lobes, seeds having 5–7 cross-furrows, and n=23 have been called var. *pediculoides* J. Howell.

P. ixodes Kellogg (p. 703) COSTA BAJA PHACELIA Ann 20–60 cm. **ST** gen erect, 0–few-branched above, densely dark-glandular-hairy. **LF** 30–180 mm; blade > petiole, gen ovate, lobed to compound, segments toothed. **FL**: pedicel < 4 mm; calyx lobes 4–6 mm, 7–11 mm in fr, subequal, oblanceolate to obovate, densely dark-glandular-hairy; corolla 8–14 mm, bell-shaped, white to rose, deciduous, scales ovate; stamens 6–9 mm, glabrous; style 5–8 mm, ± glabrous. **FR** 6–8 mm, oblong, sparsely short-hairy. **SEEDS** 10–18, 1–1.5 mm, pitted. n=11. RARE in CA. Sandy bluffs; < 100 m. s SCo (San Diego Co.); Baja CA.

P. lemmonii A. Gray Ann 7–20 cm. **ST** gen erect, simple to branched at base, glandular-puberulent. **LF** 10–40 mm; blade gen = or > petiole, ovate, toothed to slightly obtusely lobed. **FL**: pedicel ± 1 mm; calyx lobes 2–4 mm, 5–7 mm in fr, subequal, oblong to oblanceolate, short-hairy, glandular; corolla 4–6 mm, narrowly bell-shaped, tube yellow, limb white to lavender, deciduous, scales linear or 0; stamens 2–3 mm, glabrous; style 2–3 mm, glandular-puberulent. **FR** 3–4 mm, ± oblong, glandular-puberulent. **SEEDS** 30–80, ± 0.5 mm, pitted. n=22,24. Sandy washes, drying streambanks, slopes; 400–2300 m. ne PR, SNE, DMoj, n DSon; to UT.

P. leonis J. Howell (p. 707) SISKIYOU PHACELIA Ann 4–15 cm. **ST** gen erect, 0–few-branched at base, glandular-puberulent. **LVS** 10–30 mm; lowest opposite; blade linear to oblong, entire, tapered to petiole. **FL**: pedicel < 2 mm; calyx lobes 1–2.5 mm, 2–6 mm in fr, unequal, linear to oblanceolate, sparsely short-stiff-hairy; corolla 2–3 mm, widely bell-shaped, blue to lavender, deciduous, scales 0 or minute; stamens ± 2 mm, glabrous to puberulent; style ± 2 mm, glabrous. **FR** 2.5–3.5 mm, ± spheric, puberulent below; tip short-hairy. **SEEDS** 6–9, ± 1 mm, finely pitted. n=11. UNCOMMON. Sandy flats, slopes, coniferous forest; 1200–1900 m. KR.

P. linearis (Pursh) Holzinger Ann 8–60 cm. **ST** erect, simple to branched above, puberulent to short-stiff-hairy. **LF** 10–80 mm; blade linear to narrowly lanceolate, tapered to petiole, entire to 1–2-lobed near base, lobes linear. **FL**: pedicel 0.5–2 mm; calyx lobes 3–5 mm, 6–10 mm in fr, subequal, linear to narrowly oblanceolate, puberulent, stiffly ciliate; corolla 6–10 mm, widely bell-shaped, tube white to pale blue, limb gen violet, deciduous, scales linear; stamens 5–6 mm, ± short-hairy; style 5–8 mm, short-hairy. **FR** 5–8 mm, ovoid, sparsely short-hairy. **SEEDS** gen 6–15, 1–2 mm, pitted. n=11. Sandy or gravelly soils, shrubland, juniper woodland, coniferous forest; 800–2000 m. e KR, CaR, n SN, MP; to w Can, WY. ❀TRY.

P. longipes A. Gray Ann 10–40 cm. **ST** decumbent to erect, simple to branched at base, sparsely stiff-hairy, short-glandular-hairy. **LVS** ± basal, 20–140 mm; blade gen < petiole, ovate to round, ± crenate to irregularly toothed. **FL**: pedicel 3–10(30) mm; calyx lobes 3–4 mm, 4–6 mm in fr, ± linear, short-hairy, glandular; corolla 7–12 mm, widely bell-shaped, white to blue, deciduous, scales fused to filament base; stamens 10–15 mm, puberulent; style 10–15 mm, puberulent. **FR** 5–8 mm, ovoid, short-glandular-hairy; tip stiff-hairy. **SEEDS** 10–20, 1–1.5 mm, pitted. n=11. Gravelly or rocky soils, chaparral, juniper woodland, coniferous forest; 400–1900 m. WTR, SnGb, sw DMoj.

P. lyonii A. Gray (p. 707) Ann 30–120 cm. **ST** gen erect, 0–few-branched, glandular-puberulent to short-stiff-hairy. **LF** 40–150 mm; blade gen > petiole, ± ovate, deeply lobed to compound, segments crenate to finely lobed. **FL** subsessile; calyx lobes 4–5 mm, 6–8 mm in fr, oblanceolate, ± densely hairy, glandular; corolla 5–7 mm, bell-shaped, bluish, deciduous, scales ovate; stamens 4–5 mm, glabrous; style 2–4 mm. **FR** 5–7 mm, ± oblong, puberulent. **SEEDS** 7–20, 1.5–2 mm, pitted. *n*=11. Rocky slopes, canyons; < 500 m. s ChI.

P. malvifolia Cham (p. 707) Ann 20–100 cm. **ST** gen erect, gen few-branched, glandular-puberulent, densely yellow-stiff-hairy; hairs bulb-based. **LVS** 20–140 mm; blade = to > petiole, ± deltate to ± round, lower lobed to compound (segments 3, again toothed to irregularly lobed), upper gen toothed or lobed. **FL**: pedicel < 1 mm; calyx lobes 3–5 mm, 4–6 mm in fr, unequal, oblong to oblanceolate, densely yellow-bristly; corolla 5–7 mm, widely bell-shaped, cream-white, deciduous, scales ± ovate; stamens 5–10 mm, glabrous; style 8–12 mm, short-hairy. **FR** 2–3 mm, spheric, puberulent. **SEEDS** 2–6, 2–3 mm, pitted. *n*=11. Sandy or gravelly soils, coniferous forest, shrubland; < 1400 m. NCo, CW; OR.

P. marcescens J.F. Macbr. Ann 5–20 cm. **ST** gen erect, simple to branched, short-glandular-hairy, sparsely stiff-hairy. **LF** 10–50 mm; blade = to > petiole, elliptic to ovate, gen entire, sometimes 1–2-lobed at base. **FL** deciduous in fr; pedicel 1–2 mm; calyx lobes 2–3 mm, 3–4 mm in fr, subequal, linear to oblanceolate, short-hairy; corolla 4–5 mm, widely bell-shaped, purple to violet, persistent in fr, scales narrow; stamens 5–6 mm, glabrous; style 5–6 mm, short-hairy. **FR** 2.5–3.5 mm, ovoid, short-hairy. **SEEDS** (1)2, 1.5–2 mm, pitted. *n*=8. Sandy to gravelly soils, meadows, coniferous forest; 1400–2400 m. n&c SN.

P. minor (Harvey) Thell (p. 707) Ann 20–60 cm. **ST** gen erect, 0–few-branched, short-glandular-hairy, sparsely stiff-hairy. **LF** 20–110 mm; blade < to = petiole, ovate to ± round, irregularly toothed. **FL**: pedicel 10–15(20) mm; calyx lobes 5–6 mm, 6–8 mm in fr, narrowly oblong, short-hairy, glandular; corolla 10–40 mm, bell-shaped, purple, deciduous, scales fused to filament base, elongate; stamens 15–35 mm, short-glandular-hairy; style 15–40 mm, short-hairy. **FR** 7–13 mm, ovoid, beaked, puberulent; tip short-stiff-hairy. **SEEDS** 30–80, ± 1 mm, pitted. *n*=11. Open areas, burns, slopes; < 1600 m. SCo, e WTR, SnGb, SnBr, PR, w DSon; Baja CA. ❀TRY.

P. mohavensis A. Gray MOJAVE PHACELIA Ann 5–25 cm. **ST** erect, 0–few-branched, glandular-puberulent, short-stiff-hairy. **LF** 10–45 mm; blade linear to narrowly oblanceolate, entire, tapered to petiole. **FL**: pedicel 2–5 mm; calyx lobes 3–5 mm, 5–10(15) mm in fr, unequal, linear to oblanceolate, puberulent to short-hairy, glandular; corolla 5–8 mm, bell-shaped, tube base gen yellowish, limb white, aging pale blue, deciduous, scales lanceolate; stamens 5–14 mm, unequal, ± glabrous; style 5–8 mm, short-hairy. **FR** 3–5 mm, ovoid, short-stiff-hairy, gland-dotted. **SEEDS** 4–8, 1–2 mm, pitted. *n*=9. RARE. Sandy or gravelly soils, coniferous forest, dry streambeds; 1400–2500 m. SnGb, SnBr.

P. monoensis R. Halse (p. 707) MONO COUNTY PHACELIA Ann 2–12 cm. **ST** spreading to ascending, branched throughout, short-hairy, ± glandular. **LF** 8–25 mm; blade gen = petiole, ± oblong to ovate, entire to lobed. **FL**: pedicel 1–2 mm; calyx lobes 2–4 mm, 4–6 mm in fr, narrowly oblanceolate, short-hairy; corolla 2–4 mm, gen narrowly bell-shaped, yellow, persistent in fr, scales 0; stamens 1.5–3 mm, puberulent; style < 1.5 mm, puberulent. **FR** 2.5–4 mm, ovoid, puberulent. **SEEDS** < 10, 1–1.7 mm; cross-furrows 8–11. *n*=12. RARE. Clay soils, alkaline meadows, shrubland; 1900–2900 m. n SNE (Mono Co.); w NV.

P. mustelina Cov. (p. 707) DEATH VALLEY ROUND-LEAVED PHACELIA Ann 6–30 cm. **ST** decumbent to ascending, many-branched, ± short-glandular. **LF** 10–40 mm; blade = or > petiole, ± round, irregularly toothed. **FL**: pedicel 1–3(5) mm; calyx lobes 2–4 mm, 5–6 mm in fr, narrowly oblanceolate, glandular-short-hairy; corolla 6–10 mm, narrowly bell-shaped, violet to purple, deciduous, scales lanceolate; stamens 2–4 mm, short-hairy; style 3–5 mm, sparsely short-hairy. **FR** 3–4 mm, ovoid, puberulent. **SEEDS** 20–60, ± 0.5 mm, finely pitted. *n*=12. RARE. Gravelly or rocky slopes, creosote-bush scrub, pinyon/juniper woodland; 1000–2100 m. DMtns; w NV.

P. mutabilis E. Greene (p. 707) Bien or short-lived per, 10–60 cm. **ST** decumbent to erect, ± stiff-hairy, not glandular. **LF**: blade 20–180 mm, gen = petiole, lanceolate to ovate, entire to compound (then segments 3, terminal largest). **FL**: pedicels 0.5–3 mm; calyx lobes 3–5 mm, 6–10 mm in fr, ± unequal, ± linear to oblanceolate; corolla 4–6 mm, tubular to bell-shaped, white to yellowish or lavender, scales oblong; stamens 6–8 mm, exserted, hairy; style 6–8 mm, exserted. **FR** 2–3 mm, ovoid, stiff-hairy. **SEEDS** 1–4, 1.5–2.5 mm, pitted. *n*=11,22. Roadsides, ridges, open forests; 900–3100 m. KR, NCoR, CaRH, SNH, SnJt, MP; OR, NV. Closely related to *P. nemoralis*, intergrades with *P. hastata*.

P. nashiana Jepson (p. 707) CHARLOTTE'S PHACELIA Ann, 4–18 cm. **ST** ascending to erect, simple to branched at base, short-stiff-hairy; gland-tipped hairs black. **LVS** ± basal, 15–70 mm; blade gen < petiole, widely ovate to ± round, irregularly crenate to slightly lobed. **FL**: pedicel 5–10 mm; calyx lobes 3–4 mm, 5–8 mm in fr, oblong, short-hairy, glandular; corolla 10–18 mm, rotate to widely bell-shaped, tube white, throat blue with 5 ± white spots below sinuses, lobes bright blue, deciduous, scales fused to filament base, toothed; stamens 12–20 mm, short-hairy; style 10–20 mm. **FR** 7–14 mm, ovoid, beaked, short-glandular-hairy. **SEEDS** 40–80, ± 2 mm, pitted. *n*=11. RARE. Sandy to rocky, granitic slopes, gen in Joshua-tree or pinyon/juniper woodland; < 2200 m. s SNH, Teh (e slope).

P. neglecta M.E. Jones (p. 707) Ann 3–20 cm. **ST** ascending to erect, 0–few-branched, ± short-glandular. **LVS** ± basal, 10–50 mm; blade = or > petiole, ± round, wavy to crenate. **FL**: pedicel 1–8 mm, < 12 mm in fr; calyx lobes 1.5–2 mm, 2–3.5 mm in fr, narrowly ovate, densely short-hairy, glandular; corolla 4–6 mm, cream-white, funnel- to bell-shaped, deciduous, scales linear; stamens 2–4 mm, puberulent; style 2–3 mm. **FR** 3–4 mm, spheric, puberulent. **SEEDS** > 60, ± 1 mm; cross-furrows 4–7. *n*=11. Clay or alkaline soils, flats, slopes; < 1000 m. D; s NV, AZ.

P. nemoralis E. Greene Per, short-lived, 50–200 cm. **STS** ascending to erect, densely stiff-hairy, not glandular. **LVS**: blade 40–150(250) mm, < to > petiole, widely lanceolate to ovate; lowermost gen compound, lflets 3–7; uppermost often simple, entire. **FL**: pedicels < 3 mm; calyx lobes 3.5–5 mm, 5–8 mm in fr, (ob)lanceolate; corolla 3–6 mm, bell-shaped, greenish white to yellowish, scales oblong; stamens 7–9 mm, exserted, hairy; style 6–9 mm. **FR** 2–3 mm, ovoid to subspheric, stiff-hairy. **SEEDS** 1–4, 1.5–2 mm, pitted. *n*=11,22. Moist slopes, streambanks, mixed-evergreen or coniferous forest; < 800 m. NCo, NCoRO, CW (exc SCoRI); to WA.

ssp. ***nemoralis*** **ST** gen < 7 mm diam at midpoint. **LVS**: basal gen with 3 lflets. **FL**: corolla 3.5–5 mm, greenish white. *n*=11. Habitats of sp.; 50–700 m. NCoRO, CCo, SnFrB, SCoRO. ❀DRN,IRR:5,**17**&SHD:7,**14–16,22–24**.

ssp. ***oregonensis*** Heckard **ST** gen 7–10 mm diam at midpoint. **LVS**: basal gen with 3–7 lflets. **FL**: corolla 5–6 mm, yellowish. *n*=22. Habitats of sp.; < 800 m. NCo, NCoRO; to WA. Perhaps best treated as part of *P. californica*.

P. novenmillensis Munz (p. 707) NINE-MILE CANYON PHACELIA Ann 5–10 cm. **ST** ascending to erect, 0–few-branched, short-soft-hairy, sparsely glandular-puberulent. **LF** 20–80(110) mm; blade gen < petiole, (ob)lanceolate to narrowly elliptic, entire; lowermost sometimes lobed to irregularly compound. **FL**: pedicel 1–3 mm; calyx lobes 2–4 mm, 8–10 mm in fr, subequal, linear, long-ciliate; corolla 3–4 mm, bell-shaped, lavender, deciduous, scales lanceolate; stamens 2–4.5 mm, glabrous; style 4–5 mm. **FR** 2–3 mm, ovoid, short-hairy. **SEEDS** 2–4, 1.5–2 mm, pitted. RARE. Open, sandy to gravelly soils, pinyon/juniper woodland, coniferous forest; 1900–2200 m. s SNH (e slope).

P. orogenes Brand (p. 707) MOUNTAIN PHACELIA Ann 2–10 cm. **ST** erect, simple to branched, puberulent, sparsely glandular. **LVS** 5–30 mm; lower opposite; blade linear to narrowly lanceolate, entire, tapered to petiole. **FL**: pedicel 2–7 mm; calyx lobes 2–3 mm, 5–8 mm in fr, unequal, linear to oblanceolate, short-hairy; corolla 4–6 mm, widely bell-shaped, tube white to yellowish, limb violet,

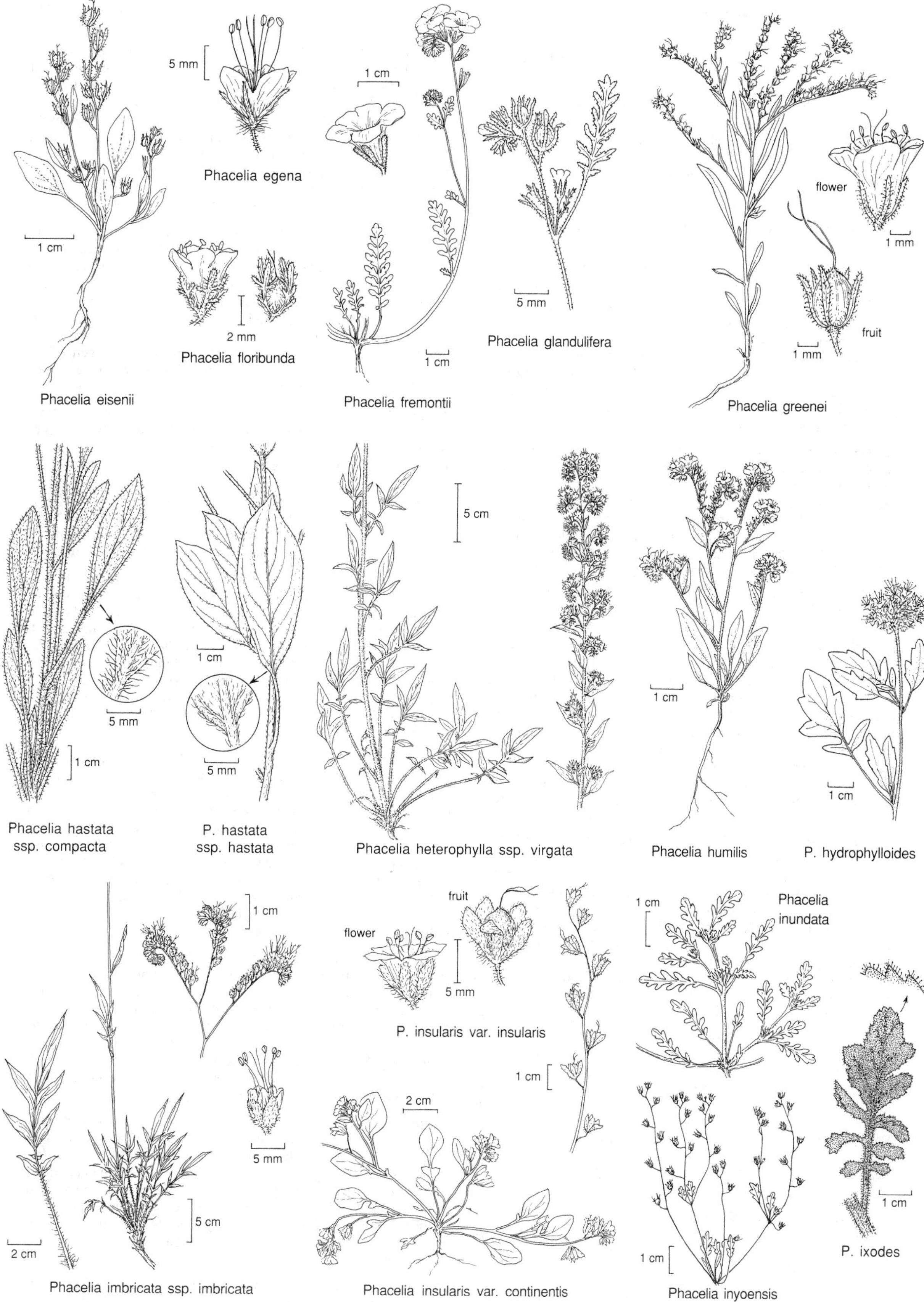

Phacelia egena

Phacelia floribunda

Phacelia eisenii

Phacelia fremontii

Phacelia glandulifera

Phacelia greenei

Phacelia hastata ssp. compacta

P. hastata ssp. hastata

Phacelia heterophylla ssp. virgata

Phacelia humilis

P. hydrophylloides

P. insularis var. insularis

Phacelia inundata

Phacelia imbricata ssp. imbricata

Phacelia insularis var. continentis

Phacelia inyoensis

P. ixodes

deciduous, scales oblong; stamens 3–5 mm, glabrous; style 3–4 mm, short-hairy. **FR** 2.5–3.5 mm, ± spheric, short-hairy. **SEEDS** 3–6, 1.5–2 mm, pitted. *n*=9. UNCOMMON. Gravelly slopes, meadow borders, coniferous forest; 2400–3400 m. s SNH (Mineral King, Tulare Co.).

P. pachyphylla A. Gray (p. 707) Ann 4–17 cm. **ST** erect, 0–few-branched, short-glandular. **LVS** ± basal, 20–50 mm; blade gen < petiole, ± round, entire to slightly crenate. **FL**: pedicel < 3 mm; calyx lobes 2–3 mm, 3–4 mm in fr, oblong, densely short-hairy, glandular; corolla 5–7 mm, funnel- to bell-shaped, violet to purple, deciduous, scales linear; stamens 2–4 mm, short-hairy; style 2–3 mm. **FR** 5–7 mm, ± spheric, short-hairy, some gland-stalked. **SEEDS** > 60, ± 1 mm; cross-furrows 6–8. *n*=11. Flats, ± alkaline soils, creosote-bush scrub; < 1000 m. D; Baja CA.

P. parishii A. Gray (p. 707) PARISH'S PHACELIA Ann 5–15 cm. **ST** ascending to erect, branched at base, glandular-puberulent. **LVS** ± basal, 8–30 mm; blade > petiole, widely elliptic to obovate, entire to barely toothed. **FL**: pedicel ± 1 mm; calyx lobes 3–5 mm, 6–8 mm in fr, unequal, ± linear to ovate, puberulent; corolla 4–6 mm, narrowly bell-shaped, tube yellow, limb lavender, deciduous, scales ± elliptic; stamens 2–4 mm, sparsely short-hairy; style 1–2 mm. **FR** 3–5 mm, ± oblong, short-hairy. **SEEDS** 20–40, 1–1.5 mm, finely pitted. RARE. Clay or alkaline soils, dry lake margins; 800–1200 m. w DMoj (nw San Bernardino Co); NV.

P. parryi Torrey Ann 10–70 cm. **ST** erect, 0–few-branched, glandular-puberulent, stiff-hairy. **LF** 10–120 mm; blade < or = petiole, oblong to ovate, irregularly toothed. **FL**: pedicel 10–20 mm; calyx lobes 4–6 mm, 6–8 mm in fr, ± linear, sparsely hairy, ± glandular; corolla 10–20 mm, rotate to widely bell-shaped, tube and lower throat white to light violet, limb violet to purple, deciduous, scales fused to filament base, square; stamens 10–20 mm, long-hairy; style 10–20 mm, short-hairy. **FR** 6–10 mm, ovoid, beaked, short-stiff-hairy, gland-dotted. **SEEDS** 40–60, ± 1 mm, pitted. *n*= 11. Open areas, burns, slopes, coastal-sage scrub, chaparral; < 2400 m. SCoRI, SW (exc ChI, w WTR); Baja CA. ❀TRY.

P. pedicellata A. Gray (p. 707) Ann 12–50 cm. **ST** erect, 0–few-branched above base, short-stiff-hairy, glandular. **LF** 20–120 mm; blade gen = petiole, ovate to round, lower compound (lflets 3–7, rounded or toothed), upper lobed to compound (segments 3, gen rounded). **FL**: pedicel 1–2 mm; calyx lobes 3–4.5 mm, 4–5.5 mm in fr, narrowly oblanceolate, sparsely hairy, ciliate, glandular; corolla 5–7 mm, bell-shaped, pink to blue, deciduous, scales round; stamens 6–8 mm, glabrous; style 6–8 mm. **FR** 3–3.5 mm, ± spheric, puberulent. **SEEDS** gen 4, ± 3 mm; back pitted; inner surface with central ridge separating 2 longitudinal grooves. *n*=11. Sandy or gravelly washes, canyons; < 1400 m. D; s NV, AZ, Baja CA.

P. peirsoniana J. Howell Ann 4–30 cm. **ST** erect, simple to branched above base, glandular-puberulent, sparsely short-stiff-hairy. **LF** 10–60 mm; blade gen < petiole, ± round, ± toothed. **FL**: pedicel 2–4 mm; calyx lobes 3–4 mm, 6–8 mm in fr, oblong, short-hairy; corolla 4–6 mm, narrowly bell-shaped, white to violet, deciduous, scales lanceolate; stamens 2–3 mm, sparsely short-hairy; style 1–2 mm, short-hairy. **FR** 4–6 mm, oblong, puberulent. **SEEDS** 20–50, ± 1 mm, pitted. *n*=12. Rocky slopes, canyons, sagebrush shrubland, pinyon/juniper woodland; 1500–2700 m. n SNE, w NV.

P. perityloides Cov. (p. 707) Per 5–40 cm. **STS** spreading to pendent, glandular; base spreading-hairy to woolly. **LF**: petiole 3–50 mm; blade 5–25 mm, ± round, irregularly toothed to lobed. **FL**: pedicel 5–30 mm; calyx lobes 3–6 mm, 4–6 mm in fr, oblong to oblanceolate; corolla 10–15 mm, narrowly funnel- to bell-shaped, tube yellowish or aging purple, limb white, scales linear; stamens 3–6 mm, unequal, incl, glabrous; style 4–6 mm, incl, short-lobed. **FR** 2.5–4 mm, narrowly ovoid, hairy. **SEEDS** 50–200, ± 0.5 mm, angular, pitted. *n*=11. Crevices on cliffs, rocky, often calcareous slopes; 600–2300 m. W&I, n&e DMoj; NV, AZ. Another var. in AZ.

var. ***jaegeri*** Munz **ST**: base spreading-hairy, not clearly woolly. **FL**: pedicels 10–30 mm in fr, reflexed; calyx lobes < 1 mm wide; corolla 12–15 mm. Habitats of sp.; 1900–2300 m. e DMtns (Clark Mtn); NV.

var. ***perityloides*** **ST**: base clearly woolly, some hairs spreading. **FL**: pedicels 5–15 mm in fr, spreading to ascending; calyx lobes 1–2 mm wide; corolla 10–12 mm. Habitats of sp.; 600–2200 m. n W&I, n DMoj.

P. phacelioides (Benth.) Brand (p. 707) MOUNT DIABLO PHACELIA Ann 5–20 cm. **ST** ascending to erect, 0–many-branched at base, puberulent, sparsely stiff-hairy. **LF** 20–80(100) mm; blade < or = petiole, elliptic to lanceolate, entire. **FL**: pedicel 1–3 mm; calyx lobes 4–6 mm, 8–11 mm in fr, ± oblong, puberulent, densely ciliate; corolla 4–6 mm, narrowly bell-shaped, white to lavender, lobes violet-streaked, deciduous, scales ovate, minutely toothed; stamens 2–3 mm, short-hairy; style ± 2 mm, short-hairy. **FR** 3.5–4 mm, ovoid, short-hairy. **SEEDS** 5–15, 1–1.5 mm, pitted. *n*=11. RARE. Open, rocky slopes; 500–1200 m. e SnFrB.

P. platyloba A. Gray (p. 707) Ann 9–45 cm. **ST** erect, simple to branched at base, puberulent, sparsely short-stiff-hairy. **LF** 15–90 mm; blade > petiole, elliptic to ovate, 1–2-compound, lflets lobed to toothed. **FL**: pedicel 1–2 mm; calyx lobes 2–4 mm, ciliate, strongly unequal, 3 gen ± lanceolate, entire, 2 ovate to ± round, gen crenate, stalked; corolla 4–5 mm, widely bell-shaped, bluish to lavender, deciduous, scales oblong; stamens 3–4 mm, glabrous; style 4–6 mm. **FR** 2–3 mm, ± oblong, densely puberulent. **SEED** gen 1, 2–3 mm, finely pitted. *n*=11. Gravelly or rocky soils, chaparral, woodland; 300–1200 m. c SNF.

P. pringlei A. Gray (p. 707) Ann 2–18 cm. **ST** erect, 0–few-branched, glabrous to glandular-puberulent below. **LVS** 7–30 mm; lower gen opposite; blade linear to narrowly lanceolate, entire, tapered to petiole. **FL**: pedicel 1–3 mm; calyx lobes 1–2 mm, 2–6 mm in fr, gen unequal, linear to oblanceolate, short-hairy, glandular; corolla 3–5 mm, ± rotate, lavender, deciduous, scales oblong; stamens 3–4 mm, papillate; style 3–4 mm, short-hairy. **FR** 2.5–3.5 mm, ± spheric, short-hairy. **SEEDS** 3–8, 1.5–2 mm, pitted. *n*=11. Open, steep slopes, ridges, coniferous forest; 1400–2700 m. KR.

P. procera A. Gray Per 50–200 cm. **ST** erect, hairy. **LF**: petiole < 40 mm; blade 50–120 mm, lanceolate to ovate, coarsely toothed to lobed. **FL**: pedicel < 3 mm, black-glandular; calyx lobes 3–5 mm, 7–8 mm in fr, linear, black-glandular; corolla 3–7 mm, ± bell-shaped, cream to greenish or brownish white, scales oblong; stamens 8–10 mm, exserted, hairy; style 6–10 mm, exserted. **FR** 6–8 mm, ovoid, short-rough-hairy, glandular. **SEEDS** gen 12–16, ± 2 mm, angled; surface net-like, pitted. *n*=11. Meadows, slopes, talus, coniferous forest; 1200–2200 m. KR, NCoRH, CaRH, n SNH; to WA, ID.

P. pulchella A. Gray var. ***gooddingii*** (Brand) J. Howell (p. 707) GOODDING'S PHACELIA Ann 5–20 cm. **ST** ascending to erect, branched throughout, glandular-puberulent. **LF** 5–40 mm; blade = or > petiole, ovate to ± round, ± toothed. **FL**: pedicel ± 2 mm; calyx lobes 4–5 mm, 6–9 mm in fr, oblanceolate, glandular-puberulent; corolla 6–12 mm, bell-shaped, tube yellow, limb lavender to violet, deciduous, scales ± lanceolate; stamens 3–5 mm, puberulent; style 4–5 mm, short-hairy. **FR** 3–5 mm, ± oblong, short-hairy. **SEEDS** (25)30–50, ± 0.5 mm, pitted. RARE in CA. Clay soils, flats; 800–1000 m. ne DMoj (Mesquite Valley, Kingston Range); also nw AZ. Other vars. in NV, s UT, n AZ.

P. purpusii Brandegee Ann 10–40 cm. **ST** gen erect, 0–few-branched, puberulent, glandular. **LF** 12–50 mm; blade > petiole, elliptic to ovate, entire to lobed. **FL**: pedicel 1–2 mm; calyx lobes 3–4 mm, 4–7 mm in fr, unequal, oblanceolate to obovate, short-hairy, ciliate; corolla 6–8 mm, bell-shaped, lavender to violet, deciduous, scales oblong; stamens 7–8 mm, papillate; style 6–9 mm, short-hairy. **FR** 3–5 mm, ovoid, short-glandular-hairy. **SEEDS** 3–7, 1.5–2 mm, pitted. *n*=9. Sandy or gravelly soils, coniferous forest; 900–2300 m. s CaRH, SN, MP.

P. quickii J. Howell (p. 707) Ann 4–18 cm. **ST** decumbent to erect, simple or branched, puberulent, glandular or not. **LF** 8–50 mm; blade linear to oblanceolate, entire, tapered to petiole. **FL**: pedicel < 1 mm; calyx lobes 2–4 mm, 4–7 mm in fr, unequal, linear to oblanceolate, short-hairy; corolla 3–5 mm, widely bell-shaped, white to blue, tardily deciduous, scales narrow; stamens 3–6 mm, glabrous; style 3–6 mm, short-hairy. **FR** 2–2.5 mm, ± spheric, short-hairy. **SEEDS** 2–4, 1–1.5 mm, pitted. *n*=8. Open granitic areas; 1000–2400 m. n&c SN.

P. racemosa (Kellogg) Brandegee Ann 4–18 cm. **ST** erect, 0–few-branched, glabrous to glandular-puberulent. **LVS** 10–40 mm;

lowest opposite; blade linear to narrowly oblanceolate, entire, tapered to petiole. **FL**: pedicel < 1 mm; calyx lobes 1–2 mm, 2–3 mm in fr, ± linear, short-hairy; corolla 2–4 mm, narrowly bell-shaped, pale blue, deciduous, scales narrow; stamens 1–2 mm, glabrous; style 1–2 mm, short-hairy. **FR** 2–3.5 mm, ± spheric, short-hairy. **SEEDS** gen 4, 1.5–2.5 mm, finely pitted. *n*=7. Gravelly to rocky slopes in coniferous forest; 1800–3000 m. CaR, n&c SN.

P. ramosissima Lehm. Per 30–150 cm. **ST** prostrate to ascending, many-branched, glabrous to densely hairy, glandular or not. **LF**: blade 40–200 mm, gen >> petiole, oblong to widely ovate, compound; lflets ± sessile, elliptic to oblong, coarsely toothed or lobed, lobes often toothed. **FL** ± sessile; calyx lobes 4–6 mm, not gen longer in fr, oblanceolate to ± spoon-shaped; corolla 5–8 mm, funnel- to bell-shaped, white to lavender, scales ovate; stamens 7–10 mm, exserted, glabrous; style 7–10 mm, exserted. **FR** 3–4 mm, ovoid, sharply bristly. **SEEDS** 2–4, 2–3 mm, pitted. *n*=11. Many habitats; < 3100 m. CA; to WA, ID, AZ. Vars. difficult, need study.

var. ***austrolitoralis*** Munz (p. 707) **ST** below infl not glandular; hairs coarse, long, stiff, spreading, bulb-based. Sand dunes, salt marshes, coastal bluffs; < 300 m. CCo, SCo, n ChI.

var. ***eremophila*** (E. Greene) J.F. Macbr. **ST** below infl gen glabrous, sometimes with few short hairs. Slopes, open places, coniferous forest; 600–2800 m. CaRH, SNH, Wrn, SNE; OR, NV. [var. *valida* M. Peck]

var. ***latifolia*** (Torrey) Cronq. **ST** below infl glandular; hairs mostly long, coarse, stiff, bulb-based, some hairs soft, spreading. Slopes, canyons, washes, flats; 50–2500 m. Teh, SnFrB, SCoR, SCo, TR, PR, n DMtns (Panamint Range); NV, AZ. [var. *suffrutescens* Parry]

var. ***montereyensis*** Munz **ST** often 100+ cm, not glandular below infl, densely puberulent; some hairs fine, long. Sand dunes, salt marshes, coastal bluffs; < 50 m. NCo, CCo, n ChI. ❀DRN, SUN:14,**15–17**,22,23,**24**.

var. ***ramosissima*** (p. 707) **ST** below infl glandular; hairs short to long, soft, spreading. Slopes, ridges, washes, meadows; 100–2800 m. KR, NCoRH, NCoRI, SN, Teh, CW, WTR, SnBr, DMtns (Panamint Range); to WA.

var. ***subglabra*** M. Peck (p. 707) **ST** gen < 60 cm, not glandular below infl, densely puberulent; some hairs fine, long. Slopes, meadows, coniferous forest; 200–3100 m. SNH, Teh, SnGb, SNE; to OR, ID.

P. rattanii A. Gray Ann 15–100 cm. **ST** gen erect, simple to branched at base, short-glandular-hairy; bristles ± white, slightly bulb-based. **LF** 10–75 mm; blade > petiole, gen ovate, irregularly toothed to slightly lobed. **FL**: pedicel ± 1 mm; calyx lobes 2–4 mm, 3–7 mm in fr, unequal, oblanceolate to obovate, short-hairy, base strongly ciliate; corolla 3–5 mm, narrowly bell-shaped, white to blue, deciduous, scales lanceolate; stamens 2–3 mm, puberulent; style 2–3 mm, glabrous. **FR** 2–3 mm, spheric, puberulent. **SEEDS** gen 2, 1.5–2 mm, pitted. *n*=11. Shaded crevices, steep slopes; < 1400 m. KR, NCoR, s CaR, SnFrB, n SCoRI.

P. rotundifolia A. Gray Ann 4–28 cm. **ST** decumbent to erect, few–many-branched, short-stiff-hairy, gen glandular. **LF** 10–40 mm; blade gen < petiole, ± round, toothed. **FL**: pedicel 1–4 mm; calyx lobes 2–4 mm, 4–6 mm in fr, narrowly oblanceolate, short-hairy; corolla 3–6 mm, narrowly bell-shaped, deciduous, tube pale yellow, limb white to violet, scales lanceolate; stamens 2–3 mm, glabrous; style 1–2 mm, sparsely short-hairy. **FR** 3.5–4.5 mm, oblong, puberulent. **SEEDS** 50–100, ± 0.5 mm, pitted. *n*=12. Rocky slopes, crevices, ledges, creosote-bush scrub & pinyon/juniper woodland; < 2000 m. W&I, D; to sw UT, AZ.

P. saxicola A. Gray Ann 5–15 cm. **ST** ascending to erect, gen many-branched, short-stiff-hairy, glandular-puberulent. **LF** 3–10 mm; blade gen = petiole, narrowly oblanceolate to ovate, entire. **FL**: pedicel 1–2 mm; calyx lobes 3–4 mm, 5–7 mm in fr, subequal, linear to narrowly oblanceolate, short-hairy; corolla 3–4 mm, narrowly bell-shaped, tube white, limb blue to violet, ± persistent in fr, scales linear; stamens 1–2 mm, ± glabrous; style 1–2 mm. **FR** 2–3 mm, ovoid, short-stiff-hairy. **SEEDS** 20–50, ± 0.5 mm, pitted. Limestone slopes, woodland; 1000–2300 m. SNE, n DMoj; s NV, nw AZ.

P. sericea (Graham) A. Gray var. ***ciliosa*** Rydb (p. 711) BLUE ALPINE PHACELIA Per 10–60 cm. **ST** ascending to erect, not glandular; hairs silky, silvery, appressed. **LF**: blade 20–65 mm, ± = petiole, lanceolate to oblong, lobed to dissected; segments entire or toothed. **FL**: pedicels 2–5 mm; calyx lobes 2–7 mm, not enlarged in fr, linear to oblong; corolla 5–8 mm, urn- to bell-shaped, dark blue to purple, persistent in fr, scales oblong to lanceolate; stamens 10–15 mm, exserted, glabrous to sparsely hairy at base; style 6–13 mm, exserted. **FR** 4–6 mm, ovoid, rough-hairy. **SEEDS** 8–18, 1–2 mm, pitted in vertical rows. *n*=11. RARE in CA. Ridges, talus slopes; 2100–2700 m. n KR (China Peak), Wrn; to ID, Colorado.

P. stebbinsii Constance & Heckard (p. 711) STEBBINS' PHACELIA Ann 10–40 cm. **ST** erect, 0–few-branched, puberulent, glandular or not. **LVS** 7–50 mm, opposite below middle; blade gen > petiole, elliptic to lanceolate, larger gen 2–6-toothed or -lobed, smaller gen entire. **FL**: pedicel < 1 mm; calyx lobes 1–3 mm, 3–4 mm in fr, narrowly oblanceolate, short-hairy, glandular; corolla 4–5 mm, rotate, white to pale blue, persistent in fr, scales lanceolate; stamens 6–7 mm, glabrous; ovules few, style 4–5 mm, glandular-puberulent. **FR** 3–4 mm, ovoid, densely short-hairy. **SEED** gen 1, 1.5–2 mm, pitted. *n*=8. RARE. Gravelly soils, meadows, coniferous forest; 1000–1600 m. n SN (El Dorado Co.).

P. stellaris Brand BRAND'S PHACELIA Ann 6–25 cm. **ST** spreading to ascending, gen branched at base, puberulent. **LVS** ± basal, 5–70 mm; blade gen > petiole, oblanceolate to ovate, deeply lobed to compound, segments rounded to obtuse. **FL**: pedicel 2–8 mm; calyx lobes 2–5 mm, 4–8 mm in fr, oblanceolate, short-stiff-hairy; corolla 3–5(7) mm, widely bell-shaped, light blue to purplish, deciduous, scales ovate to square; stamens 2–4 mm, gen glabrous; style 1–3 mm, puberulent. **FR** 4.5–6 mm, ovoid, sparsely short-hairy. **SEEDS** 8–20, 0.5–1 mm, pitted. *n*=11. RARE. Open areas, coastal-sage scrub; < 400 m. SCo; Baja CA. Some pls from near se WTR foothills intermediate to *P. douglasii* var. *d.* [*P. douglasii* var. *cryptantha* Brand]

P. suaveolens E. Greene (p. 711) Ann 5–40 cm. **ST** ascending to erect, simple to branched at base, short-hairy, glandular-puberulent. **LF** 10–75 mm; blade gen > petiole, widely elliptic to ovate, toothed to slightly lobed. **FL**: pedicel 1–2 mm; calyx lobes 4–5 mm, 6–8 mm in fr, oblanceolate, short-hairy, glandular; corolla 7–14 mm, narrowly bell-shaped, tube yellow, limb lavender to purple, deciduous, scales linear; stamens 3–6 mm, glabrous; style 3–4 mm, short-glandular-hairy. **FR** 3–5 mm, ovoid to ± oblong, short-hairy. **SEEDS** 8–16, 1–1.5 mm, pitted. Open burns, slopes, chaparral, closed-cone-pine forest; 300–1700 m. NCoRI, c SNF, SnFrB, SCoRI, nw PR. Vars. need study, may be best considered distinct spp.

var. ***keckii*** (Munz & I. M. Johnston) J. Howell SANTIAGO PEAK PHACELIA **FL**: corolla 10–14 mm. **FR** ± oblong. **SEEDS** 8–10. RARE. Open chaparral; 1000–1600 m. nw PR (Santa Ana Mtns).

var. ***suaveolens*** **FL**: corolla 7–11 mm. **FR** ovoid. **SEEDS** 10–16. *n*=12. Habitats of sp. Uncommon. NCoRI, c SNF (Ione), SnFrB, SCoRI.

P. tanacetifolia Benth. (p. 711) Ann 15–100 cm. **ST** erect, 0–few-branched, ± short-stiff-hairy, glandular-puberulent. **LF** 20–200 mm; blade > petiole, ± oblong to ovate, gen compound, lflets toothed to lobed. **FL** ± sessile; calyx lobes 4–6 mm, 6–8 mm in fr, ± linear, densely long-hairy; corolla 6–9 mm, widely bell-shaped, blue, ± persistent in fr, scales narrow; stamens 9–15 mm, glabrous; style 11–15 mm, glabrous. **FR** 3–4 mm, ± ovoid, glabrous; tip puberulent to short-hairy. **SEEDS** 1–2, 2–3 mm, wrinkled, pitted. *n*= 11. Sandy to gravelly slopes, open areas; < 2000 m. s NCoRO, c&s SNF, Teh, s ScV (Sutter Buttes), SnJV, e SnFrB, SCoR, SW (exc ChI), DMoj; s NV, AZ. ❀SUN,DRN:**7–12,14–24**.

P. tetramera J. Howell Ann 2–15 cm. **ST** spreading to ascending, branched throughout, sparsely short-hairy, ± glandular. **LF** 5–30 mm; blade gen = petiole, oblong to ovate, entire to few-toothed. **FL**: pedicel 1–4 mm; calyx lobes 4, 1.5–3 mm, 3.5–4.5 mm in fr, narrowly oblanceolate, puberulent to short-hairy; corolla 1.3–2 mm, bell-shaped, whitish, persistent in fr, lobes 4, scales 0; stamens 1–2 mm, glabrous; ovules 12–24, style < 0.5 mm, glabrous. **FR** 2.5–4 mm, oblong to ± spheric, puberulent. **SEEDS** 5–12, ± 1 mm; cross-furrows 6–9. *n*=11. Alkaline flats, washes, meadows; 1500–2400 m. SNE; to e OR, UT.

P. thermalis E. Greene (p. 711) Ann 5–45 cm. **ST** ascending to erect, gen branched at base, glandular-hairy, sparsely stiff-hairy. **LF** 12–80 mm; blade >> petiole, oblong to ovate, deeply lobed to compound, segments toothed. **FL** ± sessile; calyx lobes 3–4 mm, 5–7 mm in fr, lanceolate, short-hairy, ciliate, veins raised; corolla 3–4 mm, bell-shaped, white to pale blue, deciduous, scales oblanceolate; stamens 1–3 mm, glabrous; style ± 2 mm, glabrous. **FR** 3–4 mm, spheric, puberulent. **SEEDS** 2–4, 2–2.5 mm, pitted. *n*=11. Open clay flats; 1000–1700 m. e CaR, MP; to se OR, ID.

P. umbrosa E. Greene Ann 15–45 cm. **ST** decumbent to erect, weak, gen branched, sparsely stiff-hairy, glandular above. **LF** 25–90 mm; blade > petiole, ± ovate, gen compound, segments crenate to lobed. **FL** ± sessile; calyx lobes 3–5 mm, 5–7 mm in fr, unequal, linear to oblanceolate, long-hairy; corolla 3–6 mm, narrowly bell-shaped, pale blue to lavender, tardily deciduous, scales ovate; stamens 4–10 mm, glabrous; style 7–12 mm, glabrous. **FR** 2.5–3.5 mm, elliptic, gen sparsely short-hairy. **SEEDS** 2–4, 1.5–2 mm, pitted. Uncommon. Chaparral, oak/pine woodland; 1000–1600 m. PR; Baja CA. Relationship to *P. distans* needs further study.

P. vallicola Brand (p. 711) MARIPOSA PHACELIA Ann 6–27 cm. **ST** erect, simple to branched below middle, puberulent, sparsely stiff-hairy. **LVS** 10–35 mm; lowest opposite; blade > petiole, lanceolate to ovate, entire. **FL**: pedicel 2–6 mm; calyx lobes 3–5 mm, 7–10 mm in fr, unequal, oblong to oblanceolate, densely ciliate, sparsely glandular; corolla 4–6 mm, bell-shaped, tube lavender, limb gen purple, deciduous, scales oblong; stamens 2–3 mm, papillate; style 2–4 mm, short-hairy. **FR** 4–6 mm, ovoid, beaked, short-hairy. **SEEDS** 8–14, 1.5–2 mm, pitted. *n*=11. UNCOMMON. Open, gravelly to rocky soils, chaparral, oak/pine woodland, coniferous forest; 600–2400 m. s CaR, n&c SN.

P. vallis-mortae J. Voss (p. 711) Ann 20–60 cm. **ST** ascending to erect, simple to branched, puberulent and sparsely stiff-reflexed-hairy. **LF** 15–80 mm; blade > petiole, ± oblong, compound, lflets toothed or slightly lobed. **FL**: pedicel ± 1 mm; calyx lobes 4–6 mm, 7–10 mm in fr, linear to narrowly elliptic, long-hairy; corolla 8–15 mm, funnel- to bell-shaped, lavender to violet, deciduous, scales 0; stamens 6–12 mm, glabrous; style 5–12 mm, short-glandular-hairy. **FR** 3–4 mm, ovoid, puberulent. **SEEDS** gen 4, 2.5–3 mm, pitted. *n*=11. Sandy to rocky soils, shrubland; 600–2400 m. w SnJV, SNE, e DMoj, n DSon; to sw UT, nw AZ. Pls from w SnJV with dark-veined corolla have been called var. *heliophila* (J.F. Macbr.) J. Voss; relationships to *P. cryptantha* or *P. cicutaria* need careful study. ❀TRY.

P. viscida (Benth.) Torrey Ann 10–70 cm. **ST** erect, 0–few-branched, puberulent to short-stiff-hairy, glandular. **LF** 10–100 mm; blade > petiole, gen ovate, irregularly toothed. **FL**: pedicel 2–10 mm; calyx lobes 3–4 mm, 5–10 mm in fr, gen oblong, short-hairy, glandular; corolla 8–18 mm, rotate, tube white or purple, limb ± blue, deciduous, scales 0; stamens 8–16 mm, short-hairy; ovules many, style 8–15 mm. **FR** 8–12 mm, ovoid, beaked, puberulent; tip short-glandular-hairy. **SEEDS** 40–80, ± 1 mm, pitted. *n*= 11. Open, sandy burns, coastal sage scrub, chaparral; < 700 m. s CCo, SCoRO, SCo, ChI (exc San Clemente Island), WTR; Baja CA. ❀DRN,SUN:**14–17,22–24**.

PHOLISTOMA

Ann, fleshy. **ST** many-branched, prostrate or reclined, brittle; angles ± glabrous, bristly, or gen with hooked prickles. **LVS** simple, cauline; lower opposite; upper alternate; petioles gen winged, clasping; blade pinnately lobed, uppermost reduced, short-petioled, gen deltate, 3-lobed, with small, sharp bristles on both surfaces. **INFL** terminal, axillary, opposite lvs, or fls solitary; pedicels present. **FL**: calyx lobes hairy, bristly-ciliate; corolla rotate, lobed to middle, lobes gen hairy; stamens incl, equal, equally attached; ovary chamber 1, bristly-hairy, style 1, 2-lobed in distal 1/2. **FR** spheric; bristles stout. **SEEDS** 1–8, spheric, brown, pitted or honeycombed. (Greek: scale mouth) [Constance 1939 Bull Torrey Bot Club 66:341–352]

1. Calyx sinus appendages 0; calyx rotate below mature fr ..... ***P. membranaceum***
1′ Calyx sinus appendages present; calyx enclosing mature fr
 2. Petiole narrowly winged; style 1–3 mm ..... ***P. racemosum***
 2′ Petiole widely winged; style 4–8 mm ..... ***P. auritum***
  3. Corolla < 1 cm wide ..... var. ***arizonicum***
  3′ Corolla 1–3 cm wide ..... var. ***auritum***

P. auritum (Lindley) Lilja (p. 711) **ST** 1–15 cm. **LVS**: lower 4–16 cm, 1–8 cm wide, petiole widely winged, clasping, blade oblong to ovate-lanceolate, base cordate, tip acuminate, lobes 5–13, oblong or lanceolate, obtuse or acute, entire or 1–5-toothed. **INFL**: fls solitary or 2–6 in cymes; pedicels 1–3 cm. **FL**: calyx lobes 3–9 mm, ± lanceolate, sinus appendages 1–4 mm; corolla 3–15 mm, 5–30 mm wide, blue to purple with darker marks in throat; style 4–8 mm. **FR** 5–10 mm wide, enclosed in calyx. **SEEDS** 1–4. Ocean bluffs, talus slopes, woodlands, streambanks, canyons, desert scrub; 0–1900 m. NCoRI, s SN, c SNF, Teh, SnJV, CW, SW, ne DSon; to AZ.

var. ***arizonicum*** (M.E. Jones) Constance ARIZONA PHOLISTOMA **LF**: lobes gen 5–11, very obtuse. **FL**: corolla 3–7 mm, = calyx, < 1 cm wide. RARE in CA. Desert scrub; 300–700 m. ne DSon (Whipple Mtns); to AZ.

var. ***auritum*** FIESTA FLOWER **LF**: lobes 7–13, obtuse or acute. **FL**: corolla 7–15 mm, gen > calyx, 1–3 cm wide. *n*=9. Habitats of sp. (exc desert scrub); 0–1900 m. NCoRI, s SN, c SNF, Teh, SnJV, CW, SW. ❀DRN:**17,24**&SHD:**7–9,14–16,18–23**.

P. membranaceum (Benth.) Constance (p. 711) **ST** 5–90 cm, gen glaucous. **LVS**: lower 2–13 cm, 1–8 cm wide, petiole narrowly winged, not clasping, blade oblong to ovate, base cordate or truncate, tip obtuse, lobes 5–11, oblong, obtuse, entire or 1-toothed. **INFL**: cyme; fls gen 2–10; pedicel 5–20 mm. **FL**: calyx rotate in fr, lobes 1–3 mm, oblong, sinus appendages 0; corolla 3–6 mm, < 1 cm wide, white, gen purple spot on each lobe; style 1–2 mm. **FR** 2–4 mm wide. **SEEDS** 1–2. *n*=9. Beaches, bluffs, ravines, wooded slopes, desert washes; 40–1400 m. c&s SNF, Teh, SnJV, CW, SW, D; Baja CA.

P. racemosum (Nutt.) Constance (p. 711) **ST** 2–6 dm. **LVS**: lower 3–10 cm, 2–8 cm wide, petioles narrowly winged, clasping, blade ovate to ± deltate, base cordate or truncate, tip obtuse, lobes 5–9, obtuse or acute, entire or 3–5-toothed. **INFL**: gen cymes; fls gen 2–6; pedicels 5–30 mm. **FL**: calyx lobes 2–4 mm, ± lanceolate, sinus appendages < 2 mm; corolla 4–10 mm, 5–15 mm wide, white to blue; style 1–3 mm. **FR** 5–8 mm wide, enclosed in calyx. **SEEDS** 4–8. *n*=9. Moist shaded, areas, hillsides, ravines, ocean bluffs, coastal scrub; 0–500 m. SCo, ChI, PR; Baja CA.

ROMANZOFFIA

Per, ± scapose; herbage ± glabrous to gen sparsely soft-hairy. **ST** erect. **LVS** simple; basal long-petioled, blades reniform to round, shallowly lobed or toothed; cauline few, reduced, alternate. **INFL** loose; fls on pedicels. **FL**: corolla > calyx, bell- to funnel-shaped, white, gen yellow in center; stamens incl, subequal; ovary chambers 2, style 1, thread-like, incl, gen slightly lobed. **FR** oblong to ovoid. **SEEDS** many, ovoid, angular, brown, pitted. (Count N. von Romanzoff, promoter of Russian expedition to CA in 1816)

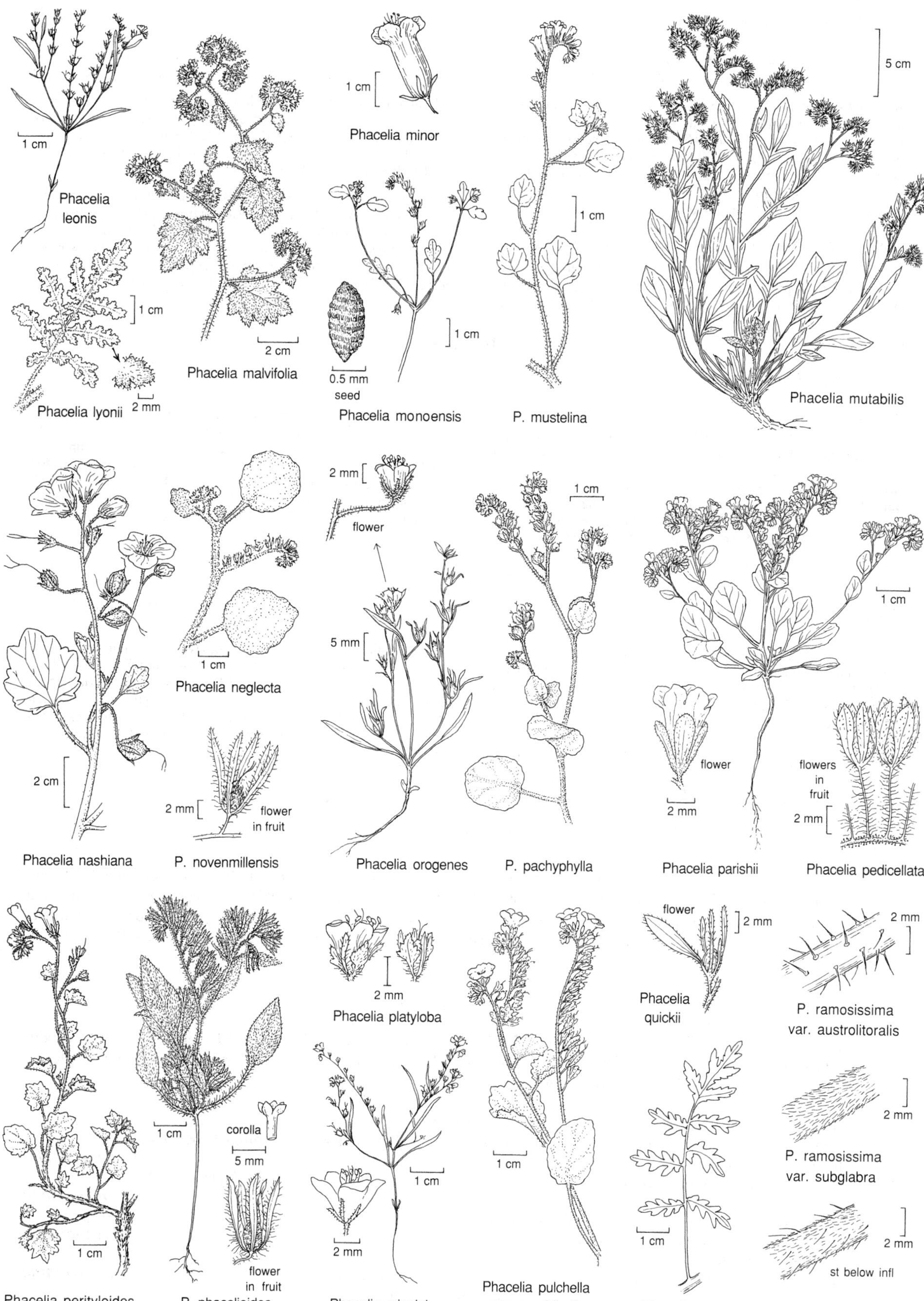
1 cm
Phacelia leonis
1 cm
2 mm
Phacelia lyonii
2 cm
Phacelia malvifolia
1 cm
Phacelia minor
1 cm
0.5 mm
seed
Phacelia monoensis
1 cm
P. mustelina
5 cm
Phacelia mutabilis
2 cm
Phacelia nashiana
1 cm
Phacelia neglecta
2 mm
flower in fruit
P. novenmillensis
2 mm
flower
5 mm
Phacelia orogenes
1 cm
P. pachyphylla
1 cm
flower
2 mm
Phacelia parishii
flowers in fruit
2 mm
Phacelia pedicellata
1 cm
Phacelia perityloides
1 cm
corolla
5 mm
flower in fruit
P. phacelioides
2 mm
Phacelia platyloba
1 cm
2 mm
Phacelia pringlei
1 cm
Phacelia pulchella var. gooddingii
flower
2 mm
Phacelia quickii
2 mm
P. ramosissima var. austrolitoralis
2 mm
P. ramosissima var. subglabra
1 cm
2 mm
st below infl
Phacelia ramosissima var. ramosissima

1. Tubers 0 (petioles widened, overlapping, forming bulbous base) .... ***R. sitchensis***
1′ Tubers brown, tomentose
 2. Infl > lvs; pedicels in fr > 1 cm .... ***R. californica***
 2′ Infl slightly, if at all > lvs; pedicels in fr < 1 cm .... ***R. tracyi***

R. californica E. Greene (p. 711) Pl 10–40 cm; tubers clustered, ovoid, brown, tomentose. **LF**: petioles 2–12 cm; base slightly widened; blade 8–45 mm wide. **INFL** > lvs, sometimes glandular; pedicels 1–3 cm in fr, slender. **FL**: calyx lobes 2–5 mm, linear-lanceolate, gen acute, glabrous, sometimes glandular; corolla 5–12 mm; ovary sparsely soft-hairy, glandular; style 4–7 mm. **FR** 6–10 mm. *n*=11. Ocean bluffs, roadbanks, wet cliffs, moist rocky areas; < 800 m. NW, n CCo, SnFrB; to WA. [*R. suksdorfii* E. Greene misapplied] ❀DRN,summerDRY:**4,5**,6,**17**&SHD:**7,14-16**,19,**20-24.**

R. sitchensis Bong. Pl 10–30 cm; tubers 0. **LF**: petioles 1–6 cm, widened, overlapping, forming bulbous base of st, gen ciliate; blades 1–4 cm wide. **INFL** > lvs, sometimes glandular; pedicels 1–3 cm in fr, slender. **FL**: calyx lobes 2–4 mm, oblong, gen obtuse, glabrous, sometimes glandular; corolla 5–11 mm; ovary sparsely soft-hairy, glandular, style 2–5 mm. **FR** 4–7 mm. *n*=11. Uncommon. Moist clefts in rocks; 1700–2100 m. KR; to AK, w Can, MT. ❀DRN,IRR:**4,5**&SHD:1,2,6,7,14–17.

R. tracyi Jepson Pl 2-12 cm, forming rounded tufts; tubers clustered, ovoid, brown, tomentose. **LF**: petioles 1–8 cm, soft-hairy; base widened; blades 10–35 mm wide, ± glabrous to soft-hairy. **INFL** compact, gen < lvs; pedicels stout, < 1 cm. **FL**: calyx lobes 2–5 mm, lanceolate, acute, soft-hairy; corolla 6–8 mm; ovary soft-hairy, style 2–3 mm. **FR** 5–8 mm. *n*=11. Uncommon. Rocky ocean bluffs; < 30 m. NCo; to WA.

TRICARDIA

1 sp. (Greek: 3 hearts, from calyx)

T. watsonii S. Watson (p. 711) THREE HEARTS Per; herbage long-soft-hairy, becoming ± glabrous; taproot woody, gen topped by a branched caudex covered by persistent petiole bases of previous years. **STS** several from root crown, erect, 5–40 cm. **LVS** simple, gen in basal rosette, petioled, 2–9 cm, 5–25 mm wide; cauline alternate, smaller upward, lower short-petioled, upper sessile, 6–20 mm, 3–15 mm wide, entire. **INFL** loose, terminal; fls pedicelled. **FL**: calyx lobes 5, very unequal, outer 3 cordate, 5–9 mm in fl, in fr 9–25 mm, wide, scarious, veiny, green to purplish, inner 2 ± 4 mm in fl, linear, enlarged in fr; corolla 4–8 mm, bell-shaped to rotate, white to cream, gen marked lavender; stamuns incl, unequal, equally attached; ovary chamber 1, style 1, 3–4 mm, tip lobed. **FR** 7–9 mm, glabrous. **SEEDS** 4–8, oblong, brown, rough. *n*=8. Sandy or gravelly desert slopes, flats, mtns, gen in shelter of shrubs; 100–2300 m. SnBr, SNE, D; to NV, UT, AZ. ❀TRY.

TURRICULA

1 sp. (Latin: little tower)

T. parryi (A. Gray) J.F. Macbr. (p. 711) POODLE-DOG BUSH Subshrub, densely glandular, sticky, ill-scented. **ST** erect, stout, densely leafy, 1–3 m, branched from base. **LVS** simple, cauline, alternate, sessile; blade 4–30 cm, lanceolate, entire or toothed, margins of upper sometimes rolled down. **INFL** terminal, branched; fls short-pedicelled, densely clustered. **FL**: calyx lobes 3–6 mm, glandular, coarsely long-hairy; corolla 10–20 mm, shallowly lobed, funnel-shaped, glandular, hairy outside, blue, lavender, or purple; stamens incl, unequal; ovary chambers 2, styles 2, 4–7 mm. **FR**: valves 4, 3–4 mm, ovoid, glandular, hairy. **SEEDS** many, oblong-ovoid, angular, shiny black, finely ridged, minutely net-sculptured. *n*=13. Gen disturbed areas, chaparral, dry granitic soils of slopes, ridges; 100–2300 m. s SN, s SnJV, s SCoRO, SW, n DMtns (Panamint Mtns); Baja CA. ❀DRN,DRY,SUN:1,2,7–9,14–24;DFCLT.

HYPERICACEAE ST. JOHN'S WORT FAMILY

Jennifer Talbot

Ann, per, shrub, tree. **LVS** simple, cauline, opposite or whorled; stipules 0; blade often with black dots or embedded clear glands. **INFL**: cyme, panicle, or fl solitary, terminal or axillary. **FL** bisexual, radial; sepals persistent, gen 5, often fused at base, overlapping; petals gen 5, free; stamens gen many, free or ± fused into 3–5 clusters; pistil 1, ovary superior, chambers 1–3, placentas gen axile, style branches 3. **FR**: capsule, gen septicidal. **SEEDS** many, small. 10 genera, 400 spp.: worldwide.

HYPERICUM

Ann, per, shrub, glabrous. **LF** sessile, ± gland-dotted. **INFL**: gen cyme, gen terminal, bracted. **FL**: sepals 5; petals 5, deciduous or persistent, yellow; anthers sometimes black dotted; ovary chambers 1 or 3, placentas 3, axile or parietal and projecting into chamber. 350 spp.: worldwide. (Greek name)

1. Petals < or = sepals; styles ± 1 mm; fr chamber 1, placentas parietal
 2. Pl prostrate to decumbent, from matted stolons, 3–25 cm; lvs 3–7 mm; fls gen 1–7 (often 1) per fl-st; common, gen montane .... ***H. anagalloides***
 2′ Pl erect, 20–60 cm; lvs 10–25 mm; fls many in compound infls; uncommon, e-c ScV, adjacent n SNF .... ***H. mutilum***
1′ Petals >> sepals; styles 3–10 mm; fr chambers 3, placentas axile
 3. Shrub, st gen > 2 m; sepal ciliate; uncommon .... ***H. canariense***
 3′ Per, st gen < 1 m; sepal margin glabrous; common
  4. Lf linear to lanceolate, gen folded, tip acute; sts many, from woody caudex .... ***H. concinnum***
  4′ Lf ± elliptic or oblong, flat, obtuse; sts few from taproot or rhizome

5. Lf elliptic; sterile axillary branches gen < 2 cm; fr 3-lobed .... ***H. formosum*** var. ***scouleri***
5′ Lf narrowly oblong; sterile axillary branches gen 2–10 cm; fr unlobed .... ***H. perforatum***

H. anagalloides Cham. & Schldl. (p. 711) TINKER'S PENNY Ann or per from matted stolons. **ST** prostrate to decumbent, 3–25 cm, slender; lower nodes rooting. **LF** 4–15 mm, elliptic to ± round, with clear to green gland-dots; base ± clasping st. **INFL**: fls 1–15. **FL**: sepals 2–4 mm, unequal, lanceolate to ovate, obtuse or acute, margin glabrous; petals 2–4 mm, ± = sepals, golden to salmon; stamens gen 15–25; styles ± 1 mm. **FR** ± 3 mm. **SEED** < 1 mm, yellow-brown. Wet places; 100–3000 m (< 1000 m in NW). NW, CaR, SN (exc Teh), CCo, SnGb, SnBr, PR, Wrn; to B.C., MT, NV. Highly variable; needs study.

H. canariense L. Shrub. **ST** gen 2–5 m. **LF** 5–7 cm, oblong-lanceolate; base tapered. **FL**: sepals ovate, acute, margin ciliate; petals 12–15 mm; styles 3–6 mm. Uncommon. Disturbed places; < 100 m. SCo; native to Canary Islands.

H. concinnum Benth. (p. 711) GOLD-WIRE Per from woody caudex. **STS** many, 15–30 cm, slender, wiry. **LF** 15–40 mm, linear to lanceolate, not clasping, gen folded, sparsely black-dotted. **INFL**: fls gen 3–9 per st. **FL**: sepals 5–9 mm, lanceolate, acute, margin black-dotted, glabrous; petals 10–15 mm, ± obovate, golden-yellow, margins black-dotted; stamens many, in 3 clusters, 4 filaments per cluster fused at base, others free; styles ± 10 mm, spreading. **FR** 6–7 mm, 3-lobed. **SEED** ± 1.5 mm, dark greenish brown. Dry, shrubby slopes; 100–1000 m. NW, CaRH, n&c SN, ScV, SnFrB.

H. formosum Kunth var. ***scouleri*** (Hook.) J. Coulter (p. 711) Per from taproot or rhizome. **STS** erect, few from base, 2–7 dm, slender; sterile axillary branches gen < 2 cm. **LF** 1–3 cm, ovate to oblong; flat; base ± clasping; margin black-dotted, lower surface inconspicuously dotted. **INFL**: fls gen 3–25 per st. **FL**: sepals 3–4 mm, oblong to ovate, gen obtuse, black-dotted, margin glabrous; petals 7–12 mm, obovate, pale to bright yellow, black-dotted; stamens many, in 3 clusters, anthers black-dotted; styles 3–5 mm. **FR** 6–7 mm, 3-lobed. **SEED** < 1 mm, brown. Springs, meadows, moist places; 100–2500 m. NW, CaRH, SN, ScV, SnFrB, SCoRI, TR, PR, GB; to w Can, Colorado.

H. mutilum L. Ann or per. **ST** ± erect, 20–60 cm.**LF** 10–25 mm, elliptic to ovate, gland-dotted. **INFL** compound; fls many. **FL**: sepals ± 3 mm, narrowly lanceolate; petals ± 2 mm. **FR** ± 4 mm. Streambanks, riparian woodland; < 300 m. e-c ScV, adjacent n SNF (Butte, Glenn cos.); expected more widely; native to e N.Am. Small pls approach robust, low elevation forms of *H. anagalloides*. Stamen number seems not to separate these 2 spp; needs study.

H. perforatum L. (p. 711) KLAMATHWEED Per from taproot. **STS** erect, many from base, 3–12 dm; sterile axillary branches gen 2–10 cm. **LF** 1.5–2.5 cm, linear to oblong; margins rolled under, black-dotted; lower surface conspicuously clear-dotted. **INFL**: fls gen 25–100 per st. **FL**: sepals 4–5 mm, lanceolate, acuminate, with black and clear dots, margin glabrous; petal 8–12 mm, ± oblong, copiously black gland-dotted, twisting after fl, bright yellow; stamens many, in 3 clusters, anthers black-dotted; styles 4–6 mm. **FR** 7–8 mm, unlobed. **SEED** ± 1 mm, brown. Pastures, abandoned fields, disturbed places; < 1500 m. NW, CaRH, n&c SN, ScV, CCo, SnFrB, PR; to e N.Am, native to Eur. May produce seed without fertilization. Seriously TOXIC to livestock; the toxin hypericin inhibits human immunodeficiency virus. NOXIOUS WEED; controlled by introduced flea beetle, *Chrysolina quadrigemina*.

JUGLANDACEAE WALNUT FAMILY

Dieter H. Wilken

Shrub, tree, gen monoecious. **LVS** gen odd-1-pinnate, alternate, deciduous; stipules 0. **INFL**: catkin, gen appearing before lvs, gen clustered; fls gen solitary in bract axils. **STAMINATE INFL** gen pendent, gen elongate, gen many-fld, on last year's twigs. **PISTILLATE INFL** gen erect; fls 1–3, at tip of new twigs. **STAMINATE FL**: sepals 3–6; petals 0; stamens 3–many; pistil 0 or vestigial. **PISTILLATE FL**: sepals 3–6; petals 0; stamens 0; pistil 1, ovary inferior, chamber 1 above, gen 2 below, styles gen 2, plumose. **FR**: nut enclosed in a ± fleshy husk, drupe-like. 7 genera, ± 60 spp.: n temp, subtrop mtns; some orn, cult for wood, nuts (*Carya* hickory, pecan; *Juglans*) [Elias 1972 J Arnold Arb 53:26–51]

JUGLANS WALNUT

Shrub, tree, gen monoecious. **ST** gen erect; bark smooth to furrowed in age, gray to brown; twig centers chambered. **STAMINATE FL**: calyx lobes gen 4, ± fused to bract. **PISTILLATE FL**: calyx lobes 4, gen ± fused to bract and bractlets, with them forming husk in fr. **FR** in clusters, gen spheric; husk leathery, strong-smelling, indehiscent yet ± deciduous. 21 spp.: temp N.Am., n temp Eur, Asia, S.Am. (Latin: walnut) [Howell 1973 Madroño 22:144] Wood used for interior finishing, furniture; source of nuts.

1. Lflets oblong to elliptic, 7–9 .... ***J. regia***
1′ Lflets lanceolate to ovate, 11–19 .... ***J. californica***
 2. Fr (incl husk) gen 2–3 cm diam .... var. ***californica***
 2′ Fr (incl husk) gen 3–3.5 cm diam .... var. ***hindsii***

J. californica S. Watson (p. 711) CALIFORNIA BLACK WALNUT Shrub, tree. **ST**: trunks 1–5, gen < 25 m; bark gray to brown; twigs brown. **LF**: lflets 11–19, 3–10 cm, lanceolate to ovate, toothed. **FR** (incl husk) 2–3.5 cm diam; woody nut shell ± grooved. Slopes, canyons, valleys; 50–900 m. s NCoRI, s ScV, n SnJV, SnFrB, SCoRO, SCo, s TR, n PR. Sometimes cult; hybrids with *J. regia* few, synthetic.

var. ***californica*** SOUTHERN CALIFORNIA BLACK WALNUT **ST**: trunks 1–5, gen < 15 m. **LF**: lflets 2–8 cm. **FR** (incl husk) gen 2–3 cm diam. UNCOMMON. Slopes, canyons; 50–900 m. SCoRO (Santa Lucia Mtns where perhaps cult), SCo, s TR, n PR (Santa Ana Mtns). ❀4–6,**7,14–24**,IRR:**8,9**,10,11.

var. ***hindsii*** Jepson NORTHERN CALIFORNIA BLACK WALNUT **ST**: trunk 1, gen 15–25 m. **LF**: lflets 5–10 cm. **FR** (incl husk) gen 3–3.5 cm diam. RARE. Canyons, valleys; 50–200 m. s NCoRI, s ScV, n SnJV, SnFrB. [*J. hindsii* (Jepson) R.E. Smith] Reported as occurring at pre-Spanish native American campsites. Formerly cult as rootstock for *J. regia* with which it hybridizes readily. ❀**4–7,14–24**,IRR:**8,9**,10,11.

J. regia L. PERSIAN or ENGLISH WALNUT Tree. **ST**: trunk 1, gen < 30 m; bark gray; twigs green to gray. **LF**: lflets 7–9, 6–12 cm, elliptic to oblong, gen entire. **FR** (incl husk) 4–5 cm diam, woody nut shell wrinkled. Persisting near abandoned habitations; < 200 m. GV, CW, SW; native to se Eur, s temp Asia. Widely cult for nuts, wood.

KOEBERLINIACEAE JUNCO FAMILY

Staria S. Vanderpool

Shrub, gen ± lfless, thorny, unscented. **ST**: branches many, smooth, rigid, interlocking. **LVS** simple, alternate, scale-like, ephemeral. **INFL**: raceme, axillary, bracted. **FL** radial, bisexual, small; sepals 4, free; petals 4, free; stamens 8, filaments flat; ovary ± stalked, spheric, stigma minute, head-like. **FR**: berry; chambers 2, each 1–2-seeded.

KOEBERLINIA ALLTHORN

The only genus. 1 sp. (C.L. Koeberlin, German clergyman, botanist, born 1794)

K. spinosa Zucc. ssp. ***tenuispina*** (Kearney & Peebles) E. Murray (p. 711) CROWN-OF-THORNS Pl < 5 m, short-spreading-hairy. **ST**: branchlets 25–70 mm, pale green, tipped by 3–6 mm, black thorns. **LF** < 2 mm. **INFL** 3–15 mm; pedicels 3–6 mm. **FL**: sepals 1–2 mm, ovate, entire, greenish white; petals 3–4 mm, 0.5–1 mm wide, short-clawed with an obovate or oblanceolate limb, white; stamens 2.8–4 mm, anthers 0.8–1 mm; ovary 1–1.2 mm, stalk 0.3–0.5 mm, style 1–1.5 mm. **FR** 2.5–3.5 mm, black. RARE in CA. Creosote-bush scrub; 1600 m. DSon (Chocolate Mtns.); sw AZ, Mex (nw Sonora). Often incl in Capparaceae. Ssp. *spinosa* not in CA.

KRAMERIACEAE RHATANY FAMILY

Beryl B. Simpson

Per, shrub, root parasite with chlorophyll. **ST** prostrate to erect, much branched. **LVS** gen simple, alternate, sessile; blade linear to ovate, hairy, sometimes glandular, tip abruptly pointed. **INFL**: fls gen solitary in axils; pedicel bracts 2. **FL** bisexual, bilateral; sepals 4–5, free, conspicuous; petals gen 5, 3 upper linear to clawed, held in ± upright "flag", 2 modified into glands flanking ovary; stamens gen 4, opening by pores; ovary superior, hairy, style slender, recurved. **FR** nut-like, bearing smooth or barbed spines. 1 genus, 17 spp.: Am, esp trop. [Simpson 1989 Fl Neotropica 49:1–109] Pollinating bees collect oils secreted by glandular petals.

KRAMERIA RHATANY

The only genus. (Possibly named for J. Kramer, 1700's, Austrian army physician)

1. Sepals cupped; claws of flag petals fused; fr spines smooth or barbed along shaft .... ***K. erecta***
1′ Sepals reflexed; flag petals free; fr spines barbed only at tip .... ***K. grayi***

K. erecta Schultes (p. 717) PIMA RHATANY, PURPLE HEATHER Shrub, ± strigose to canescent or ± silky-hairy. **ST** < 1 m; branches often ascending, tips blunt. **LF** ± linear. **FL**: buds ovate, barely curved; sepals cupped, pink; flag petal claws ± fused, blades triangular, green and pink; glandular petals pink, outer face glandular-blistered near margin. **FR** cordate, somewhat flat; spines smooth or barbs scattered. Dry, rocky ridges, slopes; < 1200 m. e PR (Santa Rosa Mtns), c&s DMoj, DSon; to NV, TX, n Mex. [*K. glandulosa* Rose & Painter; *K. parvifolia* Benth.; *K. p.* var. *imparata* J.F. Macbr.] ❀TRY.

K. grayi Rose & Painter (p. 717) WHITE RHATANY Shrub, densely canescent or silky-hairy. **ST** < 1 m; branches ± spreading, tips spiny. **LF** narrowly lanceolate. **FL**: buds curved upward; sepals reflexed, deep purple-red; flag petals free, blade oblanceolate, base green, tip pink or purple; glandular petals purple, outer face covered with blister-like glands. **FR** ± spheric; spines barbed only at tip. Dry, rocky or sandy places, esp on lime soils; < 1400 m. c&s DMoj, DSon; to NV, TX, n Mex. [*K. canescens* A. Gray] ❀TRY.

LAMIACEAE MINT FAMILY

Dieter H. Wilken, except as specifed

Ann, per, shrub, glabrous to hairy, gen aromatic. **STS** gen erect, gen 4-angled. **LVS** gen simple to deeply lobed, opposite, gen gland-dotted. **INFL**: cyme, gen clustered around st, head-like, separated by evident internodes (terminal in *Monardella*) or collectively crowded, spike-like to panicle-like (sometimes raceme or fls 2–12); subtended by lvs or bracts; fls sessile or pedicelled. **FL** gen bisexual; calyx gen 5-lobed, radial to bilateral; corolla gen bilateral, 1–2-lipped, upper lip entire or 2-lobed, ± flat to hood-like, sometimes 0, lower lip gen 3-lobed; stamens gen 4, gen exserted, paired, pairs unequal, sometimes 2, staminodes 2 or 0; ovary superior, gen 4-lobed to base, chambers 2, ovules 2 per chamber, style 1, arising from center at junction of lobes, stigmas gen 2. **FR**: nutlets 4, gen ovoid to oblong, smooth. ± 200 genera, 5500 spp.: worldwide. Many cult for herbs, oils (*Lavandula*, lavender; *Mentha*; *Ocimum*, basil; *Rosmarinus*, rosemary; *Thymus*, thyme), some cult as orn (in CA *Cedronella*, *Leonotis*, *Phlomis*). [Cantino & Sanders 1986 Syst Bot 11:163–185]

1. Fertile stamens 2; staminodes sometimes present
  2. Corolla 4-lobed, not 2-lipped; lobes subequal; fr compressed, edge corky .... **LYCOPUS**
  2′ Corolla evidently 2-lipped, lobes of upper lip very different from lobes of lower lip; fr not compressed
    3. Fertile anther sacs 1 or 2, on thread-like appendage hinged or joined to filament tip; sterile anther sac reduced, modified, or 0 .... **SALVIA**

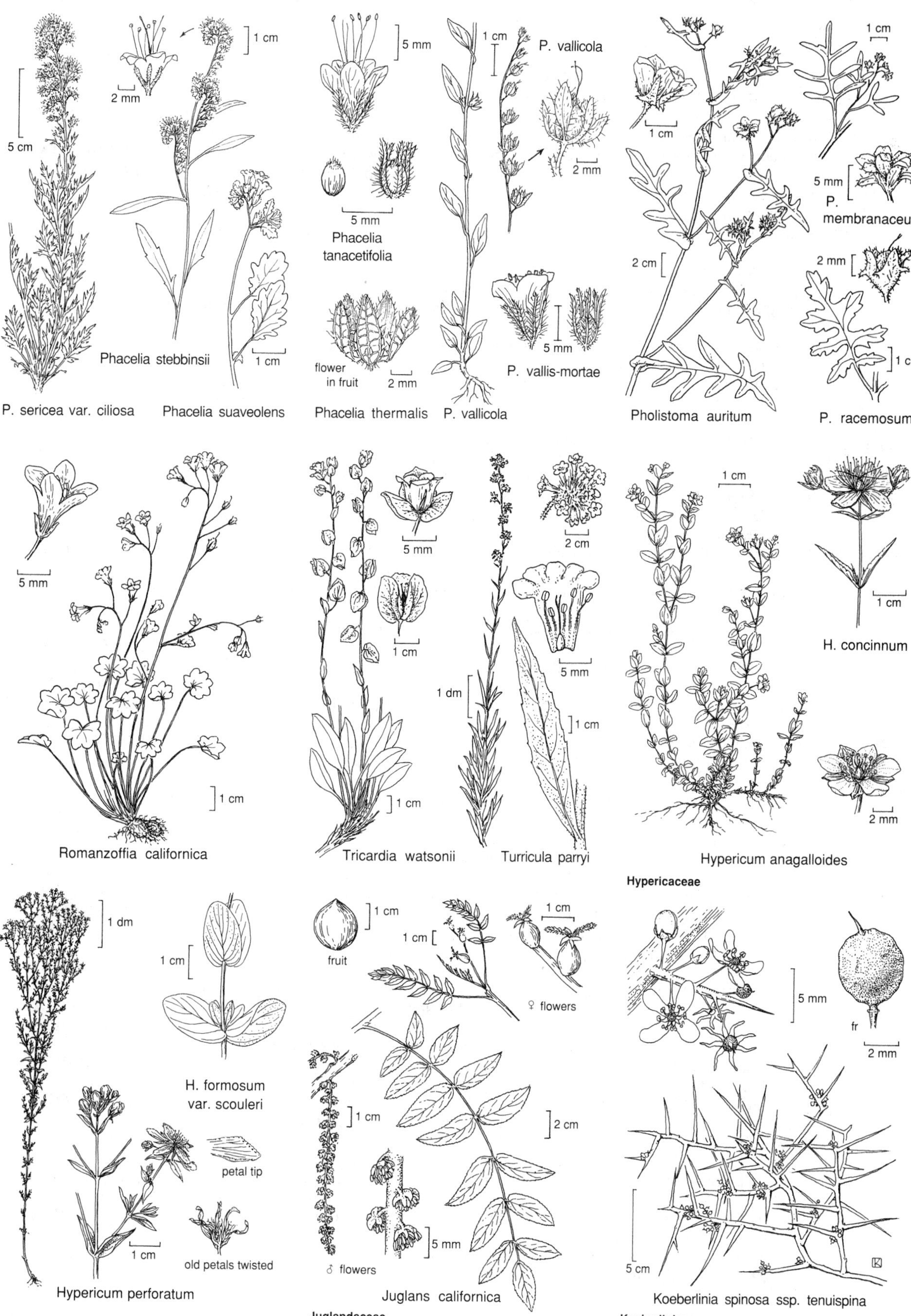

1 cm
2 mm
5 cm
Phacelia stebbinsii
1 cm
P. sericea var. ciliosa
Phacelia suaveolens
5 mm
5 mm
Phacelia tanacetifolia
flower in fruit
2 mm
Phacelia thermalis
1 cm
P. vallicola
2 mm
5 mm
P. vallis-mortae
P. vallicola
1 cm
2 cm
Pholistoma auritum
1 cm
5 mm
P. membranaceum
2 mm
1 cm
P. racemosum
5 mm
1 cm
Romanzoffia californica
5 mm
1 cm
1 cm
Tricardia watsonii
2 cm
5 mm
1 dm
1 cm
Turricula parryi
1 cm
1 cm
H. concinnum
2 mm
Hypericum anagalloides
Hypericaceae
1 dm
1 cm
H. formosum var. scouleri
petal tip
1 cm
old petals twisted
Hypericum perforatum
1 cm
fruit
1 cm
1 cm
♀ flowers
1 cm
2 cm
5 mm
♂ flowers
Juglans californica
Juglandaceae
5 mm
fr
2 mm
5 cm
Koeberlinia spinosa ssp. tenuispina
Koeberliniaceae

3′ Fertile anther sacs 2, attached to filament tip side by side
4. Shrub, rounded to mound-like, gen 5–10 dm .... **POLIOMINTHA**
4′ Ann, per, or subshrub, gen < 5 dm
5. Calyx ± radial, lobes subequal — e DMtns .... **MONARDA**
5′ Calyx ± 2-lipped, lobes of upper and lower lips unequal
6. Infl bracts ± translucent, veins ± raised, margins spiny .... [2]**ACANTHOMINTHA**
6′ Infl bracts opaque, veins not raised, margins entire or hairy, not spiny
7. Per or subshrub; infls ± evenly spaced; calyx tube swollen below middle — e DMtns .... **HEDEOMA**
7′ Ann; infls densely clustered, ± terminal, head-like; calyx tube not swollen .... [2]**POGOGYNE**
1′ Fertile stamens 4; staminodes 0
8. Corolla ± radial to slightly 2-lipped, lobes or lips equal in length
9. Infls 2+, axillary, gen subtended by 2 lvs or bracts, evenly spaced or together cylindric, spike- to panicle-like .... **MENTHA**
9′ Infl gen 1, head-like, terminal, subtended by involucre-like whorl of bracts .... **MONARDELLA**
8′ Corolla clearly bilateral, gen 1–2-lipped, lobes and lips unequal
10. Calyx obliquely funnel-shaped, flaring, margin 5–10-toothed; infl subtended by 2 lvs and forked spines .... **MOLUCCELLA**
10′ Calyx cylindric to bell-shaped, lobes 2–10; infl subtended by lvs or bracts
11. Calyx 2-lobed, lobes lip-like, entire
12. Shrub; calyx becoming bladder-like, 1–2 cm in fr .... **SALAZARIA**
12′ Per from rhizomes; calyx not becoming bladder-like, gen < 1 cm in fr — calyx back transversely ridged or with dome-like bump .... **SCUTELLARIA**
11′ Calyx 5- or 10-lobed, radial to 2-lipped (lips 2–3-lobed)
13. Calyx 10-lobed, lobe tips recurved or hooked; stamens incl in corolla tube .... **MARRUBIUM**
13′ Calyx 5-lobed, lobe tips straight; stamens exserted, hidden under upper lip to > corolla lobes
14. Stamens reclining on lower corolla lip; lip pouched; twigs densely stellate-hairy — DSon .... **HYPTIS**
14′ Stamens ascending under or ± parallel to and exceeding upper corolla lip; lip ± flat, not pouched; twigs glabrous to hairy, not stellate
15. Infl bracts ± translucent, margins spiny .... [2]**ACANTHOMINTHA**
15′ Lvs or bracts subtending infl opaque, margin glabrous to hairy
16. Nutlets fused laterally below middle; corolla 1-lipped, lip entire, ± reflexed, > other 4 lobes or lip 5-lobed, ± straight
17. Lvs crenate to deeply lobed; corolla lip 5-lobed, straight; fr ± smooth, puberulent at top .... **TEUCRIUM**
17′ Lvs entire; corolla lip entire, ± reflexed; fr irregularly ridged, puberulent to hairy throughout — stamens >> corolla lobes .... **TRICHOSTEMA**
16′ Nutlets separate to base; corolla 2-lipped, upper lip entire, notched, or lobed; lower lip 1–3-lobed
18. Subshrub or shrub
19. Corolla gen > 20 mm; upper corolla lip 4-lobed; calyx ± inflated in fr .... **LEPECHINIA**
19′ Corolla < 15 mm; upper corolla lip 2-lobed; calyx not inflated in fr .... **SATUREJA**
18′ Ann or per
20. Infls terminal, collectively ± dense, gen head-, spike-, or panicle-like, sometimes interrupted by 1–3 internodes below
21. 2 or 4 stamens clearly exceeding corolla .... **AGASTACHE**
21′ Stamens ascending under upper lip, not clearly > lower corolla lobes
22. Ann; infl lvs and bracts = or > fls — infl lvs, bracts linear to narrowly oblanceolate [2]**POGOGYNE**
22′ Per (exc *Stachys arvensis*); infl lvs or bracts gen < fls
23. Calyx 2-lipped, upper lip 3-toothed, lower lip 2-lobed, acuminate .... **PRUNELLA**
23′ Calyx ± radial, 5-lobed, lobes ± equal
24. Infl bracts white to purple, persistent; calyx < 2 mm, throat densely white-short-hairy inside .... **ORIGANUM**
24′ Infl bracts green, withering; calyx > 4 mm, throat glabrous to sparsely hairy inside
25. Calyx tube ± 15-veined; upper 2 stamens > lower 2 .... **NEPETA**
25′ Calyx tube 5–10-veined; upper 2 stamens < lower 2 .... [2]**STACHYS**
20′ Infls axillary, gen separated by evident internodes subtended by bracts or lvs
26. Ann; nutlets 3-angled, top truncate .... **LAMIUM**
26′ Per; nutlets smooth, ovoid to ± round in X-section, top ± rounded
27. Infls head-like, fls many; calyx lobe tip densely white-short-hairy .... **PYCNANTHEMUM**
27′ Infls not head-like, fls 1–12; calyx lobe tip glabrous to sparsely hairy
28. Sts gen prostrate to decumbent
29. Lf blade round to reniform, base lobed; corolla 8–25 mm .... **GLECOMA**
29′ Lf blade ovate to broadly triangular, base obtuse to rounded; corolla 3–8 mm .. ***Satureja douglasii***
28′ Sts erect
30. Most infls (exc lowest) subtended by bracts or much reduced lvs .... [2]**STACHYS**
30′ Most infls subtended by lvs
31. Calyx 7–9 mm; corolla light yellow, tube curved upward .... **MELISSA**
31′ Calyx 14–17 mm; corolla salmon, tube straight .... ***Satureja mimuloides***

ACANTHOMINTHA THORNMINT

James D. Jokerst

Ann, hairy or not, aromatic. **ST** erect, branched or not. **LF** lanceolate to obovate, entire to spiny, petioled. **INFL**: clusters, head-like, terminal and gen axillary; bracts gen scarious, veins conspicuous, margins spiny. **FL**: calyx 2-lipped, lobes spine-tipped, upper 3 acuminate, lower 2 oblong; corolla funnel-shaped, 2-lipped, white, sometimes tinged lavender or rose, throat cream, upper lip 2-lobed or entire, hooded, lower lip 3-lobed, reflexed; stamens 4, upper 2 reduced, sterile or not; style slender, lower lobe longer. **FR** smooth, ovoid. 4 spp.; s CA-FP. (Greek: thorn mint)

1. St hairs short below, conspicuously glandular above; style hairy; upper corolla lip ± = lower, 2-lobed, deeply hooded .... ***A. lanceolata***

1′ St hairs 0 to long, if glandular then inconspicuously so; style glabrous; upper corolla lip < lower, entire, shallowly hooded

 2. Upper stamens sterile; anthers glabrous; SCo, PR (sw San Diego Co.) .... ***A. ilicifolia***

 2′ Upper stamens fertile; anthers hairy; CCo, SnFrB, SCoR, WTR

  3. Anthers pink-red; margins of upper lvs not spiny; st gen unbranched; infl gen terminal; CCo, SnFrB .. ***A. duttonii***

  3′ Anthers cream; margins of upper lvs spiny; st gen branched below; infl terminal and axillary; SnFrB, SCoR, WTR .... ***A. obovata***

   4. Hairs of st and calyx 0 or short; bracts round or wider than long, cordate-clasping at base; c&s SCoRO, w&c WTR (San Luis Obispo to Ventura cos.) .... ssp. ***cordata***

   4′ Hairs of st and calyx short or long, some inconspicuously glandular; bracts gen longer than wide, obtuse or truncate at base; SCoRI (San Benito, Monterey cos.) .... ssp. ***obovata***

A. duttonii (Abrams) Jokerst SAN MATEO THORNMINT **ST** gen unbranched, < 20 cm; hairs 0 to short. **LF**: blade 8–12 mm, lance-oblong to obovate; margins of those in infl not spiny, sometimes serrate. **INFL** gen terminal; bracts 5–11 mm, ovate, green at fl, marginal spines 5, 7, or 9, 3–7 mm. **FL**: calyx 5–8 mm, hairs 0 or sparse, short; corolla 12–16 mm, white, sometimes tinged lavender, upper lip < lower, entire, shallowly hooded; upper stamens fertile, anthers short-hairy, pink-red, pollen cream; style glabrous. **ENDANGERED** US, CA. Serpentine grasslands; < 300 m. CCo, SnFrB (San Mateo Co.). [*A. obovata* Jepson ssp. *d.* Abrams]

A. ilicifolia (A. Gray) A. Gray (p. 717) SAN DIEGO THORNMINT **ST** 5–15 cm; hairs 0 to short. **LF**: blade 5–15 mm, round; margin serrate. **INFL**: bracts 7–9 mm, ovate, marginal spines 7–10, 4–8 mm. **FL**: calyx ± 5 mm; corolla 12 mm, white, lobes sometimes rose-tinged, upper lip < lower, entire, shallowly hooded; upper stamens sterile, anthers glabrous; style glabrous. **ENDANGERED** CA. Vernal pools, clay depressions on mesas, slopes in chaparral, coastal-sage scrub; < 900 m. s SCo, sw PR (sw San Diego Co); Baja CA.

A. lanceolata Curran (p. 717) SANTA CLARA THORNMINT Pl ill-smelling, soft-hairy. **ST** 10–30 cm; hairs short below, conspicuously glandular above. **LF** glandular; blades 10–20 mm, lance-oblong to ovate, upper spiny; margins entire, serrate, or spiny. **INFL**: bracts 9–12 mm, oblong, marginal spines 7–9, 10–12 mm. **FL**: calyx 12 mm; corolla 2–2.5 cm, white, sometimes pink-tipped, glandular-hairy, lips 8–10 mm, upper ± = lower, 2-lobed, deeply hooded; upper stamens fertile, anthers glabrous; style hairy. UNCOMMON. Arid, rocky, often serpentine slopes; < 1200 m. SnFrB, SCoRI.

A. obovata Jepson **ST** 4–25 cm, gen branched below; hairs 0 to sparse, short or long, some inconspicuously glandular. **LF**: blades 8–12 mm, ovate or obovate; margins of lower entire or serrate, of those in infl spiny. **INFL**: bracts 7–15 mm, shiny, straw-colored at fl, marginal spines 7, 9, or 11, 5–8 mm. **FL**: calyx 7–13 mm, hairs short or long, sometimes glandular; corolla 12–27 mm, white, lobes purple-tipped, upper lip < lower, entire, shallowly hooded; upper stamens fertile, anthers long-hairy, cream, pollen cream. n=19. Grassy slopes, oak woodland, chaparral; < 1500 m. SnFrB, SCoR, WTR. Like *A. duttonii.*

ssp. ***cordata*** Jokerst (p. 717) **ST**: hairs 0 or short. **INFL**: bracts round or wider than long, cordate-clasping at base. **FL**: calyx hairs 0 or short; anthers densely woolly. Habitats and elevations of sp. c&s SCoRO, w&c WTR (San Luis Obispo to Ventura cos.).

ssp. ***obovata*** SAN BENITO THORNMINT **ST**: hairs short or long, some inconspicuously glandular. **INFL**: bracts gen longer than wide, obtuse or truncate at base. **FL**: calyx hairs short or long, some glandular; anthers moderate woolly. RARE. Habitats and elevations of sp. SCoRI (San Benito, Monterey cos.).

AGASTACHE HORSEMINT

Deborah Engle Averett

Per, erect, gen < 1 m, aromatic. **LVS** petioled; blade ± lanceolate to triangular, margin crenate to coarsely serrate. **INFL**: spike of sessile clusters, dense; bracts 1–several at base, lf-like, lanceolate. **FL**: calyx 5-lobed, 2-lipped, turning pink before fr, lobes acuminate; corolla 5-lobed, rose to rose-purple, 2-lipped, lower longer, broader, upper 2-lobed; stamens 4, in 2 pairs (1 pair longer), exserted, anther sacs spreading; style 2-lobed, exserted. **FR** ± 2 mm, oblong, brown, smooth; with small hairs at tip. 2n=18. 22 spp.: N.Am, Mex, Asia. (Greek: many spikes)

1. Lf nearly triangular, < 5 cm (gen < 3 cm), < 3.5 cm (gen < 2 cm) wide .... ***A. parvifolia***

1′ Lf ± broadly lanceolate, 3–8 cm, 1.5–7 cm wide .... ***A. urticifolia***

A. parvifolia Eastw. (p. 717) **LF** < 5 cm, < 3.5 cm wide. **FL**: corolla rose. Uncommon. Woodlands; 1400–2200 m. CaRH, Wrn. [*A. urticifolia* ssp. *p.* (Eastw.) Vogelmann] ❀DRN, IRR:1,2.

A. urticifolia (Benth.) Kuntze (p. 717) **LF** 3–8 cm, 1.5–7 cm wide. **FL**: corolla rose to rose-purple. Common. Gen woodlands, but many habitats; 400–3000 m. NCoR, CaR, n&c SNF, SNH, SCoRO, SnBr, Wrn; to B.C., Rocky Mtns. [*A. glaucifolia* A.A. Heller] Coast Range pls gen hairier than those from farther inland. ❀DRN, IRR:**1,2,4–6**,17&SHD:**3,7**,14–16.

GLECOMA GROUND IVY

Per, glabrous to sparsely hairy, with only bisexual or only pistillate fls. **STS** prostrate to decumbent, sometimes erect, gen rooting at lower nodes. **LF** petioled; blade round to reniform, crenate to toothed. **INFLS** axillary, each subtended by lvs; fls 2–5; bracts minute or 0. **FL**: calyx 5-lobed, tube 15-veined, lobes unequal, upper >> lower; corolla 2-lipped, upper lip 2-lobed, ± flat, lower lip 3-lobed, central lobe >> lateral lobes; stamens 4, fertile, upper pair > lower; style lobes ± equal. 10 spp.: temp Eurasia. (Greek: ancient name).

G. hederacea L. **ST** 1–5 dm; hairs spreading to reflexed. **LF**: petiole gen 1–2 cm; blade 1–2 cm, ± round to reniform. **INFL**: bracts bristle-like. **FL**: calyx 3–7 mm, fine bristly to puberulent; corolla gen 15–25 mm in bisexual fl, 8–15 mm in pistillate, violet, purple-spotted. $2n$=18,36. Disturbed, often moist shaded places; < 800 m. NW, SNF, CW, SCo; widespread N.Am; native to Eur.

HEDEOMA MOCK PENNYROYAL

Ann, per, subshrub, aromatic; hairs short, spreading to recurved. **STS** decumbent to erect, branched at base. **LF** short-petioled to sessile; blade ovate to linear, entire or toothed. **INFLS** axillary at upper st nodes, each head-like, subtended by lvs; bracts minute. **FL**: calyx 2-lipped, upper lip 3-lobed, lower lip 2-lobed, lobes acuminate, sharp-pointed, tube swollen or pouched below middle; corolla 2-lipped, upper lip > lower, entire to 2-lobed, lower lip 3-lobed; stamens 2, under upper lip or exserted, staminodes minute or 0; style unequally 2-lobed. **FR**: nutlets pitted, glaucous, gelatinous when wet. 38 spp.: N.Am., S.Am. (Greek: ancient name for strongly aromatic mint) [Irving 1980 Sida 8: 218–295]

1. Lf blades linear to ± oblong; calyx lobes in fr converging, throat ± closed; sts becoming glabrous below .... ***H. drummondii***

1' Lf blades ovate to ± round; calyx lobes in fr spreading to reflexed, throat ± open; sts puberulent below .... ***H. nanum*** var. ***californicum***

H. drummondii Benth. (p. 717) Per 15–45 cm. **ST** puberulent, becoming glabrous below. **LF**: blade 5–11 mm, 1–4 mm wide, linear to ± oblong, tip gen obtuse. **INFL**: fls 3–7; pedicels 2–3.5 mm. **FL**: calyx 5–6 mm, lobes converging, throat ± closed in fr; corolla 7–9 mm, lower lip length = width. $2n$=34,36. Rocky, gravelly soils; 1400–1700 m. e DMtns (New York Mtns); to MT, NE, Mex.

H. nanum (Torrey) Briq. var. ***californicum*** W.S. Stewart (p. 717) Per 10–25 cm. **ST** puberulent. **LF**: blade 3–8.5 mm, 2–4.5 mm wide, ovate to ± round, tip acute. **INFL**: fls 3–5; pedicel 3–4.5 mm. **FL**: calyx 4.5–5.5 mm, lobes spreading or ± reflexed, throat ± open in fr; corolla 8–9 mm, lower lip length < width. $2n$=36. Rocky, often limestone outcrops; 900–2100 m. e DMtns (Providence, New York, Nopah mtns); NV, nw AZ.

HYPTIS

Ann, per, shrubs, glabrous to densely, gen stellate-hairy. **STS** erect to spreading, branched. **LVS** gen petiolate. **INFLS** axillary; each cluster ± dense, subtended by lvs or bracts. **FL**: calyx gen 5-lobed, lobes equal, acute to long-acuminate; corolla 2-lipped, upper lip 2-lobed, flat, lower lip 3-lobed, central lobe reflexed, ± pouched; stamens 4, fertile, curved, reclining on lower lip, exserted; style lobes ± equal. **FR**: nutlets angled, ridged or smooth. ± 350 spp.: warm temp, trop Am. (Greek: turned back, from lower lip position) Some spp. used for food (seeds), oil, wood, fiber.

H. emoryi Torrey (p. 717) DESERT-LAVENDER Shrub 1–3 m. **ST**: branches spreading to erect; twigs densely stellate, becoming glabrous. **LF**: petiole 3–7 mm; blade gen ovate to ± round, crenate. **INFLS**: lower subtended by lvs, upper by bracts; bracts << lvs, linear to elliptic; pedicel 1–3 mm. **FL**: calyx 4–5 mm, densely stellate, lobes ± acuminate; corolla 5–6 mm, violet. Gravelly, sandy washes, canyons, desert shrubland; < 1000 m. s DMoj, DSon; AZ, nw Mex. ❀DRN,DRYorIRR:8–14&SUN:15–17,19–24;DFCLT.

LAMIUM DEAD NETTLE

Ann, per, glabrous to hairy. **STS** decumbent to erect; base gen branched. **LVS** petioled to sessile, gen ovate, round or reniform, entire to toothed. **INFLS** terminal and axillary, each head-like, subtended by lvs. **FL**: calyx 5-lobed, lobes ± equal, gen acuminate; corolla 2-lipped, upper lip hood-like, lower lip ± 3-lobed, central lobe > lateral lobes; stamens 4, fertile, gen enclosed by upper lip, anthers gen hairy; style ± equally 2-lobed. **FR**: nutlets triangular in X-section, top truncate. 40 spp.: temp Eurasia, n Afr. (Latin: ancient name) [Bernström 1955 Hereditas 41:1–122]

1. Upper cauline lvs sessile, clasping; corolla tube glabrous inside .... ***L. amplexicaule***

1' Upper cauline lvs petioled; corolla tube puberulent below filaments .... ***L. purpureum***

L. amplexicaule L. Ann. **ST** 1–4 dm. **LVS**: middle, upper sessile; blade 1–2.5 cm, widely ovate to ± round, base truncate to lobed, margin crenate to ± lobed. **FL**: calyx 4–7 mm; corolla gen 10–18 mm, glabrous inside, red-purple. $2n$= 18. Disturbed sites, cult or abandoned fields; < 800 m. CA-FP; widespread N.Am; native to Eurasia. Cleistogamous fls have corollas < 8 mm, gen not opening.

L. purpureum L. Ann. **ST** 1–6 dm. **LVS**: middle, upper (exc uppermost) petioled; petioles 1–2 cm; blade 1–3(4) cm, ovate to ± round, base ± lobed, margin crenate to serrate. **FL**: calyx 5–7 mm; corolla 10–20 mm, puberulent inside below filaments, pink-purple. $2n$=18. Uncommon. Disturbed sites, meadows; < 300 m. NCo, CCo; widespread N.Am; native to Eur.

LEPECHINIA PITCHER SAGE

Deborah Engle Averett

Shrub, < 2 m, aromatic. **LVS** petioled, often reduced to sessile bracts upward; blade gen 4–12 cm, lanceolate to ovate, rounded to cordate at base, crenate or serrate to entire. **INFL**: raceme, open; bracts 1 per fl, lf-like, lanceolate; pedicels persistent after fr-fall. **FL**: calyx 5-lobed, barely 2-lipped, turning scarlet-purple, enlarging in fr; corolla 5-lobed, 2-lipped, white to lavender tinged, lower lip longer, upper lip 4-lobed; stamens 4, in 2 pairs, incl in throat, anther sacs spreading; style 2-lobed, incl in throat. **FR** 2–4 mm, round to oblong, black to dark brown, smooth, glabrous or with minute hairs. 2*n*=32. 40 spp.: CA, Mex, S.Am. (I.I. Lepechin, Russian botanist, 1737–1802)

1. Pedicel < 1 cm; calyx spheric in fr; fr with minute hairs .... ***L. calycina***
1′ Pedicel gen 1–4 cm (rarely shorter in *L. fragrans*); calyx bell-shaped in fr; fr glabrous
 2. Calyx lobes clearly < tube .... ***L. cardiophylla***
 2′ Calyx lobes ± = to > tube
  3. Calyx hairs gen 0.5–2 mm, lobes gen > tube .... ***L. fragrans***
  3′ Calyx hairs << 0.5 mm, lobes gen ± = tube .... ***L. ganderi***

L. calycina (Benth.) Epling (p. 717) Pl with long, branched, nonglandular hairs, sometimes with short-stalked to sessile glands. **LF** lanceolate to narrowly ovate, ± entire to crenate-serrate. **FL**: pedicel 0–1 cm; calyx inflated, gen persistent, spheric in fr, lobes < tube. **FR** hairs minute. Common. Rocky slopes; chaparral, woodland; 150–900 m. NCoR, n&c SNF, SnFrB. ❀DRN,DRY&SUN:**7,14–16**,17,**22–24**&SHD8,**9**,18,**19–21**.

L. cardiophylla Epling (p. 717) HEART-LEAVED PITCHER SAGE **ST** gen with short-stalked glands. **LF** cordate to ovate, irregularly serrate to nearly entire, gen with branched, nonglandular hairs and sessile to short-stalked glands. **FL**: pedicel gen 1–3 cm; calyx spreading at mouth, tending to drop readily in fr, with short-stalked glands, lobes < tube. **FR** glabrous. RARE. Chaparral; 600–1200 m. PR (Santa Ana Mtns). In cult.

L. fragrans (E. Greene) Epling (p. 717) FRAGRANT PITCHER SAGE Pl gen with long, branched, nonglandular hairs, sometimes with sessile or short-stalked glands (or both). **LF** deltate-lanceolate to ovate-lanceolate, serrate to entire. **FL**: pedicel gen 1–4 cm; calyx inflated at base, gen persistent in fr, lobes gen > (rarely <) tube. **FR** glabrous. UNCOMMON. Chaparral; < 1100 m. SCo, n ChI. ❀DRN,SUN,7,**15–17,24**&SHD:**21,22**&IRR:**14**,18,**19,20**.

L. ganderi Epling (p. 717) GANDER'S PITCHER SAGE Pl with short, branched, nonglandular hairs and short-stalked to sessile glands. **LF** lanceolate, serrate to entire. **FL**: pedicel gen 1–2 cm; calyx spreading, tending to drop readily in fr, lobes gen ± = tube. **FR** glabrous. RARE. Chaparral; 450–900 m. PR (Otay Mtns); Baja CA. In cult.

LYCOPUS BUGLEWEED

Per from rhizomes, glabrous or hairy. **STS** erect, branched or not. **LVS** short-petioled to sessile, gen ovate to lanceolate; margin toothed to deeply lobed or cut below middle. **INFLS** axillary, each head-like, subtended by lvs. **FL**: calyx gen 5-lobed, lobes ± equal, obtuse to short-awned; corolla slightly bilateral, not 2-lipped, gen 4-lobed, lobes ± unequal, odd lobe notched or entire; stamens 2, exserted, staminodes 2, minute, club-shaped; style exserted. **FR**: nutlets ± compressed, edge corky-thickened, truncate or rounded. 14 spp.: temp N.Am, Eurasia, 1 spp. in Australia. (Greek: wolf foot, from Grench common name) [Henderson 1962 Amer Midl Naturalist 68:95–135]

1. Calyx lobes ovate, tip obtuse to acute, ± = fr; lvs petioled if serrate .... ***L. uniflorus***
1′ Calyx lobes awl-like, tip acuminate to short-awned, gen > fr; lvs ± sessile if serrate
 2. Lvs gen short-petioled, deeply lobed to cut in lower half; fr 1–1.5 mm, top rounded .... ***L. americanus***
 2′ Lvs gen subsessile to sessile, serrate; fr 1.5–2 mm, top ± truncate .... ***L. asper***

L. americanus W.C. Barton (p. 717) Rhizomes ± slender, not thickened at tip. **STS** erect, 2–8 dm, gen glabrous; nodes short-hairy. **LF** 2–8(10) cm, short-petioled, oblong to lanceolate, lobed to cut esp in lower half, glabrous to puberulent on veins. **FL**: calyx lobes awl-like, short-awned; corolla 2–3 mm, ± = calyx, white. **FR**: nutlet 1–1.5 mm, top rounded, smooth. 2*n*=22. Moist areas, marshes, streambanks; < 1000 m. CA-FP, SNE; to B.C., e N.Am.

L. asper E. Greene (p. 717) Rhizomes thicker and tuber-like near tip. **STS** erect, 3–8(10) dm, puberulent to short-hairy. **LF** 2.5–7(9) cm, ± sessile, lanceolate to narrowly elliptic, serrate, glabrous to puberulent. **FL**: calyx lobes awl-like, acuminate to short-awned; corolla 3–5 mm, slightly > calyx, white. **FR**: nutlet 1.5–2 mm; top truncate, sometimes minutely toothed. 2*n*=22. Uncommon. Moist areas, marshes, streambanks; < 1300 m. Deltaic GV, SnFrB, GB; to w Can, Great Plains. [*L. lucidus* Benth. misapplied].

L. uniflorus Michaux (p. 717) NORTHERN BUGLEWEED Rhizomes slender, abruptly thicker and tuber-like at tip. **STS** ascending to erect, 1–5 dm, puberulent to finely strigose. **LF** 2–6(8) cm, gen short-petioled, elliptic to lanceolate, gen serrate, glabrous to sparsely puberulent. **FL**: calyx lobes ovate, obtuse to acute; corolla 2.5–4 mm, > calyx, white. **FR**: nutlet 1–2 mm, top truncate, ± finely toothed. UNCOMMON. Moist areas, marshes, near springs; < 100 m (1600–2000 m in n&c SNH). NCo, n&c SNH (Tuolumne, Plumas cos.); to B.C., e US.

MARRUBIUM HOREHOUND

Per. **STS** gen erect, gen branched, tomentose. **LVS** petioled to subsessile, gen ovate to round, crenate or toothed, **INFLS** gen axillary, each head-like, gen subtended by lvs. **FL**: calyx 10-lobed in CA, lobes spreading or recurved, sharp-pointed; corolla 2-lipped, upper lip entire to 2-lobed, lower lip 3-lobed; stamens 4, fertile, lower pair gen > upper pair, incl in tube; style incl, lobes ± equal. **FR**: nutlet top truncate. 30 spp.: Eur. (Latin: based on ancient Hebrew word for bitter juice) Some spp. cult for folk medicine, flavorings, some toxic.

M. vulgare L. **STS** ascending to erect, 1–6 dm. **LF**: petiole < blade; blade 1.5–5.5 cm, widely ovate to ± round, base rounded to ± lobed, margin crenate. **FL**: calyx 4–6 mm; teeth 10, short-soft-hairy; corolla > calyx, lips ± equal. 2*n*=34. Disturbed sites, gen overgrazed pastures; < 600 m. CA-FP; widespread worldwide; native to Eur. Formerly cult for tea, flavoring.

MELISSA BALM

Per. **STS** erect, simple or branched. **LF** petioled; blade oblong to ovate, crenate to serrate. **INFLS** axillary at upper nodes, each ± open, subtended by lvs, short-bracted. **FL**: calyx 2-lipped, upper lip > lower lip, 3-lobed, lower lip 2-lobed; corolla 2-lipped, upper lip ± entire, hood-like, erect or reflexed, lower lip 3-lobed, tube > calyx, curved upward; stamens 4, fertile, pairs subequal, incl under upper lip; style lobes unequal. **FR** ovoid, smooth. 3 spp.: Eur, w Asia. (Greek: bee)

M. officinalis L. BEE BALM **ST** branched, 2–15 dm, finely glandular hairy. **LF**: blade 2–14 cm, 1.5–7 cm wide, ovate, crenate. **INFL**: fls 4–12; subtending lvs reduced upward; pedicels 2–5 mm. **FL**: calyx 7–9 mm, tube ribbed, long-soft-hairy; corolla 8–15 mm, light yellow, becoming ± white or pinkish. Moist sites, meadows, ditches, fields; < 800 m. NW, CaRF, n SNF, CW; native to s Eur.

MENTHA MINT

Per from rhizomes, glabrous to hairy. **STS** gen ascending to erect, gen branched. **LF** petioled to sessile, elliptic, ovate, or lanceolate, toothed to lobed. **INFLS** axillary, each head-like and subtended by lvs, or collectively spike- or panicle-like and by bracts. **FL**: calyx ± radial, gen 10-veined, lobes equal or unequal; corolla ± 2-lipped, lips gen equal, upper lip notched, lower lip 3-lobed; stamens 4, ± equal, gen exserted; style lobes unequal. 25 spp.: temp. N.Am, Eurasia. (Latin: ancient name for mint) [Tucker, Harley, & Fairbrothers 1980 Taxon 29:233–255] Cult for oils, flavoring, herbs. Many cult and naturalized populations derived from hybridization, gen complexly polyploid, some sterile, reproducing vegetatively.

1. Internodes between upper infls evident, gen > 1 cm; infls subtended by ovate, elliptic, or lanceolate lvs
 2. Lvs subtending infl 1.5–3 cm, > infl diam, spreading, gen serrate . . . . ***M. arvensis***
 2′ Lvs subtending infls gen < 1 cm, gen < infl diam, reflexed, gen entire . . . . ***M. pulegium***
1′ Internodes between upper infls gen inconspicuous, gen < 6 mm; infl gen subtended by linear, lanceolate, or awl-like bracts
 3. Cauline lvs petioled, petiole 3–8 mm; infls gen > 10 mm diam, head-like, lower 2–3 separated by short internodes . . . . ***M. ×piperita***
 3′ Cauline lvs subsessile or sessile, petiole < 1 mm; infls gen < 5 mm diam, collectively spike-like
  4. Cauline lf base obtuse or rounded, tip acute or acuminate, lower surface gen glabrous or white-soft-hairy . . . . ***M. spicata***
  4′ Cauline lf base slightly lobed, tip rounded, lower surface stellate-tomentose . . . . ***M. suaveolens***

M. arvensis L. (p. 717) **ST** 1–5(8) dm, puberulent to short-hairy. **LF**: 1.5–5(8) cm; lower short-petioled; cauline gen subsessile; blade ovate to elliptic, base tapered, tip gen acute, crenate to serrate, lower surface (esp veins) short-hairy. **INFLS** axillary, each head-like, subtended by spreading lvs; bracts minute or 0. **FL**: calyx 1.5–3 mm, short-hairy; corolla 4–7 mm, white, pink, or violet; stamens > corolla lobes. 2*n*= 24,54,72,90. Moist areas, streambanks, lake shores; < 2400 m. CA-FP, GB; circumboreal. [var. *villosa* (Benth.) S.R. Stewart]. Some pls sterile; some pls naturalized from Eur.

M. ×piperita L. PEPPERMINT **ST** 3–10 dm, gen glabrous. **LF** 3–6(8) cm; petioles 3–8 mm; blade ovate to lanceolate, base tapered to slightly lobed, tip acute, gen serrate, lower surface gen glabrous. **INFLS** each head-like, clustered at upper 3–5 nodes, subtended by ovate to linear-lanceolate bracts. **FL**: calyx 2.5–4 mm, gen glabrous or lobes ciliate; corolla 3.5–6 mm, white, pink, or violet; stamens gen incl. 2*n*=66,72,84,120. Moist places, fields; < 2500 m. CA-FP, cult elsewhere; to Can, e US; naturalized from cult in Eur. Hybrid between *M. aquatica* L. & *M. spicata* L. Pls gen sterile, widely cult, spreading from rhizomes. [*M. citrata* Ehrh.]

M. pulegium L. PENNYROYAL **STS** decumbent to ascending, 1–6 dm, short-hairy. **LVS** gen 1.5–2.5 cm, reduced upward; lower short-petioled, cauline gen subsessile; blade narrowly ovate to elliptic, base tapered to obtuse, tip gen rounded, entire to finely serrate, lower surface short-hairy. **INFLS** axillary, each head-like, subtended by reflexed lvs or lf-like bracts. **FL**: calyx 2.5–4 mm, short-hairy; corolla 5–8 mm, violet to lavender; stamens > corolla lobes. 2*n*=20. Uncommon. Moist areas, ditches; < 1000 m. NW, SnJV, CW, SCo; to OR, e US; native to Eur. Oil TOXIC, has been fatal when extract ingested by humans; used as insect repellant.

M. spicata L. var. *spicata* SPEARMINT **ST** 3–10(12) dm, glabrous. **LF** gen 1–6 cm, ± sessile; blade ovate to lanceolate, base rounded to obtuse, tip acute to acuminate, gen serrate, lower surface gen glabrous. **INFLS** densely clustered, subtended by linear-lanceolate bracts, collectively spike-like. **FL**: calyx 1.5–2.5 mm, gen glabrous; corolla 3–4 mm, white, pink, or lavender; stamens exserted but not > corolla lobes. 2*n*=36,48. Moist areas, marshes, lake shores; < 1650 m. CA-FP, cult elsewhere; to e US; native to Eur. Pls with lower blade surface white-soft-hairy are var. *longifolia* L. [*M. longifolia* (L.) Hudson], native to Eur.

M. suaveolens Ehrh. **ST** 5–10 dm, soft-hairy. **LF** 1–4 cm, subsessile; blade ovate, oblong, or broadly elliptic, base slightly lobed, tip gen rounded, crenate to serrate, lower surface stellate-tomentose. **INFLS** densely clustered at upper nodes, subtended by linear or awl-like bracts, collectively spike-like. **FL**: calyx 1–1.5 mm, short-hairy; corolla 2–3 mm, white or pinkish; stamens > corolla lobes. Moist areas, ditchbanks; < 1200 m. NW, SN, CW, SCo, SNE, expected elsewhere; native to s Eur. [*M. rotundifolia* (L.) Hudson misapplied]

MOLUCCELLA SHELL FLOWER

Ann, gen glabrous. **STS** erect, gen simple. **LVS** petioled; blade crenate to lobed. **INFLS** axillary, subtended by lvs; fls 4–10; bracts spine-forked. **FL**: calyx > corolla, obliquely funnel-shaped, 5–10-toothed, becoming membranous, net-veined, persistent in fr; corolla 2-lipped, lower lip > upper, upper hood-like, lower 3–4-lobed, lobes unequal; stamens 4, ascending under upper lip, lower pair ± > upper pair; style lobes unequal. **FR**: nutlet 3-angled, top ± truncate. 2 spp.: Medit. (Latin: mistakenly named for Molucca Islands, Indonesia)

M. laevis L. **ST** < 6 dm, sometimes branched at base. **LF**: blade 1.5–4 cm, gen < petiole, ± round, palmately veined. **FL**: calyx 2–3 cm wide, gen light green, becoming tan, teeth minute; corolla 14–16 mm, white or pink, tube puberulent inside, lower lip 4-lobed, lateral lobes < central 2 lobes. Disturbed areas, roadsides; < 750 m. NW, SNF, GV, CW, SCo; to B.C., Rocky Mtns; native to s Eur.

MONARDA BEE BALM

Ann, per, gen short-hairy. **STS** erect, gen branched. **LF** petioled to sessile. **INFLS** axillary, each head-like; lower subtended by lvs; upper by bracts. **FL**: calyx 5-lobed; corolla 2-lipped, upper lip entire or ± 2-lobed, hood-like, arched, lower lip gen 2–3-lobed, central lobe gen > lateral lobes; stamens 2, ascending under upper lip, = or > upper lip; style unequally lobed. 16 spp.: N.Am. (Nicolas Monardes, Spanish physician & botanist, 1493–1588) [Scora 1967 Univ Calif Publ Bot 41:1–71] Some cult for fls, tea.

M. pectinata Nutt. (p. 717) Ann. **ST** 1.5–3.5 dm; hairs short, ± curled down. **LF** short-petioled to subsessile; blade 1.5–4 cm, gen oblong to lanceolate, entire to serrate, ± glabrous to finely strigose esp on veins. **INFL**: subtending lvs gradually reduced upward, uppermost ± 4–7 mm, ovate. **FL**: calyx tube 6–8 mm, throat densely puberulent within, lobes 2–4 mm, long-acuminate; corolla 12–25 mm, white to pink, lower lip sometimes purple-spotted. 2*n*=18,36. Washes, rocky slopes, pinyon-juniper woodland; 1150–1500 m. e DMtns (New York Mtns); to w Great Plains, n Mex. ❀TRY.

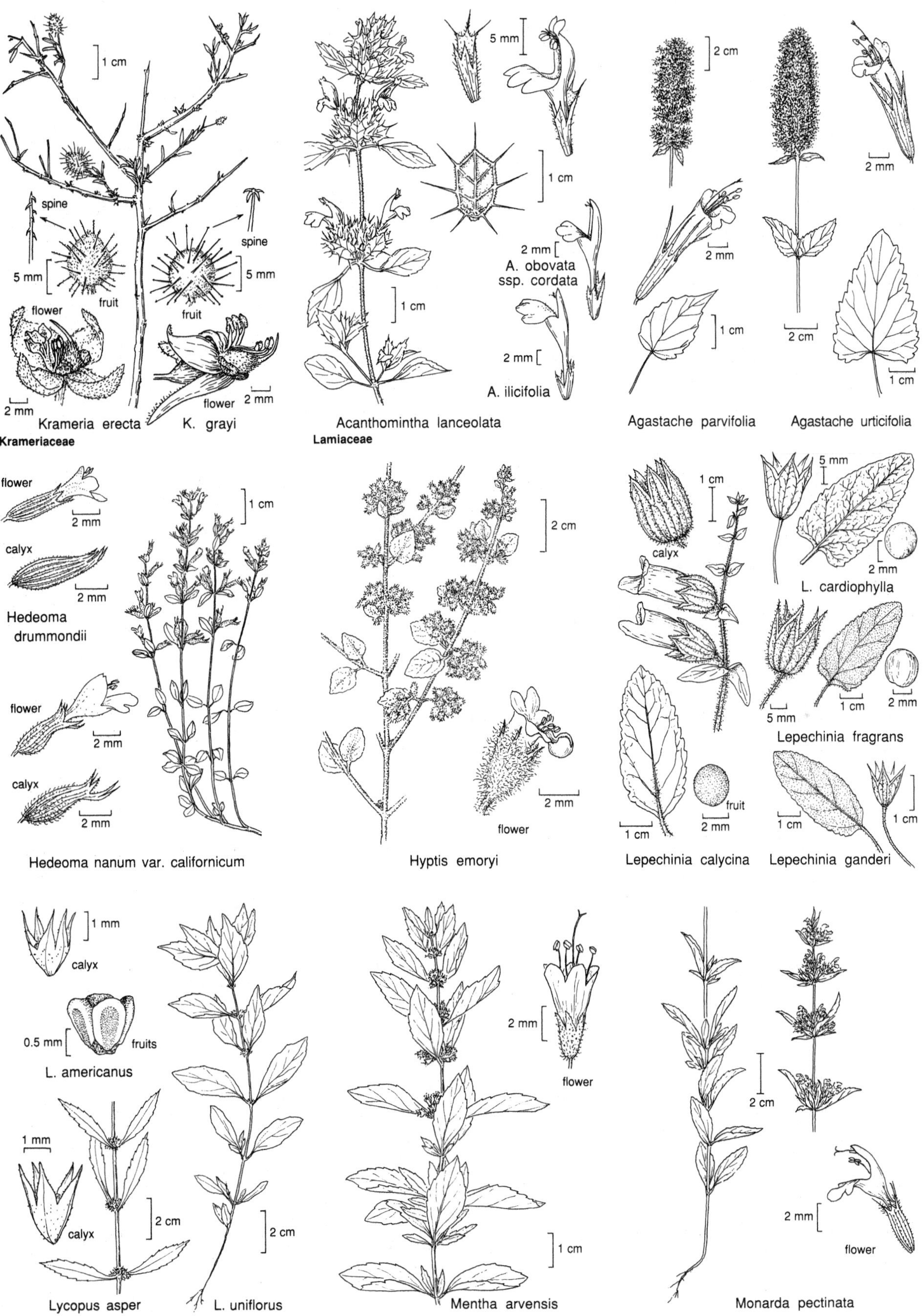
1 cm
spine
spine
5 mm
5 mm
fruit
fruit
flower
flower
2 mm
2 mm
Krameria erecta
K. grayi
Krameriaceae
5 mm
1 cm
2 mm
A. obovata
ssp. cordata
1 cm
2 mm
A. ilicifolia
Acanthomintha lanceolata
Lamiaceae
2 cm
2 mm
2 mm
1 cm
2 cm
1 cm
Agastache parvifolia
Agastache urticifolia
flower
2 mm
calyx
2 mm
Hedeoma
drummondii
1 cm
flower
2 mm
calyx
2 mm
Hedeoma nanum var. californicum
2 cm
2 mm
flower
Hyptis emoryi
1 cm
calyx
fruit
1 cm
2 mm
Lepechinia calycina
5 mm
2 mm
L. cardiophylla
5 mm
1 cm
2 mm
Lepechinia fragrans
1 cm
1 cm
Lepechinia ganderi
1 mm
calyx
0.5 mm
fruits
L. americanus
1 mm
calyx
2 cm
2 cm
Lycopus asper
L. uniflorus
2 mm
flower
1 cm
Mentha arvensis
2 cm
2 mm
flower
Monarda pectinata

MONARDELLA

James D. Jokerst

Ann, per, ± gland-dotted. **LVS** entire to serrate. **INFL**: heads 1 or more per main st, sometimes arrayed in spikes or panicles; bracts in 2–3 series (outer series (0) 1–2 pairs, ± like lvs, 0–several mm below heads, erect to reflexed; middle series 2–4 pairs, like lvs to papery or leathery, sometimes straw-colored to purple, erect in cup-like involucre to reflexed; inner series 0–few pairs, membranous, linear-lanceolate). **FL**: calyx 5-lobed, gen < 12 mm; corolla white to purple, upper lip erect, 2-lobed, lower lip recurved, 3-lobed; stamens 4; style unequally 2-lobed. ± 20 spp.; w N.Am. (Latin: small *Monarda*) [Epling 1925 Ann Missouri Bot Gard 12:1–106] Complex; hybrids common, often outnumbering non-hybrids; head width and bract orientation given for unpressed specimens. ❀often DFCLT. Many ssp., sspp., or populations have exacting soil requirements.

1. Calyx > 12 mm; corolla white to rose or red-orange, lobes 1/5–1/2 tube; per
2. Corolla 35–45 mm, red-orange to yellow, tube funnel-shaped, 1.5–3 mm wide at top, lobes ascending; calyx 20–25 mm; anthers 1.2–1.5 mm ..... ***M. macrantha***
3. St hairy; lf triangular-ovate, sparsely to densely hairy, ciliate ..... ssp. ***hallii***
3′ St glabrous to sparsely hairy; lf elliptic to ovate, ± glabrous, not ciliate ..... ssp. ***macrantha***
2′ Corolla 15–30 mm, white to cream-yellow or rose-tinged, tube cylindric, 0.5–1.5 mm wide at top, lobes spreading; calyx 12–20 mm; anthers < 1 mm ..... ***M. nana***
4. Corolla tube ± 1–1.5 mm wide at top; lf gen > 10 mm, green, glabrous to sparsely spreading-hairy
5. Corolla cream-yellow, tube 1.5–2 × calyx, lobes 7–10 mm ..... ssp. ***leptosiphon***
5′ Corolla white, tube 1.25–1.5 × calyx, lobes 6–7 mm ..... ssp. ***nana***
4′ Corolla tube ± 0.5–1 mm wide; lf gen 5–15 mm, ash-gray, densely ± appressed-puberulent
6. Calyx hairs ± dense, gen > 1 mm, wavy ..... ssp. ***arida***
6′ Calyx hairs ± sparse, gen < 1 mm, straight ..... ssp. ***tenuiflora***
1′ Calyx < 12 mm; corolla white to straw or purple, lobes > 1/2 tube; ann, per
7. Lf margin wavy, gen ± entire (less often ± serrate)
8. Lf < 10 mm, narrowly ovate to triangular-ovate; subalpine forests; s SNH ..... ***M. beneolens***
8′ Lf > 10 mm, linear to ovate; dunes, scrub; CCo, SnFrB
9. Ann; st glabrous or sparsely hairy ..... ***M. undulata***
9′ Per; st sparsely hairy to densely tomentose
10. St densely white-tomentose; head 20–30 mm wide; middle bracts 7–15 mm; lf oblanceolate to ovate, fleshy, appearing glaucous ..... ***M. crispa***
10′ St purple, ± sparsely hairy; head 10–20 mm wide; middle bracts 7–10 mm; lf linear to narrowly lanceolate, thin, green ..... ***M. frutescens***
7′ Lf margin faintly or not wavy, entire to serrate
11. Ann; heads several–many per main st, arrayed in panicles
12. Bract lateral veins ± perpendicular to midvien, areas between veins silvery-translucent ..... ***M. douglasii***
13. St slender; head 10–15 mm wide; bracts 10–15 mm, narrowly ovate ..... ssp. ***douglasii***
13′ St stout; head 20–30 mm wide; bracts 15–30 mm, widely ovate ..... ssp. ***venosa***
12′ Bract lateral veins ± parallel to midvein, areas between veins lf-like, membranous, or scarious but not silvery
14. Bracts not white; calyx lobes not white; corolla rose or purple, rarely white
15. Bracts long-hairy ..... ***M. pringlei***
15′ Bracts short-hairy
16. Bracts acuminate, not cross-veined; calyx 14–15-veined ..... ***M. breweri***
16′ Bracts acute, cross-veined; calyx 13-veined ..... ***M. lanceolata***
14′ Bracts at least partly white; calyx lobes or lobe tips or margins white; corolla gen or mostly white
17. Calyx lobe tips awl-like, recurved; bracts white throughout; stamens incl ..... ***M. leucocephala***
17′ Calyx lobe tips acute, erect; bracts partly white; stamens slightly exserted
18. Main st 1° branches above middle; bracts ovate, cross-veined; SNF ..... ***M. candicans***
18′ Main st 1° branches gen below middle; bracts gen narrowly ovate, not cross-veined; s SnJV, DMoj ..... ***M. exilis***
11′ Per; heads gen 1 per main st (or few, arrayed in spikes or racemes in *M. follettii, M. robisonii, M. siskiyouensis, M. stebbinsii*)
19. Middle bracts like lvs in texture and hairiness, greenish, gen reflexed (innermost of middle series sometimes ± scarious) — lower lf surface glandular-hairy
20. Lf narrowly to widely ovate, soft-hairy, wavy-hairy, or woolly (hairs gen > 0.5 mm exc ssp. *obispoensis*), lower surface woolly or not ..... ***M. villosa***
21. Lf ovate to widely triangular-ovate, base gen truncate; st gen densely matted-woolly (or ± glabrous) ..... ssp. ***franciscana***
21′ Lf ± narrowly ovate, base tapered to obtuse; st gen sparsely to densely spreading-hairy (or ± glabrous)
22. Some hairs branched; old pls tufted to matted; corolla white to lavender ..... ssp. ***obispoensis***
22′ Hairs unbranched; pls matted or not; corolla pink to purple
23. Pl > 50 cm; lf blade 22–50 mm; head 20–40 mm wide; outer bracts 20–30 mm ..... ssp. ***globosa***
23′ Pl < 50 cm; lf blade 10–22 mm; head 10–30 mm wide; outer bracts 8–20 mm ..... ssp. ***villosa***
20′ Lf lanceolate to narrowly ovate, glabrous, canescent, or stiff-hairy (hairs gen < 0.5 mm), lower surface not woolly
24. Pl sparsely to densely appressed- to spreading-hairy, whitish; head gen < 15 mm wide; middle bracts erect, in cup-like involucre ..... ***M. antonina***

25. Pl open, gen > 60 cm; lf greenish, not folded; hairs appressed to ascending, short, often sparse, not hiding glands of st and calyx .... ssp. ***antonina***
25′ Pl tufted, gen < 60 cm; lf whitish, often folded; hairs spreading, long, gen dense, hiding glands of st and calyx .... ssp. ***benitensis***
24′ Pl glabrous to densely and minutely canescent, green to ash-gray; head 10–30 mm wide; middle bract tips recurved
26. Inner bracts 2–many; pl glabrous to densely and minutely canescent .... ***M. sheltonii***
26′ Inner bracts 0; pl densely and minutely canescent .... ***M. siskiyouensis***
19′ Middle bracts gen not like lvs in either texture or hairiness, papery to leathery, greenish to white, straw, or purple, erect or spreading (outermost of middle series sometime like lvs, at least at tips)
27. Outer bracts present or 0; middle bracts scarious or papery, elliptic to ovate, erect, in cup-like involucre (not always apparent on pressed material)
28. Outer bract series present; middle bracts scarious or membranous, greenish to rose, sometime ± purple-tinged
29. Middle bracts long-acuminate, rose (or scarious and rose-tinged); calyx lobes hairy; corolla rose; TR .... ***M. australis***
29′ Middle bracts acute to obtuse, scarious, greenish, lavender- to rose- to purple-tinged or not; calyx lobes woolly; corolla white to purple; n CA-FP, GB .... ***M. odoratissima***
30. Outer bracts spreading, gen < 25 mm; corolla gen purple, less often lavender or white . ssp. ***odoratissima***
30′ Outer bracts reflexed, gen 25–40 mm; corolla gen white, sometimes ± purple-tinged .... ssp. ***pallida***
28′ Outer bract series 0; middle bracts papery, whitish to straw or purple, sometimed rose- or purple-tinged, less often grayish
31. St green, dark gray, or appearing glaucous; lf elliptic, narrowly ovate, or ovate; n CA .... ***M. glauca***
31′ St silvery to ash-gray with densely matted hairs; lf linear to narrowly ovate; s CA .... ***M. linoides***
32. Lf silvery or ash-gray; s SNH, Teh, e SnBr, SnJt, SNE, DMtns
33. Lf linear, oblong, or lanceolate, 10–40 mm; bracts = or > calyces .... ssp. ***linoides***
33′ Lf narrowly ovate, 10–15 mm; bracts > calyces .... ssp. ***oblonga***
32′ Lf greenish or appearing glaucous; SnGb, SnBr, PR
34. Lf gen 10–20 mm; pl rhizomed; bracts obscurely gland-dotted; SnGb, SnBr .... ssp. ***stricta***
34′ Lf 20–40 mm; pl taprooted; bracts conspicuously gland-dotted; s PR (San Diego Co.) .... ssp. ***viminea***
27′ Outer bracts present; middle bracts firm, leathery or scarious, lanceolate to ovate, reflexed to ascending, gen not in cup-like involucre
35. Lvs and lower st ± glabrous
36. Bracts and calyces conspicuously gland-dotted; head gen < 15 mm wide; bracts lanceolate or narrowly ovate .... ***M. follettii***
36′ Bracts and calyces obscurely gland-dotted; head gen > 15 mm wide; bracts narrowly ovate to ovate
37. Corolla lobes < 1/2 tube; calyx 9–12 mm .... ***M. palmeri***
37′ Corolla lobes > 1/2 tube; calyx 6–8 mm .... ***M. purpurea***
35′ Lvs and st hairy
38. Lf surfaces both greenish or ash- to dark gray, purple blotched or not, sparsely to densely hairy; heads 1–several per main st; SN, TR, DMtns
39. Upper st hairs dense, felt-like .... ***M. stebbinsii***
39′ Upper st hairs sparse to dense but not felt-like
40. Lf 5–10 mm, gen few-toothed; head 1 per main st; pl matted or tufted; s SCoRO?, e SnGb, SnJt? .... ***M. cinerea***
40′ Lf 10–40 mm, entire; heads gen several per main st; pl open; DMtns .... ***M. robisonii***
38′ Lf surfaces unequally colored, lower surface ± white- or ash-gray-canescent, felt-like, or woolly, upper surface greener, hairs 0 to dense; heads gen 1 per main st; s NCoRI, SCo, SCoRO, TR, PR
41. Lf thick, lower surface felt-like or woolly; corolla white to pale lavender .... ***M. hypoleuca***
42. St hairs 0 to sparse; lf narrowly ovate, ± flat between rolled-under margins .... ssp. ***hypoleuca***
42′ St hairs ± dense; lf linear-lanceolate, ± arched between rolled-under margins .... ssp. ***lanata***
41′ Lf thin, lower surface canescent; corolla lavender, rose, or purple .... ***M. viridis***
43. Head > 20 mm wide; bracts gen 10–20 mm; SnGb .... ssp. ***saxicola***
43′ Head 10–15(21) mm wide; bracts gen 6–10 mm; s NCoRI .... ssp. ***viridis***

M. antonina Hardham Per, rhizomed, whitish; hairs appressed to spreading, sparse to dense. **LF** narrowly ovate, often serrate. **INFL**: head gen < 15 mm wide; outer bracts reflexed or spreading, like lvs, middle bracts erect, in cup-like involucre, scarious (or their tips lf-like). **FL**: calyx hairy, purple-tipped or not; corolla lavender. Open, rocky slopes, oak woodland, chaparral; 500–900 m. c SCoR.

ssp. ***antonina*** SAN ANTONIO HILLS MONARDELLA Pl gen > 60 cm, open; hairs often sparse, appressed to ascending, short, not hiding st and calyx glands. **LF** thin, green, not folded; lower surface conspicuously gland-dotted; margin not rolled under. *n*=21. UNCOMMON. Habitats of sp. c SCoR (s Monterey Co.). Hybridizes with (and much like) *M. villosa*. ❀DRN:6,15–17&SHD:**7**,**14**,18–24&IRR:**8**,**9**,**19–23**.

ssp. ***benitensis*** (Hardham) Jokerst (p. 723) SAN BENITO MONARDELLA Pl gen < 60 cm, tufted; hairs gen dense, spreading, long, hiding st and calyx glands. **LF** thick, leathery, whitish, often folded; lower surface obscurely gland-dotted; margin rolled under. *n*=21. UNCOMMON. Serpentine barrens, chaparral; 500–900 m. c SCoRI (near New Idria, San Benito Co.). [*M. b.* Hardham]

M. australis Abrams (p. 723) Per, tufted. **ST** decumbent to erect, 10–20 cm, sparsely to densely hairy. **LF** lanceolate to narrowly ovate, entire to weakly serrate, green to ash-gray. **INFL**: outer bracts erect, like lvs, lanceolate; middle bracts erect, in cup-like involucre, long-acuminate, hairy, rose (or scarious and rose-tinged). **FL**: calyx lobes hairy; corolla 12–20 mm, rose. Montane and subalpine forests, rocky openings; 1500–3200 m. TR. [*M. odoratissima* ssp. *a.* (Abrams) Epling] Hybridizes with *M. hypoleuca* & *M. linoides* ssp. *stricta*. ❀TRY.

M. beneolens J.R. Shevock, B. Ertter & Jokerst SWEET-SMELLING MONARDELLA Per, matted to tufted, rhizomed. **ST** decumbent, 10–30 cm; hairs spreading, long, soft, some glandular. **LF** < 10 mm, narrowly to triangular-ovate; margin wavy, gen ± entire, less often ± serrate; hairs dense, long, wavy, glandular on lower surface. **INFL**: head 10–20 mm wide; outer bracts like lvs or not; middle bracts ± erect, ± in cup-like involucre, short, ovate, green to straw, rose-tinged or not. **FL**: calyx hairs dense; corolla lavender to pale rose. RARE.. Subalpine forests; 2500–3500 m. s SNH. Hybridizes with *M. odoratissima* ssp. *pallida*, *M. linoides*. ❀DRN: 15–17&SHD:**7**&IRR:1–3,8,9,**14**,18–23.

M. breweri A. Gray (p. 723) Ann, branched above base. **ST** 15–65 cm, gray-hairy. **LF** 15–45 mm, narrowly ovate; hairs short. **INFL**: head 20–30 mm wide; outer bracts 10–15 mm, widely ovate, acuminate, stiff-pointed, scarious between veins, purplish, short-hairy, veins converging at tip, cross-veins 0. **FL**: calyx 14–15-veined; corolla 12–14 mm, hairy, rose, tube exserted; stamens exserted. Oak woodland, chaparral, pinyon/juniper woodland; < 1400 m. SnFrB, SCoR, WTR. Hybridizes with *M. lanceolata* (bracts faintly cross-veined), *M. pringlei*. ❀TRY.

M. candicans Benth. Ann; 1° branches above middle of main axis, = or > main axis. **ST** purplish. **LF** lanceolate. **INFL**: head 15–25 mm wide; bracts ovate, acute or obtuse, scarious, cross-veined, veins green, margin white. **FL**: calyx lobes erect, acute, margins white-scarious; corolla 8–11 mm, white, purple-spotted or not, tube exserted; stamens slightly exserted, tissue between pollen chambers entire. Sandy or gravelly soils, oak woodland, chaparral, yellow-pine forest; < 800 m. SNF.

M. cinerea Abrams (p. 723) GRAY MONARDELLA Per, matted or tufted; hairs ± long-spreading and short-glandular. **ST** decumbent, 5–15 cm. **LF** sessile, 5–10 mm, triangular-ovate; margin sometimes faintly wavy, teeth few–several, rarely 0; lower surface ± covered by gold gland-dots. **INFL**: head 1 per main st; bracts narrowly ovate, acute, scarious, hairy, red to purple. **FL**: calyx purple, hairs ± long, short-glandular, lobes acuminate; corolla rose-purple; stamens slightly exserted. UNCOMMON. Subalpine, montane forests; > 1800 m. s SCoRO?, e SnGb, SnJt?

M. crispa Elmer (p. 723) CRISP MONARDELLA Per, low, tufted or in mounds < 3 m wide, strong-smelling. **STS** decumbent to erect, few–many, 10–50 cm, densely white-tomentose, often wavy. **LVS** clustered at nodes, 10–50 mm, oblanceolate to ovate, thick, fleshy, appearing glaucous, sparsely tomentose to woolly, margin wavy. **INFL**: head 20–30 mm wide, solitary (or several, arrayed in terminal spike); outer bracts like lvs; middle bracts 7–15 mm, ovate, papery, straw-colored or purple, soft-hairy. **FL**: calyx woolly; corolla rose-purple. RARE. Unstable coastal dunes; < 100 m. s CCo (San Luis Obispo, Santa Barbara cos.). Hybridizes with *M. frutescens* in transitional or disturbed habitats. In cult.

M. douglasii Benth. Ann, branched above base. **ST** 10–30 cm, purple, short-hairy. **LF** linear-oblong to lanceolate, hairy. **INFL**: head 10–30 mm wide; bracts 10–30 mm, narrowly to widely ovate, acuminate, midrib and lateral veins green to purple, areas between veins silvery-translucent. **FL**: calyx 15-veined, hairy, lobes tapered to rigid point; corolla 11–12 mm, deep purple; stamens exserted. Grassland, oak woodland, chaparral, serpentine outcrops; < 1100 m. NCoRO, n&c SNF, ScV, SnFrB, SCoRI.

ssp. ***douglasii*** (p. 723) **ST** slender. **INFL**: head 10–15 mm wide; bracts 10–15 mm, narrowly ovate, hairs between veins present. Uncommon. Habitats and elevations of sp. NCoRO, SnFrB, SCoRI. ❀DRN:7,14,19–24&SUN:5,15–17;DFCLT.

ssp. ***venosa*** (Torrey) Jokerst (p. 723) VEINY MONARDELLA **ST** stout. **INFL**: head 20–30 mm wide; bracts 15–30 mm, widely ovate, purple, hairs between veins sparse to 0. **RARE**. Grassland; < 400 m. n&c SNF (Butte, Tuolumne cos.), ScV (Sutter Co.).

M. exilis E. Greene Ann; 1° branches gen basal or below middle of main axis. **LF** lanceolate to narrowly ovate. **INFL**: bracts gen narrowly ovate, green or whitish, purple-tinged or not, lateral veins 0 or few, cross-veins 0, margins, tips white-scarious, abruptly acuminate. **FL**: calyx lobe tips erect, acute, white; corolla 10 mm, white; stamens slightly exserted, tissue between pollen chambers notched. Desert scrub, washes; 600–1100 m. s SnJV, DMoj. Much like *M. candicans*.

M. follettii (Jepson) Jokerst (p. 723) Per, erect, open. **ST** 30–60 cm, purple, glabrous (or puberulent or glandular-hairy above); internodes > lvs. **LF** lanceolate, entire, glabrous; lower surface purple-tinged or not. **INFL**: heads gen < 15 mm wide, 1 per main st (or several, arrayed in terminal spikes or racemes); bracts lanceolate to narrowly ovate, leathery, greenish or straw, hairy, conspicuously gland-dotted. **FL**: calyx short-hairy, conspicuously gland-dotted; corolla pink. Uncommon. Montane forests, open, rocky slopes, serpentine; 600–2000 m. n SNH (Plumas Co.). [*M. odoratissima* ssp. *glauca* (E. Greene) Epling in part, as treated by Munz]

M. frutescens (Hoover) Jokerst (p. 723) SAN LUIS OBISPO MONARDELLA Per, open. **STS** several, ± sparsely hairy, purple. **LVS** clustered at nodes, 10–50 mm, linear to narrowly lanceolate, thin, sparsely hairy, green, margin wavy. **INFL**: head gen 1 per main st, 10–20 mm wide; middle bracts 7–10 mm, narrowly ovate, papery, straw-colored or purple, outer bract series like lvs. **FL**: calyx hairy; corolla rose-purple or purple. RARE. Stabilized dunes, sandy scrub; < 200 m. s CCo (San Luis Obispo, Santa Barbara cos.). [*M. undulata* var. *f.* Hoover] Hybridizes with *M. crispa*. Threatened by coastal development. In cult.

M. glauca E. Greene (p. 723) Per. **ST** green, dark gray, or appearing glaucous, with gland-tipped bristles or not. **LF** elliptic to ovate, entire. **INFL**: outer bracts 0; middle bracts elliptic to ovate, papery, finely short-hairy, ciliate, rose or purple (or grayish), outer of middle series in cup-like involucre, sometimes lf-like at tips. **FL**: calyx lobes sparsely to densely stiff-hairy; corolla 10–20 mm, purple or red-purple. Rocky openings, sagebrush scrub to alpine forest; 1000–3500 m. KR, NCoRH, CaRH, SNH, GB. [*M. odoratissima* ssp. *g.* (E. Greene) Epling] Highly variable: small-headed, matted pls of SNH, W&I, AZ, NV, Rocky Mtns have been called *M. o.* ssp. *parvifolia* (E. Greene) Epling; pls intermediate to *M. linoides* with erect, lanceolate lvs, internodes 1.5–2 × lvs, bracts scarious below, deep rose above, of SNE, e SNH, e MP, NV, AZ, Rocky Mtns have been called *M. rubella* E. Greene. Hybridizes with *M. beneolens*, *M. o.* ssp. *pallida*, *M. linoides*. ❀TRY.

M. hypoleuca A. Gray Per, erect, open. **ST** purple, short-hairy. **LF** 20–40 mm, linear-lanceolate to narrowly ovate, thick; lower surface ± white, felt-like or woolly; upper surface green, ± glabrous to hairy. **INFL**: head 20–40 mm wide; bracts 8–12 mm, leathery or scarious, purple, tomentose, ciliate. **FL**: corolla 15–17 mm, white to pale lavender, lobes obtuse; stamens exserted. Chaparral, woodland, forest; < 1500 m. SCoRO, SCo, WTR, SnGb, PR.

ssp. ***hypoleuca*** (p. 723) **ST** glabrous to sparsely hairy. **LF** narrowly ovate, ± flat between rolled-under margins; upper surface glabrous or sparsely hairy, veins prominent. **INFL**: inner bracts ovate. Habitats and range of sp. ❀DRN,DRY:**7,14–16**,17,**24**& SHD:**8,9,18–23**.

ssp. ***lanata*** (Abrams) Munz (p. 723) FELT-LEAVED MONARDELLA **ST** ± densely hairy. **LF** linear-lanceolate, ± arched betweeen rolled-under margins; upper surface sparsely hairy to matted-hairy, veins gen not prominent. **INFL**: inner bracts elliptic. RARE. Chaparral; 300–1000 m. w PR (Orange, San Diego cos.); n Baja CA?. In cult.

M. lanceolata A. Gray (p. 723) MUSTANG MINT Ann, open, branched above. **ST** 2–5 dm, purple, glandular- or short-hairy above. **LF** 30–40 mm, ± lanceolate. **INFL**: head 5–30 mm wide; bracts 5–15 mm, acute, scarious, green, purple-tipped, short-hairy, cross-veined; outer of middle series sometimes like lvs. **FL**: calyx glabrous to scabrous, 13-veined; corolla 12–15 mm, purple; stamens exserted. n=21. Open, rocky, often disturbed sites, chaparral, woodland; < 2600 m. SN, CW, SW; Baja CA. Variable; named forms gen indistinct. Hybridizes with *M. breweri*. ❀DRN:1–3,**6–9,14–24**.

M. leucocephala A. Gray (p. 723) MERCED MONARDELLA Ann, open, branched above base, glandular-hairy. **ST** 15–20 cm, ash-gray, hairy. **LF** lanceolate or oblong, hairy; veins obscure. **INFL**: head 10–15 mm wide; bracts ovate to obovate, short-acuminate, white, scarious, cross-veins few. **FL**: calyx stiff-hairy, lobes white, tips recurved, awl-like; corolla 5–5.5 mm, white, tube barely exserted; stamens incl, upper 2 sessile. **PRESUMED EXTINCT**. Sandy soil in grasslands; 50–100 m. n SnJV (Stanislaus, Merced cos.). Last seen 1941.

M. linoides A. Gray Per, erect, open. **ST** 10–50 cm, woody below, silvery to ash-gray with densely matted hairs; internodes gen > lvs. **LF** 10–40 mm, linear to narrowly ovate, entire, greenish, silvery, or ash-gray. **INFL**: head 20–30 mm wide; outer bracts 0; middle bracts ovate, acuminate, papery, hairy, ciliate, whitish, straw-colored, rose, or purple, outer of middle series erect, in cup-like involucre, sometimes lf-like at tips. **FL**: calyx lobes stiff-hairy; corolla 12–15 mm, whitish, lavender, or pale purple; stamens exserted. *n*=21. Desert scrub, pinyon/juniper woodland, open conifer forest, subalpine; 900–3100 m. s SNH, Teh, SnGb, SnBr, e PR (and adjacent w D), SNE, DMtns; NV, Baja CA.

ssp. ***linoides*** **ST** silvery. **LF** 10–40 mm, linear to lanceolate, silvery. **INFL**: bracts = or > calyces, white to rose. Habitats, elevations of sp. s SNH, e SnBr, SnJt, SNE, DMtns; NV. Hybridizes with *M. beneolens, M. glauca*.

ssp. ***oblonga*** (E. Greene) Abrams (p. 723) FLAX-LIKE MONARDELLA **ST** silvery or ash-gray. **LF** 10–15 mm, narrowly ovate, ash-gray. **INFL**: bracts > calyces, rose-purple. RARE. Habitats of sp.; 900–2300 m. s SNH, Teh. Possibly indistinct from ssp. *l.*

ssp. ***stricta*** (Parish) Epling Pl rhizomed. **ST** greenish, sparsely hairy. **LF** gen 10–20 mm, linear to lanceolate, acute, green. **INFL**: bracts obscurely gland-dotted; middle bracts narrowly ovate, red-purple, outer of middle series like lvs, reflexed. Open, rocky slopes in montane forests; 1600–3000 m. SnGb, SnBr. Hybridizes with *M. australis*. ❀TRY.

ssp. ***viminea*** (E. Greene) Abrams (p. 723) WILLOWY MONARDELLA Pl taprooted. **ST** 10–50 cm, hairy, glaucous-green. **LF** 20–40 mm, linear to lanceolate, minutely puberulent, appearing glaucous. **INFL**: head 1 per main st (or several in ± terminal spikes); bracts greenish white, often rose-tipped, conspicuously gland-dotted, outer reflexed, like lvs or not. **FL**: corolla white to rose. **ENDANGERED** CA. Rocky washes; < 400 m. s SCo, sw PR (San Diego Co.); Baja CA. In cult.

M. macrantha A. Gray Per, low, open, tufted, rhizomed. **LF** 5–30 mm, elliptic to triangular-ovate, leathery, entire to minutely serrate, deep green, shiny. **INFL**: head 20–40 mm wide; bracts oblong-elliptic, outer series like lvs, inner series scarious, hairy, ciliate, often reddish. **FL**: calyx 20–25 mm, 2-lipped, bent, hairy, ± glandular; corolla 35–45 mm, red-orange to yellow, tube 1.5–3 mm wide at top, funnel-shaped, exserted from calyx, lobes ascending; anthers 1.2–1.5 mm, exserted. Chaparral, woodland, forest; 600– 2000 m. SCoR, TR, PR; Baja CA.

ssp. ***hallii*** Abrams (p. 723) HALL'S MONARDELLA **ST** hairy. **LF** triangular-ovate, sparsely to densely hairy, ciliate; base truncate. **FL**: calyx long-spreading-hairy. RARE. Habitats of sp. SnGb, SnBr, PR. Intermediates to ssp. *m.* common.

ssp. ***macrantha*** (p. 723) **ST** glabrous to sparsely hairy. **LF** elliptic to ovate, ± glabrous, not ciliate, base obtuse. **FL**: calyx hairy. Habitats and range of sp. ❀DRN,DRY:**15–17**,24&SHD:**7**, 14,18,22,23&IRR:8,9,19–21;DFCLT;CVS.

M. nana A. Gray Per, matted or tufted, rhizomed. **LF** 5–30 mm, ovate, entire, glabrous to hairy. **INFL**: head 10–35 mm wide; bracts 15–20 mm; outer bracts like lvs or not; middle bracts white, gen rose- or purple-tinged. **FL**: calyx 12–20 mm; corolla 15–30 mm, white or cream, sometimes rose-tinged, tube 0.5–1.5 mm wide, cylindric, lobes spreading; stamens barely exserted, anthers < 1 mm. Montane forest, chaparral; 900–2600 m. PR; Baja CA.

ssp. ***arida*** H.M. Hall **ST** < 10 cm. **LVS** crowded, 5–10 mm, ash-gray, densely ± appressed-puberulent. **FL**: calyx hairs ± dense, gen > 1 mm, wavy; corolla white to cream, rose-tinged, tube 15–20 mm, < 1 mm wide, lobes 4–6 mm, oblong. Montane desert slopes; 900–1500 m. e PR (SnJt, Santa Rosa Mtns). ❀DRY,SHD:7–14, 18–21.

ssp. ***leptosiphon*** (Torrey) Abrams (p. 723) SAN FELIPE MONARDELLA **ST** 5–30 cm; hairs 0 or spreading. **LF** 10–30 mm, ovate to round, green; hairs ± sparse, spreading. **FL**: corolla cream-yellow, tube 20–25 mm, 1.5–2 × calyx, ± 1 mm wide, lobes spreading, 7–10 mm; stamens slightly exserted. RARE. Habitats of sp.; 1200–1800 m. c PR (Palomar Mtns). In cult.

ssp. ***nana*** (p. 723) **ST** 5–30 cm; hairs 0 to sparse. **LF** gen > 10 mm, green; hairs 0 to ± sparse, spreading. **FL**: corolla white, tube 1.25–1.5 × calyx, ± 1 mm wide, lobes 6–7 mm. Habitats of sp; 1200–1600 m. PR; Baja CA.

ssp. ***tenuiflora*** (S. Watson) Abrams (p. 723) **ST** 10–15 cm; hairs short. **LF** gen 5–15 mm, ash-gray, densely ± appressed-puberulent. **FL**: calyx hairs ± sparse, gen < 1 mm, straight; corolla tube 20 mm, < 1 mm wide, lobes 6–8 mm. Habitats of sp.; 900–2600 m. n PR (incl SnJt). ❀DRN:**7**,14,**18**,19–21&SUN:**15**,**16**,17,22–24.

M. odoratissima Benth. Per, tufted or not. **ST** erect, greenish, hairy, with gland-tipped bristles or not. **LF** 5–45 mm, lanceolate to ovate, entire, sparsely to densely hairy, green to ash-gray, often purple-tinged. **INFL**: head 10–25 mm wide; outer bracts reflexed or spreading, like lvs; middle bracts erect, in cup-like involucre, lanceolate to ovate, acute to obtuse, hairy, ciliate, scarious, greenish, sometimes lavender- to rose- or purple-tinged. **FL**: calyx hairy, lobes woolly; corolla 10–20 mm, white, lavender, or purple. *n*=21. Sagebrush scrub, montane forest; 600–3100 m. KR, NCoRH, CaRH, SNH, MP, W&I; to WA, NV.

ssp. ***odoratissima*** Pl white- or gray-puberulent, gen purple-tinged. **LF** 5–20 mm; lower surface whitish or grayish; upper surface greenish. **INFL**: outer bracts spreading, < 25 mm, narrowly ovate. **FL**: calyx sparsely hairy; corolla gen purple, less often lavender or white. Uncommon. Sagebrush scrub, montane forests; 1200–2100 m. nw MP (Lava Beds National Monument, Modoc Co.); to WA. ❀TRY;DFCLT.

ssp. ***pallida*** (A.A. Heller) Epling (p. 723) Pl greenish or ash-gray. **LF** 20–45 mm, both surfaces green-glabrous to ash-gray-hairy. **INFL**: outer bracts reflexed, 25–40 mm, linear-lanceolate. **FL**: calyx woolly; corolla gen white (sometimes lavender- or purple-tinged). Montane forest, rocky slopes; 1000–3100 m. KR, NCoRH, CaRH, SNH, W&I; NV. Hybrids with *M. glauca* along SN crest have purple-tinged lvs & bracts, lavender or purple corollas; hybrids with *M. sheltonii* or *M. villosa* in NCoRI have rose corollas, spreading-soft-wavy-hairy lvs [ssp. *pinetorum* (A.A. Heller) Epling]. ❀DRN:15–17&SHD:1–3,**7**&IRR:8,9,**14**,18–23;dfclt.

M. palmeri A. Gray (p. 723) PALMER'S MONARDELLA Per, tufted, rhizomed. **ST** decumbent, 10–30 cm, purple-tinged, ± glabrous; internodes gen < lvs. **LF** 10–20 mm, lanceolate, leathery, ± glabrous, entire to weakly serrate. **INFL**: head 25–35 mm wide; outer bracts like lvs; middle bracts ovate, acute, leathery, purple, short-hairy, obscurely or not gland-dotted. **FL**: calyx 9–12 mm, 2-lipped, stiff-hairy, obscurely gland-dotted, lobes slender, red-purple; corolla 15–20 mm, red purple, slender, lobes < 1/2 tube. 2*n*= ±60. UNCOMMON. Chaparral, forest, serpentine; 200–800 m. n SCoRO (Santa Lucia Mtns). Related to *M. purpurea*. ❀DRN:**7**, 15–18,21–24&IRR:8,9,14,19,20;DFCLT.

M. pringlei A. Gray (p. 723) PRINGLE'S MONARDELLA Ann, branched near base. **ST** hairy, ash-gray. **LF** hairy. **INFL**: head 20–25 mm wide; bracts 10 mm, ovate, abruptly acuminate, long-hairy, purple above. **FL**: calyx soft-hairy, lobes awl-like, gen purple, densely hairy within; corolla 11–13 mm, rose or purple, tube exserted; stamens exserted. **PRESUMED EXTINCT**. Sandy, coastal-sage scrub; 300–400 m. e SCo (near Colton). Last seen 1921; habitat destroyed by urbanization.

M. purpurea Howell (p. 727) SISKIYOU MONARDELLA Per, erect, open, ± glabrous. **ST** 10–40 cm, dark purple, shiny; internodes gen = or > lvs. **LF** 15–30 mm, oblong-lanceolate, shiny, deep green, leathery; margin rolled under, serrate or not. **INFL**: head 15–25 mm wide, obscurely gland-dotted; outer bracts sometimes not much like lvs; middle bracts ± ovate, leathery, purple-tinged. **FL**: calyx 6–8 mm, lobes long-pointed, purple, stiff-hairy, obscurely gland-dotted; corolla 12–14 mm, exserted, rose or purple, lobes > 1/2 tube. UNCOMMON. Rocky slopes, often serpentine or related bedrock, chaparral, woodland, montane forest; 400–1400 m. KR, NCoRO, SnFrB, SCoRO; OR. [*M. odoratissima* ssp. *neglecta* (E. Greene) Epling] Hybrids with *M. villosa* (common in SnFrB and SCoRO, with lvs ovate to triangular, ± glabrous) have been called *M. villosa* var. *subglabra* Hoover; also hybridizes with *M. glauca* in KR; much like *M. follettii*. ❀DRN:6&IRR:**7**,15–17& SHD:8,9,14,18;DFCLT.

M. robisonii Epling (p. 727) ROBISON'S MONARDELLA Per, erect, open. **ST** grayish, hairs spreading, wavy. **LF** 10–40 mm, lanceolate to narrowly ovate, ash-gray, hairy, entire. **INFL**: heads 10–20 mm wide, gen several arrayed in ± terminal spikes or racemes, sometimes solitary; bracts narrowly acute, scarious, pink-tinged or not, aging brown. **FL**: calyx lobes narrowly acute; corolla pale rose. RARE. Desert scrub, pinyon/juniper woodland; 1100–1500 m. DMtns; Baja CA?. Hybridizes with *M. linoides*.

M. sheltonii Howell Per, erect, open, rhizomed, glabrous to densely and minutely canescent. **ST** purplish or not. **LF** lanceolate, entire to weakly toothed. **INFL**: head 10–30 mm wide; bracts like lvs in texture, color, hairiness; outer bracts and tips of middle bracts strongly reflexed. **FL**: corolla 10–20 mm, purple. Rocky clearings, montane forest, oak woodland, chaparral, often serpentine; 600–2000 m. KR, NCoRO, n&c SNH; OR. [*M. villosa* ssp. *sheltonii* (Torrey) Epling] Hybridizes with *M. villosa*. ❀DRN:6,15–17&SHD:**7,14**,18–24&IRR:**8,9**,19–23.

M. siskiyouensis Hardham Per, erect, open, densely and minutely canescent. **ST**: some hairs glandular, esp above. **LF** narrowly ovate, entire or weakly serrate. **INFL**: heads 10–15 mm wide, 1 per main st (or often several in ± terminal spikes or racemes); outer bracts reflexed, like lvs; middle bracts scarious, with lf-like tips, recurved; inner bracts 0. **FL**: calyx glandular-hairy; corolla purple; stamens exserted. n=19–23. Uncommon. Oak woodland, chaparral; < 900 m. s KR. ❀DRN,DRY:15–17,24&SHD:**7**,8,9,14,19–23;DFCLT.

M. stebbinsii Hardham & J. Bartel (p. 727) STEBBINS' MONARDELLA Per, matted or clumped (< 1 m diam); hairs dense, felt-like. **ST** < 40 cm. **LF** narrowly ovate, entire, ash- to dark gray, purple-blotched. **INFL**: heads < 15 mm wide, 1 per main st (or several in ± terminal spikes or racemes); bracts lanceolate; outer bracts firm, leathery, purple-tinged, soft-hairy. **FL**: calyx green, purple, or straw, lobes narrowly acute; corolla 12–18 mm, pink. n=21. RARE. Rocky serpentine slopes; 800–1100 m. n SNH (Plumas Co.). In cult.

M. undulata Benth. (p. 727) CURLY-LEAVED MONARDELLA Ann, erect, simple or branched throughout; hairs 0 or sparse. **ST** 10–50 cm, red-brown. **LVS** in axillary clusters, 10–40 mm, linear to oblanceolate, ± fleshy; margin wavy. **INFL**: head 10–30 mm wide; bracts elliptic to ovate, tips obtuse to acute, scarious or green, purple-tinged or not. **FL**: calyx lobes obtuse; corolla 14–20 mm, purple. n=21. UNCOMMON. Dunes, sandy soils in sagebrush scrub; < 300 m. CCo, SnFrB. ❀DRN,DRY:17,24&IRR:8,14–16.

M. villosa Benth. COYOTE-MINT Per, matted to erect, open, rhizomed; hairs gen > 0.5 mm, soft, wavy, or woolly, glandular and not. **LF** 10–30 mm, ± ovate, entire to serrate; base truncate to obtuse or tapered; lower surface glandular-hairy, woolly or not. **INFL**: head 20–40 mm wide; bracts 10–30 mm, reflexed; outer and middle bracts like lvs in texture, color, hairiness (or innermost of middle series ± scarious below). **FL**: calyx shaggy-glandular-hairy; corolla 10–18 mm, obtuse, purple, pink, or white. Rocky slopes, ephemeral drainages, oak woodland, chaparral, montane forest; < 1300 m. NCo, NCoR, SNF, CW. Sspp. distinct exc where they co-occur and hybrids occupy transitional habitats.

ssp. ***franciscana*** (Elmer) Jokerst (p. 727) **ST**: hairs gen densely matted-white-woolly or ± 0. **LF** ovate to widely triangular-ovate, thick, entire to serrate; lower surface white-woolly, veins sunken, base gen truncate. **INFL**: heads large, many-fld. $2n$=40,44. Coastal scrub, woodland; < 400 m. NCo, CCo, SnFrB. [var. *f.* (Elmer) Epling] ❀DRN:**17**&IRR:**15,16**&SHD:8,14,22–24.

ssp. ***globosa*** (Jepson) Jokerst Pl erect, > 50 cm, open; hairs gen sparse (0 or ± dense on upper st), unbranched. **LF**: blade 22–50 mm, base tapered to obtuse. **INFL**: head 20–40 mm wide; bracts 20–30 mm, outer reflexed. **FL**: corolla purple. Oak woodland, chaparral, openings; 200–600 m. NCoRO, SnFrB. [var. *v.* in part, as treated by Munz] ❀TRY;DFCLT.

ssp. ***obispoensis*** (Hoover) Jokerst Pl tufted to matted in age; hairs branched and not. **LF**: lower surface white-tomentose; upper surface green, hairy, base tapered to obtuse. **FL**: corolla white to lavender. $2n$=80. Chaparral, oak woodland; < 900 m. SCoR. [var. *o.* Hoover] Hybridizes with *M. hypoleuca*. ❀DRN,DRY or IRR: 14,**15–17**&SHD:7–9;often DFCLT.

ssp. ***villosa*** (p. 727) Pl erect, < 50 cm, open; hairs ± 0 to dense, wavy, soft, unbranched. **LF**: blade 10–22 mm, lower surface woolly to sparsely hairy, base obtuse. **INFL**: head 10–30 mm wide; outer bracts 8–20 mm. **FL**: corolla pink to purple. Habitats and range of sp. [ssp. *subserrata* (E. Greene) Epling] ❀DRN,DRY or IRR:**15–17,24**&SHD:**7–9,14**,19,20,**21–23**;CVS.

M. viridis Jepson Per, erect or decumbent, open. **ST** appressed-hairy. **LF** 10–20 mm, narrowly ovate, thin; lower surface ± white-canescent; upper surface green to ± gray; margin entire or minutely serrate, rolled under. **INFL**: head 10–30 mm wide; outer bracts 8–20 mm, ± like lvs, soft-hairy; inner bracts leathery or scarious. **FL**: calyx shaggy-hairy; corolla 14–17 mm, lavender to rose or purple, tube exserted. Oak woodland, chaparral, open, rocky slopes, often serpentine; < 1800 m. s NCoRI, SnGb.

ssp. ***saxicola*** (Ewan) I.M. Johnston ROCK MONARDELLA **LF** 10–60 mm, entire; lower surface coarsely ± spreading-canescent; upper surface glabrous to sparsely hairy. **INFL**: head > 20 mm wide; bracts gen 10–20 mm. **FL**: calyx 8–12 mm, slender; corolla purple to lavender, tube exserted; stamens exserted. UNCOMMON. Montane chaparral, conifer forest; 500–1800 m. SnGb. ❀DRN,DRY:**7,14–16**,17,**24**&SHD:**8,9,18–23**.

ssp. ***viridis*** Jepson (p. 727) GREEN MONARDELLA **LF** 5–30 mm, gen < 20 mm; upper surface hairy; lower surface finely appressed-canescent. **INFL**: head 10–15(21) mm wide; bracts gen 6–10 mm. **FL**: calyx 7–10 mm; corolla rose to purple, tube incl; stamens incl. UNCOMMON. Chaparral, oak woodland; 300–1000 m. s NCoRI. Hybridizes with *M. villosa*. ❀DRN,DRY:**14–16**,17&SHD:**7–9**.

NEPETA

Ann, per, glabrous to hairy. **STS** erect, gen branched. **LF** gen petioled. **INFLS** gen axillary, densely clustered at upper nodes, subtended by lvs or bracts. **FL**: calyx ± radial, ± 15-veined, lobes gen subequal; corolla 2-lipped, upper lip ± 2-lobed, < lower, hood-like, lower lip 3-lobed, central lobe >> lateral lobes; stamens 4, enclosed by or exceeding upper lip. **FR**: nutlets smooth to rough. ± 250 spp.: Eurasia, Afr. (Latin: ancient name for catnip)

N. cataria L. CATNIP Per, < 1.5 m. **ST** short-hairy to canescent. **LF** 1.5–7.5 cm; petiole < blade, blade widely lanceolate to ovate, base lobed, margin crenate to serrate, upper surface short-hairy, lower surface densely short-appressed-hairy. **INFLS** crowded at upper st nodes, head-like or collectively ± panicle-like; lower short-peduncled, subtended by lvs; upper subtended by short linear bracts. **FL**: calyx 5–6 mm, short-hairy, lobes stiff, acuminate; corolla 6–10 mm, puberulent, white, lower lip purple-spotted, central lobe minutely crenate; upper stamens > lower. $2n$=36. Moist, gen shaded areas; < 1300 m. CA-FP; to B.C., e US; native to Eurasia.

ORIGANUM

Ann, per, subshrub, glabrous to short-hairy. **STS** decumbent to erect, gen branched. **LF** petioled to sessile; blade gen ovate, entire to toothed. **INFLS** axillary, sessile or peduncled, collectively spike- or panicle-like, fl subtended by wide bract; bracts gen overlapping, white to purple. **FL**: calyx radial, 5-lobed; corolla 2-lipped, upper lip ± entire, lower 3-lobed; stamens 4, enclosed by upper lip or exserted; style lobes ± unequal. 35 spp.: Medit, w Asia. (Greek: ancient common name, mountain delight) Cult for tea, cooking herbs, essential oils (*O. dictamnus*, dittany; *O. majorana*, sweet marjoram).

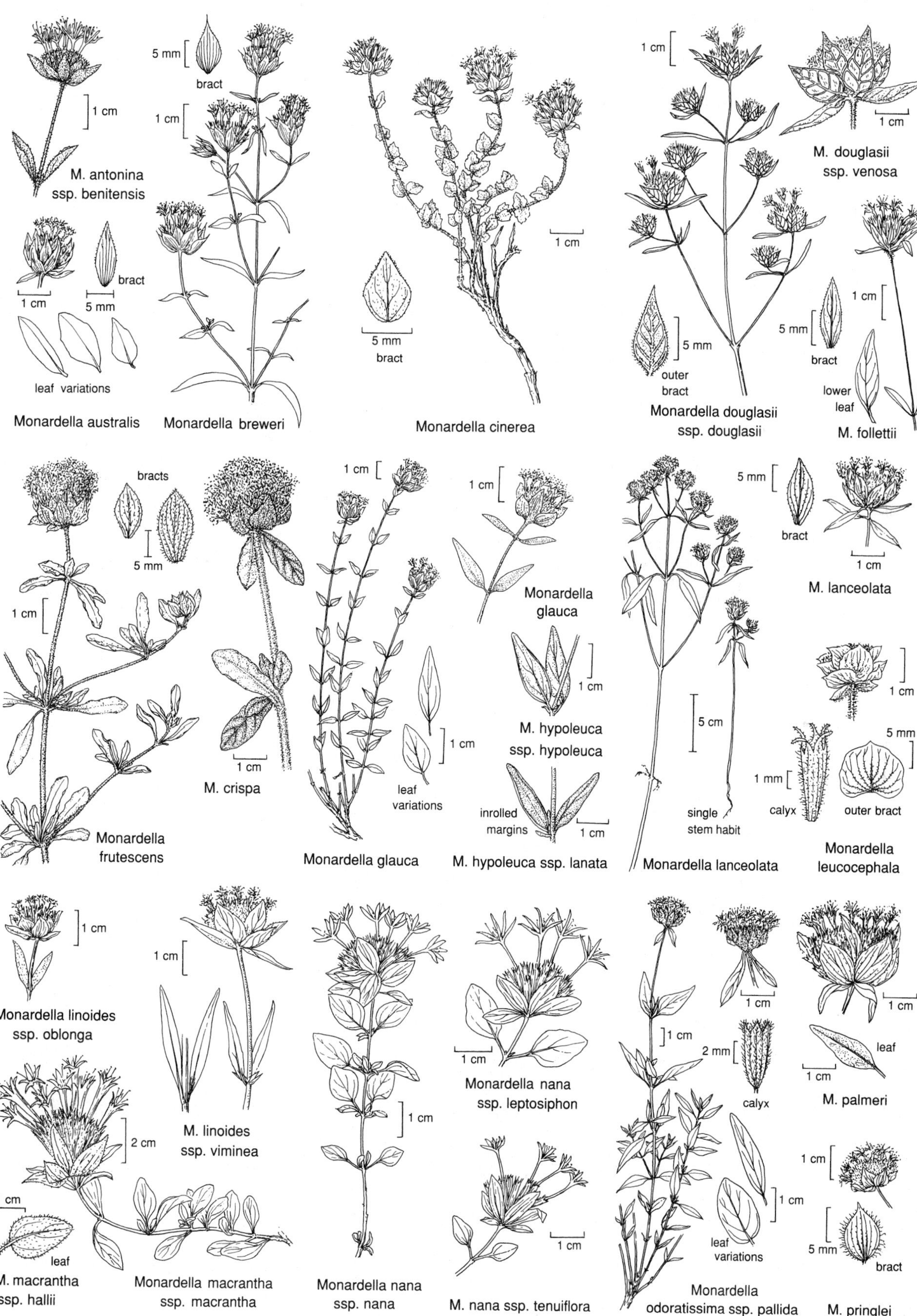

5 mm
bract
1 cm
1 cm
M. antonina
ssp. benitensis
1 cm
5 mm
bract
leaf variations
Monardella australis
Monardella breweri
1 cm
5 mm
bract
Monardella cinerea
1 cm
1 cm
M. douglasii
ssp. venosa
5 mm
outer
bract
Monardella douglasii
ssp. douglasii
1 cm
5 mm
bract
lower
leaf
M. follettii
bracts
5 mm
1 cm
Monardella
frutescens
1 cm
M. crispa
1 cm
1 cm
leaf
variations
Monardella glauca
1 cm
Monardella
glauca
1 cm
M. hypoleuca
ssp. hypoleuca
inrolled
margins
1 cm
M. hypoleuca ssp. lanata
5 cm
single
stem habit
Monardella lanceolata
5 mm
bract
1 cm
M. lanceolata
1 cm
5 mm
1 mm
calyx
outer bract
Monardella
leucocephala
1 cm
Monardella linoides
ssp. oblonga
1 cm
M. linoides
ssp. viminea
2 cm
1 cm
leaf
M. macrantha
ssp. hallii
Monardella macrantha
ssp. macrantha
1 cm
Monardella nana
ssp. nana
1 cm
Monardella nana
ssp. leptosiphon
1 cm
M. nana ssp. tenuiflora
1 cm
1 cm
2 mm
calyx
1 cm
leaf
variations
Monardella
odoratissima ssp. pallida
1 cm
leaf
1 cm
M. palmeri
1 cm
5 mm
bract
M. pringlei

O. vulgare L. OREGANO Per from long rhizomes. **ST** 5–9 dm; hairs short, curled. **LF**: petiole 3–5 mm; blade 0.5–3 cm, entire to ± serrate, glabrous to puberulent, minutely gland-dotted, margin ciliate. **INFLS** gen peduncled, collectively panicle-like; bracts 4–6 mm, light green, cream, or purple-tinged. **FL**: calyx 1–2 mm, sparsely puberulent, gland-dotted, short-bristly within; corolla 4–7 mm, white to red-purple; lower stamens slightly > upper lip. $2n$=30. Shaded sites, disturbed areas; < 500 m. NCo, SnFrB, n SCo (near Ventura); native to Medit Eur.

POGOGYNE

James D. Jokerst

Ann, hairy or not, gland-dotted, aromatic. **ST** decumbent to erect, branched or not. **LF** linear to round, entire to toothed, bristly-ciliate, short-petioled. **INFL**: clusters, head-like or interrupted, terminal and axillary, or fls solitary in axils; bracts 2 or more, bristly-ciliate. **FL**: calyx 2-lipped, lobes 5, deep, glabrous to coarse-hairy; corolla 2-lipped, lavender, purple, or white, raised area on lower lip sometimes spotted; stamens 2–4, upper 2 sometimes sterile and vestigial or 0; style hairy below stigma lobes. **FR** hairy. 7 spp.: CA (gen CA-FP); OR, Baja CA. (Greek: bearded style) [Howell 1931 Proc Cal Acad Sci 20(3): 105–128] Fl gen March–June; July in some high-elevation *P. douglasii*; June–August in *P. floribunda* (MP).

1. Corolla inconspicuous, ± incl in bracts and calyces; stamens 2–4, upper 2 sterile and vestigial or 0, lower < 2 mm; (subg. *Hedeomoides*)
  2. Infl a dense spike, from pl base to top; corolla white with purple spots .... ***P. floribunda***
  2′ Infl dense, head-like, terminal and axillary clusters or fls solitary in upper axils; corolla lavender to purple
    3. Pl inconspicuous; st prostrate to decumbent, slender; fr 1–1.3 mm; grassy areas .... ***P. serpylloides***
    3′ Pl conspicuous; st ascending to erect, stout; fr 1.5–2.5 mm; vernal pools, swales, drainages .... ***P. zizyphoroides***

1′ Corolla conspicuous, exserted from bracts and calyces; stamens 4, fertile, lower > 3.5 mm (subg. *Pogogyne*)
  4. Infl 1–3 cm wide; lower calyx lobes often 2 × upper; vernal pools, swales, n of Teh .... ***P. douglasii***
  4′ Infl < 2 cm wide; lower calyx lobes < 2 × upper; local endemics of coastal mesas, terraces (San Diego Co.), creekbeds (Monterey Co.)
    5. Calyx lobe tips flat; bract tips obtuse; corolla funnel-shaped; infl < 10 mm wide .... ***P. clareana***
    5′ Calyx lobe tips not flat; bract tips acute or acuminate; corolla bell-shaped; infl 8–20 mm wide
      6. St curved or zigzagged, thread-like, wiry-erect to lax-spreading; pl hairs dense, long, gen curved, coarse; infl 8–10 mm wide .... ***P. abramsii***
      6′ St straight, stout, rigidly erect; pl hairs 0 or sparse, short, gen straight, bristle-like; infl 10–20 mm wide .... ***P. nudiuscula***

P. abramsii J. Howell SAN DIEGO MESA MINT Pl hairs dense, long, gen curved, coarse. **ST** spreading to erect, gen 0.5–0.8 mm diam at infl base. **INFL** 8–10 mm wide; bracts often purple-tinged, tips acuminate. **FL**: calyx densely white-hairy, esp on nerves, tube 2–2.5 mm, lobes 2–5 mm, tips not flat; corolla 10–12 mm, bell-shaped, hairs sparse; style hairy 2–4 mm below stigma lobes. **FR** 1–1.5 mm. **ENDANGERED** US, CA. Vernal pools of coastal terraces; 100–200 m. s SCo (San Diego Co.) Threatened by urbanization.

P. clareana J. Howell SANTA LUCIA MINT **ST** erect, 15–25 cm, slender. **INFL** < 10 mm wide; bract tips obtuse. **FL**: calyx 4–4.5 mm, coarsely ciliate, tube 2 mm, lobe tips flat; corolla 10–15 mm, funnel-shaped, hairy outside; style hairy 2 mm below unequal stigma lobes. **FR** 1.5 mm. **ENDANGERED** CA. Intermittent creek; 300–400 m. SCoRO (Ft. Hunter-Liggett, Monterey Co., only known locality). Possibly threatened by road maintenance.

P. douglasii Benth. (p. 727) Pl robust, < 45 cm. **ST** erect. **INFL** 10–30 mm wide. **FL**: calyx tube 2–4 mm, upper lobes 2–5 mm, tips acute, lower lobes 3–8 mm, 0.75–2.5 × tube, tips lanceolate to acuminate; corolla 9–20 mm, lavender, hairy, raised area on lower lip often yellow-spotted; style hairy 2–6 mm below stigma lobes. **FR** 1–2 mm. Vernal pools, swales; < 900 m. NCoRO, NCoRI, SNF, GV, CW. Highly variable; most distinct variant is ssp. *parviflora* (Benth.) J. Howell, Douglas' pogogyne (calyx lobes 0.75–1.5 × tube; corolla 11–15 mm). Tall pls with acute calyx lobes in c&s SNF and low, branched pls of coastal bluffs in San Luis Obispo Co. may warrant recognition. Further study needed. ❀SUN,WET (winter only):**7–9,14–17**.

P. floribunda Jokerst PROFUSE-FLOWERED POGOYNE **ST** ascending to erect, branched at base. **INFL**: dense spike, from pl base to top, 10–25 mm wide. **FL**: calyx hairy, tube 2–3 mm, lobes 1.5–5 mm; corolla 4.5–6 mm, white with purple spots; stamens 2–4, upper 2 vestigial or 0; style lightly hairy 1–2 mm below stigma lobes. **FR** 2–2.5 mm. UNCOMMON. Vernal pools, seasonal lakes; 1000–1500 m. MP. Resembles *P. zizyphoroides*.

P. nudiuscula A. Gray OTAY MESA MINT Pl hairs 0 or sparse, short, gen straight, bristle-like. **ST** erect, stout, often 0.8–1 mm diam at infl base. **INFL** 10–20 mm wide; bracts green, tips acute. **FL**: calyx glabrous to sparsely hairy, tube 3–5 mm, lobes 2–5 mm, tips not flat; corolla 11–14 mm, bell-shaped, sparsely hairy; style hairy 1.5–4 mm below stigma lobes. **FR** ± 1.5 mm. **ENDANGERED** CA. Vernal pools on coastal mesas; 100–250 m. s SCo (San Diego Co.); Baja CA. Endangered by urbanization.

P. serpylloides (Torrey) A. Gray Pl inconspicuous. **ST** prostrate to decumbent, gen branched, slender. **INFL**: clusters, head-like, small, dense, terminal and axillary, or fls solitary in upper axils. **FL**: calyx tube 1–3.5 mm, lobes 1.5–4 mm; corolla 2.5–5 mm, lavender; stamens 2–4, upper 2 vestigial or 0; style sparsely hairy just below stigma lobes. **FR** 1–1.3 mm. n=19. Common. Protected grassy areas; < 1200 m. NCoR, n&c SNF, SnFrB, SCoRO. [ssp. *intermedia* J. Howell]

P. zizyphoroides Benth. (p. 727) Pl conspicuous. **ST** ascending to erect, simple or branched above, stout. **INFL**: clusters head-like, dense, terminal and axillary. **FL**: calyx tube 2.5–5 mm, lobes 1.5–6 mm; corolla 4–8 mm, lavender to purple; stamens 2–4, upper 2 vestigial; style sparsely hairy just below stigma lobes. **FR** 1.5–2.5 mm. n=19. Vernal pools, depressions; < 400 m. NCoRO, n&c SNF, GV, SnFrB; s OR. ❀STBL.

POLIOMINTHA

Shrub. **STS** spreading to erect, branched throughout, densely strigose, gen grayish. **LVS** short-petioled to subsessile; blades gen narrow, entire. **INFLS** axillary, gen subtended by lvs. **FL**: calyx ± radial, 15-veined, lobes subequal; corolla 2-lipped, lips ± equal, upper lip ± flat, lower lip 3-lobed, central lobe notched; stamens 2, ± exserted, staminodes short; style unequally lobed. 4 spp.: sw US, n Mex. (Greek: hoary white mint)

P. incana (Torrey) A. Gray (p. 727) Pl 5–10 dm, rounded or mound-like. **LF** subsessile, gen reduced upward; blade 5–18 mm, 2–4 mm wide, oblong-elliptic to narrowly linear. **INFL**: fls 2–6; pedicel 1–2 mm. **FL**: calyx densely short-hairy, tube 3–5 mm, puberulent within, lobes 1–2 mm; corolla 8–10 mm, light blue to lavender, upper lip 2–3.5 mm, lower 3–4 mm, minutely purple-dotted. Uncommon. Sandy soils, rocky slopes, 1600–1700 m. s DMoj (Cushenbury Springs); to Colorado, TX, n Mex. ❀TRY; DFCLT.

PRUNELLA SELF-HEAL

Per, glabrous to hairy, gen with bisexual fls only, sometimes with only pistillate fls. **STS** prostrate to erect, sometimes rooting at lower nodes. **LVS** basal and cauline, gen petioled; blade gen entire. **INFLS** densely clustered, collectively ± spike-like, terminal; bract gen wide, abruptly acuminate. **FL**: calyx 2-lipped, upper = lower, upper lip 3-toothed, lower 2-lobed; corolla finely hairy inside, 2-lipped, lower lip 3-lobed, upper lip ± entire, hood-like, ± enclosing stamens; stamens 4, lower pair > upper, filament minutely toothed below anther. **FR**: nutlets obovoid. 4 spp.: temp, esp Eurasia. (Latin: from early German name for pl used to treat chest pains)

P. vulgaris L. **ST** 1–5 dm, glabrous to short-hairy. **LVS**: lower petioled, petiole 5–30 mm; upper subsessile; blade 2–7 cm, gen 1–4 cm wide, ovate, elliptic, or lanceolate, base gen wedge-shaped. **INFL** 2–6.5 cm; bract margin ciliate, reddish. **FL**: calyx 7–11 mm, dark green to purplish; corolla 12–15 mm in bisexual fls, 8–11 mm in pistillate fls, bluish violet, sometimes pink or white. $2n = 28,32$. Moist areas; < 2400 m. CA-FP, SNE; circumboreal.

1. Cauline blade length gen 3 × width; sts gen decumbent to erect .... var. ***lanceolata***
1′ Cauline blade length gen 2 × width; sts gen prostrate .... var. ***vulgaris***

var. ***lanceolata*** (Barton) Fern. (p. 727) **STS** gen decumbent to erect. **LVS**: cauline blade length 3 × width. Moist areas, gen coniferous forest, woodland; < 2400 m. CA-FP, SNE; N.Am, e Asia. [var. *atropurpurea* Fern.] ❀IRR or WET:1,2,**4–7**,**15–17**,**22–24**&SHD:3,**8**,**9**,10,**14**,**18–21**;can be INV.

var. ***vulgaris*** **STS** gen prostrate, sometimes decumbent to erect. **LVS**: cauline blade length gen 2 × width. Uncommon. Moist, disturbed sites, lawns; < 200 m. NCo, CCo, SCo; e US; native to Eur. Pistillate pls with corolla < 12 mm have been called var. *parviflora* (Poiret) DC.

PYCNANTHEMUM

Per, glabrous to hairy. **STS** erect, simple or branched above middle. **LVS** petioled to subsessile. **INFLS** head-like, axillary at upper nodes and terminal, subtended by lvs or bracts; fls subsessile. **FL**: calyx 5-lobed, lobes ± equal, deltate, acute to awned; corolla 2-lipped, upper lip gen 2-lobed, ± flat, lower lip 3-lobed, lobes ± equal; stamens 4. **FR**: smooth or puberulent. 17 spp.: N.Am. (Greek: densely fld)

P. californicum Torrey (p. 727) Pl 5–10 dm. **ST** glabrous to puberulent. **LVS** > internodes, ± overlapping; petiole < 3 mm; blade 2.5–7.5 cm, gen ovate to widely lanceolate, base ± rounded, tip gen acute, entire to finely toothed, glabrous to puberulent. **INFLS** 2–5, head-like; lower subtended by ± elliptic lvs; upper subtended by linear-lanceolate bracts. **FL**: calyx 4–5.5 mm, puberulent, lobes densely short-hairy; corolla 5–6.5 mm, white, upper lip 2.5–3 mm, lower lip ± 2–3 mm, purple- or violet-spotted. $2n=40$. Moist sites, chaparral, oak woodland, coniferous forest; 500–1800 m. KR, n NCoR, CaR, SN, n ScV, TR, PR, MP. ❀IRR,SUN:1,4–6,7,8,9,**14–16**,17–24.

SALAZARIA BLADDER SAGE

1 sp. (Don Jose Salazar y Larrequi, Mexican astronomer, US-Mexican Boundary Survey)

S. mexicana Torrey (p. 727) Shrub, 5–10(15) dm, ± rounded, branched. **ST**: lateral branches spreading, rigid, tips becoming spine-like; twigs ± canescent. **LF** short-petioled to subsessile; blade 3–15(20) mm, 2–8 mm wide, gen ovate to elliptic, base rounded, margin entire, glabrous to puberulent. **INFLS** axillary at distal 3–10 nodes; fls 2; axis finely glandular-puberulent; bracts 0. **FL** 2-lipped; calyx lobes ± equal, entire, purplish, becoming 1–2 cm, bladder-like in fr; corolla 15–25 mm, upper lip ± entire, white to light violet, lower lip ± 3 lobed, violet to purple; stamens 4, gen enclosed by upper lip, lower stamen pair < upper pair, anthers ciliate. **FR**: nutlets, widely ovoid, short-stalked, tubercled. Sandy to gravelly slopes, washes, shrubland, woodland; < 1800 m. s SNE, D; to UT, TX, n MX. ❀DRN,IRR:7,**8–12**,13,**14**,18,**19–21**&SUN:15,16,22,23.

SALVIA

Deborah Engle Averett & Kurt R. Neisess

Ann, per, shrub. **LF** entire, lobed, or toothed, gen not spine-tipped. **INFL**: clusters gen many-fld, gen head-like, gen spheric, gen involucred, gen surrounding nodes in gen ± spike-like, gen interrupted panicles, or fls 1–several per lf axil. **FL**: calyx gen 2-lipped, upper lip entire or of 3 gen shallow, sometimes spine-tipped lobes, lower lip gen of 2 gen spine-tipped lobes; corolla 2-lipped, upper lip 2-lobed to entire, lower lip with 3 spreading lobes (middle often expanded); fertile stamens 2, attached in throat, anther sacs 1–2 per stamen (if 2, then separate on thread-like structure, 1 fertile, > other); style forked at tip. ± 900 spp.: ± worldwide, esp trop, subtrop Am. (Latin: to save, from medicinal use) ❀All spp. are excellent bee fodder and have edible seeds (a traditional food of native Californians).

1. Ann
 2. Corolla 15–25 mm; bract 2–5 cm; pl white-woolly .... ***S. carduacea***
 2′ Corolla 6–8 mm; bract < ± 1 cm; pl short-hairy .... ***S. columbariae***
1′ Per to shrub
 3. Lf spine-tipped
  4. Lf white-woolly, short-petioled, gen deciduous, in 1–2 pairs or 0 on margins; calyx lobes triangular, white-woolly .... ***S. funerea***

4′ Lf green-tomentose, sessile or short-petioled, persistent, in 2–7 pairs on margins; calyx lobes lanceolate, not white-woolly .... ***S. greatae***
3′ Lf not spine-tipped
5. Per; lf ± green
6. Corolla blue to pinkish purple, 6–15 mm .... ***S. verbenacea***
6′ Corolla red to salmon or yellow, 10–35 mm
7. Corolla yellow, 10–20 mm, upper lip arched; calyx white-woolly; cult, sometimes escaped ...... ***S. aethiopis***
7′ Corolla red to salmon, 25–35 mm, upper lip straight; calyx hairs short, white; native .... ***S. spathacea***
5′ Shrub to per; lf ± green to grayish white
8. Stamens incl to slightly exserted
9. Lf linear to linear-elliptic, margins rolled under, lower surface with branched hairs; heterostylous .... ***S. brandegei***
9′ Lf oblong-elliptic to obovate, margins not or only slightly rolled under, lower surface with simple hairs; not heterostylous
10. Corolla white to pale blue or lavender, rarely pale rose; fl clusters 1.6–4 cm wide; lf oblong-elliptic to obovate .... ***S. mellifera***
10′ Corolla dark blue; fl clusters 1–1.5 cm wide; lf oblanceolate to obovate .... ***S. munzii***
8′ Stamens exserted
11. Lower corolla lip > 2 × upper; hairs simple
12. Lf grayish from ± bristly hairs or gen greenish; infl gen < 5 dm
13. Erect shrub; lf linear; corolla white to pale lilac .... ***S. eremostachya***
13′ Prostrate per; lf lance-elliptic to obovate; corolla blue to lilac or purple .... ***S. sonomensis***
12′ Lf grayish velvety from minute appressed hairs; infl gen > 10 dm
14. Lf base tapered; calyx lobes barely or not spine-tipped; fls in ± spike-like clusters, these in ± raceme-like, interrupted panicles .... ***S. apiana***
14′ Lf base ± truncate to tapered; calyx lobes spine-tipped; fl clusters in ± spike-like, interrupted panicles .... ***S. vaseyi***
11′ Lower corolla lip < 2 × upper; hairs simple or branched
15. Width of lower middle corolla lobe < 1/2 length; lf puckered, teeth rounded, hairs simple or branched, moderately dense to sparse, spreading
16. Lf base ± truncate to ± cordate; lf ashy-gray, hairs branched; tissue between anther sacs glabrous .... ***S. leucophylla***
16′ Lf base tapered to ± truncate; lf yellow- to gray-green, hairs simple; tissue between anther sacs glandular-hairy
17. Bract dark, firm, < calyx; fl clusters 1–3 per fl st .... ***S. clevelandii***
17′ Bract pale, papery, >> calyx; fl clusters gen 1 per fl st .... ***S. mohavensis***
15′ Width of lower middle corolla lobe > or = length; lf not puckered, ± entire, scaly or hairs simple, very dense, appressed
18. Corolla blue-violet to rose, gen 13–23 mm; bract 10–20 mm; lf 20–50 mm .... ***S. pachyphylla***
18′ Corolla blue, rarely purple, rose, or white, gen 6–13 mm; bract 5–12(14) mm; lf 4–30 mm .... ***S. dorrii***
19. Bract, calyx lower surface with soft shaggy hairs or scaly, margin hairs long .... var. ***pilosa***
19′ Bract, calyx glabrous to scaly, margin hairs gen short
20. Lf abruptly narrowed to petiole, 6–20 mm, widest 2–13 mm from base .... var. ***dorrii***
20′ Lf tapered to petiole, 10–40 mm, widest 6–28 mm from base .... var. ***incana***

S. aethiopis L. MEDITERRANEAN SAGE Per. **ST** 30–60 cm, tomentose. **LVS** mostly basal, 5–30 cm, widely lanceolate to triangular, tomentose; teeth irregular, rounded. **INFL**: clusters 5–10-fld, in ± open panicle; bracts ± 1 cm, round, acuminate, firm. **FL**: calyx 8–15 mm, purple, white-woolly, lobes spine-tipped, upper lip deeply 3-lobed; corolla tube 10–20 mm, yellow, upper lip ± 2 × lower, arched; stamens and style exserted. **FR**: nutlet 2.5–3 mm, dark brown, shiny. 2*n*=24. Fields, roadsides; ± 500 m. MP; native to Eur. Possibly a waif from cult; more collections needed.

S. apiana Jepson (p. 727) WHITE SAGE Per, subshrub. **ST** < 1 m. **LVS** gen basal, 4–8 cm, widely lanceolate; base tapered; teeth minute, rounded; hairs dense, minute, simple, appressed. **INFL**: clusters few-fld, in ± spike-like clusters, these in ± raceme-like, interrupted panicles; bracts < to > calyx, linear-lanceolate, recurved. **FL**: calyx 8–10 mm, lobes barely or not spine-tipped, upper lip entire; corolla tube 12–22 mm, white with lavender, upper lip < 2 mm, lower lip 4–5 mm, upcurved, blocking throat; stamens and style exserted. **FR**: nutlet 2.5–3 mm, light brown, shiny. *n*=15. Common. Dry slopes, coastal-sage scrub, chaparral, yellow-pine forest; gen < 1500 m. SCo, TR, PR, w edge D; Baja CA. ❀DRN,DRY,SUN:7,8,9,11,**14–16**,17,**18–24**;also STBL.

S. brandegei Munz (p. 727) BRANDEGEE'S SAGE Shrub, > 1 m or prostrate; hairs branched. **LF** 2–6 cm, linear to linear-elliptic, puckered; upper surface glabrous, lower surface densely white-hairy; teeth small, rounded; margins rolled under. **INFL**: clusters 1.5–2 cm wide; bracts colored, ovate, tips sharp. **FL**: calyx 7–8 mm, hairs long, wavy, upper lip minutely 3-lobed; corolla tube 7–8 mm, pale blue to lavender, upper lip 3–3.5 mm, lower lip 3–4 mm; stamens ± incl. **FR**: nutlet 1.5–2.5 mm, brown, rough, oblong. *n*=15. RARE. Coastal scrub; < 200 m. n ChI (Santa Rosa Island); n Baja CA. Heterostylous. In cult.

S. carduacea Benth. (p. 727) THISTLE SAGE Ann, 1–10 dm, white-woolly. **LVS** basal, subsessile, 3–10(30) cm, oblanceolate, 1-pinnately dissected; margin wavy, short-spiny. **INFL** scapose; clusters 1.5–3 cm wide, 1–4 per fl st; bracts 2–5 cm, lanceolate, spiny. **FL**: calyx 10–17 mm, lobes spine-tipped, upper lip 3-lobed; corolla tube 15–25 mm, lavender, rarely blue or white, upper lip 2-lobed, lower lip > 2 × upper; stamens exserted. **FR**: nutlet 2.5 mm, tan to gray, flecked. *n*=16. Common. Sandy or gravelly soils; < 1400 m. Teh, SnJV, e SnFrB, SCoRI, SW, w D; n Baja CA. ❀DRN,SUN,DRY:**7–9**,**11**,14–16,**18–21**,22–24.

S. clevelandii (A. Gray) E. Greene (p. 727) Shrub, < 1.5 m, gray-green; hairs ± throughout, simple, bent downward. **LF** 2–4 cm, elliptic-ovate, puckered; base tapered to ± truncate; teeth small, rounded. **INFL**: clusters 1–3 per fl st; bracts dark, firm, < calyx. **FL**: calyx 8–10 mm, distal 1/3 to entirely purple, upper lip entire or minutely 3-lobed; corolla tube 12–18 mm, dark blue-violet, upper lip 2-lobed, 6–8 mm, lower lip 4–5 mm; stamens and style exserted. **FR**: nutlet, 2–2.5 mm, brown to gray, often mottled. 2*n*=30. Chaparral, dry slopes, coastal scrub; < 1000 m. s SCo, s PR (San Diego

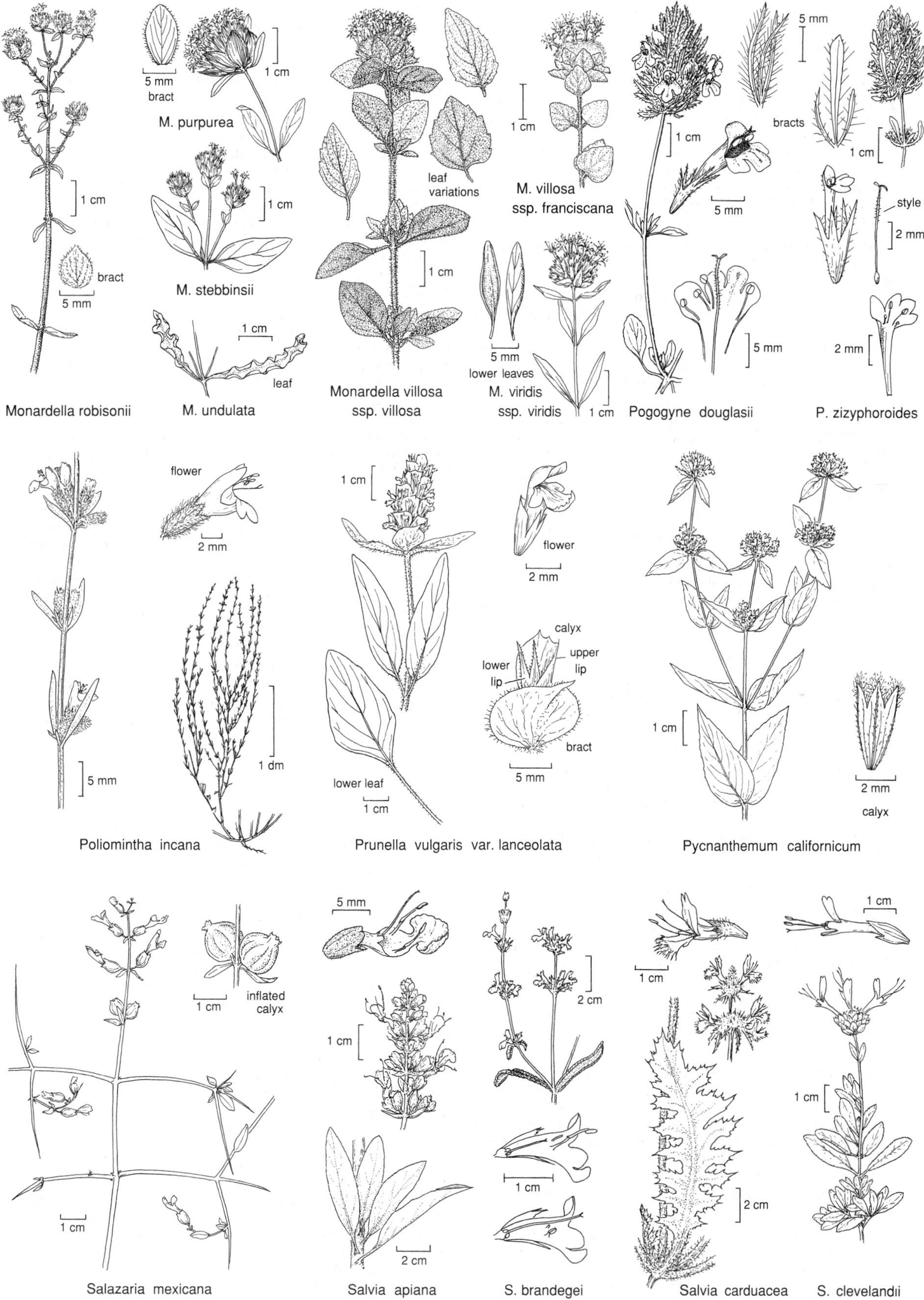

Monardella robisonii M. undulata Monardella villosa ssp. villosa M. viridis ssp. viridis Pogogyne douglasii P. zizyphoroides

Poliomintha incana Prunella vulgaris var. lanceolata Pycnanthemum californicum

Salazaria mexicana Salvia apiana S. brandegei Salvia carduacea S. clevelandii

Co.); n Baja CA. ❀DRN,SUN:7,**15–17**,18,**22–24**&IRR:8,9,**14**, **19–21**;CV.

S. columbariae Benth. (p. 733) CHIA Ann, 1–5 dm; hairs gen sparse, short. **LVS** basal, 2–10 cm, oblong-ovate, 1–2-pinnately dissected; lobes irregularly rounded, minutely bristly. **INFL** ± scapose; clusters gen 1–2 per fl st; bracts < ± 1 cm, ± round, awn-tipped. **FL**: calyx 8–10 mm, purple-tipped, upper lip unlobed but 2 (3)-awned; corolla tube 6–8 mm, pale to deep blue, upper lip entire to shallowly 2-lobed, 2–3 mm, lower lip ± 2 × upper; stamens and style exserted. **FR**: nutlet, 1.5–2 mm, tan to gray. n=13. Common. Dry, disturbed sites, chaparral, coastal-sage scrub; gen < 1200 m. CA (exc KR, CaR, n SN); to UT, AZ. [var. *ziegleri* Munz] ❀DRN,SUN,DRY:**7–10**,11,12,**14–24**.

S. dorrii (Kellogg) Abrams Shrub, spreading to mat-forming, 10–70 cm, densely white-scaly throughout. **LF** 4–30 mm, linear to spoon-shaped, ± entire. **INFL**: clusters gen 12–30 mm wide; bracts 5–12(14) mm, ± round. **FL**: calyx gen 6–11 mm, blue, purple, or rose, upper lip gen entire, rounded, lobes of lower lip acute, not spine-tipped; corolla tube gen 6–13 mm, blue, rarely purple, rose, or white, upper lip 2-lobed, 2–3 mm, < lower; stamens and style exserted. **FR**: nutlet, 1.8–3.5 mm, gray to reddish brown. $2n$=30. Common. Dry, mostly rocky places; 1000–4000 m. GB, D; to WA, ID, UT, AZ. Highly variable; vars. intergrade.

var. ***dorrii*** (Kellogg) Abrams (p. 733) **LF** 6–20 mm, widest 2–13 mm from base, abruptly narrowed to petiole. **INFL**: bract and calyx glabrous to scaly, margin hairs gen short. Dry flats, slopes; 800–3100 m. GB, n DMoj; to OR, ID, UT, AZ. [ssp. *argentea* (Rydb.) Munz; ssp. *gilmanii* (Epling) Abrams] ❀TRY.

var. ***incana*** (A. Gray) J.L. Strachan FLESHY SAGE **LF** 10–40 mm, widest 6–28 mm from base, tapered to petiole. **INFL**: bract and calyx glabrous to scaly, margin hairs gen short. UNCOMMON. Silty to rocky soils; 300–1050 m. nw CaRH (near Hornbrook, Siskiyou Co.); to WA, ID. [ssp. *carnosa* (E. Greene) Abrams; var. *c.* (E. Greene) Cronq.] CA pls may be intermediate to var. *dorrii*.

var. ***pilosa*** (A. Gray) J.L. Strachan & Rev. **LF** 4–32 mm, widest 1–15 mm from base, abruptly narrowed to petiole. **INFL**: bract and calyx lower surfaces with soft, shaggy hairs or scaly, upper surface glabrous to minutely hairy, margin hairs long. Desert slopes, washes; 900–1900 m. e slope s SNH, Teh, GB, DMoj; NV, AZ. ❀TRY;DFCLT.

S. eremostachya Jepson (p. 733) DESERT SAGE Shrub, erect, 60–80 cm, finely branched. **LF** 1.5–3.3 cm, linear, puckered; base truncate; teeth minute, rounded. **INFL**: clusters 1–3 per fl st; bracts 5–12 mm, lanceolate to ovate, rose to purple, papery. **FL**: calyx 6–9 mm, upper lip minutely 3-lobed; corolla tube white to pale lilac, 10–17 mm, upper lip 2-lobed, 1–4 mm, lower lip 4–8 mm; stamens and style exserted. **FR**: nutlet 3 mm, yellow-brown. $2n$=30. UNCOMMON. Dry, rocky, gravelly places, lower pinyon/juniper; 700–1400 m. w edge DSon; n Baja CA. Locally common.

S. funerea M.E. Jones (p. 733) DEATH VALLEY SAGE Shrub, 5–12 dm, densely branched, densely white-woolly. **LF** 9–20 mm, ± ovate, short-petioled, gen deciduous; spines 1 at tip, in 1–2 pairs or 0 on margins. **INFL**: fls gen 3 per lf axil. **FL**: calyx 4.5–6 mm, lobes 5, subequal, triangular, spine-tipped; corolla tube 12–16 mm, violet, rarely blue, upper lip 2-lobed, 2–2.5 mm, lower lip almost 2 × upper; stamens and style incl. **FR**: nutlet ± 3 mm, brown, smooth. $2n$=64. UNCOMMON. Dry washes, canyons; 0–360 m. ne DMoj (Death Valley, Amargosa & Panamint mtns).

S. greatae Brandegee (p. 733) OROCOPIA SAGE Shrub, < 1 m; hairs glandular, tangled. **LF** 9–20 mm, ± ovate, sessile or short-petioled, green-tomentose, persistent; spines 1 at tip, in 2–7 pairs on margins. **INFL**: clusters many-fld. **FL**: calyx 9–11 mm, upper lip of 3 shallow, spine-tipped lobes; corolla tube 9–11 mm, lavender to rose, upper lip 2-lobed, 2–2.5 mm, lower lip 4–5 mm; mature stamens and style exserted. **FR**: nutlet, 2–3 mm, flat, keeled, gray to brown. $2n$=±30. RARE. Alluvial slopes; 30–240 m. se DSon (Orocopia, Chocolate mtns).

S. leucophylla E. Greene (p. 733) PURPLE SAGE Shrub, prostrate to erect, < 1.5 m; hairs dense, branched. **LF** 2–8 cm, oblong-lanceolate, puckered, base ± truncate to ± cordate; teeth small, rounded; margin sometimes rolled under. **INFL**: clusters 1.5–4 cm wide; bracts ovate, < calyx. **FL**: calyx 8–12 mm, upper lip entire, acute, not spine-tipped, lower lip gen 0; corolla tube 6–13 mm, rose-lavender, upper lip 6–8 mm, slightly < lower lip; stamens and style exserted. **FR**: nutlet 2–3 mm, brown or dark gray. $2n$=30. Dry, open hills; 50–800 m. SCoRO, SCo, WTR, SnGb; Baja CA. ❀SUN,DRN:**15–17**,**22–24**&IRR:8,9,**14**,**19–21**;CVS incl GRCVR; also STBL.

S. mellifera E. Greene (p. 733) BLACK SAGE Shrub, rarely prostrate, 1–2 m; hairs simple, some glandular. **LF** 2.5–7 cm, oblong-elliptic to obovate, puckered, upper surface ± glabrous, lower surface hairy. **INFL**: clusters 1.6–4 cm wide; bracts oblong-elliptic to widely ovate, acute to spine-tipped, gen = calyx. **FL**: calyx 6–8 mm, hairs wavy, upper lip minutely 3-lobed; corolla tube 5.5–9 mm, white to pale blue or lavender, rarely pale rose, upper lip 2-lobed; stamens and style slightly exserted. **FR**: nutlet 2–3 mm, gen brown. $2n$=30. Common. Coastal-sage scrub, lower chaparral; 0–1200 m. CW, SW; n Baja CA. ❀DRN,SUN:**7**,**14–17**,18, **19–24**&IRR:8,9;CVS incl GRCVR; also STBL.

S. mohavensis E. Greene (p. 733) Shrub, gen < 1 m; hairs minute, simple. **LF** 1.5–2 cm, lance-oblong to ovate, puckered; base tapered to ± truncate; teeth small, rounded; some hairs glandular. **INFL**: clusters subspheric, gen 1 per fl st; bracts gen 1–1.5 cm, >> calyx, ovate to ± round, pale, papery. **FL**: calyx 7–12 mm, minutely glandular-hairy, sometimes tinged blue, upper lip entire to minutely 3-lobed; corolla tube 18–20 mm, tinged sky-blue, upper lip entire, slightly < lower, 4–6 mm; stamens and style exserted. **FR**: nutlet, 2–3 mm, tan to brown. n=15. Dry, rocky slopes, blackbush scrub, pinyon/juniper woodland; 300–1500 m. DMtns; AZ. Locally common. ❀TRY;DFCLT.

S. munzii Epling (p. 733) MUNZ'S SAGE Shrub, < 2.5 m. **ST**: hairs simple, nonglandular, appressed. **LF** 1.3–5 cm, oblanceolate to obovate, puckered; upper surface nearly glabrous; lower surface densely hairy. **INFL**: clusters 1–1.5 cm wide; bracts 7–30 mm, > calyx, lanceolate, lf-like. **FL**: calyx 3–6 mm, upper lip entire or minutely 3-lobed; corolla tube 7–15 mm, dark blue, upper lip 2-lobed, ± 2 mm, lower lip ± 2 × upper; stamens incl; style exserted. **FR**: nutlet, ± 1 mm, dark brown. $2n$=30. RARE in CA. Coastal-sage scrub, lower chaparral; < 800 m. s PR (San Miguel Mtns, San Diego Co.); n Baja CA. In cult.

S. pachyphylla Munz (p. 733) Shrub, prostrate, 20–80 cm, rooting at nodes, scaly. **LF** 20–50 mm, obovate to spoon-shaped; margins wavy, ± entire. **INFL**: clusters 20–50 mm wide, 2–5 per fl st; fl st ± hidden between clusters; bracts 1–2 cm, papery, ovate to oblong-elliptic, greenish to purple or rose. **FL** 8–13 mm; calyx = bracts in color, upper lip entire, obtuse to abruptly soft-pointed, 3–6 mm, lower lip 1–3 mm, lobes acuminate, not spine-tipped; corolla tube gen 13–23 mm, blue-violet to rose, upper lip entire, 4–6 mm, < lower lip; stamens and style exserted. **FR**: nutlet, 2.5–3 mm, tan to brown. $2n$=30. Dry slopes, pinyon/juniper to yellow-pine forest; 1400–2500 m. s SNH, Teh, SnBr, PR, DMtns; NV, AZ, n Baja CA. ❀TRY;DFCLT.

S. sonomensis E. Greene (p. 733) Subshrub, prostrate, mat-forming. **ST** < 4 dm; base woody. **LF** gen 3–6 cm, 5–15 mm wide, lance-elliptic to obovate, puckered; upper surface minutely hairy; lower surface white with dense, recurved hairs; teeth minute, rounded. **INFL** scapose; clusters 1–1.5 cm wide; bracts < 1 cm, lanceolate. **FL**: calyx 5–10 mm, upper lip lobes 3, minute, not spine-tipped; corolla tube 5–15 mm, blue to lilac or purple, upper lip 1–3 mm, 2-lobed, lower lip 3–7 mm; stamens and style exserted. **FR**: nutlet, 2.5 mm, oblong, brown. n=15. Chaparral, oak woodland, yellow-pine forest, dry slopes; < 2000 m. KR, NCoR, CaRF, n SNF, SCoR, SCo. Locally common. ❀DRN,DRY,SUN:**7**,14, **15–18**,19–24;CV;DFCLT;GRCVR.

S. spathacea E. Greene (p. 733) PITCHER SAGE Per, mat-like, rhizomed; hairs wavy. **LF** 8–20 cm, oblong-hastate, puckered; upper surface sparsely long-hairy; lower surface tomentose; teeth rounded. **INFL**: clusters < 6 cm wide; bracts 1.5–4 cm, green to purple. **FL**: calyx 1.5–3 cm, upper lip gen entire; corolla tube 25–35 mm, red to salmon, upper lip 7–8 mm, shallowly 2-lobed, straight, lower lip 10–12 mm; stamens and style exserted. **FR**: nutlet, 3.5–6.5 mm, brown. $2n$=30. Common. Oak woodland, chaparral, coastal-sage scrub, open or shady slopes; 0–800 m. s ScV (Solano Co.), CW, SCo, TR. ❀5;IRR:**15–17**,**22–24**&SHD:7–9, **14**,18,**19–21**;GRCVR: mow yearly.

S. vaseyi (Porter) Parish (p. 733) Subshrub, < 1.5 m. **LVS**: basal > cauline, 2–6 cm, oblong-ovate; base ± truncate to tapered; teeth minute, rounded; hairs dense, minute, appressed. **INFL**: clusters 1.5–3 cm wide; bracts < 2 cm, lanceolate. **FL**: calyx 8–14 mm, lobes spine-tipped; corolla tube 13–20 mm, white, upper lip shallowly 2-lobed, 2–4 mm, lower lip 6–9 mm; stamens and style exserted. **FR**: nutlet, 2.5–3 mm, light brown. n=15. Dry, rocky desert slopes, canyons, creosote-bush scrub; < 800 m. w edge DSon; n Baja CA. Locally common.

S. verbenacea L. Per. **ST** 3–6 dm; hairs glandular. **LVS** basal; petiole 2–10 cm; blade 3.5–9 cm, ± ovate, irregularly serrate, puckered; hairs sparse, minute. **INFL**: clusters 2–10-fld, gen 1.5–2.5 cm; bracts gen < 1/2 calyx. **FL**: calyx gen 5–8 mm, enlarging in fr, upper lip ± entire, reflexed; corolla tube 6–15 mm, blue to pinkish purple. **FR**: nutlet, ± 2 mm, brown, smooth. 2n=54. Disturbed places; probably < 200 m. e SnFrB (Alameda Co.); native to Eur. Possibly a waif from cult. More collections needed.

SATUREJA

Deborah Engle Averett

Per to shrub, decumbent to erect, < 2 m, aromatic. **LF** petioled; blade gen ovate-deltate, base truncate to tapered, margin entire to shallowly crenate-dentate, lower surface gen gland-dotted. **INFL**: fls 1–several in lf axils. **FL**: calyx 5-lobed, 2-lipped; corolla 5-lobed, white to lavender or salmon, 2-lipped, lower lip spreading, upper erect, 2-lobed; stamens 4, in 2 pairs, anther sacs spreading; style 2-lobed, exserted. **FR** 1–2 mm; surface smooth to net-like. 150 spp.: gen Medit, to Japan, Australia, also N.Am, S.Am. (Latin: savory) *S. hortensis* L. cult as herb (summer savory).

1. Per, decumbent; calyx 4–5 mm .... ***S. douglasii***
1′ Per to shrub, erect; calyx 6–17 mm
 2. Lf 5–15 mm; calyx 6–8 mm; shrub (st woody) .... ***S. chandleri***
 2′ Lf 35–80 mm; calyx 14–17 mm; per to subshrub (st base sometimes woody) .... ***S. mimuloides***

S. chandleri (Brandegee) Druce (p. 733) SAN MIGUEL SAVORY Shrub, erect, < 0.5 m. **ST** woody; bark red-brown; hairs recurved, white. **LF** 5–15 mm, 4–16 mm wide, deltate to ovate-deltate, shallowly crenate-dentate; hairs short, white. **INFL**: fls 1–6 per lf axil; pedicel 1–3 mm. **FL**: calyx 6–8 mm, bell-shaped, tinged purple, lobes ± 1 mm; corolla 4–7 mm, white to lavender. **FR** ± 1.5 mm, shiny dark brown; surface ± net-like. UNCOMMON. Rocky slopes, chaparral; 520–690 m. PR (San Miguel, Santa Ana mtns). [*Calamintha c.* Brandegee] ❀DRN,DRY:16,**17**&SHD:14,20,21, **22–24**.

S. douglasii (Benth.) Briq. (p. 733) YERBA BUENA Per, decumbent, forming mats < 1 m wide. **ST** slightly woody, sometimes rooting; hairs sparse, minute, recurved. **LF** 10–35 cm, 5–25 mm wide, ovate to ovate-triangular, shallowly crenate-dentate; hairs sparse, minute. **INFL**: fls 1(–3) per lf axil; pedicels 5–20 mm. **FL**: calyx 4–5 mm, tubular, turning purple in age, lobes ± 0.5 mm; corolla 3–8 mm, white to lavender. **FR** ± 1 mm, shiny brown, smooth. Shady places, chaparral, woodland; < 900 m. KR, NCoR, CW, TR; to B.C., ID. Dried lvs make pleasant tea. ❀DRN,SUN:**4–6**&SHD: **15–17**&IRR:2,3,7–9,14,18,19,**20–24**;GRCVR.

S. mimuloides (Benth.) Briq. (p. 733) Per to subshrub, erect, < 2 m. **ST** nonwoody, exc sometimes at base; hairs long, ± spreading. **LF** 35–80 mm, 15–55 mm wide, ± ovate to triangular, shallowly and irregularly crenate-dentate; hairs appressed, white. **INFL**: fls 1–6 per lf axil; pedicels 2–3 cm. **FL**: calyx 14–17 mm, tubular-bell-shaped, veins purple, lobes 3–5 mm; corolla 15–35 mm, salmon, often faded when dry. **FR** ± 2 mm, brown, surface smooth. Moist places, streambanks, chaparral, woodland; 400–1800 m. CCo, SCoRO, TR. ❀DRN,IRR:**5**,6,**7-9**,10,**14–17**,18,**19–24**.

SCUTELLARIA SKULLCAP

Richard G. Olmstead

Per, gen hairy, sometimes glandular, from rhizomes or tubers. **STS** erect, branched or not. **LVS** basal and cauline; lower gen petioled; cauline becoming ± sessile upward. **INFL**: fl gen 1 per lf axil (or appearing as a bracted raceme). **FL**: calyx 2-lipped, lips ± equal, enclosing nutlets, back of upper lip dome-like or transversely ridged, gen with concave depression behind ridge; corolla 2-lipped, white to violet-blue, upper lip < lower lip, ± entire, hood-like, lower lip 3-lobed; stamens 4, pairs ± equal, enclosed by upper corolla lip, anthers ciliate, lower two 1-chambered; disk below ovary gen green-yellow. **FR** gen ovoid, gen minutely papillate, brown or black. ± 300 spp.: gen temp worldwide. (Latin: tray, from calyx dome or ridge) [Olmstead 1990 Contr Univ Michigan Herb 17:223–265]

1. Corolla white to light yellow, lower lip sometimes blue- to violet-mottled
 2. Upper cauline lf blade crenate, base ± lobed; st hairs long, spreading; lower corolla lip blue-mottled .... ***S. bolanderi***
  3. Corolla 12–14 mm; lf length > 2 × width; SnBr, SnJt, s DMoj .... ssp. ***austromontana***
  3′ Corolla 15–19 mm; lf length 1–2 × width; SN, Teh, SCoRI .... ssp. ***bolanderi***
 2′ Upper cauline lf blade entire, base obtuse to wedge-shaped; st hairs short, ascending or pointed down; lower corolla lip violet-mottled or not
  4. St 1.5–4 dm, hairs ascending or upcurled; ovary disk green-yellow; lf blade ovate to oblong .... ***S. californica***
  4′ St < 2 dm, hairs pointed or curled down; ovary disk orange-red; lf blade obovate to ± diamond-shaped .... ***S. nana***
1′ Corolla blue to violet-blue, lower lip gen white-patched
 5. Fls gen > 4 in axillary or terminal racemes, gen bracted .... ***S. lateriflora***
 5′ Fls 1, axillary, subtended by lvs
  6. Back of upper calyx lip dome-like; fr brown; gen wet habitats .... ***S. galericulata***
  6′ Back of upper calyx lip transversely ridged, ± concave behind ridge; fr black; gen dry habitats
   7. St and lf hairs 1–3 mm, spreading; upper cauline lf blades gen crenate; pls with tubers .... ***S. tuberosa***
   7′ St and lf hairs << 0.5 mm, curled or appressed; upper cauline lf blades entire; rhizomes slender, tubers ± 0

8. Fl 12–22 mm; lf length 2–3 × width; corolla hairs not gland-tipped .... ***S. antirrhinoides***
8′ Fl 25–35 mm; lf length 3–7 × width; corolla hairs gland-tipped .... ***S. siphocampyloides***

S. antirrhinoides Benth. (p. 733) Pl 10–35 cm; rhizomes slender, tips ± swollen. **ST**: hairs << 0.5 mm, appressed-ascending or upcurled, sometimes gland-tipped. **LVS**: basal petioles 5–10 mm; upper cauline blades ovate to oblong, entire, base rounded to tapered, tip rounded. **FL**: pedicel 3–4.5 mm; calyx 3–4 mm, ridged; corolla 13–21 mm, violet-blue, lower lip white-patched or -mottled, inner surface long-hairy. **FR** black. Dry, rocky, often serpentine slopes, ridges, oak woodland, coniferous forest; < 2000 m. NW, CaR, n SNF, MP; to WA. Hybridizes with *S. californica* in NCoR. ❀DRN:4,5,**6**&IRR:**15–17**&SHD:**7**,8,9,**14**.

S. bolanderi A. Gray Pl 30–100 cm; rhizomes slender, tips ± swollen. **ST**: hairs 1–2 mm, spreading, often gland-tipped. **LVS**: basal petioles 2–10 mm; upper cauline blades ovate to cordate, crenate (rarely entire), base truncate to ± lobed, tip rounded. **FL**: pedicel 2–3 mm; calyx 3–5 mm, ridged; corolla 13–19 mm, white, lower lip blue-mottled, inner surface long-soft-hairy. **FR** brown to black. Gravelly soils, streambanks, oak or pine woodland; 300–2000 m. SN, SCoRI, SW, s DMoj.

ssp. ***austromontana*** Epling SOUTHERN SKULLCAP **LF** length > 2 × width. **FL**: corolla 12–14 mm. RARE. Habitat like sp.; 600–2000 m. SnBr, PR, s DMoj.

ssp. ***bolanderi*** (p. 733) **LF** length < 2 × width. **FL**: corolla 15–19 mm. Habitat like sp.; 300–1500 m. SN, SCoRI (Santa Clara Co.). ❀DRN,IRR:**7,15,16**,17,24&SHD:**8,9,14**,19–23.

S. californica A. Gray (p. 733) Pl 15–40 cm; rhizomes slender, tip ± swollen. **ST**: hairs << 0.5 mm, appressed-ascending, sometimes gland-tipped. **LVS**: basal petioles 5–10 mm; upper cauline blades ovate to oblong, entire, base obtuse to rounded, tip rounded. **FL**: pedicel 3.5–4.5 mm; calyx 3.5–5 mm, ridged; corolla 16–19 mm, white to light yellow, sometimes tinged pink or blue, lower lip not mottled. **FR** black. Open sites, shrubland, woodland; 300–2000 m. KR, NCoR, n SN. ❀DRN:6&IRR:**15–17**&SHD:1,2,**7–9**,10,**14**.

S. galericulata L. (p. 733) MARSH SKULLCAP Pl 20–80 cm; rhizomes slender. **ST** glabrous or hairs << 0.5 mm, ± descending, sometimes gland-tipped. **LVS**: basal gen 0; lower cauline petioles 1–5 mm; upper cauline blades lanceolate to narrowly oblong-ovate, entire to crenate, base truncate to ± lobed, tip ± acute. **FL**: pedicel 2–4 mm; calyx 3.5–4.5 mm, upper lip back dome-like; corolla 15–20 mm, violet-blue to blue, lower lip white-mottled, inner surface papillate. **FR** ± spheric, brown. RARE in CA. Wet sites, meadows, streambanks, coniferous forest; 1000–2100 m. n SNH (Tahoe Basin), MP (Fall River); to AK, e US, circumboreal. ❀WET:1,2,4,5,**6,15,16**&SHD:3,**7–10,14,18**,19,20.

S. lateriflora L. (p. 733) BLUE SKULLCAP Pl 20–60 cm; rhizomes slender. **ST** glabrous or hairs sparse, hairs << 0.5 mm, ascending to upcurled, gen not glandular. **LVS**: basal gen 0; lower cauline petioles 10–20 mm; upper cauline blades ovate to lanceolate, ± dentate, base rounded to truncate, tip acute. **INFL**: raceme or spike, bracted; bracts < 8 mm. **FL**: pedicel 1–3 mm; calyx 1.5–3 mm, upper lip back dome-like; corolla 6–8 mm, blue, lower lip blue, inner surface glabrous or sparsely soft-hairy. **FR** ± spheric, brown. RARE in CA. Marshes, wet meadows; < 500 m. n SnJV, SNE (Saline Valley); to B.C., e US. ❀WET:**4–9**,10,11,**14–16**,17,18,**19–23**,24.

S. nana A. Gray (p. 733) Pl < 2 dm; rhizomes thick. **ST**: hairs << 0.5 mm, ± appressed, pointed down, sometimes gland-tipped. **LVS**: basal petioles 2–5 mm; upper cauline blades obovate to diamond-shaped, entire, base obtuse to wedge-shaped, tip ± rounded. **FL**: pedicel 1–3 mm; calyx 3–4 mm, ridged; corolla 15–20 mm, white to light yellow, lower lip purple-spotted, inner surface glabrous or sparsely long-hairy; ovary disk orange-red. **FR** black. Dry, volcanic soils, shrubland; 1000–1900 m. CaR, n SN, MP; to OR, ID, NV. [*S. holmgreniorum* Cronq., Holmgren's skullcap] ❀DRN,SUN:**7,17**&IRR:1–3,14,**15,16,18**,19–21.

S. siphocampyloides Vatke Pl 20–55 cm; rhizomes slender, tips ± swollen. **ST**: hairs << 0.5 mm, appressed-ascending, gen gland-tipped. **LVS**: basal petioles 10–20 mm; upper cauline blades ovate to oblong, entire, base rounded to tapered, tip rounded. **FL**: pedicel 4–5.5 mm; calyx 3–4 mm, ridged; corolla 25–35 mm, violet-blue, lower lip white-patched, -mottled, or not, inner surface long-hairy. **FR** black. Open sites, seeps, dry streambeds, shrubland, woodland; 300–2300 m. CA-FP (exc GV). [*S. austinae* Eastw.] ❀**6**&IRR:1,2,7,8,9,**14–17**,18–24.

S. tuberosa Benth. (p. 733) Pl < 25 cm, from tubers 5–20 mm. **ST**: hairs 1–3 mm, spreading. **LVS**: basal petioles 5–20 mm; upper cauline blades ovate, entire to crenate, base obtuse to rounded, tip rounded. **FL**: pedicel 2–4 mm; calyx 4–5.5 mm, ridged or ± dome-like; corolla 13–20 mm, violet-blue, lower lip white-patched or -mottled, inner surface glabrous or long-hairy. **FR** obconic, black. Dry sites, chaparral, oak woodland, common after fires; 200–1000 m. CA-FP (exc GV); OR, Baja CA. [ssp. *australis* Epling] ❀DRN,DRY:5,**6,15–17**&SHD:**7–9,14,18–24**.

STACHYS HEDGE NETTLE

Barrett H. Anderson & Barry D. Tanowitz

Ann, per, hairy, often glandular; rhizome slender or 0. **ST** decumbent to erect, 1–25 dm. **LVS** 1.5–18 cm; lower gen petioled; upper ± sessile; blades oblong to ovate, serrate to crenate. **INFL**: spike of ± sessile clusters, gen terminal, interrupted or continuous, bracted. **FL**: calyx bell-shaped, radial to ± 2-lipped, veins 5–10, lobes 5, erect or spreading, triangular, tips sharp; corolla white, pink, red, magenta, or purple, tube narrow, with internal ring of hairs gen above base, perpendicular to oblique to tube axis, sometimes narrowed on lower surface, upper lip erect or gen parallel to tube axis, concave, entire, rarely notched, gen hairy, lower lip perpendicular to tube axis or reflexed, 3-, rarely 2-lobed, glabrous to hairy. **FR** oblong to ovoid, brown to black. ± 300 spp.: temp (exc Australia); some cult for orn. (Greek: ear of corn, from infl) [Epling 1934 Fedde Rep Sp Nov Regni Veg 80:1–75]

1. Ann; rhizome 0; uncommon .... ***S. arvensis***
1′ Per; rhizome slender; widespread
2. Corolla tube > 15 mm .... ***S. chamissonis***
2′ Corolla tube < 15 mm
3. Upper corolla lip < 2 mm; stamens barely exerted from corolla tube .... ***S. stricta***
3′ Upper corolla lip > 2 mm; stamens clearly exerted from corolla tube
4. Lvs and infl with dense, cobwebby hairs .... ***S. albens***
4′ Lvs and infl with soft or stiff hairs, sometimes dense but not cobwebby
5. Infl gen < 5 cm, with no spaces between fl clusters at maturity
6. Corolla whitish
7. St, lvs rarely glandular, slightly aromatic; st gen < 6 dm .... [3]***S. ajugoides*** var. ***ajugoides***
7′ St, lvs densely glandular, strongly aromatic; st gen > 6 dm .... ***S. pycnantha***

6′ Corolla pink to purple
8. Herbage densely glandular, strong-smelling; st gen > 6 dm ..... ***S. pycnantha***
8′ Herbage not (or not densely) glandular, not or slightly aromatic; st gen < 6 dm ..... [2]***S. ajugoides***
9. Lf base wedge-shaped; blade gen oblong; hairs often long, silky; corolla pale pink ..... [3]var. ***ajugoides***
9′ Lf base gen rounded to ± cordate; blade ovate to lanceolate; hairs rarely long, silky; corolla pink to purple ..... [2]var. ***rigida***
5′ Infl > 5 cm, with spaces between fl clusters at maturity
10. Ring of hairs in corolla tube < 2 mm from base, perpendicular to tube axis ..... ***S. bullata***
10′ Ring of hairs in corolla tube > 2 mm from base, gen oblique, rarely partially perpendicular to tube axis
11. Lvs ± sessile, longest petiole < 1 cm; MP ..... ***S. palustris*** ssp. ***pilosa***
11′ Lvs petioled, at least lower petioles > 1 cm; widespread ..... [2]***S. ajugoides***
12. Lf base wedge-shaped; blade gen oblong, hairs gen long, silky; corolla white to pale pink ..... [3]var. ***ajugoides***
12′ Lf base rounded to ± cordate; blade ovate to lanceolate, hairs rarely long, silky; corolla pink to magenta or purplish ..... [2]var. ***rigida***

S. ajugoides Benth. **ST** branched or not. **LF**: petiole 1.5–6 cm. **INFL** > 5 cm, interrupted; bracts sometimes long, leafy. **FL**: calyx tube 3–6 mm, hairs 0 or soft to stiff, sometimes glandular; corolla tube 6–11 mm, sometimes spurred or sac-like, ring of hairs > 2 mm from base, upper lip 2.5–5 mm, lower lip 4.5–8.5 mm. Gen moist places, sometimes dry hillsides, many communities; < 2500 m. CA; to B.C., Baja CA.

var. ***ajugoides*** (p. 733) **ST** often decumbent, 1–6 dm, soft-hairy. **LF**: blade 1.5–7 cm, gen oblong, hairs gen long, silky, base wedge-shaped, tip rounded. **INFL**: clusters gen 3–6-fld. **FL**: corolla white to pale pink, tube 6–8 mm, ring of hairs strongly oblique. Moist, open places often remaining wet into summer; < 1000 m. NW, CW, SW; to B.C. ❀STBL.

var. ***rigida*** Jepson & Hoover (p. 733) **ST** gen erect to ± decumbent, 6– 10 dm, ± glabrous to soft- or stiff-hairy, sometimes glandular. **LF**: blade 5–9 cm, ovate to lanceolate, glabrous or soft-hairy to densely felt-like, sometimes glandular, base rounded to ± cordate, tip acute to obtuse. **INFL**: clusters 6–16-fld. **FL**: corolla pink to magenta or purplish, tube 6–10 mm, ring of hairs strongly oblique. Moist to ± dry places; < 2500 m. CA (very uncommon in D); to WA, Baja CA. [*S. r.* Benth. incl ssp. *lanata* Epling, ssp. *quercetorum* (A.A. Heller) Epling, ssp. *rivularis* (A.A. Heller) Epling; *S. emersonii* Piper; *S. mexicana* Benth] ❀STBL.

S. albens A. Gray (p. 733) **ST** erect, 5–25 dm, often branched. **LF**: petiole < 5 cm; blade 3–15 cm, widely ovate, crenate to serrate, gen with dense, cobwebby hairs, base ± cordate, tip acute to obtuse. **INFL** 10–30 cm, ± interrupted, clusters 6–12-fld; hairs dense, cobwebby. **FL**: calyx tube 4–5.5 mm, hairs dense, cobwebby; corolla white to pinkish, tube 6–9 mm, ring of hairs > 2 mm from base, oblique, upper lip 3.5–5.5 mm, lower lip 6–8 mm. Wet, swampy to seepy places; < 3000 m. NCoR, c&s SN, Teh, GV, CCo, SW, W&I, rarely D. ❀WET or IRR:1,2,4,**5–9,14–17,19–23**,24&SHD:3,10,**18**;INV.

S. arvensis L. Ann. **ST** decumbent to erect, 1–6 dm, branched, stiff-hairy. **LF**: petiole 1–2.5 cm; blade 1.5–2.5 cm, ovate to oblong-ovate, serrate, base rounded to ± cordate, tip obtuse. **INFL** 0.5–2.0 cm, interruped; clusters 2–6-fld. **FL**: calyx tube 2.5–4 mm, upper part purple, lobes lanceolate; corolla pale purple, tube 3–6 mm, ring of hairs 0 or nor apparent. Uncommon. Moist places; < 50 m. NW, CW; native to Eur.

S. bullata Benth. (p. 733) **ST** erect, 4–8 dm, branched; hairs on angles stiff, sharp, reflexed, on sides soft, often glandular. **LF**: blade 3–18 cm, ± ovate, soft- and stiff-hairy, glandular, base ± cordate, tip obtuse. **INFL** > 5 cm, interrupted; clusters 6-fld, hairs shaggy. **FL**: calyx tube 6–7.5 mm, hairs shaggy; corolla pink, tube 7–10 mm, ring of hairs < 2 mm from base, perpendicular, upper lip 3.5–5.5 mm, lower 6–10 mm. Dry slopes near coast; < 500 m. CW, SW. Pls from n ChI might belong to undescribed ssp. ❀5:IRR: **15–17,24**&SHD:7,**8,9,14**,18,**19–23**;INV;STBL.

S. chamissonis Benth. (p. 733) **ST** erect, 1–2.5 m, simple to branched above, stiff-hairy, glandular. **LF**: petiole < 8 cm; blade 6–18 cm, ovate, crenate, soft-hairy, glandular, base cordate, tip acute. **INFL** 10–40 cm, interrupted; clusters 2–6-fld. **FL**: calyx tube 7–12 mm, soft-hairy, glandular; corolla deep magenta to purple, tube 18–24 mm, ring of hairs > 2 mm from base, perpendicular to ± oblique, upper lip 6–10 mm, lower lip 8.5–15 mm. Wet, swampy places, gen coastal; < 150 m. NCo, CCo. ❀IRR or WET:**5,15–17, 24**&SHD:**8,9,14**,18–23;INV.

S. palustris L. ssp. ***pilosa*** (Nutt.) Epling (p. 733) **ST** erect, 3–9 dm, often branched; hairs soft, spreading, ± glandular. **LF**: petiole 0–1 cm; blade 3.5–9 cm, ovate, oblong to elliptic, crenate, hairs soft, ± spreading, ± glandular, base rounded to ± cordate, tip acute to obtuse. **INFL** > 5 cm, interrupted; clusters 2–6-fld. **FL**: calyx tube 3–4.5 mm, soft- to stiff-hairy, often glandular; corolla pink to pale purple, tube 6–9 mm, ring of hairs > 2 mm from base, oblique, upper lip 3–5 mm, lower lip 5–8 mm. Moist places; 1200–1500 m. MP; to WA, w N.Am. Ssp. *palustris* naturalized in e US, native to Eur. ❀STBL.

S. pycnantha Benth (p. 733) SHORT-SPIKED HEDGE NETTLE **STS** decumbent to erect, 3–10, gen > 6 dm, often branched, soft- to stiff-hairy, glandular. **LF** strongly aromatic; petiole < 5 cm; blade 5–12 cm, ovate or lanceolate to oblong, crenate to serrate, soft- to stiff-hairy, glandular, base rounded to cordate, tip obtuse. **INFL** gen < 5 cm, ± continuous exc sometimes interrupted below; clusters 8–12-fld. **FL**: calyx tube 4–6 mm, densely soft-hairy; corolla whitish to pinkish, tube 6.5–8.5 mm, ring of hairs > 2 mm from base, oblique, upper lip 3–4 mm, lower lip 5–7 mm. UNCOMMON. Streamsides, springs, several communities; < 1100 m. NW, CW, SN (uncommon). Some ± isolated pls of SN key out here but might belong to another sp., may be ± closely associated with serpentine; more study needed. ❀TRY.

S. stricta E. Greene (p. 733) **ST** erect, 3–8 dm, branched; soft- to stiff-hairy, glandular. **LF**: petiole 1–4 cm; blade 5–15 cm, oblong to widely lanceolate, crenate, soft- to stiff-hairy, glands small, many, esp on lower surface, base ± truncate, tip obtuse. **INFL** 5–10 cm, often continuous above and interrupted below; clusters 6–12-fld. **FL**: calyx tube 4–5 mm, soft-hairy, glandular; corolla white, tube 5–7 mm, ring of hairs > 2 mm from base, oblique, upper lip < 2 mm, lower lip 3.5–7 mm. Moist, open or shady places; < 600 m. c&s NCoRI, n&c SNF, adjacent GV. ❀STBL.

TEUCRIUM

Ann, per, glabrous to short-hairy. **STS** ascending to erect, branched or not. **LVS** petioled, crenate to deeply lobed, lobes oblong. **INFLS** in CA axillary at upper nodes; fls gen 2, subtended by lvs or bracts. **FL**: calyx ± radial, 10-veined, 5-lobed, lobes subequal; corolla 1-lipped, tube split above, lip 5-lobed, ± flat, distal lobe >> lateral lobes, tip rounded, lateral lobe tips acute to obtuse; stamens 4, lower pair gen > upper; style lobes gen equal. ± 100 spp.: worldwide, esp Medit. (Greek: ancient name) [McClintock & Epling Brittonia 5:491–510]

1. Ann, 1–3 dm, simple to branched at base; pedicels < 5 mm; corolla lip 4–8 mm, white to bluish, purple-spotted .... ***T. cubense*** ssp. ***depressum***
1′ Per, 5–10 dm, gen branched throughout; pedicels 8–25 mm; corolla lip 10–17 mm, white, gen violet-streaked .... ***T. glandulosum***

T. cubense Jacq. ssp. ***depressum*** McClint. & Epling (p. 737) **ST** simple to branched at base. **LF** gen withering in fr; lower 2–4 cm, blade ovate to obovate, crenate to lobed; upper 0.5–1.5 cm, gen deeply 3-lobed. **INFL**: pedicel 1–5 mm. **FL**: calyx tube 1–3 mm, lobes 3–6 mm, bristle-tipped; corolla 7–15 mm, slightly puberulent inside, white to bluish, purple-spotted; filament glabrous. Sandy soils, washes, fields; < 400 m. DSon; to TX, Baja CA. Other 4 sspp. in s US, Caribbean, Mex, S.Am.

T. glandulosum Kellogg (p. 737) **ST** gen branched throughout. **LF** 1–4 cm, ± persistent, gen deeply 3-lobed. **INFL**: pedicel 8–25 mm. **FL**: calyx tube 2–4 mm, lobes 4–8 mm, gen acute; corolla 15–21 mm, densely puberulent inside, white, violet-streaked; filament short-hairy below middle. Uncommon. Rocky slopes, canyons; 400–500 m. ne DSon (Whipple Mtns); AZ, Baja CA. ❀TRY.

TRICHOSTEMA BLUECURLS

Harlan Lewis

Ann, shrub, strong-scented. **ST** hairy, often glandular. **LF** simple; blade linear to ovate, entire. **INFL**: cymes (racemes in *T. lanceolatum*), axillary. **FL**: calyx lobes 5, equal or uppermost 1 narrower; corolla blue or lavender, tube straight or curved upward, sometimes abruptly near throat, incl to much exserted from calyx, lobes 5, lowest a gen reflexed lip; stamens 4, attached near throat, gen much exserted, ascending between upper corolla lobes, gen arched. **FR**: nutlets 4, joined in basal ± 1/3, puberulent to hairy, irregularly ridged. ± 17 spp.: N.Am. (Greek: hair, stamen) [Lewis 1945 Brittonia 5:276–303] Ann spp. gen fl late summer or fall.

1. Shrub, < 1.5 m
 2. Corolla tube 9–14 mm; infl hairs 2–3 mm, obscuring pedicels .... ***T. lanatum***
 2′ Corolla tube 4–7 mm; infl hairs 1–2 mm, not obscuring pedicels .... ***T. parishii***
1′ Ann, < 1 m
 3. Petiole distinct, 5–15 mm
  4. Corolla tube 4–8 mm, exserted, lower lip 4–7 mm .... ***T. laxum***
  4′ Corolla tube 1.5–3 mm, incl, lower lip 2–3 mm .... ***T. simulatum***
 3′ Petiole indistinct or < 5 mm
  5. Stamens 10–20 mm; corolla tube curved abruptly upward near throat
   6. Lf blade 2–7 cm, lanceolate to narrowly ovate, length > 3 × width; corolla tube 5–10 mm ...... ***T. lanceolatum***
   6′ Lf blade 1–2 cm, ovate, length < 2 × width; corolla tube 2.5–5.5 mm .... ***T. ovatum***
  5′ Stamens 2–6 mm; corolla tube curved gradually upward
   7. Hairs on st and lvs curled or appressed; stamens 2–3 mm, ± incl or barely exserted .... ***T. micranthum***
   7′ Hairs on st and lvs straight, spreading; stamens 3–6 mm, exserted
    8. Calyx lobes < 2 × tube; calyx tube > mature nutlets; mature calyx gen red-tinged .... ***T. rubisepalum***
    8′ Calyx lobes > 2 × tube; calyx tube < mature nutlets; mature calyx green
     9. Lf blade widely elliptic, length gen < 4 × width; calyx lobes gen widest above base .... ***T. oblongum***
     9′ Lf blade elliptic, length gen > 4 × width; calyx lobes widest at base .... ***T. austromontanum***
      10. Lf ± = or < internode above .... ssp. ***austromontanum***
      10′ Lf >> internode above .... ssp. ***compactum***

T. austromontanum Harlan Lewis (p. 737) Ann < 5 dm. **ST**: short hairs appressed, long hairs spreading, some hairs glandular. **LF**: petiole indistinct or < 5 mm; blade 2–5 cm, elliptic, length > 4 × width. **FL**: calyx lobes > 2 × tube, widest at base, acute, ± equal; corolla tube 1.5–3 mm, curved gradually upward, ± = calyx, lower lip 1.8–3 mm; stamens 3–5.5 mm, exserted, arched. Uncommon. Drying margins of lakes, meadows, streams; 1000–2500 m. Teh, TR, PR, SNE; Baja CA.

ssp. ***austromontanum*** **ST** < 5 dm. **LF** < 50 mm, ± = or < internode above. *n*=14. Habitat and range of sp.

ssp. ***compactum*** Harlan Lewis HIDDEN LAKE BLUECURLS **ST** < 10 cm. **LF** < 30 mm, >> internode above. *n*=14. RARE. Montane vernal pools; 2500 m. SnJt (Hidden Lake).

T. lanatum Benth. (p. 737) WOOLLY BLUECURLS Shrub < 15 dm. **ST**: hairs short, appressed below, sometimes woolly near infl. **LF**: petiole indistinct or < 3 mm; blade 3.5–7.5 cm, linear, margin rolled under, upper surface green, glabrous, lower surface gray-hairy; gen with cluster of smaller lvs in axil. **INFL**: hairs densely woolly, 2–3 mm, obscuring pedicels, fine, blue, pink, or white. **FL**: calyx lobes = or > tube, ± equal; corolla tube 9–14 mm, ± straight, exserted, lower lip 7–12 mm; stamens 2.5–4 cm, much exserted, arched. *n*=10. Coastal scrub, chaparral; < 800 m. SCoR, SCo, WTR, SnGb, PR; Baja CA. ❀DRN,SUN,&DRY:10,14,**15**,**16**,17–21,**22–24**.

T. lanceolatum Benth. (p. 737) VINEGAR WEED Ann < 1 m. **ST**: short hairs appressed, long hairs soft, spreading, some hairs glandular. **LF**: petiole indistinct or < 4 mm; blade 2–7 cm, lanceolate to narrowly ovate, length > 3 × width, lateral veins prominent near base. **FL**: calyx lobes = to 3 × tube, uppermost often narrower; corolla tube 5–10 mm, curved abruptly upward near throat, exserted, lower lip 4–8 mm; stamens 13–20 mm, much exserted, strongly arched. *n*=7. Dry, open, gen disturbed habitats; gen < 1000 m. CA-FP; OR, Baja CA. ❀DRN,DRY,&SUN: **7–9**,10, **14**,**15**,17,**18–24**; lvs have strong odor.

T. laxum A. Gray (p. 737) TURPENTINE WEED Ann < 5 dm. **ST**: short hairs appressed, long hairs spreading, some hairs glandular. **LF**: petiole distinct, 5–15 mm; blade 3–7 cm, lanceolate to narrowly ovate. **FL**: calyx lobes ± = tube, often red-tinged, uppermost narrower; corolla tube 4–8 mm, curved abruptly upward near throat, exserted, lower lip 4–7 mm; stamens 7–16 mm, much exserted, strongly arched. *n*=7. Gravelly streambanks or sandy soil; < 1500 m. NW. ❀DRN,SUN,DRY:**7–9**,10,**14–24**;lvs have strong odor.

T. micranthum A. Gray SMALL-FLOWERED BLUECURLS Ann < 3 dm. **ST**: hairs short, appressed, some glandular. **LF**: petiole indistinct or < 5 mm; blade 2–4.5 cm, narrowly elliptic. **FL**: calyx lobes < 2 × tube, widest at base, acute, ± equal; corolla tube 1–2 mm, curved gradually upward, incl, lower lip 1.5–2 mm; stamens 2–3 mm, ± incl or barely exserted, slightly arched or ± straight. *n*=7. UNCOMMON. Dry margins of lakes, meadows, streams; 2000–2300 m. SnBr; Baja CA, AZ.

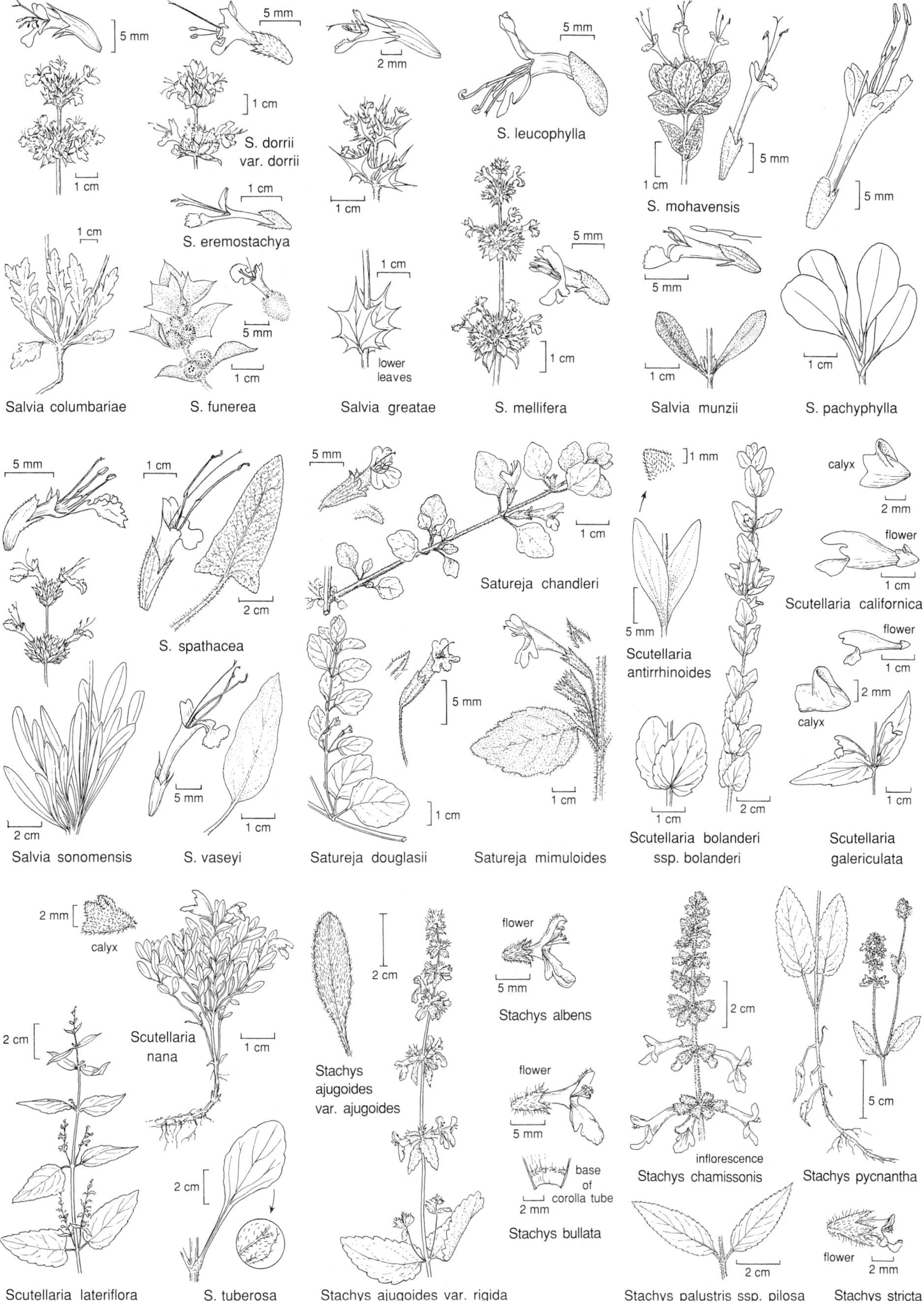

5 mm
1 cm
Salvia columbariae
5 mm
1 cm
S. dorrii var. dorrii
1 cm
S. eremostachya
5 mm
1 cm
S. funerea
2 mm
1 cm
1 cm
lower leaves
Salvia greatae
5 mm
S. leucophylla
5 mm
1 cm
S. mellifera
1 cm
5 mm
S. mohavensis
5 mm
1 cm
Salvia munzii
5 mm
1 cm
S. pachyphylla
5 mm
2 cm
Salvia sonomensis
1 cm
2 cm
S. spathacea
5 mm
1 cm
S. vaseyi
5 mm
1 cm
Satureja chandleri
5 mm
1 cm
Satureja douglasii
1 cm
Satureja mimuloides
1 mm
5 mm
Scutellaria antirrhinoides
1 cm
2 cm
Scutellaria bolanderi ssp. bolanderi
calyx
2 mm
flower
1 cm
Scutellaria californica
flower
1 cm
2 mm
calyx
1 cm
Scutellaria galericulata
2 mm
calyx
2 cm
Scutellaria lateriflora
Scutellaria nana
1 cm
2 cm
S. tuberosa
2 cm
Stachys ajugoides var. ajugoides
Stachys ajugoides var. rigida
flower
5 mm
Stachys albens
flower
5 mm
base of corolla tube
2 mm
Stachys bullata
2 cm
inflorescence
Stachys chamissonis
2 cm
Stachys palustris ssp. pilosa
5 cm
Stachys pycnantha
flower
2 mm
Stachys stricta

T. oblongum Benth. (p. 737) Ann < 5 dm. **ST**: short hairs appressed or spreading, long hairs spreading, some hairs glandular. **LF**: petiole indistinct or < 5 mm; blade 2–4 cm, widely elliptic, length < 4 × width. **FL**: calyx lobes > 2 × tube, gen widest above base, ± acuminate, ± equal; corolla tube 2–3.5 mm, curved gradually upward, ± = calyx, lower lip 2–3.5 mm; stamens 3–6 mm, exserted, arched. *n*=7. Dry margins of meadows, streams; 100–3000 m. KR, NCoR, CaR, SN, MP; to WA, ID.

T. ovatum Curran SAN JOAQUIN BLUECURLS Ann < 8 dm. **ST** woolly; hairs soft, curled or spreading, some glandular. **LF**: petiole ± indistinct; blade 1–2 cm, ovate, length < 2 × width, lateral veins prominent near base. **FL**: calyx lobes 2–3 × tube, ± equal; corolla tube 2.5–5.5 mm, curved abruptly upward near throat, exserted, lower lip 3.5–5 mm; stamens 11–16 mm, exserted, arched. *n*=7. UNCOMMON. Grassland, disturbed sites; < 300 m. s SnJV.

T. parishii Vasey (p. 737) Shrub < 12 dm. **ST**: hairs short, appressed. **LF**: petiole indistinct or < 5 mm; blade 2–6 cm, linear, margin rolled under, upper surface green, puberulent, lower surface gray-hairy; gen with cluster of smaller lvs in axil. **INFL**: hairs ± densely woolly, 1–2 mm, not obscuring pedicels, fine, blue, pink, or white, sometimes ± 0. **FL**: calyx lobes ± = tube, ± equal; corolla tube 4–7 mm, ± straight, exserted, lower lip 5–9 mm; stamens 15–25 mm, exserted, arched. *n*=10. Coastal scrub, chaparral; 600–2000 m. TR, PR; Baja CA. ❀DRN,SUN,&DRY:2,3,**7,15, 16**,17, 22–24 &IRR: 9,10,14,**18–21**.

T. rubisepalum Elmer (p. 737) HERNANDEZ BLUECURLS Ann < 5 dm. **ST**: hairs long, spreading, some short-glandular. **LF**: petiole indistinct or < 5 mm; blade lanceolate to narrowly ovate. **FL**: calyx lobes < 2 × tube, widest at base, acute, often red-tinged, uppermost narrower; corolla tube 3.5–4.5 mm, curved gradually upward, exserted, lower lip 2.5–3 mm; stamens 4–6 mm, exserted, arched. *n*= 7. UNCOMMON. Gravelly streambeds; 300–1000 m. s NCoRI (Napa Co.), c SNF (Tuolumne, Mariposa cos.), SCoRI (San Benito Co.).

T. simulatum Jepson Ann < 4 dm. **ST**: short hairs appressed, long hairs spreading, some hairs glandular. **LF**: petiole distinct, 7–15 mm; blade 2–5 cm, lanceolate to narrowly ovate. **FL**: calyx lobes ± = tube, triangular, ± equal; corolla tube 1.5–3 mm, curved slightly upward, incl, lower lip 2–3 mm; stamens 2.5–5 mm, exserted, slightly arched. *n*=7. Open, gen sandy or gravelly sites; 500–1500 m. KR, CaR, n SN, MP; s OR.

LAURACEAE LAUREL FAMILY

William J. Stone

Shrub, tree, gen evergreen, aromatic, rarely dioecious. **LVS** gen alternate, simple; surface with small pits or depressions; stipules 0. **INFL**: panicle, raceme, umbel, rarely fls solitary. **FL** bi- or unisexual, gen yellow or greenish; calyx deeply 4–6-lobed, segments in 2 series; petals 0; stamens in 3–4 whorls of 3 each, some sterile, anthers 4-celled, opening by uplifting valves; pistil 1, simple, ovary gen superior, chamber 1, ovule 1, style 1. **FR**: berry or drupe. ± 45 genera, ± 2200 spp.: widely distributed in trop, less so in temp; some cult (*Persea*, avocado, *Cinnamomum*, cinnamon, camphor).

UMBELLULARIA CALIFORNIA BAY, CALIFORNIA LAUREL, PEPPERWOOD

1 sp. (Latin: small umbrella from infl)

U. californica (Hook. & Arn.) Nutt. (p. 737) Shrub , tree, evergreen, very aromatic. **ST** < 45 m; bark greenish to reddish brown. **LF** 3–10 cm, 1.5–3 cm wide, oblong to oblong-lanceolate, shiny, smooth, deep yellow-green; petiole short. **INFL**: umbel in upper axils, simple, peduncled; fls 6–10. **FL** bisexual, yellow-green; sepals 6, 6–8 mm, oblong-ovate; stamens 9, inner 3 with 2 stalked orange glands at base. **FR**: drupe, gen solitary, 2–2.5 cm, round-ovoid, greenish, dark purple when dried, resembling an olive. Common. Canyons, valleys, chaparral; < 1600 m. NW, CaRF, SNF, SnFrB, SCoRO, scattered in TR, PR; s OR. [var. *fresnensis* Eastw.] Lf oils may produce TOXIC effects in some people. Known as Oregon myrtle in OR; used in cooking and by woodworkers. ❀4,**5,6,14–17**,IRR:1–3,**7–9,18–24**.

LENNOACEAE LENNOA FAMILY

George Yatskievych

Ann, per, root parasite lacking chlorophyll. **ST** (actually a peduncle) fleshy, underground, gen unbranched, white or ± brown. **LVS** scale-like, alternate. **INFL**: panicle, spike, or head. **FL** bisexual, ± radial; calyx lobes 4–10; corolla lobes 4–10; stamens as many as corolla lobes, epipetalous, incl; ovary superior, chambers 10–32, placentas axile, style 1, stigma lobes 5– 9. **FR**: capsule, circumscissile, hidden by persistent perianth. **SEEDS** in a ring, 1 per chamber, ± reniform, flat, brown. 2 genera, 4 spp.: sw US to n S.Am, nowhere common; some historically harvested for food. [Yatskievych & Mason 1986 Syst Bot 11:531–548]

PHOLISMA

Per. **ST** < 1.5 m. **LF** 5–25 mm, linear to triangular, glandular. **FL** 7–10 mm. **FR** ± circumscissile below middle. 3 spp.: s CA, w AZ, nw Mex. (Greek: scale, from scaly stem)

1. Infl a dense panicle or spike; calyx lobes glandular-puberulent (hairs < 0.5 mm) .... ***P. arenarium***
1′ Infl a concave head; calyx lobes appearing feathery (glandular hairs 1–1.5 mm) .... ***P. sonorae***

P. arenarium Hook. (p. 737) **ST** 3–8 dm, 1–2 cm diam. **FL**: calyx lobes linear to spoon-shaped; corolla lavender to bluish purple, margin white, exterior minutely puberulent; ovary chambers 10–20. 2*n*=36. Uncommon. Sandy soil, coastal dunes, chaparral, desert; < 1900 m. CCo, SCo, PR, D; w AZ, nw Mex. Parasitic on *Croton*, *Eriodictyon*, various shrubby Asteraceae.

P. sonorae (A. Gray) G. Yatskievych (p. 737) SAND FOOD **ST** 5–15 dm, 0.5–2 cm diam. **FL**: calyx lobes linear; corolla pink to purple, margin white, exterior glabrous; ovary chambers 12–32. 2*n*=36. RARE. Dunes, sandy areas; < 0–200 m. DSon (se Imperial Co.); w AZ, nw Mex. Parasitic on *Eriogonum*, *Tiquilia*, *Ambrosia*, *Pluchea*. [*Ammobroma s.* A. Gray] Threatened by off-road vehicles.

LENTIBULARIACEAE BLADDERWORT FAMILY

Lawrence R. Heckard

Ann, per, carnivorous, of moist or aquatic habitats. **ST** a caudex (*Pinguicula*) or filamentous with alternate or whorled branch systems in place of lvs (*Utricularia*). **LVS** 0 or in basal rosette, simple. **INFL**: raceme, spike, 1-fld or scapose. **FL** bisexual; calyx 2-lipped, gen 4–5-lobed; corolla 2-lipped, 5-lobed, spurred at base, lower lip arched and gen pouched upward, blocking throat; fertile stamens 2, epipetalous; ovary superior, chamber 1, placenta gen free-central; stigma unequally 2-lobed, ± sessile. **FR**: capsule, round, gen 2-valved (sometimes irregularly dehiscent). **SEEDS** gen many, small. 4 genera, 200 spp.: worldwide, esp trop.

1. Pl of moist ground; lvs in rosette, elliptic to ovate; corolla blue-violet or white; calyx lips 2- or 3-lobed; infl 1-fld, bracts 0 . . . . **PINGUICULA**
1′ Pl gen aquatic; "lvs" alternate, dissected into narrow segments; corolla yellow; calyx lips unlobed; infl gen several-fld, bracts present . . . . **UTRICULARIA**

PINGUICULA BUTTERWORT

Per of moist ground. **LVS** simple, in basal rosette, fleshy; margins entire, incurving over trapped prey; upper surface sticky or slimy, with stalked glands that capture small organisms, sessile glands that digest them. **INFL** 1-fld; bracts 0. **FL**: calyx 5-lobed, upper lip 3-lobed; corolla lower lip gen not blocking throat. **FR** 2-valved. ± 50 spp.: Am, Eur, Medit. (Latin: somewhat fat, from greasy lf surface) [Casper 1966 Biblioth Bot 127/128:1–209]

P. vulgaris L. ssp. ***macroceras*** (Link) Calder & Roy Taylor (p. 737) HORNED BUTTERWORT **INFLS** 1–5 per rosette, 1–2 dm. **LF** 2–5 cm, narrowly oblong to ovate. **FL**: corolla (incl spur) 2–3 cm, blue-violet (rarely white); lobes of lower lip obovate, margins overlapping, throat hairy, spur 6–9 mm. RARE in CA. Moist serpentine banks; < 500 m. n KR (n Del Norte, w Siskiyou cos.); sw OR. [*P. macroceras* Link] Not always distinct from mostly circumboreal ssp. *v.*, which has $2n$=64. A subalpine, nonserpentine race is found from e OR to AK, w Can, Japan. In cult.

UTRICULARIA BLADDERWORT

Ann, per, gen aquatic. **ST** submersed or creeping, uncoiling at tip, sometimes with claw-like appendages, with gen alternate green branch systems ("lvs") of ± linear or thread-like segments; bladders borne on st or "lvs" trap small organisms when entrance hairs triggered; terminal, overwintering buds dispersed in some spp. **LVS**: true lvs 0. **INFL**: raceme (rarely 1-fld), bracted, emergent. **FL**: calyx lips entire; corolla gen yellow, lower lip entire or 3-lobed, gen blocking throat, hairy, red-spotted, upper lip ± entire. **FR** gen circumscissile (or opening irregularly). ± 180 spp.: worldwide, esp. trop. (Latin: little bag, from bladders) [Ceska & Bell 1973 Madroño 22: 74–84] ❀Pure water ponds; DFCLT.

1. "Lf" segment margins and winter-bud scales, if present, lacking bristles ("lf" segments of *U. minor* may be bristle-tipped); corolla < 9 mm
 2. "Lf" gen 2-parted at base, each part forked again or not; ultimate segments < 6, thread-like; pedicel erect in fr; corolla lower lip ± = upper, spur conic, less wide than long; seed winged . . . . **U. gibba**
 2′ "Lf" gen 3-parted at base, each part forked 1–3 times; ultimate segments > 6, flat; pedicel recurved in fr; corolla lower lip ± 2 5 upper, spur sac-like, as wide as long; seed wingless . . . . **U. minor**
1′ "Lf" segment margins and winter-bud scales with fine, non-green bristles; corolla > 9 mm
 3. "Lf" 3-parted at base, each part unequally 1–3-forked; ultimate segments < 20, linear, flat; bladders borne on lfless sts; pedicel erect in fr; corolla lower lip ± 2 5 upper . . . . **U. intermedia**
 3′ "Lf" 1–2-parted at base, each part several times ± pinnately branched; ultimate segments > 20, thread-like; bladders borne on "lvs"; pedicel recurved in fr; corolla lower lip ± = upper . . . . **U. vulgaris**

U. gibba L. (p. 737) **STS** floating or creeping; winter buds 0; "lvs" 2–15 mm, gen 2-parted at base, each part forked again or not; ultimate segments < 6, thread-like; bladders on st and lvs. **INFL** 1–4-fld; peduncle < 15 cm, < 1 mm wide; pedicel erect in fr. **FL**: corolla 6–8 mm, lips subequal, = conic spur. **SEED** winged. $2n$=28. Uncommon. Shallow water, mud; < 1600 m. n&c CA-FP; to B.C., e N.Am, trop. [*U. fibrosa* Walter; *U. biflora* Lam.] Highly variable; sometimes cult, may be invasive; probably alien in CA.

U. intermedia Hayne (p. 737) **STS** both creeping (with bladders, without "lvs") and floating (with "lvs," gen without bladders); winter buds bristly; "lvs" dense where present, 3–15 mm, gen 3-parted at base, each part unequally 1–3-forked; ultimate segments < 20, ± linear, flat, margin bristly. **INFL** 1–5-fld; peduncle < 25 cm, < 1 mm wide; pedicel erect in fr. **FL**: corolla 10–18 mm; lower lip ± 2 × upper, slightly < cylindric, obtuse spur. **FR** rarely produced. **SEED** not seen. $2n$=44. Uncommon. Shallow water; 1200–2700 m. CaR, SNH, MP; to AK, UT, e N.Am; circumboreal.

U. minor L. (p. 737) **STS** both floating (with densely arranged "lvs" bearing some bladders) and creeping (with fewer "lvs", more bladders); winter buds glabrous; "lvs" 3–30 mm, gen 3-parted at base, each part irregularly 1–3-forked; ultimate segments gen < 16, linear, flat, sometimes bristle-tipped. **INFL** 2–9-fld; peduncle 4–15 cm, wiry, < 0.5 mm wide; pedicel recurved in fr. **FL**: corolla 6–8 mm; lower lip 2 × upper, 2 × sac-like spur, scarcely blocking throat. **SEED** unwinged. $2n$=±40. Uncommon. Shallow water; 800–2900 m. CaRH, SNH, MP; to AK, Colorado, e N.Am; circumboreal.

U. vulgaris L. (p. 737) COMMOM BLADDERWORT **STS** shallowly floating; winter buds 1–2 cm, bristly; "lvs" dense, 2–9 cm, 1–2-parted at base, each part several times unequally pinnately dissected; ultimate segments dense, 20–150, thread-like, not flat, margin bristly; bladders near "lf" base > those near tip. **INFL** 5–20-fld; peduncle 1–4 dm, stout, < 2 mm wide; pedicel recurved in fr. **FL**: corolla 1–2 cm; lower lip ± = upper, slightly > curved, cylindric, pointed spur. **SEED** unwinged. $2n$=±40. Quiet water; < 2700 m. NW, CaR, SN, SnFrB, SnBr, GB, w DMoj; to AK, e N.Am, Mex; circumboreal.

LIMNANTHACEAE MEADOWFOAM FAMILY

Robert Ornduff

Ann, glabrous to hairy. **ST** gen branched. **LVS** alternate, deeply pinnately lobed to 1–2-compound; stipules 0. **INFL**: fls solitary in axils, peduncled. **FL** gen bisexual, radial; sepals 3–5, free; petals 3–5, free, white to pink or yellow; stamens 3, 8, or 10, free, gen in 2 whorls; nectary glands at bases of outer stamens; pistils 2–5, ± free, 1-ovuled, styles fused exc sometimes at tip. **FR**: nutlets 1–5, ovoid to spheric, gen tubercled. 2 genera, 10 spp.: temp N.Am. $2n$=10 for all spp.

1. Petals gen 3, < the gen 3 sepals, tips ± entire .... **FLOERKEA**
1′ Petals 5, gen > the 5 sepals, tips toothed or jagged .... **LIMNANTHES**

FLOERKEA FALSE MERMAID

1 sp.: N.Am. (H.G. Floerke, German botanist, 1764–1835)

F. proserpinacoides Willd. (p. 737) Ann, decumbent to erect, fleshy, glabrous. **ST** < 20 cm. **LF** 1-ternate or 1-odd-pinnate, < 6 cm; lflets 3–5, < 2 cm, ± oblong. **FL**: sepals gen 3, 2–4 mm; petals gen 3, spoon-shaped, white, tips ± entire; stamens 3–6; pistils 2–3. **FR**: nutlets 2–3, 2–3.5 mm, ± spheric, wrinkled or tubercled above. Moist open places in coniferous forest or sagebrush scrub. 1500–3200 m. KR, NCoRH, CaRH, SNH, MP; to B.C., e N.Am. Fls late spring to summer.

LIMNANTHES MEADOWFOAM

Ann, decumbent to erect. **LF** gen 1-odd-pinnately lobed or compound; lobes or lflets entire to deeply lobed. **FL**: sepals 4–5; petals 4–5, gen > sepals, tips toothed or jagged; stamens 8 or 10; pistils 4–5. **FR**: nutlets 1–5, smooth or tubercled. 9 spp.: ± coastal w N.Am. (Greek: marsh fl, from habitat) [Mason 1952 Univ Calif Publs Bot 25:455–512] Sect. *Inflexae* merits critical review. Fls spring.

1. Petals curving over fr as fr matures (sect. *Inflexae*)
 2. Petals < to ± > sepals .... ***L. floccosa***
  3. Fl cup-shaped; fr papillate, gen with wide-conic tubercles .... ssp. ***californica***
  3′ Fl bell- to urn-shaped; fr with awl-shaped tubercles, also gen with bristles .... ssp. ***floccosa***
 2′ Petals > sepals
  4. Stamens 2.5–4 mm
   5. Petals aging pink; anthers ± 1 mm; PR .... ***L. gracilis*** ssp. ***parishii***
   5′ Petals aging white; anthers 0.5–1 mm; c&s SNF .... ***L. montana***
  4′ Stamens 5–6 mm .... ***L. alba***
   6. Sepals densely hairy; fr widely ridged; s NCoR, n&c SNF, GV .... ssp. ***alba***
   6′ Sepals glabrous or sparsely hairy; fr smooth or sharply tubercled; CaRF, n&c SNF, n SNH ...... ssp. ***versicolor***
1′ Petals reflexing as fr matures (exc rarely) (sect. *Reflexae*)
 7. Stamens 2–4 mm; petals 7–15 mm
  8. Lflets gen entire (lower rarely 2–3-lobed); petals 7–9 mm; c NCoRO (near Willits, Mendocino Co.) ... ***L. bakeri***
  8′ Lflets entire to 2–3-lobed; petals 8–15 mm; KR, n&c SNF .... ***L. striata***
 7′ Stamens 5–8 mm; petals 10–18 mm
  9. Lflets 3–5, entire .... ***L. vinculans***
  9′ Lflets 5–13, entire or often toothed or lobed .... ***L. douglasii***
   10. Petals yellow or yellow with white tips
    11. Petals yellow with white tips .... ssp. ***douglasii***
    11′ Petals yellow .... ssp. ***sulphurea***
   10′ Petals mostly white, not at all yellow (exc rarely when dry)
    12. Petals white, veins white or purplish; nutlets ± tubercled above or smooth; NCoR, SnFrB, SCoR .. ssp. ***nivea***
    12′ Petals white, veins rose; nutlets ridged; NCoRI, CaRF, c SNF, GV .... ssp. ***rosea***

L. alba Benth. (p. 737) **ST** < 30 cm. **LF** < 10 cm; lflets 5–9, < 15 mm, linear to ovate, entire to deeply 3-lobed. **FL** cup- to bell-shaped; sepals 7–8 mm; petals 10–15 mm, white; stamens 5–6 mm, anthers 2 mm; style 5–6 mm. **FR**: nutlets, 3–4 mm, obovoid, smooth, wrinkled, widely ridged, or sharply tubercled. Winter-wet grasslands, woodlands, edges of vernal pools, ephemeral streams; < 1800 m. s NCoR, CaRF, n&c SNF, n SNH, GV.

ssp. ***alba*** Herbage sparsely to densely hairy when young. **FL**: sepals densely hairy. **FR** widely ridged. Winter-wet grasslands, woodlands; < 1400 m. s NCoR, n&c SNF, GV. ❀TRY.

ssp. ***versicolor*** (E. Greene) C. Mason Herbage glabrous. **FL**: sepals glabrous to sparsely hairy. **FR** smooth or sharply tubercled. Winter-wet grasslands, woodlands, edges of vernal pools, ephemeral streams; < 1800 m. CaRF, n&c SNF, n SNH.

L. bakeri J. Howell (p. 737) BAKER'S MEADOWFOAM **ST** < 40 cm. **LF** < 10 cm; lflets 3–9, < 1.5 cm, elliptic to ovate, entire or lower rarely 2–3-lobed. **FL** funnel- to bell-shaped; sepals 5–7 mm; petals 7–9 mm, pale yellow with white tips; stamens 3–4 mm, anthers 1 mm; style 3–4 mm. **FR**: nutlets, 3–3.5 mm, obovoid, tubercles dense, short, wide. **RARE** CA. Vernal pools, marshy margins; < 500 m. c NCoRO (near Willits, Mendocino Co.). Possibly threatened by grazing.

L. douglasii R.Br. **ST** < 50 cm. **LF** < 25 cm; lflets 5–13, < 3 cm, entire or often toothed or lobed. **FL** cup- to bell-shaped; sepals 5–15 mm; petals 10–18 mm, white, yellow, or yellow with white tips, veins often purple, pink, or cream; stamens 5–8 mm, anthers 1–2 mm; style 5–7 mm. **FR**: nutlets, 2.5–5 mm, obovoid, smooth, tubercled at least at tip, wrinkled or not. Wet meadows, edges of vernal pools, ephemeral streams; < 1000 m. NCo, NCoR, CaRF, c SNF, GV, CCo, SnFrB, SCoR; sw OR.

ssp. ***douglasii*** (p. 743) **LF**: lflets linear to ovate, entire or irregularly toothed or lobed. **FL**: petals yellow with white tips. Wet meadows; < 700 m. NCo, NCoRO, CCo, SnFrB; sw OR. ❀winter WET or IRR,SUN:4–6,**7–9**,**14–17**,19,**20–24**.

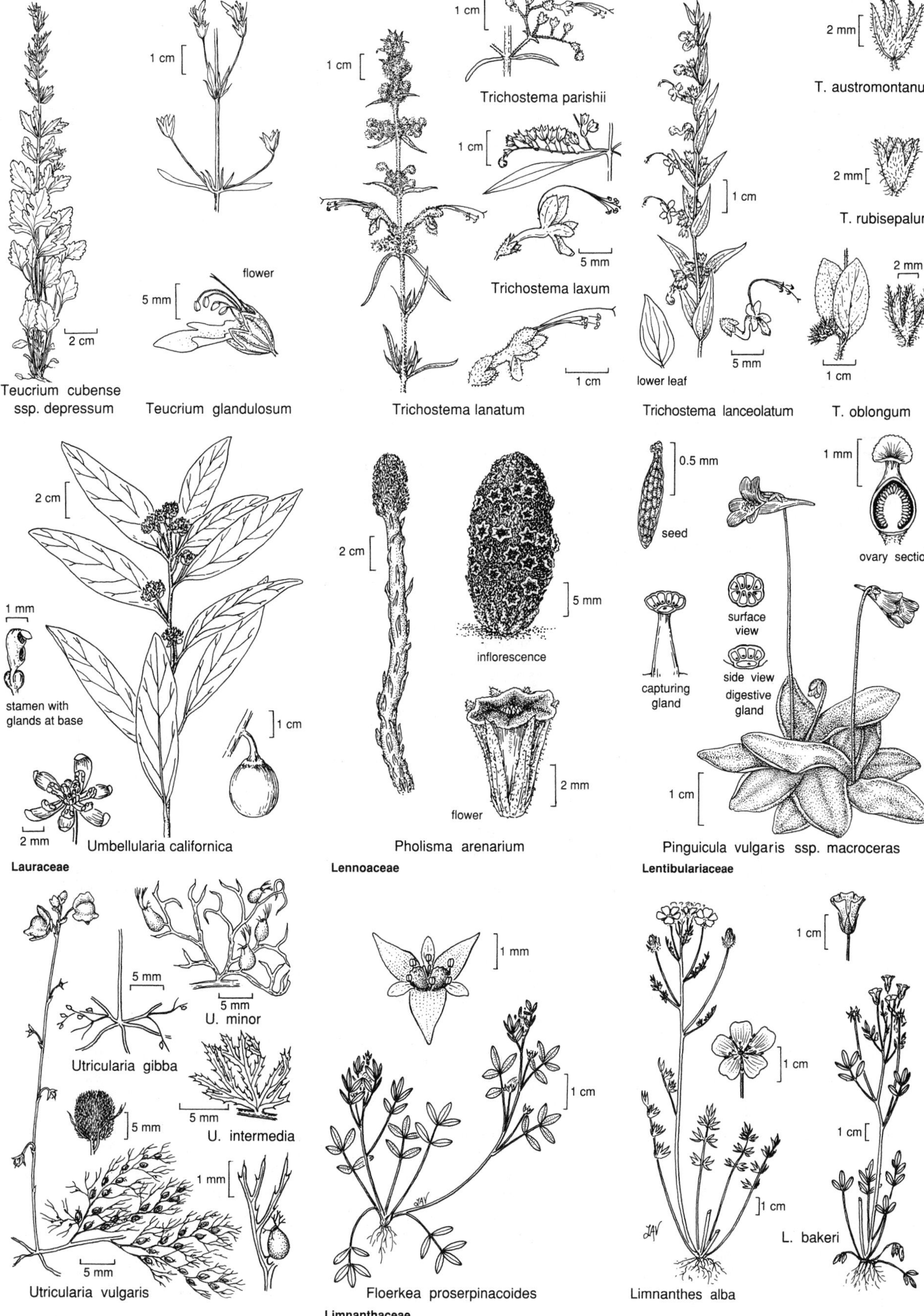
1 cm
2 cm
Teucrium cubense
ssp. depressum
1 cm
flower
5 mm
Teucrium glandulosum
1 cm
1 cm
Trichostema parishii
1 cm
5 mm
Trichostema laxum
1 cm
Trichostema lanatum
1 cm
5 mm
lower leaf
Trichostema lanceolatum
2 mm
T. austromontanum
2 mm
T. rubisepalum
2 mm
1 cm
T. oblongum
2 cm
1 mm
stamen with
glands at base
1 cm
2 mm
Umbellularia californica
Lauraceae
2 cm
5 mm
inflorescence
2 mm
flower
Pholisma arenarium
Lennoaceae
0.5 mm
seed
1 mm
ovary section
surface
view
side view
capturing
gland
digestive
gland
1 cm
Pinguicula vulgaris ssp. macroceras
Lentibulariaceae
5 mm
5 mm
U. minor
Utricularia gibba
5 mm
5 mm
U. intermedia
1 mm
5 mm
Utricularia vulgaris
1 mm
1 cm
Floerkea proserpinacoides
Limnanthaceae
1 cm
1 cm
1 cm
1 cm
L. bakeri
Limnanthes alba

ssp. ***nivea*** (C. Mason) C. Mason **LF**: lflets linear to ovate, entire to deep-lobed. **FL**: petals white, drying pale yellow at base or not, veins white or purplish. **FR**: nutlets smooth or ± tubercled above. Wet meadows, edges of vernal pools, ephemeral streams; < 1000 m. NCoR, SnFrB, SCoR. ❀winter WET or IRR,SUN:4–9,**14–17**,18,**19–24**.

ssp. ***rosea*** (Benth.) C. Mason **LF**: lflets linear to wide-ovate, entire to deeply and irregularly lobed. **FL**: petals white (sometimes drying pale pink or yellow), veins rose. **FR**: nutlets ridged. Wet meadows, edges of vernal pools; < 120 m. NCoRI, CaRF, c SNF, GV. ❀winter WET or IRR,SUN:4,5,**6–9**,**14–17**,19,**20–24**.

ssp. ***sulphurea*** (C. Mason) C. Mason POINT REYES MEADOWFOAM **LF**: lflets ovate, irregularly toothed or lobed. **FL**: petals yellow. **ENDANGERED** CA. Wet meadows of coastal prairie; < 100 m. NCo (Marin Co.), CCo (San Mateo Co.). In cult.

L. floccosa Howell (p. 743) WOOLLY MEADOWFOAM **ST** < 25 cm. **LF** < 8 cm; lflets 4–10, < 1 cm, linear to ovate, entire, toothed, or lobed. **FL** urn-, cup-, or bell-shaped; sepals 4–10 mm; petals 4.5–10 mm, < to ± > sepals, white; stamens 2.5–8.5 mm, anthers 0.4–1.5 mm; style 1.5–4 mm. **FR**: nutlets, 1–5, 3–4.5 mm, obovoid, tubercles < 1 mm or 0. UNCOMMON. Moist meadows, vernal pools; 10–400 m. KR, NCoRI (near Kelseyville, Lake Co.), CaR, ScV. In CA nutlets fall together, with calyx.

ssp. ***californica*** Arroyo SHIPPEE MEADOWFOAM **FL** cup-shaped; sepals 7.5–10 mm; petals 8–10 mm; filaments 3–7 mm, anthers ± 1–1.5 mm, dehiscing outward; style 3.5–4 mm. **FR:** nutlets, 3–5, papillate, gen with wide-conic tubercles. **ENDANGERED** CA. Vernal pool edges; < 100 m. ScV (Butte Co.).

ssp. ***floccosa*** **FL** bell- to urn-shaped; sepals 4–9 mm; petals 4.5–8.5 mm; anthers 0.4–±1 mm, gen dehiscing inward; style 1.5–3 mm. **FR:** nutlets, 1–3, with awl-shaped tubercles, also gen with bristles down-pointed. Moist meadows, vernal pools; < 400 m. KR, CaR, NCoRI (near Kelseyville, Lake Co.). [ssp. *bellingeriana* (M. Peck) Arroyo, Bellinger's meadowfoam, reported from CaRF but requires further study]

L. gracilis Howell ssp. ***parishii*** (Jepson) Beauch. PARISH'S MEADOWFOAM **ST** < 30 cm. **LF** < 8 cm; lflets < 9, < 15 mm, lanceolate to ovate, entire to 3-lobed. **FL** cup-shaped; sepals 4–8 mm; petals 8–10 mm, white, base cream or not, aging pink, tips recurved; stamens 2.5–3.5 mm, anthers ± 1 mm; style 2–3 mm. **FR**: nutlets, 3 mm, obovoid, tubercles short, wide. **ENDANGERED** CA. Wet meadows, ephemeral stream edges; 600–2000 m. PR.

L. montana Jepson **ST** < 40 cm. **LF** < 15 cm; lflets < 11, < 15 mm, linear to ovate, entire to deeply 3-lobed, lobes entire to lobed. **FL** funnel- to bell-shaped; sepals 3–6 mm; petals 7–10 mm, white, base sometimes yellowish or veins purplish, aging white; stamens 2.5–4 mm, anthers 0.5–1 mm; style 3 mm. **FR**: nutlets, 2–3 mm, obovoid, tubercles low, wide. Wet meadows, stream edges; < 1700 m. c&s SNF. ❀TRY.

L. striata Jepson **ST** < 30 cm. **LF** < 15 cm; lflets < 11, < 20 mm, linear to ovate, entire to toothed or 2–3-lobed. **FL** funnel-shaped; sepals 4–6 mm; petals 8–15 mm, white, bases yellow, veins often dark; stamens 2–4 mm, anthers 0.8–1 mm; style 3–4 mm. **FR**: nutlets, 2.5–3 mm, obovoid, smooth, wrinkled, or tubercled. Vernal pools, stream edges; < 800 m. KR, n&c SNF. Some KR pls are esp small. ❀TRY.

L. vinculans Ornd. SEBASTOPOL MEADOWFOAM **ST** < 30 cm. **LF** < 10 cm; lflets 3–5, 5–20 mm, entire, narrow-obovate. **FL** bell-shaped to rotate; sepals 6–7 mm; petals 12–18 mm, white, base drying yellowish; stamens 5–7 mm, anthers 1.5–2 mm; style 4.5–6.5 mm. **FR**: nutlets, 3–4 mm, obovoid, tubercles dense, short, wide. **ENDANGERED** CA. Wet meadows; < 300 m. s NCoRO (s Sonoma Co). Threatened by urbanization, agriculture. In cult.

LINACEAE FLAX FAMILY

Niall F. McCarten

Ann, per, shrub. **ST** gen erect, branched, glabrous, hairy, or glandular. **LVS** gen cauline, alternate, opposite, or whorled, simple, sessile, gen linear to ovate; stipules glandular or 0. **INFL**: raceme or cyme, axillary, open to dense. **FL** gen bisexual, radial, nodding in bud; sepals 4–5, free, glabrous, hairy, or margins gland-toothed; petals 4–5, free, blue, white, yellow, or pink, ephemeral; stamens 4–5, alternate petals, gen appendaged; staminodes alternate stamens or 0; ovary superior, chambers 2–5 but becoming 4–10 by growth of false septa, styles 2–5. **FR**: gen capsule. 13 genera, 300 spp.: worldwide, esp temp; some cult (*Linum usitatissimum*, flax, linseed; *L. bienne, L. grandiflorum*, orn). [Robertson 1971 J Arnold Arbor 52: 649–665]

1. Styles 5; ovary chambers 10; petals 8–25 mm; fr 5–10 mm, 10-seeded .... **LINUM**
1′ Styles 2–3; ovary chambers 4 or 6; petals gen 1–6(12) mm; fr 2–4 mm, 4- or 6-seeded
  2. Lvs gen alternate (or opposite or whorled near st base); stigmas ± = styles in width; petals with scales at base on inner surface; stipules glandular, sometimes minute; styles 2 or 3 .... **HESPEROLINON**
  2′ Lvs opposite; stigmas > styles in width; petals without scales on inner surface; stipules 0; styles 2 .... **SCLEROLINON**

HESPEROLINON DWARF FLAX

Ann. **ST** 5–50 cm. **LVS** gen alternate, gen thread-like to linear, gen flat, gen not clasping. **INFL**: cyme, gen open; pedicels thread-like, gen ascending to erect. **FL**: sepals 5, glabrous, hairy, or margins minutely gland-toothed; petals 5, gen widely spreading, 1–12 mm, yellow or white to rose, each with 3 small scales at inner base; stamens 5; staminodes 0; ovary chambers 4 or 6, styles 2–3, stigmas ± = styles in width. **FR**: surface smooth. 13 spp.: esp CA; mostly localized, mostly on serpentine. (Greek & Latin: western flax) [Sharsmith 1961 Univ Calif Publ Bot 32:235–314] Gen self-pollinated. Segregated from *Linum*.

1. Lvs lanceolate to ovate, clasping, margin clearly glandular or gland-toothed
  2. Petals yellow; lvs whorled or opposite at base, alternate above base, early deciduous, lanceolate, keeled, margin with 1 or 2 rows of gland-tipped teeth, pedicels gen 5–15 mm .... ***H. adenophyllum***
  2′ Petals white to pink; lvs whorled at base, opposite above base, persistent in fl, ovate, flat, margin with 1 row of small glands, pedicels 1–5 mm .... ***H. drymarioides***
1′ Lvs thread-like to linear, not clasping, margins minutely glandular
  3. Styles 2 (ovary chambers 4)
    4. Petals yellow .... ***H. bicarpellatum***
    4′ Petals white or light pink .... ***H. didymocarpum***

3′ Styles 3 (ovary chambers 6)
 5. Petals, anthers, and styles yellow
  6. Petals 0.5–1.5 mm, styles gen < 1 mm, incl, petals and filaments gen = sepals .... ***H. clevelandii***
  6′ Petals 2–12 mm, styles 2–10 mm, exserted, petals and filaments gen >> sepals
   7. Infl short, dense, pedicels 0.2–2 mm, petals 4–10 mm .... ***H. breweri***
   7′ Infl long, open, pedicels 2–10 mm, petals 2–5.5 mm
    8. Herbage hairy only at nodes, petals 2–3.5 mm, tip slightly notched .... ***H. serpentinum***
    8′ Herbage hairy, petals 3.5–5.5 mm, tip deeply notched .... ***H. tehamense***
 5′ Petals, anthers, and styles white to deep purple
  9. Infl short, dense, pedicels 1–8 mm
   10. Sepals glabrous outside, petals white to pink-veined .... ***H. californicum***
   10′ Sepals hairy outside, petals light to dark pink .... ***H. congestum***
  9′ Infl long, open, pedicels 1–25 mm
   11. Petals 0.5–3.5 mm, styles gen < 1 mm, incl, petals and filaments gen = sepals .... ***H. micranthum***
   11′ Petals 3–7 mm, styles 3–6 mm, gen exserted, petals and filaments gen >> sepals
    12. Pedicels straight or erect, 1–5 mm, petals 3–6 mm .... ***H. disjunctum***
    12′ Pedicels gen pendent or ± reflexed, 6–25 mm, petals 4–7 mm .... ***H. spergulinum***

H. adenophyllum (A. Gray) Small (p. 743) GLANDULAR DWARF FLAX Pl 10–50 cm. **LVS** opposite below, lanceolate, keeled, clasping; margins (and bracts) with stalked, gland-tipped teeth in 1–3 rows. **INFL**: pedicels gen 5–15 mm. **FL**: sepals 2–3 mm; petals 3–5 mm, yellow, often veined or tinged orange, fading white; stamens 4–5.5 mm, anthers yellow; ovary chambers 6, styles 3, 2.5–4 mm, yellow. RARE. Serpentine, chaparral; 150–1000 m. n&c NCoR (esp Lake, Mendocino cos.).

H. bicarpellatum (H. Sharsm.) H. Sharsm. TWO-CARPELLATE DWARF FLAX Pl 10–30 cm, gen ± glabrous. **ST** hairy above nodes. **LVS** alternate, 10–25 mm, thread-like to linear, flat, not clasping. **INFL** ± equally forked; pedicels 2–12 mm. **FL**: sepals 1.5–3 mm; petals 2–4 mm, yellow; stamens 3–5 mm, anthers yellow; ovary chambers 4, styles 2, 2–3.5 mm, yellow. *n*=17. RARE. Serpentine, chaparral; 60–1000 m. s NCoRI.

H. breweri (A. Gray) Small BREWER'S DWARF FLAX Pl 5–20 cm, ± glabrous. **LVS** alternate, linear. **INFL** short, dense; pedicels gen < 3 mm. **FL**: sepals 3–4 mm, margins gland-toothed; petals 4–10 mm, yellow; stamens 5.5–10 mm, anthers yellow; ovary chambers 6, styles 3, 3–8 mm, yellow. *n*=18. RARE. Chaparral or grassland, sometimes on serpentine; 30–700 m. s NCoRI (Napa, Solano cos.), ne SnFrB (Mount Diablo area, Contra Costa Co.).

H. californicum (Benth.) Small (p. 743) Pl 10–25 cm. **LVS** alternate, thread-like to linear; stipule glands well developed, with red exudate. **INFL**: pedicels 1–5(10) mm. **FL**: sepals 3.5–4 mm, glabrous exc minutely gland-toothed on margins; petals 4–12 mm, white or partly pink; stamens 6–11 mm, anthers white to rose; ovary chamber 6, styles 3, 4–10 mm, whitish. *n*=18. Rocky areas, chaparral, grassland, sometimes on serpentine; 30–1300 m. c&s NCoRI, s CaRF, n SNF, e SnFrB, ne SCoRI. ❀TRY.

H. clevelandii Small Pl 5–20 cm, ± glabrous. **LVS** alternate, linear. **INFL** unequally forked (main axes obvious); pedicels 5–25 mm. **FL**: sepals 1.5–2.5 mm; petals 1.5–2.5(4) mm, yellow; stamens 1.5–3 mm, anthers yellow; ovary chambers 6, styles 3, gen 0.5–1 mm, yellow. *n*=18. Chaparral margins, oak woodland, not on serpentine; 150–1400 m. c&s NCoRI, e SnFrB. Much like *H. micranthum* exc fl color. ❀TRY.

H. congestum (A. Gray) Small (p. 743) MARIN DWARF FLAX Pl gen 5–15 cm. **LVS** alternate, linear, ± glabrous; stipule glands well developed, with red exudate. **INFL** dense; pedicels 0.5–8 mm. **FL**: sepals 3–4 mm, hairy, margins minutely glandular; petals 3–8 mm, pink to rose; stamens 5.5–7 mm, anthers deep pink to purple; ovary chambers 6, styles 3, 3–5 mm, whitish. *n*=18. RARE. Serpentine, grassland; < 200 m. nw SnFrB.

H. didymocarpum H. Sharsm. (p. 743) LAKE COUNTY DWARF FLAX Pl 10–30 cm, glabrous exc just above nodes. **LVS** alternate, 10–25 mm, thread-like to linear, flat, not clasping. **FL**: sepals 2–3 mm; petals 2.5–4 mm, white or light pink; stamens 3.5–5 mm, anthers white to purplish; ovary chambers 4, styles 2, 2.5–4 mm, white. *n*=17. **ENDANGERED** CA. Serpentine, chaparral, grassland; 100–200 m. NCoRI (Big Canyon Creek, near Middletown, Lake Co.). Much like *H. bicarpellatum* exc fl color.

H. disjunctum H. Sharsm. Pl 3–30 cm, stout. **LVS** alternate, linear; stipule glands minute. **INFL**: pedicels gen 1–5 mm. **FL**: sepals 2–3 mm, glabrous exc margins minutely gland-toothed; petals 3–6 mm, white, veins often pink; stamens gen 4–6 mm, anthers pink (or rose margined white); ovary chambers 6, styles 3, 2.5–5 mm, whitish. Serpentine, chaparral; 100–1000 m. NCoRI, e SnFrB, SCoRI. Parts of geog range separated.

H. drymarioides (Curran) Small (p. 743) DRYMARIA-LIKE DWARF FLAX Pl 5–25 cm. **LVS** whorled (gen in 4's) below, alternate or subopposite above, ovate, flat, entire; surfaces and margins with stalked glands. **INFL** few-branched; pedicels 1–5(10) mm. **FL**: sepals 2.5–4 mm; petals 3–6 mm, white to pink with darker veins, reflexed; stamens 3.5–5 mm, anthers white (or purple margined white); ovary chambers 6, styles 3, 2–3.5 mm, whitish. RARE. Serpentine, chaparral or woodland; 100–1000 m. c NCoRI (Colusa, Glenn, Lake cos.).

H. micranthum (A. Gray) Small Pl 5–20 cm. **LVS** alternate, linear. **INFL**: pedicels gen 5–15(25) mm, ascending to ± pendent. **FL**: sepals 1–4 mm, sometimes very unequal, glabrous; petals 0.5–3.5 mm, white to pink or streaked rose; stamens 2–3 mm, anthers white to deep purple; ovary chambers gen 6, styles 3 (very rarely 2), 0.5–2 mm. *n*=18. Open areas, woodland margins, sometimes on serpentine; 50–2000 m. CA-FP; OR, n Baja CA. Variable; some forms very local.

H. serpentinum N. McCarten NAPA DWARF FLAX Pl 10–30 cm, gen ± glabrous. **ST** hairy above nodes. **LVS** gen alternate, 10–30 mm, thread-like to linear, flat, not clasping. **INFL** ± equally forked; pedicels 2–10 mm. **FL**: sepals 1–3 mm; petals 2–3.5 mm, yellow; stamens 3–5 mm, anthers yellow; ovary chambers 6, styles 3, 2–4 mm, yellow. *n*=17. RARE. Serpentine, chaparral; 50–800 m. s NCoRI (esp Napa & Lake cos.).

H. spergulinum (A. Gray) Small (p. 743) Pl 10–30 cm, nearly glabrous. **LVS** alternate, linear. **INFL** ± raceme-like, widely spreading; pedicels 6–25 mm, gen pendent or ± reflexed. **FL**: sepals 1.5–3.5 mm, glabrous exc margins minutely gland-toothed; petals 4–7 mm, white or pale pink, veins often darker; stamens 4–9 mm, anthers pink to red-purple margined white; ovary chambers 6, styles 3, 3.5–7 mm, whitish. *n*=18. Chaparral or woodland margins, incl serpentine; 100–1000 m. c&s NCoR, w SnFrB.

H. tehamense H. Sharsm. Pl 2–50 cm, hairy. **LVS** alternate, ± linear. **INFL**: pedicels 0.5–3 mm. **FL**: sepals 2–3 mm; petals 3.5–5.5 mm, ± 1/2 recurved, light or bright yellow, veins sometimes reddish, tip notched; stamens 4–5.5 mm, anthers yellow; ovary chambers 6, styles 3, 3–10 mm, yellow. *n*=18. RARE. Serpentine, chaparral; 100–1000 m. n&c NCoRI (Tehama, Glenn cos.).

LINUM FLAX

Ann, per. **ST** 5–90 cm. **LVS** alternate, opposite, or whorled, erect, glabrous or hairy; margins entire or gland-toothed. **INFL**: raceme or cyme. **FL**: sepals 5, margins gen translucent; petals 5, 8–25 mm; stamens 5; staminodes 5 or 0; ovary chambers 10, styles 5, free or not, stigmas > styles in width (spheric or ± elongate) or = styles in width (± linear). **FR** 5–10 mm. **SEEDS** 10, gen gelatinous when wet. ± 200 spp.: temp & subtrop, esp Medit. (Latin: flax)

1. Petals red; staminodes pink; sepals lanceolate, ciliate; stigmas ± = styles in width .... ***L. grandiflorum***
1′ Petals blue, white, yellow, or orange; staminodes white; sepals linear to ovate, margins gen glabrous or gland-toothed; stigmas > styles in width
 2. Petals yellow to orange; margins of lvs, bracts, and sepals gland-toothed; styles fused to near tip ... ***L. puberulum***
 2′ Petals blue to white; margins of lvs, bracts, and sepals gen entire, glabrous; styles free or fused only near base
  3. Stigmas ± spheric .... ***L. lewisii***
   4. Petals 6–7 mm; styles < 6 mm .... var. ***alpicola***
   4′ Petals 10–15 mm; styles gen > 6 mm .... var. ***lewisii***
  3′ Stigmas elongate
   5. Petals 8–10 mm; pedicels 5–18 mm; sepals 4–6 mm; fr 4–5 mm .... ***L. bienne***
   5′ Petals 10–15 mm; pedicels 20–25 mm; sepals 6–10 mm; fr 7–10 mm .... ***L. usitatissimum***

L. bienne Miller Per. **ST** glabrous. **LF** 5–25 mm, linear. **INFL**: pedicels 5–18 mm. **FL**: sepals 4–6 mm, glabrous, rarely ciliate; petals 8–10 mm, blue; styles free, stigmas > styles in width, elongate. **SEED** 2–3 mm. *n*=15. Grasslands, woodlands; 0–1000 m. NCo, CCo; to OR; native to Medit. [*L. angustifolium* Hudson]

L. grandiflorum Desf. Ann. **ST** glabrous. **LF** 10–20 mm, lanceolate, glaucous. **INFL**: pedicels 20–70 mm. **FL**: sepals 7–8 mm, lanceolate, ciliate; petals 15–30 mm, red; anthers blue; staminodes pink; styles free, stigmas ± = styles in width, elongate. **FR** < sepals. Uncommon. Grasslands; < 150 m. CCo; native to n Afr.

L. lewisii Pursh (p. 743) Per. **ST** glabrous. **LF** 10–20 mm, linear to lanceolate, glabrous. **INFL**: pedicels 10–30 mm. **FL**: sepals 4–6 mm, ovate; petals 6–15 mm, blue, sometimes white; styles free, stigmas ± spheric. **FR** 5–8 mm. **SEED** 3.5–4.5 mm. Dry open ridges, slopes; 400–3400 m. CA-FP, D; to AK, Rocky Mtns, TX, Mex.

var. ***alpicola*** Jepson **FL**: petals 6–7 mm; styles < 6 mm. Alpine ridges; 2000–3400 m. SNH; to UT.

var. ***lewisii*** **FL**: petals 10–15 mm; styles > 6 mm. Open slopes; 400–2000 m. CA-FP; OR. ❀DRN,SUN:1,**4–6,15–17**&IRR:2,3,**7**,8,9,**14,18–24**.

L. puberulum (Engelm.) A.A. Heller (p. 743) Per. **ST** puberulent. **LF** 5–10 mm, linear; margin gland-toothed. **INFL**: bract margins gland-toothed. **FL**: sepals 4–7 mm, < or = fr, lanceolate, margins gland-toothed; petals 10–15 mm, yellow to orange; styles fused to near tip. **FR** 3.5–4 mm. **SEED** ± 2.5 mm. *n*=15. Dry ridges; 1000–2500 m. e DMtns; to Colorado, TX. ❀TRY.

L. usitatissimum L. (p. 743) COMMON FLAX Ann. **ST** glabrous. **LF** 10–35 mm, lanceolate, entire, gen glabrous. **INFL**: pedicels 20–25 mm. **FL**: sepals 6–10 mm; petals 10–15 mm, blue or white; styles fused at base, stigmas elongate. **FR** 7–10 mm. **SEED** 3.5–4.5 mm. Disturbed areas; 0–100 m. CA–FP; to e N.Am; native to Eur.

SCLEROLINON

1 sp. (Greek & Latin: hard flax, from fr walls) [Rogers 1966 Madroño 18:181–184] Segregated from *Linum*.

S. digynum (A. Gray) C. Rogers (p. 743) Ann, glabrous. **ST** erect, 5–20 cm. **LVS** opposite, ± erect, 5–16 mm, oblong; lower lvs entire; upper lvs serrate; stipules 0. **INFL**: cyme, dense; bracts lf-like; pedicels < 3 mm. **FL**: sepals 5, 2–3 mm, unequal, margins gland-toothed; petals 5, 3–4 mm, unappendaged, yellow; stamens 5, unappendaged; ovary chambers 4, styles 2, free, 1 mm, stigmas > styles in width. **FR**: nutlets 4; surfaces rough. **SEED** ± 1 mm. *n*=8. Gen moist meadows; 1000–1800 m. KR, CaR, SN; to WA, ID.

LOASACEAE LOASA FAMILY

Barry Prigge

Ann to shrub; hairs needle-like, stinging, or rough. **LVS** alternate in CA, gen ± pinnately lobed; stipules 0. **INFL** various. **FL** bisexual, radial; sepals gen 5, gen persistent in fr; petals gen 5, free or fused to each other or to filament tube; stamens 5–many, filaments thread-like to flat, sometimes fused at base or in clusters; petal-like staminodes sometimes present; pistil 1, ovary inferior, chamber gen 1, placentas gen 3, parietal, style 1. **FR**: gen capsule (utricle). **SEEDS** 1–many. 15 genera, ± 200 spp.: esp Am (Afr, Pacific). [Ernst & Thompson 1963 J Arnold Arbor 44:138–142]

1. Subshrub; stamens 5; stigma 1; fr utricle, seed 1 .... **PETALONYX**
1′ Herbs or subshrub; stamens gen many; stigma lobes 3 or 5; fr capsule, seeds gen many
 2. Some hairs needle-like, stinging; petals fused to each other or filament tube; stigma lobes and placentas 5 .... **EUCNIDE**
 2′ Hairs not needle-like (often barbed-rough); petals free; stigma lobes and placentas 3 .... **MENTZELIA**

EUCNIDE ROCK NETTLE

Ann to subshrub; hairs gen needle-like and stinging (or barbed). **LF** widely ovate to ± round, toothed to ± lobed; base widely tapered to cordate. **INFL**: cyme, bracted. **FL**: petals fused below middle or to filament tube; stamens many, epipetalous or fused at base into short tube; ovary club-shaped to spheric, placentas 5, stigma lobes 5, gen appressed. **FR** obovoid to spheric, nodding or reflexed, clearly pedicelled, dehiscent from top by 5 valves. **SEEDS** many, < 1 mm, ± oblong, grooved or ribbed. 8 spp.: sw US, n Mex. (Greek: strongly nettle-like)

1. Ann; petals fused, tube 8–15 mm, lobes 2–5 mm, greenish; stamens epipetalous, sessile or filaments < 1.5 mm . . . . *E. rupestris*
1′ Subshrub; petals mostly free, 30–50 mm, fused at base to short filament tube, whitish to pale yellow; stamens fused at base, filaments 10–20 mm . . . . *E. urens*

E. rupestris (Baillon) H.J. Thompson & W.R. Ernst ROCK NETTLE Ann < 30 cm. **LF**: petiole < 6 cm; blade 1–8 cm, gen round, toothed to weakly lobed, green and shiny above, base widely tapered to cordate. **FL**: pedicel in fr < 2 cm, reflexed; sepals 3.5–7 mm, 2–3 mm wide, tip obtuse; corolla tube 8–15 mm, narrowly cylindric, with ring of hairs at base, gen greenish, lobes 2–5 mm, erect, green; stamens epipetalous, sessile or filaments < 1.5 mm; stigma lobes below or = anthers. **FR** 7–15 mm, cylindric. RARE in CA. Crevices, cliffs; 500–600 m. s DSon (Painted Gorge, Imperial Co.); AZ, n Mex (incl islands). [*Sympetaleia r.* (Baillon) A. Gray]

E. urens (A. Gray) C. Parry (p. 743) Subshrub 30–100 cm. **LF**: petiole < 5 cm; blade < 10 cm, gen ovate, ± irregularly toothed, gray-green and dull above, base truncate to cordate. **FL**: pedicel in fr < 1.5 cm, ascending to erect; sepals 15–22 mm, 5–7 mm wide, tip acuminate; petals ± free, 30–50 mm, spreading, whitish to pale yellow; stamens fused at base, tube 2–5 mm, filaments 10–20 mm; stigma lobes slightly above anthers. **FR** 10–20 mm, club-shaped to obconic. n=21. Cliffs, rocky slopes, washes; < 1400 m. D; to sw UT, AZ, n Mex.

MENTZELIA BLAZING STAR

Ann to shrub; hairs gen barbed-rough (smooth). **LF** linear to ovate, gen ± lobed; basal in rosettes, gen petioled; cauline gen sessile, ± reduced upward. **INFL**: gen cyme (or fl 1); bract gen 1 per fl, gen green. **FL**: sepals lanceolate to deltate, gen persistent; petals 5, free, gen yellow; stamens gen many, ± free, gen unequal, inner filaments gen thread-like; staminodes 0 or 5–many, outer often petal-like; ovary gen cylindric, placentas gen 3, style thread-like, stigma 3-furrowed or -lobed. **FR** gen tapered to base, sometimes curved. **SEEDS** gen many, grain- to prism-like (triangular in X-section), angled, or lenticular and winged (important to identification). ± 50 spp.: w US, ± trop Am. [Darlington 1934 Ann Missouri Bot Garden 21:103–226]

1. Bien or per (sometimes fls 1st year); seeds lenticular and winged or fusiform and ribbed lengthwise
 2. Seeds ± fusiform, 3-ribbed lengthwise, not winged . . . . *M. torreyi*
 2′ Seeds lenticular, winged
  3. Petal-like staminodes 5
   4. Petal tips obtuse to rounded; sandy soils . . . . *M. multiflora* ssp. ***longiloba***
   4′ Petal tips acute; gypsum-clay soils . . . . *M. pterosperma*
  3′ Petal-like staminodes 0
   5. Petals > 30 mm; fr gen > 25 mm . . . . *M. laevicaulis*
   5′ Petals < 20 mm; fr gen < 25 mm
    6. Lvs entire, narrow . . . . *M. polita*
    6′ Lvs toothed or lobed
     7. Fr > 12 mm . . . . *M. inyoensis*
     7′ Fr < 10 mm . . . . *M. oreophila*
1′ Ann; seeds grain- or prism-like, or narrowed and folded near middle
 8. Fr > 4 mm diam; seeds narrowed and folded near middle, end beak-like, ashy white; filament tip gen wide and minutely 2-toothed
  9. Petals 8, < 12 mm; filament tips acute, teeth 0 . . . . *M. reflexa*
  9′ Petals 5, gen > 15 mm; filament tips 2-toothed, anther gen on thread-like stalk between teeth
   10. Fl bract white-scarious, margin green . . . . *M. involucrata*
   10′ Fl bract green ± throughout (exc sometimes base)
    11. Upper lvs and bracts ± cordate, clasping; fr sessile, 14–25 mm, erect . . . . *M. hirsutissima*
    11′ Upper lvs and bracts not clasping; fr pedicelled, gen 9–15 mm, often reflexed
     12. Anther stalk gen < teeth; seeds narrowed at middle; e DSon . . . . *M. tricuspis*
     12′ Anther stalk gen > teeth; seeds narrowed above and below middle; c DMoj . . . . *M. tridentata*
 8′ Fr gen < 4 mm diam; seeds grain- or prism-like; filament tip ± thread-like (exc *M. micrantha*), not toothed
  13. Seeds 1-rowed above mid-ovary, prism-like, angles grooved
   14. Outer 5 filaments widened to 2-forked tip . . . . *M. micrantha*
   14′ Outer filaments ± thread-like, tip not 2-forked
    15. Basal lvs lobed; sepals 2–7 mm; petals 3–10 mm; style 3–7 mm . . . . *M. affinis*
    15′ Basal lvs entire or toothed; sepals 1–4 mm; petals 2–6 mm; style 1–3.5 mm . . . . *M. dispersa*
  13′ Seeds 2–3-rowed above mid-ovary, grain-like, angles rounded or sharp, not grooved
   16. Fl bracts gen entire; axillary frs often curved 90°+
    17. Petals gen > 8 mm
     18. Seed not folded at 1 end, angles acute, surface pointed-papillate . . . . *M. jonesii*
     18′ Seed with flap-like fold at 1 end, angles rounded, surface rounded-papillate
      19. Sepals > 8 mm; petals 12–24 mm; style 7–15 mm . . . . *M. eremophila*
      19′ Sepals < 8 mm; petals 7–15 mm; style 4–8 mm . . . . *M. nitens*
    17′ Petals gen < 8 mm
     20. Seed surface smooth . . . . *M. desertorum*
     20′ Seed surface papillate
      21. Seed angles acute, surface pointed-papillate . . . . *M. albicaulis*
      21′ Seed angles rounded, surface rounded-papillate . . . . *M. obscura*
   16′ Fl bracts toothed or lobed; axillary frs straight, erect
    22. Fl bracts green throughout, not whitish at base
     23. Petals 8–22 mm, tip obtuse to notched . . . . *M. pectinata*

23′ Petals > 20 mm, tip abruptly acuminate
24. Petals 8–17 mm wide; c&s SNF .... *M. crocea*
24′ Petals 16–33 mm wide; SnJV, e SnFrB, n SCoRI .... *M. lindleyi*
22′ Fl bracts whitish at base or ± white-scarious
25. Fl bracts wide, ± concealing fls and frs, mostly white-scarious; infl dense .... *M. congesta*
25′ Fls bracts narrow, not concealing fls and frs, whitish below middle; infl open to ± dense
26. Petals gen > 8 mm .... *M. gracilenta*
26′ Petals gen < 8 mm
27. Styles 1.5–3 mm; cauline lvs gen entire (or 2-lobed at base) .... *M. montana*
27′ Styles 3–6 mm; cauline lvs gen lobed .... *M. veatchiana*

M. affinis E. Greene (p. 743) Ann, 5–47 cm, erect. **LVS** 1–17 cm; basal toothed or lobed; cauline entire to lobed. **INFL**: bract lanceolate to obovate, entire or toothed. **FL**: sepals 2–7 mm; petals 3–10 mm, base with orange spot, tip rounded, notched, or toothed; stamens < 6.5 mm; style 3–6.5 mm. **FR** 12–28 mm, 1–3 mm wide, straight or arched < 90°. **SEED** 1–2 mm, ± 1 mm wide, prism-like, smooth; angles grooved; sides ± flat. Sandy grassland, woodland, creosote-bush scrub; < 1200 m. s SN, SnJV, se SnFrB (Mount Hamilton), SCoRI, TR, SnJt, D; AZ, Baja CA. ❀DRN,DRY, SUN:2,7,9,**10**,11,12,14–16,18–24;DFCLT.

M. albicaulis Hook (p. 743) Ann 5–42 cm. **ST** decumbent to erect, white. **LVS** 1–11 cm; lower lobed (comb-like); upper entire to lobed (comb-like). **INFL**: bract lanceolate to ovate, entire or with 2–4 small teeth or angles, base rarely faintly whitish. **FL**: sepals 1–5 mm; petals 2–7 mm, base sometimes with yellow spot; stamens 3–5 mm; style 2–5 mm. **FR** 8–28 mm, 1.5–3.5 mm wide, straight, sometimes curved < 180°. **SEED** 1–1.5 mm, grain-like, pointed-papillate; angles acute. *n*=27,36. Gravel fans, washes, shrubland to pinyon/juniper woodland; 500–2300 m. Teh, SnGb, SnBr, GB, D; to B.C., Colorado, Baja CA. Pls with *n*=27 have been called *M. mojavensis* H.J. Thompson & Joyce Roberts.

M. congesta Torrey & A. Gray (p. 743) Ann 7–40 cm. **ST** branched from base, erect, tan. **LF** 1–9 cm, entire to lobed. **INFL** dense; bract ± fused and appressed to ovary, ± entire, widely wedge-shaped, tip truncate to 3-toothed, lower 3/4 white-scarious. **FL**: sepals 1–4 mm; petals 3–9 mm, base with orange spot; stamens ± = style; style 1.5–5 mm. **FR** 5–12 mm, 2–3 mm wide, gen straight. **SEED** ± 1 mm, grain-like, ± sharp-angled, papillate, checkered or not; sides concave. Disturbed slopes, pine forest, sagebrush scrub, pinyon/juniper woodland; 1500–2700 m. SNH (e slope), Teh, WTR, SnGb, PR, SNE; to ID. [var. *davidsoniana* (Abrams) J.F. Macbr.]

M. crocea Kellogg (p. 743) Ann, 34–100 cm, erect. **LVS** 1–20 cm; basal lobed; cauline toothed or lobed. **INFL**: bract ovate to obovate, 6–10-toothed. **FL**: sepals 7–20 mm; petals 21–36 mm, base with orange spot; stamens 11–28 mm; style 20–30 mm. **FR** 20–35 mm, 3–5 mm wide, straight. **SEED** 1.5–2 mm, grain-like, minutely papillate. *n*=18. Rocky slopes, roadsides, grassland, oak/pine woodland; < 1500 m. c&s SNF. [*M. lindleyi* ssp. *c.* (Kellogg) C. Wolf]. ❀DRN,DRY,SUN:7,8,9,14–16,19–21;DFCLT.

M. desertorum (Davidson) H. J. Thompson & Joyce Roberts Ann, 10–41 cm, erect. **LVS** 1–12 cm; basal toothed or lobed; cauline entire to lobed. **INFL**: bract lanceolate to ovate, gen entire. **FL**: sepals 2–4 mm; petals 2.5–6 mm, base sometimes with orange spot; stamens = style; style 2–4 mm. **FR** 12–27 mm, 1–2.5 mm wide, straight to curved < 180°. **SEED** ± 1 mm, grain-like, rounded to slightly angled, smooth. *n*=9. Sandy flats in creosote-bush scrub; < 700 m. D; Baja CA.

M. dispersa S. Watson (p. 743) Ann, 7–48 cm, erect. **LVS** < 10 cm; basal entire to lobed; cauline entire to toothed. **INFL**: bract ovate or tapered from middle to tip, margin entire to lobed. **FL**: sepals 1–3.5 mm; petals 2–6 mm; stamens < 4.5 mm; style 1–3.5 mm. **FR** 7–25 mm, 1–2.5 mm wide, straight or curved < 30°. **SEED** 0.5–1.5 mm, 0.5–1 mm wide, prism-like; angles gen sharp, grooved; sides smooth, shiny. *n*=9,18. Sandy or rocky soils; < 2500 m. CA-FP, GB; to WA, n-c US, NM.

M. eremophila (Jepson) H.J. Thompson & Joyce Roberts (p. 743) Ann 8–43 cm. **ST** erect, tan to whitish, spreading in age. **LVS** 1–10 cm; basal lobed; cauline entire or lobed. **INFL**: bract ovate to triangular, entire or 2-toothed at base. **FL**: sepals 8–16 mm; petals 12–24 mm; stamens 3–10 mm; style 7–15 mm. **FR** 19–40 mm, 2–3.5 mm wide, gen curved < 270°. **SEED** ± 1 mm, grain-like, minutely round-papillate, ± checkered; angles rounded; tip ± folded, flap-like. *n*=9. Washes, roadsides, flats, creosote-bush scrub; 700–1100 m. s DMoj. Sometimes confused with *M. nitens*. ❀TRY.

M. gracilenta Torrey & A. Gray Ann 3–60 cm. **ST** greenish. **LVS** 2–13 cm; basal lobed (comb-like); lower cauline many-lobed, upper entire to few-lobed. **INFL**: bract gen fused to ovary base, obovate, 3–12-toothed, white below middle. **FL**: sepals 3–8 mm; petals 7–18 mm, base gen with orange spot; stamens ± = style; style 5–11 mm. **FR** gen 9–10(23) mm, 3–5 mm wide, straight or curved < 20°. **SEED** 1.5–2.5 mm, grain-like, round-papillate; angles sharp. *n*=18. Steep talus, pine/oak woodland; 200–1500 m. SCoR, WTR. [*M. ravenii* H.J. Thompson & Joyce Roberts]. ❀DRN,DRY,SUN: 2,3,7,14–16,18–21;DFCLT.

M. hirsutissima S. Watson Ann 6–31 cm. **ST** branched at base, spreading. **LF** 1–11 cm, toothed to lobed. **INFL**: bract ovate to triangular, 2–5-toothed or -lobed, ± clasping, whitish at base. **FL**: sepals 9–18 mm; petals 12–28 mm, pale yellow; stamens 4–12 mm; filaments gen minutely 2-toothed or -shouldered at tip; style 6–13 mm. **FR** 14–25 mm, 5–8 mm wide, straight. **SEED** ± 2.5 mm, ±1.5 mm wide, beaked, narrowed above and below middle, minutely papillate, ashy white. Washes, fans, slopes; creosote-bush scrub; < 600 m. DSon; Baja CA. [var. *stenophylla* (Urb. & Gilg) I.M. Johnston, hairy stickleaf]. ❀TRY.

M. involucrata S. Watson (p. 743) Ann 7–35 cm. **LVS** 2–18 cm, gen irregularly toothed. **INFL**: bracts 4–5 per fl, (ob)ovate, white-scarious, margin green, 3–10-toothed or -lobed. **FL**: sepals 7–23 mm; petals 13–62 mm, cream-yellow with orange veins; stamens 4–26 mm, filament 2-toothed at tip; style 8-30 mm. **FR** 14–22 mm, 5–10 mm wide, straight. **SEED** 2–3 mm, 2–2.5 mm wide, ± compressed, beaked, narrowed above and below middle, rough, ashy white. Washes, fans, steep slopes, creosote-bush scrub; < 900 m. D; n Mex. Pls from n DSon, with petals > 30 mm, have been called var. *megalantha* I.M. Johnston. ❀TRY.

M. inyoensis B. Prigge Per 15–40 cm. **LVS** 4–11 cm, lobed. **INFL**: bract gen linear, entire, sometimes 2-lobed at base. **FL**: sepals 4–12 mm; petals 11–18 mm, clawed; stamens 5–15 mm; outer filaments < 1.5 mm wide; style 10–13 mm. **FR** gen 12–16(25) mm, 6–8 mm wide, cylindric. **SEED** 2–3 mm, ± 2 mm wide, lenticular, faintly fine-pebbled, winged. *n*=11. Rocky slopes, canyons; 1900–2000 m. n W&I (White Mtns.); NV.

M. jonesii (Urb. & Gilg) H.J. Thompson & Joyce Roberts Ann, 10–40 cm, gen sprawling. **LF** 10–14 cm, lobed or toothed. **INFL**: bract triangular-ovate, entire. **FL**: sepals 2–10 mm; petals 6–22 mm, base sometimes with orange spot; stamens 3–10 mm, outer > inner; style 4–10 mm. **FR** 15–38 mm, 2–4 mm wide at flared tip, curved < 180° or S-shaped. **SEED** ± 1 mm, grain-like, pointed-papillate; angles acute. *n*=18,27. Washes, fans, flats, roadsides; 200–1500 m. D; NV, AZ. Closely related to *M. nitens*. ❀TRY.

M. laevicaulis (Hook.) Torrey & A. Gray (p. 743) Per 22–100 cm. **ST** erect, branching above middle. **LVS** gen lobed; basal 19–24 cm; cauline 2–10 cm. **INFL**: bract linear to lanceolate, entire, toothed at base to deeply 4–5-lobed. **FL**: sepals 15–46 mm; petals 30–80 mm; stamens 15–55 mm, outer filaments < 2 mm wide; style 2.5–7 mm. **FR** 15–44 mm, 8–13 mm wide, straight. **SEED** 3–4 mm, 2–2.5 mm wide, lenticular, winged, fine-pebbled. *n*=11. Sandy to rocky slopes, washes, roadcuts; < 2700 m. CA (exc GV, DSon); to B.C., MT, NV. ❀DRN,DRY,SUN:1,**2**,**3**,**7**,9,12,14–24; DFCLT.

M. lindleyi Torrey & A. Gray (p. 743) Ann, 10–60 cm, erect, much-branched. **LF** 2–17 cm, lobed (comb-like); sinuses toothed

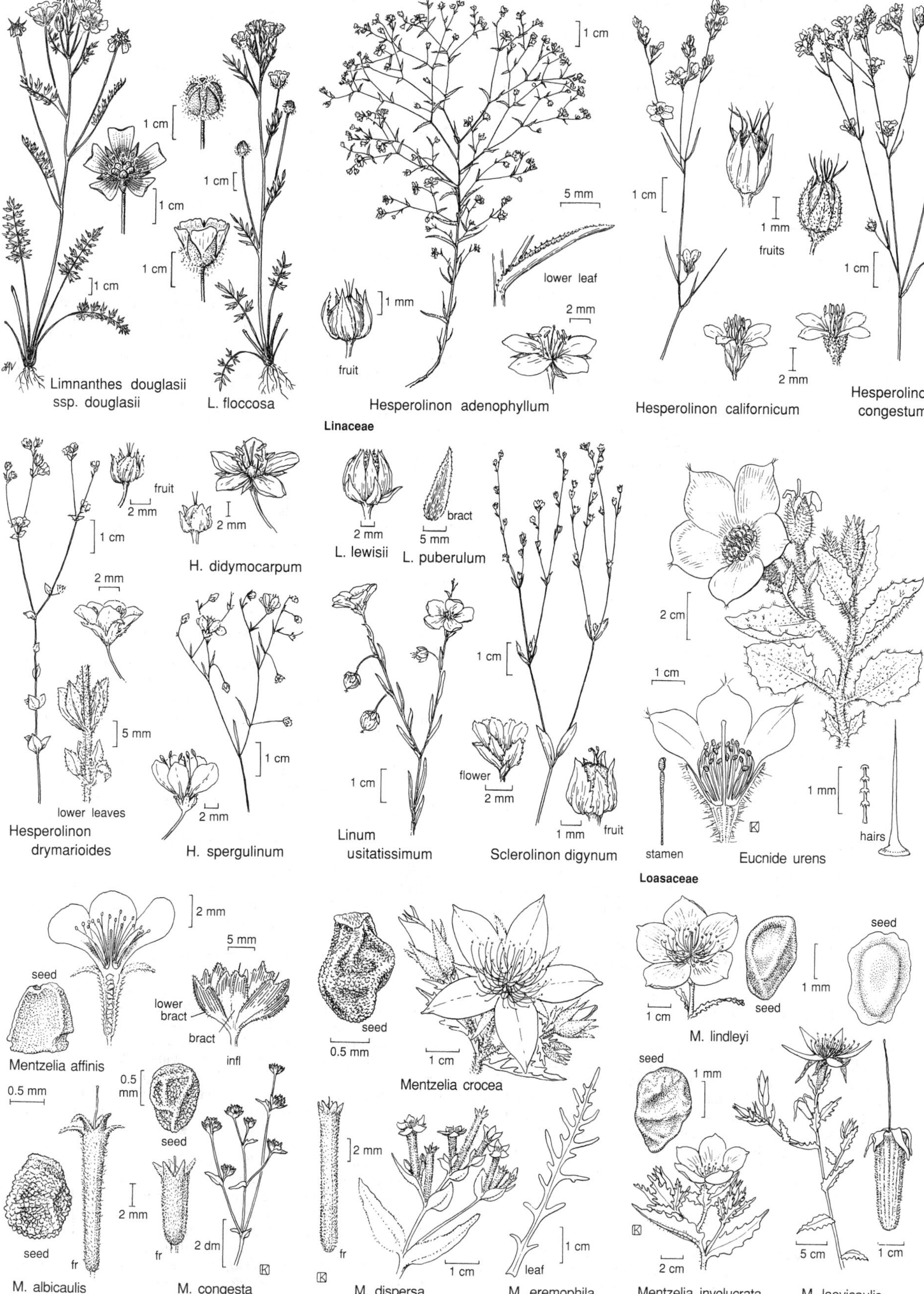
1 cm
1 cm
1 cm
1 cm
1 cm
1 cm
Limnanthes douglasii ssp. douglasii
L. floccosa
1 cm
5 mm
lower leaf
2 mm
1 mm
fruit
Hesperolinon adenophyllum
1 cm
1 mm
fruits
1 cm
2 mm
Hesperolinon californicum
Hesperolinon congestum
Linaceae
fruit
2 mm
1 cm
2 mm
H. didymocarpum
2 mm
5 mm
lower leaves
1 cm
2 mm
Hesperolinon drymarioides
H. spergulinum
2 mm
bract
5 mm
L. lewisii
L. puberulum
1 cm
Linum usitatissimum
1 cm
flower
2 mm
1 mm
fruit
Sclerolinon digynum
2 cm
1 cm
1 mm
stamen
Eucnide urens
hairs
Loasaceae
2 mm
seed
5 mm
lower bract
bract
infl
Mentzelia affinis
0.5 mm
0.5 mm
seed
2 mm
seed
fr
fr
2 dm
M. albicaulis
M. congesta
seed
0.5 mm
1 cm
Mentzelia crocea
2 mm
fr
1 cm
1 cm
leaf
M. dispersa
M. eremophila
seed
1 mm
seed
1 cm
M. lindleyi
seed
1 mm
2 cm
5 cm
1 cm
Mentzelia involucrata
M. laevicaulis

or short-lobed. **INFL**: bract ovate, toothed or lobed, green. **FL**: sepals 9–19 mm; petals 20–40 mm, golden yellow, base with orange-red spot at base; stamens 20–30 mm; anthers 0.5–1 mm; style 15–24 mm. **FR** 25–40 mm, 4–7 mm wide, gen straight. **SEED** 1–1.5 mm, grain-like, irregularly angled, minutely papillate. *n*=18. Rocky, open slopes, coastal-sage scrub, oak/pine woodland; < 800 m. SnJV, e SnFrB, n SCoRI. ❀DRN,DRY,SUN:4–6,**7**,8,**9**, 10,11,**14**,15–18,**19–24**.

M. micrantha (Hook. & Arn.) Torrey & A. Gray (p. 749) Ann 10–80 cm. **ST** erect, green. **LF** 1–18 cm, irregularly dentate or wavy. **INFL**: bract widely ovate, entire to crenate, often concealing fl. **FL**: sepals 1.5–2 mm; petals 2.5–4.5 mm; stamens ± 20, 1.5–4 mm; outer 5 filaments widened, tip 2-forked; style 2–3(5) mm. **FR** 6–13 mm, 1.5–2.5 mm wide, cylindric, gen straight (curved < 20°). **SEEDS** 5–10, pyramid-shaped, smooth; angles grooved; end walls often oblique. *n*=9. Open, often disturbed places, chaparral, oak woodland; < 1600 m. NCoRO, SnFrB, SCoRO, s ChI, WTR, SnGb, PR; Baja CA.

M. montana (Davidson) Davidson Ann, 4–48 cm, erect. **LF** < 13 cm, entire or base 2-lobed. **INFL**: bract (ob)ovate, 2–6-toothed below middle, base whitish. **FL**: sepals 1–4 mm; petals 2–7 mm, base sometimes with orange spot; stamens 2–7 mm; style 1.5–3 mm. **FR** 6–20 mm, 2–3 mm wide, sometimes cylindric. **SEED** 1–1.5 mm, ± 1 mm wide, grain-like, rough; angles sharp. *n*=18. Open slopes, flats, sagebrush scrub, coniferous forest; 1200–2600 m. CaR, SN, WTR, SnJt, W&I, n DMtns; to OR, Colorado, TX, n Mex. Intermediate to *M. congesta*, *M. veatchiana*.

M. multiflora (Nutt.) A. Gray ssp. ***longiloba*** (Darl.) Felger Bien or per, 15–100 cm, branched at base, erect. **LVS** 1–15 cm; basal crenate or shallowly lobed; cauline entire to toothed. **INFL**: bract 0 or linear to lanceolate, gen entire. **FL**: sepals 6–12 mm; petals 10–20 mm, golden yellow, tip obtuse to rounded; staminodes 5, 9–20 mm, petal-like, tip acute to serrate; stamens 3–13 mm, outer > and wider than inner; style 5–13 mm. **FR** 8–17 mm, 7–10 mm wide, cup-shaped. **SEED** 3–4 mm, 2.5–3.5 mm wide, lenticular, winged, whitish. Sandy creosote-bush scrub; < 700 m. s DMoj, DSon; AZ, n Mex. Ssp. *multiflora* e of CA. ❀TRY.

M. nitens E. Greene (p. 749) Ann 12–34 cm. **ST** glabrous to sparsely hairy, whitish. **LVS** 6–15 cm; basal toothed or lobed; cauline entire to lobed. **INFL**: bract 0 or lanceolate and entire. **FL**: sepals 3–8 mm; petals 7–15 mm, base sometimes with orange spot; stamens < or = style; style 4–8 mm. **FR** 13–26 mm, 2–3.5 mm wide, curved > 180°. **SEED** ± 1 mm, grain-like, round-papillate, checkered; end folded, flap-like; angles rounded. *n*=9. Sandy washes, slopes; 500–1800 m. n DMoj, SNE; s NV. Closely related to *M. eremophila*. ❀TRY.

M. obscura H.J. Thompson & Joyce Roberts (p. 749) Ann 8–45 cm. **LVS** toothed to lobed; basal 7–22 cm; cauline 1–7 cm. **INFL**: bract ovate, entire. **FL**: sepals 2–6 mm; petals 3–12 mm, base sometimes with faint orange spot; stamens 2–7 mm, outer > inner; style 2-6 mm. **FR** 11–31 mm, 1.5–3 mm wide, slightly flared at tip, curved 90°–250°. **SEEDS** ± 1 mm, grain-like, round-papillate; angles rounded. *n*=18. Washes, slopes, roadsides, shrubland, Joshua-tree woodland; 200–1500 m. D; to UT, AZ, n Mex. Much like *M. albicaulis* (exc seeds).

M. oreophila Darl (p. 749) Per 15–40 cm. **STS** few, 0–many branched, white, peeling. **LVS** < 9 cm, ± wavy-dentate; upper clasping. **INFL**: bract linear to lanceolate, entire. **FL**: sepals 4–11 mm, becoming inrolled; petals 7–17 mm; staminodes 5, petal-like, gen 5–8(14) mm; stamens 3–9 mm, outer > inner; style 5–9 mm. **FR** 5–10 mm, 6–11 mm wide, cup-shaped. **SEED** 2–3.5 mm, lenticular, winged, finely pebbled. *n*=11. Washes, limestone soils, talus; 900–1800 m. DMtns; to UT, w AZ. [*M. leucophylla* Brandegee misapplied] Pls with non-clasping lvs have been called *M. puberula* Darl.

M. pectinata Kellogg (p. 749) Ann 8–54 cm. **ST** erect, branched throughout, greenish. **LVS** 2–12 cm; lower gen lobed (comb-like); upper acutely lobed. **INFL**: bract often fused to ovary base, narrowly lanceolate to ovate, entire to 5-toothed. **FL**: sepals 3–13 mm; petals 8–22 mm, yellow to orange, base often with reddish or orange spot, tip obtuse to notched; stamens 4–11 mm, outer > inner; style 5–13 mm. **FR** 12–35 mm, 2–4 mm wide, straight. **SEED** 1–2 mm, grain-like, checkered or not; angles irregular, sharp; sides concave. Steep slopes, grassland, oak woodland; < 1800 m. s SNF, Teh, SCoRI. ❀DRN,DRY,SUN:7,9,14–16,19–21;DFCLT.

M. polita Nelson Per, 12–31 cm, rounded; caudex thick. **ST** white, peeling. **LF** < 7 cm, linear to lanceolate, entire. **INFL**: bract narrowly lanceolate, green, entire. **FL**: sepals 5–7 mm; petals 8–14 mm, white to pale yellow; stamens 3–7 mm, outer filaments 1–1.5 mm wide; style 5–7 mm. **FR** 5–8 mm, 5–8 mm wide, cup- or urn-shaped. **SEED** ± 3 mm, ovate, lenticular, winged, fine-pebbled. *n*= 11. Limestone, gypsum soils; 1200–1500 m. e DMtns (Clark Mtns); s NV.

M. pterosperma Eastw. Per (may fl 1st year) 6–42 cm. **LVS** 1–13 cm; basal entire to short-toothed; cauline toothed (or lobed near base). **INFL**: bract narrowly lanceolate, entire. **FL**: sepals 6–10 mm; petals 9–24 mm, tips acute; staminodes 5, petal-like, slightly < petals; stamens 4–15 mm, outer filaments 1–3 mm wide; style 6–17 mm. **FR** 8–14 mm, 6–10 mm wide, cup- to barrel-shaped. **SEED** ± 3–4 mm, elliptic in outline, lenticular, winged, fine-pebbled. *n*=11. Gypsum clay soils; 1100 m. e DMtns (Clark Mtns); to Colorado, nw AZ.

M. reflexa Cov (p. 749) Ann, 2–20 cm, rounded, branched at base. **LF** 1–10 cm, toothed. **INFL**: bract lanceolate to ovate, lobed. **FL**: sepals 5–7 mm; petals 8, 6–12 mm, pale yellow; stamens 3–8 mm, filaments slightly expanded and acute at tip; style 5–6.5 mm. **FR** 9–13 mm, 5–7 mm wide, cylindric to barrel-shaped. **SEEDS** 2–2.5 mm, ± pear-shaped, narrowed and folded near middle, minutely papillate, ashy white; tip narrow, flap-like. *n*=10. Washes, rocky slopes, roadsides; < 1200 m. DMoj.

M. torreyi A. Gray (p. 749) Per, 10–16 cm, much-branched. **LF** 2–4 cm, entire or with 2–4 linear lobes, rolled under. **INFL**: bract linear, base sometimes 2-lobed, inrolled. **FL**: sepals 3–6 mm; petals 9–15 mm, pale yellow, hairy above; stamens 7–10 mm; style 8–12 mm. **FR** 4–8 mm, 1–6 mm wide, urn-shaped. **SEED** 2–2.5 mm, ± fusiform, ± spirally 3-ribbed; 1 end acute, other truncate. Sandy or alkaline soils, slopes, shrubland, pinyon woodland; 1200–2100 m. SNE; to OR, ID.

M. tricuspis A. Gray (p. 749) Ann, 14–27 cm, spreading to erect. **LF** 2–12 cm, wavy or toothed. **INFL**: bract lanceolate to widely ovate, sessile, 2–8-toothed, base sometimes whitish. **FL**: sepals 7–17 mm; petals 11–50 mm, cream-white; stamens 7–17 mm, filaments 2-toothed at tip, teeth linear, anther stalk between teeth 1–1.5 mm (gen < teeth), thread-like; style 10–12 mm. **FR** 9–15 mm, 5–8 mm wide, cylindric to barrel-shaped, gen pedicelled, reflexed. **SEED** 2–2.5 mm, narrowed at middle, beaked, minutely papillate, whitish. *n*=10. Sandy or gravelly slopes or washes in creosote-bush scrub; 200–400 m. e DSon; to UT, AZ. ❀TRY.

M. tridentata (Davidson) H.J. Thompson & Joyce Roberts Ann, 10–25 cm, spreading to erect. **LF** 1–9 cm, wavy or toothed. **INFL**: bract lanceolate to widely ovate, sessile, 2–10-toothed, base sometimes whitish. **FL**: sepals 5–13 mm; petals 10–40 mm, cream-white; stamens 6–15 mm, filaments 2-toothed at tip, teeth linear, anther stalk between teeth 1–1.5 mm (gen > teeth), thread-like; style 9–12 mm. **FR** 9–18 mm, 5–8 mm wide, cylindric to barrel-shaped, gen pedicelled, erect to reflexed. **SEED** 1.5–2.5 mm, narrowed above and below middle, beaked, minutely papillate, whitish. *n*=10. Creosote-bush scrub; 700–1000 m. c DMoj. [*M. tricuspis* var. *brevicornuta* I.M. Johnston]. ❀TRY.

M. veatchiana Kellogg (p. 749) Ann, 3–45 cm, erect. **LVS** 1–18 cm; basal deeply lobed, sinuses or lobes toothed; cauline toothed to lobed. **INFL**: bract (ob)ovate, gen toothed, base whitish. **FL**: sepals 1–5 mm; petals 4–8 mm, (yellow-) orange; stamens 3–7 mm, outer > inner; style 3–6 mm. **FR** 8–28 mm, gen straight (or curved < 70°). **SEED** 1–2 mm, grain-like, pointed-papillate; angles acute. *n*=27. Sandy grassland, shrubland, oak/pine woodland; 400–1900 m. n SNH, Teh, SnJV, SCoRI, TR, PR, GB, D; OR, NV, AZ, Baja CA.

PETALONYX

Subshrub; hairs gen rough with whorls of fine barbs. **LF** linear to ± round, entire to toothed; base tapered to cordate. **INFL**: raceme, gen terminal; bracts 3 per fl, outer 1 > inner 2. **FL**: sepals ± deciduous; petals free or claws adherent, fused below blades; stamens gen 5, free; ovary ± ovoid, placenta 1, stigma 1. **FR**: utricle, ± ovoid, gen 5-veined or -ribbed, erect. **SEED** 1, 1.5–2.5 mm, ± fusiform, gen smooth. *n*=23 for all spp. 5 spp.: sw US, nw Mex. (Greek: petal claw) [Davis & Thompson 1967 Madroño 19:1–32]

1. Petals free; lvs linear to narrowly (ob)lanceolate . . . . . *P. linearis*
1′ Petal claws adherent below, fused above; lvs lanceolate to widely ovate
 2. Lvs petioled, cauline ± equal . . . . . *P. nitidus*
 2′ Lvs sessile, cauline reduced upwards . . . . . *P. thurberi*
  3. Hairs soft, spreading; stamens 4–7.5 mm . . . . . ssp. *gilmanii*
  3′ Hairs rough, appressed downward; stamens 6–10 mm . . . . . ssp. *thurberi*

P. linearis E. Greene Pl 15–100 cm. **LF** gen sessile, 10–25 mm, linear to narrowly (ob)lanceolate, obtuse to acute, entire to irregularly toothed. **INFL** 4–10 cm; outer bract 5–8 mm, ovate to ± round; inner bracts 3–4 mm, ovate, ± cordate, acute to notched, lobed; pedicels 1–2 mm. **FL**: petals 2–5.5 mm, free, white; stamens 3–7 mm, barely exserted; style ± 3–6 mm. Sandy or rocky canyons, gen in creosote-bush scrub; < 1000 m. se DMoj, DSon; sw AZ, nw Mex. ❀TRY.

P. nitidus S. Watson (p. 749) Pl 15–45 cm. **LF** widely tapered to base, 15–40 mm, gen ovate, acute to acuminate, serrate to coarsely few-toothed. **INFL** 3–4.5 cm; outer bract 5–13 mm, narrowly ovate; inner bracts 1–5 mm, elliptic to ovate, truncate, crenate; pedicels 1–2 mm. **FL**: petals 5–11 mm, cream-colored, claws adherent, fused in upper 1/4; stamens 7–14 mm, well exserted; style 8–15 mm. Sandy or rocky canyons in creosote-bush scrub, Joshua-tree woodland, pinyon/juniper woodland; 1000–2100 m. DMtns; to sw UT, nw AZ. ❀TRY.

P. thurberi A. Gray (p. 749) Pl < 100 cm. **LF** clasping, 4–45 mm, lanceolate to deltate-ovate, acute to acuminate, entire to few-toothed. **INFL** 1–4 cm; outer bract 4–7.5 mm, deltate-ovate; inner bracts 2–3 mm, lanceolate to ovate, tapered to cordate, acute to ± acuminate, crenate or lobed; pedicels < 1 mm. **FL**: petals 2.5–6.5 mm, cream-colored, claws adherent, fused in upper 1/5; stamens 4– 10 mm, well exserted; style 3–11 mm. Sandy or gravelly dunes, washes, canyons in creosote-bush scrub; < 1200 m. D; NV, AZ, nw Mex.

ssp. ***gilmanii*** (Munz) W.S. Davis & H.J. Thompson DEATH VALLEY SANDPAPER PLANT Hairs soft, spreading. **FL**: stamens 4–7.5 mm. RARE. Sandy washes, dunes; < 1200 m. n DMoj. [*P. g.* Munz]

ssp. ***thurberi*** Hairs rough, appressed downward. **FL**: stamens 6–10 mm. Habitats and range of sp. ❀TRY.

LYTHRACEAE LOOSESTRIFE FAMILY

Elizabeth McClintock

Ann, per, shrubs, trees. **ST** angled or cylindric. **LVS** simple, entire, gen opposite, sometimes alternate or whorled. **INFL**: raceme, spike, or panicle, terminal, or axillary clusters with 1–several fls. **FL** bisexual, gen radial; hypanthium cylindric to bell-shaped, gen membranous, persistent in fr; sepals 4–6, gen persistent, appendages 3–5 or 0, alternate sepals; petals, stamens inserted on inner hypanthium; petals 4–6 or 0, alternate sepals, deciduous; stamens gen = or 2 × petals, incl or exserted; ovary superior, chambers 2–6, style gen slender, stigma head-like. **FR**: capsule, opening by valves from top, splitting sometimes irregular or 0. **SEEDS** 3–many. ± 25 genera, 450 spp.: temp, trop, gen in wet habitats. Some orn or cult for medicine, dyes. [Graham 1964 J Arnold Arbor 45:235–250]

1. Hypanthium cylindric . . . . . **LYTHRUM**
1′ Hypanthium bell- to urn-shaped
 2. Fls in axillary clusters; fr irregularly dehiscent . . . . . **AMMANNIA**
 2′ Fls 1 per axil; fr dehiscent by 2 or 4 valves
  3. Ann; st prostrate to decumbent; lf ± petioled; hypanthium partly enclosing fr . . . . . ***Lythrum portula***
  3′ Per; st ascending to erect; lf sessile; hypanthium > fr . . . . . **ROTALA**

AMMANNIA

Ann. **ST** prostrate to erect, glabrous. **LVS** opposite, 4-ranked, sessile, linear to oblanceolate, gen with basal ear-like lobes. **INFL**: cluster, axillary; fls 3–10; bractlets 2, inconspicuous. **FL** radial; hypanthium bell- to urn-shaped, ± spheric in fr; sepals 4, appendages < or ± = sepals, thick, horn-like; petals (0)4; stamens 4(8), incl or exserted. **FR** ± spheric, irregularly dehiscent. **SEEDS** many, ± 1 mm. ± 25 spp.: temp, trop. (Paul Ammann, Germany, 1634–1691) [Howell 1985 Wasmann J Biol 43:72–74]

1. Lower peduncles > 3 mm; petals deep rose-purple; fr 3–5 mm diam . . . . . ***A. coccinea***
1′ All peduncles < 1 mm or 0; petals pale lavender; fr 4–6 mm diam . . . . . ***A. robusta***

A. coccinea Rottb. (p. 749) **ST** decumbent to erect, 1–10 dm, solitary or branched at base. **LF** 2–8 cm, 2–15 mm wide, linear to narrowly lanceolate. **INFL** compact; fls 3–5; lower peduncles 3–5 (9) mm. **FL**: hypanthium urn-shaped; petals 3–4(5) mm, obovate, deep rose-purple; stamens 4(7), exserted in fl, anthers deep yellow. **FR** 3–5 mm diam. *n*=33. Wet places, drying ponds, lake & creek margins; < 300 m. CaRF, c&s SNF, GV, SnFrB, SW, DSon.; to e US, C.Am. [*A. auriculata* Willd. misapplied]

A. robusta Heer & Regel (p. 749) **ST** decumbent to erect, 1–10 dm, solitary or branched. **LF** 1.5–8 cm, 5–15 mm wide, gen linear-lanceolate. **INFL** compact; fls 1–3(5), sessile. **FL**: hypanthium urn-shaped, often 4-ridged; appendages ± = sepals in fr; petals 3–4(8) mm, obovate, pale lavender; stamens 4(5–12), exserted in fl, anthers pale yellow. **FR** 4–6 mm diam. *n*=17. Wet places, drying pond and ditch margins; < 500 m. NCoR, s SNF, GV, CW, SCo, s ChI (Santa Catalina Island), DSon; to c US, Mex. [*A. coccinea* Rottb. misapplied]

LYTHRUM

Ann, per. **ST** prostrate to erect, sometimes 4-angled. **LVS** opposite, alternate, or whorled, linear to ovate, sessile or short-petioled. **INFL**: raceme- or spike-like; fls gen 1–2 per axil; bracts 2 per node. **FL** radial to slightly bilateral, sometimes of 2–3 forms (heterostyly); hypanthium cylindric to bell-shaped; sepals 4–6, appendages < to > sepals; petals 4–6 or 0; stamens 4–6 or 12, incl or exserted; styles < to > stamens. **FR** gen cylindric, gen dehiscent by 2 valves. **SEEDS** many, < 1 mm. ± 35 spp.: temp. (Greek: blood, from fl color) [Stuckey 1980 Bartonia 47:3–20]

1. Hypanthium bell-shaped, < 2 mm; st decumbent to erect, rooting at nodes .... ***L. portula***
1′ Hypanthium cylindric, 4–7 mm; st ascending to erect, rooting at base
 2. St branching at base; lower branches decumbent to ascending
  3. Hypanthium appendages awl-like, ± 1 mm, ± 2 × sepals .... ***L. hyssopifolium***
  3′ Hypanthium appendages deltate, ± 0.5 mm, ± = sepals .... ***L. tribracteatum***
 2′ St branching at base or not, lower gen erect
  4. Infl bracts ovate, acuminate; petals 8–14 mm; st ± puberulent .... ***L. salicaria***
  4′ Infl bracts linear, acute to obtuse; petals 2–8 mm; st glabrous
   5. Lower lvs linear-lanceolate; pedicels 1–2 mm; sepals ± = appendages; petals 4–8 mm .... ***L. californicum***
   5′ Lower lvs elliptic to oblong; fls sessile; sepals < appendages; petals 2–5 mm .... ***L. hyssopifolium***

L. californicum Torrey & A. Gray (p. 749) CALIFORNIA LOOSESTRIFE Per, heterostylous. **ST** erect, 2–6 dm, gen branching above; branches spreading to ascending, glabrous. **LVS**: lower opposite; upper alternate, 1–7 cm, linear to linear-lanceolate, ± glaucous. **INFL** terminal, ± spike-like; bracts gen linear; pedicels 1–2 mm. **FLS** of 2 style forms; hypanthium cylindric, 4–7 mm; sepals narrowly deltate, < 1 mm, ± = appendages; petals 4–8 mm, purple; stamens gen 6, incl; style incl or exserted. **FR** ovoid, ± = hypanthium. 2*n*=20. Marshes, pond and stream margins; < 2200 m. s NCoRI, SNF, s SNH, GV, CW, SW, D; to c US, n Mex. ❀INV;STBL.

L. hyssopifolium L. Ann, bien, not heterostylous. **ST** 1–6 dm, branching; lower branches prostrate to erect, gen glabrous. **LVS**: lowermost opposite, upper gen alternate, 5–30 mm, oblong to elliptic. **INFL**: spike, terminal; bracts linear; fls sessile. **FL**: hypanthium 4–6 mm, cylindric; sepals deltate, < appendages; appendages ± 1 mm, awl-like; petals 2–5 mm, pink to rose; stamens 4–6, incl; style ± exserted. **FR** ovoid, ± = hypanthium. Marshes, drying pond margins; < 1600 m. CA-FP; to e US; native to Eur.

L. portula (L.) D. Webb Ann, prostrate, not heterostylous. **ST** prostrate to decumbent, 5–25 cm, branching at base, rooting at nodes, reddish, glabrous. **LVS** opposite, 5–10 mm, oblong to obovate, fleshy. **INFL**: fl 1 per axil, sessile; bracts narrowly lanceolate. **FL**: hypanthium 1–2 mm, broadly bell-shaped; sepals deltate, ± = appendages; appendages linear; petals ± 1 mm, sometimes 0, white to rose-pink; stamens gen 6. **FR** ± spheric, > hypanthium. 2*n*=10. Drying ponds, lake margins; 1000–2200 m. n SNH; native to Eur. [*Peplis portula* L.]

L. salicaria L. PURPLE LOOSESTRIFE Per, heterostylous. **ST** erect, 5–15 dm, branching or not, ± gray-puberulent. **LVS** gen opposite or whorled (sometimes alternate above), 5–14 cm, gen lanceolate. **INFL** ± spike-like, terminal; bracts ovate, acuminate; fls ± sessile. **FLS** of 3 style forms; hypanthium cylindric, 4–5 mm; sepals deltate, < 1 mm; appendages linear, 2–3 mm; petals 8–14 mm, red-purple; stamens gen 12, incl or exserted; style incl or exserted. **FR** 3–4 mm, ovoid. Marshes, ponds, streambanks, ditches; < 1000 m. s NCo, NCoRO, n SNF, ScV, SnFrB, nw MP; worldwide; native to Eur; cult for orn. NOXIOUS WEED.

L. tribracteatum Sprengel Ann, not heterostylous. **ST** 5–30 cm, erect, often branching at base; branches ± weakly ascending, glabrous to thinly scabrous. **LVS**: lower opposite; upper alternate, 3–25 mm, gen oblong to oblanceolate. **INFL**: cluster, axillary; bracts lf-like; fls ± sessile. **FL**: hypanthium 5–6 mm, narrowly cylindric; sepals and appendages ± 0.5 mm, deltate, red-glandular; petals < 3 mm, lavender; stamens 4–6, incl; style incl. **FR** cylindric, ± = hypanthium. Wet areas, drying ponds, ditches; < 1500 m. GV, SnFrB, n MP; native to s Eur.

ROTALA

Ann, per, glabrous. **ST** ascending to erect, angled. **LVS** opposite, 4-ranked (rarely whorled). **INFL**: spike; fl 1 per axil, subtended by lf-like bract. **FL** radial; hypanthium bell-shaped; sepals gen 4, appendages 4 or 0; petals 4, tiny, < sepals; stamens gen 4, ± incl. **FR** oblong to ± spheric, dehiscent by 2–4 valves. **SEEDS** many, < 1 mm. ± 45 spp.: temp, trop. (Latin: wheel-like) [Cook 1979 Boissiera 29:7–156]

1. Sepals deltate, appendages 0; fr 2-valved .... ***R. indica***
1′ Sepals widely deltate, appendages awl-like; fr 4-valved .... ***R. ramosior***

R. indica (Willd.) Koehne **ST** decumbent to erect, gen branched, 1–3 dm. **LF** 0.5–2 cm, oblong to obovate; margins thick. **INFL**: fl 1 per axil or in short, axillary spikes. **FL**: hypanthium < 2 mm; sepals deltate; appendages 0; petals pink. **FR** oblong, gen 2-valved. Marshes, ponds; < 100 m. ScV (Butte Co.); rice-field weed in Eur; native to se Asia.

R. ramosior (L.) Koehne (p. 749) **ST** prostrate to erect, 1–4 dm, gen branched; upper branches ± 4-angled. **LF** 1–5 cm, linear to oblanceolate. **INFL**: fl 1 per axil. **FL**: hypanthium 2–5 mm; sepals broadly deltate; appendages awl-like, 2 × sepals; petals white to pink. **FR** ovoid to ± spheric, gen 4-valved. Irrigated fields, lake and pond margins, streams; < 1900 m. NCoRI, n&c SN, GV; to WA, e US, S.Am.

MALVACEAE MALLOW FAMILY

Steven R. Hill, except as specified

Ann, per, shrubs, trees, gen stellate-hairy; juice sticky; inner bark tough, fibrous. **LVS** alternate, simple, petioled; blade gen palmately veined or lobed, stipules present. **INFL** often leafy; whorl or involucre of bractlets often subtending calyx. **FL** gen bisexual, radial; calyx lobes 5, margins abutting in bud; petals 5, free (fused at base to filament tube, so falling together);

stamens many, filaments fused into a tube surrounding style, tube fused in turn to petal bases; pistil 1, ovary superior, chambers gen 5 or more, style branches, stigmas gen 1 or 2 × as many as chambers. **FR** of 5–many disk- or wedge-shaped segments, loculicidal capsule, or berry. 100 genera, 2000 spp.: worldwide, esp warm regions; some cult (e.g., *Abelmoschus*, okra; *Alcea*; *Gossypium*, cotton; *Hibiscus*, *Malvaviscus*). Mature fr important for identification.

1. Fr a capsule, chambers 5; seeds 2–many per chamber; fls showy, solitary in axils near st tips .... **HIBISCUS**
1′ Fr of 5–40 segments that gen separate from axis and each other; seeds gen 1–4 per chamber; infl various
 2. Stigmas linear, on inner side of style branches; fr segments indehiscent; seed 1 per segment
  3. Anthers borne near top of filament tube, gen clustered in 2 series; bractlets subtending calyx 0–3; infl a raceme or spike; fr segments 5–10, beak sometimes present .... **SIDALCEA**
  3′ Anthers borne along filament tube, not clustered; bractlets subtending calyx 3–9, gen forming involucre; fls 1–10 per axil; fr segments 5–40, beak 0
   4. Bractlets 3, free; gen < 1 m .... **MALVA**
   4′ Bractlets 3–9, ± fused; gen 1–4 m
    5. Bractlets 6–9; fr axis < fr segments; petal veins gen indistinct; sometimes inland .... **ALCEA**
    5′ Bractlets 3; fr axis > fr segments; petal veins gen dark, obvious; coastal .... **LAVATERA**
 2′ Stigmas head-like or obliquely squared; fr segments dehiscent or not; seeds 1 or more per segment
  6. Fr segment with 2 scarious wings at top, indehiscent, strongly net-veined below; bractlets 0; shrub; e PR, DSon .... **HORSFORDIA**
  6′ Fr segment not 2-winged, upper and lower portions sometimes different; bractlets 0–3; herb or shrub; widespread
   7. Seeds gen 2–9 per fr segment
    8. Corolla red-orange; fr segment 2-chambered (each chamber 1-seeded); fl axillary, solitary; bractlets 3, ± persistent .... **MODIOLA**
    8′ Corolla color various (rarely red-orange); fr segment 1-chambered (sometimes with a partial internal wall); infl various; bractlets 0–3, persistent or not
     9. Corolla gen yellow or orange, rarely pink to brick-red; bractlets 0; seeds 2–9 per fr segment
      10. St gen erect; fr segment not inflated, walls firm, sometimes woody; more widespread in SW, D .... **ABUTILON**
      10′ St gen trailing; fr segment greatly inflated, walls thin, papery; PR .... **HERISSANTIA**
     9′ Corolla white to rose, lavender or red-orange; bractlets gen 1–3; seeds 1–4 per fr segment
      11. Fr segment ± bur-like and ± dehiscent throughout; bractlets 3, ± persistent; stigmas gen oblique-squared, often slightly elongated; KR, CaR, MP .... **ILIAMNA**
      11′ Fr segment smooth and dehiscent at top, net-veined and indehiscent at base; bractlets 1–3, gen deciduous; stigmas gen ± symmetrical, head-like; esp D .... [3]**SPHAERALCEA**
   7′ Seed gen 1 per fr segment
    12. Bractlets 0
     13. Gen per; fr segment with smooth, dehiscent, empty upper section and strongly net-veined, indehiscent lower section; native, esp D .... [3]**SPHAERALCEA**
     13′ Gen ann; fr segment ± uniform; alien weeds, esp GV
      14. Corolla purplish; lvs ± lobed to hastate; side walls of fr segment fragile; each seed gen ± enclosed by net-veined envelope .... **ANODA**
      14′ Corolla pale yellow; lvs unlobed; side walls of fr segment firm, persistent; seeds not enveloped .... **SIDA**
    12′ Bractlets 1–3 (best seen in bud or fl)
     15. Fr segment dehiscent throughout; infl of gen dense, axillary clusters .... **MALACOTHAMNUS**
     15′ Fr segment at least partly indehiscent; infl gen raceme-like, or fls solitary in axils
      16. Fr segment with smooth, dehiscent, empty upper section and strongly net-veined, indehiscent lower section .... [3]**SPHAERALCEA**
      16′ Fr segment ± uniform, gen not strongly net-veined
       17. Ann; pl glabrous or with scattered stellate hairs; lvs wide-rounded or deeply palmately lobed; petals white to rose-purple .... **EREMALCHE**
       17′ Per; pl densely stellate-canescent or scaly; lvs wide-reniform to triangular; petals yellowish .... **MALVELLA**

ABUTILON INDIAN MALLOW

Ann, per, shrub, stellate-canescent, tomentose, or bristly-hairy. **ST** decumbent to erect. **LF**: blade crenate or toothed, cordate, lobes gen 0. **INFL**: fls solitary in axils or in leafy panicles; bractlets subtending calyx 0. **FL**: petals yellow to reddish; anthers borne at top of filament tube; stigmas head-like. **FR** ± cylindric to ± spheric; segments smooth-sided, beaked, walls firm, sometimes woody, gen not separating, dehiscent ± to base. **SEEDS** 2–9 per fr segment. 200 spp.: warm regions. (Arabic name) [Borssum Waalkes 1966 Blumea 14:159–177]

1. Ann; lf gen 10–20 cm wide; fr segments > 10, beaks 3–5 mm, spreading; alien weed .... ***A. theophrasti***
1′ Per or shrub; lf 1–5 cm wide; fr segments gen < 10, beaks 1–2 mm; native
 2. Petals 10–20 mm, orange; fr segments gen 7–10, = or < calyx .... ***A. palmeri***
 2′ Petals 3–6 mm, orange-pink to red; fr segments gen 5, >> calyx .... ***A. parvulum***

A. palmeri A. Gray (p. 749) Subshrub. **ST** 15–20 dm; hairs stellate (and either long and soft or bristly). **LF**: blade 2–5 cm wide, ± round, cordate, velvety; lobes 3, obscure. **FL**: calyx 9–15 mm, = or > mature fr; petals 10–20 mm, orange. **FR**: segments gen 7–10, ± 10 mm tall, bristly or densely soft-hairy, beaks 1.5–2 mm, ± erect. Uncommon. Dry, gen e-facing mtn slopes, creosote-bush scrub;

600–800 m. DSon, adjacent PR; AZ, n Mex. ❀DRN,DRY,SUN: 8,9,11,12,**13**.

A. parvulum A. Gray (p. 749) Per. **ST** 1–4 dm, slender, much-branched, ± decumbent from woody root, stellate-canescent. **LF**: blade 1–4.5 cm wide, ovate, cordate; hairs scattered, stellate. **FL**: calyx 3–5 mm, << mature fr; petals 3–6 mm, orange-pink to red. **FR**: segments gen 5, 7–9 mm, stellate-puberulent, beaks 1–2 mm, ± erect. Uncommon. Arid, rocky slopes, shadscale scrub; 900–1300 m. DMtns (Providence Mtns); to s Colorado, w TX, n Mex.

A. theophrasti Medikus (p. 749) VELVET-LEAF Ann. **ST** 10–20 dm. **LF**: blade 10–20 cm wide, ovate to round, cordate, velvety (densely stellate and tomentose). **FL**: petals 6–8 mm, yellow. **FR**: segments > 10, >> calyx, beaks 3–5 mm, spreading and hooked inward, with long soft hairs. Uncommon. Disturbed places; < 100 m. GV (esp ScV), SW; native to s Asia.

ALCEA HOLLYHOCK

Bien, per. **ST** gen 1–4 m, erect, leafy, unbranched. **LF**: blade ± palmately lobed. **INFL**: raceme, spike-like; bractlets subtending calyx 6–9, fused at base, < or = sepals. **FL** showy; anthers borne both at and below top of stamen tube; stigmas thread-like. **FR**: segments 18–40, falling from fr axis, indehiscent, beakless, each with an upper, empty chamber and a lower, 1-seeded chamber. 60 spp.: e Medit to c Asia. (Greek: mallow) [Zohary 1963 Israel J Bot 12:1–26]

A. rosea L. (p. 753) **ST** becoming ± glabrous. **LF**: blade gen 1–2 dm, weakly 5–7-lobed, crenate-rounded, cordate. **FL** 6–12 cm wide, gen white to red. Uncommon. Disturbed, gen urban places; esp < 100 m. SnFrB, CCo; perhaps native to Asia Minor; widely cult as orn. [*Althaea r.* (L.) Cav.]

ANODA

Ann, per. **ST** ascending to erect, sparsely bristly to tomentose. **LF**: blade shape variable within a pl, esp upper gen hastate. **INFL**: bractlets subtending calyx 0. **FL**: stigmas head-like. **FR** hemispheric to disk-like; segments 5–20, side walls fragile. **SEED** 1 per fr segment; inner seed coat net-veined, ± enveloping seed. 23 spp.: esp Mex, S.Am. (Ceylonese name) [Fryxell 1987 Aliso 11(4):485–522]

A. cristata (L.) Schldl. (p. 753) VIOLETTAS Ann. **ST** decumbent to erect, 5–10 dm, sparsely bristly. **LF**: blade 2–9 cm, lower often triangular to hastate; hairs gen unbranched. **INFL**: fls solitary in axils. **FL**: calyx spreading in fr, 20–25 mm diam, lobes acuminate, often reddish; petals 8–30 mm, purplish. **FR**: segments 9–20, dark green, bristly, beaked. 2*n*=30,60,90. Uncommon. Disturbed, ± urban places, gardens; < 800 m. GV, SNF; to e US, S.Am, Australia; perhaps native to Mex. [var. *digitata* (A. Gray) Hochr.]

EREMALCHE

David M. Bates

Ann, some pls with only pistillate fls, gen stellate-hairy. **ST** prostrate to erect. **LF**: blade toothed to lobed or parted. **INFL**: fls solitary in axils or in terminal clusters; pedicel longer in fr; bractlets subtending calyx 3, linear to thread-like. **FL**: calyx lobes > tube, acuminate; petals gen > calyx, white to purplish (drying darker); filament column incl; stigmas head-like. **FR**: segments 9–36, separating, indehiscent, unarmed, glabrous, lateral walls fragile, margins and outer wall ridged or net-veined. **SEED** 1 per fr segment. 3 spp.: sw US, nw Mex. (Greek: lonely mallow, from desert habitats)

1. Lvs ± toothed; petal bases bright purple; fr segment 2.8–3.5 mm .... ***E. rotundifolia***
1′ Lvs palmately lobed to parted; petals uniformly colored; fr segment 1.4–2 mm
 2. Petals 4–5.5 mm, ± = calyx; st prostrate to decumbent .... ***E. exilis***
 2′ Petals 5.5–25 mm, often >> calyx; st gen ± erect .... ***E. parryi***
  3. Some pls with only pistillate fls; calyx 4.5–10 mm, lobes 1.5–3.5 mm wide .... ssp. ***kernensis***
  3′ All pls with only bisexual fls; calyx 10–14 mm, lobes 2.5–5 mm wide .... ssp. ***parryi***

E. exilis (A. Gray) E. Greene (p. 753) **ST** prostrate to decumbent, < 40 cm, finely stellate-hairy. **LF** gen 1–2.5 cm wide, 3–5-lobed; lobe tips entire or 3-toothed. **INFL**: fls sometimes near st base; bractlets 3–7 mm. **FL**: calyx 4–7 mm, lobes 3–5 mm, 1.5–2.5 mm wide; petals 4–5.5 mm, white or pale pinkish purple. **FR**: segments 9–13, 1.4–1.8 mm, ± wedge-shaped in X-section, margins rounded, cushion-like, outer wall cross-ridged. 2*n*=20,40. Desert scrub; < 1500 m. D; AZ, n Baja CA. [*Malvastrum e.* A. Gray]

E. parryi (E. Greene) E. Greene **ST** ± erect, < 50 cm, often with ascending basal branches, densely stellate-hairy near tips. **LF** gen 2–5 cm wide, deeply 3–5-lobed to -parted; lobe tips ± deeply toothed. **INFL**: fls gen > lvs; bractlets 2.5–15 mm. **FL**: calyx 4.5–13 mm, lobes 3–11 mm, 1.5–5 mm wide; petals 5.5–25 mm, white to purplish. **FR**: segments 9–22, 1.5–2 mm, ± wedge-shaped in X-section, margins rounded, cushion-like, outer wall cross-ridged. Grassland, scrub, foothill woodland; 100–1300 m. c&s SNF, Teh, SnJV, e SnFrB, SCoRI, WTR. Sspp. intergrade.

ssp. ***kernensis*** (C. Wolf) D. Bates KERN MALLOW Pl with either bisexual or pistillate fls. **INFL**: bractlets 3–10 mm. **BISEXUAL FL**: calyx 5–10 mm, lobes 3.5–8 mm, 1.7–3.5 mm wide, petals 8–25 mm, sometimes white; fr segments 9–13. **PISTILLATE FL**: petals < in bisexual fls, relative to calyx length. **FR**: segments 13–19. 2*n*=20. RARE. Eroded hillsides, alkali flats with shadscale; 100–1000 m. s SnJV, s SCoRI (Kern, San Luis Obispo cos.). [*E. k.* C. Wolf, *Malvastrum k.* (C. Wolf) Munz] Threatened by agriculture, grazing, energy development.

ssp. ***parryi*** Pl with only bisexual fls. **INFL**: bractlets 7–15 mm. **FL**: calyx 10–14 mm, lobes 8–11 mm, 2.5–5 mm wide; petals 15–25 mm, purplish. **FR**: segments 14–22. 2*n*=20. Habitat and range of sp. (exc s SnJV, s SCoRI). [*Malvastrum p.* E. Greene] ❀TRY.

E. rotundifolia (A. Gray) E. Greene DESERT FIVE-SPOT **ST** erect, 8–60 cm, sometimes branched from base; hairs gen simple, bristly. **LF** 1.5–6 cm wide, round-reniform, crenate. **INFL**: fls gen > lvs; bractlets 6–10 mm. **FL**: calyx 9.5–14 mm, lobes 5.5–11 mm, 3.5–7 mm wide; petals 15–30 mm, pinkish purple, each with a bright purple basal blotch. **FR**: segments gen 25–35, 2.8–3.5 mm, wafer-like, margins sharp, outer wall net-veined. 2*n*=20. Dry desert scrub; -50–1200 m. D; NV, AZ. [*Malvastrum r.* A. Gray] ❀TRY.

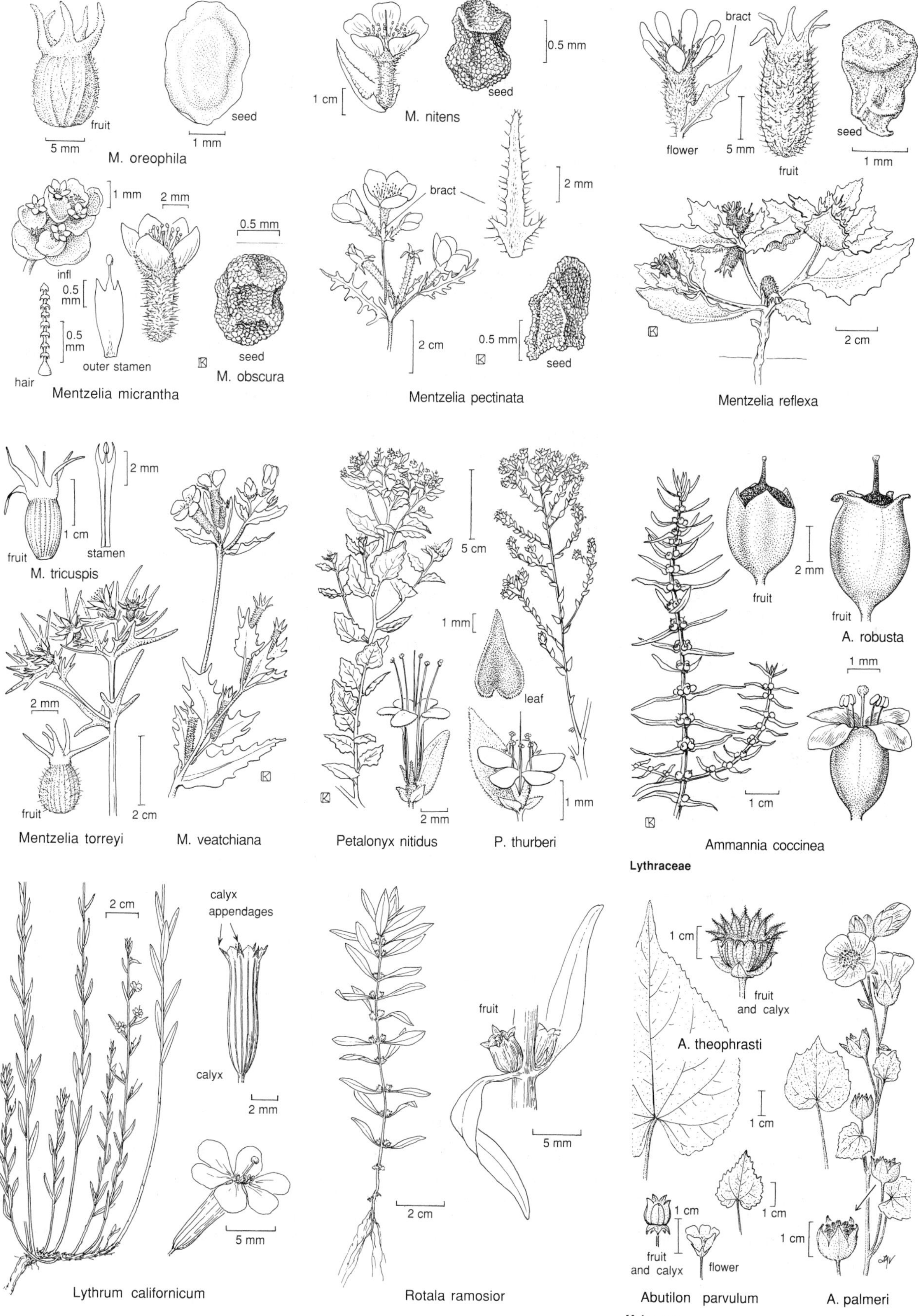
fruit
seed
5 mm
1 mm
M. oreophila
1 cm
seed
0.5 mm
M. nitens
bract
flower
5 mm
fruit
seed
1 mm
1 mm
2 mm
infl
0.5 mm
0.5 mm
hair
outer stamen
0.5 mm
seed
M. obscura
Mentzelia micrantha
bract
2 mm
2 cm
0.5 mm
seed
Mentzelia pectinata
2 cm
Mentzelia reflexa
2 mm
1 cm
fruit
stamen
M. tricuspis
2 mm
fruit
2 cm
Mentzelia torreyi
M. veatchiana
5 cm
1 mm
leaf
2 mm
1 mm
Petalonyx nitidus
P. thurberi
2 mm
fruit
fruit
A. robusta
1 mm
1 cm
Ammannia coccinea
Lythraceae
2 cm
calyx appendages
calyx
2 mm
5 mm
Lythrum californicum
fruit
5 mm
2 cm
Rotala ramosior
1 cm
fruit and calyx
A. theophrasti
1 cm
1 cm
1 cm
fruit and calyx
flower
1 cm
Abutilon parvulum
A. palmeri
Malvaceae

HERISSANTIA BLADDER MALLOW

Ann, per, gen simple and stellate hairs mixed. **ST** gen trailing. **LF** sometimes ± sessile on fl branches; blade cordate. **INFL**: fls solitary in axils; pedicels slender, jointed; bractlets subtending calyx 0. **FL**: stigmas head-like. **FR**: segments 8–15, bladdery-inflated, rounded at top, outer margin dehiscent, side walls thin, flexible, smooth. **SEEDS** gen 2–3 per segment. 5 spp.: warm and trop Am. (L.A.P. Herissant, 18th century French physician, naturalist, poet) [Fryxell 1979 J Arnold Arbor 60:316–319]

H. crispa (L.) Briz. (p. 753) CURLY ABUTILON Ann or per. **ST** < 10 dm. **LF**: blade 1–7 cm, ovate. **INFL**: pedicel gen bent down at joint. **FL**: calyx 4–7 mm, lobes ± lanceolate; petals 6–11 mm, pale yellow. **FR** 13–20 mm diam, spheric, inflated; segments bristly, silvery-shiny inside. RARE in CA. Desert scrub; < 600 m. e PR (se San Diego Co.); to se US, trop. [*Abutilon c.* (L.) Sweet]

HIBISCUS ROSE-MALLOW, HIBISCUS

Ann, per, shrubs, trees. **INFL**: fls gen solitary in axils; bractlets subtending calyx 3–many, slender, persistent, forming an involucre. **FL** gen showy; filament tube 5-toothed, anthers scattered on upper 1/2 below tip. **FR**: capsule; chambers 5. **SEEDS** 2–many per chamber. 200 spp.: warm regions. (Greek: name used by Dioscorides for marshmallow) [Fryxell 1980 Techn Bull USDA 1624:1–53]

1. Ann; petals yellow with dark basal spot; calyx inflated, bladder-like in fr .... ***H. trionum***
1' Per or subshrub; petals white to rose; calyx not inflated in fr
  2. Lf blade 1–3 cm, ovate; fls scattered; peduncle < 1 cm; sepals fused only at base; petals < 2.5 cm; seed densely silky .... ***H. denudatus***
  2' Lf blade 6–10 cm, cordate; fls near st tip; peduncle 1–8 cm; sepals fused halfway; petals 6–10 cm; seed glabrous .... ***H. lasiocarpus***

H. denudatus Benth. (p. 753) PALE FACE Subshrub, tomentose. **ST** 3–6 dm, slender. **LF**: blade 1–3 cm, ovate, finely toothed. **FL**: peduncle < 1 cm; sepals fused only at base; petals 1–2.5 cm, white to lavender, base gen purple. **FR** < calyx. **SEED** reniform, densely silky. Desert scrub of mesas, canyons; < 800 m. DSon; to w TX, n Mex.

H. lasiocarpus Cav. (p. 753) ROSE-MALLOW Per, gen clonal from rhizomes, hairy. **STS** 10–20 dm, some prostrate. **LF**: blade 6–10 cm, cordate, toothed, acuminate. **FL**: peduncle 1–8 cm; calyx bell-shaped, veiny in fr, sepals fused halfway; petals 6–10 cm, white or rose, base red. **FR** 2.5–3 cm, filling calyx. **SEED** spheric, glabrous. RARE in CA. Wet banks, marshes; < 40 m. c&s ScV, deltaic GV; c to se US. [*H. californicus* Kellogg, California hibiscus] Threatened by riverbank alteration. In cult.

H. trionum L. (p. 753) FLOWER-OF-AN-HOUR Ann, bristly-hairy. **ST** 3–6 dm, erect; some branches prostrate. **LF**: blade 2–3 cm wide, 3–5-lobed, coarsely toothed. **FL**: calyx fused ± to tip, 5-ridged or -winged, veiny, inflated and papery in fr; petals ± 2 cm, yellow, base purple-black. **FR** 1.5–2 cm, < or = calyx. **SEED** sparsely hairy. Uncommon. Disturbed places; gen < 100 m. CA-FP; native to Eur, c Afr. Widespread weed.

HORSFORDIA

Subshrub, densely stellate-tomentose or -scabrous. **ST** erect. **LF**: petiole stout; blade lanceolate to round, cordate, ± entire or irregularly fine-toothed. **INFL**: panicle or fls solitary in axils; bractlets subtending calyx 0. **FL**: petals yellow-orange to red-lavender; stigmas head-like. **FR**: segments 8–12, upper portion dehiscent, with 2 spreading, scarious wings; lower portion seed-bearing, indehiscent, firm, net-veined. **SEEDS** 1–3, white-puberulent. 4 spp.: sw US, Mex. (F.H. Horsford, Vermont, botanical collector) [Fryxell 1985 Syst Bot 10:268–272]

1. Petals pink to red-lavender, 10–21 mm; fr segments 10–12, wings >> (± 3 ×) lower part; harshly canescent .... ***H. alata***
1' Petals yellow or orange, 8–10 mm; fr segments 8–9, wings gen = lower part; velvety rusty- or yellow-hairy .... ***H. newberryi***

H. alata (S. Watson) A. Gray (p. 753) **ST** 10–40 dm, few-branched, harshly stellate-canescent. **LF**: blade 2–9 cm, slightly sticky, acute to acuminate, ± cordate, finely toothed. **FL**: pedicels 2–4 mm in fl, 6–10 mm in fr; calyx lobes 3–5 mm, wide-ovate, acuminate; petals 10–21 mm, pink to red-lavender, drying bluish. **FR**: segments 10–12, 7–8 mm; wings ± lanceolate, 3 × length of lower part. **SEED** 1 per segment. Uncommon. Rocky canyons, creosote-bush scrub, washes; 100–500 m. DSon; s AZ, n Mex. ❀TRY.

H. newberryi (S. Watson) A. Gray (p. 753) **ST** erect, 10–30 dm, branched above base; hairs velvety, yellow to rusty, stellate. **LF**: blade 3–15 cm, shallowly cordate, entire to crenate. **FL**: pedicels 5–10 mm in fl, < 20 mm in fr; calyx lobes 2–3 mm, ± triangular, acute; petals 8–10 mm, yellow or pale orange. **FR**: segments 8–9, ± 6 mm; wings ovate, ± = lower part. **SEEDS** 2–3 per segment. Uncommon. Creosote-bush scrub; 100–800 m. PR, w DSon; to sw AZ, n Mex.

ILIAMNA

Per from woody caudex, stellate-hairy. **ST** erect, branched, 6–20 dm. **LF**: blade gen 3–7-lobed. **INFL**: fls gen 1–3 in upper axils (st tip appearing as an interrupted spike); bractlets subtending calyx 3, free, linear to lanceolate, persistent. **FL** 2.5–6 cm diam; petals white to rose-purple; stigmas head-like. **FR**: segments gen 10–15, 6–10 mm, gen > calyx, ± oblong, ± bur-like, dehiscent exc for lower inside margin, attached to fr axis by strong fiber; top rounded; side walls thin, smooth, glabrous; outer surface bristly and stellate. **SEEDS** 2–4 per segment, puberulent. 7 spp.: w N.Am. (Greek: derivation uncertain) [Wiggins 1936 Contr Dudley Herb 1:213–229]

1. Upper lvs shallowly 3-lobed, 1.5–4.5 cm wide, base tapered to truncate; petals 1–3 cm; seeds 3–4 per fr segment .... ***I. bakeri***
1′ Upper lvs deeply 5–7-lobed, 8–20 cm wide, base truncate to cordate; petals 2–3 cm; seeds 2–3 per fr segment .... ***I. latibracteata***

I. bakeri (Jepson) Wiggins (p. 753) BAKER'S GLOBE MALLOW **ST** 3–7 dm, harshly stellate-hairy. **LF**: petiole 1–6 cm, stout; stipules ± 5 mm; blade 1.5–4.5 cm, shallowly 3-lobed, densely stellate-puberulent, base tapered to truncate. **INFL** gen appearing as axillary clusters; bractlets 5–8 mm. **FL**: calyx 9–12 mm; petals 1–3 cm, rose-purple. **SEEDS** 3–4 per fr segment. UNCOMMON. Mtn slopes, juniper woodland, lava beds; 1000–2500 m. CaR, MP; s-c OR. ❀TRY.

I. latibracteata Wiggins (p. 753) CALIFORNIA GLOBE MALLOW **ST** 10–20 dm; finely stellate-hairy. **LF**: petiole 5–14 cm, slender; stipules 6–10 mm; blade 8–20 cm, deeply palmately 5–7-lobed, upper surface glabrous, lower surface canescent, base truncate to cordate. **INFL** sometimes appearing as a crowded raceme; bractlets 10–14 mm. **FL**: calyx 8–10 mm; petals 2–3 cm, rose-purple. **SEEDS** 2–3 per fr segment. UNCOMMON. Conifer forests, streamsides; 500–2000 m. nw KR (Humboldt, Del Norte cos.); sw OR. ❀TRY.

LAVATERA TREE-MALLOW, MALVA ROSA

Ann, bien, per, shrub, gen stellate-hairy. **ST** gen soft-woody. **LF**: blade gen lobed. **INFL**: pedicel jointed above middle; bractlets subtending calyx 3, fused at base into an involucre. **FL** showy; petals rose-purple or white, gen dark-veined, gen reflexed in age; stigmas linear. **FR** ± disk-like; segments 6–10, smooth. **SEED** 1 per fr segment. 25 spp.: ChI, Baja CA, Medit, Australia; naturalized elsewhere. (Lavater brothers, 17th century Swiss physicians, naturalists) [Fernandes 1968 Fl Eur 2:251–253]

1. Free portion of bractlets lanceolate, petals 2.5–4.5 cm; native .... ***L. assurgentiflora***
1′ Free portion of bractlets ovate to round; petals 1–2 cm; alien
 2. Involucre > calyx; lvs densely soft-downy; st base woody .... ***L. arborea***
 2′ Involucre < calyx; lvs sparsely hairy; st base not woody .... ***L. cretica***

L. arborea L. TREE-MALLOW Bien, per, shrub, stellate-tomentose. **ST** 10–30 dm; base woody. **LF**: blade 5–20 cm, shallowly and unequally 5–9-lobed, crenate, densely soft-downy. **INFL** appearing crowded; involucre > calyx; bractlet free portions ovate to round. **FL**: calyx ± 4 mm in fl, not greatly enlarging; petals 1.5–2 cm, pale red-purple. **FR**: segments 6–9, ± glabrous, slightly net-veined. Uncommon. Disturbed places on coastal bluffs, dunes; < 100 m. n CCo, SCo; native to Eur.

L. assurgentiflora Kellogg (p. 753) MALVA ROSA, ISLAND MALLOW Per, shrub, gen glabrous. **ST** 10–40 dm. **LF**: blade 5–15 cm, 5–7-lobed, lobes triangular-ovate, toothed. **INFL** rather open; involucre < calyx, bractlet free portions lanceolate. **FL**: calyx 12–15 mm in fl, not greatly enlarging; petals 2.5–4.5 cm, rose. **FR**: segments 6–8, glabrous to hairy; outer surface smooth. RARE. Coastal bluffs; < 350 m. ChI. Variable among islands. Pls from San Clemente and Santa Catalina Islands (s ChI) have been called ssp. *glabra* Philbr., Southern island mallow. Cult as orn or windbreak on mainland (CCo, SCo) and naturalized there. In cult.

L. cretica L. Ann, bien, sparsely hairy. **ST** 10–30 dm; base not woody. **LF**: blade 4–10 cm, lower ± rounded, upper shallowly 5-lobed, crenate, truncate to ± cordate, sparsely hairy. **INFL**: loose clusters; involucre < calyx; bractlet free portions ovate to round. **FL**: calyx ± 4 mm in fl, greatly enlarging and surrounding fr; petals 1–1.6 cm, pink to lilac. **FR**: segments 7–10, glabrous to puberulent; outer surface gen ± cross-ridged. Uncommon. Disturbed places on coastal bluffs, dunes; < 50 m. CCo, SCo; native to s Eur.

MALACOTHAMNUS BUSH MALLOW

David M. Bates

Subshrubs, shrubs; hairs sparse to dense, stellate (stalked or sessile), simple, and glandular. **ST** erect; branches sometimes spreading. **LF**: blade toothed, lobes 0 or 3–7. **INFL** head-like to panicle-like, composed of axillary clusters (each a cyme) variously arrayed; clusters few–many-fld, loose to dense, congested or well separated; bractlets subtending calyx 3. **FL**: petals > calyx, gen pale pinkish purple or white (often purplish when dry); filament column incl; stigmas head-like. **FR** disk-like; segments 7–14, 2–5 mm, separating, each dehiscing into 2 valves, unarmed, smooth, top hairy. **SEED** 1 per fr segment. 11 spp.: CA, nw Mex. (Greek: soft shrub) Spp. represent major morphological variants; they are all interfertile and sometimes intergrade in areas of proximity. Variation between populations (esp in hairs, infl, and fls) is high and of ± complex pattern within most spp.

1. Bractlets subtending calyx 1–9 mm wide
 2. Fl clusters weakly defined, not subtended by bracts, arrayed in panicle-like infl — SCoR (Salinas Valley) .... ***M. abbottii***
 2′ Fl clusters dense, subtended by bracts, arrayed in head-, spike-, or panicle-like infl
  3. Infl spike- or narrowly panicle-like, clusters congested or well separated; SCoRI, PR .... ***M. aboriginum***
  3′ Infl head-like (sometimes with a few lower clusters); sw CW .... ***M. palmeri***
1′ Bractlets subtending calyx < 1 mm wide
 4. Calyx lobes < or ± = tube, length gen ± 1.5 × width; bractlets gen < half calyx length
  5. Calyx sparsely bristly (some hairs 1–3 mm); fl clusters dense, often well separated in naked, spike-like infl — PR, w DSon .... [2]***M. densiflorus***
  5′ Calyx densely hairy (not bristly); fl clusters dense or open, in spike- to panicle-like infl
   6. Lf blade 5–20 cm, thick, upper surface gen densely hairy, tawny, sinuses closed, lobes overlapping; branches gen stout; SCoRO, SCo .... ***M. davidsonii***
   6′ Lf blade 2–6(11) cm, thin, upper surface gen green, sparsely hairy, sinuses open, lobes not overlapping; branches gen slender; CW, SW .... ***M. fasciculatus***
 4′ Calyx lobes ± = to >> tube, length gen 1.5–3 × width; bractlets gen > half calyx length

7. Calyx lobes in bud joined by their margins (calyx in bud 5-angled or -winged); calyx lobes in fl widely ovate, their hairs gen tawny, coarse
8. Calyx sparsely bristly (some hairs 1–3 mm); fl clusters dense, often well separated in naked, spike-like infl; PR, w DSon . . . . [2]***M. densiflorus***
8′ Calyx densely hairy (not bristly); fl clusters dense or open, gen congested in short, basally leafy, spike-like infl; w&c TR . . . . ***M. marrubioides***
7′ Calyx lobes in bud not joined by their margins (calyx in bud not angled or winged); calyx lobes in fl narrowly triangular to ovate, their hairs white to grayish, soft
9. Petals white; lf lobes 3–5, obvious; lf upper surface becoming glabrous, dark green — s ChI . . . ***M. clementinus***
9′ Petals light purple; lf lobes 0 or obscure; lf upper surface hairy, white or green
10. Branches gen coarse, stiff; fl clusters gen densely many-fld, arrayed in spike- to panicle-like infl; lvs 4–6(11) cm; widespread . . . . ***M. fremontii***
10′ Branches slender, flexible; fl clusters loosely 3–6-fld, gen arrayed in panicle-like infl; lvs 2.5–4.5(7) cm; SCoRO . . . . ***M. jonesii***

M. abbottii (Eastw.) Kearney (p. 757) ABBOTT'S BUSH MALLOW Pl < 15 dm; hairs dense, fine, white. **ST**: branches slender. **LF**: blade < 6.5 cm, ovate, lobes 3 and shallow or 0. **INFL** panicle-like; clusters gen loosely few-fld, not subtended by conspicuous bracts; bractlets 5–8 mm, 1–9 mm wide. **FL**: calyx 9–11 mm, slightly winged in bud, lobes 6–7.5 mm, ovate, acuminate. RARE. Streambanks, often among willows; < 400 m. SCoR (Salinas River, Sargent Creek, Hesperia, all s Monterey Co.). Habitat unusual for genus. Salinas River population (type locality) probably extirpated.

M. aboriginum (Robinson) E. Greene (p. 757) INDIAN VALLEY BUSH MALLOW Pl 20–30 dm; hairs dense, gen tawny. **ST**: branches ± stout. **LF**: blade < 6(–12) cm, ovate to ± round, lobes gen 3–5, sharply angled. **INFL** spike- to panicle-like; clusters gen densely many-fld, subtended by conspicuous bracts; bractlets 6–15 mm, (1)3–9 mm wide, narrowly elliptic to ± round. **FL**: calyx 8–17 mm, angled or winged in bud, lobes 5–11 mm, ovate to ± round, acuminate. $2n$=34. RARE. Open rocky slopes; 150–1700 m. SCoRI (s San Benito, e Monterey, w Fresno cos.), PR (Laguna Mtns, San Diego Co.). PR pls have narrow bractlets, appear to be this sp.

M. clementinus (Munz & I.M. Johnston) Kearney (p. 757) SAN CLEMENTE ISLAND BUSH MALLOW Pl 4–10 dm; hairs sparse to dense, grayish, shaggy, fine. **ST**: branches from base, slender. **LF**: blade < 5(8) cm, ± round; upper surface becoming ± glabrous and dark green, lobes 3–5, sharply angled. **INFL** spike-like; clusters subsessile, loosely few-fld; bractlets 3–9 mm, gen thread-like. **FL**: calyx 5–9 mm, not angled or winged in bud, lobes free, 3.5–6.5 mm, > or >> tube, ± lanceolate-acuminate, hairs white, ± soft; petals white, fading lavender. $2n$=34. **ENDANGERED** CA, US. Rocky canyon walls; ± 250 m. s ChI (San Clemente Island). Lvs like *M. fasciculatus*; hairs and calyx like *M. fremontii*. Threatened by military activity, feral animals. In cult.

M. davidsonii (Robinson) E. Greene (p. 757) DAVIDSON'S BUSH MALLOW Pl 30–50 dm; hairs dense, appearing granular, gen tawny. **ST**: branches gen stout. **LF**: blade 5–11(20) cm, ± round, thick, wavy-margined; upper surface gen densely hairy; lobes 3–7, gen rounded, overlapping (sinuses closed). **INFL** spike- to ± panicle-like; clusters gen peduncled, congested to well separated; bractlets 1.5–4 mm, < 1 mm wide. **FL**: many abortive; calyx 5–9 mm, densely hairy, ± angled in bud; lobes 3–5 mm, ± = tube, triangular to ovate, sometimes 2–4 joined when fl open. $2n$=34. RARE. Slopes, washes; 250–700 m. c SCoRO (s Monterey, nw San Luis Obispo cos.), SCo (e San Fernando Valley, Los Angeles Co.). Intergrades with *M. fasciculatus*. In cult.

M. densiflorus (S. Watson) E. Greene (p. 757) Pl < 20 dm; hairs sparse to moderately dense, tawny; stellate hairs gen overlain by 1–few-armed, bristly hairs < 3 mm. **ST**: branches slender. **LF**: blade 3–6 cm, ovate, often leathery; lobes 0–5. **INFL** spike-like; clusters sessile, gen densely fld, well separated; bractlets 5–15 mm, < 1 mm wide. **FL**: calyx 6–14 mm, sparsely long-bristly, sometimes winged in bud, lobes 4–11 mm, = to >> tube, triangular-acute to ovate-acuminate. $2n$=34. Chaparral; 100–1100 m. PR, w edge DSon; nw Baja CA. Pls with small bractlets and calyx have been called var. *viscidus* (Abrams) Kearney, which intergrades completely with typical form. Also intergrades with *M. fasciculatus* in coastal PR.

M. fasciculatus (Torrey & A. Gray) E. Greene (p. 757) CHAPARRAL MALLOW Pl 10–50 dm; hairs sparse to dense, white to tawny. **ST**: branches gen slender. **LF**: blade 2–6(11) cm, ovate to ± round, gen thin; upper surface gen sparsely hairy; lobes 0–7, angular to rounded, not overlapping (sinuses open). **INFL** spike- to openly panicle-like; clusters gen many-fld, dense and sessile to widely spreading; bractlets 1–8 mm, < 1 mm wide. **FL**: calyx 4–11 mm, lobes 1.8–8, ± = tube, triangular to ovate, acute to short-acuminate, densely hairy, 2–3 often joined when fl open. $2n$=34. Coastal-sage scrub, chaparral; gen < 600 m (< 2450 in PR, Santa Rosa Mtns). NCoRI (Mendocino Co.), interior SnFrB, SCoRO, SW; n Baja CA. [var. *catalinensis* (Eastw.) Kearney; var. *laxiflorus* (A. Gray) Kearney; var. *nesioticus* (Robinson) Kearney (Santa Cruz Island bush mallow, **ENDANGERED** CA); var. *nuttallii* (Abrams) Kearney; *M. arcuatus* (E. Greene) E. Greene (arcuate bush mallow); *M. hallii* (Eastw.) Kearney (Hall's bush mallow); *M. mendocinensis* (Eastw.) Kearney (Mendocino bush mallow, **PRESUMED EXTINCT**); *M. parishii* (Eastw.) Kearney (Parish's bush mallow)] Highly variable, with many indistinct and intergrading local forms. Santa Cruz Island "var. *nesioticus*" is essentially indistinguishable from mainland "var. *nuttallii*"; n CA forms (e.g., "*M. mendocinensis*") tend to have smaller fls than those in s CA. Also intergrades with *M. fremontii* and *M. davidsonii* in n TR, with *M. densiflorus* in PR.

M. fremontii A. Gray (p. 757) Pl 5–20 dm; hairs ± woolly, white. **ST**: branches gen coarse, stiff. **LF**: blade 4–6(11) cm, ovate to wider than long, thin or thick, upper surface hairy, lobes obscurely 5. **INFL** spike- to panicle-like; clusters many-fld, sessile to peduncled, gen well separated; bractlets 3–15 mm, gen < 0.5 mm wide. **FL**: calyx 5.5–13 mm, not angled in bud, lobes free in bud, 3–10 mm, gen > or >> tube, gen narrowly triangular-acuminate, soft-hairy; petals light purple. $2n$=34. Chaparral to pine woodland; 60–1300 m (n CA), 450–2300 m (s CA), 1700–2800 m (SNE). NCoRH, NCoRI, SNF, Teh, e SnFrB, TR, SnJt, SNE. [ssp. *cercophorus* (Robinson) Munz; *M. helleri* (Eastw.) Kearney (Heller's bush mallow); *M. howellii* (Eastw.) Kearney; *M. orbiculatus* (E. Greene) E. Greene] Pls in NCoR with small calyces are the "*M. helleri*" form; those in SnFrB with large calyces are the "ssp. *cercophorus*" or "*M. howellii*" form; both forms also occur elsewhere. s CA pls are gen less woolly ("*M. orbiculatus*" form), may intergrade with *M. fasciculatus*. ❀DRN:6,**7**,**14**,**18**,19–23&SUN:15,**16**,**17**,24.

M. jonesii (Munz) Kearney (p. 757) JONES' BUSH MALLOW Pl < 25 dm; hairs gen velvety, close, white. **ST**: branches slender, flexible, curved. **LF**: blade 2.5–4.5(7) cm, ovate to ± diamond-shaped, thin; lobes 0 or obscurely 3–5; surfaces hairy, ash-green. **INFL** panicle-like; clusters gen 3–6-fld, open, peduncled; bractlets 2.5–8 mm, < 1 mm wide. **FL**: calyx 5–10 mm, not angled in bud, lobes free in bud, 3–7.5 mm, > or >> tube, gen narrow-triangular. $2n$=34. UNCOMMON. Open chaparral in foothill woodland; 250–830 m. SCoRO (Monterey, San Luis Obispo cos.). [*M. gracilis* (Eastw.) Kearney (slender bush mallow); *M. niveus* (Eastw.) Kearney (San Luis Obispo County bush mallow)] ❀DRN,DRY:7–9,**14**,19–23&SUN:**15–17**,**24**.

M. marrubioides (Durand & Hilg.) E. Greene (p. 757) Pl < 20 dm; hairs tawny. **ST**: branches slender. **LF**: blade 1.5–8 cm, widely ovate, thin, coarsely dentate, lobes 0–5. **INFL** spike- to panicle-like, gen leafy; clusters sessile or short-peduncled, congested or well separated; bractlets 5–13 mm. **FL**: calyx 7–15 mm, angled or slightly winged in bud, hairs dense, lobes 4.5–12 mm, > or >> tube, ovate-acuminate. $2n$=34. Chaparral, washes, hillsides; 450–1100 m. c SNF?, WTR, SnGb; nw Baja CA. ❀TRY.

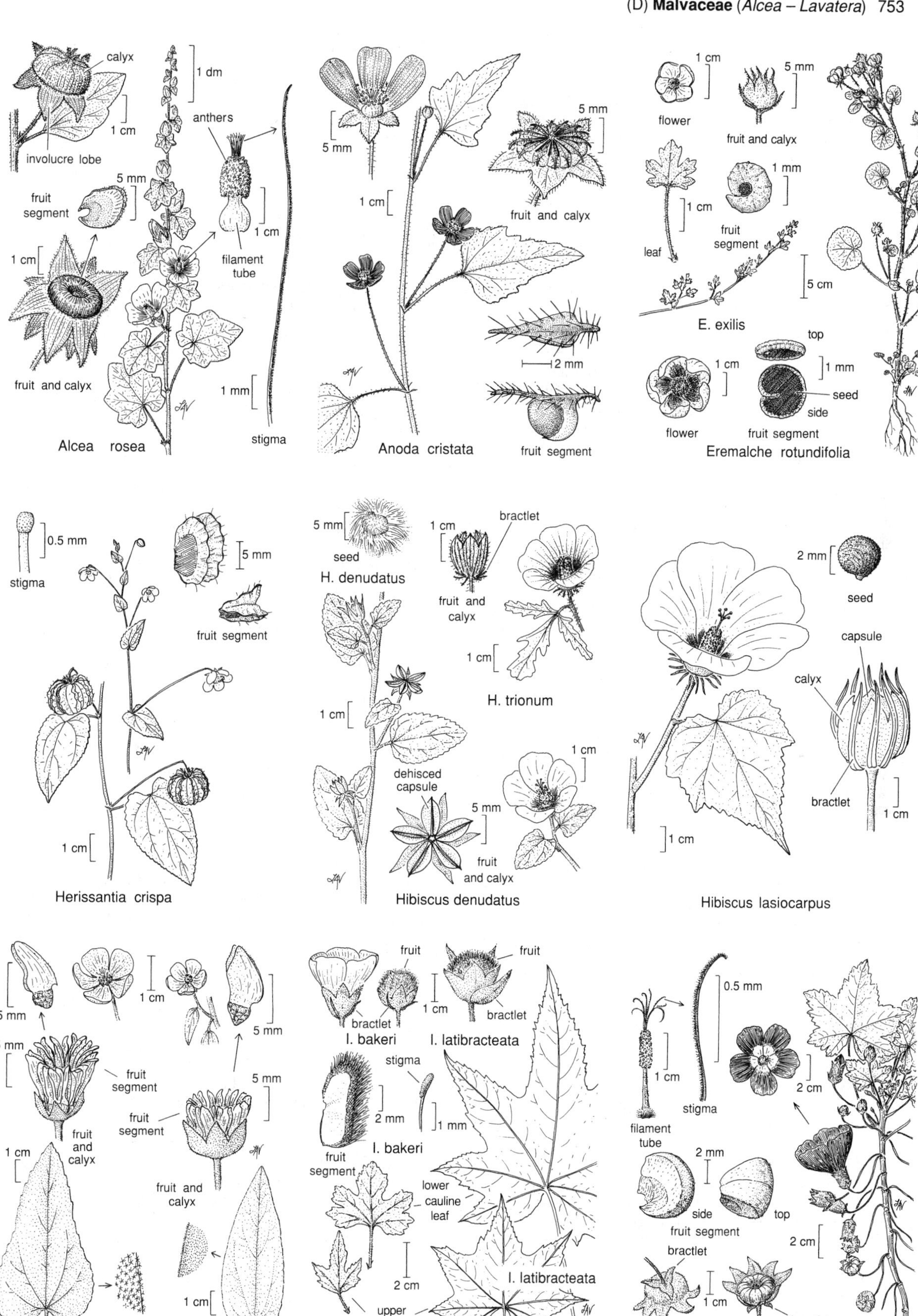
calyx
1 dm
1 cm
involucre lobe
anthers
5 mm
fruit segment
1 cm
filament tube
1 cm
fruit and calyx
1 mm
stigma
Alcea rosea
5 mm
5 mm
1 cm
fruit and calyx
2 mm
fruit segment
Anoda cristata
1 cm
5 mm
flower
fruit and calyx
1 mm
1 cm
leaf
fruit segment
5 cm
E. exilis
top
1 cm
1 mm
seed
side
flower
fruit segment
Eremalche rotundifolia
0.5 mm
stigma
5 mm
fruit segment
1 cm
Herissantia crispa
5 mm
seed
H. denudatus
1 cm
bractlet
fruit and calyx
1 cm
H. trionum
1 cm
1 cm
dehisced capsule
5 mm
fruit and calyx
Hibiscus denudatus
2 mm
seed
capsule
calyx
bractlet
1 cm
1 cm
Hibiscus lasiocarpus
5 mm
1 cm
5 mm
5 mm
fruit segment
5 mm
fruit segment
fruit and calyx
1 cm
fruit and calyx
1 cm
Horsfordia alata
H. newberryi
fruit
fruit
bractlet
1 cm
bractlet
I. bakeri
I. latibracteata
stigma
2 mm
1 mm
I. bakeri
fruit segment
lower cauline leaf
2 cm
I. latibracteata
upper cauline leaf
Iliamna bakeri
0.5 mm
1 cm
stigma
2 cm
filament tube
2 mm
side
top
fruit segment
2 cm
bractlet
1 cm
fruit and calyx
bractlet
Lavatera assurgentiflora

M. palmeri (S. Watson) E. Greene (p. 757) SANTA LUCIA BUSH MALLOW Pl < 25 dm; hairs simple and stellate, canescent or tawny. **ST**: branches stout. **LF**: blade gen < 8 cm, ± ovate, thin or thick, crenate to coarsely dentate, lobes gen 3–5. **INFL** head-like and terminal or spike-like and short; clusters dense, sessile, subtended by conspicuous bracts; bractlets 8–16(21) mm, 1–8 mm wide, ± linear to ovate. **FL**: calyx 8–19 mm, obscurely angled to winged in bud, lobes 5–15 mm, > or >> tube, ovate-acuminate. $2n$=34. UNCOMMON. Interior valleys, foothills; 30–800 m. s CCo, SCoRO (Monterey, San Luis Obispo cos). [var. *involucratus* (Robinson) Kearney (Carmel Valley bush mallow); var. *lucianus* Kearney (Arroyo Seco bush mallow)] Variable in hair types and density, bract and bractlet width, calyx size, infl form; forms intergrade; some approach *M. aboriginum*. ❀DRN,DRY:7–9,**14**,18,**19–23**& SUN:**15–17**,**24**.

MALVA MALLOW

Ann, bien, per, gen ± glabrous. **ST** prostrate to erect, gen < 1 m. **LF** petioled; blade round to reniform, gen crenate, lobes 0 or shallow. **INFL**: fls 1–several in axils; bractlets subtending calyx gen 3, free. **FL**: calyx lobes ± = tube; petals white, pink, or purple; anthers borne along filament tube, not clustered; stigmas linear, on inner side of style branches. **FR**: segments gen 6–15, indehiscent; beak 0. **SEED** 1 per segment. 100 spp.: Eur, Asia, Afr. (Greek: mallow) Some spp. reportedly TOXIC to livestock from selenium or nitrate concentration.

1. Bractlets subtending calyx lanceolate to ovate; fls gen conspicuous
  2. Petals 10–12 mm; st gen trailing or ascending, 2–6 dm ..... [2]***M. nicaeensis***
  2′ Petals 12–30 mm; st erect, 10–30 dm ..... [2]***M. sylvestris***
1′ Bractlets subtending calyx narrowly linear to lanceolate; fls gen inconspicuous
  3. Fr segments smooth or weakly ridged on back, margins round ..... ***M. neglecta***
  3′ Fr segments wrinkled and net-veined on back, margins angled or winged
    4. Petals 12–30 mm, gen bright purple or pink ..... [2]***M. sylvestris***
    4′ Petals 4–12 mm, pale pink or lilac
      5. Calyx somewhat enlarged in fr, lobes spreading or arching over fr, scarcely net-veined; petal claw gen hairy; margins of fr segment angled not winged ..... [2]***M. nicaeensis***
      5′ Calyx much enlarged in fr, lobes spreading, net-veined; petal claw glabrous; margins of fr segment winged ..... ***M. parviflora***

M. neglecta Wallr. (p. 757) COMMON MALLOW, CHEESES Ann, bien. **ST** decumbent, 2–6 dm, gen densely stellate-hairy. **LF**: blade 2–6 cm, lobes 0 (or 5–7, obscure), crenate. **INFL**: fls 3–6 per axil; pedicel in fr < 25 mm; bractlets 3–5 mm, ± broadly linear. **FL**: calyx gen 4–6 mm, not much enlarged in fr, lobes acuminate; petals 8–13 mm, pale lilac or white, claws hairy; stamen-tube hairy. **FR**: segments ± 15, puberulent, back round, smooth or weakly ridged, not net-veined, margins rounded. Disturbed places; < 3000 m. n&c CA; native to Eurasia. [sometimes identified as *M. rotundifolia* L. or *M. verticillata* L. var. *crispa* L.] The latter would key here, but has glabrous stamen-tube, 8–11 fr segments; cult for salad, likely waif.

M. nicaeensis All. (p. 757) BULL MALLOW Ann, bien. **ST** gen trailing to ascending, 2–6 dm, hairy. **LF**: blade 3–12 cm wide, lobes 5–7, shallow, ± sharp. **INFL**: fls 2–6 per axil; pedicels gen reflexed in fr; bractlets 4–5 mm, lanceolate to ovate. **FL**: calyx 4–5 mm in fl, enlarging somewhat and arching over fr, veiny, lobes gen glabrous inside; petals 10–12 mm, pink to blue-violet without darker veins, remaining blue when dry, claws gen hairy. **FR**: segments 7–9, wrinkled, net-veined, glabrous to hairy, margins sharp, not winged. Disturbed places; < 400 m. CA-FP; naturalized elsewhere, esp Mex; native to Eurasia. [not *M. borealis* Wallman]

M. parviflora L. (p. 757) CHEESEWEED, LITTLE MALLOW Ann. **ST** erect, 2–8 dm, widely branching, ± stellate near tips, glabrous below. **LF**: blade 2–8 cm wide, 5–7-angled to -lobed, teeth ± rounded. **INFL**: fls 2–4 per axil, crowded; pedicels in fr gen < 10 mm; bractlets 1–2 mm, linear. **FL**: calyx ± 3 mm in fl, enlarging and spreading in fr, net-veined, lobes widely ovate; petals 4–5 mm, white to pink, glabrous. **FR**: segments ± 11, back wrinkled, net-veined, glabrous to hairy, margins thin-winged, finely toothed. Common. Disturbed places; < 1500 m. CA-FP, D; native to Eurasia; widespread weed. Fls nearly all year.

M. sylvestris L. (p. 757) HIGH MALLOW Bien, per. **ST** gen erect, 5–30 dm; hairs sparse, simple and stellate. **LF**: blade gen 5–10 cm wide, upper 3–7-lobed, lobes ± rounded. **INFL** panicle-like, leafy; bractlets 3–7 mm, lanceolate to ovate. **FL**: calyx 6–8 mm, stellate-hairy outside, not enlarging in fr; petals 12–30 mm, notched, gen bright purple or pink, veins darker, claw hairy. **FR**: segments ± glabrous, back shallowly net-veined and wrinkled, sides wrinkled, angles sharp, not winged. Uncommon. Disturbed places; gen < 300 m. NW, CW, SCo; native to Eur; cult as orn. [ssp. *mauritiana* (L.) Boiss.]

MALVELLA ALKALI-MALLOW

Per. **ST** prostrate to decumbent. **LF**: blade gen asymmetric, gen silvery-stellate. **INFL**: pedicel ± jointed at tip, gen recurved in fr; bractlets subtending calyx 1–3, linear, sometimes deciduous. **FL**: calyx lobes ± = tube; petals stellate-hairy in bud, cream-white to yellow; stigmas head-like. **FR**: segments gen 6–10, indehiscent, slightly inflated, beak 0. **SEED** 1 per segment. 4 spp.: Am, Medit. (Greek & Latin: small mallow) [Fryxell 1974 Southw. Naturalist 19:97–103]

M. leprosa (Ortega) Krapov. (p. 757) ALKALI-MALLOW, WHITE-WEED Per. **ST** decumbent, 1–4 dm, densely white-stellate; some hairs bristly, some scale-like. **LF**: blade 1–3 mm, reniform, round, or triangular, asymmetric, toothed, margin wavy. **INFL**: fls gen 1–3 per axil; bractlets 3, 3 mm. **FL**: calyx 6–10 mm; petals 10–15 mm, cream-white to yellow. **FR**: segments 6–10, 3 mm, gen net-veined on sides. Valleys, orchards, gen in saline soil; < 1000 m. CA (esp GV); to WA, ID, TX, Mex, S.Am. NOXIOUS WEED. [*Sida l.* (Ortega) Schumann var. *hederacea* (Hook.) Schumann] TOXIC to sheep, perhaps other livestock.

MODIOLA

1 sp.: Am; introduced into Old World. (Latin: body of wheel, from fr)

M. caroliniana (L.) Don (p. 757) Per (rarely ann), bristly. **ST** decumbent, rooting at base, 1.5–5 dm. **LF**: blade 5–8 cm, 2–5 cm wide, reniform, round, or triangular, deeply toothed to 3–7-lobed, lobes (if any) ± pinnate. **INFL**: fls solitary in axils; bractlets subtending calyx gen 3, gen narrowly lanceolate. **FL**: calyx 5–6 mm, enlarging slightly in fr, lobes widely triangular-ovate, ± = tube; petals 3–8 mm, gen > calyx, red-orange; filaments borne at stamen-tube tip only; stigmas head-like. **FR**: segments 14–22, 4 mm, reniform, black, upper chamber bristly, 2-beaked, smooth-sided, lower chamber glabrous, sides wrinkled. $2n=18$. Lawns, disturbed places; gen < 400 m. CA-FP; to se US; probably native to S.Am; widely naturalized.

SIDA

Ann, per, shrub. **LF**: blade entire to palmately lobed. **INFL**: fls axillary, gen solitary or clustered; pedicels often jointed; bractlets 0. **FL**: petals gen yellow or white; stigmas head-like. **FR**: segments gen 5–15, indehiscent, often beaked, side walls firm, persistent. **SEED** 1 per fr segment, not enclosed by net-veined envelope. 150 spp.: trop, subtrop, esp Am. (Greek: name of Theophrastus for a similar pl)

S. rhombifolia L. (p. 757) Ann (less commonly per or shrub), puberulent. **ST** 3–10 dm, slender, stiff, straight. **LF**: blade 2–7 cm, 1–3 cm wide, gen diamond-shaped (narrowly oblong to ovate), base tapered, margin toothed, esp in upper 1/2, upper surface gen glabrous, lower surface puberulent, paler. **INFL**: fls solitary in axils; pedicels jointed above middle. **FL**: calyx 6–7 mm, 5–10-veined at base; petals 4–9 mm, pale yellow, > calyx. **FR**: segments 7–14, 3–4 mm, ± net-veined, beaks gen 2, prominent. Uncommon. Disturbed places, fields; gen < 300 m. GV; native to trop.

SIDALCEA CHECKER MALLOW, CHECKERBLOOM

Ann, per, sometimes from long, creeping rhizomes. **ST** gen erect or base ± decumbent. **LVS** gen mostly from near st base; lowest blades gen crenate to shallowly lobed, upper blades gen deeply lobed (gen ± compound). **INFL** gen spike- or panicle-like, gen more open in fr; bracts at pedicel base 2, gen stipule-like; bractlets subtending calyx gen 0(–3). **FL**: calyx lobes = or > tube; petals purple or rose-pink to white; stamen-tube with gen 2 series of ± fused filaments near tip; stigmas linear, on inner side of style branches. **FR**: segments gen 5–10, indehiscent, gen ± beaked, walls thin. **SEED** 1 per fr segment. ± 25 spp.: w N.Am. (Greek: combination of 2 names for mallow) [Hitchcock 1957 Univ Wash Publ Biol 18:1–96] Highly variable and difficult, with many local forms; some pls will not key with certainty. Additional work warranted.

1. Ann (exc 1 ssp. from c&s NCo; if per, upper bracts long-silky); filament series unequal; gen fl in spring
2. Upper stipules and bracts compounded into linear segments; bracts ± = calyx; lf lobes not linear ... ***S. diploscypha***
2′ Upper stipules and bracts gen simple to 2-lobed; bracts gen << calyx; lf lobes gen linear
3. Segments of upper lvs tapered-ovate, 2–5-toothed at tip; st bristly-hairy throughout; s SNF (Tulare Co.) .... ***S. keckii***
3′ Segments of upper lvs ± linear, simple (rarely 2-toothed); st gen ± glabrous below; widespread
4. Outer filaments fused into small sets; back of fr segment evenly net-veined and pitted, glabrous .... ***S. hartwegii***
4′ Outer filaments in a continuous, fused series; back of fr segment longitudinally grooved or hairy
5. Pl densely bristly-hairy above; back of fr segment ± net-veined and pitted, minutely hairy; calyx stellate-canescent .... ***S. hirsuta***
5′ Pl glabrous or sparsely hairy; back of fr segment longitudinally grooved, glabrous; calyx sparsely bristly .... ***S. calycosa***
6. Ann (base of st sometimes decumbent, rooting); bracts 2–5 mm, not silky-hairy, not obscuring calyx; petals 12–20 mm; widespread .... ssp. ***calycosa***
6′ Weak per from rhizomes; bracts 8–12 mm, silky-hairy, obscuring calyx; petals 20–25 mm; c&s NCo .... ssp. ***rhizomata***

1′ Per from fleshy roots, woody caudex, or rhizomes; filament series ± equal; gen fl in late spring or summer
7. Lvs gen evenly arrayed on st, upper and lower gen similar and lobed < halfway to base; fr segment beakless
8. Petals white or purple-tinged; most lvs shallowly lobed (grapeleaf-like); bractlets 0(1–2) ...... ***S. malachroides***
8′ Petals pink to rose-purple; lvs gen ± unlobed, often fan-shaped; bractlets 3
9. Bracts subtending pedicels involucre-like; lvs ovate; infl ± head-like; n SNF (Nevada Co.) .... ***S. stipularis***
9′ Bracts subtending pedicels stipule-like; lvs fan-shaped, rarely lobed; infl open, leafy; more widespread .... ***S. hickmanii***
10. Bracts linear, linear-lanceolate, or oblong, << calyx; bractlets < calyx
11. Pl gen 4–8 dm; calyx densely stellate-hairy, gray, marginal hairs much longer; SCoRO ...... ssp. ***hickmanii***
11′ Pl gen < 3 dm; calyx stellate-hairy but not densely so, hairs uniform in length; SnFrB .... ssp. ***viridis***
10′ Bracts gen lanceolate, ± = calyx; bractlets ± = calyx
12. Upper lvs lobed ± to base, fan-shaped; SCoRO (c San Luis Obispo Co.) .... ssp. ***anomala***
12′ Upper lvs unlobed or very shallowly lobed; SCoRO, w WTR (Santa Barbara Co.), SnBr .... ssp. ***parishii***

7′ Most lvs from near st base, lower gen ± crenate, upper gen very deeply lobed (nearly compound); fr segment gen with a short beak
13. Infl densely head- or spike-like, not 1-sided; calyces in fl gen overlapping; petals < 15 mm
14. Fr segments rough, at least on margins, gen net-veined and pitted; infl sometimes interrupted in fl, gen interrupted (or not dense) in fr; lvs often roughly bristly-hairy
15. Infl gen with evenly distributed fls; margins and sides (not back) of fr segment wrinkled and pitted; calyx hairs not obscuring surface; more widespread (sspp. key following 26′) .... [2]***S. malvaeflora***
15′ Infl with exposed st between clusters; back of fr segment net-veined and pitted; calyx hairs gen obscuring surface; s SN .... ***S. ranunculacea***
14′ Fr segments smooth, not net-veined or pitted; infl continuous, gen ± dense in fr; lvs gen sparsely soft-hairy
16. Basal lvs deeply dissected (nearly compound); fl-st with 1–2 small lvs, gen unbranched; infl elongated in fr — SnBr .... ***S. pedata***

16′ Basal lvs merely lobed; fl-st more leafy, branched; infl not much elongated in fr .... ***S. oregana***
17. Calyx (and gen infl stalk) densely stellate-canescent with minute hairs; dense part of infl gen ± 3 cm; 1° peduncles >> fr cluster; NCoR
18. Lower st < 5 mm wide, gen not stellate, with bristly hairs ± 2 mm; infl in fl gen 1.5–2.5 cm .... ssp. ***hydrophila***
18′ Lower st gen > 5 mm wide, with both minute stellate hairs and 1–1.5 mm bristly hairs; infl gen 2–5 cm in fl .... ssp. ***valida***
17′ Calyx (and gen infl stalk) with some bristly hairs 0.5–1.5 mm; dense part of infl gen > 3 cm; 1° peduncles gen ± = fr cluster; not NCoRI, s NCoRH
19. Calyx enlarged in fr to gen 8–13 mm; calyx with 1–2 mm bristly hairs; NW (Trinity, Humboldt cos.) .... ssp. ***eximia***
19′ Calyx little enlarged in fr to gen 3.5–7 mm; calyx with small stellate hairs and sometimes bristly hairs as well; more widespread .... ssp. ***spicata***
13′ Infl gen not densely head- or spike-like, sometimes 1-sided; calyces in fl gen not overlapping; if infl ± congested, petals gen > 15 mm and fr segments wrinkled
20. Fr segment narrowly wing-margined, glabrous; petals 15–35 mm; infl gen > 30 cm; s CaRF & n SNF (Butte Co.) .... ***S. robusta***
20′ Fr segment rounded or angled, not winged, often hairy; petals and infl gen shorter; more widespread
21. Infl 1-sided, loosely 3–15-fld, axis delicate, curved between fls; pl very glaucous, sparsely and minutely stellate (bristles 0); lf 1° lobes gen 5, entire on upper lvs; sts many from woody taproot, ascending; rhizome 0; CaRH, n&c SNH .... ***S. glaucescens***
21′ Pl differing in 1 or more features from 21.; widespread
22. Infl gen < 2 cm wide (incl petals), not 1-sided, fls > 10; fr segment glabrous or minutely glandular, back smooth, sides sometimes slightly wrinkled; petals 5–15(20) mm; rhizome 0; basal lvs not narrowly dissected
23. Roots fibrous, branched, not fleshy; lvs not very fleshy; infl gen ± dense; various habitats; NW, CaR, MP .... ***S. oregana*** ssp. ***oregana***
23′ Roots fleshy, simple or clustered; lvs gen fleshy; infl rather open; alkaline meadows, seeps; SW, SNE
24. Calyx minutely stellate; fr segment back slightly net-veined; SNE .... ***S. covillei***
24′ Calyx with minute stellate and longer bristly hairs; fr segment back smooth; SW .... ***S. neomexicana***
22′ Infl gen > 2 cm wide, sometimes 1-sided; fr segment obviously wrinkled, net-veined, or pitted, sometimes hairy; petals gen > 15 mm; rhizome often present; basal lvs sometimes narrowly dissected
25. Rhizome long, creeping; hairs of lower st 2–3 mm, simple; lowest lvs gen unlobed, long-petioled; fr segment back deeply pitted, stellate; SN, esp c SNH .... ***S. reptans***
25′ Pl differing in 1 or more features from 25. (if rhizome present, hairs shorter and fr segment not stellate; lvs often lobed or short-petioled; widespread)
26. Lvs (esp upper) deeply and narrowly ± 7-lobed, lobes again divided; infl loosely 3–9-fld; pl glaucous, minutely stellate only; c&s SNH, SNE .... ***S. multifida***
26′ Pl differing in 1 or more features from 26. (at least lowest lvs not so highly lobed; bristly hairs often present; widespread) .... [2]***S. malvaeflora***
27. Lvs gen crowded at st base (pl ± scapose); infl gen 30–45 cm in fr; mid-st lvs deeply lobed, unlike basal; rhizome gen 0; s SNF, SW .... ssp. ***sparsifolia***
27′ Lvs ± evenly distributed on st; infl gen 4–15 cm in fr; mid-st lvs gen like basal; rhizome sometimes present; more widespread
28. Lvs (exc lowest) nearly compound, lobes ternately divided into linear segments; st base decumbent, rooting — s NCoR, CW .... ssp. ***laciniata***
28′ Lvs (esp mid-st lvs) unlobed or at least lobes not ternately divided into linear segments; st base rooting or not
29. Calyx uniformly stellate with gen dense, minute hairs; rhizome 0 or weak; pl sometimes glaucous; lobe tips of upper lvs gen 3-toothed; infl open, sometimes 1-sided; fr segment sometimes stellate; inland; NW, CaR, SN .... ssp. ***asprella***
29′ Calyx gen both stellate and bristly (if bristles 0, stellate hairs larger); rhizome gen obvious; pl not glaucous; lf tips variable; infl open or dense, not 1-sided; fr segment gen glabrous, not stellate; gen more coastal; NCo, CW, SW
30. Fr segment lightly net-wrinkled; calyx sparsely fine-stellate and bristly; infl gen open; st base hairs simple, soft, gen ± 2 mm; SnBr .... ssp. ***dolosa***
30′ Fr segment obviously net-wrinkled; calyx hairs variable, gen denser; infl open or dense; st base hairs variable, gen more coarse; not SnBr
31. Basal lf blade gen < 2 cm; calyx and stipules gen purple; pedicel hair-like — s NCo .... ssp. ***purpurea***
31′ Basal lf blade often > 2 cm; calyx and stipules not purple; pedicel stouter
32. Pl stellate-canescent, hairs soft, felt-like, ± 1 mm; calyx densely stellate; petals gen 25–35 mm — WTR, s SCoRO .... ssp. ***californica***
32′ Pl not canescent, hairs gen few, gen coarse, spreading bristles; calyx gen sparsely hairy; petals often < 20 mm
33. Infl not very dense, fls spreading; pl gen 3–6 dm; rhizomes short; s NCo, SnFrB, CCo, SCo, n ChI .... ssp. ***malvaeflora***
33′ Infl dense, fls gen stiffly erect; pl often 5–9 dm; rhizomes long-trailing; n NCo ...... ssp. ***patula***

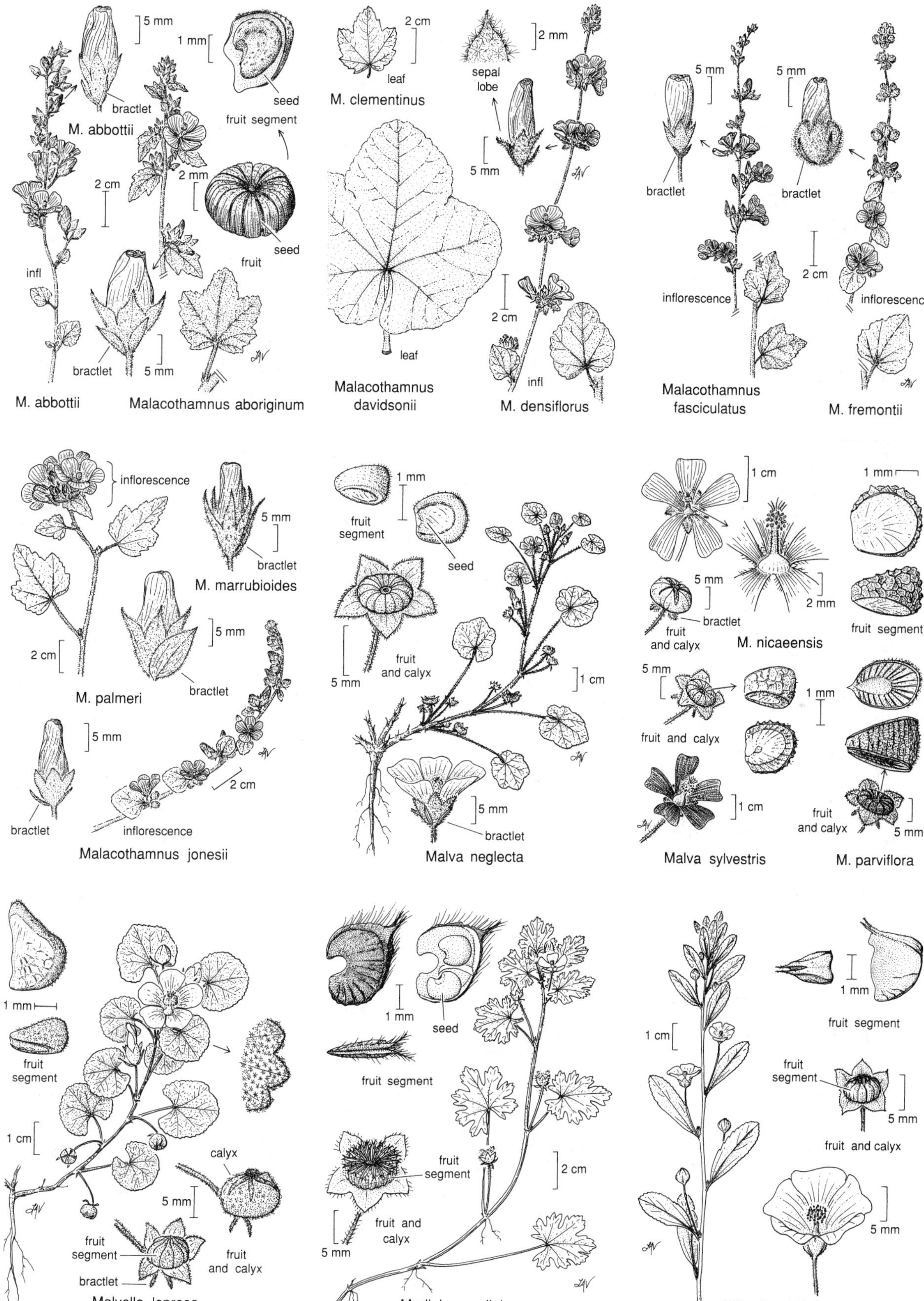

5 mm
1 mm
bractlet
seed
M. abbottii
fruit segment
2 cm
2 mm
seed
fruit
infl
bractlet
5 mm
M. abbottii
Malacothamnus aboriginum
2 cm
leaf
M. clementinus
leaf
Malacothamnus davidsonii
2 mm
sepal lobe
5 mm
2 cm
infl
M. densiflorus
5 mm
5 mm
bractlet
bractlet
2 cm
inflorescence
inflorescence
Malacothamnus fasciculatus
M. fremontii
inflorescence
5 mm
bractlet
M. marrubioides
5 mm
bractlet
2 cm
M. palmeri
5 mm
2 cm
bractlet
inflorescence
Malacothamnus jonesii
1 mm
fruit segment
seed
5 mm
fruit and calyx
1 cm
5 mm
bractlet
Malva neglecta
1 cm
1 mm
5 mm
2 mm
bractlet
fruit and calyx
M. nicaeensis
fruit segment
5 mm
1 mm
fruit and calyx
1 cm
fruit and calyx
5 mm
Malva sylvestris
M. parviflora
1 mm
fruit segment
1 cm
calyx
5 mm
fruit segment
fruit and calyx
bractlet
Malvella leprosa
1 mm
seed
fruit segment
fruit segment
2 cm
5 mm
fruit and calyx
Modiola caroliniana
1 mm
fruit segment
1 cm
fruit segment
5 mm
fruit and calyx
5 mm
Sida rhombifolia

S. calycosa M.E. Jones Ann or per from rhizome. **ST** 3–9 dm, glabrous or sparsely hairy. **LF**: lowest blades gen crenate; upper blades nearly compound with 5–11 linear to oblanceolate divisions. **INFL** dense; bracts simple or 2-lobed. **FL**: calyx 4–12 mm, often purple-tinged or scarious, ± bristly, lobes narrowly ovate, acuminate; petals 12–25 mm, pale purple (rarely white); filaments in 2 distinct continuous series. **FR**: segment 2.5–4.5 mm, back longitudinally grooved, glabrous. Marshes, wet places in woodlands or valleys; < 1200 m. NCo, NCoRO, CaRF, SNF, n SnFrB.

ssp. ***calycosa*** (p. 761) Ann. **ST** slender; base sometimes decumbent and rooting. **LF**: stipule 2–5 mm; blade gen 2–5 cm wide. **INFL**: bracts 2–5 mm. **FL**: calyx 4–7 mm; petals 12–20 mm. **FR**: segment ± 2.5 mm, back moderately grooved, sides net-veined. Wet places, esp vernal pools, swales; < 1200 m. NCoRO, CaRF, SNF, n SnFrB. ❀winter WET or IRR,DRN,SUN:**6,7**,8,9,**14–17**, 18,**19–21**,22–24.

ssp. ***rhizomata*** (Jepson) Munz (p. 761) Per from long rhizomes. **ST** fleshy. **LF**: stipule 12–18 mm; blade 2.5–10 cm wide. **INFL**: bracts 8–12 mm, silky-hairy. **FL**: calyx 6–12 mm; petals 20–25 mm. **FR**: segment ± 4.5 mm, back strongly grooved, sides lightly net-veined. Very uncommon. Marshes near coast; < 30 m. c&s NCo (Mendocino, Sonoma cos.), n CCo (Marin Co.). Easily distinguished by large, fused, ciliate bracts. ❀WET:7,14,19–23& SUN:5,**15–17,24**;GRCVR.

S. covillei E. Greene (p. 761) OWENS VALLEY CHECKERBLOOM Per; roots ± fleshy, simple to clustered; rhizome and caudex 0. **ST** 2–6 dm; lower st finely stellate to coarsely bristly, hairs finer upward. **LF**: blade fleshy, glaucous, densely stellate. **INFL** open, finely stellate. **FL**: calyx 5–8 mm, uniformly fine-stellate; petals 10–15 mm, pink-lavender. **FR**: segment ± 2.5 mm, sparsely glandular-puberulent, back net-veined, sides strongly so. $2n=20$. **ENDANGERED** CA. Alkaline flats; 1100–1300 m. SNE (Owens Valley, Inyo Co.). Threatened by lowering of water table, grazing.

S. diploscypha (Torrey & A. Gray) Benth. (p. 761) Ann. **ST** 4–6 dm, minutely stellate and sharply bristly to tip. **LF**: stipule divided into linear segments; lobes of upper lvs gen 3-toothed or -lobed, bristly. **INFL**: crowded clusters; bracts 8–12 mm, lobes 2–4, linear. **FL**: calyx 8–12 mm, lobes lanceolate, bristly; petals 20–35 mm, minutely fringed, dark pink, center sometimes darker; outer filament series fused in sets of 5–10. **FR**: segment ± 2.5 mm, glabrous, back net-veined. Grassland, open woodland, valleys; < 800 m. NCoRO, CaRF, n&c SNF, ScV, CCo, SnFrB. ❀SUN:5, 6,**7–9,14–24**.

S. glaucescens E. Greene (p. 761) Per from thick taproot and branching caudex, very glaucous, green. **ST** 2–7 dm, slender, gen ascending, not rooting; lower st minutely stellate; upper st ± glabrous. **LF**: blade glabrous or minutely stellate, 1° lobes 5, those of lower st-lvs further divided. **INFL** open, delicate, curved, gen 1-sided. **FL**: calyx 6–10 mm, finely stellate; petals 10–20 mm, pink to pink-purple. **FR**: segment 3–3.5 mm, ± inflated, obviously pitted and net-veined, back glandular-puberulent. $2n=40$. Dry grassy meadows; 900–3000 m. CaRH, n&c SNH. Intergrades with *S. multifida*, *S. malvaeflora* ssp. *asprella*. ❀IRR:1–3,7,14,18& SUN:4,5,**6,15,16,17**.

S. hartwegii Benth. (p. 761) Ann. **ST** 1.5–3 dm, slender; lower st ± glabrous to sparsely stellate. **LF**: blade segments gen 5–7, linear. **INFL** not much longer in fr; fls gen 4–6, overlapping. **FL**: calyx gen 8–10 mm, not much larger in fr, stellate-canescent; petals 18–20 mm, rose-purple; outer filament series fused in small sets. **FR**: segment 2.5–3 mm, net-veined and pitted, glabrous. Dry hillsides, often on serpentine; < 1000 m. NCoRI, CaRF, ScV, n&c SNF.

S. hickmanii E. Greene Per from woody caudex. **ST** 2–8 dm, coarsely and gen densely stellate-canescent. **LVS** ± evenly arrayed on st; blade often fan-shaped, gen wider than long, coarsely crenate to gen shallowly lobed. **INFL**: racemes many, dense or open; bractlets 3, 2–7 mm. **FL**: calyx lobes 9–18 mm; petals 12–25 mm, pale pink to pink-lavender. **FR**: segment 2–2.5 mm, gen smooth to slightly wrinkled, beakless. Chaparral, open conifer forest, sometimes on serpentine; 50–2200 m. s NCo, n CCo, SnFrB, SCoRO, w WTR, SnBr.

ssp. ***anomala*** C. Hitchc. (p. 761) CUESTA PASS CHECKERBLOOM **LVS**: upper blades fan-shaped, deeply 5–7-lobed, ± compound, segments crenate. **INFL**: bracts ± = calyx, lanceolate to widely so; bractlets ± = calyx. **RARE** CA. Closed-cone-conifer forest, gen on serpentine; 600–800 m. SCoRO (near Cuesta Pass, San Luis Obispo Co.). In cult.

ssp. ***hickmanii*** (p. 761) HICKMAN'S CHECKERBLOOM **ST** gen 4–8 dm. **INFL**: bracts << calyx, ± linear; bractlets < calyx. **FL**: calyx densely stellate, hairs longest on lobe margins. RARE. Chaparral; 500–1200 m. SCoRO (Santa Lucia Mtns, Monterey Co.).

ssp. ***parishii*** (Robinson) C. Hitchc. (p. 761) PARISH'S CHECKERBLOOM **INFL**: bracts ± = calyx, lanceolate; bractlets ± = calyx. $n=10$. **RARE** CA. Chaparral, open conifer forest; 1000–2500 m. SCoRO & WTR (Santa Barbara Co.), SnBr. Threatened by grazing, road maintenance.

ssp. ***viridis*** C. Hitchc. (p. 761) MARIN CHECKERBLOOM Pl ± sparsely stellate. **ST** gen < 3 dm. **INFL**: bracts < calyx, linear to narrowly oblong, gen ± fused at base; bractlets < calyx. **FL**: calyx hairs sparse, uniform. RARE. Dry ridges near coast; 50–400 m. s NCo (Sonoma Co.), n CCo (Marin, San Francisco, San Mateo cos.). Like ssp. *hickmanii* but gen smaller.

S. hirsuta A. Gray (p. 761) Ann. **ST** 3–8 dm, stout; lower st ± glabrous; upper st ± densely bristly. **LF**: blades ± bristly, segments of upper narrowly linear, acute. **INFL** dense. **FL**: calyx 8–10 mm, ± densely stellate-canescent and bristly; petals 13–25 mm, deep rose-pink; outer filament series continuous. **FR**: segment 3–4 mm, back wrinkled, net-veined, gen sparsely and minutely hairy, beak tip bristly. Vernally wet places, grassland, or open woodlands; < 1100 m. NCoR, CaRF, n&c SNF, GV.

S. keckii Wiggins (p. 761) KECK'S CHECKERBLOOM Ann. **ST** 1.5–3.5 dm, slender, bristly-stellate throughout. **LF**: lowest blades shallowly 7–9-lobed; lobes of upper 2–5-toothed, with tapered base. **INFL** few-fld. **FL**: calyx 8–11 mm, lobes linear-lanceolate; petals 10–20 mm, deep pink; outer filament series continuous. **FR**: segments 3–4 mm, back net-veined and pitted, gen tinted pink, gen glabrous. **PRESUMED EXTINCT**. Grassy slopes; ± 400 m. s SNF (White River region, s Tulare Co.). Like *S. diploscypha*.

S. malachroides (Hook. & Arn.) A. Gray (p. 761) Per or subshrub from woody caudex, harshly bristly and stellate throughout; fls bisexual, staminate, pistillate, or mixed. **ST** 4–15 dm, stout. **LVS** ± evenly arrayed on st; blades grapeleaf-like, ± coarsely crenate and gen shallowly lobed. **INFL**: panicle, much-branched; units head-like; bracts stipule-like; bractlets gen 0(1–2). **FL**: calyx 6–9 mm, gen purplish; petals 7–15 mm, white or purple-tinged. **FR**: segment ± 2.5 mm, glabrous or with few stellate hairs, beakless, back ridged, sides ± smooth, margins rounded. $2n=20$. RARE. Woodlands and clearings near coast; < 700 m. NCo, NCoRO, n&c CCo, SnFrB, n SCoRO; w OR. In cult.

S. malvaeflora (DC.) Benth. CHECKER MALLOW Per from gen well developed rhizomes and woody caudex. **ST** 1.5–6 dm, gen hairy; hairs variable. **LVS** variable, gen toothed or lobed; upper lvs gen much reduced. **INFL** dense to open; lowest bracts often lf-like, divided to base. **FL**: calyx 5–12 mm in fl, gen slightly enlarged in fr, gen densely stellate and bristly, bristles often on a swollen pad; petals 10–20(35) mm, bright to deep pink, gen white-veined. **FR**: segment 2.5–4 mm, gen coarsely pitted and net-veined (gen more so on sides than back). Common. Gen open, ± dry places in forest or scrub; < 2300 m. CA-FP; OR, Baja CA. Highly variable intergrading complex with many local variants. ❀SUN,DRY or IRR: 4,5,**6–9,14–24**;CVS.

ssp. ***asprella*** (E. Greene) C. Hitchc. (p. 761) Pl often gray-glaucous above. **ST** gen decumbent, rooting; lower st ± stellate and bristly. **LF**: upper and lower often similar, ± lobed, lobe tips toothed or entire; mid-st lf lobes narrowed at base. **FL**: calyx ± uniformly densely fine-stellate. **FR**: segment 3–4 mm, glandular-puberulent, somewhat stellate, net-veined and pitted on sides and back. $2n=20$, 40,60. Open woodlands, sometimes on serpentine; 100–2300 m. NW, CaR, n&c SN; sw OR. [ssp. *celata* (Jepson) C. Hitchc.; ssp. *elegans* (E. Greene) C. Hitchc., Del Norte checkerbloom; ssp. *nana* (Jepson) C. Hitchc.] Variable. Intergrades with *S. glaucescens*.

ssp. ***californica*** (Torrey & Gray) C. Hitchc. (p. 761) Pl gray, soft-hairy. **ST** gen 6–8 dm, densely and ± coarsely stellate-bristly throughout; hairs ± 1 mm. **LF**: blade stellate on both surfaces (some hairs forked), lowest gen unlobed. **FL**: calyx ± uniformly coarse-stellate (some hairs ± 1 mm). **FR**: segment 3–3.5 mm, glandular-puberulent. Coastal scrub, chaparral; < 1000 m. s SCoRO, WTR.

ssp. ***dolosa*** C. Hitchc. (p. 761) Rhizomes gen present. **ST**: lower st long-hairy (hairs ± 2 mm); upper st sometimes glabrous. **LF**: upper surfaces finely stellate. **FL**: calyx sparsely fine-stellate (longest hairs simple). **FR**: segment ± 3 mm, lightly net-veined, glabrous. Uncommon. Open pine forest; 1500–2300 m. SnBr. Like *S. neomexicana*, but rhizomes gen present, fr rougher.

ssp. ***laciniata*** C. Hitchc. **ST**: lower st ± glabrous to sparsely bristly and stellate; upper st gen more densely so. **LF** gen sparsely hairy (lower surfaces more stellate), segments of upper gen > 13, linear. **FL**: calyx finely and sparsely stellate and bristly, marginal and rib hairs longer; petals rarely white. **FR**: segment ± 3.5 mm, ± glandular-puberulent. $2n$=40. Grassland, open woodlands; gen < 700 m. s NCoRO, CW, nw SnJV. Intergrades with ssp. *malvaeflora*, ssp. *sparsifolia*. [var. *sancta* C. Hitchc.]

spp. ***malvaeflora*** (p. 761) **ST** gen 3–6 dm, trailing, rooting, gen coarsely and densely stellate and spreading-bristly. **LF** gen ± densely and coarsely hairy; mid- and upper-st lvs often similar. **INFL**: fls close but gen evenly spaced. **FL**: calyx densely stellate and coarsely bristly, bristles often on a swollen pad, hairs at base shorter, denser, ± curved and tangled, marginal hairs longer. **FR**: segment 3.5–4 mm, ± glabrous, conspicuously net-veined and pitted. $2n$=40,60. Coastal prairie, scrub, open forest; gen < 500 m. s NCo, NCoRO, CCo, SnFrB, SCo, n ChI. Intergrades with most other sspp.

ssp. ***patula*** C. Hitchc. **ST** often trailing and rooting, often 5–9 dm, sparsely bristly and ± stellate. **LF** coarsely bristly and ± stellate. **INFL** gen dense, narrow; fls stiffly erect. **FL**: calyx with fine-stellate and longer, gen forked hairs. **FR**: segment 3.5–4 mm, ± glabrous. $2n$=40. Open coastal forest; gen < 700 m. n NCo (Del Norte, Humboldt cos.); sw OR.

ssp. ***purpurea*** C. Hitchc. (p. 761) Pl ± purple-tinted (esp st base, stipules, calyx). **ST** decumbent, glabrous to sparsely bristly. **LF**: lowest < 2 cm, gen coarsely crenate, unlobed; blade ± bristly on both surfaces. **FL**: calyx sparsely fine-stellate with some coarser bristles. **FR**: segment 3–3.5 mm. $2n$=20. Uncommon. Open coastal forest; gen < 50 m. c NCo (n Sonoma, s Mendocino cos.); n CCo (San Mateo Co.).

ssp. ***sparsifolia*** C. Hitchc. Rhizomes gen 0. **ST**: hairs variable (stellate to bristly). **LVS** crowded near st base (pl appearing ± scapose); upper lvs delicate, lobed nearly to base. **INFL** very long (15)30–45 cm, narrow, often branched. **FL**: calyx densely fine-stellate, margin and rib hairs longer. **FR**: segment 2.5–3 mm, net-veined, glandular-puberulent. Moist, open places; < 2200 m. $2n$= 20,40. s SNF, SCoRO, SW; n Baja CA. Intergrades with ssp. *californica*, *S. neomexicana*.

S. multifida E. Greene (p. 761) Per from woody caudex; very gray-glaucous; rhizomes 0. **ST** 1–6 dm, appressed-stellate. **LF**: blade fleshy, deeply 7-lobed, lobes of all lvs gen deeply ternate, segments linear to oblong. **INFL** open, glabrous to finely stellate; fls gen 3–9. **FL**: calyx 7–10 mm in fl, uniformly finely stellate-hairy; petals 10–25 mm, rose-pink. **FR**: segment 3.5–4 mm, net-veined and pitted, glandular-puberulent on back. $2n$=20. Dry places in sagebrush scrub or pine forest; 2000–2500 m. SNE, c&s SNH; NV. Intergrades with *S. glaucescens*.

S. neomexicana A. Gray (p. 761) Per from clustered, fleshy roots. **ST** 2–9 dm, glabrous to sparsely hairy. **LF**: blade fleshy. **INFL** slender, loosely many-fld, glabrous to stellate. **FL**: calyx gen 5–8 mm, with few small stellate and longer hairs on swollen pads, lobes acuminate, veins gen prominent; petals 6–18 mm, rose. **FR**: segment ± 2 mm, smooth to very lightly net-veined on sides, nearly glabrous. $2n$=20. Uncommon. Alkaline springs and marshes; gen < 1500 m. SCo, PR, SnGb, SnBr; to NM, n Mex. [ssp. *thurberi* (A. Gray) C. Hitchc.]

S. oregana (Torrey & A. Gray) A. Gray Per from woody taproot. **ST** 3–15 dm, rarely rooting at base; lower st coarsely stellate to long-bristly. **LVS** ± basal; lower blades crenate to deeply lobed; upper blades ± compound, segments entire to deeply lobed. **INFL** ± spike-like, dense to open; pedicels gen 1–3 mm. **FL**: calyx gen ± 5 mm in fl, lobes lanceolate, glabrous to densely and uniformly stellate or bristly; petals 5–18 mm, rose-pink. **FR**: segment 2–3 mm, smooth to lightly net-veined and pitted. Meadows, marshes, other wet places; < 3000 m. NW, CaR, SN, GB; to WA, WY, UT. None of the vars. (within sspp.) of C. Hitchc. are recognized.

ssp. ***eximia*** (E. Greene) C. Hitchc. (p. 761) **ST** 9–12 dm, sometimes rooting at base; lower st densely bristly, hairs simple. **INFL** spike-like, very dense. **FL**: calyx ± 10 mm in fr, bristly with 1.5–2.5 mm hairs (sometimes also sparsely fine-stellate). **FR**: segment ± smooth, gen sparsely glandular-puberulent. Uncommon. Meadows; < 1200 m. n NCo, NCoRO. Intergrades with sspp. *oregana* and *spicata*. ❀WET or IRR,DRN:1–5,**6,7,14–17**,18.

ssp. ***hydrophila*** (A.A. Heller) C. Hitchc. (p. 761) **ST** 3–9 dm, gen < 5 mm diam, sometimes rooting at base; lower st bristly, hairs ± 2 mm. **INFL** much-branched; units 1.5–2.5 cm, head-like. **FL**: calyx ± 6 mm in fr, fine-stellate. **FR**: segment ± 2.2 mm, smooth, gen glabrous. $2n$=20. Uncommon. Wet soil of streambanks, meadows; 1100–2300 m. NCoRH, NCoRI. Intergrades with ssp. *valida*. ❀WET or IRR,DRN:1–5,**6,7,14–17**,18.

ssp. ***oregana*** (p. 761) **ST** 3–15 dm; lower st stellate and sometimes bristly. **INFL** open, esp in fr. **FL**: calyx gen 5–8 mm in fl (–10 mm in fr), gen densely fine-stellate to long-bristly, marginal hairs sometimes longer; petals gen 5–15 mm. **FR**: segment 2.5–3 mm, gen glabrous; back gen smooth; sides sometimes net-veined. $2n$=40,60. Moist meadows; 500–2500 m. NW, CaR, MP; to WA, WY, NV. [*S. setosa* C. Hitchc., in part] Variable; intergrades with ssp. *spicata*. ❀WET or IRR,DRN:1–5,**6,7,14–17**,18.

ssp. ***spicata*** (Regel) C. Hitchc. (p. 761) **ST** 3–8 dm; lower st gen long-bristly. **INFL** spike-like, dense, often branched. **FL**: calyx ± 6 mm in fr, gen densely bristly and stellate (or stellate only). **FR**: segment 2.5–3 mm, gen smooth to lightly net-veined, glandular-puberulent. $2n$=20,40. Meadows, streamsides; 1100–3000 m. KR, n NCoRH, CaRH, n&c SNH, GB; OR, w NV. [*S. setosa* C. Hitchc., in part, incl ssp. *s.*, Edgewood checkerbloom] Variable; most readily recognized ssp. ❀WET or IRR,DRN:1–5,**6,7,14–17**,18.

ssp. ***valida*** (E. Greene) C. Hitchc. (p. 761) KENWOOD MARSH CHECKERBLOOM **ST** 9–12 dm, gen > 5 mm wide, often rooting at base; lower st gen bristly and sometimes stellate, hairs 0.1–1.5 mm. **INFL** 2–5 cm, spike-like, dense. **FL**: calyx ± 7.5 mm in fr, fine-stellate and sparsely bristly. **FR**: segment smooth, sparsely glandular-puberulent. $2n$=20. **ENDANGERED** CA. Marsh; ± 150 m. s NCoRO (near Kenwood, Sonoma Co.). Threatened by grazing, marsh alteration. Like ssp. *hydrophila* exc in size, hairs.

S. pedata A. Gray (p. 761) BIRD-FOOTED CHECKERBLOOM Per from fleshy taproot. **ST** 2–4 dm, long-bristly, somewhat stellate near base. **LVS** (all but 1–3) basal; lobes ± ternate, segments linear to elliptic. **INFL** spike-like; upper fls crowded. **FL**: calyx 4–7 mm, uniformly finely stellate, marginal hairs longer; petals 9–12 mm, deep rose-pink, veins darker. **FR**: segment ± 2.5 mm, smooth, ± beakless. $2n$=20. **ENDANGERED** CA, US. Moist meadows in open woodlands; 1600–2500 m. SnBr (Bear Valley, Bluff Lake). Threatened by development, vehicles, grazing.

S. ranunculacea E. Greene (p. 767) Per from slender rhizomes. **ST** 2–5 dm, bristly, esp lower st. **LF**: blade fleshy, stellate to sparsely bristly. **INFL** spike-like; upper fls crowded, often in interrupted clusters; pedicels 1–3 mm. **FL**: calyx 5–9 mm, finely stellate, longest marginal bristles ± 2 mm; petals 5–15 mm, magenta-pink (drying dark purple). **FR**: segment ± 2.5 mm, ± net-veined and lightly pitted, gen sparsely fine-stellate. $2n$=20. Uncommon. Moist meadows, streambanks; 2000–3000 m. s SNF (Greenhorn Mtns, Tulare and Kern cos.). Like *S. malvaeflora* ssp. *patula*. ❀4–6, IRR:1–3,**7,14–16**,17.

S. reptans E. Greene (p. 767) Per from long rhizomes. **ST** 2–5 dm, often rooting; lower st densely long-bristly, hairs 2–3 mm. **LVS**: lowest unlobed, long-petioled, blades bristly. **INFL** ± open (some fls overlapping). **FL**: calyx 8–10 mm, finely stellate, marginal hairs longer; petals 12–20 mm, deep pink to lavender. **FR**: segment ± 3 mm, deeply net-veined and pitted, densely fine-stellate. $2n$=20. Moist meadows, dry places in pine forests; 1200–2000 m. SN (esp c SNH). ❀4,5,**6**,IRR:1–3,**7,14–16**,17,**18**,19–21;GRCVR.

S. robusta E.M. Roush (p. 767) BUTTE COUNTY CHECKERBLOOM Per from rhizomes, barely glaucous. **ST** 5–12 dm, robust; lower st densely fine-stellate; upper st glabrous. **LF**: blade upper surfaces bristly, lower surfaces stellate. **INFL** open, long (gen 30–40 cm). **FL**: calyx 10–15 mm, densely stellate; petals 15–35 mm, very pale pink (gen drying yellowish). **FR**: segment 3–3.5 mm, gen glabrous, narrowly wing-margined above, sides lightly net-veined and pitted, back less so. $2n=20$. RARE. Dry banks in chaparral, oak woodlands; 100–1600 m. s CaRF, n SNF (Butte Co.).

S. stipularis J. Howell & True (p. 767) SCADDEN FLAT CHECKERBLOOM Per from slender rhizomes. **ST** 3–6.5 dm; hairs spreading, simple, bristly. **LVS** ± evenly arrayed on st; stipules < 2 cm, ± asymmetrically ovate, cordate; blades ovate, unlobed. **INFL** ± head-like; pedicels 1–2 mm; lower bracts involucre-like; bractlets 3, linear-oblanceolate. **FL**: calyx ± 8 mm, sparsely to densely bristly; petals ± 15 mm, pink. **FR**: segment ± 2 mm, glabrous, smooth, beakless. **ENDANGERED** CA. Marsh; ± 700 m. n SNF (Scadden Flat, Grass Valley, Nevada Co.).

SPHAERALCEA GLOBEMALLOW

John C. La Duke

Ann, per, stellate-hairy. **LF** petioled; blade linear-lanceolate to triangular, entire to deeply dissected. **INFL**: panicle or raceme-like (unbranched exc for clusters in axils). **FL**: petals obovate, red-orange, white, or lavender; filament tube glabrous or stellate-hairy, anthers yellow or purple; stigmas head-like. **FR** breaking into 9–17 segments; upper part of segment dehiscent, smooth; lower part indehiscent, strongly net-veined, 1–2-seeded. **SEEDS** gray, black, or brown. ± 50 spp.: warm Am, s Afr. (Greek: globe mallow, from fr shape) [Kearney 1935 Univ Calif Publ Bot 19:1–128] Polyploidy and intermediates common.

1. Ann (obviously so) ..... ***S. coulteri***
1′ Per (perhaps in fl during first year)
 2. Lvs linear-lanceolate ..... ***S. angustifolia***
 2′ Lvs rounded, triangular, or nearly compound
  3. Lvs deeply divided (nearly compound)
   4. Sepals < 10 mm ..... ***S. grossulariifolia***
   4′ Sepals 13–15 mm ..... ***S. rusbyi*** var. ***eremicola***
  3′ Lvs triangular or rounded (often lobed, but not approaching compound)
   5. Infl an open panicle (sometimes dense in bud) ..... ***S. ambigua*** (part)
    6. Petals red-orange ..... var. ***ambigua***
    6′ Petals lavender to pink ..... var. ***rosacea***
   5′ Infl raceme-like or a dense panicle
    7. Adult lvs coarsely dentate ..... ***S. munroana***
    7′ Adult lvs entire, crenate, or finely dentate
     8. Lower lf surface markedly ridged ..... ***S. ambigua*** var. ***rugosa***
     8′ Lower lf surface ± smooth
      9. Hairs coarse; lvs gray-green; fr segment 4.5–5 mm, dehiscent part ± 60% of segment ..... ***S. emoryi*** var. ***emoryi***
      9′ Hairs soft; lvs yellow-green; fr segment 2.5–3 mm, dehiscent part < 20% of segment ..... ***S. orcuttii***

S. ambigua A. Gray APRICOT MALLOW Pl canescent. **ST** erect, 5–10 dm. **LF**: blade 15–50 mm, ± triangular, weakly 3-lobed, green or yellow-green, 3-veined, base wedge-shaped, truncate, cordate, margin crenate, wavy. **INFL**: tip not leafy. **FR**: segments 9–13, < 6 mm, < 3.5 mm wide, truncate-cylindric, dehiscent part < 3.5 mm, rounded, 60–75% of segment. **SEEDS** 2 per segment, brown, glabrous, ± hairy. Desert scrub; 150–2500 m. D; to UT, AZ, Mex.

var. ***ambigua*** (p. 767) **LF** not markedly wrinkled. **INFL** open, sometimes raceme-like; fls in clusters or solitary. **FL**: petals red-orange to apricot; filament tube 3–9 mm, hairy, anthers yellow-purple. **FR**: dehiscent portion ± 60% of segment. Habitat and range of sp. [var. *monticola* Kearney; *S. parvifolia* Nelson misapplied to CA pls] ❀DRN,SUN,DRY:2,3,**7–12**,13,**14**,15–17,**18–23**,24.

var. ***rosacea*** (Munz & I.M. Johnston) Kearney PARISH MALLOW **LF** not markedly wrinkled. **INFL** open. **FL**: petals lavender to pink; filament tube ± 11 mm, ± glabrous, anthers purple-gray. **FR**: dehiscent portion ± 60% of segment. Desert scrub; 150–800 m. D (esp DSon); AZ, Mex. ❀DRN,SUN,DRY:7,**8–14**,15–17, **18–23**,24.

var. ***rugosa*** Kearney (p. 767) **LF** markedly wrinkled. **INFL** dense. **FL**: petals red-orange; filament tube ± 5 mm, ± glabrous, anthers yellow. **FR**: dehiscent portion 75% of segment. Habitat and range of sp. ❀DRN,SUN,DRY:2,3,**7–12**,13,**14**,15–17,**18–23**,24.

S. angustifolia (Cav.) G. Don (p. 767) Pl canescent. **ST** erect, < 30 dm. **LF**: blade 15–48 mm, ± linear-lanceolate, sometimes hastately lobed, 3–5-veined, light green-gray, base tapered, margin entire to wavy-crenate. **INFL** raceme-like; tip gen leafy. **FL**: pedicel ± = calyx; petals 7–9 mm, red-orange; filament tube 5.5–7 mm, hairy, anthers yellow. **FR**: segments 9–13, 4–7 mm, 1.5–2 mm wide, truncate-conic; dehiscent part erect, 3–4 mm, ± 75% of segment. **SEEDS** 2, brown-black, hairy. Desert scrub; –6–500 m. DMoj, n DSon; to Kansas, TX, n Mex. [var. *cuspidata* A. Gray; var. *oblongifolia* (A. Gray) Shinn.; *S. emoryi* var. *nevadensis* Kearney] Relationship to *S. emoryi* merits further study. ❀TRY.

S. coulteri (S. Watson) A. Gray (p. 767) Ann; hairs few, long, soft. **ST** sprawling to erect, 1.5–15 dm, slender. **LF**: blade 15–45 mm, sometimes wider than long, ± triangular or cordate, thin, soft, gray-green, lobes 3 or 5, rounded, coarsely toothed. **INFL** gen raceme-like (fls clustered in axils); tip gen leafy. **FL**: pedicel > calyx; petals < 11 mm, salmon-orange; filament tube ± 5 mm, hairy, anthers yellow. **FR**: segments ± 15, 1.5–2 mm, 2–2.5 mm wide, ± hemispheric; dehiscent part ± 1 mm, flat, ± 30% of segment, projecting toward fr axis. **SEED** 1 per segment, brown, glabrous or ± hairy. Uncommon. Dry, sandy places; < 300 m. s DSon; AZ, Mex. ❀TRY.

S. emoryi Torrey var. ***emoryi*** (p. 767) Pl coarsely canescent. **ST** erect, < 21 dm. **LF**: blade 25–55 mm, ovate-triangular, 3-lobed, 3–5-veined, gray-green, base cordate, tip ± truncate to acute, margin crenate. **INFL** raceme-like below, compact panicle of clusters above. **FL**: calyx 6–8 mm; petals 10–12 mm, red-orange to lavender; filament column ± 6 mm, anthers yellow. **FR**: segments 10–16, 4.5–5 mm, 2.5 mm wide, truncate-conic; dehiscent part acute, < 3 mm, ± 60% of segment. **SEEDS** 1–2 per segment, brown or black. Fields, roadsides; < 600 m. s DMoj, DSon; NV, AZ, Mex. [var. *arida* (Rose) Kearney; var. *variabilis* (Cockerell) Kearney] Intergrades with *S. angustifolia*. ❀TRY.

S. grossulariifolia (Hook. & Arn.) Rydb. (p. 767) **ST** erect, 6–10 dm, white-canescent (or green or purple and ± glabrous); base woody. **LF**: blade 17–35 mm, green to gray-green, deeply 3-lobed (each lobe ± lobed or divided), 5-veined, base cordate, margin entire, lobes rounded or pointed. **INFL** raceme-like; fls in clusters or

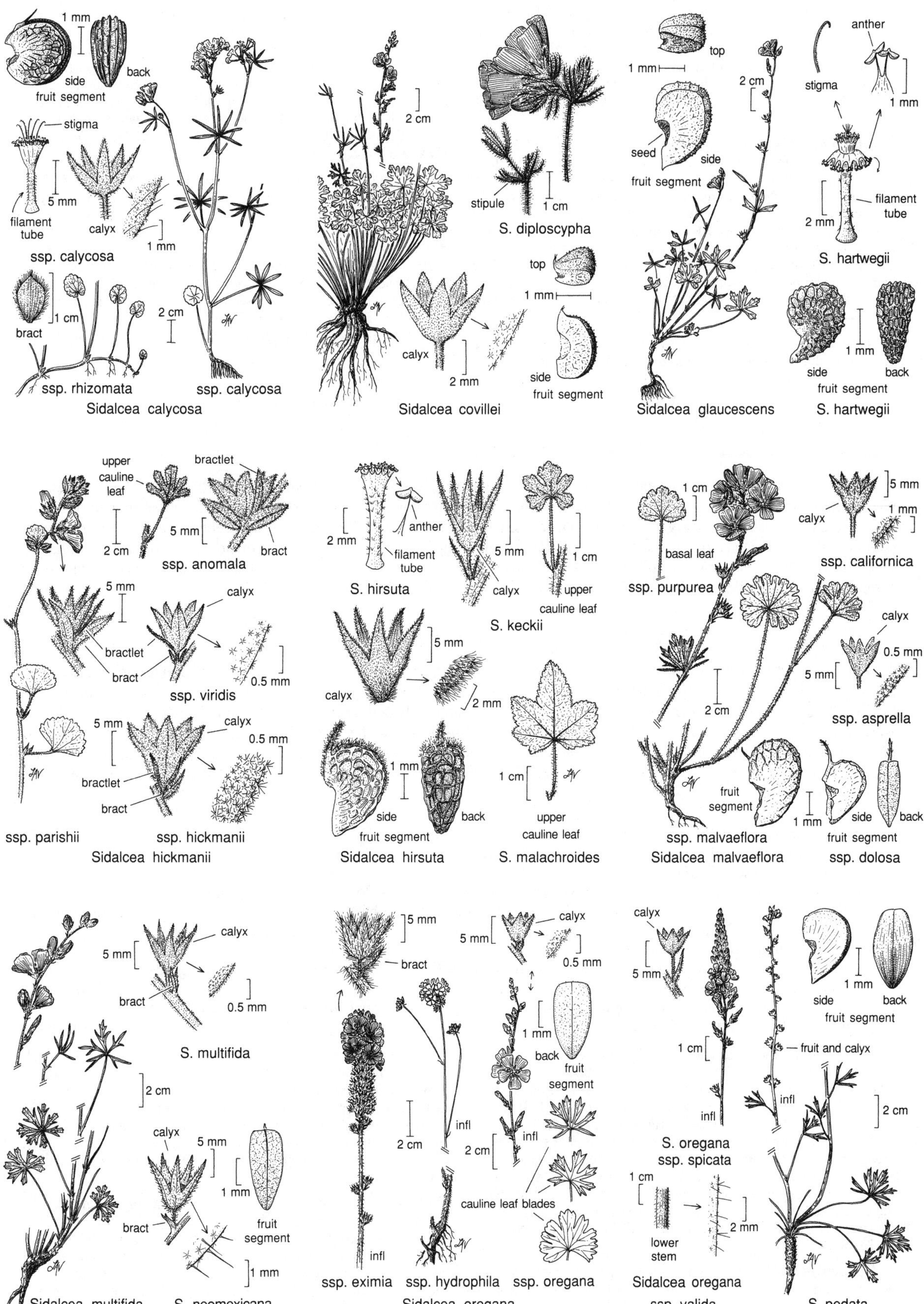
1 mm
side
back
fruit segment
stigma
5 mm
filament tube
calyx
1 mm
ssp. calycosa
bract
1 cm
2 cm
ssp. rhizomata
ssp. calycosa
Sidalcea calycosa
2 cm
stipule
1 cm
S. diploscypha
top
1 mm
calyx
2 mm
side
fruit segment
Sidalcea covillei
top
1 mm
2 cm
seed
side
fruit segment
Sidalcea glaucescens
stigma
anther
1 mm
filament tube
2 mm
S. hartwegii
1 mm
side
back
fruit segment
S. hartwegii
upper cauline leaf
bractlet
2 cm
5 mm
bract
ssp. anomala
5 mm
calyx
bractlet
bract
0.5 mm
ssp. viridis
5 mm
calyx
0.5 mm
bractlet
bract
ssp. parishii
ssp. hickmanii
Sidalcea hickmanii
2 mm
anther
filament tube
S. hirsuta
5 mm
calyx
1 cm
upper cauline leaf
S. keckii
5 mm
calyx
2 mm
1 mm
side
back
fruit segment
Sidalcea hirsuta
1 cm
upper cauline leaf
S. malachroides
1 cm
basal leaf
ssp. purpurea
5 mm
calyx
1 mm
ssp. californica
calyx
0.5 mm
5 mm
ssp. asprella
2 cm
fruit segment
1 mm
side
back
fruit segment
ssp. malvaeflora
Sidalcea malvaeflora
ssp. dolosa
calyx
5 mm
bract
0.5 mm
S. multifida
2 cm
calyx
5 mm
1 mm
fruit segment
bract
1 mm
Sidalcea multifida
S. neomexicana
5 mm
bract
calyx
5 mm
0.5 mm
1 mm
back
fruit segment
2 cm
infl
2 cm
infl
cauline leaf blades
infl
ssp. eximia
ssp. hydrophila
ssp. oregana
Sidalcea oregana
calyx
5 mm
1 cm
infl
fruit and calyx
infl
side
1 mm
back
fruit segment
2 cm
S. oregana
ssp. spicata
1 cm
lower stem
2 mm
Sidalcea oregana
ssp. valida
S. pedata

solitary; tip lfless. **FL**: pedicels < or > calyx; petals ± 11 mm, red-orange; filament tube ± 6 mm, ± hairy, anthers yellow. **FR**: segments 10–12, ± 2.5 mm, ± 2.5 mm wide, truncate-conic to spheric; dehiscent part < 1.5 mm, round, ± 60% of segment. **SEED** 1 per segment, gray, glabrous to ± hairy. Dry, volcanic soils; ± 200 m. se MP (Lassen Co.), naturalized in Santa Barbara Co.; to WA, ID, UT. ❀TRY.

S. munroana (Douglas) Spach (p. 767) Pl canescent. **ST** erect, ± 7.5 dm. **LF**: blade < 4.5 cm, triangular to weakly 5-lobed, 5-veined, green to gray-green, base truncate to tapered, margin coarsely dentate. **INFL** raceme-like; fls in clusters; tip lfless. **FL**: pedicels < calyx; petals 11–14 mm, red-orange; filament tube 7–8 mm, ± hairy, anthers yellow. **FR**: segments ± 12, 3.5–4 mm, 2.5–3 mm wide, spheric, with sharp reflexed tip, dehiscent part < 2 mm, acute, ± 55% of segment. **SEED** 1 per segment, brown, slightly hairy. Uncommon. Dry, open places; ± 2000 m. n SNH (Squaw Creek, Placer Co.); to WA, MT, WY, UT. ❀DRN,DRY,SUN:1–3, **7–9,14–24**.

S. orcuttii Rose (p. 767) CARRIZO MALLOW Taproot large. **ST** erect, 5–12 dm, ± yellow-canescent. **LF**: blade 30–50 mm, rounded to triangular, ± 3-lobed, thick, prominently 3-veined, yellow-green to pale green, base tapered to truncate, margin entire to ± wavy. **INFL** raceme-like or with branches near lfless tip; fls many per axil. **FL**: pedicel < calyx; petals 10–12 mm, red-orange; filament tube 5–6 mm, hairy, anthers yellow. **FR**: segments 12–17, 2.5–3 mm, 2–3 mm wide, ± hemispheric; dehiscent part < 1 mm, round, < 20% of segment. **SEED** 1 per segment, brown, glabrous or ± hairy. Dry, sandy, ± alkaline desert scrub; –20–900 m. s DSon; AZ, Mex. ❀TRY.

S. rusbyi A. Gray var. ***eremicola*** (Jepson) Kearney (p. 767) RUSBY'S DESERT MALLOW, PANAMINT MALLOW **ST** erect, ± 3 dm. **LF**: blade ± palmately compound, 15–20 mm, broadly ovate, 5-veined, light green, base truncate to cordate, margin entire. **INFL**: panicle; tip ± leafy. **FL**: pedicels < calyx; sepals 13–15 mm; petals < 20 mm, red-orange; filament tube ± 9 mm, hairy, anthers yellow. **FR**: segments ± 13, ± 5 mm, 2 mm wide, ± truncate-spheric; dehiscent part ± 3 mm, ± 60% of segment. **SEEDS** 1–2 per segment, black-gray, glabrous to ± hairy. RARE. Desert scrub; 1300–1500 m. n DMtns (Death Valley region, e Inyo Co.; Clark Mtns, ne San Bernardino Co.).

MARTYNIACEAE UNICORN-PLANT FAMILY

Lawrence R. Heckard

Ann, per, glandular-hairy, gen strongly scented. **LVS** simple, opposite or alternate; stipules 0; petiole long. **INFL**: raceme, terminal, bracted; bractlets 2, just below fl. **FL** bisexual; sepals 5, ± unequal; corolla 2-lipped, gen 5-lobed; stamens epipetalous, gen 2 long, 2 short, 1 vestigial; ovary superior, 1-chambered, placentas 2, parietal, each 2-lobed, style >> ovary, curved, stigma 2-lobed, flat, gen closing when touched. **FR**: capsule, drupe-like; outer layer fleshy, deciduous; inner layer ultimately exposed, woody; beak incurved, splitting to form 2 horns (claws). 3 genera, 15 spp.: gen ± trop Am; some cult. Placed by some authors in Pedaliaceae (Sesame Family). [Bretting & Nilsson 1988 Syst Bot 13:51–59]

PROBOSCIDEA UNICORN PLANT, DEVIL'S CLAW

Ann, per; taproot branched or tuberous. **ST** prostrate to spreading, gen < 1 m. **LF**: blade broadly ovate to round or triangular, palmately veined (gen palmately lobed), base cordate. **INFL**: bractlets < calyx. **FL**: calyx 1–2 cm, gen 5-lobed and split to base on lower side (or sepals free); corolla 2–5 cm, bell- to funnel-shaped, showy, tube cylindric, gen < 1 cm, bent downward, throat 10–30 mm, limb with 5 flaring lobes, throat and lower limb with colored lines ("nectar guides"). **FR**: body 5–10 cm, fusiform; surface sculptured or spiny throughout, crested with branched projections gen only along upper suture; beak (claws) 1.5–3 × body. **SEED** 8–13 mm, angled, gen black, corky. 8 spp.: Am. (Greek: beak) Dispersed by attachment of fr claw to animals.

1. Sepals free; woody fr-body surface spiny throughout (sect. *Ibicella*) ..... ***P. lutea***
1′ Sepals fused (calyx 5-lobed, cut to base on lower side); woody fr-body surface smooth or rough, lacking scattered spines (sect. *Proboscidea*)
  2. Largest lvs < 7 cm wide; corolla mostly yellow with maroon-brown flecks; fr body ± 1 cm thick, lanceolate in outline; per; taproot fusiform, tuber-like, wider than st base ..... ***P. althaeifolia***
  2′ Largest lvs gen > 7 cm wide; corolla white to pink or purplish; fr body 1.5–3 cm thick, narrowly ovate in outline; ann; taproot diam ± = st diam
    3. Infl 20–40-fld, overtopping lvs; fl fragrant; corolla upper lip lacking purplish splotch, nectar guides orange; anthers 3.8–6 mm ..... ***P. louisianica***
    3′ Infl 4–10-fld, < or barely overtopping lvs; fl not fragrant; corolla upper lip purplish splotched, nectar guides yellow; anthers 2.2–3 mm ..... ***P. parviflora***

P. althaeifolia (Benth.) Decne. (p. 767) DESERT UNICORN PLANT Per; root fusiform, tuber-like, yellow. **ST** decumbent. **LF**: blade 3–7 cm wide, broadly ovate to round or triangular, gen palmately 3–5-lobed, crenate. **INFL** 5–50-fld, gen overtopping lvs. **FL** fragrant; corolla bright yellow to orangish, maroon-brown-streaked on lower lobe and spotted in 2 rows along upper side of throat; nectar guides orange-yellow. **FR**: body ± 1 cm thick, upper crest with distal teeth sometimes extended into 2 slender, accessory beaks; sometimes also crested along lower suture. **SEED** 6–7 mm. UNCOMMON. Sandy places; < 1000 m. DSon; to TX, n Mex; Peru. [*P. altheaefolia* (Benth.) Decne.] ❀SUN,DRN,DRY:7,14–17&IRR:**8,9**,10,11, **12,13**,18–24.

P. louisianica (Miller) Thell. ssp. *louisianica* COMMON UNICORN PLANT Ann, ill-smelling. **LF**: blade gen 5–20 cm wide, broadly triangular to round, entire to shallowly indented. **INFL** 20–40-fld, overtopping lvs. **FL** musky fragrant; corolla white with purplish tinge and magenta spots in 2 rows internally along upper side of throat and scattered on base of upper lobes; upper corolla lip white to lavender-tinged; nectar guides orange; anther 3.8–6 mm. **FR**: body narrowly ovoid, 2–3 cm thick. **SEED** 7–9 mm. $2n$=30. Uncommon. Open, disturbed areas; < 500 m. esp GV (also NCoR, CW, SCo); to s-c US (where perhaps native), e US. Widely cult as novelty.

P. lutea (Lindley) Stapf (p. 767) Ann, ill-smelling. **LF**: blade 10–20 cm wide, blade ± rounded, sometimes angled, entire to dentate. **INFL** few–many-fld, gen overtopped by lvs. **FL** barely fragrant; sepals free, upper 3 narrower; corolla yellow, often with orange tinge, gen reddish dotted. **FR**: body 3–4 cm thick, ovoid; surface short-spiny. $2n$=30,32. Uncommon. Open, disturbed places; < 500 m. GV, CW, SW; native to S.Am. [*Ibicella l.* (Lindley) Eselt.]

P. parviflora (Wooton) Wooton & Standley ssp. ***parviflora*** Ann; barely ill-smelling. **LF**: blade gen 5–15 cm wide, ± broadly ovate-triangular, entire to shallowly 3–7-lobed or -toothed. **INFL** 4–10-fld, < or barely overtopping lvs. **FL** not fragrant; corolla gen white to pink, purplish mottled, throat with 2 lines of purplish spots or not; upper corolla lip purplish splotched; nectar guides yellow; anther 2–3 mm; stigma not closing with touch. **FR**: body 2–3 cm thick, narrowly ovoid. 2*n*=30. Uncommon. Disturbed, dry places; < 1000 m. SW, D; to TX, Mex. Pls from D with white seed, fr crests > 5 cm long and > 5 mm high, horns > 18 cm have been called var. *hohokamiana* Bretting: cult by sw native Americans for black basket fibers from claws. ❀TRY.

MELIACEAE MAHOGANY FAMILY

Elizabeth McClintock

Shrub, tree (rarely per); wood hard, often aromatic. **LVS** gen alternate, often clustered near st tips, gen 2-pinnate; stipules 0. **INFL**: panicle, raceme, or umbel. **FL** gen bisexual, radial; sepals gen 3–5, sometimes fused at base; petals gen 3–5, ± free (sometimes slightly fused at base or to filament tube); stamens gen 8–12, filaments gen fused; disk gen between stamens and ovary; ovary superior, chambers gen 2–5, placentas axile, style gen 1, stigma gen head-like, lobed. **FR**: gen drupe. **SEEDS** many, often winged or with an aril. ± 50 genera, 550 spp.: trop, subtrop (some temp). Timber crops, incl mahogany (*Swietenia*).

MELIA BEAD TREE

Shrub, tree. **LF** large, deciduous, petioled. **INFL**: panicle; fls many. **FL** white or purple; sepals gen 5; petals gen 5; filament tube 10–12-lobed at tip (lobes sometimes further divided), anthers 10–12; pistil surrounded by, ± = filament tube, ovary chambers 5–8, style ± as wide as ovary and stigma. ± 10 spp.: trop Asia, Australia. (Greek: ash tree, from lf shape)

M. azedarach L. CHINA BERRY, PERSIAN LILAC Tree, < 10 m. **ST**: branches broadly spreading; bark furrowed. **LF**: 1° lflets ± 5–9; 2° lflets gen 5–7 per 1° lflet, 2.5–5 cm, ovate to lanceolate, toothed. **FL**: ± purple, fragrant; sepals ± 2–3 mm; petals ± 5 mm, oblong; filament tube ± 5 mm. **FR** 10–15 mm, spheric, yellow. **SEED** 1, bony. Uncommon. Disturbed places, sometimes persisting near abandoned habitations; < 200 m. SnJV, SCo; native to se Asia, n Australia. Fast-growing, used in reforestation; fr pulp mildly toxic; seeds used for beads.

MENYANTHACEAE BUCKBEAN FAMILY

William J. Stone

Per, ± aquatic; rhizomes thick. **LVS** simple or with 3 sessile lflets, gen alternate; stipules gen 0; petiole sheathing at base. **INFL:** many types. **FL** bisexual, radial; calyx 5-lobed; corolla 5-lobed, rotate to funnel-shaped; stamens 5, alternate corolla lobes, anthers with longitudinal slits; pistil 1, ovary superior or partially inferior, chamber 1, stigma simple or 2-lobed. **FR**: capsule, septicidal. **SEEDS** few–many, smooth, shining. 5 genera, 30–40 spp.: worldwide.

1. Lflets 3; fls white or pink .... **MENYANTHES**
1′ Lvs simple; fls yellow .... **NYMPHOIDES**

MENYANTHES BUCKBEAN, BOGBEAN

1 sp. (Greek: month, for length of flowering)

M. trifoliata L. (p. 767) Pl 17–38 cm; rhizome covered with old lf bases. **ST** prostrate or fl branches ascending. **LVS** basal, alternate; lflets 3, sessile, 2–12 cm, 1–5 cm wide, oblong-obovate, ± entire; petioles 5–30 cm; stipules present as wing-margins of petiole. **INFL**: raceme on long peduncle, 2–4 dm; pedicel 5–25 mm. **FL**: calyx lobes free, 2–5 cm, oblong, spreading, tube short-conic; corolla funnel-shaped, white to pink, tube 5–8 mm, lobes 5–8 mm, spreading, gen pinkish at tip with many scaly hairs on inner surface; filaments thread-like, anthers sagittate; disk of 5 glands at base of ovary; style persistent. **FR** elliptic, almost entirely superior; chamber 1, ± indehiscent. 2*n*=54,108. Ponds, bogs, swamps, wet meadows, seeps, margins of shallow lakes; 900–3200 m. CaR, SN (exc Teh), reported from c NCo, SnFrB; to AK, Colorado, circumboreal. Lvs sometimes used in beer-making as hops substitute. ❀WET (pure or acidic water),SUN:1–3,**4–7,14–17**,18,19.

NYMPHOIDES FLOATING-HEART

Pl submersed; often with a cluster of short roots. **LVS** simple, floating, deeply cordate, gen alternate (those on fl sts opposite); petioles long. **INFL**: umbel in lf axils of fl petiole. **FL**: corolla rotate, lobes bearing a glandular appendage near base; style short or 0, stigma persistent, 2-lobed. 20 spp.: Am, Eurasia, Afr. Sometimes placed in Gentianaceae but differs in having alternate lvs. (Nymphaea-like)

N. peltata (S. Gmelin) Kuntze WATER FRINGE Pl glabrous. **ST** < 160 cm, creeping; fl sts floating. **LF** 3–10 cm wide; margin wavy or entire. **INFL**: fls 2–5 in axillary bundles. **FL**: pedicels 3–10 cm; calyx lobes oblong-lanceolate; corolla 30–40 mm diam, yellow, fringed-ciliate. 2*n*=54. Still water, slow-moving rivers; ± 1100 m. n SNH (Trout Lake, El Dorado Co.); AZ, AR, ne US; native to Eurasia, Medit. Cult in aquatic gardens.

MOLLUGINACEAE CARPET-WEED FAMILY

Wayne R. Ferren, Jr.

Ann, per, shrub, glabrous or hairy. **ST** prostrate to erect. **LVS** simple, gen basal and cauline, alternate, opposite, or whorled, rarely fleshy; stipules 0, conspicuous, or small and deciduous. **INFL**: cyme, cluster, or fl solitary, axillary. **FL** gen bisexual, small, radial; calyx persistent, sepals 4–5, gen free; corolla 0 or small; stamens 5–10, sometimes petal-like, attached to hypanthium, filaments free or fused at base; nectary a ring; ovary superior, chambers 1–10, placentas gen axile, styles 1 or 3–5, gen free. **FR**: gen capsule, gen loculicidal. **SEEDS** 1 or more per chamber, sometimes with arils. 14 genera, 95 spp.: gen trop, subtrop, esp Afr.

1. Hairy; sepals 4–7 mm; pedicel short, stout . . . . **GLINUS**
1′ Glabrous; sepals 1–2 mm; pedicel gen long, slender . . . . **MOLLUGO**

GLINUS

Ann, gen hairy. **ST** prostrate to ascending, branched from base. **LVS** alternate or appearing whorled, entire or toothed, petioled, unequal; stipules 0. **INFL**: cluster head-like, fls gen 5–10; pedicel short. **FL** bell-shaped; sepals 5, margins scarious; petals 0; stamens 3–20, free or fused in groups, outer sterile, filaments short, slender; ovary chambers 3–5, style short, stigmas 3–5. **FR** ovoid. **SEEDS** many, minute, smooth or tubercled; aril coiled around seed. 12 spp.: trop, subtrop. (Greek: sweet juice)

G. lotoides L. (p. 767) Pl hairy; hairs gen forked or stellate. **ST** 1–3 dm. **LVS** entire; petiole ± = blade; blade 0.5–3 cm, obovate to round, ± gray-green, base tapered, tip gen abruptly acute, veins gen prominent. **INFL**: fl ± sessile. **FL**: sepals 4–7 mm, oblong or lanceolate, acute, keeled, tomentose. **FR** 4 mm. **SEED** 0.6 mm, tubercled, blackish brown. Uncommon. Moist or seasonally dry margins of wetlands; < 1000 m. NCoR, GV, SnFrB, SCoRO, SCo, WTR, SnGb, PR; to AK, se US, etc.; native to Eur. *G. radiatus* (Ruiz Lopez & Pavon.) Rohrb., native to trop Am., is reported from ScV, PR: lf green, sparsely hairy, seed 0.4 mm, gen shiny red-brown.

MOLLUGO CARPET-WEED, INDIAN CHICKWEED

Ann, per, glabrous to puberulent. **ST** prostrate to ascending, slender; branches many. **LVS** gen whorled, linear to oblanceolate; stipules 0. **INFL**: umbel, cluster, or fl solitary; pedicel slender. **FL**: sepals 5, ± petal-like, persistent, midrib gen green, margins white-scarious; petals 0; stamens 3, alternate ovary chambers or 5, alternate sepals (rarely 10); ovary ± ovate, chambers 3–5, styles 3–5, linear. **FR** thin, incl in calyx. **SEEDS** many, reniform, smooth, ridged, or tubercled, red-brown; aril short. 20 spp.: trop, subtrop. (Greek: soft pl)

1. St ± erect; lvs linear, gen 1 mm wide, glaucous; infl a peduncled umbel, fls 1–several; stamens 5 . . . . . . ***M. cerviana***
1′ St prostrate; cauline lvs narrowly oblanceolate, 1–8 mm wide, not glaucous; infl a sessile cluster of stalked fls; stamens gen 3 . . . . ***M. verticillata***

M. cerviana (L.) Ser. Ann, glabrous, glaucous. **ST** ± erect, < 20 cm. **LVS** in whorls of 4–10, 3–15 mm, gen 1 mm wide, linear. **INFL**: umbel; fls 1–several; peduncle = or > slender pedicels; bracts minute, margins translucent. **FL**: sepals 5, 1–1.5 mm; stamens 5, alternate sepals, 1 mm; stigmas 3, sessile. **FR** ± round. **SEED** 0.3–0.4 mm, faintly keeled and net-sculptured, brown. $2n=18$. Uncommon. Seasonal pools, sandy washes, flats, slopes; < 1700 m. SnJt, D; to TX, Mex, trop; native to Old World.

M. verticillata L. (p. 767) Ann, mat-forming, < 50 cm diam, glabrous. **ST** prostrate, forked unequally. **LVS** in whorls of 3–6, < 40 mm, 1–8 mm wide, unequal; blade ± oblanceolate. **INFL**: sessile cluster (peduncle 0); fls 1–5; pedicel 5–15 mm, slender. **FL**: sepals 5, 1.5–2.5 mm, oblong; stamens 3–5, 3 mm, alternate ovary chambers; stigmas 3, ± sessile. **FR** ovoid, ± > sepals. **SEED** 0.6 mm, ± smooth, gen with several ridges, dark reddish brown, shiny. $2n=64$. Common. Moist, exposed, disturbed wetland margins, roadsides, fields; < 1000 m. CA-FP (exc D); N.Am; native to trop Am.

MORACEAE MULBERRY FAMILY

Elizabeth McClintock

Per, shrub, vine, tree, gen with milky juice, monoecious or dioecious. **LVS** alternate or opposite, gen simple, evergreen or deciduous, entire to lobed, petioled; stipules sometimes ± 0. **INFL**: raceme, spike, head, or fls enclosed in thickened receptacle tissue, axillary. **FL** unisexual, small, ± radial; sepals gen 4, free or fused at base; petals 0; stamens gen 4, opposite sepals; ovary gen superior, 1-chambered, style simple or 2-parted. **FR**: multiple achenes within fleshy calyces or surrounded by fleshy receptacle tissue. ±40 genera, 900–1000 spp.: trop, subtrop, some temp; many cult (*Ficus*; *Artocarpus*, breadfruit, jackfruit; *Morus*). Pollinated by insects or wind. See Cannabaceae for *Cannabis*, *Humulus*, sometimes included in this family.

1. Fls and frs surrounded by nearly closed, fleshy receptacle tissue, apparent only when infl opened; major lf veins palmate . . . . **FICUS**
1′ Fls and frs external, apparent, in dense raceme, spike, or head; major lf veins pinnate

2. Infl a ± erect, spheric head (pistillate) or umbel-like raceme (staminate); fr ± like a bumpy, yellow-green orange . . . . **MACLURA**
2′ Infl a ± pendent, catkin-like spike; fr ± like a blackberry . . . . **MORUS**

FICUS FIG

Shrub, vine, tree, sometimes growing on other pls, unarmed, monoecious. **LVS** entire or lobed, deciduous or evergreen; major veins palmate. **INFL**: fls internal, enclosed within a pear-shaped or spheric infl receptacle with a small, scale-covered opening at top. **PISTILLATE FL**: style simple. **FR**: multiple of achenes within fleshy infl receptacle. < 500 spp.: trop, subtrop. (Latin: fig) Pollination complex, mediated primarily by small wasps.

F. carica L. EDIBLE FIG Pl < 10 m. **LF**: petiole < 10 cm; blade 10–20 cm, broadly ovate to ± round, upper surface ± scabrous, lower surface hairy, major lobes gen 3–7, palmate, gen > halfway to midrib, each ± pinnately lobed. **FR** 5–8 cm, variable in color. Moist, disturbed areas, persisting near old habitations; < 800 m. n SNF, GV, SnFrB, SCo; native to Medit. Commonly cult.

MACLURA OSAGE ORANGE

Tree, gen with axillary thorns, dioecious. **LVS** alternate but often clustered with infls on short shoots, entire, deciduous; major veins pinnate. **INFL** ± erect; staminate an umbel-like raceme, several per axil, peduncled; pistillate a dense, spheric head, 1 per axil, sessile. **PISTILLATE FL**: style simple. **FR**: multiple of achenes within fleshy calyces on infl receptacle, ± like a bumpy, yellow-green orange. ± 12 spp.: warm parts of Am, Afr, Asia. (W. McClure, American geologist, 1760–1840) Wind-pollinated.

M. pomifera (Raf.) C. Schneider Tree, < 10 m; thorns 0–3 cm. **LF**: petiole 2–3 cm, slender; blade 3–12 cm, ovate to oblong-lanceolate, dark green, upper surface glabrous, lower surface soft-hairy. **FR** 8–15 cm wide, yellow-green, with hard, rough rind. Disturbed areas, abandoned wind-breaks; < 150 m. SnJV, WTR; native to s Great Plains. Widely planted; fr inedible.

MORUS MULBERRY

Tree, unarmed, monoecious or dioecious. **LVS** alternate but sometimes clustered with infls on short shoots, irregularly lobed, toothed, deciduous; major veins pinnate. **INFL** catkin-like spike, ± pendent, peduncled; staminate ephemeral. **PISTILLATE FL** pollinated by wind; style deeply 2-parted. **FR**: multiple of achenes within fleshy calyces on infl receptacle, ± like a blackberry. ± 12 spp.: temp, warm temp n hemisphere. (Latin: mulberry)

M. alba L. WHITE MULBERRY Pl 10–15 m. **LF**: petiole 5–25 mm; blade 8–20 cm, ovate-cordate, coarsely toothed, glabrous or lower surface hairy on major veins, major lobes gen 0–3, sometimes only on 1 side, gen < halfway to midrib. **FR** 1–2.5 cm, juicy, white to pinkish or red-black. Disturbed areas, moist soil, streambanks; < 1300 m. s SNF, SnJV, WTR; native to China. Widely cult; fr edible; lvs larval food of silkworm moth.

MYOPORACEAE MYOPORUM FAMILY

Elizabeth McClintock

Shrub, tree. **LVS** simple, gen alternate, evergreen, often gland-dotted, petioled; stipules 0. **INFL**: axillary clusters; fls 1-few. **FL** bisexual, radial to bilateral; calyx 5-lobed, persistent; corolla 5-lobed, sometimes 2-lipped; stamens gen 4, epipetalous, (sometimes a fifth reduced and sterile); ovary superior, chambers gen 2 (sometimes falsely 3–10), ovules gen 1–2 per chamber, style 1, stigmas 1–2. **FR**: drupe. 3–5 genera, ± 100 spp.: esp Australia (also Pacific islands, e Asia, s Afr, West Indies).

MYOPORUM

LVS alternate, entire or toothed, with often translucent glands. **FL** ± radial; corolla ± = pedicel, > calyx. ± 30 spp.: esp Australia. (Greek: from glandular pores on lvs)

M. laetum Forster f. Shrub or small tree, 3–10 m, much branched, broadly spreading; twig tips and young lvs ± bronze-green, sticky. **LF** < 10 cm, gen lanceolate, finely serrate above middle, bright green, ± fleshy; pores conspicuous. **FL**: corolla ± 10 mm diam, ± bell-shaped, white with purple spots, tube ± = lobes. **FR** 5–10 mm, ovoid, fleshy, pale to dark reddish purple. **SEED** 1. Uncommon. Disturbed, urban areas; < 200 m. SnFrB, CCo, SCo; native to New Zealand. Commonly cult near coast. TOXIC: lvs, frs may be fatal to livestock.

MYRICACEAE WAX MYRTLE FAMILY

James R. Shevock

Shrubs, trees, aromatic, evergreen or deciduous, gen monoecious or dioecious, gen with nitrogen-fixing bacteria in roots. **LVS** simple, alternate, entire to pinnately lobed, resin-dotted; stipules gen 0. **INFL**: spike, axillary, catkin-like; staminate and pistillate spikes separate. **FL** gen unisexual, small; perianth 0. **STAMINATE FL** gen subtended by 2 bractlets. **PISTILLATE FL** subtended by 2–4 bractlets; ovary superior, chamber 1, ovule 1, style 1, stigmas 2, short. **FR**: gen drupe or nut, small, rough, waxy, sometimes winged or bur-like from fused bractlets. 3 genera, ± 50 spp.: gen temp, subtrop. *Comptonia* of e N.Am, *Canacomyrica* of New Caledonia each have 1 sp. Fr of some *Myrica* are boiled to produce fragrant wax.

MYRICA

Shrubs, small trees, monoecious or dioecious (fls all unisexual). **LF** ± spicy-scented; blade unlobed, distal half gen ± sharply serrate. **FR** gen ± spheric, sometimes compressed. ± 48 spp.: temp, subtrop. (Greek: old name for a fragrant shrub)

1. Lvs glabrous, ± glossy, tip gen acute; pl evergreen; monoecious; ± coastal .... ***M. californica***
1′ Lvs sparsely puberulent, ± dull, tip gen obtuse to rounded; pl deciduous; dioecious; lower montane ... ***M. hartwegii***

M. californica Cham. & Schldl. (p. 767) WAX MYRTLE Shrub or small tree, 2–10 m, evergreen. **LF** gen ± narrowly elliptic-lanceolate, glabrous, glossy dark green, scented; tip gen sharply acute. **STAMINATE FL**: stamens 7–16, anthers much more conspicuous than subtending bractlets. **FR** few per pistillate infl, 6–8 mm diam, covered with large, irregular, dark, detaching resin globules and conspicuous whitish wax. Coastal dunes and scrub, closed-cone-pine and redwood forests; < 150 m. NCo, w KR, NCoRO, CCo, w SnFrB, SCo; to WA. ❀**5**;IRR:6,**14–17**,**22–24**&SHD:7–9,18–21; CVS.

M. hartwegii S. Watson (p. 767) SIERRA SWEET BAY Shrub, 1–2 m, deciduous. **LF** gen ± widely oblanceolate, sparsely puberulent, dull green, lighter beneath; tip gen obtuse to rounded. **STAMINATE FL**: stamens 2–4, anthers almost hidden by subtending bractlets. **FR** many per pistillate infl, gen < 2 mm diam, gen dotted with tiny, ± orange resin-glands; whitish wax inconspicuous. Streambanks, other moist places in foothills or low montane yellow-pine forest; 300–1500 m. n&c SN. ❀IRR or WET:**4**,**5**,**16**,**17**&SHD:1,**6**,**7**,14.

MYRTACEAE MYRTLE FAMILY

Elizabeth McClintock

Tree, shrub, evergreen. **LVS** simple, opposite or alternate, entire, often gland-dotted. **INFL**: umbel, panicle of umbels, or fls 1–3 in axils. **FL** gen bisexual, radial; hypanthium cylindric, deltate, ± cup-shaped, or urn-shaped; sepals, petals gen 4–5, free or sometimes fused together to form a deciduous bud cap; stamens gen many, ± showy, borne near top of hypanthium, free or in 4–5 clusters; ovary gen inferior, chambers 1–many, style 1. **FR**: gen berry or capsule. **SEEDS** gen many, small. ± 100 genera, 3000 spp.: trop, subtrop, esp s hemisphere; economically important for timber (*Eucalyptus*), spices (*Syzygium aromaticum*, cloves; *Pimenta dioica*, allspice), edible frs (*Psidium guajava*, guava; *Feijoa sellowiana*, pineapple guava), many orn. [Wilson 1960 J Arnold Arbor 41:270–278]

1. Lvs of 2 kinds, juvenile (young pls, basal sprouts) opposite, adult alternate; sepals, petals fused into deciduous bud cap .... **EUCALYPTUS**
1′ Lvs of 1 kind, opposite or alternate; sepals, petals free
  2. Lvs alternate; sepals, petals 5; fr a capsule .... **LEPTOSPERMUM**
  2′ Lvs opposite; sepals, petals 4; fr a berry .... **LUMA**

EUCALYPTUS GUM TREE

Tree, shrub. **ST** gen erect; bark persistent toward base or not, rough, otherwise gen shed, leaving trunk smooth; twigs gen round. **LVS**: juvenile opposite, often sessile, sometimes glaucous; adult alternate, petioled, gen narrower, > juvenile. **INFL**: umbel or panicle of umbels, 3–many-fld; fls rarely solitary in axils. **FL**: sepals, petals 4–5, fused into a deciduous, gen smooth bud cap; stamens gen white, yellow, pink, or red; ovary chambers 2–7, style < stamens. **FR**: capsule, woody, flat, opening at top. ± 500 spp.: Australia. (Greek: well covered, from deciduous fl bud cap) [Chippendale 1988 Fl Australia 19] Many spp. cult in CA.

1. Infl a panicle of umbels
  2. Lvs gen narrow-lanceolate, < ± 10 × longer than wide; bark not fibrous, shed .... ***E. citriodora***
  2′ Lvs gen round to ovate-lanceolate, < ± 2 × longer than wide; bark fibrous, persistent or shed .... ***E. polyanthemos***
1′ Infl 3- or more-fld umbel or fls solitary in axils
  3. Fls solitary in axils, ± sessile; fl bud 1–2.5 cm, cap centrally knobbed, warty, bluish white, waxy ..... ***E. globulus***
  3′ Fls 3 or more in a gen stalked umbel; fl bud < 1 cm, cap smooth, tan or brown, not waxy
    4. Bark persistent, deep-furrowed, hard, dark brown to ± black; stamens gen pink to red; umbel pendent .... ***E. sideroxylon***
    4′ Bark gen shed, at least above base, trunk ± smooth; stamens white; umbel not pendent
      5. Hypanthium cylindric or urn-shaped, 3–4 × bud cap; fr urn-shaped, valves incl .... ***E. cladocalyx***
      5′ Hypanthium bowl-shaped, deltate, ovoid, hemispheric, or obconic, < to ± 2 × bud cap; fr cup-shaped or ± hemispheric, valves gen exserted
        6. Umbel 3-fld
          7. Tree small, with several trunks; lvs gen juvenile, opposite, ovate, sessile, crowded, ± hiding fls .... ***E. pulverulenta***
          7′ Tree large, with 1 main trunk; lvs gen adult, alternate, lanceolate, petioled, widely spaced, not hiding fls .... ***E. viminalis***
        6′ Umbel 5–15-fld
          8. Bud cap gen hemispheric, often beaked, rarely bluntly conic, ± = hypanthium, gen < 6 mm; branches often hanging in clumps .... ***E. camaldulensis***
          8′ Bud cap gen conic-acuminate, > hypanthium, gen 6–12 mm; branches not hanging in clumps . ***E. tereticornis***

E. camaldulensis Dehnh. RED GUM, RIVER RED GUM **ST** < 25 m, ± straight; branches often hanging in clumps; bark sometimes persistent toward base, otherwise gen shed in irregular strips; trunk ± smooth, gray or tan. **LF** 6–15 cm, 1–2 cm wide, gen narrowly lanceolate. **INFL**: umbel, 5–10-fld, stalked. **FL**: hypanthium 2–3 mm, hemispheric, ± = bud cap; bud cap gen hemispheric, often beaked,

side back fruit segment 1 mm 2 cm rhizome

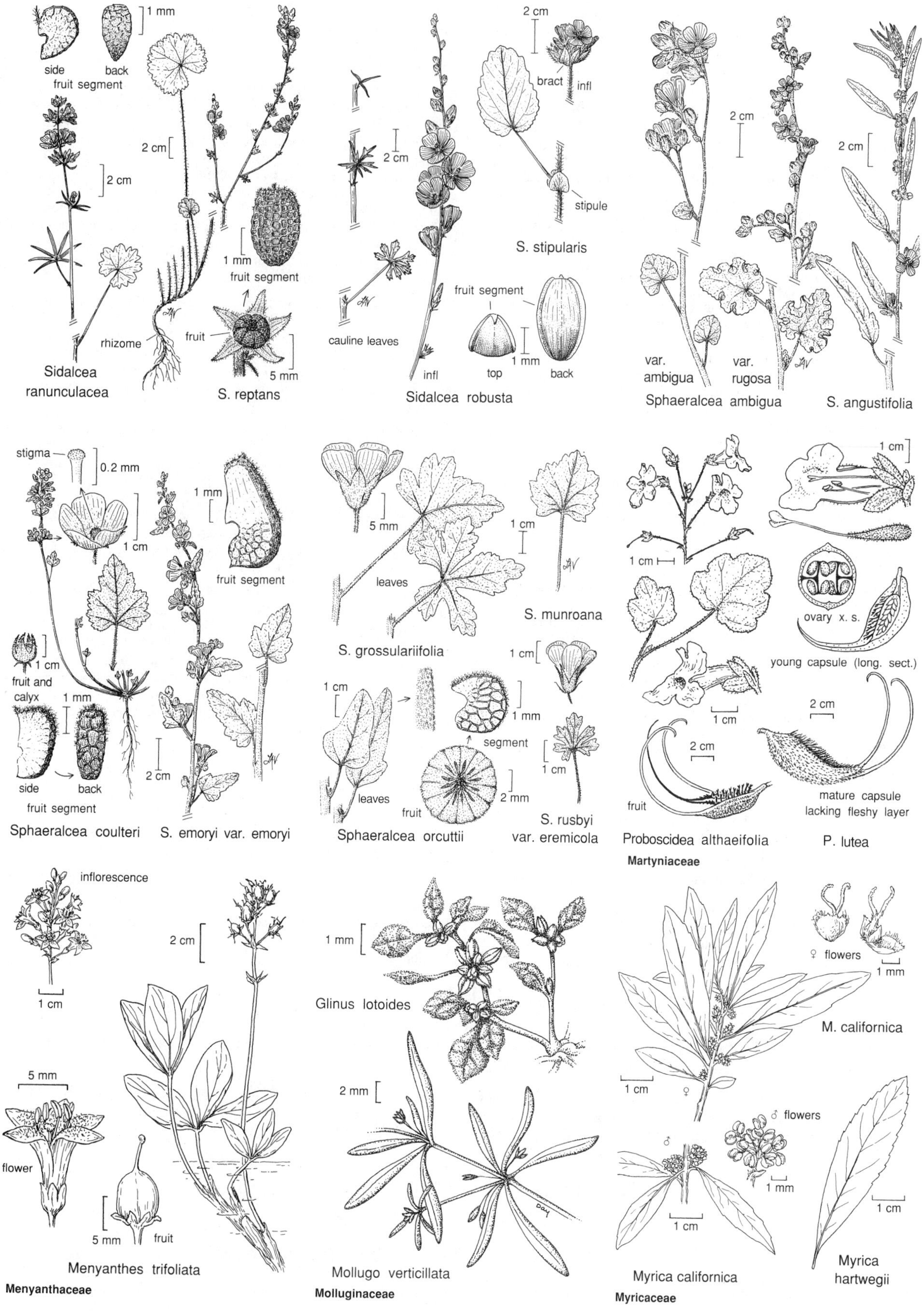

Sidalcea ranunculacea

2 cm 1 mm fruit segment fruit 5 mm

S. reptans

2 cm cauline leaves infl fruit segment top 1 mm back

Sidalcea robusta

2 cm bract infl stipule

S. stipularis

2 cm var. ambigua var. rugosa

Sphaeralcea ambigua

2 cm

S. angustifolia

stigma 0.2 mm 1 cm 1 cm fruit and calyx 1 mm side back fruit segment

Sphaeralcea coulteri

1 mm fruit segment 2 cm

S. emoryi var. emoryi

5 mm leaves

S. grossulariifolia

1 cm

S. munroana

1 cm 1 mm segment leaves fruit 2 mm

Sphaeralcea orcuttii

1 cm 1 cm

S. rusbyi var. eremicola

1 cm 1 cm 2 cm fruit

Proboscidea althaeifolia

Martyniaceae

1 cm ovary x. s. young capsule (long. sect.) 2 cm mature capsule lacking fleshy layer

P. lutea

inflorescence 1 cm 2 cm 5 mm flower 5 mm fruit

Menyanthes trifoliata

Menyanthaceae

1 mm

Glinus lotoides

2 mm

Mollugo verticillata

Molluginaceae

♀ flowers 1 mm

M. californica

1 cm ♀ ♂ flowers ♂ 1 mm 1 cm

Myrica californica

1 cm

Myrica hartwegii

Myricaceae

rarely bluntly conic; stamens white. **FR** 5–6 mm, ± hemispheric, not ribbed; valves exserted. Common. Disturbed areas; < 300 m. NCoRO, SnFrB, GV, SCoRO, WTR, SCo, n ChI (Santa Cruz Island), PR; native to Australia. Commonly cult in CA; most widely planted sp.

E. citriodora Hook. LEMON-SCENTED GUM **ST** > 20–25 m, straight, slender, graceful; bark not fibrous, shed in irregular pieces; trunk smooth, whitish. **LF** 10–20 cm, 1–2 cm wide, narrowly lanceolate, gen lemon-scented. **INFL**: panicle of 3–5-fld umbels. **FL**: hypanthium 5–6 mm, hemispheric, > bud cap; bud cap gen abruptly pointed; stamens white. **FR** < 15 mm, urn-shaped, not ribbed; valves incl. Uncommon. Disturbed coastal, urban areas; 200 m. SCo; native to ne Australia. [*E. maculata* Hook. var. *c*. (Hook.) L. Bailey] Often cult in CA.

E. cladocalyx F. Muell. SUGAR GUM **ST** 10–20 m, gen straight, graceful; bark shed in large irregular patches; trunk ± smooth, white, often mottled gray or tan. **LF** 8–15 cm, 2–3 cm wide, ± widely lanceolate. **INFL**: umbel, gen 4–16-fld, stalked, becoming panicle-like. **FL**: hypanthium < 1 cm, cylindric or urn-shaped, 3–4 × bud cap; bud cap cylindric to urn-shaped, abruptly pointed; stamens white. **FR** 1–1.5 cm, ± urn-shaped, ribbed; valves incl. Uncommon. Disturbed coastal, urban areas; 200 m. SCo; native to s Australia. [*E. corynocalyx* F. Muell.] Sometimes cult in CA. TOXIC to livestock in Australia.

E. globulus Labill. (p. 773) BLUE GUM Tree. **ST** < 45 m, straight; bark sometimes persistent toward base, otherwise shed in irregular strips; trunk smooth, bluish gray; twigs ± square or winged. **LF** 10– 20 cm, 2.5–4 cm wide, gen narrow-lanceolate, often sickle-shaped, gen aromatic. **INFL**: fls solitary in axils, ± sessile. **FL**: hypanthium < 2 cm, deltate, < bud cap; bud cap flat-hemispheric, conspicuously, centrally knobbed, warty, bluish white, waxy; stamens cream-white. **FR** > 2 cm, ± 4-ribbed, warty, glaucous, rim wide, thickened; valves ± incl. Disturbed areas; < 300 m. NCoRO, GV, SnFrB, SCoRO, SCo, n ChI (Santa Cruz Island), WTR, PR; native to se Australia. Most easily recognized (large, solitary fls, frs), most commonly cult and naturalized sp. in CA; growth rapid.

E. polyanthemos Schauer SILVER DOLLAR GUM, RED BOX **ST** < 25 m; bark fibrous, persistent or shed in flakes or irregular strips; trunk ± smooth, mottled gray or tan. **LF**: adult 5–10 cm, 1.5–5 cm wide, gen round to lanceolate-ovate, gray-green. **INFL**: panicle of 3–6-fld umbels. **FL**: hypanthium ± 3 mm, deltate, ± 2 × bud cap; bud cap conic; stamens white. **FR** 5–6 mm, pear- to bowl-shaped; valves incl. Uncommon. Disturbed coastal, urban areas; < 200 m. SnJV, SnFrB, SCo, WTR; native to Australia. Branches with juvenile, rounded, glaucous lvs used in fl arrangements.

E. pulverulenta Sims (p. 773) SILVER-LEAVED GUM, MONEY TREE **ST** < 10 m, straggling, irregularly branched; bark shed in thin strips; trunks several, ± smooth, reddish tan. **LVS**: juvenile crowded, ± hiding fls, opposite, sessile, ± clasping, 1–2 cm, broad-ovate, gray-green, very glaucous, produced by mature pls; adult gen 0, if formed then alternate, 10 cm, oblong to lanceolate, gen very glaucous, short-petioled. **INFL**: umbel, 3-fld, ± sessile. **FL**: hypanthium 5–6 mm, obconic, ± 2 × bud cap; bud cap deltate; stamens white. **FR** 5–9 mm, cup-shaped; valves gen ± incl. Uncommon. Disturbed coastal, urban areas; < 250 m. SnFrB; native to se Australia. Commonly cult in CA; branches with juvenile, rounded, silvery lvs used in fl arrangements; adult lvs gen not formed in CA.

E. sideroxylon Cunn. RED IRON BARK **ST** 7–25 m; bark persistent, deep-furrowed, hard, dark brown to ± black. **LF** 6–14 cm, 1–2 cm wide, lanceolate, dull gray-green. **INFL**: umbel, 3–7-fld, stalked, pendent. **FL**: hypanthium 4–6 mm, ovoid to hemispheric, > bud cap; bud cap conic, smooth; stamens gen pink to red. **FR** ± 1 cm, ovoid; valves incl. Uncommon. Coastal, disturbed urban areas; < 200 m. SCo; native to se Australia. Sometimes cult in CA.

E. tereticornis Smith FOREST RED GUM **ST** 10–25(50) m, straight; bark persistent toward base, otherwise gen shed in irregular sheets or flakes; trunk ± smooth, often mottled, gray or bluish gray. **LF** 8–20 cm, 1–2.5 cm wide, narrowly lanceolate, often sickle-shaped. **INFL**: umbel, 5–15-fld, stalked. **FL**: hypanthium 2–3 mm, deltate, ± 1/3 × bud cap; bud cap gen conic-acuminate, smooth; stamens white. **FR** 7–9 cm, hemispheric; valves exserted. Uncommon. Disturbed coastal, urban areas; < 200 m. SnJV, SCo, WTR; native to e Australia. Commonly cult in CA.

E. viminalis Labill. MANNA GUM **ST** 25–50 m, straight; bark sometimes persistent toward base, otherwise gen shed in long, narrow, irregular strips; trunk ± smooth, whitish or tan. **LF** 10–15 cm, 1–2.5 cm wide, lanceolate. **INFL**: umbel, 3-fld. **FL**: hypanthium 2–3 mm, ovoid, ± = bud cap; bud cap deltate-acute, smooth; stamens white. **FR** 5–7 mm, ± hemispheric; valves exserted. Uncommon. Disturbed urban areas; < 100 m. CCo; native to se Australia. Distinguished from *E. globulus, E. camaldulensis, E. tereticornis* most readily by infl; commonly cult in CA.

LEPTOSPERMUM

Shrub, small tree. **ST** erect to spreading, twisted; bark shed in long strips. **LVS** alternate; veins 0–3. **INFL**: fls 1–3 in axils. **FL**: hypanthium gen widely cup-shaped; sepals, petals 5, free; petals white, pink, or red; stamens many; ovary chambers gen 5–many, style = or < stamens. **FR**: capsule, woody, flat, opening at top. **SEEDS** many, small. ± 70 spp.: esp Australia, few New Zealand, Malay Archipelago. (Greek: slender seed)

L. laevigatum (Gaertner) F. Muell. Pl glabrous, glaucous. **ST** spreading, < 6 m; trunk ± smooth. **LF** 10–15 mm, obovate-oblong; veins 3, inconspicuous; tip blunt or with short, central point. **INFL**: fls often solitary in axils; pedicel short. **FL** 1.5–2 cm wide; petals spreading, white; stamens gen 20–30. **FR** 7–8 mm; valves 8–10. Uncommon. Disturbed coastal areas; < 100 m. CCo; native to se Australia. Commonly cult along coast in CA, sometimes for stabilization of moving sand.

LUMA

Shrub, small tree. **ST** erect; bark shed in flakes; trunk smooth. **LVS** opposite. **INFL**: cyme 3-fld or fls solitary in axils. **FL**: hypanthium ± cup-shaped; sepals, petals 4, free; stamens many; ovary chambers 2, style 6–7 mm. **FR**: berry. 4 spp.: Chile, w Argentina. (Chilean name) [Landrum 1988 Proc Cal Acad Sci 45:313–316]

L. apiculata (DC.) Burret TEMU **ST** 1.5–5(20) m, many-branched; trunk golden-brown. **LF** 2.5 cm, ovate, abruptly, sharply pointed, thinly leathery, dark green. **INFL**: fls gen solitary in axils. **FL** ± 12 mm wide; petals rounded, cream-white, pink-tinged, upper surface concave; stamens forming a dense ring, < petals. **FR** ± 12 mm wide, black. Uncommon. Cool coastal, urban areas; 100 m. CCo; native to Chile, w Argentina. [*Eugenia a.* DC.] Sometimes cult in CA.

NYCTAGINACEAE FOUR O'CLOCK FAMILY

Richard Spellenberg

Per, shrub, tree, glabrous or hairy. **ST** often forked. **LVS** opposite, sessile or petioled, pairs gen unequal; blade gen entire. **INFL** gen forked; of spikes, clusters, or umbels, each unit sometimes with a calyx-like involucre. **FL** bisexual, radial; perianth of 1 whorl, petal-like, bell- to trumpet-shaped, base hardened, tightly surrounding ovary in fr, lobes 4–5, gen notched to

± bilateral; stamens 1–many; ovary superior (appearing inferior because of hardened perianth base), style 1. **FR**: achene or nut, smooth, wrinkled, or ribbed. 30 genera, 300 spp.: warm regions, esp Am; some orn (*Bougainvillea*; *Mirabilis*, four o'clock).

1. Stigma linear, incl in perianth; infl a head or umbel; fr gen wing
 2. Fr wing thick or 0, not continuous above fr body; receptacle conic, ± smooth .... **ABRONIA**
 2′ Fr wing membranous, continuous above fr body; receptacle flat or ± conic, studded with peg-like pedicels .... **TRIPTEROCALYX**
1′ Stigma ± spheric, gen exserted; infl various; fr gen unwinged
 3. Bracts 3–5, gen fused, forming a calyx-like involucre; fls 1–many per involucre, but infl often appearing as 1 fl
  4. Fls 3 per involucre, blooming simultaneously, each subtended by an involucral bract; fr bilateral ... **ALLIONIA**
  4′ Fls 1–16 per involucre, blooming sequentially; involucre gen 5-lobed; fr radial .... **MIRABILIS**
 3′ Bracts 1–3, free, not forming an involucre; infl not fl-like
  5. Fr wing thin, membranous; perianth 30–40 mm, trumpet-shaped .... **SELINOCARPUS**
  5′ Fr wing 0; perianth < 30 mm or > 70 mm, trumpet-shaped or not
   6. St erect, > 7 mm diam; fr glabrous, ridges 10, fine, inconspicuous .... **ANULOCAULIS**
   6′ St gen trailing, gen << 5 mm diam; fr sometimes hairy, ridges or ribs 5, prominent
    7. Perianth > 70 mm, trumpet-shaped; fr glabrous, oblong, 6–10 mm .... **ACLEISANTHES**
    7′ Perianth < 30 mm; fr glabrous or hairy, club-shaped, < 3.5 mm .... **BOERHAVIA**

ABRONIA SAND VERBENA

Ann, per, gen glandular. **ST** prostrate to ascending, gen ± red. **LF** gen fleshy, petioled. **INFL**: head or umbel; fls opening together or outer first; receptacle conic, ± smooth. **FL**: perianth salverform to trumpet-shaped, gen fragrant, lobes 4–5; stamens 4–5, incl; stigma linear, incl. **FR**: body fusiform; lobe-like wings (0)2–5 (if present, prominent, opaque, thick, not continuous above fr body). 25 spp.: w N.Am. (Greek: graceful) [Galloway 1975 Brittonia 27:328–347] Closely related to *Tripterocalyx*.

1. Perianth yellow or wine-red — lf fleshy; seashore
 2. Lf blade ± as long as wide; fl yellow .... ***A. latifolia***
 2′ Lf blade longer than wide; fl wine-red .... ***A. maritima***
1′ Perianth white, pink, or magenta
 3. Sts densely tufted or in mats < 20 cm wide; montane per
  4. Infl 1–5 fld; pl mat-forming; lvs cauline, blade ± round; fr wing 0 .... ***A. alpina***
  4′ Infl many-fld; pl densely tufted; lvs basal, blade oblong to ± round; fr winged .... ***A. nana*** ssp. ***covillei***
 3′ Sts elongate, aboveground; gen desert or seashore ann or per
  5. Fr wings 0; thick, prominent angles present or not; ovary chamber extends into angle
   6. Perianth white to pale magenta; fr angles gen 5, inflated and flat on top .... ***A. turbinata***
   6′ Perianth gen bright pink; fr smooth or with few low ridges .... [2]***A. villosa*** var. ***villosa***
  5′ Fr with 2–5 thin, conspicuous wings; ovary chamber does not extend into wing
   7. Fr wings gen 2; fr outline broadly cordate .... ***A. pogonantha***
   7′ Fr wings gen 3–5; fr cordate, obconic, or ovoid
    8. Lvs thin, widely scalloped; fr beak inconspicuous, not hardened .... ***A. gracilis***
    8′ Lvs thin or thick, ± entire; fr beak evident, hardened
     9. Fr body ± smooth; immediate coast .... ***A. umbellata***
      10. Perianth light to dark magenta, infl ± hemispheric; fr wings broadly rounded .... ssp. ***umbellata***
      10′ Perianth light magenta to ± yellow; infl sparse; fr wings angled near tip .... ssp. ***breviflora***
     9′ Fr body with prominent raised veins (most evident near tip); ± inland .... ***A. villosa***
      11. Fr wings pronounced, exceeding top of fr body; fl gen> 20 mm .... var. ***aurita***
      11′ Fr wings 0 or not exceeding top of fr body; fl gen < 20 mm .... [2]var. ***villosa***

A. alpina M.K. Brandegee (p. 773) RAMSHAW MEADOWS ABRONIA Per, forming mats < 25 cm, glandular-puberulent. **LF**: petiole 1–2 cm; blade 4–9 mm, round. **INFL**: peduncle < petioles; bracts 2–3 mm, narrowly ovate; fls 1–5. **FL**: perianth white to lavender-pink, tube 10–15 mm, limb 6–9 mm wide. **FR** 3–5 mm, 5-angled, net-veined. RARE. Dry, open granitic meadows; 2400–2700 m. s SNH (Ramshaw and Templeton meadows, Tulare Co.).

A. gracilis Benth. Ann, glandular-hairy. **ST** prostrate, < 1 m. **LF**: petiole 10–35 mm; blade 20–45 cm, thin, oblong or elliptic, edges scalloped or wavy. **INFL**: peduncle 2–10 cm; bracts 6–10 mm, ovate; fls 13–24. **FL**: perianth tube 6–17 mm, ± white or pink, limb 7–16 mm wide, light magenta to purplish red. **FR** 6–11 mm, glandular-hairy; wings 5, thin, broadly rounded, extending above top of fr body, base of fl tube not forming a prominent hard beak at top of fr body. Sandy soil; < 500 m. SCo (San Diego Co.), s DSon; nw Mex. Intergrades with *A. umbellata*, *A. villosa*. ❀sand,DRN,DRY,SUN:**24**.

A. latifolia Eschsch. (p. 773) Per, fleshy, finely glandular-hairy; root deep, thick. **ST** prostrate, < 2 m; branches short, forming a dense mat. **LF**: petiole 1–6 cm; blade 2–5 cm, broadly ovate to reniform, thick. **INFL**: peduncle 1–6 cm; bracts 5–9 mm, ovate; fls 17–34. **FL**: perianth tube 9–12 mm, yellow, limb 8–13 mm wide. **FR** 8–15 mm, net-veined, glabrous or ± hairy, tapered to base and tip; wings 0 or 5. Coastal dunes and in coastal scrub; < 100 m. NCo, CCo, n SCo (Santa Barbara Co.); to B.C. Hybridizes with *A. maritima*, *A. umbellata*. ❀sand,DRN,DRY,SUN:**5,17**,24.

A. maritima S. Watson Per, densely glandular-hairy; roots spreading. **ST** prostrate, < 2 m; branches short, erect, forming thick horizontal mat. **LF**: petiole 5–30 mm; blade 5–7 cm, thick, broadly elliptic or oblong. **INFL**: peduncle 3–10 cm; bracts 7–11 mm, lanceolate or narrowly ovate; fls 10–18. **FL**: perianth tube 6–10 mm, limb 7–10 mm wide, deep wine-red. **FR** 10–14 mm, glandular-hairy near top, with 5 thick wings angled at top, tapered to base, coarsely net-veined. Coastal dunes; < 100 m. s CCo, SCo; Baja CA. Hybridizes with *A. latifolia*, *A. umbellata*. ❀sand,DRN,DRY,SUN:**17,24.**

A. nana S. Watson ssp. ***covillei*** (Heimerl) Munz Per, densely tufted. **ST** < 6 cm. **LF** glaucous; petiole 1–4 cm; blade 5–20 mm, ovate to ± round. **INFL**: peduncle < 10 cm, scapose; bracts 6–8 mm, lanceolate; fls > 6. **FL**: perianth white, tube 11–15 mm, limb 6–8 mm wide. **FR** 7–8 mm, with 5 thin wings rounded at top. Dry sandy places; 1600–2800 m. DMtns; sw NV. Pls from e DMoj have wider bracts and longer fls, approach ssp. *nana*. ❀TRY.

A. pogonantha Heimerl (p. 773) Ann, glandular-hairy. **ST** decumbent to ascending, 10–55 cm. **LF**: petiole 0.8–4 cm; blade 1.5–5.5 cm, 1–3 cm wide, ovate to oblong-ovate. **INFL**: peduncle 2–7 cm; bracts 4–9 mm, lanceolate to broadly ovate; fls 12–24. **FL**: perianth tube 1–2 cm, ± red or green, limb 6–8 mm wide, white or pink. **FR** 4–6 mm, 2–3-winged, outline cordate; wing interior spongy. Sand, desert communities; < 1550 m. s SnJV, DMoj (and adjacent margins of SN); w NV. ❀TRY.

A. turbinata S. Watson (p. 773) Ann, rarely per, glabrous or sparsely glandular-hairy. **ST** ascending to erect, < 50 cm. **LF**: petiole 1–5 cm, slender; blade 1–5 cm, broadly ovate to round. **INFL**: peduncle 3–9 cm; bracts 3–10 mm, lanceolate to ovate; fls 15–35. **FL**: perianth tube 6–18 mm, ± green or pink, limb 5–8 mm wide, white or pale magenta. **FR** 3–7 mm, as wide as long, with gen 5 wings inflated and truncate at top, glandular-hairy at top, glabrous below; wings hollow inside (some fr with 2 wings folded together). Dry, sandy soil, desert scrub; 900–2500 m. SNE; se OR, w NV. ❀TRY.

A. umbellata Lam. Ann, ± glabrous to glandular-hairy. **ST** prostrate, > 1 m. **LF**: petiole 1–6 cm, slender; blade 15–70 mm, 8–50 mm wide, ovate to ± diamond-shaped. **INFL**: peduncle 2.5–15 cm, slender, gen erect in fr; bracts 5–7 mm, lanceolate to ovate; fls 8–27. **FL**: perianth tube 6–16 mm, ± green or red, limb 6–16 mm wide, light to bright magenta. **FR** 7–12 mm, glandular-hairy at top, glabrous below, angled or with 2–5 (gen 4) wings broadly rounded above the top of fr body. Disturbed sandy areas, coastal dunes and scrub; < 100 m. NCo, CCo, SCo; to WA, Baja CA. Hybrids with *A. maritima*: ssp. *alba* (Eastw.) Munz; ssp. *platyphylla* (Standley) Munz; ssp. *variabilis* (Standley) Munz. Hybridizes also with *A. latifolia*.

ssp. ***breviflora*** (Standley) Munz PINK SAND-VERBENA **FL**: perianth tube 6–8 mm, ± green or yellow, limb 6–8 mm wide, light to bright magenta, throat surrounded by cream eyespot. **FR**: wings poorly to moderately developed, ± angled. RARE. Habitat of sp. NCo, CCo (Marin Co); s OR. Populations gen small.

ssp. ***umbellata*** **FL**: perianth tube 9–13 mm, ± green or red, limb 7–17 mm wide, light to dark magenta, throat surrounded by white eyespot. **FR**: wings poorly to well developed, often rounded at top. Habitat of sp. NCo (Sonoma Co.), CCo, SCo; Baja CA. ❀sand,DRN,DRY,SUN:**5,17,24**.

A. villosa S. Watson (p. 773) Ann, glandular-hairy. **ST** prostrate to ascending, < 80 cm. **LF**: petiole 0.5–5 cm; blade 1–5 cm, 1–4.5 cm wide, triangular-ovate to ± round. **INFL**: peduncle 2–10 cm; bracts 3–11 mm, lanceolate to narrowly ovate; fls 15–35. **FL**: perianth tube 1.3–3.5 cm, ± pink, limb 6–18 mm wide, pale to bright magenta. **FR** 5–10 mm; base of fl tube hardened as a beak on top of fr body; wings 3–5, thin, rounded or angled, or 0. Sandy places in creosote-bush or coastal-sage scrub; < 1600 m. SCo, D; s NV, sw AZ, nw Mex. Hybridizes with *A. gracilis*.

var. ***aurita*** (Abrams) Jepson **FL**: perianth tube 2–3.5 cm, limb > 1.5 cm wide. **FR**: body nearly smooth; wings wide, extending well above body. Coastal-sage scrub, chaparral, etc.; < 1600 m. c&s SCo, w DSon. ❀sand,DRN,DRY,SUN:**13,18–24**.

var. ***villosa*** **FL**: perianth tube 1.3–2 cm, limb gen < 1.5 cm wide. **FR**: body prominently wrinkled, appearing pitted; wings barely extending above fr body. Creosote-bush scrub; < 1000 m. Range of sp. (exc SCo). ❀sand,DRN,DRY,SUN:**13**.

ACLEISANTHES TRUMPETS

Per from thick taproot, minutely hairy. **ST** prostrate to ascending. **LF** sessile or petioled; blade < 8 cm, ± firm. **INFL**: bracts 1–3, free, not forming an involucre; fl gen 1. **FL** nocturnal or some cleistogamous; perianth trumpet-shaped, white, tube slender; stamens 5 (2 in cleistogamous fls), slightly exserted; stigma ± spheric, exserted slightly beyond anthers. **FR** oblong, with rounded ridges, glabrous; wings 0. 7 spp.: esp Chihuahuan Desert, ne Mex. (Greek: without closure, from absence of involucre) Pollinated by hawkmoths.

A. longiflora A. Gray (p. 773) ANGEL TRUMPETS **ST** < 1 m. **LF** petioled; blade < 4 cm, lanceolate to triangular, acute to acuminate, margins ± wavy. **FL**: perianth 7–17 cm, tube ± green or pale purple, limb 1–3 cm wide, white. **FR** 6–10 mm, oblong, ridges 5, 2 parallel grooves between adjacent ridges. Dry places, gen on limestone; 10–2500 m. e DSon (Maria Mtns, e Riverside Co.); to TX, n Mex. ❀TRY.

ALLIONIA WINDMILLS, TRAILING FOUR O'CLOCK

Ann, short-lived per, glabrous to densely glandular-hairy. **ST** trailing, < 1 m. **LF** petioled; blade < 4 cm, oval to oblong, paler below. **INFL**: involucres clustered, each involucre resembling 1 fl; bracts 3, ± 1/2 fused, hairy; fls 1 per bract, blooming together. **FL**: stamens gen 4(–7), exserted; stigma ± spheric, exserted. **FR** bilateral. 2 spp.: Am. (C. Allioni, Italian botanist, 1725–1804)

A. incarnata L. (p. 773) **FL**: perianth 3–15 mm, 3-lobed, oblique, red-purple, tube funnel-shaped, limb longest above subtending bract. **FR** 3–4.5 mm, compressed, bilateral; margin strongly incurved, entire or with 3–5 irregular teeth; outer surface convex; inner surface concave, with 2 rows of sticky glands. Creosote-bush scrub; 0–1500 m. D; to Colorado, TX, S.Am. ❀DRN,DRY,SUN: 10,11,**17,18**.

ANULOCAULIS RINGSTEM

Per from thick caudex. **ST** little branched, erect, > 7 mm diam; internodes with sticky brown ring. **LVS** few, ± on lower st half, petioled; blade oblong to round, thick. **INFL** openly branched; fls in heads, racemes, or umbel-like clusters; bracts 1–3, small, not forming an involucre. **FL**: perianth funnel-shaped; stamens 3 or 5, gen long-exserted; stigma ± spheric, exserted. **FR** finely 10-ribbed, glabrous. 5 spp.: esp Chihuahuan Desert, ne Mex. (Latin: ring stem, from sticky internodal rings)

A. annulatus (Cov.) Standley (p. 773) **ST** < 1.5 m. **LF**: blade 3–10 cm, oblong to ovate-triangular, stiff-hairy, hairs with enlarged, dark, glandular base. **INFL**: head-like umbels on long peduncles. **FL**: perianth ± 8 mm, tube ± green, hairy, limb pale pink. **FR** 4–5 mm, thick-fusiform, gray-brown. Rocky slopes, canyons; < 1200 m. ne DMoj (Death Valley region). [*Boerhavia a.* Cov.]

BOERHAVIA SPIDERLING

Ann, per. **ST** prostrate to erect; internode often with sticky area. **LF** petioled; blade 1–6 cm, paler beneath, often brown-dotted. **INFL** openly branched; unit a raceme, umbel, or head; bracts 1–3, free, not forming an involucre. **FL**: perianth < 30 mm, bell-shaped, closing by afternoon; stamens 1–5; stigma ± spheric, gen exserted. **FR** < 3.5 mm, club-shaped; ridges 4–5; wings 0. ± 30 spp.: warm regions. (H. Boerhaave, Dutch botanist, 1668–1738)

1\. Perianth red-violet; per; fr hairy .... ***B. coccinea***
1′ Perianth pale pink or white; ann; fr glabrous
 2\. Peduncle long, infl umbel-like or fl 1
  3\. Ultimate 2 cm of branches with 0–1 branches, ending in umbel-like infl; fr length 3 × width ..... ***B. intermedia***
  3′ Ultimate 2 cm of branches with 2–4 appressed side branches, each with 1–2 frs; fr length 2 × width . ***B. triquetra***
 2′ Peduncle short, infl a raceme-like spike
  4\. Bracts deciduous, << fr; fr ridges gen 5; st puberulent .... ***B. coulteri***
  4′ Bracts ± persistent, ± = fr; fr ridges gen 4; st glandular-hairy .... ***B. wrightii***

B. coccinea Miller (p. 773) SCARLET SPIDERLING Per. **ST** prostrate or sprawling, < 15 dm; glandular-hairy. **LF**: blade broadly ovate, blunt, slightly wavy. **INFL**: head. **FL** 2 mm; perianth red-violet. **FR** 2.5–3.5 mm, hairy; tip rounded; wide and smooth between ribs. Dry disturbed places; < 1000 m. s SnJV, SCo, DSon; to se US, nw S.Am. ❀DRN,DRY,SUN:**8**,**9**,11,12,**13**.

B. coulteri (Hook.) S. Watson (p. 773) Ann. **ST** decumbent to erect, < 8 dm, puberulent. **LF**: blade lanceolate to ovate-triangular, ± acute. **FL** < 1.5 mm; perianth white to pale pink. **FR** 2.5 mm, glabrous; narrow (nearly closed) and wrinkled between wide ribs; tip rounded. Gravelly hillsides, washes; < 1150 m. DSon; to AZ, Mex. *B. spicata* Choisy (fr ribs narrow, acute) is widespread in sw US, expected as agricultural weed near Mex border.

B. intermedia M.E. Jones (p. 773) Ann. **ST** ascending or erect, 2–5 dm; hairs fine, sparse. **LF**: blade broadly lanceolate or ovate, acute to obtuse. **INFL** openly branched; fls in small umbel-like clusters at branch tips. **FL** 1.5–2 mm; perianth white to pale pink. **FR** 2–2.7 mm, glabrous; wrinkled between sharp ribs; tip truncate. Gravelly washes, flats; < 1300 m. e DSon; to TX, Mex. [*B. erecta* var. *i.* (M.E. Jones) Kearney & Peebles] *B. erecta* L. (fr 3–4.5 mm) is widespread in trop Am, expected as agricultural weed near Mex border.

B. triquetra S. Watson (p. 777) Ann. **ST** branched from base, ascending to erect, < 6 dm, slender, puberulent. **LF**: blade narrowly lanceolate to oblong, obtuse to acute. **INFL** openly branched; peduncle 2–6 cm, slender; fls 1–2. **FL** 1–1.3 mm; perianth white to pale pink. **FR** 2–2.5 mm, glabrous; coarsely wrinkled between 3–5 wide, smooth, sharp ribs; tip truncate. Sandy or rocky areas; < 1600 m. e PR (n base Santa Rosa Mtns), DMtns (Little San Bernardino Mtns), DSon; w AZ, nw Mex.

B. wrightii A. Gray (p. 777) Ann. **ST** branched from base, erect, < 7 dm, glandular-hairy. **LF**: blade lanceolate to oblong-ovate, acute. **INFL** loose, spike-like. **FL** 1–2 mm, subtended by wide, reddish, persistent bracts, 2–3 mm in fr. **FR** 2–2.5 mm, nearly as wide, wrinkled between the 4 wide ridges. Dry, sandy places; < 1400 m. D: to NV, TX, Mex.

MIRABILIS FOUR O'CLOCK

Per, subshrub. **ST** repeatedly forked, decumbent to erect. **LF** gen petioled. **INFL** forked; calyx-like involucres densely clustered or solitary in axils, bell- to saucer-shaped; fls 1–16 per involucre, blooming sequentially. **FL**: perianth funnel- to bell-shaped, lobes 5; stamens 3–5, gen exserted; stigma ± spheric, gen exserted. **FR** ± round to club-shaped, smooth to 5-ribbed; wing 0. ± 60 spp.: Am, Himalayas. (Latin: wonderful) Fls open in evening, close in morning. Spp. intergrade; *Hermidium*, *Oxybaphus* sometimes segregated, but intergrade with other spp.; careful study needed. [Pilz 1978 Madroño 25:113–132]

1\. Fr strongly 5-ribbed, < 3 mm wide, gen with prominent warts or coarse wrinkles between ribs; involucre in fr enlarged, brown, papery
 2\. Lf ± sessile, ± linear
  3\. Perianth bright red, 3–4 × length of involucre in fl .... ***M. coccinea***
  3′ Perianth pale pink to magenta, ± 2 × length of involucre in fl .... ***M. linearis***
 2′ Lf obviously petioled, blade lanceolate or wider
  4\. Involucre in fl short-hairy at base, glabrous on lobes, in fr 10–15 mm .... ***M. nyctaginea***
  4′ Involucre in fl hairy throughout on outside, in fr gen < 8 mm
   5\. St branched gen near top, ascending to erect, gen glabrous (or finely strigose) in lower half; lf length > width, tip gen acute .... ***M. oblongifolia***
   5′ St much-branched from near base, decumbent to ascending, hairy or glandular throughout; lf length gen = width, tip obtuse to acute .... ***M. pumila***
1′ Fr smooth to moderately 5- or 10-ribbed or -angled, often > 3 mm wide, sometimes with low wrinkles or warts; involucre in fr little changed
 6\. Fls 3–16 per involucre, bracts either > 22 mm or free
  7\. Perianth ± 15 mm; involucral bracts 15–30 mm, free to fused ± 1/2 length .... ***M. alipes***
  7′ Perianth 40–60 mm; involucral bracts > 22 mm, fused > 1/2 length
   8\. Fr 5-angled .... ***M. greenei***
   8′ Fr not angled, often with 10 lines or low ribs .... ***M. multiflora***
    9\. Involucral bracts obtuse; fr warty, gelatinous when wet .... var. ***glandulosa***
    9′ Involucral bracts acute; fr smooth or lightly warty with 10 slender, tan ribs; not gelatinous when wet .... var. ***pubescens***
 6′ Fl 1 per involucre; bracts < 15 mm, fused
  10\. Perianth 30–50 mm; fr bluntly 5-ribbed, cross-wrinkled .... ***M. jalapa***
  10′ Perianth < 15 mm; fr not 5-ribbed
   11\. Involucre > 10 mm, lanceolate lobes > tube; perianth white; lvs ascending .... ***M. tenuiloba***
   11′ Involucre < 10 mm, ± ovate lobes < tube; perianth white to magenta; lvs widely spreading

12. Perianth pink to purple-red; lf puberulent or glandular-hairy (youngest lvs with conic-based hairs); fr gen very lightly dotted or wrinkled, sometimes smooth .... ***M. californica***
12′ Perianth white to pale pink; lf finely glandular-hairy; fr gen lightly dotted or wrinkled .... ***M. bigelovii***
13. Fr ovoid, lines 0 or inconspicuous; sts and lvs glandular-hairy .... var. ***bigelovii***
13′ Fr ± spheric, gen with 10 pale lines; sts and lvs with short, reflexed hairs, gen also ± glandular . var. ***retrorsa***

M. alipes (S. Watson) Pilz (p. 777) **ST** decumbent to erect, 2–4 dm; glaucous, glabrous, or sparsely hairy upward. **LF**: blade 2–7 cm, broadly ovate to round, fleshy, glabrous or sparsely short-hairy. **INFL**: involucre 1 per upper axil, peduncled, ± cup-shaped, glabrous; bracts gen 5–7, free to fused ± half length, 15–30 mm, broadly ovate; fl 1 per bract; pedicel fused to bract. **FL**: perianth ± 15 mm, funnel-shaped, magenta (rarely creamy white). **FR** 5.5–7 mm, elliptic, glabrous; ribs 10, slender, tan. Dry slopes, flats; 1200–2000 m. W&I, DMtns (Panamint Mtns); to w Colorado [*Hermidium a.* S. Watson] ❀TRY.

M. bigelovii A. Gray **ST** ascending to erect, < 8 dm, gen glandular-hairy. **LF**: blade 1–4 cm, ovate to ± reniform, glandular-hairy. **INFL**: involucres clustered near ends of branches, bell-shaped; bracts 5, 5–6 mm, > 1/2 fused, lobes ovate; fl 1 per involucre. **FL**: perianth 8–12 mm, widely funnel-shaped, white to pale pink (esp in w part of range). **FR** ± 3 mm, ± spheric to ovoid, lightly dotted or wrinkled, glabrous. Rocky places; < 2300 m. D; to UT, AZ, nw Mex. Vars. intergrade, esp in e D.

var. ***bigelovii*** (p. 777) **ST**: hairs glandular, spreading. **FL**: perianth white, sometimes pink. **FR** gen ± ovoid, lightly dotted, often mottled. Habitat of sp.; < 200 m. Range of sp., esp e D. [var. *aspera* (E. Greene) Munz] ❀DRN,DRY,SUN:7–10,**11–13**,14–16.

var. ***retrorsa*** (A.A. Heller) Munz **ST** scabrous above with short, reflexed hairs, often also ± glandular. **FL**: perianth white to pale pink. **FR** ± spheric, with ± prominent pale lines. Habitat of sp.; 150–2300 m. Range of sp. ❀DRN,DRY,SUN:7–10,**11–13**, 14–16.

M. californica A. Gray (p. 777) WISHBONE BUSH **ST** trailing to ascending, < 8 dm, somewhat woody, grayish when old, scabrous or ± glandular-hairy. **LF**: blade 1–3.5 cm, ovate, puberulent or glandular-hairy. **INFL**: involucres clustered near ends of branches, bell-shaped; bracts 5, 5–8 mm, > 1/2 fused, lobes ovate; fl 1 per involucre. **FL**: perianth 5–14 mm, broadly funnel-shaped, pink to purple-red (white). **FR** ± 5 mm, ovoid, gen lightly dotted or wrinkled, glabrous. Common. Grassy areas, chaparral, dunes, dry rocky areas and washes; < 1000 m. CCo, SCoR, SW, w edge D; Baja CA. [*M. laevis*, (Benth.) Curran var. *laevis* misapplied; var. *cedrosensis* (Standley) Munz] Intergrades with *M. bigelovii*. ❀DRN,DRY,SUN:**8,9,14–24**.

M. coccinea (Torrey) Benth. & Hook. **ST** ascending to erect, < 6 dm, glabrous, glaucous. **LF** ± sessile; blade 2–12 cm, linear, fleshy, glabrous. **INFL** loosely forked; involucre bell-shaped, short-hairy, enlarged and papery in fr; bracts 5, 5–8 mm in fr, ± 1/2 fused; fls 1–3 per involucre. **FL**: perianth 15–20 mm, ± salverform (tube narrowly funnel-shaped), bright red. **FR** ± 5 mm, club-shaped; ribs 5, coarsely wrinkled between ribs; hairs fine. Dry, rocky slopes, washes; 1300–1800 m. DMtns; to w NM, nw Mex. [*Oxybaphus c.* Torrey] ❀TRY.

M. greenei S. Watson **ST** ascending to erect, 4–8 dm, glabrous or sparsely hairy upward. **LF**: blade 3–10 cm, round to ovate, fleshy, glabrous or sparsely short-hairy. **INFL**: involucre 1 per upper axil, bell-shaped, ± glabrous; bracts gen 5–6, 25–40 mm, > 1/2 fused; fls 6–16 per involucre. **FL**: perianth 4–5 cm, narrowly funnel-shaped, magenta. **FR** ± 7 mm, broadly club-shaped or elliptic, bluntly 5-angled, very sparsely short-hairy. Dry slopes, flats; < 1000 m. e KR, n&c NCoRI. [*Quamoclidion g.* (S. Watson) Standley] ❀TRY.

M. jalapa L. FOUR-O'CLOCK **ST** erect, < 1 m, glabrous to sparsely short-hairy. **LF**: blade 5–14 cm, ovate to cordate. **INFL**: clusters of involucres, short-stalked, ± terminal; involucres bell-shaped; bracts 5, 5–15 mm, 1/2–3/4 fused; fl 1 per involucre. **FL**: perianth 30–50 mm, narrowly funnel-shaped, gen bright magenta (yellow, white, or variegated). **FR** 8–10 mm, ovoid, glabrous to minutely short-hairy; ribs 5, blunt, ± wrinkled or warty between ribs. Disturbed places in mild-winter areas; < 300 m. CA-FP (esp SnFrB); native to trop Am; cult as orn. **TOXIC**: intestinal irritant in fr, roots.

M. linearis (Pursh) Heimerl **ST** ascending to erect, < 10 dm, glabrous to finely strigose. **LF** ± sessile; blade 3–10 cm, linear to very narrowly lanceolate, somewhat fleshy, gray-hairy. **INFL** loosely forked; involucre cup-shaped, densely glandular-hairy, enlarged and papery in fr; bracts 5, 6–10 mm in fr, 1/2–2/3 fused; fls gen 3 per involucre. **FL**: perianth ± 10 mm, ± funnel-shaped, pale pink to magenta. **FR** ± 5 mm, club-shaped, wrinkled between the 5 ribs, hairy. Riverbeds, railroads, roadsides, gravelly places; < 300 m. SCo (Orange, Riverside cos.); native from AZ to c US, Mex; sometimes cult as orn. [*Oxybaphus l.* (Pursh) Robinson]

M. multiflora (Torrey) A. Gray (p. 777) **ST** ascending to erect, 3–8 dm. **LF**: blade 3–12 cm, round to ovate, fleshy, glandular-hairy or becoming glabrous. **INFL**: involucre 1 per upper axil, bell-shaped, ± glabrous to minutely glandular-hairy; bracts 5, 22–35 mm, 1/2–3/4 fused; fls 6 per involucre. **FL**: perianth 40–60 mm, narrowly funnel-shaped, magenta. **FR** 6–11 mm, elliptic. Dry, rocky or sandy places; < 2500 m. s SnJV, D; to w Colorado, AZ, Baja CA.

var. ***glandulosa*** (Standley) J.F. Macbr. **INFL**: involucral bracts obtuse. **FR** faintly warty, gelatinous when wet; ribs inconspicuous. Habitat of sp.; 900–2500 m. n DMoj (Inyo Co.); to w Colorado. ❀DRN,DRY,SUN:2,**10**,11.

var. ***pubescens*** S. Watson **INFL**: involucral bracts gen acute. **FR** ± smooth, 10-ribbed, not gelatinous when wet. Habitat of sp.; 50–2100 m. s SnJV, D; to sw UT, nw Mex. [*M. froebellii* (Behr) E. Greene; *M. f.* var. *glabrata* (Standley) Jepson] ❀DRN,DRY,SUN: 7,**8–10**,11–16,18,**19–23**.

M. nyctaginea (Michaux) MacMillan **ST** erect, < 10 dm, glabrous to ± hairy. **LF**: blade < 10 cm, ovate or cordate, ± glabrous. **INFL**: of umbels grouped in forked clusters; peduncles hairy; involucre cup-shaped, base short-hairy, lobes glabrous, enlarged and papery in fr; bracts 5, 10–15 mm in fr, ± 2/3 fused; fls 3–5 per involucre. **FL**: perianth ± 10 mm, broadly funnel-shaped, white or pale pink. **FR** 4–6 mm, broadly club-shaped, wrinkled or warty between 5 wide ribs, hairy. Disturbed places; < 400 m. e SCo (esp San Bernardino Co.); native to c US, n Mex. [*Oxybaphus n.* (Michaux) Sweet]

M. oblongifolia (A. Gray) Heimerl **ST** branched near top, ascending to erect, 1–5 dm; glabrous or strigose. **LF**: blade < 4 cm, narrowly 3-angled, ± finely hairy. **INFL** broadly forked; involucre cup-shaped, glandular-hairy, enlarged and papery in fr; bracts 5, ± 8 mm in fr, 1/2–2/3 fused; fls 5 per involucre. **FL**: perianth 10–12 mm, broadly funnel-shaped, magenta. **FR** 3–5 mm, club-shaped, finely warty between the 5, wide, wrinkled or warty ribs. Dry, rocky areas; 1500–2500 m. DMtns (Ivanpah & New York mtns, ne San Bernardino Co.); to Colorado, TX, n Mex. [*Oxybaphus comatus* (Small) Weath.] Variable complex, with many named forms; in need of critical study. CA pls have ± narrow lvs and branch ± at base (like *M. linearis* var. *decipiens* (Standley) Welsh of s Rocky Mtns). Exc for distribution of hairs, CA pls also resemble *M. pumila* from DMtns. ❀TRY.

M. pumila (Standley) Standley (p. 777) **ST** trailing to ascending, < 5 dm, short-hairy, often glandular. **LF**: blade 2–6 cm, triangular to broadly ovate, fleshy, hairy like sts. **INFL**: involucres in axils or narrow clusters, cup-shaped, densely glandular-hairy, enlarged and papery in fr; bracts 5, 7–8 mm in fr, ± 2/3 fused; fls 3 per involucre. **FL**: perianth 8–10 mm, broadly funnel-shaped, pale pink, hairy. **FR** ± 5 mm, club-shaped, shallowly wrinkled between 5 wide ribs. Dry, rocky places; 1400–2500 m. SnBr, SnJt, W&I, DMtns; to NV, NM, nw Mex. [*Oxybaphus p.* (Standley) Standley] ❀TRY.

M. tenuiloba S. Watson LONG-LOBED FOUR O'CLOCK **ST** trailing to erect, < 5 dm, glandular-hairy. **LF** ascending; blade 2.5–5 cm, narrowly to broadly triangular, glandular-hairy. **INFL**: involucres ± densely clustered in upper axils, narrowly bell-shaped, spreading glandular-hairy; bracts 5, 11–13 mm, < 1/2 fused, lobes

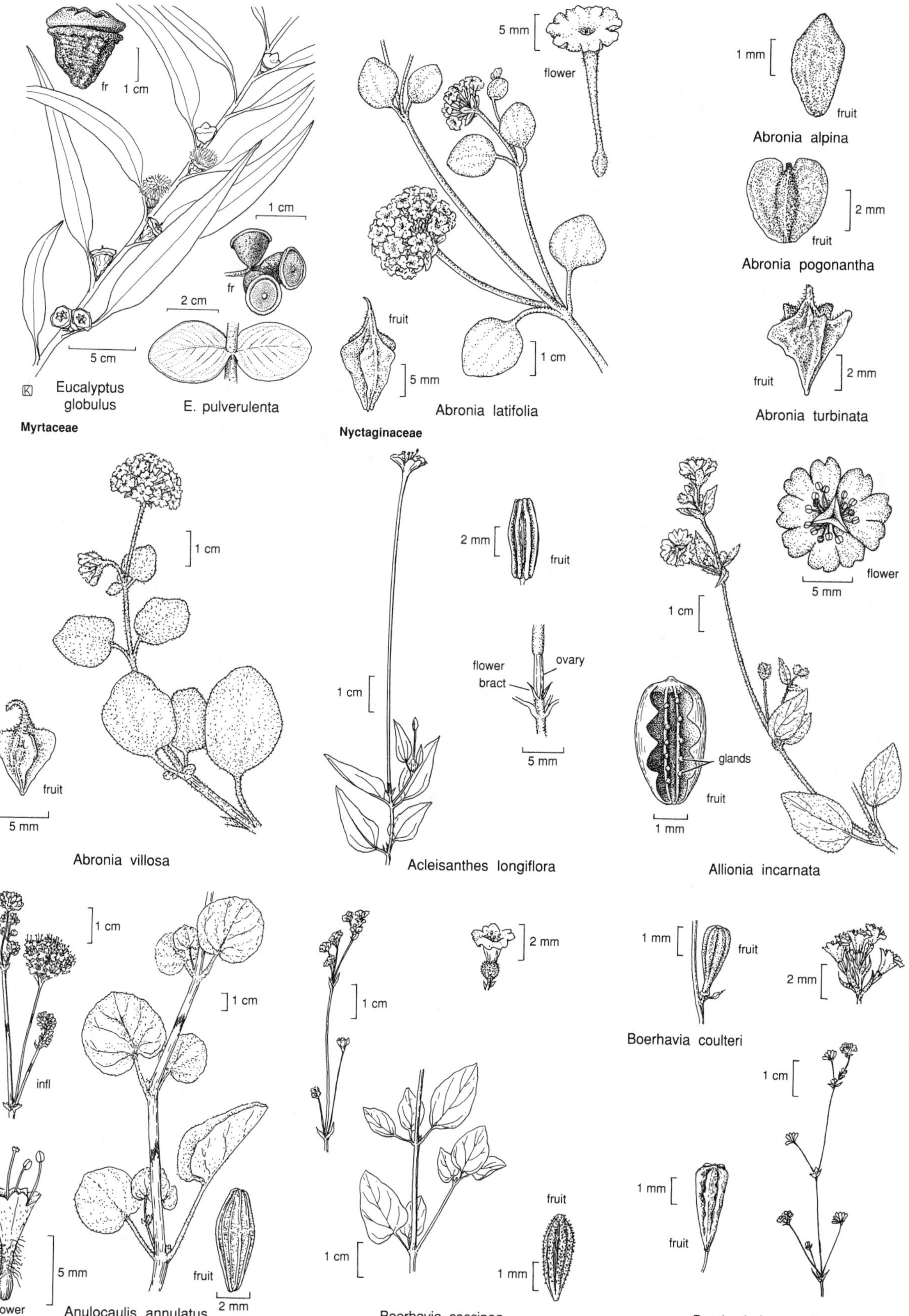
fr
1 cm
1 cm
fr
2 cm
5 cm
Eucalyptus globulus
E. pulverulenta
Myrtaceae
5 mm
flower
fruit
5 mm
1 cm
Abronia latifolia
Nyctaginaceae
1 mm
fruit
Abronia alpina
2 mm
fruit
Abronia pogonantha
fruit
2 mm
Abronia turbinata
1 cm
fruit
5 mm
Abronia villosa
2 mm
fruit
flower
bract
ovary
5 mm
1 cm
Acleisanthes longiflora
flower
5 mm
1 cm
glands
fruit
1 mm
Allionia incarnata
1 cm
1 cm
infl
5 mm
flower
fruit
2 mm
Anulocaulis annulatus
2 mm
1 cm
1 cm
fruit
1 mm
Boerhavia coccinea
1 mm
fruit
2 mm
Boerhavia coulteri
1 cm
1 mm
fruit
Boerhavia intermedia

narrowly lanceolate; fl 1 per involucre. **FL**: perianth 12–15 mm, funnel-shaped, whitish, lightly hairy. **FR** ± 5 mm, ovoid, smooth, blackish brown, glabrous. UNCOMMON. Rocky slopes in desert scrub; < 500 m. w DSon (Imperial, Riverside, San Diego cos.); Baja CA. ❀TRY.

SELINOCARPUS MOONPOD

Per, shrub. **ST** forked. **LF** sessile or petioled; blade ± fleshy. **INFL**: bracts 1–3, free, not forming an involucre; fls gen solitary in axils. **FL** nocturnal or cleistogamous; open perianth trumpet-shaped; stamens 4–8; stigma ± spheric, gen exserted. **FR**: wing thin, membranous. 8 spp.: N.Am., Afr. (Greek: parsley fruit) (Fowler & Turner 1977 Phytologia 37(3);177–208)

S. nevadensis Fowler & B. Turner (p. 777) DESERT WING-FRUIT Ann; hairs appressed and divergent, white, also glandular-puberulent. **ST** prostrate to erect, < 3 dm. **LF** petioled, < 26 mm; blade ovate to round. **FL**: perianth 30–40 mm, ± 10 mm wide, tube ± green, limb white. **FR** 5–7 mm; wings 5, 2 mm. RARE in CA. Dry, rocky areas; 1250 m. DMtns (ne Kingston Range, se Inyo Co.); to s NV, sw UT, nw AZ. [*S. diffusus* A. Gray var. *n.* Standley]

TRIPTEROCALYX

Per from large taproot. **ST** much-branched. **LF** petioled; blade < 8 cm, fleshy, margin often wavy. **INFL**: head; bracts 5–10, green, fls blooming sequentially across head; receptacle flat to ± conic, studded with peg-like pedicels. **FL**: perianth trumpet-shaped, nocturnal, lobes 4–5; stamens 3–5, incl; stigma linear, incl. **FR**: wings 3–5, wide, continuous above and below fr body, thin, transparent, conspicuously net-veined. 3 spp.: arid N.Am. (Greek & Latin: 3-winged cup, from fr) [Galloway 1977 Brittonia 27:328–347] Closely related to *Abronia*.

1. Perianth gen > 18 mm; fr wings hairy on margins and gen on veins near base; MP .... ***T. crux-maltae***
1′ Perianth gen < 18 mm; fr wings hairy only on margins; DMoj .... ***T. micranthus***

T. crux-maltae (Kellogg) Standley (p. 777) **ST** < 3 dm, glandular-sticky. **LF**: blade 2.5–7 cm, ovate to elliptic, ± glandular-hairy. **FL**: perianth (8)18–25 mm. **FR**: wings hairy on edges and often on veins near base. Sagebrush scrub; 1200–1500 m. MP (Lassen Co.); nw NV. [*Abronia c-m.* Kellogg] ❀TRY.

T. micranthus (Torrey) Hook. **ST** < 6 dm, glandular-sticky or scabrous. **LF**: blade 1–6 cm, narrowly ovate to elliptic, glabrous to glandular-hairy, hairs denser on lower surface. **FL**: perianth 6–18 mm. **FR**: wings glabrous exc on edges. Dunes; 800–2450 m. se DMoj (Kelso, San Bernardino Co.); to MT, SD, NM [*Abronia m.* Torrey] ❀TRY.

NYMPHAEACEAE WATERLILY FAMILY

William J. Stone

Per, aquatic; caudex or rhizome gen horizontal, gen large. **LVS** alternate, arising directly from caudex; petiole long; blades gen floating, submersed, or slightly emergent. **INFL**: fls solitary, axillary; peduncle long. **FLS** gen bisexual; sepals 3–many, sometimes petal-like; petals 0–many, sometimes scale- or stamen-like, inserted on receptacle or side of ovary; stamens many, spirally arranged, filaments gen broad; ovary ± compound, superior to inferior, chambers 5–many, ovules many, styles finger-like or 0. **FR** spongy, berry-like. 6 genera, ± 60 spp.: worldwide.

1. Sepals > 5, round to broad-obovate; petals ± = stamens; ovary superior, top disk-like with stigmatic lines .... **NUPHAR**
1′ Sepals 4, gen longer than wide; petals > stamens; ovary partly inferior, top with curving, finger-like stigmas .... **NYMPHAEA**

NUPHAR COW-LILY, YELLOW POND-LILY

Rhizomes prostrate, branched. **LF**: petioles gen cylindric in X-section; blade gen ovate to round, base deeply notched. **FL**: sepals 5–14, spirally arranged, persistent; petals 10–20, scale- or stamen-like, thick and oblong to thin and spoon-shaped; stamens many, gen yellow, attached to receptacle; ovary compound, superior, chambers narrowed above, top expanded into disk-like stigmatic surface, stigmas in lines radiating from center. **SEEDS** ovoid, smooth. ± 25 spp.: n hemisphere. (Arabic name) [Beal 1956 J Elisha Mitchell Sci Soc 72:317–346]

N. luteum(L.) Sibth. & Sm. ssp. ***polysepalum*** (Engelm.) E. Beal (p. 777) **LF** gen floating; blade 1–4 dm, oblong to ovate, lobes at base rounded to acute. **FL** 5–7 cm wide; sepals 7–9(12), < 5 cm, outer ovate, green, inner broadly obovate, yellow to red-tinged; petals ± = stamens, petals, stamens yellow, sometimes red-tinged; stigmatic disk 1.5–2 cm wide, entire to deeply scalloped. **SEED** 3–5 mm. Ponds, slow streams; < 2500 m. NW, n&c SN, n&c CW, MP; to AK, Rocky Mtns. [*N. p.* Englem., *Nymphaea p.* E. Greene] ❀Aquatic (fresh):1–3,**4–7**,8,9,**14–24**.

NYMPHAEA WATERLILY, WATER NYMPH

Rhizomes or tuber prostrate to erect. **LF**: blade gen floating, elliptic to round, gen deeply notched, upper surface glabrous, green, lower surface often red or purplish tinged. **FL** gen showy, fragrant; sepals gen 4, < petals, ± green; petals 12–many, white, red, blue, or yellow; stamens many, inserted on elevated part of ovary, outer filaments flat, sometimes petal-like, inner filaments linear; ovary compound, partly inferior, styles many, finger-like, spreading outward from central depression. **SEEDS** enclosed in spongy aril. ± 45 spp.: trop, n hemisphere, s Afr, Australia. (Greek: water-nymph)

1. Corolla yellow .... ***N. mexicana***
1′ Corolla white or pinkish .... ***N. odorata***

N. mexicana Zucc. YELLOW or BANANA WATERLILY Tuber often erect. **LF**: blade 10–25 cm wide, ovate to ± round. **FL** 6–10 cm, gen emergent; sepals, petals lanceolate to narrowly elliptic; petals gen ± 25, bright yellow; outer stamens gen 2–2.5 cm, anthers of inner stamens 4–6 mm; styles 7–9. **FR** 2–2.5 cm, ovoid. **SEEDS** 4–5 mm. Lakes, ponds, slow streams; < 100 m. SnJV (Merced, Madera, Fresno, Kings cos.); native to se US, Mex. NOXIOUS WEED in waterways.

N. odorata Aiton FRAGRANT or WHITE WATERLILY Rhizome prostrate, not tuber-like. **LF**: blade 5–25 cm wide, ± round. **FL** 5–15 cm, floating or ± emergent, very fragrant; sepals, petals lanceolate to ovate; petals gen > 20; outer stamens gen 3–4 mm, anthers of inner stamens 7–12 mm; styles gen 20. **FR** 2.5–3 cm, depressed spheric. **SEED** ± 2 mm. $2n$=84. Quiet waters, ponds, edges of lakes; gen < 2700 m. Widely scattered, SNH (Lake Tahoe), ScV (Butte Co.), SnBr, expected elsewhere; native to e N.Am. Cult widely for orn. NOXIOUS WEED in waterways.

OLEACEAE OLIVE FAMILY

Dieter H. Wilken

Shrub, tree, or vine, some dioecious. **LVS** alternate or opposite, deciduous or evergreen, simple to pinnately compound. **INFL** various; fl sometimes solitary. **FL** sometimes unisexual, gen radial; calyx gen minute, tube cup-shaped, lobes 4–15; petals (0)4–6, gen fused; stamens gen 2, epipetalous; pistil 1, ovary superior, chambers 2, placentas axile, ovules 2–4 per chamber, style 1, stigma gen 2-lobed. **FR**: drupe, capsule, or winged achene. **SEED** 1 per chamber. ± 25 genera, 900 spp.: ± worldwide; some cult for orn (*Forsythia*; *Jasminum*, jasmine; *Ligustrum*, privet; *Syringa*, lilac) or food (*Olea*). [Wilson & Wood 1959 J Arnold Arbor 40:369–384]

1. Lf pinnately compound, lflets (1)3–9; fr winged achene .... **FRAXINUS**
1′ Lf simple; fr capsule or drupe
 2. Lvs gen alternate; fr a deeply 2-lobed capsule .... **MENODORA**
 2′ Lvs opposite or clustered; fr a drupe
  3. Shrub, dioecious; branches short, spine-like; corolla 0; fr 5–8 mm .... **FORESTIERA**
  3′ Tree; fls bisexual; branches ± long, slender; corolla ± white; fr 9–20 mm .... **OLEA**

FORESTIERA

Shrub, gen dioecious. **LVS** simple, opposite or clustered, gen deciduous, short-petioled. **INFL**: clusters, axillary; staminate fls subsessile; pistillate fls pedicelled. **FL**: calyx minute, minutely ± 4-lobed, deciduous; corolla 0. **STAMINATE FL**: stamens 1–4; pistil vestigial. **PISTILLATE FL**: stamens 0; ovules 2 per chamber, stigma 1–2-lobed. **FR**: drupe. ± 20 spp.: Am. (Charles Le Forestier, French physician & naturalist, early 19th century)

F. pubescens Nutt. (p. 777) DESERT OLIVE Shrub 5–25 dm. **ST**: bark smooth, grayish; twigs short, spine-like, puberulent, becoming glabrous. **LF**: blade 15–40 mm, lanceolate to elliptic, leathery, entire to minutely toothed, glabrous. **INFL** gen appearing before lvs. **STAMINATE FL**: stamens 3–6 mm. **FR** 5–8 mm, elliptic in outline, purple-black, ± glaucous. $2n$=46. Streambanks, canyons, washes; 100–1800 m. s SNF, c&s SNH (e slope), Teh, e SnFrB, SCoRO (e slope), SCoRI, TR, PR, s SNE, DMoj; to Colorado, TX, n Mex. [*F. neomexicana* A. Gray] ❀1,**7,14–17,22–24**&IRR:2,3,**8–12**,13,**18–21**.

FRAXINUS ASH

Shrub or tree, gen dioecious. **ST**: bark smooth to furrowed, gen gray; twigs gen puberulent, becoming glabrous. **LVS** opposite, deciduous, odd-pinnate, petioled, gen thin, gen glabrous; lflet dark green above, pale below, gen glabrous, base and tip rounded to acute. **INFL**: clusters or panicles, axillary, often peduncled; fls pedicelled. **FL**: calyx 1–2 mm, shallowly ± 4-lobed; petals 0, 2, or 4, free or fused. **STAMINATE FL**: stamens gen 2; pistil vestigial. **PISTILLATE FL**: stamens 0; ovules 2 per chamber. **FR**: achene, winged. **SEED** gen 1. ± 65 spp.: temp N.Am, Eurasia, trop Asia. (Latin: ancient name) [Little 1952 J WA Acad Sci 42:369–380]

1. Lvs gen appearing simple (lflet 1, sometimes 3–5 on twigs); twigs gen 4-angled; fr body flat; n&e DMtns .... ***F. anomala***
1′ Lflets 3–9; twigs gen cylindric; fr body ± cylindric (exc *F. dipetala*); esp CA-FP
 2. Petals 2, white; fls bisexual, appearing with or after lvs; lflets gen serrate; fr body flat .... ***F. dipetala***
 2′ Petals 0; dioecious; fls appearing before or with lvs; lflets entire to ± serrate; fr body subcylindric
  3. Lateral lflets ± sessile; fr (incl wing) 28–50 mm, body winged from ± lower 3/4 .... ***F. latifolia***
  3′ Lateral lflets gen short-stalked; fr (incl wing) 15–30 mm, body winged from middle .... ***F. velutina***

F. anomala S. Watson (p. 777) SINGLE-LEAF ASH Shrub or tree < 6 m. **ST**: twigs gen 4-angled. **LF** 2–10 cm; petiole glandular-puberulent; lflets gen 1(–5), 1–6 cm, stalked, narrowly ovate to ± round, thick, entire to irregularly crenate. **FL** gen bisexual; petals 0. **FR** 13–25 mm, 5–10 mm wide; body flat, winged from near base. $2n$=46. Washes, rocky slopes, shrubland, pinyon/juniper woodland; 1100–2400 m. n&e DMtns; to Colorado, NM. ❀DRN,IRR:1,**2,3,7**,8–10,14–16,**18**,19–23.

F. dipetala Hook. & Arn. (p. 777) CALIFORNIA ASH Shrub or tree < 7 m. **ST**: twigs cylindric to 4-angled. **LF** 5–19 cm; petiole (sub)glabrous; lflets 3–7(9), 1–7 cm, short-stalked to subsessile, (ob)ovate to ± round, gen serrate. **FL** gen bisexual; petals 2, 2.5–4 mm, white. **FR** 20–32 mm, 5–9 mm wide; body flat, winged from near base. $2n$=46. Canyons, slopes, chaparral, oak/pine woodland; 100–1300 m. NCoR, CaRF, n SNF, c&s SN, CW, TR, PR. ❀DRN:**7,14–16**,17,**22,23**,24&IRR:**8,9,18–21**.

F. latifolia Benth (p. 777) OREGON ASH Tree < 25 m, dioecious. **ST**: twigs cylindric. **LF** 12–33 cm; petiole glabrous to puberulent; lflets 5–7, 2–10 cm, widely elliptic to narrowly ovate, entire to ± serrate, sparsely puberulent below, sessile (or lower very short-petioled). **FL**: petals 0. **FR** 28–50 mm, 5–8 mm wide; body subcylindric, winged from ± lower 3/4. 2*n*=46. Canyons, streambanks, woodland; < 1700 m. NW, CaR, SN, GV, SnFrB, MP; to B.C. ❀**4–6**&IRR:1–3,**7–9**,**14–16**,17–24.

F. velutina Torrey (p. 777) VELVET ASH Tree < 10 m, dioecious. **ST**: twigs cylindric. **LF**: 10–25 cm; petiole puberulent, often becoming glabrous; lflets (3)5–7, 3–10 cm, short-stalked, lanceolate to ovate, entire to ± serrate, sparsely puberulent to glabrous below. **FL**: petals 0. **FR** 15–30 mm, 4–8 mm wide; body subcylindric, winged from ± middle. 2*n*=46,92. Canyons, streambanks, woodland; 200–1600 m. s SN, SCo, TR, PR, s SNE, DMoj; to sw UT, TX, n Mex. [var. *coriacea* (S. Watson) Rehder] Apparently hybridizes with and difficult to separate from *F. latifolia* in s SN, w DMoj. ❀4–6,17&IRR:1,**2,3,7–11**,12,**14–16**,**18–23**,24;CVS.

MENODORA

Per to shrub, gen glabrous. **LVS** opposite to alternate, simple, gen entire, sessile to short-petioled. **INFL** appearing after lvs; axillary clusters or terminal panicle. **FL** bisexual; calyx lobes 5–15, ± linear, persistent; corolla ± rotate to funnel-shaped, lobes 4–6; ovules 2–4 per chamber, style slender, stigmatic lobes 2, ± spheric. **FR**: capsule, dehiscent by valves, circumscissile, or ± indehiscent, 2-lobed to near base. **SEEDS** 4–8. ± 25 spp.: Am, s Afr. (Greek: perhaps half-moon spear, from appearance of fr on stiff pedicel) [Steyermark 1932 Ann Missouri Bot Gard 19:87–176]

1. Shrub; branches many, spreading to ascending, branchlets short, stout, becoming spiny; corolla white, often purple- or brown-tinged; fr indehiscent .... ***M. spinescens***
1′ Per or subshrub; branches 0–few, ascending to erect, slender, not becoming spiny; corolla yellow; fr circumscissile
  2. Herbage rough-puberulent to scabrous; calyx lobes 8–11; upper lf length < 4 × width .... ***M. scabra***
  2′ Herbage ± glabrous; calyx lobes 5–8; upper lf length > 5 × width .... ***M. scoparia***

M. scabra A. Gray (p. 777) Subshrub, scabrous to rough-puberulent. **STS** 3–many from base, 11–28 cm, ascending to erect. **LVS** gen alternate, 7–20 mm, oblong to ovate, reduced and wider upward. **INFL** terminal. **FL**: calyx lobes 8–11, 3–6 mm, rough-hairy; corolla tube 4–6 mm; anthers and stigma exserted. **FR** circumscissile; lobes 5.5–8 mm. 2*n*=44. Rocky soils, canyons; 1200–1800 m. e DMtns (Clark, Eagle, New York mtns); to Colorado, TX, n Mex.

M. scoparia A. Gray Per to subshrub, ± glabrous. **STS** gen many from base, 30–60 cm, gen erect. **LVS** gen alternate, 5–25 mm, linear to obovate, reduced and narrower upward. **INFL** terminal. **FL**: calyx lobes 5–8, 3–6 mm, glabrous; corolla tube 3–5 mm; anthers and stigma exserted. **FR** circumscissile; lobes 5–7 mm. 2*n*=22. Rocky slopes, canyons; 600–2000 m. e&s DMoj, w DSon, e PR; to TX, n Mex.

M. spinescens A. Gray Shrub < 90 cm, sparsely puberulent. **ST** intricately branched; branchlets short, stout, becoming spiny. **LVS** alternate or clustered, 3–11 mm, oblong to obovate, fleshy. **INFL** axillary. **FL**: calyx lobes 5–7, 3–5 mm, sparsely rough-hairy; corolla tube 4–9 mm; anthers and stigma barely exserted. **FR** indehiscent or breaking apart irregularly; lobes 5–8 mm. Rocky slopes, canyons; 900–2300 m. SnBr (n slope), s SNE, DMtns; s NV, nw AZ. ❀TRY.

OLEA OLIVE

Shrub or tree. **LVS** simple, opposite, short-petioled, entire, leathery. **INFL**: raceme or panicle, axillary or terminal. **FL** gen bisexual; calyx 4-toothed to -lobed; corolla rotate, lobes 4. **FR**: drupe. 20 spp.: trop, warm temp Eurasia, Afr. (Greek: ancient name)

O. europaea L. Tree gen < 10 m. **ST**: bark gray, furrowed. **LF** 20–70 mm, 6–16 mm wide, densely silver-scaly below, dark green and sparsely scaly above. **FL**: corolla 2.5–4 mm, lobes > tube, margins inrolled, white. **FR** 9–20 mm, oily, green, becoming black. 2*n*=46. Gen waif, disturbed places; < 200 m. s NCoR, GV, SnFrB, SCo, n ChI (Santa Cruz Island); introduced widely; native to w Asia. Widely cult for (food and) cooking oil in Medit for ± 6000 years.

ONAGRACEAE EVENING PRIMROSE FAMILY

Warren L. Wagner, except as specified
Peter H. Raven, Family Coordinator

Ann to tree. **LVS** basal or cauline, alternate, opposite, or whorled, gen simple and toothed (to pinnately compound); stipules 0 or gen deciduous. **INFL**: spike, raceme, panicle, or fls solitary in axils; bracted. **FL** gen bisexual, gen radial, opening at dawn or dusk; hypanthium sometimes prolonged beyond ovary (measured from ovary tip to sepal base); sepals gen 4(2–7); petals gen 4 (or as many as sepals, rarely 0), often "fading" darker; stamens gen 4 or 8 (2), anthers 2-chambered, opening lengthwise, pollen gen interconnected by threads; ovary inferior, chambers gen 4 (sometimes becoming 1), placentas axile or parietal, ovules 1–many per chamber, style 1, stigma 4-lobed (or lobes as many as sepals), club-shaped, or hemispheric. **FR**: capsule, loculicidal (sometimes berry or indehiscent and nut-like). **SEEDS** sometimes winged or hair-tufted. 15 genera, ± 650 spp.: worldwide, esp w N.Am; many cult (*Clarkia, Epilobium, Fuchsia, Gaura, Oenothera*). [Munz 1965 N.Am Fl II 5:1–278]

1. Sepals persistent; hypanthium not prolonged beyond ovary; petals gen easily deciduous; moist habitats .... **LUDWIGIA**
1′ Sepals deciduous; hypanthium gen conspicuously (exc *Gayophytum*) prolonged beyond ovary; petals not gen easily deciduous; various habitats
  2. Petals and sepals 2; fr bur-like, covered with hooked hairs .... **CIRCAEA**
  2′ Petals and sepals (3)4; fr not bur-like, not covered with hooked hairs

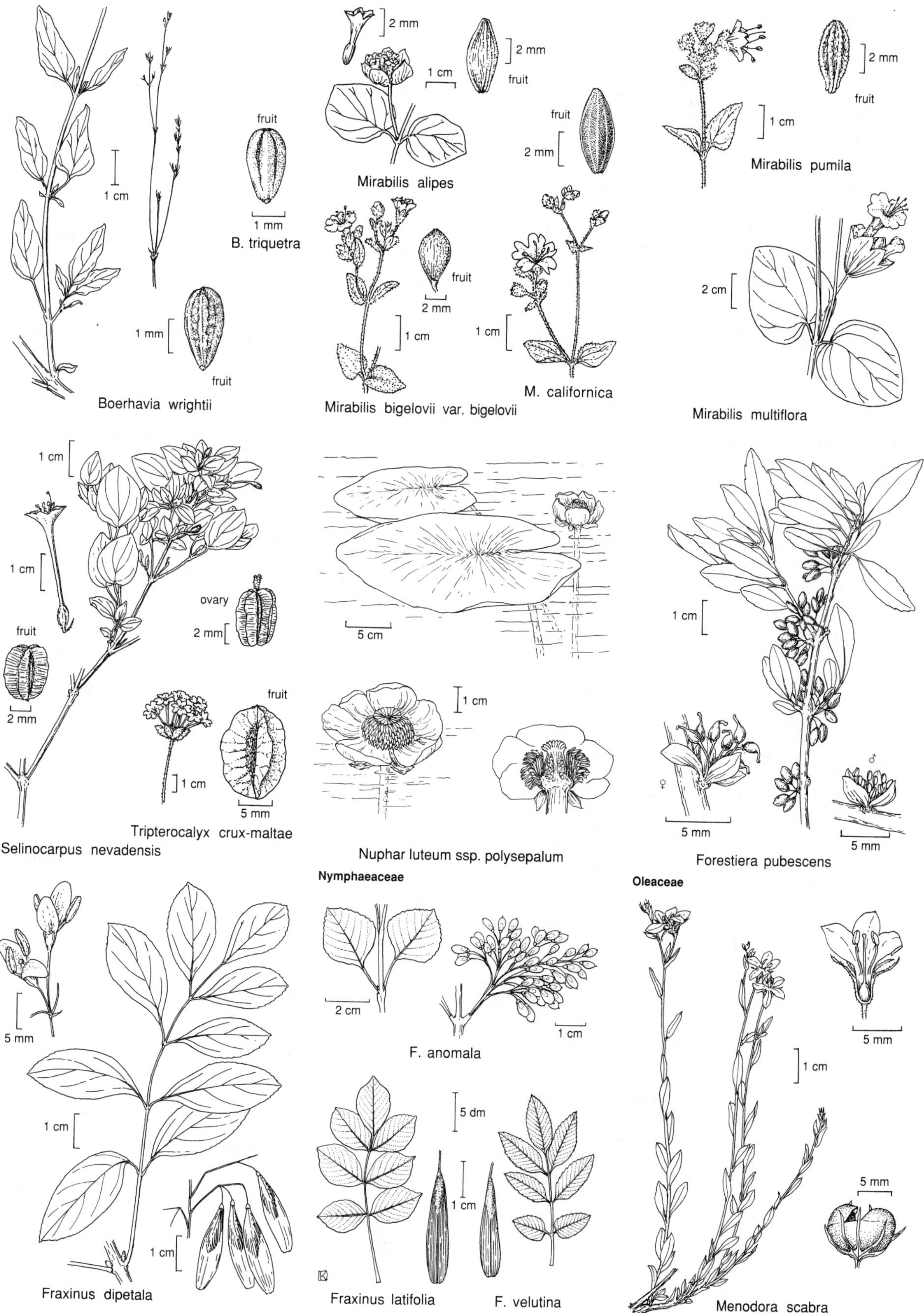
1 cm
fruit
1 mm
B. triquetra
1 mm
fruit
Boerhavia wrightii
2 mm
1 cm
2 mm
fruit
fruit
2 mm
Mirabilis alipes
fruit
2 mm
1 cm
1 cm
M. californica
Mirabilis bigelovii var. bigelovii
2 mm
fruit
1 cm
Mirabilis pumila
2 cm
Mirabilis multiflora
1 cm
1 cm
ovary
2 mm
fruit
2 mm
fruit
1 cm
5 mm
Tripterocalyx crux-maltae
Selinocarpus nevadensis
5 cm
1 cm
Nuphar luteum ssp. polysepalum
Nymphaeaceae
1 cm
5 mm
5 mm
Forestiera pubescens
Oleaceae
5 mm
1 cm
1 cm
Fraxinus dipetala
2 cm
1 cm
F. anomala
5 dm
1 cm
Fraxinus latifolia
F. velutina
5 mm
1 cm
5 mm
Menodora scabra

3. Fr indehiscent, nut-like, short
4. Fl radial; hypanthium thread-like; stamens opposite petals sterile; filament teeth 0 .... ***Clarkia heterandra***
4′ Fl bilateral; hypanthium obconic; stamens all fertile; filaments with paired teeth at base .... **GAURA**
3′ Fr dehiscent, not nut-like, long
5. Ovary chambers 2; hypanthium barely prolonged beyond ovary; st branches hair-like .... **GAYOPHYTUM**
5′ Ovary chambers 4; hypanthium well developed or not; st branches gen not hair-like
6. Seeds hair-tufted at 1 end or sepals erect at fl; pollen gen shed in groups of 4 .... **EPILOBIUM**
6′ Seeds not hair-tufted; sepals reflexed singly, in pairs, or 3 adherent and reflexed to 1 side; pollen gen shed singly
7. Stigma ± head-like or hemispheric .... **CAMISSONIA**
7′ Stigma 4-lobed
8. Anthers attached at base; petal tip often lobed or petal base clawed .... **CLARKIA**
8′ Anthers attached at middle; petal tip notched or toothed, petal base not clawed .... **OENOTHERA**

CAMISSONIA SUN CUP

Ann, per, from taproot or lateral roots. **LVS** basal, cauline, or both, alternate, simple to 2-pinnate. **INFL** bracted; spike, raceme, or fls solitary in axils. **FL** radial, gen opening at dawn (rarely at dusk); sepals 4, reflexed (sometimes 2–3 remaining adherent); petals 4, yellow, white, lavender, often with darker basal spots, gen fading purplish or reddish; stamens (4)8, longer ones opposite sepals, anthers gen attached at middle (or base), pollen grains 3-angled exc in polyploid taxa (visible with hand-lens); ovary chambers 4, stigma ± head-like or hemispheric, gen > anthers and cross-pollinated (or ± = anthers and self-pollinated). **FR** straight to coiled, gen sessile. **SEEDS** in 1–2 rows per chamber. 62 spp.: w N.Am (esp CA-FP), 1 S.Am. (L.A. von Chamisso, French-born German botanist, 1781–1838) [Raven 1969 Contr US Natl Herb 37:161–396] Polyploidy and self-pollination have predominated in evolution of genus. Previously incl in *Oenothera* ("*O.*" in synonyms).

1. Ovary tip (below hypanthium, above seeds) slender, sterile, 5–180 mm; st ± 0 (sect. *Tetrapteron*)
2. Ann; fr winged
3. Lf linear to narrowly lanceolate or oblanceolate; petals 5–18 mm; hypanthium 1.6–3.2 mm; pl densely spreading-hairy .... ***C. graciliflora***
3′ Lf narrowly oblanceolate; petals 2–3.5 mm; hypanthium 0.8–1.3 mm; pl strigose and sparsely spreading-hairy .... ***C. palmeri***
2′ Per; fr not winged
4. Lf pinnately lobed; pl hairs ± dense, spreading or appressed; hypanthium 4–8.5 mm; stigma > anthers .... ***C. tanacetifolia***
5. Pl hairs spreading, dense; pollen > 10% 4-angled .... ssp. ***quadriperforata***
5′ Pl hairs spreading or appressed; pollen < 5% 4-angled .... ssp. ***tanacetifolia***
4′ Lf not pinnately lobed (or pl ± glabrous); hypanthium 1.5–3 mm; stigma ± = or slightly > anthers
6. Lf hairs short, erect; fr subcylindric (swollen by seeds), papery .... ***C. ovata***
6′ Lf subglabrous; fr 4-angled, barely swollen by seeds, leathery .... ***C. subacaulis***
1′ Ovary without a slender, sterile tip; st present (or pl immature)
7. Seed wing thick, with club-shaped hairs; petals white, bases yellow (sect. *Chlismiella*) .... ***C. pterosperma***
7′ Seed wing 0; petals white, cream, yellow, lavender, rarely red
8. Fr pedicelled, not coiled, twisted, or wavy; seeds in 2 rows per chamber
9. Hypanthium 4.5–40 mm; pl with cauline, simple lvs (rosette 0 or poorly developed) (sect. *Lignothera*)
10. Hypanthium 18–40 mm; infl open; sepals 8–15 mm .... ***C. arenaria***
10′ Hypanthium 4.5–14 mm; infl dense; sepals 3–9 mm .... ***C. cardiophylla***
11. Hairs gen spreading (sometimes some glandular); hypanthium 4.5–12 mm .... ssp. ***cardiophylla***
11′ Hairs gen glandular (some spreading, nonglandular); hypanthium 9–14 mm .... ssp. ***robusta***
9′ Hypanthium 0.4–8 mm; pl gen with well developed basal rosette (sometimes also cauline lvs); lvs gen 1-pinnate (sect. *Chylismia*)
12. Petals lavender, bases gen yellow, lavender-dotted; rosette poorly developed .... ***C. heterochroma***
12′ Petals yellow or white (sometimes fading purple or red or bases purple- or red-dotted); rosette well developed
13. Fr gen < 2 mm wide, ± same width throughout
14. Sepals 1.5–4 mm; hypanthium 1–1.5 mm; stigma ± = anthers; infl erect, buds drooping; lflets < 30 mm .... ***C. walkeri*** ssp. ***tortilis***
14′ Sepals 5–9 mm; hypanthium 3–8 mm; stigma > anthers; infl nodding; lflets gen < 10 mm or 0 . ***C. brevipes***
15. Fl bud reflexed; petals gen fading red .... ssp. ***arizonica***
15′ Fl bud not reflexed; petals not fading red
16. Pl hairs spreading; sepals in bud with free subterminal tips; petals not red-dotted .... ssp. ***brevipes***
16′ Pl strigose (rarely also with spreading hairs below); sepals in bud without free tips; petals gen red-dotted .... ssp. ***pallidula***
13′ Fr > 2 mm wide, wider toward tip
17. Pedicel and fr becoming reflexed; petals yellow .... ***C. munzii***
17′ Pedicel and fr spreading or ascending; petals white or yellow .... ***C. claviformis***
18. Pl spreading-hairy below; free tips of sepals in bud conspicuous; petals yellow .... ssp. ***peirsonii***
18′ Pl strigose or glabrous below; free tips of sepals in bud 0 or inconspicuous (exc ssp. *funerea*); petals white or yellow

19. Petals yellow
20. Pl strigose above, also glandular or not; free tips of sepals in bud 0 or short .... ssp. ***yumae***
20′ Pl glabrous above or glandular-hairy; free tips of sepals in bud 0
21. Terminal lflet ovate to subcordate, lateral lflets few–many; fl gen opening at dawn .. ssp. ***cruciformis***
21′ Terminal lflet lanceolate, lateral lflets 0–few; fl opening at dusk .... ssp. ***lancifolia***
19′ Petals white (rarely pale yellow in ssp. *claviformis*)
22. Lateral lflets gen well developed
23. Pl strigose above, rarely glandular-hairy .... ssp. ***aurantiaca***
23′ Pl ± glabrous above (or with some glandular hairs) .... ssp. ***claviformis***
22′ Lateral lflets gen 0 or poorly developed
24. Free tips of sepals in bud conspicuous; pl gen strigose above .... ssp. ***funerea***
24′ Free tips of sepals in bud 0 or minute; pl ± glabrous above or gen with some glandular hairs .... ssp. ***integrior***
8′ Fr ± sessile (exc some *C. kernensis*), often coiled, twisted and wavy; seeds in 1 row per chamber
25. Petals white; fl opening at dusk (sect. *Eremothera*)
26. Fr of ± same width throughout; fls at lower nodes 0
27. Sepals 1.5–2.5 mm; petals 1.8–3 mm; stigma ± surrounded by anthers .... ***C. chamaenerioides***
27′ Sepals 4–6 mm; petals 3.5–7 mm; stigma exceeding anthers .... ***C. refracta***
26′ Fr base wider than tip; fls at lower nodes present or 0
28. Infl erect; petals 0.8–1.3 mm; sepals 0.8–1.8 mm .... ***C. minor***
28′ Infl nodding; petals 3–7.5 mm; sepals (2.7)4–8 mm .... ***C. boothii***
29. Rosette present at time of first fl (spring); bracts not lf-like, ± inconspicuous; pl ± glabrous (or hairs not spreading)
30. Fr 2–3.8 mm wide at base, woody, curved outward but not downward .... ssp. ***condensata***
30′ Fr 1–2.3 mm wide at base, not woody, curved outward or downward
31. Fr 1.7–2.3 mm wide at base, curved outward .... ssp. ***decorticans***
31′ Fr 1–1.6 mm wide at base, gen curved downward .... ssp. ***desertorum***
29′ Rosette gen withered by time of first fl (late spring, summer); bracts lf-like, conspicuous; pl hairs often spreading
32. Pl strigose, also with rarely spreading or glandular hairs; seeds all minutely pitted .... ssp. ***alyssoides***
32′ Pl hairs spreading, some glandular; seeds of 2 kinds, minutely pitted and coarsely papillate
33. Pl gen 15–40 cm; cauline lf lanceolate to narrowly ovate, serrate .... ssp. ***boothii***
33′ Pl 5–20 cm; cauline lf narrowly lanceolate to lanceolate, ± entire to minutely serrate .... ssp. ***intermedia***
25′ Petals yellow; fl opening near dawn
34. Pl straight, slender, erect; lf sharply pinnately lobed; petal bases ± red-flecked; fr reflexed (sect. *Eulobus*) .... ***C. californica***
34′ Pl not simultaneously straight, slender, and erect; lf entire to toothed; petal bases with 0–2 red dots; fr not reflexed
35. Lvs closely spaced at st tips; fr very flat, 5–10 mm (sect. *Nematocaulis*) .... ***C. andina***
35′ Lvs well spaced throughout; fr 4-angled or cylindric, often swollen by seeds, gen > 10 mm
36. Fr cylindric, ± swollen by seeds; fls 0 at lower nodes; lf gen linear to narrowly elliptic; seeds shiny, often < 1 mm (sect. *Camissonia*)
37. Sepals all separating when fl opens
38. Petals 8–18 mm; sepals 5–11 mm; stigma exceeding anthers .... ***C. kernensis***
39. Pl somewhat open, sparsely spreading-hairy, sparsely glandular-hairy lvs not clustered at base; pedical 0–5 mm .... ssp. ***gilmanii***
39′ Pl compact, densely spreading-hairy, sparsely glandular-hairy; lvs clustered at base; pedicel 3–15 mm .... ssp. ***kernensis***
38′ Petals 1.8–4 mm; sepals < 3.8 mm; stigma ± surrounding anthers
40. Rosette 0; lf subentire; pl ± glabrous or minutely strigose, gen sparsely glandular-hairy, rarely sparsely spreading-nonglandular-hairy .... ***C. parvula***
40′ Rosette gen ± 0; lf serrate; pl hairs gen spreading, some glandular
41. Hypanthium 1.3–3 mm; sepals 2.2–3.8 mm; fr (18)26–50 mm .... ***C. pubens***
41′ Hypanthium 0.8–1.6 mm; sepals 1.2–2 mm; fr 18–32 mm .... ***C. pusilla***
37′ Sepals gen remaining adherent in pairs when fl opens
42. Stigma exceeding anthers; sepals 3–8(12) mm; petals (3.5)5–15 mm
43. Lf lanceolate to narrowly ovate or elliptic, subentire; sepals < 4.2 mm .... [2]***C. sierrae*** ssp. ***sierrae***
43′ Lf linear to narrow-elliptic or -oblanceolate, minutely or coarsely serrate; sepals 3–8(12) mm .... ***C. campestris***
44. St gen erect; lf minutely serrate .... ssp. ***campestris***
44′ St gen decumbent; lf coarsely serrate .... ssp. ***obispoensis***
42′ Stigma ± surrounding anthers; sepals 1.6–4(5.5) mm; petals 2–7 mm
45. Lf entire (rarely with 1–2 small teeth); infl densely minutely strigose .... ***C. integrifolia***
45′ Lf minutely serrate (*C. sierrae* sometimes with only several small teeth); infl hairs glandular to spreading, sometimes minutely strigose and glandular
46. Lf lanceolate to narrowly ovate or elliptic, base obtuse or rounded .... ***C. sierrae***
47. Sepals 1.2–3 mm; petal bases not red-dotted; style 2.8–5 mm .... ssp. ***alticola***
47′ Sepals 3–4.2 mm; petal bases with 2 red dots; style 4.5–7 mm .... [2]ssp. ***sierrae***

46′ Lf linear to narrowly elliptic, base acute or long-tapered
48. Pl minutely strigose (or some hairs spreading and glandular, or sometimes coarsely spreading and nonglandular toward base); < 10% of pollen grains 4-angled .... ***C. strigulosa***
48′ Pl hairs spreading, fine (in infl often glandular); > 30% of pollen grains 4-angled
49. Sepals (3)3.8–5.5 mm; petals (4)4.5–7 mm .... ***C. lacustris***
49′ Sepals 1.6–4 mm; petals 2–4(5) mm
50. Lf gen bluish green; hypanthium 1.2–2.3 mm; pl hairs coarse; gen > 30% of pollen grains 4-angled; widespread .... ***C. contorta***
50′ Lf not (or slightly) bluish green; hypanthium ± 1.2 mm; pl hairs fine; < 10% of pollen grains 4-angled; SCoRI .... ***C. benitensis***
36′ Fr 4-angled (at least when dry, exc *C. hirtella*), not swollen by seeds; fls gen at lower (through upper) nodes; lf gen lanceolate to narrowly elliptic or narrowly ovate; seeds dull, often > 1 mm (sect. *Holostigma*)
51. Per or subshrub; ± coastal .... ***C. cheiranthifolia***
52. Per; hairs rarely silvery; style 6–9 mm; petals 6–11 mm; stigma gen ± surrounding anthers .... ssp. ***cheiranthifolia***
52′ Subshrub; hairs silvery; style 13–23 mm; petals (10)12–20 mm; stigma exceeding anthers .... ssp. ***suffruticosa***
51′ Ann or short-lived per; inland or coastal
53. Stigma exceeding anthers .... ***C. bistorta***
53′ Stigma gen ± surrounding anthers
54. Fr stout (2.8–3.5 mm wide at base), deeply grooved .... ***C. guadalupensis*** ssp. ***clementina***
54′ Fr not esp stout (0.7–2.2 mm wide at base), not deeply grooved
55. > 25% of pollen grains 4–5-angled; very localized spp.; $2n$=42
56. Infl hairs nonglandular; 25–60% of pollen grains 4–5-angled — SCoRO (Monterey Co.) .. ***C. luciae***
56′ Infl hairs both glandular and other types; 70–100% of pollen grains 4–5-angled
57. Fr 1.3–1.6 mm wide, subcylindric (when dry ± 4-angled); SCoRO (Monterey, San Luis Obispo cos.) .... ***C. hardhamiae***
57′ Fr 1.5–2 mm wide, 4-angled; ChI, PR .... ***C. robusta***
55′ < 5% of pollen grains 4–5-angled; localized or widespread spp.; $2n$=14,28
58. Fr conspicuously 4-angled, 1.8–2.2 mm wide — c&s SCo .... ***C. lewisii***
58′ Fr cylindric or weakly 4-angled when fresh, 0.7–1.8 mm wide
59. Upper petioles < 25 mm; fr < 5-coiled .... ***C. ignota***
59′ Upper petioles 0 or < 2 mm; fr straight to 3-coiled
60. Pl strigose, grayish — s SnJV, D .... ***C. pallida***
61. Petals 6.5–13 mm; hypanthium 3.8–4.2 mm; style 6.5–10.5 mm .... ssp. ***hallii***
61′ Petals 2–6.5 mm; hypanthium 1–3 mm; style 2.1–6.5 mm .... ssp. ***pallida***
60′ Pl not strigose, sometimes grayish
62. Fr 0.7–0.9 mm wide; upper lvs narrowly ovate to ovate; st ascending .... ***C. hirtella***
62′ Fr 0.8–1.8 mm wide; upper lvs narrowly lanceolate to narrowly ovate; st ± erect or decumbent
63. St gen decumbent; upper lvs gen narrowly lanceolate; infl hairs gen nonglandular .... ***C. micrantha***
63′ St ± erect; upper lvs lanceolate to narrowly ovate; infl gen with glandular hairs
64. Pl grayish with spreading hairs; hypanthium 2–3.8 mm; sepals 3.2–8 mm .... ***C. confusa***
64′ Pl with spreading hairs, but not conspicuously grayish; hypanthium 1.2–2 mm; sepals 1–2.5 mm .... ***C. intermedia***

C. andina (Nutt.) Raven (p. 785) Ann, slender, minutely strigose (infl gen more densely so). **ST** 1–15 cm, gen branched. **LVS** many, closely spaced at st tips, 10–30 mm, narrowly oblanceolate, entire. **INFL** erect. **FL**: parts rarely in 3's; hypanthium 0.8–2 mm; sepals 0.8–2 mm; petals 0.8–2.3 mm, yellow; stamens opposite petals sometimes reduced. **FR** ascending, 5–10 mm, 1–1.3 mm wide, ± straight, very flat; walls somewhat swollen by seeds. **SEEDS** in 1 row per chamber, 0.7–1.3 mm, smooth, glossy. $2n$=28,42. Seasonally moist flats, gen clay soil, sagebrush scrub to pinyon/juniper woodland; 500–2000 m. CaRH (Sierra, Plumas cos.), MP; to B.C., UT. [*O. a.* Nutt.] Self-pollinated, rarely cleistogamous.

C. arenaria (Nelson) Raven (p. 785) Ann or bushy per, erect; hairs spreading, in infl a few glandular. **ST** < 180 cm. **LF**: petiole < 60 mm; blade < 60 mm, cordate-deltate, teeth coarse or larger and smaller. **INFL** open, nodding. **FL** opening at dusk; hypanthium 18–40 mm; sepals 8–15 mm; petals 8–20 mm, yellow. **FR** 30–44 mm, ascending, cylindric, ± straight; pedicel 2–5 mm. **SEEDS** in 2 rows per chamber, 0.5–0.7 mm, brown. $2n$=14. Sandy washes, rocky slopes, desert scrub; < 0–430 m. DSon; sw AZ, n Mex (Sonora). [*O. a.* (Nelson) Raven] Gen cross-pollinated. ❀TRY.

C. benitensis Raven (p. 785) SAN BENITO EVENING PRIMROSE Ann, slender; rosette ± 0; hairs spreading, fine, in infl also glandular. **ST** erect or decumbent, 3–20 cm, peeling; branches widely spreading, wiry. **LF** 7–20 mm, very narrowly elliptic, minutely serrate. **INFL** nodding. **FL**: hypanthium ± 1.2 mm; sepals 3.2–3.5 mm, remaining adherent in pairs; petals 3.5–4 mm, yellow fading reddish, bases with 2 red dots; < 10% of pollen grains 4-angled. **FR** 15–40 mm, 0.8–1.3 mm wide, cylindric, slightly swollen by seeds, straight or wavy, subsessile. **SEEDS** in 1 row per chamber, 0.6–0.8 mm, shiny, minutely pitted. $2n$=28. **THREATENED** US. Clay or gravelly serpentine alluvial terraces; ± 600 m. SCoRI (lower Clear Creek drainage, San Benito Co.). Self-pollinated. Probably derived from *C. strigulosa*. Threatened by off-road vehicles.

C. bistorta (Torrey & A. Gray) Raven (p. 785) CALIFORNIA SUN CUP Ann or short-lived per, rosetted; strigose or hairs spreading, in infl short, erect. **ST** decumbent to ± ascending, 50–80 cm, peeling. **LVS** 12–120 mm, petioled or above gen subsessile; cauline gen lanceolate (linear), minutely dentate to subentire. **INFL** nodding. **FL**: hypanthium 2–5(7.5) mm; sepals (2.3)5–8(11) mm; petals (4.2) 7–15 mm, yellow fading reddish, bases gen with 1–2 red dots. **FR** 12–40 mm, 1.5–2.5 mm wide, ± 4-angled, gen straight or slightly wavy and twisted. **SEEDS** in 1 row per chamber, 0.9–1 mm, minutely pitted in rows, dull brownish black. $2n$=14. Sandy fields near coast or clay soils in grasslands to openings in coastal-sage scrub or chaparral; 0–600 m. SW; n Baja CA. [*O. bistorta* Torrey & A. Gray, incl var. *veitchiana* Hook.; *O. heterophylla* Hook. & Arn.] Cross-pollinated. Intergrades with *C. cheiranthifolia* ssp. *suffruticosa*.

C. boothii (Douglas) Raven Ann, gen reddish; rosette gen ± 0 (to well developed); hairs minutely strigose and spreading (some glandular, esp in infl). **ST** erect, 3–65 cm, peeling. **LVS** 20–100(130) mm, lanceolate to narrowly elliptic or narrowly ovate, sparsely minutely dentate or serrate; lower oblanceolate or not. **INFL** nodding; fls gen 0 at lower nodes. **FL** opening at dusk; hypanthium 4–8 mm; sepals (2.7)4–8 mm; petals 3–7.5 mm, gen white (red) fading reddish. **FR** 8–35 mm, 1–3.8 mm wide, cylindric exc base wider than tip, ± curved outward to very wavy and twisted, persistent, tardily dehiscent. **SEEDS** in 1 row per chamber, 1.4–2.1 mm, gen of 2 kinds (minutely pitted in rows and pale brown; coarsely papillate and dark brown). $2n$=14. Shrubby or open, dry areas, gen desert; < 2400 m. s SN, CW, GB, D; to WA, ID, UT, nw Mex. [*O. b.* Douglas] Cross-pollinated.

ssp. ***alyssoides*** (Hook. & Arn.) Raven PINE CREEK EVENING PRIMROSE Pl: hairs gen densely, minutely strigose (esp in infl, hairs rarely spreading or glandular). **ST** 3–35 cm. **LVS** 10–40 mm, lanceolate to narrowly ovate; lower oblanceolate or not. **INFL**: bracts lf-like; fls sometimes present at lower nodes. **FR** 1–1.4 mm wide, gen very wavy and twisted. **SEEDS** all alike. RARE in CA. Sandy slopes, flats, gen sagebrush scrub; 600–1700 m. MP; to NV, w UT, s ID. [*O. a.* Hook. & Arn.] Intergrades widely with sspp. *boothii, intermedia* in NV; much like ssp. *desertorum*.

ssp. ***boothii*** (p. 785) BOOTH'S EVENING PRIMROSE Pl: hairs spreading and glandular. **ST** gen 15–40 cm. **LF** 30–80 mm, lanceolate to narrowly ovate, serrate. **INFL**: bracts lf-like. **FR** 1.4–2 mm wide, gen very wavy and twisted. UNCOMMON. Sandy flats, steep loose slopes, Joshua-tree and pinyon-juniper woodlands; 900–2400 m. SNE; to WA, nw AZ. Intergrades widely with sspp. *alyssoides, intermedia* in NV.

ssp. ***condensata*** (p. 785) (Munz) Raven Pl stout, subglabrous (exc infl minutely strigose or glandular-hairy); rosette well developed. **ST** 5–20(30) cm. **LF** 25–100(130) mm, gen lanceolate to oblanceolate, subentire to minutely dentate. **INFL**: bracts inconspicuous. **FR** 2–3.8 mm wide, curved outward. Sandy slopes, washes, desert scrub; –70–1200 m. D; to s NV, s UT, w AZ, nw Mex. [*O. b.* ssp. *c.* (Munz) Munz] Intergrades widely with ssp. *desertorum*.

ssp. ***decorticans*** (Hook. & Arn.) Raven SHREDDING EVENING PRIMROSE Pl stout, subglabrous exc infl; rosette well developed. **ST** 12–65 cm. **LF** gen 20–80 mm, gen lanceolate (or lower narrowly ovate), entire to minutely dentate. **INFL**: bracts inconspicuous. **FR** 1.7–2.3 mm wide, curved outward. Open, gen steep and rocky (esp shale) slopes; 0–1850 m. s SNF, Teh, s SnJV, SnFrB, SCoRI, WTR. [*O. b.* sspp. *d.* (Hook. & Arn.) Munz & *rutila* (Davidson) Munz] Intergrades with ssp. *desertorum*.

ssp. ***desertorum*** (Munz) Raven (p. 785) Pl: rosette well developed; hairs sparse, minutely strigose (or also glandular, esp in infl). **ST** 10–35 cm. **LF** gen 10–40 mm, lanceolate to narrowly ovate (or lower oblanceolate), entire to minutely dentate. **INFL**: bracts inconspicuous. **FR** 1–1.6 mm wide, gen curved downward. Sandy or gravelly slopes, washes, gen creosote-bush scrub; 450–2000 m. s SNH, s SNE, DMoj. [*O. b.* sspp. *d.* Munz & *inyoensis* Munz] Intermediate between sspp. *condensata* & *decorticans*.

ssp. ***intermedia*** (Munz) Raven (p. 785) Pl: hairs dense, spreading (and glandular, esp in infl). **ST** 5–20 cm. **LF** gen < 25 mm, ± lanceolate (or lower oblanceolate), minutely serrate. **INFL**: bracts lf-like; fls sometimes present at lower nodes. **FR** 1–1.4 mm wide, gen curved outward or ± wavy and twisted. Sandy soils, sagebrush scrub; 1500–2150 m. SNE (ne Inyo Co.), e DMtns (Kingston Mtns); NV. [*O. b.* ssp. *i.* Munz] Intermediate between sspp. *alyssoides* and *boothii*; ± uniform.

C. brevipes (A. Gray) Raven Ann, strigose or hairs spreading. **ST** 3–75 cm. **LVS** gen basal, simple to 1-pinnate; terminal lflet < 65 mm, lateral lflets gen < 10 mm or 0. **INFL** nodding. **FL**: hypanthium 3–8 mm; sepals 5–9 mm, tips in bud gen free and subterminal; petals 3–18 mm, yellow, bases ± red-dotted; stamens ± equal. **FR** ascending to spreading, 18–92 mm, cylindric, straight or curved; valves with strong midrib; pedicel 2–20 mm. **SEEDS** in 2 rows per chamber, 1–1.5 mm. $2n$=14. Sandy or rocky slopes, washes, creosote-bush scrub, Joshua-tree woodland; –70–1800 m. D; to w&s NV, sw UT. [*O. b.* A. Gray] Cross-pollinated.

ssp. ***arizonica*** (Raven) Raven (p. 785) Pl: hairs spreading. **FL**: bud reflexed; hypanthium 3–5 mm; sepal tips in bud free or not; petals 3–8 mm, gen red-dotted, gen fading red. **FR** 18–60 mm; pedicel 2.5–5 mm. Rocky slopes and flats; 70–300 m. se DSon (Imperial Co.); sw AZ. [*O. b.* ssp. *a.* Raven] Hybridizes with *C. claviformis* sspp. *peeblesii, yumae*.

ssp. ***brevipes*** (p. 785) Pl: hairs spreading. **FL**: bud not reflexed; hypanthium 4–8 mm; petals 6–18 mm, gen not red-dotted, not fading red. **FR** 20–92 mm; pedicel 5–20 mm. Sandy slopes, washes, alluvial fans (moister than other sspp.); –70–1800 m. Range of sp. Intergrades with ssp. *pallidula*; hybridizes with *C. claviformis, C. munzii*. ❀TRY.

ssp. ***pallidula*** (Munz) Raven (p. 785) Pl strigose (hairs rarely spreading below). **FL**: bud not reflexed; hypanthium 4–5 mm; sepal tips in bud not free; petals 7–12 mm, gen red-dotted, not fading red. **FR** 20–42 mm; pedicel 2–10 mm. Dry flats, desert pavement, gen with *Larrea, Ambrosia*; 70–1100 m. D (se Inyo, ne Imperial cos.); to sw UT, nw AZ. [*O. b.* ssp. *pallidula* (Munz) Raven] ❀TRY.

C. californica (Torrey & A. Gray) Raven (p. 785) Ann, subglabrous (or lvs minutely strigose, infl sparsely glandular-hairy); rosette well developed. **ST** straight, slender, erect, 2–180 cm, ± glaucous or green. **LVS** much reduced upward, < 300 mm, narrowly elliptic, irregularly and sharply pinnately lobed. **INFL**: fls widely spaced. **FL**: hypanthium 0.6–1.5 mm, closed by fleshy disk; sepals 3.9–8 mm; petals 6–14 mm, yellow, base ± red-flecked. **FR** reflexed, 45–110 mm, cylindric (drying 4-angled), ± straight. **SEEDS** in 1 row per chamber, 1.3–1.6 mm, olive with purple spots. $2n$=14, 28. Open places in coastal-sage scrub, chaparral, desert scrub; < 1300 m. s NCoRO (Sonoma Co.), SnJV (Fresno Co.), SCoR, SW, w edge DMoj, DMtns (uncommon); sw AZ, nw Mex. [*Eulobus c.* Torrey & A. Gray; *O. leptocarpa* E. Greene] Self-pollinated. ❀SUN,DRY,DRN:**7–12,14–17**,18,**19–24**.

C. campestris (E. Greene) Raven MOJAVE SUN CUP Ann, slender; rosette gen ± 0; hairs 0, coarse, or glandular. **ST** decumbent or erect, 5–25(50) cm, peeling. **LF** 5–30 mm, linear to narrowly elliptic or narrowly oblanceolate, minutely to coarsely serrate. **INFL** nodding. **FL**: hypanthium 1.5–5 mm; sepals 3–8(12) mm, remaining adherent in pairs; petals (3.5)5–15 mm, yellow fading reddish, base with (1)2 red dots; stigma exceeding anthers. **FR** 20–43 mm, 0.7–2 mm wide, cylindric, alternately narrow and swollen by seeds, straight or wavy, subsessile. **SEEDS** in 1 row per chamber, 0.8–1.6 mm, shiny, minutely pitted. $2n$=14. Open sandy flats, desert scrub, noncoastal grasslands; 0–2000 m. SNF, GV, CW, e SW, w DMoj. [*O. c.* E. Greene] Cross-pollinated. Sspp. intergrade extensively.

ssp. ***campestris*** **ST** gen erect. **LF** linear to narrowly elliptic or narrowly oblanceolate, minutely serrate. Habitats, range of sp. [*O. c.* ssp. *parshii* (Abrams) Munz; *O. dentata* Cav. var. *johnstonii* Munz; *O. cruciata* (S. Watson) Munz misapplied] Sometimes hybridizes (hybrids seldom fertile) with *C. contorta*. ❀TRY.

ssp. ***obispoensis*** Raven **ST** gen decumbent. **LF** narrowly elliptic, coarsely serrate. Marine deposits in openings in chaparral and oak woodland; 100–500 m. SCoRO. Hybridizes with *C. contorta, C. strigulosa*. ❀TRY.

C. cardiophylla (Torrey) Raven Ann, per; rosette 0; hairs spreading, glandular or not. **ST** < 100 cm. **LF**: blade < 55 mm, ovate to rounded-cordate, irregularly dentate; petiole < 75 mm. **INFL** dense, nodding. **FL** opening at dusk; hypanthium 4.5–14 mm; sepals 3–9 mm; petals 3–12 mm, yellow or cream. **FR** 20–55 mm, ascending, cylindric, ± straight; pedicel 1–18 mm. **SEEDS** in 2 rows per chamber, 0.5–0.7 mm, brown. $2n$=14. Sandy washes, slopes, rocky walls, creosote-bush scrub; 0–1400 m. D; sw AZ, nw Mex. [*O. c.* Torrey] Gen cross-pollinated. 3 sspp., 2 in CA.

ssp. ***cardiophylla*** (p. 785) Pl: sometimes some hairs glandular. **LF** cordate. **FL**: hypanthium 4.5–12 mm; petals 3–12 mm. Habitats of sp.; 0–600 m. c&s DMoj (San Bernardino Co.), DSon; s AZ, nw Mex. ❀TRY.

ssp. ***robusta*** (Raven) Raven (p. 785) Pl: hairs both glandular and not. **LF** cordate-round. **FL**: hypanthium 9–14 mm; petals 7–11 mm. Habitats of sp.; 600–1400 m. n DMoj (Inyo Co.). [*O. c.* ssp. *r.* Raven] ❀TRY.

C. chamaenerioides (A. Gray) Raven Ann, gen reddish, glandular-hairy (infl also minutely strigose); rosette gen ± 0. **ST** erect, 8–50 cm, peeling. **LF** < 80 mm, narrowly elliptic to narrowly lanceolate, sparsely minutely dentate. **INFL** nodding. **FL** opening at dusk; hypanthium 1.6–2.3 mm; sepals 1.5–2.5 mm; petals 1.8–3 mm, white fading reddish. **FR** spreading, 35–55 mm, 0.8–0.9 mm wide, cylindric, ± straight. **SEEDS** in 1 row per chamber, 0.9–1 mm, minutely pitted in rows. 2*n*=14. Sandy slopes, flats, desert scrub; ± -50–1300 m. W&I, D; to UT, TX, nw Mex. [*O. c.* A. Gray] Self-pollinated. Related to *C. refracta*.

C. cheiranthifolia (Sprengel) Raim (p. 785) BEACH EVENING PRIMROSE Per or subshrub, short-lived, rosetted, densely strigose (rarely glabrous); hairs of infl gen erect, short. **STS** prostrate to ± ascending, < 60(130) cm, peeling. **LF** 5–50 mm, narrowly ovate to obovate, minutely serrate; cauline petioles 0–10 mm. **INFL** nodding. **FL**: hypanthium 2.1–8.5 mm; sepals 4–11.5 mm; petals 6–20 mm, yellow fading reddish, bases with 0–2 red dots. **FR** 10–25 mm, 2–2.5 mm wide, 4-angled, gen 1–2-coiled. **SEEDS** in 1 row per chamber, 1.2–1.3 mm, minutely pitted in rows, dull brownish black. 2*n*=14. Sandy slopes, flats, coastal dunes; < 100 m. NCo, CCo, SCo, ChI; sw OR, n Baja CA. [*O. c.* Sprengel incl var. *nitida* (E. Greene) Munz] Gen cross-pollinated. Sspp. intergrade on ChI.

ssp. ***cheiranthifolia*** Per; hairs rarely dense and silvery. **FL**: hypanthium 2.1–4.2(4.8) mm; petals 6–11 mm, bases with 0(–2) red dots; anthers 1–1.5 mm; style 6–9 mm. Habitats of sp. NCo, CCo, ChI; sw OR. Gen self-pollinated. ❀SUN:5,**15–17**&IRR:14,19–21,**22–24**.

ssp. ***suffruticosa*** (S. Watson) Raven Subshrub; hairs gen dense and silvery. **FL**: hypanthium 5–8.5 mm; petals (10)12–20 mm, bases with 1–2 red dots; anthers 2.2–3 mm; style 13–23 mm. Habitats of sp. SCo; n Baja CA. [*O. c.* ssp. *s.* (S. Watson) Munz] Gen cross-pollinated (± self-incompatible). Hybridizes widely with *C. bistorta*. ❀SUN:15–17,**24**&IRR:14,**19–23**.

C. claviformis (Torrey & Frémont) Raven Ann. **ST** 3–70 cm. **LVS** gen basal, gen 1-pinnate; terminal lflet 8–90 mm, lanceolate to cordate; lateral lflets < 25 mm or 0. **INFL** nodding. **FL** gen opening at dusk; hypanthium 1–6.5 mm; sepals 2–8 mm, tips in bud free and subterminal or not; petals 1.5–8 mm, yellow or white; stamens ± equal. **FR** ascending or spreading, 8–38 mm, wider to tip, straight or curved; pedicel 4–40 mm. **SEEDS** in 2 rows per chamber, 0.6–1.5 mm. 2*n*=14. Sandy or rocky slopes or washes; -70–2000 m. PR, GB, D; to OR, ID, UT, AZ, nw Mex. [*O. c.* Torrey & Frémont] Cross-pollinated; most complex, widespread sp. in genus; 11 sspp., 8 in CA.

ssp. ***aurantiaca*** (S. Watson) Raven (p. 785) Pl strigose (rarely glandular-hairy above). **LF**: terminal lflet < 30 mm, narrowly ovate; lateral lflets well developed, like terminal. **FL**: hypanthium 3–5 mm; sepal tips not free in bud (rarely inconspicuously so); petals 2.5–8 mm, white gen fading purple, bases rarely purple-dotted. Sandy flats, washes, creosote-bush scrub; -70–900 m. D; s NV, w AZ, n Baja CA. [*O. c.* ssp. *a.* (S. Watson) Raven] Intergrades extensively with sspp. *peirsonii, yumae*, and those with white petals; sometimes hybridizes with *C. brevipes, C. munzii*. ❀TRY.

ssp. ***claviformis*** Pl glabrous or strigose below, ± glabrous or glandular-hairy above. **LF**: terminal lflet < 60 mm, narrowly ovate, purple-dotted or not; lateral lflets gen large. **FL**: hypanthium 3–5.5 mm; sepal tips free in bud, conspicuous; petals 3.5–8 mm, white (pale yellow) gen fading purple, bases purple-dotted or not. Alluvial slopes, flats, creosote-bush scrub; 850–1700 m. DMoj and edges. Intergrades widely and gradually with sspp. *aurantiaca, funerea*; hybridizes with *C. brevipes* ssp. *b*. ❀TRY.

ssp. ***cruciformis*** (Kellogg) Raven (p. 785) Pl strigose or glandular-hairy below, glabrous or glandular-hairy above. **LF**: terminal lflet < 80 mm, ovate to subcordate; lateral lflets few–many. **FL** gen opening at dawn; hypanthium 2–6.5 mm; sepal tips not free in bud; petals 2.5–8 mm, yellow gen fading purple, bases gen red-dotted. Sagebrush scrub; 600–1400 m. MP (Lassen Co.); to c OR, ID, nw NV. [*O. claviformis* sspp. *c.* (Kellogg) Raven & *citrina* Raven] Outside CA, intergrades with ssp. *integrior*.

ssp. ***funerea*** (Raven) Raven (p. 785) Pl gen strigose (densely so at least below). **LF**: terminal lflet < 80 mm, ovate, gen cordate; lateral lflets gen 0. **FL**: hypanthium 3–5.5 mm; sepal tips free in bud, conspicuous, subterminal; petals 3.5–7.5 mm, white gen fading purple. Dry slopes, flats, creosote-bush scrub; -70–900 m. n DMoj (Eureka, Saline, Death valleys). [*O. c.* ssp. *f.* Raven] Intergrades with sspp. *aurantiaca, claviformis*; sometimes hybridizes with *C. munzii, C. brevipes* ssp. *b*.

ssp. ***integrior*** (Raven) Raven Pl strigose below, gen glandular-hairy or subglabrous above. **LF**: terminal lflet < 70 mm, ± ovate, base gen subcordate; lateral lflets 0–few, small. **FL**: hypanthium 3–6 mm; sepal tips not free in bud (or inconspicuously so); petals 4.5–8 mm, white fading purple, bases purple-dotted or not. Dry flats, desert scrub; 1200–2000 m. SNE; c OR, NV. [*O. c.* ssp. *i.* Raven] Intergrades with sspp. *aurantiaca, cruciformis*; hybridizes with *C. brevipes*.

ssp. ***lancifolia*** (A.A. Heller) Raven (p. 785) Pl strigose below, glabrous and glaucous above. **LF**: terminal lflet < 50 mm, lanceolate; lateral lflets 0–few, small. **FL**: hypanthium 3.5–6 mm; sepal tips not free in bud; petals 3.5–7 mm, yellow, bases gen red-dotted. Sandy soils, sagebrush scrub; 1200–1700 m. SNE. [*O. c.* ssp. *l.* (A.A. Heller) Raven]

ssp. ***peirsonii*** (Munz) Raven (p. 785) Pl spreading-hairy, rarely strigose or glandular. **LF**: terminal lflet < 90 mm, narrowly ovate; lateral lflets gen large. **FL**: hypanthium 2.5–4.5 mm; sepal tips in bud free, conspicuous; petals 4.5–7 mm, gen yellow (white). Sandy flats, creosote-bush scrub; -70–300 m. PR, DSon (Imperial Co.); n Baja CA. [*O. c.* ssp. *p.* (Munz) Raven] Intergrades with ssp. *aurantiaca*.

ssp. ***yumae*** (Raven) Raven Pl strigose (gen densely so), sometimes also glandular-hairy above. **LF**: terminal lflet < 65 mm, lanceolate; lateral lflets reduced or not. **FL**: hypanthium 2.5–4 mm; sepal tips not free in bud (or inconspicuously so); petals 4–5.5 mm, yellow, sometimes fading reddish. Dunes or sandy flats, creosote-bush scrub; 0–300 m. se DSon (se Imperial Co.); sw AZ, nw Mex. [*O. c.* ssp. *y.* Raven] Probably derived from ssp. *aurantiaca* × ssp. *peeblesii* (Munz) Raven in AZ; intergrades with ssp. *aurantiaca*.

C. confusa Raven Ann, robust, rosetted, grayish; hairs dense, spreading and strigose (infl also glandular or subglabrous). **ST** ± erect, < 60 cm. **LF** 10–50 mm; cauline lanceolate to narrowly ovate, minutely dentate, subsessile. **INFL** nodding. **FL**: hypanthium 2–3.8 mm; sepals 3.2–8 mm; petals 5–10 mm, yellow fading reddish, bases with 1–2 red dots. **FR** 13–23 mm, 0.8–1.1 mm wide, cylindric (drying 4-angled), gen 1–2-coiled. **SEEDS** in 1 row per chamber, 0.7–1.1 mm, minutely pitted in rows, dull brownish black. 2*n*=28. Dry noncoastal slopes, gen chaparral; 300–2000 m. SW. [*O. ignota* (Jepson) Munz misapplied] Self-pollinated. Probably derived from *C. hirtella* × *C. pallida* ssp. *p*.

C. contorta (Douglas) Raven Ann, slender; rosette ± 0; hairs spreading, gen coarse (only at pl base or throughout), in infl also sparsely glandular. **ST** decumbent or erect, gen 3–30 cm, wiry, peeling. **LF** gen 10–35 mm, linear to narrowly elliptic, minutely serrate, gen bluish green. **INFL** nodding. **FL**: hypanthium 1.2–2.3 mm; sepals 2.5–4 mm, remaining adherent in pairs; petals 3–5 mm, yellow fading reddish, bases with 0 or 2 red dots; > 30% of pollen grains gen 4-angled. **FR** gen 25–35 mm, 0.7–1.2 mm wide, cylindric, ± swollen by seeds, straight or wavy, subsessile. **SEEDS** in 1 row per chamber, 0.7–0.9 mm, shiny, minutely pitted. 2*n*=42. Sandy soil, slopes, flats, often disturbed, grassland, chaparral, pinyon/juniper woodland; 0–2300 m. NW, CaR, SNF, GV, CW, MP; to WA, ID, w NV. [*O. c.* Douglas; *O. cruciata* (S. Watson) Munz; *O. strigulosa* Fischer & C. Meyer misapplied] Self-pollinated. Probably derived from *C. strigulosa* × *C. campestris* ssp. *c*. (sterile hybrids formed where they co-occur).

C. graciliflora (Hook. & Arn.) Raven (p. 785) HILL SUN CUP Ann; hairs ± dense, spreading. **ST** ± 0. **LF** 10–98 mm, linear to narrowly lanceolate or narrowly oblanceolate, entire or weakly and minutely serrate. **FL** nodding; hypanthium 1.6–3.2 mm, closed by fleshy disk; sepals 4.5–8 mm; petals 5–18 mm, yellow; anthers attached at base; sterile tip of ovary 6–45 mm. **FR** 4–8 mm, 4-angled (4-winged near tip), leathery, tardily dehiscent. **SEEDS** 1.2–2 mm, papillate, tan with brown spots. 2*n*=14. Open or shrubby slopes, gen clay soils, grasslands, oak and Joshua-tree woodlands; < 800 m. NW, CaR, SNF, GV, CW, WTR; sw OR, Baja CA. [*O. g.* Hook. & Arn.] Self-pollinated.

C. guadalupensis (S. Watson) Raven ssp. ***clementina*** (Raven) Raven SAN CLEMENTE ISLAND EVENING PRIMROSE Ann, rosetted; hairs spreading, in infl also glandular. **ST** erect, 2–18(35) cm, ± fleshy, peeling. **LF** 10–38(95) mm; cauline narrowly ovate, minutely dentate or subentire, subsessile. **INFL** nodding. **FL**: hypanthium 1.6–2.4 mm; sepals 1.9–3.2 mm; petals 2.8–4.2 mm, yellow fading reddish, bases with 1 red dot. **FR** 10–18 mm, 2.8–3.5 mm wide, stout, ± 4-angled, straight or ± curved outward, deeply grooved. **SEEDS** in 1 row per chamber, 0.7–0.9 mm, minutely pitted in rows, dull brownish black. 2*n*=14. RARE. Sand dunes; 0–30 m. s ChI (San Clemente Island). [*O. g.* S. Watson ssp. *c.* Raven] Self-pollinated. Other ssp. on Guadalupe Island, Baja CA.

C. hardhamiae Raven HARDHAM'S EVENING PRIMROSE Ann, robust, rosetted; hairs spreading, in infl also glandular. **ST** erect, < 60 cm. **LF** 10–120 mm; cauline widely lanceolate to narrowly ovate, minutely dentate, subsessile. **INFL** nodding. **FL**: hypanthium 1.7–2 mm; sepals 1.8–3.2 mm; petals 2–4 mm, yellow fading reddish; 70–100% of pollen grains 4–5-angled. **FR** 13–25 mm, 1.3–1.6 mm wide, ± stout, ± cylindric (drying ± 4-angled), straight to 1-coiled. **SEEDS** in 1 row per chamber, 0.7–1.1 mm, minutely pitted in rows, dull brownish black. 2*n*=42. RARE. Sandy soil, limestone, disturbed oak woodland; 240–600 m. SCoRO (Monterey, San Luis Obispo cos.). Self-pollinated. Probably derived from *C. micrantha* × *C. intermedia.*

C. heterochroma (S. Watson) Raven (p. 785) SHOCKLEY'S EVENING PRIMROSE Ann, glandular-hairy (or subglabrous and glaucous above). **ST** 10–100 cm. **LVS** gen basal, < 70 mm, ovate; base gen cordate. **INFL** erect, longer in fr. **FL**: hypanthium 2–5 mm; sepals 1.5–3.5 mm, tips not free in bud; petals 2–6 mm, lavender, bases gen yellow, lavender-dotted. **FR** 7–13 mm, erect, club-shaped, straight; pedicel 2–5 mm. **SEEDS** in 2 rows per chamber, 1–1.2 mm, brown. 2*n*=14. Alluvial slopes, rock slides, creosote-bush scrub to pinyon/juniper woodland; 600–2100 m. SNE, n DMtns (Grapevine Mtns); NV. [*O. h.* S. Watson, incl ssp. *monoensis* (Munz) Raven] Cross-pollinated. ❀TRY.

C. hirtella (E. Greene) Raven Ann, rosetted; hairs spreading, in infl gen also glandular. **ST** ascending, < 60 cm. **LF** 10–80 mm; cauline ± ovate, minutely dentate, subsessile. **INFL** nodding. **FL**: hypanthium 1–3 mm; sepals 2.5–6 mm; petals 2–9 mm, yellow fading reddish, bases red-dotted or not. **FR** 13–20 mm, 0.7–0.9 mm wide, cylindric, gen 1–2-coiled. **SEEDS** in 1 row per chamber, 0.7–0.8 mm, minutely pitted in rows, dull brownish black. 2*n*=14. Open shrubby slopes, esp in burns; 0–2300 m. NW, SN, CW, SW; n&c Baja CA. [*O. h.* E. Greene] Self-pollinated. Rarely hybridizes with *C. ignota.*

C. ignota (Jepson) Raven Ann, rosetted, gen reddish, minutely strigose (infl glabrous, longer-hairy, or glandular). **ST** ascending or erect, < 55 cm, ± fleshy. **LF** < 60 mm, narrowly elliptic to narrowly ovate, minutely serrate; petiole < 25 mm. **INFL** nodding. **FL**: hypanthium 1.1–3 mm; sepals 2.5–5.5 mm; petals (3)4.8–8 mm, yellow fading reddish, bases red-dotted or not; < 5% of pollen grains 4–5-angled. **FR** gen 20–30 mm, 0.8–1 mm wide, cylindric (drying 4-angled), < 5-coiled. **SEEDS** in 1 row per chamber, ± 1.2 mm, minutely pitted in rows, dull brownish black. 2*n*=14. Gen clay fields, slopes in coastal-sage scrub or chaparral, sandy soils in mtns; 100–1100 m. c SNH (Madera Co.), ScV (uncommon), CW, SW (incl Santa Cruz Island); n Baja CA. [*O. i.* (Jepson) Munz] Self-pollinated.

C. integrifolia Raven KERN RIVER EVENING PRIMROSE Ann, slender, subglabrous or sparsely minutely strigose (exc infl densely strigose, grayish); rosette ± 0. **ST** decumbent or erect, gen < 30 cm, wiry, peeling. **LF** gen 10–30 mm, linear, entire; teeth 0–2, small. **INFL** nodding. **FL**: hypanthium 1.5–2.5 mm; sepals 1.6–4 mm, remaining adherent in pairs; petals gen 2–4 mm, yellow fading reddish, bases with 0–2 red dots. **FR** 45–60 mm, 0.8–1.3 mm wide, cylindric, ± swollen by seeds, straight or wavy, subsessile. **SEEDS** in 1 row per chamber, 1–1.2 mm, shiny, minutely pitted. 2*n*=28. UNCOMMON. Gen sagebrush slopes; 700–1000 m. s SNF (Kern River area, Kern Co.). Self-pollinated. Probably derived from *C. strigulosa*; now forms sterile hybrids where the two co-occur.

C. intermedia Raven Ann, rosetted; hairs dense, spreading, in infl also glandular. **ST** ± erect, < 60 cm. **LVS** 10–120 mm; cauline lanceolate to narrowly ovate, minutely dentate, subsessile. **INFL** nodding. **FL**: hypanthium 1.2–2 mm; sepals 1–2.5 mm; petals 1.5–3.5(4.5) mm, yellow fading reddish, bases with 1–2 red dots. **FR** 13–25 mm, 1.1–1.2 mm wide, cylindric (drying ± 4-angled), straight to 1-coiled. **SEEDS** in 1 row per chamber, 0.7–1 mm, minutely pitted in rows, dull brownish black. 2*n*=28. Shrubby slopes, esp burns; 300–800 m. NW, CW, SW; n Baja CA. [*O. hirta* Link and *O. micrantha* Sprengel misapplied] Self-pollinated. Probably derived from *C. hirtella* × *C. micrantha.*

C. kernensis (Munz) Raven Ann, robust; rosette gen ± 0; hairs dense, spreading (some glandular, or ± 0, esp in infl). **ST** erect, 5–30 cm. **LF** 10–38(55) mm, gen narrowly elliptic, sparsely serrate. **INFL** nodding. **FL**: hypanthium 2.2–3.8(5.5) mm; sepals 5–11 mm, free; petals 8–18 mm, yellow fading reddish, bases with 2 large red dots. **FR** 22–37 mm, 1.5–1.7 mm wide, cylindric, ± swollen by seeds, straight or wavy; pedicel 0–15 mm. **SEEDS** in 1 row per chamber, 1.1–1.2 mm, shiny, minutely pitted. 2*n*=14. Sandy slopes, flats, washes, sagebrush scrub, Joshua-tree and pinyon/juniper woodland; 700–1800 m. s SNE (esp Inyo Co.), n&w DMoj; s NV. [*O. k.* Munz] Cross-pollinated. Related to *C. parvula, C. pubens, C. pusilla.* Sspp. intergrade extensively.

ssp. ***gilmanii*** (Munz) Raven (p. 791) Pl open; hairs ± 0–few, glandular and spreading. **ST** < 30 cm. **LVS** not esp clustered at base. **FR**: pedicel 0–5 mm. Washes, slopes; 760–1800 m. Range of sp. [*O. k.* sspp. *g.* (Munz) Munz & *mojavensis* Munz] ❀TRY.

ssp. ***kernensis*** (p. 791) KERN COUNTY EVENING PRIMROSE Pl compact; hairs dense, spreading, few glandular. **ST** 5–15(22) cm. **LVS** clustered at base. **FR**: pedicel 3–15 mm. UNCOMMON. Sandy slopes, flats, gen in sagebrush scrub or Joshua-tree woodland; 850–1800 m. se SNH, w DMoj (Piute Mtns, El Paso Mtns, Grapevine Canyon, Kern Co.). Often locally abundant. ❀TRY.

C. lacustris Raven Ann, slender; rosette ± 0; hairs ± dense, spreading (glandular in infl). **ST** decumbent or erect, < 50 cm, wiry, peeling. **LF** 8–35 mm, linear to very narrowly elliptic, minutely serrate. **INFL** nodding. **FL**: hypanthium 1.7–2.7 mm; sepals 3–5.5 mm, remaining adherent in pairs; petals 4–7 mm, yellow fading reddish, bases with 0 or 2 red dots. **FR** < 45 mm, 0.8–1.3 mm wide, cylindric, ± swollen by seeds, straight or wavy, subsessile. **SEEDS** in 1 row per chamber, 0.6–0.8 mm, shiny, minutely pitted. 2*n*=28. Open grassland, on serpentine; 400–600 m, s NCoRI (c Lake Co.); not on serpentine, 200–1600 m, SNF. [*O. campestris* E. Greene, *O. cruciata* (S. Watson) Munz, and *O. dentata* Cav. misapplied] Self-pollinated. Probably derived from *C. strigulosa.*

C. lewisii Raven LEWIS' EVENING PRIMROSE Ann, rosetted; hairs spreading, in infl glandular. **ST** decumbent and branched or erect and simple, < 60 cm. **LF** 10–80 mm; cauline narrowly elliptic-lanceolate, minutely dentate, subsessile. **INFL** nodding. **FL**: hypanthium 1.5–4 mm; sepals 1.7–3.5 mm; petals 2.5–5.5 mm, yellow fading reddish, bases with 1–2 red dots; < 5% of pollen grains 4–5-angled. **FR** 13–20 mm, 1.8–2.2 mm wide, moderately stout, 4-angled, gen 1-coiled. **SEEDS** in 1 row per chamber, 0.7–0.8 mm, minutely pitted in rows, dull brownish black. 2*n*=14. RARE. Grassland, sandy or clay soils, coastal; 0–300 m. SCo, w PR; n Baja CA. [*O. micrantha* Sprengel misapplied] Self-pollinated. Related to *C. bistorta.*

C. luciae Raven Ann, rosetted; hairs dense, spreading. **ST** erect, < 60 cm. **LVS** 13–55 mm; cauline widely lanceolate to narrowly oblong, minutely dentate, sessile. **INFL** nodding. **FL**: hypanthium 2–3 mm; sepals 2.5–4.5 mm; petals 4–7 mm, yellow fading reddish, bases with 1 red dot. **FR** 15–20 mm, 1.3–2 mm wide, cylindric (drying ± 4-angled), straight to 2-coiled. **SEEDS** in 1 row per chamber, 1.3–1.5 mm, minutely pitted in rows, dull brownish black. 2*n*=42. Openings in chaparral; 300–1400 m. SCoRO. [*O. micrantha* Sprengel (incl var. *jonesii* A. Léveillé) misapplied] Self-pollinated. Probably derived from *C. hirtella* × *C. intermedia.*

C. micrantha (Sprengel) Raven Ann, rosetted; hairs dense, spreading (rarely glandular in infl). **ST** decumbent and branched or erect and simple, < 60 cm. **LF** 10–120 mm; cauline narrowly elliptic-lanceolate, minutely dentate, subsessile. **INFL** nodding. **FL**: hypanthium 1.2–2 mm; sepals 1–2.5 mm; petals 1.5–4.5 mm, yellow fading reddish, bases with 1–2 red dots. **FR** 13–25, mm, 1.1–1.8 mm wide, stout, cylindric (drying ± 4-angled), straight to 1-coiled. **SEEDS** in 1 row per chamber, 0.7–1.1 mm, minutely pitted in rows, dull brownish black. $2n$=14. Coastal dunes, beaches, sandy fields, washes; 0–300 m. w margin GV, CW, SW. [*O. m.* Sprengel; *O. hirta* Link] Self-pollinated. Previously more inclusively defined.

C. minor (Nelson) Raven NELSON'S EVENING PRIMROSE Ann, gen reddish, gen densely and minutely grayish strigose, in infl also glandular; rosette ± 0. **ST** ± erect, 3–30 cm, peeling. **LVS** 5–45 mm, ± oblanceolate; uppermost ± linear, subentire. **INFL** erect; fls from pl base. **FL** opening at dusk; hypanthium 0.5–1.9 mm; sepals 0.8–1.8 mm; petals 0.8–1.3 mm, white fading reddish. **FR** 15–25 mm, 0.8–1.2 mm wide, ± cylindric (but tapered to tip), twisted and wavy. **SEEDS** in 1 row per chamber, 1.1–1.2 mm, minutely pitted in rows. $2n$=14. UNCOMMON. Sandy slopes, flats, sagebrush scrub; ± 1200 m. ne MP (Surprise Valley, ne Modoc Co.); to WA, WY, Colorado. [*O. m.* (Nelson) Munz] Self-pollinated.

C. munzii (Raven) Raven Ann, strigose. **ST** 8–50 cm. **LVS** gen basal, 1-pinnate; terminal lflet < 60 mm, ovate; lateral lflets well developed, like terminal. **INFL** nodding. **FL**: hypanthium 2–3 mm; sepals 4–7 mm, tips not free in bud; petals 3–10 mm, yellow, bases red-dotted; stamens ± equal. **FR** 8–24 mm, wider toward tip, ± straight; pedicel (and fr) reflexed in age, 8–28 mm. **SEEDS** in 2 rows per chamber, 0.8–1.6 mm, pale brown. $2n$=14. Slopes, washes, mtns; 600–1600 m. ne DMoj; s NV. [*O. m.* Raven] Cross-pollinated. Sometimes hybridizes with *C. brevipes* ssp. *b.* and *C. claviformis* ssp. *a.*

C. ovata (Torrey & A. Gray) Raven (p. 791) SUN CUP Per, ± fleshy; taproot thick. **ST** ± 0. **LF**: blade 30–150 mm, narrowly elliptic to ovate, subentire to wavy, veins and margin with short, erect hairs; petiole 8–150 mm. **INFL** erect. **FL**: hypanthium 2–3 mm, closed by fleshy disk; sepals 11–19 mm; petals 8–23 mm, gen yellow (white); anthers attached at base; sterile tip of ovary 25–180 mm. **FR** 11–30 mm, linear-oblanceolate, ± straight or slightly curved, swollen by seeds, papery, tardily dehiscent, subsessile. **SEEDS** in 2 rows per chamber, 1.8–2.2 mm, papillate, brown. $2n$= 14. Grassy fields, gen clay soil; 0–500 m. NW, CW; sw OR. [*O. o.* Torrey & A. Gray] Cross-pollinated. ❀SUN:5&IRR:14,**15–17**, 22,23,**24**.

C. pallida (Abrams) Raven Ann, rosetted, grayish, densely strigose, in infl also glandular. **ST** decumbent and branched or erect and simple, < 60 cm. **LVS** 10–30 mm; cauline narrowly elliptic-lanceolate, subentire to minutely dentate; petiole < 2 mm. **INFL** nodding. **FL**: hypanthium 1–4 mm; sepals 1.5–8 mm; petals 2–13 mm, yellow fading reddish, bases with 1–3 red dots. **FR** 10–25 mm, 1.1–1.2 mm wide, ± 4 -angled, straight to 3-coiled. **SEEDS** in 1 row per chamber, 1–1.5 mm, minutely pitted in rows, dull brownish black. $2n$=14. Desert slopes, flats, washes, creosote-bush scrub to pinyon/juniper woodland; 30–1800 m. s SnJV (Kern Co.), n slope SnBr, D; NV, AZ. [*Sphaerostigma p.* Abrams] Gen self-pollinated. Sspp. intergrade.

ssp. ***hallii*** (Davidson) Raven **FL**: hypanthium 3.8–4.2 mm; sepals 4.8–8 mm; petals 6.5–13 mm; style 6.5–10.5 mm. Habitats of sp. n slope SnBr. [*O. h.* (Davidson) Munz] Sometimes cross-pollinated. ❀TRY.

ssp. ***pallida*** **FL**: hypanthium 1–3 mm; sepals 1.5–5.5 mm; petals 2–6.5 mm; style 2.1–6.5 mm. Habitats and range of sp. [*O. abramsii* J.F. MacBr.]

C. palmeri (S. Watson) Raven (p. 785) Ann, strigose (few hairs spreading). **ST** ± 0, peeling. **LF** 15–55 mm, narrowly oblanceolate, minutely serrate. **INFL** nodding. **FL**: hypanthium 0.8–1.3 mm, closed by fleshy disk; sepals 1.6–2.3 mm; petals 2–3.5 mm, yellow; anthers attached at base; sterile tip of ovary 5–12 mm. **FR** 5–7 mm, 4-angled, 4-winged near tip, ± straight, leathery, tardily dehiscent. **SEEDS** in 2 rows per chamber, 1.2–2 mm, papillate, tan with brown spots. $2n$=14. Desert flats, sagebrush scrub; 600–1400 m. s GV, SCoRI, WTR (Tejon Pass), PR (Jacumba), DMoj; OR, NV. [*O. p.* S. Watson] Self-pollinated.

C. parvula (Torrey & A. Gray) Raven (p. 791) Ann, slender, ± glabrous or minutely strigose (some hairs gen glandular, rarely few spreading); rosette 0. **ST** erect, < 30 cm, wiry. **LF** 10–30 mm, linear, subentire. **INFL** nodding. **FL**: hypanthium < 2 mm; sepals < 2.2 mm, free; petals < 3.6 mm, yellow fading reddish. **FR** ± 20–30 mm, 0.6–0.9 mm wide, cylindric, ± swollen by seeds, straight or wavy, subsessile. **SEEDS** in 1 row per chamber, 0.7–0.8 mm, shiny, minutely pitted. $2n$=28. Sandy soils, gen sagebrush scrub; 1000–2000 m. GB; to WA, WY, Colorado. [*O. p.* Torrey & A. Gray; *O. contorta* Douglas var. *flexuosa* (Nelson) Munz] Self-pollinated. Related to *C. kernensis, C. pubens, C. pusilla.*

C. pterosperma (S. Watson) Raven (p. 791) Ann, slender; hairs bristly, in infl also glandular. **ST** erect, 2–14 cm, peeling. **LF** 3–30 mm, narrowly lanceolate to oblanceolate, entire. **INFL** nodding. **FL**: hypanthium 1–2 mm; sepals 1.5–2.5 mm; petals 1.5–2.5 mm, white fading purplish, bases yellow. **FR** 12–28 mm, 0.6–0.8 mm wide, ascending or spreading, cylindric, ± straight, slightly swollen by seeds; pedicel 4–8 mm. **SEEDS** in 2 rows per chamber, 1–1.5 mm; wing thick; hairs club-shaped. $2n$=14. Uncommon. Well drained, gen volcanic slopes, pinyon/juniper woodland or sagebrush scrub; 1400–2400 m. s W&I (Inyo Mtns), n DMtns (Panamint Mtns); to OR, UT. [*O. p.* S. Watson] Self-pollinated.

C. pubens (S. Watson) Raven Ann; rosette gen ± 0; hairs gen glandular, some spreading. **ST** erect, < 38 cm. **LF** 15–45 mm, narrowly lanceolate, wavy-serrate. **INFL** nodding. **FL**: hypanthium 1.3–3 mm; sepals 2.2–3.8 mm, free; petals 2.2–4 mm, yellow fading reddish, bases with 1–few red dots. **FR** (18)26–50 mm, 0.8–1.2 mm wide, cylindric, ± swollen by seeds, straight or wavy; pedicel 0–2 mm. **SEEDS** in 1 row per chamber, 0.7–0.8 mm, shiny, minutely pitted. $2n$=28. Sandy soils, gen sagebrush scrub or pinyon/juniper woodland; 1000–3000 m. GB, n DMoj (scattered); w NV. [*O. p.* (S. Watson) Munz] Self-pollinated. Related to *C. kernensis, C. parvula, C. pusilla.*

C. pusilla Raven (p. 791) Ann, slender; rosette gen ± 0; hairs glandular, gen also spreading. **ST** erect, 2–22 cm. **LF** 10–30 mm, linear, minutely serrate. **INFL** nodding. **FL**: hypanthium 0.8–1.6 mm; sepals 1.2–2 mm, free; petals 1.8–3.1 mm, yellow fading reddish, bases with 2 red dots. **FR** 18–32 mm, 0.6–0.9 mm wide, cylindric, ± swollen by seeds, straight or wavy; pedicel 0–2 mm. **SEEDS** in 1 row per chamber, 0.7–0.8 mm, shiny, minutely pitted. $2n$=14. Sandy soils, gen sagebrush scrub; 760–3000 m. n slope SnBr (Cactus Flats), GB, DMoj (scattered); to WA, ID, UT. [*O. contorta* Douglas var. *flexuosa* (Nelson) Munz misapplied] Self-pollinated. Related to *C. kernensis, C. parvula, C. pubens.*

C. refracta (S. Watson) Raven (p. 791) Ann, gen reddish, sparsely minutely strigose, esp in infl also glandular; rosette ± 0. **ST** erect, 6–45 cm, peeling. **LVS** < 60 mm, narrowly elliptic to narrowly lanceolate (lowest gen oblanceolate), sparsely minutely dentate. **INFL** nodding. **FL** opening at dusk; hypanthium 4–6 mm; sepals 4–6 mm; petals 3.5–7 mm, white fading reddish; 70–100% of pollen grains 4–5-angled. **FR** 20–50 mm, 0.7–1 mm wide, cylindric, straight or wavy. **SEEDS** in 1 row per chamber, 0.9–1.5 mm, minutely pitted in rows. $2n$=14. Sandy slopes, flats, desert scrub; –30–1300 m. D; to s NV, sw UT, w AZ. [*O. r.* S. Watson; *O. deserti* M.E. Jones] Cross-pollinated. ❀TRY.

C. robusta Raven (p. 791) Ann, rosetted; hairs spreading, in infl also glandular. **ST** erect, < 60 cm. **LVS** 10–80 mm; cauline narrowly elliptic-lanceolate, minutely dentate, subsessile. **INFL** nodding. **FL**: hypanthium 1.8–3.7 mm; sepals 2.6–4.2 mm; petals 3.2–7 mm, yellow fading reddish, bases with 1–2 red dots. **FR** 14–25 mm, 1.5–2 mm wide, moderately stout, 4-angled, gen 1-coiled. **SEEDS** in 1 row per chamber, 0.9–1.2 mm, minutely pitted in rows, dull brownish black. $2n$=42. Coastal-sage scrub, chaparral, gen in disturbed places; 0–300 m. ChI (Santa Cruz, San Clemente, Santa Catalina islands) PR; n Baja CA. [*O. micrantha* Sprengel misapplied] Self-pollinated. Probably derived from *C. lewisii* × *C. intermedia.*

C. sierrae Raven Ann, slender; rosette gen ± 0; hairs spreading, coarse, in infl also glandular. **ST** decumbent to ascending, < 15 cm, peeling. **LF** 5–18 mm, lanceolate to narrowly ovate or elliptic; teeth few or minute. **INFL** nodding. **FL**: hypanthium 1–2.2 mm; sepals 1.2–4.2 mm, remaining adherent in pairs; petals 2.2–7 mm, yellow fading reddish, bases with 0 or 2 red dots. **FR** 20–30 mm, 0.5–0.7 mm wide, cylindric, ± swollen by seeds, straight or wavy, sessile. **SEEDS** in 1 row per chamber, 0.8–1.6 mm, shiny, minutely pitted. $2n$=14. Granite outcrops, ponderosa-pine or foothill-pine/blue-oak forests; 500–2350 m. c SNF (Madera, Mariposa cos.), s SNH (ne Fresno Co.). Related to *C. campestris.*

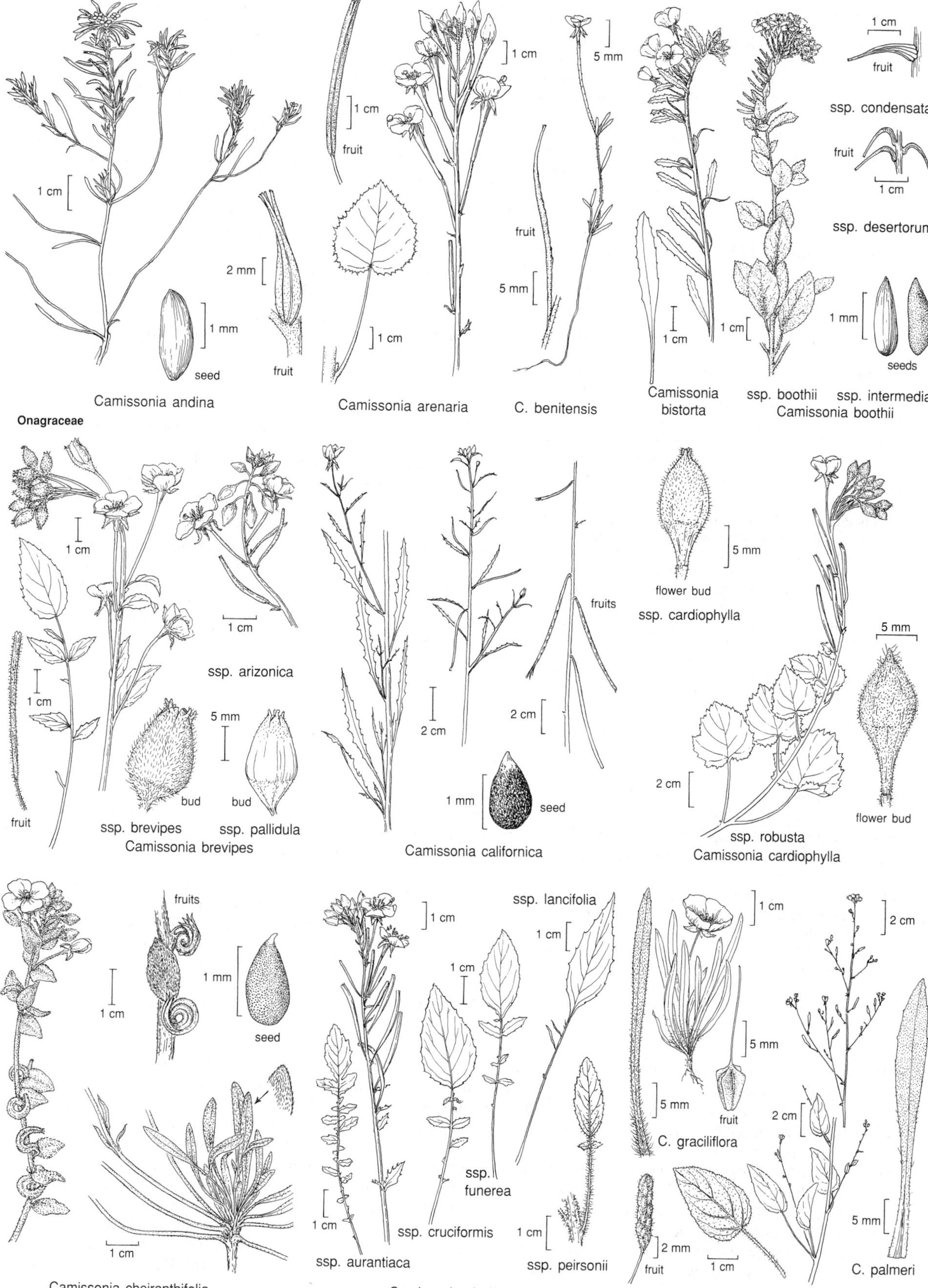

1 cm
2 mm
1 mm
seed
fruit
Camissonia andina
Onagraceae
fruit
1 cm
1 cm
1 cm
Camissonia arenaria
5 mm
fruit
5 mm
C. benitensis
1 cm
Camissonia bistorta
1 cm
ssp. boothii
1 cm
fruit
ssp. condensata
fruit
1 cm
ssp. desertorum
1 mm
seeds
ssp. intermedia
Camissonia boothii
1 cm
1 cm
1 cm
ssp. arizonica
5 mm
bud
bud
fruit
ssp. brevipes
ssp. pallidula
Camissonia brevipes
fruits
2 cm
2 cm
1 mm
seed
Camissonia californica
5 mm
flower bud
ssp. cardiophylla
5 mm
2 cm
flower bud
ssp. robusta
Camissonia cardiophylla
fruits
1 cm
1 mm
seed
1 cm
Camissonia cheiranthifolia
1 cm
ssp. lancifolia
1 cm
1 cm
1 cm
ssp. funerea
ssp. cruciformis
1 cm
ssp. aurantiaca
ssp. peirsonii
Camissonia claviformis
1 cm
2 cm
5 mm
5 mm
fruit
C. graciliflora
2 cm
2 mm
fruit
1 cm
5 mm
C. palmeri
Camissonia heterochroma

ssp. ***alticola*** Raven (p. 791) MONO HOT SPRINGS EVENING PRIMROSE **LF** lanceolate; teeth several, small. **FL**: hypanthium 1–2.2 mm; sepals 1.2–3 mm; petals 2.2–4 mm, not red-dotted; style 2.8–5 mm. RARE. Shallow soil on granite outcrops, ponderosa-pine forest; 2000–2350 m. s SNH (ne Fresno Co.). Self-pollinated. Pls from Merced Lake (± 2200 m, e Mariposa Co.) are similar, warrant further study.

ssp. ***sierrae*** (p. 791) **LF** narrowly ovate to elliptic, minutely serrate. **FL**: hypanthium 1.3–2.2 mm; sepals 3–4.2 mm, bases with 2 red dots; petals 4–7 mm; style 4.5–7 mm. Ponderosa-pine or foothill-pine/blue-oak forests; 500–1300 m. c SNF (Madera, Mariposa cos.). Cross- or self-pollinated. May be locally abundant.

C. strigulosa (Fischer & C. Meyer) Raven (p. 791) Ann, slender, minutely strigose (hairs glandular or not, toward base also coarse, spreading); rosette ± 0. **ST** decumbent or erect, < 50 cm, wiry, peeling. **LF** 8–35 mm, linear to very narrowly elliptic, minutely serrate. **INFL** nodding. **FL**: hypanthium 1.6–2.7 mm; sepals 1.6–4 mm, remaining adherent in pairs; petals 2.1–4.5 mm, yellow fading reddish, bases with 0–2 red dots. **FR** 15–45 mm, 0.8–1.3 mm wide, cylindric, ± swollen by seeds, straight or wavy, subsessile. **SEEDS** in 1 row per chamber, 0.6–0.8 mm, shiny, minutely pitted. $2n$=28. Open sandy soils of dunes, grassland, desert scrub; 0–2100 m. s edge s SNH, Teh, CW, SW, n ChI (Santa Rosa Island), w DMoj; n Baja CA. [*O. s.* (Fischer & C. Meyer) Torrey & A. Gray; *O. dentata* Cav. misapplied] Self-pollinated. Related to S.Am *C. dentata*; hybridizes with *C. campestris* ssp. *obispoensis*, *C. contorta*, *C. integrifolia*, *C. kernensis* ssp. *k.*

C. subacaulis (Pursh) Raven (p. 791) Per, ± fleshy, ± glabrous; taproot thick. **ST** ± 0. **LF**: blade 20–220 mm, lanceolate to narrowly elliptic, subentire to irregularly pinnately lobed, rarely sparsely and minutely strigose; petiole 10–120 mm. **FL** erect; hypanthium 1.5–3 mm, closed by fleshy disk; sepals 4.1–13 mm; petals 5–16 mm, yellow; anthers attached at base; ovary with sterile tip 15–80 mm. **FR** 11–28 mm, linear-ovoid, 4-angled, barely swollen by seeds, ± straight or slightly curved, leathery; pedicel 0–10 mm. **SEEDS** in 2 rows per chamber, 1.3–1.9 mm, pitted, pale brown. $2n$=14. Moist meadows, gen clay soils; 450–2600 m. SN, GB; to WA, MT, Colorado. [*O. s.* (Pursh) A.O. Garrett incl var. *taraxacifolia* (S. Watson) Jepson] Gen cross-pollinated. ❀TRY.

C. tanacetifolia (Torrey & A. Gray) Raven Per; taproot woody, new shoots from lateral roots; hairs gen ± dense (sparse), short, spreading or appressed. **ST** 0. **LF** 65–320 mm, narrowly elliptic, deeply and irregularly pinnately lobed; petiole 10–80 mm. **INFL** erect. **FL**: hypanthium 4–6.5(8.5) mm, closed by fleshy disk; sepals 5.5–13 mm; petals 8–23 mm, yellow; anthers attached at base; sterile tip of ovary 14–55 mm. **FR** 7–25 mm, long-tapered, swollen by seeds, leathery, ± straight or slightly curved, disintegrating irregularly. **SEEDS** in 2 rows per chamber, 1.5–2 mm, pitted in rows, pale brown. Open fields, moist slopes, clay soils; 700–2500 m. CaR, SN, GB; to WA, ID, NV. [*O. t.* Torrey & A. Gray] Cross-pollinated.

ssp. ***quadriperforata*** Raven SIERRA VALLEY EVENING PRIMROSE Pl: hairs dense, spreading. **FL**: > 10% of pollen grains 4-angled. $2n$=42. UNCOMMON. Clay flats, sagebrush scrub; 1300–1500 m. n SNH (Plumas, Sierra cos.), MP (Lassen Co.). Locally common. ❀TRY.

ssp. ***tanacetifolia*** (p. 791) Pl: hairs gen dense (sparse), spreading or appressed. **FL**: < 5% of pollen grains 4-angled. $2n$= 14,28. Habitats and range of sp. [*C. breviflora* (Torrey & A. Gray) Raven misapplied: not in CA] ❀SUN:**15–17**&IRR:1–3,7,14,18–24.

C. walkeri (Nelson) Raven ssp. ***tortilis*** (Jepson) Raven Ann, short-lived per; hairs spreading, in infl also glandular or bristly. **ST** 10–60 cm. **LVS** gen basal, simple to 2-pinnate, 30–220 mm, gen purple-dotted; lateral lflets < 30 mm. **INFL** erect; buds drooping. **FL**: hypanthium 1–1.5 mm; sepals 1.5–4 mm, gen purple-dotted, in bud tips free, subterminal; petals 3–6 mm, yellow. **FR** ascending or spreading, 10–45 mm, 1.2–1.8 mm wide, cylindric, straight or curved; valves gen twisted; pedicel 5–30 mm. **SEEDS** in 2 rows per chamber, 0.6–1.2 mm. $2n$=14. Rocky places near cliffs, along ephemeral streams, creosote-bush scrub to pinyon/juniper woodland; 600–1800 m. n DMoj (Inyo, ne San Bernardino cos.); to s NV, sw UT. [*O. scapoidea* Torrey & A. Gray var. *t.* Jepson; *O. w.* (Nelson) Raven ssp. *t.* (Jepson) Raven] Self-pollinated. Ssp. *w.* outside CA.

CIRCAEA ENCHANTER'S NIGHTSHADE

Peter C. Hoch

Per from tuber-tipped rhizomes or stolons. **LVS** opposite, petioled, entire to toothed. **INFL**: raceme or panicle, gen bracted. **FL** biradial; sepals 2, often reflexed; petals 2, erect; stamens 2, pollen yellow, grains shed singly; ovary chambers 1–2. **FR** indehiscent, gen club-shaped, hooked-hairy. **SEEDS** 1 per chamber, adhering to inner fr wall. 8 spp.: n hemisphere. (Greek: Circe, the enchantress) [Boufford 1982 Ann Missouri Bot Gard 69:804–994] Often self-pollinated.

C. alpina L. ssp. ***pacifica*** (Asch. & Magnus) Raven (p. 791) **ST** 1–5 dm, simple, erect, slender, cylindric, gen densely strigose. **LF** 3–11 cm, ovate to ± round, glabrous or ± hairy; base round to ± cordate; tip acute; petiole 1.5–5 cm. **INFL**: raceme, erect, densely strigose and glandular; pedicel in fr 2–5 mm. **FL**: hypanthium 0.3–0.5 mm; sepals 1–2 mm, white, reflexed; petals 1–1.5 mm, white; stamens ± = pistil; stigma 2-lobed. **FR** ± 2 mm, 1-chambered. $2n$= 22. Cool, moist, coniferous forest; < 2700 m. NW, CaR, n SNF, SNH, SnFrB, SnBr, Wrn; to w Can, MT, NM. ❀WET:1,2,6&SHD:4,5,7,14–16;DFCLT.

CLARKIA

Harlan Lewis

Ann. **ST** prostate to erect, < 1.5 m, glabrous, often glaucous, or puberulent; hairs rarely long and spreading. **LF** simple, pinnately veined; petiole < 4 cm or 0; blade 1–10 cm, linear to elliptic or ovate, entire or shallow-toothed, glabrous or sparsely puberulent. **INFL**: spike, raceme; bracts lf-like; axis in bud straight or recurved at tip, in fl ± straight; buds erect or not. **FL**: hypanthium obconic to cup-shaped or long, slender, gen with ring of hairs within; sepals 4, gen fused to tip in bud, reflexed at least at base, staying fused at least at tip, in 4's or 2's, or all coming free; corolla bowl-shaped to rotate, petals 5–60 mm, often lobed or clawed, lavender or pink to dark red, pale yellow, or white, often spotted, flecked, or streaked with red, purple, or white; stamens 8, in 2 like or unlike series, or 4, filaments cylindric to wider above, subtended by ciliate scales or gen not, anthers attached at base, pollen white or yellow to blue-gray, lavender, or reddish; ovary 4-chambered, glabrous or not, cylindric, fusiform, or wider above, gen shallowly to deeply 4- or 8-grooved, stigma lobes 4, gen prominent. **FR**: gen capsule, elongate, rarely short, indehiscent, nut-like. **SEEDS** gen many, rarely 1–2, 0.5–2 mm, angled, crested or not, brown, gray or mottled. ± 41 spp.: w N.Am, 1 S.Am. (Captain William Clark, 1770–1838, of Lewis & Clark Expedition) [Lewis & Lewis 1955 UC Publ Bot 20:241–392] Self-compatible or -pollinated or outcrossed; on herbarium specimens, curvature of infl axis in bud gen reliable, pollen color gen not.

1. Fr 2–3 mm, nut-like, indehiscent, 1–2-seeded .... ***C. heterandra***
1′ Fr > 5 mm, not nut-like, dehiscent, many-seeded
2. Petal blade lobed above
3. Stamens 4; hypanthium slender, > 10 mm
4. Petal length ± = width, outer lobes wider than middle .... ***C. breweri***
4′ Petal length ± 2 × width, lobes ± equally wide .... ***C. concinna***
5. Stigma beyond anthers; petals 15–30 mm .... ssp. ***concinna***
5′ Stigma not beyond anthers; petals 10–20 mm
6. Petal lobes prominent, separated by deep sinuses .... ssp. ***automixa***
6′ Petal lobes short, not separated by deep sinuses .... ssp. ***raichei***
3′ Stamens 8; hypanthium not slender, < 7 mm
7. Petal lobes 2, a tooth between .... ***C. xantiana***
8. Petals 6–12 mm; stigma not beyond anthers .... ssp. ***parviflora***
8′ Petals 12–20 mm; stigma beyond anthers .... ssp. ***xantiana***
7′ Petal lobes 2, no tooth between .... ***C. biloba***
9. Petals bright pink to magenta, length gen > 1.5 × width .... ssp. ***australis***
9′ Petals lavender to pink or purplish pink, length gen not > 1.5 × width
10. Petal lobes gen 1/5–1/2 petal .... ssp. ***biloba***
10′ Petal lobes gen < 1/5 petal .... ssp. ***brandegeae***
2′ Petal blade not or inconspicuously lobed above (see also *C. biloba* ssp. *brandegeae*)
11. Petal claw 2-lobed or -toothed, not entire
12. Petals 6–14 mm; stigma not beyond anthers
13. Petal gen spotted, blade unlobed above .... ***C. rhomboidea***
13′ Petal unspotted, blade inconspicuously 3-lobed above .... ***C. stellata***
12′ Petals gen > 12 mm; stigma beyond anthers
14. Infl axis in bud recurved < 3 nodes above open fls; petal length 1.4–1.6 × width .... ***C. mildrediae***
14′ Infl axis in bud straight 4 or more nodes above open fls, recurved at tip; petal length 1.6–3 × width
15. Fl bud fusiform, tip acute .... ***C. borealis***
16. Seed 1.8–2.5 mm; CaRF .... ssp. ***arida***
16′ Seed 1.5–1.8 mm; KR .... ssp. ***borealis***
15′ Fl bud obovate, tip obtuse
17. Petal length 1.6–2 × width — Butte Co. .... ***C. mosquinii***
18. Lf elliptic to ovate .... ssp. ***mosquinii***
18′ Lf linear-lanceolate .... ssp. ***xerophylla***
17′ Petal length 1.9–3 × width
19. Lf linear-lanceolate .... ***C. australis***
19′ Lf elliptic to ovate .... ***C. virgata***
11′ Petal claw ± 0 or entire
20. Infl axis in bud recurved at tip, in fl ± straight; buds pendent
21. Petals white to pale cream, not darker-flecked or -spotted, fading pink, gen < 10 mm .... ***C. epilobioides***
21′ Petals darker than white to pale cream, at least on margin, or darker-flecked or -spotted, gen > 10 mm
22. Anthers alike, tips acute or slender; pollen yellow to ± white
23. Ovary, immature fr 8-grooved; fr not wider above
24. Ovary hairs ± sparse; pedicel 5–15 mm .... ***C. arcuata***
24′ Ovary hairs dense; pedicel < 3 mm .... ***C. lassenensis***
23′ Ovary, immature fr 4-, rarely 8-grooved; fr gen wider above .... ***C. gracilis***
25. Petals 6–23 mm, spot 0, base darker or not; stigma not beyond anthers .... ssp. ***gracilis***
25′ Petals 20–40 mm, with a red spot near middle or red at base; stigma beyond anthers
26. Petals with a red spot near middle .... ssp. ***sonomensis***
26′ Petals red at base
27. Petals gen 3–4 cm; s CaRF .... ssp. ***albicaulis***
27′ Petals gen 2–3 cm; NCoRI .... ssp. ***tracyi***
22′ Anthers in 2 unlike series, obtuse at tip, inner gen << outer; inner pollen gen cream, outer gen blue-gray or lavender
28. Petals 6–12 mm; stigma not beyond anthers
29. Petals pink, gen darker-flecked .... ***C. modesta***
29′ Petals pale pink shading to ± white near base, purple-flecked .... ***C. similis***
28′ Petals 10–35 mm; stigma beyond anthers
30. Petals oblanceolate; corolla rotate .... ***C. lingulata***
30′ Petals fan-shaped; corolla bowl-shaped
31. Ovary, immature fr 8-grooved; petals gen white-streaked .... ***C. dudleyana***
31′ Ovary, immature fr 4-grooved; petals darker near tip, shading to white near middle, with a darker line or zone near base
32. Width of outer filaments ± 2 × inner; ring of hairs in hypanthium below tip .... ***C. cylindrica***
33. Ovary, fr wider above .... ssp. ***clavicarpa***
33′ Ovary, fr cylindric .... ssp. ***cylindrica***
32′ Width of all filaments ± equal; ring of hairs in hypanthium ± at tip
34. Fr beak < 3 mm or 0 .... ***C. lewisii***
34′ Fr beak 7–15 mm .... ***C. rostrata***

20′ Infl axis in bud gen ± straight; buds erect or reflexed
35. Fl buds reflexed
36. Corolla bowl-shaped; petals ± not clawed
37. Seeds brown; not Monterey Co. . . . . . **C. bottae**
37′ Seeds gray; n SCoRO (Monterey Co.) . . . . . **C. jolonensis**
36′ Corolla rotate; petals clawed
38. Petal claw < blade . . . . . **C. delicata**
38′ Petal claw = or > blade
39. Calyx, ovary sparsely to densely puberulent and with longer, straight, spreading hairs < 3 mm . . . . . **C. unguiculata**
39′ Calyx, ovary sparsely to densely puberulent without longer, straight, spreading hairs < 3 mm
40. Lf not glaucous, bright green . . . . . **C. exilis**
40′ Lf glaucous, gray-green or reddish
41. Sepals gen dark red-purple; stigma beyond anthers . . . . . **C. springvillensis**
41′ Sepals green, red-tinged or not; stigma beyond anthers or not . . . . . **C. tembloriensis**
42. Petals ± 10 mm wide; stigma not beyond anthers . . . . . ssp. **calientensis**
42′ Petals gen < 10 mm wide; stigma beyond anthers or not . . . . . ssp. **tembloriensis**
35′ Fl buds erect
43. Ovary, immature fr 4-grooved
44. Petals with a ± red spot or zone at base
45. Petals 10–30 mm; stigma beyond anthers . . . . . **C. rubicunda**
45′ Petals 5–13 mm; stigma not beyond anthers . . . . . **C. franciscana**
44′ Petals with a red spot or marks near middle . . . . . **C. amoena**
46. St decumbent; infl internode gen < subtending fl . . . . . ssp. **amoena**
46′ St erect; infl internode gen > subtending fl . . . . . ssp. **huntiana**
43′ Ovary, immature fr 8-grooved
47. Stigma not beyond anthers; petals gen 5–15 mm
48. St decumbent to prostrate; lf oblanceolate, elliptic, or obovate, tip gen obtuse
49. Petal 5–11 mm, spot 0 . . . . . **C. davyi**
49′ Petal 10–15 mm, spot reddish, above base . . . . . **C. prostrata**
48′ St erect; lf linear to ovate, tip gen acute
50. Immature fr length > 9 × width; sepals staying fused in 4's . . . . . **C. affinis**
50′ Immature fr length < 8 × width; sepals staying fused in 2's or all coming free . . . . . **C. purpurea** ssp. **quadrivulnera**
47′ Stigma gen beyond anthers; petals 10–60 mm
51. Ovary, immature fr fusiform, grooves 8, 4 deeper . . . . . **C. amoena** ssp. **whitneyi**
51′ Ovary, immature fr gen ± cylindric, grooves 8, equal
52. Petals ± without a red or purple spot
53. Petals dark reddish purple
54. Sepals fused to tip in bud . . . . . [2]**C. speciosa** ssp. **speciosa**
54′ Sepals free at tip in bud . . . . . [2]**C. williamsonii**
53′ Petals pink or lavender to purple or purplish red, below often cream or lighter
55. Infl open; lf linear to narrow-lanceolate . . . . . **C. speciosa** ssp. **immaculata**
55′ Infl dense; lf wide-lanceolate to elliptic or ovate . . . . . [2]**C. purpurea** ssp. **purpurea**
52′ Petals gen with a red or purple spot
56. Petal spot near middle or below
57. Infl dense; ovary gen > internode above . . . . . **C. speciosa** ssp. **nitens**
57′ Infl open; ovary gen < internode above . . . . . **C. speciosa**
58. Branches on well developed pls few, many-fld . . . . . ssp. **polyantha**
58′ Branches on well developed pls many, few-fld . . . . . [2]ssp. **speciosa**
56′ Petal spot above middle
59. Lf lanceolate to elliptic or ovate, length gen < 5 × width; infl dense
60. Hypanthium 10–15 mm, veined; petals 20–25 mm . . . . . **C. imbricata**
60′ Hypanthium 2–10 mm, ± not veined; petals 10–25 mm . . . . . [2]**C. purpurea** ssp. **purpurea**
59′ Lf linear to narrow-lanceolate, length > 5 × width; infl open
61. Sepals fused to tip in bud . . . . . **C. purpurea** ssp. **viminea**
61′ Sepals free at tip in bud . . . . . [2]**C. williamsonii**

C. affinis Harlan Lewis & M. Lewis (p. 791) **ST** erect, < 8 dm, puberulent. **LF**: petiole < 3 mm or 0; blade 1.5–7 cm, linear to narrowly lanceolate. **INFL**: axis in bud straight; buds erect. **FL**: hypanthium 1.5–4 mm; sepals staying fused in 4's; corolla bowl-shaped, petals 5–15 mm, obovate, pale pink to dark wine-red, often purple-flecked or -marked; stamens 8, anthers alike; ovary 8-grooved, length > 9 × width, stigma not beyond anthers. **FR** 1.5–3 cm; beak 3–7 mm, slender. *n*=26. Openings in woodland, chaparral; < 500 m. NCoRI, c SNF, SnFrB, SCoR, WTR. ❀TRY.

C. amoena (Lehm.) Nelson & J.F. Macbr. FAREWELL-TO-SPRING **ST** decumbent to erect, < 1 m, puberulent. **LF**: petiole < 1 cm; blade 1–6 cm, linear to lanceolate. **INFL** open or dense; axis in bud straight; buds erect. **FL**: hypanthium 3–10 mm; sepals staying fused in 4's; corolla bowl-shaped, petals 1.5–6 cm, obovate to fan-shaped, pink to lavender, rarely white, red-spotted or -marked near middle; stamens 8, anthers alike; ovary cylindric with 4 grooves (or fusiform with 8), stigma beyond anthers. Gen open, drying places; < 700 m. NCo, KR, NCoRO, n SnFrB; to B.C. Intermediates among sspp. frequent.

ssp. ***amoena*** (p. 791) GODETIA **ST** decumbent. **INFL** dense; internode gen < subtending fl. **FL**: petals 2–3.5 cm; ovary cylindric, grooves 4. *n*=7. Uncommon. Open slopes, bluffs on or near coast; < 100 m. NCo. ❀SUN:**4–6**,**15–17**&IRR:7,8,9,**14**,19,**20–24**;CVS.

ssp. ***huntiana*** (Jepson) Harlan Lewis & M. Lewis **ST** erect. **INFL** open; internode gen > subtending fl. **FL**: petals 1.5–3 cm; ovary cylindric, grooves 4. *n*=7. Openings in forest, woodland, gen not on coast; < 700 m. KR, NCoRO, n SnFrB; OR. ❀SUN:**4–6, 15–17**&IRR:7,8,9,**14**,19,**20–24**.

ssp. ***whitneyi*** (A. Gray) Harlan Lewis & M. Lewis (p. 791) WHITNEY'S FAREWELL-TO-SPRING **ST** decumbent. **INFL** dense. **FL**: petals gen 3–6 cm; ovary fusiform, grooves 8, 4 deeper. *n*=7. Uncommon (reportedly only 1 wild population remains). Coastal scrub, open; < 100 m. NCo (formerly n of Fort Bragg to Shelter Cove, Humboldt, Mendocino cos.). In cult.

C. arcuata (Kellogg) Nelson & J.F. Macbr. **ST** erect, < 8 dm; hairs 0 or scattered, spreading. **LF** sessile, 1.5–6 cm, linear to narrow-lanceolate or oblanceolate. **INFL**: axis in bud recurved at tip; buds pendent; pedicels 5–15 mm. **FL**: hypanthium 3–7 mm; sepals staying fused in 4's, puberulent, hairs spreading, some short, glandular; corolla bowl-shaped, petals 1–3 cm, fan-shaped, pinkish lavender, lighter below, often with a dark reddish spot at base; stamens 8, anthers alike; ovary 8-grooved, gen ± sparsely glandular-puberulent, stigma beyond anthers. **FR** 1–3.5 cm; beak < 7 mm. *n*=7. Openings in woodland, chaparral, drying soils incl serpentine; < 1000 m. CaRF, n&c SNF. ❀TRY.

C. australis E. Small SMALL'S SOUTHERN CLARKIA **ST** erect, < 1 m, puberulent. **LF**: petiole 1–3 cm; blade 2–5 cm, linear-lanceolate. **INFL**: axis in bud straight 4 or more nodes above open fls, recurved at tip; buds pendent, obovate, tip obtuse. **FL**: hypanthium 2–4 mm; sepals all coming free; corolla rotate, petals 8–15 mm, lavender-purple, reddish purple-mottled or -spotted, length 2.2–3 × width, claw 2-lobed, blade ± diamond-shaped; stamens 8, subtended by ciliate scales, anthers alike, pollen blue-gray; ovary 4-grooved, stigma beyond anthers. **SEED** 1–1.5 mm. *n*=5. RARE. Yellow-pine forest; 800–1500 m. c SN (Tuolumne, Madera, Mariposa cos.).

C. biloba (Durand) Nelson & J.F. Macbr. **ST** erect, < 1 m, puberulent. **LF**: petiole < 1.5 cm; blade 2–6 cm, linear to lanceolate. **INFL**: axis in bud recurved at tip; buds pendent. **FL**: hypanthium 1–4 mm; sepals staying fused in 4's, pink or purplish red; corolla rotate to bowl-shaped, petals 1–2.5 cm, narrowly wedge- to fan-shaped, lavender to bright pink or magenta, often red-flecked, lobes 2, tooth between 0; stamens 8, outer anthers lavender, inner smaller, paler; ovary 8-grooved, stigma beyond anthers. Dry sites; 300–1000 m. n&c SNF, e SnFrB. Sspp. intergrade.

ssp. ***australis*** Harlan Lewis & M. Lewis (p. 791) MARIPOSA CLARKIA **LF** linear to narrow-lanceolate. **FL**: petals bright pink to magenta, length gen > 1.5 × width, lobes gen 1/5–1/3 petal. *n*=8. RARE. Chaparral, woodland; 300–500 m. c SNF (Merced River drainage, Tuolumne, Mariposa cos.).

ssp. ***biloba*** (p. 791) **LF** narrow-lanceolate. **FL**: petals purplish pink, length gen not > 1.5 × width, lobes gen 1/5–1/2 petal. *n*= 8. Foothill woodland, serpentine or not; < 1000 m. n&c SNF, e SnFrB. ❀DRY:1,6,**15–17**;IRR:**7,22–24**&SHD:8,9,**14,19–21**.

ssp. ***brandegeae*** (Jepson) Harlan Lewis & M. Lewis (p. 791) **LF** lanceolate. **FL**: petals lavender, length gen not > 1.5 × width, lobes gen < 1/5 petal, sometimes obscure. *n*=8. Foothill woodland: < 500 m. n SNF. ❀TRY.

C. borealis E. Small **ST** erect, < 1 m, puberulent. **LF**: petiole 1.5–5 cm; blade 2–5 cm, elliptic to ovate. **INFL**: axis in bud straight 4 or more nodes above open fls, recurved at tip, buds pendent, fusiform, tip acute. **FL**: hypanthium 2–4 mm; sepals all coming free; corolla rotate, petals 13–19 mm, lavender-pink, often dark-flecked, length 1.6–2 × width, claw 2-lobed, blade triangular to semicircular; stamens 8, subtended by ciliate scales, anthers alike, pollen blue-gray; ovary 4-grooved, stigma beyond anthers. **SEED** 1.5–2.5 mm. Foothill woodland, forest margin; 400–800 m. KR, CaRF.

ssp. ***arida*** E. Small NORTHERN CLARKIA **SEED** 1.8–2.5 mm. *n*=7. RARE. Foothill woodland; 500 m. CaRF (known only from type, s Shasta Co.).

ssp. ***borealis*** NORTHERN CLARKIA **SEED** 1.5–1.8 mm. *n*=7. UNCOMMON. Habitats, elevations of sp. KR (e Trinity, w Shasta cos.).

C. bottae (Spach) Harlan Lewis & M. Lewis (p. 791) PUNCHBOWL GODETIA **ST** erect, < 1 m, glabrous, glaucous, rarely puberulent. **LF**: petiole < 5 mm; blade 3–10 cm, narrow-lanceolate to lanceolate. **INFL**: axis in bud straight; buds reflexed. **FL**: hypanthium 2–3 mm; sepals staying fused in 4's; corolla bowl-shaped, petals 1.5–3 cm, fan-shaped, pale lavender to pinkish lavender, often white toward base, gen red-flecked; stamens 8, outer anthers lavender, inner smaller, paler; stigma beyond anthers. **SEED** brown. *n*=9. Openings in chaparral, woodland, coastal scrub; < 1000 m. SCoRO, SCo, WTR, nw PR. [*C. deflexa* (Jepson) Harlan Lewis & M. Lewis] ❀SUN:6,**7**,8,9,**14–24**.

C. breweri (A. Gray) E. Greene BREWER'S CLARKIA, FAIRY FANS **ST** decumbent or erect, < 2 dm, glabrous or sparsely puberulent. **LF**: petiole < 2 cm; blade 2–5 cm, linear to lanceolate. **FL**: hypanthium 20–35 mm, slender, ring of hairs within 0; sepals staying fused in 4's, not petal-like, green to magenta; corolla rotate, petals 1.5–2.5 cm, pink, length ± = width, lobes 3, middle longer, narrower, linear to oblanceolate; stamens 4, filaments wider above; stigma beyond anthers. *n*=7. UNCOMMON. Openings in woodland, chaparral; < 1000 m. se SnFrB (Mount Hamilton), SCoRI. ❀DRN,SUN,DRY:**7**,8,9,**14**,15–17,19–24.

C. concinna (Fischer & C. Meyer) E. Greene RED RIBBONS **ST** erect, < 4 dm, glabrous or puberulent. **LF**: petiole 5–25 mm; blade 1–4.5 cm, lanceolate to elliptic or ovate. **FL**: hypanthium 10–25 mm, slender, ring of hairs within 0; sepals staying fused in 4's, below petal-like, thin above, red; corolla rotate, petals 1–3 cm, bright pink, gen white-streaked, length ± 2 × width, lobes 3, ± equally wide, middle = or > others, gen oblanceolate; stamens 4, filaments cylindric; stigma beyond anthers or not. *n*=7. Mixed-evergreen forest, woodland, coastal scrub; < 1500 m. NW, n SNF, CCo, SnFrB. ❀SUN:**15–17,24**;part SHD:**7–9,14,19–23**.

ssp. ***automixa*** R.N. Bowman SANTA CLARA RED RIBBONS **FL**: sepals staying fused only near tip; petals 10–20 mm, lobes prominent, separated by deep sinuses; stigma not beyond anthers. *n*=7. RARE. Woodland; < 1500 m. s SnFrB (foothills, Santa Clara Valley, s Alameda Co.).

ssp. ***concinna*** (p. 795) **FL**: sepals staying fused only near tip; petals 15–30 mm, lobes prominent, separated by deep sinuses; stigma beyond anthers. *n*=7. Habitats, elevations of sp. NW, n SNF, e SnFrB (Oakland Hills, Mount Diablo).

ssp. ***raichei*** G. Allen, V.S. Ford, & Gottlieb RAICHE'S RED RIBBONS **FL**: sepals staying fused in upper 2/3; petals 10–20 mm, lobes short, not separated by deep sinuses; stigma not beyond anthers. *n*=7. RARE. Exposed sites; < 100 m. n CCo (known only from type locality near Tomales, Marin Co.).

C. cylindrica (Jepson) Harlan Lewis & M. Lewis **ST** erect, < 6 dm, puberulent or glabrous. **LF**: petiole < 5 mm; blade 1–6 cm, linear to narrow-lanceolate. **INFL**: axis in bud recurved at tip; buds pendent. **FL**: hypanthium 2–7 mm, ring of hairs below tip; sepals staying fused in 4's, reddish purple; corolla bowl-shaped, petals 1–3.5 cm, fan-shaped, base bright purplish red, middle ± white, above purple to pinkish lavender, middle and above often reddish purple-flecked; stamens 8, outer filaments ± 2 × wider, outer anthers lavender, inner smaller, paler; ovary 4-grooved, stigma beyond anthers. **FR** 2–5 cm, cylindric or wider above; beak 3–5 mm. Open grassland, woodland, chaparral; < 1000 m. c&s SNF, Teh, SCoR, WTR. Sspp. intergrade.

ssp. ***clavicarpa*** W. Davis (p. 795) **FL**: hypanthium length > 2× width; ovary, fr wider above. *n*=9. Habitats and elevations of sp. c&s SNF, Teh. ❀TRY.

ssp. ***cylindrica*** (p. 795) **FL**: hypanthium length ± = width; ovary, fr cylindric. *n*=9. Habitats and elevations of sp. SCoR, WTR. ❀SUN:**15–17,24**or part SHD:7,8,9,**14**,18,**19–23**.

C. davyi (Jepson) Harlan Lewis & M. Lewis **ST** prostrate to decumbent, < 9 dm, sparsely puberulent. **LF** ± sessile; blade 1–2.5 cm, oblanceolate to wide-elliptic or obovate, tip gen obtuse. **INFL**: axis in bud straight; buds erect. **FL**: hypanthium 2–5 mm; sepals staying fused in 2's or all coming free; corolla bowl-shaped, petals 5–11 mm, fan-shaped to obovate, lavender-pink shading in middle to white or pale yellow, spot 0; stamens 8, anthers alike; ovary 8-grooved, stigma not beyond anthers. *n*=17. Coastal grassland, bluffs; < 100 m. NCo, n CCo, n ChI (Santa Rosa Island).

C. delicata (Abrams) Nelson & J.F. Macbr. (p. 795) DELICATE or CAMPO CLARKIA **ST** erect, < 7 dm; glabrous, ± glaucous above, gen puberulent below. **LF**: petiole < 1 cm; blade 1.5–4 cm, lanceolate to elliptic or ovate. **INFL**: axis in bud straight; buds reflexed. **FL**: hypanthium ± 2 mm; sepals staying fused in 4's; corolla rotate, petals 8–12 mm, oblanceolate to obovate, rose-lavender to pale pink, claw < blade, tapered, entire; stamens 8, outer anthers orange-red, inner smaller, paler; ovary 8-grooved, stigma not beyond anthers. *n*=18. RARE in CA. Oak woodland, chaparral; < 1000 m. s PR (San Diego Co.); n Baja CA. Probably from hybrids between *C. unguiculata, C. epilobioides.*

C. dudleyana (Abrams) J.F. Macbr. (p. 795) **ST** erect, < 7 dm, puberulent. **LF**: petiole < 1 cm; blade 1.5–7 cm, narrowly lanceolate to lanceolate. **INFL**: axis in bud recurved at tip; buds pendent. **FL**: hypanthium 1–3 mm; sepals staying fused in 4's, gen pink or purplish red; corolla bowl-shaped, petals 1–3 cm, fan-shaped, lavender pink, gen white-streaked, often red-flecked; stamens 8, outer anthers lavender, inner smaller, paler; ovary 8-grooved, stigma beyond anthers. *n*=9. Openings in woodland, chaparral, yellow-pine forests; < 1500 m. c&s SNF, Teh, TR. ❀TRY.

C. epilobioides (Nutt.) Nelson & J.F. Macbr. **ST** erect, < 7 dm, sparsely puberulent. **LF**: petiole < 7 mm; blade 15–25 mm, linear to narrowly lanceolate or oblanceolate. **INFL**: axis in bud recurved at tip; buds pendent. **FL**: hypanthium 0.5–1.5 mm; sepals staying fused in 4's or 2's, often red; corolla bowl-shaped, petals 5–10 mm, obovate, white to pale cream, fading pink, flecks 0; stamens 8, outer anthers larger; ovary cylindric, not obviously grooved, stigma not beyond anthers. *n*=9. Gen shady sites in woodland, chaparral; < 1000 m. CW, SW, ChI (Santa Cruz, Santa Catalina islands); to AZ, n Baja, CA.

C. exilis Harlan Lewis & Vasek SLENDER CLARKIA **ST** erect, < 1 m, glabrous, glaucous. **LF**: petiole < 5 mm or 0; blade 1–6 cm, lanceolate to narrow-elliptic, glabrous, bright green. **INFL**: axis in bud straight; buds reflexed. **FL**: hypanthium 1–3 mm; sepals staying fused in 4's, gen green, sparsely to densely puberulent; corolla rotate, petals 5–15 mm, lavender, pink, or white, often with dark purplish spot, claw = or > blade, slender, blade gen diamond-shaped; stamens 8, outer anthers gen reddish, inner smaller, paler; ovary 8-grooved, hairs as on sepals, stigma ± not beyond anthers. **FR** 1.5–3 cm, ± 2 mm wide. *n*=9. UNCOMMON. Woodland; < 1000 m. s SNF, Teh.

C. franciscana Harlan Lewis & Raven (p. 795) PRESIDIO CLARKIA **ST** erect, < 4 dm, puberulent. **LF**: petiole < 5 mm; blade 1–3.5 cm, narrowly lanceolate. **INFL**: axis in bud straight; buds erect. **FL**: hypanthium 1–3 mm; sepals staying fused in 4's; corolla bowl-shaped, petals 5–13 mm, wedge-shaped, ± truncate at tip, lavender-pink shading to white near middle, base bright reddish purple; stamens 8, anthers alike; ovary 4-grooved, stigma not beyond anthers. *n*=7. **ENDANGERED** CA. Serpentine soil; ± 50 m. SnFrB (Presidio, San Francisco; Oakland hills). In cult.

C. gracilis (Piper) Nelson & J.F. Macbr. **ST** erect, < 9 dm; hairs 0 to dense. **LF**: petiole < 1 cm; blade 2–7 cm, linear to narrow-lanceolate. **INFL**: axis in bud recurved at tip; buds pendent. **FL**: hypanthium 1.5–10 mm; sepals staying fused in 4's; corolla bowl-shaped, petals 8–40 mm, obovate to fan-shaped, pinkish lavender, often lighter toward base, spot 0 or 1 near middle or base, red; stamens 8, anthers alike; ovary 4-, rarely 8-grooved, puberulent, stigma beyond anthers or not. **FR** 2.5–5 cm, gen wider above; beak slender, < 10 mm. *n*=14. Open, ± dry sites; < 1500 m. NW, CaRF, n SNF, ScV, SnFrB, MP; to WA. From hybrids between *C. amoena* and a probably extinct relative of *C. lassenensis.*

ssp. ***albicaulis*** (Jepson) Harlan Lewis & M. Lewis WHITE-STEMMED CLARKIA **FL**: petals gen 3–4 cm, pinkish lavender to light purple shading to white near middle, base red; stigma beyond anthers. *n*=14. RARE. (known only from 3–4 small populations). Foothill woodland; ± 500 m. s CaRF (Butte Co.). Very local.

ssp. ***gracilis*** (p. 795) **FL**: petals 6–23 mm, ± pink to lavender, spot 0, base darker or not; stigma not beyond anthers. *n*=14. Common. Openings in woodland, forest; < 1500 m. NW, n SNF, ScV, SnFrB, MP; to WA. ❀TRY.

ssp. ***sonomensis*** (C. Hitchc.) Harlan Lewis & M. Lewis (p. 795) **FL**: petals 2–4 cm, pinkish lavender shading to white below, with red spot near middle; stigma beyond anthers. *n*=14. Openings in woodland, forest; < 500 m. KR, NCoR, n SnFrB; s OR. ❀TRY.

ssp. ***tracyi*** (Jepson) Abdel-Hameed & R. Snow **FL**: petals gen 2–3 cm, pinkish lavender, base red; stigma beyond anthers. *n*= 14. Uncommon. Gen serpentine soil; 100–500 m. NCoRI. ❀SUN:7,**14–16,19–24.**

C. heterandra (Torrey) Harlan Lewis & Raven (p. 795) **ST** erect, < 6 dm, glandular-puberulent. **LF**: petiole 5–20 mm; blade 2–8 cm, lanceolate to ovate. **INFL**: axis in bud straight. **FL**: hypanthium 1–2 mm; sepals staying fused in 4's; corolla rotate, petals 5–6 mm, elliptic, pink; stamens 8, inner anthers smaller, sterile; stigma not beyond anthers. **FR** 2–3 mm, nut-like, indehiscent. **SEEDS** 1–2. *n*=9. Shady sites, woodland, yellow-pine forest; 500–1700 m. KR, SN, s SCoRO, TR. [*Heterogaura h.* (Torrey) Cov.]

C. imbricata Harlan Lewis & M. Lewis (p. 795) VINE HILL CLARKIA **ST** erect, < 6 dm, glabrous or sparse-puberulent. **LF**: petiole 0–2 mm; blade 2–2.5 cm, lanceolate. **INFL** dense; axis in bud straight; buds erect. **FL**: hypanthium 10–15 mm, veined; sepals all coming free; corolla bowl-shaped, petals 2–2.5 cm, fan-shaped, lavender shading paler near middle, with large wedge-shaped purplish red spot above; stamens 8, anthers alike; ovary 8-grooved, stigma beyond anthers. *n*=8. **ENDANGERED** CA. Clearings, roadsides, perhaps chaparral; ± 50 m. NCoRO (Sonoma Co.). In cult.

C. jolonensis D. Parnell JOLON CLARKIA **ST** erect, < 6 dm, glabrous, glaucous. **LF**: petiole < 5 mm; blade 3–5 cm, narrow-lanceolate to lanceolate. **INFL**: axis in bud straight; buds reflexed. **FL**: hypanthium 2–3 mm; sepals staying fused in 4's; corolla bowl-shaped, petals 1–2 cm, fan-shaped, pale lavender to pinkish lavender, gen red-flecked; stamens 8, outer anthers lavender, inner smaller, paler; stigma beyond anthers. **SEED** gray. *n*=9. UNCOMMON. Dry woodland; ± 500 m. n SCoRO (near Jolon, Monterey Co.).

C. lassenensis (Eastw.) Harlan Lewis & M. Lewis (p. 795) **ST** erect, < 9 dm, puberulent. **LF**: petiole < 1 cm; blade 2–5 cm, linear to narrow-lanceolate. **INFL**: axis in bud recurved at tip; buds pendent; pedicels < 3 mm. **FL**: hypanthium 3–5 mm; sepals staying fused in 4's; corolla bowl-shaped, petals 8–16 mm, obovate to fan-shaped, pinkish lavender shading lighter toward reddish purple base; stamens 8, anthers alike; ovary 8-grooved, hairs dense, stigma not beyond anthers. **FR** 2.5–4 cm, beak ± 2 mm. *n*=7. Woodland, conifer forest; 500–2000 m. KR, NCoRI, CaR, n SN, MP; OR, NV.

C. lewisii Raven & D. Parnell (p. 795) LEWIS' CLARKIA **ST** erect, < 5 dm, puberulent or glabrous. **LF**: petiole < 7 mm; blade 2–5 cm, narrowly lanceolate to lanceolate. **INFL**: axis in bud recurved at tip; buds pendent. **FL**: hypanthium 1.5–4 mm, ring of hairs at tip; sepals staying fused in 4's, pink to reddish purple; corolla bowl-shaped, petals 1–3 cm, fan-shaped, pinkish lavender shading to white near middle, base purplish red or with a red line, often reddish purple-flecked; stamens 8, filaments alike, outer anthers lavender, inner smaller, paler; ovary 4-grooved, stigma beyond anthers. **FR** 1.5–7 cm, cylindric; beak 0–3 mm. *n*=9. UNCOMMON. Coastal scrub, woodland, chaparral; < 300 m. CCo, n SCoR (Monterey, San Benito cos.). [Misapplied to *C. bottae* (Spach) Harlan Lewis & M. Lewis] ❀SUN:**17** or part SHD:**7,14–16**,19–21,**22–24.**

C. lingulata Harlan Lewis & M. Lewis (p. 795) MERCED CLARKIA **ST** erect, < 6 dm, puberulent. **LF**: petiole < 1.5 cm; blade 2–6 cm, linear to narrowly lanceolate. **INFL**: axis in bud recurved at tip; buds pendent. **FL**: hypanthium 1–4 mm; sepals staying fused in 4's, pink or purplish red; corolla rotate, petals 1–2 cm, oblanceolate, bright pink, red-flecked or not; stamens 8, outer anthers lavender, inner smaller, paler; ovary 8-grooved, stigma beyond anthers. *n*=9. **ENDANGERED** CA. Open chaparral, steep n-facing slopes; 400–450 m. c SNF (2 sites, Merced River Canyon, Mariposa Co.). Derived from *C. biloba.*

C. mildrediae (A.A. Heller) Harlan Lewis & M. Lewis (p. 795) MILDRED'S CLARKIA **ST** erect, < 1 m, puberulent. **LF**: petiole 1.5–4 cm; blade 3–6 cm, elliptic to ovate. **INFL**: axis in bud gen

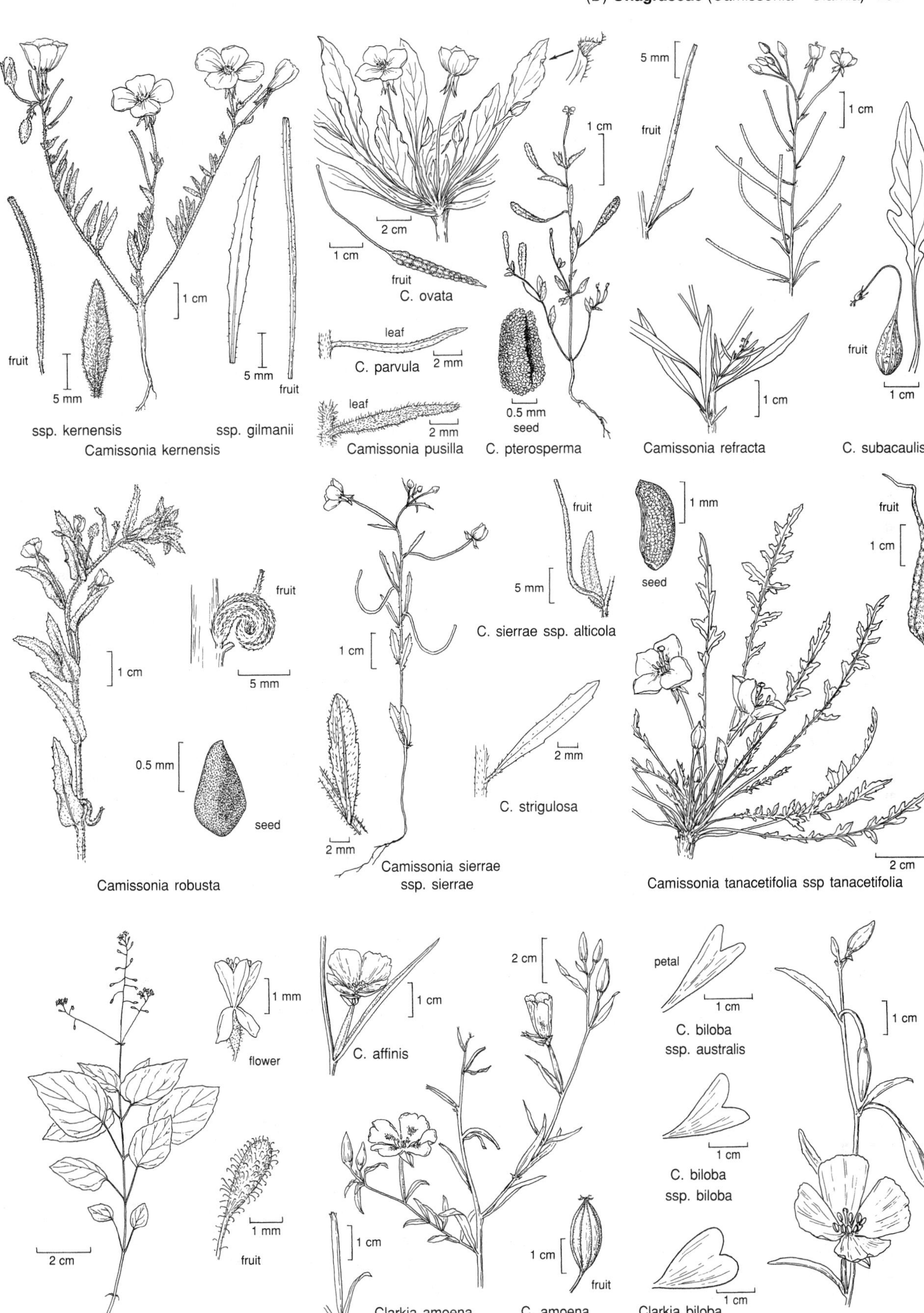
fruit
5 mm
1 cm
5 mm
fruit
ssp. kernensis
ssp. gilmanii
Camissonia kernensis
2 cm
1 cm
fruit
C. ovata
leaf
C. parvula
2 mm
leaf
2 mm
Camissonia pusilla
1 cm
0.5 mm
seed
C. pterosperma
5 mm
fruit
1 cm
1 cm
Camissonia refracta
fruit
1 cm
C. subacaulis
fruit
1 cm
5 mm
0.5 mm
seed
Camissonia robusta
1 cm
2 mm
Camissonia sierrae
ssp. sierrae
fruit
5 mm
C. sierrae ssp. alticola
2 mm
C. strigulosa
1 mm
seed
fruit
1 cm
2 cm
Camissonia tanacetifolia ssp tanacetifolia
1 mm
flower
1 mm
fruit
2 cm
Circaea alpina ssp. pacifica
1 cm
C. affinis
2 cm
1 cm
1 cm
fruit
Clarkia amoena
ssp. amoena
C. amoena
ssp. whitneyi
petal
1 cm
C. biloba
ssp. australis
1 cm
C. biloba
ssp. biloba
1 cm
Clarkia biloba
ssp. brandegeae
1 cm
Clarkia bottae

recurved < 3 nodes above open fls; buds pendent, tips acute. **FL**: hypanthium 2–3 mm; sepals gen all coming free; corolla rotate, petals 1.5–2 cm, lavender to purplish red, often darker-flecked or -spotted, length 1.4–1.6 × width, claw 2-lobed, blade triangular to semicircular; stamens 8, subtended by ciliate scales, anthers alike; ovary 4-grooved, stigma beyond anthers. *n*=7. UNCOMMON. Yellow-pine forest; 500–1600 m. s CaR, n SN (Feather River drainage). ❀DRN:4,5,**6,15–17**&SHD:**7**,14.

C. modesta Jepson **ST** erect, < 7 dm, puberulent. **LF**: petiole < 15 mm; blade 2–4 cm, linear to narrowly lanceolate or elliptic. **INFL**: axis in bud recurved at tip; buds pendent. **FL**: hypanthium 1–3 mm; sepals staying fused in 4's; corolla gen rotate, petals 8–12 mm, oblanceolate to diamond-shaped, pink, gen darker-flecked; stamens 8, outer anthers larger; ovary 8-grooved, stigma not beyond anthers. *n*=8. Shady places in woodland; < 1000 m. NCoRI, c&s SNF, SnFrB, SCoR.

C. mosquinii E. Small **ST** erect, < 1 m, puberulent. **LF**: petiole 1–3 cm; blade 2–5 cm, linear-lanceolate to ovate or elliptic. **INFL**: axis in bud straight 4 or more nodes above open fls, recurved at tip; buds pendent, obovate, tip obtuse. **FL**: hypanthium 2–4 mm; sepals all coming free; corolla rotate, petals 10–20 mm, lavender-purple, often reddish purple-spotted, length 1.6–2 × width, claw 2-lobed, blade ± diamond-shaped; stamens 8, subtended by ciliate scales, anthers alike, pollen blue-gray; ovary 4-grooved, stigma beyond anthers. **SEED** < 1 mm. Dry, rocky places, probably foothill woodland; 185 & ± 500 m. n SNF (ne Butte. Co.). 2 sspp. total.

ssp. ***mosquinii*** MOSQUIN'S CLARKIA **LF** elliptic to ovate. *n*=6. **PRESUMED EXTINCT** (last seen 1959). Habitat of sp.; ± 500 m. n SNF (known only from type, in 1959, Feather River Canyon near Pulga, border with CaR).

ssp. ***xerophylla*** E. Small ENTERPRISE CLARKIA **LF** linear-lanceolate. *n*=6. **PRESUMED EXTINCT** (last seen 1967). Habitat of sp., s-facing slope; 185 m. n SNF (known only from type, now under Lake Oroville).

C. prostrata Harlan Lewis & M. Lewis (p. 795) **ST** prostrate to decumbent, < 5 dm, sparse-puberulent. **LF** ± sessile; blade 1–2.5 cm, elliptic to oblanceolate, tip gen obtuse. **INFL**: axis in bud straight; buds erect. **FL**: hypanthium 3–7 mm; sepals gen staying fused in 2's; corolla bowl-shaped, petals 1–1.5 cm, fan-shaped to obovate, lavender-pink shading to pale yellow below, spot above base, reddish purple; stamens 8, anthers alike; ovary 8-grooved, stigma not beyond anthers. *n*=26. Coastal bluffs in grassland, closed-cone-pine forest; < 100 m. s CCo (San Luis Obispo Co.). Possibly from hybrids between *C. davyi*, *C. speciosa*. ❀4,5;IRR: **15–17,22–24**.

C. purpurea (Curtis) Nelson & J.F. Macbr. **ST** decumbent to erect, < 1 m, glabrous, glaucous or not, to densely puberulent, rarely some hairs longer. **LF**: petiole 0–2 mm; blade 1.5–7 cm, linear or narrow-oblanceolate to elliptic or ovate. **INFL** open or dense; axis in bud straight; buds erect. **FL**: hypanthium 2–10 mm; sepals staying fused in 2's or all coming free; corolla bowl-shaped, petals 5–25 mm, fan-shaped, obovate or elliptic, lavender or pale pink to purple or dark wine-red, often with red or purple spot near middle or above or below; stamens 8, anthers alike; ovary 8-grooved, length < 8 × width, stigma beyond anthers or not. **FR** 1–3 cm; beak 0–2 mm. *n*=26. Open, grassy or shrubby places; < 500 m. CA-FP; to WA, AZ, Baja CA. Sspp. intergrade extensively.

ssp. ***purpurea*** (p. 795) **LF** 1.5–4.5 cm, widely lanceolate to elliptic or ovate. **INFL** dense. **FL**: petals 10–25 mm, lavender to purple or purplish red, often lighter below, often with a darker spot above; stigma beyond anthers or not. *n*=26. Uncommon. Grassland; < 100 m. GV, CCo; OR. ❀DRN,SUN:**7–9,14–17**,19,**20–24**.

ssp. ***quadrivulnera*** (Douglas) Harlan Lewis & M. Lewis (p. 795) FOUR-SPOT **ST** erect. **LF** 1.5–5 cm, linear to lanceolate. **INFL** gen open. **FL**: petals < 15 mm, gen ± 10 mm, lavender to purple or dark wine-red, often with a purple spot near middle or above; stigma not beyond anthers. *n*=26. Common. Habitats, elevations, and range of sp. Highly variable. ❀DRN,SUN:**7–9, 14–17,19–24**.

ssp. ***viminea*** (Douglas) Harlan Lewis & M. Lewis **LF** 3–7 cm, linear to narrowly lanceolate. **INFL** open. **FL**: petals 15–25 mm, lavender to purple, gen with a darker spot above, darker below or not; stigma beyond anthers. *n*=26. Habitats and elevations of sp. Much of CA-FP; OR. ❀TRY.

C. rhomboidea Douglas (p. 795) **ST** erect, < 1 m, puberulent. **LF**: petiole 5–25 mm; blade 1–6 cm, lanceolate to elliptic or ovate. **INFL**: axis in bud recurved at tip; buds pendent. **FL**: hypanthium 1–3 mm; sepals all coming free; corolla rotate, petals 7–14 mm, pinkish lavender, often darker-flecked, claw 2-lobed, blade lanceolate to widely ovate or diamond-shaped; stamens 8, subtended by ciliate scales, anthers alike, pollen blue-gray; ovary 4-grooved, stigma not beyond anthers. *n*=12. Common. Yellow-pine forest, woodland; < 2500 m. CA-FP; ± w US, Baja CA. From hybrids between spp. related to *C. mildrediae*, *C. virgata*. ❀DRN:1–4,**7–9,14**,18,19,**20–23**&SUN:**15–17,24**.

C. rostrata W. Davis (p. 795) BEAKED CLARKIA **ST** erect, < 6 dm, puberulent. **LF**: petiole < 10 mm; blade 1–6 cm, lanceolate. **INFL**: axis in bud recurved at tip; buds pendent. **FL**: hypanthium ± 2 mm, ring of hairs ± at tip; sepals staying fused in 4's, pink to purplish red; corolla bowl-shaped, petals 1–2.5 cm, fan-shaped, pinkish lavender shading to white near middle, often darker-flecked, base reddish purple; stamens 8, filaments alike, outer anthers lavender, inner smaller, paler; ovary 4-grooved, stigma beyond anthers. **FR** 1–3 cm; beak 7–15 mm. *n*=9. RARE. Oak/pine woodland; ± 500 m. c SNF (Hell Hollow near Bear Valley, Mariposa Co.), SnJV (Merced River drainage, probably ephemeral populations).

C. rubicunda (Lindley) Harlan Lewis & M. Lewis (p. 795) **ST** decumbent to erect, < 1.5 m, puberulent. **LF**: petiole < 1 cm; blade 1–4 cm, lanceolate to elliptic. **INFL** open or dense; axis in bud straight; buds erect. **FL**: hypanthium 4–10 mm; sepals staying fused in 4's; corolla bowl-shaped, petals 1–3 cm, obovate to fan-shaped, rosy-pink to lavender, base bright red or purplish red; stamens 8, anthers alike; ovary 4-grooved, cylindric, stigma beyond anthers. *n*=7. Openings, woodland, forest, chaparral near coast; < 500 m. CCo, SnFrB. [ssp. *blasdalei* (Jepson) Harlan Lewis & M. Lewis] ❀SUN,DRYorIRR:7–9,**14–17,22–24**;CVS.

C. similis Harlan Lewis & Ernst **ST** erect, < 9 dm, puberulent. **LF**: petiole < 8 mm; blade 2–4 cm, narrow-lanceolate to elliptic. **INFL**: axis in bud recurved at tip; buds pendent. **FL**: hypanthium 1.5–2 mm; sepals staying fused in 4's; corolla rotate to bowl-shaped, petals 6–10 mm, oblanceolate to diamond-shaped or obovate, pale pink shading to ± white near base, purple-flecked, fading pink; stamens 8, outer anthers larger; ovary very shallowly 8-grooved, stigma not beyond anthers. *n*=17. Gen shady sites, woodland, chaparral; < 1000 m. SCoRI (uncommon), SW. Probably from hybrids between *C. epilobioides*, *C. modesta*.

C. speciosa Harlan Lewis & M. Lewis **ST** decumbent to erect, < 6 dm, puberulent. **LF**: petiole 0–5 mm; blade 1–6 cm, linear to lanceolate. **INFL** open or dense; axis in bud straight; buds erect. **FL**: hypanthium 5–15 mm; sepals gen all coming free; corolla bowl-shaped, petals 1–2.5 cm, fan-shaped, lavender to deep wine-red often shading to white or pale yellow at base, spot 0 or dark reddish, near or below middle; stamens 8, anthers alike; ovary 8-grooved, stigma beyond anthers. Woodlands, sandy coastal hills; < 700 m. c&s SNF, s CCo, SCoR. Sspp. intergrade.

ssp. ***immaculata*** Harlan Lewis & M. Lewis (p. 795) PISMO CLARKIA **ST** decumbent. **LF** linear to narrowly lanceolate. **INFL** open. **FL**: petals reddish lavender shading to pale yellow or white below, spot 0. *n*=9. **RARE** CA. Sandy hills near coast; < 100 m. s CCo (± Pismo to Edna, San Luis Obispo Co.). In cult.

ssp. ***nitens*** (Harlan Lewis & M. Lewis) Harlan Lewis (p. 795) **ST** erect. **LF** lanceolate. **INFL** dense. **FL**: petals lavender, shading to pale yellow below or pale yellow throughout, gen with a bright purplish red spot near or below middle; ovary gen > internode above. *n*=9. Foothill woodland; < 700 m. c&s SNF. [*C. nitens* Harlan Lewis & M. Lewis] ❀TRY.

ssp. ***polyantha*** Harlan Lewis & M. Lewis (p. 795) **ST** erect; branches on well developed pls few, many-fld. **LF** linear to narrow-lanceolate. **INFL** open. **FL**: petals purple or lavender, gen lighter toward base, with purplish red spot near middle; ovary gen < internode above. *n*=9. Foothill woodland; < 700 m. s SNF. Intergrades with ssp. *nitens*. ❀TRY.

ssp. ***speciosa*** (p. 795) **ST** decumbent to erect; branches on well developed pls many, few-fld. **LF** linear to narrowly lanceolate. **INFL** open. **FL**: petals dark purplish red to lavender, gen white or pale yellow toward base, or pale yellow throughout, gen with red spot near middle; ovary gen < internode above. *n*=9. Woodland; < 500 m. SCoR. Petal color variable in San Luis Obispo Co. ❀DRN,DRY,SUN:7,8,9,**14–17,19–24**.

C. springvillensis Vasek (p. 795) SPRINGVILLE CLARKIA **ST** erect, < 1 m, glabrous, glaucous. **LF**: petiole 0–5 mm; blade 2–9 cm, lanceolate, glabrous, glaucous. **INFL**: axis in bud straight; buds reflexed. **FL**: hypanthium 3–4 mm; sepals staying fused in 4's, gen dark red-purple, sparsely to densely puberulent; corolla rotate, petals ± 15 mm, claw ± = or > blade, slender, blade ± diamond-shaped, lavender-pink, gen with dark purplish spot near base; stamens 8, outer anthers red, inner smaller, paler; ovary 8-grooved, hairs as on sepals, stigma beyond anthers. *n*=9. **ENDANGERED** CA. Woodland; ± 500 m. s SNF (Springville, Tulare Co.).

C. stellata Mosq. **ST** erect, < 1 m, puberulent. **LF**: petiole 5–30 mm; blade 1–5 cm, elliptic to ovate. **INFL**: axis in bud recurved < 3 nodes above open fls; buds pendent. **FL**: hypanthium 1.5–2 mm; sepals all coming free; corolla rotate, petals 6–8 mm, lavender-purple, not dark-flecked or -spotted, claw 2-lobed, blade obovate, inconspicuously 3-lobed; stamens 8, subtended by ciliate scales, anthers alike, pollen yellow; ovary 4-grooved, stigma not beyond anthers. *n*=7. Uncommon. Conifer forest; 1000–1500 m. CaRH, n SNH. ❀TRY.

C. tembloriensis Vasek **ST** erect, < 8 dm. **LF**: petiole 0–5 mm; blade 2–7 cm, lanceolate, glaucous, gray-green. **INFL**: axis in bud straight; buds reflexed. **FL**: hypanthium 2–3 mm; sepals staying fused in 4's, green, red-tinged or not, sparsely to densely puberulent; corolla rotate, petals 1–2.5 cm, lavender-pink, claw ± = or > blade, slender, blade ± diamond-shaped, spot purplish or 0; stamens 8, outer anthers lavender to red, inner smaller, paler; ovary 8-grooved, hairs as on sepals, stigma beyond anthers or not. **FR** 1.5–3 cm, ± 3 mm wide. Dry grassland, scrub; 100–500 m. s SNF, SnJV, SCoRI. Hybrids between sspp. have low fertility.

ssp. ***calientensis*** (Vasek) Holsinger VASEK'S CLARKIA **FL**: petals ± 10 mm wide; stigma not beyond anthers. *n*=9. RARE. Grassland; ± 500 m. s SNF (Caliente Hills, Kern Co.) [*C. calientensis* Vasek]

ssp. ***tembloriensis*** (p. 795) **FL**: petals gen < 10 mm wide; stigma beyond anthers or not. *n*=9. Habitats, elevations, range of sp. ❀TRY.

C. unguiculata Lindley (p. 795) **ST** erect, < 1 m, glabrous, glaucous. **LF**: petiole 0–1 cm; blade 1–6 cm, lanceolate to elliptic or ovate. **INFL**: axis in bud straight; buds reflexed. **FL**: hypanthium 2–5 mm; sepals staying fused in 4's, green to dark red, sparsely to densely puberulent and with longer, straight, spreading hairs < 3 mm; corolla rotate, petals 1–2.5 cm, lavender-pink to salmon or dark reddish purple, claw = or > blade, slender, blade triangular or diamond-shaped to ± round; stamens 8, outer anthers red, inner smaller, paler; ovary 8-grooved, hairs as on sepals, stigma beyond anthers. *n*=9. Common. Woodland; < 1500 m. NCoR, SNF, Teh, SnFrB, SCoR, SCo, WTR, PR. ❀SUN,DRYorIRR:1–5,**6,15–17**, or part SHD:**7–9,14,18–24**;CVS.

C. virgata E. Greene (p. 795) SIERRA CLARKIA **ST** erect, < 1 m, puberulent. **LF**: petiole 1.5–5 cm; blade 2–5 cm, elliptic to ovate. **INFL**: axis in bud straight 4 or more nodes above open fls, recurved at tip; buds pendent, obovate, tip obtuse. **FL**: hypanthium 2–4 mm; sepals all coming free; corolla rotate, petals 8–15 mm, lavender-purple, reddish purple-mottled or -spotted, length 1.9–2.7 × width, claw 2-lobed, blade ± diamond-shaped; stamens 8, subtended by ciliate scales, anthers alike, pollen blue-gray; ovary 4-grooved, stigma beyond anthers. **SEED** 1–1.5 mm. *n*=5. UNCOMMON. Yellow-pine forest, foothill woodland; 400–1100 m. n&c SN (Eldorado to Tuolumne cos.). Sterile hybrids with *C. australis*.

C. williamsonii (Durand & Hilg.) Harlan Lewis & M. Lewis (p. 795) **ST** erect, < 1 m, puberulent. **LF**: petiole 0–10 mm; blade 2–7 cm, linear to narrowly lanceolate. **INFL** open; axis in bud straight; buds erect. **FL**: hypanthium 7–13 mm; sepals staying fused in 2's or all coming free, free at tip in bud; corolla bowl-shaped, petals 1–3 cm, fan-shaped, lavender, white near middle, with purple spot above, rarely uniformly wine-red; stamens 8, anthers alike; ovary 8-grooved, stigma beyond anthers. *n*=9. Foothill woodland, yellow-pine forest; 400–2000 m. n&c SNF. [f. *incerta* (Jepson) Harlan Lewis & M. Lewis] ❀SUN:6,**17**&IRR:1–3,**7, 14–16,18**,19–24.

C. xantiana A. Gray **ST** erect, < 8 dm, glabrous, glaucous. **LF**: petiole 0–2 mm; blade 2–6 cm, linear to lanceolate, **INFL**: axis in bud straight; buds reflexed. **FL**: hypanthium 2–5 mm; sepals staying fused in 4's; corolla rotate, petals 6–20 mm, lavender to reddish purple, clawed, blade lobes 2, a slender, 1–3 mm tooth between, upper petals gen with white-surrounded spot; stamens 8, outer anthers lavender to purple, inner anthers smaller, paler; ovary 8-grooved, stigma beyond anthers or not. Dry slopes; 500–2000 m. s SN, Teh, WTR.

ssp. ***parviflora*** (Eastw.) Harlan Lewis **FL**: petals 6–12 mm, lavender-pink or white; stigma not beyond anthers. *n*=9. Uncommon. Dry slopes; 1000–1500 m. s SNF (Kern River drainage).

ssp. ***xantiana*** (p. 795) **FL**: petals 12–20 mm, lavender to reddish purple; stigma beyond anthers. *n*=9. Foothill woodland; 500–2000 m. s SN (esp Kern River drainage), Teh, WTR. ❀TRY.

EPILOBIUM FIREWEED, WILLOW HERB

Peter C. Hoch

Ann to subshrub. **LVS** gen opposite below (or clustered in axils), gen ± fine-toothed; veins gen obscure. **INFL**: gen raceme, bracted. **FL** radial or ± bilateral; sepals 4, erect; petals 4, gen notched; stamens 8, anthers attached at middle, pollen grains gen shed in 4's, gen cream-yellow; ovary chambers 4, stigma gen club-like. **FR** straight, cylindric to club-like. **SEEDS** gen in 1 row per chamber, gen with white, deciduous hair-tuft. 171 spp.: worldwide exc trop, deserts. (Greek: upon pod, from inferior ovary) [Raven 1976 Ann Missouri Bot Gard 63:326–340] Incl *Boisduvalia*, *Zauschneria*. Most taxa polyploid; many with anthers ± = stigma self-pollinated; many hybrids.

1. Lvs alternate; hypanthium 0; petals entire; stamens subequal, in 1 whorl; pollen bluish gray, grains shed singly (sect. *Chamaenerion*)
  2. Bracts << cauline lvs, ± linear; lf veins conspicuous; petioles 2–7 mm; sepals 7–16 mm; petals 10–25 mm .... ***E. angustifolium*** ssp. ***circumvagum***
  2′ Bracts little reduced, lanceolate; lf veins obscure; petioles 0–3 mm; sepals 10–24 mm; petals 13–32 mm .... ***E. latifolium***

1′ Lvs opposite at least at base; hypanthium 0.3–34 mm; petals notched; stamens in 2 unequal whorls; pollen cream, grains gen shed in 4's
  3. Ann; lvs opposite only near base; lower st peeling; seed hair-tuft 0 (sect. *Boisduvalia*)
    4. Fr wall tough, ± indehiscent below; seeds in 2 rows per chamber
      5. Fr 8–12 mm, sharply 4-angled .... ***E. cleistogamum***
      5′ Fr 5–8 mm, cylindric .... ***E. pygmaeum***

4′ Fr wall flexible, dehiscent to base; seeds in 1 row per chamber
6. Fr not beaked, central axis remaining intact .... *E. densiflorum*
6′ Fr beaked, central axis disintegrating
7. Sepals 2–6 mm, petals 4–11 mm; fr 14–19 mm; seeds in 1 staggered row, 1.8–2 mm .... *E. pallidum*
7′ Sepals 0.7–2 mm, petals 1–3 mm; fr 8–13 mm; seeds in 4 rows, 1–1.5 mm .... *E. torreyi*
3′ Ann to subshrub; lvs gen opposite up to infl; lower st gen not peeling; seeds hair-tufted
8. Hypanthium 17–34 mm; corolla red-orange, ± bilateral (sect. *Zauschneria*)
9. Matted per 1–2 dm; lvs not clustered; lower lvs white-canescent; NW .... *E. septentrionale*
9′ Clumped per to subshrub 1–9 dm; lvs clustered or not, mostly not white-canescent; widespread CA-FP .... *E. canum*
10. Lvs linear to lanceolate, often clustered; per to subshrub 2–9 dm .... ssp. *canum*
10′ Lvs widely lanceolate to ovate, not clustered; per 1–5 dm .... ssp. *latifolium*
8′ Hypanthium gen 0.4–8(16) mm; corolla white to purple (sect. *Epilobium*)
11. Ann; st peeling below; lvs ± narrowly lanceolate, upper often clustered
12. Pl 2–20 dm, ± glandular; hypanthium 1.5–16 mm, sepals 2–8 mm, petals 2–15 mm; seed 1.4–2.7 mm .... *E. brachycarpum*
12′ Pl 0.5–4 dm, nonglandular; hypanthium 0.4–1 mm, sepals 1.5–4 mm, petals 1.8–5 mm; seed 0.6–1.2 mm
13. Lvs gen folded along midrib, upper clustered; infl dense; petals 1.8–2.5 mm; seeds 0.6–0.9 mm, papillate .... *E. foliosum*
13′ Lvs flat, not clustered; infl open; petals 2–5 mm; seeds 0.8–1.2 mm, netted .... *E. minutum*
11′ Per from caudex; st not peeling; lvs (narrowly) lanceolate to ovate, not clustered
14. Stigma 4-lobed; sepals (3)6–15 mm; petals (6)10–24 mm, pink to rose-purple or cream
15. Petals cream-yellow; seed hairs reddish, persistent; fr 35–75 mm .... *E. luteum*
15′ Petals pink to rose-purple; seed hairs white, deciduous; fr 8–45 mm
16. St 4–10 dm; lf 30–90 mm, veins reddish, conspicuous; seed 0.9–1.3 mm, ridged .... *E. oreganum*
16′ St < 4 dm; lf 6–45 mm, veins obscure; seed 1.4–3.4 mm, papillate
17. Petals 6–10 mm; seeds 2 per chamber (8 per fr); fr 8–16 mm .... *E. nivium*
17′ Petals 11–24 mm; seeds > 8 per chamber; fr 20–45 mm
18. Pl 1–4 dm, nonglandular; lf 20–45 mm, petiole 2–6 mm; hypanthium 1–1.8 mm; seed 2.5–3.4 mm .... *E. rigidum*
18′ Pl < 2.5 dm, sometimes glandular (esp ovaries); lf 6–26 mm, petiole 0–3 mm; hypanthium 2–6 mm; seed 1.4–2.1 mm
19. Pedicel in fr 2–3 mm; lf 6–20 mm, tip round to obtuse, petiole 0–3 mm; pl ± glaucous . *E. obcordatum*
19′ Pedicel in fr 6–25 mm; lf 13–26 mm, tip ± acute; petiole 0; pl not glaucous .... *E. siskiyouense*
14′ Stigma entire, club- or head-like; sepals 1–7.5 mm; petals 1.6–14 mm, white to rose-purple
20. St ± erect, with offset rosettes or fleshy underground shoots
21. Lf veins ± obscure; st with fleshy shoot; seed papillate
22. Pedicel in fr 10–40 mm; infl ± nodding; lf 5–45 mm .... *E. halleanum*
22′ Pedicel in fr 0–5 mm; infl erect; lf 10–65 mm .... *E. saximontanum*
21′ Lf veins conspicuous; st with rosette or fleshy shoot; seed ridged .... *E. ciliatum*
23. Petals 2–6(9) mm, white to pink; lf lanceolate, reduced in open infl — st with basal rosette . ssp. *ciliatum*
23′ Petals 4–15 mm, pink to rose-purple; lf lanceolate to ovate, little reduced on crowded infl
24. St with fleshy shoot; infl long .... ssp. *glandulosum*
24′ St with leafy rosette; infl ± flat-topped .... ssp. *watsonii*
20′ Sts ± ascending, clumped or cespitose; leafy or thread-like stolons sometimes tipped by fleshy bulblets
25. Pl subglabrous (exc scattered infl hairs)
26. Pl < 3.5 dm, matted; stolons thread-like; pedicels 20–65 mm; infl open (bracts << internodes); lf tip round to obtuse .... *E. oregonense*
26′ Pl < 8.5 dm, clumped; stolons short, scaly; pedicels 5–25 mm; infl crowded (bracts ± = internodes); lf tip subacute .... *E. glaberrimum*
27. St < 3.5 dm; lf 10–35 mm, lanceolate to narrowly ovate; fr 20–55 mm .... ssp. *fastigiatum*
27′ St 2–8.5 dm; lf 20–70 mm, ± narrowly lanceolate; fr 45–75 mm .... ssp. *glaberrimum*
25′ Pl variously hairy, often in decurrent lines on st
28. St densely glandular-hairy; petals 2–3 mm, white .... *E. howellii*
28′ St hairy, nonglandular; petals 2–9 mm, white or pink
29. Stolons thread-like, tipped by fleshy bulblets; lf ± narrowly lanceolate; seed 1.4–2.2 mm, hairs persistent
30. Lf densely minutely strigose; infl minutely strigose and glandular .... *E. leptophyllum*
30′ Lf minutely strigose only on margins and midrib; infl minutely strigose, nonglandular .... *E. palustre*
29′ Stolons short, leafy; lf gen lanceolate or wider; seeds 0.7–2.1 mm, hair-tuft gen deciduous
31. Pl cespitose, < 2.2 dm; lf 8–28 mm; fr 17–40 mm
32. Infl ± nodding; pedicel in fr 10–55 mm; seed 0.7–1.4 mm, netted — hair-tuft persistent .... *E. anagallidifolium*
32′ Infl erect; pedicel in fr 2–18 mm; seed 1.5–2.1 mm, papillate .... *E. clavatum*
31′ Pl loosely clumped, 1–5 dm; lf 15–55 mm; fr 40–100 mm
33. Pedicel in fr 5–15 mm; fr 40–65 mm; seed 0.9–1.2 mm, papillate; petals gen pink to rose-purple .... *E. hornemannii*
33′ Pedicel in fr 20–45 mm; fr 50–100 mm; seed 1.1–1.6 mm, netted; petals gen white ... *E. lactiflorum*

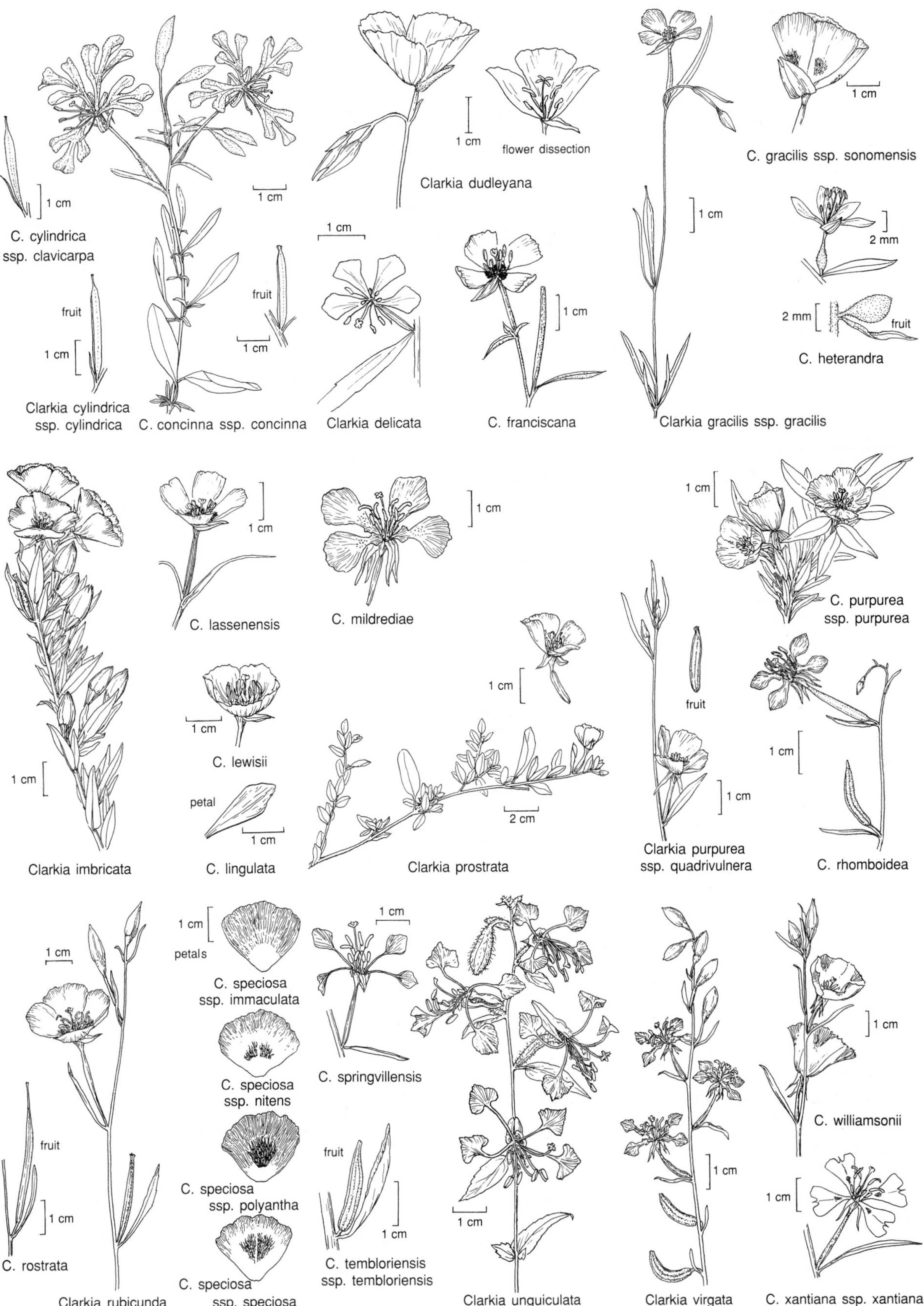
1 cm
C. cylindrica
ssp. clavicarpa
fruit
1 cm
Clarkia cylindrica
ssp. cylindrica
1 cm
C. concinna ssp. concinna
fruit
1 cm
1 cm
flower dissection
Clarkia dudleyana
1 cm
Clarkia delicata
1 cm
C. franciscana
1 cm
Clarkia gracilis ssp. gracilis
1 cm
C. gracilis ssp. sonomensis
2 mm
2 mm
fruit
C. heterandra
1 cm
Clarkia imbricata
1 cm
C. lassenensis
1 cm
C. lewisii
petal
1 cm
C. lingulata
1 cm
C. mildrediae
1 cm
2 cm
Clarkia prostrata
fruit
1 cm
Clarkia purpurea
ssp. quadrivulnera
1 cm
C. purpurea
ssp. purpurea
1 cm
C. rhomboidea
1 cm
fruit
1 cm
C. rostrata
Clarkia rubicunda
1 cm
petals
C. speciosa
ssp. immaculata
C. speciosa
ssp. nitens
C. speciosa
ssp. polyantha
C. speciosa
ssp. speciosa
1 cm
C. springvillensis
fruit
1 cm
C. tembloriensis
ssp. tembloriensis
1 cm
Clarkia unguiculata
1 cm
Clarkia virgata
1 cm
C. williamsonii
1 cm
C. xantiana ssp. xantiana

E. anagallidifolium Lam. (p. 801) Per < 2 dm, cespitose, ascending to erect, ± strigose in decurrent lines; stolons short. **LVS** 8–25 mm; basal spoon-shaped to oblong, glabrous; upper elliptic to lanceolate, sparsely strigose, obtuse; petiole 1–6 mm. **INFL** ± nodding, sometimes glandular. **FL**: hypanthium 0.6–1.4 mm; sepals 1.5–5 mm; petals 2.5–9 mm, pink to rose-purple. **FR** 17–36 mm, subglabrous; pedicel 10–55 mm. **SEED** 0.7–1.4 mm, netted; hairs persistent. $2n$=36. Moist alpine slopes, meadows, streambanks; 2100–3700 m. KR, CaRH, SNH, SnBr, n W&I; circumboreal. ❀TRY.

E. angustifolium L. ssp. ***circumvagum*** Mosq. (p. 801) FIREWEED Per < 30 dm, gen strongly colonial, ± glabrous to densely strigose above. **LVS** alternate, 15–200 mm, lanceolate; midrib strigose below; veins ± conspicuous; petiole 2–7 mm. **INFL** dense, gen canescent; bracts small, linear. **FL** nodding in bud; hypanthium 0 (exc as greenish disk); sepals 7–16 mm; petals 10–25 mm, gen deep pink to magenta, entire; stamens subequal, < pistil, maturing before stigma, pollen bluish gray, shed singly; stigma 4-lobed. **FR** 40–100 mm, gray-hairy; pedicel 7–20 mm. **SEED** 1–1.3 mm, fusiform, irregularly netted; hairs persistent. $2n$=72. Common. Open places, gravel bars, roadsides, esp after fires; < 3300 m. NCo, KR, NCoRO, CaRH, SNH, SnBr, ne DMtns; circumboreal. Ssp. *angustifolium* ($2n$=36), farther n & higher, might be expected in CA. ❀SUN:**4–6**&IRR:**1–3**,**7**,**14–18**,19–21;STBL;INV;[CVS non CA].

E. brachycarpum C. Presl (p. 801) Ann 2–20 dm, glabrous and peeling below, strigose and gen glandular-hairy above. **LVS** gen early deciduous, 10–50 mm, linear to narrowly elliptic, acuminate, gen folded along midrib, ± glabrous; petiole 0–4 mm. **FL**: hypanthium 1.5-8(16) mm; sepals 2–8 mm; petals 2–15 mm, white to rose-purple; stamens < or = pistil; stigma sometimes 4-lobed. **FR** 15–35 mm, glabrous or glandular; pedicel 3–20 mm. **SEED** 1.4–2.7 mm, papillate. $2n$=24. Common. Dry open woodland, grassland, roadsides; < 3300 m. CA-FP (exc ChI), MP; to B.C., SD, NM, also e Can; introduced in s S.Am. [*E. paniculatum* Torrey & A. Gray incl vars. *laevicaule* (Rydb.) Munz & *tracyi* (Rydb.) Munz] Highly variable, esp fl size. KR pls with large fls, pollen shed singly, have been called *E. p.* var. *jucundum* (Rydb.) Trel. ❀SUN,DRN:**4–6**&IRR:1–3,**7**,8,9,**14–24**.

E. canum (E. Greene) Raven CALIFORNIA FUCHSIA, ZAUSCHNERIA Per (clumped with basal scaly shoots) to subshrub 1–9 dm, ± densely spreading-hairy and gen glandular. **LF** subsessile, 5–50 mm, linear to ovate, green to grayish, sometimes strongly toothed. **FL** red-orange; hypanthium 20–34 mm; sepals 8–15 mm; petals 8–17 mm; stamens << pistil; stigma 4-lobed. **FR** 20-35 mm, ± beaked, hairy; pedicels 0–2 mm. **SEED** 1.5–2.3 mm, low-papillate. Dry slopes, ridges; < 3000 m. CA-FP (exc NCo, NCoRH); to OR, WY, NM, n Mex. Hummingbird-pollinated. Sspp. intergrade, esp in s CA; ssp. *garrettii* (Nelson) Raven in UT, WY.

ssp. ***canum*** (p. 801) Per or subshrub 2–9 dm, rarely glandular. **LVS** often clustered, linear to lanceolate, gen grayish. $2n$=30, 60. Habitats and range of sp. (exc SNH, GB); < 1500 m. [*Zauschneria c.* E. Greene; *Z. californica* C. Presl ssp. *californica*] Variable. ❀SUN,DRN:4,**5**,**6**,**15–17**&IRR:2,**7–10**,11,**14**,**18–24**; rather INV;decid.GRNCVR;CVS;also STBL.

ssp. ***latifolium*** (Hook.) Raven (p. 801) Per 1–5 dm, gen glandular. **LVS** opposite, widely lanceolate to ovate, gen green. $2n$=60. Habitats of sp.; 500-3000 m. KR, CaR, c&s SNF, SNH, Teh, SnJV, TR; to sw OR, w NM, nw Mex. [*Z. californica* ssp. *l.* (Hook.) Keck]. ❀SUN,DRN:**4–6**,**17**&IRR:1–3,7,8–10,**14–16**,**18–24**;rather INV;CVS;some formsGRNCVR(deciduous);also STBL.

E. ciliatum Raf. Per < 19 dm, loosely clumped, with basal rosettes or fleshy bulblets, gen strigose in lines or spreading-hairy. **LF** 1–15 cm, narrowly lanceolate to ovate; veins conspicuous; petiole 0–8 mm. **INFL** densely strigose, ± spreading-hairy, gen glandular. **FL**: hypanthium 0.5–2.6 mm; sepals 2–7.5 mm; petals 2–14 mm, white to rose-purple; stamens < or = pistil; stigma club- or head-like. **FR** 15–100 mm, hairy; pedicels 0–30 mm. **SEED** 0.8–1.9 mm, ridged. $2n$=36. Abundant. Disturbed places, moist meadows, streambanks, roadsides; < 4100 m. ± CA; most of N.Am, e Asia, s S.Am, introduced in Australasia, Eur, w Asia.

ssp. ***ciliatum*** (p. 801) Rosettes well developed. **LF** lanceolate. **INFL** openly branched, not leafy. **FL**: petals 2–6(9) mm, white to pink. Common. Habitats and range ± of sp. [*E. adenocaulon* Hausskn. incl vars. *holosericeum* (Trel.) Munz, *occidentale* Trel., & *parishii* (Trel.) Munz]

ssp. ***glandulosum*** (Lehm.) P. Hoch & Raven Basal offshoots fleshy, dense, rarely leafy. **LF** gen ovate, little reduced upward. **INFL** long, dense, leafy. **FL**: petals 4–14 mm, pink to rose-purple. Often ± shaded streambanks, seeps, meadows; < 3500 m. KR, NCoR, CaR, SNH, TR, Wrn; to AK, ne N.Am; introduced to n Eur. [*E. brevistylum* Barbey var. *b.*; *E. glandulosum* Lehm.] Variable; larger-fld pls cross-pollinated.

ssp. ***watsonii*** (Barbey) P. Hoch & Raven (p. 801) Pl robust; rosettes well developed. **LF** lanceolate to ovate. **INFL** dense, leafy, flat-topped. **FL**: petals 5–14 mm, pink to rose-purple. Moist streamsides, ± disturbed coastal sites; < 350 m. NCo, CCo; to B.C. [*E. watsonii* Barbey incl var. *franciscanum* (Barbey) Jepson] Distinctive.

E. clavatum Trel. (p. 801) Per < 2.2 dm, cespitose, ascending, ± strigose (esp above); stolons wiry, scaly. **LF** 12–28 mm, ± elliptic to ovate, obtuse to subacute; petiole 0–3 mm. **INFL** erect, sparsely glandular. **FL**: hypanthium 0.6–2 mm; sepals 2.5–4.5 mm; petals 3.5–6 mm, pink to rose-purple; stigma widely club-like. **FR** 20–40 mm, gen sparsely hairy; pedicel 2–18 mm. **SEED** 1.5–2.1 mm, papillate. $2n$=36. Subalpine scree, rocky streambanks; 1200–3600 m. KR, CaRH, c SNH, Wrn; to AK, MT, Colorado.

E. cleistogamum (Curran) P. Hoch & Raven (p. 801) Ann < 3.2 dm, peeling below, sometimes with prostrate basal branches, ± glabrous to spreading-hairy. **LVS** opposite only near base, subsessile, 20–45 mm, linear to narrowly elliptic; upper lvs hairy. **INFL** leafy, hairy and glandular. **FL** gen cleistogamous; hypanthium 0.5–1 mm; sepals 1–3 mm; petals 2–5 mm, white to pale pink; stigma sometimes barely 4-lobed. **FR** 8–12 mm, sharply 4-sided, tough, sessile, dehiscent only near tip; central axis 0. **SEEDS** in 2 rows per cell, 1 mm, netted, glabrous. $2n$=30. Vernal pools, clay flats; 0–300 m. SNF, GV, SCoRO. [*Boisduvalia c.* Curran]

E. densiflorum (Lindley) P. Hoch & Raven (p. 801) Ann < 15 dm, simple or branched above, peeling below, hairy or strigose. **LVS** opposite only near base, subsessile, 10–85 mm, narrowly lanceolate; upper lvs hairy. **INFL** leafy, ± dense, sometimes glandular. **FL**: hypanthium 1.5–4 mm; sepals 2–9 mm; petals 3–10 mm, white to rose-purple; stigma irregularly 4-lobed. **FR** 4–11 mm, cylindric, pliable, sessile; axis persistent. **SEED** 1.2–2 mm, netted, glabrous. $2n$=20. Streambanks, outwashes; < 2600 m. CA-FP (exc SCo, ChI); to B.C., MT, n Baja CA. [*Boisduvalia d.* (Lindley) S. Watson incl vars. *pallescens* Suskd. & *salicina* (Torrey & A. Gray) Munz] Highly variable.

E. foliosum (Torrey & A. Gray) Suksd. (p. 801) Ann < 4 dm, ± peeling below, ± strigose. **LVS** gen alternate below, clustered above, 5–30 mm, sublinear to lanceolate, gen folded along midrib; petiole 3–12 mm. **INFL** crowded, densely strigose. **FL**: hypanthium 0.4–0.8 mm; sepals 1.5–2.5 mm; petals 1.8–3 mm, white. **FR** 12–20 mm, sparsely hairy; pedicel 2–5 mm. **SEED** 0.6–0.9 mm, low-papillate. $2n$=32. Dry, open, disturbed areas, roadsides; 20–1500(2400) m. KR, NCoR, CaRH, n SNF, SNH, ScV, CCo, SnFrB, SCoRO, WTR, SnGb; to B.C., ID, AZ, Mex.

E. glaberrimum Barbey Per < 8.5 dm, clumped, ± glabrous, glaucous; stolons short, scaly. **LF** subsessile, 10–70 mm, narrowly lanceolate to narrowly ovate, clasping. **FL**: hypanthium 0.7–2.3 mm; sepals 1.6–7.5 mm; petals 2.5–12 mm, gen pink to rose-purple (white). **FR** 20-75 mm, sometimes sparsely hairy; pedicel 5–25 mm. **SEED** 0.7–1.2 mm, papillate. $2n$=36. Gravel bars, scree, roadsides, moist rocky areas; 600–3800 m. KR, NCoRH, CaRH, SNH, TR, PR, Wrn, n W&I; to w Can, MT, WY, n Mex.

ssp. ***fastigiatum*** (Nutt.) P. Hoch & Raven (p. 801) **ST** < 3.5 dm, gen simple. **LF** 10–35 mm, lanceolate to narrowly ovate. **FL**: sepals 1.6–4 mm; petals 2.5–7 mm, pink to rose-purple. **FR** 20–55 mm; pedicel 5–17 mm. **SEED** 0.8–1.3 mm. Habitats of sp.; 1200–3800 m. KR, NCoRH, CaRH, SNH, Wrn; to w Can, MT, n Mex. [var. *f.* (Nutt.) Trel.]

ssp. ***glaberrimum*** (p. 801) **ST** 2–8.5 dm, gen branched above. **LF** 20–70 mm, (narrowly) lanceolate. **FL**: sepals 3–7.5 mm;

petals 5–12 mm. **FR** 45–75 mm; pedicel 5–25 mm. **SEED** 0.7–1 mm. Well drained, gravelly soils, streambanks, roadsides; 600–3000 m. KR, NCoRH, CaRH, SNH, TR, PR, Wrn, n W&I; to w Can, MT, WY, n Mex.

E. halleanum Hausskn. (p. 801) Per < 6 dm, gen sparsely strigose with minutely strigose lines; winter bulblets small, underground. **LF** 5–45 mm, ovate below (narrower above), ciliate, obtuse to subacute; petioles 0–1.5 mm. **INFL** ± nodding, gen ± glandular. **FL**: hypanthium 0.5–1.7 mm; sepals 1.2–2.8 mm; petals 1.6–5.5 mm, white to pink. **FR** 24–60 mm, sometimes hairy; pedicel 10–40 mm. **SEED** 1.1–1.6 mm, papillate. $2n$=36. Moist meadows, streambanks; 600–3200 m. KR, NCoR, CaR, SNH, SnFrB, TR, PR, MP; to w Can, SD, Colorado, AZ. [*E. brevistylum* var. *ursinum* (Trel.) Jepson; *E. pringleanum* Hausskn. incl var. *tenue* (Trel.) Munz]

E. hornemannii Reichb. ssp. ***hornemannii*** (p. 801) Per 1–4.5 dm, loosely clumped, ± strigose in lines (esp above); stolons short, leafy. **LF** 15–55 mm, lanceolate to ovate (narrower above), gen ± glabrous, obtuse to acute; petiole 0–8 mm. **INFL** glandular. **FL**: hypanthium 1–2.2 mm; sepals 2–4.5 mm; petals 3–9 mm, (white) pink to rose-purple. **FR** 40–65 mm, hairy; pedicel 5–15 mm. **SEED** 0.9–1.2 mm, papillate. $2n$=36. Moist meadows, streambanks; 1500–3400 m. KR, NCoRH, CaRH, SNH, SnBr, SnJt, Wrn, n W&I; ± circumboreal.

E. howellii P. Hoch Per < 2 dm, loosely clumped; stolons short, thread-like, minutely leafy. **ST** densely glandular. **LF** sessile, 4–20 mm, round to lanceolate (narrower above), sparsely strigose; tip round to obtuse. **INFL** glandular (ovaries sparsely so). **FL**: bud gen nodding; hypanthium 0.4–0.8 mm; sepals 1.5–2 mm; petals 2–3 mm, white; stigma head-like. **FR** 35–45 mm, nearly glabrous; pedicel 25–40 mm. **SEED** 0.8–1.1 mm, papillate. $2n$=36. UNCOMMON. Wet meadows, mossy seeps; 2000–2700 m. n&c SNH (Yuba Pass, Sierra Co.).

E. lactiflorum Hausskn. (p. 801) Per 1.5–5 dm, gen clumped, minutely strigose in lines (esp above); stolons short, leafy. **LF** 20–55 mm, narrowly ovate to narrowly lanceolate (narrower above), ± ciliate, obtuse to acute; petioles ± winged, 0–12 mm. **INFL** nodding to erect, glandular. **FL**: hypanthium 1–2.2 mm; sepals 2–3.6 mm; petals 3–6.8 mm, white (pink). **FR** 50–100 mm, hairy; pedicel 20–45 mm. **SEED** 1.1–1.6 mm, netted. $2n$=36. Moist meadows, streambanks, talus; 1400–3300 m. KR, CaRH, SNH, n W&I; to AK, MT, Colorado, also e N.Am, n Eurasia.

E. latifolium L. (p. 801) Per < 7 dm, clumped from woody caudex, ± glaucous to densely strigose. **LVS** alternate, 10–100 mm, elliptic to widely lanceolate, glabrous or hairy on margins and veins, subobtuse to acute; petiole 0–3 mm. **INFL** gen ± strigose; bracts lf-like. **FL**: bud nodding; hypanthium 0 (exc as greenish disk); sepals 10–24 mm; petals 13–32 mm, (white) pink to rose-purple, entire; stamens subequal, ± = pistil, maturing before stigma; pollen grayish, shed singly; stigma 4-lobed. **FR** 30–105 mm, subglabrous to gray-hairy; pedicel 5–18 mm. **SEED** 1.3–2.4 mm, irregularly netted; hairs persistent. $2n$=36(CA),72. Gravel bars, talus slopes, glacial outwashes; 2500–3100 m. n&c SNH; ± circumboreal. ❀DRN,IRR:1,2,4,6,7,14–16;DFCLT.

E. leptophyllum Raf. Per 1–10 dm, ± densely strigose; stolons thread-like, tipped with fleshy, winter bulblets. **LF** subsessile, 20–75 mm, linear to narrowly elliptic. **INFL** ± glandular. **FL**: hypanthium 0.8–1.5 mm; sepals 2.5–4.5 mm; petals 3.5–7 mm, white to pink. **FR** 35–80 mm, gray-hairy; pedicel 1–4 mm. **SEED** 1.5–2.2 mm, papillate; hairs persistent. $2n$=36. Uncommon. Boggy meadows, damp places; 2000 m. s SNE; to B.C., MT, Colorado; native to e N.Am.

E. luteum Pursh (p. 801) YELLOW WILLOWHERB Per 1–8 dm, loosely clumped, ± strigose in decurrent lines; stolons short and leafy or compact. **LF** 25–80 mm, spoon-shaped to ovate, acute to acuminate; veins conspicuous; petiole 1–3 mm. **INFL** nodding in bud, densely glandular and minutely strigose. **FL**: hypanthium 1.2–3 mm; sepals 10–12 mm; petals 12–22 mm, cream-yellow; stamens < pistil; stigma 4-lobed. **FR** 35–75 mm, hairy; pedicels 10–22 mm. **SEED** 1–1.2 mm, netted; hairs reddish, persistent. $2n$=36. RARE in CA. Moist streambanks, montane meadows; ± 1500 m. n KR; to AK.

E. minutum Lehm. (p. 801) Ann < 4 dm, subglabrous to strigose, ± peeling below. **LVS** alternate to opposite, 5–20 mm, narrowly elliptic to spoon-shaped, subglabrous; petiole 0–5 mm. **INFL** strigose. **FL**: hypanthium 0.5–1 mm; sepals 2–4 mm; petals 2–5, white or pink; stigma sometimes barely lobed. **FR** 15–25 mm, subglabrous; pedicels 1–7 mm. **SEED** 0.8–1.2 mm, netted. $2n$=26. Dry, open, disturbed areas, vernal pools, often after fire; 100–1600 m. NW, CaR, n&c SN, ScV, CW, MP; to B.C., MT, w NV.

E. nivium Brandegee (p. 801) SNOW MOUNTAIN WILLOWHERB Subshrub 1–2.5 dm, from woody caudex, grayish strigose. **LVS** sometimes clustered, 8-18 mm, elliptic-lanceolate, densely spreading-hairy; petiole 0–3 mm. **INFL** densely spreading-hairy. **FL**: hypanthium 5–8 mm; sepals 3–5 mm; petals 6–10 mm, rose-purple; stamens < pistil; stigma 4-lobed. **FR** 8–16 mm, subfusiform, glandular; pedicels 2–4 mm. **SEEDS** 2 per chamber, 1.5–2.4 mm, ± obovoid, low-papillate. $2n$=30. RARE. Dry talus, shaly slopes, 1500–2500 m. s NCoRH. In cult.

E. obcordatum A. Gray (p. 801) Per < 1.5 dm, cespitose from woody caudex, ± subglabrous, ± glaucous. **LF** 6–20 mm, elliptic to nearly round; tip obtuse to round; petiole 0–3 mm. **FL**: hypanthium 3.3–6 mm; sepals 5–15 mm; petals 13–24 mm, pink to rose-purple; stamens < pistil; stigma 4-lobed. **FR** 20–40 mm, widely club-like, densely glandular; pedicel 3 mm. **SEED** 1.4–2.1 mm, low-papillate. $2n$=36. Rocky places; 1900–4000 m. CaRH, SNH, MP; to e OR, c ID, ne NV. ❀DRN,IRR:1,2,4–7,14–17;DFCLT.

E. oreganum E. Greene (p. 801) OREGON FIREWEED Per 4–10 dm from leafy, basal shoots, ± glabrous, ± glaucous. **LF** 30–90 mm, narrowly lanceolate to narrowly ovate; veins reddish, conspicuous; petiole 1–3 mm. **INFL** sparsely strigose. **FL**: hypanthium 2.4–3 mm; sepals 6–10 mm; petals 10–15 mm, pink to rose-purple; stamens < pistil; stigma 4-lobed. **FR** 25–45 mm, hairy and glandular; pedicel 3–6 mm. **SEED** 0.9–1.3 mm, papillate-ridged. $2n$=36. RARE in CA. Bogs, small streams, ditches; 500–1600 m. KR, NCoRO; sw OR. [*E. exaltatum* E. Drew] Often confused with *E. ciliatum*, esp ssp. *glandulosum*.

E. oregonense Hausskn. (p. 801) Per < 4 dm, often matted, delicate, (sub)glabrous, ± purplish above; stolons sprawling, thread-like, with minute rounded lvs. **LF** sessile, clasping, 5–25 mm, round (below) to linear (above); tip obtuse to round. **INFL** very open (bracts << internodes), sparsely glandular. **FL**: hypanthium 0.8–1.5 mm; sepals 1–4 mm; petals 2–8 mm, white to rose-purple; stigma sometimes cylindric. **FR** 20–50 mm; pedicel 20–65 mm, gen > fr. **SEED** 1–1.4 mm, papillate. $2n$=36. ± boggy areas; 1700–3600 m. KR, CaRH, SNH, SnBr, SnJt, Wrn; to WA, w MT, WY.

E. pallidum (Eastw.) P. Hoch & Raven (p. 801) Ann 1–6 dm, peeling below, strigose or spreading-hairy. **LVS** opposite only near base, subsessile, 10–50 cm, narrowly elliptic, glabrous (upper hairy). **INFL** crowded, sometimes glandular; bracts = lvs. **FL**: hypanthium 1–2.5 mm; sepals 2–6 mm; petals 4–11 mm, rose-purple; stamens < or = pistil; stigma 4-lobed. **FR** 14–19 mm, subcylindric, beaked, pliable, sessile, dehiscent to base; septa and axis disintegrating. **SEEDS** in 1 staggered row, 1.8–2 mm, netted, glabrous. $2n$=20. Streambanks, moist slopes; 60–1800 m. CaR, n SNF, ScV, MP; to sw OR, ID. [*Boisduvalia p.* Eastw.; *B. macrantha* A.A. Heller]

E. palustre L. Per < 8 dm, ± strigose (esp above); stolons thread-like, sparsely leafy, tipped with compact, fleshy bulblets. **LF** subsessile, 15–70 mm, sublinear to elliptic. **INFL** ± nodding to erect, densely strigose, nonglandular. **FL**: hypanthium 0.6–1.8 mm; sepals 1.4–4.5 mm; petals 2–9 mm, white (pink); stigma sometimes cylindric. **FR** 30–90 mm, hairy; pedicel 15–60 mm. **SEED** 1.4–2.2 mm, finely papillate; hairs persistent. $2n$=36. Wet meadows, seeps, boggy areas, disturbed wet areas; ± 2200 m. c SNH (Grass Lake, El Dorado Co.); circumboreal.

E. pygmaeum (Speg.) P. Hoch & Raven (p. 801) Ann < 5.5 dm, gen with decumbent basal branches, minutely strigose or spreading-hairy above, glabrous and peeling below. **LVS** opposite only near base, subsessile, 8–35 mm, narrowly lanceolate to narrowly ovate. **INFL** crowded; bracts narrowly ovate or wider. **FLS** gen cleistogamous; hypanthium 0.3–1 mm; sepals 0.7–2 mm; petals 1–3 mm, pink; stigma sometimes barely 4-lobed. **FR** sessile, 5–8 mm, cylindric, tough, ± indehiscent below; axis disintegrating. **SEEDS** in 2 rows per chamber, 1–1.3 mm, netted, glabrous. $2n$=30. Vernal pools, clay mud flats; < 2500 m. CA-FP (exc ChI, e TR), MP; to w Can, ND, UT, n Baja CA. [*Boisduvalia glabella* (Nutt.) Walp. incl var. *campestris* (Jepson) Jepson]

E. rigidum Hausskn. (p. 801) SISKIYOU MOUNTAINS WILLOWHERB Per 1–4 dm, cespitose from woody caudex, sometimes decumbent (shoots wiry), glabrous and glaucous to densely strigose or hairy (esp above), nonglandular. **ST** peeling below. **LF** 20–45 mm, gen ± (ob)ovate; petiole 2–6 mm. **FL**: hypanthium 1–1.8 mm; sepals 9–15 mm; petals 16–20 mm, pink; stamens < pistil; stigma 4-lobed. **FR** 20–35 mm, hairy, club-like; pedicel 8–12 mm. **SEED** 2.5–3.4 mm, papillate. $2n$=36. UNCOMMON. Dry, open places, dry stream beds, sometimes on serpentine-like soils; 150–1200 m. nw KR (Del Norte Co.); sw OR. ❀DFCLT.

E. saximontanum Hausskn. (p. 801) Per < 6 dm, from fleshy, scaly, underground shoots, ± strigose in decurrent lines. **LF** subsessile, gen clasping, 10–65 mm, narrowly elliptic to (ob)ovate. **INFL** glandular. **FL**: hypanthium 0.8–1.4 mm; sepals 1.2–3.5 mm; petals 2.2–7 mm, white (pink). **FR** 20–70 mm, hairy; pedicels 0–5 mm. **SEED** 1–1.8 mm, low-papillate. $2n$=36. Moist montane meadows, streambanks, ± disturbed roadsides; 1400–3500 m. c&s SNH, SNE; to e OR sw Can (also e Can), MT, NM.

E. septentrionale (Keck) Raven (p. 801) HUMBOLDT COUNTY FUCHSIA Per < 2 dm, matted, wiry, scaly below, from ± woody caudex, sometimes decumbent, densely white-canescent. **LVS** ± sessile, 10–25 mm, lanceolate to ovate; upper bract-like, ± green, glandular and spreading-hairy. **INFL** densely glandular. **FL** red-orange; hypanthium 17–23 mm; sepals 8–12 mm; petals 13–15 mm; stamens << pistil; stigma 4-lobed. **FR** 20–25 mm, ± beaked, hairy; pedicels 0–2 mm. **SEED** 1.9–2.5 mm, low-papillate. $2n$=30. UNCOMMON. Dry, sandy or rocky ledges; 100–1800 m. NCo, NCoRO. [*Zauschneria septentrionalis* Keck; *E. canum* ssp. *s.* (Keck) Raven] Bird-pollinated. ❀DRN,IRR:**5,17**&SHD:7,14, **15,16**,19–24.

E. siskiyouense (Munz) P. Hoch & Raven (p. 801) SISKIYOU FIREWEED Subshrub 1–2.5 dm, clumped, with scaly shoots from woody caudex, ± glabrous to ± densely strigose and glandular. **LF** sessile, 13–26 mm, widely lanceolate to ovate, ± sparsely strigose. **INFL** ± densely glandular. **FL**: hypanthium 2–4 mm; sepals 5–10 mm; petals 11–20 mm, pink to rose-purple; stamens < pistil; stigma 4-lobed. **FR** 25–45 mm, hairy; pedicel 6–25 mm. **SEED** 1.4–1.9 mm, papillate. $2n$=36. RARE in CA. Scree, serpentine ridges; 1700–2500 m. n KR; sw OR. [*E. obcordatum* A. Gray ssp. *s.* Munz & var. *laxum* Jepson]. In cult.

E. torreyi (S. Watson) P. Hoch & Raven (p. 801) Ann 1–6.2 dm, grayish spreading-hairy, peeling below. **LVS** opposite only near base (where glabrous), subsessile, 5–45 mm, ± linear-lanceolate, hairy. **INFL** glandular. **FLS** gen cleistogamous; hypanthium 0.4–1 mm; sepals 0.7–2 mm; petals 1–3 mm, pink or white; stigma rarely 4-lobed. **FR** sessile, 8–13 mm, cylindric, beaked, flexible; axis disintegrating. **SEED** 1–1.5 mm, netted; hairs 0. $2n$=18. Streambanks, moist slopes; < 2600 m. NW (exc NCo), CaR, SN, GV, SnFrB, SCoRO, MP, sw DMoj; to B.C., ne NV. [*Boisduvalia stricta* (A. Gray) E. Greene]

GAURA

Ann, bien, per, from woody caudex, rhizome, or taproot. **LVS** basal and cauline, alternate, sessile; margin gen wavy-dentate. **INFL**: spike, terminal, bracted. **FL** gen bilateral, opening at dusk or dawn; sepals gen 4, gen widely opening; petals gen 4, white or yellow, often fading reddish or purplish; stamens 8, filaments gen with paired teeth at base, anthers attached at middle; ovary chambers gen 4 (in fr 1), stigma deeply lobed, gen elevated above anthers (pl then cross-pollinated). **FR** indehiscent, ± erect, nut-like, gen 4-angled or -winged; walls woody; base stalk-like or not. **SEEDS** gen 3–4, gen 2–3 mm, ovoid, gen flat-sided, yellowish to pale brown. 21 spp.: temp N.Am (esp TX), C.Am. (Greek: proud, from showy fls of some) [Raven & Gregory 1972 Mem Torrey Bot Club 23:1–96]

1. Fr ± linear; stalk-like base 2–8 mm, slender (sect. *Stipogaura*) . . . . ***G. sinuata***
1′ Fr fusiform to ovoid; stalk-like base 0–3 mm, thick if present
  2. Sepals 2–3.5 mm, barely opening; ann; stalk-like base 0 (sect. *Schizocarya*) . . . . ***G. parviflora***
  2′ Sepals 5–14 mm, widely opening; per; stalk-like base short, thick
    3. Fr base > 1/2 diam of widest part, fr becoming wider ± gradually; caudex woody, branched below ground (sect. *Campogaura*) . . . . ***G. coccinea***
    3′ Fr base ± 1/4 diam of widest part, fr with a conspicuous, abrupt bulge near middle; rhizomed (sect. *Xenogaura*) . . . . ***G. drummondii***

G. coccinea Pursh (p. 801) WILD HONEYSUCKLE, LINDA TARDE Per, gen minutely strigose and with long spreading hairs, or ± glabrous; caudex woody, branched below ground. **ST** 10–120 cm. **LF** 10–70 mm, linear to narrow-elliptic, entire to coarsely wavy-serrate. **INFL**: bracts 2–5 mm. **FL**: hypanthium 4–13 mm; sepals 5–10 mm; petals 3–8 mm. **FR** erect or spreading, 4–9 mm, 4-angled; stalk-like base short, thick, > 1/2 diam of widest part. **SEED** 1.5–3 mm. $2n$=14,42,56. Dry slopes, gen limestone, Joshua-tree or pinyon/juniper woodland; 900–1600 m. DMtns (naturalized in Teh, SW); to w Can, c US, Mex. [var. *glabra* (Lehm.) Torrey & A. Gray] Though native, may become a NOXIOUS WEED. ❀TRY;INV.

G. drummondii (Spach) Torrey & A. Gray (p. 801) Per, forming large mats, rhizomed; minutely strigose and gen with long, spreading hairs. **STS** 1–several, 2–6(12) cm. **LF** 5–95 mm, narrowly lanceolate to elliptic, subentire to shallowly and coarsely wavy-dentate. **INFL**: bracts 2–8 mm. **FL**: hypanthium 4–14 mm; sepals 7–14 mm; petals 6–10 mm. **FR** erect, 7–13 mm, ovate, 4-angled; stalk-like base short and thick, ± 1/4 diam of widest part, with a conspicuous, abrupt bulge near middle. **SEEDS** 2–8, 2–2.5 mm. $2n$=28. Cult fields; < 300 m. CW, SW; native c TX to c Mex. [*G. odorata* Lagasca misapplied] NOXIOUS WEED.

G. parviflora Douglas (p. 801) LIZARD-TAIL, VELVET WEED Ann, densely short-glandular-hairy, sparsely long-spreading-hairy (lvs also minutely strigose). **ST** erect, 20–200(300) cm; branches gen 0 or below infl. **LVS**: cauline 20–125 mm, narrowly elliptic to narrowly ovate, slightly wavy-dentate. **INFL**: bracts 1.5–5.5 mm. **FL**: hypanthium 1.5–5 mm; sepals 2–3.5 mm, barely opening; petals 1.5–3 mm, stigma ± surrounding anthers. **FR** gen reflexed, 5–11 mm, fusiform, 4-angled, 8-ribbed below; stalk-like base 0. **SEED** 2–3 mm. $2n$=14. Cult fields, pastures, waste places, streambanks; < 400 m. SnFrB, SW; probably native to c US, naturalized widely. [var. *lachnocarpa* Weath.] Self-pollinating.

G. sinuata Ser. (p. 801) WAVY-LEAVED GAURA Per, forming large mats, rhizomed; hairs gen sparse or 0. **ST** 20–60 cm, branched, sparsely minutely strigose and with long, spreading hairs. **LF** 10–110 mm, linear to narrow-oblanceolate, slightly wavy-dentate. **INFL**: bracts 1–5 mm. **FL**: hypanthium 2.5–5 mm; sepals 7–14 mm; petals 7–14.5 mm. **FR** erect, 4–16 mm, ± linear, narrowly 4-winged; stalk-like base 2–8 mm, slender, tapered. **SEEDS** 1–4, 2–3 mm. $2n$=28. Light sandy loam of cult fields; < 1000 m. GV, CW, SW, DMoj; native to OK, TX; widely naturalized, esp se US. [*G. villosa* Torrey (incl var. *mckelveyae* Munz) misapplied] NOXIOUS WEED, limited by self-sterility.

GAYOPHYTUM

Harlan Lewis

Ann. **ST** gen erect, < 1 m, slender; hairs 0 to dense, rarely glandular. **LVS** cauline, alternate (or lowest subopposite), entire, petioled or not, narrow-lanceolate. **INFL**: fls axillary, pedicelled or not, opening at dawn. **FL**: hypanthium inconspicuous; sepals 4, staying fused in 2's or all coming free; petals 4, 0.5–8 mm, white, with 1–2 yellow or greenish spots at base, fading pink or red; stamens 8, those opposite sepals larger, pollen ± yellow; ovary chambers 2, stigma gen not beyond anthers, gen touching them, gen ± spheric. **FR**: capsule, ± cylindric or flat; valves 4, all gen coming free, gen equal. **SEEDS** few–many, gen all maturing, gen appressed to septum, alternate or subopposite between chambers, in each chamber gen in 1 row and gen not overlapped, 0.5–2.3 mm, ovoid, glabrous or hairy, brown or gray mottled with brown; appendages 0. ± 9 spp.: w N.Am, 2 S.Am. (C. Gay, French author of Flora of Chile, 1800–1873) [Lewis & Szweykowski 1964 Brittonia 16:343–391] Self-compatible; taxa with petals < 3 mm self-pollinated.

1. Petals 3–8 mm; stigma beyond anthers (see also *G. heterozygum*)
 2. Petals 4–8 mm; SN (s of Placer Co.) .... ***G. eriospermum***
 2′ Petals 3–7 mm; CaR, SnBr, Wrn, n SNH (Plumas Co.) .... ***G. diffusum*** ssp. ***diffusum***
1′ Petals 0.5–3 mm; stigma not beyond anthers
 3. Seeds ± 50% aborted; fr very irregular, lumpy .... ***G. heterozygum***
 3′ Seeds all maturing; fr not irregular, slightly to very knobby
  4. Fr flat; pl gen branched only below, 2° branches 0–few
   5. Lateral fr valves staying attached; petals 0.8–1.5 mm; larger stamens ± 0.7 mm .... ***G. humile***
   5′ All fr valves coming free; petals 1.3–1.8 mm; larger stamens 0.8–1.5 mm .... ***G. racemosum***
  4′ Fr ± not flat; pl branched only above to throughout, 2° branches gen many
   6. Seeds in each chamber overlapped, gen in 2 rows
    7. Petals 0.7–1.5 mm; pedicel gen > fr .... ***G. ramosissimum***
    7′ Petals 1.5–3 mm; pedicel gen < fr .... ***G. diffusum*** ssp. ***parviflorum***
   6′ Seeds in each chamber not overlapped, in 1 row
    8. Seeds > 9, subopposite; fr slightly knobby
     9. Branches ± throughout, gen 2–8 nodes between .... ***G. decipiens***
     9′ Branches at base or not, many above, gen 0–1 node between .... ***G. diffusum*** ssp. ***parviflorum***
    8′ Seeds (1)3–9, gen alternate; fr very knobby
     10. Seeds 6–9; pedicel gen < fr .... ***G. diffusum*** ssp. ***parviflorum***
     10′ Seeds (1)3–5; pedicel ± = fr .... ***G. oligospermum***

G. decipiens Harlan Lewis & J. Szweykowski **ST** < 50 cm; branches ± throughout, gen 2–8 nodes between. **LVS** 1–3 cm, gen ± reduced above. **INFL**: 1st fl gen 1–5 nodes above base. **FL**: petals 1.1–1.8 mm; larger stamens 0.8–1.5 mm; ovary hairy. **FR** 6–8 mm, > pedicel, ± not flat, slightly knobby. **SEEDS** 10–25, subopposite, glabrous to dense-puberulent. $2n=14$. Pinyon/juniper woodland, pine or fir forest; 1800–4200 m. SNH, SnGb, SnBr, Wrn, W&I, DMtns (Panamint Mtns); ± w US.

G. diffusum Torrey & A. Gray **ST** < 60 cm; branches at base or not, gen forked above. **LVS** 1–6 cm, gen reduced above. **INFL**: 1st fl 1–20 nodes above base. **FL**: petals 1.2–7 mm; larger stamens 0.9–6 mm; ovary hairy, stigma beyond anthers or not, hemispheric or not. **FR** 3–15 mm, sessile or gen > pedicel, cylindric, slightly to very knobby. **SEEDS** 3–18, alternate or subopposite, in each chamber sometimes in 2 rows and overlapped, glabrous to densely puberulent. $2n=28$. Common. Open montane forest, sagebrush scrub; 800–3700 m. CA; w US to B.C., Baja CA. Complex from several $2n=14$ spp.; sspp. may intergrade locally.

ssp. ***diffusum*** **FL**: petals 3–7 mm; larger stamens 2–6 mm; stigma beyond anthers, hemispheric. Uncommon in CA. Habitats, elevations of sp. CaR, SnBr, Wrn, n SNH (Plumas Co.); range of sp., esp nw US.

ssp. ***parviflorum*** Harlan Lewis & J. Szweykowski (p. 801) **FL**: petals 1.2–3 mm; larger stamens 0.9–2 mm; stigma not beyond anthers, ± spheric. **FR**: seeds in each chamber sometimes in 2 rows and overlapped. Common. Habitats, elevations of sp. NW, CaR, SN, TR, PR, GB. Variable; most small-fld pls assigned by Munz, others to *G. nuttallii* belong here; may occur with any member of genus.

G. eriospermum Cov. **ST** < 1 m; branches 0 near base, forked above. **LVS** 2–7.5 cm, much reduced above. **INFL**: 1st fl gen 10–20 nodes above base. **FL**: petals 4–8 mm; larger stamens 3–7 mm; ovary hairy, stigma beyond anthers, hemispheric. **FR** 4–16 mm, > pedicel, slightly knobby. **SEEDS** 4–12, alternate, glabrous to densely puberulent. $2n=14$. Open montane forest; 100–3000 m. SN (s of Placer Co.).

G. heterozygum Harlan Lewis & J. Szweykowski (p. 801) **ST** < 80 cm; branches gen 0 at base, forked above. **LVS** 1.5–6 cm, much reduced above. **INFL**: 1st fl gen 10–20 nodes above base. **FL**: petals gen 2–3 mm; larger stamens 2–4 mm, pollen ± 50% aborted; ovary hairy, stigma hemispheric or not. **FR** 6–15 mm, > pedicel, very irregular, lumpy. **SEEDS** 2–10 maturing, ± 50% aborted, irregularly alternate, glabrous to densely puberulent. $2n=14$. Open montane forest; 800–3000 m. KR, NCoRH, CaRH, SNH, Teh, SCoRO, TR, PR, Wrn; to WA, NV. Gen self-pollinated; stable hybrid, probably between *G. eriospermum, G. oligospermum.*

G. humile A.L. Juss. (p. 801) Pl herbage gen glabrous. **ST** < 30 cm; branches 0 or gen below, not forked, tips ± glandular-puberulent or not. **LVS** 1–2.5 cm, little reduced above. **INFL**: 1st fl 1–3 nodes above base. **FL**: petals 0.8–1.5 mm; larger stamens ± 0.7 mm; ovary glabrous or glandular-puberulent. **FR** 8–17 mm, sessile, ± like lvs, flat, grooved along midline, smooth or slightly knobby; lateral 2 valves staying attached, wider. **SEEDS** 24–50, ascending, subopposite, glabrous. $2n=14$. Drying margins of wet sites, snow beds; 800–3000 m. NW, CaR, SN, SCoR, TR, Wrn; ± w US, S.Am. [*G. nuttallii* Torrey & A. Gray]

G. oligospermum Harlan Lewis & J. Szweykowski Pl herbage gen glabrous. **ST** < 70 cm; branches gen 0 at base, forked above, tips sometimes barely hairy. **LVS** 1.5–6 cm, much reduced above. **INFL**: 1st fl gen 10–20 nodes above base. **FL**: petals 1.5–2.5 mm; larger stamens 1.2–1.5 mm; ovary puberulent. **FR** 4–9 mm, ± = pedicel, very knobby. **SEEDS** (1)3–5, alternate, glabrous. $2n=14$. Open pine forest; 1300–2800 m. TR, PR.

G. racemosum Torrey & A. Gray **ST** < 40 cm; branches gen below, not forked. **LVS** 1–2.5 cm, not much reduced above. **INFL**: 1st fl 1–3 nodes above base. **FL**: petals 1.3–1.8 mm; larger stamens 0.8–1.5 mm; ovary hairy or not. **FR** 10–15 mm, > pedicel, flat, ± grooved along midline, smooth or slightly knobby; lateral 2 valves not attached, wider. **SEEDS** 10–35, appressed or above ascending, subopposite, glabrous or densely puberulent. $2n=28$. Wet meadows, drying margins; 1000–4000 m. CaRH, SNH, Teh, WTR, Wrn; to WA, Alberta, Colorado. Probably from hybrids between *G. decipiens, G. humile.*

G. ramosissimum Torrey & A. Gray Pl herbage ± glabrous. **ST** < 50 cm; branches throughout, forked exc at base. **LVS** 1–4 cm, much reduced above. **INFL**: 1st fl 5–15 nodes above base. **FL**: petals 0.7–1.5 mm; larger stamens ± 0.5 mm; ovary glabrous or puberulent. **FR** 3–9 mm, gen < pedicel, cylindric, slightly knobby. **SEEDS** 10–30, in each chamber in 2 rows, overlapped, glabrous. 2*n*=14. Sagebrush scrub; 500–3000 m. CaRH, n SN, GB; ± w US.

LUDWIGIA FALSE LOOSESTRIFE, WATER PRIMROSE

Peter C. Hoch

Ann to subshrub, sometimes floating or rooting at nodes. **LVS** alternate to opposite, simple; stipules gen deciduous. **INFL**: spike; fls 1 per bract. **FL** radial; hypanthium 0; sepals 4–5(7), persistent; petals (0)4–5(7), white to yellow; stamens 4 or 10 (12), pollen gen shed singly in CA; stigma club-shaped to spheric. **FR** dehiscing irregularly; wall thick or thin. **SEEDS** free or embedded in fr wall. 82 spp.: ± worldwide. (C.G. Ludwig, German botanist & physician, 1709–1773) [Raven 1963 Reinwardtia 6:327–427] Many polyploids.

1. Lvs opposite; sepals 4; petals 0 or 4, 1–3 mm, soon deciduous; stamens 4; frs 1–10 mm, erect; seeds free from fr wall (sect. *Dantia*)
  2. Petals 0; sepals 1–2 mm; fr 1–5 mm; pedicels 0–0.5 mm; ovary with 4 green stripes . . . . ***L. palustris***
  2′ Petals 1–3 mm; sepals ± 2–5 mm; fr 4–10 mm; pedicels 0–3 mm; ovary not striped . . . . ***L. repens***

1′ Lvs alternate; sepals 5(6); petals 5(6), 7–30 mm; stamens 10(12) in 2 unequal sets; fr 10–40 mm, reflexed; seeds embedded in fr wall (sect. *Oligospermum*)
  3. St gen ascending or erect (sometimes floating or creeping); bractlets obovate; petioles 3–23 mm; sepals (8)13–19 mm; petals (15)20–30 mm . . . . ***L. hexapetala***
  3′ St gen floating or creeping; bractlets deltate; petioles 3–60 mm; sepals 3–12 mm; petals 7–24 mm . . . . ***L. peploides***
    4. Pls spreading-hairy; lf tip glandular; fr 25–40 mm . . . . ssp. ***montevidensis***
    4′ Pls gen subglabrous; lf tip not glandular; fr 10–25 mm . . . . ssp. ***peploides***

L. hexapetala (Hook. & Arn.) Zardini, Gu, & Raven (p. 807) Per or subshrub, sometimes floating or rooting at nodes. **ST** 2–20 dm, prostrate to erect, simple or branched above, glabrous to spreading-hairy. **LVS** 1–11 cm, alternate; blade narrowly elliptic to widely obovate, entire, subglabrous. **FL**: sepals 5(6), 8–19 mm; petals 5(6), 15–30 mm; stamens 10(12) in 2 unequal sets, anthers 1.5–4 mm. **FR** reflexed, falling with 10–60 mm pedicel; body 12–30 mm, cylindric, tapered to pedicel, ± hairy. **SEED** 1.2–1.5 mm, embedded in inner fr wall. 2*n*=80. Swamps, lake margins; < 300 m. NCo, s NCoRO, CCo, SnFrB, SCo; to WA, se US, s S.Am, introduced into Eur. [*Jussiaea uruguayensis* Cambess.]

L. palustris (L.) Elliott (p. 807) Per, matted or floating. **ST** 1–5 dm, prostrate or ascending, rooting at nodes, well branched, subglabrous. **LVS** < 5 cm, opposite; blade narrowly elliptic to subovate, entire, subglabrous. **FL**: sepals 4, 1.1–2 mm; petals 0; stamens 4, anthers 0.2–0.4 mm. **FR** erect; pedicel 0–0.5 mm; body (1.5)2–5 mm, ± oblong, minutely strigose, with 4 green stripes. **SEED** 0.5–0.7 mm, free of fr wall. 2*n*=16. Roadside ditches, wet meadows, pond margins; < 1000 m. w NW, n SN, c SNH, SnJV, CCo, SnFrB, SCo; to B.C., e US, n S.Am, introduced ± worldwide. [vars. *americana* (DC.) Fern. & Griscom & *pacifica* Fern. & Griscom] Highly variable, weedy.

L. peploides (Kunth) Raven Per, matted, floating, or creeping. **ST** 1–30 dm, prostrate to erect, simple or branched. **LVS** < 10 cm, alternate, ± clustered; blade oblong to round, subentire, glabrous to spreading-hairy above. **FL**: sepals 5(6), 3–12 mm; petals 5(6), 7–24 mm; stamens 10(12) in 2 unequal sets, anthers 0.5–2.2 mm. **FR** reflexed; pedicel 6–90 mm; body cylindric, ± 5-angled, hard, subglabrous to spreading-hairy. **SEED** 1–1.5 mm, embedded in inner fr wall. 2*n*=16. Ditches, streambanks, lakeshores; < 900 m. NCo, NCoRO, SNF, GV, CCo, SnFrB, SCo, WTR, DMoj; to OR, se US, S.Am, Eurasia, Australia. [*Jussiaea repens* L. misapplied] May be serious wetland or agricultural weed.

ssp. *montevidensis* (Spreng.) Raven Pl ± spreading-hairy (exc st base), ± sticky when fresh. **LF**: tip glandular; lower petioles (5)8–16 mm. **FR**: body 25–40 mm. 2*n*=16. Lake shores, streambanks; < 500 m. s NCo, n&c SNF, GV, SCo; to se US, Eur, Australia; native to s S.Am. [*J. r.* var. *m.* (Spreng.) Munz misapplied]

ssp. ***peploides*** (p. 807) Pl (sub)glabrous. **LF**: tip gen not glandular; lower petioles 3–8 mm. **FR**: body 10–25 mm. 2*n*=16. Ditches, shores, streambanks; < 900 m. NCo, NCoRO, SNF, GV, CCo, SnFrB, SCo, WTR, sw DMoj; to OR, TX, S.Am, introduced into Australia. [*J. r.* var. *p.* (Kunth) Griseb. misapplied]

L. repens Forster Per, matted. **ST** 1–3 dm, decumbent, rooting at nodes, ± branched, subglabrous. **LVS** opposite, < 5 cm; blade narrowly elliptic to ± round, entire, subglabrous to densely and minutely strigose. **FL**: sepals 4, 1.8–5 mm; petals 4, 1–3 mm, yellow; stamens 4, anthers 0.4–0.9 mm, pollen shed ± in 4's. **FR** erect; pedicel 0–3 mm; body 4–10 mm, oblong to narrowly obconic, sometimes ± hairy. **SEED** 0.6–0.8 mm, free of fr wall. 2*n*=32. Muddy or sandy streambanks, ponds, ditches; < 900 m. SnBr, PR, sw DMoj; to se US, Caribbean, n Mex, introduced into s Asia, Japan. [*L. natans* Elliott var. *stipitata* Fern. & Griscom]. ❀INV.;STBL.

OENOTHERA EVENING PRIMROSE

Ann, bien, per, gen from taproot. **LVS** basal or cauline, alternate, gen pinnately toothed to lobed. **INFL**: spike, raceme-like, or fls in axils of upper, reduced lvs. **FL** radial, gen opening at dusk; sepals 4, reflexed in fl (sometimes 2–3 remaining adherent); petals 4, yellow, white, rose or ± purple, gen fading orangish to purplish, tip notched or toothed; stamens 8, anthers attached at middle; ovary chambers 4, stigma deeply lobed, gen > anthers and cross-pollinated (or ± = anthers and self-pollinated). **FR** cylindric to 4-winged, straight to curved, gen sessile (base sometimes seedless, stalk-like). **SEEDS** in gen 2 (1–3) rows per chamber, or clustered. 119 spp.: Am, some widely naturalized. (Greek: wine-scented) [Dietrich & Wagner 1988 Syst Bot Monogr 24:1–91] Many spp. self-pollinated; some of these have chromosome peculiarities (ring of 14 in meiosis) and ± 50% pollen fertility; they yield genetically ± identical offspring; noted with abbreviation "PTH" (permanent translocation heterozygote).

1. Petals white or rose to ± purple; seeds in 1 row per chamber or clustered (fr not tubercled) or in 2 rows per chamber (fr tubercled)
  2. Fr much wider toward tip; petals white, rose, or rose-purple; seeds clustered
    3. Hypanthium 4–8 mm; petals 5–10 mm; fr base tapered (sect. *Hartmannia*) . . . . ***O. rosea***

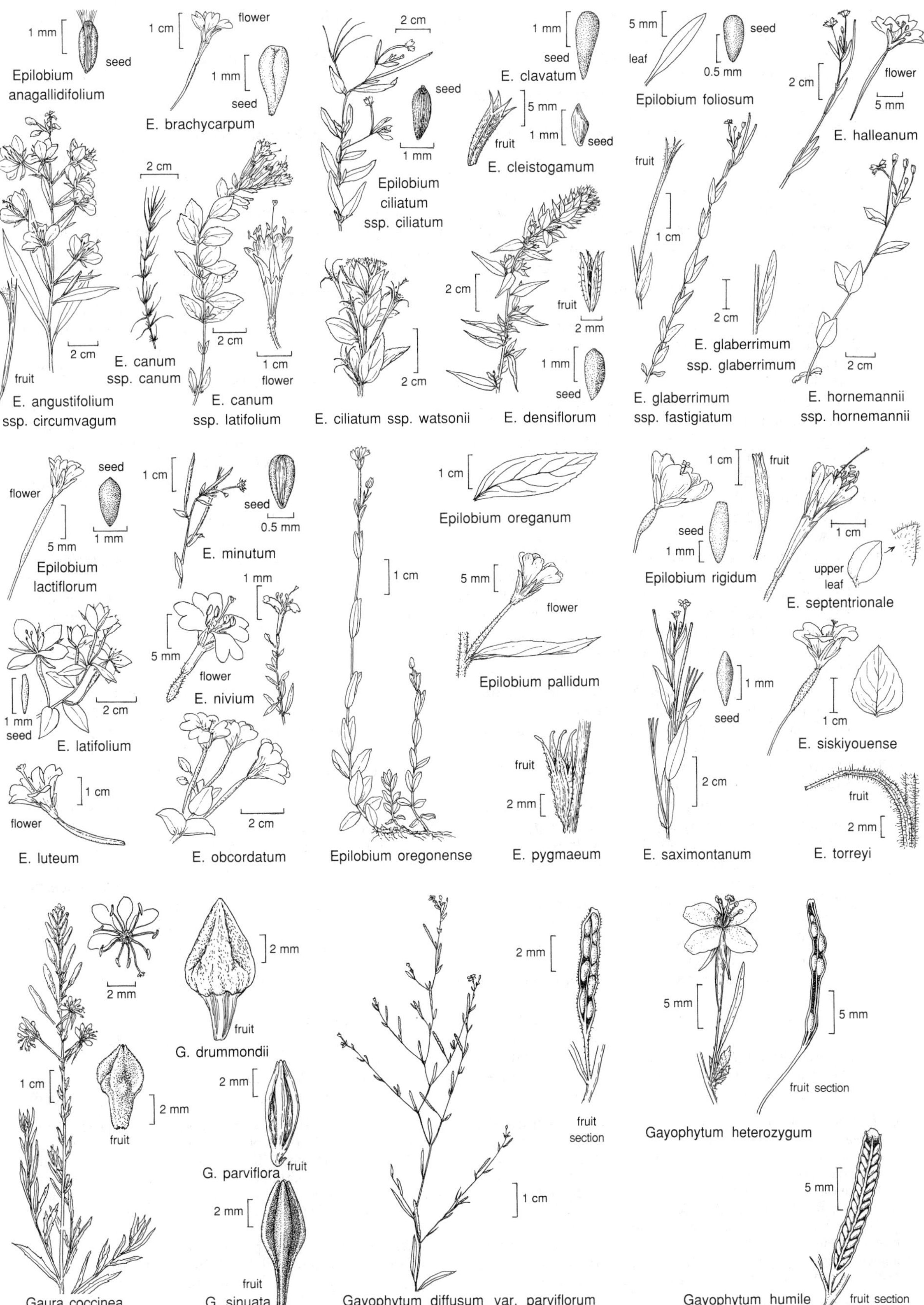
1 mm
seed
Epilobium anagallidifolium
1 cm
flower
1 mm
seed
E. brachycarpum
2 cm
seed
1 mm
Epilobium ciliatum ssp. ciliatum
1 mm
seed
E. clavatum
5 mm
1 mm
fruit
seed
E. cleistogamum
5 mm
leaf
seed
0.5 mm
Epilobium foliosum
2 cm
flower
5 mm
E. halleanum
2 cm
fruit
E. angustifolium ssp. circumvagum
2 cm
E. canum ssp. canum
2 cm
1 cm
flower
E. canum ssp. latifolium
2 cm
E. ciliatum ssp. watsonii
2 cm
fruit
2 mm
1 mm
seed
E. densiflorum
fruit
1 cm
2 cm
E. glaberrimum ssp. glaberrimum
E. glaberrimum ssp. fastigiatum
2 cm
E. hornemannii ssp. hornemannii
seed
flower
5 mm
1 mm
Epilobium lactiflorum
1 cm
seed
0.5 mm
E. minutum
1 cm
Epilobium oreganum
1 cm
fruit
seed
1 mm
Epilobium rigidum
1 cm
upper leaf
E. septentrionale
1 mm
5 mm
flower
E. nivium
1 cm
5 mm
flower
Epilobium pallidum
1 mm
seed
2 cm
E. latifolium
1 mm
seed
2 cm
1 cm
E. siskiyouense
1 cm
flower
E. luteum
2 cm
E. obcordatum
Epilobium oregonense
fruit
2 mm
E. pygmaeum
E. saximontanum
fruit
2 mm
E. torreyi
2 mm
2 mm
fruit
G. drummondii
1 cm
2 mm
fruit
2 mm
fruit
G. parviflora
2 mm
fruit
G. sinuata
Gaura coccinea
2 mm
fruit section
1 cm
Gayophytum diffusum var. parviflorum
5 mm
5 mm
fruit section
Gayophytum heterozygum
5 mm
Gayophytum humile
fruit section

3′ Hypanthium 10–23 mm; petals 25–40 mm; fr base cylindric (sect. *Xylopleurum*) .... ***O. speciosa***
2′ Fr narrower toward tip (exc slightly wider in *O. stricta*); petals white; seeds in rows
4. St not peeling; bud erect; fr 4–9 mm wide, tubercled; seeds in 2 rows per chamber, with an obvious cavity on 1 side (sect. *Pachylophus*) .... ***O. caespitosa***
5. Fr 10–34 mm, lanceolate to elliptic-ovate, gen S-shaped, stalk-like base 0–1 mm; petals fading rose to purple; margin of seed-cavity lobed .... ssp. ***crinita***
5′ Fr 25–68 mm, cylindric, ± straight, stalk-like base 0–55 mm; petals fading lavender to pink; margin of seed-cavity entire .... ssp. ***marginata***
4′ St peeling; bud nodding; fr 1.5–5 mm wide, not tubercled; seeds in 1 row per chamber, cavity 0 (sect. *Anogra*)
6. Per; new shoots gen from lateral roots; free sepal tips in bud 0 or < 1 mm .... ***O. californica***
7. St hairs dense, short, appressed and long, spreading, wavy; roots fleshy; new rosettes at st tips . ssp. ***eurekensis***
7′ St hairs 0 (to dense, short, appressed and long, spreading); roots not fleshy; new rosettes not forming at st tips
8. Cauline lvs gen ± pinnately lobed; pl ± grayish green; .... ssp. ***avita***
8′ Cauline lvs subentire to deeply wavy-dentate; pl green to slightly grayish .... ssp. ***californica***
6′ Ann or per; new shoots from lateral roots 0; free sepal tips in bud 0–1 mm (pl ann or short-lived per) or 1–9 mm (pl per) .... ***O. deltoides***
9. Grayish green per; free sepal tips in bud 1–9 mm .... ssp. ***howellii***
9′ Green ann or short-lived per; free sepal tips in bud 0–1 mm
10. Upper lvs pinnately lobed; st gen < 1 dm; petals 15–25(30) mm .... ssp. ***piperi***
10′ Upper lvs subentire to ± dentate, rarely pinnately lobed; st gen 2–10 dm; petals 15–40 mm
11. Spreading hairs ± 2 mm above; fr base 3–5 mm wide .... ssp. ***cognata***
11′ Spreading hairs 1–1.5 mm above; fr base 2–3 mm wide .... ssp. ***deltoides***
1′ Petals yellow; seeds gen in 2 rows per chamber (fr not tubercled, but hairs sometimes with red blister-like bases)
12. St erect; bien; fls in gen dense spikes; seeds angled, irregularly pitted (sect. *O.* subsect. *Oenothera*)
13. Hypanthium 60–135 mm .... ***O. longissima***
13′ Hypanthium 20–55 mm
14. Petals 25–52 mm; stigma well above anthers; pollen ± all fertile or ± 50% sterile (sterile grains small and shriveled at 10×)
15. Lf crinkled; seeds and pollen ± 50% sterile .... ***O. glazioviana***
15′ Lf ± flat, ± not crinkled; seeds and pollen ± all fertile .... ***O. elata***
16. Sepals green or red-flushed; red blister-like hair bases ± 0; anthers 8–15 mm .... ssp. ***hirsutissima***
16′ Sepals red-flushed; red blister-like hair bases obvious; anthers 12–23 mm .... ssp. ***hookeri***
14′ Petals 7–25(30) mm; stigma ± not above anthers; pollen ± 50% sterile
17. Petals 13–23 mm, gen < sepals; coastal .... ***O. wolfii***
17′ Petals 7–25(30) mm, ± = sepals; gen not near coast
18. Spike dense (internodes in fr < fr); longest hairs gen without red blister-like bases; petals 10–25(30) mm; sepals gen green or yellowish (reddish) .... ***O. biennis***
18′ Spike ± open (internodes in fr gen > fr); longest hairs gen with red blister-like bases; petals 7–20 mm; sepals often marked reddish .... ***O. villosa*** ssp. ***strigosa***
12′ St ± 0, decumbent, or erect; ann or per (exc *O. stricta* bien); fls few, in upper axils; seeds not both angled and irregularly pitted
19. Fr winged; seeds obliquely wedge-shaped (sect. *Lauvaxia*) .... ***O. flava*** ssp. ***flava***
19′ Fr not winged (sharply angled in *O. xylocarpa*); seeds ± round to oblanceolate
20. Fr lanceolate to ovate, 4–11 mm wide; seed 2.4–3.5 mm, ± wrinkled; pollen ± all fertile
21. Petals fading deep salmon-red; fr gen ± twisted; 1 side of seed with 2 small ridges; c&s SNH (sect. *Contortae*) .... ***O. xylocarpa***
21′ Petals fading reddish purple to orange; fr not twisted; 1 side of seed with thick, U-shaped area surrounding a groove; D (sect. *Eremia*) .... ***O. primiveris***
22. Lf grayish green; petals (22)29–40 mm; stigma above anthers .... ssp. ***bufonis***
22′ Lf gen green; petals 6–25(28) mm; stigma ± not above anthers .... ssp. ***primiveris***
20′ Fr cylindric (sometimes slightly wider toward tip), 2–4 mm wide; seed 1–1.8 mm, pitted; pollen ± 50% sterile
23. Bud erect; petal base gen with a red spot; fr slightly wider toward tip (sect. *O.* subsect. *Munzia*) .... ***O. stricta*** ssp. ***stricta***
23′ Bud nodding or curved upward; petal base unspotted; fr tapered to tip
24. Bud curved upward; sepal free tips in bud 0.3–3 mm (sect. *O.* subsect. *Raimannia*) .... ***O. laciniata***
24′ Bud nodding; sepal free tips in bud 0.1–1 mm (sect. *O.* subsect. *Nutantigemma*) .... ***O. pubescens***

O. biennis L. COMMON EVENING PRIMROSE Bien, rosetted; hairs glandular (esp infl), spreading (blister-like bases gen 0), and minutely strigose. **ST** erect, 3–20 dm. **LVS**: cauline 5–20 cm, oblanceolate to elliptic, subentire to dentate, ± lobed toward base. **INFL**: spike, dense. **FL**: hypanthium (20)25–40 mm; sepals 12–20(28) mm, green or yellowish, rarely reddish, free tips in bud, 1.5–3 mm; petals 10–25(30) mm, yellow fading duller or orange. **FR** 20–40 mm, 4–6 mm wide, narrowly lanceolate, ± straight. **SEED** 1–2 mm, angled, irregularly pitted. $2n$=14. Disturbed places; gen < 300 m. NW, CW, SW; native to c&e N.Am; ± worldwide weed.

O. caespitosa Nutt. FRAGRANT EVENING PRIMROSE Per, rosetted; caudex woody, new shoots gen from lateral roots; hairs glandular and sometimes also coarse and non-glandular. **ST** sprawling, < 2 dm, or ± 0. **LF** 1.7–36 cm, oblanceolate to narrowly elliptic, gen irregularly dentate to lobed. **INFL**: fls in axils. **FL**: hypanthium 30–165 mm; sepals 16–50 mm, tips in bud not free; petals 16–56 mm, white. **FR** 10–68 mm, 4–9 mm wide, cylindric to elliptic-ovate, tubercled. **SEED** obovate to ± triangular, papillate or netted, 1 side with a cavity sealed by a depressed, gen splitting membrane. $2n$=14,28. Open desert scrub, pinyon/juniper woodland, coniferous and bristlecone-pine forests; 1100–3400 m. MP (likely), SNE, D; w US. Cross-pollinated. 5 intergrading sspp., 2 in CA.

ssp. ***crinita*** (Rydb.) Munz CAESPITOSE EVENING PRIMROSE Pl loosely to densely cespitose. **FL**: hypanthium 30–85 mm; petals fading rose to purple. **FR** 10–34 mm, lanceolate to elliptic-ovate, gen S-shaped; stalk-like base 0–1 mm. **SEED** 2.9–3.5 mm; cavity margin lobed. UNCOMMON. Calcium soils in bristlecone-pine forest, pinyon/juniper woodland, desert scrub; 1150–3370 m. Range ± of sp. [var. *c.* (Rydb.) Munz] 2 intergrading forms differ in elevation, habit, lf size, petal color; more study needed. ❀TRY.

ssp. ***marginata*** (Hook. & Arn.) Munz (p. 807) Pl loosely cespitose. **FL**: hypanthium 65–165 mm; petals fading lavender to pink. **FR** 25–68 mm, cylindric, ± straight; stalk-like base 0–55 mm. **SEED** 2.2–3.4 mm; cavity margin entire. Rocky or sandy sites in granite, limestone, or sandstone soils, pinyon/juniper woodland to pine forest; < 2400 m. Range ± of sp. [var. *m.* (Hook. & Arn.) Munz] Variable. ❀DRN,DRYorIRR,SUN:1–3,**7–14,18–21**.

O. californica (S. Watson) S. Watson Per, rosetted when young, glabrous to densely minutely strigose, also sometimes with longer, spreading hairs; roots gen not fleshy, gen new shoots from laterals. **ST** decumbent or ascending, 1–8 dm, peeling. **LVS**: cauline 1–6 cm, lanceolate or deltate-ovate, entire to pinnately lobed. **INFL**: fls in upper axils; buds nodding. **FL**: hypanthium 20–40 mm; sepals 15–30 mm, free tips in bud 0–1 mm; petals 15–35 mm, white fading pink. **FR** 20–80 mm, 2–3 mm wide, cylindric, straight or curved. **SEEDS** in 1 row per chamber, 1.4–3 mm, obovate, smooth. $2n$=14,28. Sandy or gravelly areas, dunes, desert scrub to pinyon/juniper or ponderosa-pine woodlands; < 2500 m. CW, SW, SNE, D; to sw UT, nw AZ, nw Baja CA. Cross-pollinated.

ssp. ***avita*** Klein (p. 807) Pl ± grayish green; hairs dense, short, appressed and also longer, spreading. **ST**: new rosettes not formed at tips. **LVS**: cauline oblong to lanceolate, gen ± pinnately lobed. **FR** 20–80 mm. $2n$=14. Sandy or gravelly areas, desert scrub to pinyon/juniper or ponderosa-pine woodlands; 800–2500 m. SNE, D; ± sw US. [*O. a.* (Klein) Klein] ❀TRY.

ssp. ***californica*** (p. 807) Pl green to slightly grayish; hairs 0, or dense, short, appressed, also sometimes long, spreading. **ST**: new rosettes not formed at tips. **LVS**: cauline oblong to lanceolate, subentire to deeply wavy-dentate. **FR** 30–55 mm. $2n$=28. Sandy or gravelly areas, open, coastal-sage scrub, chaparral, oak woodland; < 1900 m. CW (San Luis Obispo Co. s), SW, s DMtns (Little San Bernardino Mtns); nw Baja CA. [*O. c.* var. *glabrata* Munz] Intergrades with ssp. *avita*. ❀TRY.

ssp. ***eurekensis*** (Munz & Roos) Klein (p. 807) EUREKA DUNES EVENING PRIMROSE Pl: roots fleshy; hairs dense, short, appressed, also long, spreading, wavy. **ST**: new rosettes formed at tips. **LVS**: cauline deltate-ovate, entire to dentate. **FR** 30–70 mm. $2n$=14. **RARE** CA, **ENDANGERED** US. Dunes, gen with *Psorothamnus polydenius*; 900–1200 m. n-most DMoj (Eureka Valley, ne Inyo Co.). [*O. avita* (Klein) Klein ssp. *e.* (Munz & Roos) Klein] Populations few, large.

O. deltoides Torrey & Frémont DEVIL'S LANTERN, LION-IN-A-CAGE, BASKET EVENING PRIMROSE Ann, per, loosely rosetted; hairs 0, curly, or straight, also sometimes glandular. **ST** decumbent or erect, 2–10 dm, stout, spongy, peeling. **LVS**: cauline 2–15 cm, ± diamond-shaped-obovate to oblanceolate, subentire to pinnately lobed. **INFL**: fls in upper axils; buds nodding. **FL**: hypanthium 20– 40 mm; sepals 8–30 mm, free tips in bud 0–9 mm; petals 10–40 mm, white fading pink. **FR** 20–60(80) mm, 1.5–4(5) mm wide, cylindric, gen curved, ± twisted. **SEEDS** in 1 row per chamber, 1.5– 2 mm, obovate, smooth. $2n$=14. Sandy, often dunes; < 1800 m. SnJV, ne SnFrB, MP, D; to OR, UT, AZ, nw Mex. Gen cross-pollinated. 5 sspp., 4 in CA.

ssp. ***cognata*** (Jepson) Klein (p. 807) Ann, short-lived per; hairs spreading, ± 2 mm above. **ST** gen 2–4(6) dm, branched from base. **LVS**: upper subentire to wavy-dentate, rarely pinnately lobed. **FL**: bud tip obtuse; sepal tips not free; petals 25–40 mm. **FR**: base 3–5 mm wide. Sandy soils, grassland; < 700 m. SnJV, w DMoj. Sometimes self-pollinated. ❀TRY.

ssp. ***deltoides*** (p. 807) Ann; hairs spreading, 1–1.5 mm above, also minutely strigose or not. **ST** gen 2–10 dm, ± branched from base. **LVS**: upper dentate. **FL**: bud tip obtuse to acute; sepal free tips 0–1 mm; petals 15–38 mm. **FR**: base 2–3 mm wide. Sandy soils, incl dunes; gen < 1100 m. D; s NV, w AZ, nw Mex. ❀TRY.

ssp. ***howellii*** (Munz) Klein (p. 807) ANTIOCH DUNES EVENING PRIMROSE Per, grayish green; hairs spreading, 1–3 mm, wavy, also minutely strigose, also sometimes glandular. **ST** gen 4–8 dm, ± branched throughout. **LVS**: upper wavy-lobed. **FL**: bud tip obtuse; free sepal tips 1–9 mm; petals 20–40 mm. **FR**: base 3–4 mm wide. **ENDANGERED** CA, US. Sandy bluffs, dunes; < 100 m. ne SnFrB (Antioch, Contra Costa Co.). Populations with few to ± 150 pls at any one time. In cult.

ssp. ***piperi*** (Munz) Klein (p. 807) Ann; hairs wavy or curly, 1.5–3 mm above. **ST** gen < 1 dm, gen simple or few-branched from base. **LVS**: upper pinnately lobed. **FL**: bud tip bluntly acute; free sepal tips 0–1 mm; petals 15–25(30) mm. **FR** base 3–5 mm wide. Sand, incl dunes, sagebrush scrub; 850–1800 m. MP; c OR, w NV. ❀TRY.

O. elata Kunth Bien, short-lived per, densely minutely strigose and (esp in infl) glandular; hairs also long, appressed to spreading, sometimes with red, blister-like bases. **ST** erect, 4–25 dm. **LVS**: cauline 4–25 cm, oblanceolate to lanceolate or elliptic, gen dentate to subentire. **INFL**: spike. **FL**: hypanthium 20–48(55) mm; sepals 27–48 mm, free tips in bud 1–7 mm; petals 25–52 mm, yellow fading reddish orange. **FR** 20–65 mm, 4–7 mm wide, narrowly lanceolate, ± straight. **SEED** 1–2 mm, angled, irregularly pitted. $2n$=14. Moist places; 0–2800 m. CA; w N.Am to C.Am. Gen cross-pollinated. 3 sspp., 2 in CA.

ssp. ***hirsutissima*** (S. Watson) W. Dietr. (p. 807) **ST** 10–25 dm. **FL**: sepals green or red-flushed, hairs glandular or not, also spreading and sometimes with red, blister-like bases, free tips in bud 3–6 mm; anthers 8–15 mm. Moist places, gen inland; < 2800 m. CA; w US, nw Mex. [*O. hookeri* Torrey & A. Gray sspp. *angustifolia* (Gates) Munz, *grisea* (Bartlett) Munz, & *venusta* (Bartlett) Munz] Several intergrading forms. ❀SUN:4–6,**14–17**,&IRR:1–3,**7–12**,13,**18–24**;INV,STBL.

ssp. ***hookeri*** (Torrey & A. Gray) W. Dietr. & W.L. Wagner (p. 807) **ST** < 8 dm. **FL**: sepals red-flushed, hairs glandular and also spreading with conspicuous, red, blister-like bases, free tips in bud 1–5 mm; anthers 12–23 mm. Moist, coastal, slightly inland, sandy bluffs; < 200 m. CW, SW. [*O. hookeri* Torrey & A. Gray ssp. *montereyensis* Munz] ❀IRRorWET:7,**14,18–24**&SUN:5,6,**15–17**; INV,STBL.

O. flava (Nelson) A.O. Garrett ssp. ***flava*** (p. 807) Per, ± rosetted, green; hairs glandular, also gen minutely strigose; taproot fleshy. **ST** ± 0. **LF** 3–36 cm, oblanceolate to oblong, irregularly pinnately lobed, ± fleshy. **INFL**: fls in axils. **FL**: hypanthium 24–80 (150) mm; sepals 10–34 mm, free tips in bud 1–2(5) mm; petals 10–30(38) mm, yellow fading pale orange. **FR** 10–40 mm, 4–7 mm wide, narrowly ovate to elliptic; wings 2–6 mm wide. **SEED** 1.8–2.6 mm, obliquely wedge-shaped, minutely beaded, narrow-winged distally and along 1 margin. $2n$=14. Drying depressions, streambanks, gen clay soils, sagebrush scrub to pinyon/juniper woodland; 900–1600 m. CaR, n SNH, MP; to s Can, Colorado, Mex. Gen self-pollinated. ❀TRY.

O. glazioviana Micheli (p. 807) Bien, rosetted, densely minutely strigose; hairs also long, spreading, gen with red blister-like bases, in infl also glandular. **ST** erect, 5–15 dm. **LVS**: cauline 5–15 cm, elliptic to lanceolate, crinkled, dentate to subentire. **INFL**: spikes. **FL**: hypanthium 35–50 mm; sepals 28–45 mm, free tips in bud 5–8 mm; petals 35–50 mm, yellow fading reddish orange. **FR** 20–35 mm, 5–6 mm wide, narrowly lanceolate, ± straight. **SEED** 1.3–2 mm, angled, irregularly pitted, ± 50% sterile. $2n$=14. Disturbed places; ± < 500 m. NW, CW, SW. [*O.* × *erythrosepala* Borbás] Commonly cult; naturalized occurrences scattered. Gen cross-pollinated. PTH. Possibly derived in Eur from hybrids between two N.Am spp.

O. laciniata Hill Ann, short-lived per, rosetted, minutely strigose; hairs above gen also long, spreading, and also glandular. **ST** decumbent to erect, 0.5–5 dm. **LVS**: cauline 2–10 cm, narrowly oblanceolate to ± elliptic or oblong, subentire to pinnately lobed. **INFL**: fls in upper axils; buds curved upward. **FL**: hypanthium 12–35 mm; sepals 5–15 mm, free tips in bud 0.3–3 mm; petals 5–22 mm, yellow fading orange. **FR** 20–50 mm, 2–4 mm wide, cylindric. **SEED** 1–1.8 mm, ± spheric or widely elliptic, pitted. $2n$=14.

Open, gen sandy, gen urban, disturbed places; gen < 500 m. CW, SW, DMoj; native to e US, widely naturalized, but occurrences scattered. Self-pollinated. PTH.

O. longissima Rydb. Bien, rosetted, minutely strigose; hairs also gen long, spreading, with red, bristle-like bases, sometimes some glandular. **ST** erect, 6–30 dm. **LVS**: cauline 5–22 cm, narrowly oblanceolate to ± elliptic, subentire to dentate. **INFL**: spike. **FL**: hypanthium 60–135 mm; sepals 23–47 mm, free tips in bud 2–6 mm; petals 28–65 mm, yellow fading reddish orange. **FR** 25–55 mm, 4–9 mm wide, narrowly lanceolate, ± straight. **SEEDS** 1–2 mm, angled, irregularly pitted. 2*n*=14. Seasonally moist places in creosote-bush scrub, pinyon/juniper woodland; 1000–1700 m. e DMtns (New York Mtns); to Colorado, AZ. [ssp. *clutei* (Nelson) Munz] Gen cross-pollinated. ❀TRY.

O. primiveris A. Gray Ann, rosetted, minutely strigose, in infl gen glandular; hairs also coarse, with red, blister-like bases or not. **ST** gen 0 (sometimes erect or ascending, < 3.5 dm). **LF** 4–28 cm, oblanceolate, wavy-dentate to 1–2-pinnately lobed. **INFL**: fls in axils. **FL**: hypanthium 20–72 mm; sepals 7–30 mm, free tips in bud 0; petals 6–40 mm, yellow fading reddish orange to purple. **FR** 10–60 mm, 4–8 mm wide, lanceolate to ovate, straight, curved, or S-shaped. **SEED** 3–3.5 mm, irregularly obovate to oblanceolate, papillate, 1 side coarsely wrinkled in distal 1/2, other side with thick, U-shaped area forming groove and small cavity at tip. 2*n*=14. Sandy flats, low hills, dune margins, arroyos; 30–1400 m. D; ± sw US, n Mex. Self- or cross-pollinated. 2 intergrading sspp.

ssp. ***bufonis*** (M.E. Jones) Munz (p. 807) **LF** grayish green. **FL**: petals (22)29–40 mm. Habitats of sp. D; to UT, w AZ, nw Mex. Gen cross-pollinated. ❀TRY.

ssp. ***primiveris*** **LF** gen green. **FL**: petals 6–25(28) mm. Uncommon. Habitats of sp. D; to TX, nw Mex. [ssp. *caulescens* (Munz) Munz] Self-pollinated. ❀TRY.

O. pubescens Sprengel (p. 807) Ann, bien, rosetted, minutely strigose; hairs also long, spreading, and gen some glandular. **ST** decumbent to erect, 0.5–8 dm. **LVS**: cauline 2–8 cm, narrow-oblanceolate to lanceolate or elliptic, pinnately lobed to subentire. **INFL**: fls in upper axils; buds nodding. **FL**: hypanthium 15–50 mm; sepals 5–25 mm, free tips in bud 0.1–1 mm; petals 5–35 mm, yellow fading orange. **FR** 20–45 mm, 2–4 mm wide, cylindric. **SEED** 1–1.5 mm, subspheric, pitted. 2*n*=14. Open places; ± 600 m. c DMoj (Newberry Springs, San Bernardino Co.); to NM, S.Am. [*O. laciniata* Hill ssp. *p.* (Sprengel) Munz] PTH. CA pls possibly introduced from AZ.

O. rosea L'Hér. (p. 807) Per, rosetted when young, minutely and sometimes densely strigose (lower st sometimes also long-hairy). **ST** decumbent to ascending, 1–5(10) dm. **LVS**: cauline 1.5–4 cm, oblanceolate to elliptic, gen subentire to wavy-dentate. **INFL**: fls in upper axils. **FL**: hypanthium 4–8 mm; sepals 5–8 mm, free tips in bud 1 mm; petals 5–10 mm, rose to rose-purple. **FR** much wider toward tip; body 8–10 mm, 3–4 mm wide; valves each with median rib; stalk-like base 5–20 mm, ribbed, tapered. **SEEDS** clustered in each chamber, 0.5–0.8 mm, obliquely oblanceolate, finely granular-papillate. 2*n*=14. Disturbed places; gen < 500 m. WTR; native to TX, Mex, S.Am. Self-pollinated. PTH.

O. speciosa Nutt. (p. 807) Per, rosetted when young, forming large patches from woody caudices and rhizomes, minutely strigose (some hairs also longer). **ST** weakly ascending to erect, 1–5 dm. **LVS**: cauline 2.5–8 cm, oblanceolate to ± elliptic, subentire to wavy-lobed. **INFL**: fls in upper axils; st tip nodding. **FL**: hypanthium 10–23 mm; sepals 15–30 mm, free tips in bud 1–4 mm; petals 25–40 mm, white fading pink or rose-purple. **FR** 10–25 mm, widening upward (to 3–5 mm), 8-ribbed; stalk-like base (4)8–15 mm, 1.5–2 mm wide, cylindric. **SEEDS** clustered in each chamber, 1–1.5 mm, obliquely oblanceolate, finely granular-papillate. 2*n*= 14,28,42. Disturbed places; gen < 500 m. WTR; native NM to c US, c Mex. [var. *childsii* (L. Bailey) Munz] Commonly cult. Cross-pollinated.

O. stricta Link ssp. *stricta* Ann, bien, rosetted, minutely strigose; hairs also long, spreading, some glandular. **ST** gen erect (decumbent), 2.5–10 dm. **LVS**: cauline 6–10 cm, very narrowly elliptic-lanceolate, gen slightly wavy, serrate. **INFL**: fls in upper axils; buds erect. **FL**: hypanthium 20–45 mm; sepals 14–20 mm; petals 15–25 mm, yellow fading reddish orange, bases gen with 1 red spot. **FR** 30–40 mm, 3–4 mm wide, ± cylindric (slightly wider toward tip). **SEED** 1.4–1.8 mm, widely elliptic, pitted. 2*n*=14. Uncommon. Gen moist, disturbed places; gen < 500 m. SCoRO; native to Chile. PTH.

O. villosa Thunb. ssp. ***strigosa*** (Rydb.) W. Dietr. & Raven (p. 807) Bien, rosetted, minutely strigose, esp in infl also glandular; hairs also long, spreading, gen with red, blister-like bases. **ST** erect, 5–20 dm. **LVS**: cauline 10–30 cm, lanceolate or elliptic, entire to minutely dentate. **INFL**: spike, open, few-fld; internodes in fr gen > fr. **FL**: hypanthium 25–40 mm; sepals 9–18 mm, often marked reddish, free tips in bud 0.5–2.5 mm; petals 7–20 mm, yellow fading duller to pale orange. **FR** 20–35 mm, 4–7 mm wide, narrowly lanceolate, ± straight. **SEED** 1–2 mm, angled, irregularly pitted. 2*n*=14. Moist openings in forests; esp 500–2000 m. KR, NCoRI (Tehama Co.), CaRH (Plumas Co.), MP; to sw Can, c US. [*O. s.* (Rydb.) Mackenzie & Bush; *O. biennis* L. and *O. hookeri* Torrey & A. Gray misapplied] PTH. ❀TRY.

O. wolfii (Munz) Raven, W. Dietr. & W. Stubbe (p. 807) WOLF'S EVENING PRIMROSE Bien, rosetted, densely minutely strigose; many hairs also with red, blister-like bases, some glandular. **ST** erect, 5–10 dm. **LVS**: cauline 5–18 cm, narrowly lanceolate to elliptic, wavy-dentate, upper dentate. **INFL**: spike. **FL**: hypanthium 30–46 mm; sepals 17–28 mm, free tips in bud erect, 1–3 mm; petals 13–23 mm, yellow fading reddish orange. **FR** 30–48 mm, 5–8 mm wide, narrowly lanceolate, ± straight. **SEED** 1–2 mm, angled, irregularly pitted. 2*n*=14. RARE. Coastal sand, incl dunes, bluffs, roadsides, gen moist places (perhaps also inland); < 100 m. n NCo, w KR (Del Norte, Humboldt cos.), e KR (Carrville, ± 800 m, Trinity Co.); OR. [*O. hookeri* Torrey & A. Gray ssp. *w.* Munz] PTH. In cult.

O. xylocarpa Cov. (p. 807) Per, rosetted, grayish green; hairs short (on fl long, coarse), erect; taproot fleshy. **ST** 0. **LF**: main segment ± round to oblanceolate, 26–62 mm; lobes few, gen small. **INFL**: fls in axils. **FL**: hypanthium 27–45(55) mm; sepals 25–30 mm, tips not free in bud; petals 25–38 mm, yellow fading deep salmon-red. **FR** 35–90 mm, 7–11 mm wide, narrowly lanceolate, curved, twisted, wrinkled. **SEEDS** in 1 row (or near base 2 rows) per chamber, 2.4–3.2 mm, obovate; tip gen truncate, coarsely wrinkled, papillate; 1 side with 2 small ridges. 2*n*=14. Gravelly to pumice meadows, Jeffrey-pine or lodgepole-pine/fir forests; 2200–3100 m. c&s SNH; w NV. Gen cross-pollinated. Locally common. ❀TRY;DFCLT.

OROBANCHACEAE BROOM-RAPE FAMILY

Lawrence R. Heckard

Ann, per, non-green root parasites; roots modified into absorptive structures; pl an erect, fleshy, mostly underground st (peduncle) with terminal infl. **LF**: true lvs 0. **INFL**: spike, raceme, or panicle; bracts alternate, scale-like. **FL** bisexual; calyx cylindric or cup-shaped, lobes 0–5, persistent; corolla ± 2-lipped, lobes gen 5; stamens 4, epipetalous in 2 pairs (sometimes a 5th vestigial); ovary superior, chamber 1, placentas gen 2–4, parietal, simple or lobed, stigma gen 2–4-lobed, gen bowl- to funnel-shaped. **FR**: capsule, loculicidal; valves 2–4. **SEEDS** many, small, angled; surface netted. 14 gen, 200 spp.: esp n temp. Sometimes placed in Scrophulariaceae. [Thieret 1971 J Arnold Arbor 52:404–432]

1. Infl glabrous; bracts widely ovate to obovate (width ± = length), densely overlapping; corolla with ring of hairs at base of stamens . . . . **BOSCHNIAKIA**
1′ Infl hairy; bracts ± lanceolate (width << length), little or not overlapping; corolla gen lacking ring of hairs at base of stamens . . . . **OROBANCHE**

BOSCHNIAKIA GROUND-CONE

Per, glabrous. **STS** gen simple, ann, 1–many from corm-like thickening surrounding host root. **INFL** spike-like; bracts densely overlapping, ± ovate; pedicels gen < 2 mm; bractlets on pedicel 0–3, narrow. **FL**: calyx cup-shaped, teeth 0–5 (variable on a pl), tip acute or tapered; corolla with ring of hairs in upper tube at base of stamens, upper lip entire or indented, lower lip 3-lobed; upper filament and anther hairy. **FR** 2–4-valved; placenta 1 per valve. 3 spp.: nw NA, ne Asia. (Boschniaki, Russian botanist) [Gilkey 1945 OR State Monogr Bot 9]

1. Infl < 3 cm diam; largest bracts gen acute, << 10 mm at widest point; corolla 10–15 mm . . . . *B. hookeri*
1′ Infl > 3 cm diam; largest bracts gen obtuse to rounded, > 10 mm at widest point; corolla 15–20 mm . *B. strobilacea*

B. hookeri Walp. (p. 807) SMALL GROUND-CONE Pl 7–12 cm. **INFL** 3–6 cm, < 3 cm diam, purplish to pale yellow; lower bracts 10–12 mm, ovate to narrowly obovate, tip obtuse to acute. **FL**: calyx cup 2–3 mm, teeth 2–4 mm, deltate; corolla 10–15 mm, gen pale, lips erect, 3–4 mm (upper > lower). **SEED** ± 1.5 mm. RARE in CA. Open woods, shrubby places, gen on *Gaultheria shallon*; < 300 m. NCoRO (1 site, Humboldt Co.), SnFrB (Mount Tamalpais, Marin Co.); to B.C. CA pls are among the most distinct from *B. strobilacea* in their small size and pale color.

B. strobilacea A. Gray (p. 807) CALIFORNIA GROUND-CONE Pl 10–30 cm. **INFL** gen 7–18 cm, 3–6 cm diam, gen reddish brown to dark purplish; lower bracts 15–20 mm, ovate to widely obovate, margin gen pale, tip obtuse to rounded. **FL**: calyx cup 2–4 mm, teeth 3–7 mm, narrowly deltate; corolla 15–20 mm, gen purplish (lobe margins pale), lips 5–6 mm, ± equal, lower lip gen spreading. **SEED** ± 2 mm. Open woods, chaparral, on *Arctostaphylos* or *Arbutus*; < 3000 m. CA-FP (exc s SW); s OR; widely scattered, most common NW, TR.

OROBANCHE BROOM-RAPE

Ann, per, gen glandular-puberulent above; root attachment sometimes tuber-like. **ST** simple or branched. **INFL** gen ± spike-like (lower fls often short-pedicelled or on short branches), gen dense; fls gen > 20; bracts gen lanceolate to deltate (wider on peduncle); bractlets 0 or 2. **FL**: calyx lobes gen 4–5; corolla glandular-puberulent (hairs short and tack-shaped or long-stalked), gen lacking ring of hairs at stamen bases, upper lip erect to reflexed, gen 2-lobed, lower lip 3-lobed, spreading, yellow-lined; anthers glabrous to hairy; stigma lobes 2, spreading or peltate. **FR** 2-valved; placentas gen 2 or 4, often lobed. **SEED** < 0.7 mm. 140 spp.: worldwide, esp Medit. (Greek: vetch strangler, from parasitic habit) [Heckard 1973 Madrono 22:41–70]

1. Fls 1–20; pedicels long, scapose; bractlets on pedicel 0 (sect. *Gymnocaulis*)
2. Fls gen 5–20; st (peduncle + infl axis) > 5 cm; bracts > 6; upper pedicels < st; corolla lobes glabrous to soft-hairy, hairs colorless . . . . *O. fasciculata*
2′ Fls gen 1–3; st (peduncle + infl axis) gen < 5 cm; bracts gen < 6; upper pedicels > st; corolla lobes minutely ciliate, hairs violet . . . . *O. uniflora*
1′ Fls gen > 20; pedicels 0 or short, not scapose; bractlets on pedicel 2
3. Branches at st base many, slender, yellow; infl open, fls becoming well separated; calyx gen 4-lobed, cut deepest on upper side; agricultural weed (sect. *Trionychon*) . . . . *O. ramosa*
3′ Branches at st base few, stout; infl dense, gen remaining so; calyx 5-lobed, cut ± equally or deepest on lower side; native (sect. *Nothaphyllon*)
4. Infl and fls ± dark purple (lighter in *O. pinorum*); calyx gen 5–11 mm (–12 mm in *O. cooperi*)
5. Calyx divided to base on lower side; hairs of infl and corolla papillate, not glandular; root attachment rounded, coral-like; gen on *Adenostoma fasciculatum* . . . . *O. bulbosa*
5′ Calyx divided ± equally; most hairs of infl and corolla glandular, not papillate; root attachment not rounded and coral-like (exc *O. pinorum*, sometimes *O. cooperi*); on various shrubs and herbs
6. Infl and corolla pale purple-tinged; corolla with a ring of hairs at base of filaments; root attachment rounded, coral-like . . . . *O. pinorum*
6′ Infl and corolla dark purple; corolla lacking ring of hairs; root attachment branched (rarely rounded, coral-like in *O. cooperi*)
7. Infl 4–5 cm wide; corolla 18–32 mm, lips 5–10 mm; on herbs of Asteraceae; D . . . . *O. cooperi*
7′ Infl 2–3 cm wide; corolla 12–18 mm, lips 3–5 mm; on shrubs, gen in chaparral . . . . *O. valida*
8. Corolla 14–18 mm, hairy outside, densely so at sinuses, hairs 0.4–0.7 mm; filament base and anther hairy . . . . ssp. *howellii*
8′ Corolla 12–14 mm, puberulent outside, weakly so to glabrous at sinus, hairs ± 0.1 mm; filament base and anther glabrous . . . . ssp. *valida*
4′ Infl and fls gen buff to pinkish, corolla lips white to pink or lavender with darker veins (purple in *O. californica* ssp. *c.*); calyx gen 10–20 mm
9. Corolla 15–30 mm, lips 4–10 mm, upper erect, lower spreading
10. Infl 3–4 cm, branched, forming a convex or ± flat-topped cluster; anthers densely hairy throughout; GB, adjacent CaR, SNH, n DMtns . . . . *O. corymbosa*
10′ Infl > 4 cm, of long, gen unbranched, ± raceme-like units; anthers glabrous or hairy along dehisced margin
11. Upper corolla lobes acute, tips mostly pointed; bracts lanceolate, veins 3–5, inconspicuous; fls short-pedicelled throughout; CA-FP (exc c&s SW) . . . . *O. vallicola*

11′ Upper corolla lobes obtuse, tips rounded to blunt; bracts ± ovate, veins > 5, conspicuous; only lower fls pedicelled; SW, D . . . . *O. parishii*

12. Corolla 15–24 mm, lips 4–6 mm; calyx lobes gen 7–10 mm; CCo, SCo, ChI . . . . ssp. ***brachyloba***

12′ Corolla 20–25 mm, lips 6–8 mm; calyx lobes 10–16 mm; TR, PR, DMoj . . . . ssp. ***parishii***

9′ Corolla 20–50 mm, lips 10–14 mm, widely flaring (see also *O. corymbosa* for pls of SNH, GB) . *O. californica*

13. Corolla lips gen purple; calyx lobes, pedicels, and bracts purplish, drying dark purple; coastal dunes and hills, NCo, n&c CCo . . . . ssp. ***californica***

13′ Corolla lips buff to yellowish to pinkish or purple-tinged; calyx lobes, pedicels, and bracts pale to pink-tinged, drying brown; CA-FP, inland (exc ssp. *grandis*)

14. Corolla tube ± stout, > 4 mm wide at constriction, abruptly expanded to form hump-back throat; lips moderately recurved — s SNH, Teh, TR, PR, s SNE . . . . ssp. ***feudgei***

14′ Corolla tube slender below, < 4 mm at constriction, gradually expanded; lips widely recurved

15. Corolla 35–50 mm, lower lobes narrowly ovate, > 5 mm wide . . . . ssp. ***grandis***

15′ Corolla 20–35 mm, lower lobes narrowly triangular to oblong, < 5 mm wide

16. Pl 4–10 cm; st below infl < 5 cm; infl ± flat- or convex-topped cluster < 5 cm; moist or vernally wet places . . . . ssp. ***grayana***

16′ Pl 10–35 cm; st below infl gen > 6 cm; infl elongate, 5–20 cm; dry places . . . . ssp. ***jepsonii***

O. bulbosa G. Beck (p. 807) Pl 8–30 cm, dark purplish aboveground, white-papillate, not glandular. **ST** arising from round, coral-like root attachment, stout, thickened at base, bulb-like, with overlapping scales. **INFL** sometimes pyramid-shaped. **FL**: calyx 6–10 mm, tube cut nearly to base on lower side, lobes narrowly triangular, gen < or = tube; corolla 10–18 mm, yellowish to purplish, lobes 2–4 mm, obtuse to acute; anthers glabrous to sparsely hairy; stigma 2-lobed, margins recurved. 2*n*=48. Openings in chaparral, gen on *Adenostoma fasciculatum*; < 1700 m. s NCoRO, NCoRI, SNF (Eldorado Co. s), SCoRO, SW; Baja CA.

O. californica Cham. & Schldl. Pl 4–35 cm, pale to dark purple aboveground, glandular-puberulent. **STS** 1 or clustered, slender to stout, branched below or throughout. **INFL** long or branched and ± flat- to convex-topped; pedicels 0–4 cm, shorter upward. **FL**: calyx 12–20 mm, gen pale to pinkish, lobes linear-triangular, >> tube; corolla 20–50 mm, purplish or white to rose with darker veins, moderately glandular-puberulent, lips 10–14 mm, gen widely flaring; upper lip obtuse to rounded; anthers woolly; placentas 4, stigma lobes 2, triangular, margins recurved. Gen dry, ± rocky soils, on herbs, gen Asteraceae; < 2500 m. CA-FP; to B.C., Baja CA. [*O. grayana* G. Beck misapplied] Sspp. highly variable, intergrading.

ssp. ***californica*** Pl 5–27 cm, gen (dark) purple aboveground. **ST** gen branched. **INFL** < 8 cm, ± head-like or round-topped. **FL**: corolla 22–45 mm, throat abruptly wider than tube, 8–10 mm wide, lips purplish, paler. 2*n*=48. Uncommon. Sandy or heavy soils of coastal bluffs, gen on *Grindelia*; < 150 m. NCo, n&c CCo. [*O. grayana* G. Beck vars. *nelsonii* Munz & *violacea* (Eastw.) Munz] Intergrades with ssp. *grandis* in Monterey and San Luis Obispo cos.

ssp. ***feudgei*** (Munz) Heckard (p. 807) Pl 10–30 cm. **ST** gen branched above, stout. **INFL** ± flat- or round-topped; branches gen < 9 cm. **FL**: corolla 25–35 mm, tube ± stout, abruptly expanded above sinus, > 4 mm wide at constriction, forming ± hump-back throat 8–10 mm wide, lips moderately recurved, whitish to yellowish, purple-tinged and veined. 2*n*=48. Dry washes, mtn slopes, flats, on *Artemisia tridentata*; 700–2500 m. s SNH, Teh, TR, PR, s SNE; Baja CA. [*O. grayana* G. Beck var. *f.* Munz]

ssp. ***grandis*** Heckard Pl 8–30 cm. **ST** gen branched above, stout. **INFL** gen < 5 cm, round-topped or not. **FL**: corolla 35–50 mm, pinkish or pale brownish red, veins dark, tube < 4 mm wide at constriction, throat 9–10 mm wide, lips widely recurved, lower lobes narrowly ovate. Uncommon. s CCo, s SCoRO (Santa Barbara Co.), c SCo (Los Angeles Co.), n ChI (Santa Rosa Island).

ssp. ***grayana*** (G. Beck) Heckard Pl 4–10 cm. **ST** branched or not, 1–5 cm. **INFL** < 5 cm, convex to ± flat-topped, few-fld. **FL**: corolla 20–30 mm, tube gradually widening to 5–8 mm, lips whitish to yellowish to pale purplish with lavender veins. 2*n*=48. Uncommon. Moderately moist meadows, stream margins, gen on *Erigeron, Aster*; gen 1200–1800 m. NCoRH, n&c SNH, SnFrB (s Sonoma Co., ± 60 m), MP; to WA. [*O. grayana* G. Beck var. *g.*]

ssp. ***jepsonii*** (Munz) Heckard Pl 10–35 cm. **ST** gen branched, > 6 cm. **INFL** 5–20 cm. **FL**: corolla 25–40 cm, whitish or pinkish to pale yellowish brown, tube gradually widened to 5–8 mm, veins gen purplish. 2*n*=48. Uncommon. Gen dry flats, slopes, gen on herbs and shrubs of Asteraceae; < 2200 m. KR, NCoR, CaRH, SNF, GV, SnFrB, SCoR. [*O. grayana* var. *j.* Munz] Pls < 12 cm with flat- to round-topped infl on *Heterotheca villosa* in SCoRO have been called ssp. *condensa* Heckard. Stout pls from s CCo, s SCoRO, c SCo, n ChI that key here are an undescribed ssp.

O. cooperi (A. Gray) A.A. Heller (p. 813) Pl 10–40 cm, gen dark purplish aboveground, glandular-puberulent; root attachment sometimes a coral-like thickening. **STS** simple or branched, often forming large clumps, stout, little enlarged at base. **INFL** 4–5 cm wide; lower pedicels < 5 cm, upper 0. **FL**: calyx 8–12 mm, lobes > tube, triangular, acuminate; corolla 18–32 mm, purplish, hairs long-stalked, gen glandular, tube lacking ring of hairs, lips 5–10 mm, upper lobes 6–10 mm, > lower, obtuse; anthers gen hairy; stigma lobes 2, thin, recurved. 2*n*=24,48,72. Sandy flats, washes, on Asteraceae (gen *Ambrosia, Hymenoclea, Encelia*) (weed on tomatoes, DSon, in 1960's); < 500 m. D; to UT, AZ, Baja CA. [*O. ludoviciana* vars. *c.* (A. Gray) G. Beck & *latiloba* Munz] An undescribed form (probably best a ssp.), 2*n*=96, with smaller, shorter-lobed corolla and peltate, & bowl-shaped stigma occurs on same hosts, over range of sp.

O. corymbosa (Rydb.) Ferris (p. 813) Pl 3–17 cm, pale to purple-tinged aboveground, glandular-puberulent. **ST**: branches clustered, stout, gen thickened at base. **INFL** 3–4 cm, branched, gen round-topped, few-fld; lower pedicels 5–30 mm below, upper 0. **FL**: calyx 12–18 mm, lobes > tube, linear-triangular; corolla 20–30 mm, sparsely short-glandular, lips 5–8 mm, purplish to pink, veins darker, upper lobes rounded; anthers woolly; stigma 2-lobed, peltate. 2*n*=48,96. Openings in sagebrush scrub, gen on *Artemisia tridentata*; 1200–2800 m. n CaR, SNH, GB, n DMtns (Panamint Mtns); to B.C., MT, UT. [*O. californica* var. *co.* (Rydb.) Munz] Closely related to *O. californica*, esp ssp. *feudgei*.

O. fasciculata Nutt. (p. 813) CLUSTERED BROOM-RAPE **STS** 1 or clustered, 5–20 cm, branched or not. **INFL**: raceme, ± flat-topped, gen 5–20-fld; bracts > 6, glandular-puberulent; pedicels 3–15 cm, shorter upwards; bractlets 0. **FL**: calyx lobes 3–7 mm, gen < tube, deltate, gen ± acuminate; corolla 15–30 mm, curved, becoming erect, yellow to purple-tinged, lobes rounded to narrowly acute; anthers gen hairy; stigma 2-lobed, recurved. 2*n*=48. Dry, gen ± bare places, gen on shrubs (esp *Artemisia, Eriodictyon, Eriogonum*); < 3300 m. CA-FP, GB, DMtns; to Yukon, c N.Am, n Mex. [var. *franciscana* D.B. Achey; var. *lutea* (C. Parry) D.B. Achey] Pls intermediate to *O. uniflora*, gen on *Galium*, are scattered in CA-FP (esp w WTR, SNF, NCoR), s OR.

O. parishii (Jepson) Heckard Pl 5–26 cm, ± yellowish white. **ST** gen simple, stout, glandular-puberulent. **INFL**: bracts narrowly ovate, with > 5 conspicuous, parallel veins. **FL**: calyx 10–20 mm, ± narrowly triangular, pale; corolla 15–25 mm, buff to pinkish, lips 4–8 mm, lobes rounded, veins reddish; anthers glabrous to hairy. Bare, sandy to rocky soils, gen on shrubs; < 2800 m. SNH, Teh, CCo, SW, W&I, DMtns; Baja CA. [*O. californica* Cham. & Schldl. var. *p.* Jepson]

ssp. ***brachyloba*** Heckard (p. 813) SHORT-LOBED BROOM-RAPE Pl 5–18 cm. **INFL** 3–8 cm. **FL**: calyx lobes 7–10 mm; corolla 15–24 mm, lips 4–6 mm, erect or slightly spreading; anthers gen glabrous; stigma lobes gen narrow, recurved. 2*n*=96. RARE. Sandy soil near ocean, on shrubs such as *Isocoma Menziesii*; < 300 m. CCo, SCo, ChI; Baja CA.

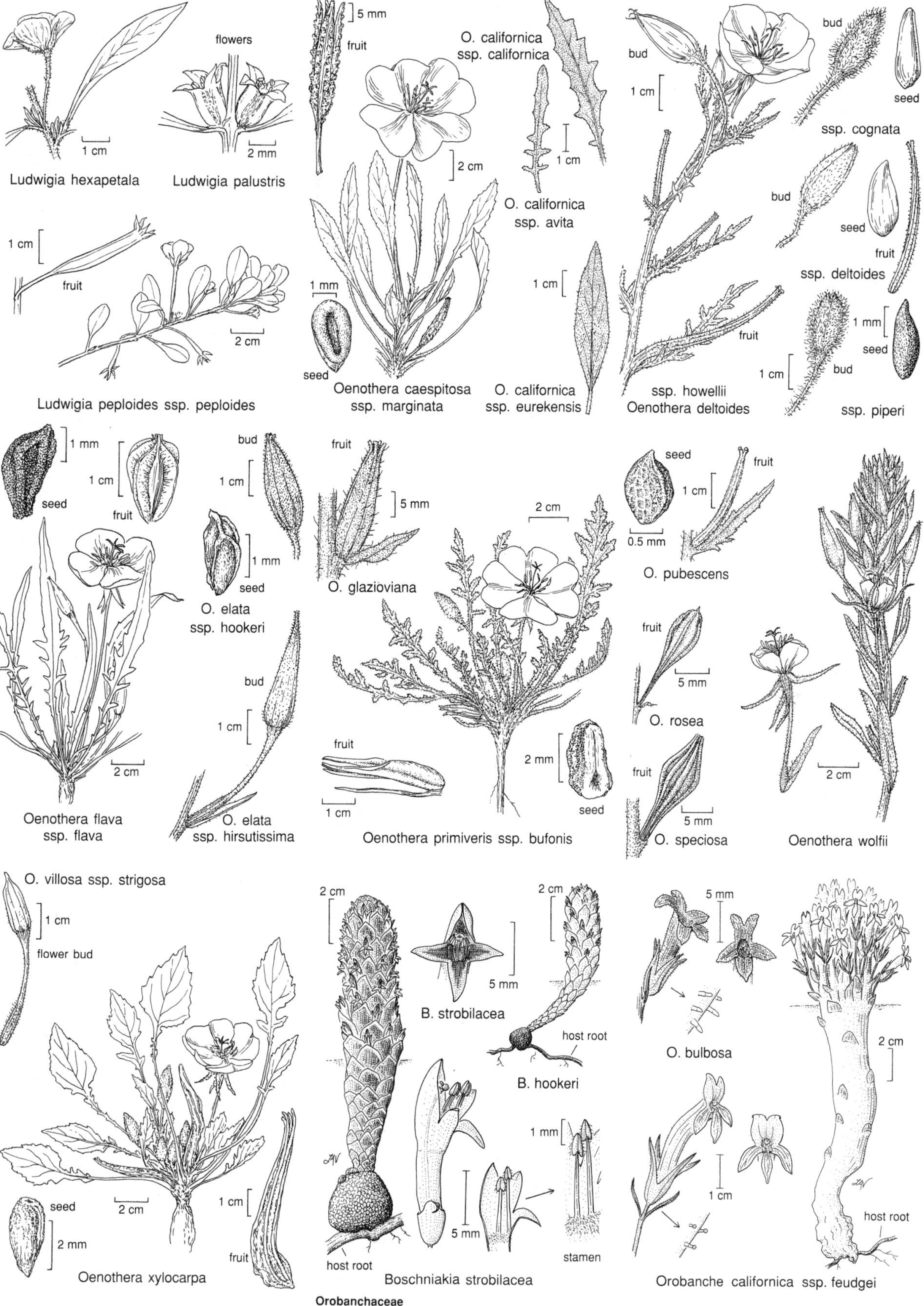

Orobanchaceae

ssp. ***parishii*** Pl 15–26 cm. **INFL** 5–14 cm. **FL**: calyx lobes 10–16 mm; corolla 20–25 mm, lips 6–8 mm, spreading; anthers glabrous or hairy; stigma lobes wide, spreading. 2*n*=48. Uncommon. Openings in chaparral, scrub, gen on shrubs; < 2800 m. s SNH, Teh, SW, W&I, DMtns; Baja CA. [*O. californica* var. *p.* Jepson] Separation from *O. ludoviciana* Nutt. var. *arenosa* (Suksd.) Cronq. blurred in GB.

O. pinorum Hook. (p. 813) Pl 10–30 cm, glandular-puberulent aboveground. **ST** slender above; base gen enlarged, with many overlapping bracts; root attachment roundish, coral-like. **INFL** at first dense, gen becoming open; lower pedicels 2–6 mm, upper 0. **FL**: calyx 5–8 mm, lobes ± = tube, triangular-acuminate; corolla 12–20, yellowish, hairy in a ring at stamen bases, lips erect, lobes tinged pale purple; anthers glabrous or sparsely hairy; stigma lobes 2, recurved. 2*n*=48. Uncommon. Rocky, open forest slopes, on *Holodiscus* spp. (not known on conifers); < 2100 m. NW, CaRH, SnFrB; to WA, ID; NM.

O. ramosa L. Pl 10–60 cm, yellowish, glandular-puberulent. **ST**: branches many from near base, slender. **INFL** open; lower pedicels short, upper 0. **FL**: calyx divided more deeply on top, lobes gen 4; corolla 10–15 mm, tube whitish, throat and lobes pale blue or lilac; anthers gen glabrous; stigma 2-lobed. 2*n*=24. Near fields; < 50 m. n SnJV, SnFrB, SCoRI, SCo; native of Eur. NOXIOUS (controlled) weed on tomatoes; may persist.

O. uniflora L. (p. 813) NAKED BROOM-RAPE **ST** 0.5–5 cm. **INFL**: gen raceme; fls gen 1–3; bracts gen < 6, gen glabrous; pedicels 3–12 cm, scapose; bractlets 0. **FL**: calyx lobes gen 4–8 mm, > tube, narrowly triangular; corolla 12–35 mm, ± horizontal, purplish to yellowish, lobes gen rounded, finely ciliate; anthers gen hairy; stigma lobes 2, margins recurved. 2*n*=36,48,±70. Gen moist places, on herbs, esp *Sedum*, Saxifragaceae, Asteraceae; < 3100 m. NW, CaR, SN, ScV (Sutter Buttes), CW, SCo, n ChI, Wrn; to Yukon, e N.Am. [var. *minuta* (Suksd.) D.B. Achey; var. *sedi* (Suksd.) D.B. Achey; ssp. *occidentalis* (E. Greene) Ferris] Variable. Pls < 20 cm, with deep violet corollas 25–35 mm, of n&c CA to Can have been called var. *purpurea* (A.A. Heller) D.B. Achey.

O. valida Jepson Pl 6–35 cm, dark purplish, glandular-puberulent. **ST** gen 1; branches (if any) slender to stout, base enlarged or not. **INFL** 2–3 cm wide. **FL**: calyx 5–11 mm, lobes subequal, linear-triangular; corolla 12–18 mm, lips 3–5 mm, acute, purple; anthers glabrous or hairy; stigma bowl-shaped, slightly 2-lobed. Rocky soils in chaparral, on shrubs; < 2000 m. NCoR, WTR, SnGb.

ssp. ***howellii*** Heckard & L.T. Collins HOWELL'S BROOM-RAPE Pl 6–20 cm; glandular hair stalks 0.2–0.4 mm, gen 3-celled. **ST** slender, gen enlarged at base. **FL**: corolla 14–18 mm, densely hairy outside at sinuses, hairs 0.4–0.7 mm; folds of lower throat puberulent; filament base and anther hairy. 2*n*=48. UNCOMMON. Volcanic and serpentine slopes, open chaparral, gen on *Garrya*; 200–1700 m. s NCoRH, c&s NCoRI.

ssp. ***valida*** (p. 813) ROCK CREEK BROOM-RAPE Pl 10–35 cm; glandular hair stalks ± 0.1 mm, gen 2-celled. **ST** gen stout, not enlarged at base. **FL**: corolla 12–14 mm, sparsely hairy outside at sinuses, hairs ± 0.1 mm; folds of lower throat glabrous; filament base and anther glabrous. 2*n*=48. RARE. Decomposed granite, on various chaparral shrubs; 1250–2000 m. c WTR (Topatopa Mtns), SnGb.

O. vallicola (Jepson) Heckard (p. 813) Pl 8–40 cm, pale to pinkish, glandular-puberulent. **ST** stout; base gen thickened, single or branched at base and above. **INFL** > 4 cm, sometimes branched from base into raceme-like units; bracts narrowly triangular, veins 3–5, inconspicuous; pedicels short. **FL**: calyx 8–15 mm, pale or pinkish; corolla 17–30 mm, yellowish to pinkish, lips 5–9 mm, upper lobes triangular with pointed tip, veins gen reddish; anthers glabrous to hairy; stigma 2-lobed, margins recurved. 2*n*=48. Uncommon. Woodlands, forest openings, gen on *Sambucus*; < 300 m. NCoR, CaRF, GV, CW, SCo (Los Angeles Co.). [*O. californica* vars. *californica* & *claremontensis* Munz misapplied] Difficult to separate from *O. californica* ssp. *jepsonii*.

OXALIDACEAE OXALIS FAMILY

Robert Ornduff

Ann to tree. **LVS** compound (palmate, pinnate, or lflet 1), alternate, often ± basal in rosettes or in clusters at st or rhizome tips, gen petioled; stipules gen 0; lflets gen sessile. **INFL**: cyme, sometimes umbel- or raceme-like, or fls solitary, gen in axils; peduncle bracted. **FL** gen bisexual, radial; sepals 5, free or fused at base; petals 5, free or fused above base; stamens 10 or 15, fused below, of 2 lengths; pistil 1, ovary superior, chambers 3–5, placentas axile, styles 1–5, gen ± free. **FR**: gen capsule, loculicidal. **SEEDS** gen with aril. 8 genera, 575 spp.: esp temp. Often heterostylous.

OXALIS

Ann, per, shrub; roots fibrous or woody; bulbs, tubers, or rhizomes often present. **ST** sometimes 0 or very short. **LF** petioled: stipules 0 or small; lflets 3, gen ± obcordate in CA, gen entire, gen green. **FL**: petals clawed; stamens 10; ovary chambers 5, styles 5, free, erect or curved. **FR** cylindric to spheric, explosively dehiscent. **SEEDS** flat, often ridged; aril translucent. ± 480 spp.: esp temp. (Greek: sour) [Eiten 1963 Amer Midl Nat 69:257–309; Lourteig 1975 Phytologia 42:57–197] Gen heterostylous; many (esp aliens in CA exc *O. laxa*) fine orn; some noxious weeds; contained oxalates may be toxic to livestock.

1. Petals yellow
 2. Ann; fr < 5 mm .... ***O. laxa***
 2′ Per; fr > 6 mm or 0
  3. Bulbs present; lvs in loose, ± basal rosette .... ***O. pes-caprae***
  3′ Bulbs 0; lvs cauline
   4. Petals > 12 mm; NCo .... ***O. suksdorfii***
   4′ Petals < 12 mm; NCo and elsewhere
    5. Petals gen < 8 mm; st rooting at nodes; taproot ± fleshy; disturbed urban places .... ***O. corniculata***
    5′ Petals gen 8–12 mm; st not rooting at nodes; taproot ± woody; native of ± undisturbed places .... ***O. albicans***
     6. St hairs 0 or gen curled .... ssp. ***californica***
     6′ St hairs slightly curved .... ssp. ***pilosa***
1′ Petals white to pink, red, or purple
 7. Bulbs 0

8. Sepals with 2 orange tubercles at tip; rhizome woody, not creeping; disturbed urban places .... ***O. rubra***
8′ Sepals without tubercles; rhizome fleshy, creeping; ± undisturbed forests
9. Fls solitary .... ***O. oregana***
9′ Fls 3–9 in umbel-like cyme .... ***O. trilliifolia***
7′ Bulbs present
10. St elongate
11. St glabrous, axils often with bulblets; petiole 2–7 cm .... ***O. incarnata***
11′ St hairy, axils without bulblets; petiole gen < 0.5 cm .... ***O. hirta***
10′ St 0 or very short
12. Fls in umbel-like cyme .... ***O. latifolia***
12′ Fls solitary .... ***O. purpurea***

O. albicans Kunth Per; taproot ± woody; bulbs 0. **ST** < 40 cm, ± trailing, not rooting at nodes, glabrous or hairy. **LVS** cauline; petiole < 7 cm; lflets < 1.5 cm. **INFL** 1–3-fld; pedicel < 2 cm. **FL**: sepals < 6 mm, lanceolate; petals gen 8–12 mm, yellow. **FR** 6–18 mm, cylindric. Coastal grassland & scrub, chaparral; < 700 m. NCo, NCoRO, CCo, SnFrB, SCoRO, SCo, ChI, WTR; to TX, n Mex. Fls spring, summer; confused with, perhaps same sp. as *O. corniculata*.

ssp. ***californica*** (Abrams) Eiten **ST**: hairs 0 or gen curled. Coastal scrub, chaparral; < 700 m. SCo, ChI; Baja Ca.

ssp. ***pilosa*** (Nutt.) Eiten (p. 813) **ST**: hairs slightly curved. Habitats of sp.; < 200 m. Range of sp. (exc ChI).

O. corniculata L. (p. 813) Per; taproot ± fleshy; bulbs 0. **ST** creeping, rooting at nodes, < 30 cm, ± hairy. **LVS** cauline; petiole < 7 cm; lflets < 2 cm, often maroon. **INFL**: cyme, ± umbel-like or not, 2–5-fld; pedicel < 1 cm. **FL**: sepals < 4.5 mm; petals gen < 8 mm, yellow. **FR** 6–25 mm, cylindric, ± angled. Abundant. Lawns, gardens; gen < 2500 m. CA; probably native to Old World. Fls ± all year. Pernicious urban weed. Possibly toxic in quantity to sheep.

O. hirta L. Per; bulbs < 3 cm, pale brown. **ST** gen branched, < 30 cm, hairy; axils without bulblets. **LVS** cauline; petiole gen < 0.5 cm; lflets < 1.5 cm, linear to obovate or shallow-obcordate, lower surface hairy. **INFL**: fls solitary; peduncle < 7 cm, bracts 2, near tip. **FL**: sepals < 8 mm, ± lanceolate, hairy; petals < 4 cm, rosy-purple to white. **FR** 0 in CA. Waste places, esp near gardens; < 100 m. CCo, SCo; native to s Afr.

O. incarnata L. Per; rhizome slender; bulbs < 2 cm, pale brown, beaked. **ST** erect, branched, < 30 cm, glabrous; axils often with bulblets. **LVS** cauline, in whorl-like clusters of < 10, lower sometimes opposite; petiole 2–7 cm; lflets < 1.5 cm. **INFL**: fls solitary; peduncle < 7 cm; bracts 2, near middle. **FL**: sepals < 6 mm, oblong, acute, tips with 2 orange tubercles; petals < 2 cm, white to pale pink. **FR** 0 in CA. Shady, disturbed, gen urban places; < 200 m. NCo, SnFrB; native to s Afr. Cult as orn.

O. latifolia Kunth Per; bulblets at tips of short stolons; bulbs < 25 mm, scaly. **ST** 0 or very short. **LVS** basal; petiole < 30 cm; lflets ± 4.5 cm, hairs 0 or sparse. **INFL** umbel-like, < 15-fld; peduncle < 14 cm. **FL**: sepals < 4 mm, oblong to lanceolate, tips with 2 brownish orange tubercles; petals < 1.5 cm, pink. **FR** 0 in CA. Gardens, fields, orchards; < 100 m. NCo, CCo, SCoRO; native to NM, Mex. [*O. martiana* Zucc.] Fls late winter to summer. Weed of n SCoR (Salinas Valley) ± matching this description may be *O. vallicola* (Rose) Kunth.

O. laxa Hook. & Arn. Ann; roots fibrous; bulbs 0. **ST** erect, < 20 cm, hairy. **LVS** cauline; petiole < 5 cm; lflets < 13 mm, hairs on both surfaces sparse. **INFL** raceme-like, 6–14-fld; peduncle < 15 cm; pedicels recurved in fr. **FL**: sepals < 4 mm, linear to lanceolate; petals < 12 mm, oblong, yellow. **FR** < 5 mm, ovoid to ± spheric. Disturbed sites; < 1000 m. NCo, c SNF; native to Chile. CA pls may be *O. micrantha* Bertero.

O. oregana Nutt. (p. 813) REDWOOD SORREL Per; rhizome creeping, fleshy, scaly; bulbs 0. **ST** 0 or very short. **LVS** < 10, clustered at rhizome tips; petiole < 20 cm, hairy; lflets < 4 cm, lower surface often purplish, upper often marbled. **INFL**: fls solitary; peduncle < 15 cm; bracts 2, near peduncle tip. **FL**: sepals < 1 cm, lanceolate; petals < 2.5 cm, white to deep pink. **FR** < 12 mm, ± ovoid. Moist conifer forests; < 1000 m. NCo, NCoRO, CCo, SnFrB; to WA. Fls spring to summer. ❀SHD:**17**&IRR:**4–7**,8,9,**14–16**,**20–24**;GRCVR,CVS.

O. pes-caprae L. (p. 813) BERMUDA BUTTERCUP Per; bulbs < 2.5 cm, pale brown. **ST** mostly underground, vertical, short, often with bulblets, whitish. **LVS** < 40, in loose ± basal rosette at enlarged st tip; petiole < 12 cm; lflets < 3.5 cm, often spotted, lower surface hairy. **INFL** umbel-like, < 20-fld; peduncle < 30 cm. **FL**: sepals < 7 mm, lanceolate to oblong, tips often with 2 orange or yellow tubercles; petals < 2.5 cm, yellow; filaments < 7 mm, hairy. **FR** 0 in CA. Disturbed areas, grasslands; < 500 m. NCo, NCoRO, CCo, SnFrB, SCoRO, SCo; native to s Afr. [*O. cernua* Thunb.] Cult as orn. Fls late fall to early summer. Pernicious urban weed. Possibly toxic in quantity to sheep.

O. purpurea L. Per; bulbs ovoid, dark brown, gummy. **ST** 0 or very short. **LVS** < 8, in basal rosette; petiole < 5 cm; lflets < 4 cm, hairy on lower surface, drying with dark spots, streaks. **INFL**: fls solitary; peduncle 1–11 cm, bracts 2, at or below peduncle middle. **FL**: sepals < 1 cm, ± lanceolate, hairy; petals < 2.5 cm, purple to red, salmon, or white. **FR** 0 in CA. Waste places, esp near gardens; < 100 m. NCo, SnFrB; native to s Afr. [*O. variabilis* Jacq.] Fls late winter to early spring. Cult as orn.

O. rubra A. St. Hil. Per; rhizome woody, not creeping; bulbs 0. **ST** 0 or very short. **LVS** basal; petiole < 30 cm; lflets < 2 cm. **INFL** umbel-like, < 10-fld, in umbel-like assemblages or not; peduncle > petiole. **FL**: sepals < 5 mm, broad-lanceolate, tips with 2 orange tubercles; petals purplish rose to white. **FR** < 8 mm, ovoid. Waste, disturbed places, esp near gardens; < 50 m. NCo, SnFrB; native to Brazil. Cult as orn.

O. suksdorfii Trel. (p. 813) Per, taprooted, often stoloned; bulbs 0. **ST** < 25 cm, ± trailing to erect, ± hairy. **LVS** cauline; petiole < 5 cm; lflets < 2 cm. **INFL** umbel-like or not, 1–3-fld; peduncle < 10 cm. **FL**: sepals < 8 mm, obtuse; petals 12–20 mm, yellow. **FR** 1–1.5 cm, oblong. Dry, shrubby or wooded areas; < 700 m. NCo; to WA. Fls summer. ❀TRY.

O. trilliifolia Hook. (p. 813) Per; rhizome creeping, thick, scaly; bulbs 0. **ST** 0 or very short. **LVS** clustered at rhizome tips; petiole < 30 cm, hairy; lflets < 4 cm. **INFL** umbel-like, 3–9-fld; peduncle < 25 cm. **FL**: sepals < 5 mm, lanceolate; petals < 1.3 cm, white or pinkish. **FR** < 3 cm, linear. Dense conifer forests; < 1500 m. NCo, KR, NCoRO; to WA. Fls summer. ❀TRY

PAEONIACEAE PEONY FAMILY

Fosiée Tahbaz

Per, subshrubs; roots gen in fleshy clusters. **STS** 1–several from root crown. **LVS** basal and cauline, alternate, large, deeply (gen ternately) dissected to compound, ± fleshy; stipules 0. **INFL**: fls solitary or few in a cluster, terminal. **FL**: sepals gen 5, free, leathery, persistent; petals gen 5–10; stamens many, maturing outward from fl axis; pistils 2–5, simple, coarse, thick-walled, surrounded at base by lobed nectary-disk. **FR**: follicles, 2–5, thick-walled, subtended by persistent calyx. **SEEDS** large, gen several per follicle, gen black, with an aril.

PAEONIA PEONY

The only genus. ± 30 spp.: Eurasia, 2 spp. in w US; many cult. (Greek: Paeon, physician to the gods)

1. Lf segment tips gen obtuse or rounded; bases of 1° lf divisions narrowed, stalk-like; st 20–40 cm; lvs gen 5–8 per st, glaucous, esp beneath; petals rounded, < inner sepals .... ***P. brownii***
1′ Lf segment tips gen long-acute; bases of 1° lf divisions little narrowed; st 35–75 cm; lvs gen 7–12 per st, paler green beneath; petals elliptic, > inner sepals .... ***P. californica***

P. brownii Hook. (p. 813) **ST** 20–40 cm, gen simple. **LVS** 5–8 per st, glaucous (esp beneath); 1° divisions 3–6 cm, 2–5 cm wide, narrowed and stalk-like at base; ultimate segments ± elliptic, bases tapered, tips gen obtuse or rounded. **FL**: petals 8–13 mm, < inner sepals, rounded, maroon to bronze, margins yellowish or greenish; filaments 3–5 mm, anthers 2–4 mm. **FR**: follicles 2–4 cm. **SEEDS** ± 11 mm, cylindric. *n*=5. Open, dry pine forest and scrubland; 200–3000 m. KR, NCoR, CaR, n&n-c SN, se SnFrB (Mount Hamilton); to B.C., WY, NV. ❀TRY;DFCLT.

P. californica Torrey & A. Gray (p. 813) **ST** 35–75 cm, gen branched. **LVS** 7–12 per st, dark green above, paler beneath, not very glaucous; 1° divisions 3–9 cm, 1–6 cm wide, bases slightly narrowed; ultimate segments ± widely linear or oblong, bases barely or not tapered, tips gen long-acute. **FL**: petals 15–25 mm, > inner sepals, elliptic, deep blackish red, margins pink; filaments 5–8 mm, anthers 3–7 mm. **FR**: follicles 3–4 cm. **SEEDS** ± 16 mm, ± curved. *n*=5. Chaparral, coastal scrub; 200–1500 m. SCoR, TR, PR. ❀DRN,DRY:**15–17**,24&SHD:**7**,9,**14**,18–23.

PAPAVERACEAE POPPY FAMILY

Curtis Clark

Ann to small tree; sap often colored, often milky. **LVS** basal, cauline, or both, gen toothed, lobed, or dissected; cauline gen alternate; stipules 0. **INFL**: cyme, raceme, or panicle (terminal), or fl solitary. **FL** bisexual, gen radial; sepals 2–4, sometimes shed ± at fl; petals gen 4 or 6 (or more), sometimes in 2 unlike pairs; stamens 4–many; ovary gen 1, superior, chamber gen 1, stigma lobes 0–many, ovules 1–many. **FR**: gen capsule, dehiscent by valves or pores, gen septicidal. 40 genera, 400 sp.: n temp, n trop, s Afr; some cult (*Papaver, Dicentra, Eschscholzia*). Petal length incls any spur or pouch. *Hunnemannia fumariifolia* Sweet (*Eschscholzia*-like garden per with free sepals) an uncommon waif in CA. *Corydalis, Dicentra, Fumaria* formerly treated in Fumariaceae.

1. Fl bilateral or biradial; petals 4, 1 or both outer ones spurred or pouched at base; lvs dissected — cauline 0 or alternate (Fumarioideae)
  2. Fl biradial; outer 2 petals alike, both pouched at base .... **DICENTRA**
  2′ Fl bilateral; outer 2 petals not alike, upper petal spurred at base
    3. Fr a capsule, ± oblong, several–many-seeded .... **CORYDALIS**
    3′ Fr a nutlet, spheric, 1-seeded .... **FUMARIA**
1′ Fl radial; petals 4, 6, or more, not spurred or pouched at base; lvs entire to dissected (Papaveroideae)
  4. Subshrub to small tree
    5. Petals 4, yellow; lf entire or minutely toothed .... **DENDROMECON**
    5′ Petals 6, white; lf deeply lobed .... **ROMNEYA**
  4′ Herb, sometimes woody at base
    6. Lvs entire to minutely toothed, cauline opposite; petals gen 6
      7. Stamens 4–6, 8–16, or many; fr a capsule, not composed of separating segments .... **MECONELLA**
      7′ Stamens > 12; longitudinal segments (carpels) of fr 9–18, separating, linear, narrowed between seeds, breaking into 1-seeded units .... **PLATYSTEMON**
    6′ Lvs toothed to dissected (exc entire in *Canbya*), cauline alternate; petals 4, 6, or more
      8. Pl spiny-armed; lvs gen cauline .... **ARGEMONE**
      8′ Pl unarmed; lvs basal, cauline, or both
        9. Hairs gen ± 5–15 mm; lvs gen basal — ne DMoj .... **ARCTOMECON**
        9′ Hairs 0–3 mm; lvs basal, cauline, or both
          10. Lf entire; petals 6; stamens 6–9 — small ann; D .... **CANBYA**
          10′ Lf toothed to dissected; petals gen 4; stamens 6–many
            11. Ovary and fr < 3 × longer than wide
              12. Stigma disk-like; style 0; sap milky .... **PAPAVER**
              12′ Stigma head-like; style 1, slender; sap clear .... **STYLOMECON**
            11′ Ovary gen > 3 × (fr > 4 ×) longer than wide
              13. Lf deeply pinnately lobed; receptacle tip narrow — probably naturalized .... **GLAUCIUM**
              13′ Lf dissected into ± linear segments; receptacle tip enlarged, cupped around ovary base .... **ESCHSCHOLZIA**

ARCTOMECON

Per; hairs gen ± 5–15 mm; taproot stout; sap colorless. **LVS** gen basal, obovate or wedge-shaped; teeth rounded. **INFL**: fls solitary, terminal. **FL**: bud nodding; sepals 2–3, long-hairy; petals 4 or 6, free, obovate, white or yellow, gen persistent after pollination; stamens many, free; placentas 3–6, style 1, stigma lobes 3–6. **FR** ovate to oblong, dehiscent from tip. **SEEDS** few, oblong, wrinkled, black. 3 spp.: sw US. (Greek: bear poppy, from long hairs)

A. merriamii Cov. (p. 813) WHITE BEAR POPPY Per 2–5 dm. **ST** glaucous. **LF** 25–75 cm. **FL**: sepals 15–20 mm; petals 25–40 mm, white. **FR** 25–35 mm. **SEEDS** 1.5–2 mm. RARE in CA. Rocky slopes; 900–1400 m. ne DMoj; s NV. *A. californica* Torrey & Frémont (petals yellow) in sw NV, undocumented in CA.

ARGEMONE PRICKLY POPPY

Ann, per, armed; sap yellow or orange, milky. **LVS** gen cauline, oblanceolate to ovate, toothed or deeply pinnately lobed. **INFL**: fls solitary, terminal. **FL**: sepals 2(3), prickly, with pointed appendage below tip, shed at fl; petals 4 (or 6), free, obovate, crinkled, white, shed after pollination; stamens 12–many, free; placentas 4–6, style 0 or 1, stigma lobes 4–6. **FR** ovate to lanceolate, prickly, dehiscent at tip by slits. **SEEDS** many, 1–2.5 mm, round to ovate, net-ridged, brown or black. ± 30 spp.: N.Am, S.Am, Hawaii. (Greek: ocular cataract, supposedly remedied by sap of other pls with this name)

1. Sap orange; lf gen less prickly on upper surface; fr 25–30 mm; stamens 100–120; seed 1–1.5 mm; SNE, DMoj .... ***A. corymbosa***

1′ Sap gen yellow; lf ± equally prickly on both surfaces; fr 35–55 mm; stamens 150–250; seed 2–2.5 mm; CW, SW, GB, nw DMtns .... ***A. munita***

A. corymbosa E. Greene Per 4–8 dm; sap orange. **LF** 8–15 cm, prickly on margins and veins, gen less so on upper surface. **FL**: petals 20–35 mm; stamens 100–120. **FR** 25–30 mm. **SEED** 1–1.5 mm. Dry slopes, flats; 400–1100 m. SNE, DMoj. ❀TRY.

A. munita Durand & Hilg. (p. 813) CHICALOTE Ann or per 6–15 dm; sap gen yellow (rarely red in NCoRH). **LF** 5–15 cm, ± equally prickly on margins and veins of both surfaces. **FL**: petals 25–40 mm; stamens 150–250. **FR** 35–55 mm. **SEED** 2–2.5 mm. Open areas; 70–3000 m. NW (exc NCo), CW, SW, GB, DMtns; n Baja CA. [sspp. *argentea* Ownbey, *rotundata* (Rydb.) Ownbey, & *robusta* Ownbey, prickly poppy] TOXIC but not gen eaten. ❀DRN,SUN,DRY:1–3,**7**,8,**9**,10,11,**14**,15–17,**18–21**,22–24.

CANBYA

Ann; sap colorless. **LVS** ± basal, linear-oblong, entire. **INFL**: fls solitary, terminal. **FL**: sepals 3; petals 6, free, elliptic, white, persistent after pollination; stamens 6–9, free; placentas 3, style 0, stigma lobes 3, linear, radiating from below ovary top. **FR** ovate, dehiscent from tip. **SEEDS** many, shiny, brown. 2 spp.: CA, OR, NV. (W.M. Canby, Delaware botanist, 1831–1904)

C. candida C. Parry (p. 817) Pl 10–30 mm, tufted, ± glabrous. **LF** 5–9 mm, fleshy. **INFL**: peduncle 10–20 mm. **FL**: petals 3–4 mm, closing over fr. **FR** 2–2.5 mm. **SEED** 0.6 mm. Sandy places; 600–1200 m. w DMoj, adjacent SN. ❀TRY.

CORYDALIS

Ann to per, glabrous, glaucous; sap colorless. **LVS** deeply pinnately dissected. **INFL**: raceme or panicle. **FL** bilateral; sepals 2, shed at fl; petals 4, yellow or white to pink, persistent after pollination, outer 2 petals free, not alike, keeled (upper spurred at base), inner 2 petals adherent at tips, oblanceolate, crested on back; stamens 6, ± fused in 2 sets, opposite outer petals; ovary lanceolate, placentas 2, style 1, stigma lobes 4–8. **FR** gen cylindric to oblong, dehiscent from tip. **SEEDS** several–many, 2–2.5 mm, round-reniform, smooth or rough, black. ± 100 spp.: n hemisphere, s Afr (some orn). (Greek: crested lark)

1. Petals yellow; ann or bien; spur 4–5 mm .... ***C. aurea***

1′ Petals white to pink, becoming purple-tipped; per; spur 12–16 mm .... ***C. caseana*** ssp. ***caseana***

C. aurea Willd. (p. 817) Ann or bien, 10–40 cm, glaucous. **LVS** several–many, 3–18 cm. **INFL**: raceme, few-fld. **FL**: petals 9–11 mm, yellow, spur 4–5 mm. **FR** 18–25 mm, cylindric. Open areas; 1500–2300 m. GB; to AK, e US, n Mex. ❀TRY.

C. caseana A. Gray ssp. ***caseana*** (p. 817) SIERRA CORYDALIS, FITWEED Per 50–100 cm, glaucous. **LVS** several, 15–35 cm. **INFL**: raceme or panicle, dense, narrow. **FL**: petals 10–12 mm, white to pink, becoming purple-tipped, spur 12–16 mm. **FR** 10–15 mm, oblong. UNCOMMON. Damp, shaded areas; 2400–5400 m. CaRH, n SNH, s MP. Other sspp. in OR, Rocky Mtns. TOXIC, eaten by naive livestock. ❀TRY.

DENDROMECON TREE POPPY

Shrub to small tree, evergreen; sap colorless. **LVS** entire or minutely toothed. **INFL**: fls solitary, terminal. **FL**: receptacle funnel-shaped around ovary base; sepals 2, shed at fl; petals 4, free, obovate or wedge-shaped, yellow, satiny, shed after pollination; stamens many, free; placentas 2, style 0, stigma lobes 2, flat. **FR** cylindric, dehiscent from base. **SEEDS** many, 2–3 mm, obovate, smooth, brown or black with pale outgrowth near stalk. 2 spp.: CA, Baja CA. (Greek: tree poppy)

1. Lf 1.5–3 × longer than wide, tip gen rounded, margin entire; ChI .... ***D. harfordii***

1′ Lf 3–8 × longer than wide, tip acute, margin minutely toothed; mainland .... ***D. rigida***

D. harfordii Kellogg CHANNEL ISLAND TREE POPPY Tree or shrub, 2–6 m. **LF** 3–8 cm, 15–45 mm wide (1.5–3 × longer than wide), tip gen rounded, margin entire. **FL**: petals 20–30 mm. **FR** 7–10 cm. **SEED** 25–30 mm. UNCOMMON. Shrubby slopes; < 600 m. ChI. [*D. rigida* ssp. *h.* (Kellogg) Raven] Pls with paler lvs from s ChI (extant on Santa Catalina Island) have been called var. *rhamnoides* (E. Greene) Munz [= *D. rigida* ssp. *rhamnoides* (E. Greene) Thorne, island tree poppy] ❀SUN,DRN,DRY:7,**8**,**9**,**14–17**,18,**19–24**;CV.

D. rigida Benth. (p. 817) BUSH POPPY Shrub 1–3 m. **LF** 2.5–10 cm, 7–25 mm wide (3–8 × longer than wide), tip acute, margin minutely toothed. **FL**: petals 2–3 cm. **FR** 5–10 cm. **SEED** 20–25 mm. Dry slopes, washes, esp recent burns; < 1800 m. s NW, CaRF, SNF, CW, SW; n Baja CA. ❀SUN,DRN,DRY:4–6,**7–9**,10–12,**14–16**,17,**18–24**;CV.

DICENTRA

Per, glabrous, sometimes glaucous; sap colorless. **LVS** gen basal (sometimes some cauline), deeply dissected. **INFL**: raceme, panicle, or fls solitary. **FL** biradial; sepals 2, shed just after fl; petals 4, white to yellow, sometimes persistent, outer 2 free, lanceolate, alike, both pouched at base, inner 2 adherent at tips, oblanceolate, ± crested on back; stamens 6, ± fused in 2 sets opposite outer petals; ovary cylindric to long-conic, placentas 2, style 1, stigma lobes 2. **FR** ovate or long-conic, dehiscent from tip. **SEEDS** few, 1–2 mm, oblong to reniform, smooth, rough, or netted, black. 16 spp.: N.Am, Asia; some orn. (Greek: twice spurred, from outer petals)

1. Lvs basal and cauline; st 50–160 cm; corolla shed after pollination; fl erect
 2. Petals yellow, 12–16 mm, outer recurved to middle .... ***D. chrysantha***
 2′ Petals whitish to cream, 20–25 mm, outer recurved at tips .... ***D. ochroleuca***
1′ Lvs basal; st < 45 cm; corolla persistent after pollination; fl nodding
 3. Infl 4+-fld; inner petals conspicuously crested; pl rhizomed
  4. Base of central stamen of each set aligned with lateral 2; NW, CaR, SNH, n CW; < 2200 m .... ***D. formosa***
  4′ Base of central stamen of each set bulging out from lateral 2; s SNH; 2200–3000 m .... ***D. nevadensis***
 3′ Infl 1–3-fld; inner petals not or obscurely crested; pl tubered
  5. Recurved tips of outer petals < bodies; infl 1–3-fld .... ***D. pauciflora***
  5′ Recurved tips of outer petals > bodies; infl 1-fld .... ***D. uniflora***

D. chrysantha (Hook. & Arn.) Walp. (p. 817) GOLDEN EARDROPS Pl 50–160 cm, glaucous, taprooted. **LVS** basal and cauline, 2-pinnately dissected, 15–30 cm. **INFL**: panicle, 2–5 dm, many-fld. **FL** erect; petals 12–16 mm, yellow, outer recurved to middle. **FR** 15–25 mm, long-conic. Dry slopes, burns, disturbed areas; < 1800 m. NW, CW, SN, SW; n Baja CA. TOXIC to livestock, esp when abundant after burns. ❀DRN,DRY,SUN:7,8,9,**14**,15–17,**18–21**,22–24;DFCLT.

D. formosa (Haw.) Walp. (p. 817) BLEEDING HEART Pl 20–45 cm, sometimes glaucous, rhizomed. **LVS** 2-ternately dissected, 20–50 cm. **INFL** 4+-fld. **FL** nodding; petals 14–18 mm, rose-purple to whitish, not drying black; base of central stamen of each set aligned with lateral 2. **FR** 14–20 mm, long-conic. Damp, shaded areas; < 2200 m. NW, CaR, SNH, n CW; to B.C. Small pls from nw KR with bluish glaucous lvs and light petals have been called ssp. *oregana* (Eastw.) Munz, Oregon bleeding heart. ❀DRN,**4,5**&SHD:**6**&IRR:1,**7**,8,9,**14–24**.

D. nevadensis Eastw. TULARE COUNTY BLEEDING HEART Pl 20–45 cm, sometimes glaucous, rhizomed. **LVS** 2-ternately dissected, 20–50 cm. **INFL**: 4+-fld. **FL** nodding; petals 13–15 mm, cream, pink-tipped, drying black; base of central stamen of each set bulging out from lateral 2. **FR** 14–20 mm, long-conic. UNCOMMON. Gravel bars; 2200–3000 m. s SNH (Sequoia & Kings Canyon National Parks, Tulare Co.). ❀DRN,IRR:1;DFCLT.

D. ochroleuca Engelm. Pl 50–160 cm, glaucous, taprooted. **LVS** basal and cauline, 3-pinnately dissected, 15–30 cm. **INFL** many-fld, 20–50 cm. **FL** erect; petals 20–25 mm, whitish to cream, purple-tipped, outer recurved at tips. **FR** 15–25 mm, long-conic. Dry slopes, disturbed areas; < 800 m. s CW, SW. ❀DRN,DRY,SUN:7–9,14–24;DFCLT.

D. pauciflora S. Watson Pl 3–7 cm, sometimes glaucous, tubered. **LVS** 1–3, 2–3-ternately dissected, 4–6 cm. **INFL** 1–3-fld. **FL** nodding; petals 18–20 mm, white to pink or lavender, recurved tips of outer petals < bodies. **FR** ± 12 mm, ovate. Gravelly areas; 1200–3000 m. KR, n&s SNH. ❀TRY;DFCLT.

D. uniflora Kellogg (p. 817) STEER'S HEAD Pl 3–7 cm, sometimes glaucous, tubered. **LVS** 1–3, 2–3-ternately dissected, 4–6 cm. **INFL** 1-fld. **FL** nodding; petals ± 15 mm, white to pink or lavender, recurved tips of outer petals > bodies. **FR** ± 12 mm, ovate. Gravelly or rocky areas; 1600–3700 m. KR, NCoRH, CaRH, SNH; to WA, UT. ❀DRN,IRR:1–3;DFCLT.

ESCHSCHOLZIA

Ann, per; taproot sap colorless or clear orange. **LVS** basal and gen some cauline, ± linear-dissected. **INFL**: cyme, 1–many-fld. **FL**: receptacle funnel-shaped, tip cupped around ovary base, sometimes spreading-rimmed below petals; sepals 2, fused, shed as a unit at fl; petals gen 4 (exc doubled fls), free, obovate or wedge-shaped, gen yellow to orange (white or pink), shed after pollination leaving crown-like membrane (formerly called "inner rim"); stamens 12–many, free; placentas 2, style 0, stigma lobes 4–8, spreading, linear. **FR** cylindric, dehiscent from base. **SEEDS** many, 1–2 mm, round to ovate, net-ridged, bur-like, or pitted, tan, brown, or black. 12 spp.: w N.Am. (J.F. Eschscholtz, Russian naturalist, 1793–1831)

1. Per; receptacle rim 0.5–5 mm; cotyledons gen 2-lobed .... [2]***E. californica***
1′ Ann; receptacle rim gen 0–0.3 mm; cotyledons gen entire
 2. Receptacle rim 0.5–5 mm; cotyledons sometimes 2-lobed .... [2]***E. californica***
 2′ Receptacle rim 0–0.3 mm; cotyledons entire
  3. Lvs gen basal; lf segments acute; petals yellow; fl bud glabrous
   4. Fl bud gen nodding; seed minutely pitted; petals (10)12–25 mm; D .... ***E. glyptosperma***
   4′ Fl bud erect; seed with prominent discontinuous ridges; petals 7–12 mm; SNF, GV .... ***E. lobbii***
  3′ Lvs basal and cauline; lf segments acute or obtuse; petals yellow or orange; fl bud glabrous or hairy
   5. Fl bud gen hairy, nodding; petiole gen hairy
    6. Petals yellow, bases sometimes orange-spotted, 10–15(20) mm; receptacle < 1.5 mm wide ..... ***E. hypecoides***
    6′ Petals orange or deep yellow throughout, 15–40 mm; receptacle > 1.5 mm wide ... ***E. lemmonii*** ssp. ***lemmonii***
   5′ Fl bud glabrous, nodding or erect; petiole gen glabrous
    7. Fl bud gen nodding; SNE, D
     8. Lf segments widened to tip, gen obtuse, < 2 mm; lvs grayish or bluish green; petals 3–26 mm; seeds gen oblong to elliptic .... ***E. minutiflora***
     8′ Lf segments not widened to tip, acute, > 1.5 mm; lvs bright green or yellow-green; petals 8–30 mm; seeds gen round .... ***E. parishii***
    7′ Fl bud gen erect; not D

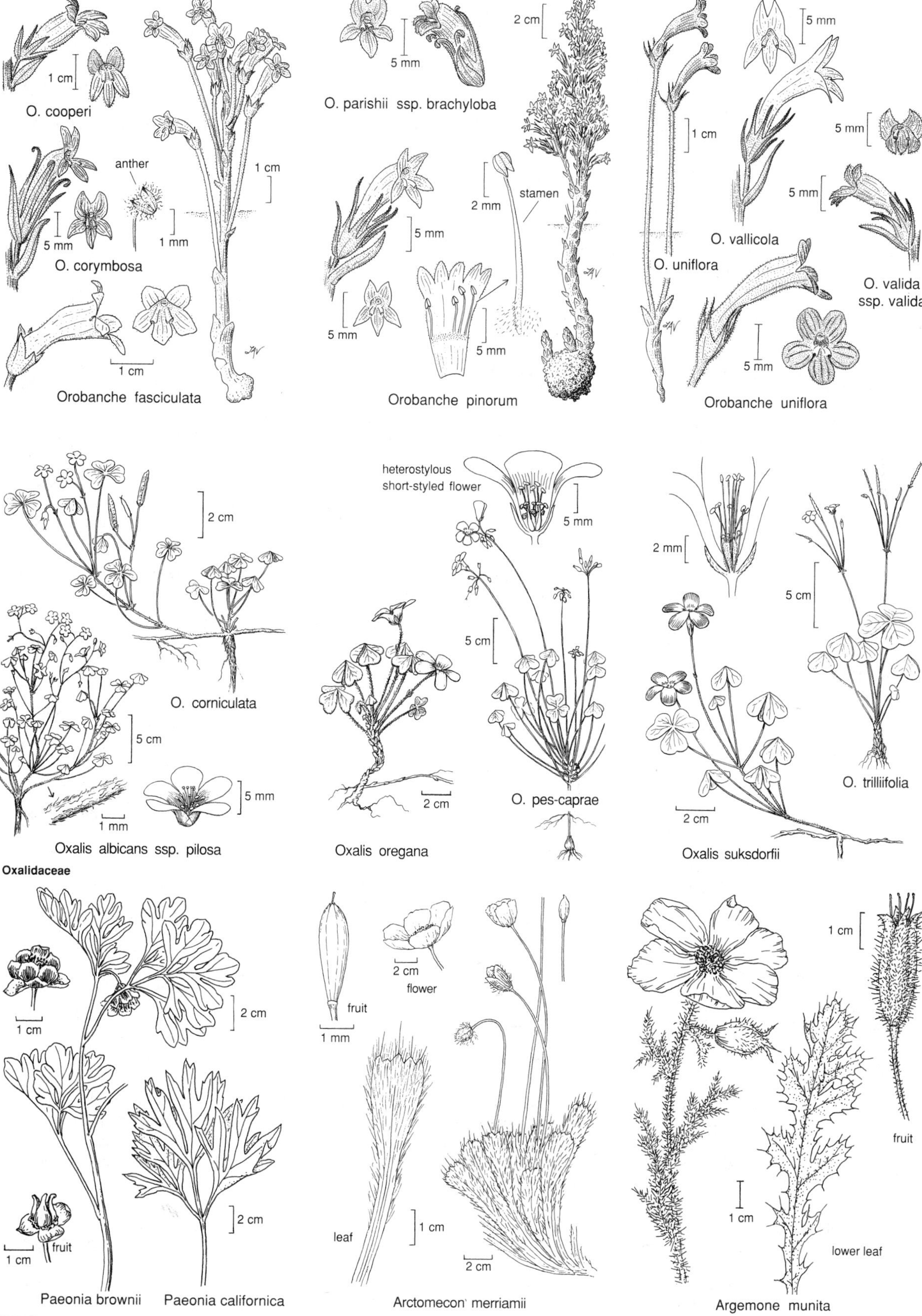

O. cooperi
1 cm
anther
1 cm
5 mm
1 mm
O. corymbosa
1 cm
Orobanche fasciculata
5 mm
O. parishii ssp. brachyloba
2 cm
2 mm
stamen
5 mm
5 mm
5 mm
Orobanche pinorum
5 mm
1 cm
5 mm
5 mm
O. vallicola
O. uniflora
O. valida
ssp. valida
5 mm
Orobanche uniflora
2 cm
O. corniculata
5 cm
1 mm
5 mm
Oxalis albicans ssp. pilosa
Oxalidaceae
heterostylous
short-styled flower
5 mm
5 cm
2 cm
O. pes-caprae
Oxalis oregana
2 mm
5 cm
O. trilliifolia
2 cm
Oxalis suksdorfii
1 cm
2 cm
2 cm
1 cm
fruit
Paeonia brownii
Paeonia californica
Paeoniaceae
fruit
1 mm
2 cm
flower
leaf
1 cm
2 cm
Arctomecon merriamii
Papaveraceae
1 cm
fruit
1 cm
lower leaf
Argemone munita

9. Receptacle obconic, < 2.5 mm wide
10. Fl bud long-pointed gen > 1/4 bud; pl appearing compact from short lower lf segments; mainland . . . . **E. caespitosa**
10′ Fl bud blunt or short-pointed < 1/4 bud; pl appearing open from ± long lower lf segments; ChI . . . . **E. ramosa**
9′ Receptacle barrel-shaped, > 2 mm wide
11. Petals 20–40 mm, orange or deep yellow; sw Teh, n WTR . . . . **E. lemmoni** ssp. **kernensis**
11′ Petals 3–15 mm, yellow; NCoRI, SnFrB, SCoRI, e SCoRO . . . . **E. rhombipetala**

E. caespitosa Benth. (p. 817) Ann, erect, tufted, 5–30 cm, glabrous, sometimes slightly glaucous. **LF**: segments obtuse or acute. **FL**: bud erect, glabrous, tip long-pointed gen > 1/4 bud; receptacle obconic; petals 10–25 mm, yellow, bases sometimes orange-spotted. **FR** 4–8 cm. **SEEDS** 1.5–2.4 mm wide, elliptic to obovate, net-ridged, brown to black. $2n=12$. Open chaparral; 0–1500 m. CA-FP (exc ChI); sw OR. ❀DRN,SUN:7,8,9,11,**14–16**,17,**18–24**.

E. californica Cham. (p. 817) CALIFORNIA POPPY Ann (or per from heavy taproot), erect or spreading, 5–60 cm, glabrous, sometimes glaucous. **LF** segments obtuse or acute. **FL**: bud erect, acute to long-pointed, glabrous, sometimes glaucous; receptacle obconic; petals 20–60 mm, yellow, bases gen orange-spotted. **FR** 3–9 cm. **SEED** 1.5–1.8 mm wide, round to elliptic, net-ridged, brown to black. $2n=12$. Grassy, open areas; 0–2000 m. CA-FP, w SNE, DMoj; to s WA, NV, NM, nw Baja CA. Highly variable (> 90 taxa have been described). Ann pls with entire cotyledons from DMtns have been called ssp. *mexicana* (E. Greene) C. Clark; large pls from s SNF have been called *E. procera* E. Greene, Kernville poppy. CA state fl. TOXIC but rarely eaten. ❀SUN:1–5,**6–12,14–24**&IRR:**13**;CVS.

E. glyptosperma E. Greene (p. 817) Ann, erect, 5–25 cm, glabrous, sometimes glaucous. **LVS** basal; segments acute. **INFL**: fl 1. **FL**: bud gen nodding, acute, glabrous, sometimes glaucous; receptacle obconic; petals (10)12–25 mm, yellow. **FR** 4–7 cm. **SEED** 1.2–1.8 mm wide, round, minutely pitted, tan to brown. $2n=14$. Desert washes, flats, slopes; 50–1500 m. D; to s NV, sw UT, w AZ. ❀TRY.

E. hypecoides Benth. SAN BENITO POPPY Ann, erect, 5–30 cm, sparsely hairy. **LF**: segments gen obtuse. **FL**: bud nodding, pointed, gen hairy; receptacle barrel-shaped; petals 10–15(20) mm, yellow, base sometimes orange-spotted. **FR** 3–7 cm. **SEED** 1–1.3 mm wide, round to elliptic, net-ridged, brown. $2n=12$. UNCOMMON. Grassy areas in woodland, chaparral; 500–1500 m. SCoRI.

E. lemmonii E. Greene Ann 5–30 cm, glabrous to sparsely hairy. **LF**: segments gen obtuse. **FL**: bud pointed; receptacle barrel-shaped; petals orange or deep yellow. **FR** 3–7 cm. **SEED** 1.3–1.8 mm wide, net-ridged, brown. $2n=12$. Open grasslands; 200–1000 m. s-most SNF, w Teh, e SCoRO, SCoRI, n WTR.

ssp. ***kernensis*** (Munz) C. Clark TEJON POPPY Pl erect. **FL**: bud gen erect, glabrous; petals 20–40 mm. **SEED** round to elliptic. RARE. Habitats and elevations of sp. sw Teh, n WTR. [*E. caespitosa* ssp. *k.* Munz]

ssp. ***lemmonii*** Pl spreading or erect. **FL**: bud nodding, gen hairy; petals 15–40 mm. **SEED** elliptic. Habitats and elevations of sp. s-most SNF, w Teh, e SCoRO, SCoRI. ❀DRN,DRY,SUN:7,**8**,**9**,11,**14–21**.

E. lobbii E. Greene (p. 817) FRYING PANS Ann, erect, 5–15 cm, glabrous. **LVS** basal; segments acute. **INFL**: fl 1. **FL**: bud erect, acute, glabrous; receptacle obconic; petals 7–12 mm, yellow. **FR** 4–7 cm. **SEED** 1.4–2 mm wide, round to elliptic, with discontinuous ridges, brown. $2n=12$. Open fields, grasslands; 0–600 m. SNF, GV. ❀DRN,DRY,SUN:**7–9**,11,**14–24**.

E. minutiflora S. Watson (p. 817) Ann, erect or spreading, 5–35 cm, glabrous, gray- or blue-glaucous. **LF**: segments obtuse. **FL**: bud nodding, short-pointed, glabrous, sometimes glaucous; receptacle obconic; petals 3–26 mm, yellow, base sometimes orange-spotted. **FR** 3–6 cm. **SEED** 1–1.4 mm wide, gen oblong to elliptic, net-ridged, brown to black. Desert washes, flats, slopes; 0–2000 m. $2n=12,24,36$. SNE, D; to s NV, sw UT, w AZ, nw Mex. Variable. Pls from n&c DMoj with petals 6–18 mm & $2n=24$ have been called ssp. *covillei* (E. Greene) C. Clark; pls from w DMoj (ne Kern Co.) with 10–26 mm & $2n=12$ have been called ssp. *twisselmannii* C. Clark & Faull. [*E. parishii* has been misapplied to the pls called ssp. *t.*] ❀TRY.

E. parishii E. Greene Ann, erect, 5–30 cm, glabrous, bright green to yellow-green, sometimes glaucous. **LF**: segments > 1.5 mm, not widened to tip, acute. **FL**: bud nodding, long-pointed, glabrous, sometimes glaucous; receptacle obconic; petals 8–30 mm, yellow. **FR** 5–7 cm. **SEED** 1–1.4 mm wide, gen round, net-ridged, tan to brown. $2n=12$. Desert slopes, hillsides; 0–1200 m. s DMoj, DSon; nw Mex. ❀TRY.

E. ramosa E. Greene ISLAND POPPY Ann, erect, 5–30 cm, glabrous, sometimes glaucous. **LF**: segments obtuse. **FL**: bud erect, glabrous, sometimes glaucous, tip blunt or short-pointed < 1/4 bud; receptacle obconic; petals 5–20 mm, yellow, base sometimes orange-spotted. **FR** 4–7 cm. **SEED** 1.4–1.6 mm wide, elliptic, net-ridged, brown. $2n=24$. UNCOMMON. Open places, esp chaparral; 0–300 m. ChI; Guadalupe Island, Mex. [*E. elegans* E. Greene misapplied]

E. rhombipetala E. Greene DIAMOND-PETALED CALIFORNIA POPPY Ann, erect, 5–30 cm, glabrous. **LF**: segments gen obtuse. **FL**: bud erect or nodding, blunt or short-pointed, glabrous, sometimes glaucous; receptacle barrel-shaped; petals 3–15 mm, yellow. **FR** 4–7 cm. **SEED** 1.3–1.8 mm wide, round, net-ridged, black. RARE (possibly extinct). Fallow fields, open places; 0–500 m. Formerly known from 7 sites in NCoRI, e SnFrB, e SCoRO, SCoRI; last known from Contra Costa Co.

FUMARIA

Ann, gen glabrous; sap colorless. **ST** branched, 20–60 cm in CA. **LVS** cauline, 2–6 cm in CA, finely pinnately dissected. **INFL**: raceme, narrow in CA. **FL** bilateral; sepals 2, shed after fl; petals 4, cream to purple, persistent after pollination, outer 2 petals free, not alike (upper spurred at base), inner 2 petals adherent at tips, oblanceolate, crested on back; stamens 6, ± fused in 2 sets opposite outer petals; ovary subspheric, style 1, deciduous, stigma dot-like. **FR**: nutlet, 2 mm in CA, spheric. **SEED** 1. ± 15 spp.: Eurasia. (Latin: smoky, perhaps from odor of fresh roots)

1. Petals 5–7 mm, purple; fr top not crested, depressed on both sides of stigma . . . . **F. officinalis**
1′ Petals 3–4 mm, cream, inner purple-tipped; fr top ± crested . . . . **F. parviflora**

F. officinalis L. FUMITORY **FL**: petals 5–7 mm, purple. **FR**: top not crested, depressed on both sides of stigma. Disturbed places; < 1000 m. SCo; native to Eur.

F. parviflora Lam. **FL**: petals 3–4 mm, cream, inner purple-tipped. **FR** top ± crested. Disturbed places; < 1000 m. CCo, SCo; native to Eur.

GLAUCIUM

Bien; taproot sap clear yellow. **LVS** basal and cauline, deeply pinnately lobed, glaucous; 1° lobes entire to lobed. **INFL**: fls solitary, terminal or axillary. **FL**: sepals 2, shed at fl; petals 4, free, obovate, yellow to orange, shed after pollination; stamens many, free; ovary chambers 1–2, placentas 2, style 0, stigma lobes 2, convex. **FR**: dehiscent from tip, linear. **SEEDS** many, ovate to reniform, with shallow depressions, dark brown. 6 spp.: Medit. (Greek: glaucous)

G. flavum Crantz SEA POPPY Pl 50–100 cm, glabrous or sparsely hairy. **LF** 5–15 cm. **FL**: petals 2–3 cm. **FR** 15–20 cm. **SEED** 1.5 mm. Uncommon. Disturbed areas, where ± naturalized; < 1000 m. CA-FP; native to Eur.

MECONELLA

Ann; sap colorless. **LVS** basal (or some cauline, opposite), gen entire, glabrous to rough-hairy. **INFL**: fls solitary, gen terminal and axillary. **FL**: sepals (2)3; petals (4 or) 6, free, obovate to oblong, white or yellow, shed after pollination; stamens 4–6, 8–16, or many, free; placentas 3, style 0, stigma lobes 3, ± erect, ovate. **FR** linear or obovate, dehiscent from tip. **SEEDS** many, 0.5–1 mm, ovate to reniform, smooth, shiny, brown or black. 4 spp.: w N.Am. (Greek & Latin: little poppy)

1. Lvs basal; fr obovate ..... ***M. linearis***
1′ Lvs basal and cauline; fr linear
  2. Stamens 8–16, in 2 series; petals 5–10 mm ..... ***M. californica***
  2′ Stamens 4–6, in 1 series; petals 2–4 mm ..... ***M. denticulata***

M. californica Torrey (p. 817) Pl 10–20 cm, glabrous. **LVS** basal and cauline, 5–25 mm, linear to oblanceolate. **FL**: petals 5–10 mm, white; stamens 8–16, in 2 series. **FR** 15–35 mm, linear, twisted. **SEED** 0.6 mm, obovate. Open, rocky areas; < 1000 m. CaRF, SNF, n CCo, SnFrB. ❀SUN,DRN,DRY:7,9,14,**15**,**16**,17, 19–24.

M. denticulata E. Greene Pl 5–30 cm, glabrous. **LVS** basal and cauline, 10–40 mm, linear to (ob)ovate, entire to minutely toothed. **FL**: petals 2–4 mm, white; stamens 4–6, in 1 series, anthers 1 mm, = filaments. **FR** 20–30 mm, linear, twisted. **SEED** 0.5 mm, obovate-reniform. Shaded canyons; < 1000 m. w CW, w SW. [*M. oregana* Nutt. var. *d.* (E. Greene) Jepson]

M. linearis (Benth.) Nelson & J.F. Macbr. Pl 5–25 cm, spreading-hairy. **LVS** basal, 25–70 mm, linear. **INFL**: fls terminal. **FL**: petals 6–15 mm, cream, base yellow; stamens many. **FR** 10–14 mm, obovate. **SEED** 0.4 mm, reniform-obovate. Grassy areas, washes; < 1000 m. w NW, s SNF, CW, w SW; s OR. [*Hesperomecon l.* (Benth.) E. Greene] ❀TRY.

PAPAVER

Ann, per; sap white or yellow, milky. **LVS** basal and cauline, deeply pinnately lobed, glabrous or hairy, sometimes glaucous. **INFL**: fls solitary, terminal. **FL**: bud nodding; sepals 2, shed at fl; petals 4, free, obovate to wedge-shaped, white to red or purple, persistent after pollination; stamens many, free; placentas 4–20, style 0, stigma disk-like, lobes 4–20. **FR** oblong to round, dehiscent by pores beneath stigma. **SEEDS** many, < 0.7 mm, reniform, net-ridged, brown or black. ± 50 spp.: esp Eurasia. (Latin: poppy) *P. apulum* Ten. var. *micranthum* (Boreau) Fedde, *P. dubium* L., *P. hybridum* L. perhaps in CA as waifs.

1. Cauline lf base wide, clasping; fr 3–5 cm, round ..... ***P. somniferum***
1′ Cauline lf base narrow, not clasping; fr 1–2 cm, narrowly obovate to round
  2. Peduncle hairs appressed; lf segments linear to lanceolate ..... ***P. argemone***
  2′ Peduncle hairs 0 or spreading; lf segments oblong to lanceolate
    3. Petals 10–20 mm; fr narrowly obovate; native fire- or disturbance-follower ..... ***P. californicum***
    3′ Petals 20–40 mm; fr widely obovate to round; alien from cult ..... ***P. rhoeas***

P. argemone L. Ann 20–50 cm, stiff-hairy. **LF** 3–12 cm. **FL**: petals 2 cm, pale scarlet, base dark-spotted. **FR** 1.5–2 cm, narrowly obovate. Uncommon. Disturbed areas; < 500 m. CA-FP; native to Eurasia; cult. Perhaps only a waif in CA.

P. californicum A. Gray (p. 817) FIRE POPPY Ann 30–60 cm, glabrous to shaggy-hairy. **LF** 3–9 cm. **FL**: petals 10–20 mm, brick-red, base greenish spotted. **FR** 1–1.6 cm, narrowly obovate. Burns, disturbed areas, open woodland; < 800 m. CW, SW. ❀DRN:7–9, **14**,15,16,18,**19–24**.

P. rhoeas L. CORN POPPY Ann 30–80 cm, hairy. **LF** 3–15 cm. **FL**: petals 20–40 mm, white (sometimes red-marked), red, or purple. **FR** 1–2 cm, widely obovate to round. Disturbed areas, fallow fields; < 500 m. CA-FP; native to Eurasia; cult. Perhaps only a waif in CA.

P. somniferum L. OPIUM POPPY Ann 60–120 cm, glaucous, glabrous or sparsely hairy. **LF** 10–20 cm; bases of cauline lvs wide, clasping st. **FL**: petals 4–6 cm, white to red or purple. **FR** 3–5 cm, round. Disturbed areas, fallow fields; < 500 m. CA-FP; native to Eurasia; cult. Perhaps only a waif in CA. TOXIC, source of morphine & other alkaloids.

PLATYSTEMON

1 sp. (Greek: wide stamen)

P. californicus Benth. (p. 817) CREAM CUPS Ann 10–30 cm, shaggy-hairy; sap colorless. **LVS** basal and cauline, opposite, 2–8 cm, linear to lanceolate or narrowly oblong, entire. **INFL**: fls solitary, terminal; peduncle 10–20 cm. **FL**: sepals 3, hairy; petals 6, free, elliptic, white to yellowish, 8–16 mm, persistent after pollination; stamens > 12, free, filaments flattened; longitudinal ovary segments (carpels) gen 9–18, fused, separating in fr. **FR**: segments 10–16 mm, linear, narrowed between seeds, breaking into 1-seeded units. **SEED** 1 mm, elliptic to reniform, smooth, brown. Open grasslands, sandy soils, burns; < 1000 m. CA-FP, w D; to OR, UT, AZ, Baja CA. Highly variable. [vars. *crinitus* E. Greene, *horridulus* E. Greene, *nutans* Brandegee, & *ornithopus* (E. Greene) Munz] Pls of s ChI (Santa Barbara Island) have been called var. *ciliatus* Dunkle, Santa Barbara Island cream cups. ❀DRN,DRY,SUN:6, **7–9**,10–12,**14–24**.

ROMNEYA MATILIJA POPPY

Subshrub or shrub, 100–250 cm, woody below; sap clear, bitter; rhizomes creeping. **LVS** cauline, gray-green-glaucous; lobes 3–5, deep, lanceolate to ovate. **INFL**: fls solitary, terminal. **FL**: sepals 3; petals 6, free, obovate, crinkled, white, shed after pollination; stamens many, free; chambers 1–12, placentas 7–12, style 0, stigma lobes 7–12, flat. **FR**: oblong to ovoid, bristly, dehiscent from tip. **SEEDS** many, 1.3–1.5 mm, ovate. 2 spp.: CA, Baja CA. (T. Romney Robinson, Irish astronomer, early 1800's)

1. Sepals glabrous; petals 60–100 mm; lf 5–20 cm ..... ***R. coulteri***
1′ Sepals appressed-hairy; petals 40–80 mm; lf 3–10 cm ..... ***R. trichocalyx***

R. coulteri Harvey (p. 823) COULTER'S MATILIJA POPPY **LF** 5–20 cm. **FL**: sepals glabrous; petals 60–100 mm. **FR** 3–4 cm. **SEED** papillate, dark brown. UNCOMMON. Dry washes, canyons; < 1200 m. e SCo, PR. Largest fls of any pl native to CA. ❀DRN, SUN:**4,5**,6,**7–9**,10–12,**14–24**;CV;INV.

R. trichocalyx Eastw. HAIRY MATILIJA POPPY **LF** 3–10 cm. **FL**: sepals appressed-hairy; petals 40–80 mm. **FR** 2.5–3.5 cm. **SEED** smooth, ± light brown. UNCOMMON. Dry washes, canyons; < 1100 m. w SCo, PR; n Baja CA. ❀DRN;SUN:**4,5**,6,7,**8,9**,10–12,**14–24**;INV.

STYLOMECON

1 sp. (Greek: style poppy)

S. heterophylla (Benth.) G.C. Taylor (p. 823) WIND POPPY Ann 30–60 cm; sap yellow, clear. **LVS** basal and cauline, 2–12 cm, deeply pinnately lobed, glabrous to hairy, petioled; 1° lobes toothed or lobed. **INFL**: fls solitary, axillary; peduncle 5–20 cm. **FL**: sepals 2, 4–10 mm, shed at fl; petals 4, free, 10–20 mm, wedge-shaped, orange-red, base purple-spotted, shed after pollination; stamens many, free; placentas 4–11, style 1, slender, stigma head-like, lobes 4–11. **FR** 1–2 cm, obconic to obovate, dehiscent by flaps below tip. **SEEDS** many, 0.4 mm, reniform, roughly net-ridged, brown or black. Grassy areas, openings in chaparral; < 1200 m. s NW, s SNF, SnJV, CW, SW; n Baja CA. ❀DRN,DRY:4–6,**7–9,14–24**; CV.

PHILADELPHACEAE MOCK ORANGE FAMILY

Charles F. Quibell

Shrub, subshrub. **ST** < 3 m, gen erect; bark gen peeling as thin sheets or narrow strips. **LVS** simple, opposite, deciduous or not, ± hairy; stipules 0; blade ± round to narrowly elliptic, entire or toothed. **INFL**: cyme, raceme, panicle, or fl 1, terminal or axillary, gen bracted. **FL** bisexual, radial; sepals 4–7, free, spreading or erect; petals 4–7, free, ± round to narrowly elliptic, gen white; stamens 10–12 in 2 whorls or many and clustered, filament base linear or wide and flat; pistil 1, ovary superior to 2/3 inferior, chambers 2–8, ovules 1–2 or many per chamber, placentas axile or parietal, styles 3–8, free or fused at base. **FR**: capsule, loculicidal or septicidal; styles persistent or not. **SEEDS** gen many, small to minute. 7 genera, 130 spp.: temp, subtrop n hemisphere; some cult for orn (*Carpenteria, Deutzia, Philadelphus*) Sometimes treated as Saxifragaceae or Hydrangeaceae.

1. Fls < 8 mm wide, odorless; sepals 1.5–2 mm, ± deciduous; petals 3–6 mm
 2. Lf blade 8–16 mm, 3–6 mm wide; fr cylindric, styles persistent; trunks or main sts erect; DMtns ..... **FENDLERELLA**
 2′ Lf blade 15–40 mm, 10–30 mm wide; fr depressed-spheric, styles deciduous; main sts prostrate; n coastal mtns ..... **WHIPPLEA**
1′ Fls gen > 15 mm wide, fragrant; sepals 3–10 mm, persistent; petals 5–25 mm
 3. Stamens 10, alternating long and short; filament base wide, flat; stigma terminal, < 0.5 mm ..... **JAMESIA**
 3′ Stamens > 20, ± equal, filament base ± linear; stigma linear, elongated along style, > 1 mm
  4. Fls 3–6 cm wide; lvs 1-veined from base, persistent, leathery; ovary ± superior; petals ± round — c&s SNF (Fresno Co.) ..... **CARPENTERIA**
  4′ Fls 1.5–3 cm wide; lvs 3-veined from base, deciduous; ovary partly to completely inferior; petals oblong to elliptic ..... **PHILADELPHUS**

CARPENTERIA TREE-ANEMONE

1 sp. (William M. Carpenter, Louisiana botanist, 1811–1848)

C. californica Torrey (p. 823) Shrub < 3 m. **ST**: bark ± gray, peeling as thin, wide sheets. **LVS** persistent, leathery; petiole 0.5–1 cm; blade 4–10 cm, 1–2.5 cm wide, narrowly elliptic, 1-veined, margin gen rolled under, entire or ± toothed, upper surface green, glabrous, lower surface pale, short-tomentose. **INFL**: cyme or short raceme, terminal, open; fls 3–13, bracts < lvs, sessile. **FL** 3–6 cm wide, fragrant; sepals 5–7, 8–10 mm, gen glabrous, inner surface white-tomentose at tip; petals 5–8, 15–25 mm, ± round, white; stamens many, clustered, filaments linear; ovary ± superior, chambers 5–8, ± incomplete; placentas axile, ovules many; style 1, branches 5–7 above middle, 3–6 mm; stigma linear along style branch. **FR** 0.6–1.2 cm wide, depressed-spheric, weakly septicidal. **SEEDS** many, fusiform, red-brown. $2n$=20. RARE. Streambanks, chaparral, oak woodland; 450–1000 m. c&s SNF (between Kings & San Joaquin Rivers, Fresno Co.). Pls appear to be fire-resistant, stump-sprouters; seedling establishment not observed in nature. In cult.

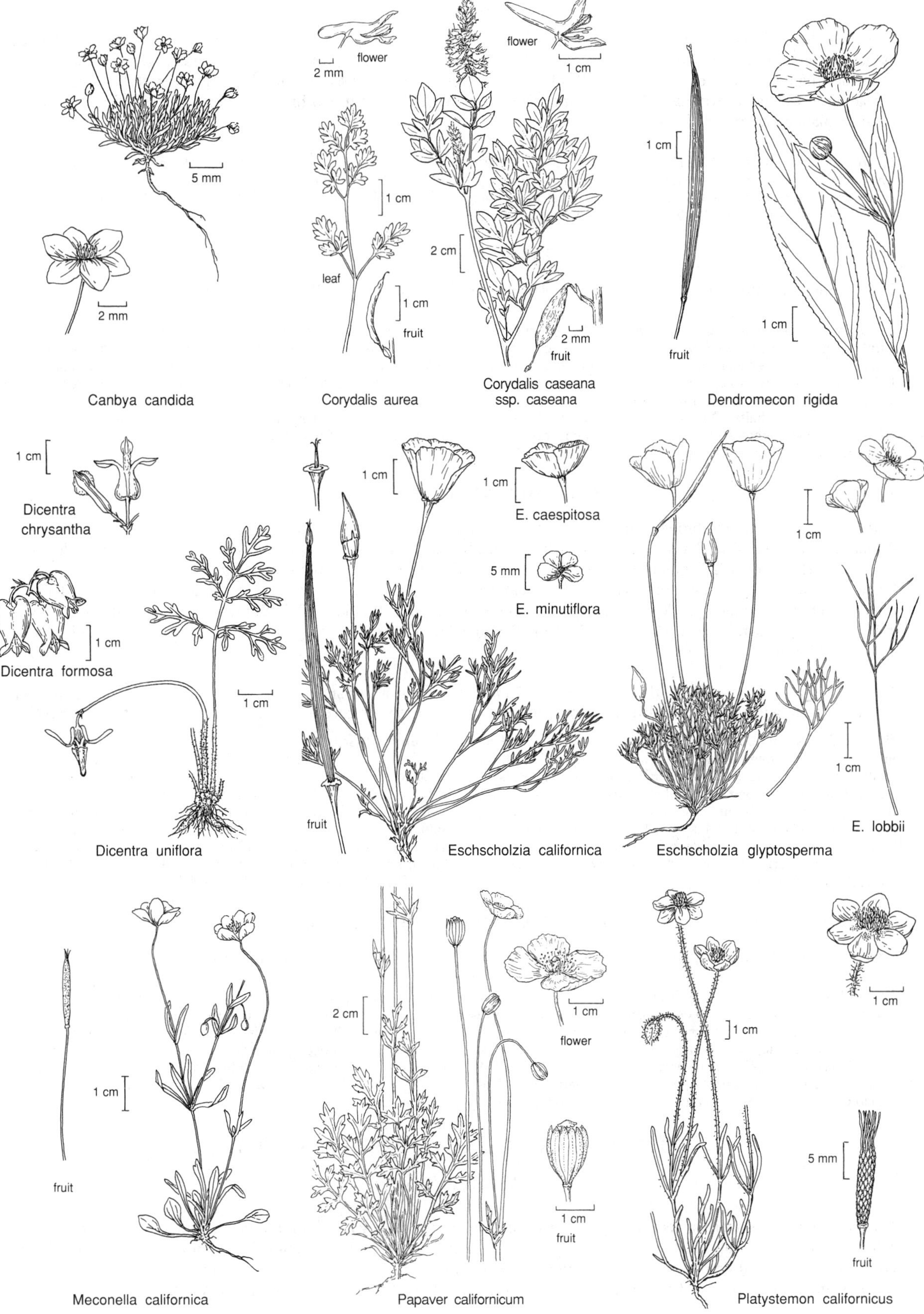

5 mm
2 mm
Canbya candida
flower
2 mm
1 cm
leaf
1 cm
fruit
Corydalis aurea
flower
1 cm
2 cm
2 mm
fruit
Corydalis caseana ssp. caseana
1 cm
fruit
1 cm
Dendromecon rigida
1 cm
Dicentra chrysantha
1 cm
Dicentra formosa
1 cm
Dicentra uniflora
1 cm
fruit
1 cm
E. caespitosa
5 mm
E. minutiflora
Eschscholzia californica
1 cm
1 cm
E. lobbii
Eschscholzia glyptosperma
1 cm
fruit
Meconella californica
2 cm
1 cm
flower
1 cm
fruit
Papaver californicum
1 cm
1 cm
5 mm
fruit
Platystemon californicus

FENDLERELLA

Shrub < 8 dm. **ST**: bark whitish, peeling as thin sheets or strips; twigs strigose. **LVS** deciduous, leathery, ± sessile. **INFL**: cymes clustered, terminal, dense to open; fls (1)3–11. **FL**: odorless; sepals 5; petals 5, white; stamens 10, alternating long and short, filament base wide, flat; ovary half-inferior, chambers 3, placentas axile, ovule 1 per chamber, styles 3, persistent, spreading in fr, stigma terminal. **FR** ± cylindric, septicidal. **SEED** fusiform, red-brown. 3 spp.; sw US, n Mex. (August Fendler, plant collector, 1813–1883)

F. utahensis (S. Watson) A.A. Heller (p. 823) YERBA DESIERTO **LF**: blade 8–16 mm, 3–6 mm wide, ovate to elliptic, strigose, 3-veined from base, margin entire, ± rolled under. **INFL** 12–18 mm, short-peduncled. **FL**: sepals ± 1.5 mm, linear-lanceolate, sparsely strigose; petals 3–4 mm, oblong-obovate. **FR** ± 4 mm. **SEED** ± 2 mm. UNCOMMON. Limestone soils, crevices, slopes; 1300–2800 m. DMtns; to Colorado, n Mex. ❀TRY;DFCLT.

JAMESIA CLIFFBUSH

1 sp. (Edwin P. James, naturalist, 1797–1861) [Holmgren & Holmgren 1989 Brittonia 41:335–350]

J. americana Torrey & A. Gray var. ***rosea*** C. Schneider (p. 823) Shrub < 1 m; herbage gen densely hairy. **ST**: bark gen gray, peeling as narrow strips. **LVS** deciduous; petiole 2–6 mm; blade 1.5–4 cm, 1–2 cm wide, widely ovate to ± round, pinnately veined, margin toothed, upper surface green, finely strigose, lower surface pale gray-white, densely hairy. **INFL**: cyme, terminal; fls (1)3–11. **FL** 1.2–1.5 cm wide, slightly fragrant; sepals 5, 3–4 mm, gray-strigose; petals 5, 5–8 mm, elliptic to obovate, gen pink; stamens 10, alternating long and short, filament base wide, flat; ovary half-inferior, chambers 3–5, 1 in fr, placentas parietal, ovules many, styles 3–5, > sepals, persistent, spreading in fr, stigma terminal. **FR** 1–1.3 cm, conic, septicidal. **SEEDS** many, fusiform, brown. Rocky slopes, cliffs; 2250–3700 m. c&s SNH, W&I, n DMtns (Grapevine Mtns); w NV. [var. *californica* (E. Small) Jepson] Other vars. in GB, Rocky Mtns. ❀TRY;DFCLT.

PHILADELPHUS MOCK ORANGE

Shrub < 3 m. **ST**: bark red-brown, aging gray, peeling as narrow strips or narrow rectangles; twigs glabrous to hairy. **LVS** deciduous, petioled; blade 3-veined from base, ± glabrous to hairy, margin entire to toothed. **INFL**: raceme, panicle, or fl one, terminal, ± open. **FL** fragrant; sepals 4–5, glabrous to hairy; petals 4–5, white; stamens gen many, clustered, filaments linear, fused at base, ovary inferior to half-inferior, chambers 4–5, placentas axile, ovules many, style 1, branches gen 4 above middle, stigma linear along style branch. **FR** becoming woody, gen loculicidal. **SEEDS** many, gen fusiform, gen brown. ± 65 spp.: temp Amer, Eurasia. (Greek: for Ptolemy Philadelphus, Greek king of Egypt, 309–247 B.C.) [Hu 1956 J Arnold Arbor 37:15–90]

1. Fl 2–3 cm wide; infl raceme, fls 6+; lf blade 25–75 mm, lower surface sparsely strigose .... ***P. lewisii***
1' Fl 1–1.5 cm wide; fls 1–3; lf blade 8–25 mm, lower surface densely strigose .... ***P. microphyllus***

P. lewisii Pursh (p. 823) WILD MOCK ORANGE **LF**: petiole 3–8 mm; blade 25–75 mm, 20–40 mm wide, ovate, margin entire to toothed, flat. **INFL**: raceme; fls 6+. **FL** 2–3 cm wide; sepals 4–7 mm; petals 8–12 mm, obovate to oblong. $2n$=26. Slopes, canyons, forest openings; < 1500 m. NW, CaR, SN; to B.C., MT. [ssp. *californicus* (Benth.) Munz] Pls from NW with clearly toothed lvs have been called ssp. *gordonianus* (Lindley) Munz. Variation in lvs, fl size needs critical study. ❀SUN:**4–6,17**&IRR:1–3,**7–9**,10,**14–16,18–24**;CVS.

P. microphyllus A. Gray (p. 823) LITTLELEAF MOCK ORANGE **LF**: petiole 5–18 mm; blade 8–25 mm, 3–8 mm wide, narrowly ovate to elliptic, margin entire, ± rolled under. **INFL**: fls 1–3. **FL** 1–1.5 cm wide; sepals 3–5 mm, petals 6–8 mm, widely elliptic. Rocky slopes, cliffs; 1200–2750 m. e PR (San Jacinto, Santa Rosa mtns), W&I, e DMtns; to TX, n Mex. [ssp. *stramineus* (Rydb.) C. Hitchc.] Pls with lf blades 8–12 mm, rough-hairy above have been called ssp. *pumilus* (Rydb.) C. Hitchc. [Hitchcock 1943 Madroño 7:35–56] ❀DRN,IRR:1–3,**7**,10,14–16,**18**.

WHIPPLEA MODESTY, YERBA DE SELVA

1 sp. (Lieutenant A.W. Whipple, 1818–1863, Commander, Pacific Railroad Expedition of 1853–1854)

W. modesta Torrey (p. 823) Subshrub. **STS**: main prostrate, lateral and terminal decumbent to erect; bark gray-brown, peeling as narrow strips. **LVS** persistent, ± sessile; blade 1.5–4 cm, 1–3 cm wide, ovate to elliptic, coarsely strigose, upper surface hairs ascending, bulbous-based. **INFL**: cyme or raceme, dense; fls 4–12. **FL** 4–6 mm wide, odorless; sepals 4–6, 1.5–2 mm, narrowly oblong, ± glabrous; petals 4–6, 3–6 mm, obovate, white; stamens 8–12, ± equal, filament base wide, flat; ovary half-inferior, chambers 4–5, incomplete, placentas ± axile, ovule 1 per chamber, styles 4–5, stigma terminal. **FR** depressed-spheric, separating into segments. **SEEDS** 1 per chamber, plump, honeycomb-pitted. Slopes, coniferous forest; < 1400 m. NW, w SnFrB (Santa Cruz Mtns), n SCoRO; to WA. ❀DRN:**4–6**&SHD,IRR:**7**,14,**15–17**;GRCVR.

PHYTOLACCACEAE POKEWEED FAMILY

William J. Stone

Per, shrub, tree. **LVS** alternate, entire; stipule 0. **INFL**: raceme or panicle. **FLS** bi- or unisexual, radial; calyx gen deeply 4–5-lobed, persistent; petals gen 0; stamens 4–many, free or fused at base, anthers opening by longitudinal slits; pistils 1 and compound or free and simple, ovary gen superior, ovules 1 per pistil or chamber, styles same number as pistils or chambers (or 0), stigmas linear or thread-like. **FR**: berry; chambers gen many. ± 17 genera, 100 spp.: N.Am, S.Am.

PHYTOLACCA

Per, tall, stout. **LVS** large, petioled, ovate to ovate-lanceolate. **FLS** small; sepals 5, petal-like; stamens 5–30; pistil 1. **FR**: chambers 5–12. **SEED** 1 per chamber, vertical. ± 35 ssp.: trop & subtrop, esp Am. (Greek & Latin: plant, crimson lake, from fr color)

P. americana L. (p. 823) POKEWEED, POKEBERRY, PIGEONBERRY Pl ill-smelling; roots 10–15 cm, thick, fleshy, carrot-like, white. **STS** 1–several from crown, 1–3 m, fleshy, bright purple at maturity. **LF** 8–30 cm, 3–12 cm wide, glabrous; petiole 1–5 cm. **INFL**: raceme erect to nodding, 5–20 cm; pedicels with bracts. **FL** bisexual, 5–6 mm; sepals equal, white or pinkish; stamens 10. **FR** 4–6 mm high, 7–10 mm wide, shiny purplish black. **SEED** 2.5–3 mm, flat, black. $2n$=36. Uncommon. Disturbed areas, gardens, roadsides; gen < 1000 m. CA-FP; native to e US. *P. heterotepala* H. Walter (sepals unequal, stamens 13–20) reported from San Francisco, probably not naturalized.

PITTOSPORACEAE PITTOSPORUM FAMILY

Elizabeth McClintock

Tree, shrub, woody vine, gen evergreen. **LVS** simple, alternate, sometimes crowded at branch tips, often leathery, gen entire, gen petioled. **INFL**: panicle, cluster, or fls solitary. **FL** gen bisexual, radial; sepals 5, free, sometimes fused toward base; petals 5, free, erect or spreading, or weakly adherent ± into a tube; stamens 5; ovary superior, chambers gen 2–3, style 1, stigma gen head-like. **FR**: capsule, berry. **SEEDS** several, often in pulp. 9 genera, ± 200 spp.: warm temp, trop, Old World, esp Australia, New Zealand, e Asia; some orn, esp *Pittosporum*, in CA.

1. Tree, shrub; corolla white or dark red to purple-black; fr a dehiscent capsule .... **PITTOSPORUM**
1′ Woody vine; corolla blue; fr an indehiscent berry .... **SOLLYA**

PITTOSPORUM

Tree, shrub. **LF**: tip acute or rounded. **INFL**: panicle, cluster, or fls solitary, terminal or axillary. **FL** gen functionally unisexual; petals gen adherent below, free, spreading above; anthers free; ovary chambers 2–3. **FR**: capsule, woody, 2–3 valved; pulp resinous. **SEEDS** sticky. ± 150 spp.: warm parts of Australia, New Zealand, Pacific Islands, e Asia, Afr. (Greek: pitch, seed, from resinous seed coating) Some spp. have medicinal and poisonous properties; saponins in *P. crassifolium*.

1. Corolla white; fl fragrant; seed ± red
 2. Shrub, small tree; lf obovate, leathery, tip gen rounded, margin turned under .... ***P. tobira***
 2′ Tree; lf widely lanceolate, papery, tip acuminate, margin ± wavy .... ***P. undulatum***
1′ Corolla dark red to purple black; fl not fragrant; seed purple-black or black
 3. Lf oblong to obovate, leathery, margin turned under, lower surface densely hairy .... ***P. crassifolium***
 3′ LF oblong to oblong-ovate, papery, margin ± wavy, lower surface glabrous .... ***P. tenuifolium***

P. crassifolium Cunn. Shrub, small tree. **ST** < 9 m; branchlets densely hairy. **LF** gen 5–7 cm, oblong to obovate, leathery; lower surface densely gray-hairy; margin turned under; petiole < 2 cm. **INFL**: cluster, terminal. **FL** not fragrant; petals ± 1 cm, dark red to purple-black. **FR** 2–3 cm, round or ovoid, densely hairy, gen 3-valved. **SEED** purple-black. Uncommon. Disturbed urban areas; < 200 m. CCo, expected elsewhere; native to New Zealand.

P. tenuifolium Gaertner Tree. **ST** < 8 m; branchlets lightly hairy. **LF** gen 2–3 cm, oblong to oblong-ovate, papery; lower surface gen paler, glabrous; margin ± wavy; petiole 3–10 mm. **INFL**: cluster, terminal or axillary. **FL** not fragrant; petals 10–12 mm, dark red to purple-black. **FR** 10–12 mm, ± round, glabrous, 2-valved. **SEED** black. Uncommon. Disturbed urban areas; < 200 m. CCo (San Francisco Peninsula), expected elsewhere; native to New Zealand.

P. tobira (Thunb.) Aiton JAPANESE PITTOSPORUM, MOCK ORANGE Shrub, small tree. **ST** < 8 m; branchlets hairy when young. **LF** 5–10 cm, obovate, leathery, glabrous; margin turned under; petiole 5–15 mm. **INFL**: cluster, umbel-like, terminal. **FL** fragrant; petals 10–12 mm, white. **FR** 10–12 mm, ovate, densely hairy, 3-valved. **SEED** ± black. Disturbed urban areas, expected elsewhere; < 200 m. SCo; native to Japan.

P. undulatum Vent. VICTORIAN BOX, MOCK ORANGE Tree. **ST** < 15 m; branchlets hairy, becoming glabrous. **LF** 5–15 cm, widely lanceolate, papery, glabrous; tip acuminate; margin ± wavy; petiole 5–25 mm. **INFL**: cluster, umbel-like, terminal. **FL** fragrant; petals 10–15 mm, white. **FR** 10–15 mm, ± round, glabrous, 2-valved. **SEED** reddish. Uncommon. Disturbed areas; < 200 m. CCo, SCo; native to se Australia.

SOLLYA

Woody vine. **INFL**: cyme, terminal or axillary, few-fld; fls rarely solitary. **FL**: petals free, spreading, blue; anthers adherent into cone around style; ovary chambers 2. **FR**: berry; pulp sticky. **SEEDS** sticky. 3 spp.: w Australia. (R.H. Solly, English botanist, 1778–1850)

S. heterophylla Lindley **LF** 3–5 cm, oblong to narrowly ovate; petiole < 0.5 cm. **INFL** pendent. **FL**: petals 8–10 mm, ovate; style < 5 mm. **FR** 2.5 cm, ± black. Uncommon. Disturbed, urban areas; < 200 m. CCo, SCo; native to w Australia. [*S. fusiformis* (Labill.) Briq.]

PLANTAGINACEAE PLANTAIN FAMILY

Lauramay T. Dempster

Ann, per, gen scapose. **LVS** gen basal, simple, with longitudinal ribs; stipules 0. **INFL**: spike, gen terminal, gen dense; fls few–many, each subtended by 1 bract. **FL** unisexual or bisexual, gen radial; calyx deeply 4-lobed, lobes gen overlapping, persistent, margin mostly scarious; corolla salverform or cylindric, lobes 4, spreading to erect, scarious, persistent, colorless; stamens 2–4, alternate corolla lobes, epipetalous; ovary superior, 2–4-chambered, ovules several per chamber, style 1, stigma long, hairy. 3 genera, ± 270 spp.: worldwide, esp temp; some weedy, some (esp *P. afra*, psyllium) cult for laxative.

PLANTAGO PLANTAIN

Ann, per. **ST** decumbent to erect. **FL** gen bisexual; corolla radial, sometimes bilateral. **FR**: capsule, circumscissile ± at or below middle. **SEEDS** 2–many, gelatinous when wet. ± 250 spp.; worldwide. (Latin: sole of foot)

1. Lvs cauline; infl branched, spikes gen in terminal, cyme-like clusters, peduncles axillary .... ***P. indica***
1′ Lvs in basal rosette; infl scapose, with 1 terminal spike
 2. Per
  3. Pl with clustered fibrous roots — spike 3–30 cm, infl 5–60 cm incl peduncle
   4. Lf widely elliptic to ± cordate, narrowed abruptly to petiole; corolla lobes spreading .... ***P. major***
   4′ Lf elliptic-oblanceolate, tapered gradually to petiole; corolla lobes gen enclosing fr, together beak-like (spreading in staminate fls) .... ***P. subnuda***
  3′ Pl taprooted
   5. Lf deeply pinnately lobed .... ***P. coronopus***
   5′ Lf entire or lobes few and shallow
    6. Spike 2–8 cm, ± ovoid, dense (becoming cylindric in fr) .... ***P. lanceolata***
    6′ Spike (2)8–18 cm, narrowly cylindric, gen ± loose near base
     7. Lf petioled; blade oblanceolate to narrowly elliptic-ovate; corolla glabrous; inland (KR, n CaR, SNE) .... ***P. eriopoda***
     7′ Lf not petioled; blade linear or narrowly oblanceolate; corolla tube short-hairy outside; coastal (NCo, CCo, SnFrB, n ChI) .... ***P. maritima***
 2′ Ann (sometimes bien)
  8. Lf deeply cut, lobes ascending, sharp; spike nodding, becoming erect or wavy in fr — fr gen with 3 seeds .... ***P. coronopus***
  8′ Lf ± entire; spike erect
   9. Corolla lobes lanceolate, acute
    10. Lf petioled; blade oblanceolate to obovate, margin entire or ± toothed .... ***P. virginica***
    10′ Lf not petioled; blade thread-like to narrowly oblanceolate, margin gen entire
     11. Lf narrowly oblanceolate; infl dense; corolla lobes ± 2 mm .... ***P. truncata*** ssp. ***firma***
     11′ Lf thread-like to linear; infl ± loose; corolla lobes ± 0.5 mm
      12. Corolla gen with 1 erect and 3 spreading lobes; seeds 4–28 .... ***P. elongata***
      12′ Corolla lobes gen erect; seeds 4 .... ***P. pusilla***
   9′ Corolla lobes round-ovate, gen obtuse (± acute in *P. patagonica*)
    13. Bracts linear, 1–12 × sepals, some or all visible
     14. Bracts gen 3–12 × sepals; pl drying dark green, glabrous or ± hairy .... ***P. aristata***
     14′ Bract gen 1–2 × sepals; pls drying light green, softly hairy throughout .... ***P. patagonica***
    13′ Bracts ovate to round, gen < or = sepals, ± hidden in infl
     15. Herbage gen sparsely hairy; bracts gen < sepals; seeds not shiny; grassland, woodland .... ***P. erecta***
     15′ Herbage ± densely silky-hairy; bracts ± = sepals; seeds shiny; mostly dry places .... ***P. ovata***

P. aristata Michaux (p. 823) Ann, glabrous to ± hairy. **LF** 5–15 cm, linear; margin entire. **INFLS** gen 1–6, 9–35 cm incl peduncle; spike 1–13 cm, dense; bract 3–12 × sepals, linear, ascending, clearly > fls. **FL**: corolla lobes 1.5–2.3 mm, round-ovate, 3 gen spreading, 1 erect, tips obtuse; stamens 4. **SEEDS** 2, 2–2.5 mm. $2n$=20. Waste places; < 200 m. NCo, GV, SCo (expected elsewhere); native to e US.

P. coronopus L. (p. 823) Ann or bien from taproot, coarsely hairy. **LF** 4–25 cm, narrowly lanceolate, gradually tapered to base, pinnately lobed; lobes ascending, acute. **INFLS** gen 1–6, 5–50 cm incl peduncle; spike 2–22 cm, narrowly cylindric, dense, nodding, becoming erect or often wavy in fr; bract 1–2 × sepals, exserted, base round, tips long-acuminate. **FL**: corolla lobes ± 1 mm, 3 gen spreading, 1 gen erect, ovate-lanceolate; stamens 4. **SEEDS** gen 3, ± 1 mm. Coastal bluffs, salt marshes, trampled places (where weedy), chaparral, grassy flats; < 300 m. NCo, n SNF (Amador Co.), ScV, CCo, SCo (Ventura Co.), s ChI (Santa Catalina Island); native to Eur.

P. elongata Pursh (p. 823) Ann, glabrous or sparsely hairy. **LF** 3–10 cm, linear to thread-like, entire, sometimes few-toothed. **INFLS** few–many, 2–18 cm incl peduncle; spike 0.5–5 cm, dense or often ± loose; bract ± = sepals, appressed to sepals, round-ovate, keel wide, fleshy, margin translucent, not exserted. **FL**: corolla lobes 0.5 mm, gen 3 spreading, 1 erect, ovate-lanceolate; stamens 2. **SEEDS** 4–28, ± 1.5 mm, slender. $2n$=12,36. Saline and alkaline places, beaches, vernal pools; < 200 m. CA-FP (exc CaR, SN); to B.C. [*P. bigelovii* A. Gray ssp. *californica* (E. Greene) I.J. Bassett] Pls with 5 seeds per fr have been called ssp. *pentasperma* I.J. Bassett; pls with fr 2–3 mm, 4–9 seeds, $2n$=20, have been called *P. bigelovii* A. Gray.

P. erecta E. Morris (p. 823) Ann, ± silky or long-hairy; hairs scattered. **LF** 3–13 cm, thread-like to narrowly oblanceolate, entire or teeth few, small. **INFLS** 1–many, 3–30 cm incl peduncle; spike 0.5–3 cm, head-like to short-cylindric, dense, hairy; bract 0.4–0.5 × sepals, ovate, not exserted. **FL**: corolla lobes 2–2.7 mm, spreading or reflexed, round-ovate, tips obtuse; stamens 4. **SEEDS** 2, 2–2.5 mm. 2*n*=20. Sandy, clay, or serpentine soils, grassy slopes and flats, open woodland; < 700 m. CA-FP; to OR, Baja CA. [ssp. *rigidior* Pilger; *P. hookeriana* Fisher & C. Meyer var. *californica* (E. Greene) I. Poe & *P. bigelovii* ssp. *c.* (E. Greene) I.J. Bassett, misapplied] ❀TRY.

P. eriopoda Torrey (p. 823) Per, from stout caudex, taprooted below, gen glabrous exc peduncle. **LF** 5–25 cm, oblanceolate to narrowly elliptic-ovate, tapered gradually to wide petiole, margin gen ± wavy or minutely dentate. **INFLS** gen 1–6, 10–55 cm incl peduncle; spike 8–18 cm, narrowly cylindric, becoming arched, dense, loose near base; bract ± = sepals, appressed to sepals, widely ovate, not exserted. **FL**: corolla lobes ± 1 mm, spreading, ovate-lanceolate; stamens 4. **SEEDS** 2–4, ± 2 mm. 2*n*=24. Moist and alkaline places; < 1000 m. KR, n CaR, SNE (Mono Co.); to Can, c US.

P. indica L. Ann, ± coarsely hairy. **ST** 2–6 cm, branched, leafy. **LVS** cauline, opposite (appearing clustered at nodes), 2–4.5 cm, thread-like or linear; margin entire. **INFLS** axillary; spikes gen in terminal, cyme-like clusters, 1–2 cm, round to elliptic in outline, dense; peduncle 3–7 cm; lower bracts > fls, tips long, slender; upper bracts gen = sepals, round-ovate, obtuse, enclosing fls. **FL**: corolla lobes ± 2 mm, spreading, narrowly ovate; stamens 4. **SEEDS** 2, 2.5 mm. Sandy waste places; < 200 m. CCo, SnFrB, SCo; native to Eurasia. Often used for bird seed and poultry feed.

P. lanceolata L. (p. 823) ENGLISH PLANTAIN Per from stout caudex, taprooted, short-hairy. **LF** 5–25 cm, lanceolate to lance-oblong, tapered gradually to petiole; margin gen finely dentate. **INFLS** few–many, 20–80 cm incl peduncle; spike 2–8 cm, ± ovoid, becoming cylindric, dense; bract ± = sepals, widely ovate, acute, not exserted. **FL**: corolla lobes 2–2.5 mm, spreading, ovate, tips ± acute; stamens 4. **SEEDS** 1–2, ± 3 mm. 2*n*=12,24,96. Common. Weed of waste places, lawns, roadsides; < 1600 m. CA-FP; native to Eur.

P. major L. (p. 827) COMMON PLANTAIN Per (rarely ann), gen glabrous exc peduncle; caudex short, with fibrous roots. **LF** blade 5–18 cm, widely elliptic to somewhat cordate, narrowed ± abruptly to petiole; margin entire or ± finely dentate, rough to touch. **INFLS** gen 3–7, 5–60 cm incl peduncle; spike gen 3–20 cm, linear-cylindric, becoming ± loose; bract ± = sepals, ovate, not exserted. **FL**: corolla lobes 0.5 mm, spreading, ovate-lanceolate; stamens 4. **SEEDS** 5–16, < 1 mm. 2*n*=6,12,24. Disturbed areas; < 2200 m. CA-FP (exc SN), GB; to e US; native to Eur. Highly variable; named vars. (incl *pilgeri* Domin & *scopulorum* Fries & S.P. Broberg) indistinct.

P. maritima L. (p. 827) Per, gen glabrous exc peduncle; caudex stout; taproot long, thick below. **LF** 3–15 cm, linear to narrowly oblanceolate, gradually tapered to base, entire or sparsely dentate. **INFLS** few–many, gen 8–30 cm incl peduncle; spike 2–10 cm, narrowly cylindric, gen dense, looser near base; bract ± = sepals, widely ovate, not exserted. **FL**: corolla lobes ± 1 mm, 3 gen spreading, 1 erect, ovate-lanceolate; stamens 4. **SEEDS** 2–3, 2 mm. Coastal bluffs, wet and saline places; < 150 m. NCo, CCo, SnFrB, n ChI; to AK, ne N.Am. Tall, narrow-lvd pls from wet places have been called var. *juncoides* (Lam.) A. Gray; low pls with spreading or ascending peduncles from dry places or coastal bluffs have been called var. *californica* (Fern.) Pilger. ❀DRN,IRR:**14,22–24**& SUN:5,**15–17**.

P. ovata Forsskal Ann, ± densely silky-hairy. **LF** 2–17 cm, linear or rarely oblong, entire or teeth few, minute. **INFLS** few–many, gen 2–27 cm incl peduncle; spike 0.5–3.5 cm, gen short-cylindric, dense, woolly; bract ± = sepals, ovate to round, not exserted. **FL**: corolla lobes 1.3–2.8 mm, spreading, round-ovate, tips obtuse; stamens 4. **SEEDS** 2, 2–2.5 mm. 2*n*=8. Sandy or gravelly soils, creosote-bush scrub, Joshua-tree woodland, sagebrush scrub, coastal strand; < 1400 m. CW, SW, SNE, D; to UT, TX, Baja CA; also Medit. [*P. insularis* Eastw. incl var. *fastigiata* (E. Morris) Jepson] May be alien, naturalized very early from Medit.

P. patagonica Jacq. (p. 827) Ann, ± densely hairy. **LF** 2–10 cm, linear or very narrowly oblanceolate, entire. **INFLS** often many, 2–18 cm incl peduncle; spike 1–6 cm, nearly cylindric, a little wider at base, dense, ± densely woolly; bract 1–2 × sepals, linear, lower exserted. **FL**: corolla lobes 1.5–2 mm, spreading (or 1 erect), ovate, tips ± acute; stamens 4. **SEEDS** 2, ± 3 mm. 2*n*=20. Sandy, rocky or grassy slopes, pinyon/juniper or Joshua-tree woodland, chaparral; 500–2200 m. SnBr, SnJt, DMoj; to B.C., n-c US, TX, Mex; also Argentina. [*P. purshii* Roemer & Shultes var. *oblonga* (E. Morris) Shinners] ❀TRY.

P. pusilla Nutt. Ann, glabrous or sparsely hairy. **LF** 1–10 cm, thread-like, gen entire. **INFLS** 1–many, 2–19 cm incl peduncle; spike 0.5–8 cm, slender, ± loose; bract ± = sepals, lanceolate, not exserted. **FL**: corolla lobes 0.5 mm, gen erect, lanceolate, tips acute; stamens 2. **SEEDS** 4, 0.7–1.3 mm. 2*n*=12. Sandy soils; < 100 m. SCo (San Diego Co.); native to e US.

P. subnuda Pilger (p. 827) Per, nearly glabrous to ± hairy; caudex stout; fibrous roots below. **LF** 12–40 cm, elliptic-oblanceolate, tapered to ± wide petiole; margin entire or finely dentate. **INFLS** gen 1–5, 9–50 cm incl peduncle; spike 8–30 cm, cylindric, dense, becoming loose at base; bract ± = sepals, elliptic-lanceolate, not exserted. **FL**: corolla lobes ± 3 mm, mostly erect in pistillate fls (spreading in staminate fls), lanceolate, tips acute, together beak-like; stamens 4. **SEEDS** 3, ± 1.6 mm. Coastal bluffs and marshes, wet meadows, ditches; < 300 m. NCo, CCo, SnFrB, SCo, n ChI; to c Mex. [*P. hirtella* Kunth var. *galeottiana* (Decne.) Pilger]

P. truncata Cham. ssp. *firma* (Kunze) Pilger (p. 827) Ann, finely long-hairy. **LF** 1–6 cm, narrowly oblanceolate, tapered gradually to base, gen entire, sometimes with few small teeth. **INFLS** 1–few, 3–7 cm incl peduncle; spike 0.5–2.5 cm, dense; bract ± = sepals, ovate-lanceolate, not exserted. **FL**: corolla lobes ± 2 mm, erect in pistillate fls, lanceolate, tips acute, forming sharp beak; stamens 4. **SEEDS** 2, ± 1.8 mm. Moist meadows and ditches; < 300 m. ScV, SnFrB; native to Chile.

P. virginica L. Ann, bien, finely long-hairy. **LF** 2–12 cm, widely oblanceolate or obovate, tapered gradually to petiole, entire or slightly toothed. **INFLS** 1–many, 3–33 cm incl peduncle; spike 1–14 cm, dense, becoming loose below; bract ± = sepals, lanceolate to narrowly elliptic, not exserted. **FL**: corolla lobes 2–3 mm, curved upward in pistillate fls (spreading in staminate fls), lanceolate, tips acute; stamens 4. **SEEDS** 2, 1.5–2 mm, pale brown. 2*n*=24. Disturbed places; < 1000 m. CaRF (Shasta Co.), SCo; native to c&e U.S. Pls with seeds 2.5–3 mm and red-brown have been called *P. rhodosperma* Decne.

PLATANACEAE PLANE TREE, SYCAMORE FAMILY

Elizabeth McClintock

Tree, gen monoecious, wind-pollinated. **ST**: branches irregular below, spreading to erect above; bark irregularly colored, scale-like, peeling; twigs dense-hairy. **LVS** simple, alternate, deciduous, gen palmately 3- or 5-lobed, -veined; stipules gen lf-like, free or fused around st, shed before lvs; petiole at base dilated, hollow, ± covering bud; blade dense-hairy, glabrous in age, hairs stellate. **INFL**: heads 1–6, ± evenly spaced on axis, spheric, many-fld, sessile or on pendent peduncles, gen unisexual; staminate breaking apart in age; pistillate persistent; bracts subtending heads, fls. **FLS** unisexual; calyx cup-shaped, scale-like, entire or 3–6(8)-lobed or sepals ± free. **STAMINATE FL**: petals 3–6, fleshy or scale-like, minute or vestigial; stamens 3–6(8), alternate petals, anthers subsessile, axis above anther expanded, disc-like; pistils vestigial. **PISTILLATE FL**: petals 3–6, minute, or gen 0; staminodes often 3–4; pistils (3)5–9, ovaries superior, 1-chambered, gen 1-ovuled, style 1. **FR**: spheric head of small, hairy, basally bristly achenes; style persistent. 1 genus, ± 8 spp.: n temp; some orn, esp *P.* ×*acerifolia* (Aiton) Willd., London plane tree; some cult for wood, veneer. [Ernst 1963 J Arnold Arbor 44: 206–210]

PLATANUS

(Greek: probably broad, from lvs)

P. racemosa Nutt. (p. 827) WESTERN SYCAMORE **ST** 10–35 m; base < 1 m wide; bark smooth, pale. **LF**: stipules 2–3 cm; petiole 3–8 cm; blade ± 10–25 cm, ± round, lobes 3 or 5, acute to acuminate, entire. **INFL**: heads 3–5, ± 1 cm. **FR**: heads 2–3 cm, sessile or not. $2n=42$. Common. Streamsides, canyons; < 2000 m. c&s SNF, Teh, GV, CW, SW; Baja CA. Pls in PR, with some pistillate infls with peduncles 5–15 mm have been called var. *wrightii* (S. Watson) L. Benson. ❀IRR,SUN:1–3,7–9,**11**,14–17,**18–21**,22–24; susceptible to sycamore anthracnose.

PLUMBAGINACEAE LEADWORT FAMILY

Elizabeth McClintock

Ann, per, subshrub, or vine. **LVS** simple, gen in a basal rosette (sometimes cauline), entire or lobed. **INFL**: panicle, head, raceme, or cyme, gen ± scapose. **FL** bisexual, radial, gen small; calyx tubular, often membranous or partly scarious, lobes 5; petals 5, nearly free to ± fused, clawed, ± twisted together; stamens 5, opposite petals, sometimes epipetalous; ovary superior, gen 5-lobed or -ribbed, chamber and ovule 1, styles 5, sometimes fused. **FR**: utricle, achene, or capsule, ± enclosed in persistent calyx. ± 12 genera, ± 400 spp.: ± worldwide, esp Medit, w&c Asia; some cult as orn (*Limonium* used as dried fl). [Channell & Wood 1959 J Arnold Arbor 40:391–397]

1. Infl a head on an unbranched, scapose peduncle; lf linear .... **ARMERIA**
1′ Infl a panicle, branched ± from pl base; lf oblanceolate to obovate (wings of infl axes with linear appendages in *L. sinuatum*) .... **LIMONIUM**

ARMERIA THRIFT, SEA-PINK

Per, rhizomed. **LVS** many, often linear, entire, sessile; veins 1–7, parallel. **INFL**: head, on unbranched, scapose peduncle; involucre sheathing, recurved, other bracts subtending individual fls. **FL**: calyx 10-ribbed; styles 5, fused at base, hairy. ± 35 spp.: Eur, Medit, w coast N.Am, w coast S.Am. (Latin, from Old French: name for a dianthus with spheric heads) [Porsild 1955 Natl Mus Canada Bull 135:170–174]

A. maritima (Miller) Willd. ssp. ***californica*** (Boiss.) Pors. (p. 827) **LF** 6–12 cm, 1–2.5 mm wide, flat, glabrous. **INFL** < 2 cm wide; peduncle < ± 30 cm. **FL**: calyx < corolla, both gen pink; calyx tube ± hairy on ribs, lobe tips short, sharp. Ocean bluffs, ridges, coastal strand, sand, exposed grasslands; < 200m. NCo (common), CCo, SCo (uncommon). [var. *c.* (Boiss.) G. Lawr.] ❀DRN:4–6& IRR:7,14,**15–17**,**21–24**.

LIMONIUM SEA-LAVENDER, MARSH-ROSEMARY

Per (rarely ann); rhizome ± woody. **LVS** few–several, oblanceolate to obovate, entire or lobed, gen petioled. **INFL**: panicle, branched ± from pl base, often ending in (but ± open between) 1-sided, spike-like clusters; involucre 0, other bracts subtending individual fls and along axes. **FL**: calyx gen 5–10-ribbed; styles 5, ± free. ± 150 spp.: ± worldwide, often in saline soils. (Greek: meadow, from habitat of many spp.) [Kunkel & Sunding 1967 Cuad de Bot Mus Canar 2:9–18]

1. Lf pinnately lobed; infl axis wings 3–5, very narrow, rough-hairy .... ***L. sinuatum***
1′ Lf ± entire or wavy-margined; infl axis wings 0 or 2, wide, glabrous
  2. Infl with some non-fl branches .... ***L. otolepis***
  2′ Infl without non-fl branches
    3. Infl > 1 m, axes winged .... ***L. arborescens***
    3′ Infl < 1 m, axes not winged
      4. Lf blade obovate to oblong, base tapered; calyx lobes white .... ***L. californicum***
      4′ Lf blade round to wide-ovate, base ± truncate; calyx lobes blue-purple .... ***L. perezii***

L. arborescens (Brouss.) Kuntze **LF**: blade 5–12 cm, < petiole, ovate to oblong-ovate, entire, base tapered. **INFL** > 1 m; axes glabrous, wings 2, wide, lf-like, veiny, not prolonged at branch points. **FL**: calyx tube 5–6 mm, glabrous, lobes blue-purple; corolla lobes white. Uncommon. Disturbed urban areas, coastal lagoons; ± 10 m. SCo; native to Canary Islands.

L. californicum (Boiss.) A.A. Heller (p. 827) WESTERN MARSH-ROSEMARY **LF**: blade 5–15 cm, 1.5–6 cm wide, obovate to oblong, ± entire or wavy-margined, ± thick, leathery, base tapered; petiole gen < blade, stout, very narrowly winged. **INFL** < 35 cm; axes glabrous, not winged. **FL**: calyx tube 4–5 mm, ribs gen ± hairy, lobes ± 0.5 mm, white; corolla lobes blue, slightly > calyx. Common. Coastal strand, salt marshes, beaches, bays; < 50 m. NCo, CCo, SCo; n Baja CA. [var. *mexicanum* (S.F. Blake) Munz] ❀WET or IRR (brackish marsh,SUN:5,15,16,**17**,21,**22–24**; also STBL.

L. otolepis (Schrenk) Kuntze **LVS** in basal rosette dead before fls appear; blade 3–8 cm, 1.5–3 cm wide, < or > petiole, obovate to oblong, ± entire, base tapered; lvs on lower infl axis < 3 cm, ± round, sessile, clasping st. **INFL** 40–80 cm; branches slender, most without fls; axes glabrous, not winged. **FL**: calyx tube ± 1 mm, lobes ± 0.5 mm, white. Uncommon. Disturbed, coastal, urban areas, esp salt marshes, roadsides; < 100 m. SnFrB, SCo; native to w&c Asia. [*L. perfoliatum* (Boiss.) Kuntze misapplied]

L. perezii (Stapf) Hubb. **LF**: blade 4–15 cm, < or = petiole, round to wide-ovate, ± entire, base ± truncate. **INFL** 15–45 cm; axes glabrous, not winged. **FL**: calyx lobes blue-purple; corolla lobes white. Uncommon. Disturbed coastal areas, cliffs, sand dunes, roadsides; < 100 m. SCoRO, SCo; native to Canary Islands.

L. sinuatum (L.) Miller **LF**: blade 3–12 cm, ± obovate, pinnately lobed, base tapered; petiole < or > blade, winged. **INFL** 30–50 cm;

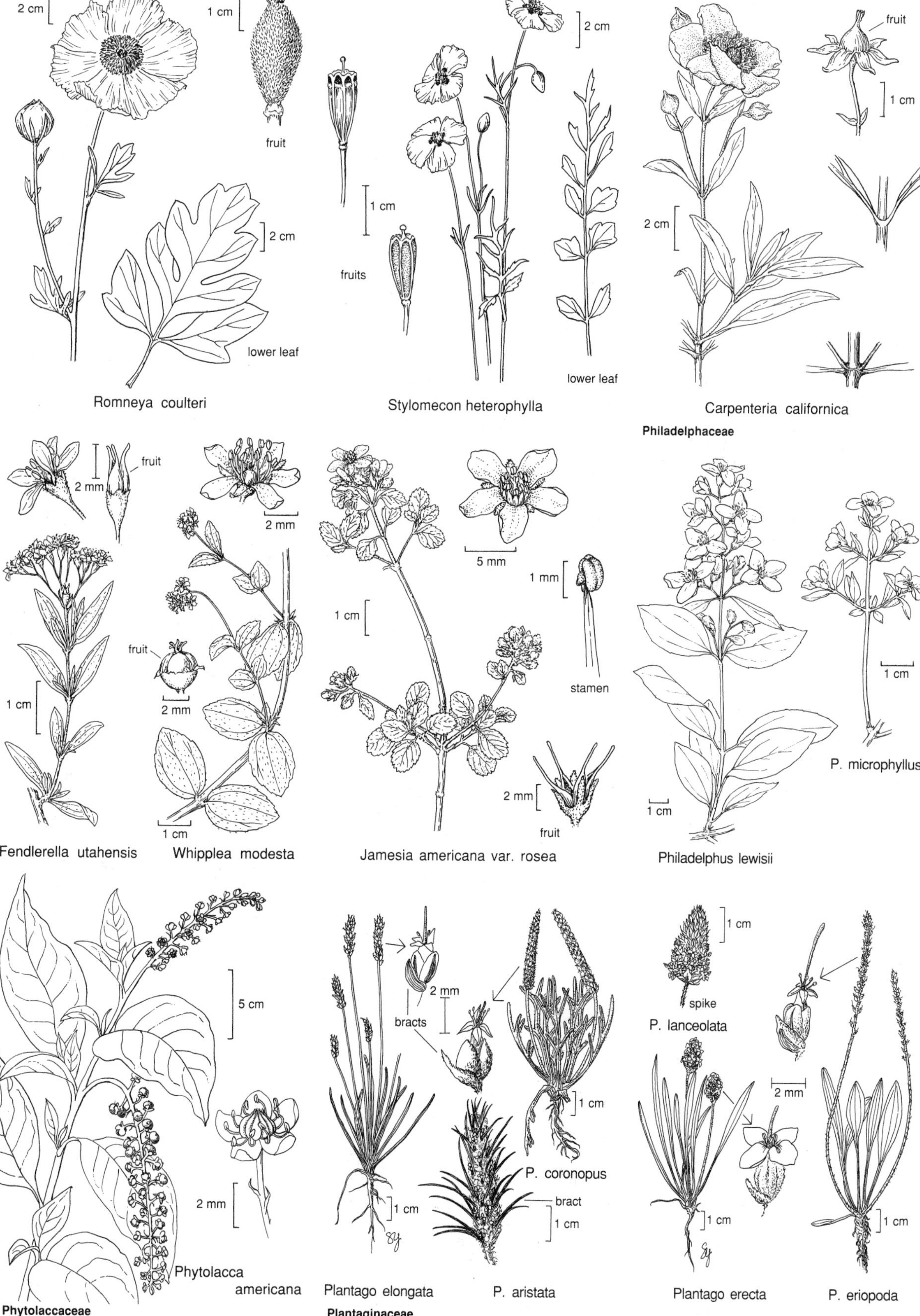

2 cm
1 cm
fruit
1 cm
fruits
2 cm
2 cm
lower leaf
lower leaf
Romneya coulteri
Stylomecon heterophylla
fruit
1 cm
2 cm
Carpenteria californica
Philadelphaceae
fruit
2 mm
2 mm
fruit
2 mm
1 cm
1 cm
Fendlerella utahensis
Whipplea modesta
5 mm
1 mm
1 cm
stamen
2 mm
fruit
Jamesia americana var. rosea
1 cm
1 cm
P. microphyllus
Philadelphus lewisii
5 cm
2 mm
Phytolacca americana
Phytolaccaceae
2 mm
bracts
1 cm
P. coronopus
bract
1 cm
1 cm
Plantago elongata
P. aristata
Plantaginaceae
1 cm
spike
P. lanceolata
2 mm
1 cm
1 cm
Plantago erecta
P. eriopoda

axes rough-hairy; wings 3–5, very narrow, each with a wider, ± linear, lf-like appendage, gen at branch points. **FL**: calyx lobes blue; corolla lobes pale yellow. Uncommon. Disturbed coastal areas, vacant lots, old fields, roadsides; < 300 m. SnFrB, SCoRO, SCo; native to Medit, w Asia.

POLEMONIACEAE PHLOX FAMILY

Robert W. Patterson, Family Editor

Ann, per, shrub, vine. **LVS** simple or compound, cauline (or most in basal rosette), alternate or opposite; stipules 0. **INFL**: cymes, heads, or fls solitary. **FL**: calyx gen 5-ribbed, ribs often connected by translucent membranes that are gen torn by growing fr; corolla gen 5-lobed, radial or bilateral, salverform to bell-shaped, throat often well defined; stamens gen 5, epipetalous, attached at same or different levels, filaments of same or different lengths, pollen white, yellow, blue, or red; ovary superior, chambers gen 3, style 1, stigmas gen 3. **FR**: capsule. **SEEDS** 1–many, gelatinous or not when wet. 19 genera, 320 spp.: Am, n Eur, n Asia; some cult (*Cantua*, *Cobaea* (cup-and-saucer vine), *Collomia*, *Gilia*, *Ipomopsis*, *Linanthus*, *Phlox*). [Grant 1959 Nat Hist Phlox Family, M. Nijhof]

1. Lvs pinnately compound .... **POLEMONIUM**
1′ Lvs simple, entire to deeply lobed
2. Lvs 0 (exc cotyledons); involucral bracts fused near base; filaments 0 .... **GYMNOSTERIS**
2′ Lvs present; involucral bracts (if present) not fused; filaments present
3. Calyx membrane 0; calyx sinus pleated to expanded in fr .... **COLLOMIA**
3′ Calyx membrane present between ribs; calyx sinus not pleated or expanded in fr
4. Lvs opposite, gen cauline
5. Stamens attached at different levels .... **PHLOX**
5′ Stamens attached at same level
6. Lvs pinnately lobed, lobes spine-tipped; fl parts gen in 6's .... [2]**LEPTODACTYLON**
6′ Lvs palmately lobed or entire, lobes sometimes sharp but not spine-tipped; fl parts in 5's ..... **LINANTHUS**
4′ Lvs alternate, sometimes in basal rosette
7. Lvs deeply palmately lobed .... [2]**LEPTODACTYLON**
7′ Lvs not deeply palmately lobed
8. Lvs and calyx lobes bristle-tipped
9. Corolla radial; lf bristles in 2's or 3's .... **LANGLOISIA**
9′ Corolla bilateral; lf bristles solitary .... **LOESELIASTRUM**
8′ Lvs and calyx lobes not bristle-tipped
10. Infl a dense, spiny-bracted head; calyx lobes gen unequal
11. Infl woolly .... **ERIASTRUM**
11′ Infl gen hairy, not woolly .... **NAVARRETIA**
10′ Infl not a dense, spiny-bracted head; calyx lobes equal
12. St scapose; lvs abruptly reduced above well developed basal rosette .... [3]**GILIA**
12′ St not clearly scapose; lvs ± gradually reduced above basal lvs
13. Fls densely clustered, collectively head-like, ± subsessile
14. Lower lvs 1–2-pinnately lobed, axis and lobes ± linear, not clearly flat; corolla blue or throat yellow, purple-spotted .... [3]**GILIA**
14′ Lower lvs entire or 1-pinnately to -palmately toothed to lobed, axis and lobes narrowly elliptic to oblanceolate, ± flat; corolla white or light yellow .... [2]**IPOMOPSIS**
13′ Fls not densely clustered, 1 per node or few-fld cyme, collectively raceme- or panicle-like; fls gen pedicelled
15. Upper lvs ± palmately lobed, distal lobe gen > lateral lobes; lf and sepal tips gen rounded at 20× (if pointed, not translucent); seeds ± dark brown to black .... **ALLOPHYLLUM**
15′ Upper lvs entire or pinnately lobed, distal lobe gen = or > lateral lobes; lf and sepal tips ± sharp-pointed at 20 ×; seeds light brown to grayish
16. Corolla gen bell- or funnel-shaped; throat color clearly different from tube color .... [3]**GILIA**
16′ Corolla gen salverform; tube color gen uniform (red, pink, lavender, or white) — corolla lobes gen mottled .... [2]**IPOMOPSIS**

ALLOPHYLLUM

Alva G. Day

Ann, hairy, ± glandular; glands minute. **ST** erect, leafy, gen branched. **LVS** alternate, dark green, gen deeply pinnately lobed, ± palmately lobed upward; lobes linear to lanceolate, blunt-tipped, central lobe widest. **INFL**: fls gen in clusters. **FL**: calyx lobes blunt-tipped, tube narrowly membranous between ribs, ribs translucent below in fr, membrane splitting; corolla funnel-shaped, radial or bilateral, throat narrow, tapered, lobes narrowly obovate; stamens attached in tube; stigmas 3. **FR** spheric, < calyx; valves gen falling. **SEEDS** 1–3 per chamber, concave, black or brown, gelatinous when wet; ends rounded. 4 spp.: w N.Am. (Greek: other leaf) [Grant & Grant 1955 El Aliso 3:93–110]

1. Widest lvs or lobes < 3 mm wide, linear to narrowly lanceolate; corolla dark blue-purple; st glands short-stalked .... ***A. gilioides***

2. Fls 4–8 in ± dense clusters .... ssp. ***gilioides***
2′ Fls loosely paired or in 3's, not clustered .... ssp. ***violaceum***
1′ Widest lvs or lobes 3–15 mm wide, lanceolate; corolla not dark blue-purple; st glands gen long-stalked
3. Corolla ± 2-lipped, lobes > 1/2 tube + throat; longest stamens well exserted .... ***A. glutinosum***
3′ Corolla not 2-lipped; lobes < 1/3 tube + throat; longest stamens exserted or incl
4. Lower lvs 3–13-lobed; corolla 8–22 mm, tube reddish purple, lobes pink .... ***A. divaricatum***
4′ Lower lvs entire or coarsely toothed; corolla < 11 mm, white .... ***A. integrifolium***

A. divaricatum (Nutt.) A.D. Grant & V. Grant (p. 827) Pl with skunk-like odor. **ST** < 60 cm; glands long- or short-stalked. **LF**: lobes 3–13, 4–8 mm wide, ± lanceolate. **INFL** open or dense; fls 2–8 in clusters; pedicels elongating in fr. **FL**: corolla 8–22 mm, tube reddish purple, lobes 2–4 mm, pink; stamens unequal, incl or longest exserted; style incl or exserted. **SEEDS** 1–3 per chamber, brown. $2n$=16,18. Sandy areas, chaparral, woodlands; 300–1800 m. KR, NCoRI, s CaR, SNF, SnFrB, SCoR, TR.

A. gilioides (Benth.) A.D. Grant & V. Grant **ST** < 40 cm, puberulent; hairs gland-tipped in infl. **LF**: lobes 0–11, linear to narrowly lanceolate. **INFL** open or dense; fls 2–8 in clusters. **FL**: corolla < 10 mm, dark blue-purple, lobes 1–3 mm; stamens nearly equal, incl to slightly exserted; style incl. **SEEDS** 1 per chamber, black. Open, sandy, gen damp or grassy areas; 200–2900 m. NCoR, CaR, SN, SnFrB, SCoR, TR, PR, DMtns (Panamint Mtns); s OR, w NV, AZ, Baja CA.

ssp. ***gilioides*** (p. 827) **ST** < 40 cm. **LVS** basal, cauline; lower pinnately 5–11-lobed, lobes 2–4 mm wide. **INFL** dense; fls 4–8 in clusters; pedicels gen not elongating in fr. **FL**: corolla 6–10 mm. Common. Habitat and range of sp.; 200–1900 m.

ssp. ***violaceum*** (A.A. Heller) Day **ST** gen ± 15 cm, slender. **LVS** ± cauline, entire or 3–7-lobed; lobes 0.5–3 mm wide. **INFL** open; fls solitary or 2–3 in clusters; pedicels gen elongating in fr. **FL**: corolla 5–8 mm. Habitat and range of sp. (exc not in CaR, OR); 1200–2900 m. [*A. violaceum* (A.A. Heller) A.D. Grant & V. Grant]

A. glutinosum (Benth.) A.D. Grant & V. Grant (p. 827) Pl with skunk-like odor. **ST** 10–60 cm; glands long-stalked, conspicuous. **LVS** 7–21-lobed; lobes of basal lvs 1–2 mm wide, of upper lvs 3–11 mm wide, lanceolate to elliptic. **INFL** ± open; fls 2–8 in clusters; pedicels elongating slightly in fr. **FL**: corolla bilateral, ± 2-lipped, 6–11 mm, pale blue to lavender, lobes 1/2 to = tube + throat; longest stamens, style well exserted, curving up from lower lip. **SEEDS** 1–3 per chamber, brown. $2n$=18. Rocky or sandy, often moist areas in sun or shade; 30–1600 m. SCoRO, s ChI (Santa Catalina Island), TR, PR; n Baja CA.

A. integrifolium (Brand) A.D. Grant & V. Grant **ST** 10–25 cm, gen unbranched; glands long-stalked, conspicuous. **LVS** ± cauline; lower entire or coarsely toothed; upper gen 3-lobed; largest lobes 5–15 mm wide. **INFL** ± open; fls 2–4 in clusters; pedicels elongating in fr. **FL**: corolla 6–11 mm, lobes 1–2 mm, white to pale blue; stamens, style incl. **SEEDS** 1–2 per chamber, brown. $2n$=16,18. Common. Open, rocky or sandy, gen moist areas; 1300–2700 m. CaR, SNH, TR; w NV.

COLLOMIA

Dieter H. Wilken

Ann, per. **ST** hairy, glandular. **LVS** alternate, simple, entire to gen pinnately lobed, linear to ovate; basal short-petioled; cauline sessile. **INFL**: heads or clusters, terminal (or fls 1–3 in axils). **FL**: calyx green, becoming straw-colored, membranous in age, gen bell-shaped, sinuses in fr pleated to expanded; corolla salverform to funnel-shaped. **FR** ovoid to elliptic. **SEEDS** 3 per chamber, oblong, gen gelatinous, brown. $2n$=16. 15 spp.: N.Am; also in s S.Am. (Greek: glue, from wet seed surface) Self-compatible; ann spp. self-pollinating; per spp. gen cross-pollinating. [Wilken et al. 1982 Biochem Syst Ecol 10:239–243]

1. Per
2. St < 5 cm; lvs gen basal, lobed; corolla purple .... ***C. larsenii***
2′ St > 10 cm; lvs cauline, toothed; corolla red-orange .... ***C. rawsoniana***
1′ Ann
3. St gen simple, erect; infl head-like, terminal
4. Corolla yellow to orange, gen > 15 mm (< 5 mm in cleistogamous fls); pollen gen blue .... ***C. grandiflora***
4′ Corolla white to pink, < 15 mm; pollen gen white .... ***C. linearis***
3′ St branched, branches spreading; fls in axillary and terminal clusters
5. Lvs 1–2-pinnately lobed .... ***C. heterophylla***
5′ Lvs entire to toothed
6. Lower cauline lvs elliptic, 3-toothed; corolla throat yellow .... ***C. diversifolia***
6′ Lower cauline lvs linear to lanceolate, entire; corolla throat violet to pink
7. Stigma and at least 1 anther exserted, stamens attached at same level .... ***C. tinctoria***
7′ Stigma incl; anthers in throat (not truly exserted), stamens attached at different levels .... ***C. tracyi***

C. diversifolia E. Greene (p. 827) SERPENTINE COLLOMIA Ann. **ST** erect, branched diffusely. **LVS**: basal and lower cauline 3-toothed; upper cauline entire, glandular-puberulent. **INFL**: clusters, terminal and axillary; fls 7–15. **FL**: calyx 10–12 mm, lobes acuminate; corolla 9–12 mm, tube purple, throat yellow, lobes violet. UNCOMMON. Rocky to gravelly serpentine areas; 300–600 m. NCoRI, ne SnFrB (Contra Costa Co.). ❀TRY.

C. grandiflora Lindley Ann. **ST** erect, simple or branched above in robust pls. **LVS**: basal lanceolate, toothed; cauline lanceolate to linear, entire, gen glabrous above, glaucous, slightly glandular below. **INFL**: head, terminal; fls sessile. **FL**: calyx 7–10 mm, lobes lanceolate; corolla 15–30 mm (< 5 mm in cleistogamous fls), yellow to orange; pollen gen blue. Open areas; 600–2500 m. CA-FP; to B.C., ID, Colorado, AZ; naturalized in Eur. Fully cleistogamous pls have single sts < 10 cm with 3–7 fls at tip. ❀SUN, DRN:1–5,**6**,**7**,8,**9**,**10**,**14**–**24**.

C. heterophylla Hook. Ann. **ST** erect, branched diffusely. **LVS**: lower 1–2-lobed; upper lobed to entire, gen hairy, slightly glandular. **INFL**: clusters, terminal and axillary; fls 7–25. **FL**: calyx 5–8 mm, lobes narrowly lanceolate; corolla 10–14 mm, white to pink. Sandy to gravelly, open areas; < 1300 m. KR, NCoR, CaR, SNF, SnFrB, SCoR; to B.C., ID. ❀STBL.

C. larsenii (A. Gray) Payson (p. 827) TALUS COLLOMIA Per, gen cespitose, from slender rhizomes. **ST** gen branched. **LF** 1–2-pinnately or -palmately lobed; lobes linear-oblong, glandular, hairy, narrowed at base. **INFL**: clusters, terminal; fls 6–9. **FL**: calyx 5–9 mm; corolla 10–15 mm, light to deep purple. RARE in CA. Volcanic talus; 3000–3500 m. CaRH (Mt. Lassen); to WA. [*C. debilis*

(S. Watson) E. Greene var. *l.* (A. Gray) Brand] Distinct from *C. debilis* of n WA, Rocky Mtns.

C. linearis Nutt. (p. 827) Ann. **ST** erect, gen simple, branched above in robust pls. **LVS**: basal lanceolate, toothed; cauline lanceolate to linear, entire, gen glabrous above, glaucous, slightly glandular below. **INFL**: bracted head, terminal; fls 7–20, sessile. **FL**: calyx 4–7 mm; corolla 8–15 mm, white or pink; pollen gen white. Open areas; 600–3300 m. KR, CaR, SN, SnBr, MP; to AK, e N.Am, AZ; naturalized elsewhere.

C. rawsoniana E. Greene (p. 827) FLAMING TRUMPET Per. **ST** erect, 1–6 dm. **LVS** cauline, 3–8 cm, ovate to elliptic, coarsely toothed, glandular, hairy. **INFL** terminal; fls 3–7. **FL**: calyx 8–12 mm; corolla 25–40 mm, red-orange; stamens, style slightly exserted; pollen blue. RARE. Shaded areas near streams in woodlands; 1000–2200 m. c SNF (Mariposa, Madera cos.). In cult.

C. tinctoria Kellogg (p. 827) Ann. **ST** 2–8 cm, branched diffusely. **LF** linear-lanceolate, glandular, entire. **INFL** axillary; fls gen 2–3. **FL**: calyx 5–7 mm, lobes long tapered, awned; corolla 8–11 mm, slender, tube maroon to violet, lobes pink; stamens attached at same level, 1–2 anthers exserted, stigma exserted. Gravelly to rocky, open areas; 600–2800 m. KR, NCoRH, CaRH, n&c SNH, WTR, MP; to WA, ID, NV.

C. tracyi H. Mason (p. 827) TRACY'S COLLOMIA Ann. **ST** 2–8 cm, openly branched. **LF** linear-lanceolate, glandular, entire. **INFL** axillary; fls gen 2–3. **FL**: calyx 6–8 mm; corolla 10–20 mm, gen white to lavender; stamens attached at different levels, upper anthers in throat; stigmas below anthers. UNCOMMON. Rocky, gravelly, or sandy areas; 300–2100 m. KR, n NCoRH. Much like *C. tinctoria*, but does not intergrade.

ERIASTRUM

Robert W. Patterson

Ann, per, glabrous to woolly. **ST** gen erect. **LVS** cauline, alternate, pinnately lobed or simple; lobes gen linear or lanceolate. **INFL**: heads, bracted, gen densely woolly; bracts lf-like; fls sessile. **FL**: calyx lobes unequal; corolla radial or bilateral, funnel-shaped to salverform; stamens equal or not, attached at same level, anthers gen sagittate, pollen white or blue; style incl or exserted. **SEEDS** 1–few per chamber. 13 spp.: w N.Am. (Greek, woolly star) [Harrison 1972 Brigham Young Univ Sci Bull, Biol Ser 16:1–26]

1. Per ..... ***E. densifolium***
  2. Lvs few-lobed or entire, gen glabrous; coastal dunes ..... ssp. ***densifolium***
  2′ Lvs variously lobed, densely hairy; gen away from immediate coast
    3. Lvs strongly recurved, blade wider at base than at tip, lobes spine-tipped — DMoj ..... ssp. ***mohavense***
    3′ Lvs not recurved, blades gen equally wide at base, tip, lobes not spine tipped
      4. Corolla 25–32 mm; pls densely woolly — e SCo (Santa Ana River) ..... ssp. ***sanctorum***
      4′ Corolla < 20 mm; pls gen hairy but not densely white-woolly
        5. Bracts gen 5–9-lobed; fls gen 15–20 ..... ssp. ***austromontanum***
        5′ Bracts gen 1–5-lobed; fls gen < 15 ..... ssp. ***elongatum***

1′ Ann
  6. Corolla 15–23 mm
    7. Corolla bilateral, stamens unequal, bent toward lower lip ..... ***E. eremicum***
    7′ Corolla radial, stamens equal
      8. Stamens attached in corolla sinus ..... ***E. pluriflorum***
      8′ Stamens attached in base of corolla throat ..... ***E. virgatum***
  6′ Corolla < 15 mm
    9. Corolla lobes bright yellow ..... ***E. luteum***
    9′ Corolla lobes blue or white
      10. Stamens extended to tip of corolla lobes or beyond; corolla lobes bright blue ..... ***E. sapphirinum***
      10′ Stamens not extended as far as tip of corolla lobes; corolla lobes pale blue or white
        11. Stamens unequal
          12. Stamens attached at top of throat; calyx 5–6 mm ..... ***E. diffusum***
          12′ Stamens attached at base of throat or lower; calyx 6–10 mm ..... ***E. wilcoxii***
        11′ Stamens equal
          13. Filaments 2–3 mm
            14. Bracts, calyces extended beyond rest of infl; s CA, Baja CA, gen not D ..... ***E. filifolium***
            14′ Bracts, calyces not extended beyond rest of infl; SNE, D ..... ***E. sparsiflorum***
          13′ Filaments < 2 mm
            15. Lf lobes gen > 3; corolla lobes blue, tube, throat yellow ..... ***E. abramsii***
            15′ Lf lobes 1–3; corolla white or pale blue
              16. Corolla > calyx; NCoRI ..... ***E. brandegeae***
              16′ Corolla = calyx; SCoRI ..... ***E. hooveri***

E. abramsii (Elmer) H. Mason Ann. **ST** 5–15 cm. **LF** 10–45 mm; lobes gen 2–9, thread-like, woolly. **FL**: corolla salverform, tube 3–4 mm, yellow, lobes 2–3 mm, blue or white; stamens attached in lower throat or upper tube, incl. Dry slopes, flats; < 1200 m. NCoRI, SnFrB, SCoRI.

E. brandegeae H. Mason (p. 827) BRANDEGEE'S ERIASTRUM Ann. **ST** 5–30 cm. **LF** 10–25 mm; lobes gen 2–6, thread-like, woolly. **FL**: corolla salverform or slightly bilateral, white or pale blue, tube 4–6 mm, lobes 3–5 mm; stamens equal, attached in lower throat or upper tube, incl. **RARE** CA. Volcanic soils; 800–1000 m. n&c NCoRI. [*E. tracyi* H. Mason] Threatened by grazing, vehicles.

E. densifolium (Benth.) H. Mason (p. 837) Per. **ST** erect or spreading, nearly glabrous to woolly. **LF** 10–50 mm; lobes gen 2–16, glabrous to woolly. **FL**: corolla funnel-shaped, blue or white, tube 15–35 mm, lobes 5–15 mm; stamens equal, attached in throat or upper tube, exserted. Dunes, dry river beds, open slopes; 0–2900 m. s SN, s CW, SW, DMoj; Baja CA.

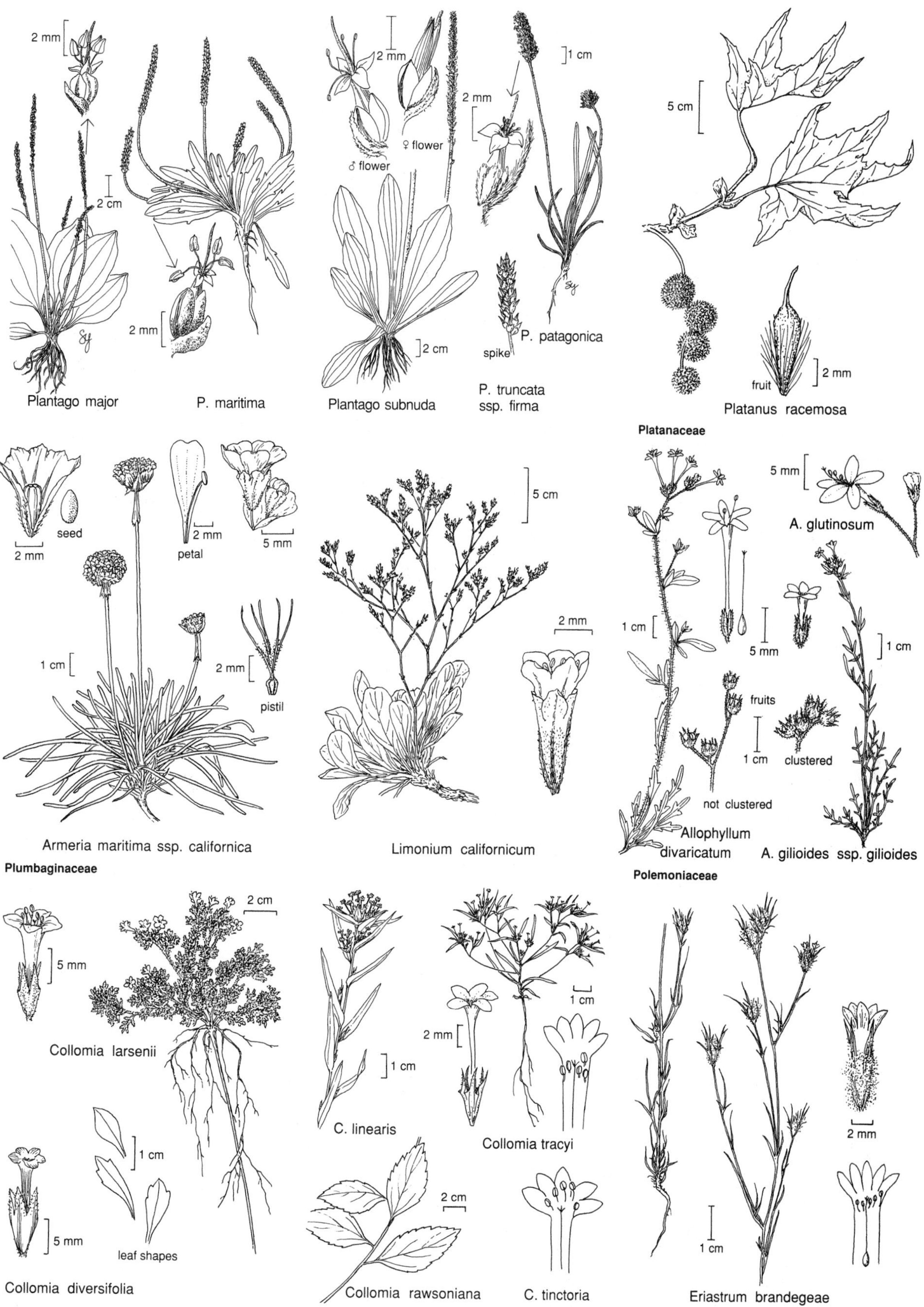
2 mm
2 cm
2 mm
Plantago major
P. maritima
2 mm
♂ flower
♀ flower
2 cm
Plantago subnuda
1 cm
2 mm
spike
P. patagonica
P. truncata
ssp. firma
5 cm
fruit
2 mm
Platanus racemosa
Platanaceae
2 mm
seed
2 mm
petal
5 mm
1 cm
2 mm
pistil
Armeria maritima ssp. californica
Plumbaginaceae
5 cm
2 mm
Limonium californicum
5 mm
A. glutinosum
1 cm
5 mm
1 cm
fruits
1 cm
clustered
not clustered
Allophyllum
divaricatum
A. gilioides ssp. gilioides
Polemoniaceae
5 mm
2 cm
Collomia larsenii
1 cm
5 mm
leaf shapes
Collomia diversifolia
1 cm
C. linearis
1 cm
2 mm
Collomia tracyi
2 cm
Collomia rawsoniana
C. tinctoria
2 mm
1 cm
Eriastrum brandegeae

ssp. ***austromontanum*** (Craig) H. Mason **LF**: lobes 2–15, slightly woolly. **INFL** terminal; bracts gen 5–9-lobed; fls gen 15–20. **FL**: corolla tube narrow. 2*n*=14. Dry, open slopes; 1300–2900 m. s SN, w SCoR, TR, PR. ❀TRY.

ssp. ***densifolium*** **LF** gen glabrous, gen entire. **FL**: corolla tube broad. 2*n*=14. Coastal dunes; ± 0 m. s CCo. ❀TRY;DFCLT.

ssp. ***elongatum*** (Benth.) H. Mason **LF**: lobes gen 2–8, 5 mm, slightly to moderately woolly. **INFL** terminal, axillary; bracts gen 1–5-lobed; fls gen < 15. **FL**: corolla tube 14–18 mm. 2*n*=14. Dry places; < 1800 m. s SCoRO, SW, w edge D; Baja CA. ❀DRN, SUN:7–12,14,**19–24**.

ssp. ***mohavense*** (Craig) H. Mason **LF**: axis 1–4 mm wide, strongly recurved; blade base wider than tip; lobes 2–8, spine-tipped, densely woolly. **FL**: corolla tube 15–17 mm. 2*n*=14. Sandy slopes, flats; 800–2800 m. SNE, w DMoj. In cult.

ssp. ***sanctorum*** (Milliken) H. Mason (p. 837) SANTA ANA RIVER WOOLLYSTAR **LF**: lobes gen 2–6, densely woolly. **FL**: corolla tube > 30 mm. 2*n*=14. **ENDANGERED** US, CA. Gravelly river beds; < 500 m. e SCo (along Santa Ana River, sw San Bernardino Co.) Threatened by habitat alteration. In cult.

E. diffusum (A. Gray) H. Mason Ann. **ST** erect or spreading, 3–20 cm. **LF** 10–25 mm; lobes gen 2–4 near base, nearly glabrous to woolly. **FL**: corolla gen funnel-shaped, tube 3–5 mm, blue or yellow, lobes 3–5 mm, light blue or cream; stamens attached in upper throat, unequal, exserted. 2*n*=14. Open, areas, sandy soil; < 2000 m. D; to Colorado, NM, TX. Despite name, not always openly or diffusely branched.

E. eremicum (Jepson) H. Mason ssp. ***eremicum*** (p. 837) Ann. **ST** 5–30 cm. **LF** 5–55 mm; lobes 2–6, nearly glabrous to woolly. **FL**: corolla gen bilateral, tube 4–9 mm, blue or yellow, throat gen yellow, lobes 4–8 mm, light to dark blue; stamens attached in lower throat or upper tube, unequal, bent toward lower corolla lip, exserted. 2*n*=14. Open areas in sandy soils; < 1800 m. D; to NV, UT, AZ. Ssp. *yageri* (M.E. Jones) H. Mason occurs in AZ.

E. filifolium (Nutt.) Wooton & Standley Ann. **ST** 4–40 cm. **LF** 20–40 mm, thread-like, gen maturing red-brown, nearly glabrous to woolly, entire or 2 lobes at base. **INFL**: bracts, calyces extended beyond infl. **FL**: corolla salverform, tube 4 mm, yellow, throat yellow, lobes 3 mm, blue; stamens attached in lower throat, equal, exserted, filaments 2–3 mm. Dry slopes; < 1700 m. SCo; Baja Ca.

E. hooveri (Jepson) H. Mason (p. 837) HOOVER'S ERIASTRUM Ann. **ST** 3–15 cm. **LF** 5–25 mm, thread-like, nearly glabrous to woolly, entire or 2-lobed. **FL**: corolla salverform, white, tube 2–3 mm, lobes 2–3 mm; stamens attached in lower throat or upper tube, anthers placed ± at opening. RARE. Drying grassy areas; < 170 m. s&e SCoR.

E. luteum (Benth.) H. Mason YELLOW-FLOWERED ERIASTRUM Ann. **ST** 2–25 cm. **LF** 20–40 mm, thread-like, woolly, entire or 2-lobed at base. **FL**: corolla salverform or slightly bilateral, bright yellow, tube 3 mm, lobes 5–6 mm; stamens attached in lower throat, equal, exserted. UNCOMMON. Drying slopes; < 1000 m. SCoR (Monterey, San Luis Obispo cos.).

E. pluriflorum (A.A. Heller) H. Mason (p. 837) MANY-FLOWERED ERIASTRUM Ann. **ST** 2–32 cm. **LF** 20–50 mm, thread-like, woolly, entire or 2–10-lobed. **FL**: corolla salverform, tube 12–16 mm, white or yellow, throat yellow, lobes 4–9 mm, bright blue; stamens attached in sinuses, gen equal, exserted. 2*n*=14. UNCOMMON. Chaparral, woodlands, pine forests; < 2000 m. c&s SNF, SnJV, SnFrB, e SCoR, w DMoj. [ssp. *sherman-hoytae* (Craig) H. Mason] ❀TRY.

E. sapphirinum (Eastw.) H. Mason (p. 837) Ann. **ST** 5–40 cm; hairs minute, glandular. **LF** 5–55 mm, thread-like, nearly glabrous to woolly, entire or 2-lobed near base. **FL**: corolla funnel-shaped, bilateral from unequal sinuses, tube 5–6 mm, blue or yellow, throat yellow, lobes 5–10 mm, bright blue; stamens attached in lower throat or upper tube, equal, exserted, gen extended beyond corolla lobes. 2*n*=14. Many communities; 700–2700 m. s CA-FP, w D. [sspp. *ambiguum* (M.E. Jones) H. Mason, *dasyanthum* (Brand) H. Mason, *gymnocephalum* (Brand) H. Mason] ❀TRY.

E. sparsiflorum (Eastw.) H. Mason Ann. **ST** 5–30 cm. **LF** 5–35 mm, thread-like, woolly, entire or 2-lobed near base. **FL**: corolla salverform or narrow funnel-shaped, tube 4–6 mm, yellow, throat yellow, lobes 2–3 mm, bright blue, pink, yellow, or cream; stamens attached in lower throat or upper tube, equal, slightly exserted or incl. 2*n*=14. Desert slopes; < 2400 m. s SNH, Teh, SNE, DMtns; OR, NV, Baja CA. [ssp. *harwoodii* (Craig) H.K. Harrison]

E. virgatum (Benth.) H. Mason VIRGATE ERIASTRUM Ann. **ST** 3–40 cm. **LF** 20–50 mm, thread-like, woolly, entire or 2-lobed near base. **FL**: corolla salverform, tube 7–12 mm, yellow, throat yellow, lobes 7–10 mm, bright blue; stamens attached in base of throat, gen equal, exserted. UNCOMMON. Sandy soils; < 500 m. s CCo, n SCoRO (Monterey Co.).

E. wilcoxii (Nelson) H. Mason Ann. **ST** 3–30 cm. **LF** 15–30 mm, thread-like, densely woolly, entire or 2–4-lobed near base. **FL**: corolla gen salverform, tube 5–8 mm, yellow, throat yellow, lobes 4–5 mm, blue; stamens attached in lower throat or upper tube, unequal, filaments 1–5 mm, exserted or opening. 2*n*=14. Desert flats, slopes; < 2700 m. SNE, n DMtns; to ID, WY.

GILIA

Alva G. Day

Ann, per, gen erect. **ST** glabrous, hairy, glandular, or cobwebby. **LVS** simple, gen alternate; basal gen in rosette, toothed, pinnately lobed, or entire; cauline gen reduced; lf tips, calyx lobes acute, acuminate, or needle-like. **INFL**: fls solitary or clustered, 1–many in axils of bracts. **FL**: calyx membranous between ribs, membrane splitting or expanding; corolla > calyx, lobes gen ovate. **FR** gen ovoid; chambers 3, valves separating from top. **SEEDS** 3–many, brown, gen gelatinous when wet. ± 70 spp.: w N.Am, S.Am. (Felipe Gil, 18th century Spanish botanist)

Key to Groups

1. Cobwebby (fine, kinky) hairs in lf axils or elsewhere; st ± scapose; basal lvs gen in rosette .... **Group 1**
1′ Cobwebby hairs 0; st ± scapose or not; basal rosette present or 0 .... **Group 2**

Group 1: Cobwebby hairs present; st ± scapose; basal rosette gen present (sect. *Arachnion*)

1. Calyx, pedicel densely glandular (rarely cobwebby); fls, frs gen in clusters; calyx ribs wider than membrane .... ***G. brecciarum***
  2. Stamens long-exserted; corolla 10–20 mm, upper throat white with yellow spots .... ssp. ***neglecta***
  2′ Stamens slightly exserted; corolla 5–11 mm, upper throat ± purple
    3. Branches spreading to nearly erect; corolla throat yellow with purple veins .... ssp. ***brecciarum***
    3′ Branches long-decumbent; corolla throat gen all purple .... ssp. ***jacens***
1′ Calyx, pedicel barely or not glandular; fls, frs gen not in clusters; calyx ribs narrower than membrane
  4. Cauline lvs clasping or expanded at base
    5. Corolla 6–12 mm, 1–2 × calyx

6. St green (not glaucous), slightly cobwebby below middle; upper lvs gen deeply toothed .... ***G. modocensis***
6′ St glaucous, glabrous below middle; upper lvs gen not deeply toothed
7. Stamens, style exserted to middle of corolla lobes; pollen blue; gen s of TR .... ***G. diegensis***
7′ Stamens, style exserted to base of corolla lobes; pollen white; gen n of TR .... ***G. sinuata***
5′ Corolla 10–35 mm, 2–7 × calyx .... ***G. latiflora***
8. St cobwebby below middle
9. Corolla tube stout, throat > 7 mm wide, corolla gen 3 × lobes .... ssp. [2]***davyi***
9′ Corolla tube slender, throat < 7 mm wide, corolla 4–6 × lobes .... ssp. ***elongata***
8′ St glabrous and glaucous below middle
10. Fls in loose clusters; corolla purple to mid-throat or higher .... ssp. [2]***davyi***
10′ Fls not in clusters; corolla purple to base of throat
11. Calyx 2–4 mm; corolla 9–16 mm .... ssp. ***cuyamensis***
11′ Calyx 4–5 mm; corolla 15–22 mm .... ssp. ***latiflora***
4′ Cauline lvs not clasping or expanded at base
12. Calyx glabrous or cobwebby (not glandular)
13. Stamens = or > corolla lobes; calyx membrane gen purple or purple-spotted
14. Basal lvs densely cobwebby; calyx membrane gen dull purple-spotted; SNH, w GB .... ***G. salticola***
14′ Basal lvs glabrous or slightly cobwebby; calyx membrane bright purple-spotted; DMoj .... ***G. aliquanta***
15. Corolla lobes = tube + throat .... ssp. ***aliquanta***
15′ Corolla lobes gen 1/2 tube + throat .... ssp. ***breviloba***
13′ Stamens < corolla lobes; calyx membrane gen not colored
16. Basal lf lobes spreading; corolla throat yellow or yellow-spotted
17. Corolla 4–7 mm; throat yellow-spotted, < or = lobes; calyx lobes short-pointed .... ***G. clokeyi***
17′ Corolla 7–12 mm; throat yellow, = or > lobes; calyx lobes acuminate .... ***G. ophthalmoides***
16′ Basal lf lobes gen ascending; corolla throat blue in upper 1/2, yellow below .... ***G. ochroleuca***
18. St glabrous and glaucous below infl
19. Corolla 8–14 mm; style > stamens .... ssp. ***bizonata***
19′ Corolla 4–6 mm; style < or = stamens .... ssp. ***ochroleuca***
18′ St gen cobwebby or glandular below infl (if glabrous, not glaucous)
20. Pl tall; basal lvs entire to gen 1-pinnate; corolla pale; s SnBr, PR .... ssp. ***exilis***
20′ Pl low, spreading; basal lvs 2-pinnate; corolla bright; SnGb, Teh .... ssp. ***vivida***
12′ Calyx glandular (rarely cobwebby in young pls)
21. Lobes of basal lvs toothed on both sides
22. Stamens < corolla lobes; corolla throat bluish in upper 1/2, yellow in lower 1/2 .... ***G. cana***
23. Adjacent pedicels widely spreading
24. Corolla throat conic, tube 2–4 × calyx .... ssp. ***speciformis***
24′ Corolla throat cup-shaped, tube gen < 2 × calyx .... ssp. ***triceps***
23′ Adjacent pedicels narrowly spreading
25. Fr narrowly ovoid, gen < 5 mm; basal lvs cobwebby-matted .... ssp. ***cana***
25′ Fr widely ovoid, gen > 5 mm; basal lvs cobwebby, not matted
26. Corolla tube = throat; stamens unequal, longest well exserted .... ssp. ***bernardina***
26′ Corolla tube 3–6 × throat; stamens equal, slightly exserted .... ssp. ***speciosa***
22′ Stamens gen > corolla lobes; corolla throat yellow or white .... ***G. leptantha***
27. Corolla tube 2–4 × calyx
28. Corolla lobes 1–3 mm wide; SnBr .... ssp. ***leptantha***
28′ Corolla lobes 3–5 mm wide; s SN .... ssp. ***purpusii***
27′ Corolla tube 1–2 × calyx
29. Corolla throat conic; stamens > corolla lobes .... ssp. ***pinetorum***
29′ Corolla throat cup-shaped; stamens < corolla lobes .... ssp. ***transversa***
21′ Lobes of basal lvs linear, entire or toothed on 1 side
30. Infl showy; corolla 10–22 mm, unspotted below lobes .... ***G. tenuiflora***
31. Corolla tube stout, lobes 4–7 mm wide
32. Corolla throat > lobes; mainland .... ssp. ***amplifaucalis***
32′ Corolla throat < lobes; ChI .... ssp. ***hoffmannii***
31′ Corolla tube slender, lobes 2–4 mm wide
33. Branches decumbent; calyx densely glandular; stamens exserted to base of lobes .... ssp. ***arenaria***
33′ Branches spreading, calyx slightly glandular; longest stamens exserted beyond middle of lobes .... ssp. ***tenuiflora***
30′ Infl not showy; corolla 3–12 mm, gen spotted below lobes (*G. inconspicua* complex)
34. Longest stamens exserted > mid-lobe
35. Lf lobes ascending; corolla 3–4 × lobes; GB .... ***G. inconspicua***
35′ Lf lobes spreading; corolla 2–3 × lobes; s SNF, Teh .... ***G. interior***
34′ Longest stamens exserted < mid-lobe
36. Fr 2 × calyx, valves not detaching; lf lobes < 1 mm wide, ascending .... ***G. minor***
36′ Fr = calyx, valves detaching; lf lobes 1–2 mm wide, spreading
37. Branches decumbent; corolla without spots, lobe tips obtuse .... ***G. malior***
37′ Branches erect to spreading; corolla with purple spots below lobes, lobe tips acute
38. Lobes of basal lvs 2–3 mm; fr 3–4 mm; SCoR .... ***G. austro-occidentalis***
38′ Lobes of basal lvs 3–11 mm; fr 5–6 mm; DMoj .... ***G. transmontana***

Group 2: Cobwebby hairs 0; st ± scapose or not; basal rosette sometimes 0

1. Sepals gen fused below middle or at base; pollen yellow (sect. *Giliastrum*)
 2. Principal lvs sharp-toothed, holly-like; pl > 10 cm; corolla pink
  3. Ann; lf teeth acuminate .... ***G. latifolia***
  3′ Per; lf teeth needle-like .... ***G. ripleyi***
 2′ Principal lvs not sharp-toothed, not holly-like; pl < 9 cm; corolla yellow or white
  4. Lf linear; corolla yellow .... ***G. filiformis***
  4′ Lf oblong, lanceolate, or narrowly oblanceolate; corolla mostly white
   5. Pl < 3 cm, st not glandular; lf 2–5 mm; calyx membrane ciliate; DSon .... ***G. maculata***
   5′ Pl > 3 cm, st glandular; lf 6–20 mm; calyx membrane not ciliate; GB, DMoj
    6. Corolla throat spotted, tube exserted from calyx .... ***G. campanulata***
    6′ Corolla throat not spotted, tube incl in calyx
     7. Lower lvs white-hairy; corolla 4–7 mm; seeds 10–15 per chamber .... ***G. inyoensis***
     7′ Lower lvs gen glabrous; corolla < 3 mm; seeds 1 per chamber .... ***G. tenerrima***

1′ Sepals gen fused to middle or higher; pollen blue, yellow, or white
 8. Pl glandular-puberulent; upper lvs gen entire
  9. Lvs gen cauline; infl not scapose; pollen blue; seeds gelatinous when wet (sect. *Kelloggia*)
   10. Corolla 6–8 mm, lobes bluish white; branches, lvs ascending .... ***G. capillaris***
   10′ Corolla 5–20 mm, lobes pink; branches, lvs spreading
    11. Calyx glandular; corolla tube, throat yellow with reddish streaks .... ***G. sinistra***
     12. Corolla 9–20 mm; style exserted .... ssp. ***pinnatisecta***
     12′ Corolla 5–9 mm; style incl .... ssp. ***sinistra***
    11′ Calyx glabrous; corolla tube yellow, throat purple or yellow .... ***G. leptalea***
     13. Corolla 8–15 mm, throat 2–5 mm, yellow, sometimes with short purple marks .... ssp. ***bicolor***
     13′ Corolla 13–21 mm, throat 6–8 mm, purple .... ssp. ***leptalea***
  9′ Lvs in basal rosette; infl gen ± scapose; pollen yellow or white; seeds not gelatinous when wet (sect. *Giliandra*)
   14. Lower lvs gen 2-pinnately lobed; corolla tube glandular-puberulent, lobes wavy-margined .. ***G. hutchinsifolia***
   14′ Lower lvs gen 1-pinnately lobed or toothed; corolla tube glabrous, lobes not wavy-margined
    15. Corolla tube thread-like, exserted from calyx
     16. Corolla lobes truncate and short-pointed; throat wider than tube .... ***G. leptomeria***
     16′ Corolla lobes 3-toothed; throat narrower than tube .... ***G. triodon***
    15′ Corolla tube stout, incl or slightly exserted from calyx
     17. Corolla 2–4 mm, throat gen incl; pedicels in fr spreading or recurved; fr spheric, gen < calyx .... ***G. micromeria***
     17′ Corolla 4–7 mm, throat exserted; pedicels in fr nearly erect; fr ovoid, gen > calyx
      18. Lf axis strap-shaped; basal lvs gen without sand grains on upper surface .... ***G. lottiae***
      18′ Lf axis linear; basal lvs dotted with sand grains on both surfaces .... ***G. subacaulis***
 8′ Pl hairy on lf or st; upper lvs lobed or toothed
  19. Lf lobes gen not linear; fls not in heads or clusters; pedicel glands large, gen flat-topped, subsessile (sect. *Saltugilia*)
   20. Calyx gland-dotted
    21. Hairs on basal lvs translucent, straight or curved, gland-tipped; spots 0 in corolla .... ***G. scopulorum***
    21′ Hairs on basal lvs white, sharply bent, non-glandular; purple spots below each corolla lobe ...... ***G. stellata***
   20′ Calyx gen glabrous
    22. Corolla 5–13 mm, lavender or white
     23. Stamens < corolla lobes; corolla lobes 1–1.5 × throat, throat not purple-marked .... ***G. australis***
     23′ Stamens gen > corolla lobes; corolla lobes 2–3 × throat, throat purple-marked below each lobe .... ***G. caruifolia***
    22′ Corolla 10–36 mm, gen pink .... ***G. splendens***
     24. Corolla tube 7–18 mm, 2–5 × calyx .... ssp. ***grantii***
     24′ Corolla tube 4–10 mm, 1–2 × calyx .... ssp. ***splendens***
  19′ Lf lobes ± linear; fls gen in heads or clusters; pedicel glands, if present, minute, spheric, stalked (sect. *Gilia*)
   25. Infl axis gen glandular; corolla throat yellow, sometimes purple-spotted
    26. Corolla throat not spotted
     27. Corolla 6–9 mm; throat cup-shaped; lvs 1–2-pinnate .... ***G. angelensis***
     27′ Corolla 8–14 mm, throat conic; lvs 2–3-pinnate .... ***G. nevinii***
    26′ Corolla throat purple-spotted
     28. Corolla 8–19 mm, throat well exserted from calyx .... ***G. tricolor***
      29. Fls in loose clusters or solitary; corolla 7–13 mm .... ssp. ***diffusa***
      29′ Fls gen in dense clusters; corolla 10–19 mm .... ssp. ***tricolor***
     28′ Corolla 7–11 mm, throat partly incl in calyx
      30. Calyx in fr 5–8 mm, gen < or = fr .... ***G. clivorum***
      30′ Calyx in fr 8–13 mm, gen >> fr .... ***G. millefoliata***
   25′ Infl axis gen glabrous or tomentose; corolla throat neither yellow nor spotted
    31. Fls gen in hemispheric heads or clusters; corolla throat gen > tube; stamens < or = corolla lobes .... ***G. achilleifolia***

32. Corolla 10–20 mm; heads 8–25-fld; pedicels 1–2 mm .... ssp. ***achilleifolia***
32′ Corolla 5–10 mm; clusters 2–7-fld; pedicels 1–30 mm .... ssp. ***multicaulis***
31′ Fls in spheric heads; corolla throat < or = tube; stamens = or > corolla lobes .... ***G. capitata***
33. Corolla 5–8 mm, lobes linear or narrowly oblong; calyx lobes acute, erect
34. Heads glabrous or sparsely hairy to glandular at base
35. Calyx membrane colorless; corolla lobes < 1 mm wide; seeds gen 1–6 .... ssp. ***capitata***
35′ Calyx membrane blue-violet; corolla lobes 1–2 mm wide; seeds 10–25 .... ssp. ***pacifica***
34′ Heads tomentose at base
36. Heads 12–25 mm wide; corolla pale blue-violet; SN .... ssp. ***mediomontana***
36′ Heads 20–35 mm wide; corolla bright blue-violet; NCo .... ssp. ***tomentosa***
33′ Corolla 7–13 mm, lobes oblong; calyx lobes acuminate, tips recurved
37. Heads sparsely hairy to glabrous at base; fls short-pedicelled; s CA .... ssp. ***abrotanifolia***
37′ Heads tomentose at base; fls gen sessile; esp n&c CA
38. Corolla bright blue-violet; pl with skunk-like odor — n CCo .... ssp. ***chamissonis***
38′ Corolla pale blue-violet; pl not with skunk-like odor
39. Infl 10–20 mm wide; corolla lobes ± 2 mm wide; SNF .... ssp. ***pedemontana***
39′ Infl 20–30 mm wide; corolla lobes ± 3 mm wide; GV, SCoRI .... ssp. ***staminea***

G. achilleifolia Benth. **ST** 6–70 cm, glabrous or lightly hairy below, slightly glandular in infl or throughout. **LVS**: basal in cluster, 1–2-pinnate, 5–10 cm; axis linear, lobes 3–25 mm, ± 2 mm wide, linear, spreading, gen curving, glabrous or finely white-hairy in axils or at base of lobes. **INFL**: heads, clusters, or fls solitary. **FL**: calyx lobes acute; corolla lavender or white, lobes spreading or semi-erect; stamens, style gen exserted. **SEEDS** ± 10–18. Open or shaded, gen grassy places in sandy or rocky soil; 60–1200 m. s NCoRI (Solano Co.), SnFrB, SCoR.

ssp. ***achilleifolia*** (p. 837) **ST**: branches spreading. **INFL**: head, hemispheric; fls 8–25; pedicels 1–2 mm. **FL**: corolla 10–20 mm, 2–3 × calyx, lavender; stamens < lobes; stigmas = or > anthers. $2n=18$. Habitat of sp. SnFrB, SCoR. Sometimes garden escape elsewhere. ❀SUN,DRN:7,8,9,**14–16**,17–21,**22–24**.

ssp. ***multicaulis*** (Benth.) V. Grant & A.D. Grant (p. 837) **ST** trailing to erect. **INFL**: clusters; fls 1–7; pedicels 1–30 mm. **FL**: corolla 5–10 mm, 1–2 × calyx, lavender or white; stamens, style slightly exserted; stigmas touching anthers. $2n=18$. Habitat and range of sp. Often grows with ssp. *achilleifolia*.

G. aliquanta A.D. Grant & V. Grant **ST**: branches spreading from base, 8–16 cm, glabrous or slightly cobwebby below, glandular in infl. **LVS**: basal in semi-erect cluster, 1–3 cm, pinnate; axis 1–3 mm wide; lobes 2–7 mm, gen > axis width. **INFL** loose; pedicels 1–11 mm in unequal pairs. **FL**: calyx 3–5 mm, glabrous or cobwebby, bright purple or purple-spotted, membrane puckered between ribs; corolla 6–12 mm, tube purple, lobes lavender, obovate, tips rounded; stamens unequal, exserted; style exserted. **FR** = or > calyx, widely ovoid. Rocky slopes, washes; 700–1600 m. s SN, n SnGb, n SnBr, DMoj; s NV.

ssp. ***aliquanta*** (p. 837) **FL**: corolla tube incl in calyx, throat spreading, tube + throat = lobes, lobes widely obovate; longest stamens gen > corolla lobes; stigmas touching or > anthers. $2n=18$. Habitat of sp.; 700–1300 m. Range of sp.

ssp. ***breviloba*** A.D. Grant & V. Grant (p. 837) **FL**: corolla tube exserted from calyx, throat narrow, tapered, tube + throat > 2 × lobes, lobes narrowly obovate; stamens < lobes; stigmas touching anthers. Uncommon. Habitat of sp.; 800–1600 m. DMoj; s NV.

G. angelensis V. Grant **ST**: branches ± erect from base, 7–70 cm, with short, translucent hairs below. **LVS**: basal in ± erect cluster, 1–2-pinnate, glabrous or short-hairy; axis and lobes linear. **INFL**: cluster, loose; fls 1–10; pedicels yellow-glandular. **FL**: calyx 3–5 mm, gen short-hairy, lobes acute, membrane gen blue, width < ribs; corolla 6–9 mm, tube incl in calyx, throat cup-shaped, yellow, not spotted, lobes lavender or white; stamens, style exserted. **FR** 4–5 mm, < calyx. **SEEDS** 18–30. $2n=18$. Open, sandy or rocky, gen grassy areas; 200–1900 m. e SnFrB (Mount Hamilton Range), SCoRI, SW; Baja CA.

G. australis (H. Mason & A.D. Grant) V. Grant & A.D. Grant **ST**: branches ± erect, 20–45 cm, hairy below. **LVS**: basal in loose rosette, 2-pinnate, 2–7 cm, translucent-hairy; axis linear; lobes short-acuminate; upper linear, entire or lobed at base. **INFL** loose; pedicels 5–20 mm in unequal pairs, glands flat-topped. **FL**: calyx gen glabrous, green with purple marks or not, rib width < membrane, lobes acuminate; corolla 5–10 mm, 2–3 × calyx, lavender to white, tube incl, throat yellow-spotted, lobes 1–1.5 × throat; stamens, style exserted. **FR** 4–6 mm, 1–2 × calyx, narrowly ovoid. **SEEDS** 7–13 per chamber. $2n=18$. Open, sandy areas; 600–1200 m. TR, PR, DMtns. Like *G. splendens* but fls smaller, paler, self-pollinated.

G. austro-occidentalis (A.D. Grant & V. Grant) A.D. Grant & V. Grant **ST** 10–30 cm; branches spreading from base, cobwebby near base, glandular above. **LVS**: basal in rosette, 1-pinnate, glabrous or cobwebby; axis linear; lobes 2–3 mm, linear, entire or toothed, spreading. **INFL** clustered in fl, loose in fr; pedicels unequal. **FL**: calyx glandular or cobwebby, green or purplish, lobes acuminate, rib width = membrane; corolla 6–8 mm, tube purple, throat yellow, lobes 1–3 mm, lavender, purple-spotted below; stamens unequal, longest exserted to middle of corolla lobes. **FR** 3–4 mm, detaching. **SEEDS** 24–30. $2n=18$. Sandy flats, interior and coastal valleys; 600–1300 m. SCoR. Like *G. minor, G. interior*.

G. brecciarum M.E. Jones Pl sometimes with skunk-like odor. **ST** 8–35 cm, densely cobwebby near base. **LVS**: basal in semi-erect cluster, 2–5 cm, 2-pinnate; lobes > axis width; upper cauline lobes gen finger-like, middle lobe widest. **INFL**: fls (and gen frs) clustered; pedicel glands dense, minute, black, stalked. **FL**: calyx lobes acute, spreading, ribs thick, densely glandular, width > membrane, membrane gen purple-tinged. **FR** 4–7 mm, widely ovoid. Open, sandy places; 500–2300 m. s SN, Teh, SCoRI, TR (n edges), w DMoj, DMtns; se OR, NV, UT. Sspp. intergrade somewhat in s CA.

ssp. ***brecciarum*** (p. 837) **ST**: branches spreading. **FL**: corolla 7–11 mm, tube incl, purple, throat exserted, narrowly V-shaped, purple below, yellow and purple-veined above, lobes pinkish lavender with white base; stamens, style slightly exserted. $2n=18$. Sandy flats in open woodlands or shrubland; 1200–2300 m. s SNH, Teh, n edges SnGb, GB; to se OR, NV, UT.

ssp. ***jacens*** (A.D. Grant & V. Grant) Day **ST**: branches long-decumbent, densely cobwebby. **FL**: corolla 5–7 mm, tube incl, purple, throat gen incl, purple (or upper throat yellow and purple-veined); stamens, style slightly exserted. $2n=18$. Low, sandy or clay flats; 500–2200 m. Teh, SCoRI, WTR. [*G. j.* A.D. Grant & V. Grant]

ssp. ***neglecta*** A.D. Grant & V. Grant (p. 837) **ST** spreading to erect. **FL**: corolla showy, 10–20 mm, tube purple, gen exserted, throat widely V-shaped, lower half purple, white above, yellow spotted, lobes white, tips lavender; stamens, style long-exserted. $2n=18$. Sandy desert flats; 650–2100 m. c&s SNH, Teh, w DMoj, w DMtns. [ssp. *argusana* A.D. Grant & V. Grant] ❀TRY.

G. campanulata A. Gray **ST** 3–12 cm; branches spreading, gland-dotted. **LVS**: basal few, rosette 0; lower cauline narrowly oblanceolate, entire or 3–7-toothed, with white, jointed hairs; upper cauline spreading or recurved, entire, gland-dotted. **INFL**: fls 2 per st; pedicels 3–11 mm, thread-like, glandular. **FL**: calyx lobes acuminate, free nearly to base, membrane-margined to near tip; corolla 7–9 mm, 2–3 × calyx, bell-shaped, yellow with white lobes, 2 purple lines below lobes, tube = 1 mm, throat > lobes; stamens attached near base, incl; style incl. **FR** 2–3 mm, < calyx. **SEEDS** 7–8 per chamber, gelatinous when wet. $2n=18$. Open, sandy flats; 900–2100 m. SNE; NV. ❀TRY.

G. cana (M.E. Jones) A.A. Heller Pl sometimes with skunk-like odor. **ST**: branches 1–several from base, 9–32 cm, cobwebby near base, gen glandular above. **LVS**: basal in rosette, 1–2-pinnate, cobwebby; axis < 3 mm wide; lobes gen ascending, toothed on both sides, teeth short-pointed or acuminate. **INFL** spreading, showy; fls many. **FL**: calyx gland-dotted, lobes acute to acuminate; corolla tube purple, throat bluish in upper half, yellow below, lobes pink; stamens exserted, to below middle of corolla lobes; style > stamens. **FR** < calyx, ovoid to spheric; valves detaching. Open, gravelly or sandy flats or washes; 800–3100 m. SnBr, s SNE, DMoj; sw NV; nw AZ. ❀TRY.

ssp. ***bernardina*** A.D. Grant & V. Grant **LVS**: basal densely cobwebby. **INFL**: fls in 2's or 3's; pedicels very unequal. **FL**: calyx cobwebby to slightly glandular; corolla 17–23 mm, tube exserted, throat widely conic, lobes 4–6 mm wide; stamens unequal, longest exserted. **FR** 5–9 mm, widely ovoid. **SEEDS** 4–6 per chamber. 2*n*=18. Habitat of sp.; 800–1500 m. n SnBr, sw edge DMoj. Some populations intergrade with *G. leptantha* ssp. *transversa*.

ssp. ***cana*** (p. 837) **LVS**: basal gen cobwebby-matted. **INFL**: pedicels ascending, narrowly spreading, very unequal. **FL**: calyx cobwebby or glandular; corolla 14–26 mm, tube well exserted, throat narrowly conic, lobes 2–3 mm wide; stamens gen unequal, longest exserted to above middle of corolla lobes. **FR** 3–6 mm. **SEEDS** 2–4 per chamber. 2*n*=18. Rocky, sandy, or granitic slopes; 1800–3100 m. e slope s SNH, s SNE (Coso Mtns). [*G. latiflora* ssp. *cosana* A.D. Grant & V. Grant]

ssp. ***speciformis*** A.D. Grant & V. Grant (p. 837) **LVS**: basal 2–5 cm, densely cobwebby. **INFL**: fls 4–8 per bract; pedicels slender, in spreading, subequal pairs. **FL**: corolla 15–29 mm, tube 2–4 × calyx, throat narrowly conic, lobes 3–9 mm wide; stamens equal, exserted to below middle of corolla lobes. **FR** 5–8 mm, widely ovoid to spheric. **SEEDS** 4–6 per chamber. Gen basalt gravel or sand; 800–1200 m. e DMtns; sw NV, AZ.

ssp. ***speciosa*** (Jepson) A.D. Grant & V. Grant (p. 837) **LVS**: basal 2–4 cm, densely cobwebby. **INFL**: fls 4–8 per bract; pedicels slender, spreading, unequal, nearly glabrous. **FL**: calyx glabrous or glandular; corolla 20–32 mm, tube 3–4 × throat, throat short, narrow, lobes 4–9 mm wide; stamens equal, exserted to base of lobes. **FR** 5–9 mm. **SEEDS** 10–20 per chamber. 2*n*=18. Habitat of sp.; 200–1200 m. e slope s SNH. Locally abundant. Intergrades with ssp. *cana*.

ssp. ***triceps*** (Brand) A.D. Grant & V. Grant (p. 837) **LVS**: basal 1–5 cm, slightly cobwebby. **INFL**: fls many; pedicels slender, spreading, nearly equal. **FL**: corolla 8–23 mm, tube 1–2.5 × calyx, throat cup-shaped, lobes ± 3 mm wide; stamens exserted to below middle of corolla lobes. **FR** 4–7 mm, widely ovoid to spheric. **SEEDS** 4–6 per chamber. 2*n*=18. Sandy flats, gen on limestone; 800–1600 m. n&w DMtns; sw NV, w AZ.

G. capillaris Kellogg (p. 837) **ST**: branches ascending, 4–20 cm, glandular-puberulent. **LVS** cauline, ascending, 1–2 mm wide, linear to narrowly lanceolate, entire. **INFL**: fls 1–3 per st. **FL**: calyx glandular-puberulent, fused to middle, membrane splitting to base in fr, lobes linear, acuminate, translucent at tip; corolla 6–8 mm, 2 × calyx, tube yellow, glandular-puberulent, throat, lobes white or bluish white; stamens unequal, attached in upper throat, slightly exserted; style < stamens, slightly exserted. **FR** < calyx, splitting to base, detaching. **SEEDS** 2 per chamber. 2*n*=18. Open, wet, gravelly areas, meadows, streamsides, or snow pockets; 1500–3100 m. KR, CaRH, SNH, WTR, SnBr, PR, Wrn; to OR, ID, UT.

G. capitata Sims **ST** branched, 10–90 cm, leafy below infl, glabrous, glandular, or with white hairs. **LVS**: lower 1–2-pinnate, lobes 5–20 mm, axis and lobes ± 1 mm wide, linear, end lobe gen wider, axils and midribs gen with white hairs; upper cauline reduced or lobes finger-like. **INFL**: head, terminal, spheric; fls 50–100. **FL**: calyx expanded in fr, lobes acute to acuminate, membrane width > ribs; corolla tube incl, throat gen exserted, lobes linear to oblong; stamens and style exserted; stigmas < 2 mm. **FR** < calyx, ovoid to spheric. 2*n*=18. Open, gen sandy to rocky areas; < 2100 m. NW, CaR, SN, SnJV, SnFrB, SCoR, TR, PR; to B.C., ID, Baja CA.

ssp. ***abrotanifolia*** (E. Greene) V. Grant (p. 837) **ST** 20–80 cm, glabrous to glandular. **LF** 1–2-pinnate. **INFL** 17–35 mm wide, loosely 25–50-fld; pedicels 1–2 mm. **FL**: calyx ribs hairy, green or brown, membrane colorless or bluish, lobes acuminate, tips recurved; corolla 8–12 mm, pale blue-violet or white, lobes 2–3 mm wide, oblong. **FR** detaching. **SEEDS** 9–18. 2*n*=18. Sandy, loamy slopes; < 1900 m. s SN, Teh, SCoR, ChI, TR, PR; Baja CA. ❀3, **7–9**,**14–16**,17,**18–24**.

ssp. ***capitata*** (p. 837) **ST** 20–80 cm, glabrous, glandular, or hairy. **LF** 2-pinnate. **INFL** 14–40 mm wide; pedicels < 1 mm. **FL**: calyx gen glabrous, membrane colorless, lobes acute, gen erect; corolla 6–8 mm, pale blue-violet, lobes linear, < 1 mm wide; stigmas < 0.5 mm. **FR** spheric, detaching. **SEEDS** 1–6. 2*n*=18. Rocky slopes; 100–1900 m. NW, CaR; to B.C., ID. Also in SnFrB, probably as garden escape. ❀SUN:1–3,**4–6**,**17**&IRR:7,8,9,**14–16**,**18–24**.

ssp. ***chamissonis*** (E. Greene) V. Grant DUNE GILIA Pl gen with skunk-like odor. **ST**: branches spreading, 15–70 cm, stout, glandular. **LVS**: basal in rosette, fleshy, glandular, 2-pinnate. **INFL** 25–35 mm wide; base densely tomentose; fls sessile. **FL**: calyx tomentose, ribs green, membrane purplish, lobes acuminate, tips recurved; corolla 9–10 mm, bright blue-violet, lobes ± 3 mm wide, oblong. **FR** ovoid, not detaching. **SEEDS** 10–25. 2*n*=18. Coastal sandhills; < 60 m. n CCo. ❀SUN,DRN,IRR:5,7–9,**14–17**,19–21, **22–24**.

ssp. ***mediomontana*** V. Grant **ST** simple or branches ± erect, 15–40 cm, leafy below middle, glabrous or glandular. **LVS** in loose basal rosette, 1–2-pinnate. **INFL** 12–25 mm wide; base tomentose; fls sessile. **FL**: calyx hairy, membrane colorless, lobes acute, erect to recurved; corolla 5–8 mm, pale blue-violet to white, lobes ± 1 mm wide, linear. **FR** ovoid to spheric, detaching. **SEEDS** 3–10. 2*n*=18. Open, rocky slopes; 900–2100 m. n&c SN. ❀TRY.

ssp. ***pacifica*** V. Grant **ST** 25–50 cm. **LF** 2-pinnate; axis and lobes ± 1 mm wide. **INFL** 12–40 mm wide; pedicels 0–3 mm. **FL**: calyx glabrous to slightly hairy or glandular, ribs green, membrane blue-violet, lobes erect, acute; corolla 6–8 mm, pale to bright blue-violet, lobes 1–2 mm wide, linear. **FR** obovoid, detaching. **SEEDS** 10–25. 2*n*=18. Coastal bluffs, slopes; gen < 300 m. n&c NCo; OR. ❀TRY.

ssp. ***pedemontana*** V. Grant **ST** 30–90 cm, glabrous or glandular. **LF** 1–2-pinnate. **INFL** 10–20 mm wide; base tomentose; fls sessile. **FL**: calyx densely hairy, membrane colorless, lobes acuminate, recurved; corolla 7–11 mm, pale blue-violet, lobes ± 2 mm wide, oblong. **FR** not detaching. **SEEDS** 6–15. 2*n*=18. Open, rocky slopes; 60–1500 m. SNF. ❀SUN,DRN:**7**,**8**,**9**,**14–24**.

ssp. ***staminea*** (E. Greene) V. Grant (p. 837) **ST** 10–60 cm, glabrous or glandular. **LVS** in loose basal rosette, gen 1-pinnate. **INFL** 20–30 mm wide; base tomentose; fls sessile. **FL**: calyx hairy, membrane bluish or colorless, lobes acuminate, tips recurved; corolla 7–13 mm, pale blue-violet, lobes ± 3 mm wide, oblong. **FR** gen not detaching. **SEEDS** 10–25. 2*n*=18. Sandhills, flats; < 100 m. SnJV, SCoRI; AZ. Intergrades with ssp. *pedemontana*. ❀DRN, SUN:7,**8**,**9**,10,**14–24**.

ssp. ***tomentosa*** (Brand) V. Grant **ST** 10–70 cm, stout, gen hairy. **LF** 2-pinnate. **INFL** 20–35 mm wide, 1–several in cluster; base tomentose; fls sessile. **FL**: calyx densely hairy, membrane colorless, lobes acute, tips slightly recurved; corolla 7–11 mm, bright blue-violet, lobes 1–2 mm wide, linear. **FR** detaching. **SEEDS** 3–10. 2*n*=18. Sea bluffs; < 30 m. NCo. Intergrades with ssp. *capitata* in ne SnFrB. ❀TRY.

G. caruifolia Abrams CARAWAY-LEAVED GILIA **ST**: branches spreading, 12–100 cm, glabrous or hairy below, glandular in infl; glands flat-topped. **LVS** with translucent hairs; basal and lower cauline in semi-erect rosette, 2–8 cm, finely 2–3-pinnate, axis linear, lobes acuminate or short-pointed. **INFL** loose; pedicels 1–11 mm in unequal pairs, slender. **FL**: calyx gen glabrous, purple-blotched; corolla 7–13 mm, lavender, throat < lobes, yellow, lobes purple-marked at base; stamens, style long-exserted. **FR** 1.2–2 × calyx. **SEEDS** 12–20. 2*n*=18. UNCOMMON. Open areas in woodlands, chaparral, sandy soil; 1400–2300 m. PR; Baja CA.

G. clivorum (Jepson) V. Grant (p. 837) **ST**: branches spreading or erect, gen many, 6–30 cm, leafy, hairy below, glandular or hairy in infl. **LVS**: basal 2–6 cm, 1–2-pinnate, axis and lobes narrowly linear, ascending. **INFL**: clusters in fl, loose in fr; fls 2–5. **FL**: calyx 4–5 mm, hairy or glandular, membrane gen purple, rib width > membrane; corolla 6–8 mm, tube, throat gen incl, yellow, purple-spotted, lobes light blue or white; stamens and style slightly exserted. **FR** < calyx. **SEEDS** 24–36. $2n$=36. Common. Open, grassy areas; < 1100 m. s NCoR, CW, ChI, WTR; also in AZ.

G. clokeyi H. Mason (p. 837) **ST** 8–17 cm, cobwebby below middle, glandular above. **LVS** 1–3 cm, gray-green, cobwebby; basal in rosette, 1-2-pinnate, lobes short-pointed, spreading; upper cauline palmate, middle lobe widest. **INFL** loose; pedicels paired, unequal, spreading, thread-like. **FL**: calyx 2–4 mm, glabrous or cobwebby in early fls, lobe short-pointed, tip thick; corolla 4–6 mm, tube gen incl, throat exserted, base yellow, lobes 1–3 mm, acute, white, gen blue-streaked; stamens, style slightly exserted. **FR** 3–6 mm, spheric, detaching. $2n$=18. Open, rocky slopes or sandy washes, gen on limestone; 400–2000 m. e DMoj, DMtns; to Colorado, NM.

G. diegensis (Munz) A.D. Grant & V. Grant **ST** 10–40 cm, glabrous, glaucous below middle, glandular above. **LVS**: basal in prostrate rosette, 1–7 cm, strap-shaped, toothed or pinnate, lobes spreading, cobwebby; cauline shorter, clasping, lanceolate, serrate. **INFL** cluster in fl, loose in fr; pedicels unequal. **FL**: calyx 3–5 mm; corolla 8–12 mm, tube incl or slightly exserted, purple, lower throat purple, upper yellow, lobes 2–5 mm, lavender or white; stamens unequal, exserted to mid-lobe, pollen blue; style gen = or > stamens. **FR** 4–7 mm, < calyx. **SEEDS** many. $2n$=18. Sandy areas, open forest or shrubland; 500–2200 m. SnGb, SnBr, PR; Baja CA.

G. filiformis A. Gray Pl glabrous or sparsely gland-dotted. **ST**: branches few–many, 5–15 cm, thread-like, glaucous. **LVS** 1–3 cm, linear, entire; basal rosette 0 or ephemeral; cauline spreading. **INFL**: fls gen 2 per st; pedicels 2–16 mm. **FL**: calyx 2–4 mm, tube 1–2 mm, lobes acuminate, margins membranous to near tip, spreading in fr; corolla 4–7 mm, yellow, lobes 3–5 mm, oblong; stamens attached near base; stamens, style well exserted, < corolla lobes. **FR** 2–4 mm. **SEEDS** many, gelatinous when wet. $2n$=18. Open, rocky canyon slopes or washes; 300–1800 m. DMoj; to UT, AZ.

G. hutchinsifolia Rydb. Pl densely glandular-puberulent (studded with sand grains), glandular-hairy below, with skunk-like odor. **ST** branched from base or above, 5–40 cm. **LVS**: lower in semi-erect cluster, 2–8 cm, 1-2-pinnate, axis linear, lobes > 2 × axis width; upper linear, entire. **INFL**: fls gen 2 per st; pedicels unequal. **FL**: calyx 3–4 mm, lobes linear, membrane attached well below tip, U-shaped between lobes in fr; corolla 7–14 mm, tube well exserted, glandular-puberulent, throat green-spotted, yellow above, lobes 2–4 mm, gen wavy-margined, lavender; stamens, style slightly exserted. **FR** 3–6 mm, = or > calyx. **SEEDS** many, not gelatinous when wet. $2n$=18. Sandy or gravelly flats, slopes, dunes; 400–1800 m. GB, DMoj; to UT, AZ. ❀TRY.

G. inconspicua (Smith) Sweet (p. 837) **ST**: branches ascending or spreading, 8–32 cm, cobwebby below infl, black-glandular above. **LVS**: basal in rosette, 1-pinnate, cobwebby, lobes 2–10 mm, = or > axis width, linear (or rounded and short-pointed), entire or toothed, ascending. **INFL**: cluster in fl, loose in fr; fls 2–4 per st; pedicels unequal. **FL**: calyx glabrous or gland-dotted or cobwebby in early fls; corolla 6–11 mm, 3–4× lobes, tube and throat gen yellow, lobes lavender with purple spot at base; stamens and style exserted; longest stamens > middle of corolla lobe. **FR** 5–8 mm, < calyx, oblong-ovoid; valves detaching. **SEEDS** 4–6 per chamber. $2n$=36. Rocky or sandy sagebrush slopes, washes; 1200–2000 m. GB; to WA, ID, NV.

G. interior (H. Mason & A.D. Grant) A.D. Grant **ST**: branches spreading, 6–13 cm, cobwebby below infl, glandular above. **LVS**: basal in rosette, 1-pinnate, cobwebby; axis 1–2 mm wide; lobes entire or toothed, 1–2 × axis width, spreading. **INFL**: fls 1–2 per st; pedicels 2–20 mm, unequal. **FL**: calyx glandular; corolla 6–11 mm, 2–3 × lobes, tube purple, throat yellow, purple-spotted, lobes 3–5 mm, lavender; stamens and style exserted, longest stamens > middle of corolla lobe. **FR** 3–6 mm, < or = calyx, narrowly ovoid. **SEEDS** 3–4 per chamber. $2n$=18. Common. Rocky slopes in open woodland or shrubland; 700–1700 m. s SNF, Teh.

G. inyoensis I.M. Johnston (p. 837) **ST**: branches spreading, 3–10 cm, finely glandular, white-hairy below. **LVS** white-jointed-hairy; basal few, 4–8 mm, oblanceolate to obovate, entire or toothed; cauline 3–6 mm, upper entire, spreading or recurved. **INFL**: fls 2 per st; pedicels 4–8 mm, thread-like, spreading. **FL**: calyx 2–4 mm, tube short, membrane bordering lobes to near tip; corolla 4–7 mm, tube and throat yellow, incl in calyx, lobes white, obovate; stamens, style slightly exserted. **FR** 2–4 mm. **SEEDS** 10–15 per chamber, gen not gelatinous. Common. Open, sandy flats in pine forest or sagebrush shrubland; 1900–2600 m. s SNH, SNE; NV. [var. *breviuscula* Jepson]

G. latiflora (A. Gray) A. Gray BROAD-FLOWERED GILIA Pl ± scapose. **LVS**: basal in prostrate rosette, 2–7 cm, cobwebby, strap-shaped, toothed or lobed, lobes spreading; cauline shorter, clasping, entire or lobed at base, tapered. **INFL** gen loose; pedicels unequal. **FL** showy, fragrant; calyx slightly glandular, or early fls cobwebby; corolla 10–35 mm, tube purple, upper throat, base of lobes white, tips lavender; stamens exserted, < lobes; style > stamens. **FR** gen > calyx. Gen deep sandy soils; 700–1500 m. SCoRI, DMoj and w margins. Forms showy displays. Sspp. variable within populations, intergrading where ranges overlap.

ssp. ***cuyamensis*** A.D. Grant & V. Grant CUYAMA GILIA **ST** 6–25 cm, glabrous, glaucous below middle. **FL**: calyx 2–4 mm; corolla 9–16 mm, tube well exserted, throat tapered or widely expanded, purple at base. **FR** 3–6 mm. $2n$=18. UNCOMMON. Sandy flats, pinyon/juniper woodland, lower river valleys; 600–2000 m. s SCoRI, n WTR.

ssp. ***davyi*** (Milliken) A.D. Grant & V. Grant (p. 837) **ST** 10–30 cm, glabrous, glaucous below middle, or cobwebby. **INFL**: cluster in fl, loose in fr. **FL**: calyx 4–7 mm; corolla 18–24 mm, lobes 1/3 tube + throat, tube exserted, throat gen long-tapered, purple to middle or higher. **FR** 6–9 mm. Common. Open, sandy flats; 700–1200 m. SCoRI, s DMoj and w margins. [*G. latiflora* ssp. *excellens* (Brand) A.D. Grant & V. Grant] ❀DRN,SUN:**7–12,14–16**,17,**18–23**,24

ssp. ***elongata*** A.D. Grant & V. Grant (p. 837) **ST** 13–32 cm, cobwebby below middle. **FL**: calyx 4–6 mm; corolla 21–35 mm, 4–6 × lobes, tube > 3 × calyx, white-veined, throat narrowly tapered, yellow, white. **FR** 6–9 mm. Open, sandy slopes; 700–1200 m. w DMoj (Rand & El Paso ranges, Black Rock Hills). ❀TRY.

ssp. ***latiflora*** (p. 837) **ST** 10–33 cm, glabrous, glaucous below middle. **INFL** loose. **FL**: calyx 4–5 mm; corolla 15–22 mm, tube slightly exserted, white-veined, throat widely expanded, yellow at base. **FR** 6–9 mm. $2n$=18. Common. Open, sandy flats; 700–1100 m. sw DMoj, n base SnGb & SnBr. ❀TRY.

G. latifolia S. Watson BROAD-LEAVED GILIA Pl strong-scented. **ST** 10–30 cm, glandular-hairy below; branches spreading. **LVS**: basal in rosette, petioled, ovate, holly-like, appressed-hairy; upper lvs reduced, needle-like. **INFL**: cluster, loose, gland-dotted; fls many; pedicels spreading, longer in fr. **FL**: calyx fused in lower 1/2, gland-dotted, lobes fine-pointed; corolla 7–11 mm, tube white, lobes bright pink above, whitish or dull pink below; stamens attached in lower throat, unequal, longest slightly exserted, pollen white. **FR** equal or > calyx, ovoid. **SEEDS** many, not gelatinous. $2n$=36. Common. Rocky slopes, washes; < 1800 m. D; to UT, AZ. Like *G. ripleyi*.

G. leptalea (A. Gray) E. Greene Pl glabrous or slightly glandular-puberulent; glands black. **ST**: branches spreading. **LVS** cauline, linear to narrowly elliptic, gen entire; lower rarely pinnately lobed. **INFL** loose, spreading; pedicels thread-like. **FL**: calyx glabrous, lobes tapered, fine-pointed, membrane in fr splitting between lobes; corolla funnel-shaped, tube gen exserted, lobes ovate-acute, pink; stamens unequally attached in throat, style and longest stamens well exserted. **FR** < calyx; valves detaching. **SEEDS** 1–3 per chamber. Common. Open, rocky areas in forest or meadows; 900–3100 m. CaR, SN; OR. Forms large showy populations. Sspp. intergrade.

ssp. ***bicolor*** H. Mason & A.D. Grant (p. 839) **ST** 4–26 cm. **FL**: corolla 8–15 mm, throat 2–5 mm, < 2 × lobes, tube, throat yellow, sometimes with short purple lines below lobes. Habitat of sp.; 1600–3100 m. SNH. A small form (1–4 cm, corolla ± 6 mm) grows in rocky meadows, c&s SNH, 2200–3100 m. Differs from *G. capillaris* by pink corolla lobes, exserted stamens, more spreading habit.

ssp. ***leptalea*** (p. 839) **ST** 8–33 cm. **FL**: corolla 13–21 mm, throat 6–8 mm, 2–3 × lobes, tube yellow, throat purple, veins yellow. $2n=18$. Habitat of sp.; 900–2100 m. CaR, s SNF, n SNH; OR. ❀TRY.

G. leptantha Parish **ST** spreading or erect, cobwebby below middle, glandular above. **LVS**: basal in rosette, 1-pinnate, cobwebby, axis 1–2 mm wide, linear, lobes toothed on both sides, teeth short-pointed or acuminate. **INFL** clustered or loose with 2–6 fls above a bract; pedicels unequal, barely spreading. **FL**: calyx slightly glandular or cobwebby in early fls; corolla tube yellow or purple with yellow veins, throat gen yellow, lobes pink to lavender; stamens unequal, longest gen > corolla lobes. Open, rocky or sandy areas; 700–2800 m. s SNH, SCoR, n slope SnGb, SnBr.

ssp. ***leptantha*** (p. 839) **ST** spreading or erect, 15–45 cm. **FL**: corolla 13–23 mm, tube 2–4 × calyx, narrow, flared to narrow throat, lobes 1–3 mm wide, narrowly oblong; longest stamens > lobes. **FR** 3–4 mm, ovoid. $2n=18$. Open, rocky soil in forest, streambanks; 1500–2300 m. SnBr. ❀TRY.

ssp. ***pinetorum*** A.D. Grant & V. Grant (p. 839) **ST**: lower branches long-decumbent, 5–25 cm. **LVS**: basal in compact rosette, densely cobwebby. **FL**: corolla 9–14 mm, tube 1–1.5 × calyx, throat conic, > tube, lobes 3–4 mm wide, obovate; longest stamens > lobes. **FR** 4–5 mm, spheric. $2n=18$. Bare summits, open, rocky or sandy areas with pines; 1500–2800 m. SCoR, WTR (Mt. Pinos).

ssp. ***purpusii*** (Milliken) A.D. Grant & V. Grant **ST** < 60 cm. **FL**: corolla 13–23 mm, tube 2–4 × calyx, slender, throat, lobes 3–5 mm wide, gen narrowly oblong; longest stamens well exserted, < corolla lobes. **FR** 3–6 mm, ovoid. $2n=18$. Common. Open, rocky or sandy soil in forest or chaparral; 700–2600 m. s SN. Intergrades with ssp. *pinetorum* in s SNF, Teh. ❀DRN:1–3,**7**,14–16,18–23.

ssp. ***transversa*** A.D. Grant & V. Grant **ST** 16–40 cm, densely glandular above rosette. **LVS** basal, lower cauline 2–8 cm. **FL**: glandular or cobwebby; corolla 13–17 mm, tube 1–1.5 × calyx, throat widely expanding, lobes 3–6 mm wide, obovate; stamens unequal, longest well exserted, < corolla lobes; style > stamens. **FR** 3–8 mm, < or = calyx, ovoid. $2n=18$. Rocky or sandy soil, gen near streams; 900–1800 m. n slope SnGb, SnBr. ❀TRY.

G. leptomeria A. Gray (p. 839) GREAT BASIN GILIA **ST**: branches spreading, 1–several, 7–23 cm, thread-like, glandular-puberulent. **LVS**: basal in rosette, 1–6 cm, lanceolate or strap-shaped, toothed or lobed, lobes rounded and short-pointed, gen < axis width; cauline linear, entire. **INFL**: fls 1–3 per st. **FL**: calyx 2–3 mm, lobe tips thickened, membrane U-shaped; corolla 4–7 mm, tube 1.5–3 × calyx, thread-like, purple, throat yellow, lobes truncate and short-pointed, white, outside purple; stamens, style slightly exserted, pollen white. **FR** 3–5 mm, = calyx, narrowly ovoid. **SEEDS** many, not gelatinous. $2n=34,36$. Common. Open, sandy or rocky areas; 800–2100 m. GB; to OR, ID, Colorado, NM.

G. lottiae Day (p. 839) **ST** 5–43 cm, glandular-puberulent; branches gen spreading from base. **LVS**: basal 2–11 cm, in flat rosette, widely strap-shaped, fleshy, serrate or pinnate, lobes spreading; cauline linear, entire. **INFL** clustered, nodding, becoming erect. **FL**: calyx lobe short; corolla 6–7 mm, tube slightly exserted, stout, < throat, yellow, lobes lanceolate to ovate, pink to white; stamens, styles barely exserted. **FR** 3–5 mm, < 1.5 × calyx, ovoid, tip pointed. $2n=50$. Sandy soils, sagebrush scrub; 400–2100 m. GB, e DMtns (Clark Mtns); to WA, ID, UT.

G. maculata Parish (p. 839) LITTLE SAN BERNARDINO MOUNTAINS GILIA Pl hairy. **ST** 1–3 cm. **LF** 2–5 mm, entire, gen oblong or oblanceolate. **INFL**: cluster, dense; fls ± sessile. **FL**: calyx tube < 1 mm, lobe margins widely membranous, membrane ciliate; corolla 3–5 mm, white, lobes 1–2 mm, wavy, recurved, red-spotted at base; stamens, style slightly exserted; pollen yellow. **FR** < calyx, widely ovoid. **SEEDS** 6–16, not gelatinous. $2n=18$. RARE. Sandy flats; 900–1100 m. s DMtns (Little San Bernardino Mtns). [*Linanthus m.* (Parish) Milliken] Threatened by development.

G. malior Day & V. Grant **ST** 5–20 cm; lower branches decumbent, spreading, cobwebby below infl, gland-dotted above. **LVS**: lower in basal rosette, 1-pinnate, axis, lobes 1–2 mm wide, linear, cobwebby, lobes spreading, sometimes toothed; upper palmate to entire above. **INFL**: cluster in fl, loose in fr; pedicels unequal. **FL**: calyx gen slightly glandular; corolla 6–11 mm, tube and throat purple or partly yellow above, throat tapered, lobes lavender, obovate, tip obtuse; stamens, style slightly exserted. **FR** = calyx; tip rounded; valves detaching. $2n=36$. Open, sandy, rocky flats; 700–2000 m. Teh, s SnJV, SCoRI, SNE, DMoj; NV, OR. Intermediate between *G. minor, G. aliquanta*.

G. micromeria A. Gray (p. 839) **ST**: branches many, spreading, 4–14 cm, thread-like, slightly puberulent. **LVS**: basal in rosette, 1–6 cm, pinnate, lobes spreading, length 0.5–1 × axis width; upper linear, entire. **INFL** loose, nodding in bud; pedicels spreading or recurved. **FL**: calyx ± 2 mm, membrane attached near lobe tips, V-shaped; corolla ± 3 mm, tube gen incl, throat pale yellow, lobes oval, entire, white or lavender; stamens, style slightly exserted; pollen white. **FR** ± 3 mm, ± = calyx, spheric. **SEEDS** 2–8 per chamber, not gelatinous. $2n=18$. Uncommon. Rocky or sandy soil; 1200–1670 m. GB (Modoc, Inyo cos.); to OR, Colorado.

G. millefoliata Fischer & C. Meyer (p. 839) Pl densely glandular, skunk-like odor. **ST** 8–30 cm; main st short; branches gen long-decumbent. **LVS** ± fleshy; basal in rosette, 1–2-pinnately lobed, lobes 2–5 mm; upper shorter, palmate. **INFL**: clusters; fls 2–6; pedicels 2–5 mm. **FL**: calyx 4–6 mm, glandular, membrane purple or colorless, narrower than ribs; corolla 8–11 mm, tube yellow, throat incl, with paired purple spots, lobes < 2 mm wide, obovate; stamens, style slightly exserted to base of lobes; pollen white. **FR** 7–9 mm, < calyx. **SEEDS** 25–50. $2n=18$. Stabilized coastal dunes; < 10 m. NCo; s OR. Formerly in San Francisco. ❀TRY.

G. minor A.D. Grant & V. Grant (p. 839) **ST**: lower branches decumbent, 6–20 cm, cobwebby below infl. **LVS**: basal in rosette, 1-pinnate, axis and lobes < 1 mm wide, linear, cobwebby, lobes ascending; upper glandular. **INFL** clustered in fl, loose in fr; pedicels very unequal. **FL**: calyx glandular or cobwebby in early fls; corolla 4–8 mm, ± 2 × calyx, tube incl, purple, throat purple or yellow with purple-veins, tapered, lobes oval-acute, lavender; stamens, style slightly exserted. **FR** 2 × calyx, narrowly ovoid; valves partly separating, not detaching. $2n=18$. Firm sand, gen at base of shrubs; 300–1100 m. SCoR, DMoj; AZ.

G. modocensis Eastw. **ST**: branches gen several, spreading from below, 10–34 cm, slightly cobwebby or glandular near base. **LVS**: basal in rosette, strap-shaped, pinnate, lobes short, spreading, toothed; cauline clasping, gen deeply lobed. **INFL** clustered in fl, loose in fr; pedicels glandular; glands black. **FL**: calyx gland-dotted, rib width > membrane, lobes erect or spreading in fr; corolla 7–10 mm, tube incl, purple, throat yellow, or midveins purple, lobes 2–3 mm, lavender; stamens, style, slightly exserted. **FR** < or = calyx. $2n=36$. Open, rocky areas, pinyon/juniper woodland; 400–2300 m. Teh, WTR, SnBr, GB, DMtns; OR, ID, NV. Decumbent pls in n WTR (Mt. Pinos) with lower lvs 2-pinnate and densely cobwebby have been called *G. tetrabreccia* A.D. Grant & V. Grant.

G. nevinii A. Gray (p. 839) NEVIN'S GILIA **ST**: branches ± erect, 10–40 cm, leafy, white- or translucent-hairy. **LVS**: basal hairy, 2–3-pinnate, axis and lobes < 1 mm wide. **INFL** glandular. **FL**: calyx 5–6 mm, longer in fr, lobes acute, ribs wider than membrane; corolla 8–14 mm, lavender with narrow, yellow throat, lobes 2–3 mm, short-pointed; stamens slightly exserted, pollen blue. **FR** 6–8 mm, < calyx, narrowly ovoid. $2n=36$. UNCOMMON. Rocky, grassy slopes, coastal canyons; 20–400 m. ChI; Baja CA (Guadalupe Island). ❀SUN,DRN:7–9,**14–17,19–24**.

G. ochroleuca M.E. Jones **ST**: branches gen spreading from below, 5–30 cm, glabrous or cobwebby, gen glandular in infl. **LVS**: lower in rosette, 1–2-pinnate or entire, axis, lobes gen linear, ascending; cauline palmate; infl lvs gen entire. **INFL**: pedicels thread-like, widely spreading, gen in subequal pairs, glandular. **FL**: calyx 2–4 mm, ribs and lobes linear, membrane colorless; corolla tube purple, throat well expanded, blue in upper half, yellow below, lobes pink to white; stamens slightly exserted. **FR** 2–5 mm; valves detaching. **SEEDS** 3–15. Open, gen sandy places; 200–2500 m. SCoRI, SnGb, SnBr, PR, w DMoj; Baja CA.

ssp. ***bizonata*** A.D. Grant & V. Grant Pl gray-green. **ST** glabrous below calyces, gen glabrous, glaucous below infl. **LF** cobwebby or glabrous, 1–2-pinnate; lobes 1–15 mm. **FL**: calyx glabrous, rarely cobwebby, lobes acute; corolla 8–14 mm, tube gen incl, < throat; stamens < style. **FR** spheric. $2n=18$. Common. Flats; 300–2200 m. SCoRI, TR (n slope), w DMoj.

ssp. ***exilis*** (A. Gray) A.D. Grant & V. Grant (p. 839) Pl yellow-green. **ST**: branches sub-erect, gen cobwebby below. **LF** cobwebby, entire or 1-pinnate. **INFL** glabrous or glandular. **FL**: calyx slightly cobwebby, lobes acute to acuminate; corolla 8–11 mm, tube slightly exserted, > throat; style slightly exserted; stigmas near or just above anthers. **FR** spheric to ovoid. 2*n*=18. Common. Flats; 200–1800 m. s SnBr, PR; Baja CA. Small-fld race occurs in se PR.

ssp. ***ochroleuca*** (p. 839) Pl herbage gray-green. **ST**: branches spreading, < 30 cm, glabrous, glaucous below infl, glabrous beneath fls. **LF** slightly cobwebby, 1–2-pinnate, lobes 3–10 mm, ascending; infl lvs palmate. **FL**: calyx glabrous; corolla 4–6 mm, tube incl, throat = tube; style slightly exserted, stigmas near anthers. **FR** spheric. 2*n*=18. Desert sand; 700–1500 m. w DMoj.

ssp. ***vivida*** (A.D. Grant & V. Grant) A.D. Grant & V. Grant (p. 839) Pl dark gray-green. **ST** gen glabrous or cobwebby below middle. **LF** densely cobwebby-matted or tufted, 2–3-pinnate; axis 2–4 mm wide; lobes linear or reduced to teeth. **INFL**: longer pedicel of a pair 3–6 × shorter pedicel. **FL**: calyx glabrous, cobwebby or finely glandular, lobes acute to acuminate; corolla 9–14 mm, tube 1–1.5 × calyx, throat slightly > tube, colors vivid; stamens < style. **FR** spheric. 2*n*=18. Common. Rocky or sandy soil, pine forests; 1800–2500 m. Teh, SnGb. [*G. leptantha* ssp. *v.* A.D. Grant & V. Grant] ❀TRY.

G. ophthalmoides Brand (p. 839) **ST**: branches sub-erect, 15–30 cm, densely cobwebby below middle, glandular above. **LVS**: lower in rosette, densely cobwebby, 1–2-pinnate, axis linear, lobes spreading, gen toothed (or lobed) on both sides. **INFL**: pedicels thread-like, unequal. **FL**: calyx glabrous, lobes acuminate, reddish; corolla 7–12 mm, tube 1.5–2 × calyx, slender, purple, throat narrow, long-tapered, bright yellow, lobes 1–3 mm, pink; stamens, style slightly exserted. **FR** 4–6 mm, gen < calyx; valves detaching. **SEEDS** 3–4 per chamber. 2*n*=36. Open, rocky soil, gen pinyon/juniper woodland; 1100–2600 m. W&I, DMtns; to Colorado, AZ. ❀TRY.

G. ripleyi Barneby (p. 839) RIPLEY'S GILIA Per. **ST** 10–30 cm, densely glandular-hairy. **LVS**: basal, lower in cluster, 3–6 cm, glandular-hairy, obovate, veins raised beneath, ending in needle-like teeth; upper < 1 cm, gland-dotted, needle-like. **INFL** loose; fls many, gland-dotted. **FL** opening in evening; calyx 4–6 mm, tube 2 mm, lobes acuminate, margins membranous to near tip; corolla 7–10 mm, tube 1/2 corolla, white, throat white, lobes 3–5 mm, ovate, both surfaces pink; stamens attached near base, incl or longest slightly exserted; style slightly exserted. **FR** 3–4 mm. **SEEDS** many, not gelatinous. 2*n*=18. RARE in CA. Limestone cliffs; 900–1400 m. n DMtns (Inyo Co.); NV (where rare).

G. salticola Eastw. (p. 839) **ST**: branches spreading, 4–20 cm, densely cobwebby. **LVS**: basal in rosette, densely cobwebby, 1–2-pinnate, axis and lobes 2–3 mm wide, entire or toothed. **INFL** loose-clustered, black-gland-dotted. **FL**: calyx glabrous or cobwebby, lobes acuminate, outcurved in fr, membrane purple-spotted or colorless; corolla 6–15 mm, tube incl, yellow, throat yellow, lobes 3–6 mm, gen > throat, bright pinkish lavender; stamens, style exserted, > lobes. **FR** 4–6 mm, < calyx. **SEEDS** 10–15. 2*n*=18. Open, volcanic or granitic areas; 1500–2700 m. n&c SNH, w MP, nw SNE (Mono Co.); w NV. [*G. leptantha* Parish ssp. *s.* (Eastw.) A.D. Grant & V. Grant]

G. scopulorum M.E. Jones (p. 839) **ST**: branches ± erect, 10–30 cm, glandular-hairy. **LVS** glandular-hairy; hairs translucent; lower lvs in semi-erect rosette, 1–2-pinnate, axis linear below, winged above, merging with lobes, lobes toothed, teeth short-pointed; upper minute, 2-toothed. **INFL** loose; pedicels thread-like, spreading in unequal pairs, gland-dotted, glands flat-topped, short-stalked. **FL**: calyx slightly glandular, ribs green, width > membrane; corolla 9–17 mm, 2–4 × calyx, tube purple, throat < tube, yellow, lobes ovate, acute, lavender to pink; stamens, style slightly exserted. **FR** 5–6 mm, < or = calyx, widely ovoid. **SEEDS** many. 2*n*=18,36. Semi-shaded, rocky ravines; < 1200 m. s SNE, e DMoj, DSon; to UT, AZ.

G. sinistra M.E. Jones **ST**: branches spreading, leafy, densely glandular-puberulent; glands black or yellowish. **LVS** spreading, linear or narrowly lanceolate, entire or lobed, if lobed then lower pinnate, upper palmate, 3–5-lobed, middle lobe much longer. **INFL** loose; fls 2–3 above a lf; pedicels unequal, thread-like, longer in fr. **FL**: calyx glandular-puberulent, lobes linear, acute, rib margins reddish, membrane finally splitting; corolla narrowly funnel-shaped, tube and throat yellow with reddish streaks, lobes narrowly oblong, bright pink, tips rounded; stamens unequal, exserted, unequally attached in upper throat; pollen blue. **FR** < or = calyx, oblong. Open, chaparral or forest, serpentine or red volcanic soil; 300–2200 m. KR, NCoR, CaR, SN, SnBr, Wrn; NV, OR, Colorado.

ssp. ***pinnatisecta*** (H. Mason & A.D. Grant) Day (p. 839) **ST** 12–50 cm. **LVS**: lower in semi-erect rosette, pinnate, lobes 7–9, linear; upper palmate. **FL**: corolla 9–20 mm, tube 1.5–2.5 × calyx, slender, lobes 3–5 mm; stamens, style exserted, > mid-lobe. **SEEDS** 3–6 per chamber. 2*n*=18. Habitat of sp.; 300–2200 m. NCoRI. [*G. leptalea* ssp. *p.* H. Mason and A.D. Grant] ❀TRY.

ssp. ***sinistra*** (p. 839) **ST** 7–30 cm. **LVS**: lower gen few, entire; upper spreading, entire or palmate. **FL**: corolla 5–9 mm, tube incl to 1.5 × calyx, lobes 1–3 mm; stamens incl or longest exserted, style incl. **SEEDS** 2–4 per chamber. 2*n*=18. Habitat of sp.; 1000–2200 m. KR, NCoR, CaR, SN, SnBr, Wrn; to OR, Colorado. [*G. capillaris* Kellogg, in part]

G. sinuata Benth. (p. 837) **STS** 1 or several spreading from base, 13–34 cm, glabrous, glaucous below middle. **LVS**: basal in prostrate rosette, cobwebby, pinnate, lobes spreading, entire or toothed; cauline reduced, clasping with expanded base, toothed or entire, tapering to tip. **INFL** glandular; fls in clusters; pedicels longer in fr, unequal, 1–9 mm. **FL**: calyx 3–5 mm; corolla 7–12 mm, tube exserted, purple, white-veined, lobes 2–3 mm, lavender, pink, or white; stamens, style slightly exserted; pollen white. **FR** 4–7 mm. 2*n*=36. Open, sandy flats; 150–1800 m. GB, DMoj; to WA, ID, Colorado, AZ.

G. splendens H. Mason & A.D. Grant SPLENDID GILIA **ST** gen 1; branches several, ± erect, 10–80 cm, glabrous or hairy near base. **LVS**: lower in semi-erect rosette, finely-divided, 2-pinnate, axis linear, lobes well spaced, ascending, hairs shiny, translucent. **INFL** loose; fls 8–10 on main branches; pedicels in unequal pairs, glandular, glands large, flat-topped. **FL**: calyx glabrous, ribs purplish, linear, width < membrane, lobes acuminate to long-tapered; corolla trumpet-shaped, tube exserted, throat tapered, lobes obovate, acute; stamens exserted, < corolla lobes, gen < style. **FR** 4–8 mm, 2 × calyx, narrowly ovoid. **SEEDS** 30–70. Chaparral or forest, rocky soil; 300–2400 m. SCoR, TR, SnJt. ❀TRY.

ssp. ***grantii*** (Brand) V. Grant & A.D. Grant (p. 839) **FL**: corolla 20–36 mm, bright pink, tube 10–18 mm, sometimes throat yellow. 2*n*=18. Habitat of sp.; 800–2200 m. SnGb, SnBr.

ssp. ***splendens*** (p. 839) **FL**: corolla 10–23 mm, gen pink, tube < 10 mm. 2*n*=18. Habitat and range of sp.

G. stellata A.A. Heller (p. 839) STAR GILIA Pl hairy below; hairs white, sharply bent. **ST**: branches gen several from base, 10–40 cm, gland-dotted above. **LVS**: basal in rosette, grayish, 1–2-pinnate, axis linear, lobes toothed on both sides, teeth short-pointed; upper cauline minute, entire or 2-toothed. **INFL** loose; fls 4–8 per st; pedicels spreading, in unequal pairs, glabrous or gland-dotted. **FL**: calyx hairy or glandular, glands black, stalked; corolla funnel-shaped, tube incl or slightly exserted, throat yellow with purple spots, lobes pink or white; stamens, style slightly exserted. **FR** 5–7 mm, widely ovoid. **SEEDS** several per chamber. 2*n*=18. Common. Sandy desert flats, washes; < 1700 m. D; to UT, AZ, Baja CA.

G. subacaulis Rydb. (p. 839) **ST** 5–30 cm, glandular-puberulent; branches gen spreading from base. **LVS** basal 1–7 cm, in ± erect cluster, glandular-puberulent, pinnate, axis narrow, lobes widely obtuse to pointed; cauline entire, linear. **INFL**: cluster, nodding, becoming erect. **FL**: calyx lobe short; corolla 4–7 mm, tube incl, stout, throat > tube, well-expanded, yellow-spotted, lobes ovate, white, lavender-streaked below; stamens exserted; styles gen exserted. **FR** 3–5 mm, 1.5–2 × calyx, ovoid; tip pointed. 2*n*=16. Sandy soils, sagebrush scrub, pinyon/juniper woodland; 500–2500 m. DMoj; to OR, WY, AZ. Closely related to *G. lottiae*.

G. tenerrima A. Gray (p. 839) **ST** spreading, 5–22 cm, thread-like, glandular-hairy below, leafy. **LVS** 1–2 cm, glabrous, linear to narrowly lanceolate, entire or 1–2-lobed; basal few. **INFL** loose; fls 2–3 per st; pedicels gland-dotted, spreading or recurved in fr. **FL**:

calyx tubular below, lobe margins membranous to tip; corolla 2–3 mm, 2 × calyx, tube, throat white, incl, lobes bluish; stamens, style slightly exserted. **FR** 1–3 mm, spheric; valves spreading, detaching. **SEED** 1 per chamber, gelatinous when wet. 2*n*=36. Creeks, rocky canyon slopes; 500–2800 m. SNH, SNE; to OR, WY, UT.

G. tenuiflora Benth. **ST**: branches gen several, spreading from below, glabrous or cobwebby or glandular at base. **LVS**: basal gen in cluster or rosette, cobwebby on upper surface or in axils or nearly glabrous, 1–2-pinnate, lobes gen linear, spreading, axis linear or strap-shaped. **INFL** loose or ± clustered; pedicels unequal. **FL**: calyx glandular or cobwebby; corolla 3–4 × calyx, tube purple, throat gen tapered, partly or all purple, lobes bright pinkish lavender, white at base; stamens unequal, exserted. Sandy coastal or flood plains; < 1400 m. CCo, SCoR, n ChI.

ssp. ***amplifaucalis*** A.D. Grant & V. Grant (p. 839) **ST** 16–24 cm, stout, glabrous or cobwebby near base, slightly glandular above. **INFL** densely clustered, showy. **FL**: calyx cobwebby or slightly glandular, rib width < or = membrane; corolla 13–22 mm, tube stout, 1–2 × calyx, throat widely expanded, purple in lower 1/2, lobes 5–7 mm wide; stamens, style well exserted. **FR** 5–8 mm, < calyx. Sandy soil of dry creeks, floodplains, slopes; 39–900 m. SCoRI. ❀TRY.

ssp. ***arenaria*** (Benth.) A.D. Grant & V. Grant (p. 839) SAND GILIA **ST**: lower branches decumbent, 6–17 cm, gen densely glandular or base cobwebby. **LVS**: basal in prostrate rosette, serrate or 1-pinnate, lobes short. **FL**: calyx densely glandular, rib width > membrane; corolla 10–14 mm, tube 1–2 × calyx, slender, lobes 3–4 mm wide; stamens, style exserted to base of lobes. **FR** 5–6 mm, 1–2 × calyx. 2*n*=18. **THREATENED** US. Coastal sand dunes; < 30 m. c CCo (Monterey Bay). Intergrades with ssp. *tenuiflora* near Salinas River mouth.

ssp. ***hoffmannii*** (Eastw.) A.D. Grant & V. Grant (p. 839) HOFFMANN'S SLENDER-FLOWERED GILIA **ST** 6–12 cm, stout, leafy, glabrous near base, densely glandular above. **LVS** cobwebby in axils; basal, lower lvs in rosette or not, 1-pinnate, lobes 1–4 mm. **INFL** clustered (looser in fr). **FL**: calyx glandular or cobwebby, ribs wider than membrane; corolla 18–20 mm, throat < lobes, lobes 4–6 mm wide, obovate, tube stout, 1.5–2.5 × calyx; stamens well exserted; style slightly exserted, stigmas below longest stamens. **FR** 6–7 mm, < calyx. RARE. Coastal sandhills; < 30 m. n ChI (Santa Rosa Island).

ssp. ***tenuiflora*** (p. 839) **ST** 15–40 cm, glabrous or slightly cobwebby; branches spreading from base. **LVS** cobwebby in axils; basal in semi-erect cluster, 2–6 cm, 1–2-pinnate, lobes 1–2 mm wide. **INFL** glandular. **FL**: calyx slightly glandular or cobwebby, rib width < membrane; corolla 11–22 mm, tube slender, 1.5–3 × calyx, throat entirely purple or with pale yellow spots, lobes 2–4 mm wide; longest stamens > mid-lobe, < style. **FR** 4–6 mm, gen = calyx. 2*n*=18. Sand hills, flood plains, dry river beds; 100–1400 m. CCo, SCoR. ❀TRY.

G. transmontana (H. Mason & A.D. Grant) A.D. Grant & V. Grant **ST**: branches ascending or spreading, 10–32 cm, slightly cobwebby below middle. **LVS** ± cobwebby; basal in rosette, 1-pinnate, axis and lobes linear, lobes 3–11 mm, spreading, entire or few-toothed. **INFL** loose; fls 2–3 per st; pedicels unequal, glandular. **FL**: calyx glabrous or slightly glandular; corolla 5–8 mm, tube incl, purple, lower throat yellow, upper throat white with purple spots below lobes, lobes > throat, ovate, acute, lavender; stamens, style slightly exserted. **FR**: tip pointed or round; valves detaching. 2*n*=36. Rocky or sandy desert slopes or washes; 600–2000 m. DMoj; to UT.

G. tricolor Benth. BIRD'S EYES **ST**: branches many from base, spreading, 10–38 cm, leafy, glabrous or white-hairy below. **LVS** hairy in axils and on upper surfaces; basal in loose cluster, 1–2-pinnate; upper palmate, lobes linear, entire. **INFL** glandular; glands black, minute, slender-stalked. **FL**: calyx ribs flat, green, wider than membrane, lobes acute or acuminate, erect or spreading; corolla 2–3 × calyx, tube incl, yellow, throat exserted, yellow with purple spots below lobes, lobes spreading, blue-violet at tip, paler below; stamens attached near corolla sinuses, unequal; stamens, style slightly exserted. **SEEDS** 16–36. Open, grassland, hills, valleys; < 1200 m. NCoR, SNF, SnJV, SCoR. Sspp. intergrade.

ssp. ***diffusa*** (Congdon) H. Mason & A.D. Grant (p. 845) **INFL** loose; pedicels 15–35 mm. **FL**: calyx 3–4 mm, membrane purple or colorless; corolla 7–13 mm, lobes 3–6 mm wide, colors gen pale, throat spots in pairs. **FR** 3–6 mm. 2*n*=18. Habitat of sp.; 90–1200 m. SNF, SnJV, SCoRI.

ssp. ***tricolor*** (p. 845) **INFL**: fls 2–5 in dense clusters (sometimes loose); pedicels < 1–5 mm, longer in fr. **FL**: calyx 4–7 mm, membrane purple; corolla 10–19 mm, lobes 4–8 mm wide, colors bright, spots merged in a ring. **FR** 5–6 mm. 2*n*=18. Habitat of sp.; < 1200 m. SNF, NCoR, SnJV, SCoR. ❀SUN:4–6,**7,14–17,22–24**&IRR:8–13,**18–21**.

G. triodon Eastw. (p. 845) Pl glandular-puberulent. **ST** gen 1 branched above; branches spreading, 5–13 cm, thread-like. **LVS**: basal in rosette, 5–20 mm, oblanceolate to obovate, toothed or lobed, lobes spreading, short-pointed, length < axis width; upper lvs linear, entire. **INFL** loose; fls 1–3 per st. **FL**: calyx 2–3 mm, lobes short-pointed, tips thickened, membrane U-shaped; corolla 5–7 mm, tube 1.5–2 × calyx, thread-like, purple, throat narrower than tube, lobes 3-toothed, tips white, base yellow; stamens, style slightly exserted; pollen white. **FR** 3–4 mm, narrowly ovoid. **SEEDS** many, not gelatinous. 2*n*=18. Open, sandy or rocky areas, sagebrush shrubland, juniper woodland; 1200–1700 m. e DMtns (Clark Mtns); to Colorado, NM. Like *G. leptomeria* exc corolla throat, lobes, lf shape.

GYMNOSTERIS

Dieter H. Wilken

Ann. **ST** erect, gen solitary, glabrous. **LVS** 0. **INFL** head-like; bracts lf-like, in involucre; fls 1–5. **FL**: calyx urn-shaped, scarious, abruptly awned, puberulent; corolla salverform, white to lavender, throat gen yellow; stamens attached in throat, filaments 0. **FR** ovoid. **SEEDS** 1–5 per chamber, angled. 2 spp.: w US. (Greek, naked st) [Wherry 1944 Am Mid Nat 31: 216–231] Self-pollinated.

G. parvula A.A. Heller (p. 837) **ST** < 7 cm. **INFL**: bracts 2–13 mm, lanceolate to ovate, glabrous. **FL**: calyx < 4 mm; corolla < 7 mm, tube = calyx, lobes gen oblong, acuminate, white or tinged pink. Gravelly, sandy areas, gen in meadows or wet depressions in shrublands; 2400–3700 m. n&c SNH, SNE; to OR, WY, Colorado. *G. nudicaulis* (Hook. & Arn.) E. Greene, with corolla 8–12 mm, occurs in w NV, may be expected in SNE.

IPOMOPSIS SCARLET GILIA

Dieter H. Wilken

Ann, per. **ST** gen branched at base. **LVS** alternate, simple, gradually smaller upward, entire to pinnately or palmately lobed; lobes gen small-pointed at tip. **INFL**: clusters, lateral and 1-sided or terminal and open to head-like, rarely solitary and pedicelled in lower axils. **FL**: calyx gen bell-shaped, tube and sinuses membranous, lobes gen small-pointed at tip; corolla bell-shaped or salverform, radial or bilateral, white to red or lavender. **SEEDS** slender, angled, slightly winged, white to light brown. 30 spp.: w N.Am, se U.S., s S.Am. (Greek, striking appearance) [Grant & Wilken 1988 Bot Gaz 149:443–449] Per spp. cross-pollinated, ann spp. gen self-pollinated. Distinguished from *Gilia* by infl, lf morphology, chromosome number, flavonoid chemistry.

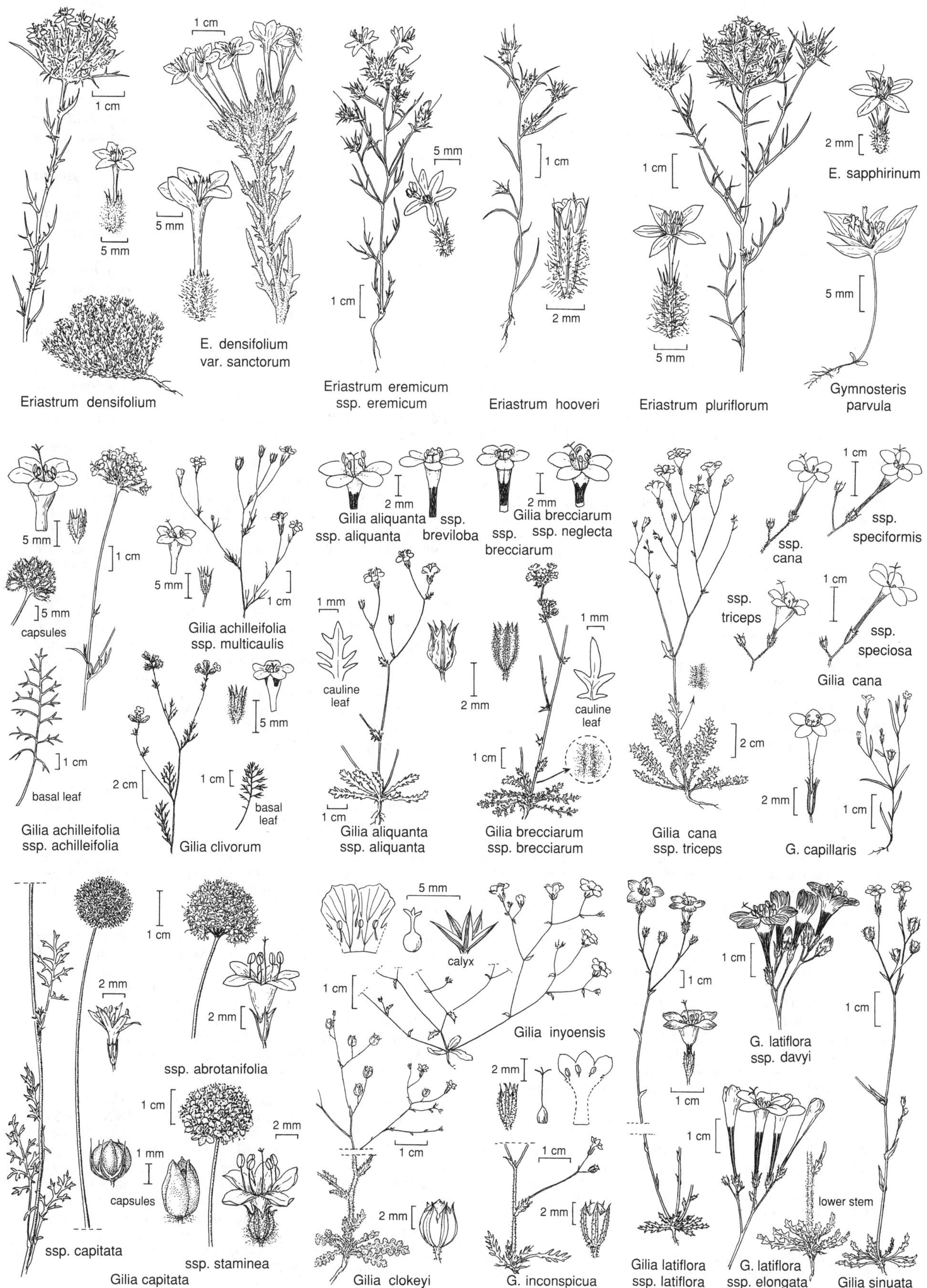
1 cm
1 cm
5 mm
5 mm
Eriastrum densifolium
E. densifolium
var. sanctorum
5 mm
1 cm
Eriastrum eremicum
ssp. eremicum
1 cm
2 mm
Eriastrum hooveri
1 cm
5 mm
Eriastrum pluriflorum
2 mm
E. sapphirinum
5 mm
Gymnosteris
parvula
5 mm
1 cm
5 mm
capsules
1 cm
basal leaf
Gilia achilleifolia
ssp. achilleifolia
5 mm
1 cm
Gilia achilleifolia
ssp. multicaulis
5 mm
2 cm
1 cm
basal
leaf
Gilia clivorum
2 mm
Gilia aliquanta
ssp. aliquanta
ssp.
breviloba
2 mm
Gilia brecciarum
ssp.
brecciarum
ssp. neglecta
1 mm
cauline
leaf
2 mm
1 cm
Gilia aliquanta
ssp. aliquanta
1 mm
cauline
leaf
1 cm
Gilia brecciarum
ssp. brecciarum
2 cm
Gilia cana
ssp. triceps
1 cm
ssp.
cana
ssp.
speciformis
ssp.
triceps
1 cm
ssp.
speciosa
Gilia cana
2 mm
1 cm
G. capillaris
1 cm
2 mm
ssp. abrotanifolia
2 mm
1 cm
1 mm
capsules
2 mm
ssp. capitata
ssp. staminea
Gilia capitata
5 mm
calyx
1 cm
Gilia inyoensis
1 cm
2 mm
Gilia clokeyi
2 mm
1 cm
2 mm
G. inconspicua
1 cm
1 cm
1 cm
Gilia latiflora
ssp. latiflora
1 cm
G. latiflora
ssp. davyi
1 cm
lower stem
G. latiflora
ssp. elongata
1 cm
Gilia sinuata

1. Infl of lateral clusters on 1 side of axis
  2. Calyx lobes long-tapered, hairy; corolla gen white to lavender, tube 25–45 mm .... ***I. tenuituba***
  2′ Calyx lobes deltate to acuminate, sparsely hairy; corolla gen red, tube 10–30 mm
    3. Corolla tube 10–20 mm; anthers incl in tube; DMtns .... ***I. arizonica***
    3′ Corolla tube 20–30 mm; anthers incl in throat or exserted; widespread .... ***I. aggregata***
      4. Pollen blue; cauline lf lobes blunt to rounded .... ssp. ***bridgesii***
      4′ Pollen gen white or yellow; cauline lf lobes acute .... ssp. ***formosissima***
1′ Infl not 1-sided; terminal, open, and few-fld, or head-like and bracted, or fls solitary in lower axils
  5. Corolla 15–28 mm, red .... ***I. tenuifolia***
  5′ Corolla gen < 10 mm, white to deep pink
    6. Per .... ***I. congesta***
      7. Lf pinnately lobed .... ssp. ***congesta***
      7′ Lf palmately lobed
        8. Lf lobes gen 3; st < 1 dm; alpine .... ssp. ***montana***
        8′ Lf lobes gen 5; st 1–3 dm; montane .... ssp. ***palmifrons***
    6′ Ann
      9. Corolla deep pink or white mottled with pink; stamens exserted .... ***I. effusa***
      9′ Corolla gen white; stamens gen incl
        10. Lvs entire to slightly toothed, linear to elliptic; lower fls 1–3 in axil .... ***I. depressa***
        10′ Lvs gen pinnately lobed, gen oblanceolate; fls terminal .... ***I. polycladon***

I. aggregata (Pursh) V. Grant (p. 845) Per, dying after flowering once. **ST** erect, glabrous or glandular, slightly hairy. **LVS**: basal 3–5 cm, pinnately 9–11-lobed, withered at fl; cauline 5–7-lobed, glabrous to puberulent. **INFL** 1-sided; clusters lateral, compact; fls 1–7. **FL**: calyx lobes deltate to acuminate; corolla tube 20–30 mm, salverform, gen red with yellow mottling in throat and bases of lobes, lobes acute to acuminate; stamens attached at different levels, exserted; style exserted. Openings in shrublands, woodlands; 1100–3300 m. KR, CaRH, n&c SNH, MP; to B.C., Colorado, Mex. 8 sspp., ssp. *aggregata* in Rocky Mtns.

ssp. ***bridgesii*** (A. Gray) V. Grant & A.D. Grant (p. 845) **LF**: lobes blunt to rounded. **INFL**: fls 1–5. **FL**: calyx lobes 1–2 mm, deltate, lacking small point at tip; 1–4 stamens exserted, pollen blue. Gen upper montane, subalpine; 1900–3300 m. n&c SNH. Hybridizes with *I. tenuituba* near Sonora Pass, where pls highly variable. ❀TRY;DFCLT.

ssp. ***formosissima*** (E. Greene) Wherry **LF**: lobes acute. **INFL**: fls 3–7. **FL**: calyx lobes 3–4 mm, narrowly acuminate; stamens exserted, pollen white, light yellow, or slightly bluish. Shrublands, montane; 1100–2500 m. KR, CaRH, n&c SNH, MP; to WA, Colorado, n Mex. In e SNH this ssp. gen occurs lower than ssp. *bridgesii*. Intergrades with ssp. *bridgesii* in n&c SNH. ❀DRN:1–4,6,15–17&IRR:7,14,18;DFCLT.

I. arizonica (E. Greene) Wherry (p. 845) Per, dying after flowering once. **ST** erect, glabrous or glandular, slightly hairy. **LVS**: basal 3–5 cm, pinnately 7–11-lobed; cauline hairy. **INFL** 1-sided; fls 5–13 on upper third of axis. **FL**: calyx lobes 1–3 mm, acuminate, glabrous to glandular-puberulent; corolla 11–20 mm, salverform, red, lobes 5–10 mm, flared to reflexed; stamens attached at same level, incl; style incl. Open, sandy to rocky areas in canyons; 1500–3100 m. DMtns; s NV, n AZ. [*I. aggregata* ssp. *a.* (E. Greene) V. Grant & A.D. Grant] ❀TRY.

I. congesta (Hook.) V. Grant Per. **ST** decumbent to erect, 1–3 dm, glabrous to densely puberulent. **LF** 1–4 cm, gen hairy, entire or pinnately to palmately 3–5-lobed. **FL**: calyx 3–5 mm; corolla gen > calyx, tube yellow, lobes gen oblong, white; stamens attached at same level, exserted; style incl. Dry, open shrublands, woodlands, alpine; 1200–3700 m. KR, n CaR, n&c SNH, MP, W&I; to OR, SD, Colorado. 7 sspp., esp GB.

ssp. ***congesta*** **ST** 1–2 dm, branched upward. **LVS**: lower pinnately 3–5-lobed, hairy. Valleys, basins; 1200–2150 m. n CaR, MP, n SNE; to OR, w Great Plains. ❀TRY.

ssp. ***montana*** (Nelson & Kenn.) V. Grant Pl cespitose. **ST** < 1 dm. **LF** gen palmately 3-lobed, hairy. Alpine; 2100–3700 m. GB; to OR, NV.

ssp. ***palmifrons*** (Brand) Day (p. 845) **ST** 1–3 dm. **LF** palmately 5-lobed, sparsely hairy. Gen woodlands, shrublands of foothills, montane; 1500–2750 m. GB; to OR, ID. Intergrades with ssp. *montana* in n CaR, MP.

I. depressa (M.E. Jones) V. Grant Ann. **ST** decumbent, < 1 dm, glandular-puberulent. **LVS**: cauline < 2 cm, linear to elliptic, acute, entire to toothed; upper crowded below infl. **INFL** compact, terminal; bracts lf-like, hairy to canescent; fl 1–3 per axil. **FL**: calyx lobes acuminate, canescent; corolla 5–7 mm, lobes < 2 mm, white; stamens attached at same level, unequal, slightly exserted; style incl. Sandy soils of gentle slopes, flats; 1000–1550 m. n DMoj; to UT.

I. effusa (A. Gray) Moran BAJA CALIFORNIA IPOMOPSIS Ann. **ST** < 3 dm, sparsely to moderately puberulent. **LVS**: basal < 3 cm, oblong to ovate, pinnately 2–10-lobed; cauline entire to 1-pinnately lobed, lobes linear. **INFL** terminal, open; fls 1–8. **FL**: calyx reddish to purplish; corolla 8–12 mm, slightly bilateral, tube < 3 mm, throat 1–2 mm, lobes 4–7 mm, oblong, notched, deep pink to white with pink mottling; stamens, style exserted. RARE in CA. Sandy soils, desert washes; < 100 m. se PR (Pinto Wash, sw Imperial Co.); Baja CA (where in montane forest).

I. polycladon (Torrey) V. Grant (p. 845) Ann. **ST** decumbent to prostrate, < 1 dm, glandular-puberulent. **LVS** < 2 cm, entire to toothed or pinnately 5–7-lobed; basal, cauline crowded below infl, terminal lobe = axis width, glabrous above, puberulent below. **INFL** head-like, terminal; bracts oblanceolate, lobed, pointed at tip. **FL**: corolla 3–6 mm, salverform, lobes < 2 mm; stamens attached at same level, filaments = anthers; style incl. **FR** < 5 mm, ellipsoid. Sandy, gravelly soils; 900–2150 m. SNE, DMoj; to TX, n Mex.

I. tenuifolia (A. Gray) V. Grant (p. 845) SLENDER-LEAVED IPOMOPSIS Per. **ST** 1–4 dm, sparsely puberulent. **LVS**: basal 5–35 mm, pinnately 3–5-lobed, lobes remote, linear; cauline entire. **INFL** open, terminal or in upper axils; fls 1–7. **FL**: calyx lobes < 3 mm; corolla 15–28 mm, bell-shaped, slightly bilateral, tube 5–10 mm, throat 6–11 mm, lobes 5–7 mm, oblong, notched, red with white mottling on lobes, throat; stamens, style exserted. RARE in CA. Gravelly to rocky slopes, canyons; 100–1200 m. se PR (s San Diego Co); Baja CA. [*Loeselia t.* A. Gray]

I. tenuituba (Rydb.) V. Grant (p. 845) Per, dying after flowering once. **ST** erect, glabrous or glandular, slightly hairy. **LVS**: basal 3–6 cm, pinnately 9–17-lobed, withered at fl; cauline gen puberulent. **INFL** 1-sided; fls 3–7, lower well spaced on axis. **FL**: calyx lobes tapered, glandular, hairy; corolla tube 25–45 mm, salverform, lobes white to pink or lavender, slightly speckled at base; stamens attached at different levels, incl, pollen white to yellow, rarely blue; style incl to slightly exserted. Gravelly to rocky slopes; 2400–3050 m. n-c SNH, e MP, n SNE; to Colorado. [*I. aggregata* ssp. *attenuata* (A. Gray) V. Grant & A.D. Grant misapplied to CA pls] Hybridizes with *I. aggregata* in SNH. ❀TRY.

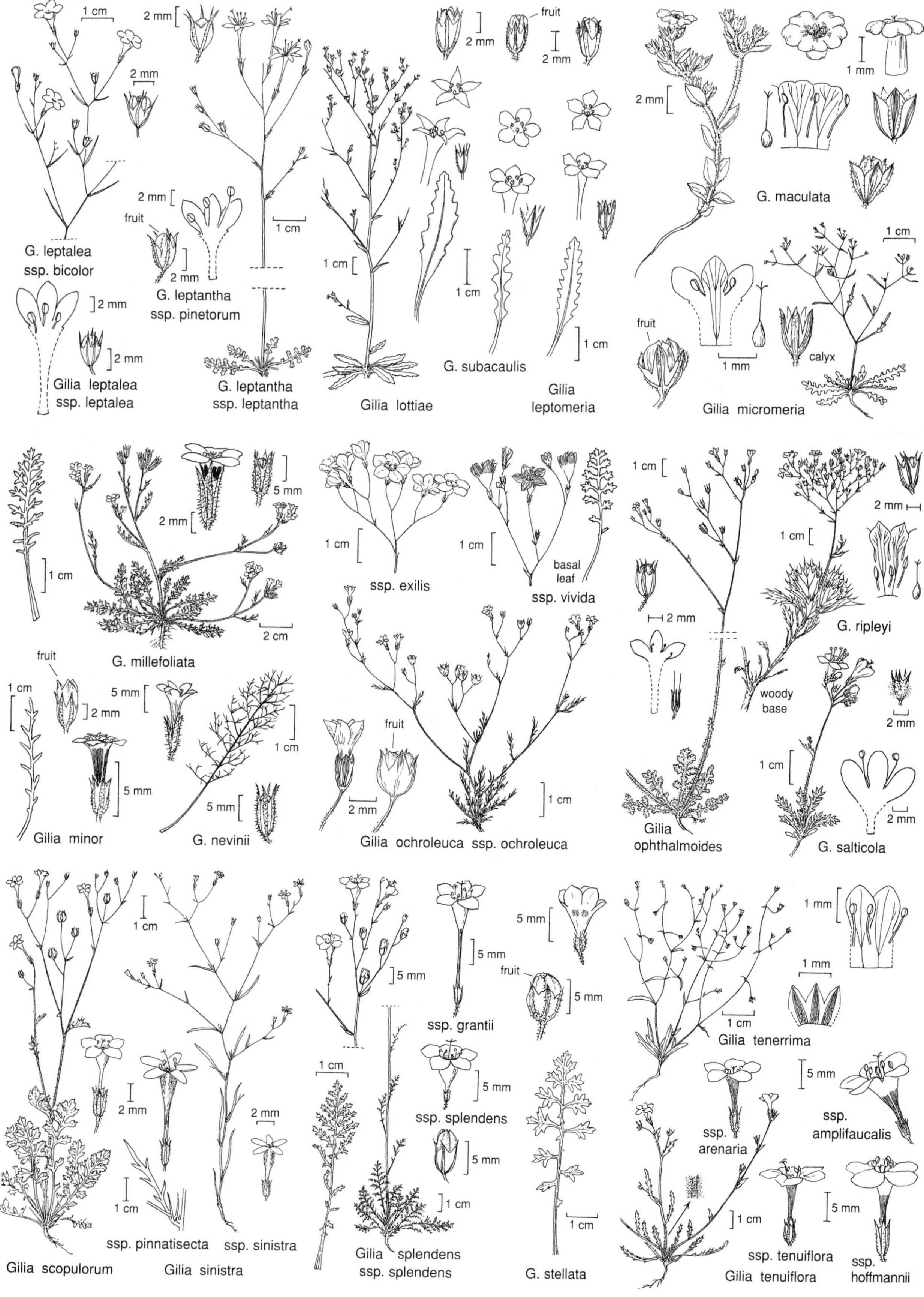
G. leptalea
ssp. bicolor
Gilia leptalea
ssp. leptalea
G. leptantha
ssp. pinetorum
G. leptantha
ssp. leptantha
Gilia lottiae
G. subacaulis
Gilia
leptomeria
G. maculata
Gilia micromeria
calyx
fruit
G. millefoliata
Gilia minor
G. nevinii
ssp. exilis
ssp. vivida
basal
leaf
Gilia ochroleuca ssp. ochroleuca
Gilia
ophthalmoides
G. ripleyi
woody
base
G. salticola
Gilia scopulorum
ssp. pinnatisecta
ssp. sinistra
Gilia sinistra
ssp. grantii
ssp. splendens
Gilia splendens
ssp. splendens
G. stellata
Gilia tenerrima
ssp.
arenaria
ssp.
amplifaucalis
ssp. tenuiflora
Gilia tenuiflora
ssp.
hoffmannii
1 cm
2 mm
5 mm
1 mm
2 cm

LANGLOISIA

Steven L. Timbrook

1 sp. (Rev. A.B. Langlois, Louisiana botanist) [Timbrook 1986 Madroño 33:157–174]

L. setosissima (Torrey & A. Gray) E. Greene Ann, bristly, gen hairy; hairs branched, nonglandular. **ST** erect, gen naked below, leafy above. **LVS** alternate, simple, linear or oblanceolate, 3–5-toothed at tip, each tooth with 1 bristle; basal teeth of upper lvs reduced to cluster of 2–3 bristles. **INFL**: clusters, terminal, head-like; bracts lf-like; pedicels 0-short. **FL**: calyx lobes equal, bristle-tipped; corolla funnel-shaped; stamens attached at or below sinuses, equal, exserted, pollen white to blue; style exserted. **FR** oblong-lanceolate, triangular in X-section; outer wall of valve flat. **SEEDS** gelatinous when wet. Dry, gen sandy places; gen < 1800 m. SNE, D; to NV, AZ, n Mex; also in e OR, w ID. Self-compatible; gen cross-pollinated. Sspp. intergrade.

1. Corolla lobes 1/2 to nearly = tube, purple dotted; stamens > 3 mm ..... ssp. ***punctata***
1′ Corolla lobes 1/3–1/2 tube, unmarked or streaked with purple, seldom dotted; stamens < 3 mm ..... ssp. ***setosissima***

ssp. ***punctata*** (Cov.) S. Timbrook (p. 851) LILAC SUNBONNET **FL**: corolla white to light blue with purple dots, lobes 1/2 to nearly = tube, gen 2 yellow spots in middle of each; stamens > 3 mm. Common. Desert washes, flats, slopes, gravelly to sandy soils; < 1800 m. SNE, DMoj; to NV. Smaller-fld populations in w ID, e OR. [*L. p.* (Cov.) Goodd.] ❀TRY.

ssp. ***setosissima*** (p. 851) BRISTLY LANGLOISIA **FL**: corolla lavender to blue, unmarked or with purple lines, lobes 1/3–1/2 tube; stamens < 3 mm. Common. Desert washes, flats, slopes, gravelly to sandy soils; gen < 1800 m. SNE, D; to NV, AZ, n Mex.

LEPTODACTYLON

Robert W. Patterson & Paul A. Meyers

Per, open or cespitose. **ST** decumbent to erect. **LVS** cauline with clustered axillary lvs, alternate or opposite, simple, deeply lobed; lobes linear, gen spine-tipped, palmate or pinnate. **INFL** gen terminal; fls gen sessile. **FL**: calyx membrane wider than ribs, lobes 4–6, linear; corolla funnel-shaped or salverform, lobes 4–6; stamens attached at same level, anthers at throat, pollen yellow; style incl. 7 spp.: w N.Am. (Greek: narrow finger) [Gordon-Reedy 1989 Madroño 37:28–42]

1. Lvs opposite, pinnately 3-lobed ..... ***L. jaegeri***
1′ Lvs alternate, palmately lobed
  2. Pl gen 3–10 dm; corolla salverform, lobes 9–18 mm ..... ***L. californicum***
  2′ Pl gen 1–3 dm; corolla funnel-shaped, lobes 7–10 mm ..... ***L. pungens***

L. californicum Hook. & Arn. (p. 845) PRICKLY PHLOX **ST** 3–10 dm. **LF**: lobes 3–12 mm. **FL** open during day; calyx lobes equal; corolla salverform, gen pink, lobes elliptic obovate to round; stamens attached in mid-tube. Scrublands, forests, coastal strand; 0–1500 m. s SCoRO, n&c SW. 4 sspp. have been separated from ssp. *californicum* by hair morphology: ssp. *brevitrichomum* Gordon-Reedy; ssp. *glandulosum* (Eastw.) H. Mason; ssp. *leptotrichomum* Gordon-Reedy; ssp. *tomentosum* Gordon-Reedy. ❀DRN,DRY:7, 9,14,18–21&SUN:5,15–17,22–24;DFCLT.

L. jaegeri (Munz) Wherry (p. 845) SAN JACINTO PRICKLY PHLOX Pl cespitose, 2–10 cm; st, lvs, calyx lobes glandular-hairy. **LVS** gen opposite, pinnately 3-lobed; lobes 10–15 mm, spine-tipped. **INFL** crowded in upper axils. **FL**: calyx lobes gen 6, gen unequal; corolla funnel-shaped, white, tube + throat 17–30 mm, lobes gen 6, 7–9 mm, oblanceolate; stamens gen 6, attached at throat. RARE. Dry rocky areas; 2900–3000 m. SnJt.

L. pungens (Torrey) Rydb. (p. 845) Pl gen hairy, glandular or not. **ST** 1–3 dm. **LVS** alternate, 3–7-lobed; middle lobe gen longest, spine-tipped. **FL** gen opening in evening; calyx lobes gen unequal, spine-tipped, membrane extended along lobes; corolla funnel-shaped, tube + throat 7–15 mm, lobes 7–10 mm, obovate, white or pink with purplish shading on outer surface; stamens attached at throat. $2n$=18. Open, rocky areas in montane, subalpine forests, alpine fell-fields; 1700–4000 m. CA; to B.C., Rocky Mtns. [spp. *hallii* (Parish) H. Mason; spp. *hookeri* (Douglas) Wherry; spp. *pulchriflorum* (Brand) H. Mason] Some proposed sspp. sort well elsewhere, but not in CA. Further study warranted. ❀DRN,DRY, SUN:1–3,6,7,14–18.

LINANTHUS

Robert W. Patterson

Ann, per. **ST** gen erect, gen branched from base. **LVS** cauline, opposite, entire or palmately 3–9-lobed; lobes linear to narrowly lanceolate or spoon-shaped. **INFL** head-like, open, or fl solitary; bracts lf-like; fls sessile or pedicelled. **FL**: calyx tubular, or lobes nearly free, bordered by translucent membrane; corolla funnel-shaped, salverform, or bell-shaped; stamens attached at same level, pollen yellow. 41 spp.: w N.Am, Chile. (Greek: flax flower) [Patterson 1977 Madroño 24:36–48]

1. Per (sect. *Siphonella*)
  2. Lf gen entire; ne PR ..... ***L. floribundus*** ssp. ***hallii***
  2′ Lf gen palmately lobed; SW, incl ne PR
    3. Lf lobes gen 3; pedicels gen 1–6 mm ..... ***L. floribundus***
      4. Lf and calyx hairy ..... ssp. ***floribundus***
      4′ Lf and calyx glabrous ..... ssp. ***glaber***
    3′ Lf lobes gen 5; pedicels 0 or obscure
      5. Lf lobes linear-lanceolate; corolla tube gen > calyx; seeds > 2 mm; SNH ..... ***L. pachyphyllus***
      5′ Lf lobes gen linear; corolla tube < calyx; seeds < 2 mm; not SNH ..... ***L. nuttallii***
        6. Lf and calyx sparsely hairy; NW, MP ..... ssp. ***nuttallii***
        6′ Lf, calyx densely hairy
          7. Lf lobes 3–7 mm; NCoRH ..... ssp. ***howellii***
          7′ Lf lobes 5–10 mm; SNE ..... ssp. ***pubescens***

1′ Ann
8. Corolla salverform, tube >> calyx, < 1 mm wide; infl a dense bracted head (sect. *Leptosiphon*)
9. Calyx membrane gen as wide as ribs
10. Bract margins densely white-ciliate; infl gen spheric
11. Corolla tube gen < 25 mm, lobes 2–4 mm . . . . . **L. ciliatus**
11′ Corolla tube gen > 25 mm, lobes 5–8 mm . . . . . **L. montanus**
10′ Bract margin not white-ciliate; infl gen not spheric
12. Lf lobes oblanceolate; corolla lobes gen truncate or slightly notched at tip; corolla tube white; s SNH . . . . . **L. oblanceolatus**
12′ Lf lobes linear or linear-lanceolate; corolla lobes rounded; corolla tube darker than lobes; more widespread s CA, incl s SNH
13. Lobes of bracts free; TR, sw DMoj . . . . . **L. breviculus**
13′ Lobes of bracts connected by a translucent membrane; s SN . . . . . **L. nudatus**
9′ Calyx membrane << ribs or obscure
14. Corolla tube < 10 mm, < 2 × calyx length; corolla lobes white, throat dark purple . . . . . **L. serrulatus**
14′ Corolla tube > 10 mm, 2–6 × calyx length; corolla lobes variously colored, if white then throat not purple
15. Corolla yellow; lf lobes needle-like, gen ascending . . . . . **L. acicularis**
15′ Corolla tube or lobes (not both) yellow; lf lobes linear-lanceolate or oblanceolate, gen spreading
16. Stigma < 1 mm . . . . . **L. bicolor**
16′ Stigma > 1 mm
17. Calyx gen hairy only on lobe margins; NCoR, SnFrB . . . . . **L. androsaceus**
17′ Calyx gen hairy throughout; widespread . . . . . **L. parviflorus**
8′ Corolla not both salverform and in dense bracted heads
18. Corolla 15–30 mm; infl a dense, bracted head (sect. *Pacificus*) . . . . . **L. grandiflorus**
18′ Corolla < 15 mm, or if > infl not a dense, bracted head
19. Corolla gen open only in evening (if open in daytime, then limb > 10 mm wide); calyx membrane > ribs; corolla lacking red marks near throat (sect. *Linanthus*)
20. Calyx glandular-hairy . . . . . **L. jonesii**
20′ Calyx sometimes hairy but not glandular
21. Calyx hairy on inner surface; gypsum soils . . . . . **L. arenicola**
21′ Calyx not hairy on inner surface; gen not gypsum soils
22. Lf not lobed; corolla tube without hairy pads where filaments attach . . . . . **L. bigelovii**
22′ Lf lobed; corolla tube with hairy pads where filaments attach . . . . . **L. dichotomus**
19′ Corolla open in daytime, limb < 10 mm wide; calyx membrane gen narrower than ribs (if wider, then corolla with red marks near throat)
23. Fls gen sessile, corolla gen glabrous inside (if otherwise, lvs not lobed) (sect. *Dianthoides*)
24. Calyx lobes free to base, margins translucent; corolla tube very short
25. St long, thread-like, openly branched at base; calyx ribs with purple marks at base . . . . . **L. bellus**
25′ St not thread-like, compactly branched; calyx ribs without purple marks near base
26. Corolla bell-shaped, 5–6 mm, white with 2 red marks at base of lobes . . . . . **L. demissus**
26′ Corolla funnel-shaped, 6–12 mm, blue-purple or cream with purple kidney-shaped crest at base of lobes . . . . . **L. parryae**
24′ Calyx fused by membranes; corolla tube conspicuous
27. Lvs not lobed; tips of corolla lobes finely toothed . . . . . **L. dianthiflorus**
27′ Lvs deeply lobed; tips of corolla lobes entire
28. Calyx membrane much wider than ribs; SnGb . . . . . **L. concinnus**
28′ Calyx membrane as wide as ribs
29. Corolla 10–15 mm, tube slightly exserted from calyx; SnBr . . . . . **L. killipii**
29′ Corolla 15–25 mm, tube long-exserted from calyx; PR . . . . . **L. orcuttii**
23′ Fls gen on thread-like pedicels (if sessile, then corolla with hairy ring inside tube); lf palmate, lobes very short (sect. *Dactylophyllum*)
30. Fls sessile . . . . . **L. lemmonii**
30′ Fls on thread-like pedicels
31. Corolla tube and filaments glabrous; corolla gen < calyx
32. St gen branched above base or unbranched; corolla tube << lobes; stamens exserted; seeds 1 per chamber . . . . . **L. harknessii**
32′ St branched at base; corolla tube ± = lobes; stamens incl; seeds > 1 per chamber . . . . . **L. pygmaeus**
33. Corolla white; mainland . . . . . ssp. **continentalis**
33′ Corolla blue; ChI . . . . . ssp. **pygmaeus**
31′ Corolla tube, filaments, or both hairy; corolla gen > calyx
34. Filaments gen hairy at base (if glabrous, then attached at hairy ring inside corolla)
35. Corolla 2–4 mm; hairy ring gen present inside corolla tube, tube = lobes . . . . . **L. septentrionalis**
35′ Corolla 4–15 mm; hairy ring 0 inside corolla tube, base of filaments hairy, tube << lobes
36. St diffusely branched at base; calyx lobes as wide as membrane; corolla veins obscure . . . . . **L. filipes**
36′ St branched above base; calyx lobes narrower than membrane; corolla veins conspicuous . . **L. liniflorus**
34′ Filaments glabrous, attached above hairy ring
37. Corolla tube gen < lobes . . . . . **L. aureus**
38. Corolla lobes bright yellow . . . . . ssp. **aureus**

38′ Corolla lobes cream .... ssp. ***decorus***
37′ Corolla tube gen > lobes
39. Stigmas 2–4 mm; corolla throat widely widened, 3 mm; SnFrB, SCoR .... ***L. ambiguus***
39′ Stigmas 1 mm; corolla throat not widened, < 2 mm; SnFrB, NCoR
40. Corolla tube white, pink, or lilac, = calyx .... ***L. bolanderi***
40′ Corolla tube maroon, > 1.5 × calyx .... ***L. rattanii***

L. acicularis E. Greene Ann, hairy. **ST** 3–15 cm. **LF**: lobes 3–10 mm, needle-like. **INFL** head-like; fls sessile. **FL**: calyx 6–8 mm, lobes needle-like, membrane obscure; corolla salverform, tube 10–20 mm, thread-like, lobes 4–6 mm, lanceolate, yellow; stamens exserted; stigmas 2–4 mm. 2*n*=18. Grassy areas, woodland, chaparral; < 700 m. NCo, NCoR, SnFrB. Needle-like lvs, yellow corolla tube differentiate this from yellow-fld s CA populations of *L. parviflorus*. ❀TRY.

L. ambiguus (Rattan) E. Greene SERPENTINE LINANTHUS Ann. **ST** 10–20 mm, thread-like. **LF**: lobes 3–5 mm, narrowly lanceolate. **INFL**: fl solitary; peduncle 5–25 mm, thread-like. **FL**: calyx 3–6 mm, gen hairy, membrane gen = calyx length; corolla funnel-shaped, tube 4–6 mm, purple, wide hairy band inside near throat, throat violet, lobes 4–6 mm, lanceolate, pink with yellow base; stamens exserted; stigmas 2–4 mm. 2*n*=18. UNCOMMON. Grassy areas on serpentine soil; < 1000 m. SnFrB, SCoRI. ❀TRY.

L. androsaceus (Benth.) E. Greene Ann, hairy. **ST** decumbent to erect, 5–30 cm. **LF**: lobes 8–30 mm, narrowly oblanceolate to linear. **INFL** head-like; bracts ciliate, outer surface gen glabrous; fls sessile. **FL**: calyx 6–8 mm, margin hairy, outer surface gen glabrous, membrane obscure; corolla salverform, tube 15–28 mm, pink or lavender, throat gen violet at base, yellow above, lobes 5–10 mm, lanceolate, pink; stamens attached in upper throat. Open or shaded areas in woodlands, chaparral; < 1200 m. 2*n*=18. NCoR, SnFrB. Many variants often placed here are synonyms of *L. parviflorus*. ❀TRY.

L. arenicola (M.E. Jones) Jepson & V. Bailey SAND LINANTHUS Ann. **ST** 1–8 cm, glabrous to slightly hairy. **LF**: lobes 3–12 mm, upper surface hairy, lower surface glabrous. **INFL**: cyme. **FL** opening in evening; calyx 4–5 mm, lobes unequal, membrane 2/3 calyx length; corolla 5–7 mm, funnel-shaped, light yellow with purple throat; stamens attached in lower throat. RARE in CA. Saline flats in gypsum soils; 800–1400 m. DMoj; NV.

L. aureus (Nutt.) E. Greene (p. 845) Ann. **ST** thread-like, glabrous or hairy, sometimes glandular. **LF**: lobes 3–6 mm, linear. **INFL**: fl solitary; peduncle 5–15 mm, thread-like. **FL**: calyx 4–6 mm, membrane wider than ribs; corolla funnel-shaped, tube 3–5 mm, ring of hairs inside and outside, lobes 5–7 mm, oblanceolate; stamens attached in throat; stigmas 3–4 mm, exserted. Desert flats; < 2000 m. D; to NM, Baja CA. Both sspp. occur in gen same range but rarely occur together.

ssp. ***aureus*** **FL**: corolla tube, lobes bright yellow, throat gen brighter yellow or maroon. 2*n*=18. Habitat and range of sp. ❀TRY.

ssp. ***decorus*** (A. Gray) H. Mason **FL**: corolla tube, lobes white or cream, throat maroon. 2*n*=18. Habitat and range of sp. ❀TRY.

L. bellus (A. Gray) E. Greene (p. 845) DESERT BEAUTY Ann. **STS** many, thread-like, gen glabrous. **LF**: lobes 2–4 mm, linear. **INFL**: fl solitary, sessile; bracts several. **FL**: calyx 3–4 mm, tube obscure, purple band at base, lobes bordered by membrane; corolla widely funnel-shaped, tube 2 mm, purple, throat yellow, lobes 5–10 mm, widely lanceolate, pink with paired basal purple spots; stamens attached in base of throat. 2*n*=18. RARE in CA. Desert chaparral areas in sandy soils; 1000–1400 m. se PR (se San Diego Co.); Baja CA.

L. bicolor (Nutt.) E. Greene Ann, hairy. **ST** 5–16 cm. **LF**: lobes 3–10 mm, middle lobe narrowly oblanceolate to spoon-shaped, lateral lobes similar or linear. **INFL** head-like, gen 1 fl open at a time in each infl; fls sessile. **FL**: calyx 6–8 mm, membrane obscure; corolla salverform, tube 12–22 mm, reddish, throat yellow, lobes 2–3 mm, rounded or truncate, pink or white; stamens attached in throat, exserted; stigmas < 1 mm. 2*n*=18. Common. Open, grassy areas, chaparral, woodlands, etc.; gen < 1700 m. NCoR, SNF, SnFrB, SCoR, s ChI, WTR. Distinguished from *L. acicularis, L. androsaceus, L. parviflorus* by its small corolla, extremely short stigmas. ❀TRY.

L. bigelovii (A. Gray) E. Greene Ann, glabrous. **ST** erect, 5–20 cm. **LF** simple, 10–30 mm, linear. **INFL**: cyme. **FL** opening in evening; calyx 8–12 mm, membrane much wider than ribs; corolla funnel-shaped, tube 4–5 mm, lobes 7–8 mm, cream or white, purplish shading on back of lobes; stamens attached in upper tube. 2*n*=18. Deserts, dry areas; gen < 1700 m. SCoRI, WTR, D.

L. bolanderi (A. Gray) E. Greene Ann. **ST** 5–20 cm, thread-like, glabrous or hairy, sometimes glandular. **LF**: lobes 2–5 mm, linear. **INFL**: fls solitary; peduncles 5–25 mm, thread-like. **FL**: calyx 4–6 mm, membrane 1/2 to nearly = calyx length, gen hairy, glandular; corolla funnel-shaped, tube 3–15 mm, incl or exserted beyond calyx, white, pink, or purple, ring of hairs on inner surface, throat yellow, lobes 2–3 mm, white or pink; stamens attached in throat, exserted; stigmas 1 mm. 2*n*=18. Drying areas in woodlands, chaparral; 200–1700 m. NCoR, n SNF, SnFrB; OR. [*L. bakeri* H. Mason]

L. breviculus (A. Gray) E. Greene Ann, hairy. **ST** 10–25 cm. **LF**: lobes 3–10 mm, linear. **INFL** head-like; fls sessile. **FL**: calyx 7–8 mm, membranes as wide as ribs; corolla salverform, tube 15–25 mm, maroon, glabrous, throat purple, lobes 4–6 mm, white, pink, or blue; anthers gen incl. 2*n*=18. Deserts, dry montane areas; < 2400 m. SnGb, SnBr, DMoj. ❀TRY.

L. ciliatus (Benth.) E. Greene (p. 845) WHISKER BRUSH Ann, hairy. **ST** 2–30 cm. **LF**: lobes 5–20 mm, linear. **INFL** head-like, spheric; bracts white-ciliate; fls sessile. **FL**: calyx 7–10 mm, membranes wider than ribs, ± 1/2 calyx length; corolla salverform, tube 10–25 mm, white or pink, hairy on outside, throat yellow, lobes 2–4 mm, obovate or truncate at tip, light to deep pink, gen with darker pink or red spot at base; anthers gen incl. 2*n*=18. Common. Open or wooded areas; < 3000 m. CA-FP. ❀DRN:1–5,7–9,14–24; DFCLT.

L. concinnus Milliken (p. 845) SAN GABRIEL LINANTHUS Ann, glandular-hairy. **ST** 1–12 cm. **LVS** opposite or ± alternate; lobes 8–15 mm, linear, gen connected at base (lvs appearing pinnately lobed). **INFL** crowded; fls 3–7. **FL**: calyx 10 mm, membrane wider than ribs; corolla funnel-shaped, white, tube 1–2 mm, throat 6–8 mm, lobes 5–10 mm, widely lanceolate, entire or toothed at tips, with 2 dark purple marks at base; stamens incl. RARE. Dry rocky slopes; 1700–2800 m. SnGb.

L. demissus (A. Gray) E. Greene (p. 845) Ann. **ST** decumbent, 2–10 cm, hairy or glandular. **LF**: lobes 6–10 mm, hairy or glabrous. **INFL**: clusters, terminal, bracted; fls gen sessile. **FL**: calyx 3–4 mm, tube obscure, lobes membrane margined; corolla bell-shaped, tube 1–2 mm, white or yellow green, throat, 4–6 mm, white or light yellow, lobes 2–3 mm, white with 2 purple lines at base; stamens incl. 2*n*=18. Limestone soils, desert pavement, sandy areas; < 1700 m. DMoj; to UT, AZ.

L. dianthiflorus (Benth.) E. Greene (p. 845) Ann. **ST** erect or spreading, 5–12 cm, hairy. **LF** 5–20 mm, thread-like or linear-oblong, glabrous, entire. **INFL** leafy; fls 1–few, gen sessile. **FL**: calyx 10–16 mm, gen hairy, membrane wider than ribs; corolla funnel-shaped, tube 4–6 mm, light pink, throat 5–10 mm, yellow, lobes 8–12 mm, toothed at tip, pink or white with purple marks at base; stamens incl. Open areas; < 1300 m. SCo, ChI, WTR, PR. [ssp. *farinosus* (Brand) H. Mason] ❀TRY.

L. dichotomus Benth. (p. 845) EVENING SNOW Ann, glabrous. **ST** 5–20 cm, glaucous. **LF**: lobes 3–7, 10–22 mm, linear. **INFL**: cyme. **FL** gen opening in evening; calyx 8–14 mm, membrane much wider than ribs; corolla funnel-shaped, tube 7–10 mm, purple, throat white or cream, lobes 10–16, white with light purple shading on back; stamens attached in lower tube, incl, filaments dilated into hairy pad at base. 2*n*=18. Common. Drying open areas, esp serpentine; < 1700 m. CA-FP, D; AZ, NV. On pls from

SnFrB s, fls open in evening; on pls from SnFrB n [ssp. *meridianus* (Eastw.) H. Mason], fls open during daylight. ❀DRN,DRY,SUN: **7**,8–12,14–24;fragrant;DFCLT.

L. filipes (Benth.) E. Greene Ann. **ST** 5–20 cm, thread-like, hairy. **LF**: lobes 3–7 mm, linear. **INFL**: fls solitary; peduncles 4–12 mm, thread-like. **FL**: calyx 2–3 mm, membrane 2/3 calyx length, hairy; corolla funnel-shaped, tube 1–2 mm, white or pink, throat yellow, lobes 2–3 mm, pink or sometimes white; stamens slightly exserted, hairy at base; stigmas exserted. 2*n*=18. Open areas in woodlands; < 1300 m. NCoR, CaRF, SNF.

L. floribundus (A. Gray) Milliken Per, hairy or glabrous. **LF**: lobes 8–20 mm. **INFL**: clusters, bracted; pedicels 0–15 mm. **FL**: calyx 7–9 mm, membrane obsolete; corolla funnel-shaped, tube 5–9 mm, white, throat yellow, lobes 5–9 mm, white; stamens incl or slightly exserted. Open, wooded areas, desert canyons; < 2300 m. SnGb, SnBr, SnJt, PR; Baja CA. Gen self-pollinated. [*L. nuttallii* ssp. *f.* (A. Gray) Munz] A related tetraploid, *L. melingii* (Wiggins), is restricted to high mtns of Baja CA.

ssp. ***floribundus*** St, lvs, calyx hairy. 2*n*=18. Open, wooded areas; < 2300 m. Range of sp. (but not found with ssp. *glaber*). ❀DRN,SUN:2–6,**15–16**,17&IRR:**7,14,18**,19–24.

ssp. ***glaber*** R. Patterson St, lvs, calyx glabrous. 2*n*=18. Open, wooded areas; < 2100 m. SnBr, PR; Baja CA. [ssp. *glabrus* R. Patterson] ❀TRY.

ssp. ***hallii*** (Jepson) H. Mason SANTA ROSA MOUNTAINS LINANTHUS **LF** entire. UNCOMMON. Desert canyons; 1000–2000 m. e PR (Santa Rosa Mtns). Further study needed to understand relationships.

L. grandiflorus (Benth.) E. Greene (p. 845) LARGE-FLOWER LINANTHUS Ann. **ST** hairy, branched above. **LF**: lobes 10–30 mm, at acute angle to st, linear, hairy or glabrous. **INFL**: head-like, terminal; fls sessile. **FL**: calyx 10–14 mm, hairy, membrane wider than ribs, 2/3 calyx length; corolla funnel-shaped, tube 5–6 mm, white, ring of hairs on inner surface, throat yellow, lobes 10–15 mm, white or pink; stamens incl. 2*n*=18. UNCOMMON. Open, grassy flats, gen in sandy soil; < 1200 m. NCo, CCo, SnFrB. ❀SUN,DRN:**4–6,17**&IRR:7,8,9,**14–16**,18,**19–24**.

L. harknessii (Curran) E. Greene (p. 845) Ann. **ST** 5–15 cm, thread-like, gen glabrous. **LF**: lobes 5–15 mm, thread-like, glabrous or hairy. **INFL**: fls solitary, bracted; peduncles 5–20 mm, thread-like. **FL**: calyx 1–3 mm, membranes 1/2 calyx length , glabrous; corolla funnel-shaped, white or pale blue, gen incl in calyx; stamens incl. 2*n*=18. Open flats; 1000–3200 m. NCoR, CaR, SNH, GB; to WA. [ssp. *condensatus* H. Mason, Plaskett Meadows linanthus]

L. jonesii (A. Gray) E. Greene Ann, glandular-hairy. **ST** 3–15 cm. **LF** 10–20 mm, linear, curved outward, entire. **INFL**: cyme. **FL** opening in evening; calyx 7–8 mm, glandular, membrane wider than ribs; corolla funnel-shaped, tube 5–6 mm, white, throat yellow, lobes 3–5 mm, white or cream-yellow, light purple shading on back; stamens incl. Sandy flats, washes; < 900 m. D; AZ, Baja CA.

L. killipii H. Mason (p. 845) BALDWIN LAKE LINANTHUS Ann, hairy. **ST** 5–15 cm. **LF**: lobes 3–10 mm, linear. **INFL** terminal; fls gen sessile. **FL**: calyx 6–7 mm, membrane wider than ribs, extended along lobes; corolla funnel-shaped, tube 4–5 mm, white, throat 3–4 mm, yellow, lobes 3–4 mm, white with purple mark at base; stamens incl. 2*n*=18. RARE. Dry, open areas in pinyon/juniper woodland; 1700–2400 m. c SnBr (Baldwin Lake). Threatened by development, vehicles.

L. lemmonii (A. Gray) E. Greene Ann, hairy. **ST** 5–15 cm, gen glandular. **LF**: lobes 2–5 mm, linear. **INFL** terminal, bracted; fls sessile. **FL**: calyx 4–5 mm, membrane 1/2 calyx length; corolla funnel-shaped, tube 1–3 mm, light yellow, ring of hairs on upper inner surface, throat yellow or maroon, lobes 2–3 mm, yellow or cream; stamens incl. 2*n*=18. Dry, open areas gen away from coast, chaparral, woodlands, deserts; < 1900 m. SW, w DSon; Baja CA.

L. liniflorus (Benth.) E. Greene Ann. **ST** 10–50 cm, glabrous or hairy. **LF**: lobes 1–3 cm, linear. **INFL**: cyme or panicle; peduncles 1–3 cm. **FL**: calyx 3–5 mm, hairy, membrane wider than ribs, extended along lobes; corolla funnel-shaped, white, tube 1–2 mm, throat wide, 2–4 × tube, lobes 8–10 mm, veins gen purple; stamens exserted, filaments hairy at base. 2*n*=18. Woodlands, open areas, common on serpentine soil, deserts; < 1700 m. n&c CA-FP, w DMoj; to WA. [ssp. *pharnaceoides* (Benth.) H. Mason]

L. montanus (E. Greene) E. Greene MUSTANG CLOVER Ann, hairy. **ST** 10–60 cm. **LF**: lobes 20–30 mm, linear. **INFL**: heads, spheric, terminal; bracts white-ciliate; fls sessile. **FL**: calyx ± 10 mm, hairy-ciliate, membrane wider than ribs, 1/2 calyx length; corolla salverform, tube 25–30 mm, maroon, hairy outside, throat yellow, lobes 5–8 mm, white or pink with purple spot at base; stamens incl. 2*n*=18. Dry, open areas of woodlands; 300–1700 m. c&s SNF. ❀DRY,SUN:1–6,**17**&IRR:**7**,8,9,**14–16,18–24**.

L. nudatus E. Greene (p. 845) Ann, hairy. **ST** 5–30 cm. **LF**: lobes 3–12 mm, linear, gen fused by membrane for 1/3–1/2 their length, often glandular. **INFL** head-like, terminal; fls sessile. **FL**: calyx 5 mm, membrane > 1/2 calyx length; corolla salverform, tube 10–12 mm, white, hairy outside, throat yellow, lobes 3–4 mm, white or pink; stamens exserted. 2*n*=18. Open areas in woodlands, chaparral; 600–2400 m. s SNH, Teh. ❀TRY.

L. nuttallii (A. Gray) Milliken (p. 845) Per, hairy. **STS** many, 10–20 cm. **LF**: lobes gen 5, linear. **INFL**: cluster, terminal, bracted; fls gen sessile. **FL**: calyx 8–9 mm, membrane gen obscure, 1/2 calyx length; corolla funnel-shaped, tube incl in calyx, white or light yellow, throat yellow, lobes 4–5 mm, white; stamens incl or slightly exserted; stigmas slightly exserted. Dry flats, openings in forest, sometimes on serpentine; 500–3500 m. KR, NCoRO, NCoRH, CaR, GB; to B.C., Rocky Mtns, n Mex. Gen self-pollinated.

ssp. ***howellii*** T.W. Nelson & R. Patterson MOUNT TEDOC LINANTHUS **LF**: lobes 3–7 mm, grayish, densely hairy. 2*n*=18. RARE. Open Jeffrey-pine forest on serpentine soil; 1500–1800 m. NCoRH.

ssp. ***nuttallii*** **LF**: lobes 5–10 mm, green, nearly glabrous to moderately hairy. 2*n*=18. Open areas in forests; 500–2200 m. KR, NCoRO, CaRH, Wrn; to B.C., Rocky Mtns, n Mex.

ssp. ***pubescens*** R. Patterson **LF**: lobes 5–10 mm, densely hairy. 2*n*=18. Dry flats, openings in forests; 2800–3500 m. SNE; to NV.

L. oblanceolatus (Brand) Jepson SIERRA NEVADA LINANTHUS Ann, hairy. **ST** 2–12 cm. **LF**: lobes 5–15 mm, oblanceolate. **INFL** head-like, terminal; fls sessile. **FL**: calyx ± 5 mm, membrane 1/2 calyx length; corolla salverform, tube 8–12 mm, white, throat yellow, hairy outside, lobes ± 2 mm, truncate or notched at tip, white; stamens incl. UNCOMMON. Open flats near meadows; 2800–3700 m. s SNH.

L. orcuttii (C. Parry & A. Gray) Jepson (p. 845) ORCUTT'S LINANTHUS Ann. **ST** 5–10 cm, sparsely puberulent. **LF**: lobes 5–12 mm, linear, hairy. **INFL**: clustered or fls solitary, bracted; fls sessile. **FL**: calyx 6–10 mm, membrane < 1/2 calyx length, extended along lobes; corolla funnel-shaped, tube 5–15 mm, stout, pink or white, throat yellow, lobes 5–8 mm, pink, blue, or white with purple kidney-shaped mark at base; stamens incl. RARE. Openings in chaparral, pine forests; 1300–2000 m. PR; Baja CA.

L. pachyphyllus R. Patterson Per, hairy. **ST** 10–20 cm. **LF**: lobes gen 5, 10–16 mm, linear. **INFL**: cluster, terminal, bracted; fls gen sessile. **FL**: calyx 9–10 mm, densely hairy, membrane gen obscure, 1/2 calyx length; corolla funnel-shaped, tube 12–15 mm, white or pale yellow, throat yellow, lobes 6–9 mm, white; stamens gen exserted; stigmas slightly exserted. 2*n*=36. Open, wooded areas; 1700–2500 m. SNH. Gen self-pollinated. Distinguished from *L. nuttallii* by gen larger features, range, chromosome number.

L. parryae (A. Gray) E. Greene Ann. **ST** decumbent or very short-erect concealed by lvs, 2–10 cm, glandular-hairy. **LF**: lobes 5–15 mm, linear, hairy. **INFL** crowded leafy; fls sessile. **FL**: calyx 6–8 mm, tube obscure, membrane extended along lobes; corolla funnel-shaped, white or blue-purple, tube 1 mm, throat 1–2 mm, lobes 8–12 mm, gen jagged at tip, dark purple kidney-shaped arch at base; stamens incl. 2*n*=18. Sandy, open, flat areas; < 2000 m. s SNF, s SnJV, SCoRI, WTR, DMoj. Some populations have pls with both corolla colors. ❀TRY.

L. parviflorus (Benth.) E. Greene (p. 851) Ann, hairy. **ST** decumbent to erect, 5–25 cm. **LF**: lobes 8–25 mm. **INFL** head-like, terminal; fls sessile. **FL**: calyx 6–8 mm, membrane obscure, 1/2 calyx length; corolla salverform, tube 20–35 mm, thread-like, maroon, pink, or yellow, throat yellow, lobes 5–6 mm, oblong or lanceolate, white, yellow, pink, blue, or purple, often with red marks at base; stamens, style gen exserted, stigmas 3–6 mm. $2n$=18. Abundant. Open or wooded areas, many pl communities; < 1200 m. CA-FP. [*L. androsaceus* sspp. *croceus* (Milliken) H. Mason, *laetus* H. Mason, *luteolus* (E. Greene) H. Mason, *luteus* (Benth.) H. Mason, *micranthus* (Steudel) H. Mason; *L. plaskettii* Eastw.] Highly variable, esp corolla color. Differentiated from *L. androsaceus* by hairy calyx, thread-like corolla tube. Detailed study needed before infraspecific taxa can be recognized. ❀DRN,SUN:5,6,7,8,9,**14–17**,18,**19–24**.

L. pygmaeus (Brand) J. Howell Ann. **ST** 2–10 cm, puberulent. **LF**: lobe 2–6 mm, thread-like. **INFL** openly forked; peduncles 4–15 mm, thread-like. **FL**: calyx 3–5 mm, membrane 2/3 calyx length; corolla funnel-shaped, white or blue, tube 2–3 mm, lobes 1–2 mm; anthers incl. Dry, open areas; < 1700 m. CA-FP.

ssp. ***continentalis*** Raven **FL**: corolla white. Dry, open areas; < 1700 m. $2n$=18. SNF, GV, CW, SW (exc ChI).

ssp. ***pygmaeus*** **FL**: corolla lavender-blue. Dry, open areas; < 500 m. s ChI (San Clemente); Baja CA (Guadalupe Islands)

L. rattanii (A. Gray) E. Greene (p. 851) RATTAN'S LINANTHUS Ann. **ST** 5–20 cm, thread-like, gen hairy, glandular. **LF**: lobes 2–5 mm, linear. **INFL**: fls solitary; peduncles 5–25 mm, thread-like. **FL**: calyx 4–6 mm, membrane 1/2 to = calyx length, gen hairy, glandular; corolla funnel-shaped, tube 10–15 mm, exserted beyond calyx, maroon, ring of hairs inside, throat yellow, lobes 2–3 mm, white; stamens attached in throat, exserted; stigmas 1 mm. UNCOMMON. $2n$=18. Drying areas in conifer forest; 1700–2000 m. NCoRH.

L. septentrionalis H. Mason Ann. **ST** 5–30 cm, thread-like, glabrous or hairy. **LF**: lobes 5–20 mm, thread-like. **INFL**: fls solitary, bracted; peduncles 10–25 mm, thread-like. **FL**: calyx 1–3 mm, membrane 2/3 calyx length, extended along lobes; corolla funnel-shaped, tube 2–3 mm, white, ring of hairs at or above stamen attachment, throat yellow, lobes 1–2 mm, white or pale blue; stamens slightly exserted. Common. Sagebrush shrublands, pinyon/juniper woodlands; 2000–3000 m. GB; to w Can, Colorado.

L. serrulatus E. Greene MADERA LINANTHUS Ann. **ST** 5–18 cm, puberulent. **LF**: lobes 4–10 mm, linear, hairy. **INFL** head-like, terminal; fls sessile. **FL**: calyx 5–7 mm, sparsely hairy, membrane obscure, < 1/2 calyx length; corolla funnel-shaped, tube 7–8 mm, dark purple, throat wide, 4–6 mm, yellow, lobes 5–7 mm, white; stamens incl. UNCOMMON. Open areas, woodland, chaparral; 300–1300 m. s SNF.

LOESELIASTRUM

Steven L. Timbrook

Ann, bristly and gen soft; hairs unbranched. **ST** erect, gen naked below, leafy above. **LVS** alternate, simple, 1–4 cm, linear to oblanceolate, toothed; teeth with 1 bristle. **INFL**: bracts lf-like; pedicels glandular. **FL**: calyx lobes equal, bristle-tipped, glandular-hairy; corolla 2-lipped, white to deep pink, upper lip gen 3-lobed with maroon arches above throat, lower lip 2-lobed, less marked; stamens attached at or below sinuses, unequal, curved, exserted, pollen yellow; style exserted. **FR** 2–5 mm, ovoid, 3-lobed in X-section; outer wall of valve indented between walls separating chambers. **SEEDS** gelatinous when wet. 2 spp.: sw US, n Mex. (Latin: like *Loeselia*) [Timbrook 1986 Madroño 33:157–174] Self-compatible; gen cross-pollinated.

1. Corolla 11–21 mm, upper lip 3/4 to > tube; longer stamens > upper lip; calyx (exc bristles) 1/2–3/4 corolla tube .... ***L. matthewsii***
1′ Corolla 8–15 mm, upper lip gen 1/2–3/4 tube; longer stamens < upper lip; calyx (exc bristles) > 3/4 corolla tube .... ***L. schottii***

L. matthewsii (A. Gray) S. Timbrook (p. 851) DESERT CALICO **FL**: corolla strongly bilateral, white to deep rose-purple, upper lip 5–11 mm, 3/4 to > tube, lobe bases with bright maroon arches and white blotch, lower lip 4–7 mm, lobe tips gen truncate, 3-toothed or notched, gen with inward directed projections in sinuses on either side of middle lobe of upper lip; longer stamens > upper lip. Common. Desert washes, flats, slopes, sandy to gravelly soils; gen < 1800 m. SNE, DMoj, w DSon (e San Diego Co); NV. [*Langloisia m.* (A. Gray) E. Greene] ❀TRY.

L. schottii (Torrey) S. Timbrook (p. 851) **FL**: corolla weakly to moderately bilateral, gen white to pink, upper lip 3–7 mm, gen 1/2–3/4 tube, lobe bases maroon-streaked, lower lip 2–5 mm, lobe tips gen acute; longer stamens < upper lip. Common. Desert washes, flats, slopes, sandy to gravelly soils; gen < 1800 m. sw SnJV, SNE, D; to sw UT, w AZ, n Mex. Weakly bilateral forms are self-pollinated. [*Langloisia s.* (Torrey) E. Greene]

NAVARRETIA

Alva G. Day

Ann, gen erect; branches spreading or ascending, hairy, glandular or puberulent. **LVS** simple, alternate, gen deeply pinnately lobed or entire. **INFL**: head; bracts pinnately to palmately toothed or lobed, spine-tipped; fls sessile or subsessile. **FL**: calyx membranous between ribs, lobes 4–5, entire or toothed, unequal, spine-tipped; corolla lobes 4–5; stigmas 2 or 3. **FR** gen ovoid, chambers 1–3. **SEEDS** 1–many per chamber, free or stuck together, brown, gelatinous when wet. ± 30 spp.: w N.Am, also in Argentina, Chile. (F. Navarrete, Spanish physician, 1700's)

1. Calyx ribs strap-shaped, wider at base than membrane; lobes of bract tip 7–many; fr dehiscing from base upward, not to tip (sect. *Mitracarpium*)
  2. Axis of upper lvs and often of bracts wider above middle
    3. Central st < or = branches; axis white-hairy .... ***N. jaredii***
    3′ Central st gen > branches; axis puberulent
      4. Bracts gen hairy and puberulent, axis of inner bracts wider below middle, gen not above .... [2]***N. pubescens***
      4′ Bracts puberulent, axis wider below and above middle .... ***N. setiloba***
  2′ Axis of upper lvs and bracts linear
    5. Corolla throat lacking spots

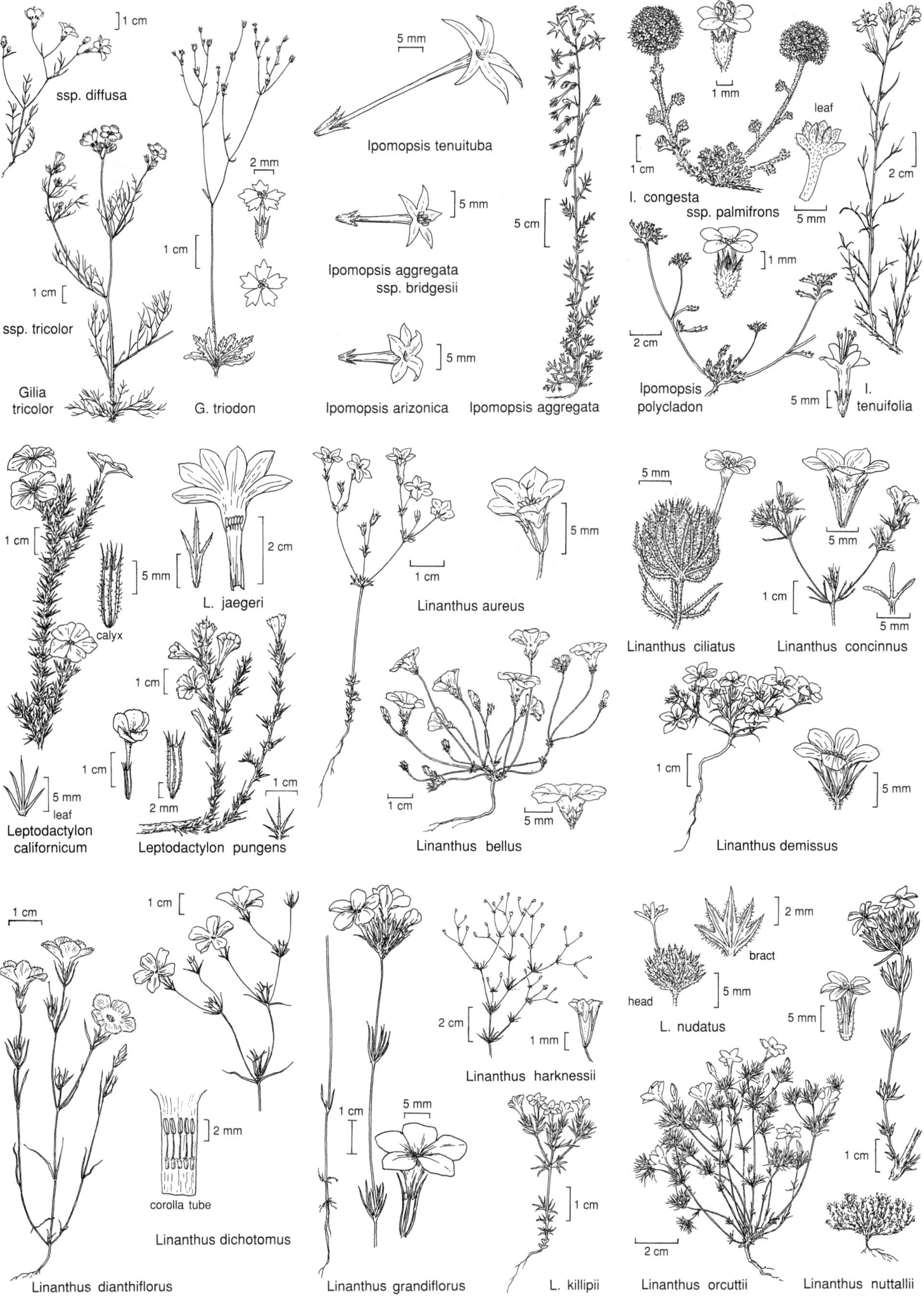
1 cm
ssp. diffusa
1 cm
ssp. tricolor
Gilia tricolor
2 mm
1 cm
G. triodon
5 mm
Ipomopsis tenuituba
5 mm
Ipomopsis aggregata ssp. bridgesii
5 mm
Ipomopsis arizonica
5 cm
Ipomopsis aggregata
1 mm
1 cm
leaf
I. congesta ssp. palmifrons
5 mm
1 mm
2 cm
Ipomopsis polycladon
2 cm
5 mm
I. tenuifolia
1 cm
calyx
5 mm
2 cm
L. jaegeri
1 cm
1 cm
2 mm
1 cm
5 mm
leaf
Leptodactylon californicum
Leptodactylon pungens
1 cm
Linanthus aureus
5 mm
1 cm
5 mm
Linanthus bellus
5 mm
Linanthus ciliatus
5 mm
1 cm
5 mm
Linanthus concinnus
1 cm
5 mm
Linanthus demissus
1 cm
1 cm
2 mm
corolla tube
Linanthus dichotomus
Linanthus dianthiflorus
1 cm
5 mm
Linanthus grandiflorus
2 cm
1 mm
Linanthus harknessii
1 cm
L. killipii
2 mm
bract
head
5 mm
L. nudatus
5 mm
2 cm
Linanthus orcuttii
1 cm
Linanthus nuttallii

6. Corolla bright blue with purple throat . . . . . [2]***N. pubescens***
6′ Corolla yellow or white (sometimes lobe tips blue)
7. Infl puberulent, slightly or not hairy; corolla lobes 4 . . . . . ***N. cotulifolia***
7′ Infl densely white-hairy, corolla lobes gen 5 . . . . . ***N. eriocephala***
5′ Corolla throat with purple or brown spots
8. Corolla lobes purple; fr 8-valved . . . . . ***N. jepsonii***
8′ Corolla lobes white or yellow; fr 4-valved
9. Corolla white; fr 1–2-seeded . . . . . ***N. heterandra***
9′ Corolla yellow; fr 2–9-seeded . . . . . ***N. nigelliformis***
10. Corolla 12–16 mm; branches ascending; herbage dark green . . . . . ssp. ***nigelliformis***
10′ Corolla 9–11 mm; branches decumbent; herbage gray-green . . . . . ssp. ***radians***
1′ Calyx ribs tapered, narrower at base than membrane; lobes of bract tip gen 0–few; fr dehiscing from tip downward (if dehiscing upward, then only when wet)
11. Pl gen not glandular; seeds stuck together until wet; fr translucent, sticking to seeds, opening at base when wet (sect. *Navarretia*)
12. Axis of inner bracts concave-expanded at base; corolla lobes ovate
13. Bract tips with 3 spreading spines; corolla lobes with 1 vein entering base; stigmas 2 . . . . . ***N. intertexta***
14. Bracts and calyces densely white-hairy; pl erect, loosely branched, 8–28 cm . . . . . ssp. ***intertexta***
14′ Bracts and calyces sparsely white-hairy; pl spreading, densely branched, < 10 cm . . . . . ssp. ***propinqua***
13′ Bract tips without 3 spreading spines; corolla lobes with 3 veins entering base; stigmas 3
15. Calyx and bracts glabrous; bract tips entire; calyx lobes entire . . . . . ***N. subuligera***
15′ Calyx and bracts appressed-hairy; bract tips many-lobed; calyx lobes toothed . . . . . ***N. tagetina***
12′ Axis of inner bracts widely membranous-winged at base, not concave-expanded; corolla lobes linear or narrowly ovate (*N. leucocephala* group)
16. Mature infl a head (fls sessile); pls prostrate
17. Corolla 12–21 mm, tube long-exserted; involucral lvs lobed near base . . . . . ***N. myersii***
17′ Corolla 7–9 mm, tube slightly exserted; involucral lvs with many lobes above and below . . . . . ***N. prostrata***
16′ Mature infl a dense cyme (fls subsessile or short-pedicelled); pls gen not prostrate
18. Calyx tube hairy; seeds 5–25; s CA; Baja CA . . . . . ***N. fossalis***
18′ Calyx tube glabrous or slightly hairy; seeds 2–7; n&c CA . . . . . ***N. leucocephala***
19. Corolla tube exserted, throat 2–3 mm wide, lobes narrowly ovate . . . . . ssp. ***leucocephala***
19′ Corolla tube incl, throat 1–2 mm wide, lobes linear
20. Infl 2–20-fld; outer bracts gen with 3–5 entire or forked lobes . . . . . ssp. ***pauciflora***
20′ Infl 10–60-fld; outer bracts gen with 7 or more 3–4-branched lobes
21. Corolla < or = calyx; some calyx lobes gen toothed; CaR, SN, GB . . . . . ssp. ***minima***
21′ Corolla = or > calyx; calyx lobes gen entire; NCoR, w ScV
22. St gen erect, branches ascending; corolla white . . . . . ssp. ***bakeri***
22′ St low, branches spreading; corolla blue — rare . . . . . ssp. ***plieantha***
11′ Pl gen glandular; seeds not stuck together; fr opaque, not sticking to seeds, dehiscing downward from tip to base
23. Outer bracts with short or linear axis and linear lobes, terminal lobe ± as wide as laterals (sect. *Masonia*)
24. Bracts pinnately lobed
25. Corolla yellow; pl puberulent, not glandular-hairy . . . . . ***N. breweri***
25′ Corolla lavender; pl puberulent and glandular-hairy . . . . . ***N. peninsularis***
24′ Bracts palmately lobed
26. Corolla 8–10 mm; stamens and style exserted . . . . . ***N. prolifera***
27. Corolla lobes yellow . . . . . ssp. ***lutea***
27′ Corolla lobes blue to purple . . . . . ssp. ***prolifera***
26′ Corolla 4–7 mm; stamens and style incl . . . . . ***N. divaricata***
28. Corolla white with pink lobes; longest bract 5–10 mm . . . . . ssp. ***divaricata***
28′ Corolla purple; longest bract 7–20 mm . . . . . ssp. ***vividior***
23′ Outer bracts with broad axis and tapered lobes (if bracts palmate then middle lobe much wider than laterals) (sect. *Aegochloa*)
29. Tip of outer bracts 3-lobed, often hooked
30. Lf lobes gen pointed tipward (basal lobes spreading); bract lobes pointed tipward, tips not hooked . . . . . ***N. atractyloides***
30′ Lf and bract lobes spreading; bract tips gen hooked . . . . . ***N. hamata***
31. Corolla pale colored; bracts hairy, not glandular, 3 lobes of hook equal . . . . . ssp. ***parviloba***
31′ Corolla bright colored; bracts gland-dotted or glandular-hairy, 3 lobes of hook gen unequal
32. Corolla tube incl, throat wide . . . . . ssp. ***hamata***
32′ Corolla tube exserted, throat narrow . . . . . ssp. ***leptantha***
29′ Tip of outer bracts neither 3-lobed nor hooked
33. Bract bases not concave, lobes many, gen forked on back; stamens attached in tube
34. Corolla light blue, 5–7 mm; outer bracts with middle lobe ovate-acuminate . . . . . ***N. mellita***
34′ Corolla dark blue, 9–12 mm; outer bracts with middle lobe lanceolate, long-tapered . . . . . ***N. squarrosa***
33′ Bracts or their bases concave, lobes few, not forked on back; stamens attached in throat
35. Outer bracts much longer than wide
36. Infl lvs thread-like, lobed only at base . . . . . ***N. filicaulis***
36′ Infl lvs strap-shaped or lanceolate, lobed at base and above . . . . . ***N. viscidula***

35′ Outer bracts little longer than wide
37. Corolla > calyx, purple; stamens, style exserted .... *N. heterodoxa*
37′ Corolla = calyx, lavender or white; stamens, style incl .... *N. rosulata*

N. atractyloides (Benth.) Hook. & Arn. (p. 851) **ST** 5–29 cm. **LF** strap-shaped to lanceolate, pinnate; cauline lf axis 2–6 mm wide; basal lobes spreading; upper lobes unequal, pointed tipward. **INFL**: bracts gland-dotted or glandular-hairy, recurved, base widely clasping, lobes at tip gen 3, unequal, pointed tipward. **FL**: calyx lobes entire or toothed; corolla 8–9 mm, incl in head, gen purple (white), tube red-veined, lobes < 2 mm; stamens, style slightly exserted; stigmas 3. **FR** < calyx, dehiscing from tip. **SEEDS** 2–6 per chamber. $2n$=18. Open, rocky or sandy areas; 90–900 m. NCoR, SnFrB, SCoR, ChI, TR, PR; OR, Baja CA. Like *N. hamata*, but not skunk-like odor, bract tips not hooked. [*N. hamata* ssp. *foliacea* (E. Greene) H. Mason] ❀DRN,SUN,IRR:5,7–9,**14–17,19–24**.

N. breweri (A. Gray) E. Greene (p. 851) Pl gen densely branched, as wide as high. **ST** 3–8 cm, white-puberulent, brown or red. **LF** gray-green, pinnate; axis < 0.5 mm wide; lobes needle-like, entire or forked. **INFL**: outer bracts like lvs but axis wider, shorter. **FL**: calyx 6–9 mm, sinus U-shaped, lobes needle-like, hairy inside; corolla 6–7 mm, yellow, lobes 1 mm; stamens, style exserted; stigmas 3, minute, appearing as 1. **FR** obovoid, dehiscing from tip. **SEEDS** 1–2 per chamber. Open, wet areas, meadows, streamsides; 1200–3300 m. SNH, GB; to WA, ID, Colorado, AZ. ❀TRY.

N. cotulifolia (Benth.) Hook. & Arn. Pl glandular-puberulent. **ST** 3–31 cm, gen branched, leafy; internodes < lvs. **LF** 2-pinnate; base hairy; axis and lobes < 0.5 mm wide, linear; lobes clustered, pointed tipward. **INFL**: heads leafy at base; bracts pinnate; lower lobes forked; bracts and calyces hairy at middle. **FL**: calyx ribs 4, strap-shaped, lobes entire; corolla 8–11 mm, > head, pale yellow, tube thread-like, lobes 4, ± 3 mm; stamens, style exserted beyond lobes; stigmas 2. **FR**: chamber 1; valves 4, dehiscing in lower half. **SEED** 1. Heavy soils; < 50 m. NCoRI, ScV, SnFrB, SCoRI. ❀TRY.

N. divaricata (A. Gray) E. Greene MOUNTAIN NAVARRETIA **ST** 2–10 cm; branches paired or whorled, 2–5 below heads, spreading, slender, brown or purple, puberulent or hairy. **LVS**: basal thread-like, glabrous, entire; cauline pinnate, lobes thread-like, abruptly pointed. **INFL**: bracts and calyces white-hairy at middle, outer bracts palmate, middle lobe > lateral. **FL**: corolla incl or exserted, lobes < 2 mm; stamens, style incl; stigmas 3, minute. **FR** 2–4 mm, dehiscing from tip to base. Common. Open, gravelly (often volcanic) areas; 300–2600 m. NW, CaR, SN, s SCoRO (Santa Barbara Co.), MP; to WA, ID.

ssp. ***divaricata*** (p. 851) **INFL**: longest bracts 5–10 mm, bract tips gen glabrous. **FL**: corolla 4–5 mm, < or = calyx, white with throat yellow, lobes pink-tipped (corolla rarely purple). **SEEDS** 1–3 per chamber. Habitat and range of sp.

ssp. ***vividior*** (Jepson & V. Bailey) H. Mason (p. 851) **INFL**: longest bracts 7–20 mm, bract lobes glandular-puberulent to tip. **FL**: corolla 5–7 mm, exserted, dark blue-violet. **SEEDS** 6–8 per chamber. Clay or volcanic soil; 600–1900 m. NW, CaR, SN, MP; OR. ❀TRY.

N. eriocephala H. Mason (p. 851) HOARY NAVARRETIA **ST** gen branched, 5–25 cm, white-puberulent. **LF** 2-pinnate; axis and lobes thread-like; upper lobes clustered, pointed tipward. **INFL** compact, white-hairy, like woolly ball; bract lobes puberulent, tips exposed, spreading. **FL**: calyx ribs 4–5, strap-shaped, wider at base than membrane, hairy at middle, longest lobes toothed; corolla 8–12 mm, slightly > head, white, blue-tipped, throat gen lacking spots, lobes 4–5, 2–3 mm; stamens unequal, longest exserted; style exserted; stigmas 2. **FR** obovoid; chamber 1; valves 4, dehiscing in lower half. **SEED** 1. UNCOMMON. Heavy soil of seasonally wet flats; < 400 m. n&c SNF, e ScV. Intergrades somewhat with *N. heterandra*.

N. filicaulis (A. Gray) E. Greene (p. 851) Pl glandular-puberulent. **ST** 7–18 cm; branches zig-zag, slender. **LF** 1–3 cm, 1-pinnately lobed near base; axis and lobes thread-like; lobes 0 or 2–6. **INFL**: bracts gland-dotted, palmate, base wide and concave; outer bracts lf-like. **FL**: calyx 3–4 mm; corolla 5–7 mm, purple, tube thread-like, exserted, lobes 1–2 mm, narrowly ovate; stamens attached in lower throat; stamens, style exserted beyond corolla lobes; stigmas 2, minute. **FR** 2–4-valved, dehiscing from tip to base. **SEEDS** 1–2 per chamber. Open areas, chaparral, woodlands, in gravel or clay; 300–900 m. CaRF, n&c SNF.

N. fossalis Moran Pl spreading, gen not prostrate. **ST** 1–15 cm, glabrous or puberulent; hairs recurved. **LF** glabrous, pinnate; axis and lobes linear; lobes few, entire or forked. **INFL** 1–2 cm wide; fls clustered; longest bracts gen < 2 × head, hairy below middle, membranous-winged at base, gen ciliate; lobes few, 2–4-branched. **FL**: calyx ribs hairy, slightly glandular at base, narrower than membrane at base, lobes glabrous, membrane truncate, ciliate; corolla 4–7 mm, exserted, white, lobes linear, < 1 mm wide; stamens, style slightly exserted; stigmas 2, minute. **FR**: chambers 2, translucent, sticking to seeds until wet. **SEEDS** 5–25, separating when wet. $2n$= 18. RARE. Vernal pools, ditches; 30–1300 m. s SCoRO (San Luis Obispo Co.), SW; Baja CA.

N. hamata E. Greene Pl gen with skunk-like odor. **ST** 8–30 cm, glandular-puberulent. **LF** pinnate; axis linear to widely lanceolate; lobes spreading, not hooked; tip 3-lobed, ± hooked. **INFL**: bracts widely clasping, outer lanceolate, recurved, hooked at tip, pinnate, lobes spreading. **FL**: calyx lobes entire or toothed; corolla gen purple or pinkish; stigmas 3. **FR** < calyx, dehiscing from tip to base. **SEEDS** gen 2–10 per chamber. Dry, sandy or rocky places in coastal or inland chaparral; < 1100 m. SnFrB, SCoRO, SW; Baja CA. Sspp. intergrade.

ssp. ***hamata*** (p. 851) **LF**: cauline axis linear or lanceolate. **INFL** gen terminal; bracts gland-dotted or glandular-hairy, lobes of hook gen unequal. **FL**: calyx lobes entire or toothed; corolla 10–15 mm, bright pink or purple, tube incl, throat wide, lobes 3–5 mm. $2n$=18. Habitats of sp. TR, PR. Some pls from TR have wider lvs, axillary infl, are much like ssp. *parviloba* exc for larger, brighter fls.

ssp. ***leptantha*** (E. Greene) H. Mason (p. 851) **LF**: axis and lobes linear. **INFL** terminal; bracts gland-dotted or glandular-hairy, lobes of hook unequal. **FL**: calyx lobes gen entire; corolla tube thread-like, exserted, purple, throat narrow, lobes 2–3 mm, pink. Habitats of sp., on coast; < 700 m. SCo, ChI; Baja CA.

ssp. ***parviloba*** Day (p. 851) **LF**: axis linear to widely lanceolate, narrowest below hook, lobes of hook equal, widely spreading, middle lobe recurved. **INFL** axillary and terminal; bracts long-hairy, not glandular, hooks like those of lvs. **FL**: calyx lobes entire or largest 2-toothed; corolla 8–14 mm, incl in head, tube white, lobes 2–3 mm, pale blue or lavender. Open, sandy areas, often on sandhills; < 1000 m. SnFrB (Santa Cruz Mtns), SCoRO.

N. heterandra H. Mason TEHAMA NAVARRETIA Pl gen wider than tall. **ST** decumbent, 3–11 cm, puberulent. **LF** 2-pinnate; axis linear; lobes linear, needle-like, widely spreading. **INFL** compact, densely hairy in center; bract tips spreading, often red, glabrous. **FL**: calyx ribs 4–5, strap-shaped, hairy above middle, longest lobes toothed; corolla 6–8 mm, white, throat with purple spots below lobes, lobes 4–5; stamens unequal, exserted, < lobes; style incl; stigmas 2. **FR** obovoid; chamber 1; valves 4, dehiscing in lower half. **SEEDS** 1–2. UNCOMMON. Heavy soil, vernal pools, wet or drying flats; < 400 m. NCoRI, CaRF, w ScV, e SnFrB, SCoRI, MP; OR.

N. heterodoxa (E. Greene) E. Greene (p. 851) Pl with skunk-like odor. **ST** 8–24 cm, glandular-puberulent. **LVS** pinnate; basal lf axis and lobes thread-like; cauline lf recurved at tip, axis wider, lobes acuminate. **INFL**: bracts widely ovate, palmate, glands sessile, lobe length < axis width, middle lobe acuminate, tip recurved; fls many. **FL**: calyx 6–8 mm, lobes long-hairy; corolla 6–11 mm, > bracts, purple, lobes 2 mm, narrowly oblong; stamens, style exserted, stigmas 3, minute. **FR** dehiscing from tip to base. **SEEDS** 4–5 per chamber. Open, rocky, gen serpentine slopes; 100–900 m. SnFrB.

N. intertexta (Benth.) Hook. **ST** puberulent or hairy below heads, hairs reflexed, appressed or spreading. **LF** 1–2-pinnate, glabrous or white-hairy near base; axis and lobes needle-like, spreading at tip. **INFL**: bracts and calyces white-hairy about middle;

bracts pinnate, lobes needle-like, forked, 3 lobes at tip spreading; inner bracts clasping, base expanded-concave, narrowly membrane-margined. **FL**: calyx ribs tufted-hairy within sinuses, lobes sometimes toothed, membrane V-shaped; corolla gen white, lobes ovate, 1 vein entering base; stamens, style exserted, stigmas 2. **FR**: chambers 1–2, translucent below middle, sticking to seeds until wet. **SEEDS** 8–10, dark brown, pitted. Open, wet areas, meadows, vernal pools; < 2300 m. KR, NCoR, CaR, SN, GV, SnFrB, SCoR, MP; to B.C., MT, WY, Colorado, AZ, Baja CA.

ssp. ***intertexta*** (p. 851) **ST** 8–28 cm. **INFL** densely white-hairy; bracts 10–20 mm, 3 terminal lobes equal. **FL**: corolla > calyx, blue or white. Habitat of sp.; < 1800 m. KR, NCoR, CaR, SNF, GV, SnFrB, SCoR; OR. ❀TRY.

ssp. ***propinqua*** (Suksd.) Day (p. 851) Pl densely branched, gen wider than high. **ST** 3–10 cm. **INFL** sparsely hairy; bracts 5–10 mm, terminal lobe > laterals. **FL**: corolla = calyx, white. Habitat of sp.; 800–2300 m. CaR, SNH, MP; to B.C., MT, WY, Colorado, AZ, Baja CA. [*N. p.* Suksd.]

N. jaredii Eastw. (p. 851) PASO ROBLES NAVARRETIA **ST** 3–20 cm, puberulent or white-hairy. **LVS**: lower glabrous, 1–2-pinnate, axis linear, lobes needle-like; upper hairy below middle, tip toothed along wide axis. **INFL**: bracts hairy, gland-dotted, tip of outer bracts toothed along wide axis. **FL**: calyx ribs strap-shaped, lobes sometimes toothed; corolla 7–11 mm, exserted, tube and throat white, glandular-puberulent, lobes 2–4 mm, blue; stamens, style exserted; stigmas 2. **FR**: chamber 1; valves 4, dehiscing in lower half. **SEED** 1. Open, grassy, serpentine areas; 200–500 m. SCoRI, n SW. [*N. mitracarpa* E. Greene ssp. *j.* (Eastw.) H. Mason]

N. jepsonii Jepson (p. 851) JEPSON'S NAVARRETIA **ST** 10–15 cm, reddish, densely white-puberulent. **LF** glabrous or slightly puberulent, 1–2-pinnate; axis and lobes needle-like. **INFL**: bracts dark red, coarsely white-hairy below, lobes gland-dotted. **FL**: calyx coarsely white-hairy below, ribs strap-shaped, wider at base than membrane, lobes sometimes toothed; corolla 9–11 mm, tube and throat white, lobes 2–3 mm, purple with darker basal spot; stamens and style exserted; stigmas 2. **FR**: chamber 1; valves 8, dehiscing in lower half. **SEED** 1. UNCOMMON. Open, grassy or clay flats, serpentine areas; 400–800 m. NCoRI. ❀TRY.

N. leucocephala Benth. **ST** puberulent or hairy below heads; hairs recurved. **LVS** 1–2-pinnate; lobes linear; lower lvs glabrous; upper hairy at base. **INFL**: outer bracts lf-like, lobes needle-like, 2–4-branched at base, lobes near tip 2, shorter, entire, pointed tipward; inner bracts simpler, base membranous-winged, ciliate, width of wing > midrib. **FL**: calyx ribs glabrous or slightly hairy, tapered, gland-dotted below, membrane truncate, ciliate; corolla 4–10 mm, lobes with 1 vein entering base; stamens exserted; stigmas 2, minute. **FR** translucent, adhering to seeds until wet. **SEEDS** separating when wet. Vernal pools; < 2200 m. NCoRI, CaR, n SNH, GV, GB; to WA, NV. Some sspp. intergrade. Variable with water level and duration.

ssp. ***bakeri*** (H. Mason) Day BAKER NAVARRETIA **ST** 2–10 cm, gen erect; branches ascending. **INFL**: longer bracts < 2 × head, lobes 2–4-branched on back. **FL**: calyx ribs sometimes slightly hairy; corolla white to blue, tube incl, throat 1–2 mm wide, lobes linear, exserted; style exserted. **SEEDS** 2–4. UNCOMMON. Habitat of sp.; < 1700 m. NCoRI, w ScV. Intermediate between sspp. *leucocephala* and *plieantha*. [*N. b.* H. Mason] ❀TRY.

ssp. ***leucocephala*** (p. 851) **ST** 2–22 cm. **INFL**: longer bracts > 2 × head, lobes 3–4-branched on back. **FL**: calyx ribs gen glabrous; corolla white, tube exserted, lobes narrowly ovate; style exserted. **SEEDS** 4–7. 2*n*=18. Habitat of sp.; < 500 m. GV; OR. Prostrate pls with spreading branches occur in Butte Co. Intergrades with ssp. *minina* in n SNF, CaR.

ssp. ***minima*** (Nutt.) Day **ST** 2–8 cm. **INFL** 1–2 cm wide, bracts < 2 × head, lobes 3–4-branched on back. **FL**: calyx lobes gen toothed; corolla white, gen incl, throat < 2 mm wide, lobes linear; style incl or slightly exserted. **SEEDS** 1–6. 2*n*=18. Habitat of sp.; < 2200 m. CaR, n SN, GB; to WA, UT. [*N. m.* Nutt.]

ssp. ***pauciflora*** (H. Mason) Day FEW-FLOWERED NAVARRETIA **ST** 1–4 cm. **INFL** ± 1 cm wide, 2–20-fld; longest bracts 1.5–2 × head, bract lobes few, entire or forked; fls in 2–4 clusters within head. **FL**: corolla white to blue, tube incl, lobes linear. **SEEDS** 2. RARE. Habitat of sp.; 400–600 m. s NCoRI (Lake, Napa cos.). [*N. p.* H. Mason] Intergrades rarely with ssp. *plieantha*.

ssp. ***plieantha*** (H. Mason) Day (p. 851) MANY-FLOWERED NAVARRETIA **ST** 1–3 cm, spreading, many-branched. **INFL** 10–60-fld; bracts < 2 × head, lobes 2–4-branched on back; fls short-pedicelled in clusters within head. **FL**: calyx ribs lightly glandular at base, puberulent at middle; corolla blue, tube incl, lobes linear. **SEEDS** 2–3. **ENDANGERED** CA. Habitat of sp.; 800–950 m. s NCoR (Lake, Sonoma cos.). [*N. p.* H. Mason]

N. mellita E. Greene (p. 855) Pl glandular-hairy. **ST** 10–20 cm. **LVS**: basal pinnate, axis and lobes linear; cauline axis shorter, wider. **INFL**: outer bracts palmate, base expanded, middle lobe ovate, acuminate, > head, lateral lobes linear to lanceolate, forked on back. **FL**: corolla 5–7 mm, > calyx, throat white, lobes light blue; stamens attached in tube, incl; style incl, stigmas 3. **FR** dehiscing from tip to base. **SEEDS** gen 5 per chamber. Open, wet, sandy or gravelly areas, chaparral; 300–1200 m. NCoRI, SCoRI, c SNF (Tuolumne Co.). ❀TRY.

N. myersii P.S. Allen & Day (p. 851) PINCUSHION NAVARRETIA **ST** < 20 mm, glabrous or puberulent; hairs recurved. **LVS** gen radiate from base of central head, 4–8 cm, pinnate, axis linear, base membrane-winged, lobes few, in lower half, linear. **INFL** gen 1, sessile (or peduncle prostrate), 1–2 cm wide, 5–60-fld; outer bracts lf-like, >> head, basal wings hairy; inner bracts < head, basal wings ciliate; fls sessile. **FL**: calyx ribs long-hairy, gland-dotted below, tapered, lobes linear, glabrous, membrane truncate, ciliate; corolla 12–21 mm, 2–4 × calyx, white, tube thread-like, throat cup-shaped, lobes 2.5–3 mm, oblong-ovate; stamens, style exserted, stigmas 2, minute. **FR** translucent, sticking to seeds until wet. **SEEDS** 2–4, separating when wet. Vernal pools; 20–90 m. c SNF, c GV (4 sites: Sacramento, Amador, Merced cos.).

N. nigelliformis E. Greene **ST** 9–32 cm, puberulent; hairs white, recurved. **LF** 2-pinnate; axis and lobes linear, < 1 mm wide, spreading (or upper clustered, pointing tipward). **INFL**: bracts pinnate, axis and lobes linear, hairy below middle. **FL**: calyx ribs strap-shaped, hairy near middle, longest lobes toothed; corolla yellow, throat with paired, purple or brown spots below lobes; stigmas 2, minute. **FR**: chamber 1; valves 4, dehiscing in lower half. Vernal pools, clay depressions; 100–1000 m. SNF, GV, SCoRI. ❀TRY.

ssp. ***nigelliformis*** (p. 851) Pl herbage dark green. **ST**: branches spreading-ascending. **INFL** slightly white-hairy in center; bract tips glandular-puberulent. **FL**: corolla 12–16 mm, exserted from head, lobes 3–4 mm; stamens and style > lobes. **SEEDS** 4–9. Habitat and range of sp. ❀TRY.

ssp. ***radians*** (J. Howell) Day (p. 851) Pl herbage light gray-green. **ST**: branches decumbent (pl wider than high). **INFL** densely white-hairy in center; bract tips gen glabrous. **FL**: corolla 9–11 mm, incl in head, lobes 1–2 mm; stamens exserted < corolla lobes; style incl. **SEEDS** 2–8. Habitat of sp.; 200–1000 m. SCoRI. Like *N. heterandra*. Further study needed.

N. peninsularis E. Greene (p. 851) Pl glandular-puberulent. **ST** 3–25 cm. **LF** 1–3 cm, hairy near base, pinnate; axis and lobes linear, < 1 mm wide; lobes 3–5 mm, needle-like, simple or forked, middle lobe longest. **INFL**: outer bracts 10–15 mm, lf-like, axis wider, shorter. **FL**: calyx lobes needle-like; corolla 6–9 mm, slightly > calyx, lavender, lobes 1–2 mm; stamens unequal, longest exserted; style incl, stigmas 3. **FR** dehiscing from tip to base. **SEEDS** 6–8 per chamber. Wet areas in open forest; 1500–2300 m. Teh, TR, PR; Baja CA. Like *N. breweri*.

N. prolifera E. Greene **ST**: branches paired or whorled, 2–4 from below heads, 6–16 cm, spreading, slender, brown, gen glabrous. **LVS**: basal entire, thread-like, glabrous; cauline pinnate, axis and lobes thread-like. **INFL**: bracts (and calyces) densely white-hairy at middle, glandular-puberulent above, lobe tips glabrous, needle-pointed; outer bracts palmate, middle lobe < 2 × longest lateral. **FL**: calyx lobes entire; corolla 8–10 mm, gen > calyx, lobes ovate, 1–3 mm; stamens, style exserted; stigmas 3. **FR** dehiscing from tip to base. **SEEDS** 3 per chamber. Dry, rocky flats near drainage channels; 600–1400 m. SNF.

ssp. ***lutea*** (Brand) Mason YELLOW BUR NAVARRETIA **FL**: corolla yellow throughout. UNCOMMON. Habitat of sp.; 900–1400 m. n SNF (El Dorado Co.). ❀TRY.

ssp. ***prolifera*** **FL**: corolla tube and throat yellow, lobes blue or purple. Habitat of sp.; 600–1300 m. SNF.

N. prostrata A. Gray Pl prostrate with central head and radiating lvs and sts bearing heads. **ST** gen lfless and glabrous exc below heads, hairs recurved. **LF** > 2 × head diam, glabrous, 1–2-pinnately lobed from base to tip, axis 1–4 mm wide, lobes linear, pointed tipward. **INFL**: bracts like lvs but shorter, hairy below, basal wing wide, membranous, ciliate, lower lobes 2–3-branched on back. **FL**: calyx ribs hairy, tapered, tube slightly glandular at base, lobes glabrous, often 3-toothed; corolla 7–9 mm, blue to white, lobes linear, < 1 mm wide; stamens and style exserted, stigmas 2, minute. **FR** translucent, sticking to seeds until wet. **SEEDS** 5–25, separating when wet. Alkaline floodplains, vernal pools; < 700 m. w SnJV (Merced Co.), SCoRI, c SCo (Los Angeles Co.), PR (Santa Rosa Plateau). ❀TRY.

N. pubescens (Benth.) Hook. & Arn. (p. 855) Pl glandular. **ST** 14–33 cm; internodes gen < lvs, tan to red-brown. **LF** 2-pinnate, axis and lobes linear; lower lvs glabrous, lobes clustered; upper lvs puberulent or hairy, tip linear or wider, many-toothed. **INFL** gen glandular hairy at middle of bracts (and calyces), puberulent above; bracts pinnate; outer bracts 1.5–2 × head, axis narrow or wider above, toothed; inner bracts shorter, base clasping, tip narrow, toothed. **FL**: calyx ribs strap-shaped to base, gen 2 lobes toothed; corolla 10–16 mm, > calyx, tube and throat purple, lobes bright blue; stamens attached in mid-throat, exserted; style exserted, stigmas 2, minute. **FR** dehiscing in lower half; chamber 1. **SEEDS** 1–2. Open, slopes, gravel or clay; < 1100 m. SNF, NCoR, SnFrB, SCoRI; s OR. [*N. mitracarpa* E. Greene] ❀SUN,DRY:7,8,9,**14–24**.

N. rosulata Brand (p. 851) MARIN COUNTY NAVARRETIA Pl glandular-puberulent; odor skunk-like. **ST** 6–13 cm. **LVS**: lower pinnate, axis and lobes linear; upper with wider base, ovate-acuminate tip. **INFL**: outer bracts 7–8 mm, lobe length < axis width. **FL**: calyx 6–7 mm, lobes glandular-hairy; corolla = calyx, lavender to white, throat narrow, lobes ± 1 mm, narrowly oblong; stamens attached midway in tube-throat, incl; stigmas 3. **FR** dehiscing from tip to base. **SEEDS** 4–6 per chamber, bullet-shaped. RARE. Rocky, serpentine areas; 200–600 m. s NCoRI (Napa Co.), n SnFrB (Marin Co). [*N. heterodoxa* ssp. *r.* (Brand) H. Mason]

N. setiloba Cov. (p. 855) PIUTE MOUNTAINS NAVARRETIA Pl puberulent, glandular. **ST** 10–20 cm. **LF** 2-pinnate; axis linear below middle, wider and toothed above; lobes linear, forked. **INFL**: outer bracts 1.5–2 × head, wider and toothed above middle; inner bracts clasping at base, narrowest at middle, toothed above. **FL**: calyx ribs widely strap-shaped, lobes < tube, hairy, longest sometimes toothed; corolla 10–11 mm, mostly white; stamens exserted; style exserted, stigmas 2, minute. **FR** obovoid; chamber 1. **SEEDS** gen 1 (2). RARE. Depressions in clay or gravelly loam; 500–2100 m. s SNH, Teh (Tulare, Kern cos.).

N. squarrosa (Eschsch.) Hook. & Arn. (p. 855) SKUNKWEED Pl spreading, long-hairy, glandular; odor skunk-like. **ST** 10–60 cm. **LVS**: lower 1–2-pinnate, axis and lobes linear; upper with axis shorter, wider, lobes narrowly lanceolate, gen forked on back, spiny. **INFL**: outer bracts lf-like, < lvs, middle lobe lanceolate, wider than lateral. **FL**: calyx 7–10 mm, lobes narrowly lanceolate; corolla 9–12 mm, > calyx, deep blue, lobes 2 mm; stamens attached in tube, incl; style incl, stigmas 3. **FR** dehiscing from tip to base. **SEEDS** 8–11 per chamber. Common. Open, wet, gravelly flats, slopes; < 800 m. NCoR, n SNF (Sacramento, Amador cos.) SnFrB, SCoR; to B.C. ❀SUN:**4–6**&IRR:7,**14–17**,**19–24**.

N. subuligera E. Greene (p. 851) AWL-LEAVED NAVARRETIA Pl gen purple. **ST** 4–16 cm, puberulent; hairs recurved. **LF** pinnate near base, glabrous; axis and lobes linear. **INFL**: outer bracts pinnate, glabrous, basal margins membranous, ciliate, lobes short, well separated, tip awl-shaped, glabrous; inner bracts wider and concave at base. **FL**: calyx 7–8 mm, ribs tapered, membrane V-shaped, ciliate; corolla 6–7 mm, white, lobes with 3 veins entering base; stamens and style incl, stigmas 3. **FR** translucent, adhering to seeds until wet. **SEEDS** 8–10, separating when wet. UNCOMMON. Open, rocky, wet places; 150–1100 m. NCoRI, n SNF (Amador Co.), ScV; OR.

N. tagetina E. Greene MARIGOLD NAVARRETIA **ST** 7–30 cm, recurved-hairy near heads, glabrous or puberulent below. **LF** 2-pinnate, hairy near base; lobes needle-like, spreading, **INFL**: outer bracts lf-like, > head, tip short-toothed; inner bracts wider and concave at base, margins membranous, ciliate. **FL**: calyx 7–8 mm, ribs tapered, ciliate within at sinuses, lobes toothed; corolla 9–11 mm, pale blue, lobes with 3 veins entering base; stamens attached in upper throat, exserted; style incl, stigmas 3. **FR** translucent, adhering to seeds until wet. **SEEDS** 1–2 per chamber, separating when wet. Common. Open, grassy flats, vernal pools; 30–600 m. n&c SNF, NCoRI, GV; to WA. ❀SUN:4–6,**7–9**,**14–17**,18,**19–24**.

N. viscidula Benth. (p. 855) Pl gland-dotted, glandular-hairy. **ST** 3–24 cm; branches widely spreading. **LVS** pinnate; lobes or teeth pointed tipward; lower lvs 2–3 cm, axis and lobes thread-like; upper lvs gen strap-shaped, > 1 mm wide. **INFL**: outer bracts lf-like, gen > head; inner bracts shorter, palmate, middle lobe longest, tip acuminate or toothed. **FL**: calyx lobes lanceolate, hairy, tips glandular; corolla 9–16 mm, 2 × calyx, purple or red-purple; stamens attached in lower throat, exserted; style exserted, stigmas 3. **FR** dehiscing from tip to base. **SEEDS** 1–3 per chamber. Open, sandy or clay flats, near pool, marsh, or meadow; 100–800 m. NCoR, SNF, SnFrB. Small-fld race in SNF has been called ssp. *purpurea* (Brand) H. Mason, is being studied. ❀SUN:4–7,**14–17**, 19–24.

PHLOX

Robert W. Patterson & Dieter H. Wilken

Ann, per. **ST** prostrate to erect, or tufted to cushion-like. **LVS** cauline, opposite, simple, sessile, linear-lanceolate to elliptic, entire. **FL**: corolla salverform; stamens attached at different levels, some stamens unequal. ± 60 spp.: Am, Siberia. (Greek: flame, ancient name for *Lychnis* of Caryophyllaceae) [Cronquist, A. 1984. in Intermountain Flora, V.4.]

1. Ann (sect. *Microsteris*) .... ***P. gracilis***
1′ Per
 2. Stigmas > style; stamens very short, anthers well below corolla throat (sect. *Protophlox*) .... ***P. speciosa***
  3. Pl glabrous .... ssp. ***nitida***
  3′ Pl ± puberulent; calyx sometimes glandular .... ssp. ***occidentalis***
 2′ Stigmas << style; stamens long, at least some anthers at corolla throat
  4. Pl erect, branches open, not matted; fls gen 3–15 per infl (sect. *Phlox*)
   5. Lower lvs elliptic, 1–4 cm wide, glabrous; st decumbent, flowering branches erect .... ***P. adsurgens***
   5′ Lower lvs linear-lanceolate to lanceolate, < 1 cm wide, gen hairy; st gen ± erect from woody base
    6. Styles 5–8 mm, < calyx; lvs coarsely hairy .... ***P. hirsuta***
    6′ Styles 11–40 mm, gen > calyx; lvs variously hairy
     7. Corolla tube gen ± 40 mm; SnBr .... ***P. dolichantha***
     7′ Corolla tube 12–33 mm; SNE, DMtns .... ***P. stansburyi***
  4′ Pl ± matted to tightly cushion-like; fls gen 1–5 per infl (sect. *Microphlox*)

8. Pl tightly cushion-like (sometimes difficult to determine on dried specimens)
9. Lvs coarsely ciliate; st not appearing square in X-section .... ***P. condensata***
9′ Lvs not coarsely ciliate; st appearing square in X-section .... ***P. muscoides***
8′ Pl ± matted, not tightly cushion-like
10. Pl growing from long rhizomes with widely dispersed leafy clusters .... ***P. dispersa***
10′ Pl growing from short rhizomes with densely clumped sts
11. Calyx membrane keeled at base .... ***P. austromontana***
11′ Calyx membrane not keeled
12. Glandular hairs present
13. Glandular throughout herbage; GB .... ***P. douglasii*** var. ***rigida***
13′ Glandular only on calyces; SNH, SNE .... ***P. pulvinata***
12′ Glandular hairs 0
14. Lvs hairy only at base .... ***P. diffusa***
14′ Lvs hairy throughout .... ***P. hoodii*** var. ***canescens***

P. adsurgens A. Gray (p. 855) Per. **ST** decumbent, 1–3 dm; branches erect, open. **LF** 1–3 cm, elliptic-ovate, glabrous. **INFL**: fls few; peduncles 5–20 mm. **FL**: calyx 10–13 mm, lobes > tube, glandular-hairy; corolla bright pink, tube 12–20 mm, lobes gen rounded. $2n=14$. Open, wooded areas, mixed evergreen forest, montane coniferous forest; 500–2000 m. NCoR; OR. ❀DRN:**4**, **5**&IRR,SHD:1,2,**6**,**7**,14,**15–17**,18,24.

P. austromontana Cov. (p. 855) Per, ± matted. **LF** 10–15 mm, lanceolate; upper surface hairy; lower surface gen glabrous. **INFL** terminal; fls solitary; peduncle short. **FL**: calyx membrane keeled; corolla white to pink or lavender, tube 11–14 mm, lobes gen rounded. Dry, rocky areas, pinyon/juniper woodland, coniferous forest, sagebrush scrub; 1500–2700 m. SnGb, SnBr, PR; to w Colorado, AZ, Baja CA. [*P. diffusa* ssp. *subcarinata* Wherry] ❀TRY; DFCLT.

P. condensata (A. Gray) E. Nelson (p. 855) Per, tightly cushion-like. **LF** 3–5 mm, lanceolate, coarsely ciliate; upper surface gen concave; lower surface with 2 elongate grooves. **INFL** terminal; fls solitary, sessile. **FL**: calyx 5–6 mm, gen glandular-puberulent; corolla white or pale pink, tube 8–10 mm. $2n=28$. Dry, open, rocky areas, esp limestone, travertine; 2000–4000 m. SNH, SnBr, SNE; NV. [*P. covillei* E. Nelson; *P. caespitosa* Nutt. var. *condensata* A. Gray] ❀TRY;DFCLT.

P. diffusa Benth. (p. 855) Per, ± matted, ± glabrous to hairy, not glandular. **ST** decumbent. **LF** 10–15 mm, linear-lanceolate, not sharp-tipped. **INFL** terminal; fls solitary; peduncle short. **FL**: calyx 8–10 mm, hairy, membrane not keeled; corolla white to pink or blue, tube 9–13 mm. $2n=28$. Dry, open areas; 1100–3600 m. CA (exc SW, D); w N.Am. [*P. azurea* G. Smith] ❀DRN,SUN:1,2–6, 15–17&IRR:3,7,14,18–21;DFCLT.

P. dispersa Sharsm. (p. 855) HIGH SIERRA PHLOX Per, ± matted; rhizome creeping, slender; hairs short, gland-tipped. **LF** 5–10 mm, linear-lanceolate, leathery, sharp-tipped. **INFL** terminal; fls few, sessile. **FL**: calyx ± 7 mm, sinus membrane not keeled; corolla white, tube ± 10 mm, lobes irregularly margined. UNCOMMON. Dry flats of loose granite; gen 3600–4200 m. s SNH (Tulare, Inyo cos.).

P. dolichantha A. Gray (p. 855) BEAR VALLEY PHLOX Per. **ST** erect, nearly glabrous to glandular-hairy; branches open. **LF** 2–5 cm, linear-lanceolate. **INFL** terminal; fls few; peduncle slender. **FL**: calyx 10–12 mm, glandular-puberulent; corolla white or pink, tube > 40 mm; style 25–30 mm. RARE. Open areas in forests; 2000–2700 m. SnBr. Threatened by urbanization.

P. douglasii Hook. ssp. ***rigida*** (Benth.) Wherry Per, ± matted, glandular-hairy. **LF** 4–8 mm, linear-lanceolate, sharp-tipped. **INFL** terminal; fls 1–3, sessile. **FL**: calyx 7–9 mm; corolla pink or pale lavender to white, tube ± 10 mm; style 4–6 mm. Dry areas, sagebrush scrub, juniper woodland; 1500–2000 m. GB; to s WA. ❀TRY;DFCLT.

P. gracilis E. Greene (p. 855) Ann; upper parts gen glandular-hairy. **ST** decumbent and highly branched to simple and erect, < 20 cm. **LVS** opposite (uppermost alternate), 10–30 mm, oblanceolate to lanceolate. **FL**: calyx 5–10 mm; corolla tube 8–12 mm, yellowish, lobes 1–2 mm, gen truncate or notched, bright pink to white. $2n=14$. Dry to moist areas; < 3300 m. CA; to B.C., MT, Colorado, Mex; also in S.Am. [*Microsteris g.* (Hook.) E. Greene, incl ssp. *humilis* (E. Greene) V. Grant]

P. hirsuta E. Nelson (p. 855) YREKA PHLOX Per, densely coarse-hairy. **ST** erect; branches open. **LF** 10–20 mm, ± lanceolate. **INFL**: fls few; peduncles 5–25 mm. **FL**: calyx 8–12 mm, membranes keeled; corolla pink to white, tube 12–15 mm; style 5–8 mm. **ENDANGERED** CA. Dry serpentine talus, open Jeffrey pine/incense-cedar forest; 1000–1500 m. n CaRH (Siskiyou Co.).

P. hoodii Richardson ssp. ***canescens*** (Torrey & A. Gray) Wherry Per, ± matted. **ST** glabrous. **LF** tapered, nearly glabrous exc base densely white-woolly. **INFL** terminal; fls solitary, sessile. **FL**: calyx 7–8 mm, woolly near base of lobes; corolla white to lilac, tube 10–12 mm. Open, rocky areas, sagebrush scrub, pinyon/juniper woodland; 1500–2700 m. n SNH, GB; MT, UT, n AZ. [ssp. *lanata* (Piper) Munz] ❀TRY;DFCLT.

P. muscoides Nutt. (p. 855) MOSS PHLOX Per, tightly cushion-like. **LF** 3–5 mm, closely overlapping, 4-ranked (appearing square in X-section), completely concealing st. **INFL**: fls solitary, sessile. **FL**: calyx ± 5 mm; corolla white to lilac, tube 7–10 mm. RARE in CA. Open, rocky areas; 1400–2700 m. CaRH (Mount Lassen), nw MP (ne Siskiyou Co.); to OR, w Great Plains, Colorado. [*P. bryoides* Nutt.]

P. pulvinata (Wherry) Cronq. Per, cespitose. **LF** 4–10 mm, lanceolate, coarsely ciliate; upper surface gen flat. **INFL** terminal; fls solitary, sessile. **FL**: calyx 7–8 mm, gen glandular-puberulent; corolla white or pale pink, tube 8–10 mm. Dry, rocky areas, subalpine forest, alpine fell-fields; 3300–4300 m. SNH, SNE; to ID, Colorado. [*P. caespitosa* Nutt. var. *p.* (Wherry) Cronq.]

P. speciosa Pursh (p. 855) Per. **ST** erect; branches open. **LF** 1–5 cm, linear-lanceolate. **INFL** terminal; lower parts with leafy bracts; fls few; peduncles 3–20 mm, slender. **FL**: calyx 7–10 mm; corolla bright pink to white, tube 10–15 mm, lobes obcordate to deeply 2-lobed; stamens very short, anthers well hidden in corolla tube; style 2–4 mm, stigmas > style. Rocky, wooded slopes; 500–2400 m. KR, NCoR, CaR, SN; to B.C., MT.

ssp. ***nitida*** (Suksd.) Wherry Pl glabrous. Habitat of sp. KR, NCoR, CaR; to WA. [var. *nitida* Suksd.] ❀TRY;DFCLT.

ssp. ***occidentalis*** (Torrey) Wherry Pl puberulent (sometimes nearly glabrous), sometimes glandular, esp in calyx. Habitat and range of sp. ❀TRY;DFCLT.

P. stansburyi (Torrey) A.A. Heller (p. 855) Per. **ST** openly branched from base, gen growing up through shrubs. **LF** 1–3 cm, linear-lanceolate, > internodes, glandular-hairy to long-soft-hairy. **INFL**: fls few; peduncles 5–25 mm. **FL** 8–12 mm; calyx glandular-hairy, membrane gen keeled; corolla pink to white, tube 12–33 mm; style nearly = corolla tube. Dry areas in sagebrush scrub, pinyon/juniper woodland; 1700–3000 m. SNE, DMtns; to UT, NM. [*P. longifolia* Nutt. var. *puberula* E. Nelson; *P. viridis* E. Nelson var. *compacta* (Brand) Wherry] ❀TRY;DFCLT.

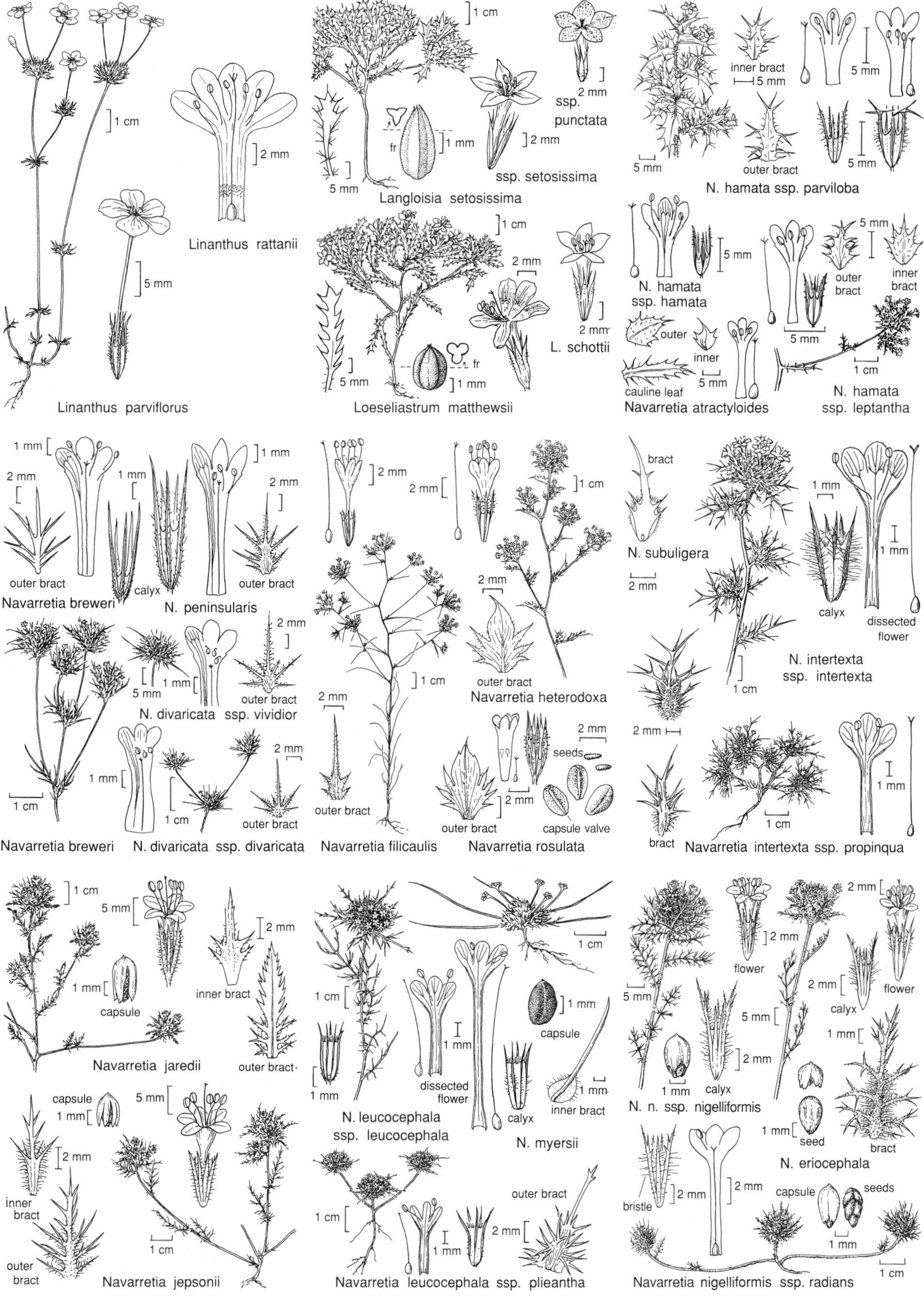
Linanthus parviflorus
Linanthus rattanii
Langloisia setosissima
ssp. punctata
ssp. setosissima
Loeseliastrum matthewsii
L. schottii
N. hamata ssp. parviloba
inner bract
outer bract
N. hamata ssp. hamata
outer bract
inner bract
cauline leaf
Navarretia atractyloides
N. hamata ssp. leptantha
outer bract
Navarretia breweri
calyx
N. peninsularis
N. divaricata ssp. vividior
Navarretia breweri
N. divaricata ssp. divaricata
Navarretia filicaulis
outer bract
Navarretia heterodoxa
seeds
capsule valve
Navarretia rosulata
bract
N. subuligera
calyx
dissected flower
N. intertexta ssp. intertexta
bract
Navarretia intertexta ssp. propinqua
inner bract
capsule
outer bract
Navarretia jaredii
capsule
inner bract
outer bract
Navarretia jepsonii
dissected flower
N. leucocephala ssp. leucocephala
capsule
calyx
inner bract
N. myersii
outer bract
Navarretia leucocephala ssp. plieantha
flower
calyx
N. n. ssp. nigelliformis
seed
N. eriocephala
bristle
capsule
seeds
Navarretia nigelliformis ssp. radians

POLEMONIUM

Dieter H. Wilken

Ann, per. **STS** decumbent to erect, 1–10, 1–10 dm, glandular, hairy. **LVS** pinnately compound, alternate; basal petioled; cauline sessile upward; lflets entire to divided, glabrous to glandular. **INFL** open to head-like. **FL**: calyx bell-shaped, membranous in age but not separated into membrane and ribs, glandular; corolla rotate to funnel-shaped, white to blue or purple; stamens attached at same level, filaments hairy at base. **FR** ovoid to spheric. **SEEDS** 3–10 per chamber, gen 1–3 mm, elliptic to ovate in outline, slightly gelatinous when wet, brown to black. ± 20 spp.: Am, Eurasia. (Greek: perhaps from Polemon, Athenian philosopher, or polemos, strife or war) [Grant 1989 Bot Gaz 150:158–169] Pers gen cross-pollinated, anns self-pollinated.

1. Ann; fls solitary, axillary .... ***P. micranthum***
1′ Per; fls clustered gen at st tips
 2. Lflets deeply 3–5-lobed (appearing ± whorled); infl head-like
  3. Stamens and style exserted, > corolla tube; petiole base membranous .... ***P. chartaceum***
  3′ Stamens and style incl, < corolla tube; petiole base green, not membranous .... ***P. eximium***
 2′ Lflets entire; infl gen open
  4. St erect, < 10 dm; lvs cauline
   5. Corolla gen purple to flesh-pink; fr 6–8 mm .... ***P. carneum***
   5′ Corolla lobes gen blue; fr 3–4 mm .... ***P. occidentale***
  4′ St cespitose, < 3 dm; lvs ± basal
   6. Terminal lflet ± fused with distal pair; corolla 8–15 mm wide; montane .... ***P. californicum***
   6′ Terminal lflet gen free from distal pair; corolla 5–8 mm wide; alpine .... ***P. pulcherrimum***
    7. Corolla white, throat yellow; pl hairy .... var. ***pilosum***
    7′ Corolla blue to purple; pl glandular .... var. ***pulcherrimum***

P. californicum Eastw. (p. 855) Per, cespitose; rhizome gen short. **ST** decumbent to erect, < 3 dm; hairs soft. **LVS** 5–20 mm; lower lflets 15–25; upper lflets 11–19, lanceolate to ovate, ± glabrous; terminal lflet ± fused with most distal pair. **INFL** open; pedicels < 10 mm. **FL**: calyx 4–8 mm, lobes > tube, acute; corolla 7–15 mm wide, bell-shaped, lobes >> tube, tube white, throat blue or yellow, lobes blue to blue-violet; stamens and style exserted. $2n=36$. Open to shaded areas in woodlands; 1600–3100 m. KR, CaRH, n&c SNH; to WA. Self-incompatible. Intergrades with *P. pulcherrimum* at high elevations in SNH. ❀DRN:4–6&IRR, SHD:1–3,7,14–18;DFCLT.

P. carneum A. Gray Per, minutely hairy. **STS** decumbent to erect, 1–3, 5–10 dm. **LVS** cauline, smaller upward; lflets 7–21, 2–4 cm, lanceolate to ovate. **INFL**: cluster, open, umbel-like; fls 3–7; pedicels 2–12 mm. **FL**: calyx 8–20 mm, lobes = tube; corolla 10–25 mm, rotate to bell-shaped, throat + lobes > tube, flesh-pink to purple; style >> stamens. **FR** 6–8 mm. Moist to dry, open areas; < 1800 m. w NW, CCo, SnFrB; to WA. ❀SHD,DRY:5,**6**,**15**,**16**,17.

P. chartaceum H. Mason (p. 855) MASON'S SKY PILOT Per, hairy. **STS** erect, 3–8, < 3 dm. **LVS** basal, narrowly oblong in outline; petioles sheathing, membranous at base, brownish; lflets > 13, < 5 mm, deeply 3–5-lobed; lobes gen elliptic, glandular, appearing as whorled lflets. **INFL** head-like; pedicels < 5 mm. **FL**: calyx 6–8 mm, lobes = tube, rounded; corolla 10–14 mm, funnel-shaped, tube > lobes, white, throat blue or yellow, lobes blue; stamens and style exserted. UNCOMMON. Rocky slopes, talus, 1800–4200 m. KR, SNE (Sweetwater Mtns), W&I. Probably related to both *P. eximium* (c&s SNH) and *P. elegans* E. Greene (WA, B.C.); warrants further study, esp because of interrupted range. ❀DFCLT.

P. eximium E. Greene (p. 855) SKY PILOT Per, hairy. **ST** erect, < 4 dm. **LVS** gen basal, narrowly oblong in outline; petioles sheathing, green, not membranous; lflets > 11, < 6 mm, deeply 3–5-lobed; lobes gen oblanceolate, glandular, appearing as whorled lflets. **INFL** head-like; pedicels < 6 mm. **FL**: calyx 5–10 mm, lobes ± = tube, oblong, rounded; corolla 11–15 mm, funnel-shaped, tube > lobes, white, throat and lobes blue; stamens and style incl. Rocky outcrops, talus; 3000–4200 m. c&s SNH. ❀DFCLT.

P. micranthum Benth. (p. 855) Ann. **ST** prostrate to erect; hairs soft. **LVS** cauline; petioles << blades; lflets 5–15, < 10 mm, ovate to elliptic. **INFL**: fls solitary in axils. **FL**: calyx 3–7 mm, lobes = tube, acute, hairy; corolla rotate, < 5 mm, white to bluish, lobes > tube; stamens and style incl. Open, seasonally wet areas, gen among shrubs; 600–1800 m. KR, CaRH, n SNH, s SnJV, se SCoRO, n WTR, MP; to B.C., MT; also in S.Am.

P. occidentale E. Greene (p. 855) Per; rhizomes short. **ST** erect, 5–10 dm. **LVS** cauline, 1–4 dm; lower with 13–17 lflets; upper with 9–13 lflets; lflets elliptic to lanceolate, green above, glaucous below, glabrous. **INFL** open; pedicels < 5 mm. **FL**: calyx 4–10 mm, lobes = tube, acute; corolla 10–15 mm, bell-shaped, gen blue, throat gen yellow, lobes >> tube; stamens incl; style exserted. **FR** < 4 mm. Moist areas, meadows, woodlands; 900–3300 m. KR, SN, SnBr; to B.C., Colorado. [*P. caeruleum* L. ssp. *amygdalinum* (Wherry) Munz] ❀DRN:4–6&IRR,SHD:1–3,7,14–18.

P. pulcherrimum Hook. Per, cespitose, gen from slender rhizome. **STS** decumbent to erect, 4–10, < 3 dm; hairs soft. **LVS** ± basal, abruptly smaller upward; lflets 9–21, 4–8 mm, ovate to round, glandular, hairs small. **INFL** crowded or head-like; pedicels < 10 mm. **FL**: calyx 4–8 mm, lobes = tube; corolla 5–8 mm wide, rotate to bell-shaped, tube white, throat often yellow, lobes = tube, blue to white; stamens and style ± incl. Subalpine to alpine talus; 2400–3700 m. KR, NCoRH, n&c SNH, MP; to AK, WY.

var. ***pilosum*** (E. Greene) Brand Pl gen < 1 dm; herbage hairy. **FL**: corolla white, throat yellow. Uncommon. Volcanic talus; 2700–3000 m. CaRH, MP; to WA.

var. ***pulcherrimum*** Pl gen 1–2 dm; herbage sparsely hairy and glandular. **FL**: corolla blue to purple. Talus; 2400–3700 m. KR, n&c SNH; to AK, WY. ❀TRY;DFCLT.

POLYGALACEAE MILKWORT FAMILY

Thomas L. Wendt

Ann, per, shrub, tree, vine, some non-green, dependent on fungi for nutrition; hairs unbranched. **LVS** simple, gen alternate (rarely opposite or whorled); veins pinnate; margin gen ± entire; stipules gen 0. **INFL**: raceme, spike, or panicle. **FL** bisexual, gen bilateral (appearing ± pea-fl-like) or ± radial; sepals 5, free or fused, lateral (inner) pair often larger and petal-like (wings); petals 5 or 3, individually fused to stamen tube, ± similar or different with 1 lower keel petal, 2 strap-like upper petals, and 0 or 2 small lateral petals; stamens 3–10, ± fused, tube open on upper side; ovary chambers 1–8, ovule 1 per chamber, style 1 or 0. **FR**: capsule, drupe, or nut, sometimes winged. **SEED** often with aril. 18 genera, 800 spp.: esp trop, subtrop; very few cult. [Blake 1924 N Amer Fl 25:305–379]

POLYGALA MILKWORT

Ann, per, shrub, tree, vine; roots gen with wintergreen odor. **INFL**: raceme or spike, sometimes grouped and panicle-like. **FL** bilateral; lateral 2 sepals enlarged as wings; petals 3 or 5, keel petal often with cylindric beak or fringed crest at tip; stamens 6–8, anthers dehiscent at tip, appearing 1-chambered; nectary disc or gland present; ovary chambers 2, stigma 2-lobed. **FR**: capsule. **SEED** fusiform or ovoid, black, gen hairy, gen with prominent white aril on one end. ± 500 spp.: trop, temp. (Greek: much milk, some Eur spp. said to increase milk flow in cows) [Wendt 1979 J Arnold Arbor 60:504–514]

1. Infl thornless
  2. Fls 2.5–5 mm, ± cleistogamous, keel-petal beakless .... [2]***P. californica***
  2′ Some fls > 7 mm, opening, keel-petal with a beak
    3. Beak of keel-petal oblong, gen notched or contorted, ± 0.7–1 mm diam near tip; aril glabrous; fl 9–14.5 mm (sometimes 2.5–5 mm cleistogamous fls also present near base) .... [2]***P. californica***
    3′ Beak of keel-petal linear, entire, ± 0.2 mm diam near tip; aril hairy; fl 7–14 mm, of one type .... ***P. cornuta***
      4. Fl 8.5–14 mm; wings densely puberulent; upper sepal gen acute to acuminate .... var. ***cornuta***
      4′ Fl 7–11.2 mm; wings glabrous, ciliate, or puberulent near tip; upper sepal gen rounded to obtuse ... var. ***fishiae***

1′ Infl thorn-tipped
  5. Fls 2.5–5.3 mm; wings cream-colored or greenish; twigs densely hairy
    6. Pedicels, outer sepals, and lvs with spreading hairs .... ***P. acanthoclada***
    6′ Pedicels glabrous; outer sepals glabrous, ciliate, or sparsely hairy near tip; lf hairs incurved or appressed .... ***P. intermontana***
  5′ Fls 6–13.5 mm; wings pink; twigs glabrous or puberulent
    7. Beak of keel-petal with 1–2 prominent notches on lower side due to excess tissue; seeds more hairy at aril end .... ***P. heterorhyncha***
    7′ Beak of keel-petal entire or slightly jagged; seeds evenly hairy .... ***P. subspinosa***

P. acanthoclada A. Gray (p. 855) Subshrub, shrub; hairs spreading. **ST** weak and sprawling to erect, gen not highly branched, < 10 dm; twig hairs dense, white. **LF** 5–25 mm, oblanceolate to narrowly elliptic or obovate, hairy. **INFL** thorn-tipped; fls 1–15; pedicels 1.5–5.8 mm, hairy. **FL** 3–5.3 mm; outer sepals hairy, wings cream-colored; beak of keel-petal 0 or minute. **FR** 4–6 mm. n=9. Desert scrub, Joshua-tree or pinyon/juniper woodlands, often loose, sandy or gravelly soils; 900–1700 m. DMoj (Eagle and New York Mtns, Lucerne Valley); to s UT, AZ.

P. californica Nutt. (p. 855) Per, often from rhizomes. **ST** gen decumbent, 0.5–3.5 dm. **LF** 7–60 mm, ovate to obovate. **INFL** thornless. **FL** 9–14.5 mm; sepal wings ciliate, sometimes puberulent near tip, pink or rarely white; beak of keel-petal 1.2–3 mm, lower side notched or contorted (rarely entire), 0.7–1 mm diam near tip, yellow or white when pollen shed; ± cleistogamous fls (2.5–5 mm) often present in separate, gen basal racemes. **FR** 4.5–10.5 mm incl prominent stalk, thin-textured, green; veins prominent. **SEED** 3.5–6 mm incl hairs; aril glabrous. n=9. Exposed slopes, chaparral, forests; 10–1400 m. NW, CW, n ChI; se OR. ❀5,**17**;SHD:**15**, **16**&IRR:7,14,22–24.

P. cornuta Kellogg Per, shrub, often from rhizomes. **ST** prostrate to erect, < 25 dm. **LF** 10–65 mm, linear to ovate. **FL** 7–14 mm; wings ciliate, glabrous to puberulent, cream or greenish to pink; beak of keel-petal 0.5–2.5 mm, entire, ± 0.2 mm diam near tip, dull rose to green when pollen shed. **FR** 5.9–10 mm incl stalk, thick-textured, dark yellow-brown; veins obscure. **SEED** 4.5–7.3 mm incl hairs; aril with shaggy hairs. n=9. Chaparral, open forest, exposed slopes; 100–2100 m. KR, CaR, SN, s SCoRO, TR, PR. Vars. intergrade in s SCoRO, WTR.

var. ***cornuta*** (p. 855) Per, subshrub. **ST** prostrate to erect, 1–10 dm. **LF** often ± 2 × as long as wide. **FL** 8.5–14 mm; upper sepal acute or acuminate (rarely rounded), outer sepals and wings cream or greenish to pink in bud, wings densely puberulent. n=9. Chaparral, pine forests; 100–2100 m. KR, CaR, SN. ❀17;SHD:15,16&IRR:1,2,7,14,18.

var. ***fishiae*** (C. Parry) Jepson (p. 855) FISH'S MILKWORT Shrub, often forming thickets < 2 m diam. **ST** decumbent to erect, 6–25 dm. **LF** > 2 × as long as wide. **FL** 7–11.2 mm; upper sepal rounded or obtuse (rarely acute), outer sepals (at least at tip) and wings dark pink in bud, wings ciliate, glabrous, or puberulent near tip. n=9. UNCOMMON. Chaparral, oak woodland; 100–1100 m. s SCoRO, TR, PR; n Baja CA. [var. *pollardii* Munz, Pollard's milkwort] ❀TRY.

P. heterorhyncha (Barneby) T. Wendt (p. 861) NOTCH-BEAKED MILKWORT Per, subshrub, forming thorny mats 1–2 dm high, < 7 dm diam. **ST** glabrous to puberulent, ± glaucous. **LF** 4–20 mm, ovate, elliptic, or obovate, often ± scabrous; base tapered to rounded. **INFL** strongly thorn-tipped. **FL** 7.5–13.5 mm; sepal wings pink; beak of keel-petal 1.4–4 mm, prominently notched on lower side, yellow. **FR** 4.2–7.8 mm incl stalk; veins prominent. **SEED** 3–4.4 mm, incl hairs; seed body widely elliptic to round in X-section, hairier near aril end; aril glabrous. n=18. RARE. Desert scrub; 900–1600 m. DMtns (Funeral Mtns, Inyo Co); s NV. [*P. subspinosa* var. *h.* Barneby]

P. intermontana T. Wendt (p. 861) Subshrub, shrub, stiffly branched, ± open, < 10 dm high (or forming thorny mats 1.5–5 dm high, gen < 10 dm diam). **ST**: twig hairs dense, white, appressed to irregularly ascending. **LF** 3–25 mm, linear to obovate; hairs incurved or appressed. **INFL** thorn-tipped; fls 1–7; pedicels 2.5–9 mm, glabrous. **FL** 2.5–5.2 mm; outer sepals glabrous or ciliate, sometimes with incurved hairs near tip, wings cream or greenish; beak of keel-petal 0 or minute. **FR** 3.5–5.8 mm. n=9. Pinyon/juniper woodland; 2600–2700 m. SNE (n Mono Co); to UT, n AZ. [*P. acanthoclada* var. *intricata* Eastw.]

P. subspinosa S. Watson (p. 861) SPINY MILKWORT Per, shrub. **ST** gen < 2.5 dm, glabrous or puberulent, sometimes glaucous. **LF** 4–31 mm, obovate or elliptic, not at all scabrous; base long-tapered. **INFL** thorn-tipped, often weakly so. **FL** 6–13 mm; sepal wings pink; beak of keel-petal 1–3 mm, entire or slightly jagged, yellow or greenish. **FR** 5.5–10 mm incl stalk; veins prominent. **SEED** 3.3–4.9 mm, incl hairs; seed body elliptic in X-section, gen evenly hairy throughout; aril glabrous. n=9,18. RARE in CA. Desert scrub, volcanic mesas; 1400–1500 m. MP (Lassen Co); to UT, sw Colorado, nw NM, n AZ.

POLYGONACEAE BUCKWHEAT FAMILY

James C. Hickman

Ann to trees, some dioecious. **ST**: nodes often swollen. **LVS** simple, basal or cauline, alternate, opposite, or whorled, gen entire; stipules 0 or obvious and fused into a gen scarious sheath around st. **INFL**: small cluster, axillary or arrayed in cymes or panicles; involucres sometimes subtending 1–many fls. **FL** gen bisexual, small, ± radial; perianth gen 5–6-lobed, base ± tapered, often jointed to pedicel; stamens 2–9, often in 2 whorls; ovary superior, styles gen 3, gen fused at base. **FR**: achene, gen enclosed by persistent perianth, gen 3-angled, ovoid, and glabrous. 50 genera, 1100 spp.: worldwide, esp n temp; some cult for food (*Fagopyrum*; *Rheum*, rhubarb; *Rumex*, sorrel) or orn (*Antigonon*, coral-vine; *Muehlenbeckia*; *Polygonum*). [Ronse Decraene & Akeroyd 1988 Bot J Linn Soc 98:321–371; Reveal et al. 1989 Phytologia 66(2–4):83–414] Treatments of the 15 eriogonoid genera are based on the monographic work of James L. Reveal, who is gratefully acknowledged.

1. Stipule obvious, sheathing st; lvs cauline (or some basal); node ± swollen; fls often in clusters but not involucred (Polygonoideae)
 2. Perianth lobes 5
  3. Fl unisexual; perianth in fr woody, spiny — uncommon ann weed ..... **[2]EMEX**
  3′ Fl bisexual; perianth in fr fleshy or membranous
   4. Herb or shrub, st not twining nor vine-like; native or alien
    5. Lf blade triangular-cordate; fr much exserted — ann with hollow st; naturalized seed crop, SCo ..... **FAGOPYRUM**
    5′ Lf blade lanceolate to round, not triangular-cordate; fr not or little exserted ..... **[4]POLYGONUM**
   4′ Vine, st twining or twisting, gen woody; alien garden escape or weed
    6. Fl ± 2 mm, perianth in fr fleshy, appearing berry-like ..... **MUEHLENBECKIA**
    6′ Fl 4–10 mm, perianth in fr dry, ± keeled to winged ..... **[4]POLYGONUM**
 2′ Perianth lobes 6 or 4 (sometimes 5 in staminate fl of *Emex*)
  7. Twining vine, sometimes woody; midrib of outer perianth lobes in fr ± keeled to winged ..... **[4]POLYGONUM**
  7′ Not twining, not woody; midrib of outer perianth lobes in fr not keeled or winged
   8. Fl parts in 2's
    9. Lvs ± basal, reniform; fr conspicuously winged all around; alpine per ..... **OXYRIA**
    9′ Lvs cauline, lanceolate; fr not winged; lowland ann ..... **[4]POLYGONUM**
   8′ Fl parts in 3's
    10. Outer perianth lobes becoming coarse spines in fr, inner remaining small; uncommon ann weed ..... **[2]EMEX**
    10′ Outer perianth lobes in fr inconspicuous, inner enlarged, covering fr; common, gen per ..... **RUMEX**
1′ Stipule 0; lvs gen basal (pl scapose); node gen not swollen; fls gen in involucres of ± fused bracts (Eriogonoideae)
 11. Lvs cauline, opposite, often deeply notched, appearing 2-lobed; involucral bracts in fr large, scarious, 2-winged, net-veined — sprawling ann ..... **PTEROSTEGIA**
 11′ Lvs gen basal or alternate, not deeply notched so as to appear 2-lobed; involucre ± tubular (often spiny) or 0
  12. Involucral bracts, if any, essentially free
   13. Shrub ..... **DEDECKERA**
   13′ Ann
    14. Lf blade obovate to round, petiole obvious; perianth yellow; stamens 9
     15. Perianth glabrous to sparsely hairy; involucral bracts 0; lf-like bracts 3 per node ..... **GILMANIA**
     15′ Perianth hairy, esp at base; involucral bracts 5, linear-lanceolate; lf-like bracts opposite ..... **GOODMANIA**
    14′ Lf blade linear to oblanceolate, petiole often indistinct; perianth white, pink, or green (sometimes yellow-tinged); stamens sometimes 3 or 6
     16. Involucre 0; perianth involucre-like, leathery, sparsely hairy, lobes hooked; stamens 3 ..... **LASTARRIAEA**
     16′ Involucral bracts and perianth lobes blunt or straight, bracts woolly; fls 2–many per involucre, perianth texture softer
      17. Involucral bracts 3, awns straight; perianth lobes densely woolly, tips long-acuminate; fls 2 per involucre, greenish; stamens 6 or 9 ..... **HOLLISTERIA**
      17′ Involucral bracts many, blunt and unawned; perianth lobes glabrous, tips round; fls many per involucre, pink or yellow-tinged; stamens 3 ..... **NEMACAULIS**
  12′ Involucral bracts obviously fused (even if only at base), forming a funnel- or tube-like involucre
   18. Involucral bracts (lobes or teeth on involucral tube), if any, acute to rounded ..... **ERIOGONUM**
   18′ Involucral bracts awned
    19. Involucre ± funnel-shaped, narrowly to widely flaring; awns 4–36; fls 2–20 per involucre
     20. Involucral awns 4–36, sharp; infl bracts 3 per node or fused into one 3-lobed bract ± surrounding st ..... **OXYTHECA**
     20′ Involucral awns 5, dull; infl bract 1 per node, deeply 3-lobed, on 1 side of st — uncommon, SCoR ..... **ARISTOCAPSA**
    19′ Involucre cylindric or bell-shaped, flaring or not; awns 3–6; fls 1–3 per involucre
     21. Bracts 2–3 per infl node; fl 1 per involucre (involucre thus might be confused with an outer perianth whorl) ..... **CHORIZANTHE**
     21′ Bract 1 per infl node; fls gen 2–3 per involucre

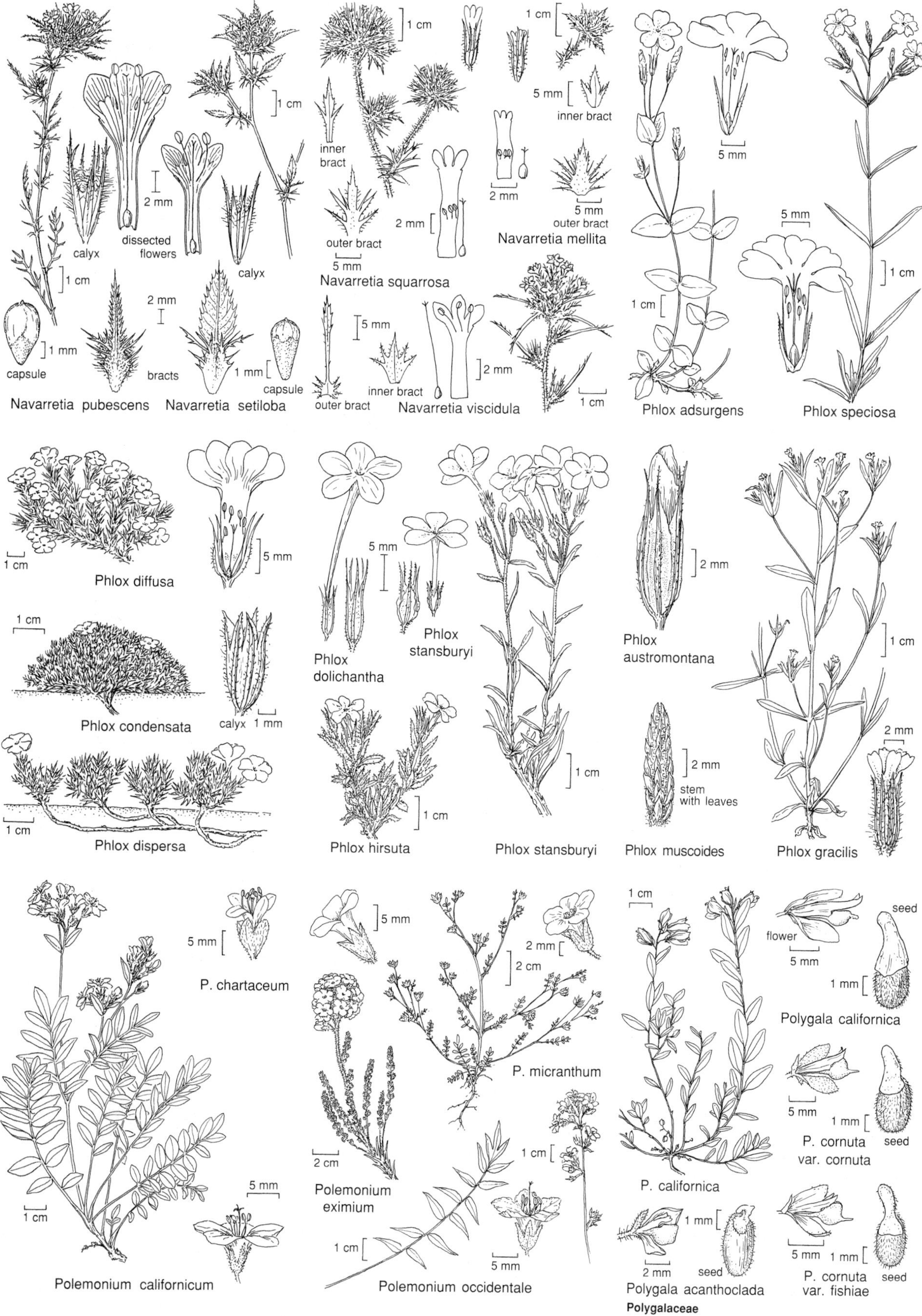

Polygalaceae

22. Involucre awned at base and tip
  23. Awns at involucre base 3; fls 2 per involucre ..... **CENTROSTEGIA**
  23′ Awns at involucre base 6; fls 3 per involucre ..... **DODECAHEMA**
22′ Involucre awned at tip only
  24. Bracts at infl nodes large, conspicuous; fl hairy ..... **MUCRONEA**
  24′ Bracts at infl nodes small, obscure; fl glabrous ..... **SYSTENOTHECA**

ARISTOCAPSA

1 sp. (Latin: awned box, from involucre) [Reveal & Hardham 1989 Phytologia 66:83–88]

A. insignis (Curran) Rev. & Hardham (p. 861) INDIAN VALLEY SPINE-FLOWER Ann, spreading, 2–10 cm, glandular. **LVS** basal; stipule 0; petiole 0; blade 3–15 mm, oblanceolate, glabrous. **INFL** open, 3–10 cm diam; bract 1 per node, 1–4 mm, 3-lobed, awns 0.2–2 mm; involucre 1 per node, ± sessile, 3–5 mm, narrowly funnel-shaped, teeth 5, awns 1–3 mm, straight, spreading; fls 4–6 per involucre. **FL**: perianth 1.5–2 mm, white to pink or rose, sparsely hairy, lobes 6; stamens 9. **FR** 1.5–1.8 mm, brown. *n*=14. UNCOMMON. Foothill woodland; 300–600 m. SCoRI (Monterey, San Luis Obispo cos.). [*Chorizanthe i.* Curran; *Centrostegia i.* (Curran) A.A. Heller]

CENTROSTEGIA

1 sp. (Greek: spurred cover, from involucre) [Reveal 1989 Phytologia 66:199–220]

C. thurberi Benth. (p. 861) THURBER'S SPINEFLOWER Ann, spreading, 3–30 cm, glandular. **LVS** basal; stipule 0; ± petioled; blade 5–40 mm, oblanceolate, glabrous. **INFL** 6–50 cm diam; bract 1 per node, 1–10 mm, 3-lobed, awns 1–2 mm; involucre 1 per node, sessile, 2–8 mm, cylindric above, swollen at base, upper awns 5, < 1 mm, basal awns 3, < 2 mm, curved; fls 2 per involucre. **FL**: perianth 2–3.5 mm, white to pink, hairy, lobes 6; stamens 9. **FR** 2–2.5 mm, brown. *n*=19. Common. Desert scrub, other dry sandy places; 300–2400 m. s SnJV, e SCoRI, TR, SNE, D; to NV, UT, AZ, nw Mex. [*Chorizanthe t.* (Benth.) S. Watson]

CHORIZANTHE SPINEFLOWER

Ann, per, glabrous or hairy, sometimes glandular. **ST** gen scapose (made up of infl axes). **LVS** basal (rarely some cauline); stipule 0; blade gen ± oblanceolate. **INFL** open or of few heads, sometimes 1-sided; bracts gen opposite, lf-like to scale-like; involucres 1–several per axil, sessile, tube cylindric to bell-shaped, gen ± cross-ridged or net-veined, bracts (and ribs) 3–6, awns straight or hooked; fls 1–2 per involucre. **FL**: perianth white to red or yellow, lobes 6, entire to fringed or toothed; stamens 3–9. **FR** 1.5–4.5 mm, gen ± brown, glabrous. 50 spp.: temp w N.Am, sw S.Am. (Greek: divided fl, from perianth) [Reveal & Hardham 1989 Phytologia 66(2):98–198]

1. Involucral bracts and awns (or ribs) 3–5
 2. Involucral bracts and awns 3
  3. Involucral tube cylindric, prominently cross-ridged, glabrous; stamens 6; D ..... *C. corrugata*
  3′ Involucral tube 3-angled, ± bell-shaped or flaring, smooth to moderately cross-ridged, hairy; stamens 9; SW, SNE, D
   4. Pl erect; awns of infl and involucral bracts straight — SNE, D ..... *C. rigida*
   4′ Pl prostrate to decumbent; infl bracts unawned, involucral awns hooked
    5. Involucral tube very narrowly flaring, smooth, ± brown; SCo ..... *C. orcuttiana*
    5′ Involucral tube bell-shaped, thinly hairy, reddish; PR ..... [2]*C. polygonoides* var. *longispina*
 2′ Involucral bracts and awns 4–5
  6. Awns straight; perianth 2.5–3.5 mm, glabrous; stamens attached at base of perianth tube; DMoj ..... *C. spinosa*
  6′ Awns hooked; perianth 1.5–2.5 mm, hairy; stamens attached near top of perianth tube
   7. Involucral tube cylindric, 3–4.5 mm, largest bract ± lf-like; perianth yellow, thinly hairy; GB, DMoj, n WTR ..... *C. watsonii*
   7′ Involucral tube bell-shaped, 1.5–2.5 mm, bracts not lf-like; perianth white to rose, densely hairy; CA-FP ..... *C. polygonoides*
    8. Involucre (incl awn) 3–4 mm; PR ..... [2]var. *longispina*
    8′ Involucre (incl awn) ± 5 mm; n CA ..... var. *polygonoides*
1′ Involucral bracts and awns (or ribs) 6
 9. Involucral bract margin scarious or membranous
  10. Involucral bract margin membranous, continuous across sinuses (involucre thus mimicking a fused corolla); true cauline lvs (no infl or involucre in axil) gen present; gen inland CA-FP
   11. Cauline lvs (and lower bracts) alternate; involucre ± woolly or becoming glabrous, bracts ± equal; perianth 1.5–3 mm, densely hairy — CA-FP ..... *C. membranacea*
   11′ Whorl of true cauline lvs present near mid-st in most pls; involucre bristly, bracts equal or not; perianth 3.5–5 mm, sparsely hairy
    12. 3 involucral bracts >> alternating 3, membrane purple; SCoR ..... *C. douglasii*
    12′ 6 involucral bracts ± equal, membrane white; foothills around n&e GV ..... [2]*C. stellulata*
  10′ Involucral bract margin scarious, not continuous across sinuses (exc sometimes between pairs of erect bracts); true cauline lvs gen 0; ± coastal CW, NCo (exc *C. stellulata*) (*C. pungens* complex)
   13. Involucral awns straight
    14. Pl gen decumbent; involucral bracts spreading, awns brown; perianth 3–4.5 mm; c NCo ..... *C. howellii*
    14′ Pl gen ± erect; involucral bracts erect, awns ivory or straw-colored; perianth 4–6 mm; n CCo ..... [2]*C. valida*

13′ Involucral awns hooked
15. Perianth glabrous, lobes white, tube yellow .... ***C. diffusa***
15′ Perianth hairy, lobes same color as tube
16. Pl gen erect; whorl of true cauline lvs often present; involucral tube 3–4 mm, bracts clearly margined
17. Pl decumbent to erect, soft-hairy; involucre thinly hairy; perianth 2.5–4 mm; n CCo, s SnFrB .. ***C. robusta***
17′ Pl strictly erect, coarse-hairy; involucre bristly; perianth 4–5 mm; foothills around n GV .... [2]***C. stellulata***
16′ Pl prostrate to ascending; cauline lvs 0; involucral tube 1.5–3 mm, bract margins sometimes indistinct
18. Involucral bracts obviously white- to purple-margined, tube 2–3 mm; stamens 9 .... ***C. pungens***
18′ Involucral bract margins thin, ± indistinct; involucral tube 1.5–2.5 mm; stamens 3–9
19. Perianth lobes barely exserted, entire to jagged (central tooth, if any, gen not greatly predominant); stamens 3–9; s CCo .... [2]***C. angustifolia***
19′ Perianth lobes well exserted, with an abrupt central tooth (if teeth > 1, central greatly predominant); stamens 9; n CCo, SnFrB .... [2]***C. cuspidata***
9′ Involucral bract margin not scarious or membranous
20. Involucral awns unequal, longest gen > 2 × length of next longest in some or all involucres
21. Longest involucral awn hooked, ± 2 × length of next longest — NW, CW, s SN .... ***C. clevelandii***
21′ Longest involucral awn straight, gen 3–5 × length of next longest
22. Longest involucral awn ± 5 × length of next longest — e CW, s SN, n WTR .... ***C. uniaristata***
22′ Longest involucral awn ± 3 × length of next longest
23. Involucres (incl awns) gen 5–7 mm, scattered in small clusters; SCoRO .... ***C. rectispina***
23′ Involucres gen 8–10 mm, clumped in large clusters; SCoRI .... ***C. ventricosa***
20′ Involucral awns equal or of alternating lengths, longest gen < 2 × length of next longest
24. Inner perianth lobes either rather deeply fringed or notched (may require dissection of involucre to see)
25. Outer perianth lobes fringed along sides; SW .... ***C. fimbriata***
26. Fringe segments, esp lobe tip, wide, ± lobe-like; s SCo, w PR .... var. ***fimbriata***
26′ Fringe segments, incl lobe tip, thin, not lobe-like; e PR .... var. ***laciniata***
25′ Outer perianth lobes entire or tip notched or fringed; SCoR (*C. palmeri* complex)
27. Perianth lobes white to pinkish
28. Inner perianth lobes notched, 5–6 mm; pl yellowish green; WTR (Sierra Madre Mtns, Santa Barbara Co.) .... ***C. blakleyi***
28′ Inner perianth lobes fringed, 4–5 mm; pl reddish or greenish; SCoRO .... ***C. obovata***
27′ Perianth lobes red to purple
29. Outer perianth lobes entire, erect; on serpentine; SCoRO .... ***C. palmeri***
29′ Outer perianth lobes ± notched, spreading to recurved; not on serpentine; SCoR .... ***C. biloba***
30. Outer perianth lobes deeply notched; SCoR .... var. ***biloba***
30′ Outer perianth lobes shallowly notched; SCoRI .... var. ***immemora***
24′ Inner perianth lobes entire, with a central tooth, or shallowly jagged (may require dissection to see)
31. Perianth lobes all equal in length and width
32. Pl erect; infl breaking at nodes; stamens 3, attached near top of perianth tube .... ***C. brevicornu***
33. Lvs ± linear; gen D .... var. ***brevicornu***
33′ Lvs oblanceolate; SNE .... var. ***spathulata***
32′ Pl prostrate to decumbent; infl not breaking at nodes; stamens 3–9, attached at base of perianth tube; w CA-FP
34. Perianth lobes entire; pl thinly hairy, yellowish green, when fresh; filaments fused at base; SW .... ***C. procumbens***
34′ Perianth lobes ± jagged or with a central tooth; pl densely hairy, brown or gray; filaments free; CCo
35. Perianth lobes barely exserted, entire to jagged (central tooth, if any, gen not greatly predominant); stamens 6–9; s CCo .... [2]***C. angustifolia***
35′ Perianth lobes well exserted, with an abrupt central tooth (if teeth > 1, central greatly predominant); stamens 9; n CCo, SnFrB .... [2]***C. cuspidata***
31′ Perianth lobes unequal, inner gen < and narrow than outer
36. Most involucral awns straight; perianth lobes sometimes ± jagged at tip
37. Pl decumbent; involucre 1.5–2 mm, bracts spreading; SCo .... ***C. parryi*** var. ***fernandina***
37′ Pl ± erect; involucral tube 3–4.5 mm, bracts erect; s NCo .... [2]***C. valida***
36′ Involucral awns hooked; perianth lobes gen entire
38. Pl grayish, canescent; involucral tube 1.5–2 mm — TR .... ***C. parryi*** var. ***parryi***
38′ Pl reddish or greenish, thinly hairy; involucral tube 2.5–6 mm (*C. staticoides* complex)
39. Lower bracts ± linear, not very lf-like, deciduous
40. Perianth well exserted; involucres loosely clustered; e PR .... ***C. leptotheca***
40′ Perianth moderately exserted; involucres densely clustered; s CW, SW .... ***C. staticoides***
39′ Lower bracts ± oblanceolate, ± lf-like, persistent
41. Involucral tube 3.5–5 mm, cylindric; not on serpentine .... ***C. xanti***
42. Involucre densely hairy; e SnBr, n SnJt .... var. ***leucotheca***
42′ Involucre thinly hairy; s SNF, SCoRI, n border TR, SNE .... var. ***xanti***
41′ Involucral tube gen 2–3 mm, ± bell-shaped; on serpentine or not
43. Infl clusters ± open; involucral tube gen 2.5–3 mm; stamens 9; serpentine; s SCoRO .... ***C. breweri***
43′ Infl clusters head-like; involucral tube gen 2-2.5 mm; stamens 6; non-serpentine; n ChI ... ***C. wheeleri***

C. angustifolia Nutt. **ST** prostrate to decumbent, 3–10 cm, soft-hairy. **LF**: blade 5–50 mm. **INFL**: involucral tube 1.5–2.5 mm, cylindric, bracts 6, margins sometimes thinly scarious, pink, awns ± equal, hooked. **FL**: perianth 2–3 mm, barely exserted, white to rose, hairy, lobes ± jagged at tip (central tooth, if any, gen not predominant); stamens 3–9. *n*=20. Sandy places; < 200 m. s CCo (San Luis Obispo, Santa Barbara cos.). [var. *eastwoodiae* Goodman] Closely related to *C. pungens*.

C. biloba Goodman **ST** erect, 5–40 cm. **LF**: blade 10–50 mm. **INFL**: involucral tube 4–6 mm, urn-shaped, strigose, bracts 6, 1–2 mm, erect to spreading, awns hooked. **FL**: perianth 4.5–6 mm, sparsely hairy, tube white to yellow, lobes red to purple, outer lobes spreading, notched, inner lobes erect, fringed; stamens 9. Uncommon. Non-serpentine dry woodland, chaparral; 200–800 m. SCoR. Closely related to *C. palmeri*.

var. ***biloba*** **FL**: outer perianth lobes deeply notched, sometimes jagged. *n*=20. Habitat of sp.; 200–700 m. e SCoRO, w SCoRI (e SCoRI near Parkfield Grade, w Fresno Co.). [*C. palmeri* S. Watson var. *b.* (Goodman) Munz] ❀TRY.

var. ***immemora*** Rev. & Hardham HERNANDEZ SPINE-FLOWER **FL**: outer perianth lobes shallowly notched. *n*=18–22. Very uncommon. Habitat of sp.; 600–800 m. e SCoRI (se San Benito, adjacent Monterey cos). Reasonable candidate for legal protection. ❀TRY.

C. blakleyi Hardham (p. 861) BLAKLEY'S SPINEFLOWER **ST** gen ascending, 5–15 cm, yellowish green, spreading-hairy. **LF**: blade 5–25 mm. **INFL**: involucral tube 3–4.5 mm, urn-shaped, thinly hairy, bracts 6, longest < 2 × length of next longest and with straight or curved awn, other awns hooked. **FL**: perianth 5–6 mm, white to pinkish, sparsely hairy, lobes erect, deeply notched; stamens 9. *n*=±19. UNCOMMON. Chaparral; 600–1600 m. WTR (n slope, Sierra Madre Mtns, Santa Barbara Co.). Closely related to *C. palmeri*. ❀TRY.

C. brevicornu Torrey BRITTLE SPINEFLOWER **ST** ± erect, 5–50 cm, thinly hairy. **LF**: blade 10–40 mm, ± linear to oblanceolate. **INFL** breaking at nodes; involucral tube 3–5 mm, cylindric, green, thinly strigose, bracts 6, awns hooked. **FL**: perianth 2–4 mm, gen white, glabrous, lobes equal, ± linear, entire; stamens 3, attached near top of perianth tube. Common. Desert scrub, sagebrush scrub, juniper woodland; < 3000 m. SNE, D; to se OR, s ID, w UT, AZ, nw Mex.

var. ***brevicornu*** (p. 861) **ST** 5–50 cm. **LF**: blade 15–40 mm, ± linear; petiole indistinct. **INFL**: involucral tube prominently ribbed. *n*=19–21,23. Desert scrub; < 2300 m. s SNE, D; s NV, UT, AZ, nw Mex.

var. ***spathulata*** (Rydb.) C. Hitchc. **ST** 5–30 cm. **LF**: blade 10–20 mm, oblanceolate; petiole distinct. **INFL**: involucral tube obscurely ribbed. *n*=19. Sagebrush scrub, pinyon/juniper woodland; 700–3000 m. SNE; to se OR, s ID. [ssp. *s.* (Rydb.) Munz]

C. breweri S. Watson (p. 861) BREWER'S SPINEFLOWER **ST** decumbent to ascending, 3–35 cm, reddish, thinly hairy. **LF**: blade 5–20 mm, gen ovate, thinly hairy to tomentose below. **INFL**: lower bracts ± oblanceolate, ± lf-like, persistent, awns straight; involucres in ± open clusters, tube gen 2.5–3 mm, ± bell-shaped, hairs slender, curly, bracts 6, awns hooked. **FL**: perianth 3–3.5 mm, white to red, hairy, lobes unequal, gen entire; stamens 9. *n*=19. RARE. Chaparral, foothill woodland, on serpentine; < 800 m. SCoRO (sw San Luis Obispo Co.). Closely related to *C. staticoides*.

C. clevelandii C. Parry (p. 861) **ST** ± decumbent, 2–10 cm, grayish, hairy. **LF**: blade 5–20 mm. **INFL**: involucral tube 3–3.5 mm, urn-shaped, densely hairy, bracts 6, longest gen ± 2 × length of next longest, awns hooked. **FL**: perianth 2.5–3 mm, white, sparsely hairy; perianth lobes entire to slightly notched or jagged; stamens 3. *n*=21. Common. Chaparral, dry woodland; 400–2100 m. NW, CW, s SN, Teh, n WTR.

C. corrugata (Torrey) Torrey & A. Gray (p. 861) **ST** erect, 3–50 cm, thinly tomentose. **LF**: blade 5–20 mm, ovate to rounded. **INFL**: involucral tube 3–4 mm, cylindric, green to tan, prominently cross-ridged, ± glabrous, bracts 3, 2–4.5 mm, gen lanceolate, awns hooked. **FL**: perianth 2–2.5 mm, white, thinly hairy; stamens 6, attached near top of perianth tube. *n*=19. Common. Dry soils in desert scrub; –70–920 m. D; s NV, w AZ, nw Mex.

C. cuspidata S. Watson (p. 861) **ST** ± decumbent, 5–50 cm, soft-hairy. **LF**: blade 5–50 mm. **INFL**: involucral tube 1.5–3 mm, cylindric, bracts 6, margins sometimes thinly scarious, white to pink, awns ± equal, hooked or straight. **FL**: perianth 2–3 mm, well exserted, white to rose, hairy, lobes with an abrupt, predominating, central tooth; stamens 9. *n*=20. Sandy places; < 250 m. n CCo, SnFrB. Pls from Pt. Reyes to Bodega Head tend to have woollier involucre, no scarious bract margin, and straight involucral awns; they may be called var. *villosa* (Eastw.) Munz. Closely related to *C. pungens*.

C. diffusa Benth. **ST** prostrate to erect, 3–15 cm, soft-hairy. **LF**: blade 3–20 mm. **INFL**: involucral tube 2–2.5 mm, narrowly bell-shaped, bracts 6, of 2 lengths, scarious margins thin (n end of range) to conspicuous (s end of range), white, extending nearly full length of awn, awns hooked. **FL**: perianth 2.5–3 mm, glabrous, tube lemon-yellow, lobes white, entire to jagged or with a minute central tooth; stamens 3–9. *n*=20. Coastal scrub and forest; < 800 m. CCo, w SnFrB, w SCoRO. Closely related to *C. pungens*.

C. douglasii Benth. (p. 861) DOUGLAS' SPINEFLOWER **ST** erect, 10–50 cm, soft-hairy. **LF**: basal blades 5–40 mm; whorl of cauline lvs gen present near mid-st. **INFL**: bract awns straight; involucral tube 3–5 mm, cylindric, hairs long, ± soft, bracts 6 (3 larger), membranous margins purple, extending fully across sinuses (involucre appears ± equilaterally triangular from above), awns hooked. **FL**: perianth 3.5–4.5 mm, white to rose, slightly hairy, lobes with a minute tooth or shallow notch; stamens 9. *n*=20. UNCOMMON. Foothill woodland, pine forest; 200–1600 m. e SCoRO, SCoRI. ❀TRY.

C. fimbriata Nutt. FRINGED SPINEFLOWER **ST** decumbent to erect, 10–50 cm, hairy, glandular. **LF**: blade 5–35 mm, gen obovate, thinly hairy. **INFL**: involucral tube 4–7 mm, cylindric, hairy, bract 6, awns straight. **FL**: perianth 6–10 mm, glabrous, tube gen yellow, lobes gen rose, fringed, esp along sides; stamens 9. Common. Chaparral, coastal scrub; < 1600 m. PR; n Baja CA. The vars. grow mixed in s San Diego Co., with some individuals not placeable.

var. ***fimbriata*** (p. 861) **FL**: perianth 6–8 mm, fringe segments on lobe sides widely linear and lobe-like, lobe tip irregular in outline, ± oblong, rounded. *n*=20(22–23). Habitat of sp. w PR; n Baja CA.

var. ***laciniata*** (Torrey) Jepson (p. 861) **FL**: perianth 8–10 mm, fringe segments on lobe sides narrowly linear and ± hair-like, lobe tip long-acuminate. *n*=20. Habitat of sp. e PR; n Baja CA.

C. howellii Goodman (p. 861) HOWELL'S SPINEFLOWER **ST** ± decumbent, 3–10 cm, soft-hairy. **LF**: blade 10–30 mm, gen obovate. **INFL**: involucral tube 3–4 mm, cylindric, hairy, bracts 6, spreading, scarious margins white, not continuous across sinuses, awns straight, brown. **FL**: perianth 3–4.5 mm, white to rose, hairy, lobes jagged or with a minute central tooth; stamens 9. *n*=40. **THREATENED** CA. Sand dunes; < 15 m. c NCo (c Mendocino Co.). Threatened by recreation, alien pls. Closely related to *C. pungens*.

C. leptotheca Goodman **ST** ± erect, 5–35 cm, reddish, thinly hairy. **LF**: blade 5–30 mm, oblong to narrowly ovate, thinly hairy above, tomentose below. **INFL**: lower bracts ± linear, not very lf-like, soon deciduous, awns straight; involucres loosely clustered, tube 3–4 mm, cylindric, smooth, hairs slender, curly, bracts 6, awns hooked. **FL**: perianth 4.5–6 mm, well exserted, rose to red, hairy, lobes unequal, gen entire; stamens 9. *n*=19. Chaparral or pine forest, often on granite; 300–1900 m. e PR; n Baja CA. Closely related to and hard to distinguish from *C. staticoides*. ❀TRY.

C. membranacea Benth. (p. 861) PINK SPINEFLOWER **ST** erect, 10–100 cm, ± woolly. **LF**: basal blade 1–5 cm, ± linear; some cauline, alternate. **INFL**: involucres few in axils of bract-like lvs below, in dense heads above, tube 3–4 mm, urn-shaped, woolly to ± glabrous, bracts 6, ± equal, membranous margin white, shallower in sinuses, awns hooked. **FL**: perianth 1.5–3 mm, white to rose, densely hairy; stamens 9. *n*=19–21,40–42. Common. Grassland, chaparral, foothill woodland, etc.; < 1600 m. NW, CaR, SNF, Teh, GV, CW, n WTR; s OR. ❀DRN,DRY,SUN:7,8,**9**,10,14–24.

C. obovata Goodman **ST** prostrate to erect, 5–40 cm, reddish or greenish gray, spreading-hairy. **LF**: blade 5–25 mm. **INFL**: involucral tube 3–4 mm, urn-shaped, hairy, bracts 6, erect to spreading, awns hooked, exc often the longest. **FL**: perianth 4–5 mm, white to pinkish, sparsely hairy, outer lobes spreading, obovate to round, often slightly notched, inner lobes erect, fringed; stamens 9. *n*=19, 20. Chaparral, dry woodlands, in sand or calcareous soil; < 1300 m. s SCoRO. Closely related to *C. palmeri*. ❀TRY.

C. orcuttiana C. Parry (p. 861) ORCUTT'S SPINEFLOWER **ST** prostrate, 1–15 cm, yellowish, hairy. **LF**: blade 5–15 mm, narrowly oblanceolate, thinly hairy. **INFL**: involucral tube ± 2 mm, 3-angled, bracts 3, ± 2 mm, thick, spreading, awns hooked. **FL**: perianth 1.5–1.8 mm, incl, yellow, densely hairy; stamens 9, attached near top of perianth tube. *n*=38–40. **ENDANGERED** CA, **PRESUMED EXTINCT**. Coastal scrub; 60–125 m. s SCo (Del Mar to Point Loma, San Diego Co.). Last known habitat has been developed.

C. palmeri S. Watson (p. 861) **ST** ± erect, 5–40 cm, reddish, hairy. **LF**: blade 10–30 mm. **INFL**: involucral tube 3.5–4 mm, urn-shaped, thinly hairy, bracts 6, erect to spreading, awns hooked. **FL**: perianth 4–5 mm, gen glabrous, tube white to yellow, lobes erect, red to purple, outer ± round, entire, inner fringed; stamens 9. *n*=19, 20. Uncommon. Serpentine outcrops; < 700 m. SCoRO (w Monterey, w San Luis Obispo cos.). Isolated populations show local differences. ❀DRN,DRY,SUN:7,9,14–17,20–24;DFCLT.

C. parryi S. Watson **ST** prostrate to ascending, 2–30 cm, strigose. **LF**: blade 5–40 mm, oblanceolate to oblong. **INFL**: involucral tube 1.5–2 mm, urn-shaped, canescent, bracts 6, awns straight or hooked. **FL**: perianth 2.5–3 mm, white, sparsely hairy, lobes equal or not, gen nearly entire; stamens 9. Uncommon. Sandy places, gen in coastal or desert scrub; 200–1200 m. c&e SCo, e TR, nwedge DSon.

var. ***fernandina*** (S. Watson) Jepson (p. 861) SAN FERNANDO VALLEY SPINEFLOWER **INFL**: involucral awns straight. **FL**: perianth lobes ± equal. **PRESUMED EXTINCT**. Habitat of sp.; 200–350 m. SCo, e WTR, SnGb. Extirpated from Los Angeles Basin; most likely to be found near Elizabeth Lake, n edge e WTR or SnGb.

var. ***parryi*** **INFL**: involucral awns hooked. **FL**: perianth lobes unequal. Habitat of sp.; 300–1200 m. c&e SCo, e TR, nw edge DSon.

C. polygonoides Torrey & A. Gray KNOTWEED SPINEFLOWER **ST** prostrate, 1–15 cm, greenish or reddish, soft-hairy. **LF**: blade 3–10 mm, thinly hairy. **INFL**: involucral tube 1.5–2.5 mm, bell-shaped, 3-angled, often reddish, thinly hairy, prominent bracts 3, 1.5–3 mm, spreading, awns hooked (0–2 smaller bracts between larger), none lf-like. **FL**: perianth 1.5–2 mm, white to rose, densely hairy; stamens 9, attached near top of perianth tube. *n*=20. Diverse sandy to gravelly places; < 1400 m. CA-FP; n Baja CA.

var. ***longispina*** (Goodman) Munz (p. 861) Pl gen reddish. **INFL**: involucre (incl awns) gen 3–4 mm, tube 1.5–2 mm, bracts sometimes only 3, 2–3 mm incl awns. Chaparral; < 1400 m. PR; n Baja CA. ❀TRY.

var. ***polygonoides*** (p. 861) Pl gen greenish. **INFL**: involucre (incl awns) gen ± 5 mm, tube 2–3 mm, bracts 1.5–2 mm incl awns. Habitat of sp. NW, CaR, n SNF, ScV, CW.

C. procumbens Nutt. (p. 861) PROSTRATE SPINEFLOWER **ST** prostrate to decumbent, 2–25 cm, yellowish green, thinly hairy. **LF**: blade 5–40 mm. **INFL**: involucral tube 1.5–3 mm, cylindric to narrowly bell-shaped, bracts 6, gen spreading, awns recurved to hooked, **FL**: perianth 1.7–3 mm, white or yellow, hairy, lobes equal, entire; stamens 9, attached at base of perianth tube, filaments fused at base, hairy. *n*=19–23. Common. Coastal scrub, chaparral; < 800 m. c&s SCo, s TR, w PR; Baja CA. [var. *albiflora* Goodman UNCOMMON, var. *mexicana* Goodman; *C. jonesiana* Goodman] Highly variable; populations vary in fl color.

C. pungens Benth. (p. 861) **ST** prostrate to erect, 5–50 cm, grayish or reddish, soft-hairy. **LF**: blade 5–70 mm. **INFL**: involucral tube 2–3 mm, cylindric, bracts 6, scarious margins conspicuous, white to purple, awns ± equal, hooked. **FL**: perianth 2–3.5 mm, white to rose, hairy, lobes ± jagged at tip; stamens 9. *n*=20. Coastal strand, scrub, etc.; < 450 m. n&c CCo, SnFrB. The earliest name in a variable and confusing complex with many local forms (see key). More erect pls with pink to purple involucral margins have been called var. *hartwegiana* Rev. & Hardham.

C. rectispina Goodman (p. 861) STRAIGHT-AWNED SPINEFLOWER **ST** ± decumbent, 3–25 cm, grayish, hairy. **LF**: blade 5–20 mm. **INFL**: involucre (incl awns) gen 5–7 mm, tube 2–3 mm, urn-shaped, densely hairy, bracts 6, longest ± 3 × length of next longest, with a straight awn, awns of others hooked. **FL**: perianth 3.5–4 mm, sparsely hairy, tube yellow, outer lobes spreading, white, often slightly notched, inner lobes erect, yellow, often jagged; stamens 9. *n*=20. RARE. Chaparral, dry woodland; 200–600 m. SCoRO (s Monterey, San Luis Obispo cos.).

C. rigida (Torrey) Torrey & A. Gray (p. 861) SPINY-HERB **ST** erect, densely branched, 2–15 cm, greenish, hairy. **LF** long-petioled; blade gen 5–10 mm, oblanceolate to elliptic, hairy; rosette withering early. **INFL**: main bracts like lvs but larger (blade gen 10–35 mm) and wider, later bracts many, spine-like, < 30 mm; involucral tube 2–3 mm, funnel-shaped, bracts 3, 2–10 mm (1 much the longest), lanceolate, awns straight, **FL**: perianth < 2 mm, incl, yellow, densely hairy; stamens 9, attached near top of perianth tube. *n*=19,20. Common. Desert scrub; < 2000 m. SNE, D; to sw UT, w AZ, nw Mex. Dead spiny skeletons persist.

C. robusta C. Parry (p. 861) ROBUST SPINEFLOWER **ST** decumbent to erect, 10–50 cm, grayish, soft-hairy. **LF**: blade 10–50 mm. **INFL**: involucral tube 2.5–4 mm, cylindric, thinly hairy, bracts 6, scarious margins white to pinkish, awns hooked. **FL**: perianth 2.5–4 mm, white to rose, hairy, lobes jagged or with central tooth; stamens 9. UNCOMMON. Coastal sand, scrub; < 300 m. n-c CCo, sw SnFrB (s San Mateo, Santa Cruz, n Monterey cos., extirpated in Alameda Co.). Closely related to *C. pungens*. More erect pls with pink involucral margins have been called var. *hartwegii* (Benth.) Rev. & R. Morgan.

C. spinosa S. Watson (p. 861) MOJAVE SPINEFLOWER **ST** prostrate to ascending, 3–40 cm, grayish, thinly hairy. **LF**: blade 3–20 mm, oblong. **INFL**: bracts gen 3 per node, ± lanceolate, awns straight; involucral tube 2–2.5 mm, urn-shaped, bracts gen 5, one >> others (to 10 mm, ± lanceolate), awns straight. **FL**: perianth 2.5– 3.5 mm, gen white, glabrous; stamens 9, attached at base of perianth tube. **FR** black. *n*=22. UNCOMMON. Desert scrub; 6–1300 m. w DMoj.

C. staticoides Benth. (p. 871) TURKISH RUGGING **ST** ascending to erect, 5–60 cm, greenish or reddish, thinly hairy. **LF**: blade 5–80 mm, oblong to narrowly ovate, thinly hairy to glabrous above, tomentose below. **INFL**: lower bracts ± linear, not very lf-like, soon deciduous, awns straight; involucres densely clustered, tube 3–5 mm, cylindric, nearly glabrous to curly-hairy, bracts 6, awns hooked. **FL**: perianth 3–5 mm, rose to red, hairy, lobes unequal, gen entire; stamens 9. *n*=19–21. Common. Coastal scrub, chaparral, desert scrub; < 2000 m. c&s CCo, SCoRO, SW (exc e PR). [var. *brevispina* Goodman; var. *elata* Goodman; var. *latiloba* Goodman; ssp. *chrysacantha* (Goodman) Munz (RARE: coastal Orange, San Diego cos.)] Highly variable; some forms, incl "ssp. *chrysacantha*," seem to be environmentally induced. Named entities are only the central forms of a large, intergrading complex. More study warranted.

C. stellulata Benth. **ST** erect, 5–30 cm, hairy. **LF**: basal blade 5–20 mm, narrowly lanceolate to oblanceolate; cauline in a whorl near mid-st and sometimes alternate. **INFL**: bract awns straight; involucres in dense heads, tube 3–4 mm, cylindric, ± bristly, bracts 6, ± equal, membranous margin white, shallower in (or not continuous through) sinuses (appearing star-like from above), awns hooked. **FL**: perianth 4–5 mm, creamy-white to rose, slightly hairy, lobes ± deeply notched; stamens 9. *n*=19–22. Uncommon. Dry foothill woodland; < 900 m. NCoRI, CaRF, SNF, e SnFrB.

C. uniaristata Torrey & A. Gray (p. 871) ONE-AWNED SPINEFLOWER **ST** decumbent to ascending, 2–25 cm, grayish or reddish hairy. **LF**: blade 5–20 mm. **INFL**: involucral tube 2–3 mm, urn-shaped, hairy, bracts 6, longest ± 5 × length of next longest, with a straight awn, other awns hooked. **FL**: perianth 2–3 mm, white, sparsely hairy, lobes gen with 1–3 minute teeth; stamens 3. *n*=40. Sandy to gravelly, talus or clay places; 800–2000 m. e CW, s SN, n WTR.

C. valida S. Watson (p. 871) SONOMA SPINEFLOWER **ST** ± erect, 10–30 cm, soft-hairy. **LF**: blade 10–50 mm, gen ± obovate. **INFL**: involucral tube 3–4.5 mm, cylindric, thinly hairy, bracts 6, erect, reddish, scarious margins thin or indistinct, not continuous across sinuses, white, awns straight, ivory to straw-colored. **FL**: perianth 4–6 mm, white to rose, hairy, lobes unequal, jagged or with a minute central tooth; stamens 9. RARE. Sandy soils in coastal grassland; < 300 m. n CCo (Point Reyes Peninsula, Marin Co.). One extant population known; threatened by cattle. Closely related to *C. pungens*.

C. ventricosa Goodman (p. 871) **ST** decumbent to ± erect, 5–50 cm, reddish, hairy. **LF**: blade 5–40 mm. **INFL**: involucral tube 4–4.5 mm, urn-shaped, much wider at base, thinly hairy, bracts 6, spreading, longest ± 3 × length of next longest, with a straight awn, other awns hooked. **FL**: perianth 4–4.5 mm, sparsely hairy, tube gen white, lobes red to maroon, often ± jagged; stamens 9. n=20–22. Uncommon. Serpentine outcrops; 500–1000 m. SCoRI (s San Benito, se Monterey, ne San Luis Obispo, w Fresno cos.). [*C. palmeri* S. Watson var. *v.* (Goodman) Munz] Closely related to *C. palmeri*. ❀DRN,DRY,SUN:7,9,14,16,17,20–24;DFCLT.

C. watsonii Torrey & A. Gray WATSON'S SPINEFLOWER **ST** ± erect, 2–15 cm, canescent. **LF**: blade 3–20 mm, gen thinly tomentose. **INFL**: involucral tube 3–4.5 mm, cylindric, green, hairy, bracts 5, largest 2–6 mm, ± lf-like, other four 1–2 mm, awns hooked. **FL**: perianth 1.5–2.5 mm, yellow, thinly hairy; stamens 9, attached near top of perianth tube. Desert and sagebrush scrub and woodland; 300–2200 m. sw SnJV, n edge TR, GB, w DMoj; to NV, sw UT, nw AZ.

C. wheeleri S. Watson (p. 871) WHEELER'S SPINEFLOWER **ST** ± erect, 5–25 cm, reddish, thinly hairy. **LF**: blade 5–20 mm, gen oblong, thinly hairy above, tomentose below. **INFL**: lower bracts ± oblanceolate, ± lf-like, persistent, awns straight; involucres in dense heads, tube gen 2–2.5 mm, ± bell-shaped, hairs coarse, recurved, bracts 6, awns hooked. **FL**: perianth 2.5–3 mm, white to red, gen glabrous, lobes unequal, gen entire; stamens 6. n=19. UNCOMMON. Non-serpentine chaparral; < 600 m. n ChI (Santa Cruz, Santa Rosa Islands). Closely related to *C. staticoides*. Stamen number is unique in the complex. ❀TRY.

C. xanti S. Watson **ST** ± erect, 3–30 cm, reddish, thinly hairy. **LF**: blade 3–15 mm, oblong to ovate, thinly hairy above, densely tomentose below. **INFL**: lower bracts ± oblanceolate, ± lf-like, persistent, awns straight; involucres ± loosely clustered, tube 3.5–5 mm, cylindric, smooth, hairs slender, curly, bracts 6, awns hooked. **FL**: perianth 4.5–6 mm, rose to red, hairy, lobes unequal, gen entire; stamens 9. Foothill and pinyon/juniper woodland, desert scrub; 100–1600 m. s SN, Teh, s SCoRI, n&e TR, SnJt, SNE (s Mono Co.). Closely related to *C. staticoides*.

var. ***leucotheca*** Goodman **INFL**: involucre densely white-hairy. Uncommon. Desert scrub; 300–1200 m. e SnBr, n SnJt. ❀TRY.

var. ***xanti*** (p. 871) **INFL**: involucre thinly hairy. n=19,21. Common. Habitat and range of sp., exc desert scrub in e SnBr, SnJt. ❀TRY.

DEDECKERA

1 sp. (Mary C. DeDecker, e CA botanist, 1909–) [Reveal & Howell 1976 Brittonia 28:245–251]

D. eurekensis Rev. & J. Howell (p. 871) JULY GOLD Shrub, 2–10 dm, 5–20 dm diam, minutely hairy. **LVS** cauline; stipule 0; petiole 2–5 mm; blade 7–15 mm, oblanceolate. **INFL** 1–8 cm; bracts 2–5 per node, free; most nodes with 1 ± sessile fl and 1–several branches; branches and bracts reduced and fls more per node upward; true involucre 0. **FL**: perianth gen 2–4 mm, yellowish, lobes 6; stamens 9. **FR** 2–3.5 mm, light reddish brown, puberulent at tip. n=14. **RARE** CA. Limestone outcrops; 1220–2200 m. W&I, DMtns (Last Chance, Panamint Mtns).

DODECAHEMA

1 sp. (Greek: twelve awns) [Reveal & Hardham 1989 Phytologia 66:83–88]

D. leptoceras (A. Gray) Rev. & Hardham (p. 871) SLENDER-HORNED SPINE-FLOWER Ann, spreading, 3–10 cm. **LVS** basal; stipule 0; petiole 0; blade 10–60 mm, linear-oblanceolate, glabrous. **INFL** 1–5 dm diam; bract 1 per node, 3–6 mm, 3-lobed; involucre ± sessile, 2–4 mm, cylindric, glandular, lobes 6, lobe awns 2–3 mm, awns at base 6, 1–2 mm; fls 3 per involucre. **FL**: perianth 1.2–2 mm, white to pink, hairy, lobes 6; stamens 9. **FR** 1.7–2 mm, dark brown to black. n=±17. **ENDANGERED** CA, US. Alluvial sand in coastal scrub; 200–700 m. c&e SCo, adjacent foothills of TR, PR. [*Centrostegia l.* A. Gray; *Chorizanthe l.* (A. Gray) S. Watson] Threatened by development, off-road vehicles.

EMEX

Ann, monoecious, glabrous. **ST** decumbent to erect, 3–8 dm. **LVS** cauline, alternate, petioled; stipules fused, sheathing st above nodes; blade base ± cordate. **STAMINATE INFL**: panicle, axillary. **PISTILLATE INFL**: cluster, axillary, small. **STAMINATE FL**: perianth lobes 5; stamens 4–6. **PISTILLATE FL**: perianth lobes 6. **FR** enclosed in woody perianth tube that has 3 major spines. 2 spp.: Medit, s Afr, Australia. (Latin: from *Rumex*)

1. Perianth in fr ± 10 mm, inner lobes sharply spine-tipped . . . . ***E. australis***
1′ Perianth in fr < 8 mm, inner lobes not spine-tipped . . . . ***E. spinosa***

E. australis Steinh. THREE-CORNERED JACK **FR**: perianth bell-shaped, indented at base; sculpture coarse, irregular, of a few dot-like indentations, inner lobes ± round. Uncommon. Coastal areas, esp dunes; < 200 m. CCo, expected elsewhere; native to s Afr, Australia. NOXIOUS WEED.

E. spinosa (L.) Campdera (p. 889) DEVIL'S THORN **FR**: perianth ± urn-shaped, sculpture fine, regular, netted or ladder-like, inner lobes ± elliptic. 2n=20. Uncommon. Disturbed places; < 500 m. SCo, expected elsewhere; native to Medit. NOXIOUS WEED.

ERIOGONUM WILD BUCKWHEAT

Ann to shrub. **LVS** gen ± basal (clustered on low sts or cauline), petioled, gen ± tomentose below (often shedding above); stipule 0. **INFL** openly cyme-like, umbel-like, or head-like, gen ± scapose; bracts (any whorled, lf-like structures on infl) 3–many per node, lf-like to scale-like; involucres gen 1 per node, gen ± obconic, lobes (or short teeth) gen 3–10, gen erect; fls gen many per involucre, pedicelled. **FL**: perianth white, yellow, or red, lobes 6, gen ± oblong to obovate; stamens 9. **FR** brown to black, glabrous to hairy. ± 250 spp.: N.Am. (Greek: woolly knees, from hairy nodes of some) [Reveal 1989 Phytologia 66:295–414] Largest genus in CA; apparently currently differentiating; many taxa ± indistinct. Better habitat data needed. Many are excellent bee fodder. ❀ Most are attractive and easy to grow with good drainage.

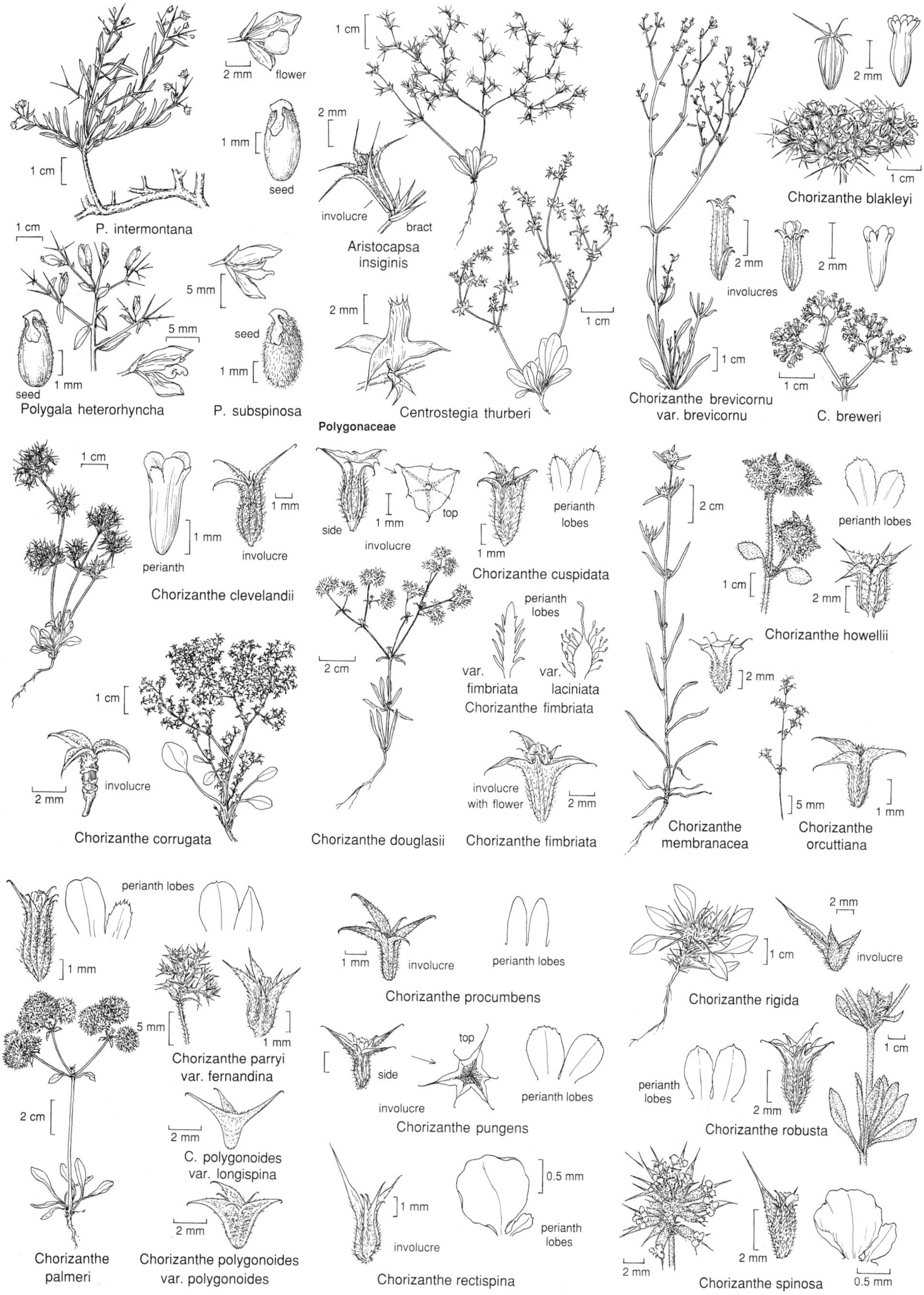

2 mm flower
1 mm seed
1 cm
P. intermontana

1 cm
5 mm
5 mm
seed
1 mm
seed 1 mm
Polygala heterorhyncha

P. subspinosa

1 cm
2 mm
involucre
bract
Aristocapsa insiginis

2 mm
1 cm
Centrostegia thurberi

Polygonaceae

2 mm
2 mm
involucres
2 mm
1 cm
Chorizanthe brevicornu var. brevicornu

1 cm
Chorizanthe blakleyi

1 cm
C. breweri

1 cm
1 mm
perianth
1 mm
involucre
Chorizanthe clevelandii

1 cm
2 mm
involucre
Chorizanthe corrugata

side
1 mm
top
involucre
2 cm
Chorizanthe douglasii

1 mm
perianth lobes
Chorizanthe cuspidata

perianth lobes
var. fimbriata
var. laciniata
Chorizanthe fimbriata

involucre with flower
2 mm
Chorizanthe fimbriata

2 cm
2 mm
Chorizanthe membranacea

1 cm
perianth lobes
2 mm
Chorizanthe howellii

5 mm
1 mm
Chorizanthe orcuttiana

perianth lobes
1 mm
2 cm
Chorizanthe palmeri

5 mm
1 mm
Chorizanthe parryi var. fernandina

2 mm
C. polygonoides var. longispina

2 mm
Chorizanthe polygonoides var. polygonoides

1 mm
involucre
perianth lobes
Chorizanthe procumbens

side
top
involucre
perianth lobes
Chorizanthe pungens

1 mm
involucre
0.5 mm
perianth lobes
Chorizanthe rectispina

1 cm
2 mm
involucre
Chorizanthe rigida

perianth lobes
2 mm
1 cm
Chorizanthe robusta

2 mm
2 mm
0.5 mm
Chorizanthe spinosa

Key to Groups

1. Ann
 2. Involucre ± ribbed (or angled), gen either sessile or erect ..... **Group 1**
 2′ Involucre unribbed, unangled, gen either stalked or reflexed ..... **Group 2**
1′ Per to shrub
 3. Perianth base ± stalk-like, ± as wide as pedicel and jointed to it; bracts gen lf-like (± reduced above), 2–10 per node ..... **Group 3**
 3′ Perianth base not stalk-like; bracts gen scale-like, 3 per node ..... **Group 4**

Group 1: Ann; involucre ± ribbed (gen either sessile or erect)

1. Involucres appearing terminal (or few lateral); infl ± repeatedly forked (or ± evenly branched) or umbel-like; infl axes often short
 2. Perianth ± 1 mm, yellow to cream; SNE, w DMoj
  3. Perianth cream; n SNE ..... ***E. ampullaceum***
  3′ Perianth yellow; w DMoj, s SNE ..... ***E. mohavense***
 2′ Perianth ± 2 mm, white to red; NW, CW
  4. Infl axes glabrous; involucres sessile (directly subtended by bracts)
   5. Involucre ± 3 mm, ± cylindric, teeth 5; NCoRI, SnFrB ..... ***E. luteolum*** var. ***caninum***
   5′ Involucre ± 4 mm, bell-shaped, teeth 8; SCoRI ..... ***E. nortonii***
  4′ Infl axes gen ± hairy (if glabrous, lower involucres stalked)
   6. Involucre glabrous — infl axes subglabrous; s SnFrB, n SCoRI ..... ***E. argillosum***
   6′ Involucre tomentose
    7. Lvs basal; SCoRI — styles 0.1–0.3 mm ..... ***E. eastwoodianum***
    7′ Some lvs cauline (or lowest bracts quite lf-like); SCoRI, SnFrB
     8. Involucres sessile throughout; ne SnFrB; probably extinct ..... ***E. truncatum***
     8′ Lower involucres stalked; SCoRI — styles 0.7–1 mm
      9. Perianth lobes smooth ..... ***E. temblorense***
      9′ Perianth lobes papillate when fresh ..... ***E. vestitum***
1′ Most involucres appearing lateral; infl cyme-like, ± unevenly branched; axes sometimes long, wand-like
 10. Perianth ± hairy (magnification needed)
  11. Perianth densely long-hairy, 2–2.5 mm; NW ..... ***E. dasyanthemum***
  11′ Perianth minutely hairy, ± 1 mm; DMtns ..... ***E. puberulum***
 10′ Perianth (sub)glabrous (sometimes minutely glandular)
  12. Involucre < 1.5 mm
   13. Infl axes ± tomentose
    14. Perianth lobes widened at base; SnBr, PR; rare ..... ***E. foliosum***
    14′ Perianth lobes ± fan-shaped, narrowed at base; GB, D; common
     15. Perianth yellow to red; infl branches gen curved ± inward ..... ***E. nidularium***
     15′ Perianth white to pink or pale yellow; infl branches gen curved ± outward ..... ***E. palmerianum***
   13′ Infl axes glabrous
    16. Perianth yellow, ± 1 mm — SNE, DMoj ..... ***E. brachyanthum***
    16′ Perianth white to reddish, 1–2 mm
     17. Perianth 1–1.5 mm; e CW, WTR ..... ***E. elegans***
     17′ Perianth 1–2 mm; SW, GB, w D ..... ***E. baileyi***
      18. Infl glabrous; SW, GB, D; common ..... var. ***baileyi***
      18′ Infl tomentose; GB; uncommmon ..... var. ***praebens***
  12′ Involucre 2–7 mm
   19. Lf blade eliiptic to (ob)lanceolate; infl axes gen tomentose
    20. Involucre 4–5 mm, teeth minute ..... ***E. roseum***
    20′ Involucre gen 2–3.5 mm, teeth ± prominent
     21. Infl axes ± upcurved; petiole often winged — SCoR, TR ..... ***E. cithariforme***
     21′ Infl axes straight; petiole slender ..... ***E. gracile***
      22. Infl gen tomentose; widespread ..... var. ***gracile***
      22′ Infl ± glabrous; PR ..... var. ***incultum***
   19′ Lf blade ± round; infl axes glabrous or woolly above nodes
    23. Infl axes gen woolly above nodes; KR, CaR, n SNH, MP ..... ***E. vimineum***
    23′ Infl axes glabrous; widespread
     24. Some lvs ± cauline (or lowest bracts lf-like and scale-like) ..... ***L. luteolum*** var. ***luteolum***
     24′ Lvs basal; bracts scale-like
      25. Involucre 2–2.5 mm — CW, nw WTR ..... ***E. covilleanum***
      25′ Involucre 3–7 mm
       26. Involucre 3–4 mm; pl 5–40 cm; SW, SNE, D ..... ***E. davidsonii***
       26′ Involucre 4–7 mm; pl 40–100 cm; n WTR, SnBr, PR ..... ***E. molestum***

Group 2: Ann; involucre unribbed (gen either stalked or reflexed)

1. Some lvs cauline (or lower bracts quite lf-like)
 2. Lf linear, < 2 mm wide, rolled under — CA-FP, GB ..... ***E. spergulinum***

3. Infl prostrate to ascending, gen 2–5 cm, axes glabrous — s SNH ..... var. ***pratense***
3′ Infl erect, 5–40 cm, axes glandular
4. Perianth ± 2 mm; widespread, incl SN ..... var. ***reddingianum***
4′ Perianth ± 3 mm; SN ..... var. ***spergulinum***
2′ Lf wider than linear, > 2 mm wide, flat
5. Perianth covered with hooked hairs
6. Fls 2 per involucre ..... ***E. hirtiflorum***
6′ Fls 4–6 per involucre ..... ***E. inerme***
5′ Perianth glabrous, glandular, or soft-wavy-hairy
7. Perianth soft-wavy-hairy — adjacent to s SnJV ..... [2]***E. ordii***
7′ Perianth glabrous or glandular
8. Stamens long-exserted; widespread c&s CA ..... ***E. angulosum***
8′ Stamens incl; some widespread
9. Fls hidden by dense, cottony hairs; s SNF, w SnJV ..... ***E. gossypinum***
9′ Fls not hidden; widespread, incl SnJV
10. Outer perianth lobes flat, wavy-margined — s SN, SnJV, SCoRI, WTR, w DMoj ..... ***E. gracillimum***
10′ Outer perianth lobes with a pouch-like bulge or cup
11. Outer perianth lobes cupped at base; GB, s SNH, TR ..... ***E. maculatum***
11′ Outer perianth lobes cupped at tip; Teh, SnJV, SCoRI, WTR, w DMoj ..... ***E. viridescens***
1′ Lvs basal (at most a few ± cauline on lowermost st)
12. Lf glabrous to coarsely hairy or hairy in patches (not tomentose); perianth often yellow (sometimes white to red)
13. Perianth glabrous — e-c SN, SNE ..... ***E. esmeraldense*** var. ***esmeraldense***
13′ Perianth subglabrous to hairy
14. Perianth subglabrous to sparsely soft-hairy
15. Infl axes sparsely hairy — adjacent to s SnJV ..... [2]***E. ordii***
15′ Infl axes subglabrous or sparsely glandular
16. Perianth white, ± 2 mm; PR, sw DMtns ..... ***E. apiculatum***
16′ Perianth pink to red, ± 0.5 mm; s SN, TR, PR, W&I ..... ***E. parishii***
14′ Perianth densely and often coarsely hairy
17. Perianth whitish; infl glandular throughout — ne DMoj ..... ***E. glandulosum***
17′ Perianth yellow; infl glandular only at base or nodes
18. Involucre teeth gen 4(5); CW, D ..... ***E. trichopes***
19. Pl erect, tall; involucre gen 1–1.5 mm; SCoRI, n WTR ..... var. ***hooveri***
19′ Pl spreading, low; involucre < 1 mm; D ..... var. ***trichopes***
18′ Involucre teeth gen (4)5; D
20. Infl axes slender, nodes glandular; e DMoj; uncommon ..... ***E. contiguum***
20′ Main infl axes wider below nodes, gen glabrous; s GB, D; common (pls gen 1st-year-fl per: see also Group 4) ..... ***E. inflatum*** var. ***inflatum***
12′ Lf ± tomentose on one or both surfaces; perianth gen white to reddish (sometimes yellow)
21. Outer perianth lobes ± cordate at base (magnification needed)
22. Involucres erect
23. Infl flat-topped, gen < 30 cm; ne DMoj ..... ***E. bifurcatum***
23′ Infl narrow, 20–120 cm; D ..... ***E. deflexum*** var. ***rectum***
22′ Involucres reflexed
24. Infl axes glandular — SNE, DMoj ..... ***E. brachypodum***
24′ Infl axes glabrous (exc sometimes very base)
25. Perianth yellow
26. Involucre sessile; infl ± umbrella-like; SNE ..... ***E. hookeri***
26′ Involucre stalks spreading or recurved; infl ± open; D ..... [2]***E. thomasii***
25′ Perianth white
27. Infl pagoda-like, in several tiers; involucre ± 1 mm; perianth base swollen; n DMoj ..... ***E. rixfordii***
27′ Infl not tiered; involucre ± 2 mm; perianth base not swollen; D, GB ..... ***E. deflexum*** (part)
28. Involucre 2–3 mm, stalk 3–15 mm; main infl axes often wider below nodes; Teh, TR, n&w DMoj ..... var. ***baratum***
28′ Involucre 1–2 mm, stalk 0–5 mm; infl axes slender; SNE, D ..... var. ***deflexum***
21′ Outer perianth lobes obtuse to truncate at base
29. Perianth minutely hairy
30. Perianth white to red, ± hair-tufted inside — e PR, s DMoj, DSon ..... ***E. thurberi***
30′ Perianth yellow (if whitish, not hair-tufted inside)
31. Outer perianth lobe bases swollen — D ..... [2]***E. thomasii***
31′ Outer perianth lobe bases not swollen
32. Involucre minutely glandular; SCoRI, WTR, GB, D ..... ***E. pusillum***
32′ Involucre glabrous; D ..... ***E. reniforme***
29′ Perianth glabrous
33. Involucres ± sessile; ne DMoj ..... ***E. hoffmannii***
34. Lf margin flat; pl spreading, < 50 cm; perianth lobes narrowly spoon-shaped ..... var. ***hoffmannii***
34′ Lf margin strongly wavy; pl erect, < 100 cm; perianth lobes narrowly ovate ..... var. ***robustius***
33′ Involucres on ± recurved or reflexed stalks; GB, DMtns
35. Infl glandular ± throughout; W&I, DMtns ..... ***E. eremicola***

35′ Infl glabrous (exc sometimes involucre stalks); GB
36. Outer perianth lobes wavy-margined .... ***E. cernuum***
37. Involucre stalk 2–25 mm .... var. ***cernuum***
37′ Involucre stalk 0–2 mm .... var. ***viminale***
36′ Outer perianth lobes flat
38. Involucre obconic, ± 1.5 mm .... ***E. deflexum*** var. ***nevadense***
38′ Involucre bell-shaped, ± 2 mm .... ***E. nutans***

Group 3: Per to shrub; perianth base ± stalk-like, jointed to pedicel; lower bracts gen ± lf-like

1. Perianth base ± indistinctly stalk-like, often ± wider than pedicel and ± funnel-shaped or sharply 3-angled
2. Infl axis gen > 20 cm, stout, erect; infl dense, ball-like, branches hidden; involucres gen many
3. Perianth bright yellow to rose; lf blade narrowly elliptic to ovate, largest often < 10 mm wide; n KR .... ***E. hirtellum***
3′ Perianth cream to pale yellow; lf blade gen ± round, largest gen > 15 mm wide; c&s SN, W&I, DMtns .... ***E. latens***
2′ Infl axis gen < 5(40) cm, slender, often prostrate, branches gen visible (or 0); involucres gen 1–5 per node or cluster
4. Bracts whorled near middle of infl axis; perianth glabrous, ± yellow; e KR .... ***E. alpinum***
4′ Bracts at branching nodes only; perianth glabrous or hairy, white to cream or rose; CA
5. Perianth hairy; bracts 2 per node; CaRH .... ***E. pyrolifolium***
5′ Perianth glabrous; bracts 3–5 per node; CA
6. Infl dense; involucres terminal; bracts lf-like; stalk-like perianth base very short, not angled; n CA .. ***E. lobbii***
6′ Infl open; most involucres lateral; bracts scale-like; stalk-like perianth base long, sharply 3-angled; s CA .... ***E. saxatile***
1′ Perianth base distinctly stalk-like, ± as wide as pedicel
7. Dioecious; staminate and pistillate pls obviously different
8. Lf upper surface shedding as lf ages, becoming olive-green — e KR, CaRH, n&c SNH .... ***E. marifolium***
8′ Lf upper surface remaining tomentose
9. Lf blade 5–15 mm wide; petiole 7–30 mm; KR .... ***E. diclinum***
9′ Lf blade 3–7 mm wide; petiole 3–10 mm; SNH .... ***E. incanum***
7′ Not dioecious (or staminate and pistillate pls not different)
10. Perianth hairy
11. Involucre lobes short, erect — NCoRI, n&c SNF .... [2]***E. tripodum***
11′ Involucre lobes long, spreading or reflexed
12. Infl branches (and involucres) 2 or more
13. Perianth 4–5 mm; NCoRI, SNF .... [2]***E. tripodum***
13′ Perianth 5–9 mm; CaRH, n SNH, MP .... [2]***E. sphaerocephalum***
14. Perianth cream or pale yellow, gen ± conspicuously hairy .... var. ***halimioides***
14′ Perianth bright yellow; gen ± inconspicuously hairy .... var. ***sphaerocephalum***
12′ Infl unbranched (involucre 1 per infl axis)
15. Infl axis bractless — GB .... ***E. cespitosum***
15′ Infl axis with whorl of bracts near middle
16. Fr ± glabrous; s SNH .... ***E. twisselmannii***
16′ Fr tip hairy; CaRH, MP, n SNH
17. Infl unbranched; involucre lobes reflexed, appressed to tube .... ***E. douglasii***
17′ Infl gen branched; involucre lobes gen spreading, not appressed to tube (see 13′ for vars.) .... [2]***E. sphaerocephalum***
10′ Perianth (sub)glabrous
18. Lf silky-hairy ± on both surfaces — c NCoRO .... ***E. kelloggii***
18′ Lf tomentose to woolly (rarely glabrous) esp below, not silky-hairy
19. Involucre lobes ± erect
20. Bracts whorled near middle of infl axis
21. Pl densely matted; s KR, n NCoR .... ***E. libertini***
21′ Pl loosely matted or openly branched; SN (see 30′ for vars.) .... [2]***E. prattenianum***
20′ Some bracts ± immediately subtending involucres
22. Perianth white or pink — s SNH .... ***E. polypodum***
22′ Perianth yellowish, often becoming red
23. Perianth pale yellow in early fl
24. Infl axis < 10 cm; se NCoR .... ***E. nervulosum***
24′ Infl axis gen 20–40 cm; s KR, CaR, n SN .... ***E. ursinum***
23′ Perianth bright yellow in early fl
25. Inner perianth lobes wider than outer lobes; fr 2.5–3 mm; not on serpentine, WTR .... ***E. crocatum***
25′ Inner and outer perianth lobes similar; fr 3.5–5.5 mm; serpentine, NW
26. Lf blade narrowly elliptic to oblong, margin rolled under; KR .... ***E. congdonii***
26′ Lf blade ovate to round, margin flat; n NW .... ***E. ternatum***
19′ Involucre lobes obviously reflexed
27. Infl axis with whorled bracts near middle

28. Infl branched; involucres several
29. Infl rays gen 5–8; bracts subtending rays gen << bracts near mid-axis, gen sharply reflexed; Wrn .... ***E. heracleoides***
29′ Infl rays gen 1–5; bracts subtending rays ± = bracts near mid-axis, gen ascending to spreading; widespread, incl Wrn (see 33. for vars.) .... [2]***E. umbellatum***
28′ Infl unbranched; involucre 1
30. Pl densely matted; e KR .... ***E. siskiyouense***
30′ Pl loosely branched, spreading; SN .... [2]***E. prattenianum***
31. Lf blade becoming subglabrous above; c&s SNH .... var. ***avium***
31′ Lf blade remaining tomentose above; n&c SN .... var. ***prattenianum***
27′ Infl axis bractless near middle
32. Lf blade lanceolate to ovate, largest gen > 30 mm wide .... ***E. compositum***
32′ Lf blade elliptic to round, largest gen < 20 mm wide .... [2]***E. umbellatum***
33. Infl a strictly simple umbel (obviously bracted only subtending rays, rays unbranched)
34. Perianth whitish, pale yellow, pink, or red, gen strongly striped
35. Lf blade ± round, very densely tomentose (± felt-like) below; perianth ± bright red — SnGb, SnBr .... var. ***minus***
35′ Lf blade ± elliptic, ± thinly tomentose below; perianth cream to reddish brown
36. Perianth gen cream to pale yellow; Wrn, n W&I .... var. ***dichrocephalum***
36′ Perianth gen pink to reddish brown; s W&I, n DMtns .... [2]var. ***versicolor***
34′ Perianth bright yellow, gen weakly striped
37. Low dense mat; infl gen spreading; uncommon
38. Lf blade ± narrowly elliptic, 2–6 mm wide; s SNH .... var. ***covillei***
38′ Lf blade ± ovate to round, 5–10 mm wide; KR, n NCoRH, CaRH .... var. ***humistratum***
37′ Spreading open mat or subshrub; infl ± erect; common
39. Subshrub; lf blade thinly tomentose below; SN (e slope), GB, nw DMoj .... var. ***nevadense***
39′ Spreading mat; lf blade densely tomentose below; e KR, CaR, n SNH, MP .... [2]var. ***polyanthum***
33′ Infl at least somewhat compound (some rays branched, or bracts whorled near mid-axis or mid-ray)
40. Lf blade ± glabrous on both surfaces
41. Some infl rays again branched, bractless exc at branching nodes — s SNH, e Teh, SNE .... var. ***chlorothamnus***
41′ Infl rays unbranched, with whorl of bracts near middle
42. Perianth whitish or cream, 4–7 mm; Wrn .... var. ***glaberrimum***
42′ Perianth bright yellow, 7–10 mm; n SNH .... var. ***torreyanum***
40′ Lf blade hairy at least below
43. Perianth 7–12 mm; shrub, 50–200 cm — KR, n SNH .... var. ***speciosum***
43′ Perianth 3–8 mm; low matted herb or subshrub
44. Perianth pink to reddish brown, conspicuously dark-striped; s W&I, n DMtns .... [2]var. ***versicolor***
44′ Perianth whitish to yellow; widespread
45. Perianth whitish or cream; e DMtns .... var. ***juniporinum***
45′ Perianth yellow; widespread
46. Lf blade sparsely hairy on both surfaces — e SN, TR, SNE, DMtns .... var. ***subaridum***
46′ Lf blade densely tomentose, at least below
47. Lf blade remaining densely white-tomentose above— s NCoR, SnFrB, n SCoR ... var. ***bahiiforme***
47′ Lf blade finally shedding above, becoming ± glabrous
48. Infl subglabrous to thinly tomentose
49. Lf blade gen finely toothed or wavy-margined; n NW .... var. ***argus***
49′ Lf blade margin gen entire, flat; SN .... var. ***furcosum***
48′ Infl ± densely tomentose
50. Infl open (rays gen > 4 cm, involucres well separated); TR, SnJt .... var. ***munzii***
50′ Infl ± compact (rays gen < 3 cm, involucres crowded); n CA .... [2]var. ***polyanthum***

Group 4: Per to shrub; perianth base not stalk-like; bracts gen scale-like

[note: longest key, but 16 or fewer decisions will key any taxon]

1. St jointed, breaking into short-cylindric sections; fls 150–200 per involucre — ne DMoj; uncommon (subg. *Clastomyelon*) .... ***E. intrafractum***
1′ St not jointed; fls 7–100 per involucre (esp subg. *Eucycla*)
2. Pl cushion-like, matted, or cespitose
3. Inner and outer perianth lobes distinctly different (outer round, inner narrower, often notched)
4. Perianth balloon-like from bulging outer perianth lobes; DMtns; uncommon .... ***E. gilmanii***
4′ Perianth not balloon-like; CA-FP, GB; common
5. Infl cyme- or umbel-like .... ***E. strictum***
6. Perianth bright yellow — esp MP .... [2]var. ***anserinum***
6′ Perianth ± white
7. Lf blade gen 5–8 mm, gen ± round, both surfaces white-woolly — KR .... var. ***greenei***
7′ Lf blade gen 10–30 mm, gen ± elliptic, gen tomentose below, ± shedding and brownish above .... var. ***proliferum***

5′ Infl head-like (involucres in 1 cluster per axis)
8. Perianth bright yellow
9. Infl axis 5–20 cm .... ***E. ovalifolium*** var. ***ovalifolium***
9′ Infl axis gen 1–5 cm .... ***E. strictum*** [2]var. ***anserinum***
8′ Perianth white to cream or purplish .... ***E. ovalifolium*** (most)
10. Lf blade 5–60 mm; infl axis gen 5–20 cm; perianth gen 4–6 mm
11. Lf distinctly brown-margined; perianth white to cream .... var. ***eximium***
11′ Lf not brown-margined; perianth white to purplish .... var. ***purpureum***
10′ Lf blade gen 2–6 mm; infl axis gen < 4 cm; perianth gen 2–3 mm
12. Involucre 2–4 mm; CaRH, SNH, W&I; not on limestone .... var. ***nivale***
12′ Involucre 5–7 mm; ne SnBr; limestone .... var. ***vineum***
3′ Inner and outer perianth lobes similar
13. Perianth ± yellow
14. Involucre glandular — lf blade ± oblanceolate
15. Lf blade 10–25 mm; involucre teeth 5; n SNE, 1500–2500 m .... ***E. beatleyae***
15′ Lf blade 4–15 mm; involucre teeth 5–8; s CaRH, n&c SNH, n SNE, 2300–4000 m .... ***E. rosense***
14′ Involucre glabrous or hairy but not glandular
16. Lf blade 4–12 mm, ± oblanceolate; infl gen 5–15 cm; MP .... ***E. prociduum***
16′ Lf blade 10–35 mm, gen ± ovate; infl 10–40 cm; GB .... ***E. ochrocephalum***
17. Infl axis ± tomentose; n SNE .... var. ***alexanderae***
17′ Infl axis glabrous or glandular; MP .... var. ***ochrocephalum***
13′ Perianth white (yellowish) to rose
18. Infl axis hairy or glandular
19. Perianth and fr long-hairy; DMtns .... ***E. shockleyi*** var. ***shockleyi***
19′ Perianth gen ± glabrous or glandular; fr glabrous
20. Infl cyme-like, branches long; most involucres sessile, lateral — SN, CW, TR (see 64′ for vars.) .... [4]***E. wrightii***
20′ Infl ± head-like; most involucres appearing terminal
21. Infl branched (but branches often ± hidden); s SNH .... ***E. breedlovei*** var. ***breedlovei***
21′ Infl head-like, ± unbranched; c&s SNH, Teh, TR, SNE
22. Infl axis glandular; involucre flexible; c&s SNH, W&I .... ***E. gracilipes***
22′ Infl axis tomentose; involucre rigid; c&s SN, TR, SNE (see 24′ for vars.) .... [2]***E. kennedyi***
18′ Infl axis glabrous
23. Infl cyme-like, branches long; most involucres sessile, lateral; SN, CW (see 64′ for vars.) .... [4]***E. wrightii***
23′ Infl ± head-like; c&s SN, TR, SNE
24. Lf coarsely hairy, ± rusty .... ***E. breedlovei*** var. ***shevockii***
24′ Lf ± soft-tomentose, gen white .... [2]***E. kennedyi***
25. Lf oblong, 2–6 mm; c&s SN, SNE
26. Lf gray to rusty; involucres 2.5–3.5 mm, tomentose; perianth 2.5–3.5 mm; se-most SNF ... var. ***pinicola***
26′ Lf white; involucres 1.5–2 mm, subglabrous; perianth 2–2.5 mm; SNH (e slope), SNE, nw DMtns .... var. ***purpusii***
25′ Lf oblanceolate to elliptic, 2–12 mm; TR
27. Lf gen 6–10 mm, tip gen acute; infl axis 8–15 cm; involucre 2.5–4 mm .... var. ***austromontanum***
27′ Lf 2–5 mm, tip ± rounded; infl axis 1–12 cm; involucre 1–2.5 mm
28. Infl axis gen 1–2 cm; 2600–3500 m .... var. ***alpigenum***
28′ Infl axis 4–12 cm; 1500–2600 m .... var. ***kennedyi***
2′ Pl ± erect, not matted (exc some *E. microthecum* & *E. wrightii*)
29. Subshrub or shrub
30. Involucres solitary, 5–6 mm; fl 4–5 mm; n SCoRO .... ***E. butterworthianum***
30′ Involucres solitary or not, 1.5–3 mm; fl 2–4 mm; CA
31. Gen most involucres sessile, lateral; infl branches long
32. Branches fragile; involucres lateral esp near infl tips; GB .... [2]***E. nummulare***
32′ Branches stout; involucres lateral throughout infl; CA-FP, D (see 64′ for vars.) .... [4]***E. wrightii***
31′ Most involucres terminal; infl branches ± short
33. Involucres clustered at nodes (pl of ChI if solitary)
34. Lvs ± clustered, narrow; widespread, esp inland c&s CA .... ***E. fasciculatum***
35. Lf oblanceolate, slightly rolled under, ± tomentose above .... var. ***polifolium***
35′ Lf linear, tightly rolled under, subglabrous above
36. Lf light yellow-green; perianth gen hairy inside — D .... var. ***flavoviride***
36′ Lf dark (or gray-) green; perianth glabrous or hairy outside
37. Pl gen decumbent; involucre and perianth gen glabrous — CCo .... var. ***fasciculatum***
37′ Pl gen erect; involucre and perianth (outside) gen ± hairy .... var. ***foliolosum***
34′ Lvs borne singly, narrow or wide; ± coastal
38. Infl branches few; gen mainland
39. Perianth hairy; s CCo, n SCo, n ChI .... ***E. cinereum***
39′ Perianth glabrous; CCo, SCo .... ***E. parvifolium***
38′ Infl branches many; ChI
40. Lf 2–5 cm, linear, rolled under .... ***E. arborescens***
40′ Lf 2–10 cm, ± elliptic to ovate, flat .... ***E. giganteum***

41. Lf blade narrowly elliptic to oblong — San Clemente Island .... var. ***formosum***
41′ Lf blade gen ± ovate
42. Pl < 100 cm; infl dense or open; Santa Barbara Island .... var. ***compactum***
42′ Pl gen 150–350 cm; infl open; Santa Catalina Island .... var. ***giganteum***
33′ Involucres solitary (1 per node); mainland
43. Infl with definable main axes (± unevenly branched, side branches smaller)
44. Infl ± tiered or pagoda-shaped; outer perianth lobes narrowed to base — Teh, D .... ***E. plumatella***
44′ Infl gen ± funnel-shaped, not tiered; outer perianth lobe bases ± cordate .... [2]***E. heermannii***
45. Infl axes deeply grooved — limestone cliffs, DMtns .... var. ***sulcatum***
45′ Infl axes round
46. Infl axes thinly tomentose; e DMtns .... var. ***floccosum***
46′ Infl axes ± glabrous; widespread
47. Pl gen 10–50 cm, low, spreading; lvs gen oblanceolate to spoon-shaped
48. Fl 1.5–2.5 mm; limestone cliffs, DMtns .... var. ***argense***
48′ Fl 2.5–3 mm; gravel, esp SNE, n DMtns .... var. ***humilius***
47′ Pl gen 50–200 cm, erect, ± rounded; lvs gen ± lanceolate
49. Lf 5–15 mm, both surfaces remaining ± tomentose; infl gen dense; s CA-FP .... var. ***heermannii***
49′ Lf 15–40 mm, gen shedding above; infl gen ± open; s SCoRI .... var. ***occidentale***
43′ Infl axis not evident above first node, branches ± even
50. Perianth and fr hairy
51. Perianth yellow; s DSon .... ***E. deserticola***
51′ Perianth white; KR .... ***E. pendulum***
50′ Perianth and fr glabrous
52. Large, ± erect shrub; lf blade elliptic to oblong, 10–30 mm
53. Infl branches stout, ± spreading (see 44′ for vars.); widespread .... [2]***E. heermannii***
53′ Infl open, branches fragile; GB .... [2]***E. nummulare***
52′ Low, spreading shrub or subshrub; lf blade linear to narrowly elliptic, 3–20 mm
54. Lf 4–6 mm, linear, strongly rolled under; e DMtns .... ***E. ericifolium***
54′ Lf 3–25 mm, gen ± oblanceolate, often ± flat; SN, TR, GB, D (vars. difficult) .... ***E. microthecum***
55. Largest lvs gen 15–25 mm, nearly flat
56. Perianth yellowish; hairs gen ± white or gray; e SN, GB .... var. ***ambiguum***
56′ Perianth white to pink or red-brown; hairs often pale red-brown; SnBr, s SNE, DMoj
57. Upper lvs gen largest, nearly flat; infl dense, flat-topped; perianth gen brownish; SnBr .... var. ***corymbosoides***
57′ Upper lvs smaller, strongly rolled under; infl ± open, not very flat-topped; perianth white to pink .... var. ***simpsonii***
55′ Largest lvs gen 3–15 mm, gen distinctly rolled under
58. Pl gen 20–60 cm; esp GB, n DMtns
59. Lf gen 8–12 mm, gen barely rolled under .... var. ***laxiflorum***
59′ Lf gen 5–8(15) mm, strongly rolled under .... var. ***panamintense***
58′ Pl 4–15 cm; SNH, e TR, s W&I (var. *alpinum* complex)
60. Infl ± open, main axes gen 2–4 cm; lf gen ± densely tomentose above; s W&I, rare .... var. ***lapidicola***
60′ Infl compact, main axes gen < 1 cm; lf gen thinly hairy above; c SNH, e TR, n SNE
61. Lf oblanceolate; perianth white to rose; c SNH, s SNE .... var. ***alpinum***
61′ Lf elliptic to ovate; perianth ± brownish; SnGb, SnBr .... var. ***johnstonii***
29′ Per (at most somewhat woody above caudex)
62. Most involucres sessile, lateral; infl branches long, ± wand-like
63. Involucre 6–7 mm; pl 60–180 cm — CW, SW .... ***E. elongatum*** var. ***elongatum***
63′ Involucre 2–5 mm; pl often < 50 cm
64. Lf blade oblong to round; W&I, DMtns .... ***E. panamintense***
64′ Lf blade linear to elliptic; CA .... [4]***E. wrightii***
65. Matted per or subshrub; esp SN, TR
66. Pl densely matted, 1–6 cm; lf blade 1–2 mm; involucre ± 2 mm; s SN; rare .... var. ***olanchense***
66′ Pl loosely matted, 10–30 cm; lf blade 5–12 mm; involucre 1.5–4 mm; SN, SNE, CW, TR, SnJt; common .... var. ***subscaposum***
65′ Shrub or subshrub, not matted; esp D, NW
67. Lf woolly; infl axes stout, rigid, densely gray-woolly — DSon; uncommon .... var. ***nodosum***
67′ Lf tomentose (esp below); infl gen slender, glabrous to white-, red- or green-tomentose
68. Lf blade gen 2–10 mm, strongly rolled under; base of petiole fully ringing st; PR .. var. ***membranaceum***
68′ Lf blade gen 5–30 mm, ± flat; base of petiole not fully ringing st; NW, DMoj
69. Lf blade 15–30 mm; involucre 2–3 mm; perianth 3–4 mm; NW .... var. ***trachygonum***
69′ Lf blade gen 5–15 mm; involucre 2–2.5 mm; perianth 2.5–3.5 mm; DMoj .... var. ***wrightii***
62′ Most involucres terminal; infl branches either ± short or ± equal
70. Lf blade < 20 mm; ± round; n SNF — infl axis glabrous .... ***E. apricum***
71. Infl erect .... var. ***apricum***
71′ Infl prostrate .... var. ***prostratum***
70′ Lf blade gen > 20 mm; often not round; widespread
72. Lf erect, blade 40–150 mm, lanceolate .... ***E. elatum***

73. Infl axes and involucres glabrous; GB, SN (e slope) .... var. ***elatum***
73′ Infl axes and involucres hairy; n CA-FP, MP .... var. ***villosum***
72′ Lf spreading, blade 20–100 mm, ± ovate to reniform
74. Infl axes gen tomentose; NCo, CCo .... ***E. latifolium***
74′ Infl axes gen glabrous (if tomentose, not coastal)
75. Involucre < 1.5 mm; perianth densely white-hairy; GB, D (sometimes ann; see also Group 2) .... ***E. inflatum***
76. Infl slender; uncommon; s D .... var. ***deflatum***
76′ Infl main axes wider below major nodes; common; GB, D .... var. ***inflatum***
75′ Involucre 2–7 mm; perianth (sub)glabrous; CA
77. Involucre 5–7 mm; ChI .... ***E. grande***
78. Perianth pink to red; involucre 5–7 mm; pl 15–50 cm — n ChI .... var. ***rubescens***
78′ Perianth white; involucre 4–6 mm; pl 10–220 cm
79. Pl 50–220 cm; involucre 5–6 mm; ChI .... var. ***grande***
79′ Pl 10–25 cm; involucre 4–5 mm; s ChI .... var. ***timorum***
77′ Involucre 2–5 mm; mainland .... ***E. nudum***
80. Infl and involucre tomentose
81. Involucre 1(2) per node, 4–6 mm; perianth glabrous, 3–4 mm; SnFrB .... var. ***decurrens***
81′ Involucres 3–6 per node, 3–5 mm; perianth hairy, 1.5–4 mm; widespread
82. Perianth white, rose, or yellow, hairy; infl spreading; lvs gen 10–40 mm; NW, CaR, s SN, MP .... var. ***oblongifolium***
82′ Perianth white, densely hairy; infl erect; lvs gen 30–55 mm; s SNF .... var. ***regirivum***
80′ Infl (and gen involucre) glabrous
83. Some lvs ± cauline, margin ± strongly curled
84. Lf densely woolly below, gray-tomentose above; involucres 5–10 per cluster — s SNF .... var. ***murinum***
84′ Lf tomentose below, ± glabrous above; involucres 1–5 per cluster
85. Infl slender; perianth white to pink; NW, CW, common .... var. ***auriculatum***
85′ Infl main axes inflated; perianth pale yellow; SCoRI, uncommon .... var. ***indictum***
83′ Lvs basal, margins gen nearly flat
86. Involucre gen 1(2) per cluster — perianth white; TR, PR .... var. ***pauciflorum***
86′ Involucres (1)2–10 per cluster
87. Perianth hairy, yellow or white
88. Perianth white; lf becoming ± glabrous above; n CA-FP, SN .... var. ***pubiflorum***
88′ Perianth bright yellow; lf ± tomentose above; s CA-FP, DMoj .... var. ***westonii***
87′ Perianth gen subglabrous, white
89. Infl gen a head of 3–6 involucres — c&s SNH, 3000–3800 m .... var. ***scapigerum***
89′ Infl few–many-branched; involucres 1–10 per cluster
90. Infl few-branched, involucres ± 5–10 per head — n NCo .... var. ***paralinum***
90′ Infl many-branched, open, involucres 1–6 per cluster
91. Lf gen 10–20 mm; involucres 1–3 per node; SN, 1000–3200 m .... var. ***deductum***
91′ Lf gen 20–50 mm; involucres 2–5 per node; n CA-FP, < 2300 m .... var. ***nudum***

E. alpinum Engelm. (p. 871) TRINITY BUCKWHEAT [Group 3] Per from thick woody caudex, 3–8 cm; mats 3–15 cm diam. **LVS** ± basal; blade 10–30 mm, round, tomentose (or shedding above). **INFL** head-like; axis often ± prostrate, slender, tomentose; bracts ± lf-like, whorled near mid-axis; involucre 1, 3–6 mm, bell-shaped, tomentose, teeth 6–12. **FL**: perianth 3–8 mm, yellow to rose, glabrous, base funnel-shaped, indistinctly stalk-like. **FR** 4–5 mm, glabrous. **ENDANGERED** CA. Rocky slopes, ridges; 2200–2900 m. e KR (Mount Eddy area, s Siskiyou, ne Trinity cos.).

E. ampullaceum J. Howell (p. 871) [Group 1] Ann 5–30 cm. **LVS** basal; blade 5–15 mm, ± round, tomentose (esp below). **INFL** cyme-like, erect, slender, ± repeatedly forked; bracts scale-like; involucres sessile (few lateral), ± erect, 1.5–2 mm, ± ribbed, teeth 5. **FL**: perianth 1–1.5 mm, cream. **FR** 1–1.3 mm, glabrous. Uncommon. Sand; 1700–2200 m. n SNE; w NV.

E. angulosum Benth. (p. 871) [Group 2] Ann 10–90 cm. **LVS** basal and cauline on lower st; blade 5–40 mm, ± lanceolate, tomentose (esp below). **INFL** cyme-like; axes gen grooved or angled; bracts ± lf-like below (reduced upward), and scale-like; involucres on slender stalks, 1.5–3 mm, bell-shaped, unribbed, sparsely hairy, teeth 5, rounded. **FL**: perianth 1.5–1.8 mm, white to rose, minutely glandular, lobes ± cupped, inner narrower. **FR** 1–1.3 mm, glabrous. Common. Sand or clay; 150–1900 m. c&s SNF, Teh, SnJV, CW, TR, s PR, w DMoj. ❀TRY.

E. apiculatum S. Watson (p. 871) [Group 2] Ann 20–90 cm, ± glandular. **LVS** basal; blade 5–40 mm, ± oblanceolate, coarsely hairy and glandular. **INFL** cyme-like, slender, unevenly branched; involucres on thread-like stalks, ± 1.5 mm, unribbed, ± glabrous, teeth 4. **FL**: perianth 1.5–2.5 mm, ± white, short-white-hairy, outer lobes abruptly soft-pointed. **FR** ± 1.5 mm, glabrous. Uncommon. Granite sand; 1100–2700 m. c&s PR (Santa Rosa, Palomar, Cuyamaca mtns), sw DMtns (Little San Bernardino Mtns).

E. apricum J. Howell [Group 4] Per 8–25 cm, 10–25 cm diam, compact. **LVS** ± basal; blade 3–10 mm, ± round, tomentose below. **INFL** cyme-like, glabrous; branches ± short and equal; bracts scale-like; involucres sessile, gen terminal, 2–2.5 mm, glabrous, teeth 5. **FL**: perianth 2–3 mm, white, glabrous, stalk-like base 0. **FR** 2.5–3 mm, glabrous. Red clay of Ione Formation; ± 100 m. n SNF (w Amador Co.). Both vars. threatened by vehicles, mining.

var. ***apricum*** (p. 871) IONE BUCKWHEAT **INFL** erect. *n*=20. **ENDANGERED** CA. Habitat and range of sp. In cult.

var. ***prostratum*** R. Myatt IRISH HILL BUCKWHEAT **INFL** prostrate. **ENDANGERED** CA. Habitat and range of sp.

E. arborescens E. Greene (p. 871) [Group 4] Shrub 60–200 cm, 50–300 cm diam. **LVS** cauline, borne singly ± near st tips; blade 20–50 mm, ± linear, rolled under, tomentose (esp below). **INFL** cyme- to head-like, dense; branches many; bracts lf-like ± throughout; involucres gen short-stalked, 2–3 mm, bell-shaped, hairy, teeth 5–7. **FL**: perianth 2–4 mm, white to pinkish, ± hairy, stalk-like base 0. **FR** 2.5–3.5 mm, glabrous. *n*=20. Uncommon. Dry slopes; < 150 m. n ChI (exc San Miguel Island). ❀DRN:5,**14–17**,**24**&IRR:7,**8**,**9**,18,**19–23**; also STBL.

E. argillosum J. Howell (p. 871) CLAY-LOVING BUCKWHEAT [Group 1] Ann 8–60 cm. **LVS** basal; blade 5–50 mm, ± elliptic, ± tomentose (esp below). **INFL** cyme-like, subglabrous; lower bracts lf-like and scale-like, reduced upward; involucres slender-stalked (few lateral), 2–3 mm, ribbed, glabrous, teeth 5. **FL**: perianth 1–2 mm, white to rose, glabrous. **FR** 2–2.5 mm, glabrous. UNCOMMON. Bare clay (sometimes on serpentine); 150–800 m. s SnFrB, n SCoRI.

E. baileyi S. Watson [Group 1] Ann 10–50 cm. **LVS** basal; blade 5–20 mm, ± round, tomentose. **INFL** cyme-like, ± unevenly branched; bracts scale-like; involucres sessile (many lateral), 1–1.5 mm, ± ribbed, teeth 5. **FL**: perianth 1–2 mm, white to pink, subglabrous or minutely glandular. **FR** 1–1.5 mm, glabrous. Common. Sand or gravel; 650–2900 m. SNF, Teh, s SCoR, TR, GB, w DMoj; to WA, ID, w UT.

var. ***baileyi*** (p. 871) **INFL** glabrous. Common. Habitats and range of sp.

var. ***praebens*** (Gand.) Rev. **INFL** tomentose. Uncommon. Sand; 1300–2900 m. GB; w NV. [var. *divaricatum* (Gand.) Rev.]

E. beatleyae Rev. (p. 871) BEATLEY'S BUCKWHEAT [Group 4] Per 8–15 cm; mats 10–50 cm diam. **LVS** ± basal; blade 10–25 mm, ± oblanceolate, densely tomentose. **INFL** head-like; axis erect, glandular; bracts scale-like, ± hidden; involucres 3–6 per head, 3–4 mm, bell-shaped, rigid, glandular and thinly hairy, teeth 5. **FL**: perianth 2.5–4 mm, pale to reddish yellow, subglabrous, stalk-like base 0. **FR** 2.5–3.5 mm, glabrous. UNCOMMON. Dry volcanic outcrops; 1500–2500 m. n SNE; NV. Much like *E. ochrocephalum*.

E. bifurcatum Rev. (p. 871) FORKED BUCKWHEAT [Group 2] Ann 5–40 cm. **LVS** basal; blade 5–30 mm, round, tomentose (esp below). **INFL** cyme-like, 30–150 cm diam, spreading, flat-topped; bracts scale-like; involucres slender-stalked (esp lower), 2–2.5 mm, unribbed, glabrous, teeth 5. **FL**: perianth 1.5–2 mm, white to reddish, outer lobe bases cordate, inner lobes narrower. **FR** 2–2.5 mm, glabrous. *n*=20. RARE. Sand; 700–800 m. ne DMoj (s Inyo, ne San Bernardino cos.); s NV.

E. brachyanthum Cov. [Group 1] Ann 5–40 cm. **LVS** basal; blade 5–20 mm, ovate to round, tomentose (esp below). **INFL** ± cyme-like, densely short-branched, gen rounded, glabrous; bracts scale-like; involucres sessile (most lateral), ± 1 mm, angled, glabrous, teeth 5. **FL**: perianth ± 1 mm, yellow, glabrous, inner lobes narrower. **FR** ± 1 mm, glabrous. Sand; 600–2000 m. SNE, DMoj; NV. Locally common. ❀STBL.

E. brachypodum Torrey & A. Gray (p. 871) [Group 2] Ann 5–50 cm, 20–100 cm diam, minutely glandular. **LVS** basal; blade 10–50 mm, round-cordate, tomentose (esp below). **INFL** cyme-like, stout, gen ± flat-topped; bracts scale-like; involucres on slender, reflexed stalks, 1–2.5 mm, minutely glandular, unribbed, teeth 5. **FL**: perianth 1–2.5 mm, white to reddish, outer lobes gen bulged and cordate at base, inner narrower. **FR** 1–2 mm, glabrous. *n*=20. Sand or gravel; 150–2400 m. SNE, DMoj; to sw UT, nw AZ. Locally common. ❀STBL.

E. breedlovei (J. Howell) Rev. [Group 4] Per gen < 10 cm; mats 8–20 cm diam. **LVS** ± basal on branched caudex; blade 2–10 mm, elliptic, white-woolly below, gray-tomentose above. **INFL** densely umbel-like, sometimes ± compound; bracts scale-like; involucres ± few, slender-stalked, 2.5–4 mm, rigid, angled, teeth 7–9, becoming incurved. **FL**: perianth 2–4 mm, whitish to reddish, ± hairy, stalk-like base 0. **FR** 2–3 mm, glabrous. Granite or limestone; 1700–2500 m. s SNH. Much like *E. kennedyi*.

var. ***breedlovei*** (p. 871) BREEDLOVE'S BUCKWHEAT **INFL** dense, densely glandular; involucre teeth gen acute, inconspicuously scarious-margined. RARE. Metamorphic limestone; 2300–2500 m. s SNH (Piute Mtn, Kern Co.).

var. ***shevockii*** J. Howell (p. 871) THE NEEDLES BUCKWHEAT **INFL** dense to open, gen ± glabrous; involucre teeth rounded, obviously scarious-margined. UNCOMMON. Granite; 1700–2500 m. s SNH (The Needles, Baker Point, Little Kern River gorge, Kern & Tulare cos.).

E. butterworthianum J. Howell (p. 871) BUTTERWORTH'S BUCKWHEAT [Group 4] Subshrub to shrub 10–30 cm, 10–30 cm diam. **LVS** cauline; blade 5–20 mm, narrowly elliptic, ± rolled under, white-woolly below, gray-tomentose above, becoming reddish. **INFL**: involucres sessile, solitary, 5–6 mm, ± ribbed, teeth 5. **FL**: perianth 4–5 mm, brownish yellow to rose, glabrous, stalk-like base 0. **FR** ± 3 mm, glabrous. **RARE** CA. Dry sandstone; 650–700 m. n SCoRO (The Indians, Santa Lucia Mtns, c Monterey Co.).

E. cernuum Nutt. NODDING BUCKWHEAT [Group 2] Ann 5–60 cm, ± slender. **LVS** basal; blade 5–25 mm, ± round, white-tomentose (esp below). **INFL** cyme-like, glabrous, glaucous; bracts scale-like; involucres gen on reflexed stalks, 1–2 mm, unribbed, glabrous, teeth 5. **FL**: perianth 1–2 mm, white to pinkish, glabrous, outer lobes ± wavy-margined. **FR** 1.5–2 mm, glabrous. Sand or gravel; 1300–3300 m. c SNH (e slope), GB; to WA, w&c Can, c US, NM.

var. ***cernuum*** (p. 871) **INFL**: involucre stalk 2–25 mm, ascending, spreading, or reflexed. Uncommon. Sand or gravel; 2100–3300 m. c SNH (e slope), W&I; range of sp. outside CA.

var. ***viminale*** (S. Stokes) Rev. **INFL**: involucre stalk 0–2 mm, reflexed. Uncommon. Sand; 1300–2200 m. GB; to se OR, w UT.

E. cespitosum Nutt. (p. 871) [Group 3] Per from much-branched woody caudex, < 10 cm; mats 20–70 cm diam; some fls staminate. **LVS** ± basal on caudex branches; blade 2–5(15) mm, ± elliptic, rolled under, tomentose. **INFL**: bracts 0; involucre 1 per axis, 2–3.5 mm, lobes 6–9, long, reflexed, tomentose. **FL**: perianth 2.5–5 mm (staminate fl) or 6–10 mm (bisexual fl), yellow to reddish, hairy, stalk-like base short, distinct. **FR** 3.5–5 mm; tip sometimes slightly hairy. Sand or gravel; 1200–3100 m. GB; to OR, MT, Colorado. Locally common. ❀TRY;DFCLT.

E. cinereum Benth. (p. 871) [Group 4] Shrub 60–150 cm, 100–200 cm diam, gray-tomentose. **LVS** cauline; blade 15–30 mm, ± ovate, wavy-margined, tomentose (esp below). **INFL** compound; units head-like; bracts ± lf-like; involucres 3–10 per head, 3–5 mm, ± hairy, teeth 5. **FL**: perianth 2.5–3 mm, white to pinkish, densely long-hairy, stalk-like base 0. **FR** 2–2.5 mm, glabrous. *n*=40. Beaches, coastal bluffs; < 500 m. s CCo, w SCo, n ChI (Santa Rosa Island). Locally commmon. ❀DRN,DRY:**14**,19–21,**22,23**&SUN:5,**15–17,24**; also STBL.

E. cithariforme S. Watson (p. 875) [Group 1] Ann 20–50 cm. **LVS** basal (few barely cauline); blade 3–30 mm, lanceolate to round, wavy-margined, tomentose (esp below). **INFL** cyme-like; axes gen ± tomentose; bracts scale-like; branches few, upcurved; involucres sessile (many lateral), ± cylindric, 2.5–3 mm, ribbed, sometimes hairy, teeth 5. **FL**: perianth 1.5–2 mm, white to rose, glabrous. **FR** 1–1.5 mm, glabrous. Sand; 500–2400 m. SCoR, TR. [*E. gracile* Benth. var. *c.* (S. Watson) Munz] Common intermediates to *E. molestum* (lvs gen round-reniform) have been called var. *agninum* (E. Greene) Rev. [*E. gracile* Benth. var. *polygonoides* (S. Stokes) Munz] ❀STBL.

E. compositum Benth. var. ***compositum*** (p. 875) [Group 3] Per from woody caudex, erect, 20–70 cm. **LVS** ± basal; blade 30–250 mm, lanceolate to deltate, ± woolly below, ± thinly tomentose above. **INFL**: umbel, often compound; axis gen stout, hollow; bracts lf-like, reduced upward; involucres 1 per ultimate ray, 6–10 mm, lobes 5–10, long, gen spreading. **FL**: perianth 4–6 mm, cream to yellow, glabrous, stalk-like base very long, distinct. **FR** 5–6 mm; tip slightly hairy. *n*=20. Gravel, often on serpentine; < 2500 m. NW, CaR; to WA, ID. Locally common. ❀DRN,SUN:1–3,6,**7**,8,9,**14–24**.

E. congdonii (S. Stokes) Rev. (p. 875) CONGDON'S BUCKWHEAT [Group 3] Shrub, low, spreading, 5–40 cm, 30–60 cm diam. **LVS** cauline, ± clustered near st tips; blade 5–20 mm, ± oblanceolate, ± woolly below. **INFL** umbel-like, slender; bracts ± lf-like, subtending 1(3) ray(s); involucres 1 per ray, 5–7 mm, tomentose, lobes 6–8. **FL**: perianth 4–6 mm, yellow, glabrous, stalk-like base short, distinct. **FR** 4–5.5 mm; tip slightly hairy. UNCOMMON. Serpentine outcrops; 1000–2200 m. KR. [*E. ternatum* Howell var. *c.* (S. Stokes) J. Howell] ❀TRY.

E. contiguum (Rev.) Rev. (p. 875) REVEAL'S BUCKWHEAT [Group 2] Ann 3–30 cm, spreading. **LVS** basal; blade 4–20 mm, round-cordate, sparsely and coarsely hairy. **INFL** cyme-like, dense, slender; nodes glandular; bracts scale-like; involucres on ± erect, thread-like stalks, 1–2 mm, unribbed, glabrous, teeth gen 5. **FL**: perianth 1–2.5 mm, yellow, coarsely yellow-hairy. **FR** 1.5–2 mm, glabrous. *n*=16. RARE in CA. Sandy flats; 50–1000 m. e DMoj (se Inyo, ne San Bernardino cos.); sw NV.

E. covilleanum Eastw. (p. 875) [Group 1] Ann 10–40 cm. **LVS** basal; blade 3–15 mm, round to reniform, tomentose (esp below). **INFL** cyme-like, unevenly branched; bracts scale-like; involucres sessile, most lateral, 2–2.5 mm, ribbed, glabrous, teeth 5, ± white-margined. **FL**: perianth 2–2.5 mm, white to rose or pale yellow, ± glabrous. **FR** ± 2 mm, glabrous. *n*=12. Uncommon. Sand or gravel; 300–1600. c&s CW, WTR. ❀TRY.

E. crocatum Davidson (p. 875) CONEJO BUCKWHEAT [Group 3] Subshrub to shrub 30–50 cm, 30–100 cm diam, spreading. **LVS** cauline, dense; blade 10–35 mm, widely ovate, densely white-woolly. **INFL** umbel-like, often compound; bracts ± strap-like, ± hidden; involucre 1 per ultimate ray (central involucre often rayless), 3–4 mm, widely bell-shaped, woolly, teeth 5–6. **FL**: perianth 5–6 mm, bright yellow to red, gen subglabrous, inner lobes wider, stalk-like base very long, distinct. **FR** 2.5–3 mm, glabrous, unwinged. *n*=20. **RARE** CA. Dry rocky slopes; 50–150 m. s WTR (nw Santa Monica Mtns, Ventura Co.). In cult.

E. dasyanthemum Torrey & A. Gray (p. 875) [Group 1] Ann 5–60 cm. **LVS** basal; blade 5–20 mm, round, tomentose (esp below). **INFL** cyme-like, open; lower bracts lf- and scale-like, reduced upward; involucre sessile, ± erect, 3–4 mm, cylindric, ribbed, sometimes hairy between ribs, teeth 5. **FL**: perianth 2–2.5 mm, white to rose, ± densely long-hairy. **FR** 1.5–2 mm, glabrous. *n*=12. Uncommon. Sand; 150–600 m. c&s NCoR.

E. davidsonii E. Greene [Group 1] Ann 5–40 cm. **LVS** basal; blade 10–20 mm, round to reniform, wavy-margined, tomentose (esp below). **INFL** cyme-like, unequally branched; bracts scale-like; involucres sessile (most lateral), 3–4 mm, cylindric, ribbed, glabrous, teeth 5. **FL**: perianth 1.5–2 mm, white to pink or red, glabrous, inner lobes narrower, shorter. **FR** 2–2.5 mm, glabrous. Volcanic or granitic soils; 200–2800 m. SW, SNE, DMoj; to sw UT, n AZ, n Baja CA. [*E. molestum* S. Watson var. *d.* (E. Greene) Jepson] ❀DRN,DRY,SUN:1–3,**7**,8–10,**14–24**.

E. deflexum Torrey FLAT-TOPPED BUCKWHEAT [Group 2] Ann 5–200 cm. **LVS** ± basal; blade 10–40 mm, round to reniform, white-woolly below, tomentose above. **INFL** cyme-like, ± unevenly branched, gen ± widely spreading, glabrous; bracts scale-like; involucres on gen reflexed stalks, 1–3 mm, ± unribbed, teeth 5. **FL**: perianth 1–3 mm, white to pinkish, glabrous, outer lobes gen cordate at base, inner ± narrow. **FR** 1.5–3 mm, glabrous. *n*=20 (all vars.). Sandy desert scrub; < 2900 m. Teh, s SnJV, TR, SNE, D; to w UT, w NM, nw Mex.

var. ***baratum*** (Elmer) Rev. (p. 875) **INFL** sometimes narrow; main axes often stout or wider below main nodes; involucre stalk 3–15 mm, reflexed; involucre 2–3 mm. **FL**: perianth 1.5–2 mm, outer lobe bases truncate. Dry slopes; 900–2900 m. Teh, s SnJV, TR, n&w DMoj; s NV. [ssp. *b.* (Elmer) Munz] Locally common. ❀TRY.

var. ***deflexum*** (p. 875) **INFL**: axes ± slender; involucre stalk 0–5 mm, reflexed; involucre 1–2 mm. **FL**: perianth 1–2 mm, outer lobe bases cordate. Sand; < 2000 m. D; to s UT, AZ, nw Mex. Locally common. ❀TRY.

var. ***nevadense*** Rev. **INFL**: axes ± slender; involucre stalk 0–5 mm, reflexed; involucre ± 1.5 mm. **FL**: perianth 1–2 mm, outer lobe bases truncate. Sand; 1100–2200 m. SNE; to w UT. Locally common. Intergrades with var. *deflexum* in n DMoj. ❀TRY.

var. ***rectum*** Rev. **INFL** sometimes narrow; axes slender to stout; involucre 1.5–2 mm, erect, stalk 0–5 mm. **FL**: perianth 1–2 mm. Uncommon. Sand; < 1000 m. s DMoj, DSon. [*E. insigne* S. Watson misapplied to CA pls]

E. deserticola S. Watson (p. 875) [Group 4] Shrub 60–150 cm, 100–300 cm diam, openly branched. **LVS** cauline; blade 5–15 mm, ± ovate, densely tomentose. **INFL** cyme-like, open, fragile; bracts scale-like; involucres gen sessile (many lateral), 1.5–2 mm, gen ± hairy, teeth 4, very short. **FL**: perianth 2.5–3.5 mm, yellow, hairy, stalk-like base 0, lobe tips truncate. **FR** 3–4 mm, gen hairy. Uncommon. Sand; −65–100 m. s DSon (Imperial Co.); sw AZ, nw Sonora.

E. diclinum Rev. (p. 875) JAYNE'S CANYON BUCKWHEAT [Group 3] Subshrub 5–20 cm, dioecious; loose mats 10–50 cm diam. **LVS** clustered on low sts; blade 5–20 mm, elliptic to ovate, ± yellow-gray-woolly. **INFL** umbel- to ± head-like; axis ± tomentose; bracts lf-like, often ± hidden among fls; involucres sessile, 2.5–4 mm (larger in pistillate pls), tomentose, lobes 4–6. **FL**: perianth 2–3 mm (staminate fl) or 3.5–8 mm (pistillate fl), yellow to red, glabrous, stalk-like base short, distinct. **FR** 3–4 mm; tip slightly hairy. UNCOMMON. Serpentine; 1700–2400 m. KR (Siskiyou, Trinity cos.); sw OR. Closely related to *E. marifolium*.

E. douglasii Benth. var. ***douglasii*** (p. 875) [Group 3] Per from densely branched woody caudex, 3–10 cm; mats 5–40 cm diam. **LVS** clustered at branch-nodes and st tips; blade 5–15 mm, oblanceolate, white-tomentose. **INFL** appearing head-like; bracts lf-like, whorled near mid-axis (subtending solitary ray); involucre 1 per head, 2.5–3 mm, lobes 6–10, long, strongly reflexed, tomentose. **FL** perianth 5–8 mm, yellow, hairy, stalk-like base distinct. **FR** 3–4.5 mm; tip slightly hairy. Uncommon. Sand or gravel; 1200–2500 m. e CaRH, n SNH, MP; to WA, ID, NV. Closely related to *E. cespitosum*. ❀TRY.

E. eastwoodianum J. Howell EASTWOOD'S BUCKWHEAT [Group 1] Ann 10–50 cm. **LVS** basal; blade 5–30 mm, ± round, densely tomentose (esp below). **INFL** cyme-like, ± evenly branched; bracts scale-like; involucres stalked (some upper involucres sessile, lateral), 2–2.5 mm, ± ribbed, tomentose, teeth 5. **FL**: perianth 1.5–2 mm, white, glabrous. **FR** 1.6–2 mm, glabrous. *n*=17. UNCOMMON. Sand or barren clay; 500–1000 m. s SCoRI (Fresno, Monterey cos.). Not very distinct from *E. temblorense*, *E. vestitum*.

E. elatum Benth. [Group 4] Per 40–150 cm, 5–15 cm diam. **LVS** basal; blade gen 40–250 mm, ± lanceolate, gen tomentose (esp below). **INFL** cyme-like; bracts scale-like or barely lf-like; involucres 1–5 per cluster, gen short-stalked, 2.5–4 mm, teeth 5. **FL**: perianth 2.5–3 mm, white to red, base hairy, not stalk-like. **FR** 3.5–4 mm, glabrous. Sand or gravel; 600–3100 m n CA; to WA, ID, c NV.

var. ***elatum*** (p. 875) **INFL**: axes and involucres glabrous. *n*=20. Common. Habitats and range of sp. ❀DRN,SUN:1–3,6,**7**,14,**15–17**,18–21.

var. ***villosum*** Jepson **INFL**: axes and involucres long-hairy. Uncommon. Habitats of sp.; 700–2200 m. KR, CaR, n SN, MP; s OR, w NV. ❀TRY.

E. elegans E. Greene (p. 875) [Group 1] Ann 10–50 cm. **LVS** basal; blade 3–20 mm, ± round-cordate, wavy-margined, tomentose. **INFL** cyme-like, open; bracts scale-like; involucres sessile (most lateral), 1–1.5 mm, ribbed, teeth 5. **FL**: perianth 1–1.5 mm, white to pink, glabrous. **FR** 0.8–1 mm, glabrous. Uncommon. Sand or gravel; 200–1200 m. SnFrB, SCoR, WTR. ❀TRY.

E. elongatum Benth. var. ***elongatum*** (p. 875) [Group 4] Per 60–180 cm. **LVS** basal and some ± cauline; blade 10–30 mm, ± elliptic, wavy-margined, tomentose. **INFL** cyme-like, open; bracts scale-like (lowest sometimes lf-like); branches long, wand-like, tomentose; involucres sessile (most lateral), gen 6–7 mm, tomentose, teeth 5. **FL**: perianth 2.5–3 mm, white, glabrous, stalk-like base 0. **FR** 2–3 mm, glabrous. 2*n*=34. Common. Dry places; 50–1900 m. c&s CW, SW; Baja CA. Other vars. in Mex. ❀DRN, SUN:**7**,**14**,15,16,18,**19–24**.

E. eremicola J. Howell & Rev. (p. 875) WILD ROSE CANYON BUCKWHEAT [Group 2] Ann 8–25 cm, ± spreading. **LVS** basal; blade 3–25 mm, ± round, white-woolly below, tomentose above. **INFL** cyme-like, open, finely glandular; bracts scale-like; involucres on reflexed stalks (at least below), 1–2 mm, unribbed, teeth 5. **FL**: perianth 2–2.5 mm, white, yellow, or red, glabrous. **FR** 2–2.3 mm, glabrous. RARE. Sand or gravel; 2200–3100 m. s W&I (Inyo Mtns), n DMtns (Panamint Mtns).

E. ericifolium Torrey & A. Gray var. ***thornei*** Rev. & Henrickson THORNE'S BUCKWHEAT [Group 4] Subshrub 4–15 cm, 10–45 cm diam. **LVS** cauline; blade 4–6 mm, linear, rolled under, tomentose

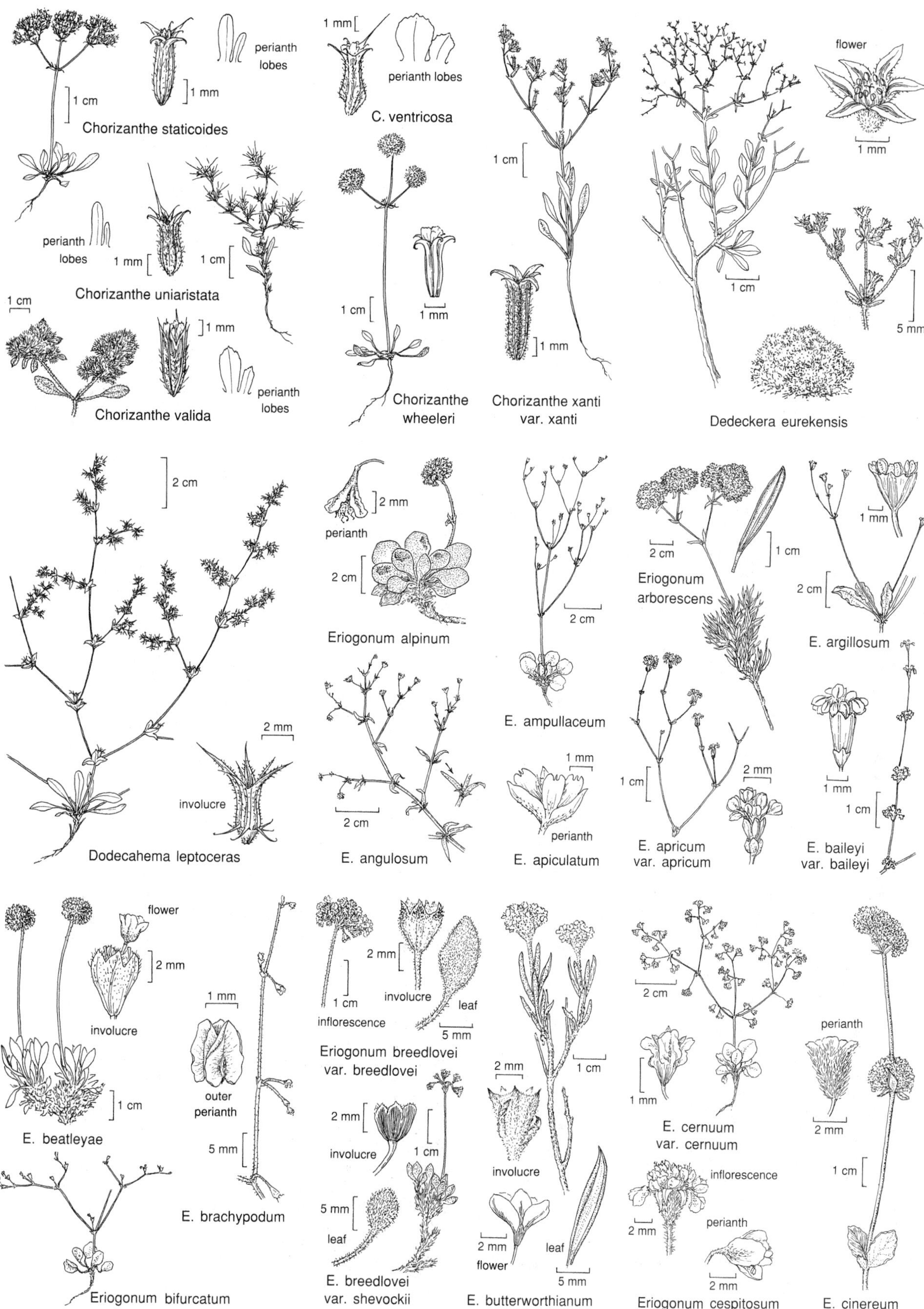

perianth lobes
1 cm
1 mm
Chorizanthe staticoides
perianth lobes
1 mm
1 cm
Chorizanthe uniaristata
1 cm
1 mm
perianth lobes
Chorizanthe valida
1 mm
perianth lobes
C. ventricosa
1 cm
1 mm
Chorizanthe wheeleri
1 cm
1 mm
Chorizanthe xanti var. xanti
flower
1 mm
1 cm
5 mm
Dedeckera eurekensis
2 cm
2 mm
involucre
Dodecahema leptoceras
2 mm
perianth
2 cm
Eriogonum alpinum
2 cm
E. angulosum
2 cm
E. ampullaceum
1 mm
perianth
E. apiculatum
2 cm
1 cm
Eriogonum arborescens
1 cm
2 mm
E. apricum var. apricum
1 mm
2 cm
E. argillosum
1 mm
1 cm
E. baileyi var. baileyi
flower
2 mm
involucre
1 cm
E. beatleyae
1 mm
outer perianth
5 mm
E. brachypodum
Eriogonum bifurcatum
2 mm
1 cm
inflorescence
involucre
leaf
5 mm
Eriogonum breedlovei var. breedlovei
2 mm
involucre
1 cm
5 mm
leaf
E. breedlovei var. shevockii
2 mm
1 cm
involucre
2 mm
flower
leaf
5 mm
E. butterworthianum
2 cm
1 mm
E. cernuum var. cernuum
inflorescence
2 mm
perianth
2 mm
Eriogonum cespitosum
perianth
2 mm
1 cm
E. cinereum

(esp below). **INFL** ± cyme-like, ± dense; bracts scale-like; involucres sessile or short-stalked, 1.5–2 mm, glabrous, teeth 5. **FL**: perianth 1.5–2 mm, white to pink, glabrous, stalk-like base 0. **FR** 2 mm, glabrous. **ENDANGERED** CA. Copper-rich gravel; ± 1800 m. e DMtns (New York Mtns). Other vars. in AZ. Very difficult to distinguish from *E. microthecum* complex.

E. esmeraldense S. Watson var. ***esmeraldense*** [Group 2] Ann 5–50 cm. **LVS** basal; blade 5–40 mm, obovate to round, sparsely and coarsely hairy. **INFL** cyme-like, gen spreading, glabrous, glaucous; bracts scale-like; involucres on slender, spreading to reflexed stalks, 0.8–1.8 mm, unribbed, glabrous, teeth 5. **FL**: perianth 1–3 mm, white to red, glabrous. **FR** 1.4–2.5 mm, glabrous. Sand; 1500–3200 m. c SNH (e slope), SNE; NV. Locally common. Another var. in c NV.

E. fasciculatum Benth. CALIFORNIA BUCKWHEAT [Group 4] Shrub 10–200 cm, 50–300 cm diam. **LVS** cauline, clustered at nodes; blade 6–18 mm, linear to oblanceolate, leathery, ± rolled under, ± white-tomentose below. **INFL**: head or umbel of heads, gen ± canescent; lowest bracts lf-like, upper scale-like; involucres gen several per head, 2.5–4 mm, angled, teeth 5. **FL**: perianth 2.5–3 mm, white to pinkish, stalk-like base 0. **FR** 1.8–2.5 mm, glabrous. Common. Dry slopes, washes, canyons in scrub; < 2300 m. c&s CA-FP, SNE, D; to UT, AZ, nw Mex.

var. ***fasciculatum*** Pl gen decumbent. **LF** linear, tightly rolled under, dark (or gray-) green, glabrous above. **INFL**: involucre 3–4 mm, gen glabrous. **FL**: perianth ± glabrous. *n*=20. Coastal scrub, often on bluffs; < 400 m. CCo. ❀DRN:5,**15–17**&IRR:7–9,**14**, **19–23**;GRNCVR;CVS.

var. ***flavoviride*** Munz & I.M. Johnston Pl ± rounded. **LF** linear, tightly rolled under, light (or yellow-) green, glabrous above. **INFL**: involucre 2–3 mm, gen glabrous. **FL**: perianth gen hairy inside. *n*=20. Dry slopes, washes; 200–1300 m. D; Baja CA. ❀TRY.

var. ***foliolosum*** (Nutt.) Abrams (p. 875) Pl ± rounded. **LF** linear, tightly rolled under, dark (or gray-) green, sometimes thinly tomentose above. **INFL**: involucre 3–4 mm, hairy. **FL**: perianth hairy outside. *n*=40. Dry slopes, canyons; < 1600 m. CW, SW; nw Mex. Important honey plant, planted more widely. ❀DRN,SUN: 5,**7**,**14–24**; CVS,GRNCVR; also STBL.

var. ***polifolium*** (A. DC.) Torrey & A. Gray (p. 875) Pl ± rounded. **LF** oblanceolate, slightly rolled under, ± tomentose above. **INFL** ± tomentose; involucre 2.5–3.5 mm, hairy. **FL**: perianth hairy. *n*=20. Dry slopes, washes; 60–2300 m. s SN, SCoRI, e SCo, TR, e PR, SNE, D; to sw UT, w AZ, nw Mex. ❀DRN,DRY,SUN: **7**,8–11,**14–16**,17,**18–24**;STBL.

E. foliosum S. Watson (p. 875) LEAFY BUCKWHEAT [Group 1] Ann 5–10 cm, 10–50 cm diam. **LVS** ± basal; blade 5–15 mm, oblong to round, tomentose (esp below). **INFL** cyme-like, open, tomentose; lowest bracts sometimes lf-like, upper scale-like; involucres sessile (most lateral), ± 1 mm, ± hairy, teeth 5. **FL**: perianth ± 1 mm, white to rose, glabrous, outer lobes widest near base, inner narrower, longer. **FR** ± 1 mm, glabrous. RARE. Sand; 1200–2200 m. SnBr (Bear Valley), PR (scattered); n Baja CA. [var. *hastatum* (Wiggins) Rev.]

E. giganteum S. Watson [Group 4] Shrub 30–350 cm, 50–300 cm diam, often rounded. **LVS** cauline, ± clustered at st tips and below infls; blade 20–100 mm, narrowly elliptic to ovate, leathery, white-woolly (esp below). **INFL** cyme-like, gen very dense; bracts ± lf-like, smaller upward; involucres crowded, ± sessile, 3–5 mm, bell-shaped, tomentose, teeth 5. **FL**: perianth 2–4 mm, white to rose, hairy, stalk-like base 0. **FR** 2–3.5 mm, glabrous. Dry rocks; < 450 m. s ChI. Hybridizes with *E. arborescens*, *E. fasciculatum* where cult on mainland; vars. indistinct.

var. ***compactum*** Dunkle SANTA BARBARA ISLAND BUCKWHEAT Pl < 100 cm. **LF**: blade 25–60 mm, ± ovate. **INFL** dense or open; involucres 3–5 mm. **FL**: perianth 2–2.5 mm. **FR** 2–2.5 mm. **RARE** CA. Dry rocks; < 300 m. s ChI (Santa Barbara Island). In cult.

var. ***formosum*** M.K. Brandegee (p. 875) SAN CLEMENTE ISLAND BUCKWHEAT Pl < 250 cm. **LF**: blade 50–80 mm, narrowly elliptic to oblong. **INFL** dense or open; involucre 4–5 mm. **FL**: perianth 3–4 mm. **FR** 3–3.5 mm. RARE. Dry rocks; < 100 m. s ChI (San Clemente Island). In cult.

var. ***giganteum*** SANTA CATALINA ISLAND BUCKWHEAT Pl gen 150–350 cm. **LF**: blade 30–100 mm, ovate. **INFL** open; involucre 3–4 mm. **FL**: perianth 2–2.5 mm. **FR** 2–2.5 mm. *n*=20. Uncommon. Dry slopes, ridges; < 450 m. s ChI (Santa Catalina Island). ❀DRN,SUN:5,**14–17**,**24**&IRR:7,**8**,**9**,18,**19–23**.

E. gilmanii S. Stokes (p. 875) GILMAN'S BUCKWHEAT [Group 4] Per from a shredding woody caudex, 1–5 cm; mats < 20 cm diam. **LVS** ± rosetted on low sts; blade 2–4 mm, elliptic, tomentose. **INFL** dense, gen ± head-like; bracts scale-like; involucres gen several per head, ± sessile, gen 3–4 mm, rigid, teeth 5. **FL**: perianth ± balloon-like, 3.5–5 mm, reddish, glabrous, lobes 3–5 mm, outer shorter, ± round, cupped, inner narrow, stalk-like base 0. **FR** 2.5–3 mm, glabrous. UNCOMMON. Gravel slopes; 1800–2000 m. n DMtns (Panamint, Last Chance mtns).

E. glandulosum (Nutt.) Benth. (p. 875) [Group 2] Ann 5–40 cm, ± widely spreading. **LVS** basal; blade 5–15 mm, widely elliptic to round, sparsely and coarsely hairy. **INFL** cyme-like, flat-topped, glandular ± throughout; bracts scale-like; involucres on slender, ± reflexed stalks, ± 1 mm, unribbed, glabrous, teeth 5. **FL**: perianth 1–2 mm, white to pink, stiffly white-hairy, lobes ± widely lanceolate. **FR** 1–1.3 mm, glabrous. Sand or gravel; 800–1600 m. n&ne DMtns; sw NV. [*E. carneum* (J. Howell) Rev.] Locally common.

E. gossypinum Curran (p. 875) COTTONY BUCKWHEAT [Group 2] Ann 5–20 cm. **LVS** basal and seemingly cauline; blade 3–40 mm, ± oblanceolate, ± white-tomentose. **INFL** cyme-like, forked; bracts lf- and scale-like below, reduced upwward; involucres on slender spreading stalks, ± 3 mm, unribbed, subglabrous outside, cottony inside (hairs obscuring fls), teeth 5. **FL**: perianth ± 1.5 mm, white to rose, glandular. **FR** 1.3–1.5 mm, glabrous. *n*=20. UNCOMMON. Clay hills; 100–500 m. s SNF (Greenhorn Mtns), sw SnJV.

E. gracile Benth. [Group 1] Ann 20–60 cm. **LVS** basal and barely cauline; blade 10–60 mm, ± elliptic or (ob)lanceolate, wavy-margined, tomentose (esp below). **INFL** cyme-like, unevenly branched; bracts lf-like below, scale-like above; involucres sessile (most lateral), ± 2–3 mm, ± ribbed, ± hairy, teeth 5. **FL**: perianth 1–2 mm, white to pink or yellow, glabrous. **FR** 1–1.5 mm, glabrous. *n*=11. Sand; < 1900 m. CA–FP (exc w SW); nw Mex. Highly variable; young pls much like *E. cithariforme*.

var. ***gracile*** (p. 875) **INFL**: axis gen tomentose. **FL**: perianth white or yellow. Common. Habitats and range of sp.

var. ***incultum*** Rev. **INFL**: axis ± glabrous or thinly hairy at base. **FL**: perianth white to pale yellow. Uncommon. Sand; 1200–1900 m. PR. Populations scattered.

E. gracilipes S. Watson (p. 875) [Group 4] Per 5–10 cm; mats 5–20 cm diam. **LVS** clustered on low sts; blade 5–20 mm, ± elliptic, white-tomentose. **INFL** head-like; axis glandular; bracts scale-like; involucres 5–7 per head, sessile, 2–3 mm, flexible, ± hairy, teeth 5. **FL**: perianth 2–3 mm, white to rose, glabrous, stalk-like base 0. **FR** 2–2.5 mm, glabrous. Uncommon. Dry granite; 2900–4000 m. c&s SNH (e slope), n W&I (White Mtns); w NV. ❀TRY;DFCLT.

E. gracillimum S. Watson (p. 875) [Group 2] Ann 5–50 cm. **LVS** basal and seemingly cauline; blade 5–40 mm, ± elliptic, white-tomentose (esp below), margins wavy, ± rolled under. **INFL** cyme-like, slender; bracts lf- and scale-like below, reduced upward; axes gen ± grooved; involucres on slender, spreading stalks, ± 2 mm, bell-shaped, nearly unangled, finely glandular, teeth 5. **FL**: perianth ± 2 mm, white to rose, subglabrous, margins gen ± wavy. **FR** ± 1 mm, glabrous. *n*=20. Common. Clay to gravel; 500–1300 m. s SN, SCoR, TR. ❀STBL.

E. grande E. Greene [Group 4] Per; mats 10–220 cm. **LVS** basal and sometimes cauline low on sts; blade 20–100 mm, ± oblong-ovate, strongly wavy-margined, woolly (esp below). **INFL** cyme- or umbel-like, gen stout; bracts ± scale-like; involucres 1–3 per cluster, sessile, 5–7 mm, angled, sometimes hairy, teeth 5–8. **FL**: perianth 2–4 mm, white to red, glabrous, stalk-like base 0. **FR** 2.5–3 mm, glabrous. Dry cliffs; < 200 m. ChI; nw Baja CA. [*E. latifolium* ssp. *g.* (E. Greene) S. Stokes] var. ***grande*** ISLAND BUCKWHEAT Pl 50–220 cm. **LF**: blade 20–100 mm, greenish above. **INFL**: involucre 5–6 mm, gen narrowly obconic. **FL**: perianth white. *n*=20. UNCOMMON. Habitats of sp. ChI (Santa Cruz, Anacapa, Santa Catalina, San Clemente Islands). ❀DRN,SUN:5,15,16,**17,24**&IRR:8,9,**14,19–23**.

var. ***rubescens*** (E. Greene) Munz SAN MIGUEL ISLAND BUCKWHEAT Pl 15–50 cm. **LF**: blade 20–60 mm, greenish above. **INFL**: involucre 5–7 mm, ± widely bell-shaped. **FL**: perianth light pink to red. UNCOMMON. Habitats of sp. n ChI. ❀DRN,SUN: 5,**15–17,23–24**&IRR:**14**,18–20,**21–22**;CVS.

var. ***timorum*** Rev. (p. 875) SAN NICOLAS ISLAND BUCKWHEAT Pl 10–25 cm. **LF**: blade 15–45 mm, ± gray above. **INFL**: involucre 4–5 mm, bell-shaped. **FL**: perianth white. **ENDANGERED** CA. Dry cliffs; < 50 m. s ChI (San Nicolas Island).

E. heermannii Durand & Hilg. [Group 4] Shrub 10–200 cm, 10–250 cm diam, gen rounded. **LVS** cauline; blade 5–40 mm, ± linear to spoon-shaped, gen ± tomentose. **INFL** cyme-like, gen dense; bracts scale-like; axes many, rigid, ± widely spreading; involucres sessile, 1–2 mm, glabrous, teeth 5. **FL**: perianth 1.5–4 mm, white or yellowish to rose, glabrous, outer lobes ± obovate-cordate, inner ± oblanceolate, stalk-like base 0. **FR** 2–2.5 mm, glabrous. Dry rocks, washes; 600–2800 m. c&s SNH (e slope), Teh, s SCoR, TR, SNE, esp DMoj; to UT, n AZ. Highly variable complex.

var. ***argense*** (M.E. Jones) Munz Pl 10–45 cm, 20–80 cm diam, ± low, spreading. **LF**: blade 5–10 mm, oblanceolate to spoon-shaped. **INFL**: axes round, glabrous. **FL**: perianth 1.5–2.5 mm, white. Uncommon. Limestone outcrops, cliffs; 1200–2800 m. DMtns; s NV.

var. ***floccosum*** Munz CLARK MOUNTAIN BUCKWHEAT Pl 30–60 cm, 40–80 cm diam. **LF**: blade 5–15 mm, ± linear, rolled under. **INFL**: axes round, thinly tomentose. **FL**: perianth 2–3 mm, yellowish white. UNCOMMON. Dry slopes, washes; 900–2400 m. e DMtns; s NV. ❀TRY.

var. ***heermannii*** (p. 877) Pl 50–150 cm, 60–200 cm diam. **LF**: blade 5–15 mm, gen ± lanceolate, gen thinly tomentose. **INFL**: axes round, glabrous. **FL**: perianth 2.5–4 mm, yellowish white. Gravel; 1000–1900 m. s SNH (Kern Co.), Teh, n TR, se SCoRO (se San Luis Obispo Co.). ❀TRY.

var. ***humilius*** (S. Stokes) Rev. Pl 30–70 cm, 50–120 cm diam, low, spreading. **LF**: blade 8–15 mm, oblanceolate to spoon-shaped. **INFL** ± open to dense; axes round, glabrous. **FL**: perianth 2.5–3 mm, white. Gravel; 1000–2000 m. c&s SNH (e slope), SNE, n DMtns; NV.

var. ***occidentale*** S. Stokes WESTERN HEERMANN'S BUCKWHEAT Pl 100–200 cm, 100–250 cm diam. **LF**: blade 15–40 mm, ± lanceolate, gen ± glabrous above. **INFL** ± open; axes round, glabrous. **FL**: perianth 3–4 mm, white to pink. UNCOMMON. Clay or shale slopes; 600–1000. ± c SCoRI (San Benito, e Monterey cos.).

var. ***sulcatum*** (S. Watson) Munz & Rev (p. 877). Pl 10–40 cm, 15–60 cm diam. **LF**: blade 4–12 mm, narrowly (ob)lanceolate to spoon-shaped. **INFL**: axes deeply grooved, minutely scabrous. **FL**: perianth 1.5–2 mm, yellowish white. Uncommon. Limestone outcrops; 700–2700 m. DMtns; to sw UT, nw AZ.

E. heracleoides Nutt. var. ***heracleoides*** (p. 877) [Group 3] Per 5–40 cm; loose mats < 10 m diam. **LVS** clustered on low sts; blade 10–50 mm, ± oblanceolate, tomentose (esp below). **INFL** umbel-like, sometimes compound; bracts lf-like, gen whorled near mid-axis (reflexed bracts also subtend umbels); rays gen 5–8; involucre gen 1 per ultimate umbel ray, 3–4.5 mm, lobes 5–12, long, reflexed. **FL**: perianth gen 3–6 mm, gen ± cream, ± glabrous, stalk-like base long, distinct. **FR** 2–5 mm, glabrous or tip thinly hairy. Sand or gravel slopes; 1200–2800 m. Wrn; to B.C., MT, Colorado. Close to *E. umbellatum*. ❀DRN,SUN,IRR:1–3,**7,14–18**,19–21.

E. hirtellum J. Howell & Bacigal. (p. 877) KLAMATH MOUNTAIN BUCKWHEAT [Group 3] Per from stout woody caudex, 10–40 cm; mats 20–40 cm diam, ± glabrous. **LVS** ± basal; blade 5–20 mm, narrowly elliptic to ovate, yellow-green. **INFL** ± ball-like (branches and bracts ± hidden); axis stout; involucres sessile, 5–6 mm, lobes 5–6, erect. **FL**: perianth 3–4 mm, bright yellow or rose, hairy, stalk-like base very short, indistinct. **FR** 2–3 mm; tip slightly hairy. RARE. Serpentine slopes; 1400–1900 m. n KR (e Del Norte, Siskiyou cos).

E. hirtiflorum S. Watson (p. 877) [Group 2] Ann 5–15 cm. **LVS** basal, ± sessile, 5–50 mm, ± oblanceolate, glabrous, ciliate. **INFL** cyme-like, open, ± forked, ± glandular; bracts lf-like below, scale-like above; involucres gen on slender stalks, ± 1 mm, subglabrous, 2-fld, teeth 4. **FL**: perianth ± 1 mm, white to reddish, hooked-bristly. **FR** ± 1 mm, glabrous. Sand or gravel; gen 500–2000 m. NW, SN, n CW, TR. Widely scattered. ❀TRY.

E. hoffmannii S. Stokes [Group 2] Ann 10–100 cm. **LVS** ± basal; blade 10–50 mm, ± round, tomentose (esp below). **INFL** cyme-like, open; bracts scale-like; main axes gen dominant; involucres ± sessile (many lateral), 1–2 mm, unribbed, teeth 5. **FL**: perianth 1.5 mm, white to reddish, glabrous; stamen incl, glabrous. **FR** 2 mm, glabrous. *n*=20. Dry slopes; 300–1700 m. n DMtns (Death Valley region, Inyo Co.).

var. ***hoffmannii*** HOFFMANN'S BUCKWHEAT Pl 10–50 cm, spreading. **LF**: blade flat. **FL**: perianth lobes narrowly spoon-shaped. UNCOMMON. Dry slopes; 1000–1700 m. n DMtns (w slope Panamint Mtns).

var. ***robustius*** S. Stokes ROBUST HOFFMANN'S BUCKWHEAT Pl 30–100 cm, erect. **LF**: blade strongly wavy-margined. **FL**: perianth lobes narrowly ovate. UNCOMMON. Dry slopes; 300–750 m. ne DMtns (Black, Funeral mtns).

E. hookeri S. Watson (p. 877) [Group 2] Ann 10–60 cm. **LVS** basal; blade 10–50 mm, ± round, white-tomentose (esp below). **INFL** cyme-like, gen umbrella-like; bracts scale-like; involucres sessile, reflexed, 1–2 mm, widely bell-shaped, unribbed, teeth 5. **FL**: perianth ± 2–4 mm, yellow or reddish, outer lobes round, bases ± cordate, inner lobes narrower. **FR** 2–2.5 mm, glabrous. *n*=20. Uncommon. Sand or gravel; 1000–2500 m. SNE; to ID, WY, Colorado, NM.

E. incanum Torrey & A. Gray [Group 3] Per from densely branched woody caudex, 3–20 cm, dioecious; mats 10–50 cm diam. **LVS** clustered on low sts; petiole 3–10 mm; blade 5–15 mm, ± elliptic, ± densely white-, yellow-, or gray-tomentose or -woolly. **INFL** head- or umbel-like, gen ± tomentose; bracts ± lf-like; involucres sessile, gen dense, 2–3 mm, tomentose, teeth 5–8. **FL**: perianth 2–3 mm (staminate) or 4–6 mm (pistillate), yellow to red, glabrous, stalk-like base distinct. **FR** 3–3.5 mm; tip slightly hairy. Sand; 2100–3700 m. SNH; also OR. Intergrades with *E. marifolium* in c SNH. ❀TRY;DFCLT.

E. inerme (S. Watson) Jepson (p. 877) [Group 2] Ann 5–50 cm, glandular. **LVS** basal; blade 3–30 mm, lanceolate to spoon-shaped, sessile, glabrous, ciliate. **INFL** cyme-like, slender, ± forked; bracts lf-like, smaller upward; involucres gen on short stalks, 1.5–2 mm, ± bell-shaped, unribbed, often minutely hairy, lobes 4. **FL**: perianth ± 1.5 mm, pink to red, white-hooked-bristly. **FR** 1.5–1.9 mm, glabrous. Sand; 600–2100 m s NCoR, c&s SN, Teh, CW, WTR. [var. *hispidulum* Goodman]

E. inflatum Torrey & Frémont DESERT TRUMPET [Group 4] Ann or per 10–150 cm. **LVS** basal; blade 5–50 mm, oblong to reniform, ± coarsely short-hairy. **INFL** cyme-like, glabrous, glaucous; main axes gen wider below nodes; bracts scale-like; involucres on erect or spreading, thread-like stalks, 1–1.5 mm, teeth 5. **FL**: perianth 1–3 mm, yellow, coarsely white-hairy, lobes narrowly ovate. **FR** 2–2.5 mm, glabrous. *n*=16. Dry sand or gravel; –15–2000 m. GB, D; to Colorado, NM, nw Mex.

var. ***deflatum*** I.M. Johnston Per, long-lived, not fl 1st year, 50–150 cm. **INFL**: branches many, slender. Habitats of sp.; < 1100 m. s DMoj, DSon; s NV, w AZ, nw Mex. Young pls much like var. *inflatum*. ❀TRY.

var. ***inflatum*** (p. 877) Ann or per, often fl 1st year, 2–80 cm. **INFL**: branches few–many; main axes wider below nodes. Common. Habitats and range of sp. Nearly identical to *E. trichopes* var. *hooveri* of e CW. ❀DRN,DRY,SUN:1–3,7–9,**10–13**,14,19–24.

E. intrafractum Cov. & C. Morton (p. 877) JOINTED BUCKWHEAT [Group 4] Per 60–150 cm. **LVS** basal; blade 25–70 mm, oblong-ovate, ± woolly. **INFL** ± cyme-like; axes few, stout, brittle, finally breaking into napkin-ring-like segments; bracts ± lf-like at first node, scale-like above; branches 2–3, long, wand-like; involucres gen 3 per node, sessile, 2.5–3 mm, bell-shaped, splitting as fls mature, teeth 5. **FLS** many per involucre; perianth 1.5–3 mm, yellow to red, bristly, lobes oblanceolate, stalk-like base 0. **FR** 2–2.5 mm; tip bristly. *n*=20. UNCOMMON. Dry limestone; 1100–1600 m. n DMtns (Grapevine, Panamint mtns).

E. kelloggii A. Gray (p. 877) KELLOGG'S BUCKWHEAT [Group 3] Per from spreading woody caudex, 4–9 cm; loose mats 20–50 cm diam. **LVS** clustered on low sts; blade 4–10 mm, ± oblanceolate, silvery-silky (esp below). **INFL** head-like; bracts lf-like, whorled near mid-axis; involucre 1 per head, 4–6 mm, lobes 6–8, short, erect. **FL**: perianth 5–7 mm, whitish to rose, glabrous, stalk-like base short, distinct. **FR** 4–5 mm; tip slightly hairy. **ENDANGERED** CA. Serpentine soil; 1000–1200 m. c NCoRO (Red Mtn, Mendocino Co.). In cult.

E. kennedyi S. Watson [Group 4] Per 5–15 cm; cushions 10–50 cm diam. **LVS** densely clustered on low sts; blade 2–12 mm, elliptic to oblong, often rolled under, tomentose. **INFL** head-like, gen ± tomentose; bracts scale-like, subtending involucres; involucres 3–8, sessile, 1.5–4 mm, rigid, ± hairy, teeth 5–7. **FL**: perianth 1.5–3.5 mm, white to rose, glabrous, lobes ± widely elliptic, stalk-like base 0. **FR** 2–4 mm, glabrous. Dry gravel; 1200–3600 m. c&s SN, TR, SNE, nw DMtns.

var. ***alpigenum*** (Munz & I.M. Johnston) Munz & I.M. Johnston (p. 877) SOUTHERN ALPINE BUCKWHEAT **LF**: blade 2–4 mm, oblanceolate to elliptic, tip rounded. **INFL**: axis gen 1–2 cm; involucre 1–2 mm. **FL**: perianth 1.5–2.5 mm. **FR** 1.8–2 mm. UNCOMMON. Dry granitic gravel; 2600–3500 m. SnGb, SnBr.

var. ***austromontanum*** Munz & I.M. Johnston (p. 877) SOUTHERN MOUNTAIN BUCKWHEAT **LF**: blade gen 6–10 mm, oblanceolate to elliptic, tip acute. **INFL**: axis 8–15 cm; involucre 2.5–4 mm. **FL**: perianth 2–3 mm. **FR** 3.5–4 mm. RARE. Dry gravel in yellow-pine forest; 1900–2100. n WTR (Mount Pinos), SnBr (Bear Valley). In cult.

var. ***kennedyi*** **LF**: blade 2–5 mm, oblanceolate to elliptic, tip rounded. **INFL**: axis 4–12 cm; involucre 1.5–2.5 mm. **FL**: perianth 1.5–2.5 mm. **FR** 2–2.5 mm. Uncommon. Dry gravel or rocks; 1500–2600 m. n WTR (Mount Pinos), SnBr. ❀DRN,SUN:1–3,7, 15–18;DFCLT.

var. ***pinicola*** Rev. KERN BUCKWHEAT **LF**: blade 3–5 mm, oblong. **INFL**: axis 5–13 cm; involucre 2.5–3.5 mm. **FL**: perianth 2.5–3.5 mm. **FR** 2.5–3 mm. RARE. Dry ridges; 1500–1700 m. se-most SNF (s of Cache Peak, se Kern Co.)

var. ***purpusii*** (Brandegee) Rev. **LF**: blade 2.5–6 mm, oblong. **INFL**: axis 4–10 cm; involucre 1.5–2 mm, glabrous. **FL**: perianth 2–2.5 mm. **FR** 2.5–3 mm. Uncommon. Dry granite sand; 1200–2500 m. c&s SN (e slope), SNE, nw DMtns (Argus, Coso mtns).

E. latens Jepson [Group 3] Per from thick woody caudex, 10–50 cm; mats 10–20 cm diam. **LVS** basal; blade 10–35 mm, gen ± round, ciliate, subglabrous, green. **INFL** ball-like (branches and bracts ± hidden); axis stout, ± glabrous; involucres gen many per head, 6–8 mm, lobes 5–8, ± spreading. **FL**: perianth 3–6 mm, cream to pale yellow, stalk-like base indistinct, angled, hairy. **FR** 3–5 mm, glabrous. Uncommon. Granite sand; 2500–3400 m. c&s SNH (e slope), W&I, n DMtns (Panamint Mtns); w NV. ❀DRN, SUN,IRR:1–3,7,14,**15–17**,18.

E. latifolium Smith (p. 877) COAST BUCKWHEAT [Group 4] Per 10–70 cm, 5–20 cm diam. **LVS** basal and cauline on lower st, 15–50 mm, oblong to ovate, white-tomentose (esp below), margins often wavy. **INFL** head- to umbel-like, sometimes compound; axes gen ± tomentose; involucres 3–20+ per head, 3.5–6 mm, gen tomentose, teeth 5–8. **FL**: perianth 3–3.5 mm, white to red, glabrous, stalk-like base 0. **FR** 3–4 mm, glabrous. *n*=20. Common. Coastal bluffs, scrub; < 150 m. NCo, n&c CCo; to WA. Glabrous pls have been called *E. oblongifolium* Benth. ❀DRN,SUN:4,5,**17**&IRR: 14,**15,16,22–24**.

E. libertini Rev. (p. 877) DUBAKELLA MOUNTAIN BUCKWHEAT [Group 3] Per from compact woody caudex, 5–30 cm; mats 30–50 cm diam. **LVS** clustered on low sts; blade 5–20 mm, ± elliptic, tomentose (esp below). **INFL** head-like; axis slender, thinly tomentose; bracts lf-like, whorled at mid-axis; involucre 1 per head, 4–8 mm, tomentose, teeth 5–8. **FL**: perianth 5–8 mm, yellow, glabrous, stalk-like base distinct. **FR** 4–5 mm; tip slightly hairy. UNCOMMON. Serpentine outcrops; 1200–1600 m. s KR, n NCoRH, n NCoRI (Trinity, Shasta, Tehama cos.). ❀TRY.

E. lobbii Torrey & A. Gray var. ***lobbii*** (p. 877) [Group 3] Per from thick, sparsely branched, woody caudex, gen 3–15 cm, 5–40 cm diam. **LVS** basal; blade 10–50 mm, ± round, woolly below, ± tomentose above. **INFL** densely umbel-like (sometimes compound); axis often ± prostrate, slender, tomentose; bracts lf-like, subtending branches; involucre 1 per ultimate ray, 4–12 mm, lobes 6–10, long, ± spreading. **FL**: perianth 5–7 mm, white to rose, dark-striped, glabrous, stalk-like base very short, ± indistinct. **FR** 5–8 mm, glabrous. Common. Open, rocky slopes and ridges; 1400–3500 m. NW, CaR, SN, GB; OR, w NV. Another var. in w NV. ❀DRN,SUN:1–3,6,7,14–18;DFCLT.

E. luteolum E. Greene [Group 1] Ann 5–60 cm. **ST** prostrate to erect, 5–60 cm. **LVS** basal and often some cauline; blade 5–50 mm, gen round, wavy-margined, densely tomentose (esp below). **INFL** cyme-like; lowest bracts often lf-like, upper scale-like; involucres gen sessile (some lateral), ± 3 mm, ± cylindric, ribbed, gen glabrous, teeth 5. **FL**: perianth gen 2–3 mm, white to rose or yellow, glabrous; stamens ± incl, glabrous. **FR** 1–2 mm, glabrous. *n*=12. Open serpentine or granite; < 2300 m. n CA-FP; s OR. Vars. merge.

var. ***caninum*** (E. Greene) Rev. (p. 877) TIBURON BUCKWHEAT **INFL** prostrate to ascending, 5–30 cm, ± evenly branched; few involucres lateral. **FL**: perianth ± 2 mm, white to rose. RARE. Open serpentine; < 500 m. c NCoRI (Colusa Co.), n CCo, n SnFrB (Marin, formerly Alameda cos.). [*E. c.* (E. Greene) Munz]

var. ***luteolum*** (p. 877) **INFL** erect, 20–60 cm, unevenly branched; many involucres lateral. **FL**: perianth ± 3 mm, white or pale yellow to red. Common. Serpentine or granite sand or gravel; 100–2400 m. NW, CaR, n&c SN, SnFrB. [var. *pedunculatum* (S. Stokes) Rev. = *E. p.* S. Stokes; var. *saltuarium* Rev.] Variable; no characters known reliably segregate other taxa. Intergrades with *E. vimineum* in ne NW, CaR. ❀DRN,SUN:1–3,**7**,14,**15–17**,18–21, **22–24**.

E. maculatum A.A. Heller (p. 877) [Group 2] Ann 5–40 cm. **LVS** basal (some seemingly cauline); blade 5–40 mm, lanceolate to obovate, densely white-tomentose (esp below); margin sometimes wavy or rolled under. **INFL** cyme-like, open; axes ± angled, ± tomentose; lower bracts lf- and scale-like, reduced upward; involucres on thread-like, spreading stalks, 1–2 mm, cup-shaped, unribbed, glandular, teeth 5. **FL**: perianth 1–2.5 mm, white to red or yellow, striped rose to purple, glandular, outer lobes wider, base ± cupped, inner lobes longer. **FR** 1–1.5 mm, glabrous. *n*=20. Common. Gravel to clay soils; 100–2500 m. s SN, TR, GB; to WA, UT, Baja CA. ❀TRY.

E. marifolium Torrey & A. Gray (p. 877) [Group 3] Subshrub 10–40 cm, loose mats 20–70 cm diam, dioecious. **LVS** clustered on low sts; blade 3–30 mm, elliptic to ovate, woolly below, finally shedding and olive-green above. **INFL** umbel- or ± head-like; axis slender, gen thinly tomentose; bracts lf-like, subtending rays; central ray often shorter; involucres terminal, 2–3 mm, gen ± tomentose, lobes 5–6, short, erect. **FL**: perianth 1–2 mm and dull yellow (staminate) or 4–7 mm and bright yellow to red (pistillate), glabrous, stalk-like base distinct. **FR** 3.5–5 mm; tip slightly hairy. 2*n*=32. Common. Sand; 1000–3400 m. e KR, CaRH, n&c SNH; s OR, nw NV. Intergrades with *E. incanum* in c SNH. ❀TRY;DFCLT.

E. microthecum Nutt. [Group 4] Subshrub or shrub 5–150 cm, 60–160 cm diam. **LVS** cauline; lower ± clustered; blade 3–25 mm, gen ± oblanceolate, gen ± rolled under, tomentose (esp below). **INFL** cyme-like, gen ± flat-topped; bracts scale-like; involucres terminal, 1.5–4 mm. **FL**: perianth 1.5–4 mm, white to rose or yellow, glabrous, stalk-like base 0. **FR** 1.5–3 mm, glabrous. Dry places; 1100–3300 m. SN, SW, GB, D; to WA, MT, WY, Colorado, NM. [Reveal 1971 Brigham Young Univ Sci Bull Biol Ser 13(1):1–45] Highly variable; vars. intergrading, often indistinct, differentiated esp by geog.

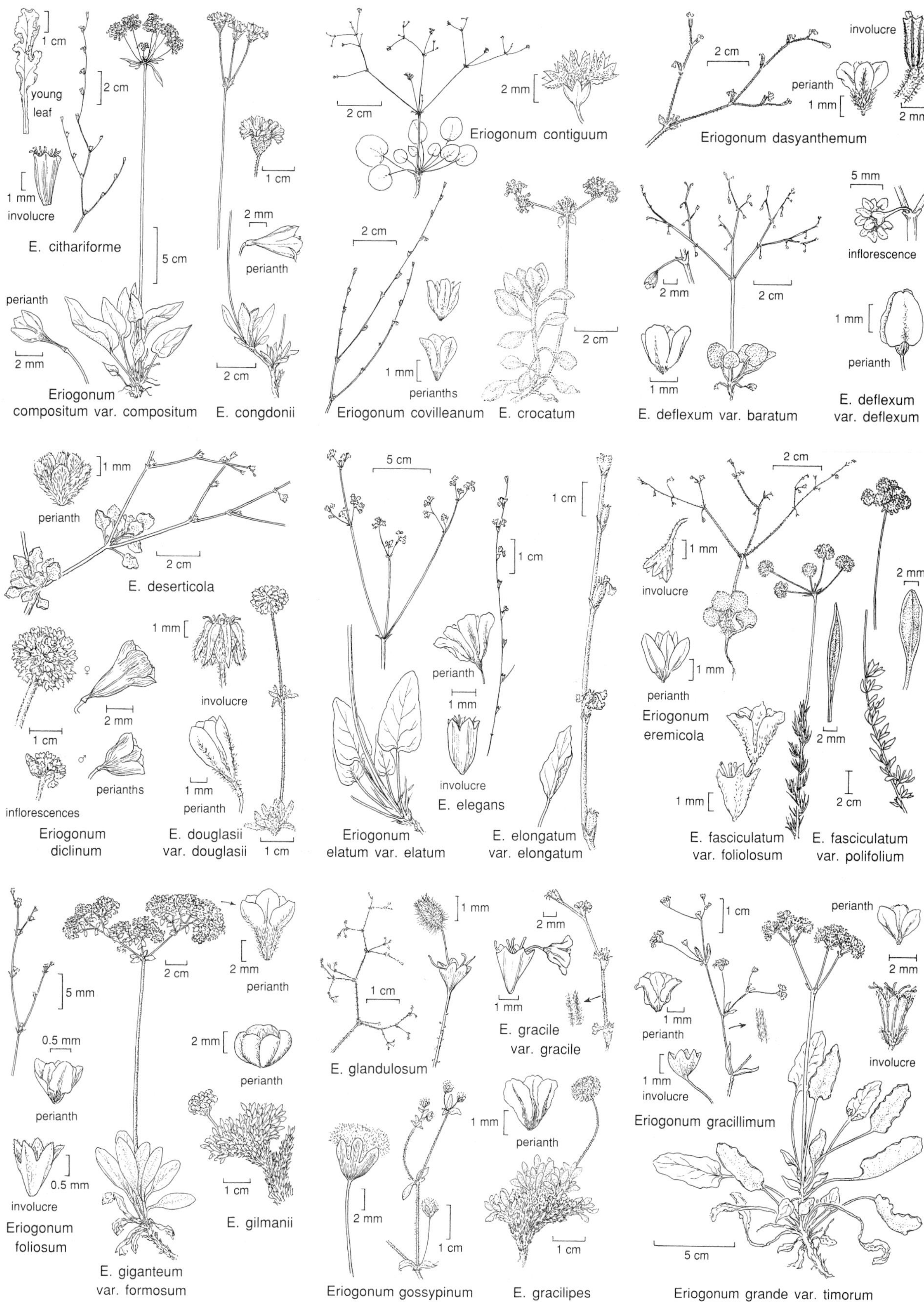
young leaf
1 cm
2 cm
1 mm
involucre
E. cithariforme
5 cm
perianth
2 mm
Eriogonum compositum var. compositum
1 cm
2 mm
perianth
2 cm
E. congdonii
2 cm
2 mm
Eriogonum contiguum
2 cm
1 mm
perianths
Eriogonum covilleanum
2 cm
E. crocatum
2 cm
perianth
1 mm
involucre
2 mm
Eriogonum dasyanthemum
2 mm
2 cm
1 mm
E. deflexum var. baratum
5 mm
inflorescence
1 mm
perianth
E. deflexum var. deflexum
1 mm
perianth
2 cm
E. deserticola
1 cm
2 mm
inflorescences
perianths
Eriogonum diclinum
1 mm
involucre
1 mm
perianth
E. douglasii var. douglasii
1 cm
5 cm
Eriogonum elatum var. elatum
perianth
1 mm
involucre
E. elegans
1 cm
1 cm
E. elongatum var. elongatum
2 cm
1 mm
involucre
1 mm
perianth
Eriogonum eremicola
1 mm
E. fasciculatum var. foliolosum
2 mm
2 mm
2 cm
E. fasciculatum var. polifolium
5 mm
0.5 mm
perianth
0.5 mm
involucre
Eriogonum foliosum
2 cm
E. giganteum var. formosum
2 mm
perianth
2 mm
perianth
1 cm
E. gilmanii
1 cm
E. glandulosum
1 mm
2 mm
1 cm
Eriogonum gossypinum
2 mm
1 mm
E. gracile var. gracile
1 mm
perianth
1 cm
E. gracilipes
1 cm
1 mm
perianth
1 mm
involucre
Eriogonum gracillimum
perianth
2 mm
involucre
5 cm
Eriogonum grande var. timorum

var. ***alpinum*** Rev. Pl 4–10 cm. **LF**: blade 3–9 mm, oblanceolate, gen not very rolled under, thinly hairy above. **INFL**: main branches gen < 1 cm. **FL**: perianth white to rose. Uncommon. Rocks; 2500–3300 m. c SNH (Alpine Co.), n SNE.

var. ***ambiguum*** (M.E. Jones) Rev. (p. 877) Pl 5–50 cm. **LF**: blade 8–25 mm, largest nearly flat. **FL**: perianth yellowish. Common. Rocks; 1100–3000 m. SN (e slope), GB; NV. ❀TRY.

var. ***corymbosoides*** Rev. SAN BERNARDINO BUCKWHEAT Pl 30–60 cm. **LF**: blade 8–25 mm, upper gen largest, gen ± oblong, ± flat. **INFL** dense, flat-topped. **FL**: perianth brownish. RARE. Rocks; 1800–2900 m. SnBr.

var. ***johnstonii*** Rev. JOHNSTON'S BUCKWHEAT Pl 6–13 cm. **LF**: blade 5–10 mm, elliptic to ovate, thinly hairy above. **INFL**: main branches gen < 1 cm. **FL**: perianth brownish. RARE. Rocks; 2600–2900 m. SnGb, SnBr.

var. ***lapidicola*** Rev. (p. 877) INYO MOUNTAINS BUCKWHEAT Pl 5–15 cm. **LF**: blade 3–7 mm, gen ± densely tomentose above. **INFL**: main branches gen 2–4 cm. **FL**: perianth whitish to rose or orange. RARE. Rocks; 2600–3100 m. s W&I (Inyo Mtns); NV.

var. ***laxiflorum*** Hook. Pl 10–50 cm. **LF**: blade 5–25 mm, gen nearly flat. **FL**: perianth white to pink. Rocks; 1200–3100 m. SN (e slope), GB; range of sp. outside CA.

var. ***panamintense*** S. Stokes (p. 877) PANAMINT MOUNTAINS BUCKWHEAT Pl 30–60 cm. **LF**: blade gen 5–8(15) mm, strongly rolled under. **FL**: perianth brownish. RARE. Rocks; 1900–2800 m. s W&I (Inyo Mtns), n DMtns (Panamint Mtns).

var. ***simpsonii*** (A. DC.) Rev. Pl 10–150 cm. **LF**: blade 5–25 mm, upper smaller, strongly rolled under. **INFL** ± open, not very flat-topped. **FL**: perianth white to pink. Sand or gravel; 1300–2200 m. s SNE, DMoj; to Colorado, NM. ❀TRY.

E. mohavense S. Watson (p. 877) [Group 1] Ann 5–30 cm. **LVS** basal; blade 4–20 mm, oblong to round, wavy-margined, tomentose (esp below). **INFL** cyme-like, evenly branched, slender, subglabrous; involucres sessile (few lateral), 1.5–2 mm, glabrous, ribbed, teeth 5. **FL**: perianth < 1 mm, yellow, gen glabrous. **FR** 1–1.3 mm, glabrous. Uncommon. Sand; 700–1300 m. DMoj.

E. molestum E. Greene (p. 877) [Group 1] Ann 20–100 cm. **LVS** basal; blade 10–30 mm, round to reniform, tomentose (esp below). **INFL** cyme-like; branches ± few, ± upcurved; bracts scale-like; involucres sessile (most lateral), 4–7 mm, cylindric, ribbed, teeth 5. **FL**: perianth 1.5–2 mm, white to pink, glabrous. **FR** 2–2.5 mm, glabrous. Uncommon. Granite sand; 1000–2200 m. n WTR, SnBr, PR.

E. nervulosum (S. Stokes) Rev. (p. 877) SNOW MOUNTAIN BUCKWHEAT [Group 3] Per from spreading woody caudex, 5–15 cm; mats 5–30 cm diam. **LVS** ± rosetted on low sts; blade 4–10 mm, elliptic to round, woolly below, finally shedding above. **INFL** umbel-like; axis slender, tomentose; bracts ± lf-like; involucres sessile, 3–6 mm, lobes 6–8, short, erect. **FL**: perianth 4–6 mm, pale yellow to reddish, glabrous, stalk-like base short, distinct. **FR** 4.5–5 mm, glabrous. RARE. Serpentine outcrops; 300–2100 m. s NCoRH, c&s NCoRI. In cult.

E. nidularium Cov. (p. 877) [Group 1] Ann 5–30 cm. **LVS** basal; blade 5–20 mm, ± round. **INFL** cyme-like, densely and ± unevenly branched, often curved ± inward; axis thinly tomentose; bracts scale-like; involucres sessile (most lateral), ± 1 mm, ribbed, teeth 5. **FL**: perianth gen 2–3 mm, yellow to red, glabrous, lobes fan-shaped. **FR** ± 1 mm, glabrous. Common. Sand or gravel flats, washes; 300–2300 m. GB, D; to se OR, s ID, UT, AZ. ❀STBL.

E. nortonii E. Greene (p. 877) PINNACLES BUCKWHEAT [Group 1] Ann 5–30 cm, ± ascending, glabrous. **LVS** basal (or some seemingly cauline); blade 5–15 mm, round to reniform, wavy-margined, tomentose below, finally glabrous above. **INFL** cyme-like (or compound umbel-like); bracts lf- and scale-like below, reduced upward; involucres sessile (immediately subtended by bracts), 3–4 mm, ribbed, teeth 8. **FL**: perianth 1–2 mm, white to rose, glabrous. **FR** ± 1 mm, glabrous. RARE. Sand; 300–700 m. n SCoR (Gabilan, e Santa Lucia Mtns, San Benito and Monterey cos.).

E. nudum Benth. [Group 4] Per, 10–200 cm. **LVS** basal or ± spaced on lower st; blade 10–70 mm, oblanceolate to ovate, gen tomentose (esp below). **INFL** cyme- to head-like; axes glabrous, sometime wider below major nodes; bracts scale-like (lowest sometimes lf-like); involucres 1–10 per cluster, 2–5 mm, teeth 5–8. **FL**: perianth 1.5–4 mm, white, yellow, or red, subglabrous, stalk-like base 0. **FR** 1.5–3.5 mm, glabrous. Abundant. Dry open places; < 3800 m. CA (exc ChI); to WA, NV, nw Mex. Vars. difficult, intergrading.

var. ***auriculatum*** (Benth.) Jepson (p. 881) **LVS** cauline on lower st, 30–70 mm; margins strongly wavy. **INFL** slender, glabrous; involucres 1–3 per node. **FL**: perianth white to pink, gen glabrous. $2n$=80. Common. Rocks or gravel; 50–1600 m. NW, CW. ❀DRN,DRY,SUN:4–7,14,**15–17**.

var. ***decurrens*** (S. Stokes) M. Bowerman (p. 881) BEN LOMOND BUCKWHEAT **LVS** cauline on lower st, 10–30 mm; margins gen wavy. **INFL** slender, tomentose; involucre 1(2) per node, 4–6 mm, tomentose. **FL**: perianth 3–4 mm, white, glabrous. RARE. Sand; 50–800 m. SnFrB (near Mount Diablo, Contra Costa Co.; Ben Lomond Hills, Santa Cruz Co.)

var. ***deductum*** (E. Greene) Jepson **LVS** basal, gen 10–20 mm; margins gen flat. **INFL** slender, glabrous; involucres gen 1–3 per node. **FL**: perianth 2.5–3.5 mm, white, glabrous. Open places; 1000–3200 m. SN; w NV. Intergrades with var. *nudum*.

var. ***indictum*** (Jepson) Rev. **LVS** cauline on lower st, 10–60 mm; margins strongly curled. **INFL** very stout, glabrous; main axes inflated; involucre gen 1 per node, 4–5 mm, glabrous. **FL**: perianth 2.5–3 mm, pale yellow, glabrous. $2n$=80. Uncommon. Barren slopes; 150–800 m. e SCoRI. Intergrades with var. *auriculatum*. ❀DRN,DRY,SUN:**7–9,14–17,19–24**.

var. ***murinum*** Rev. (p. 881) MOUSE BUCKWHEAT **LVS** cauline on lower st, 15–35 mm, densely white-woolly below, gray-tomentose above; margins slightly wavy. **INFL** slender, glabrous; involucres 5–10 per node, 4–6 mm, glabrous. **FL**: perianth 3–4 mm, white, hairy. RARE. Dry sandy slopes; 500–800 m. s SNF (Kaweah River drainage, Tulare Co.).

var. ***nudum*** (p. 881) **LVS** basal, gen 20–50 mm; margins ± flat. **INFL** much-branched, slender, glabrous; involucres gen 2–5 per node, 3–5 mm, ± glabrous. **FL**: perianth 2–3 mm, white, glabrous. n=20. Abundant. Sand or gravel; 70–2300 m. NW, CaR, SN, SnFrB; to WA, w NV. Highly variable; intergrades with all other vars. Not on immediate coast. ❀DRN,DRY,SUN:1–6,**7, 14–16**,17,24.

var. ***oblongifolium*** S. Watson **LVS** basal, gen 10–40 mm; margins flat to slightly wavy. **INFL** slender, spreading, tomentose; involucres 3–6 per node, 3–5 mm, tomentose. **FL**: perianth 3–4 mm, white to rose or yellow, hairy. n=20. Common. Dry open places; 150–1900 m. NW, CaR, n SN; s OR, w NV. ❀DRN,DRY, SUN:1–3,**7**,8,9,**14–17**,18.

var. ***paralinum*** Rev. (p. 881) DEL NORTE BUCKWHEAT **LVS** basal, 10–20 mm; margins flat. **INFL** sparsely branched, ± slender, glabrous; involucres gen 5–10 per node, 3–5 mm, glabrous. **FL**: perianth 2–4 mm, white, glabrous. RARE in CA. Open places along immediate coast; < 80 m. n NCo (Del Norte Co.); sw OR. Much like var. *nudum* but isolated from it.

var. ***pauciflorum*** S. Watson **LVS** basal, 15–30 mm, glabrous above; margins flat. **INFL** slender, glabrous; involucre gen 1(2) per node, 5–7 mm, gen glabrous. **FL**: perianth 2–2.5 mm, white, glabrous or hairy. Uncommon. Dry slopes, flats; 1100–2800 m. TR, PR; nw Mex. Intergrades with var. *westonii*.

var. ***pubiflorum*** Benth. **LVS** basal, 10–40 mm, becoming ± glabrous above; margins flat. **INFL** slender, glabrous; involucres 3–10 per node, 2–5 mm, gen glabrous. **FL**: perianth 2.5–3 mm, white or pale yellow, obviously hairy. n=20. Common. Dry flats, slopes; 50–2000 m. NW, CaR, SN, ScV, SnFrB; s OR, w NV. Intergrades with vars. *oblongifolium* & *westonii*. ❀SUN,DRN:1,2,4–6,**7,14–17,24**.

var. ***regirivum*** Rev. & J. Stebb. (p. 881) KINGS RIVER BUCKWHEAT **LVS** cauline on lower st, gen 30–55 mm, densely tomentose below; margins slightly wavy. **INFL** slender, erect, tomentose; involucres 3–6 per node, 3–4 mm, thinly tomentose. **FL**: perianth 1.5–3 mm, white, densely hairy. n=20. RARE. Limestone slopes; 150–300 m. s SNF (Kings River near Pine Flat Reservoir, Fresno Co.).

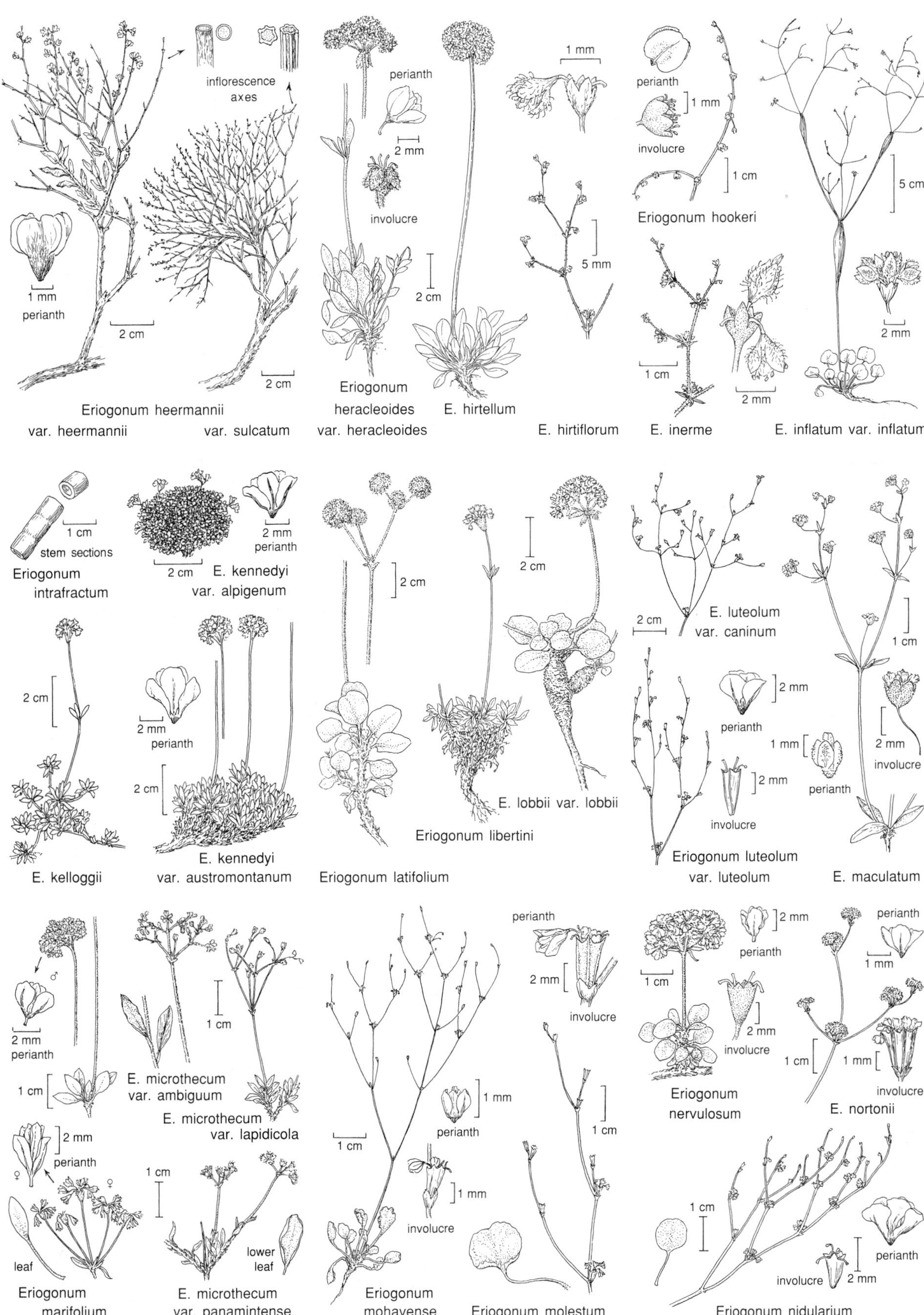

inflorescence axes
perianth
1 mm
2 cm
2 cm
Eriogonum heermannii
var. heermannii
var. sulcatum
perianth
2 mm
involucre
2 cm
Eriogonum
heracleoides
var. heracleoides
E. hirtellum
1 mm
5 mm
E. hirtiflorum
perianth
1 mm
involucre
1 cm
Eriogonum hookeri
1 cm
2 mm
E. inerme
5 cm
2 mm
E. inflatum var. inflatum
1 cm
stem sections
Eriogonum
intrafractum
2 cm
E. kennedyi
var. alpigenum
2 mm
perianth
2 cm
2 cm
2 cm
E. kelloggii
2 mm
perianth
2 cm
E. kennedyi
var. austromontanum
Eriogonum latifolium
Eriogonum libertini
E. lobbii var. lobbii
2 cm
E. luteolum
var. caninum
2 mm
perianth
2 mm
involucre
Eriogonum luteolum
var. luteolum
1 mm
perianth
1 cm
2 mm
involucre
E. maculatum
♂
2 mm
perianth
1 cm
E. microthecum
var. ambiguum
1 cm
E. microthecum
var. lapidicola
2 mm
perianth
♀
♀
leaf
Eriogonum
marifolium
1 cm
lower
leaf
E. microthecum
var. panamintense
1 cm
1 mm
perianth
1 mm
involucre
Eriogonum
mohavense
perianth
2 mm
involucre
1 cm
Eriogonum molestum
2 mm
perianth
1 cm
2 mm
involucre
Eriogonum
nervulosum
perianth
1 mm
1 cm
1 mm
involucre
E. nortonii
1 cm
perianth
involucre
2 mm
Eriogonum nidularium

var. ***scapigerum*** (Eastw.) Jepson (p. 881) **LVS** basal, 10–20 mm; margins flat. **INFL**: gen simple head, slender, glabrous; involucres 3–6 per head, 2–3 mm, gen glabrous. **FL**: perianth 2–3 mm, white, glabrous. Uncommon. Alpine slopes; 3000–3800 m. c&s SNH. Perhaps only a dwarf form of var. *deductum*.

var. ***westonii*** (S. Stokes) J. Howell **LVS** basal, 10–30 mm, often tomentose above; margins ± flat. **INFL** slender or stout, glabrous; involucres 1–4 per node, 3–5 mm, gen glabrous. **FL**: perianth 2.5–3 mm, yellow, obviously hairy. *n*=20. Common. Dry places; 300–2800 m. s SN, Teh, s CW, TR, s SNE, w DMoj. Intergrades with var. *pubiflorum*. ❀TRY.

E. nummulare M.E. Jones (p. 881) [Group 4] Shrub 15–100 cm, 30–150 cm diam. **LVS** cauline; blade 10–30 mm, gen ± oblanceolate, tomentose (esp below). **INFL** cyme-like, open; axes thick, fragile; bracts scale-like; involucres sessile (some lateral, esp near branch tips), 2–3 mm, teeth 5. **FL**: perianth 1.5–3 mm, white, glabrous, stalk-like base 0. **FR** 2–3 mm, glabrous. Sand; 1100–2600 m. s MP, SNE; to w UT. [*E. kearneyi* Tidestrom, Kearney's buckwheat, incl var. *monoense* (S. Stokes) Rev.]

E. nutans Torrey & A. Gray (p. 881) NODDING BUCKWHEAT [Group 2] Ann 5–30 cm, reddish. **LVS** basal; blade 5–20 mm, round, densely tomentose (esp below). **INFL** cyme-like, open; bracts scale-like; involucres on slender, ± reflexed stalks, ± 2 mm, bell-shaped, subglabrous, teeth 5. **FL**: perianth 2–3 mm, white to red, glabrous, outer lobes ± widely elliptic, inner narrower. **FR** ± 2 mm, glabrous. *n*=40. UNCOMMON. Sand or gravel; ± 3000 m. e MP (ne Lassen Co.), n SNE (e Mono Co.); to se OR, w UT. Scattered but expected more widely. Grayish pls from NV [var. *glabrum* Rev., glabrous nodding buckwheat] may be roadside waifs in SNH.

E. ochrocephalum S. Watson [Group 4] Per 10–40 cm; mats 5–15 cm diam. **LVS** clustered on low sts; blade 10–35 mm, lanceolate to obovate, tomentose. **INFL** head-like; involucres 5–8 per head, 3.5–5 mm, ± bell-shaped, rigid, thinly tomentose, ribbed, teeth 6–8. **FL**: perianth 2–3 mm, yellow, glabrous, lobes oblong, stalk-like base 0. **FR** 1.5–3 mm, glabrous. Uncommon. Dry clay or rocky slopes; 1200–2400 m. GB; to sw OR, s ID, w NV. Vars. not very distinct. Much like *E. beatleyae*.

var. ***alexanderae*** Rev. (p. 881) ALEXANDER'S BUCKWHEAT **INFL**: axis tomentose; involucre gen tomentose. RARE in CA. Shale or gravel; 1300–2100 m. n SNE (e Mono Co.); w NV.

var. ***ochrocephalum*** (p. 881) **INFL**: axis glabrous or glandular; involucre gen ± sparsely hairy. Uncommon. Dry volcanic or clay soils; 1200–2400 m. MP; OR, w NV. ❀TRY;DFCLT.

E. ordii S. Watson (p. 881) [Group 2] Ann 5–70 cm, erect, gen glabrous. **LVS** basal; blade 20–80 mm, narrowly oblanceolate to obovate, subglabrous to hairy in patches. **INFL** cyme-like, slender; lowest bracts gen lf-and scale-like, reduced upward; involucres on thread-like stalks, 1–1.5 mm, glabrous, teeth 4. **FL**: perianth 1–2.5 mm, white to red, soft-hairy. **FR** ± 2 mm, glabrous. Uncommon. Barren clay; 15–1500 m. s SNF, Teh, SCoRI, n WTR, sw DSon?; nw AZ?

E. ovalifolium Nutt. [Group 4] Per 15–35 cm; mats 3–40 cm diam. **LVS** ± basal on dense caudex; petiole long, slender; blade 2–60 mm, elliptic to ± round, tomentose to woolly. **INFL** gen head-like, < 30 cm; bracts scale-like, ± hidden; involucres 1–15 per head, 2–6.5 mm, rigid, hairy, teeth 5. **FLS** sometimes unisexual; perianth 2.5–7 mm, white to purple or yellow, glabrous, outer lobes obovate to round, inner narrower, gen notched, stalk-like base 0. **FR** 2–3 mm, glabrous. *n*=20. Common. Dry sand or gravel; 1200–4100 m. KR, CaR, SNH, SnBr, GB; to s Can, MT, Colorado, NM. Vars. ± intergrade.

var. ***eximium*** (Tidestrom) J. Howell (p. 881) BROWN-MARGINED BUCKWHEAT **LF**: blade 5–20 mm, gen elliptic, distinctly brown-margined. **INFL**: axis gen 5–10 cm; involucres 3–10, 4–6 mm. **FL**: perianth gen 4–6 mm, white to cream. UNCOMMON. Granite sand; 1800–3400 m. n SNH (El Dorado, Alpine cos.); w-c NV. Intergrades with var. *nivale*. ❀TRY;DFCLT.

var. ***nivale*** (Canby) M.E. Jones (p. 881) **LF**: blade 2–8 mm, gen round, ± unmargined. **INFL**: axis < 5 cm; involucres 3–5, 2–4 mm. **FL**: perianth gen 2–3 mm, white to cream. Common. Alpine sand or gravel; 1500–4100 m. CaRH, SNH, W&I; to B.C., w UT. ❀DRN,SUN:1–3,**7,15–17**.

var. ***ovalifolium*** (p. 881) **LF**: blade 10–60 mm, gen ± obovate, ± unmargined. **INFL**: axis 4–20 cm; involucres 3–15, 4–6.5 mm. **FL**: perianth gen 4–6 mm, yellow. Common. Sand or gravel; 1200–2900 m. SNH (e slope), GB; to OR, MT, Colorado. Intergrades with and difficult to distinguish from *E. strictum* var. *anserinum*. ❀DRN,SUN:1–3,**7,15–17**,22–24.

var. ***purpureum*** (Nelson) Durand **LF**: blade 5–20 mm, gen ± obovate, ± unmargined. **INFL**: axis 4–20 cm; involucres 3–15 per cluster, 4–6.5 mm. **FL**: perianth gen 4–6 mm, white to cream or purplish. Common. Sand or gravel; 1200–2800 m. KR, CaRH, SNH (e slope), GB; range of sp. [var. *ovalifolium* misapplied] ❀TRY.

var. ***vineum*** (Small) Jepson CUSHENBURY BUCKWHEAT **LF**: blade gen 6–8 mm, gen round, unmargined. **INFL**: axis 3–6 cm; involucres 2–5, 5–7 mm. **FL**: perianth gen 2–3 mm, white to cream. RARE. Limestone; 1500–2100 m. ne SnBr (Cushenbury Canyon). Threatened by mining, vehicles. Isolated and distinct. In cult.

E. palmerianum Munz (p. 881) [Group 1] Ann 5–30 cm. **LVS** basal; blade 5–15 cm, gen round, tomentose (esp below). **INFL** cyme-like, ± open; branches few, tips often outcurved; involucres sessile (many lateral), ± 1 mm, thinly hairy, ± angled, teeth 5. **FL**: perianth 1.5–2 mm, white to pink or pale yellow, glabrous. **FR** ± 1.7 mm, glabrous. Sand or gravel; 600–3400 m. SNE, D; to Colorado, NM. Locally common. ❀TRY.

E. panamintense C. Morton (p. 881) [Group 4] Per 15–50 cm, 10–40 cm diam. **LVS** basal; blade 10–40 mm, oblong to round, ± densely tomentose. **INFL** cyme-like, slender, ± unevenly branched; bracts gen lf- and scale-like below, reduced upward; involucres sessile (many lateral), 2–5 mm, tomentose, teeth 5. **FL**: perianth 3–5 mm, white or brownish, glabrous, stalk-like base 0. **FR** 2.5–3 mm, glabrous. Open gravel or rocks; 1500–2900 m. W&I, DMtns; NV. [*E. rupinum* Rev.] Pls with only scale-like bracts have been called var. *mensicola* (S. Stokes) Rev.

E. parishii S. Watson (p. 881) [Group 2] Ann 10–50 cm, densely and finely branched. **LVS** basal; blade 20–60 mm, ± oblanceolate, coarsely hairy. **INFL** cyme-like, open, slender, glandular above nodes; bracts scale-like; involucres on thread-like, spreading stalks (many seemingly lateral), ± 0.5 mm, glabrous, unribbed, teeth 4. **FL**: perianth ± 0.5 mm, pink to red, subglabrous. **FR** 1–1.3 mm, glabrous. *n*=20. Common. Granite sand; 1200–3200 m. s SN, TR, PR, n W&I (White Mtns); AZ, n Baja CA. ❀TRY.

E. parvifolium Smith (p. 881) [Group 4] Shrub 30–100 cm, 50–200 cm diam. **LVS** cauline; blade 5–30 mm, lanceolate to ± round (or folded under, appearing ± triangular), thick, ± woolly below, glabrous above, becoming ± brownish. **INFL** head- to cyme-like; bracts lf-like below, gen scale-like above; involucres 2–7 per node or head, 2.5–4 mm, gen subglabrous, teeth 5. **FL**: perianth 2.5–3 mm, white to pinkish or greenish yellow, glabrous, stalk-like base 0. **FR** 2.5–3 mm, glabrous. 2*n*=40. Dunes, seabluffs; < 700 m. CCo, SCo. [vars. *lucidum* (S. Stokes) Rev. & *paynei* (Munz) Rev.] ❀DRN,SUN:5,**15–17,24**&IRR:**14,21–23**.

E. pendulum S. Watson (p. 881) WALDO BUCKWHEAT [Group 4] Subshrub or shrub 20–50 cm, 20–80 cm diam, open. **LVS** cauline, ± clustered near st tips; blade 15–50 mm, ± oblanceolate, white-woolly (esp below). **INFL** cyme-like, open; branches few; bracts lf-like, reduced upward; involucres gen ± stalked above bracts, 3.5–5 mm, white-woolly, teeth 5–8. **FL**: perianth 3–7 mm, white, woolly, stalk-like base 0. **FR** 3–5 mm, hairy. RARE in CA. Open serpentine; 400–1000 m. nw KR (n Del Norte Co.); sw OR.

E. plumatella Durand & Hilg. (p. 881) [Group 4] Shrub 30–60 cm, 30–60 cm diam. **LVS** cauline; blade 6–15 mm, gen obovate, ± densely white-tomentose. **INFL** cyme-like, open, appearing ± layered; main axes obvious; branches spreading-recurved; bracts ± scale-like; involucres sessile, ± 2 mm, glabrous, teeth 5. **FL**: perianth 2–2.5 mm, white to pale yellow, glabrous, outer lobes spreading, obovate, inner erect, narrower, stalk-like base 0. **FR** 2.5–3 mm, glabrous. Common. Dry slopes, washes; 300–1800 m. D; to sw UT, nw AZ. [var. *jaegeri* (Munz & I.M. Johnston) Munz] ❀TRY.

E. polypodum Small (p. 881) TULARE COUNTY BUCKWHEAT [Group 3] Per from densely branched woody caudex, 5–15 cm; mats 10–50 cm diam. **LVS** clustered on low sts; blade 2–10 mm, ovate, densely tomentose (esp below). **INFL** head- to umbel-like, slender, tomentose; bracts ± lf-like; involucres ± densely clustered, sessile, 2–4 mm, lobes 5–7, ± spreading. **FL**: perianth 2.5–3.5 mm, chalky-white or red-striped, glabrous, stalk-like base short, distinct. **FR** 3–3.5 mm, glabrous. UNCOMMON. Open granite sand in subalpine forest; 2400–3500 m. s SNH (esp Tulare Co.).

E. prattenianum Durand [Group 3] Subshrub 10–50 cm, 30–50 cm diam, low, open. **LVS** clustered on low sts; blade 5–20 mm, elliptic to obovate, ± woolly (esp below). **INFL** head-like, slender, thinly hairy; bracts lf-like, whorled near mid-axis (sometimes 0); involucres 3–5 mm, ± tomentose, lobes 8–10, erect to reflexed. **FL**: perianth 3–8 mm, yellow, glabrous, stalk-like base ± long, distinct. **FR** 4–5 mm, glabrous. Outcrops; 800–2600 m. SN.

var. ***avium*** Rev. & J.R. Shevock (p. 881) KETTLE DOME BUCKWHEAT Pl 10–30 cm. **LF**: blade becoming subglabrous above. **FL**: perianth gen 3–6 mm. UNCOMMON. Granite; 1200–2600 m. c&s SNH (Madera, Fresno cos.).

var. ***prattenianum*** Pl 10–50 cm, spreading. **LF**: blade tomentose above. **FL**: perianth gen 4–8 mm. Uncommon. Volcanics; 800–2600 m. n&c SN. ❀DRN,SUN:1–3,7,14,**15,16**,18.

E. prociduum Rev. (p. 881) PROSTRATE BUCKWHEAT [Group 4] Per from densely branched caudex, gen 5–15 cm; mats 10–30 cm diam. **LVS** clustered on low sts; blade 4–20 mm, ± oblanceolate, tomentose. **INFL** head-like, glabrous; bracts scale-like; involucres 4–6, sessile, ± 3 mm, bell-shaped, rigid, gen thinly hairy, teeth 5. **FL**: perianth 2–3 mm, bright yellow, glabrous, stalk-like base 0. **FR** 2–2.5 mm, glabrous. RARE. Barren, light-volcanic slopes; 1300–2500 m. MP; s-c OR, nw NV.

E. puberulum S. Watson (p. 881) DOWNY BUCKWHEAT [Group 1] Ann 2–30 cm. **LVS** basal; blade 5–15 mm, obovate to round, sparsely hairy. **INFL** cyme-like, minutely silky; lowest bracts ± strap-like, upper scale-like; involucres sessile (most lateral), ± 1 mm, barely angled, minutely silky, lobes 4, deep. **FL**: perianth 1–1.5 mm, white to red, minutely hairy. **FR** 1–1.5 mm, glabrous. RARE in CA. Gravelly pinyon/juniper woodland; 1300–2900 m. n DMtns (Cottonwood Mtns, Death Valley region, Inyo Co.); to sw UT.

E. pusillum Torrey & A. Gray (p. 881) [Group 2] Ann 5–30 cm. **LVS** basal; blade gen 5–15 mm, gen round, white-woolly (esp below). **INFL** cyme-like, slender, ± glabrous; bracts scale-like; involucres on slender straight stalks, ± 2 mm, cup-shaped, unribbed, minutely glandular, teeth 5. **FL**: perianth 1–2.5 mm, yellow becoming reddish, glandular, lobes ± obtuse at base. **FR** ± 0.7 mm, glabrous. *n*=16. Common. Sand or gravel; 150–2600 m. SCoRI, WTR, GB, D; to OR, ID, NV. ❀STBL.

E. pyrolifolium Hook. (p. 881) [Group 3] Per from thick woody caudex, 3–20 cm, 5–30 cm diam. **LVS** ± basal; blade 10–40 mm, ± round, glabrous to tomentose; petiole ± brown-woolly. **INFL** head- to umbel-like, gen ± prostrate, slender; bracts gen 2 subtending rays or head, ± strap-like; involucres terminal on gen stout rays, 4–6 mm, glabrous or hairy, teeth 4–5. **FL**: perianth 4–6 mm, white to rose, hairy, stalk-like base very short, ± indistinct. **FR** 4–5 mm; tip hairy. *n*=20. Sand or pumice flats; 1700–3200 m. CaRH; to WA, MT. Locally common. ❀DRN,SUN,IRR:1–3,7, 15,16;DFCLT.

E. reniforme Torrey & Frémont (p. 881) [Group 2] Ann 5–40 cm. **LVS** basal; blade 5–20 mm, round-reniform, ± densely white-tomentose. **INFL** cyme-like, spreading, slender, ± glabrous, glaucous; bracts scale-like; involucres on slender, ± curved stalks, ± 2 mm, unribbed, glabrous, glaucous, teeth 5. **FL**: perianth ± 1 mm, yellow, becoming reddish, minutely glandular, outer lobes wider, ± obtuse. **FR** ± 1 mm, glabrous. *n*=16. Common. Sand or gravel; < 1500 m. D; s NV, w AZ, nw Mex. ❀STBL.

E. rixfordii S. Stokes (p. 881) RIXFORD'S BUCKWHEAT [Group 2] Ann 15–40 cm. **LVS** basal; blade 5–30 mm, round-cordate, densely white-tomentose (esp below). **INFL** cyme-like, of pagoda-like tiers of branches, stout, glabrous, glaucous; bracts scale-like; involucres ± sessile, reflexed, ± 1 mm, unribbed, glabrous, teeth 5. **FL**: perianth ± 1.5 mm, white to reddish, outer lobes narrowly ovate, bases ± cordate. **FR** 1.5 mm, glabrous. *n*=20. UNCOMMON. Sand or gravel; 150–1600 m. ne DMoj (Death Valley); sw NV.

E. rosense Nelson & Kenn. (p. 881) [Group 4] Per gen < 10 cm; mats 5–20 cm diam, ± glandular throughout. **LVS** clustered on low sts; blade 4–15 mm, ± oblanceolate, coarsely tomentose (esp below). **INFL** head-like; bracts scale-like; involucres 3–5, 3–3.5 mm, rigid, sparsely hairy, teeth 5–8. **FL**: perianth 2–3 mm, bright yellow to reddish, stalk-like base 0. **FR** 1.5–2 mm, glabrous. *n*=20. Uncommon. Dry granite or volcanic outcrops; 2300–4000 m. s CaRH (Lassen Peak), n&c SNH, n SNE; w NV. [*E. anemophilum* E. Greene misapplied]

E. roseum Durand & Hilg. (p. 881) [Group 1] Ann 10–80 cm. **LVS** basal; blade 10–30 mm, oblanceolate to elliptic, tomentose (esp below). **INFL** cyme-like; branches few, wand-like; bracts lf- and scale-like below, reduced upward; involucres sessile (most lateral), 4–5 mm, tomentose, teeth 5. **FL**: perianth ± 2 mm, white to pink or yellow, glabrous. **FR** ± 2 mm, glabrous. *n*=9. Common. Sand, gravel, or rocks; < 2200 m. NW, SNF, Teh, CW, TR; s OR. ❀TRY.

E. saxatile S. Watson (p. 881) [Group 3] Per 10–20(40) cm; mats 5–20 cm diam. **LVS** clustered on low sts; blade 3–25 mm, elliptic to round, densely tomentose. **INFL** cyme-like, ± stout, ± tomentose; involucres sessile (most lateral), 2–4 mm, ± tomentose, teeth 5–6. **FL**: perianth 3–7 mm, white to rose or yellowish, glabrous, outer lobes narrower, stalk-like base long, sharply 3-angled, ± indistinct. **FR** 3.5–4 mm, glabrous, winged. *n*=20. Common. Decomposed granite or volcanic rocks; 900–3400 m. s SN, CW, SW, SNE, DMtns; s NV. ❀DRN,DRY,SUN:1–3,**7**,14, **15–18**,19–24.

E. shockleyi S. Watson var. ***shockleyi*** (p. 881) SHOCKLEY'S BUCKWHEAT [Group 4] Per gen < 6 cm; mats 5–100 cm diam. **LVS** clustered on low sts; blade 2–12 mm, oblanceolate to elliptic, tomentose. **INFL** head-like, ± erect; bracts scale-like; involucres 2–6 per head, 2–6 mm, bell-shaped, ± tomentose, teeth 5–10. **FL**: perianth 2.5–4 mm, white or yellowish, densely soft-hairy toward base, stalk-like base 0. **FR** ± 3 mm, hairy. UNCOMMON. Dry limestone gravel in pinyon/juniper woodland; 1700–2700 m. SNE, n DMtns (Last Chance Mtns); to s ID, Colorado, NM.

E. siskiyouense Small (p. 881) SISKIYOU BUCKWHEAT [Group 3] Per from densely branched woody caudex, 5–20 cm; mats 10–50 cm diam. **LVS** clustered on low sts; blade 3–8 mm, elliptic to round, gray-tomentose (esp below), ± yellow-green above. **INFL** appearing head-like, slender, ± tomentose; bracts lf-like, whorled near mid-axis; involucre 1 per head, 3–4 mm, lobes 6–10, long, reflexed, tomentose. **FL**: perianth 4–6 mm, yellow, glabrous, stalk-like base distinct. **FR** 4.5–5 mm, glabrous. UNCOMMON. Serpentine slopes; 1700–2500 m. e KR (Mount Eddy, Scott Mtns). ❀DRN,SUN:1–3,7,14,**15–17**,18.

E. spergulinum A. Gray [Group 2] Ann 5–40 cm. **LVS** basal; blade 3–30 mm, linear (< 2 mm wide), ciliate. **INFL** cyme-like, prostrate to erect, slender; involucres on ± erect, thread-like stalks, 0.5–1 mm, unribbed, glabrous, teeth 4. **FL**: perianth 1–3 mm, white, striped darker. **FR** ± 2 mm, glabrous. Dry sand (esp granitic); 1200–3500 m. NCoRH, CaRH, SN, n SCoRO, TR; to OR, ID, NV. Vars. indistinct, intergrading.

var. ***pratense*** (S. Stokes) J. Howell **INFL** prostrate to ascending, gen 2–5 cm; axes not glandular. **FL**: perianth ± 1.5 mm. Uncommon. Granite sand; 2500–3500 m. s SNH.

var. ***reddingianum*** (M.E. Jones) J. Howell **INFL** erect, 8–40 cm; axes glandular. **FL**: perianth ± 2 mm. Common. Sand or gravel; 1400–3500 m. NCoRH, CaRH, SN, n SCoRO, TR; to OR, ID, NV.

var. ***spergulinum*** (p. 881) **INFL** erect, 10–40 cm; internodes glandular. **FL**: perianth ± 3 mm. Uncommon. Sand; 1200–3000 m. SN.

E. sphaerocephalum Benth. [Group 3] Subshrub or shrub 5–40 cm, 30–60 cm diam. **LVS** cauline; blade 10–40 mm, ± oblanceolate, densely tomentose (esp below). **INFL** head- or umbel-like (sometimes compound), sometimes hairy; bracts lf-like, subtending rays (or whorled near mid-axis or mid-ray); involucre 1 per ray, 3–4 mm, lobes 6–10, long, gen spreading, thinly hairy. **FL**: perianth 5–9 mm, pale yellow, glabrous or hairy, stalk-like base very long, distinct. **FR** 3–4 mm; tip slightly hairy. Common. Sagebrush shrublands; 900– 2300 m. CaRH, n SNH, MP; to WA, ID, w NV.

var. ***halimioides*** S. Stokes (p. 885) **FL**: perianth cream or pale yellow, gen ± conspicuously hairy. *n*=20. Common. Habitat and range of sp. ❀TRY.

var. ***sphaerocephalum*** (p. 885) **FL**: perianth bright yellow, gen ± obscurely hairy. Uncommon. Habitat of sp.; 1200–2300 m. Range of sp. (exc CaRH).

E. strictum Benth. [Group 4] Per 10–35 cm; mats 10–40 cm diam. **LVS** ± basal; blade 5–40 mm, elliptic to ovate, ± woolly (esp below). **INFL** gen umbel- to cyme-like, erect, slender, gen ± tomentose; involucres sessile or terminal, 2.5–6 mm, rigid, sometimes hairy, teeth 5. **FL**: perianth 3–5 mm, white to yellow, becoming reddish, glabrous, outer lobes shorter, round, inner narrower, stalk-like base 0. **FR** ± 3.5 mm, glabrous. Sagebrush shrublands, rocky places; 400–2600 m. KR, n NCoRO, NCoRH, CaR, ne SNH, MP; to WA, MT, NV.

var. ***anserinum*** (E. Greene) R. Davis **LF**: blade 5–20 mm, ovate. **INFL** gen 1–5 cm. **FL**: perianth yellow. Sagebrush shrublands; 400– 2600 m. CaR, ne SN, MP; to OR, s ID, w NV. Perhaps composed of intermediates between var. *proliferum* and *E. ovalifolium*. ❀TRY.

var. ***greenei*** (A. Gray) Rev. (p. 885) GREENE'S BUCKWHEAT **LF**: blade 5–10 mm, ovate, densely white-woolly. **INFL** 3–5 cm. **FL**: perianth white. UNCOMMON. Rocky places; 800–2100 m. KR. ❀DRN,SUN,DRY:1–3,**7**,14–18.

var. ***proliferum*** (Torrey & A. Gray) Rev. (p. 885) **LF**: blade 10–30 mm, gen widely elliptic. **INFL** 5–15 cm. **FL**: perianth white to rose or purple. Common. Sagebrush shrublands, rocky openings in forest; 400–2600 m. KR, n NCoRO, NCoRH, CaR, ne SNH, w MP; to WA, MT, n NV. [ssp. *p.* (Torrey & A. Gray) S. Stokes] ❀DRN,SUN,DRY:1–3,**7,14–18**.

E. temblorense J. Howell & Twisselm. (p. 885) TEMBLOR BUCKWHEAT [Group 1] Ann 10–80 cm. **LVS** basal (or some seemingly cauline); blade 15–40 mm, elliptic to obovate, densely tomentose. **INFL** cyme-like; axes tomentose; bracts lf- and scale-like below, reduced upward; involucres (at least lowest) on slender stalks, 2–2.5 mm, bell-shaped, ribbed, tomentose, teeth 5. **FL**: perianth ± 2 mm, white, dark-striped, glabrous. **FR** 2–3 mm, glabrous. *n*=17. UNCOMMON. Barren clay in grassland, sandstone outcrops; 300–1000 m. s SCoRI (e Monterey, e San Luis Obispo, w Kern cos.). Not very distinct from *E. eastwoodianum, E. vestitum*.

E. ternatum Howell (p. 885) TERNATE BUCKWHEAT [Group 3] Per 10–30 cm; mats 30–50 cm diam. **LVS** clustered on low sts; blade 10–15 mm, ovate to round, densely brown-tomentose, finally shedding above, margin flat. **INFL** umbel-like; axis thinly tomentose; bracts strap-like; involucre gen 1 per ray, sessile, 5–8 mm, tomentose, lobes 5–8, short, erect. **FL**: perianth 6–8 mm, bright yellow, glabrous, stalk-like base short, distinct. **FR** 3.5–5 mm; tip slightly hairy. UNCOMMON. Serpentine outcrops; 500–2200 m. n KR, s NCoRO (nw Sonoma Co.), n NCoRH; s OR. ❀DRN, SUN,IRR:1–3,**6,7**,14,**15,16,18**.

E. thomasii Torrey (p. 885) [Group 2] Ann 5–30 cm. **LVS** basal; blade 5–20 mm, round-cordate, densely tomentose (esp below). **INFL** cyme-like, open, ± glabrous, glaucous; bracts scale-like; involucres on slender, spreading stalks, ± 1 mm, cup-shaped, unribbed, glabrous, teeth 5. **FL**: perianth 0.5–1 mm (2 mm in fr), yellow becoming white or rose, minutely hairy, outer lobes sometimes swollen-cordate at base. **FR** ± 1 mm, glabrous. *n*=20. Sand or gravel; –61–1300 m. D; s NV, w AZ, nw Mex. Locally common. ❀STBL.

E. thurberi Torrey (p. 885) [Group 2] Ann 5–40 cm. **LVS** basal; blade 8–45 mm, elliptic to round, tomentose (esp below). **INFL** cyme-like, ascending; axes glandular; involucres on slender stalks, ± 1.5 mm, unribbed, minutely glandular, teeth 5. **FL**: perianth ± 1.5 mm, white becoming reddish, minutely glandular, lobes ± hair-tufted inside, outer lobes ± fan-shaped, bases ± obtuse. **FR** ± 0.7 mm, glabrous. Sand or gravel; 100–1300 m. e PR, s DMoj, DSon; AZ, nw Mex. ❀STBL.

E. trichopes Torrey (p. 885) [Group 2] Ann 10–180 cm. **LVS** basal; blade 5–40 mm, oblong to reniform, gen ± wavy-margined, glabrous to thinly hairy. **INFL** cyme-like; base ± hairy; axes often wider below main nodes; bracts scale-like; involucres on thread-like stalks, 0.5–2.5 mm, glabrous, teeth gen 4. **FL**: perianth 1–2.5 mm, yellow becoming reddish, coarsely and minutely white-hairy, lobes ovate. **FR** ± 2 mm, glabrous. *n*=16. Open clay, sand, gravel, or serpentine; –60–1300 m. SCoRI, n WTR, D; to UT, NM, nw Mex.

var. ***hooveri*** Rev. Pl erect, 40–180 cm, bright green. **INFL**: main axes often ± inflated; involucre gen 1–2.5 mm, stalks 10–40 mm. **FR** 2–2.5 mm. Clay, often of serpentine origin; 400–1200 m. SCoRI, n WTR. Nearly identical to 1st-year-fl *E. inflatum* var. *i* of D.

var. ***trichopes*** Pl ± decumbent, 10–60 cm, yellow-green. **INFL**: axes gen slender; involucre < 1 mm, stalks 5–20 mm. **FR** 1–1.5 mm. Sand or gravel; –60–1300 m. D; to sw UT, sw NM, nw Mex. ❀TRY.

E. tripodum E. Greene (p. 885) TRIPOD BUCKWHEAT [Group 3] Subshrub 25–50 cm, 30–60 cm diam, spreading. **LVS** clustered on low sts; blade 15–25 mm, oblanceolate, woolly below, tomentose or shedding above, margin rolled under. **INFL** umbel- or head-like, sometimes compound; rays long, slender, thinly hairy; bracts lf-like, subtending rays (if ray 1, appearing whorled near mid-axis); involucre 1 per ray, 3–4 mm, bell-shaped, woolly, lobes 6–10, gen short and erect. **FL**: perianth 4–5 mm, yellow, hairy, stalk-like base distinct. **FR** 2.5–3 mm; tip slightly hairy. UNCOMMON. Serpentine outcrops; 200–1600 m. NCoRI, n&c SNF. ❀DRN,SUN:**7**, 14,**15–17**.

E. truncatum Torrey & A. Gray (p. 885) MOUNT DIABLO BUCKWHEAT [Group 1] Ann 10–60 cm. **LVS** basal; blade 10–70 mm, oblong to obovate, densely tomentose (esp below). **INFL** umbel-like, ± compound; bracts ± identical to lvs below, scale-like above; involucres sessile, 2.5–4 mm, ribbed, tomentose, teeth 5. **FL**: perianth ± 2 mm, white to rose, glabrous. **FR** ± 2 mm, glabrous. **PRESUMED EXTINCT**. Sand; 300–600 m. Formerly ne SnFrB (Alameda and Contra Costa cos., esp e slope Mount Diablo), deltaic ScV (Suisun, Solano Co.). Last seen in 1940.

E. twisselmannii (J. Howell) Rev. TWISSELMANN'S BUCKWHEAT [Group 3] Per from woody caudex, 10–20 cm; mats loose, 30–40 cm diam. **LVS** clustered on low sts; blade 5–10 mm, oblanceolate to elliptic, tomentose (esp below). **INFL** head-like, ± tomentose; bracts lf-like, whorled near mid-axis; involucre 1 per head, 2.5–5 mm, widely bell-shaped, ± tomentose, lobes 6–9, long, reflexed. **FL**: perianth 4–6 mm, pale yellow, hairy, stalk-like base distinct. **FR** 5–5.5 mm; tip gen slightly hairy. **RARE** CA. Outcrops; 2500–2700 m. s SNH (The Needles and Slate Mtn, Tulare Co.).

E. umbellatum Torrey SULFUR FLOWER [Group 3] Per to shrub, 10–200 cm, 10–200 cm diam. **LVS** clustered on low sts; blade 3–40 mm, gen ± elliptic, gen densely tomentose (esp below). **INFL** umbel- to head-like, erect, slender; bracts lf-like, subtending rays (if ray 1, appearing whorled near mid-axis), rarely alternate on main axis); involucre gen 1 per ray, 1–6 mm, ± tomentose, lobes 6–12, long, reflexed. **FL**: perianth 2.5–12 mm, gen ± yellow becoming reddish (cream to purple), glabrous, lobes ± obovate, stalk-like base long, distinct. **FR** 2–5 mm, glabrous. Abundant. Dry, open, often rocky places; 200–3700 m. CA; to w Can, Colorado, NM. Extremely variable and difficult. Many vars. intergrade, best dispositions unclear; more study needed.

var. ***argus*** Rev. **LF**: blade margin often finely wavy. **INFL** compound, gen thinly tomentose; axis often with whorled or alternate bracts near middle; rays often branched again. **FL**: perianth 3–8 mm, bright yellow. Serpentine or other rocks; 1100–2300 m. KR, c NCoRH; sw OR. Some pls are ± glabrous with flat lvs.

var. ***bahiiforme*** (Torrey & A. Gray) Jepson **LF**: both blade surfaces densely tomentose. **INFL** compound, tomentose; rays often long, often with whorled bracts near middle, often branched again. **FL**: perianth 5–8 mm, bright yellow. Uncommon. Serpentine or other rocks; 700–2200 m. s NCoR, SnFrB, n SCoR. ❀DRN,DRY,SUN:1,2,**7**,14,**15–17**.

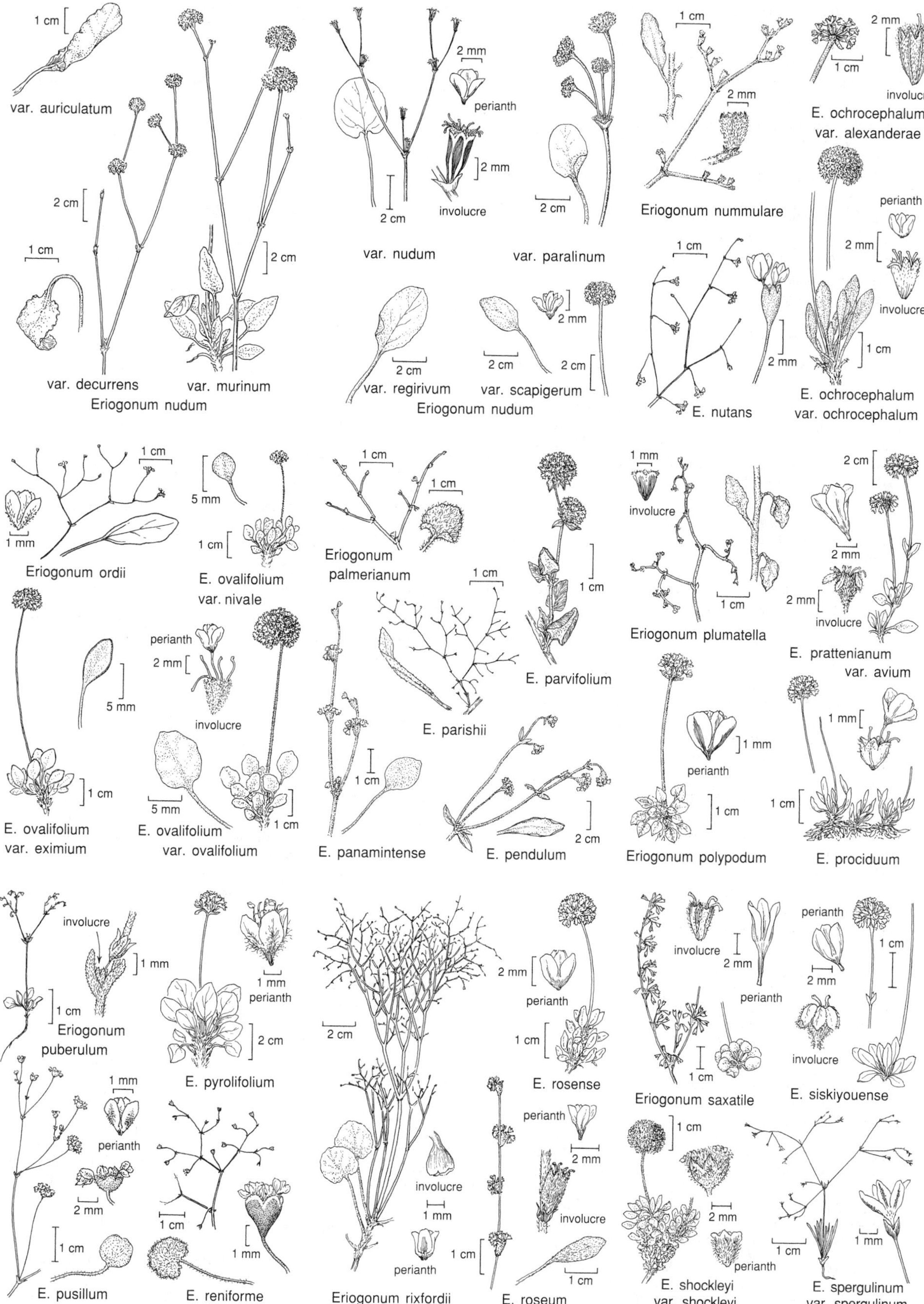
1 cm
var. auriculatum
2 cm
1 cm
2 cm
var. decurrens
var. murinum
Eriogonum nudum
2 mm
perianth
2 mm
involucre
2 cm
2 cm
var. nudum
var. paralinum
2 cm
var. regirivum
2 cm
2 mm
2 cm
var. scapigerum
Eriogonum nudum
1 cm
2 mm
Eriogonum nummulare
1 cm
2 mm
E. nutans
2 mm
1 cm
involucre
E. ochrocephalum
var. alexanderae
perianth
2 mm
involucre
1 cm
E. ochrocephalum
var. ochrocephalum
1 cm
1 mm
Eriogonum ordii
5 mm
1 cm
E. ovalifolium
var. nivale
1 cm
1 cm
Eriogonum
palmerianum
1 cm
1 cm
E. parvifolium
1 mm
involucre
1 cm
Eriogonum plumatella
2 cm
2 mm
2 mm
involucre
E. prattenianum
var. avium
5 mm
1 cm
E. ovalifolium
var. eximium
perianth
2 mm
involucre
5 mm
1 cm
E. ovalifolium
var. ovalifolium
E. parishii
1 cm
E. panamintense
2 cm
E. pendulum
1 mm
perianth
1 cm
Eriogonum polypodum
1 mm
1 cm
E. prociduum
involucre
1 mm
1 cm
Eriogonum
puberulum
1 mm
perianth
2 cm
E. pyrolifolium
2 cm
2 mm
perianth
1 cm
E. rosense
involucre
2 mm
perianth
1 cm
Eriogonum saxatile
perianth
1 cm
2 mm
involucre
E. siskiyouense
1 mm
perianth
2 mm
1 cm
E. pusillum
1 cm
1 mm
E. reniforme
involucre
1 mm
perianth
Eriogonum rixfordii
perianth
2 mm
involucre
1 cm
1 cm
E. roseum
1 cm
2 mm
perianth
E. shockleyi
var. shockleyi
1 cm
1 mm
E. spergulinum
var. spergulinum

var. ***chlorothamnus*** Rev. Gen shrub, glabrous. **LF**: blade gen ± oblanceolate. **INFL** compound; some rays again branched. **FL**: perianth 3–6 mm, bright yellow. Uncommon. Sagebrush shrublands; 1600–2900 m. s SNH, SNE, nw DMoj. Intergrades with var. *subaridum*.

var. ***covillei*** (Small) Munz & Rev. Low mat. **LF**: blade 2–6 mm wide, ± narrowly elliptic, sometimes remaining densely tomentose above. **INFL** simple, gen spreading. **FL**: perianth 2–5 mm, bright yellow. Uncommon. Subalpine rocks; 2600–3600 m. c&s SNH (esp Tulare, Inyo cos.). Intergrades with var. *nevadense*.

var. ***dichrocephalum*** Gand. **LF**: blade ± elliptic, gen subglabrous below. **INFL** simple, subglabrous. **FL**: perianth 4–8 mm, whitish to pale yellow, gen conspicuously striped. Sand or gravel; 1500–3700 m. Wrn, W&I; to OR, MT, WY. ❀TRY.

var. ***furcosum*** Rev. (p. 885) Low shrub. **LF**: blade elliptic or oblanceolate, margin gen entire, flat. **INFL** compound, thinly tomentose; rays often with whorled bracts near middle or branched again. **FL**: perianth 5–8 mm, bright yellow. Common. Sand or gravel; 1200–3000 m. SNH; w NV. [var. *stellatum* (Benth.) M.E. Jones (OR to MT) missaplied] ❀DRN,DRY,SUN:1–3,7,14,**15–17**.

var. ***glaberrimum*** (Gand.) Rev. GREEN BUCKWHEAT Pl glabrous. **LF**: blade ± oblanceolate. **INFL** ± compound; rays unbranched, most with whorled bracts near middle. **FL**: perianth 4–7 mm, whitish to cream. RARE. Sand or gravel; 1600–2300 m. Wrn; OR. [var. *aureum* (Gand.) Rev., golden buckwheat, misapplied to CA pls]

var. ***humistratum*** Rev. (p. 885) MOUNT EDDY BUCKWHEAT Low dense mat. **LF**: blade gen 5–10 mm, ± ovate to round. **INFL** simple, gen spreading; rays often hidden in fls. **FL**: perianth 3–7 mm, bright yellow. UNCOMMON. Rocks, gen on serpentine; 1700–2800 m. KR (esp Mount Eddy), n NCoRH (Yolla Bolly Mtns), n CaRH (Mount Shasta). ± identical to var. *hausknechtii* (Dammer) M.E. Jones as it occurs on Mount Hood, OR. Intergrades with var. *polyanthum*. ❀DRN,SUN:1,2,6,**7,15–17**,24& IRR:14,18.

var. ***juniporinum*** Rev. JUNIPER BUCKWHEAT Subshrub. **INFL** compound, ± tomentose; rays gen branched again. **FL**: perianth 4–6 mm, whitish to pale yellow. RARE in CA. Pinyon/juniper woodlands, sagebrush shrublands; 1300–2500 m. e DMtns (e San Bernardino Co.); NV. Like var. *subaridum*.

var. ***minus*** I.M. Johnston (p. 885) ALPINE SULFUR-FLOWERED BUCKWHEAT Dense mat. **LF**: blade ± round, very densely tomentose (± felt-like) below. **INFL** simple, often head-like. **FL**: perianth 2.5–6 mm, ± bright red (or yellow when young), gen conspicuously striped. UNCOMMON. Open gravel slopes; 1800–3000 m. SnGb, SnBr.

var. ***munzii*** Rev. Subshrub. **INFL** compound, open, ± densely tomentose; rays often branched again. **FL**: perianth 3–8 mm, bright yellow. Open rocks or among shrubs; 1500–2900 m. TR, SnJt.

var. ***nevadense*** Gand. (p. 885) Subshrub. **LF**: blade ± thinly tomentose below. **INFL** simple, gen thinly tomentose. **FL**: perianth 4–7 mm, bright yellow. Common. Jeffrey-pine forest, sagebrush shrublands; 1500–3200 m. SN (e slope), GB, nw DMoj; s OR, c NV. [var. *umbellatum* (Rocky Mtns) misapplied] ❀TRY.

var. ***polyanthum*** (Benth.) M.E. Jones (p. 885) Spreading mat. **LF**: blade elliptic to round. **INFL** gen simple (bracts sometimes whorled near mid-axis), ± erect; rays gen < 2 cm. **FL**: perianth 4–7 mm, bright yellow. Common. Open forest, among rocks or shrubs; 200–3000 m. KR, CaR, n SN, MP; OR. [var. *umbellatum* (Rocky Mtns) misapplied; var. *hausknechtii* (Dammer) M.E. Jones, "Warner Mtns buckwheat" (OR, WA) missaplied to pls from Wrn at 3000 m that are intermediate to var. *humistratum*] Smaller pls on serpentine in KR have been called var. *goodmanii* Rev. Intergrades esp with vars. *nevadense* & *speciosum*. ❀DRN,SUN:1,2, 4–6,**7,15–17**,24&IRR:3,8,9,**14,18–23**;CV.

var. ***speciosum*** (Drew) S. Stokes Shrub, robust. **INFL** compound, often > 5 cm diam; rays often branched again. **FL**: perianth 7–12 mm, bright yellow. Uncommon. Rocks, esp serpentine; 300–1600 m. KR, n SNF (Butte Co.); OR. Intergrades with var. *polyanthum*. ❀DRN,SUN:1,2,4–6,**15–17**&IRR:**7,14**.

var. ***subaridum*** S. Stokes (p. 885) Subshrub. **LF**: blade narrowly elliptic or oblanceolate. **INFL** compound; rays gen again branched (or with small bracts whorled near middle). **FL**: perianth 3–7 mm, bright yellow. Common. Sagebrush shrublands, rocks; 1300–2800 m. se SN, TR, SNE, DMtns; to Colorado, AZ. Closely related to var. *chlorothamnus*. For serpentine pls from e KR that key here, see var. *argus*. ❀TRY.

var. ***torreyanum*** (A. Gray) M.E. Jones (p. 885) DONNER PASS BUCKWHEAT Pl glabrous. **INFL** ± compound; rays unbranched, gen with whorled bracts near middle. **FL**: perianth 7–10 mm, bright yellow. RARE. Rocky meadows, outcrops; 2100–2400 m. n SNH (Sierra, Nevada, Placer cos.). Much like var. *glaberrimum*.

var. ***versicolor*** S. Stokes Mat, gen 10–40 cm diam. **INFL** simple or rays sometimes branched again. **FL**: perianth 3–6 mm, pink to reddish brown, conspicuously dark-striped. Uncommon. Sagebrush shrublands; 1800–3200 m. s W&I (Inyo Mtns), DMtns; s NV.

E. ursinum S. Watson (p. 885) [Group 3] Per from woody caudex 5–40 cm; loose mats 30–60 cm diam. **LVS** clustered on low sts; blade 8–25 mm, elliptic to ovate, yellow-brown-woolly below, gen shedding above. **INFL** umbel-like, compound, tomentose; bracts ± strap-like, subtending branches and involucres; involucres gen several per main ray, ± 4 mm, lobes 6–8, short, erect. **FL**: perianth 4–6 mm, pale yellow, ± glabrous, stalk-like base distinct. **FR** 3–3.5 mm; tip slightly hairy. Uncommon. Sand or gravel; 400–2800 m. s KR, s CaR, n SN. ❀DRN,DRY,SUN:1,2,**7,15–17**.

E. vestitum J. Howell (p. 885) IDRIA BUCKWHEAT [Group 1] Ann 5–50 cm. **LVS** basal; blade 10–50 mm, oblong to obovate, tomentose (esp below). **INFL** cyme-like; axes ± densely tomentose; bracts lf- and scale-like below (sometimes whorled near mid-axis), reduced upward; involucres sometimes ± clustered (few lateral), 1.5–2 mm, ribbed, tomentose, teeth 5. **FL**: perianth ± 2 mm, white, glabrous, minutely papillate. **FR** 2–2.5 mm, glabrous. n=17. UNCOMMON. Open clay; 400–700 m. ne SCoRI (Merced, San Benito, Fresno cos.). Not very distinct from *E. eastwoodianum*, *E. temblorense*.

E. vimineum Benth. (p. 885) WICKER BUCKWHEAT [Group 1] Ann 5–30 cm. **LVS** basal; blade 5–20 mm, ovate to round, tomentose (esp below). **INFL** cyme-like, open, glabrous to thinly hairy; bracts scale-like (lowest sometimes lf-like); involucres sessile (many lateral), 2–4 mm, nearly cylindric, ribbed, gen glabrous, teeth 5. **FL**: perianth 2–2.5 mm, white to rose or pale yellow, striped darker, glabrous, outer lobes larger, obovate. **FR** 2–2.5 mm, glabrous. 2n=24. Common. Volcanic or granite sand or gravel; 400–2400 m. ne KR, CaR, n SNH, MP; to WA, ID, n NV. Much like *E. luteolum*. ❀DRN,SUN:1,2,4,5,**6,7,14–17**,18.

E. viridescens A.A. Heller (p. 885) [Group 2] Ann 5–30 cm. **LVS** basal and seemingly cauline on lower st, 5–30 mm, gen ± elliptic, white-tomentose (esp below). **INFL** cyme-like, open, ± tomentose; axes gen round; bracts lf- and scale-like below, reduced upward; involucres on spreading, slender stalks, 2–3 mm, cup-shaped, unribbed, minutely glandular, teeth 5. **FL**: perianth 1–2 mm, white to rose, minutely glandular, outer lobes obovate, ± cupped at tip. **FR** ± 1 mm, glabrous. n=20. Sand, gravel, or clay; < 1700 m. SnJV, SCoR, TR, w DMoj. Locally common. ❀TRY.

E. wrightii Benth. [Group 4] Per to shrub, 15–100 cm, 10–150 cm diam. **LVS** cauline or clustered on low sts; blade 1–30 mm, linear to widely elliptic, gen densely tomentose (esp below). **INFL** cyme- to head-like; branches long, wand-like, glabrous to densely woolly; bracts scale-like; involucres sessile (most lateral), 1–4 mm, glabrous to densely woolly, teeth 5. **FL**: perianth 1.5–4 mm, white to pink or rose, glabrous, lobes obovate, stalk-like base 0. **FR** 1–3 mm, glabrous. Dry gravel or rocks; 50–3500 m. CA-FP, D; to s NV, TX, Mex. Variable; vars. intergrade.

var. ***membranaceum*** Jepson (p. 885) Subshrub 20–50 cm, 30–60 cm diam, gen thinly tomentose. **LF**: petiole base fully ringing st; blade 2–10 mm, strongly rolled under. **INFL** open; branches stout; involucre 2–3 mm. **FL**: perianth 3–4 mm. **FR** 2.5–3 mm. Common. Dry gravel or rocks; 300–2200 m. SW; nw Mex. ❀DRN,SUN:7,8–11,**14–16,18–21**,22–24.

var. ***nodosum*** (Small) Rev. (p. 885) Shrub 30–100 cm, 30–150 cm diam, densely gray-woolly. **LF**: blade 8–12 mm, ± flat. **INFL** open; branches stout; involucre 1.5–2.5 mm. **FL**: perianth 3–4 mm. **FR** 2.5–3 mm. Uncommon. Dry gravel or rocks; 150–1600 m. e PR, s-most DMoj, w DSon; n Baja CA.

var. ***olanchense*** (J. Howell) Rev. (p. 885) OLANCHA PEAK BUCKWHEAT Per 1–6 cm; mats 5–30 cm diam, dense, gen thinly tomentose. **LF**: blade 1–2 mm, ± flat. **INFL** ± head-like, gen slender; involucres gen 2–3 per head, ± 2 mm. **FL**: perianth ± 2 mm. **FR** 1.5–2 mm. RARE. Dry gravel or rocks; ± 3500 m. s SNH (Olancha Peak, Tulare Co.). Perhaps only a stunted form of var. *subscaposum*.

var. ***subscaposum*** S. Watson Subshrub 5–30 cm; mats 10–50 cm diam, loose, glabrous to tomentose. **LF**: blade 5–12 mm, ± flat. **INFL** dense to open, gen slender; involucre gen 2–4 mm. **FL**: perianth 2–3 mm. **FR** 2–2.5 mm. $2n$=34. Common. Dry gravel or rocks; 200–3400 m. SN, CW, TR, e PR. Intergrades with vars. *olanchense*, *trachygonum*. ❀DRN,SUN:1–3,**7**,14,**15**,**16**,17,**18–21**,22–24.

var. ***trachygonum*** (Benth.) Jepson (p. 885) Subshrub 5–50 cm, 10–50 cm diam, gen densely tomentose. **LF**: blade 15–30 mm, ± flat. **INFL** open, stout; involucre 2–3 mm. **FL**: perianth 3–4 mm. **FR** 2.5–3 mm. Common. Dry gravel; 50–800 m. NW, s CaR, n SNF. ❀DRN,SUN:1,2,**7**,14,**15**,**16**,17,22–24.

var. ***wrightii*** Subshrub 15–50 cm, 10–50 cm diam, gen tomentose. **LF**: blade 5–15 mm, ± flat. **INFL** open, slender; involucre 2–2.5 mm. **FL**: perianth ± 3 mm. **FR** 2.5–3 mm. Uncommon. Dry gravel or rocks; 30–2300 m. DMoj; to TX, n Mex.

FAGOPYRUM BUCKWHEAT

Ann, heterostylous, ± glabrous; fls bisexual and staminate. **ST** erect, 2–8 dm, hollow. **LVS** cauline, alternate, petioled; stipules fused, sheathing st above nodes; blade triangular-cordate to ± hastate. **INFL**: short spike or ± flat-topped panicle. **FL**: perianth lobes 5; stamens 8. **BISEXUAL FL**: styles 3, either short or long, stigmas 3, head-like. **FR** >> calyx, sharply 3-angled. 15 spp.: esp temp Eurasia. (Latin + Greek: beech wheat, from beechnut-like, edible fr)

F. esculentum Moench (p. 889) Pl becoming tinged with red. **ST** < 6 dm. **LF** < 7 cm; blade ± triangular, gen longer than wide. **INFLS** in most axils, dense. **FL**: perianth 3–4 mm, white; nectaries 8, yellow. **FR** 5–6 mm, shiny brown. n=8. Fields, disturbed places; < 500 m. SCo; native to Eurasia. [*F. sagitattum* Gilib. invalid] Cult as grain, honey crop. TOXIC to livestock.

GILMANIA

1 sp. (M. French Gilman, CA botanist, 1871–1944) [Reveal 1989 Phytologia 66:236–245]

G. luteola (Cov.) Cov. (p. 885) GOLDEN CARPET Ann, ± prostrate, 3–20 cm diam, thinly hairy. **LVS** basal (bracts appear as cauline lvs); stipule 0; petiole < 3 cm; blade 1–1.5 cm, ± obovate, ± glabrous. **INFL** 5–30 cm diam; bracts clustered at nodes, lf-like; involucre 0; fls several–many per bract axil; pedicels 2–7 mm. **FL**: perianth 1–2 mm, yellow, thinly hairy, lobes 6; stamens 9. **FR** 1.5–2 mm, brownish. RARE. Barren alkaline scrub; < 500 m. DMoj (Death Valley, Inyo Co.).

GOODMANIA

1 sp. (George J. Goodman, Oklahoma botanist, 1904–) [Reveal & Ertter 1976 Brittonia 28:427–429]

G. luteola (C. Parry) Rev. & B. Ertter (p. 885) Ann, spreading, 1–8 cm, thinly hairy. **LVS** basal; stipule 0; petiole 3–20 mm; blade 2–6 mm, ± round, lower surface tomentose, upper surface thinly hairy. **INFL** 2–10 cm diam; bracts lf-like to linear and awned; involucral bracts 5, free, 3–8 mm, unequal, linear, awned; fls several–many per involucre. **FL**: perianth 0.8–1 mm, yellow, base hairy, lobes 6; stamens 9. **FR** 1–1.2 mm, brownish. n=20. Uncommon. Grassland, alkaline and desert scrub; < 2200 m. s SnJV, SNE, w DMoj; NV [*Oxytheca l.* C. Parry]

HOLLISTERIA

1 sp. (Col. William W. Hollister, CA pioneer, 1818–1886) [Reveal 1989 Phytologia 66:199–220]

H. lanata S. Watson (p. 889) Ann, spreading, 3–20 cm, tomentose. **LVS** basal; stipule 0; petiole 0; blade 15–60 mm, gen linear, thinly hairy. **INFL** 5–50 cm diam; bracts 3–4 per node, awned, the largest 7–20 mm, ± elliptic, others 2–5 mm, linear; involucral bracts 3–4, free, 1.5–2 mm, awned; fl 1 per involucre. **FL**: perianth 1.5–2 mm, yellowish, woolly, lobes 6, outer 3 narrower; stamens 6–9. **FR** 1.7–2 mm, brown to dark black. n=21. UNCOMMON. Grassland in ± clay soils; 15–975 m. s SnJV, SCoRI.

LASTARRIAEA

Ann, hairy. **LVS** basal; stipule 0; petiole 0; blade linear. **ST** prostrate to ascending. **INFL** open, brittle; bracts gen 5 per node, awned, awn tips hooked; true involucre 0; fl 1 per node (more at st tips), sessile. **FL**: perianth greenish, thinly hairy, lobes 6, awned, awn tips hooked; stamens 3. **FR** brownish, lanceolate in outline. 3 spp.: w N.Am., S.Am. (J. V. Lastarria, Chile, 1817–1888) [Reveal 1989 Phytologia 66:199–220]

L. coriacea (Goodman) Hoover (p. 889) Pl 2–15 cm. **LF** 5–30 mm, ciliate. **INFL** 5–50 cm diam; bracts gen 5–10 mm, awns 0.5–2.5 mm. **FL**: perianth 2–3.5 mm incl awns. **FR** 2.5–3 mm. n=20 (30). Common. Coastal scrub, chaparral, sandy to gravelly soils; < 900 m. SN, GV, CW, SW; nw Mex. [*L. chilensis* Gay ssp. *californica* Gross; *Chorizanthe lastarriaea* C. Parry var. *c.* (Gross) Goodman; *C. coriacea* Goodman]

MUCRONEA

Ann, glandular. **LVS** basal; stipule 0; petiole indistinct; blade oblanceolate. **INFL** open; bract 1 per node, 5–25 mm, lobes 3, awned; involucre gen 1 per node, sessile, cylindric, lobes 2–4, awned, awns straight, often unequal; fl gen 1 per involucre. **FL**: perianth white to pink, hairy, lobes 6; stamens 6–9. **FR** 2–3 mm, lanceolate in outline, brown to black. 2 spp.: CA. (Latin: sharp-pointed, from bracts) [Reveal 1989 Phytologia 66:199–220]

1. Bracts on 1 side of st; perianth lobes entire .... ***M. californica***
1′ Bracts encircling st; perianth lobes gen jagged .... ***M. perfoliata***

M. californica Benth. (p. 889) CALIFORNIA SPINEFLOWER Pl 3–50 cm; non-glandular hairs ± sparse. **LF** gen 10–30 mm. **INFL**: bracts on 1 side of st, gen 5–12 mm, gen lobed > halfway to base; involucre 1 per node (2 smaller involucres often on a short axillary branch), 2–9 mm, ± radial, lobes gen 3, awns gen ± 2 mm; fls 1–2 per involucre. **FL**: perianth lobes entire. *n*=19. Coastal scrub, chaparral, in sandy soils; < 1400 m. s CW, SW. [*Chorizanthe c.* (Benth.) A. Gray; *C. c.* var. *suksdorfii* J.F. Macbr.]

M. perfoliata (A. Gray) A.A. Heller (p. 889) PERFOLIATE SPINEFLOWER Pl 2–30 mm; non-glandular hairs nearly 0. **LF** gen 30–60 mm. **INFL**: bracts encircling st, gen 10–25 mm, gen lobed < halfway to base; involucre 1 per node, 3–8 mm, asymmetric or ± bilateral, lobes 4, awns < 1.5 mm; fl 1 per involucre. **FL**: perianth lobes gen jagged. *n*=19. Chaparral, Joshua-tree woodland, etc., in sandy soils; 150–1900 m. s SNF, Teh, SnJV, SCoRI, n WTR, w DMoj. [*Chorizanthe p.* A. Gray] ❀TRY.

MUEHLENBECKIA

Vine-like shrubs, high-climbing or in dense twisted masses if no support is available, monoecious. **ST** much-branched, < 10 m, slender, wiry. **LVS** cauline, alternate, petioled; stipules fused, closely sheathing st above nodes, delicate; blade linear to round, ± thick. **INFL** axillary, open or dense; staminate and pistillate fls mixed. **STAMINATE FL**: perianth lobes 5; stamens 8. **PISTILLATE FL**: perianth lobes 5; style < 1 mm, stigmas 3. **FR** enclosed in fleshy perianth, appearing berry-like; achene 3-angled, ovoid. 15 spp.: s temp. (H.G. Muehlenbeck, Alsatian physician, 1798–1845)

1. Lvs 0.5–1 cm, blade nearly round .... ***M. complexa***
1′ Lvs 5–10 cm, blade narrowly lanceolate, ± hastate .... ***M. hastatula***

M. complexa Meissner MAIDENHAIR VINE **ST** ± twisted, not twining, dark brown or reddish. **LF** gen < internode; blade sometimes with 2 shallow sinuses near base, leathery. **INFL** < 1 cm, spike-like. **FL** 3–5 mm; perianth greenish, becoming waxy-white in fr. **FR** 4–5 mm. Disturbed urban areas; < 500 m. NCo, SnFrB; native to New Zealand. Cult as orn.

M. hastatula (Smith) I.M. Johnston **ST** twining when young; greenish brown. **LF** > internode. **INFL** 1–4 cm, panicle-like. **FL** 1–2 mm; perianth green or brown. **FR** ± 4 mm. Disturbed urban areas; < 500 m. SnFrB, expected elsewhere; native to S.Am. Cult as orn; invasive, difficult to eradicate.

NEMACAULIS

1 sp. (Latin: thread st) [Reveal & Ertter 1980 Madroño 27:101–109]

N. denudata Nutt. WOOLLY-HEADS Ann, prostrate to erect, 4–40 cm, woolly. **LVS** basal; stipule 0; petiole indistinct; blade 5–80 mm, linear to oblanceolate. **INFL** open, 20–80 cm diam; branches wiry; bracts 3 per node, 1–5 mm, gen linear; involucral bracts many, 1–4 mm, oblanceolate, upper surface very woolly; fls 5–30 per involucre. **FL**: perianth 0.8–1.5 mm, greenish white to dark red, glabrous or glandular, lobes 6; stamens 3. **FR** ± 1 mm, brown to black. Coastal strand, desert scrub, etc., in sandy soils; < 400 m. SW, DSon; AZ, nw Mex.

1. Pl prostrate to decumbent; involucral bracts gen red with white wool; fls 12–30 per involucre; coastal .. var. ***denudata***
1′ Pl ascending to erect; involucral bracts gen brownish with tawny wool; fls 5–12 per involucre; coastal and inland .... var. ***gracilis***

var. ***denudata*** (p. 889) **FLS** gen visible in surrounding wool; outer perianth lobes gen widely ovate. Coastal beaches; < 100 m. SCo; nw Mex. ❀TRY.

var. ***gracilis*** (p. 889) Goodman & L. Benson **FLS** gen obscured by surrounding wool; outer perianth lobes gen linear. Habitat and range of sp. ❀TRY.

OXYRIA MOUNTAIN SORREL

1 sp. (Greek: sour, from acidic taste)

O. digyna (L.) Hill (p. 889) Per, glabrous, becoming red-tinged; caudex thick, scaly. **ST** erect, < 5 dm. **LVS** ± basal, alternate, petioled; stipules fused, loosely sheathing st above nodes, breaking; blade << petiole, reniform, fleshy, sour-tasting. **INFL**: panicle, erect, ± open. **FL** nodding; perianth lobes 4, < 2 mm; stamens 6; styles 2. **FR** flat, conspicuously winged. 2*n*=14. Alpine rock crevices, talus; 1800–4000 m. CA; circumboreal. ❀DRN,IRR:1; DFCLT.

OXYTHECA

Ann, glandular, glaucous. **LVS** basal; stipule 0; petiole indistinct; blade linear to obovate, hairy. **INFL** open; bracts gen 3 per node, gen linear-lanceolate, gen fused at base, awned; involucre 1 per node, gen stalked, narrowly funnel-shaped to bell-shaped, teeth 3–36, awned; fls 2–20 per involucre. **FL**: perianth white to rose or pale yellow, hairy, lobes 6; stamens 9. **FR** golden brown to dark brown. w N.Am., s S.Am. (Latin: sharp box, from involucre) [Ertter 1980 Brittonia 32:70–102]

1. Involucral awns 5; fls incl; each perianth lobe again lobed or fringed (sect. *Neoxytheca*)
 2. Involucral bracts >> half fused; involucre very widely funnel-shaped, laterally flattened, white-margined; perianth lobes irregularly fringed; e PR .... ***O. emarginata***
 2′ Involucral bracts << half fused; involucre funnel-shaped, not flattened or white-margined; perianth lobes with 3 ± regular lobes
  3. Fl 1–2 mm; s SNH, TR, SnJt .... ***O. caryophylloides***
  3′ Fl 2.5–4 mm; SnGb, SnBr, PR, w DMtns .... ***O. trilobata***
1′ Involucral awns (3)4 or > 7; fls exserted; perianth lobes gen entire

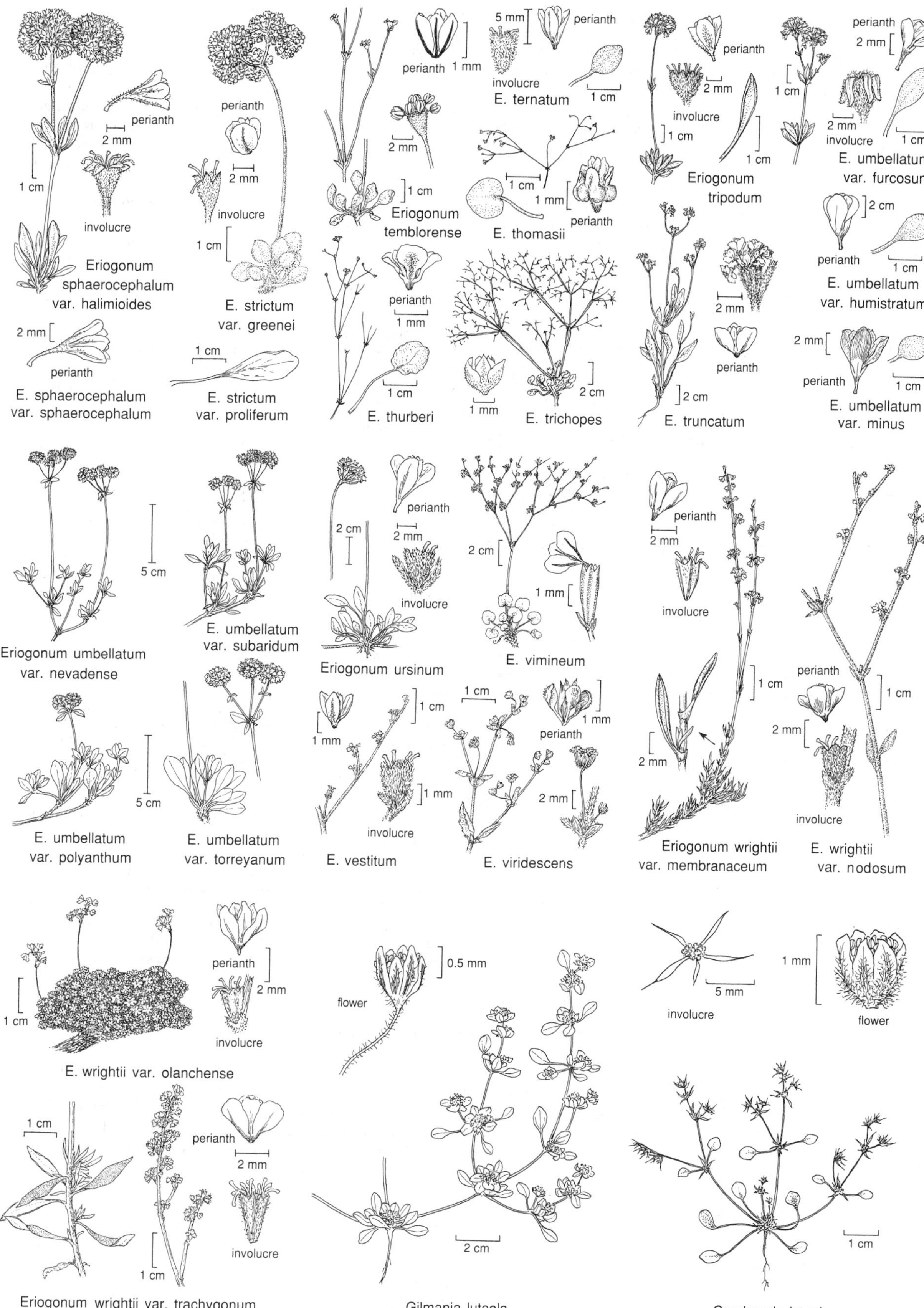
perianth
2 mm
1 cm
involucre
Eriogonum sphaerocephalum var. halimioides
2 mm
perianth
E. sphaerocephalum var. sphaerocephalum
perianth
2 mm
involucre
1 cm
E. strictum var. greenei
1 cm
E. strictum var. proliferum
perianth 1 mm
2 mm
1 cm
Eriogonum temblorense
perianth
1 mm
1 cm
E. thurberi
5 mm
perianth
involucre
E. ternatum
1 cm
1 cm
1 mm
perianth
E. thomasii
1 mm
2 cm
E. trichopes
perianth
2 mm
involucre
1 cm
1 cm
Eriogonum tripodum
2 mm
perianth
2 cm
E. truncatum
perianth
2 mm
1 cm
2 mm
involucre
1 cm
E. umbellatum var. furcosum
2 cm
perianth
1 cm
E. umbellatum var. humistratum
2 mm
perianth
1 cm
E. umbellatum var. minus
5 cm
Eriogonum umbellatum var. nevadense
E. umbellatum var. subaridum
5 cm
E. umbellatum var. polyanthum
E. umbellatum var. torreyanum
2 cm
perianth
2 mm
involucre
Eriogonum ursinum
2 cm
1 mm
E. vimineum
1 cm
1 mm
involucre
1 mm
E. vestitum
1 cm
1 mm
perianth
2 mm
E. viridescens
perianth
2 mm
involucre
1 cm
2 mm
Eriogonum wrightii var. membranaceum
perianth
2 mm
involucre
1 cm
E. wrightii var. nodosum
1 cm
perianth
2 mm
involucre
E. wrightii var. olanchense
1 cm
perianth
2 mm
involucre
1 cm
Eriogonum wrightii var. trachygonum
0.5 mm
flower
2 cm
Gilmania luteola
5 mm
involucre
1 mm
flower
1 cm
Goodmania luteola

4. Involucral awns gen 4 (mostly sect. *Oxytheca*)
5. Bracts fully fused and completely surrounding st; involucres sessile — s SnJV, GB, DMoj .... ***O. perfoliata***
5′ Bracts fused at base only; involucres nearly sessile or ± stalked
6. Lf blade obovate, ciliate, surface nearly glabrous; n SnBr — on limestone ...... [2]***O. parishii*** var. ***goodmaniana***
6′ Lf blade oblanceolate, ± hairy; GB
7. Stalks of lower involucres 2–15 mm; bract awns < 0.5 mm; GB .... ***O. dendroidea*** ssp. ***dendroidea***
7′ Stalks of lower involucres 0.5–2 mm throughout; bract awns > 1 mm; se W&I .... ***O. watsonii***
4′ Involucral awns 7–36 (4–5 in var. *goodmaniana*) (sect. *Acanthoscyphus*); TR .... ***O. parishii***
8. Involucral awns 7–16, reddish, longest 3–4 mm; WTR .... var. ***abramsii***
8′ Involucral awns 4–36, ivory-colored, longest gen 2–5 mm; TR
9. Involucral awns 4–5, longest 2–3 mm; n SnBr, on limestone, RARE .... [2]var. ***goodmaniana***
9′ Involucral awns 7–36, longest 3–5 mm; TR, not on limestone, more common
10. Involucral awns 7–10; RARE, e SnBr .... var. ***cienegensis***
10′ Involucral awns 10–36; TR .... var. ***parishii***

O. caryophylloides C. Parry (p. 889) CHICKWEED OXYTHECA Pl 10–25 cm. **LF** 1–8 cm, gen oblanceolate, strigose, glandular. **INFL** sparsely glandular at nodes; involucre funnel-shaped, glabrous or glandular, involucral bracts 5, 4–7 mm, fused only at base, awns 0.3–1 mm; fls 2–3 per involucre. **FL**: perianth 1–2 mm, greenish white to reddish, each lobe with 3 ± regular lobes at tip. **FR** 1.2–1.5 mm. *n*=20. UNCOMMON. Sandy soils in pine forest; 1200–2600 m. s SNH (Tulare Co.), TR, SnJt.

O. dendroidea Nutt. ssp. ***dendroidea*** (p. 889) Pl 5–40 cm. **LF** 1–4.5 cm, gen narrowly oblanceolate, hairy, sparsely glandular. **INFL** sparsely glandular; bract awns 0–0.5 mm; branches sometimes recurved; involucre narrowly funnel-shaped, glabrous, involucral bracts 4, 1–2 mm, < half fused, awns 0.5–3 mm; fls 2–6 per involucre. **FL**: perianth 1–2 mm, white to pink, strigose, lobes entire. **FR** 1.5–2 mm. *n*=20. Dry sandy places in scrub or pine forest: 1200–2200 m. GB; to WA, WY, NV.

O. emarginata H.M. Hall (p. 889) WHITE-MARGINED OXYTHECA Pl 3–30 cm. **LF** 1.5–7.5 cm, gen oblanceolate, sparsely strigose and glandular. **INFL** sparsely glandular; bracts linear to ovate; involucre very widely funnel-shaped, laterally flattened, glabrous, white-margined; involucral bracts 5, 4–8 mm, >> half fused, awns 1–1.5 mm; fls 3–6 per involucre. **FL**: perianth 2–5 mm, white to pink, lobes irregularly fringed. **FR** 1.8–2 mm. *n*=20. UNCOMMON. Dry places, open scrub or woodlands; 1200–2500 m. e PR (SnJt, Santa Rosa Mtns, Riverside Co.).

O. parishii C. Parry Pl 5–60 cm. **LF** 1–7 cm, oblanceolate to widely obovate; margin ciliate. **INFL** glandular on lower half of internodes; involucre funnel-shaped, glabrous; involucral bracts 4–36, 1.5–2 mm, gen > half fused (exc awns), awns gen 2–5 mm; fls 3–20 per involucre. **FL**: perianth 2–2.5 mm, white to pink, gen strigose; lobes entire. **FR** 1.7–2 mm. Dry, rocky or sandy soils; 1300–2500 m. s SCoRO, TR.

var. ***abramsii*** (McGregor) Munz (p. 889) **INFL**: involucral awns 7–16, dark red, longest 3–4 mm. Uncommon. Shale to sandy places; 1700–2000 m. s SCoRO (San Rafael Mtns), WTR (Topatopa Mtns, Mt. Pinos). [*O. a.* McGregor; *O. p.* ssp. *a.* (McGregor) Munz]

var. ***cienegensis*** B. Ertter (p. 889) CIENEGA SECA OXYTHECA **INFL**: involucral awns 7–10, ivory, longest 3–5 mm. RARE. Sandy, granitic places; 2170–2450 m. e SnBr (Cienega Seca Creek, Coon Creek).

var. ***goodmaniana*** B. Ertter (p. 889) CUSHENBURY OXYTHECA **INFL**: involucral awns 4–5, ivory, longest 2–3 mm. RARE. Limestone talus; 1300–2300 m. n SnBr (Greenlead Mine, Holcomb Valley, Cushenbury Canyon). [*O. watsonii* of most CA references] Threatened by limestone mining.

var. ***parishii*** (p. 889) **INFL**: involucral awns 10–36, ivory, longest 3–5 mm. *n*=20. Sand to gravelly places; 2000–2500 m. TR.

O. perfoliata Torrey & A. Gray (p. 889) Pl 6–20 cm, nearly glabrous or sparsely glandular. **LF** 1–6 cm, oblanceolate to oblong, ciliate. **INFL** glandular on lower half of internodes; bracts fused around node, fused unit 1–2.5 cm diam; involucre sessile, narrowly funnel-shaped, involucral bracts 4, 2–5 mm, < half fused, awns 2–3 mm; fls 5–10. **FL**: perianth 1.5–2.5 mm, white to greenish yellow, hairy, lobes entire. **FR** 1.5–2 mm. *n*=20. Common. Creosote-bush or pinyon scrub, sandy to rocky soils; 200–2000 m. s SnJV, GB, DMoj; to NV, sw UT, nw AZ.

O. trilobata A. Gray (p. 889) Pl 7–50 cm. **LF** 1–9 cm, gen linear to oblanceolate, sparsely strigose and glandular. **INFL** sparsely glandular near base of internodes; involucre widely funnel-shaped, sparsely glandular, involucral bracts 5–6, << half fused, awns sometimes 3, 0.3–2 mm; fls 3–10 per involucre. **FL**: perianth 2.5–4 mm, white to pink, hairy and glandular, lobes deeply divided into 3 ± regular, overlapping lobes. **FR** 1.2–2 mm. *n*=20. Common. Sandy places; 700–2800 m. SnGb, SnBr, PR, w DMtns (Little San Bernardino Mtns); Baja CA.

O. watsonii Torrey & A. Gray (p. 889) WATSON'S OXYTHECA Pl 5–25 cm. **LF** 0.7–5 cm, oblanceolate to obovate, gen sparsely strigose and glandular. **INFL** glandular at or near nodes; bract awns 1–3 mm; involucre narrowly funnel-shaped, glabrous, involucral bracts 4, 1.5–2 mm, ± half fused (exc awns), awns 2.5–3 mm; fls 2–7 per involucre. **FL**: perianth 1–1.5 mm, greenish white to pink, hairy; lobes entire. **FR** 1–1.5 mm. *n*=20. RARE in CA. Sandy places; 1200–2000 m. se W&I (Santa Rosa Hills, Inyo Co.); NV. Name misapplied to *O. parishii* var. *goodmaniana*.

POLYGONUM KNOTWEED, SMARTWEED

Ann, per, shrub, vine. **ST** prostrate to erect, or climbing, or floating, < 3 m. **LVS** gen cauline, alternate, sessile or petioled; stipules fused, sheathing st above nodes, gen scarious or membranous; blade sometimes obviously jointed to stipule sheath. **INFL**: unit a 1–8-fld cluster, these arrayed singly or in head-like to open panicles. **FL**: perianth lobes gen 5; stamens 3–8, filaments gen wider at base. **FR** gen ovoid, 3-angled, sometimes round, flat, indented; shiny to dull, brown to black. ± 300 spp.: worldwide, esp n temp. (Greek: many knees, from swollen nodes of some spp.) [Ronse Decraene & Ackeroyd 1988 Bot J Linn Soc 98:321–371] Segregate genera (e.g., *Bistorta, Fallopia, Persicaria*) are sometimes recognized.

1. 3 outer perianth lobes ± keeled to widely winged, esp in fr — uncommon aliens (subg. *Fallopia*)
2. Lf gen 3–10 cm, ovate-sagittate; vine (sect. *Fallopia*)
3. Woody per, perianth bright pink .... ***P. baldschuanicum***
3′ Non-woody ann, perianth greenish white .... ***P. convolvulus***
2′ Lf gen 10–30 cm, widely ovate to round; stout per from rhizome; (sect. *Reynoutria*)
4. St gen 1–2 m; lf blade 10–15 cm, ± round, tip abruptly pointed; perianth ± white .... ***P. cuspidatum***
4′ St gen 2–3 m; lf blade 15–30 cm, widely ovate, acuminate; perianth ± green .... ***P. sachalinense***

1′ 3 outer perianth lobes not keeled or winged, even in fr
5. Lf gen 5–35 cm, petioled (exc sometimes *P. davisiae*), blade gen lanceolate or wider; if ann, st gen 50–200 cm; filaments thread-like (subg. *Persicaria*)
6. St unbranched, erect; gen subalpine; lvs ± basal, blade gen oblong, strap-shaped; perianth lobes nearly free (sect. *Bistorta*) .... ***P. bistortoides***
6′ St branched, gen ± sprawling; lower elevations; lvs cauline, blade rarely oblong; perianth lobes fused to nearly free
7. Lf widely lanceolate to round
8. Infl head-like, ± spheric, dense; perianth fused ± 1/3 its length (sect. *Cephalophilon*) .... ***P. capitatum***
8′ Infl a panicle or appearing as a small, dense cluster; perianth fused << 1/3 its length (sect. *Aconogonon*)
9. Infl an open, ± lfless panicle, fls > 50; st erect, > 1 m, stout; some lvs > 10 cm
10. Montane native; st from thick caudex; lf blade base round to tapered, basal lobes 0 .... ***P. phytolaccifolium***
10′ North coastal alien; st from extensive rhizome; lf blade base tapered, often with 2 small, round lobes .... ***P. polystachyum***
9′ Infl a small axillary cluster, fls 2–25; st prostrate to decumbent, < 1 m; lvs < 8 cm
11. Petioles < 1 cm; lf blade gen lanceolate, surfaces ± glabrous to scabrous, yellow-green, margin scabrous, tip acute; fls 2–5 per axil .... ***P. davisiae***
11′ Some petioles > 1 cm; lf blade gen ovate to round, surfaces ± glabrous to velvety, dark green becoming reddish, tip gen ± round, margin glabrous to soft-hairy; fls 5–25 per axil .... ***P. newberryi***
7′ Lf ± narrowly lanceolate (exc in *P. amphibium*) — infl spike-like; perianth fused ± 1/3 its length (sect. *Persicaria*)
12. Perianth gland-dotted; infl ± open, slender; fr 2–3-angled
13. Perianth margins gen ± red; fr dark brown, dull, bitter-tasting; ann .... ***P. hydropiper***
13′ Perianth margins gen white; fr black, shiny, not bitter-tasting; per from rhizome .... ***P. punctatum***
12′ Perianth not gland-dotted; infl open or dense; fr gen 2-angled or ± flat
14. Perianth greenish to light pink; infl spikes 2–6 mm wide
15. Perianth gen light pink, opening, veins at fruiting inconspicuous; fr 3-angled; per from extensive rhizome .... ***P. hydropiperoides***
15′ Perianth gen greenish, remaining closed, veins at fruiting prominent, forked, recurved; fr flat, gen indented on both sides; ann .... ***P. lapathifolium***
14′ Perianth deep pink; infl spikes 7–20 mm wide
16. Ann; ± terrestrial; st ± erect (lower nodes sometimes rooting); lf narrowly lanceolate
17. Stipule (incl margin) glabrous; peduncle glandular; fr ± 3.5 mm, round, nearly flat to indented .... ***P. pensylvanicum***
17′ Stipule strigose, margin bristly; peduncle glabrous; fr ± 2.5 mm, widely ovate, often 3-angled .... ***P. persicaria***
16′ Per from rhizome; aquatic or on shorelines; st gen floating or decumbent; lf widely lanceolate to elliptic .... ***P. amphibium***
18. St terrestrial or emergent in fl; lf blade lanceolate, tip ± acuminate; infl > 4 cm .... var. ***emersum***
18′ St floating in fl; lf blade oblong, tip round to acute; infl < 3.5 cm .... var. ***stipulaceum***
5′ Lf gen 1–5 cm, ± sessile, gen linear to lanceolate (to round in *P. minimum*); ann, subshrub, or shrub, st gen 3–35 cm; filaments much wider at base (subg. *Polygonum*)
19. Shrub or subshrub
20. St ± erect (gnarled at base), fl branches erect, whip-like; foothills or low mtns .... ***P. bolanderi***
20′ St prostrate to ascending throughout; coast or high mtns
21. Subshrub; stipule ± 20 mm, persistent; coastal dunes .... ***P. paronychia***
21′ Shrub, gnarled; stipule ± 5 mm, ± falling with lf; high mtns .... ***P. shastense***
19′ Ann
22. St ± round, obviously and ± regularly 8–16 ribbed (if st ± angled, ribs obvious between angles); gen alien (sect. *Polygonum*)
23. Fl clusters scattered, some or all in axils of full-sized lvs; st prostrate to erect
24. Fr 3.5–4.5 mm, olive-brown, shiny — uncommon, SnFrB salt marshes, possibly native ..... [2]***P. marinense***
24′ Fr 2–3 mm (late-season fr sometimes larger), dark brown, gen ± dull
25. Perianth ± 2.5 mm, exserted from stipule; abundant .... ***P. arenastrum***
25′ Perianth ± 2 mm, ± enclosed in stipule; uncommon .... ***P. prolificum***
23′ Fl clusters ± restricted to branch tips, in axils of bracts; st erect
26. All but lowest bracts < fls or falling early, so infl appears raceme-like and restricted to st tips; perianth margin pink; fr < 2 mm, brown, very shiny .... ***P. argyrocoleon***
26′ Most bracts > fls, so infl clusters appear axillary and less restricted to st tips; perianth green or yellowish (margin rarely pink); fr > 2 mm, not very shiny
27. Perianth ± green, ± incl in stipule; fr 2–2.5 mm, brown, ± shiny .... ***P. patulum***
27′ Perianth gen dark yellowish (esp margin), exserted from stipule; fr ± 3 mm, dark brown, ± dull .... ***P. ramosissimum***
22′ St ± sharply angled (esp just below nodes), ribs between angles obscure or 0; native
28. Lvs firm, ± spiny, uniformly linear, joint with stipule obscure; styles free, bases hard, persistent in fr (sect. *Duravia*, in part)
29. Pl dense, often cushion-like, 1–7 cm; stipule deeply fringed, segments gen curled .... ***P. parryi***
29′ Pl open, 2–40 cm; stipule bristly or shallowly fringed, segments not curled
30. Stipule prominent, silvery, widely lanceolate, ± obscuring lvs, margin fringed .... ***P. bidwelliae***

30′ Stipule not at all obscuring lvs, silvery or golden, lanceolate, margin bristly .... *P. californicum*
28′ Lvs soft, not spiny, linear to round (gen lanceolate), joint with stipule obvious; styles fused, falling in fr as a unit (sect. *Monticola*)
31. Fr 1.5–2.5 mm, light brown to nearly black, dull to somewhat shiny (the darkest dull) [sspp. intergrade] .... *P. polygaloides*
32. Bract margin white; fr brown to nearly black, gen dull (lanceolate if shiny)
33. Infl ± 7–9 mm wide; bracts ascending to spreading, shorter, rounder upward, white margin wide, gen reflexed; fr < 2 mm, ovate in outline, nearly black, dull; widespread .... ssp. ***confertiflorum***
33′ Infl ± 4–6 mm wide; bracts ± appressed, uniformly lance-elliptic, white margin narrow, gen not reflexed; fr 2–2.5 mm, lanceolate in outline, brown, gen ± shiny; RARE, MP .... [2]ssp. ***esotericum***
32′ Bract margin green; fr brown, gen ± shiny
34. Bracts lance-elliptic, stiffly appressed; fr 2–2.5 mm, lanceolate in outline, brown, shiny to ± dull; RARE, MP .... [2]ssp. ***esotericum***
34′ Bracts linear-lanceolate, soft, spreading; fr ± 1.5 mm, ± widely ovate in outline, light brown, ± shiny; widespread .... ssp. ***kelloggii***
31′ Fr 2–5 mm, shiny, black or olive brown
35. Fr olive-brown; salt marshes, SnFrB; uncommon .... [2]*P. marinense*
35′ Fr black; widely distributed; ± common
36. Lvs elliptic to round, persistent, regularly and widely spaced below, barely smaller upward, crowded at st tips; st reddish .... *P. minimum*
36′ Lowest lvs linear to oblanceolate or narrowly elliptic, gen falling early, reduced, more widely spaced upward; st ± green [sspp. intergrade extensively] .... *P. douglasii*
37. Perianth 3–5 mm, opening partly or widely; infl open or dense; fls and fr gen spreading
38. Fl clusters scattered; perianth white, opening widely .... ssp. ***majus***
38′ Fl clusters crowded at st tips; perianth ± pink, opening partly .... ssp. ***spergulariiforme***
37′ Perianth gen 2–3 mm, gen remaining closed; infl ± open; fls and fr erect or becoming reflexed
39. Fr erect, 2–3 mm; enfolding perianth not stalked at base; lvs uniformly lance-elliptic, gradually smaller upward .... ssp. ***johnstonii***
39′ Fr reflexed, 2–4 mm; enfolding perianth gen with minute, curved stalk at base; lvs linear to elliptic, abruptly smaller upward
40. Fr ± 2 mm; pl funnel-shaped from ascending basal branches; lower lvs ± elliptic, persistent, appearing crowded .... ssp. ***austiniae***
40′ Fr 3–4 mm; pl ± open, not funnel-shaped; lower lvs linear to oblanceolate, falling early, ± widely spaced .... ssp. ***douglasii***

P. amphibium L. WATER SMARTWEED Per from rhizome, terrestrial, emergent, or floating. **ST** prostrate to erect, rooting at nodes, stout. **LF** < 35 cm, petioled; stipule veiny, brown, fracturing, bristles 0 exc in terrestrial sts with flared stipules; blade widely lanceolate, oblong, or ovate, base tapered to cordate, tip acuminate to round. **INFL** 1–15 cm, dense. **FL**: perianth 4–6 mm, deep pink; filaments thread-like. **FR** 2.5–3 mm, round, flat, brown, shiny. $2n$=96,98. Common. Shallow lakes, streams, shores; < 3000 m. CA-FP, w DMoj; to e N.Am, Eurasia. Variable, in part due to environment. Vars. intergrade.

var. ***emersum*** Michaux (p. 889) KELP Rhizome coarse. **ST** gen erect, terrestrial or emersed in fl. **LF**: blade lanceolate, ± hairy, tip ± acuminate. **INFL** 4–10 cm, narrowly cylindric or conic. Habitat of sp.; < 1500 m. Range of sp. [*P. coccineum* Muhlenb.] Prostrate, non-fl, terrestrial sts hairier. NOXIOUS WEED.

var. ***stipulaceum*** Coleman (p. 889) Rhizome slender. **ST** prostrate, gen floating in fl. **LF**: blade < 10 cm, gen oblong, ± glabrous, tip round to acute. **INFL** < 3.5 cm, ± ovoid. Habitat of sp.; < 3000 m. CA-FP. Shore growth form has spreading hairs, flared stipules. ❀Water or WET,SUN:**1–9**,**14**,15–24.

P. arenastrum Boreau (p. 893) COMMON KNOTWEED, DOORWEED Ann or weak per. **ST** prostrate to erect, < 1.5 m, round, ribbed. **LVS** gen 0.5–2 cm, gen smaller upward and on side branches, jointed to stipule sheath, sessile; blade linear to narrowly elliptic. **INFL**: fls 2–8 in nearly all axils. **FL**: perianth 2–3.2 mm, fused 30–50% its length, green, margin white or pink, base in fr often prominently veiny. **FR** 2–3 mm, at maturity gen 2 sides convex, 1 ± concave, dark brown, ± dull. $2n$=40,60. Abundant. Disturbed places; < 2500 m. CA; to e N.Am, native to Eur. [*P. montereyense* Brenckle] Highly variable, tolerant of trampling. *P. aviculare* L. (perianth 3.1–4.9 mm, fr 2.8–3.8 mm: Styles 1962 Watsonia 5:177–214) is very closely related, apparently undocumented in CA.

P. argyrocoleon Kunze (p. 893) Ann. **ST** erect, < 6 dm, round, ribbed. **LF** gen 2–4 cm, sessile; blades linear to lanceolate, reduced gradually upward, then abruptly at infl. **INFL**: branches raceme-like, restricted to st tips, ± 15 cm; fls 4–8 per axil; bracts < fls or falling early. **FL**: perianth ± 2 mm, exserted from stipule, margin red. **FR** < 2 mm, widely ovoid, 3 sides ± equal, brown, shiny. $2n$=40. Fields, disturbed places; < 1000 m. CA (esp s); native to sw Asia.

P. baldschuanicum Regel Woody vine, ± glabrous. **ST** climbing, < 6 m. **LF** gen 3–10 cm, petioled; blade narrowly ovate, base cordate or sagittate. **INFL** much branched, open. **FL**: perianth 5–8 mm, flaring, bright pink, midribs of outer 3 lobes winged. **FR** ± 2 mm, black, shiny. Uncommon. Disturbed places; < 500 m. SnFrB, SCo, expected elsewhere; native to c Asia.

P. bidwelliae S. Watson (p. 893) BIDWELL'S KNOTWEED Ann. **ST** erect, 2–20 cm, ± angled; ribs 0 or obscure. **LVS** 4–8 mm, crowded above, sessile; stipules ± obscuring lvs and fls, silvery, shallowly fringed; blade obscurely jointed to stipule, < 2 cm, linear, spine-tipped, ± recurved, veins 3. **FL** 1 per upper axil, ± sessile; perianth ± 2.5 mm, pink; styles free. **FR** 2.5 mm, ovoid-elliptic, brown, shiny. UNCOMMON. Thin volcanic soils; 60–1200 m. CaR, ne ScV (Butte, Tehama, Shasta cos.). ❀SUN,DRN:**7**,8,**9**,10, 15,**16**,17–21.

P. bistortoides Pursh (p. 893) WESTERN BISTORT Per from short, bulb-like rhizome. **ST** erect, unbranched, gen 20–70 cm. **LVS** ± basal or reduced rapidly upward, gen 15–40 cm, petioled; blade oblong or strap-shaped to narrowly lanceolate, leathery, ± glaucous, esp below. **INFL** long peduncled, dense, ± oblong. **FL**: perianth 4–5 mm, white or pink, lobes nearly free; stamens exserted, filaments thread-like. **FR** 3–4.5 mm, ± obovoid, brown, shiny. Common. Wet mtn meadows; 1500–3000 m. CA-FP (uncommon in coastal freshwater marshes; 0–20 m. NCo, SnFrB); to AK, e N.AM. Variable. ❀WET,SUN:**1–6**,7,15–18;INV.

P. bolanderi Brewer (p. 893) Shrub. **ST** gnarled below; twigs erect, < 6 dm, barely woody, slender, whip-like. **LVS** crowded at branch tips and in axillary clusters, sessile; stipule margin deeply fringed; blade obscurely jointed to stipule, < 2 cm, linear, spine-tipped, some ± recurved, veins 3. **FLS** 1–2 in upper axils, sessile; perianth ± 3 mm, white to pink, midribs dark; styles free. **FR** ± 3 mm, ovoid-elliptic, brown, shiny. Uncommon. Open, rocky places; 300–1500 m. NW (esp Napa Co.), CaR, n&c SN. ❀TRY STBL.

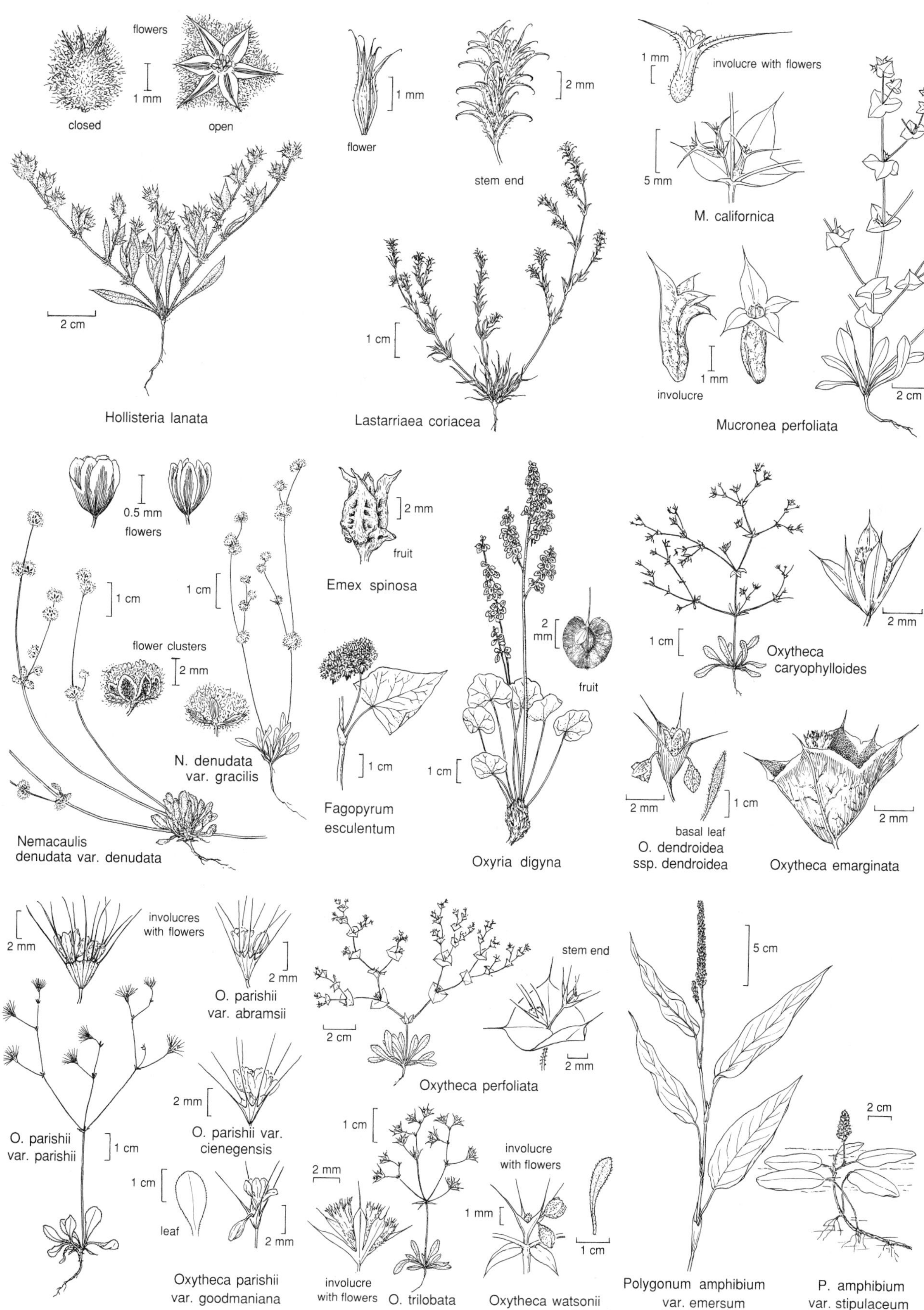
flowers
1 mm
closed
open
2 cm
Hollisteria lanata
1 mm
flower
2 mm
stem end
1 cm
Lastarriaea coriacea
1 mm
involucre with flowers
5 mm
M. californica
1 mm
involucre
2 cm
Mucronea perfoliata
0.5 mm
flowers
1 cm
1 cm
flower clusters
2 mm
N. denudata
var. gracilis
Nemacaulis
denudata var. denudata
2 mm
fruit
Emex spinosa
1 cm
Fagopyrum
esculentum
2
mm
fruit
1 cm
Oxyria digyna
1 cm
Oxytheca
caryophylloides
2 mm
2 mm
1 cm
basal leaf
O. dendroidea
ssp. dendroidea
2 mm
Oxytheca emarginata
2 mm
involucres
with flowers
2 mm
O. parishii
var. abramsii
O. parishii
var. parishii
1 cm
2 mm
O. parishii var.
cienegensis
1 cm
leaf
2 mm
Oxytheca parishii
var. goodmaniana
2 cm
stem end
2 mm
Oxytheca perfoliata
1 cm
2 mm
involucre
with flowers
O. trilobata
involucre
with flowers
1 mm
1 cm
Oxytheca watsonii
5 cm
Polygonum amphibium
var. emersum
2 cm
P. amphibium
var. stipulaceum

P. californicum Meissner (p. 893) Ann. **ST** erect, 4–40 cm, ± angled; ribs 0 or obscure. **LF** < 3 cm, sessile; stipule fringe ± bristly; blade obscurely jointed to stipule, gen < 2 cm, linear, some ± recurved, ± weakly spine-tipped, veins 3. **FL** 1 in upper axils, sessile; perianth ± 2.5 mm, white to pink; styles free. **FR** 2 mm, narrowly elliptic, brown, shiny. Open places (incl serpentine); 40–1200 m. NW, CaR, SN, GV, sw MP; to WA. Variable.

P. capitatum D. Don Per from woody caudex (sometimes ann), wavy-hairy, tinged red with age. **ST** prostrate, gen < 50 cm. **LF** < 4 cm, petioled; stipule sheath loose; blade elliptic to round, upper surface with dark inverted "V", base ± cordate. **INFL** 1–2 cm wide, ± spheric. **FL**: perianth ± 2 mm, bright pink; filaments thread-like. **FR** 1.5 mm, brown, shiny. $2n=22$. Disturbed urban places in mild-winter areas near coast; < 500 m. NW, CW, SW; native to Himalayas.

P. convolvulus L. BLACK BINDWEED Ann vine. **ST** climbing, < 6 m. **LF** 3–15 cm, petioled; blade ± ovate-sagittate, tip acuminate. **INFL** axillary, gen < 15 cm, open, raceme-like. **FL**: perianth ± 4 mm, gen not opening, greenish white, midribs of 3 outer lobes ± keeled. **FR** 2.5 mm, widely ovoid, black, ± dull. $2n=20,40$. Uncommon. Disturbed places; < 1000 m. CA-FP; to e N.Am; native to Eur.

P. cuspidatum Siebold & Zucc. (p. 893) JAPANESE KNOTWEED Per from spreading rhizome, dioecious. **ST** erect, 1–3 m. **LF** petioled; blade gen 10–15 cm, widely ovate to wider than long, tip abruptly pointed. **INFL**: open panicle, in most upper axils; branches < 15 cm, longer in fr, ± drooping. **FL**: perianth 2–8 mm, ± white, 3 outer lobes keeled in fl, winged in fr; fertile stamens exserted. **FR** ± 4 mm, brown, shiny. $2n=44,88$. Uncommon. Disturbed places; < 1000 m. n CA-FP (esp s NCoR, SnFrB); to e N.Am; native to Japan. NOXIOUS WEED.

P. davisiae A. Gray (p. 893) Per from woody caudex. **ST** decumbent to ascending, < 4 dm. **LF** < 5 cm; stipule sheath oblique, open above axil; petiole 0–1 cm; blade gen lanceolate (to round), yellow-green, gen scabrous, often glaucous, base tapered to round, tip gen acute. **INFL**: clusters; fls 2–5 per axil. **FL**: perianth 3 mm, fused << 1/3 its length, yellow, green, or purple; filaments thread-like. **FR** 3–4 mm, ± oblong, light brown, shiny. Talus, rocky sites of snow accumulation; 1500–2800 m. KR, NCoRH, CaRH, n&c SNH; s OR. Intergrades with *P. phytolaccifolium*. ❀TRY.

P. douglasii E. Greene Ann. **ST** gen erect, 3–80 cm, ± angled; ribs 0 or obscure. **LF** < 8 cm, sessile; blade linear, elliptic, or oblanceolate, smaller upward, vein 1. **INFL** 5–20 cm, open or dense. **FL**: perianth 2–5 mm, opening or not, pink or white, margins and midribs gen red or green; stamens gen 8 (3–8, may vary within pl). **FR** 2–5 mm, black, shiny. $2n=40$. Common. Open, slopes, dry meadows; < 3500 m. CA (esp mtns, coast); to Can, e N.Am. Variable intergrading complex; sspp. maintained in mixed populations merge over geog range. Gen self-pollinating. Homopteran insect parasites shorten st, broaden lvs, enlarge and sterilize fls in infected pls. Further study warranted. Treatment improved by E. Neese.

ssp. ***austiniae*** (E. Greene) E. Murray Pl ± funnel-like in shape. **ST**: branches ascending from base, gen < 20 cm; internodes ± equal. **LVS** appearing ± basal and crowded, persistent, ± elliptic, ± sharply acute, abruptly reduced to scale-like bracts. **INFL** spike-like, open. **FL**: perianth ± 2 mm, closed. **FR** 2 mm, reflexed; enfolding perianth with a minute, curved stalk at base. Uncommon. Sagebrush plains; ± 1500 m. MP; to OR, MT, WY. [var. *a.* (E. Greene) M.E. Jones]

ssp. ***douglasii*** (p. 893) **ST** decumbent at high elevation, 3–80 cm. **LVS** often falling before fls appear, < 8 cm, linear to oblanceolate, abruptly reduced to scale-like bracts. **INFL** spike-like, open. **FL**: perianth 3–4 mm, gen closed (< 2.5 mm if open). **FR** 3–4 mm, reflexed; enfolding perianth with a minute, curved stalk at base. Common. Drying places; 1000–3000 m. CA; to Can, e N.Am. [var. *latifolium* (Engelm.) E. Greene]

ssp. ***johnstonii*** (Munz) J. Hickman (p. 893) **ST** 5–50 cm. **LVS** persistent, < 3 cm, all elliptic, gradually smaller upward. **INFL** ± open, ± leafy. **FL**: perianth ± 2 mm, closed or ± open; fertile stamens 3–8. **FR** 2–3 mm, erect; perianth stalk 0. Dry forests, open rocky places; 1200–2800 m. KR, CaR, SN, TR, PR; to WA, ID. Many pls are intermediate to ssp. *douglasii*, but both sspp. may grow together without local intergradation. Rocky Mtn pls called *P. sawatchense* Small have been placed here but may warrant separate status.

ssp. ***majus*** (Meissner) J. Hickman (p. 893) **ST** 15–60 cm. **LVS** often falling before fls appear, < 6 cm, ± linear. **INFL** spike-like, open (denser than ssp. *douglasii*). **FL**: perianth 3–5 mm, widely open, white, midribs dark. **FR** 3.5–5 mm, gen spreading, narrowly ovoid; perianth stalk 0. Uncommon. Dry plains, meadows (incl serpentine); 500–2000 m. NCoR, CaR, SN, SnBr?, w MP; to B.C., MT. [*P. m.* (Meissner) Piper] Self-incompatible.

ssp. ***spergulariiforme*** (Meissner) J. Hickman (p. 893) **ST** ascending to erect, 5–50 cm. **LVS** often falling before fls appear, < 6 cm, gen ± linear. **INFL**: fl clusters crowded at branch tips. **FL**: perianth 3–5 mm, gen partly open, appearing cylindric, gen bright pink, midribs dark. **FR** 3–4 mm, gen spreading, narrowly ovoid; perianth stalk 0. Open, rocky places (incl serpentine); 10–2000 m. NW, CaR, n SN, SnFrB, w MP; to B.C. [*P. s.* Meissner] Self-incompatible. Many intermediates to sspp. *douglasii* or *majus* range e to Rocky Mtns.

P. hydropiper L. MARSHPEPPER, WATERPEPPER Ann, ± glabrous. **ST** ascending to erect, < 1 m. **LF** < 12 cm; stipule margin bristly; petiole short; blade lanceolate. **INFL** gen drooping; branches gen ± 6 cm, densely spike-like. **FL**: perianth ± 3 mm, gland-dotted, ± green, margins gen ± red; filaments thread-like. **FR** 2–3-angled, ± round, dark brown, dull, bitter tasting. $2n=20,22$. Uncommon. Wet places; < 1500 m. n&c CA-FP; native to Eur.

P. hydropiperoides Michaux (p. 893) WATERPEPPER Per from extensive, slender rhizomes. **ST** ascending to erect, < 1 m. **LF** gen ± 10 cm, petioled; stipule strigose, margin bristly; blade linear to lanceolate, strigose (esp margin, midrib). **INFL**: branches spike-like, ± erect, gen 6–10 cm, 2–5 mm wide; bract margin bristly. **FL**: perianth ± 3 mm, light pink; filaments thread-like. **FR** ± 2 mm, 3-angled, dark brown, shiny. $2n=40$. Wet banks; < 1500 m. CA-FP; to Mex, e N.AM. [var. *asperifolium* E. Stanford] Often confused with *P. persicaria*.

P. lapathifolium L. (p. 893) WILLOW WEED Ann. **ST** ascending to erect, < 1.5 m. **LF** < 20 cm, petioled; stipule veiny, esp at base, ± glabrous, clear brown; blade ± lanceolate, often hairy below. **INFL**: branches spike-like, 3–8 cm, ± 5 mm wide, gen drooping, dense. **FL**: perianth ± 2 mm, greenish to pink, lobes gen 4, veins in fr prominent, those of outer lobes forked, hooked near tips; filaments thread-like; styles 2. **FR** ± 2 mm, round, flat, indented on both sides, brown, shiny. $2n=22$. Moist places; < 1500 m. Common. CA; to e N.Am. Highly variable, some forms are induced by environment. Merits study.

P. marinense T. Mert. & Raven (p. 893) MARIN KNOTWEED Ann. **ST** prostrate to erect, < 4 dm, ± round; ribs obvious. **LVS** gen 1–3 cm, ± sessile, gradually smaller upward; blade narrowly oblanceolate to elliptic. **INFL**: fls in axils of most lvs. **FL**: perianth ± 3.5 mm, green, margin white. **FR** 3.5–4.5 mm, narrowly ovoid, olive-brown, very shiny, exserted. $2n=60$. UNCOMMON. Coastal salt marshes; < 10 m. SnFrB (esp Marin Co.). Related to *P. aviculare*, taxonomic status uncertain: possibly = *P. robertii* Lois.; if so, alien, native to w Medit. Endangered by salt marsh development. Merits immediate study.

P. minimum S. Watson (p. 893) Ann. **ST** 2–30 cm, slender, ± angled; ribs 0 or obscure, ± red-brown. **LVS** well spaced below, crowded at st tips, persistent, < 2 cm, barely smaller upward, sessile; stipule lobes lanceolate; blade lance-elliptic to ± round, vein 1. **INFL**: fls 1–3 per axil from near base. **FL**: perianth 2 mm, white; stamens 5 or 8. **FR** ± 2 mm, black, shiny. Common. Meadows, open rocky places; 1500–3300 m. NW, CaR, SN; temp montane w N.Am.

P. newberryi Small Per from woody caudex. **ST** decumbent to ascending, < 4 dm. **LF** < 8 cm; some petioles > 1 cm; blade lanceolate to round, glabrous to velvety, dark green becoming reddish, base often cordate, tip ± round. **INFL**: fls 5–25 per axil. **FL**: perianth ± 3 mm, greenish with some red; filaments thread-like. **FR** 4–4.5 mm, obovoid, brown, shiny. Volcanic slopes, esp pumice; 1500–2400 m. CaRH (esp Mount Shasta), n&c SNH, MP; to WA. Glabrous forms are intermediate to (possibly hybrids with) *P. phytolaccifolium*. ❀DRN,SUN:1,2.

P. paronychia Cham. & Schldl. (p. 893) Subshrub. **ST** prostrate to ascending, < 1 m. **LVS** crowded at branch tips, sessile; stipule ± 20 mm, persistent, becoming frayed, joint with blade obvious; blade linear to oblanceolate, margin rolled under, midrib on lower surface a flat, sharp-toothed ridge. **INFL**: fls crowded in upper axils. **FL**: perianth 6–10 mm, white or pink, lobes ± lanceolate. **FR** 4–5 mm, black, shiny. $2n=28$. Coastal dunes, scrub; 0–50 m. NCo, CCo; to B.C. ❀DRN,SUN:4,5,15,**16,17,24**&SHD:**14**&IRR:19–21,**22,23**;GRCVR (Good bee plant).

P. parryi E. Greene (p. 893) Ann, compact, often cushion-like. **ST** 1–7 cm, angled; ribs 0 or obscure. **LVS** 5–20 mm, smaller upward, sessile; stipule deeply fringed, segments fine, gen weak and ± curled, joint with blade obscure; blade linear, spine-tipped, veins 3. **INFL**: fl 1 in most axils, sessile. **FL**: perianth 1.5 mm, ± white; styles free, minute. **FR** gen ± 1.5 mm, widely ovoid, brown, ± shiny. Vernally moist, open, sandy or rocky places; 500–2000 m. NW, CaR, SN, SnFrB, PR; to WA. Some pls from sw SnFrB (Santa Cruz Co.) have straight stipule fringe (like *P. californicum*); styles in fr obvious; fr ± 2.5 mm, less shiny.

P. patulum M. Bieb. Ann. **ST** erect, < 6 dm, ± round; ribs obvious. **LVS** < 4 cm, sessile; blade linear to lanceolate, gradually smaller upward then abruptly reduced to bracts, often falling early. **INFL**: fl clusters in most upper lf axils; fls ± sessile, enclosed by stipule. **FL**: perianth 2.5 mm, greenish, margin gen white. **FR** 2–2.5 mm, widely ovoid, brown, very shiny. $2n=20$. Uncommon. Salt marshes; < 10 m. Deltaic GV, SnFrB; native to e Eur. Closely related to *P. aviculare* and *P. argyrocoleon* (neither found in salt marshes); applicability of this name to CA pls remains uncertain.

P. pensylvanicum L. PINKWEED Ann. **ST** ascending to erect, < 1 m. **LF** < 20 cm, petioled; stipule glabrous, margin without bristles; blade gen narrowly lanceolate. **INFL** spike-like, dense, not interrupted; peduncle glandular; bracts glabrous. **FL**: perianth gen 4–5 mm, pink. **FR** ± 3.5 mm, round in outline, gen nearly flat on 1 side, indented on other, dark brown, shiny. Moist disturbed places, drying ponds; ± 30 m. e ScV, expected elsewhere; native to e US, where large frs are important waterfowl food. Possibly planted.

P. persicaria L. (p. 893) LADY'S THUMB Ann. **ST** ± erect, < 1 m, sometimes ± swollen above nodes. **LF** < 20 cm, petioled; stipule strigose, margin bristly; blade narrowly lanceolate, gen with dark central spot, gen gland-dotted below. **INFL** spike-like, dense, < 3 cm, gen interrupted below; peduncles glabrous; bracts ± strigose-bristly. **FL**: perianth 2 mm, deep pink (rarely white). **FR** ± 2 mm, ± flat to equally 3-angled, brown to black, shiny. $2n=22,40,44$. Moist urban places; < 1500 m. Common. CA; native to Eur. [*P. fusiforme* E. Greene] Variable. Hybrids with *P. lapathifolium* have narrower infl and ± flat, brown frs. Often confused with *P. hydropiperoides*.

P. phytolaccifolium Meissner (p. 893) Per from a large, fleshy caudex. **ST** erect, stout, > 1 m. **LF** 10–20 cm, petioled; blade lanceolate to ovate, ± glabrous, base round to tapered, tip acuminate. **INFL** 15–30 cm, open, fls many; lower branches in lf axils. **FL**: perianth ± 4 mm, ± white; stamens exserted, filaments thread-like. **FR** 4–6 mm, exserted, ovoid, brown, shiny. Meadows, moist rocky places; 1500–2800 m. KR, NCoRH, n&c SNH, MP; to WA, ID, NV. [misspelled *P. phytolaccaefolium*] *P. alaskanum* Small (AK) is intermediate to Old World *P. alpinum* All. ❀IRR,DRN:1,2.

P. polygaloides Meissner Ann. **ST** 1–30 cm, slender, ± angled; ribs 0 or obscure. **LVS** < 4 cm, sessile, joint with stipule obvious; lower blades linear to narrowly lanceolate. **INFL** 1–7 cm, ± crowded at st tips; bracts narrowly lanceolate to round, margin green or white, sometimes reflexed. **FL**: perianth opening or not, 1.5–2.5 mm, white to red, midribs ± green; fertile stamens 3, 5, or 8, varying within pls. **FR** 1–2 mm, light brown to nearly black, shiny to sculptured in lines and dull. Common. Vernally moist places; 100–3300 m. CA; to B.C., MT, Colorado, AZ. Highly variable intergrading complex centered in MP, 1500 m. Sspp. may grow together, sometimes without intergrading. CA sspp. self-pollinating. Bracts may change dramatically with age; immature pls may not key.

ssp. ***confertiflorum*** (Piper) J. Hickman (p. 893) **ST** 2–15 cm. **INFL** gen 1–2 cm, gen 8–10 mm wide; bracts 4–10 mm, spreading to ascending, shorter, rounder upward, tip ± pointed, white margin ± 1/2 total bract width, ± reflexed. **FL**: perianth 2 mm, white to red; anthers gen 3. **FR** < 2 mm, ovate in outline, nearly black, deeply sculptured, dull. Common. Vernal pools, wet meadows; gen 500–1900 m. NW, CaR, n SN, MP; to WA, MT, WY. [*P. c.* Piper] Intergrades with all other sspp., esp ssp. *kelloggii*; in n MP possible intermediates to ssp. *polygaloides* (which otherwise does not occur in CA) are larger, with bracts larger, whiter, rounder, flatter, more clustered.

ssp. ***esotericum*** (Wheeler) J. Hickman (p. 893) MODOC COUNTY KNOTWEED **ST** 5–12 cm. **INFL** 2–7 cm, gen 5–7 mm wide; bracts 4–6 mm, barely smaller upward, ± appressed, uniformly lance-elliptic, white margin 0 or narrow, not reflexed. **FL**: perianth 2 mm, white; anthers gen 8 or 5. **FR** 2–2.5 mm, lanceolate in outline, brown, smooth to sculptured, ± shiny. RARE. Vernal pools, other seasonally wet places; ± 1500 m. MP (near Goose Lake, Modoc Co.; Sierra Valley, s Plumas Co.). [*P. e.* Wheeler] Intermediates to ssp. *confertiflorum* occur more widely in MP, to OR.

ssp. ***kelloggii*** (E. Greene) J. Hickman (p. 893) **ST** 1–15 cm. **INFL** 3–15 mm, 5–15 mm wide; bracts 3–10 mm, smaller upward, spreading, all linear-lanceolate, margin green. **FL**: perianth 1.5 mm, white to red; anthers 3. **FR** 1.5 mm, widely ovoid, light brown, ± shiny. Common. Mtn meadows, seeps, rainpools; 1500–3300 m. NW, CaR, SN, TR, PR; to B.C., MT, Colorado, AZ. [*P. k.* E. Greene]

P. polystachyum Meissner HIMALAYAN KNOTWEED Per from extensive rhizome. **ST** erect, stout, gen > 1 m. **LF** 10–20 cm, petioled; blade lanceolate, densely soft-hairy, esp below, base tapered or ± cordate, often with 2 small basal lobes. **INFL** 20–35 cm, open; fls many; lower branches in lf axils. **FL**: perianth 5–7 mm, white; filaments thread-like. **FR** apparently not produced in CA. Uncommon. Moist disturbed places, marshes; 0–500 m. NCo (esp Del Norte, Humboldt cos.), n CCo; native to s-c Asia. NOXIOUS WEED. Cult as orn.

P. prolificum (Small) Robinson Ann. **ST** ± erect, much branched, < 4 dm, round, ribbed. **LF** gen 1–2 cm, sessile; blade linear-oblong, rough-veiny beneath, bluish green, often covered with whitish fungus, smaller upward; lower lvs falling early. **INFL**: fl clusters scattered in upper lf axils; fls ± sessile. **FL**: perianth 2 mm, incl in stipule. **FR** 2 mm, widely ovoid, brown, ± dull. $2n=20,60$. Wet salty places; 100–2000 m. n SNH (Lake Tahoe Area), SnFrB (Napa Co.), expected elsewhere; native to e N.Am. Closely related to *P. aviculare*.

P. punctatum Elliott (p. 893) Ann or per from slender rhizome. **ST** decumbent to erect, < 1 m. **LF** < 15 cm, ± sessile; stipule margin entire or bristly; sessile or petiole short; blade narrowly lanceolate. **INFL**: branches gen 5–8 cm, slender, interrupted below. **FL**: perianth gen green, dotted with green glands, margin gen white; filaments thread-like. **FR** 2–3-angled, black, shiny. Common. Shallow water, shores; < 1500 m. CA; to WA, e N.Am, S.Am. Highly variable. Planted as waterfowl food.

P. ramosissimum Michaux (p. 897) Ann, yellowish. **ST** erect, much branched, < 1 m, round, ribbed. **LF** gen 1–5 cm, sessile; blade linear-lanceolate, smaller upward; lower lvs falling early. **INFL**: fl clusters in most upper lf axils, exserted from stipules. **FL**: perianth 3 mm, ± yellow (rarely pink), esp margin. **FR** 3 mm, dark brown, ± dull. $2n=20,60$. Uncommon. Disturbed places; < 500 m. n CA-FP; probably native to e N.Am.

P. sachalinense Maxim. (p. 897) GIANT KNOTWEED, SACALINE Per from spreading rhizome, monoecious or dioecious. **ST** erect, gen 2–3 m. **LF** 15–35 cm, petioled; blade ± ovate, base cordate, tip acuminate. **INFL**: panicles in upper lf axils, open, ± 10 cm; branches spreading. **STAMINATE FL**: perianth 6–7 mm, pale green, keels 3 at base; stamens exserted. **PISTILLATE FL**: perianth 12–15 mm, ± white, widely 3-winged throughout. **FR** 3–4 mm, narrowly ovoid, brown, shiny, retained in perianth. $2n=44,66$. Uncommon. Disturbed places; < 500 m. n CA-FP; native to Japan. NOXIOUS WEED. Forms dense infestations. Cult as orn.

P. shastense A. Gray (p. 897) Shrub. **ST** gnarled, prostrate to ascending, < 4 dm. **LF** < 2 cm, sessile; stipule ± 5 mm, base persistent, membranous upper part ± falling with lf, joint with blade obvious; blade lanceolate to elliptic, 2-ridged above, grooved on both sides of midrib below. **INFL**: fl clusters in upper lf axils but not at branch tips. **FL**: perianth 5–9 mm, pink or white, ribs green, lobes ± round; styles free. **FR** 3–4 mm, narrowly ovoid, brown, shiny. Rocky or gravelly slopes; 2100–3400 m. CaRH, SNH; OR, NV.

PTEROSTEGIA

1 sp. (Greek: winged cover, from involucre) [Reveal 1989 Phytologia 66:228–235]

P. drymarioides Fischer & C. Meyer (p. 897) Ann, prostrate or sprawling, (mats < 12 dm diam), thinly hairy, monoecious. **LVS** cauline, opposite, petioled; stipule 0; blade 3–20 mm, gen fan-shaped, entire to deeply notched and appearing 2-lobed. **INFL**: fls 1–2 per node, 1 or both gen staminate; only pistillate fls involucred; involucre growing with fr to 1–3 mm, net-veined, cream to pink or rose, wings 2, outer surface rounded, inner hollowed. **FL**: perianth 0.9–1.2 mm, pale yellow to pink or rose, sparsely hairy, lobes 5–6; stamens 6. **FR** ± 1.5 mm, brownish. *n*=14. Common. Shady, often moist places; < 1600 m. CA; to s OR, s NV, sw UT, w AZ, nw Mex.

RUMEX DOCK

Ann, bien, per, gen from stout taproot, some dioecious, glabrous. **ST** gen erect, gen unbranched below infl, < 2 m, ± ridged, gen red-brown in fr; nodes ± swollen. **LVS** gen ± basal, alternate, petioled; stipules fused, sheathing st above nodes, fracturing; blade < 50 cm. **INFL**: bracted clusters gen arrayed in erect panicles. **FL** gen bisexual, < 3 mm, gen green; perianth lobes 6, persistent, outer 3 in fr ± inconspicuous, inner 3 in fr enlarged, hardened, ± veiny, covering fr, midrib often expanded into a tubercle; stamens 6; stigmas 3, fringed. **FR** brown, shiny. (Latin: sorrel) [Mitchell 1978 Brittonia 30:293–296] Hybrids common. Mature inner perianth lobes gen required for identification. Some cult for vegetable greens. TOXIC in quantity to livestock; seldom eaten.

1. Dioecious (rarely some fls bisexual); st erect, slender, gen < 4 dm; perianth ± red in fl, tubercles 0 in fr (subg. *Acetosa*)
  2. Lvs gen sagittate to hastate; st single from slender rhizome; inner perianth lobes ± = fr faces .... ***R. acetosella***
  2′ Lvs linear to lanceolate; sts clustered on taproot; inner perianth lobes >> fr faces .... ***R. paucifolius***

1′ Fls gen bisexual; st decumbent to erect, gen stout, gen > 7 dm; perianth gen green in fl, tubercles various or 0 in fr (subg. *Rumex*)
  3. Lvs cauline; main st prostrate to ± erect, gen with smaller branches at most nodes, branches often producing fls after main st; native (sect. *Axillares*)
    4. Inner perianth lobes > 12 mm; MP .... ***R. venosus***
    4′ Inner perianth lobes < 6 mm; widespread .... ***R. salicifolius***
      5. Tubercles 0 or inconspicuous; inner perianth lobes triangular, with short, triangular teeth near base .... var. ***denticulatus***
      5′ Tubercles 1 or 3 per fl, ± obvious; inner perianth lobes lanceolate to round, teeth gen 0
        6. 1 tubercle obvious, width > 1/3 width of inner perianth lobe
          7. Infl dense; lf ± thick, some oblong to elliptic; coastal .... var. ***crassus***
          7′ Infl ± open; lf thin, linear to lanceolate; coastal or in mtns .... var. ***salicifolius***
        6′ 3 tubercles obvious, ± equal, width < or > 1/3 width of inner-perianth lobe
          8. Tubercles ovate to nearly round, > 1/3 width of inner perianth lobe .... var. ***transitorius***
          8′ Tubercles lanceolate, gen << 1/3 width of inner perianth lobe
            9. Some lvs oblong to elliptic; GB — shores, ± salty lake beds .... var. ***lacustris***
            9′ Lvs linear to lanceolate; esp SN .... var. ***triangulivalvis***
  3′ Lvs basal and cauline (basal rosette often withered when fr mature); st ± erect, gen unbranched below infl; native or alien (sect. *Rumex*)
    10. Tubercles 0 or inconspicuous — margin of inner perianth lobe entire
      11. Lf blade fleshy, base tapered; inner perianth lobe 8–15 mm .... ***R. hymenosepalus***
      11′ Lf blade ± leathery, base truncate to cordate; inner perianth lobe 4–10 mm .... ***R. occidentalis***
    10′ Tubercles 1 or 3 per fl, obvious
      12. Margin of inner perianth lobe entire
        13. Inner perianth lobe 2–3 mm, tubercle 1/3 width of lobe or wider; infl open, interrupted
          14. Per; tubercle ± 1/2 width of perianth lobe; widespread, alien .... ***R. conglomeratus***
          14′ Ann; tubercle ± 1/3 width of perianth lobe; gen D, native .... [2]***R. violascens***
        13′ Inner perianth lobe 5–6 mm, tubercle << 1/3 width of lobe; infl dense, continuous
          15. Tubercle ovate to round; abundant .... ***R. crispus***
          15′ Tubercle ± lanceolate; uncommon .... ***R. orbiculatus***
      12′ Margin of inner perianth lobe with 0.5–4 mm teeth
        16. Ann (sometimes bien)
          17. Inner perianth lobe 4–5 mm, teeth ± 1.5 mm; infl interrupted throughout; alien — uncommon .. ***R. dentatus***
          17′ Inner perianth lobe 2–3.5 mm, teeth 0.5–4 mm; infl dense, not interrupted near tip; native
            18. Widest inner perianth lobe narrowly ovate, teeth 2–4 mm, fine; infl dense, leafy; widespread .... ***R. maritimus***
            18′ Widest inner perianth lobes widely ovate or ± equilaterally triangular, teeth < 1 mm, or obscure; infl ± open, nearly lfless; esp D .... [2]***R. violascens***
        16′ Per
          19. Largest lf 20–75 cm, blade lance-ovate, ± cordate; inner perianth lobe narrowly ovate-triangular .... ***R. obtusifolius***
          19′ Largest lf < 20 cm (if larger, blade narrowly lanceolate, tapered); inner perianth lobe widely ovate-triangular to round-cordate
            20. Infl branches spreading; teeth of inner perianth lobe slender to base, longest ± 1.5 mm .... ***R. pulcher***
            20′ Infl branches ascending; teeth of inner perianth lobe coarse, wider at base, longest 0.5–1 mm

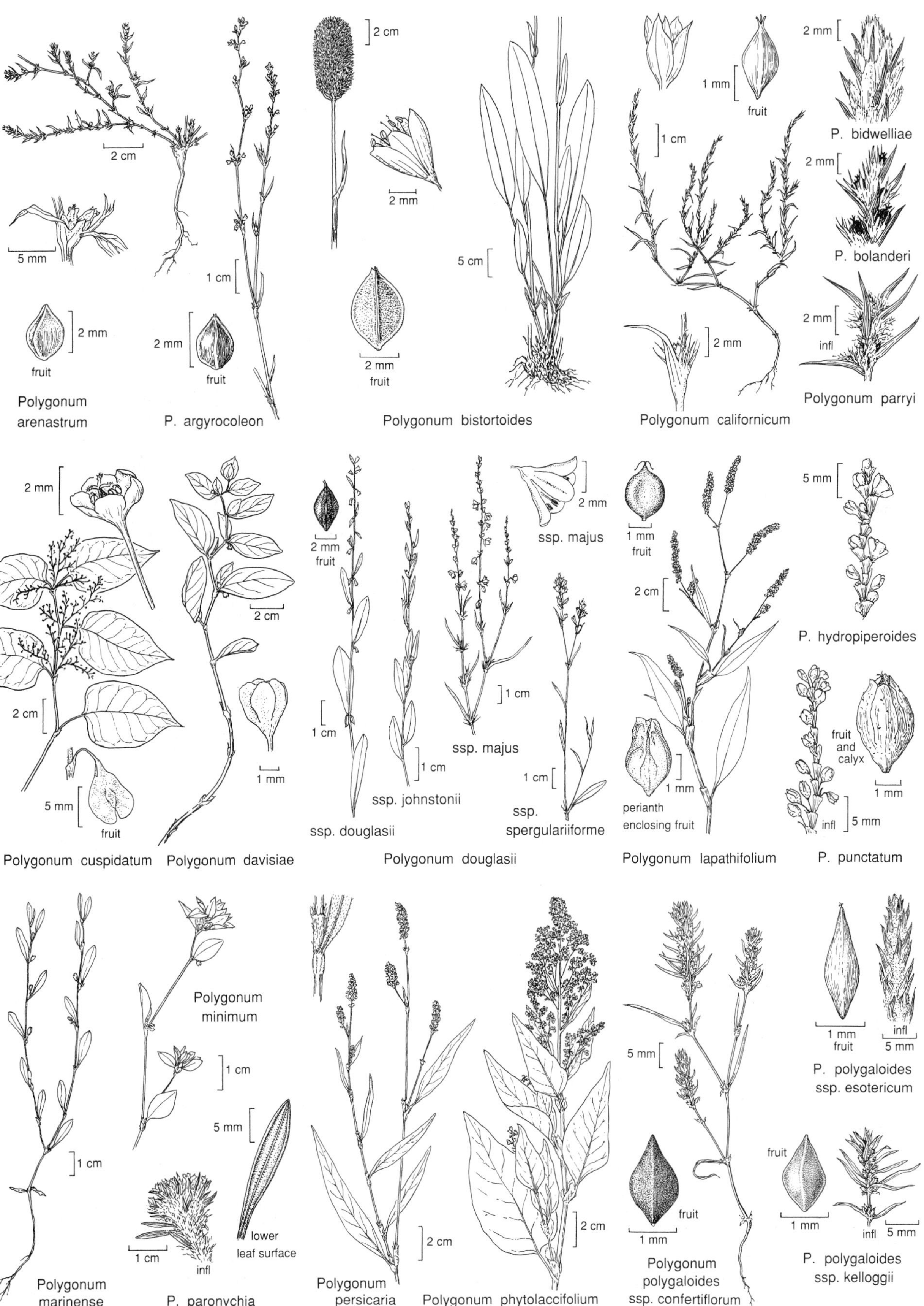
2 cm
5 mm
2 mm
fruit
Polygonum arenastrum
1 cm
2 mm
fruit
P. argyrocoleon
2 cm
2 mm
5 cm
2 mm
fruit
Polygonum bistortoides
1 mm
fruit
1 cm
2 mm
Polygonum californicum
2 mm
P. bidwelliae
2 mm
P. bolanderi
2 mm
infl
Polygonum parryi
2 mm
2 cm
2 cm
5 mm
fruit
Polygonum cuspidatum
2 cm
1 mm
Polygonum davisiae
2 mm
fruit
1 cm
ssp. douglasii
1 cm
ssp. johnstonii
1 cm
ssp. majus
2 mm
ssp. majus
1 cm
ssp. spergulariiforme
Polygonum douglasii
1 mm
fruit
2 cm
1 mm
perianth enclosing fruit
Polygonum lapathifolium
5 mm
P. hydropiperoides
fruit and calyx
1 mm
infl
5 mm
P. punctatum
1 cm
Polygonum marinense
Polygonum minimum
1 cm
5 mm
1 cm
infl
lower leaf surface
P. paronychia
2 cm
Polygonum persicaria
2 cm
Polygonum phytolaccifolium
5 mm
1 mm
fruit
Polygonum polygaloides ssp. confertiflorum
1 mm
fruit
infl
5 mm
P. polygaloides ssp. esotericum
fruit
1 mm
infl
5 mm
P. polygaloides ssp. kelloggii

21. Inner perianth lobe > 5 mm, round-cordate, longest teeth ± 0.5 mm, each much wider at base . . . . **R. kerneri**

21′ Inner perianth lobe < 5 mm, ± triangular-cordate, longest teeth ± 1 mm, each somewhat wider at base . . . . **R. stenophyllus**

R. acetosella L. (p. 897) SHEEP SORREL Per from slender rhizome, dioecious, gen clonal. **ST** < 4 dm, slender. **LVS** gen ± basal, 2–10 cm; blade gen sagittate to hastate. **INFL** ± open. **FL**: perianth ± red; inner lobe ± 2 mm, ± = (and sometimes firmly adherent to) fr faces, smooth, teeth 0, tubercles 0. 2*n*=14,32,40,42. Moist, ± disturbed places; 0–3000 m. Common. CA-FP; native to Eur. [*R. angiocarpus* Murb.] Gen polyploid.

R. conglomeratus Murray (p. 897) Per. **ST** < 15 dm, slender. **LVS** basal and cauline, < 30 cm; blade lance-ovate, base cordate to ± tapered, basal margin ± curled. **INFL** open, interrupted, leafy. **FL**: inner perianth lobe 2–3 mm, ovate, teeth 0, tubercles 1–3, ± 1/2 width of lobe, oblong. **FR** 1.5–2 mm. 2*n*=18,20. Common. Moist places; < 1500 m. CA-FP; native to Eur.

R. crispus L. (p. 897) CURLY DOCK Per. **ST** < 15 dm, stout. **LVS** basal and cauline, < 50 cm; blade lanceolate, margin strongly curled, esp near base. **INFL** dense, narrow, ± leafy. **FL**: inner perianth lobe ± 5 mm, ovate to round, base cordate, teeth 0, tubercles 3, ovate, unequal, << 1/3 width of lobe. **FR** 2 mm. 2*n*=60. Abundant. Disturbed places; < 2500 m. CA; N.Am, native to Eurasia. Variable.

R. dentatus L. Ann, bien. **ST** < 7 dm, slender. **LVS** basal and cauline, < 15 cm; blade lanceolate, base cordate to tapered, margin ± curled. **INFL** open, interrupted, leafy below. **FL**: inner perianth lobe 4–5 mm, narrowly ovate, teeth ± basal, ± 1.5 mm, tubercles 3, lanceolate, ± equal. **FR** 2 mm. Uncommon. Wet places, rice fields; < 30 m. n SnJV (possibly the only US sites); native to Eurasia. [ssp. *klotzschianus* (Meissner) Rech.f.]

R. hymenosepalus Torrey (p. 897) CANAIGRE, WILD-RHUBARB Per from cluster of tuber-like roots. **ST** 6–12 dm, stout, ± fleshy, smooth. **LVS** basal and cauline, < 6 dm, fleshy; blade lanceolate to oblong, base tapered, tip acuminate, margin curled. **INFL** dense. **FL**: inner perianth lobe 8–15 mm, round-cordate, ± pink, teeth 0, tubercles 0. **FR** 4–7 mm. 2*n*=40,100. Dry sandy places; 0–2000 m. SW, DMoj; to WY, TX, Baja CA. ❀DRN,SUN:7–14**16**,17–24.

R. kerneri Borbas Per. **ST** < 15 dm, ± slender. **LVS** basal and cauline, < 25 cm; blade oblong to lanceolate, margin ± curled, fine-toothed. **INFL** interrupted, esp below. **FL**: inner perianth lobe ± 6 mm, round-cordate, teeth ± 0.5 mm, their bases very wide, tubercles 3, oblong, ± equal. **FR** 3 mm. Uncommon. Wet places near coast; < 50 m. CCo (San Luis Obispo Co.), SCo (Santa Barbara Co.); native to se Eur.

R. maritimus L. (p. 897) GOLDEN DOCK Ann, bien. **ST** sometimes branched, < 10 dm, slender, or stout and hollow. **LVS** gen cauline, < 20 cm; blade linear to ovate, base cordate to tapered, margin ± curled. **INFL** dense, interrupted below, leafy. **FL**: inner perianth lobe 2–3.5 mm, narrowly ovate, teeth variable, gen 2–4 mm, fine, tubercles 3, ± equal, linear and inconspicuous to ovate-round and as wide as lobes. **FR** 1–2 mm. 2*n*=40. Wet, ± salty places; 0–2000 m. CA; to AK, e N.Am, S.Am, Eurasia. [*R. fueginus* Philippi; *R. persicarioides* L.] Highly variable in size, color, lvs, inner perianth lobes (all traits of which vary within fls); apparently inseparable into geog races. ❀TRY STBL.

R. obtusifolius L. (p. 897) BITTER DOCK Per. **ST** < 15 dm, stout. **LVS** basal and cauline, < 75 cm; blade lance-ovate, cordate, flat, entire. **INFL** open, interrupted below, ± lfless. **FL**: inner perianth lobe 3–5 mm, narrowly ovate-triangular, teeth gen ± 0.5 mm (in some fls 1–2 mm or 0), tubercle 1 (or 3, unequal). **FR** 2–3 mm. 2*n*=40. Moist places; < 1500 m. CA; native to w Eur. [ssp. *agrestis* (Fries) Danser] Hybridizes with *R. crispus*, other Eur spp.

R. occidentalis S. Watson (p. 897) WESTERN DOCK Per. **ST** 5–20 dm, stout, ± red. **LVS** basal and cauline, < 60 cm, ± leathery; blade lanceolate, base truncate to cordate, margin ± curled. **INFL** dense above, ± interrupted below. **FL**: inner perianth lobe 4–10 mm, round to ovate, cordate, often pink, teeth 0, tubercles 0. **FR** 3–4 mm. 2*n*=140,200. Uncommon. Wet, ± salty places; 0–1500 m. CA; to AK, e N.Am. [*R. fenestratus* E. Greene] Variable.

R. orbiculatus A. Gray Per. **ST** < 16 dm, stout, deeply ridged, ± red. **LVS** basal and cauline, < 5 dm; blade lanceolate, base tapered, margin ± curled. **INFL** dense. **FL**: inner perianth lobe ± 5 mm, round-cordate, teeth 0, tubercles 3, << 1/3 width of lobe, ± lanceolate, ± equal. **FR** 4 mm. 2*n*=160. Uncommon. Bog; 1100 m. n SNH (one site known near Taylorsville, Plumas Co.); native to e N.Am.

R. paucifolius S. Watson Per, dioecious (rarely some fls bisexual). **STS** clustered on taproot, sometimes branched, < 4(7) dm, slender. **LVS** gen ± basal, 2–9 cm; blade linear to lanceolate. **INFL** ± open. **FL**: perianth ± red, inner lobe 3–4 mm, >> fr faces, teeth 0, tubercles 0. **FR** ± 1.3 mm. 2*n*=14,28. Moist places; 1500–4000 m. CaR, SN, MP, W&I; to B.C., Alberta, Colorado. [var. *gracilescens* Rech.f.] ❀TRY STBL.

R. pulcher L. (p. 897) FIDDLE DOCK Per. **ST** < 1 m, slender. **LVS** basal and cauline, < 12 cm; blade lanceolate, basal cordate, often narrowed below middle, forming a "fiddle" shape. **INFL** interrupted; branches spreading. **FL**: inner perianth lobe 3–5 mm, ovate, teeth slender to base, some 1.5 mm, tubercles 3, ovate, unequal or inconspicuous. **FR** 2–3 mm. 2*n*=20,40. Common. Disturbed places; gen < 1500 m. CA; native to Medit. Variable.

R. salicifolius J.A. Weinm. WILLOW DOCK Per. **ST** prostrate to ± erect, < 1 m; branches from most nodes often producing fls later than main axis. **LVS** cauline, gen crowded below, 3–20 cm; blade linear to ovate, margin entire, flat or ± curled. **INFL** dense to open. **FL**: inner perianth lobe 2–5 mm, narrowly ovate to ± round, teeth gen 0, tubercles variable. **FR** 2–3 mm. Common. Moist places; 0–3500 m. CA; to AK, Rocky Mtns, Mex. [Hickman 1984 Madroño 31:249–252] Highly variable, even within pls; intergrading complex, warrants detailed study.

var. ***crassus*** (Rech.f.) J. Howell **ST** prostrate to decumbent. **LF**: blade ± thick, leathery, gen lanceolate, some oblong, ovate, or elliptic. **INFL** dense, continuous, gen 7–15 cm. **FL**: inner perianth lobe 2.5–4 mm, tubercle 1, > half as wide as lobe. 2*n*=20. Coastal dunes, bluffs, marshes; < 100 m. NCo, CCo, SCo, more common n; to WA. [*R. c.* Rech.f.] ❀IRR,DRN,SUN:**5,16,17,24**.

var. ***denticulatus*** Torrey (p. 897) **ST** ascending to ± erect. **LF**: blade linear to lanceolate. **INFL** ± open, interrupted, 15–30 cm. **FLS** gen both bi- and unisexual on 1 pl; inner perianth lobe 3–4 mm, ± triangular, teeth near base, short, irregular, tubercles inconspicuous or 0. 2*n*=20. Moist places; 0–3500 m. CA (esp montane, coastal); to Yukon, MT, Colorado, n Mex. [*R. californicus* Rech.f.; *R. utahensis* Rech.f.] Except for tubercles, very much like var. *salicifolius*. ❀TRY STBL.

var. ***lacustris*** (E. Greene) J. Hickman (p. 897) **ST** ± decumbent if terrestrial, erect and emersed if aquatic. **LF**: blade gen widely lanceolate, some elliptic or oblong, ± thickened. **INFL** dense, continuous, ± 5 cm. **FL**: inner perianth lobe 2–5 mm, tubercles 3, gen << 1/3 width of lobe, narrowly lanceolate, ± equal. Beds, shores of ± salty lakes; 1000–2500 m. GB (esp MP); s OR, NV. [*R. l.* E. Greene] Growth form variable with water level: emergent pls gen unbranched, lfless below initial water level. ❀TRY STBL.

var. ***salicifolius*** (p. 897) **ST** decumbent to ± erect. **LF**: blade linear to lanceolate. **INFL** open, ± interrupted, 15–30 cm. **FL**: inner perianth lobe 2–3 mm, tubercle 1, > 1/3 width of lobe. 2*n*=20. Moist places, esp coastal, montane; 0–3000 m. CA-FP; NV, Baja CA. Coastal forms intergrade with var. *crassus*. ❀TRY STBL.

var. ***transitorius*** (Rech.f.) J. Hickman (p. 897) **ST** ± erect. **LF**: blade lanceolate to oblong or elliptic. **INFL** open or dense, ± continuous, 5–15 cm. **FL**: inner perianth lobe ± 3 mm, tubercles 3, > 1/3 width of lobe, ± equal, ovate to oblong. 2*n*=20. Moist places; 0–2000 m. CA-FP (esp NCo, SN, CCo); to AK. [*R. t.* Rech.f.] Variable. ❀TRY STBL.

var. ***triangulivalvis*** (Danser) C. Hitchc. (p. 897) **ST** ± erect. **LF**: blade linear to lanceolate. **INFL** dense, ± continuous, gen 15–30 cm. **FL**: inner perianth lobe 3–5 mm, tubercles 3, < 1/3 width of lobe, ± equal, lanceolate. $2n$=20. Many habitats; 300–2500 m. CA (esp SN); to B.C., e N.Am, Mex, naturalized in Eur. [*R. t.* (Danser) Rech.f.] ❀TRY STBL.

R. stenophyllus Ledeb. (p. 897) Per. **ST** < 6(20) dm. **LVS** basal and cauline, 8–30 cm; blade narrowly lanceolate to oblong, base tapered, margin curled. **INFL** < 60 cm, ± dense, ± interrupted below. **FL**: inner perianth lobe 3–5 mm, ± triangular-cordate, teeth ± 1 mm, each somewhat wider at base, tubercles 3, oblong, ± equal. **FR** ± 2 mm. $2n$=20,40,60. Uncommon. Wet, ± salty places; ± 25 m. n SnJV (Gustine, Merced Co.); native to Eurasia.

R. venosus Pursh (p. 897) WINGED DOCK, WILD-BEGONIA Per. **ST** decumbent to ascending, 1–4 dm; branches from most nodes often producing fls later than main axis. **LVS** cauline, 2–20 cm; blade lanceolate to ovate, light green, base gen tapered. **FL**: inner perianth lobe 12–18 mm, round-cordate, pink or red, teeth 0, tubercles 0. **FR** 5–7 mm. $2n$=40. Dry, sandy places; 1200–1800 m. MP; to WA, e N.Am.

R. violascens Rech.f. (p. 897) MEXICAN DOCK Ann, bien. **ST** < 1 m, gen stout (appearing per). **LVS** basal and cauline, gen 6–10 cm; blade lance-linear to elliptic, base cordate to tapered, margin ± curled. **INFL** ± open, interrupted below, 10–20 cm, nearly lfless. **FL**: inner perianth lobe ± 3 mm, widely ovate to ± equilaterally triangular, teeth < 1 mm, sometimes obscure, tubercles 3, ovoid, unequal. **FR** ± 2 mm. Uncommon. Wet, ± salty places; < 500 m. s ScV, SnJV, PR, D; to TX, Mex. Variable.

SYSTENOTHECA

1 sp. (Greek: tapering case, from involucre) [Reveal & Hardham 1989 Phytologia 66:83–88]

S. vortriedei (Brandegee) Rev. & Hardham (p. 897) VORTRIEDE'S SPINE-FLOWER Ann, spreading, 2–15 cm, glandular. **LVS** basal; stipule 0; petiole indistinct; blade 20–50 mm, tongue-shaped to oblanceolate, glabrous. **INFL** open, 5–30 cm diam; bract 1 per node, 1–5 mm, deeply 3–4-lobed, awned, lower bracts encircling st; involucre 1 per node, gen 4–5 mm, cylindric and box-like, lobes 4, awned; fls 1 each of 2 types per involucre. **FLS**: 1 bisexual, exserted, perianth gen reddish; 1 sterile or pistillate, incl, perianth white; in both types perianth glabrous, lobes 6, tips notched; stamens 9. **FR** 2–2.5 mm, dark brown to black. UNCOMMON. Woodland, on sand or serpentine; 700–1600 m. SCoRO (Santa Lucia Mtns, Monterey, San Luis Obispo cos.) [*Chorizanthe v.* Brandegee; *Centrostegia v.* (Brandegee) Goodman] ❀DRN,SUN: **7**,14–17.

PORTULACACEAE PURSLANE FAMILY

Ann or per, gen fleshy. **STS** gen glabrous. **LVS** simple, alternate or opposite, sometimes stipuled. **INFL** various. **FL** bisexual, radial; sepals gen 2(–8), free or fused at base; petals 3–18, free or ± fused; stamens 1–many, free or inserted on corolla; ovary superior or partly inferior, chamber 1, placenta free-central or basal; styles 2–8, gen fused at base. **FR**: capsule, circumscissile or 2–3-valved. **SEEDS** 1–many, gen black, gen shiny. ± 20 genera, ± 400 spp.: gen temp Am, Australia, s Afr; some cult (*Lewisia, Portulaca, Calandrinia*). [Bogle 1969 J Arnold Arbor 50:566–598] Family description and key to genera by Dieter H. Wilken & Walter A. Kelley.

1. Fr circumscissile
 2. Sepals free, 2–8; ovary superior; fr circumscissile near base .... **LEWISIA**
 2′ Sepals fused at base, 2-lobed; ovary partly inferior; fr circumscissile near middle .... **PORTULACA**
1′ Fr 2–3-valved, dehiscing from tip
 3. Infl umbel or panicle of head-like clusters, dense; sepals in fr gen round to reniform, ± scarious [2]**CALYPTRIDIUM**
 3′ Infl cyme, raceme, panicle, or fls solitary, gen open; sepals in fr gen not round or reniform, gen not scarious
  4. Style 2-branched (or stigmas 2, sessile); petals fused at base, tube expanding in fr, cap-like; fr 2-valved .... [2]**CALYPTRIDIUM**
  4′ Styles 3, sometimes fused at base; petals ± free, tube not expanding in fr, not cap-like; fr gen 3-valved
   5. Cauline lvs alternate
    6. Infl not 1-sided; petals gen red (exc *C. ambigua*); ovules many; seeds 6–40 per fr .... **CALANDRINIA**
    6′ Infl gen 1-sided; petals white to pink; ovules gen 3; seeds 1–3 per fr .... [2]**MONTIA**
   5′ Cauline lvs opposite
    7. Lvs mostly basal; cauline lvs gen 2, free or fused (2-lobed to disk- or cup-like) .... **CLAYTONIA**
    7′ Lvs ± gradually reduced upward; cauline lvs gen 4+, not fused .... [2]**MONTIA**

CALANDRINIA

Walter A. Kelley

Ann, per, ± fleshy, glabrous or glaucous. **STS** several–many, prostrate to erect, 3–45 cm. **LVS** simple, alternate; blade linear to spoon-shaped, cylindric or flat. **INFL**: raceme or panicle; bracts scarious or lf-like. **FL**: sepals 2, overlapping, persistent in fr; petals 3–7, gen 5, red or white; stamens 3–15; style 3-branched. **FR**: capsule, 3-valved from tip. **SEEDS** many, ovate to ± elliptic, gen black, smooth, finely tuberculate, or with a fine, net-like pattern, short-hairy or not. 150 spp.: w Am, Australia. (J.L. Calandrini, Switzerland, born 1703) [Kelley 1973 MS Thesis, CA State Univ, Northridge]

1. Petals white; fls in umbel-like clusters .... ***C. ambigua***
1′ Petals gen red; fls not in umbel-like clusters
 2. Pl glaucous; sepals gen purple-veined; seed at 10× dull gray, at 30× black, with short, white hairs .... ***C. maritima***
 2′ Pl not glaucous; sepals not purple-veined; seed at 10× shiny black, at 30× black, glabrous
  3. Fr gen > calyx by 3+ mm; seed at 30× finely tuberculate .... ***C. breweri***
  3′ Fr gen not > calyx by 3+ mm; seed at 30× with very fine, net-like pattern .... ***C. ciliata***

C. ambigua (S. Watson) Howell (p. 901) Ann, glabrous. **ST** spreading to erect. **LF** 1.5–6 cm, linear to ± spoon-shaped, ± cylindric. **INFL**: panicle of umbel-like clusters, compact; bracts scarious; pedicels 1–3 mm, straight in fr. **FL**: sepals 2–5 mm, glabrous, margins white-scarious; petals 3–5, 2–5 mm, white; stamens 5–10. **FR** < calyx. **SEEDS** 6–15, 1–2 mm wide, ± elliptic to ovate, at 10× shiny black, smooth. Common. Sandy to silty soil; < 1000 m. D; AZ, nw Mex.

C. breweri S. Watson (p. 901) Ann, ± glabrous. **ST** prostrate to ascending. **LF** 2–8 cm, ± ovate to spoon-shaped, flat. **INFL**: raceme, elongate; bracts lf-like; pedicels 6–20 cm, gen curved in fr. **FL**: sepals 4–6 mm, glabrous to ± ciliate; petals gen 5, 3–5 mm, red; stamens 3–6. **FR** gen > calyx by 3 mm or more. **SEEDS** 10–15, 1–2 mm wide, ± elliptic, at 10× shiny black, at 30× black, glabrous, finely tuberculate. Uncommon. Sandy to loamy soil, disturbed places, burns; < 1200 m. NCoR, c SNF, SnFrB, SCoRO, SCo, WTR; Baja CA. Like *C. ciliata*. ❀TRY.

C. ciliata (Ruiz López & Pavón) DC. (p. 901) RED MAIDS Ann, ± glabrous. **ST** spreading. **LF** 1–10 cm, linear to oblanceolate, flat. **INFL**: raceme, elongate; bracts lf-like; pedicels 4–25 mm, gen straight in fr. **FL**: sepals 2.5–8 mm, glabrous to ciliate; petals gen 5, 4–15 mm, red; stamens 3–15. **FR** not > calyx by 3 mm or more. **SEEDS** 10–20, 1–2.5 mm wide, elliptic, at 10× shiny black, at 30× black, glabrous, with a very fine, net-like pattern. $2n=24$. Common. Sandy to loamy soil, grassy areas, cult fields; < 2200 m. CA-FP, w MP, s SNE; to NM, C.Am; also in nw S.Am. [var. *menziesii* (Hook.) J.F. Macbr.; *C. micrantha* Schldl.] Variable vegetatively; uniform in fl, fr, and seed. ❀SUN:4–6,**7–9,14–24**&IRR:1–3,**10–12**,13.

C. maritima Nutt. SEASIDE CALANDRINIA Ann, glaucous. **ST** spreading or ascending. **LF** 1–6 cm, obovate to spoon-shaped, flat. **INFL**: panicle, gen above lvs; bracts scarious; pedicels 5–15 mm, ± straight in fr. **FL**: sepals 3–5 mm, glaucous, gen purple-veined; petals gen 5, 3–6 mm, red-purple; stamens gen 5. **FR** gen > calyx. **SEEDS** 20–40, 0.5–1 mm wide, ± elliptic, at 10× dull gray, at 30× black, with short, white hairs. UNCOMMON. Sandy soil, sea bluffs; < 300 m. SCo, ChI; Baja CA. ❀DRN,DRY,SUN:17,24; DFCLT.

CALYPTRIDIUM PUSSYPAWS

Dieter H. Wilken & Walter A. Kelley

Ann, per, ± fleshy, from taproot or fibrous roots, gen glabrous. **STS** gen several, gen spreading to ascending. **LVS** in basal rosette or basal and cauline, simple, oblanceolate to spoon-shaped. **INFL**: raceme, panicle, or umbel, scapose, leafy, or bracted; fls gen on 1 side of axis, deciduous or persistent in fr; bracts gen < sepals, scarious. **FL**: sepals 2, ovate to reniform, gen scarious or scarious-margined, persistent in fr; petals 2–4, minute, < sepals, tips adherent and cap-like in fr, falling as 1 unit; stamens 1–3; style 0 or 1, thread-like, stigmas gen 2. **FR**: capsule, gen translucent, 2-valved, gen compressed, oblong to ± round. **SEEDS** 1–many, black, gen shiny. 8 spp.: w Am. (Greek: cap, from petals in fr) [Hinton 1975 Brittonia 27:197–208; Thomas 1956 Leafl W Bot 8:9–11] Observation of fl, seeds requires 20× magnification.

1. Style thread-like; fr widely elliptic to ± round
 2. Infl terminal, 1 per rosette; bracts subtending infl ± round, ± = sepals .... ***C. umbellatum***
 2′ Infl axillary, 2+ per rosette; bracts subtending infl ovate to deltate, < sepals
  3. Per; taproot slender to thick; infl umbel; anthers pink to rose .... ***C. monospermum***
  3′ Ann; roots ± fibrous; infl panicle of head-like clusters; anthers yellow .... ***C. pulchellum***
1′ Style 0, stigmas sessile; fr ovate to narrowly oblong
 4. Fls pedicelled, pedicels 1–3 mm; petals 2 or 4
  5. Petals 2, ± 1 mm; sepals ± round to reniform — stamen 1 .... ***C. roseum***
  5′ Petals 4, 2–3 mm; sepals ovate to ± reniform
   6. Sepals ovate; ± fleshy; fls persistent in fr .... ***C. pygmaeum***
   6′ Sepals ± reniform, scarious; fls deciduous in fr .... ***C. quadripetalum***
 4′ Fls subsessile; petals gen 3
  7. Fr > 2 × sepal length .... ***C. monandrum***
  7′ Fr < 2 × sepal length .... ***C. parryi***
   8. Fls persistent in fr; seeds fine-tubercled throughout; TR, s SNH .... var. ***parryi***
   8′ Fls deciduous in fr; seeds smooth throughout or margin fine-tubercled; SnFrB, n SCoRI, SNE
    9. Outer sepal ovate to ± round, margin membranous or narrowly scarious; SnFrB, n SCoRI .... var. ***hesseae***
    9′ Outer sepal reniform, margin clearly white-scarious; SNE .... var. ***nevadense***

C. monandrum Nutt. (p. 901) Ann, 1.5–18 cm; taproot slender. **STS** spreading to decumbent, leafy. **LVS** basal and cauline; basal 1–5 cm, withering in fr; cauline 1–2(4) cm. **INFL**: raceme or panicle, 1–4 cm, gen open, axillary; bracts narrowly ovate; fls subsessile, gen persistent in fr. **FL**: sepals 1–2 mm, ovate to deltate, fleshy, becoming membranous, narrowly white-margined; petals 3, 1–3 mm, pink to reddish; stamen gen 1; stigmas sessile. **FR** 3–6 mm, oblong. **SEEDS** 4–10. Sandy soils, open sites, burned areas, shrubland; < 2150 m. s SN, Teh, s SnJV, SCoRI, SW, SNE, D; NV, AZ. ❀TRY.

C. monospermum E. Greene (p. 901) Per, < 50 cm; caudex short, thick; taproot slender to thick. **STS** gen spreading to ascending, scapose to leafy. **LVS**: basal rosette 1; basal lvs 1.5–6 cm; cauline lvs 0.8–3 cm. **INFL**: umbel, gen compound, ± open, simple and dense in small pls, axillary, 2+ per rosette, 1–10 cm diam; bracts ovate to deltate; fls subsessile to short-pedicelled, persistent in fr. **FL**: sepals 3–8 mm, ± round, clearly scarious; petals 4, 3–7 mm, rose to white; stamens 3, pink to rose; style ± 1 mm, thread-like. **FR** 2–3.5 mm, widely elliptic to ± round. **SEEDS** 1–3. Open, sandy or gravelly soils, coniferous forest; 600–3200 m. KR, NCoRH, CaR, SN, TR, SNE; s OR, NV, Baja CA. Hybridizes and intergrades with *C. umbellatum* in SN.

C. parryi A. Gray Ann, 2–11 cm; taproot slender. **STS** spreading to ascending, leafy. **LVS** basal and cauline, 1–3 cm; basal withering in fr. **INFL**: raceme or panicle, open to dense, 1–3.5 cm, axillary; bracts ovate to elliptic; fls subsessile, persistent to deciduous in fr. **FL**: sepals 2–5 mm, equal to unequal (outer > and wider than inner), ovate, round, or reniform, scarious to membranous; petals 3, 1.5–3 mm, gen white; stamens gen 3; stigmas sessile. **FR** 3–7 mm, ovate to oblong. **SEEDS** 10–15. Open areas, chaparral, oak or pinyon/juniper woodland, coniferous forest; 700–3350 m. s SNH, s SnFrB, n SCoRI, TR, SNE; w NV, AZ. (1 other var. in s AZ, n Mex.)

var. ***hesseae*** J.H. Thomas (p. 901) SANTA CRUZ MOUNTAINS PUSSYPAWS **INFL**: fl deciduous in fr. **FL**: outer sepal ovate to ± round, margin narrowly white-scarious to membranous. **SEED**: margin fine-tubercled. UNCOMMON. Chaparral, oak woodland; 700–1100 m. s SnFrB (Mount Hamilton, Santa Cruz Mtns), n SCoRI.

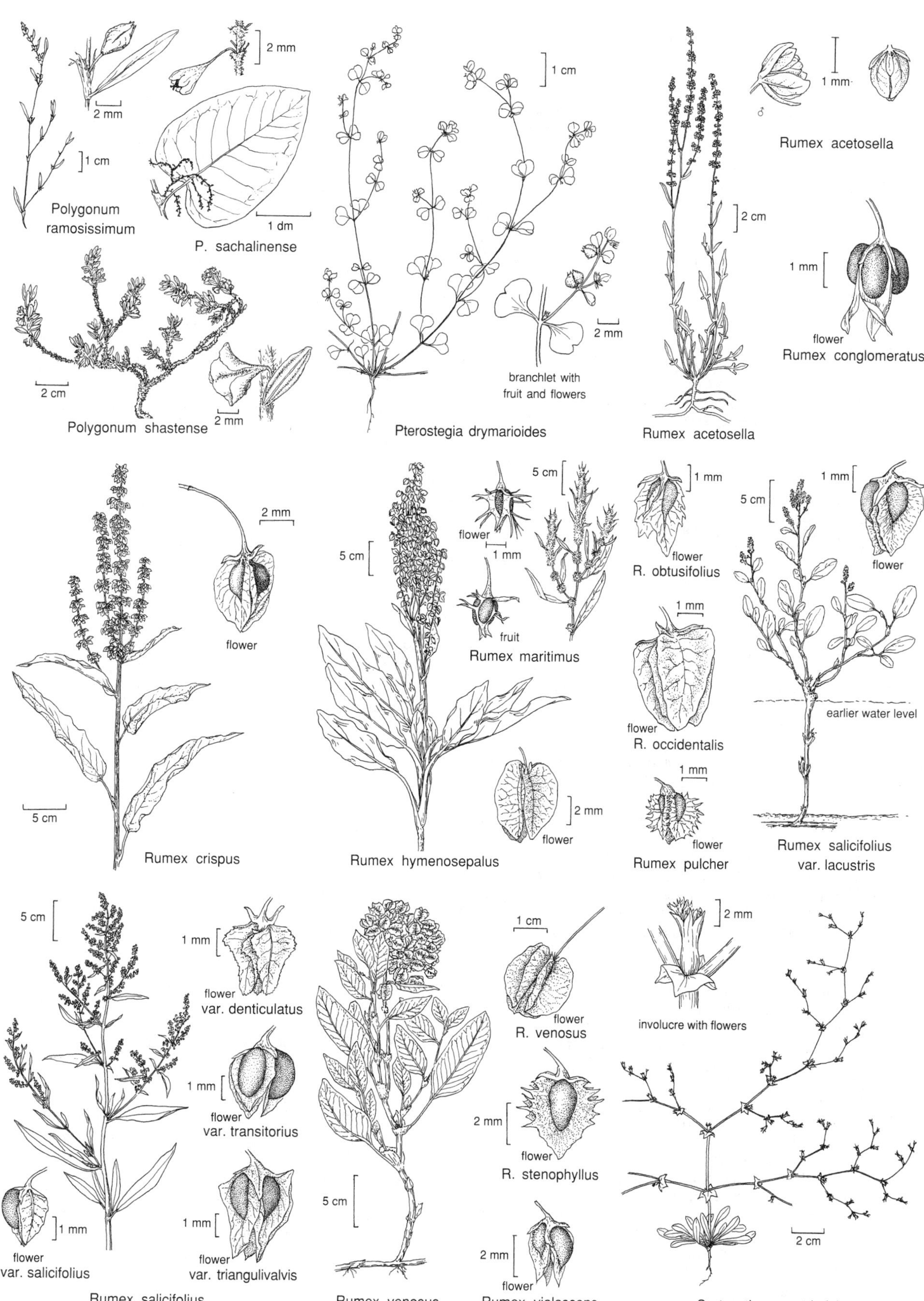

2 mm
2 mm
1 cm
Polygonum ramosissimum
1 dm
P. sachalinense
1 cm
2 mm
branchlet with fruit and flowers
Pterostegia drymarioides
2 cm
Polygonum shastense
2 mm
1 mm
Rumex acetosella
2 cm
Rumex acetosella
1 mm
flower
Rumex conglomeratus
2 mm
flower
5 cm
Rumex crispus
5 cm
5 cm
flower
1 mm
fruit
Rumex maritimus
2 mm
flower
Rumex hymenosepalus
1 mm
flower
R. obtusifolius
1 mm
flower
R. occidentalis
1 mm
flower
Rumex pulcher
5 cm
1 mm
flower
earlier water level
Rumex salicifolius var. lacustris
5 cm
1 mm
flower
var. denticulatus
1 mm
flower
var. transitorius
1 mm
flower
var. triangulivalvis
1 mm
flower
var. salicifolius
Rumex salicifolius
5 cm
Rumex venosus
1 cm
flower
R. venosus
2 mm
flower
R. stenophyllus
2 mm
flower
Rumex violascens
2 mm
involucre with flowers
2 cm
Systenotheca vortriedei

var. ***nevadense*** J. Howell **INFL**: fl deciduous in fr. **FL**: outer sepal ± reniform, margin white-scarious. **SEED**: margin fine-tubercled. Pinyon/juniper woodland; 1550–2500 m. SNE; w NV.

var. ***parryi*** (p. 901) **INFL**: fl gen persistent in fr. **FL**: outer sepal ± reniform, margin clearly scarious. **SEED** fine-tubercled throughout. Coniferous forest, alpine; 1400–3350 m. s SNH (Fresno, Inyo cos.), TR.

C. pulchellum (Eastw.) Hoover MARIPOSA PUSSYPAWS Ann, 2–7 cm; roots fibrous. **STS** spreading to ascending, leafy. **LVS**: basal rosette 1; basal lvs 0.5–2 cm, cauline 0.2–1.5 cm. **INFL**: panicle of head-like clusters, open, 1–4 cm, axillary, gen 2+ per rosette; bracts ovate to deltate; fls subsessile, gen persistent in fr. **FL**: sepals 3–4 mm, ± round, clearly scarious; petals 4, ± 3 mm, rose; stamens 3, yellow; style < 1 mm, thread-like. **FR** 1.5–2.5 mm, widely elliptic to ± round. **SEEDS** 1–2. RARE. Sandy soils, decomposed granite, oak woodland; 400–1100 m. c SN (s Mariposa, Madera, n Fresno cos.).

C. pygmaeum Rydb. Ann, 0.5–8 cm; taproot slender or roots fibrous. **STS** spreading to erect, leafy. **LVS** basal and cauline, 5–1.5 cm, gen persistent in fr. **INFL**: raceme or panicle, ± dense, 0.5–3 cm, axillary; bracts ovate to ± round; fls persistent in fr; pedicels 1–3 mm. **FL**: sepals 2–4 mm, ovate, fleshy, becoming membranous, margin sometimes white; petals 4, 2–3 mm, white; stamens gen 3; stigmas sessile. **FR** 3–5 mm, ± ovate. **SEEDS** 5–9. Sandy to gravelly soils, coniferous forest; 2100–3500 m. s SNH, SnBr.

C. quadripetalum S. Watson (p. 901) FOUR-PETALED PUSSYPAWS Ann, 1.5–13 cm; taproot slender or roots fibrous. **STS** spreading to erect, leafy. **LVS** basal and cauline, 0.5–6 cm, gen persistent in fr. **INFL**: raceme or panicle, ± dense, 0.5–4 cm, axillary; bracts ovate to ± round; fls deciduous in fr; pedicel 1–2 mm. **FL**: sepals 4–6 mm, round to reniform, scarious; petals 4, 2–3 mm, white to pink; stamens 1–3; stigmas sessile. **FR** 3–4 mm, lanceolate to ovate. **SEEDS** 5–14. UNCOMMON. Sandy or gravelly areas, gen serpentine; 500–2000 m. NCoRH, NCoRI. ❀TRY;DFCLT.

C. roseum S. Watson Ann, 1.5–10 cm; taproot slender or fibrous. **STS** spreading to ascending, leafy. **LVS** basal and cauline, 0.5–4 cm, gen persistent in fr. **INFL**: raceme or panicle, open, 1–4 cm, axillary; bracts ovate to elliptic; fl persistent to ± deciduous in fr; pedicels 1–3 mm. **FL**: sepals 2–3 mm, ± round to reniform, white-margined; petals 2, ± 1 mm, white; stamen 1; stigmas sessile. **FR** 2–3 mm, ovate to oblong. **SEEDS** 5–11. Gravelly soils, coniferous forest, sagebrush shrublands; 1500–3750 m. East slope SNH, W&I; to ID, NV.

C. umbellatum (Torrey) E. Greene (p. 901) Per, < 6 dm; caudex short, thick; taproot slender to thick. **STS** gen spreading to ascending, scapose. **LVS**: basal rosettes gen 2+; lf 1.5–7 cm. **INFL**: umbel, gen compound, simple in small pls, 1–7 cm diam, terminal, gen 1 per rosette, dense; bracts subtending infl < or = sepals, ± round; fls subsessile to short-pedicelled, persistent in fr. **FL**: sepals 3–8 mm, ± round, clearly scarious; petals 4, 3–8 mm, white; stamens 3, yellow to red; style ± 1 mm, thread-like. **FR** 2–3 mm, widely elliptic to ± round. **SEEDS** 1–8. Open, sandy to rocky soils, coniferous forest, alpine; 1500–4300 m (100–200 m in SnFrB). KR, NCoRH, CaR, SN, sw SnFrB (Santa Cruz Mtns), GB; to MT, w WY. Alpine pls from SNH with simple umbels, persistent lvs < 1 cm have been called var. *caudiciferum* (A. Gray) Jepson. ❀TRY;DFCLT.

CLAYTONIA

Kenton L. Chambers

Ann or per, from stolon, rhizome, tuber, or taproot, glabrous, ± fleshy. **LVS** entire; basal 0–many, rosetted; cauline gen 2, gen opposite, free to fully fused into ± 2-toothed disk or cup surrounding st. **INFL**: raceme, terminal, 1-sided; pedicels reflexed, becoming erect in fr. **FL**: petals 5, pink or white; stamens 5, epipetalous; ovary chamber 1, placentas basal, style 1, stigmas 3. **FR**: capsule; valves 3, margins rolling inward and forcibly expelling seeds. **SEEDS** 3–6, gen black, gen clearly appendaged. 28 spp.: N.Am, e Asia. (John Clayton, colonial Am botanist, born 1686) [Miller 1978 Syst Bot 3:322–341] Some spp. formerly placed in *Montia*.

1. Per; fl-sts from thick, taprooted caudex or spheric tuber
 2. Basal lvs many, clustered at caudex tips; cauline lvs linear to narrowly oblanceolate; alpine; SNH, Wrn .... ***C. megarhiza***
 2′ Basal lvs 0–few, from spheric tuber; cauline lvs ± linear to ovate; montane to alpine; widespread
  3. Cauline lvs sessile (if short-petioled, blade linear-lanceolate); tuber fibrous-rooted only .... ***C. lanceolata***
  3′ Cauline lvs with petiole ± = ovate blade; tuber often from slender taproot — CaRH, GB .... ***C. umbellata***
1′ Ann or per from rhizomes; caudex gen 0 or short; taproot 0 or slender
 4. Per; rhizomes long, gen branched, often with bulb-like winter buds; petals >> sepals
  5. Pl < 15 cm; rhizomes fleshy, pale, bulb-like buds 0; bract 1, below lowest fl; petals pinkish ...... ***C. nevadensis***
  5′ Pl 15–60 cm; rhizomes slender, often brownish, with bulb-like buds at nodes; bracts 0 or 1 per fl; petals gen white
   6. Bracts 0; basal lf blade ovate, base truncate to cordate, petiole linear; cauline lvs ovate, sessile .... ***C. cordifolia***
   6′ Bracts 1 per fl; basal lf blade elliptic to oblanceolate, tapered, petiole indistinct; cauline lvs oblanceolate to obovate, gen short-petioled .... ***C. palustris***
 4′ Ann (or per with rhizomes 0 to very short); petals ± = to >> sepals
  7. Bracts 1 per 1–4 fls; petals 4–12 mm; lf blades oblanceolate to ovate, abruptly tapered to petiole (if blade ± linear, pl per, some lf bases bulb-like)
   8. Per; fls gen 1–2 per bract; cauline lvs free, sometimes ± lanceolate; petals 6–12 mm; stigmas maturing well after anthers (cross-pollinated); widespread .... ***C. sibirica***
   8′ Ann; fls often 2–4 per bract; cauline lvs widely ovate, bases often ± fused on 1 side; petals 4–5 mm; stigmas maturing ± with anthers (self-pollinated); NCo .... ***C. washingtoniana***
  7′ Bract gen 1, subtending lowest fl; petals 1–8 mm; lf blades variable, bulb-like bases 0
   9. Seeds dull black, ± tubercled at 20×, appendage brownish, fleshy or shrunken, filling notch or not
    10. Lvs ± prostrate; pl cushion-shaped, < 6 cm wide, gen not glaucous; cauline lvs widely ovate, free; petals 5–8 mm, pink; seed appendage filling notch; KR, NCoRO, NCoRH .... ***C. saxosa***
    10′ Lvs decumbent to erect; pl loose or tufted, often > 6 cm wide, gen glaucous; cauline lvs linear if free; petals 1–8 mm, pink or white; seed appendage minute; widespread
     11. Petals 5–8 mm; styles maturing well after anthers (cross-pollinated); infl many-fld, long-stalked above cauline lf-pair; NCoR, CW .... ***C. gypsophiloides***
     11′ Petals 2–5 mm; styles maturing ± with anthers (self-pollinated); infl few–many-fld, stalk 0 to long; widespread .... ***C. exigua***

12. Cauline lvs free or fused and appearing 2-lobed, lobes 2–30 mm; petals 2–5 mm; widespread . . ssp. ***exigua***
12′ Cauline lvs fused around st into a round or squarish disk 3–15 mm wide; petals 2 mm; KR, NCoR, SnFrB, SCoRO . . . . ssp. ***glauca***
9′ Seeds shiny black, smooth at 20×, appendage white, fleshy, filling notch or slightly projecting (*C. perfoliata* complex)
13. Basal lvs linear or blade > 3 × longer than wide, tapered gradually to petiole; petals 1–6 mm . . . ***C. parviflora***
14. Petals 4–6 mm; stigmas maturing well after anthers (cross-pollinated); cauline lf-pair disk-like — SNF, Teh . . . . ssp. ***grandiflora***
14′ Petals gen < 4 mm; stigmas maturing ± with anthers (self-pollinated); cauline lvs free or fused and disk-like
15. Cauline lf-pair ± disk-like; widespread, often weedy . . . . ssp. ***parviflora***
15′ Cauline lvs linear, free or ± fused on 1 side; s SN, SCoR, TR, PR, SNE, DMtns . . . . ssp. ***viridis***
13′ Basal lf blades < 3 × longer than wide (exc seedlings), base abruptly tapered to cordate; petals gen < 5 mm
16. Basal lvs gen many, prostrate to spreading, smaller inward, blades ± elliptic to widely deltate; cauline lf-pair often unequally fused; petals minute . . . . ***C. rubra***
17. Basal lf blades widely elliptic to obovate, base wedge-shaped; infl ± sessile; only petioles often red; nw CA, GB . . . . ssp. ***depressa***
17′ Basal lf blades widely diamond-shaped to deltate, base ± truncate; infl ± stalked; petioles or whole pl often red; widespread . . . . ssp. ***rubra***
16′ Basal lvs few–many, spreading to erect, ± equal; cauline lvs gen equally fused; petals minute to > sepals, white or pinkish . . . . ***C. perfoliata***
18. Basal lf blades abruptly small-pointed, often wider than long, base truncate; cauline lf disk abruptly 2-pointed; esp SW . . . . ssp. ***mexicana***
18′ Basal lf blades obtuse to acute, gen longer than wide, base wedge-shaped; cauline lf disk teeth gen obtuse or acute; widespread, often weedy . . . . ssp. ***perfoliata***

C. cordifolia S. Watson Per; caudex short, 2–10 mm diam, horizontal, brownish; rhizome long, with bulb-like buds; stolons 0. **ST** 10–40 cm, erect. **LVS**: basal 2–25 cm, blade 1–9 cm, widely ovate, base truncate or cordate, tip obtuse to acute, petiole linear; cauline 1–4 cm, free, widely ovate, sessile, obtuse to acute. **INFL** stalked, open; bracts 0; fls 3–12. **FL**: sepals 3–5 mm; petals 8–13 mm, white. **FR** 4–5 mm. **SEED** 1.5–2 mm, round, shiny. $2n$=10,20. Gen shaded swamps, streambanks, seeps, wet meadows; 1200–2300 m. KR, NCoRH, n SNH; to B.C., MT, UT, NV. [*Montia c.* (S. Watson) Pax & K. Hoffm.] ❀IRR or WET:**4**&SHD:1,**5–7**,14,**15–17**;GRNCVR.

C. exigua Torrey & A. Gray Ann, ± glaucous. **ST** 1–15 cm, erect or spreading. **LVS**: basal < 12 cm, ± linear; cauline < 8 cm, linear and free, crescent-shaped and ± fused on 1 side, or fully fused and disk-like. **INFL** sessile to long-stalked; fls 3–15, lowest bracted. **FL**: sepals 1.5–3 mm; petals 2–5 mm, white or pinkish; stamens maturing ± with stigmas. **FR** 1.5–2.5 mm. **SEED** 1–1.5 mm, elliptic, dull; appendage minute. Dry or moist, sometimes disturbed bare clay to sandy soils, often on serpentine; < 2000 m. NW, SNF, GV, CW, TR, PR; to B.C. Self-pollinating.

ssp. ***exigua*** (p. 901) **LVS**: cauline pair free to ± fused on 1 side. **FL**: petals 2–5 mm, white or pinkish. $2n$=32,48. Habitats and range of sp. [*Montia spathulata* Howell (illegitimate) incl vars. *e.* (Torrey & A. Gray) Robinson, *rosulata* (Eastw.) J. Howell, & *tenuifolia* (Torrey & A. Gray) Munz] ❀DRN,SUN:1,2,4,**5–7**,8,9,**14–17**,18–24.

ssp. ***glauca*** (Torrey & A. Gray) John M. Miller & Chambers (p. 901) **LVS**: cauline pair fully fused into round or squarish disk 3–15 mm wide. **FL**: petals 2 mm, white. $2n$=16. Habitats of sp. KR, NCoR, SnFrB, SCoRO. [*Montia perfoliata* forma *g.* (Torrey & A. Gray) J. Howell] ❀TRY.

C. gypsophiloides Fischer & C. Meyer (p. 901) Ann. **ST** 3–25 cm, spreading to erect. **LVS**: basal 2–15 cm, linear; cauline < 6 cm, linear and free, fused on 1 side, or fully fused into 2-toothed disk. **INFL** 2–15 cm, long-stalked; fls 3–30, lowest fl bracted. **FL**: sepals 2–3 mm; petals 5–8 mm, gen pinkish; stamens maturing well before stigmas. **FR** 1.5–2 mm. **SEED** 1–1.5 mm, elliptic, dull; appendage minute. $2n$=16. Moist, bare, often stony places, in sun or shade, often on serpentine; < 1300 m. NCoR, SnFrB, SCoR (Santa Lucia, Temblor mtns). [*Montia g.* (Fischer & C. Meyer) Howell; *M. perfoliata* var. *nubigena* (E. Greene) Jepson] Cross-pollinated. ❀DRN:5,**15,16**,17&SHD:**7,9,19–24**.

C. lanceolata Pursh (p. 901) WESTERN SPRING BEAUTY Per; caudex 0; tuber 1–3 cm wide, brownish, spheric; rhizomes and stolons 0. **ST** 5–15 cm, erect. **LVS**: basal 0–2, 5–8 cm, blade 1–3 cm, elliptic, base wedge-shaped, tip acute, petiole thread-like; cauline 1–7 cm, ± linear to ovate, gen ± sessile. **INFL** gen short-stalked or sessile, 1-bracted at base; fls 3–15. **FL**: sepals 3–7 mm; petals 5–12 mm, white or pinkish (base sometimes yellow). **FR** 3.5–4.5 mm. **SEED** 2–2.5 mm, round, shiny. $2n$=16,24,32, many other numbers. Gravelly woodlands, meadows; 1500–2600 m. KR, NCoRH, CaRH, n&c SNH, SnGb, MP; to w Can, MT, NM. [vars. *peirsonii* Munz & I.M. Johnston, Peirson's spring beauty, & *sessilifolia* Torrey (Nelson)] Much variability apparently environmental; needs study. ❀TRY;DFCLT.

C. megarhiza (A. Gray) S. Watson (p. 901) FELL-FIELDS CLAYTONIA Per; caudex long, 5–30 mm diam, vertical, brownish, scaly with withered lf bases; rhizomes and stolons 0. **ST** 1–6 cm, spreading. **LVS**: basal 2–10 cm, crowded, blade 1–3 cm, widely elliptic to obovate, base tapered, tip obtuse; cauline 1–3 cm, free, ± linear, acute. **INFL** ± sessile, dense, bracted throughout; fls 2–6. **FL**: sepals 4–6 mm; petals 5–9 mm, white or pinkish. **FR** 3.5–4.5 mm. **SEED** 2–2.5 mm, round, shiny. $2n$=24. Subalpine and alpine gravel, talus, crevices; 2600–3300 m. RARE in **CA**. n&c SNH, Wrn; to w Can, MT, WY, Colorado, NM. [var. *bellidifolia* (Rydb.) C. Hitchc.; *C. b.* Rydb.] In cult.

C. nevadensis S. Watson (p. 901) Per; caudex long, 1–3 mm diam, horizontal, white or yellowish, continuous with fleshy, much-branched rhizomes; bulb-like offshoots and stolons 0. **ST** 2–10 cm, spreading or erect. **LVS**: basal 2–15 cm, blade 1–5 cm, elliptic to widely ovate, base wedge-shaped, tip gen obtuse, petiole linear, often buried; cauline < 2 cm, free, ovate, sessile, obtuse. **INFL** short-stalked or sessile, gen dense; bracts 0–1; fls 2–8. **FL**: sepals 3–8 mm; petals 6–10 mm, pinkish. **FR** 3–4 mm. **SEED** 1.5–2 mm, round, shiny, smooth. $2n$=14. Subalpine streams, springs, melting snowbeds, in gravel or sand; 2200–3500 m. KR, CaRH, SNH, SNE; OR.

C. palustris Swanson & Kelley (p. 901) MARSH CLAYTONIA Per; caudex short, 2–5 mm diam, horizontal, white; rhizomes and stolons slender, with bulb-like buds. **ST** 10–60 cm. **LVS**: basal 8–30 cm, blade 2–8 cm, narrowly elliptic to oblanceolate, base tapered gradually to petiole, tip obtuse to acute, petiole linear; cauline 2–9 cm, free, oblanceolate to widely elliptic, sessile or tapered to winged petiole, obtuse to acute. **INFL** stalked, open, bracted throughout; fls 5–18. **FL**: sepals 3–4 mm; petals 5–9 mm, gen white (pinkish). **FR** 2–3 mm. **SEED** 1.5–1.8 mm, round, dull. $2n$=12. UNCOMMON. Marshy meadows, springs, streambanks, in sun or shade; 1000–2500 m. KR, CaRH, n&c SNH. [*Montia sibirica* (L.) Howell var. *heterophylla* (Torrey & A. Gray) Robinson misapplied] n KR pls may intergrade with *C. sibirica* L.

C. parviflora Hook. Ann. **ST** 1–30 cm, spreading to erect. **LVS**: basal 1–18 cm, linear to narrowly oblanceolate, blade 1–7 cm, > 3 × longer than wide, gradually tapered to petiole, obtuse to acute; cauline free (< 6 cm, linear) or disk-like (< 5 cm diam, round or squarish). **INFL** stalked or sessile, dense or open, 1-bracted at base; fls 3–40. **FL**: sepals 1.5–4 mm; petals 2–6 mm, white or pinkish. **FR** 1.5–4 mm. **SEEDS** 1.2–2.3 mm, ovate to round, shiny, smooth. Vernally moist, often disturbed places in sun or shade; < 2300 m. CA-FP, GB, DMtns; to B.C., MT, n Mex.

ssp. ***grandiflora*** John M. Miller & Chambers (p. 901) **LVS**: cauline pair fused, disk-like. **INFL** stalked, open. **FL**: sepals 1.5–2.5 mm; petals 4–6 mm; stamens maturing well before stigmas. **SEED** 1.2–1.5 mm. 2*n*=12. Habitats of sp.; 150–1200 m. SNF, Teh. Cross-pollinating.

ssp. ***parviflora*** (p. 901) **LVS**: cauline pair fused, disk-like. **INFL** stalked or sessile, open or dense. **FL**: sepals 1.5–4 mm; petals 2–6 mm; stamens maturing ± with stigmas. **SEED** 1.2–2.3 mm. 2*n*=24,36,48. Habitats and range of sp. [*Montia perfoliata* forma *pa.* (Hook.) J. Howell & var. *utahensis* (Rydb.) Munz] Variable; intergrades with other members of *M. perfoliata* complex. Self-pollinating.

ssp. ***viridis*** (Davidson) John M. Miller & Chambers (p. 901) Herbage green. **LVS**: cauline free, linear, often curved, spreading or erect. **FL**: sepals 1.5–2 mm; petals 2–3.5 mm; stamens maturing ± with stigmas. **SEED** 1.2–1.5 mm. 2*n*=24,36. Shrub- or woodland, sometimes dry; < 1800 m. s SN, SCoR, TR, PR, SNE, DMtns; to n Mex. [*Montia spathulata* var. *v.* Davidson] Intergrades with ssp. *p.*, *C. rubra*. Self-pollinating.

C. perfoliata Willd. MINER'S LETTUCE Ann. **ST** 1–40 cm, spreading to erect. **LVS**: basal 1–25 cm, blade < 4 cm, < 3 × longer than wide, elliptic to reniform, tip rounded to acute, petiole linear; cauline pair fused, disk-like, < 10 cm diam, round or squarish (or free on 1 side). **INFL** stalked or sessile, open or dense, 1-bracted at base; fls 5–40. **FL**: sepals 1.5–5 mm; petals 2–6 mm, white or pinkish. **FR** 1.5–4 mm. **SEED** 1.2–2.7 mm, ovate to round, shiny, smooth. Common. Vernally moist, often shady or disturbed sites; < 2000 m. CA-FP, GB, DMtns; to B.C., MT, C.Am. Highly variable; sspp. difficult because of environmental plasticity, genetic mixing among polyploids, and geog overlap of distinct, self-pollinating forms.

ssp. ***mexicana*** (Rydb.) John M. Miller & Chambers (p. 907) **LVS**: basal round-deltate to reniform, tip short-pointed; cauline pair gen 2-angled, angles short-pointed. 2*n*=12,24,36. Habitats of sp. NW, CW, SW; AZ, Mex. Intergrades with ssp. *perfoliata*, *C. parviflora*, *C. rubra*. ❀TRY.

ssp. ***perfoliata*** (p. 907) **LVS**: basal elliptic to round-deltate, tip obtuse to acute; cauline pair gen round or ± obtuse-angled. 2*n*=24,36,48,60. Habitats and range (exc s AZ to C.Am) of sp. [*Montia p.* (Donn) Howell] Polyploids are derived from hybridization with ssp. *mexicana*, *C. parviflora*, *C. rubra*. ❀DRN:4&SHD:**5–7,9,10,14–24**&IRR:1,2,8; grown or collected as a salad plant; rather INV.

C. rubra (Howell) Tidestrom Ann. **ST** 1–15 cm. **LVS**: basal 1–8 cm, blade < 2 cm, elliptic to widely deltate, base truncate to wedge-shaped, tip gen obtuse, petiole linear; cauline pair fused or partly free on 1 side, < 4 cm wide, gen round or with 2 squarish corners. **INFL** sessile, ± dense, 1-bracted at base; fls 3–30. **FL**: sepals 1.5–3 mm; petals 2–3.5 mm, white or pinkish. **FR** 2–3 mm. **SEED** 1–2 mm, elliptic, shiny, smooth. Vernally moist dunes, coniferous forest, shrubland, in sun or shade; < 2500 m. NW, CaRH, SNH, Teh, w CW, GB; to B.C., SD, Colorado. Intergrades with *C. parviflora*, *C. perfoliata*.

ssp. ***depressa*** (A. Gray) John M. Miller & Chambers (p. 907) **LVS**: basal elliptic, base wedge-shaped; cauline fused. 2*n*=24,36. Open, often disturbed sites, shrubland; < 2500 m. NCo, KR, n CCo, SnFrB, GB; to B.C., SD. [*Montia perfoliata* var. *d.* (A. Gray) Jepson]

ssp. ***rubra*** (p. 907) **LVS**: basal diamond-shaped or deltate, widest below middle, base often truncate; cauline ± free (at least on 1 side). 2*n*=12. Habitats and range (exc NCo, CCo) of sp. [*Montia perfoliata* var. *d.* (A. Gray) Jepson misapplied]

C. saxosa Brandegee (p. 907) Ann. **ST** 1–3 cm, spreading to erect. **LVS**: basal crowded, prostrate, < 2 cm, oblong to obovate, sessile, tip obtuse; cauline < 1.5 cm, free, ovate, sessile, obtuse. **INFL** sessile, dense; bracts 0; fls 2–10. **FL**: sepals 2–3 mm; petals 5–8 mm, pinkish. **FR** 2–3.5 mm. **SEED** 1.3–2 mm, elliptic, dull. 2*n*=16. Open, rocky places, gen on serpentine; 800–2150 m. KR, NCoRO, NCoRH. [*Montia s.* (Brandegee) Robinson] ❀DFCLT.

C. sibirica L. (p. 907) CANDY FLOWER (Ann) gen per; caudex short, < 1 cm diam, vertical, brownish; rhizomes and stolons short, sometimes forming offset rosettes; taproot slender. **ST** 5–60 cm, spreading to erect. **LVS**: basal 3–30 cm, blade 1–8 cm, oblanceolate to deltate, base truncate to long-tapered, tip obtuse to long-tapered, some petioles often with bulb-like base; cauline 1–8 cm, free, lanceolate to ovate, sessile, gen acute. **INFL** simple or branched, stalked, open, bracted throughout; fls gen 10–20. **FL**: sepals 3–6 mm; petals 6–12 mm, gen pinkish; stamens maturing well before stigmas. **FR** 2.5–3.5 mm. **SEED** 1.5–2 mm, round to elliptic, shiny or dull. 2*n*=12,24,36. Gen shady moist woodlands, streambanks, marshes; < 1300 m. NW, CaRH, SnFrB; to AK, MT. [*Montia s.* (L.) Howell] ❀DRN,SHD,IRR:**4,5,7,14–17**.

C. umbellata S. Watson Per from oblong to spheric, brownish tuber 1–5 cm diam; caudex, rhizomes, and stolons gen 0; taproot gen 2–5 mm diam. **ST** 5–25 cm, erect, mostly underground. **LVS**: basal 0–3, 5–25 cm, blade 1–3 cm, widely elliptic to ovate, base acute or truncate, tip obtuse, petiole mostly underground; cauline 1.5–5 cm, blade 1–3 cm, widely elliptic, obtuse, abruptly tapered to linear petiole. **INFL** ± sessile, dense, 1–2-bracted at base; fls 2–12. **FL**: sepals 4–7 mm; petals 6–12 mm, pinkish. **FR** 5–6 mm. **SEED** 2.5–3 mm, round, shiny. 2*n*=16. Uncommon. Talus slopes, stony flats, crevices; 1900–3500 m. KR, GB; OR, NV.

C. washingtoniana (Suksd.) Suksd. (p. 907) Ann. **ST** 6–40 cm, spreading to erect. **LVS**: basal 4–30 cm, blade 1–5 cm, gen widely diamond-shaped to ovate or deltate, base wedge-shaped to truncate, tip gen acute; cauline 1–4 cm, free or fused < 5 mm on 1 side, widely ovate to ± round, sessile, obtuse. **INFL** stalked or sessile, open, bracted throughout; fls 5–25, clustered. **FL**: sepals 2.5–4 mm; petals 4–5 mm, pinkish in CA; stamens maturing ± with stigmas. **FR** 2.5–3 mm. **SEED** 1.5–2 mm, elliptic, shiny or dull. 2*n*=24. Uncommon. Gen shady moist woodlands, streambanks, seeps; < 150 m. NCo; to B.C. Fertile derivative of *C. sibirica* × *C. perfoliata*. Self-pollinating.

LEWISIA

Lauramay T. Dempster

Per, gen from short, thick, ± branched taproot, topped by short, sometimes very thick caudex at or below ground level, sometimes from spheric corm. **ST**: aerial parts restricted to infl. **LVS** gen in basal rosette, simple, entire or not; base wide; margin gen ± translucent. **INFL** ± scapose; sts 1–many, gen lfless but bracted, sometimes disjointing in age, 1–many-fld. **FL**: sepals 2–8, free, persistent; petals 4–18, variously colored, overlapping in bud; stamens 5–many; styles 2–8, fused at base, stigmas 2–8, thread-like. **FR**: capsule, translucent, spheric or ovoid, circumscissile near base. **SEEDS** 2–many, dark, gen shiny, smooth or finely tuberculate. ± 20 spp.: w N.Am. (Captain Meriwether Lewis, 1774–1809, of Lewis & Clark Expedition) [Elliott 1966 Bull Alpine Gard Soc 34] ❀DRN&IRR:pots and rock gardens only;DRY when dormant;DFCLT.

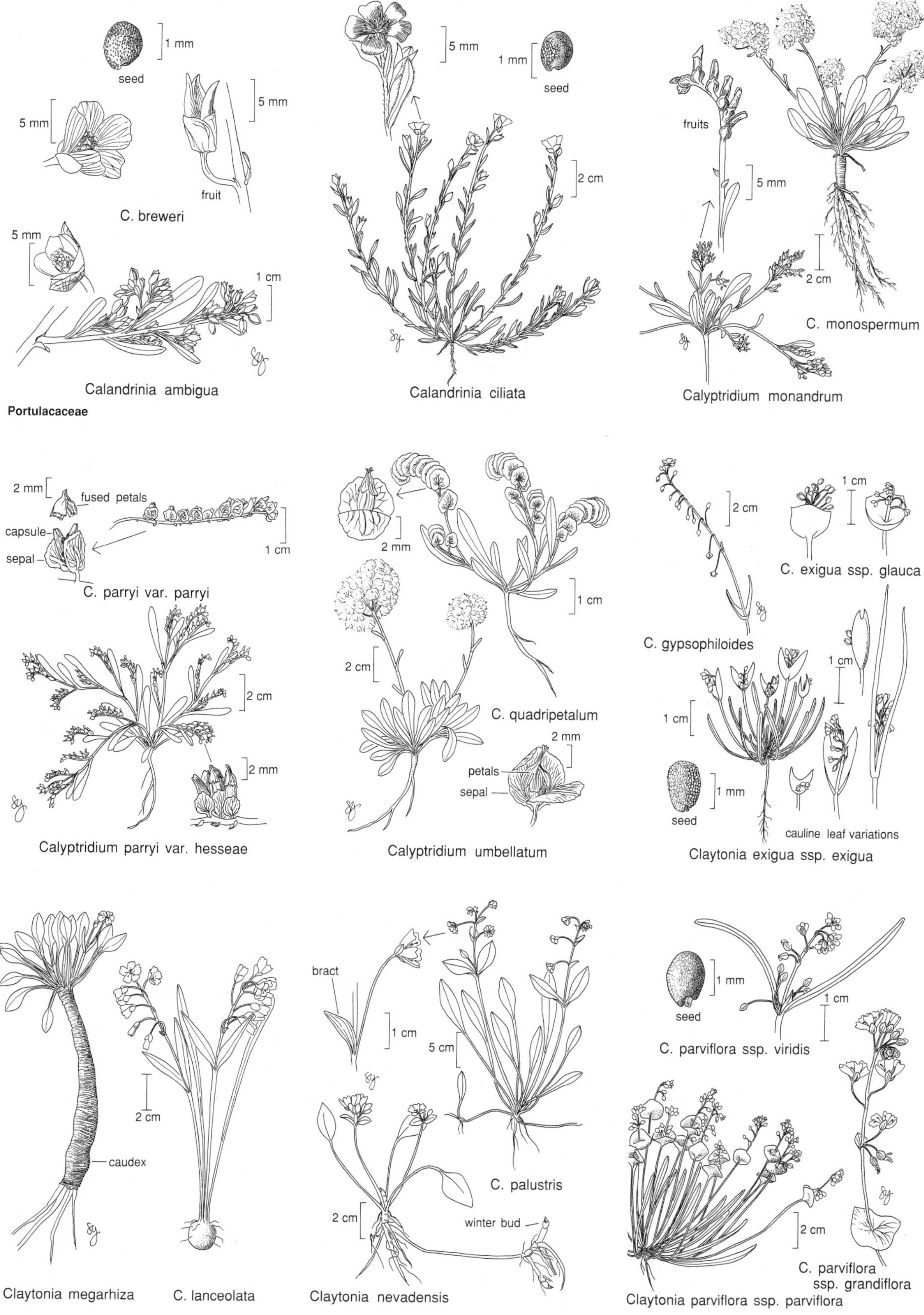
seed
1 mm
5 mm
5 mm
fruit
C. breweri
5 mm
1 cm
Calandrinia ambigua
Portulacaceae
5 mm
1 mm
seed
2 cm
Calandrinia ciliata
fruits
5 mm
2 cm
C. monospermum
Calyptridium monandrum
2 mm
fused petals
capsule
sepal
1 cm
C. parryi var. parryi
2 cm
2 mm
Calyptridium parryi var. hesseae
2 mm
1 cm
2 cm
C. quadripetalum
2 mm
petals
sepal
Calyptridium umbellatum
2 cm
1 cm
C. exigua ssp. glauca
C. gypsophiloides
1 cm
1 cm
1 mm
seed
cauline leaf variations
Claytonia exigua ssp. exigua
2 cm
caudex
Claytonia megarhiza
C. lanceolata
bract
1 cm
5 cm
C. palustris
2 cm
winter bud
Claytonia nevadensis
1 mm
seed
1 cm
C. parviflora ssp. viridis
2 cm
C. parviflora ssp. grandiflora
Claytonia parviflora ssp. parviflora

1. Sepals entire, petal-like, becoming scarious in age; st disjointing near middle, leaving ring of scarious bracts (sect. *Lewisia*)
  2. Sepals 2; bracts ovate, 2–4 in ring — fl 10–15 mm .... ***L. disepala***
  2′ Sepals 6–8; bracts awl-like, 5–many in ring .... ***L. rediviva***
1′ Sepals entire or not, not petal-like; st not disjointing, bracts gen 2 or alternate on st, not in ring (sect. *Oreobroma*)
  3. Sepals 2 but seemingly 4 due to 2 sepal-like bracts immediately below calyx; sts 1-fld
    4. Sepal margins entire — SnBr, PR .... ***L. brachycalyx***
    4′ Sepal margins gland-toothed .... ***L. kelloggii***
  3′ Sepals 2, bracts well below; sts 1–many-fld
    5. Lvs few, cauline; sts from small, spheric corm — fls 1–10 in loose cluster .... ***L. triphylla***
    5′ Lvs many, mostly in basal rosette; sts from stout, elongated root + caudex
      6. St gen 1–2-fld; fls gen not or slightly exserted from lvs
        7. Sepal margins gland-toothed; root + caudex elongate, widest at top, tapered or branched below; bracts ± ovate, immediately or well below fls, margins ± gland-toothed; ± geog restricted — sts often 2-fld
          8. Petals ± 8 mm, tips irregular, not glandular, not acuminate; root + caudex very stout, rosette dense; bracts in 1–2 irregular pairs, above middle of st; c&s SNH, SNE (White Mtns) ***L. glandulosa***
          8′ Petals 15–18 mm, tips glandular-acuminate; root + caudex stout, rosette loose; bracts in 1–2 pairs, near middle of st; n SNH .... ***L. longipetala***
        7′ Sepal margins entire or not, gen not glandular (exc rarely in *L. pygmaea*); root + caudex fusiform, corm-like, or elongate, widest at top, tapered or branched below; bracts gen lanceolate, well below fls, margins gen entire, not glandular; widespread
          9. Sepal tips ± acute, gen spreading, margins gen ± entire, gen not glandular; fls 8–15 mm; st 1-fld; root + caudex fusiform or corm-like .... ***L. nevadensis***
          9′ Sepal tips ± round, not spreading, margins ± jagged, rarely glandular; fls 5–10 mm; st rarely 2-fld; root + caudex gen elongate, widest at top, tapered or branched below .... ***L. pygmaea***
      6′ St gen 2–many-fld; fls well exserted from lvs (exc sometimes in *L. oppositifolia*)
        10. Lf ± linear, flat or cylindric .... ***L. leana***
        10′ Lf narrowly oblanceolate or wider, flat
          11. Fl gen 6–8 mm — st 4–7 × lvs, several–many-fld
            12. Lvs finely to coarsely dentate, ± spoon-shaped, tip round, truncate, or notched — KR, CaRH, n SNH .... ***L. cantelovii***
            12′ Lvs entire, narrowly oblanceolate to obovate, tip acute to blunt
              13. Lvs 3–8 cm, basal; KR .... ***L. columbiana***
              13′ Lvs 10–20 cm, basal and a few cauline; c SNH .... ***L. congdonii***
          11′ Fl 8–17 mm
            14. St 1.5–3 × lvs, 3–11-fld; pedicels > fls
              15. Sepal margins jagged-toothed, not glandular; KR .... ***L. oppositifolia***
              15′ Sepal margins gland-toothed; NCoRH .... ***L. stebbinsii***
            14′ St 3–7 × lvs, 12–many-fld; pedicels gen < fls; KR, CaRH .... ***L. cotyledon***
              16. Lf margins entire; > 1300 m .... var. ***cotyledon***
              16′ Lf margins not entire; < 2100 m
                17. Lf margins dentate .... var. ***heckneri***
                17′ Lf margins wavy .... var. ***howellii***

L. brachycalyx A. Gray (p. 907) SHORT-SEPALED LEWISIA Root + caudex short, stout, tapered or often branched below. **LVS** several–many, in spreading rosette, 2–8 cm, oblanceolate, ± fleshy, entire, tapered to petiole; tip blunt. **INFL**: sts many, 1–3.5 cm, 1-fld; fls incl in lvs; bracts 2, closely below and ± like sepals. **FL**: sepals 2 (but seemingly 4 due to bracts), 1/3–1/2 × corolla, obovate-obtuse to round, entire; petals 5–9, 2–2.5 cm, oblong, white or pinkish, tip blunt or notched; stamens 9–15; stigmas 5–8. *n*=10. UNCOMMON. Wet meadows, open forest; 1400–2300 m. SnBr, PR; to UT, NM, Baja CA.

L. cantelovii J. Howell (p. 907) CANTELOW'S LEWISIA Root + caudex short, stout, tapered abruptly to narrow root. **LVS** many, in spreading rosette, < 8 cm, spoon-shaped, finely to coarsely dentate, tapered to broad petiole; tip round, truncate, or notched. **INFL**: sts 1–several, 7–45 cm, each with a broad, lax, gen many-fld panicle; fls well exserted from lvs; bracts among fls and few below, minute, lance-ovate, gland-toothed; pedicels ± = fls, ± thread-like. **FL**: sepals 2, ± 1/2 × corolla, ± round, margin gland-toothed; petals 5–6, 4.5–9 mm, obovate, pale pink with deeper veins, tip round; stamens 5–6; stigmas 3. *n*=14. RARE. Granite cliff faces, rocky outcrops; 400–1300 m. KR, CaRH, n SNH. [*L. cantelowii* J. Howell] Pls with smaller infl, deeper lf serrations have been called *L. serrata* Heckard & Stebb. (Saw-toothed lewisia; RARE).

L. columbiana (Howell) Robinson Root + caudex short, thick, tapered to slender root. **LVS** many, in dense rosette, 3–8 cm, narrowly oblanceolate, fleshy, entire, tapered to base; tip blunt. **INFL**: sts several, 12–30 cm, each with a spreading, many-fld panicle; fls exserted from lvs; bracts among fls and few below, gland-toothed; pedicels ± = fls, slender. **FL**: sepals 2, ± 1/4 × corolla, roundish, gland-toothed; petals 7–10, 6–11 mm, ovate, rose, pink, or white with pink lines, tip notched; stamens 5–6; stigmas 2–3. *n*=15. Granite slopes, cliffs; 2100–2300 m. KR (Trinity Co.); to B.C., ID. CA pls under this name are probably hybrid *L. leana* × *L. cotyledon*.

L. congdonii (Rydb.) J. Howell (p. 907) CONGDON'S LEWISIA Root + caudex short, stout. **LVS** few–many in rosette and few cauline, 8–30 cm, narrowly oblanceolate to obovate, entire, tapered to long, slender petiole; tip acute or obtuse. **INFL**: sts 1–several, 2–6 dm, each with a loose, several–many-fld panicle; fls well exserted from lvs; bracts among fls and below, awl-like, gland-toothed; pedicels ascending, < fls, slender. **FL**: sepals 2, 1/4–1/3 × corolla, roundish, margin gland-toothed; petals 6–7, ± 8 mm, lanceolate, rose, tip jagged; stamens 5; stigmas 3. *n*=±12. **RARE** CA. Granite outcrops; 500–2800 m. c SN (Fresno, Mariposa cos.?). Reported from foothill woodland by Munz.

L. cotyledon (S. Watson) Robinson CLIFF MAIDS Root + caudex thick at top, ± slender, branched below. **LVS** many, in rosette, 3–9 cm, ovate or spoon-shaped, fleshy, entire or not, tapered to base; tip round. **INFL**: sts gen 1–6, 10–30 cm, each with a ± flat-topped, 12–many-fld panicle; fls exserted from lvs; bracts among fls and 2–4 below, gland-toothed; pedicels gen < fls. **FL**: sepals 2, ± 1/3 × corolla, round or truncate, margin gland-toothed; petals gen 7–13, ± 15 mm, oblanceolate or obovate, white or pink with darker veins, tip ± notched; stamens 6–9; stigmas 3–4. *n*=14. Rocky, sandy slopes, subalpine forests; 150–2300 m. KR, CaRH; OR. Following vars. not clearly distinct, overlap in range. Hybrids with *L. leana* reported (see *L. columbiana*).

var. ***cotyledon*** (p. 907) **LF**: margin entire. Rocky, sandy areas, alpine slopes, subalpine forests; 1300–2300 m. KR, CaRH?; to OR, ID.

var. ***heckneri*** (C. Morton) Munz (p. 907) HECKNER'S LEWISIA **LF**: margin dentate. *n*=14. RARE. Rocky cliffs, slopes; 300–2100 m. KR (n Trinity, e Humboldt cos.).

var. ***howellii*** (S. Watson) Jepson (p. 907) HOWELL'S LEWISIA **LF**: margin wavy. *n*=14. UNCOMMON. Rock outcrops, canyon walls near oak woodland; 150–900 m. CaRH (Shasta Co.), KR; OR?.

L. disepala Rydb. (p. 907) YOSEMITE LEWISIA Caudex short, roots much branched, some fibrous. **LVS** many, in rosette, 0.5–1.5 cm, linear or club-shaped, fleshy, entire, tapered to base; tip blunt. **INFL**: sts few, < 1 cm, 1-fld, disjointing near middle, leaving ring of 2–4 scarious, ovate bracts; fls exserted from lvs. **FL**: sepals 2, 0.7 × corolla, petal-like, scarious in age, ± round, entire; petals 5–8, 1–1.5 cm, obovate, pink, tip round; stamens 15; stigmas 2. RARE. Granitic sand; 1900–3500 m. c&s SNH. [*L. rediviva* var. *yosemitana* M.K. Brandegee; not *L. yosemitana* Jepson]

L. glandulosa (Rydb.) Dempster (p. 907) Root + caudex elongate, gen widest at top, tapered or branched below. **LVS** many, in dense rosette, gen 2–10 cm, thread-like to narrowly lanceolate, entire, gen persistent after withering, tapered to fleshy, expanded base; tip obtuse. **INFL**: sts many, 1–3.5 cm, 1- or often 2-fld; fls not or barely exserted from lvs; bracts in 1–2 irregular pairs, above middle of st, ± ovate, gland-toothed. **FL**: sepals 2, ± 1/2 × corolla, roundish or truncate, margin reddish gland-toothed; petals 6–8, ± 8 mm, obovate, white, pink, or reddish, tip irregular, gen acuminate; stamens 6; stigmas 4. Granite sand, rock cracks, wet meadows; 3000–4000 m. c&s SNH, SNE. [*L. pygmaea* ssp. *g.* (Rydb.) Ferris]

L. kelloggii M.K. Brandegee (p. 907) Root + caudex short, thick, often branched below. **LVS** many, in dense rosette, 1.5–9 cm, spoon-shaped or obovate, leathery, entire, abruptly narrowed to long petiole; tip obtuse. **INFL**: sts several, 1–4 cm, 1-fld; fls often exserted from lvs; bracts 2, closely below and like sepals. **FL**: sepals 2 (but seemingly 4 due to bracts), ± 1/2 × corolla, oblong-lanceolate, minutely gland-toothed; petals 5–9, 15–30 mm, obovate, white or pink, tip blunt, ± notched; stamens 8–20; stigmas 3–6. Decomposed granite, slate; 1900–2900 m. n&c SNH; ID. Pls in n part of CA range (e.g., Sierra Co.) are larger, with fls more deeply colored than farther s or in ID.

L. leana (Porter) Robinson (p. 907) QUILL-LEAVED LEWISIA Root slender, woody; caudices 1–many. **LVS** many, in dense rosette, 1.5–4 cm, ± linear, flat or cylindric, entire, ± fleshy, tapered to narrow base; tip blunt. **INFL**: sts 1–several, 8–24 cm, each with a loose, rounded, many-fld panicle; fls well exserted from lvs; bracts among fls and below, lanceolate, entire or increasingly gland-toothed above; pedicels ± = fls. **FL**: sepals 2, ± 1/3 × corolla, ovate or roundish, ± gland-toothed; petals 6–8, 5–6.5 mm, obovate, white, pink, or purple, tip gen round; stamens 4–5; stigmas 2. *n*=14. Sandy, rocky places, pine forest; 1300–3300 m. KR, NCoRH, SNH (Fresno Co.).

L. longipetala (Piper) Clay LONG-PETALED LEWISIA Root + caudex ± slender, often branched below. **LVS** many, in loose rosette, 3–6 cm, linear to narrowly oblanceolate, thin, entire, tapered to base; tip blunt. **INFL**: sts 3–8, ± = lvs, each 1–2-fld; fls not or slightly exserted from lvs; bracts in 1–2 pairs, ± near middle of st, ± ovate, gland-toothed; pedicels ± > fls. **FL**: sepals 2, < 1/3 × corolla, fan-shaped, margin red-gland-toothed; petals ± 8, 15–18 mm, oblanceolate, rose, tipped with a gland; stamens 5–8; stigmas ± 5. RARE. *n*=11. Crevices in granitic rocks; 2500–2800 m. n SNH (Nevada, Eldorado cos.) [*L. pygmaea* ssp. *l.* (Piper) Ferris]

L. nevadensis (A. Gray) Robinson (p. 907) Root + caudex short, fusiform or corm-like. **LVS** few–many, in ± loose rosette, 3–13 cm, thread-like to narrowly lanceolate, entire, tapered to base; tip obtuse. **INFL**: sts several–many, 2.5–9 cm, each gen 1-fld; fls incl in to barely exserted from lvs; bracts 2, below middle of st, 6–18 mm, erect, lanceolate, margins entire, translucent. **FL**: sepals 2, 1/2–2/3 × corolla, widely ovate, entire to ± jagged, tips acute, spreading; petals 6–10, 9–15 mm, ± obovate, white or faintly pink, tips blunt or ± pointed; stamens 6–15; stigmas 3–6. *n*=±28. Grassy meadows, moist gravel flats, open forest; 1300–3000 m. KR, NCoRH, CaRH, SNH, TR; to WA. Intergrading with *L. pygmaea* to Rocky Mtns.

L. oppositifolia (S. Watson) Robinson (p. 907) OPPOSITE-LEAVED LEWISIA Root + caudex small, gen ± fusiform. **LVS** few–many, in loose rosette, sometimes few, reduced on lower part of st, 5–11 cm, narrowly linear-oblanceolate, entire, tapered to slender base; tip blunt. **INFL**: sts 1–3, 6–20 cm, each with a 2–5-fld, ± umbel-like cluster; fls rarely incl in lvs; bracts among fls and below, lanceolate, ± toothed, not glandular; pedicels > fls. **FL**: sepals 2, ± 1/3 × corolla, round or truncate, jagged-dentate, not glandular; petals 8–11, 12–17 mm, ovate, white or pink, tips blunt or ± jagged; stamens ± 12–21; stigmas 3–5. RARE. Moist places in open pine forest; 300–1200 m. KR (n Del Norte Co.); sw OR. CA pls low, perhaps hybrid with *L. nevadensis*. Threatened by logging, mining.

L. pygmaea (A. Gray) Robinson Root + caudex gen largest at top, tapered below or short-fusiform or corm-like. **LVS** gen several, in rosette, 2–8 cm, thread-like to linear-lanceolate, fleshy, entire, tapered to expanded base; tip blunt. **INFL**: sts several–many, 1–5 cm, each gen 1-fld (less often several-fld); fls gen incl in lvs; bracts 2, at or below middle of st, ± widely lanceolate, entire or dentate. **FL**: sepals 2, ± 1/2 × corolla, ± ovate, margin ± jagged or toothed, rarely glandular, tip pointed, rounded, or truncate; petals 6–9, 5–10 mm, obovate, white, pink, or red, often striped, tip ± jagged; stamens ± 8; stigmas ± 4. *n*=±33. Rocky slopes, wet granite sand or gravel, moist meadows, along streams; 1700–4500 m. KR, NCoRH, CaRH, SNH, WTR (Mt. Pinos), SnBr, Wrn; to B.C., Rocky Mtns. [*L. sierrae* Ferris] Ill-defined, probably of hybrid origin; intergrades and apparently hybridizes with *L. nevadensis*, *L. glandulosa*, sometimes *L. triphylla*.

L. rediviva Pursh (p. 907) BITTER ROOT Root + caudex short, thick, ± expanded above; roots radiating, fleshy. **LVS** many, in rosette, 0.5–5 cm, linear, thick, entire, barely tapered at base; tip blunt. **INFL**: sts several–many, 2–6 cm, each 1-fld, disjointing near middle, leaving ring of 5–many scarious, awl-like bracts; fls exserted from lvs. **FL**: sepals 6–8, ± 3/4 × corolla, petal-like, scarious in age, widely obovate, entire; petals 10–19, 12–25 mm, obovate-oblong, white or pink, tip obtuse-notched; stamens 40–47; stigmas 6–8. *n*=14. Rocky, sandy ground, talus, serpentine, clay, granite, shale, open woodlands and sagebrush shrublands with pine, oak, or juniper; 60–3000 m. KR, NCoRH, NCoRI, SN, SnFrB, SCoRI, TR, SnJt, MP, SNE; to Rocky Mtns. Roots once food for native Americans, bitter if taken after fl. Smaller pls of TR, PR, SNE have been called var. *minor* (Rydb.) Munz.

L. stebbinsii R. Gankin & W.R. Hildreth STEBBINS' LEWISIA Root + caudex slender. **LVS** 5–15, in rosette, < 9 cm, oblanceolate to obovate, fleshy, entire or ± wavy, tapered to long petiole; tip obtuse. **INFL**: sts 1–9, ± prostrate, 7–14 cm, each with a 3–11-fld, ± umbel-like cluster; fls gen exserted from lvs; bracts among fls and several below, often paired, gland-toothed; pedicels > fls, slender. **FL**: sepals 2, 1/3–1/2 × corolla, widely ovate, margins red-gland-toothed; petals 7–10, 8–15 mm, oblanceolate-obovate, white at base, pink above, tips blunt, ± jagged; stamens 10–12; stigmas 3–4. RARE. Open, gravelly places; 1600–1950 m. NCoRH (se Trinity, ne Mendocino cos.).

L. triphylla (S. Watson) Robinson (p. 907) Roots fibrous from small, deep, spheric corm. **LVS** 2–5, cauline, 1–6 cm, linear or thread-like, entire, tapered to base; tip obtuse. **INFL**: sts 1–several, gen 2–7 cm aboveground, each with a loose, gen 1–10-fld cluster; fls incl in or exserted from lvs; bracts among fls, ± lanceolate, entire. **FL**: sepals 2, ± 1/2 × corolla, ovate, entire; petals 5–9, 4–7 mm, oblong-oblanceolate, white or pink, tips rounded; stamens 3–5; stigmas 3–5. Moist sandy or gravelly slopes, grassy meadows, open conifer forest; 1300–3400 m. KR, NCoRH, CaRH, SNH, Wrn. Gen distinct from closely related *L. pygmaea*; some intermediates may be found.

MONTIA

Ann or per, glabrous, ± fleshy, sometimes ± aquatic, sometimes matted. **LVS**: cauline > 2, alternate or opposite, entire. **INFL**: raceme, 1-sided; lowest fl gen bracted; pedicels recurved, erect in fr. **FL**: petals (3–)5, equal or 2 larger; stamens (3–)5, filaments fused to corolla-base; ovary chamber 1, placentas basal, style 1, stigmas 3. **FR**: capsule; valves 3, margins rolling inward and forcibly expelling seeds. **SEEDS** 1–3, gen black, smooth to tubercled, fleshy-appendaged or not. 12 spp.: Am, Siberia, Australia. (Giuseppe Monti, Italian botanist, 1682–1760) [McNeill 1975 Can J Bot 53:789–809] Sometimes divided into 9 genera.

1. Lvs opposite, not much reduced upwards
 2. Per with stolon-like sts bearing pink bulblets; petals 5–9 mm, >> sepals .... *M. chamissoi*
 2′ Ann; sts often matted; petals 1–2 mm, ± = sepals .... *M. fontana*
1′ Lvs all alternate, reduced upwards or not
 3. Per; basal lvs on short, caudex-like st; petals 7–15 mm — fl-st with axillary, easily detached, fleshy bulblets .... *M. parvifolia*
 3′ Ann; lvs cauline, gradually reduced upwards; petals < 6.5 mm
  4. Lvs petioled; blades wide, base truncate or wedge-shaped; seeds with projecting fleshy appendage 0.2–0.4 mm .... *M. diffusa*
  4′ Lvs ± linear; seed appendage 0 or minute, not projecting
   5. Sepals 3–7 mm; st 5–25 cm; lower st lvs 2–10 cm; seeds 1.4–2.5 mm .... *M. linearis*
   5′ Sepals < 2.5 mm; st 1–9 cm; lower st lvs 0.5–4.5 cm; seeds 0.7–1.2 mm
    6. Infls both terminal and from axils of st lvs; pls tufted, not matted or rooting at nodes .... *M. dichotoma*
    6′ Infls lateral, each from membranous bract opposite a st lf; pls moss-like, matted, rooting at nodes . *M. howellii*

M. chamissoi (Sprengel) E. Greene (p. 907) TOAD LILY Per 2–30 cm, from leafy stolons, erect to prostrate or floating, often tufted or matted; rhizomes scaly, with pink, overwintering bulblets. **LVS** opposite, 5–50 mm, oblanceolate. **INFLS** gen few, some axillary; fls 2–8. **FL**: sepals 1.5–2 mm, spheric; petals 5–9 mm, equal, pink or white; stamens 5. **FR** 1–1.5 mm. **SEED** 1–1.3 mm; tubercles low, rounded; appendage flattened. $2n=22$. Wet, sandy or loamy soil, seeps, wet meadows; 1100–3700 m. KR, NCoRH, CaR, SNH, TR, PR, GB; to AK, WY, NM, also c Can, n-c US.

M. dichotoma (Nutt.) Howell (p. 907) Ann 2–9 cm, erect, often appearing tufted. **LVS** alternate, 5–40 mm, linear. **INFLS** few–many, terminal; fls 4–16. **FL** often cleistogamous; sepals 1.5–2.5 mm, widely obovate, obtuse to truncate; petals = or slightly > sepals, ± unequal, white; stamens 3. **FR** 1.4–1.8 mm. **SEED** 0.8–1.2 mm, ± smooth; appendage 0. $2n=14$. Uncommon. Moist depressions, grassland, open woodland; 1000–1600 m. KR, CaR, MP; to B.C., MT. Self-pollinating.

M. diffusa (Nutt.) E. Greene (p. 907) Ann 5–20 cm, erect, gen diffusely branched. **LVS** basal and cauline, alternate; blade 8–50 mm, < or = petiole, lanceolate to deltate, tip acute to obtuse, base wedge-shaped to cordate. **INFLS** gen many, terminal; fls 3–10. **FL**: sepals 1.8–3.5 mm, widely obovate, obtuse to truncate; petals 3–4.5 mm, equal, pink or white; stamens 5. **FR** 2.3–3 mm. **SEED** 1.3–1.5 mm, gen ± smooth (or minutely netted); appendage protruding. $2n=16$. Uncommon. Often disturbed or burned coniferous forest or mixed woodland; < 1000 m. NW, n SN, SnFrB; to B.C.

M. fontana L. (p. 907) WATER CHICKWEED, BLINKS Ann 1–30 cm, prostrate to erect, sometimes floating, often tufted or matted, rooting from lower nodes. **LVS** opposite, ± sessile, 3–20 mm, linear to widely oblanceolate; tip acute to obtuse; base tapered. **INFLS** many, some axillary; fls 1–8, lowest 1–2-bracted. **FLS** often cleistogamous; sepals 1–2 mm, round, truncate; petals 1–2 mm, ± unequal, white; stamens 3. **FR** 1–2 mm. **SEED** 0.5–1.2 mm, gen ± rough with acute tubercles; appendage round or flat. $2n=18,20$. Common. Ponds, streams, vernal pools, seeps, ditches; < 3200 m. CA (exc D); to AK, e N.Am, ± worldwide. Highly variable. [*M. hallii* (A. Gray) E. Greene; *M. verna* Necker] Gen self-pollinated. Pls with seed densely acute-tubercled have been called ssp. *chondrosperma* (Fenzl) Walters; pls with seed smooth on sides, tubercled on keel have been called ssp. *amporitana* Sennen; pls from s SNH, TR, SNE, with seed smooth have been called ssp. *variabilis* Walters [*M. funstonii* Rydb., Funston's montia].

M. howellii S. Watson (p. 907) Ann 1–9 cm, matted, rooting from lower nodes. **LVS** alternate 5–25 mm, linear or ± oblanceolate. **INFLS** few–many, lateral, sessile; base bracted opposite lf; fls 2–6. **FLS** gen cleistogamous; sepals 0.8–1.5 mm, widely ovate, obtuse; petals 0–0.7 mm, ± unequal, white; stamens 3. **FR** 0.9–1.2 mm. **SEED**: 0.7–0.9 mm, smooth; appendage minute, not protruding. Uncommon. Vernally wet sites, often on compacted soil; < 400 m. NW; to B.C. Sometimes mistaken for *M. fontana* or *M. dichotoma*. Self-pollinated.

M. linearis (Hook.) E. Greene (p. 907) Ann 4–25 cm, erect, often much-branched. **LVS** alternate, 10–100 mm, linear. **INFLS** few–many, terminal; fls 3–14. **FL**: sepals 3–7 mm, widely ovate, obtuse to truncate; petals 4–6.5 mm, ± unequal, white or pale pink; stamens 3. **FR** 3–4 mm. **SEED** 1.4–2.5 mm, gen ± smooth (or minutely netted); appendage 0. $2n=24,28$. Moist grassland, shrubland, open woodland, fields; < 2300 m. KR, NCoR, CaR, n&c SN, SnFrB, TR, PR, MP; to w Can, MT.

M. parvifolia (DC.) E. Greene (p. 907) Per 5–40 cm; caudex branched, matted, creeping, becoming erect and densely leafy at tips. **LVS** alternate; basal 15–60 mm, very fleshy, narrowly oblanceolate to widely ovate, often sessile; lvs of stolons smaller, much reduced upward, axils gen with easily detached bulblets. **INFLS** terminal; 1+ fls bracted; fls 2–12. **FL**: sepals 2–3.5 mm, obovate, obtuse; petals 7–15 mm, equal, pink or white; stamens 5. **FR** 2–3 mm. **SEED** 1–1.5 mm, gen ± smooth (or minutely netted); appendage 0-minute. $2n=22,44$. Moist, rocky slopes, cliffs, creekbanks, forests; < 2600 m. NW, CaR, SN, CCo, SnFrB; to AK, MT. Highly variable; reproduces by bulblets as well as seeds. ❀IRR,DRN:1, 2,**4–7**,14,**15–17**.

PORTULACA

Walter A. Kelley

Ann, ± fleshy. **STS** several–many, spreading to erect. **LVS** simple, alternate or opposite, upper 2–5 forming involucre; blade linear, ovate, or spoon-shaped, flat or cylindric. **INFL**: fls solitary or in clusters at st tips. **FL**: sepals 2, fused at base, lower portion fused to ovary and persistent in fr; petals 4–6, inserted on calyx, yellow; stamens 4–20; ovary ± inferior, 1-chambered; style 3–6-branched. **FR**: capsule, circumsissle. **SEEDS** many, reniform, ± tuberculate. 100 spp.: warm regions worldwide. (Probably Latin: small gate or door, from capsule lid) [Matthews & Levins 1985 Sida 11(1):45–61]

1. Lf axils with many long (to 8 mm) hairs; involucre hairy ..... ***P. halimoides***
1′ Lf axils glabrous, scarious, or with few hairs; involucre glabrous ..... ***P. oleracea***

P. halimoides L. (p. 907) **ST** spreading to ascending, 1–6 cm. **LF** 3–15 mm, ± linear, ± cylindric. **INFL**: fls in clusters of 2–10 at st tips. **FL**: sepals 1–2.5 mm, gen reddish, not keeled; petals 1–2.5 mm, yellow, drying reddish; stamens 4–15; style 3–5-branched. **FR** 1–3 mm wide. **SEED** 0.5–0.8 mm wide, black or metallic silver. Uncommon. Sandy washes, flats; 1000–1200 m. DMtns (Little San Bernardino Mtns, New York Mtns); to TX, n Mex. [*P. parvula* A. Gray] CA pls reported as *P. mundula* I.M. Johnston (Desert portulaca, UNCOMMON) appear to belong here; additional study needed.

P. oleracea L. (p. 907) COMMON PURSLANE **ST** spreading. **LF** 5–30 mm, ovate to spoon-shaped, gen flat. **INFL**: fls solitary or in clusters of 2–5 at st tips. **FL**: sepals 3–5 mm, green or reddish, gen keeled; petals 3–5 mm, yellow; stamens 5–20; style 4–6-branched. **FR** 3–8 mm wide. **SEED** 0.6–1 mm wide, dark brown to black. $2n=36,54$. Disturbed soil; < 1400 m. CA; native to Eur. [*P. retusa* Engelm.]

PRIMULACEAE PRIMROSE FAMILY

Anita F. Cholewa & Douglass M. Henderson

Ann, per, subshrub, glabrous to glandular-hairy. **LVS** simple, basal or cauline, alternate, opposite, or whorled, sessile or petioled; stipules 0. **INFL** sometimes scapose. **FL** bisexual, radial; parts gen in 4's or 5's; calyx deeply lobed, often persistent; corolla lobes spreading to reflexed; stamens epipetalous, opposite corolla lobes; ovary gen superior, 1-chambered, placenta basal or free-central, style 1, stigma head-like. **FR**: capsule, circumscissile or 2–6-valved. **SEEDS** small, few–many. ± 25 genera, 600 spp.: esp n hemisphere; several orn (*Cyclamen*, *Dodecatheon*, *Primula*). [Channell & Wood 1959 J Arnold Arbor 40:268–288]

1. Ovary partly inferior; infls terminal racemes ..... **SAMOLUS**
1′ Ovary superior; infls gen not terminal racemes
 2. Infl scapose, ending in an umbel
  3. Corolla lobes reflexed, tips acute to obtuse ..... **DODECATHEON**
  3′ Corolla lobes spreading or erect, tips notched or obcordate
   4. Corolla lobes 1–3 mm ..... **ANDROSACE**
   4′ Corolla lobes 5–10 mm ..... **PRIMULA**
 2′ Fls gen in lf axils
  5. Lvs alternate (exc lowest); corolla < calyx ..... **CENTUNCULUS**
  5′ Lvs opposite or whorled; corolla, if present, = or > calyx
   6. Lvs mainly in a single whorl at st tip ..... **TRIENTALIS**
   6′ Lvs distributed along st
    7. Corolla 0, calyx petal-like; fls solitary and ± sessile, in lf axils ..... **GLAUX**
    7′ Corolla present, calyx not petal-like; fls solitary and pedicelled or in racemes or panicles, in lf axils
     8. Corolla salmon or blue; fr circumscissile; ann ..... **ANAGALLIS**
     8′ Corolla yellow; fr with valves; per ..... **LYSIMACHIA**

ANAGALLIS PIMPERNEL

Ann, per, low, spreading. **ST** diffusely branched. **LVS** cauline, opposite or whorled, ± sessile, gen entire. **INFL**: fls solitary in axils, pedicelled. **FL**: parts in 5's; calyx = or < corolla, persistent; filaments free or fused at base, hairy, anthers oblong; ovary superior, spheric, style thread-like. **FR** circumscissile, spheric. 20 spp.: gen Eurasia. (Greek: pimpernel)

A. arvensis L. (p. 911) POOR-MAN'S WEATHERGLASS, SCARLET PIMPERNEL Ann, glabrous. **ST** 5–40 cm. **LF** sessile; blade 5–20 mm, ovate to elliptic. **FL**: pedicel 1–3 cm, recurved in fr; calyx lobes 3–5 mm, lanceolate; corolla 7–11 mm wide, salmon to blue. $2n=40$. Common. Disturbed places, ocean beaches; gen < 1000 m. CA-FP; to e N.Am; native to Eur. TOXIC to livestock, humans. *A. linifolia* L. (per; corolla > 15 mm wide) reported from Siskiyou, Monterey cos.; apparently a waif.

ANDROSACE ROCK-JASMINE, FAIRY-CANDELABRA

Ann, per, gen < 12 cm. **LVS** in basal rosette. **INFL**: umbel 1 per scapose peduncle, terminal, subtended by involucral bracts. **FL**: parts in 5's; calyx tube scarious; corolla salverform, tube narrowed at top, lobes obcordate or notched at tip; stamens incl, filaments ± 0 or short, anthers oblong; ovary superior, spheric, style short. **FR** 5-valved, spheric. ± 100 spp.: n temp, arctic, esp Asia. (Greek: uncertain sea-plant) [Robbins 1944 Amer Midl Nat 32:137–163]

1. Calyx < 2(3) mm; lf blade abruptly narrowed to petiole ..... ***A. filiformis***
1′ Calyx gen > 3 mm; lf blade tapered to petiole
 2. Calyx lobes awl-like, tips stiffly acute ..... ***A. elongata*** ssp. ***acuta***
 2′ Calyx lobes widely lanceolate to triangular, tips acute to obtuse
  3. Involucral bracts ± ovate, gen ± 0.7–1.7 mm wide; corolla < calyx; peduncles gen 1–few ..... ***A. occidentalis*** var. ***simplex***
  3′ Involucral bracts ± lanceolate, gen ± < 0.5 mm wide; corolla = or > calyx; peduncles 1 to gen several ..... ***A. septentrionalis*** ssp. ***subumbellata***

A. elongata L. ssp. ***acuta*** (E. Greene) G. Robb. (p. 911) Ann, 2–8 cm, hairy; peduncles gen solitary. **LF** 5–20 mm, linear-lanceolate, tapered to petiole, acute to acuminate, entire or finely dentate, ciliate. **INFL**: involucre bracts 2.5–5 mm, < 1 mm wide, ovate to lanceolate, acute to acuminate; pedicels 0.6–6.4 cm. **FL**: calyx 3.5–5 mm, hairy, lobes gen = tube, awl-like, tips stiffly acute, reddish; corolla < calyx, white. 2*n*=40. Dry grassy slopes; < 1200 m. SnJV, SnFrB, SCoRI, SCo; OR, Baja CA. [*A. occidentalis* Pursh var. *a.* (E. Greene) Jepson]

A. filiformis Retz. (p. 911) SLENDER-STEMMED ANDROSACE Ann, 3–12 cm, glabrous or ± glandular-hairy; peduncles gen several. **LF** 3–20 mm, ovate to ± triangular, abruptly narrowed to petiole, finely dentate. **INFL**: involucre bracts 0.8–1.2 mm, < 0.5 mm wide; pedicels 1–4 cm. **FL**: calyx < 3, gen 2 mm, glabrous to sparsely glandular-hairy, lobes < to ± = tube; corolla > calyx, white. UNCOMMON. Meadows; 1800 m. CaRH (s slope Willow Creek Mtn, n Siskiyou Co.); sporadic to WA, Rocky Mtns; also in Eurasia.

A. occidentalis Pursh var. ***simplex*** (Rydb.) St. John (p. 911) Ann, 3–7 cm, hairy; peduncles gen 1–few. **LF** 5–15 mm, lanceolate-elliptic, tapered to petiole, entire to finely dentate. **INFL**: involucre bracts 2–5 mm, gen ± 0.7–1.7 mm wide, narrowly ovate to ovate; pedicels 0.5–3 cm. **FL**: calyx 3.6–6 mm, hairy, lobes gen = tube, widely lanceolate to triangular, tips obtuse to acute; corolla < calyx, white. 2*n*=20. Gen moist sites; 1700 m. n SNH (Emigrant Gap, Placer Co.); to B.C., c US.

A. septentrionalis L. ssp. ***subumbellata*** G. Robb. (p. 911) Ann or weak per, 1–6 cm, hairy; peduncles 1 to gen several. **LF** 5–20 mm, linear-lanceolate, tapered to petiole, entire to finely dentate. **INFL**: involucre bracts 1.7–3 mm, gen ± < 0.5 mm wide, linear-lanceolate to lanceolate; pedicels 0.5–5 cm. **FL**: calyx (2.5)3–4 mm, glabrous or puberulent at base, tube > lobes, scarious between ridges, lobes widely lanceolate to triangular, gen reddish, tips acute to obtuse; corolla = or > calyx, white. Dry, rocky sites; 2700–3600 m. c&s SNH, SnBr, SNE; to B.C., Rocky Mtns. [ssp. *puberulenta* (Rydb.) G. Robb.]

CENTUNCULUS CHAFFWEED

Ann. **LVS** cauline, alternate or lowest opposite, ± sessile, entire. **INFL**: fls solitary in axils, ± sessile. **FL** tiny; parts in 4's or 5's; calyx > corolla, persistent; corolla tube short, urn-shaped, lobes acute, spreading; filaments short, free, glabrous; ovary superior, style very short. **FR** circumscissile, spheric. ± 3 spp.: temp, trop. (Latin: patchwork)

C. minimus L. (p. 911) **ST** ascending, 3–10 cm. **LF** 2–5 mm, oblanceolate to widely obovate. **FL**: parts gen in 4's; calyx ± 3 mm; corolla < calyx, pink; filaments wider at base. 2*n*=22. Vernal pools, moist places; < 950 m. NCo, n SNF (Calaveras Co.), CCo, SnFrB, SCo; to B.C., e N.Am, Eur, S.Am.

DODECATHEON SHOOTING STAR

Per, glabrous or glandular-hairy; roots fleshy-fibrous. **LVS** basal. **INFL**: umbel, 1 per scapose peduncle, terminal, few–many-fld, subtended by bracts. **FL** nodding; parts in 4's or 5's; sepals reflexed, later erect, persistent; corolla tube short, lobes reflexed; stamens exserted, filaments very short, wide, often fused, anthers erect, ± lanceolate, surrounding style; ovary superior, style slender, ± exserted from anthers. **FR** ± 5-valved or circumscissile, ovate to spheric. ± 14 spp.: gen N.Am. (Greek: 12 gods, presumably the Olympians) [Thompson 1953 Contr Dudley Herb 4:73–154] Polyploid group; spp. sometimes intergrade.

1. Stigma enlarged, >> style in width; tissue at base of anther wrinkled, dark maroon to black; filaments < 1.5 mm, free or joined by a thin membrane but not fused into a tube
  2. Corolla tube gen covering anther bases; anthers 5, tips acute; pl densely glandular-hairy .... ***D. redolens***
  2' Corolla tube not covering anther bases; anthers gen 4, tips truncate to obtuse; pl glabrous to glandular-hairy
    3. Lf linear to linear-oblanceolate, 2–20 cm; st glabrous; infl glabrous to sometimes ± glandular-hairy; fr with valves .... ***D. alpinum***
    3' Lf linear-oblanceolate to oblanceolate, 9–50 cm; st glandular-hairy; infl glandular-hairy; fr circumscissile or with valves, sometimes both on same pl .... ***D. jeffreyi***

1' Stigma not much enlarged, ± = style in width; tissue at base of anther smooth or wrinkled, yellow or dark; filaments 1–4 mm, fused into a tube, sometimes < 1.5 mm and free
  4. Tissue at anther base smooth (often longitudinally wrinkled upon drying); fr circumscissile or with valves
    5. Corolla lobe gen 9–14 mm, magenta to lavender; roots white, without bulblets; fr with valves .... ***D. pulchellum***
    5' Corolla lobe gen 5–9 mm, magenta to white; roots reddish, with many bulblets; fr circumscissile .... [2]***D. subalpinum***
  4' Tissue at anther base transversely wrinkled; fr circumscissile
    6. Lf blade gen tapered gradually to petiole, length gen > 2.5 × width (if blade narrowed abruptly, filament tube < 1.5 mm and yellow to dark, smooth)
      7. Anther 5–9 mm; filament 0.5–1.5 mm, free or fused into a tube; corolla lobe gen 7–20 mm ...... ***D. conjugens***
      7' Anther 3–4 mm; filament 2–3.5 mm, fused into a tube; corolla lobe gen 5–9 mm .... [2]***D. subalpinum***
    6' Lf blade gen narrowed abruptly to petiole, length gen < 2.5 × width; filament tube 1–4 mm, dark, gen wrinkled
      8. Lf blade length gen < 2 × width; filament tube gen < 3 mm wide; bulblets present among roots at flowering .... ***D. hendersonii***
      8' Lf blade length gen > 2 × width; filament tube gen 3–4 mm wide; bulblets 0 .... ***D. clevelandii***
        9. Tissue at base of anther completely yellow; filament tube lacking yellow or white spot below each anther, sometimes entirely yellow .... ssp. ***clevelandii***
        9' Tissue at base of anther maroon to black; filament tube sometimes with yellow or white spot below each anther
          10. Filament tube lacking yellow or white spot below each anther .... ssp. ***insulare***
          10' Filament tube with yellow or white spot below each anther
            11. Anther dark maroon or purple, tip obtuse to notched .... ssp. ***patulum***
            11' Anther gen yellow, tip acute to obtuse .... ssp. ***sanctarum***

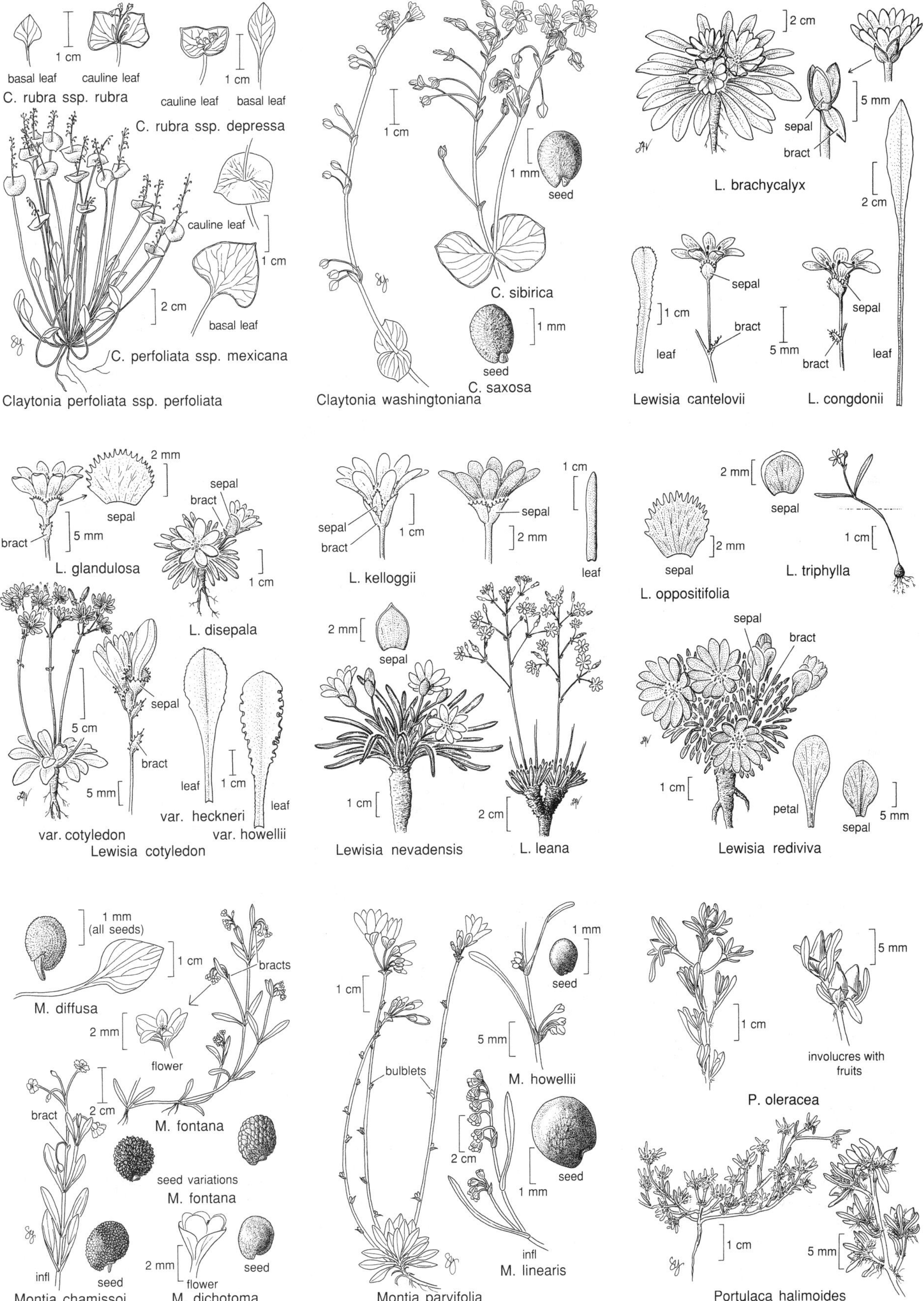
basal leaf
cauline leaf
1 cm
C. rubra ssp. rubra
cauline leaf
basal leaf
1 cm
C. rubra ssp. depressa
cauline leaf
1 cm
2 cm
basal leaf
C. perfoliata ssp. mexicana
Claytonia perfoliata ssp. perfoliata
1 cm
Claytonia washingtoniana
1 mm
seed
C. sibirica
1 mm
seed
C. saxosa
2 cm
5 mm
sepal
bract
L. brachycalyx
2 cm
1 cm
leaf
sepal
bract
5 mm
Lewisia cantelovii
sepal
bract
leaf
L. congdonii
2 mm
sepal
bract
5 mm
L. glandulosa
sepal
bract
1 cm
L. disepala
5 cm
sepal
bract
5 mm
leaf
1 cm
leaf
var. cotyledon
var. heckneri
var. howellii
Lewisia cotyledon
sepal
bract
1 cm
sepal
2 mm
1 cm
leaf
L. kelloggii
2 mm
sepal
1 cm
Lewisia nevadensis
2 cm
L. leana
2 mm
sepal
sepal
2 mm
L. oppositifolia
1 cm
L. triphylla
sepal
bract
1 cm
petal
sepal
5 mm
Lewisia rediviva
1 mm
(all seeds)
1 cm
M. diffusa
bracts
2 mm
flower
2 cm
bract
M. fontana
seed variations
M. fontana
infl
seed
Montia chamissoi
2 mm
flower
seed
M. dichotoma
1 cm
bulblets
Montia parvifolia
1 mm
seed
5 mm
M. howellii
2 cm
seed
1 mm
infl
M. linearis
5 mm
1 cm
involucres with
fruits
P. oleracea
1 cm
5 mm
Portulaca halimoides

D. alpinum (A. Gray) E. Greene (p. 911) ALPINE SHOOTING STAR Pl gen glabrous. **LF** 2–20 cm; blade linear to linear-oblanceolate, narrowed gradually to petiole. **INFL** 1–10-fld. **FL**: parts gen in 4's; corolla lobes 8–16 mm, magenta to lavender; filaments free, 0.5 mm, black, anthers 4.7–8.5 mm, tissue at base wrinkled, black; stigma enlarged. **FR** with valves. *n*=22. Boggy meadows, streambanks; 2400–3400 m. KR, NCoRH, SNH, n WTR (Mt. Pinos), SnBr, SnJt, Wrn, SNE; to OR, UT, AZ. [ssp. *majus* H.J. Thompson] ❀bog and pure water:1–4,6;DFCLT.

D. clevelandii E. Greene Pl glandular-hairy. **LF** 1–18 cm; blade gen oblanceolate, gen narrowed abruptly to petiole, margin dentate to subentire. **INFL** 1–16-fld. **FL**: parts in 5's; corolla lobes 6–25 mm, magenta to white, lobes 6–25 mm; filament tube 2.5–4 mm, anthers 3–5 mm, tissue at base transversely wrinkled, yellow to black; stigma not enlarged. **FR** circumscissile. Grassy slopes, flats; gen < 600 m for all sspp. n&c SNF, c SNH, GV, CW, SW; Baja CA. Extremes (sspp. here) gen segregated by geog but may intergrade.

ssp. ***clevelandii*** (p. 911) **INFL** 1–16-fld. **FL**: filament tube lacking yellow or white spot below each anther, tissue at base of anther yellow. *n*=22. Grassy slopes, flats. SCo, SnBr, SnJt, PR: Baja CA. ❀DRN,DRY:15–17,24&SHD:7–9,14,18–23; DFCLT.

ssp. ***insulare*** H.J. Thompson **INFL** 5–9-fld. **FL**: filament tube lacking yellow or white spot below each anther, tissue at base of anther maroon to black. *n*=22. Grasslands, woodlands. SCoRO, ChI; Guadalupe Island, Mex. ❀DRN,DRY:**15–17,24**&SHD:**7–9, 14,18–23**.

ssp. ***patulum*** (E. Greene) H.J. Thompson **INFL** 1–6-fld. **FL**: filament tube with yellow or white spot below each anther, tissue at base of anthers maroon to black, anthers gen dark purple, tip obtuse to notched. *n*=22,44. Moist places, often on serpentine or in subalkaline sites. n&c SNF, c SNH, GV, SnFrB, SCoRI. ❀DRY:7–9, 14–24.

ssp. ***sanctarum*** (E. Greene) Abrams **INFL** 3–7-fld. **FL**: filament tube with yellow or white spot below each anther, tissue at base of anther maroon to black, anther gen yellow, sometimes dark, tip acute to obtuse. *n*=22,33,44. Woodlands. SnFrB, SCoR, WTR. ❀DRN,DRY:**16,17**,24&SHD:**7**,8,9,**14,15**,18–23;DFCLT.

D. conjugens E. Greene (p. 911) Pl glabrous. **LF** 3–14 cm; blade linear-oblanceolate to obovate, narrowed gradually, sometimes abruptly, to petiole, entire. **INFL** 1–7-fld. **FL**: parts in 5's; corolla lobes 7–20 mm, magenta to white; filaments free or fused into a tube 0.5–1.5 mm, anthers 5–9 mm, tissue near base transversely wrinkled, dark maroon to black; stigma not much enlarged. **FR** circumscissile. *n*= 22. Moist slopes, meadows, often with sagebrush shrublands; 1200–1900 m. MP; to w Can, WY. ❀WET:1–4,6;DFCLT.

D. hendersonii A. Gray (p. 911) MOSQUITO BILLS, SAILOR CAPS Pl glabrous to glandular-hairy; root with rice-like bulblets. **LF** 2–16 cm; blade elliptic to ovate or obovate, gen narrowed abruptly to petiole, entire to ± toothed. **INFL** 3–17-fld. **FL**: parts in 4's or 5's, even on same pl; corolla lobes 6–23 mm, magenta to deep lavender or white; filament tube 1–3 mm, anthers 3–5 mm, tissue near base transversely wrinkled, dark maroon to black; stigma not much enlarged. **FR** circumscissile. *n*=22,33,66. Gen in shady sites; < 1900 m. NW (exc NCo), CaR, SNF, GV, SnFrB, n SCoRI, SnBr; to s B.C., ID. [ssp. *cruciatum* (E. Greene) H.J. Thompson; ssp. *parvifolium* (Knuth) H.J. Thompson, *D. hansenii* (E. Greene) H.J. Thompson] Highly variable; may hybridize with *D. clevelandii*. ❀DRN,DRY:15–17,24&SHD:**7**,8,9,**14,18**,19–23.

D. jeffreyi Van Houtte SIERRA SHOOTING STAR Pl (exc st) glabrous to densely glandular-hairy (st glandular-hairy). **LF** 9–50 cm; blade gen oblanceolate, gen narrowed gradually to petiole, entire to crenate. **INFL** 3–18-fld. **FL**: parts in 4's or 5's; corolla lobes 10–25 mm, magenta to lavender or white; filaments free or partly fused, gen < 1.5 mm, anthers 6.5–11 mm, tissue near base strongly wrinkled to ± smooth, gen dark; stigma enlarged. **FR** circumscissile or with valves. 2*n*=42,43,44, 66,86. Moist to dry meadows, streambanks; 600–3000 m. NW (exc NCo), CaR, SN; to AK, MT. [ssp. *pygmaeum* (H.M. Hall) H.J. Thompson] Highly variable, esp in CaR, OR, WA. Intergrades with *D. alpinum*. ❀IRR:1–4,6,7; DFCLT.

D. pulchellum (Raf.) Merr. (p. 911) Pl glabrous. **LF** 4–25 cm; blade oblanceolate to ovate, gen narrowed ± gradually to petiole, gen entire. **INFL** 2–15-fld. **FL**: parts in 5's; corolla lobes gen 9–14 mm, magenta to lavender; filament tube 1.5–3.5 mm, dark maroon to black, anthers 3–5.5 mm, tissue near base smooth to longitudinally wrinkled (when dry), maroon to black; stigma not much enlarged. **FR** with valves. 2*n*=44,88. Wet meadows; 1200–2200 m. GB; to AK, e US, Mex. [ssp. *monanthum* (E. Greene) H.J. Thompson] The name *D. pauciflorum* E. Greene may have priority. ❀bog and pure water:1–4,6;DFCLT.

D. redolens (H.M. Hall) H.J. Thompson (p. 911) Pl densely glandular-hairy. **LF** 20–40 cm; blade oblanceolate, narrowed gradually to petiole, entire. **INFL** 5–10-fld. **FL**: parts in 5's; corolla tube covering anther bases, lobes 15–25 mm, magenta to lavender; filaments free, gen < 1 mm, anthers 7–11 mm, tissue near base transversely wrinkled, dark maroon to black; stigma enlarged. **FR** with valves. Moist sites; 2400–3600 m. c&s SNH, SnGb, SnBr, SnJt, SNE; to NV, w UT. ❀WET:1–4,6;DFCLT.

D. subalpinum Eastw. Pl glabrous, reddish exc lf blades; root reddish, with bulblets. **LF** 3–10 cm; blade oblanceolate, gradually tapered to petiole, entire. **INFL** 1–5-fld. **FL**: parts in 5's; corolla lobes 5–9 mm, magenta to white; filament tube 2–3.5 mm, anthers 3–4 mm, tissue near base transversely wrinkled to smooth, maroon to black; stigma not much enlarged. **FR** circumscissile. 2*n*=66. Moist places, often shaded; 2100–4000 m. c&s SNH. ❀WET, SHD:1–4,6;DFCLT.

GLAUX SEA-MILKWORT

1 sp.: n temp, arctic. (Greek: bluish green)

G. maritima L. (p. 911) Per, fleshy, erect, tufted. **ST** 5–40 cm. **LVS** cauline, opposite, sessile, 4–20 mm, linear to oblong, entire. **INFL**: fls solitary in axils, ± sessile. **FL** 3–4 mm; parts in 5's; calyx corolla-like, lobes ± 2 × tube, ovate or oblong, white to reddish or lavender; corolla 0; stamens 5, alternate calyx lobes, filaments awl-like, anthers cordate; ovary superior. **FR** 5-valved, < calyx, ovoid. **SEEDS** few, elliptic in outline, ± flat. 2*n*=30. Coastal salt marsh, saline meadows. NCo, KR, deltaic GV, CCo, SCo, e-most MP, SNE; to e N.Am, Eurasia. ❀TRY:STBL.

LYSIMACHIA LOOSESTRIFE

Per. **LVS** cauline, opposite or whorled, sessile or petioled, lanceolate to widely ovate, gland-dotted, entire. **INFL**: raceme, panicle, or fl solitary. **FL**: parts gen in 5's or 6's; corolla > calyx, yellow; filaments free or fused at base; ovary superior. **FR** 5–6-valved, ovate. 100 spp.: gen n temp. (Greek: loose dagger) [Ray 1956 Illinois Biol Monogr 24:1–160]

1. Lf petioled; fls solitary in axils; corolla lobes 7–13 mm; st creeping . . . . ***L. nummularia***
1' Lf sessile; fls in racemes in axils; corolla lobes 3–5 mm; st erect . . . . ***L. thyrsiflora***

L. nummularia L. (p. 911) MONEYWORT **ST** often rooted at nodes, sometimes branched, 20–50 cm, glabrous. **LF** 1.5–2.5 cm, ovate to widely so; base truncate to rounded; petiole 2–4 mm. **FL**: calyx lobes 5–13 mm; corolla lobes cordate-ovate; filaments free or ± fused at base, erect, wider at base, glandular. **FR** < calyx. $2n$=36. Moist meadows; ± 1000 m. n SNH (near Quincy, Plumas Co.); e US; native to Eur.

L. thyrsiflora L. (p. 911) TUFTED LOOSESTRIFE **ST** erect, 20–80 cm, unbranched, glabrous. **LF** 5–12 cm, lanceolate, sessile; base long-tapered. **FL**: calyx lobes 1–4 mm; corolla yellow, often with purple dots, lobes linear, gen with small tooth in each sinus; filaments free, erect, thread-like throughout, glabrous. **FR** ± > calyx, black dotted. $2n$=40,54. Wet places; 800–1300 m. KR, CaR, c SNF (Calaveras Co.), n SNH (Plumas Co.); to AK, e N.Am, Eurasia. ❀TRY.

PRIMULA PRIMROSE

Per, rhizomed, sometimes stoloned. **LVS** basal or crowded on branches near ground, sessile. **INFL**: umbel 1 per scapose peduncle, terminal, subtended by bracts. **FL**: parts in 5's; calyx tube angled; corolla funnel-shaped or salverform, lobes spreading or erect, entire or notched at tip; stamens incl, filaments short, anthers oblong, obtuse; ovary superior. **FR** 5-valved, elliptic to ovate. **SEEDS** many, peltate, dotted. 200 spp.: gen n temp. (Latin: diminutive of first, from early fl) Spp. often heterostylous.

P. suffrutescens A. Gray (p. 911) SIERRA PRIMROSE Subshrub. **ST** creeping, branched; base woody. **LF** 15–35 mm, ± spoon-shaped, glabrous; base tapered; tip gen rounded, dentate. **INFL** glandular-hairy; peduncle 3.5–12 cm; pedicels 2–several, < 1.5 cm. **FL**: calyx 4–8 mm, lobes lanceolate; corolla magenta with yellow throat, tube 7–10 mm, = lobes. **FR** ± = calyx, ovoid. $2n$=44. Gen in rock crevices; gen 2000–4200 m. KR, SNH, SNE. ❀DRN,IRR: 1–4,6;DFCLT.

SAMOLUS WATER-PIMPERNEL, BROOKWEED

Per, erect, glabrous. **LVS** cauline, alternate, petioled or ± sessile, entire. **INFL**: raceme, terminal. **FL** small; parts in 5's; corolla > calyx, white, lobes opposite stamens; stamens 5, alternate 5 stamen-like appendages; ovary partly inferior. **FR** 5-valved, spheric. ± 10 spp.: ± worldwide. (Celtic: from curative properties)

S. parviflorus Raf. (p. 911) **ST** 1.5–4 dm. **LF** 2–5 cm; blade obovate to oblong-lanceolate, gen tapered to winged petiole, tip obtuse to rounded. **INFL**: pedicels ascending to spreading, 1–2 cm, with 1 bractlet. **FL**: calyx 1–2 mm, lobes short, triangular, tooth-like; corolla ± 1.5 mm wide. **FR** ± 2.5 mm wide. $2n$=26. Moist sites; gen < 1300 m. SnJV, CCo, SnFrB, SCoR, SCo, n ChI, WTR; to B.C., e N.Am, Mex, S.Am. *S. ebracteatus* Kunth (lower 4 lvs >> 5 cm; bractlet 0), from Baja CA and Clark Co., NV, expected in extreme se CA. ❀TRY:STBL.

TRIENTALIS STARFLOWER

Per, low, glabrous; roots tuber-like. **ST** erect, simple. **LVS** at base of st scale-like, others well developed, mainly crowded near st tip. **INFL**: fls solitary in axils of few uppermost lvs; pedicels slender. **FL**: parts gen in 5's to 7's; sepals ± free, persistent; corolla spreading, ± flat, = or > calyx, white or pinkish to rose, lobes free ± to base; filaments fused at base, slender, anthers oblong. **FR** 5-valved, spheric. **SEEDS** few. 3 spp.: N.Am, n Eurasia. (Latin: 1/3 foot, from height of pl)

1. Corolla gen white; pedicels > lvs; some cauline lvs scattered below main cluster .... ***T. arctica***
1' Corolla gen pinkish to rose; pedicels gen < lvs; cauline lvs all crowded at st tip .... ***T. latifolia***

T. arctica Hook. (p. 911) ARTIC or NORTHERN STARFLOWER **ST** 5–20 cm. **LF** 15–50 mm, 12–16 mm wide, elliptic to obovate. **FL**: corolla 12–16 mm wide. $2n$=±170. RARE in CA. Bogs, other wet areas; gen < 10 m. NCo (near Crescent City, Del Norte Co.); to AK, w Can., ID.

T. latifolia Hook. (p. 911) **ST** 5–30 cm. **LF** 25–90 mm, 10–50 mm wide, ovate to obovate. **FL**: corolla 8–15 mm wide. Shaded places, esp woodlands; < 1400 m. NCo, NCoR, CaRH (Mount Shasta), SN, SnJV, SnFrB, SCoR; to B.C. ❀TRY.

PUNICACEAE POMEGRANATE FAMILY

Elizabeth McClintock

Shrub, small tree, glabrous, sometimes thorny. **LVS** simple, opposite, entire, deciduous. **INFL**: axillary clusters; fls 1–5. **FL** bisexual, radial; hypanthium bell-shaped to cylindric, ± leathery; sepals 5–8, persistent, appearing as hypanthium lobes; petals 5–8, crumpled in bud (and often in fl); stamens many, inserted along hypanthium tube; ovary inferior, chambers gen 3–8, irregular, style 1, stigma 1, head-like. **FR**: berry, ± spheric, leathery-rinded, divided into irregular sections. **SEEDS** many, embedded in juicy pulp. 1 genus, ± 1 sp.: Medit, ne Afr, to Himalayas. Seeds edible.

PUNICA POMEGRANATE

1 variable sp. (Latin: "Apple of Carthage")

P. granatum L. **ST** < 5 m. **LF**: blade 2–8 cm, oblong or ovate-lanceolate, >> petiole. **FL** 3–5 cm wide; hypanthium ± brownish red; petals coral-red. **FR** gen 8–10 cm wide, bright orange-red. **SEEDS** ruby-red. Uncommon. Disturbed ground; < 500 m. s SnJV, s CCo, n SCo; native se Eur to Himalayas. Long, widely cult in warm areas, often naturalized; fermented seeds source of grenadine; dwarf and double-fld forms cult as orn; persistent calyx on fr inspired King Solomon's crown (hence others).

RAFFLESIACEAE RAFFLESIA FAMILY

George Yatskievych

Per, st or root parasites lacking chlorophyll, gen monoecious or dioecious. **ST** reduced to thread-like tissue underground or inside host st. **LF** scale-like. **INFL**: short spike or fl solitary. **FL** gen unisexual, radial, fleshy; sepals 4–16, often fused at base; petals 0; stamens 5–many, fused to style axis forming a column that is gen expanded at tip into a disk with stigmatic areas or stamens along or under margin; ovary ± inferior, chambers 1–several, placentas parietal. **FR** various, gen fleshy. **SEEDS** many, minute. ± 8 genera ± 50 spp.: worldwide, esp trop. Poorly known taxonomically. *Rafflesia* spp. have world's largest fls (1 m diam); *Pilostyles* fls are < 2 mm diam.

PILOSTYLES

Per, st parasites. **ST** appearing 0. **LVS** reduced to bracts subtending fl. **INFL**: fl solitary. **FL** minute; sepals 4–5. **STAMINATE FL**: anthers many, sessile on column under margin of disk. **PISTILLATE FL**: ovary inferior, chamber 1, stigmas in ring along disk margin. **FR**: fleshy capsule. 20 spp.: trop Am, Afr, Australia, sw Asia (esp Iran). (Latin: hair pillar, from the central column) Only fls and bracts are visible on surface of host st.

P. thurberi A. Gray (p. 915) THURBER'S PILOSTYLES **FL** < 2 mm, brown or maroon; bracts 4–7, 1–1.5 mm, overlapping, round or ovate; sepals like bracts; disk < 1 mm diam; stamens in ring of ± 3 rows. UNCOMMON. Open desert scrub; 0–300 m. s DSon (Riverside, San Diego, Imperial cos.); to NV, AZ, TX, Mex. Parasitic on *Psorothamnus*, esp *P. emoryi*.

RANUNCULACEAE BUTTERCUP FAMILY

Dieter H. Wilken, except as specified

Ann, per, sometimes aquatic. **LVS** gen basal and cauline, gen alternate, simple or compound; petioles at base gen flat, sometimes sheathing or stipule-like. **INFL**: cyme, raceme, panicle, or fls solitary. **FL** gen bisexual, radial; sepals gen 5, free, early deciduous or withering in fr, gen green; petals 0–many, free; stamens gen 10–many; pistils 1–many, ovary superior, chamber 1, style 1, gen ± persistent in fr as beak, ovules 1–many. **FR**: achene, follicle, berry, or utricle-like, 1–many-seeded. ± 60 genera, 1700 spp.: worldwide, esp n temp, trop mtns; many orn (*Adonis, Aquilegia, Clematis, Consolida, Delphinium, Erianthis, Helleborus*), some highly TOXIC (*Aconitum, Actaea, Delphinium, Ranunculus*). [Duncan & Keener 1991 Phytologia 70:24–27]

1. Fl bilateral; sepals not alike, uppermost spurred or hooded, gen > others
  2. Upper sepal hooded, gen enclosing upper 2 petals, stamens, not spurred .... **ACONITUM**
  2' Upper sepal spurred, ± flat or curved but not hooded, gen not enclosing petals, stamens
    3. Ann, gen from taproot; petal 1; pistil 1 .... **CONSOLIDA**
    3' Per, gen from ± fibrous or fleshy roots; petals 4; pistils 3(–5) .... **DELPHINIUM**

1' Fl radial; sepals ± alike
  4. Sepal spurs 1–3 mm; pl gen tufted; lvs basal, simple, thread-like to narrowly oblanceolate .... **MYOSURUS**
  4' Sepal spurs 0 (petal spurs (4)10–42 mm in *Aquilegia*); pl not tufted; lvs gen basal and cauline, gen dissected or compound (or basal, simple, lanceolate to reniform)
    5. Pistil 1; fr berry — lf 1–3-ternate, segments 2–9 cm, lateral lanceolate to ovate .... **ACTAEA**
    5' Pistils 2–many; fr achene, follicle, or utricle-like
      6. Ovary with 2+ ovules; fr a follicle
        7. Petals spurred .... **AQUILEGIA**
        7' Petals 0 or not spurred
          8. Lf simple .... **CALTHA**
          8' Lf 1–3-ternate
            9. Pl from slender, yellow rhizomes or stolons, scapose; sepals 5–8, petal-like, linear, petals 5–7, blade linear .... **COPTIS**
            9' Pl from slender to fusiform, clustered, fleshy roots, not scapose; sepals gen 5, petal-like, oblong to obovate, petals gen 0 .... **ISOPYRUM**
      6' Ovule 1; fr an achene (utricle-like in *Kumlienia*)
        10. Cauline lvs gen opposite or in 1–2 whorls
          11. Per; basal lvs rosetted, cauline gen 2–3, in 1–2 whorls; infl terminal; sepals gen 5–8 .... **ANEMONE**
          11' Gen woody vine; basal lvs 0, cauline gen many, gen opposite; infl axillary; sepals gen 4 .... **CLEMATIS**
        10' Cauline lvs alternate or 0
          12. Perianth parts in 2 whorls, sepals gen green to yellowish, petals 3–many, white to yellow or orange
            13. Petals yellow to orange, bases bluish to purplish; anthers blue to purplish; lf gen 2–3-pinnately dissected .... **ADONIS**
            13' Petals gen white to yellow throughout; anthers yellow; lf simple to 1-palmately dissected, 1-ternate, or 1-pinnate .... **RANUNCULUS**
          12' Perianth parts in ± 1 whorl (inner << outer, gland-like in *Kumlienia*), sepals gen green to white or purplish, petal-like or not, petals 0 or 5–12, yellow or green

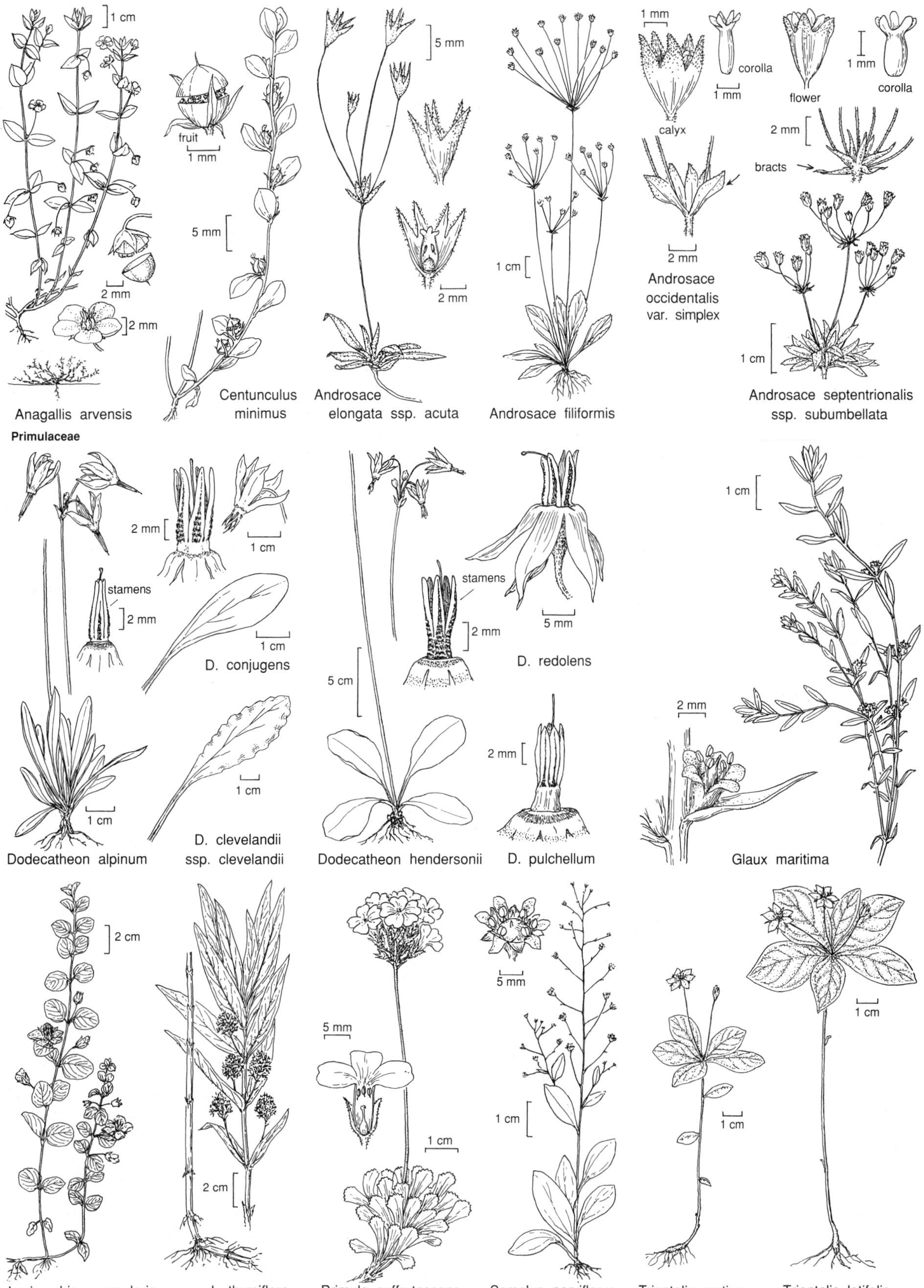
1 cm
2 mm
2 mm
Anagallis arvensis
fruit
1 mm
5 mm
Centunculus minimus
5 mm
2 mm
Androsace elongata ssp. acuta
1 cm
Androsace filiformis
1 mm
corolla
1 mm
calyx
2 mm
Androsace occidentalis var. simplex
flower
1 mm
corolla
2 mm
bracts
1 cm
Androsace septentrionalis ssp. subumbellata
Primulaceae
2 mm
1 cm
stamens
2 mm
1 cm
D. conjugens
1 cm
1 cm
Dodecatheon alpinum
D. clevelandii ssp. clevelandii
stamens
2 mm
5 cm
Dodecatheon hendersonii
5 mm
D. redolens
2 mm
D. pulchellum
1 cm
2 mm
Glaux maritima
2 cm
Lysimachia nummularia
2 cm
L. thyrsiflora
5 mm
1 cm
Primula suffrutescens
5 mm
1 cm
Samolus parviflorus
1 cm
Trientalis arctica
1 cm
Trientalis latifolia

14. Lf gen 1–4-ternate, segments wedge- or fan-shaped or ± round; fl unisexual or bisexual . . . . **THALICTRUM**
14′ Lf simple, crenate to deeply palmately lobed; fl bisexual
 15. Lf crenate to palmately lobed, lobes rounded to obtuse, entire to toothed or lobed; infl scapose, fls 1(–3), terminal (or lateral); sepals 6–10 mm, flat — petals 2–3 mm, gland-like . . . . **KUMLIENIA**
 15′ Lf deeply palmately lobed, lobes ± wedge-shaped, toothed above middle; infl not scapose, fls 5+ in terminal panicle; sepals 2.5–5 mm, blade cup-like . . . . **TRAUTVETTERIA**

ACONITUM MONKSHOOD

Per from rhizome or tuber; roots fibrous or fleshy. **STS** 1–few, gen erect, gen simple. **LF** gen palmately lobed; deep lobes 3–7, toothed to lobed; cauline gradually reduced upward. **INFL**: gen raceme, terminal, bracted; pedicels spreading to upcurved. **FL**: bilateral; sepals 5, petal-like, upper 1 > others, hooded, gen enclosing upper 2 petals and stamens, tip gen rounded to beaked; petals 2–5, upper 2 clawed, blades gen inflated, tip spurred, lower 3 << sepals, scale-like, or 0; pistils gen 3. **FR**: follicles. **SEED** angled or winged, dark brown to black. ± 100 spp.: temp N.Am, Eurasia. (Greek: ancient name) [Brink 1980 Amer J Bot 67:263–273] Most spp. highly TOXIC, causing death in livestock, humans.

A. columbianum Nutt. (p. 915) Pl 3–15(20) dm. **ST** erect, less often reclining or twining above; upper axils (incl infl) with deciduous bulblets or not. **LF** 3–17 cm, 5–14 cm wide; deep lobes 3–5, wedge- to diamond-shaped, toothed to irregularly cut or lobed above middle. **INFL** 5–55 cm, open. **FL**: sepals deep bluish purple or white to yellow-green, upper 10–15(20) mm, beak 3–8 mm, lateral 8–18 mm, round to reniform, lower 7–12 mm, lanceolate to ovate; upper petals blue to whitish, spur < blade, lower petals 0. **FR** glabrous to puberulent, glandular or not. n=8. Streambanks, moist areas, meadows, coniferous forest; 600–2900 m. NW (exc NCo), CaR, SNH, MP; to B.C., SD, NM. [*A. geranioides* E. Greene, *A. leibergii* E. Greene] Pls with bulblets (KR, n SNH) have been called var. *howellii* (Nelson & J.F. Macbr.) C. Hitchc. [*A. hanseni* E. Greene, *A. viviparum* E. Greene] ❀WET,DRN:1–3,**4–6**,7, 14–18;DFCLT.

ACTAEA BANEBERRY

Per from stout, branched caudex. **STS** 1–few, ascending to erect, branched to not. **LVS** 1–4(5), gen 1–3(4)-ternate, lowest scale-like, fibrous, sheathing. **INFL**: raceme, axillary or terminal. **FL** radial; sepals 3–5, petal-like, early deciduous; petals 4–10, rarely 0, oblanceolate to spoon-shaped; pistil 1, placentas 2, ovules several. **FR**: berry. 5–6 spp.: temp N.Am, Eurasia. (Greek: ancient name, from wet habitat and similarity to *Sambucus* lvs) Fr TOXIC to humans.

A. rubra (Aiton) Willd. (p. 915) BANEBERRY Pl (2)4–8 dm. **ST** few-branched above, sparsely puberulent. **LVS** 10–70 dm; lower 2–3-ternate, upper 1–2-ternate; segments 2–9 cm, toothed to irregularly cut, lateral lanceolate to ovate, terminal widely ovate to ± round. **INFL**: pedicels spreading to ascending, 5–8 mm, in fr 6–15 mm. **FL**: sepals 2–3 mm, whitish or purplish green; petals ± = sepals, white; stamens 4–7 mm. **FR** 5–10 mm, red or white, shiny. n= 8. Deep soils, moist, open to shaded sites, mixed-evergreen or coniferous forest; < 2800 m. NW, CaR, SN (exc Teh), SnFrB, SCoRO, SnBr; to AK, ne US. [ssp. *arguta* (Nutt.) Hultén] ❀SHD:**4–6**&IRR:1–3,**7**,14,15,**16**,**17**,18.

ADONIS

Ann, per, gen from taproot. **STS** 1–few, erect, branched to not. **LVS** 2–many-pinnately dissected, cauline; segments gen linear. **INFL**: raceme or fls solitary, terminal or axillary. **FL** radial; sepals 5, gen ± greenish; petals 5–20, white to yellow or orange; pistils many. **FR**: achene, rough to wrinkled or ridged, beaked. ± 20 spp.: temp Eurasia; some cult as orn. (Greek: Adonis of mythology, from whose blood the pl allegedly grew)

A. aestivalis L. Ann 3–7 dm, glabrous. **LVS** gen 2–3-pinnately dissected, 2–7 cm. **FL**: receptacle in fr 1.5–3 cm; sepals 5–8 mm, ovate to oblong; petals (6)8–10, 8–15 mm, oblong to oblanceolate, yellow to orange, bases bluish to purplish; anthers blue to purplish. **FR** 4–5 mm, 3.5–5 mm wide, with minute, lateral ridge below middle; back keeled; keel base short-toothed; beak 1–1.5 mm, straight. Disturbed sites, fields, open pine forest; 1200–1400 m. MP; to WA, ID, also e US; native to Eur.

ANEMONE ANEMONE

Per from stout, simple to branched caudex, rhizome, or tuber. **STS** 1–several, erect, gen simple. **LVS** simple to 1-ternate, blade or lflet toothed to dissected; basal lvs rosetted, petioled, in fl or fr withered or persistent; cauline lvs gen 2–3, in 1–2 whorls, petiole 0 to short. **INFL** terminal; peduncles 1–5, erect, 1-fld, in fr elongated. **FL** radial; receptacle in fr elongated; sepals 5–8(10), petal-like; petals 0; pistils many, styles in fr gen persistent, gen glabrous to puberulent. **FR**: achenes, densely clustered. ± 100 spp.: temp worldwide. (Greek: fl shaken by wind) Some spp. cult for orn. Pls with long, plumose styles sometimes separated as *Pulsatilla*.

1. Sepals 18–30 mm; styles in fr 20–30(35) mm, plumose . . . . *A. occidentalis*
1′ Sepals 5–25 mm; styles in fr < 5 mm, glabrous to puberulent
 2. Basal lf gen 1; cauline lvs simple or 1-ternate, blade or lflet crenate to lobed; pl from rhizome; fr glabrous to puberulent
  3. Cauline lvs simple, ± sessile; sepals 15–25 mm . . . . *A. deltoidea*
  3′ Cauline lvs gen 1-ternate, petioles 8–40 mm; sepals 5–20 mm . . . . *A. oregana*
 2′ Basal lvs few–many; cauline lvs dissected; pl from tuber atop slender caudex or from branched caudex; fr densely woolly or silky

4. Pl from tuber atop slender caudex; fr cluster oblong to elliptic, (15)20–30 mm; e DMtns .... ***A. tuberosa***
4′ Pl from branched caudex; fr cluster gen ovoid to spheric, 10–20 mm; KR, CaR, SN
 5. Segments of basal lvs linear, gen < 1.5 mm wide; cauline petioles gen short; sepals white to bluish .... ***A. drummondii***
 5′ Segments of basal lvs oblong to lanceolate, gen 2–4 mm wide; cauline petioles gen ± 0; sepals yellowish to purplish .... ***A. multifida***

A. deltoidea Hook. (p. 915) Pl 10–25 cm; rhizome slender. **ST** gen 1; hairs 0 to sparse. **LVS**: basal gen 1, 1-ternate, petiole 10–15 cm, short-stiff-hairy, lflets 3–6 cm, ovate, crenate or sharp-toothed to lobed; cauline simple, 3–8 cm, ovate, toothed to lobed, petioles ± 0. **INFL**: fl 1. **FL**: sepals gen 5, 15–25 mm, white, glabrous. **FR**: cluster ± spheric, 9–12 mm, glabrous to puberulent; styles < 1 mm. Open to shaded sites, coniferous forest; 200–2000 m. n NCo, KR, NCoRH; to WA. ❀SHD,IRR,DRN:1–5,**6**,7,14,**15–17**,18.

A. drummondii S. Watson (p. 915) Pl 10–25(30) cm; caudex branched. **STS** 1–several; hairs gen soft. **LVS**: blades 2–5 cm, dissected, segments linear, gen < 1.5 mm wide; basal few–many, petioles 2–10 cm, soft-hairy; cauline petioles gen short. **INFL**: fls 1–2 (3). **FL**: sepals 5–8, 8–20 mm, white to bluish, lower surface soft-hairy. **FR**: cluster ovoid to spheric, 10–15 mm, densely woolly; styles 1–4 mm. *n*=24. Rocky slopes, coniferous forest, alpine; 1200–3350 m. KR, CaR, SN; to AK, ID. ❀TRY;DFCLT.

A. multifida Poiret (p. 915) GLOBE ANEMONE Pl (5)15–50 cm; caudex branched. **STS** 1–several; hairs soft. **LVS** dissected; basal few–many, petioles 3–12 cm, soft-hairy, blades 2–9 cm, segments lanceolate to oblong, gen 2–4 mm wide; cauline segments 2–5(7) mm wide, petioles gen ± 0. **INFL**: fls 1–3. **FL**: sepals 5–8, 6–12 mm, yellowish to purplish, lower surface soft-hairy. **FR**: cluster gen ovoid to spheric, 10–20 mm, woolly to densely silky; styles 1–2 mm. *n*=16. RARE in CA. Open, gravelly or rocky slopes, subalpine; 1800–2750 m. KR (Marble Mtns, Siskiyou Co.), n SNH (The Dardanelles, Alpine Co.); to AK, e Can, Colorado, also Chile. In cult.

A. occidentalis S. Watson (p. 915) Pl 20–75 cm; caudex branches 0–few. **STS** 1–few; hairs soft, in age ± 0 exc dense at nodes. **LVS**: blade 3.5–8 cm, dissected, segments linear, 1–2 mm wide; basal few, petioles 3–14 cm, soft-hairy; cauline petioles 0 to short. **INFL**: fls gen 1. **FL**: sepals 5–8, persistent in fr, 18–30 mm, white to purplish, lower surface soft-hairy. **FR**: cluster spheric, 20–40 mm, densely woolly to soft-hairy; styles 20–30(35) mm, plumose. Open, rocky slopes, alpine; 1200–3100 m. KR, CaR, SN; to B.C., MT. ❀TRY;DFCLT.

A. oregana A. Gray (p. 915) Pl 10–25 cm; rhizome thick. **ST** gen 1; hairs 0. **LVS**: basal gen 1, deeply 3-lobed to 1-ternate, in fl withered or persistent, petiole 2–6.5 cm, subglabrous, lobes or lflets ± ovate, irregularly toothed, deeply lobed or not; cauline gen 1-ternate, petioles 8–40 mm. **INFL**: fl 1. **FL**: sepals gen 5, 5–20 mm, white, reddish, blue, or purplish, ± glabrous. **FR**: cluster ovoid to spheric, 8–15 mm, puberulent; styles < 1.5 mm. Shaded sites, coniferous forest; 100–1600 m. NW, CaR, n SN; to WA. [*A. quinquefolia* L. vars. *grayii* (Behr & Kellogg) Jepson, *o.* (A. Gray) Robinson, *minor* (Eastw.) Munz]. Sepal color, stamen number varies with geography, ecology, needs further study. ❀SHD,IRR,DRN:4,5,**6**,7,15–17.

A. tuberosa Rydb. (p. 915) Pl 12–35 cm; tuber atop slender caudex. **STS** 1–few, glabrous. **LVS** dissected; segments oblong to ovate, 4–8 mm wide; basal few, petioles 5–8 cm, glabrous; cauline petioles ± 0 to short. **INFL**: fls 1–5. **FL**: sepals 5–8, 6–14 mm, reddish, lower surface soft-hairy. **FR**: cluster oblong to elliptic, (15) 20–30 mm, densely woolly; styles 1–2 mm. Rocky slopes, ledges; 900–1900 m. e DMtns; to UT, NM. ❀TRY;DFCLT.

AQUILEGIA COLUMBINE

Per from thick, simple to branched caudex. **STS** 1–few, ascending to erect, branched to not, scapose to not, glabrous to glandular-hairy. **LVS**: basal 1–3-ternate, petiole gen long; cauline 0–few, gen much reduced, deeply 3-lobed to 1–2-ternate, petiole short to ± 0; segments gen wedge-shaped to obovate, upper surface green to pale green, lower surface pale green to glaucous. **INFL**: few-fld raceme or fl solitary, terminal; axis and pedicels glabrous to glandular; fls often pendent. **FL** radial; sepals 5, petal-like, spreading to slightly reflexed; petals 5, gen with spur projecting between sepals; pistils gen 5. **FR**: follicles, glabrous to glandular. **SEEDS** smooth, shiny, brown to black. ± 70 spp.: temp N.Am, Eurasia. (Derivation uncertain, perhaps Latin: eagle, from spurs, or water-drawer, from habitats) [Munz 1946 Gentes Herb 7:1–150] Many spp. and hybrids cult as orn; natural hybrids common.

1. Fl erect; sepals and petals gen cream to pink .... ***A. pubescens***
1′ Fl nodding; sepals red, petals gen red to yellowish
 2. Petal spur ± outcurved, mouth > 90° to fl axis, elliptic to triangular; petal blade 0 .... ***A. eximia***
 2′ Petal spur ± straight to ± incurved, mouth ± 90° to fl axis, ± round; petal blade 1–8 mm .... ***A. formosa***

A. eximia Planchon (p. 915) Pl 20–160 cm. **LVS**: basal and lower cauline 2–3-ternate, petioles 4–30 cm, segments 8–35(50) mm; upper cauline gen simple to deeply 3-lobed. **INFL**: fl nodding, in fr erect. **FL**: sepals 10–28 mm, red; petal blade 0, spur 12–35 mm, ± outcurved, tube red, tip 2–4 mm wide, mouth 6–10 mm wide, elliptic to triangular, > 90° to fl axis, red to yellowish; stamens 10–25 mm. 2*n*=14. Seeps, moist ravines, often serpentine soils, mixed-evergreen or coniferous forest; 100–1800 m. NCoR, SnFrB, SCoR, w WTR. ❀IRR:**4–6**&SHD:**7,14–17**,18,**19–24**.

A. formosa Fischer (p. 915) Pl 20–80(150) cm. **LVS**: basal and lower cauline 2–3-ternate, petioles 5–30(40) cm, segments 7–45 (130) mm; upper cauline gen simple to deeply 3-lobed. **INFL**: fl nodding, in fr erect. **FL**: sepals 12–20(25) mm, red; petal blade 1–8 mm, yellow, spur (4)10–23 mm, ± straight to ± incurved, tube red, tip 1.5–4 mm wide, mouth 4–8 mm wide, ± round, ± 90° to fl axis, red to yellowish; stamens 10–18 mm. 2*n*=14. Streambanks, seeps, moist places, chaparral, oak woodland, mixed-evergreen or coniferous forest; < 3300 m. CA-FP (exc GV, SCo, ChI), GB, DMtns; to AK, MT, Baja CA. [vars. *hypolasia* (E. Greene) Munz, *pauciflora* (E. Greene) H.S. Boothman, *truncata* (Fischer & C. Meyer) Baker; forma *anomala* J. Howell] Lf variation needs study; pls with pale green to glaucous, gen 3-ternate lvs (DMtns), have been called *A. shockleyi* Eastw. ❀IRR,DRN:**4–6,15–17,24**&SHD:1–3,**7**,8,9,**14, 18–23**.

A. pubescens Cov. (p. 915) Pl 15–50 cm. **LVS**: basal and lower cauline 1–2-ternate, petioles 5–25 cm, segments 10–25(40) mm; upper cauline gen deeply 3-lobed to 1-ternate. **INFL**: fl erect. **FL**: sepals 10–24 mm, cream to pink; petals cream to pink exc spur mouth ± cream only, blade 8–15 mm, spur 20–42 mm, gen ± straight, tip 1–2.5 mm wide, mouth 6–10 mm wide, round, ± 90° to fl axis; stamens 12–21 mm. Open, gen rocky slopes, shrubland, subalpine forest, alpine; 2600–3650 m. SNH. Variable, hybridizes with *A. formosa* in SNH. ❀IRR,DRN:**4–6**,14–17&SHD:1–3,7,8, 9,18–24;DFCLT.

CALTHA MARSH MARIGOLD

Per from short caudex, gen fleshy, glabrous. **STS** 1–few. **LVS** simple, mostly basal, gen cordate to reniform, entire to sharp-toothed; basal petioles > blades. **INFL**: cyme or fl solitary, terminal or axillary; bracts lf-like. **FL** radial; sepals 5–12, petal-like, white to yellow; petals 0; pistils 5–many. **FR**: follicles, sessile to short-stalked, gen beaked. **SEED** gen smooth, brown to black. ± 10 spp.: temp N.Am, Eurasia. (Greek: ancient name, from bowl-shaped fl)

C. leptosepala DC. var. ***biflora*** (DC.) G. Lawson (p. 915) Pl 8–36 cm. **ST** gen < 5 cm, simple. **LVS**: petioles 3–25 cm, blades 2–9 cm wide, subentire to crenate. **INFL**: peduncle 4–32 cm, gen > lvs, 1(2)-fld. **FL**: sepals 5–11, 8–20 mm, oblong to elliptic. **FR**: follicles 3–8, 7–18 mm; beak gen straight, erect. Marshes, pond margins, streambanks, coniferous forest; 900–3300 m. KR, CaR, SNH, MP; to AK. [*C. howellii* (Huth) E. Greene] ❀WET:**4**,**5**,17&SHD:**1**,**2**,3,**6**,7,15,16.

CLEMATIS VIRGIN'S BOWER

Frederick B. Essig

Gen woody vine, sometimes dioecious. **LVS** gen 1–2-pinnate, cauline, gen opposite; petiole gen twining; lflets ovate to lanceolate, often irregularly 2–3-lobed, coarsely toothed. **INFL**: panicle to 1-fld, axillary or terminal. **FL** radial; sepals gen 4, free, petal-like, gen lanceolate, white to cream (or brightly colored elsewhere); petals 0; stamens many, free; pistils gen many, simple. **FR**: many achenes, each gen with an elongate, feathery style. 250 spp.: worldwide. (Greek: twig) Worldwide revision badly needed but CA spp. distinct.

1. Infl several–many-fld, appearing June-September .... ***C. ligusticifolia***
1′ Infl 1–3-fld, appearing January-June
 2. Fr body hairy; sepal hairy on both surfaces .... ***C. lasiantha***
 2′ Fr body glabrous; sepal hairy only on lower surface .... ***C. pauciflora***

C. lasiantha Nutt. (p. 915) PIPESTEMS **LF**: lflets 3–5, ± 3-lobed, toothed, largest on pl gen 3–5 cm. **INFL**: gen 1-fld, axillary. **FL**: sepals 10–21 mm, both surfaces hairy; stamens or staminodes 50–100, 7–13 mm, << sepals; pistils 75–100. **FR**: body hairy. Hillsides, chaparral, open woodlands; < 2000 m. SNF, CW, SW. Fls January-June. ❀DRN,DRY:**7**,**14–17**,**22-24**&SHD:**8**,**9**,**18–21**.

C. ligusticifolia Nutt. (p. 915) VIRGIN'S BOWER, YERBA DE CHIVA **LF**: lflets 5–15, irregularly lobed or toothed, largest on pl 2–8 cm. **INFL** several–many-fld, axillary. **FL**: sepals 6–10 mm, both surfaces hairy; stamens 25–40, 5–9 mm, ± = sepals; pistils 35–65. **FR**: body hairy. Along streams, wet places; < 2400 m. CA; to B.C., SD, NM, nw Mex. Fls June-September. ❀IRR:**1–7**,**14–17**,**22–24**&SHD:**8**,**9**,**18–21**.

C. pauciflora Nutt. (p. 915) ROPEVINE **LF**: lflets 3–9, ± 3-lobed, toothed, largest on pl gen 1–3 cm. **INFL** 1–3-fld, axillary. **FL**: sepals 6–12 mm, only lower surface hairy; stamens 30 –50, 6– 12 mm, ± = sepals; pistils 25–50. **FR**: body glabrous. Dry chaparral; < 1300 m. SW, DMtns (Little San Bernardino Mtns); Baja CA. Fls January-June. Pls ± intermediate to *C. lasiantha* occur in SW. ❀DRN,DRY:**7**,**14–17**,**22–24**&SHD:**8**,**9**,**18–21**.

CONSOLIDA

Ann gen from taproot. **ST** gen 1, erect, branched to not. **LVS** palmately or palmate-pinnately deeply lobed or dissected; cauline merging into bracts upward. **INFL**: raceme, terminal; pedicels spreading to ascending. **FL** bilateral; sepals 5, petal-like, white to blue or violet, uppermost spurred; petal 1, in color = sepals, tip gen lobed, arched over stamens, base with spur enclosed in uppermost sepal spur; pistil 1. **FR**: follicle. **SEED** brown to black, minutely scaly in rows. ± 40 spp.; Medit Eur, w Asia. (Latin: consolidate, from reported ability to heal wounds) [Keener 1976 Castanea 41:12–20] Segregated from *Delphinium*.

C. ambigua (L.) P. Ball & Heyw. Pl ± puberulent throughout exc st glabrous below. **ST** 45–100 cm. **LVS** short-petioled; segments 14–21, < 2 mm wide, linear. **INFL** (3)8–21-fld; pedicels 5–16 mm. **FL**: sepals white, pink, or blue, lateral 8–15 mm, spur 12–18 mm. **FR** 9–18 mm. **SEED** brown. $2n$=16. Disturbed soils, waste places, often a waif; < 300 m. NCoRI, n&c SNF, GV, SnFrB, SCo; to e US; native to Eur. [*Delphinium ajacis* L. misapplied]

COPTIS GOLDTHREAD

Per from slender rhizome or stolon, gen glabrous. **ST** short, simple, stout, scaly. **LVS** gen 1–2-ternate, basal, petioled. **INFL** scapose, gen 1–3-fld, umbel-like in fr. **FL** bisexual (or some staminate); radial; sepals 5–8, petal-like, early deciduous; petals 5–7, blade gen linear to elliptic, often with a gland at junction with claw; pistils 5–15, short-stalked. **FR**: follicles, stalked, glabrous to puberulent. **SEED** smooth to wrinkled. ± 10 spp.: temp N.Am, e Asia. (Greek: cut, from lvs) Petals sometimes considered to be modified staminodes.

C. laciniata A. Gray (p. 921) Pl 11–24 cm; rhizome or stolon yellow. **LVS** 3–8, gen 1-ternate; lflets ovate to triangular, terminal stalked or not, lobes gen 3, very deep, irregularly toothed to cut. **INFL**: peduncle 8–19 cm, < lvs in fl, > lvs in fr; pedicels 1.5–3 cm, in fr elongated. **FL**: sepals 5–9 mm, linear; petals 4–7 mm, claw ± thread-like, blade linear. **FR**: follicles 6–12; stalks 4–6 mm, bodies 7–12 mm, glabrous, walls papery, slightly translucent. Wet sites, seeps, streambanks, coniferous forest; < 1000 m. n NCo, w KR; to WA. ❀WET or IRR:**4**,**5**,17&SHD:**6**,7,15,16.

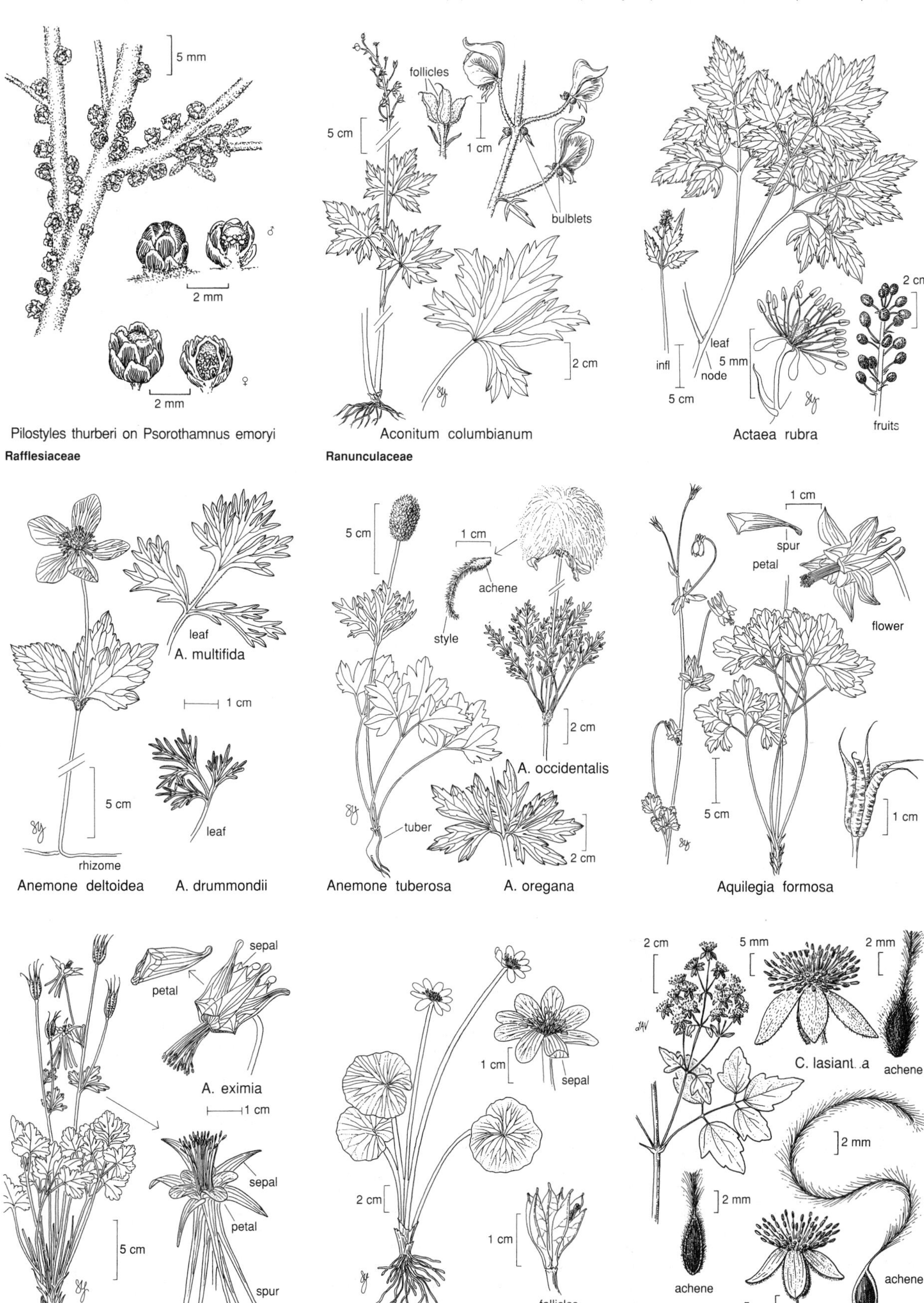
5 mm
♂
2 mm
2 mm
♀
Pilostyles thurberi on Psorothamnus emoryi
Rafflesiaceae
follicles
5 cm
1 cm
bulblets
2 cm
Aconitum columbianum
Ranunculaceae
2 cm
leaf
infl
5 mm
node
5 cm
fruits
Actaea rubra
leaf
A. multifida
1 cm
5 cm
leaf
rhizome
Anemone deltoidea
A. drummondii
5 cm
1 cm
achene
style
2 cm
A. occidentalis
tuber
2 cm
Anemone tuberosa
A. oregana
1 cm
spur
petal
flower
5 cm
1 cm
Aquilegia formosa
sepal
petal
A. eximia
1 cm
sepal
petal
5 cm
spur
Aquilegia pubescens
1 cm
sepal
2 cm
1 cm
follicles
Caltha leptosepala var. biflora
2 cm
5 mm
2 mm
C. lasiant..a
achene
2 mm
2 mm
achene
achene
5 mm
Clematis
ligusticifolia
C. pauciflora

DELPHINIUM LARKSPUR

Michael J. Warnock

Per; root gen < 10 cm, ± fibrous or fleshy; buds gen obscure. **ST** gen 1, erect, gen unbranched; base gen ± as wide as root, gen firmly attached to root, gen ± reddish or purplish. **LVS** simple, basal and cauline, petioled; blades gen palmately lobed, deep lobes gen 3–5, gen < 6 mm wide, gen also lobed; lower lvs gen dry, often 0 in fl; cauline merging into bracts upward. **INFL**: raceme or somewhat branched, terminal; fls gen 10–25; pedicels gen ± spreading. **FL** bilateral; sepals 5, petal-like, gen spreading, gen ± dark blue, uppermost spurred; petals 4, << sepals, upper 2 with nectar-secreting spurs enclosed in uppermost sepal, lower 2 clawed, with blades gen 4–8 mm, notched, gen ± perpendicular to claws, gen colored like sepals, gen obviously hairy; pistils 3(–5). **FR** aggregate of 3(–5) erect follicles, gen 2.5–4 × longer than wide. **SEED** dark brown to black, often appearing white, gen winged when immature, gen without inflated collar; coat cell margins gen straight. (Latin: dolphin, from bud shape) [Lewis & Epling 1954 Brittonia 8:1–22] Hybrids common, esp in disturbed places. Root length here includes coarse but not thread-like parts. Most spp. highly TOXIC, attractive and causing many deaths to cattle, less often to horses, sheep. ❀Exc as noted, successful in cult only within natural range and habitat. Lowland ssp.: DRY. Upland spp.: winter chilling required.

1\. Pl with all the following: sepals pink to deep rosy-pink; pedicel hairs glandular, yellow; fr erect; pedicels ± ascending; angle between claw and blade on lower petals > 130° ..... ***D. purpusii***
1′ Pl with at least one exception to the characters in 1.
 2\. Lf blade base tapered ..... ***D. uliginosum***
 2′ Lf blade base cordate, truncate, or rounded
  3\. Sepals ± red or yellow; angle between claw and blade of lower petals 140–180°, blade hairs 0 or obscure to naked eye
   4\. Seed without inflated collar; fr ± straight; pls of ± dry sites; SCoRI & s — sepals red ..... ***D. cardinale***
   4′ Seed with inflated collar at widest end: fr curved; pls of moist sites; SCoRO & n
    5\. Sepals bright yellow, not red — n CCo ..... ***D. luteum***
    5′ Sepals scarlet to orange-red, rarely yellow but not brightly so ..... ***D. nudicaule***
  3′ Sepals not red or yellow (sometimes maroon or pink); blade of lower petal angled 60–130° to claw, its hairs apparent to naked eye
   6\. Pl with all the following: infl ± 1-sided; fl gen on and after 15 June; > 1900 m; st gen < 40 cm; pedicels ascending; seed winged, without inflated collar at widest end; root < 5 cm — n&c SNH, Wrn ..... ***D. depauperatum***
   6′ Pl with two or more exceptions to the characters in 6.
    7\. St base gen much narrower than adjacent root, attachment not firm; seed with inflated collar at widest end, unwinged; infl in mid-fl gen with < 6 fls on main axis or oldest open-fld pedicels gen > 25 mm and > 2 × youngest; fr curved
     8\. Sepals ± maroon; pls within populations gen dissimilar, often ± infertile
      9\. Lf segments 5 or fewer, > 5 mm wide ..... ***D. nudicaule* × *decorum***
      9′ Lf segments > 5, gen < 5 mm wide ..... ***D. nudicaule* × *depauperatum***
     8′ Sepals not maroon; pls within populations gen similar, gen ± fertile
      10\. Most lvs on upper 1/3 of st; st gen > 50 cm; fls on main infl axis gen > 15 — lf lobe > 1 cm at widest point
       11\. Sepals pinkish to light blue ..... ***D. trolliifolium* × *nudicaule***
       11′ Sepals dark blue
        12\. Lf lobe tips crenate; n CCo, n SnFrB (Sonoma Co.) —pedicels ± glandular ..... ***D. bakeri***
        12′ Lf lobe tips sharply, irregularly cut; NCo, NCoRO (n of Sonoma Co.) ..... ***D. trolliifolium***
      10′ Most lvs on lower 1/3 of st; st gen < 50 cm; fls on main infl axis often < 20
       13\. Root > 10 cm, fibrous, not clustered, not spheric — lf lobes gen > 7; st glabrous; NW
        14\. Sepals blue — lvs ± fleshy; talus, 1100–2600 m ..... ***D. antoninum***
        14′ Sepals pinkish
         15\. Population with some blue but not red fls; gen > 1500 m ..... ***D. antoninum* × *nudicaule***
         15′ Population with some red but not blue fls; < 1500 m ..... ***D. nudicaule* × *trolliifolium***
       13′ Root < 10 cm, clustered, ± spheric
        16\. Sepals dark blue-purple (often faded, mottled on herbarium specimens), puberulent outside, gen not reflexed; lower st hairy; lower petal blade gen > 6 mm — spur 13–20 mm; st < 45 cm ..... ***D. decorum***
         17\. Lf with gen 5 lobes that extend > 50% to petiole; < 200 m ..... ssp. ***decorum***
         17′ Lf with 6 or more lobes that extend > 50% to petiole; > 700 m ..... ssp. ***tracyi***
        16′ Sepals gen bright blue to white or pink (gen not faded, not mottled on herbarium specimens), gen glabrous, often reflexed (if sepals purplish and puberulent outside and not reflexed, then lower st ± glabrous or lower petal blade < 6 mm)
         18\. Lf with 6 or more lobes that extend > 50% to petiole, these < 7 mm wide; pedicels glabrous or puberulent; SNH & n ..... ***D. nuttallianum***
         18′ Lf with 5 or fewer lobes that extend > 50% to petiole (if > 5, then pedicels puberulent, pls of s SN and s), these often > 7 mm wide
          19\. Terminal lf lobe widest above middle; pedicel spreading at 70–90° angle; lf lobes gen 5; SN ..... ***D. gracilentum***

19′ Terminal lf lobe widest near middle; pedicel spreading at gen < 70° angle; lf lobes 3–many; not SN .... ***D. patens***
20. Lobes of lower lvs gen > 1.5 cm wide; basal, lower cauline lvs gen divided < 80% to petiole — SCoRO, SW (exc ChI) .... ssp. ***hepaticoideum***
20′ Lobes of lower lvs gen < 1.5 cm wide; basal (if present), lower cauline lvs gen divided > 80% to petiole
21. Pedicels puberulent, gen glandular; SW .... ssp. ***montanum***
21′ Pedicels gen glabrous; not SW
22. Hairs of lower petals white .... ssp. ***patens***
22′ Hairs of lower petals yellow .... ***D. patens × decorum***
7′ St base gen ± as wide as adjacent root, attachment firm; seed without inflated collar at widest end, winged or not; infl in mid-fl with gen > 6 fls on main axis or oldest open-fld pedicel gen < 25 mm or < 2 × youngest; fr ± straight
23. Root > 15 cm, gen with enlarged, whitish (brown on herbarium specimens) buds; sts gen > 1 m, often 2 or more per root system
24. Fls rare after 30 June; sepals ± lavender or greenish white; < 1200 m .... ***D. californicum***
25. Sepals ± lavender; main infl axis puberulent; upper petals ± hairy .... ssp. ***californicum***
25′ Sepals greenish white; main infl axis glabrous; upper petals ± glabrous .... ssp. ***interius***
24′ Fls rare before 1 July; sepals gen bluish (if ± lavender or greenish white, then > 1200 m), sometimes canescent
26. Lvs present on lower 20% of st in fl
27. Fl spur 9–12 mm; sepals white to light blue .... ***D. inopinum***
27′ Fl spur 11–22 mm; sepals dark blue .... ***D. polycladon***
26′ Lvs gen 0 on lower 20% of st in fl
28. Sepals purplish blue, outer surface midline not lighter; lower st glaucous; KR, SNH, SnGb, SnBr .... ***D. glaucum***
28′ Sepals ± bright blue, outer surface midline lighter due to hairs; lower st puberulent; Wrn .... ***D. stachydeum***
23′ Root < 15 cm (if larger, then enlarged buds 0); sts gen < 1 m, gen 1 per root system
29. Some hairs of lower petioles > 0.5 mm, white, straight, spreading — seeds unwinged
30. Seeds finely prickly, fuzzy to naked eye; lateral sepals 7–13 mm; fls gen > 12 on main infl axis .... ***D. hansenii***
31. Sepals violet-purple to maroon — SNF, SnJv .... ssp. ***ewanianum***
31′ Sepals dark blue-purple to white or pink
32. Basal lvs gen 0 in fl; cauline lvs 3 or more; SN, CaRF, NCoRI, ScV .... ssp. ***hansenii***
32′ Basal lvs present but sometimes dry in fl (thus 0 on herbarium specimens); cauline lvs gen 2 or fewer; s SN, Teh .... ssp. ***kernense***
30′ Seeds not prickly, not fuzzy to naked eye; lateral sepals 10–25 mm; fls gen < 12 on main infl axis
33. Hairs on margin of lower petal blades 0; sepal spur gen downcurved for > 3 mm at tip — CCo (c Monterey Co.) .... ***D. hutchinsoniae***
33′ Hairs on margin of lower petal blades present; sepal spur straight or downcurved for < 3 mm at tip .... ***D. variegatum***
34. Pl of mainland — sepals gen dark royal blue, rarely white or lavender .... ssp. ***variegatum***
34′ Pl of islands
35. Sepals white to light blue .... ssp. ***kinkiense***
35′ Sepals bright blue .... ssp. ***thornei***
29′ Hairs of lower petioles 0–0.5 mm, or curved, or both (if some > 0.5 mm and straight, then pl of NW or SnFrB and seeds winged)
36. Inner lobes of lower petals hairier than outer; lower st ± striate — lateral sepals < 16 mm; seeds winged; sepals not reflexed; pedicels ascending at < 45° angle; NW, CW, GV, PR .... ***D. hesperium***
37. Lateral sepals < or = 4 mm wide; PR .... ssp. ***cuyamacae***
37′ Lateral sepals > 4 mm wide; n of TR
38. Sepals dark blue-purple .... ssp. ***hesperium***
38′ Sepals white to pinkish or light blue .... ssp. ***pallescens***
36′ Lobes of lower petals equally as hairy; lower st not striate
39. Seed coat cell margins wavy (barely visible at 10× when held up to light); fr gen < or = 3× longer than wide; desert, grassland (sometimes with scattered trees), shrubland; GV, s SNF, Teh, TR, SNE, D
40. Sepals rarely reflexed; st gen > 60 cm; grasslands, open woodlands — sepals white or light pink to light blue
41. Sepals light blue .... ***D. parryi × D. gypsophilum*** ssp. ***parviflorum***
41′ Sepals white to light pink .... ***D. gypsophilum***
42. Lower petal blades 5–8 mm; lateral sepals gen > 10 mm; pedicels gen > 1 cm apart on infl axis .... ssp. ***gypsophilum***
42′ Lower petal blades 3–4 mm; lateral sepals gen < 10 mm; pedicels gen < 1 cm apart on infl axis .... ssp. ***parviflorum***
40′ Sepals, esp lateral, often reflexed; st gen < 60 cm; deserts or shrublands, rarely grasslands
43. Pl of fine, alkaline soil; sepals gen light blue; lower petals white; GV .... ***D. recurvatum***

43′ Pl of often coarse, not very alkaline soil; sepals ± sky or dark blue to white or pink; lower petals same color as sepals; D and adjacent areas ..... ***D. parishii***

44. Sepals ± sky blue, lateral sepals reflexed ..... ssp. ***parishii***

44′ Sepals dark blue or white to pink or blue but not sky blue, lateral sepals not reflexed

45. Fls rare before 20 May; sepals white to pink or blue; WTR ..... ssp. ***pallidum***

45′ Fls rare after 20 May; sepals dark blue; w DSon ..... ssp. ***subglobosum***

39′ Seed coat cell margins straight (barely visible at 10× when held up to light); fr gen > 3× longer than wide; shrubland in GB; chaparral, woodland, or forest in CW, SW, CaR, SN

46. Green lvs gen present on lower 20% of st in fl; lower st and petioles ± glabrous

47. Root > 7 cm; lobes of lower lvs < 4 mm wide; lateral sepals rarely reflexed; not in broadleafed woodlands; ne SNH, GB ..... ***D. andersonii***

47′ Root < 7 cm; lobes of lower lvs gen > 4 mm wide; lateral sepals reflexed; gen in broadleafed woodlands; SCoRO, WTR ..... ***D. umbraculorum***

46′ Green lvs gen 0 on lower 20% of st in fl (if present, then lower st, or petioles, or both with short, curled hairs) — CW, SW, Teh

48. Lobes of lower lvs gen > 5 mm wide; > 10 km inland; > 400 m ..... ***D. umbraculorum* × *parryi***

48′ Lobes of lower lvs < 5 mm wide (or pl from < 10 km inland or < 400 m) ..... ***D. parryi***

49. Basal lvs gen 0 in fl

50. Lateral sepals gen > 16 mm ..... ssp. ***blochmaniae***

50′ Lateral sepals < 16 mm ..... ssp. ***parryi***

49′ Basal lvs gen present in fl

51. Pl from > 700 m ..... ssp. ***purpureum***

51′ Pl from < 700 m

52. Sepals gen reflexed ..... ssp. ***eastwoodiae***

52′ Sepals gen spreading ..... ssp. ***maritimum***

D. andersonii A. Gray (p. 921) Root gen > 10 cm, distally branched. **ST** 20–90 (gen 30–60) cm; base often narrower than root, but firmly attached to root, ± glabrous. **LVS** mostly on lower 1/2 of st, ± glabrous; lobes 7–30, < 4 mm wide. **INFL** cylindric; pedicels 8–68 mm, 5–25 cm apart, ± S-shaped, glabrous to puberulent. **FL**: sepals dark blue, lateral 9–16 mm, spur 11–18 mm. **FR** 17–32 mm, gen > 4 × longer than wide. **SEED**: coat smooth, shiny, ± translucent, inflated. $2n$=16. Talus, dry soils in sagebrush scrub; 1300–2700 m. ne SNH, GB; to se OR, c ID, s UT. [ssp. *cognatum* (E. Greene) Ewan; *D. decorum* var. *nevadense* S. Watson; not *D. menziesii* DC.] ❀DRN,DRY;DFCLT.

D. antoninum Eastw. (p. 921) ANTHONY PEAK LARKSPUR Root gen > 15 cm. **ST** 7–60 (gen < 30) cm; base narrower than root, not firmly attached to root, glabrous. **LVS** mostly on lower 1/3 of st, ± fleshy, ± glabrous; lobes 3–15. **INFL**: fls 3–25; pedicels 6–32 mm, 5–25 mm apart, gen puberulent. **FL**: sepals blue, lateral 11–13 mm, spur 12–16 mm; lower petal blades 3–5 mm, hairier on inner lobes. **FR** 14–22 mm, curved. **SEED** with inflated collar at widest end, otherwise ± bumpy. Uncommon. Talus; 1100–2600 m. KR, NCoRH. Hybridizes with *D. decorum* ssp. *tracyi*, *D. nudicaule*. ❀DRN,DRY;DFCLT.

D. bakeri Ewan (p. 921) BAKER'S LARKSPUR **ST** 45–100 (gen > 55) cm, ± glabrous; base narrower than root, not firmly attached to root. **LVS** mostly on upper 2/3 of st; basal ± glabrous; lobes > 1 cm at widest, tips crenate. **INFL**: pedicels 8–91 mm, > 10 mm apart, ± glandular. **FL**: sepals dark blue, lateral 9–11 mm, spur 9–13 mm; lower petal blades hairier on inner lobes. **FR** 18–20 mm, ± curved. **SEED** with inflated collar at widest end, otherwise smooth, shiny. **RARE** CA (possibly extinct). Coastal scrub; 100–300 m. n SnFrB, n CCo, (s Sonoma Co.).

D. californicum Torrey & A. Gray Root gen > 15 cm, distally branched; buds prominent (exc on herbarium specimens). **ST** often 2 or more per root system, 60–220 (gen > 100) cm, gen puberulent. **LF**: lobes 3–15, tips ± sharply cut. **INFL** gen branched; fls gen > 50; lowermost bracts ± lf-like; pedicels 5–65 mm, 5–25 mm apart, puberulent. **FL**: sepals forward-pointing, ± lavender to greenish white, lateral 6–11 mm, spur 7–14 mm; lower petal blades 3–5 mm. **FR** 11–16 mm. **SEED** with ± overlapping scales. Chaparral, open woodlands; 0–1000 m. CCo, SnFrB.

ssp. ***californicum*** (p. 921) COAST or CALIFORNIA LARKSPUR **LF** puberulent on lower surface and margins. **INFL**: main axis puberulent. **FL**: sepals ± lavender, densely puberulent, lateral 7–11 mm, spur 7–14 mm; upper petals ± hairy. $2n$=16. Gen slopes in dense chaparral, w side of coast ranges; 0–1000 m. CCo, w SnFrB. ❀DRN,IRR:15–17&SHD:14.

ssp. ***interius*** (Eastw.) Ewan (p. 921) HOSPITAL CANYON LARKSPUR **LF** glabrous. **INFL**: main axis ± glabrous; pedicel tips often puberulent. **FL**: sepals greenish white, hairs gen near tips only, lateral 6–8 mm, spur 7–10 mm; upper petals ± glabrous. UNCOMMON. Gen slopes in open woodlands, e side of coast ranges; 300–800 m. e SnFrB. ❀DRN,DRY;DFCLT.

D. cardinale Hook. (p. 921) CARDINAL or SCARLET LARKSPUR Root gen > 15 cm, distally branched. **ST** 30–270 cm, gen curled-puberulent, base sometimes narrower than root, but firmly attached to root. **LVS**: basal present in fl or not, ± glabrous, lobes 5–27. **INFL**: pedicels 15–55 mm, 6–80 (gen > 15) mm apart, puberulent. **FL**: sepals red, gen ± forward-pointing, lateral 11–15 mm, spur 15–24 mm; lower petals flattened, blades 4–5 mm, with hairs few, short, yellow, obscure to naked eye. **FR** 12–18 mm, ± straight. **SEED** bumpy. $2n$=16. Slopes (often talus) in chaparral; 300–1500 m. SCoR, SW; Baja CA. Hybrids with *D. parryi* (and man-made hybrids with *D. hutchinsoniae*) have been called *D. inflexum* Davidson. Gen hummingbird pollinated. ❀DRN,DRY:15–17&SHD:7,14,18–23.

D. decorum Fischer & C. Meyer YELLOWTINGE LARKSPUR Roots clustered, ± spheric or not. **ST** 7–45 cm, narrower than root, not firmly attached to root, lower st gen hairy. **LVS** mostly basal, ± glabrous on upper surface, gen ± puberulent on lower surface and margins; lobes 3–15, sometimes > 6 mm at widest. **INFL**: fls 2–20; pedicels 10–63 mm, 10–25 mm apart, gen puberulent. **FL**: sepals gen not reflexed, dark blue-purple, esp veins (gen faded, mottled on herbarium specimens), puberulent outside, lateral 11–24 mm, spur 13–20 mm; lower petal blades 6–11 mm, gen hairier on inner lobes. **FR** 9–20 mm, ± curved. **SEED** with inflated collar at widest end, otherwise bumpy. Grasslands, open chaparral, meadows; 0–2300 m. NW, SnFrB.

ssp. ***decorum*** (p. 921) COAST LARKSPUR **ST** 8–35 cm, ± erect. **LF** with gen 5 lobes that extend > 50% to petiole. **INFL** puberulent. **FL**: lateral sepal 12–24 mm, spur 13–19 mm. Grasslands, open coastal chaparral; 0–200 m. NCo, SnFrB. Hybridizes with *D. luteum*, *D. nudicaule*, *D. patens*, *D. trolliifolium*. ❀DRN, DRY:15–17&SHD:7,14.

ssp. ***tracyi*** Ewan TRACY'S LARKSPUR **ST** 7–45 cm, prostrate to erect. **LF** with 6 or more lobes that extend > 50% to petiole. **INFL** gen glabrous. **FL**: lateral sepal 11–18 mm, spur 13–20 mm. $2n$=16. Meadows in coniferous forests; 700–2300 m. KR, NCoRO, NCoRH. Hybridizes with *D. antoninum*. ❀DRN,DRY; DFCLT.

D. depauperatum Torrey & A. Gray Root < 5 cm. **ST** 7–60 (gen < 40) cm; base ± glabrous, gen not reddish. **LVS** mostly basal (cauline much reduced), ± glabrous; lobes 5–10. **INFL** ± 1-sided; fls 4–22; pedicels ascending, 7–73 mm, 15–50 mm apart, puberulent, often glandular. **FL**: lateral sepals 11–13 mm, spur 12–16 mm. **FR** 9–16 mm. **SEED** winged, shiny, ± bumpy. Moist meadows; 1900–2600 m. n&c SNH, Wrn; OR, NV. [*D. diversifolium* ssp. *harneyense* Ewan] May hybridize with *D. nuttallianum*. ❀DRN, DRY;DFCLT.

D. glaucum S. Watson (p. 921) MOUNTAIN LARKSPUR Root gen > 15 cm, distally branched; buds gen prominent (exc on herbarium specimens). **ST** gen 2 or more per root system, 80–300 (gen > 150) cm, lower glabrous, glaucous. **LVS** gen glabrous; lobes sharply cut at tips, often > 6 mm at widest. **INFL** gen branched; fls gen > 50; lower bracts lf-like; pedicels 10–48 mm, 3–25 mm apart, glabrous to puberulent. **FL**: sepals ± forward-pointing to spreading, purplish blue, lateral 8–14(21) mm, spur 10–19 mm. **FR** 9–20 mm. **SEED** bumpy. $2n=16$. Wet thickets, streamsides; 1600–3200 m. KR, SNH, SnGb, SnBr; to AK, NV. May hybridize with *D. polycladon*. ❀WET,DRN:1,4–6&SHD:2,3.

D. gracilentum E. Greene (p. 921) Root ± spheric, diffuse-fibrous. **ST** 15–80 (gen < 50) cm, narrower than root, not firmly attached to root, ± glabrous, exc sometimes near base. **LVS** mostly on lower 1/3 of st, ± glabrous; lobes gen 5, gen > 6 mm at widest, gen rounded at tips, terminal widest above middle. **INFL**: pedicels 15–45 mm, gen > 10 mm apart, glabrous to glandular-puberulent. **FL**: sepals reflexed, blue to white or pink, glabrous, lateral 7–13 mm, spur 9–14 mm; lower petal blades 3–5 mm, hairier on inner lobes. **FR** 8–16 mm, curved. **SEED** with inflated collar at widest end, otherwise gen bumpy. Coniferous forest; 150–2700 m. SN. [*D. pratense* Eastw. (meadow larkspur); *D. patens* ssp. *greenei* (Eastw.) Ewan] ❀DRN,DRY;DFCLT.

D. gypsophilum Ewan Root sometimes > 15 cm, distally branched. **ST** 30–150 (gen 60–100) cm; base sometimes narrower than root but firmly attached to root, gen glabrous, glaucous. **LF**: margins gen puberulent. **INFL** gen branched; pedicels 5–25 mm, 3–50 mm apart, glabrous. **FL**: sepals gen white to pink, lateral 7–19 mm, spur 7–15 mm; lower petal blades 3–8 mm. **FR** 9–18 mm, gen < 3 × longer than wide. **SEED**: coat cell margins wavy. Grassland, oak woodland; 150–1200 m. s SNF, Teh, SnJV, SCoR.

ssp. ***gypsophilum*** (p. 921) GYPSUM-LOVING LARKSPUR **ST** 50–150 cm. **INFL**: pedicels gen > 1 cm apart. **FL**: sepals gen white, lateral 10–19 mm, spur 10–15 mm; lower petal blades 5–8 mm, white. $2n=16,32$. UNCOMMON. Slopes in grassland, open oak woodland; 150–1200 m. s SNF, Teh, SnJV, SCoRI. ❀DRN, DRY;DFCLT.

ssp. ***parviflorum*** Harlan Lewis & Epling SMALL-FLOWERED GYPSUM-LOVING LARKSPUR **ST** 30–140 cm. **INFL**: pedicels gen < 1 cm apart. **FL**: sepals white to pinkish, lateral 7–10 mm, spur 7–11 mm; lower petal blades 3–4 mm, white or yellow. $2n=16$. UNCOMMON. Open oak woodland; 200–350 m. SCoRO. ❀DRN, DRY;DFCLT.

D. hansenii (E. Greene) E. Greene HANSEN'S LARKSPUR **ST** 25–180 (gen 40–80) cm; base puberulent to hairy. **LF** hairy, esp on lower surface; lobes 3–18; petioles hairy. **INFL**: fls often > 25; pedicels 3–57 mm, gen < 8 mm apart, ± ascending, puberulent. **FL**: sepals spreading to forward-pointing, lateral 7–13 mm, spur 6–16 mm; lower petal blades gen hairier on inner lobes. **FR** 8–20 mm, often < 3 × longer than wide. **SEED** finely prickly. Oak woodland or open chaparral; 60–3000 m. NCoRI, CaRF, SNF, c&s SNH, Teh, GV.

ssp. ***ewanianum*** M.J. Warnock EWAN'S LARKSPUR **ST** 25–130 cm; base gen puberulent. **LVS**: basal gen few in fl. **INFL**: pedicels 6–25 mm apart. **FL**: sepals violet-purple to maroon, lateral 8–12 mm, spur 6–16 mm. $2n=32$. UNCOMMON. Oak woodland; 60–600 m. SNF, SnJV. ❀DRN,DRY;DFCLT.

ssp. ***hansenii*** (p. 921) **ST** 40–180 cm; base gen puberulent or hairy. **LVS**: basal gen 0 in fl. **INFL**: pedicels < 10(–25) mm apart. **FL**: sepals dark blue-purple to white or pink, lateral 7–13 mm, spur 8–13 mm. $2n=16,32$. Oak woodland; 150–3000 m. NCoRI, CaRF, SNF, c&s SNH, ScV. [var. *arcuatum* E. Greene] ❀DRN,DRY; DFCLT.

ssp. ***kernense*** (Davidson) Ewan **ST** 34–110 cm; base puberulent. **LVS** gen basal but dry in fl. **INFL**: pedicels 7–25 (gen > 10) mm apart. **FL**: sepals white to dark blue-purple, lateral 7–13 mm, spur 8–16 mm. $2n=16,32$. Open oak woodland, chaparral; 800–1900 m. s SN, Teh. ❀DRN,DRY;DFCLT.

D. hesperium A. Gray WESTERN or COAST LARKSPUR **ST** 10–120 (gen 40–80) cm, hairy to ± glabrous below; base seldom reddish, gen prominently veined. **LF**: upper surface ± glabrous; lower surface puberulent, prominently veined; lobes 3–14. **INFL**: fls 5–100; pedicels ascending, 6–75 mm, 3–50 mm apart, puberulent. **FL**: sepals white to pinkish or dark blue-purple, lateral 7–16 mm, spur 8–18 mm; lower petal blades 3–8 mm, hairier on inner lobes. **FR** 8–18 mm, sometimes < 3 × longer than wide. **SEED** winged, otherwise smooth. Grassland, open woodlands; 10–1500 m. NW, CaRF, ScV, SnFrB, PR.

ssp. ***cuyamacae*** (Abrams) Harlan Lewis & Epling CUYAMACA LARKSPUR **ST**: base puberulent. **INFL**: fls gen > 25; pedicels gen < 8 mm apart. **FL**: sepals spreading-forward-pointing, dark blue-purple, lateral 7–10 mm, gen < 4 mm wide, spur 8–12 mm; lower petal blades 3–5 mm. $2n=16$. **RARE** CA. Grassland, yellow-pine forest; ± 1500 m. c PR.

ssp. ***hesperium*** (p. 921) **ST**: base ± glabrous to hairy. **INFL**: fls gen < 30; pedicels gen > 8 mm apart. **FL**: sepals spreading, dark blue-purple, lateral 8–16 mm, > 4 mm wide, spur 10–18 mm; lower petal blades 5–8 mm. $2n=16$. Oak woodland, w slope coast ranges; 10–1100 m. NW, SnFrB. ❀DRN,DRY;DFCLT.

ssp. ***pallescens*** (Ewan) Harlan Lewis & Epling **ST**: base gen puberulent, rarely hairy. **INFL**: fls gen < 25; pedicels > 8 mm apart. **FL**: sepals spreading, white to pinkish or light blue, lateral 10–15 mm, > 4 mm wide, spur 10–17 mm; lower petal blades 4–7 mm. $2n=16$. Oak woodland, e slope coast ranges; 20–1000 m. NCoRI, SCoRI, CaRF, ScV, SnFrB. Hybridizes with *D. parryi*, *D. uliginosum*. ❀DRN,DRY:7,14–16.

D. hutchinsoniae Ewan (p. 921) HUTCHINSON'S LARKSPUR **ST** 25–100 cm; base puberulent. **LF**: lobes 3–17, 2° lobes spreading; petiole hairs gen spreading. **INFL**: fls gen < 10 per main axis; pedicels ± ascending, 8–40 mm, 10–25 mm apart, puberulent. **FL**: lateral sepals 12–24 mm, spur 11–19 mm, gen downcurved for > 3 mm at tip; lower petal blades 5–10 mm, hairs few but more common on inner lobes, 0 on margin. **FR** 9–21 mm; veins gen blue. **SEED** winged, otherwise smooth. RARE. Coastal prairie, chaparral, forest; 0–400 m. c CCo (c Monterey Co). See *D. cardinale*.

D. inopinum (Jepson) Harlan Lewis & Epling (p. 921) UNEXPECTED LARKSPUR Root gen > 15 cm, distally branched; buds apparent (exc on herbarium specimens). **ST** often 2 or more per root system, 80–150 (gen 100–140) cm, glabrous, gen glaucous. **LVS** mostly on lower third of st. **INFL**: fls gen > 25; pedicels 5–25 mm, 6–25 mm apart, glabrous. **FL**: sepals white to light blue, often forward-pointing, lateral 8–12 mm, spur 9–12 mm; lower petal blades 3–5 mm. **FR** 12–20 mm. **SEED** winged, otherwise smooth. RARE. Open conifer forest, rock outcrops; 2200–2800 m.

D. luteum A.A. Heller (p. 927) YELLOW or GOLDEN LARKSPUR Root gen > 15 cm, distally branched. **ST** 20–55 cm; base narrower than root, not firmly attached to root, ± glabrous. **LVS** mostly ± basal, ± fleshy, gen ± glabrous; lobes > 6 mm at widest or not. **INFL** gen branched; pedicels 8–68 mm, 8–50 mm apart, puberu-lent. **FL**: sepals bright yellow, ± forward-pointing, lateral 11–16 mm, spur 13–20 mm; lower petals flattened, blades 3–4 mm, ± glabrous. **FR** 11–14 mm, ± curved. **SEED** with inflated collar at widest end, otherwise smooth. **RARE** CA. Moist sites, cliffs, coastal grassland, chaparral; 0–50 m. n CCo (Marin, Sonoma cos.). Hybridizes with *D. decorum*, *D. nudicaule*. In cult.

D. nudicaule Torrey & A. Gray (p. 927) RED or ORANGE LARKSPUR Root gen > 15 cm, distally branched. **ST** 15–125 (gen < 50) cm; base narrower than root, not firmly attached to root, gen glabrous. **LVS** mostly on lower 30% of st, ± glabrous; lobes 3–10, > 6 mm at widest or not. **INFL**: pedicels 15–80 mm, 7–50 mm apart, glabrous to glandular-puberulent. **FL**: sepals scarlet to orange-red, rarely dull yellow (maroon to magenta in hybrids), forward-pointing, lateral 8–16 mm, spur 12–34 mm; lower petals flattened, blades 2–3 mm, ± glabrous. **FR** 13–26 mm, curved. **SEED** with inflated collar at widest end, otherwise smooth. $2n=16$. Common.

Moist talus, wooded, rocky slopes: 0–2600 m. NW, CaR, n&c SN, CW (exc SCoRI), nw MP; sw OR. Gen hummingbird pollinated; hybridizes with most other larkspurs in its range. ❀DRN,DRY:5 &SHD:1,2,**7**,14,**15–17**.

D. nuttallianum Walp. DWARF, MEADOW, SLIM, or SONNE'S LARKSPUR Root ± spheric, clustered or not. **ST** 5–100 (gen < 50) cm; base narrower than root, not firmly attached to root, glabrous to puberulent. **LVS** mostly on lower 20% of st, ± glabrous; lobes 7–25 (6 or more extending > 60% to petiole), < 7 mm wide. **INFL**: fls rarely > 12; pedicels 7–75 mm, 15–50 mm apart, glabrous or puberulent. **FL**: sepals gen ± reflexed, gen ± glabrous, lateral 8–18 mm, spur 8–20 mm. **FR** 7–17 mm, ± curved. **SEED** with inflated collar at widest end, otherwise smooth, shiny. $2n=16$. Open woodlands, sage-brush scrub, meadow edges, streamsides; 300–3300 m. KR, CaR, SNH, MP; to OR, ID, NV. [*D. sonnei* E. Greene] Hybridizes with *D. depauperatum*, *D. nudicaule*, *D. polycladon*; extremely difficult, variable complex. ❀DRN,DRY;DFCLT.

D. parishii A. Gray Root often > 15 cm, branched. **ST** 17–100 (gen < 60) cm; base often narrower than root, but firmly attached to root, glabrous to puberulent. **LVS** mostly basal or mostly cauline in fl, glabrous to puberulent; lobes 3–20, gen < 6 mm at widest. **INFL**: fls 6–75; pedicels 3–48 mm, 3–25 mm apart, glabrous to puberulent. **FL**: sepals dark blue to white or pink, lateral 7–13 mm, spur 8–14 mm; lower petal blades 3–6 mm. **FR** 8–21 mm, gen < 3 × longer than wide. **SEED** ± winged; coat inflated; cell margins wavy. Desert scrub, with scattered trees or not; 300–2500 m. Teh, TR, SNE, D; to sw UT, AZ, n Baja CA.

ssp. ***pallidum*** (Munz) M.J. Warnock PALE-FLOWERED LARKSPUR **ST** 27–95 cm. **LVS** gen basal, cauline much reduced; lobes 3–7, often > 6 mm at widest. **INFL**: pedicels 4–17 mm apart. **FL**: sepals gen spreading, white to pink or blue, lateral 6–11 mm, 2–4 mm wide, spur 7–13 mm; lower petal blades 3–4 mm. **FR** 11– 14 mm. $2n=16$. Uncommon. Sagebrush scrub, chaparral; 900–1900 m. WTR. ❀DRN,DRY;DFCLT.

ssp. ***parishii*** PARISH'S or DESERT LARKSPUR **ST** 17–100 cm. **LVS**: basal 3–5-lobed but gen 0 in fl; cauline 3–15-lobed. **INFL**: pedicels 8–25 mm apart. **FL**: sepals ± sky blue, lateral reflexed, 8–12 mm, 3–6 mm wide, spur 8–15 mm; lower petal blades 3–6 mm. **FR** 9–21 mm. $2n=16$. Desert scrub, juniper woodland; 300–2500 m. Teh, e TR, SNE, D; to sw UT, AZ, n Baja CA. ❀DRN,DRY; DFCLT.

ssp. ***subglobosum*** (Wiggins) Harlan Lewis & Epling **ST** 19–78 cm. **LVS**: basal gen present in fl (cauline much reduced); lobes 7–12. **INFL**: pedicels 8–17 mm apart. **FL**: sepals spreading, dark blue, lateral 9–13 mm, 5–7 mm wide, spur 12–14 mm; lower petal blades 4–6 mm. **FR** 8–11 mm. UNCOMMON. Chaparral or desert scrub; 600–1300 m. w DSon; n Baja CA. ❀DRN,DRY;DFCLT.

D. parryi A. Gray Root sometimes > 10 cm. **ST** 15–110 (gen < 80) cm, gen curled-puberulent, esp below. **LVS** gen curled-puberulent; basal present or 0 in fl; lobes 5–27, gen < 6 mm at widest. **INFL**: fls 3–60; pedicels 5–68 mm, 8–50 mm apart, ± ascending, gen puberulent. **FL**: sepals reflexed or not, lateral 9–25 mm, spur 8–21 mm; lower petal blades 3–10 mm. **FR** 10–19 mm. **SEED** winged, otherwise ± bumpy. Chaparral, open woodlands; 0–2600 m. Teh, CW, SW; n Baja CA.

ssp. ***blochmaniae*** (E. Greene) Harlan Lewis & Epling (p. 927) DUNE or BLOCHMAN'S LARKSPUR Root < 10 cm. **LVS**: basal gen 0 in fl; lobes of cauline 5–15. **FL**: sepals gen reflexed, lateral 16–25 mm, spur 11–16 mm; lower petal blades 7–10 mm, paler than sepals (esp on herbarium specimens). $2n=16$. RARE. Coastal chaparral, sand; 0–200 m. s CCo.

ssp. ***eastwoodiae*** Ewan EASTWOOD'S LARKSPUR Root < 10 cm. **LVS** basal and cauline in fl, mostly on lower third of st; lobes 5–15. **FL**: sepals gen reflexed, lateral 11–20 mm, spur 11–17 mm; lower petal blades 6–9 mm. Uncommon. Coastal chaparral, grassland, serpentine; 100–500 m. s CCo, SCoRO (San Luis Obispo Co?). ❀DRN,DRY;DFCLT.

ssp. ***maritimum*** (Davidson) M.J. Warnock Root < 10 cm. **LVS** basal and cauline in fl; lobes 5–10, often > 6 mm at widest. **FL**: sepals gen spreading, lateral 9–20 mm, spur 8–21 mm; lower petal blades 4–11 mm. $2n=16$. Coastal chaparral; 0–300 m. CCo, SCo, ChI; n Baja CA. ❀DRN,DRY;DFCLT.

ssp. ***parryi*** PARRY'S LARKSPUR Root 5–20 cm. **LVS**: basal gen 0 in fl; lobes of cauline 7–27. **FL**: sepals gen spreading, lateral 9–15 mm, spur 8–15 mm; lower petal blades 3–8 mm. $2n=16$. Chaparral, oak woodland; 200–1700 m. CW, SW; n Baja CA. [ssp. *seditiosum* (Jepson) Ewan] Hybrids with *D. cardinale* have been called *D. inflexum* Davidson; also hybridizes with *D. gypsophilum*, *D. hesperium*, *D. umbraculorum*. ❀DRN,DRY:15–24.

ssp. ***purpureum*** (Harlan Lewis & Epling) M.J. Warnock MOUNT PINOS LARKSPUR Root gen > 10 cm. **LVS** basal and cauline in fl, mostly on lower third of st; lobes 3–20. **FL**: sepals gen reflexed, purplish or dark blue, lateral 7–11 mm, spur 10–13 mm; lower petal blades 3–5 mm. **FR** 10–15 mm. $2n=16$. UNCOMMON. Sagebrush scrub, dry chaparral; 1000–2600 m. Teh, WTR. [*D. parishii* ssp. *p.* Harlan Lewis & Epling] May be threatened by development. ❀DRN,DRY;DFCLT.

D. patens Benth. SPREADING LARKSPUR Root round to diffuse-fibrous. **ST** 10–90 (gen < 50) cm; base gen narrower than root, not firmly attached to root, ± glabrous. **LVS** mostly on lower third of st, ± glabrous; lobes 3–10, terminal widest near middle. **INFL**: pedicels 10–78 mm, gen > 10 mm apart, glabrous to glandular. **FL**: sepals reflexed, lateral 7–20 mm, spur 8–18 mm; lower petal blades 3–8 mm. **FR** 12–23 mm, ± curved. **SEED** with inflated collar at widest end, otherwise smooth, shiny. Grassland, woodland, forest; 80–2800 m. s NCoR, GV, SnFrB, SCoR, SW.

ssp. ***hepaticoideum*** Ewan (p. 927) **ST** 25–80 cm. **LVS** basal and lower cauline gen divided < 80% to petiole; lobes 3–5, at least those of lower lvs gen > 15 mm wide. **INFL**: pedicels gen glabrous. **FL**: lateral sepal 11–17 mm, spur 10–18 mm; lower petal blades 5–8 mm. Riparian woodlands; 300–1300 m. SCoRO, SW (exc ChI). ❀DRN,DRY,SHD;DFCLT.

ssp. ***montanum*** (Munz) Ewan **ST** 30–70 cm. **LVS**: basal 0 in fl; cauline gen divided > 80% to petiole; lobes 5–10, < 10 mm wide. **INFL**: pedicels puberulent, gen glandular. **FL**: lateral sepal 7–11 mm, spur 8–14 mm; lower petal blades 3–6 mm. Open conifer forest; 1500–2800 m. TR, PR. ❀DRN,DRY,SHD;DFCLT.

ssp. ***patens*** **ST** 10–90 cm. **LVS** basal and lower cauline divided > 80% of way to petiole; lobes 3–5, gen < 15 mm wide. **INFL**: pedicels gen glabrous. **FL**: lateral sepal 9–20 mm, spur 10–15 mm; lower petal blades 4–6 mm, inner lobes gen hairier. $2n=16$. Grassland, open woodlands; 80–1100 m. s NCoR, GV, SnFrB, n SCoR. Hybridizes with *D. decorum*, *D. nudicaule*, *D. variegatum*. ❀DRN,DRY,SHD;DFCLT.

D. polycladon Eastw. HIGH MOUNTAIN LARKSPUR Root gen > 15 cm, distally branched; buds prominent (exc on herbarium specimens). **ST** 15–160 (gen 80–120) cm, glabrous, often 2 or more per root system. **LVS** mostly on lower third of st, glabrous; lobes often > 6 mm at widest. **INFL** often ± 1-sided; fls 3–35; pedicels 10–150 mm, ± S-shaped, 10–80 mm apart, glabrous to puberulent. **FL**: lateral sepals 12–18 mm, spur 11–22 mm. **FR** 13–20 mm. **SEED** ± striate. $2n=16$. Streamsides, wet talus; 2200–3600 m. SNH, W&I. Hybridizes with *D. nuttallianum*, *D. glaucum*. ❀DRN,DRY; DFCLT.

D. purpusii Brandegee (p. 927) PURPUS', ROSE-FLOWERED, or KERN COUNTY LARKSPUR Root often > 15 cm, distally branched. **ST** 30–120 cm; base gen narrower than root, not firmly attached to root, ± glabrous. **LVS** mostly on lower half of st, ± puberulent; lobes > 6 mm at widest or not. **INFL** narrow; pedicels 5–48 mm, 8–50 mm apart, ± ascending, hairs glandular, yellow. **FL**: sepals gen reflexed, pink to deep pink, lateral 10–16 mm, spur 10–19 mm; lower petal blades 3–4 mm, ± glabrous, angled > 130° to claw. **FR** 11–29 mm. **SEED** shiny; coat inflated, ± clear, winged. $2n=16$. UNCOMMON. Talus, cliffs: 300–1300 m. s SN. ❀DRN,DRY, SHD;DFCLT.

D. recurvatum E. Greene VALLEY or RECURVED LARKSPUR Root > 15 cm or not. **ST** 18–85 (gen < 60) cm; base often narrower than root, but firmly attached to root, ± glabrous. **LVS** ± glabrous; basal gen >> cauline; lobes 3–11. **INFL**: pedicels 10–56 mm, 7–25 mm apart, ± glabrous. **FL**: sepals gen reflexed, gen light blue, lateral

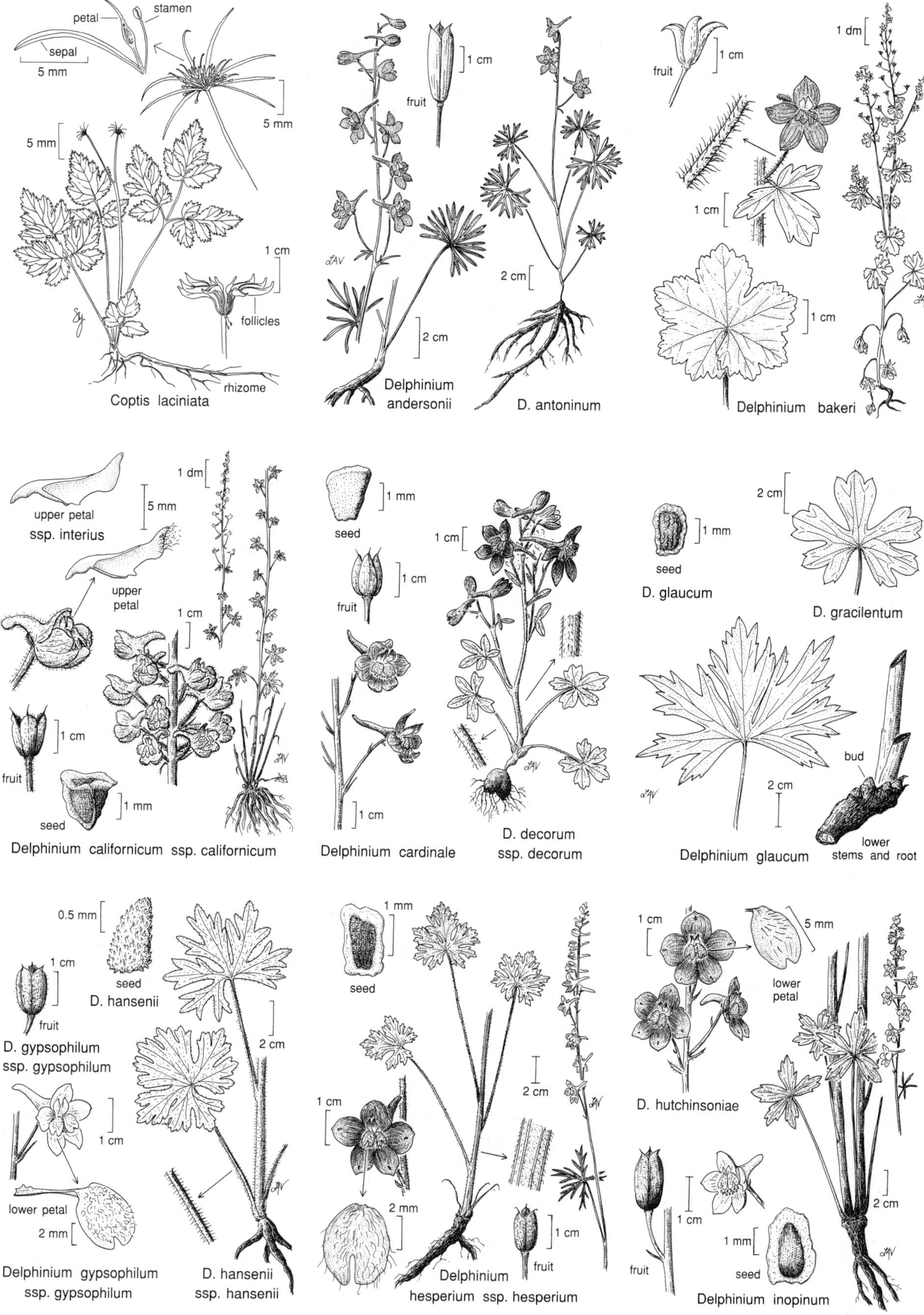

petal
stamen
sepal
5 mm
5 mm
5 mm
1 cm
follicles
rhizome
Coptis laciniata
1 cm
fruit
2 cm
Delphinium andersonii
2 cm
D. antoninum
fruit
1 cm
1 dm
1 cm
1 cm
Delphinium bakeri
upper petal
ssp. interius
5 mm
1 dm
upper petal
1 cm
1 cm
fruit
seed
1 mm
Delphinium californicum ssp. californicum
seed
1 mm
1 cm
1 cm
fruit
1 cm
Delphinium cardinale
D. decorum ssp. decorum
seed
1 mm
D. glaucum
2 cm
D. gracilentum
2 cm
bud
lower stems and root
Delphinium glaucum
0.5 mm
seed
D. hansenii
1 cm
fruit
D. gypsophilum ssp. gypsophilum
2 cm
1 cm
lower petal
2 mm
Delphinium gypsophilum ssp. gypsophilum
D. hansenii ssp. hansenii
1 mm
seed
2 cm
1 cm
2 mm
1 cm
fruit
Delphinium hesperium ssp. hesperium
1 cm
5 mm
lower petal
D. hutchinsoniae
1 cm
2 cm
1 mm
fruit
seed
Delphinium inopinum

11–16 mm, spur 10–18 mm; lower petals white. **FR** 8–21 mm, gen < 3 × longer than wide. **SEED** winged; coat cell margins wavy. 2*n*=16. RARE. Poorly drained, fine, alkaline soils in grassland, *Atriplex* scrub; 30–600 m. GV. In cult.

D. stachydeum (A. Gray) Tidestrom SPIKED or ROCKY MOUNTAIN LARKSPUR Root gen > 15 cm, distally branched; buds ± prominent (exc on herbarium specimens). **STS** often > 1 per root system, 40–180 (gen > 100) cm, puberulent. **LVS** mostly on lower half but not in lower 1/5 of st; lobes 3–18. **INFL** gen branched; fls gen > 30; pedicels 10–30 mm, 4–25 mm apart, puberulent. **FL**: sepals ± bright blue, paler along median line due to hairs, lateral 9–13 mm, spur 11–16 mm. **FR** 10–15 mm. **SEED** ± striate. RARE. Coniferous forest edges, sagebrush scrub; 2300–2600 m. Wrn; to se WA, s ID, nw UT.

D. trolliifolium A. Gray (p. 927) COW POISON **ST** 40–180 (gen 60–120) cm; base gen narrower than root, not firmly attached to root, glabrous to hairy. **LF** glabrous; lobes > 1 cm at widest, tips sharply, irregularly cut. **INFL**: pedicels 7–96 mm, 4–50 mm apart, glabrous to hairy. **FL**: lateral sepals 8–21 mm, spur 10–23 mm. **FR** 15–34 mm, gen > 4 × longer than wide, curved. **SEED** with inflated collar at widest end, otherwise smooth. 2*n*=16. Oak woodland, coastal chaparral; 30–1100 m. NCo, n&c NCoRO. Hybridizes with *D. decorum*, *D. nudicaule*. ❀DRN:**4**,5,17&SHD:**6**,16.

D. uliginosum Curran (p. 927) SWAMP or BOG LARKSPUR **ST** 8–70 cm; base narrower than root, but firmly attached to root, ± glabrous. **LVS**: basal, ± fleshy, glabrous, blade gen divided < 50% to petiole, base tapered. **INFL**: fls 5–45; pedicels ascending, 3–104 mm, puberulent. **FL**: lateral sepals 9–14 mm, spur 10–14 mm; lower petals hairier on inner lobes. **FR** 10–18 mm. **SEED** ± winged; surface bumpy. 2*n*=16. UNCOMMON. Streamsides, chaparral, grassland, on serpentine; 400–600 m. s NCoRI. Very local. Hybridizes with *D. hesperium*. ❀DRY;DFCLT.

D. umbraculorum Harlan Lewis & Epling UMBRELLA LARKSPUR **ST** 40–85 cm; base ± glabrous, often narrower than root, but firmly attached to root. **LVS** basal and cauline, ± glabrous; lobes 3–10, those of lower lvs > or = 4 mm wide. **INFL**: pedicels ± ascending, 6–73 mm, 10–50 mm apart, glabrous or puberulent. **FL**: lateral sepals reflexed, 10–16 mm, spur 11–14 mm. **FR** 15–19 mm. **SEED** winged, otherwise smooth. 2*n*=16. UNCOMMON. Moist oak forest; 400–1600 m. SCoRO, WTR. Hybridizes with *D. parryi*. ❀DRY,SHD:14–17.

D. variegatum Torrey & A. Gray (p. 927) ROYAL LARKSPUR **ST** 14–85 (gen < 50) cm; base hairy. **LVS** mostly on lower third of st; lobes 3–15, gen overlapping; petioles hairy. **INFL** gen branched; fls 4–25; pedicels ± ascending, 6–74 mm, 10–25 mm apart, gen puberulent. **FL**: lateral sepal 10–25 mm, spur 10–19 mm, straight or downcurved for < 3 mm at tip; lower petal blades 4–11 mm, hairier on inner lobes, margins. **FR** 9–19 mm; veins gen colored. **SEED** winged. Grassland, open woodlands; 0–800 m. NCo, NCoR, CaR, SNF, GV, SnFrB, SCoR, ChI.

ssp. ***kinkiense*** (Munz) M.J. Warnock SAN CLEMENTE ISLAND LARKSPUR **INFL**: fls on main axis gen < 12. **FL**: sepals light blue to white, lateral 11–18 mm; lower petal blades 4–9 mm. RARE. Coastal grassland; 0–500 m. s ChI (San Clemente Island). [*D. kinkiense* Munz] In cult.

ssp. ***thornei*** Munz THORNE'S ROYAL LARKSPUR **INFL**: fls on main axis gen < 16. **FL**: sepals bright blue, lateral 17–21 mm; lower petal blades 6–11 mm. RARE. Grassland, oak woodland; 0–500 m. ChI (San Clemente Island).

ssp. ***variegatum*** **INFL**: fls on main axis gen < 10. **FL**: sepals dark royal blue, rarely white or lavender, lateral 10–25 mm; lower petal blades 4–11 mm. 2*n*=16,32. Grassland, open oak woodland; 20–800 m. NCo, NCoR, CaR, SNF, GV, SnFrB, SCoR. Pls with large fls [forma *superbum* Ewan] occur throughout range (most common SnFrB & n). ❀DRN,DRY;DFCLT.

ISOPYRUM

Per from clustered, slender to fusiform or ± spheric, fleshy roots, glabrous. **ST** ascending to erect; branches 0–few. **LVS** 1–3-ternate; lflet upper surface green, lower surface pale green to glaucous, segments gen wedge-shaped; basal 0–few, petioles gen > blade; cauline petioles short to ± 0. **INFL**: cyme, flat-topped, to fls solitary in CA, terminal or axillary. **FL** bisexual, radial; sepals gen 5, petal-like; petals gen 0; stamens 10–many; pistils 2–many, stalk-like base 0 or short. **FR**: follicles, glabrous; veins obvious, stalk-like base 0 to obvious, curved or not; beak straight to recurved. **SEED** brown, smooth to wrinkled. ± 30 spp.: temp N.Am, Eurasia. (Greek: ancient name, from grain-like fr) [Calder & Taylor 1963 Madroño 17: 69–76]

1. Fr stalk-like base 0; filaments > 15, ± thread-like; sepals 7–10 mm; lf segment lobes 2–9 mm wide .. ***I. occidentale***
1' Fr stalk-like base 0.5–2 mm; filaments ± 10, flat; sepals 3.5–5.5 mm; lf segment lobes 0.5–4 mm wide .... ***I. stipitatum***

I. occidentale Hook. & Arn. (p. 927) Pl 8–25 cm. **STS** 1–3, erect, gen simple. **LF** 1–2-ternate, 3–12 cm; segment lobes 2–3, gen < 1/2 segment length. **FL**: sepals 3–6 mm wide, white to pink; stamens 2–5.5 mm; pistils 3–5. **FR** 10–12 mm. Shaded slopes, chaparral, oak woodland, coniferous forest; 400–1700 m. NCoRI, SN, CW (exc CCo), n WTR. ❀SHD,DRN:7,15,16;DFCLT.

I. stipitatum A. Gray (p. 927) Pl 5–15 cm. **STS** (1)3–7, decumbent to erect, gen simple. **LF** 1–2-ternate, 4–11 cm; segment lobes gen 3 (middle segment sometimes entire), gen > 1/2 segment length. **FL**: sepals 1.5–2 mm wide, white; stamens 2–3 mm; pistils 3–5. **FR** 5–8 mm. Shaded slopes, chaparral, mixed-evergreen forest, oak woodland; 600–1400 m. NW, CaR, n SN, se SnFrB (Mount Hamilton Range), MP; s OR. SnFrB pls grow with and are much like *I. o.* ❀SHD,DRN:7,15,16;DFCLT.

KUMLIENIA

Per from short, stout caudex, gen ± glabrous; roots clustered, fleshy. **LVS** simple, basal, rosetted. **INFL** scapose, 1(–3)-fld; peduncle long. **FL** bisexual, radial; sepals 5–6, petal-like; petals 5–12, < sepals, gland-like; pistils many, style persistent in fr, beak-like. **FR** utricle-like, splitting tardily from base along 2 lines; wall thin, slightly transparent between veins; beak tapered, straight, tip hooked. **SEED** 1, cylindric. 2 spp.: w N.Am. (T.L. Kumlien, Swedish naturalist, mentor of E.L. Greene, 19th century) [Greene 1886 Bull Calif Acad Sci 1:337–338]

K. hystricula (A. Gray) E. Greene (p. 927) Pl 8–25 cm. **LF**: petiole 2–14 cm; blade 1.5–3 cm, reniform to round, crenate to palmately lobed, lobes rounded to obtuse, entire to toothed or lobed. **INFL**: peduncle reclining or decumbent. **FL**: receptacle glabrous, not elongated in fr; sepals 6–10 mm, white, glabrous; petals 2–3 mm, ± = stamens, narrowly oblanceolate, yellow or green. **FRS** in head-like cluster; body 3–4 mm, lanceolate, hairs sparse, short, stiff. Wet places, among rocks, streambanks, gen coniferous forest; 300–2300 m. SN (exc Teh). [*Ranunculus h.* A. Gray] ❀DRN, IRR:1,2,7,15,16;DFCLT.

MYOSURUS MOUSE-TAIL

Ann from fibrous roots, gen glabrous. **STS** 1–many, ascending to erect, slender, gen tufted. **LVS** simple; basal, ± sessile, thread-like to narrowly oblanceolate, entire. **INFL** scapose; peduncle 1-fld, in fr gen > 1 cm. **FL** bisexual, radial; receptacle in fr much elongated, cylindric; sepals 5–7, spurred, white to green, fading brownish; petals 0 or 3–5, white to greenish or yellowish, gen early deciduous; stamens 5–many; pistils many. **FR**: achenes, glabrous to puberulent; keel on outer surface, in depression or not, beak (continuation of keel) ± ascending to erect-appressed. 10–15 spp.: temp Am, Eurasia, New Zealand. (Greek: mouse tail, from receptacle in fr) [Campbell 1952 El Aliso 2:389–403; Stone 1959 Evolution 13:151–174] Fr needed for identification.

1. Fr beak gen erect to erect-appressed, gen < 0.5 mm, gen << or = body; stamens gen 10 .... ***M. minimus***
1′ Fr beak ± ascending, 0.3–2 mm, gen > body; stamens gen 5
  2. In fr, infl < lvs, peduncle < 1 cm; sepal spur < 1 mm .... ***M. sessilis***
  2′ In fr, infl > lvs, peduncle gen > 1 cm; sepal spur 1–3 mm
    3. Fr body ± compressed laterally, longer than wide, keel not in a depression; KR, CaR, SN (exc Teh), ScV, SnBr, MP .... ***M. apetalus***
    3′ Fr body not compressed, ± as long as wide, keel in a depression; e DMtns .... ***M. cupulatus***

M. apetalus C. Gay (p. 927) Pl 1.5–12 cm. **LF** 1–6 cm, thread-like to linear. **INFL** in fr > lvs. **FL**: receptacle in fr 4–15 mm; sepals 1–2.5 mm, spur 1–3 mm; petals 0 or 5, ± = sepals; stamens gen 5. **FR**: body ± compressed laterally, longer than wide; keel not in a depression; beak ± ascending, 0.3–2 mm. Wet places, vernal pools, marshes, shrubland; 1300–3100 m. KR, CaR, SN (exc Teh), ScV, SnBr, MP; to B.C., MT, Colorado, Chile. [*M. aristatus* Benth., *M. minimus* L. ssp. *montanus* G.R. Campbell]

M. cupulatus S. Watson (p. 927) Pl 2–14 cm. **LF** 1–7 cm, linear to narrow-oblanceolate. **INFL** in fr > lvs. **FL**: receptacle in fr 5–40 mm; sepals 1.5–3 mm, spur 1–2.5 mm; petals 5, 1.5–2.5 mm; stamens gen 5. **FR**: body not compressed, ± as long as wide; keel in a depression; beak ± ascending, 0.5–1 mm. Wet places, shrubland, pinyon/juniper woodland; 1000–1700 m. e DMtns; to NM.

M. minimus L. (p. 927) Pl 2–12 cm. **LF** 0.5–6 cm, thread-like to narrowly oblanceolate. **INFL** in fr < to > lvs. **FL**: receptacle in fr 1–40 mm; sepals 1–3 mm, spur ± = sepal; petals 5, 1.5–3 mm; stamens gen 10. **FR**: body ± compressed laterally or not; keel in a slight depression or not; beak gen erect to erect-appressed, gen < 0.5 mm. Wet places, vernal pools, marshes; < 1500 m. NCoR, CaRF, SNF, GV, SnFrB, s SCoRO, SCo, SnJt, MP; to B.C., e US, Eurasia. [var. *filiformis* E. Greene, ssp. *major* (E. Greene) G.R. Campbell] Pls from GV, SCo with infl < or = lvs have been called ssp. *apus* (E. Greene) G.R. Campbell, (little mousetail), hybridizes with *M. sessilis* in GV.

M. sessilis S. Watson (p. 927) Pl 1–10 cm. **LF** 1–7 cm, linear. **INFL** in fr < lvs; peduncle in fr < 1 cm. **FL**: receptacle in fr 10–30 mm; sepals 1.5–2.5 mm, spur < 1 mm; petals 3–5, ± = sepals; stamens gen 5. **FR** ± compressed laterally; keel not in a depression; beak ± ascending, ± 1 mm. Vernal pools, grassland; < 150 m. GV; s OR. [*M. minimus* var. *sessiliflorus* (Huth) G.R. Campbell] Sometimes confused with *M. m.*

RANUNCULUS BUTTERCUP

Ann, per, sometimes from stolons or caudices, terrestrial or aquatic; roots gen fibrous. **ST** prostrate to erect. **LVS** basal and gen cauline, gen reduced upwards, gen glabrous; petiole base flat, stipule-like or not; basal and lower cauline petioles gen long; blades simple to dissected or compound, entire to toothed. **INFL**: cyme, axillary or terminal, 1–few-fld. **FL** radial; sepals gen 5, gen early deciduous, gen glabrous, gen green to yellowish; petals gen 5, gen > sepals, gen white to yellow, shiny; nectar gland near petal base, pocket-like or with flap-like scale; anthers yellow; pistils gen many. **FR**: achene, gen compressed, beaked, gen glabrous; walls thick. ± 250 spp.: temp world-wide, trop mtns; some orn. (Latin: (Pliny) little frog, from gen wet habitats)

1. Pl gen aquatic, st gen submersed or floating (sometimes stranded); petals white, white and yellow, or yellow
  2. Most lvs floating or emergent, simple, entire .... ***R. hydrocharoides***
  2′ Most or all lvs submersed, gen 2–6-dissected (if floating or emergent, lvs, if any, sometimes simple, 3-lobed)
    3. Floating or emergent lvs gen 0 or like submersed, 3–6-dissected, segments thread-like, entire ..... ***R. aquatilus***
      4. Petiole below gen flat, wide, above gen thread-like; lf gen = or < internode; pedicel ± straight in fr; frs 10–30 .... var. ***capillaceus***
      4′ Petiole gen flat, wide ± throughout; lf gen << internode; pedicel recurved in fr; frs 10–50 ..... var. ***subrigidus***
    3′ Floating or emergent lvs gen 0 or unlike submersed, ± 3-lobed to 1-ternate, lobes or lflets linear to obovate, entire to lobed
      5. Petals yellow throughout; sepals 4.5–8 mm; fr body sides smooth or slightly papillate on sides .... ***R. flabellaris***
      5′ Petals white, bases yellow or not; sepals 1.5–5 mm; fr body cross-ridged
        6. Receptacle in fr hairy; frs 15–many, bodies 1–2 mm; submersed lvs 3–6-dissected; per .... ***R. aquatilus*** var. ***hispidulus***
        6′ Receptacle in fr glabrous; frs 2–6, bodies 2–3 mm; submersed lvs gen 2–3-dissected; ann .... ***R. lobbii***
1′ Pl gen terrestrial (if aquatic st gen emergent); petals gen yellow (sometimes white or reddish throughout)
  7. Basal lf entire to crenate or lobed << 1/2 way to midrib or blade base
    8. St erect to gen prostrate or reclining, often rooting at lower nodes, stolon-like or from stolons
      9. Basal lf blade linear to lanceolate, base gen tapered .... ***R. flammula***
      9′ Basal lf blade elliptic to reniform, base gen rounded or lobed
        10. Lf gen crenate; fr cluster cylindric, body striate to ridged on sides; major roots slender, not narrowed to st .... ***R. cymbalaria*** var. ***saximontanus***
        10′ Lf ± entire; fr cluster spheric, body smooth to striate on sides; major roots thick, narrowed to st .... ***R. gormanii***
    8′ St decumbent to gen erect, not rooting at nodes, not stolon-like, not from stolons

11. Petals 1–3, 1–3 mm, ± = sepals; fr body 0.9–1.5 mm
12. Sepals 3; upper cauline lf lanceolate to ovate; fr body smooth to finely net-like .... ***R. bonariensis*** var. ***trisepalus***
12′ Sepals 5; upper cauline lf linear to oblong; fr body rough to minutely papillate .... ***R. pusillus***
11′ Petals 5–many, 3–15 mm, gen > sepals; fr body 1–2.5 mm
13. Sepals 1–3.5 mm, ± spreading; petals 3–6.5 mm; cauline lf blade ovate to cordate, entire to barely toothed .... ***R. populago***
13′ Sepals 3–7 mm, ± reflexed; petals 5–15 mm; cauline lf blade linear to lanceolate, entire or deeply 2–3-lobed
14. Upper cauline lf linear to narrowly lanceolate, entire; sts gen 3–5 from base, gen branched from base .... ***R. alismifolius***
15. Basal lf blade 2–4 cm, gen ovate .... var. ***alismellus***
15′ Basal lf blade 4–12 cm, lanceolate to oblong .... var. ***alismifolius***
14′ Upper cauline lf 0 or oblong to ovate, entire or gen deeply 2–3-lobed; sts gen 1 from base, sometimes few-branched above .... ***R. glaberrimus***
16. Basal lf elliptic to obovate; upper cauline lf 0 or with terminal lobe gen > lateral .... var. ***ellipticus***
16′ Basal lf ovate to round; upper cauline lf 0 or with lobes gen ± equal .... var. ***glaberrimus***
7′ Basal lf gen lobed > 1/2 way to midrib or blade base to dissected or compound
17. Lf blades gen glabrous, petiole hairs very sparse or gen 0
18. Petals white, red-tinged or not; fr body 8–15 mm, very compressed, sides 6–10 mm wide, margin wing-like; pl scapose .... ***R. andersonii***
18′ Petals yellow; fr body < 7 mm, compressed to plump, sides < 4 mm wide, margin ± thick, not wing-like (but back sometimes keeled); pl scapose or gen not
19. Fr body 2–7 mm, beak 2–4 mm, cluster spheric
20. Basal lf toothed to deeply 3-lobed; fr body sides spiny, back very keeled; ann, bien .... [2]***R. muricatus***
20′ Basal lf gen 1-ternate; fr body sides smooth, back slightly keeled; per .... ***R. orthorhyncus*** var. ***bloomeri***
19′ Fr body 1–2 mm, beak < 1 mm, cluster ovoid to subcylindric
21. Ann; sepals 2–3 mm; petals 2–5 mm; fls many, axillary, terminal; st hollow below .... ***R. sceleratus***
21′ Per; sepals 3.5–7 mm; petals 5–12 mm; fls few, terminal; st solid .... ***R. eschscholtzii***
22. Basal lf with 3–5(7) lobes > 1/2 way to blade base, middle lobe with 1 tooth or lobe or gen entire, tips obtuse to rounded .... var. ***oxynotus***
22′ Basal lf with 3 lobes > 1/2 way to blade base, middle lobe with 3 teeth or lobes, tips acute var. ***suksdorfii***
17′ Lf blades gen soft-hairy to tomentose, petiole hairs gen > sparse
23. Pl scapose; lf 2–3-dissected, segments linear, ± entire; fr body < beak, finely tomentose or glabrous, with 2 lateral bulges .... ***R. testiculatus***
23′ Pl not scapose; lf entire, toothed, or deeply lobed to 1-compound, lobes or lflets oblong to round, ± entire to deeply lobed or cut; fr body gen > beak, sometimes puberulent but not tomentose, lenticular to ovoid, without bulges
24. Fr body sides papillate or spiny, puberulent or glabrous; ann, bien
25. Sepals ± 1–2 mm; petals 1–2 mm or 0; fr body 1–3 mm, sides papillate, hairs curved
26. Petals gen fewer than 5; basal lf deeply 3–5-lobed to 1-ternate, lobes or lflets oblanceolate to wedge-shaped, toothed to lobed; upper cauline lf 3-lobed to 1-ternate .... ***R. hebecarpus***
26′ Petals gen 5; basal lf coarsely toothed to deeply 3-lobed, lobes ± semicircular, toothed; upper cauline lf entire to deeply 3-lobed .... ***R. parviflorus***
25′ Sepals 3.5–7 mm; petals 5–8 mm; fr body 4–7 mm, sides spiny, hairs 0
27. Basal lf ± ovate to wedge-shaped; upper cauline lf lobes linear to lanceolate or narrow-oblanceolate; frs 4–10; fr keel gen few-spined .... ***R. arvensis***
27′ Basal lf cordate to reniform; upper cauline lf lobes wide-oblanceolate; frs gen 10–20; fr keel not spiny .... [2]***R. muricatus***
24′ Fr body sides ± smooth, not papillate (exc sometimes in *R. sardous*), gen glabrous; per (ann, bien in *R. sardous*)
28. St prostrate to ascending, rooting at lower nodes, stolon-like or not .... ***R. repens***
28′ St ± prostrate to erect, not rooting at nodes, not stolon-like
29. Petals 2–8 mm, = or 1–2.5 mm > sepals
30. Basal lf ovate to deltate, gen 1-ternate, lflets stalked; sepals 3–6 mm; petals 4–8 mm; fr beak ± straight or slightly curved .... ***R. macounii***
30′ Basal lf gen cordate, deeply 3-lobed; sepals 1.5–3 mm; petals 2–5.5 mm; fr beak curved .... ***R. uncinatus***
29′ Petals 4–19 mm, gen 3+ mm > sepals
31. Caudex bulb-like; sepals gen spreading, hairs 0-sparse — NCo (waif) .... ***R. bulbosus***
31′ Caudex 0 or not bulb-like; sepals gen puberulent to hairy, gen reflexed (exc in *R. acris*)
32. Receptacle, esp in fr, puberulent to hairy
33. Petals 7–22, blade gen oblong to obovate, length gen 2 × width .... [2]***R. californicus***
33′ Petals 5–8, blade gen widely obovate to round, length gen < 2 × width
34. Basal lf cordate to gen ovate, lflets gen 5; fr body 2–4 mm, beak 2–4 mm, ± straight .... ***R. orthorhynchus*** var. ***orthorhynchus***
34′ Basal lf ovate to gen cordate, lflets gen 3; fr body 2–3 mm, beak < 0.5 mm, ± straight to curved .... ***R. sardous***
32′ Receptacle, esp in fr, gen glabrous
35. Basal lf pentagon-shaped, deeply lobed; sepals spreading .... ***R. acris***
35′ Basal lf ovate to cordate or round, deeply lobed to compound; sepals gen reflexed

36. Fr body 3.5–5.5 mm, sides 3–4.5 mm wide .... ***R. canus***
36′ Fr body 2–4 mm, sides 1.5–3.5 mm wide
 37. Petals gen 7–22; fr beak gen 0.5–1 mm, gen curved .... [2]***R. californicus***
 37′ Petals gen 5–6; fr beak gen 1.5–2.5 mm, curved or ± straight .... ***R. occidentalis***

R. acris L. (p. 927) Per 15–45(60) cm. **ST** erect, gen few-branched above base, gen hairy. **LVS** hairy; basal and lower cauline petioles 4–16 cm, blades 2.5–6 cm, pentagon-shaped, with 3 lobes > 1/2 way to base, of these, lateral and often terminal deeply 2–3-lobed, these again lobed or toothed; upper cauline lvs deeply lobed, toothed. **FL**: receptacle glabrous; sepals 4–7 mm, soft-hairy; petals 7.5–15 mm, 5–10 mm wide. **FRS** 18–many; cluster spheric; body 2–2.5 mm, sides ± 2 mm wide, smooth, back keeled above middle; beak ± 0.5 mm, curved. 2*n*=14. Waste places, fields; < 100 m. n NCo (waif); to AK, e US, Caribbean; native to Eur.

R. alismifolius Benth. Per 4–50 cm; roots ± thick, fleshy. **STS** decumbent to erect, gen 3–5 from base, gen branched from base, glabrous. **LVS**: basal and lower cauline petioles 2–7 cm, blades 2–12 cm, ± oblong to ovate, entire, base gen tapered; upper cauline lvs linear to narrowly lanceolate, entire. **FL**: receptacle glabrous; sepals 3–5 mm, ± reflexed, glabrous to puberulent; petals 5–7 in CA, 5–15 mm, 2–5.5 mm wide. **FRS** 12–many; cluster spheric; body 1.5–2.5 mm, sides 1–2 mm wide, smooth to sparsely puberulent, back rounded; beak 0.5–1 mm, ± straight. Wet places, streambanks, meadows, coniferous forest; 1300–3600 m. KR, NCoRH, CaRH, SNH, SnBr, SnJt, MP, n W&I; to B.C., ID, WY, NV. Vars. intergrade, difficult.

var. ***alismellus*** A. Gray (p. 927) Pl 4–20(30) cm. **LF**: basal blade 2–4 cm, gen ovate. **FL**: petals 5–7 mm. Habitats of sp.; 1400–3600 m. KR, NCoRH, CaRH, SNH, SnBr, SnJt, n W&I; to WA, NV. ❀WET or IRR,DRN:1,2,7&SUN:4–6,15–17.

var. ***alismifolius*** (p. 927) Pl 15–50 cm. **LF**: basal blade 4–12 cm, oblong to lanceolate. **FL**: petals 6–10(15) mm. 2*n*=16. Habitats of sp.; 1300–2300 m. KR, NCoRH, CaRH, n&c SNH, MP; to B.C., ID, WY. [vars. *hartwegii* (E. Greene) Jepson & *lemmonii* (A. Gray) L. Benson] ❀TRY;DFCLT.

R. andersonii A. Gray (p. 927) Per 8–18 cm, scapose, glabrous. **LVS** basal; petioles 3–6 cm, blades 2–3 cm, cordate to reniform, 1-ternate, lflets gen 2-dissected, segments lanceolate to oblong. **FL**: receptacle puberulent; sepals 6.5–8 mm, persistent, white, red-tinged or not; petals 10–21 mm, 8–15 mm wide, white, red-tinged or not. **FRS** 14–many; cluster spheric; body 8–15 mm, very compressed, sides 6–10 mm wide, smooth, margin wing-like; beak ± 0.5 mm, ± straight. Rocky slopes, gravelly soils, shrubland, pinyon/juniper woodland; 900–2300 m. SNH (e slope), MP, W&I, n DMtns (Grapevine, Panamint Mtns); to OR, ID, NV. ❀TRY; DFCLT.

R. aquatilus L. Per (5)20–80 cm, aquatic, often mat-forming. **ST** submersed or floating, branched throughout, rooting at lower nodes, glabrous. **LVS** cauline, most or all submersed; submersed blades 10–40 mm, 3–6-dissected, cordate to reniform, segments thread-like, petioles < 1.5 cm; floating or emergent gen 0 or few, like submersed or not. **FL** floating or emergent; receptacle hairy; sepals 2–5 mm; petals 4–10(14) mm, 1–2.5 mm wide, white, base yellow or not. **FRS** 10–50; cluster spheric; body 1–2 mm, sides 1–1.5 mm, with transverse, broken, wavy ridges, glabrous to sparsely puberulent, back ± rounded; beak < 0.5 mm, ± straight. Ponds, lake margins, marshes, rivers; < 2900 m. CA-FP (exc ChI), GB; to AK, e N.Am, Mex. Vars. intergrade, difficult to separate; var. *a.* native to Eurasia. ❀WET,SUN:1–3,**4–7**,8,9,14,**15–18**,19–24.

var. ***capillaceus*** (Thuill.) DC. (p. 927) **LVS** gen = or < internodes; submersed lvs 4–6-dissected, segments thread-like, petioles below gen flat, wide, above gen thread-like; floating or emergent gen 0 or like submersed. **INFL**: pedicel ± straight in fr. **FRS** 10–30. Habitats, elevations of sp. CA-FP (exc ChI), GB; to AK, e N.Am, Mex. [var. *harrisii* L. Benson]

var. ***hispidulus*** E. Drew (p. 927) **LVS**: submersed 3–6-dissected, segments thread-like, petioles below gen flat, wide, above gen thread-like; floating or emergent gen 0 or unlike submersed, ± 3-lobed to 1-ternate, lobes or lflets linear to obovate, entire to few-toothed or -lobed. Habitats of sp.; < 1700 m. NW, CaR, SN (exc Teh), CCo, SnFrB, PR, MP; to AK, MT.

var. ***subrigidus*** (Drew) Breitung **LVS** gen << internodes; submersed lvs 3–5-dissected, segments thread-like, petioles gen flat, wide ± throughout; floating or emergent gen 0 or like submersed. **INFL**: pedicel recurved in fr. **FRS** 10–50. Habitats of sp.; < 1800 m. n CaR, CCo, n SNH, SnJt, MP; to B.C., e N.Am, Mex. [*R. subrigidus* Drew] Much like var. *caprillaceus*.

R. arvensis L. (p. 927) Ann or bien 10–40 cm; roots 3–4.5 mm wide above, tapered. **ST** gen erect; branches 0–few; hairs 0-sparse. **LVS**: hairs 0 or sparse, appressed; basal and lower cauline petioles 1–4.5 cm, blades 1.5–6 cm, ± ovate to wedge-shaped, entire to deeply lobed or dissected, lobes or segments 3, entire to few-toothed or -lobed; upper cauline lvs gen deeply 3-lobed, lobes linear to (ob)lanceolate. **FL**: receptacle glabrous, tip puberulent; sepals 3.5–6 mm, short-hairy; petals 6–8 mm, 2.5–3.5 mm wide. **FRS** 4–10; cluster spheric; body 4–6 mm, sides ± 3 mm wide, spiny; keel thick, few-spined; beak 2–3 mm, curved. Waste places, fields, ditchbanks; < 1000 m. KR, NCoRO, KR, CaR, SNF, SCoRO, expected elsewhere; to WA, ID, UT, also e US; native to Eur.

R. bonariensis Poiret var. ***trisepalus*** (Gill) Lourt. (p. 927) Ann 4–20(30) cm. **ST** decumbent to erect, gen branched below middle; hairs 0–sparse. **LVS** lanceolate to ovate, gen glabrous; basal and lower cauline petioles 1.5–8.5 cm, blades 0.6–2 cm, gen entire; upper cauline lvs gen smaller. **FL**: receptacle glabrous; sepals 3, 2–3 mm; petals 2–3, 2–3 mm, ± 1 mm wide. **FRS** 9–many; cluster conic to ovoid; body 1–1.5 mm, ± 1 mm wide, ± plump, smooth to finely net-like, back rounded; beak minute. Clay soils, vernal pools; < 1200 m. n&c SNF, e GV (exc s SnJV). [*R. alveolatus* Carter] Other vars. in Chile, Argentine.

R. bulbosus L. Per 20–30(50) cm; caudex 10–20 mm wide, bulb-like. **ST** gen erect, few-branched; hairs below sparse, rough, above ± 0. **LVS** gen hairy; basal and lower cauline petioles 8–15 cm, blades 1.5–2.5 cm, gen cordate, deeply 3-lobed to 1-ternate, lobes or lflets reniform to wedge-shaped, toothed or lobed; upper cauline lvs 3-lobed, lobes ± oblong, entire to toothed. **FL**: receptacle puberulent; sepals 5.5–8 mm, gen spreading, hairs 0–sparse; petals 8–15 mm, 6–10 mm wide. **FRS** 12–many; cluster spheric; body 2.5–3 mm, sides 2–2.5 mm wide, smooth, back keeled, whitish; beak < 0.5 mm, curved. 2*n*=16. Meadows, roadside ditches; < 100 m. n NCo (waif); to B.C., e US; native to Eur.

R. californicus Benth. (p. 929) Per (11)18–70 cm. **ST** ± prostrate to erect, branched to not, ± glabrous to hairy. **LVS** glabrous to hairy; basal and lower cauline petioles 3–20 cm, blades 1.5–8 cm, ovate, cordate, or round, gen 1-ternate (terminal lflet deeply 3-lobed) to -pinnate, lflets 3 or 5, gen wedge-shaped, toothed to deeply cut; upper cauline lvs very deeply lobed, lobes linear to oblong, toothed. **FL**: receptacle gen glabrous; sepals 3.5–7 mm, hairs short, stiff; petals gen 7–22, 5–13 mm, 2–6 mm wide. **FRS** 5+, cluster spheric to wide-ovoid; body gen 2–3 mm, sides 1.5–2.5 mm wide, ± smooth, glabrous or puberulent, back not keeled; beak gen 0.5–1 mm, gen curved. Grassland, oak woodland, mixed-evergreen or coniferous forest; < 2400 m. CA-FP; s OR, Baja CA. [vars. *austromontanus* L. Benson, *cuneatus* E. Greene, *gratus* Jepson & *rugulosus* (E. Greene) L. Benson] Intergrades complexly with *R. canus*, *R. occidentalis*. ❀DRN:**4–6,17**&part SHD:**7,22–24**&IRR:2,3,8, 9,**14–16,18–21**; dry in summer.

R. canus Benth. (p. 929) Per 11–65 cm. **ST** gen erect, gen few-branched, hairy. **LVS** appressed-hairy; basal and lower cauline petioles 2–12 cm, blades 3–6.5 cm, cordate, gen deeply 3-lobed, lobes wedge-shaped, toothed to deeply dissected; upper cauline lvs deeply 3-lobed, lobes linear to narrowly lanceolate, toothed. **FL**: receptacle glabrous; sepals 4.5–7 mm, hairs dense, short; petals 5–17 (23), 8–13 mm, 3–9 mm wide. **FRS** 11–many; cluster spheric; body 3.5–5.5 mm, sides 3–4.5 mm wide, smooth, glabrous or puberulent, back keeled; beak ± 0.5–1 mm, ± straight, tip curved. Gen clay soils, grassland, oak/pine woodland, coniferous forest; < 2300 m. s NCoRI, SNF, GV, e SnFrB, SCoR, TR, PR; s OR, Baja CA. [vars. *laetus* (E. Greene) L. Benson & *ludovicianus* (E. Greene) L. Benson] Intergrades complexly with *R. californicus*. ❀TRY.

R. cymbalaria Pursh var. ***saximontanus*** Fern. (p. 929) Per 3–20(30) cm, often scapose, gen from stolons. **ST** gen erect, simple; hairs 0–sparse. **LVS**: petioles 1.5–8 cm, blades 0.5–2.5 cm, ovate to reniform, gen crenate. **FL**: receptacle glabrous to puberulent; sepals 2–5 mm; petals 5–8, 4–8 mm, 2–3 mm wide. **FRS** many; cluster cylindric; body 1–2 mm, sides ± 1 mm wide, striate to ridged, glabrous, back rounded; beak 0.5–1 mm, ± straight. 2*n*=16. Meadows, streambanks, marshes, pond margins; 200–3200 m. NW (exc NCo), CaR, SN, e SCo, TR, PR, GB, DMoj; to w Can, SD, Mex, also S.Am. Other vars. in Rocky Mtns, Asia. ❀WET:1,**2**,3,**7**,8, 10,11,14,18–23&SUN:4,5,**6**,**15**,**16**,17,24.

R. eschscholtzii Schldl. Per 5–25 cm, scapose or not. **ST** gen erect, glabrous; branches 0–few, at base. **LVS**: basal and lower cauline (if present) petioles 1–8 cm, blades 1–3 cm, round to reniform, with 3–5(7) lobes > 1/2 way to base, lobes entire to few-toothed or -lobed; upper cauline lvs 0 or 1–2, sessile, deeply 3-lobed to 1-ternate, lobes or lflets oblong to elliptic, entire. **FL**: receptacle glabrous; sepals 3.5–7 mm, hairs 0-sparse; petals 5–12 mm, 5–9 mm wide. **FRS** 17–many; cluster narrowly ovoid to subcylindric; body 1.5–2 mm, ± plump, sides ± 1 mm wide, glabrous, back rounded; beak 0.5–1 mm, ± straight. 2*n*=56. Meadows, rocky slopes, ledges; 1800–4000 m. KR, CaR, SNH, SnBr, SnJt, GB; to AK, MT, Colorado, AZ. Var. *e.* in alpine, arctic w N.Am but probably not in CA.

var. ***oxynotus*** (A. Gray) Jepson (p. 929) **LVS**: basal and lower cauline (if present) with 3–5(7) lobes > 1/2 way to blade base, middle lobe with 1 tooth or lobe or gen entire, tips obtuse to rounded. Habitats and elevations of sp. CaR, SNH, SnBr, SnJt, GB; NV. Scapose pls in SNH with lf bases not persistent doubtfully belonging to var. *e.* ❀TRY;DFCLT.

var. ***suksdorfii*** (A. Gray) L. Benson (p. 929) **LVS**: basal and lower cauline (if present) with 3 lobes > 1/2 way to blade base, middle lobe with 3 teeth or lobes, tips acute. Meadows, rocky slopes; 1700–2200 m. KR; to WA, MT, WY. ❀TRY;DFCLT.

R. flabellaris Raf. (p. 929) Per < 70 cm, submersed or terrestrial. **STS** thick, floating or prostrate, branched throughout, rooting at lower nodes, glabrous. **LVS** cauline, most or all submersed; petioles 0.5–4 cm; blades 1.5–8 cm, round to reniform, submersed 3–5-dissected, segments thread-like; floating or emergent gen 0 or unlike submersed, 3-lobed, lobes linear to oblong. **FL**: receptacle puberulent; sepals 4.5–8 mm; petals 5–8, 7–13 mm, 4–8.5 mm wide. **FRS** many; cluster ovoid to spheric; body 1.5–2 mm, sides 1–1.5 mm wide, smooth or slightly papillate; keel thick; beak 1–1.5 mm, ± straight. Shallow water, ponds, marshes, streams; 600–1900 m. KR, n SNH, MP; to B.C., e&s US. ❀TRY.

R. flammula L. (p. 929) Per (3)10–45 cm. **ST** prostrate to reclining, often stolon-like, branched, glabrous. **LVS**: basal and lower cauline petioles 2–8 cm, blades 2–6 cm, linear to lanceolate, entire, base gen tapered; upper cauline lvs smaller. **FL**: receptacle glabrous; sepals 2–5 mm, glabrous to puberulent; petals 3–6 mm, 2–7 mm wide. **FRS** 5–many, cluster spheric; body 1.5–2 mm, sides ± 1 mm wide, gen smooth, back faintly keeled; beak < 1 mm, ± straight. 2*n*=32. Wet meadows, marshes, pond margins, streambanks, coniferous forest; < 2300 m. NW, CaR, n&c SN, CCo, SnFrB, SnBr; to AK, e N.Am. [vars. *ovalis* (Bigelow) L. Benson and *samolifolius* (E. Greene) L. Benson] ❀TRY.

R. glaberrimus Hook. Per 5–26 cm, scapose or not; roots fleshy. **STS** gen 1 from base, decumbent to erect, glabrous; branches gen 0–few, above. **LVS** gen basal; basal and lower cauline (if present) petioles 2–8 cm, blades 2–5 cm, elliptic to round, gen tapered at base, entire to 3-lobed; upper cauline lvs 0 or oblong to ovate, gen sessile, entire or gen deeply 2–3-lobed. **FL**: receptacle glabrous; sepals 4–7 mm, reflexed, hairs 0-sparse; petals gen 5, 6–15 mm, 4–8 mm wide. **FRS** many; cluster spheric; body ± 2 mm, sides ± 1.5 mm wide, glabrous to sparsely puberulent, back faintly keeled; beak ± 0.5 mm, ± straight. Open areas, meadows, rocky soils in coniferous forest, juniper woodland, sagebrush scrub, shrubland; 900–3500 m. CaR, n SNH, MP, n SNE; to B.C., SD, NM.

var. ***ellipticus*** (E. Greene) E. Greene (p. 929) **LVS**: basal and lower cauline (if present) elliptic to obovate, gen entire; upper cauline lvs entire or gen deeply 2–3-lobed, terminal lobe gen > lateral. Coniferous forest, shrubland; 900–3500 m. n SNH (e slope), MP, n SNE; to B.C., NE, NM. ❀DRN,IRR:1–3,7,14,18&SUN:6,15, 16;DFCLT.

var. ***glaberrimus*** (p. 929) **LVS**: basal and lower cauline (if present) ovate to round, entire to coarsely crenate; upper cauline lvs entire or gen deeply 2–3-lobed, lobes ± equal. Juniper woodland, sagebrush scrub; 900–3300 m. CaR, n SNH, MP; to B.C., SD, NM. ❀DRN,IRR,SUN:1–3,6,7,14–16,18;DFCLT.

R. gormanii E. Greene (p. 929) Per 5–20 cm; major roots thick, narrowed to st. **ST** prostrate to ascending, rooting at lower nodes, stolon-like or not, few-branched, glabrous. **LVS** gen basal; basal and lower cauline (if present) petioles 2–7 cm, blades 0.5–4 cm, elliptic to widely ovate, ± entire; upper cauline lvs 0 or smaller. **FL**: receptacle glabrous; sepals 1.5–3 mm; petals 5–7, 3.5–6 mm, 1.5–3 mm wide. **FRS** 5–15; cluster spheric; body 1.5–2 mm, sides ± 1.5 mm wide, smooth to striate, back faintly keeled; beak 0.5–1 mm, curved. Wet meadows, pond margins, coniferous forest; 1300–3300 m. KR; OR.

R. hebecarpus Hook. & Arn. (p. 929) Ann (2)8–30 cm. **ST** erect, slender, soft-hairy; branches 0–few. **LVS** soft-hairy; basal and lower cauline petioles 1.5–8 cm, blades 0.5–2 cm, gen cordate, deeply 3–5-lobed to 1-ternate, lobes or lflets oblanceolate to wedge-shaped, toothed to lobed; upper cauline lvs 3-lobed to 1-ternate, lobes or lflets entire to toothed. **FL**: receptacle glabrous; sepals 1–2 mm, hairs short, soft; petals 0–3(5), 1–2 mm, < 1 mm wide. **FRS** 3–11; cluster spheric; body ± 2 mm, sides ± 2 mm wide, papillate, puberulent, hairs curved or hooked at tip, back keeled; beak ± 0.5 mm, ± straight to curved. Shaded areas, chaparral, oak, pine, or juniper woodland; < 1300 m. NCoRI, CaR, SN, ScV (Sutter Buttes), CW, SW (exc n ChI), MP; to WA, ID, Baja CA. ❀TRY.

R. hydrocharoides A. Gray (p. 929) FROG'S-BIT BUTTERCUP Per 5–25 cm, aquatic or terrestrial, scapose or not. **ST** submersed, floating, or prostrate to decumbent, glabrous; branches 0–few. **LVS** most floating or emergent; basal and lower cauline (if present) petioles 2–10(15) cm, blades simple, 0.6–3 cm, ovate to ± cordate, entire; upper cauline lvs 0 or much smaller. **FL**: receptacle glabrous; sepals 2–4 mm; petals 5–8, 2.5–6 mm, 1–3 mm wide. **FRS** 9–many; cluster spheric; body 1.5–4 mm, plump, ± 1 mm wide, sides smooth, back rounded; beak 1–1.5 mm, ± straight. RARE in CA. Marshes, small streams; 1100–2000 m. s SNE (Owens Valley); to NM, Mex. Last seen in 1874.

R. lobbii (Hiern) A. Gray (p. 929) LOBB'S AQUATIC BUTTERCUP Ann 20–80 cm, aquatic. **ST** submersed or floating, rooting at lower nodes, glabrous; branches few, upper often emergent. **LVS**: submersed blades 0.8–1.5 cm, gen 2–3-dissected, segments thread-like; floating or emergent lvs gen 0 or unlike submersed, blades 0.5–1 cm, gen reniform, flat, deeply 3-lobed to 1-ternate, lobes or lflets elliptic, entire to toothed. **FL**: receptacle glabrous; sepals 1.5–3 mm; petals 4–6 mm, 1–2 mm wide, white. **FRS** 2–6; cluster spheric; body 2–3 mm, ± plump, ± 1.5 mm wide, sides with transverse, broken, wavy ridges, back faintly keeled; beak minute. UNCOMMON. Shallow water, vernal pools, oak woodland, mixed-evergreen or redwood forest; < 300 m. NCo, s NCoR, CCo, SnFrB; to B.C. ❀TRY;DFCLT.

R. macounii Britton (p. 929) Per (sometimes fl 1st year) 20–65 cm; roots thick, fleshy. **ST** prostrate to erect, branched, gen rough-hairy. **LVS**: hairs on blades 0–sparse, on petioles dense; basal and lower cauline petioles (5)8–16 cm, blades 5–8(12) cm, ovate to deltate, gen 1-ternate, lflets stalked, deeply lobed, toothed; upper cauline lvs gen deeply 3-lobed, toothed. **FL**: receptacle puberulent; sepals 3–6 mm; petals 4–8 mm, 3–6 mm wide. **FRS** 20–many; cluster spheric; body 2–3 mm, sides 1.5–2 mm wide, smooth, back keeled; beak 1–1.5 mm, ± straight or slightly curved. 2*n*=42. Marshes, wet meadows, shrubland; 1400–1800 m. MP; to AK, MI, AZ. ❀TRY.

R. muricatus L. (p. 929) Ann or bien (5)15–50 cm. **ST** decumbent to erect, glabrous, branches 0–few. **LVS** glabrous or hairy; basal and lower cauline petioles 3–15 cm, blades 1–3.5(5) cm, cordate to reniform, toothed to deeply 3-lobed, lobes widely oblanceolate, toothed; upper cauline lvs slightly smaller. **FL**: receptacle hairs short, stiff; sepals 4–7 mm, glabrous to sparsely puberulent; petals 5–8 mm, 2.5–3.5 mm wide. **FRS** gen 10–20; cluster spheric; body

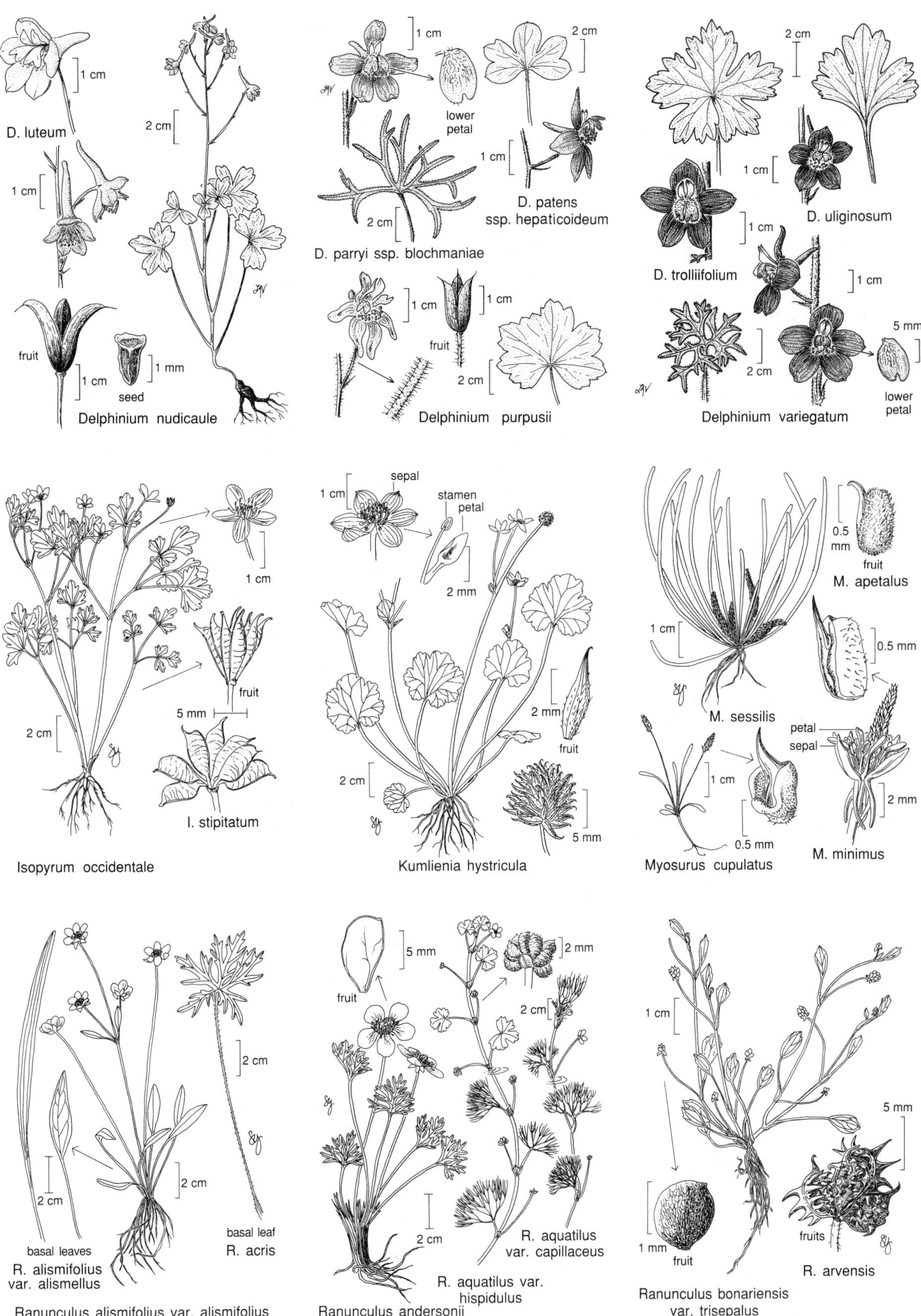

D. luteum
1 cm
2 cm
fruit
seed
1 mm
Delphinium nudicaule
1 cm
lower petal
2 cm
D. patens ssp. hepaticoideum
D. parryi ssp. blochmaniae
fruit
Delphinium purpusii
2 cm
D. uliginosum
D. trolliifolium
5 mm
lower petal
Delphinium variegatum
sepal
stamen
petal
2 mm
1 cm
fruit
I. stipitatum
Isopyrum occidentale
5 mm
Kumlienia hystricula
0.5 mm
M. apetalus
M. sessilis
petal
sepal
Myosurus cupulatus
M. minimus
basal leaves
R. alismifolius var. alismellus
basal leaf
R. acris
Ranunculus alismifolius var. alismifolius
R. aquatilus var. capillaceus
R. aquatilus var. hispidulus
Ranunculus andersonii
fruits
R. arvensis
Ranunculus bonariensis var. trisepalus

5–7 mm, sides 2.5–4 mm wide, spiny, margin thick, back very keeled, keel spines 0; beak 2–2.5 mm, curved. $2n$=48, 64. Wet fields, ditches, vernal pools; < 700 m. NW, CaR, SNF, GV, SnFrB, SCoRO; to WA, se US, native to Eur.

R. occidentalis Nutt. (p. 929) Per 10–60 cm. **ST** gen erect, few-branched, gen hairy. **LVS** gen soft-hairy; basal and lower cauline petioles 2–11 cm, blades 1.5–6 cm, ovate to cordate, gen deeply 3-lobed, lobes toothed to deeply cut; upper cauline lvs smaller. **FL**: receptacle glabrous; sepals 3–8 mm, hairs short; petals gen 5– 6, 4–15 mm, 2–8 mm wide. **FRS** 10–22; cluster spheric; body 2–4 mm, sides 1.5–3.5 mm wide, smooth, sparsely puberulent or glabrous, back keeled; beak gen 1.5–2.5 mm, curved or ± straight. Meadows, flats, open woodland or forest; 100–2200 m. CA-FP (exc GV, SW); to AK. [vars. *dissectus* L. Henderson, *eisenii* (Kellogg) A. Gray, *howellii* E. Greene, *rattanii* A. Gray, *ultramontanus* E. Greene] Intergrades complexly with *R. californicus*. ❀DRN: **4–6,17**& part SHD:**7,15,16,22–24**&IRR:2,3,8,9,**14,18–21**; dry in summer.

R. orthorhynchus Hook. Per 15–50(85) cm; roots ± thick, tapered. **ST** ascending to erect, branched throughout, glabrous to hairy. **LVS** glabrous to hairy; basal and lower cauline petioles 6–20 (25) cm, blades 3.5–10 cm, ovate to cordate, gen 1-ternate or -pinnate, lflets 3–7, coarsely toothed; upper cauline lvs smaller. **FL**: receptacle hairs short, stiff; sepals 6–10 mm, reflexed, hairs 0 to short, soft; petals 5–8, 10–20 mm, 4–7 mm wide. **FRS** 4–many; cluster spheric; body 2–4 mm, sides 2–3 mm wide, smooth; keel weak; beak 2–4 mm, ± straight. n=16. Meadows, wet, open areas, shrubland, woodland, or forest; < 2300 m. CA-FP (exc s SnJV, SCoR, SW), MP; to AK, MT, WY. Other vars. in AK, B.C.

var. ***bloomeri*** (S. Watson) L. Benson (p. 929) Pl 15–45 cm. **LVS**: basal and lower cauline ovate to cordate, glabrous, lflets ovate to ± round. **FL**: petals 10–17 mm, 4–8 mm wide, slightly obcordate. Clay soils, wet meadows, flats; < 300 m. NCo, NCoR, deltaic GV, CCo, SnFrB; OR. ❀SUN,IRRorWET:**4,5**,7–9,**14–17**,19–23, **24**.

var. ***orthorhynchus*** (p. 929) Pl 15–50 cm. **LVS**: basal and lower cauline cordate to gen ovate, hairy, lflets ovate to wedge-shaped. **FL**: petals 10–19 mm, 4–10 mm wide, wide-ovate to round. Meadows, open areas, gen coniferous forest; 300–2300 m. NW, CaR, SN, w ScV, n SnJV, n SnFrB; to B.C. [vars. *hallii* Jepson, *platyphyllus* A. Gray] NW, w ScV, n SnJV, n SnFrB; to B.C., MT, WY. ❀TRY.

R. parviflorus L. Ann 8–40 cm. **ST** erect, few-branched throughout; hairs sparse, soft. **LVS** soft-hairy; basal and lower cauline petioles 1.5–10 cm, blades 0.5–2.5 cm, cordate to round, coarsely toothed to deeply 3-lobed, lobes ± semicircular, toothed; upper cauline lvs entire to deeply 3-lobed. **FL**: receptacle glabrous; sepals ± 1 mm, hairs dense, short; petals 1–2 mm, < 1 mm wide. **FRS** 10–many; cluster spheric; body 1.5–3 mm, sides 1–2 mm wide, papillate, hairs on protuberances, curved or hooked at tip; keel thick, weak; beak ± 0.5 mm, curved. Waste areas, wet fields; < 1200 m. NCo, w KR, s NCoRO, CCo (Monterey), expected elsewhere; c&e US; native to Eur.

R. populago E. Greene (p. 939) Per 8–30 cm; roots ± fleshy above, tapered. **ST** gen erect, few-branched, glabrous. **LVS**: basal and lower cauline petioles 3–8(12) cm, blades 1.5–4.5 cm, ovate to cordate, entire to barely toothed; upper cauline lvs smaller. **FL**: receptacle glabrous or puberulent; sepals 1–3.5 mm; petals 3–6.5 mm, 1.5–3 mm wide. **FRS** 7–many; cluster spheric; body 1–2 mm, ± plump, 1–1.5 mm wide, smooth; keel faint; beak 0.5–1 mm, curved. Wet meadows, pond margins, streambanks; 1000–1900 m. KR, CaR, n SNH; to WA, ID.

R. pusillus Poiret (p. 939) Per 8–50 cm, emergent or terrestrial. **ST** decumbent to erect, slender, glabrous; branches 0–few. **LVS**: basal and lower cauline petioles 1–5 cm, blades 0.6–5 cm, narrowly lanceolate to ovate, entire; upper cauline lvs linear to oblong, entire. **FL**: receptacle glabrous; sepals 1–2 mm; petals 1–3, 1–2 mm, < 1 mm, gland pocket-like. **FRS** 18–many; cluster spheric to ovoid; body ± 1 mm, sides < 1 mm wide, rough to minutely papillate, back faintly keeled; beak minute. Shallow water, wet meadows, ditches; < 500 m. NCo, NCoRO, n&c SNF, SnFrB; also e&s US.

R. repens L. Per 10–60 cm. **ST** prostrate to ascending, rooting at lower nodes, stolon-like or not, branched; hairs 0 to rough. **LVS** glabrous to hairy; basal and lower cauline petioles 4–20 cm, blades 1.5–8 cm, ovate to cordate, gen 1-ternate, lflets coarsely lobed to toothed; upper cauline lvs smaller. **FL**: receptacle puberulent; sepals 5–7 mm, hairs short, soft; petals 8–14 mm, 5–10 mm wide. **FRS** 17–many; cluster spheric; body 2–3 mm, sides 1.5–2 mm wide, smooth, back faintly keeled; beak ± 1 mm, curved. $2n$=32. Waste areas, ditches, wet fields; < 200 m. NCo, CaR, c&s SNF, CCo, SnFrB, n SCo, expected elsewhere; to AK, e N.Am; native to Eur. Pls with weakly stoloned, erect sts, many petals, probably escaped from cult, have been called var. *pleniflorus* Fern.

R. sardous Crantz Ann or bien 18–50 cm. **STS** erect, few-branched; hairs 0 to sparse. **LVS** hairy; basal and lower cauline petioles 2.5–10 cm, blades 1.5–5 cm, ovate to gen cordate, gen 1-ternate, lflets gen 3, toothed, terminal often > lateral; upper cauline lvs deeply 3-lobed to 1-ternate. **FL**: receptacle hairs short; sepals 3–6 mm, reflexed, hairs short; petals gen 5, 7–10 mm, 5–8.5 mm wide. **FRS** 12–many; cluster spheric; body 2–3 mm, sides 2–2.5 mm wide, smooth or low-papillate esp near margin; keel thick; beak < 0.5 mm, ± straight or curved. Waste areas, as waif; < 300 m. n NCo; to WA, e N.Am; native to Eur.

R. sceleratus L. (p. 939) Ann 15–50 cm; roots ± fleshy. **ST** erect, simple to branched, hollow below; hairs 0–sparse. **LVS**: basal and lower cauline petioles 3–9(12) cm, blades 1.5–3 cm, cordate to reniform, deeply 3-lobed to 1-ternate, lobes or lflets toothed to deeply lobed; upper cauline lvs ± sessile. **FL**: receptacle gen puberulent; sepals 2–3 mm; petals 2–5 mm, 1–2 mm wide. **FRS** many; cluster ovoid to subcylindric; body 1–2 mm, sides < 1 mm wide, smooth or transversely ridged, back slightly keeled; beak < 1 mm. $2n$=32. Shallow water, lake or pond margins, streambanks; < 2000 m. CaR, GV, MP; to AK, MN, AZ, to e N.Am, Eur. [var. *multifidus* Nutt.]

R. testiculatus Crantz (p. 939) Ann 1–6(10) cm, scapose. **ST** decumbent to erect; branches 0–few, at base. **LVS** basal, tomentose; petioles < 3 cm, blades 0.5–3 cm, ovate, 2–3-dissected, segments linear, ± entire. **FL**: receptacle tomentose; sepals 2–8 mm, persistent in fr, tomentose; petals 2–5, 3–8 mm, 1–4.5 mm wide, early deciduous. **FRS** many; cluster ovoid to spheric; body 1.5–2.5 mm, plump, with 2 lateral bulges, finely tomentose or glabrous; beak 3–4 mm, spine-like, ± straight. Waste areas, overgrazed pastures, shrubland; 1400–2500 m. CaR, s SNH, GB, expected elsewhere; to WA, Colorado; native to Eurasia. [*Ceratocephalus t.* (Crantz) Roth] TOXIC to livestock.

R. uncinatus D. Don (p. 939) Per (sometimes fl 1st year) 15–60 cm. **ST** erect, slender; branches 0–few, above middle, hairs 0–sparse. **LVS**: basal and lower cauline petioles 5–16 cm, hairs gen soft, blades 2–9 cm, gen cordate, deeply 3-lobed, lobes toothed, hairs gen appressed; upper cauline lvs gen smaller. **FL**: receptacle glabrous; sepals 1.5–3 mm, hairs short; petals 4–5, 2–5.5 mm, 1–2.5 mm wide, gland with flap-like scale. **FRS** 5–many; cluster 2–2.5 mm, spheric, sides 1.5–2 mm wide, smooth, back faintly keeled; beak 1–2 mm, curved. Wet, gen shaded areas, mixed-evergreen or coniferous forest; 200–2400 m. KR, NCoR, CaR, SN, n SnFrB, SnBr, MP; to AK, MT, NM, AZ. Pls with fr body stiff-hairy, esp on margin, have been called var. *parviflorus* (Torrey) L. Benson.

THALICTRUM MEADOW-RUE

Per from caudex or rhizomes, dioecious or fls bisexual, gen glabrous. **STS** 1–few, gen erect; branches 0 or few. **LVS** gen 1–4-ternate, basal or basal and cauline, gen reduced upwards, petioled; segments wedge-shaped, fan-shaped, or ± round; upper surface gen green; lower surface pale green. **INFL**: raceme or panicle, axillary or terminal, gen erect; pedicels gen erect in fr;

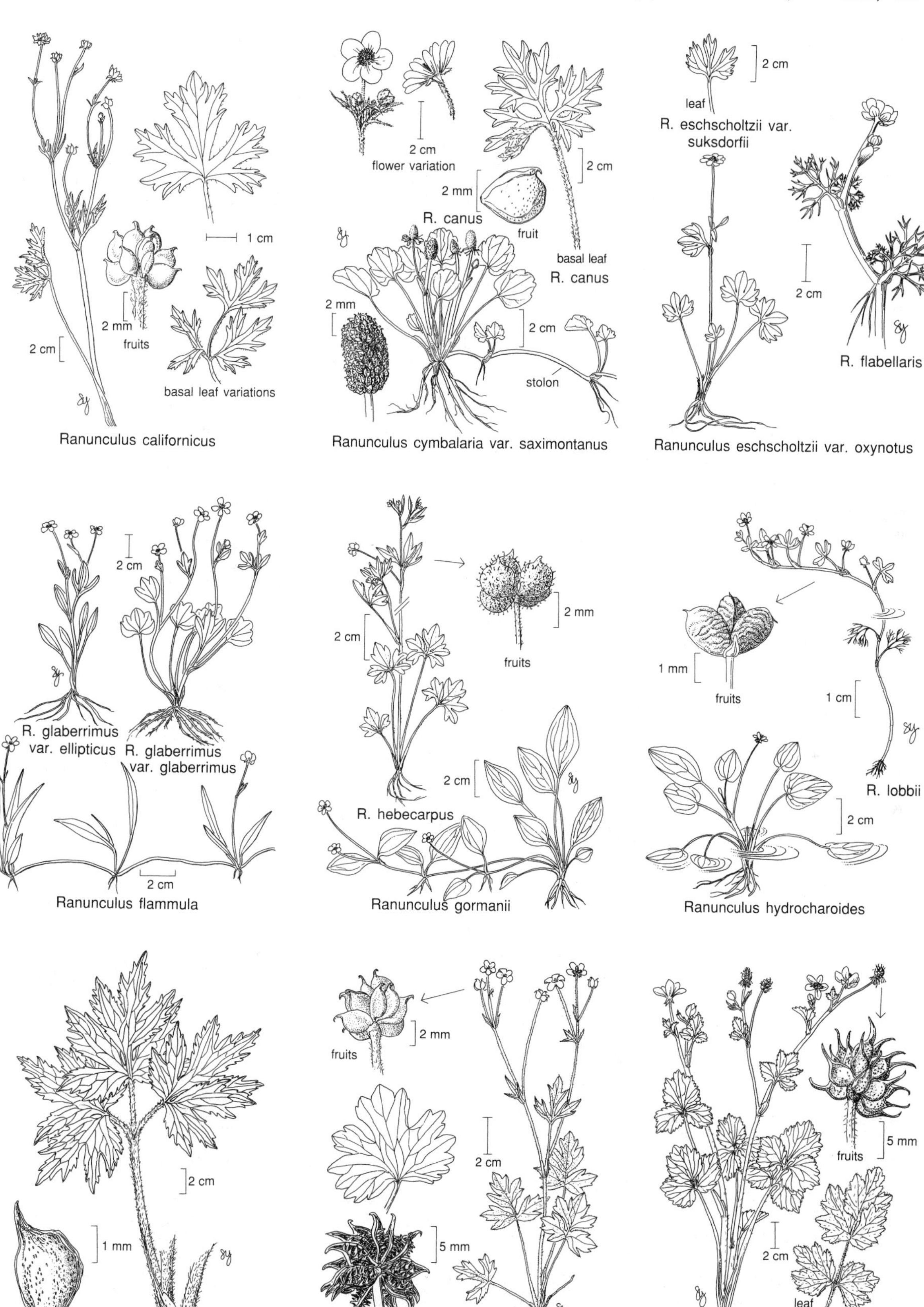

1 cm
2 mm
fruits
2 cm
basal leaf variations
Ranunculus californicus
2 cm
flower variation
2 cm
2 mm
R. canus
fruit
basal leaf
R. canus
2 mm
2 cm
stolon
Ranunculus cymbalaria var. saximontanus
2 cm
leaf
R. eschscholtzii var. suksdorfii
2 cm
R. flabellaris
Ranunculus eschscholtzii var. oxynotus
2 cm
R. glaberrimus var. ellipticus
R. glaberrimus var. glaberrimus
2 cm
Ranunculus flammula
2 mm
fruits
2 cm
2 cm
R. hebecarpus
Ranunculus gormanii
1 mm
fruits
1 cm
R. lobbii
2 cm
Ranunculus hydrocharoides
2 cm
1 mm
fruit
Ranunculus macounii
2 mm
fruits
2 cm
5 mm
fruits
R. muricatus
Ranunculus occidentalis
5 mm
fruits
2 cm
leaf
R. orthorhynchus var. orthorhynchus
Ranunculus orthorhynchus var. bloomeri

bracts simple to 1-ternate. **FL** radial; sepals 4–5, gen green, petal-like or not, often early deciduous; petals 0; stamens 8–many, gen > sepals, anthers gen narrowly oblong, tip gen abruptly pointed, filaments gen thread-like; pistils 2–20. **FR**: achenes, compressed laterally to not, ribbed or veined, beaked. ± 80 spp.: temp N.Am, Eurasia, Afr; some orn, medicinal. (Greek: name given by Dioscorides, Greek physician-botanist) [Boivin 1944 Rhodora 46:337–377,391–445,453–487]

1. Pl 5–15(20) cm; lvs gen basal, 1.5–6 cm; infl raceme; pedicel recurved in fr — alpine .... ***T. alpinum***
1' Pl 40–200 cm; lvs basal, cauline, 4–46 cm; infl gen panicle; pedicel ± erect in fr
2. Fl bisexual; anther tip obtuse, filament flat; fr body gen semi-circular to crescent-shaped .... ***T. sparsiflorum***
2' Fl unisexual; anther tip abruptly pointed, filament thread-like; fr body obliquely ovate to fusiform or ± circular to obovate
3. Fr mostly reflexed, body slightly compressed laterally, ribs gen 3 per side, ± straight; n NCo, nw KR .... ***T. occidentale***
3' Fr spreading to ascending, body compressed laterally throughout or only near margins, ribs 1–3 per side, ± curved; CA-FP (exc GV, ChI), GB .... ***T. fendleri***
4. Fr body ± compressed laterally throughout, side with 2–3 ribs, 0 veins, obliquely ± ovate to ± obovate; lower surface of upper lvs gen finely glandular-puberulent (at 20×) .... var. ***fendleri***
4' Fr body compressed laterally gen only near margins, side with gen 1 rib, several wavy veins, obliquely and ± widely obovate to ± circular; lower side of upper lvs gen glabrous .... var. ***polycarpum***

T. alpinum L. (p. 939) Pl 5–15(20) cm; fls bisexual. **LVS** gen basal, 1.5–6 cm; segments 4–10 mm, glabrous; upper surface dull green; lower surface glaucous, tip rounded. **INFL**: raceme, scapose; pedicels recurved in fr. **FL**: sepals 1.5–2.5 mm, purplish; stamens 7–12(15). **FRS** 1–6, ± pendent (due to recurved pedicel); body 2–3.5 mm, slightly compressed laterally, side obliquely ± ovate to ± obovate, with 2–3 curved ribs. $2n$=14. Meadows, often moist, gravelly soils, gen alpine; 2900–3700 m. n&c SNH, n W&I; to AK, NM, arctic Eurasia. ❀IRR,DRN,SUN:**1**,2–7,15,16,18; DFCLT.

T. fendleri A. Gray Pl 60–200 cm, gen dioecious. **LVS** basal and cauline, 7–46 cm; segments 8–20 mm, glabrous to finely glandular-puberulent, tip acute to rounded. **INFL**: panicle, leafy to bracted above. **FL**: sepals 2–5 mm, greenish white to purplish; stamens 15–28. **FRS** 7–20, spreading to ascending; body 4–8 mm, side with 1–3 ± curved ribs, 0 or several wavy veins. Moist, open to shaded places, woodland, forest; < 3200 m. CA-FP (exc GV, ChI), GB; to WA, WY, TX, n Mex. Some pls in NCoR have some bisexual fls; vars. in CA difficult, need study.

var. ***fendleri*** (p. 939) **LF**: lower surface (esp upper lvs) gen finely glandular-puberulent (at 20 ×). **FR**: body ± compressed laterally throughout, side obliquely ± ovate to ± obovate, with 2–3 ribs, 0 veins. $2n$=28,56,70. Habitats of sp.; 900–3200 m. KR, CaR (very uncommon), SN, SnFrB, SCoRO, TR, PR, GB; to OR, WY, TX, n Mex. ❀DRN:**4–6**&IRR:**17**&SHD:1–3,7,9,**14–16,18**,19–21,**22–24**.

var. ***polycarpum*** Torrey (p. 939) **LF**: lower surface gen glabrous. **FR**: body compressed laterally gen only near margins, side obliquely and ± widely obovate to ± circular, with gen 1 rib, several veins. $2n$=28. Habitats of sp.; < 1000 m (1500 in SN). NCo, w KR, NCoR, SN (very uncommon), CW, TR, w PR; to WA. [*T. p.* (Torrey) S. Watson] ❀DRN:**4–6**&IRR:**15–17**&SHD:**7**,9,**14**,19–21,**22–24**.

T. occidentale A. Gray (p. 939) Pl 40–100 cm, dioecious. **LVS** basal and cauline, 6–40 cm; segments 11–35 mm, glabrous to finely glandular-puberulent, tip obtuse to rounded. **INFL**: panicle, leafy to bracted above. **FL**: sepals 2–5 mm, greenish white to purplish; stamens 15–30. **FRS** (3)6–14, mostly reflexed; body 4–6 mm, slightly compressed laterally, side obliquely and narrowly ovate to narrowly fusiform, with gen 3, ± straight ribs. Moist, shaded places, coniferous forest; < 200 m. n NCo, nw KR; to B.C., WY. [var. *palousense* H. St. John] ❀TRY.

T. sparsiflorum Fischer & C. Meyer Pl 60–180 cm; fls bisexual. **LVS** basal (few) and cauline, 4–30 cm; segments 12–20 mm, finely glandular-puberulent, tip obtuse to rounded. **INFL**: panicle, gen leafy. **FL**: sepals 2.5–4 mm, greenish white; stamens 10–20, anthers ovate to oblong, tip obtuse, filaments flat. **FRS** 6–22, ± spreading; body 4–6 mm, strongly compressed laterally, side gen semi-circular to crescent-shaped, veined or weakly ribbed; beak 1–1.5 mm, straight. $2n$=42. Uncommon. Moist places, streambanks, coniferous forest; 1400–3300 m. CaR, c&n SN, SnBr, SnJt, n SNE (Sweetwater, White mtns); to AK, Colorado, Asia. ❀TRY.

TRAUTVETTERIA

Per from rhizomes; roots often clustered. **ST** gen 1, erect, simple. **LVS** few, round to reniform, deeply palmately lobed. **INFL**: panicle, ± flat-topped, terminal. **FL** bisexual, radial; sepals 3–7, petal-like; petals 0; stamens many; pistils 10+, ovule 1, style in fr persistent, ± hooked or coiled. **FR**: achenes; wall papery, shiny, veined or ribbed. 2 spp.: temp N.Am, e Asia. (E.R. von Trautvetter, Russian botanist, 1809–1889)

T. caroliniensis (Walter) Vail var. ***occidentalis*** (A. Gray) C. Hitchc. (p. 939) Pl 3–9 dm. **ST** gen glabrous. **LVS**: blade upper surface green, lower surface paler; basal 1–2, petiole 15–45 cm, lobes 5–10, ± wedge-shaped, toothed above middle; cauline petiole 0–15 cm. **INFL**: fls 5+; bracts < 2 cm; pedicels 4–10 mm. **FL**: sepals 2.5–5 mm, early deciduous, blade ovate, cup-like, greenish white; stamens 5–10 mm, filaments flat, wider than anthers or not. **FR** 2.5–4.5 mm, ± round. Moist, shaded places, streambanks; 1100–1800 m. KR, CaR (uncommon), c&n SNH; to B.C., MT, NM. [*T. grandis* Torrey & A. Gray] Other var. in e US. ❀SHD, DRN,IRRorWET:1,**4–7**,14,**15–17**.

RESEDACEAE MIGNONETTE FAMILY

Thomas F. Daniel

Ann, bien, per, shrub. **LVS** simple, alternate; stipules small, tooth- or gland-like; blade entire to deeply lobed. **INFL**: raceme, terminal, spike-like. **FL** gen bisexual, small, asymmetric, 1 per bract; sepals 2–8; petals 0–8; disk sometimes present; stamens 3–50+, gen on disk, anthers 2-chambered; pistils gen ± compound, 2–7-parted, gen open at top, ovary superior, sessile or short-stalked, gen 1-chambered, stigmas beak-like. **FR**: capsule, gaping at top, or berry. **SEEDS** few–many, reniform. 6 genera, 70 spp.: n&e hemispheres, esp Medit. [Abdallah & de Wit 1978 Meded Landbouwhogeschool 78(14):99–416]

1. Petals 2; stamens 3; disk 0; fr 2–3 mm .... **OLIGOMERIS**
1′ Petals 4–6; stamens > 10; disk present; fr 3–11 mm .... **RESEDA**

OLIGOMERIS

Ann, per. **LF** sessile, entire. **FL**: sepals 2–6, margins white; petals 2 (rarely more), entire to shallowly lobed; disk 0; stamens 3–12; stigmas 3–5. **FR**: capsule. 3 spp.: w N.Am, e hemisphere. (Greek: few parts)

O. linifolia (M. Vahl) J.F. Macbr. (p. 939) Ann, ± fleshy, glabrous. **ST** erect, < 45 mm. **LF** 8–45 mm, 0.5–2 mm wide, linear to ± oblanceolate. **INFL**: bracts triangular to awl-like. **FL** 1–2 mm; sepals 4; petals 2, whitish; stamens 3; pistil barely compound, 8-lobed. **FR** 2–3 mm, 4-parted, depressed-spheric. **SEEDS** black, shiny. n=24. Rocky slopes, open dunes, ocean bluffs, roadsides, alkaline places; < 850 m. SW, D; to TX, Mex; also in Eurasia.

RESEDA MIGNONETTE

Ann, per, shrub. **LF** petioled or not; blade entire to deeply lobed. **FL**: sepals 4–8, margins gen white; petals 4–8, base gen dilated, limb gen lobed; stamens 10–40, on prominent disk; stigmas 3–5. **FR**: capsule. 55 spp.: esp Eur, Medit; cult as orn and naturalized widely. (Latin: to calm, from supposed sedative property)

1. Sepals and petals 4; lf entire .... ***R. luteola***
1′ Sepals and petals 5–6; lf entire or lobed
  2. Lf pinnately lobed, lobes 4–11 per side; petals alike, ± equally 3-lobed; fr 4-parted, with persistent filaments .... ***R. alba***
  2′ Lf entire or ternately or pinnately lobed, lobes 0–3 per side; petals unlike, entire to unequally 6-lobed; fr 3-parted, without persistent filaments
    3. Lf entire to ternately lobed, lobe 0–1 per side; bracts persistent in fr; pedicel 4–12 mm; stamens 20–25; seed rough, 2–2.2 mm .... ***R. odorata***
    3′ Lf pinnately lobed, lobes 1–3 per side; bracts not persistent in fr; pedicel 2–4.5 mm; stamens 14–17; seed smooth, 1.5–1.7 mm .... ***R. lutea***

R. alba L. WHITE MIGNONETTE Per, glabrous. **ST** < 10 dm. **LF** pinnately lobed; lobes 4–11 per side. **INFL**: bracts 3.5–9 mm, persistent in fr; pedicel 2–8 mm. **FL**: sepals 5–6; petals 5–6, 4.5–6 mm, whitish, alike, ± equally 3-lobed; stamens 11–13, filaments persistent in fr. **FR** erect, 4-parted. **SEED** 1.2–1.3 mm, rough. n=10; 2n=20,40. Waste ground, fields, roadsides; < 100 m. CW, SW (gen coastal); native to Medit.

R. lutea L. (p. 939) YELLOW MIGNONETTE Per, glabrous. **ST** < 7 dm. **LF** pinnately lobed; lobes 1–3 per side. **INFL**: bracts 1.5–2 mm, not persistent in fr; pedicel 2–4.5 mm. **FL**: sepals 6; petals 6, 2–3 mm, yellow, unlike, unequally 2–3-lobed; stamens 14–17, not persistent in fr. **FR** erect, 3-parted. **SEED** 1.5–1.7 mm, smooth. 2n=24,48. Creekbeds; 400–750 m. NCoRI (s Lake Co); native to Medit.

R. luteola L. DYER'S ROCKET Bien, glabrous. **ST** < 10 dm. **LF** entire. **INFL**: bracts 2–3.5 mm, persistent in fr; pedicel 1–2.5 mm. **FL**: sepals 4; petals 4, 2–4 mm, yellowish, unlike, irregularly lobed; stamens 20–25, filaments persistent in fr. **FR** erect, 3-parted. **SEED** 0.8–1 mm, smooth. 2n=24,26,28. Waste ground, fields, roadsides; < 100 m. NCo, NCoR, CCo, s SCo (San Diego); native to Old World. Source of yellow dye used from Neolithic time.

R. odorata L. GARDEN MIGNONETTE Ann, glabrous or sparsely hairy. **ST** < 6 dm. **LF** entire to ternately lobed. **INFL**: bracts 2–6 mm, persistent in fr; pedicel 4–12 mm. **FL**: sepals 6; petals 6, 2.5–4.5 mm, whitish, unlike, upper deeply lobed, lower very reduced; stamens 20–25, filaments not persistent in fr. **FR** gen pendent, 3-parted. **SEED** 2–2.2 mm, rough. n=6; 2n=12,14. Disturbed ground; < 100 m. CCo, SCo; native to Medit; possibly only waif in CA.

RHAMNACEAE BUCKTHORN FAMILY

John O. Sawyer, Jr. (except *Ceanothus*)

Shrub, vine, tree, gen erect, often thorny. **LVS** simple, gen alternate, often clustered on short-shoots, gen petioled, gen stipuled; blade often 1–3-ribbed from base. **INFL**: cyme, panicle, or fls solitary in axils. **FL** gen bisexual, radial; hypanthium subtending, surrounding, or partly fused to ovary; sepals 4 or 5; petals 0, 4, or 5, clawed; stamens 4 or 5, alternate sepals, attached to hypanthium top, each gen fitting into a petal concavity; ovary superior or partly inferior, chambers 2–5, each 1–2-ovuled, style lobes or parts 1–3. **FR**: capsule, drupe. 55 genera, 900 spp.: esp trop, subtrop; some cult (*Ceanothus*; *Colletia*, anchor-plant; *Gouania*; *Phylica*; *Rhamnus*; *Ventilago*; *Ziziphus*). [Brizicky 1965 J Arnold Arbor 45:439–463]

1. Fls showy; petals, sepals, pedicels blue, pink, or white .... **CEANOTHUS**
1′ Fls inconspicuous; petals, if any, green or white; sepals and pedicels green or gray
  2. Fr a capsule
    3. Hypanthium not splitting with capsule; local, SCo, PR .... **ADOLPHIA**
    3′ Hypanthium splitting with capsule; uncommon, D .... **COLUBRINA**
  2′ Fr a drupe
    4. Fr of 2–4 separate stones; widespread .... **RHAMNUS**
    4′ Fr of 1 stone; DSon, e PR
      5. Petals 0; hypanthium below and not surrounding ovary .... **CONDALIA**
      5′ Petals 5; hypanthium surrounding base of ovary .... **ZIZIPHUS**

ADOLPHIA

Shrub. **ST**: branches opposite, rigid, green; twigs spreading, jointed at base, thorn-tipped. **LVS** opposite, sometimes clustered in axils, early deciduous; stipules ± triangular, black, shiny; blade entire, 1–3-ribbed from base. **INFL**: small axillary clusters. **FL**: hypanthium bowl-shaped, 10-ribbed, free from ovary, in fr papery, not splitting; sepals 5, triangular, persistent; petals 5, spoon-shaped, white; stamens 5; ovary superior, chambers 3, each 1-ovuled, style parts 3. **FR**: capsule, spheric. 2 spp.: SCo, PR, TX, Mex. (Adolphe Brongniart, student of Rhamnaceae, 1801–1876)

A. californica S. Watson (p. 939) SPINESHRUB, CALIFORNIA ADOLPHIA Shrub < 1 m. **ST**: twigs stout, glabrous to puberulent. **LF**: petiole 2–4 mm; blade 3–9 mm, oblong, base wedge-shaped, tip obtuse to acute. **FL**: hypanthium 3 mm wide, 2 mm deep; petals 2 mm. **FR** 5 mm. RARE in CA. Coastal scrub, chaparral; < 300 m. SCo, PR; Mex. Local.

CEANOTHUS CALIFORNIA-LILAC

Clifford L. Schmidt

Shrub, small tree, prostrate to erect, thorny or not. **ST**: branches gen arranged as lvs. **LVS** alternate or opposite, deciduous or evergreen, petioled; blade 1–3-ribbed from base, margin entire or not. **INFL**: gen panicle-like aggregations of umbel-like, 3-fld clusters. **FL** gen < 5 mm; hypanthium surrounding fleshy disk below ovary base, in fr thick, not splitting; sepals gen 5, lanceolate-deltate, incurved, colored like petals, persistent; petals gen 5, hooded, white to deep blue; stamens gen 5, opposite petals; ovary superior, 3-lobed, chambers 3, each 1-ovuled, style parts 3. **FR**: capsule, ± spheric, 3-valved. **SEEDS** 3, ± 3 mm, 1 surface convex. 45 spp.: N.Am, esp w. (Greek: thorny pl) [Rensselaer & McMinn 1942 Ceanothus Santa Barbara Bot Gard 1–308] Hybridization common (named hybrids not recognized here); hybrid forms may not key adequately.

1. Lvs alternate, gen thin, evergreen or deciduous, stomata on lower surface not in cavities; stipules thin, deciduous; infl raceme- or panicle-like; fr without horns but often with crests or ridges (subg. *Ceanothus)*
 2. Lf gen 3-ribbed from blade base (if not, basal vein-pair more prominent than others)
  3. Twigs rigid, gen thorny; lf glaucous; branches gen gray
   4. Pl ascending, < 1.5 m; lf < 3 cm; gen > 1000 m .... ***C. cordulatus***
   4' Pl erect, > 1.5 m; lf 1.5–6 cm; < 1800 m
    5. Lf 2.5–6 cm; fr warty, not sticky; KR, NCoRO, SnFrB; < 900 m .... ***C. incanus***
    5' Lf 1.5–4 cm; fr sticky, not warty; SNF, SnFrB, SCoR, WTR, SnBr, SnJt, PR, n Baja CA; < 1800 m .... ***C. leucodermis***
  3' Twigs flexible, gen not thorny; lf not glaucous; branches gen not gray
   6. Lf entire, ± deciduous
    7. Lf gen > 2 cm; pl erect, < 4 m; infl < 15 cm; fl gen white or blue; fr sticky, valve crests minute .... [2]***C. integerrimus***
    7' Lf gen < 2 cm; pl ascending, < 1.2 m; infl < 7 cm; fl pale to deep blue; fr not sticky, valve crests ± 0 .... [2]***C. parvifolius***
   6' Lf variously toothed, deciduous or evergreen
    8. Lvs deciduous; twigs glabrous; fl white .... ***C. sanguineus***
    8' Lvs evergreen; twigs variously hairy; fl white or pale to dark blue
     9. Fl white — twigs puberulent, becoming smooth, brown; lf aromatic, upper surface ± shiny, glabrous .... ***C. velutinus***
      10. Lf lower surface light green, glabrous; near coast .... var. ***hookeri***
      10' Lf lower surface gray, canescent; inland .... var. ***velutinus***
     9' Fl gen pale to dark blue, rarely ± white
      11. Twigs round, not striate
       12. Lf lower surface tomentose
        13. Lf 3–8 cm, margin serrate, few-gland-toothed or not; twigs brown, with soft, short hairs; pl < 7 m; ChI .... ***C. arboreus***
        13' Lf 1.5–3 cm, margin gland-toothed; twigs ± rusty-tomentose; pl < 3 m; mainland .... [2]***C. tomentosus***
       12' Lf lower surface ± hairy, esp on veins, but not tomentose .... ***C. oliganthus***
        14. Twigs ± red-brown, hairy; lf upper surface hairy; infl gen interrupted, < 5 cm; fr lobed, crested, hairy or not; SCoR, TR, PR; < 1300 m .... var. ***oliganthus***
        14' Twigs gen gray-green, ± glabrous to puberulent; lf upper surface ± glabrous; infl continuous, < 4 cm; fr barely lobed or crested, glabrous; gen coastal TR to NCoR; < 1000 m .... var. ***sorediatus***
      11' Twigs angled, striate
       15. Lf margin turned under
        16. Lf blade ovate, lower surface gray-green, short-tomentose; infl continuous, < 5 cm; fr sticky when young, at maturity black, shiny .... ***C. griseus***
        16' Lf blade oblong to elliptic, lower surface pale green, densely cobwebby; infl interrupted, < 15 cm; fr not sticky when young, at maturity not shiny, not black .... [2]***C. parryi***
       15' Lf margin not or only slightly turned under
        17. Lf lower surface veins not raised; fr not glandular-sticky; PR .... ***C. cyaneus***
        17' Lf lower surface veins very raised; fr glandular-sticky; NW, CW .... ***C. thyrsiflorus***
 2' Lf gen 1-ribbed from blade base, basal vein-pair more prominent than others or not
  18. Lf entire or rarely minutely toothed, ± glabrous
   19. Lvs evergreen, thick, rarely minutely toothed

20. Fl white; fr 5–7 mm, valves glandular-crested; pl < 3.5 m; branches gray-green to red-brown ..... ***C. palmeri***
20′ Fl pale blue to ± white; fr 4–5 mm, valves barely crested; pl < 6 m; branches olive-green .... ***C. spinosus***
19′ Lvs ± deciduous, thin, entire or rarely minutely toothed toward tip
21. Lf gen > 2 cm; pl erect, < 4 m; infl < 15 cm, clusters gen stalked; fl gen white or blue; fr sticky, valves minutely crested .... [2]***C. integerrimus***
21′ Lf gen < 2 cm; pl ascending, < 1.2 m; infl < 7 cm, clusters gen sessile; fl pale to deep blue; fr not sticky, valves not crested .... [2]***C. parvifolius***
18′ Lf gen toothed, at least lower surface gen hairy
22. Lf margin ± turned under
23. Lf upper surface glandular-papillate, margin minutely gland-toothed
24. Pl ± prostrate; twigs angled, lower side gray-tomentose — coastal, grassy hills; CCo ...... ***C. hearstiorum***
24′ Pl ascending to erect; twigs round, gen densely hairy .... ***C. papillosus***
25. Pl ascending to erect, < 5 m; lf blade 1–2 cm wide .... var. ***papillosus***
25′ Pl ± ascending, < 1.8 m; lf blade < 1 cm wide .... var. ***roweanus***
23′ Lf upper surface glabrous or hairy, rarely glandular-papillate, margin gland-toothed or not
26. Twigs angled, tomentose; lf upper surface not furrowed on veins, glabrous, margin not gland-toothed; infl < 15 cm .... [2]***C. parryi***
26′ Twigs round, variously hairy; lf upper surface furrowed on veins, puberulent, margin gen gland-toothed; infl < 4 cm — sts ascending
27. Lf linear or narrowly oblong to elliptic, tip truncate to notched, upper surface rarely glandular-papillate; pl densely branched, < 1.5 m; infl < 3 cm .... ***C. dentatus***
27′ Lf ovate to round, tip not truncate or notched, upper surface not glandular-papillate; pl moderately branched, gen < 3.5 m; infl < 4 cm .... ***C. impressus***
22′ Lf margin not turned under
28. Pl < 0.3 m, forming mat; lf shape variable on single pls, upper surface blue-green .... ***C. diversifolius***
28′ Pl < 3 m, ascending to erect, not forming mat; lvs ± constant in shape on individual pls, not blue-green
29. Pl < 3 m; twigs gray-brown to reddish, rusty-tomentose, not glandular; infl panicle-like; immature fr sticky .... [2]***C. tomentosus***
29′ Pl < 2 m; twigs gray or pale green, gen glandular, not rusty-tomentose; infl ± raceme-like; immature fr not sticky
30. Twigs gray, glandular-hairy or not; petiole < 8 mm; lf blade upper surface dull green; fl pale blue; pl ascending, < 1 m — c SNF to CaRF, NCoRI .... ***C. lemmonii***
30′ Twigs pale green, glandular-hairy; petiole < 3 mm; lf blade upper surface gen dark green; fl blue; pl ascending to erect, gen < 2 m .... ***C. foliosus***
31. Lf blade upper surface ± glabrous, lower glabrous to densely hairy; pl ascending, gen < 1 m; NCoRO, SnFrB, PR .... var. ***foliosus***
31′ Lf blade upper and lower surfaces hairy; pl ascending or erect, < 2 m; NCoRO, SnFrB, SCoR
32. Lf blade upper surface dull green, glandular-hairy, lower gray, densely hairy throughout; pl erect, < 2 m; SnFrB, SCoR .... var. ***medius***
32′ Lf blade upper surface dark green, sparse-hairy, lower pale green, sparse-hairy on veins; pl ascending, < 0.8 m; NCoRO .... var. ***vineatus***
1′ Lvs gen opposite (at least some alternate in *C. verrucosus, C. megacarpus*), thick, evergreen, stomata on lower surface in cavities; stipules thick, at least bases persistent, gen corky; infl raceme-like; fr gen with horns, with intervening ridges or not (subg. *Cerastes*)
33. Lvs on single pl alternate or alternate and opposite
34. Lvs alternate, blades round to deltate-obovate; fr gen 5 mm wide — SCo (San Diego Co.) ...... ***C. verrucosus***
34′ Lvs on single pl alternate and sometimes opposite, blades elliptic to obovate; fr 8–12 mm wide . ***C. megacarpus***
35. Lvs gen opposite, < 4 cm; fr horns 0 or minute; ChI .... var. ***insularis***
35′ Lvs gen alternate, < 2.6 cm; fr horns prominent; CCo, SCoRO, s ChI .... var. ***megacarpus***
33′ Lvs opposite
36. Fl gen white
37. Lf blade lower surface canescent to tomentose throughout, margin gen turned under — twigs white to rusty-tomentose .... ***C. crassifolius***
37′ Lf blade lower surface ± sparse-hairy on veins, canescent between, margin gen not turned under
38. Lf upper surface flat or ± cupped .... ***C. greggii***
39. Lf < 2 cm, blade round or broad-elliptic to -obovate, upper surface yellow-green, margin sharp-toothed, gen near tip, or entire .... var. ***perplexans***
39′ Lf < 1.5 cm, blade elliptic-ovate to oblanceolate, upper surface gray-green, margin entire to toothed .... var. ***vestitus***
38′ Lf upper surface gen flat or ± cupped
40. Lf blade margin entire, upper surface dull green, flat; twigs glabrous; fr 4–6 mm, horns slender or minute — fl sometimes pale blue to lavender .... [2]***C. cuneatus*** var. ***cuneatus***
40′ Lf blade margin gen toothed, upper surface dark- or yellow-green, cupped or not; twigs puberulent; fr 6–9 mm, horns not slender, not minute
41. Lf blade upper surface dark-green, gen flat; twigs rusty-puberulent; fr 6–9 mm, horns rough; SnFrB .... ***C. ferrisae***
41′ Lf blade upper surface yellow-green, ± cupped; twigs gray-brown, puberulent; fr ± 6 mm, horns wrinkled; NCoRI, NCoRH .... ***C. jepsonii*** var. ***albiflorus***
36′ Fl blue, lavender, or pale or white with blue tinge

42. St prostrate, rooting at nodes, or ascending
43. Lf blade upper surface dull
44. Sts ± ascending or not; lf not ± erect, < 2 cm, lf blade upper surface gen puberulent; fl blue; fr horns slender, near top .... ***C. fresnensis***
44′ Sts prostrate; lf ± erect, ± 1 cm, blade upper surface ± glabrous; fl white with blue tinge; fr horns not slender, on side, or 0 .... ***C. roderickii***
43′ Lf blade upper surface gen shiny
45. Lf 2–5 cm — lf margin dentate; fl deep blue to violet; fr 3.5–4.5 mm, horns minute, ridges 0 .... ***C. gloriosus*** var. ***gloriosus***
45′ Lf gen < 2.5 cm
46. Lf ± as long as wide, margin entire or coarse-toothed
47. Fr 4–5 mm, horns minute, not wrinkled, ridges 0; lf < 1.5 cm; s SnFrB, CCo (near Hazard Canyon, San Luis Obispo Co.) .... [2]***C. cuneatus*** var. ***rigidus***
47′ Fr 7–9 mm, horns wrinkled, not minute, ridges present; lf gen < 2 cm; s SNH (Kern Plateau) .... [2]***C. pinetorum***
46′ Lf longer than wide, margin gen toothed
48. Lf blade < 5 mm wide; twigs puberulent .... ***C. pumilus***
48′ Lf blade > 5 mm wide; twigs puberulent or glabrous
49. Lf blade lower surface white-woolly, margin turned under; twigs glabrous — CCo (San Luis Obispo Co.) .... ***C. maritimus***
49′ Lf blade lower surface not white-woolly but hairy on veins, canescent between, margin not turned under; twigs puberulent
50. Lf < 2 cm, petiole < 1 mm; fr gen 5 mm, horns slender, not wrinkled, ridges minute — NCoRO (Sonoma Co.) .... ***C. confusus***
50′ Lf < 3.2 cm, petiole < 3 mm; fr 5–9 mm, horns wrinkled, not slender, ridges prominent .... ***C. prostratus***
42′ St ± prostrate to erect, not rooting at nodes
51. Lf often holly-like, marginal teeth spine-like, at least near tip
52. Fr ± 6 mm, horns thick, wrinkled, ridges prominent — fl blue to violet; lf < 3 cm, gen ± cupped; pl ± ascending, < 6 dm .... ***C. jepsonii*** var. ***jepsonii***
52′ Fr gen 3–5 mm, horns slender, not wrinkled, ridges 0 or minute
53. Lf < 1.5 cm, flat; pl erect — fr 5 mm; fl blue to lavender .... ***C. sonomensis***
53′ Lf 1.5–2.5 cm, flat or not; pl ascending to erect
54. Pl ascending, weakly arched; lf blade ± flat, margin not wavy, ± turned under, tip ± truncate .... ***C. divergens***
54′ Pl ± erect, rigid; lf blade ± cupped, margin wavy, not turned under, tip obtuse .... ***C. purpureus***
51′ Lf gen not holly-like, margin gen dentate, teeth gen not spine-like
55. Fr 7–9 mm, horns prominent, wrinkled, ridges present — pl ± prostrate to erect; lf margins ± turned under or not .... [2]***C. pinetorum***
55′ Fr 3–6 mm, horns slender or minute, not wrinkled, ridges 0
56. Lf margin entire or toothed near tip, blade narrowly oblanceolate to round-obovate
57. Lvs 2–7 mm, lower surface glabrous; fls pale blue, sometimes pink — n PR (Riverside Co.) .... ***C. ophiochilus***
57′ Lvs 5–50 mm, lower surface minutely canescent to hairy on veins; fls gen white, sometimes bluish
58. Branches thick, stiff, straight; blade ± cupped, obovate to round-obovate, margin toothed toward tip; fl light to dark blue — s SnFrB, CCo (near Hazard Canyon, San Luis Obispo Co.) .... [2]***C. cuneatus*** var. ***rigidus***
58′ Branches slender, flexible, arched; blade ± flat, narrowly oblanceolate to round-obovate, margin gen entire; fl lavender, pale blue, or white
59. Lvs closely clustered, blades narrowly oblanceolate .... ***C. cuneatus*** var. ***fascicularis***
59′ Lvs not closely clustered, blades gen round-obovate .... [2]***C. cuneatus*** var. ***cuneatus***
56′ Lf margin dentate, blade round to widely elliptic or oblong
60. Pl gen ± prostrate, < 5 dm — twigs slender; lf 1–2 cm; SnFrB (Point Reyes, Marin Co.) .... ***C. gloriosus*** var. ***porrectus***
60′ Pl gen erect, sometimes ascending, < 2 m
61. Lf 15–40 mm; st flexible, arched; n SnFrB, NCoRO .... ***C. gloriosus*** var. ***exaltatus***
61′ Lf 6–18 mm; st stiff, straight; SnFrB (Bolinas Ridge, sw Marin Co.) .... ***C. masonii***

C. arboreus E. Greene FELTLEAF CEANOTHUS Pl erect, < 7 m. **ST**: twigs round, hairy, brown, becoming gray. **LVS** alternate, evergreen, < 8 cm; stipules deciduous; petiole < 25 mm; blade widely ovate to elliptic, 3-ribbed from base, thin, tip acute to obtuse, margin serrate, few-gland-toothed or not, upper surface dull green, puberulent, lower pale green, canescent-short-tomentose. **INFL** panicle-like, < 15 cm. **FL** pale blue. **FR** < 8 mm, rough, 3-lobed; valves ± crested. $2n=24$. Dry, shrubby slopes; < 300 m. ChI. ± glabrous pls from n ChI have been called var. *glabra* Jepson. ❀DRN:**15–17,24**&IRR:7–9,**14,19–23**;CVS.

C. confusus J. Howell (p. 943) RINCON RIDGE CEANOTHUS Pl prostrate to decumbent, < 1.2 m wide. **ST** rooting at nodes or not; twigs angled, red-brown, ± puberulent, becoming gray-brown. **LVS** opposite, evergreen, flat; stipules persistent; petiole < 1 mm; blade < 2 cm, oblanceolate or elliptic to obovate, tip obtuse, margin ± turned under, toothed above middle, teeth gen 3–5, coarse, sharp, upper surface green, ± glabrous, lower surface paler, veins gen hairy, canescent between veins. **INFL** raceme-like, < 2 cm. **FL** blue, lavender, or purple. **FR** ± 5 mm; horns near top, slender, not wrinkled; ridges minute. $2n=24$. RARE. Dry, shrubby slopes; < 500 m. s NCoR (esp Sonoma Co.). [*C. divergens* ssp. *c.* (J. Howell) Abrams]. Closely related to *C. prostratus*. In cult.

C. cordulatus Kellogg MOUNTAIN WHITETHORN Pl spreading to ascending, < 1.5 m, thorny. **ST**: twigs round, yellow-green, puberulent, becoming light gray. **LVS** alternate, evergreen, < 3 cm; stipules deciduous; petiole < 6 mm; blade ovate to elliptic, 3-ribbed from base, tip acute to obtuse, margin gen entire, upper surface light to gray-green, glabrous to puberulent, lower paler, glabrous to puberulent. **INFL** panicle-like, < 4 cm. **FL** white. **FR** < 5 mm, rough, 3-lobed; valves ± crested. 2*n*=24. Rocky ridges, open pine forests; 900–2900 m. KR, NCoR, SN, SnGb, SnBr, SnJt; OR, NV, n Baja CA. ❀DRN,DRY,SUN:1–3,6,7,14–16,18;STBL.

C. crassifolius Torrey (p. 939) HOARYLEAF CEANOTHUS Pl erect, open, < 3.5 m. **ST**: twigs round, white- to rusty-tomentose, becoming gray to brown. **LVS** opposite, evergreen, < 3 cm; stipules persistent; petiole 2–6 mm; blade ± ovate, 1-ribbed from base, firm, tip obtuse, margin turned under or not, gen with few shallow teeth, upper surface olive-green, glabrous, minutely rough, lower canescent to tomentose. **INFL** raceme-like, < 3 cm. **FL** white. **FR** 6–8 mm; horns on shoulder; ridges 0. 2*n*=24. Dry ridges, slopes; < 1100 m. SCoRO (Santa Barbara Co.), TR, PR; n Baja CA. Less hairy pls with ± flat lf margins in n of range have been called var. *planus* Abrams; pls < 1 m, with upper lf surface tomentose in Otay Mtns, San Diego Co., may be hybrid, *C. ×otayensis* McMinn. ❀DRN,DRY,SUN:**7**,**8**,**9**,14–17,**18–24**;STBL.

C. cuneatus (Hook.) Nutt. Pl prostrate to erect, < 3 m. **ST**: twigs round, gray to brown, gen glabrous, becoming gray. **LVS** opposite, clustered or not, evergreen, < 3.2 cm, flat; stipules persistent; petiole ± 0–3 mm; blade oblong, oblanceolate, or obovate to roundish, 1-ribbed from base, firm, tip obtuse to notched, margin entire or with spine-like teeth, upper surface dull to shiny green, ± glabrous, lower surface ± hairy on veins. **INFL** raceme-like, < 2.5 cm. **FL** white, pale blue, blue, or lavender. **FR** 4–6 mm, gen with horns near top; ridges 0 or minute. Dry fans, slopes, ridges; < 1800 m. CA-FP (exc GV); OR, n Baja CA.

var. ***cuneatus*** (p. 939) BUCK BRUSH Pl gen erect, rarely prostrate, < 3 m. **ST**: twigs gen light gray, becoming gray. **LF** < 3.2 cm; petiole < 3 mm; blade oblanceolate to obovate or rounded, entire, upper surface gen dull green, glabrous, lower minutely canescent. **FL** white or pale blue to lavender. **FR** 4–6 mm; horns minute, some slender; ridges 0. 2*n*=24. Common. Dry, rocky slopes, ridges, incl sands, serpentine; < 1800 m. Range of sp. [var. *dubius* J. Howell; var. *submontanus* (Rose) McMinn; *C. ramulosus* (E. Greene) McMinn var. *r.*] ❀DRN,DRY,SUN:**7–9**,**14–16**,17,**18–24**;CVS;STBL.

var. ***fascicularis*** (McMinn) Hoover Pl erect, 6–15 dm. **ST**: twigs brown, becoming dark brown. **LVS** clustered, < 2 cm; petiole < 3 mm; blade narrowly oblanceolate, margin entire or few-toothed near tip, upper surface gen dull green, glabrous, lower minutely canescent. **FL** blue. **FR** 4–5 mm; horns gen minute; ridges 0. Chaparral, coastal sandy mesas; < 400 m. s CCo (Santa Barbara, San Luis Obispo cos.). [C. ramulosus (E. Greene) McMinn var. *f.* McMinn] ❀DRN,DRY,SUN:**7–9**,**14–17**,18,**19–24**.

var. ***rigidus*** (Nutt.) Hoover MONTEREY CEANOTHUS Pl ± prostrate to erect, < 1.5 m, much branched. **ST**: twigs brown, tomentose, becoming dark gray. **LF** < 1.5 cm; petiole < 2 mm; blade oblanceolate to round-obovate, margin toothed above middle or entire, upper surface gen ± cupped, dark green, glabrous, lower paler, puberulent between veins. **FL** bright to pale blue. **FR** gen 5 mm; horns minute, on sides; ridges 0. 2*n*=24. UNCOMMON. Sandy hills, flats, closed-cone-pine forests; < 200 m. s SnFrB, CCo (near Hazard Canyon, San Luis Obispo Co.). [*C. r.* Nutt., incl var. *albus* Roof] ❀DRN,DRY,SUN:8,9,**14–17**,19–21,**22–24**;CVS.

C. cyaneus Eastw. LAKESIDE CEANOTHUS Pl erect, < 5 m. **ST**: twigs angled, green, glabrous to puberulent, with scattered, brownish glands, becoming gray-green. **LVS** alternate, evergreen, < 4.5 cm; stipules deciduous; petiole < 6 mm; blade ovate-elliptic, 3-ribbed from base, tip acute to obtuse, margin glandular-serrate to ± entire, upper surface green, glabrous, lower pale green, puberulent on veins. **INFL** panicle-like, < 18 cm. **FL** bright blue. **FR** gen 4 mm, smooth, shiny, shallowly 3-lobed; valve crests 0 or minute. 2*n*=24. RARE. Dry, shrubby slopes; < 400 m. s PR (San Diego Co.). In cult.

C. dentatus Torrey & A. Gray (p. 941) Pl ± ascending, < 1.5 m, densely branched. **ST**: twigs round, dark gray, hairy, not changing color. **LVS** alternate, evergreen, gen clustered, < 2 cm; stipules deciduous; petiole 2 mm; blade elliptic to narrowly oblong or linear, 1-ribbed from base, tip gen truncate or notched, margin turned under, appearing ± crenate, ± glandular-papillate, upper surface dark green, gen hairy, lower pale green, hairy. **INFL** raceme-like, < 3 cm. **FL** deep blue. **FR** ± 4 mm; lobes shallow; valves minutely crested. Coastal sandy hills, slopes, flats; < 1500 m. CCo, SCoRO. [var. *floribundus* (Hook.) Trel.] ❀DRN:7,14&SUN:**15–17**.

C. divergens C. Parry (p. 943) CALISTOGA CEANOTHUS Pl ascending to erect < 2 m. **ST**: twigs angled, red-brown, with whitish bloom or not, glabrous, becoming gray to red-brown. **LVS** opposite, evergreen, < 2.5 cm; stipules persistent; petiole < 2 mm; blade oblong to obovate, 1-ribbed from base, firm, tip ± truncate or not, margins ± turned under with spine-like teeth, upper surface ± flat, surface green, glabrous, lower surface hairy on veins, minutely canescent between. **INFL** raceme-like, < 3 cm. **FL** deep blue to purple. **FR** gen 6 mm; ridges minute. 2*n*=24. RARE. Dry, shrub-covered, rocky, volcanic slopes; < 500 m. NCoR. In cult.

C. diversifolius Kellogg PINE MAT Pl ± prostrate, < 3 dm, forming mats. **ST**: twigs round, yellow-green, hairy, becoming red-brown. **LVS** alternate, evergreen, < 5.5 cm; stipules deciduous; petiole < 12 mm; blade round, ovate, obovate, elliptic, or ± oblong, gen 1-ribbed from base, thin, tip gen obtuse, margin serrate, glands few, upper surface bluish green, hairy, lower paler, tomentose. **INFL** raceme-like, < 3 cm. **FL** blue to ± white. **FR** gen 4 mm, smooth; valves crested toward tip. Draws, flats under oaks, pines; 900–1800 m. KR, NCoRH, CaRH, SNH, Teh. ❀DRN:4–6,**17**&IRR:1,2&SHD:**7**,14,**15**,**16**,18;GRCVR.

C. ferrisae McMinn (p. 941) COYOTE CEANOTHUS Pl erect, < 2 m. **ST**: twigs round, rusty-puberulent, becoming gray. **LVS** opposite, evergreen, < 3.2 cm; stipules persistent; petiole < 3 mm; blade round to widely elliptic, 1-ribbed from base, firm, tip obtuse, margin toothed above middle or entire, upper surface flat or ± cupped, dark green, glabrous, lower canescent between veins. **INFL** raceme-like, < 3 cm. **FL** white. **FR** 6–9 mm; horns near top, rough; ridges 0. RARE. Serpentine slopes; < 300 m. se SnFrB (Santa Clara Co.). In cult.

C. foliosus C. Parry Pl ± prostrate to erect, < 2 m. **ST**: twigs round, ± pale green, often glandular-hairy, becoming gray. **LVS** alternate, evergreen, < 2 cm; stipules deciduous; petiole < 3 mm; blade narrowly oblong to widely elliptic or obovate, gen 1-ribbed or faintly 3-ribbed from base, tip obtuse, margin gen minutely gland-toothed, upper surface dull to dark green, puberulent, lower paler, hairy or not. **INFL** gen raceme-like, 3–8 cm. **FL** blue. **FR** gen 4 mm, smooth, faintly lobed; valves ± crested. Dry slopes; < 1500 m. NCoRO, SnFrB, SCoR, PR.

var. ***foliosus*** WAVYLEAF CEANOTHUS Pl ± ascending, gen < 1 m. **LF**: blade oblong to widely elliptic, upper surface dark green, ± glabrous, lower hairs few, scattered. 2*n*=24. Openings on dry slopes; < 1500 m. NCoRO, SnFrB, PR. ❀DRN,SUN:3,7,14,**15**,**16**,17,18–21;CV.

var. ***medius*** McMinn LA CUESTA CEANOTHUS Pl erect, < 2 m. **LF**: blade narrowly to widely elliptic, upper surface dull green, glandular-hairy, lower gray, densely hairy. Chaparral succeeding burned broad-lvd-evergreen forest; < 650 m. SnFrB, SCoR. ❀DRN,DRY,SUN:7,14,**15–17**,18–21.

var. ***vineatus*** McMinn VINE HILL CEANOTHUS Pl ± prostrate to ascending, < 8 dm. **LF**: blade widely elliptic to obovate, margin glands few, upper surface dark green, hairs sparse, lower paler, hairs scattered on veins, fewer between. UNCOMMON. Dry, rolling hills; < 300 m. NCoRO. ❀DRN,DRY,SUN:14,**15–17**.

C. fresnensis Abrams FRESNO MAT Pl gen prostrate, < 6 m wide. **ST** rooting at nodes; twigs round, gen red-brown, short-tomentose, not changing color. **LVS** opposite, evergreen, gen < 2 cm; stipules persistent; petiole gen 1–2 mm; blade oblanceolate to obovate or elliptic, 1-ribbed from base, firm, tip ± obtuse to notched, margin entire, ± toothed at tip or not, upper surface dull green, gen puberulent, lower canescent between veins. **INFL** raceme-like, < 3 cm. **FL** blue. **FR** 5–6 mm; horns slender, near top. UNCOMMON. Dry ridges in coniferous forest; 900–2000 m. c SN. Variable due to hybridization. ❀DRN,DRY:1,7;DFCLT.

C. gloriosus J. Howell Pl prostrate to erect, < 2 m. **ST**: twigs angled, green, red-brown, or brown, ± strigose or short-tomentose, becoming gen red-brown. **LVS** opposite, evergreen, < 5 cm; stipules persistent; petiole < 4 mm; blade round to widely elliptic or oblong, 1-ribbed from base, firm, tip obtuse, truncate, or notched, margin dentate, teeth spine-like or not, upper surface dark green, glabrous, lower hairy on veins, canescent between. **INFL**: raceme-like, < 2.5 cm. **FL** deep blue to violet. **FR** gen 3.5–4.5 mm; horns minute, near top; ridges 0. Sandy places, coastal bluffs, shrubby slopes, ridges; < 500 m. NCo, NCoRO, CCo, SnFrB.

var. ***exaltatus*** J. Howell GLORY BRUSH Pl erect, < 2 m. **ST**: twigs angled, dark, gen strigose. **LF** 1.5–4 cm. 2*n*=24. Shrubby slopes, ridges, chaparral, coniferous forest; < 500 m. NCoRO, n SnFrB. ❀DRN,SUN:**15–17**&IRR:**14**,18,**19–24**;CV.

var. ***gloriosus*** GLORY MAT, POINT REYES CEANOTHUS Pl prostrate or decumbent, < 3 dm, < 3 m wide. **ST**: twigs angled, red-brown, ± strigose. **LF** 2–5 cm. 2*n*=24. UNCOMMON. Sandy places, coastal bluffs, closed-cone-pine forest; < 500 m. s NCo, n CCo (Marin Co.). ❀DRN:**15–17**&IRR:14,19–21,**22–24**;CVS.

var. ***porrectus*** J. Howell MOUNT VISION CEANOTHUS Pl sprawling, < 5 dm. **ST**: twigs angled, green, minutely tomentose. **LF** 1–2 cm. 2*n*=24. RARE. Closed-cone-pine forest; < 300 m. SnFrB (Point Reyes). In cult.

C. greggii A. Gray Pl erect, < 2 m. **ST**: twigs round, gray, tomentose, not changing color. **LVS** opposite, evergreen; stipules persistent; petiole < 3 mm; blade esp variable in shape on individual pls, 1-ribbed from base, firm, tip obtuse to truncate, margin entire to dentate, upper surface flat or ± cupped, both surfaces gray-canescent. **INFL** raceme-like, < 2 cm. **FL** white. **FR** 3–5 mm; horns near middle or 0; ridges 0. Dry slopes; 300–2300 m. Teh, TR, PR, W&I, DMtns; to UT, TX, n Mex.

var. ***perplexans*** (Trel.) Jepson (p. 941) CUPLEAF CEANOTHUS **LF** < 2 cm; blade round to widely elliptic or obovate, sharp-toothed (gen near tip) or entire, upper surface yellow-green. 2*n*=24. Chaparral, pinyon/juniper woodland, yellow-pine forest; < 2100 m. SnBr, SnJt, PR; n Baja CA. ❀DRN,SUN:2,3,**7**,10,**18**&IRR:11, 14,**19–21**;STBL.

var. ***vestitus*** (E. Greene) McMinn (p. 941) MOJAVE CEANOTHUS **LF** < 1.5 cm; blade elliptic-ovate to oblanceolate, entire to toothed, upper surface gray-green. Desert slopes, Joshua-tree and pinyon/juniper woodlands, sagebrush scrub; < 2300 m. Teh, WTR, W&I, DMtns; to UT, AZ. ❀DRN,SUN:2,3,**7**,10,15–17&IRR:11, 14,**19–21**;STBL.

C. griseus (Trel.) McMinn CARMEL CEANOTHUS Pl prostrate to erect, gen < 2.5 m. **ST**: twigs angled, greenish brown, puberulent, not changing color. **LVS** alternate, evergreen, < 6 cm; stipules deciduous; petiole 5–10 mm; blade broad- to round-ovate, 3-ribbed from base, tip obtuse, margin serrate, turned under, wavy near base, upper surface dark green, glabrous, lower gray-green, short-tomentose. **INFL** raceme-like, dense, < 6 cm. **FL** blue. **FR** gen 4 mm, shallowly 3-lobed, glandular-sticky when young, black, shiny when mature; valve crest 0. Coastal scrub, closed-cone-pine forests; < 200 m. NCo, NCoRO, CCo, SnFrB, SCoRO. ± prostrate pls in CCo (Yankee Point, Monterey Co.) have been called var. *horizontalis* McMinn. ❀DRN:5,**15–17**,**22–24**&IRR or part SHD:7–9, **14**,**19–21**;CVS;GRCVR.

C. hearstiorum Hoover & Roof (p. 941) HEARST'S CEANOTHUS Pl ± prostrate or sprawling, some fl branches ascending. **ST**: twigs angled, light-green, densely hairy, becoming red-brown. **LVS** alternate, evergreen, < 2.5 cm; stipules deciduous; petiole < 2 mm; blade elliptic to ± rectangular, 1-ribbed from base, tip gen truncate, margin ± turned under, minutely gland-toothed, upper surface gen ± cupped, dark green, glandular-hairy, lower gray-tomentose. **INFL** ± raceme-like, < 5 cm. **FL** blue. **FR** ± 4 mm, smooth, 3-lobed; valve crest 0. **RARE** CA. Coastal, low, grassy hills; < 200 m. CCo (near Arroyo de la Cruz, San Luis Obispo Co.). In cult.

C. impressus Trel. (p. 941) Pl ascending, gen < 3.5 m, < 7 m wide, densely branched. **ST**: twigs round, dark brown, puberulent, not changing color. **LVS** alternate, evergreen, < 2.5 cm; stipules deciduous; petiole < 3 mm; blade ovate to ± round; 1-ribbed from base, tip obtuse, margin strongly turned under, appearing ± crenate, gen minutely toothed, glandular or not, upper surface furrowed along veins, dark green, puberulent, lower gray-hairy. **INFL** raceme-like, < 4 cm. **FL** blue. **FR** gen 4 mm, deeply 3-lobed; valves crested. Dry, sandy mesas, slopes; < 200 m. CCo. Pls around Nipomo Mesa, San Luis Obispo Co. (gen taller, lvs larger, light green, infl longer) have been called var. *nipomensis* McMinn. ❀DRN,SUN:5,7,**15–17**,**22–24**&IRR:8,9,**14**,**19–21**;CVS.

C. incanus Torrey & A. Gray COAST WHITETHORN Pl erect, < 4 m, gen thorny. **ST**: twigs round, gray-glaucous, puberulent, not changing color. **LVS** alternate, evergreen, < 6 cm; stipules deciduous; petiole < 12 mm; blade elliptic- to round-ovate, 3-ribbed from base, firm, tip obtuse, margin entire, rarely ± minutely serrate, surfaces glabrous to puberulent, upper gray-green, lower paler. **INFL** panicle-like, < 7 cm. **FL** white. **FR** < 5 mm, rough, shallowly 3- lobed at top; valve crest 0. 2*n*=24. Open, moist flats, slopes; < 900 m. KR, NCoRO, SnFrB. ❀DRN,SUN:5,**15–17**,**22–24**&IRR:7,8,9, **14**,18,**19–21**;STBL.

C. integerrimus Hook. & Arn. (p. 941) DEER BRUSH Pl ascending-erect, < 4 m. **ST**: twigs round, yellow- to pale-green, glabrous to ± strigose, not changing color. **LVS** alternate, deciduous, < 8 cm; stipules deciduous; petiole < 15 mm; blade lanceolate, elliptic or ± oblong to widely ovate, 1–3-ribbed from base, thin, tip acute to obtuse, margin gen entire or minutely dentate toward tip, upper surface light green, ± glabrous to puberulent, lower paler, ± glabrous to hairy. **INFL** ± raceme-like, < 15 cm, clusters gen stalked. **FL** white or blue, rarely pink. **FR** 4–5 mm, sticky; top ± depressed; valves minutely crested. 2*n*=24. Dry slopes, ridges; 300–2100 m. KR, NCoR, CaR, SN, SnFrB, SCoR, TR, PR; to WA. Several poorly defined vars. have been recognized: var. *i.* (lf blade oblong to elliptic-oblong, gen 1-ribbed from base; fl white or blue; SnFrB, SCoR); var. *californicus* (Kellogg) Benson (lf blade elliptic or lanceolate to oblong-ovate, 3-ribbed from base, upper surface gen glabrous, lower glabrous or hairy; fl white; KR, NCoR, CaR, SN); var. *macrothyrsus* (Torrey) Benson (lf blades oblong to ovate, gen 3-ribbed from base, gen hairy on both surfaces; fl white to blue; KR, CaRH, to WA); var. *puberulus* (E. Greene) Abrams (lf blade elliptic or lanceolate to oblong-ovate, 3-ribbed from base, gen hairy on both surfaces; fl white; TR, PR). ❀DRN:4–6,**15**,**16**,17&IRR: 1–3,7,14,18–21.

C. jepsonii E. Greene MUSK BRUSH **ST**: twigs gen round, gray-brown, puberulent, not changing color. **LVS** opposite, evergreen, < 3 cm; stipules persistent; petiole < 3 mm; blade ovate to elliptic, 1-ribbed from base, firm, tip obtuse, margin, ± turned under, with spine-like teeth, upper surface gen ± cupped, yellow-green, glabrous, lower canescent between veins. **INFL** raceme-like, < 2 cm. **FL**: odor musky. **FR** ± 6 mm; horns prominent, wrinkled; ridges present. 2*n*=24. Dry, shrubby, rocky slopes, serpentine; < 600 m. NCoR, SnFrB.

var. ***albiflorus*** J. Howell (p. 941) Pl erect, < 1 m. **FL** white. **FR** ± oblong. Habitats and elevations of sp. NCoRH, NCoRI. ❀DRN,DRY:7,15,16;DFCLT.

var. ***jepsonii*** (p. 941) Pl ± ascending, < 6 dm. **FL** blue to violet. **FR** spheric. Habitats of sp. NCoRO, SnFrB. ❀DRN,DRY:7, 14–16;DFCLT.

C. lemmonii C. Parry Pl ± ascending, < 1 m. **ST**: twigs round, green, hairy, glandular or not, becoming gray. **LVS** alternate, evergreen, < 3 cm; stipules deciduous; petiole < 8 mm; blade elliptic to ± oblong, 1- or ± 3-ribbed from base, tip acute to obtuse, margin minutely gland-toothed, upper surface dull green, ± glabrous to strigose, lower paler, densely hairy. **INFL** ± raceme-like, dense, < 5 cm. **FL** pale blue. **FR** < 4 mm, smooth, 3-lobed, depressed at top; valves prominently crested. 2*n*=24. Open, wooded slopes; 350–1000 m. NCoRI, CaRF, n&c SNF. ❀DRN,DRY:**7**,14–17;DFCLT.

C. leucodermis E. Greene (p. 941) CHAPARRAL WHITETHORN Pl erect, < 4 m, thorny. **ST**: twigs round, gray-glaucous, puberulent, becoming glaucous. **LVS** alternate, evergreen, < 4 cm; stipules deciduous; petiole 2–3 mm; blade ovate to elliptic-oblong, 3-ribbed from base, tip acute to obtuse, margin entire to minutely serrate, glandular or not, both surfaces gray-glaucous, glabrous to puberulent. **INFL** gen raceme-like, < 11 cm; clusters stalked. **FL** pale blue to white. **FR** < 6 mm, sticky, 3-lobed, depressed at top; valve crests 0. 2*n*=24. Dry, rocky or sandy slopes; < 1800 m. SNF, SnFrB, SCoR, TR, PR; n Baja CA. Variable in lvs, infl. ❀DRN,DRY:**7**, 14–17,**18**.

C. maritimus Hoover (p. 941) MARITIME CEANOTHUS Pl decumbent to ascending, < 1 m. **ST**: twigs round, gray to red-brown, glabrous, becoming gray. **LVS** opposite, evergreen, firm, < 2 cm; stipules persistent; petiole < 2 mm; blades narrowly to widely obovate or oblong, 1-ribbed from base, firm, tip abruptly pointed, truncate, or notched, margin turned under, toothed above middle, upper surface shiny green, glabrous, lower white-woolly. **INFL** raceme-like, < 2 cm. **FL** deep blue to whitish. **FR** 6 mm; horns minute, erect; ridges 0. **RARE** CA. Coastal hills, bluffs; < 150 m. CCo (Hearst Ranch, n San Luis Obispo Co.). In cult.

C. masonii McMinn MASON'S CEANOTHUS Pl ± ascending to erect, < 2 m. **ST**: twigs round, dark, puberulent. **LVS** opposite, evergreen, 0.6–1.8 cm; stipules present; petiole < 4 mm; blade widely elliptic to round, 1-ribbed from base, tip obtuse to truncate, margin dentate, teeth spine-like or not, upper surface dark green, glabrous, lower surface hairy on veins, canescent between. **INFL** raceme-like, < 2.5 cm. **FL** deep blue to violet. **FR** ± 3.5 mm; horns near top minute; ridges 0. $2n$=24. **RARE** CA. Dry, rocky, shrubby slopes; < 500 m. SnFrB (Bolinas Ridge, sw Marin Co.). Closely related to *C. gloriosus*.

C. megacarpus Nutt. Pl < 4 m. **ST**: twigs round, gray to red-brown, ± tomentose, becoming gray. **LVS** evergreen; stipules persistent, warty; petiole < 7 mm; blade elliptic to obovate, 1-ribbed from base, firm, tip gen notched to truncate, margin entire, upper surface flat, dull-green, glabrous, lower canescent between veins. **INFL** raceme-like, < 2 cm. **FL** white or pale lavender exc disk; ovary dark. **FR** 8–12 mm; ridges 0. Dry, shrubby slopes, canyons near coast; < 750 m. CCo, SCoRO, ChI.

var. ***insularis*** (Eastw.) Munz (p. 943) ISLAND CEANOTHUS Pl erect. **LVS** gen opposite, < 4 cm. **FR**: horns gen 0 or minute, on side or near tip. UNCOMMON. Habitats of sp.; < 600 m. ChI. [*C. i.* Eastw.] ❀DRN,DRY,SUN:7,**14–17,19–24**.

var. ***megacarpus*** (p. 943) BIGPOD CEANOTHUS Pl ± ascending or erect, < 4 m. **LVS** gen alternate, < 2.6 cm. **FR**: horns prominent, near tip. $2n$=24. Habitats and elevations of sp. CCo, SCoRO, s ChI (Catalina Island). Pls with whip-like, ± pendent branches (Santa Barbara, Ventura cos.) have been called var. *pendulus* McMinn. ❀DRN,DRY,SUN:7,**14–24**.

C. oliganthus Nutt. Pl erect, often tree-like, < 3 m. **ST**: twigs round, warty or smooth, not changing color. **LVS** alternate, evergreen, < 4 cm; stipules deciduous; petiole < 8 mm; blade ovate, elliptic, to elliptic-oblong, 3-ribbed from base, tip obtuse or acute, margin minutely gland-toothed, upper surface dark green, lower paler, hairy, esp on veins. **INFL** panicle- or raceme-like. **FR** ± 4 mm, gen sticky, 3-lobed; valves ± crested. Dry, shrubby slopes; < 1300 m. NCoR, SnFrB, SCoR, TR, PR.

var. ***oliganthus*** (p. 943) **ST**: twigs ± red-brown, hairy. **LF**: blade upper surface hairy. **INFL** gen interrupted, < 5 cm. **FL** gen deep blue or purple. **FR** smooth or rough, lobed, crested, hairy or not. Habitats of sp.; < 1300 m. SCoR, TR, PR. Some pls in s CA (fl pale blue; fr very wrinkled, stickier, hairy) have been called var. *orcuttii* (C. Parry) Jepson. ❀DRN,DRY:7,**14–17**,18,**19–24**.

var. ***sorediatus*** (Hook. & Arn.) Hoover JIM BRUSH **ST**: twigs gen gray-green, ± glabrous to puberulent. **LF**: blade upper surface ± glabrous. **INFL** continuous, < 4 cm. **FL** pale to deep blue. **FR** smooth, barely lobed or crested, glabrous. $2n$=24. Habitats of sp.; < 1000 m. NCoR, SnFrB, SCoR, TR. [*C. s.* Hook. & Arn.] ❀DRN,DRY:7–9,**14–17**,18,**19–24**.

C. ophiochilus S. Boyd, T. Ross, & L. Arnseth VAIL LAKE CEANOTHUS Pl erect, 0.3–2 m. **ST**: twigs ± round, red-brown, glabrous, becoming ashy gray. **LVS** opposite, clustered or not, evergreen, < 7 mm; stipules persistent; petiole < 1 mm; blade narrowly oblanceolate to obovate, 1-ribbed near base, thick, firm, tip rounded to acute, margin gen entire; upper surface convex, surface green, glabrous, lower surface pale green. **INFL** raceme-like, gen < 1 cm. **FL** pale blue, sometimes pinkish. **FR** 3–3.5 mm, horns gen 0; smooth. RARE. Rocky, n-facing slopes, ridges, chaparral; ± 600 m. n PR (near Vail Lake, Riverside Co.). Closely related to *C. greggii*.

C. palmeri Trel. PALMER CEANOTHUS Pl ascending to erect, < 3.5 m. **ST**: twigs round, gray-green, glabrous, becoming red-brown. **LVS** alternate, gen evergreen, < 4 cm; stipules deciduous; petiole < 8 mm; blade oblong to oblong-ovate, ± 1-ribbed from base, firm, thick, tip rounded to notched, margin entire, upper surface light green, glabrous, lower paler, glabrous exc sometimes ± hairy on midrib. **INFL** panicle-like, < 12 cm. **FL** white. **FR** 5–7 mm, smooth, 3-lobed; valves glandular-crested. Dry slopes, chaparral, yellow-pine forest; 100–1800 m. c SNF, SnJt, PR; n Baja CA. Pls in c SNF (Amador, Eldorado cos.) at 100–450 m have more flexible lvs, ± smaller fr. ❀DRN,DRY:1–3,**7,18**.

C. papillosus Torrey & A. Gray **ST**: twigs round, olive green, gen densely hairy, becoming dark. **LVS** alternate, ± crowded, evergreen, < 5 cm; stipules deciduous; petiole < 6 mm; blade linear or elliptic to oblong-elliptic, 1-ribbed from base, margin minutely gland-toothed, ± turned under, upper surface dark green, ± hairy, glandular-papillate, lower paler, gen densely hairy. **INFL** gen raceme-like, < 5 cm. **FL** deep blue. **FR** ± 3 mm, smooth, shallowly 3-lobed at top; valve crests minute. Open, dry, ± wooded or shrubby slopes; < 1200 m. SnFrB, SCoR, WTR, PR. Pls with lf upper surface ± glabrous, barely or not glandular-papillate have been called *C.* ×*regius* (Jepson) McMinn.

var. ***papillosus*** WARTLEAF CEANOTHUS Pl ascending-erect, < 5 m. **LF**: blade 1–2 cm wide, tip obtuse to truncate, gen not notched. $2n$=24. Open, dry, ± wooded slopes; 0–900 m. SnFrB, SCoRO. ❀DRN,DRY:5,**15–17,22–24**& part SHD;**7,14,19–21**.

var. ***roweanus*** McMinn (p. 943) Pl ascending-erect or ± prostrate, < 1.8 m. **LF**: blade < 1 cm wide, tip truncate or notched. Open, dry, ± shrubby slopes; 600–1200 m. SCoR, WTR, nw PR (Orange Co.). ❀DRN,DRY:**5,15–17,22–24**& part SHD;7,**14**,18, **19–21**;CV.

C. parryi Trel. Pl erect, < 5 m. **ST**: twigs angled, gray or red-brown, tomentose, becoming dark. **LVS** alternate, evergreen, < 4.5 cm; stipules deciduous; petiole < 10 mm; blade oblong or ± elliptic, gen 1-ribbed from base, tip obtuse, margin entire to minutely gland-toothed, turned under, upper surface dark green, glabrous, lower paler, tomentose. **INFL** panicle-like, < 15 cm. **FL** deep blue. **FR** 3–4 mm, smooth, ± 3-lobed; valve crests 0. $2n$=24. Wooded canyons, slopes; < 750 m. NCoR. ❀DRN:**7**,14,**15**&SUN:**16**,17.

C. parvifolius (S. Watson) Trel. Pl ascending, < 12 dm. **ST**: twigs round, olive-green to red, glabrous, becoming red. **LVS** alternate, deciduous, < 2 cm; stipules deciduous; petiole < 5 mm; blade oblong-elliptic to elliptic, 3-ribbed ± from base, sometimes weakly so, tip acute to obtuse, margin entire, minutely toothed near tip or not, upper surface green, glabrous, lower paler, rarely hairy. **INFL**: raceme-like, < 7 cm; clusters gen sessile. **FL** pale to deep blue. **FR** 4–5 mm, smooth, 3-lobed; valve crests ± 0. Moist wooded slopes, flats; 1350–2100 m. SNH. ❀DRN,IRR:1–3,7.

C. pinetorum Cov. (p. 943) KERN CEANOTHUS Pl ± prostrate to erect, < 1.2 m, < 8 m wide. **ST** rooting at nodes or not; twigs round, brown, glabrous or puberulent, becoming gray-white to brown. **LVS** opposite, gen clustered, evergreen, gen < 2 cm; stipules persistent; petiole < 2 mm; blade oblong to round-elliptic, 1-ribbed from base, firm, tip obtuse, margin coarse-toothed, ± turned under or not, upper surface green, gen glabrous, lower minutely canescent between veins. **INFL** raceme-like, < 2 cm. **FL** whitish to blue. **FR** 7–9 mm; horns near top, wrinkled; ridges present. Uncommon. Dry rocky slopes, open pine forests; 1600–2700 m. s SNH (Kern Plateau). ❀DRN,IRR, part SHD:1,7.

C. prostratus Benth. (p. 943) MAHALA MAT Pl prostrate, < 2.5 m wide. **ST** rooting at nodes or not; twigs angled, red-brown, ± puberulent, becoming gray-brown. **LVS** opposite, evergreen, < 3.2 cm; stipules persistent; petiole < 3 mm; blade oblanceolate to obovate, 1-ribbed from base, firm, tip obtuse, margin gen 3–9-toothed at tip or above middle; upper surface flat, surface green, ± glabrous, lower surface paler, veins gen hairy, canescent between. **INFL** raceme-like, < 2 cm. **FR** light to deep blue, lavender, or purple. **FR** 5–9 mm; horns ± on sides, not slender, wrinkled; ridges prominent. $2n$=24. Open flats, coniferous forest; 900–2200 m. KR, NCOR, CaRH, n&c SNH, MP; to WA, w NV. Pls with large lvs and frs have been called var. *laxus* Jepson; pls with small, sometimes cupped, wavy-margined lvs and small frs from NCoR have been called var. *occidentalis* McMinn. ❀DRN,DRY:4–6,15,16&SHD;1–3,14,18; DFCLT,GRCVR.

C. pumilus E. Greene SISKIYOU MAT Pl prostrate to decumbent, < 2.5 m wide. **ST** rooting at nodes or not; twigs angled, red-brown, puberulent, becoming gray-brown. **LVS** opposite, evergreen, < 1.5 cm; stipules persistent; petiole < 2 mm; blade oblanceolate to obovate-oblong, 1-ribbed from base, firm, tip truncate, margin gen 3-toothed at tip, ± turned under, upper surface flat or ± cupped, green, glabrous, lower paler, veins gen hairy, canescent between. **INFL** raceme-like, < 2 cm. **FL** white to blue or lavender. **FR** < 5 mm; horns on side, minute; ridges 0. 2*n*=24. Gen serpentine, dry slopes, open flats in chaparral or coniferous forest; 600–1800 m. KR, NCoRO, NCoRH; sw OR. ❀DRN,DRY:4–6,15–17& part SHD: 1–3,**7**,14,18;DFCLT;GRCVR.

C. purpureus Jepson (p. 943) HOLLYLEAF CEANOTHUS Pl ascending to erect, < 2 m. **ST**: twigs angled, red-brown, with whitish bloom or not, glabrous, becoming gray to red-brown. **LVS** opposite, evergreen, < 2.5 cm; stipules persistent; petiole < 2 mm; blade ± round to widely elliptic, 1-ribbed from base, firm, tip obtuse, margin wavy with spine-like teeth, upper surface ± cupped, surface green, glabrous, lower surface hairy on veins, minutely canescent between. **INFL** raceme-like, < 3 cm. **FL** deep blue to purple. **FR** gen 5 mm; horns at top, slender; ridges 0. 2*n*=24. UNCOMMON. Dry, shrub-covered, rocky, volcanic slopes; < 500 m. NCoRI. ❀DRN,DRY:**15,16**,17&partSHD:**7,14**,19–24.

C. roderickii W. Knight (p. 943) PINE HILL CEANOTHUS Pl prostrate, < 3 m wide. **ST** rooting at nodes; twigs round, red-brown, puberulent, becoming gray-brown. **LVS** opposite, evergreen, suberect, ± 1 cm; stipules persistent; petiole 1–2 mm; blade oblanceolate, 1-ribbed from base, firm, tip obtuse to notched, margin entire or toothed near tip, upper surface dull green, ± glabrous, lower canescent between veins. **INFL** raceme-like, short. **FL** white with blue tinge. **FR** gen 5 mm; horns stubby, on side, or 0; ridges 0. **RARE** CA. Dry, stony, soils from serpentine or similar rocks; 300–600 m. n SNF (Pine Hill, w El Dorado Co.). Suberect lvs ± unique in genus. Threatened by development. In cult.

C. sanguineus Pursh REDSTEM CEANOTHUS Pl erect, < 3 m. **ST**: twigs round, greenish, glabrous, becoming red to purple. **LVS** alternate, deciduous, < 10 cm; stipules deciduous; petiole < 25 mm; blade ovate to ovate-elliptic or obovate, 3-ribbed from base, thin, tip acute to obtuse, margin minutely gland-toothed, upper surface green, ± glabrous, lower paler, gen hairy, esp on veins. **INFL** panicle-like, < 12 cm. **FL** white. **FR** ± 4 mm, smooth, 3-lobed; valve crests minute or 0. Dry forest floor; ± 1200 m. KR; to B.C., MT. ❀DRN, part SHD,IRR:1–3,7;STBL.

C. sonomensis J. Howell (p. 939) SONOMA CEANOTHUS Pl erect, < 1 m. **ST**: twigs round, gray to brown, glabrous, not changing color. **LVS** opposite, evergreen, < 1.5 cm, flat; stipules persistent; petiole ± 0; blade obovate to rounded, 1-ribbed from base, tip notched, margin with 3–5 ± spine-like teeth, upper surface shiny green, glabrous, lower surface gray, short-tomentose. **INFL** raceme-like, < 2.5 cm. **FL** blue or lavender. **FR** 5 mm; horns slender; ridges minute. 2*n*=24. RARE. Chaparral, in sand, serpentine, volcanic soils; < 800 m. NCoRO (Hood Mtn Range, Sonoma & Napa cos.). Closely related to *C. cuneatus*. In cult.

C. spinosus Torrey & A. Gray GREENBARK CEANOTHUS Pl erect, ± tree-like, < 6 m, gen thorny. **ST**: twigs round, ridged, olive-green, glabrous, not changing color. **LVS** alternate, evergreen, < 5 cm; stipules deciduous; petiole < 8 mm; blade elliptic to oblong, 1-ribbed from base, firm, thick, tip obtuse to notched, margin gen entire, upper surface green, glabrous, lower paler, glabrous or hairy, esp on midrib. **INFL** panicle-like, < 15 cm. **FL** pale blue to ± white. **FR** 4–5 mm, barely lobed, smooth; valve crests minute or 0. Dry slopes; < 900 m. SCoRO, WTR, PR; n Baja CA. ❀DRN,DRY, SUN:7–9,**14–17**,18,**19–24**.

C. thyrsiflorus Eschsch. (p. 943) BLUE BLOSSOM Pl prostrate to erect, < 6 m. **ST**: twigs angled, green, glabrous, becoming darker. **LVS** alternate, evergreen, < 5 cm; stipules deciduous; petiole < 12 mm; blade oblong-ovate to widely elliptic, 3-ribbed from base, tip obtuse to acute, margin ± gland-toothed, upper surface dark green, glabrous, lower paler, sparse-hairy on veins, puberulent between or not. **INFL** raceme- or panicle-like, < 8 cm. **FL** light to deep blue. **FR** ± 3 mm, glandular-sticky, slightly lobed; valve crests 0. 2*n*=24. Wooded slopes, canyons; < 600 m. NCo, KR, NCoRO, SnFrB, SCoRO; sw OR. Prostrate pls from c&s NCo have been called var. *repens* McMinn. ❀DRN:6,**15–17,24**& part SHD,IRR:7–9,**14**,19, **20–23;CVS;GRCVR.**

C. tomentosus C. Parry Pl erect, < 3 m. **ST**: twigs round, ± rusty-tomentose, becoming gray-brown or reddish. **LVS** alternate, evergreen, < 3 cm; stipules deciduous; petiole < 5 mm; blade ovate to elliptic, 3- or 1-ribbed from base, tip obtuse, margin gland-serrate, upper surface dark green, finely hairy, lower light or dark tomentose. **INFL** panicle-like, < 9 cm. **FL** blue to ± white. **FR** ± 4 mm, sticky when young, shallowly lobed; valves crested. 2*n*=24. Dry, shrubby slopes, scattered; < 1500 m. n&c SN, SCo, SnBr, PR; n Baja CA. Pls with lf margin glandular-dentate, lower surface gray-green, hairy esp on veins, not tomentose, from < 1100 m in SCo, SnBr, PR, n Baja CA, have been called var. *olivaceous* Jepson. ❀DRN:7,**15,16**,17,19,**20–24**&IRR:**14**,18.

C. velutinus Hook. **ST**: twigs round, brown, ± puberulent, becoming dark brown. **LVS** alternate, evergreen, aromatic, < 8 cm; stipules deciduous; petiole < 18 mm; blade widely elliptic to ovate-elliptic, 3-ribbed from base, tip obtuse, margin gland-toothed, upper surface dark green, ± shiny, glabrous, lower paler. **INFL** panicle-like, < 12 cm. **FL** white. **FR** 3–4 mm, 3-lobed. Open, wooded slopes; < 3000 m. KR, NCoRO, CaRH, SNH, n SnFrB, Wrn; to B.C., SD.

var. ***hookeri*** M. Johnston (p. 943) Pl tree-like or not, < 6 m. **LF**: lower surface glabrous. **FR** ± rough; valve crests minute. Near coast; < 900 m. KR, NCoRO, n SnFrB; to B.C. [var. *laevigatus* (Hook.) Torrey & A. Gray] ❀DRN:5,**6,15–17**&IRR:**7**,14.

var. ***velutinus*** (p. 943) TOBACCO BRUSH Pl ascending-erect, < 2 m; crown rounded. **LF**: lower surface canescent. **FR** ± smooth; valve crests minute or 0. Inland; 1000–3000 m. Range of sp. (exc NCoRO, n SnFrB). Pls with lf < 3 cm, entire to minutely gland-toothed, upper surface not shiny, lower ± silky, in KR, CaRH, Wrn, SNH, have been called *C.* ×*lorenzenii* (Jepson) McMinn (possibly hybrid with *C. cordulatus*). ❀DRN:1,2,6,7,15–17;DFCLT.

C. verrucosus Nutt. WART-STEMMED CEANOTHUS Pl erect, < 3 m. **ST**: twigs angled, gray-brown, minutely hairy, becoming gray. **LVS** alternate, evergreen, < 1.5 cm; stipules persistent, warty; petiole < 3 mm; blade round- to deltate-obovate, 1-ribbed from base, firm, tip gen notched, margin gen entire, upper surface flat or ± cupped, dark green, glabrous, minutely rough, lower canescent between veins. **INFL** raceme-like, < 2 cm. **FL** white exc disk; ovary dark. **FR** gen 5 mm; horns 0 or minute, on side; ridges 0. 2*n*=24. RARE in CA. Dry hills, mesas; < 300 m. s SCo (San Diego Co.); n Baja CA. Threatened by development. In cult.

COLUBRINA

Tree, shrub. **ST**: branches opposite and alternate, rigid; twigs spreading, thorn-tipped or not, gen hairy. **LVS** in part clustered on short-shoots, evergreen or deciduous; stipules deciduous; blade 3–5-ribbed from base, entire but gen with round, marginal glands. **INFL** umbel-like, few-fld; axis sometimes thorn-tipped. **FL**: hypanthium filled with nectar in fl, lower part adhering to developing fr, upper deciduous; sepals 5; petals 5, oblanceolate, = sepals; stamens 5; ovary chambers 3, each 1-ovuled, styles 3. **FR**: capsule, shallowly 3-valved. 31 spp.: warm places, worldwide. (Latin: from French for serpent tree) [Johnston 1971 Brittonia 23:2–53]

C. californica I.M. Johnston (p. 947) LAS ANIMAS COLUBRINA Shrub < 3 m. **LVS** deciduous; blade 12–30 mm, oblong to obovate, dull gray-green, hairs silky, denser on lower surface, base rounded or wedge-shaped, tip rounded to ± notched, sometimes with a small point. **INFL** 5–10 mm, 3–12-fld, very dense; pedicel 1–2 mm, 2–4 mm in fr. **FLS** appearing after rain; hypanthium 3 mm wide. **FR** 8– 10 mm, persistent 3–6 months. UNCOMMON. Creosote-bush scrub; < 1000 m. DSon; AZ, Mex. Local. ❀DRN,DRY, SUN:8,9,**10–12**,13,14,19–23.

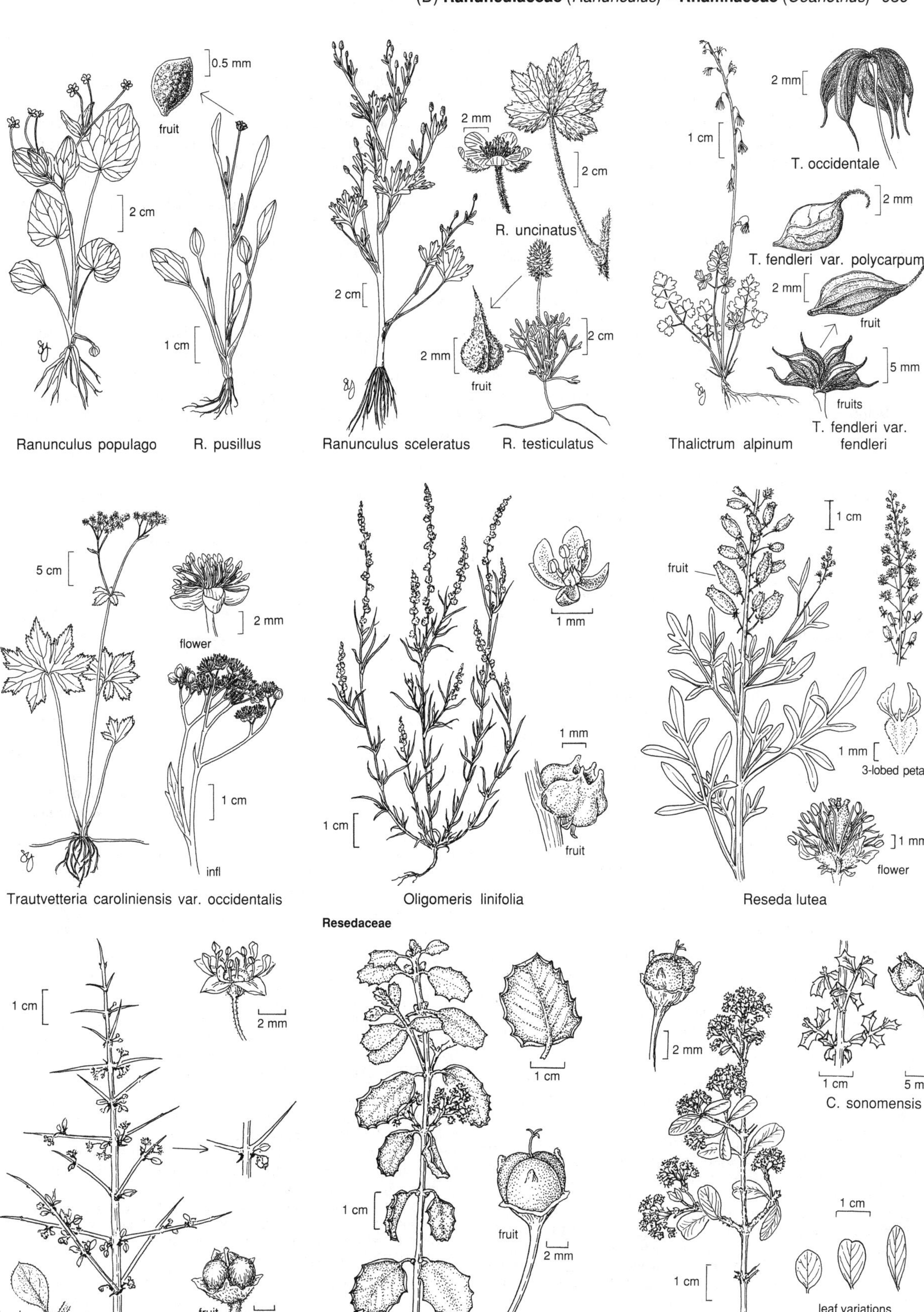
0.5 mm
fruit
2 cm
1 cm
Ranunculus populago
R. pusillus
2 mm
2 cm
R. uncinatus
2 cm
2 mm
fruit
2 cm
Ranunculus sceleratus
R. testiculatus
2 mm
1 cm
T. occidentale
2 mm
T. fendleri var. polycarpum
2 mm
fruit
5 mm
fruits
Thalictrum alpinum
T. fendleri var. fendleri
5 cm
2 mm
flower
1 cm
infl
Trautvetteria caroliniensis var. occidentalis
1 mm
1 mm
fruit
1 cm
Oligomeris linifolia
1 cm
fruit
1 mm
3-lobed petal
1 mm
flower
Reseda lutea
Resedaceae
1 cm
2 mm
2 mm
fruit
2 mm
Adolphia californica
Rhamnaceae
1 cm
1 cm
fruit
2 mm
Ceanothus crassifolius
2 mm
1 cm
5 mm
C. sonomensis
1 cm
1 cm
leaf variations
Ceanothus cuneatus var. cuneatus

CONDALIA

Shrub. **ST**: branches alternate, rigid, 3-ranked; twigs spreading, thorn-tipped. **LVS** clustered on short-shoots, deciduous; stipules deciduous; petioles ± 0 or short; blade obovate, 1-ribbed from base, entire. **INFL**: fls solitary or in clusters on short-shoots. **FL**: hypanthium below, not surrounding ovary, becoming flat after fl, persistent below fr; sepals 5, deciduous; petals gen 0; stamens 5; ovary spheric, strongly narrowed at base, chambers 2, each 1-ovuled, style 1. **FR**: drupe, stone 1. **SEED** tightly held in stone. 18 spp.: arid Am. (A. Condal, Spanish physician, 1745–1804) [Johnston 1962 Brittonia 14:332–368]

C. globosa I.M. Johnston var. ***pubescens*** I.M. Johnston (p. 947) **ST** < 4 m; bark smooth, gray; twigs 3–13 cm, thorn-tipped, pale olive or purple, short-hairy. **LVS** in clusters of 2–7; blade 3–12 mm, slightly thickened; stipules brown. **INFL** 1–8-fld. **FL**: hypanthium 1–1.5 mm, olive or purple, short-hairy; sepals 1 mm, olive; stamens < sepals; pistil purple. **FR** 3–5 mm, black, juicy, bitter or sweet. Uncommon. Creosote-bush scrub; < 1000 m. DSon; AZ, Mex. ❀DRN,DRY,SUN:**10,12**,14–16,**22–24**&IRR:8,9,11,13,**19–21**.

RHAMNUS BUCKTHORN

Shrub, small tree. **ST**: branches alternate, flexible; twigs sometimes thorn-tipped. **LVS** sometimes clustered on short-shoots, deciduous or evergreen, petioled; stipules deciduous; blade 1-ribbed from base, entire or not. **INFL**: umbel or fls solitary, axillary. **FL** bisexual or unisexual, gen < 3 mm; hypanthium at base fused to, developing around ovary in fr, above base deciduous; sepals 4 or 5; petals 0, 4, or 5; stamens 4 or 5; ovary appearing partly inferior, chambers 2–4, each 1–2-ovuled, style lobes 2–4. **FR**: drupe, 2–4-stoned. 125 spp.: temp, few trop. (Greek: name for pls of this genus) [Wolf 1938 Rancho Santa Ana Bot Gard Monogr 1] Some of value in medicine or as dyes.

1. Terminal bud covered with scales; petals gen 0; style exserted (subg. *Rhamnus*)
 2. Lvs deciduous; sepals 5; fr black, 3-stoned ..... ***R. alnifolia***
 2′ Lvs evergreen; sepals 4; fr red, 2-stoned (*R. crocea* complex)
  3. Lf blade < 15 mm, flat; branches spreading ..... ***R. crocea***
  3′ Lf blade > 15 mm, lower side concave; branches gen ascending
   4. Lf blade elliptic; small tree; ChI ..... ***R. pirifolia***
   4′ Lf blade ovate to round; shrub; mainland
    5. Branches stiff, twigs glabrous to densely hairy; CA-FP, DMtns ..... ***R. ilicifolia***
    5′ Branches flexible, twigs densely hairy; PR ..... ***R. pilosa***
1′ Terminal bud not covered with scales; petals present; style incl (subg. *Frangula*)
 6. Lvs thin, deciduous
  7. Fr 3-stoned; lf blade 50–150 mm, gen > 80 mm ..... ***R. purshiana***
  7′ Fr 2-stoned; lf blade 15–80 mm ..... ***R. rubra***
 6′ Lvs thick, gen evergreen (*R. californica* complex)
  8. Lf blade ± glabrous on lower side ..... ***R. californica***
   9. Fr 2-stoned; blade dark green on upper side, bright green or yellow on lower ..... ssp. ***californica***
   9′ Fr 3-stoned; blade ± equally yellow on both sides ..... ssp. ***occidentalis***
  8′ Lf blade tomentose on lower side ..... ***R. tomentella***
   10. Lf margin entire or with blunt teeth
    11. Lf blade widely elliptic; s KR, NCoRI ..... ssp. ***crassifolia***
    11′ Lf blade narrowly elliptic; widespread ..... ssp. ***tomentella***
   10′ Lf margin with sharp teeth
    12. Lf margin dentate to dentate-serrate; c&s SN, Teh, TR, PR, SnJt ..... ssp. ***cuspidata***
    12′ Lf margin serrate to ± entire; DMtns ..... ssp. ***ursina***

R. alnifolia L'Her. ALDER BUCKTHORN Shrub < 2 m. **ST**: bark gray; twigs brown; terminal bud scales 5 mm. **LVS** deciduous; petiole 4–12 mm; blade 45–100 cm, ovate to elliptic, thin enough to transmit light, base acute to obtuse, tip acute, margin irregularly toothed, sides glabrous to puberulent, veins prominent, arched. **INFL** 1–3-fld; pedicels 2–10 mm. **FL** unisexual; hypanthium 1 mm; sepals 5; petals 0; style exserted. **FR** 3-stoned, ± 8 mm, black. Uncommon. Wet meadow edges of lodgepole-pine forest; 1800 m. n SNH (Nevada, Placer cos.); to Can, n Rocky Mtns, n US. ❀WET,DRN:1,2,7,15,16&SUN:4,6.

R. californica Eschsch. (p. 947) CALIFORNIA COFFEEBERRY Shrub < 5 m. **ST**: bark bright gray or brown; twigs glabrous to finely hairy; terminal bud not covered with scales. **LVS** evergreen; petiole 3–10 mm; blade 20–80 mm, ovate to elliptic, thick, base acute to rounded, tip acute, rounded, or truncate, margin serrate to entire, sometimes rolled under, both surfaces ± glabrous. **INFL** 5–60-fld; pedicels < 20 mm. **FL** bisexual; hypanthium 1–2 mm; sepals 5; petals 5; style incl. **FR** 10–15 mm, black. Coastal-sage scrub, chaparral, woodlands, forests; < 2300 m. NW, CW, SW; sw OR. Sspp. intergrade in intermediate habitats.

ssp. ***californica*** **ST**: twigs red. **LF**: blade narrowly to widely elliptic, base and tip acute, upper surface gen dark green, lower bright green or yellow, veins prominent. **FR** 2-stoned. $2n=24$. Non-serpentine; < 2000 m. NW, CW, SW (incl Santa Cruz Island). ❀DRN, DRY:**14–17,22–24**&SHD,IRR:**7**,8,9,18–21;CVS,incl.GRCVRS.

ssp. ***occidentalis*** (J. Howell) C. Wolf **ST**: twigs brown. **LF**: blade ovate to elliptic, base rounded, tip acute, rounded, or truncate, both surfaces ± equally yellowish, veins not prominent. **FR** 3-stoned. Serpentine; < 2300 m. KR, NCoRO, n SNF; sw OR. ❀DRN,DRY:**14–17,22–24**&SHD&IRR:**7**,8,9,18–21.

R. crocea Nutt. (p. 947) SPINY REDBERRY Shrub < 2 m. **ST**: bark gray; branches many, spreading, stiff, sometimes thorn-tipped, rooting, glabrous; terminal bud scales 3 mm. **LVS** evergreen; petiole 1–4 mm; blade 10–15 mm, obovate, thick, base acute to rounded, tip rounded, margin sharply toothed or entire, surfaces glabrous, veins not prominent. **INFL** 1–6-fld; pedicels 1–6 mm. **FL** unisexual; hypanthium 2 mm; sepals 4; petals 0; style exserted. **FR** 2-stoned, 6 mm, red. Coastal-sage scrub, chaparral, woodlands; < 1000 m. NCoRO (Sonoma, Mendocino cos.), CW, SW; Baja CA. ❀DRN,DRY:**7,14–17,20–24**&SHD:**18,19**.

R. ilicifolia Kellogg (p. 947) HOLLY-LEAF REDBERRY Shrub < 4 m. **ST**: bark gray; branches ascending, stiff; twigs glabrous to densely hairy; terminal bud scales 3 mm. **LVS** evergreen; petiole 2–10 mm; blade 20–40 mm, ovate to round, thick, base and tip rounded, margin spiny, surfaces glabrous or hairy, lower concave,

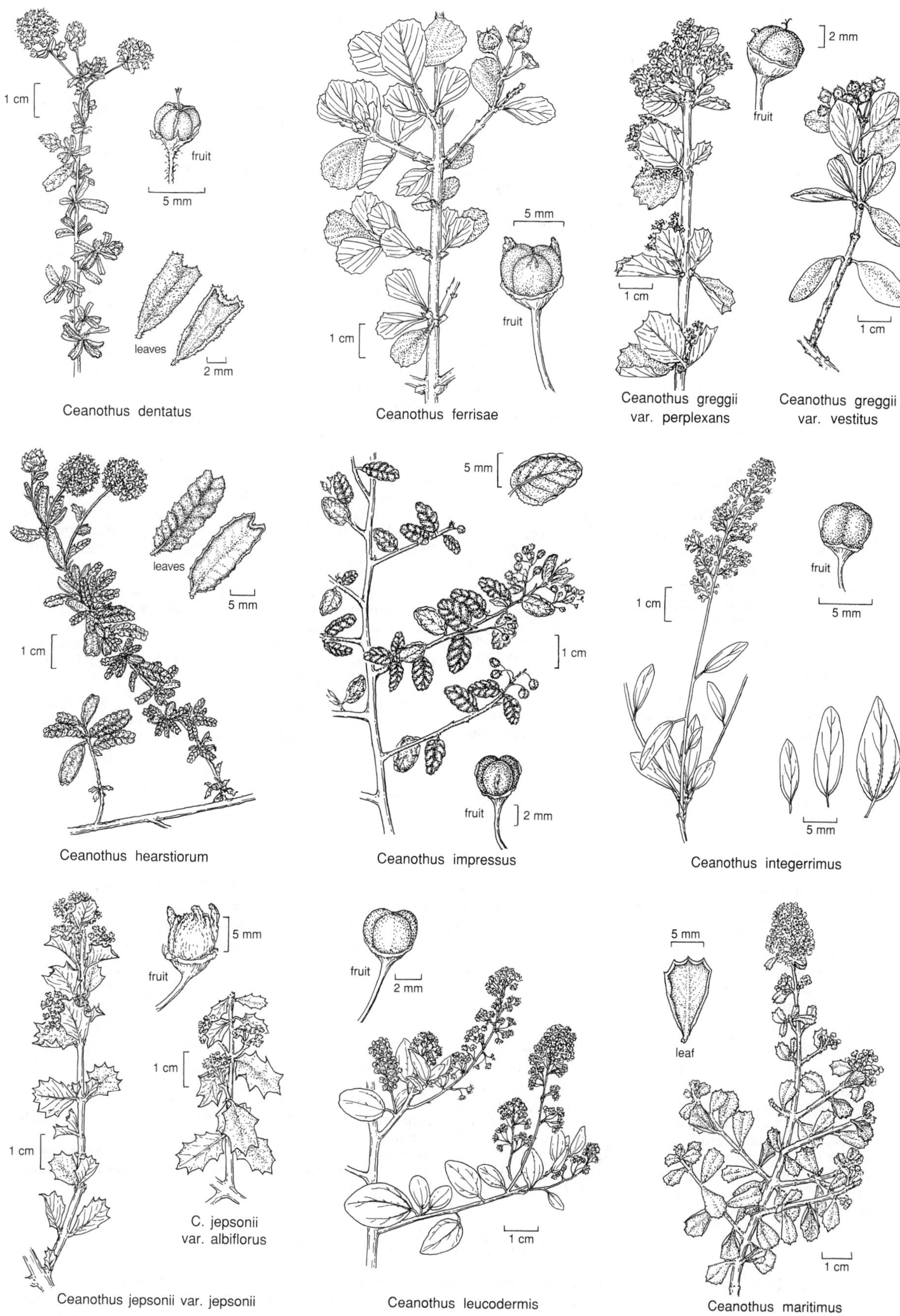
1 cm
fruit
5 mm
leaves
2 mm
Ceanothus dentatus
5 mm
fruit
1 cm
Ceanothus ferrisae
2 mm
fruit
1 cm
1 cm
Ceanothus greggii var. perplexans
Ceanothus greggii var. vestitus
leaves
5 mm
1 cm
Ceanothus hearstiorum
5 mm
1 cm
fruit
2 mm
Ceanothus impressus
1 cm
fruit
5 mm
5 mm
Ceanothus integerrimus
5 mm
fruit
1 cm
1 cm
C. jepsonii var. albiflorus
Ceanothus jepsonii var. jepsonii
fruit
2 mm
1 cm
Ceanothus leucodermis
5 mm
leaf
1 cm
Ceanothus maritimus

veins not prominent. **INFL** 1–6-fld; pedicels 2–4 mm. **FL** unisexual; hypanthium 2 mm; sepals 4; petals 0; style exserted. **FR** 2-stoned, 8 mm, red. $2n=24$. Chaparral, montane forests; < 2000 m. CA-FP, DMtns; AZ, Baja CA. [*R. crocea* ssp. *i.* (Kellogg) C. Wolf] ❀DRN,DRY:**7,14–17**,20–23&SHD:**18,19**.

R. pilosa (Trel.) Abrams (p. 947) Shrub < 2 m. **ST**: bark gray; branches spreading, flexible; twigs densely hairy; terminal bud scales 3 mm. **LVS** evergreen; petiole 2–5 mm; blade 15–20 mm, ovate to round, thick, base, tip rounded, margin prickly, surfaces densely hairy, veins not prominent. **INFL** 1–6-fld; pedicels 2–4 mm. **FL** unisexual; hypanthium 2 mm; sepals 4; petals 0 or 4; style exserted. **FR** 2-stoned, 6 mm, red. Uncommon. Chaparral; 300–700 m. PR; Baja CA. [*R. crocea* ssp. *p.* (Trel.) C. Wolf] ❀DRN, DRY,SUN:7,**14**, 15,**16,17**,20–23,**24**.

R. pirifolia E. Greene (p. 947) ISLAND REDBERRY Tree < 10 m. **ST**: bark gray; branches ascending; twigs purple; terminal bud scales 3 mm. **LVS** evergreen; petiole 5–10 mm; blade 20–40 mm, elliptic, thick, base rounded, tip acute to rounded, with a small point, margin toothed to entire, surfaces glabrous, lower concave, veins prominent. **INFL** 1–6-fld; pedicels 3–6 mm. **FL** unisexual; hypanthium 2 mm; sepals 4; petals 0; style exserted. **FR** 2-stoned, 8 mm, red. Coastal-sage scrub, chaparral. ChI; Mex (Guadalupe Island). [*R. crocea* ssp. *p.* (E. Greene) C. Wolf] ❀DRN,DRY,SUN: 7,**14**,15,**16,17**,20–23, **24**.

R. purshiana DC. (p. 947) CASCARA Tree or shrub, < 12 m. **ST**: bark gray; twigs red to brown; terminal bud not covered with scales, hairs brown. **LVS** deciduous; petiole 6–23 mm; blade 50–150 mm, broadly elliptic to obovate, thin, base obtuse or tapered, tip obtuse to truncate, margin irregularly toothed to entire, surfaces sparsely hairy to glabrous, veins prominent. **INFL** < 25-fld; pedicels < 25 mm. **FL** bisexual; hypanthium 3 mm; sepals 5; petals 5; stamens 5; style incl. **FR** 3-stoned, 10 mm, black. Coniferous forests, chaparral; < 2000 m. NCo, KR, NCoRO, NCoRH, CaR, n&c SN; to B.C., MT. Pls with wedge-shaped lvs in KR, CaR, n&c SN assignable with difficulty to var. *annonifolia* (E. Greene) Jepson. Bark yields cathartic drugs; bark and fr TOXIC in excess, esp to children. ❀**4,5**&IRR:1–3,**6,16**, **17**&SHD:7,14,**15**.

R. rubra E. Greene (p. 947) SIERRA COFFEEBERRY Shrub < 2 m. **ST**: bark red to bright gray; twigs red to gray; terminal bud not covered with scales, hairy. **LVS** deciduous; petiole 2–12 mm; blade 15–80 mm, narrowly elliptic to obovate, thin, green or gray, base, tip acute to rounded, margin finely toothed to entire, surfaces glabrous to finely hairy, veins not prominent. **INFL** 6–15-fld; pedicels 1–12 mm. **FL** bisexual; hypanthium 2 mm; sepals 5; petals 5; style incl. **FR** 2-stoned, 12 mm, black. Chaparral, montane forests, sagebrush steppe; 1000–2200 m. KR, NCoRH, CaR, n&c SNH, MP; NV. Several sspp. distinguished with difficulty: sspp. *rubra* (lf green; n SNH), grading into sspp. *modocensis* C. Wolf (lvs on short-shoots; MP); sspp. *nevadensis* (Nelson) C. Wolf (fr pear-shaped; w NV); sspp. *obtusissima* (E. Greene) C. Wolf (lf obovate; KR, CaR); sspp. *yosemitana* C. Wolf (lf finely hairy; c SNH). ❀DRN,SUN:1–5,**6,15**,16&IRR:**7**.

R. tomentella Benth. HOARY COFFEEBERRY Shrub < 6 m. **ST**: bark gray or red; twigs velvety; terminal buds not covered with scales. **LVS** evergreen; petiole 3–10 mm; blade 20–70 mm, ovate to narrowly elliptic, thick, base acute to rounded, tip acute, rounded, or with a small point, margin entire to toothed, sometimes rolled under, lower surface tomentose, veins prominent. **INFL** 5–60-fld; pedicels < 20 mm. **FL** bisexual; hypanthium 1–2 mm; sepals 5; petals 5. **FR** 3-stoned, 10–15 mm, black. Chaparral, woodlands; < 2300 m. KR, NCoRH, NCoR, CaRF, SN, ScV, SnFrB, SCoR, SW, DMtns; to NM, Baja CA. Several sspp. distinguishable, but intermediates are common.

ssp. ***crassifolia*** (Jepson) J.O. Sawyer (p. 947) **ST**: twigs gray, tomentose. **LF**: blade 30–100 mm, widely elliptic, upper surface white-tomentose, tip obtuse, margin entire or with blunt teeth. Chaparral, woodlands; < 1000 m. s KR, NCoRI [*R. californica* ssp. *crassifolia* (Jepson) C. Wolf] Similar pls in SW best considered ssp. *tomentella*. ❀DRN,DRY,SUN:7,**14–17**,20–24.

ssp. ***cuspidata*** (E. Greene) J.O. Sawyer (p. 947) **ST**: twigs red, hairs of 2 lengths. **LF**: blade 20–60 mm, elliptic, upper surface green, tip abruptly narrowed to a point, margin dentate to dentate-serrate. Chaparral, montane woodlands; 600–2300 m. c&s SN, Teh, TR, nw PR, SnJt. [*R. californica* ssp. *cuspidata* (E. Greene) C. Wolf] Pls at high elevations sometimes deciduous. ❀DRN,DRY, SUN:2,3,**7,14**,15–21.

ssp. ***tomentella*** **ST**: twigs gray, tomentose. **LF**: blade 30–70 mm, narrowly elliptic, upper surface green, tip acute, margin entire or teeth blunt. Chaparral, woodlands; < 1000 m. s KR, NCoR, CaRF, SNF, ScV, SnFrB, SCoR, SW; Baja CA. [*R. californica* Eschsch. ssp. *t.* (Benth.) C. Wolf] ❀DRN,DRY,SUN:**15–17,24**& SHD:**14**&IRR:**7**,8,9,18–21,**22,23**.

ssp. ***ursina*** (E. Greene) J.O. Sawyer (p. 947) **ST**: twigs gray, tomentose. **LF**: blade 30–70 mm, elliptic or ovate, upper surface green, tip acute to rounded, margin serrate to ± entire. Woodlands; 1200–2100 m. DMtns (Clark, New York, Providence Mtns); to NV, NM. [*R. californica* ssp. *ursina* (E. Greene) C. Wolf] ❀DRN, DRY,SUN:2,3,**7,14**,15–21.

ZIZIPHUS

Tree, shrub, vine. **ST**: branches alternate, flexible, sometimes 2–3-ranked; twigs thorn-tipped. **LVS** in part clustered on short-shoots, deciduous or evergreen, petioled; stipules sometimes unequal spines; blade ovate to oblong, 1–3-ribbed from base, ± entire to serrate. **INFL**: cyme or small panicle. **FL**: hypanthium surrounding base of ovary; sepals 5; petals 5, < or = sepals; stamens 5; ovary broadly attached at base, chambers 2, each 1-ovuled, styles 2. **FR**: drupe, stone 1. 100 spp.: gen trop. (Latin: reason for application obscure) [Johnston 1963 Amer J Bot 50:1020–1027]

1. Fr 7–10 mm, blue-black, not beaked .... ***Z. obtusifolia*** var. ***canescens***
1′ Fr 10–25 mm, brown, beaked .... ***Z. parryi*** var. ***parryi***

Z. obtusifolia (Torrey & A. Gray) A. Gray var. ***canescens*** (A. Gray) M. Johnston (p. 947) GRAYTHORN **ST** < 3 m; bark gray; twigs 1–8 cm, thorn-tipped, hairs short, dense, white. **LVS** deciduous; blade 2–20 mm, ovate or oblong, firm, gray, margin entire or teeth 2–10, glandular; stipules brown. **INFL** 10–30-fld. **FL**: hypanthium 1.5–2 mm, olive, glabrous to tomentose; sepals = petals, yellowish; stamens < petals; pistil olive. **FR** juicy. Uncommon. Creosote-bush scrub; < 1000 m. DSon; AZ, Mex. [*Condalia lycioides* (A. Gray) Weberb. var. *c.* (A. Gray) Trel.] ❀DRN,DRY,SUN:8,9, **10–12**,13,14,16,19–23.

Z. parryi Torrey var. ***parryi*** (p. 947) LOTEBUSH **ST** < 4 m; bark gray to brown; twigs 1.3–3 cm, thorn-tipped, with 1 node, 1 short-shoot, glabrous. **LVS** deciduous; blade 10–25 mm, elliptic to obovate, membranous, olive, margin ± entire; stipules brownish, membranous. **INFL** 2–5-fld. **FL**: hypanthium 2–2.2 mm, purplish green, glabrous; sepals < petals, yellow; stamens < petals; pistil brownish. **FR** dry. Uncommon. Chaparral; 500–1500 m. w edge DSon; Mex. [*Condalia p.* (Torrey) Weberb.] ❀DRN,DRY,SUN: 8,9,**10–12**,13,14,19–23.

ROSACEAE ROSE FAMILY

Ann to tree. **LVS** simple to pinnately to palmately compound, gen alternate; stipules free to fused, persistent to deciduous. **INFL**: cyme, raceme, panicle, or fls solitary. **FL** gen bisexual, radial; hypanthium free or fused to ovary, saucer- to funnel-shaped, often with bractlets alternate with sepals; sepals gen 5; petals gen 5, free; stamens (0)5–many, pistils (0)1–many,

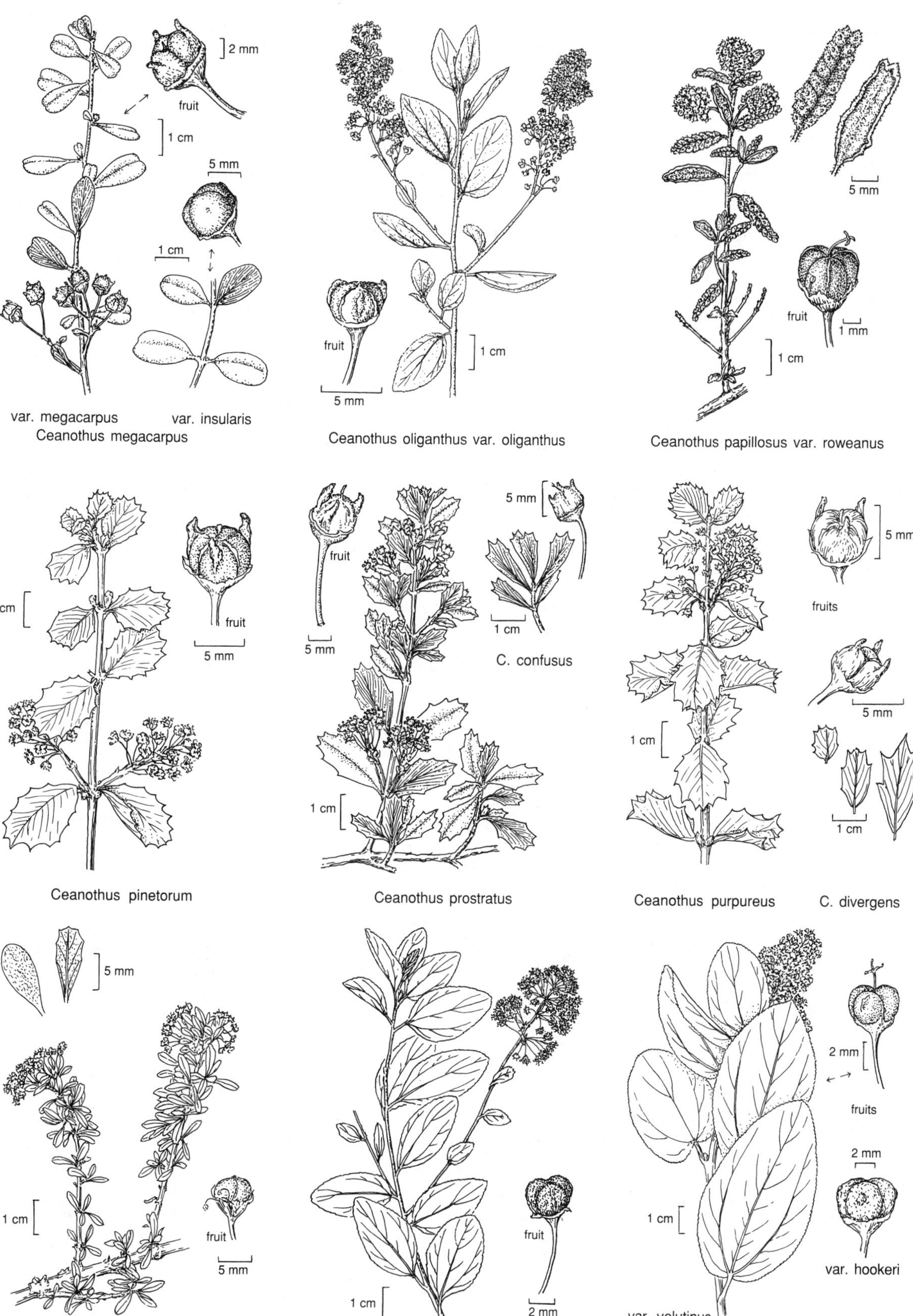
2 mm
fruit
1 cm
5 mm
1 cm
var. megacarpus
var. insularis
Ceanothus megacarpus
fruit
1 cm
5 mm
Ceanothus oliganthus var. oliganthus
5 mm
fruit
1 mm
1 cm
Ceanothus papillosus var. roweanus
1 cm
fruit
5 mm
Ceanothus pinetorum
fruit
5 mm
5 mm
1 cm
C. confusus
1 cm
Ceanothus prostratus
5 mm
fruits
5 mm
1 cm
1 cm
Ceanothus purpureus
C. divergens
5 mm
1 cm
fruit
5 mm
Ceanothus roderickii
fruit
1 cm
2 mm
Ceanothus thyrsiflorus
2 mm
fruits
2 mm
1 cm
var. hookeri
var. velutinus
Ceanothus velutinus

simple or compound; ovary superior to inferior, styles 1–5. **FR**: achene, follicle, drupe, pome, or blackberry- to raspberry-like. **SEEDS** gen 1–5. 110 genera, ± 3000 spp.: worldwide, esp temp. Many cult for orn and fr, esp *Cotoneaster*, *Fragaria*, *Malus*, *Prunus*, *Pyracantha*, *Rosa*, and *Rubus*. [Robertson 1974 J Arnold Arbor 55:303–332,344–401,611–662] Family description, key to genera by Barbara Ertter and Dieter H. Wilken.

1. Ann or per
 2. Pl 1–2 m; lvs 2–3-pinnate; fls unisexual (dioecious); fr a follicle ..... **ARUNCUS**
 2′ Pl gen < 1 m; lvs simple, ternate, palmate, or 1-pinnate; fls gen bisexual; fr an achene
  3. Petals 0; sepals gen 4; frs 1(–3) per fl, inside ± urn-shaped hypanthium
   4. Lf palmately lobed; infl axillary, hidden by sheathing stipules; hypanthium herbaceous in fr; ann ... **APHANES**
   4′ Lf pinnate; infl spike or head, not hidden; hypanthium ± hardened in fr; gen per
    5. Hypanthium ± prickly, not angled; stamens 2 or 4 ..... **ACAENA**
    5′ Hypanthium not prickly, 4-angled; stamens 1–many ..... **SANGUISORBA**
  3′ Petals and sepals gen 5; frs 1–many per fl; hypanthium shallow to ± obconic
   6. Sts long-trailing to decumbent; infl ± scapose; — lvs mostly basal, ternate, segments linear ...... [2]**LUETKEA**
   6′ Sts gen erect; infl gen not scapose (if so, sts not long-trailing or decumbent)
    7. Hypanthium obconic, rim hooked-prickly; infl ± raceme — bractlets 0 ..... **AGRIMONIA**
    7′ Hypanthium gen shallow, not prickly; infl cyme or fls solitary
     8. Bractlets 0; achenes fine-hairy, fleshy-coated — lvs palmate-lobed to ternate ..... [3]**RUBUS**
     8′ Bractlets gen present; achenes gen glabrous (exc *Geum*), not fleshy-coated
      9. Style tapered to fr, hooked or ± hairy — lvs 1-pinnate ..... **GEUM**
      9′ Style jointed to fr, not hooked, not hairy
       10. Lf (sub)palmate to ternate
        11. Receptacle in fr strawberry-like (enlarged, red, fleshy); stolons gen present; petals white or yellow; lvs ternate
         12. Petals yellow; fls solitary, axillary, on leafy stolons; bractlets gen wider than sepals .... **DUCHESNEA**
         12′ Petals white; fls 1–several; infl ± umbel-like, from among basal lvs; stolons not leafy; bractlets narrower than sepals ..... **FRAGARIA**
        11′ Receptacle in fr not strawberry-like; stolons 0; petals ± yellow; lvs ternate to (sub)palmate
         13. Stamens 10–25; petals gen >> 1 mm; lflets 3–7, gen > 3-toothed ..... [2] **POTENTILLA**
         13′ Stamens 5; petals ± 1 mm; lflets 3, 3-toothed — low, matted per, > 1900 m ..... **SIBBALDIA**
       10′ Lf 1-pinnate
        14. Hypanthium cup-like, ± flat-bottomed; filaments gen ± flat, often forming a tube; petals ± white to pinkish
         15. Stamens 10; lflets 2–15 per side; CA-FP, MP ..... **HORKELIA**
         15′ Stamens 20; lflets 15–35 per side; c&s SNH, SNE ..... **HORKELIELLA**
        14′ Hypanthium ± shallow (if cup-like, not flat-bottomed); filaments gen thread-like, not flat, not forming a tube; petals white to yellow, sometimes red
         16. Pl gen hanging clump in vertical rock crevices — ± resin-scented ..... [2]**IVESIA**
         16′ Pl not hanging clump, gen not in vertical rock crevices
          17. Lf gen ± cylindric; lflets 4–80 per side; pistils 1–8(20); stamens 5–20(40); petals 1–5(7) mm, linear to obovate or round ..... [2]**IVESIA**
          17′ Lf gen ± flat; lflets 2–8(13) per side; pistils gen > 10; stamens 20–25; petals (2)4–20 mm, obcordate to elliptic, sometimes ovate ..... [2]**POTENTILLA**

1′ Subshrub, shrub, or tree
 18. Lvs very deeply lobed (gen to midrib) or compound
  19. Ovary inferior, chambers 1–5; styles 1–5; fr pome ..... **SORBUS**
  19′ Ovary superior, chamber 1; style 1; fr achene or follicle
   20. Pistils gen 1–5 per fl
    21. Lvs opposite; primary lflets 5–14 cm; infl conspicuously flat-topped; bark peeling in strips; ChI ..... [2]**LYONOTHAMNUS**
    21′ Lvs alternate; primary lflets or lobes gen < 5 cm; infl not conspicuously flat-topped; bark gen not peeling in strips; mainland
     22. Lf 3–9-lobed; fls gen solitary; styles in fr 2–6 cm, plumose ..... [4]**PURSHIA**
     22′ Lvs 1–3-pinnate or -ternate; infl raceme or panicle; styles in fr gen << 1 cm, not plumose
      23. Sts decumbent to prostrate; pl < 2 dm; lvs ternate, not strong-smelling — infl ± scapose ..... [2]**LUETKEA**
      23′ Sts ascending to erect; pl > 2 dm; lvs pinnate, glandular, strong-smelling
       24. Pistil 1; ovule 1; fr achene; lvs 2–3-pinnate ..... **CHAMAEBATIA**
       24′ Pistils 4–5; ovules gen 2+; fr follicle; lvs (1)2-pinnate ..... **CHAMAEBATIARIA**
   20′ Pistils gen 10–many per fl
    25. Petals yellow; hypanthium bractlets 5; style attached below fr tip ..... ***Potentilla fruticosa***
    25′ Petals white to red; hypanthium bractlets 0; style terminal
     26. Pl not prickly; lvs pinnately lobed, rusty-scaly below, margins rolled under; style plumose in fr; e DMtns ..... **FALLUGIA**
     26′ Pl gen ± prickly; lvs pinnately to palmately compound, not rusty-scaly, margins not gen rolled under; style short-hairy to ± glabrous; mostly CA-FP, GB
      27. Lvs pinnate; hypanthium urn-shaped; fr gen enclosed in fleshy hypanthium ..... **ROSA**
      27′ Lvs palmate; hypanthium shallow; fr raspberry- or blackberry-like ..... [3]**RUBUS**

18′ Lvs simple to lobed (but not deeply so)
28. Ovary inferior, chambers gen 2–5; fr a pome
29. Sts armed with thorns
30. Lvs ± ovate, deciduous, ± lobed above middle, margin plane, gen clearly toothed; fr purple to black .... **CRATAEGUS**
30′ Lvs oblanceolate, ± evergreen, margin gen rolled under, 0–few-toothed; fr orange to red .. **PYRACANTHA**
29′ Sts gen not armed
31. Lf gen sharp-toothed from base to tip
32. Infl many-fld, flat-topped; petals 2–4 mm; fr 5–10 mm diam .... **HETEROMELES**
32′ Infl few-fld, ± umbel-like; petals 10–15 mm; fr 10+ mm diam .... **MALUS**
31′ Lf entire, minutely gland-toothed, or toothed from above middle
33. Lvs ± clustered on short lateral branches, blade tapered to base, ± subsessile .... **PERAPHYLLUM**
33′ Lvs evenly distributed on branches and twigs, blade and petiole well defined
34. Petals ascending to erect; lvs deciduous; blade glabrous to sparsely tomentose, pale green below, tip ± rounded; fr gen blue-black .... **AMELANCHIER**
34′ Petals ± spreading; lvs ± evergreen; blade gray- to white-tomentose, esp below, tip obtuse to acute; fr gen red .... **COTONEASTER**
28′ Ovary superior (sometimes hidden in funnel- or urn-shaped hypanthium), chamber gen 1; fr not a pome
35. Lf veins palmate
36. Lf blade 3(5)-lobed from above middle, gen wedge-shaped, margin entire, rolled under; fls solitary on branch tips .... [4]**PURSHIA**
36′ Lf blade gen 3–7-lobed from below middle, not wedge-shaped, lobes crenate to toothed, margin ± plane; fls several–many, clustered
37. Petals 2.5–4 mm; pistils 1–5 (if 3–5, fused below middle); infl umbel-like; fr follicle ... **PHYSOCARPUS**
37′ Petals 4–30 mm; pistils many; infl ± cyme or raceme; fr blackberry- or raspberry-like .... [3]**RUBUS**
35′ Lf veins pinnate
38. Pl matted, scapose; lf rosettes many — gen on limestone .... **PETROPHYTON**
38′ Erect shrubs, not scapose; lf rosettes 0
39. Lvs opposite
40. Shrub gen < 2 dm; petals 0; pistil 1; fr achene; SNE, D .... **COLEOGYNE**
40′ Tree; petals 5; pistils 2; fr follicle; ChI .... [2]**LYONOTHAMNUS**
39′ Lvs alternate
41. Pistil gen 1 per fl
42. Hypanthium cup- to urn-shaped, deciduous in fr; style thread-like, stigma ± disc-like; fr drupe **PRUNUS**
42′ Hypanthium ± funnel-shaped to obconic, persistent and ± enclosing fr; style tapered to stigma; fr achene
43. Hypanthium 1–1.5 mm; petals 1.5–2 mm; infl panicle, fls many; lvs linear, margin not rolled .... **ADENOSTOMA**
43′ Hypanthium 2–14 mm; petals 0 or 6–8 mm; fls 1–12, clustered; lvs linear to round, margin gen rolled under
44. Petals 0; lf lobes 0, margin entire to short .... **CERCOCARPUS**
44′ Petals 6–8 mm, white to cream-colored; lf lobes 3–9, gen entire .... [4]**PURSHIA**
41′ Pistils gen 2–6 per fl
45. Lf margin gen entire; infl raceme, pendent; dioecious; fr drupe .... **OEMLERIA**
45′ Lf lobed or margin toothed; entire; infl umbel-like, panicle, or fl 1, spreading to erect; bisexual; fr achene or follicle
46. Lf gen widest above middle, lobes 3–9, margin gen entire, ± strongly rolled under; fl solitary [4]**PURSHIA**
46′ Lf gen widest at or below middle, not clearly lobed, margin gen toothed, plane; fls gen 3–many
47. Infl umbel-like, fls gen 3–5; petals 4–6 mm; stamens ± 50, sepals ± toothed — shaded limestone slopes, CaR .... **NEVIUSIA**
47′ Infl gen panicle, fls many; petals 1.5–2 mm; stamens gen < 50, sepals not toothed
48. Petals gen white; hypanthium saucer-shaped; lf blade ± glabrous both surfaces; stigma ± 2-lobed; fr achene .... **HOLODISCUS**
48′ Petals pink to rose; hypanthium obconic to bell-shaped; lf blade ± hairy, esp below; stigma head-like; fr follicle .... **SPIRAEA**

ACAENA

Barbara Ertter

Per, nonglandular. **LVS** gen ± basal, odd-1-pinnate; lflets ± evenly toothed to lobed. **INFL**: dense spike or head. **FL**: hypanthium ± cone-shaped; bractlets 0; sepals gen 4; petals 0; stamens 2 or 4, opposite sepals; pistils 1(2), ovaries superior, continuous to style at top, stigma many-branched, exserted from hypanthium. **FR**: hypanthium ± hardened, encasing elliptic achenes; prickles gen 4–many, barbed. 50–150 spp.: esp s hemisphere. (Greek: thorn, from fr) [Yeo 1973 Bot Soc British Isles Conf Rep 13:193–221]

1. Infl a head; prickles 4, on fr top, 10–15 mm; pl mat-forming from stolons .... ***A. novae-zelandiae***
1′ Infl ± a spike, interrupted below; prickles > 4, scattered on fr, gen 1–3 mm; pl from ± woody, branched caudex .... ***A. pinnatifida*** var. ***californica***

A. novae-zelandiae Kirk (p. 947) BIDDY-BIDDY **ST**: fl-st 10–20 cm. **LF** 2–6 cm in CA; stipules lflet-like; lflets 3–5 per side, 5–15 mm, elliptic-oblanceolate, evenly toothed < 1/4 to midvein. **INFL**: head ± 10 mm diam (30–35 mm in fr), spheric. **FL**: sepals ± 1.5 mm, elliptic-ovate; stamens 2, pale. **FR**: hypanthium body 2.5–4 mm, obconic; prickles 4, on fr top, 10–15 mm. Gen ± disturbed areas; < 200 m. NCo, CCo, n SCo; native to New Zealand. Cult. NOXIOUS WEED. [*A. anserinifolia* (Forster & Forster f.) Druce misapplied] At least 2 other spp. cult, perhaps naturalized in CCo: *A. pallida* (Kirk) H.H. Allan (lflets 1–3 cm, hypanthium body in fr 5–6 mm, prickles 8–15 mm); Australian sp. undescribed (lflets 1–3 cm, hypanthium body in fr 4–6 mm, prickles 6–10 mm).

A. pinnatifida Ruiz Lopez & Pavon var. ***californica*** (Bitter) Jepson (p. 947) **ST**: fl-st 10–60 cm. **LF** 3–12 cm; stipules 0; lflets 5–8 per side, 4–15 mm, ± oblanceolate-ovate, lobed nearly to midvein into 3–8 linear-lanceolate segments. **INFL**: spike 10–20 mm diam, ± oblong. **FL**: sepals 2.5–3.5 mm, elliptic-lanceolate; stamens gen 4, purple-black. **FR**: hypanthium body 4–7 mm, obovate, ± ridged; prickles > 4, scattered, gen 1–3 mm. Coastal grassland, open, rocky slopes; 50–400 m. s NCo, CCo, w SnFrB. [*A. c.* Bitter] S.Am pls (var. *p.*) variable. ❀DRN:4,5,**15–17**&IRR:14,19–24.

ADENOSTOMA

Margriet Wetherwax

Shrub, ± resinous. **ST**: bark shredding. **LVS** simple, alternate or clustered, evergreen, short-petioled, stipuled or not, ± linear, stiff. **INFL**: panicle, terminal, 4–15 cm; bracts gen < calyx, ± lanceolate. **FL**: hypanthium 1–1.5 mm, ± = sepals, obconic, persistent, enclosing fr; calyx lobes 1–1.5 mm; corolla lobes ± round, cream to white; stamens 10–15 in clusters alternate petals; ovary superior, chamber 1, ovules 1–2. **FR**: achene. 2 spp.: CA, Baja CA. (Greek: glandular mouth, from hypanthium glands)

1. Lvs clustered; hypanthium throat glandular; pedicels 0 . . . . ***A. fasciculatum***
1′ Lvs alternate; hypanthium throat glandless; pedicels 0–4 mm . . . . ***A. sparsifolium***

A. fasciculatum Hook. & Arn. (p. 947) CHAMISE Shrub < 4 m, burled, much-branched. **ST**: trunk bark gray-brown; twigs glabrous to hairy. **LVS** 4–10 mm, clustered; stipules minute. **INFL** 4–12 cm, dense; bracts linear to lanceolate, green; pedicels 0. **FL:** hypanthium strongly 10-ribbed, throat glandular; corolla lobes ± 1.5 mm; ovary obliquely truncate. 2*n*=18. Dry slopes, ridges, chaparral; < 1600 m. s NCoR, SN, CW, SW; Baja CA. Pls from sw PR, sw SCo, with ± puberulent twigs, lvs < 6 mm, have been called var. *obtusifolium* S. Watson. ❀DRN,SUN,DRY:**7**,8,9,**14–16**,17,**18–24**;CV:GRNCVR.

A. sparsifolium Torrey (p. 947) RED SHANK Shrub or tree-like, 2–6 m. **ST**: trunk bark red-brown; twigs glabrous. **LVS** 4–15 mm, not clustered; stipules 0. **INFL** < 15 cm, open; bracts narrowly ovate, papery; pedicels 0–4 mm. **FL**: hypanthium 5-ribbed, nonglandular; corolla lobes ± 2 mm; ovary truncate. 2*n*=18. Dry slopes, flats, chaparral; 300–2000 m. s CCo, s SCoRO, TR, PR; Baja CA. ❀DRN,SUN,DRY:7–9,**14–16**,17,**18–21**,22–24.

AGRIMONIA

Barbara Ertter

Per, finely glandular. **STS** 1–several from rhizome, erect. **LVS** odd-1-pinnate; lflets evenly toothed, often alternately large and small. **INFL**: raceme or spike-like. **FL**: hypanthium cup-shaped to obconic; bractlets 0; sepals 5; petals 5; stamens 5–15; pistils 2, ovaries superior, continuous to style at top. **FR**: hypanthium hard, encasing achenes, rimmed with spreading prickles. ± 20 spp.: gen n temp. (Greek: eye disease, from former use as cure) [Kline & Sorensen 1990 Taxon 39:512–515]

A. gryposepala Wallr. (p. 951) COMMON AGRIMONY **ST** gen 50–140 cm. **LVS**: largest gen 10–25 cm; main lflets 2–5 per side, largest 4–12 cm, elliptic to obovate, slightly < terminal. **INFL** 2–4 dm, gen 10–50-fld; pedicels gen 1–10 mm, recurved in fr. **FL**: sepals 1–4 mm; petals 2–4 mm, ± oblong, yellow. **FR**: hypanthium 3–5 mm, obconic, ridged, prickles 1–4 mm, hooked; sepals incurved. 2*n*=56. Moist places, gen in woodlands; 150–1700 m. NW, CaRH, n SN, SnBr, PR; to e N.Am, C.Am.

AMELANCHIER SERVICE-BERRY

Dieter H. Wilken

Shrub or small tree. **ST**: bark gray- to red-brown; twigs gen short. **LVS** alternate or clustered, simple, deciduous; stipules deciduous. **INFL**: racemes or clusters; fls 3–16+. **FL**: hypanthium bell- to urn-shaped; sepals persistent; petals ascending to erect, white; stamens ± 10–20; ovary inferior, 2–5-chambered, styles 2–5. **FR**: pome, berry-like, gen spheric, blue-black. ± 10 spp.: temp N.Am, Eurasia, n Afr. Fr of some spp. used by native Americans for food. (Latin: from old French common name) [Jones 1946 Illinois Biol Mongr 20(2):1–126] Variation in w N.Am needs further study.

1. Styles gen 2–4; lf upper surface in fr gen puberulent to sparsely and finely tomentose (glabrous in n DMtns) . . . . ***A. utahensis***
1′ Styles gen 5; lf upper surface in fr gen glabrous . . . . ***A. alnifolia***
  2. Ovary top glabrous; petals 6–12 mm; lf lower surface glabrous; fr 7–10 mm diam . . . . var. ***pumila***
  2′ Ovary top tomentose; petals 11–15 mm; lf lower surface sometimes sparsely hairy; fr 10–14 mm diam . . . . var. ***semiintegrifolia***

A. alnifolia (Nutt.) Nutt. Shrub 1–8 m. **ST**: twigs glabrous to tomentose. **LF**: blade 9–50 mm, 8–45 mm wide, elliptic to round, gen serrate above middle. **FL**: petals 3–4 mm wide, oblong to oblanceolate; styles (4)5. **FR** 7–14 mm diam. Open shrubland, coniferous forest; 50–2600 m. NCo, KR, n NCoRO, n&c SNH; to AK, n-c US, NM. 3–4 other vars. in nw US, Rocky Mtns.

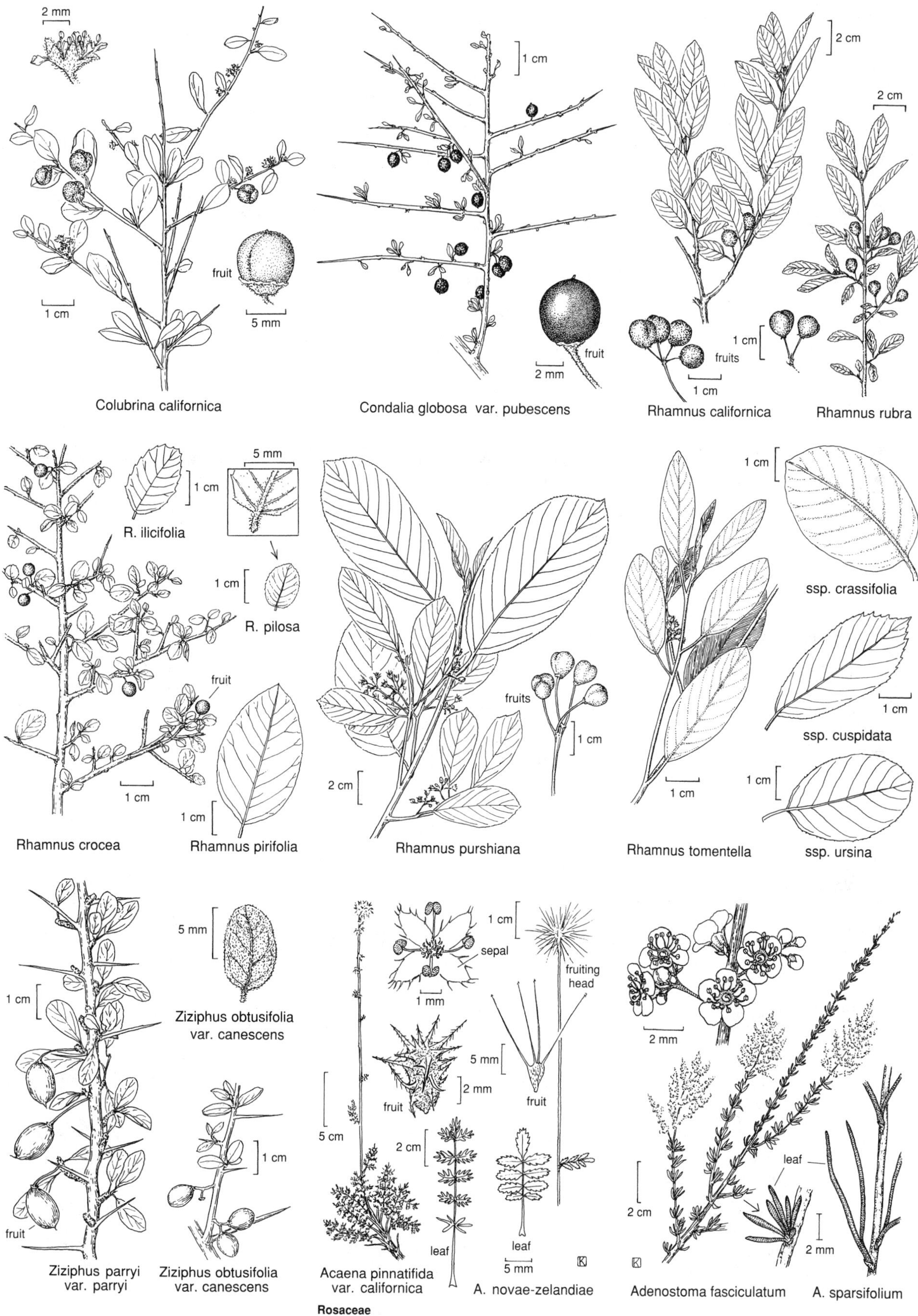
2 mm
fruit
1 cm
5 mm
Colubrina californica
1 cm
fruit
2 mm
Condalia globosa var. pubescens
2 cm
2 cm
1 cm
fruits
1 cm
Rhamnus californica
Rhamnus rubra
1 cm
R. ilicifolia
5 mm
1 cm
R. pilosa
fruit
1 cm
1 cm
Rhamnus crocea
Rhamnus pirifolia
fruits
1 cm
2 cm
Rhamnus purshiana
1 cm
ssp. crassifolia
1 cm
ssp. cuspidata
1 cm
1 cm
Rhamnus tomentella
ssp. ursina
5 mm
1 cm
Ziziphus obtusifolia
var. canescens
1 cm
fruit
Ziziphus parryi
var. parryi
Ziziphus obtusifolia
var. canescens
1 cm
sepal
1 mm
fruiting
head
5 mm
2 mm
fruit
fruit
5 cm
2 cm
leaf
leaf
5 mm
Acaena pinnatifida
var. californica
A. novae-zelandiae
2 mm
leaf
2 cm
2 mm
Adenostoma fasciculatum
A. sparsifolium

Rosaceae

var. ***pumila*** (Nutt.) Nelson (p. 951) Shrub 1–4 m. **LF**: blade 9–50 mm, 8–45 mm wide, glabrous, lateral veins 12–18. **INFL** 1–5 cm; fls 3–9. **FL**: petals 6–12 mm; ovary top glabrous. **FR** 7–10 mm diam. Open, often moist shrubland, coniferous forest; 1400–2600 m. n&c SNH (e slope); to MT, Colorado. [*A. pumila* Nutt.] ❀DRN:4,5,**6**&IRR:1,**2**,**7**,14,**15–17**&SHD:3,8–10,**18**,**19**,20,21.

var. ***semiintegrifolia*** (Hook.) C. Hitchc. (p. 951) Shrub or tree-like, 2–8 m. **LF**: blade 25–45 mm, 20–40 mm wide, lower surface glabrous to sparsely hairy, upper surface glabrous, lateral veins 16–24. **INFL** 3–7 cm; fls 5–16. **FL**: petals 11–15 mm; ovary top tomentose. **FR** 10–14 mm diam. $2n$=34. Open coniferous or mixed-evergreen forest; 50–2500 m. NCo, KR, n NCoRO; to AK, ID. [*A. florida* Lindley] ❀DRN:4,**5**,**6**,**17**&IRR:1–3,7,8,9,**14–16**,**18**,19–21.

A. utahensis Koehne (p. 951) UTAH SERVICE-BERRY Shrub 1–4 m. **ST**: twigs glabrous to white-tomentose. **LF**: blade 13–45 mm, 10–45 mm wide, entire to serrate from below middle, dull green below, darker above, glabrous to minutely tomentose; lateral veins 12–24. **INFL** 1–4 cm; fls 3–8. **FL**: petals 5–11 mm, 1–4 mm wide, elliptic to wedge-shaped; ovary top glabrous to tomentose, styles 2–4(5). **FR** 4–10 mm diam. Open, rocky slopes, shrubland, pinyon/juniper woodland, coniferous forest; 200–3400 m. NW, CaR, SN, CW, SW, SNE, DMtns; to OR, MT, TX, Baja CA. Variable; pls from n DMtns gen < 2 m, with twigs and lvs glabrous, have been called ssp. *covillei* (Standley) Clokey; pls with petals > 9 mm, lf blades with < 18 lateral veins, have been called *A. pallida* E. Greene. ❀DRN:**2**,4,5,**6**,**15–17**&IRR:**1**,3,7,8–10,**14**,**18**,19–21.

APHANES

Barbara Ertter

Ann, inconspicuous, nonglandular. **ST** spreading to erect. **LVS** palmately lobed. **INFL**: fls several per lf axil, ± hidden by sheathing stipules. **FL**: hypanthium ± urn-shaped; bractlets 0–4; sepals 4; petals 0; stamen gen 1; pistil gen 1, ovary superior, jointed to style near base. **FR**: hypanthium encasing achene. 10–20 spp.: worldwide, esp Medit. (Greek: unseen, from hidden fls) [Rothmaler 1937 Feddes Repert 42:164–173]

A. occidentalis (Nutt.) Rydb. (p. 951) **ST** gen 2–10 cm. **LF** gen 3–12 mm; stipules widely ovate, deeply few-toothed; petiole (between stipule, blade) gen 1–5 mm; blade gen 2–5 mm, ± round, lobes 3, > 2/3 to base, again toothed or lobed. **FL** 0.5–2 mm; bractlets 0–0.5 mm; sepals 0.2–0.6 mm. **FR**: achene ± 1 mm, ovoid. Seasonally moist grassland, chaparral, woodland; 30–1200 m. NW, SNF, ScV, CW, SW; to WA, Baja CA. [*Alchemilla o.* Nutt.] Highly variable, with several ± separable forms (see illustrations). Relation to spp. of Medit unclear. n NCo pls with more hidden fls resemble *A. microcarpa* (Boiss. & Reuter) Rothm. of e N.Am, Eur. *A. arvensis* L. (larger pls) weedy in OR, WA.

ARUNCUS GOAT'S BEARD

Margriet Wetherwax

1 sp.: n temp N.Am, Eur. (Greek: goat's beard)

A. dioicus (Walter) Fern. var. ***pubescens*** (Rydb.) Fern. (p. 951) Per, gen dioecious; rhizome stout. **ST** < 2 m, glabrous. **LF** < 60 cm, 2–3-pinnate, petioled; stipules 0; lflets 3–15 cm, ovate, acute, 2-serrate, ± hairy. **INFL**: panicle, < 50 cm; fls gen sessile, staminate > pistillate. **FL**: hypanthium saucer-shaped; sepals 5, 1–2 mm; petals 1–2 mm, oblong to (ob)ovate, white to pale yellowish. **STAMINATE FL**: stamens 15–30, exserted. **PISTILLATE FL**: pistils 3 (4–5), styles short. **FR**: follicle, 3–5 mm, ± cylindric, reflexed, septicidal. **SEED** ± 2 mm. $2n$=18. Moist streambanks, mixed-evergreen or coniferous forest; < 1800 m. NCo, KR, n NCoRO, CaR; to AK. [*A. vulgaris* Raf.] Var. *d.* in e N.Am. ❀WET or IRR,DRN:**4–6**&SHD:7,14,**15–17**;rather INV.

CERCOCARPUS MOUNTAIN-MAHOGANY

Richard Lis

Shrub or small tree, evergreen. **ST**: trunk < 80 cm diam; bark gen gray to reddish brown; twigs short. **LVS** gen clustered, simple; stipules often deciduous; blade ± thin to leathery, entire to toothed, upper surface gen ± glabrous. **INFL**: clusters; fls 1–12. **FL**: hypanthium funnel-like, tube persistent in fr, rim cup-like, deciduous; petals 0; stamens 10–45, in 2–3 rows on hypanthium rim, anthers glabrous or hairy; pistil 1, free from hypanthium tube, ovary superior, 1-ovuled, style terminal, persistent in fr, straight or becoming twisted, plumose. **FR**: achene, cylindric, hairy, incl in hypanthium. 13 spp.: w N.Am, Mex. (Greek: tailed fr) [Lis 1992 Int J Pl Sci 153:258–272]

1. Lf blades entire; anthers glabrous
  2. Lf blade gen 3–10 mm, margin inrolled to midrib . . . . ***C. intricatus***
  2′ Lf blade 10–40 mm, margin plane to inrolled (but not to midrib) . . . . ***C. ledifolius***
    3. Lf oblanceolate to elliptic-lanceolate, glabrous to sparsely woolly below, midrib & veins visible . . . . var. ***intermontanus***
    3′ Lf narrowly lanceolate, densely woolly below, midrib and veins obscure . . . . var. ***ledifolius***
1′ Lf blades toothed to crenate; anthers hairy
  4. Lf ± leathery, woolly below; hypanthium densely white-tomentose — s ChI (Santa Catalina Island) . . . . ***C. traskiae***
  4′ Lf ± thin, not leathery, ± sparsely hairy below; hypanthium glabrous to sparsely hairy
    5. Lf ± glabrous below; hypanthium rim 2–5 mm diam . . . . ***C. minutiflorus***
    5′ Lf hairy below; hypanthium rim 4–7 mm diam . . . . ***C. betuloides***
      6. Lf blade with 4–7 lateral veins, 1–4 cm . . . . var. ***betuloides***
      6′ Lf blade with 6–12 lateral veins, 2–8 cm
        7. Fls (and frs) gen 5–12 per cluster; lf blade gen ovate, teeth ± deltate . . . . var. ***blancheae***
        7′ Fls (and frs) 1–3 per cluster; lf blade gen obovate, long, narrowly triangular . . . . var. ***macrourus***

C. betuloides Torrey & A. Gray Shrub or small tree 2–8 m; branches spreading to erect. **LF**: petiole 1–10 mm; blade 1–8 cm, widely elliptic to obovate, ± thin, ± entire to serrate, sparsely hairy below; lateral veins 4–10. **INFL**: fls 1–12. **FL**: hypanthium glabrous to sparsely hairy; stamens 25–45, anthers hairy; style incl, stigma straight. **FR** 8–12 mm, strigose; style 5–10 cm. Chaparral, pine/oak woodland, coniferous forest; < 2500 m. NW, CaR, SN, CW, ChI, TR, PR, MP; sw OR, AZ, Baja CA.

var. ***betuloides*** (p. 951) BIRCH-LEAF MOUNTAIN-MAHOGANY **LF**: petiole 1–7 mm; blade 1–4 cm, obovate to ± round, finely toothed to serrate; lateral veins 4–7. **INFL**: fls 1–3; pedicel 1–4 mm, 2–6 mm in fr. **FL**: hypanthium 5–8 mm, 8–14 mm in fr, rim 5–6 mm diam. **FR**: style 5–9 cm. $2n$=18. Dry, rocky slopes, chaparral; < 2500 m. Range of sp. (exc ChI). ❀SUN,DRY:1–3,5,**7**,8–10,**14–21**,22–24;also STBL.

var. ***blancheae*** (C. Schneider) Little (p. 951) ISLAND MOUNTAIN-MAHOGANY **LF**: petiole 6–10 mm; blade 2.5–6 cm, widely elliptic to ovate, serrate; lateral veins 6–10. **INFL**: fls gen 5–12; pedicel 2–4 mm, 5–12 mm in fr. **FL**: hypanthium 7–9 mm, 9–14 mm in fr, rim 5–7 mm diam. **FR**: style 3–6 cm. UNCOMMON. Chaparral; < 600 m. ChI (exc San Clemente Island), s WTR. [ssp. *blancheae* (C. Schneider) Thorne] ❀5,**14–17,22–24**&IRR:7–9, **19–21**.

var. ***macrourus*** (Rydb.) Jepson (p. 951) **LF**: petiole 5–10 mm; blade 2–8 cm, oblanceolate to diamond-shaped, serrate; lateral veins 5–12. **INFL**: fls 1–3; pedicel 3–4 mm, 8–12 mm in fr. **FL**: hypanthium 7–9 mm, 10–16 mm in fr, rim 4–7 mm diam. **FR**: style 7–10 cm. Coniferous forest; 800–1800 m. KR, CaR, MP; sw OR. ❀5,6,**15–17**&IRR:1,2,**7**,**14**,19–24.

C. intricatus S. Watson (p. 951) Shrub 1–3 m, intricately branched. **LF**: petiole 0–1 mm; blade gen 0.3–1 cm, linear, thick-leathery, entire, inrolled to midrib, glabrous to gray-white-hairy below. **INFL**: fls 2–3, sessile. **FL**: hypanthium 3–5 mm in fl and fr, rim 1–3 mm diam; stamens 10–15, anthers glabrous; style slightly exserted, stigma straight. **FR** 5–7 mm; style 1–2 cm. Dry, rocky outcrops, slopes, pinyon/juniper woodland; 1400–3000 m. s SNH, SNE, DMtns; to Colorado, AZ. ❀DRN,DRY,SUN:1–3,**7**,9,**10**,11, **14**,15,16,**18**,19–24;DFCLT.

C. ledifolius Nutt. CURL-LEAF MOUNTAIN-MAHOGANY Shrub or tree 1–10 m, much-branched. **LF**: petiole 0–6 mm; blade 1–4 cm, narrowly (ob)lanceolate, resinous, entire, inrolled or plane, glabrous to densely woolly below. **INFL**: fls 1–3, axillary, ± sessile. **FL**: stamens 15–25, anthers glabrous; style exserted, stigma straight. **FR** 4–13 mm. Deep soils, rocky slopes; 1200–2800 m. KR, NCoRH, CaRH, SNH, Teh, TR, n PR, GB, DMtns; to WA, MT, AZ, Baja CA. Hybridizes with *C. intricatus*.

var. ***intermontanus*** N. Holmgren (p. 951) Pl < 8 m; trunk < 8 dm diam. **LF**: petiole < 6 mm; blade 1–4 cm, ± elliptic-lanceolate, glabrous to sparsely woolly below, midrib and veins visible. **FL**: hypanthium 5.5–9 mm, 6–10 mm in fr, rim 4–5 mm diam. **FR** 8–13 mm; style 4–7 cm. 2n=18. Pinyon/juniper woodland, sagebrush scrub; 1500–2800 m. KR, NCoRH, CaRH, SNH, Teh, TR, n PR, GB, DMtns; to WA, WY, AZ, Baja CA. [Hybrids with *C. intricatus* are var. *intercedens* C. Schneider] ❀DRN,SUN:**1**,4–6,15–17& IRR:**2,3,7**,8–10,**14,18–21**,22,23.

var. ***ledifolius*** (p. 951) Pl < 3 m; trunk < 2 dm. **LF**: petiole 1–3 mm; blade 1–1.5 cm, narrowly lanceolate, densely white-woolly below, midrib and veins obscure. **FL**: hypanthium 2–5 mm, 4–7 mm in fr, rim 3–5 mm diam. **FR** 4–8 mm; style 2.4–3.5 cm. Uncommon. Steep slopes, open pine forest; 1200–2500 m. s SNH; to e OR, sw MT, n WY, n UT. ❀DRN,SUN:15–17&IRR:**1,2**,3,**7,14, 18**,19–21.

C. minutiflorus Abrams (p. 951) Shrub 2–5 m, much-branched. **LF**: petiole 1–5 mm; blade 1–2.5 cm, ± widely (ob)ovate, ± thin, ± entire to serrate, ± glabrous below; lateral veins 3–6. **INFL**: fls gen 1(–3), axillary; pedicel 1–3 mm, 4–8 mm in fr. **FL**: hypanthium 5–8 mm, 10–11 mm in fr, glabrous or sparsely hairy, rim 2–5 mm diam; stamens 15–25, anthers sparsely hairy; style exserted, stigma ± hook-like. **FR** 8–13 mm; style 3–7 cm. Chaparral; < 1300 m. s PR (San Diego Co.); n Baja CA. ❀DRN,DRY,SUN:3,7,**14**,15, 16,**18–23**,24;STBL.

C. traskiae Eastw. (p. 951) CATALINA ISLAND MOUNTAIN-MAHOGANY Tree 3–8 m; branches spreading to erect. **LF**: petiole 2–10 mm; blade 2–6 cm, elliptic to (ob)ovate, thick-leathery, ± serrate or crenate, white-woolly below; lateral veins 6–10. **INFL**: fls 4–10; pedicel 0–5 mm, 5–7 mm in fr. **FL**: hypanthium 7–14 mm, white-tomentose, rim 4–7 mm diam; stamens 20–40, anthers hairy; style exserted, stigma ± hook-like. **FR** 10–13 mm; style 4–6 cm. **ENDANGERED** CA. Dry, rocky soils; 100–250 m. s ChI (Santa Catalina Island). [*C. betuloides* var. *t.* (Eastw.) Dunkle] 7 plants surviving in 1990. In cult.

CHAMAEBATIA

Thomas J. Rosatti

Shrub, strong-smelling, evergreen, gen stellate-hairy, glandular. **LF** odd-2–3-pinnate; stipules entire; segments gen thick, gland-tipped, smaller and larger sometimes alternating, lobes or teeth 0–3(5). **INFL**: panicle or raceme, 3–10 cm. **FL**: hypanthium densely hairy inside near base; bractlets 0; sepals lanceolate; petals spreading, ± round, white; stamens many; pistil 1, ovary superior, style base hairy. **FR**: achene, obovoid, leathery. 2 spp.: CA, Baja CA. (Greek: low bramble) [Armstrong 1980 Madroño 27:111]

1. Lf gen elliptic, gland-tips gen sessile or sunken; 300–700 m, s PR .... ***C. australis***
1′ Lf gen obovate, gland-tips gen stalked; 600–2200 m, CaR, SN .... ***C. foliolosa***

C. australis (Brandegee) Abrams (p. 951) **LF** gen elliptic; gland-tips gen sessile or sunken. **ST**: bark ± black. **LF** 3–8 cm, gen 2-pinnate. **FL**: hypanthium gen ± 3 mm; sepals ± 3 mm; petals 4–5 mm; ovary glabrous. Dry slopes, chaparral; 300–700 m. s PR; n Baja CA. Uncommon. Some populations threatened by clearing for avocado orchards. ❀SUN,DRN:7,**14–16**,17,**19–23**,24.

C. foliolosa Benth. (p. 951) MOUNTAIN MISERY **LF** gen obovate; gland-tips gen stalked. **ST**: bark dark brown. **LF** 2–10 cm, gen 3-pinnate. **FL**: hypanthium ± 4–5 mm; sepals ± 4 mm; petals 6–8 mm; ovary base hairy. Coniferous forests; 600–2200 m. CaR, SN. ❀DRN,SHD:1,2,17&IRR:3,**7**,14–16,18;DFCLT;GRNCVR;may be INV.

CHAMAEBATIARIA FERN BUSH, DESERT SWEET

Thomas J. Rosatti

1 sp. (Greek: *Chamaebatia*-like)

C. millefolium (Torrey) Maxim. (p. 951) Shrub < 20 dm, densely branched, strong-smelling, gen stellate-hairy, glandular. **LF** odd-(1)2-pinnate, 2–8 cm, oblong; stipules entire; 1° lflets 31–43; 2° lflets (0)11–35, < 2 mm, sessile, sometimes with a sessile gland-tip, lobes or teeth 0–3(5). **INFL**: panicle or raceme, 3–10 cm. **FL**: bractlets on hypanthium 0; sepals 3–5 mm, lanceolate; petals ± 5 mm, ± round, white; pistils 4–5, ovaries superior, ± fused below, styles free. **FR**: follicles 3–5 mm, leathery, dehiscent on inner suture and upper half of outer suture. **SEEDS** few, 2.5–3.5 mm, narrowly fusiform, yellowish. Dry, rocky sagebrush scrub, pinyon/juniper woodland, pine forest; 900–3400 m. KR, CaR, SN (e slope), GB, ne DMtns; to OR, WY, AZ. ❀DRN,SUN:1,**2,3,7,14–16**,17,**18**,19–21.

COLEOGYNE BLACKBUSH

Thomas J. Rosatti

1 sp. (Greek: sheath female, from hypanthium enclosing pistil)

C. ramosissima Torrey (p. 951) Shrub 3–20 dm, much-branched, ± strigose, thorny. **LVS** in opposite clusters, 5–15 mm, linear-oblanceolate, ± thick, entire; stipules persistent. **INFL**: fls solitary at st tips; bract (closely subtending fl) ± linear, base 2-lobed. **FL** gen yellow (sepals often reddish outside); hypanthium bell-shaped, leathery, sheath at top enclosing pistil, 4–5 mm; sepals 4, 7–8 mm, erect, elliptic, persistent, inner 2 widely scarious-margined; petals 0; stamens 30–40; pistil 1, ovary superior, style lateral, long-hairy esp below, persistent. **FR**: achene, ± 3–4 mm, ± crescent-shaped, brown, glabrous. Dry, open slopes, creosote-bush scrub, pinyon/juniper woodland; 600–1600 m. SNE, D (esp DMtns); to Colorado, NM.

COTONEASTER

Thomas J. Rosatti

Shrub or tree, unarmed. **LVS** stipuled, petioled, subevergreen in CA, entire. **INFL**: fls solitary or clustered at branch tips. **FL**: bractlets on hypanthium 0; sepals persistent; stamens ± 20; ovary inferior, 2–5-chambered, styles 2–5, free. **FR**: pome, drupe-like, ± 6.5 mm in CA; stones 2–5. ± 50 spp.: e hemisphere; many orn. (Latin: quince-like, possibly from lf shape) [Gladkova 1968 Bot Zhur 53:1263–1273]

1. Petals erect, ± pink to rose (at least at base); stones gen 3; lf gen yellowish, or grayish tomentose below ***C. franchetii***
1′ Petals spreading, white; stones 2; lf white-tomentose below .... ***C. pannosa***

C. franchetii Boiss. Shrub 15–30 dm. **LF** 15–35 mm, (ob)ovate, gen yellowish or grayish tomentose below, ± hairy above, becoming ± glossy and sometimes glabrous. **FL**: petals erect, ± pink to rose (at least at base). **FR** (orange-) red; stones gen 3. Disturbed places near dwellings; gen < 500 m. n CW, expected elsewhere; native to China.

C. pannosa Franchet Shrub 15–37 dm. **LF** 15–40 mm, elliptic to ± ovate, white-tomentose below, dull and ± glabrous above. **FL**: petals spreading, white. **FR** red; stones gen 2. Disturbed places, mixed-evergreen forest; gen < 1000 m. NCo, n CW, c WTR, expected elsewhere; native to sw China.

CRATAEGUS HAWTHORN

Thomas J. Rosatti

Shrub or tree, gen thorny. **LVS** stipuled, petioled, deciduous, toothed to lobed. **INFL**: racemes at branch tips. **FL**: hypanthium cup- to urn-shaped, bractlets 0; sepals becoming reflexed; petals gen white; stamens 5–25; ovary inferior, styles 1–5, free. **FR**: pome, drupe-like, purple-black and glabrous in CA; stones 1–5, gen 1-seeded. ± 300 spp.: n temp. (Greek: fl thorn, used by Theophrastus?; or strength, from strong wood or thorns) [Brunsfeld & Johnson 1990 Madroño 37:274–282] *C. erythropoda* may occur in e D; *C. monogyna* Jacq. escaped from or persistent beyond cult, perhaps naturalized. Many hybrids and morphologically identifiable clones.

1. Stamens (5–)10; lvs on short lateral branches lobed above middle, tips truncate; older thorns 13–18 mm ***C. douglasii***
1′ Stamens (12–)20; lvs on short lateral branches unlobed, tips rounded; older thorns 8–12 mm .... ***C. suksdorfii***

C. douglasii Lindley (p. 951) **ST**: older thorns 13–18 mm. **LF** 2–9 cm; blade base wedge-shaped, entire, margin above base toothed, gen lobed, tips truncate. **FL** 14–16 mm wide when dry; sepals entire to ± gland-toothed; stamens (5–)10; ovary often ± hairy, styles gen 5. **FR** gen many per infl, 9–10 mm wide when fresh; stone ± 5 mm, outer face rounded, plump, sides shallowly pitted. 2*n*=34. Streamsides, forest, meadows, grassland, sagebrush scrub; 700–1700 m. MP; to AK, e-c N.Am. Some pls may = *C. columbiana* Howell (hairier, styles gen 2–4, older thorns 2–7 cm, sepals strongly gland-toothed). ❀SUN:**4–6**,15–17&IRR:1,**2**,**7**,10,14,18.

C. suksdorfii (Sarg.) Kruschke (p. 951) **ST**: older thorns 8–12 mm. **LF** 2–9 cm; blade base wedge-shaped, entire, margin above base toothed, gen unlobed, tips rounded. **FL** 12–13 mm wide when dry; sepals entire to ± gland-toothed; stamens (12–)20; ovary gen glabrous, styles gen 5. **FR** gen few per infl, 7–8 mm wide when fresh; stone ± 4.5 mm, outer face keeled, narrow, sides deeply pitted. 2*n*=34. Moist streamsides, lakesides, forest; gen < 1500 m. KR, NCoR, CaR, n SnFrB; to B.C., MT. ❀**4–6**&IRR:1–3,**7**,8,9,**14**,**15–17**,18–24;may be INV.

DUCHESNEA

Barbara Ertter

Per, stoloned; glands not evident. **LVS** basal and on stolons, ternate; lflets often double-toothed. **INFL**: fl solitary in axils. **FL**: hypanthium shallowly cup-shaped; bractlets 5; sepals 5; petals 5, yellow; stamens 20, filaments slender; pistils many, ovaries superior, style slender, jointed below fr tip. **FR** ± strawberry-like; receptacle enlarged, red, completely covered with achenes. 2 spp.: se Asia. (A.N. Duchesne, French botanist, 1747–1827)

D. indica L. MOCK-STRAWBERRY Pl forming ground-cover, sparsely hairy. **LVS**: basal lf petiole 2–20 cm, central lflet 1–6 cm, short-stalked, elliptic-ovate. **INFL**: pedicel 2–13 cm. **FL**: sepals 4–10 mm; bractlets = sepals but wider, obovate, 3-toothed; petals 4–8 mm, obovate-oblong. **FR**: receptacle ± 1 cm wide, spheric-ovoid, ± pithy, tasteless; achenes 1–1.5 mm, red. *n*=42. Disturbed urban areas; gen < 500 m. CA-FP; to e US; native to India. Cult.

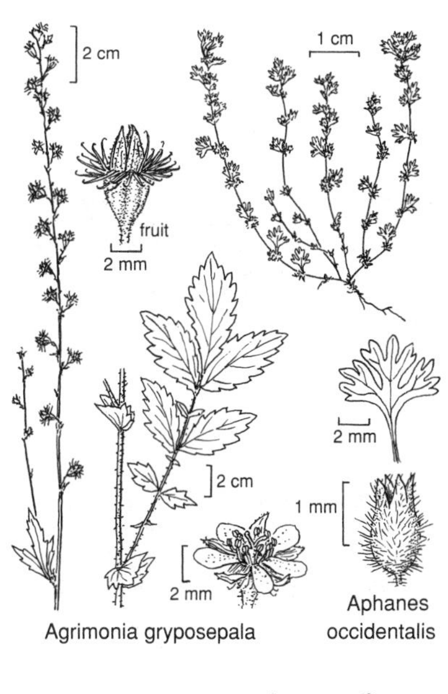

Agrimonia gryposepala

Aphanes occidentalis

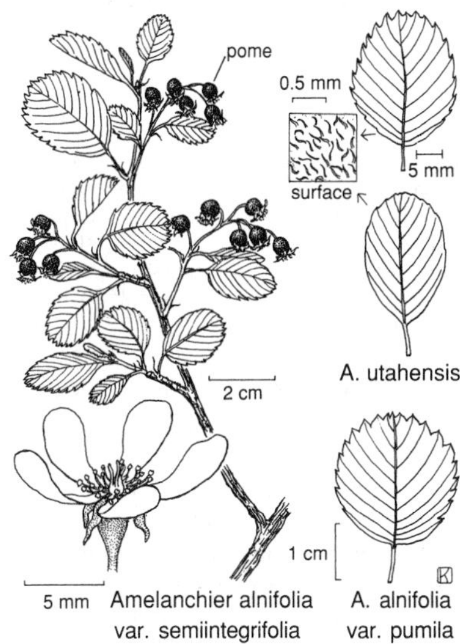

Amelanchier alnifolia var. semiintegrifolia

A. alnifolia var. pumila

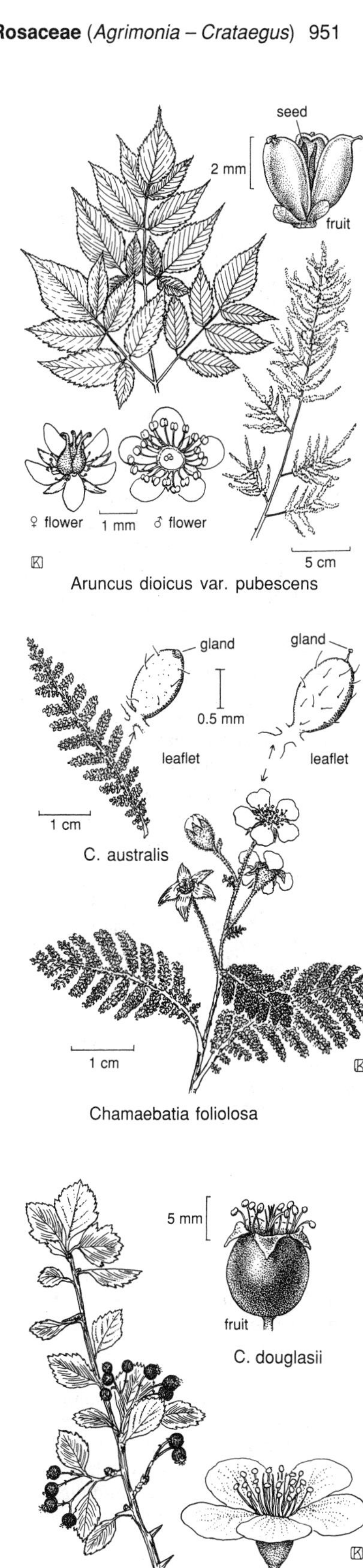

Aruncus dioicus var. pubescens

Chamaebatia foliolosa

Crataegus suksdorfii

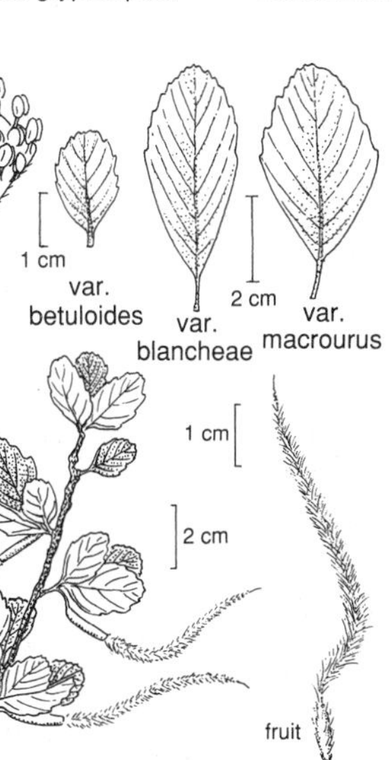

Cercocarpus betuloides

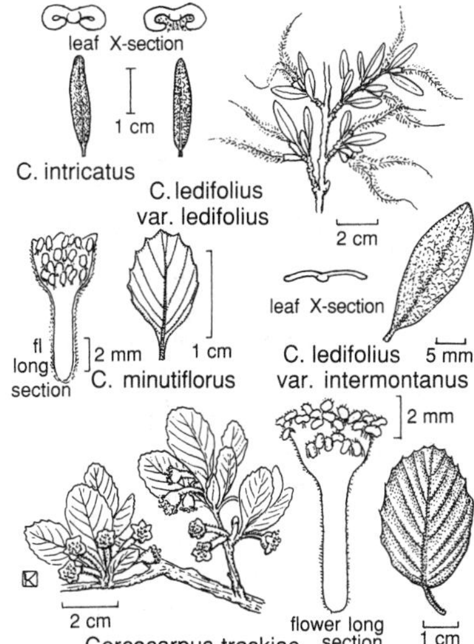

Cercocarpus traskiae

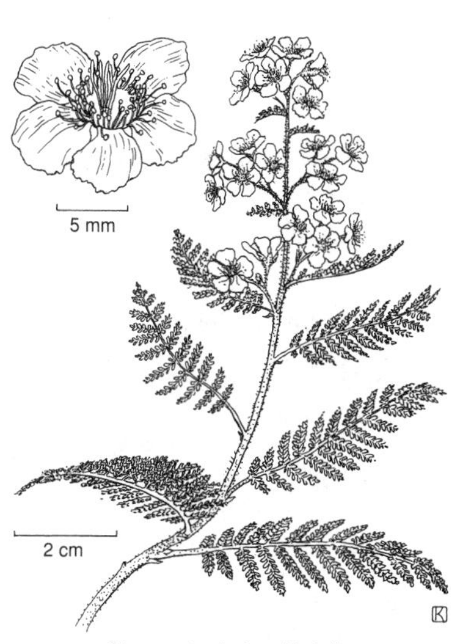

Chamaebatiaria millefolium

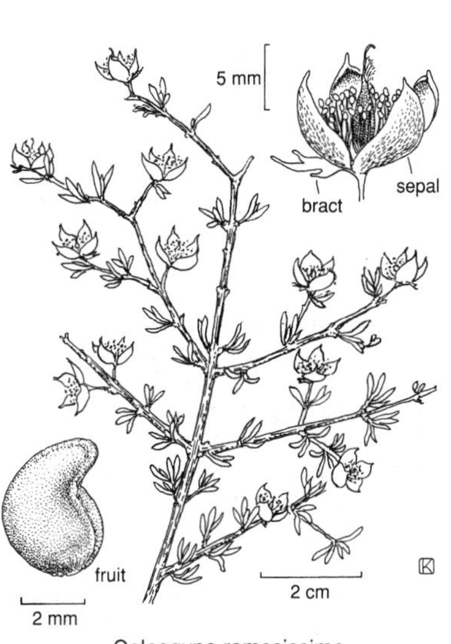

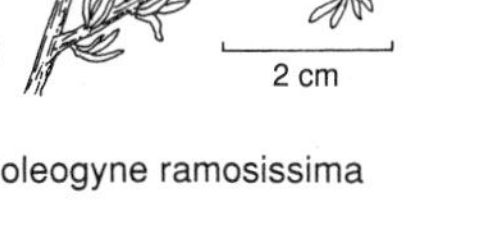

Coleogyne ramosissima

FALLUGIA APACHE PLUME

Margriet Wetherwax

1 sp. (V. Fallugi, Italian abbot)

F. paradoxa (D. Don) Endl. (p. 959) Shrub < 2 m, ± erect. **ST** much-branched; bark grayish white-tomentose, peeling. **LVS** alternate to clustered, 7–15 mm, ovate to wedge-shaped, lobed; lobes 3–7, deeply pinnate, linear, obtuse, rolled under, densely hairy above, rusty-scaly below; stipules lanceolate, deciduous. **INFL** terminal; fls 1–3; bracts 5, linear, alternate sepals. **FLS**: hypanthium hemispheric, silky-hairy; sepals 5–8 mm, ovate, acute to long-acuminate, tomentose; petals 10–25 mm, ± round, white; stamens many; pistils many, ovary superior, chamber 1, style 1, persistent, 30–50 mm, plumose, purplish. **FR**: achene, 3–5 mm, silky-hairy. $2n$=18. Dry, ± rocky slopes in pinyon/juniper woodland; 1000–2200 m. e DMtns; to Colorado, w TX, n Mex. ❀DRN,SUN:1,2,7–9,**14, 18–23**,24&DRY:5,**15,16**,17.

FRAGARIA STRAWBERRY

Barbara Ertter

Per from short rhizomes and lfless stolons, ± nonglandular. **LVS** basal, 1-ternate; lflet teeth gen simple. **INFL**: cyme, ± umbel-like, open, 1–several-fld; pedicels recurved in fr. **FL**: hypanthium shallow; bractlets 5; sepals 5; petals 5, ± obovate, gen white; stamens 20–35, filaments ± flat; pistils many, ovaries superior, jointed to stout style on side. **FR**: receptacle enlarged, fleshy, red, incompletely covered with achenes. 15–30 spp.: gen n temp. (Latin: fragrant) [Hancock & Bringhurst 1981 Amer J Bot 68:1–5] Hair orientation and pl size have been used to define sspp. but seem to have no taxonomic significance, at least in CA. All spp. intergrade.

1. Lflets sparsely hairy above; central lflet acute to obtuse, teeth gen 12–21, sharp or obtuse, central tooth < to > adjacent ones; bractlets often 2-lobed — infl often > lvs; < 2000 m .... ***F. vesca***
1′ Lflets gen glabrous above; central lflet rounded to truncate, teeth gen < 13, ± obtuse, central tooth << adjacent ones; bractlets unlobed
  2. Lf leathery, central lflet stalk 1–10 mm; petals (8)10–18 mm; infl < or > lvs; ± coastal, < 200 m .... ***F. chiloensis***
  2′ Lf thin, central lflet stalk 1–3 mm; petals 4–9 mm; infl < lvs; mtns, gen 1200–3300 m .... ***F. virginiana***

F. chiloensis (L.) Duchesne BEACH STRAWBERRY Often dioecious. **ST** gen 5–20 cm. **LF** leathery; petiole gen 2–20 cm; central lflet stalk 1–10 mm, blade 10–60 mm, obovate, densely hairy below, gen glabrous above, rounded to truncate, teeth gen 7–11, above middle, rounded-obtuse, central tooth << adjacent ones. **INFL** < or > lvs. **FL** gen 20–40 mm wide; bractlets unlobed; sepals 6–10 mm; petals 8–18 mm. **FR**: receptacle 10–20 mm; achene 1.5–2 mm, smooth. n=28. Ocean beaches, grassland; < 200 m. NCo, CCo; to AK, also coastal S.Am, Hawaii. Cult and possible escape in SCo. [ssp. *pacifica* Staudt] ❀DRN:**4,5**&IRR:7,**15–17,22–24**&SHD:8, 9,**14**,19–21;GRNCVR;CVS.

F. vesca L. (p. 959) WOOD STRAWBERRY **ST** gen 3–30 cm. **LF** thin; petiole gen 3–12 cm; central lflet stalk < 2 mm, blade 15–70 mm, widely elliptic-obovate, acute to obtuse, teeth gen 12–21, below and above middle, sharp or obtuse, central tooth < to > adjacent ones; lflets sparsely hairy above, hairier below. **INFL** often >> lvs. **FLS** bisexual, gen ± 15 mm wide; bractlets often 2-lobed; sepals gen 4–8 mm; petals gen 5–8 mm. **FR**: receptacle ± 10 mm; achene ± 1.5 mm. n=7. Gen partial shade in forests; 30–2000 m. NW, CaR, SN, CW, SnBr, PR; to e N.Am, Baja CA, also Eur. [ssp. *californica* (Cham. & Schldl.) Staudt] Pls intermediate to *F. chiloensis* or *F. virginiana* (esp in KR) have been called *F. crinita* Rydb. ❀**4–6**SHD,IRR:1,2,**7**,8,9,**14–24**;DRNCVR;CVS non CA.

F. virginiana Duchesne (p. 959) MOUNTAIN STRAWBERRY Often dioecious. **ST** 2–12 cm. **LF** thin; petiole gen 1–12 cm; central lflet stalk 1–3 mm, blade gen 15–60 mm, obovate, rounded to truncate, teeth gen 7–11, above middle, ± obtuse, central tooth << adjacent ones; lflets gen glabrous above, ± hairy below. **INFL** < lvs. **FL** ± 10–20 mm wide; bractlets unlobed; sepals 3–6 mm; petals 4–9 mm. **FR**: receptacle ± 10 mm; achene ± 1.5 mm. n=28. Meadows, forest openings; gen 1200–3300 m. KR, CaRH, SNH, MP; to e N.Am. [ssp. *platypetala* (Rydb.) Staudt] Pls of c SNH (Yosemite Valley) are unusually large. ❀DRN:**4–6**&IRR:1–3,**7**,14–16,18; GRNCVR.

GEUM

Barbara Ertter

Per; glands inconspicuous. **ST** ascending to erect. **LVS** gen basal, odd-1-pinnate; lflets lobed, unevenly toothed, often alternately large and small. **INFL**: gen cyme, open. **FL**: hypanthium shallow; bractlets gen 5; sepals 5; petals 5; stamens > 20; pistils many, ovaries superior, continuous to style at top. **FR**: achene ± flat; style long, persistent. 40–50 spp.: gen n temp, arctic. (Latin: ancient name)

1. Style in fr 20–35 mm, plumose, not hooked; fls 1–3, nodding, ± cup-shaped; petals cream or tinged pink, persistent; bractlets 5–14 mm .... ***G. triflorum***
1′ Style in fr 4–10 mm, not plumose, hooked; fls gen 3–10, erect, rotate; petals yellow, deciduous; bractlets < 3 mm
  2. Terminal lflet of basal lf gen lobed nearly to base; pl drying blackish green; head in fr (incl styles) ± 15 mm wide, pedicel 1–1.5 mm diam; petals ± ovate; uncommon .... ***G. aleppicum***
  2′ Terminal lflet of basal lf lobed < 3/4 to base; pl drying greenish; head in fr (incl styles) gen 10–15 mm diam, pedicel ± 0.8 mm diam; petals obovate-round; common .... ***G. macrophyllum***

G. aleppicum Jacq. Pl tufted, drying blackish green. **ST** 40–100 cm. **LF** 15–40 cm; main lflets 1–3 per side, largest nearly = lobes of terminal lflet; terminal lflet 5–10 cm, gen 3-lobed nearly to base, sharply toothed. **INFL** gen 3–10-fld; pedicels straight. **FL** rotate; bractlets < 3 mm, linear-lanceolate; sepals reflexed, 4–7 mm; petals 4–8 mm, ± ovate, unnotched, yellow, deciduous. **FR**: body 4–5 mm; style 4–6 mm, hooked below gen deciduous tip, hairs 0 to sparse. n=21. Very uncommon. Meadows; 450–1500 m. n CaRH (Mount Shasta), MP (Susanville); to n&e N.Am, Eurasia. [*G. strictum* Aiton]

G. macrophyllum Willd. (p. 959) BIGLEAF AVENS Pl tufted, drying greenish. **ST** gen 20–100 cm. **LF** gen 10–45 cm; main lflets gen 2–4 per side, gen << terminal lflet; terminal lflet gen 8–10 cm, ± bluntly cordate-reniform, 3-lobed < 3/4 to base, irregularly toothed. **INFL** gen 3–10-fld; pedicels straight. **FL** rotate; bractlets gen ± 2 mm, linear, often some missing; sepals reflexed, 3–5 mm; petals 3–7 mm, obovate-round, shallowly notched, yellow, deciduous. **FR**: achene body 2.5–3.5 mm; style 2.5–5 mm, hooked below gen deciduous tip, glabrous to glandular. *n*=21. Meadows, streambanks; < 3300 m. NW, CaRH, SNH, SnBr, GB; to n&e N.Am. Vars. indistinct in CA. Some pls from s SNH approach *G. aleppicum*. ❀**4–6**&IRR:1,2,**7**,**15–17**&SHD:3,**14**,**18–24**;rather INV.

G. triflorum Pursh (p. 959) OLD MAN'S WHISKERS Pl in patches, rhizomed, ± gray-green. **ST** gen 20–50 cm. **LF** gen 5–20 cm; lflets wedge-shaped, gen 2–3-lobed ± 1/2 to base, lobes deeply few-toothed, main lflets 3–6 per side, largest 1–3 cm, ± = terminal lflet **INFL** 1–3-fld; pedicels curved (straight in fr). **FL** ± cup-shaped bractlets 5–14 mm, linear-oblanceolate; sepals erect, 6–12 mm; petals 7–13 mm, ± elliptic, outcurved, cream or pink-tinged, persistent. **FR**: achene body 2.5–5 mm; style < 35 mm incl gen persistent tip, not hooked, plumose. *n*=21. Dry meadow edges, sagebrush scrub, open yellow-pine forest; 1300–3200 m. c KR (Marble Mtns), CaRH, n&c SNH, GB; to n&e N.Am. [*G. canescens* (E. Greene) Munz; *G. ciliatum* Pursh] Vars. indistinct in CA. ❀1,**2**,6&IRR:3, **7**,**14–16**,17,**18–21**,22–24;GRNCVR.

HETEROMELES CHRISTMAS BERRY, TOYON

Dieter H. Wilken

1 sp. (Greek: different apple)

H. arbutifolia (Lindley) Roemer (p. 959) Shrub or small tree gen < 5 m. **ST**: trunk bark grayish; twigs puberulent. **LVS** simple, evergreen, short-petioled; blade 4–11 cm, elliptic to oblong, leathery, sharply toothed, shiny dark green above, dull and paler below. **INFL**: panicle, terminal, ± open, flat-topped; fls many. **FL**: hypanthium 2–3 mm, ± obconic; sepals 1–2 mm; petals 2–4 mm, white; stamens 10 in pairs opposite sepals; pistil 1, ovary ± inferior, 2–3-chambered, styles 2–3. **FR**: pome, 5–10 mm diam, bright red, pulp mealy. **SEEDS** 3–6, compressed, brown. Chaparral, oak woodland, mixed-evergreen forest; < 1300 m. CA-FP; Baja CA. [var. *macrocarpa* (Munz) Munz] ❀DRN,SUN:**7**,8,9,**14–24**&DRY:5, **17**;CVS include yellow-berried form.

HOLODISCUS

Richard A. Lis

Shrub, ± hairy. **ST** 3–60 dm; bark reddish, in age gray, shredding. **LVS** simple, alternate, 0.3–12 cm, thin to leathery, toothed; base truncate to gen ± wedge-shaped; lower surface strongly veined; stipules 0; petiole distinct or not. **INFL**: panicle, ± terminal, dense, 2.5–25 cm, ± conic, many-fld, persistent; pedicel slender, bractlets 1–3. **FL**: hypanthium 3–5 mm wide, saucer-shaped, prominent nectary-disk below inner rim; sepals 5, 1–2 mm; petals 5, 1.5–2 mm, ± ovate, gen white; stamens 15–20, wider at base; pistils 5, ovaries superior, 2-ovuled, hairs dense, bristle-like, persistent in fr, style 1 mm, persistent, stigma ± 2-lobed. **FR**: achenes 5, 1–1.5 mm, often with sessile glands. 5 spp.: w N.Am, C.Am, n S.Am. (Greek: whole disk) Spp. highly variable; lvs of peg-like sts best for identification; complexity in c SNH evidently from local climatic variation, hybridization.

1. Lvs on peg-like sts gen 2–4 cm, elliptic, others 2–12 cm, ovate to wide-ovate; petiole gen distinct; infl gen not mixed with lvs .... ***H. discolor***

1′ Lvs on peg-like sts gen 0.3–2 cm, obovate, others 1–3 cm, ovate to ± round; petiole gen indistinct; infl gen mixed with lvs .... ***H. microphyllus***

 2. Lf puberulent to glabrous on both surfaces, hairs much longer on margins and veins .... var. ***glabrescens***

 2′ Lf ± long-hairy on 1 or both surfaces, hairs not greatly longer on margins and veins .... var. ***microphyllus***

H. discolor (Pursh) Maxim. (p. 959) OCEANSPRAY Pl 1.5–6 m, ± open. **ST** hairy; glands 0; peg-like sts not very predominant. **LVS** of peg-like sts gen 2–4 cm, elliptic, others 2–12 cm, ovate to wide-ovate; teeth gen below middle of blade, gen of 2 sizes; petiole gen distinct, gen not winged. **INFL** gen not mixed with lvs, 10–25 cm, 5–25 cm wide. **FL**: sepal outer surface hairs gen sparse to dense, short, inner surface gen glabrous. Moist woodland edges, rocky slopes; < 1800 m. NW, SNH, GV (Sutter Buttes), CW, ChI, WTR, SnGb (300–1300 m), PR; to B.C., MT, Colorado, TX, Mex. [var. *delnortensis* Ley; var. *dumosus* (S. Watson) Dippel; var. *franciscanus* (Rydb.) Jepson; *H. boursieri* (Carrière) Rehder] ❀DRN:**4–6**, **15–17**&SHD:**18**,**24**&IRR:1–3,**7**,8,9,**14**,**19–23**;CV.

H. microphyllus Rydb. Pl 0.3–1 m, ± dense. **ST**: peg-like sts very predominant. **LVS** of peg-like sts gen 0.3–2 cm, obovate, others 1–3 cm, ovate to ± round; teeth of 1 size, those of lvs on peg-like sts near tip, rarely to middle, those of others gen to middle; petiole gen indistinct, winged or not. **INFL** gen mixed with lvs, 2–8 cm, 1.5–5 cm wide. **FL**: sepal lower surface glabrous to densely short- to long-hairy, upper surface glabrous. Rocky places, outcrops; 600–4000 m. SNH (2700–4000 m), SnGb (1800–3300 m), SnBr, SnJt, SNE, DMtns; to OR, ID, WY, Colorado, AZ, Baja CA. Lvs of peg-like sts from pls of wet places more often ovate to round than obovate.

var. ***glabrescens*** (Greenman) Ley (p. 959) **ST** glabrous to puberulent; glands 0 or gen not obscured by hairs. **LF** glabrous to puberulent on both surfaces; hairs much longer on margins and veins; glands not obscured by hairs. **FL**: sepals glabrous to puberulent, glandular. Habitats of sp.; 600–2900 m. NCoRH (Mendocino Co.), MP; to OR, UT. ❀TRY.

var. ***microphyllus*** (p. 959) ROCK SPIRAEA **ST** hairy to long-hairy; glands 0 or obscured by hairs. **LF** ± long-hairy on 1 or both surfaces; hairs not greatly longer on margins and veins; glands 0 or often obscured by hairs. **FL**: sepals hairy to long-hairy; glands 0. Habitats of sp.; 1200–4000 m. Range of sp. [var. *sericeus* Ley] ❀TRY.

HORKELIA

Barbara Ertter

Per, gen ± glandular, gen resinous-smelling; caudex gen branched. **ST** gen ascending to erect. **LVS** gen basal, odd-1-pinnate, gen ± flat; cauline alternate, reduced upward; uppermost lateral lflets gen ± fused with terminal. **INFL**: cyme, open or of dense clusters; pedicels gen straight. **FL**: hypanthium a ± flat-bottomed cup, width ± 2 × length; bractlets 5, gen 2/3 sepals;

sepals 5, often reflexed; petals 5, gen ± = sepals, blunt, white; stamens 10, filaments flat, often forming a tube; pistils 2–many, ovaries superior, styles jointed below fr tip, ± thicker at base. **FR**: achene. 19 spp.: w N.Am. (J. Horkel, German pl physiologist, 1769–1846) Data apply to basal lvs, pressed hypanthia.

1. Fls gen in head-like clusters; pl gen rosetted or tufted; NW, CaR, SN, MP — bractlets < 0.5 mm wide; hypanthium gen 2–5 mm wide; lower pedicels often reflexed in fr; > 300 m
 2. Lateral lflets 2–4 per side, entire to ± 3-toothed; petals linear to obovate; fr ± 2 mm; 300–2500 m
  3. St densely glandular above, ascending to erect; petals gen widely obovate, 2.5–4 mm — nw KR ..... ***H. congesta*** ssp. ***nemorosa***
  3′ St ± glabrous to sparsely glandular above, often decumbent; petals linear to widely oblanceolate, 1.5–4 mm ..... ***H. tridentata***
   4. Petals ± widely oblanceolate, often < sepals; hypanthium 2.5–5 mm wide, inner wall hairy (exc in NCoRH); KR, NCoR, n SNH ..... ssp. ***flavescens***
   4′ Petals gen narrowly oblanceolate, gen = or > sepals; hypanthium gen 2–3.5 mm wide, inner wall gen glabrous; CaR, SN, MP ..... ssp. ***tridentata***
 2′ Lateral lflets 3–15 per side, > 3-toothed; petals ± wedge-shaped; fr 1–1.8 mm; > 1000 m ..... ***H. fusca***
  5. Lateral lflets gen divided ± 3/4 to base into oblanceolate lobes, often > 8 per side, ± crowded; filaments gen < 0.5 mm, gen wider than long — s CaRH ..... [2]ssp. ***tenella***
  5′ Lateral lflets gen toothed < 1/2 to base, gen 4–8 per side, separated; filaments often > 0.5 mm, longer than wide
   6. Petals 4–6 mm; lflets gen 10–20 mm, toothed < 1/4 to base, gen ± obovate; anthers ± 0.6 mm; style 1–1.5 mm; fr ± 1.8 mm; Wrn ..... ssp. ***capitata***
   6′ Petals 2–4 mm; lflets gen 5–15 mm, toothed 1/4–1/2 to base, ± round to wedge-shaped; anthers ± 0.4 mm; style ± 1 mm; fr ± 1.2 mm; n CA, SNH ..... ssp. ***parviflora***
1′ Fls gen either in ± flat-topped clusters or infl open; pl often matted; CA-FP, W&I
 7. Pedicels ± recurved in fr; infl open, often ± 1-sided
  8. Petals 4–7 mm; lflets toothed 1/4 to midvein; pistils gen > 20; n&c SNF (esp Ione Formation) ..... ***H. parryi***
  8′ Petals 2–3 mm; lflets lobed > 1/2 to base; pistils < 5; SnBr ..... ***H. wilderae***
 7′ Pedicels erect in fr; infl open or dense, not 1-sided
  9. Filaments gen < 0.5 mm, gen wider than long; style ± 1 mm — petals ± wedge-shaped; s CaRH ..... [2]***H. fusca*** ssp. ***tenella***
  9′ Filaments > 0.5 mm, gen longer than wide; style 1–4 mm
   10. Pl gen rosetted from simple or few-branched caudex; resinous odor indistinct; lateral lflets gen < 10-lobed; fr > 2 mm; often ± serpentine clay; esp KR
    11. Lflets 5–10 per side, 5–25 mm, lobed > 3/4 to base; st base hairs ± 2 mm; infl open to dense; hypanthium 3.5–5 mm wide; petals ± obovate, cream; pistils 5–15; style 2.5–4 mm; esp e KR ..... ***H. daucifolia***
    11′ Lflets ± 15 per side, 2–10 mm, gen lobed ± 1/2 to base; st base hairs ± 1 mm; infl open; hypanthium 2–3 mm wide; petals obcordate, often pink-tinged; pistils 2–6; style ± 1.5–2 mm; w KR ..... ***H. sericata***
   10′ Pl gen tufted or matted from a few–many-branched caudex; resinous odor gen obvious; lateral lflets often > 10-toothed or -lobed; fr often < 2 mm; gen non-serpentine; gen not KR (exc *H. hendersonii*)
    12. Lateral lflets pinnately veined, ± elliptic, ± evenly > 10-toothed — hypanthium width 4–8 mm, > 2 × length; bractlets ± 2–3 mm wide; CW, SW
     13. Lflets 1–3 per side, < distinct terminal lflet, teeth gen double, > 20; petals ± round, ± 5 mm wide; 400–1300 m; PR, n Baja CA ..... ***H. truncata***
     13′ Lflets 5–12 per side, ± = gen indistinct terminal lflet, teeth gen single, < 20; petals obovate or narrower, 1.5–4 mm wide; < 700 m; CCo, SCoRO, SCo ..... ***H. cuneata***
      14. Hairs ± all glandular; infl open; hypanthium inner rim ± glabrous; filament base 0.5–2 mm wide ..... ssp. ***puberula***
      14′ Hairs mostly glandless; infl open to dense; hypanthium inner rim gen hairy; filament base 0.5–1 mm wide
       15. Infl gen ± open; hairs ± spreading, moderate; glands ± obvious ..... ssp. ***cuneata***
       15′ Infl dense; hairs ascending to appressed, dense esp on lflets; glands hidden, if any ..... ssp. ***sericea***
    12′ Lateral lflets ± palmately veined at base, wedge-shaped or ± round-ovate, ± unevenly few–many-toothed or -lobed
     16. Hypanthium width 4–10 mm, 1–2 × length; bractlets ovate, gen ± 2 mm wide, ± = sepals; lateral lflets round to ovate, 5–60 mm, gen ± separated, gen > 20-toothed or -lobed; st ± leafy; fr ± 1 mm ..... ***H. californica***
      17. Lflets 3–5 per side, gen unlobed, 15–60 mm, elliptic to ovate — hypanthium inner wall glabrous; sepals not red-mottled; style 2–3 mm; CW ..... ssp. ***frondosa***
      17′ Lflets 4–9 per side, ± lobed, 5–40 mm, ovate to round
       18. Lateral lflets gen few-lobed ± 1/2 to base, 10–40 mm; sepals red-mottled inside; bractlets gen toothed; inner wall of hypanthium ± hairy; filaments gen 1.5–3 mm; style gen 3–4 mm; NCo, CCo ..... ssp. ***californica***
       18′ Lateral lflets lobed 1/2–3/4 to base, 5–25 mm; sepals not red-mottled; bractlets gen entire; inner wall of hypanthium glabrous; filaments gen 0.5–1.5 mm; style gen 2–3 mm; NCoR, SNF, SnFrB, SnBr ..... ssp. ***dissita***
     16′ Hypanthium width gen < 6 mm, gen 2+ × length; bractlets gen lanceolate, < 2 mm wide (exc. *H. yadonii*), < sepals; lateral lflets round or wedge-shaped, gen < 15 mm, often crowded, < 20-toothed or -lobed; st lvs few, ± reduced; fr > 1 mm

19. St < 25 cm; bractlets gen < 0.5 mm wide; lvs 3–10 cm; fr 1.5–2.5 mm; > 2000 m; KR, s SNH, W&I
20. Hypanthium densely hairy on inner rim; sepals 3.5–6 mm, 2+ × hypanthium length; longer hairs ± 1 mm, soft — ne KR ..... ***H. hendersonii***
20′ Hypanthium ± hairy on inner wall; sepals gen 2–4 mm, < 2 × hypanthium length; longer hairs ± 0.5 mm, stiff
21. Lflets gen 10–14 per side, gen 2–4 mm; pistils 10–20; W&I ..... ***H. hispidula***
21′ Lflets gen 4–6 per side, gen 3.5–8 mm; pistils 5–12; s SNH ..... ***H. tularensis***
19′ St 10–70 cm; bractlets gen > 0.5 mm wide; lvs 3–20 cm; fr gen < 1.5 mm; < 2800 m; NCo, NCoR, CW, SW
22. St decumbent to ascending, gen 10–30 cm; pl matted, odor strong; infl ± dense; lvs gen 4–10 cm; lflets 5–10 per side, ± crowded — hairs dense; NCo, CCo ..... ***H. marinensis***
22′ St ascending to erect, 10–70 cm; pl tufted or matted, odor variable; infl gen ± open; lvs gen 3–20 cm; lflets 6–16 per side, separated or crowded
23. Bractlets 1–2 mm wide, lanceolate to ovate; anthers ± 1 mm — hypanthium gen 3–6 mm wide; st hairs spreading; SCoRO ..... ***H. yadonii***
23′ Bractlets gen 0.5–1 mm wide, linear or elliptic to lanceolate; anthers 0.5–1 mm
24. St and petiole hairs gen spreading; pl often green
25. Lflets toothed ± 1/3 to base; hypanthium inner wall glabrous; PR ..... ***H. clevelandii***
25′ Lflets divided > 1/2 to base into linear to oblanceolate lobes; hypanthium inner wall hairy; NCo, NCoRO, SnFrB ..... ***H. tenuiloba***
24′ St and petiole hairs gen ascending to appressed; pl gen grayish — hypanthium inner wall ± hairy
26. Lf 3–8 cm; pistils < 20; style 1–2 mm; NCoRI ..... ***H. bolanderi***
26′ Lf 4–20 cm; pistils gen > 20; style gen 2–3 mm; TR ..... ***H. rydbergii***

H. bolanderi A. Gray BOLANDER'S HORKELIA Pl ± matted, grayish. **ST** 10–30 cm; hairs ascending to appressed. **LF** 3–8 cm; lflets ± 7 per side, gen ± crowded, 4–10 mm, wedge-shaped to obovate, ± 5-toothed ± 1/3 to base, densely hairy. **INFL** ± dense to open, several–many-fld; pedicels gen 2–4 mm. **FL**: hypanthium width 2–4 mm, > 2 × length, inner wall ± hairy; bractlets 0.5–1 mm wide, ± elliptic; sepals ± 3–4 mm; petals 3–5.5 mm, oblong to oblanceolate; filaments 0.5–2 mm, bases 0.3–0.4 mm wide, anthers ± 0.5 mm; pistils gen 10–20, styles 1–2 mm. **FR** ± 1.2 mm. UNCOMMON. Edges of vernally wet places in pine forest; 450–1100 m. NCoRI (Colusa, Lake, Mendocino? cos.). ❀IRR,DRN:1,**7**, **14–16**,**19–24**&SUN:4–6,**17**;GRNCVR.

H. californica Cham. & Schldl. (p. 959) Pl clumped, green. **ST** 10–120 cm. **LF** 5–40 cm; lflets 3–9 per side, gen ± separated, 5–60 mm, ± ovate to round, unevenly toothed and lobed (teeth gen > 20), ± hairy. **INFL** open, of few–many separate fls and few-fld clusters; pedicels gen 1–20 mm. **FL**: hypanthium width 4–10 mm, 1–2 × length; bractlets ± 4–6 mm, gen ± 2 mm wide, ovate, often toothed; sepals gen 4–6 mm; petals 3–8 mm, gen ± oblanceolate to elliptic; filaments 0.5–3 mm, bases 0.2–1.5 mm wide, anthers 0.8–1.8 mm; pistils gen > 50, styles 2–4 mm. **FR** ± 1 mm. Open scrub, n slopes, streambanks; < 1600 m. NCo, NCoR, SNF, CW, SnBr. Often confused with *Potentilla glandulosa* (terminal lflets distinct). Sspp. intergrade. [Ertter 1992 Phytologia 71:420–422]

ssp. ***californica*** **LF** 8–40 cm; sheathing bases ± hairy; lflets 4–9 per side, 10–40 mm, toothed, few-lobed ± 1/2 to base; terminal lflet 10–40 mm. **FL**: hypanthium inner wall ± hairy; bractlets gen toothed; sepals red-mottled inside; filaments gen 1.5–3 mm, bases 0.5–1 mm wide; styles gen 3–4 mm. *n*=28. Grassy openings, edges of coastal scrub, esp n slopes; < 400 m. NCo, CCo. ❀DRN,IRR:7,**14–17**,19–23,**24**&SUN:**5**.

ssp. ***dissita*** (Crum) B. Ertter **LF** gen 5–25 cm; sheathing bases gen glabrous; lflets gen 7–9 per side, 5–25 mm, unevenly toothed, lobed 1/2–3/4 to base; terminal lflet 10–40 mm. **FL**: hypanthium inner wall glabrous; bractlets often entire; sepals not red-mottled inside; filaments gen 0.5–1.5 mm, bases gen < 0.5 mm wide; styles gen 2–3 mm. 2*n*=28. Shady meadow edges, seasonal streams, open chaparral; 50–1600 m. NCoR, SNF, SnFrB, SnBr. [*H. elata* (E. Greene) Rydb.] ❀IRR:**5**,**7**,**15–17**&SHD:1,**14**,19–24;GRNCVR.

ssp. ***frondosa*** (E. Greene) B. Ertter (p. 959) **LF** 10–40 cm; sheathing bases ± hairy; lflets gen 3–4 per side, 15–60 mm, ± evenly double-toothed < 1/4 to base, gen unlobed; terminal lflet 20–90 mm. **FL**: hypanthium inner wall glabrous; sepals not red-mottled inside; filaments 1–2 mm, bases 0.5–1.5 mm wide; styles 2–3 mm. Coastal scrub, canyons, poison-oak thickets; 10–400 m. CW. [*H. frondosa* (E. Greene) Rydb.] ❀DRN,IRR:**5**,**17**&SHD:14,**15**,**16**, 22–24.

H. clevelandii (E. Greene) Rydb. Pl tufted, green to grayish. **ST** 10–50 cm; hairs spreading. **LF** 5–18 cm; lflets 6–12 per side, gen ± separated, gen 5–12 mm, wedge-shaped to ± round, 5–10-toothed ± 1/3 to base, ± densely hairy. **INFL** gen open, ± 5–30-fld; pedicels gen 1.5–6 mm. **FL**: hypanthium width 2–4 mm, ± 2 × length, inner wall glabrous; bractlets 0.5–1.5 mm wide, gen lanceolate; sepals ± 3–4 mm; petals 3–6 mm, widely oblanceolate; filaments 0.5–1.5 mm, bases ± 0.5 mm wide, anthers 0.5–1 mm; pistils gen 10–50, styles ± 1.5 mm. **FR** ± 1.2 mm. Meadows, under pines, on granite; 1200–2500 m. PR; n Baja CA (where some possibly distinct). [*H. bolanderi* ssp. *c.* (E. Greene) Keck] ❀IRR:1–3,**7**,**15–17**&SHD: **14**,**18**,19–24;GRNCVR.

H. congesta Hook. ssp. ***nemorosa*** Keck JOSEPHINE HORKELIA Pl tufted or rosetted, ± grayish green, ± odorless; caudex 0–few-branched. **ST** 15–30 cm, densely glandular above. **LF** 4–8 cm; lflets 2–5 per side, separated, 5–12 mm, ± elliptic, < 5-toothed < 1/4 to base, very hairy. **INFL**: clusters 1–several, head-like, gen 5–15-fld; pedicels gen ± 2 mm. **FL**: hypanthium width gen 3–4.5 mm, < or = 2 × length; bractlets < 0.5 mm wide, linear; sepals 2–4.5 mm; petals 2.5–4 mm, gen widely obovate; filaments 0.5–2 mm, bases 0.2–0.5 mm wide, anthers ± 0.5 mm; pistils ± 10, styles 2–3 mm. **FR** ± 2 mm. RARE in CA. Vernally moist, rocky clay, gen serpentine; 300–800 m. nw KR (Del Norte Co.); sw OR.

H. cuneata Lindley (p. 959) Pl matted, green or grayish. **ST** gen 20–70 cm. **LF** gen 10–30 cm; lflets 5–12 per side, ± separated, gen 10–25 mm, ± = terminal, ± elliptic, pinnately veined, evenly ± 10–15-toothed gen < 1/3 to midvein, ± glabrous to densely hairy. **INFL** open, of several–many separate fls or < 10-fld clusters; pedicels gen 1–30 mm. **FL**: hypanthium width 4–7 mm, > 2 × length; bractlets ± 2 mm wide, ovate; sepals 4–6 mm; petals 4–8 mm, 1.5–4 mm wide, ± oblanceolate to obovate; filaments 1–3 mm, bases 0.5–2 mm wide, anthers ± 1 mm; pistils gen 30–60, styles gen 2–3 mm. **FR** 1.5–1.8 mm. Old dunes, open chaparral; < 700 m. CCo, SCoRO, SCo. Sspp. intergrade.

ssp. ***cuneata*** Hairs glandular and nonglandular, spreading. **INFL** ± open; clusters several–many, few-fld; pedicels gen 1–12 mm. **FL**: hypanthium inner rim gen ± hairy; filament bases 0.5–1 mm wide. *n*=14. Old dunes, coastal sandhills; gen < 400 m. CCo, SCoRO, SCo. ❀DRN:5,**15–17**&IRR:7,**14**,22,23,**24**;GRNCVR.

ssp. ***puberula*** (E. Greene) Keck Hairs sparse, ± all glandular. **INFL** open, of many ± separate fls; pedicels gen 5–30 mm. **FL**: hypanthium inner rim ± glabrous; filament bases gen 0.5–2 mm wide. Dry, sandy, coastal chaparral; gen 70–700 m. SCoRO, SCo (esp foothill edge of Los Angeles Basin). More inland than other sspp. ❀TRY.

ssp. ***sericea*** (A. Gray) Keck KELLOGG'S HORKELIA Hairs dense, ± nonglandular, often appressed. **INFL** dense to ± open;

clusters several–many, several-fld; pedicels gen 1–12 mm. **FL**: hypanthium inner rim very hairy; filament bases 0.5–1 mm wide. *n*= 14. RARE. Old dunes, coastal sandhills; gen < 200 m. CCo. Remaining pls less distinct from ssp. *cuneata* than those formerly near San Francisco. Threatened by coastal development. In cult.

H. daucifolia (E. Greene) Rydb. (p. 959) Pl rosetted, ± grayish; caudex gen simple; odor indistinct. **ST** 15–30 cm; hairs ± 2 mm at st base. **LF** 5–15 cm; lflets 5–10 per side, ± crowded, 5–25 mm, divided > 3/4 to base into 2–6 linear to oblanceolate lobes, hairs many. **INFL** dense to open, ± flat-topped, 5–25-fld; pedicels gen 3–9 mm. **FL**: hypanthium width 3.5–5 mm, > 2 × length; bractlets ± 0.5 mm wide, linear; sepals 4–6 mm; petals 4–6 mm, ± obovate, cream drying yellowish; filaments 1–2.5 mm, bases 0.5–1.5 mm wide, anthers ± 0.7 mm; pistils 5–15, styles 2.5–4 mm. **FR** ± 2.5 mm. Dry open places, often on serpentine clay; 500–1650 m. w KR, CaRH (esp Shasta, Scott valleys), n NCoRI?; sw OR. [ssp. *latior* Keck] *Potentilla d.* E. Greene var. *indicta* Jepson (Tehama Co.) resembles OR phase, which may be distinct. ❀TRY.

H. fusca Lindley Pl gen tufted (± matted), green to grayish. **ST** gen 10–60 cm. **LF** gen 4–15 cm; lflets 3–15 per side, separated to ± crowded, gen 4–20 mm, narrowly wedge-shaped to ± round, ± 5–10-toothed to lobed, hairs sparse to dense. **INFL**: clusters 1–several, gen ± head-like, gen 5–30-fld; pedicels gen 1–3 mm. **FL**: hypanthium width gen 2–3.5 mm, ± 1–2 × length; bractlets < 0.5 mm wide, linear; sepals gen 2–3 mm; petals 2–6 mm, ± wedge-shaped; filaments 0.1–1.5 mm, bases 0.2–1 mm wide, anthers gen ± 0.5 mm; pistils gen 10–20, styles 0.8–1.5 mm. **FR** 1–1.8 mm. Dry meadow edges, open forest, volcanic or granitic soils; 1000–3300 m. e KR, CaRH, SNH, GB; to WA, WY, NV. Sspp. need study.

ssp. ***capitata*** (Lindley) Keck **LF** 10–15 cm; lflets 4–7 per side, separated, gen 10–20 mm, gen ± obovate, ± 10-toothed < 1/4 to base, hairs gen sparse. **INFL**: clusters gen 10–30-fld. **FL**: petals 4–6 mm; filaments 0.5–1.5 mm, gen longer than wide, anthers ± 0.6 mm; styles 1–1.5 mm. **FR** ± 1.8 mm. 2*n*=28. Habitats of sp.; 1800–2300 m. Wrn; to WA, ID. ❀TRY;DFCLT.

ssp. ***parviflora*** (Nutt.) Keck (p. 959) **LF** gen 4–15 cm; lflets gen 4–8 per side, separated, gen 5–15 mm, wedge-shaped to ± round, ± 5-toothed 1/4–1/2 to base, hairs sparse to dense. **INFL**: clusters gen 5–20-fld. **FL**: petals 2–4 mm; filaments 0.2–1 mm, gen longer than wide, anthers ± 0.4 mm; styles ± 1 mm. **FR** ± 1.2 mm. *n*=14. Habitats of sp.; 1000–3300 m. e KR, n CaRH, SNH, MP (exc Wrn), n SNE; to OR, ID, NV. [ssp. *pseudocapitata* (Rydb.) Keck] ❀DRN,IRR:1–7,15–18;DFCLT.

ssp. ***tenella*** (S. Watson) Keck **LF** gen 4–15 cm; lflets gen 8–15 per side, ± crowded, gen 5–15 mm, widely obovate, gen divided ± 3/4 to base into 5–15 oblanceolate lobes, hairs sparse. **INFL**: clusters gen 5–10-fld. **FL**: petals gen 2–4 mm; filaments gen < 0.5 mm, gen wider than long, anthers ± 0.5 mm; styles ± 1 mm. **FR** 1.2–1.5 mm. 2*n*=28. Gen lodgepole-pine forest; 1200–2200 m. s CaRH. [*Potentilla douglasii* E. Greene var. *tenuisecta* (Rydb.) Crum misapplied] ❀DRN,IRR:1,16,17;DFCLT.

H. hendersonii Howell HENDERSON'S HORKELIA Pl matted, grayish. **ST** < 20 cm. **LF** 3–8 cm, gen ± cylindric; lflets 5–10 per side, crowded, 4–10 mm, wedge-shaped, ± 5-toothed ± 1/2 to base, hairs dense. **INFL** ± dense, ± flat-topped, 5–20-fld; pedicels 1–7 mm. **FL**: hypanthium width 2–4 mm, ± 2 × length, inner rim densely hairy; bractlets < 0.5 mm wide, linear; sepals 3.5–6 mm; petals gen 3–4 mm, linear to narrowly oblanceolate; filaments ± 2 mm, bases ± 0.5 mm wide, anthers ± 0.5 mm; pistils ± 10–15, styles 2–3 mm. **FR** ± 2 mm. RARE. Dry granitic flats; 2000–2300 m. ne KR; sw OR (Mount Ashland). In cult.

H. hispidula Rydb. WHITE MOUNTAINS HORKELIA Pl matted, gen grayish. **ST** < 25 cm. **LF** 3–10 cm, gen ± cylindric; lflets gen 10–14 per side, crowded, gen 2.5–4 mm, divided > 3/4 to base into 3–7 oblanceolate lobes, hairs many. **INFL** ± dense or open, gen 3–15-fld; pedicels gen 1–6 mm. **FL**: hypanthium width 3–4 mm, ± 2 × length, inner wall ± hairy; bractlets < 0.5 mm wide, linear to lanceolate; sepals 2.5–4 mm; petals 3–5 mm, oblong to oblanceolate; filaments 0.5–2 mm, bases ± 0.5 mm wide, anthers ± 0.5 mm; pistils 10–20, styles ± 2 mm. **FR** ± 1.5 mm. RARE. Dry flats; 3000–3400 m. n W&I (White Mtns). In cult.

H. marinensis (Elmer) Crum POINT REYES HORKELIA Pl matted, grayish; odor very strong. **ST** decumbent to ascending, gen 10–30 cm. **LF** gen 4–10 cm; lflets gen 5–10 per side, ± crowded, 7–12 mm, gen ± wedge-shaped, ± 5–10-toothed ± 1/3–1/2 to base, hairs dense. **INFL** ± dense; clusters indistinct, 5–10-fld; pedicels gen 1–6 mm. **FL**: hypanthium width 4–5 mm, > 2 × length, inner wall hairy; bractlets ± 1 mm wide, lanceolate; sepals 3–6 mm; petals gen 4–6 mm, oblong to oblanceolate; filaments 1–2.8 mm, bases ± 0.5 mm wide, anthers ± 0.7 mm; pistils gen 20–30, styles 2–4 mm. **FR** 1.5–2 mm. 2*n*=56. RARE. Sandy coastal flats; ± 15–350 m. c NCo (Fort Bragg), n CCo (Point Reyes to Santa Cruz). In cult.

H. parryi E. Greene PARRY'S HORKELIA Pl openly matted, green. **ST** gen 10–30 cm. **LF** 5–10 cm; lflets 3–6 per side, separated, 5–15 mm, ± obovate, 5–10-toothed 1/4 to base, hairs sparse. **INFL** open, ± 5–10-fld; pedicels 5–15 mm, ± recurved in fr. **FL**: hypanthium width ± 3–4 mm, > 3 × length; bractlets 0.5–1.5 mm wide, lanceolate; sepals 4–6 mm; petals 4–7 mm, obovate to elliptic; filaments gen 1–3 mm, bases ± 1 mm wide, anthers 0.6–1 mm; pistils gen 20–50, styles ± 2 mm. **FR** ± 1.5 mm. 2*n*=28. RARE. Open chaparral; 80–900 m. n&c SNF (esp Ione Formation). Reported once from limestone. In cult.

H. rydbergii Elmer Pl ± matted, grayish green. **ST** 10–70 cm; hairs ascending to ± appressed. **LF** 4–20 cm; lflets 7–14 per side, separated or crowded, 5–15 mm, ± wedge-shaped, ± 5–10-toothed ± 1/3 to base, hairs gen dense. **INFL** ± open, gen 8–40-fld; pedicels gen 2–8 mm. **FL**: hypanthium width gen 2.5–4 mm, ± > 2 × length, inner wall ± hairy; bractlets 0.5–1 mm wide, ± lanceolate; sepals 3–5 mm; petals 4–5.5 mm, oblong to oblanceolate; filaments 0.5–2 mm, bases 0.5–1 mm wide, anthers 0.6–1 mm; pistils gen 20–50, styles gen 2–3 mm. **FR** ± 1.2 mm. 2*n*=28. Meadows, streambanks, under pines; 1200–2800 m. TR. [*H. bolanderi* A. Gray ssp. *parryi* (S. Watson) Keck] ❀DRN,IRR:1–3,**7,14–18**,19–24.

H. sericata S. Watson (p. 959) HOWELL'S HORKELIA Pl ± tufted, green to silvery; caudex few-branched; odor indistinct. **ST** 15–50 cm; hairs at base ± 1 mm. **LF** 3–15 cm, often ± cylindric; lflets ± 15 per side, crowded, 2–10 mm, gen divided ± 1/2 to base into < 10 elliptic lobes, hairs many. **INFL** open; fls many, ± separate; pedicels gen 1–6 mm. **FL**: hypanthium width 2–3 mm, 1–2 × length; bractlets ± 0.5 mm wide, linear-lanceolate; sepals 2.5–4 mm; petals 3–5 mm, narrowly obcordate, often pink-tinged; filaments 0.5–1.5 mm, bases < 0.5 mm wide, anthers ± 0.5 mm; pistils 2–6, styles ± 1.5–2 mm. **FR** 2–2.7 mm. *n*=14. UNCOMMON. Dry, rocky serpentine clay, open chaparral or pine forest; 60–1200 m. w KR; sw OR. [*Potentilla howellii* E. Greene] Variation in hairiness, stipule lobing needs study. ❀DRN,IRR,SUN:**7**,14,**15–17**,18–21.

H. tenuiloba (Torrey) A. Gray THIN-LOBED HORKELIA Pl loosely matted, gen ± green. **ST** < 40 cm; hairs ± spreading. **LF** 5–15 cm, often ± cylindric; lflets gen 8–15 per side, ± crowded, 3–10 mm, divided > 1/2 to base into 3–8 linear to oblanceolate lobes, hairs sparse to dense. **INFL** dense or ± open, few–many-fld; pedicels gen 1–6 mm. **FL**: hypanthium width 3–4.5 mm, > 2 × length, inner wall ± hairy; bractlets 0.5–1 mm wide, linear-lanceolate; sepals 3–5 mm; petals ± 4 mm, oblanceolate; filaments ± 1.5–2 mm, bases ± 0.5 mm wide, anthers ± 0.5 mm; pistils 10–25, styles ± 2 mm. **FR** ± 1.5 mm. 2*n*=28. UNCOMMON. ± sandy soils, open chaparral; 50–500 m. c&s NCo, c&s NCoRO, nw SnFrB. ❀DRN,IRR:5,7,**14–17**,19–24;GRNCVR.

H. tridentata Torrey (p. 959) Pl rosetted or tufted, ± gray, ± odorless; caudex 0–few-branched. **ST** decumbent to erect, < 45 cm, gen ± glabrous. **LF** 3–12 cm; lflets 2–5 per side, separated, 5–30 mm, gen elliptic to oblong, ± 3-toothed < 1/4 to base (rarely entire), at least lower surface ± densely hairy. **INFL**: clusters 1–many, ± head-like, gen 3–40-fld; pedicels gen 1–6 mm. **FL**: hypanthium width 2–5 mm, 1–2 × length; bractlets < 0.5 mm wide, linear; sepals gen 1.5–3 mm; petals 1.5–4 mm, linear to widely oblanceolate; filaments 1–2 mm, bases gen ± 0.2–0.5 mm wide, anthers 0.2–0.5 mm; pistils 5–15, styles 1–2.5 mm. **FR** 1.5–2.5 mm. Dry, open coniferous forest; 300–2500 m. KR, NCoR, CaR, SN, MP; s OR. Sspp. intergrade extensively. ❀DRN,IRR,SUN: 1–3,7,14–18;DFCLT.

ssp. ***flavescens*** (Rydb.) Keck **ST** decumbent. **INFL**: cluster gen 1. **FL** 6–10 mm wide; hypanthium width 2.5–5 mm, inner wall hairy (exc in NCoRH); petals often < sepals, ± widely oblanceolate; filaments 1–2 mm. **FR** 2–2.5 mm. Often on ± serpentine; 750–2000 m. KR, NCoR, n SNH (esp Plumas Co.).

ssp. ***tridentata*** **ST** decumbent to ± erect. **INFL**: clusters 1–many. **FL** 4–7 mm wide; hypanthium width gen 2–3.5, inner wall gen glabrous; petals gen = or > sepals, gen narrowly lanceolate; filaments gen 0.5–1 mm. **FR** 1.5–2 mm. n=14. Granitic or volcanic soils; 300–2500 m. CaR, SN, MP; s OR.

H. truncata Rydb. (p. 959) RAMONA HORKELIA Pl tufted, green. **ST** 20–60 cm. **LF** 4–13 cm; lflets 1–3 per side, separated, 10–30 mm, oblong to obovate, gen > 20-toothed < 1/4 to midvein, ± glabrous; terminal lflet largest, gen unlobed. **INFL** open, 5–20-fld; pedicels gen 4–20 mm. **FL**: hypanthium width 5–8 mm, > 2 × length; bractlets ± 2–3 mm wide, widely ovate; sepals gen 3.5–5.5 mm; petals 5–7 mm, ± 5 mm wide, ± round; filaments 1–2 mm, bases 0.5–2 mm wide, anthers ± 1 mm; pistils 50–80, styles ± 2 mm. **FR** ± 1.3 mm. RARE. Dry red clay, open chaparral; 400–1300 m. PR; n Baja CA. In cult.

H. tularensis (J. Howell) Munz (p. 959) KERN PLATEAU HORKELIA Pl ± matted, grayish. **ST** < 25 cm. **LF** 2–10 cm, often ± cylindric; lflets gen 4–6 per side, ± crowded, gen 3.5–8 mm, divided > 3/4 to base into 5–8 ± oblanceolate lobes, hairs dense. **INFL** ± dense to open, gen 3–15-fld; pedicels gen 2–7 mm. **FL**: hypanthium width 2.5–4.5 mm, ± 2 × length, inner wall hairy; bractlets gen < 0.5 mm wide, linear; sepals ± 3–4.5 mm; petals ± 3–4 mm, oblong to oblanceolate; filaments 1–2 mm, bases ± 0.5 mm wide, anthers ± 0.8 mm; pistils 5–12, styles ± 2 mm. **FR** ± 2–2.5 mm. RARE. Dry, rocky balds, flats; 2350–2850 m. s SNH (Bald Mtn area, Tulare Co.). In cult.

H. wilderae Parish (p. 959) BARTON FLATS HORKELIA Pl ± rosetted, green; caudex gen simple. **ST** gen decumbent, gen 10–25 cm. **LF** 4–10 cm; lflets 4–7 per side, separated, 3–10 mm, ± obovate, divided 1/2–3/4 to base into 5–15 lobes, hairs ± sparse. **INFL** open, gen 5–15-fld; pedicels 3–15 mm, ± recurved in fr. **FL**: hypanthium width ± 2–3 mm, ± 2 × length; bractlets gen < 0.5 mm wide, lanceolate to ovate; sepals ± 2 mm; petals 2–3 mm, oblanceolate to oblong; filaments 0.5–1 mm, bases 0.5–1 mm wide, anthers ± 0.4 mm; pistils < 5, styles ± 1–1.5 mm. **FR** ± 2 mm. RARE. Edge of chaparral, under pines; 2200–3000 m. SnBr (Barton Flats area).

H. yadonii B. Ertter SANTA LUCIA HORKELIA Pl tufted to matted, grayish green. **ST** gen 20–60 cm; hairs spreading. **LF** gen 6–20 cm; lflets 7–16 per side, separated to ± crowded, gen 4–15 mm, wedge-shaped to round, ± 3–20-toothed or -lobed, often deeply notched, hairs gen dense. **INFL** ± open, gen 5–10-fld; pedicels gen 1–7 mm. **FL**: hypanthium width gen 3–6 mm, ± > 2 × length, inner wall ± hairy; bractlets 1–2 mm wide, lanceolate to ovate; sepals gen 4–6 mm; petals 3–5 mm, oblanceolate to elliptic; filaments 1–2 mm, bases 0.5–1 mm wide, anthers ± 1 mm; pistils gen > 20, styles 2.5–3 mm. **FR** ± 1.5 mm. UNCOMMON. Sandy meadow edges, seasonal streambeds in chaparral or foothill-pine woodland; 350–1900 m. SCoRO. Sometimes confused with *H. cuneata* ssp. *sericea*, *H. rydbergii*, or *H. tenuiloba*. ❀IRR:7,8,9,**14–17**,**19–24**; GRNCVR.

HORKELIELLA

Barbara Ertter

Per, tufted, ± glandular; odor ± resinous. **ST** ascending to erect, 15–50 cm. **LVS** gen basal, odd-pinnately compound; lflets many, ± overlapped, toothed to palmately divided, segments ± oblanceolate. **INFL**: ± cyme; pedicels straight. **FL**: hypanthium cup-like, flat-bottomed; bractlets 5; sepals 5, often reflexed; petals 5, white or pinkish, midvein often reddish; stamens 20, filaments ± flat, often an erect tube; pistils many, ovaries superior, style jointed below fr tip, ± rough, thick. **FR**: achene, ± 1.5 mm. 2 spp.: CA. (Latin: small *Horkelia*) Fl ± like that of *Horkelia*.

1. Filament opposite sepal center > adjacent ones; nonglandular hairs ± 0 to sparse, not obscuring glands; lflet lobes (or 2° lflets) gen 5–10; c SNH (e slope), w SNE . . . . ***H. congdonis***
1′ Filament opposite sepal center < adjacent ones; nonglandular hairs ± 0 to dense, often obscuring glands; lflet lobes gen < 5; s SNH (w slope) . . . . ***H. purpurascens***

H. congdonis (Rydb.) Rydb. (p. 959) Pl green, gen glandular-sticky. **LVS**: basal gen 8–25 cm, lflets 25–35 per side, gen 2–6 mm, 5–10-lobed; cauline 2–7. **INFL** ± dense; pedicels gen ± 2 mm. **FL** ± 15 mm wide; sepals 4–8 mm; petals 3–6 mm, oblong-oblanceolate; anther ± 1 mm, filaments 1.5–3 mm, central longest; style 2–5 mm. Meadows in sagebrush flats; 1500–2900 m. c SNH (e slope), w SNE. [*Ivesia purpurascens* (S. Watson) Keck ssp. *c.* (Rydb.) Keck] ❀DRN,IRR,SUN:1–3,**7**,**15–16**,17.

H. purpurascens (S. Watson) Rydb. Pl often grayish, often not sticky. **LVS**: basal gen 7–17 cm, lflets 15–30 per side, gen 3–10 mm, lobes gen < 5; cauline 2–4. **INFL** open at least in fr; pedicels 1–15 mm. **FL** 10–15 mm wide; sepals 3–6 mm; petals 3–7 mm, ± oblanceolate; anthers 0.5–1 mm, filaments 1–2.5 mm, central shortest; style ± 3–4 mm. Partial shade at granitic meadow edges in conifer forest; 1400–2900 m. s SNH (w slope). [*Ivesia p.* (S. Watson) Keck ssp. *p.*] ❀DRN,IRR:**15–17**&partSHD:1–3,**7**.

IVESIA

Barbara Ertter

Per, glandular; odor resinous. **LVS** gen basal, odd-1-pinnate, gen ± cylindric; cauline reduced; lflets gen overlapped, gen divided ± to base. **INFL**: cyme. **FL**: hypanthium shallow or deep; bractlets (0)5, gen < sepals; sepals gen 5; petals gen 5, acute to rounded; stamens gen < or = 20; pistils 1–many, ovaries superior, style jointed below fr tip, base ± rough-thickened. **FR**: achene. 30 spp.; w N.Am. (E. Ives, Yale Univ. pharmacologist, 1779–1861) [Ertter 1989 Syst Bot 14:231–244] Lf and lflet data are for basal lvs.

1. Lf ± flat, lflets toothed or lobed gen < 3/4 to base, gen ± separate — pls hanging clumps on vertical rock; infl open
  2. Stamens 15–40 — s SNH, SnBr, PR, SNE, DMtns . . . . ***I. saxosa***
  2′ Stamens 5–10
    3. Bractlets 5; MP, adjacent n SNH . . . . ***I. baileyi***
      4. Petals pale yellow; bractlets ± = sepals . . . . var. ***baileyi***
      4′ Petals white; bractlets ± < 1/2 sepals . . . . var. ***beneolens***
    3′ Bractlets gen 0; DMtns
      5. Hypanthium width < to = length, receptacle stalked in pistil-bearing portion; n DMtns . . . . ***I. arizonica***
      5′ Hypanthium width > 2 × length; receptacle not stalked; e DMtns . . . . ***I. patellifera***

1′ Lf ± cylindric, lflets gen lobed ± to base, overlapped
6. Stamens 10–20 (if 10, cauline lvs > 3)
7. Lflets < 8 per side; sts hanging to ± matted from few–many-branched caudex; rock outcrops
8. Lflet lobes < 4; petals white; SnJt .... ***I. callida***
8′ Lflet lobes 3–6; petals yellow; e DMtns .... ***I. jaegeri***
7′ Lflets > 10 per side; sts decumbent to erect from 0–few-branched caudex; meadows, sandy flats
9. Lflets 15–20 per side, sheathing lf bases glabrous; stamens 10–16, filaments ± < 1 mm; styles < 1.5 mm
10. Petals 4(5), light yellow; infl not red-tinged; s SNH .... ***I. campestris***
10′ Petals 5, white or pink-tinged; infl often red-tinged; c&s SNH .... ***I. unguiculata***
9′ Lflets > 20 per side, sheathing lf bases gen ± strigose; stamens 15–20; filaments 1–4 mm; styles ± > 2 mm
11. Infl of separate fls or gen < 5-fld clusters; pedicels gen > 5 mm; lflets gen 30–80 per side
12. Lf mousetail-like, lflets indistinct, obscured by dense, silvery hairs, lobes < 1.5 mm; stamens 15; pistil 1; granitic sand — SNH, TR, SnJt .... ***I. santolinoides***
12′ Lf not mousetail-like, lflets overlapped but distinct, not obscured by hairs, lobes > 2 mm; stamens 20; pistils > 2; meadows
13. Herbage hairs ± 0 or appressed, < 1 mm; petals ± obovate; alkali meadows; GB .... ***I. kingii*** var. ***kingii***
13′ Herbage hairs spreading, < 4 mm; petals ± oblanceolate; gen on serpentine clays; KR .... ***I. pickeringii***
11′ Infl of gen > 5-fld clusters; pedicels gen < 3 mm (exc lowest); lflets 20–35 per side
14. St 10–20 cm, ± decumbent; lvs < 8 cm, cauline ± 2; filaments ± flat; SnBr .... ***I. argyrocoma***
14′ St gen > 20 cm, decumbent to erect; lvs > 8 cm, cauline > 3; filaments not flat; n SNH, MP
15. Petals white; hypanthium length = or > width; hairs of st base ± spreading, 2–4 mm .... ***I. sericoleuca***
15′ Petals yellow; hypanthium length < or = width; hairs of st base ascending, < 2 mm .... ***I. aperta***
16. Petals 2–3 mm, oblanceolate; filaments 1–1.5 mm; n SNH (exc Dog Valley), s MP .... var. ***aperta***
16′ Petals gen 4–7 mm, ± obovate; filaments 2–4 mm; n SNH (Dog Valley, e Sierra Co.) .... var. ***canina***
6′ Stamens 5–10; cauline lvs 1–2
17. Bractlets > sepals — KR (Castle Crags) .... ***I. longibracteata***
17′ Bractlets ± 1/2 sepals
18. Infl gen ± open between separate fls or ± loose clusters; pl ± matted from much-branched caudex; pedicel gen ± S-shaped in fr; sheathing lf bases strigose
19. Stamens 10; pistils 10–30 — c&s SNH .... [2]***I. pygmaea***
19′ Stamens 5; pistils gen < 5
20. Petals white to pale yellowish, linear, ± 1 mm; pl densely hairy, glands not obvious; 1500–1700 m; MP .... ***I. paniculata***
20′ Petals yellow, oblanceolate, ± 2 mm; pl moderately hairy, glands many; 2700–4000 m; n&c SNH, W&I .... ***I. shockleyi*** var. ***shockleyi***
18′ Infl of 1–few gen head-like, 5–20-fld clusters (becoming more open in *I. webberi*); pl gen rosetted from 0–few-branched caudex; pedicels straight in fr; sheathing lf-bases hairy or not
21. Cauline lvs 2, ± opposite; lflets 4–8 per side, lobes linear to lanceolate; 1500–1900 m — s-most MP, adjacent n SNH .... ***I. webberi***
21′ Cauline lvs gen 1 or alternate; lflets > 8 per side, lobes oblanceolate to ± round; > 1800 m
22. Lf mousetail-like, lflets indistinct, < 1 mm, hidden by dense, silvery hairs; petals 1–2 mm — c&s SNH .... ***I. muirii***
22′ Lf not mousetail-like, lflets overlapped but distinct, gen > 1 mm, ± glabrous to ± hairy; petals > 2 mm
23. Stamens 10; sheathing lf-bases gen ± strigose; hypanthium length < 1/2 width — c&s SNH ... [2]***I. pygmaea***
23′ Stamens 5; sheathing lf-bases not strigose (± glabrous or glandular); hypanthium length > 1/2 width
24. Hypanthium length = or > width; petals narrowly oblanceolate; pistils gen 2–4; fr ± 2 mm, mottled brown .... ***I. gordonii***
24′ Hypanthium length ± < width; petals ± obovate; pistils gen 5–15; fr 1–1.5 mm, not mottled .... ***I. lycopodioides***
25. Lflet lobes ± 1 mm, ± round, ± glabrous; petals 2–3 mm — rocky areas; n&c SNH, SNE .... ssp. ***lycopodioides***
25′ Lflet lobes > 1 mm, narrowly oblanceolate to obovate, often ± hairy; petals gen > 3 mm
26. Lflet lobes 2–8 mm, ± glabrous to sparsely hairy, bristle-tip 0–0.5 mm; wet meadows; c&s SNH .... ssp. ***megalopetala***
26′ Lflet lobes 1–3 mm, moderately to densely hairy, bristle-tip gen 0.5–1 mm; vernally moist, rocky areas; W&I, c&s SNH .... ssp. ***scandularis***

I. aperta (J. Howell) Munz Pl tufted, greenish or white-hairy; caudex 0–few-branched. **ST** decumbent to erect, 15–45 cm. **LF** 10–20 cm; sheathing bases densely strigose; lflets 20–35 per side, lobes < 5, 3–15 mm, elliptic to oblanceolate; cauline lvs 3–8. **INFL** open; clusters many, 10–20 mm wide, ± head-like, ± 5–20-fld; pedicels gen < 3 mm (exc lowest), straight. **FL** 5–15 mm wide; hypanthium length < or = width; petals 2–7 mm, oblanceolate to obovate, yellow, < to > sepals; stamens gen 20; pistils 2–7. **FR** 2–3 mm, smooth, brown. Dry, rocky meadows, gen volcanic soils; 1500–2300 m. n SNH, s MP; w NV.

var. ***aperta*** (p. 963) SIERRA VALLEY IVESIA **ST** gen ascending to erect. **INFL**: clusters gen > 10-fld. **FL**: hypanthium 2–3 mm wide; petals 2–3 mm, oblanceolate; filaments 1–1.5 mm; styles ± 2.5 mm. RARE. Habitats of sp.; 1500–2300 m. Range of sp. (exc Dog Valley). With *I. sericoleuca* near Beckwith, Sierra Valley. In cult.

var. ***canina*** B. Ertter (p. 963) DOG VALLEY IVESIA **ST** decumbent to ascending. **INFL**: clusters gen < 10-fld. **FL**: hypanthium 4–5 mm wide; petals gen 4–7 mm, ± obovate; filaments 2–4 mm; styles 3–4 mm. UNCOMMON. Habitats of sp.; 1600–2000 m. n SNH (Dog Valley, e Sierra Co.)

I. argyrocoma (Rydb.) Rydb. SILVER-HAIRED IVESIA Pl rosetted or tufted, silvery-hairy; caudex gen simple. **ST** ± decumbent,

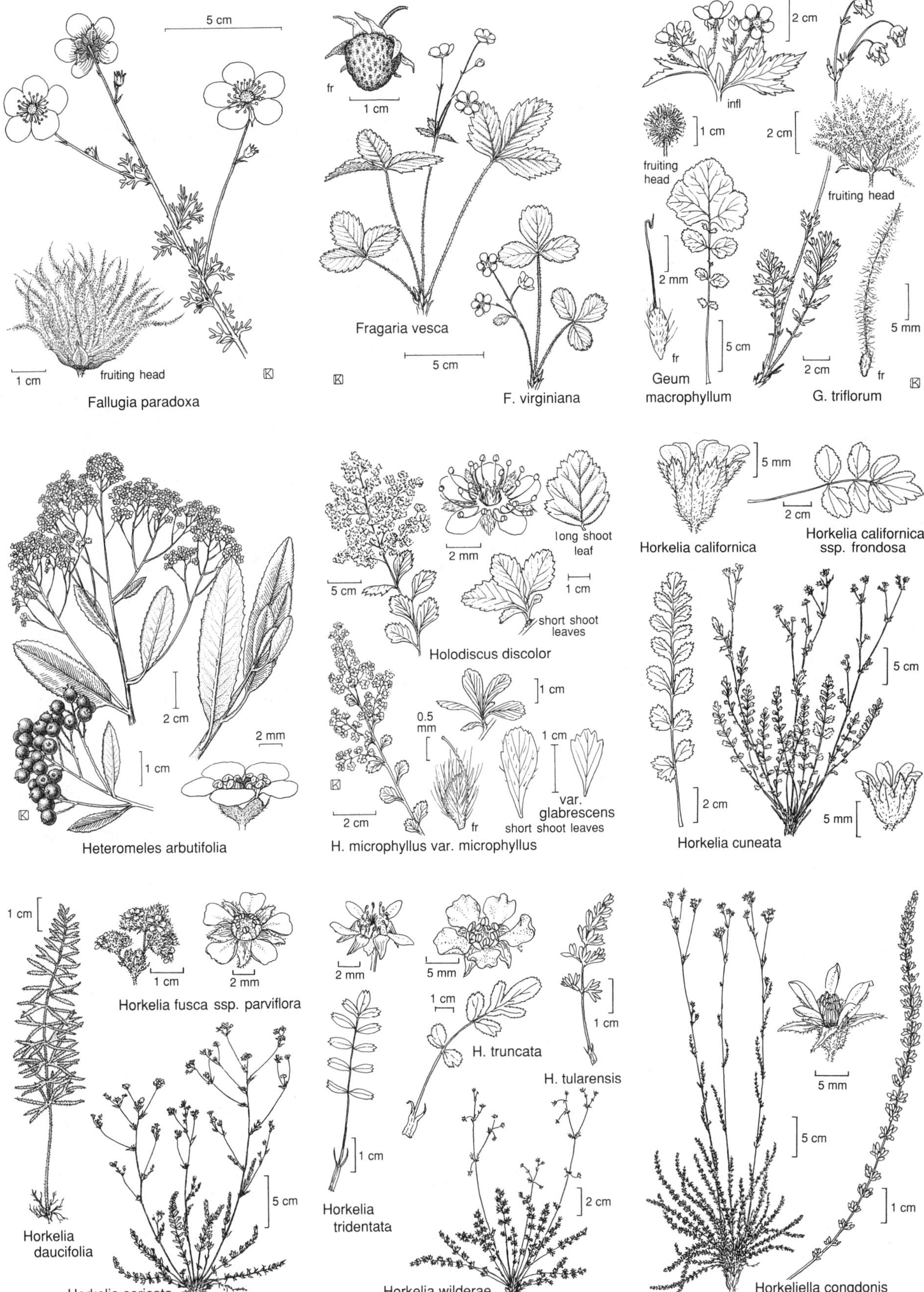

5 cm
1 cm
fruiting head
Fallugia paradoxa
fr
1 cm
Fragaria vesca
5 cm
F. virginiana
2 cm
infl
1 cm
fruiting head
2 cm
fruiting head
2 mm
fr
5 cm
Geum macrophyllum
2 cm
5 mm
fr
G. triflorum
2 cm
1 cm
2 mm
Heteromeles arbutifolia
5 cm
2 mm
long shoot leaf
1 cm
short shoot leaves
Holodiscus discolor
1 cm
0.5 mm
1 cm
var. glabrescens
2 cm
fr
short shoot leaves
H. microphyllus var. microphyllus
5 mm
Horkelia californica
2 cm
Horkelia californica ssp. frondosa
5 cm
2 cm
5 mm
Horkelia cuneata
1 cm
1 cm
2 mm
Horkelia fusca ssp. parviflora
Horkelia daucifolia
5 cm
Horkelia sericata
2 mm
5 mm
1 cm
H. truncata
1 cm
H. tularensis
1 cm
Horkelia tridentata
2 cm
Horkelia wilderae
5 mm
5 cm
1 cm
Horkeliella congdonis

10–20 cm. **LF** gen 4–8 cm; sheathing bases densely strigose; lflets 25–35 per side, lobes ± 3, gen 2–3 mm, elliptic to obovoid; cauline lvs ± 2. **INFL**: clusters 1–several, 10–20 mm wide, loosely head-like, < 20-fld; pedicel < 3 mm (exc lowest), ± straight. **FL** ± 10 mm wide; hypanthium length = width; petals 2–4 mm, obovate, white, > sepals; stamens 20, filaments ± flat (unique in *Ivesia*); pistils 4–8. **FR** 2–2.5 mm, smooth, brown. RARE. Pebble plains; 2000–2300 m. SnBr; n Baja CA (where perhaps distinct).

I. arizonica (J. Howell) B. Ertter var. ***arizonica*** (p. 963) YELLOW PURPUSIA Pls hanging clumps, green. **ST** 5–10 cm. **LF** gen 5–10 cm, flat; sheathing bases gen glabrous; lflets 2–4 per side, separated, 5–15 mm, ± round, ± evenly toothed or lobed < 3/4 to base; cauline lvs 1–3. **INFL** open, gen < 10-fld; pedicels 5–30 mm, often ± S-shaped in fr. **FL** 5–10 mm wide; hypanthium length 1–2 × width; bractlets gen 0; petals 2–3 mm, oblanceolate to elliptic, yellow, ± = sepals; stamens 5; pistils 2–10; receptacle stalked in pistil-bearing portion (unique in *Ivesia*). **FR** 1.5–2 mm, ± ridged, pale. UNCOMMON. Limestone crevices; 1200–3100 m. n DMtns (Inyo Co.); s NV, nw AZ. Var. *saxosa* (Brandegee) B. Ertter [*Purpusia s.* Brandegee], rock purpusia, (NV) has white petals.

I. baileyi S. Watson Pls hanging clumps, green. **ST** 5–20 cm. **LF** gen 3–10 cm, ± flat; sheathing bases ± glabrous; lflets 2–6 per side, ± separated, ± 4–10 mm, ± round, toothed ± < 1/2 to base; cauline lvs 1–2. **INFL** open, often > 10-fld; pedicels gen 2–12 mm, S-shaped in fr. **FL** 4–8 mm wide; hypanthium length < 1/2 width; petals 1.5–2 mm, ± oblanceolate, pale yellow or white, < sepals; stamens 5; pistils 1–6. **FR** ± 1.5 mm, pale. Volcanic crevices; 1600–2600 m. MP, adjacent SNH; to se OR, s ID, n NV.

var. ***baileyi*** (p. 963) BAILEY'S IVESIA **LF**: sheathing base ± glabrous. **INFL**: pedicels 2–12 mm. **FL**: bractlets ± = sepals; petals pale yellow. **FR** smooth. RARE in CA. Habitats and elevations of sp. s MP, adjacent SNH; nw NV.

var. ***beneolens*** (Nelson & J.F. Macbr.) B. Ertter OWYHEE IVESIA **LF**: sheathing base glabrous. **INFL**: pedicels 6–12 mm. **FL**: bractlets ± < 1/2 sepals; petals white. **FR** ± ridged. RARE in CA. Habitats of sp.; ± 2150 m. Wrn; to se OR, s ID, n NV.

I. callida (H.M. Hall) Rydb. (p. 963) TAHQUITZ IVESIA Pl hanging to ± matted, green; caudex branched. **ST** gen 2–15 cm. **LF** gen 1–7 cm; sheathing bases ± hairy; lflets ± 6 per side, lobes < 4, 2–7 mm, oblanceolate to elliptic; cauline lvs ± 2. **INFL** ± open, < 15-fld; pedicels gen 5–15 mm, ± S-shaped in fr. **FL** ± 7 mm wide; hypanthium length < 1/2 width; petals ± 3 mm, obovate, white, ± = sepals; stamens 20; pistils 4–8. **FR** ± 1.5 mm, smooth, pale. **RARE** CA. Granite crevices; ± 2450 m. SnJt.

I. campestris (M.E. Jones) Rydb. FIELD IVESIA Pl rosetted, green to white-hairy; caudex simple. **ST** ascending, ± 10–30 cm. **LF** 5–15 cm; sheathing bases glabrous; lflets gen 15–20 per side, lobes 3–5, 2–10 mm, oblanceolate; cauline lvs 3–4. **INFL**: clusters 1–few, 10–20 mm wide, loosely head-like, gen < 15-fld; pedicels < 4 mm, straight. **FL** < 10 mm wide; hypanthium length ± 1/2 width; petals 4(5), (4 unique in *Ivesia*), 3–4 mm, narrowly obovate, light yellow, > sepals; stamens gen 12–16; pistils 4–20. **FR** 1–1.5 mm, smooth, light brown. UNCOMMON. Meadow edges; 2200–3100 m. s SNH. ❀DRN,IRR:1–3,7,15–18;DFCLT.

I. gordonii (Hook.) Torrey & A. Gray (p. 963) Pl tufted, green; caudex 0–few-branched. **ST** ascending to erect, 5–20 cm. **LF** gen 3–8 cm; sheathing bases ± glandular; lflets gen 10–16 per side, overlapped but distinct, lobes 4–8, gen 2–5 mm, oblanceolate to obovate; cauline lf gen 1. **INFL**: cluster gen 1, 10–30 mm wide, head-like, 10–20-fld; pedicels gen < 3 mm, straight. **FL** ± 5 mm wide; hypanthium length = or > width; petals 2–3 mm, narrowly oblanceolate, yellow; stamens 5; pistils gen 2–4. **FR** ± 2 mm, ± smooth, mottled brown. Open, dry, rocky ridges, slopes; 1800–3500 m. e KR (Mount Eddy), NCoRH, n&c SNH, Wrn, n SNE (Sweetwater Mtns); to WA, WY, Colorado. ❀TRY;DFCLT.

I. jaegeri Munz & I.M. Johnston JAEGER'S IVESIA Pl hanging to ± matted, green; caudex branched. **ST** 3–15 cm. **LF** 2–10 cm; sheathing bases sparsely hairy; lflets 4–6 per side, lobes 3–6, 2–5 mm, oblanceolate to obovate; cauline lvs ± 2. **INFL** open, gen < 10-fld; pedicels 6–30 mm, ± S-shaped in fr. **FL** 5–10 mm wide; hypanthium length < 1/2 width; petals gen 2 mm, narrowly oblanceolate, yellow, < or = sepals; stamens 20; pistils 3–8. **FR** 1–2 mm, ± ridged, pale. RARE. Limestone crevices; 2100–3600 m. e DMtns (Clark Mtn); sw NV.

I. kingii S. Watson var. ***kingii*** ALKALI IVESIA Pl rosetted, glabrous or short-appressed-hairy, glaucous or not; caudex gen simple. **ST** decumbent to ascending, 15–40 cm. **LF** 7–15 cm; sheathing bases gen strigose; lflets 30–50 per side, overlapped but distinct, lobes < 4, 2–6 mm, oblanceolate to obovate; cauline lvs 4–13. **INFL** open; clusters gen < 10, 10–20 mm wide, loosely head-like, gen < 5-fld; pedicels 3–25 mm, straight. **FL** ± 10 mm wide; hypanthium length ± 1/2 width; petals 3–5 mm, ± obovate, white, > sepals; stamens 20; pistils 2–6. **FR** 2–2.5 mm, smooth, light brown. RARE in CA. Moist alkaline clay; 1200–2000 m. n SNE (Mono Co.); to UT. Var. *eremica* (Cov.) B. Ertter, Ash Meadows mousetails, w NV, has branched caudex, denser hairs, more tightly overlapped lflets.

I. longibracteata B. Ertter (p. 963) CASTLE CRAGS IVESIA Pl tufted, green; caudex 0–few-branched. **ST** ascending to erect, 3–12 cm. **LF** gen 2–4 cm, ± flat; sheathing bases ± ciliate; lflets 4–6 per side, lobes 2–7, 2–6 mm, ± oblanceolate; cauline lvs 1–3. **INFL** loosely head-like, 10–20 mm wide, < 15-fld; pedicels 2–6 mm, straight. **FL** ± 8 mm wide; hypanthium length < 1/2 width; bractlets > sepals (unique in *Ivesia*); petals 1.5–2.5 mm, ± linear, pale yellowish, ± = sepals; stamens 5; pistils 6–11. **FR** 1–1.5 mm, ± veined, pale. RARE. Granite crevices; 1200–1400 m. e KR (Castle Crags, Shasta Co.).

I. lycopodioides A. Gray CLUB-MOSS IVESIA Pl rosetted, green; caudex gen simple. **ST** decumbent to erect, 3–30 cm. **LF** 1–15 cm; sheathing bases ciliate; lflets 10–35 per side, overlapped but distinct, lobes 4–10, 1–8 mm, oblanceolate to round; cauline lf gen 1. **INFL**: cluster gen 1, 10–20 mm wide, ± dense or head-like, gen < 15-fld; pedicels gen < 5 mm, straight. **FL** 8–15 mm wide; hypanthium length ± < width; petals 2–5 mm, ± obovate, yellow, > sepals; stamens 5; pistils gen 5–15. **FR** 1–1.5 mm, smooth, pale. Wet meadows to alpine rocks; 2300–4000 m. SNH, SNE. Sspp. intergrade.

ssp. ***lycopodioides*** **ST** 3–15 cm. **LF** 1–7 cm; lobes ± 1 mm, ± round, ± glabrous, bristle-tip gen 0. **FL**: hypanthium length ± 1/2 width; petals 2–3 mm, gen < 2 mm wide, obovate; filaments ± 1 mm; styles 1–2 mm. $2n=28$. Rocky areas; 3000–4000 m. n&c SNH, n SNE (Sweetwater Mtns). ❀DRN,IRR,SUN:1,16,17; DFCLT.

ssp. ***megalopetala*** (Rydb.) Keck (p. 963) **ST** 10–30 cm. **LF** 4–15 cm; lobes 2–8 mm, ± oblanceolate, ± glabrous to sparsely hairy, bristle-tip 0–0.5 mm. **FL**: hypanthium length 1/2–1 × width; petals 3–5 mm, gen > 2 mm wide, widely obovate; filaments 1–2 mm; styles 2.5–3 mm. Wet meadows; 2300–3700 m. c&s SNH. ❀DRN,IRR,SUN:1,**2**,4–6,**7**,14,**15**,**16**,17,18.

ssp. ***scandularis*** (Rydb.) Keck **ST** gen 5–15 cm. **LF** gen 3–8 cm; lobes 1–3 mm, ± obovate, moderately to densely hairy, bristle-tip gen 0.5–1 mm. **FL**: hypanthium length ± < 1/2 width; petals 3–5 mm, gen > 2 mm wide, widely obovate; filaments ± 1.5 mm; styles 2–3 mm. Vernally moist, open, rocky areas; 3000–4000 m. c&s SNH, W&I.

I. muirii A. Gray (p. 963) Pl tufted, silvery; caudex gen simple. **ST** ascending to erect, gen 5–15 cm. **LF** 2–5 cm, mousetail-like; sheathing bases densely strigose; lflets 25–40 per side, indistinct, lobes ± 3, < 1 mm, obovate to round; cauline lf 1, gen ± bract-like. **INFL**: clusters 1–few, 10–15 mm wide, head-like, gen 10–20-fld; pedicels ± < 2 mm, straight. **FL** ± 5 mm wide; hypanthium length 1/2–1 × width; petals 1–2 mm, narrowly oblong to oblanceolate, pale yellow, ± = sepals; stamens 5; pistils 1–4. **FR** ± 2 mm, smooth, gray spotted red. Rocky areas; 2900–4000 m. c&s SNH. Apparently hybridizes with *I. lycopodioides*.

I. paniculata T. Nelson & J. Nelson (p. 963) ASH CREEK IVESIA Pl matted, grayish green; caudex much-branched. **ST** prostrate in fr, gen 5–15 cm. **LF** gen 2–5 cm; sheathing bases densely strigose; lflets gen 8–15 per side, lobes gen > 5, < 2 mm, obovate; cauline lf 1. **INFL** ± open; clusters 1–many, 10–20 mm wide, ± loose, ± 10–20-fld; pedicels 2–6 mm, ± S-shaped in fr. **FL** ± 5 mm wide; hypanthium length ± 1/2 width; petals ± 1 mm, linear, white to pale yellowish, < sepals; stamens 5; pistils 1–3. **FR** 1–1.5 mm, smooth, brown. RARE. Shallow, rocky soil, open sagebrush; 1500–1700 m. MP (Ash Valley, n-c Lassen Co.). Related to *I. rhypara* B. Ertter & Rev. of OR, NV (lflet lobes 2–4, gen > 2 mm). In cult.

I. patellifera (J. Howell) B. Ertter (p. 963) KINGSTON MOUNTAINS IVESIA Pls hanging clumps, green. **ST** 10–20 cm. **LF** gen 5–12 cm, flat; sheathing bases glabrous; lflets 2–3 per side, separated, 5–20 mm, ± round, evenly toothed or lobed < 1/2 to base, cauline lvs 0–2. **INFL** open, few–many-fld; pedicels 5–30 mm, gen ± S-shaped in fr. **FL** 5–10 mm wide; hypanthium length < 1/2 width; bractlets gen 0; petals 2–3 mm, narrowly oblanceolate, yellow, < sepals; stamens 5–10; pistils 4–10. **FR** 1.5–2 mm, ± ridged, pale. RARE. Granite crevices; 1400–2100 m. e DMtns (Kingston Mtns). [*Potentilla p.* J. Howell, Kingston Mtns cinquefoil]

I. pickeringii A. Gray PICKERING'S IVESIA Pl tufted, grayish, long-spreading-hairy; caudex 0–few-branched. **ST** ascending to erect, 30–50 cm. **LF** 10–20 cm; sheathing bases strigose; lflets 35–50 per side, overlapped but distinct, lobes gen 3–5, 2–5 mm, oblanceolate to obovate; cauline lvs 5–10. **INFL** open; fls many, gen separate; pedicels 2–10 mm, straight. **FL** ± 10 mm wide; hypanthium length ± = width; petals 3–5 mm, ± oblanceolate, white or pink-tinged, > sepals; stamens 20; pistils 2–4. **FR** 2.5–3 mm, smooth, dark brown. $2n=28$. RARE. Wet, rocky meadows, gen on serpentine clay; 800–1500 m. c KR. In cult.

I. pygmaea A. Gray Pl ± matted, green; caudex branched. **ST** decumbent to erect, gen 3–15 cm. **LF** 1–10 cm; sheathing bases gen ± strigose; lflets gen 10–15 per side, lobes gen 5–8, gen 1–3 mm, ± widely oblanceolate; cauline lf 1. **INFL** dense to ± open, < 30 mm wide, gen < 10-fld; pedicels 2–10 mm, straight. **FL** ± 10 mm wide; hypanthium length ± < 1/2 width; petals 2–4 mm, widely oblanceolate, yellow, < or = sepals; stamens 10; pistils 10–30. **FR** 1–1.5 mm, smooth, pale. Rocky (granitic) places; 2800–4000 m. c&s SNH.

I. santolinoides A. Gray (p. 963) MOUSETAIL IVESIA Pl tufted, silvery; caudex 0–several-branched. **ST** erect, gen 15–40 cm. **LF** mousetail-like, gen 4–10 cm; sheathing bases densely strigose; lflets 60–80 per side, indistinct, lobes 1–5, < 1.5 mm, obovate to round; cauline lvs 1–3. **INFL** open; fls many, separate; pedicels 5–30 mm, straight. **FL** gen 5–8 mm wide; hypanthium length 1/2 width; petals ± 2 mm, 2 × sepals, obovate to round, white; stamens 15; pistil 1. **FR** ± 2 mm, smooth, mottled gray-brown. $n=14$. Bare places, sandy, granite ledges; 2700–3600 m. SNH, TR, SnJt. ❀DRN,SUN:1–3,16–18;DFCLT.

I. saxosa (E. Greene) B. Ertter (p. 963) Pls hanging clumps, green. **ST** 5–30 cm. **LF** gen 5–15 cm, flat; sheathing bases ± glabrous; lflets 2–4 per side, separated, gen 5–15 mm, ± round, shallowly and evenly toothed to unevenly lobed ± to base; cauline lvs 2–4. **INFL** open, few–many-fld; pedicels 7–30 mm, gen ± S-shaped in fr. **FL** 7–10 mm wide; hypanthium length < 1/2 width; petals 2–4 mm, oblanceolate to obovate, yellow, < sepals; stamens 15–40; pistils 3–20. **FR** 1–1.5 mm, ± ridged, pale. Granitic or volcanic crevices; 900–3000 m. s SNH, SnBr, PR, SNE, DMtns; n Baja CA. [*Potentilla s.* E. Greene incl ssp. *sierrae* Munz] Stamen number, lf lobing, hairiness variations apparently not useful taxonomically.

I. sericoleuca (Rydb.) Rydb. (p. 963) PLUMAS IVESIA Pl tufted, greenish to white-hairy; caudex 0–few-branched. **ST** gen decumbent to ascending, 15–45 cm. **LF** 10–20 cm; sheathing bases densely strigose; lflets 20–35 per side, lobes 1–4, 3–15 mm, elliptic to oblanceolate; cauline lvs 3–8. **INFL** open; clusters ± 5–20, 10–20 mm wide, ± head-like, ± 5–10-fld; pedicels < 3 mm (exc lowest), straight. **FL** 10–15 mm wide; hypanthium length ± 1–2 × width; petals 4–7 mm, widely obovate to obcordate, white, > sepals; stamens gen 20; pistils 2–7. **FR** 2–3 mm, smooth, brown. $n=14$. RARE. Dry, gen volcanic meadows; 1500–2200 m. n SNH, s MP (Sierra Valley). Pls from Truckee River drainage gen less hairy, infl more open, fl smaller. Related to *I. pityocharis* B. Ertter (NV, flatter hypanthium, 8–20 pistils).

I. shockleyi S. Watson var. ***shockleyi*** Pl matted, green; caudex much-branched. **ST** spreading, < 15 cm. **LF** gen 2–8 cm; sheathing bases strigose; lflets gen 5–10 per side, lobes 3–5, 1–5 mm, oblanceolate to obovate; cauline lf 1, gen ± bract-like. **INFL** open, gen < 10-fld; pedicels 4–10 mm, S-shaped in fr. **FL** 5–10 mm wide; hypanthium length 1/2 width; petals ± 2 mm, oblanceolate, yellow, < or = sepals; stamens 5; pistils 2–5(6). **FR** ± 2 mm, smooth, light brown. Rocky areas; 2700–4000 m. n&c SNH, W&I; se OR, c&n NV. *I. cryptocaulis* (Clokey) Keck (sw NV) has open, buried caudex. ❀DRN,IRR,SUN:1–3,16,17;DFCLT.

I. unguiculata A. Gray YOSEMITE IVESIA Pl rosetted, silvery- to gray-hairy; caudex gen simple. **ST** ascending, gen 10–30 cm. **LF** 7–15 cm; sheathing bases glabrous; lflets gen 15–25 per side, lobes 3–8, gen 3–5 mm, linear to oblanceolate; cauline lvs 3–6. **INFL**: clusters 1–several, ± 10 mm wide, head-like, gen < 10-fld; pedicels < 3 mm, straight. **FL** 6–9 mm wide; hypanthium length ± = width; petals 3–4 mm, ± obovate, white or pink-tinged, > sepals; stamens 10–15; pistils 3–9. **FR** ± 1.5 mm, smooth, light brown. RARE. Meadows; 1500–2500 m. c&s SNH. In cult.

I. webberi A. Gray WEBBER'S IVESIA Pl rosetted, green; caudex gen simple. **ST** decumbent to ascending, gen 5–15 cm. **LF** 3–7 cm; sheathing bases strigose; lflets 4–8 per side, lobes 5–12, 3–10 mm, linear to lanceolate; cauline lvs 2, ± opposite (unique in *Ivesia*). **INFL**: cluster 1, 15–50 mm wide, head-like (open in fr), 5–15-fld; pedicels gen 2–5 mm, straight. **FL** ± 10 mm wide; hypanthium length 1/2 width; petals gen 2–3 mm, oblanceolate, yellow, < sepals; stamens 5; pistils ± 5. **FR** 2.5 mm, ± smooth, light brown, mottled darker. RARE. Rocky clay in sagebrush flats; 1500–1900 m. n SNH (Dog Valley, e Sierra Co.), s MP (Sierra Valley); w NV.

LUETKEA

Thomas J. Rosatti

1 sp. (Count F.P. Luètke, Russian sea captain, 1797–1882)

L. pectinata (Pursh) Kuntze (p. 963) Subshrub 5–15 cm, ± prostrate. **LVS** ± sessile, unstipuled; lower lvs linear; upper lvs 5–15 mm, ± 2-ternately dissected; segments linear, pointed, ribbed below, grooved above. **INFL**: raceme, ± scapose, 1–7 cm, narrow; bracts linear to ternately dissected. **FL**: hypanthium hemispheric, bractlets 0; sepals ± 2 mm, ovate; petals ± 3 mm, ± obovate, white; stamens ± 20; ovaries 4–6, free, superior. **FR**: follicles 4–6, ± 4 mm, leathery, dehiscent along both sutures. **SEEDS** > 1, ± fusiform, flat, smooth. Moist slopes, often near snow, coniferous forests; 1800–2800 m. KR, CaRH; to AK. ❀DRN,IRR:4–6&SHD: 1,2,7,15,16;DFCLT.

LYONOTHAMNUS CATALINA IRONWOOD

Dieter H. Wilken

1 sp. (Greek: Lyon's shrub, W.S. Lyon, early resident of Los Angeles)

L. floribundus A. Gray Tree, evergreen. **ST**: trunk bark gray to reddish brown, peeling in strips. **LVS** opposite; stipules deciduous; petiole 1–3 cm; blade dark, shiny green above, ± grayish below. **INFL**: panicle, gen flat-topped; bractlets 1–3. **FL**: hypanthium bell-shaped, tomentose; sepals persistent; petals ± round, white; stamens ± 15; pistils 2, ovary superior, style stout, stigma ± head-like. **FR**: follicles 2, 3–4 mm, woody. **SEEDS** 1–4, ± 2 mm, compressed, brownish. Rocky slopes, canyons, oak woodland, chaparral; 20–500 m. ChI. Sspp. hybridize in cult.

1. Lf gen compound .... ssp. ***asplenifolius***
1' Lf gen simple .... ssp. ***floribundus***

ssp. ***asplenifolius*** (E. Greene) Raven (p. 963) FERN-LEAVED CATALINA IRONWOOD Pl 4–12 m. **LF** 9–21 cm, palmately to pinnately compound; 1° lflets gen 3–7, 5–14 cm, linear; 2° lflets gen 20–30, 4–12 mm, subopposite. $2n$=54. RARE. Habitats of sp.; 20–500 m. ChI (Santa Cruz, Santa Rosa, San Clemente islands). [var. *a.* (E. Greene) Brandegee] Threatened by grazing. In cult.

ssp. ***floribundus*** (p. 963) SANTA CATALINA ISLAND IRONWOOD Pl 5–15 m. **LF** 7–17 cm, simple, linear to oblong, gen entire (crenate to irregularly lobed). $2n$=±48. RARE. Habitats of sp.; 100–500 m. s ChI (Santa Catalina Island). Threatened by feral animals. Seedling lvs often compound. In cult.

MALUS APPLE

Thomas J. Rosatti

Tree or shrub, sometimes thorny. **LVS** gen toothed (lobed). **INFL**: clusters at ends of short lateral branches. **FL**: bractlets on hypanthium 0; stamens many; ovary inferior, chambers (2–)5, 2-ovuled, styles (2–)5, ± fused at base. **FR**: pome. ± 25 spp.: n temp. (Classical name of apple) Often incl in *Pyrus*, pear.

1. Some main lvs gen lobed; fr 10–15 mm, oblong, yellow to gen purple-red or -black; native .... ***M. fusca***
1' Main lvs unlobed; fr gen > 30 mm, round, ± red; escape from cult .... ***M. sylvestris***

M. fusca (Raf.) C. Schneider (p. 963) OREGON CRAB APPLE Shrub or small tree. **LF** 3–12 cm, gen widely lanceolate; teeth gen sharp; petiole 15–50 mm; some main lvs gen lobed. **FL**: petals white. **FR** 10–15 mm, oblong, yellow to gen purple-red or -black; pedicel 20–30 mm. n=17. Moist, open coniferous forest; < 800 m. NCo, NCoRO, CaRF, n CCo; to AK.

M. sylvestris Miller APPLE Tree. **LF** 5–11 cm, elliptic to widely ovate; teeth gen blunt; petiole < 20 mm; main lvs unlobed. **FL**: petals white or pink. **FR** gen > 30 mm, round, ± red; pedicel 10–25 mm. $2n$=34,51. Disturbed places; gen < 1000 m. n CA; native to Eurasia.

NEVIUSIA SNOW-WREATH

Barbara Ertter

Shrub, ± strigose. **LVS** alternate, deciduous; blade shallowly lobed, toothed; stipules linear. **INFL** umbel-like. **FL**: hypanthium flat, bractlets 0; sepals 5–6, ± toothed; petals 0 or 5–6; stamens 50–100+, filaments showy, white, ± expanded; pistils 2–6, ovary superior, strigose, chamber 1, ovule 1. **FR**: achene, wall ± soft. 2 spp.: CA, se US. (Reverend R.D. Nevius, 1827–1913)

N. cliftonii J.R. Shevock, B. Ertter, & D. Taylor (p. 967) Pl 5–25 dm, erect. **ST**: branches slender. **LF**: petiole gen 4–10 mm; blade 20–60 mm, ovate to cordate, lobes sharply toothed, sparsely strigose. **INFL**: fls gen 3–5; pedicels 1–3 cm. **FL**: sepals 4–6 mm, ± obovate; petals 4–6 mm, clawed, white; stamens ± 50, filaments 4– 5 mm. **FR** 3–4 mm. Uncommon. Shaded, n-facing, limestone slopes; 300–500 m. CaR (near Lake Shasta).

OEMLERIA OSO BERRY

Thomas J. Rosatti

1 sp. (derivation unknown) [Landon 1975 Taxon 24:200]

O. cerasiformis (Hook. & Arn.) J.W. Landon (p. 967) Shrub or small tree, 1–5 m, ± dioecious. **LVS** deciduous, 5–13 cm, elliptic to narrowly obovate, gen entire, ± rolled under, paler below, glabrous above; stipules soon deciduous; petiole 5–15 mm. **INFL**: racemes on short lateral branches, 3–10 cm, pendent, bracted; bractlets 1–2 at pedicel tip. **FLS** fragrant (gen some bisexual); hypanthium ± 5 mm; petals clawed, white. **STAMINATE FL**: hypanthium persistent; petals spreading, > pistillate petals; stamens 15, prominent. **PISTILLATE FL**: most of hypanthium deciduous; petals erect; ovaries gen 5, prominent, free, superior, styles deciduous. **FR**: drupes 1–5 per pistillate fl, 5–15 mm, bean-shaped, blue-glaucous. n=8. Shaded coniferous forest, chaparral; < 1700 m. NW, w CaR, w SN, ScV (Sutter Buttes), s-c SnJV, w CW, sw WTR; to B.C. [*Osmaronia c.* (Hook. & Arn.) E. Greene] ❀partSHD:4,**5**,**6**,**15–17**&IRR:**7**,8,9,**14**,18–24.

PERAPHYLLUM WILD CRAB APPLE

Thomas J. Rosatti

1 sp. (Greek: very leafy)

P. ramosissimum Nutt. (p. 967) Shrub 1–3 m, much-branched. **LVS** ± clustered on short lateral branches, deciduous, 2–4 cm; stipules soon deciduous; petiole 0–short; blade ± oblanceolate, pointed, strigose, entire to minutely gland-toothed. **INFL**: fls 1–3, pedicelled, on short lateral branches. **FL**: hypanthium ± funnel-shaped, bractlets 0; sepals spreading to reflexed, 3–5 mm, hairy at least inside, persistent; petals spreading, 6–8 mm, ± obovate, white to rose; stamens ± 15–20, free; ovary 1, inferior, 4(6)-chambered, styles 2(3). **FR**: pome, 8–10 mm, spheric, yellowish to reddish. Dry washes, sagebrush scrub, pinyon/juniper woodland, pine forest; 1200–2500 m. e CaRH, n&c SNH (e slope), GB; to OR, Colorado, NM. ❀DRN,SUN:1,**2**,15–17&IRR:**3**,**7**,14,**18**,19–21.

PETROPHYTON ROCK SPIRAEA

Thomas J. Rosatti

Shrub, matted, scapose. **LVS** crowded, evergreen, gen ± oblanceolate, entire. **INFL** ± spike-like. **FL**: bractlets on hypanthium 0; sepals persistent; petals white; stamens 20–40; pistils gen 5, simple, ovary superior, hairy, styles thread-like. **FR**: follicles, dehiscing along both sutures. **SEEDS** 1–several, linear. ± 4 spp.: w N.Am. (Greek: rock pl)

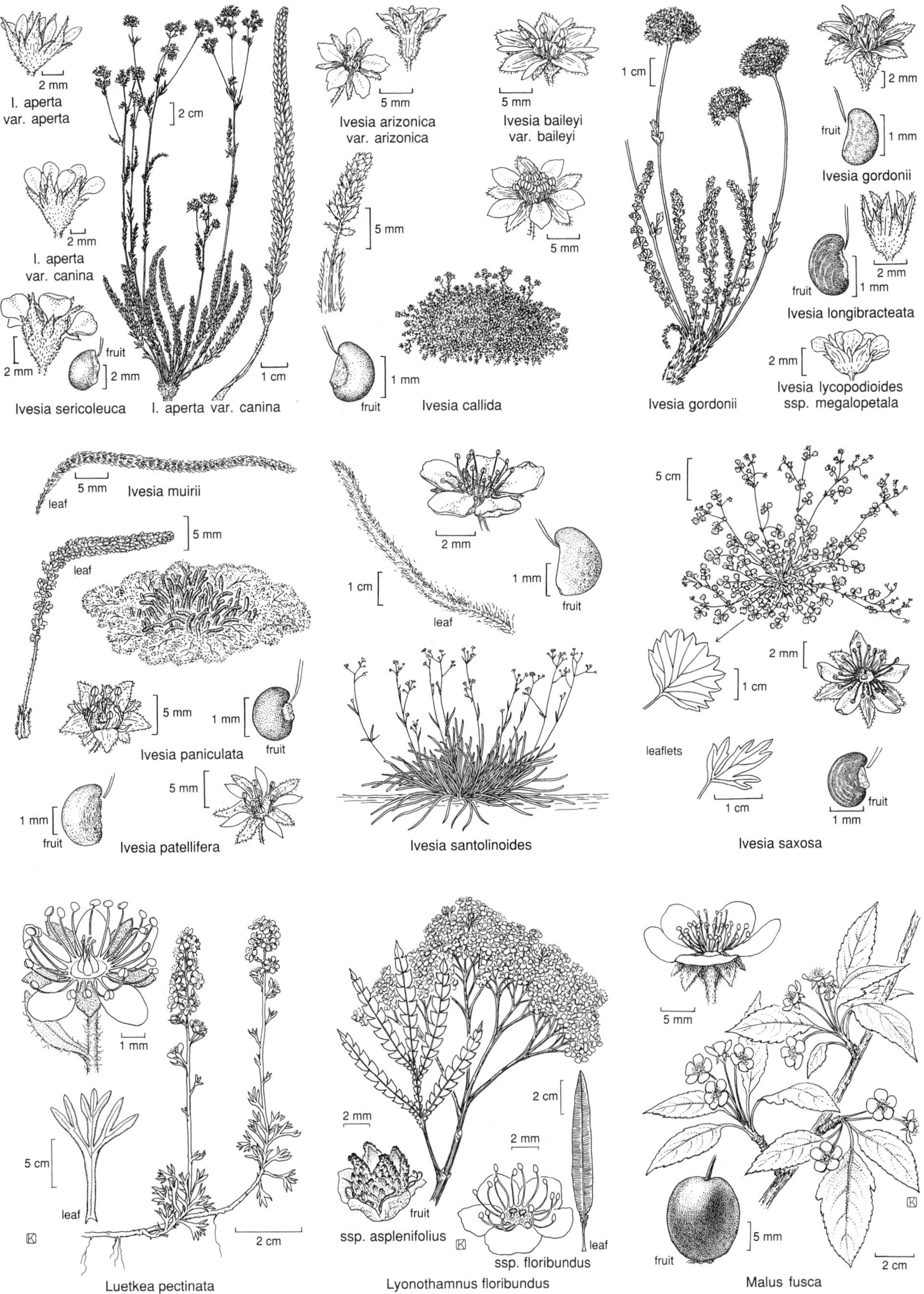
2 mm
I. aperta
var. aperta
2 cm
2 mm
I. aperta
var. canina
2 mm
fruit
2 mm
1 cm
Ivesia sericoleuca
I. aperta var. canina
5 mm
Ivesia arizonica
var. arizonica
5 mm
Ivesia baileyi
var. baileyi
5 mm
5 mm
1 mm
fruit
Ivesia callida
1 cm
Ivesia gordonii
2 mm
fruit
1 mm
Ivesia gordonii
2 mm
fruit
1 mm
Ivesia longibracteata
2 mm
Ivesia lycopodioides
ssp. megalopetala
5 mm
Ivesia muirii
leaf
5 mm
leaf
5 mm
1 mm
fruit
Ivesia paniculata
5 mm
1 mm
fruit
Ivesia patellifera
2 mm
1 cm
leaf
1 mm
fruit
Ivesia santolinoides
5 cm
2 mm
1 cm
leaflets
1 cm
fruit
1 mm
Ivesia saxosa
1 mm
5 cm
leaf
2 cm
Luetkea pectinata
2 mm
fruit
ssp. asplenifolius
2 mm
2 cm
leaf
ssp. floribundus
Lyonothamnus floribundus
5 mm
5 mm
fruit
2 cm
Malus fusca

P. caespitosum (Nutt.) Rydb. (p. 967) Pl 3–8 dm wide; rosettes many. **STS** very stout. **LF** 1–3-veined below. **INFL** 4–14 cm; peduncle 3–10 cm, bracted. **FL**: sepals ± 1.5 mm, narrowly ovate, acute; petals ± 1.5 mm, gen obtuse; style ± 3 mm. **FR** ± 2 mm. **SEEDS** 1–2, ± 1.5 mm, linear to obovoid, brown, smooth. Limestone soils, pinyon/juniper woodland, coniferous forest; 1200–3000 m. s SNH, W&I, DMtns; to Rocky Mtns.

1. Lf 10–18 mm, sparsely hairy; s SNH .... ssp. ***acuminatum***
1′ Lf 5–12 mm, densely hairy; W&I, DMtns .. ssp. ***caespitosum***

ssp. ***acuminatum*** (Rydb.) Munz **LF** 10–18 mm, sparsely hairy. Limestone cliffs, coniferous forest; 1200–2300 m. s SNH. ❀TRY;DFCLT.

ssp. ***caespitosum*** **LF** 5–12 mm, densely hairy. Limestone ledges, rocks, often in pinyon/juniper woodland; 1500–3000 m. W&I, DMtns; to Rocky Mtns. ❀DRN,SUN:1,2,7,14–16;DFCLT.

PHYSOCARPUS NINEBARK

Thomas J. Rosatti

Shrub, gen ± stellate-hairy. **LVS** stipuled, petioled, deciduous; blade ovate to ± round in CA, gen palmately 3–7-lobed, lobes crenate to serrate. **INFL** ± umbel-like, bracted. **FL**: hypanthium bell-shaped, bractlets 0; sepals persistent; petals rounded, white; stamens 20–40; pistils 1–5, free or ± fused, ovary superior, style thread-like, stigma head-like. **FR**: follicles, inflated, often opening along both sutures. **SEEDS** 2–4, ovoid; coat hard, shiny. ± 10 spp.: N.Am, Asia. (Greek: bladdery fr)

1. Lf blade 5–20 mm, petiole 5–10 mm; pistil gen 1; stamens alternately short and long; W&I, n DMtns .. ***P. alternans***
1′ Lf blade 30–140 mm, petiole 5–20+ mm; pistils 3–5, fused at base; stamens ± equal; CA-FP (exc GV) ***P. capitatus***

P. alternans (M.E. Jones) J. Howell (p. 967) Pl 5–15 dm. **LF**: petiole 5–10 mm; blade 5–20 mm, gen densely hairy, lobes 3–7, gen crenate. **FL**: sepals ± 3 mm; petals 3–4 mm; stamens alternately short and long; pistil gen 1. **FR** hairy. Dry, rocky pinyon/juniper woodland; 1800–3100 m. W&I, n DMtns; to UT. [sspp. *annulatus* J. Howell and *panamintensis* J. Howell] ❀TRY;DFCLT.

P. capitatus (Pursh) Kuntze (p. 967) Pl 10–25 dm. **LF**: petiole 5–20+ mm; blade 30–140 mm, often ± glabrous; lobes gen 3–5, gen serrate (if 0, lf coarsely ± 9–11-toothed). **FL**: sepals 2.5–3 mm; petals 2.5–3 mm. **FR** glabrous to ± hairy. *n*=9. Moist banks, n-facing slopes, coniferous forests; < 1400 m. CA-FP (exc GV); to AK, MT, UT. ❀**4–6,17**&IRR,partSHD:1,2,7,8,9,**14–16,18–24**;CV:decid.GRCVR.

POTENTILLA CINQUEFOIL

Barbara Ertter

Ann to shrub; odor resinous or 0. **LVS** gen basal, odd-1-pinnate to 1-palmate or 1-ternate; lflets ± toothed or lobed, terminal gen ± = lateral. **INFL**: cyme, gen ± open; pedicels gen ± straight. **FL**: hypanthium ± shallow; bractlets 5; sepals gen 5, ± triangular; petals gen 5, gen = or > sepals, gen ± widely obcordate, gen yellow; stamens gen 20; pistils gen many, styles gen jointed near tip. **FR**: achene. 200–500 spp.: n temp. (Latin: diminutive of powerful, for reputed medicinal value) [Clausen, Keck, & Hiesey 1940 Carn Inst Wash Pub 520:26–195] *P. anglica* Laicharding is a waif from cult: st trailing; pedicels 2–8 cm, slender; sepals and petals gen 4; lflets 3–5.

1. Shrub; fr densely hairy .... ***P. fruticosa***
1′ Herb; fr glabrous (often among receptacle hairs)
 2. Petals dark red, narrowly ovate, << sepals — pl openly matted from stolons, often floating; bogs, marshes .... ***P. palustris***
 2′ Petals yellow to white, elliptic to obcordate, < to > sepals
  3. Fls solitary from stolons; lf pinnate, main lateral lflets alternating with reduced lflets .... ***P. anserina***
   4. Lf 3–15 cm, gen densely hairy above; pedicel gen 2–7 cm; 1200–2500 m; s SNH, SnBr, GB ..... ssp. ***anserina***
   4′ Lf 3–75 cm, ± glabrous above; pedicel gen 5–30 cm; gen < 150 m; NCo, CCo, SCo .... ssp. ***pacifica***
  3′ Fls in cymes (stolons 0); lf pinnate, palmate, or ternate; reduced lflets 0
   5. Styles attached below middle of fr, widest near middle — lf pinnate, terminal lflet > others ..... ***P. glandulosa***
    6. Styles 1.5–2.5 mm, ± slender; fls gen not opening fully; sheathing lf base gen strigose
     7. St gen 5–20 cm; hairs sparse, some glandular; terminal lflet gen 5–15 mm, < 10-toothed; SnGb .. ssp. ***ewanii***
     7′ St 20–50 cm; hairs gen ± dense, ± nonglandular; terminal lflet gen 15–30 mm, > 10-toothed; n CA .... ssp. ***globosa***
    6′ Styles gen < 1.5 mm, gen swollen; fl opening fully; sheathing lf base often glabrous
     8. Petals ± elliptic-(ob)ovate; sepals ± = or > petal, gen ± obtuse; st and pedicel hairs gen glandular; lflet gen double-toothed; gen < 2600 m
      9. Petals ± elliptic-ovate, 3.5–6 mm wide, cream to pale yellow; infl appearing ± leafy (bracts ± 1/2 subtended branches); gen < 1200 m .... ssp. ***glandulosa***
      9′ Petals narrowly elliptic-obovate, 2–3.5 mm wide, yellow; infl not appearing very leafy (bracts < 1/2 subtended branches); gen 500–2600 m .... ssp. ***reflexa***
     8′ Petals ± obovate to round; sepals < petals, gen ± acute; st and pedicel hairs glandular or not; lflet gen single-toothed; > 900 m
      10. St gen 5–20 cm; lflet teeth < 10; sheathing lf base gen appressed-hairy — gen glandular above; rocky areas, 2400–3800 m .... ssp. ***pseudorupestris***
      10′ St gen > 20 cm; lflet teeth > 10; sheathing lf base gen glabrous
       11. Petals ± yellow; infl branches spreading (angle gen 20–40°), sometimes glandular above; NW, CaR, n SNH, MP .... ssp. ***ashlandica***
       11′ Petals cream; infl branches ascending (angle gen 10–30°), nonglandular above; CaR, SN, TR, SnJt, SNE

12. St-base hairs < 1 mm, glandless; st gen 20–60 cm; lflet teeth gen single; 1800–3700 m .. ssp. ***nevadensis***
12′ St-base hairs gen 2–3 mm, glandular; st 40–90 cm; lflet teeth gen double; 1200–2200 m .... ssp. ***hansenii***
5′ Styles attached just below tip of fr, slender throughout or widest near base
13. Styles < or ± = 1 mm, tapered from rough-thickened base to tip
14. Per; basal lvs gen present in fl; rocky alpine barrens, > 2700 m — lflets densely hairy, at least below
15. Lf ± pinnate; lflets 5–11, lobed ± 2/3 to midvein .... ***P. pensylvanica***
15′ Lf gen ± palmate; lflets ± 5, lobed > 3/4 to midvein .... ***P. pseudosericea***
14′ Ann (or short-lived per); basal lvs often withered or fallen in fl; gen moist or disturbed areas, < 3100 m
16. Lf pinnate, lflets 9–17; petals > sepals, cream, ± obcordate — lflets lobed > 1/2 to widvein; MP .... [2]***P. newberryi***
16′ Lf gen ± palmate or ternate, lflets 3–7; petals < sepals, yellow fading white, gen elliptic to obovate
17. St-base hairs gen ± dense, glandular or not, < 1 mm, spreading or not; petals < 2.5 mm; hypanthium gen 2–4 mm wide; fr smooth, < 1 mm, whitish or light brown; native
18. St hairs glandular and not; lflets 3, ± obovate; petals oblanceolate-elliptic .... ***P. biennis***
18′ St hairs glandless; lflets 3–5, oblanceolate to elliptic; petals obovate-elliptic .... ***P. rivalis***
17′ St-base hairs sparse, glandless, < 2 mm, spreading; petals > 3 mm; hypanthium 3–10 mm wide; fr veined, ± 1–1.5 mm, ± brown; alien
19. Lflets gen 3; petals 3–4 mm; anthers ± 0.3 mm; ann or short-lived per .... ***P. norvegica***
19′ Lflets gen 6–7; petals 6–9 mm; anthers ± 1 mm; short-lived per .... [2]***P. recta***
13′ Styles > 1 mm (exc *P. recta*), gen slender — petals ± obcordate, > sepals
20. Lflets 3; infl gen < 7-fld — st ± ascending to erect, gen < 30 cm (see also *Fragaria, Sibbaldia*)
21. Lflets lobed ± 1/2 to base, sparsely glandular; styles ± rough-thickened at base; KR .... ***P. cristae***
21′ Lflets not lobed, ± nonglandular; styles slender throughout; KR, CaRH, SNH
22. Teeth of central lflet gen > 7, gen uneven, ± 1/4 to base; pedicel 10–20 mm; pistils > ± 20; st ± glabrous or sparsely spreading-hairy; KR, CaRH, SNH .... ***P. flabellifolia***
22′ Teeth of central lflet 7, even, 1/4–1/2 to base; pedicel 10–40 mm; pistils < ± 20; st sparsely strigose; SNH .... ***P. grayi***
20′ Lflets > 3; infl often > 7-fld
23. Lf pinnate to subpalmate (see also *Horkelia, Horkeliella, Ivesia*)
24. Ann to short-lived per; petals cream; styles 1–1.5 mm; fr veined — lflet lobes oblanceolate; receding shorelines; MP .... [2]***P. newberryi***
24′ Per; petals yellow; styles > 1.5 mm; fr gen smooth
25. Lf densely white-tomentose below, green and strigose above; styles ± 1.5–2 mm — alpine barrens; n W&I .... ***P. morefieldii***
25′ Lf surfaces similarly hairy (but hairs often denser below); styles 2–3 mm
26. Basal lf gen > 1/2 st; pedicel gen ± recurved in fr; < 2000 m
27. Lateral lflets toothed < 1/2 to base; pistils ± 5; st 20–50 cm — SCo, brackish meadows, presumed extinct .... ***P. multijuga***
27′ Lateral lflets toothed ± 1/2 or lobed ± fully to base; pistils gen > 10; st 5–45 cm
28. Lflet ± 4-toothed ± 1/2 to base; pistils gen ± 10; pl ± subglabrous; c CCo, s NCoRO?, < 100 m .... ***P. hickmanii***
28′ Lflet 3–10-lobed > 1/2 to base; pistils 10–30; pl sometimes hairy; CaRH, GB, adjacent n SNH, 900–2000 m .... ***P. millefolia***
26′ Basal lf < 1/2 st; pedicel ± straight in fr; 1100–3700 m .... ***P. drummondii***
29. Petiole gen < blade, cottony-hairy; lflets 3–7 per side, ± overlapped, palmately lobed > 1/2 to base — gen white-hairy .... ssp. ***breweri***
29′ Petiole gen > blade, glabrous to shaggy-hairy; lflets gen 2–4 per side, sometimes separated, pinnately toothed ± 1/2 to midvein, sometimes split to base
30. Lflets gen gray-hairy, ± overlapped; petiole gen densely straight- or shaggy-hairy ...... ssp. ***bruceae***
30′ Lflets green, glabrous to ± hairy, gen separated; petiole glabrous or strigose .... ssp. ***drummondii***
23′ Lf palmate
31. St ± prostrate or hanging, gen < 25 cm; infl not flat-topped, pedicels often ± recurved in fr; lflet teeth < 10 — > 1800 m, s CA-FP, W&I
32. Lflets densely white-tomentose below, green and strigose above; infl gen 1–5-fld — ± 3200 m; W&I .... ***P. concinna***
32′ Lflet surfaces ± equally strigose; infl gen > 5-fld
33. Pl hanging from rock crevices; frs gen 5–20, ± smooth; pedicels gen > 15 mm; SnJt .... ***P. rimicola***
33′ Pl rosetted to tufted in sandy soil; frs gen > 15, ± veined; pedicels gen < 15 mm; s SNH, SnBr, SnJt .... ***P. wheeleri***
31′ St ± ascending, gen 10–100 cm; infl flat-topped, pedicels ± straight; lflet teeth often > 10
34. Styles ± 1 mm, gen > 50; basal lvs gen fallen or withered in fl; st short- and long-spreading-hairy; fr veined; bractlets ± = sepals; alien .... [2]***P. recta***
34′ Styles gen 1.5–2 mm, < 50; basal lvs gen prominent in fl; st glabrous or hairs of 1 length; fr smooth; bractlets ± 1/2–2/3 sepals; native
35. Lflet base gen untoothed; st 10–40 cm, base glabrous; anthers 0.4–0.9 mm; fr 1.5 mm; 2600–3500 m; c&s SNH .... ***P. diversifolia*** var. ***diversifolia***
35′ Lflet gen toothed or lobed ± throughout; st 20–100 cm, base hairy; anthers 0.6–1.6 mm; fr 1–1.5 mm; 120–3500 m; widespread .... ***P. gracilis***

36. Lflet lobed > 3/4 to midvein; st and petiole hairs gen appressed
37. Longest petiole gen 5–10 cm; central lflet 2–6 cm, lobes often narrowest at base, entire; petals 5–8 mm; widespread . . . . . var. ***elmeri***
37′ Longest petiole gen 10–25 cm; central lflet 5–9 cm, lobes gen widest at base, often few-toothed; petals 7–10 mm; MP . . . . . var. ***flabelliformis***
36′ Lflet toothed ± 1/2 or less to midvein; st and petiole hairs spreading to appressed
38. Lf surfaces similarly hairy, not tomentose; petals gen 4–7 mm; longest petiole gen 5–15 cm; widespread, 800–3500 m . . . . . var. ***fastigiata***
38′ Lf surfaces very different, gen tomentose below, dark and ± glabrous above; petals 7–10 mm; longest petiole 5–25 cm; nw KR, n NCoRO, 120–1100 m . . . . . var. ***gracilis***

P. anserina L. (p. 967) Pl tufted from stolons, nonglandular. **LF** pinnate, 3–50(75) cm; main lflets 5–10 per side (reduced lflets alternating), 10–50 mm, ± elliptic to oblanceolate, toothed 1/2–1/3 to midvein, densely hairy (at least below). **INFL**: fls solitary from stolon nodes. **FL**: hypanthium 4–7 mm wide; petals 7–20 mm; filaments 1–3.5 mm, anthers ± 1 mm; styles ± 2 mm, slender. **FR** ± 2 mm, rough, dark red-brown. $2n$=28,42. Wet, often brackish areas; 0–2500 m. NCo, s SNH, CCo, SCo, SnBr, GB; circumboreal. [Rousi 1965 Ann Bot Fenn 2:47–112]

ssp. ***anserina*** **LF** 3–15 cm, gen densely hairy above; main lflets 10–25 mm. **INFL**: pedicel gen 2–7 cm. **FL**: petals 7–10 mm; pistils 10–50. $2n$=28,42. Habitats of sp.; 1200–2500 m. s SNH, SnBr, GB; circumboreal. [var. *sericea* Hayne] ❀TRY.

ssp. ***pacifica*** (Howell) Rousi **LF** 3–50(75) cm, ± glabrous above; main lflets 10–50 mm. **INFL**: pedicel gen 5–30 cm. **FL**: petals 8–20 mm; pistils gen 20–200. $2n$=28. Wetlands; gen < 150 m. NCo, CCo, SCo; coastal N.Am, Asia. [*P. egedii* Wormsk. var. *grandis* (Torrey & A. Gray) J. Howell] ❀IRR or WET:7–9,**14,19–23**&SUN:**4–6,15–17,24**;GRNCVR;INV; also STBL.

P. biennis E. Greene (p. 967) Ann or bien, taprooted, glandular. **ST** ascending to erect, 10–70 cm, ± spreading-hairy. **LF** ternate; basal often withered in fl; cauline gen 2–10 cm, lflets 3, 5–30 mm, ± obovate, toothed ± 1/4 to midvein, ± hairy. **INFL** few–many-fld. **FL**: hypanthium 2–4 mm wide; petals 1–2.5 mm, < sepals, oblanceolate-elliptic; stamens 10, filaments 0.5–1 mm, anthers ± 0.2 mm; style ± 0.7 mm, tapered from rough-thickened base. **FR** ± 0.6 mm, smooth, whitish. Moist shores; 1500–3100 m. SNH, TR, GB, DMtns; w N.Am.

P. concinna Richardson (p. 967) ALPINE CINQUEFOIL Pl tufted from branched caudex; glands 0 or hidden. **ST** prostrate to decumbent, 3–15 cm, ± ascending-hairy. **LVS** palmate; basal gen 1.5–5 cm, lflets gen 5, central lflet 5–20 mm, narrowly obovate, distal 1/2 few-toothed ± 1/2 to midvein, densely white-tomentose below, ± green and strigose above. **INFL** gen 1–5-fld; pedicels often ± recurved in fr. **FL**: hypanthium 2.5–5 mm wide; petals 2.5–5.5 mm; filaments 1–2.5 mm, anthers 0.6–0.9 mm; styles 1.5–2.5 mm, slender. **FR** 1.5–2 mm, smooth, light brown. RARE in CA. Rocky meadows, ridges; ± 3170 m. n W&I (White Mtns); w N.Am. [var. *divisa* Rydb.; *P. beanii* Clokey] Pls intermediate to *P. gracilis* var. *pulcherrima* (Lehm.) Fern. (otherwise unknown from CA) also in n W&I.

P. cristae W. Ferlatte & Strother (p. 967) CRESTED POTENTILLA Pl loosely matted from branched caudex, glandular. **ST** erect, 5–20 cm, spreading-hairy. **LVS** ternate; basal 1.5–9 cm, lflets 3, 5–20 mm, ± round, few-lobed ± 1/2 to base, unevenly toothed, sparsely hairy. **INFL** gen < 7-fld. **FL**: hypanthium 3.5–6 mm wide; petals 3–5.5 mm; filaments 0.7–1.7 mm, anthers ± 0.5 mm; styles ± 1.3–2 mm, ± slender exc ± rough-thickened base. **FR** 1–1.5 mm, smooth, crested, light brown. $2n$=42. RARE. Seasonally moist, often serpentine-like gravels, talus; 1800–2800 m. KR.

P. diversifolia Lehm. var. ***diversifolia*** (p. 967) Pl tufted from few-branched caudex; glands 0 or hidden. **ST** ascending, gen 10–40 cm, base glabrous. **LVS** palmate; basal gen 4–13 cm, lflets gen 5, gen untoothed near base, distally toothed ± 1/4 to midvein, glabrous (exc margins, veins), central lflet gen 15–40 mm, oblanceolate. **INFL** gen < 20-fld. **FL**: hypanthium 3–4 mm wide; petals 6–8 mm; filaments 1–2.5 mm, anthers 0.4–0.9 mm; styles ± 2 mm, slender. **FR** 1.5 mm, smooth, pale brown. $2n$=82-± 101. Gen ± rocky, moist areas; 2600–3500 m. c&s SNH; w N.Am. [*P. glaucophylla* Lehm.] Intergrades with *P. gracilis*. ❀TRY.

P. drummondii Lehm. Pl ± tufted from few-branched caudex; glands gen 0 or hidden. **ST** decumbent to ± erect, gen 10–60 cm, glabrous to cottony. **LVS** pinnate to subpalmate; basal gen 3–25 cm, lflets 2–9 per side, separated or overlapped, 5–50 mm, oblanceolate to obovate, toothed to lobed, glabrous to densely hairy. **INFL** gen < 20-fld. **FL**: hypanthium 3–6 mm wide; petals 5–10 mm; filaments 1–4 mm, anthers 0.7–1 mm; styles 2–3 mm, slender. **FR** 1.5–2 mm, smooth, ± brown. $2n$=64–108,129. Meadows, rocks; 1100–3700 m (highest in SN). NW, CaR, SN, GB; w N.Am. May hybridize with *P. diversifolia*, *P. gracilis*, *P. wheeleri*; sspp. poorly defined.

ssp. ***breweri*** (S. Watson) B. Ertter (p. 967) **LVS**: basal pinnate; petiole gen < blade, cottony; lflets 3–7 per side, ± overlapped, gen 5–20 mm, palmately lobed > 1/2 to base, gen white-hairy. $2n$= ± 72–102. Habitats of sp.; 1500–2300 m (KR) or 2700–3700 m. c KR, SNH, Wrn; to WA, NV. [*P. b.* S. Watson (var. *viridis* Jepson misapplied); *P. millefolia* var. *algida* Jepson] ❀IRR,DRN,SUN:**1**,2,3,**6,7,15–17**,18.

ssp. ***bruceae*** (Rydb.) Keck (p. 967) **LVS**: basal subpalmate; petiole gen > blade, gen densely straight- or shaggy-hairy; lflets ± 3 per side, ± overlapped, gen 15–25 mm, pinnately toothed ± 1/2 to midvein (sometimes also split to base), gen gray-hairy. $2n$=±64–98,129. Habitats of sp.; 1200–3700 m. KR, CaRH, SNH, Wrn; OR. ❀TRY.

ssp. ***drummondii*** (p. 967) **LVS**: basal pinnate to subpalmate; petiole gen > blade, glabrous or strigose; lflets gen 2–4 per side, gen separated, 10–50 mm, pinnately toothed ± 1/2 to midvein (often also split to base), green, glabrous to ± hairy. $2n$=±64–108. Meadows; 1110–3000 m. KR, n&c SNH; to AK. ❀IRR,DRN,SUN:**1,6,7,15–18**.

P. flabellifolia Hook. (p. 967) FAN-FOIL Pl loosely clustered from openly branched caudex, ± nonglandular. **ST** ascending to erect, gen 10–30 cm, glabrous or sparsely spreading-hairy. **LVS** ternate; basal gen 3–12 cm, lflets 3, gen 10–30 mm, widely obovate, gen unevenly > 7-toothed ± 1/4 to midvein, ± glabrous. **INFL** gen 1–5-fld. **FL**: hypanthium gen 3–4.5 mm wide; petals 6– 10 mm; filaments gen 2–3 mm, anthers 0.6–1.5 mm; styles ± 2 mm, slender. **FR** gen 1.2 mm, gen ± smooth, brown or reddish. n=14. Moist meadows; 1700–3700 m. KR, CaRH, SNH; to B.C., MT. ❀WET:**1–3,7**,14,18,19–21&SUN:**4–6,15–17**.

P. fruticosa L. (p. 967) SHRUBBY CINQUEFOIL Shrub < 1 m; glands 0 or hidden. **LF** ± pinnate, 1–3 cm; lflets 2–3 per side, 5–20 mm, linear to narrowly elliptic, entire, ± hairy. **INFL**: fls 1–2 at end of twigs. **FL**: hypanthium 5–8 mm wide; petals 5–10 mm; stamens 20–25, filaments 2–3 mm, anthers ± 1 mm; styles ± 2 mm, attached near middle of fr, narrowly club-shaped. **FR** ± 1.5 mm, densely hairy (unique in CA). n=7,14. Meadows, rocks; 2000–3600 m. KR, CaRH, SNH, Wrn, W&I; circumboreal. ❀DRN,SUN:**4–6**&IRR:1,7,15–17&WET:2,3,14,18–24;DFCLT;CV.

P. glandulosa Lindley Pl gen ± tufted from loosely branched caudex; glandular hairs often many. **ST** ± erect, 5–90 cm, spreading-hairy. **LVS** pinnate; basal 3–30 cm, terminal lflet largest (0.5–12 cm), lateral gen 3–5 per side, ± obovate, toothed 1/4–1/2 to midvein, ± hairy. **INFL** gen 2–30-fld. **FL**: hypanthium 3–6 mm wide; petals 3–10 mm, gen widely ovate, yellow to white; stamens ± 25, filaments 1–3.5 mm (longest opposite sepals), anthers 0.6–1.2 mm; styles ± 1–2.5 mm, attached below middle of fr, ± fusiform and rough. **FR** ± 1 mm, smooth or ± ridged, golden to reddish brown. n=7. Common. Many habitats; < 3800 m. CA-FP (exc GV), GB; w N.Am. Often confused with *Horkelia*; despite much work, sspp. remain poorly defined.

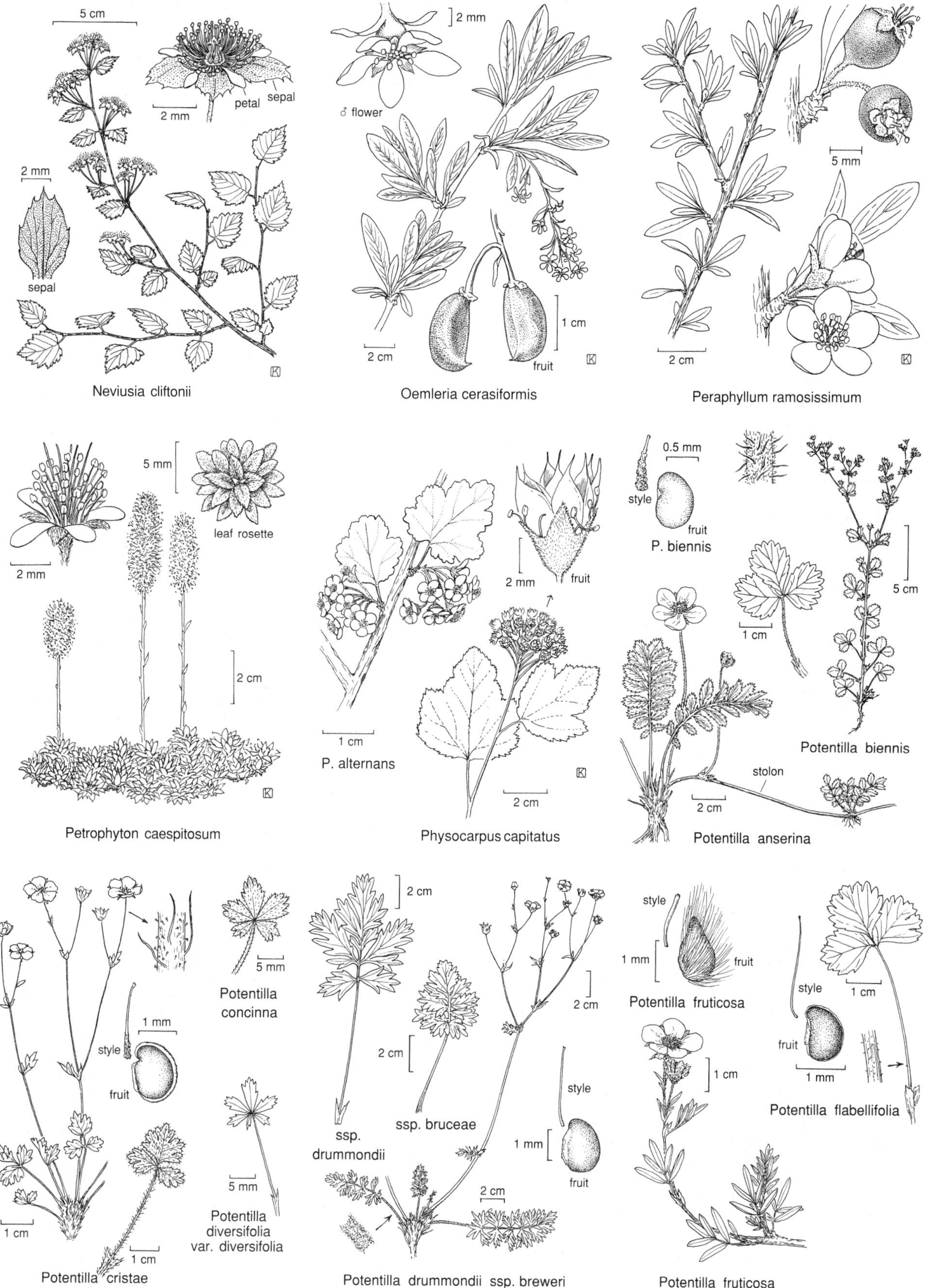

5 cm
2 mm
petal
sepal
2 mm
sepal
Neviusia cliftonii
2 mm
♂ flower
1 cm
2 cm
fruit
Oemleria cerasiformis
5 mm
2 cm
Peraphyllum ramosissimum
5 mm
leaf rosette
2 mm
2 cm
Petrophyton caespitosum
2 mm
fruit
1 cm
P. alternans
2 cm
Physocarpus capitatus
0.5 mm
style
fruit
P. biennis
5 cm
1 cm
Potentilla biennis
stolon
2 cm
Potentilla anserina
5 mm
Potentilla concinna
1 mm
style
fruit
1 cm
1 cm
Potentilla cristae
5 mm
Potentilla diversifolia var. diversifolia
2 cm
2 cm
2 cm
ssp. drummondii
ssp. bruceae
style
1 mm
fruit
2 cm
Potentilla drummondii ssp. breweri
style
1 mm
fruit
Potentilla fruticosa
1 cm
Potentilla fruticosa
style
1 cm
fruit
1 mm
Potentilla flabellifolia

ssp. ***ashlandica*** (E. Greene) Keck (p. 971) **ST** gen 20–60 cm, gen nonglandular-hairy. **LVS**: basal gen 5–20 cm; sheathing base gen glabrous; terminal lflet gen 20–35 mm; lateral lflet teeth ± 10, gen single. **INFL**: branch angle gen 20–40°. **FL**: bractlets ± 1 mm wide; petals gen 5–10 mm, > sepals, ± yellow; styles ± 1 mm. Moist places; 900–2600 m. NW, CaR, n SNH, MP; sw OR. [var. *austiniae* Jepson] Incls most yellow-fld pls formerly in ssp. *nevadensis*. ❀TRY.

ssp. ***ewanii*** Keck (p. 971) EWAN'S CINQUEFOIL **ST** gen 5–20 cm, sparsely ± glandular-hairy. **LVS**: basal gen 3–10 cm, sheathing base gen strigose, terminal lflet gen 5–15 mm; lateral lflet teeth < 10, gen single. **INFL**: branch angle gen ± 50°. **FL** not fully opening; bractlets < 0.5 mm wide; petals 3–4 mm, ± = sepals, yellow; styles 1.5–2.5 mm. RARE. Edges of seeps, small waterways; 1900–2400 m. SnGb (Mount Islip area). [*P. peirsonii* Munz misapplied]

ssp. ***glandulosa*** (p. 971) **ST** gen 20–80 cm, densely glandular-hairy. **LVS**: basal gen 10–25 cm, sheathing base glabrous to glandular, terminal lflet gen 20–60 mm, lateral lflet teeth > 10, gen double. **INFL**: branch angle gen 30–60°. **FL**: bractlets 1–3 mm wide; petals 4.5–7 mm, ± = sepals, ± ovate, cream to pale yellow; styles ± 1–1.5 mm. Gen ± shady or cleared slopes; gen < 1200 m. NW, SNF, CW, SW; to B.C., ID [ssp. *typica* Keck] ❀SUN:**4–6**&IRR:7,8–10,**14–24**.

ssp. ***globosa*** Keck (p. 971) **ST** gen 20–50 cm, nonglandular, gen densely hairy. **LVS**: basal gen 5–12 cm, sheathing base strigose, terminal lflet gen 15–30 mm, lateral lflet teeth > 10, gen double. **INFL**: branch angle gen 20–40°. **FL** not fully opening; bractlets ± 1 mm wide; petals ± 4 mm, gen < sepals, cream; styles 1.5–2 mm. Dry rocky slopes; 1400–2500 m. KR, NCoRH, n CaRH; sw OR. [*P. rhomboidea* Rydb.]

ssp. ***hansenii*** (E. Greene) Keck **ST** 40–90 cm, glandular-hairy below. **LVS**: basal gen 10–25 cm, sheathing base gen glabrous, terminal lflet gen 30–60 mm, lateral lflet teeth > 10, gen double. **INFL**: branch angle gen 10–30°. **FL**: bractlets gen ± 1 mm wide; petals 4–7 mm, > sepals, cream; styles ± 1 mm. Moist meadows; 1200–2200 m. CaR, SN. ❀TRY.

ssp. ***nevadensis*** (S. Watson) Keck (p. 971) **ST** gen 20–60 cm, nonglandular, short-hairy. **LVS**: basal gen 5–20 cm, sheathing base gen glabrous, terminal lflet gen 20–40 mm, lateral lflet teeth gen 10–20, gen single. **INFL**: branch angle gen 10–20°. **FL**: bractlets gen 0.5–1 mm wide; petals 4–8 mm, > sepals, cream; styles ± 1 mm. Gen ± moist, often rocky places; 1800–3700 m. SNH, TR, SnJt, SNE; NV. ❀IRR,DRN:4–6&SHD:1,**2**,**3**,**7**,14,**15–17**.

ssp. ***pseudorupestris*** (Rydb.) Keck **ST** gen 5–20 cm, gen glandular-hairy. **LVS**: basal gen 3–8 cm, sheathing base gen appressed-hairy, terminal lflet gen 5–20 mm; lateral lflet teeth < 10, gen single. **INFL**: branch angle gen ± 40°. **FL**: bractlets ± 1 mm wide; petals 4–7 mm, > sepals, cream to pale yellow; styles ± 1–1.5 mm. Rocky areas; 2400–3800 m. NW, CaRH, SNH, Wrn, W&I; to MT. [*P. p.* Rydb.] CA pls possibly separable; further study needed.

ssp. ***reflexa*** (E. Greene) Keck (p. 971) **ST** gen 20–80 cm, glandular-hairy. **LVS**: basal gen 7–15 cm, sheathing base glabrous to glandular, terminal lflet gen 15–40 mm, lateral lflet teeth > 10, gen double. **INFL**: branch angle gen 20–40°. **FL**: bractlets ± 1 mm wide; petals 3–5 mm, < sepals, ± widely oblong (gen wider near tip), yellow; styles ± 1 mm. Moist or ± shaded places; gen 500–2600 m. NW, CaRH, SN, TR, PR, Wrn; OR, NV. ❀TRY.

P. gracilis Hook. Pl tufted from short, thick rhizome; glands gen 0 or hidden. **ST** ± ascending, gen 20–100 cm, strigose to spreading-hairy. **LVS** palmate; basal gen 6–30 cm, lflets gen 5–7, central lflet ± oblanceolate, toothed or lobed ± throughout, hairy esp below. **INFL** few–many-fld. **FL**: hypanthium gen 3–5 mm wide; petals gen 4–10 mm; filaments 1.5–2.5 mm, anthers 0.6–1.6 mm; styles 1.5–2 mm, slender. **FR** 1–1.5 mm, smooth, light brown. $2n$=52–109. Common. Meadows, open forests; 120–3500 m. NW (exc sw), CaR, SNH, Teh, TR, PR, GB; w N.Am. Variation complex; many but not all pls are assignable to the following extremes.

var. ***elmeri*** (Rydb.) Jepson (p. 971) Hairs gen appressed. **ST** gen 20–50 cm. **LVS**: basal central lflet 20–60 mm, gen tomentose below, lobed > 3/4 to midvein, lobes often narrowest at base, entire. **FL**: petals 5–8 mm. n=21. Dry meadows; 1280–3050 m. SNH, Teh, TR, SnJt, SNE; w N.Am. [*P. pectinisecta* Rydb.; *P. flabelliformis* Lehm. var. *inyoensis* Jepson]

var. ***fastigiata*** (Nutt.) S. Watson (p. 971) Hairs spreading to appressed. **ST** gen 20–50 cm. **LVS**: basal central lflet 20–60 mm, ± equally hairy above and below, toothed < 1/2 to midvein, teeth widest at base. **FL**: petals gen 4–7 mm. $2n$=52–109. Common. Gen open forests, dry meadows; 800–3500 m. Range of sp. [ssp. *nuttallii* (Lehm.) Keck; vars. *glabrata* (Lehm.) C. Hitchc. & *permollis* (Rydb.) C. Hitchc.]

var. ***flabelliformis*** (Lehm.) Torrey & A. Gray (p. 971) Hairs gen appressed. **ST** gen 50–100 cm. **LVS**: basal central lflet 50–90 mm, tomentose below, sparsely hairy above, lobed > 3/4 to midvein, lobes gen widest at base, often few-toothed. **FL**: petals 7–10 mm. $2n$=± 56–65. Moist or wet meadows; 1050–1600 m. MP; to sw Can. Similar pls sporadic elsewhere (e.g., c SNH, Yosemite Valley). [*P. f.* Lehm.] ❀TRY.

var. ***gracilis*** (p. 971) Hairs gen spreading. **ST** gen 40–90 cm. **LVS**: basal central lflet 30–70 mm, gen tomentose below, ± glabrous and dark above, toothed ± 1/2 to midvein. **FL**: petals 7–10 mm. Meadows; 120–1100 m. nw KR, n NCoRO; to B.C. ❀DRN,IRR:**7**,**14–16**,**19–24**&SUN:**4–6**,**17**.

P. grayi S. Watson (p. 971) Pl ± rosetted or tufted from short rhizome or taproot, nonglandular. **ST** ± ascending, gen 10–20 cm, sparsely strigose. **LVS** ternate; basal gen 2–6 cm, lflets 3, central lflet 10–25 mm, ± obovate, evenly 7-toothed 1/4–1/2 to midvein, ± subglabrous. **INFL** gen 1–5-fld. **FL**: hypanthium 2.5–4 mm wide; petals gen 4–7 mm; filaments 1–3 mm, anthers ± 0.6 mm; styles 1– 2.5 mm, slender. **FR** 1.2–1.5 mm, smooth, light brown. Meadows; 2000–2800 m. SNH. [*P. flabellifolia* Hook. var. *g.* (S. Watson) Jepson] May hybridize with *P. flabellifolia*.

P. hickmanii Eastw. (p. 971) HICKMAN'S CINQUEFOIL Pl rosetted from thick taproot, nonglandular. **ST** prostrate to decumbent, 5–45 cm, ± glabrous. **LVS** pinnate; basal 4–30 cm, lflets 4–7 per side, separated or overlapped, 5–25 mm, wedge-shaped, ± 4-toothed gen ± 1/2 to base, ± subglabrous. **INFL** gen < 10-fld; pedicels gen ± recurved in fr. **FL**: hypanthium 3–6 mm wide; petals 6–10 mm; filaments 1.5–4 mm, anthers ± 1 mm; pistils gen ± 10, styles 2–3 mm, slender. **FR** ± 2 mm, smooth, ± tan. **ENDANGERED** CA. Vernally wet meadows, open pine forests; < 100 m. n&c CCo, s NCoRO?. 1 protected population remains. s NCoRO (Solano Co.) pls possibly extinct, may warrant recognition.

P. millefolia Rydb. (p. 971) Pl rosetted from thick taproot, sometimes glandular. **ST** prostrate to decumbent, 5–20 cm, spreading- to appressed-hairy. **LVS** pinnate; basal 2–15 cm, lflets 5–13 per side, overlapped, 5–20 mm, narrowly 3–10-lobed > 1/2 to base, ± glabrous to hairy. **INFL** gen < 10-fld; pedicels gen ± recurved in fr. **FL**: hypanthium 3–6 mm wide; petals 4–8 mm; filaments gen 2–3.5 mm, anthers ± 1 mm; pistils 10–30, styles 2–3 mm, slender. **FR** 1.5–2 mm, smooth, ± tan. Vernally wet meadows; 900–2000 m. CaRH, GB, ne-most SNH; OR, NV. Variable; further study needed. Spreading-hairy pls have been called var. *klamathensis* (Rydb.) Jepson. ❀DRN,IRR,SUN:1–3,6,7,15–18;DFCLT.

P. morefieldii B. Ertter (p. 971) MOREFIELD'S CINQUEFOIL Pl tufted from few-branched caudex; glands 0 or hidden. **ST** prostrate to decumbent, 5–15 cm, ± ascending-hairy. **LVS** pinnate; basal 2–6 cm, lflets 2–4 per side, ± overlapped, 5–20 mm, ± oblanceolate, few-lobed ± 3/4 to midvein, often ± asymmetrically split ± to base, white-tomentose below, ± green and strigose above. **INFL** gen 5–15-fld; pedicels often ± recurved in fr. **FL**: hypanthium 2.5–5 mm wide; petals 4–6 mm; filaments ± 1–2 mm, anthers 0.5–1 mm; styles ± 1.5–2 mm, ± slender. **FR** ± 1.8 mm, smooth or ± ridged, pale brown. RARE. Low areas in alpine calcareous (or granite?) rocks; 3300–4000 m. n W&I (n White Mtns, Mono Co.).

P. multijuga Lehm. (p. 971) BALLONA CINQUEFOIL Pl rosetted from thick taproot, ± nonglandular. **ST** prostrate to decumbent, 20–50 cm. **LVS** pinnate; basal 10–20 cm, lflets 5–8 per side, 10–20 mm, ± obovate, 5-toothed < 1/2 to base. **INFL** gen < 10-fld; pedicels gen ± recurved in fr. **FL**: hypanthium 4–6 mm wide; petals 4.5–10 mm; filaments 1.5–4.5 mm, anthers ± 1 mm; pistils ± 5, styles 2–3 mm, slender. **FR** 1.8 mm, smooth, ± tan. **PRESUMED EXTINCT**. Brackish marshes; ± 0 m. c SCo (Ballona Marsh, Los Angeles Co.).

P. newberryi A. Gray (p. 971) NEWBERRY'S CINQUEFOIL Ann to short-lived per, rosetted from taproot, inconspicuously glandular. **ST** prostrate to decumbent, gen 10–40 cm, spreading- to ascending-hairy. **LVS** pinnate; basal gen 2–10 cm, lflets 4–8 per side, overlapped, 3–10 mm, 2–8-lobed > 1/2 to midvein, lobes oblanceolate, ± hairy. **INFL** many-fld; pedicels gen ± recurved in fr. **FL**: hypanthium 2.5–5 mm wide; petals 3–5.5 mm, cream; filaments 1–1.5 mm, anthers ± 0.5 mm; styles 1–1.5 mm, ± tapered from rough-thickened base. **FR** 1–1.5 mm, veined, brown. RARE in CA. Receding shorelines; 1300–2200 m. MP; to WA, NV.

P. norvegica L. (p. 971) Ann to short-lived per from taproot, nonglandular. **ST** ascending to erect, 10–70 cm; hairs spreading, sparse and long below, denser and shorter above. **LVS** gen ternate; basal often withered or fallen in fl; cauline gen 3–12 cm, lflets gen 3, 15–50 mm, oblanceolate, toothed ± 1/3 to midvein, ± hairy. **INFL** several–many-fld. **FL**: hypanthium 4–10 mm wide; petals 3–4 mm, < sepals; stamens 15–20, filaments 0.5–2 mm, anthers ± 0.3 mm; styles ± 0.8 mm, tapered from rough-thickened base. **FR** ± 1 mm, veined, light brown. n=28,35; $2n$=56,63,70. Moist, disturbed areas; < 2300 m. c SNH, ScV, SNE; N.Am, native to Eurasia. [ssp. *monspeliensis* (L.) Asch. & Graebn.]

P. palustris (L.) Scop. (p. 971) Pl openly matted from stolons, often floating, sparsely glandular. **ST**: fl-sts ascending, 20–50 cm, ± glabrous to ± strigose. **LVS** cauline, pinnate to subpalmate, 5–20 cm; lflets 2–3 per side, terminal 15–70 mm, oblanceolate-elliptic, distal 1/3 toothed ± 1/4 to midvein, ± glabrous to sparsely strigose. **INFL** gen < 10-fld. **FL**: hypanthium 5–10 mm wide; petals 2–6 mm, << sepals, narrowly ovate, dark red; stamens 20–25, filaments 1.5–2.5 mm, anthers 1–1.5 mm; styles 0.8–2 mm, ± slender. **FR** 1–1.5 mm, glossy, light brown. $2n$=28,42–64. Bogs, marshes; < 2400 m. NCo, CaRH, n SNH, ScV, Wrn; circumboreal. ❀WET,SUN: 1,**2**,3,**4–7**,8,9,14,**15–18**,19–21.

P. pensylvanica L. (p. 971) Pl tufted from simple or branched caudex, inconspicuously glandular. **ST** ± ascending, 8–25 cm, short-spreading- and long-ascending-hairy. **LVS** ± pinnate; basal gen 3–10 cm, lflets 2–5 per side, 8–25 mm, ± oblanceolate, ± evenly 10-lobed ± 2/3 to midvein, densely hairy below, gen greener and less hairy above. **INFL** ± 5–10-fld. **FL**: hypanthium 3.5–5.5 mm wide; petals ± 4 mm; filaments 0.5–1.5 mm, anthers ± 0.6 mm; styles ± 1 mm, tapered from rough-thickened base. **FR** ± 1 mm, ± smooth, pale brown. $2n$=28. Alpine rocky barrens; 2700–3800 m. c SNH, W&I; N.Am, Eurasia. Highly variable outside CA; w N.Am pls have been called var. *strigosa* Pursh.

P. pseudosericea Rydb. (p. 971) Pl tufted or matted from ± branched caudex; glands 0 or hidden. **ST** decumbent to ascending, gen 5–10 cm, ± strigose. **LVS** gen ± palmate; basal gen 2–4 cm, lflets ± 5, central lflet gen 5–15 mm, ± obovate, 3–10-lobed > 3/4 to midvein, tomentose below, densely white-strigose above. **INFL** gen ± 5-fld. **FL**: hypanthium 3–4 mm wide; petals ± 3 mm; filaments 0.5–1 mm, anthers ± 0.4 mm; styles ± 1 mm, tapered from rough-thickened bases. **FR** ± 1 mm, ± smooth, pale brown. Rocky flats, slopes; 3200–4300 m. c&s SNH, n SNE (Sweetwater Mtns), W&I.

P. recta L. (p. 971) Pl gen ± tufted from taproot or branched caudex, sparsely glandular. **ST** ± ascending, 20–60 cm, spreading-hairy; base with short and long hairs. **LVS** palmate; basal gen fallen or withered in fl; cauline 8–15 cm, lflets gen 6–7, central lflet 30–80 mm, oblanceolate, toothed ± 1/2 to midvein, sparsely hairy. **INFL** many-fld. **FL**: hypanthium 3–6 mm wide; petals 6–9 mm; stamens ± 25, filaments 1–2 mm, anthers ± 1 mm; styles ± 1 mm, stubby. **FR** 1–1.5 mm, prominently veined, brown. $2n$=28,42. Gen disturbed areas; 150–1500 m. CaRH, ScV, SnFrB; N.Am; native to Eurasia. Probably closer to Eurasian *P. norvegica* than to *P. gracilis*, with which it is most often confused.

P. rimicola (Munz & I.M. Johnston) B. Ertter (p. 971) CLIFF CINQUEFOIL Pl hanging, taprooted, ± glandular. **ST** gen 5–20 cm, spreading- to ascending-hairy. **LVS** palmate; basal 2–4 cm, lflets 5, central lflet 10–30 mm, ± obovate, distal 1/3 few-toothed ± 1/4 to midvein, ± strigose. **INFL** gen 5–20-fld; pedicels gen > 15 mm, often ± recurved in fr. **FL**: hypanthium 2–3 mm wide; petals gen 4–7 mm; filaments 1–2.5 mm, anthers 0.5–1 mm; pistils gen 5–20, styles 1.5–2.5 mm, slender. **FR** ± 1.5 mm, ± smooth, red-tipped. RARE in CA. Granite crevices; 2400–2800 m. SnJt; n Baja CA. [*P. wheeleri* var. *r.* Munz & I.M. Johnston]

P. rivalis Nutt. (p. 971) Ann or bien from taproot; glands 0 or hidden. **ST** gen ascending, gen 10–50 cm, spreading- to ascending-hairy. **LVS** ternate, subpalmate, or ± subpinnate; basal gen withered or fallen in fl; cauline gen 2–12 cm, lflets 3–5, gen 10–40 mm, oblanceolate to elliptic, toothed ± 1/3 to midvein, ± hairy. **INFL** many-fld. **FL**: hypanthium 2.5–3.5 mm wide; petals 1.5–2 mm, < sepals, obovate-elliptic; stamens ± 15, filaments 0.5–1 mm, anthers ± 0.3 mm; styles 0.5–1, tapered from rough-thickened bases. **FR** ± 0.8 mm, smooth, light brown. Moist, ± disturbed areas; 40–2100 m. s NCoRO, CaRH, SN, GV, CW, SnBr, MP; w&c N.Am. [var. *millegrana* (Engelm.) S. Watson]

P. wheeleri S. Watson (p. 971) Pl rosetted to tufted from 0–few-branched caudex, inconspicuously glandular. **ST** prostrate to decumbent, 2–25 cm, ± appressed-hairy below, spreading-hairy above. **LVS** palmate; basal 2–10 cm, lflets 5, central lflet 5–25 mm, wedge-shaped, distal 1/3 few-toothed ± 1/4 to midvein, ± densely strigose. **INFL** gen > 5-fld; pedicels gen < 15 mm, often ± recurved in fr. **FL**: hypanthium 3–4 mm wide; petals 3–6 mm; filaments 1–2 mm, anthers ± 0.5 mm; styles gen 1.5–2 mm, ± slender. **FR** 1–1.5 mm, ± veined, pale. Sandy, ± moist flats; 1800–3500 m. s SNH, SnBr, SnJt; n Baja CA. Further study needed.

PRUNUS

Dieter H. Wilken

Shrub or tree. **ST**: bark gray to red-brown. **LVS** gen alternate, simple, gen glabrous; stipules deciduous. **INFL**: raceme or umbel-like cluster, often on short branchlets. **FL**: hypanthium cup- to urn-shaped; sepals spreading to reflexed; stamens 15+, gen in 2+ whorls; pistil 1, ovary superior, chamber 1, ovules 2, style 1, stigma subspheric. **FR**: drupe, gen ovoid to spheric. ± 400 spp.: temp N.Am, Eurasia, n Afr; many cult for wood, orn, edible fr; some persisting near human habitation (*P. armeniaca*, apricot; *P. avium*, sweet cherry; *P. cerasus*, sour cherry; *P. domestica*, plum; *P. laurocerasus*, laurel cherry; *P. lusitanica*, portugal laurel; *P. mahaleb*; *P. persica*, peach). Seeds of many spp. ± TOXIC from production of hydrocyanic acid.

1. Infl a raceme, fls 10–many; frs on naked infl axis
 2. Raceme short, flat-topped; fls 3–12 .... [2]***P. emarginata***
 2′ Raceme longer than wide; fls gen 15–many
  3. Pedicel 5–8 mm; petals 4–7 mm; fr 6–14 mm, pulp thick; lf margin finely serrate; lvs deciduous .... ***P. virginiana*** var. ***demissa***
  3′ Pedicel 1–5 mm; petals 1–3 mm; fr 12–22 mm, pulp thin; lf margin entire or spiny-serrate; lvs evergreen .... ***P. ilicifolia***
   4. Lf margin spiny-serrate, ± wavy, blade widely ovate to round; petiole 4–10 mm; mainland .... ssp. ***ilicifolia***
   4′ Lf margin gen entire, flat, blade gen ovate; petiole 10–25 mm; ChI .... ssp. ***lyonii***
1′ Infl a cluster, often umbel-like, fls 1–12; frs from short, leafy branchlet or lf axil
 5. Twigs flexible to ± stiff, gen not becoming spine-like; ovary and fr gen glabrous (exc *P. dulcis*)
  6. Fls subsessile, pedicel in fr gen < 3 mm; sepal back glabrous, margin tomentose; fr densely puberulent, gray-green, pulp leathery .... ***P. dulcis***

6′ Pedicel 3–15 mm; sepals glabrous to evenly puberulent; fr gen glabrous, yellow to purple
7. Pedicel 3–12 mm; infl raceme-like; fr 7–14 mm; some or all lf blades widest above middle ..... [2]*P. emarginata*
7′ Pedicel 10–15 mm; infl umbel-like; fr 15–30 mm; some or all lf blades widest at or below middle
8. Petiole 15–20 mm; lf blade 45–80 mm; gen tree near human habitation .... *P. cerasifera*
8′ Petiole 4–15 mm; lf blade 25–50 mm; gen shrub, native .... *P. subcordata*
5′ Twigs rigid, gen becoming spine-like; ovary and fr gen densely puberulent
9. Lf blade ovate to round, 7–22 mm wide, base obtuse to ± cordate .... *P. fremontii*
9′ Lf blade narrowly elliptic to oblanceolate, 1–7 mm wide, base tapered
10. Petals reddish, 5–9 mm; lf blade 4–7 mm wide, finely serrate .... *P. andersonii*
10′ Petals white to yellowish, 2–4 mm; lf blade 1–3 mm wide, gen entire .... *P. fasciculata*
11. Lf surface puberulent; inland .... var. *fasciculata*
11′ Lf surface glabrous to low-papillate; coastal .... var. *punctata*

P. andersonii A. Gray (p. 971) DESERT PEACH Shrub < 2 m. **ST** much-branched; twigs rigid, becoming spine-like. **LVS** gen clustered, deciduous; petiole 0–7 mm; blade 9–30 mm, elliptic to oblanceolate, finely serrate, base tapered, tip gen acute. **INFL**: fls 1–2; pedicel 4–7 mm. **FL**: sepals puberulent inside; petals 5–9 mm, reddish. **FR** 10–14 mm, subspheric, densely puberulent, red-orange; pulp dry. Rocky slopes, flats, shrubland, coniferous forest; 900–2600 m. SNH (e slope), GB, n DMtns (Last Chance Range); w NV. ❀DRN,SUN:1,2,**3**,**7**,8–10,**18–21**,22–23&DRY:**14**,15–17.

P. cerasifera Ehrh. CHERRY PLUM Tree < 6 m. **LVS** deciduous; petiole 15–20 mm; blade 45–80 mm, elliptic to ovate, finely serrate, sometimes purplish. **INFL**: fls 1–3; pedicel 10–15 mm. **FL**: hypanthium and sepals glabrous; petals 7–14 mm, white to reddish. **FR** 15–30 mm, glabrous, yellow to dark red; pulp fleshy. Roadsides, chaparral; < 200 m. SnFrB; native to se Eur. Variable; may hybridize with other cult spp.

P. dulcis (Miller) D. Webb ALMOND Tree < 7 m. **LVS** deciduous; petiole 8–15 mm; blade 25–95 mm, oblong to lanceolate, crenate-serrate. **INFL**: fls 1–3, subsessile; pedicels in fr < 3 mm. **FL**: hypanthium glabrous; sepals tomentose near margin; petals 12–15 mm, pink, becoming white. **FR** 25–40 mm, densely puberulent, gray-green; pulp leathery, splitting to reveal stone. $2n$=16. Canyons, slopes, grassland as waif; < 200 m. GV, e SnFrB, s SCoRO (expected elsewhere); probably native to w Asia, n Afr. [*P. amygdalus* Batsch]

P. emarginata (Hook.) Walp. (p. 971) BITTER CHERRY Shrub or tree 1–10 m, often forming dense thickets. **LVS** gen clustered, deciduous; petiole 3–12 mm; blade 20–65 mm, gen elliptic to obovate, crenate-serrate, base obtuse to tapered. **INFL**: raceme, ± flat-topped; fls 3–12; pedicel 3–12 mm. **FL**: hypanthium and sepals glabrous to puberulent; petals 4–8 mm, white. **FR** 7–14 mm, ovoid to spheric, glabrous, red to purple; pulp ± fleshy. Rocky slopes, canyons, chaparral, mixed-evergreen or coniferous forest; < 2800 m. CA-FP (exc GV, ChI), MP; to B.C., MT, WY. ❀DRN:**4–6**&IRR:1,**2**,3,**7**,8–10,14,**15–18**,19–23.

P. fasciculata (Torrey) A. Gray (p. 971) DESERT ALMOND Shrub 1–2 m. **ST** much-branched; twigs rigid, becoming ± spine-like. **LVS** gen clustered, deciduous, subsessile; blade 7–15 mm, narrowly oblanceolate, gen entire, base tapered, tip acute to obtuse. **INFL**: fls 1–3, subsessile. **FL**: sepals glabrous to sparsely puberulent; petals 2–4 mm, white to yellowish. **FR** 8–15 mm, ovoid to spheric, densely puberulent, gray to red-brown; pulp dry. Slopes, canyons, washes, shrubland, woodland; < 2200 m. s SNH, Teh, s CCo, s SCoRI, n TR, e PR, D; to UT, Baja CA.

var. ***fasciculata*** **LF**: surface puberulent. Creosote-bush scrub, Joshua-tree or pinyon/juniper woodland; 700–2200 m. s SNH, Teh, s SCoRI, n TR, D; to UT, Baja CA. ❀DRN,SUN,DRY:1,2,**3**,**7**,8–11,**14**,**18**,19–23.

var. ***punctata*** Jepson **LF**: surface glabrous to minutely gland-dotted or low-papillate. Sandy soils, shrubland, oak woodland; < 200 m. s CCo, e WTR. ❀DRN,SUN:**7**,**14-16**,**19-24**&IRR:8,9,**18**.

P. fremontii S. Watson (p. 971) DESERT APRICOT Shrub 1–4 m. **ST** much-branched; twigs rigid, becoming spine-like. **LVS** gen clustered, deciduous; petiole 2–7 mm; blade 12–30 mm, ovate to round, serrate, base obtuse to ± cordate, tip obtuse to rounded. **INFL**: fls 1–5; pedicel 1–3 mm. **FL**: petals 4–8 mm, white. **FR** 8– 15 mm, ovoid to spheric, puberulent, yellowish; pulp dry. Rocky slopes, canyons, shrubland, pinyon/juniper woodland; 200–1200 m. e PR, w DSon; Baja CA. ❀DRN:3,7,8,9,**14**,**18**,**19**&SUN:**15**,**16**,**20–24**.

P. ilicifolia (Nutt.) Walp. Shrub or tree < 15 m. **LVS** evergreen; petiole 4–25 mm; blade 16–120 mm, ovate to round, entire or spiny-serrate, base truncate to ± cordate, tip acute to rounded. **INFL**: raceme; fls many; pedicel 1–5 mm. **FL**: petals 1–3 mm, white. **FR** 12–25 mm, ovoid to spheric, glabrous, red to blue-black; pulp fleshy. Canyons, slopes, shrubland, woodland; < 1600 m. s NCoR, CW, SW; Baja CA.

ssp. ***ilicifolia*** (p. 977) ISLAY, HOLLY-LEAFED CHERRY, Pl < 9 m. **LF**: petiole 4–10 mm; blade widely ovate to round, margin spiny-serrate, ± wavy. **INFL**: pedicel 2–5 mm. **FR** 12–18 mm, gen red. Habitats and range (exc ChI) of sp. ❀DRN,SUN or part SHD:5,**7**,**14–24**&IRR:8–10.

ssp. ***lyonii*** (Eastw.) Raven (p. 977) CATALINA CHERRY Pl < 15 m. **LF**: petiole 10–25 mm; blade gen ovate, gen entire, flat. **INFL**: pedicel 1–3 mm. **FR** 15–25 mm, gen blue-black. Chaparral, woodland; < 600 m. ChI; Baja CA (mainland). [*P. lyonii* (Eastw.) Sargent] ❀DRN,SUN or part SHD:**14–17**,18,**22–24**&IRR:**7**,8,9,**19–21**.

P. subcordata Benth. (p. 977) Shrub or tree-like, < 8 m. **ST**: branches stiff. **LVS** deciduous; petiole 4–15 mm; blade 25–50 mm, elliptic to widely ovate, finely serrate, base obtuse to ± cordate, tip obtuse to rounded. **INFL**: fls 1–7; pedicel 10–15 mm. **FL**: hypanthium and sepals glabrous to puberulent; petals 5–10 mm, white. **FR** 15–25 mm, ovoid to subspheric, glabrous to puberulent, yellow to dark red; pulp fleshy. Mixed-evergreen or coniferous forest; 100–1900 m. NW, CaR, SN, CW, MP; s OR. ❀DRN,SUN or part SHD:5,6,**15**,**16**&IRR:1–3,**7**,8,9,**14**,**18**,19–21.

P. virginiana L. var. ***demissa*** (Nutt.) Torrey (p. 977) WESTERN CHOKE-CHERRY Shrub or small tree < 6 m. **LVS** deciduous; petiole 10–25 mm; blade 50–100 mm, elliptic to obovate, finely serrate, base rounded to ± cordate, tip gen acute. **INFL**: raceme; fls many; pedicel 5–8 mm. **FL**: petals 4–7 mm, white. **FR** 6–14 mm, ovoid to spheric, glabrous, dark red to black; pulp ± fleshy. $2n$=16. Rocky slopes, canyons, shrubland, oak/pine woodland, coniferous forest; 100–2900 m. CA-FP (exc coast, GV), GB; to B.C., c US, TX, n Mex. [var. *melanocarpa* (Nelson) Sarg.] Var. *virginiana* of e N.Am, gen a tall tree, sometimes cult. ❀DRN,SUN or part SHD: **4–6**,**14–17**&IRR:1,**2**,3,**7**,8–10,**18–20**,21–24;STBL; rather INV.

PURSHIA ANTELOPE BUSH

Thomas J. Rosatti

Shrub or small tree. **LVS** ± clustered on short lateral branches, mostly deciduous, gen deeply lobed, ± strongly rolled under, gen with ± sunken glands above; bases persistent, overlapping, sheathing st. **INFL**: fls solitary on side-branch tips. **FL**: hypanthium ± funnel-shaped, sometimes partly glandular, bractlets 0; stamens ± 25; pistils 1–5(12), simple, styles persistent, ± hairy. **FR**: achene, ± fusiform to oblong. ± 5 spp.: w N.Am. (Frederick T. Pursh, N.Am flora author, 1774–1820) [Koehler & Smith 1981 Madroño 28:13–25; Henrickson 1986 Phytologia 60:468]

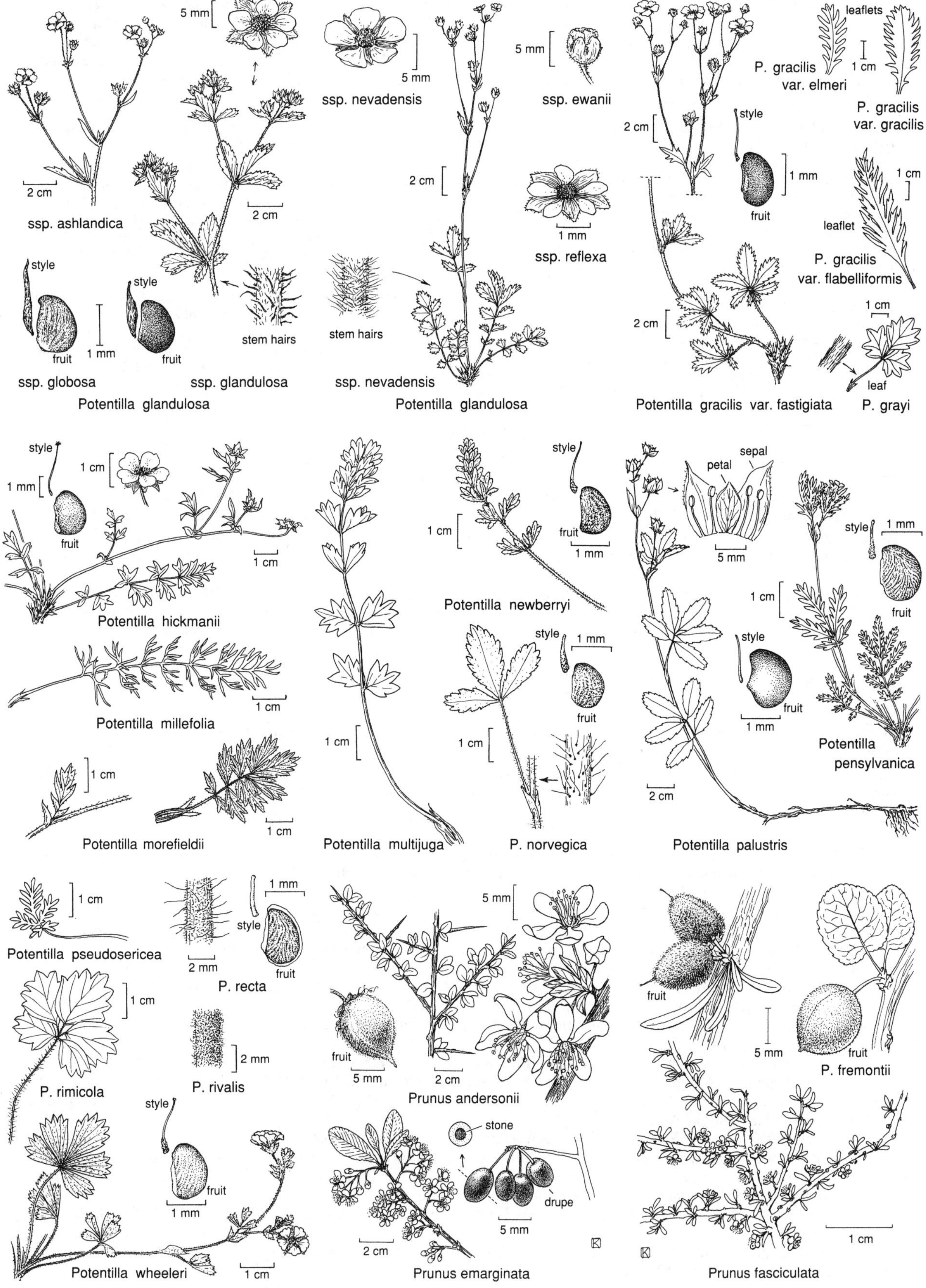
5 mm
2 cm
ssp. ashlandica
2 cm
style
style
1 mm
fruit
fruit
stem hairs
ssp. globosa
ssp. glandulosa
Potentilla glandulosa
5 mm
ssp. nevadensis
5 mm
ssp. ewanii
2 cm
1 mm
ssp. reflexa
stem hairs
ssp. nevadensis
Potentilla glandulosa
leaflets
P. gracilis
var. elmeri
1 cm
P. gracilis
var. gracilis
style
2 cm
1 mm
fruit
1 cm
leaflet
P. gracilis
var. flabelliformis
1 cm
2 cm
leaf
Potentilla gracilis var. fastigiata
P. grayi
style
1 cm
1 mm
fruit
1 cm
Potentilla hickmanii
1 cm
Potentilla millefolia
1 cm
1 cm
Potentilla morefieldii
style
1 cm
fruit
1 mm
Potentilla newberryi
style
1 mm
fruit
1 cm
1 cm
Potentilla multijuga
P. norvegica
sepal
petal
5 mm
style
1 mm
1 cm
fruit
style
fruit
1 mm
Potentilla
pensylvanica
2 cm
Potentilla palustris
1 cm
Potentilla pseudosericea
1 mm
style
2 mm
fruit
P. recta
1 cm
2 mm
P. rimicola
P. rivalis
style
fruit
1 mm
Potentilla wheeleri
1 cm
5 mm
fruit
5 mm
2 cm
Prunus andersonii
stone
drupe
5 mm
2 cm
Prunus emarginata
fruit
5 mm
fruit
P. fremontii
1 cm
Prunus fasciculata

1. Pistils 4–5(12); styles in fr 2–6 cm, plumose; lf lobes (3)5–9, from just below middle *P. mexicana* var. *stansburyana*
1′ Pistils 1(–3); styles in fr < 1 cm, canescent at least below tip; lf lobes 3(–5), gen from above middle ... *P. tridentata*
2. Lf upper surface sparsely nonglandular-hairy, sessile or sunken glands few–many; twig hairs mostly glandular .... var. *glandulosa*
2′ Lf upper surface densely nonglandular-hairy, sessile or sunken glands 0–few; twig hairs mostly nonglandular .... var. *tridentata*

P. mexicana (D. Don) Welsh var. ***stansburyana*** (Torrey) Welsh (p. 977) Shrub 3–30 dm. **LF**: lobes (3)5–9, from just below middle, entire to again lobed. **FL**: hypanthium ± 5 mm; sepals 4–6 mm, ± ovate; petals 6–8 mm, widely ovate, white to cream; pistils 4–5(12). **FR** glabrous to becoming so; styles 2–6 cm, plumose. n= 9. Dry Joshua-tree or pinyon/juniper woodland; 1100–2500 m. W&I, DMtns; to Colorado, NM, n Mex. [*Cowania m.* D. Don var. *stansburiana* (Torrey) Jepson] Hybridizes with *P. tridentata* var. *glandulosa*. ❀DRN,SUN:1,2,**3,7**,9,**10**,11,**14,18–21**&DRY:15,16, 22,23.

P. tridentata (Pursh) DC. Shrub 10–50 dm. **LF**: lobes 3(–5), gen from above middle, gen entire. **FL**: hypanthium ± 3–4 mm; sepals ± 3 mm, ± oblong; petals 6–8 mm, ± obovate, cream to yellow; pistils 1(–3). **FR** canescent; style < 1 cm, canescent at least below tip. Dry sagebrush scrub, chaparral, Joshua-tree or pinyon/juniper woodland, coniferous forest; 700–3400 m. KR, NCoRH, CaR, SNH (e slope), Teh, n TR and e PR (D edge), GB, DMtns; to B.C., MT, NM. Vars. intergrade.

var. ***glandulosa*** (Curran) M.E. Jones (p. 977) Pl greenish; twig hairs mostly glandular. **LF** 5–10 mm, sparsely tomentose and greenish, sessile or sunken glands few–many above. $2n$=18. Chaparral, Joshua-tree or pinyon/juniper woodland; 700–3000 m. c&s SNH (e slope), Teh, n TR & e PR (D edge), SNE, DMtns; NV, AZ. [*P. glandulosa* Curran] ❀DRN,SUN:1,2,**3,7**,9–11&DRY:**14**,15, 16,**18**,19–21.

var. ***tridentata*** (p. 977) Pl grayish; twig hairs mostly nonglandular. **LF** 5–30 mm, ± densely white-tomentose, sessile or sunken glands 0–few above. $2n$=18. Sagebrush scrub, coniferous forest, juniper woodland; 900–3400 m. KR, NCoRH, CaR, SNH (e slope), GB; to B.C., MT, NM. ❀DRN,SUN:1,**2,3,7**&DRY:**14**, 15,16,18–21.

PYRACANTHA FIRETHORN

Thomas J. Rosatti

Shrub, ± evergreen; thorns often leafy or branched. **LF**: stipules soon deciduous; petiole short. **INFL**: racemes or panicles at ends of short lateral branches. **FL** bisexual; bractlets on hypanthium 0; sepals persistent; petals whitish; stamens 20, fused at base; ovary half-inferior, chambers 5. **FR**: pome, ± spheric, open at top; stones 5, free, 1–2-seeded. ± 6 spp.: Medit, Asia. (Greek: fire thorn, from fr color, thorns)

P. angustifolia (Franchet) C. Schneider Pl < 4 m, prostrate to gen erect, ± gray-tomentose (sometimes becoming glabrous). **LF** 25–50 mm, gen ± oblanceolate, gen rolled under; teeth 0–few, gen distal, inconspicuous. **INFL** ± several-fld. **FR** 6–8 mm, yellow-orange to red. Disturbed areas, abandoned fields, roadsides; < 200 m. s NCoRO, CCo, SnFrB, SCo, expected elsewhere; native to China.

ROSA

Barbara Ertter

Shrub to vine, often thicket-forming, gen prickly. **LVS** gen odd-pinnately compound; stipules gen attached to petiole, gen gland-margined. **INFL**: gen ± cyme or fls solitary. **FL**: hypanthium urn-shaped; bractlets 0; sepals 5, often with long expanded tip; petals gen 5 (exc cultivars), gen pink in CA (white to red or yellow); stamens gen >> 20; pistils gen many, ovaries superior, jointed to gen hairy styles. **FR**: bony achenes enclosed in fleshy, gen reddish hypanthium (hip). 100+ spp.: gen n temp. (Latin: ancient name) Spp. hybridize freely; other non-natives established locally.

1. Lflets << 1 cm, toothed 1/2 to base; hypanthium densely prickly — rare PR .... ***R. minutifolia***
1′ Lflets > 1 cm, toothed < 1/4 to base; hypanthium glabrous to stalked-glandular
2. Sepal with toothed lateral lobes; st gen ± olive-green; prickles compressed side-to-side, curved; alien
3. Lvs and sepals glandless; terminal lflet ± ovate .... ***R. canina***
3′ Lvs and sepals glandular; terminal lflet gen elliptic to ± widely obovate .... ***R. eglanteria***
2′ Sepal body gen entire (tip sometimes toothed); st gen ± grayish or brownish; prickles compressed or slender, not curved (exc *R. californica*); native
4. Sepals deciduous in fr; hypanthium gen 2–3 mm wide at fl; pistils < 10; sepal tip gen << body; petals ± 10 mm — lflets glabrous, fls 1–3 .... ***R. gymnocarpa***
4′ Sepals persistent; hypanthium gen 3–7 mm wide at fl; pistils gen > 10 (exc some *R. bridgesii*); sepal tip often > body; petals gen 10–25 mm
5. Dwarf ± rhizomatous shrubs gen 1–10 dm; lflet tip often ± truncate — lf margins gen ± glandular, double-toothed
6. Hypanthium stalked-glandular — prickles gen ± slender, often many; fls gen 2–5; NW, CW, 150–1550 m .... ***R. spithamea***
6′ Hypanthium gen glabrous
7. Prickles few, base thick; terminal lflet gen widely obovate; fl 1(–5); CaR, SN, 900–2500 m ..... ***R. bridgesii***
7′ Prickles gen many, both slender and base thick; terminal lflet gen ± elliptic; fls gen 1–5; CW, gen < 300 m .... ***R. pinetorum***
5′ Open or thicket-forming shrubs gen > 5 dm; lflet tip not truncate
8. Prickles gen ± slender, ± straight, few–many; lvs and sepals glandless; fls 1–5(10); CaR, SNH, SnGb, SnBr, GB, DMtns .... ***R. woodsii*** var. ***ultramontana***
8′ Prickles gen ± thick-based, often curved, gen ± few (exc *R. nutkana*); lvs and sepals often glandular; fls often > 5; CA-FP (exc SNH)

9. Prickles gen curved, not contrasting with st; sepals often glandless, tip gen ± = body; pedicel gen ± hairy . . . . **R. californica**

9′ Prickles ± straight, gen whitish, contrasting with dark st; sepals gen ± glandular, tip gen > body; pedicel glabrous to glandular, gen not hairy

10. Fl gen 1(–4); sepal tip gen toothed; lf ± glandular; hypanthium gen 5–7 mm wide at fl; pedicel sometimes glandular; petals gen 18–25 mm; w NW; < 700 m . . . . **R. nutkana** var. **nutkana**

10′ Fls gen 2–10; sepal tip entire; lf glandless; hypanthium gen 3–6 mm wide at fl; pedicel gen glabrous; petals ± 15–20 mm; NW, CaR; < 2100 m . . . . **R. pisocarpa**

R. bridgesii Crépin (p. 977) Dwarf shrub gen 1–8 dm, rhizomed. **ST** dark brownish; prickles few, gen paired, base thick, straight. **LF**: lflets gen ± hairy and glandular; terminal lflet gen 10– 30 mm, gen widely obovate (elliptic), tip gen ± truncate, margins gen double-toothed, glandular. **INFL** 1(–5)-fld; pedicels gen ± 10– 15 mm, glabrous to glandular. **FL**: hypanthium gen 2.5–4 mm wide at fl, gen glabrous (glandular), neck gen 1.5–2 mm wide; sepals gen glandular, margins entire, tip gen < body, entire; petals 10–15 mm; pistils ± 5–20. **FR** 6–10 mm wide. Open forests, rocky areas; 900–2500 m. CaR, SN; OR?. [*R. gymnocarpa* var. *pubescens* S. Watson] ❀TRY.

R. californica Cham. & Schldl. (p. 977) CALIFORNIA ROSE Shrub 8–25 dm, often thicket-forming. **ST** gray-brown; prickles gen few, compressed, gen strongly curved. **LF**: lflets ± hairy, sometimes glandular; terminal lflet gen 15–60 mm, ± elliptic, tip ± obtuse, margins single- or double-toothed, glandular or not. **INFL** 1– 20-fld; pedicels gen ± 5–20 mm, gen ± hairy, glandless. **FL**: hypanthium 3–5 mm wide at fl, glabrous or sparsely hairy, neck 2–5 mm wide; sepals glandular or not, entire, tip gen ± = body, entire; petals gen 10–20 mm; pistils > 10. **FR** 8–20 mm wide. *n*=14. Gen ± moist areas, esp streambanks; < 1600 m. CA-FP (exc CaRH, SNH); s OR, n Baja CA. Needs study. ❀SUN or part SHD:5,6, **15**,**16**,17&IRR:2,**7–9**,10,**14**,**18–24**;INV;also STBL.

R. canina L. DOG ROSE Shrub or thicket-forming, gen 8–40 dm. **ST** olive-green; prickles ± few, gen compressed, curved. **LF**: lflets (sub)glabrous; terminal lflet gen 15–40 mm, ± ovate, tip ± acute, margins ± single-toothed, sparsely glandular. **INFL** gen 1–5-fld; pedicels gen 10–20 mm, gen glabrous. **FL**: hypanthium gen 4–5 mm wide at fl, glabrous, neck 2–3 mm wide; sepals ± glandless, margins with toothed lateral lobes, tip gen ± = body, ± toothed; petals 15–30 mm; pistils > 10. **FR** 10–20 mm wide; sepals often deciduous. 2*n*=35. Gen ± dry open areas; 100–900 m. NW, n SN; to WA, ID, also e US; native to Eurasia.

R. eglanteria L. (p. 977) SWEET-BRIER Thicket-forming, 8–30 dm. **ST** ± olive-green; prickles gen ± few, compressed, curved. **LF**: lflets ± hairy and glandular; terminal lflet gen 10–35 mm, elliptic to ± widely obovate, tip obtuse, margins double-toothed, glandular. **INFL** gen 1–8-fld; pedicels gen ± 10 mm, glandular. **FL**: hypanthium gen 4–5 mm wide at fl, gen ± glabrous, neck ± 3 mm wide; sepals glandular, with toothed lateral lobes, tip gen ± = body, gen toothed; petals gen 10–20 mm; pistils ± 10–20. **FR** ± 15 mm wide; sepals often deciduous. 2*n*=35. Gen ± dry, often disturbed open sites; 30–1100 m. NW, SNF, CW; to e NAm; native to Eur.

R. gymnocarpa Nutt. (p. 977) WOOD ROSE Loose shrub ± 5–20 dm. **ST** grayish brown; prickles few–many, slender, ± straight. **LF**: lflets glabrous; terminal lflet ± 10–30 mm, ± widely elliptic, tip ± obtuse, margins ± double-toothed, glandular. **INFL** 1(–3)-fld; pedicels gen ± 15–30 mm, gen stalked-glandular. **FL**: hypanthium gen 2–3 mm wide at fl, glabrous, neck ± 2 mm wide; sepals glandular or not, entire, tip gen << body, entire; petals gen ± 10 mm; pistils < 10. **FR** 5–12 mm wide; neck and sepals deciduous. *n*=7. Common. Gen in forests, shrublands; 30–2000 m. NW, CaR, n&c SN, CW, PR, MP; to BC, MT. Pls of c PR, nwNW need further study. ❀part SHD,DRN:4,**5**,**6**,**15–17**&IRR:1,2,**7**,8,9,**14**,18–24.

R. minutifolia Engelm. (p. 977) SMALL-LEAVED ROSE Dense shrub ± 3–10 dm. **ST** gray; prickles many, slender, straight. **LF**: lflets hairy; terminal lflet ± 3–6 mm, ± obovate, tip ± obtuse, margins toothed ± 1/2 to midvein, ± glandless. **INFL** gen 1-fld; pedicels ± 2–10 mm, hairy, glandless. **FL**: hypanthium ± 3 mm wide at fl, densely prickly, neck ± 2 mm wide; sepals glandless, with toothed lateral lobes, tip gen ± = body, toothed; petals ± 10–20 mm; pistils gen ± 10. **FR** ± 5 mm wide. RARE. Chaparral; ± 160 m. s PR (Otay Mesa); n Baja CA. In cult.

R. nutkana Presl var. ***nutkana*** (p. 977) NOOTKA ROSE Shrub, loose to thicket-forming, gen 5–20 dm. **ST** gen ± black; prickles few–many, gen ± compressed, ± straight. **LF**: lflets sparsely hairy and glandular; terminal lflet ± 15–50 mm, ± widely elliptic, base gen ± rounded, tip ± obtuse, margins ± single- to double-toothed, ± glandular. **INFL** gen 1(–4)-fld; pedicels gen ± 10–20 mm, glabrous or glandular. **FL**: hypanthium gen 5–7 mm wide at fl, glabrous or glandular, neck 3–6 mm wide; sepals gen glandular, entire, tip gen > body, toothed; petals gen 18–25 mm; pistils > 10. **FR** 10–15 mm wide. *n*=21. Gen ± moist flats; < 700 m. w NW; to AK. [vars. *muriculata* (E. Greene) G. Jones and *setosa* G. Jones] May hybridize with coastal spp. to n ChI. ❀part SHD or SUN:**4–6**& IRR:7,**14–17**,19–24;INV;also STBL.

R. pinetorum A.A. Heller PINE ROSE Dwarf shrub, gen < 10 dm, ± rhizomed. **ST** gray-brown; prickles gen many, both slender and ± thick-based, straight. **LF**: lflets glabrous to hairy; terminal lflet ± 10–40 mm, gen ± elliptic, tip ± obtuse, margins ± single- or double-toothed, ± glandular. **INFL** gen 1–5-fld; pedicels gen 10–30 mm, glabrous to glandular. **FL**: hypanthium gen ± 4 mm wide at fl, glabrous, neck ± 3 mm wide; sepals gen ± glandular, entire, tip gen ± = body, entire or toothed; petals ± 15–20 mm; pistils ± 10–20. **FR** ± 12 mm wide. Pine woodlands, canyons; gen < 300 m. w-c CW. Possibly hybrids of *R. spithamea*, *R. gymnocarpa*, or others; further study essential. ❀DRN,part SHD:4,5&IRR:7,14,**15–17**, 19–24; also STBL.

R. pisocarpa A. Gray (p. 977) CLUSTER ROSE Shrub to thicket, gen 5–25 dm. **ST** gen ± black; prickles few, gen ± thick-based, straight. **LF**: lflets (sub)glabrous; terminal lflet ± 15–40 mm, ± obovate-elliptic, tip ± obtuse, margins single-toothed, gen glandless. **INFL** gen 2–10-fld; pedicels ± 10–20 mm, gen glabrous, glandless. **FL**: hypanthium gen 3–6 mm wide at fl, glabrous, neck 2–3 mm wide; sepals gen glandular, margins entire, tip gen > body, entire; petals ± 15–20 mm; pistils gen > 10. **FR** ± 10 mm wide. *n*=7. Gen ± moist or shady areas; 30–2100 m. NW, CaR; to BC. Intergrades with *R. californica*, *R. woodsii*, *R. nutkana*. ❀IRR,DRN:**4–6**& part SHD:1,7,14–17;DFCLT.

R. spithamea S. Watson (p. 977) GROUND ROSE Dwarf shrub, gen < 5 dm, rhizomed. **ST** gray-brown; prickles few–many, both slender and thick-based, ± straight. **LF**: lflets 2–4 per side, (sub) glabrous; terminal lflet ± 10–30 mm, ± widely elliptic (obovate), tip ± obtuse to ± truncate, margins ± double-toothed, glandular. **INFL** 1–10-fld; pedicels gen 5–15 mm, gen ± glandular. **FL**: hypanthium gen 4–5 mm wide at fl, glandular, neck ± 3–4 mm wide; sepals gen glandular, entire, tip gen ± = body, entire; petals ± 10–15 mm; pistils ± 10–20. **FR** ± 7–10 mm wide. Open forest, chaparral, esp after fire; gen 150–1550 m. NW, CW. [var. *sonomensis* (E. Greene) Jepson] s CW pls possibly varietally distinct. ❀DRN,IRR:**7**,14& SUN:4–6,**15**,**16**,17;also STBL;GRNCVR.

R. woodsii Lindley var. ***ultramontana*** (S. Watson) Jepson (p. 977) INTERIOR ROSE Shrub, loose to thicket-forming, gen 5–30 dm. **ST** gray- or red-brown; prickles few–many, gen ± slender, gen ± straight. **LF**: lflets (sub)glabrous; terminal lflet ± 10–40 mm, ± elliptic, tip ± obtuse, margins single-toothed, glandless. **INFL** gen 1–5-fld; pedicels gen ± 10–20 mm, gen ± glabrous, glandless. **FL**: hypanthium gen 3–5 mm wide at fl, glabrous, neck 2–4 mm wide; sepals glandless, gen entire (or with simple, linear lobes), tip gen ± = body, entire; petals gen 15–20 mm; pistils gen > 10. **FR** 5–12 mm wide. Gen ± moist areas; 800–3400 m. CaRH, SNH, SnGb, SnBr, GB, DMtns; to B.C., MT, NV. [vars. *glabrata* (Parish) Cole & *gratissima* (E. Greene) Cole] Some DMtns pls have thick-based, curved prickles. Var. *woodsii* in c US. ❀DRN,SUN:4,5,**6** &IRR: **15–17**&part SHD:1,**2**,3,**7**,**14**,18–24; rather INV; also STBL.

RUBUS

Barbara Ertter

Per to shrub, often bramble-forming, often prickly, prostrate or clambering to erect. **LVS** gen palmately lobed to compound; lflets often stalked, toothed. **INFL** various. **FL**: hypanthium shallow; bractlets 0; sepals 5, gen reflexed, tips gen linear; petals 5; stamens gen >> 20; pistils few–many, ovaries superior, jointed to slender to club-shaped styles. **FR**: aggregate of sweet, fleshy-coated achenes (drupelets) that gen separate jointly from receptacle (raspberry-like) or separate jointly with part of fleshy receptacle (blackberry-like). 200–700 spp.: worldwide, esp n temp, Andes. (Latin: ancient name for bramble) Coastal forms often have smaller, hairier, rounder lvs; hybrids and other escapes from cult expected.

1. Lf ± 5-lobed — shrub, large-fld, unarmed; fr finely hairy ..... ***R. parviflorus***
1' Lf, gen 3-lobed or compound, sometimes ± unlobed
  2. Main sts 5-angled, stout; lflets 5 on large lvs; fls often > 10; escaped from cult — prickles gen wide-based, ± curved (0)
    3. Lflets deeply lobed or again compound — infl ± panicle ..... ***R. laciniatus***
    3' Lflets toothed
      4. Infl a raceme (bracts often lf-like); lflets gen widest below middle, green below; sepal tips 1–8 mm ..... ***P. pensilvanicus***
      4' Infl a panicle; lflets gen widest above middle, white below; sepal tips gen ± 1 mm
        5. Prickles many; common ..... ***R. discolor***
        5' Prickles 0; uncommon ..... ***R. ulmifolius*** var. ***inermis***
  2' Main sts round, slender to stout; lflets seldom > 3; fls gen 1–10 — native
    6. Pl ± prostrate; pistils 5–15; prickles often 0 — fr puberulent; > 750 m
      7. Stipules linear; lf compound, white below; st not rooting at nodes — prickly ..... [2]***R. glaucifolius***
      7' Stipules ± ovate; lf often simple, green below; st rooting at nodes — KR, NCoRH
        8. Prickles 0; petals white ..... ***R. lasiococcus***
        8' Prickly; petals gen dull purple to pink (white) ..... ***R. nivalis***
    6' Pl decumbent to erect; pistils gen > 10; prickles gen present
      9. Lf green below; < 1500 m
        10. Mains sts erect; petals ± red; fr yellow to red; fl bisexual; nw CA ..... ***R. spectabilis***
        10' Main sts clambering; petals white; fr black; fls gen unisexual; common, CA-FP ..... ***R. ursinus***
      9' Lf white below; < 2400 m
        11. Pl prostrate to decumbent; st 2–3 mm diam; pistils < 10; infl prickles few, ± slender, straight [2]***R. glaucifolius***
        11' Pl arched to erect, st 4–10 mm diam; pistils gen > 15; infl prickles many, gen ± wide-based, gen curved ..... ***R. leucodermis***

R. discolor Weihe & Nees HIMALAYAN BLACKBERRY Arched bramble. **ST** 5–15 mm diam, 5-angled; prickles many, ± wide-based, gen ± curved. **LF** compound; stipules linear; petiole ± 3–9 cm; lflets 3–5, gen widest above middle, sharply toothed, white below; longest lflet stalk ± 10–40 mm; longest lflet blade ± 5–11 cm. **INFL**: panicle, many-fld, nonglandular. **FL**: sepal tips gen ± 1 mm; petals 10–15 mm, obovate, white to pinkish; pistils > 15. **FR** blackberry-like, ± oblong, black, ± glabrous. $2n$=28. Common. Disturbed moist areas, roadsides, fencerows; < 1600 m. CA-FP; to B.C.; native to Eurasia. [*R. procerus* Mueller] Favored by rats for food, shelter.

R. glaucifolius Kellogg (p. 977) Pl prostrate to decumbent, not rooting at nodes. **ST** 2–3 mm diam, round; prickles few, ± slender, gen ± straight. **LF** compound; stipules linear; petiole ± 2–5 cm; lflets 3(–5), ± ovate, shallowly lobed, irregularly toothed, white below; longest lflet stalk ± 5–20 mm; longest lflet blade ± 2–5 cm. **INFL** gen ± glandular; fls 1–few. **FL**: sepal tips 1–3 mm; petals 4–6 mm, ± obovate, white; pistils gen < 10. **FR** hemispheric, red, puberulent. Open, often moist forest; 750–2300 (highest in s SNH). NW, SN, PR; OR. Barely glandular pls in c PR have been called var. *ganderi* (L. Bailey) Munz, Cuyamaca raspberry. ❀TRY.

R. laciniatus Willd. CUT-LEAVED BLACKBERRY Clambering to arched brambles. **ST** 3–10 mm diam, 5-angled; prickles many, ± wide-based, gen ± curved. **LF** 1–2-compound; stipules linear; petiole ± 3–12 cm; lflets 3–5, ± ovate, deeply lobed to again compound, sharply toothed, ± green below; longest lflet stalk ± 0–40 mm; longest lflet blade ± 2–6 cm. **INFL**: panicle, sometimes glandular; fls few–many. **FL**: sepal tips 2–8 mm; petals 5–7 mm, ± obovate, pinkish to white; pistils gen > 15. **FR** blackberry-like, ± ovoid, glabrous, black. Gen ± moist, disturbed areas; 60–1900 m. NW, CaR, SN, SnFrB, PR; to B.C., e N.Am; Eur cultivar.

R. lasiococcus A. Gray (p. 977) Prostrate ground-cover, rooting at nodes. **ST** 1–3 mm diam, round; prickles 0. **LF** ovately 3-lobed to ternately ± compound; stipules ± ovate; petiole ± 2–5 cm; blade ± 2–4 cm, finely toothed, green below. **INFL** 1–2-fld, nonglandular. **FL**: sepal tips ± 1 mm; petals 5–8 mm, obovate-round, white; pistils gen < 15. **FR** ± hemispheric, red, densely puberulent. Open forest; 1100–2000 m. KR, n NCoRH; to B.C. [*R. pedatus* Smith misapplied in CA] ❀DRN:**4–6**&IRR,SHD:1,2,7,14–17;DFCLT;GRNCVR.

R. leucodermis Torrey & A. Gray (p. 977) BLACKCAP RASPBERRY Arched to erect brambles. **ST** 4–10 mm diam, round; prickles many, slender to wide-based, straight to ± curved (esp in infl). **LF** compound; stipules linear; petiole ± 1–5 cm; lflets 3(–7), ± ovate, shallowly lobed, irregularly toothed, white below; longest lflet stalk ± 5–20 mm; longest lflet blade ± 2–6 cm. **INFL**: ± cyme, few-fld, glandular or not. **FL**: sepal tips 1–3 mm; petals 3–5 mm, oblanceolate-elliptic, white; pistils gen > 15. **FR** raspberry-like, ± spheric, red-purple to black, puberulent. $2n$=14. Gen ± rocky, esp moist areas; 40–2400 m. CA-FP (exc coast, GV); to B.C., MT, NV. Glandular pls (TR, PR) have been called var. *bernardinus* (E. Greene) Jepson; simple-lvd pls (KR) have been called var. *trinitatis* A. Berger. ❀**4–6**&IRR:1,2,7,8,9,**14–17**,18–24.

R. nivalis Douglas (p. 977) SNOW DWARF BRAMBLE Pl prostrate, rooting at nodes. **ST** 1–2 mm diam, round; prickles ± slender, ± curved. **LF** gen simple; stipules ± ovate; petiole 1–4 cm; blade ± 3–6 cm, ± cordate, gen irregularly lobed, toothed, green below. **INFL** 1–2-fld, nonglandular. **FL**: sepal tips ± 2 mm; petals 4–6 mm, lanceolate-elliptic, gen dull purple to pink (white); pistils gen < 10. **FR** ± hemispheric, red, puberulent. UNCOMMON. Moist semi-shade; ± 1250 m. nw KR (Del Norte Co.); to B.C., ID.

R. parviflorus Nutt. (p. 977) THIMBLEBERRY Pl erect, woody. **ST** 3–15 mm diam, round; prickles 0. **LF** simple; stipules lanceolate; petiole ± 2–12 cm; blade ± 5–15 cm, ± cordate, ± 5-lobed < 1/2 to base, finely toothed, green below. **INFL**: ± cyme, few-fld, densely glandular. **FL**: sepal tips 2–6(20) mm; petals 15–30 mm, elliptic-ovate, gen white (pink); pistils > 15. **FR** raspberry-like, ± hemispheric, red, puberulent. Gen moist shade; < 2500 m. CA-FP (exc GV), MP; to AK, e Can, NM. Hairier coastal pls have been called var. *velutinus* (Hook. & Arn.) E. Greene. ❀**4–6**,**17**&IRR:1,2,7,14,**15**,**16**,18;CVS;may be INV.

R. pensilvanicus Poiret (p. 977) Arched to erect brambles, ± nonglandular. **ST** 5–10 mm diam, 5-angled; prickles many, ± wide-based, gen ± curved. **LF** compound; stipules linear; petiole ± 3–9 cm; lflets 3–5, ± ovate-elliptic, sharply toothed, green below; longest lflet stalk ± 10–30 mm; longest lflet blade ± 6–12 cm. **INFL**: raceme, gen 5–10-fld; bracts often lf-like. **FL**: sepal tips 1–8 mm; petals 10–20 mm, obovate-round, white; pistils gen > 15. **FR** blackberry-like, ± oblong, black, glabrous. Disturbed areas; < 1500 m. CA-FP; native to e N.Am. [*R. almus* (L. Bailey) L. Bailey] Glandular *R. allegheniensis* Porter of e N.Am a scattered waif in CA.

R. spectabilis Pursh (p. 977) SALMONBERRY Pl ± erect, woody, ± thicket-forming. **ST** 3–15 mm diam, round; prickles (0) few, slender, ± straight. **LF** gen compound; stipules linear; petiole ± 1–9 cm; lflets gen 3, ± ovate, shallowly lobed, irregularly toothed, green below; longest lflet stalk ± 5–30(50) mm; longest lflet blade ± 3–10 (17) cm. **INFL** 1–2(4)-fld, nonglandular. **FL**: sepal tips < 2 mm; petals 10–15 mm, obovate-elliptic, ± red; pistils gen > 15. **FR** ± raspberry-like, ± ovoid, yellow to red, glabrous. *n*=7. Moist places, esp streamsides; < 1400 m. NCo, nw KR, n NCoRO, n CCo, w SnFrB; to AK. Hairier coastal pls have been called var. *franciscanus* (Rydb.) J. Howell. ❀IRR:**4,5**&SHD:**6**,7,**15–17**;CV;STBL; may be INV.

R. ulmifolius Schott var. *inermis* (Willd.) W.O. Focke Arched brambles. **ST** 5–10 mm diam, 5-angled; prickles 0. **LF** compound; stipules linear; petiole ± 2–6 cm; lflets 3–5, ± obovate-oblong, finely toothed, white below; longest lflet stalk ± 5–30 mm; longest lflet blade ± 3–8 cm. **INFL**: ± panicle, many-fld, nonglandular. **FL**: sepal tips ± 1 mm; petals 10–15 mm, ± obovate, white to pink; pistils gen > 15. **FR** blackberry-like, ± ovoid, black, glabrous. Uncommon. Gen moist, ± disturbed areas; < 500 m. s NW, ScV, CW, n SW; Eur cultivar.

R. ursinus Cham. & Schldl. (p. 977) CALIFORNIA BLACKBERRY Pl clambering, gen dioecious. **ST** 2–10 mm diam, round; prickles slender, ± straight. **LF** gen 3(7)-lobed or compound; stipules linear; petiole ± 2–5 cm; blade ± 2–10 cm, irregularly toothed, green below; longest lflet stalk 0–30 mm; longest lflet blade elliptic-ovate to ± cordate. **INFL**: ± cyme, few-fld, often glandular. **FL** gen unisexual; sepal tips 0–2 mm; petals 5–25 mm, ± elliptic, white; pistils gen > 10. **FR** blackberry-like, oblong to spheric, black, gen glabrous. *n*=21,28,42. Common. Gen moist places, shrubland, streamsides; < 1500 m. CA-FP; to B.C., ID. [var. *sirbenus* (L. Bailey) J. Howell; *R. macropetalus* Hook.; *R. vitifolius* Cham. & Schldl. incl var. *eastwoodianus* (Rydb.) Munz] Variable, esp hairs. An ancestor of loganberry, boysenberry. ❀SUNorSHD:**4–6**,**17**&IRR:1,2,**7**,8,9,**14–16**,18,**19–24**;also STBL;INV.

SANGUISORBA

Barbara Ertter

Ann or per, sometimes monoecious; hairs 0 or partitioned, not glandular. **LVS** alternate, odd-1-pinnate; lflets gen evenly toothed or lobed. **INFL**: spikes 1–many, head-like; peduncles long. **FLS** bisexual (or upper pistillate, lower staminate); hypanthium urn-shaped; bractlets 0; sepals gen 4; petals 0; stamens 0–many; pistils 0–3, ovaries superior, continuous to style at top, stigma gen ± shrub-like, exserted. **FR**: hypanthium hard, 4-angled, enclosing achenes. ± 25 spp.: n temp, arctic. (Latin: blood-absorbing, from styptic properties) [Nordberg 1966 Opera Bot 11(2):1–103]

1. Lflets lobed > 2/3 to midvein; ann or bien; basal lvs withered at fl; sepals green .... ***S. occidentalis***
1′ Lflets toothed < 1/3 to midvein; per; basal lvs present at fl; sepals green or purplish
  2. Largest lflet blade gen 5–20 mm, on stalk gen 1–4 mm, teeth gen < 15; stamens many, filaments thread-like; open areas; CA-FP (exc SNH) .... ***S. minor*** ssp. ***muricata***
  2′ Largest lflet blade 25–50 mm, on stalk 3–25 mm, teeth gen > 15; stamens 2–4, filaments not thread-like; bogs, streams; NCo, KR, NCoRO .... ***S. officinalis***

S. minor Scop. ssp. *muricata* Briq. (p. 983) GARDEN BURNET Per, tufted, taprooted. **ST** erect, gen 2–7 cm. **LVS**: basal present at fl, largest gen 4–20 cm; lflets 4–10 per side, largest blade gen 5–20 mm, round-oblong, stalk gen 1–4 mm, teeth gen < 15, < 1/3 to midvein. **INFL** 7–30 mm, 6–20 mm wide, ovoid-spheric, ± 5–30-fld. **FL**: sepals 3–6 mm, elliptic, green or purplish; stamens many, filaments thread-like. **FR** ± 5 mm; angles equally short-winged; faces with raised bumpy network. *n*=14,28. Open, esp disturbed areas; 30–1600 m. CA-FP (exc SNH); to e US; native to Eur. Often used in seeding mixtures after fires and in pastures.

S. occidentalis Nutt. (p. 983) WESTERN BURNET Ann or bien, taprooted. **ST** gen ascending to erect, gen 10–70 cm. **LVS**: basal withered at fl; largest cauline gen 3–12 cm; lflets 4–7 per side, largest blade 5–20 mm, ± sessile, ± obovate-elliptic, lobes < 15, > 2/3 to midvein, linear. **INFL** 5–35 mm, 5–10 mm wide, cylindric-ovoid, ± 10–50-fld. **FL**: sepals 2–3 mm, ovate, green; stamens gen 2, filaments thread-like. **FR** 2–4 mm, barely winged; faces wrinkled. *n*=7. Open, esp disturbed areas; 700–1900 m. KR, CaRH, n&c SNH, n CCo, PR, MP; to B.C., MT. *S. annua* (Hook.) Torrey & A. Gray (c US) has winged fr with smooth faces. ❀TRY.

S. officinalis L. (p. 983) GREAT BURNET Per; rhizome thick, creeping. **ST** erect, gen 50–140 cm. **LVS**: basal present at fl, largest gen 20–40 cm; lflets 3–6 per side, largest blade 25–50 mm, ovate-oblong, stalk 3–25 mm, teeth gen > 15, < 1/3 to midvein. **INFL** gen 12–20 mm, 7–10 mm wide, ± elliptic-ovoid, > 20-fld. **FL**: sepals 2–3.5 mm, elliptic-ovate, dark purplish; stamens 2–4, filaments not thread-like. **FR** 2.5–3.5 mm, barely winged; faces ± smooth. *n*=14, 21,28. RARE in CA. Bogs, streams, often on serpentine; 120–1400 m. c NCo, nw KR, n NCoRO; to AK. CA pls may be ssp. *microcephala* (Presl) Calder & Roy Taylor [*S. m.* Presl]. In cult.

SIBBALDIA

Barbara Ertter

Per, low, ± matted; caudex branched. **LVS** gen basal, gen 1-ternate in CA. **INFL**: cyme. **FL**: hypanthium shallow; bractlets gen 5; sepals gen 5; petals gen 5; stamens 4–10; pistils few–many, ovaries superior, jointed to slender style on side. **FR**: achene. ± 20 spp.: n hemisphere, esp Himalayan. (R. Sibbald, Scottish naturalist, physician, 1641–1722) [Dixit & Panigrahi 1981 Proc Indian Acad Sci 90:253–272]

S. procumbens L. (p. 983) Pl hairs appressed, gen ± sparse. **ST** 2–15 cm, ± spreading. **LF**: petiole 1–7 cm; lflets 5–25 mm, ± wedge-shaped, teeth gen 3, at tip. **INFL**: pedicels gen 3–10 mm, straight. **FL**: bractlets linear, < sepals; sepals 2–4 mm, triangular; petals ± 1 mm, widely oblanceolate, yellow; stamens 5. **FR** ± 1 mm, smooth, brown, often retained in disintegrating fl. *n*=7. Moist rocky areas; 1900–3700 m (lowest in n). KR, CaRH, SNH, SnBr, Wrn, SNE; to ne N.Am, Eurasia. ❀TRY;DFCLT.

SORBUS MOUNTAIN ASH

Thomas J. Rosatti

Shrub or tree. **LVS** odd-1-pinnate in CA, petioled, deciduous; lflets gen toothed, terminal sometimes partly fused to uppermost lateral(s). **INFL**: panicle, many-fld. **FL**: sepals 5; petals 5; stamens 15–20; ovary gen inferior, styles gen free. **FR**: pome; chambers 1–5, 1–2-seeded. ± 80 spp.: n temp. (Latin: ancient name) [Jones 1939 J Arnold Arbor 20:1–43] Intergrading: native spp. might all = *S. sitchensis*.

1. St-bud hairs dense, whitish, evenly distributed; lflets 9–17; alien .... ***S. aucuparia***
1' St-bud hairs (0) gen sparse, whitish or red-brown, gen ± marginal; lflets 7–13; native
 2. Lflet toothed gen only above middle; styles 2–3 mm .... ***S. sitchensis*** var. ***grayi***
 2' Lflet toothed from below middle (or near base); styles 1–2.5 mm
  3. Pedicel glabrous; lflets 7–9(11), 2–4 cm; st-buds 3–8(11) mm, hairs red-brown; petals 3–4 mm; pl 1–2 m .... ***S. californica***
  3' Pedicel sparsely hairy; lflets 9–13, 2–7 cm; st-buds 7–15 mm, hairs whitish; petals 5–6 mm; pl 1–5 m ***S. scopulina***
   4. Lflets 9–11, gen oblong (ovate); stipules often ± persistent; esp w of CaR summit .... var. ***cascadensis***
   4' Lflets (9)11–13, gen lanceolate; stipules gen not persistent in fr; esp e of CaR summit .... var. ***scopulina***

S. aucuparia L. Tree < 15 m; st-bud hairs dense, evenly distributed, whitish. **LF**: lflets 9–17, 3–5 cm, oblong-lanceolate. **INFL**: pedicel hairs gen dense. **FL**: petals ± 4 mm, ± round. **FR** 9–11 mm, bright red. **SEEDS** ± 4 mm, ovate, ± flat, light brown. 2*n*=34. Disturbed places, gen near dwellings; gen < 1700 m. SnFrB, SnBr (expected elsewhere); native to Eur.

S. californica E. Greene (p. 983) Shrub 1–2 m; st-buds 3–8(11) mm, sticky, hairs gen sparse, marginal, red-brown. **LF**: lflets 7–9 (11), 2–4 cm, 1–2.5 cm wide, oblong-ovate. **INFL**: pedicel glabrous. **FL**: petals 3–4 mm, ± round. **FR** 7–10 mm, bright red. **SEEDS** ± 4 mm, ovate, barely flat, light brown. Moist coniferous forests; 1500–3400 m. KR, NCoRH, CaR, SNH, SnFrB, Wrn?; w NV. Intergrades with *S. scopulina*. ❀IRR,DRN,SHD:1,2,4,5,**6**, 7,15,16;DFCLT.

S. scopulina E. Greene Shrub 1–5 m; st-buds 7–15 mm, sticky, hairs gen sparse, ± marginal, whitish. **LF**: lflets 9–13, 2–7 cm, lanceolate, oblong, or ovate. **INFL**: pedicel sparsely hairy. **FL**: petals 5–6 mm, ovate to ± round. **FR** 8–10 mm, orange to bright red. **SEEDS** 3.5–4 mm, ovate to oblong, flat, brown. Canyons, wooded slopes, moist places, coniferous forests; 1200–2800 m. KR, NCoR, CaR, SNH, MP; to B.C., MT, NM. Vars. intergrade.

var. ***cascadensis*** (G. Jones) C. Hitchc. (p. 983) **LF**: lflets 9–11, 2.5–7 cm, 1–3 cm wide, gen oblong (ovate); stipules often ± persistent. **FL**: petals ± round. **FR** bright red. **SEEDS** ± 4 mm, ovate, dark brown. Habitats of sp.; 1200–2100 m. KR, n NCoR, CaR, n SNH, MP; to B.C. [*S. c.* G. Jones] ❀IRR,DRN:1,4–6,15, 16&SHD:2,7;DFCLT.

var. ***scopulina*** **LF**: lflets (9)11–13, 2–6 cm, gen lanceolate (oblong); stipules gen not persistent in fr. **FL**: petals ovate. **FR** orange to bright red. **SEEDS** 3.5–4 mm, oblong, light brown. Habitats and elevations of sp. KR, NCoR, CaR, SNH, MP; to B.C., Rocky Mtns. ❀IRR,DRN:1,4–6,15–17&SHD:2,7;DFCLT.

S. sitchensis K.F. Roemer var. ***grayi*** (Wenzig) C. Hitchc. (p. 983) Shrub 1–3 m; st-buds 5–7 mm, hairs gen sparse, marginal or ± evenly distributed, red-brown. **LF**: lflets 7–11, 2–6 cm, 8–25 mm wide, ovate to oblong, gen toothed only above middle. **INFL**: pedicel glabrous or ± red-brown-hairy. **FL**: petals 3–4 mm, ovate. **FR** 7–10 mm, red. **SEEDS** ± 3.5 mm, ovate, barely flat, dark brown. Montane areas; 600–3000 m. c KR (Salmon Mtns); to B.C. [*S. occidentalis* (S. Watson) E. Greene] ❀IRR,DRN,SHD:1,2,4,5,**6**, 15,16;DFCLT.

SPIRAEA

Thomas J. Rosatti

Shrub, unarmed. **LVS** elliptic to (ob)ovate, gen serrate, deciduous; stipules gen 0; petiole 0–short. **INFL**: raceme or panicle, many-fld, bracted; bractlet gen 1 at top of pedicel, gen linear. **FL**: hypanthium obconic to bell-shaped; sepals spreading to erect; petals spreading, white to rose; stamens 15–many; pistils 5, opposite petals, free or fused at base, surrounded by hypanthium, ovaries superior, styles ± terminal, beak-like in fr, stigmas head-like. **FR**: follicles 5, dehiscent along inner suture and top of outer. **SEEDS** ± fusiform; coats membranous. ± 50 spp.: n temp. (Greek: shrub) [Henrickson 1985 Aliso 11:199–211]

1. Infl gen wider than long, ± flat-topped; lf glabrous to sparsely hairy, surfaces similar; sepals spreading to erect .... ***S. densiflora***
1' Infl longer than wide, not flat-topped; lf tomentose and lighter below; sepals mostly soon reflexed ..... ***S. douglasii***

S. densiflora Torrey & A. Gray (p. 983) Pl 2–9 dm, glabrous to finely and sparsely hairy. **LF** gen 1–7 cm; petiole < 3 mm; blade surfaces similar. **INFL** gen wider than long, ± flat-topped. **FL**: hypanthium 2–2.5 mm; sepals ± 1 mm, spreading to erect; petals ± 1.5 mm, rose. Moist, rocky areas incl serpentine, coniferous forests; 600–3400 m. KR, CaR, SNH; to B.C. Pls with young herbage finely hairy have been called ssp. *splendens* (Baumann) Abrams. ❀IRR,DRN:1,**4–6,17**&SHD:2,7,14,**15,16**,18–21.

S. douglasii Hook. (p. 983) Pl 10–20 dm, ± tomentose. **LF** gen 3–12 cm; petiole < 10 mm; blade darker, nearly glabrous above. **INFL** longer than wide, not flat-topped; **FL**: hypanthium ± 1 mm; sepals 0.5–1 mm, mostly soon reflexed; petals ± 1.5 mm, pink to rose. *n*=18. Moist areas, coniferous forests; < 2000 m. n NCo, KR, CaR, n SNH, w MP; to B.C. ❀IRR:1,**4–6,15–17**&SHD:2,**7**,8,9, **14**,18–24;may be INV.

RUBIACEAE MADDER FAMILY

Lauramay T. Dempster

Ann, per, shrub, vine, tree. **LVS** gen opposite, entire; stipules gen on st, sometimes lf-like (then lvs apparently whorled and stipules considered lvs), adjacent pairs sometimes fused. **INFL**: cyme, panicle, cluster, or fl solitary, gen terminal and ± axillary. **FL** gen bisexual; calyx gen ± 4-lobed, sometimes 0; corolla gen radial, 4-lobed; stamens epipetalous, alternate

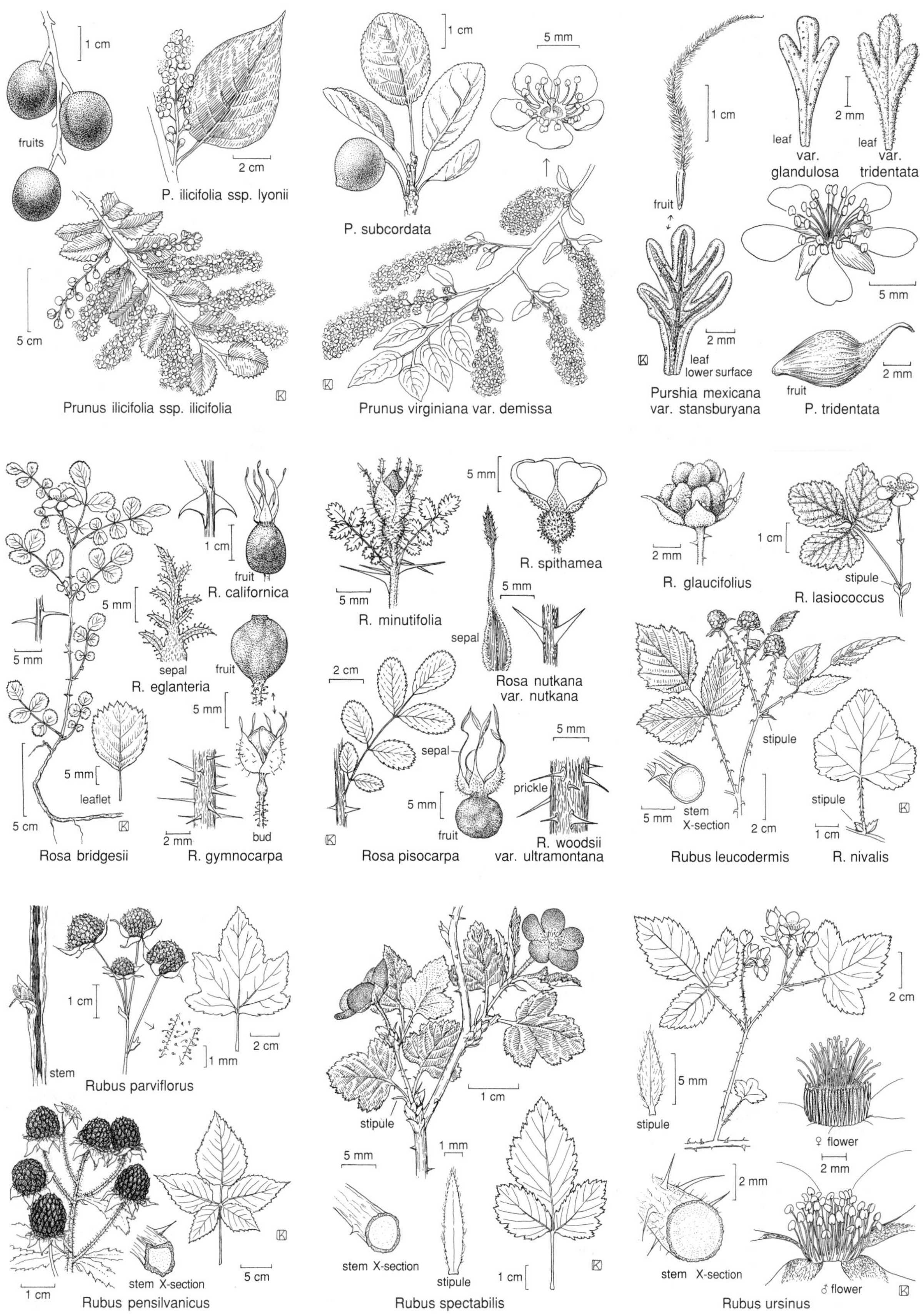

fruits
1 cm
2 cm
5 cm
P. ilicifolia ssp. lyonii
Prunus ilicifolia ssp. ilicifolia
1 cm
5 mm
P. subcordata
Prunus virginiana var. demissa
1 cm
fruit
2 mm
leaf
var.
glandulosa
var.
tridentata
5 mm
2 mm
leaf
lower surface
Purshia mexicana
var. stansburyana
fruit
P. tridentata
5 mm
1 cm
fruit
R. californica
5 mm
sepal
R. eglanteria
fruit
5 mm
5 mm
leaflet
5 cm
2 mm
bud
Rosa bridgesii
R. gymnocarpa
5 mm
R. spithamea
5 mm
R. minutifolia
5 mm
sepal
Rosa nutkana
var. nutkana
2 cm
sepal
5 mm
fruit
5 mm
prickle
Rosa pisocarpa
R. woodsii
var. ultramontana
2 mm
R. glaucifolius
1 cm
stipule
R. lasiococcus
stipule
5 mm
stem
X-section
2 cm
stipule
1 cm
Rubus leucodermis
R. nivalis
1 cm
2 cm
1 mm
stem
Rubus parviflorus
1 cm
stem X-section
5 cm
Rubus pensilvanicus
1 cm
stipule
5 mm
1 mm
stem X-section
stipule
1 cm
Rubus spectabilis
2 cm
5 mm
stipule
♀ flower
2 mm
2 mm
stem X-section
♂ flower
Rubus ursinus

corolla lobes, gen incl; ovary gen inferior, chambers gen 2 or 4, style 1, ± fused if 2. **FR**: 2 or 4 nutlets or a berry, drupe, or capsule. ± 500 genera, 6000 spp.: world-wide, esp trop; many cult (incl *Coffea*, coffee; *Cinchona*, quinine; many orn). [Dempster 1979 Fl CA 4(2):1–47]

1. Large shrub; infl a dense, spheric head ..... **CEPHALANTHUS**
1′ Herb, subshrub; infl not a spheric head
 2. Lvs opposite
  3. Fls sessile, 1–several per node ..... **DIODIA**
  3′ Fls on long pedicels; infl terminal, open, few-fld ..... **KELLOGGIA**
 2′ Lvs whorled
  4. Infl a dense spike with overlapping bracts; calyx 0 ..... **CRUCIANELLA**
  4′ Infl not a spike; calyx present or 0
   5. Fls pedicelled, gen in clusters, sometimes 1; involucre 0; calyx 0 ..... **GALIUM**
   5′ Fls sessile, in few-fld heads; involucral bracts ± free, ± = lvs; calyx teeth 6 ..... **SHERARDIA**

CEPHALANTHUS BUTTON BUSH

Shrub, small tree. **LVS** opposite or in whorls of 3; stipules not lf-like. **INFL**: head, dense, spheric, peduncled. **FL**: calyx 4-toothed; corolla narrowly funnel-shaped; style 1, thread-like, exserted, stigma ± spheric. **FR**: 2–4 nutlets, wider at top. ± 17 spp.: warm-temp Am., Asia, s Afr. (Greek: head fl)

C. occidentalis L. var. ***californicus*** Benth. (p. 983) CALIFORNIA BUTTON WILLOW Pl 2–10 m. **ST** when young round, reddish, glabrous. **LF** 7–20 cm, elliptic or ovate, petioled, becoming glabrous. **INFL** 3–3.5 cm; peduncle 2.5–5 cm. **FL**: corolla white or yellowish, tube slender, lobes obtuse; stigma exserted to 4 mm. Lake, stream edges; 3–1000 m. NCoRI, SNF, ScV, SnJV (common); to TX, Mex. ❀SUN&WETorIRR:**7–9**,10–12,**14–24**.

CRUCIANELLA CROSS-WORT

Ann. **LVS** whorled below, opposite (without stipules) in infl; stipules lf-like. **INFL**: spike; bracts overlapping, 3 per fl. **FL**: calyx 0; corolla tubular, lobes 4–5, erect; style divided above, stigmas head-like. **FR**: 2 nutlets, wider at top. 35–40 spp.: Medit to c Asia. (Latin: little cross, presumably from corolla lobes)

C. angustifolia L. Pl erect or ± decumbent, grass-like. **ST** 12–30 cm, 4-angled. **LVS** in whorls of 4, 8–25 mm, linear. **INFL** 2.5–7.5 cm, dense, narrow, 4-angled; bracts scarious, rib green, tip sharp to touch. **FL**: corolla white or yellow. **FR** glabrous. $2n$=22. Oak woodlands; 200–600 m. CaRF, n SNF SCoRI; native to Eur.

DIODIA BUTTONWEED

Ann. **LVS** opposite; stipules fused to lf bases, small, fringed. **INFL**: fls 1–several in axils, sessile. **FL**: corolla funnel-shaped, lobes gen 4. **FR**: 2 nutlets; calyx teeth persistent. 50 spp.: warm Am, Afr. (Greek: thoroughfare, from habitats)

D. teres Walter ROUGH BUTTONWEED, POOR JOE **ST** erect, simple or branched, 7–25 cm, 4-angled. **LF** 2.5 cm, lanceolate, tip sharp; petiole and blade base transparent, fringed, forming a cup ± enclosing infl. **FL**: calyx lobes 4, stiff, unequal; corolla 3 mm, white or pink. **FR**: hairs stiff. Sandy places; 50 m. DSon (sw Imperial Co); to e US, Mex.

GALIUM BEDSTRAW, CLEAVERS

Ann, per, sometimes ± shrubby, often ± dioecious, glabrous or hairy, often scabrous. **ST** when young 4-angled. **LVS** in whorls of 4 or more, incl lf-like stipules. **INFL**: panicle, or axillary clusters of 1–many fls. **FL** bisexual or unisexual (with sterile stamens or pistils); calyx 0; corolla gen rotate, sometimes ± bell-shaped, gen greenish, fading yellow or white, sometimes reddish, lobes gen 4; ovary 2-lobed, styles 2, ± fused basally. **FR**: 2 nutlets or 1 berry. ± 400 spp.: worldwide, esp temp. (Greek: milk, from use of some spp. in its curdling) Hairiness of ovary and fr gen ± equal on a single pl; staminate pls often identified only by association with pistillate.

1. Ann
 2. Corolla lobes gen 3
  3. Lvs in whorls of 2–4; fls gen solitary in axils ..... ***G. bifolium***
  3′ Lvs in whorls of 4–6; fls several in axils ..... [2]***G. trifidum***
   4. Pl weak, straggling ..... [2]var. ***pacificum***
   4′ Pl tufted or matted, dwarfed ..... [2]var. ***pusillum***
 2′ Corolla lobes gen 4
  5. Nutlets sausage-shaped, with unevenly distributed, hooked hairs ..... ***G. murale***
  5′ Nutlets ± spheric, glabrous or with evenly distributed, hooked hairs, sometimes with hair-like tubercles
   6. Lvs in whorls of 2–4; fr with hooked hairs ..... ***G. proliferum***
   6′ Lvs in whorls of 5–8; fr with or without hooked hairs
    7. Pl erect, very slender; lf < 7 mm
     8. Fr glabrous; lf seldom or little reflexed ..... ***G. divaricatum***
     8′ Fr gen with hooked hairs; lf eventually reflexed ..... ***G. parisiense***
    7′ Pl gen climbing or decumbent, coarse; lf 13–32 mm

9. Fr with hooked hairs, the pair of nutlets < 4 mm wide; pedicel slender, spreading .... ***G. aparine***
9′ Fr with hair-like tubercles, without hairs, the pair of nutlets > 4 mm wide; pedicel stout, recurved .... ***G. tricornutum***
1′ Per or shrub
10. Corolla gen 3-lobed; fr 2 nutlets, spheric, smooth, hard .... [2]***G. trifidum***
11. Pl weak, straggling .... [2]var. ***pacificum***
11′ Pl tufted or matted, dwarfed .... [2]var. ***pusillum***
10′ Corolla gen 4-lobed; fr 2 nutlets, various, or a berry
12. Fr hairs hooked
13. Lvs in whorls of 4 .... ***G. oreganum***
13′ Lvs in whorls of 6 .... ***G. triflorum***
12′ Fr hairs 0 or not hooked
14. Fls bisexual
15. Lvs in whorls of 4 .... ***G. boreale*** ssp. ***septentrionale***
15′ Lvs in whorls of 5–12
16. Corolla yellow .... ***G. verum***
16′ Corolla white
17. Pls forming dense mats in lawns; sts 7–15 cm .... ***G. saxatile***
17′ Pls ascending or decumbent, not forming mats; sts 30–120 cm
18. Corolla bell-shaped; ovaries with short, upwardly curved hairs .... ***G. mexicanum*** var. ***asperulum***
18′ Corolla rotate; ovaries glabrous or with small tubercles
19. Lf gen < 15 mm, obovate to oblanceolate .... ***G. mollugo***
19′ Lf gen 25–60 mm, narrowly elliptic .... ***G. schultesii***
14′ Fls gen unisexual, either anthers or ovaries small, sterile — lvs in whorls of 4 (exc *G. hardhamiae*, sometimes *G. clementis*)
20. Fr a berry, hairiness like that of herbage
21. Lvs in whorls of 6 .... ***G. hardhamiae***
21′ Lvs in whorls of 4 (sometimes 6 in *G. clementis*)
22. Pl low, gen not woody above ground
23. Lf ± bristle-like, sharp to touch, ± thick; dry areas .... ***G. andrewsii***
24. Pl dense, internode < lf, and a main st not apparent, and lf gen awl-like, keeled, ascending; herbage not hairy .... ssp. ***andrewsii***
24′ Pl elongate, internode > lf, or a main st apparent, or lf wider than awl-like, flat, or spreading; herbage hairy or not
25. Herbage sometimes hairy; lf often flat, lanceolate, spreading; pl open, a main st apparent .... ssp. ***gatense***
25′ Herbage gen not hairy; lf flat or ± keeled, narrower than lanceolate, gen ascending; pl ± dense, a main st ± not apparent .... ssp. ***intermedium***
23′ Lf linear to elliptic, not sharp to touch, gen thin (exc *G. ambiguum* var. *siskiyouense*); moist areas
26. Lf ± linear, gen ciliate .... ***G. ambiguum***
27. Lf gen thin, ± finely hairy, margins ± parallel; ovary, fr gen glabrous or velvety .... var. ***ambiguum***
27′ Lf often thick, gen glabrous or with marginal bristles, margins gen convex; ovary, fr densely hairy .... var. ***siskiyouense***
26′ Lf ovate, elliptic, oblanceolate, or seeming linear, not ciliate (or hairs at least not restricted to margin)
28. Lf seeming linear, margin strongly turned down; SCoRO .... ***G. clementis***
28′ Lf gen ovate, elliptic, or oblanceolate, margin not or slightly turned down; widespread
29. Lf with few hairs, 6–10 mm; st gen glabrous .... ***G. muricatum***
29′ Lf ± hairy, 3–20 mm; st gen hairy .... [2]***G. californicum***
30. Pl open; mainland .... [2]ssp. ***californicum***
30′ Pl dense; n ChI .... ssp. ***miguelense***
22′ Pl often climbing, ± woody above ground
31. Herbage gen with minute prickles; woody st often long, slender, climbing or sprawling
32. Lf often obtuse or rounded but tip often with a short point, gen not sharp to touch .... ***G. porrigens***
33. Lf broadly ovate to oblong .... var. ***porrigens***
33′ Lf ± linear .... var. ***tenue***
32′ Lf gen tapered, tip often sharp to touch
34. Herbage loosely sprawling, not climbing high, gen remaining green; SCoRO .... ***G. californicum*** ssp. ***maritimum***
34′ Herbage densely tangled, climbing high, maturing red; SCo, s ChI, WTR, PR .... ***G. nuttallii***
35. Herbage gen glabrous; s ChI .... ssp. ***insulare***
35′ Herbage with tiny, sharp, curved hairs; SCo, WTR, PR .... ssp. ***nuttallii***
31′ Herbage gen without prickles; woody st gen short
36. Lf ± glabrous, exc sometimes margin, or puberulent
37. Lf sharp to touch at tip, 1-veined; SCoRO, WTR .... ***G. cliftonsmithii***
37′ Lf not sharp to touch at tip, ± 3-veined; n of SCoRO, WTR

38. Lf lance-ovate to elliptic, sessile (at least broadly attached); infl branchlets ± stout; corolla gen red ..... [2]***G. bolanderi***

38′ Lf broadly elliptic to ovate, ± petioled; infl branchlets thread-like; corolla gen yellowish — CaRF, SN, n SCoRI ..... ***G. sparsiflorum***

39. Lf gen glabrous, ovate, ± leathery ..... ssp. ***glabrius***

39′ Lf puberulent, elliptic or sometimes ± ovate, gen limp ..... ssp. ***sparsiflorum***

36′ Lf hairy

40. Woody sts ± stout, ± stiff; montane

41. Pl sprawling widely on low shrubs; lf gen broadly elliptic; SnGb ..... ***G. grande***

41′ Pl tufted or ± climbing; lf gen elliptic; farther n than SnGB ..... [2]***G. bolanderi***

40′ Woody sts wiry, slender; gen coastal, insular, sometimes montane ..... [2]***G. californicum***

42. Lf gen 4–6 mm, ± fleshy, abruptly petioled; SCoRO ..... ssp. ***luciense***

42′ Lf 4–25 mm, not fleshy, gen not petioled

43. Lf 5–8 × longer than wide; n SNF ..... ssp. ***sierrae***

43′ Lf 2–4 × longer than wide

44. Herbage gen with dense, soft, fine hairs, or sometimes nearly glabrous — SCoRO, WTR, SnGB, n ChI ..... ssp. ***flaccidum***

44′ Herbage with ± sparse, scattered hairs

45. Hairs ± coarse; SnFrB, NCoRO, NCoRI ..... [2]ssp. ***californicum***

45′ Hairs fine; SnJt ..... ssp. ***primum***

20′ Fr 2 nutlets, hairs unlike those of herbage, long, straight, (short, curved in *G. buxifolium*, *G. jepsonii*)

46. Pl with staminate, pistillate, and sometimes bisexual fls

47. Herb; st gen slender, flexible, woody; mainland

48. Lf ovate to orbicular, not sharp at tip; infl narrow, open between dense clusters ..... ***G. parishii***

48′ Lf linear to oblanceolate, acute or sharp at tip; infl open ..... ***G. wrightii***

47′ Erect shrub; st rigid, stout, woody; ChI

49. Ovary hairs short, curved, ascending; n ChI ..... ***G. buxifolium***

49′ Ovary hairs long, straight, spreading (or nearly 0); s ChI ..... ***G. catalinense***

50. St, lf hairs sparse, gen very short, straight or, if longer, curved but not curly; ChI (San Clemente Island) ..... ssp. ***acrispum***

50′ St, lf hairs ± dense, short, curly; ChI (Santa Catalina Island) ..... ssp. ***catalinense***

46′ Pl with staminate or pistillate fls only

51. Ovary, fr with short, curved, ascending hairs; corolla bell-shaped, divided halfway to base or less ..... ***G. jepsonii***

51′ Ovary, fr with long, straight, spreading hairs; corolla rotate or, if bell-shaped, divided > halfway to base (exc *G. hilendiae* ssp. *kingstonense*)

52. Lf ± linear, gen widest above middle

53. Pedicel gen > fr; upper nodes >> lower; fr hairs < fr ..... ***G. johnstonii***

53′ Pedicel gen < fr; upper nodes ± = lower; fr hairs = or > fr ..... ***G. angustifolium***

54. Corolla gen hairy (sometimes glabrous in ssp. *jacinticum*)

55. St hairy, sometimes scabrous

56. St hairs ± as long as those of lf; st ridges gen narrower than surfaces between; SnGb ..... ssp. ***gabrielense***

56′ St hairs shorter than those of lf; st ridges gen wider than surfaces between; s SNH ..... ssp. ***onycense***

55′ St ± glabrous

57. Infl pyramid-shaped, many-fld, much-branched — pl 35–50 cm; DSon (Borrego Desert) ..... ssp. ***borregoense***

57′ Infl narrow, few-fld, few-branched

58. Pl 17–35 cm; lf 11–26 mm; SnJt ..... ssp. ***jacinticum***

58′ Pl 6–16 cm; lf gen 2–10 mm; SnGb, SnBr ..... ssp. ***nudicaule***

54′ Corolla gen glabrous or not more hairy than st or lf

59. Woody st internode often < lf; pl gen glabrous; n ChI ..... ssp. ***foliosum***

59′ Woody st internode gen > lf; pl hairy or not

60. Pl tall or low, ± stout, glabrous to white-hairy; lvs not deciduous; widespread in hills, mtns, but not in D ..... ssp. ***angustifolium***

60′ Pl tall, very slender, gen glabrous; lvs deciduous; DMtns ..... ssp. ***gracillimum***

52′ Lf gen ovate, gen widest below middle

61. Pl woody at base and above

62. St flexible, few-branched; lf obtuse to acute but not sharp at tip; TR, s SNH ..... ***G. hallii***

62′ St rigid, much-branched; lf sharp at tip; D ..... ***G. stellatum*** var. ***eremicum***

61′ Pl woody only at base

63. Corolla gen bell-shaped or cupped at base

64. Herbage glabrous

65. Lvs at each node ± equal, lanceolate to ovate, tapered to a pointed tip; DMtns ..... ***G. argense***

65′ Lvs at each node ± unequal, the 2 larger ovate to orbicular, rounded abruptly to a pointed tip; n SNH, GB ..... [2]***G. multiflorum***

64′ Herbage hairy, sometimes scabrous

66. Lf rounded abruptly to a pointed tip, gen ± arched ..... [2]***G. multiflorum***

66′ Lf tapered to a pointed tip, gen not arched .... ***G. hilendiae***
67. Pl gen tall, stiff; internode 2–6 × lf; herbage ± hairy, gen scabrous; lf tip ± sharp to touch .... ssp. ***carneum***
67′ Pl gen ± low, flexible; internode 1–4.5 × lf; herbage with long hairs; lf tip not sharp
68. Corolla throat wide; lf ovate to ± round .... ssp. ***hilendiae***
68′ Corolla throat slender; lf lanceolate, ovate, or elliptic .... ssp. ***kingstonense***
63′ Corolla gen rotate, sometimes shallowly bell-shaped
69. Corolla hairy — SNE, DMtns
70. Herbage hairy .... ***G. munzii***
70′ Herbage glabrous below hairy infl
71. Pl flexible; lf 8–23 mm, gen lanceolate to ovate, tip acute but not sharp to touch .... ***G. magnifolium***
71′ Pl stiff; lf 2–10 mm, ovate, tip sharp to touch .... ***G. matthewsii***
69′ Corolla glabrous or velvety with magnification
72. Fr, incl hairs, 3–4 mm, well exserted from lvs; c&s SNH, SNE, DMtns .... ***G. hypotrichium***
73. Corolla shallowly bell-shaped, reddish — herbage gray with ± dense, fine hairs; pl < 9 cm .... ssp. ***tomentellum***
73′ Corolla gen rotate, greenish, yellow, or pink-tinged
74. Pl 12–23 cm, not dense .... ssp. ***inyoense***
74′ Pl 2.5–15 cm, ± dense (ssp. *subalpinum* often open)
75. Herbage velvety with magnification .... ssp. ***hypotrichium***
75′ Herbage hairy without magnification, not velvety .... ssp. ***subalpinum***
72′ Fr, incl hairs, 5–8 mm, gen not much exserted from lvs; n SNH (n of Lake Tahoe), NCoRH
76. Herbage grayish; lf broadly ovate to round .... ***G. grayanum***
77. Pl 5–23 cm .... var. ***grayanum***
77′ Pl 2.5–14 cm, ± dwarfed .... var. ***nanum***
76′ Herbage gen green; lf lanceolate to elliptic or obovate
78. Lf lanceolate to elliptic .... ***G. serpenticum***
79. Corolla often ± cup-shaped; lf tip gen flat; KR .... ssp. ***scotticum***
79′ Corolla rotate; lf tip often reflexed; Wrn .... ssp. ***warnerense***
78′ Lf ovate to obovate or nearly round .... ***G. glabrescens***
80. Lf ovate to obovate or round, tip sometimes blunt; KR .... ssp. ***glabrescens***
80′ Lf ovate, tip gen slender, acute; Wrn .... ssp. ***modocense***

G. ambiguum Wight YOLLA BOLLY BEDSTRAW Per, low, matted, not woody above ground, dioecious, hairy. **ST** 5–15 cm. **LVS** in whorls of 4, 6–16 mm, ± linear; tip ± acute; margin gen ciliate. **STAMINATE INFL**: few-fld clusters. **PISTILLATE INFL**: fls gen solitary in axils. **FL**: corolla rotate, yellowish. **FR**: berry. Light shade of conifers, oaks; 150–2100 m. KR, NCoRO, NCoRH, n SNF; s OR.

var. ***ambiguum*** (p. 983) **LF** gen thin, ± finely hairy; margins ± straight. **FR** (and ovary in fl) gen glabrous or velvety. $2n$=22,44. NCoRH, n SNF (Eldorado Co).

var. ***siskiyouense*** Ferris (p. 983) **LF** often thick, gen glabrous or with marginal bristles; margins gen convex. **FR** (and ovary in fl) densely hairy. $2n$=66. KR (Del Norte Co), NCoRO (uncommon); s OR. ❀IRR:1&SUN:5,17.

G. andrewsii A. Gray PHLOX-LEAVED BEDSTRAW Per, low, tufted, matted, dioecious, green or silvery. **ST** 5–22 cm. **LVS** in whorls of 4, 4–11 mm, ± bristle-like, sharp to touch, ± thick; tip with 1 persistent hair. **STAMINATE INFL**: few-fld clusters. **PISTILLATE INFL**: fls solitary in axils. **FL**: corolla rotate, yellowish. **FR**: berry; hairs 0. Dry slopes, ridges, chaparral, or open woodlands; 220–2580 m. CW, SW; Baja CA. Sspp. difficult to distinguish but biologically valid.

ssp. ***andrewsii*** (p. 983) Pl low, dense, a main st not apparent, glabrous. **ST** 5–15 cm. **LVS** > internodes, hiding st, awl-like, ± keeled, ascending. $2n$=22. High chaparral, open woodland, gen serpentine or sandy-loam soil; 250–2580 m. SCoR, SnBr, PR; mtns of Baja CA. ❀DRN:**15,16**,17&SHD:3,**14,18**.

ssp. ***gatense*** (Dempster) Dempster & Stebb. Pl sometimes ± erect, open; a main st apparent, sometimes hairy. **ST** < 22 cm. **LVS** often < internodes, often lanceolate, flat, sometimes spreading. $2n$=88. Dry, rocky places in serpentine soil, chaparral or open oak/pine woodland; 220–1450 m. SnFrB, SCoRI. [var. *g.* Dempster]

ssp. ***intermedium*** Dempster & Stebb. Pl gen low, ± dense; a main st sometimes apparent, glabrous. **ST** 6–20 cm. **LVS** sometimes < internodes, ± lanceolate, gen ± flat, sometimes spreading. $2n$=44. High chaparral, open oak/pine woodland, in various soils, incl serpentine; 300–700 m. SCoRO, WTR, SnGb.

G. angustifolium Nutt. (p. 983) NARROW-LEAVED BEDSTRAW Per, low, tufted or with elongate sts, woody at least at base, dioecious, glabrous to hairy. **ST** 6–100 cm. **LVS** in whorls of 4, ± strap-shaped, 3-veined. **INFL**: panicle, individual clusters with few–many fls. **FL**: corolla rotate, gen yellowish. **FR**: nutlets, gen > pedicel; hairs dense, long, straight, spreading. Mtn areas where roots sheltered; 15–2650 m. s SN, Teh, SCoR, SW, DMtns, DSon; Baja CA.

ssp. ***angustifolium*** Pl glabrous or hairy. **ST** 15–100 cm; ridges narrower than surfaces between. **LF** < 27 mm. **INFL** many-fld, gen open. **FL**: corolla not hairier than herbage. $2n$=22,44. Cliffs, canyons, protected places on hillsides; 15–2650 m. s SN, Teh, SCoRO, s ChI; n Baja CA.

ssp. ***borregoense*** Dempster & Stebb. BORREGO BEDSTRAW Pl glabrous. **ST** 35–50 cm, slender; ridges ± covering surfaces between. **LF** < 8 mm. **INFL** many-fld, ± dense. **FL**: corolla hairs longer than those of herbage. RARE. Among boulders; 350–1250 m. DSon (Palm Canyon, Hellhole Canyon, Pinyon Mtn Valley; San Diego Co.).

ssp. ***foliosum*** (Hilend & Howell) Dempster & Stebb. Pl glabrous exc lf margin. **ST** 30–60 cm; nodes close together, prominent after lf-fall; ridges ± as wide as surfaces between. **LF** gen 3–11 mm. **INFL** many-fld, ± dense. **FL**: corolla glabrous or puberulent. $2n$=22. Rocky slopes; 30–60 m. n ChI. [var. *f.* Hilend. & Howell]

ssp. ***gabrielense*** (Munz & I.M. Johnston) Dempster & Stebb. (p. 983) SAN ANTONIO CANYON BEDSTRAW Pl tufted, woody at base, ± hairy. **ST** 6–30 cm; ridges narrower than surfaces between. **LF** gen 2–14 mm. **INFL** narrow, few-fld, ± dense. **FL**: corolla yellow or reddish, bristly on outside. $2n$=44. UNCOMMON. Slopes, ridges, open forest, high chaparral; 1200–2650 m. SnGb (near San Antonio Canyon; Los Angeles, San Bernardino cos.). [*G. g.* Munz & I.M. Johnston] ssp. ***gracillimum*** Dempster & Stebb. Pl very slender throughout, gen glabrous exc lvs scabrous. **ST** ± 40 cm, woody; ridges ± as wide as surfaces between. **LVS** gen 4–15 mm, ephemeral. **INFL** few-fld, ± open. **FL**: corolla hairs gen 0. **FR**: nutlets, < 2 mm, > hairs. 2*n*=22. Shaded places among granite boulders in canyons, on outcrops; 130–1550 m. DMtns (Providence, Little San Bernardino mtns).

ssp. ***jacinticum*** Dempster & Stebb. Pl ± low, gen glabrous or lvs ± hairy. **ST** 17–35 cm; ridges ± as wide as surfaces between. **LF** gen 11–26 mm. **INFL** narrow, few-fld, ± open. **FL**: corolla hairs scattered. 2*n*=66. Open mixed forest; 1350–2100 m. w SnJt.

ssp. ***nudicaule*** Dempster & Stebb. Pl low, gen glabrous or lvs ± hairy; rhizome creeping. **ST** 6–16 cm; ridges ± as wide as surfaces between. **LF** gen 2–10 mm. **INFL** narrow, few-fld, open. **FL**: corolla gen red, sometimes yellow, hairy outside. 2*n*=22. Steep slopes, open mixed forest; 1650–2650 m. SnGb, SnBr.

ssp. ***onycense*** (Dempster) Dempster & Stebb. ONYX PEAK BEDSTRAW Pl slender; ± scabrous, grayish green; base woody. **ST** 12–30 cm; ridges often hiding surfaces between. **LF** gen 5–10 mm. **INFL** ± narrow, few-fld, open. **FL**: corolla gen pink, hairy outside. 2*n*=22. UNCOMMON. Granite outcrops, open oak/pine woodland; 950–2300 m. s SNH (Onyx Peak area, e Kern Co).

G. aparine L. (p. 983) GOOSE GRASS Ann, climbing or prostrate, or sometimes low, erect; herbage adhesive by small, hooked prickles. **ST** 3–9 dm, weak, brittle. **LVS** in whorls of 6–8, 13–31 mm; lowest petioled, ± round; upper sessile, ± narrowly oblanceolate. **INFL**: fls few on branchlets in most axils. **FL** bisexual; corolla rotate, whitish. **FR**: nutlets; hairs many, short, hooked. 2*n*=20,22, 42,44,63,64,66,±86,88. Grassy, half-shady places, weedy in gardens; 30–1500 m. CA-FP; to AK, e Coast, perhaps native to Eur. Small pls with slender, pointed lvs, yellowish corollas, have been called *G. spurium* L.

G. argense Dempster & Ehrend. Per, erect, 20–50 cm, stiff, open, dioecious, glabrous. **LVS** in whorls of 4, 8–17 mm, lanceolate to ovate, tapered to sharp tip. **INFL**: clusters terminal on branchlets in upper, ± lfless part of pl. **FL**: corolla ± bell-shaped to rotate, cream. **FR**: nutlets; hairs long, straight, spreading. 2*n*=44. Loose, stony slopes; 1250–2400 m. DMtns (Argus, Nelson ranges; Inyo Co.).

G. bifolium S. Watson (p. 987) LOW MOUNTAIN BEDSTRAW Ann, erect, gen 5–15 cm, glabrous; branches few or 0. **LVS** in whorls of 4, in 2 unequal pairs, or uppermost often opposite, 10–21 mm, lanceolate or narrowly elliptic. **INFL**: fls gen solitary on slender branchlets; pedicels nodding. **FL** bisexual; corolla white, lobes gen 3, ovate, ascending. **FR**: nutlets; hairs short, hooked. Open coniferous forest, gravelly slopes, meadows; 1500–3700 m. KR, NCoRI, SNH, WTR, MP; to B.C., MT, Colorado.

G. bolanderi A. Gray (p. 987) BOLANDER'S BEDSTRAW Per, tufted or ± climbing, woody at base, dioecious, glabrous, scabrous, or gray-hairy. **ST** gen 15–25 cm. **LVS** in whorls of 4, in 2 ± unequal pairs, 6–27 mm, narrowly to broadly elliptic, tapered at base, acute or abruptly soft-pointed at tip, obscurely 3-veined. **STAMINATE INFL**: axillary clusters. **PISTILLATE INFL**: fl solitary in axils. **FL**: corolla rotate, gen maroon, sometimes yellowish. **FR**: berry, sometimes hairy. 2*n*=66. Open forest, lush chaparral; 150–2600 m. KR, CaRF, SNF, Teh, n SCoRI. [*G. pubens* A. Gray]

G. boreale L. ssp. ***septentrionale*** (Roemer & Schultes) Iltis NORTHERN BEDSTRAW Per, erect, 3–6 dm, ± glabrous to puberulent. **LVS** in whorls of 4, 13–31 mm, linear to broadly lanceolate, ± leathery, 3-veined. **INFL**: panicle, terminal; fls many. **FL** bisexual; corolla rotate, white. **FR**: nutlets; hairs gen present, curved, not hooked. 2*n*=66. Wet places; 15–2000 m. KR; to AK, NJ, e Asia. ❀WET:1–3,**4**,5–7.

G. buxifolium E. Greene (p. 987) BOX BEDSTRAW Shrub, erect, 3–6 dm, ± dioecious, glabrous, scabrous, or sparsely hairy. **LVS** in whorls of 4, 8–30 mm, elliptic to broadly oblanceolate or obovate, rounded, sometimes with a small point at tip, ± leathery; petiole swollen at base. **INFL**: dense, leafy clusters from nodes of woody sts. **FL** bisexual or unisexual; corolla rotate, white. **FR**: nutlets; hairs curved toward tip. **RARE** CA. Rocky bluffs, slopes; 10–400 m. n ChI. Similar to *G. catalinense*.

G. californicum Hook. & Arn. CALIFORNIA BEDSTRAW Per or sometimes ± woody above ground, low, tufted, or ± climbing, dioecious, ± hairy. **ST** 5–90 cm. **LVS** in whorls of 4, ± petioled, 3–20 mm, ovate to elliptic; tip acute to obtuse. **STAMINATE INFL**: clusters; fls few. **PISTILLATE INFL**: fls gen solitary in axils. **FL**: corolla rotate, yellowish. **FR**: berry, ± hairy. Moist, ± shaded sites, open slopes, forests, canyons, bluffs; 15–1700 m. NCo, n SNF, SnFrB, n ChI, TR, PR.

ssp. ***californicum*** (p. 987) Pl open, in mats or tufts (forest, chaparral) or tangled masses (sea cliffs), gen not woody; hairs ± coarse. **ST** 8– 32 cm. **LF** 6–18 mm, ovate to elliptic; tip round or abruptly acute, not sharp to touch. 2*n*=132. Dense shade to open places; 30–1350 m. NCoRO, NCoRI, SnFrB.

ssp. ***flaccidum*** (E. Greene) Dempster & Stebb. (p. 987) Pl ± open, not woody; hairs gen dense, soft, fine, sometimes 0. **ST** 7–30 cm. **LF** 6–25 mm, gen elliptic; tip obtuse or round, with short point. 2*n*=88. Open or dense non-coastal woodlands; 30–1500 m. SCoRO, WTR, SnGb, n ChI.

ssp. ***luciense*** Dempster & Stebb. CONE PEAK BEDSTRAW Pl low, spreading mat, not woody, grayish hairy. **ST** 5–16 cm. **LF** gen 4–6 mm, elliptic or ± obovate; tip acute, not sharp to touch. 2*n*=44. RARE. Pine, oak forests; 1100–1300 m. SCoRO (n Santa Lucia Mtns).

ssp. ***maritimum*** Dempster & Stebb. (p. 987) Pl loosely sprawling to low-climbing, ± woody; hairs on sts, lf margins ± stout, straight or curved. **ST** < 90 cm, wiry. **LF** 4–8 mm, ovate; tip gradually acute, gen sharp to touch. **FL**: corolla hairs gen 0. **FR**: hairs gen 0. 2*n*=88. Coastal bluffs, canyons; 15–60 m. SCoRO.

ssp. ***miguelense*** (E. Greene) Dempster & Stebb. SAN MIGUEL ISLAND BEDSTRAW Pl low, prostrate or climbing, dense, woody; hairs ± stout, curved. **ST** 15–45 cm; internodes short. **LF** 2–8 mm, broadly ovate to round, leathery; margin ± turned down; tip acute to round, sometimes with small point. UNCOMMON. Grassy slopes; 30–60 m. n ChI. [var. *m.* (E. Greene) Jepson]

ssp. ***primum*** Dempster & Stebb. CALIFORNIA BEDSTRAW Pl low, weak, tufted, or decumbent, not woody; hairs ± sparse, soft. **ST** 9–16 cm. **LF** 4–12 mm, elliptic; tip acute to round. 2*n*=22. RARE. Shade, lower edge of pine belt; 1350–1700 m. SnJt.

ssp. ***sierrae*** Dempster & Stebb. ELDORADO BEDSTRAW Pl weak, slender, in loose tufts, not woody; hairs many, straight, soft. **ST** 7–14 cm. **LF** < 20 mm, 5–8 × longer than wide; tip acute, with a weak, terminal hair. 2*n*=88. **RARE** CA. Open pine, oak forests, chaparral; 100–500 m. n SNF (Eldorado Co.).

G. catalinense A. Gray (p. 987) Shrub, erect, < 12 dm. **ST** stout; rigid, nodes enlarged; lf bases persistent. **LVS** in whorls of 4, 13–25 mm, lanceolate to oblanceolate, gen 1-veined; tip gen obtuse or round; petiole short, gen swollen at base. **INFL**: clusters, axillary, dense, leafy. **FL** bisexual or unisexual; corolla rotate, whitish. **FR**: nutlets; hairs long, straight, spreading, or ± 0. Rocky slopes, cliffs; 5–300 m. s ChI.

ssp. ***acrispum*** Dempster SAN CLEMENTE ISLAND BEDSTRAW Pl gen scabrous; hairs sparse, very short, or, if longer, curved but not curly. 2*n*=22. **ENDANGERED** CA. s ChI (San Clemente Island).

ssp. ***catalinense*** SANTA CATALINA BEDSTRAW Pl not scabrous; hairs ± dense, short, curly. 2*n*=22. UNCOMMON. s ChI (Santa Catalina Island).

G. clementis Eastw. (p. 987) SANTA LUCIA BEDSTRAW Per, low, matted, dioecious, gray-hairy. **ST** 8–13 cm. **LVS** gen in whorls of 4, sometimes 6, 2–7 mm, seeming linear; margins turned down; tip tapered to ± obtuse. **STAMINATE INFL**: few-fld clusters in upper axils. **PISTILLATE INFL**: fl solitary in upper axils. **FL**: corolla rotate, yellowish. **FR**: berry, hairy. 2*n*=22. UNCOMMON. N-facing slopes, open woodlands; 1130–1780 m. SCoRO (n Santa Lucia Mtns).

G. cliftonsmithii (Dempster) Dempster & Stebb. SANTA BARBARA BEDSTRAW Per, ± climbing with slender, woody sts, dioecious, shiny, coarse, ± prickly. **ST** 3–6 dm. **LVS** in whorls of 4, 7–15 mm, ovate to elliptic, 1-veined; tip acute, with a sharp hair. **INFL**: few-fld clusters on branchlets. **FL**: corolla rotate, yellowish. **FR**: berry; hairs 0. 2*n*=± 188. UNCOMMON. Light shade, upper parts of canyons to ocean; 200–1220 m. SCoRO, WTR.

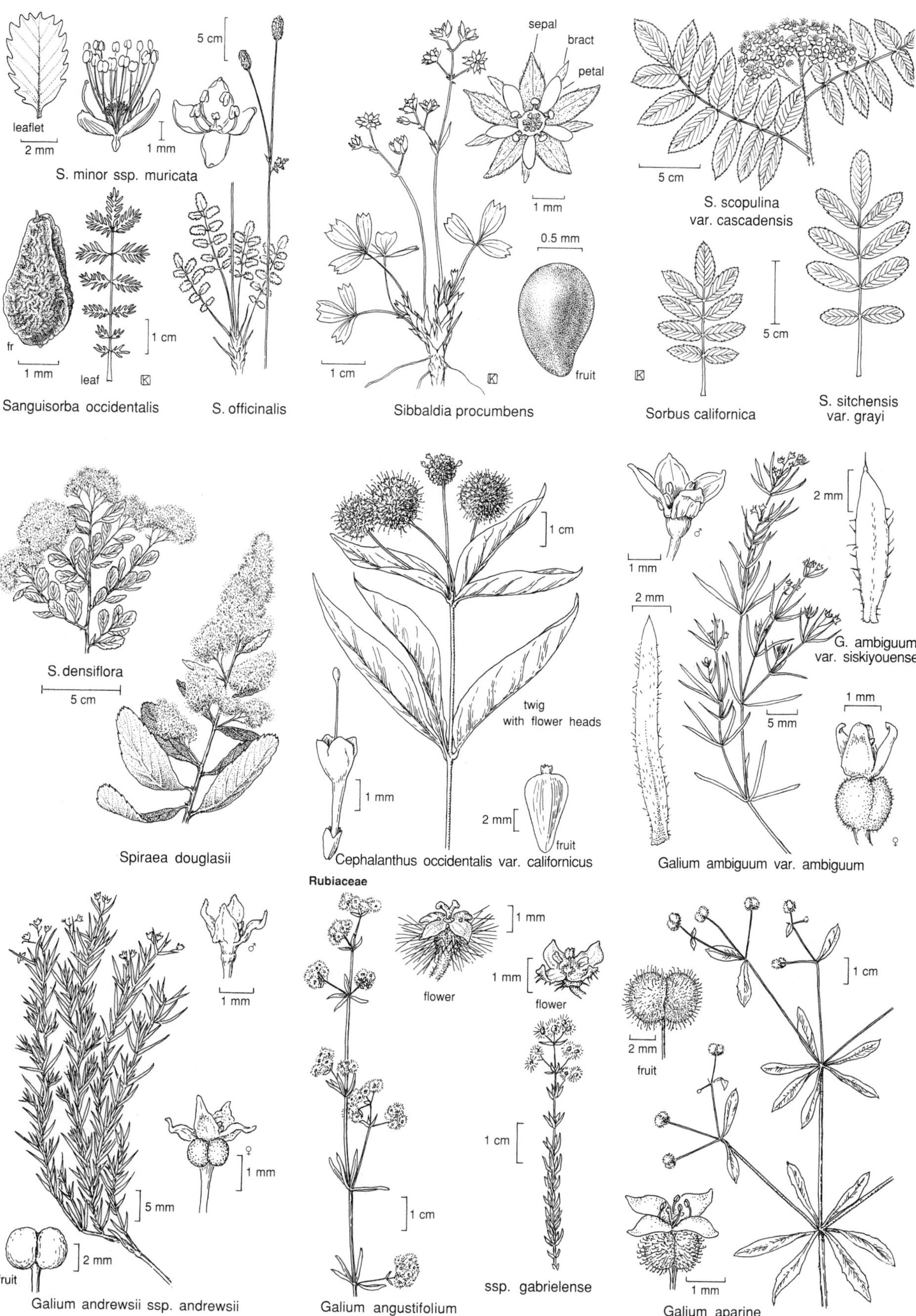
leaflet
2 mm
1 mm
5 cm
S. minor ssp. muricata
fr
1 mm
leaf
1 cm
Sanguisorba occidentalis
S. officinalis
sepal
bract
petal
1 mm
0.5 mm
1 cm
fruit
Sibbaldia procumbens
5 cm
S. scopulina
var. cascadensis
5 cm
Sorbus californica
S. sitchensis
var. grayi
S. densiflora
5 cm
Spiraea douglasii
1 cm
twig
with flower heads
1 mm
2 mm
fruit
Cephalanthus occidentalis var. californicus
1 mm
2 mm
2 mm
G. ambiguum
var. siskiyouense
5 mm
1 mm
Galium ambiguum var. ambiguum
Rubiaceae
1 mm
1 mm
5 mm
2 mm
fruit
Galium andrewsii ssp. andrewsii
1 mm
flower
1 mm
flower
1 cm
1 cm
ssp. gabrielense
Galium angustifolium
1 cm
2 mm
fruit
1 mm
Galium aparine

G. divaricatum Lam. LAMARCK'S BEDSTRAW Ann, erect, < 30 cm; glabrous to short-hairy; branches spreading. **ST** very slender. **LVS** in whorls of 5–8, seldom or little reflexed, < 7 mm, lanceolate to oblanceolate. **INFL**: panicle, terminal, open. **FL** bisexual; corolla white, lobes ± erect. **FR**: nutlets; hairs 0. 2*n*=44. Fields, slopes; 10–700 m. NCoRO, n&c SNF, SnFrB; native to Medit. Similar to *G. parisiense*.

G. glabrescens (Ehrend.) Dempster & Ehrend. Per, erect, 8–31 cm, dioecious, gen green, ± glabrous; base woody. **LVS** in whorls of 4, ovate or obovate. **INFL**: narrow, leafy; fls few on short, axillary branchlets little exserted from lvs. **FL**: corolla gen rotate, yellowish or reddish. **FR**: nutlets; hairs long, straight, spreading. Gravelly slopes, talus; 1600–2800 m. KR, Wrn.

ssp. ***glabrescens*** **ST** 12–24 cm. **LF** ovate or obovate, tip sometimes blunt. Habitats and elevations of sp. KR.

ssp. ***modocense*** Dempster & Ehrend. MODOC BEDSTRAW **ST** 8–31 cm. **LF** ovate, tip acute. RARE. Habitats and elevations of sp. Wrn.

G. grande McClatchie SAN GABRIEL BEDSTRAW Per, sprawling, climbing, or forming large clones, woody, dioecious, grayish hairy. **ST** gen < 60 cm, slender, rooting at nodes. **LVS** in whorls of 4, 5–12 mm, elliptic, petioled; tip acute to abruptly soft-pointed. **STAMINATE INFL**: clusters, axillary; fls few. **PISTILLATE INFL**: fl solitary in axils. **FL**: corolla rotate, yellowish. **FR**: berry, velvety. 2*n*= ± 220. RARE. Open broad-leafed forest, chaparral; 450–1500 m. SnGb (local).

G. grayanum Ehrend. (p. 987) GRAY'S BEDSTRAW Per, tufted or matted, dioecious, densely velvety. **ST** 2.5–23 cm. **LVS** in whorls of 4, 4–17 mm, broadly ovate to obovate or orbicular; tip obtuse or abruptly soft-pointed, not sharp to touch. **INFL**: narrow, leafy; fls few on ascending branchlets. **FL** unisexual; corolla rotate but ± cupped at base, yellowish or reddish. **FR**: nutlets; hairs long, straight, brownish. 2*n*=22. Rocky slopes, ridges, open forests; 1950–3500 m. KR, NCoRH, CaRH, n SNH; NV.

var. ***grayanum*** Pl 5–23 cm. **LF** 5–17 mm. Rocky slopes, ridges; 1950–3500 m. KR, NCoRH, CaRH, n SNH.

var. ***nanum*** Dempster & Ehrend. Pl 2.5–14 cm, dwarfed. **LF** 4–9 mm. Very open fir forest; 1950–2500 m. NCoRH; NV (Mt. Rose).

G. hallii Munz & I.M. Johnston (p. 987) NODDING BEDSTRAW Per, climbing or sprawling, dioecious, grayish hairy; base woody. **ST** 3–6 dm, pliant. **LVS** in whorls of 4, 6–10 mm, broadly ovate to elliptic, 3-veined; petiole expanded at base, tip obtuse to acute but not sharp to touch. **INFL**: clusters, axillary, drooping, leafy. **FL**: corolla rotate, cream. **FR**: nutlets; hairs long, straight, white, tawny, or bluish. 2*n*=22. N-facing slopes, canyon bottoms; 820–2350 m. s SNH, TR. ❀DRN,SHD,IRR:1–3,7.

G. hardhamiae Dempster (p. 987) HARDHAM'S BEDSTRAW Per, low, matted, dioecious; hairs few. **ST** < 30 cm; internodes >> lvs. **LVS** in whorls of 6, 1–3 mm, lanceolate to ovate, fleshy, bright green; margin strongly rolled down; petiole distinct, colorless. **INFL** long, leafy; fls few in axils. **FL**: corolla rotate, yellow, green, or pinkish. **FR**: berry; hairs ± 0. 2*n*=22. RARE. Serpentine soil with Sargent cypress; 400–950 m. SCoRO (Santa Lucia Range).

G. hilendiae Dempster & Ehrend. HILEND'S BEDSTRAW Per, dioecious, hairy, sometimes scabrous; base woody. **ST** 13–38 cm. **LVS** in whorls of 4, lance-ovate to orbicular. **INFL**: panicle, ± terminal, leafy. **FL**: corolla ± bell-shaped, gen ± pink. **FR**: nutlets; hairs long, straight. Rocky places; 1200–3400 m. DMoj; c NV.

ssp. ***carneum*** (Hilend & J. Howell) Dempster & Ehrend. PANAMINT MOUNTAINS BEDSTRAW **ST** ± erect, gen 13–38 cm, stiff; internodes 2–6 × > lvs. **LF** lanceolate to ovate, tip tapered to acute or sharp to touch. **FL**: corolla throat open. 2*n*=44. UNCOMMON. Rocky slopes, open flats; 1650–3400 m. DMtns (Panamint Mtns). [*G. munzii* var. *c.* Hilend. & Howell]

ssp. ***hilendiae*** Pl hairs many, long. **ST** 13–33 cm, not stiff; internodes 1.5–4.5 × > lvs. **LF** ovate to ± round; tip tapered to acute but not sharp to touch. **FL**: corolla throat open. Dry canyons, rocky places; 1300–2700 m. DMoj.

ssp. ***kingstonense*** (Dempster) Dempster & Ehrend. KINGSTON MOUNTAIN BEDSTRAW Pl often matted; hairs many, long. **ST** < 35 cm, slender, weak; internodes 2–4 × > lvs. **LF** lance-elliptic to ovate; tip obtuse or acute but not sharp to touch. **FL**: corolla pink, bell-shaped, throat slender, ± = lobes. 2*n*=44. RARE. Rocky places; 1200–2100 m. DMtns (Kingston Mtns, ne San Bernardino Co.); c NV. [*G. munzii* var. *k.* Dempster]

G. hypotrichium A. Gray ALPINE BEDSTRAW Per, low or ± dwarfed, dioecious; base ± woody. **ST** 2.5–23 cm. **LVS** in whorls of 4, ± crowded, ovate to round, gen ± fleshy; tip obtuse or acute but not sharp to touch. **INFL** leafy; fls few on branchlets. **FL**: corolla rotate or ± bell-shaped. **FR**: nutlets; hairs long, straight, yellowish. Talus, rocky places; 1950–4200 m. c&s SNH, SNE, DMtns.

ssp. ***hypotrichium*** (p. 987) Pl dense, velvety with magnification. **ST** 2.5–12 cm. **LF** 3–8 mm, ovate to nearly round. **FL**: corolla ± rotate, pale yellow or pink-tinged. 2*n*=22,44. Ridges, talus; 3000–4200 m. c&s SNH, SNE (White, Sweetwater Mtns). Tetraploids (n of Sonora Pass) have been called ssp. *ebbettsense* Dempster & Ehrend. but differ little from diploids (s of Sonora Pass).

ssp. ***inyoense*** Dempster & Ehrend. Pl ± open, scabrous. **ST** 12–23 cm. **LF** gen 6–11 mm, broadly ovate to round. **FL** ± rotate, ± white. 2*n*=44. Talus, rocky washes in pinyon belt; 1950–3200 m. s SNH.

ssp. ***subalpinum*** (Hilend & J. Howell) Ehrend. Pl often ± open, hairy, sometimes scabrous. **ST** < 15 cm. **LF** 5–10 mm, ovate to round. **FL**: corolla ± rotate, ± white. 2*n*=44. Rocks, talus, gen above timber line; 2650–3880 m. s SNH.

ssp. ***tomentellum*** Ehrend. TELESCOPE PEAK BEDSTRAW Pl ± dense, gray; hairs dense, fine. **ST** < 9 cm. **LF** 5–9 mm, ovate. **FL**: corolla shallowly bell-shaped, reddish. 2*n*=22. RARE. Talus; 3300–3550 m. DMtns (Telescope Peak, Panamint Mtns).

G. jepsonii Hilend & J. Howell (p. 987) JEPSON'S BEDSTRAW Per, erect, 8–16 cm, dioecious, ± glabrous exc lf margin. **STS** in small clumps. **LVS** in whorls of 4, dense near base, sparse above, 6–15 mm, broadly linear. **INFL**: panicle, terminal, ± lfless, open. **FL**: corolla bell-shaped, divided halfway to base or less, cream, lobe tips often pink. **FR**: nutlets; hairs ascending, short, curved, not hooked. 2*n*=22. UNCOMMON. Open woodlands; 2000–2500 m. SnGb, SnBr.

G. johnstonii Dempster & Stebb. JOHNSTON'S BEDSTRAW Per, erect, 18–35 cm, dioecious, glabrous exc lf margin; base woody. **ST**: upper internodes 2–5 × > lvs. **LVS** in whorls of 4, 14–30 mm, gen linear. **INFL**: clusters, terminal on ascending branchlets; fls few. **FL**: corolla rotate or ± bell-shaped, yellowish. **FR**: nutlets, gen < pedicel; hairs ± spreading, < fr, not hooked. 2*n*=66. UNCOMMON. Open mixed forest; 1650–2300 m. SnGb, SnBr. [*G. angustifolium* var. *pinetorum* Munz & I.M. Johnston]

G. magnifolium (Dempster) Dempster Per, erect, 10–45 cm, dioecious, glabrous exc upper bracts; base woody. **LVS** in whorls of 4, 8–23 mm, gen lanceolate to ovate; tip tapered to acute. **INFL**: panicle, open; fls solitary or in clusters in axils. **FL**: corolla rotate, whitish, with long hairs outside. **FR**: nutlets; hairs few, short, straight. Rocky slopes in juniper belt; 800–2000 m. DMtns (Clark Mtn); NV, UT.

G. matthewsii A. Gray (p. 987) MATTHEWS' BEDSTRAW Per, erect, 13–30 cm, open, dioecious, glabrous exc upper bracts; base woody. **STS** gen many, slender, tangled or not; internodes 2–7 × > lvs. **LVS** in whorls of 4, 2–10 mm, ovate, leathery; tip sharp. **INFL**: panicle; fls in terminal clusters on branches. **FL**: corolla rotate, yellowish, with long hairs outside. **FR**: nutlets; hairs long, straight. 2*n*=22. Dry slopes, washes; 1100–3000 m. DMtns.

G. mexicanum Kunth var. ***asperulum*** (A. Gray) Dempster (p. 987) Per, spreading or reclining, glabrous or hairs few, short, reflexed. **ST** 45–90 cm. **LVS** gen in whorls of 5–8, 13–25 mm, lanceolate to narrowly elliptic; tip tapered or acuminate. **INFL**: panicle, open, leafy; fls few; pedicel, branchlet thread-like. **FL** bisexual; corolla bell-shaped, white, > ovary; ovary top-shaped. **FR**: nutlets, but ± fleshy; hairs few, short, curved upward. 2*n*=22,44. Wet places near streams; 550–2500 m. KR, NCoRO, c SNH; to WA, w MT, UT.

G. mollugo L. HEDGE BEDSTRAW Per, erect, 3–12 dm, sometimes glabrous. **ST** stout, ± swollen at nodes. **LVS** in whorls gen of 8, oblanceolate to obovate; tip abruptly soft-pointed. **INFL**: panicle, many-fld. **FL** bisexual; corolla rotate, white; stamens exserted. **FR**: nutlets, small, few; hairs 0. $2n$=22,44. Disturbed areas; gen < 1000 m. NCoRO, SnFrB; native to Eur.

G. multiflorum Kellogg (p. 987) KELLOGG'S BEDSTRAW Per, erect, 15–35 cm, dioecious, glabrous or hairy; base woody. **LVS** in whorls of 4, in 2 unequal pairs; the larger 8–15 mm, broadly ovate, 3-veined, base round, tip ± obtuse but often with a small point. **INFL**: panicle of few-fld clusters, terminal. **FL**: corolla ± bell-shaped, whitish or pinkish. **FR**: nutlets; hairs long, straight. $2n$=22. Rocky places among sagebrush; 1300–2900 m. n SNH, GB; NV. [f. *hirsutum* (A. Gray) Ehrend.; *G. watsonii* (A. Gray) A.A. Heller]

G. munzii Hilend & J. Howell (p. 987) MUNZ'S BEDSTRAW Per, erect, 10–30 cm, dioecious; hairs coarse; base woody. **ST** flexible. **LVS** in whorls of 4, in 2 unequal pairs; the larger 6–19 mm, gen ovate, ± 3-veined, tip tapered, ± sharp to touch. **INFL**: panicle, open; clusters axillary, few-fld. **FL**: corolla rotate, whitish or pink, hairy outside. **FR**: nutlets; hairs long, straight. $2n$=44. UNCOMMON. Cool n- or e-facing slopes, shady canyon bottoms; 1100–2250 m. DMtns; to s UT.

G. murale (L.) All. (p. 989) TINY BEDSTRAW Ann, erect or spreading, glabrous exc lf margin and tip. **ST** 1–6.5 cm. **LVS** in whorls of 4–6, 1–3 mm, obovate to oblanceolate; tip often with a slender hair. **INFL**: fls 1–2 in axils, ± sessile. **FL** bisexual; corolla 1 mm, green, fading yellow or white, lobes ascending, < half as long as ovary, ovate, obtuse at tip. **FR**: nutlets, each sausage-shaped; hairs unevenly distributed, hooked. $2n$=44. Locally abundant. Damp, mossy places, undergrowth on grassy hillsides; 20–650 m. c SNF, SnFrB; native to Eur.

G. muricatum Wight (p. 989) HUMBOLDT BEDSTRAW Per, low, matted, dioecious, glabrous to sparsely hairy. **ST** 7–20 cm. **LVS** in whorls of 4, 6–10 mm, elliptic to obovate, ± leathery, shiny, obtuse, petioled; tip abruptly soft-pointed. **INFL** open; clusters in upper axils; fls few. **FL**: corolla rotate, yellowish. **FR**: berry, velvety. $2n$= 22,44. Shady, damp forest; 50–1100 m. NCoRO; to s OR.

G. nuttallii A. Gray SAN DIEGO BEDSTRAW Per, climbing, woody, dioecious; herbage densely tangled, becoming dark red. **ST** 6–15 dm, slender. **LVS** in whorls of 4, 3–8 mm, linear to triangular-ovate, gen leathery; tip sharp to touch. **STAMINATE INFL**: axillary clusters. **PISTILLATE INFL**: fl 1 in axils. **FL**: corolla rotate, gen ± reddish. **FR**: berry; hairs 0. $2n$=22. Chaparral, groves of pine or *Lyonothamnus*; 3–500 m. SCo, s ChI, WTR, PR; Baja CA.

ssp. ***insulare*** Ferris NUTTALL'S ISLAND BEDSTRAW Pl gen glabrous. UNCOMMON. Chaparral, groves of pine or *Lyonothamnus*; 3–400 m. s ChI.

ssp. ***nuttallii*** Pl with tiny, sharp, curved hairs. Chaparral; 3–500 m. SCo, WTR, PR; to Baja CA.

G. oreganum Britton OREGON BEDSTRAW Per, erect, 20–30 cm, glabrous exc lf margin and veins. **LVS** in whorls of 4, 20–40 mm, broadly ovate or elliptic, 3-nerved. **INFL**: clusters, terminal on upper branches; fls few. **FL** bisexual: corolla rotate, yellowish or greenish. **FR**: nutlets; hairs long, hooked. UNCOMMON. Open coniferous forest; 1500 m. KR (Del Norte Co.); w OR, WA.

G. parishii Hilend & J. Howell (p. 989) PARISH'S BEDSTRAW Per, erect or matted, gen velvety; base woody. **ST** 5–40 cm. **LVS** in whorls of 4, in 2 unequal pairs, 1–9 mm, ovate to orbicular, gen 3-veined. **INFL** open between dense clusters; pedicel ± 0. **FL** bisexual or unisexual (sometimes on individual pls); corolla rotate, gen red or pink, sometimes yellow. **FR**: nutlets; hairs long, straight. $2n$=22. Steep slopes, talus, at base of boulders; 1675–3400 m. SnGb, SnBr, SnJt, DMtns; s NV.

G. parisiense L. WALL BEDSTRAW Ann, erect, 15–25 cm; hairs short. **ST** slender. **LVS** in whorls of 6, 4–6 mm, lanceolate to oblanceolate, gen becoming reflexed. **INFL**: panicle, open, few-fld; pedicel thread-like. **FL** bisexual; corolla rotate basally, whitish or purplish, lobes erect. **FR**: nutlets, ± elongate; hairs short, hooked, rarely 0. $2n$=44,66. Warm, dry, gen rocky soil; 160–2200 m. KR, NCoRO, SN, SnFrB; native to Medit.

G. porrigens Dempster CLIMBING BEDSTRAW Per, climbing, woody, dioecious, scabrous, clinging with tiny prickles. **ST** 6–15 dm, slender. **LVS** in whorls of 4, 2–18 mm; tip round or obtuse; terminal hair weak, ephemeral. **STAMINATE INFL**: axillary clusters. **PISTILLATE INFL**: fls gen solitary in axils. **FL**: corolla rotate, yellowish or often ± reddish. **FR**: berry; hairs 0. $2n$=22. Among shrubs in chaparral or forest; 150–1060 m. CA-FP; s OR, n Baja CA.

var. ***porrigens*** (p. 989) **LF** ± widely ovate to oblong. Habitats of sp.; < 1060 m. CA-FP (gen w of GV); to s OR, n Baja CA. [*G. nuttallii* var. *ovalifolium* Dempster] ❀DRN,SHD:3,**7**,**14–16**,**18**.

var. ***tenue*** (Dempster) Dempster (p. 989) **LF** ± linear. Habitats of sp.; 150–1050 m. NCoRI, CaRF, SNF, SCoRI; s OR. [*G. nuttallii* var. *t.* (Dempster) Dempster]

G. proliferum A. Gray DESERT BEDSTRAW Ann, erect, 5–30 cm, ± hairy, often scabrous. **LVS** in whorls of 2–4, 3–10 mm, gen lanceolate, sessile (lowermost broad, petioled); tip obtuse or abruptly soft-pointed. **INFL**: fls gen solitary in axils, subsessile. **FL** bisexual; corolla rotate, yellowish. **FR**: nutlets; hairs short, hooked. Very uncommon. Rocky banks, limestone ledges; 1100–1400 m. DMtns.

G. saxatile L. Per, dense, matted, glabrous exc lf tip. **ST** 7–15 cm. **LVS** in whorls gen of 6–8, 6–11 mm, obovate to oblanceolate. **INFL**: clusters. **FL** bisexual; corolla rotate, white. **FR**: nutlets; hairs 0; tubercles fine. $2n$=22,44. Lawns; 30–100 m. SnFrB; native to Eur.

G. schultesii Vest Per, ± erect, 30–120 cm, glabrous or lf margin ± scabrous. **LVS** in whorls of 6–8, 25–60 mm, narrowly elliptic; tip obtuse, with small point. **INFL**: panicle, terminal, open; fls many. **FL**: corolla rotate, whitish. **FR**: nutlets; hairs 0. Probably hillsides; probably ± 500 m. SnFrB (Santa Clara Co.); native to Eur.

G. serpenticum Dempster Per, erect, 7–33 cm, tufted, little branched, dioecious, ± puberulent; base woody. **LVS** in whorls of 4, < 15 mm, linear to lanceolate or elliptic. **INFL**: panicle of clusters, terminal, narrow, leafy, on erect or ascending branchlets. **FL**: corolla whitish. **FR**: nutlets; hairs long, straight, yellowish. Steep slopes, meadows, open pine forests; 1000–2750 m. KR, Wrn; to WA, ID, n NV.

ssp. ***scotticum*** Dempster & Ehrend. SCOTT MOUNTAIN BEDSTRAW **LF**: tip gen flat. **FL**: corolla ± cupped. $2n$=22. RARE. Steep slopes in open pine forest; 1000–2000 m. KR.

ssp. ***warnerense*** Dempster & Ehrend. WARNER MOUNTAINS BEDSTRAW **LF**: tip often reflexed. **FL**: corolla rotate. $2n$=22. RARE. Steep slopes, about rocks, meadows: 1450–2750 m. Wrn.

G. sparsiflorum Wight SEQUOIA BEDSTRAW Per, erect, 20–50 cm, loosely tufted, dioecious; base woody. **LVS** in whorls of 4, 6–25 mm, broadly elliptic to ovate or round, 3-veined, tip abruptly soft-pointed. **STAMINATE INFL**: panicle open, leafy; clusters many-fld. **PISTILLATE INFL**: fls gen solitary in axils. **FL**: corolla rotate, gen yellowish. **FR**: berry, sometimes hairy. Shady places, often in forest; 350–2300 m. CaRF, SN, n SCoRI.

ssp. ***glabrius*** Dempster & Stebb. Pl sometimes reddish. **LF** gen ovate, ± leathery; hairs 0. Shady places; 350–1600 m. CaRF, n SCoRI.

ssp. ***sparsiflorum*** (p. 989) Pl green. **LF** gen elliptic, sometimes ± ovate, thin; puberulent under magnification. $2n$=22. Shady places in forest; 800–2300 m. SN.

G. stellatum Kellogg var. ***eremicum*** Hilend & J. Howell (p. 989) Shrub, ± erect, 30–75 cm, dioecious, scabrous. **ST** stout, brittle; branches many, spreading. **LVS** in whorls of 4, gen 4–8 mm, lanceolate to needle-like, gray-green; tip sharp to touch. **INFL**: panicle, axillary, leafy; fls few–many. **FL**: corolla rotate, whitish. **FR**: nutlets; hairs dense, long, straight, white. $2n$=22,44. Rocky slopes; 130–1600 m. D; to NV, NM, nw Mex.

G. tricornutum Dandy ROUGH CORN BEDSTRAW Ann, sprawling or trailing, dense in fl, open in fr. **ST** 10–35 cm, stout, ± glabrous. **LVS** in whorls of 6–8, 12–19 mm, linear to narrowly oblanceolate; margin thickened, with short, recurved prickles. **INFL**: clusters, axillary; fls few; peduncle < lf; pedicel recurved. **FL** bisexual; corolla rotate, white. **FR**: nutlets, the pairs > 4 mm wide; hairs 0; tubercles stout. $2n$=44. Grain fields; 100–2000 m. c SNF, GV.

G. trifidum L. (p. 989) Per, rarely ann, minutely scabrous. **ST** 10–50 cm, slender, weak, tangled. **LVS** in whorls of 4–6, 4–19 mm, linear to elliptic or narrowly ovate-obovate, petioled; tip rounded. **INFL**: few-fld clusters; pedicel slender. **FL** bisexual; corolla rotate, white or pinkish, gen 3-lobed. **FR**: nutlets, spheric, hard, smooth, black when dry; hairs 0. Wet places; < 2800 m. CA exc D; to AK, Mex.

var. ***pacificum*** Wieg. Pl straggling. **ST** 10–50 cm. $2n$=24. Wet places; < 2400 m. CA exc D; to AK, Mex. [*G. cymosum* Wieg.] ❀WET:**4–6,17,24**&SHD:2,3,7,8,9,**14–16,18–23.**

var. ***pusillum*** A. Gray. Pl dwarfed, tufted or matted. **ST** 5–10 cm. $2n$=24. Montane meadows; 1700–2800 m. KR, CaRH, SNH, SnBr, SnJt, MP; to AK, Rocky Mtns.

G. triflorum Michaux (p. 989) SWEET-SCENTED BEDSTRAW Per, gen decumbent, radiating from center, glabrous to ± scabrous. **ST** 20–45 cm. **LVS** in whorls of 6, gen horizontal regardless of st orientation, 6–38 mm, ovate to obovate; tip acute to acuminate, sometimes with small point. **INFL**: clusters, axillary, peduncled; fls 2–3. **FL** bisexual; corolla rotate, cream. **FR**: nutlets; hairs hooked, soft, white or brown. $2n$=22,44,66. Damp, shady forests; 10–3000 m. KR, NCoRO, SN, SnFrB, SnBr, MP; to AK, e N.Am, Greenland, Mex, Eur, Asia.

G. verum L. LADIES' BEDSTRAW, YELLOW BEDSTRAW Per, erect or decumbent, ± glabrous. **ST** 10–25 cm. **LVS** in whorls of 6–12, 12–22 mm, thread-like or linear. **INFL**: panicle, terminal; fls many. **FL** bisexual; corolla rotate, yellow. **FR**: nutlets, small; hairs 0. $2n$=22, 44. Lawns; 30–60 m. SnFrB; native to Eur, Asia.

G. wrightii A. Gray WRIGHT'S BEDSTRAW Per, erect or straggling, ± dioecious, glabrous exc just above base; base woody. **ST** 15–50 cm. **LVS** in whorls of 4, < 8 mm, gen linear to oblanceolate, attachment broad; tip acute or ± sharp to touch. **INFL**: panicle, terminal, open. **FL** bisexual or unisexual on individual pls; corolla rotate, gen red or pink. **FR**: nutlets; hairs few, long, straight. $2n$=22. RARE in CA. Shady canyons among rocks; 1600–2000 m. DMtns (Clark Mtn); to NM, w TX, Mex (Coahuila, Sonora, Baja CA).

KELLOGGIA

Per from slender rhizome. **LVS** opposite; stipules ± scale-like, fused basally. **INFL**: cyme, terminal, open; fls few; pedicel long, slender. **FL**: calyx 4–5-lobed; corolla funnel-shaped, 4–5-lobed; styles 2, fused below. **FR**: 2 nutlets, wider at top. 2 spp.: w N.Am, China. (Dr. Albert Kellogg, pioneer CA botanist, 1813–1887)

K. galioides Torrey (p. 989) **ST** erect, 15–40 cm, ± 4-angled. **LVS** often clustered in axils, 19–38 mm, lanceolate to narrowly ovate. **FL**: calyx inconspicuous; corolla pink or white, tube slender, throat short, lobes 4–5, lanceolate. **FR**: hairs hooked. Somewhat open places in coniferous forests; 1100–3000 m. KR, NCoRH, CaRH, SNH, SnBr, SnJt, MP; to WA, WY, UT, AZ.

SHERARDIA

1 sp. (Dr. Wm. Sherard, patron of Dillenius and friend of John Ray, 1659–1728)

S. arvensis L. FIELD MADDER Ann, matted. **ST** decumbent, often much branched from base, 7–16 cm, 4-angled. **LVS** in whorls of 5–6, 4–13 mm, lanceolate or oblanceolate; margin thick; tip acute or sharp to touch; stipules lf-like. **INFL**: head, axillary, involucred; fls gen 2–3, gen incl in involucre; involucral bracts ± free, ± = lvs, gen < peduncle; pedicel ± 0. **FL**: calyx teeth 6, ± dissected at tips, persistent; corolla salverform, gen 4-lobed, pink or lavender; styles 2, thread-like, fused below. **FR**: 2 nutlets; hairs soft. $2n$=22. Pastures, lawns; 10–600 m. NCoRO, CaRF, n SNF, SnFrB, SCoRO, SCo; native to Medit.

RUTACEAE RUE FAMILY

James R. Shevock

Per, shrubs, trees, very aromatic, sometimes thorny. **LVS** gen alternate, simple to pinnately compound (sometimes reduced to spines), prominently oil-gland-dotted; stipules 0. **INFL**: cyme, raceme, or fls solitary, gen bracted. **FL** gen bisexual, gen strongly aromatic; sepals gen 5, free or fused at base, gen persistent; petals gen 5, free or fused at base, gen whitish or greenish; stamens gen 2–4 × petal number; ovary gen superior, gen lobed, chambers gen 4–5, ovules gen many. **FR**: berry, drupe, winged achene, or capsule, gen aromatic. **SEEDS** gen oily. ± 150 genera, ± 1500 spp.: esp trop, warm temp, esp s Afr, Australia; used or cult for food (*Citrus*, 50 spp.), perfume, medicine, timber, orn (*Choisya, Skimmia*, etc). Some TOXIC: oils may promote localized sunburn or produce dermatitis.

1. Lf simple, entire
  2. Lvs opposite; ovary chamber 1; fr a berry .... **CNEORIDIUM**
  2' Lvs alternate; ovary chambers 2; fr a capsule .... **THAMNOSMA**

1' Lf dissected or compound
  3. Shrub or small tree; petals white, entire; fr round, winged, achene-like .... **PTELEA**
  3' Per; petals yellow, wavy to fringed; fr a capsule .... **RUTA**

CNEORIDIUM

1 sp.: s CA, Baja CA. (Latin: resembling *Cneorum*, an old world genus)

C. dumosum (Nutt.) Baillon (p. 989) BUSHRUE Shrub, gen < 1.5 m, glabrous, evergreen, **ST** intricately branched. **LVS** opposite, simple, 1–2.5 cm, ± linear, entire, glabrous. **INFL**: cluster, gen axillary; fls 1–3. **FL**: calyx 4-lobed, 1–1.5 mm; petals 4(5), 5–6 mm, obovate, white; stamens 8, longest opposite sepals, filaments wider at base; ovary sessile, chamber 1, ovules 2, style short, flat, from near ovary base, stigma head-like. **FR**: berry, 5–6 mm diam, reddish brown, spheric, resin-dotted. **SEEDS** 1–2, 5–6 mm, ± spheric, dark brown. Mesas, bluffs near coast; < 800 m. s SCo, s ChI (San Clemente Island); Baja CA. ❀DRN:14–18&IRR:8,9,**19–24**.

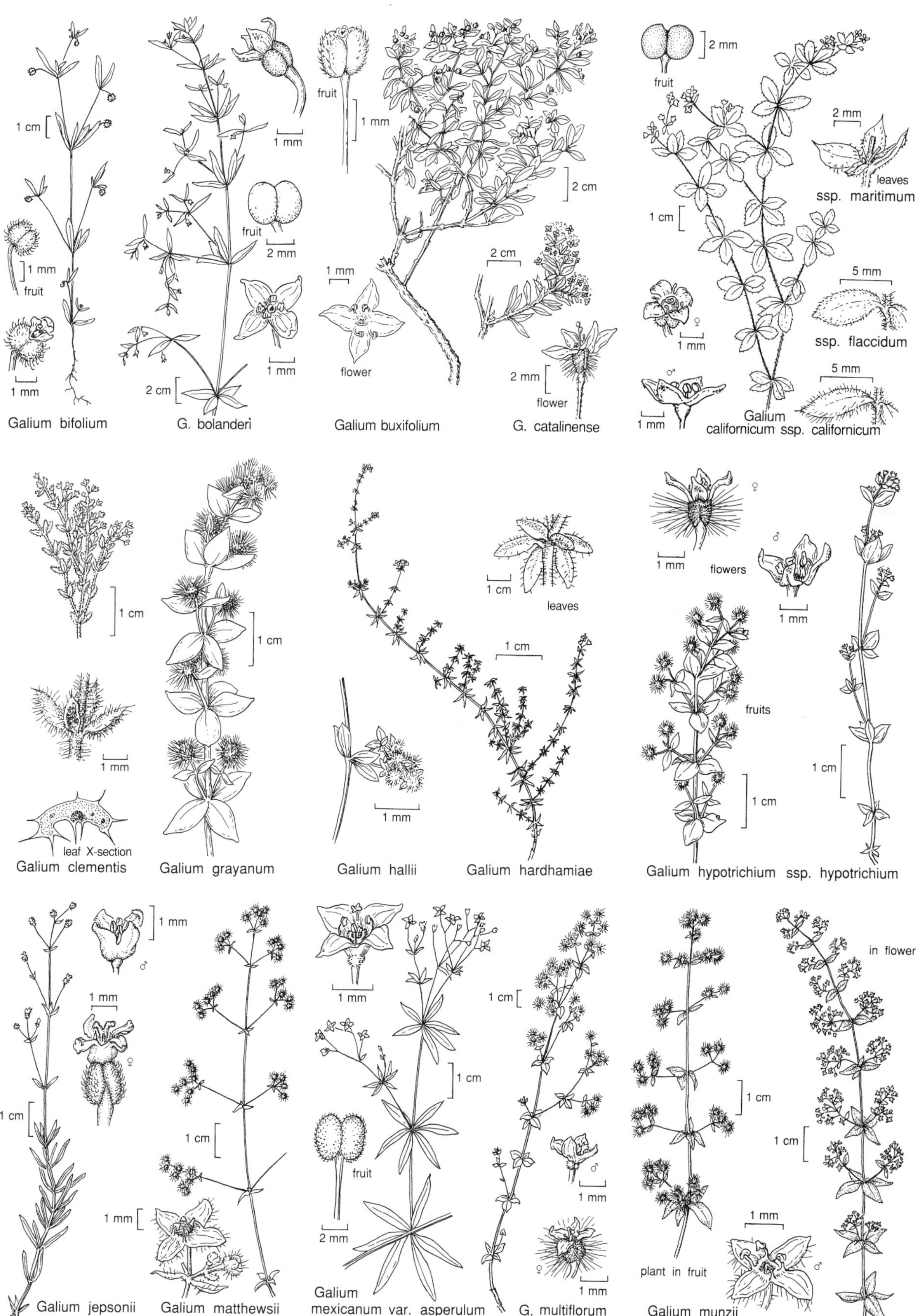
1 cm
1 mm
fruit
1 mm
Galium bifolium
1 mm
fruit
2 mm
1 mm
2 cm
G. bolanderi
fruit
1 mm
2 cm
1 mm
flower
Galium buxifolium
2 cm
2 mm
flower
G. catalinense
2 mm
fruit
2 mm
leaves
ssp. maritimum
1 cm
5 mm
1 mm
ssp. flaccidum
5 mm
1 mm
Galium
californicum ssp. californicum
1 cm
1 mm
leaf X-section
Galium clementis
1 cm
Galium grayanum
1 cm
leaves
1 cm
1 mm
Galium hallii
Galium hardhamiae
1 mm
flowers
1 mm
fruits
1 cm
1 cm
Galium hypotrichium ssp. hypotrichium
1 mm
1 mm
1 cm
1 cm
1 mm
Galium jepsonii
Galium matthewsii
1 mm
1 cm
fruit
2 mm
Galium
mexicanum var. asperulum
1 cm
1 mm
1 mm
G. multiflorum
in flower
1 cm
1 cm
1 mm
plant in fruit
Galium munzii

PTELEA HOP TREE

Shrub, small tree, gen with bisexual and pistillate fls. **LVS** pinnately compound; lflets 3–5, ± sessile, entire to finely serrate. **INFL** panicle-like. **FL**: sepals 4–5; petals 4–5, greenish white; stamens 4–5, filaments hairy on inner side; ovary chambers 2, style short, stigmas 2. **FR** dry, indehiscent, ± flat, gland-dotted; wing prominent, round. **SEEDS** 2. 3 spp.: US, Mex. (Greek: elm, from similar fr)

P. crenulata E. Greene (p. 989) Pl gen < 5 m. **LVS** deciduous; lflets 3, 2– 7 cm, lanceolate to obovate, glabrous above, ± hairy beneath; petiole 2–5 cm. **INFL** ± flat-topped. **FL**: sepals minute; petals 4–5 mm, fragrant. **FR** 1–2 cm, ± straw-colored; wing ± notched at both ends; style persistent. Scrub, woodland; gen < 700 m. NCoRI, CaRF, SNF, SnFrB. ❀**7,14–17**;IRR:8,9,18,**19–21**, 22–24.

RUTA

Per, subshrub. **LVS** pinnately or ternately compound. **INFL**: panicle or cluster, erect, terminal. **FL**: petals 4–5, wavy to fringed; stamens 8 or 10 in 2 series; ovary chambers 4–5. **FR**: capsule, 4–5 lobed at top, sometimes indehiscent. **SEEDS** several, angled. 7 spp.: nw Afr islands, Medit, sw Asia. (Latin: the classical name) Cult for orn, flavoring, medicine.

R. chalepensis L. Pl glaucous, puberulent. **ST** 4–8 dm. **LF** 2–3-pinnate, gen 10–20 cm, oblong in outline; segments gen 1–1.5 cm, narrowly elliptic, entire. **FL**: petals 6–8 mm, yellow, margins in-rolled, fringed. **FR**: lobes pointed. Uncommon. Disturbed places; gen < 500 m. CA-FP (esp near coast); native to Medit. *R. graveolens* L. may also be naturalized (lf ovate in outline; petal margin wavy, not fringed).

THAMNOSMA TURPENTINE-BROOM

Subshrub or shrub. **LF** very small, ephemeral. **INFL**: panicle (raceme-like or fls scattered along sts). **FL** bisexual; calyx persistent, 4-lobed; petals 4, erect in fl; stamens 8, in 2 series; ovary stalked, deeply 2-lobed, style thread-like. **FR**: capsule, deeply 2-lobed, leathery, opening at tip. 6 spp.: sw N.Am, s Afr. (Greek: bush odor)

T. montana Torrey & Frémont (p. 989) **ST** 3–6 dm, broom-like, yellowish green, thickly covered with blister-like glands, gen lfless. **FL**: sepals ± 2 mm, ± round, greenish; petals 8–12 mm, ± elliptic, ± leathery, purplish, tip rolled out; style ± exserted, ovules 8–9 per chamber. **FR**: lobes ± 5 mm thick, ± spheric. **SEEDS** 1–3, ± 4 mm, reniform, whitish. Dry slopes; < 700 m. D; to NM, Mex. ❀DRN, DRY:8–13,18–21&SUN:7,14–16,22–24;DFCLT.

SALICACEAE WILLOW FAMILY

Shrub, tree, gen dioecious (rarely monoecious). **ST**: trunk < 40 m; wood soft; bark smooth, bitter; buds scaly. **LVS** simple, alternate, deciduous; stipules gen deciduous, often large. **INFL**: catkin, gen appearing before lvs; each fl subtended by disk or 1–2 nectary glands and 1 bract. **FL**: perianth 0. **STAMINATE FL**: stamens 1–many. **PISTILLATE FL**: pistil 1, ovary superior, chamber 1, stigma lobes 2–4. **SEEDS** many; hairs fine, white, cottony. **FR**: capsule; valves 2–4. 2 genera, 340 spp.: gen temp (exc Australia, Malay Archipelago) moist places; many cult. Hybridization common; identification often difficult. Family description, key to genera by John O. Sawyer, Jr.

1. Lf length not >> (often ± =) width; catkin pendent, bract cut into narrow segments; staminate fl with 8–many stamens; pistillate fl with cup-like disk; winter bud scales > 3 .... **POPULUS**
1′ Lf length gen >> width; catkin erect, bract entire; staminate fl with 1–8 stamens; pistillate fl with 1–2 glands; winter bud scale 1 .... **SALIX**

POPULUS POPLAR, COTTONWOOD, ASPEN

John O. Sawyer, Jr.

Tree. **ST**: trunk < 40 m; young bark smooth, pale yellow-green to gray; older bark furrowed, brown to gray; twigs with swellings below lf scars; winter bud gen resinous, scales > 3. **LVS** gen glabrous (juvenile lvs may differ from adult lvs); blade 3–11 cm, elliptic to triangular, veins pinnate or subpalmate, tip gen elongate. **INFL**: catkin pendent, 3–8 cm; bract cut into narrow segments; fls sessile, on a cup- or saucer-like disk. **STAMINATE FL**: stamens 8–60. **PISTILLATE FL**: style short, stigmas 2–3(4), large, scalloped to 2-lobed. **FR** spheric to conic; valves 2–3(4), 3–12 mm. 40 spp.: n hemisphere. (Latin: name for pls of this genus)

1. Lf blade white-tomentose below .... ***P. alba***
1′ Lf blade glabrous or hairy below
  2. Lf blade lanceolate .... ***P. angustifolia***
  2′ Lf blade ovate, nearly round, or triangular
    3. Lf blade triangular, margin coarsely scalloped .... ***P. fremontii*** spp. ***fremontii***
    3′ Lf blade ovate to nearly round, margin finely scalloped
      4. Lf blade narrowly to widely ovate, 3–7 cm; petiole 1/3–1/2 blade length, round in X-section .... ***P. balsamifera*** ssp. ***trichocarpa***
      4′ Lf blade widely ovate to nearly round, gen 2–4 cm, petiole 2/3 to = blade length, laterally compressed in X-section .... ***P. tremuloides***

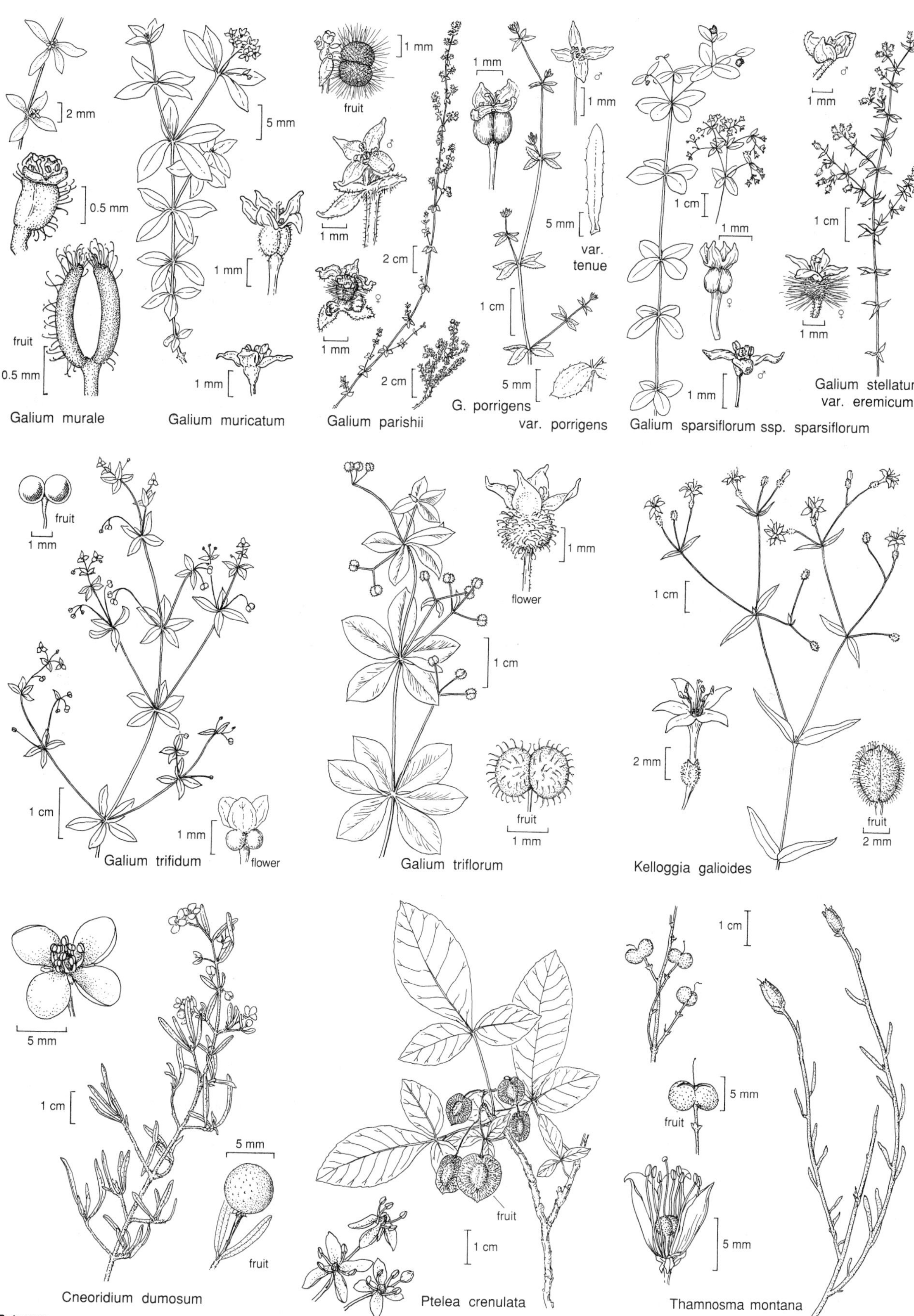
2 mm
0.5 mm
fruit
0.5 mm
Galium murale
5 mm
1 mm
1 mm
Galium muricatum
1 mm
fruit
♂
1 mm
2 cm
♀
1 mm
2 cm
Galium parishii
1 mm
♂
1 mm
5 mm
var.
tenue
1 cm
5 mm
G. porrigens
var. porrigens
1 cm
1 mm
♀
♂
1 mm
Galium sparsiflorum ssp. sparsiflorum
♂
1 mm
1 cm
♀
1 mm
Galium stellatum
var. eremicum
fruit
1 mm
1 cm
1 mm
Galium trifidum
flower
1 mm
flower
1 cm
fruit
1 mm
Galium triflorum
1 cm
2 mm
fruit
2 mm
Kelloggia galioides
5 mm
1 cm
5 mm
fruit
Cneoridium dumosum
fruit
1 cm
Ptelea crenulata
1 cm
5 mm
fruit
5 mm
Thamnosma montana

P. alba L. WHITE POPLAR **ST**: crown wide; trunk < 20 m; twigs and winter buds white-tomentose. **LF**: petiole 1/3–1/2 blade length; blade 3–9 cm, 3–5-lobed, base rounded to slightly cordate, tip acute, margin entire to toothed, upper surface blue-green, glossy, lower white-tomentose. Disturbed places near settlements; 600–1800 m. KR, CaR, GB, expected elsewhere; native to c Eur, c Asia. Persisting primarily by clonal root-sprouting.

P. angustifolia James (p. 999) NARROW-LEAVED COTTONWOOD **ST**: crown slender; trunk < 15 m; twigs glabrous; winter buds very gummy. **LF**: petiole 1/3 blade length, lower side round, upper widely channeled; blade 4–11 cm, lanceolate, base wedge-shaped, tip acute or tapered, margin finely scalloped, surfaces yellow-green, glabrous. $2n=38$. RARE in CA. Streamsides; 1200–1800 m. e SN, SnBr; to OR, Rocky Mtns, n Mex. [×*P. acuminata* Rydb. misapplied] In cult.

P. balsamifera L. spp. ***trichocarpa*** (Torrey & A. Gray) Brayshaw (p. 999) BLACK COTTONWOOD **ST**: crown wide; trunk < 30 m; twigs brown, becoming gray; winter buds finely ciliate, very resinous, fragrant when opening. **LF**: petiole 1/3–1/2 blade length, lower side round, upper channeled, with pair of glands at junction with blade; blade 3–7 cm, narrowly to widely ovate, base round to cordate, tip acute to tapered, margin finely scalloped, upper surface green, lower glaucous, often stained with brown resin. $2n=38$. Scattered. Alluvial bottomlands, streamsides; < 2800 m. CA-FP, GB; to AK, n Rocky Mtns, UT, n Baja CA. [*P. t.* Torrey & A. Gray] [Brayshaw 1965 Canad Field-Naturalist 79:91–95] ❀IRR or WET,SUN:1,**2–7**,8,9,14,**15–18**,19–23,**24**; susceptible to galls.

P. fremontii S. Watson ssp. ***fremontii*** (p. 999) ALAMO or FREMONT COTTONWOOD **ST**: crown wide; trunk < 20 m; twigs yellow, becoming gray, glabrous to hairy; winter buds resinous. **LF**: petiole 1/2 to = blade length, laterally compressed near blade; blade 3–7 cm, deltate, base slightly cordate to flat, tip ± tapered, margin coarsely scalloped, surfaces yellow-green, glabrous to hairy, often stained with milky resin. Scattered. Alluvial bottomlands, streamsides; < 2000 m. CA (exc MP); to c Rocky Mtns, n Mex. [var. *arizonica* (Sarg.) Jepson; var. *macdougalii* (Rose) Jepson] [Eckenwalder J Arnold Arbor 58:193–208] ❀IRR or WET, SUN:1,**2,3**, 4–6,**7–24**; susceptible to mistletoe; INV;STBL.

P. tremuloides Michaux (p. 999) QUAKING ASPEN **ST**: crown slender; trunk < 15 m, highly clonal; twigs greenish white, glabrous; winter buds shiny. **LF**: petiole 2/3 to = blade length, laterally compressed; blade 2–4(7) cm, widely ovate or wider, base rounded to cordate, tip tapered, margin finely scalloped, surfaces glabrous, upper green, lower glaucous. $2n=38$. Streamsides, moist openings and slopes in montane and subalpine forests, woodlands, sagebrush steppe; 1800–3000 m. KR, NCoRH, CaR, SNH, SnBr, GB; to AK, e N.Am, Mex. ❀IRR,SUN,DRN:**1,2**,3,**4–7**,14–17,**18**,19–21;INV; also STBL.

SALIX WILLOW

George W. Argus

Shrub, tree, dioecious; bud scale 1, not sticky, margins gen fused (or free, overlapping). **ST**: twigs gen flexible and not glaucous. **LF**: blade linear to widely obovate, entire to toothed, gen ± hairy. **INFL**: dense catkin emerging before, with, or after lvs, sessile or on a short leafy shoot; bract subtending each fl. **FL**: perianth 0. **STAMINATE FL**: stamens 1–8. **PISTILLATE FL**: ovary stalked or sessile, style 1 or 0, stigmas 2, each sometimes 2-lobed; nectaries 1–several, gen rod-like, gen between infl axis and fl. **FR**: valves 2. ± 400 spp.: ± worldwide, esp n temp, arctic. (Latin: ancient name) [Argus 1986 Syst Bot Monog 9:1–170; Dorn 1976 Canad J Bot 54:2769–2789] Difficult, highly variable. Not all specimens will key easily; sprout shoots and other extreme forms are not incl in keys, may require field comparison for identification. Studies of variation, hybridization needed.

1. Pl vegetative (without reproductive structures) .... **Key 1**
1′ Pl with reproductive structures
  2. Pl pistillate .... **Key 2** (p. 992)
  2′ Pl staminate .... **Key 3** (p. 994)

Key 1 (vegetative plants)

1. Dwarf alpine shrub, < 0.2 m, forming mats by layering or rhizomes
  2. Young lvs hairy; mature lf blade 20–40 mm, not strongly net-like below, puberulent on both surfaces; pl forming colonies by adventitious rooting of sts .... ***S. arctica***
  2′ Young lvs glabrous; mature lf blade 6–22 mm, strongly net-like below, glabrous on both surfaces; pl forming rhizomatous mats .... ***S. reticulata*** ssp. ***nivalis***
1′ Low shrub to tree, not alpine, > 0.2 m, not mat-forming
  3. Tree; margins of bud scale free and overlapping
    4. Lf blade not glaucous below; twigs yellowish or yellow-gray .... ***S. gooddingii***
    4′ Lf blade glaucous below; twigs yellow- or red-brown .... ***S. laevigata***
  3′ Shrub or tree; margins of bud scale fused
    5. Petiole with glands at base of blade
      6. Tree, 7–25 m, alien; twigs pendent or erect, brittle at base; petioles tomentose or silky
        7. Sts gen erect (pendent); twigs yellow, gray-brown, or red-brown; twigs and young lvs densely silky; petioles silky; stipules acute at tip .... ***S. alba***
        7′ Sts pendent ± to ground; twigs yellow- to gray-brown; twigs and young lf blade sparsely long-soft-wavy-hairy, becoming glabrous; petioles tomentose; stipules acuminate at tip .... [2]***S. babylonica***
      6′ Shrub or small tree, 0.6–10 m, native; twigs erect, flexible or brittle at base; petioles glabrous or long-soft-wavy-hairy
        8. Lf blade tomentose or silky, tip acute; stipules not strongly glandular; twigs dull or shiny; shrub 0.4–6 m .... [2]***S. eastwoodiae***
        8′ Lf blade becoming glabrous, tip ± long-acuminate (rarely acute); stipules glandular; twigs highly glossy; tall shrub or small tree 1–10 m .... ***S. lucida***
    5′ Petiole lacking glands at base of blade
      9. Lf blade not glaucous below
        10. Lf blade margins entire or sparsely fine-dentate with well separated teeth; pl clonal by root-sprouting; petiole glabrous or silky, 2–6 mm

11. Lf blade linear (length 10–23 × width); young lvs silky .... [2]*S. exigua*
11′ Lf blade narrowly elliptic to narrowly obovate (length 2.8–8.5 × width; young lvs long-soft-wavy-hairy, tomentose, or velvety
12. Petiole glabrous; twigs gray-brown; lf tip acute, both surfaces becoming glabrous to moderately tomentose or long-soft-wavy-hairy .... *S. melanopsis*
12′ Petiole hairy; twigs red-brown; lf tip acuminate, both surfaces densely silky or long-soft-wavy-hairy .... *S. sessilifolia*
10′ Lf blade margins entire to finely serrate; pl not clonal; petiole soft-shaggy-hairy, tomentose, or velvety, 3–17 mm
13. Lf blade margins strongly rolled under; young lvs woolly or silky .... [2]*S. sitchensis*
13′ Lf blade margins flat or slightly rolled under; young lvs soft-shaggy-hairy, tomentose, woolly, or silky
14. Young lvs shaggy-hairy; mature lf blade glabrous or soft-shaggy-hairy below, shiny to highly glossy above .... *S. boothii*
14′ Young lvs densely tomentose, silky, or woolly; mature lf blade tomentose, long-soft-wavy-hairy, woolly-tomentose, silky, or becoming glabrous below and dull above .... [2]*S. eastwoodiae*
9′ Lf blade glaucous below or surface obscured by hair
15. Lf blade linear (length 10–23 × width); young lvs silky; clonal by root-sprouting .... [2]*S. exigua*
15′ Lf blade narrowly elliptic or wider; not clonal by root-sprouting
16. Tree, 7–25 m, alien; twigs pendent ± to ground; stipule tip ± long-acuminate .... [2]*S. babylonica*
16′ Shrub or tree, 0.2–10 m, native; twigs decumbent or erect; stipule tip rounded to acute
17. Lf blade margins strongly rolled under
18. Lf blade densely woolly or silky below, surface obscured by white hair, dull above; margin of first lvs on shoot glandular-dotted to finely serrate .... [2]*S. sitchensis*
18′ Lf blade silky, tomentose, or woolly below, often with mixture of white and rusty hairs, glaucous below, shiny above; margin of first lvs on shoot entire .... *S. scouleriana*
17′ Lf blade margins flat or slightly rolled under
19. Lf blade highly glossy above
20. Lf blade moderately to densely shaggy, tomentose, or woolly below
21. Twigs gen glaucous; lf blade gen woolly or shaggy-hairy below, 18–55 wide, length 1.6–4.2 × width .... *S. hookeriana*
21′ Twigs not glaucous; lf blade gen tomentose below, 6–32 mm wide, length 3.2–9.6 × width .... [8]*S. lasiolepis*
20′ Lf blade glabrous (or soon becoming so) to sparsely silky below
22. Petiole tomentose or velvety; lf blade 35–125 mm; twig hairs spreading; lf blade glands slightly on upper surface of blade; gen 0–2500 m .... [8]*S. lasiolepis*
22′ Petiole sparsely soft-shaggy-hairy; lf blade 20–54 mm; twig hairs appressed or curved toward st; lf blade glands at margin of blade; 2500–4000 m .... *S. planifolia* ssp. *planifolia*
19′ Lf blade dull to shiny above
23. Lf blade glabrous, downy, or soft-shaggy-hairy below
24. Lf blade with white hairs, sometimes mixed with rusty hairs
25. Stipules gen lf-like, sometimes vestigial; young lvs and lf blade gen short-tomentose, sometimes densely silky; 0–2500 m .... [8]*S. lasiolepis*
25′ Stipules 0 or vestigial on first lvs, lf-like on later lvs; young lvs and lf blade gen densely silky; 1400–3600 m
26. Lf blade 32–74 mm, persistently long-(often white)-hairy; petiole 3–9 mm; stipules gen 0 or vestigial .... [5]*S. geyeriana*
26′ Lf blade 44–102 mm, becoming ± glabrous or sparsely short-(gen rusty)-hairy; petioles 5–16 mm; stipules vestigial on first lvs, lf-like on later .... [3]*S. lemmonii*
24′ Lf blade with white hairs only
27. Stipules 0 or vestigial on first lvs, lf-like on later lvs
28. Young lvs and lf blades gen densely silky; twigs gen glaucous; 1450–3600 m ...... [5]*S. geyeriana*
28′ Young lvs and lf blades gen short-tomentose (sometimes densely silky); twigs not glaucous; 0–2500 m .... [8]*S. lasiolepis*
27′ Stipules lf-like
29. Lf blade gen with ridge-like, netted veins below, blade length 2–3.8 × width, margin entire to shallowly crenate; twigs long-soft-wavy-hairy; young lvs densely silky — n MP ... [2]*S. bebbiana*
29′ Lf blade smooth below, blade length 2.2–9.6 × width, margin gen finely serrate (sometimes entire in *S. lasiolepis*); twigs tomentose or velvety, gen becoming glabrous; young lvs gen glabrous or very sparsely hairy (densely silky in *S. lasiolepis*)
30. Lf base wedge-shaped to acute; blade shiny above; twigs tomentose, becoming glabrous .... [8]*S. lasiolepis*
30′ Lf base acute to rounded or ± cordate; blade dull above; twigs glabrous to sparsely long-soft-wavy-hairy
31. Twigs yellowish or grayish, smooth .... *S. lutea*
31′ Twigs gen red-brown, peeling
32. Lf blade lanceolate to elliptic (length 2.9–6.4 × width), base acute to rounded (rarely cordate); stipule tip acute .... *S. ligulifolia*
32′ Lf blade lanceolate to ovate (length 2.4–4.5 × width), base gen cordate; stipule tip rounded .... *S. prolixa*

23′ Lf blade long-soft-wavy-hairy, tomentose, silky, or velvety below
33. Twigs glaucous
34. Lf blade dull above
35. Lf blade very densely white-silky below (some hairs sometimes rusty); young lvs short-silky; twigs brittle at base .... [2]***S. drummondiana***
35′ Lf blade sparsely to moderately densely white-silky below; young lvs long-silky; twigs flexible at base .... [3]***S. orestera***
34′ Lf blade shiny above
36. Lf blade hairier below than above (but midrib glabrous below), gen densely white-silky to -tomentose; young lvs short-silky .... [2]***S. drummondiana***
36′ Lf blade equally hairy above and below or becoming glabrous (midrib below gen hairy like blade), hairs white, rusty, or mixed; young lvs silky
37. Lf blade 32–74 mm, persistently long-(gen white)-hairy (sometimes mixed with rusty); petiole 3–9 mm; stipules gen 0 or vestigial .... [5]***S. geyeriana***
37′ Lf blade 44–102 mm, becoming glabrous or short-(gen rusty)-hairy; petiole 5–16 mm; stipules vestigial on first lvs, lf-like on later lvs .... [3]***S. lemmonii***
33′ Twigs not glaucous
38. Young lvs sparsely or moderately hairy; petiole puberulent; twigs long-soft-wavy-hairy — n MP .... [2]***S. bebbiana***
38′ Young lvs very densely hairy; petiole tomentose, silky, or velvety; twigs soft-shaggy-hairy, tomentose, silky, velvety, or becoming glabrous
39. Lf blade ± widely elliptic or widely obovate (length 1.3–2.8 × width); petiole velvety (or tomentose); young lvs velvety (or silky); lf tip obtuse to rounded .... ***S. delnortensis***
39′ Lf blade ± narrowly elliptic or narrowly obovate (length 2.8–7.3 × width); petiole tomentose (or silky or velvety in *S. jepsonii*); young lvs silky; lf tip acute to acuminate (or obtuse in *S. jepsonii*)
40. Subalpine shrub < 1 m; lf blade 10–95 mm; 2100–4000 m
41. Lf blade 10–20 mm, shiny above; petiole silky; young lvs very densely hairy; rare, e slope c SNH (Mono Co.) .... ***S. brachycarpa*** ssp. ***brachycarpa***
41′ Lf blade 35–95 mm, dull above; petiole soft-shaggy-hairy; SNH .... [3]***S. orestera***
40′ Shrub > 1 m; lf blade 32–144 mm; 0–2800 m (± 3600 m in *S. geyeriana, S. lemmonii, S. jepsonii*)
42. Lf blade with white and rusty hair
43. Stipules gen lf-like (or vestigial); young lvs and lf blade gen short-tomentose, sometimes densely silky; 0–1900 m .... [8]***S. lasiolepis***
43′ Stipules 0 or vestigial on first lvs, gen lf-like on later lvs; young lvs and lf blade gen densely silky; 1450–3600 m
44. Lf blade 32–74 mm, persistently long-(often white)-hairy; petiole 3–9 mm; stipules gen 0 or vestigial .... [5]***S. geyeriana***
44′ Lf blade 44–110 mm, becoming glabrous or short-(gen rusty)-hairy; petiole 5–16 mm; stipules vestigial on first lvs, lf-like on later lvs .... [3]***S. lemmonii***
42′ Lf blade with white hairs only
45. Petiole silky or velvety
46. Lf blade 32–74 mm, gen densely silky on both surfaces; stipules gen 0; petiole velvety .... [5]***S. geyeriana***
46′ Lf blade 35–144 mm, gen becoming glabrous or hairier below; stipules gen lf-like
47. Lf blade densely silky below, dark green above; margin of first lvs on shoot gland-dotted .... ***S. jepsonii***
47′ Lf blade sparsely or moderately hairy below, not dark green above; margin of first lvs on shoot glandless .... [8]***S. lasiolepis***
45′ Petiole tomentose or soft-shaggy-hairy
48. Lf blade silky on both surfaces; stipule tip acute .... [3]***S. orestera***
48′ Lf blade tomentose below, long-soft-wavy-hairy or tomentose above; stipule tip rounded
49. Lf blade gen densely tomentose below; twigs gen velvety; on serpentine — hybridizes with *S. lasiolepis* .... ***S. breweri***
49′ Lf blade gen sparsely or moderately tomentose; twigs gen sparsely tomentose; not always on serpentine .... [8]***S. lasiolepis***

Key 2 (pistillate plants)

1. Ovary glabrous
2. Fl bract deciduous after fl, tawny, sparsely hairy, tip often ragged
3. Margins of bud scale free and overlapping; tree
4. Lf blade not glaucous below; beak of ovary abruptly tapered to style .... [2]***S. gooddingii***
4′ Lf blade glaucous below; beak of ovary gradually tapered to style .... ***S. laevigata***
3′ Margins of bud scale fused; shrub or tree
5. Petiole with glands near base of blade; pl not clonal; stigmas persistent after fl
6. Tree 10–25 m; stipules not conspicuously glandular; fr 3.5–5 mm; twigs golden-yellow, gray-, or red-brown; alien .... ***S. alba***
6′ Tree or shrub 1–10 m; stipules conspicuously glandular; fr 6–11 mm; twigs yellow-brown; native ... ***S. lucida***

5′ Petiole glands 0; pl spreading clonally by root-sprouting; stigmas deciduous after fl
7. Fr hairy, at least when young; fl bract tip acute to obtuse; young lvs silky .... [2]***S. exigua***
7′ Fr glabrous; fl bract tip rounded; young lvs soft-shaggy-hairy .... ***S. melanopsis***
2′ Fl bract persistent after fl, tawny to black, gen hairy, tip entire
8. Tree, alien; twigs pendent ± to ground; nectary > stalk .... ***S. babylonica***
8′ Shrub or tree, native; twigs erect; nectary < stalk
9. Fl bract dark brown, widest at round tip, hairs dense, short, gen wavy, ± equal; twigs tomentose, brittle at base; hairs of young lvs white and sometimes also rusty .... [2]***S. lasiolepis***
9′ Fl bract not as above in all ways; twigs variously hairy, flexible at base (exc some *S. hookeriana*); hairs of young lvs white only
10. Lf blade not glaucous below
11. Catkins gen appearing before the lvs, sessile; lf blade glabrous, becoming so, or soft-shaggy-hairy and highly glossy above; ovary gen glabrous .... ***S. boothii***
11′ Catkins appearing with the lvs, on distinct leafy shoots; lf blade tomentose or silky and dull to shiny above; ovary gen silky (sometimes glabrous) .... [3]***S. eastwoodiae***
10′ Lf blade glaucous below
12. Stipules 0 or vestigial on early lvs, 0 or prominent on later lvs; fl bract hairs straight or wavy
13. Style 0.6–1.4 mm; fr 6–7 mm; twigs thinly glaucous; lf blade length 1.6–3.9 × width, soft-shaggy or woolly below; fl bract with long, straight hairs, tip sometimes pointed .... [3]***S. hookeriana***
13′ Style 0.1–0.6 mm; fr 2.5–5.5 mm; twigs not glaucous; lf blade length 2.2–9.6 × width, tomentose or silky below (but becoming glabrous); fl bract with short, wavy hairs, tip gen widest at round tip .... [2]***S. lasiolepis***
12′ Stipules prominent on all lvs; fl bract hairs gen curly (rarely straight)
14. Twigs yellow or gray-yellow, smooth, not peeling; twigs glabrous or sparsely soft-shaggy-hairy ... ***S. lutea***
14′ Twigs red-brown, sometimes yellow-brown, gen peeling; twigs glabrous or long-soft-wavy-hairy
15. Ovary on stalk 0.9–2.5 mm; lf blade lanceolate to elliptic (length 2.9–6.4 × width), base acute to rounded (rarely cordate); stipule tip acute .... ***S. ligulifolia***
15′ Ovary on stalk 1.3–4.2 mm; lf blade lanceolate to ovate (length 2.4–4.5 × width), base gen cordate; stipule tip rounded .... ***S. prolixa***
1′ Ovary hairy
16. Dwarf alpine shrub, forming mats by adventitious rooting of sts or by rhizomes
17. Catkins > 20-fld; fl bract hairy; nectary 0.6–1.2 mm; style 0.4–1.6 mm; lf veins not strongly net-like below .... ***S. arctica***
17′ Catkins < 20-fld; fl bract glabrous; nectary ± 0.2 mm; style 0.1–0.4 mm; lf veins strongly net-like below .... ***S. reticulata*** ssp. ***nivalis***
16′ Low to tall shrub or tree; not forming mats
18. Fl bract deciduous, tawny, sparsely to moderately hairy
19. Tree, not clonal by root-sprouting; margins of bud scale free and overlapping; lf margins entire to finely serrate; petiole glands present; stigmas persistent .... [2]***S. gooddingii***
19′ Slender shrub, clonal by root-sprouting; margins of bud scale fused; lf margins entire or finely dentate with well separated teeth; petiole glands 0; stigmas deciduous after fl
20. Lf blade linear (length 10–23 × width); ovary silky or soft-shaggy-hairy; style 0–0.2 mm; outer nectary present .... [2]***S. exigua***
20′ Lf blade narrowly elliptic, narrowly obovate or elliptic (length 3–8.5 × width); ovary densely long-soft-wavy-hairy; style 0.2–0.4 mm; outer nectary 0 .... ***S. sessilifolia***
18′ Fl bract persistent, brown or tawny, gen moderately hairy
21. Catkins appearing with the lvs, on distinct, leafy shoot
22. Lf blade not glaucous below or surface obscured by hair
23. Lf blade densely silky or silky-woolly below, dark green and sparsely silky (but becoming glabrous) above; twigs gen brittle at base
24. Lf blade 8–25 mm wide, length 3–7.3 × width, margins flat or slightly rolled under; nectary 0.3–0.6 mm; twigs highly brittle at base; CaRH, SNH; 1000–3400 m .... [3]***S. jepsonii***
24′ Lf blade 17–40 mm wide, length 2–4 × width, margins strongly rolled under; nectary 0.6–0.8 mm; twigs flexible or somewhat brittle at base; NW, CCo, SnFrB, SCoRO, w WTR; 0–400 m (exc 1800–2500 m in n NW) .... [4]***S. sitchensis***
23′ Lf blade sparsely to densely tomentose or silky and gray-green on both surfaces; twigs flexible at base
25. Lvs on catkin shoots (and first lvs on vegetative twigs) distinctly finely serrate with long teeth .... [3]***S. eastwoodiae***
25′ Lvs on catkin shoots (and first lvs on vegetative twigs) entire or gland-dotted
26. Lf blade sparsely to moderately woolly-tomentose or silky, becoming glabrous below
27. Lf blade shaggy-(sometimes appressed)-hairy below, base rounded or cordate (rarely acute); nectary 0.5–0.8 mm; SNH, KR, CaRH .... [3]***S. eastwoodiae***
27′ Lf blade silky below, base acute; nectary 0.8–1 mm; SNH .... [2]***S. orestera***
26′ Lf blade densely silky to woolly below
28. Lf blade 8–25 mm wide, length 3–7.3 × width, margins flat or slightly rolled under; nectary 0.3–0.6 mm; twigs highly brittle at base; CaRH, SNH; 1000–3400 m .... [3]***S. jepsonii***
28′ Lf blade 17–40 mm wide, length 2–4 × width, margins strongly rolled under; nectary 0.6–0.8 mm; twigs flexible or somewhat brittle at base; NW, CCo, SnFrB, SCoRO, w WTR; 0–400 m (exc 1800–2500 m in n NW) .... [4]***S. sitchensis***

22′ Lf blade glaucous below
29. Ovary on stalk 0–0.3 mm; subalpine shrub 0.2–0.5 m .... ***S. brachycarpa*** ssp. ***brachycarpa***
29′ Ovary on stalk 0.4–5 mm; alpine, subalpine, or montane shrub or tree > 0.5 m
30. Young lvs with rusty or white and rusty hairs
31. Catkin subspheric, 9–20 mm; style 0.1–0.4 mm; fr 4–5.2 mm; fl bract pale; lvs gen persistently silky on both surfaces; petioles 3–9 mm; stipules 0 or vestigial .... [4]***S. geyeriana***
31′ Catkin cylindric, 14–50 mm; style 0.3–0.8 mm; fr 5–6 mm; fl bract dark; lvs silky (but becoming glabrous) on both surfaces; petiole 5–16 mm; stipules lf-like (exc vestigial on early lvs) .... [2]***S. lemmonii***
30′ Young lvs with white hairs
32. Style 0.1–0.4 mm; ovary on stalk 1–5 mm
33. Catkin 25–60 mm; twigs shaggy, not glaucous; petiole puberulent; lf blade dull above; stipules lf-like; ovary hairs gray, stalk 2–5 mm; 1000–1400 m .... [2]***S. bebbiana***
33′ Catkin 9–15 mm; twigs tomentose or velvety, gen glaucous; petiole velvety; lf blade shiny above; stipules vestigial; ovary hairs white or white and rusty, stalk 1–2.8 mm; 1500–3600 m ... [4]***S. geyeriana***
32′ Style 0.4–1 mm; ovary on stalk 0.4–2 mm
34. Style 0.4–0.6 mm; ovary short-silky; twigs highly brittle at base; lf blade densely silky below; petiole silky or velvety .... [3]***S. jepsonii***
34′ Style 0.6–1 mm; ovary long-silky; twigs flexible at base; lf blade sparsely to moderately silky below; petiole tomentose .... [2]***S. orestera***
21′ Catkins appearing before (sometimes just before) the lvs, sessile or subtended by greenish bracts
35. Fl bract tawny or light rose
36. Fl bract sparsely hairy; ovary sparsely or moderately silky, long-beaked; style 0.1–0.4 mm
37. Catkin 25–60 mm; twigs shaggy, not glaucous; petiole puberulent; lf blade dull above; stipules lf-like; ovary hairs gray, stalk 2–5 mm; 1000–1400 m .... [2]***S. bebbiana***
37′ Catkin 9–15 mm; twigs tomentose or velvety, gen glaucous; petiole velvety; lf blade shiny above; stipules vestigial; ovary hairs white or white and rusty, stalk 1–2.8 mm; 1500–3600 m ..... [4]***S. geyeriana***
36′ Fl bract moderately to densely hairy; ovary densely hairy, short-beaked; style 0.4–0.8 mm
38. Ovary tomentose, stalk 0–0.4 mm; nectary gen thread-like; lf margins slightly rolled under, blade sparsely or moderately tomentose below; stipule tip rounded; on serpentine .... ***S. breweri***
38′ Ovary silky, stalk 0.4–1.2 mm; nectary short, rod-like; lf margins strongly rolled under, blade very densely woolly or silky below; stipule tip acute; not on serpentine .... [4]***S. sitchensis***
35′ Fl bract brown, black, or 2-colored
39. Twigs strongly glaucous .... ***S. drummondiana***
39′ Twigs not glaucous or weakly so
40. Young lvs with white hairs
41. Catkin stout; ovary stalk 1–2 mm .... [3]***S. hookeriana***
41′ Catkin slender; ovary stalk 0–1.2 mm
42. Catkin 15–40 mm; ovary stalk 0–0.3 mm; lf blade densely tomentose below; on serpentine, KR .... ***S. delnortensis***
42′ Catkin 40–95 mm; ovary stalk 0.4–1.2 mm; lf blade woolly or silky below; NW, CW (exc SCoRI), w WTR .... ***S. sitchensis***
40′ Young lvs with white or white and rusty hairs
43. Stigma 0.2–0.4 mm; catkin < 45 mm
44. Catkin 9–15 mm; style 0.1–0.4 mm; fr 4–5.2 mm .... [4]***S. geyeriana***
44′ Catkin 13–50 mm; style 0.3–0.8 mm; fr 5–6 mm .... [2]***S. lemmonii***
43′ Stigma 0.5–1.2 mm; catkin 30–140 mm in *S. hookeriana*, 10–45 mm in *S. planifolia*, 15–65 mm in *S. scouleriana*
45. Young lvs sparsely hairy; ovary stalk 0.3–0.8 mm; fr < 6 mm .... ***S. planifolia*** ssp. ***planifolia***
45′ Young lvs densely hairy; ovary stalk 0.8–2.3 mm; fr > 6 mm
46. Style 0.6–1.4 mm; stigma 0.5–0.7 mm .... [3]***S. hookeriana***
46′ Style 0.3–0.6 mm; stigma 0.7–1.2 mm, slender .... ***S. scouleriana***

Key 3 (staminate plants)

1. Stamens > 2 per fl
2. Margins of bud scale fused; anthers 0.7–1 mm .... ***S. lucida***
2′ Margins of bud scale free and overlapping; anthers 0.4–0.6 mm
3. Lf blade not glaucous below; twigs yellowish or yellow-gray; margins of lvs on catkin shoots finely serrate; young lf hairs white .... ***S. gooddingii***
3′ Lf blade glaucous below; twigs yellow- or red-brown; margins of lvs on catkin shoots entire; young lf hairs white or white and rusty .... ***S. laevigata***
1′ Stamens 1 or 2 per fl
4. Stamen 1 per fl
5. Lf blade 8–25 mm wide, length 3–7.3 × width; fl bract sparsely hairy; filaments hairy; twigs highly brittle at base; CaRH, SNH; 1000–3400 m .... [2]***S. jepsonii***
5′ Lf blade 17–40 mm wide, length 2–4 × width; fl bract moderately hairy; filaments glabrous; twigs flexible or somewhat brittle at base; NW, CCo, SnFrB, SCoRO, w WTR; 0–400 m (exc 1800–2500 m in n NW) .... ***S. sitchensis***
4′ Stamens 2
6. Dwarf alpine shrub, < 0.2 m, forming mats by rhizomes or adventitious rooting of sts

7. Catkin on leafy shoot 9–18 mm; fl bract hairy; young lvs long-soft-wavy-hairy; only axillary nectary present . . . . **S. arctica**
7′ Catkin on leafy shoot 2.1–4 mm; fl bract glabrous; young lvs glabrous; both axillary and outer nectary present . . . . **S. reticulata** ssp. **nivalis**
6′ Low to tall shrub or tree, > 0.2 m, not mat-forming
8. Both axillary and outer nectary present; catkins appearing with or after lvs, on leafy shoots; alien or native
9. Tree; twigs brittle at base; alien
10. Twigs erect or weakly pendent, gen yellow, sometimes gray- or red-brown; nectaries 2; twigs and young lvs densely silky . . . . **S. alba**
10′ Twigs pendent ± to ground, gray- to yellow-brown; nectaries 1–several; twigs and young lvs glabrous or sparsely silky . . . . **S. babylonica**
9′ Shrub; twigs flexible at base; native
11. Low subalpine shrub, 0.2–0.5 m, not clonal by root-sprouting; catkin 4–5 mm on leafy shoot ± 1 mm; anthers 0.3–0.4 mm . . . . **S. brachycarpa** ssp. **brachycarpa**
11′ Medium or tall riparian shrub, 0.8–7 m, clonal by root-sprouting; catkin 20–70 mm on leafy shoot 5–110 mm; anthers 0.5 to 1.2 mm
12. Lf blade linear; young lvs silky . . . . **S. exigua**
12′ Lf blade narrowly elliptic to narrowly obovate; young lvs silky or long-soft-wavy-hairy
13. Petiole tomentose or becoming glabrous; fl bract glabrous to sparsely hairy; lf blade becoming glabrous or sparsely to moderately tomentose below, tip acute; twigs gray-brown . . . . **S. melanopsis**
13′ Petiole silky; fl bract moderately hairy; lf blade sparsely to very densely silky (or long-soft-wavy-hairy below), tip acuminate; twigs red-brown . . . . **S. sessilifolia**
8′ Only axillary nectary present; catkins appearing before or with lvs, sessile or on leafy shoots; native
14. Catkins appearing with lvs, on distinct, leafy shoots
15. Hairs of young lvs rusty (or white and rusty)
16. Catkin 6–12 mm; fl bract tawny to light brown; anthers 0.4–0.5 mm . . . . [4]**S. geyeriana**
16′ Catkin 13–35 mm; fl bract brown to 2-colored; anthers 0.5–0.9 mm . . . . [2]**S. lemmonii**
15′ Hairs of young lvs white only
17. Filaments hairy
18. Lf blade not glaucous below; lvs on catkin shoots finely serrate with long teeth (rarely only gland-dotted) . . . . [2]**S. eastwoodiae**
18′ Lf blade glaucous below or surface obscured by hair; lvs on catkin shoots entire or gland-dotted
19. Stipules 0 or vestigial
20. Catkin 9–15 mm; anthers 0.4–0.5 mm; lf blade shiny above; petiole velvety . . . . [4]**S. geyeriana**
20′ Catkin 12–25 mm; anthers 0.6–1 mm; lf blade dull above; petiole tomentose . . . . [2]**S. orestera**
19′ Stipules lf-like
21. Lf blade densely silky below; petiole silky or velvety; twigs very brittle at base . . . . [2]**S. jepsonii**
21′ Lf blade sparsely to moderately tomentose or silky below; petiole puberulent or tomentose; twigs flexible at base
22. Fl bract tawny; twigs moderately to densely long-soft-wavy-hairy; petiole puberulent; 1200–1400 m . . . . [2]**S. bebbiana**
22′ Fl bract brown; twigs sparsely soft-shaggy-hairy or silky, becoming glabrous; petiole tomentose; 2100–4000 m . . . . [2]**S. orestera**
17′ Filaments glabrous
23. Stipules 0 or vestigial, stipule scars 0 or minute . . . . [4]**S. geyeriana**
23′ Stipules lf-like, stipule scars distinct
24. Lf blade not glaucous below . . . . [2]**S. eastwoodiae**
24′ Lf blade glaucous below
25. Twigs yellowish to yellow-gray, not peeling; anthers 0.7–1.2 mm . . . . **S. lutea**
25′ Twigs gen red-brown sometimes yellow-brown, peeling; anthers 0.5–0.8 mm
26. Lf blade margin entire to crenate; fl bract 1.2–3.2 mm, hairs wavy . . . . [2]**S. bebbiana**
26′ Lf blade gen serrate; fl bract 0.8–1.6 mm, hairs curly
27. Lf blade lanceolate to elliptic, base acute to rounded; stipule tip acute . . . . **S. ligulifolia**
27′ Lf blade lanceolate to ovate, base gen cordate; stipule tip rounded . . . . **S. prolixa**
14′ Catkins appearing before lvs, sessile or subtended by greenish bracts
28. Fl bracts gen tawny
29. Catkin 20–32 mm; fl bracts densely hairy; twigs not glaucous; hairs of young lvs white only; on serpentine; 300–1300 m . . . . **S. breweri**
29′ Catkin 6–12 mm; fl bracts sparsely hairy; twigs gen glaucous; hairs of young lvs gen white and rusty; not on serpentine; 1500–3600 m . . . . [4]**S. geyeriana**
28′ Fl bracts gen dark brown to blackish
30. Twigs densely glaucous . . . . **S. drummondiana**
30′ Twigs not glaucous (or thinly so)
31. Twigs gen glabrous or becoming so
32. Nectary short, square . . . . [2]**S. scouleriana**
32′ Nectary distinctly rod-like
33. Young lvs gen glabrous . . . . **S. planifolia** ssp. **planifolia**
33′ Young lvs gen densely hairy

34. Fl bracts widest at widely rounded tip, gen densely hairy; 0–2800 m . . . . [2]*S. lasiolepis*
34′ Fl bracts widest at middle, sparsely to moderately hairy; 1450–3500 m . . . . [2]*S. lemmonii*
31′ Twigs gen persistently (sometimes densely) hairy
35. Anthers 0.4–0.6 mm; fl bracts gen obscured by hairs . . . . [2]*S. lasiolepis*
35′ Anthers 0.6–1.2 mm; fl bracts sparsely to moderately hairy
36. Twigs brittle at base; on serpentine . . . . *S. delnortensis*
36′ Twigs brittle or flexible at base; not on serpentine
37. Catkins 30–60 mm, stout; twigs brittle at base; coastal dunes and meadows, gen < 100 m . . . . *S. hookeriana*
37′ Catkins 15–45 mm, slender; twigs flexible at base; upland forest to subalpine meadows, 90–3400 m
38. Nectary rod-like; lvs on catkin shoots gland-dotted or finely serrate; subalpine meadows, lake- and stream-shores, 1700–3200 m . . . . *S. boothii*
38′ Nectary square; lvs on catkin shoots entire; upland forests, meadows, 90–3400 m . . . [2]*S. scouleriana*

S. alba L. WHITE WILLOW Tree < 25 m. **ST** erect, spreading, or pendent; twigs brownish to golden-yellow, sometimes ± brittle at base, silky, becoming glabrous. **LVS**: stipules lf-like; petiole with glands; young lvs silky; mature blade 63–115 mm, lanceolate to elliptic, acute (base acute or wedge-shaped), serrate, silky, becoming glabrous, lower surface glaucous, upper surface dull. **INFL** appearing with lvs, 30–60 mm, on leafy shoot 3–25 mm; fl bract tawny, glabrous or sparsely hairy, tip rounded; pistillate bract deciduous after fl. **STAMINATE FL**: stamens 2. **PISTILLATE FL**: ovary glabrous, style ± 0.2 mm, stalk 0.2–0.8 mm. $2n$=76. Disturbed places, gen near settlements; gen < 1000 m. CA; to e N.Am; native to Eur. Mostly cult as orn, but many CVS and hybrids ± naturalized. Var. *vitellina* (L.) Stokes (twigs bright yellow or golden, spreading) and hybrids with *S. fragilis* most common.

S. arctica Pallas (p. 999) ARCTIC WILLOW Shrub < 0.1 m, trailing, rooting. **ST**: twigs brownish, sparsely soft-shaggy-hairy. **LVS**: stipules vestigial; young lvs long-soft-hairy; mature blade 20–40 mm, ± elliptic, acuminate (base acute to rounded), entire, sparsely soft-shaggy-hairy, becoming glabrous, lower surface glaucous. **INFL** appearing with lvs, 9–55 mm, on leafy shoot 5–55 mm; fl bract tawny or dark brown. **STAMINATE FL**: stamens 2. **PISTILLATE FL**: ovary silky, style 0.4–1.6 mm, stalk 0.3–0.8 mm. $2n$=76,114. Alpine tundra; 2300–4000 m. CaRH (Lassen Peak), c&s SNH; to n Can, NV. [*S. anglorum* Cham. vars. *antiplasta* C. Schneider & *petraea* Andersson] Pls with narrow, sharp-pointed lvs approach *S. cascadensis* Cockerell of WA and n; need study. ❀WET:1–3,7,18&SUN:4–6,15,16;DFCLT.

S. babylonica L. (p. 999) WEEPING WILLOW Tree < 20 m. **ST** pendent ± to ground; twigs gray- to yellow-brown, long-soft-wavy-hairy, becoming glabrous, brittle at base. **LVS**: petiole gen with glands; young lvs glabrous or sparsely silky; mature blade 70–140 mm, linear-lanceolate, acuminate (base wedge-shaped), gen finely sharp-serrate, silky, becoming glabrous, lower surface glaucous. **INFL** appearing with lvs, 18–35 mm, on leafy shoots 4–10 mm (± 0.3 mm in staminate); fl bract tawny. **STAMINATE FL**: stamens 2. **PISTILLATE FL**: ovary glabrous, stigma shrub-like, style ± 0.2 mm, stalk 0–0.3 mm. $2n$=76. Disturbed places, around settlements; gen < 1000 m. CA; to se US; native to Asia. Mostly cult as orn, but many CVS and hybrids ± naturalized, incl hybrids with *S. alba*.

S. bebbiana Sarg. (p. 999) GRAY WILLOW Shrub, small tree, < 10 m. **ST**: twigs spreading widely, red-brown, soft-shaggy-hairy. **LVS**: young lvs hairy; mature blade 20–80 mm, elliptic to narrowly obovate, abruptly acute (base obtuse to acute), entire to crenate, tomentose or silky on both surfaces, becoming glabrous. **INFL** appearing with or just before lvs, 6–60 mm, on leafy shoots 3–15 mm (± 0 in staminate); fl bract tawny. **STAMINATE FL**: stamens 2. **PISTILLATE FL**: ovary long-beaked, silky, style 0.1–0.4 mm, stalk 2–5 mm. $2n$=38. RARE in CA. Streams, lake shores; 1200–1400 m. n MP (Lower Klamath Lake, Siskiyou Co.; Goose Lake, Modoc Co.); to AK, e N.Am, NM. In cult.

S. boothii Dorn (p. 999) BOOTH'S WILLOW Shrub < 6 m. **ST**: twigs brownish, glabrous to shaggy-hairy. **LVS**: young lvs shaggy-hairy; mature blade 26–102 mm, lanceolate to widely elliptic, acuminate (base rounded to acute), entire to finely serrate, soft-shaggy-hairy, becoming glabrous, lower surface not glaucous, upper surface shiny. **INFL** appearing before or with lvs, 10–70 mm, sessile or on leafy shoots < 15 mm (margins of shoot-lvs gland-dotted or finely serrate); fl bract gen dark brown, hairs wavy. **STAMINATE FL**: stamens 2. **PISTILLATE FL**: ovary glabrous, style 0.3–1 mm, stalk 0.5–2.5 mm. $2n$=76. Uncommon. Wet subalpine meadows, shores; 1700–3200 m. KR, CaRH, n SNH, Wrn; to w Can, Colorado. [*S. pseudocordata* Andersson; *S. myrtillifolia* Andersson misapplied]

S. brachycarpa Nutt. ssp. ***brachycarpa*** (p. 999) SHORT-FRUITED WILLOW Shrub < 0.5 m. **ST**: twigs brownish, silky, becoming glabrous. **LVS**: stipules gen vestigial; young lvs densely silky; mature blade 10–20 mm, ± elliptic, acute (base rounded to acute), entire, surfaces silky, lower glaucous. **INFL** appearing with lvs, densely-fld, 4–14 mm, ± spheric, on leafy shoots < 4 mm; fl bract tawny. **STAMINATE FL**: stamens 2. **PISTILLATE FL**: ovary very densely silky, style 0.6–1.2 mm, stalk 0–0.3 mm. $2n$=38. RARE in CA. Subalpine meadows (esp on limestone); 3200–3500 m. c SNH (Mono Co); to n&e Can, Colorado.

S. breweri Bebb (p. 999) BREWER'S WILLOW Shrub < 4 m. **ST**: twigs yellow-brown, velvety or silky-tomentose. **LVS**: stipules vestigial or lf-like; young lvs silky; mature blade 58–144 mm, lanceolate to widely elliptic or oblanceolate, acute (base rounded to acute), entire to shallowly wavy-toothed, slightly rolled under, lower surface glaucous, gen densely tomentose, midrib glabrous, upper surface sparsely tomentose. **INFL** appearing before lvs, 20–85 mm, sessile or on leafy shoots < 5 mm; fl bract tawny to light rose. **STAMINATE FL**: stamens 2. **PISTILLATE FL**: ovary very densely tomentose, style 0.4–0.8 mm, stalk 0–0.4 mm, nectary gen thread-like. Serpentine streamsides; 300–1300 m. NCoR, SnFrB, n&c SCoR. Hybrids with *S. lasiolepis* in Yolo Co. have lvs less hairy, fr long, slender, stalked, becoming glabrous. Like *S. delnortensis* but has shorter petioles, flexible twig bases. ❀IRR or WET, SUN:4,**5**,6–9,14–16,**17**,22–24.

S. delnortensis C. Schneider (p. 999) DEL NORTE WILLOW Shrub < 2 m. **ST**: twigs yellowish brown, brittle at base, tomentose or velvety, becoming glabrous. **LVS**: young lvs velvety or silky; mature blade 53–102 mm, elliptic to widely obovate, obtuse or abruptly pointed (base obtuse to acute), entire, slightly rolled under, lower surface glaucous, densely tomentose, upper surface sparsely tomentose. **INFL** appearing before lvs, 15–40 mm, sessile or on leafy shoots < 5 mm; fl bract tawny to pinkish. **STAMINATE FL**: stamens 2. **PISTILLATE FL**: ovary silky, style 0.6–1.2 mm, stalk 0–0.3 mm. RARE in CA. Serpentine streamsides; 90–500 m. nw KR (near Gasquet, Del Norte Co.); sw OR. Like *S. breweri*; may hybridize with *S. lasiolepis*. In cult.

S. drummondiana Hook. (p. 999) DRUMMOND'S WILLOW Shrub < 5 m. **ST**: twigs brownish, brittle at base, glaucous, hairy, becoming glabrous. **LVS**: young lvs white- or white-and-rusty-silky; mature blade 40–85 mm, lanceolate, elliptic, or oblanceolate, acute or acuminate (base acute or wedge-shaped), entire to shallowly crenate, slightly rolled under, lower surface glaucous, gen obscured by dense, silky, white (or white and rusty) hairs, upper surface becoming glabrous. **INFL** appearing before lvs, 15–110 mm, sessile; fl bract brown to black. **STAMINATE FL**: stamens 2. **PISTILLATE FL**: ovary silky, style 0.5–1.5 mm, stalk 0.3–2 mm. $2n$=38,57,76. Streamsides, subalpine red-fir forests; 2200–3000 m. c&s SNH; to w Can, NV, NM. [var. *subcoerulea* (Piper) C. Ball] Hybrids with *S. jepsonii* have long, slender stigmas, dark bracts; like *S. geyeriana*. ❀WET:1–6,18.

S. eastwoodiae A.A. Heller (p. 999) SIERRA WILLOW Shrub < 4 m. **ST**: twigs violet to brownish, glaucous or not, shaggy-hairy, becoming glabrous. **LVS**: petiole gen with glands; young lvs hairy;

mature blade 27–99 mm, lanceolate to widely elliptic or oblanceolate, acute (base rounded to acute), entire to finely short-slender-serrate or gland-dotted, surfaces hairy, becoming glabrous, lower not glaucous. **INFL** appearing with lvs, 16–45 mm, on leafy shoots 2–20 mm; margin of shoot lvs finely short-slender-serrate or gland-dotted; fl bract brown or tawny. **STAMINATE FL**: stamens 2. **PISTILLATE FL**: ovary silky, style 0.6–1.5 mm, stalk 0.2–1.6 mm. $2n=76$. Alpine and subalpine meadows, streams, talus; 1600–3800 m. KR, CaRH, SNH; to OR, MT, WY. [*S. commutata* Bebb (incl var. *denudata* Bebb) misapplied] ❀WET:1–6,18.

S. exigua Nutt. (p. 999) NARROW-LEAVED WILLOW Shrub < 7 m, clonal by root-sprouting; **ST**: twigs brownish, silky or long-soft-wavy-hairy, becoming glabrous. **LVS**: petiole very short; young lvs silky; mature blade 50–124 mm, linear, acuminate (base wedge-shaped), entire to sharply serrate (teeth well separated), surfaces silky, lower glaucous or not. **INFL** appearing with or after lvs, sometimes branched, 22–70 mm, on leafy shoots 5–110 mm; fl bract tawny; pistillate bract deciduous after fl. **STAMINATE FL**: stamens 2. **PISTILLATE FL**: ovary glabrous, silky, or sparsely soft-shaggy-hairy, stigma deciduous after fl, style 0–0.2 mm, stalk 0.2–1 mm. $2n=38$. Common. Streamsides, marshes, wet ditches; < 2700 m. CA; to AK, e N.Am, AZ, Mex. [var. *stenophylla* (Rydb.) C. Schneider; *S. hindsiana* Benth. incl vars. *leucodendroides* (Rowlee) C. Ball and *parishiana* (Rowlee) C. Ball] Pls with spreading hairs on lvs and twigs, slender stigmas 0.6–1 mm, and ± entire lvs from throughout CA have been called *S. hindsiana*; these features vary independently; the type of *S. h.* does not share them all. Such forms may be derived from *S. e.* × *S. sessilifolia*; further study needed. ❀WET,SUN:1–5,**6–24**;STBL.

S. geyeriana Andersson (p. 999) GEYER'S WILLOW Shrub < 5 m. **ST**: twigs yellowish to brownish, gen glaucous, tomentose or velvety, becoming glabrous, sometimes brittle at base. **LVS**: stipules 0 or vestigial; petiole short; young lvs silky; mature blade 32–74 mm, lanceolate, elliptic, or oblanceolate, acuminate (base wedge-shaped), entire, flat, surfaces ± persistently white- or white-and-rusty-silky, lower glaucous. **INFL** appearing with or just before lvs, 6–20 mm, sessile or on leafy shoots < 8 mm, subspheric; fl bract tawny to brown. **STAMINATE FL**: stamens 2. **PISTILLATE FL**: ovary white- or white-and-rusty-silky, style 0.1–0.4 mm, stalk 1–2.8 mm. $2n=38$. Subalpine streams, meadows; 1450–3600 m. s CaRH, n SNH, s SNH (esp Kern Plateau), SnBr, MP; to B.C., MT, NM. [var. *argentea* (Bebb) C. Schneider] Like *S. lemmonii*; hybrids occur in Sierra and Lassen cos. *S. drummondiana* has lower lf surface densely silky, midrib glabrous, margins slightly rolled under. ❀WET:**1–3**,4,5,**6,7**;STBL.

S. gooddingii C. Ball (p. 999) GOODDING'S BLACK WILLOW Tree < 30 m. **ST**: twigs yellowish, velvety or soft-shaggy-hairy, becoming glabrous, sometimes brittle at base; bud scale margins free, overlapping. **LVS**: petiole with glands; young lvs hairy; mature blade 67–130 mm, narrowly lanceolate or narrowly ovate, acuminate (base acute), finely serrate, surfaces glabrous or becoming so, lower not glaucous. **INFL** appearing with lvs, 22–65 mm, on leafy shoot 4–25 mm; fl bract tawny; pistillate bract deciduous after fl. **STAMINATE FL**: stamens 4–8. **PISTILLATE FL**: ovary glabrous or hairy, style 0.1–0.3 mm, stalk 1.2–3.2 mm. $2n=38$. Common. Streamsides, marshes, seepage areas, washes, meadows; gen < 500 m (–23–1600 m). NCoRI, CaRF, SNF, GV, SCo, PR, D (esp GV, D); to TX, Mex. [var. *variabilis* C. Ball] ❀WET:5,6,**7–9**,10, 11,**12–14**,15–17,**18–24**;STBL.

S. hookeriana Hook. (p. 1001) COASTAL WILLOW Shrub, small tree, < 8 m. **ST**: twigs brownish, gen thinly glaucous, densely tomentose or soft-shaggy-hairy, becoming glabrous, brittle at base. **LVS**: young lvs white- or white-and-rusty-hairy; mature blade 46–113 mm, elliptic to widely ovate, entire to coarsely crenate, rounded to acuminate (base acute), rarely glabrous, lower surface glaucous, shaggy-hairy, upper surface highly glossy, tomentose (rarely glabrous). **INFL** appearing before lvs, 30–140 mm, sessile or on leafy shoots < 10 mm; fl bract dark brown. **STAMINATE FL**: stamens 2. **PISTILLATE FL**: ovary glabrous or tomentose, style 0.6–1.4 mm, stalk 1–2 mm. $2n=57$. Coastal dunes, floodplains, meadows; < 100 m. NCo (< 100 m), n NCoRO (500–1000 m); to AK. [*S. piperi* Bebb] Glabrous and densely tomentose forms intergrade, may occur in same population. Higher elevation glabrous form (Humboldt Co.) warrants study. May hybridize with *S. lasiolepis*. ❀IRR or WET:**4,5**,6,7,**14–17**,19–21,**22–24**;STBL.

S. jepsonii C. Schneider (p. 1001) JEPSON'S WILLOW Shrub < 3 m. **ST**: twigs brownish, tomentose or silky, becoming glabrous, brittle at base. **LVS**: young lvs silky; mature blade 55–103 mm, ± lanceolate, acute (both ends), entire, lower surface glaucous, densely silky, upper surface sparsely so. **INFL** appearing with lvs, 13–60 mm, on leafy shoots 1.5–7 mm; fl bract tawny or brown. **STAMINATE FL**: stamens 2 or 1. **PISTILLATE FL**: ovary silky, narrowly ovoid, style 0.4–0.6 mm, stalk 0.4–1.2 mm. Margins of lakes and streams, wet meadows; 1000–3400 m. CaRH, SNH. [*S. sitchensis* vars. *angustifolia* Bebb and *ralphiana* (Jepson) Jepson] Like *S. sitchensis* but smaller, lvs narrower (variable), stamens 2 vs 1 (variable), and ovary narrower. May be derived from *S. drummondiana* × *S. sitchensis*. ❀TRY;STBL.

S. laevigata Bebb (p. 999) RED WILLOW Tree < 15 m. **ST**: twigs red- to yellow-brown, hairy, becoming glabrous, brittle at base; bud scale margins free, overlapping. **LVS**: stipules vestigial on early lvs, lf-like later; petiole gen with glands; young lvs glabrous or white- or white-and-rusty-hairy; mature blade 67–150 mm, lanceolate to widely elliptic, acuminate (base rounded to acute), indistinctly finely crenate, surfaces glabrous, lower glaucous, upper shiny to highly glossy. **INFL** appearing with or after lvs, 35–110 mm, on leafy shoots 3–35 mm; fl bract tawny; pistillate bract deciduous after fl. **STAMINATE FL**: stamens 5. **PISTILLATE FL**: ovary glabrous, style 0.2–0.6 mm, stalk 1–3.4 mm. $2n=38$. Common. Riverbanks, seepage areas, lake shores (subalkaline or brackish), canyons, ditches; 0–1700 m. CA; s OR, n NV, AZ, Mex, n C.Am. [var. *araquipa* (Jepson) C. Ball]

S. lasiolepis Benth. (p. 1001) ARROYO WILLOW Shrub, small tree, < 10 m. **ST**: twigs yellowish to brownish, tomentose, long-soft-wavy-hairy, or velvety, becoming glabrous, gen brittle at base. **LVS**: young lvs hairy; mature blade 35–125 mm, ± lanceolate-elliptic to oblanceolate, acute (base wedge-shaped), entire to irregularly serrate, slightly rolled under, gen white- or white-and-rusty-tomentose, becoming glabrous, lower surface glaucous, upper surface shiny. **INFL** appearing before lvs, 15–70 mm, sessile or on leafy shoots < 10 mm; fl bract dark brown, ± densely straight- or wavy-hairy. **STAMINATE FL**: stamens 2. **PISTILLATE FL**: ovary glabrous, style 0.1–0.6 mm, stalk 0.5–2.4 mm. $2n=76$. Abundant. Shores, marshes, meadows, springs, bluffs; < 2800 m. CA; to WA, ID, TX, Mex. [vars. *bracelinae* C. Ball and *sandbergii* (Rydb.) C. Ball; *S. lutea* Nutt. var. *nivaria* Jepson; *S. tracyi* C. Ball, Tracy's willow] Highly variable. Coastal pls with lvs widely obovate, lower surface densely woolly or coarsely veined, tip gen obtuse to rounded have been called var. *bigelovii* (Torrey) Bebb. NCo populations suggest intergradation with *S. hookeriana*. ❀WET,SUN: 1–5,**6–9**,10–13,**14–24**;STBL;INV.

S. lemmonii Bebb (p. 1001) LEMMON'S WILLOW Shrub < 4 m. **ST**: twigs yellowish to brownish, gen thinly glaucous, gen glabrous, sometimes brittle at base. **LVS**: stipules lf-like (vestigial on first lvs); young lvs silky; mature blade 44–110 mm, oblong-lanceolate to elliptic, acuminate (base acute), entire or finely serrate, (white-or) white-and-rusty-silky, becoming glabrous, lower surface glaucous. **INFL** appearing with or just before lvs, 13–50 mm, sessile or on leafy shoots < 6 mm; fl bract tawny to dark brown. **STAMINATE FL**: stamens 2. **PISTILLATE FL**: ovary silky, style 0.3–0.8 mm, stalk 1.2–2.1 mm. $2n=76$. Streams, wet meadows, burns in subalpine pine forests; 1400–3500 m. e KR, CaRH, SNH, SnBr, MP; to B.C., ID, NV. See *S. geyeriana*. ❀WET or IRR,SUN:**1–3**, 4,5,**6,7**;STBL.

S. ligulifolia (C. Ball) C. Ball (p. 1001) STRAP-LEAFED WILLOW Shrub < 8 m. **ST**: twigs brownish, glabrous or long-shaggy-hairy. **LVS**: stipule tip acute; young lvs glabrous or hairy; mature blade 60–133 mm, lanceolate to elliptic, acuminate (base acute to rounded or cordate on vigorous lvs), margin gland-dotted to serrate, surfaces glabrous or sparsely long-soft-wavy-hairy, lower surface glaucous. **INFL** appearing with or just before lvs, 12–38 mm, sessile or on leafy shoots < 5 mm; fl bract brown, hairs curly. **STAMINATE FL**: stamens 2. **PISTILLATE FL**: ovary glabrous, style 0.2–0.6 mm, stalk 0.9–2.5 mm. $2n=38$. Rocky streams; 1100–2500 m. CaRH, SNH, Wrn; to ID, WY, NM. Lf margins often incorrectly called "entire". ❀TRY;STBL.

S. lucida Muhlenb. SHINING WILLOW Shrub, tree, < 10 m. **ST**: twigs brownish, glabrous or soft-shaggy-hairy, sometimes brittle at base. **LVS**: stipules glandular; petiole with glands; young lvs glabrous or white- or white-and-rusty-hairy; mature blade 53–170 mm, lanceolate, long-acuminate (base acute), finely serrate, glabrous or becoming so, upper surface shiny. **INFL** appearing with lvs, 20–90 mm, on leafy shoots 8–45 mm; fl bract tawny; pistillate bract deciduous after fl. **STAMINATE FL**: stamens 3–5. **PISTILLATE FL**: ovary glabrous, style 0.2–0.6 mm, stalk 0.8–2 mm. $2n$=76. Wet places; < 3200 m. CA; to AK, w Can, NV.

1. Lf blade not glaucous below, with stomata above .... ssp. ***caudata***
1′ Lf blade glaucous below, at least largest lacking stomata above .... ssp. ***lasiandra***

ssp. ***caudata*** (Nutt.) E. Murray Wet meadows, lakeshores; 1500–3200 m. SNH, SnBr, MP; to BC, NV. [*S. c.* (Nutt.) A.A. Heller var. *bryantiana* C. Ball and N.F. Bracelin] ❀TRY.

ssp. ***lasiandra*** (Benth.) E. Murray (p. 1001) Common. Wet meadows, shores, seepage areas; 0–2700 m. Range of sp. (less common in s CA, D). [*S. lasiandra* Benth. incl vars. *abramsii* C. Ball and *lancifolia* (Andersson) Bebb] ❀WET:1–5,**6–10**,11–13,**14–24**;STBL.

S. lutea Nutt. (p. 1001) YELLOW WILLOW Shrub, small tree, < 7 m. **ST**: twigs yellow-gray; bark smooth, glabrous or hairy. **LVS**: young lvs glabrous or silky; mature blade 42–116 mm, lanceolate to oblanceolate, acuminate (base acute), entire to finely crenate, glabrous to sparsely hairy, lower surface glaucous. **INFL** appearing just before or with lvs, 15–75 mm, sessile or on leafy shoots < 8 mm; fl bract brown. **STAMINATE FL**: stamens 2. **PISTILLATE FL**: ovary glabrous, style 0.2–0.6 mm, stalk 1–3.4 mm. $2n$=38. Creek margins, wet meadows; 900–3100 m. c&s SNH (esp e slope), SnBr, SnJt, GB, w DMoj; to w&c Can, NE. [var. *watsonii* (Bebb) Jepson] ❀WET,SUN:**1–3**,4–6,**7–9**,10–12,**14**,15–17,**18–21**,22–24;STBL.

S. melanopsis Nutt. (p. 1001) DUSKY WILLOW Shrub < 4 m, clonal by root-sprouting. **ST**: twigs brownish, silky, becoming glabrous. **LVS**: petiole very short; young lvs long-soft-wavy-hairy; mature blade 30–85 mm, very narrowly elliptic to narrowly obovate, acute (both ends), entire to sparsely spiny-serrate, hairy, becoming glabrous, lower surface not glaucous. **INFL** appearing with or after lvs, sometimes branched, 19–50 mm, on leafy shoots 4–70 mm; fl bract tawny or light brown; pistillate bract deciduous after fl. **STAMINATE FL**: stamens 2. **PISTILLATE FL**: ovary glabrous (rarely sparsely soft-shaggy-hairy), stigmas deciduous after fl, style 0.2–0.4 mm, stalk 0.2–0.7 mm. Streambanks, often among rocks; 800–2700 m. KR, SNH, MP; to w Can, Colorado, NV. [var. *bolanderiana* (Rowlee) C. Schneider] ❀TRY;STBL.

S. orestera C. Schneider (p. 1001) GRAY-LEAFED SIERRA WILLOW Shrub < 2 m. **ST**: twigs yellowish to brownish, glaucous or not, silky or short-shaggy-hairy, becoming glabrous. **LVS**: young lvs silky; mature blade 35–95 mm, lanceolate to oblanceolate, acute (both ends), entire, silky, lower surface glaucous or obscured by hairs. **INFL** appearing with lvs, 12–55 mm, on leafy shoots 3–15 mm; fl bract dark brown. **STAMINATE FL**: stamens 2. **PISTILLATE FL**: ovary silky, style 0.6–1 mm, stalk 0.8–2 mm. Alpine and subalpine meadows, wet places; 2100–4000 m. SNH; w NV. [*S. commutata* Bebb var. *rubicunda* Jepson misapplied] Possibly derived from *S. eastwoodiae* × *S. lemmonii* (should be studied near Kaiser Pass, ne Fresno Co.). Does not hybridize with *S. eastwoodiae* where they occur together. ❀TRY;STBL.

S. planifolia Cham. ssp. ***planifolia*** (p. 1001) TEA-LEAFED WILLOW Shrub < 1 m. **ST**: twigs brownish, glabrous or silky. **LVS**: stipules 0 or vestigial; young lvs glabrous or white- or white-and-rusty-silky; mature blade 20–54 mm, ± elliptic, acute (both ends), entire to finely serrate, glabrous or silky, lower surface glaucous, upper surface glossy. **INFL** appearing before lvs, 10–45 mm, sessile, rarely on leafy shoots < 3 mm; fl bract brown to black. **STAMINATE FL**: stamens 2. **PISTILLATE FL**: ovary white- or white-and-rusty-silky, style 0.5–1.4 mm, stalk 0.3–0.8 mm. $2n$=76, 57. Subalpine meadows, riverbanks; 2500–4000 m. c SNH; to n&e N.Am, NM. [*S. phylicifolia* L. var. *monica* (Bebb) Jepson] ❀TRY.

S. prolixa Andersson (p. 1001) MACKENZIE'S WILLOW Shrub < 5 m. **ST**: twigs brownish, glabrous or long-shaggy-hairy. **LVS**: stipule tip rounded, young lvs glabrous or hairy; mature blade 50–150 mm, lanceolate to ovate, acuminate (base gen cordate), glabrous, margin gland-dotted to serrate, lower surface glaucous. **INFL** appearing with lvs, 25–85 mm, sessile or on leafy shoots < 5 mm; fl bract brown, hairs curly. **STAMINATE FL**: stamens 2. **PISTILLATE FL**: ovary glabrous, style 0.3–1 mm, stalk 1.3–4.2 mm. Streams; 1600–2000 m. KR, CaRH, n SNH; to n Can, UT. [*S. mackenzieana* (Hook.) Andersson; *S. cordata* Muhlenb. misapplied] ❀TRY;STBL.

S. reticulata L. ssp. ***nivalis*** (Hook.) Löve et al. (p. 1001) SNOW WILLOW Shrub < 0.1 m, mat-forming. **ST**: twigs brownish, glabrous. **LVS**: stipules 0; young lvs glabrous; mature blade 6–22 mm, ± elliptic, acute to rounded (both ends), entire, gen glabrous, lower surface glaucous, upper surface shiny. **INFL** appearing with lvs, 7–30 mm, on leafy shoots 2–10 mm; fl bract tawny to pink; pistillate fls 4–17 per catkin. **STAMINATE FL**: stamens 2. **PISTILLATE FL**: ovary silky, style 0.1–0.4 mm, stalk 0–0.8 mm. $2n$=38. RARE in CA. Alpine cirques; 3100–3500 m. c SNH (near Mount Dana); to w Can, Colorado. [*S. nivalis* Hook.] In cult.

S. scouleriana Hook. (p. 1001) SCOULER'S WILLOW Shrub, slender tree, < 10 m. **ST**: twigs brownish, hairy or becoming glabrous. **LVS**: petiole gen velvety; young lvs hairy; mature blade 29–80 mm, oblanceolate to narrowly elliptic, rounded or acute (base wedge-shaped), entire or irregularly serrate, strongly rolled under, lower surface glaucous, sparsely white-and-rusty-tomentose or densely white-woolly, upper surface shiny, becoming glabrous (exc midrib). **INFL** appearing before lvs, 15–65 mm, gen sessile; fl bract dark brown or 2-colored. **STAMINATE FL**: stamens 2. **PISTILLATE FL**: ovary silky, style 0.3–0.6 mm, stalk 0.8–2.3 mm. $2n$=76,114. Common. Dry to moist forests, meadows, springs, swamps; 90–3400 m. NW, CaRH, SN, n CCo, SnGb, SnBr, SnJt, MP; to AK, c Can, NM, Mex. [var. *coetanea* C. Ball] ❀IRR or WET,SUN:**1–3**,4–6,**7**,8–10,14,**15–18**,19–24;STBL.

S. sessilifolia Nutt. (p. 1001) SANDBAR WILLOW Shrub < 5 m, clonal by root-sprouting. **ST**: twigs brownish, persistently silky, long-soft-wavy-hairy, or becoming glabrous. **LVS**: petiole very short; young lvs hairy; mature blade 40–120 mm, ± elliptic to narrowly obovate, acuminate (base wedge-shaped), sparsely spiny-serrate, silky, lower surface not glaucous. **INFL** appearing with or after lvs, branched, 20–70 mm, on leafy shoots 20–125 mm; fl bract tawny or brown, tip rounded or acute; pistillate bract deciduous after fl. **STAMINATE FL**: stamens 2. **PISTILLATE FL**: ovary long-soft-wavy-hairy, stigmas deciduous after fl, style 0.2–0.4 mm, stalk 0.2–0.7 mm. Streambanks, sandbars; gen < 200 m. NCo, w KR; to B.C. [*S. parksiana* C. Ball, Del Norte coast willow] ❀WET:**4–7**,14,**15–17**,19–24;STBL.

S. sitchensis Bong. (p. 1001) SITKA WILLOW Shrub, small tree, < 7 m. **ST**: twigs brownish, silky or long-soft-wavy-hairy, flexible or ± brittle at base. **LVS**: young lvs hairy; mature blade 52–110 mm, lanceolate to widely obovate, acute or pointed (base wedge-shaped), entire to finely serrate, strongly rolled under near base, lower surface not glaucous, densely short-silky to densely silky-woolly and gen obscured, upper surface dark green, becoming glabrous. **INFL** appearing just before or with lvs, 26–95 mm, on leafy shoots 3–20 mm, rarely sessile; fl bract tawny or brown. **STAMINATE FL**: stamen 1. **PISTILLATE FL**: ovary silky, widely ovoid, style 0.4–0.8 mm, stalk 0.4–1.2 mm. $2n$=38. Common. Tidal swamps, marshes, springs, streambeds; 0–400 m (1800–2500 m in Siskiyou, Humboldt cos). NW, CW (exc SCoRI), w WTR; to AK, MT. [*S. coulteri* Andersson] ❀WET:1,2,**4–7**,**14–17**,18–24; STBL.

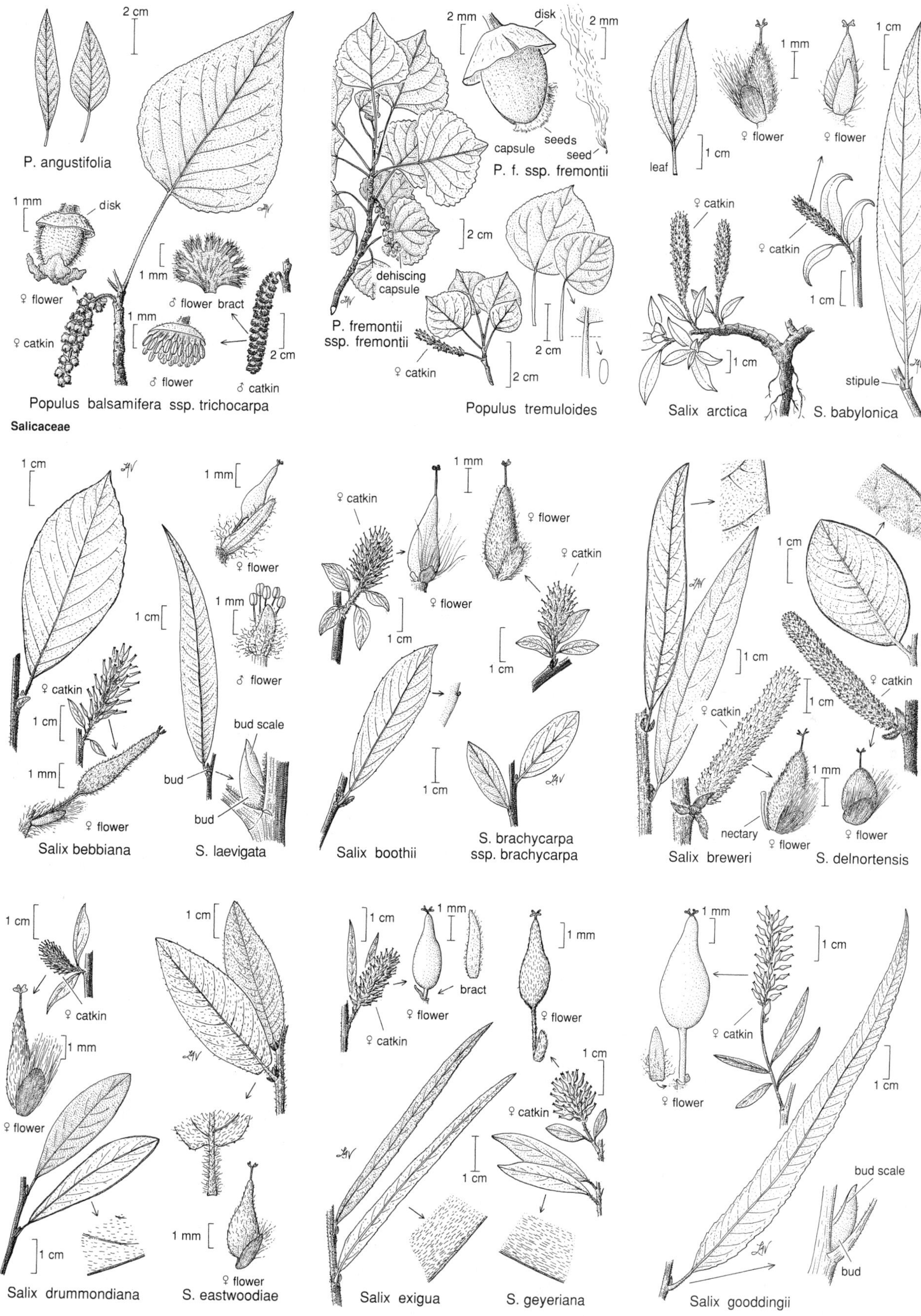
2 cm
P. angustifolia
1 mm
disk
♀ flower
♀ catkin
1 mm
♂ flower bract
1 mm
♂ flower
2 cm
♂ catkin
Populus balsamifera ssp. trichocarpa
Salicaceae
2 mm
disk
2 mm
capsule
seeds
seed
P. f. ssp. fremontii
2 cm
dehiscing capsule
P. fremontii ssp. fremontii
♀ catkin
2 cm
2 cm
Populus tremuloides
1 mm
♀ flower
♀ flower
1 cm
leaf
1 cm
♀ catkin
♀ catkin
1 cm
1 cm
stipule
Salix arctica
S. babylonica
1 cm
1 mm
♀ flower
1 mm
♂ flower
1 cm
♀ catkin
1 cm
1 mm
♀ flower
bud scale
bud
bud
Salix bebbiana
S. laevigata
1 mm
♀ catkin
♀ flower
♀ flower
1 cm
♀ catkin
1 cm
1 cm
Salix boothii
S. brachycarpa ssp. brachycarpa
1 cm
1 cm
♀ catkin
1 cm
♀ catkin
1 mm
nectary
♀ flower
♀ flower
Salix breweri
S. delnortensis
1 cm
♀ catkin
1 mm
♀ flower
1 cm
1 cm
1 mm
♀ flower
Salix drummondiana
S. eastwoodiae
1 cm
1 mm
bract
♀ flower
♀ catkin
1 mm
♀ flower
1 cm
♀ catkin
1 cm
Salix exigua
S. geyeriana
1 mm
1 cm
♀ catkin
♀ flower
1 cm
bud scale
bud
Salix gooddingii

SANTALACEAE SANDALWOOD FAMILY

William J. Stone

Herbs, shrubs, trees, green root-parasites. **LVS** entire; stipules 0. **INFL**: gen small cymes in upper axils. **FL**: calyx lobes 4–5 parted, fused side-to-side in bud, persistent; petals 0; stamens opposite sepals, inserted on fleshy disk; ovary superior to fully inferior, chamber 1, ovules 2–4, suspended from top of free-central placenta, style 1, thread-like, stigma head-like. **FR**: drupe or nut. **SEED** 1, spheric or ovoid. ± 26 genera, 250 spp.: gen trop.

COMANDRA BASTARD TOAD-FLAX

Per, subshrub, smooth; rhizome extensive, horizontal. **ST** 7–40 cm, green, blue-green, or grayish, striate. **LVS** alternate, ± sessile. **FL** subtended by bractlet; calyx tube bell- or urn-shaped, lobes (4-)5, ovate to oblong; anther base with a tuft of hair. **FR** drupe-like, crowned by persistent calyx. **SEED** spheric. 4 ssp.: 3 Am, 1 Eur. (Greek: hair, man, for hairy attachment of stamens) [Piehl 1965 Mem Torrey Bot Club 22(1):1–97]

C. umbellata (L.) Nutt. (p. 1001) ssp. ***californica*** (Rydb.) Piehl Sub-shrub. **STS** many from rhizome, 10–40 cm, leafy, glaucous, branched. **LF** 15–55 mm, lanceolate, acute to sharp-pointed, sometimes paler below. **FL** 2–3.5 mm; sepals lanceolate to narrowly ovate, whitish; anthers ± 0.5 mm. **FR** 5–7.5 mm, oblong-ovoid; calyx tube sometimes forms "neck" above fr. Gen dry, ± rocky areas; 300–3000 m. KR, CaR, SN, SNE, n DMoj?; to B.C., sw NV, nw AZ. [*C. californica* Rydb.] Intergrades with ssp. *pallida* (A. DC.) Piehl in OR, WA, NV, AZ. ❀TRY.

SARRACENIACEAE PITCHER-PLANT FAMILY

William J. Stone

Per, gen from slender rhizome or short caudex, carnivorous; roots poorly developed. **LVS**: basal rosette, spreading to erect, modified into tubular "pitcher" sometimes containing digestive fluids. **INFL**: fl gen 1, on long peduncle. **FL** bisexual, radial, nodding; sepals gen 5, overlapping, gen clawed; petals 5 or 0; stamens many; pistil 1, ovary superior, chambers gen 5, sometimes incomplete above, placentas axile (or parietal above), style 1, entire to 4–6 lobed or umbrella-like, stigma terminal or under tips of style lobes. **FR**: capsule, loculicidal; valves gen 5. **SEEDS** many, club-like, often winged. 3 genera, 15 spp.: CA, OR, e N.Am, n S.Am, esp bogs, streamsides.

1. Lf widest at top, with obvious, translucent-dotted hood .... **DARLINGTONIA**
1′ Lf widest at middle, hood 0 .... **SARRACENIA**

DARLINGTONIA CALIFORNIA PITCHER PLANT, COBRA LILY

1 sp. (William Darlington, Philadelphia botanist, 1782–1863)

D. californica Torrey (p. 1007) **LF** gen 1–6 dm, tubular, greenish yellow, with stiff, reflexed hairs inside, enlarged upward into a translucent-dotted hood with 2 yellow- to purple-green appendages at front border; lf opening downward-facing, under hood. **INFL** < 1 m; bracts obvious, yellow. **FL**: sepals 4–6 cm, oblong to oblanceolate, yellow-green, veins purple; petals 2–4 cm, narrowly ovate, dark purple, heavily veined; stamens 12–15 in 1 whorl; ovary tip truncate or concave, style 2–3 mm, deeply 5-lobed, stigmas 5. **FR** 2.5–4.5 cm, ovoid. **SEED** ± 2 mm, papillate, light reddish brown. $2n$=30. UNCOMMON. Seeps, boggy places with running water, gen on serpentine; 60–2200 m. KR, CaRH?, n SNH (c Plumas, Sierra, Nevada cos.); w OR. Lvs produce no digestive enzymes; breakdown of insects is bacterial. ❀ WET, bog and pure water DRN:1,**2**,3,**4**,**5**,**17** &SHD: 6,7,14,**15**,**16**; DFCLT.

SARRACENIA PITCHER PLANT

LF tubular, curved, enlarged upward, upward-facing opening partly covered by a sometimes umbrella-like flap; hollow hood 0. **INFL**: fl subtended by 3 bracts. **FL**: sepals 5; petals 5; ovary chambers 5, style tip expanded into large, umbrella-like, peltate disk. 8 spp.: se US. (M.S. Sarrazin, Quebec physician and collector, 1659–1734) *S. rubra* was planted in NCo (Mendocino Co.) and may persist.

S. purpurea L. (p. 1007) **LVS** many, nearly erect, < 20 cm, widest at middle, top open or partly covered by reniform flap, with reflexed hairs on inside of flap and tube. **INFL** 20–60 cm. **FL**: sepals 2–6 cm, ovate to rhombic, dark purple-red outside, pale green inside; petals 2–6 cm, obovate, tapered to a short claw, purple-red on both surfaces. $2n$=26. Uncommon. Seeps, marshes, bogs; 1000 m. n SNH (Butterfly Valley, Plumas Co.), likely eradicated at that site but possibly naturalized elsewhere; native to e N.Am. Cult as novelty. Lvs contain digestive fluids.

SAURURACEAE LIZARD'S-TAIL FAMILY

Elizabeth McClintock

Per, from rhizomes or stolons. **LVS** simple, alternate; stipules joined to petiole. **INFL**: spike, dense, many-fld, terminal, sometimes subtended by petal-like involucre bracts and so resembling a single fl. **FLS** small, bisexual; perianth 0; stamens 6, 8 (or 3); ovary superior but sometimes embedded in infl axis, compound, 1-chambered or carpels fused only at base, styles 3–4, distinct. **FR**: capsule, ± fleshy, dehiscent at tip or ± berry-like. **SEEDS** many or 1, spheric or ovate. 5 genera, 7 spp.: e Asia, N.Am. [Wood 1971 J Arnold Arb 52:479–485]

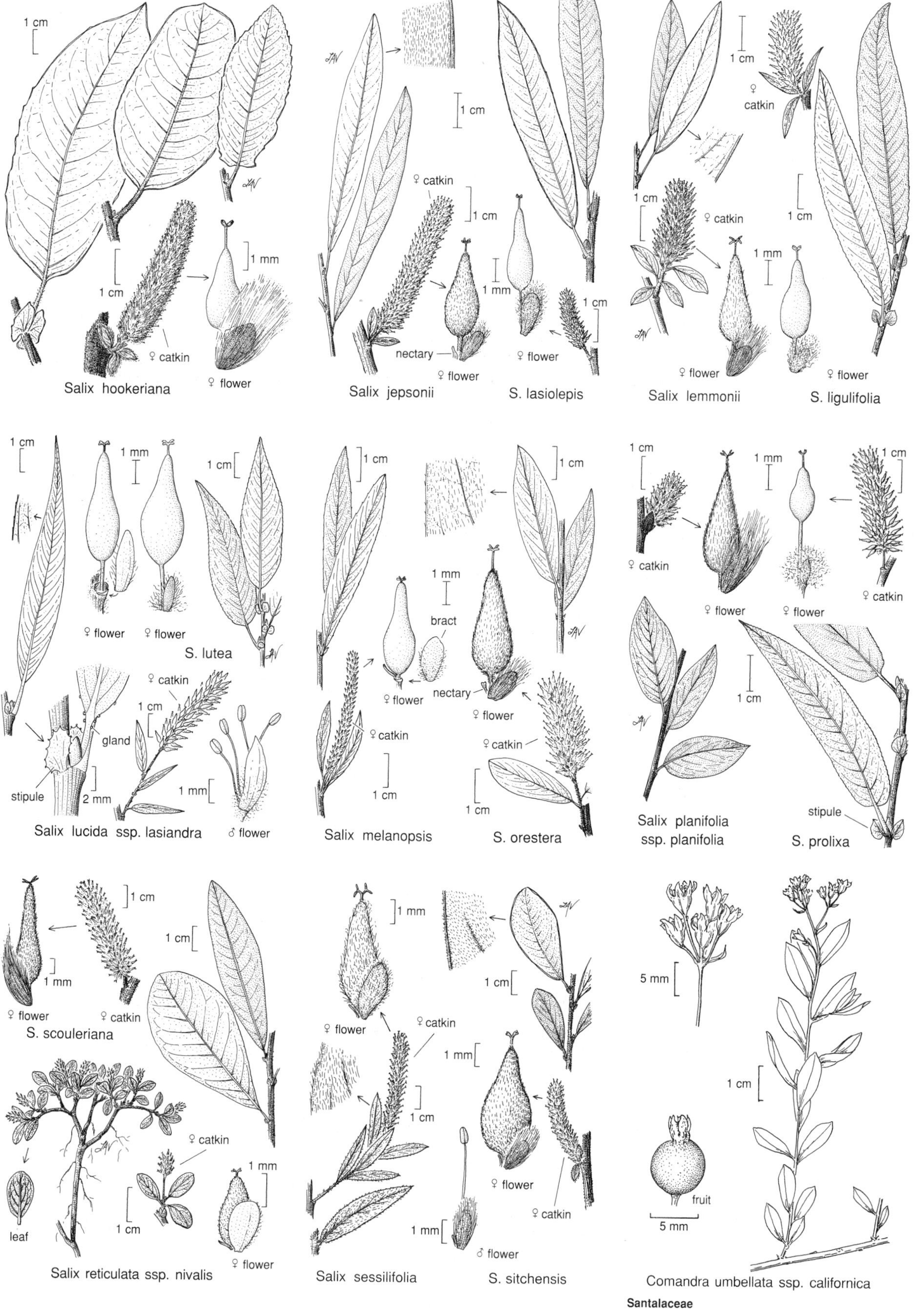

Santalaceae

ANEMOPSIS

1 sp. (Greek: anemone-like, from infl) [Howell 1971 Wasmann J Biology 29:97–100]

A. californica (Nutt.) Hook. & Arn. (p. 1007) YERBA MANSA Rhizome creeping, thick, woody. **ST** 15–50 cm, gen naked, hollow, glabrous or hairy. **LVS**: basal several, blade 5–15 cm, elliptic to oblong or base sometimes cordate, petiole 10–20 cm; cauline few, ovate, gen subsessile to clasping, sometimes subtending 1–3 short-petioled lvs. **INFL** 1.5–4 cm, conic; involucre bracts 5–8, 1–3 cm, petal-like, white, often tinged reddish. **FL** (exc lowermost) subtended by a 5–6 mm white bract; stamens appearing to arise from infl axis. Common. Saline or alkaline soil, wet or moist areas, seeps, springs; 75–1700 m. sw ScV, SnJV, SnFrB, SCoR, SCo, n&s ChI, PR, DMoj; to Utah, w TX, nw Mex. [*A. c.* var. *subglabra* Kelso] Pls aromatic, once used to treat diseases of skin, blood. ❀IRR or WET:3,6,**7–10,14–24**&SHD:**11,12**;deciduous GRCVR.

SAXIFRAGACEAE SAXIFRAGE FAMILY

Patrick E. Elvander

Per or subshrub from caudex or rhizome, gen ± hairy. **ST** often ± leafy on lower half, rarely trailing and leafy throughout. **LVS** gen simple, basal or sometimes cauline, gen alternate, gen petioled; veins ± palmate. **INFL**: panicle, gen ± scapose. **FL** gen bisexual, gen radial; hypanthium free to ± fused to ovary; calyx lobes gen 5; petals gen 5, free, gen clawed, gen white; stamens gen 5 or 10; pistils 2 and simple or 1 and compound (chambers 1–2, placentas 2–4, axile or parietal), ovary superior to inferior, sometimes more superior in fr, styles gen 2. **FR**: 2 follicles or 2–4-valved capsule. **SEEDS** gen many, small. 40 genera, 600 spp.: esp n temp, arctic, alpine; some cult (*Bergenia*, *Darmera*, *Heuchera*, *Saxifraga*, *Tellima*, *Tolmiea*). [Soltis 1988 Syst Bot 13:64–72]

1. Fl solitary, showy; staminodes 5, toothed to fringed; styles inconspicuous; placentas 4, parietal (Parnassioideae) .... **PARNASSIA**
1' Fls 2 or more or not showy; staminodes 0; styles gen obvious; placentas 2–3, parietal or axile (Saxifragoideae)
 2. Fl inconspicuous; sepals 4; petals 0; stamens 8 .... **CHRYSOSPLENIUM**
 2' Fl ± conspicuous; sepals 5; petals 4–5; stamens 3, 5, or 10
  3. Stamens 3; petals 4; hypanthium bilateral, open on 1 side .... **TOLMIEA**
  3' Stamens 5 or 10; petals gen 5; hypanthium gen radial, closed
   4. Lf blade gen > 1 dm wide, peltate; infl appearing before lvs, 7–20 dm .... **DARMERA**
   4' Lf blade gen << 1 dm wide, not peltate; infl not appearing before lvs (or, in *Jepsonia*, < 3 dm)
    5. Stamens 5
     6. Ovary superior
      7. Petals thread-like; calyx ± bilateral; rhizome scaly, bulblets 0 .... **BENSONIELLA**
      7' Petals triangular, long-tapered; calyx radial; caudex with bulblets, scales 0 .... **BOLANDRA**
     6' Ovary > half inferior, at least in fl
      8. Lvs basal (at most 3 cauline and reduced); ovary chamber 1, placentas 2, parietal
       9. Infl a panicle; petals ± entire or 0 .... **HEUCHERA**
       9' Infl a raceme; petals 3–7-lobed .... **MITELLA**
      8' Lvs both basal and cauline; ovary chambers 2, placentas 2, axile
       10. Lf blade ± shallowly lobed; fls > 15; infl ± round-topped, branches gen > 3, with fls on 1 side .... **BOYKINIA**
       10' Lf blade deeply lobed; fls < 15; infl ± flat-topped, branches gen < 3, with fls on all sides .... **SUKSDORFIA**
    5' Stamens 10
     11. Ovary chamber 1, placentas parietal
      12. Styles and placentas 3 .... **LITHOPHRAGMA**
      12' Styles and placentas 2
       13. Infl a raceme; fr valves equal, narrow; petal lobes ± 5–7, linear .... **TELLIMA**
       13' Infl a panicle; fr valves unequal, wide; petal lobes 0 .... **TIARELLA**
     11' Ovary chambers 2 (rarely 3 in *Saxifraga*), placentas axile
      14. Infl appearing before lvs; caudex corm-like; some heterostylous .... **JEPSONIA**
      14' Infl appearing with lvs; caudex 0 or not corm-like; not heterostylous
       15. Lf blade not jointed to nor falling before petiole; styles free throughout .... **SAXIFRAGA**
       15' Lf blade jointed to and falling before petiole; styles gen fused at base, at least in fl .. **SAXIFRAGOPSIS**

BENSONIELLA

1 sp. (G.T. Benson, Stanford botanist, 1896–1928)

B. oregana (Abrams & Bacigal.) C. Morton (p. 1007) BENSONIELLA Rhizome scaly; bulblets 0. **LVS** basal; petiole 2–15 cm, hairs dense, long, brown; blade 4–20 cm, round-ovate, base cordate, unevenly crenate, ± glabrous above, veins hairy below. **INFL**: raceme, 20–40 cm, spike-like, glandular; bracts 0; pedicels < 1 mm. **FL**: hypanthium free of ovary; calyx ± bilateral, lobes 1.5–2.5 mm, obovate; petals ± = calyx lobes, thread-like, ephemeral; stamens 5, > calyx lobes; pistil 1, ovary superior, chamber 1, placentas 2, parietal. **FR**: capsule, opening before seeds mature. **RARE** CA. Wet meadows, bogs; > 1000 m. KR; sw OR. [*Bensonia o.* Abrams & Bacigal.]

BOLANDRA

Caudex scaleless, bearing bulblets. **LVS** basal and cauline, reduced and merging into bracts upward; blade ± round, base cordate to reniform, lobes 3–7, deep, teeth sharp. **INFL** panicle-like; fls few; bracts sessile, lower often clasping, lf-like. **FL**: hypanthium free from ovary; stamens 5; pistil 1, ovary superior, chambers 2, placentas 2, axile. **FR**: capsule. 2 spp.: WA, OR, CA. (H.N. Bolander, CA botanist, 1831–1897) [Gornall & Bohm 1985 Bot J Linn Soc 90:1–71]

B. californica A. Gray (p. 1007) SIERRA BOLANDRA **LF**: petiole 3–10 cm; blade 1–5 cm wide. **INFL** 15–60 cm, glandular; pedicels 0.5–3 mm. **FL**: calyx lobes 3–6 mm, elliptic; petals 4–7 mm, triangular, long-tapered, green, margin purple; stamens << calyx lobes. 2*n*=14. UNCOMMON. Rock crevices, wet cliffs; > 2000 m. n&c SNH (Eldorado to Mariposa cos.).

BOYKINIA

Pl glandular; rhizome scaly; bulblets 0. **LVS** basal and cauline, reduced and merging into bracts upward; stipules inconspicuous to lf-like; blade round to ovate, base cordate to reniform, sometimes truncate, obtuse, or tapered, primary lobes 3–many, very shallow to deep, teeth sharp-tipped. **INFL** gen 1-sided; bracts sessile to short-petioled, lower lf-like. **FL**: hypanthium partly fused to ovary; petals ephemeral; stamens 5; pistil 1, ovary > half inferior, chambers 2, placentas 2, axile. **FR**: capsule. 8 spp.: N.Am, Asia. (Dr. S. Boykin, Georgian naturalist, 1786–1848) [Gornall & Bohm 1985 Bot J Linn Soc 90:1–71]

1. Petal gen < 3 mm, ± = calyx lobe; stipules inconspicuous .... ***B. rotundifolia***
1' Petal > 3 mm, >> calyx lobe; stipules obvious
 2. Stipule ± 1 cm or more, gen toothed; infl dense; petal gen > 5 mm .... ***B. major***
 2' Stipule ± 2–4 mm, bristly; infl open; petal gen < 5 mm .... ***B. occidentalis***

B. major A. Gray **LF** 9–50 cm; stipules conspicuous, green, entire or teeth gen sharp-tipped; petiole 5–35 cm; blade < 20 cm wide, divided > 1/2 way to base, lobes and teeth ± straight-sided. **INFL** 3–10 dm, ± flat-topped, dense. **FL**: calyx lobes 2–5 mm, triangular to elliptic; petals 5–7 mm, elliptic to round. Shaded, moist meadows, banks; < 2500 m. KR, CaR, SN; to WA, MT. ❀WET:**4,5,17** &SHD:1,2,**3,6–9,14–17**; GRCVR.

B. occidentalis Torrey & A. Gray (p. 1007) **LF** 6–45 cm; stipules slight expansion of petiole base, gen green, brown-bristled; petiole 3–30 cm; blade < 12 cm wide, divided ± 1/4–1/3 way to base, lobes and teeth ± round-sided. **INFL** 3–6 dm, tapered, ± open. **FL**: calyx lobes 1–3 mm, triangular, tapered; petals 3–4 mm, obovate. 2*n*=14. Shady wet banks; < 1500 m. NW, n&c SNF, CW, WTR; to B.C. [*B. elata* (Nutt.) E. Greene] ❀WET&DRN:**4,5,17** &SHD:2,**3,6–9, 14–24**.

B. rotundifolia C. Parry ROUND-LEAVED BOYKINIA **LF** 10–30 cm; stipules inconspicuous, gen brown-bristled; petiole 5–18 cm; blade to 18 cm wide, divided < 1/8 way to base, lobes and teeth ± round-sided. **INFL** 3–10 dm, tapered, dense. **FL**: calyx lobes 2–3 mm, triangular; petals 2–3 mm, obovate. 2*n*=26. UNCOMMON. Streambanks; < 2000 m. s SCoRO, TR, PR. ❀WET&DRN:**4,5, 17**&SHD:**6,7,14–24**;GRCVR.

CHRYSOSPLENIUM GOLDEN SAXIFRAGE

Pl glabrous; caudex without scales or bulblets. **ST** prostrate to ascending, rooting at nodes. **LVS** cauline, opposite to sometimes alternate; blade round to ovate, base gen truncate, teeth round. **INFL**: fl solitary, axillary. **FL**: hypanthium fused to ovary; calyx lobes 4, petal-like; petals 0; stamens 8; pistil 1, ovary inferior to half inferior, chamber 1, placentas 2, parietal. **FR**: capsule. 15 spp.: n temp, s S.Am. (Greek: gold spleen)

C. glechomifolium Nutt. (p. 1007) **LF** 7–30 mm; petiole < 10 mm; blade 5–20 mm wide. **FL**: calyx lobes 1.5–2 mm, elliptic; stamens ± 1 mm; nectaries 2, surrounding ovary, each with 4 lobes alternate stamens. 2*n*=18. Uncommon. Shady wet areas; < 200 m. NCo; to B.C.

DARMERA

1 sp. (Karl Darmer, German horticulturist, 19th century) [Schmid & Turner 1977 Madroño 24:68–74]

D. peltata (Torrey) (p. 1007) Voss INDIAN RHUBARB, UMBRELLA PLANT Rhizome < 5 cm thick, fleshy, scaly; bulblets 0. **LVS** basal, < 1.5 m; blade peltate, ± 10 dm, ± round, lobes deep, teeth irregular. **INFL** appearing before lvs, 3–15 dm, ± flat-topped; bracts gen scale-like or 0. **FL**: hypanthium minute, free from ovary; calyx lobes reflexed, 3–4 mm, elliptic; petals sessile, 5–7 mm, obovate, white to pink; stamens 10, 3–4 mm, filaments tapered; pistils 2, simple, ovaries superior. **FR** 2 follicles, 8–12 mm. 2*n*=34. Uncommon. Rocky streambanks; < 2000 m. KR, CaR, SN; sw OR. [*Peltiphyllum p.* (Torrey) Engl.] ❀WET:**1–7,17** &SHD:**14–16**,18–20.

HEUCHERA ALUMROOT

Rhizome scaly; bulblets 0. **LVS** basal, sometimes a few cauline; blade ovate, base cordate to reniform, lobes and teeth shallow, irregular. **INFL** gen raceme-like; bracts gen scale-like. **FL** radial to ± bilateral; hypanthium partly fused to ovary; calyx lobes equal or not; petals 0 or 5, gen equal; stamens 5, gen equal; pistil 1, ovary > half inferior, chamber 1, placentas 2, parietal. **FR**: capsule. 50 spp.: N.Am. (J.H. von Heucher, German professor of medicine, 1677–1747) [Rosendahl, Butters, & Lakela 1936 Minn Stud Pl Sci 2:1–180] A very difficult genus, highly variable at many levels and needing much additional research.

1. Styles and most or all stamens incl in full fl
 2. Petals gen 0; infl very narrow, spike-like .... ***H. cylindrica*** var. ***alpina***
 2' Petals present, ± conspicuous; infl more open, panicle or raceme-like
  3. Fl (exc petals) gen < 3 mm, radial; calyx lobes equal, yellow with pink tips .... ***H. duranii***

3′ Fl (exc petals) gen > 3.5 mm, ± bilateral; calyx lobes ± unequal, light pink to red, sometimes with green tips (*H. cespitosa* complex)

4. Glands of infl ± sessile; lf blade gen < 15 mm, gen broadly ovate, ± deeply lobed ..... ***H. abramsii***

4′ At least some glands of infl on hairs; lf blade gen > 15 mm, gen round to reniform, shallowly lobed

5. Stamens gen < half length of calyx lobes; part of hypanthium free from ovary ± 0.7 mm on short side, ± 1.5 mm on long side; PR (San Diego Co.) ..... ***H. brevistaminea***

5′ Stamens ± = calyx lobes; part of hypanthium free from ovary > 0.8 mm on short side, > 1.8 mm on long side; s CA n of San Diego Co.

6. Calyx lobes on long side of hypanthium minute; SnJt ..... ***H. hirsutissima***

6′ Calyx lobes all well developed; s SN, s SCoRO, SnGb

7. Stamens equal, ± incl; calyx lobes unequal; SnGb ..... ***H. elegans***

7′ Stamens unequal, longest barely exserted; calyx lobes ± equal; s SN, s SCoRO ..... ***H. cespitosa***

1′ Styles and all stamens exserted in full fl

8. Fl radial; calyx lobes equal; part of hypanthium fused to ovary gen >> free part

9. Styles 2–4 mm, exserted; hypanthium broadly obconic, base tapered; infl gen open

10. Hypanthium and calyx lobes together 3.5–4.5 mm; lf blade 6–18 cm wide; n ChI ..... ***H. maxima***

10′ Hypanthium and calyx lobes together 1–3 mm; lf blade 2–7 cm wide; mainland ..... ***H. micrantha***

9′ Styles gen < 1.8 mm, barely exserted; hypanthium bell-shaped, base truncate; infl gen dense

11. Lf blade gen round, base ± truncate; petiole 2–5 cm, hairs short, stiff; > 1000 m ..... ***H. merriamii***

11′ Lf blade gen ovate, base cordate; petiole 7–20 cm, hairs long, soft; < 500 m ..... ***H. pilosissima***

8′ Fl ± bilateral; calyx lobes gen unequal; part of hypanthium fused to ovary < long side of free part

12. Petals 2–3 mm, > 0.3 mm wide, oblanceolate, unequal; hypanthium strongly oblique ..... ***H. parishii***

12′ Petals 3–6 mm, gen << 0.3 mm wide, thread-like to narrowly oblanceolate, equal; hypanthium slightly oblique ..... ***H. rubescens***

13. Infl gen ± 1-sided; part of hypanthium fused to ovary gen longer than wide in fl, ± obconic; styles either smooth or strongly papillate throughout; s SN, PR, SNE, D

14. Styles strongly papillate throughout; s SN, SNE, D ..... var. ***alpicola***

14′ Styles smooth; PR ..... var. ***versicolor***

13′ Infl rarely 1-sided; part of hypanthium fused to ovary gen as wide as or wider than long in fl, broadly obconic to spheric; styles papillate only near base; n&c SN

15. Infl open, conic; petals thread-like; n SN ..... var. ***glandulosa***

15′ Infl dense, narrow, cylindric; petals oblanceolate; c SN ..... var. ***rydbergiana***

H. abramsii Rydb. ABRAMS' ALUMROOT **LF**: blade gen < 15 mm, ± deeply 5-lobed. **INFL** 5–15 cm, dense; glands sessile. **FL**: part of hypanthium fused to ovary < long side of free part; calyx lobes unequal, gen red-purple; stamens < calyx lobes, equal, incl. UNCOMMON. Dry, rocky areas; 2800–3500 m. SnGb. [*H. rubescens* var. *a.* (Rydb.) M.G. Stewart] Otherwise like *H. cespitosa.* ❀DRN:5,6&IRR,SHD:2,3,7,14,15,**16**,18.

H. brevistaminea Wiggins MOUNT LAGUNA ALUMROOT **INFL** 18–25 cm. **FL**: part of hypanthium fused to ovary > long side of free part; calyx lobes slightly unequal, gen red-purple; stamens << calyx lobes, equal incl. RARE. Dry, steep, rocky areas; 1500–2000 m. PR (Laguna Mtns, San Diego Co.). Otherwise like *H. cespitosa.* Styles and placentas rarely 3 per fl.

H. cespitosa Eastw. **LF**: petiole 1–6 cm; blade 15–40 mm, round to reniform, shallowly 5-lobed. **INFL** 10–36 cm, narrow, open, glandular. **FL** ± bilateral; part of hypanthium fused to ovary 1–2.2 mm, ± = long side of free part, together with calyx lobes 4–7 mm; calyx lobes ± equal, pink-red with green tips; petals > calyx lobes, obovate-oblanceolate, unequal; stamens gen < calyx lobes, unequal, longest barely exserted; mature styles > 1.5 mm, incl. Uncommon. Rocky areas; 1900–2300 m. s SN (Kern, Tulare cos.), s SCoRO (Big Pine Mtn). [*H. rubescens* var. *c.* (Eastw.) M.G. Stewart] Otherwise like *H. abramsii*, *H. brevistaminea*, *H. elegans*, *H. hirsutissima*, forming a complex much in need of monographic study.

H. cylindrica Hook. var. ***alpina*** S. Watson **LF**: petiole 1–9 cm, glandular to stiff-hairy; blade gen < 3 cm wide, broadly ovate, moderately 3–5-lobed, gen truncate to ± cordate. **INFL** spike-like, 8.5–55 cm, dense, ± short-glandular. **FL** ± radial; part of hypanthium fused to ovary 2–2.5 mm, < to > free part, together with calyx lobes 6–8 mm; calyx lobes cream-white to green, sometimes with pink tips; petals gen 0; stamens < calyx lobes, incl; mature styles gen < 1 mm, incl. Rocky banks, slopes; 1000–3000 m. MP; to WA, ID, NV. 2 other vars. from nw US & B.C., with lf blades gen > 3 cm, gen strongly cordate, occur with var. *a.* in se OR and may be discovered in MP: var. *cylindrica* (petiole densely glandular and stiff-hairy) and var. *glabella* (Torrey & A. Gray) W.E. Wheelock (petiole glabrous to sparsely glandular).

H. duranii Bacigal. (p. 1007) DURAN'S ALUMROOT **LF**: petiole 2.5–5 cm; blade 1–2 cm, round-reniform, shallowly 5–9-lobed. **INFL** 14–20 cm, dense toward tips, ± short-glandular. **FL** radial; part of hypanthium fused to ovary 0.8 mm, < free part, together with calyx lobes 2–2.5 mm; calyx lobes yellow with pink tips; petals 2.5–3 mm, ± = calyx lobes, lanceolate to oblanceolate; stamens < calyx lobes, incl; mature styles < 1 mm, incl. RARE in CA. Rocky areas; 2500–3500 m. W&I (White Mtns); w NV. May be inseparable from *H. parvifolia* Nutt. In cult.

H. elegans Abrams URN-FLOWERED ALUMROOT **INFL**: bracts few, small, lf-like. **FL**: part of hypanthium fused to ovary < long side of free part; calyx lobes unequal; stamens < calyx lobes, equal, ± incl. UNCOMMON. Rocky areas; 1500–2500 m. SnGb. [*H. rubescens* var. *e.* (Abrams) Jepson] Otherwise like *H. cespitosa.* ❀IRR&DRN:**2,4–6,15–17,24**&SHD:**3,7**,10,14,18–23.

H. hirsutissima C. Rosend., F.K. Butters, & Lakela SHAGGY-HAIRED ALUMROOT **FL**: part of hypanthium fused to ovary << long side of free part; calyx lobes on long side of hypanthium minute; stamens ± = calyx lobes, equal, incl. UNCOMMON. Rocky areas; 2200–3500 m. SnJt. Otherwise like *H. cespitosa.* ❀DRN&IRR: **2–6,15–17**&SHD:**7,14,18–21**.

H. maxima E. Greene ISLAND ALUMROOT **LF**: petiole 8–20 cm; blade 6–18 cm, round, broadly 7–9-lobed. **INFL** 45–60 cm, narrow, ± open between fl clusters, glandular; bracts few, lf-like. **FL** radial; part of hypanthium fused to ovary 2.5–3.5 mm, >> free part, together with calyx lobes 3.5–4.5 mm; calyx lobes white to pink; petals 3.5–6 mm, 3 × calyx lobes, oblong to lanceolate; stamens > calyx lobes, ± exserted; mature styles 3–4 mm, exserted. RARE. Cliffs in canyons; < 500 m. n ChI. In cult; CVS:hybrids with *H. sanguinea* of AZ, n Mex.

H. merriamii Eastw. **LF**: petiole 2–5 cm; blade 2–5 cm, round, shallowly 5–7-lobed. **INFL** 7–23 cm, narrow, ± open between fl clusters, glandular. **FL** radial; part of hypanthium fused to ovary 1.5–2.5 mm, > free part, together with calyx lobes (2-) 3–5 mm; calyx lobes equal, pink; petals ± 1.5 mm, 2 × calyx lobes, oblanceolate; stamens > calyx lobes, exserted; mature styles < 1.8 mm, barely exserted. Uncommon. Dry, rocky areas; 1500–2500 m. KR;

sw OR. Otherwise like *H. pilosissima*. Pls with fls 2–3 mm, slightly more exserted stamens and styles may be called *H. pringlei* Rydb., are possibly of hybrid origin. ❀DRN&WET:**4,5**,17&SHD:**1–3,6,7**,15,16.

H. micrantha Lindley (p. 1007) **LF**: petiole 3–30 cm, gen glandular; blade 20–80 mm, broadly ovate to oblong, 5–7-lobed, gen hairy. **INFL** 1–10 dm, wide to narrow, open, glandular; bracts on peduncle few, small, lf-like. **FL** radial; part of hypanthium fused to ovary 0.5–2.5 mm, gen > free part, together with calyx lobes 1–3 mm; calyx lobe tips green to red; petals 3–5 mm, 2 × calyx lobes, oblanceolate, sometimes ± pink; stamens > calyx lobes, exserted; mature styles 2–4 mm, exserted. $2n=14$. Moist, rocky banks and cliffs; < 2500 m. NW, CaR, SN, CW (exc SCoRI); to B.C., ID. Highly variable. Pls from SN, CaR with petioles minutely glandular, upper surface of lf blade glabrous may be called var. *erubescens* (A. Braun & C. Bouch) C. Rosend. Pls from NW, w CW with petioles densely glandular, upper surface of lf blade hairy may be called var. *pacifica* C. Rosend., F.K. Butters, & Lakela [var. *hartwegii* (W.E. Wheelock) C. Rosend.] ❀IRR&DRN:**2,4–6,17,24**&SHD:**1,3,7**, 8–10, **14–16,18–23**; GRCVR; CVS.

H. parishii Rydb. PARISH'S ALUMROOT **LF**: petiole 1–10 cm; blade 5–40 mm, broadly ovate to round-reniform, shallowly 5-lobed. **INFL** 5–27 cm, narrow, ± dense, glandular-hairy. **FL** ± bilateral; hypanthium inflated on longer side, part fused to ovary ± 1.4 mm, ± = free part, together with calyx lobes 3.5–6 mm; calyx lobes ± equal, pink with green tips; petals 2–3 mm, > calyx lobes, unequal, oblanceolate; stamens > calyx lobes, ± unequal, exserted; mature styles > 1.5 mm, exserted. UNCOMMON. Rocky places; 1500–3800 m. SnBr. [*H. rubescens* var. *p.* Jepson] Possibly of hybrid origin, involving gen *H. hirsutissima* and *H. rubescens*. Smaller pls with equal stamens from higher elevations have been called *H. alpestris* C. Rosend.

H. pilosissima Fischer & C. Meyer **LF**: petiole 7–20 cm; blade 45–90 mm, ovate, deeply 5–7-lobed. **INFL** 15–55 cm, ± dense, glandular; bracts on peduncle few, small, lf-like. **FL** radial; part of hypanthium fused to ovary ± 1.5 mm, >> free part, together with calyx lobes 2.5–4.5 mm; calyx lobes pink to red; petals 3–5 mm, ± 2 × calyx lobes, oblanceolate, white to pink; stamens > calyx lobes, exserted; mature styles 1–1.5 mm, slightly exserted. $2n=14$. Ocean bluffs, shaded slopes; < 500 m. NCo, CCo, SCoRO. ❀DRN&IRR:**5,15–17**&SHD:14,18–21,**22–24**.

H. rubescens Torrey **LF**: petiole 1–15 cm; blade 8–60 mm, broadly ovate to ± round, ± deeply 5–9-lobed. **INFL** 7–55 cm, often 1-sided, open or dense, cylindric to conic, glandular. **FL** ± bilateral; part of hypanthium fused to ovary 0.9–2.5 mm, ± = long side of free part, together with calyx lobes 3–6 mm; calyx lobes gen unequal, whitish to pink-red with green tips, becoming redder; petals 3–6 mm, > calyx lobes, narrowly oblanceolate to thread-like; stamens > calyx lobes, exserted; mature styles gen >> 1.5 mm, exserted. Dry, rocky areas; 1500–4000 m. SN, PR, GB (exc n MP), DMtns; to OR, ID, Colorado, TX, n Mex. Highly variable. Closely related to *H. parishii* and *H. mexicana* of Mex; hybridizes with other spp. Vars. intergrade. Needs monographic study. ❀DRN:**4,5**&IRR:**1, 2**,15–17,24 &SHD:**3,6**,7,14,18–23.

var. ***alpicola*** Jepson **INFL**: ± open, ± narrow, ± 1-sided. **FL**: hypanthium gen ± as long or longer than wide; part of hypanthium fused to ovary together with calyx lobes gen 4–6 mm; petals oblanceolate; styles strongly papillate throughout. Habitat of sp. s SN, SNE, DMtns. [var. *pachypoda* (E. Greene) C. Rosend., F.K. Butters, & Lakela]

var. ***glandulosa*** Kellogg **INFL**: open, conic. **FL**: petals thread-like. Habitat and elevations of sp. n SN; NV. Otherwise like var. *rydbergiana*.

var. ***rydbergiana*** C. Rosend, F.K. Butters, & Lakela **INFL**: dense, narrow, cylindric, rarely 1-sided. **FL**: hypanthium gen wider than long; part of hypanthium fused to ovary together with calyx lobes 3–4 mm; petals oblanceolate; styles papillate only at base. Habitat and elevations of sp. c SN.

var. ***versicolor*** (E. Greene) M.G. Stewart SAN DIEGO COUNTY ALUMROOT **INFL**: ± open, ± conic, gen 1-sided. **FL**: hypanthium gen ± as long as or longer than wide; part of hypanthium fused to ovary together with calyx lobes 3–5 mm; petals ± oblanceolate; styles smooth. RARE in CA. Habitat and elevations of sp. PR (Cuyamaca Peak, San Diego Co.); to TX, n Mex. [*H. leptomeria* E. Greene; *H. l.* var. *peninsularis* C. Rosend., F.K. Butters, & Lakela] In cult.

JEPSONIA

Caudex corm-like, ovoid or flat, lobed, without scales or bulblets. **LVS** gen 1–3, basal; blade ± round, base cordate to reniform, lobes and teeth shallow. **INFL**: cyme, appearing before lvs; bracts scale-like. **FL**: hypanthium free of ovary, gen truncate at base; stamens 10; pistil 1, ovary superior, chambers 2, placentas 2, axile. **FR**: capsule. 3 spp.: CA. (W.L. Jepson, CA botanist, 1867–1946) [Ornduff 1969 Brittonia 21:286–298] Spp. heterostylous, fls blooming in autumn (most members of family bloom in spring).

1. Lf gen 1; fls gen < 4; hypanthium gen > 2 × calyx lobes; SCo, PR, Baja CA .... ***J. parryi***
1′ Lvs gen 2–3; fls gen > 4; hypanthium < 1.5 × calyx lobes; ChI or c SNF
 2. Petals withering; peduncle red-pink; seeds pale brown; c SNF .... ***J. heterandra***
 2′ Petals persistent but not withering; peduncle ± green; seeds dark brown; ChI .... ***J. malvifolia***

J. heterandra Eastw. (p. 1007) Caudex often flat, branched. **LVS** gen 2–3. **INFL**: fls gen 4–17; peduncle 5–23 cm, red-pink, often drying tan. **FL**: hypanthium 1.5–3 mm, gen > calyx lobes, truncate to tapered at base; calyx lobes 1.3–2 mm, pink; petals gen 3.5–6 mm, persistent, withering, veins deep pink. **FR** gen ± green or rose, ± red-striped. **SEED** pale brown. $2n=14$. Uncommon. Crevices, esp in slate-like rock, slopes; 50–500 m. c SNF. [*J. parryi* var. *h.* (Eastw.) Jepson] ❀DRN&SHD:**7**,14–17.

J. malvifolia (E. Greene) Small (p. 1007) ISLAND JEPSONIA Caudex often flat, branched. **LVS** gen 2–3. **INFL**: fls gen 4–17; peduncle 5–25 cm, ± green, often drying brown. **FL**: hypanthium 1–2 mm, gen ± = calyx lobes, gen truncate at base; calyx lobes 1–2 mm, yellow-green to pink; petals gen 3–3.5 mm, persistent but not withering, veins red. **FR** yellow-green, gen tan-striped. **SEED** dark brown. $2n=14$. UNCOMMON. Rocky outcrops, clay slopes; < 1000 m. ChI; Guadalupe Island, Mex. ❀DRN&DRY:17,**24**&SHD:15–16, 20–23.

J. parryi (Torrey) Small (p. 1007) Caudex ovoid, unbranched. **LF** gen 1. **INFL**: fls gen < 4; peduncle 3–28 cm, brown. **FL**: hypanthium 2.5–4 mm, >> calyx lobes, gen truncate at base; calyx lobes 0.8–2 mm, ± green; petals gen 3.5–6 mm, veins tan or ± purple. **FR** green or tan, brown-striped. **SEED** dark brown. $2n=14$. Uncommon. Shrubby, rocky slopes; < 1000 m. SCo, PR; Baja CA. ❀DRN&DRY:17,**24**&SHD:**7,14–16**,18–21,**22,23**.

LITHOPHRAGMA WOODLAND STAR

Rhizome slender, scaleless, bearing bulblets. **LVS** basal and cauline, reduced, sometimes opposite, more deeply lobed upward; blade round, base cordate to reniform, ± lobed, gen toothed. **INFL**: raceme; bracts scale-like or 0. **FL**: hypanthium gen partly fused to ovary; petals gen lobed or toothed; stamens 10; pistil 1, ovary superior to ± inferior, chamber 1, placentas 3, parietal, styles 3. **FR**: capsule, valves 3. 12 spp.: w N.Am. (Greek: rock hedge, from habitats) [Taylor 1965 U Calif Publs Bot 37:1–122] Generic names ending in "phragma" are considered of neuter, not feminine, gender.

1. Basal lvs compound or deeply lobed (> halfway to blade base), lflets or primary lobes 3; petals white to pink
  2. Hypanthium long-obconic, part fused to ovary gen > free part; ovary > half inferior; petals 7–16 mm, gen obovate, 3-lobed ..... ***L. parviflorum***
    3. Hypanthium ± 2 × longer than wide ..... var. ***parviflorum***
    3′ Hypanthium 3–4 × longer than wide ..... var. ***trifoliatum***
  2′ Hypanthium spheric to bell-shaped, part fused to ovary gen < free part; ovary < half inferior; petals 3–7 mm, gen ovate, gen 4–7-lobed
    4. Infl 40–60 cm; s ChI ..... ***L. maximum***
    4′ Infl 8–30 cm; mainland
      5. Petal lobes deep, gen 4–5; infl often with bulblets in axils of lower bracts; seed spiny ..... ***L. glabrum***
      5′ Petal lobes shallow, 5–7; infl without bulblets; seed smooth ..... ***L. tenellum***
1′ Basal lvs ± shallowly lobed (< halfway to blade base), primary lobes 3–5; petals white
  6. Hypanthium ± long-obconic; ovary gen > half inferior
    7. Cauline lvs 1–3, alternate; petals 3-lobed at tip; seed smooth ..... ***L. affine***
    7′ Cauline lvs 2, opposite; petals entire to shallowly toothed; seed spiny ..... ***L. cymbalaria***
  6′ Hypanthium spheric to bell-shaped; ovary < half inferior
    8. Petals obovate, lobes 3; hypanthium bell-shaped, base truncate; infl often with bulblets in axils of upper bracts ..... ***L. heterophyllum***
    8′ Petals ovate-elliptic, entire or lobes or teeth > 3; hypanthium spheric to bell-shaped, base tapered to round; infl bulblets gen 0
      9. Petals spreading, entire or ± evenly toothed or lobed; hypanthium spheric; ovary partly inferior; lf teeth ± round ..... ***L. bolanderi***
      9′ Petals ± erect, ± unevenly lobed (gen with teeth near base); hypanthium broadly bell-shaped; ovary superior; lf teeth ± sharp ..... ***L. campanulatum***

L. affine A. Gray (p. 1007) **LF**: basal blade ± 3–5-lobed, teeth ± sharp-tipped. **INFL** 10–60 cm; fls 3–15; pedicels 3–10 mm. **FL**: hypanthium obconic, ± inflated above, part fused to ovary ± = free part; petals 5–13 mm, ovate-elliptic, 3-lobed at tip; ovary > half inferior. **SEED** smooth. $2n$=14,21,28,35. Open, grassy slopes; < 2000 m. KR, NCoR, c SNF, SnFrB, SCoR, s ChI, TR, PR; sw OR, Baja CA. [*ssp. mixtum* R.L. Taylor, *L. tripartita* (E. Greene) E. Greene] Intermediates to *L. parviflorum* common in n CA. ❀DRN&DRY:4,5,**6**,**17**,24&SHD:**7**,**9**,**14–16**,18–23.

L. bolanderi A. Gray (p. 1007) **LF**: basal blade ± 3–5-lobed, teeth ± round. **INFL** 20–80 cm; fls 5–25; pedicels 1.5–4 mm. **FL**: hypanthium spheric, part fused to ovary < free part; petals 4–7 mm, ovate-elliptic, entire or ± evenly toothed or lobed; ovary < half inferior. **SEED** spiny. $2n$=14,28,35,42. Open slopes; < 1500 m. NCoR, CaRF, SNF, SnFrB, SnGb. [*L. scabrella* (E. Greene) E. Greene]

L. campanulatum J. Howell (p. 1007) **LF**: basal blade ± 3–5-lobed, teeth ± sharp. **INFL** 15–55 cm; fls 2–11; pedicels 1.5–3 mm. **FL**: hypanthium broadly bell-shaped, part fused to ovary << free part; petals 3–7 mm, ovate-elliptic, unevenly lobed (gen with teeth near base); ovary superior. **SEED** spiny. Shady, well drained slopes; < 2500 m. KR, NCoR, CaR, n SN; sw OR.

L. cymbalaria Torrey & A. Gray **LF**: basal blade shallowly 3-lobed, teeth 0; cauline 2, opposite. **INFL** 10–35 cm; fls 2–8; pedicels 4–10 mm. **FL**: hypanthium ± long-obconic, part fused to ovary ± = free part; petals 4–8 mm, ovate-elliptic, entire to shallowly toothed; ovary gen > half inferior. **SEED** spiny. $2n$=14+. Shady moist areas; < 2500 m. SnFrB, SCoR, WTR, SnGb, n ChI.

L. glabrum Nutt. (p. 1007) **LF**: basal blade deeply 3-lobed or ± palmately compound (lflets 3), lobes and lflets lobed, teeth ± sharp-tipped. **INFL** 8–25 cm; fls 1–7; pedicels 3–6 mm; lower bracts often with axillary bulblet. **FL**: hypanthium spheric to bell-shaped, part fused to ovary gen < free part; petals 3–7 mm, ovate, deeply 3–5-lobed; ovary < half inferior. **SEED** spiny. $2n$=14,28. Dry, gravelly places; < 3500 m. KR, CaRH, SNH, MP; to B.C., SD, Colorado. [*L. bulbiferum* Rydb.]

L. heterophyllum (Hook. & Arn.) Torrey & A. Gray **LF**: basal blade ± 3–5-lobed, teeth gen round. **INFL** 15–50 cm; fls 3–12; pedicels 0.5–2 mm; upper bracts often with axillary bulblets. **FL**: hypanthium bell-shaped, base truncate, part fused to ovary << free part; petals 5–12 mm, obovate, 3-lobed; ovary superior. **SEED** spiny. $2n$=14. Shaded slopes; < 1500 m. NCoR, SnFrB, SCoR, w WTR. ❀DRN&DRY:4,5,**6**,**17**,24&SHD:**7**,**9**,**14–16**,18–23.

L. maximum Bacigal. SAN CLEMENTE ISLAND WOODLAND STAR **LF**: basal blade palmately compound, lflets 3, lobed, teeth sharp. **INFL** 40–60 cm; fls 6–25; pedicels ± 1 mm. **FL**: hypanthium spheric to bell-shaped, part fused to ovary gen < free part; petals 3.5–4.5 mm, ovate, ± 5-lobed; ovary < half inferior. **SEED** spiny. **ENDANGERED** CA. Steep, moist, n-facing canyon slopes; < 400 m. s ChI (San Clemente Island). Threatened by feral goats, erosion.

L. parviflorum (Hook.) Torrey & A. Gray (p. 1007) **LF**: basal blade deeply 3-lobed to ± palmately compound, lobes lobed, teeth ± sharp-tipped. **INFL** 10–50 cm; fls 4–14; pedicels 3–7 mm. **FL**: hypanthium long-obconic, part fused to ovary > free part; petals 7–16 mm, gen obovate, 3-lobed; ovary > half inferior. **SEED** smooth. $2n$=14,21,28,35. Open areas; < 3000 m. KR, NCoR, SN, CaRF, SnFrB, SCoR, WTR, MP; to B.C., SD.

var. ***parviflorum*** **FL** not fragrant; hypanthium ± 2 × longer than wide; petals gen white. **SEED** 0.5–0.6 mm. Common. Habitat and range of sp.

var. ***trifoliatum*** (Eastw.) Jepson **FL** fragrant; hypanthium 3–4 × longer than wide; petals pink. **SEED** 0.6–0.8 mm. $2n$=28. Habitat of sp.; < 600 m. CaRF, SNF. [*L. t.* Eastw.] Seeds gen sterile.

L. tenellum Nutt. **LF**: basal blade deeply 3-lobed, lobes lobed, teeth ± round. **INFL** 8–30 cm; fls 3–12; pedicels 3–10 mm. **FL**: hypanthium spheric to bell-shaped, part fused to ovary < free part; petals 3–7 mm, ovate, 5–7-lobed, white to pink; ovary < half inferior. **SEED** smooth. $2n$=14,35. Dry areas; < 3000 m. CaR, n SN, MP, SnGb, SnBr; to B.C., Colorado, NM. [*L. breviloba* Rydb.; *L. rupicola* E. Greene]

MITELLA BISHOP'S CAP

Rhizome scaly; bulblets 0. **LVS** basal, sometimes a few cauline or arising from rhizome; blade ovate to round, base ± cordate, ± lobed, gen toothed. **INFL**: raceme or spike, ± 1-sided; bracts gen scale-like. **FL**: hypanthium ± fused to ovary; petals 5, pinnately lobed, not clawed; stamens gen 5; pistil 1, ovary gen > half inferior, chamber 1, placentas 2, parietal. **FR**: capsule, ± circumscissile. 12 spp.: temp to arctic N.Am, Asia. (Latin: small cap, from fr)

1. Stamens opposite petals ..... ***M. pentandra***
1′ Stamens alternate petals

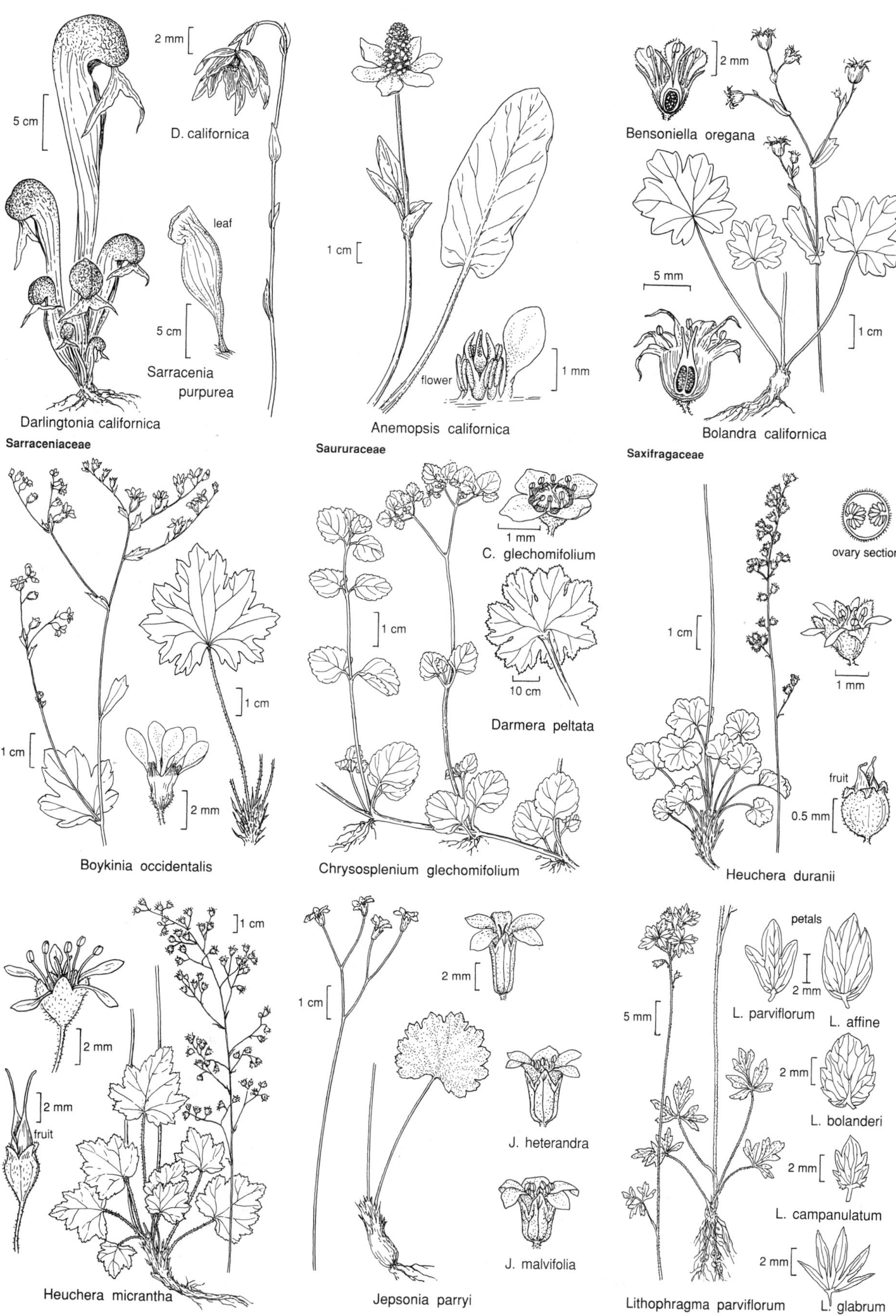
2 mm
5 cm
D. californica
leaf
5 cm
Sarracenia purpurea
Darlingtonia californica
Sarraceniaceae
1 cm
flower
1 mm
Anemopsis californica
Sauruaceae
2 mm
Bensoniella oregana
5 mm
1 cm
Bolandra californica
Saxifragaceae
1 cm
1 cm
2 mm
Boykinia occidentalis
1 mm
C. glechomifolium
1 cm
10 cm
Darmera peltata
Chrysosplenium glechomifolium
ovary section
1 cm
1 mm
fruit
0.5 mm
Heuchera duranii
1 cm
2 mm
2 mm
fruit
Heuchera micrantha
1 cm
2 mm
J. heterandra
J. malvifolia
Jepsonia parryi
petals
5 mm
2 mm
L. parviflorum
L. affine
2 mm
L. bolanderi
2 mm
L. campanulatum
2 mm
Lithophragma parviflorum
L. glabrum

2. Cauline lvs 1–3; infl blooming top to bottom . . . . *M. caulescens*
2′ Cauline lvs gen 0 (sometimes 1); infl blooming bottom to top
 3. Petal lobes gen 5 or more, pinnate; styles each 2-lobed at tip; hypanthium wider than long, bell- or saucer-shaped
  4. Lf blade ± round, 3–8 cm wide, lobes indistinct, 7–11, teeth ± round; pedicel gen > 1.5 mm; petiole glabrous to ± hairy; montane, > 1500 m . . . . *M. breweri*
  4′ Lf blade ovate-elliptic, gen < 4 cm wide, lobes distinct, ± 5, teeth ± sharp; pedicel gen < 1.5 mm; petiole densely hairy; ± coastal n CA, < 1000 m . . . . *M. ovalis*
 3′ Petal lobes 3, ± palmate at lobe tip; styles unlobed; hypanthium longer than wide, cup-shaped
  5. Lf blade ovate, lobes ± deep, gen ± entire; petal lobes narrowly triangular; sometimes 1 lf cauline . . . . *M. diversifolia*
  5′ Lf blade ± round, lobes ± shallow, teeth gen round; petal lobes ± elliptic; lvs basal . . . . *M. trifida*

M. breweri A. Gray (p. 1013) **LVS** basal; petiole glabrous to ± hairy; blade 3–8 cm wide, ± round, base gen slightly reniform, lobes 7–11, shallow, teeth ± round. **INFL** 1–3 dm, blooming bottom to top; pedicels gen 1.5–2.5 mm. **FL**: hypanthium 3–3.5 mm wide, saucer-shaped; petals yellow-green, lobes 5–9, gen opposite, linear; stamens alternate petals; styles each 2-lobed at tip. $2n$=14. Moist slopes; 1500–3500 m. KR, CaRH, SNH; to B.C., MT, NV. ❀IRR &DRN:**1,2**,4,**5,6,17**&SHD:3,7,14,**15,16**;GRCVR.

M. caulescens Nutt. **LVS** basal and a few cauline; petiole glabrous to ± hairy; blade 2–7 cm wide, ± round, lobes 3–7, teeth sharp. **INFL** 1.5–4.5 dm, blooming top to bottom; pedicels 2–8 mm. **FL**: hypanthium 2.5–4 mm wide, saucer-shaped; petals yellow-green, lobes 4–7, alternate, linear; stamens alternate petals; styles unlobed. $2n$=14. Uncommon. Wet shaded areas; < 1700 m. NCo, KR; to B.C., MT.

M. diversifolia E. Greene **LVS** basal and sometimes 1 cauline; petiole glabrous to sparsely hairy; blade 3–6 cm wide, ovate, lobes 5–7, ± deep, gen entire. **INFL** 2–5 dm, blooming bottom to top; pedicels 2–7 mm. **FL**: hypanthium 2.5–3.5 mm wide, cup-shaped; petals light pink, lobes 3, at tip, narrowly triangular; stamens alternate petals; styles unlobed. Uncommon. Moist woodlands, streambanks; 500–2000 m. KR; to WA.

M. ovalis E. Greene **LVS** basal or single from rhizome; petiole densely hairy; blade 1.5–5 cm wide, ovate-elliptic, lobes 5–9, teeth ± sharp, shallow. **INFL** 1–3 dm, blooming bottom to top; pedicels 0.5–1.5 mm. **FL**: hypanthium 2.5–3.5 mm wide, saucer-shaped; petals yellow-green, lobes 4–7, alternate, linear; stamens alternate petals; styles each 2-lobed at tip. Wet woodlands, shaded banks; < 1000 m. NCo, NCoRO, n CCo; to B.C. ❀DRN&WET:**4,5,17**& SHD:**6,15,16**;GRCVR.

M. pentandra Hook. **LVS** basal; petiole gen glabrous; blade 1.5–8 cm wide, ovate, lobes 5–9, teeth sharp. **INFL** 1–4 dm, blooming bottom to top; pedicels 2–7 mm. **FL**: hypanthium 3–5 mm wide, saucer-shaped; petals green, lobes 5–10, gen opposite, linear; stamens opposite petals; styles unlobed or each shallowly 2-lobed at tip. $2n$=14. Streambanks, wet meadows; 1500–2500 m. KR, NCoRH, CaRH, SNH; to AK, Colorado. ❀WET:1–3,**4–6, 16,17**&SHD:7,**15**.

M. trifida Graham **LVS** basal; petiole glabrous to sometimes sparsely hairy; blade 2–8 cm wide, ± round, lobes 5–7, shallow, teeth gen round. **INFL** 1.5–4.5 dm, blooming bottom to top; pedicels 0.5–2 mm. **FL**: hypanthium 2.5–3.5 mm wide, cup-shaped; petals white or ± purple, lobes 3, at tip, elliptic; stamens alternate petals; styles unlobed. Uncommon. Wet shaded slopes; 900–2000 m. KR, NCoRO; to B.C., Alberta. ❀IRR&DRN:4,**5,17**& SHD:**2**,3,**6**,14,**15,16**.

PARNASSIA GRASS-OF-PARNASSUS

Pl glabrous; caudex without scales or bulblets. **LVS** basal; blade ovate, entire, base tapered to cordate or reniform. **INFL**: fl solitary; bract gen 1, sessile, lf-like. **FL**: hypanthium minute, free of ovary; stamens 5, staminodes 5, toothed to fringed or divided; pistil 1, ovary superior, chamber 1, placentas 4, parietal, styles inconspicuous, stigmas 4. **FR**: capsule, valves 4. **SEED** ± winged. 25 spp.: n temp, arctic. (Mount Parnassus, Greece)

1. Petal margin fringed near base
 2. Staminode lobe tips spheric; calyx lobes entire; lf blade base tapered . . . . *P. cirrata*
 2′ Staminode lobe tips flat, round; calyx lobes irregularly toothed; lf blade base reniform . . . . *P. fimbriata*
1′ Petal margin entire
 3. Bract attached above middle of peduncle; staminode lobes > 13 . . . . *P. californica*
 3′ Bract attached below middle of peduncle; staminode lobes < 12 . . . . *P. parviflora*

P. californica (A. Gray) E. Greene (p. 1013) **LF** 4–14 cm; blade 2–5 cm, ovate-lanceolate, base tapered. **INFL** 15–47 cm; bract above middle of peduncle, elliptic. **FL**: calyx lobes 3–7 mm, elliptic, entire, ± reflexed in fr; petals 8–20 mm, round-ovate, entire, ephemeral; staminodes 5–9 mm, lobes > 13, tips spheric. **FR** 8–12 mm. Wet banks, meadows; < 4000 m. KR, NCoRI, CaRH, SNH, SCoR; s OR, NV. [*P. palustris* L. var. *c.* A. Gray] ❀WET&DRN: **4,5,17**&SHD:14,**15,16**.

P. cirrata Piper **LF** 3–18 cm; blade 1–6 cm, round-ovate, base tapered. **INFL** 17–43 cm; bract above middle of peduncle, ovate, base cordate, clasping. **FL**: calyx lobes 3–7 mm, elliptic, ± entire, reflexed in fr; petals 8–15 mm, ovate to ± elliptic, fringed below; staminodes 3–6 mm, lobes < 15, ± equal, tips spheric. **FR** 5–13 mm. Uncommon. Wet places; < 3000 m. SnGb, SnBr; Mex.

P. fimbriata König (p. 1013) **LF** 8–16 cm; blade 2–4 cm, round, base reniform. **INFL** 10–40 cm; bract above middle of peduncle, ovate, base cordate, clasping. **FL**: calyx lobes 3–6 mm, elliptic, jagged to fringed below, gen spreading to ascending; petals 7–13 mm, obovate, fringed below; staminodes 3–4 mm, lobes < 9, tips flat, round. **FR** 5–10 mm. $2n$=36. Uncommon. Wet banks, meadows, rocky seeps; < 4000 m. KR, n SNH, Wrn; to AK, Alberta, Colorado. ❀WET,DRN&SHD:**1,2**.

P. parviflora DC **LF** 2–7 cm; blade 1–3.5 cm, ± ovate, base gen ± truncate. **INFL** 10–35 cm; bract below middle of peduncle, ± elliptic. **FL**: calyx lobes 4–7 mm, elliptic, entire, gen erect to spreading; petals 5–13 mm, ovate-elliptic, entire; staminodes 3.5–6 mm, lobes < 12, tips spheric. **FR** 4–13 mm. Uncommon. Rocky seeps; < 3000 m. SNE, W&I; to e Can. [*P. palustris* L. var. *p.* (DC) J. Boivin] Probably hybridizes with *P. californica* in CA.

SAXIFRAGA SAXIFRAGE

Pl gen ± hairy, often glandular; caudex or rhizome gen not woody, gen scaly. **LF**: blade oblanceolate to round, base tapered to reniform, margin entire or toothed. **INFL**: fls 1–many; bracts scale-like. **FL**: hypanthium free or ± fused to ovary; petals sometimes spotted; stamens 10, filaments gen flat; pistil 1 (chambers 2, placentas 2, axile) or 2 (each with 1 chamber, 1 marginal placenta), ovary superior to ± inferior (sometimes more superior in fr), styles 2, free throughout. **FR**: capsule or 2 follicles. ± 400 spp.: esp cool n temp. (Latin: rock-breaking) [Elvander 1984 Syst Bot Monogr 3:1–44]

1. Stolons conspicuous; fl bilateral (3 upper petals spotted, << 2 lower) .... ***S. stolonifera***
1′ Stolons 0; fl radial or ± bilateral (3 upper petals spotted, ± = 2 lower in *S. ferruginea*)
 2. Lvs cauline on ± trailing st
  3. Fls in few-fld racemes or panicles .... ***S. nuttallii***
  3′ Fls at tip of long peduncle, 1–2 or in head-like clusters
   4. Lf blade 3–5-lobed at tip; fls 1–2 .... ***S. cespitosa***
   4′ Lf blade entire; fls in head-like clusters .... ***S. tolmiei***
 2′ Lvs basal, sometimes a few smaller and cauline
  5. Lf blade gen 5–8 mm .... ***S. rivularis***
  5′ Lf blade gen > 15 mm
   6. Filaments wider above than below
    7. Lf blade base truncate to tapered; ovaries fused only at base; fr 2 follicles
     8. Infl open; sepals reflexed; petal spots 2; filaments gen much-widened above middle .... ***S. marshallii***
     8′ Infl dense toward tips; sepals spreading; petal spots 0; filaments ± widened exc near tip .... ***S. rufidula***
    7′ Lf blade base gen cordate to reniform; ovaries fused throughout; fr a capsule
     9. Lf blade teeth irregular, rounded, base gen cordate; petiole hairy; petals not falling early, ovate to elliptic, spots 0; capsule dark purple .... ***S. mertensiana***
     9′ Lf blade teeth regular, sharp, base gen reniform; petiole ± glabrous; petals ephemeral, round to elliptic, spots 2; capsule tip purple, base yellow-banded .... ***S. odontoloma***
   6′ Filaments wider below than above
    10. Ovary superior in fl and fr
     11. Infl bulblets 0; ovaries < half-fused; petal spots 0 .... ***S. howellii***
     11′ Infl bulblets axillary to lower bracts; ovaries > half-fused; some petals spotted
      12. All petals spotted; lf blade linear-elliptic, entire or teeth minute or sparse .... ***S. bryophora***
      12′ 3 upper petals spotted; lf blade obovate, teeth gen coarse .... ***S. ferruginea***
    10′ Ovary ± half-inferior in fl, superior in fr
     13. Lf blade linear to oblanceolate, gen > 10 cm, long-tapered to an indistinct petiole; bogs and marshes .... ***S. oregana***
     13′ Lf blade elliptic, obovate, or ovate, gen < 10 cm, short-tapered to a distinct petiole; many habitats but not bogs or marshes
      14. Lf blade shallowly toothed; infl ± open, often 1-sided; styles on fr gen > 1.5 mm .... ***S. californica***
      14′ Lf blade ± entire; infl open between ± head-like clusters, not 1-sided; styles on fr gen < 1 mm
       15. Petals gen < 2 mm, sepals ± reflexed to spreading .... ***S. nidifica*** var. ***nidifica***
       15′ Petals gen > 2 mm, sepals ± erect to spreading
        16. Infl ± glabrous to sparsely glandular; lvs ± glabrous; > 1500 m .... ***S. aprica***
        16′ Infl densely glandular; lvs ± hairy; < 1000 m .... ***S. integrifolia***

S. aprica E. Greene (p. 1013) Caudex gen with bulblets. **LF** 10–45 mm; petiole 5–15 mm; blade obovate to elliptic, base tapered, entire or teeth minute. **INFL** ± 5–7 cm; 1 head-like cluster at tip and sometimes 1–2 more nearby. **FL**: sepals ± spreading to erect, ovate; petals 1.8–3 mm, > sepals, gen ovate; filaments widened toward base; nectaries disk-like, lobed; ovary > half inferior in fl. **FR**: ± follicles. $2n$=20. Rocky, wet alpine meadows, near snowbeds or meltwater; 1600–3500 m. KR, CaRH, SNH; sw OR, w NV. [not *S. nivalis* L.] Pls from KR and s OR are robust, pollen often sterile. ❀WET,DRN&SHD:**1**,**2**,17.

S. bryophora A. Gray (p. 1013) Rhizome < 1 mm diam; bulblets 0. **LF** 1–4 cm, ± fleshy, sessile; blade linear-elliptic, base tapered, entire or teeth minute or sparse. **INFL** 3–25 cm, open; all but uppermost bracts gen with axillary bulblets. **FL**: sepals reflexed, < petals, ± elliptic; petals 3–5 mm, triangular to ovate, 2-spotted near base; filaments slightly widened toward base; nectaries 0; ovary superior. **FR**: capsule. Sandy meadows, ledges; 1600–4500 m. KR, CaRH, SNH.

S. californica E. Greene (p. 1013) Caudex gen with long slender rhizomes and bulblets. **LF** 4–10 cm; petiole 2–5 cm; blade ovate, base tapered, teeth shallow, ± round or sharp. **INFL** 15–35 cm, often 1-sided, ± open. **FL**: sepals gen reflexed, elliptic; petals 2.5–4.5 mm, >> sepals, elliptic to round; filaments widened throughout or only toward base; nectaries disk-like, lobed; ovary ± half-inferior in fl. **FR**: ± follicles. Moist, shady places; < 1200 m. CA-FP; sw OR, nw Baja CA. Variable. Pls with coarse lf-teeth, spreading sepals have been called *S. fallax* E. Greene. ❀DRN&DRY:4,5,**17**&SHD:**6**,**7**,9,14,**15**,**16**.

S. cespitosa L. Caudex slender, somewhat woody; bulblets 0. **ST** short, trailing, ± woody. **LVS** cauline, gen crowded, 5–10 mm, sessile; blade obovate, 3–5-lobed at tip. **INFL** 2–5 cm; peduncle long, gen with 1 lf-like bract near middle, gen 1–2 fls at tip. **FL**: sepals erect to spreading, << petals, elliptic; petals 3–5 mm, elliptic to obovate, ephemeral; filaments slightly widened toward base; nectaries inconspicuous or 0; ovary inferior in fl. **FR**: capsule. $2n$=52,56,60,63,65,80. RARE in CA. Damp rocky places; > 2000 m. KR (Marble Mtns); to arctic, e N.Am, circumboreal. [var. *emarginata* (Small) C. Rosend.] Extremely variable, with many poorly defined vars. In cult.

S. ferruginea Graham Caudex slender, gen producing short rhizomes; bulblets 0. **LF** 1.5–6 cm, ± fleshy, ± sessile; blade obovate, teeth gen coarse, sharp. **INFL** 5–40 cm, open; lower bracts gen with axillary bulblets. **FL** ± bilateral; sepals reflexed, < petals, ovate; petals 3.5–5 mm, ovate to triangular, upper 3 with 1 spot; filaments thread-like; nectaries inconspicuous; ovary superior. **FR**: capsule. $2n$=38. Wet banks, gravel; 1500–2500 m. KR, s NCoRH; to AK, MT. Pls with axillary bulblets have been called var. *macounii* Engl. & Irmscher.

S. howellii E. Greene HOWELL'S SAXIFRAGE Caudex slender, gen producing short rhizomes; bulblets 0. **LF** 2–6 cm; petiole 1–4 cm; blade oblong to ovate, base tapered, teeth coarse, round or sharp. **INFL** 5–20 cm, open to dense; fls 5–15. **FL**: sepals reflexed, < petals, elliptic to ovate; petals 2.5–4.5 mm, ± obovate; filaments gen widened toward base; nectaries not seen; ovary superior. **FR**: ± follicles. RARE in CA. Moist ledges, crevices; < 900 m. KR; sw OR.

S. integrifolia Hook. Caudex with short rhizomes, sometimes with bulblets. **LF** 2–7 cm; petiole 5–40 mm; blade ± ovate, base tapered, entire or teeth minute. **INFL** 12–35 cm; gen 1 head-like cluster at tip, sometimes 1–3-branched, ± open below. **FL**: sepals gen spreading to rarely reflexed, < petals, ovate; petals 2–4 mm, obovate; filaments ± thread-like; nectaries disk-like, lobed; ovary > half inferior in fl. **FR**: ± follicles. 2*n*=38. Uncommon. Vernally moist meadows; < 1000 m. KR, NCoR, ScV; to B.C., ID, NV. Several intergrading taxa have been recognized, incl some in CA.

S. marshallii E. Greene Caudex gen producing rhizomes; bulblets 0. **LF** 5–15 cm; petiole 3–10 cm; blade ovate or elliptic, base truncate to tapered, teeth coarse, sharp to round. **INFL** 10–40 cm, open. **FL**: sepals reflexed, gen < petals, ovate to elliptic; petals 2.5–4.5 mm, elliptic or obovate, 2-spotted, ephemeral; filaments gen much-widened above middle; nectaries band-like; ovary superior. **FR**: ± follicles. Mossy rocks, cliffs; < 1500 m. NCo, KR, NCoRO; sw OR.

S. mertensiana Bong. Caudex with bulblets. **LF** 4–20 cm; petiole 2–15 cm; blade ± round, base gen cordate, teeth coarse, irregular, ± round. **INFL** 15–30 cm, open. **FL**: sepals reflexed, < petals, ovate to elliptic; petals 4–5 mm, ovate to elliptic; filaments gen club-shaped; nectaries not seen; ovary superior. **FR**: capsule. 2*n*=48. Mossy rocks, cliffs; < 2500 m. NCo, KR, NCoRO, uncommon in n&c SN; to AK, MT. ❀DRN&DRY:**4,5,17**&SHD:**2,6,15,16**&IRR:3,7,14.

S. nidifica E. Greene var. ***nidifica*** Caudex with rhizomes and gen bulblets. **LF** 3–10 cm; petiole 1.5–5 cm; blade ovate to triangular, base gen tapered, entire or teeth minute. **INFL** 10–50 cm, gen open between ± head-like clusters. **FL**: sepals ± reflexed to spreading, < to = petals, triangular to ovate; petals 1–2 mm, narrowly elliptic to round; filaments slightly widened near base; nectaries disk-like, lobed; ovary > half-inferior in fl. **FR**: ± follicles. 2*n*=38. Open wet meadows, slopes; 500–3000 m. KR, NCoR, CaR, SNH, MP; to B.C., MT, NV. Variable. Intergrades with var. *claytoniifolia* (Canby) P. Elvander in e OR, WA.

S. nuttallii Small Rhizome < 1 mm diam; bulblets 0. **ST** 5–25 cm, trailing. **LVS** cauline, 3–20 mm, larger along middle of st; petiole 1–5 mm; blade obovate to elliptic, gen 3-lobed to sometimes entire at tip. **INFL**: raceme or panicle; fls few. **FL**: sepals erect, << petals, triangular; petals 3–6 mm, elliptic; filaments thread-like; nectaries not seen; ovary ± inferior. **FR**: capsule. **SEEDS** spiny. RARE in CA. Wet, shaded cliffs, ledges; < 700 m. NCo (Del Norte Co.); sporadically to WA.

S. odontoloma Piper (p. 1013) Caudex producing rhizomes; bulblets 0. **LF** 4–40 cm; petiole 2–30 cm, base gen ± expanded, sheathing, membranous; blade ± round, base gen reniform, teeth coarse, sharp. **INFL** 20–50 cm, open. **FL**: sepals reflexed, gen ± = petals, ovate to elliptic; petals 3–4.5 mm, round to elliptic, 2-spotted, ephemeral; filaments club-shaped; nectaries band-like; ovary superior. **FR**: capsule. 2*n*=48. Wet meadows, ledges; > 1500 m. KR, NCoRH, CaRH, SNH, SnBr; to B.C., MT, NM. [*S. aestivalis* Fischer; not *S. arguta* D. Don]

S. oregana J. Howell Caudex > 15 cm, thick, fleshy, sometimes branched; bulblets 0. **LF** 7–25 cm; petiole ± indistinct; blade linear to oblanceolate, base tapered, entire or sharp-toothed. **INFL** 25–125 cm, dense toward tips of main axis and branches, otherwise open. **FL**: sepals reflexed, gen < petals, ovate to triangular; petals 2–4 mm, linear to elliptic; filaments widened exc at tip; nectaries disk-like, lobed; ovary > half-inferior in fl. **FR**: ± follicles. 2*n*=38, 72. Bogs, marshes; 150–2500 m. KR, CaRH, SN, MP; to WA, ID, NV. Variable. Like *S. pensylvanica* L. of e US. ❀WET,DRN&SHD:**1–7**,15–17.

S. rivularis L. Caudex fragile; bulblets 0. **LVS** basal and cauline, < 3 cm, smaller upward; petiole 5–25 mm; blade broader than long, base ± shallowly reniform, lobes 3–5, deep, ± round. **INFL**: raceme; fls gen few. **FL**: sepals erect, gen < petals, elliptic to ovate; petals 2–6 mm, elliptic to ± obovate; filaments thread-like, slightly widened at base; nectaries not seen; ovary >> half inferior, more superior in fr. **FR**: capsule. Uncommon. Moist crevices, shaded rocky areas; 3500–4500 m. c&s SNH; circumboreal, sporadic to arctic, Colorado, AZ. Glabrous pls from Rocky Mtns have been called *S. debilis* Engelm.

S. rufidula (Small) Macoun Caudex with short rhizomes or bulblets. **LF** 4–12 cm; petiole 4–8 cm; blade ovate-elliptic, base tapered to truncate, teeth gen ± round. **INFL** 5–20 cm, dense toward tips. **FL**: sepals spreading, < petals, ovate; petals 1.5–3.5 mm, ovate; filament widened exc near tip; nectaries band-like; ovary superior. **FR**: ± follicles. 2*n*=20,38. RARE in CA. Moist rocky areas; ± 2000 m. KR (Marble Mtns, Siskiyou Co.); c OR, WA. [*S. aequidentata* (Small) C. Rosend.]

S. stolonifera Meerb. STRAWBERRY-GERANIUM Caudex branched, gen producing stolons; bulblets 0. **LF** 8–14 cm; petiole 5–8 cm; blade ± round, base shallowly cordate to truncate, teeth irregular, ± round. **INFL** ± 3–4 dm, open. **FL** bilateral; sepals reflexed, < petals, ovate; petals unequal, lower two 10–15 mm, >> upper 3 gen with 1–3 spots; filaments gen widened toward tips; nectaries band-like; ovary superior. **FR**: ± follicles. 2*n*=36,54. Moist, shady places; < 200 m. s NCoRO, expected more widely; native to China, Japan. [*S. sarmentosa* L.] Cult as orn.

S. tolmiei Torrey & A. Gray (p. 1013) Pl ± glabrous; caudex slender, ± woody; bulblets 0. **ST** trailing, ± woody. **LVS** cauline, gen crowded, 8–15 mm, fleshy, sessile; blade elliptic or obovate, entire. **INFL**: head-like cluster; peduncle 3–12 cm; bracts gen 1–3, lf-like. **FL**: sepals spreading to ascending, < petals, ± ovate; petals 2.5–5 mm, linear to oblanceolate; filaments gen much-widened exc near base and tip; nectaries not seen; ovary ± superior. **FR**: capsule. **SEED** winged. 2*n*=30. Alpine tundra, fell-fields; > 2000 m. KR, NCoRH, CaRH, SNH; to AK. ❀WET&DRN:**1**,2–5,17&SHD:14–16.

SAXIFRAGOPSIS

1 sp. (Greek: like Saxifraga)

S. fragarioides (E. Greene) Small (p. 1013) Pl ± glandular; caudex thick, woody, branched, scaly; bulblets 0. **ST** thick, woody, trailing. **LVS** basal, a few cauline, reduced, merging into sessile, linear bracts upward; petiole 1.5–4 cm, base expanded, membranous; blade 1.5–4 cm, obovate, jointed to petiole and falling from it, base tapered, teeth coarse. **INFL**: panicle, 10–25 cm, open below and between ± dense clusters; fls > 40. **FL**: hypanthium partly fused to ovary; calyx lobes 2–3 mm, spreading to reflexed, elliptic to ovate; petals 2–3 mm, obovate; stamens 10; pistil 1, ovary ± half inferior, chambers 2, placentas 2, axile, styles 2, fused at base at least in fl. **FR**: capsule. Uncommon. Rock crevices; 1500–3000 m. KR; sw OR. [*Saxifraga f.* E. Greene] ❀DRN:4,5&IRR:**6,17**&SHD:**1–3,7,15,16**.

SUKSDORFIA

Caudex scaleless, bearing bulblets. **LVS** basal and cauline, reduced and merging into linear bracts upward; petiole base expanded, membranous, entire, ± sheathing, ± stipule-like; blade ovate, base cordate to reniform, primary lobes 3, deep, teeth coarse, round, with sharp tips. **INFL**: cyme, open or sometimes a few dense clusters. **FL**: hypanthium partly fused to ovary; stamens 5; pistil 1, ovary > half inferior, chambers 2, placentas 2, axile. **FR**: capsule. 2 spp.: w N.Am. (W.N. Suksdorf, WA botanist & collector, 1850–1932) [Gornall & Bohm 1985 Bot J Linn Soc 90:1–71]

S. ranunculifolia (Hook.) Engl. (p. 1013) **LF**: blade 1–4 cm wide, ± glabrous. **INFL** 1–3 dm; hairs yellow-tipped; pedicels gen < 3 mm. **FL**: calyx lobes 1–3 mm, elliptic to triangular; petals 2–4 mm, elliptic to obovate. Uncommon. Moist rocky slopes; 1500–2500 m. KR, CaRH, n SNH (Plumas Co.); to B.C., MT.

TELLIMA FRINGE CUPS

1 sp. (anagram of Mitella)

T. grandiflora (Pursh) Lindley (p. 1019) Pl hairy, with some glands; rhizome scaly; bulblets 0. **LVS** basal and cauline, 5–40 cm, reduced upward; stipules ± 5 mm, membranous, ± sheathing, ciliate to toothed; petiole 3–30 cm; blade 2–10 cm, ovate, base cordate, clasping above, lobes shallow, teeth sharp. **INFL**: spike-like raceme, 40–100 cm, gen 1-sided; fls many; bracts scale-like; pedicels 2–5 mm. **FL** fragrant or not; hypanthium partly fused to ovary; calyx lobes 2–3 mm, ± elliptic; petals 3–7 mm, lobes ± 5–7, linear, ± green-white to rose, ephemeral; stamens 10; pistil 1, ovary > half inferior, chamber 1, placentas 2, parietal. **FR**: capsule; valves 2, equal. $2n=14$. Moist slopes; < 2000 m. NW, CaR, n SN, n&c CW; to AK, MT. ❀IRR&DRN:**4,5,17**&SHD:2,3,**6,7**,8,9,**14–16**,18–24; GRCVR.

TIARELLA SUGAR-SCOOP, FOAMFLOWER

Pl hairy; glands few; rhizome scaly; bulblets 0. **LVS** simple or compound, basal and cauline, reduced upward; stipules minute on cauline lvs, membranous, with marginal bristles; blade ovate, base cordate, lobes 3–5, ± deep, or lflets 3, teeth sharp. **INFL** gen raceme-like. **FL**: hypanthium minute, free of ovary; petals thread-like; stamens 10; pistil 1, ovary superior, chamber 1 below, 2 above, placentas 2, parietal. **FR**: capsule; valves 2, unequal. 2 spp.: N.Am, Asia. (Greek: small tiara) [Kern 1966 Madroño 18:152–160]

T. trifoliata L. LACE FLOWER **LF** 2–20 cm; stipules on basal lvs < 1 cm, on cauline lvs < 1 mm; petiole > blade; blade 3–12 cm wide, teeth sharp. **INFL** 15–40 cm; glandular hairs dense. **FL**: calyx lobes 1.5–2.5 mm; petals 3–4 mm. UNCOMMON. Moist shady banks; < 2500 m. NW, n CCo, SnFrB; to AK, MT. Vars. intergrade.

1. Lf compound, lflets 3 .... var. ***trifoliata***
1′ Lf simple, lobes 3–5 .... var. ***unifoliata***

var. ***trifoliata*** (p. 1019) **LF** gen 3–9 cm wide; lflets stalked, shallowly lobed, teeth sharp. Habitat of sp.; ± 1500 m. KR (Humboldt-Trinity Co. line); to AK, MT.

var. ***unifoliata*** (Hook.) Kurtz (p. 1019) **LF** gen 9–12 cm wide; lobes ± deep, teeth gen round, tips sharp. $2n=14$. Habitat and range of sp. [*T. u.* Hook.] ❀IRR&DRN:**4,5,17**&SHD:1–3,**6**,7,14, **15,16**;GRCVR.

TOLMIEA PIG-A-BACK PLANT

1 sp. (W.F. Tolmie, Hudson's Bay Co. physician, Fort Vancouver, 1812–1886)

T. menziesii (Pursh) Torrey & A. Gray (p. 1019) Pl ± hairy, glandular; rhizome scaly; bulblets 0. **LVS** basal and cauline, 8–40 cm, reduced upward; stipules 3–8 mm, elliptic, lf-like, with irregular teeth and short bristles; blade 3–10 cm wide, ovate, base gen cordate, sometimes bearing plantlets, lobes shallow, teeth sharp. **INFL**: raceme, 10–80 cm; bracts scale-like. **FL** bilateral; hypanthium free of ovary; calyx cleft between lower 2 lobes, upper 3 larger; petals 4, 8–12 mm, thread-like, brown-purple; stamens 3, opposite upper 3 sepals; pistil 1, ovary superior, chamber 1, placentas 2, parietal. **FR**: capsule, opening by 2 equal valves. **SEED** spiny. $2n=28$. Moist banks, cliffs; < 1800 m. NW, n CCo, SnFrB; to AK. ❀WET&DRN:4,**5,17**&SHD:**6,7**,8,9,**14–16**,20–23,**24**;GRCVR;CVS.

SCROPHULARIACEAE FIGWORT FAMILY

Lawrence R. Heckard, Family Coordinator

Ann to shrubs, gen glandular, some green root-parasites. **ST** gen round. **LVS** gen alternate, simple, gen ± entire; stipules gen 0. **INFL**: spike to panicle, gen bracted, or fls 1–2 in axils. **FL** bisexual; calyx lobes gen 5; corolla gen strongly bilateral, gen 2-lipped (upper lip gen 2-lobed, lower lip gen 3-lobed); stamens gen 4 in 2 pairs, gen incl, a 5th (gen uppermost) sometimes present as a staminode; pistil 1, ovary superior, chambers gen 2, placentas axile, style 1, stigma lobes gen 2. **FR**: capsule, gen ± ovoid, loculicidal or septicidal. **SEED**: coat sculpture often characteristic. ± 200 genera, 3000 spp.: ± worldwide; some cult as orn (e.g., *Antirrhinum, Mimulus, Penstemon*) or medicinal (*Digitalis*). Key to genera by Elizabeth Chase Neese & Margriet Wetherwax.

1. Lvs alternate, not all basal
 2. Stamens 5; corolla ± radial, 5-lobed; fr septicidal .... **VERBASCUM**
 2′ Stamens 2 or 4; corolla ± radial to 2-lipped; fr loculicidal or dehiscing by terminal slits or pores
  3. Corolla base neither spurred nor sac-like; fr loculicidal
   4. Corolla upper lobe not forming a beak or hood
    5. Corolla 40–60 mm, long-bell-shaped, gen pink or white; stamens 4; stigma lobes 2, flat .... **DIGITALIS**
    5′ Corolla < 15 mm, ± rotate, gen blue; stamens 2; stigma head-like .... [2]**VERONICA**
   4′ Corolla upper lip forming a beak or hood
    6. Corolla hood rounded; anther sacs 2 per stamen, equal; lvs gen toothed (or divisions gen > 7) **PEDICULARIS**
    6′ Corolla beak ± straight; anther sacs 1 per stamen or unequal; lvs entire or divisions 3–7
     7. Beak tip closed (opening directed downward); stigma unexpanded (dot-like); seed attached at side
      8. Calyx sheath-like (sometimes ± notched); bracts grading from lf-like to calyx-like; lower corolla lip 3-lobed; infl various; gen ± low elevation .... **CORDYLANTHUS**

8′ Calyx unequally 4-lobed; bract 1 per fl; lower corolla lip 3-toothed; infl spike-like; montane .... **ORTHOCARPUS**
7′ Beak tip open; stigma expanded (head-like or 2-lobed); seed attached at base
9. Anther sacs 2 per stamen; lower corolla lip highly reduced, sometimes 3-pouched .... **CASTILLEJA**
9′ Anther sacs 1 per stamen; lower corolla lip < 6 mm or strongly 3-pouched .... **TRIPHYSARIA**
3′ Corolla base spurred or sac-like; fr dehiscent by terminal slits or pores
10. Lvs palmately veined, long-petioled, entire to dentate
11. Corolla base spurred; lvs round to reniform, entire to lobed .... **CYMBALARIA**
11′ Corolla base sac-like; lvs gen ± hastate (if round to reniform, then irregularly bristly-dentate) .... **MAURANDYA**
10′ Lvs not palmately veined, ± sessile, gen entire
12. St twining or decumbent
13. Pl glabrous (exc base); pedicels twining or U-shaped; lvs lanceolate to ovate .... [2]**ANTIRRHINUM**
13′ Pl hairy; pedicels straight; lvs ovate to hastate .... [2]**KICKXIA**
12′ St erect
14. Corolla base spurred; pl gen glabrous .... [2]**LINARIA**
14′ Corolla base sac-like; pl gen hairy (at least infl)
15. Stamens 4, anther sacs 2 per stamen; fr asymmetric, dehiscing by 1–2 pores near tip ... [2]**ANTIRRHINUM**
15′ Stamens 2, anther sacs 1 per stamen; fr symmetric, dehiscent by irregular slit at tip .... **MOHAVEA**
1′ Lvs basal (or some cauline lvs opposite or whorled)
16. Stamens 2, staminodes 0 or 2
17. Lvs basal, long-petioled; infl scapose .... **SYNTHYRIS**
17′ Lvs cauline, sessile to short-petioled; infl not scapose
18. Corolla ± radial, gen 4-lobed, tube < lobes; calyx gen 4-lobed
19. Shrub; fr compressed parallel to septum; alien .... **HEBE**
19′ Ann or per; fr compressed perpendicular to septum; native .... [2]**VERONICA**
18′ Corolla 2-lipped, 5-lobed, tube > lobes; calyx gen 5-lobed
20. Anther sacs of each stamen separated, parallel; corolla tube 4-angled .... **GRATIOLA**
20′ Anther sacs of each stamen touching, not parallel; corolla tube cylindric
21. Upper lvs scale-like; herbage ± glandular .... **DOPATRIUM**
21′ Upper lvs ovate; herbage glabrous .... **LINDERNIA**
16′ Stamens 4, staminode 0 or 1
22. Calyx tube well developed
23. Pl aquatic; lvs all basal; stolons present .... **LIMOSELLA**
23′ Pl of drier habitats; lvs opposite; stolons gen 0
24. Stigma lobes 2, flat; calyx tube gen prominently ribbed or pleated .... **MIMULUS**
24′ Stigma head-like; calyx tube not prominently ribbed or pleated
25. Stamens exserted .... **TONELLA**
25′ Stamens incl in corolla
26. Stamens in central lobe of lower corolla lip .... **COLLINSIA**
26′ Stamens in upper corolla lip
27. Upper corolla lobes pink, lower white; calyx lobes unequal; fr gen ± wider than long ..... **BELLARDIA**
27′ Corolla yellow or purple; calyx lobes ± equal; fr longer than wide .... **PARENTUCELLIA**
22′ Calyx segments ± free
28. Corolla spurred at base of tube
29. St decumbent; fls solitary in axils .... [2]**KICKXIA**
29′ St erect; infl a terminal raceme .... [2]**LINARIA**
28′ Corolla not spurred
30. Staminode 0, filaments 4
31. Lvs in whorls of 3; fls > 2 cm, red .... **GALVEZIA**
31′ Lvs opposite; fls < 2 cm, not red
32. Pl gen aquatic; lvs spoon-shaped to round, entire; corolla white to pink .... **BACOPA**
32′ Pl gen of dry habitats; lvs ± lanceolate, toothed; corolla violet to purple .... **STEMODIA**
30′ Staminode 1, filaments 4–5
33. St square; staminode attached near base of upper lip .... **SCROPHULARIA**
33′ St round; staminode attached near base of corolla throat
34. Fertile filament bases glabrous, attached to corolla at different levels .... **PENSTEMON**
34′ Fertile filament bases densely hairy, attached to corolla at 1 level
35. Shrub; anthers glabrous; seeds angled .... **KECKIELLA**
35′ Per; anthers woolly; seeds winged .... **NOTHOCHELONE**

ANTIRRHINUM SNAPDRAGON

David M. Thompson

Ann, per, glabrous to hairy. **ST** vine-like, ascending or erect, often clinging by twining pedicels or branchlets. **LVS** gen opposite below, alternate above, gen reduced upward; veins pinnate. **INFL**: raceme or fls solitary in axils. **FLS** often cleistogamous; uppermost calyx lobe gen largest; corolla tube of opening fls truncate or with rounded sac-like extension at base,

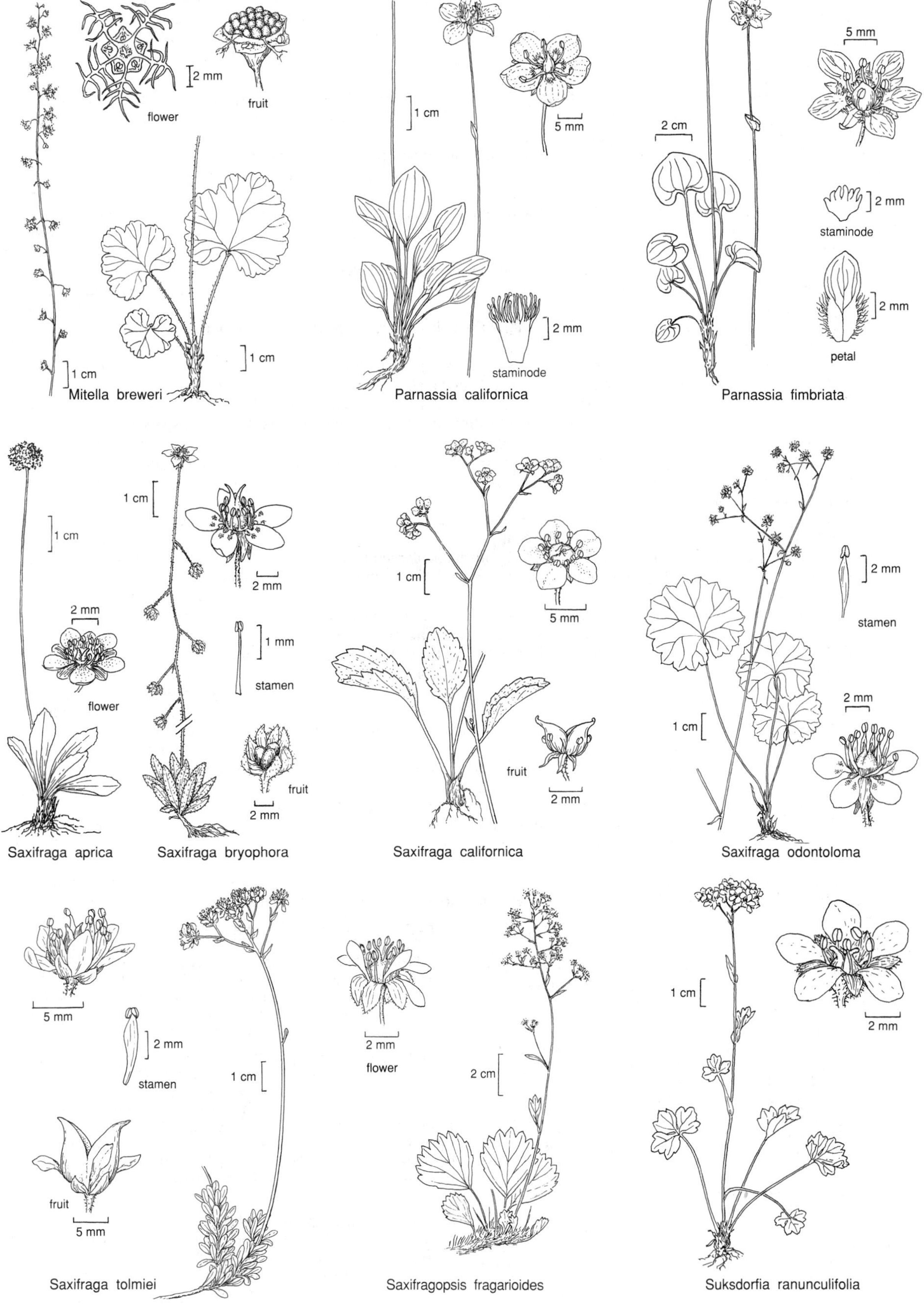
flower
2 mm
fruit
1 cm
1 cm
Mitella breweri
1 cm
5 mm
2 mm
staminode
Parnassia californica
2 cm
5 mm
2 mm
staminode
2 mm
petal
Parnassia fimbriata
1 cm
2 mm
flower
Saxifraga aprica
1 cm
2 mm
1 mm
stamen
fruit
2 mm
Saxifraga bryophora
1 cm
5 mm
fruit
2 mm
Saxifraga californica
2 mm
stamen
1 cm
2 mm
Saxifraga odontoloma
5 mm
2 mm
stamen
1 cm
fruit
5 mm
Saxifraga tolmiei
2 mm
flower
2 cm
Saxifragopsis fragarioides
1 cm
2 mm
Suksdorfia ranunculifolia

lower lip base gen swollen and closing mouth; staminode 0. **FR** ovoid to spheric; chambers gen dehiscent by 1–2 pores near tip, lower chamber gen larger, upper sometimes indehiscent. **SEEDS** many, gen with tubercles or netted ridges. 36 spp.: w N.Am, w Medit. (Greek: nose-like, from corolla shape) [Thompson 1988 Syst Bot Monogr 22:1–142]

1. Pedicels 3–10 cm, twining or with U-shaped hook
 2. Corollas of opening fls yellow and gold; fr fragile, opening by irregular bursting on sides .... ***A. filipes***
 2′ Corollas of opening fls lavender to blue-purple; fr firm, dehiscent by 2 slits at tip .... ***A. kelloggii***
1′ Pedicels < 2.6 cm, not twining, not hooked
 3. Lowest fls not subtended by branchlets; fls all opening; ann to per; pls self-supporting
  4. Lower calyx lobes > or = corolla tube; seed smooth, 1 side flattened, with raised, rough, irregular border .... ***A. orontium***
  4′ Lower calyx lobes < to << corolla tube; seed with netted ridges or a cup-shaped wing
   5. Corolla > 23 mm .... ***A. majus***
   5′ Corolla < 23 mm
    6. Fls in terminal racemes; corolla pale pink to red, 13–18 mm; often per from caudex
     7. Pl hairy .... ***A. multiflorum***
     7′ Pl glabrous .... ***A. virga***
    6′ Fls solitary in lf axils; corolla gen white, cream, or pale lavender, often with violet or purple veins or marks, 7–14 mm; ann
     8. Pedicel in fr bending down; fr chambers equal; seed wing large, cup-shaped .... ***A. cyathiferum***
     8′ Pedicel in fr not changing orientation; fr chambers unequal; seed wing 0
      9. Corolla white, veins violet; swollen lower lip base with dense, cylindric hairs .... ***A. cornutum***
      9′ Corolla white to pale lavender, veins not contrasting in color; swollen lower lip base with sparse, sphere-tipped hairs .... ***A. leptaleum***
 3′ Lowest fls and often others gen subtended by branchlets; fls cleistogamous or opening; ann, rarely bien; pls often not self-supporting, climbing or sprawling
  10. Calyx lobes ± equal
   11. Pl hairy only in infl; upper fr chamber indehiscent .... ***A. coulterianum***
   11′ Pl hairy throughout; both fr chambers dehiscent .... ***A. nuttallianum***
    12. Hairs very dense, very fine, of mixed lengths, with ± unenlarged tips .... ssp. ***nuttallianum***
    12′ Hairs sparse to moderately dense, coarse, of ± uniform length, with much-enlarged tips ...... ssp. ***subsessile***
  10′ Calyx lobes unequal
   13. Fls cleistogamous or opening; corollas of opening fls 5–7 mm, white, veins violet .... ***A. kingii***
   13′ Fls opening; corollas 8–20 mm, white to tan, cream and pink, or lavender
    14. Corolla lower lip base not swollen, not closing mouth, throat floor with 2 longitudinal folds ...... ***A. ovatum***
    14′ Corolla lower lip base swollen, closing mouth, throat floor without folds
     15. Corollas white to tan, throat floor expanded at mouth, lower lip lobes reflexed, poorly developed .... ***A. subcordatum***
     15′ Corolla lavender, throat uniformly narrowed at mouth, lower lip lobes thrust forward, erect or spreading, conspicuous .... ***A. vexillo-calyculatum***
      16. Lower st glabrous or with only nonglandular hairs; corollas 11–17 mm — SnFrB and nearby .... ssp. ***vexillo-calyculatum***
      16′ Lower st glandular-hairy, with or without nonglandular hairs; corollas 8–14 mm
       17. Branchlets subtending fls with 1 lf at lowest node; nw CA .... ssp. ***breweri***
       17′ Branchlets subtending fls gen with 2 lvs at lowest node; SN .... ssp. ***intermedium***

A. cornutum Benth. (p. 1019) Ann, hairy. **ST** erect, self-supporting. **LVS** not reduced upward. **INFL**: fls solitary; pedicels gen 1–2 mm, subtending branchlets 0. **FL**: calyx lobes equal; corolla 9–11 mm, white, veins violet, lower lip base rounded, with dense cylindric hairs. **FR**: upper chamber indehiscent. *n*=16. Uncommon. Dry stream margins, disturbed areas, often on serpentine; 0–1220 m. NCoRI, w CaR, n ScV. ❀DRN,SUN:5,7,8–10,**14–16**,17–21,**22–24**.

A. coulterianum Benth. (p. 1019) Ann, glabrous below infl, hairy in infl. **ST** erect but weak, often clinging to other pls or debris. **LVS**: basal rosette often present (unique in genus). **INFL**: raceme, terminal; pedicels 1–5 mm, lowest gen subtended by twining branchlets. **FL**: calyx lobes equal; corolla 9–12 mm, white to lavender; lower stamens gen exserted. **FR**: upper chamber indehiscent. *n*=15. Among shrubs in deserts, gen on burns elsewhere; 0–1700 m. s SCoRO, SW (exc ChI); n Baja CA. ❀DRN,DRY:**7,14**,15–17,**18–24**&part SHD:**8,9**,10–12;also STBL.

A. cyathiferum Benth. (p. 1019) DEEP CANYON SNAPDRAGON Ann, hairy. **ST** erect, self-supporting. **LVS** not reduced upward. **INFL**: fls solitary; pedicels 1–6 mm, in fr bending down, subtending branchlets 0. **FL**: calyx lobes equal; corolla 8–9 mm, cream and purple, veins purple. **FR**: chambers equal, opening by irregular bursting near tip. **SEED**: wing large, cup-shaped. *n*=13. RARE in CA. Washes, rocky slopes; 0–800 m. w DSon (Deep Canyon, Riverside Co.); s AZ, nw Mex.

A. filipes A. Gray (p. 1019) Ann, glabrous. **ST** vine-like, climbing. **INFL**: fls solitary; pedicels 3–10 cm, twining, subtending branchlets gen 0. **FL**: calyx lobes equal; corolla of opening fls 10–13 mm, yellow and gold with maroon flecks on lower lip. **FR**: chambers equal, fragile, opening by irregular bursting on sides. **SEED**: ridges 4–6, thick, parallel. *n*=15. On shrubs, debris, gen in washes; 0–1400 m. D; to sw UT, w AZ, nw Mex.

A. kelloggii E. Greene (p. 1019) Ann, glabrous. **ST** ascending to vine-like, often clinging to other pls or debris. **INFL**: fls solitary; pedicels 3–9 cm, twining or with U-shaped hook, subtending branchlets gen 0. **FL**: calyx lobes equal; corolla of opening fls 10–14 mm, lavender to deep blue-purple, veins darker. **FR**: chambers equal, firm, dehiscent by 2 slits at tip. **SEED**: tubercles scattered, block-like. *n*=15. Disturbed areas, esp burns; 0–1300 m. s NCoRO, CW, SW; nw Baja CA. ❀SUN or part SHD:5,**15–17,22–24**&IRR: **7,14**,18,**19–21**.

A. kingii S. Watson (p. 1019) Ann, gen glabrous below infl (infl sometimes sparsely hairy). **ST** erect but weak, often clinging to other pls or debris. **INFL**: raceme or fls solitary; upper pedicels gen 1–4 mm, lowest gen subtended by twining branchlets. **FL**: calyx lobes unequal; corolla of opening fls 5–7 mm, white, veins violet. *n*=15. Uncommon. Washes, scree; 500–2300 m. SNE, DMtns (esp ne San Bernardino Co.); to se OR, w UT, nw AZ.

A. leptaleum A. Gray (p. 1019) Ann, hairy. **ST** erect, self-supporting. **LVS** not reduced upward. **INFL**: fls solitary; pedicels 1–2 mm, subtending branchlets 0. **FL**: calyx lobes equal; corolla 7–10 mm, white to pale lavender, veins not contrasting in color, lower lip base angled, sparse, sphere-tipped hairs gen hidden in mouth. **FR**: upper chamber indehiscent. *n*=16. Uncommon. Small washes, shallow ditches, disturbed areas; 300–2100 m. SN (esp SNF). [*A. cornutum* Benth. var. *l.* (A. Gray) Munz] ❀TRY.

A. majus L. Ann, per, glabrous below infl (infl gen hairy). **ST** erect, self-supporting. **INFL**: raceme, terminal; pedicels 1–10 mm, subtending branchlets 0. **FL**: calyx lobes equal; corolla 25–45 mm, color various. *n*=8,16. Open, disturbed areas; 0–1200 m. CA-FP; native to Medit. Cult as orn. Waif.

A. multiflorum Pennell (p. 1019) Ann, per, hairy. **ST** erect, self-supporting, often from woody caudex. **LVS** alternate (on seedlings, basal lvs opposite). **INFL**: raceme, terminal; pedicels 2–4(10) mm, subtending branchlets 0. **FL**: calyx lobes unequal; corolla 13–18 mm, pale pink to red with tan-brown withered area on lower lip. *n*=16. Rocky or disturbed areas, burns; 0–1700 m. c SNF (Calaveras Co.), s SnFrB, SCoR, n ChI, TR. [*A. glandulosum* Lindley] ❀DRN,DRY,SUN:**7**,8–10,**14–24**.

A. nuttallianum Benth. Ann, rarely bien, hairy. **ST** erect but weak, often clinging to other pls or debris. **INFL**: fls solitary; pedicels gen 2–18 mm, lowest gen subtended by twining branchlets, fl branchlets, or both. **FL**: calyx lobes ± equal; corolla of opening fls 7–12 mm, lavender to blue-purple with 1–2 blue-veined white blotches on lower lip base and gold hairs in mouth (hair color unique in CA). *n*=16. Dunes, rocky or disturbed places; 0–1400 m. s CCo, SW; Baja CA. Sspp. hybridize near coast. ❀TRY.

ssp. ***nuttallianum*** Hairs very dense, very fine, of mixed lengths, tips ± unenlarged. **LVS** opposite at lowest 5–8 nodes of main st. **INFL**: fls not present at lowest few 1-lvd nodes; upper pedicels gen > 6 mm. **FL**: whitish blotch on lower corolla lip uninterrupted. **SEED**: ridges entire, longitudinal. Rocky areas, gen inland areas, esp burns; 0–1300 m. s SCo, SnBr, PR; n Baja CA.

ssp. ***subsessile*** (A. Gray) D. Thompson (p. 1019) Hairs sparse to moderately dense, coarse, of ± uniform length, tips much enlarged. **LVS** opposite at lowest 2–5 nodes of main st. **INFL**: fls at all 1-lvd nodes; upper pedicels gen < 6 mm. **FL**: whitish blotch on lower corolla lip gen interrupted by lavender. **SEED**: ridges broken, fragments longitudinal or unpatterned. Stabilized coastal dunes, rocky or disturbed areas; 0–1400 m. s CCo, SCo, ChI; s Baja CA, also in AZ.

A. orontium L. (p. 1019) Ann, gen glabrous below infl (infl ± hairy). **ST** erect, self-supporting. **INFL** raceme-like; bracts ± lf-like; pedicels 1–4 mm, subtending branchlets 0. **FL**: calyx lobes unequal, gen > corolla tube; corolla 10–15 mm, pink. **FR**: upper chamber indehiscent. **SEED** gen smooth; 1 side slightly keeled, other side flat, with raised, rough, irregular border. *n*=8. Open, disturbed areas; 0–1000 m. CA-FP; native to Medit. Waif.

A. ovatum Eastw. (p. 1019) OVAL-LEAVED SNAPDRAGON Ann, hairy. **ST** erect but weak, often clinging to other pls or debris. **INFL** raceme-like or fls solitary; pedicels 2–5 mm, lowest gen subtended by twining branchlets. **FL**: calyx lobes very unequal; corolla 17–20 mm, cream and pink, lower lip base not swollen, not closing mouth, throat floor with 2 longitudinal folds (unique in genus). *n*=16. RARE. Heavy, adobe-clay soils on gentle, open slopes, also disturbed areas; 200–1000 m. s SnJV (esp w Kern, e San Luis Obispo cos.), s SCoRI. Abundant every 20–50 years.

A. subcordatum A. Gray (p. 1019) DIMORPHIC SNAPDRAGON Ann, non glandular-hairy below infl (infl glandular-hairy). **ST** erect but weak, often clinging to other pls or debris. **INFL** raceme-like or fls solitary; pedicels 1–3 mm, lowest subtended by twining branchlets. **FL**: calyx lobes very unequal; corolla 13–17 mm, white to tan, throat floor expanded at mouth, lower lip lobes reflexed, poorly developed. *n*=15. RARE. Gentle, open slopes on serpentine, often under shrubs; 300–800 m. n&c NCoRI. In cult.

A. vexillo-calyculatum Kellogg Ann, glabrous to hairy below infl (infl gen glandular-hairy). **ST** erect but weak, often clinging to other pls or debris. **INFL** raceme-like or fls solitary; pedicels 1–4 mm, lowest subtended by twining branchlets, fl branchlets, or both. **FL**: calyx lobes unequal; corolla 8–17 mm, lavender, veins often vaguely darker, throat uniformly narrowed, curved upward at mouth, lower lip lobes conspicuous, thrust forward, erect or spreading. *n*=15. Gravelly lower slopes of rockslides, disturbed areas, often on serpentine; 0–2000 m. KR, NCoR, c CaRH, n&c SN, ScV (Sutter Buttes), SnFrB, n SCoRI; sw OR. Sspp. intergrade in NCoR.

ssp. ***breweri*** (A. Gray) D. Thompson **ST**: lower st hairs all glandular. **INFL**: branchlets subtending pedicels with 1 lf at lowest node. **FL**: corolla 8–12 mm, some veins dark. Habitats and elevations of sp. KR, n NCoRO, NCoRH, c CaRH; sw OR. [*A. b.* A. Gray]

ssp. ***intermedium*** D. Thompson **ST**: lower st hairs glandular, often also longer (< 3 mm), tapered, non glandular. **INFL**: branchlets subtending pedicels gen with 2 lvs at lowest node. **FL**: corolla 10–14 mm, some veins dark. Habitats of sp.; 100–1400 m. n&c SN.

ssp. ***vexillo-calyculatum*** (p. 1019) **ST**: lower st glabrous or hairs sparse, < 3 mm, tapered, non glandular. **INFL**: branchlets subtending fls with 2 lvs at lowest node. **FL**: corolla 11–17 mm, no veins dark. Habitats of sp.; 0–1200 m. s NCoR, ScV (Sutter Buttes), SnFrB, n SCoRI. ❀DRN,DRY,SUN:**7**,**8**,**9**,**14–16**,17,**18–24**.

A. virga A. Gray (p. 1019) TALL SNAPDRAGON Per, glabrous. **ST** erect, self-supporting, from woody caudex. **LVS** alternate (on seedlings, lowest opposite). **INFL**: raceme, terminal, loose; pedicels 2–6 mm, subtending branchlets 0. **FL**: calyx lobes equal; corolla 14–17 mm, pink, lips often withering, turning brown. *n*=16. UNCOMMON. Openings in chaparral, rocky areas, often on serpentine; 200–2000 m. s NCoRH, s NCoRI.

BACOPA WATER-HYSSOP

John L. Strother

Ann, ± per. **ST** prostrate to erect, gen < 6 dm. **LVS** cauline, opposite, gen < 4 cm, gen narrowly ± obovate to ± round. **INFL**: fls 1–3 per lf axil, sessile or pedicelled. **FL**: sepals 4–5; corolla (3–)5-lobed, gen white to pink, throat yellow; stamens (2–)4(–5); ovary subtended by nectary, stigma weakly 2-lobed. **FR**: capsule, spheric, loculicidal or septicidal. **SEEDS** > 30, 0.1–0.3 mm. ± 100 spp.: trop, warm temp. (Presumed to be aboriginal name) [Barrett & Strother 1978 Syst Bot 3:408–419]

1. Lf narrowly ± obovate, vein 1; pedicel bractlets 2, near tip ..... ***B. monnieri***
1' Lf ovate or obovate to ± round, veins > 6, palmate; pedicel bractlets 0
  2. Sepals 4(5), outer narrowly ovate to elliptic, gen ciliate; lf 10–23 mm, 8–12 mm wide ..... ***B. repens***
  2' Sepals 5, outer ovate to ± round, not ciliate; lf 12–36 mm, 14–20 mm wide
    3. Pedicel 15–51 mm, gen >> lf; corolla 10–14 mm ..... ***B. eisenii***
    3' Pedicel 5–18 mm, gen < lf; corolla 5–8 mm ..... ***B. rotundifolia***

B. eisenii (Kellogg) Pennell (p. 1019) **ST** prostrate to ascending. **LF** 12–34 mm, obovate to ± round; veins > 6, palmate. **INFL**: pedicel 15–51 mm, gen >> lf, stout; bractlets 0. **FL**: sepals 5, outer 4.3–6.6 mm, widely ovate to ± round, not ciliate; corolla 10–14 mm. 2*n*=56. Rice fields, muddy places, wet soil, or floating; < 100 m (< 1200 in SNE). GV, s CW, SNE; NV. ❀TRY.

B. monnieri (L.) Wettst. (p. 1019) **ST** prostrate to ascending. **LF** gen 5–25 mm, narrowly ± obovate; vein 1. **INFL**: pedicel gen > lf; bractlets 2, near tip. **FL**: sepals 5, outer 5–7 mm, lanceolate to ovate, not ciliate. 2*n*=64. Wet soil or in shallow water; < 100 m. DSon (e Riverside Co.); to s US; native to trop. Recent alien, probably naturalized.

B. repens (Sw.) Wettst. **ST** prostrate to ascending. **LF** 10–23 mm, gen obovate; veins > 6, palmate. **INFL**: pedicel 0–25 mm, very slender; bractlets 0. **FL**: sepals 4(5), outer 2.2–4.6 mm, narrowly ovate to elliptic, gen ciliate. 2*n*=28. Agricultural weed, wet soil or floating; < 100 m. ScV (Butte Co); Mex, Caribbean, C.Am, S.Am; probably native to Caribbean, C.Am, or both.

B. rotundifolia (Michaux) Wettst. **ST** prostrate to ascending. **LF** 12–36 mm, ± round; veins > 6, palmate. **INFL**: pedicel 5–18 mm, gen < lf, stout; bractlets 0. **FL**: sepals 5, outer 3.1–5.3 mm, ovate to ± round, not ciliate; corolla 5–8 mm. Wet soil or floating; < 100 m. 2*n*=56. GV; to ID, e US, AZ, Mex; native to c US. [*B. nobsiana* H. Mason]

BELLARDIA

Margriet Wetherwax

Ann, hairy. **ST** erect, gen simple. **LVS** opposite above, sessile, toothed. **INFL**: raceme, spike-like, bracted. **FL**: calyx 4-lobed; corolla 2-lipped, upper lip hood-like, lower lip >> upper, 3-lobed, throat with 2 ridges; stamens 4, in 2 pairs, incl, anthers hairy, awned at base; stigma club-shaped. **FR** ovoid, loculicidal. **SEEDS** many, small, ridged. 2 spp.: Medit. (C.A.L. Bellardi, Italian botany professor, Turin, 1740–1826)

B. trixago (L.) All. Pl glandular-hairy. **ST** erect, 15–70 cm. **LF** 15–90 mm, linear to lanceolate, coarsely crenate-dentate. **INFL**: bracts lf-like, reduced upwards, becoming ovate, cordate, entire. **FL**: calyx 8–10 mm, lobes 1–1.5 mm, unequal, triangular; corolla 20–25 mm, pale purple, lower lip white. **FR** ± 7mm. **SEED** 0.5–1 mm, ± oblong. 2*n*=24. Disturbed places; < 200 m. NCoRO, CCo, SnFrB; native to Medit.

CASTILLEJA INDIAN PAINTBRUSH, OWL'S-CLOVER

T.I. Chuang & Lawrence R. Heckard

Ann to subshrub, green root-parasites. **LVS** sessile, entire to dissected. **INFL** spike-like; bracts becoming shorter, wider, more lobed than lvs, tips gen colored. **FL**: calyx gen unequally 4-lobed, gen colored like bract tips; corolla upper lip beak-like, tip open, lower lip gen reduced, 3-toothed to -pouched; anther sacs 2, unequal; stigma entire to 2-lobed, gen exserted. **FR** loculicidal, ± ovoid, ± asymmetric. **SEED** gen ± brown, attached at base; coat netted, net-like walls sometimes aligned ladder-like. ± 200 spp.: esp w N.Am. (Domingo Castillejo, Spanish botanist) [Chuang & Heckard 1991 Syst Bot 16:644–666] Highly variable within & between populations. Hybridization & polyploidy common; polyploid forms may have separate ranges or be ± identifiable within populations by minor characters. Biologically consistent taxa very difficult to define. ❀TRY with host; usually DFCLT.

1. Infl not very reddish (gen green, white, yellow, or purplish); fl ± short (gen bee-pollinated); lower lip 3-pouched, teeth gen ± erect, white or green to lavender, upper lip << 3 × lower (subg. *Colacus*)
 2. Per
  3. Calyx sinuses much deeper in front and back than on sides (sect. *Pallescentes*)
   4. Herbage glandular; corolla 17–22 mm, lower lip 5–7 mm, teeth yellow-green, ± triangular; st gen simple; bract lobes ± acute .... ***C. lemmonii***
   4′ Herbage hairy, nonglandular; corolla 13–16 mm, lower lip 2 mm, teeth dark green, incurved; st much-branched; bract lobes truncate or rounded .... ***C. praeterita***
  3′ Calyx sinuses subequal (sect. *Pilosae*)
   5. Pl ± white-woolly
    6. Pl 8–30 cm; lvs 20–60 mm; hairs simple; KR, CaRH, MP .... ***C. arachnoidea***
    6′ Pl 8–15 cm; lvs 5–20 mm; hairs branched; KR .... ***C. schizotricha***
   5′ Pl puberulent to coarsely hairy
    7. Pl ashy-puberulent; bract tips gen dusty red or purplish; hairs dense, partitioned, much-branched; seed coat walls irregularly netted — seed coat tight-fitting; SnBr .... ***C. cinerea***
    7′ Pl not ashy-puberulent; bract tips gen yellow-green; hairs ± stiff, glandular and not, ± unbranched; seed coat walls with ladder-like thickenings
     8. Bract lobes acuminate, green-margined; seed coat loose-fitting .... ***C. nana***
     8′ Bract lobes gen rounded (acute), white-margined; seed coat tight-fitting .... ***C. pilosa***
 2′ Ann (sect. *Oncorhynchus*)
  9. Bracts green throughout
   10. Lvs and bracts entire .... ***C. campestris***
    11. Lvs and bracts linear, thin, flexible; bracts < or = fls; corolla light to bright yellow; lower anther sac 1/4–1/3 upper .... ssp. ***campestris***
    11′ Lvs and bracts lanceolate, thick, ± brittle; bracts > fls; corolla deep yellow to orange; lower anther sac ± 1/2 upper .... ssp. ***succulenta***
   10′ Lvs entire or lobed; bracts lobed
    12. Corolla beak densely white-hairy — SnBr, n&c PR .... ***C. lasiorhyncha***
    12′ Corolla beak puberulent
     13. Lower corolla lip pouches 2–4 mm wide, ± 2 mm deep; stigma well incl .... ***C. tenuis***
     13′ Lower corolla lip pouches 4–10 mm wide, 3–6 mm deep; stigma gen ± exserted
      14. Corolla 13–22 mm, pouches 4–8 mm wide; CaR, n&c SN, MP .... ***C. lacera***
      14′ Corolla gen 20–28 mm, pouches 8–10 mm wide; NCoR .... ***C. rubicundula***

15. Corolla yellow .... ssp. ***lithospermoides***
15′ Corolla white, turning pinkish .... ssp. ***rubicundula***
9′ Bracts tipped purplish, yellowish, or whitish
16. Corolla beak hooked, densely shaggy-hairy; filaments puberulent .... ***C. exserta***
17. Upper bracts gen 5–7 mm wide, lobes < 2 mm, tipped pale lavender; infl appearing alternately light- and dark-banded; NCo, n&c CCo .... ssp. ***latifolia***
17′ Upper bracts < 5 mm wide, lobes > 2 mm; infl not banded
18. Corolla white to yellow or rose to purple-red; widespread .... ssp. ***exserta***
18′ Corolla bright rose-red exc orange-tipped lower lip; w DMoj .... ssp. ***venusta***
16′ Corolla beak straight, puberulent; filaments glabrous
19. Infl 10–20 mm wide; corolla linear, pouches ± 2 mm wide; stigma well incl
20. Bract lobes 3, tipped white or pale yellow; seed < 1 mm, coat shallowly netted .... ***C. attenuata***
20′ Bract lobes 3–5, tipped pink or purplish red; seed 1–1.5 mm, coat ± deeply netted .... ***C. brevistyla***
19′ Infl 20–40 mm wide; corolla wider distally, pouches gen 3–5 mm wide; stigma gen ± exserted
21. St ± stiffly spreading-hairy and glandular; infl pale — n&c SNF .... ***C. lineariloba***
21′ St subglabrous, not glandular; infl bright (yellow to purplish red)
22. Branches gen many; lf oblong, > 3 mm wide; seed coat tight-fitting; ± coastal .... ***C. ambigua***
23. Bracts tipped yellow .... ssp. ***ambigua***
23′ Bracts tipped rose or purplish red
24. Pl ± fleshy, branches 0 or few from mid-st; n NCo, salt marshes .... ssp. ***humboldtiensis***
24′ Pl not fleshy, much-branched from base; c CCo, bluffs, grasslands .... ssp. ***insalutata***
22′ Branches 0 or few from mid-st; lf linear-lanceolate, < 3 mm wide; seed coat loose-fitting; gen inland .... ***C. densiflora***
25. Calyx 5–10 mm; lower corolla lip widened abruptly, pouches as deep as long .... ssp. ***gracilis***
25′ Calyx 8–20 mm; lower corolla lip widened gradually, pouches longer than deep
26. Infl rose-purple .... ssp. ***densiflora***
26′ Infl white to pale yellow .... ssp. ***obispoensis***
1′ Infl ± red (yellow-orange); fl long (gen hummingbird-pollinated); lower lip of small, incurved, gen dark green (red) teeth, upper lip > 3 × lower (subg. *Castilleja*)
27. Ann; st simple; wet places — lvs and bracts entire, ± narrowly lanceolate .... ***C. minor***
28. Corolla 15–20(30) mm; pl shaggy-hairy (most hairs glandular) .... ssp. ***minor***
28′ Corolla 25–35 mm; pl puberulent (many hairs nonglandular) .... ssp. ***spiralis***
27′ Per; sts gen clustered; gen drier places (exc *C. miniata*)
29. Pl ± densely white-woolly or ash-gray with branched hairs
30. Herbage ± green, calyces white-woolly; calyx sinuses deeper on sides than in back and front ... ***C. plagiotoma***
30′ Whole pl densely white-woolly or ash-gray; calyx sinuses deeper in back and front than on sides
31. Lvs oblong to obovate, entire, tip rounded; infl ± yellow; bracts entire or 3-toothed — n ChI ...... ***C. mollis***
31′ Lvs linear to lanceolate, upper sometimes 3-lobed, tip acute to obtuse; infl gen red; bracts gen 3–7-lobed
32. Pl white-felty with long, interwined, slightly branched hairs; lvs entire — n ChI ... ***C. lanata*** ssp. ***hololeuca***
32′ Pl white-woolly or ash-gray with shorter, much-branched or stellate hairs; upper lvs sometimes 3-lobed
33. Infl pale yellow-green; pl ash-gray; hairs matted, short, stellate; bracts 5–7-lobed — s ChI ...... ***C. grisea***
33′ Infl bright red to bright yellow; pl white-woolly or ash-gray; hairs much-branched; bracts gen entire (0–5-lobed)
34. Pl white-woolly; calyx ± undivided on sides .... ***C. foliolosa***
34′ Pl ash-gray; calyx divided ± 1/4 on sides .... [2]***C. pruinosa***
29′ Pl gen not white-woolly or ash-gray from branched hairs
35. Calyx divided 2/3 in front, < 1/3 in back, lobes gen curved upward; corolla (incl lower lip) gen curved out through front calyx sinus
36. Pl gen glabrous; lvs linear, rolled upward; bracts narrowly 3-lobed, < calyx; corolla beak sparsely puberulent; CaR, TR, GB, D .... ***C. linariifolia***
36′ Pl puberulent; lvs lanceolate, not rolled; bracts gen entire, > calyx; corolla beak ± densely shaggy-puberulent; CA-FP .... ***C. subinclusa***
37. Corolla yellow-orange; calyx lobes clearly curved upward, sides ± entire to divided < 1/5 .. ssp. ***franciscana***
37′ Corolla reddish; calyx lobes barely curved upward, sides divided at least 1/5 .... ssp. ***subinclusa***
35′ Calyx divided ± 1/3–1/2 in front and back, lobes not curved upward; corolla gen not curved out through front calyx sinus (lower lip incl)
38. Hairs branched and dense (or puberulent) .... [2]***C. pruinosa***
38′ Hairs gen unbranched (or ± 0), puberulent to bristly (esp *C. affinis* may have some branched hairs)
39. Pl glandular-puberulent to -sticky below infl
40. Calyx divided slightly > 1/2 in front; corolla beak 1/2 tube; stigma clearly 2-lobed .... ***C. parviflora***
40′ Calyx gen divided 1/3–1/2 in front; corolla beak 1–2 × tube; stigma entire or barely notched
41. Branches gen crowded below spike; lvs gen crowded; seed coat deeply netted; coastal .... ***C. wightii***
41′ Branches few below spike; lvs gen ± well spaced; seed coat shallowly netted; inland
42. Pl glandular but not sticky; lvs entire or shallowly lobed, not wavy; calyx lobes obtuse to rounded — nw KR .... ***C. hispida*** ssp. ***brevilobata***
42′ Pl very glandular-sticky; lvs entire to deeply lobed, margin wavy; calyx lobes gen acute ... ***C. applegatei***
43. St 10–25 cm; lf lobes 0–3; calyx 13–15 mm, divided ± 1/4 on sides; subalpine — SNH ... ssp. ***pallida***
43′ St 30–80 cm; most lvs entire; calyx 12–25 mm, gen divided < 1/5 on sides; gen below subalpine

44. Calyx lobes obtuse to rounded .... ssp. ***martinii***
44′ Calyx lobes acute
45. Calyx 12–18 mm; c&s SNH .... ssp. ***disticha***
45′ Calyx 16–25 mm; n CA .... ssp. ***pinetorum***
39′ Pl gen not glandular below spike (exc barely in some *C. affinis* or *C. latifolia*)
46. Hairs gen bristly or shaggy; lvs ± fleshy, oblong to rounded, gen entire, tips truncate-rounded; bracts 0–3-lobed
47. Corolla 20–30 mm, beak 10–15 mm; n&c CCo .... ***C. latifolia***
47′ Corolla 30–40 mm, beak 15–25 mm; c NCo .... ***C. mendocinensis***
46′ Hairs gen puberulent or 0 (bristly); lvs gen not fleshy, linear to lanceolate, entire or lobed, tips acute to obtuse; bracts entire to deeply lobed
48. Lvs gen 3–5-lobed; calyx divided 1/4–1/3 in back and front; gen sagebrush or desert scrub ***C. angustifolia***
48′ Lvs gen 0–3(5)-lobed; calyx divided ± 1/2 in back and front; gen CA-FP
49. Pl gen ± bristly-puberulent; lvs sometimes 5-lobed; dry places most of CA-FP .... ***C. affinis***
50. Infl gen bright red(-orange) (rarely yellow), 30–50 mm wide; variable, common .... ssp. ***affinis***
50′ Infl gen yellow (rarely pink or orangish), 15–25 mm wide; relatively uniform, rare — s NCoRI, nw SnFrB, on serpentine .... ssp. ***neglecta***
49′ Pl glabrous to sparsely puberulent; lvs entire; moist or dry places
51. Bract and calyx lobes obtuse or rounded; seed coat deeply netted, tight-fitting; dry coastal bluffs, NCo .... ***C. affinis*** ssp. ***litoralis***
51′ Bract and calyx lobes acute; seed coat shallowly netted, loose-fitting; moist places, NCo and elsewhere
52. Infl gray-puberulent; s NCo (Point Reyes), probably alien, presumed extinct .... ***C. chrymactis***
52′ Infl (sub)glabrous; widespread .... ***C. miniata***
53. St slender; infl pinkish to yellow-orange; serpentine bogs, rare; nw KR .... ssp. ***elata***
53′ St stout; infl gen bright red to orange-red; common, widespread .... ssp. ***miniata***

C. affinis Hook. & Arn. Per 15–60 cm, few-branched with short axillary shoots, (yellow-) green becoming purplish, ± glabrous to bristly, gen nonglandular. **LF** 30–80 mm, linear to narrowly oblong; lobes 0–5, tips obtuse or rounded. **INFL** 5–30 cm; bracts 17–25 mm, lobes 0–5, bright red to yellow (gen lighter upward). **FL**: calyx 15–35 mm, divided ± 1/3–1/2 in back and front, ± 1/4–1/3 on sides, long-nonglandular- and short-glandular-hairy, lobes acute to rounded; corolla 18–40 mm, beak ± 1–1.5 × tube, back gen shaggy-hairy, margins reddish or yellowish, lower lip 2–3 mm, green to dark violet; stigma slightly notched. **FR** 10–15 mm. **SEED** 1.5–2 mm; coat deeply netted, tight-fitting, most walls ladder-like. Sea bluffs, dry places in chaparral; < 1200 m. n&c NCo, n&s NCoRO, n CaRF, SNF, CW, SW; OR, n Baja CA. Many forms, some geog isolated; hybridizes with other spp.

ssp. ***affinis*** (p. 1019) Pl ± bristly. **LF** 30–80 mm, ± lanceolate; lobes 0–5. **INFL** 30–50 mm wide, gen bright red to orange-red (yellow). **FL**: calyx 20–35 mm; corolla 25–40 mm. $2n$=48,72, 96. Chaparral, coastal scrub; < 1200 m. c NCo (Mendocino Co.), n NCoRO (Humboldt Co.), s NCoRO, n CaRF, SNF, CW, SW; Baja CA. [ssp. *insularis* (Eastw.) Munz] Pls of n CCo with ± fleshy lvs, inflated calyx, and less exserted corolla beak have been called *C. inflata* Pennell. Dune pls of s CCo, n SCo with some branched hairs may represent past hybridization with diploid *C. mollis* of n ChI.

ssp. ***litoralis*** (Pennell) Chuang & Heckard (p. 1019) Pl ± glabrous or puberulent. **LF** 30–80 mm, ± oblong, gen entire. **INFL** gen 30–50 mm wide, bright red to orange-red. **FL**: calyx 20–25 mm; corolla 25–40 mm. $2n$=120,144. Sea bluffs; < 100 m. n&c NCo; OR. [*C. l.* Pennell; *C. wightii* ssp. *l.* (Pennell) Munz]

ssp. ***neglecta*** (E.M. Zeile) Chuang & Heckard TIBURON INDIAN PAINTBRUSH Pl ± bristly. **LF** 20–40 mm, ± lanceolate; lobes 0–5. **INFL** 15–25 mm wide, yellow (pinkish). **FL**: calyx 15–20 mm; corolla 18–22 mm. RARE. $2n$=72. Open serpentine slopes; < 300 m. s NCoRI (Napa Co.), nw SnFrB (Marin Co.). [*C. n.* E.M. Zeile]

C. ambigua Hook. & Arn. JOHNNY-NIP Ann 10–30 cm, gen much-branched and decumbent, puberulent. **LF** 10–50 mm, lanceolate to ovate; lobes 0–5. **INFL** 3–12 cm, 3–4 cm wide, often dense; bracts 15–25 mm, oblong to ovate, tipped white to rose-purple, lobes 3–9, central lobe gen rounded. **FL**: calyx 12–20 mm, divided 1/2 in front and on sides, 2/3 in back, lobes linear; corolla 14–25 mm, pale yellow or rose-purple, beak 4–5 mm, straight, puberulent, lower lip 3–4 mm, pouches 3–7 mm wide, 1–2 mm deep, gen purple-dotted at base, teeth 1–3 mm; stigma gen ± exserted. **FR** 8–12 mm. **SEED** ± 1–2 mm; coat shallowly netted, tight-fitting. $2n$=24 (all sspp.). Coastal bluffs, salt marshes, grassland; < 150 m. NCo, s NCoR, n&c CCo; to B.C. Highly variable and difficult; many local, ± ecological forms; needs more study. [*Orthocarpus castillejoides* Benth.]

ssp. ***ambigua*** (p. 1023) Pl not fleshy. **INFL** gen ± yellow; bract tips acute or rounded. **FL**: corolla yellow, lower lip teeth ± 1 mm. **SEED** ± 1 mm. Coastal bluffs, grassland; < 100 m. Range of sp.

ssp. ***humboldtiensis*** (Keck) Chuang & Heckard (p. 1023) HUMBOLDT BAY OWL'S-CLOVER Pl ± fleshy, few-branched, ascending. **INFL** gen pink to rose-purple; bract tips rounded to truncate. **FL**: corolla pink or rose-purple, lower lip teeth 2–3 mm. **SEED** 2–2.5 mm. Salt marshes; ± 0 m. n NCo (Humboldt Bay), n CCo (Point Reyes). [*O. c.* var. *h.* Keck] Threatened by coastal development.

ssp. ***insalutata*** (Jepson) Chuang & Heckard (p. 1023) Pl much-branched, not fleshy. **INFL** ± rose-purple; bract tips acute or rounded. **FL**: corolla ± rose-purple, lower lip teeth 1.5–2 mm. **SEED** ± 1 mm. Grassy coastal bluffs; < 100 m. c CCo (Monterey, San Luis Obispo cos.). [*O. c.* var. *i.* Jepson]

C. angustifolia (Nutt.) G. Don (p. 1023) DESERT INDIAN PAINTBRUSH Per 15–45 cm, few-branched, gray-green, ± bristly, nonglandular. **LF** 20–70 mm, linear-lanceolate; lobes 0–5, widely spreading. **INFL** 4–15 cm; bracts 20–30 mm, lobes 3–5, bright red to yellowish orange (violet). **FL**: calyx 15–25 mm, divided 1/4–1/3 in back and front, ± 1/7 on sides, long-nonglandular- and short-glandular-hairy, lobes obtuse to rounded; corolla 20–35 mm, beak ± = tube, yellowish green, back puberulent, margins reddish, lower lip 2–3 mm, dark green, incl; stigma 2-lobed. **FR** 10–15 mm. **SEED** 1.5-2 mm; coat deeply netted, most walls ladder-like. $2n$= 24,48. Dry sagebrush scrub, pinyon/juniper woodland; 1000–3000 m. ne SnBr, GB, w DMoj; to OR, MT, WY, Colorado, NM. [*C. martinii* Abrams ssp. *ewanii* (Eastw.) Munz ($2n$=24)] Type from ID; earliest epithet in the widespread *C. chromosa* Nelson complex. More study needed.

C. applegatei Fern. Per 10–80 cm, few-branched, green to dusty, gen very short-glandular-sticky-hairy and long-nonglandular-hairy. **STS** gen ± few. **LF** 20–70 mm, ± lanceolate, gen wavy-margined; lobes 0–3. **INFL** 5–20 cm; bracts 15–25 mm, lobes 0–7, bright red to yellowish. **FL**: calyx 13–25 mm, divided 1/3–1/2 in back and front, ± 1/8 on sides, lobes acute to rounded; corolla 20–40 mm, beak ± = tube, back puberulent, margins reddish, lower lip 1–3 mm, dark green, incl or exserted; stigma slightly 2-lobed, well

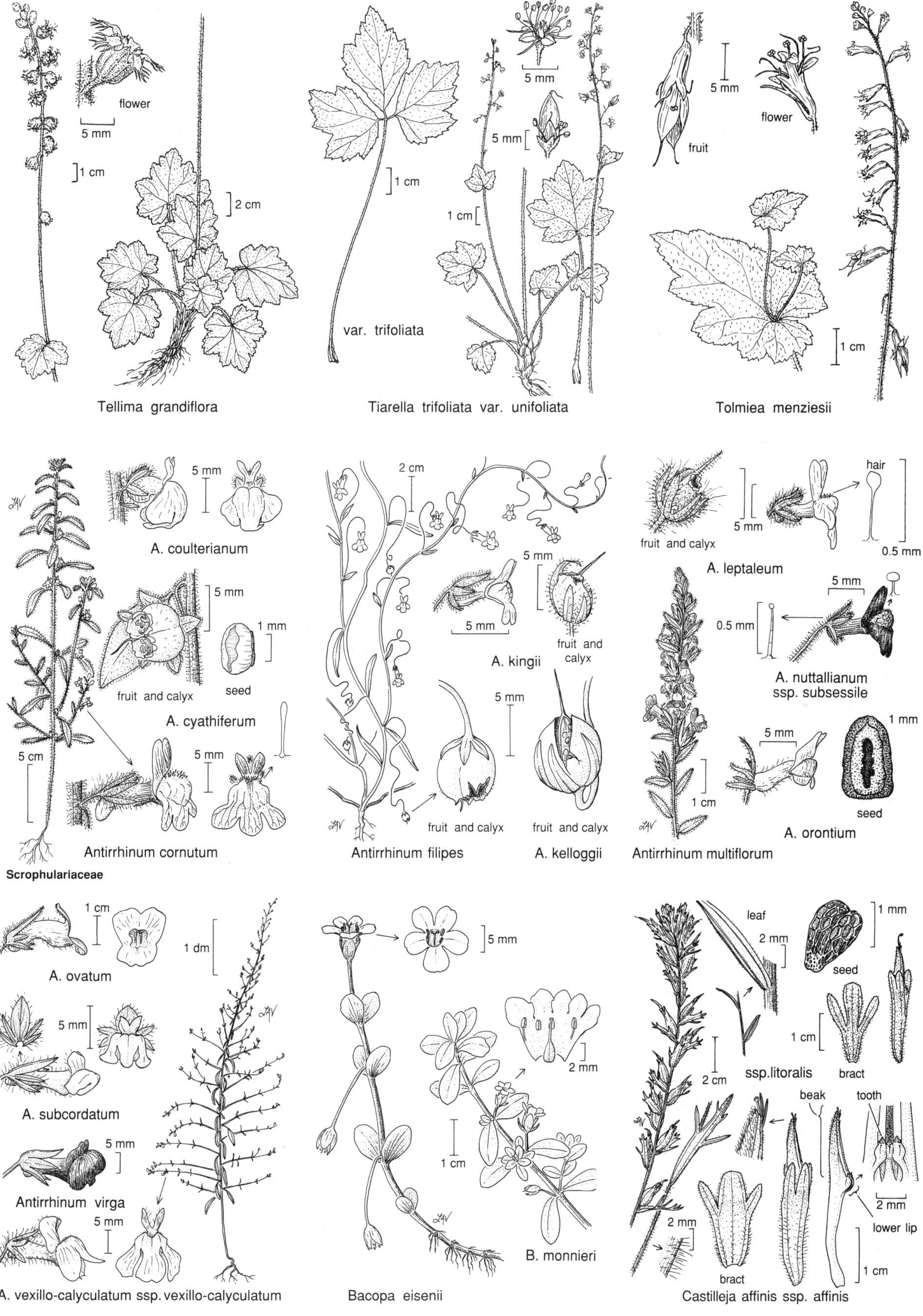
flower
5 mm
1 cm
2 cm
Tellima grandiflora
1 cm
5 mm
5 mm
1 cm
var. trifoliata
Tiarella trifoliata var. unifoliata
fruit
5 mm
flower
1 cm
Tolmiea menziesii
5 mm
A. coulterianum
5 mm
1 mm
fruit and calyx
seed
A. cyathiferum
5 cm
5 mm
Antirrhinum cornutum
2 cm
5 mm
5 mm
A. kingii
fruit and calyx
5 mm
fruit and calyx
fruit and calyx
Antirrhinum filipes
A. kelloggii
hair
5 mm
fruit and calyx
0.5 mm
A. leptaleum
5 mm
0.5 mm
A. nuttallianum
ssp. subsessile
1 mm
5 mm
1 cm
seed
A. orontium
Antirrhinum multiflorum
Scrophulariaceae
1 cm
A. ovatum
1 dm
5 mm
A. subcordatum
5 mm
Antirrhinum virga
5 mm
A. vexillo-calyculatum ssp. vexillo-calyculatum
5 mm
2 mm
1 cm
B. monnieri
Bacopa eisenii
leaf
2 mm
1 mm
seed
1 cm
2 cm
ssp.litoralis
bract
beak
tooth
2 mm
lower lip
2 mm
bract
1 cm
Castilleja affinis ssp. affinis

exserted. **FR** 8–15 mm. **SEED** 1–1.5 mm; coat shallowly netted, loose-fitting, side walls ladder-like. Dry open forest or scrub; 300–3600 m. CA; to OR, ID, NV, n Baja CA. Highly variable complex (other sspp. outside CA), unique in combination of glandular-sticky herbage and wavy lf margins.

ssp. ***disticha*** (Eastw.) Chuang & Heckard (p. 1023) **ST** 30–80 cm. **LF**: lobes 0–3. **FL**: calyx 12–18 mm, divided > 1/3 in back and front, ± 1/6 on sides, lobes acute; corolla 25–35 mm, beak 14–18 mm. 2*n*= 24. Open coniferous forest; 2000–3000 m. c&s SNH. [*C. d.* Eastw.] Intergrades with ssp. *pinetorum*.

ssp. ***martinii*** (Abrams) Chuang & Heckard (p. 1023) **ST** 30–80 cm. **LF**: lobes 0–3. **FL**: calyx 15–25 mm, divided ± 1/3 in back and front, ± 1/12 on sides, lobes ovate, obtuse or rounded; corolla 25–40 mm, beak 12–18 mm. 2*n*=24,48,72. Dry chaparral, open yellow-pine forest, sagebrush scrub; 300–2800 m. NCoRI, e SnFrB, SCoRI, SW, SNE; n Baja CA, s NV. [*C. m.* Abrams incl var. *clokeyi* (Pennell) N. Holmgren; *C. roseana* Eastw.] Hybrids with *C. angustifolia* [*C. chromosa*] with 2*n*=48,72 from SnBr have been called *C. montigena* Heckard, Heckard's Indian paintbrush.

ssp. ***pallida*** (Eastw.) Chuang & Heckard (p. 1023) **STS** many, 10–25 cm. **LF**: lobes gen 3. **FL**: calyx 13–15 mm, divided > 1/3 in back and front, ± 1/4 on sides, lobes acute; corolla 16–22 mm, beak 7–10 mm. 2*n*=48. Dry rocky slopes and flats, red-fir forest to alpine barrens; 2400–3600 m. SNH. [*C. breweri* Fern. incl var. *p.* Eastw.]

ssp. ***pinetorum*** (Fern.) Chuang & Heckard (p. 1023) **ST** 30–60 cm. **LF**: lobes 0–3. **FL**: calyx 16–25 mm, divided 1/3–1/2 in back and front, 1/8–1/5 on sides, lobes lanceolate, acute; corolla 25–35 mm, beak 11–16 mm. 2*n*=24,48. Open coniferous forests, dry sagebrush scrub; 700–2700 m. KR, NCoRH, CaR, SN, GB, n DMtns; to OR, ID.

C. arachnoidea Greenman (p. 1023) Per 8–30 cm, white-woolly; hairs simple. **LF** 20–60 mm, linear-lanceolate; lobes 0–5. **INFL** 3–12 cm; bracts 10–25 mm, dull yellow to rusty red, cobwebby-woolly, lobes gen 3–5, rounded. **FL**: calyx 12–18 mm, subequally divided ± 1/2, lobes linear-lanceolate; corolla 12–20 mm, beak 3–5 mm, lower lip 2–4 mm, pouches shallow, pale green or purplish red, teeth pale yellow; stigma ± exserted, notched, dark. **FR** 8–12 mm. **SEED** ± 1 mm; coat shallowly netted, loose-fitting. 2*n*=24. Open summits, dry rocks; 1700–3000 m. KR, s CaRH, MP; sw OR. [ssp. *shastensis* (Eastw.) Pennell; *C. payneae* Eastw.]

C. attenuata (A. Gray) Chuang & Heckard (p. 1023) VALLEY TASSELS Ann 10–50 cm, nonglandular, spreading-hairy. **LF** 20–80 mm, ± linear; lobes 0–3. **INFL** 3–30 cm, 1–3 cm wide; bracts 15–35 mm, lobes 3, ± lanceolate, tips white or pale yellow. **FL**: calyx 10–20 mm, divided ± 1/2 in front, 1/3 on sides, 3/4 in back; corolla 10–25 mm, linear, beak 4–5 mm, straight, puberulent, lower lip 3–4 mm, pouches ± 2 mm wide, 1–1.5 mm deep, whitish, purple-dotted; stigma well incl. **FR** 7–11 mm. **SEED** < 1 mm; coat shallowly netted, loose-fitting. 2*n*=24. Grassland; gen < 1200 m. CA-FP; to B.C., n Baja CA; also c Chile. [*Orthocarpus a.* A. Gray]

C. brevistyla (Hoover) Chuang & Heckard (p. 1023) Ann 6–40 cm, nonglandular, sparsely hairy. **LF** 20–60 mm, ± linear; lobes 0–3. **INFL** 5–25 cm, 1–2 cm wide; bracts 10-20 mm, tipped pink or purplish red, lobes 3–5, ± linear. **FL**: calyx 15–20 mm, divided < 1/3 in front, 1/3 on sides, 2/3 in back; corolla 15–28 mm, linear, beak 4–6 mm, straight, puberulent, lower lip 3–5 mm, pouches gen 2 mm wide, 1–1.5 mm deep, purplish red or blotched pink; stigma well incl. **FR** 8–12 mm. **SEED** 1–1.5 mm; coat ± deeply netted, loose-fitting. 2*n*=48. Grassland; < 1200 m. s SNF, s SCoRI. Probable hybrid: *C. attenuata* × *C. exserta*.

C. campestris (Benth.) Chuang & Heckard Ann 10–30 cm, ± glabrous. **LF** 15–40 mm, linear-lanceolate, entire. **INFL** 3–15 cm, 2–3 cm wide; bracts 10–25 mm, linear-lanceolate, entire, ± green. **FL**: calyx ± 7 mm, divided 1/3 in front, 1/4 on sides, 1/2 in back, ± hairy; corolla 15–25 mm, pale yellow to orange, beak 5–6 mm, straight, lower lip 4–5 mm, pouches 6–10 mm wide, 3–4 mm deep; stigma ± exserted. **FR** 5–7 mm. **SEED** ± 7 mm; coat shallowly netted, tight-fitting. 2*n*=24. Vernal pools, moist places; < 2300 m. s NCoRI, CaR, n SN, GV, MP; s OR. [*Orthocarpus c.* Benth.]

ssp. ***campestris*** (p. 1023) **LF** narrowly linear, thin, flexible. **INFL**: bracts gen < fls. **FL**: corolla light to bright yellow; lower anther sac 1/4–1/3 upper. Habitats and range of sp.

ssp. ***succulenta*** (Hoover) Chuang & Heckard (p. 1023) SUCCULENT OWL'S-CLOVER **LF** lanceolate, thick, ± brittle. **INFL**: bracts > fls. **FL**: corolla gen deep yellow to orange (pale yellow); lower anther sac 1/2 upper. RARE. Habitats of sp.; < 750 m. s SNF, e SnJV. [*O. c.* ssp. *s.* Hoover] Threatened by urbanization, agriculture.

C. chrymactis Pennell Per 30–100 cm, few-branched, green, puberulent to strigose, nonglandular. **LF** 30–60 mm, ± lanceolate, gen entire; tip acute. **INFL** 8–10 cm, gray-puberulent; bracts 25–35 mm, lobes 3–7, purplish red, tips acute. **FL**: calyx 20–25 mm, divided 1/2 in back and front, 1/4 on sides, ± bristly, lobe tips acute; corolla 25–30 mm, beak ± = tube, back puberulent, lower lip ± 1 mm, dark green; stigma 2-lobed. **FR** 9–11 mm. n CCo (Point Reyes); native to AK. [*C. leschkeana* J. Howell] 1 CA pl; probably accidental introduction from AK. Part of *C. miniata* complex.

C. cinerea A. Gray (p. 1023) ASH-GRAY INDIAN PAINTBRUSH Per 5–15 cm, densely ashy-puberulent, stiff-hairy. **LF** 10-20 cm, linear-lanceolate; lobes gen 0(3). **INFL** 3–6 cm; bracts 12–20 mm, dusty red or purplish, densely branched-hairy, lobes 3–5, truncate or rounded. **FL**: calyx 15–20 mm, subequally divided 1/3–1/2, lobes linear, densely branched-hairy; corolla 15–18 mm, incl, beak 4–5 mm, pale yellowish, lower lip ± 2 mm, greenish, pouches shallow, teeth minute, incurved; stigma ± exserted, dark. **FR** 6–10 mm. **SEED** ± 1 mm; coat ± deeply netted, tight-fitting, walls irregularly net-thickened. RARE. Dry sagebrush scrub; 1800–2800 m. SnBr. Threatened by grazing, development, vehicles.

C. densiflora (Benth.) Chuang & Heckard Ann 10–40 cm, subglabrous. **LF** 20–80 mm; lobes 0–3, linear-lanceolate. **INFL** 3–25 cm, 2.5–4 cm wide; bracts 10–25 mm, tipped white or purplish, lobes 3–5, linear. **FL**: calyx 5–20 mm, divided 1/2 in front and on sides, 2/3 in back, lobes ± linear, tip wider; corolla 10–25 mm, pale yellow, pink, or purplish red, beak 5–6 mm, straight, puberulent, lower lip 4–5 mm, pouches 4–6 mm wide, 2–3 mm deep; stigma ± exserted, ± 2-lobed. **FR** 7–10 mm. **SEED** ± 0.5 mm; coat shallowly netted, loose-fitting. 2*n*=24 (all sspp.). Grassland; < 1400 m. NCoR, c SNF, s CCo, SCoR, SW; n Baja CA. [*Orthocarpus d.* Benth.] Highly variable; many local forms; more study needed.

ssp. ***densiflora*** (p. 1023) **INFL** gen rose-purple (cream); bracts = fls. **FL**: corolla ± incl, lower lip widened gradually, pouches longer than deep, teeth ± 2 mm. Grassland; < 800 m. NCoR, c SNF, SCoR.

ssp. ***gracilis*** (Benth.) Chuang & Heckard (p. 1023) **INFL** rose-purple; bracts < fls. **FL**: corolla exserted, lower lip widened abruptly, pouches gen shorter than deep, teeth gen < 1 mm. Grassland; < 1400 m. SCoR, SW; n Baja CA. [*O. d.* var. *g.* (Benth.) Keck]

ssp. ***obispoensis*** (Keck) Chuang & Heckard **INFL** white to pale yellow; bracts = fls. **FL**: corolla ± incl, lower lip widened gradually, pouches longer than deep, teeth ± 2 mm. Coastal grassland; < 100 m. s CCo (San Luis Obispo Co.). [*O. d.* var. *o.* Keck]

C. exserta (A.A. Heller) Chuang & Heckard PURPLE OWL'S-CLOVER Ann 10–45 cm, glandular-puberulent, stiff-hairy. **LF** 10–50 mm; lobes 5–9, ± thread-like. **INFL** 2–20 cm, 2–4 cm wide; bracts 10–25 mm, white to purplish red, lobes 5–9 (lowest pair often again 2–4-lobed), ± linear (tips wider). **FL**: calyx 10–22 mm, divided 1/2 in front, 1/3 on sides, 2/3 in back; corolla 12–30 mm, beak 6–7 mm, shaggy-hairy, tip hooked, lower lip 4–6 mm, pouches 3–8 mm wide, 3–4 mm deep; filaments puberulent; stigma ± incl. **FR** 10–15 mm. **SEED** 1–2 mm; coat deeply netted, loose-fitting. Open fields, grassland; < 1600 m. NW, SNF, GV, CW, SW, w DMoj; AZ, nw Mex. [*Orthocarpus purpurascens* Benth.] Highly variable; hybridizes with *C. attenuata*, *C. densiflora*, *C. lineariloba*. Very showy in spring.

ssp. ***exserta*** (p. 1023) **INFL**: upper bracts gen < 5 mm wide, lobes > 2 mm, tipped white, pale yellow, or purplish red. **FL**: corolla colored like bracts. 2*n*=24. Habitats and range of sp. [*O. p.* incl var. *pallidus* Keck] ssp. ***latifolia*** (S. Watson) Chuang & Heckard (p. 1023) **ST**: branches many, decumbent. **INFL** banded alternately light and dark; upper bracts gen 5–7 mm wide, lobes < 2 mm, tipped pale lavender. **FL**: corolla colored like bracts. 2*n*=24. Coastal bluffs, dunes; < 200 m. NCo, n&c CCo. [*O. p.* var. *l.* S. Watson]

ssp. ***venusta*** (A.A. Heller) Chuang & Heckard **INFL**: upper bracts gen < 5 mm wide, lobes > 2 mm, tipped deep pink or purple. **FL**: corolla pouch orange-tipped. Dry sand, washes; 600–900 m. w DMoj. [*O. p.* var. *ornatus* Jepson]

C. foliolosa Hook. & Arn. (p. 1023) WOOLLY INDIAN PAINTBRUSH Per or subshrub 30–60 cm, felt-like with white to gray, much-branched hairs. **ST** much-branched, with short, axillary shoots. **LF** 10–50 mm, ± linear; lobes 0–3, tips obtuse. **INFL** 3–20 cm; bracts 15–25 mm, lobes 0–5, orange-red (yellow-green). **FL**: calyx 15–18 mm, divided 1/3–2/5 in back and front, entire or barely notched on sides, swollen in fr; corolla 18–25 mm, beak ± = tube, exserted, back puberulent, margins pale, lower lip 2 mm, dark green, incl; stigma club-shaped, slightly 2-lobed. **FR** 10–15 mm. **SEED** 1.5–2 mm; coat deeply netted, most walls ladder-like. 2*n*=24. Dry, open, rocky slopes, edges of chaparral; < 1800 m. NCoR, SNF, CW, s ChI (Santa Catalina Island); n Baja CA. ❀DRN,DRY,SUN:7,14–17, 22–24;DFCLT.

C. grisea Dunkle (p. 1023) SAN CLEMENTE ISLAND INDIAN PAINTBRUSH Per or subshrub 40–60 cm, ash-gray, densely stellate-hairy. **ST** openly branched, with short leafy axillary shoots. **LF** 10–50 mm, ± linear; lobes 0–3. **INFL** 3–10 cm; bracts 10–20 mm, lobes 5–7, pale yellow-green. **FL**: calyx 10–20 mm, divided 1/3 in back and front, entire or slightly notched on sides, swollen in fr; corolla 15–25 mm, beak 1/4–1/3 tube, dull yellow, back puberulent, margins pale yellow, lower lip 2 mm, dark green; stigma club-shaped, well exserted. **FR** 10–12 mm. **SEED** 1–1.5 mm; coat deeply netted, most walls ladder-like. 2*n*=24. RARE. Coastal bluffs; 300 m. s ChI (San Clemente Island). [*C. hololeuca* E. Greene ssp. *g.* (Dunkle) Munz]

C. hispida Benth. ssp. ***brevilobata*** (Piper) Chuang & Heckard (p. 1023) SHORT-LOBED INDIAN PAINTBRUSH Per 10–50 cm, green or yellowish, long-nonglandular- and short-glandular-hairy or glandular-puberulent. **LF** 15–60 mm, lanceolate to ovate; lobes 0–7, tips rounded to acute. **INFL** 5–20 cm, open below; bracts 15–45 mm, lobes gen 3–5, bright red to yellow. **FL**: calyx 15–30 mm, divided ± 1/3 in back and front, 1/5–1/4 on sides, soft-hairy, lobe tips obtuse to rounded; corolla 20–35 mm, beak ± = tube, yellowish green, back puberulent, margins reddish, lower lip 1–2 mm, dark green, incl; stigma slightly 2-lobed. **FR** 8–13 mm. **SEED** 1–2 mm; coat shallowly netted, loose-fitting. 2*n*=24. UNCOMMON. Dry, open serpentine, or forest edge; 200–1700 m. nw KR; sw OR. [*C. b.* Piper] *C. h.* highly variable, widespread, intergrades with *C. angustifolia* and *C. miniata* outside CA; complex needs study.

C. lacera (Benth.) Chuang & Heckard (p. 1023) Ann 10–40 cm, glandular-puberulent and spreading-hairy. **LF** 10–50 mm, linear-lanceolate; lobes 0–5. **INFL** 3–15 cm, 2–3 cm wide; bracts 10–20 mm, ovate, green, lobes 3–7, linear-lanceolate. **FL**: calyx 8–13 mm, divided 1/2 in front and on sides, 2/3 in back, stiffly hairy and short-glandular; corolla 13–22 mm, deep yellow, beak 4–6 mm, straight, puberulent, lower lip 3–5 mm, pouches 4–8 mm wide, 3–6 mm deep, gen purple-dotted at base; stigma ± exserted, ± 2-lobed. **FR** 5–8 mm. **SEED** < 1 mm; coat ± deeply netted, loose-fitting. 2*n*=22, 24. Grassland; 200–2800 m. CaR, n&c SN, MP; s OR. [*Orthocarpus l.* Benth.]

C. lanata A. Gray ssp. ***hololeuca*** (E. Greene) Chuang & Heckard (p. 1023) WHITE-FELTED INDIAN PAINTBRUSH Per or subshrub 30–100 cm, much-branched, white or gray, felt-like with dense, white-woolly, long, interwined, slightly branched hairs. **ST** with short axillary shoots. **LF** 10–50 mm, linear, entire; tip obtuse. **INFL** 3–10 cm; bracts 15–20 mm, lobes 3, very deep, gen orange-red. **FL**: calyx 15–18 mm, divided 2/5 in back and front, entire or slightly notched on sides, swollen in fr; corolla 20–25 mm, beak ± = tube, well exserted, back glandular-puberulent, margins pale, lower lip 2–3 mm, dark green, incl; stigma club-shaped, slightly 2-lobed. **FR** ± 10 mm. **SEED** 1–1.5 mm; coat deeply netted, most walls ladder-like. RARE. Coastal scrub; 20–300 m. n ChI (all islands). [*C. h.* E. Greene] Ssp. *lanata* (n Mex to TX) larger-fld. In cult.

C. lasiorhyncha (A. Gray) Chuang & Heckard (p. 1023) SAN BERNARDINO MOUNTAINS OWL'S-CLOVER Ann gen 10–20(40) cm, glandular-puberulent, spreading-hairy. **LF** 10–30 mm, linear-lanceolate; lobes 0–3. **INFL** 2–15 cm, 2–3 cm wide, ± open; bracts 8–20 mm, lobes 3–5, linear-lanceolate, green. **FL**: calyx 8–12 mm, divided 1/3 in front and on sides, 1/2 in back; corolla 14–22 mm, yellow, beak ± 7 mm, straight, densely white-hairy, lower lip ± 5 mm, pouches 4–8 mm wide, 3–4 mm deep, teeth ± 2 mm; stigma incl. **FR** 6–9 mm. **SEED** ± 1 mm; coat ± deeply netted, loose-fitting. 2*n*=24. RARE. Meadows, flats, open forest; 1300–2300 m. SnBr, n&c PR. [*Orthocarpus l.* A. Gray]

C. latifolia Hook. & Arn. (p. 1023) MONTEREY INDIAN PAINTBRUSH Per or subshrub 30–60 cm, gray-green becoming purplish, ± bristly, mostly nonglandular. **ST** with many, short, axillary shoots. **LF** 5–20 mm, oblong to rounded, ± fleshy; lobes 0–3, tips truncate-rounded. **INFL** 5–20 cm; bracts 15–20 mm, widely wedge-shaped to widely obovate, lobes 0–3, bright red to yellow, central lobe tip truncate to rounded. **FL**: calyx 15–25 mm, divided ± 1/3–1/2 in back, more deeply in front, < 1/8 on sides, long-nonglandular- and short-glandular-hairy, lobes obtuse to rounded; corolla 20–30 mm, beak ± = or > tube, back shaggy-puberulent, margins reddish, lower lip 2 mm, dark green, incl; stigma club-shaped, ± entire. **FR** 12–20 mm. **SEED** 2–2.5 mm; coat deeply netted, most walls ladder-like. 2*n*=24. UNCOMMON. Coastal dunes, scrub; < 100 m. n&c CCo. Scattered but locally abundant in good years.

C. lemmonii A. Gray (p. 1023) Per 10–20 cm, ± spreading-hairy and glandular. **LF** 20–40 mm, linear or lanceolate; lobes 0–3. **INFL** 3–12 cm; bracts 10–15 mm, lobes 3–5, ± acute, purplish red. **FL**: calyx 16–18 mm, divided 1/2–2/3 in back and front, ± 1/8 on sides, lobes gen rounded; corolla 16–20 mm, beak 7–9 mm, pale yellow, lower lip 5–7 mm, yellow-green, pouches shallow, short, teeth whitish or purplish, erect; stigma ± 2-lobed. **FR** 7–9 mm. **SEED** 1–1.5 mm; coat shallowly netted, loose-fitting. 2*n*=24. Moist meadows; 2100–3500 m. CaRH, SNH. [*C. culbertsonii* E. Greene]

C. linariifolia Benth. (p. 1025) Per 30–100 cm, few-branched, yellowish to gray-green, gen becoming purplish, glabrous to slightly puberulent. **LF** 20–80 mm, gen linear, rolled upward; lobes 0–3. **INFL** 5–20 cm, open below; bracts 15–30 mm, lobes gen 3, narrow, < calyx, bright red to yellow. **FL**: calyx 20–35 mm, divided 1/3 in back, 2/3 in front, ± 1/8 on sides, puberulent, lobes acute; corolla 25–45 mm, beak ± = tube, yellow-green, back puberulent, margins red, lower lip 2–3 mm, dark green; stigma slightly 2-lobed. **FR** 10–15 mm. **SEED** 1.5–2 mm; coat shallowly netted, loose-fitting, side walls ladder-like. 2*n*=24,48. Dry plains, rocky slopes, sagebrush shrubland or pinyon/juniper woodland; 1000–3000 m. CaR, TR, GB, s edge DMoj; to OR, MT, NM.

C. lineariloba (Benth.) Chuang & Heckard (p. 1025) Ann 15–45 cm, spreading-hairy and glandular. **LF** 20–75 mm, linear-lanceolate; lobes 0–7. **INFL** 20–50 cm, 2.5–4 cm wide; bracts 12–25 mm, lobes 5–7, linear, tipped white, yellowish, or pale purple. **FL**: calyx 15–25 mm, subequally divided ± 2/3; corolla 15–30 mm, whitish, yellowish, or ± rose, beak 4–5 mm, straight, puberulent, lower lip 3–4 mm, pouches 4–5 mm wide, ± 2 mm deep, purple-dotted at base; stigma exserted, ± 2-lobed. **FR** 7–9 mm. **SEED** ± 1 mm; coat ± deeply netted, loose-fitting. 2*n*=20. Grassland; < 1600 m. SNF. [*Orthocarpus l.* Benth.]

C. mendocinensis (Eastw.) Pennell (p. 1025) MENDOCINO COAST INDIAN PAINTBRUSH Per, decumbent to ascending, 40–60 cm, much-branched, gray-green, shaggy-bristly, nonglandular. **ST** with leafy axillary shoots. **LF** 5–20 mm, oblong to rounded, ± fleshy, gen ± entire; tip truncate-rounded. **INFL** 5–20 cm; bracts 15–20 mm, widely wedge-shaped to widely obovate, lobes 0–3, bright red to orange-red, central lobe wide; tip truncate-rounded. **FL**: calyx 20–25 mm, divided 1/2 in back and front, < 1/8 on sides, shaggy-hairy (some hairs glandular); corolla 30–45 mm, beak ± = tube, back shaggy-puberulent, margin reddish, lower lip 2 mm, dark green, ± incl; stigma club-shaped, entire. **FR** 15–20 mm. **SEED** 2–2.5 mm; coat deeply netted, most walls ladder-like. RARE. Coastal scrub; < 100 m. c NCo (Mendocino Co.). [*C. latifolia* ssp. *m.* Eastw.] Related to *C. affinis* ssp. *litoralis*. Threatened by coastal development.

C. miniata Hook. Per 40-80 cm, few-branched, green or becoming purplish, glabrous, to long-soft-hairy above. **LF** 30–60 mm, ± lanceolate, entire; tip acute. **INFL** 3–15 cm; bracts 15–35 mm, lobes 0–5, bright red to yellowish. **FL**: calyx 10–30 mm, divided 1/2–2/3 in back, more deeply in front, ± 1/4 on sides, lobes acute; corolla 15–40 mm, beak ± = tube, yellow-green, back puberulent, margins red, lower lip 1–2 mm, dark green, slightly exserted; stigma slightly 2-lobed. **FR** 6–12 mm. **SEED** 1.5–2 mm; coat shallowly netted, loose-fitting, inner walls membranous, persistent. Moist places, esp meadows, bogs; < 3500 m. NW, CaR, SNH, c-w CW, SW, GB; to AK, Rocky Mtns.

ssp. ***elata*** (Piper) Munz (p. 1025) SISKIYOU INDIAN PAINTBRUSH **ST** slender. **INFL** pinkish to yellow-orange. **FL**: calyx 9–17 mm; corolla 15–25 mm. **FR** 6–10 mm. $2n=24$. RARE in CA. Bogs, often near serpentine; < 1400 m. nw KR (Del Norte, Siskiyou cos.); sw OR. [*C. e.* Piper]

ssp. ***miniata*** (p. 1025) **ST** stout. **INFL** gen bright red (orange-red). **FL**: calyx 15–30 mm; corolla 20–40 mm. **FR** 10–12 mm. $2n=24,48,72,96,120$. Common. Wet montane meadows, streambanks; gen 1500–3500 m. Range of sp. Pls from lowland s NCoRO (Pitkin Marsh, Sonoma Co., ± 240 m) with yellow infl have been called *C. uliginosa* Eastw., Pitkin Marsh Indian paintbrush. ❀DRN,IRR,SUN,with host:1,2,4–7,14–17;DFCLT.

C. minor (A. Gray) A. Gray Ann, ± simple, ± slender, 30–150 cm, green to grayish, variously hairy. **LF** 40–100 mm, linear-lanceolate, entire. **INFL** 10–40 cm, narrow, open below; bracts 20–50 mm, lf-like, entire, tips red, ± long-tapered; pedicels < 10 mm below (0 above). **FL**: calyx 14–28 mm, divided 2/3–3/4 in back and front, ± 1/8 on sides, soft-hairy, lobes narrow, acute; corolla 15–35 mm, beak < tube, back ± puberulent, margins pale, lower lip 2–3 mm, yellowish, exserted from calyx, spreading; stigma slightly 2-lobed. **FR** 10–15 mm. **SEED** 1–1.5 mm; coat ± deeply netted, most walls ladder-like. $2n=24$. Wet places; < 2300 m. NCoR, SnJV, c&s SNF, CCo, n SnFrB, SCo, GB; to WA, NM.

ssp. ***minor*** (p. 1025) Hairs long, soft, mostly glandular. **FL**: corolla 15–20(30) mm. Alkaline marshes; 1200–2200 m. GB; to WA, NM. [*C. exilis* Nelson]

ssp. ***spiralis*** (Jepson) Chuang & Heckard Hairs short-glandular-puberulent, long-glandular, and long-nonglandular. **FL**: corolla 25–35 mm. Wet places; < 2300 m. NCoR, c&s SNF, SnJV, CCo, n SnFrB, SCo. [*C. stenantha* A. Gray incl ssp. *spiralis* (Jepson) Munz]

C. mollis Pennell (p. 1025) SOFT-LEAVED INDIAN PAINTBRUSH Per or subshrub, ± prostrate, 30–40 cm, openly branched, woolly, intertwined-long-branched-hairy and short-simple-glandular-hairy, gray(-green). **ST** with short axillary shoots. **LF** 10–30 mm, ± ovate or oblong, entire; tip rounded. **INFL** 3–8 cm; bracts 15–20 mm, obovate, fleshy, tip rounded or 3-toothed, yellow(-green). **FL**: calyx 16–18 mm, divided ± 1/2 in back and front, ± 1/8 on sides, lobes ovate, acute; corolla 17–18 mm, beak gen < tube, yellowish green, slightly exserted, back densely puberulent, margins pale, lower lip 2 mm, green; stigma slightly 2-lobed, dark green, slightly exserted. **FR** 16 mm. **SEED** 2 mm; coat deeply netted, most walls ladder-like. $2n=24$. RARE. Coastal dunes; < 20 m. n ChI (Santa Rosa, e San Miguel islands). [Heckard et al. 1991 Madroño 38:141–142]

C. nana Eastw. (p. 1025) Per 5–25 cm, spreading-hairy, mostly nonglandular. **LF** 10–35 mm, linear-lanceolate; lobes 0–5. **INFL** 3–13 cm; bracts 15–30 mm, lobes 3–5, acuminate, yellow-green or purplish, green-margined. **FL**: calyx 12–20 mm, subequally divided ± 1/2, lobes linear-lanceolate; corolla 15–20 mm, beak 4–6 mm, pale yellow blotched purplish, lower lip 3–5 mm, pouches shallow; stigma ± exserted, dark. **FR** 8–12 mm. **SEED** 1–1.5 mm; coat shallowly netted, loose-fitting, with ladder-like thickenings. Dry, ± alpine barrens; 2400-3700 m. SNH.

C. parviflora Bong. (p. 1025) Per 15–40 cm, ± glabrous to long-nonglandular- and short-glandular-hairy. **LF** 15–50 mm, ± lanceolate to oblong; lobes 0–5. **INFL** 3–15 cm; bracts 15–25 mm, lobes 3–5, gen bright red (yellowish). **FL**: calyx 12–20 mm, divided ± 1/2 in back, more in front, 1/4–1/3 on sides, lobes acute to notched; corolla 15–28 mm, beak ± 1/2 tube, yellowish green, back sparsely puberulent, margins reddish, lower lip 1–2 mm, dark green, incl; stigma clearly 2-lobed, obvious, blackish. **FR** 6–9 mm. **SEED** 1–1.5 mm; coat shallowly netted, loose-fitting. $2n=24$. Montane to alpine meadows; 1500–3000 m. CaRH, c&s SNH; to AK. CA pls have been called *C. peirsonii* Eastw. but are like typical *C. parviflora* of n OR to AK; needs further study.

C. pilosa (S. Watson) Rydb. (p. 1025) Per 8–35 cm, often decumbent, spreading-hairy, nonglandular. **LF** 10–50 mm, linear-lanceolate; lobes 0–3. **INFL** 3–20 cm; bracts 10–30 mm, lobes 3–5, pale green or purplish, white-margined, central lobe truncate or rounded. **FL**: calyx 10–20, subequally divided ± 1/2, lobes linear-lanceolate; corolla 13–22 mm, beak 4–6 mm, pale yellow-green, lower lip 3–5 mm, pouches shallow; stigma ± exserted, dark. **FR** 8–12 mm. **SEED** ± 1 mm; coat deeply netted, tight-fitting, walls ladder-like. $2n=24$. Dry sagebrush scrub to alpine barrens; 1200–3400 m. n&c SNH, GB; to e OR, c ID, w WY. [ssp. *jussellii* (Eastw.) Munz; *C. psittacina* (Eastw.) Pennell] Highly variable; forms intergrade; needs further study.

C. plagiotoma A. Gray (p. 1025) MOJAVE INDIAN PAINTBRUSH Per 30–60 cm, gray-green, becoming ± maroon, puberulent (esp lvs); hairs branched. **LF** 20–50 mm, ± linear; lobes 3–5. **INFL** 3–20 cm; bracts 13–20 mm, white-woolly lobes 3–5, central lobe wide, truncate, green. **FL**: calyx 12–18 mm, pale yellow, white-woolly, divided 1/8 in back, 1/4 in front, ± 1/2 on sides, front lobes ± 2 mm > back lobes; corolla 12–20 mm, beak ± = tube, incl, yellowish, back puberulent, margins pale, lower lip 1 mm, pale green; stigma barely exserted, head-like. **FR** ± 10 mm. **SEED** 1–1.5 mm; coat deeply netted, tight-fitting, most walls ladder-like. UNCOMMON. Dry sagebrush scrub, pinyon woodland; 300–2500 m. s SN, s SnJV, SCoRI, TR, DMoj.

C. praeterita Heckard & Bacigal. (p. 1025) Per 10–45 cm, much-branched, stiff-hairy, nonglandular (exc infl). **LF** 30–50 mm, ± linear; lobes 0–3. **INFL** 8–14 cm, 1.5–2 cm wide; bracts 15–25 mm, pale green, tipped lemon yellow or pale red, lobes 3(5), central lobe truncate to rounded. **FL**: calyx 14–18 mm, divided ± 1/2 in back and front, 1/8 on sides, lobes acute to rounded; corolla 13–16 mm, incl or exserted, beak ± 5 mm, puberulent, margins yellowish or dark purple, lower lip 2 mm, pouches narrow, teeth green, incurved; stigma ± exserted, ± 2-lobed. **FR** 8–10 mm. **SEED** 1–1.5 mm; coat shallowly netted, loose-fitting. $2n=24$. Dry places, esp near *Artemisia rothrockii* meadows; 2200–3400 m. s SNH.

C. pruinosa Fern. (p. 1025) Per or subshrub 30–80 cm, few-branched, ± gray; hairs gen dense, branched. **LF** 20–80 mm, ± lanceolate; lobes 0–3, tips obtuse. **INFL** 3–20 cm; bracts 10–25 mm, lobes 0–5, gen bright red to orange-red. **FL**: calyx 13–20 mm, divided 1/3–1/2 in back and front, 1/5–1/4 on sides, lobes obtuse to ± acute; corolla 25–32 mm, beak 1–2 × tube, back puberulent, margins reddish, lower lip 1–2 mm, dark green, incl or ± exserted; stigma unlobed, club-shaped. **FR** 8–15 mm. **SEED** 1.5–2 mm; coat deeply netted, loose-fitting, side walls ladder-like, inner walls membranous, splitting. $2n=48$. Dry, open serpentine or forest edge; 500–2200 m. NW, CaR, n&c SN, SnGb, MP; s OR. Pls with ± wider calyx lobes from SnGb that key here have been called *C. gleasonii* Elmer, Mount Gleason Indian paintbrush, probably a polyploid derivative of *C. affinis* ssp. *a.* × *C. foliolosa*. Highly variable and confusing complex; needs further study.

C. rubicundula (Jepson) Chuang & Heckard CREAM SACS Ann 20–70 cm, spreading-hairy and glandular. **LF** 20–80 mm, lanceolate; lobes 0–7. **INFL** 5–15 cm, 3–4 cm wide, dense; bracts 15–30 mm, ovate, green, lobes 5–9, ± lanceolate. **FL**: calyx 8–10 mm, subequally divided 1/2; corolla 20–28 mm, white, pink, or yellow, beak 5–7 mm, straight, puberulent, lower lip 4–6 mm, pouches 8–10 mm wide, 4–6 mm deep, gen purple-dotted at base; stigma = beak, ± 2-lobed. **FR** 7–10 mm. **SEED** < 1 mm; coat ± deeply netted, loose-fitting. $2n=24$. Open grassland; < 900 m. NCoR; sw OR. [*Orthocarpus r.* Jepson]

ssp. ***lithospermoides*** (Benth.) Chuang & Heckard (p. 1025) **FL**: corolla yellow. Habitats and elevations of sp. NCoR; sw OR. [*O. l.* Benth. var. *l.*]

ssp. ***rubicundula*** **FL**: corolla white, becoming pinkish. Habitats and elevations of sp. s NCoRI. [*O. l.* Benth. var. *bicolor* (A.A. Heller) Jepson]

2 mm
habit
5 cm
habit
2 mm
ssp. insalutata
ssp. humboldtiensis
1 mm
tooth
5 cm
5 mm
Castilleja ambigua ssp. ambigua
bracts
ssp. pallida
ssp. disticha
2 cm
5 mm
calyx
ssp. martinii
1 mm
5 mm
bract
ssp. pinetorum
Castilleja applegatei
flower
C. angustifolia
2 cm
0.5 mm
seed
bract
5 mm
flower
C. attenuata
hairs
2 mm
0.5 mm
seed
C. brevistyla
5 mm
bract
calyx
corolla
Castilleja arachnoidea
5 mm
0.5 mm
seed
bract
2 mm
0.5 mm
hair
1 cm
C. cinerea
1 cm
bract
anthers
1 cm
1 mm
ssp. succulenta
ssp. campestris
Castilleja campestris
5 mm
bracts
calyx
5 mm
corolla
ssp. gracilis
2 cm
2 cm
0.5 mm
seed
5 mm
bract
flower
Castilleja densiflora ssp. densiflora
2 mm
anther
1 mm
5 mm
2 cm
bract
calyx
corolla
5 mm
upper bract
ssp. exserta
ssp. latifolia
Castilleja exserta
1 cm
bract
1 dm
5 mm
1 cm
5 mm
2 mm
bract
Castilleja foliolosa
C. hispida ssp. brevilobata
C. grisea
top view
5 mm
5 mm
bract
Castilleja lacera
1 cm
5 mm
bract
C. lanata ssp. hololeuca
bract
5 mm
C. lasiorhyncha
5 mm
leaf
bract
flower
Castilleja latifolia
1 cm
bract
2 mm
C. lemmonii

C. schizotricha Greenman (p. 1025) SPLIT-HAIR INDIAN PAINT-BRUSH Per 8–15 cm, ± white-woolly with branched hairs. **LF** 5–20 mm, linear-lanceolate; lobes 0–3. **INFL** 3–8 cm; bracts 10–20 mm, pinkish to dusty red, lobes 3, obtuse to rounded. **FL**: calyx 13–18 mm, subequally divided ± 1/2; corolla 15–20 mm, tomentose, distal 1/2 purplish red, beak 4–5 mm, lower lip ± 4 mm, pouches shallow; stigma ± exserted, ± notched, dark. **FR** 8–10 mm. **SEED** ± 1 mm; coat shallowly netted, loose-fitting. Decomposed granite or marble; 1500–2300 m. KR (Siskiyou Co.); sw OR. [*C. arachnoidea* ssp. *s.* (Greenman) Pennell]

C. subinclusa E. Greene Per 30–120 cm, (gray-) green, becoming purplish, glabrous to long-hairy and short-glandular. **ST** with short axillary shoots. **LF** 30–80 mm, ± lanceolate; lobes 0–3. **INFL** 6–40 cm, open below; bracts 20–60 mm, gen entire, tips bright red to orange-red; pedicels < 10 mm (0 above). **FL**: calyx 20–32 mm, divided 1/6–1/3 in back, 2/3 in front, < 1/3 on sides, glandular-puberulent and long-hairy, lobes acute to rounded; corolla 30–50 mm, beak ± = tube, yellow-green, margins rose-pink, beak densely shaggy-puberulent, lower lip 2–3 mm, exserted from calyx, dark green or purplish; stigma unlobed. **FR** 10–15 mm. **SEED** ± 2 mm; coat deeply netted, tight-fitting, most walls ladder-like. Coastal scrub or chaparral; < 2200 m. s NCo, SNF, Teh, n CCo, w SnFrB, SCoRI, WTR, SnGb, s PR; n Baja CA. Range discontinuous.

ssp. ***franciscana*** (Pennell) Chuang & Heckard (p. 1025) **INFL** strongly 2-colored. **FL**: calyx divided < 1/5 on sides, lobes linear to narrowly deltate, obviously bent upward; corolla yellow-orange. $2n$=24,48,72, 96. Coastal scrub; < 100 m. s NCo (s Mendocino, Sonoma cos.), n CCo (to Santa Cruz Co.), w SnFrB. [*C. f.* Pennell] Closely related to more inland or more southern ssp. *s.* ❀DRN,SUN,IRR,with host:15–17;DFCLT.

ssp. ***subinclusa*** (p. 1025) **INFL** ± uniformly colored. **FL**: calyx divided 1/5–1/3 on sides, lobes long-tapered, barely bent upward; corolla ± red. $2n$=±72. Open chaparral; < 2200 m. SNF, Teh, SCoRI, WTR, SnGb, s PR; n Baja CA. [*C. jepsonii* Bacigal. & Heckard]

C. tenuis (A.A. Heller) Chuang & Heckard (p. 1025) Ann 10–45 cm, spreading-hairy, glandular above. **LF** 10–40 mm, linear-lanceolate; lobes 0–3(5). **INFL** 5–25 cm, 1–3 cm wide; bracts 10–30 mm, ovate, green, lobes 3–7, ± lanceolate. **FL**: calyx 8–12 mm, divided 1/3 in front and on sides, 1/2 in back; corolla 12–20 mm, white or yellow, beak 4–5 mm, straight, puberulent, lower lip 3–4 mm, pouches 2–4 mm wide, ± 2 mm deep; stigma well incl. **FR** 6–9 mm. **SEED** ± 1 mm; coat ± deeply netted, loose-fitting. $2n$=24,48. Moist flats, meadows; 1000–2800 m. NCoR, CaR, SNH, MP, rare in SnBr and PR; to AK. [*Orthocarpus hispidus* Benth.]

C. wightii Elmer (p. 1025) Per 30–80 cm, much-branched, (yellow-)green tinged purplish, densely long-bristly and sticky-puberulent. **ST** with many short axillary shoots below infl. **LF** 20–60 mm, gen crowded, lanceolate to ± ovate; lobes 0–3. **INFL** 5–20 cm; bracts 10–25 mm, lobes 3, bright red to yellow. **FL**: calyx 15–28 mm, divided 1/3–1/2 in back and front, 1/8–1/4 on sides, lobes acute or rounded; corolla 20–30 mm, beak 1–2 × tube, back shaggy-hairy, margins reddish to yellowish, lower lip 2 mm, dark green, incl; stigma slightly 2-lobed. **FR** 10–15 mm. **SEED** 2 mm; coat deeply netted, most walls ladder-like. $2n$=24,48. Coastal scrub; < 300 m. c&s NCo, n CCo, SnFrB. ❀DRN,SUN, IRR, with host:15–17;DFCLT.

COLLINSIA

Elizabeth Chase Neese

Ann, often glandular, sometimes brown-staining. **LVS** opposite; lower petioled. **INFL** bracted, often interrupted; fls 1-many in lf axils. **FL**: calyx 5-lobed; corolla ± pea-like, gen glabrous outside, tube short, throat ± angled to tube, ± pouched on upper side, lips gen ± = throat, upper lobes 2, ± reflexed, gen paler, lower lobes 3, lateral spreading, central lobe keeled, enclosing stamens and style; stamens 4, attached unequally near throat base; staminode gland-like. **FR** septicidal and loculicidal (valves 2-lobed). **SEEDS** gen few, ± oblong, gen plump; inner surface ± hollow. ± 18 spp.: N.Am, esp CA. (Zaccheus Collins, 1764–1831, Philadelphia botanist) Late-season fls gen atypically small.

1. Fls crowded in whorls; lower pedicels < calyx
 2. Corolla ± uniformly purple; upper lip with winged appendages > 1 mm near base .... ***C. greenei***
 2′ Upper lip or entire corolla paler; upper lip wingless
  3. Upper filaments with curved basal spur > 1 mm projecting into pouch; corolla throat as wide as long; calyx lobes gen sharp-tipped .... ***C. heterophylla***
  3′ Filaments ± unspurred; corolla throat longer than wide; calyx lobes gen ± blunt-tipped
   4. Upper lip < 1/2 lower
    5. Lf glabrous below; infl ± head-like; corolla upper lobes ± 1 mm, dry, brownish, lateral lobes glabrous, throat barely angled to tube, pouch low; infl glandless; NCo .... ***C. corymbosa***
    5′ Lf hairy below; infl gen of several whorls; corolla upper lobes >> 1 mm, red-banded, lateral lobes gen hairy, throat strongly angled to tube, pouch prominent; infl densely glandular, staining; widespread ***C. tinctoria***
   4′ Upper lip ± = lower in fully open fls
    6. Corolla bluish, not drying veiny, upper lobes notched, lateral lobes and tip of middle lobe gen hairy; lvs thin, entire to toothed; SnBr, PR .... ***C. concolor***
    6′ Corolla gen white to ± purplish, gen drying veiny, upper lobes ± toothed, often back-to-back, lower lip gen glabrous; lvs thick, crenate; widespread .... ***C. bartsiifolia***
     7. Corolla 15–20 mm; seed < 1 mm; esp n CA .... var. ***bartsiifolia***
     7′ Corolla 9–14 mm; seed ± 1.5 mm; esp s CA .... var. ***davidsonii***
1′ Fls solitary or whorled, not very crowded; lower pedicels > calyx
 8. Corolla ± uniformly dark purple, upper lip ± 1/2 lower, with winged appendages > 1 mm near base .... ***C. greenei***
 8′ Upper lip or entire corolla paler, upper lip ± = lower, wingless
  9. Lvs gen ± lanceolate-deltate, gen coarsely toothed; lowest pedicels 1–2 per node, >> calyx, upper 3+ per node, shorter — n&c CCo, n SCoRO .... ***C. multicolor***
  9′ Lvs linear to ovate, gen entire or crenate; pedicels ± equal
   10. Infl ± glandless
    11. Calyx lobes gen blunt-tipped .... ***C. parryi***
    11′ Calyx lobes sharp-tipped
     12. Upper filaments and middle corolla lobe glabrous; calyx ± = fr, lobes not widely spreading; seeds plump, oblong, unwinged

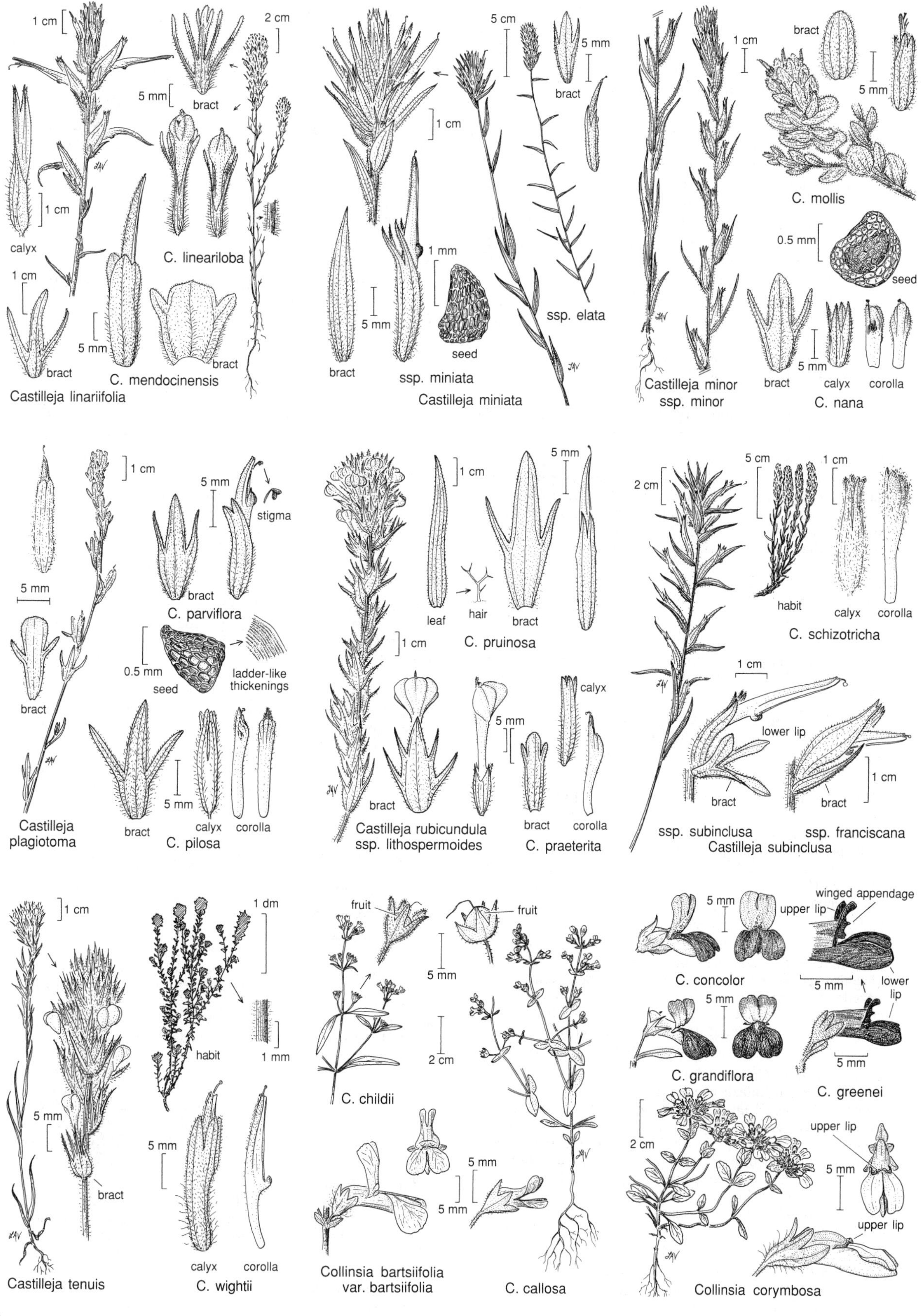

1 cm
2 cm
5 mm
bract
1 cm
calyx
1 cm
C. lineariloba
5 mm
bract
bract
C. mendocinensis
Castilleja linariifolia
5 cm
5 mm
bract
1 cm
1 mm
5 mm
seed
ssp. elata
bract
ssp. miniata
Castilleja miniata
1 cm
bract
5 mm
C. mollis
0.5 mm
seed
5 mm
Castilleja minor
ssp. minor
bract
calyx
corolla
C. nana
1 cm
5 mm
stigma
5 mm
bract
C. parviflora
0.5 mm
seed
ladder-like
thickenings
bract
5 mm
Castilleja
plagiotoma
bract
calyx
corolla
C. pilosa
5 mm
1 cm
leaf
hair
bract
1 cm
C. pruinosa
calyx
5 mm
bract
Castilleja rubicundula
ssp. lithospermoides
bract
corolla
C. praeterita
2 cm
5 cm
1 cm
habit
calyx
corolla
C. schizotricha
1 cm
lower lip
1 cm
bract
bract
ssp. subinclusa
ssp. franciscana
Castilleja subinclusa
1 cm
1 dm
habit
1 mm
5 mm
bract
5 mm
Castilleja tenuis
calyx
corolla
C. wightii
fruit
fruit
5 mm
2 cm
C. childii
5 mm
5 mm
Collinsia bartsiifolia
var. bartsiifolia
C. callosa
5 mm
upper lip
winged appendage
C. concolor
5 mm
lower
lip
5 mm
C. grandiflora
5 mm
C. greenei
2 cm
upper lip
5 mm
upper lip
Collinsia corymbosa

13. Corolla 8–15 mm, main lobes gen 2–6 mm wide, obovate, throat strongly angled to tube . . . *C. grandiflora*
13′ Corolla 4–8 mm, main lobes ± 1 mm wide, oblong, throat barely angled to tube . . . . *C. parviflora*
12′ Upper filaments and middle corolla lobe hairy; calyx gen >> fr, lobes widely spreading in fr; seeds flat, roundish, ± winged — fr gen red-blotched . . . . *C. sparsiflora*
14. Largest corollas > 13 mm . . . . var. *arvensis*
14′ Corollas < 13 mm
15. Corolla 5–8 mm, throat barely angled to tube, pouch hidden by calyx . . . . var. *collina*
15′ Corolla 7–13 mm, throat strongly angled to tube, pouch evident . . . . var. *sparsiflora*
10′ Infl conspicuously glandular
16. Pedicels spreading to ascending in fr; upper most bracts > 2 mm
17. Middle and upper lvs oblong to ovate, gen < 6 × longer than wide
18. Pl fleshy; upper lvs clasping; seeds 6–8, ± 2 mm; calyx lobes ± = fr . . . . *C. callosa*
18′ Pl not fleshy; upper lf bases tapered; seeds 2, ± 3 mm; calyx lobes > fr . . . . *C. childii*
17′ Middle and upper lvs ± linear, gen > 6 × longer than wide
19. Corolla gen 8–12 mm, >> 2 × calyx, strongly angled to tube, pouch prominent . . . . *C. linearis*
19′ Corolla < 8 mm, gen < 2 × calyx, barely angled to tube, pouch ± hidden by calyx . . . . *C. rattanii*
16′ Some pedicels strongly reflexed in fr; upper most bracts 0–2 mm . . . . *C. torreyi*
20. Corollas < 7 mm . . . . var. *wrightii*
20′ Some corollas > 7 mm
21. Middle corolla lobe << lateral lobes . . . . var. *brevicarinata*
21′ Middle corolla lobe ± = lateral lobes
22. Lvs elliptic to ovate (gen 2–5 × longer than wide) . . . . var. *latifolia*
22′ Lvs ± linear (gen > 6 × longer than wide) . . . . var. *torreyi*

C. bartsiifolia Benth. Pl 7–35 cm. **LF** gen 1–4 cm, ± oblong, thickish, obtuse, crenate, rolled under, gen finely hairy. **INFL** interrupted, ± finely glandular or shaggy; whorls dense; pedicels < calyx. **FL**: calyx lobes ± blunt; corolla glabrous outside, gen whitish to pinkish lavender (purplish), veiny when pressed, throat longer than wide, hairy inside, lips ± equal; upper lobes ± oblong, toothed, often back-to-back, lateral lobes obovate, notched; upper filaments hairy, spur 0–0.5 mm. **SEEDS** many, ± plump. *n*=7. Open sandy places; < 1300 m. NW, CaRF, SNF, CW, w DMoj. Variable; sometimes intergrades with *C. corymbosa, C. heterophylla, C. tinctoria*; more study needed.

var. ***bartsiifolia*** **FL**: corolla 15–20 mm, gen white to pale lavender. **SEED** < 1 mm. Habitats of sp; gen < 600 m. NW, CaRF, SNF, CW. ❀TRY.

var. ***davidsonii*** (Parish) V. Newsom **FL**: corolla 9–14 cm, gen pale lavender to purple. **SEEDS** ± 1.5 mm. Habitats of sp.; 500–1300 m. s SNF, SCoRI, w DMoj. Pls intermediate to *C. tinctoria* have been called *C. stricta* E. Greene. ❀TRY.

C. callosa Parish Pl 4–25 cm, stout, fleshy. **LVS** gen < 3 cm, oblong to ovate, obtuse, gen entire, ± rolled under; middle and upper lvs clasping. **INFL** open, glandular; bracts > 2 mm; fls 1–3 per node, pedicels > calyx, not reflexed. **FL**: calyx in fr gen > 5 mm wide, base ± truncate, lobes ± = fr; corolla gen 7–9 mm, lavender-blue. **SEEDS** 6–8, ± 2 mm. Disturbed rocky slopes, open chaparral, sagebrush scrub, pinyon/juniper or pine woodland; 1000–2300 m. s SNH, Teh, TR, nw DMtns.

C. childii A. Gray Pl 8–35 cm. **LF** oblong to (ob)lanceolate, gen flat, entire or finely toothed; base tapered. **INFL** open, densely glandular; bracts > 2 mm; pedicels 2–5 per node, > calyx, not reflexed. **FL**: calyx in fr < 5 mm wide, base rounded, lobes > fr; corolla 6–8 mm, whitish or pale lavender; upper lip ± = lower. **SEEDS** 2, ± 3 mm. Shaded slopes, open oak and mixed coniferous woodlands; 1000–2200 m. c&s SN, Teh, SCoRO, TR, PR.

C. concolor E. Greene Pl 15–40 cm. **LF** narrowly oblong to widely lanceolate, thin, gen entire (to toothed), subglabrous. **INFL** interrupted, finely hairy to shaggy, gen finely glandular; whorls dense; pedicels gen < calyx. **FL**: calyx gen long-hairy, lobes widely acute, ciliate; corolla 11–15 mm, blue to bluish purple, upper lip evenly purple-dotted in a triangular white area near base, lateral lobes obovate, notched, gen sparsely hairy, middle lobe gen sparsely hairy at tip, throat longer than wide, hairy inside; upper filaments hairy, spur 0–0.5 mm. **SEEDS** many, ± round, flat. *n*=7. Openings and margins of chaparral, oak or pinyon/juniper woodland; 300–1700 m. SnBr, PR; Baja CA. ❀TRY.

C. corymbosa Herder ROUND-HEADED CHINESE HOUSES Pl 5–25 cm. **ST**: branches gen long, decumbent, gen reddish, densely scaly-hairy. **LF** lanceolate to ovate, thickish, obtuse, crenate, subglabrous to finely gray-hairy. **INFL** gen head-like, glandless; pedicels < calyx. **FL**: calyx ± hairy, lobes blunt; corolla 14–22 mm, gen whitish, throat longer than wide, barely angled to tube, pale lavender, gen undotted, hairy inside, pouch low, smoothly arched, upper lobes reflexed ± 1 mm, brownish, dry, papery, lateral lobes spoon-shaped; filaments hairy, spur 0. **SEEDS** many, ± cupped. *n*=7. RARE. Coastal sand; 0–20 m. NCo (scattered), formerly n CCo, where transitional to *C. bartsiifolia*.

C. grandiflora Lindley Pl 6–35 cm. **LF** narrowly oblong to lanceolate, subentire. **INFL** open, glabrous to finely glandular or scaly-hairy; pedicels 1–4 per node, gen > calyx, ascending to reflexed. **FL**: calyx lobes acuminate, ± = fr; corolla 8–15 mm, bluish, upper lip pale at center, main lobes widely obovate, notched, middle lobe glabrous, throat strongly angled to tube, longer than wide, pouch prominent, angular; filament spur 0. **SEEDS** gen 4. *n*=7. Gravelly or grassy margins of coniferous woodlands; 400–1600 m. w NW; to B.C. [var. *pusilla* A. Gray] ❀TRY.

C. greenei A. Gray Pl 10–30 cm, gen purplish, ± candelabra-like. **LF** narrowly (ob)lanceolate to ovate, entire to toothed, obtuse. **INFL** interrupted, glandular; pedicels 1–5 per node, ± = calyx. **FL**: calyx lobes blunt, > fr; corolla gen 10–15 mm, ± evenly dark purplish; upper lip ± 1/2 lower, with forward-projecting, 2-lobed, 1–1.5 mm winged appendages near base. **SEEDS** 2–4, wafer-like, ± cupped. *n*=7. Open chaparral or coniferous forest, serpentine slopes; 300–2500 m. NW. ❀TRY.

C. heterophylla Buist (p. 1031) CHINESE HOUSES Pl 10–50 cm. **LF** lanceolate-deltate, toothed (often deeply lobed in seedlings). **INFL** interrupted, glabrous to hairy, ± glandular; whorls dense; pedicels < calyx. **FL**: calyx lobes gen acute, glabrous to shaggy; corolla 10–20 cm, gen glabrous outside, throat hairy inside, strongly angled to tube, as wide as long, pouch prominent, squarish, upper lip white to lavender or tipped dark violet, wine-spotted and gen reddish lined near base, lower lip whitish to rose-purple, middle lobe gen with darker red tip; upper filaments hairy, spur 1–2 mm, curved into pouch. **SEEDS** many, ovate, ± flat. *n*=7. Common. Shady places; < 1000 m. CA-FP; n Baja CA. Pls with very short upper lip have been called var. *austromontana* Newsom. May hybridize with *C. bartsiifolia, C. multiflora*, and *C. tinctoria*; more study needed. ❀DRN:**5,6,15–17,22–24**&SHD:**7,14,18–21**&IRR:8–11.

C. linearis A. Gray (p. 1031) Pl 10–40 cm. **LF** linear to narrowly oblanceolate, rolled under. **INFL** open, finely scaly-hairy and spreading-glandular; bracts > 2 mm; pedicels gen 1–3(5) per node, > calyx, ascending. **FL**: calyx lobes blunt, ± = fr; corolla gen 8–12 (15) mm, >> 2 × calyx, white to blue-purple, throat strongly angled to tube, as wide as long, pouch squarish, prominent. **SEEDS** ± plump, gen unwinged. *n*=7. Open coniferous forest; 200–2000 m. KR, c SNH (near Wawona, Mariposa Co.); s OR. ❀TRY.

C. multicolor Lindley & Paxton (p. 1031) SAN FRANCISCO COLLINSIA Pl gen 30–60 cm. **ST** loosely branched, weak. **LVS**: middle and upper lanceolate-deltate, clasping, coarsely toothed. **INFL** ± glandular-clammy, lowermost pedicels 1–2 per node, >> calyx, upper ones 3+, ± crowded, ± = calyx. **FL**: calyx lobes acute; corolla 12–18 mm, throat longer than wide, pouch rounded, not prominent, upper lip whitish, not or faintly dotted and lined, lower lip lavender to bluish purple, lateral lobes obovate, notched, middle lobe sometimes sparsely hairy; upper filaments hairy, spur 0–0.5 mm. **SEEDS** 8+, ± plump. UNCOMMON. Moist, ± shady scrub, forests; < 250 m. n&c CCo, n SCoRO [*C. franciscana* Bioletti] ❀TRY.

C. parryi A. Gray (p. 1031) Pl 10–40 cm. **ST** finely hairy. **LVS** ± lanceolate, obtuse, entire to crenate; upper clasping. **INFL** ± glandless; pedicels gen 1–3(5) per node, gen >> calyx. **FL**: calyx lobes bluntish, gen ciliate, ± = fr; corolla 4-10 mm, blue-violet to lavender (white), glabrous; upper filaments sparsely short-spreading-hairy. **SEEDS** 8–12, sometimes slightly winged. *n*=7. Open chaparral, sagebrush scrub, mixed woodland; 500–1600 m. c SCoRO, TR. [*C. antonina* Hardham incl ssp. *purpurea* Hardham, San Antonio collinsia]

C. parviflora Lindley (p. 1031) BLUE-EYED MARY Pl 3–40 cm. **LF** ± linear-lanceolate, obtuse, gen rolled under. **INFL** open, leafy, glabrous to sparsely and finely glandular; pedicels > calyx, gen 1 per node below, 3–5 per node above, often reflexed. **FL**: calyx > 2/3 corolla, ± = fr, lobes sharply acute to acuminate; corolla 4–8 mm, throat barely angled to tube, tube and throat white, narrowed to lips, pouch angular, ± hidden by calyx, upper lip whitish or blue-tipped, main lobes ± 1 mm wide, oblong, gen blue (purplish). **SEEDS** gen 4. *n*=7. Common. Moist, ± shady places in mtns; 800–3500 m. CA-FP, Wrn; to B.C., e Can, Colorado. ❀TRY.

C. rattanii A. Gray (p. 1031) Pl 8–40 cm. **LF** rolled under, ± linear, gen entire (crenate), gen finely hairy and gray-green above, purplish and subglabrous below. **INFL** open, scaly-hairy and spreading-glandular; pedicels 1–3(5) per node, ascending, > calyx; bracts > 2 mm. **FL**: calyx lobes ± blunt; corolla 4–8 mm, < 2 × calyx, gen purplish lavender (white), glabrous or keel finely glandular, throat barely angled to tube, pouch ± hidden by calyx. **SEEDS** 2–6, plump to ± flat and narrowly winged. *n*=7. Open coniferous forest; 100–1500 m. NW; to WA.

C. sparsiflora Fischer & C. Meyer Pl 5–30 cm. **LF** gen linear to oblong, entire. **INFL** open, glabrous to finely hairy, nonglandular; pedicels 1–2(3) per node, gen >> calyx. **FL**: calyx lobes sharply acute to long-tapered, gen ciliate, gen >> fr; corolla gen lavender to purple (white), middle lobe sparsely long-hairy near tip; upper filaments (sparsely) short-spreading-hairy. **FR** spheric; top red-blotched, revealed by spreading calyx lobes. **SEEDS** 4–12, roundish, flat, gen cupped, winged. Grassy disturbed areas, old fields, chaparral; < 1200 m. NW, CaR, SNF, GV, CW. Variable; more study needed.

var. ***arvensis*** (E. Greene) Jepson **FL**: corolla 12–20 mm, throat strongly angled to tube, pouch prominent. Drying meadows, old fields, rocky grassy slopes; < 700 m. s NCoR. ❀SUN:7,**14–17**.

var. ***collina*** (Jepson) V. Newsom **FL**: corolla 5–8 mm, throat barely angled to tube, pouch hidden by calyx. Disturbed grassy fields, roadbanks, open chaparral and foothill woodland; 100–1200 m. SNF, GV, SnFrB, SCoR. Widespread intermediates to var. *sparsiflora* have been called var. *bruceae* (M.E. Jones) V. Newsom.

var. ***sparsiflora*** (p. 1031) **FL**: corolla 7–13 mm, throat ± angled to tube, pouch gen ± evident. *n*=7. Grassy, sometimes disturbed places, chaparral; < 1000 m. NW, CaR, n&c SNF, ScV, n CW. ❀SUN,DRN:**7**,8,9,**14–17**,**19–24**.

C. tinctoria Benth. (p. 1031) Pl 20–60 cm, robust. **LF** gen lanceolate-deltate, ± entire to serrate, (sub)glabrous above, densely hairy below, gen strongly mottled. **INFL** interrupted, glandular-clammy, staining; whorls dense; pedicels gen crowded, < calyx. **FL**: calyx >> fr, lobes deep, tips blunt; corolla 12–20 mm, gen white to yellowish or pale lavender (purple), throat strongly angled to tube, pouch projecting backward 2–4 mm from tube, upper lip < 1/2 lower, reflexed portion gen 2–5 mm, red-banded, lateral lobes gen red-dotted, gen long-hairy on upper surface, middle lobe glandular, hairy; upper filaments hairy. **SEEDS** gen 4, flat, gen slightly winged. *n*=7. Rocky, dry mixed woodland, coniferous forest; 600–2500 m. e NW, CaR, SN. Intergrades with *C. bartsiifolia*, *C. heterophylla*. ❀DRN:**15–17**,22–24&SHD:1,2,**7**,**14**,19–21.

C. torreyi A. Gray Pl 5–25 cm. **LF** linear to ovate, entire to toothed, glabrous. **INFL** open, densely glandular, staining; bracts much reduced upward; pedicels gen >> calyx, some sharply reflexed in fr, S-shaped. **FL**: calyx lobes subacute to blunt, ± = fr; corolla 4–10 mm, throat angled to tube, as wide as long, throat and lower lip gen blue (lavender purplish), upper lip whitish or pale lavender. **SEEDS** 2. Coniferous forest, often in sandy, granitic soil; 800–4000 m. KR, NCoRH, CaRH, SNH, TR, MP.

var. ***brevicarinata*** V. Newsom **LF** gen linear to elliptic (> 5 × longer than wide). **FL**: corolla gen 6–9 mm, middle lobe 1–2 mm < lateral lobes. Habitats of sp.; 1800–3000 m. s SNH (rare in n SNH, Placer Co.), SnGb.

var. ***latifolia*** V. Newsom **LF** elliptic to ovate (gen 2–5 × longer than wide). **FL**: corolla gen 6–9 mm, middle lobe ± = lateral. Habitats of sp.; 1000–2500 m. KR, NCoRH, CaRH, MP.

var. ***torreyi*** (p. 1031) **LF** ± linear (gen > 6 × longer than wide). **FL**: corolla gen 6–9 mm, middle lobe ± = lateral. *n*=21. Habitats of sp.; 1000–3000 m. NCoRH, CaRH, SNH, MP.

var. ***wrightii*** (S. Watson) I.M. Johnston **LF** linear to elliptic (gen > 5 × longer than wide). **FL**: corolla gen 4–6 mm, middle lobe ± = lateral. Habitats of sp.; 800–4000 m. NCoRH, CaRH, SNH, TR.

CORDYLANTHUS BIRD'S-BEAK

T.I. Chuang & Lawrence R. Heckard

Ann, green root-parasites, gen much-branched. **LVS** sessile, 0–11-lobed. **INFL**: spike (subtended by outer bracts), or fls solitary (each subtended by outer bracts) but often clustered; outer bracts ± lf-like; inner bract calyx-like (formerly confused with calyx). **FL**: calyx gen divided to base in front, sheath-like, tip gen entire or shallowly notched; corolla ± club-shaped, upper lip beak-like, enclosing anthers and style, tip closed, lower lip ± = upper, pouched, middle lobe gen tightly rolled under; stamens gen 4, anther sacs 1–2 per stamen, unequal; style bent near tip, stigma unexpanded, ± exserted downward from closed beak tip. **FR** loculicidal. **SEEDS** gen 10–20, attached at side; coat netted or ridged, tight-fitting. 18 spp.: w N.Am. (Greek: club-shaped fl) [Chuang & Heckard, 1986 Syst Bot Monogr 10:1–105] Close to *Orthocarpus*, distinguished by infl and calyx. Gen fls late summer.

1. Lateral lf lobes 8–10, paired; calyx << 1/2 corolla, divided > 1/2; anther sacs 2, widely separated (subg. *Dicranostegia*) . . . . ***C. orcuttianus***

1′ Lf entire to ± palmately 3–7-lobed; calyx gen entire to divided ± 1/3 in back, ± = corolla; anther sacs 1 (or if 2, ± overlapping)

2. Lf ± oblong to lanceolate, entire; infl a spike, 20–150 mm; middle lobe of lower corolla lip erect, not rolled under; saline or alkaline places (subg. *Hemistegia*)

3. Stamens 4; inner bract entire or notched . . . . ***C. maritimus***

4. Inner bract gen entire; seeds 25–40, 1–1.5 mm; inland . . . . ssp. ***canescens***

4′ Inner bract notched; seeds 10–20, 2–3 mm; gen coastal

5. St gen much-branched, upper branches > central spike; seeds ± 2 mm; SCo ..... ssp. ***maritimus***

5′ St few-branched, branches < or = central spike; seeds 2–3 mm; NCo, CCo ..... ssp. ***palustris***

3′ Stamens 2; inner bract 3–7-lobed

6. Lf 1–2 mm wide; inner bract 3-lobed; style puberulent ..... ***C. tecopensis***

6′ Lf 3–8 mm wide; inner bract 3–7-lobed; style glabrous

7. Pl soft-hairy (or becoming glabrous), longest hairs < 1 mm; seed coat deeply netted, wavy-crested . ***C. palmatus***

7′ Pl (esp infl) stiff-hairy, longest hairs > 1 mm; seed coat netted, not wavy-crested ..... ***C. mollis***

8. Pl bristly; st much-branched from near base; infl gen 2–6 cm; corolla pouch and tube sparsely tomentose; seed 1–1.5 mm; GV ..... ssp. ***hispidus***

8′ Pl gen ± soft-hairy; st few-branched from middle; infl 5–15 cm; corolla pouch and tube densely tomentose; seed 2–3 mm; n CCo ..... ssp. ***mollis***

2′ Lf thread-like to linear, entire to subpalmately 7-lobed; infl a spike < 20 mm or fls solitary (but often clustered); middle lobe of lower corolla lip tightly rolled under; nonsaline places (subg. *Cordylanthus*)

9. Corolla > 3 × longer than wide; seed (shallowly) netted; esp GB and D in sagebrush or juniper scrub

10. Calyx divided to base; corolla ± dusty yellow ..... ***C. ramosus***

10′ Calyx tube 1–4 mm; corolla gen pink to maroon (yellowish)

11. Stamens 2; calyx tip divided < 1/3 — MP ..... ***C. capitatus***

11′ Stamens 4; calyx tip shallowly notched

12. Pl canescent, not glandular-sticky; outer bracts 3–7-lobed ..... ***C. eremicus***

13. Outer bracts sparsely scabrous, tips slightly wider, with maroon thickening ..... ssp. ***eremicus***

13′ Outer bracts distinctly bristly, tips acute, unmarked ..... ssp. ***kernensis***

12′ Pl densely glandular-sticky; outer bracts gen 3- or 5-lobed

14. Infl a short spike; corolla lips ± equal; inner bract pinnately lobed; s MP, SNE ..... ***C. kingii*** ssp. ***helleri***

14′ Fls solitary, scattered; lower corolla lip << upper; inner bract gen entire; e DMtns ..... ***C. parviflorus***

9′ Corolla 2–3 × longer than wide; seed finely wavy-striate; esp CA-FP in foothill woodland

15. Outer bracts fan-shaped, shallowly 3–7-lobed; inner bract and calyx tips rough-papillate; corolla < 10 mm, pouch sides yellow — serpentine, NCoR ..... ***C. pringlei***

15′ Outer bracts entire to deeply 3-lobed; inner bract scabrous to puberulent; corolla 10–20 mm, pouch sides ± white

16. Anther sac 1 (+ bearded vestige) per stamen; calyx tube ± 1 mm; outer bracts 3-lobed, << fl — SW ***C. nevinii***

16′ Anther sacs 2 per stamen; calyx gen divided to base (exc *C. rigidus*); outer bracts entire or 3-lobed, gen = or > fl

17. Pl ± bristly (exc ssp. *littoralis*); fls solitary but densely clustered in 5–15-fld heads; calyx tube < 2 mm; corolla maroon in U-shape on lower side ..... ***C. rigidus***

18. Outer bract << fl, gen 5–10 mm, middle lobe oblong — sw SN ..... ssp. ***brevibracteatus***

18′ Outer bract gen > fl, 15–20 mm, middle lobe linear to lanceolate

19. Middle lobe of outer bract linear, tip wider, notched, maroon-thickened, marginal bristles > 2 × lobe width — SW ..... ssp. ***setigerus***

19′ Middle lobe of outer bract gen ± lanceolate, tip not wider (or ± truncate and green-, whitish-, or faintly maroon-thickened) marginal bristles ± = lobe width

20. Pl puberulent or soft-hairy; outer bract weakly ciliate and puberulent, tip of middle lobe tapered; dunes, c&s CCo ..... ssp. ***littoralis***

20′ Pl bristly; outer bracts scabrous, tip of middle lobe parallel-sided or widening; esp chaparral and woodland, c&s SN, c&s CW, WTR ..... ssp. ***rigidus***

17′ Pl soft-hairy or becoming glabrous; fls solitary but gen in ± loose 1–7-fld clusters; calyx divided to base; corolla streaked or blotched maroon

21. St decumbent, branches ascending < 15 cm; outer bracts 3-lobed near base — serpentine, ne SnFrB ..... ***C. nidularius***

21′ St erect, branches ascending gen > 20 cm; outer bracts entire or 3-lobed in upper 2/3

22. Main st densely long-soft-hairy; tips of outer bracts (or middle lobes) > 1.5 mm wide, longest hairs 2–3 mm; corolla barely marked maroon ..... ***C. pilosus***

23. Outer bracts entire (tip wider, obtuse to 2-notched); NCoR, n SnFrB ..... ssp. ***pilosus***

23′ Outer bracts 3-lobed; CaRF, n&c SNF

24. Outer bracts lobed in upper 1/3, lateral lobes < 3 mm; corolla = or > calyx and inner bract ssp. ***hansenii***

24′ Outer bracts lobed in lower 1/2; lateral lobes > 3 mm; corolla < calyx and inner bract ..... ssp. ***trifidus***

22′ Main st ± glabrous to sparsely stiff-hairy or ± glandular; tips of outer bracts (or middle lobes) < 1.5 mm wide, longest hairs < 2 mm; corolla heavily blotched maroon ..... ***C. tenuis***

25. St glabrous or minutely puberulent; lvs (or lobes) thread-like, tightly folded up, gen ± glabrous; serpentine outcrops — s NCoR

26. Outer bracts entire; sts minutely and sparsely glandular, esp below fls ..... ssp. ***brunneus***

26′ Outer bracts 3-lobed; sts ± glabrous (not glandular below fls) — sw NCoRO ..... ssp. ***capillaris***

25′ St gen puberulent and stiff-hairy (or sticky); lvs (or lobes) linear to ± oblong, flat to channeled, puberulent and gen glandular; gen not serpentine

27. Fls 3–7 in dense clusters; outer bracts gen 3-lobed, densely soft-hairy; longest hairs of inner bract > 2 mm; s SNF ..... ssp. ***barbatus***

27′ Fls 1–6 in loose clusters; outer bracts entire or 3-lobed, ± stiff-hairy; longest hairs of inner bract < 1 mm

28. Outer bracts entire; SN ..... ssp. ***tenuis***

28′ Outer bracts 3-lobed; n CA
29. Bracts and calyx yellow-green, puberulent, weakly glandular; fls gen 3–6 per cluster — CaRH . . . . ssp. ***pallescens***
29′ Bracts and calyx green or maroon-tinged, glandular, puberulent and with long, stiff hairs; fls 1–3 per cluster . . . . ssp. ***viscidus***

C. capitatus Benth. (p. 1031) YAKIMA BIRD'S-BEAK Pl 10–50 cm, glaucous-green or gray-purple, densely glandular- and non-glandular-hairy. **LF** 20–40 mm, linear. **INFL**: spike, 15–20 mm, 2–5-fld, head-like; outer bracts 4–7, 10–20 mm, lobes 3 in lower 1/2, linear-lanceolate; inner bract 12–18 mm. **FL**: calyx 10–15 mm, tube 2–4 mm, tip divided ± 1/3; corolla 10–20 mm, maroon, yellow-tipped, throat 4–6 mm wide; stamens 2, anther sac 1. **SEEDS** 4–6, 2–2.5 mm, ± reniform, shallowly netted, ± smooth between nets. 2*n*=26. RARE in CA. Open coniferous forest, juniper woodland; 1800–2000 m. Wrn; to c WA, ID.

C. eremicus (Cov. & C. Morton) Munz Pl 10–80 cm, (yellow-)green, often tinged red, canescent. **LF** 10–40 mm, thread-like or linear, entire to 5-lobed. **INFL**: spike, 15–25 mm, 3–14-fld; outer bracts 3–7, 5–20 mm, lobes 3–7; inner bract 10–18 mm. **FL**: calyx 10–18 mm, tube 1–3 mm; corolla 10–20 mm, purplish, pinkish, or yellow-green, often yellow-tipped, throat base blotched maroon, pouch 4–6 mm wide, soft-white-hairy; stamens 4, anther sacs 2. **SEED** 1.5–2 mm, ± ovoid, deeply netted, pale brown. 2*n*=26. Dry rocks; gen 1800–3000 m. s SNH, n SnBr, n DMtns.

ssp. ***eremicus*** (p. 1031) **INFL**: outer bracts ± sparsely scabrous, tips wider, ± maroon-thickened. **FL**: calyx tube 2–3 mm; corolla lavender to pinkish, blotched purple. **SEED** papillate between nets. Sagebrush scrub, pinyon/juniper woodland; gen 1800–2800 m. n SnBr (near Cushenbury, 1000 m), n DMtns. [*C. ramosus* ssp. *e.* (Cov. & C. Morton) Munz; *C. bernardinus* Munz]

ssp. ***kernensis*** Chuang & Heckard (p. 1031) INYO BIRD'S-BEAK **INFL**: outer bracts distinctly bristly, tips not wider, unthickened, uncolored. **FL**: calyx tube 1 mm; corolla gen pinkish. **SEED** smooth between nets. UNCOMMON. Open Jeffrey-pine or juniper forest; 2100–3000 m. s SNH (Kern Plateau, Tulare, and Inyo cos.).

C. kingii S. Watson ssp. ***helleri*** (Ferris) Chuang & Heckard (p. 1031) Pl 10–60 cm, gray-green, tinged red, densely glandular-sticky. **LF** gen 10–25 mm, linear. **INFL**: spike, 20–30 mm, 1–4-fld; outer bracts 3–6, 10–15 mm, gen 3-lobed; inner bract 10–15 mm, 3–5-lobed, glandular. **FL**: calyx 15–20 mm, tube 2.5 mm; corolla 15–25 mm, rosy-lavender to dull purple-red or -yellow, hairy; stamens 4, anther sacs 2. **SEED** 2–2.5 mm, ± ovoid, deeply netted, pale brown, papillate between nets. 2*n*=26. Open pinyon/juniper woodland or sagebrush scrub; 1300–3100 m. s MP, SNE; NV. [*C. h.* (Ferris) J.F. Macbr.] Ssp. *k.* in NV, UT.

C. maritimus Benth. Pl 10–40 cm, gray-green, glaucous, often tinged purple and salt-encrusted, gen ± short-hairy. **LF** 5–25 mm, ± linear-lanceolate, entire. **INFL**: spike, 20–90 mm, many-fld; outer bract lf-like; inner bract 15–30 mm. **FL**: calyx 15–25 mm; corolla 15–25 mm, white to cream, puberulent, lips pale to brownish or purplish red, middle lobe of lower lip erect; stamens 4, anther sacs 2 (lower pair) or 1 (upper pair). **SEEDS** 10–40, 1–3 mm, ± reniform, deeply netted, dark brown. 2*n*=30. Coastal salt marshes (< 10 m), inland alkaline flats (1200–1900 m). n NCo, n CCo, SCo, SNE; to OR, UT, n Baja CA.

ssp. ***canescens*** (A. Gray) Chuang & Heckard (p. 1031) **ST**: branches gen many, ± erect, upper gen > central spike. **INFL** dense; inner bract gen entire. **SEEDS** 25–40, 1–1.5 mm. Inland alkaline flats; 1200–1900 m. SNE; to s OR, UT. [*C. c.* A. Gray] ❀STBL.

ssp. ***maritimus*** (p. 1031) SALT MARSH BIRD'S-BEAK **ST**: branches gen many, decumbent to ascending, gen > central spike. **INFL** loose or dense; inner bract gen notched. **SEEDS** 15–20, ± 2 mm. RARE. Coastal salt marshes; < 10 m. SCo; n Baja CA.

ssp. ***palustris*** (Behr) Chuang & Heckard (p. 1031) POINT REYES BIRD'S-BEAK **ST**: branches 0–few, ascending, < or = central spike. **INFL** dense; inner bract notched. **SEEDS** 10–20, 2–3 mm. RARE. Coastal salt marshes; < 10 m. n NCo (Humboldt Co.), n CCo (Marin, Sonoma cos.); sw OR.

C. mollis A. Gray Pl 10–40 cm, gray-green, often tinged purple, glandular-puberulent and long-nonglandular-hairy. **LF** 10–25 mm, ± oblong, entire to 7-lobed. **INFL**: spike, 20–150 mm; outer bract lf-like; inner bract 15–25 mm, ± pinnately 3–7-lobed. **FL**: calyx 15–20 mm; corolla 15–20 mm, whitish, ± densely tomentose, middle lobe of lower lip erect; stamens 2 (lower pair vestigial), anther sacs 2; style glabrous. **SEEDS** 20–30, 1–3 mm, ± reniform, deeply netted, dark brown. 2*n*=28 (both sspp.). Coastal and inland salt marshes; < 10 m. c&s GV, n CCo.

ssp. ***hispidus*** (Pennell) Chuang & Heckard (p. 1031) HISPID BIRD'S-BEAK Pl bristly. **ST**: branches many, from near base, spreading. **INFL** gen 20–60 mm. **FL**: corolla sparsely tomentose. **SEED** 1–1.5 mm. RARE. Saline marshes and flats; < 10 m. c&s GV (Solano, Merced, Kern cos.). [*C. h.* Pennell]

ssp. ***mollis*** (p. 1031) SOFT BIRD'S-BEAK Pl gen ± soft-hairy. **ST**: branches few–many, from middle, ± ascending. **INFL** 50–150 mm. **FL**: corolla densely tomentose. **SEED** 2–3 mm. RARE. Coastal salt marshes; < 10 m. n CCo.

C. nevinii A. Gray (p. 1031) Pl 20–80 cm, gen gray-green, tinged red, densely glandular-puberulent and long-soft-hairy. **LF** 5–30 mm, linear. **INFL**: fls solitary, 1–3 in loose clusters, often inverted; outer bracts 1–3, 5–10 mm, 3-lobed, tips wider, sometimes notched, cream; inner bract 10–15 mm, tip scabrous. **FL**: calyx 10–15 mm, tube ± 1 mm; corolla 12–18 mm, white, yellow-tipped, pouch 5–8 mm wide, purple-veined; stamens 4, anther sac 1 + bearded appendage. **SEED** 2 mm, finely wavy-striate, dark brown. 2*n*=28. Dry, open Jeffrey-pine/oak forest; 1700–2500 m. s SNH (Piute Mtns), TR, PR; w AZ, n Baja CA.

C. nidularius J. Howell (p. 1031) MOUNT DIABLO BIRD'S-BEAK Pl < 15 cm, gray-green, maroon-tinged, glandular-puberulent and long-nonglandular-hairy. **ST** decumbent, openly branched. **LF** 10–30 mm, narrowly linear. **INFL**: fls solitary, 1–3 in ± dense clusters; outer bracts 2–3, 10–15 mm, 3-lobed near base, segments linear; inner bract ± 15 mm. **FL**: calyx 13–16 mm; corolla ± 15 mm, white, lower lip and throat purple-lined; stamens 4, anther sacs 2. **SEEDS** 7–10, 1.5–2 mm, ± ovoid, finely wavy-striate, dark brown. 2*n*=28. **RARE** CA. Dry, open serpentine in chaparral; 600–800 m. ne SnFrB (e slope Mount Diablo).

C. orcuttianus A. Gray (p. 1031) ORCUTT'S BIRD'S-BEAK Pl 10–50 cm, green or red-tinged, ± stiff-hairy. **LF** 20–80 mm, pinnately dissected; lateral lobes 8–10, paired, ± linear. **INFL**: spike, 20–100 mm, dense; outer bracts lf-like; inner bracts 15–25 mm, 3–7-lobed; lobes linear, ascending. **FL**: calyx ± 6 mm, gen divided > 1/2; corolla 18–25 mm, white, yellow-tipped; stamens 2 (upper pair staminodes), anther sacs 2, widely separated. **SEEDS** 30–40, 1–1.5 mm, ± reniform, deeply netted, wavy-crested, dark brown. 2*n*=32. RARE in CA. Coastal scrub; < 350 m. s SCo (sw San Diego Co.); n Baja CA. Threatened by urbanization. Fls spring-summer.

C. palmatus (Ferris) J.F. Macbr. (p. 1031) PALMATE BIRD'S-BEAK Pl 10–30 cm, gray-green, ± glandular and soft-hairy (longest hairs < 1 mm). **LF** 7–20 mm, ± oblong, entire to 5-lobed. **INFL**: spike, 50–150 mm, dense; outer bract lf-like; inner bract 15–20 mm, 3–7-lobed. **FL**: calyx ± 15 mm; corolla 15–20 mm, whitish, finely puberulent, sides often ± pale lavender, middle lobe of lower lip erect; stamens 2 (upper pair vestigial); style glabrous. **SEED** 2.5–3 mm, ± reniform, deeply netted, wavy-crested, ± dark-brown. 2*n*=42. **ENDANGERED** CA, US. Alkaline flats; < 60 m. GV (Colusa, Yolo, Alameda, San Joaquin, Madera, Fresno cos.). [ssp. *carnulosus* (Pennell) Munz] Threatened by agriculture, urbanization.

C. parviflorus (Ferris) Wiggins (p. 1031) PURPLE BIRD'S-BEAK Pl 20–60 cm, gray-green, tinged red, glandular-sticky and long-hairy. **LF** 5–30 mm, linear. **INFL**: fls solitary, 1–4 per loose cluster; outer bract gen 1, 5–15 mm, 3–5-lobed, tips obtuse, ciliate, densely glandular; inner bract 10–12 mm. **FL**: calyx 10–15 mm, tube 1–1.5 mm; corolla 15–20 mm, pink to lavender, pouch ± 7 mm wide, often dark-veined; stamens 4, anther sacs 2. **SEED** 1.5–2 mm, ± ovoid, shallowly netted, dark brown, densely papillate between nets. $2n$=26. RARE in CA. Dry sagebrush scrub or pinyon/juniper/Joshua-tree woodland; 700–2200 m. e DMtns (New York, Providence mtns); to se NV, sw UT, nw AZ, also s-c ID.

C. pilosus A. Gray Pl 20–120 cm, (gray-)green, gen tinged purple, densely puberulent (partly glandular) and long-soft-hairy. **LF** 10–40 mm, linear to lanceolate. **INFL**: fls solitary, 1–3 in loose clusters; outer bracts 1–4, 15–20 mm, linear, entire or 3-lobed, tips > 1.5 mm wide, angled to slightly 3-lobed, often ivory-thickened; inner bract 15–22 mm, tip abruptly pointed to notched. **FL**: calyx 15–20 mm; corolla 15–20 mm, whitish, yellow-green-tipped, pouch 5–8 mm wide, ± lightly marked maroon; stamens 4, anther sacs 2. **SEED** 1.5–2.5 mm, ± ovoid, finely wavy-striate, dark brown. $2n$=28 (all sspp.). Open foothill woodland, chaparral, serpentine; < 1300 m. NCoR, s CaR, n&c SNF, SnFrB. Variable; close to *C. tenuis*.

ssp. ***hansenii*** (Ferris) Chuang & Heckard (p. 1031) **INFL**: outer bracts ± linear, 3-lobed in upper 1/3, lobes < 3 mm, tips ± truncate. **FL**: corolla = or > calyx and inner bract. Open foothill woodland; < 1000 m. s CaR, n SNF. [*C. h.* (Ferris) J.F. Macbr.]

ssp. ***pilosus*** (p. 1031) **INFL**: outer bracts entire, tip wider, obtuse to 2-notched. **FL**: corolla < or = calyx and inner bract. Habitats of sp.; gen < 700 (1300) m. NCoR, SnFrB. [ssp. *diffusus* (Pennell) Munz]

ssp. ***trifidus*** (Robinson & Greenman) Chuang & Heckard (p. 1031) **INFL**: outer bracts ± lanceolate, 3-lobed in lower 1/2, lobes > 3 mm, tips truncate to obtuse. **FL**: corolla < calyx and inner bract. Foothill woodland; gen < 1100 m. n&c SNF.

C. pringlei A. Gray (p. 1031) Pl 30–150 cm, green, lightly maroon-tinged, glabrous or puberulent. **LF** 10–40 mm, thread-like. **INFL**: fls solitary, 1–4 in head-like clusters; outer bracts 1–3, 5–8 mm, fan-shaped, shallowly 3–7-lobed; inner bracts 8–10 mm, tip rough-papillate. **FL**: calyx 8–10 mm; corolla 8–9 mm, yellowish mottled maroon on top; stamens 4, anther sacs 2. **SEEDS** 4–6, 2.5–3 mm, ± ovoid, finely wavy-striate, dark brown. $2n$=28. Open dry serpentine in chaparral and mixed-evergreen forest; 400–1700 m. NCoRH, NCoRI.

C. ramosus Benth. (p. 1031) Pl 10–90 cm, gray-green or tinged red, ± canescent. **LF** 10–40 mm, ± thread-like, entire to 5-lobed. **INFL**: fls solitary; clusters spike-like, 15–25 mm, 3–7-fld; outer bract 1 per fl, 10–20 mm, entire to 7-lobed, sometimes bristly, segments ± thread-like; inner bract 10–20 mm, entire. **FL**: calyx 10–15 mm; corolla 10–20 mm, dusty yellow, often marked maroon, ± puberulent; stamens 4, anther sacs 2. **SEED** ± 2 mm, ± ovoid, deeply netted, light brown. $2n$=24. Rocky to alkaline soils in sagebrush scrub; 1200–2800 m. n CaRH (Shasta Valley), GB; to OR, WY, Colorado. [ssp. *setosus* Pennell]

C. rigidus (Benth.) Jepson Pl 30–150 cm, yellow-green or tinged red, gen ± stiff-hairy, gen not glandular. **LF** 10–40 mm, ± linear, often inrolled. **INFL**: fls solitary, gen 5–15 in loose, head-like clusters; outer bract gen 1 per fl, 5–20 mm, gen 3-lobed; inner bract 14–20 mm, often ± bristly. **FL**: calyx 10–20 mm, tube 1–2 mm; corolla 12–20 mm, yellowish, lower side marked maroon in U-shape, pouch 3–10 mm wide, white; stamens 4, anther sacs 2. **SEED** 1.5–2 mm, ± ovoid, wavy-striate, dark brown. $2n$=28. Many habitats; < 2500 m. c&s SNF, Teh, CW (exc n SnFrB), SW; n Baja CA. Highly variable, intergrading. Many geog races, variably distinct; 5 most discrete (1 in Baja CA) considered sspp.

ssp. ***brevibracteatus*** (A. Gray) Munz **INFL** 2–12-fld, 10–20 mm wide; outer bract gen 5–10 mm (< fl), lobed in upper 1/2, ± long-bristly, middle lobe oblong, tip slightly wider, unthickened, uncolored. **FL**: corolla 8–17 mm. Granitic openings in Jeffrey-pine/pinyon/juniper forest or sagebrush shrubland; 1000–2500 m. s SNH (Kern Plateau, Tulare and Kern cos.).

ssp. ***littoralis*** (Ferris) Chuang & Heckard (p.1035) SEASIDE BIRD'S-BEAK Pl puberulent to soft-hairy. **INFL** 5–8-fld, 10–20 mm wide; outer bract 10–20 mm, lobed in upper 1/2, middle lobe ± lanceolate, tip tapered. **FL**: corolla 15–20 mm. **ENDANGERED** CA. Dunes; 0–200 m. c CCo (s Monterey Bay and Peninsula), s CCo (w Santa Barbara Co.). [*C. l.* (Ferris) J.F. Macbr.]

ssp. ***rigidus*** (p. 1035) **INFL** 5–15-fld, 20–40 mm wide; outer bract 10–20 mm, lobed in lower half, ± short-bristly, middle lobe linear to oblong, tip tapered or wider and notched. **FL**: corolla 12–20 mm. Open foothill woodland, chaparral margins, coniferous forest; 150–2300 m. c&s SN, CW (exc n SnFrB), WTR. [*C. ferrisianus* Pennell; *C. littoralis* ssp. *platycephalus* (Pennell) Munz]

ssp. ***setigerus*** Chuang & Heckard. (p. 1035) **INFL** 5–13-fld, 20–30 mm wide; outer bract 10–20 mm, lobed in lower 1/2, densely long-bristly (hairs purple-based), middle lobe ± linear, tip ± wider, notched, thickened, purple. **FL**: corolla 13–17 mm. Open coastal-sage scrub, chaparral, oak woodland, coniferous forest; < 1900 m. SW (exc WTR); n Baja CA. [*C. filifolius* Benth. misapplied]

C. tecopensis Munz & Roos (p. 1035) TECOPA BIRD'S-BEAK Pl 10–60 cm, grayish or tinged purple, sparsely puberulent, glaucous. **LF** 5–15 mm, linear-lanceolate, entire. **INFL**: spike, 20–150 mm, loose; outer bract lf-like; inner bract 10–15 mm, lobed near middle. **FL**: calyx 10–13 mm; corolla 10–15 mm, pale lavender, densely puberulent, middle lobe of lower lip erect; stamens 2 (upper pair vestigial); style puberulent. **SEEDS** 8–10, 2–3 mm, ± reniform, deeply netted, light brown. $2n$=28. RARE. Alkaline meadows and flats; 100–900 m. s SNE, n DMoj; w NV.

C. tenuis A. Gray Pl 20–120 cm, (gray- or yellow-)green or tinged maroon, glabrous to glandular-sticky. **ST** ± wiry. **LF** 10–60 mm, ± linear. **INFL**: fls solitary, 1–7 in loose clusters; outer bracts 1–4, 5–20 mm, linear, entire or 3-lobed, tips gen ± wider (< 1.5 mm) and thickened, longest hairs < 2 mm; inner bract 10–20 mm. **FL**: calyx 10–20 mm; corolla 10–20 mm, 4–8 mm wide, whitish, ± yellow-tipped, heavily blotched maroon; stamens 4, anther sacs 2. **SEEDS** 6–16, 1.5–2.5 mm, ± ovoid or angled, finely wavy-striate, dark brown. $2n$=28. Open woodland or forest, serpentine outcrops; < 2500 m. NW, CaR, n&c SN; s OR. Highly variable. Close to *C. pilosus*.

ssp. ***barbatus*** Chuang & Heckard (p. 1035) FRESNO COUNTY BIRD'S-BEAK **ST** gen puberulent and long-hairy, sometimes glandular. **LF** linear to ± oblong. **INFL** dense, 3–7-fld; outer bracts gen 3-lobed; inner bract bearded, hairs opaque, longest > 2 mm. **FL**: corolla 15–18 mm. UNCOMMON. Open mixed forest; 1300–2000 m. s SNF (Fresno Co.).

ssp. ***brunneus*** (Jepson) Munz (p. 1035) SERPENTINE BIRD'S-BEAK **ST** finely glandular-puberulent, esp below fls. **LF** thread-like, gen folded or channeled. **INFL** loose, 1–4-fld; outer bracts entire, thread-like; inner bract sparsely glandular-puberulent. **FL**: corolla 12–15 mm. UNCOMMON. Serpentine in mixed-evergreen forest, chaparral; gen < 900 m. s NCoR (Lake, Sonoma, Napa cos.).

ssp. ***capillaris*** (Pennell) Chuang & Heckard (p. 1035) PENNELL'S BIRD'S-BEAK **ST** ± glabrous. **LF** thread-like, gen folded. **INFL** loose, 1–4-fld; outer bracts 3-lobed, thread-like; inner bract tip gen puberulent. **FL**: corolla 14–16 mm. RARE. Serpentine in chaparral; ± 200 m. sw NCoRO (near Occidental, Sonoma Co).

ssp. ***pallescens*** (Pennell) Chuang & Heckard (p. 1035) PALLID BIRD'S-BEAK **ST** gen puberulent and long-hairy, gen nonglandular. **LF** linear. **INFL** loose, gen 3–6-fld; outer bracts 3-lobed; inner bract hairs < 1 mm. **FL**: corolla 10–15 mm. RARE. Open volcanic alluvium; 1200 m. CaRH (near Black Butte, Siskiyou Co.). [*C. p.* Pennell] 1 large population.

ssp. ***tenuis*** **ST** gen puberulent (esp > 1600 m) to glandular-sticky (esp < 900 m). **LF** linear, entire. **INFL** loose, 1–3-fld; outer bracts entire; inner bract hairs < 1 mm. **FL**: corolla 10–20 mm. Open coniferous forests, foothill woodland; 300–2500 m. SN. [*C. pilosus* ssp. *bolanderi* (A. Gray) Munz] Pls from c KR (Scott Mtns) that key here may be derived from ssp. *viscidus*.

ssp. ***viscidus*** (Howell) Chuang & Heckard **ST** densely glandular-sticky, puberulent, and long-hairy. **LF** linear. **INFL** loose, 1–3-fld; outer bracts 3-lobed; inner bract hairs < 1 mm. **FL**: corolla 10–18 mm. Open yellow-pine forest on serpentine; 400–1600 m. KR, NCoRH, CaR, n SN; OR. [*C. v.* (Howell) Pennell] Close to ssp. *tenuis*.

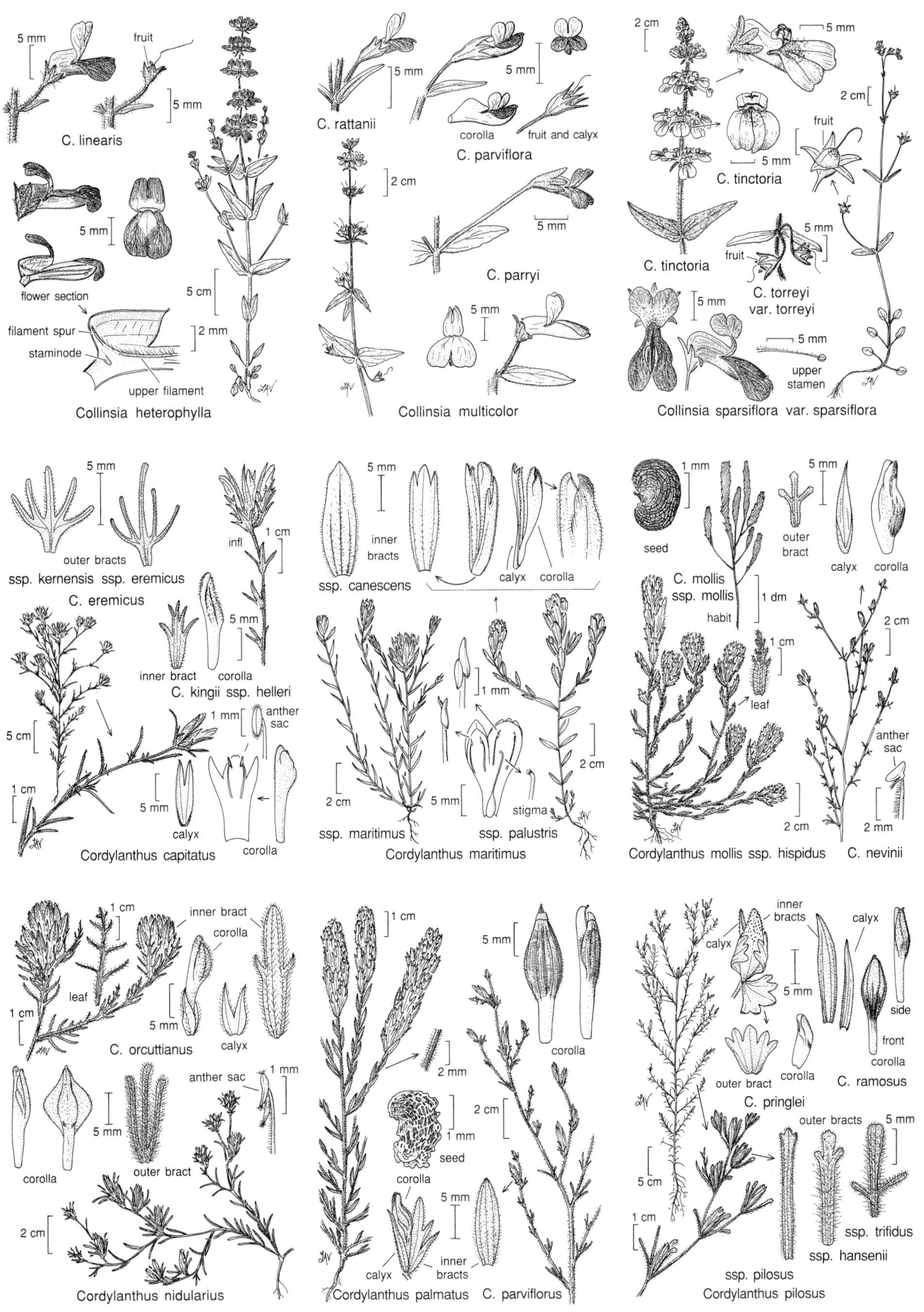

5 mm
fruit
5 mm
C. linearis
5 mm
flower section
filament spur
staminode
upper filament
5 cm
2 mm
Collinsia heterophylla
5 mm
C. rattanii
5 mm
corolla
fruit and calyx
C. parviflora
2 cm
5 mm
C. parryi
5 mm
Collinsia multicolor
2 cm
5 mm
5 mm
C. tinctoria
C. tinctoria
2 cm
fruit
5 mm
fruit
C. torreyi
var. torreyi
5 mm
5 mm
upper
stamen
Collinsia sparsiflora var. sparsiflora
5 mm
outer bracts
ssp. kernensis ssp. eremicus
C. eremicus
infl
1 cm
5 mm
inner bract
corolla
C. kingii ssp. helleri
5 cm
1 mm
anther
sac
1 cm
5 mm
calyx
corolla
Cordylanthus capitatus
5 mm
inner
bracts
ssp. canescens
calyx
corolla
1 mm
2 cm
5 mm
stigma
2 cm
ssp. maritimus
ssp. palustris
Cordylanthus maritimus
1 mm
seed
C. mollis
ssp. mollis
habit
1 dm
5 mm
outer
bract
calyx
corolla
2 cm
1 cm
leaf
anther
sac
2 cm
2 mm
Cordylanthus mollis ssp. hispidus
C. nevinii
1 cm
inner bract
corolla
leaf
1 cm
5 mm
C. orcuttianus
calyx
anther sac
1 mm
5 mm
corolla
outer bract
2 cm
Cordylanthus nidularius
1 cm
5 mm
2 mm
2 cm
1 mm
seed
corolla
5 mm
corolla
calyx
inner
bracts
Cordylanthus palmatus
C. parviflorus
inner
bracts
calyx
calyx
5 mm
side
front
corolla
outer bract
corolla
C. ramosus
C. pringlei
5 cm
outer bracts
5 mm
1 cm
ssp. trifidus
ssp. hansenii
ssp. pilosus
Cordylanthus pilosus

CYMBALARIA

David M. Thompson

Ann, per, glabrous to hairy. **ST** decumbent or vine-like. **LVS** gen opposite below, alternate above, ± round to reniform, entire to palmately lobed; veins palmate. **INFL**: fls solitary in lf axils. **FL**: calyx lobes 5, deep, ± unequal; corolla tube with conic or cylindric spur at base, lower lip base swollen, closing mouth. **FR** spheric; chambers dehiscent by several slits radiating from tip. **SEEDS** many, gen ridged or tubercled. *n*=7. ± 8 spp.: esp Medit. (Latin: round lvs)

C. muralis Gaertner, Meyer, & Scherb. (p. 1035) KENILWORTH IVY **LF** 1–3 cm wide; lobes 5–9, moderately shallow, rounded to triangular, often abruptly pointed. **FL**: calyx 2–2.5 mm; corolla 9–15 mm, pale lilac to violet, spur 1–3 mm, cylindric, lower lip base yellowish. **FR** ± 4 mm, glabrous, pedicel growing away from light. *n*=7. Rock walls, shady, disturbed areas; 0–1000 m. NCo, CCo, SCo; native to Medit. Cult as orn.

DIGITALIS FOXGLOVE

Margriet Wetherwax

Bien, per. **ST** erect. **LVS** basal and cauline, alternate; lowermost in a rosette. **INFL**: raceme, 1-sided, bracted. **FL** nodding; calyx deeply 5-lobed; corolla ± bilateral, long-bell-shaped, lowest 3 lobes forming a prominent lip; stamens 4, in 2 pairs, incl; stigmas 2, flat. 30 spp.: Eur (esp Medit), w&c Asia; some cult as orn or as source of the cardiac glycoside digitalis, a medically important heart stimulant. (Latin: finger, from corolla shape)

D. purpurea L. Gen bien. **ST** < 18 dm, simple, gray-tomentose and glandular, esp upward. **LF** 10–30 cm; petiole winged; blade lanceolate to ovate, margins crenate to dentate, upper surface green and soft-hairy, lower surface gray-tomentose. **INFL**: pedicel 6–25 mm, tomentose; fls many. **FL**: calyx lobes < 1.8 cm, lanceolate to ovate; corolla 4–6 cm, white to pink-purple with darker spots on lower inside surface, lobes ciliate, sparsely long-hairy inside; stamens, style incl. **FR** ± 12 mm, ovoid. **SEEDS** many, ± 0.5 mm. 2*n*=56. Acid soil in open woodlands, disturbed places; < 1000 m. NCo, NCoRO, n SNF, SnFrB, SCoRO; to B.C.; native to w Eur, nw Afr. All parts TOXIC, unpalatable to livestock; lvs may be mistaken for comfrey or sage — tea and salad containing them have been fatal to humans.

DOPATRIUM

Margriet Wetherwax

Ann, fleshy, sparsely glandular. **ST** erect. **LVS** opposite, fleshy. **INFL**: fls solitary in axils; lower bracts lf-like, upper scale-like. **FL**: calyx 5-lobed; corolla 2-lipped, yellow or pale blue to lavender, upper lip erect, 2-lobed, lower lip 3-lobed (middle lobe > lateral); fertile stamens 2, sterile stamens 2. **FR** spheric, loculicidal. 7 spp: Afr, Asia, Australia. (Latin: aboriginal name)

D. junceum (Roxb.) Buch.-Ham. Aquatic; upper parts emergent. **ST** 10–30 cm. **LVS** few, fleshy, 2–5 cm, oblong, obtuse. **INFL**: upper bracts 0.5–2 mm; pedicels 4–11 mm, sparsely glandular, spreading in fr; submersed fls sessile. **FL**: calyx 2–5 mm, lobes = tube, obtuse; corolla 4–7 mm, pale blue to lavender, tube slightly > calyx; submersed fls cleistogamous, 2–3 mm. **FR** ± 4 mm, spheric. **SEEDS** netted. Uncommon. Rice fields; < 100 m. ScV (Butte Co.); native to Asia, Australia.

GALVEZIA

David M. Thompson

Per, shrub, glabrous to hairy. **ST** erect, arching, or pendent, often much-branched. **LVS** opposite or whorled in 3's, entire; veins pinnate. **INFL**: raceme. **FL**: calyx lobes 5, entire, equal; corolla gen red, tube with sac-like extension at base, lower lip base ± swollen, gen not fully closing mouth; stamens often exserted; stigma exserted. **FR** ovoid to spheric; chambers dehiscent by 1–2 pores near tip. **SEEDS** many; ridges thin, netted. 5 spp.: w N.Am, w S.Am, Galapagos Islands. (José Gálvez, Spanish administrator, 1720's–1787)

G. speciosa (Nutt.) A. Gray (p. 1035) SHOWY ISLAND SNAPDRAGON Pl spreading, < 1 m, < 2 m wide. **ST** arching, often ± pendent. **LVS** thickish, whorled (opposite on seedlings). **INFL**: pedicels 1–2 cm. **FL**: calyx lobes 7–10 mm; corolla 2–2.5 cm, bright red, lower lip base ± closing mouth. **FR** 6–7 mm, ovoid, asymmetric; lower chamber larger. *n*=15. RARE. Rocky cliffs, canyons; 0–900 m. s ChI; Guadalupe Island, Mex. In cult.

GRATIOLA HEDGE-HYSSOP

Margriet Wetherwax

Ann, ± glandular. **ST** erect or ascending. **LVS** opposite, entire to dentate, sessile. **INFL**: raceme; bractlets beneath calyx present or 0. **FL**: sepals 5, gen free; corolla tube 4-angled or -sided, upper lip entire or 2-lobed, lower 3-lobed; fertile stamens 2, anther sacs of each stamen separated, parallel; stigmas 2, flat. **FR** 4-valved. **SEEDS** many, wingless. ± 20 spp.: temp, trop mtns. (Latin: grace or favor, from supposed medicinal quality)

1. Pedicel with 2 sepal-like bractlets below calyx; corolla 2–3 × calyx ..... ***G. neglecta***
1' Pedicel without bractlets; corolla ± = calyx
  2. Lvs and sepals long-tapered; sepals 8–11 mm, equal; corolla throat yellow, limb white ..... ***G. ebracteata***
  2' Lvs and sepals truncate; sepals 4–6 mm, unequal; corolla yellow (only 3 lower lobes white) ..... ***G. heterosepala***

G. ebracteata Benth. (p. 1035) **ST** < 15 cm, glabrous. **LF** 5–25 mm, lanceolate, long-tapered, slightly clasping at base, dentate toward tip. **INFL**: pedicels 10–23 mm, stout, erect; bractlets 0. **FL**: sepals 8–11 mm, equal, free, long-tapered; corolla ± 6 mm, throat yellow, limb white, upper lobes 1/2 fused. **FR** 4–5 mm, spheric. Wet, muddy places; < 2000 m. n&c CA-FP; to B.C., MT.

G. heterosepala H. Mason & Bacigal. (p. 1035) BOGGS LAKE HEDGE-HYSSOP **ST** < 10 cm, glabrous below, glandular-puberulent upward. **LF** < 20 mm, not clasping, truncate; lower linear-lanceolate, becoming shorter, oblong (upper 2–5 mm). **INFL** glandular-puberulent; pedicels 10–23 mm, slender, erect; bractlets 0. **FL**: sepals 4–6 mm, unequally fused, truncate; corolla 6–8 mm, yellow (exc lower 3 lobes white), lobes of upper lip ± entire; stigmas 2, lobes unequal. **FR** 4–6 mm, spheric. **ENDANGERED** CA. Shallow water, margins of vernal pools; < 1200 m. NCoRI, c SNF, ScV, MP.

G. neglecta Torrey (p. 1035) **ST** < 30 cm. **LVS** 5–50 mm, lanceolate to obovate, round-clasping at base, entire to dentate toward tip. **INFL**: pedicels 10–20 mm, slender, spreading; bractlets sepal-like, = or > calyx. **FL**: sepals 5–6 mm, equal, free, long-tapered; corolla ± 10 mm, tube white or yellowish, limb small, purple, lobes of upper lip entire. **FR** ± 5 mm, spheric. Wet, muddy places, vernal pools, sandbars; < 2000 m. c NCo (Mendocino Co.), w KR (Humboldt Co.), CaRH, n&c SNH, MP.

HEBE

Elizabeth McClintock

Shrub. **LVS** simple, opposite, 4-ranked, sessile or nearly so, evergreen. **INFL**: gen raceme, gen axillary, dense, many-fld. **FL** bisexual, gen bilateral; calyx gen 4-lobed; corolla often 4-lobed, often ± blue, tube short, lobes spreading; stamens 2, exserted; staminodes 0; style exserted. **FR**: capsule, septicidal. **SEEDS** few–many, gen flat, smooth. ± 100 spp.: esp New Zealand, also S.Am, se Australia, New Guinea; many cult as orn. (Greek: Hebe, goddess of youth, cup-bearer to the gods, wife of Hercules)

1. Corolla dark blue-lavender; lf 2–6 cm ..... ***H. ×franciscana***
1′ Corolla red-magenta to -purple; lf 5–10 cm ..... ***H. speciosa***

H. ×franciscana (Eastw.) Souster Shrub < 1 m. **LF** 1.5–3 cm wide, elliptic, ± thin. **INFL** 4–6 cm. **FL**: corolla ± 1 cm wide. Uncommon. Disturbed urban areas, ocean bluffs; < 100 m. SnFrB; native region unknown. Hybrid between *H. elliptica* (Forster f.) Pennell and *H. speciosa*.

H. speciosa (Cunn.) Cockayne & Allan Shrub 1–2 m. **LF** 2.5–5 cm wide, wide-oblong, ± thick to nearly fleshy. **INFL** 6–8 cm. **FL**: corolla ± 8 mm wide. Disturbed urban areas; < 100 m. SnFrB; native to New Zealand.

KECKIELLA

Noel H. Holmgren

Subshrub or shrub. **ST** wand-like to much-branched. **LVS** drought-deciduous, ± opposite or in 3's, gen short-petioled. **INFL**: panicle or spike-like; bracts small. **FL**: calyx lobes 5, ± equal; corolla not purplish (whitish to red), short-glandular outside, upper lip hooded, lobes short, rounded, external in bud, lower lip rounded, lobes often reflexed; filaments densely nonglandular-hairy at base, anthers small, glabrous, anther sacs gen spreading flat at dehiscence; staminode well developed, glabrous to densely bearded; nectary a disk; stigma head-like. **FR** ovoid, septicidal and sometimes also loculicidal at tip. **SEEDS** many, irregularly angled or ± winged. 7 spp.: esp CA. (David D. Keck, CA botanist, 1903–) [Straw 1966 Brittonia 18:80–95] Segregated from *Penstemon*.

1. Corolla pink to red, tube 16–25 mm, throat indistinct
 2. Lvs opposite or subopposite
  3. Lf base gen subcordate; anther sacs 1.1–1.5 mm ..... ***K. cordifolia***
  3′ Lf base wedge-shaped; anther sacs 0.9–1 mm ..... ***K. corymbosa***
 2′ Lvs whorled in 3's (sometimes a few opposite) ..... ***K. ternata***
  4. Calyx glandular, 3.8–5.5 mm ..... var. ***septentrionalis***
  4′ Calyx glabrous, 4.5–7.2 mm ..... var. ***ternata***
1′ Corolla white, yellow, or brown-purple, short tube and distinct throat together 4–11 mm
 5. Corolla 11–16 mm, tube + throat > upper lip
  6. Infl a panicle; fls clearly pedicelled; anther sac valves spreading widely; staminode densely hairy; young sts glabrous, glaucous ..... ***K. lemmonii***
  6′ Infl spike-like; fls ± sessile; anther sac valves barely spreading; staminode glabrous near tip; young sts densely short-hairy, green ..... ***K. rothrockii***
   7. Corolla 13–15 mm, sparsely long-hairy; lvs green, becoming glabrous ..... var. ***jacintensis***
   7′ Corolla 10–12 mm, becoming glabrous; lvs canescent ..... var. ***rothrockii***
 5′ Corolla (12)15–23 mm, tube + throat < upper lip
  8. Staminode densely hairy; corolla yellow; anther sacs 1.1–1.8 mm ..... ***K. antirrhinoides***
   9. Calyx 3–6 mm, lobes widely ovate, tips obtuse to acute; pl finely hairy; anther sacs 1.4–1.8 mm ..... var. ***antirrhinoides***
   9′ Calyx 5.5–9 mm, lobes lanceolate, tips acute to acuminate; pl canescent; anther sacs 1.1–1.5 mm var. ***microphylla***
  8′ Staminode glabrous near tip; corolla white, cream, or rose-tinged; anther sacs 0.6–0.8 mm ..... ***K. breviflora***
   10. Calyx glandular ..... var. ***breviflora***
   10′ Calyx glabrous ..... var. ***glabrisepala***

K. antirrhinoides (Benth.) Straw **ST** spreading to erect, 6–25 dm; young sts canescent (rarely glabrous). **LVS** ± opposite, in axillary clusters on older sts; blade 5–20 mm, (ob)lanceolate to narrowly (ob)ovate, base tapered, margin gen entire. **INFL** finely short-hairy and sparsely glandular. **FL**: corolla 15–23 mm, yellow (drying blackish), tube + widely expanded throat 6–10 mm, upper lip 8–15 mm; anther sacs 1.1–1.8 mm; staminode densely yellow-hairy, exserted. Scrub, woodland; 100–1600 m. SnBr, PR, D; AZ, Mex.

var. ***antirrhinoides*** (p. 1035) Pl finely hairy. **FL**: calyx 3–6 mm, lobes widely ovate, tips obtuse to acute; anther sacs 1.4–1.8 mm. Chaparral, oak forest; 100–1600 m. SnBr, PR; Mex. ❀DRN, SUN:**15,16**,17,24,or part SHD:**7,14,18–23**&IRR:3,8,9,11,12.

var. ***microphylla*** (A. Gray) N. Holmgren (p. 1035) Pl canescent. **FL**: calyx 5.5–9 mm, lobes lanceolate, tips acute to acuminate; anther sacs 1.1–1.5 mm. Juniper/pinyon woodland, Joshua-tree scrub; 400–1800 m. D; AZ. [ssp. *m.* (A. Gray) Straw] ❀TRY.

K. breviflora (Lindley) Straw **ST** 5–20 dm; young sts glabrous, glaucous. **LVS** opposite, subsessile; main lvs 10–40 mm, ± (ob)lanceolate, gen serrate. **INFL** glabrous to sticky-hairy. **FL**: calyx 4–8 mm, lobes lanceolate to ovate, acuminate; corolla 12–18 mm, white or cream, lobes sometimes rose-tinged, lined purplish or pinkish, tube + throat 4–8 mm, upper lip 8–12 mm; anther sacs 0.6–0.8 mm; staminode glabrous, slightly exserted. $2n=16$. Rocky slopes, forest, chaparral; < 2700 m. NCoRI, SN, ScV, SCoRO, WTR; w NV.

var. ***breviflora*** **FL**: calyx glandular. Habitats and elevations of sp. SN, SCoRO, WTR. ❀SUN,DRN:1–3,**7**,8,9,**14–16**,17,**18–23**,24.

var. ***glabrisepala*** (Keck) N. Holmgren **FL**: calyx glabrous. Habitats and elevations of sp. NCoRI, SN, ScV; w NV. [ssp. *g.* (Keck) Straw] ❀TRY.

K. cordifolia (Benth.) Straw (p. 1035) **ST** often spreading, < 30 dm; young sts glabrous to short-hairy. **LVS** ± opposite; blade 20–65 mm, ovate, base rounded, truncate, or cordate, margin gen shortly 3–11-toothed. **INFL** glandular and stiffly hairy, gen more densely hairy than herbage. **FL**: calyx 7–13 mm, lobes ± lanceolate; corolla 31–43 mm, red to reddish orange, tube 18–25 mm, throat indistinct, upper lip 11–21 mm; anther sacs 1.1–1.5 mm; staminode densely yellow-hairy, incl. $2n=16$. Chaparral, forest; < 1200 m. s SCoRO, SW; n Baja CA. ❀DRN,SUN:**17,24** or part SHD:7,**14–16,18–23**&IRR:8,9. IRR in all zones reduces summer dormancy. A yellow-fl. variant is also grown.

K. corymbosa (Benth.) Straw (p. 1035) Shrub 3–6 dm, < 1 m wide. **ST**: young sts glabrous or hairy. **LVS** opposite; blade 10–35 mm, (ob)lanceolate to narrowly (ob)ovate, base wedge-shaped, margin entire to 3–5-toothed. **INFL**: pedicels (and calyces) glandular-hairy and densely coarse-hairy. **FL**: calyx 6.4–11 mm, lobes ± lanceolate; corolla 22–40 mm, pink to red, tube 17–22 mm, throat indistinct, upper lip 9–15 mm; anther sacs 0.9–1 mm; staminode densely yellow-hairy, incl. Rocky slopes in coniferous or hardwood forests, rarely in chaparral; 100–1600 m. NW, n&c CW. ❀DRN, SUN:**15,16**,17,or partSHD:7,**14**,19–23.

K. lemmonii (A. Gray) Straw (p. 1035) **ST** 5–15 dm; young sts glabrous, glaucous, wand-like. **LVS** opposite; blade 10–65 mm, lanceolate to ovate, base rounded to widely wedge-shaped, margin gen 2–12-toothed. **INFL** glandular-hairy. **FL**: calyx 3.5–6.7 mm, lobes ± lanceolate; corolla 11–15 mm, purplish brown, lower lip pale yellow, brown-purple-lined, tube + expanded throat 5.5–9.5 mm, upper lip 2.6–6 mm; anther sacs 0.6–0.9 mm; staminode densely yellow-hairy, exserted. Rocky slopes, coniferous and mixed forests, chaparral; 200–1900 m. NW, n SN; OR. ❀DRN, part SHD:**7,14–16** or SUN:4–6,17.

K. rothrockii (A. Gray) Straw Pl wide, low, 3–6 dm. **ST** densely short-hairy; young sts green. **LVS** subopposite or in 3's, subsessile; main lvs 5–16 mm, (ob)lanceolate to widely obovate, entire or finely serrate towards tip. **INFL** short-hairy. **FL**: calyx 4–7 mm, sometimes glandular, lobes lanceolate; corolla 11–16 mm, brownish yellow, pale yellow, or cream, purple- or reddish brown-lined, tube + expanded throat 7–11 mm, upper lip 4–5 mm; anther sacs 0.8–1.1 mm, barely spreading at dehiscence; staminode glabrous distally, ± incl. Sagebrush steppe, juniper/pinyon woodland; 1900–3200 m. SnJt, SNE, DMtns; NV.

var. ***jacintensis*** (Abrams) N. Holmgren **LF** green, becoming glabrous. **FL**: corolla 13–15 mm, sparsely hairy. Habitats and elevations of sp. SnJt. [ssp. *j.* (Abrams) Keck] ❀TRY.

var. ***rothrockii*** (p. 1035) **LF** canescent. **FL**: corolla 10–12 mm, becoming glabrous. Habitats and elevations of sp. SNE, DMtns; NV. ❀TRY.

K. ternata (Torrey) Straw **ST** spreading to erect, < 25 dm; young sts glabrous, glaucous, wand-like. **LVS** gen whorled in 3's (or opposite); blade 15–60 mm, linear to narrowly (ob)lanceolate, acutely tapered at both ends, often folded and curved, margin gen 5–11-toothed. **INFL** glabrous to glandular-hairy. **FL**: calyx 3.7–7.2 mm, lobes lanceolate to ovate; corolla 21–31 mm, red, tube 16–24 mm, throat indistinct, upper lip 5.5–9.5 mm; anther sacs 0.8–1.1 mm; staminode densely hairy, incl. Juniper/pinyon woodland, chaparral, forest; 300–2700 m. Teh, TR, PR; n Baja CA.

var. ***septentrionalis*** (Munz & I.M. Johnston) N. Holmgren (p. 1035) **INFL** glandular-hairy. **FL**: calyx 3.8–5.5 mm. Mixed-hardwood forest, chaparral; ± 1500 m. Teh, WTR. [ssp. *s.* (Munz & I.M. Johnston) Straw] ❀TRY.

var. ***ternata*** (p. 1035) **INFL** glabrous. **FL**: calyx 4.5–7.2 mm. Habitats and elevations of sp. SnGb, SnBr, PR; n Baja CA. ❀DRN,part SHD:1–3,**7,14,18–21**or SUN:**15,16**,17,**22,23**,24.

KICKXIA FLUELLIN

David M. Thompson

Ann in CA, hairy. **ST** gen decumbent, much-branched. **LVS** alternate (lowest sometimes opposite), ± entire; veins pinnate. **INFL**: fls solitary in axils. **FL**: calyx lobes 5, ± equal; corolla tube with conic or cylindric spur at base, lower lip base swollen, closing mouth. **FR** ± spheric; chambers equal, circumscissile. **SEEDS** many, gen honeycombed. ± 30 spp.: esp Eur, Medit. (J.J. Kickx, Belgian professor, 1842–1887) Potentially pernicious weeds.

1. Lvs ovate to hastate below, hastate to sagittate upward . . . . ***K. elatine***
1' Lvs narrowly to widely ovate or subcordate . . . . ***K. spuria***

K. elatine (L.) Dumort. (p. 1041) **FL**: calyx lobes lanceolate, not enlarging in fr; corolla 7–15 mm, yellowish or bluish, upper lip violet. **FR** 4–4.5 mm. $n=9$. Disturbed, open places; 0–1000 m. CA-FP; native to Eur. Waif.

K. spuria (L.) Dumort. (p. 1041) **FL**: calyx lobes ovate, enlarging, base in fr ± cordate; corolla 10–15 mm, yellow, upper lip deep purple. **FR** 3–5 mm. $n=9$. Disturbed, open places; 0–1000 m. CA-FP; native to Eur. Waif.

LIMOSELLA MUDWORT

Margriet Wetherwax

Ann, sometimes partly submersed, with stolons. **LVS** basal, erect; stipules gen present. **INFL** scapose; fl 1. **FL**: calyx bell-shaped, lobes 5, ovate, acute; corolla radial, bell-shaped, lobes 5, upper surfaces sparsely papillate; stamens 4; style gen subterminal, stigmas gen fused, head-like. **FR** elliptic to spheric, chambers 2 below, 1 above. **SEEDS** many, minute. ± 15 ssp.: worldwide. (Latin: mud seat, from habitat)

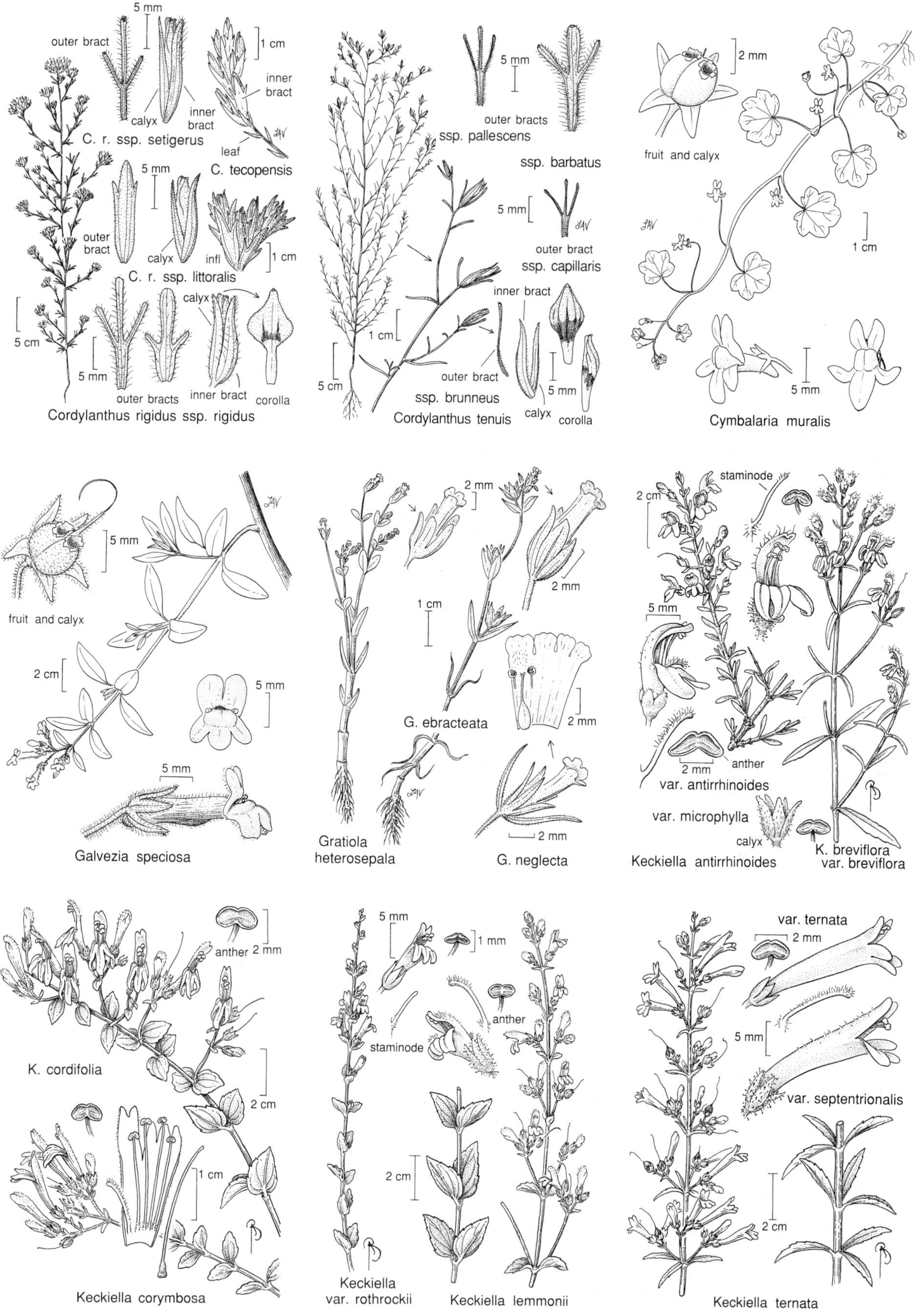

5 mm
outer bract
1 cm
inner bract
calyx
inner bract
C. r. ssp. setigerus
leaf
C. tecopensis
5 mm
outer bract
calyx
infl
1 cm
C. r. ssp. littoralis
calyx
5 cm
5 mm
outer bracts
inner bract
corolla
Cordylanthus rigidus ssp. rigidus
5 mm
outer bracts
ssp. pallescens
ssp. barbatus
5 mm
outer bract
ssp. capillaris
inner bract
1 cm
outer bract
5 mm
5 cm
ssp. brunneus
Cordylanthus tenuis
calyx
corolla
2 mm
fruit and calyx
1 cm
5 mm
Cymbalaria muralis
5 mm
fruit and calyx
2 cm
5 mm
5 mm
Galvezia speciosa
2 mm
2 mm
1 cm
G. ebracteata
2 mm
2 mm
Gratiola heterosepala
G. neglecta
staminode
2 cm
5 mm
2 mm
anther
var. antirrhinoides
var. microphylla
calyx
Keckiella antirrhinoides
K. breviflora var. breviflora
anther
2 mm
K. cordifolia
2 cm
1 cm
Keckiella corymbosa
5 mm
1 mm
anther
staminode
2 cm
Keckiella var. rothrockii
Keckiella lemmonii
var. ternata
2 mm
5 mm
var. septentrionalis
2 cm
Keckiella ternata

1. Lvs awl-like, cylindric; style > ovary, slender .... ***L. subulata***
1′ Lvs flat, blade linear to ovate; style < or = ovary, stout
 2. Lf linear to ± spoon-shaped, petiole not differentiated; corolla lobes rounded .... ***L. acaulis***
 2′ Lf blade spoon–shaped to ovate, petiole clearly differentiated; corolla lobes acute .... ***L. aquatica***

L. acaulis Sessé & Mociño (p. 1041) Cespitose, gen mat-forming. **LF**: 1– 6 cm, 0.5–2 mm wide, flat, linear to ± spoon-shaped; petiole gen not differentiated; stipules gen ear-shaped, transparent. **INFL**: pedicels < 2/3 lf length, spreading in fr. **FL**: calyx lobes ± = tube; corolla 2–3 mm, white to lavender, lobes rounded; stamens attached at 1 level; style 0.5–1 mm, erect to slightly curved. **FR** 3–5 mm, spheric. Wet, muddy places; < 3300 m. SNH, GV, SnFrB, SW, MP; to NM, Mex. ❀TRY.

L. aquatica L. (p. 1041) Tufted. **LF**: petiole 3–10(30) cm, gen clearly differentiated from blade; blade 5–30 mm, flat, spoon-shaped to ovate; stipules 0 or sheathing lf base, transparent. **INFL**: pedicels << petioles, spreading in fr. **FL**: calyx lobes 2 × tube; corolla ± 2 mm, white to pinkish, outside sometimes blue, lobes acute; stamens attached at different levels; style 0.2–0.4 mm, stout, ± curved, stigma slightly 2-lobed. **FR** 3–5 mm, elliptic to spheric. $2n=40$. Wet, muddy, periodically flooded places; < 3200 m. CaR, SNH, CCo, SnFrB, SnBr, MP; to AK, e N.Am, Eurasia. ❀TRY.

L. subulata Ives (p. 1041) Tufted. **LF** 1–3 cm (petiole not differentiated), linear, awl-like to cylindric, obtuse; stipules 0 or tapered. **INFL**: pedicels < lvs, curved in fr. **FL**: calyx lobes < to = tube; corolla ± 3 mm, lobes rounded, white to lavender-blue; stamens attached at 1 level; style > ovary, < 2 mm, head-like, straight to slightly curved, subterminal. **FR** spheric. $2n=20$. Muddy or sandy intertidal flats; < 10 m. Deltaic GV (Contra Costa, Sacramento cos.); s B.C., native to e coast N.Am, Eur.. Poorly known in w N.Am.

LINARIA TOADFLAX

Margriet Wetherwax

Ann to per, gen glabrous. **ST** ascending to erect, simple or branched at base. **LVS** gen opposite or whorled (or upper alternate), sessile, simple, linear to ovate, gen wider on non-fl shoots, entire to dentate, pinnately veined. **INFL**: spike or raceme, terminal; bracts reduced, alternate. **FL**: calyx lobes 5, deep, ± equal; corolla 5-lobed, 2-lipped, lower side of tube spurred at base, lower side of throat swollen, ± hairy, ± closing corolla below lips; stamens 4, in 2 pairs, incl; stigma small, head-like or lobes 2, flat. **FR** ± spheric, opening by slits into chambers near tip. **SEEDS** many, flat and winged or pyramid-like and ± ridged. ± 100 spp.: esp Eurasia; many cult. (Latin: flax, from flax-like lvs of some) Corolla length incls spur.

1. Corolla yellow
 2. Ann; st 5–20 cm, trailing; corolla 15–20 mm .... ***L. supina***
 2′ Per; st 30–120 cm, ascending to erect; corolla 20–50 mm
  3. Lf lanceolate to ovate; corolla 35–50 mm; seed ± pyramid-like, ridged .... ***L. genistifolia*** ssp. ***dalmatica***
  3′ Lf ± linear; corolla 18–32 mm; seed flat, winged .... ***L. vulgaris***
1′ Corolla gen not yellow
 4. Per; lf linear to oblanceolate .... ***L. purpurea***
 4′ Ann; lf linear to lance-linear
  5. Corolla spur gen strongly curved
   6. Corolla violet-purple, throat swelling obvious, orange .... ***L. bipartita***
   6′ Corolla violet to blue, throat swelling ± obscure, ± white-ridged .... ***L. canadensis***
  5′ Corolla spur ± straight
   7. Corolla 20–38 mm, purplish, throat swelling with small whitish patch (CVS highly variably colored) .... ***L. maroccana***
   7′ Corolla 15–18 mm, purple, throat swelling yellow, purple-veined .... ***L. pinifolia***

L. bipartita Willd. Ann, glabrous, glaucous. **ST** 10–30 cm. **LF** 30–50 mm, linear. **FL**: calyx 2.5–6 mm; corolla 18–20 mm, violet-purple, throat swelling orange; stigma lobes 2, flat. **FR** ± 2 mm. **SEED** minute, pyramid-like, ridged. Disturbed places; gen < 500 m. SW; native to Medit.

L. canadensis (L.) Dum.-Cours. (p. 1041) BLUE TOADFLAX Ann or bien, gen glabrous, ± glaucous. **ST** 10–60 cm, slender, with decumbent non-fl shoots. **LF** 5–25 mm, narrowly linear, obtuse. **INFL**: raceme, dense in fl, open in fr, ± glandular-puberulent; pedicels 1.5–6 mm, > bracts. **FL**: calyx ± 3 mm, lobes linear-lanceolate, acute; corolla 10–24 mm, violet to blue, lips spreading, lower lip >> upper, throat swelling ± obscure, white-ridged, spur curved, slender. **FR** ± 3 mm. **SEED** 0.5 mm, ± pyramid-like, ± tubercled. $2n=12$. Sand or gravel; < 1800 m. NCo, NCoR, SNF, GV, CW, WTR, PR. Pls with larger fls, seeds densely tubercled have been called var. *texana* (Scheele) Pennell. ❀SUN,DRN,IRR:1,2,5,6,**7–9**,10,**14–24**.

L.genistifolia (L.) Miller ssp. *dalmatica* (L.) Maire & Petitm. (p. 1041) DALMATIAN TOADFLAX Per, glabrous. **ST** erect, 20–120 cm, branched. **LVS** crowded, < 60 mm, lanceolate to ovate, acute to long-tapered, ± clasping, rigid. **INFL** dense to open; pedicels 1–13 mm, ± = bracts. **FL**: calyx 2–12 mm, lobes linear to triangular-ovate; corolla 20–50 mm, yellow, lower lip closing throat, densely white- to orange-hairy. **FR** 3–7 mm. **SEED** ± 1.2 mm, ± pyramid-like, ridged. $2n=12$. Disturbed places, pastures, fields; gen < 1000 m. CA-FP; widely naturalized elswehere; native to Medit. [*L. d.* (L.) Miller] NOXIOUS WEED.

L. maroccana Hook. f. Ann. **ST** < 50 cm. **LF** 20–40 mm, linear. **INFL** ± glandular-puberulent. **FL**: calyx 3–6 mm, lobes lanceolate; corolla 20–38 mm, gen purple, throat swelling with small whitish patch (CVS variably colored, some like *L. bipartita*); spur gen ± straight; stigma lobes 2, flat. **FR** 3–5 mm. **SEED** ± 1 mm, pyramid-like, ridged, black. $2n=12$. Roadsides, disturbed places; gen < 500 m. CW, SW; native to Medit. Garden escape and component of “wildflower” seed mixes.

L. pinifolia (Poiret) Thell. Ann. **ST** 80–100 cm, with decumbent non-fl shoots. **LF** 20–40 mm, linear (elliptic on non-fl shoots). **INFL** ± glandular-puberulent. **FL**: calyx 3–5 mm, lobes lanceolate; corolla 15–18 mm, purple, throat swelling yellow, veins purple; stigma lobes 2, flat. **FR** ± 4 mm. **SEED** ± 1 mm, pyramid-like, ridged. $2n=12$. Disturbed places; gen < 500 m. s CW; native to Medit. Garden escape.

L. purpurea (L.) Miller Per. **ST** 30–70 cm. **LF** 20–50 mm, linear. **FL**: calyx ± 2.5 mm; corolla 15–18 mm, lavender-purple, throat swelling darker. **FR** ± 3 mm. **SEED** ± 1 mm, ± pyramid-like, ridged. Disturbed places; gen < 500 m. CW, SW; native to Medit. Garden escape.

L. supina (L.) Chaz. Ann. **ST** 5–20 cm, simple. **LVS** whorled, 5–30 mm, ± linear. **INFL** glandular-hairy; pedicels < 6 mm, < bracts.

FL: calyx 3–7 mm, corolla 15–20 mm, pale yellow, sometimes violet-tinged. **FR** 3–7 mm. **SEED** < 3 mm, flat, winged. $2n=12$. Disturbed places; gen < 500 m. s CW; native to Medit. Garden escape; perhaps only a waif in CA.

L. vulgaris Miller (p. 1041) BUTTER-AND-EGGS Per. **ST** 30–100 cm, ascending to erect, simple or not. **LVS** crowded, 25–50 mm, linear, glabrous to sparsely soft-shaggy-hairy. **INFL**: raceme, dense, ± glandular-hairy; pedicel 1–4 mm. **FL**: calyx lobes ± 3 mm, unequal; corolla 18–32 mm, yellow, lower lip orange-hairy. **FR** 9–12 mm. **SEED** 1.5 mm, flat, winged. $2n=12$. Disturbed places; gen < 1000 m. CA-FP; widely naturalized elsewhere; native to Medit.

LINDERNIA FALSE PIMPERNEL

Margriet Wetherwax

Ann, bien. **ST** openly branched. **LVS** opposite, entire to finely dentate. **FL**: calyx radial, sepals 5, ± free; corolla 2-lipped, upper lip erect, 2-lobed, lower lip > upper, spreading, 3-lobed, throat with 2 yellow, hairy ridges; fertile stamens 2 (lower stamen pair antherless, forming corolla throat ridges and with free, forked, filament-like tips); stigmas 2, flat. ± 50 spp.: esp trop, subtrop. (Franz B. von Lindern, German physician, botanist, 1682–1755)

L. dubia (L.) Pennell Ann, glabrous. **ST** < 38 cm, spreading to erect. **LF** 1–37 mm; blade lanceolate to ovate, entire to finely dentate, tapered to round-clasping at base. **INFL**: pedicels 0.5–28 mm. **FL**: sepals < 7 mm, linear; corolla 7–10 mm, white to bluish or lavender. **FR** < or = sepals. **SEED** length 1.5–3 × width, yellow. Wet places; < 1600 m. n&c CA-FP, SNE; N.Am, S.Am. Vars. unclearly differentiated.

1. Pedicels gen >> subtending lvs; lvs gen much reduced above .... var. ***anagallidea***
1′ Pedicels < or slightly > subtending lvs; lvs slightly reduced above .... var. ***dubia***

var. ***anagallidea*** (Michaux) Cooperr. (p. 1041) **LF** 1–24 mm, gen much reduced above; blade ovate to elliptic, ± entire, nearly all widest and round-clasping at base. **INFL**: pedicels 3–28 mm, gen >> subtending lvs (lowermost sometimes shorter). **FL**: corolla 7–9 mm, white to lavender. **FR** > calyx. **SEED** length < 2 × width, brownish yellow. Habitats and elevations of sp. SN, GV; to WA, e US, Mex. [*L. a.* (Michaux) Pennell]

var. ***dubia*** (p. 1041) **LF** 1–37 mm, slightly reduced above; blade lanceolate, lanceolate to ovate; lower lvs (or all) tapered at base. **INFL**: pedicels 0.5–27 mm, < or slightly > subtending lvs. **FL**: corolla 9–10 mm, pale lavender, darker distally. **FR** < or = calyx. **SEED** length 2–3 × width, pale yellow. Streamsides; 100–700 m. KR, SN, GV, SnFrB, SNE; to WA, e US, Mex.

MAURANDYA

David M. Thompson

Per, glabrous to hairy. **ST** prostrate, erect, pendent, or climbing by twining sts or petioles. **LVS** gen alternate (lowest opposite on seedlings), entire to irregularly bristly-dentate; veins palmate. **INFL**: fls solitary in lf axils. **FL**: calyx lobes 5, ± equal; corolla tube with sac-like extension at base, tube-throat floor gen with 2 longitudinal folds, lower lip base gen not swollen, gen not closing mouth. **FR** ovoid to ± spheric, often oblique; chambers 1–2, equal or not, each gen dehiscent by 2–3 pores or irregularly near tip. **SEEDS** many, ± tubercled, pitted, or winged. 20 spp.: sw US, Mex. (C.P. Maurandy, botany professor, Cartagena, Spain, ± 18th century) [Elisens 1985 Syst Bot Monogr 5:1–97]

1. Pl glabrous; lf hastate to sagittate, entire; sts or petioles twining; corolla ± reddish to violet .... ***M. antirrhiniflora*** ssp. ***antirrhiniflora***
1′ Pl hairy; lf ± round to reniform, irregularly bristly-dentate; st erect to pendent, neither sts nor petioles twining; corolla yellowish to white .... ***M. petrophila***

M. antirrhiniflora Willd. ssp. ***antirrhiniflora*** (p. 1041) VIOLET TWINING SNAPDRAGON Pl glabrous; sts or petioles twining. **LF** hastate to sagittate, entire; **INFL**: pedicel 12–47 mm. **FL**: calyx lobes entire; corolla ± reddish to violet, tube-throat 13–17 mm, floor without longitudinal folds, lower lip swollen at base, ± closing throat. **FR**: chambers unequal. $2n=24$. RARE in CA. Desert flats, washes; 0–2600 m. e DMtns (Providence Mtns, e San Bernardino Co.); to TX, Mex. In cult.

M. petrophila Cov. & C. Morton ROCK LADY Pl hairy; neither sts nor petioles twining. **ST** erect to pendent. **LF** ± round to reniform, irregularly bristly-dentate. **INFL**: pedicel 1–4 mm. **FL**: calyx lobes irregularly bristly-dentate; corolla tube-throat 20–24 mm. **FR**: chamber 1 (septum incomplete). **RARE** CA. Limestone crevices of canyons; 1200–1400 m. n DMoj (Titus, Fall canyons, Death Valley region, Inyo Co.).

MIMULUS MONKEYFLOWER

David M. Thompson

Ann to shrub, glabrous to hairy. **ST** gen erect. **LVS** opposite, gen ± sessile, gen toothed or gen entire, reddish or gen green. **INFL**: raceme, bracted, or fls gen 2 per axil. **FL** sometimes cleistogamous; calyx gen green, lobes 5, gen << tube, equal or not, gen uppermost largest; corolla gen deciduous, white to red, maroon, purple, gold or yellow, limb width measured at widest point looking into fl, lower lip base sometimes swollen, ± closing mouth, tube-throat floor gen with 2 longitudinal folds; pollen chambers spreading; placentas 2, axile or parietal; stigma lobes gen lf-like, gen incl. **FR** gen ovoid to fusiform, gen upcurved if elongate, gen ± fragile, loculicidal near tip (sometimes hard, indehiscent); chambers 1–2. **SEEDS** many, gen < 1 mm, ovoid, yellowish to dark brown. ± 100 spp.: w N.Am, Chile, e Asia, s Afr, New Zealand, Australia. (Latin: little mime or comic actor, from face-like corolla limb of some) [Grant 1924 Ann Missouri Bot Gard 11:99–388]

1. Pedicel gen > calyx; corolla deciduous (exc *M. breweri*), not salverform; placentas axile, fused, not parted by fr dehiscence (subg. *Mimulus*)
2. Per; corolla lavender to purple or orange to red — corolla tube-throat > 2 cm

3. Lf pinnately veined; corolla lavender to purple, mouth nearly closed by swollen lower lip base (sect. *Mimulus*) .... ***M. ringens***

3′ Lf palmately veined; corolla lavender to purple or orange to red, mouth open (sect. *Erythranthe*)

4. Corolla orange to red, anthers, stigma exserted but arched in upper lip .... ***M. cardinalis***

4′ Corolla lavender to purple, anthers, stigma incl .... ***M. lewisii***

2′ Gen ann; corolla color various, incl yellow (if per, corolla yellow)

5. Calyx in fr strongly and asymmetrically swollen, uppermost lobe longest, lowest 2 gen upcurved; corolla yellow, lower lip base gen swollen, ± closing mouth (sect. *Simiolus*)

6. Bracts at nodes subtending fls fused around st completely, forming ± circular disk, glaucous ... ***M. glaucescens***

6′ Bracts or lvs at nodes subtending fls petioled or fused around st only at their bases, not forming circular disk, not glaucous

7. Lvs or at least some ± pinnately lobed or dissected into narrow segments .... ***M. laciniatus***

7′ Lvs ± entire or ± crenate, not pinnately lobed or dissected (but base often irregularly dissected or small-lobed)

8. Bract or lf pairs at nodes subtending fls linear to lanceolate, not fused at base .... ***M. nudatus***

8′ Bract or lf pairs at nodes subtending fls ovate to cordate or round, sometimes fused at base around st

9. Fls gen > 5 in a definite, bracted raceme; ann or per; corolla tube-throat 2–40 mm; fls cleistogamous or opening .... ***M. guttatus***

9′ Fls 1–5 per st, solitary in axils of upper lvs, not in bracted raceme; per; corolla tube-throat 17–45 mm; fls opening .... ***M. tilingii***

5′ Calyx in fr not (or symmetrically) swollen, lobes gen ± equal or lowest 2 longest, not upcurved; corolla yellow or not, lower lip base not swollen, not closing mouth

10. Calyx lobes ± = tube; pl hairs dense, long, soft, wavy (sect. *Mimuloides*) .... ***M. pilosus***

10′ Calyx lobes << tube; pl glabrous, puberulent, or hairs ± straight (sect. *Paradanthus*)

11. Per; stolons or rhizomes present

12. Calyx glabrous; lf palmately 3-veined .... ***M. primuloides***

13. Lvs linear to oblanceolate, ± erect, some ± basal but not in rosettes .... ssp. ***linearifolius***

13′ Lvs oblong to obovate, gen ± spreading in ± distinct rosettes .... ssp. ***primuloides***

12′ Calyx ± glabrous to hairy; lf pinnately 5- or more veined

14. Corolla tube-throat 28–40 mm, tube funnel-shaped, gen > 4 mm wide .... ***M. dentatus***

14′ Corolla tube-throat 15–26 mm, tube ± cylindric, 2–4 mm wide exc at mouth .... ***M. moschatus***

11′ Ann; stolons and rhizomes 0

15. Calyx tube ± hairy

16. Lf palmately veined; calyx 8–10(13 in fr) mm; corolla pinkish to white .... ***M. parishii***

16′ Lf pinnately veined or ± so; calyx 3–8 mm; corolla yellow

17. Corolla lobe bases without large blotches; calyx lobes ± acute; pedicel in fr not growing away from light .... ***M. floribundus***

17′ Corolla lobe bases each with 1 large, central, maroon-purple blotch; calyx lobes rounded; pedicel in fr growing away from light .... ***M. norrisii***

15′ Calyx tube glabrous to puberulent

18. Calyx in fr greatly swollen, membranous; corolla overall lavender or rose to whitish or yellow

19. Upper lvs sessile; corolla lavender or rose to whitish or yellow

20. Corolla tube-throat 7–16 mm, limb < 14 mm wide; drained upland areas .... ***M. inconspicuus***

20′ Corolla tube-throat 7–11 mm, limb 4–6 mm wide; vernally wet depressions .... ***M. latidens***

19′ Upper lvs petioled; corolla yellow (rarely whitish with pinkish border in *M. pulsiferae*)

21. Corolla tube-throat 3–6.5 mm .... ***M. breviflorus***

21′ Corolla tube-throat 7–13 mm .... ***M. pulsiferae***

18′ Calyx in fr slightly enlarging but not or slightly swollen, ± corky or membranous; corolla purple, lavender, yellow, or partly white

22. Corolla lobes yellow or white

23. Lower calyx lobes > upper .... ***M. alsinoides***

23′ Lower calyx lobes ± = upper

24. Corolla tube-throat 4–10 mm

25. Corolla tube-throat 6–10 mm; calyx lobes ciliate .... [2]***M. rubellus***

25′ Corolla tube-throat 4–6 mm; calyx lobes glabrous .... ***M. suksdorfii***

24′ Corolla tube-throat 10–19 mm

26. White corolla lobes 0–4; calyx in fr ± corky; pedicel in fr thickened, ± rigid .... ***M. bicolor***

26′ White corolla lobes 0; calyx in fr membranous; pedicel in fr thin, flexible .... [2]***M. montioides***

22′ Corolla lobes, or some, maroon, purple, rose, or lavender

27. Corolla tube-throat gen 2–10 mm (if 8–10 mm, then calyx lobes ciliate, corolla pink to lavender)

28. Corolla tube-throat 2–3.5 mm and deciduous (or 4.5–8 mm and persistent); calyx lobes puberulent but not ciliate

29. Calyx 3.5–9 mm, > fr; corolla tube-throat > 4.5–8 mm, persistent .... ***M. breweri***

29′ Calyx 2–3.5 mm, < fr; corolla tube-throat 2–3.5 mm, deciduous .... ***M. exiguus***

28′ Corolla tube-throat 5–10 mm, deciduous; calyx lobes ± glabrous or ciliate

30. Calyx lobes ± glabrous .... ***M. androsaceus***

30′ Calyx lobes ciliate

31. Corolla limb 5–9 mm wide .... ***M. gracilipes***

31′ Corolla limb 3–4 mm wide .... [2]***M. rubellus***

27′ Corolla tube-throat > 8 mm (if 8–10 mm, calyx lobes not ciliate, some or all corolla lobes maroon or purple)
32. Upper corolla lip 4-lobed, lower lip of 1 large, notched, yellow lobe .... ***M. shevockii***
32′ Upper corolla lip of 2 lobes, lower of 3 ± equal, partly yellow or entirely purple lobes, or corolla not lipped
33. Calyx gen with red-brown spots, in fr with wide corky ribs, lobes acute to long-tapered . ***M. filicaulis***
33′ Calyx gen without spots, in fr membranous, lobes gen obtuse to truncate
34. Lvs linear, gen 1–3 mm wide; pedicel 3–25 mm .... [2]***M. montioides***
34′ Lvs (at least lower) ovate, > 3 mm wide; pedicel 10–65 mm
35. Corolla not lipped, tube-throat 10–22 mm, lobes equally colored; upper lvs often ± linear .... ***M. palmeri***
35′ Corolla strongly 2-lipped, tube-throat 8–13 mm, upper lip darker; upper lvs ovate to elliptic .... ***M. purpureus***

1′ Pedicel gen < calyx; corolla persistent, (exc *M. pictus*; placentas parietal, parted at least at tip by fr dehiscence (or fr hard, indehiscent) (subg. *Schizoplacus*)
36. Shrub, subshrub, or per from woody caudex; calyx 2–4 cm; corolla tube-throat 25–60 mm; lf edges often rolled under; main lf axils often with clusters of smaller lvs (sect. *Diplacus*)
37. Shrub or subshrub; lf upper surface glabrous; calyx not swollen at base; corolla gen not yellow; fr splitting only along upper suture .... ***M. aurantiacus***
37′ Per from woody caudex; lf upper surface hairy; calyx gen swollen at base; corolla yellow; fr 4-valved at tip .... ***M. clevelandii***
36′ Ann, rarely per; calyx gen < 2 cm, corolla tube-throat gen < 25 mm; lf edges gen ± flat; main lf axils not leafy
38. Corolla radial, salverform, maroon and ± white, throat floor without longitudinal folds (sect. *Mimulastrum*)
39. Corolla limb ± solid maroon at base, maroon veins appearing << 0.5 mm wide, fading into white border .... ***M. mohavensis***
39′ Corolla limb ± white ± throughout, maroon veins with much wider maroon border, so appearing > 0.5 mm wide .... ***M. pictus***
38′ Corolla bilateral, gen not maroon and white, throat floor gen with 2 longitudinal folds — fls sometimes all cleistogamous
40. Fr hard, often oblique at base, indehiscent while st alive; gen fls March–April (later in vernally wet depressions or on desert cliffs) (sect. *Oenoe*)
41. Sts < 1 cm, tightly tufted; lf ± ciliate at least in basal 1/2, often ± linear; vernally wet depressions
42. Corolla tube-throat 5–10 mm .... ***M. pygmaeus***
42′ Corolla tube-throat 20–60 mm
43. Corolla tube-throat glabrous outside; lowest corolla lobe magenta with a large, dark purple spot at base .... ***M. angustatus***
43′ Corolla tube-throat puberulent outside; lowest corolla lobe gold with scattered, small, dark dots at base .... ***M. pulchellus***
41′ Sts gen > 1 cm, not tightly tufted but often branched; lf gen not ciliate (or, if so, hairs not restricted to margin), gen not linear; vernally wet depressions or not
44. Corolla lobes each with 1 large, dark spot at center of base; fls opening
45. Corolla pink to white; limestone crevices .... ***M. rupicola***
45′ Corolla magenta to purple; vernally wet depressions .... ***M. tricolor***
44′ Corolla without large, dark spots; fls often cleistogamous
46. Corolla tube-throat gen < 2 cm (if fls cleistogamous then fr angular, lf dull purplish, petiole ciliate)
47. Lf ± ciliate on petiole, blade base; fr 4–8.5 mm .... ***M. congdonii***
47′ Lf not ciliate; fr 6–11 mm .... ***M. latifolius***
46′ Corolla tube-throat 2–4.5 cm (if fls cleistogamous then fr ± ovoid, lf shiny green, petiole glabrous)
48. Corolla with upper lip >> lower, tube abruptly widened to long throat, throat longitudinally striped within; fls often cleistogamous .... ***M. douglasii***
48′ Corolla with both lips well developed, often funnel-like but tube-throat not abruptly widened above, not striped within; fls opening
49. Corolla tube-throat ± 2 × calyx, upper lip lobes magenta to red-purple .... ***M. kelloggii***
49′ Corolla tube-throat ± = calyx, upper lip lobes white .... ***M. traskiae***
40′ Fr fragile, symmetric at base, gen promptly dehiscent; fls gen May–September (often earlier in deserts) (sect. *Eunanus*)
50. All fl nodes with 1 fl or fr developing
51. Corolla throat floor densely hairy; SNF .... ***M. viscidus***
51′ Corolla throat floor glabrous to sparsely puberulent; coastal and sw mtns, adjacent D
52. Corolla tube-throat 10–23 mm, limb 8–26 mm wide .... ***M. fremontii***
52′ Corolla tube-throat 7–10 mm, limb 4–7 mm wide .... ***M. rattanii***
50′ At least some fl nodes (gen upper) with 2 fls or frs developing
53. Stigma ± exserted, lobes rounded
54. Upper lf tips sharply acuminate; corolla magenta .... ***M. cusickii***
54′ Upper lf tips rounded to acute; corolla ± magenta, lavender-purple, violet-purple, or yellow
55. Corolla tube-throat 7–12 mm; fr 4–6 mm .... ***M. jepsonii***
55′ Corolla tube-throat 9–19 mm; fr 6–12 mm
56. Lvs and calyx hairy; corolla ± magenta or yellow .... ***M. mephiticus***
56′ Lvs and calyx ± puberulent; corolla ± magenta to purplish

57. Calyx mouth in fr strongly oblique ..... ***M. johnstonii***
57′ Calyx mouth in fr slightly oblique ..... ***M. nanus***
53′ Stigma incl (if ± exserted, stigma lobes acute)
58. Corolla tube-throat 6–8 mm, limb 3–5 mm wide ..... ***M. leptaleus***
58′ Corolla tube-throat > 9 mm, limb > 5 mm wide
59. Calyx 10–25 mm, corolla yellow ..... ***M. brevipes***
59′ Calyx < 9 mm (or corolla pink to magenta or maroon)
60. Calyx lobes spreading, long-tapered (or upper lobe rounded, ± with small point at tip); SNE, D
61. Calyx puberulent, often purplish throughout, upper lobe rounded, ± with small point at tip, fr incl ..... ***M. parryi***
61′ Calyx hairy, sometimes purplish on ribs, upper lobe long-tapered, fr exserted ..... ***M. bigelovii***
62. Lower internodes gen > upper; lf elliptic, sometimes acuminate but not abruptly so ..... var. ***bigelovii***
62′ Lower internodes ± = upper; lf gen wide-ovate to ± round, abruptly sharp-pointed ..... var. ***cuspidatus***
60′ Calyx lobes converging, ± erect or ± spreading, obtuse to long-tapered; CA-FP
63. Calyx swollen, esp in fr
64. Corolla limb and throat floor glabrous or puberulent ..... ***M. bolanderi***
64′ Corolla limb and throat floor long-hairy ..... ***M. constrictus***
63′ Calyx not swollen
65. Calyx not ribbed, uniformly green, puberulent, lobes obtuse to acute; corolla light rose to magenta, throat floor with 2 gold stripes extending from folds onto lower lip base ..... ***M. torreyi***
65′ Calyx dark-ribbed, puberulent or ± hairy, lobes acute to long-tapered; corolla almost white to dark magenta or yellow, throat often yellowish within but floor without gold stripes
66. Corolla almost white to dark magenta with darker central marks on lobes, throat dark-spotted ..... ***M. layneae***
66′ Corolla dark magenta or yellow, lobes unmarked, throat gen dark-lined ..... ***M. whitneyi***

M. alsinoides Benth. (p. 1041) Ann, 0.5–15 cm, puberulent. **LF**: petiole 0.5–20 mm; blade 3–18 mm, ovate to round. **FL**: pedicel 7–27 mm; calyx 4–8 mm, minutely puberulent, upper 3 lobes ± 0.5 mm, acute to acuminate, lower 2 lobes ± 1 mm, widely rounded; corolla yellow, gen with large, reddish spot on lower lip base, tube-throat 7–11 mm; placentas axile. **FR** 5–7 mm. Moist to wet places, often on moss-filled crevices of rock faces; < 860 m. KR; to B.C.

M. androsaceus E. Greene (p. 1041) Ann, 0.5–9 cm, minutely puberulent. **LF** 3–13 mm, ± lanceolate to oblong or ovate, narrowly sheathing st. **FL**: pedicel 7–27 mm, ± spreading (in fr ascending at tips); calyx 3.5–6.5 mm, tube gen minutely puberulent, lobes equal, 0.2–1 mm, ± glabrous; corolla red-purple, tube-throat 5–8 mm, limb 4–6 mm wide; placentas axile. FR 4–5 mm. Uncommon. Moist runoff areas on gentle slopes; < 2100 m. NCoRI, Teh, CW, WTR, SnBr, e PR (Santa Rosa Mtns, Riverside Co.).

M. angustatus (A. Gray) A. Gray (p. 1041) Ann. **STS** < 1 cm, tightly tufted, glabrous. **LF** 5–36 mm, ± linear, at least basal 1/2 ciliate. **FL**: pedicel 0–1 mm; calyx 6–14 mm, hairy, lobes unequal, 1–4 mm, obtuse, often ciliate; corolla persistent, magenta-purple, lowest lobe magenta with large, dark purple spot at base, tube-throat 20–60 mm, glabrous outside; placentas parietal. **FR** 2–4 mm, ± oblique-ovoid, hard, indehiscent. n=9. Vernally wet depressions; 250–1200 m. c NCoRO (Longvale, Mendocino Co.), s NCoRI, s CaRF, n SNF, SnJV (Pinehurst, Fresno Co.).

M. aurantiacus Curtis (p. 1041) Subshrub, shrub, glabrous to hairy. **ST** 10–150 cm; main lf axils often with clusters of smaller lvs. **LF** 20–80 mm, linear to obovate; edges gen rolled under; upper surface glabrous, often sticky. **FL**: pedicel 3–30 mm; calyx 20–37 mm, not swollen at base, glabrous to hairy, lobes unequal, 3–10 mm, acute to acuminate; corolla persistent, white to buff, yellow, orange, or red, tube-throat 25–60 mm; placentas parietal. **FR** 12–20 mm, splitting only along upper suture. n=10. Common. Rocky hillsides, cliffs, canyon slopes, disturbed areas, borders of chaparral, open forest; < 1600 m. CA-FP. [sspp. *australis* (McMinn) Munz & *lompocensis* (McMinn) Munz] Highly complex, with many intergrading, hybridizing, local forms; the most distinct have been called *M. aridus* (Abrams) A.L. Grant, low bush monkeyflower (corolla yellow; se PR, San Diego Co., nw Baja CA); *M. bifidus* Pennell incl ssp. *fasciculatus* Pennell (corolla lobes deeply notched; SNF, SCoRO); *M. flemingii* Munz (corolla red; n ChI, San Clemente Island); *M. longiflorus* (Nutt.) A.L. Grant incl ssp. *calycinus* (Eastw.) Munz & var. *rutilus* A.L. Grant (calyx hairy; s SCoRO, SW); *M. puniceus* (Nutt.) Steudel (corolla red; sw SW, incl Santa Catalina Island). The name *M. parviflorus* Lindley has been misapplied to *M. flemingii* Munz. ❀DRN,IRR fall through spring & DRY when dormant:**7**,8,9,**14**,18,**19–23**&SUN:4,5,**15–17**,**24**;CVS.

M. bicolor Benth. (p. 1041) Ann, 4–27 cm, densely puberulent. **LF** 4–30 mm, linear to oblanceolate. **FL**: pedicel 6–30 mm; calyx 5.5–12 mm, ± corky in fr, ± dotted red, puberulent, lobes equal, 1–3.5 mm, acute; corolla tube-throat 10–19 mm, lobes each 2-lobed, upper lip white or yellow, lower lip yellow, often white on sides; placentas axile. **FR** 4–8 mm. Moist places, gen on clay soils along level drainage areas; 360–2100 m. se KR, s CaR, SN. ❀IRR or WET,SUN:1–3,6,**7**,8–10,**14–24**.

M. bigelovii (A. Gray) A. Gray (p. 1041) Ann, 2–25 cm, densely hairy. **LF** 7–35 mm, elliptic to round, often abruptly acuminate. **FL**: pedicel 1–4 mm; calyx 6–13 mm, sometimes purplish but only on ribs, hairy, lobes spreading, unequal, 1.5–6 mm, long-tapered; corolla persistent, rose to dark magenta, tube-throat 12–22 mm, throat floor gold, mouth gen with 2 lateral, dark maroon patches; placentas parietal. **FR** 7–13 mm, exserted. n=8. Rocky desert slopes, margins of washes; 120–2300 m. s SNE, D; to sw UT, w AZ. Variable; vars. intergrade.

var. ***bigelovii*** **ST**: lower internodes gen > upper. **LVS** elliptic, sometimes acuminate but not abruptly so, upper narrower, longer-tapered than lower. Habitats of sp.; 120–1700 m. D; s NV, w AZ. ❀DRN,SUN:6,**7**,**14**,15–17&IRR:**8–12**,13,**18–24**.

var. ***cuspidatus*** A.L. Grant **ST**: lower internodes ± = upper (all reduced in severe drought). **LVS** gen widely ovate to ± round, abruptly sharp-pointed, upper lvs as wide or wider, often more abruptly sharp-pointed than lower lvs. Habitats and elevations of sp. s SNE, n D; to sw UT, nw AZ. [var. *panamintensis* Munz; *M. spissus* A.L. Grant] ❀TRY.

M. bolanderi A. Gray (p. 1041) Ann, 2–90 cm, gen densely hairy. **LVS** 5–60 mm, oblanceolate to ovate; lowest lvs glabrous, upper densely hairy. **FL**: pedicel 2–5 mm; calyx 7–27 mm, wide-ribbed, swollen (esp in fr), densely hairy, lobes often ± spreading in fr, unequal, 1–9 mm, acute to acuminate; corolla persistent, magenta, throat floor folds white, tube-throat 12–30 mm, throat floor and limb glabrous or puberulent; placentas parietal. **FR** 7–20 mm. n=8. Burns, openings in chaparral, disturbed areas; gen 300–1700 m. NCoR, SN (common in c SNF), CW, WTR. [*M. platylaemus* Pennell] Very small pls much like *M. rattanii*. ❀DRN,SUN:**7**,8–10, **14–24**.

M. breviflorus Piper (p. 1041) Ann, 2–17 cm, minutely puberulent. **LF** petioled, 3–28 mm, elliptic, not sheathing or clasping st. **FL**: pedicel 3–16 mm, ascending in fr; calyx 2–6 mm in fl, 3.5–9.5 mm and swollen in fr, gen glabrous, lobes equal, < 1 mm, acute, gen ciliate; corolla yellow, tube-throat 3–6.5 mm; placentas axile. **FR** 3–6.5 mm. Uncommon. Moist places; 1500–2200 m. n SNH (El Dorado, Alpine, Amador cos.), MP; to WA, ID, NV.

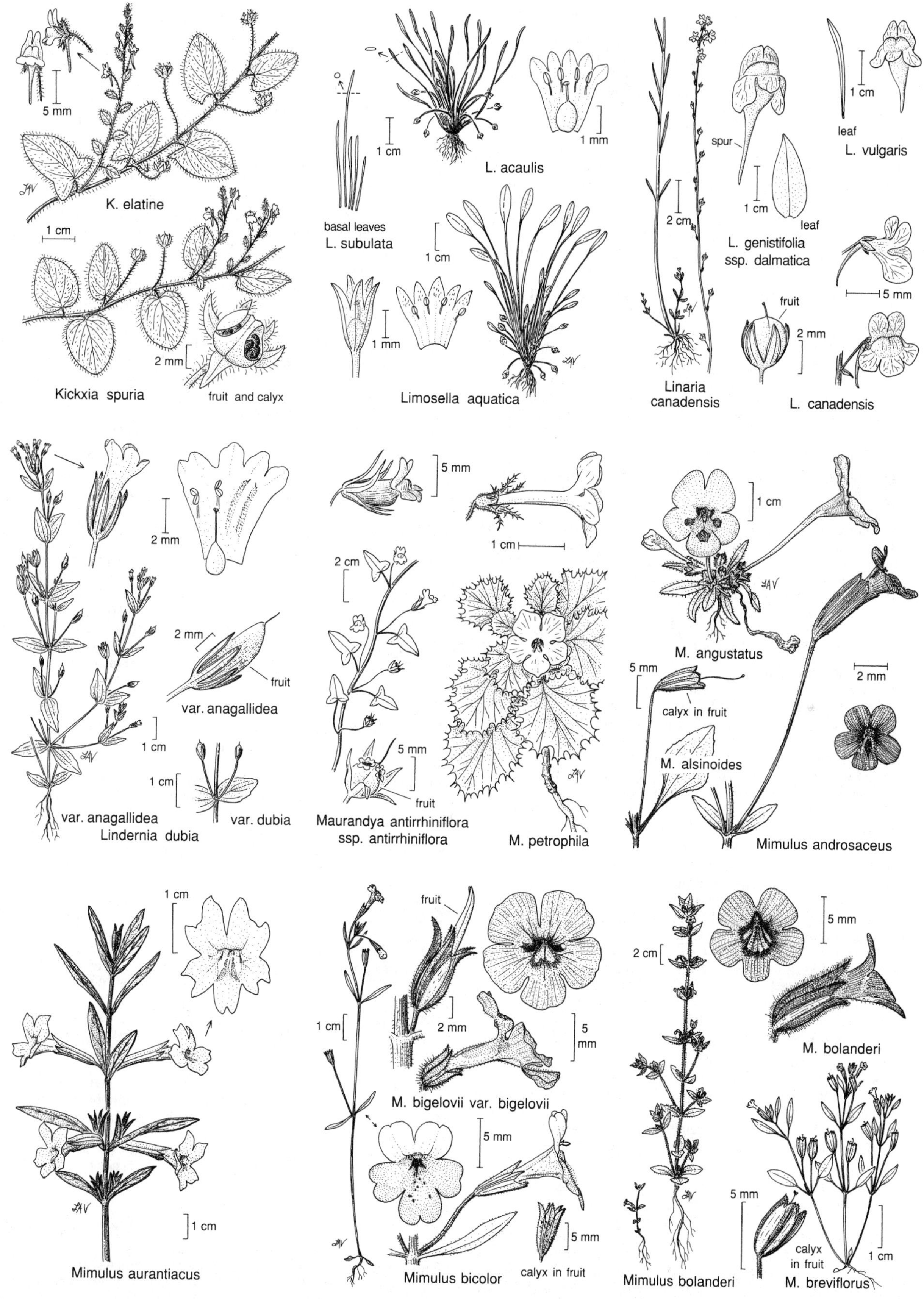

5 mm
K. elatine
1 cm
2 mm
Kickxia spuria
fruit and calyx
1 cm
basal leaves
L. subulata
1 mm
L. acaulis
1 cm
1 mm
Limosella aquatica
2 cm
spur
1 cm
leaf
L. genistifolia
ssp. dalmatica
1 cm
leaf
L. vulgaris
5 mm
fruit
2 mm
Linaria
canadensis
L. canadensis
2 mm
2 mm
fruit
var. anagallidea
1 cm
1 cm
var. anagallidea
Lindernia dubia
var. dubia
5 mm
1 cm
2 cm
5 mm
fruit
Maurandya antirrhiniflora
ssp. antirrhiniflora
M. petrophila
1 cm
M. angustatus
5 mm
calyx in fruit
M. alsinoides
2 mm
Mimulus androsaceus
1 cm
1 cm
Mimulus aurantiacus
fruit
1 cm
2 mm
5
mm
M. bigelovii var. bigelovii
5 mm
5 mm
Mimulus bicolor
calyx in fruit
2 cm
5 mm
M. bolanderi
Mimulus bolanderi
5 mm
calyx
in fruit
1 cm
M. breviflorus

M. brevipes Benth. (p. 1045) Ann, 5–80 cm, gen densely hairy. **LVS** 7–90 mm, lanceolate to obovate; lowest glabrous. **FL**: pedicel 2–10 mm; calyx 10–25 mm, hairy, lobes acuminate, very unequal, upper lobes 4–13 mm, lower 1–5 mm; corolla persistent, yellow, tube-throat 15–30 mm; placentas parietal. **FR** 8–14 mm. *n*=8. Bare areas on slopes in chaparral, esp after disturbance, incl fire; 30–1800 m. SW; Baja CA. ❀DRN,SUN:7–10,**14–16**,17,**19–24**.

M. breweri (E. Greene) Cov. (p. 1045) Ann, 3–21 cm, ± hairy. **LF** 3–35 mm, ± linear. **FL**: pedicel 2–15 mm, ascending in fr; calyx 3.5–9 mm, puberulent along ridges, lobes equal, 1–2 mm, puberulent; corolla persistent, rose to lavender, tube-throat 4.5–8 mm; placentas axile. **FR** 3.5–6.5 mm. Common. Moist places near seepage areas, streams; gen 1200–3400 m. NW, CaR, SNH, SnBr, SnJt, MP; to B.C., MT, WY, UT. ❀WET,SUN:**1**,2–5,**6**,**7**,8–10,**14–18**,19–21,**22–24**.

M. cardinalis Benth. (p. 1045) Per, rhizomed, hairy. **ST** 25–80 cm, often decumbent or ascending. **LVS** 20–80 mm, oblong to obovate, palmately 3–5-veined; upper clasping st. **FL**: pedicel 50–80 mm; calyx 20–30 mm, hairy, lobes equal, 4–5 mm, acute to acuminate; corolla orange to red, tube-throat 40–50 mm, upper lip arched forward, lower lip reflexed; anthers, stigma exserted but arched in upper lip; placentas axile. **FR** 16–18 mm. *n*=8. Moist to wet places along streams, seepage areas; < 2400 m. Gen CA-FP; to se OR, sw UT, w-c NV, w NM, n Baja CA. ❀WET or IRR:1–3,**7–9**,10,11,**14,18–23**&SUN:4,5,**6**,**15–17**,**24**;CV.

M. clevelandii Brandegee (p. 1045) CLEVELAND'S BUSH MONKEYFLOWER Per from woody caudex. **ST** 30–90 cm, hairy; main lf axils often with clusters of smaller lvs. **LF** 20–100 mm, lanceolate to oblong; edges gen rolled under; upper surface hairy. **FL**: pedicel 3–4 mm; calyx 20–25 mm, gen swollen at base, narrowed above ovary, hairy, lobes unequal, 6–9 mm, acute to acuminate; corolla persistent, yellow, tube-throat 35–40 mm; placentas parietal. **FR** 10–12 mm, splitting into 4 parts at tip. *n*=10. UNCOMMON. Disturbed areas, open borders of woodlands, chaparral; 1000–2000 m. PR. Hybridizes with *M. aurantiacus* at lower elevations. ❀DRN, some IRR:7–10,**14**,18,**19–23**&SUN:5,15–17,**24**.

M. congdonii Robinson (p. 1045) Ann, < 10 cm, puberulent. **LF**: blade 8–32 mm, oblanceolate to elliptic, dull purplish green, base and petiole ± ciliate. **FL**: pedicel 2–5 mm; calyx 5–14 mm, hairy on ribs, lobes unequal, 0.2–1 mm, obtuse; corolla persistent, magenta, tube-throat 8–30 mm (or fl cleistogamous), lobes equal or not, 1–4 mm; lower 2 anthers and stigma exserted; placentas parietal. **FR** 4–8.5 mm, angled, laterally compressed, oblique, hard, indehiscent. *n*=9. Disturbed areas or seepage, runoff areas on slopes, gen granitic soils; 120–1100(1700) m. SNF, uncommon in s NCoR, SnFrB, single localities in SCoRO (Monterey Co.), WTR (Ventura Co.), PR (San Diego Co.).

M. constrictus (A.L. Grant) Pennell (p. 1045) Ann, 2–24 cm, hairy. **LF** 5–32 mm, narrowly elliptic to obovate. **FL**: pedicel 0.5–3 mm; calyx 7–12 mm, ribbed, swollen (esp in fr), coarsely hairy, slightly oblique at mouth, lobes erect or ± spreading, ± equal, 1–4 mm, acuminate; corolla persistent, pale lavender to dark magenta, tube-throat 13–22 mm, throat pale with variable, radiating, dark maroon marks, throat floor and limb long-hairy; placentas parietal. **FR** 8–12 mm. *n*=8. Slopes, roadbanks of decomposing granite; 750–2200 m. s SNF, Teh, n WTR. Intergrades with *M. whitneyi* in n, *M. johnstonii* in s. ❀TRY.

M. cusickii (E. Greene) Rattan (p. 1045) Ann, 3–24 cm, densely puberulent to hairy. **LVS** 15–45 mm, ovate to obovate, at least upper lvs sharply acuminate. **FL**: pedicel 1–3 mm; calyx 10–17 mm, densely puberulent, lobes ± unequal, 1.5–5 mm, acuminate; corolla persistent, magenta with throat floor folds gold at mouth, tube-throat 20–28 mm; stigma and gen lower anthers exserted; placentas parietal. **FR** 10–17 mm. *n*=8. Steep, unstable canyon slopes, scree; gen 600–1600 m. s Wrn; to WA, ID. ❀TRY.

M. dentatus Benth. (p. 1045) Per, rhizomed, ± hairy. **ST** 15–40 cm, ascending. **LF**: petioles 0–12 mm; blade 15–70 mm, ovate, pinnately 5+-veined. **FL**: pedicel 20–50 mm; calyx 9–18 mm, ± glabrous, ribs ± hairy, lobes ± equal, 2–7 mm, ± ciliate; corolla yellow, tube-throat 28–40 mm, tube funnel-shaped, gen > 4 mm wide; placentas axile. **FR** 8–9 mm. Coastal streambanks, gen in partial shade; < 400 m. n NCo (Del Norte, Humboldt cos.); to WA. ❀IRR or WET,SHD:**4,5**,6,7,14,**15–17**,19–23,**24**.

M. douglasii (Benth.) A. Gray (p. 1045) Ann, 0.3–4 cm, ± hairy. **LF**: blade 5–28 mm, ovate to obovate, shiny green. **FL**: pedicel 2–4 mm; calyx 8–14 mm, hairy, lobes unequal, 0.5–2 mm, obtuse; corolla persistent, tube-throat 20–41 mm (or fl cleistogamous), limb magenta, tube abruptly widened to long throat (which is boldly mottled, striped gold and purple), upper lobes 4–5 mm, lower lobes < 1 mm; anthers and stigma nearly exserted; placentas parietal. **FR** 3–7 mm, oblique-ovoid, hard, indehiscent. *n*=9. Gentle slopes, bare, weathered clay, serpentine or granitic soils; 45–1200 m. NW, CaRF, SNF, CW; sw OR. ❀DRN,SUN:**4–7,14– 17**,19–24.

M. exiguus A. Gray (p. 1045) SAN BERNARDINO MOUNTAINS MONKEYFLOWER Ann, 2–10 cm, minutely puberulent. **LF** 2–8 mm, narrowly elliptic to narrowly ovate, not sheathing st. **FL**: pedicel 6–20 mm, ascending in fr; calyx 2–3.5 mm, ± glabrous, lobes equal, 0.5–1 mm, acute, ± minutely puberulent; corolla lavender, tube-throat 2–3.5 mm; placentas axile. **FR** 3–4 mm, nearly spheric, > calyx. RARE. Gentle slopes, along streamlets, runoff areas in clay soils; 1800–2300 m. SnBr; n Baja CA.

M. filicaulis S. Watson (p. 1045) SLENDER-STEMMED MONKEYFLOWER Ann, 4–30 cm, densely puberulent. **LF** 8–23 mm, linear to oblanceolate. **FL**: pedicel 9–25 mm; calyx 5–9 mm, with wide corky ribs in fr, gen with red-brown spots, puberulent, lobes equal, 1–2 mm, acute to long-tapered; corolla not lipped, rose- purple with dark, red-purple throat marks extending slightly onto upper 4 lobes, yellow from throat folds also on lower lobe, tube-throat 13–17 mm, lobes all notched; placentas axile. **FR** 5.5–6.5 mm. *n*=8. RARE. Disturbed, moist loamy soil, often in partial shade; 1200–1750 m. c SNH (Tuolumne, Mariposa cos.). [*M. biolettii* Eastw.]

M. floribundus Lindley (p. 1045) Ann, hairy, often ± slimy. **ST** 3–50 cm, often decumbent or ± climbing. **LF**: petiole 0–20 mm; blade 5–45 mm, lanceolate to ovate, base gen rounded to cordate, veins ± pinnate. **FL**: pedicel 5–30 mm, not reflexed in fr; calyx 3–8 mm, ± hairy, lobes equal, 1–2 mm, ± acute; corolla yellow, tube-throat 6–15 mm; placentas axile. **FR** 4–7 mm. *n*=16. Crevices, seeps around granite outcrops, near streams; < 2500 m. CA-FP (esp c&s SNF); to B.C., SD, n Mex. [ssp. *subulatus* A.L. Grant; *M. arenarius*, *M. dudleyi* A.L. Grant] Many minor, ± indistinct forms (if corolla tube-throat > 15 mm, see *M. moschatus*). ❀DRN,SUN:4–6&IRR:1–3,**7**,8–10,**14–24**.

M. fremontii (Benth.) A. Gray (p. 1045) Ann, 1–20 cm. **LVS** 2–30 mm, gen narrowly elliptic (to obovate); lowest glabrous, upper hairy. **INFL**: fl 1 per node. **FL**: pedicel 1–4 mm; calyx 5–14 mm, wide-ribbed, swollen, often reddish, often ± whitish hairy, lobes ± equal, 0.5–3.5 mm, acute to acuminate; corolla persistent, magenta to red-purple (rarely yellow), ± darker at mouth, tube-throat 10–23 mm, throat floor ± glabrous, folds gold; placentas parietal. FR 7–13 mm. *n*= 8. Sandy, disturbed areas among shrubs, often on banks, benches along streams; 75–2100 m. SCoR, SW, s DMoj; n Baja CA. [*M. subsecundus* A. Gray, one-sided monkeyflower] ❀TRY.

M. glaucescens E. Greene (p. 1045) SHIELD-BRACTED MONKEYFLOWER Ann, 6–80 cm, ± glabrous, glaucous. **LF**: petiole 0–60 mm; blade 5–70 mm, ovate to ± round. **INFL**: raceme, gen > 5-fld; bracts fused around st completely, forming ± circular disks 5–45 mm wide. **FL**: pedicel 5–35 mm; calyx 7–25 mm, asymmetrically swollen in fr, ± glabrous, lobes unequal, lowest 2 upcurved in fr; corolla yellow, tube-throat 15–35 mm; placentas axile. **FR** 6–12 mm. *n*=14. UNCOMMON. Seepage areas; < 600 m. s CaRF, adjacent n SNF. ❀IRR:6,7,8,9,**14–17**,**19–24**.

M. gracilipes Robinson (p. 1045) SLENDER-STALKED MONKEYFLOWER Ann, 2–8 cm, puberulent. **LF** 3–13 mm, elliptic to ovate. **FL**: pedicel 8–25 mm; calyx 4–6 mm, ± puberulent, lobes equal, 0.5–2 mm, obtuse to rounded, ± with small point at tip, densely ciliate; corolla pink to lavender, 2-lipped, tube-throat 7–10 mm, limb 5–9 mm wide; placentas axile. **FR** 4–5 mm. UNCOMMON. Appearing gen after fire or other disturbance on decomposed granite; 500–1300 m. c SNF (Mariposa, Fresno cos.).

M. guttatus DC. (p. 1045) Ann or rhizomed per, 2–150 cm, glabrous to hairy. **LVS** abruptly reduced to sessile bracts; petioles 0–95 mm; blades 4–125 mm, ovate to round, often crenate, base often irregularly small-lobed or dissected. **INFL**: raceme, gen > 5-fld;

bracts ovate to cordate, fused at base or not, not glaucous. **FL** cleistogamous or opening; pedicel 10–80 mm; calyx 6–30 mm, asymmetrically swollen in fr, glabrous to hairy, lobes unequal, lowest 2 upcurved in fr; corolla yellow, tube-throat 2–40 mm; placentas axile. **FR** 5–12 mm. *n*=14,15,16,24,28. Common. Wet places, gen terrestrial, sometimes emergent or floating in mats; < 2500 m. CA; to AK, w Can, Rocky Mtns, n Mex. [sspp. *arenicola* Pennell, *arvensis* (E. Greene) Munz, *litoralis* Pennell, & *micranthus* (A.A. Heller) Munz; *M. glabratus* Kunth ssp. *utahensis* Pennell, Utah monkeyflower; *M. microphyllus* Benth., small-lvd monkeyflower; *M. nasutus* E. Greene; *M. whipplei* A.L. Grant, Whipple's monkeyflower] Exceedingly complex: local populations may be unique but their forms intergrade over geog or elevation; variants not distinguished here. ❀WET or IRR:1–3,**7–9**,10–12,**14**,**18–23**&SUN:**4–6**,**15–17**,**24**;occas.INV.

M. inconspicuus A. Gray (p. 1045) SMALL-FLOWERED MONKEYFLOWER Ann, 3–30 cm, glabrous; large pls branched at base and above. **LVS** 6–40 mm, widely elliptic; lowest in basal rosette, ± petioled; others sessile, widely attached. **FL**: pedicel 5–25 mm; calyx 5–9 mm (6–11 mm, greatly and symmetrically swollen in fr), gen glabrous, lobes equal, 0.5–1.5 mm; corolla gen rose to lavender, tube-throat 7–16 mm, limb < 14 mm wide, lobe tips gen deeply notched; placentas axile. **FR** 4–9 mm, ± stalked above receptacle. UNCOMMON. Near hillside streams or seeps in partial shade; 160–2000 m. SNF (s of El Dorado Co.). [*M. acutidens* E. Greene, Kings River monkeyflower; *M. grayi* A.L. Grant, Gray's monkeyflower]

M. jepsonii A.L. Grant (p. 1045) Ann, 1–10 cm, finely puberulent. **LF** 3–14 mm, ± linear to narrowly elliptic or oblanceolate; lower surface purplish; tip rounded. **FL**: pedicel 0.5–2 mm; calyx 2–6 mm, finely puberulent, lobes ± unequal, 0.5–1.7 mm, acute to acuminate; corolla persistent, lavender-purple, tube-throat 7–12 mm, throat floor with 2 gold stripes surrounded by and often dotted with deeper purple; stigma exserted; placentas parietal. **FR** 4–6 mm. *n*=8. Pine forest openings, gen granitic soils; 1200–2400 m. CaRH, n SNH, w MP; s OR.

M. johnstonii A.L. Grant (p. 1045) Ann, 3–20 cm, gen finely whitish puberulent. **LF** 7–30 mm, oblanceolate to obovate; tip gen acute. **FL**: pedicel 1–5 mm; calyx 6–11 mm, membranous, reddish, finely puberulent, mouth strongly oblique in fr, lobes erect to spreading, unequal, 1–3 mm, ± acuminate; corolla persistent, magenta, tube-throat 9–15 mm, throat floor gold, mouth gen with 2 lateral, dark maroon patches; stigma ± exserted; placentas parietal. **FR** 7–12 mm. *n*=8. Roadbanks, disturbed areas, esp scree; gen 1280–2920 m. SnGb, SnBr. Intergrades with *M. constrictus* in WTR.

M. kelloggii (E. Greene) A. Gray (p. 1045) Ann, 1–31 cm, hairy. **LF** 6–40 mm, elliptic to obovate; lower surface often purplish. **FL**: pedicel 2–6 mm; calyx 8–16 mm, puberulent, lobes ± unequal, 0.5–1.5 mm, obtuse; corolla persistent, magenta to red-purple, deepening toward spotted-gold throat, tube-throat 20–45 mm, ± 2 × calyx, gradually wider and funnel-like toward tip, upper lobes 4–5 mm, lower lobes 2–3 mm; lower anthers and stigma exserted; placentas parietal. **FR** 6–12 mm, oblique-cylindric, slightly curved, hard, indehiscent. *n*=9. Bare areas, unstable or disturbed, steep slopes in scree or soil; 50–1600 m. NW, CaRF, n&c SNF; sw OR. ❀DRN,IRR,SUN:6,**7**,8–10,**14–17**,19–24.

M. laciniatus A. Gray (p. 1045) CUT-LEAVED MONKEYFLOWER Ann, 3–38 cm, glabrous to sparsely hairy. **LVS**: petioles 0–35 mm; blades 3–55 mm, oblanceolate to ± ovate, at least some ± pinnately lobed or dissected into narrow segments. **INFL**: raceme, gen > 5-fld; bracts ± lanceolate or pinnately lobed, bases long-tapered or petioled. **FL**: pedicel 10–45 mm; calyx 3–15 mm, asymmetrically swollen in fr, ± glabrous, lobes unequal, lowest 2 upcurved in fr; corolla yellow, tube-throat 4–15 mm (or fl cleistogamous); placentas axile. **FR** 3–8 mm. *n*=14. UNCOMMON. Seeps on granite outcrops; > 900 m. SNH. ❀IRR,SUN:**7**,14,**15–17**,19–24.

M. latidens (A. Gray) E. Greene (p. 1045) Ann, 10–27 cm, branched only near base. **LVS** 4–25 mm, gen lanceolate to ± ovate; lowest ± petioled, in basal rosette; others sessile, widely attached. **FL**: pedicel 6–30 mm; calyx 6–7 mm (9–12 mm, strongly swollen, pleated, membranous in fr), ± glabrous, lobes equal, 1–2 mm; corolla pale pink to white, rarely yellowish, tube-throat 7–11 mm, limb 4–6 mm wide, lobes entire; placentas axile. **FR** 7–8 mm, ± stalked. Vernally wet depressions; < 900 m. NCoRI, GV, CW, e SCo (Menifee Valley, w Riverside Co.); Baja CA. Scattered.

M. latifolius A. Gray Ann, 1–10 cm, puberulent. **LF** not ciliate; petiole 1–4 mm; blade 6–39 mm, elliptic to ovate. **FL**: pedicel 1.5–4 mm; calyx 7.5–13 mm, puberulent, lobes unequal, 0.5–3 mm, obtuse; corolla persistent, magenta to purple, tube-throat 12–19 mm, upper lip 3–4 mm, lower lip ± 1 mm; placentas parietal. **FR** 6–11 mm, oblique, angled, hard, indehiscent. Very uncommon. Rocky places; < 150 m. n ChI (Santa Cruz Island); Guadalupe Island, Mex. [*M. brandegei* Pennell, Santa Cruz Island monkeyflower]

M. layneae (E. Greene) Jepson (p. 1045) Ann, 2–28 cm, ± hairy. **LF** 6–25 mm, narrowly elliptic to obovate. **FL**: pedicel 1–3 mm; calyx 5–11 mm, dark-ribbed, ± hairy, lobes ± equal, 1–3 mm, acute to acuminate; corolla persistent, almost white to dark magenta with darker central marks on lobes, tube-throat 13–21 mm, throat gen whitish with dark spots, lips spreading, ± equal; placentas parietal. **FR** 6–10 mm. *n*=8. Bare or disturbed areas, often serpentine or granitic crevices or soils; gen 200–2000 m. NW, CaR, SN. [*M. brachiatus* Pennell, serpentine monkeyflower] Intermediates with *M. nanus* (NW) and *M. mephiticus* (s SN) have yellow at corolla mouth. ❀TRY.

M. leptaleus A. Gray (p. 1045) Ann, 0.5–14 cm, puberulent. **LF** 5–25 mm, linear to oblanceolate. **FL**: pedicel gen 0.5–1 mm; calyx 3–6 mm, puberulent, lobes ± unequal, 0.8–2 mm, acute; corolla persistent, magenta or white, tube-throat 6–8 mm, throat gaping, whitish with dark dots, limb 3–5 mm wide; placentas parietal. **FR** 4–6 mm. *n*=8. Granitic soils or sand near outcrops, disturbed areas; 2000–3400 m. SNH; w NV (Mount Rose, Washoe Co.). Intermediates with *M. whitneyi* in s SNH often have yellowish fls.

M. lewisii Pursh (p. 1047) Per, rhizomed, hairy. **ST** 25–80 cm. **LF** 20–70 mm, oblong to elliptic, gen clasping st, palmately 3–5-veined. **FL**: pedicel gen 30–70 mm; calyx 15–25 mm, hairy, lobes equal, 4–6 mm, acute; corolla lavender to purple, tube-throat 30–50 mm, lips spreading; placentas axile. **FR** 13–14 mm. *n*=8. Streambanks, seeps; 1200–3100 m. CaRH, SN; to w Can, MT, WY, UT. ❀IRR or WET,DRN:**1**,2,3,7,14,18–21&SUN:4–6,15–17;DFCLT.

M. mephiticus E. Greene (p. 1047) Ann, 1–15 cm, hairy. **LF** 6–30 mm, ± linear to narrowly lanceolate; tip obtuse to rounded. **FL**: pedicel 2–7 mm; calyx 5–11 mm, hairy, lobes ± unequal, 1–3 mm, acute to long-tapered; corolla persistent, ± magenta or yellow (often mixed in a population), tube-throat 11–19 mm, lower lip base with 3 maroon lines and maroon spots at mouth; stigma and sometimes anthers ± exserted; placentas parietal. **FR** 6–11 mm. *n*=8. Bare, sandy or gravelly areas, often around granite outcrops; gen 1500–3500 m. SN, s MP, SNE; NV. [*M. coccineus* Congdon; *M. densus* A.L. Grant] Corolla more often magenta (gradually darker), pl more tufted, anthers more often exserted from low to high elevations.

M. mohavensis Lemmon (p. 1047) MOJAVE MONKEYFLOWER Ann, 2–10 cm, ± puberulent. **LF** 7–27 mm, narrowly elliptic, red-purple. **FL**: pedicel 2–5 mm; calyx 7–15 mm, enlarged in fr, red-purple, ± puberulent along veins, lobes spreading, ± unequal, 2–4 mm, acuminate, ciliate; corolla radial, salverform, persistent, tube-throat 9–15 mm, tube-throat, limb at base ± solid maroon, veins maroon, fading into white border, appearing << 0.5 mm wide; placentas parietal. **FR** 8–13 mm. *n*=7. RARE. Gravelly banks of desert washes; 600–1000 m. DMoj (near Barstow, San Bernardino Co.).

M. montioides A. Gray (p. 1047) Ann, 1–18 cm, puberulent. **LF** 4–32 mm, linear (rarely > 3 mm wide). **FL**: pedicel 3–25 mm; calyx 3.5–8.5 mm, minutely puberulent, membranous in fr, lobes equal, 0.5–2 mm, obtuse to rounded, with small point at tip, ± glabrous to ciliate; corolla yellow, purple, or yellow with purple upper lip, tube-throat 10–16 mm, lobes often each 2-lobed; placentas axile. **FR** 3–6 mm. Disturbed areas along streamlets, gen in granitic soils; > 1800 m. c&s SN (s of Madera Co.); near Carson City, NV. [*M. barbatus* E. Greene; *M. discolor* A.L. Grant] Complex; corolla colors mixed in some areas, uniform in others; pls with purple corollas at lower elevation much like *M. palmeri*. ❀TRY.

M. moschatus Lindley (p. 1047) MUSK MONKEYFLOWER Per, rhizomed, ± glabrous to densely slimy-hairy, often musk-scented. **ST** 5–30 cm, prostrate to ascending. **LF**: petiole 0–15 mm; blade 10–60 mm, oblong to ovate, pinnately 5+-veined. **FL**: pedicel 10–50 mm; calyx 8–12 mm, ± glabrous to hairy, lobes equal, 2–5 mm; corolla yellow, tube ± cylindric, 2–4 mm wide (exc mouth), tube-throat 15–26 mm, throat floor deeply grooved; placentas axile. **FR** 4–9 mm. *n*=16. Common. Seeps, streambanks, often in partial shade; < 2900 m. CA-FP; to B.C., Rocky Mtns; naturalized in ne US, Chile, Eur. [var. *moniliformis* (E. Greene) Munz] Variable. ❀WET or IRR:**1**,2,3,**7**,8–10,**14–18**,19–24&SUN:**4–6**;occas. INV.

M. nanus Hook. & Arn. (p. 1047) Ann, 1–10 cm, ± puberulent. **LF** 7–25 mm, oblanceolate to obovate; lower surface ± purple; tip obtuse to rounded. **FL**: pedicel 1–4 mm; calyx 6–10 mm, puberulent, lobes ± unequal, 1–4 mm, acute to ± long-tapered, mouth in fr slightly oblique; corolla persistent, magenta to purplish, tube-throat 11–19 mm, throat floor gen with 2 gold stripes surrounded by and often dotted deeper magenta; stigma ± exserted; placentas parietal. **FR** 6–12 mm. *n*=8. Sandy runoff areas above streamlets or in bare openings among shrubs; gen 1000–2300 m. NW, CaRH, GB; to WA, MT, WY. Intermediates with *M. mephiticus* (e GB), with *M. layneae* in NW. ❀WET,DRN,SHD:1–3,**4–7**,14,**15–17**,18,24.

M. norrisii Heckard & J.R. Shevock (p. 1047) KAWEAH MONKEYFLOWER Ann, gen 3–15 cm, ascending, hairy. **LF** 15–45 mm; blade ovate, ± tapered to a petiole, veins pinnate. **FL**: pedicel 10–50 mm; calyx 3.5–6 mm, hairy, ribs wide, purplish, lobes equal, 1–2 mm, rounded; corolla yellow, tube-throat 15–30 mm, lobe bases each with 1 large, central, maroon-purple blotch; placentas axile. **FR** 4–6 mm, growing away from light, tardily dehiscent. *n*=16. RARE. Marble crevices; 600–1300 m. s SNF (Kaweah River drainage, Tulare Co.).

M. nudatus E. Greene (p. 1047) BARE MONKEYFLOWER Ann, 4–30 cm, ± glabrous. **LF**: petiole 0–30 mm, abruptly reduced upward; blade 3–25 mm, lanceolate to ovate below, narrower upward; pairs at nodes not fused. **INFL**: fls solitary in upper axils, gen > 5 per st. **FL**: pedicel 10–50 mm; calyx 4–15 mm, asymmetrically swollen in fr, glabrous, lobes unequal, lowest 2 upcurved in fr; corolla yellow, tube-throat 10–20 mm; placentas axile. **FR** 5.5–9.5 mm. UNCOMMON. Seeps on serpentine outcrops; 250–700 m. s NCoRI (Lake, Napa cos.). ❀IRR or WET,DRN,SUN:**7**,14–17; DFCLT.

M. palmeri A. Gray (p. 1047) Ann, 1–28 cm, ± puberulent. **LF** 3–28 mm, ± linear to ovate (< 12 mm wide). **FL**: pedicel 10–65 mm; calyx 4–10 mm, glabrous to puberulent, lobes equal, 0.5–1.5 mm, rounded to truncate, with small point at tip, glabrous to densely ciliate; corolla purple with variable marks, not lipped, tube-throat 10–22 mm; placentas axile. **FR** 3.5–9 mm. Sandy washes, disturbed areas; < 2100 m. s SNF, Teh (and adjacent w DMoj), n SCoRO, TR, PR; n Baja CA. [*M. diffusus* A.L. Grant, Palomar monkeyflower] Intergrades with *M. montioides* in SN, TR. ❀TRY.

M. parishii E. Greene (p. 1047) Ann, 3–85 cm, hairy. **LF** 8–75 mm, oblanceolate to ovate, veins 3, palmate. **FL**: pedicel 15–18 mm; calyx 8–10 (13 in fr) mm, hairy, lobes equal, 1–2 mm; corolla pinkish to white, tube-throat 9–14 mm; placentas axile. **FR** 6–10 mm. Uncommon. Wet, sandy streamsides; < 2100 m. s SN, SW (and adjacent w D), n&e DMtns (Granite, New York, Panamint mtns); n Baja CA.

M. parryi A. Gray (p. 1047) Ann, 1–17 cm, densely puberulent. **LF** 3–26 mm, ± linear to oblanceolate; tip rounded to acute. **FL**: pedicel 1–5 mm; calyx 6–12 mm, membranous, often purplish throughout, puberulent, lobes spreading, upper lobe 1.5–3 mm, rounded, tip ± small-pointed, other lobes gen 0.5–2 mm, acuminate; corolla persistent, magenta, tube-throat 11–18 mm; placentas parietal. **FR** 5.5–10 mm, incl. *n*=8. Very uncommon. Steep hillsides, along washes; 1200–2600 m. W&I (esp Inyo Mtns); to sw UT, nw AZ. ± common outside CA, where corollas yellow.

M. pictus (E. Greene) A. Gray (p. 1047) CALICO MONKEYFLOWER Ann, 2–38 cm, hairy. **ST** 4-angled. **LF** 7–45 mm, oblanceolate to obovate. **FL**: pedicel 2–6 mm; calyx 6–13 (–18 in fr) mm, hairy, lobes ± unequal, 1–4 mm, obtuse; corolla radial, salverform, gen deciduous, tube dark, brownish maroon, tube-throat 6.5–18 mm, limb ± white ± throughout, veins maroon with much wider maroon border, so appearing > 0.5 mm wide; placentas parietal. **FR** 7–17 mm. RARE. Bare, sunny areas around shrubs, rock outcrops on granitic soils; 100–1300 m. s SNF, Teh. In cult.

M. pilosus (Benth.) S. Watson (p. 1047) Ann, 2–35 cm, densely long-soft-wavy-hairy. **LF** 10–30 mm, lanceolate to oblong. **FL**: pedicel gen 10–15 mm; calyx 6–7 mm, densely hairy, lobes unequal, 3–4 mm, ± = tube, obtuse; corolla yellow, tube-throat 7–8 mm; placentas axile. **FR** 4–7 mm. Moist, sandy areas, esp running or dry streamlets, disturbed areas; < 2600 m. Mostly CA-FP; to WA, UT, AZ, Baja CA. ❀IRR,DRN,SUN:1–6,**7**,8–10,**14–18**,19–24.

M. primuloides Benth. Per; rhizomes or stolons forming mats of ± distinct rosettes or tufted pls; forming bulblets in autumn. **ST** 0.5–12 cm, glabrous. **LF** 7–50 mm, linear to obovate; upper surface glabrous to densely long-hairy; veins 3, palmate. **FL**: pedicel 10–120 mm, stiffly erect; calyx 5–12 mm, glabrous, lobes equal, 0.5–1.5 mm; corolla yellow, tube-throat 8–20 mm, base of each lower lip lobe gen with reddish spot; placentas axile. **FR** 6–7 mm. Wet meadows, seeps, streamsides; 600–3400 m. NW, CaR, SN, SnBr, SnJt, GB; to WA, NV.

ssp. ***linearifolius*** (A.L. Grant) Munz (p. 1047) **ST** 4–12 cm, tufted. **LVS** ± erect, some ± basal but not in rosettes, all intermingled with those of nearby sts, 15–50 mm, linear to oblanceolate. **FL**: calyx 7–12 mm. Habitats of sp.; 600–2200 m. KR. ❀WET or IRR,DRN:**1**,2,3,**7**,14,18&SUN:**4–6**,15–17.

ssp. ***primuloides*** (p. 1047) **ST** 0.5–4 cm. **LVS** gen ± spreading in ± distinct rosettes, 7–35 mm, oblong to obovate. **FL**: calyx 5–9 mm. *n*=17. Habitats & range of sp. Smaller, hairier pls sometimes found with others (e.g., Echo Summit, El Dorado Co.) have been called var. *pilosellus* (E. Greene) Smiley; such variation ± continuous. ❀WET or IRR,DRN:**1**,2,3,7,14,18&SUN:**4–6**,15–17; DFCLT.

M. pulchellus (E. Greene) A.L. Grant (p. 1047) PANSY MONKEYFLOWER Ann, < 1 cm, tightly tufted, puberulent. **LF** 8–35 mm, ± linear, ciliate at least on basal 1/2. **FL**: pedicel 0–2 mm; calyx 7–15 mm, hairy, lobes unequal, 1–3 mm, obtuse, ± ciliate; corolla persistent, lavender to purple exc lower lip, or at least lowest lobe gold, lowest lobe with scattered, small, dark dots at base, tube-throat 20–40 mm, puberulent outside; placentas parietal. **FR** 3–5.5 mm, oblique-ovoid, hard, indehiscent. *n*=9. UNCOMMON. Vernally wet depressions; 600–2000 m. n&c SNF (Calaveras, Tuolumne, Mariposa cos.).

M. pulsiferae A. Gray (p. 1047) Ann, 2–21 cm, puberulent. **LVS**: petioles 1–6 mm; blades 3–20 mm; basal in ± rosette, ovate; cauline elliptic. **FL**: pedicel 6–24 mm, ascending; calyx 3.5–7.5 mm, 5–12 mm in fr, minutely puberulent, lobes ± equal, 0.5–1.2 mm, acute; corolla yellow (limb often whitish with pinkish border in n NW), tube-throat 7–13 mm; placentas axile. **FR** 5–10 mm. Moist places along streams, runoff areas; 500–1900 m. NW, CaR, MP; to WA.

M. purpureus A.L. Grant (p. 1047) PURPLE MONKEYFLOWER Ann, 0.5–7 cm, puberulent. **LF** 3–16 mm, elliptic to ovate. **FL**: pedicel 15–56 mm; calyx 4–8 mm, tube puberulent, lobes equal, 0.5–1 mm, rounded to truncate, with small point at tip, glabrous; corolla tube-throat 8–13 mm, lips 2, strong, upper lip red-purple, lower lip rose, limb 8–10 mm wide; placentas axile. **FR** 4–7 mm. RARE. Along streamlets on open, gentle slopes; 1900–2300 m. SnBr; Baja CA. Vars. do not warrant recognition. Threatened by development, vehicles.

M. pygmaeus A.L. Grant (p. 1047) EGG LAKE MONKEYFLOWER Ann, < 1 cm, tightly tufted, puberulent. **LF** 2–15 mm, oblanceolate to narrowly elliptic, puberulent, ciliate on basal 1/2. **FL**: pedicel 0–0.5 mm; calyx 4–8 mm, ± hairy, lobes unequal, 0.5–1.5 mm, rounded, sometimes ± ciliate; corolla persistent, yellow, tube-throat 5–10 mm; placentas parietal. **FR** 2–4 mm, ovoid, hard, indehiscent. RARE. Vernally wet depressions; 1200–1700 m. s CaRH, n SNH (Lake Almanor region, Plumas Co.), MP (Egg Lake, Modoc Co., w of Eagle Lake, Lassen Co.); s OR.

M. rattanii A. Gray (p. 1047) Ann 1–18 cm. **LVS** 3–46 mm, narrowly elliptic to obovate; lowest glabrous; upper hairy. **INFL**: fl 1 per node. **FL**: pedicel 1–3 mm; calyx 5–9 mm, wide-ribbed, swollen, hairy, lobes ± equal, 0.7–2 mm, acute; corolla persistent, rose to

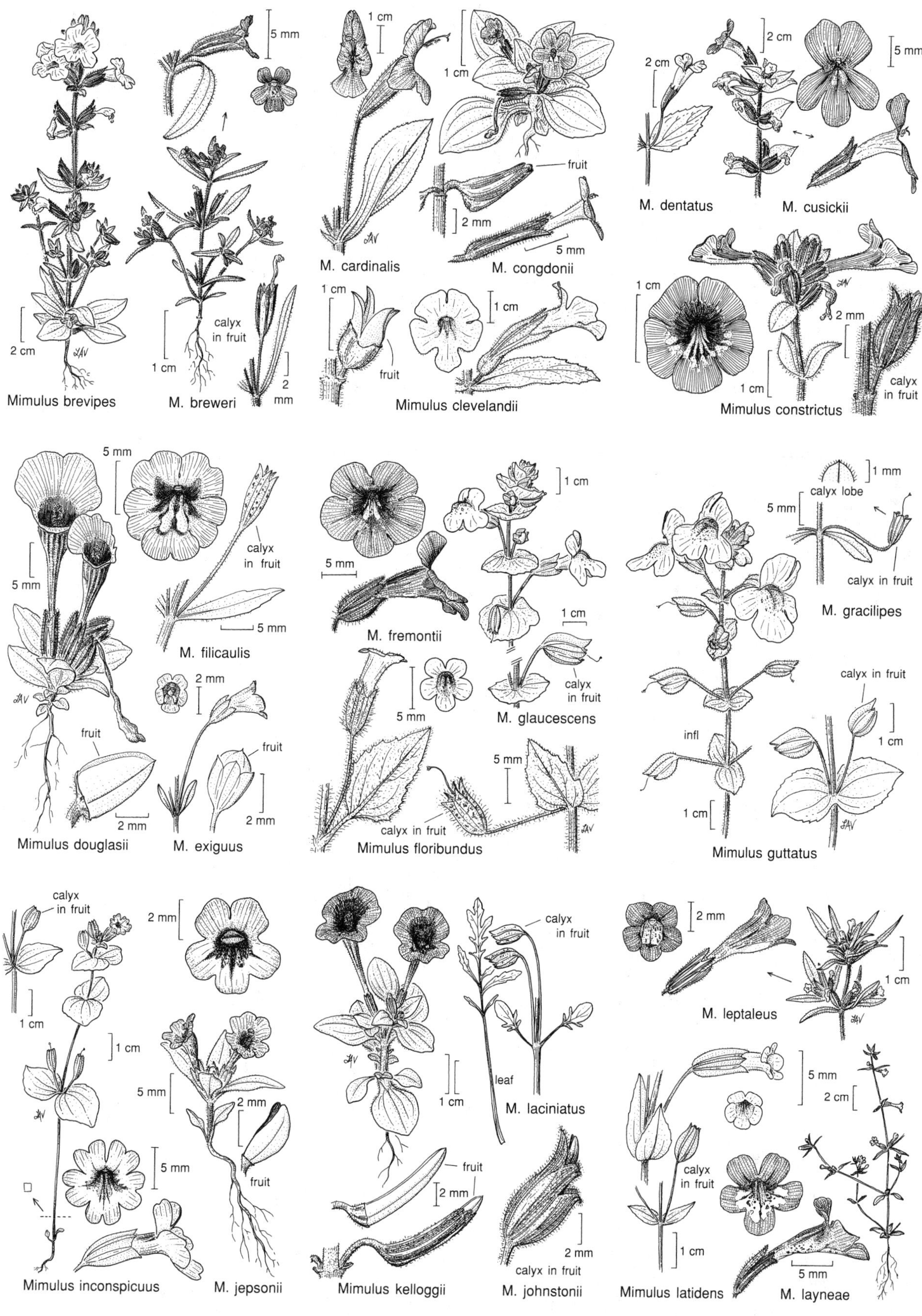
5 mm
2 cm
1 cm
calyx
in fruit
2
mm
Mimulus brevipes
M. breweri
1 cm
1 cm
fruit
2 mm
5 mm
M. cardinalis
M. congdonii
1 cm
1 cm
fruit
Mimulus clevelandii
2 cm
2 cm
5 mm
M. dentatus
M. cusickii
1 cm
2 mm
1 cm
calyx
in fruit
Mimulus constrictus
5 mm
5 mm
calyx
in fruit
5 mm
M. filicaulis
2 mm
fruit
fruit
2 mm
2 mm
Mimulus douglasii
M. exiguus
1 cm
5 mm
M. fremontii
1 cm
5 mm
calyx
in fruit
M. glaucescens
5 mm
calyx in fruit
Mimulus floribundus
1 mm
calyx lobe
5 mm
calyx in fruit
M. gracilipes
calyx in fruit
1 cm
infl
1 cm
Mimulus guttatus
calyx
in fruit
2 mm
1 cm
1 cm
5 mm
2 mm
fruit
5 mm
Mimulus inconspicuus
M. jepsonii
calyx
in fruit
leaf
1 cm
M. laciniatus
fruit
2 mm
2 mm
calyx in fruit
Mimulus kelloggii
M. johnstonii
2 mm
1 cm
M. leptaleus
5 mm
2 cm
calyx
in fruit
1 cm
5 mm
Mimulus latidens
M. layneae

magenta, tube-throat 7–10 mm, throat floor ± glabrous, folds yellow at mouth; placentas parietal. **FR** 7–11 mm. *n*=8. Sandy, open places, esp around sandstone outcrops or on burns, other disturbed areas; 90–1250 m. NCoRI, CW. [ssp. *decurtatus* (A.L. Grant) Pennell, Santa Cruz County monkeyflower]

M. ringens L. (p. 1047) Per, rhizomed, glabrous. **ST** 20–130 cm, often ascending, 4-angled. **LF** 25–80 mm, lanceolate to narrowly oblong, often ± clasping st; veins pinnate. **FL**: pedicel 20–35 mm; calyx 10–16 mm, ± puberulent, lobes equal, 2–6 mm, acute, minutely ciliate at base; corolla lavender to purple, tube-throat 20–30 mm, lower lip base swollen, nearly closing mouth; placentas axile. **FR** 10–12 mm. *n*=8,11,12. Wet places; < 200 m. ne SnJV (near La Grange, Stanislaus Co.); mostly c Can to e US; also ID, Colorado. Probably introduced in CA.

M. rubellus A. Gray (p. 1047) Ann, 2–32 cm, minutely puberulent. **LVS** 3–31 mm, lanceolate to ovate; lower ± petioled; upper sessile, narrowly sheathing st. **FL**: pedicel 5–18 mm, ascending in fr; calyx 4.5–9 mm, gen glabrous, ribs brownish, lobes equal, 0.5–1.5 mm, rounded, often with small point at tip, ciliate; corolla yellow or lavender-magenta, tube-throat 6–10 mm, limb 3–4 mm wide; placentas axile. **FR** 3.5–7 mm. Gen in and along washes; 800–2600 m. W&I, DMtns; to WY, NM, n Baja CA.

M. rupicola Cov. & A.L. Grant (p. 1047) DEATH VALLEY MONKEYFLOWER Per, 1–17 cm, puberulent. **LF** 18–60 mm, oblanceolate, not ciliate. **FL**: pedicel 1–3 mm; calyx 8–17.5 mm, puberulent, lobes unequal, 1.5–7 mm, long-tapered; corolla persistent, pinkish to white with 1 large, magenta-purple spot at center of each lobe base, tube-throat 17–35 mm; placentas parietal. **FR** 3–8 mm, ovoid-oblong, slightly curved, hard, indehiscent. **SEEDS** few, 1–2 mm. *n*=8. UNCOMMON. Limestone crevices; 300–1800 m. n DMtns (Cottonwood, Funeral, Grapevine, Last Chance, n Panamint ranges).

M. shevockii Heckard & Bacigal. (p. 1047) KELSO CREEK MONKEYFLOWER Ann, 2–12 cm, minutely puberulent. **LF** 3–10 mm, lanceolate to ovate, clasping st; pairs at nodes fused or not. **FL**: pedicel 10–22 mm; calyx 4–7 mm, with reddish spots or solid red, ± puberulent, lobes equal, 0.5–1 mm, gen rounded; corolla tube-throat 8–12 mm, upper lip appearing 4-lobed (lateral pair small), maroon-purple, lower lip appearing as 1 lobe, notched, yellow with maroon dots at base; anthers and stigma exserted; placentas axile. **FR** 5–6 mm. *n*=16. RARE. Alluvial fans, dry streamlets, gen granitic soils; 900–1300 m. s SNF (Kelso Creek, Cortez, Cyrus canyons, Kern Co.).

M. suksdorfii A. Gray (p. 1051) Ann, 0.5–10 cm, minutely puberulent. **LVS** 4–23 mm, ± linear to ovate; upper sessile, narrowly sheathing st. **FL**: pedicel 2–8 mm, spreading and ± S-curved in fr; calyx 3–5.5 mm, glabrous, lobes equal, 0.3–1 mm, rounded or with small point at tip; corolla yellow, tube-throat 4–6 mm; placentas axile. **FR** 2.5–5 mm. Moist, gen clay soils in ± full sun; 1100–4000 m. CaRH, SNH, WTR, SnBr, SnJt, n DMtns (Grapevine Mtns, Inyo Co.); to WA, MT, WY, Colorado, AZ.

M. tilingii Regel (p. 1051) Per, 2–35 cm, rhizomed, glabrous to ± hairy. **LF**: petiole 0–25 mm; blade 5–30 mm, elliptic to ± round; pairs at nodes not fused. **INFL**: fls 1–5 per st, solitary in axils of upper lvs, not in bracted raceme. **FLS** opening; pedicel 10–90 mm; calyx 7–25 mm, asymmetrically swollen in fr, glabrous to puberulent, lobes unequal, lowest 2 upcurved in fr; corolla yellow, tube-throat 17–45 mm; placentas axile. **FR** 5–10 mm. *n*=14,15,24,25, 28. Seeps, streamsides, wet meadows; 1400–3400 m. NW, CaRH, SNH, SnBr, SnJt, W&I; to AK, MT, Colorado. Intergrades with *M. guttatus* in some areas. ❀WET or IRR:**1–3**,**7**,14,**18**,19–23&SUN: 4,5,**6**,15–17.

M. torreyi A. Gray (p. 1051) Ann, 4–38 cm, puberulent. **LF**: petioles poorly defined, 0–10 mm, reduced upward; blade 6–40 mm, gen oblanceolate or elliptic. **FL**: pedicel 1–3 mm; calyx 5–10 mm, membranous, uniformly green, puberulent, lobes slightly unequal, 0.5–2 mm, acute to obtuse; corolla persistent, light rose to magenta, with 2 gold stripes with dark magenta margins extending onto lower lip base from throat floor folds, gen without other marks, tube-throat 9–18 mm, lower lip gen ± projecting forward, upper lip ± reflexed; placentas parietal. **FR** 6–11 mm. *n*=10. Bare or disturbed, open areas; 600–2000 m. s CaRH, SN. ❀TRY.

M. traskiae A.L. Grant SANTA CATALINA ISLAND MONKEYFLOWER Ann, 8–12 cm, puberulent. **LF**: petiole 1–12 mm; blade 12–41 mm, ovate, not ciliate. **FL**: pedicel 3–5 mm; calyx 18–21 mm, puberulent on ridges, lobes unequal, 2–4 mm, acute; corolla persistent, tube-throat 20–23 mm, ± = calyx, red-purple, upper lip 4–5 mm, white, lower lip < 1 mm, red-purple; placentas parietal. **FR** ± 8 mm, ± oblique-oblong, hard, probably indehiscent. **PRESUMED EXTINCT**. Habitat unknown. s ChI (Avalon, Santa Catalina Island).

M. tricolor Lindley (p. 1051) Ann, 1–14 cm, densely puberulent. **LF** 8–45 mm, oblanceolate, not ciliate. **FL**: pedicel 1–3 mm; calyx 10–23 mm, puberulent, lobes unequal, 1–5 mm, obtuse; corolla persistent, magenta to purple with whitish throat stained with yellow, mottled with dark dots, and with 1 large, maroon spot at center of each lobe base, tube-throat 15–50 mm, puberulent outside; placentas parietal. **FR** 5–8 mm, oblique-ovoid, slightly compressed laterally, hard, indehiscent. *n*=9. Vernally wet depressions; gen < 600 m (< 1400 m in MP). s NCoRO (s Sonoma Co.), NCoRI, e GV, sw MP; OR. ❀TRY;DFCLT.

M. viscidus Congdon (p. 1051) Ann, 2–37 cm, hairy. **LVS** (exc basal) 4–45 mm, oblanceolate to oblong; lowest gen glabrous. **INFL**: fl 1 per node. **FL**: pedicel 1–4 mm; calyx 6–15 mm, wide-ribbed, swollen, densely hairy, lobes ± unequal, 1–4 mm, acute; corolla persistent, lavender or magenta with darker, central marks on lobe bases, tube-throat 10–20 mm, throat folds ± gold, fading toward mouth, throat floor densely hairy; placentas parietal. **FR** 6–11 mm, often indehiscent. *n*=8. Burns, openings in chaparral, disturbed areas; 90–1300 m. SNF (El Dorado Co. to n Tulare Co.). ❀TRY.

M. whitneyi A. Gray (p. 1051) Ann, 1–14 cm, puberulent. **LF** 7–23 mm, ± linear to narrowly elliptic. **FL**: pedicel 1–3 mm; calyx 4–8 mm, dark-ribbed, puberulent, lobes ± equal, 1–3 mm, acute; corolla persistent, dark, magenta or yellow, tube-throat 13–18 mm, throat with gen 5 dark longitudinal lines, 2 reddish brown lateral, longitudinal stripes, these ± obscured in magenta fls; placentas parietal. **FR** 6–10 mm. *n*=8. Disturbed areas, roadbanks, exposed runoff areas, granitic soils; gen 1500–3400 m. s SNH. Intergrades with *M. constrictus* in s of range at lower elevations; with *M. leptaleus* & *M. mephiticus* in ne of range at higher elevations. ❀TRY.

MOHAVEA

David M. Thompson

Ann, hairy. **ST** erect; branches 0 or few. **LVS** alternate, widely lanceolate, entire; veins pinnate. **INFL** crowded; fls solitary in axils. **FL**: calyx lobes 5, ± equal; corolla tube with sac-like extension at base, lips flaring, ± fan-shaped, lower lip base swollen, closing mouth; stamens 2, incl, staminodes 2. **FR** obliquely ovoid, fragile; chambers dehiscent by 1–2 large pores near tip. **SEEDS** ovate, flat, smooth; wing incurved, ± cup-shaped. 2 spp.: sw US, n Mex. (Mojave River, where first collected by John Frémont)

1. Corolla dark yellow; fl 15–20 mm . . . . ***M. breviflora***
1′ Corolla yellowish to white; fl 25–35 mm . . . . ***M. confertiflora***

M. breviflora Cov. (p. 1051) Pl 5–20 cm. **INFL**: pedicel 2–5 mm. **FL** 15–20 mm; corolla dark yellow, lower lip lobed to within 2–3 mm of swollen base, maroon-spotted only on swollen base. **FR** 8–10 mm. *n*=15. Gravelly desert slopes, washes; –100–1400 m. n&e DMoj; s NV, nw AZ. ❀TRY;DFCLT.

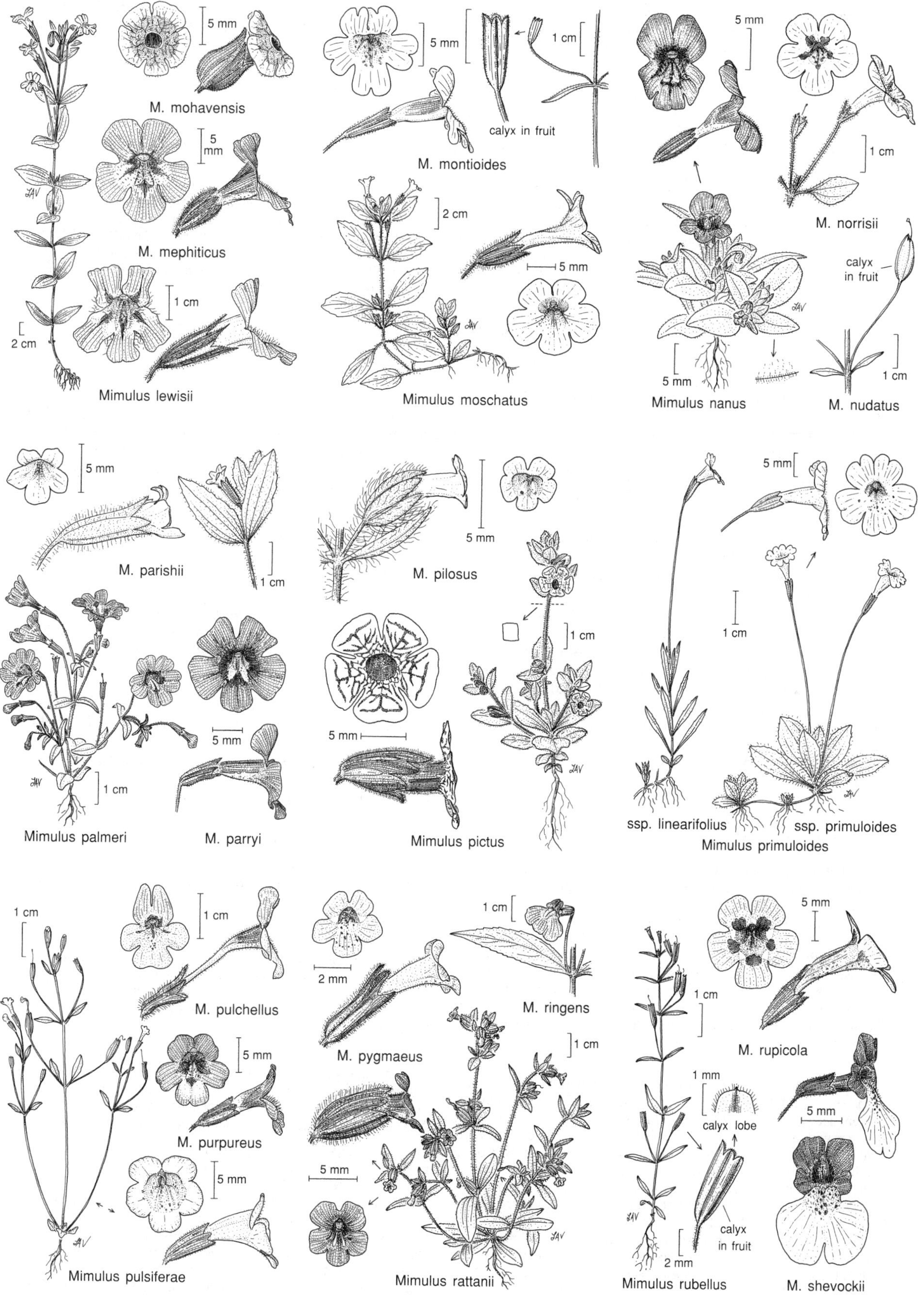

5 mm
M. mohavensis
5 mm
M. mephiticus
1 cm
2 cm
Mimulus lewisii
5 mm
calyx in fruit
1 cm
M. montioides
2 cm
5 mm
Mimulus moschatus
5 mm
1 cm
M. norrisii
calyx in fruit
5 mm
1 cm
Mimulus nanus
M. nudatus
5 mm
M. parishii
1 cm
5 mm
1 cm
Mimulus palmeri
M. parryi
5 mm
M. pilosus
1 cm
5 mm
Mimulus pictus
5 mm
1 cm
ssp. linearifolius
ssp. primuloides
Mimulus primuloides
1 cm
1 cm
M. pulchellus
5 mm
M. purpureus
5 mm
Mimulus pulsiferae
2 mm
1 cm
M. ringens
M. pygmaeus
1 cm
5 mm
Mimulus rattanii
5 mm
1 cm
M. rupicola
1 mm
calyx lobe
5 mm
calyx in fruit
2 mm
Mimulus rubellus
M. shevockii

M. confertiflora (Benth.) A.A. Heller (p. 1051) GHOST FLOWER Pl 10–40 cm. **INFL**: pedicel 5–10 mm. **FL** 25–35 mm; corolla yellowish to white, lower lip lobed to within 6–8 mm of swollen base, maroon-spotted on swollen base and limb. **FR** 10–12 mm. *n*= 15. Gravelly or sandy desert slopes, washes; 0–1100 m. D (exc Inyo Co.); s NV, w AZ, nw Mex. ❀TRY;DFCLT.

NOTHOCHELONE

Noel H. Holmgren

1 sp. (Greek: false turtle, from corolla like that of *Chelone*, turtlehead) Segregated from *Penstemon*.

N. nemorosa (Lindley) Straw (p. 1051) Per from caudex; hairs often pointing downward. **STS** erect, 3–10 dm. **LVS** cauline, opposite, short-petioled; larger blades 4–14 cm, lanceolate to ovate, base cordate or rounded, tip acute to acuminate, margin coarsely toothed. **INFL**: panicle, glandular, few-fld; bracts small. **FL**: calyx lobes 5, ± equal, 5–13 mm, narrowly lanceolate to ovate; corolla 26–33 mm, pink to maroon, lower side often paler, glandular outside, glabrous inside, upper lip 2.4–5 mm, shallowly 2-lobed, lower lip 5–9 mm; filament hairs longest at base, anthers ± 1 mm, long-woolly, anther sacs spreading ± flat at dehiscence; staminode coarsely bearded, incl; nectary a disk; stigma head-like. **FR** 10–15 mm. **SEEDS** flat, widely winged, 2–3.5 mm. 2*n*=30. Rocky places in open Douglas-fir, yellow-pine, and mixed-evergreen forests; 1000–1400 m. KR, n CaRH; to B.C. ❀TRY.

ORTHOCARPUS

T.I. Chuang & Lawrence R. Heckard

Ann, green root parasites. **LVS** sessile, entire to 3-lobed. **INFL**: spike; bracts entire to 5-lobed, tips gen colored. **FL**: calyx 4-lobed, deepest sinus in back; corolla club-shaped, upper lip beak-like, tip closed, enclosing anthers and style, lower lip shorter, ± 3-pouched, gen 3-toothed; stamens 4, anther sacs 2, unequal; style and stigma slender. **FR** loculicidal, ± ovoid, gen ± notched. **SEEDS** gen 8–15, often ± curved, ± keeled, attached at side; coat netted or ridged, tight-fitting. 9 spp.: w N.Am. (Greek: straight fr) [Chuang & Heckard 1992 Syst Bot 17:560–582] Close to *Cordylanthus*; other spp. formerly placed here are in *Castilleja* (Owl's-clovers) or *Triphysaria* (Johnny-tuck). ❀TRY with host; DFCLT.

1. Bracts grading into upper lvs, uniformly ± green (or uppermost purple-tinged), all 3 lobes triangular-lanceolate; infl densely glandular-puberulent
  2. Corolla white to purplish, gen 15–20 mm; beak tip strongly hooked; lower corolla lip strongly pouched, teeth 0 ..... ***O. bracteosus***
  2′ Corolla golden-yellow, 10–15 mm; beak tip obscurely hooked; lower corolla lip moderately pouched, teeth 3, incurved ..... ***O. luteus***
1′ Bracts differing abruptly from upper lvs, tips purplish or white; lateral 2–4 lobes much narrower than ovate or oblong central lobe; infl glabrous or sparsely glandular-puberulent
  3. Beak with conspicuous, cylindric, hooked tip — KR, CaR
    4. Corolla 10–12 mm ..... ***O. imbricatus***
    4′ Corolla 25–30 mm — presumed extinct ..... ***O. pachystachyus***
  3′ Beak straight, without cylindric, hooked tip ..... ***O. cuspidatus***
    5. Corolla 16–25 mm; beak 7–8 mm, gen 1–4 mm > lower lip; lower lip pouches 4–5 mm deep — NW, CaR ..... ssp. ***cuspidatus***
    5′ Corolla 9–18 mm; beak 3–6 mm, < 1 mm > lower lip; pouches 1–3 mm deep
      6. Corolla 14–18 mm, well exserted, beak 5–6 mm, pouches ± 3 mm deep; NW, CaR ..... ssp. ***copelandii***
      6′ Corolla 9–14 mm, ± hidden by bract, beak 3–4 mm, pouches 1–2 mm deep; SN, GB ..... ssp. ***cryptanthus***

O. bracteosus Benth. (p. 1051) Pl 10–40 cm, minutely scabrous, glandular-puberulent, gen becoming ± purple-tinged. **ST** gen simple, slender. **LVS** 15–35 mm, ± linear; lower entire; upper 3-lobed to middle. **INFL** 3–20 cm, densely puberulent; bracts grading into upper lvs, 10–20 mm, ± ovate, 2 lateral lobes below middle, central lobe ± lanceolate, 3–5 mm wide. **FL**: calyx 6–10 mm, divided 2/3 in back, 1/2 on front and sides; corolla 12–20 mm, gen rose (to white), exserted, lips ± equal, beak 4–6 mm, tip with prominent, downward-projecting, glabrous, cylindric hook, lower lip pouched, teeth 0; stigma barely exserted. **FR** 5–7 mm. **SEED** light brown. 2*n*=30. Moist meadows; 500–2000 m. NW, CaR, MP; to B.C.

O. cuspidatus E. Greene Pl 10–40 cm, puberulent to scabrous, sparsely glandular, becoming ± purple-tinged. **ST** simple to much-branched, gen slender. **LVS** 10–50 mm, ± lanceolate; lower entire; upper with 3 deep, linear lobes. **INFL** 2–10 cm, dense; bracts differing abruptly from lvs, 10–20 mm, ± ovate, with 2 narrow, ± basal lobes, central lobe 7–15 mm wide, tip abrupty pointed, uppermost purplish pink on distal 1/3. **FL**: calyx 7–10 mm, divided 2/3 in back, 1/2 in front, 1/4 on sides; corolla 10–25 mm, exserted or not, lips purplish pink, densely puberulent, beak nearly straight, 4–10 mm, 0–4 mm > lower lip, lower lip ± pouched, teeth 1–2 mm, triangular, densely puberulent; stigma well incl. **FR** 6–8 mm. **SEED** dark brown. 2*n*=28. Open slopes or sagebrush; 700–3200 m. NW, CaR, SN, GB; to OR, n-c US.

ssp. ***copelandii*** (Eastw.) Chuang & Heckard (p. 1051) **FL**: corolla 14–18 mm, exserted, beak 5–6 mm, 0–1 mm > lower lip, pouches ± 3 mm deep; upper medial anther sac ± 1.3 mm. Open, grassy to rocky slopes; 1300–2300 m. NW, CaR; sw OR. Intergrades with ssp. *cuspidatus*.

ssp. ***cryptanthus*** (Piper) Chuang & Heckard (p. 1051) **FL**: corolla 9–14 mm, ± hidden by bract, beak 3–5 mm, 0–1 mm > lower lip, pouches 1–2 mm deep; upper medial anther sac 0.7–1 mm. Drying meadows, open sagebrush; 1500–3200 m. SN, GB; to OR, n-c US.

ssp. ***cuspidatus*** (p. 1051) SISKIYOU MOUNTAINS ORTHOCARPUS **ST** gen stout. **FL**: corolla 16–25 mm, exserted, beak 7–8 mm, gen 1–4 mm > lower lip, pouches 4–5 mm deep; upper medial anther sac 1.5–2 mm. UNCOMMON. Open, grassy to rocky slopes; 700–2200 m. NW, CaR; sw OR. Intergrades with ssp. *copelandii*.

O. imbricatus S. Watson (p. 1051) Pl 10–35 cm, ± puberulent, green. **ST** simple or branched above. **LF** 20–50 mm, entire, ± lanceolate, 3-veined. **INFL** 2–8 cm; bracts differing abruptly from lvs, 10–15 mm, strongly net-veined, with 2 small, lateral lobes from near base, central lobe widely oblong, 5–10 mm wide, distal 1/3 ± rose. **FL**: calyx 5–7 mm, hairy, divided 3/4 in back and front, 1/4 on sides; corolla 10–12 mm, nearly hidden by bract, ± puberulent, tube ± rose, beak ± 4 mm, ± 0.5 mm > lower lip, tip strongly hooked, cylindric, lower lip moderately pouched, yellowish, teeth 0.5 mm, triangular, densely puberulent; stigma well incl. **FR** ± 5 mm. **SEEDS** 3–5, dark brown. 2*n*=28. Mtn meadows, grassy or rocky slopes; 1600–2400 m. NW, CaR; to B.C.

O. luteus Nutt. (p. 1051) Pl 10–40 cm, glandular- and longer-nonglandular hairy, yellow-green, often becoming ± purple-tinged. **ST** gen simple, slender. **LF** 15–50 mm, ± linear, entire or upper deeply 3-lobed. **INFL** 5–20 cm, densely glandular-puberulent; bracts grading into upper lvs, 10–20 mm, ± green, 2 lateral lobes below middle, narrowly triangular, central lobe ± lanceolate, 2–5 mm wide. **FL**: calyx 5–8 mm, divided 3/4 in back, 1/3 in front, 1/4 on sides; corolla 10–15 mm, golden-yellow, exserted, puberulent (esp beak), lips ± equal, beak 2–4 mm, tip minute, downward-projecting, cylindric, lower lip moderately pouched, teeth 0.5 mm, blunt, incurved; stigma incl. **FR** 5–7 mm. **SEEDS** 20–35, yellowish to dark brown. 2*n*=28. Moist fields, sagebrush scrub, mtn meadows; 1500–3000 m. GB; to B.C., n-c US, NM.

O. pachystachyus A. Gray (p. 1051) SHASTA ORTHOCARPUS Pl 10–20 cm, scabrous, sparsely soft-hairy, and (partly glandular-) puberulent. **ST** simple, stout. **LF** 30–50 mm; lobes of upper lvs 3–5, ± linear. **INFL** 5–10 cm; bracts differing abruptly from upper lvs, 20–30 mm, widely ovate, prominently net-veined, gen with 4 narrow lateral lobes below middle, central lobe ± ovate, 7–10 mm wide, upper margin pink-lavender. **FL**: calyx 15–20 mm, divided 2/3 in back, 1/2 in front, 1/4 on sides; corolla 25–30 mm, ± rose above, beak ± 10 mm, ± 3 mm > lower lip, glabrous, strongly curved, tip with 1 mm hook, lower lip slightly pouched, puberulent, teeth rounded; stigma barely incl. **FR** 5–7 mm. **PRESUMED EXTINCT**. Meadows?; ± 850 m. e KR, adjacent w CaR (c Siskiyou Co.). Last seen in 1913.

PARENTUCELLIA

Margriet Wetherwax

Ann, sticky-hairy, gen erect and unbranched in CA. **LVS** gen opposite above, sessile. **INFL**: raceme, spike-like, terminal; bracts lf-like. **FL**: calyx 4-lobed, radial; corolla 2-lipped, upper lip entire or notched, forming hood, lower lip > upper, 3-lobed; stamens 4, incl in upper lip, anthers hairy, awned at base. **FR** cylindric. **SEEDS** many, minute, smooth. 4 spp.: w Eur, Medit, c Asia. (Tomaso Parentucelli, early Renaissance librarian to Cosimo de' Medici in Florence; as humanist Pope Nicholas V, 1447–1455, established Vatican library and gardens)

1. Corolla reddish purple, persistent; calyx lobes ± 1/2 tube length . . . . ***P. latifolia***
1′ Corolla yellow, deciduous; calyx lobes ± = tube . . . . ***P. viscosa***

P. latifolia (L.) Caruel (p. 1051) **ST** < 30 cm. **LF** 4–12 mm, triangular-lanceolate, dentate. **FL**: calyx 6–12 mm, lobes ± 1/2 tube length, triangular-lanceolate; corolla persistent, 8–10 mm, reddish purple, with 2 yellow, partly free appendages near lower throat. **FR** = calyx, glabrous. Uncommon. Pastures; ± 100 m. NCoRO, n CCo; native to Eur.

P. viscosa (L.) Caruel (p. 1051) **ST** < 50 cm. **LF** 20–40 mm, lanceolate, serrate. **FL**: calyx 10–16 mm, lobes ± = tube, lanceolate; corolla deciduous, ± 20 mm, yellow, with 2 ridges near lower throat. **FR** slightly > calyx tube, hairy. Damp, grassy places, gen near coast; gen < 100 m. NCo, NCoRO, CCo, SCo; OR; native to Eur.

PEDICULARIS LOUSEWORT

Linda Ann Vorobik

Per, green root parasites. **STS** decumbent to erect, gen 1–several from gen short caudex. **LVS** alternate, gen ± basal, gen < infl, crenate to divided, gen reduced upward; petiole gen < blade. **INFL**: raceme, spike-like; bracts (at least lower) gen ± like upper lvs; pedicels 1–6 mm. **FL**: calyx lobes (2,4) gen 5, uppermost gen shortest (all gen < tube), lateral fused in pairs; corolla white or yellow to red or purple, upper lip hood-, beak-, or trunk-like, lower lip 3-lobed, narrow to fan-shaped, central lobe gen smallest; fertile stamens 4, gen glabrous, anthers gen incl; stigma head-like, gen exserted. **FR** loculicidal, gen ± ovoid, asymmetric, opening mostly on upper side. **SEED** smooth or netted. ± 500 spp.: cool wet n temp, circumboreal, S.Am. (Latin: lice, from belief that ingestion by stock promoted lice infestation) [Macior 1977 Bull Torrey Bot Club 104: 148–154]

1. Lf simple, ± toothed; calyx lobes gen 2 (4 in some pls); fr ± lanceolate in outline
 2. Upper corolla lip hooded, not beaked; lf crenate; sts gen simple . . . . ***P. crenulata***
 2′ Upper corolla lip extended in a long, down-curved beak; lf serrate to dentate; sts gen branched above . ***P. racemosa***
1′ Lf deeply lobed to compound; calyx lobes 5; fr ovoid
 3. Lower cauline lf segments gen 3–5(7), ovate, entire to toothed, terminal largest; basal lvs 0 at fl . . . . ***P. howellii***
 3′ Lower cauline lf segments > 5, serrate to divided, terminal segment not largest; basal lvs gen present at fl (exc *P. bracteosa*)
  4. Corolla not club-like, upper lip with a long narrow, beak, lower lip ± fan-like
   5. Corolla white to yellowish, upper lip 4–5.5 mm, beak curved downward, not very trunk-like . . . . ***P. contorta***
   5′ Corolla pink to purple, upper lip 3.5–18 mm, beak curved upward, ± like an elephant's trunk
    6. Infl densely hairy; beak 3–5 mm; lower lip ± fan-like . . . . ***P. attollens***
    6′ Infl gen glabrous; beak 6–13 mm; outer lobes of lower lip appearing like ears . . . . ***P. groenlandica***
  4′ Corolla ± club-like, upper lip hooded, not beaked; lower lip not fan-like
   7. Corolla deep red to red-purple (or yellow to orange), lower lip ± 1/4 upper . . . . ***P. densiflora***
   7′ Corolla yellowish or purplish, lower lip gen >> 1/4 upper
    8. St 30–90 cm; infl > lvs; basal lvs gen 0 at fl; corolla yellowish . . . . ***P. bracteosa*** var. ***flavida***
    8′ St < 30 cm; infl gen < or = lvs; basal lvs present at fl; corolla yellowish or purplish
     9. Corolla yellowish (sometimes purple-tipped), hairy; filaments hairy . . . . ***P. semibarbata***
     9′ Corolla purplish, glabrous; filaments glabrous
      10. Corolla 30–42 mm, pale purple, darker-tipped, lower lip > 1/2 upper, its lobes rounded, wavy-margined; lf margin wavy, white, thickened . . . . ***P. centranthera***
      10′ Corolla 17–24 mm, purple to purplish red, lower lip < upper, its lobes pointed, not wavy-margined; lf margin not wavy, white, or thickened . . . . ***P. dudleyi***

P. attollens A. Gray (p. 1051) LITTLE ELEPHANT'S HEAD **ST** 6–60 cm, tomentose above. **LVS**: basal 3–20 cm, ± linear; segments 17–41, linear, toothed. **INFL** 2–30 cm; bracts tomentose, < fls. **FL**: calyx 4–6 mm, tomentose; corolla 4.5–7 mm, light pink to purple, marked darker, glabrous, upper lip 3.5–6 mm, trunk-like, curved upward, beak 3–5 mm, lower lip 3–5 mm, ± fan-like; anthers ± 1 mm, bases obtuse. **FR** 6–10 mm. **SEED** 2.5–4 mm, finely netted. Wet meadows, streamsides, bogs; 1500–3900 m. KR, CaR, SNH, MP, W&I; OR. ❀very DFCLT.

P. bracteosa Benth. var. ***flavida*** (Pennell) Cronq. (p. 1051) **ST** 30–110 cm, sparsely hairy above. **LVS** ± ovate; segments singly to doubly toothed; basal lvs gen 0 at fl (8–25 cm, 5–21-segmented, petiole ± = blade); cauline lvs 3–15 cm, lower 13–25-segmented, upper dentate. **INFL** 4–17 cm, glabrous; lower bracts > fls. **FL**: calyx 7–13 mm, lobes minutely ciliate; corolla 15–24 mm, ± club-like, light greenish to yellow, upper lip 8–10 mm, deeply hooded, lower lip 4–6 mm; anthers 1.5–3 mm, bases acute to obtuse. **FR** 9–10 mm. **SEED** 2.5–3.5 mm, netted. Uncommon. Moist upper montane coniferous forest; 1200–2300 m. KR; OR. [*P.f.* Pennell]

P. centranthera A. Gray (p. 1051) DWARF LOUSEWORT **ST** 4–10 cm, ± glabrous; caudex long. **LVS**: basal 6–20 cm, > infl, ± lanceolate; segments 13–25, oblong, crenate, dentate, or lobed. **INFL** 3–10 cm; lower bracts < or = fls. **FL**: calyx 15–22 mm, ciliate or long-hairy; corolla 30–42 mm, ± club-like, pale purple, darker-tipped, glabrous, upper lip 10–18 mm, hooded, lower lip 9–15 mm, lobes ± equal, widely rounded; anthers 3.5–5 mm, exserted, bases acuminate. **FR** 10–13 mm. **SEED** 3–4.5 mm, shallowly netted. Sagebrush scrub, alluvial fans; 1300–1500 m. RARE in CA. se MP (e Lassen Co.); to OR, Colorado, NM.

P. contorta Benth. (p. 1051) CURVED-BEAK LOUSEWORT **ST** 15–40 cm, subglabrous. **LVS**: basal 3–18 cm, oblong-lanceolate; segments 25–41, linear, entire to toothed. **INFL** 5–28 cm; bracts gen < fls, ciliate. **FL**: calyx 6–10 mm, ciliate, sometimes hairy; corolla 10–13 mm, white to pale yellow, upper lip 4–5.5 mm, beaked, curved downward, not very trunk-like, lower lip 5–8 mm, ± fan-like; anthers 2.5–3 mm, bases acute. **FR** 6–10 mm. **SEED** 2–2.5 mm, finely netted. UNCOMMON. Bogs, meadows, streamsides, moist, open montane coniferous forest; 1600–2400 m. KR; to w Can, MT, WY, UT.

P. crenulata Benth. (p. 1051) SCALLOP-LEAVED LOUSEWORT **ST** 12–40 cm, tomentose above. **LVS** basal and cauline, 3–11 cm, ± linear; margins ± doubly crenate, thick, rolled under, wavy, white. **INFL** 2–12 cm, gen simple; bracts gen < fls. **FL**: calyx 8–12 mm, tomentose, lobes 2(4); corolla 20–26 mm, ± club-like, white in CA, glabrous, upper lip 11–15 mm, hooded, lower lip 7–12 mm, ± fan-like, margins ± wavy, lobes ± equal; anthers 2–3 mm, bases acute. **FR** 10–20 mm, lanceolate in outline. **SEED** 1.5–2 mm, smooth. Wet meadows, streambanks; 2100–2300 m. RARE in CA. c SNH (Convict Creek, Mono Co); to NV, WY, Colorado. [forma *candida* J.F. Macbr.]

P. densiflora Hook. (p. 1051) INDIAN WARRIOR Pl softly to coarsely brown-hairy. **ST** 6–55 cm. **LVS**: basal 5–28 cm, oblong-lanceolate, segments 13–41, ± linear to ovate, doubly toothed to lobed. **INFL** 4–12 cm; lower bracts > fls. **FL**: calyx 8–15 mm, gen hairy, lobes ± equal; corolla 23–36 mm, straight, club-like, deep red or purple to yellow, gen minutely hairy, upper lip 8–17 mm, hooded, lower lip 2–4 mm, lobes ± equal; anthers 2–3 mm, bases acute. **FR** 8–13 mm. **SEED** 2.5–4.5 mm, netted. Dry chaparral, oak/pine or yellow-pine forests; < 2100 m. NW, CaR, SNF, Teh, CW, SW; s OR. [ssp. *aurantiaca* Sprague] ❀very DFCLT.

P. dudleyi Elmer (p. 1057) DUDLEY'S LOUSEWORT Pl gen ± hairy. **ST** 10–30 cm. **LVS**: basal 3–26 cm, > infl, oblong-lanceolate, segments 15–25, oblong to ovate, doubly toothed to lobed. **INFL** 2–6 cm; lower bracts > fls. **FL**: calyx 10–14 mm, tomentose; corolla 17–24 mm, ± club-like, light pink to purple, marked darker, glabrous, upper lip 9–11 mm, hooded, lower lip 4–7 mm, lobes ± equal; anthers 1.5–2.5 mm, gen incl, bases acute. **FR** 8–13 mm. **SEED** 3.5–5 mm, netted. RARE. Coastal chaparral or forest; < 350 m. CW (exc SCoRO). Widely scattered. Pls from c CCo (Arroyo de la Cruz, San Luis Obispo Co.) warrant further study (smaller, lvs < infl, anthers often exserted with bases somewhat acuminate); also like *P. semibarbata* but filaments glabrous.

P. groenlandica Retz. (p. 1057) ELEPHANT'S HEAD Pl ± glabrous. **ST** 8–80 cm. **LVS**: basal 3–25 cm, ± lanceolate; segments 25–51, linear to oblong, toothed. **INFL** 1–30 cm; lower bracts ± = fl. **FL**: calyx 3.5–6 mm, lobes ± equal, densely short-ciliate; corolla 8–15 mm, light pink to red-purple, glabrous, tube 4–10 mm, upper lip 7–18 mm, hood curved downward, trunk-like beak 6–13 mm, curved upward, lower lip 3–7 mm, lateral lobes resembling elephant ears; anthers 1.5–2.5 mm, bases acute. **FR** 6–9 mm. **SEED** 2.5–4 mm, netted. Wet meadows, streamsides, bogs; 1000–3600 m. KR, CaR, SNH; to AK, e N.Am., NM. CA pls have been considered var. *surrecta* Piper. ❀very DFCLT.

P. howellii A. Gray (p. 1057) HOWELL'S LOUSEWORT **ST** 6–45 cm, glabrous to puberulent; caudex long. **LVS**: basal gen 0 at fl (4–20 cm, widely lanceolate, segments 7–17, ovate, entire-dentate to lobed); cauline lvs 2–9 cm, ovate, lobes or segments 0–11, entire to dentate, terminal >> lateral. **INFL** closely subtended by 2–3 lvs, 1.5–4 cm; bracts < fls, ± deltate, hairy. **FL**: calyx 6–7 mm, hairy; corolla 7–9 mm, sickle-shaped, whitish to purplish, marked darker, glabrous, upper lip 5–6 mm curved, beak ± 4 mm, tapered, lower lip 2–3 mm, ± incl in calyx; anthers 1–2 mm, bases acute. **FR** 7–9 mm. **SEED** 1.5–2 mm, finely netted. UNCOMMON. Dry ridges, open red-fir forest; 1500–1900 m. n KR; s OR.

P. racemosa Hook. (p. 1057) LEAFY LOUSEWORT **STS** few-many, ± decumbent, branched above, 12–80 cm, subglabrous. **LVS** cauline, 2–10 cm, ± narrowly lanceolate, singly to doubly toothed. **INFL** 1–5 cm; lower bracts = or > fls. **FL**: calyx 4.5–8 mm, glabrous, lobes 2(4); corolla 10–16 mm, whitish, yellowish, or purplish, glabrous, upper lip 5–7.5 mm, beak strongly incurved, lower lip 5–9 mm, fan-like; anthers 1.5–2.5 mm, bases acute. **FR** 10–16 mm, lanceolate in outline. **SEED** 1.5–3 mm, smooth. Open coniferous forests; 900–2300 m. KR, CaR, n SNH, SNE; to w Can, WY, NM. Pls e of CA-FP with linear lvs and whitish to yellowish corollas have been called ssp. *alba* Pennell. ❀very DFCLT.

P. semibarbata A. Gray (p. 1057) **ST** mostly underground, 3–20 cm, sparsely tomentose; caudex long. **LVS**: basal 5–22 cm, gen > infl, ± lanceolate, segments 11–25, lanceolate to ovate, toothed to lobed. **INFL** 3–12 cm; bracts hairy, lower >> fls. **FL**: calyx 8–10 mm, ciliate; corolla 15–24 mm, ± club-shaped, yellowish, tinged red or tipped purple, hairy, upper lip 7–10 mm, hooded, lower lip 5–7 mm, lobes ± equal, rounded; anthers 2–2.5 mm, exserted, bases acuminate, filaments hairy. **FR** 6.5–10 mm. **SEED** 3.5–4.5 mm, smooth. Dry ridgetops, coniferous woodlands, often with red fir; 1500–3500 m. NCoRH, CaRH, SN, SnGb, SnBr, PR; NV. [ssp. *charlestonensis* (Pennell & Clokey) Clokey] ❀very DFCLT.

PENSTEMON BEARDTONGUE

Noel H. Holmgren

Per to shrub. **LVS** gen opposite, entire to toothed; upper sessile. **INFL**: panicle or raceme; bracts gen small. **FL**: calyx lobes 5, ± equal; corolla tube ± cylindric or lower side expanded, ± 2-lipped, gen pink or blue to purple (some red, yellow, or white), upper lip 2-lobed, external in bud; anther sacs 2, gen spreading ± flat at dehiscence; staminode attached near base of corolla tube, well developed, gen hairy on upper side; nectaries 2, at bases of upper stamens; stigma head-like. **FR**: capsule, septicidal and sometimes also loculicidal at tip. **SEEDS** gen many, irregularly angled. 250 spp.: N.Am., esp w US. (Latin & Greek: almost thread, from stamen-like staminode) [Holmgren 1984 In Cronquist et al. Intermtn fl 4:370–457] Largest genus of fl pls endemic to N.Am. See also *Keckiella*, *Nothochelone*.

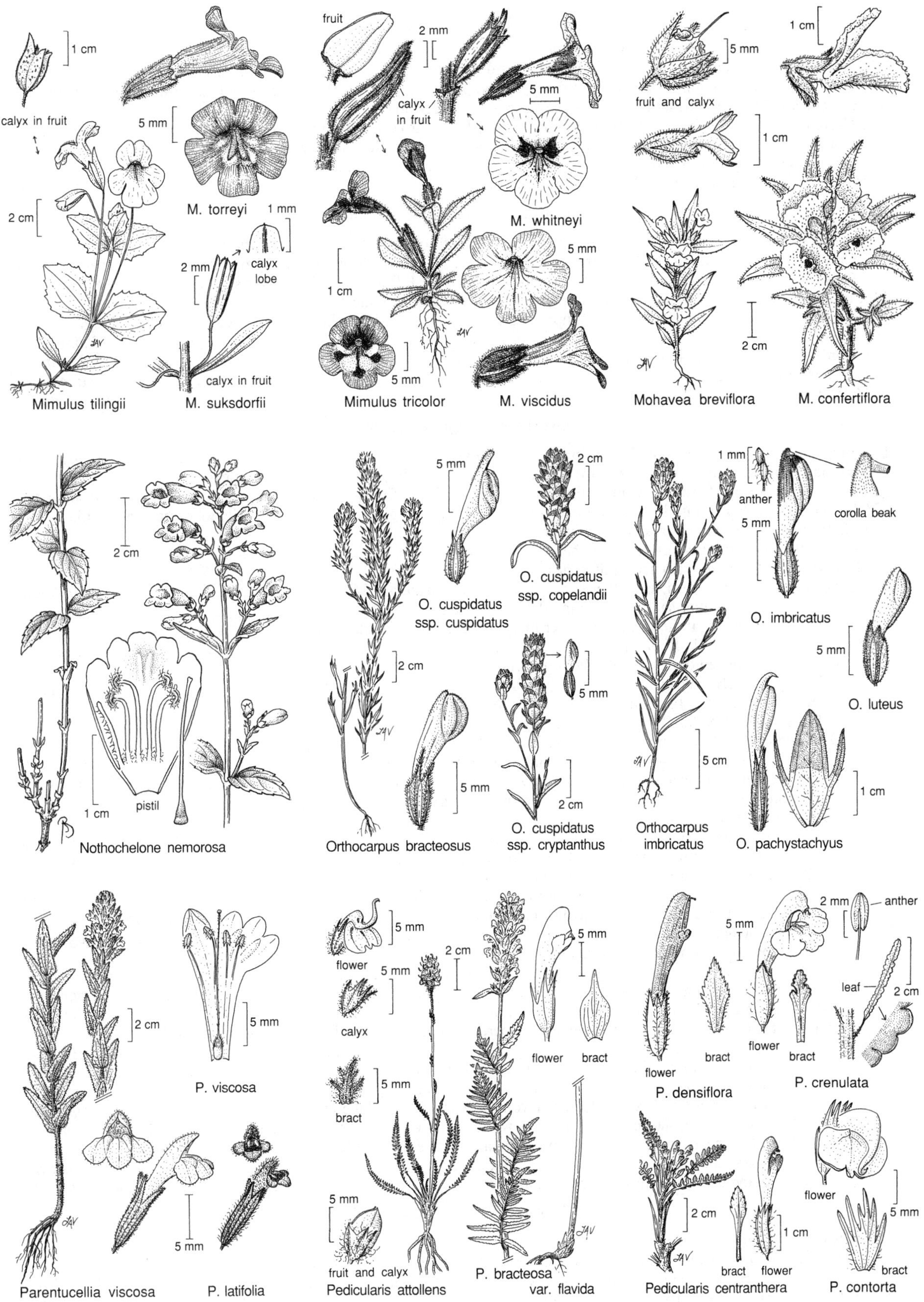

calyx in fruit
1 cm
5 mm
2 cm
M. torreyi
1 mm
2 mm
calyx lobe
calyx in fruit
Mimulus tilingii
M. suksdorfii
fruit
2 mm
calyx in fruit
5 mm
M. whitneyi
1 cm
5 mm
5 mm
Mimulus tricolor
M. viscidus
5 mm
fruit and calyx
1 cm
1 cm
2 cm
Mohavea breviflora
M. confertiflora
2 cm
1 cm
pistil
Nothochelone nemorosa
5 mm
2 cm
O. cuspidatus ssp. copelandii
O. cuspidatus ssp. cuspidatus
2 cm
5 mm
5 mm
2 cm
Orthocarpus bracteosus
O. cuspidatus ssp. cryptanthus
1 mm
anther
corolla beak
5 mm
O. imbricatus
5 mm
O. luteus
5 cm
1 cm
Orthocarpus imbricatus
O. pachystachyus
2 cm
5 mm
P. viscosa
5 mm
Parentucellia viscosa
P. latifolia
5 mm
flower
2 cm
5 mm
calyx
5 mm
bract
5 mm
flower
bract
5 mm
fruit and calyx
Pedicularis attollens
P. bracteosa var. flavida
5 mm
flower
bract
P. densiflora
2 mm
anther
leaf
2 cm
flower
bract
P. crenulata
2 cm
1 cm
bract
flower
Pedicularis centranthera
flower
5 mm
bract
P. contorta

Key to Groups

1. Anther sacs not dehiscing full length (opening only partly)
  2. Anther sacs dehiscing at distal end (proximal portion indehiscent); per (subg. *Habroanthus*) .... **Group 1**
  2′ Anther sacs dehiscing at proximal end (leaving distal portion indehiscent); subshrubs (subg. *Saccanthera*) **Group 2**
1′ Anther sacs dehiscing full length, valves gen spreading widely apart
  3. Corolla mouth closed by arched floor, hiding densely hairy throat interior (subg. *Cryptostemon*) ... ***P. personatus***
  3′ Corolla mouth open, exposing glabrous interior of throat (corolla floor frequently hairy)
    4. Anthers densely woolly; shrubs (subg. *Dasanthera*) .... **Group 3**
    4′ Anthers glabrous; per or subshrubs (subg. *Penstemon*) .... **Group 4**

Group 1 (subg. Habroanthus)

1. Corolla blue or blue-violet, tube ± abruptly expanded into throat, 6–13 mm wide at widest point
  2. Calyx 3–5 mm; corolla throat 6–8 mm wide when pressed, floor hairy; anther sacs 1.6–2.2 mm ..... ***P. pahutensis***
  2′ Calyx 4–13 mm; corolla throat 7.5–13 mm wide when pressed, floor glabrous; anther sacs 1.8–3 mm ***P. speciosus***
1′ Corolla red, tube gradually expanding into throat, 4–9 mm wide at widest point
  3. Corolla 30–40 mm, distinctly 2-lipped, upper lip projecting, hood-like, short-lobed, lower lip reflexed, longer-lobed .... ***P. labrosus***
  3′ Corolla 24–33 mm, nearly radial, lobes ± equal, projecting .... ***P. eatonii***
    4. Pl glabrous throughout .... var. ***eatonii***
    4′ Pl short-hairy .... var. ***undosus***

Group 2 (subg. Saccanthera)

1. Corolla red to orange, lips long, lower lip strongly reflexed .... ***P. rostriflorus***
1′ Corolla lavender, bluish, or purplish, lips short, lower lip projecting or spreading, not reflexed
  2. Lvs sharply serrate; filaments white-hairy distally; corolla lobes finely ciliate .... ***P. venustus***
  2′ Lvs entire (weakly serrate in *P. parvulus*, *P. purpusii*); filaments glabrous distally; corolla lobes not finely ciliate
    3. Staminode hairy; anthers glabrous on inner margins
      4. Corolla 15–20 mm; lvs glabrous; anther sacs 1–1.3 mm .... ***P. gracilentus***
      4′ Corolla 24–34 mm; lvs hairy; anther sacs 1.4–1.9 mm
        5. Lvs well distributed on sts; anther sacs 1.7–1.9 mm, dehiscing < 1/3 length .... ***P. papillatus***
        5′ Lvs gen basal; anther sacs 1.4–1.7 mm, dehiscing for ± 1/2 length .... ***P. scapoides***
    3′ Staminode glabrous; anthers hairy on inner margins near filament attachment
      6. Lvs gen basal; corolla floor hairy .... ***P. caesius***
      6′ Lvs well distributed on sts; corolla floor glabrous
        7. Herbage densely canescent-hairy; pl 0.5–2 dm .... ***P. purpusii***
        7′ Herbage glabrous to finely hairy; pl gen > 2 dm
          8. Infl glandular, peduncles widely spreading
            9. Corolla 13–22 mm; anther sacs 1.2–2.2 mm, dehiscing for 2/5–3/5 length
              10. Lvs thread-like, 0.4–1.5 mm wide; anther sacs 1.2–1.4 mm .... ***P. filiformis***
              10′ Lvs linear to (ob)lanceolate, 2–9 mm wide; anther sacs 1.5–2.2 mm .... ***P. roezlii***
            9′ Corolla 21–38 mm; anther sacs 1.7–3.2 mm, dehiscing for 3/5–4/5 length
              11. Herbage glaucous, glabrous .... ***P. neotericus***
              11′ Herbage green, hairy, lvs sometimes glabrous .... ***P. laetus***
                12. Calyx 8–14 mm, lobes narrowly lanceolate .... var. ***leptosepalus***
                12′ Calyx 4.7–9 mm, lobes gen lanceolate to ovate
                  13. Corolla throat open, lips ± spreading; anther sacs dehiscing for 2/3–3/5 length .... var. ***laetus***
                  13′ Corolla throat somewhat narrowed at mouth, lips ± projecting; anther sacs dehiscing for 3/5–4/5 length .... var. ***sagittatus***
          8′ Infl glabrous, peduncles appressed-ascending
            14. Upper lvs widest near base, gen cordate-clasping; herbage gen glabrous, glaucous
              15. Corolla 14–20 mm; anther sacs 1.4–2 mm .... ***P. parvulus***
              15′ Corolla 20–35 mm; anther sacs 1.8–3.5 mm .... ***P. azureus***
                16. Calyx 5–7.5 mm, lobes ± ovate, long-tapered tips 1.5–3.2 mm; lvs linear, 4–9 cm, 2–7.5 mm wide .... var. ***angustissimus***
                16′ Calyx 3–6.5 mm, lobes (ob)lanceolate to ± round, pointed tips 0.2–1.5 mm; lvs (ob)lanceolate, 1–6.5 cm, 5–17 mm wide .... var. ***azureus***
            14′ Upper lvs tapered to base; herbage glabrous or hairy, rarely glaucous .... ***P. heterophyllus***
              17. Lvs narrowly (ob)lanceolate, 2–7.5 mm wide, rarely with axillary lf clusters .... var. ***purdyi***
              17′ Most lvs linear (some narrowly (ob)lanceolate), 0.5–5 mm wide, gen with axillary lf clusters
                18. Pl short-hairy .... var. ***australis***
                18′ Pl glabrous, sometimes lower sts short-hairy .... var. ***heterophyllus***

Group 3 (subg. Dasanthera)

1. Fl sts gen < 1.2 dm
  2. Corolla blue-violet or blue-purple; staminode densely hairy; lvs green, glabrous .... ***P. davidsonii***
  2′ Corolla rose, lavender, or violet; staminode glabrous to sparsely hairy; lvs glaucous, gen puberulent ... ***P. rupicola***
1′ Fl sts 1.2–3 dm — corolla rose-red to rose-purple .... ***P. newberryi***

3. Corolla 27–35 mm, throat well expanded, 7.5–12 mm wide when pressed, floor ± long-wavy-hairy; anthers incl ..... var. ***berryi***

3′ Corolla 20–30 mm, throat moderately expanded, 4.7–7.5 mm wide when pressed, floor short-hairy; anthers, at least longer pair, exserted

4. Corolla rose-red; lvs of fl-st < basal lvs ..... var. ***newberryi***

4′ Corolla dark rose-purple; lvs of fl-st barely reduced ..... var. ***sonomensis***

Group 4 (subg. Penstemon)

1. Herbage hairy

2. Herbage glandular-hairy; corolla lobes glandular-hairy inside

3. Herbage densely glandular-hairy; fl-sts erect, gen > 4 dm, stout ..... ***P. sudans***

3′ Herbage sparsely glandular-hairy; fl-sts gen decumbent at base, gen < 4 dm ..... ***P. deustus***

4. Lvs lanceolate, gen < 8 mm wide; calyx gen < 4 mm, lobes lanceolate, acute; corolla 10–15 mm var. ***pedicellatus***

4′ Lvs ovate to ± round, 7–28 mm wide; calyx gen > 4 mm, lobes narrowly lanceolate, sharply acute or long-tapered; corolla 9–11 mm ..... var. ***suffrutescens***

2′ Herbage nonglandular-hairy; corolla lobes glabrous inside or corolla floor nonglandular-hairy

5. Infl glabrous; calyx 1.5–4.5 mm; corolla 6–10 mm

6. Lvs gen all cauline, linear or narrowly lanceolate, gen folded lengthwise and arching-recurved ..... [2]***P. cinicola***

6′ Lvs best developed on basal sterile sts, these (ob)lanceolate to obovate, cauline lvs ascending, rarely folded lengthwise, not arching-recurved (for vars. see 29) ..... [2]***P. procerus***

5′ Infl hairy, gen glandular; calyx 3–12 mm; corolla 10–30 mm

7. Calyx 3–6 mm

8. Corolla 20–25 mm, floor arched, closing off throat, hiding stamens and densely hairy interior .. ***P. personatus***

8′ Corolla 10–18 mm, floor not closing off throat, stamens and glabrous throat interior exposed

9. Lvs at base of pl 15–75 mm, 2–32 mm wide, petioled (upper lvs sessile) ..... ***P. humilis***

9′ Lvs similar throughout, 8–20 mm, 0.8–6 mm wide, petioled

10. Pl ± erect, not mat-forming, sts 1–3 dm; calyx lobes ovate ..... ***P. californicus***

10′ Pl low, gen mat-forming, sts 0.5–1.5 dm; calyx lobes lanceolate ..... ***P. thompsoniae***

7′ Calyx 6–12 mm

11. Corolla gradually expanded into throat, 4–6 mm wide when pressed; staminode ± incl, not coiled at tip

12. Anther sacs 0.6–0.7 mm, valves spreading widely; corolla 13–17 mm ..... ***P. calcareus***

12′ Anther sacs 1.1–1.4 mm, valves barely spreading; corolla 15–20 mm ..... ***P. monoensis***

11′ Corolla abruptly and widely expanded into throat, 7–12 mm wide (4–5 mm in P. barnebyi) when pressed; staminode exserted, coiled at tip

13. Corolla 10–14 mm, throat 4–5 mm wide when pressed ..... ***P. barnebyi***

13′ Corolla 18–28 mm, throat 7–12 mm wide when pressed ..... ***P. janishiae***

1′ Herbage (below infl) glabrous

14. Infl glabrous; lf margin gen entire

15. Corolla 17–34 mm

16. Anther sacs 0.6–1.2 mm; corolla floor glabrous

17. Corolla gen < 17 mm, moderately expanded, lavender, magenta, or violet ..... [2]***P. patens***

17′ Corolla > 17 mm, cylindric or narrowly funnel-shaped, red

18. Corolla lobes projecting, not spreading, glabrous; lvs cordate-clasping ..... ***P. centranthifolius***

18′ Corolla lobes spreading, glandular; lvs not cordate-clasping ..... ***P. utahensis***

16′ Anther sacs 1.2–2.4 mm; corolla floor glabrous or hairy

19. Upper lf pairs fused around st, sharply serrate

20. Corolla 19–22 mm, bright pink; anther-sacs 1.3–1.6 mm ..... ***P. clevelandii*** var. ***connatus***

20′ Corolla 24–34 mm, bluish purple or blue; anther-sacs 1.8–2.4 mm ..... ***P. spectabilis*** var. ***spectabilis***

19′ Lf pairs free (or barely fused in some hybrids), entire or weakly serrate

21. Lvs lanceolate to ovate, 10–35 mm wide; staminode incl, glabrous; common hybrid .. (see *P. centranthifolius*)

21′ Lvs linear to narrowly lanceolate, 3–9 mm wide; staminode exserted, densely hairy ..... ***P. fruticiformis***

22. Corolla 22–24 mm, throat 8–10 mm wide when pressed, glandular outside; calyx lobes ovate, gen 5–7.5 mm ..... var. ***amargosae***

22′ Corolla 24–28 mm, throat 10–14 mm wide when pressed, glabrous outside; calyx lobes widely ovate to ± round, gen 4.5–6 mm ..... var. ***fruticiformis***

15′ Corolla 6–17 mm

23. Openly branched shrub; lvs narrowly linear, gen rolled inward, gen < 1.4 mm wide ..... ***P. thurberi***

23′ Per, fl-sts gen little-branched; lvs gen lanceolate to ovate (linear in *P. cinicola*), flat to folded, 1–22 mm wide

24. Lvs thick; anther sacs 1–1.2 mm; corolla floor glabrous ..... [2]***P. patens***

24′ Lvs ± thin; anther sacs 0.3–1 mm; corolla floor hairy

25. Corolla 9–17 mm (if < 10 mm anther sacs 0.5–0.8 mm)

26. Pl 8–12 cm; upper cauline lvs ovate to ± round, 10–20 mm; anther sacs < 0.5 mm; calyx 2.5–3 mm ..... ***P. tracyi***

26′ Pl 15–60 cm; upper cauline lvs (ob)lanceolate, 15–70 mm; anther sacs 0.5–1 mm; calyx 3–5 mm

27. St base gen buried in sand; lower cauline lvs scale-like ..... ***P. albomarginatus***

27′ St base aboveground; lower cauline lvs well developed ..... ***P. rydbergii*** var. ***oreocharis***

25′ Corolla 6–10 mm; anther sacs 0.3–0.5 mm

28. Lvs gen all cauline, linear to narrowly lanceolate, gen folded lengthwise and arching-recurved . [2]***P. cinicola***

28′ Lvs best developed on basal, sterile sts, (ob)lanceolate to obovate, cauline lvs ascending, not arching-recurved . . . . ***P. procerus***
29. Pl 10–40 cm; upper cauline lvs 10–45 mm, 3–10 mm wide . . . . var. ***brachyanthus***
29′ Pl 5–12 cm; upper cauline lvs gen 5–20 mm, 1–5.5 mm wide . . . . var. ***formosus***
14′ Infl glandular-hairy; lf margin toothed (exc *P. heterodoxus*)
30. Upper lf pairs fused around st
31. Anther sacs 0.8–1.3 mm, valves spreading flat
32. Corolla gen > 20 mm, throat 5–9 mm wide when pressed; lower lip strongly lined; anther sacs 1.1–1.3 mm . . . . ***P. pseudospectabilis***
32′ Corolla gen < 20 mm, throat 4–6 mm wide when pressed, lower lip lacking prominent guide-lines; anther sacs 0.8–1.1 mm . . . . ***P. stephensii***
31′ Anther sacs 1.3–2.4 mm, valves hardly spreading (exc *P. bicolor*)
33. Corolla 18–27 mm; anther sacs 1.3–2 mm . . . . ***P. bicolor***
33′ Corolla 24–34 mm; anther sacs 1.8–2.4 mm
34. Corolla pale pink to lavender-pink; staminode densely hairy, exserted . . . . ***P. palmeri***
34′ Corolla bluish purple or blue; staminode glabrous, incl . . . . ***P. spectabilis*** var. ***subviscosus***
30′ Lvs free, not fused around st
35. Corolla 10–18 mm
36. Lvs ± toothed, upper cordate-clasping; upper side of corolla abruptly expanded . . . . ***P. anguineus***
36′ Lvs entire, upper not cordate-clasping; upper side of corolla moderately expanded . . . . ***P. heterodoxus***
37. Pl 5–20 cm; infl of 1(2) fl cluster . . . . var. ***heterodoxus***
37′ Pl gen > 20 cm; infl gen of > 2 fl clusters
38. Infl strongly glandular . . . . var. ***cephalophorus***
38′ Infl moderately glandular . . . . var. ***shastensis***
35′ Corolla 18–35 mm
39. Lvs thin; upper side of corolla tube ± abruptly expanded; calyx 6–12 mm . . . . ***P. rattanii***
40. Calyx lobes ovate, obtuse or rounded, calyx 6–7 mm . . . . var. ***kleei***
40′ Calyx lobes lanceolate, acute or long-tapered, calyx 6–12 mm . . . . var. ***rattanii***
39′ Lvs thick; corolla gradually expanded or lower side ± abruptly expanded; calyx 3–8.5 mm
41. Lvs linear to narrowly lanceolate, 1.5–7 mm wide . . . . ***P. incertus***
41′ Lvs lanceolate to widely ovate, 10–60 mm wide
42. Corolla gradually expanded, throat 4.5–9 mm wide when pressed
43. Corolla 24–30 mm; anther sacs 1.2–1.6 mm; main upper lvs 45–90 mm; common hybrid . . . . (see *P. centranthifolius*)
43′ Corolla 18–24 mm; anther sacs 0.8–1.3 mm; main upper lvs 20–55 mm . . . . ***P. clevelandii***
44. Lvs entire to moderately serrate; staminode glabrous or sparsely hairy . . . . var. ***clevelandii***
44′ Lvs sharply serrate; staminode densely hairy . . . . var. ***mohavensis***
42′ Corolla strongly expanded, throat 6–20 mm wide when pressed
45. Staminode incl, glabrous; anther-sacs 1.3–1.9 mm . . . . ***P. floridus***
46. Corolla 21–27 mm, expanded gradually to throat, 6–11 mm wide when pressed, mouth perpendicular to tube, tube 6–8 mm . . . . var. ***austinii***
46′ Corolla 24–30 mm, ± abruptly expanded to throat, 10–16 mm wide when pressed, mouth oblique to tube, tube 7–12 mm . . . . var. ***floridus***
45′ Staminode exserted, hairy; anther-sacs 1.6–2.2 mm . . . . ***P. grinnellii***
47. Pl 10–60 cm; young branches light green; corolla 22–30 mm, whitish, tinged pink or lavender . . . . var. ***grinnellii***
47′ Pl 45–85 cm; young branches glaucous; corolla 26–35 mm, tinged (blue-) violet . . var. ***scrophularioides***

P. albomarginatus M.E. Jones (p. 1057) WHITE-MARGINED BEARDTONGUE Per 15–35 cm, glabrous; crown buried in sand. **LVS**: lower-most scale-like; upper 15–50 mm, oblanceolate, white-margined, entire or weakly dentate. **INFL** glabrous. **FL**: calyx 3–5 mm, lobes ovate-elliptic, white-margined, finely serrate; corolla 13–17 mm, pink to purple, glabrous (exc floor hairy); anther sacs ± 0.9–1 mm, spreading flat; staminode glabrous. $2n=16$. RARE. Loose desert sand, gen on stabilized dunes; 700–900 m. DMoj; NV, AZ. In cult.

P. anguineus Eastw. (p. 1057) Per 20–90 cm, glabrous below infl. **LVS** thin, ± serrate; basal lvs lanceolate to narrowly ovate; middle cauline lvs narrowly elliptic, slightly narrowed in middle, widest at sessile, rounded or truncate base; upper cauline lvs 10–70 mm, lanceolate to ovate, cordate-clasping. **INFL** glandular. **FL**: calyx 5–8 mm, lobes narrowly lanceolate; corolla 13–18 mm, abruptly expanded throat on upper side, blue to purple-violet, glandular outside, floor gen long-hairy; anther sacs 1–1.3 mm, dehiscing full length, valves barely spreading; staminode sparsely hairy. $2n=16$. Open, gen logged areas of ± coniferous forest; 600–2000 m. NW; OR. ❀DRN,SUN:**4-6**,or part SHD,IRR:1,2,**7,14-17**,18-24;DFCLT.

P. azureus Benth. Per 20–70 cm, with woody branches below, glabrous. **LVS** gen cauline; upper lvs widest at gen cordate-clasping base, entire. **INFL** glabrous. **FL**: corolla 20–35 mm, (lavender-)blue, glabrous; anther sacs 1.8–3.5 mm, dehiscing across common tip 1/2–2/3 their length, inner margins hairy; staminode glabrous. Gen ± moist woodland, open forest; 300–2500 m. NW, CaR, n&c SN; OR.

var. ***angustissimus*** A. Gray (p. 1057) **LF** 40–90 mm, linear. **FL**: calyx 5–7.5 mm, lobes ovate, long-tapered tip 1.5–3.2 mm. $2n=48$. Habitats of sp; 300–700 m. NCoRH, n&c SN. [ssp. *an.* (A. Gray) Keck] ❀SUN,DRN,DRY:**7,14–17**,19–21.

var. ***azureus*** (p. 1057) **LF** 10–65 mm, lanceolate. **FL**: calyx 3–6.5 mm, lobes (ob)lanceolate to round, pointed tip 0.2–1.5 mm. $2n=48$. Habitats of sp; 500–2500 m. NW, CaR, n&c SNH; OR. ❀DRN,SUN:1,2,6,**7**,14,**15,16**,17.

P. barnebyi N. Holmgren (p. 1057) BARNEBY'S BEARDTONGUE Per 6–30 cm; hairs short, backward-pointing. **LVS** 20–75 mm; basal lvs well developed; upper cauline lvs lanceolate, (sub)entire. **INFL** glandular. **FL**: calyx 4.5–12 mm, lobes ± lanceolate; corolla 10–14 mm, abruptly expanded to throat on lower side, 4–5 mm wide when pressed, violet, blue distally, glandular outside, throat white, dark-lined, floor ± yellow-hairy; anther sacs 0.7–0.9 mm, spreading flat; staminode much exserted, densely orange-yellow-hairy. RARE in CA. Limestone gravel or silt in sagebrush scrub or juniper/pinyon woodland; 1500–2500 m. W&I; NV.

P. bicolor (Brandegee) Clokey & Keck Per < 150 cm, glabrous. **LVS** thick; upper cauline 4–11 cm, ovate, bases fused around st, sharply serrate. **INFL** glandular. **FL**: calyx 4–6 mm, lobes ± ovate; corolla 18–27 mm, throat 6–11 mm wide when pressed, cream to magenta, strongly lined, glandular outside, floor long-whitish-hairy; anther sacs 1.3–2 mm, spreading flat; staminode incl, densely yellow-hairy. Gravelly or rocky soils, creosote-bush or blackbush scrub, Joshua-tree woodland; 700–1500 m. ne DMtns (Castle Mtns); NV. ❀TRY;DFCLT.

P. caesius A. Gray (p. 1057) Per 20–80 cm, subscapose, glabrous, glaucous, with woody branches below. **LVS** gen basal; blade 15–43 mm, widely obovate to round, entire, petioled. **INFL** glandular. **FL**: calyx 3.7–7.5 mm, lobes lanceolate or ovate; corolla 15–23 mm, purplish blue, glandular outside, floor hairy; anther sacs 1.3–1.6 mm, dehiscing across common tip slightly > 1/2 their length, inner margin glabrous; staminode glabrous. Rocky ridges and slopes in open coniferous forests and alpine communities; 1800–3400 m. s SNH, SnGb, SnBr. ❀TRY;DFCLT.

P. calcareus Brandegee (p. 1057) LIMESTONE BEARDTONGUE Per 7–25 cm; hairs fine, backward-pointing, ashy (short, densely glandular on upper st and infl). **LVS**: upper cauline 20–60 mm, widely lanceolate, entire to shallowly dentate. **FL**: calyx 6.5–8.5 mm, lobes ± lanceolate; corolla 13–17 mm, cylindric to funnel-shaped, bright pink to rose-purple, glandular outside, throat 4–6 mm wide when pressed, floor nearly glabrous; anther sacs 0.6–0.7 mm, spreading flat; staminode densely yellow-hairy, ± incl. RARE in CA. Limestone crevices, rocky slopes in juniper/pinyon woodland, Joshua-tree scrub; 1200–1600 m. n DMtns.

P. californicus (Munz & I.M. Johnston) Keck (p. 1057) CALIFORNIA PENSTEMON Per 10–30 cm, spreading to ascending; hairs appressed, flat, scale-like, backward-pointing. **LF** 7–15 mm, linear to narrowly oblanceolate, entire, petioled. **FL**: calyx 4–8.5 mm, lobes ovate, hairy as on lvs, (or also sparsely glandular); corolla 14–18 mm, (blue-) purple, white inside, dark-lined, sparsely glandular outside, floor sparsely hairy; anther sacs 1–1.3 mm, dehiscing full length; staminode hairy. $2n=16$. RARE. Sandy soils, yellow-pine forest or juniper/pinyon woodland; 1200–2300 m. PR; Mex.

P. centranthifolius (Benth.) Benth. (p. 1057) SCARLET BUGLER Per 30–120 cm, glabrous, glaucous. **LVS** thick; middle cauline gen largest, 40–100 mm, lanceolate to ovate, cordate-clasping, entire. **FL**: calyx 3.5–6.5 mm, lobes ovate to ± round; corolla 20–30 mm, cylindric, lobes projecting, not spreading, bright red, glabrous (incl floor); anther sacs 0.8–1.2 mm, spreading flat; staminode glabrous. $2n=16$. Dry, open or wooded places, gen in chaparral or oak woodland; < 1800 m. NCoR, n SNF, GV (margins), CW, SW; Mex. Common hybrids with *P. spectabilis* have been called *P. ×parishii*. ❀SUN,DRY,DRN:7,9,**14–21**,22,23;may be DFCLT.

P. cinicola Keck (p. 1057) ASH BEARDTONGUE Per 15–40 cm, glabrous or short-hairy below. **LVS** mostly cauline, 30–60 mm, linear, entire, folded lengthwise, arching-recurved. **INFL** glabrous. **FL**: calyx 1–2.5 mm, lobes widely obovate, tips gen jagged-toothed; corolla 6–9 mm, blue-purple, glabrous exc hairy floor; anther sacs 0.3–0.5 mm, valves spreading widely; staminode yellow-hairy. $2n=16,32$. UNCOMMON. Dry, rocky, igneous soils, in sagebrush openings of montane forests; 1250–2200 m. MP; OR.

P. clevelandii A. Gray Per 30–70 cm, much-branched, glabrous. **LF** thick, 20–60 mm, ± ovate, entire to serrate. **INFL** glabrous or glandular. **FL**: calyx 3.5–6.5 mm, lobes ovate to ± round; corolla 18–24 mm, gradually expanded into throat, 4.5–8 mm wide when pressed, pink, magenta, or reddish purple, unlined, glandular outside and inside (floor lacks nonglandular hairs); anther sacs dehiscing full length; staminode incl, glabrous to densely hairy. Rocky hillsides, rock crevices in creosote-bush scrub, juniper/pinyon woodland, chaparral; 400–1500 m. PR, s DMtns, w DSon; Mex.

var. ***clevelandii*** (p. 1057) **LVS** entire to moderately serrate; upper cauline lf bases free, cordate-clasping to rounded. **FL**: anther sacs 0.8–1.3 mm, spreading flat; staminode glabrous or sparsely hairy. $2n=16$. Habitats and elevations of sp. PR, w DSon; Mex. Variable; hybridizes with *P. centranthifolius*. ❀DRN,DRY:7,14–16,**18–21**,22,23.

var. ***connatus*** Munz & I.M. Johnston (p. 1057) SAN JACINTO BEARDTONGUE **LVS** sharply serrate; upper cauline lf bases fused around st. **FL**: anther sacs 1.3–1.6 mm, dehiscing full length, valves barely spreading; staminode ± hairy. UNCOMMON. Habitats and elevations of sp. e PR (e slope SnJt, Santa Rosa mtns). [ssp. *c.* (Munz & I.M. Johnston) Keck] Distinctive; perhaps should be recognized as sp. ❀DRN,DRY,SUN:3,7,14,15,18–21;DFCLT.

var. ***mohavensis*** (Keck) McMinn (p. 1057) **LVS** sharply serrate; upper cauline lf bases free, widely wedge-shaped. **FL**: anther sacs 0.9–1.1 mm, valves spreading but not flat; staminode densely hairy. Habitats of sp. s DMtns (Little San Bernardino, Granite mtns). [ssp. *m.* Keck]

P. davidsonii E. Greene var. ***davidsonii*** (p. 1057) Subshrub < 10 cm, mat-forming, short-hairy. **LVS** ± basal (much reduced upward); main blades 5–30 mm, elliptic to (ob)ovate, (sub)entire, glabrous, green. **INFL** glandular; pedicels short. **FL**: calyx 7–13 mm, lobes linear to lanceolate; corolla 20–36 mm, blue-violet to blue-purple, floor ± white-shaggy-hairy; anther sacs 0.9–1.3 mm, valves spreading flat, white-woolly; staminode incl, densely pale yellow-hairy. $2n=16$. Montane to alpine outcrops, talus; 2000–3600 m. KR, CaRH, SNH, Wrn, n SNE (Sweetwater Mtns); OR. ❀DRN, IRR:15–17&SHD:1–6,7,14–21;DFCLT.

P. deustus Lindley Per gen < 40 cm, subglabrous to glandular. **LVS** mostly basal, dentate; cauline lvs 10–50 mm. **INFL** ± glandular. **FL**: corolla 8–15 mm, cream-white, dark-lined, glandular outside and on floor; anther sacs 0.5–0.7 mm, valves spreading flat; staminode glabrous or sparsely hairy distally. $2n=16$. Open, rocky, gen volcanic places; 600–3000 m. KR, NCoRH, CaR, SNH, MP; to WA, MT, WY, UT.

var. ***pedicellatus*** M.E. Jones (p. 1059) **LF** lanceolate. **FL**: calyx 2.5–5 mm, lobes lanceolate, acute; corolla 10–15 mm. Sagebrush scrub, juniper/pinyon woodland, yellow-pine and montane forests; 900–3000 m. CaR, SNH, MP; OR, NV, UT. [ssp. *heterander* (Torrey & A. Gray) Pennell & Keck] ❀DRN,DRY,SUN:1, **2,3,7**,14,**15–17**,18–21.

var. ***suffrutescens*** L. Henderson (p. 1059) **LF** ovate to ± round. **FL**: calyx 3.5–6.5 mm, lobes narrowly lanceolate, sharply acute or long-tapered; corolla 9–11 mm. Open forest; 600–2200 m. KR, NCoRH, s CaRH, n SNH; OR. [var. *savagei* L. Henderson] ❀TRY.

P. eatonii A. Gray (p. 1059) Per 40–100 cm. **LVS**: cauline 30–90 mm, widely lanceolate to ovate, entire. **FL**: calyx 3.5–6 mm, lobes ovate; corolla 24–33 mm, cylindric, obscurely 2-lipped, lobes subequal, barely spreading, scarlet, glabrous; anther sacs 1.4–2.8 mm, dehiscing in distal 1/2–3/4, short-hairy on sides; staminode glabrous to sparsely hairy at tip. $2n=16$. Dry sagebrush scrub, juniper/pinyon woodland, yellow-pine forest; 1500–2800 m. SnBr, DMtns; to Colorado, NM. ❀SUN,DRY:1–3,**7**,10,15–21.

var. ***eatonii*** Pl glabrous. Habitats and range of sp. (exc SnBr).

var. ***undosus*** M.E. Jones Pl short-hairy. Habitats and range of sp. [ssp. *u.* (M.E. Jones) Keck]

P. filiformis (Keck) Keck (p. 1059) THREAD-LEAVED BEARDTONGUE Per 20–50 cm, woody-branched below, finely backward-pointing-hairy. **LVS** narrowly linear, sometimes glabrous; short basal lvs densely clustered; cauline lvs 20–70 mm, gen ± 0.5 mm wide, tightly rolled outward, entire. **INFL** glandular. **FL**: calyx 3.4–5.7 mm, lobes lanceolate; corolla 13–16 mm, blue, glandular outside, floor glabrous; anther sacs 1.2–1.4 mm, dehiscing across common tip to 1/2 their length, inner margins short-hairy; staminode glabrous. $2n=16$. RARE. Open, rocky places among shrubs or yellow pines; ± 450 m. e KR (n Trinity, nw Shasta cos.). In cult.

P. floridus Brandegee Per 50–120 cm, glabrous, glaucous. **LVS** thick; upper cauline lvs lanceolate or ovate, cordate-clasping, gen dentate (upper-most sometimes subentire). **INFL** glandular. **FL**: calyx 4.2–6.2 mm, lobes ovate; corolla 21–30 mm, throat narrowed towards mouth, rose-pink, strongly lined, glandular outside and inside (floor lacks nonglandular hairs); anther sacs 1.3–1.9 mm, spreading flat; staminode incl, glabrous. $2n=16$. Gravely washes, canyon floors, in sagebrush scrub and juniper/pinyon woodland; 1000–2400 m. SNE, n DMtns; NV. Vars intergrade in s White Mtns.

var. ***austinii*** (Eastw.) N. Holmgren (p. 1059) **FL**: corolla 21–27 mm, gradually expanded into throat, 6–11 mm wide when pressed, tube 6–8 mm, mouth perpendicular to tube. Habitats and elevations of sp. n DMtns; NV. [ssp. *a*. (Eastw.) Keck] ❀TRY; DFCLT.

var. ***floridus*** (p. 1059) **FL**: corolla 24–30 mm, ± abruptly expanded into throat, 10–16 mm wide when pressed, tube 7–12 mm, mouth oblique to tube. Habitats of sp.; 1600–2400 m. SNE; NV. ❀TRY;DFCLT.

P. fruticiformis Cov. Shrub 30–60 cm, much-branched below, gen wider than tall. **ST**: young sts glabrous, gen glaucous. **LVS** thick, 25–65 mm; upper lvs ± narrowly lanceolate, (sub)entire, gen folded lengthwise or rolled inward. **INFL** glabrous. **FL**: corolla pale pink to whitish, limb sometimes ± lavender, strongly lined, floor shaggy-hairy; anther sacs, 1.6–2.1 mm, dehiscing full length, valves barely spreading; staminode exserted, densely hairy. Gravelly washes, canyon floors in creosote-bush scrub, juniper/pinyon woodland; 1000–1800 m. s SNE, n DMtns; w NV.

var. ***amargosae*** (Keck) N. Holmgren (p. 1059) DEATH VALLEY BEARDTONGUE **FL**: calyx 4.5–7.5 mm, lobes ovate; corolla 22–24 mm, throat 8–10 mm wide when pressed, glandular-hairy outside. UNCOMMON. Creosote-bush scrub; 1000–1200 m. ne DMtns (Kingston Mtns); w NV. [ssp. *a*. Keck]

var. ***fruticiformis*** (p. 1059) **FL**: calyx 4.5–6.5 mm, lobes widely ovate to ± round; corolla 24–28 mm, throat 10–14 mm wide when pressed, glabrous exc on floor. $2n=16$. Habitats and range of sp. (exc NV). ❀TRY;DFCLT.

P. gracilentus A. Gray (p. 1059) Per 25–65 cm, glabrous, woody-branched below. **LVS** gen cauline; upper lvs 40–100 mm, linear to narrowly lanceolate, entire. **INFL** glandular (esp pedicels). **FL**: calyx 4–6 mm, lobes lanceolate; corolla 15–20 mm, red- to blue-purple, floor glabrous to shaggy-hairy; anther sacs 1–1.3 mm, dehiscing across common tip to 1/2 their length, inner margins glabrous; staminode yellow-hairy. $2n=16$. Sagebrush scrub, juniper woodland, yellow-pine to subalpine forest; 1000–3000 m. CaR, n SNH, MP; OR, NV. ❀DRN,DRY,SUN:1-3,7,14-16,18-21;DFCLT.

P. grinnellii Eastw. Per 10–85 cm, ± glabrous. **LVS** 50–90 mm, ± (ob)lanceolate, thick, gen folded lengthwise and arching-recurved, subentire to dentate. **INFL** glandular. **FL**: calyx 4.7–8.5 mm, lobes ± ovate; corolla abruptly expanded into throat, 10–18 mm wide when pressed, strongly lined, glandular outside and inside, floor long-hairy; anther sacs 1.6–2.2 mm, dehiscing full length, valves barely spreading; staminode exserted, densely yellow- or golden-hairy. $2n=16$. Chaparral, foothill and juniper/pinyon woodland, montane forest; 500–2700 m. s SN, SW (exc SCo, ChI) Hybrids with *P. centranthifolius* have been called *P. ×dubius* Davidson.

var. ***grinnellii*** (p. 1059) Pl 10–60 cm; young branches light green. **FL**: corolla 22–30 mm, whitish, pink- or lavender-tinged. Habitats of sp.; 900–2700 m. SW (exc SCo, ChI). [*P. hians* I.M. Johnston] ❀DRN,DRY,SUN:1–3,**7,14–21**,22,23.

var. ***scrophularioides*** (M.E. Jones) N. Holmgren Pl 45–85 cm; young branches glaucous. **FL**: corolla 26–35 mm, ± violet. Habitats of sp.; 500–2500 m. s SN, WTR. [ssp. *s*. (M.E. Jones) Munz; *P. peirsonii* Munz & I.M. Johnston] ❀DRN,DRY,SUN:1–3,**7**,14,**15–19**,may be DFCLT.

P. heterodoxus A. Gray Per 5–65 cm, mat-forming, ± glabrous. **LVS** entire, sometimes folded lengthwise; basal many; cauline narrowly lanceolate to ovate, sometimes clasping. **INFL** glandular. **FL**: calyx 2.5–6 mm, lobes narrowly oblong to obovate; corolla 10–16 mm, cylindric to moderately expanded, deep blue-purple, glandular outside, floor yellow-brown-hairy; anther sacs 0.5–1 mm, dehiscing full length but barely spreading; staminode moderately yellow-hairy. Montane to alpine slopes, meadows, scree; 1100–3900 m. KR, CaRH, SN, W&I; NV.

var. ***cephalophorus*** (E. Greene) N. Holmgren (p. 1059) Pl 15–35 cm. **LF** 20–70 mm, 6–14 mm wide. **INFL** gen 2–6 clusters, strongly glandular. Subalpine and alpine habitats of sp.; 2100–3300 m. s SNH. [ssp. *c*. (E. Greene) Keck]

var. ***heterodoxus*** (p. 1059) Pl 5–20 cm. **LF** 5–50 mm, 2–10 mm wide. **INFL** of 1(2) clusters, glandular. $2n=16$. Montane and sulapine habitats of sp.; 2000–3900 m. KR, SN, W&I; NV. ❀DRN,IRR:1–3,**7**,14&SUN:15,16;usually DFCLT.

var. ***shastensis*** (Keck) N. Holmgren SHASTA BEARDTONGUE Pl 15–65 cm. **LF** 30–140 mm, 6–20 mm wide. **INFL** gen of 2–6 fl clusters, moderately glandular. $2n=32$. UNCOMMON. Montane meadows; 1100–2400 m. CaRH. [*P. s*. Keck]

P. heterophyllus Lindley Per 25–150 cm, woody-branched below. **LVS** gen cauline, 20–95 mm, linear to oblanceolate, (sub)entire; base tapered. **INFL** glabrous to short-hairy. **FL**: calyx 4.2–8 mm, lobes lanceolate-acuminate to (ob)ovate; corolla 23–40 mm, magenta to blue, glabrous; anther sacs 2.2–3 mm, dehiscing across common tip to 1/2–2/3 their length, inner margins long-hairy; staminode glabrous. Grassland, chaparral, forest openings; 50–1600 m. NW, CaRF, n SNF, ScV, CW, SW.

var. ***australis*** Munz & I.M. Johnston (p. 1059) Pl short-hairy, bearing axillary lf clusters at lower nodes. **LF** 0.5–4 mm wide, linear to narrowly (ob)lanceolate. Habitats and elevations of sp. CW, SW. [ssp. *a*. (Munz & I.M. Johnston) Keck] ❀DRN,SUN,DRY:7, **15–21**,22,23;may be DFCLT.

var. ***heterophyllus*** (p. 1059) Pl glabrous (lower sts rarely short-hairy), gen bearing axillary lf clusters at lower nodes. **LF** 0.5–5 mm wide, linear to narrowly (ob)lanceolate. Habitats and elevations of sp. NW, CW, WTR. ❀SUN,DRN,DRY:2,**7**,14,**15–17**, 18–24;may be DFCLT.

var. ***purdyi*** (Keck) McMinn (p. 1059) Pl short-hairy (exc sometimes lf tips, calyces); axillary lf clusters 0. **LF** 2–7.5 mm wide, narrowly (ob)lanceolate. Habitats and elevations of sp. NCoR, CaRF, n SNF, ScV, SnFrB, SCoRI. [ssp. *p*. Keck] ❀SUN or part SHD,DRN,DRY:2,4–6,**7,14–17**,18–24;CV;may be DFCLT.

P. humilis A. Gray var. ***humilis*** (p. 1059) Per 5–35 cm, gen mat-forming, ± short-(sometimes ashy-)hairy. **LVS** entire; basal lvs many, 15–75 mm, (ob)ovate, petioled; cauline lvs lanceolate to obovate, sessile, clasping. **INFL** glandular. **FL**: calyx 3–6 mm, lobes lanceolate to ovate; corolla 11–15 mm, cylindric narrowly funnel-shaped, blue with lighter floor, dark-lined, glandular outside, floor ± yellow- or white-hairy; anther sacs 0.5–0.8 mm, dehiscing full length, valves barely spreading; staminode orange- to yellow-hairy. $2n=16$. Open montane to subalpine forests, sagebrush scrub, juniper/pinyon woodland; 1500–3000 m. CaRH, n&c SNH, GB; to OR, WY, Colorado. Small-fld, small-lvd pls from n CA, s OR, nw NV have been called *P. cinereus* Piper, gray beardtongue, which intergrades fully with *P. h.*

P. incertus Brandegee (p. 1059) Shrub 20–100 cm, rounded; young sts glabrous, glaucous. **LVS** thick; largest lvs at mid st, 40–70 mm, linear to narrowly lanceolate, gen rolled inward, entire. **INFL** glandular. **FL**: calyx 4.5–7.3 mm, lobes ovate, glabrous to glandular; corolla 23–32 mm, throat 8–12 mm wide when pressed, violet to purple (limb bluish), unlined, glandular outside, glabrous inside exc small hair patch on floor; anther sacs 1.8–2.4 mm, dehiscing full length, valves barely spreading; staminode incl, densely hairy. Gen sandy soil along washes, canyon slopes, in sagebrush scrub, Joshua-tree and juniper/pinyon woodlands; 1000–1700 m. s SNH, Teh, SnBr, nw DMtns. ❀TRY;DFCLT.

P. janishiae N. Holmgren (p. 1059) Per 8–25 cm; hairs short, backward-pointing. **LVS** 20–60 mm; basal lvs well developed; upper cauline lvs lanceolate, entire to toothed. **INFL** glandular. **FL**: calyx 6–13 mm, lobes lanceolate; corolla 18–28 mm, abruptly expanded to throat on lower side, 7–12 mm wide when pressed, pink to dull purple (lobes sometimes bluish), dark-lined, glandular outside, floor white- to pale-yellow-hairy; anther sacs 0.8–1.2 mm, spreading flat; staminode much exserted, densely orange-yellow-hairy. Gen igneous-clay soils in sagebrush scrub, juniper/shrub savanna, yellow-pine forest; 1300–2300 m. MP; to OR, ID, NV. [*P. miser* A. Gray misapplied] ❀TRY;DFCLT.

P. labrosus (A. Gray) Hook.f. (p. 1059) Per 30–70 cm, glabrous. **LVS**: cauline 30–85 mm, linear, gen rolled inward, entire.

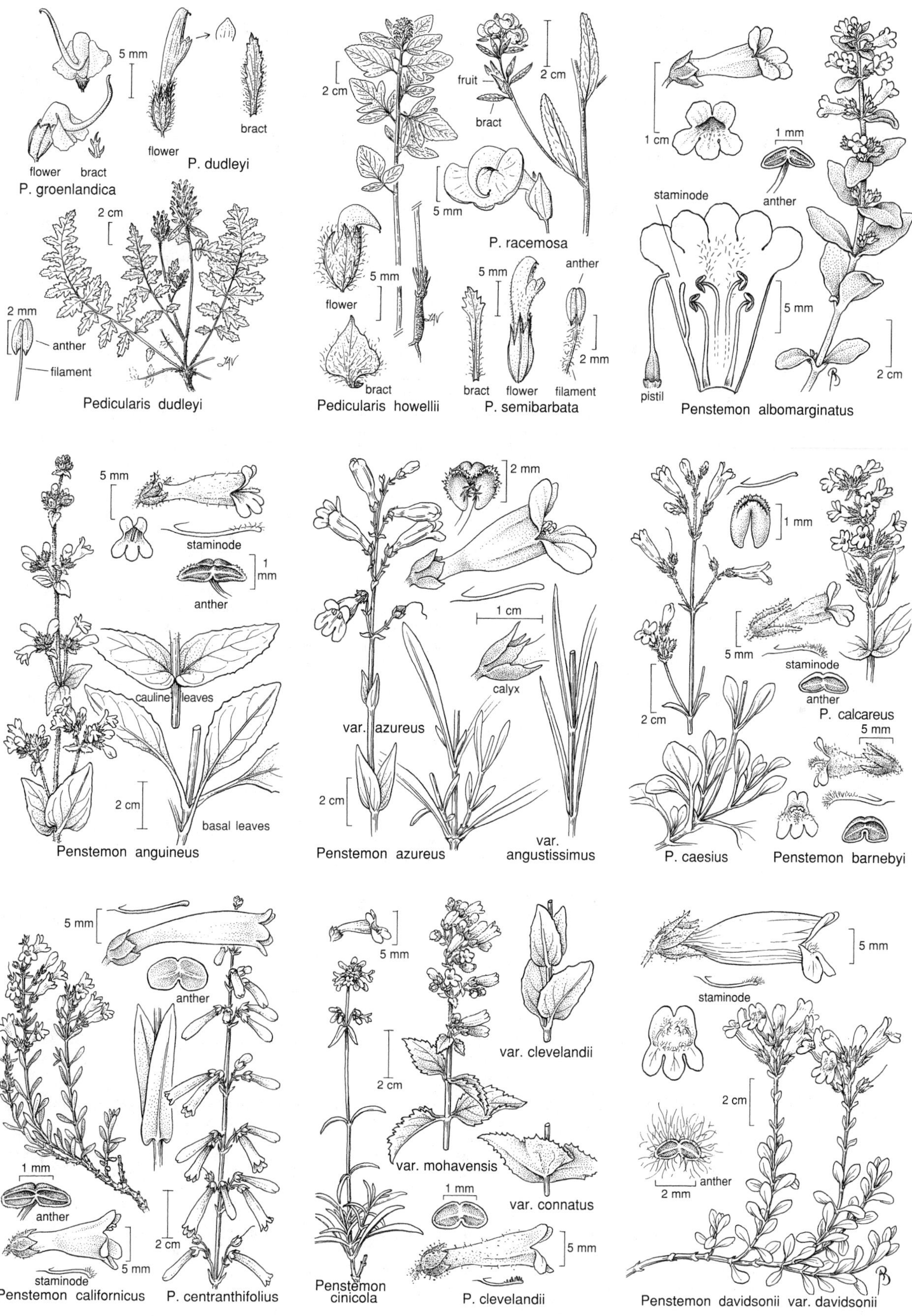
5 mm
flower
bract
P. groenlandica
flower
P. dudleyi
bract
2 cm
2 mm
anther
filament
Pedicularis dudleyi
2 cm
5 mm
flower
bract
Pedicularis howellii
fruit
bract
2 cm
5 mm
P. racemosa
5 mm
anther
2 mm
bract
flower
filament
P. semibarbata
1 cm
1 mm
anther
staminode
5 mm
2 cm
pistil
Penstemon albomarginatus
5 mm
staminode
1
mm
anther
cauline leaves
2 cm
basal leaves
Penstemon anguineus
2 mm
1 cm
calyx
var. azureus
2 cm
Penstemon azureus
var.
angustissimus
1 mm
5 mm
staminode
anther
P. calcareus
2 cm
5 mm
P. caesius
Penstemon barnebyi
5 mm
anther
1 mm
anther
5 mm
staminode
2 cm
Penstemon californicus
P. centranthifolius
5 mm
2 cm
var. clevelandii
var. mohavensis
1 mm
var. connatus
5 mm
Penstemon
cinicola
P. clevelandii
5 mm
staminode
2 cm
anther
2 mm
Penstemon davidsonii var. davidsonii

INFL glabrous. **FL**: calyx 3.5–5.5 mm, lobes ovate; corolla 30–40 mm, ± cylindric, upper lip forming hood, lower lip strongly reflexed, (orange-)red (rarely yellow), glabrous; anther-sacs 1.5–2 mm, dehiscing in distal 2/3, glabrous; staminode glabrous. 2*n*=16. Juniper/pinyon woodland, pine and mixed hardwood forests; 1500–2400 m. TR, PR; Mex. ❀SUN,DRN,DRY or IRR:1–6,**7**, 14,**15–21**,22,23;CVS.

P. laetus A. Gray Per 15–75 cm, gen hairy, woody-branched below. **LVS** gen cauline; upper lvs 15–100 mm, linear to lanceolate, entire. **INFL** glandular. **FL**: corolla 21–35 mm, (violet-) blue, short-glandular outside, floor glabrous; anther sacs dehiscing across common tip to 4/5 their length, inner margins long-hairy; staminode glabrous. 2*n*=16. Dryish open scrub or partial shade of coniferous forests; 400–2500 m. KR, CaR, SN, WTR; OR.

var. ***laetus*** **LF** 2–11 mm wide. **FL**: calyx 4.7–9 mm, lobes ± widely lanceolate; corolla throat relatively open, lips ± spreading; anthers 1.8–2.3 mm, dehiscing 2/3–3/5. Foothill to montane forests; 400–2500 m. SN, WTR. Most s SN pls have larger calyx lobes. ❀DRN:1–3,**7**,14&SUN:**15–17**,18–21.

var. ***leptosepalus*** A. Gray **LF** 4.5–22 mm wide. **FL**: calyx 8–14 mm, lobes narrowly lanceolate; corolla throat relatively open, lips ± spreading; anthers 2–2.7 mm, dehiscing 2/3–3/5. Coniferous forest; 400–1600 m. s CaR, n SN. [ssp. *l.* (A. Gray) Keck] ❀TRY.

var. ***sagittatus*** (Keck) McMinn (p. 1059) **LF** 1.5–8.5 mm wide. **FL**: calyx 5–8.5 mm, lobes ± widely lanceolate; corolla throat slightly narrowed at mouth, lips projecting; anthers 2.1–2.8 mm, narrowly sagittate, dehiscing 3/5–4/5. Sagebrush scrub, evergreen forests; 600–1600 m. KR, CaR; OR. [ssp. *s.* Keck] ❀DRN,IRR: **7**,14&SUN:**4–6,15,16**,17

P. monoensis A.A. Heller (p. 1059) Per 7–30 cm; hairs dense, ashy, backward-pointing. **LVS** gen cauline, 50–120 mm, ± (widely) lanceolate, entire to toothed. **INFL** glandular. **FL**: calyx 8–11 mm, lobes gen lanceolate (ovate); corolla 15–20 mm, cylindric to narrowly funnel-shaped, glandular outside, pink to red-violet, throat 4–6 mm wide when pressed, floor white to pale pink to white, sparsely hairy; anther sacs 1.1–1.4 mm, dehiscing full length, valves barely spreading; staminode ± incl, yellow-hairy. Sandy and gravely washes and hills, sagebrush scrub, Joshua-tree and juniper/pinyon woodland; 1300–1800 m. W&I. ❀TRY;DFCLT.

P. neotericus Keck (p. 1059) PLUMAS COUNTY BEARDTONGUE Per 25–80 cm, glabrous, glaucous, woody-branched below. **LVS** gen cauline; upper 30–85 mm, lanceolate, entire. **INFL** glandular. **FL**: calyx 4–7 mm, lobes lanceolate to ovate; corolla 23–38 mm, blue to pink-purple, glandular outside, floor white, glabrous; anther sacs 2.4–3.2 mm, dehiscing across common tip 2/3–4/5 their length, inner margins white-hairy; staminode glabrous. 2*n*=32. UNCOMMON. Gen in volcanic soils of scrub, open forest; 1000–2200 m. CaR, n SN.

P. newberryi A. Gray Subshrub 12–30 cm, mat-forming; hairs gen short, backward-pointing (0). **LVS** gen basal, gen reduced upward, short-petioled; main blades, 10–40 mm, (ob)ovate, finely serrate. **INFL** glandular. **FL**: calyx 7–12 mm, lobes ± lanceolate; corolla floor hairy on ridges; anther sacs 0.8–1.2 mm, spreading flat, woolly; staminode incl, pale-yellow-hairy distally. 2*n*=16. Outcrops, talus; 700–3500 m. NW, CaR, SNH, n Teh; OR, NV.

var. ***berryi*** (Eastw.) N. Holmgren **LVS** much reduced upward. **FL**: corolla 27–35 mm, rose-red, throat 7.5–12 mm wide when pressed, floor long-curly-hairy; stamens incl, anthers sacs 1.2–1.7 mm. Habitats of sp.; 1600–2300 m. KR; OR. [ssp. *berryi* (Eastw.) Keck] ❀DRN,IRR:1,2,7&SUN:4–6,15–18;DFCLT.

var. ***newberryi*** (p. 1061) **LVS** much reduced upward. **FL**: corolla 20–30 mm, rose-red, throat 4.7–7.5 mm wide when pressed, floor short-hairy; 2–4 stamens exserted, anther sacs 0.8–1.3 mm. Habitats of sp.; 1300–3500 m. NCoRH, CaRH, SNH, n Teh; NV. ❀DRN,IRR:1,2,7,14&SUN:4–6,15–19;DFCLT.

var. ***sonomensis*** (E. Greene) Jepson SONOMA BEARDTONGUE **LVS** little reduced upward. **FL**: corolla 22–30 mm, dark rose-purple, throat 5.5–6.5 mm wide when pressed, floor short-hairy; 2–4 stamens exserted, anther sacs ± 1.2 mm. UNCOMMON. Habitats of sp.; 700–1300 m. NCoR (peaks of Lake, Napa, Sonoma cos.). [ssp. *s.* (E. Greene) Keck] ❀DRN,IRR:2,7,14–16&SUN,DRY:4–6,17;DFCLT;CVS.

P. pahutensis N. Holmgren (p. 1061) PAHUTE BEARDTONGUE Per 15–35 cm, glabrous, sometimes glaucous. **LVS**: cauline 30–100 mm, linear to narrowly lanceolate, entire. **INFL** glabrous. **FL**: calyx 3–5 mm, lobes widely ovate; corolla 17–30 mm, bluish lavender, throat 6–8 mm wide when pressed, floor yellow- or white-hairy; anther sacs 1.6–2.2 mm, ± S-shaped, dehiscing in distal 2/3, glabrous; staminode densely golden-yellow-hairy. 2*n*=16. RARE in CA. Sagebrush scrub, juniper/pinyon woodland; 1900–2300 m. ne DMtns (Grapevine Mtns); NV.

P. palmeri A. Gray var. ***palmeri*** (p. 1061) Per 50–200 cm, glabrous, glaucous. **LVS** thick; upper cauline lvs 40–120 mm, triangular-clasping, ± dentate. **INFL** glandular. **FL** fragrant; calyx 4–7 mm, lobes ± ovate; corolla 25–32 mm, abruptly expanded into throat, 12–20 mm wide when pressed, pale (lavender-) pink, strongly lined, glandular outside and inside, floor sparsely long-whitish hairy; anther sacs 1.8–2.2 mm, dehiscing full length, valves barely spreading; staminode exserted, densely long-spreading-yellow-hairy. 2*n*=16,32. Washes, roadsides, canyon floors, in creosote-bush scrub to juniper/pinyon woodland; 1100–2300 m. e DMoj; to UT, AZ. [*P. bryantae* Keck] ❀DFCLT.

P. papillatus J. Howell (p. 1061) INYO BEARDTONGUE Per 20–40 cm, ashy-hairy, woody-branched below. **LVS** gen cauline, 25–40 mm, narrowly elliptic to oblanceolate, gen widest at base, cordate-clasping, entire. **INFL** glandular. **FL**: calyx 7–12 mm, lobes narrowly lanceolate; corolla 24–30 mm, blue-violet, glandular outside, floor glabrous; anthers sacs 1.7–1.9 mm, dehiscing only across common tip, inner margins glabrous; staminode pale-yellow-hairy. UNCOMMON. Rocky openings of juniper/pinyon woodland and montane forest communities; 2000–2700 m. c&s SNH, SNE.

P. parvulus (A. Gray) Krautter (p. 1061) Per 15–30 cm, loosely matted, glabrous, glaucous, woody-branched below. **LVS** gen cauline, 10–45 mm, lanceolate to ovate, clasping, (sub)entire. **INFL** glabrous. **FL**: calyx 3–5 mm, lobes lanceolate to widely obovate; corolla 14–20 mm, blue (-violet), glabrous; anther sacs 1.4–2 mm, dehiscing across common tip 1/2–2/3 their length, inner margins hairy; staminode glabrous. 2*n*=32 (s SNH). Rocky, open foothill and montane forests; 500–2200 m. KR, s SNH; OR. Some n CA and OR pls intergrade with *P. azureus* var. *azureus*. ❀DRN,IRR, SUN:1,2,4–7,14–17;DFCLT.

P. patens (M.E. Jones) N. Holmgren (p. 1061) Per 15–40 cm, glabrous, glaucous. **LVS** thick; basal lvs well developed; cauline lvs 25–90 mm, lanceolate, entire. **INFL** glabrous. **FL**: calyx 3–7 mm, lobes widely ovate; corolla 13–20 mm, lavender to magenta, glabrous; anther sacs 1–1.2 mm, dehiscing full length, valves barely spreading; staminode minutely orange- to yellow-hairy. 2*n*=16. Sagebrush scrub, juniper/pinyon woodland, yellow-pine forest; 1900–3000 m. c&s SNH, s SNE; s NV. [*P. confusus* M.E. Jones ssp. *patens* (M.E. Jones) Keck] ❀TRY.

P. personatus Keck (p. 1061) CLOSE-THROATED BEARDTONGUE Per 30–50 mm, gen glabrous (short-hairy), glaucous. **LVS** cauline; lower-most scale-like; largest (at mid st) short-petioled, blade 30–50 mm, ovate, (sub)entire. **INFL** glandular. **FL**: calyx 5–6 mm, lobes widely lanceolate; corolla 20–25 mm, blue-purple, glabrous to glandular outside, densely hairy throughout inside, arched floor closing throat; anther sacs 1.2–1.4 mm, dehiscing full length, valves barely spreading; staminode incl, coarsely hairy. RARE. Yellow-pine, montane forests; 1500–1800 m. n SN (Butte, Plumas cos.).

P. procerus Graham Per 5–40 cm, gen matted, glabrous (exc some lower sts). **LVS** entire; basal lvs many, oblanceolate; upper cauline lvs < basal, narrowly lanceolate to narrowly ovate. **INFL** glabrous. **FL**: calyx 1.5–4.5 mm, lobes (ob)ovate; corolla 6–10 mm, blue-purple, ± glabrous, floor whitish, white- to yellow-hairy; anther sacs 0.3–0.5 mm, widely spreading; staminode glabrous or sparsely orange-yellow-hairy. Montane meadows to alpine barrens; 1200–3600 m. NW, SN; to AK, MT, WY, Colorado.

var. ***brachyanthus*** (Pennell) Cronq. Pl 10–40 cm. **LVS**: upper 10–45 mm, 3–10 mm wide. Habitat of sp (exc alpine); 1200–2400 m. KR, NCoRH; OR. [ssp. *b.* (Pennell) Keck] ❀DRN,IRR: 2,4–6,**7**,14,**15–17**.

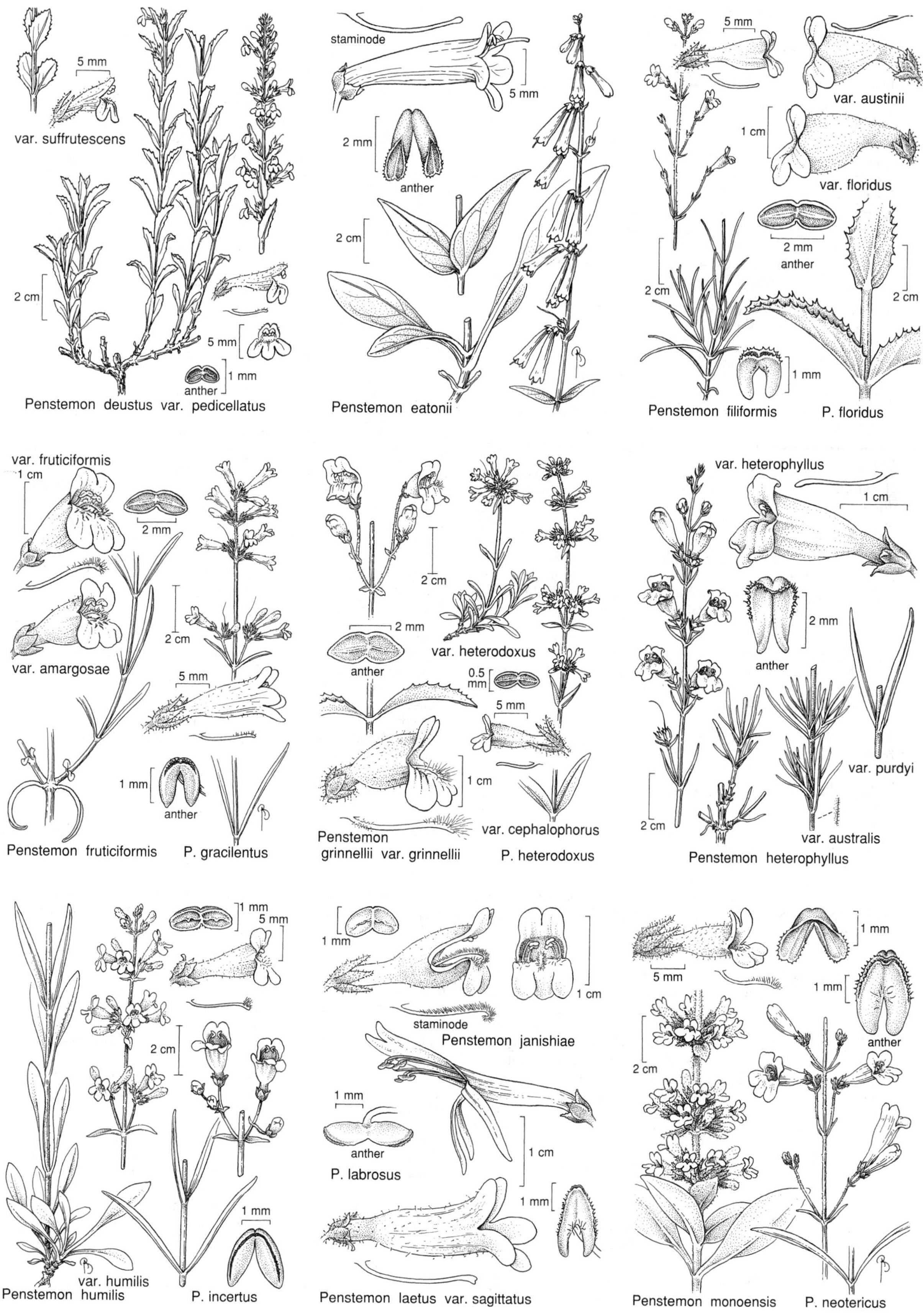
var. suffrutescens
5 mm
2 cm
5 mm
1 mm
anther
Penstemon deustus var. pedicellatus
staminode
5 mm
2 mm
anther
2 cm
Penstemon eatonii
5 mm
var. austinii
1 cm
var. floridus
2 mm
anther
2 cm
2 cm
1 mm
Penstemon filiformis
P. floridus
var. fruticiformis
1 cm
2 mm
var. amargosae
2 cm
5 mm
1 mm
anther
Penstemon fruticiformis
P. gracilentus
2 cm
2 mm
anther
var. heterodoxus
0.5 mm
5 mm
1 cm
var. cephalophorus
Penstemon grinnellii var. grinnellii
P. heterodoxus
var. heterophyllus
1 cm
2 mm
anther
var. purdyi
2 cm
var. australis
Penstemon heterophyllus
1 mm
5 mm
2 cm
1 mm
var. humilis
Penstemon humilis
P. incertus
1 mm
1 cm
staminode
Penstemon janishiae
1 mm
anther
P. labrosus
1 cm
1 mm
Penstemon laetus var. sagittatus
1 mm
5 mm
1 mm
anther
2 cm
Penstemon monoensis
P. neotericus

var. ***formosus*** (Nelson) Cronq. (p. 1061) Pl 5–12 cm. **LVS**: upper 5–20 mm, 1–5.5 mm wide. 2*n*=32. Habitats of sp. (exc montane); 2500–3600 m. KR, n&c SNH; OR, NV. [ssp. *f.* (Nelson) Keck] SN pls have lvs folded lengthwise. Vars. separated elevationally in KR. ❀DRN,IRR:1,2,6,7,14–16;DFCLT.

P. pseudospectabilis M.E. Jones (p. 1061) Shrub 30–100 cm. **ST**: young sts glabrous, glaucous. **LVS** ± thin; upper cauline lvs 30–90 mm, widely triangular-ovate, ± serrate, bases fused around st; upper lf pairs ± disk-like. **INFL** gen glandular. **FL**: calyx 3.5–7.5 mm, lobes ovate; corolla 17–25 mm, gradually expanded to throat, 5–9 mm wide when pressed, reddish pink, strongly lined, glandular outside and inside (floor lacks nonglandular hairs); anther sacs 1.1–1.3 mm, spreading flat; staminode incl, glabrous. 2*n*=16. Gravelly or rocky desert washes, canyon floors, in creosote-bush scrub, juniper/pinyon woodland; 100–1400 m. s DMtns, ne DSon; AZ. ❀DRN,DRY,SUN:10,12–21;DFCLT.

P. purpusii Brandegee (p. 1061) SNOW MOUNTAIN BEARDTONGUE Per 5–20 cm, densely canescent or glaucous, woody-branched below. **LVS**: cauline 10–30 mm, ± (ob)lanceolate, gen folded lengthwise, (sub)entire. **INFL** glandular. **FL**: calyx 6–10 mm, lobes ± lanceolate; corolla 27–31 mm, blue-violet, glandular outside, floor glabrous; anther sacs 2.4–2.6 mm, dehiscing across common tip 2/3–3/4 their length, inner margins hairy; staminode glabrous. UNCOMMON. Rocky ridges, peaks, open slopes in montane forest; 1500–2400 m. w KR, NCoRH. ❀DRN:1,2,7,14&SUN:15–17; DFCLT.

P. rattanii A. Gray Per 25–120 cm, ± glabrous. **LVS**: main cauline lvs thin, 25–140 mm, narrowly elliptic, slightly narrowed just above rounded to truncate base (where widest), shallowly toothed, ± = basal. **INFL** glandular. **FL**: corolla 20–30 mm, abruptly expanded to throat, gen on upper side, blue-violet, whitish within, glandular outside, floor ± hairy; anther sacs 1.3–1.7 mm, dehiscing full length, valves barely spreading; staminode yellow-hairy. 2*n*=16. Open places in ± coastal forest; < 1200 m. w NW, n CW; OR.

var. ***kleei*** (E. Greene) A. Gray SANTA CRUZ MOUNTAINS BEARDTONGUE **FL**: calyx 6–7 mm, lobes ovate. Redwood, hardwood forests; 400–1000 m. SnFrB. [ssp. *k.* (E. Greene) Keck] ❀SHD, DRY:7,14 or SUN,IRR:**15–17**;may be DFCLT.

var. ***rattanii*** (p. 1061) **FL**: calyx 6–12 mm, lobes lanceolate. Grassy slopes in grand-fir/sitka-spruce, redwood, mixed-evergreen forests; 10–1200 m. n NCo, w KR, n NCoRO; OR. ❀TRY.

P. roezlii Regel (p. 1061) Per 15–55 cm, hairy, woody-branched below. **LVS** gen cauline; upper lvs 15–70 mm, linear to (ob)lanceolate, gen folded lengthwise, entire. **INFL** glandular. **FL**: calyx 3.5–6 mm, lobes ± linear to narrowly ovate; corolla 14–22 mm, ± purplish blue, glandular outside, floor glabrous; anther sacs 1.5–2.2 mm, dehiscing across common tip 1/2–3/5 their length, inner margins long-hairy; staminode glabrous. 2*n*=16. Dryish sagebrush or juniper scrub, coniferous forest; 700–2500 m. KR, CaR, n&c SNH, MP; OR, NV. [*P. laetus* ssp. *r.* (Regel) Keck] ❀DRN,IRR,SUN: 1–3,7,14–16;DFCLT.

P. rostriflorus Kellogg (p. 1061) Per 30–100 cm, glabrous or sts finely hairy below; clumps large, woody-branched. **LVS** gen cauline, 20–70 mm, linear to lanceolate, entire. **INFL** glandular. **FL**: calyx 4–8 mm, lobes lanceolate to narrowly ovate; corolla 22–33 mm, upper lip forming hood over anthers, lower lip strongly reflexed, (orange-) red sparsely glandular outside, floor glabrous; anther sacs 1.8–2.5 mm, dehiscing across common tip 1/4–1/3 their length, glabrous on sides; staminode glabrous. 2*n*=16. Dry sagebrush or Joshua-tree scrub, juniper/pinyon woodland, montane forest; 1600–2800 m. c&s SNH, SW, SNE, DMtns; to Colorado, NM. [*P. bridgesii* A. Gray]

P. rupicola (Piper) Howell (p. 1061) Subshrub 5–14 cm, matted, spreading-hairy (exc some lvs). **LVS** gen basal (reduced upward); main blades 7–20 mm, widely (ob)ovate to ± round, (sub)entire, glaucous; petiole 1.5–6 mm. **INFL** glandular. **FL**: calyx 4–10 mm, lobes ± lanceolate; corolla 25–38 mm, lavender or violet, floor glabrous or sparsely and finely hairy; stamens incl, anther sacs 1–1.4 mm, spreading flat, woolly; staminode gen sparsely long-hairy distally. 2*n*=16. Outcrops, rocky slopes in Douglas-fir forest; 1100–2300 m. KR; to WA. In CA, hybridizes readily with *P. davidsonii*, *P. newberryi*. ❀DRN,IRR:4–6,15–17&SHD:1,2,7;DFCLT.

P. rydbergii Nelson var. ***oreocharis*** (E. Greene) N. Holmgren (p. 1061) Per 20–60 cm, gen glabrous (or finely hairy on sts below). **LVS** entire; basal and lower cauline lvs well developed, oblanceolate; upper cauline lvs 25–70 mm, lanceolate. **FL**: calyx 3–5 mm, lobes narrowly oblong to (ob)ovate; corolla 9–14 mm, blue to purple, glabrous (exc white- to yellow-hairy floor); anther sacs 0.5–0.8 mm, dehiscing full length; staminode densely golden-yellow hairy. 2*n*=16. Moist meadows, streambanks, gen in montane to subalpine forests; 1200–3100 m. CaR, SNH, GB; OR, NV. [*P. o.* E. Greene] ❀DRN,IRR,SUN:1,2,4–6,**7**,14–16;DFCLT.

P. scapoides Keck (p. 1061) Per 15–60 cm, subscapose, woody-branched below. **ST** glabrous, glaucous. **LVS** gen dense, in basal mat, entire, densely hairy; basal lvs 1.5–6 mm, ovate to ± round; cauline lvs few, narrowly linear to narrowly oblanceolate. **INFL** glandular. **FL**: calyx 3–5 mm, lobes oblong to widely ovate; corolla 25–34 mm, pale lavender to purple or blue, glandular outside, floor yellow-hairy; anther sacs 1.4–1.7 mm, dehiscing across common tip 1/2 their length, inner margins glabrous; staminode pale yellow-hairy. 2*n*=16. Sagebrush, juniper-pinyon, and bristlecone-pine woodland; 2000–3200 m. W&I, n DMtns (Last Chance Mtns). ❀TRY;DFCLT.

P. speciosus Lindley (p. 1061) Per 5–60 cm, short-hairy on sts. **LVS**: upper cauline 35–90 mm, lanceolate, clasping, sometimes folded lengthwise, entire, gen glabrous. **INFL** glabrous to short-hairy, rarely glandular. **FL**: calyx 4–13 mm, lobes lanceolate to ovate; corolla 25–37 mm, ± abruptly expanded into throat, 7–13 mm wide when pressed, sky-blue (gen white inside), glabrous; anther sacs 1.8–3 mm, ± S-shaped, dehiscing in distal 2/3, glabrous; staminode glabrous or tip hairy. 2*n*=16. Open sagebrush scrub to subalpine forest; 1200–3300 m. KR, CaR, SNH, WTR, SnGb, MP, n DMtns; to WA, ID, UT. [ssp. *kennedyi* (Nelson) Keck] ❀DRN, IRR,SUN:1–3,7,14–16;DFCLT.

P. spectabilis Thurber (p. 1061) Per 80–120 cm, glabrous, green or glaucous. **LVS** ± thin; upper cauline lvs 35–100 mm, lanceolate to widely ovate, pairs fused at base, gen folded lengthwise and recurved at tip, ± serrate. **FL**: calyx 3.5–6.3 mm, lobes ovate to ± round; corolla 24–34 mm, blue (-purple), throat 8–14 mm wide when pressed, pale violet to lavender, whitish within, glandular outside and on throat roof, floor gen glabrous; anther sacs 1.8–2.4 mm, dehiscing full length but barely spreading; staminode incl, glabrous. Gravelly and sandy slopes, banks of washes, coastal-sage scrub, chaparral, oak woodland; 100–2400 m. TR, PR; Mex. Vars. hybridize in e TR.

var. ***spectabilis*** **INFL** glabrous. 2*n*=16. Habitats of sp.; 100–1100 m. SnGb, SnBr, PR; Mex. ❀DRN,SUN:**7**,14–17,**18–23**;DFCLT.

var. ***subviscosis*** (Keck) McMinn **INFL** glandular. Habitats of sp.; 500–2400 m. TR. [ssp. *subviscosus* Keck] ❀DRN,DRY, SUN:2,3,7,14,**15,16**,18–21;DFCLT.

P. stephensii Brandegee (p. 1065) STEPHENS' BEARDTONGUE Shrub 30–150 cm, glabrous. **LVS** ± thin; upper cauline lvs 25–50 mm, triangular-ovate, pairs fused at base, finely and sharply serrate. **INFL** glabrous to glandular. **FL**: calyx 3–5.2 mm, lobes ovate to ± round; corolla 15–20 mm, rose to magenta, unlined, throat 4–6 mm wide when pressed, glandular outside and inside (floor lacks nonglandular hairs); anther sacs 0.8–1.1 mm, spreading flat; staminode incl, glabrous. RARE. Rocky slopes, washes, rock crevices in creosote-bush scrub, juniper/pinyon woodland; 1200–1500 m. DMtns.

P. sudans M.E. Jones (p. 1065) Per < 70 cm, sticky throughout. **LVS** gen cauline, 30–60 mm, ovate, serrate. **FL**: calyx 3.5–5.5 mm, lobes lanceolate to ovate; corolla 9–11 mm, cream-white, dark-lined; anther sacs 0.5–0.7 mm, spreading widely; staminode gen glabrous. Open, rocky, igneous soils in sagebrush scrub, yellow-pine and montane forests; 1200–1700 m. s CaRH, MP; NV. [*P. deustus* ssp. *s.* (M.E. Jones) Pennell & Keck] ❀TRY;DFCLT.

P. thompsoniae (A. Gray) Rydb. (p. 1065) Per 5–15 cm, ± prostrate, gen matted; hairs appressed, backward-pointing, scale-like,

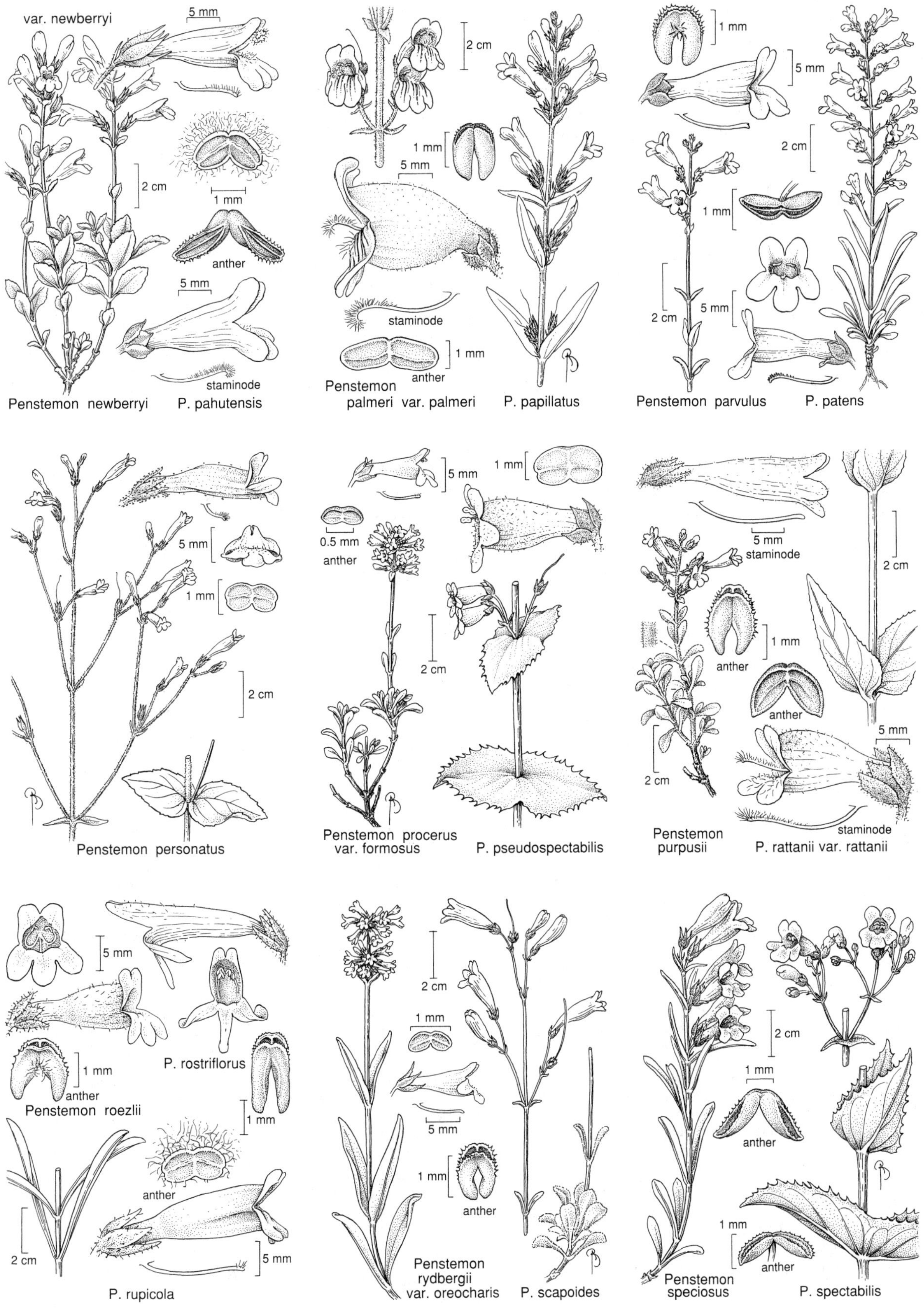
var. newberryi
5 mm
2 cm
1 mm
anther
5 mm
staminode
Penstemon newberryi
P. pahutensis
2 cm
1 mm
5 mm
staminode
1 mm
anther
Penstemon
palmeri var. palmeri
P. papillatus
1 mm
5 mm
2 cm
1 mm
2 cm
5 mm
Penstemon parvulus
P. patens
5 mm
1 mm
2 cm
Penstemon personatus
5 mm
1 mm
0.5 mm
anther
2 cm
Penstemon procerus
var. formosus
P. pseudospectabilis
5 mm
staminode
2 cm
anther
1 mm
anther
5 mm
2 cm
staminode
Penstemon
purpusii
P. rattanii var. rattanii
5 mm
P. rostriflorus
1 mm
anther
Penstemon roezlii
1 mm
anther
2 cm
5 mm
P. rupicola
2 cm
1 mm
5 mm
1 mm
anther
Penstemon
rydbergii
var. oreocharis
P. scapoides
2 cm
1 mm
anther
1 mm
anther
Penstemon
speciosus
P. spectabilis

ashy. **LF** 5–20 mm, ± obovate, entire, petioled. **FL**: calyx 4–7.5 mm, lobes lanceolate, some hairs gen glandular; corolla 10–18 mm, violet to blue, glandular outside, floor pale yellow-hairy; anther sacs 0.8–1.3 mm, dehiscing full length but barely spreading; staminode densely orangish- to golden-yellow-hairy. White calcareous soils in juniper/pinyon woodland; 1500–2700 m. e DMtns (New York, Clark mtns); to UT, AZ. ❀TRY;DFCLT.

P. thurberi Torrey (p. 1065) THURBER'S BEARDTONGUE Shrub 20–80 cm, ± round, glabrous. **LVS** cauline, narrowly linear; main blades 10–45 mm, gen rolled upward, nearly round in X-section, entire. **INFL** glabrous. **FL**: calyx 2.4–4.5 mm, lobes ovate; corolla 8–15 mm, ± funnel-shaped (lips obliquely spreading), lavender, rose, or blue-purple, glabrous outside, floor hairy; anther sacs 0.7–0.9 mm, spreading flat; staminode glabrous. $2n=16$. UNCOMMON. Sandy and gravelly slopes and mesas, in chaparral, creosote-bush or Joshua-tree scrub, juniper/pinyon woodland; 500–1200 m. PR, DMoj; to NV, NM, Mex.

P. tracyi Keck (p. 1065) TRACY'S BEARDTONGUE Per 8–12 cm, glabrous. **LVS** gen basal, obovate, (sub)entire; cauline lvs 10–20 mm, (ob)ovate to ± round, base widely wedge-shaped. **INFL** gen glabrous. **FL**: calyx 2.5–3 mm, lobes ovate, glabrous or glandular-ciliate; corolla 11–13 mm, glabrous (exc densely long-hairy floor); anther sacs ± 0.4 mm, spreading flat; staminode tip hairy-tufted. RARE. Exposed outcrops; 2000–2100 m. s KR (n Trinity Co.). In cult.

P. utahensis Eastw. (p. 1065) Per 15–50 cm, glabrous, glaucous. **LVS** thick, gen folded lengthwise; basal lvs well developed; upper cauline lvs 15–55 mm, ± lanceolate, entire. **INFL** glabrous. **FL**: calyx 2.5–5 mm, lobes widely ovate; corolla 17–25 mm, cylindric to funnel-shaped (limb slightly oblique, lobes spreading), pink to red (or purplish), glandular outside and inside; anther sacs 0.6–1.1 mm, spreading flat; staminode glabrous or tip minutely hairy. Sagebrush scrub, juniper/pinyon woodland; 1200–2500 m. e DMtns (Kingston, New York mtns); to UT, AZ. ❀DRN,SUN, DRY:2,3,7,10,14–16,18–21;DFCLT.

P. venustus Lindley (p. 1065) Shrub 30–80 cm, spreading, gen glabrous (or sts short-hairy). **LVS** stiff, cauline; largest (at mid-st) 50–120 mm, lanceolate, gen sharply serrate. **INFL** glabrous. **FL**: calyx 3.4–6.2 mm, lobes ovate to lanceolate; corolla 22–38 mm, lavender to purple, glabrous (exc lobes short-ciliate); filaments and inner margins of anthers long-white-hairy, anther sacs 1.5–2 mm, dehiscing across common tip; staminode long-white-hairy. $2n=64$. Dry, rocky, exposed places; 900–2200 m. s CaRH (sw Lassen Co.); native to OR, WA, ID. Naturalized from cult.

SCROPHULARIA FIGWORT

Margriet Wetherwax

Per or shrub, long-soft-shaggy- and short-stout-glandular-hairy. **ST** square. **LVS** opposite, 4-ranked, lanceolate to triangular-ovate, petioled. **INFL**: panicle; pedicels > bracts; bracts lanceolate, long-tapered. **FL**: calyx lobes 5, obtuse to rounded; corolla 5-lobed, 2-lipped, greenish yellow to blackish, tube urn-shaped to spheric, lowest lobe gen recurved; fertile stamens 4, incl, anther sacs 1, staminode attached near base of upper lip, appressed to upper side of corolla; style slender, gen reflexed at fl, stigma head-like; nectary at base of ovary, fleshy. **FR** septicidal. **SEEDS** many, oblong-ovoid, ridged. ± 150 spp.: n temp, esp Eurasia. (Latin: scrofula, a disease supposedly cured by some spp.)

1. Shrub; infl long-soft-wavy-hairy, sticky; staminode ± 0; s ChI ..... ***S. villosa***
1′ Per; infl glandular-puberulent or short-hairy; staminode well developed; CA (exc ChI)
  2. Staminode fan-shaped, wider than long; KR, CaR ..... ***S. lanceolata***
  2′ Staminode club-shaped to obovate, longer than wide; gen CA-FP
    3. Corolla urn-shaped, dark maroon, upper 1/2 blackish ..... ***S. atrata***
    3′ Corolla spheric, yellowish green to dark maroon
      4. Corolla distinctly 2-colored; lf base wedge-shaped to triangular, narrowed to petiole; SN, SNE ..... ***S. desertorum***
      4′ Corolla not distinctly 2-colored; lf base truncate to cordate, not narrowed to petiole; CA-FP ..... ***S. californica***
        5. Lf blade slightly crenate or serrate to coarsely dentate; w NW, w CW, WTR ..... ssp. ***californica***
        5′ Lf blade deeply dissected, sections serrate; e NW, s SN, e CW, TR ..... ssp. ***floribunda***

S. atrata Pennell (p. 1065) BLACK-FLOWERED FIGWORT Per. **ST** 10–12 dm. **LF**: blade 6–10 cm, ovate, acute, dentate, teeth rounded to widely acute; petiole 2–7 cm. **INFL** long, glandular-puberulent; branches widely spreading. **FL**: calyx lobes ± 3 mm, ovate, rounded; corolla 9–11 mm, urn-shaped, upper half blackish, lower half dark maroon, lowest lobe spreading to reflexed; staminode ± oblong, width << limb of upper lip, dark maroon. **FR** 6–8 mm, ovoid, obtuse or acute. UNCOMMON. Calcareous (sometimes diatomaceous) soils; < 500 m. s SCoRO (Santa Barbara, s San Luis Obispo cos.). ❀DRN,IRR,SUN:14,**15–17**,18–21,**22–24**.

S. californica Cham. & Schldl. CALIFORNIA FIGWORT Per, minutely glandular. **ST** 8–12 dm. **LF**: blade 7–19 cm, ovate to ± triangular, truncate to cordate, acute, crenate to deeply dissected; petiole 3–7 cm. **INFL** narrow, compact to widely spreading. **FL**: calyx 3–4 mm, lobes obtuse to long-tapered; corolla 8–12 mm, upper half brown-maroon, lower half paler or yellowish green, lowest lobe recurved; staminode club-shaped to obovate, brown to light maroon, longer than wide; nectary inconspicuous. **FR** 6–8 mm, conic-ovoid. Common. Moist places, roadsides, chaparral; < 1500 m. CA-FP; to B.C.

ssp. ***californica*** (p. 1065) **LF**: blade ± crenate to coarsely dentate. Habitats and elevations of sp. w NW, w CW, WTR; to B.C. ❀IRR:**4–6,15–17,24**&SHD:7,8,9,**14,18–23**;may be INV.

ssp. ***floribunda*** (E. Greene) Shaw (p. 1065) **LF**: blade deeply dissected, sections serrate. Habitats and elevations of sp. e NW, s SN, e CW, SCo. ❀TRY.

S. desertorum (Munz) Shaw (p. 1065) Per, minutely glandular. **STS** clustered, 7–12 dm. **LF**: blade 10–13 cm, lanceolate, base wedge-shaped, narrowly acute, serrate; petiole 7–10 cm. **INFL** long. **FL**: calyx lobes 2–3 mm, triangular-ovate, obtuse to rounded; corolla 7–9 mm, ± spheric, upper half maroon, lower half cream, lowest lobe recurved; staminode club-shaped, longer than wide, light maroon, with 2 nectar drops gen conspicuous at base. **FR** 5–9 mm, ovoid, long-tapered. Dry slopes; 1000–3000 m. SN, SNE, n DMtns; NV. ❀DFCLT.

S. lanceolata Pursh (p. 1065) Per, minutely glandular. **STS** clustered, 8–15 dm. **LF**: blade 9–14 cm, lanceolate to ovate, base wedge-shaped, truncate or cordate, long-tapered, serrate, gen glabrous; petiole 2–4 cm. **INFL** long, narrow, long-soft-wavy-glandular-hairy. **FL**: calyx lobes 2–3 mm, ovate, rounded; corolla 8–14 mm, spheric to urn-shaped (throat slightly narrowed), greenish brown, lowest lobe recurved; staminode fan-shaped, wider than long, yellow-green. **FR** 6–8 mm, conic-ovoid. Sandy roadsides, woodland; 1400–2800 m. KR, CaR; to WA, ID, MT, Colorado. ❀TRY.

S. villosa Pennell (p. 1065) SANTA CATALINA FIGWORT Shrub, glandular-puberulent. **ST** 12–18 dm. **LF**: blade 10–15 cm, ovate, truncate to cordate, long-tapered, sharply dentate; petiole 3–5 cm. **INFL** long, narrowly branched, long-soft-wavy-glandular-hairy. **FL**: calyx lobes 3 mm, triangular-ovate, acute; corolla 8–11 mm, spheric to urn-shaped, blackish maroon, lowest lobe recurved; staminode 0 or awl-like vestige, maroon. **FR** 6–8 mm, conic. RARE. Canyons, coastal scrub, chaparral; < 500 m. s ChI (Santa Catalina, San Clemente islands).

STEMODIA

Margriet Wetherwax

Per. **STS** 1–many, ascending to erect. **LVS** gen opposite, clasping, serrate. **INFL** raceme; bracts 2 below calyx. **FL**: sepals 5, free; corolla 5-lobed, 2-lipped, violet to purple, tube 4-angled, upper lip 2-lobed, arched, lower lip reflexed; stamens 4, in 2 pairs, anther sacs well separated by thick tissue; stigma lobes 2, flat. **FR** ± cylindric, loculicidal. **SEEDS** many, netted. ± 20 spp.: trop Am. (Latin: from "stemodiacra" — stamens with 2 tips)

S. durantifolia (L.) Sw. (p. 1065) Pl glandular-hairy. **ST** 10–50 cm. **LVS** 2–3 per node, 2–4 cm (smaller upward), lanceolate, widest at base, ± dentate, ± sessile. **INFL** open; pedicels 2–12 mm; bracts ± = sepals. **FL**: sepals 5–7 mm, lanceolate, long-tapered; corolla 7–10 mm. **FR** ± 5 mm, ovoid-cylindric. Wet sand or rocks, drying river beds; < 300 m. SnJt, DSon; AZ, Mex. ❀TRY.

SYNTHYRIS KITTENTAILS

Margriet Wetherwax

Per, rhizomed. **ST**: aerial part scapose. **LVS** basal, petioled; blade ovate to round, cordate, serrate to shallowly pinnately lobed. **INFL**: raceme; bracts < lvs, sessile, alternate. **FL**: sepals 4, free; corolla bell-shaped to rotate, blue to lavender, lobes 4; stamens 2, exserted; stigmas fused, minutely head-like. **FR** loculicidal, plump, entire to notched. **SEEDS** 2–many per chamber, flat or with incurved margins. $2n=24$. 14 spp.: w N Am. (Greek: closing door, from fr valves adhering to placenta)

1. Corolla 3–5 mm, ± rotate; lvs glabrous; seeds > 2 per chamber; Wrn . . . . ***S. missurica***
1′ Corolla 6–9 mm, bell-shaped; lvs ± long-shaggy-hairy; seeds 2 per chamber; NW, SnFrB . . . . ***S. reniformis***

S. missurica (Raf.) Pennell (p. 1065) **LF**: blade nearly round, cordate, glabrous, palmately veined. **INFL** ± hairy; fls gen many; peduncle < 60 cm, erect in fr; pedicel in fr 2–6 mm, < or = bracts; bracts ± oblanceolate. **FL**: sepals 3–4 mm, linear to oblong, not ciliate; corolla 3–5 mm, ± rotate, blue to purplish, lobes > tube; style 4–5 mm. **FR** 5 mm, ± round, slightly notched, not ciliate. **SEEDS** > 2 per chamber, flat, thinly winged, yellow-brown. Moist cliffs, grassy forest openings; 2000–2500 m. Wrn; to ne WA, n ID. ❀IRR,DRN,SHD:1,2,4–7,14–17;DFCLT.

S. reniformis (Douglas) Benth. (p. 1065) SNOW QUEEN **LF**: blade ovate to reniform, cordate, upper surface soft-shaggy-hairy. **INFL** ± soft- to rough-shaggy-hairy; fls gen few; peduncle < 15 cm, curved to reclining in fr; pedicel in fr 7–10 mm, > 2 × bracts; bracts ovate. **FL**: sepals ± 4.5 mm, ovate, ciliate; corolla 6–9 mm, bell-shaped, blue or lavender to white, lobes < tube; style 5–8 mm. **FR** 2–4 mm, 6–8 mm wide; lobes widely spreading, ciliate. **SEEDS** 2 per chamber, brown; margins thick, incurved. Grassy places; < 1500 m. NCo, NCoRO, SnFrB; to w WA. Pls from NCoRO (s Humboldt, Mendocino, Marin cos.) with lf blade longer than wide have been called var. *cordata* A. Gray. ❀IRR,DRN:**4–6**, **17**&SHD:**7**,14,**15**,**16**.

TONELLA

Margriet Wetherwax

Ann. **ST** erect, slender, branched. **LVS** opposite, entire to 3-lobed. **INFL**: raceme, bracted. **FL**: calyx deeply 5-lobed; corolla 2-lipped, upper lip 2-lobed, lower lip 3-lobed (middle lobe wider than lateral); stamens 4, equal, exserted, filaments hairy; stigmas fused. **FR** spheric to ovoid, loculicidal. **SEEDS** large, wingless. 2 spp: N.Am. (Derivation unknown)

T. tenella (Benth.) A.A. Heller (p. 1069) **ST** ascending, 10–30 cm. **LF** 10–15 mm; upper surface soft-shaggy-hairy; lower lvs petioled, ovate to round, becoming sessile upward, entire to deeply 3-lobed. **INFL** minutely glandular-hairy distally; bracts lanceolate, subtending 1–3 fls; pedicels 8–15 mm. **FL**: calyx < 3 mm, lobes < 2 mm, acute to obtuse, minutely ciliate; corolla 2–2.5 mm, upper lobes < lower, white proximally, blue or violet distally. **FR** 2–2.5 mm. **SEEDS** 1 per chamber, < 1.5 mm. Moist, shaded places; < 1600 m. KR, NCoR, CaR, n SNH, ScV, SnFrB, SCoRI; to WA.

TRIPHYSARIA

T.I. Chuang & Lawrence R. Heckard

Ann, green root-parasites. **LVS** sessile, ± linear; upper finely divided. **INFL**: spike; bracts gen lf-like. **FL**: calyx ± equally 4-lobed; corolla club-shaped (exc *T. pusilla*), lips nearly equal, upper lip beak-like, tip open, lower lip deeply 3-pouched, 3-toothed, throat abruptly indented forming a fold; stamens 4, anther sac 1; stigma gen slightly enlarged, entire to slightly 2-lobed. **FR** loculicidal, ± ovoid. **SEEDS** many, 0.8–1.2 mm, ovoid, attached at base; coat netted, gen tight-fitting. 5 spp.: w N.Am. (Greek: 3 bladders, from lower lip pouches) [Chuang & Heckard 1991 Syst Bot 16:644–666] Hybrids common. Related to *Castilleja* and *Orthocarpus* but isolated reproductively. ❀TRY with host; DFCLT.

1. Pl gen << 15 cm, decumbent-branched from base; corolla 4–7 mm, beak hooked, pouches ± 1 mm deep . . . . ***T. pusilla***
1′ Pl gen > 15 cm, ± ascending-branched; corolla 8–25 mm, beak not hooked, pouches gen 1.5–4 mm deep
 2. Stamens exserted; pouch depth > 1.5 × length . . . . ***T. floribunda***
 2′ Stamens ± incl; pouch depth < 1.5 × length
  3. Pl green- or yellow-brown, gen glabrous; corolla yellow or white, beak yellowish . . . . ***T. versicolor***
   4. Corolla yellow . . . . ssp. ***faucibarbata***
   4′ Corolla white . . . . ssp. ***versicolor***
  3′ Pl purplish, gen puberulent; corolla sulphur-yellow or white to rose, beak dark purple

5. Corolla 8–15 mm, ± = bract, pouches 1–2 mm deep .... ***T. micrantha***
5′ Corolla 10–25 mm, > bract, pouches 3–4 mm deep .... ***T. eriantha***
 6. Corolla yellow .... ssp. ***eriantha***
 6′ Corolla white, becoming rose-pink .... ssp. ***rosea***

T. eriantha (Benth.) Chuang & Heckard BUTTER-AND-EGGS, JOHNNY-TUCK Pl gen 10–35 cm, purplish, puberulent and glandular- and nonglandular-hairy. **LF** 10–50 mm, 3–7-lobed. **INFL** 2–15 cm, dense above; bracts 5–18 mm, 3–5-lobed. **FL**: calyx 8–13 mm, divided 1/4–1/2; corolla 10–25 mm, sulphur-yellow or white (beak dark purple), tube slender, densely puberulent, ± 2 × calyx, lower lip ± = beak, pouches 3–4 mm deep; stamens incl. **FR** 5–8 mm, oblong. **SEEDS** 30–50, dark brown. $2n$=22. Common. Grassland, foothills, coastal fields; < 1300 m. CA-FP; sw OR.

ssp. ***eriantha*** (p. 1069) **FL**: corolla sulphur-yellow. Grassland, foothills; < 1300 m. CA-FP; sw OR. [*Orthocarpus e.* Benth. incl var. *gratiosus* Jepson & Tracy]

ssp. ***rosea*** (A. Gray) Chuang & Heckard **FL**: corolla white, gen becoming rose-pink. Coastal fields, bluffs; < 100 m. NCo, CCo, SCo. [*O. e.* Benth. var. *r.* A. Gray]

T. floribunda (Benth.) Chuang & Heckard (p. 1069) SAN FRANCISCO OWL'S-CLOVER Pl 10–30 cm, yellow-brown, ± glabrous or sparsely stiff-hairy. **LF** 10–40 mm, 5–9-lobed. **INFL** gen 1–5 cm, dense; bracts 5–12 mm, 3–7-lobed, ± glabrous. **FL**: calyx 4–6 mm, divided 1/4–1/2; corolla 10–14 mm, creamy white, tube slender, glabrous, ± 2 × calyx, lower lip ± = beak, pouches ± 2 mm deep; stamens exserted. **FR** 4–5 mm. **SEEDS** 20–30, dark brown. $2n$= 22. RARE. Coastal grassland, serpentine slopes; < 100 m. n CCo, w SnFrB. [*Orthocarpus f.* Benth.]

T. micrantha (A. Gray) Chuang & Heckard (p. 1069) Pl 5–15 cm, gen dark purplish, glabrous below, puberulent and glandular- and nonglandular-hairy above. **LF** 7–25 mm, 0–5-lobed. **INFL** 3–6 cm, slender, ± open; bracts 8–15 mm, 3–5-lobed, long-hairy. **FL**: calyx 6–7 mm, divided 1/3–2/3, hairy; corolla 8–15 mm, yellow (beak dark purple), lower lip 1 mm < beak, marked dark purple, pouches 1–2 mm deep; stamens ± incl; stigma slightly 2-lobed. **FR** 4–5 mm. **SEEDS** 50–80, dark brown; coat loose-fitting. $2n$=22. Grassland; < 1000 m. c&s SNF, SnJV, SCoR. [*Orthocarpus erianthus* var. *m.* (A. Gray) Jepson]

T. pusilla (Benth.) Chuang & Heckard (p. 1069) Pl 5–20 cm, yellow-brown or purplish, finely spreading-hairy. **ST** decumbent-branched from base; main axis obscure. **LF** 5–30 mm, 3–9-lobed. **INFL** 4–15 cm, ± from st base; bracts 5–12 mm, 3–11-lobed (sometimes 2-pinnate). **FL**: calyx 5–7 mm, divided 1/3–2/3, hairy; corolla 4–7 mm, gen purplish (yellow), beak hooked, pouches ± 1 mm deep; stamens incl. **FR** 4–6 mm, round, compressed. **SEEDS** 20–30, dark brown. $2n$=22. Grassland; < 400 m. NCoR, c SNF, CW; to B.C. [*Orthocarpus p.* Benth.] Self-pollinating.

T. versicolor Fischer & C. Meyer Pl 10–60 cm, green to yellow-brown, ± glabrous. **LF** 20–80 mm, 5–9-lobed. **INFL** 5–20 cm, dense above; bracts 8–18 mm, 3–5-lobed, lobes lanceolate. **FL**: calyx 5–10 mm, divided ± 1/3; corolla 12–22 mm, white or yellow, tube >> calyx, slender, densely puberulent, beak yellowish, lower lip ± 1 mm < beak, pouches 2–4 mm deep, margins purple-dotted; stamens incl. **FR** 6–9 mm. **SEEDS** 30–50, dark brown. $2n$=22 (both sspp.). Grassland; < 500 m. NCo, NCoR, CCo; sw OR.

ssp. ***faucibarbata*** (A. Gray) Chuang & Heckard **FL**: corolla yellow. Habitats and elevations of sp. NCoR. [*Orthocarpus f.* A. Gray]

ssp. ***versicolor*** (p. 1069) **FL**: corolla white fading rose. Habitats of sp.; < 100 m. NCo, CCo; sw OR. [*O. f.* A. Gray var. *albidus* Keck]

VERBASCUM MULLEIN

Margriet Wetherwax

Bien in CA (lf rosette 1st year, stout fl-st 2nd year). **ST** erect, simple or branched just below infl. **LVS** basal and cauline, alternate, reduced upward. **INFL**: raceme or panicle, bracted. **FL**: calyx ± radial, deeply 5-lobed; corolla ± radial, rotate, 5-lobed; stamens 5, lowest 2 filaments > upper 3, all or upper hairy; stigmas fused, head-like. **FR**: capsule, septicidal. **SEEDS** small, wingless, many. ± 360 spp.: Eurasia. (Latin: from root for bearded) Lvs used medicinally in cigarets for asthmatics.

1. Lf glabrous or ± hairy; filaments all densely purple-hairy; fr glabrous to glandular-puberulent
 2. Lf glabrous; lower pedicels 10–15(25) mm, 1 per node .... ***V. blattaria***
 2′ Lf ± hairy; lower pedicels < 10 mm, 1–4 per node .... ***V. virgatum***
1′ Lf densely woolly-tomentose; upper 3 filaments white- or yellow-hairy, lower 2 glabrous to ± hairy; fr densely tomentose
 3. Lf not decurrent; infl openly branched .... ***V. speciosum***
 3′ Lf decurrent; infl dense, gen unbranched .... ***V. thapsus***

V. blattaria L. (p. 1069) MOTH MULLEIN Bien, glabrous below. **ST** 30–120 cm, sometimes branched just below infl. **LVS**: basal 4–25 cm, oblanceolate, crenate, glabrous, short-petioled; cauline 2–12 cm, elliptic to ovate, cordate, dentate, sessile. **INFL**: raceme, terminal, glandular distally; fls 1 per node; bracts < pedicels, linear to lanceolate; pedicels 10–15(25) mm. **FL**: calyx 5–8 mm, lobes linear-lanceolate; corolla 25–30(40) mm wide, yellow or white (purple); filaments all densely purple-hairy. **FR** 6–8 mm, spheric. $2n$= 18,30,32. Disturbed areas; < 1600 m. CA-FP; native of Eurasia. Specific epithet derived from supposed function of lvs in repelling cockroaches.

V. speciosum Schrader. SHOWY MULLEIN Bien, densely stellate-hairy. **ST** 100–150 cm. **LVS** 30–40 cm, oblanceolate, entire, not decurrent, petioled. **INFL**: panicle; fls 5–9 per node; bracts < pedicels, ovate; pedicels 5–10 mm. **FL**: calyx 3–6 mm, lobes linear; corolla 20–30 mm wide, yellow; filament hairs white. **FR** 3–7 mm, ovoid-oblong. Disturbed areas; < 500 m. SnFrB; native to Eurasia.

V. thapsus L. (p. 1069) WOOLLY MULLEIN Bien, densely stellate-hairy. **ST** 30–200 cm, gen simple. **LVS**: basal 8–50 cm, oblanceolate, gen entire, short-petioled; cauline 5–30 cm, lanceolate, gen entire, long-decurrent. **INFL**: raceme, dense; bracts 12–18 mm; pedicels < 2 mm, gen fused to st. **FL**: calyx 7–9 mm, lobes lanceolate; corolla 15–25(30) mm wide, yellow; upper 3 filaments white- or yellow-hairy, lower 2 glabrous to sparsely hairy. **FR** 7–10 mm, ovoid. $2n$=32,34,36. Disturbed areas; < 2200 m. CA-FP; native to Eurasia.

V. virgatum Stokes WAND MULLEIN Gen non-glandular bristly; hairs ± branched. **ST** 60–120 cm, branched just below infl. **LVS**: basal 10–30 cm, obovate, crenate to dentate, petioled; cauline 7–15 cm, lanceolate, cordate, crenate, sessile. **INFL**: raceme, open, glandular; fls 1–4 per node; bracts < 8 mm, ovate; pedicels < or = bract. **FL**: calyx 5–8(9) mm, lobes lanceolate; corolla ± 25 mm wide, yellow. **FR** 7–8 mm, spheric, glandular-puberulent, sometimes also branched- or stellate-hairy. $2n$=32,64,66. Disturbed areas; < 300 m. NCoRO, c SnJV, SnFrB, SW; native to Eurasia.

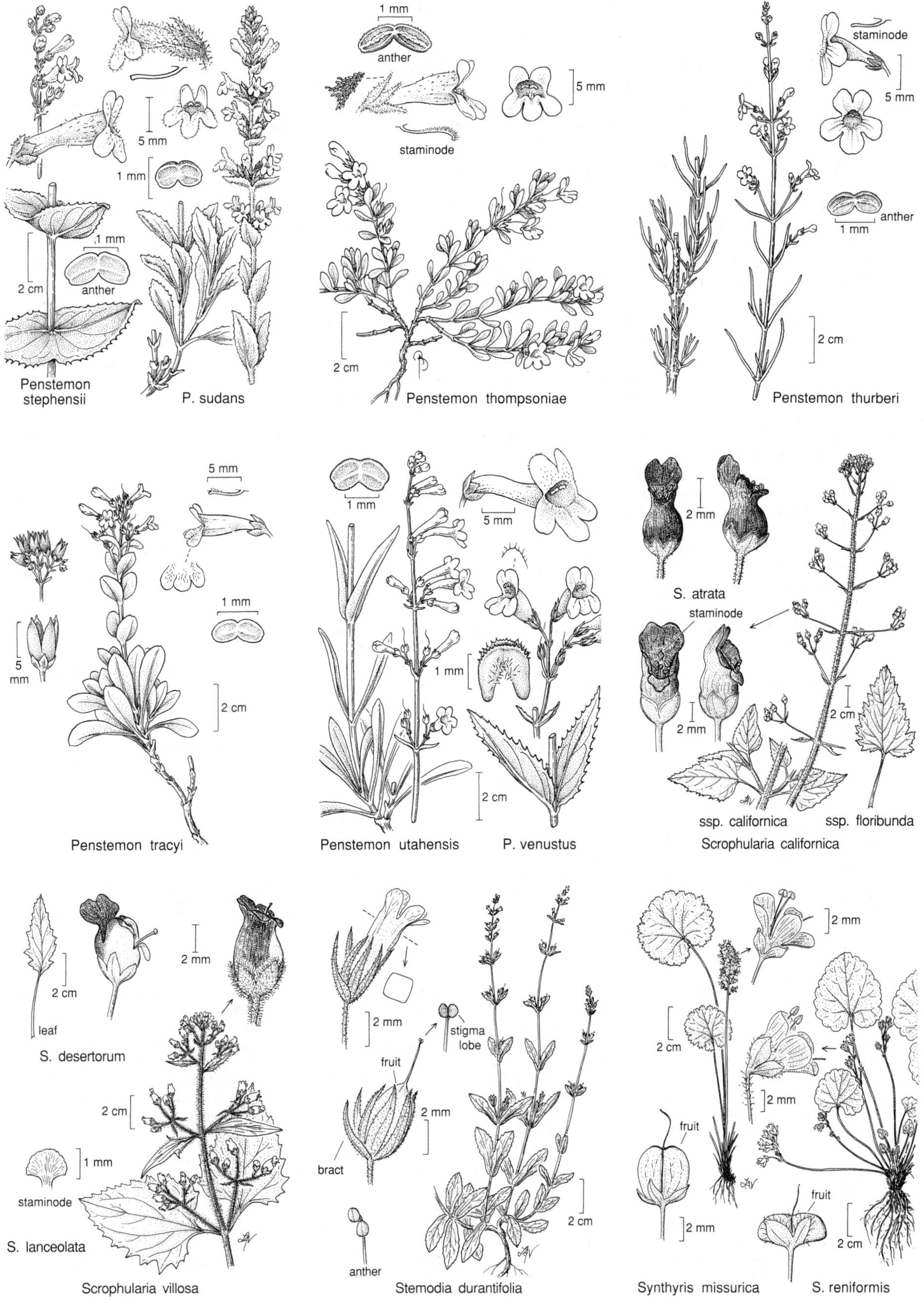

1 mm
5 mm
2 cm
anther
Penstemon stephensii
P. sudans
anther
staminode
Penstemon thompsoniae
staminode
Penstemon thurberi
Penstemon tracyi
Penstemon utahensis
P. venustus
S. atrata
staminode
ssp. californica
ssp. floribunda
Scrophularia californica
leaf
S. desertorum
staminode
S. lanceolata
Scrophularia villosa
stigma lobe
fruit
bract
anther
Stemodia durantifolia
fruit
Synthyris missurica
S. reniformis

VERONICA SPEEDWELL, BROOKLIME

Margriet Wetherwax

Ann, per. **ST** erect or prostrate. **LVS** opposite. **INFL**: raceme, terminal or axillary, or fls solitary in axils; bracts small, alternate. **FL**: sepals gen 4(5), ± free, gen unequal; corolla ± rotate, 4-lobed, upper lobe wide (formed by fusion of upper pair), blue or violet to white; stamens 2, exserted; stigma head-like. **FR**: capsule, flattened perpendicular to septum, gen obcordate, loculicidal and septicidal. ± 250 spp.: n temp, esp Eurasia. (Possibly named for Saint Veronica)

1. Racemes axillary, gen opposite; herbage gen glabrous (exc *V. chamaedrys*); per
2. Lvs petioled
3. Lf lanceolate to ± ovate, obtuse to acute; pedicel 5–10 mm; widespread ..... ***V. americana***
3′ Lf obovate, rounded; pedicel 3–6 mm; SNE (near Bridgeport, Mono Co.) ..... ***V. beccabunga***
2′ Lvs sessile (lower rarely short-petioled)
4. Racemes alternate; lf linear-lanceolate (length gen >> 5 × width); fr deeply notched ..... ***V. scutellata***
4′ Racemes opposite; lf lanceolate to ovate (length < 5 × width); fr entire or slightly notched
5. Lf lanceolate (length 3–5 × width); corolla 3–5 mm, pink; fr notched 0.1–0.3 mm ..... ***V. catenata***
5′ Lf elliptic to ovate (length 1.5–3 × width); corolla 5–10 mm, lavender to blue; fr rounded to obcordate
6. Pl glabrous; lf ± entire; fr rounded, notched < 0.1 mm; corolla pale lavender, violet-lined ..... ***V. anagallis-aquatica***
6′ Pl hairy; lf crenate to dentate; fr obcordate; corolla bright blue with white center ..... ***V. chamaedrys***
1′ Most racemes terminal, or fls solitary in axils; herbage gen hairy; ann or per
7. Rhizomed per
8. Fls solitary in axils — mat-forming in lawns, SnFrB ..... ***V. filiformis***
8′ Fls in ± distinct racemes
9. Style 6–9 mm, > fr; corolla 8–13 mm
10. Lf hairy; corolla 8–10 mm, pale blue to lavender-rose; KR ..... ***V. copelandii***
10′ Lf gen glabrous; corolla 10–13 mm, deep blue; CaR, SNH ..... ***V. cusickii***
9′ Style 0.8–3 mm, gen < fr; corolla 6–10 mm
11. Style 2–3 mm, fr gen < calyx, wider than long, st decumbent ..... ***V. serpyllifolia*** ssp. ***humifusa***
11′ Style 0.8–1.3 mm, fr > calyx, longer than wide, st erect nearly from base ..... ***V. wormskjoldii***
7′ Fibrous- or tap-rooted ann
12. Pedicel 0.5–2 mm, < calyx; seeds many, flat, smooth
13. Lf triangular to ovate, crenate to serrate; sepals unequal (outer pair longer); corolla blue to violet ... ***V. arvensis***
13′ Lf oblong to spoon-shaped, entire to slightly serrate; sepals ± equal; corolla whitish ..... ***V. peregrina*** ssp. ***xalapensis***
12′ Pedicel 4–30 mm, gen > calyx; seeds 1–12 per chamber, cup-shaped, outer surface rough
14. Pedicel 15–30 mm; lf crenate to serrate; style 2–3 mm; fr lobes spreading ..... ***V. persica***
14′ Pedicel 4–15 mm; lf palmately 3–9 lobed; style 0.6–1.6 mm; fr lobes ± parallel (or fr ± spheric)
15. Sts prostrate; fls solitary in axils; fr ± spheric, barely notched; seeds 1–2 per chamber; style 0.6–0.9 mm ..... ***V. hederifolia***
15′ Sts ± erect; infl a terminal raceme; fr obcordate; seeds 6–12 per chamber; style 1.2–1.6 mm ..... ***V. triphyllos***

V. americana (Raf.) Schwein. (p. 1069) AMERICAN BROOKLIME Per, rhizomed, glabrous. **ST** gen decumbent, rooting at lower nodes, branched, 5–60(100) cm. **LF** 5–50 mm, lanceolate to ovate, ± serrate, acute to obtuse, petioled. **INFL** axillary; fls 10–25; bracts linear; pedicels 5–10(13) mm. **FL**: sepals ± 3 mm, oblanceolate, acute; corolla 7–10 mm, violet-blue, dark-lined; style 1.5–3(4)mm. **FR** 3–4 mm, ± round, entire to barely notched. **SEED** 0.5 mm, flat. $2n=36$. Common. Moist to wet soil, springs, slow streams, meadows, lakeshores; < 3300 m. CA-FP.

V. anagallis-aquatica L. WATER SPEEDWELL Per, rhizomed, glabrous. **ST** gen decumbent, rooting at lower nodes, simple to many-branched from base, 10–60(100) cm. **LF** 20–80 mm, elliptic to ovate, clasping to cordate, entire to serrate, light green, sessile (exc lowest ± short-petioled). **INFL** axillary, glabrous to ± glandular-puberulent; fls gen > 30; bracts linear-lanceolate; pedicels 4–8 mm, upcurved. **FL**: sepals 3–5.5 mm, lanceolate to elliptic; corolla 5–10 mm, pale lavender-blue, violet-lined; style 1.5–3 mm. **FR** 2.5–4 mm, at least as wide, rounded, barely notched. **SEED** 0.5 mm, flat. $2n=18,36$. Wet meadows, streambanks, slow streams; < 2500 m. CA-FP; widely naturalized in N.Am, S.Am; native to Eur.

V. arvensis L. Ann, hairy; roots fibrous. **ST** prostrate to erect, ± branched, 5–40 cm. **LF** 2–15 mm, ovate, truncate to cordate, crenate to serrate; lower ± short-petioled. **INFL** terminal; bracts < pedicels; pedicels 0.5–2 mm. **FL**: sepals 3–4 mm, unequal (outer pair larger), lanceolate; corolla 2–3 mm, blue to violet; style 0.4–1.0 mm. **FR** 3–4 mm, flat, ciliate; notch 0.5–0.8 mm. **SEED** 1 mm, flat, smooth. $2n=14,16,18$. Meadows, lawns, gardens; < 1500 m. KR, NCoR, SNF, GV, WTR; widely naturalized; native to Eurasia.

V. beccabunga L. Per, rhizomed, glabrous. **ST** gen prostrate to ascending, rooting at nodes, 15–55 cm. **LF** 10–50 mm, obovate, widely obtuse, ± crenate, petioled. **INFL** axillary; fls 6–15; bracts small; pedicels 3–5 mm. **FL**: sepals 2.5–3.5 mm, unequal, oblanceolate; corolla 5–7 mm, blue to purple; style 1.5–2 mm. **FR** ± 3 mm, ± spheric; notch 0.2–0.3 mm. **SEED** 0.5 mm, flat. $2n=16,18,36$. Wet meadows, slow streams; < 2000 m. SNE (near Bridgeport); to se US; native to Eurasia.

V. catenata Pennell CHAIN SPEEDWELL Per, rhizomed, glabrous. **ST** ascending to erect, rooting at lower nodes, branched, 10–60 (100) cm. **LF** 25–90 mm, lanceolate, clasping, acute, ± entire, dark green, sessile. **INFL** axillary; fls 15–25; bracts linear-lanceolate; pedicels 3–7 mm, ± straight, spreading. **FL**: sepals ± 3 mm, lanceolate to ovate, acute to obtuse; corolla 4–5 mm, pale pink; style < 2 mm. **FR** 3–4 mm, obcordate to rounded; notch < 0.3 mm. **SEED** 0.5 mm, flat. $2n=36$. Wet meadows, slow streams; < 2500 m. CA-FP; widespread US, adjacent Can; native to Eur. [*V. comosa* Richter] Sterile hybrids with *V. anagallis-aquatica* have been found in mixed populations.

V. chamaedrys L. Per, rhizomed, ± hairy. **ST** decumbent to ascending, rooting at lower nodes, 7–40 cm. **LF** 10–30 mm, ovate, crenate, obtuse, sessile (exc lower ± short-petioled). **INFLS** axillary, opposite (lower often ± alternate); bracts 3–7 mm; pedicels 6–7 mm. **FL**: sepals 4–6 mm, ± equal, linear to lanceolate; corolla ± 10 mm, bright blue, center white; style 4–5 mm. **FR** < calyx, wider than long, obcordate, ciliate. $2n=16,32$. Lawns; < 100 m. SnFrB; native to Eur.

V. copelandii Eastw. (p. 1069) COPELAND'S SPEEDWELL Per, rhizomed, shaggy-hairy, ± glandular. **ST** ascending, branched, 5–12 cm. **LF** 5–35 mm, oblong to elliptic, acute to obtuse, entire, sessile. **INFL** terminal; pedicels 6–8 mm. **FL**: sepals gen 5, 2–3 mm, unequal, elliptic; corolla 8–10 mm, pale blue to lavender; style ± 7 mm, > fr. **FR** longer than wide, barely notched. UNCOMMON. Subalpine meadows, slopes; ± 2500 m. e KR (Trinity, Siskiyou cos.). ❀IRR,DRN:1–7,15–18;DFCLT.

V. cusickii A. Gray (p. 1069) CUSICK'S SPEEDWELL Per, rhizomed. **ST** ascending, branched, 10–15(20) cm, glandular-hairy. **LF** 5–25 mm, elliptic to ovate, acute to obtuse, entire, gen glabrous, sessile. **INFL** terminal, glandular-hairy; pedicels 3–9 mm. **FL**: sepals 2–3 mm, ± unequal, lanceolate; corolla 10–13 mm, deep blue; style 6–9 mm, > fr. **FR** 5–6 mm, longer than wide, deeply notched, glandular-hairy. UNCOMMON. Forest openings, meadows, < 3000 m. CaR, n&c SNH; to WA. ❀IRR,DRN,SUN:1–7,15–18; DFCLT.

V. filiformis Sm. Per, rhizomed, mat-forming, hairy. **ST** ± prostrate, freely rooting at nodes, < 50 cm. **LF** 5–10 mm (reduced upward), reniform, petioled. **INFL**: fls solitary in axils; pedicels < 4 cm, > subtending lf, reflexed in fr. **FL**: sepals 3–5 mm, oblanceolate to elliptic, obtuse; corolla 10–15 mm, pale lavender-blue; style ± 3 mm. **FR** (rarely produced) 3–4 mm; keeled midline long-hairy; lobes ± parallel. **SEED** ± 1.5 mm, ± cup-shaped. Lawns; < 100 m. SnFrB; native to Asia Minor.

V. hederifolia L. Ann, taprooted, ± hairy. **ST** ± prostrate, branched at base, 10–60 cm. **LF** 5–15 mm, ovate to round, 3–5-lobed, ± truncate at base, obtuse, petioled. **INFL**: fls solitary in axils; pedicels gen < subtending lf. **FL**: sepals 4–5 mm, ovate, truncate at base, acute, ciliate; corolla, 4–9 mm, pale bluish; style 0.6–0.9 mm. **FR** ± 4 mm, ± spheric, glabrous; notch < 0.3 mm. **SEEDS** 1–2 per chamber, < 3 mm, cup-shaped. Lawns; < 100 m. ScV; widely naturalized in ne US; native to Eur.

V. peregrina L. ssp. ***xalapensis*** (Kunth) Pennell (p. 1069) PURSLANE SPEEDWELL Ann, taprooted, gen glandular-hairy. **ST** erect, gen branched, 5–30 cm. **LF** 5–25 mm, linear to spoon-shaped, entire to ± serrate; lower ± petioled. **INFL** terminal, open; bracts lanceolate, > pedicels; pedicels 0.5–2 mm. **FL**: sepals 3–6 mm, ± equal, lanceolate; corolla 2–3 mm, whitish; style 0.1–0.4 mm. **FR** 3–4 mm, obovate; notch 0.2–0.5 mm. **SEED** 0.5 mm, flat. 2*n*=52. Moist places; < 3100 m. CA-FP; to w Can, Mex, S.Am. Glabrous ssp. *p.* is widespread in e N.Am, Eur.

V. persica Poiret PERSIAN SPEEDWELL Ann, taprooted, hairy. **ST** prostrate, simple or branched, 5–60 cm. **LF** 5–25 mm, ovate, truncate, dentate to crenate, acute to obtuse, short-petioled. **INFL** terminal, open; pedicels 15–30 mm, recurved in fr. **FL**: sepals 4–7 mm, lanceolate to ovate; corolla 8–12 mm, blue, purple-lined, center white; style 2–3 mm. **FR** much wider than long; lobes spreading; notch 0.7–1.2 mm. **SEEDS** 5–11 per chamber, ± 2 mm, cup-shaped, rough. 2*n*=28. Lawns, fields; < 1100 m. CA-FP; N.Am; native to Asia Minor.

V. scutellata L. (p. 1069) MARSH SPEEDWELL Per, rhizomed, ± glabrous. **ST** decumbent to erect, slender, 10–60 cm, gen glabrous. **LF** sessile, 20–40 mm, linear-lanceolate, ± entire, often purple-tinged. **INFLS** alternate, open; fls 5–20; bracts < 1/2 pedicels, linear; pedicels 6–16 mm, spreading at right angles in fr. **FL**: sepals ± 3 mm, ± equal, ovate; corolla 5–7 mm, white or blue, purple-lined. **FR** 3–4 mm, longer than wide; notch 0.4–1 mm. **SEED** ± 1.5 mm, flat. 2*n*=18. Wet meadows, ponds; < 2200 m. CaR, SN, SnBr; to B.C., e US; Eur.

V. serpyllifolia L. ssp. ***humifusa*** (Dickson) Syme (p. 1069) Per, rhizomed, ± hairy. **ST** erect at tips only, 5–30 cm. **LF** 10–25 mm, elliptic to widely ovate, obtuse, entire to crenate. **INFL** terminal, glandular-hairy; pedicels 2.5–7 mm. **FL**: sepals 2.5–4 mm, ± equal, oblong to ovate; corolla 6–7 mm, bright blue; style 2–3 mm. **FR** 2.8–3.7 mm, wider than long, ± glandular-hairy; notch 0.3–0.8 mm. **SEED** 0.5 mm, flat. 2*n*=14,28. Moist streambanks, lakeshores, meadows; 1700–3000 m. CaR, SN, SnBr; to AK, ne US, NM; also in S.Am, Eurasia. [var. *h.* Dickson] Ssp. *s.* less hairy, pedicels shorter, fls and fr smaller; uncommon lawn weed in w CA-FP; native to Eur.

V. triphyllos L. Ann, taprooted, glandular-hairy. **ST** erect, simple or branched from base, 5–20 cm. **LF** 4–12 mm, sessile; lobes 3–7, palmate, spoon-shaped to oblong, obtuse; lowest lvs ovate, crenate. **INFL** terminal open; lower bracts like lvs, upper oblanceolate, < or = pedicels. **FL**: sepals 4–5 mm, ovate; corolla 3–4 mm, deep blue; style ± 1.5 mm. **FR** 5–7 mm, obcordate. **SEEDS** 6–12 per chamber, cup-shaped. Uncommon. 2*n*=14. Gravelly flats; < 1000 m. KR; native to Eur.

V. wormskjoldii Roemer & Schultes (p. 1069) AMERICAN ALPINE SPEEDWELL Per, rhizomed, long-wavy-hairy. **ST** decumbent to erect, rooting at nodes, gen simple, 10–25 (40) cm. **LF** 20–40 mm, lanceolate to elliptic, acute to obtuse, entire to crenate, sessile. **INFL** terminal, dense to interrupted, glandular- to sticky-hairy; bracts linear to lanceolate; pedicels 2–6 mm. **FL**: sepals 3.5–5.5 mm, lanceolate; corolla 6–10 mm, deep blue; style 0.8–1.3 mm. **FR** 4.5–6.5 mm, longer than wide; notch 0.1–0.5 mm. **SEED** ± 1 mm, flat. 2*n*=18,36. Moist alpine meadows, streambanks, lakeshores; 2600–3500 m. CaR, SN, MP; to AK, ne US, n Mex. [*V. alpina* L. var. *alterniflora* Fern.] ❀TRY.

SIMAROUBACEAE QUASSIA or SIMAROUBA FAMILY

Elizabeth McClintock

Shrub, tree, gen dioecious. **ST** often thorny; bark often bitter. **LVS** gen alternate, simple, entire, or pinnately compound with subentire lflets. **INFL**: panicle, raceme, or fls solitary. **FL** inconspicuous; sepals gen 5, gen fused at base, gen erect; petals gen 5, free, gen spreading; stamens gen 10–15, gen inserted on a disk, filaments often with a basal scale; pistils gen 2–5, ovaries superior, 1-ovuled (if pistil 1, chambers gen 2–5, 1-ovuled), styles free or partly fused. **FR**: winged achene, drupe, berry, or capsule. ± 25 genera, ± 150 spp.: trop, warm temp; some cult. [Brizicky 1962 J Arnold Arbor 43:173–186] Bark, lvs used in medicine.

1. Tree, unarmed; lf large, pinnately compound, lflets with 2–4 teeth near base; fr a winged achene .... AILANTHUS
1′ Shrub or small tree, thorny; gen appearing ± lfless (lf scale-like, simple, entire); fr a drupe .... CASTELA

AILANTHUS

Tree, unarmed, ± dioecious, with a few bisexual fls. **LVS** deciduous, gen ± odd-pinnate, ill-smelling when crushed. **INFL**: large panicle, terminal. **FL**: calyx lobes 5–6; petals 5–6; stamens 10–12; ovaries very compressed, adherent near middle, styles ± free but twisted around one another. **FR**: winged achene, 2–5 per fl, ± pendent. ± 15 spp.: e Asia, ne Australia. (Moluccan: sky tree)

A. altissima (Miller) Swingle (p. 1069) TREE OF HEAVEN Pl < 20 m, rapidly growing; young parts ± glandular-puberulent. **LF** 3–9 dm; lflets 13–25, 8–13 cm, lanceolate, base gen ± truncate, with 2–4 teeth, each with a large gland on lower surface. **INFL** 10–20 cm. **FR** < 5 cm, linear or oblong, seed near middle. Common. Disturbed urban areas, waste places; < 1250 m. CA-FP; native to e Asia. Cult as street tree, spreading by seeds, invasive roots. Common near old Chinese habitations, esp SNF.

CASTELA

Shrub or small tree, appearing ± lfless, dioecious; thorns large, branched. **LF** simple, sometimes scale-like. **FL**: calyx lobes 4–8; petals 4–8; stamens 8–24; ovaries 4–8, adherent near middle, style bases fused, tips spreading. **FR**: drupe, dry, 4–8 per fl, spreading. ± 15 spp.: sw & s-c US, to S.Am. (René R.L. Castel, French botanist, poet, editor, opera librettist, 1759–1832) [Moran & Felger 1968 Trans San Diego Soc Nat Hist 15:31–40]

C. emoryi (A. Gray) Moran & Felger (p. 1075) CRUCIFIXION THORN Pl gen < 1 m, intricately branched; young parts densely puberulent. **LF** scale-like, entire, ephemeral and rarely seen. **INFL**: panicle, much branched, 2.5–5 cm, stiff. **FL** 6–8 mm diam. **FR** ± 6 mm, flat-topped; base ± rounded, sometimes persisting several years. **SEED** 1. RARE in CA. Dry, gravelly washes, slopes, plains; ± 650 m. D; AZ, nw Mex. [*Holacantha e.* A. Gray] Two other desert pls have same common name.

SIMMONDSIACEAE JOJOBA FAMILY

William J. Stone

Shrub, evergreen, dioecious, much branched, unusual secondary growth. **ST**: bark smooth. **LVS** opposite, simple, small, leathery; base jointed; stipules 0. **INFL**: staminate fls in axillary clusters; pistillate fls gen solitary. **FL**: small, radial; sepals gen 5, overlapping, becoming larger in female, disk 0; corolla 0; stamens 8–12, free, anthers elongate with longitudinal slits; ovary superior, chambers 3, styles 3, stigmas long, feathery. **FR**: capsule, loculicidal. **SEED** 1. 1 genus: sw U.S., Mex. Sometimes placed in Buxaceae.

SIMMONDSIA JOJOBA, GOAT-NUT, PIG-NUT

1 sp. (F.W. Simmonds, English botanist, died exploring Trinidad in 1804)

S. chinensis (Link) C. Schneider (p. 1075) **ST** 1–2 m; young growth ± hairy; branches stiff. **LF** 2–4 cm, oblong-ovate, dull green, ± canescent-puberulent, subsessile. **INFL**: peduncles 3–10 mm. **FL**: sepals in staminate fls 3–4 mm, greenish, in pistillate fl becoming 10–20 mm. **FR** < 2.5 mm, nut-like, ovoid, tough, leathery, obtusely 3-angled. **SEED** large, contains liquid wax. $2n$=26. Locally common. Arid areas; < 1500 m. w PR, SnJt, DSon; n Mex. [*S. californica* Nutt.] Important as forage pl, seed wax is source of substitute for sperm oil, fr edible. ❀DRN:**8**,**9**,10,11,**12**,13,**19–21**&SUN:**14**,18,**22–24**&DRY:7,15–17;CVS.

SOLANACEAE NIGHTSHADE FAMILY

Michael Nee

Ann to shrub. **LVS** gen simple, gen alternate, gen petioled; stipules 0; blade entire to deeply lobed. **INFL** various. **FL** bisexual; calyx lobes gen 5; corolla ± radial, cylindric to rotate, lobes gen 5; stamens 5, alternate corolla lobes; ovary superior, gen 2-chambered, style 1. **FR**: berry or capsule, 2–5-chambered. 75 genera, 3000 spp.: worldwide, esp ± trop; many alien weeds in CA; many cult for food, drugs, or orn (potato, tomato, peppers, tobacco, petunia); many TOXIC.

1. Fr a capsule
  2. Fr wall prickly ..... **DATURA**
  2′ Fr wall not prickly
    3. Calyx lobes > tube, wing-like at base in fr ..... **NICANDRA**
    3′ Calyx lobes < or, if > tube, not wing-like at base in fr
      4. Corolla narrowly urn-shaped; seeds flat, narrowly winged ..... **ORYCTES**
      4′ Corolla funnel-shaped; seeds angled but not flat, not winged
        5. Fls in racemes or panicles (only lower bracts lf-like) ..... **NICOTIANA**
        5′ Fls solitary in axils of lvs or lf-like bracts ..... **PETUNIA**

1′ Fr a berry
  6. Corolla salverform to funnel-shaped (urn-shaped in *Salpichroa*)
    7. Shrub, thorny ..... **LYCIUM**
    7′ Per or shrub, unarmed
      8. Erect shrub; corolla narrowly funnel-shaped, red ..... **CESTRUM**
      8′ Decumbent or climbing per; corolla urn-shaped, white or greenish ..... **SALPICHROA**
  6′ Corolla ± rotate to shallowly bell-shaped
    9. Anthers opening by pores or short slits near tip ..... **SOLANUM**
    9′ Anthers opening by slits from tip to near base
      10. Anthers gen > filaments
        11. Anthers adherent; lvs ± odd-1–2-pinnate ..... **LYCOPERSICON**
        11′ Anthers free; lvs entire to deeply lobed ..... **SOLANUM**
      10′ Anthers gen < filaments
        12. Calyx in fr not bladder-like; corolla tube tomentose between filament bases ..... **CHAMAESARACHA**
        12′ Calyx in fr bladder-like; corolla tube hairy but not tomentose between filament bases ..... **PHYSALIS**

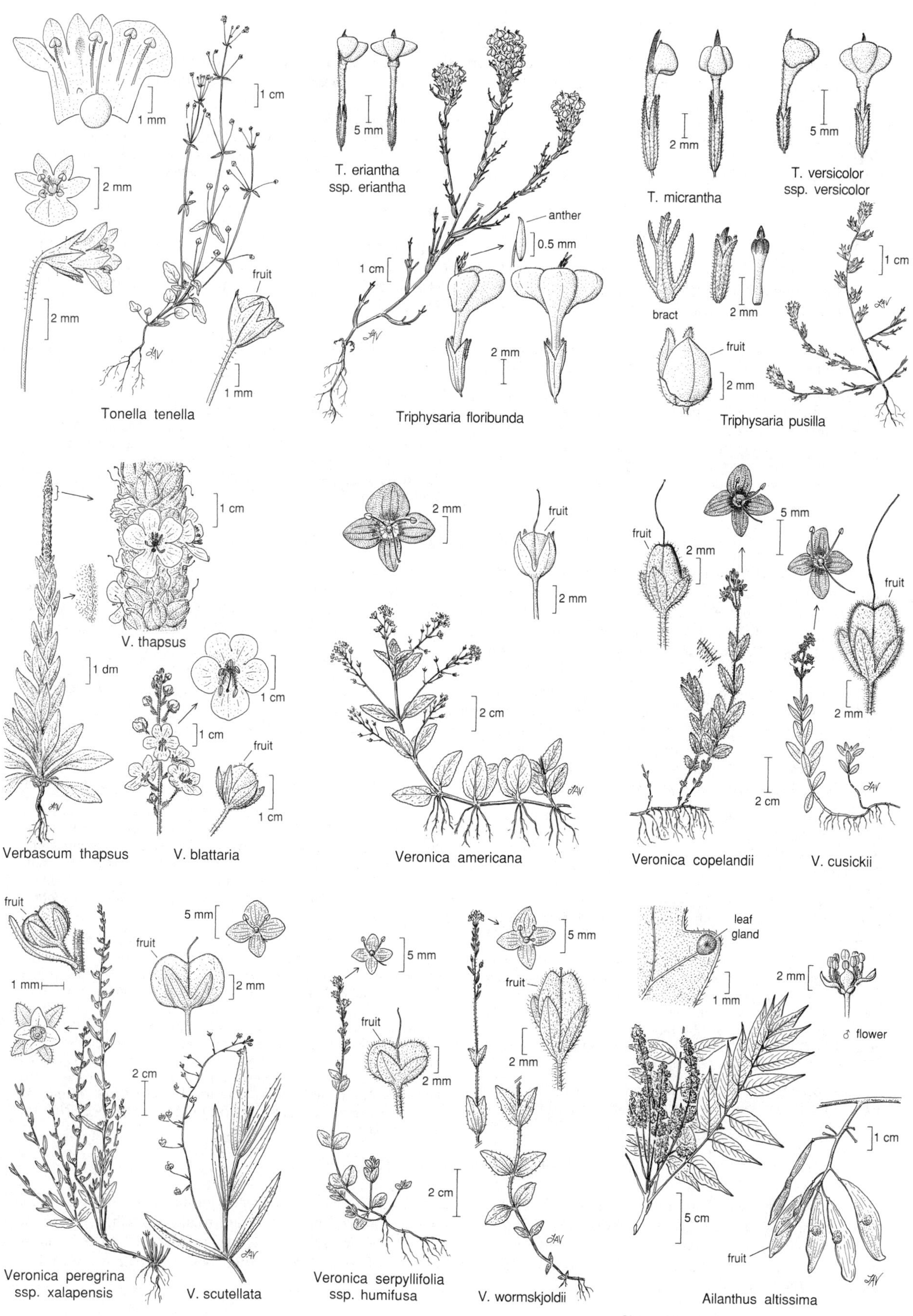

Simaroubaceae

CESTRUM

Shrub; hairs simple or branched. **LF** entire, glabrous to hairy. **INFL**: panicle. **FL**: corolla ± salverform; stamens incl, filaments sometimes unequal. **FR**: berry, sometimes ± dry. 250 spp.: trop Am. (Greek: derivation unknown)

C. fasciculatum (Schldl.) Miers Shrub < 3 m; hairs soft, ± dense, esp on axes. **ST** often purple. **LF** 4–13 cm, ± ovate, acuminate. **INFL** < 10 cm, ± head-like, ± dense, 5–10-fld. **FL**: calyx 8–9 mm; corolla 15–25 mm, purplish red, lobes 2–2.5 mm. **FR** 15 mm, red, white-spongy inside. **SEEDS** ± 10, 2.5–4 mm, dark brown. Disturbed areas; < 1000 m. n NCo, SnFrB, expected elsewhere; native to Mex. Other spp. (esp *C. aurantiacum* Lindley with yellow-orange fls) widely cult, may become naturalized.

CHAMAESARACHA

Per; hairs ± scale-like. **ST** decumbent, branched. **LF** entire to ± deeply pinnately lobed. **INFL**: cluster, axillary, 1–5-fld. **FL**: calyx in fr ± enlarged, open at top; corolla rotate, tomentose near base between stamens; anthers free, gen < filaments, opening by slits; style 1. **FR**: berry, spheric, partly enclosed by calyx. **SEEDS** ± flat, reniform. ± 9 spp.: esp sw US, Mex. (Greek: low *Saracha*, a S.Am genus in family) [Averett 1973 Rhodora 75:325–365]

1. Lf ± linear to lanceolate, ± entire to ± deeply lobed; corolla 10–15 mm wide; e DMoj .... ***C. coronopus***
1′ Lf ovate, entire; corolla 15–25 mm wide; CaR, SNH, GB .... ***C. nana***

C. coronopus (Dunal) A. Gray (p. 1075) **STS** many from base, 10–50 cm. **LF** 20–65 mm, 1–10 mm wide. **INFL**: pedicels ± 1 cm, in fr < 2 cm, reflexed. **FL**: calyx 3–5 mm, in fr 5–10 mm, lobes in fr 2 mm; corolla dirty or greenish white; filaments 3 mm. **FR** 5–8 mm wide. **SEEDS** 3 mm. *n*=12,24,36. Dry, clay soil; ± 1500 m. e DMtns (New York Mtns); to UT, TX, n Mex.

C. nana (A. Gray) A. Gray (p. 1075) **STS** 1–several from base, 5–25 cm. **LF** 15–50 mm. **INFL**: pedicels 8–18 mm, in fr < 3 cm, recurved. **FL**: calyx ± 5 mm, in fr < 10 mm, lobes in fr 2–3 mm; corolla whitish; filaments 5 mm. **FR** 1–1.2 cm wide. **SEEDS** 1.5–2 mm. *n*=12. Sandy soils, slopes, coniferous forest; 1500–2800 m. CaR, SNH, GB; OR, NV. [*Leucophysalis nana* (A. Gray) Averett] ❀DRN,DRY:1–3,7,14,18&SUN:15–17;GRNCVR.

DATURA JIMSON WEED, THORN-APPLE

Ann to subshrub, ± glabrous or hairs simple, ill-smelling. **LF** entire to deeply lobed. **INFL**: fls solitary in branch forks. **FL**: calyx circumscissile near base, leaving a ± rotate collar in fr; corolla funnel-shaped, white or purplish, lobes 5(10); filaments attached below middle of corolla tube; ovary 2- or 4-chambered. **FR**: capsule, leathery or ± woody, gen prickled; valves 2–4, or indefinite. **SEEDS** ± flat, black, brown, grayish brown, or tan. ± 13 spp.: warm regions, esp Mex. (Hindu: ancient name) All spp. HIGHLY TOXIC; several orn, some source of drugs.

1. Corolla 15–20 cm; calyx 8–12 cm, ribbed; seeds tan .... ***D. wrightii***
1′ Corolla 4.5–16 cm; calyx 2–9 cm, angled or winged; seeds black or grayish brown
 2. Fr nodding, prickles many, weak; st grayish hairy .... ***D. discolor***
 2′ Fr erect, prickles few, strong; st glabrous to sparsely hairy
  3. Fr prickles 8–22 mm; calyx 20–25 mm .... ***D. ferox***
  3′ Fr prickles 3–10 mm; calyx 35–45 mm .... ***D. stramonium***

D. discolor Bernh. Ann < 5 dm. **ST** grayish hairy. **LF** 6–12 cm, 4–10 cm wide, widely ovate, coarsely toothed. **FL** erect; calyx 5–9 cm, 5-winged toward base, lobes 1–1.5 cm; corolla 10–16 cm, glabrous, white with purple markings in tube, lobes shallow; filaments 4.5 cm, anthers 5–6 mm; style 10–14 cm. **FR** 4-valved, nodding, 35 mm wide, puberulent; prickles < 2 cm. **SEEDS** 3–3.5 cm, coarsely wrinkled, black with a white outgrowth near attachment scar. *n*=12. Sandy, gravelly soils, washes; < 500 m. DSon; Mex.

D. ferox L. Ann < 6 dm. **ST** ± glabrous. **LF** 10–20 cm, gen deeply wavy-lobed. **FL** erect; calyx 2–2.5 cm, lobes 2–4 mm; corolla 4.5–5 cm, glabrous, white or pale blue-purple, lobes 2–4 mm, tips short, narrow; filaments 2 cm, anthers 3 mm; style 2.5 cm. **FR** 25 mm wide; prickles few, 8–22 mm, very stout. *n*=12. Disturbed areas; < 200 m. n SNF (Amador Co.), n SnJV; native to Mex, s S.Am. Hybridizes with *D. stramonium*.

D. stramonium L. JIMSON WEED Ann < 10 dm. **ST** ± glabrous. **LF** 5–15 cm, 4–10 cm wide, ovate, coarsely toothed to shallowly lobed. **FL** erect; calyx 3.5–4.5 cm, angled toward base, lobes 5–7 mm; corolla 6–9 cm, glabrous, white or pale blue-purple, lobes 8–10 mm, spreading, tips long, narrow; filaments 22–25 mm; anthers 3.5–5 mm; style 4–6 cm. **FR** 4-valved, erect, 25–35 mm wide, glabrous; prickles 3–10 mm, upper > lower. **SEEDS** 3–3.5 mm, black. *n*=12. Sandy soils, open, often disturbed areas; < 1500 m. CA-FP (exc KR, CaR, SNH); to e US; native to Mex. [var. *tatula* (L.) Torrey]

D. wrightii Regel (p. 1075) Ann or per 5–15 dm. **ST** whitish puberulent. **LF** 7–20 cm, ovate, entire or coarsely lobed. **FL** erect to nodding; calyx 8–12 cm, ribbed toward base, lobes ± 2 cm; corolla 15–20 cm, puberulent, white, lobes 1–2 cm, tips long, narrow; filaments 13–15 cm, anthers 12–15 mm; style 15–18 cm. **FR** irregularly valved, nodding, 25–30 mm wide, puberulent; prickles 5–12 mm. **SEEDS** 5 mm, flat, tan; margin grooved. *n*=12. Sandy or gravelly open areas; < 2200 m. NCoRI, c&s SNF, Teh, GV, CW, SW; to UT, TX, Mex. Sometimes cult for showy fls; may have been introduced by early Spanish; may be the same as *D. inoxia* J.S. Miller [*D. meteloides* A.DC.], native to Mex. ❀DRN,SUN:**7–10, 14–24**&IRR:11–13;occas.INV.

LYCIUM BOX THORN

Shrub 1–4 m, gen with leafy thorns, glabrous, hairy, or glandular. **LVS** alternate or clustered, entire, small, fleshy. **INFL**: clusters; fls 1–several. **FL**: calyx cylindric to bell-shaped, lobes 2–5; corolla funnel-shaped or rotate, whitish, greenish, or purplish, lobes 4–5; stamens attached at different levels in corolla. **FR**: berry, 2-chambered, dry to fleshy. **SEEDS** 2–many. ± 100 spp.: warm, dry areas worldwide. (Latin: Lycia, ancient country of Asia Minor) [Hitchcock 1932 Ann Missouri Bot Gard 19:179–374]

1. Seeds 2; lf ± round in X-section ..... ***L. californicum***
1′ Seeds several–many; lf ± flat to elliptic in X-section
 2. Calyx lobes 1/2 to > tube (or gen > 2 mm)
  3. Fr firm, greenish purple or yellow to orange
   4. Herbage glandular-puberulent; fr yellow to orange, with 2 cross-grooves above middle ..... ***L. cooperi***
   4′ St glabrous, lvs glaucous; fr greenish purple, not cross-grooved ..... ***L. pallidum*** var. ***oligospermum***
  3′ Fr soft, red
   5. Filaments hairy at base, attached near middle of corolla tube; s ChI, w DSon ..... ***L. brevipes*** var. ***hassei***
   5′ Filaments glabrous, attached near top of corolla tube; s ChI ..... ***L. verrucosum***
 2′ Calyx lobes < 2/3 tube (or gen < 2 mm)
  6. Central sts curved to arched, often unarmed; alien of waste places ..... ***L. barbarum***
  6′ Sts gen stiffly spreading to ascending, gen thorny; natives
   7. Corolla lobes 1/3 to = tube ..... ***L. brevipes*** var. ***brevipes***
   7′ Corolla lobes < 1/3 tube
    8. Pl glandular
     9. Calyx 4–8 mm, lobes triangular, 1–2 mm ..... ***L. fremontii***
     9′ Calyx 2.5–5 mm, lobes oblong-ovate, 2–4 mm ..... ***L. parishii***
    8′ Pl ± glabrous
     10. Corolla lobe margin glabrous to finely straight-ciliate; lf ± linear-oblanceolate ..... ***L. andersonii***
     10′ Corolla lobe margin woolly-ciliate; lf narrowly oblanceolate to obovate ..... ***L. torreyi***

L. andersonii A. Gray (p. 1075) Pl ± glabrous; branches stiffly spreading to erect. **LF** 3–15 mm, ± linear, ± elliptic in X-section. **FL**: calyx 1.5–3 mm, cup-shaped, lobes (2)4–5, ± 0.8 mm; corolla narrowly funnel-shaped, whitish, often violet-tinged, tube 5–10 mm, lobes 1–1.5 mm; stamens ± incl to slightly exserted, unequal, attached 1/3 from base. **FR** 3–8 mm, red or orange, juicy. **SEEDS** many. Gravelly or rocky slopes, washes; < 1900 m. s Teh, s SnJV, SCo, n WTR, PR, SNE, D; to UT, NM, nw Mex. [var. *deserticola* (C. Hitchc.) Jepson] ❀STBL.

L. barbarum L. MATRIMONY VINE Pl glabrous; thorns 0 or short; branches curved or arched. **LF** 2–6 cm, (ob)lanceolate. **FL**: calyx 3–4 mm, bell-shaped, lobes gen 2–3, 1–2 mm; corolla funnel-shaped, lavender to purple, fading tan, tube 4–6 mm, ± abruptly expanded to throat, lobes spreading, < tube; stamens ± equal, attached at middle of tube, hair-tufted at base. **FR** 1 cm, fleshy, bright red. **SEEDS** 10–20. Waste places, fields; < 1500 m. s SN, GV, SCo, MP, expected elsewhere; formerly cult; native to Eurasia. [*L. halimifolium* J.S. Miller] Pls with corolla tube < lobes and ± hidden by calyx are *L. chinense* J.S. Miller, waif in ScV, c SNF.

L. brevipes Benth. Pl gen glandular-puberulent; branches spreading. **LF** 5–15 mm, ± obovate. **FL**: calyx 2–6 mm, bell-shaped, lobes 2–4(6); corolla funnel-shaped, tube 6–10 mm, lavender to whitish, lobes 3–5 mm, ovate; stamens exserted, attached ± at middle of tube. **FR** ± 10 mm, red. **SEEDS** many. Coastal bluffs, slopes; < 600 m. SCo, s ChI, w DSon; nw Mex. ❀SUN,DRN:7,**8, 9,14**,15,16,18,**19–24**&IRR:**12,13**;also STBL.

var. ***brevipes*** **FL**: calyx lobes 1/3 to = tube, triangular to linear. Habitats and elevations of sp. s ChI (San Clemente Island), w DSon; nw Mex. Intergrades with *L. torreyi* in w DSon.

var. ***hassei*** (E. Greene) C. Hitchc. SANTA CATALINA ISLAND DESERT-THORN **FL**: calyx lobes 1–3 × tube, narrowly obovate. **PRESUMED EXTINCT**. Habitats of sp.; < 300 m. s ChI (Santa Catalina, San Clemente islands). [*L. h.* E. Greene] Reports from SCo are cult pls.

L. californicum Nutt. (p. 1075) Pl glabrous; branches rigidly spreading. **LF** 3–10 mm, linear to oblanceolate. **FL**: calyx 3 mm, bell-shaped, lobes 2–4(5), ± 0.8 mm, widely triangular; corolla 4–6 mm, ± rotate, sometimes reflexed, white, purple-tinged, lobes = tube, ovate; stamens exserted, attached near base of tube. **FR** 3–6 mm, red, firm. **SEEDS** 2, oblong. Coastal bluffs, coastal-sage scrub; < 150 m. s SCo, ChI; Baja CA. ❀DRN,SUN:**14–17,19–24**&IRR:8,9;also STBL.

L. cooperi A. Gray (p. 1075) Pl glandular-puberulent; branches rigidly ascending to erect, leafy. **LF** 1–3 cm, oblanceolate to obovate. **FL**: calyx 8–15 mm, narrowly bell-shaped, lobes 4–5, 1.5–3 mm, 1/2 to = tube; corolla narrowly funnel-shaped, greenish white, lavender-veined, outside glabrous or puberulent, tube 9–12 mm, lobes ovate-triangular; stamens barely exserted, attached ± at middle of tube. **FR** 5–9 mm, yellow to orange, with 2 cross-grooves above middle. **SEEDS** several. Sandy to rocky flats, washes; < 2000 m. SNH (e slope), s SnJV, SNE, DMoj (incl w margins); to UT, AZ. ❀STBL.

L. fremontii A. Gray (p. 1075) Pl glandular-hairy; branches spreading to ascending. **LF** 10–25 mm, narrowly obovate. **FL**: calyx 4–8 mm, cylindric, lobes 5, 1–2 mm, < tube, triangular; corolla 10–15 mm, narrowly funnel-shaped, violet or whitish, purple-veined, lobes < 1/3 tube; stamens mostly incl, unequal, attached in lower 1/3 of tube. **FR** 6–8 mm, red, juicy. **SEEDS** many. Alkaline soils, flats; < 500 m. PR (e slope), s DSon; s AZ, nw Mex. ❀DRN,SUN:**14**,15,16,18,**19–24**&IRR:**8,9**,10,11,**12,13**;also STBL.

L. pallidum Miers var. ***oligospermum*** C. Hitchc. (p. 1075) Pl very thorny, glabrous; branches many, spreading to ascending. **LF** 10–50 mm, oblong to narrowly obovate, glaucous. **FL**: calyx 5–8 mm, bell-shaped, lobes 5, ± = tube; corolla narrowly funnel-shaped, greenish white, veins purple, tube 8–12 mm; stamens exserted, attached at middle of tube. **FR** 8–10 mm, fleshy, firm, greenish purple. **SEEDS** 5–7. Flats, washes, slopes; < 1200 m. DMoj, n DSon; s NV. Var. *p.* (corolla tube 12–18 mm, seeds 15+) expected in e DMoj. ❀TRY.

L. parishii A. Gray PARISH'S DESERT-THORN Pl glandular-hairy; intricately spreading-branched. **LF** 5–30 mm, oblanceolate. **FL**: calyx 2.5–5 mm, bell-shaped, lobes 5, 2–4 mm, ± = tube, oblong-ovate; corolla narrowly funnel-shaped, purple, tube 2.5–6 mm, lobes ± 1 mm, ovate; stamens slightly exserted, attached ± at middle of tube. **FR** 4–6 mm, red. **SEEDS** 7–12. RARE in CA. Sandy to rocky slopes, canyons; < 1000 m. e SCo, w DSon; AZ, nw Mex.

L. torreyi A. Gray Pl nearly glabrous (exc lf clusters hair-tufted); twigs slender, ± climbing. **LF** 1–5 cm, ± oblanceolate. **FL**: calyx 2.5–4.5 mm, lobes 5, 1–1.5 mm; corolla 8–15 mm, narrowly funnel-shaped, greenish lavender or whitish, lobes 3–4 mm, lanceolate to ovate; stamens slightly exserted, attached at middle of tube. **FR** 6–10 mm, red or orange, juicy. **SEEDS** 8–30. Washes, streambanks; < 700 m. s DMoj, DSon; to UT, TX, n Mex. ❀TRY.

L. verrucosum Eastw. SAN NICOLAS ISLAND DESERT-THORN Pl hairy; branches spreading; thorns thick. **LF** 5–10 mm, narrowly obovate. **FL**: calyx 4 mm, bell-shaped, lobes ± 3–4, 2–3.5 mm, ± = tube, lanceolate; corolla 8–10 mm, lavender, tube cylindric, lobes narrowly obovate; stamens barely exserted, attached at top of corolla tube. **FR** reddish. **SEEDS** many. **PRESUMED EXTINCT**. Habitat unknown. s ChI (San Nicolas Island). Known only from type specimen; perhaps a form of *L. brevipes*.

LYCOPERSICON TOMATO

Ann or per, sticky-glandular, aromatic. **LF** irregularly odd-1–2-pinnate. **INFL**: raceme; pedicels jointed. **FL** often nodding; parts gen 5; calyx divided nearly to base; corolla reflexed, yellow; anthers > filaments, cone-like around style, tapered, sterile at tips; stigma exserted from anther cone. **FR**: berry, green to red, juicy. **SEEDS**: coat gelatinous. ± 6 spp.: w S.Am., C.Am. (Greek: wolf peach, from supposed toxic properties) [Muller 1940 USDA Misc Publ 382]

1. Herbage green, hairs << 3.5 mm; lflets 3–10 cm, 10–45 mm wide; fr glabrous .... ***L. esculentum***
1′ Herbage gray-green, hairs < 0.5 mm; lflets mostly ± 3.5 cm, ± 10 mm wide; fr puberulent .... ***L. peruvianum***

L. esculentum L. TOMATO Pl erect or reclining, often fleshy. **LF** 10–20 cm; lflets 3–10 cm, 10–45 mm wide. **INFL** few-fld; peduncle 1.5–4 cm; bracts gen 0. **FR** 3–12 cm wide, spheric, compressed, or pear-shaped, yellow-green to red. Waste areas, abandoned fields, roadsides; < 100 m. GV, SnFrB, SCo; native to S.Am.

L. peruvianum (L.) Miller Ann or short-lived per from thick taproot, spreading to reclining. **LF** 5–12 cm; lflets ± 3.5 cm, ± 10 mm wide, often folded or very narrow. **INFL** few–many-fld; peduncle 3–15 cm; bracts lf-like, simple. **FR** 1.5 cm wide, spheric, green. Waste areas, disturbed sites; < 100 m. s SCo; native to S.Am.

NICANDRA

1 sp. (Nicander, botanical poet, Asia Minor, 2nd–1st centuries B.C.)

N. physalodes (L.) Gaertner Ann < 2 m. **ST** strongly 5-ribbed, hollow. **LF** 8–30 cm; blade toothed to lobed. **INFL**: fls solitary in axils. **FL**: calyx lobes wing-like at base, enlarging and ± enclosing fr; corolla 15–30 mm, widely bell-shaped, shallowly lobed, pale blue, tube base with ring of hairs; filaments 4–5 mm; ovary on orange disc, 3–5-chambered, style 5–7 mm. **FR:** capsule 1–2 cm, spheric. **SEEDS** many, brown. Fields; < 200 m. SCo; native to S.Am. Probable waif in CA.

NICOTIANA TOBACCO

Ann to small tree. **LVS** entire. **INFL**: raceme or panicle, terminal. **FL**: calyx 5-lobed, ± enlarging, not fully enclosing fr; corolla gen radial, gen funnel-shaped to salverform; stamens 5, equal or 1 smaller. **FR**: capsule. **SEEDS** many, minute, angled. ± 60 spp.: gen Am; *N. tabacum* L. widely cult. (J. Nicot, 1530–1600, said to have introduced tobacco to Eur) [Goodspeed 1954 Chron Bot 16:1–536] Seriously TOXIC to livestock.

1. Shrub or small tree; herbage glabrous, glaucous .... ***N. glauca***
1′ Ann or per; herbage gen glandular-hairy
  2. Corolla 50–75 mm; cauline lvs gen elliptic; basal lvs tapered to winged petiole .... ***N. sylvestris***
  2′ Corolla 15–50 mm; cauline lvs gen lanceolate; basal lvs petioled
    3. Cauline lvs clasping; per, base often slightly woody; corolla open during day .... ***N. obtusifolia***
    3′ Cauline lvs not clasping; ann; corolla closed during day
      4. Cauline lvs gen ± sessile (exc lowest)
        5. Corolla tube 15–20 mm, limb 8–10 mm wide; filaments attached ± at 1 level below middle of corolla tube; calyx lobes very unequal .... ***N. clevelandii***
        5′ Corolla tube 25–50 mm, limb 20–50 mm wide; filaments attached at various levels above middle of corolla tube; calyx lobes ± unequal .... ***N. quadrivalvis***
      4′ Cauline lvs petioled
        6. Calyx dark-striped, lobes narrowly lanceolate, unequal, gen > tube, not pock-marked; pl densely glandular .... ***N. acuminata*** var. ***multiflora***
        6′ Calyx not striped, lobes triangular, ± equal, < tube, pock-marked; pl glabrous to ± sparsely glandular .... ***N. attenuata***

N. acuminata Hook. var. *multiflora* (Philippi) Reiche Ann 5–15 dm, densely glandular. **LVS** 5–25 cm, petioled; basal ovate; cauline reduced. **INFL**: bracts < 50 mm, linear. **FL**: calyx 15–20 mm, dark-striped, lobes > or ± = tube, unequal, narrowly lanceolate; corolla ± salverform, greenish white, tube 30–40 mm, limb 10–20 mm wide; stamens unequal, filaments attached near middle of tube. **FR** 10–12 mm. Open, sandy or gravelly areas; < 1600 m. CA-FP (exc SW), MP; to WA, NV; native to S.Am.

N. attenuata Torrey (p. 1075) Ann 5–15 dm, ± glandular. **LVS** 3–10 cm, petioled; basal elliptic to ovate; cauline gradually reduced. **INFL**: bracts < 30 mm, linear. **FL**: calyx 6–10 mm, pock-marked, lobes < tube, ± equal, triangular, acute; corolla ± salverform, greenish, pink-tinged, tube 20–27 mm, limb 4–8 mm wide, white; stamens unequal, filaments attached below middle of tube. **FR** 8–12 mm. Open, well drained slopes; 200–2800 m. CA (exc coast); to B.C., MT, NM, nw Mex. ❀TRY.

N. clevelandii A. Gray (p. 1075) Ann 2–6 dm, slender, ± glandular. **LVS** 3–18 cm; lowermost lvs petioled, elliptic to ovate; upper lvs sessile, lanceolate. **INFL**: bracts < 30 mm, linear to lanceolate. **FL**: calyx 8–10 mm, lobes 4, very unequal (1 > tube), linear; corolla ± salverform, greenish white, tube 15–20 mm, limb 8–10 mm wide; stamens unequal, filaments attached below middle of corolla tube. **FR** 4–6 mm. Gen sandy washes, shrubland; < 500 m. SCo, ChI (Santa Catalina, Santa Cruz islands), DSon; AZ, nw Mex.

N. glauca Graham TREE TOBACCO Gen tree < 600 dm, gen glabrous, gen glaucous; wood soft. **LF** 5–21 cm, petioled, gen ± ovate. **INFL**: bracts < 5 mm, linear. **FL**: calyx ± 10 mm, lobes < tube, ± unequal, triangular; corolla 30–35 mm, ± cylindric, yellow; stamens ± equal, filaments attached below middle of tube. **FR** 7–15 mm. Open, disturbed flats or slopes; < 1100 m. NCoRI, c&s SNF, GV, CW, SW, D; to s US, Mex, Afr, Medit; native to S.Am.

N. obtusifolia Martens & Galeotti (p. 1075) Per from woody base, 2–8 dm, glandular. **LVS** 2–10 cm; lower lvs short-petioled, (ob)ovate; upper lvs ± narrowly ovate, clasping. **INFL**: bracts < 20 mm, linear to lanceolate. **FL**: calyx 10–15 mm, lobes ± = tube, ± equal, narrowly triangular; corolla ± funnel-shaped, greenish or dull white, tube and throat 15–26 mm, limb 8–10 mm wide; stamens unequal, filaments attached near base of tube. **FR** 8–10 mm. Gravelly or rocky washes, slopes; < 1600 m. s SNE, D; to UT, TX, Mex. [*N. trigonophylla* Dunal]

N. quadrivalvis Pursh Ann 3–20 dm, glandular. **LVS** 4–15 cm; lower lvs short-petioled, elliptic to ovate; upper lvs sessile, reduced. **INFL**: bracts < 35 mm, ± linear. **FL**: calyx 15–20 mm, 10-ridged, lobes narrowly lanceolate, 1 sometimes > tube; corolla ± salverform, white with green or violet tinge, tube 25–50 mm, limb 20–50 mm wide; stamens unequal, filaments attached at different levels above middle of tube. **FR** 15–20 mm. Open, well-drained washes, slopes; < 1500 m. CA-FP, uncommon in SNE and DMoj; to WA, c US. [*N. bigelovii* (Torrey) S. Watson] Pls from SCo with corolla limb > 50 mm wide have been called *N. b.* var. *wallacei* A. Gray. Widely cult by w Am native people. ❀DRN,SUN:**7–10,14–24**.

N. sylvestris Speg. & O. Comes Per 1–1.5 dm, glandular-puberulent. **LVS** 5–30 cm; lower lvs elliptic to ovate, tapered to winged petiole; upper lvs ± clasping. **INFL**: bracts < 40 mm, linear to lanceolate. **FL**: calyx 8–15 mm, lobes < tube, ± equal, triangular; corolla narrowly funnel-shaped, white to cream, tube and throat 50–75 mm, limb 15–20 mm wide; stamens unequal, filaments attached below middle of tube. **FR** 12–16 mm. Open, disturbed areas, fields; < 200 m. n CCo, SnFrB; native to S.Am. Pls with long-tapered sepal lobes ± = tube are *N. alata* Link & Otto, cult for orn, waif in fields.

ORYCTES

1 sp. (Greek: a digger)

O. nevadensis S. Watson (p. 1075) NEVADA ORYCTES Ann 5–20 cm, branched from slender taproot, leafy, sticky; some hairs scale-like. **LF** 1–3 cm, linear to ovate, entire to shallowly lobed, drying ± wavy; petiole 5–10 mm, narrowly winged. **INFL**: umbels in axils, few-fld. **FL**: calyx 2–3 mm, < 10 mm in fr; corolla 5–8 mm, narrowly urn-shaped, purplish, lobes small; stamens attached at base of tube, unequal, some slightly exserted; style = stamens. **FR**: capsule, 2-valved, 6–7 mm wide, spheric. **SEEDS** 10–15, round, flat; body 2 mm wide; wing 0.5 mm wide. RARE in CA. Sandy soils, dunes; 1200–1500 m. s SNE (Inyo Co.); w NV. Seriously threatened by grazing.

PETUNIA

Ann or per, sticky-glandular. **ST**: main branches from base, with ± long internodes. **LVS** subopposite near fls, entire. **INFL**: fls solitary in axils. **FL**: calyx divided nearly to base, lobes ± lf-like, esp in fr; corolla funnel-shaped, 5-lobed; stamens ± equal or 1 short, 2 medium, 2 long. **FR**: capsule. **SEEDS** many, minute, spheric to angled. ± 30 spp.: gen S.Am; some cult for orn, sometimes waifs. (Native Am: petun, name for tobacco)

P. parviflora A.L. Juss. (p. 1075) **ST** prostrate to decumbent, < 4 dm, rooting at nodes; axillary branches short, leafy. **LF** 5–14 mm, ± oblanceolate, fleshy, ± subsessile. **FL**: calyx lobes 3–6 mm, < 11 mm in fr, elliptic to narrowly obovate; corolla 4–6 mm, purplish, tube whitish. **FR** 2–3 mm. **SEEDS** pale brown. Open washes, dry streambeds; < 1300 m. CW, SCo, n ChI (Santa Rosa Island), PR, DSon; to sw UT, se US, Mex; also S.Am; waif elsewhere in US.

PHYSALIS GROUND-CHERRY

Ann or rhizomed per; hairs sometimes branched. **LVS** sometimes ± opposite, entire to pinnately lobed. **INFL**: fls 1–few per axil, pedicelled. **FL**: calyx 5-lobed, enlarged and persistent in fr; corolla rotate to widely bell-shaped, yellowish, often dark-spotted inside; stamens 5, filaments inserted on hairy band in corolla tube, anthers free, gen < filaments, opening by slits; style gen straight. **FR**: berry. **SEEDS** many, 2–2.5 mm, ± spheric to reniform. ± 85 spp.: Am, Eurasia, Australia. (Greek: bladder, from calyx in fr) [Sullivan 1985 Syst Bot 10:426–444] Some spp. cult for edible or orn fr. Unripe fr often TOXIC. Needs further study in w US.

1. Fl erect; corolla purple; style curved ..... ***P. lobata***
1′ Fl gen nodding; corolla yellow; style straight
  2. Lf with branched hairs; per
    3. Corolla with 5 purple-brown spots inside ..... ***P. hederifolia*** var. ***fendleri***
    3′ Corolla greenish or deep yellow inside, unspotted ..... ***P. viscosa***
  2′ Lf glabrous or with unbranched hairs; ann or per
    4. Per or subshrub from rhizomes
      5. Anthers bluish ..... ***P. longifolia***
      5′ Anthers yellow
        6. Pedicel > fl or calyx in fr ..... ***P. crassifolia***
        6′ Pedicel < fl or calyx in fr ..... ***P. hederifolia*** var. ***palmeri***
    4′ Ann from taproot
      7. Pl ± densely hairy ..... ***P. pubescens***
        8. Lf 5–9 cm, ± thick, teeth many ..... var. ***grisea***
        8′ Lf 3–5 cm, thin, teeth 0–few ..... var. ***integrifolia***
      7′ Pl glabrous to sparsely hairy
        9. Corolla with 5 dark-purple spots inside; anthers twisted after opening ..... ***P. philadelphica***
        9′ Corolla not dark-spotted inside; anthers not clearly twisted after opening
          10. Corolla pale yellow with dark yellow center, 15–23 mm wide; lf deeply toothed; anthers 3 mm ..... ***P. acutifolia***
          10′ Corolla evenly yellow throughout, 7–8 mm wide; lf entire to ± weakly toothed; anthers 1–2 mm ..... ***P. lancifolia***

P. acutifolia (Miers) Sandw. (p. 1079) Ann 2–10 dm, branched; hairs simple, short, appressed. **LF** 4–12 cm, lanceolate to ± ovate, tapered to base, teeth < 7 mm, prominent, slender. **INFL**: pedicels 15–25 mm, in fr < 40 mm. **FL**: calyx 3–4.5 mm, in fr 20–25 mm, spheric, with 10 equal veins; corolla 15–23 mm wide, rotate, pale yellow with darker yellow center; anthers 3 mm, yellow and blue-green. Waste places, roadsides; < 200 m. s SnJV, SCo, DSon; to TX, n Mex. [*P. wrightii* A. Gray]

P. crassifolia Benth. (p. 1079) Per or subshrub < 8 dm; hairs simple, dense, short, gen glandular. **ST** often zigzag, ridged. **LF** 1–3 cm, gen ovate, fleshy, entire or ± wavy; petiole ± = blade. **INFL**: pedicels 15–30 mm, in fr > fls. **FL**: calyx 4–7 mm, in fr 20–25 mm, weakly angled; corolla 15–20 mm diam, widely bell-shaped, yellow; anthers 2–3 mm, yellow. Gravelly to rocky flats, washes, slopes; < 1300 m. PR, s SNE, D; NV, AZ, n Mex. [var. *versicolor* (Rydb.) Waterf.; *P. greenei* Vasey & Rose, Greene's ground-cherry] ❀TRY;DFCLT.

P. hederifolia A. Gray Per or subshrub 1–8 dm, from fleshy rhizome; hairs sometimes branched or glandular. **LF** 2–4 cm, ovate, entire to coarsely toothed, gen gray-green; base tapered or subcordate. **INFL**: pedicels 3–5(10) mm, in fr < 15 mm. **FL**: calyx 6–7 mm, in fr 20–30 mm, ± spheric, with 10 green veins; corolla 10–15 mm, widely bell-shaped, yellow, gen with 5 purple-brown spots at base inside. Gravelly to rocky slopes; 700–1800 m. s PR, s SNE, DMtns; to TX, n Mex.

var. ***fendleri*** (A. Gray) Cronq. Hairs gen branched, nonglandular. *n*=12. Habitats of sp; 900–1800 m. s SNE (incl Inyo Mtns), DMtns; to TX, n Mex. [var. *cordifolia* (A. Gray) Waterf.; *P. f.* A. Gray]

var. ***palmeri*** (A. Gray) C. Hitchc. Hairs unbranched, many glandular. *n*=12. Habitats of sp.; 700–1600 m. s PR, DMtns; to UT, AZ, Baja CA.

P. lancifolia Nees Ann < 8 dm; hairs simple, few, minute, appressed. **LF** 3–13 cm, lanceolate, tapered to base, entire or teeth < 3 mm. **INFL**: pedicels 10–25 mm, in fr < 40 mm. **FL**: calyx 2–3 mm, in fr 20–25 mm, 10-veined; corolla 7–8 mm wide, widely bell-shaped, yellow; anthers 1–2 mm, blue-tinged. Wet places, fields, waste places; < 200 m. GV, SnFrB, SCoRI, DSon; to TX, c Mex; native to S.Am. [*P. angulata* L. var. *l.* (Nees) Waterf.]

P. lobata Torrey LOBED GROUND-CHERRY Per < 5 dm, decumbent to spreading, few-branched, glabrous to minutely papillate. **LF** 1–7 cm, lanceolate to ovate, entire to lobed, tapered to base. **INFL**: pedicels 3–4.5 mm, not gen longer in fr. **FL**: calyx 3–4.5 mm, in fr 15–20 mm; corolla 15–20 mm wide, rotate, purple, tube white inside; anthers 1–2.5 mm, yellowish. *n*=11,22. RARE in CA. Granitic soils, dry lake margins; 500–800 m. se DMoj, ne DSon; to KS, TX, n Mex.

P. longifolia Nutt. Per 2–6 dm, from rhizome; hairs simple, few, appressed. **LF** 4–7 cm, lanceolate to ± elliptic, petioled, gen entire (irregularly toothed). **INFL**: pedicels 6–15 mm, in fr < 20 mm. **FL**: calyx 9–10 mm, in fr 25–35 mm, 10-veined; corolla 15–20 mm wide, bell-shaped, yellow with dark purple spots or veins at base; anthers 3–4 mm, bluish. *n*=12. Waste places, fields; < 1000 m. n CaRH (Shasta Valley) as waif; native to c&e US. [*P. subglabrata* K. K. MacKenzie & Bush misapplied to CA pls of this and possibly other spp.]

P. philadelphica Lam. TOMATILLO Ann < 10 dm; hairs 0 or sparse, simple, and glandular. **LF** gen 4–8 cm, ± ovate, entire to toothed. **INFL**: pedicels 3–8 mm, not longer in fr, < calyx. **FL**: calyx 4.5–6 mm, in fr 20–30+ mm, green with 10 purple veins; corolla 8–15 mm wide, ± rotate, yellow with 5 dark purple spots inside, tube hairy inside; anthers 2.5–3 mm, twisted after opening, yellow and blue-green. *n*=12. Waste places, cult fields, roadsides; < 700 m. SNF, GV, CW, SCo, PR; to e US; native to Mex. [*P. ixocarpa* Hornem.] Widely cult for food.

P. pubescens L. Ann < 8 dm; hairs simple, ± dense, spreading, most with small glands. **LF** 3–9 cm, widely ovate to ± cordate, entire to coarsely toothed. **INFL**: pedicels 3–12 mm, in fr ± longer. **FL**: calyx 5 mm, in fr 20–40 mm, sharply 5-angled with ribs between; corolla ± 10 mm wide, widely bell-shaped, yellow, tube with 5 dark spots inside; anthers 1.5–2 mm, blue. Waste places, cult fields; < 1500 m. s NCoRO, SnJV, D, expected elsewhere; native to e US.

var. *grisea* Waterf. **LF** 5–9 cm, ± thick, often drying reddish brown; teeth many. *n*=12. Habitats and elevations of sp. s SNE, D; native to c&e US. [*P. pruinosa* L.; *P. neomexicana* Rydb. misapplied to CA pls] Often cult.

var. *integrifolia* (Dunal) Waterf. **LF** 3–5 cm, thin, drying ± translucent; teeth 0–few. *n*=12. Habitats and elevations of sp. s NCoRO, SnJV, expected elsewhere; native to e US.

P. viscosa L. GRAPE GROUND-CHERRY Per 1–4 dm, few-branched, from rhizome, forming colonies; hairs branched, few, small, nonglandular. **LF** 3–5 cm, ovate, entire to irregularly wavy-toothed; base tapered or obtuse. **INFL**: pedicels 5–15 mm, in fr < 25 mm. **FL**: calyx 8–9 mm, in fr 20–30 mm, 10-ribbed; corolla 14–17 mm wide, bell-shaped, yellow with darker yellow or yellow-green center, not spotted; anthers 4 mm, yellow, filling corolla throat. Waste places, fields, roadsides as waif; < 300 m. *n*=12. SCo; native to c US, e Mex. [var. *cinerascens* (Dunal) Waterf.; *P. mollis* Nutt.] NOXIOUS WEED.

SALPICHROA

Per from rhizome. **LF** fleshy, entire. **INFL**: fls solitary in axils. **FL**: calyx deeply lobed; corolla lobes small; stamens ± incl, anthers converging toward style, oblong; style 1, thread-like, stigma head-like. **FR**: berry, juicy. **SEEDS** many, compressed. 15 spp.: S.Am, esp Andes. (Greek: trumpet color)

S. origanifolia (Lam.) Baillon Pl < 2 m, decumbent to climbing, hairy. **LF** 10–35 mm; petiole 5–10 mm; blade widely elliptic to ovate, ± glabrous. **INFL**: pedicels 3–14 mm. **FL**: calyx lobes 2–3 mm; corolla urn-shaped, white or greenish, with a densely woolly ring inside, lobes 2–3 mm, triangular; anthers 1.5–2 mm, slightly exserted; disc below ovary brick-red. **FR** 10–15 mm, ovoid, white or pale yellow, ill-smelling. *n*=12. Disturbed places; < 100 m. NCo, w CW, SCo; to TX, n Mex, Eur; native to S.Am. [*S. rhomboidea* (Gillies & Hook.) Miers]

SOLANUM NIGHTSHADE

Ann to shrub or vine, often glandular, sometimes prickly. **LVS** alternate to subopposite, often unequal, entire to deeply pinnately lobed. **INFL**: panicle or umbel-like, often 1-sided. **FL**: calyx ± bell-shaped; corolla ± rotate, white to purple; anthers free, > filaments, oblong or tapered, opening by 2 pores or short slits near tip; ovary 2-chambered, style 1, stigma head-like. **FR**: berry, gen spheric (or dry, capsule-like). **SEEDS** many, compressed, gen reniform. ± 1500 spp.: worldwide, esp trop Am. (Latin: quieting, from narcotic properties) [Symon 1981 J Adelaide Bot Gard 4:1–367] Many cult for food (incl potato, *S. tuberosum*), orn; many TOXIC.

1. Some or all hairs stellate; anthers tapered, opening only by terminal pores; pl often prickly
  2. Stellate hairs scale-like, rays fused at center .... ***S. elaeagnifolium***
  2′ Stellate hairs not scale-like, rays free above stalk
    3. 1 anther > other 4; calyx tightly enclosing fr .... ***S. rostratum***
    3′ Anthers ± equal; calyx not tightly enclosing fr
      4. Lf densely white-tomentose or velvety below
        5. Calyx not prickly; infl branches 2–4; frs several–many, 7–15 mm wide .... ***S. lanceolatum***
        5′ Calyx prickly; infl branches 0; fr 1, 35–50 mm wide .... ***S. marginatum***
      4′ Lf sparsely tomentose below, not clearly white or velvety

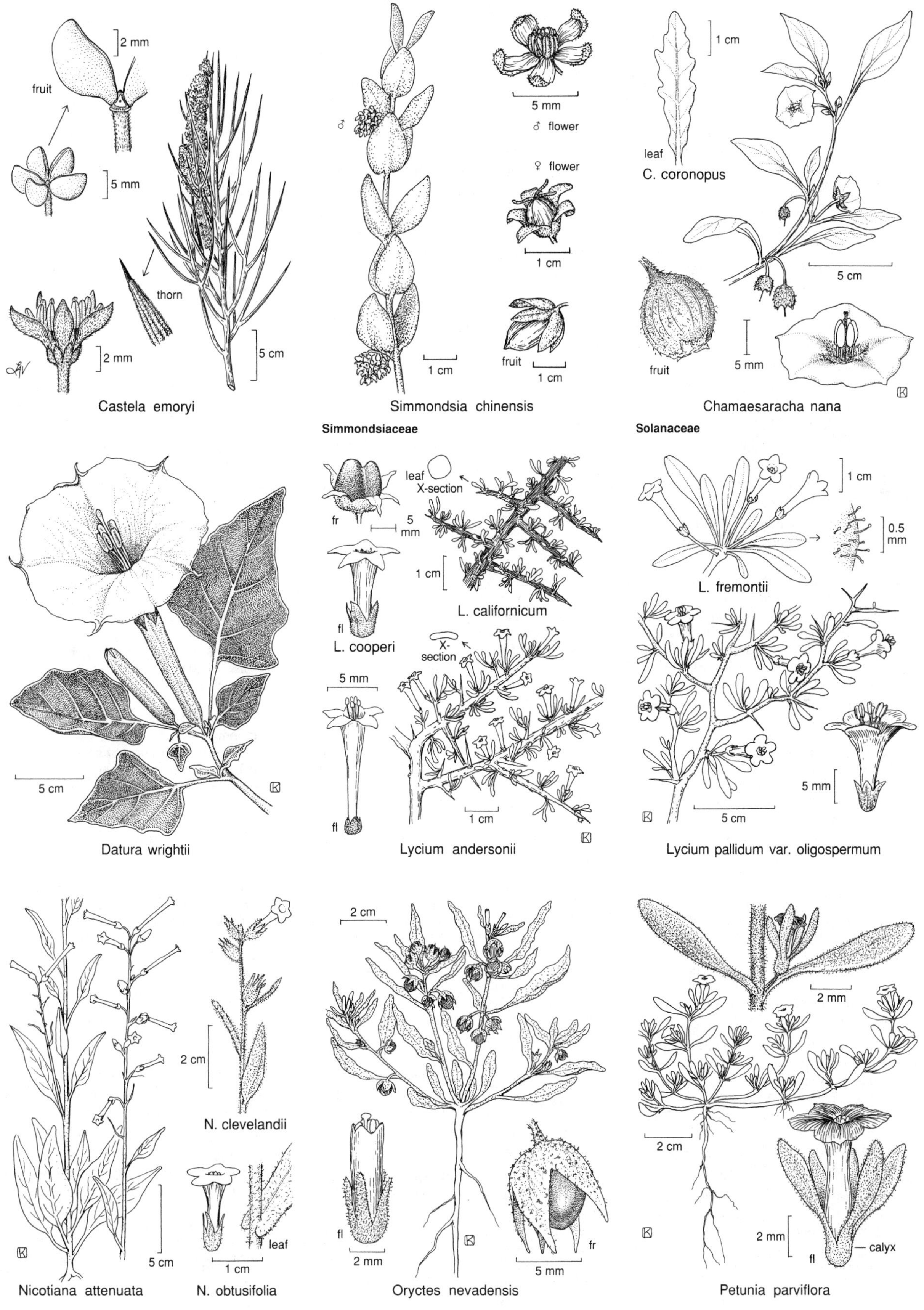
2 mm
fruit
5 mm
thorn
2 mm
5 cm
Castela emoryi
♂
5 mm
♂ flower
♀ flower
1 cm
fruit
1 cm
1 cm
Simmondsia chinensis
Simmondsiaceae
1 cm
leaf
C. coronopus
5 cm
fruit
5 mm
Chamaesaracha nana
Solanaceae
5 cm
Datura wrightii
leaf
X-section
fr
5
mm
1 cm
L. californicum
fl
L. cooperi
X-
section
5 mm
fl
1 cm
Lycium andersonii
1 cm
0.5
mm
L. fremontii
5 mm
5 cm
Lycium pallidum var. oligospermum
2 cm
N. clevelandii
5 cm
1 cm
leaf
Nicotiana attenuata
N. obtusifolia
2 cm
fl
2 mm
5 mm
fr
Oryctes nevadensis
2 mm
2 cm
2 mm
fl
calyx
Petunia parviflora

6. Lf deeply pinnately lobed, lobes wavy to again lobed; ann; fr red .... ***S. sisymbriifolium***
6′ Lf ± entire to wavy-lobed, teeth or lobes ± entire; per from rhizome; fr yellow
7. Corolla white or pale violet; stellate hairs fine, gen sessile; infl gen raceme-like; fr 8–20 mm .. ***S. carolinense***
7′ Corolla gen purple; stellate hairs coarse, some stalked; infl gen branched; fr > 20 mm diam ... ***S. dimidiatum***
1′ Hairs simple or branched, not stellate; anthers oblong, opening by terminal pores that become short slits; pl gen not prickly
8. Lf very deeply lobed .... ***S. triflorum***
8′ Lf entire to shallowly lobed (or deeply 2-lobed below middle)
9. Corolla weakly lobed, lobes < tube + throat, gen spreading
10. Shrub 2–4 m; lf ± entire to deeply lobed; fr orange-red .... ***S. aviculare***
10′ Per or subshrub gen < 1 m; lf ± entire to 1–2-lobed at base; fr green to purplish
11. Upper st densely branched-hairy .... ***S. umbelliferum***
11′ Upper st glabrous or mostly simple-hairy
12. St ± glabrous, clearly ridged; lf sessile or gradually tapered to base .... ***S. parishii***
12′ St gen soft- or glandular-hairy, smooth or weakly ridged; lf base obtuse to subcordate
13. Lf gen 6–10 cm, oblong to ovate; fr gen 15–25 mm; ChI .... ***S. wallacei***
13′ Lf 2–7 cm, lanceolate to ovate; fr 10–15 mm; mainland .... ***S. xanti***
9′ Corolla lobes deep (= or > tube + throat), often reflexed
14. Infl axis gen forked, umbel-like clusters gen 2
15. Corolla violet to purple, each lobe with 2 shiny green spots at base; pl ± woody, ± climbing ... ***S. dulcamara***
15′ Corolla white to pale violet, lobes not spotted; pl not woody, gen reclining .... ***S. furcatum***
14′ Infl umbel- or raceme-like, axis not forked
16. Calyx enlarged and enclosing fr base; fr yellowish or green .... ***S. sarrachoides***
16′ Calyx not enlarged, not enclosing fr base; fr black
17. Anthers gen 2.5–4 mm .... ***S. douglasii***
17′ Anthers 1.4–2.2 mm
18. Seeds 1–1.5 mm; hairs ± appressed or curved, glands 0 .... ***S. americanum***
18′ Seeds 1.5–2.5 mm; hairs ± spreading or curved, some glandular .... ***S. nigrum***

S. americanum Miller (p. 1079) Ann to subshrub 3–8 dm, ± glabrous or hairs short, curved or ± appressed. **LF** 2–15 cm, ovate, entire to coarsely wavy-toothed. **INFL** umbel- or ± raceme-like. **FL**: calyx 1–2 mm, lobes in fr recurved; corolla 3–6 mm wide, deeply lobed, white; anthers 1.4–2.2 mm; style 2.5–4 mm. **FR** 5–8 mm diam, greenish or black. **SEED** 1–1.5 mm. *n*=12. Open, often disturbed places; < 1000 m. CA-FP, DMoj (uncommon); to Can, e US, Mex. [*S. nodiflorum* Jacq.] Much like *S. nigrum*; may be early introduction from S.Am.

S. aviculare Forst. f. Shrub 2–4 m, gen glabrous. **LF** 10–30 cm, entire to deeply lobed. **INFL** gen raceme-like. **FL**: calyx 3–4 mm; corolla 30–40 mm wide, shallowly lobed, bluish violet; anthers ± 4 mm; style 7–10 mm. **FR** 10–15 mm diam, orange-red. **SEED** ± 1.5 mm. Uncommon. Open, gen disturbed places; < 100 m. NCo, n CCo; native to Australasia. [*S. laciniatum* Ait. misapplied to CA pls] Urban SCo waifs with densely hairy herbage and corollas 11–15 mm wide are *S. mauritianum* Scop.

S. carolinense L. (p. 1079) CAROLINA HORSE-NETTLE Per 2–9 dm, from rhizome, little-branched; prickles slender, spreading; hairs stellate. **LF** 5–15 cm, ovate, ± entire to wavy-lobed, teeth or lobes entire. **INFL** raceme-like. **FL**: calyx 5–6 mm, lobes ± = tube; corolla 20–30 mm wide, gen pale violet (white); anthers 7–9 mm, ± equal. **FR** 10–15 mm diam, yellow. **SEED** ± 2.5 mm. *n*=12. Uncommon. Disturbed places, fields, roadsides; < 200 m. s NCo-RO, n SN, GV, SCo; native to c&e US, n Mex. NOXIOUS WEED.

S. dimidiatum Raf. TORREY'S NIGHTSHADE Per 3–8 dm, from rhizome, little-branched; prickles slender, yellowish; hairs stellate, some short-stalked. **LF** 8–15 cm, ovate, wavy to 5–7-lobed, teeth or lobes gen entire. **INFL**: branches 2–3. **FL**: calyx 8–13 mm; corolla 30–50 mm wide, gen purple (white); anthers 8–12 mm, ± equal. **FR** 25–30 mm diam, yellow. **SEED** ± 4 mm. *n*=12. Uncommon. Disturbed places; < 200 m. GV, SCo; native to c US. [*S. torreyi* A. Gray] NOXIOUS WEED.

S. douglasii Dunal Per or subshrub < 20 dm, much-branched; hairs gen simple, < 1 mm, ± curved, white. **LF** 1–9 cm, ovate, entire to coarsely and irregularly toothed. **INFL** gen umbel-like. **FL**: calyx ± 2 mm; corolla ± 10 mm wide, deeply lobed, white with greenish spots at base; anthers 2.5–4 mm; style 4–5 mm, puberulent below. **FR** 6–9 mm diam. **SEED** 1.2–1.5 mm. *n*=12. Dry shrubland, woodland; < 1000 m. s NCo, Teh, CW, SW, DMoj; n Mex.

S. dulcamara L. Per to subshrub gen < 10 dm, climbing, ± glabrous or soft-hairy. **LF** 5–12 cm, ± cordate to deeply 1–2-lobed near base. **INFL**: axis forked; clusters gen 2, umbel-like. **FL**: calyx 3–4 mm; corolla 8–12 mm wide, deeply lobed, purple or violet, lobe base with 2 green spots. **FR** 8–12 mm diam, ± ovoid, red. **SEED** 2–3 mm. *n*=12. Moist disturbed places, marshes; < 1000 m. CCo, SnFrB, MP; to Can, e US; native to n Eurasia.

S. elaeagnifolium Cav. (p. 1079) WHITE HORSE-NETTLE Per < 10 dm, from rhizome, forming colonies, gen prickly, densely scaly-stellate (hair rays fused at center). **LF** 2–15 cm, oblong, entire to barely lobed, ± yellowish. **INFL** raceme-like. **FL**: calyx 5–8 mm, lobes ± = tube; corolla 20–30 mm wide, purple or blue; anthers 8–10 mm, ± equal. **FR** 8–15 mm diam, orange, often persistent. **SEED** 2.5–3 mm. *n*=12. Common. Dry, disturbed places, fields; < 1200 m. CA (exc NCo, KR, GB); to S.Am; native to c US, n Mex. NOXIOUS WEED.

S. furcatum Dunal Per 5–12 dm, weak or ± reclining; hairs simple, sparse, ± curved. **LF** 3–7 cm, ovate, ± entire or irregularly few-toothed. **INFL**: axis forked; clusters 2, umbel-like; **FL**: calyx 2.5–3 mm; corolla 10–18 mm wide, white to pale violet; anthers 3–3.5 mm. **FR** 5–6 mm, dark purple. **SEED** ± 2 mm. *n*=36. Open, often disturbed places; < 200 m. NCo, CCo, SnFrB; OR, Baja CA; native to S.Am. [*S. gayanum* (E.J. Remy) Phil.f. misapplied to CA pls]

S. lanceolatum Cav. LANCELEAF NIGHTSHADE Shrub 8–50 dm, sparsely prickly, stellate, glandular. **LF** 3–15 cm, gen lanceolate, entire to lobed. **INFL**: branches 2–4; peduncle 1–3 cm. **FL**: calyx 5–8 mm, lobes ± = tube, tips slender, dark; corolla 25–30 mm wide, blue-purple; anthers 6–8 mm, gen ± equal. **FR** 7–15 mm diam, yellow-orange. Disturbed places; < 200 m. GV, SnFrB, SCo; native to Mex, C.Am. NOXIOUS WEED.

S. marginatum L. WHITE-MARGINED NIGHTSHADE Shrub < 2 m, prickly, densely stellate-hairy. **LF** 10–18 cm, widely ovate, wavy-lobed, densely white-tomentose below. **INFL** raceme-like; lowermost fl bisexual, reflexed, others staminate. **FL**: calyx ± 10 mm, prickly; corolla 15–30 mm wide, white; anthers 6–7 mm. **FR** 35–50 mm diam, tough, yellow, glabrous. **SEED** ± 3 mm. *n*=12. Disturbed places; gen < 1000 m. CCo, SCo; native to Afr. NOXIOUS WEED.

S. nigrum L. BLACK NIGHTSHADE Ann to subshrub 3–8 dm; hairs ± spreading or arched, some glandular. **LF** 4–7 cm, ovate, entire to coarsely wavy-toothed. **INFL** raceme-like. **FL**: calyx 2–3 mm; corolla 10 mm wide, deeply lobed, white; anthers 2 mm; style 3–5 mm. **FR** 6–8 mm diam, black. **SEED** ± 2 mm. *n*=36. Disturbed places; < 200 m. NCo, n CCo, SnFrB (expected elsewhere); to WA, e US; native to Eurasia. Much like *S. americanum.*

S. parishii A.A. Heller (p. 1079) Per to subshrub < 10 dm, much-branched, glabrous to sparsely hairy. **ST** angled to ribbed. **LF** 2–7 cm, lanceolate to elliptic, entire to wavy, tapered to base. **INFL** ± umbel-like; pedicels 13–18 mm, > peduncle. **FL**: calyx 4 mm; corolla 17–22 mm wide, gen blue-purple (white), lobes << tube; anthers 3.5–5 mm. **FR** 7–10 mm diam. **SEED** ± 2 mm. Dry chaparral, oak/pine woodland, pine forest; < 2000 m. CA-FP (exc GV), MP; s OR, Baja CA. May hybridize with *S. xanti*; needs study. ❀DRN,SUN:1–3,7,14–24;DFCLT.

S. rostratum Dunal (p. 1079) BUFFALO BERRY Ann 1–7 dm, densely yellow-prickly, stellate. **LF** 5–15 cm, deeply lobed. **INFL** raceme-like. **FL**: calyx enclosing fr, prickly; corolla 23–35 mm wide, yellow; anthers unequal (4 straight and 6–8 mm, 1 curved and 10–14 mm); style curved. **FR** 9–12 mm diam, ± dry. **SEED** 2–2.5 mm. *n*=12. Disturbed places, roadsides, fields; < 500 m. s NCoRO, CaRF, GV, CCo, s SCoRO, SCo; to e US, Mex; native to Great Plains.

S. sarrachoides Sendtner Ann 1–9 dm, decumbent, sticky; hairs spreading, some glandular. **LF** 2–6 cm, ovate, entire to irregularly toothed or shallowly lobed. **INFL** gen raceme-like, few-fld. **FL**: calyx 2–2.5 mm, in fr 4–6 mm, enclosing fr base; corolla 3–5 mm wide, white, lobes > tube; anthers 1.5–2 mm. **FR** 6–7 mm diam, yellowish. **SEED** ± 2 mm, yellow. *n*=12. Disturbed areas; gen < 1000 m. NCo, KR, CaRF, GV, w CW, SCo, n ChI (Santa Cruz Island), SNE; to Can, e US, Mex; native to S.Am.

S. sisymbriifolium Lam. Ann 5–15 dm, ± sticky, yellow-prickly, stellate. **LF** 4–15 cm, deeply lobed, lobes wavy-toothed to again lobed. **INFL** raceme-like. **FL**: calyx 7–9 mm, slightly enlarged in fr; corolla 25–30 mm wide, very pale blue to white; anthers 10 mm. **FR** 10–20 mm diam, red. **SEED** ± 2 mm, orange-yellow. Disturbed places, roadsides; < 100 m. GV (expected elsewhere); native to S.Am.

S. triflorum Nutt. Ann 1–5 dm, decumbent; hairs ± curved or spreading, sometimes glandular. **LF** 2–5 cm, oblong to ovate, very deeply lobed. **INFL** umbel-like; fls 2–3. **FL**: calyx 2.5–3 mm, enlarged in fr, enclosing fr base; corolla 7–9 mm wide, white, lobes ± = to > tube; anthers ± 3 mm. **FR** 8–12 mm diam, green. **SEEDS** many, 2.5–3 mm, yellow. *n*=12. Dry shrubland, juniper woodland; 100–2300 m. s SNH (e slope), SCo, GB, n DMoj; to c US; native to S.Am.

S. umbelliferum Eschsch. (p. 1079) Per to subshrub gen < 10 dm, much-branched; hairs gen dense, branched. **LF** 1–4 cm, ± elliptic to (ob)ovate, gen entire. **INFL** ± umbel-like, sometimes forked; pedicels 12–15 mm, > peduncle. **FL**: calyx 3–3.5 mm; corolla 16–25 mm wide, lavender to blue-purple, lobes < tube, each with 2 greenish spots at base; anthers 3–5 mm. **FR** 12–14 mm diam. **SEED** ± 2 mm. Shrubland, mixed-evergreen forest, woodland; < 1600 m. NW (uncommon), deltaic GV, CW, SW; AZ, Baja CA. [vars. *glabrescens* Torrey and *incanum* Torrey] May hybridize with *S. parishii, S. xanti*; needs study. ❀DRN,IRR:**7**,8–10,**14,19–23**&SUN:**15–17,24**.

S. wallacei (A. Gray) Parish (p. 1079) WALLACE'S NIGHTSHADE Per to subshrub < 10 dm, much-branched, gen densely soft-hairy, often glandular. **LF** gen 6–10 cm, ovate to oblong, entire to ± wavy; base rounded to subcordate. **INFL**: branches gen 2. **FL**: calyx 4.5–6 mm; corolla 30–40 mm wide; anthers 4.5–5.5 mm. **FR** 15–25 mm diam. **SEED** 1.5–2 mm. UNCOMMON. Canyons, chaparral; < 800 m. ChI; Guadalupe Island, Mex. [var. *clokeyi* (Munz) McMinn (*S. clokeyi* Munz), island nightshade] ❀DRN:5,**17**&SHD:7,**14**,15,16,**24**&IRR:8,9,**19–23**.

S. xanti A. Gray (p. 1079) Per to subshrub 4–9 dm, much-branched, ± hairy, sometimes glandular. **LF** 2–7 cm, lanceolate to ovate, ± entire to 1–2-lobed at base; base obtuse to subcordate. **INFL** umbel-like, sometimes branched. **FL**: calyx 4–5 mm; corolla 15–30 mm wide, dark blue or lavender, lobes < tube; anthers 4–5.5 mm. **FR** 10–15 mm diam, greenish. **SEED** 1.5–2 mm. Shrubland, oak/pine woodland, coniferous forest; < 2700 m. CA-FP (exc CaR, GV), n DMtns; Baja CA. [vars. *hoffmannii* Munz (Hoffmann's nightshade), *intermedium* Parish, *montanum* Munz, *obispoense* (Eastw.) Wiggins (San Luis Obispo nightshade); *S. tenuilobatum* Parish (narrow-lvd nightshade)] Variation complex; may hybridize with *S. parishii, S. umbelliferum.* ❀DRN:2,3,7,9,14,18–21&SUN:15,16,**17,22–24**;CV.

STAPHYLEACEAE BLADDERNUT FAMILY

James R. Shevock

Per, shrubs, trees, some monoecious or dioecious. **LVS** gen opposite, pinnately compound; lflets gen toothed; stipules 0 or paired. **INFL**: panicle or raceme, drooping. **FLS** small, radial; sepals 5, free or fused at base, often petal-like; petals 5, free; stamens 5, alternate petals, often attached to nectary-disk; ovary superior, chambers 2–4, each with 1–12 ovules in 2 rows. **FR**: gen inflated capsule with open top, sometimes follicles, drupe, or berry. **SEEDS** gen 1–2 per chamber. 5 genera, ± 50 spp.: n temp, Asia, C.Am, S.Am.; some cult for showy fr (*Staphylea*) or used for timber (*Turpinia*).

STAPHYLEA BLADDERNUT

Per, shrub, trees. **LVS** deciduous, opposite; lflets gen 3. **INFL**: panicle. **FL**: petals white or pink. **FR**: capsule, inflated, bladdery, deeply 3-lobed, pendent. **SEEDS** spheric, light brown. 10 spp.: n temp. (Greek: cluster, from infl)

S. bolanderi A. Gray (p. 1079) SIERRA BLADDERNUT Shrub, small tree, 2–6 m, glabrous. **LF**: lflets 3, 2.5–6 cm, widely ovate to round, finely serrate. **INFL** gen appearing before or with lvs. **FL**: sepals 8–10 mm, white; petals 10–12 mm, white; stamens well exserted. **FR** 2.5–5 cm, prominently horned, greatly inflated. **SEEDS** 5–7 mm, ± obovoid, light brown, smooth. Uncommon. Wooded or shrubby slopes; 300–1400 m. KR, CaR, SN. ❀DRN, IRR:1,6,**15–17**&SHD:**7**,14,18.

STERCULIACEAE CACAO FAMILY

R. David Whetstone & T.A. Atkinson

Per to tree; hairs stellate (or scale-like, peltate). **LVS** cauline, alternate, simple or palmately compound, evergreen, petioled; stipules gen deciduous. **INFL**: gen complex clusters, cymes, or fls solitary (in axils, opposite lvs, or on a spur branch); whorl of bracts often subtends calyx (esp if petals 0). **FL** bisexual, radial; sepals 5, gen fused at base; petals 0 or 5, clawed, sometimes fused to filament tube; stamens 5 (sometimes alternate 5 staminodes), filaments fused below into tube; ovary superior, sometimes on a stalk that may be fused to filament tube, chambers gen 5, style 1. **FR**: capsule. 60 genera, 700 spp.: gen trop, subtrop; some cult for orn (*Fremontodendron*) or for drugs and food (*Cola*; *Theobroma*, chocolate). [Brizicky 1966 J Arnold Arbor 47:60–74]

1. Lf unlobed, serrate; fl gen < 3 mm wide; calyx not showy; petals with a long, thread-like claw; ovary and fr stalked . . . . **AYENIA**
1′ Lf 3-lobed, lobes often again lobed, otherwise entire; fl > 25 mm wide; calyx showy; petals 0; ovary and fr sessile . . . . **FREMONTODENDRON**

AYENIA

Shrub; taproot stout. **ST** erect, much-branched at base, 1–4 dm; twig hairs stellate or 0. **LF** ovate-obovate, unlobed, serrate. **INFL**: fls 1(–2) in axils. **FL** gen < 3 mm wide; sepals ± spreading, narrowly ovate; petal claw thread-like, coiled, limb ± obcordate, incurved, parachute-like, sinus with an anther below and a stalked, gland-like appendage above; filament tube ± cup- or urn-shaped at top, surrounding ovary, with 5 short, stalked anthers bent outward and downward (each stalk inserted in sinus, thereby attached to petal), staminodes 5, < stamens; ovary (and fr) stalked above receptacle. **FR**: chambers 1-seeded. ± 50 spp.: warm Am. (Louis de Noailles, Duc d'Ayen, 1739–1777) [Cristobal 1960 Opera Lilloana 4:1–230]

A. compacta Rose (p. 1079) AYENIA **FL**: sepals ± 1.5 mm; petals 2–3 mm, claws ± 2 mm. **FR** ± 5 mm, spheric, straw-colored, with ± cylindric, purplish protuberances. **SEED** black. RARE in CA. Dry, rocky canyons; < 500 m. e DMtns (Providence Mtns), w&c DSon (incl Eagle Mtns); Baja CA. [*A. californica* Jepson]

FREMONTODENDRON FREMONTIA, FLANNELBUSH

Shrub, small tree. **ST** decumbent to erect, 3–8 m; inner bark gelatinous; twig hairs very dense, stellate. **LF** often on a spur, ± ovate, gen with 3 main and few–many 2° lobes, otherwise entire; hairs gen denser on lower surface. **INFL**: fls gen solitary, opposite a lf or on a spur; fl subtended by gen 3 sepal-like bracts. **FL** > 25 mm wide, showy; sepals spreading, widely ovate to ± round, upper surfaces pitted between raised, hard, fused basal margins, tip awned or not; petals 0; filaments thick, fleshy, tube surrounding and ± = ovary, < style; ovary (and fr) sessile. **FR** 2–4 cm, ovoid, bristly; chambers 2–3 seeded. 2 spp.; CA, AZ, Mex. (John C. Frémont, explorer in West, 1813–1890) [Kelman 1991 Syst Bot 16:3–20]

1. Sepal pits glabrous; pl erect, unbranched near ground; stipules ± 4–4.5 mm . . . . ***F. mexicanum***
1′ Sepal pits with silky hairs ± 1 mm; pl erect and branched near ground or decumbent; stipules ± 2 mm . . . . ***F. californicum***
2. Pl erect, taller than wide; fl ± 35–60 mm wide; sepals yellow, margins sometimes reddish . . . . . ssp. ***californicum***
2′ Pl decumbent, much wider than tall; fl ± 35 mm wide; sepals orange, coppery, or reddish . . . . ssp. ***decumbens***

F. californicum (Torrey) Cov. Pl erect and branched near ground or decumbent. **LF** 1–5 cm; blade palmately to pinnately lobed, soft to leathery; stipules ± 2 mm. **FL** ± 35–60 mm wide; sepals yellow, margins sometimes reddish (or orange, coppery, or reddish), pits silkyhairy. Chaparral, oak/pine woodland, rocky ridges; ± 400–2000 m. CA-FP; AZ, Baja CA.

ssp. ***californicum*** (p. 1079) Pl erect, 2–5 m, taller than wide. **LF**: blade palmately to pinnately lobed, rarely entire. **FL** < 60 mm wide; sepals yellow, margins sometimes reddish. Common. Chaparral, oak-pine woodland; ± 400–2000 m. Range of sp. [sspp. *crassifolium* (Eastw.) J.H. Thomas, *napense* (Eastw.) Munz, *obispoense* (Eastw.) Munz] Highly variable; some variation induced by habitat; described taxa gen indistinct. ❀DRN,DRY,SUN:4–6,**7–9**,10–12,**14–24**;CVS.

ssp. ***decumbens*** (R. Lloyd) Munz PINE HILL FLANNELBUSH Pl decumbent, < 1 m, much wider than tall. **LF**: blade palmately lobed. **FL** ± 35 mm wide; sepals orange, coppery, or reddish. **RARE** CA. Rocky ridges; ± 2000 m. n SNH (Pine Hill, El Dorado Co.). [*F. decumbens* R. Lloyd] In cult.

F. mexicanum Davidson MEXICAN FLANNELBUSH Pl erect and unbranched near ground, < 7 m. **LF** 25–50 mm; blade palmately lobed, thickly leathery; stipules ± 4–4.5 mm. **FL** ± 60 mm wide; sepals gen orange, reddish toward bases, pits glabrous. **RARE** CA. Chaparral, canyons; 300–1000 m. PR (Orange, San Diego, sw Imperial cos.); Baja CA. In cult.

STYRACACEAE STORAX FAMILY

James R. Shevock

Shrubs, small trees, often stellate- or peltate-hairy; bark resinous. **LVS** simple, alternate, gen entire; stipules 0. **INFL**: panicle, raceme, cyme, or fls solitary. **FL** gen bisexual, radial, gen showy, fragrant; sepals 4–5, fused, persistent; petals 2–5(7), fused below, tube short; stamens gen 2–4 × petal number, gen epipetalous in 1 series; ovary superior or inferior, chambers 2–5. **FR**: gen capsule or drupe. **SEEDS** gen 1–2 per chamber. ± 10 genera, ± 150 spp.: e&w US to S.Am; also in Medit, e Asia; some cult as orn (*Styrax, Halesia*); some bark resins are medicinal.

STYRAX SNOWDROP BUSH

Shrubs, trees. **LVS** gen deciduous. **INFL**: gen raceme, drooping. **FL**: calyx truncate; corolla white, fragrant; stamens 10–16, fused at base. **FR**: gen capsule, ± spheric. **SEED** gen 1 per chamber. ± 120 spp.: esp se Asia; some with medicinal resins. (Greek: name of Theophrastus for pl providing gum storax) [Gonsoulin 1974 Sida 5:191–258]

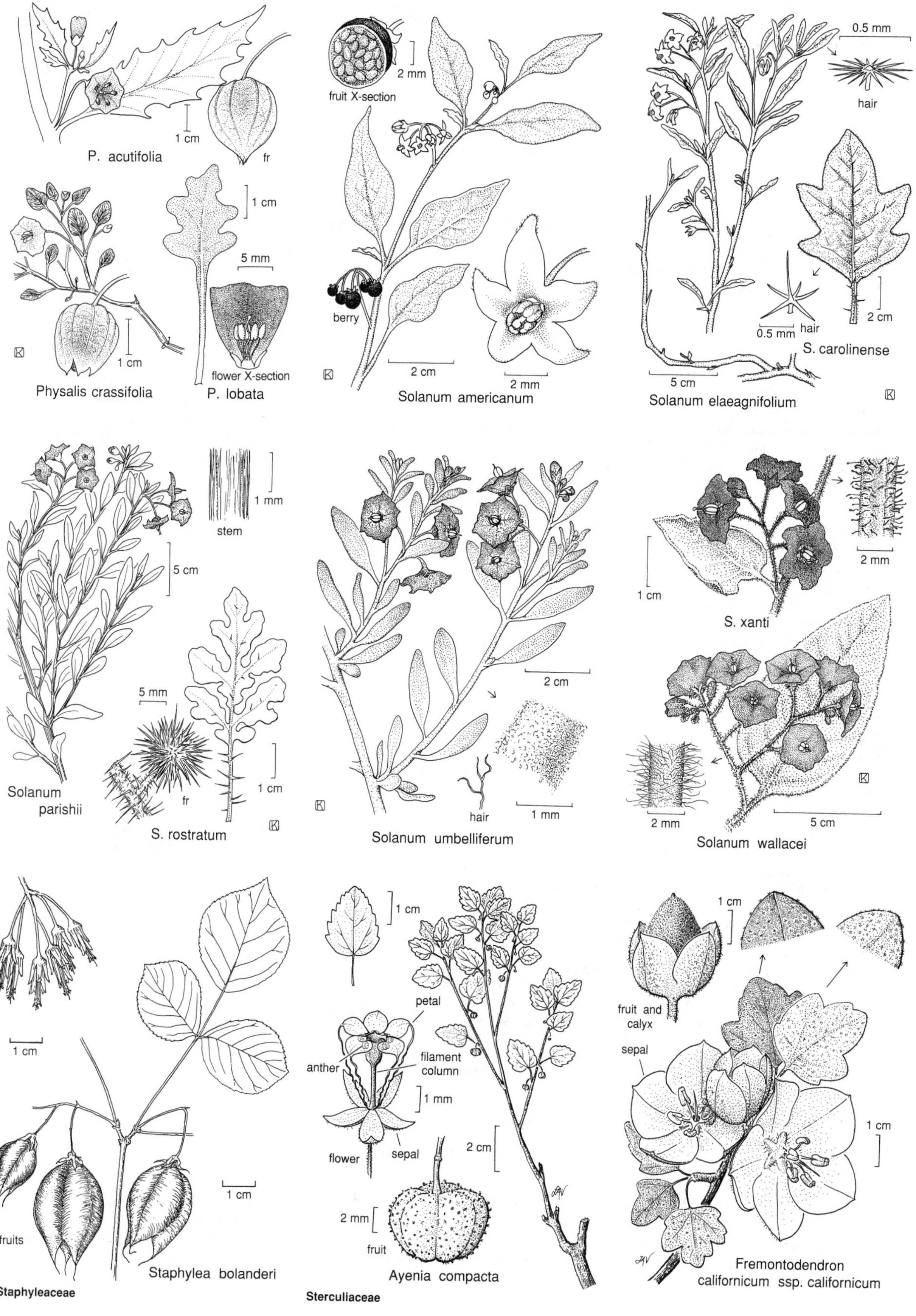
P. acutifolia
1 cm
fr
Physalis crassifolia
1 cm
P. lobata
1 cm
5 mm
flower X-section
fruit X-section
2 mm
berry
2 cm
2 mm
Solanum americanum
0.5 mm
hair
0.5 mm
hair
2 cm
S. carolinense
5 cm
Solanum elaeagnifolium
1 mm
stem
5 cm
5 mm
fr
1 cm
Solanum parishii
S. rostratum
2 cm
hair
1 mm
Solanum umbelliferum
2 mm
1 cm
S. xanti
2 mm
5 cm
Solanum wallacei
1 cm
1 cm
fruits
Staphylea bolanderi
Staphyleaceae
1 cm
petal
anther
filament column
1 mm
flower
sepal
2 cm
2 mm
fruit
Ayenia compacta
Sterculiaceae
1 cm
fruit and calyx
sepal
1 cm
Fremontodendron californicum ssp. californicum

S. officinalis L. var. ***redivivus*** (Torrey) H. Howard (p. 1087) Shrub 1–4 m. **LF**: petiole 3–10 mm, variably hairy; blade 2–7 cm, obovate to ± round, upper surface gen glabrous, lower surface paler, stellate-hairy, base often ± cordate. **INFL**: cluster, few-fld; peduncle 6–12 mm. **FL**: calyx teeth obscure, unequal; corolla 12–18 mm, 4–10-lobed, white, showy. **FR** gen 12–14 mm. **SEED** spheric-ovoid, light brown, smooth. Uncommon. Dry places in chaparral, woodland; gen < 1500 m. NCoRH, NCoRI, CaR, SN, SCoRO, w WTR, SnBr, PR. [var. *californicus* (Torrey) Rehd.; var. *fulvescens* (Eastw.) Munz & I.M. Johnston] Hair density and color variable throughout range. ❀6,**7**,8–10,**14,18–21**;SUN:4,**15,16**,17,**22–24**.

TAMARICACEAE TAMARISK FAMILY

Dieter H. Wilken

Shrubs, trees, much-branched, often in saline habitats. **ST**: trunk bark rough. **LVS** alternate, sessile, entire. **INFL**: racemes or spikes; bracts scale-like. **FL**: sepals 4–6, gen free, overlapping; petals 4–5, overlapping, gen attached below nectary; stamens 4–10, attached to disk-like nectary; ovary 1-chambered, placentas parietal or basal, ovules 2–many, styles 2–5. **FR**: capsule, loculicidal. **SEEDS** many, hairy. ± 5 genera, 100 spp.: Eurasia, Afr, esp Medit.

TAMARIX TAMARISK

STS green, glabrous; twigs jointed, slender, often drooping. **LVS** on twigs, gen overlapping, awl- to scale-like, gen excreting salt. **INFLS** gen in panicle-like clusters on current or previous year's twigs. **FL**: sepals gen 5, persistent; petals gen 5, deciduous to ± persistent, white to reddish; stamens gen 5, filaments alternate or confluent with nectar disk lobes; nectar disk 4–5-lobed; placentas basal, styles 3. **SEEDS** hairy-tufted. (Latin: Tamaris River, Spain) [Baum 1967 Baileya 15:19–25] Invasive weeds with deep roots that lower water table, esp along streams, irrigation canals. Most CA spp. cult for orn, windbreaks; some may hybridize.

1. Twig-lvs not overlapping, strongly clasping, abruptly pointed; sepals ± round ..... ***T. aphylla***
1′ Twig-lvs overlapping, not clasping, acute to long-acuminate; sepals elliptic to ovate
  2. Sepals, petals, and stamens gen 4 ..... ***T. parviflora***
  2′ Sepals, petals, and stamens gen 5
    3. Nectar disk lobes longer than wide, confluent with filaments ..... ***T. gallica***
    3′ Nectar disk lobes wider than long, alternate filaments
      4. Lvs oblong to narrowly lanceolate, acute; sepals entire ..... ***T. chinensis***
      4′ Lvs ovate, acute to acuminate; sepals minutely toothed ..... ***T. ramosissima***

T. aphylla (L.) Karsten (p. 1087) ATHEL Tree < 12 m. **LVS** on twigs not overlapping, ± 2 mm, strongly clasping, abruptly pointed. **INFL**: spike 2–6 cm; bract clasping, triangular, tip acuminate. **FL**: sepals 1–1.5 mm, ± round, tip obtuse, entire, outer < inner; petals ± 2 mm, oblong to elliptic; nectar disk lobes wider than long; stamens alternate disk lobes. Uncommon. Washes, roadsides; < 200 m. SnJV, e SCo, D; to TX; native to India, Afr.

T. chinensis Lour. Tree < 10 m. **LF** 1.5–3 mm, oblong to narrowly lanceolate, acute. **INFL**: raceme 1.5–7 cm, dense; bract linear to lanceolate, acute to acuminate; pedicel ± 1 mm. **FL**: sepals 0.5–1.5 mm, ovate, acute; petals 1.5–2 mm, oblong to elliptic; nectar disk lobes wider than long; stamens alternate disk lobes. Uncommon. Canyons, riverbanks, roadsides; < 200 m. SCo, D; to s Can, Mex; native to Asia. Possibly = *T. ramosissima*.

T. gallica L. Shrub or tree < 8 m. **LF** 1.5–2 mm, linear to narrowly lanceolate, acute. **INFL**: spike 2–5 cm; bract oblong to lanceolate, acute to acuminate. **FL**: sepals 0.5–1 mm, ovate, acute, outer slightly < inner; petals 1.5–2 mm, elliptic to ovate; nectar disk lobes longer than wide, confluent with filaments. Uncommon. Washes, flats, roadsides; < 300 m. s NCoR, SnJV, SnFrB, SCo, n DMoj (Death Valley); to TX; native to s Eur. Pls with outer sepals narrower than inner, petals sometimes > 2 mm, have been called *T. africana* Poiret; native to Afr.

T. parviflora DC. (p. 1087) Shrub or tree 1.5–5 m. **LF** 2–2.5 mm, ± linear, long-acuminate. **INFL**: spike 1–4 cm; bract > pedicel, triangular, tip obtuse to blunt. **FL**: sepals 4, 1–1.5 mm, elliptic to ovate, entire to finely toothed, outer slightly > and wider than inner, acute to abruptly pointed, inner obtuse; petals 4, ± 2 mm, ± oblong; stamens gen 4; nectar disk 4-lobed, lobes longer than wide, confluent with filaments. Common. Washes, slopes, sand dunes, roadsides; < 800 m. s NCoR, s SNF, Teh, GV, CCo, SnFrB, SCoRI, SCo, SNE, D; to ne N.Am, Mex; native to se Eur. [*P. tetrandra* Pallas misapplied]

T. ramosissima Ledeb. (p. 1087) Shrub or tree < 8 m. **LF** 1.5–3.5 mm, ovate; tip acute to acuminate. **INFL**: spike 1.5–7 cm; bract > pedicel, triangular, acuminate. **FL**: sepals 0.5–1 mm, ± ovate, obtuse to acute; petals 1–2 mm, elliptic to oblanceolate; nectar gland lobes wider than long; stamens alternate disk lobes. Common. Washes, streambanks, ditches; < 800 m. Teh, SnJV, SCo, WTR, SNE, D; to c US, AZ; native to e Asia. [*T. pentandra* Pallas] Possibly = *T. chinensis*.

THYMELAEACEAE MEZEREUM FAMILY

Robert J. Berman

Shrubs, trees. **ST** extremely pliable; bark smooth, leather-like. **LVS** simple, alternate or opposite; stipules 0; blade entire. **INFL**: cluster, raceme, umbel, or fl solitary. **FL** bisexual; calyx corolla-like, shallowly 4-lobed; corolla 0 or inconspicuous; stamens 4 and alternate calyx lobes or 8, inserted on calyx; disk-like nectary often present; pistil 1, ovary superior, chamber 1, ovule 1. **FR**: drupe or nut. 50 genera, 500 spp.: worldwide, esp Australia, trop Afr. Some cult (*Daphne*).

DIRCA LEATHERWOOD

Shrub. **ST** < 3 m; bark smooth, dark brown. **LVS** deciduous. **INFL** appearing before lvs, axillary and terminal clusters; fls 2–3, pendent. **FL**: calyx yellow; stamens 8, well exserted; style slender, > stamens. **FR**: drupe. 2 spp.: CA, e N.Am. (Greek: a fountain in Thebes)

D. occidentalis A. Gray (p. 1087) WESTERN LEATHERWOOD **LF**: petiole 3–5 mm; blade 3–7 cm, broadly elliptic to obovate. **FR** 8–10 mm, broadly elliptic to oblong in outline, green to light brown, infrequently produced. UNCOMMON. Moist slopes in partial shade; 50–300 m. SnFrB. ❀DRN&SHD**17**&IRR14,15,**16**,23,24; DFCLT.

TROPAEOLACEAE NASTURTIUM FAMILY

Elizabeth McClintock

Ann, per, ± fleshy, gen ± glabrous; roots sometimes tuber-like. **ST** prostrate, often twining by petioles. **LVS** gen simple, gen alternate, peltate; blade entire, lobed, or dissected; stipules often 0; petiole gen >> blade. **INFL**: fl gen solitary on long, axillary peduncle. **FL** showy, bisexual, gen bilateral; sepals 5, uppermost with a gen long nectar spur; petals gen 5, clawed, upper 2 unlike lower 3; stamens 8, in 2 whorls, unequal; ovary superior, chambers 3, placentas axile, style 1, >> ovary, 3-lobed, stigmas 3. **FR** separating into 3 gen nut-like, 1-seeded segments. 3 genera, 92 spp.: Mex to S.Am; some cult.

TROPAEOLUM NASTURTIUM

FL bilateral; nectar spur conspicuous. **FR** segments nut-like. ± 90 spp.: Mex to Chile, c Argentina. (Greek: trophy, from shield-like lvs) Edible tubers of *T. tuberosum* Ruiz López & Pavón are sold in Andean markets.

T. majus L. GARDEN NASTURTIUM Ann or per. **ST** > 1 m. **LF**: petiole 5–25 cm; blade 3–12 cm wide, round to ± reniform, veins palmate. **INFL**: peduncle < 20 cm. **FL** 2.5–6 cm diam; sepals > 2 cm, ovate, ± tan, nectar spur > 3 cm; petals > 4.5 cm, gen orange, claws of lower 3 > 1.5 cm, fringed near top. **FR** > 1.5 cm, > 2 cm wide; segments broadly ovate, deeply lobed. Uncommon. Cool, often moist, shaded ravines, intermittent streams; < 300 m. NCo, SnFrB, SCo, s ChI; native to S.Am. Abundantly cult.

ULMACEAE ELM FAMILY

Dieter H. Wilken

Shrub or tree, often monoecious. **LVS** simple, gen alternate, gen 2-ranked, short-petioled; stipules deciduous; blade base often oblique, veins pinnate. **INFL**: cyme, clustered, axillary; fls 1–few; bracts 0. **FL** radial; sepals 4–6, free to fused; corolla 0; stamens 4–6, opposite sepals; ovary superior, chamber 1, ovule 1, style branches gen 2. **FR**: drupe or winged nutlet. ± 15 genera, ± 150 spp.: temp to trop; some cult for orn (*Celtis*, *Ulmus*, *Zelkova*), some used for wood, fibers (esp *Ulmus*). [Elias 1970 J Arnold Arbor 51: 18–40]

1. Fr a spheric drupe; lf blade entire to serrate, lower surface yellowish green, clearly net-veined .... **CELTIS**
1′ Fr a compressed, winged nutlet; lf blade serrate (sometimes doubly so), lower surface pale green, not clearly net-veined .... **ULMUS**

CELTIS HACKBERRY

Shrub or tree; fls often bisexual and unisexual on same pl. **ST**: trunk bark deeply furrowed. **LF**: blade entire to serrate, base oblique. **INFL** gen stalked. **FL**: sepals gen 5–6, gen free to near base; stamens 4–6, exserted; style branches elongate, spreading to reflexed. **FR**: drupe, ± spheric, persistent. **SEED** 1, spheric. ± 60 spp.: esp trop and dry temp.

C. reticulata Torrey NET-LEAF HACKBERRY Pl 1–8 m. **ST**: trunk bark corky, checkered between furrows; twigs puberulent. **LF**: blade 2–8 cm, lanceolate to ovate, leathery, entire to serrate, lower surface yellowish green, clearly net-veined, base obtuse to ± cordate, tip obtuse to acuminate, scabrous. **FL**: pedicel in fr 4–15 mm. **FR** 5–12 mm diam, brownish to purple; pulp thin. Uncommon. Canyons, seeps, washes; 500–1700 m. s SNF, Teh, SnBr, PR, s SNE, e DMtns; to WA, KS, TX, n Mex. Sometimes also cult, persisting (ScV, SNE, D). ❀SUN,DRN:4–6,**14–17**,**22–24**&IRR: 1,**2**,**3**,**7–13**,**18–21**.

ULMUS ELM

Shrub or tree. **ST**: trunk bark scaly to furrowed, gray to brown. **LF** serrate (or doubly so), base gen oblique, obtuse to cordate. **INFL** sessile to short-stalked. **FL** bisexual; calyx gen bell-shaped, lobes 5–9; stamens 5–9, exserted; style branches spreading. **FR**: nutlet, compressed, clearly winged. ± 20 spp.: n temp; some widely cult (e.g. *U. americana*, american elm, *U. ×hollandica*, dutch elm).

1. Lf blade sometimes doubly serrate, lower surface or veins short-hairy .... ***U. minor***
1′ Lf blade gen singly serrate, lower surface gen glabrous (or vein axils puberulent)
 2. Fls appearing after lvs (in late summer to fall); calyx lobes deep; fr wing opaque, brown .... ***U. parvifolia***
 2′ Fls appearing before lvs (in spring); calyx lobes short, unequal; fr wing ± translucent, whitish .... ***U. pumila***

U. minor Miller SMOOTH-LEAVED ELM, ENGLISH ELM Tree gen < 10 m. **ST**: twigs puberulent, becoming glabrous. **LF**: blade 4–8 cm, elliptic to widely ovate, margin sometimes doubly serrate, upper surface glabrous to ± scabrous, lower surface ± glabrous (exc veins short-hairy) to short-soft-hairy. **INFL**: fls appearing before lvs in spring. **FL**: calyx lobes short, ± equal. **FR** 10–20 mm, 8–12 mm wide; body 3–5 mm wide; wing translucent, whitish. $2n=28$. Waste places, canyons, as waif; < 600 m. n&c SNF, SnJV, SnFrB, SCo; to e US; native to s Eur, n Afr. [*U. procera* Salisb.] Cult; gen reproducing by root-sprouts.

U. parvifolia Jacq. CHINESE ELM Shrub or tree gen < 8 m. **ST**: twigs glabrous to puberulent. **LF**: blade 2–7 cm, elliptic to ovate, margin gen singly serrate, gen glabrous. **INFL**: fls appearing after lvs in late summer to fall. **FL**: calyx deeply lobed. **FR** 8–12 mm, 5–10 mm wide; body 3–4.5 mm wide; wing opaque, brown. Waste places, washes, as waif; < 900 m. s SNF, Teh, s SnJV, SnFrB, SCo; to e US; native to e Asia. Cult; gen reproducing by root-sprouts.

U. pumila L. SIBERIAN ELM Tree gen < 10 m. **ST**: twigs glabrous to sparsely puberulent. **LF**: blade 3–9 cm, narrowly elliptic to lanceolate, margin gen singly serrate, glabrous (lower surface vein axils puberulent). **INFL**: fls appearing before lvs in spring. **FL**: calyx lobes short, unequal. **FR** 10–15 mm, 10–12 mm wide; body 2–4 mm wide; wing translucent, whitish. Waste places, roadsides, washes; < 1800 m. SNF, GV, SNE, n DMoj; to c US; native to c Asia. Cult; reproducing by seeds and root-sprouts.

URTICACEAE NETTLE FAMILY

Dennis W. Woodland

Ann to (soft-wooded) trees, glabrous or stinging-hairy, monoecious or dioecious. **LVS** alternate or opposite, gen stipuled, petioled, often with embedded crystals. **INFL** various, axillary. **FLS** gen unisexual, small, greenish; sepals gen 4–5, free to fused; petals 0. **STAMINATE FL**: stamens gen 4–5, opposite sepals, in bud incurved, then springing out. **PISTILLATE FL**: ovary 1, superior, chamber 1, style 0–1, stigma 1, gen hair-tufted. **FR**: gen achene (drupe). 50 genera, 700 spp.: worldwide; some cult (*Boehmeria*; *Pilea*, clearweed) [Miller 1971 J Arnold Arb 52:40–68] Wind-pollinated.

1. Stinging hairs present; lvs opposite
  2. Pistillate sepals 2–4, ± equal, fused ± to tip ..... **HESPEROCNIDE**
  2′ Pistillate sepals 4, ± free, outer 2 < inner 2 ..... **URTICA**
1′ Stinging hairs 0; lvs alternate
  3. Lf entire, crystals gen elongate; fls solitary ..... **SOLEIROLIA**
  3′ Lf entire or toothed, crystals round; fls in cymes
    4. Lf toothed; stipules (or scars) present; per, subshrub ..... **BOEHMERIA**
    4′ Lf entire; stipules (or scars) 0; gen ann ..... **PARIETARIA**

BOEHMERIA FALSE NETTLE

Per to shrub, not stinging-hairy. **LVS** gen opposite, ± ovate, toothed, 3-veined from base; base round to cordate; tip obtuse to acuminate; crystals round. **INFL** spike- or panicle-like. **STAMINATE FL**: sepals 4, ± free; stamens 4. **PISTILLATE FL**: sepals 2–4, fused ± to tip, enclosing ovary. **FR** flat to lenticular. ± 50 spp.: esp trop, subtrop. (G.R. Boehmer, Saxony, 1723–1803)

B. nivea (L.) Gaudich. Per or subshrub 1–3 m, monoecious. **LVS** alternate, scabrous above, white-tomentose below. **INFL** panicle-like, gen < petioles; lower staminate; upper pistillate. $n=14$. Roadsides, waste places; < 200 m. GV, SW; native to e Asia; cult for fiber.

HESPEROCNIDE WESTERN NETTLE

Ann, erect, slender, stinging-hairy, monoecious. **LVS** opposite, toothed. **INFL** gen head-like, with both staminate and pistillate fls. **STAMINATE FL**: sepals 4, ± free; stamens 4. **PISTILLATE FL**: sepals 2–4, ± equal, fused ± to tip. **FR** lenticular, enclosed by ovate, sac-like, membranous calyx. 2 spp.: CA, Hawaii. (Greek: western nettle) [Woodland et al. 1976 Can J Bot 54:374–383]

H. tenella Torrey (p. 1087) **ST** < 5 dm. **LF** 4–40 mm, ovate, thin, bluntly serrate; stipules small; petiole ± < blade, slender; crystals gen elongate, rounder in fully exposed pls. **INFL** < petiole, ± round. **FL**: calyx 1–1.5 mm. **FR** ovate. Chaparral, oak woodland, coastal scrub; < 1000 m. NCoRI, CW, SW; n Baja CA.

PARIETARIA PELLITORY

Ann or per, not stinging-hairy. **LVS** alternate, 1–8 cm, lanceolate to round, entire; stipules 0; crystals round. **INFL** head-, spike-, or panicle-like, gen few-fld; fls subtended by involucre of 1–3 bracts. **FL**: sepals 4, fused below. **STAMINATE FL**: stamens 4. **FR** ovoid, shiny. 20–30 spp.: worldwide. (Latin: wall, from habitat of some) [Hinton 1969 Sida 3:293–297]

1. Fr ± black, tip acute; per; alien ..... ***P. judaica***
1′ Fr tan to ± brown or reddish brown, tip obtuse; ann; native
  2. Lf blade ± linear to ovate, base ± tapered, lowest veins gen from midrib; fr sometimes hidden between calyx lobes ..... ***P. pensylvanica***
  2′ Lf blade round to ovate, base gen truncate to cordate, lowest veins often from basal margin; fr hidden between calyx lobes ..... ***P. hespera***
    3. Lf blade ± round, width = length; calyx lobes spreading to recurved, acuminate ..... var. ***californica***
    3′ Lf blade ovate, wider than long; calyx lobes erect, acute ..... var. ***hespera***

P. hespera B.D. Hinton Ann 2–55 cm, decumbent to erect, sometimes matted. **LF**: blade 5–20 mm, ovate to round, base gen truncate to cordate, tip notched to short-acuminate, lowest veins often from basal margin. **FL**: calyx lobes gen 2–3 mm, acute or acuminate. **FR** hidden between calyx lobes, 0.9–1.3 mm, ovate, tan to ± brown, tip obtuse. Chaparral, deserts, dry woodlands, roadsides, often moist shade, sandy or rocky soils; < 1400 m. SnFrB, SW, D; to UT, NM, nw Mex. [*P. floridana* Nutt. misapplied]

var. ***californica*** B.D. Hinton (p. 1087) **LF**: blade widely ovate to round, length = width. **FL**: calyx lobes recurved to spreading, acuminate, yellow-brown to light reddish brown. Uncommon. Habitats and elevations ± of sp. SnFrB, s CA; n Baja CA.

var. ***hespera*** (p. 1087) **LF**: blade ovate, wider than long. **FL**: calyx lobes erect, acute, dark reddish brown. ± habitats and range of sp.

P. judaica L. Per 1–8 dm, decumbent to erect. **LF** 11–90 mm, narrowly lanceolate to widely ovate, base tapered to round; tip acuminate to long-tapered. **FL**: calyx lobes 2–3.5 mm. **FR** 1–1.5 mm, 0.5–0.9 mm wide, ± black, tip acute. Waste places; gen < 100 m. CCo, SCo; native to Eurasia, n Afr. Often invasive in coastal urban settings.

P. pensylvanica Willd. (p. 1087) Ann < 6 dm, decumbent to erect. **LF**: blade 7–90 mm, linear-lanceolate to ovate, base tapered to obtuse, tip gen acuminate to long-tapered, lowest veins gen from midrib. **FL**: calyx lobes ± erect, 1.5–2 mm, acute, dark reddish brown. **FR** hidden between calyx lobes or not, 0.9–1.5 mm, 0.5–1 mm wide, ovate, ± (reddish) brown, tip obtuse. $2n$=16. Dry ledges, slopes, shady waste areas; < 2500 m. CA-FP; US.

SOLEIROLIA BABY'S TEARS

1 sp. (Capt. J.F. Soleirol, collector esp in Corsica, 1791–1863)

S. soleirolii (Req.) Dandy Ann 2–30 cm, ± mat-forming, not stinging-hairy. **LVS** alternate, 3–8 mm, oblong to ± round, oblique, entire; crystals gen elongate. **INFL**: fls solitary, subtended by 3 winged bractlets; lower fls pistillate, upper fls staminate. **STAMINATE FL**: sepals 4, ± free; stamens 4. **PISTILLATE FL**: sepals fused. **FR**: achene, 0.8–1 mm, ovoid, shiny, enclosed by calyx and bractlets. $2n$=20. Damp, shaded waste places, gardens; gen < 100 m. CCo, SnFrB, SCo, s ChI; Baja CA; native to Corsica, Sardinia.

URTICA STINGING NETTLE

Ann to shrub, weak, stinging-hairy or not, monoecious or dioecious. **LVS** opposite, lanceolate to cordate, toothed, prominently 3–5-veined from base; crystals round to elongate. **INFL** head-, raceme-, or panicle-like. **STAMINATE FL**: sepals 4, ± free, green, sharp-bristly; stamens 4. **PISTILLATE FL**: sepals 4, ± free, outer 2 < inner 2. **FR** lenticular to deltate, enclosed by 2 inner sepals. ± 50 spp.: esp temp. (Latin: to burn, from stinging hairs) [Woodland 1982 Syst Bot 7:282–290]

1. Ann 1–6 dm; lf blade gen < 40 mm; infl gen ± head-like, often < petiole, with staminate and pistillate fls; fr deltate .... ***U. urens***
1′ Per 10–30 dm; lf blade gen > 40 mm; infl spike-, raceme-, or panicle-like, gen > petiole, with staminate or pistillate fls; fr ovate .... ***U. dioica***
  2. St and lower lf surface with no or moderate nonstinging hairs; st green; NW, n CCo .... ssp. ***gracilis***
  2′ St and lower lf surface with moderate to dense nonstinging hairs; st gray-green; CA-FP (± exc NW) .... ssp. ***holosericea***

U. dioica L. Per 10–30 dm, from rhizome, ± erect. **LF**: blade gen > 40 mm, narrowly lanceolate to widely ovate, base tapered to cordate. **INFL** spike-, raceme-, or panicle-like, 1–7 cm, gen > petiole, with only staminate or pistillate fls. **FR** ovate. Streambanks, margins of deciduous woodlands, moist waste places; < 3000 m. CA-FP; US, Can, n Mex, Eurasia. Ssp. *d.* dioecious, native to Eurasia; naturalized in N.Am.

ssp. ***gracilis*** (Aiten) Selander (p. 1087) AMERICAN STINGING NETTLE Gen monoecious. **ST** 10–25 dm, green; nonstinging hairs (st and lower lf surface) 0 to moderate. $2n$=26,52. Habitats and elevations of sp. NW, n CCo; US, Can. [*U. californica* E. Greene; *U. lyallii* S. Watson]

ssp. ***holosericea*** (Nutt.) Thorne (p. 1087) HOARY NETTLE Gen monoecious. **ST** 10–30 dm, gray-green; nonstinging hairs (st and lower lf surface) moderate to dense. $2n$=26. Habitats of sp.; < 3000 m. CA-FP (± exc NW); w US, n Mex. [*U. h.* Nutt.; *U. serra* Blume misapplied].

U. urens L. DWARF NETTLE (p. 1087) Ann 1–6 dm, from slender taproot, monoecious; nonstinging hairs 0 to moderate. **LF**: blade gen < 40 mm, elliptic to ± ovate, base tapered to rounded. **INFL** gen ± head- (spike-) like, 5–25 mm, often < petiole, with staminate and pistillate fls. **FR** 1.5–2.5 mm, deltate. $2n$=24,26,52. Waste areas, gardens, orchards; gen < 3000 m. CA-FP (± exc NW, CaRH); native to Eur.

VALERIANACEAE VALERIAN FAMILY

Lauramay T. Dempster (except as specified)

Ann, per, sometimes strongly scented; odor gen disagreeable. **LVS** simple, pinnately lobed, or compound; petioles sometimes sheathing; basal ± whorled; cauline opposite, petioled to sessile. **INFL**: cyme, panicle, or head-like, gen ± dense. **FLS** gen bisexual; calyx fused to ovary top, limb 0 or highly modified (if present, lobes gen 5–15, coiled inward, becoming plumose, pappus-like, spreading in fr); corolla radial to 2-lipped, lobes gen 5, throat gen > lobes, >> tube, base gen spurred or swollen, tube slender, long or short; stamens gen 1–3, epipetalous; ovary inferior, chamber gen 1 (sometimes 3 but 2 empty or vestigial). **FR**: achene, smooth, ribbed, or winged. ± 17 genera, 300 spp.: gen temp, worldwide exc Australia. Some spp. cult (*Centranthus*), some medicinal (*Valeriana*). [Ferguson 1965 J Arnold Arbor 46:218–225]

1. Per; calyx lobes rolled inward, becoming plumose, pappus-like in fr
  2. Stamen 1; corolla gen purplish red, long-spurred; cauline lvs entire .... **CENTRANTHUS**
  2′ Stamens 3; corolla white or pink, throat swollen at base, spur 0; cauline lvs pinnately lobed .... **VALERIANA**
1′ Ann; calyx gen 0
  3. St simple or unequally branched; ovary 1-chambered; corolla gen spurred .... **PLECTRITIS**
  3′ St equally and repeatedly forked; ovary 2–3-chambered; corolla throat ± swollen at base, spur 0 .... **VALERIANELLA**

CENTRANTHUS

Ann, per (in CA). **STS** 1–many; base sometimes woody. **LVS** gen cauline, simple, entire, toothed, or lobed. **INFL**: cyme, clustered, dense, terminal or axillary. **FL**: calyx lobes 5–15, coiled inward, becoming plumose, spreading and persistent in fr; corolla ± funnel-shaped, lobes ± unequal, spreading, tube long, slender, long-spurred; stamen 1. **FR** ± compressed; inner surface 1-veined, outer surface 3-veined. 12 spp.: Medit. (Greek: spurred fl)

C. ruber (L.) DC. RED VALERIAN Pl glabrous, glaucous; base gen woody. **ST** decumbent to erect, simple or branched, 3–9 dm, hollow. **LVS** 5–8 cm; lower petioled; upper sessile; blades widely oblong to elliptic-lanceolate, acute to rounded, entire. **INFL**: fls many. **FL**: corolla 14–18 mm, gen purplish red, sometimes lavender or white; spur 2 × ovary length. **FR** 3–4 mm, glabrous. $2n$=14. Disturbed urban places, rock or wall crevices, roadsides; < 200 m. s NCo, GV, CCo, SnFrB, cult elsewhere; native to Medit Eur.

PLECTRITIS

Fred R. Ganders

Ann, glabrous. **ST** gen erect, 5–80 cm, gen angled, simple or few-branched. **LVS** simple, basal and cauline, opposite, gen entire; basal short-petioled, spoon-shaped; cauline gen sessile, oblong, ovate, or obovate. **INFL** clustered, head-like or interrupted spike, terminal. **FL**: calyx 0; corolla 2-lipped to ± radial and funnel-shaped, white to dark pink, tube gen spurred at base; stamens 3. **FR**: achene; body ± triangular, 2–4 mm, strongly winged or not, wings lateral, wide, ± glabrous to densely hairy. 5 spp.: w N.Am, sw S.Am. (Greek: spur) [Morey 1962 PhD thesis, Stanford U] Self-compatible; large-fld taxa cross- and self-pollinated, small fld taxa self-pollinated. Wing shape, color, hairiness vary in some spp.

1. Spur gen < 1/2 corolla tube length (sometimes ± 0), slender, tip gen enlarged; fr winged or not, wing margins not thickened; convex side of fr body gen keeled, not grooved lengthwise
 2. Corolla 1.5–3.5 mm, white to very pale pink .... ***P. brachystemon***
 2′ Corolla 4–9.5 mm, pale to dark pink .... ***P. congesta***

1′ Spur gen 1/2+ corolla tube length, slender or blunt, tip not enlarged; fr winged, wing margins thickened; convex side of fr body grooved lengthwise, not keeled
 3. Corolla white to pale pink, ± radial, lobes ± equal, not red-spotted, spur thick, blunt; fr glabrous to hairy near tip or along margins .... ***P. macrocera***
 3′ Corolla pink to dark pink, 2-lipped, lower lip with 2 red spots, spur slender, pointed; fr hairs in vertical bands .... ***P. ciliosa***
  4. Corolla 3.5–8.5 mm, spur > ovary .... ssp. ***ciliosa***
  4′ Corolla 1.5–3.5 mm, spur gen < ovary .... ssp. ***insignis***

P. brachystemon Fischer & C. Meyer (p. 1087) **FL**: corolla 1.5–3.5 mm, uniformly white or pale pink, ± radial to slightly 2-lipped; spur slender, tip gen enlarged or spur reduced to minute swelling; undehisced anthers < 0.7 mm. **FR**: convex side of body keeled lengthwise, winged or not; wings hairy near tip or near margins, margin thin, not grooved lengthwise. $2n$=32. Common. Coastal bluffs, open, partly shaded slopes; < 1850 m. NW, CaR, n SN, GV, CW, SCo; to B.C. [*P. anomala* (A. Gray) Suksd. incl var. *gibbosa* (Suksd.) Dyal; *P. aphanoptera* (A. Gray) Suksd.; *P. congesta* var. *major* (Fischer & C. Meyer) Dyal; *P. magna* (E. Greene) Suksd. incl var. *nitida* (A.A. Heller) Dyal; *P. samolifolia* (DC.) Hoeck incl var. *involuta* (Suksd.) Dyal misapplied]

P. ciliosa (E. Greene) Jepson **FL**: corolla 1.5–8.5 mm, pink or dark pink, 2-lipped, lower lip with 2 red spots; spur slender, pointed. **FR**: convex side of body grooved lengthwise, winged; wing hairs in vertical bands esp near body, wing margin thick, gen grooved lengthwise. $2n$=32. Common. Open, partly shaded slopes; < 1700 m. CA-FP (exc NCo); to WA, AZ, n Baja CA.

ssp. ***ciliosa*** **FL**: corolla 3.5–8.5 mm; spur 1/2+ corolla length, > ovary. Habitat of sp.; < 1100 m. CA-FP (exc NCo, SW); to WA. [*P. californica* (Suksd.) Dyal var *c.*; *P. macroptera* (Suksd.) Rydb. var. *m.*] ❀TRY.

ssp. ***insignis*** (Suksd.) D. Morey (p. 1087) **FL**: corolla 1.5–3.5 mm; spur < 1/2 corolla length, gen < ovary. Habitat and range of sp. [var. *davyana* (Jepson) Dyal; *P. californica* var. *rubens* (Suksd.) Dyal; *P. macroptera* (Suksd.) Rydb. var. *patelliformis* (Suksd.) Dyal]

P. congesta (Lindley) A. DC. SEA BLUSH **FL**: corolla 4–9.5 mm, pale pink to dark pink, 2-lipped, lower lip uniformly colored; spur slender, tip gen enlarged; indehisced anthers > 0.7 mm. **FR** winged or not; convex side of body keeled lengthwise; wings hairy near tip or near margins, margin thin, not grooved lengthwise. $2n$=32. Coastal bluffs, open, partly shaded slopes; 100–900 m. NW, CW; to B.C. ❀TRY.

P. macrocera Torrey & A. Gray (p. 1087) **FL**: corolla 2–3.5 mm, uniformly white or pale pink, ± radial; spur thick, blunt. **FR** winged; convex side of body grooved lengthwise; wings ± glabrous to evenly hairy, margin thick, grooved lengthwise. $2n$=32. Common. Open, partly shaded slopes; < 2000 m. CA-FP (exc SW); to B.C., MT, UT. [var. *collina* (A.A. Heller) Dyal; var. *grayi* (Suksd.) Dyal; var. *mamillata* (Suksd.) Dyal; *P. eichleriana* (Suksd.) A.A. Heller; *P. jepsonii* (Suksd.) Burtt Davy]

VALERIANA VALERIAN

Ann, per (in CA) from rhizome or short underground caudex, glabrous to soft-hairy. **ST** gen erect, 1–several. **LVS**: basal simple or pinnately lobed, tapered to petiole; cauline subsessile to ± clasping, pinnately lobed, distal lobe gen > others. **INFL**: cyme, clustered, ± dense to open, terminal or axillary. **FL**: calyx lobes 5–15, gen ± coiled inward, becoming plumose, spreading and persistent in fr; corolla ± funnel-shaped, white or pink, lobes ± equal, throat >> tube, sometimes swollen near base, tube slender, sometimes obscured by swollen throat; stamens 3; ovary ± 1-chambered. **FR** gen compressed, gen 6-veined vertically. ± 200 spp.: temp worldwide exc Australia. (Latin: strength, from use in folk medicine, or after Valerian, a Roman emperor) [Meyer 1951 Ann Missouri Bot Gard 38:377–503]

1. Corolla throat slightly > lobes; lvs simple or compound, most basal
 2. Corolla throat wide, base strongly swollen, oblique; lvs gen short-hairy; pls gen with bisexual fls . . . ***V. californica***
 2′ Corolla throat tapered to tube, base barely swollen; lvs gen glabrous; some pls with bisexual fls, some with pistillate fls — Wrn . . . . ***V. occidentalis***
1′ Corolla throat ± 2 × lobes; lvs gen compound, basal and cauline
 3. Lvs mostly basal and simple; W&I . . . . ***V. pubicarpa***
 3′ Lvs basal and cauline, mostly lobed or compound; NW . . . . ***V. sitchensis***
  4. Lflets not lobed, fine-crenate to fine-dentate; fr 5–6 mm, lanceolate to ± oblong in outline; < 1200 m . . . . ssp. ***scouleri***
  4′ Lflets irregularly lobed or sparsely toothed; fr ± 4 mm, ovoid or narrowly so; 1500–2200 m . . . . . . ssp. ***sitchensis***

V. californica A.A. Heller (p. 1087) Pls gen with bisexual fls, ± glabrous to short-hairy. **ST** 2.5–5 dm. **LVS** 2–13 cm; basal simple or compound, sometimes deeply 3-lobed, lateral lobes < terminal lobe, blade or terminal lobe ovate-obovate; cauline gen 3–9-lobed, lobe margin entire, finely dentate or few-toothed, terminal lobe gen rounded. **FL**: corolla 4–5 mm, cream-white, lobes slightly < throat. **FR** 4–6 mm, ± ovoid. Moist places, coniferous forest; 1500–3300 m. KR, NCoRH, CaR, SNH, MP; OR, NV. [*V. capitata* Link ssp. *c.* (A.A. Heller) F. Meyer] ❀IRR,DRN:4–6,15–17&SHD:1–3,7,18; DFCLT.

V. occidentalis A.A. Heller (p. 1087) Pls with bisexual or pistillate fls, gen glabrous; nodes, petiole base, sometimes sinuses between lobes short-hairy. **ST** 3–7.5 dm. **LVS** 5–30 cm; basal gen simple, blade ovate to round, sometimes deeply 3-lobed, lateral lobes < terminal lobe; cauline deeply lobed or compound, lobes or lflets 3–7, margin entire, fine-crenate, or fine-dentate, terminal lobe gen obtuse. **FL**: corolla 3–4 mm, white, lobes slightly < throat. **FR** 3–5 mm, ovoid. Moist places, coniferous forest; 1500–1800 m. Wrn; to OR, MT, Colorado. ❀IRR,DRN:4,5,**6**,15,17&SHD:1,2, 7;DFCLT.

V. pubicarpa Rydb. Pls gen with bisexual fls, gen glabrous. **ST** 1.1–7 dm. **LVS** 4–30 cm; basal gen simple, blade elliptic to spoon-shaped; cauline few, reduced, simple or few-lobed, margin ± entire. **FL**: corolla ± 6 mm, white, sometimes pinkish, throat ± 2 × lobe length. **FR** 3.5–5 mm, lanceolate in outline. Moist, rocky slopes, coniferous forest; 2100–3200 m. W&I; to OR, MT, UT.

V. sitchensis Bong. Pls gen with bisexual fls, gen glabrous. **ST** 1.1–7 dm; nodes short-hairy. **LVS** 3–20 cm, basal or cauline; basal simple, lobed, or compound; cauline deeply lobed to compound, lobes or lflets gen 3–7, lanceolate to round. **FL**: corolla 5–8 mm, white or pinkish, throat ± 2 × lobe length. **FR** 4–6 mm, ovoid to ± oblong in outline. Moist places; < 2200 m. NW; to AK, MT, Colorado.

ssp. ***scouleri*** (Rydb.) F. Meyer (p. 1093) **ST** 1–6 dm. **LVS** mostly basal, < 15 cm; lflets ovate to round, margin ± entire, fine-crenate, or fine-dentate; cauline reduced. **FR** 5–6 mm, lanceolate to ± oblong in outline. Moist cliffs, streambanks; < 1200 m. NCo, KR, n NCoRO; to B.C. ❀IRR or WET,DRN:**4–6,15–17**&SHD:7, 14,19–24.

ssp. ***sitchensis*** (p. 1093) **ST** 3–7 dm. **LVS** mostly cauline, 4–30 cm; lflets gen lobed or coarsely toothed. **FR** ± 4 mm, ovoid or narrowly so. Moist places, meadows; 1500–2200 m. KR; to AK, MT. ❀WET or IRR,DRN:1,**4–6**,7,15–17;DFCLT.

VALERIANELLA CORN SALAD

Ann. **ST** erect, equally and repeatedly forked. **LVS** cauline, gen simple, entire or toothed. **INFL**: cymes, dense, terminal, gen paired, subtended by involucre-like ring of bracts, peduncled. **FL**: calyx gen 0; corolla funnel-shaped, lobes unequal, throat swollen near base; stamens 3; ovary 2–3-chambered, 1 chamber fertile, 2 chambers empty (sometimes fused into 1). **FR** ± compressed, grooved vertically. ± 80 spp.: Eurasia, n Afr. (Latin: diminutive of *Valeriana*) [Dyal 1938 Rhodora 40:185–212]

V. locusta (L.) Betcke **ST** 1–4 dm, sparsely hairy; hairs pointed down. **LVS** 0.5–3 cm; lower petioled; upper ± sessile; blade obovate to narrowly oblong, entire, upper sometimes dentate. **FL**: corolla 1.5–2 mm, white, lobes bluish. **FR** 2–3 mm; fertile chamber flanked by thickened, corky wall on 1 side, by empty chambers on opposite side; groove between sterile chambers narrow, shallow. $2n=16$. Moist, gen shaded sites; < 1400 m. KR, NCoRO, n SN, n ScV, SnFrB; to WA, ID; native to Eur. [*V. olitoria* (L.) Pollich] Pls from n SNF (Amador Co.) without thickened corky wall and with deep, wide groove between sterile chambers are *V. carinata* Lois., native to Eur. Both spp. cult for edible lvs.

VERBENACEAE VERVAIN FAMILY

Dieter H. Wilken

Ann to tree, gen hairy. **LVS** cauline, opposite, gen toothed; stipules 0. **INFL**: raceme, spike, or head, gen elongated in fr; bract gen 1 per fl. **FL** bisexual; calyx gen 4–5-toothed; corolla 4–5-lobed, radial to bilateral, salverform to 2-lipped; stamens 4–5, epipetalous (if 4, gen in unequal pairs); ovary superior, 2- or 4-lobed, gen 2-chambered, style 1, often with 2 unequal lobes, only 1 stigmatic, lateral. **FR**: 2 or 4 nutlets, drupe-like, or capsule. ± 90 spp., ± 1900 spp.: esp Am trop. Some cult (*Clerodendron*, *Lantana*, *Verbena*, *Vitex*); some weedy worldwide (*Lantana*); some used for wood (*Tectona*, teak).

1. Ann to per
 2. Calyx 2–4-toothed, ± compressed; corolla ± 2-lipped; nutlets 2 . . . . **PHYLA**
 2′ Calyx 5-toothed, cylindric; corolla gen radial; nutlets 4 . . . . **VERBENA**
1′ Shrub
 3. Lvs entire; corolla ± rotate, lobes leathery; coastal salt marshes . . . . **AVICENNIA**
 3′ Lvs crenate to toothed; corolla salverform to ± 2-lipped, lobes thin; dry places
  4. Infl longer than wide; corolla 2.5–3.5 mm, white . . . . **ALOYSIA**
  4′ Infl wider than long (head); corolla 8–12 mm, yellow to purple . . . . **LANTANA**

ALOYSIA

Shrub, strong-smelling. **LF**: blade lanceolate to ovate. **INFL**: raceme or spike. **FL**: calyx 4-toothed; corolla 5-lobed, ± 2-lipped; stamens 4, in unequal pairs; ovary 2-chambered, ovules 2, style unlobed, stigma ± spheric, terminal. **FR**: nutlets 2. ± 35 spp.: Am. Some used for food flavoring, tea. (Maria Luisa Teresa, Queen of Spain or Luis Antonio Bourbon, Prince of Asturias, Spain, both 18th century).

A. wrightii Abrams (p. 1093) OREGANILLO Shrub < 2 m, ± rounded. **ST**: branches many; twigs brown, angles white. **LF**: petiole < 4 mm; blade 4–17 mm, ovate to ± round, crenate, lower surface densely hairy. **INFL**: spike 1.5–6 cm; axis densely and finely tomentose; bract slightly < calyx, lanceolate. **FL**: calyx 1.5–3 mm, puberulent; corolla 2.5–3.5 mm, white, lobes rounded, upper 2 larger. **FR** < 2 mm. Uncommon. Rocky, often limestone, slopes, Joshua-tree or pinyon/juniper woodland; 900–1600 m. s&e DMtns; to TX, n Mex. [*Lippia w.* Torrey invalid] ❀TRY;DFCLT.

AVICENNIA RED MANGROVE

Shrub, tree; roots branches spreading, exposed at low tide. **LVS** persistent, petioled; blade thick, entire. **INFL** head-like, clusters terminal; bractlets 3, gen < calyx, outer 1 > inner 2. **FL**: sepals 5, fused at base; corolla 4-lobed, radial, rotate; stamens 4, attached at 1 level; ovary gen 1-chambered, ovules gen 4 per chamber, style not lobed, stigma ± spheric, terminal. **FR**: capsule, 2-valved. ± 15 spp.: temp, subtrop worldwide, gen coastal marshes, lagoons. (Avicenna, Persian physician and philosopher, 980–1037)

A. marina (Forskal) Vierh. var. *resinifera* (Forster f.) Bakh. Shrub < 2.5 m. **ST**: trunk < 10 cm. **LF** 4–11 cm, elliptic to ovate, leathery; upper surface ± shiny, lower felt-like. **INFL**: clusters gen 3 per branch; fls 4–9 per cluster; peduncles 2–4 cm; bractlets ± ovate. **FL**: perianth lobes ± ovate, leathery, puberulent outside. **FR** ± 2 cm diam, ovoid. Coastal salt marsh; ± 0 m. SCo (Mission Bay, San Diego Co.); native to New Zealand, Australasia. First planted in 1960's, becoming weedy, extirpation attempted, but pls persistent in 1990. [Moran 1980 Madroño 27:143]

LANTANA

Per or shrub, strong-smelling. **STS** prostrate to erect, branched. **LF**: blades oblong to ovate. **INFL**: spike or head, not elongated in fr; involucre bracts many, inner bracts 1 per fl. **FL**: calyx ± cylindric, tip truncate to 4–5-toothed; corolla 4–5-lobed, ± salverform; stamens 4, in unequal pairs; ovary 2-chambered, ovule 1 per chamber, stigma lateral. **FR**: nutlets 2, ± spheric, drupe-like. ± 150 spp.: trop, esp Am (few Asia, Afr). (Latin: ancient name for *Viburnum*, which this resembles)

1. Corolla yellow, becoming red or orange; involucre bracts oblong to lanceolate .... ***L. camara***
1' Corolla rose-red to purple; involucre bracts ovate .... ***L. montevidensis***

L. camara L. **INFL**: involucre bracts 2–3 mm wide, oblong to lanceolate. **FL**: corolla 8–11 mm, yellow, becoming orange or reddish. Disturbed slopes, abandoned fields, urban weed; < 200 m. SCo as waif; naturalized ± worldwide; perhaps native to Caribbean.

L. montevidensis (Sprengel) Briq. **INFL**: involucre bracts 3.5–5 mm wide, ovate. **FL**: corolla 8–12 mm, rose-red to purple. Disturbed slopes, roadsides, urban weed; < 200 m. SCo as waif; naturalized ± worldwide; native to S.Am.

PHYLA

Per, gen mat-like. **ST**: central gen stolon-like; branches decumbent to erect, glabrous or ± strigose. **LVS** opposite or clustered, strigose to appressed-hairy; hairs forked. **INFL**: spike, ± spheric, becoming cylindric in fr, dense; bracts ovate to wedge-shaped. **FL**: calyx ± compressed, 2–4-toothed; corolla ± 2-lipped, tube gen > calyx; stamens 4, in unequal pairs; ovary 2-chambered, ovules 2, style lobes 2, stigma lateral. **FR**: nutlets 2. ± 15 spp.: warm temp, subtrop Am. (Greek: clan or tribe, from clustered fls) ❀IRR or WET:7,**8,9**,10–12,**13,14**,18,**19–23**&SUN:4–6,**15–17,24**;turf-like GRNCVR; fls attract bees.

1. Lf widest below middle, 15–25 mm wide, teeth 11–21 .... ***P. lanceolata***
1' Lf gen widest at or above middle, 5–10 mm wide, teeth 5–11 .... ***P. nodiflora***
  2. Lf gen 4–5 × longer than wide, narrowly wedge-shaped .... var. ***incisa***
  2' Lf gen 2–4 × longer than wide, oblanceolate to elliptic .... var. ***nodiflora***

P. lanceolata (Michaux) E. Greene (p. 1093) **ST**: internodes gen 3–10 cm; branches 15–50 cm. **LF**: blade 25–60 mm, lanceolate to ovate; margin serrate from below middle, teeth 11–21. **INFL** 7–18 mm; peduncle 4–9 cm. **FL**: corolla white or pale blue to purplish. 2*n*=32. Wet places, marshes, ditches; < 400 m. GV, CCo, SnFrB, SCo, D; to e N.Am, n Mex. [*Lippia l.* Michaux]

P. nodiflora (L.) E. Greene **ST**: internodes gen < 4 cm; branches gen < 15 cm. **LF**: blade 5–30 mm, ± ovate to wedge-shaped; margin gen serrate from above middle; teeth 5–11. **INFL** 6–10 mm; peduncle 1.5–9 cm. **FL**: corolla white to reddish. 2*n*=36. Gen wet places; < 400 m. NW (exc KR, NCoRH), GV, CCo, SnFrB, SCo, PR, DSon; to e N.Am, n Mex.

var. ***incisa*** (Small) Mold. (p. 1093) **LF** gen 4–5 × longer than wide; blade narrowly wedge-shaped. Wet places, pond margins, ditches; < 400 m. SnJV, SCo, DSon; to c US, n Mex. [*Lippia i.* (Small) Tidestrom]

var. ***nodiflora*** (p. 1093) **LF** gen 2–4 × longer than wide; blade elliptic to oblanceolate. Wet places, ditches, fields; < 300 m. NW (exc KR, NCoRH), GV, CCo, SnFrB, SCo, ChI (Santa Cruz, Santa Catalina islands), PR, DSon; warm temp, trop ± worldwide. [*Lippia n.* vars. *canescens* (Kunth) Kuntze, *reptans* (Kunth) Kuntze] Densely matted pls with lvs < 1 cm, widely naturalized from S.Am, have been called *L. n.* var. *rosea* (D. Don) Munz]

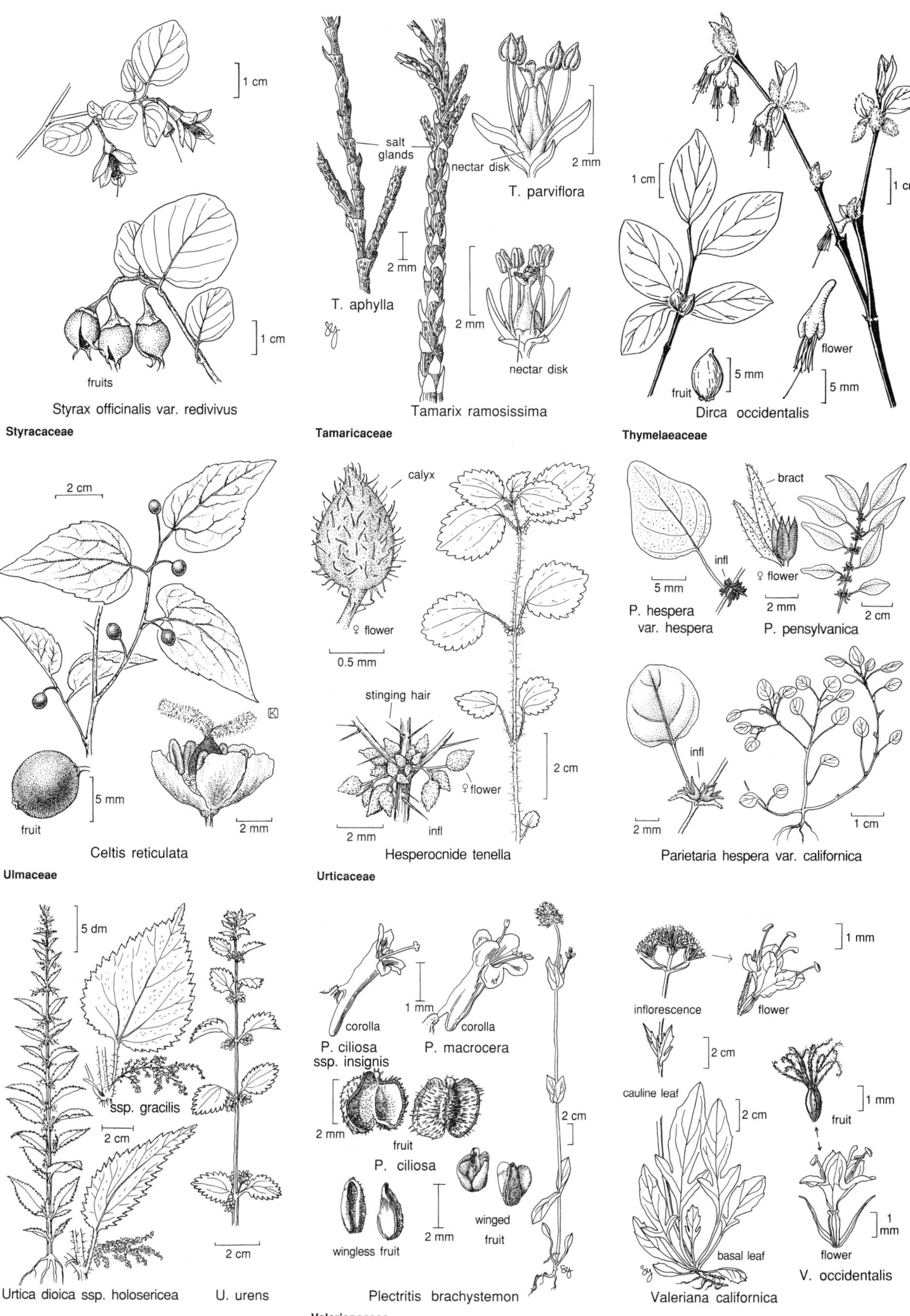
1 cm
fruits
1 cm
Styrax officinalis var. redivivus
Styracaceae
salt glands
nectar disk
2 mm
T. parviflora
2 mm
T. aphylla
2 mm
nectar disk
Tamarix ramosissima
Tamaricaceae
1 cm
1 cm
flower
5 mm
fruit
5 mm
Dirca occidentalis
Thymelaeaceae
2 cm
5 mm
fruit
2 mm
Celtis reticulata
Ulmaceae
calyx
♀ flower
0.5 mm
stinging hair
♀ flower
2 cm
2 mm
infl
Hesperocnide tenella
Urticaceae
bract
infl
♀ flower
5 mm
2 mm
2 cm
P. hespera var. hespera
P. pensylvanica
infl
2 mm
1 cm
Parietaria hespera var. californica
5 dm
ssp. gracilis
2 cm
2 cm
Urtica dioica ssp. holosericea
U. urens
1 mm
corolla
corolla
P. ciliosa ssp. insignis
P. macrocera
2 mm
fruit
P. ciliosa
2 cm
wingless fruit
2 mm
winged fruit
Plectritis brachystemon
Valerianaceae
1 mm
inflorescence
flower
2 cm
cauline leaf
2 cm
1 mm
fruit
1 mm
basal leaf
flower
V. occidentalis
Valeriana californica

VERBENA

Ann to shrub. **STS** often 4-angled; hairs gen short, stiff. **LVS** reduced upward; blade entire to pinnately lobed. **INFL**: spikes, often in panicle-like clusters, gen terminal, gen very long in fr. **FL**: calyx 5-ribbed, 5-toothed, hairs gen strigose or appressed; corolla 4–5-lobed, ± radial to bilateral and 2-lipped; stamens 4, paired; ovary 4–chambered, ovules 4, style 1, lobes 2, 1 tooth-like, 1 with ± spheric stigma. **FR**: nutlets 4, gen oblong. ± 250 spp.: temp, trop Am, Medit Eur. (Latin: ancient name) [Umber 1979 Syst Bot 4:72–102; Barber 1982 Syst Bot 7:433–456]

1. Corolla 8–17 mm, limb 6–10 mm wide; calyx 5–9.5 mm
 2. Lvs 3–5-lobed; calyx hairs spreading ..... ***V. gooddingii***
 2′ Lvs 1–2-pinnately dissected; calyx hairs appressed ..... ***V. tenuisecta***
1′ Corolla 1.5–6 mm, limb < 5 mm wide; calyx 2.5–4 mm
 3. Fl bract 4–8 mm, 2–3 mm > calyx; most sts prostrate to decumbent ..... ***V. bracteata***
 3′ Fl bract 1–4.5 mm, gen < calyx; most sts ascending to erect
  4. Most cauline lf blades 1–2-lobed (lobes below middle 8–25 mm deep), toothed
   5. Herbage sparsely strigose; calyx minutely strigose ..... ***V. menthifolia***
   5′ Herbage and calyx short-spreading-(gen soft-)hairy ..... [2]***V. lasiostachys***
    6. Lf surface grayish green above, hairs gen soft; nutlet scar brownish at 20× ..... var. ***lasiostachys***
    6′ Lf surface green above, hairs gen stiff; nutlet scar white-papillate at 20× ..... var. ***scabrida***
  4′ Most cauline lf blades gen unlobed (exc some *V. hastata*), teeth 2–8 mm
   7. Lower and mid-cauline petioles gen 1–2.5 cm, sometimes narrowly winged
    8. Petioles, infl axis, and calyx short-spreading-(gen soft-) hairy ..... [2]***V. lasiostachya*** (see 6. for vars.)
    8′ Petioles, infl axis, and calyx strigose or scabrous
     9. Frs gen overlapping (axis with 8–10 frs per cm); upper surface of lvs rough-puberulent to slightly scabrous ..... ***V. hastata***
     9′ Frs not overlapping, esp below (axis gen with < 8 frs per cm); upper surface of lvs strigose to scabrous ..... ***V. scabra***
   7′ Lower and mid-cauline petiole gen 0 (blade often clasping)
    10. Frs not overlapping (esp below middle); fl bracts 3.5–4 mm; calyx 4–4.5 mm; spikes 1–5 per cluster ..... ***V. californica***
    10′ Frs overlapping; fl bracts 3–3.5 mm; calyx 3–3.5 mm; spikes gen 5–17 per cluster
     11. Lf base cordate to truncate, gen clasping; spikes ± 5–6 mm diam in fr ..... ***V. bonariensis***
     11′ Lf base ± acute, subsessile; spikes 3–4 mm diam in fr ..... ***V. litoralis***

V. bonariensis L. (p. 1093) Ann or bien 50–150+ cm. **STS** 1–few, erect, glabrous to sparsely rough-hairy; angles smooth to scabrous. **LF** gen clasping, 7–15 cm, elliptic to lanceolate, coarsely serrate, scabrous; base cordate to truncate. **INFL**: spikes (3)8–17 per cluster, 15–125 mm in fr, ± 5–6 mm diam, dense; fl bract 3–3.5 mm. **FL**: calyx 3–3.5 mm; corolla 5–6 mm, white or purplish. **FR** 1–1.5 mm. Disturbed, often wet places, fields, roadsides; < 200 m. s ScV, n SnJV, n SnFrB; to s US; native to S.Am.

V. bracteata Lagasca & J.D. Rodriguez (p. 1093) Ann or bien 8–30 cm. **STS** few–many from base, prostrate to decumbent; hairs sparse, spreading. **LF** 1–3(6) cm, ± oblanceolate, coarsely serrate to lobed, rough-hairy; base tapered to ± flat petiole. **INFL**: spikes 1–3 per st branch, 2–10 cm in fr, 6–10 mm diam, gen dense; fl bract 4–8 mm. **FL**: calyx 2–4 mm; corolla 4–5 mm, white to lavender or blue. **FR** 1–2 mm. $2n$=14. Open, disturbed places, pond or lake margins; < 2200 m. CaR, GV, SCoR, SW, GB, D; to B.C., e N.Am, n Mex.

V. californica Mold. (p. 1093) CALIFORNIA VERVAIN Bien or per, 30–75 cm. **STS** gen 1–3, decumbent to erect, ± canescent. **LF** sessile or ± clasping, 2–9 cm, elliptic to oblanceolate, entire to irregularly and obtusely toothed, ± canescent, tapered to truncate base. **INFL**: spikes 1–3(5) per st branch, 10–24 cm in fr, 5–10 mm diam; open in fr (frs not overlapping); fl bract 3.5–4 mm. **FL**: calyx 4–4.5 mm; corolla 2.5–3.5 mm, violet to purplish. **FR** ± 2 mm. RARE. Wet places, seeps, gen serpentine soils, pine/oak woodland; 300–400 m. c SNF (Tuolumne Co.).

V. gooddingii Briq. (p. 1093) Per (sometimes fl 1st year) 10–45 cm. **STS** 3–10+ from base, decumbent to erect; hairs soft, spreading. **LF** 1–4 cm, lanceolate to ovate, obtusely toothed, short-soft-hairy; base 3–5-lobed, tapered to ± flat petiole. **INFL**: spike gen 1 per st branch, 2–6 cm in fr, 10–15 cm diam, dense; fl bract 4.5–6 mm. **FL**: calyx 6–9.5 mm, hairs spreading; corolla 8–14 mm, purplish blue. **FR** 2–3 mm. $2n$=30. Sandy soils, washes, rocky slopes; 1200–2000 m. e DMoj, ne DSon; to UT, N Mex. [var. *nepetifolia* Tidestrom; *Glandularia g.* (Briq.) Solbrig] ❀DRN,SUN:**10**&IRR:**7–9,11–16**,17,**18–23**,24.

V. hastata L. Bien or per 35–150 cm. **STS** 1–2, erect, sparsely strigose. **LF** 9–15 cm, lanceolate, serrate, rough-puberulent, sparsely short-bristly, or ± scabrous; base acute; petiole 1–2.5 cm. **INFL**: spikes 1–8 per cluster, 3–15 cm, 4–5 mm diam, dense; fl bract 2–3 mm. **FL**: calyx 2.5–3 mm; corolla 2.5–5 mm, blue to violet. **FR** ± 2 mm. $2n$=14. Wet places, ditches, marshes; < 1300 m. GV, CCo, SnFrB, MP; to B.C., e N.Am, AZ. [var. *scabra* Mold.] ❀WET,SUN:1–3,**7–10,14–16**,18,**19–23**,24.

V. lasiostachys Link (p. 1093) Per 35–80 cm. **STS** 1–5+, decumbent to erect; hairs short, spreading. **LF** 4–10 cm, ± ovate, coarsely toothed, gen deeply 1–2-lobed near base, soft-hairy to scabrous; petiole < 2 cm, winged. **INFL**: spikes 1–3 per st, 7–25 cm, gen open below (frs not overlapping); fl bract 3–4.5 mm. **FL**: calyx 2.5–4 mm; corolla 2.5–5 mm, blue to purple. **FR** 1–2 mm. Open, dry to wet places; < 2500 m. CA-FP; OR, Baja CA. Vars. need further study.

var. ***lasiostachys*** **LF**: upper surface grayish green, gen ± soft-hairy, esp near base. **FR**: nutlet scar brownish, ± rough at 20× magnification. $2n$=14. Habitats and range of sp. [*V. abramsii* Mold.] ❀SUN:6,**15–17**&IRR:1,2,**7–9,14,18–24**.

var. ***scabrida*** Mold. **LF**: upper surface green, gen scabrous. **FR**: nutlet scar white-papillate at 20× magnification. Habitats of sp.; < 1500 m. s NW, s SNF, Teh, CW, SCo, ChI (exc San Clemente Island), WTR, w PR; Baja CA. [*V. robusta* E. Greene] ❀SUN:**15, 16**,17&IRR:**7–9,14,18–24**.

V. litoralis Kunth Bien or per 40–150+ cm. **STS** 1–few, erect, glabrous to sparsely short-bristly; angles smooth to scabrous. **LF** 3–10 cm, elliptic to lanceolate, irregularly serrate, sparsely short-bristly; base ± acute; petiole ± 0. **INFL**: spikes (1)3–11 per cluster, 3–7 cm in fr, 3–4 mm diam, dense to open below; fl bract 3–3.5 mm. **FL**: calyx 3–3.5 mm; corolla 4–5.5 mm, violet to purplish. **FR** 1–1.5 mm. $2n$=28,56. Disturbed places, fields; < 200 m. n&c SNF, ScV, n SnJV; to s US, Pacific; native to C.Am, S.Am. [*V. brasiliensis* Vell.]

V. menthifolia Benth. Bien or per 30–75 cm. **STS** 1–3 from base, ascending to erect, gen sparsely strigose. **LF** 2–4(6) cm, ovate, gen deeply 1–2-lobed near base, coarsely serrate, sparsely strigose; base tapered to ± flat petiole. **INFL**: spikes 1–3 per st,

6–30 cm in fr, < 0.5 mm diam, open (frs not overlapping); fl bract 2–3 mm. **FL**: calyx 2.5–3 mm; corolla 2–3 mm, purple. **FR** 1–1.5 mm. Open, gen dry places, shrubland; < 300 m. SCo, w PR; to TX, n Mex.

V. scabra M. Vahl Bien or per 40–100+ cm. **STS** 1–3 from base, gen erect, strigose to scabrous. **LF** 4–10 cm, lanceolate to ovate, toothed, strigose to scabrous; petiole 1–2.5 cm. **INFL**: spikes 1–5 per st, 4–12 cm in fr, < 0.5 mm diam, open (frs not overlapping); fl bract 1–2 mm. **FL**: calyx 2–2.5 mm; corolla 1.5–2.5 mm, blue to lavender. **FR** 1–1.5 mm. Wet places, marshes; < 300 m. s SCo; to se US, n Mex.

V. tenuisecta Briq. Ann to per 15–60 cm. **STS** 1–many from base, decumbent to erect, glabrous to sparsely strigose. **LF** 1–3.5 cm, ± ovate, 1–2-pinnately dissected, strigose. **INFL**: spike gen 1 per st branch, 3–8 cm in fr, 10–15 mm wide, dense; fl bract 3.5–9 mm. **FL**: calyx 5–7 mm; corolla 9–14 mm, white to purple. **FR** 2–3 mm. Dry, disturbed places, abandoned fields (as waif); < 300 m. SCo, SnJV; to s US; native to S.Am. [*V. tenera* Sprengel]

VIOLACEAE VIOLET FAMILY

R. John Little

Ann to shrub or vine (gen per in CA). **LVS** basal, cauline, or both, gen alternate, entire to compound; stipules gen small. **INFL**: head, raceme, panicle, or fls solitary; peduncle bractlets 2. **FL** gen bisexual, gen bilateral; sepals 5, free to slightly fused, gen persistent; petals 5, free, lowest gen spurred or pouched at base; stamens gen 5, alternate petals, filaments short, wide, anthers surrounding ovary, adherent or fused, often with nectaries at base, often with membranous appendage at tip; ovary superior, chamber 1, placentas 3, parietal, ovules gen many, style 1. **FR**: gen capsule, 3-valved, gen explosively dehiscent. **SEEDS** gen appendaged. 15 genera, 600 spp.: gen temp, worldwide; some cult as orn; some Eur spp. medicinally useful as emetics, diuretics, purgatives. [Brizicky 1961 J Arnold Arbor 42:321–333]

VIOLA VIOLET

Ann or per < 35 cm, glabrous to hairy. **LF** entire to compound. **INFL**: fl gen solitary, axillary. **FL** bilateral; sepals subequal, appendaged at base; petals unequal, lowest spurred or pouched at base, lateral 2 equal, gen spreading, often hairy near base, upper 2 equal, erect; lower 2 stamens with nectaries projecting into spur. **FR**: capsule, ovoid to oblong. (Latin: ancient name) [Clausen 1964 Madroño 17:173–197] Cleistogamous fls gen present. Seeds often dispersed by ants that feed on seed-appendages.

1. Stipule lf-like, ± = lf, pinnately lobed; gen ann; alien
 2. Petals gen < sepals, mostly cream, sometimes with some blue-violet on upper 4, spur ± = sepal appendages ..... ***V. arvensis***
 2′ Petals gen > sepals, mainly blue-violet (rarely yellow), spur > sepal appendages — sometimes per ..... ***V. tricolor***
1′ Stipule not lf-like, << lf, entire or toothed; per; gen native
 3. Lf dissected to compound
  4. Cleistogamous fls present, sometimes exclusively; lf blade gen wider than long ..... ***V. sheltonii***
  4′ Cleistogamous fls 0; lf blade gen longer than wide
   5. Upper 2 petals golden-yellow on front, brown to very dark on back ..... ***V. douglasii***
   5′ Upper 2 petals dark red-violet
    6. Lower 3 petals bluish to white, bases with yellow area; st glabrous to puberulent ..... ***V. beckwithii***
    6′ Lower 3 petals cream, bases with deep yellow area; st glabrous ..... ***V. hallii***
 3′ Lf gen ± entire to deeply lobed (seldom dissected)
  7. Petals white to blue or purple, without yellow
   8. St branched; lvs basal and cauline ..... ***V. adunca***
   8′ St 0; lvs basal
    9. Petals white, dark-veined or not
     10. Lf blade elliptic to ovate, longer than wide — nw KR, 100–500 m ..... ***V. primulifolia*** ssp. ***occidentalis***
     10′ Lf blade ovate to reniform, gen as wide as or wider than long
      11. Rhizome < 5 mm diam; spur white; lf blade thin, not leathery, (sub)entire; native; mtns ..... ***V. macloskeyi***
      11′ Rhizome 5–20 mm diam; spur violet; lf blade thick, leathery, ± toothed; alien; SnFrB, SCo ..... [2]***V. odorata***
    9′ Petals blue (sometimes very pale) to purple
     12. Lf glabrous; rhizome thin; NCo ..... ***V. palustris***
     12′ Lf ± glabrous or finely hairy; rhizome thick; KR, SnFrB, SCo, SnBr, Wrn, SNE
      13. Pl stoloned; lf finely hairy; alien; SnFrB, SCo ..... [2]***V. odorata***
      13′ Pl not stoloned; lf ± glabrous; native; KR, SnBr, Wrn, SNE ..... ***V. sororia*** ssp. ***affinis***
  7′ Petals yellow to ± orange, or white with yellow base and spur
   14. Petals white inside, with gen yellow base and spur, lower 3 veined purple-red
    15. Lf blade base gen tapered; basal blade midvein 10–40 mm; pl glabrous ..... ***V. cuneata***
    15′ Lf blade base gen truncate to deeply cordate; basal blade midvein 10–60 mm; pl finely hairy ..... ***V. ocellata***
   14′ Petals yellow to ± orange inside, gen veined dark brown to purple
    16. Sts few, erect, unbranched below; basal lvs 0–2; cauline lvs and peduncles restricted to upper st
     17. Upper 2 petals yellow outside; cauline lf ovate-cordate to widely reniform; moist places ..... ***V. glabella***
     17′ Upper 2 petals purple to brown-purple outside; cauline lf diamond-shaped to reniform, entire to dissected; dry places ..... ***V. lobata***
      18. Cauline lf diamond-shaped to gen deltate, entire to toothed ..... ssp. ***integrifolia***
      18′ Cauline lf ± ovate to gen reniform, palmately lobed or dissected, segments 3–12, gen entire ..... ssp. ***lobata***

16′ Sts 0–many, prostrate to erect, branched below; basal lvs 0–6; cauline lvs and peduncles on lower and upper st

19. Pl stoloned; lf evergreen, leathery, gen purple-spotted on 1 or both sides — coastal forests . ***V. sempervirens***

19′ Pl not stoloned; lf not evergreen, not leathery, not purple-spotted

20. Basal lvs 0; cleistogamous fls 0; petals orange-yellow — fr glabrous .... ***V. pedunculata***

20′ Basal lvs 1–6; cleistogamous fls present; petals yellow to lemon-yellow

21. Basal blade 1.5–6 × longer than wide, 15–80 mm, linear to obovate; cauline blade 2–10 × longer than wide

22. Pl white-tomentose — n&c SNH .... ***V. tomentosa***

22′ Pl glabrous to densely hairy (not white-tomentose)

23. Lf gen entire .... ***V. bakeri***

23′ Lf toothed, jagged, or wavy (seldom entire)

24. Cauline lf blade 3–25 mm wide, linear to ± ovate; peduncle 20–90 mm .... ***V. pinetorum***

25. Pl canescent, 4–10 cm; lvs 3–10 mm wide, basal 40–95 mm, cauline 30–80 mm — s SNH ssp. ***grisea***

25′ Pl glabrous to hairy, 6–22 cm; lvs 4–25 mm wide, basal 50–200 mm, cauline 30–150 mm .... ssp. ***pinetorum***

24′ Cauline lf blade 14–45 mm wide, elliptic to ± ovate; peduncle 40–180 mm .... ***V. praemorsa***

26. Cauline lf blade much longer than wide, gen irregularly toothed or wavy, base gen tapered .... ssp. ***linguifolia***

26′ Cauline lf blade slightly longer than wide, gen regularly toothed, base ± truncate to subcordate .... ssp. ***praemorsa***

21′ Basal blade gen 1–1.5 × longer than wide, 10–50 mm, ovate to round; cauline blade gen < 2.5 × longer than wide

27. Lf canescent above; pl gen 6–13 cm; peduncle 30–100 mm; uncommon — SNE, DMoj .... ***V. aurea***

27′ Lf glabrous to hairy (not canescent) above; pl 1–35 cm; peduncle 15–170 mm; widespread .. ***V. purpurea***

28. Pl 1–12 cm; st gen buried, little branched or elongated by late summer; lower petal (incl spur) gen < 11 mm

29. Cauline lf gen entire; pl 1–12 cm .... ssp. ***integrifolia***

29′ Cauline lf gen crenate to deeply toothed; pl 4–11 cm .... ssp. ***venosa***

28′ Pl 4–35 cm; st gen not buried, clearly branched and elongated above by late summer; lower petal (incl spur) gen > 11 mm

30. Cauline lf deeply toothed .... ssp. ***mohavensis***

30′ Cauline lf entire, jagged, shallowly toothed, or lobed

31. Lf gen green tinged purple, gen glabrous above; lower petal (incl spur) gen < 13 mm; peduncle gen < 13 cm; 500–2500 m, gen montane forest .... ssp. ***purpurea***

31′ Lf gen gray-green (± without purple), puberulent; lower petal (incl spur) gen 13+ mm, or peduncle > 13 cm, or both; 400–1800 m, gen foothill communities .... ssp. ***quercetorum***

V. adunca Smith (p. 1093) WESTERN DOG VIOLET Pl 6–30 cm. **STS** clustered on thin, much-branched rhizomes, branched, compact, gen elongating late-season, gen hairy. **LVS** simple, basal, cauline; petiole 5–70 mm; blade 5–40 mm, round-ovate, gen crenate, ± thickish, glabrous to hairy, base truncate to cordate, tip obtuse. **INFL**: peduncle < 100 mm. **FL**: petals pale to deep violet, lowest (incl spur) 8–16 mm, lower 3 white at base, purple-veined, lateral 2 white-bearded, often curved or hooked at tip. **FR** 6–11 mm, glabrous. n=10. Damp banks, meadow edges in forests; 6–3500 m. CA-FP (gen mtns); to AK, Colorado, e N.Am. [var. *oxyceras* (S. Watson) Jepson] Pls in c SNH (Yosemite) with petals white (exc inside bases violet-tinged) and spur violet have been called var. *kirkii* V. Duran. ❀IRR,DRN:2,3,**7**,14,18–24&SUN:**1,4–6,15–17**.

V. arvensis Murray Ann 5–35 cm. **ST** prostrate to erect, much-branched, puberulent. **LVS** simple, cauline; stipules like and nearly = lf, pinnately lobed; petiole ± 1 cm; blade 15–30 mm, 3–10 mm wide, narrowly lanceolate to ± ovate, coarsely crenate-serrate. **INFL**: peduncle < 80 mm. **FL**: sepals < 12 mm, appendages < 4 mm, ear-like, ± = petal spur; petals gen < sepals, mainly cream (or tinged blue-violet on upper 4), lowest (incl spur) 7–15 mm. **FR** 5–9 mm, glabrous. 2n=34. Lawns, gardens, waste places; < 900 m. CaR, SNF, GV, waif from cult elsewhere; widespread N.Am; native to Eur.

V. aurea Kellogg (p. 1093) Pl 6–13 cm. **ST** decumbent or erect from woody taproot, gray-tomentose. **LVS** simple, canescent; blade base tapered to truncate; basal 1–6, petiole 40–70 mm, blade 12–50 mm, oblong to round, crenate to shallowly and irregularly serrate, tip obtuse; cauline petiole 14–55 mm, blade 15–37 mm, ± lanceolate to ovate, gen dentate-serrate. **INFL**: peduncle 30–100 mm, gen canescent. **FL**: petals yellow, lowest (incl spur) 8–13 mm, lower 3 veined dark brown, lateral 2 with thick hairs, at least upper 2 purple or brown outside. **FR** ± 6 mm, puberulent. n=6. Uncommon. Dry, sandy slopes; 1000–1800 m. SNE, DMoj; w NV.

V. bakeri E. Greene (p. 1093) Pl 3–30 cm. **ST** from erect, woody taproot. **LVS** simple, basal and cauline, gen entire; petiole 1–8 cm; cauline blade ± 55 mm, ± lanceolate, thin, hairs 0 or on margins, veins, or throughout, base tapered, tip obtuse or acute. **INFL**: peduncle 15–75 mm. **FL**: petals yellow, lowest (incl spur) 6–14 mm, lower 3 veined brownish, lateral 2 scaly-puberulent near base, upper 2 sometimes purple or brown outside. **FR** 5–9 mm, glabrous to puberulent. n=24. Openings in coniferous forests; 1300–2700 m. KR, SNH; to WA. [ssp. *shastensis* M. Baker; ssp. *grandis* M. Baker & J. Clausen illegitimate] ❀DRN:1,4–6&SHD,IRR:2,3,7,15–17; DFCLT.

V. beckwithii Torrey & A. Gray (p. 1093) GREAT BASIN VIOLET Pl 6–22 cm. **STS** clustered on short, erect rhizome, glabrous to puberulent. **LVS** basal and cauline, 1-ternate; petiole < 80 mm; blade < 50 mm, deltate to ovate, ± fleshy, lflets dissected, segments ± linear to lanceolate. **INFL**: peduncle 20–120 mm. **FL**: lowest petal (incl spur) 10–18 mm, lower 3 bluish to white, veined dark red-violet, base with yellow area, upper 2 dark red-violet. **FR** 7–12 mm, glabrous. Dry places, often among shrubs; 900–2700 m. n SNH, MP; to OR, ID, UT. [ssp. *glabrata* M. Baker] Cleistogamous fls 0. ❀DRN:1,6,15–17&SHD:2,3,7,18;DFCLT.

V. cuneata S. Watson (p. 1093) Pl 6–25 cm, glabrous. **ST** prostrate to erect from erect or spreading rhizome, glabrous. **LVS** simple; blade base gen tapered; basal blade 10–40 mm, ovate to round, veined purple; cauline smaller, petiole 3–200 mm. **INFL**: peduncle 10–110 mm, very thin. **FL**: petals white with gen yellow base and spur, gen with deep red-violet patches or veins on back, lowest petals (incl spur) 8–14 mm, lower 3 veined purple-red, lateral 2 bearded with club-shaped hairs, with purple eye-spot near base, upper 2 with purplish base or not. **FR** 5–8.5 mm, glabrous to ± minutely scabrous, purplish. Uncommon. Open pine or oak forests, often serpentine; 600–2200 m. NW; sw OR. ❀DRN:1,4–6,17&SHD:2,7,15,16;DFCLT.

V. douglasii Steudel (p. 1093) DOUGLAS VIOLET Pl 4–20 cm. **STS** erect, clustered on short, erect rhizome. **LVS** basal and cauline, odd-2-pinnate (dissected or compound); petiole < 60 mm; blade < 50 mm, ovate, 1° divisions 3–7, segments 1–5 mm wide, linear-elliptic, glabrous to puberulent. **INFL**: peduncle 20–125 mm. **FL**: petals golden-yellow, veined dark, lowest (incl spur) 8–21 mm, upper 2 brown to very dark outside. **FR** 5–12 mm, glabrous. *n*=12,24. Vernally moist, grassy slopes and flats, often serpentine; 150–2300 m. NW, CaRH, n&c SNF, SnJV, CW, SnGb, SnBr, PR; n Baja CA. Cleistogamous fls 0. ❀DRN:1–3,**7**,8,9,14,18&SUN:15–17;DFCLT.

V. glabella Nutt. (p. 1093) STREAM VIOLET Pl 3–34 mm. **ST** erect from scaly, spreading rhizome, glabrous or hairy. **LVS** simple; basal 0–2, petiole 25–270 mm, blade 2–13 cm wide, gen wider than long; cauline only near st tip, ovate to widely reniform, ± entire to crenate or serrate, thin, tips gen acute. **INFL**: peduncles only from upper axils, 20–80 mm. **FL**: petals deep yellow, lowest (incl spur) 6–16 mm, lower 3 veined purple, lateral 2 bearded. **FR** 7–17 mm, glabrous. *n*=12. Damp, wet, or shady places in forests, streambanks, etc; 0–2600 m. NW, CaR, SN, CW, Wrn. [*V. californica* M. Baker] ❀IRR or WET,DRN:**4–6,17**&SHD:1–3,7,14,**15,16**,18; occas. INV.

V. hallii A. Gray (p. 1093) Pl 7–22 cm. **STS** clustered on short, erect rhizome, glabrous. **LVS** basal and cauline, ternate; petiole < 70 mm; blade < 65 mm, ovate to deltate, ± fleshy, lflets dissected, segments 1–7 mm wide, ± lanceolate. **INFL**: peduncle 3–11 cm. **FL**: lowest petal (incl spur) 10–17 mm, lower 3 cream, veined dark red-violet, bases with deep yellow area, upper 2 dark red-violet. **FR** 4–12 mm, glabrous. *n*=36. Open forests, grassy hills, chaparral, serpentine, gravelly soil; 500–2100 m. NW; sw OR. Cleistogamous fls 0. ❀DRN,DRY,SHD:1,5,7,15–17;DFCLT.

V. lobata Benth. PINE VIOLET Pl 5–46 cm. **ST** erect from spreading or erect rhizome. **LVS** simple; basal 0–2, petiole 50–215 mm; cauline only near st tip, petiole gen 10–50 mm, blade 15–150 mm wide, reniform, deltate, ovate, or diamond-shaped, entire to dissected, segments entire to toothed, glabrous or hairy, sometimes glaucous. **INFL**: peduncles only from upper axils, 20–130 mm. **FL**: petals yellow, all or lower 3 veined purple-brown toward base, lowest (incl spur) 8–19 mm, deep yellow, lateral 2 yellow-bearded, upper 2 purple to brown-purple outside. **FR** 6–16 mm, glabrous. *n*=6. Dry, shaded or open woodlands; 150–2300 m. NW, CaR, SN, PR; s OR, n Baja CA.

ssp. ***integrifolia*** (S. Watson) R.J. Little (p. 1095) **LVS**: cauline blade diamond-shaped to gen deltate, entire to toothed. Uncommon. Habitats of sp.; 400–1700 m. Range of sp. exc Baja CA. [var. *i.* S. Watson]

ssp. ***lobata*** (p. 1095) **LVS**: cauline blade gen wider than long, gen reniform to ± ovate or deltate, palmate-lobed or -dissected, lobes or segments 3–12, fine- or coarse-serrate to gen entire. Habitats of sp.; 150–2300 m. Range of sp. [ssp. *psychodes* (E. Greene) Munz] ❀DRN,DRY,SHD:1,4–7,14–18;DFCLT.

V. macloskeyi F. Lloyd (p. 1095) Pl 2–14 cm from rhizome < 5 mm diam, forming dense patches by late-season stolons. **ST** 0. **LVS** basal, simple, glabrous to hairy; petiole 5–100 mm; blade 10–55 mm, ovate-cordate to round-reniform, entire or obscurely toothed, thin, tip gen obtuse. **INFL**: peduncle 20–150 mm, sometimes reddish. **FL**: petals white, lowest (incl spur) 6–12 mm, lower 3 veined purple, lateral 2 bearded or not. **FR** 5–9 mm, glabrous. *n*=12. Wet meadows, seeps, banks; 1000–3400 m. n&c CA-FP (mtns), SnBr, SnJt; to Can, ne US, NV. Glabrous pls with deeply crenate lvs have been called ssp. *pallens* (Banks) M. Baker. ❀WET,DRN:1,2,4,5, **6**,15–17&SHD:7,1,8;DFCLT.

V. ocellata Torrey & A. Gray (p. 1095) WESTERN HEART'S EASE Pl 5–37 cm. **ST** prostrate to erect from gen erect rhizome, finely hairy. **LVS** simple; blade base truncate to gen deeply cordate; basal blade midvein 10–60 mm, lower ovate-cordate; cauline smaller, petiole 3–90 mm. **INFL**: peduncle 10–100 mm. **FL**: petals white with yellow base and spur, lowest (incl spur) 8–15 mm, with yellow patch near base, lower 3 veined red-purple, lateral 2 bearded with yellow club-shaped hairs, purple eye-spot near base, at least upper 2 deep red-violet outside. **FR** 5–8 mm, minutely scabrous, green. *n*=6. Rocky or grassy banks, thickets, often on serpentine; 150–1100 m. NW, CaRF, n&s CW; sw OR. ❀IRR,DRN:4,**5**&SHD:6, 7,15,16,**17**.

V. odorata L. ENGLISH VIOLET Pl 6–24 cm from rhizome 5–20 mm diam, spreading by long, leafy stolons, finely hairy. **ST** 0. **LVS** basal, simple; petiole 20–170 mm; blade 15–60 mm, widely ovate-cordate, ± toothed, thick, leathery, base cordate, tip obtuse. **INFL**: peduncle 20–150 mm. **FL**: petals deep violet with white base or white with violet spur, lowest (incl spur) < 20 mm. **FR** 5–8 mm, puberulent, purple. *n*=10. Gen disturbed places; < 200 m. SnFrB, SCo; e US; native to Eurasia. Uncommonly naturalized.

V. palustris L. (p. 1095) MARSH VIOLET Pl 5–22 cm from thin, creeping rhizome, forming late-season stolons, glabrous. **ST** 0. **LVS** basal, simple; stipules reddish; petiole 20–170 mm; blade 20–65 mm, ovate to round, crenate, base cordate, tip obtuse. **INFL**: peduncle < 210 mm. **FL**: sepal appendages 1–2 mm, ear-like; petals acute, almost white to pale blue or violet, lowest (incl spur) 8–17 mm, lower 3 violet-veined at base, lateral 2 ± bearded. **FR** 6–7 mm, glabrous. *n*=24. Uncommon. Swampy, shrubby places; < 150 m. NCo; to AK, e N.Am, Colorado; also in Eur. ❀TRY.

V. pedunculata Torrey & A. Gray (p. 1095) JOHNNY-JUMP-UP Pl 5–39 cm. **ST** gen decumbent to erect from deep, spongy rhizome with many fleshy roots, branched, thin, puberulent. **LVS** cauline, simple; petiole 20–65 mm; blade 10–55 mm, deltate-ovate to cordate, crenate to serrate, glabrous to hairy, tip acute to obtuse. **INFL**: peduncle < 200 mm. **FL**: petals orange-yellow, lowest (incl spur) 10–20 mm, lower 3 veined dark brown, lateral 2 bearded, upper 2 red-brown outside; style 2.9 mm. **FR** 5–11 mm, glabrous. *n*=6. Open, grassy slopes, hillsides, chaparral, oak woodlands; 0–1000 m. NCoRO, NCoRI, SnFrB, CW, SW; n Baja CA. Cleistogamous fls 0. Pls from SCoRI (w San Benito Co.) with lvs smaller, thinner, narrower, petals yellow, style ± 2 mm have been called ssp. *tenuifolia* M. Baker & J. Clausen. ❀DRN,DRY in summer, SUN:5,**7**,8,9, 14–24;DFCLT.

V. pinetorum E. Greene Pl 4–22 cm, rosette-forming or not, glabrous to hairy. **ST** prostrate to erect from woody taproot. **LVS** simple, linear to ovate, gen ± toothed or wavy-margined; blade base tapered, tip acute; basal 4–20 cm, thin or not; cauline 30–150 mm, gen 4 × width. **INFL**: peduncle 20–90 mm. **FL**: lowest petal (incl spur) 5–11 mm, lemon-yellow, lower 3 petals veined purple-brown, lateral 2 bearded, upper 2 ± purple-brown outside. **FR** 4–7 mm, puberulent. *n*=6. Dry places under shrubs or conifers to barren alpine rock; 1400–3400 m. CaR, SN, WTR, SnBr, SnJt. Variable; needs study. ❀TRY;DFCLT.

ssp. ***grisea*** (Jepson) R.J. Little (p. 1095) Pl 4–10 cm, canescent. **LVS** 3–10 mm wide; basal 40–95 mm; cauline 30–80 mm. **INFL**: peduncle < 70 mm. Dry mtn peaks and slopes; 1500–3400 m. Uncommon. s SNH. [*V. purpurea* ssp. *xerophyta* M. Baker & J. Clausen in part]

ssp. ***pinetorum*** (p. 1095) Pl 6–22 cm, glabrous to hairy. **LVS** 4–25 mm wide; basal 50–200 mm; cauline 30–150 mm. **INFL**: peduncle 30–90 mm. Gen dry soil, often under pines; 1400–3100 m. CaR, SN, WTR, SnBr, SnJT. [*V. purpurea* ssp. *mesophyta* M. Baker & J. Clausen & ssp. *xerophyta* M. Baker & J. Clausen in part]

V. praemorsa Douglas ASTORIA VIOLET Pl 6–30 cm. **ST** appearing 0 early, gen later ascending to erect from short, erect rhizome. **LVS** simple; petiole < 200 mm; basal blade 20–85 mm, gen widely (ob)ovate; cauline blade 20–100 mm, elliptic to gen ± ovate, ± irregularly toothed or wavy-margined, hairs 0 to dense, base tapered to subcordate, tip acute or obtuse. **INFL**: peduncle 40–180 mm. **FL**: petals yellow, lowest (incl spur) 12–20 mm, deep yellow, lower 3 veined brown-purple, lateral 2 bearded, upper 2 sometimes purple or brown outside. **FR** 6–11 mm, glabrous to hairy. 2*n*=36, 48. Moist to dry soil, grassy slopes, meadows, yellow-pine forests; 500–2400 m. NW, CaR n&c SNH, Wrn; to sw Can, MT, Colorado. Complex; needs study. ❀TRY.

ssp. ***linguifolia*** (Torrey & A. Gray) M. Baker & J. Clausen (p. 1095) **LVS**: cauline blade < 100 mm, ± elliptic, gen irregularly toothed or wavy-margined, base gen tapered. 2*n*=36,48. Habitats and ± range of sp. [sspp. *arida* M. Baker, *major* (Hook.) M. Baker, and *oregona* M. Baker] ssp. ***praemorsa*** (p. 1095) **LVS**: cauline blade < 75 mm, gen widely ovate, gen regularly toothed, base ± truncate to subcordate. 2*n*=36. Habitats and ± range of sp.; 500–1700 m.

V. primulifolia L. ssp. ***occidentalis*** (A. Gray) L.E. McKinney & R.J. Little (p. 1095) WESTERN BOG VIOLET Pl 8–19 cm from spreading or erect rhizomes, producing thin late-season stolons, glabrous. **ST** 0. **LVS** basal, simple; petiole 30–110 mm; blade 15–70 mm, elliptic to widely ovate, crenate, base tapered, tip acute or rounded. **INFL**: peduncle 40–165 mm. **FL**: petals white, lowest (incl spur) 10–14 mm, lower 3 veined purple, lateral 2 heavily bearded. **FR** 5–8 mm, glabrous. *n*=12. RARE. *Darlingtonia* marshes, bogs; 100–500 m. nw KR (near Gasquet, Del Norte Co.); sw OR. [*V. lanceolata* L. ssp. *o.* (A. Gray) N. Russell] Cleistogamous fls 0. In cult.

V. purpurea Kellogg Pl 1–35 cm. **ST** appearing ± early, sometimes elongating, ascending to erect from woody taproot, gen hairy. **LVS** simple, entire to toothed, often purplish (esp below), ± hairy; basal 1–5, petiole 20–145 mm, blade 10–50 mm, ovate to round, tapered to cordate at base, often fleshy; cauline blade < basal, lanceolate to ovate, base tapered to cordate. **INFL**: peduncle 15–170 mm. **FL**: petals deep lemon-yellow, lowest (incl spur) 6–17 mm, lower 3 veined purple-brown, lateral 2 bearded, upper 2 purplish outside. **FR** 5–12 mm, puberulent. n=6,12. Chaparral, dry forest, timberline communities, sagebrush or desert scrub; 400–3100 m. CA-FP, GB; to WA, WY, Colorado, AZ, n Baja CA. Variable; sspp. intergrade; needs study. ❀TRY;DFCLT.

ssp. ***integrifolia*** M. Baker & J. Clausen Pl 1–12 cm. **LVS**: basal erect to spreading, blade often fleshy, entire, wavy, or shallowly and irregularly dentate, base tapered to subcordate; cauline blade ± lanceolate to ovate, gen entire. **INFL**: peduncle gen < 70 mm. *n*=6. Dry, coniferous forests to timberline, in sandy to rocky soil, incl serpentine; 1200–2700 m. NW, CaR, n&c SNH; sw OR.

ssp. ***mohavensis*** (M. Baker & J. Clausen) J. Clausen (p. 1095) Pl 8–22 cm. **LVS**: basal gen erect, blade regularly dentate, base tapered to truncate; cauline blade deeply toothed, base tapered, tip acute. **INFL**: peduncle < 140 mm. *n*=6. Desert or sagebrush scrub, dry yellow-pine forest; 1000–2400 m. TR, PR, SNE; sw NV. [*V. aurea* ssp. *m.* M. Baker & J. Clausen] Variable.

ssp. ***purpurea*** (p. 1095) MOUNTAIN VIOLET Pl 4–22 cm. **LVS** gen green, tinged purple, gen glabrous above; basal gen erect, blade base gen truncate to cordate; cauline blade toothed. **INFL**: peduncle gen < 130 mm. *n*=6. Open montane forests, scrub; 500–2500 m. NW, CaR, SN, CW, SW; sw OR. [ssp. *dimorpha* M. Baker & J. Clausen]

ssp. ***quercetorum*** (M. Baker & J. Clausen) R.J. Little (p. 1095) Pl 8–35 cm. **LVS** gen gray-green (or barely purplish), puberulent; basal erect, blade base gen tapered; cauline blade toothed. **INFL**: peduncle often > 130 mm. *n*=12. Dry foothill slopes, in grass or shrubs; 400–1800 m. NW, CaR, SN, CW, SW; sw OR. [*V. q.* M. Baker & J. Clausen] Similar to ssp. *p.* when dried.

ssp. ***venosa*** (S. Watson) M. Baker & J. Clausen (p. 1095) Pl 4–11 cm. **LVS**: basal erect to spreading, blade often fleshy, base tapered to subcordate; cauline blade gen crenate to deeply toothed, base gen tapered. **INFL**: peduncle < 70 mm. *n*=6. Dense shade of forests or shrublands; 1300–3100 m. GB; to WA, WY, Colorado. [sspp. *atriplicifolia* (E. Greene) M. Baker & J. Clausen & *geophyta* M. Baker & J. Clausen]

V. sempervirens E. Greene (p. 1095) EVERGREEN VIOLET Pl 2–15 cm; stolons with scattered lvs, at nodes rooting, rosette-forming. **ST** erect from scaly rhizome. **LVS** basal and cauline, simple, leathery, evergreen; hairs 0 to sparse, esp on veins; petiole 10–160 mm; blade 10–40 mm, gen wider than long, gen purple-spotted on 1 or both surfaces, crenate, base cordate, tip obtuse. **INFL**: peduncle 50–100 mm. **FL**: petals lemon-yellow, lowest (incl spur) 8–17 mm, very wide, lower 3 veined purple, lateral 2 bearded. **FR** < 7 mm, purple-mottled. *n*=12,24. Coastal forests; 30–1400 m. NW, CW; to B.C. ❀SHD,IRR,DRN:**4,5**,6,14–16,**17**,24;DFCLT;GRNCVR.

V. sheltonii Torrey (p. 1095) Pl 3–27 cm. **STS** clustered on short, erect rhizome. **LVS** basal and cauline, ternate; petiole < 210 mm; blade < 70 mm, gen ± reniform, gen glabrous, lflets dissected, segments 1–10 mm wide, ± linear to obovate, gen glabrous. **INFL**: peduncle < 180 mm. **FL**: petals deep lemon-yellow, lowest (incl spur) 7–18 mm, lower 3 veined brown-purple, lateral 2 bearded with club-shaped hairs, upper 2 brown-purple outside. **FR** 6–8 mm, glabrous to puberulent. *n*=6. Fir, pine, or oak woodlands, rich or gravelly soil; 800–2500 m. NW, n SNH, SCoR, n WTR, SnBr, nw PR, Wrn; to WA, Colorado. ❀TRY;DFCLT.

V. sororia Willd. ssp. ***affinis*** (Le Conte) R.J. Little (p. 1095) LE-CONTE VIOLET Pl 3–25 cm from short thick rhizome with long fibrous roots, ± glabrous. **ST** 0. **LVS** basal, simple; petiole 50–230 mm; blade 10–45 mm, gen widely ovate to reniform, toothed, tip acute to obtuse. **INFL**: peduncle 50–250 mm. **FL**: petals deep blue-violet, lowest (incl spur) 10–19 mm (spur ± 3 mm, as wide as long), lower 3 dark-veined, white-bearded. **FR** 5–10 mm, glabrous. *n*=27. Shady, moist ground; 300–2300 m. KR, SnBr, Wrn, SNE. [*V. nephrophylla* E. Greene] ❀IRR,DRN:**4–6,15–17**&SHD:1–3,**7**,8, 9,**14**,18–24;occas. INV.

V. tomentosa M. Baker & J. Clausen (p. 1095) WOOLLY VIOLET Pl 5–10 cm, white-tomentose. **ST** ascending or erect from shallow or deep, woody taproot. **LVS** simple, gen elliptic to narrowly ovate; basal petiole 10–60 mm, blade 15–50 mm, (sub)entire; cauline blade 20–40 mm, gen entire. **INFL**: peduncle 10–40 mm. **FL**: petals yellow, lowest (incl spur) 6–9 mm, veined dark brown, lateral 2 sometimes with short, dark brown veins near base, short-bearded, upper 2 purple or brown outside. **FR** < 5 mm, hairy. *n*=6. RARE. Dry, gravelly places; 1500–2000 m. n-c SNH. Cleistogamous fls 0.

V. tricolor L. WILD PANSY Ann (per) 3–20 cm. **ST** angled, slightly winged. **LVS** cauline, simple; stipules lf-like, nearly = lf, pinnately lobed; petiole 7–33 mm; blade < 20 mm wide, oblong to ovate, coarsely toothed. **INFL**: peduncle < 90 mm. **FL**: sepals < 12 mm, appendages < 4 mm, ear-like, lanceolate, < petal spur; petals mostly blue-violet (rarely yellow), lowest (incl spur) 10–15 mm. 2*n*=26. Uncommon. Disturbed places; < 1000 m. CCo, SnFrB; native of Eur; cult, profusely self-seeding.

VISCACEAE MISTLETOE FAMILY

Frank G. Hawksworth & Delbert Wiens

Per, shrub, gen ± green, parasitic on aboveground parts of woody pls, dioecious or monoecious. **ST** brittle; 2° branches gen many. **LVS** simple, entire, opposite, 4-ranked, with blade or lvs scale-like (then each pair gen fused). **INFL**: spikes or open cymes, gen axillary, sometimes terminal; bracts opposite, 4-ranked, scale-like, each pair fused. **FL** unisexual, radial, 2–4 mm; perianth parts in gen ± 1 series. **STAMINATE FL**: perianth parts 3–4(7); anthers gen ± sessile, opposite and gen on perianth parts. **PISTILLATE FL**: perianth parts gen 2–4; ovary inferior, 1-chambered, style unbranched, stigma ± obscure. **FR**: berry, shiny, gelatinous. **SEEDS** 1(–2), without thickened coat. 7 genera, ± 450 spp.: trop, gen n temp. [Kuijt 1982 J Arnold Arbor 63:401–410] Sometimes incl in Loranthaceae; parasitic on pls in many other families. Frs gen dispersed by birds or seeds explosively ejected. All parts of most members may be TOXIC.

1. St gen < 20 cm, ± angled, at least when young, yellow, straw, yellow-green, olive-green, green, brown, reddish, purple; lf < 1 mm, scale-like; berry ± compressed-spheric, 2-colored (1 color below, 1 above), not white, not pink, not reddish, explosive, pedicel short, recurved; pistillate perianth parts 2; anthers ± 1-chambered; on *Abies, Pinus, Pseudotsuga, Tsuga*, rarely *Picea* .... **ARCEUTHOBIUM**

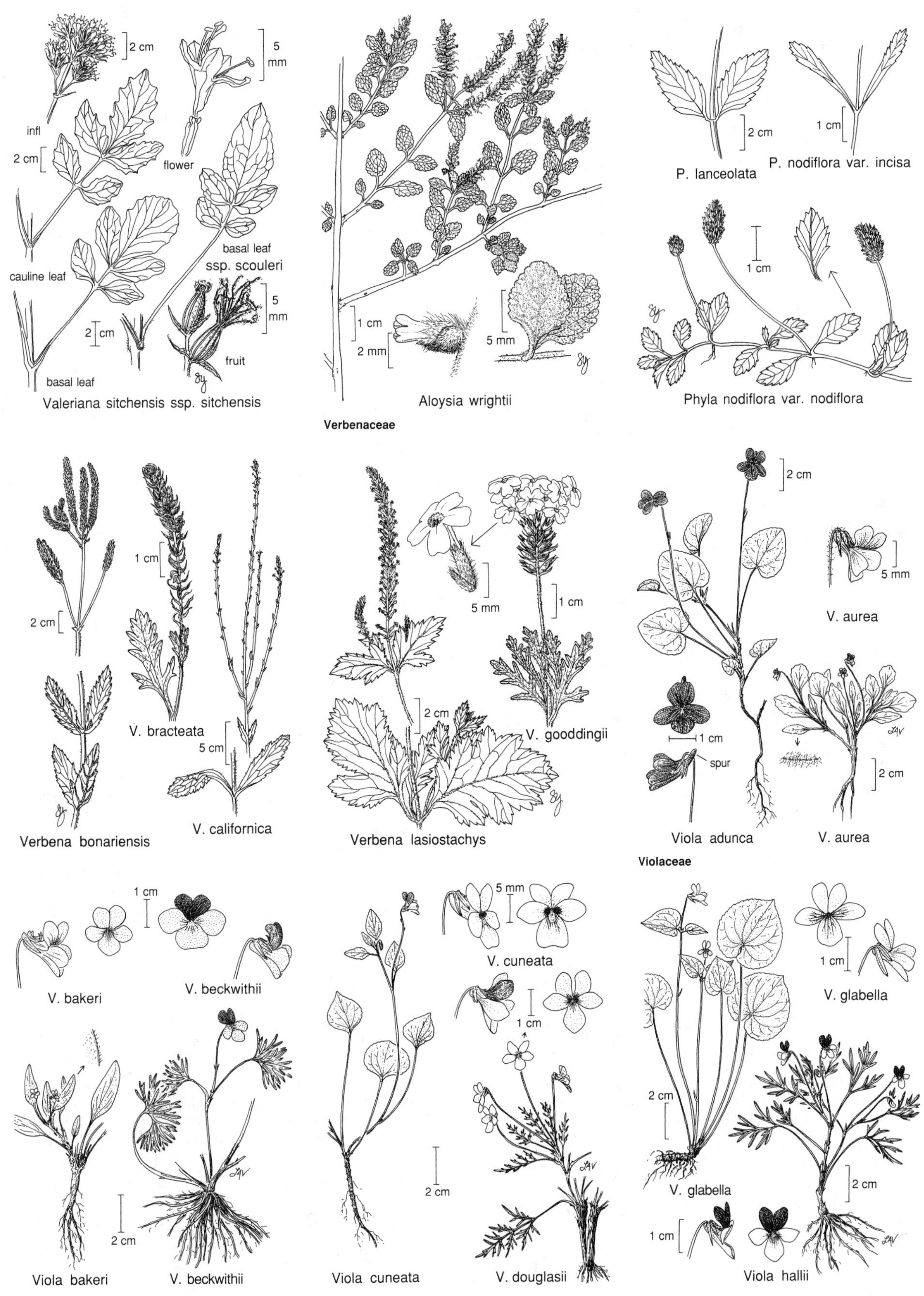
2 cm
5 mm
infl
2 cm
flower
cauline leaf
basal leaf
ssp. scouleri
5 mm
2 cm
fruit
basal leaf
Valeriana sitchensis ssp. sitchensis
1 cm
2 mm
5 mm
Aloysia wrightii
Verbenaceae
2 cm
1 cm
P. lanceolata
P. nodiflora var. incisa
1 cm
Phyla nodiflora var. nodiflora
1 cm
2 cm
V. bracteata
5 cm
Verbena bonariensis
V. californica
5 mm
1 cm
2 cm
V. gooddingii
Verbena lasiostachys
2 cm
5 mm
V. aurea
1 cm
spur
2 cm
Viola adunca
V. aurea
Violaceae
1 cm
V. bakeri
V. beckwithii
2 cm
Viola bakeri
V. beckwithii
5 mm
V. cuneata
1 cm
2 cm
Viola cuneata
V. douglasii
1 cm
V. glabella
2 cm
V. glabella
2 cm
1 cm
Viola hallii

1′ St gen > 20 cm, rounded, green, yellow-green, gray-green, less often reddish; lf blade, 5–70 mm, or scale-like, < 1 mm; berry ± spheric, 1-colored, white, pink, or reddish, not explosive, pedicel short, ± straight, or 0; pistillate perianth parts gen 3–4; anthers 2-several–chambered; on woody dicots or *Abies, Calocedrus, Cupressus, Juniperus*, reportedly *Pinus* (*P. monophylla*)
2. Perianth parts gen 3, pistillate persistent; infl many-fld, open or ± interrupted spikes; fls sunken into axis; berry 3–6 mm; lf with blade, gen 5–47 mm, or scale-like, < 1 mm; anthers 2-chambered; on esp *Abies, Acacia, Adenostoma, Alnus, Arctostaphylos, Calocedrus, Cercidium, Cupressus, Fraxinus, Juglans, Juniperus, Larrea* (rarely), *Olneya, Platanus, Populus, Prosopis, Quercus, Rhus, Robinia, Salix, Umbellularia*, reportedly on *Pinus* (*P. monophylla*) .... **PHORADENDRON**
2′ Perianth parts gen 4, pistillate gen deciduous; infl few-fld, dense cymes; fls not sunken into axis; berry 6–10 mm; lf with blade, gen 50–70 mm; anthers several-chambered; on esp *Acer, Alnus, Betula, Crataegus, Malus, Populus, Robinia, Salix, Ulmus* .... **VISCUM**

ARCEUTHOBIUM DWARF MISTLETOE

Per, shrub, glabrous, dioecious. **ST** gen < 20 cm, ± angled, at least when young, yellow, straw, yellow-green, olive-green, green, brown, reddish, purple; 2° branches gen not whorled, gen in ± 1 plane. **LF** scale-like. **INFL**: spikes, many-fld, open or ± interrupted, short-peduncled; fls gen opposite, 4-ranked, less often whorled. **STAMINATE FL**: perianth parts 3–4(7); anthers ± 1-chambered. **PISTILLATE FL**: perianth parts 2, persistent. **FR** gen 2–5 mm, ± compressed-spheric, 2-colored (1 color below, 1 above), dispersed by explosion, seeds projected < 15 m; pedicel short, recurved. ± 45 spp.: temp and trop n hemisphere. (Greek: juniper, life) Most important timber pathogens, causing annual loss of many millions of dollars; most spp. cause abnormal branching (witches' brooms) in hosts. [Hawksworth & Wiens 1972 USDA Handbook No. 401]

1. St gen << 2 cm, 2° branches rare; on *Pseudotsuga menziesii*, rarely associated *Abies* .... ***A. douglasii***
1′ St > 2 cm, 2° branches many; on *Abies, Pinus, Tsuga*, rarely *Picea breweriana*
2. 2° branches whorled, in > 1 plane; on *Pinus contorta* ssp. *murrayana*, rarely associated pines .... ***A. americanum***
2′ 2° branches not whorled, in ± 1 plane; on conifers other than *Pinus contorta* ssp. *murrayana*
3. On *Abies, Tsuga*, associated *Pinus monticola* (5 lvs per bundle), rarely *Picea breweriana*
4. St 6–15(22) cm, yellow to yellow-green; on *Abies* .... ***A. abietinum***
4′ St 3–13 cm, green to reddish green; on *Tsuga*, associated *Pinus monticola*, rarely *Picea breweriana* [2]***A. tsugense***
5. St 3–9 cm, on *Tsuga mertensiana*, associated *Pinus monticola*, rarely *Picea breweriana* .... ssp. ***mertensianae***
5′ St 3–13 cm; on gen *Tsuga heterophylla* .... ssp. ***tsugense***
3′ On *Pinus* (1–5 lvs per bundle), less often *Tsuga*, rarely *Picea breweriana*
6. On *Pinus edulis* (gen 2 lvs per bundle), *P. monophylla* (gen 1 lf per bundle), expected on *P. quadrifolia* (gen 4 lvs per bundle) .... ***A. divaricatum***
6′ On pines other than *Pinus edulis, P. monophylla, P. quadrifolia* with 2–3 or 5 lvs per bundle, less often on *Tsuga*, rarely on *Picea breweriana*
7. On pines with 5 lvs per bundle or *Tsuga*, rarely *Picea breweriana*
8. St 3–4 cm, 2° branches dense; on *Pinus albicaulis, P. flexilis*, rarely *P. balfouriana* ssp. *b., P. monticola* .... ***A. cyanocarpum***
8′ St > 4 cm, 2° branches open; on *Pinus lambertiana, P. monticola, Tsuga*, rarely *Picea breweriana*
9. On *Pinus lambertiana*; 600–2000 m; KR, NCoR, CaR, SN, TR, PR .... ***A. californicum***
9′ On *Pinus monticola, Tsuga*, rarely *Picea breweriana*; 0–2500 m; NW, CaRH
10. St 5–10 cm, dark brown to reddish; on *Pinus monticola* .... ***A. monticola***
10′ St 3–13 cm, green to reddish green; on *Tsuga*, associated *Pinus monticola*, rarely *Picea breweriana* .... [2]***A. tsugense*** sspp. see (5., 5′)
7′ On pines with 2–3 lvs per bundle
11. On *Pinus muricata, P. radiata*; NCo, CCo .... ***A. littorum***
11′ On pines other than *P. muricata, P. radiata*; NW, CaR, SN, CW, TR, PR, MP
12. On *Pinus jeffreyi, P. ponderosa*, less often *P. coulteri*; 50–2400 m; KR, NCoR, CaR, SN, TR, PR, MP .... ***A. campylopodum***
12′ On *Pinus attenuata, P. sabiniana*, rarely *P. coulteri*; 100–1200 m; NW, CaR, SN, CW, WTR
13. On *Pinus sabiniana*, rarely *P. coulteri*; NW, CaR, SN, CW, WTR .... ***A. occidentale***
13′ On *Pinus attenuata*; KR .... ***A. siskiyouense***

A. abietinum (Engelm.) Hawksw. & Wiens FIR DWARF MISTLETOE **ST** 6–15(22) cm, 2–6 mm wide at base, yellow to yellow-green. **FL** gen August-September. **SEED** mature September–October. n=14. Common. Fir forests, on *Abies concolor, A. grandis, A. magnifica*; < 2400 m. KR, NCoR, CaRH, SNH, TR, PR; to s WA, s UT, NM, n Mex. [*A. campylopodum* f. *a.* (Engelm.) Gill] 2 "special forms," 1 on *Abies concolor, A. grandis* (0–2200 m; range of sp.), 1 on *A. magnifica* (1500–2400 m; KR, NCoR, CaRH, SNH, s OR).

A. americanum Engelm. LODGEPOLE-PINE DWARF MISTLETOE **ST** 6–18(25) cm, 1–2 mm wide at base; staminate yellow, pistillate green; 2° branches whorled, in > 1 plane. **FL** April–May. **SEED** mature August–September. n=14. Lodgepole pine forests, on *Pinus contorta* ssp. *murrayana*, rarely associated pines; 1300–2500 m. CaRH, SNH; to B.C., Colorado.

A. californicum Hawksw. & Wiens SUGAR-PINE DWARF MISTLETOE **ST** 8–12 cm, 2–3 mm wide at base, yellow to green. **FL** gen July-August. **SEED** mature August–October. n=14. Mixed conifer forests, on *Pinus lambertiana*; 600–2000 m. KR, NCoR, CaR, SN, TR, PR.

A. campylopodum Engelm. (p. 1095) WESTERN DWARF MISTLETOE **ST** 8–14 cm, 3–6 mm wide at base, olive-green to yellow. **FL** August–November. **SEED** mature October–December. n=14. Common. Pine woodlands, forests, on *Pinus jeffreyi, P. ponderosa*, less often *P. coulteri*; 50–2400 m. KR, NCoR, CaR, SN, TR, PR, MP; to WA, NV, NM, n Baja CA. [f. *blumeri* (Nelson) Gill; f. *microcarpum* (Engelm.) Gill, both misapplied.]

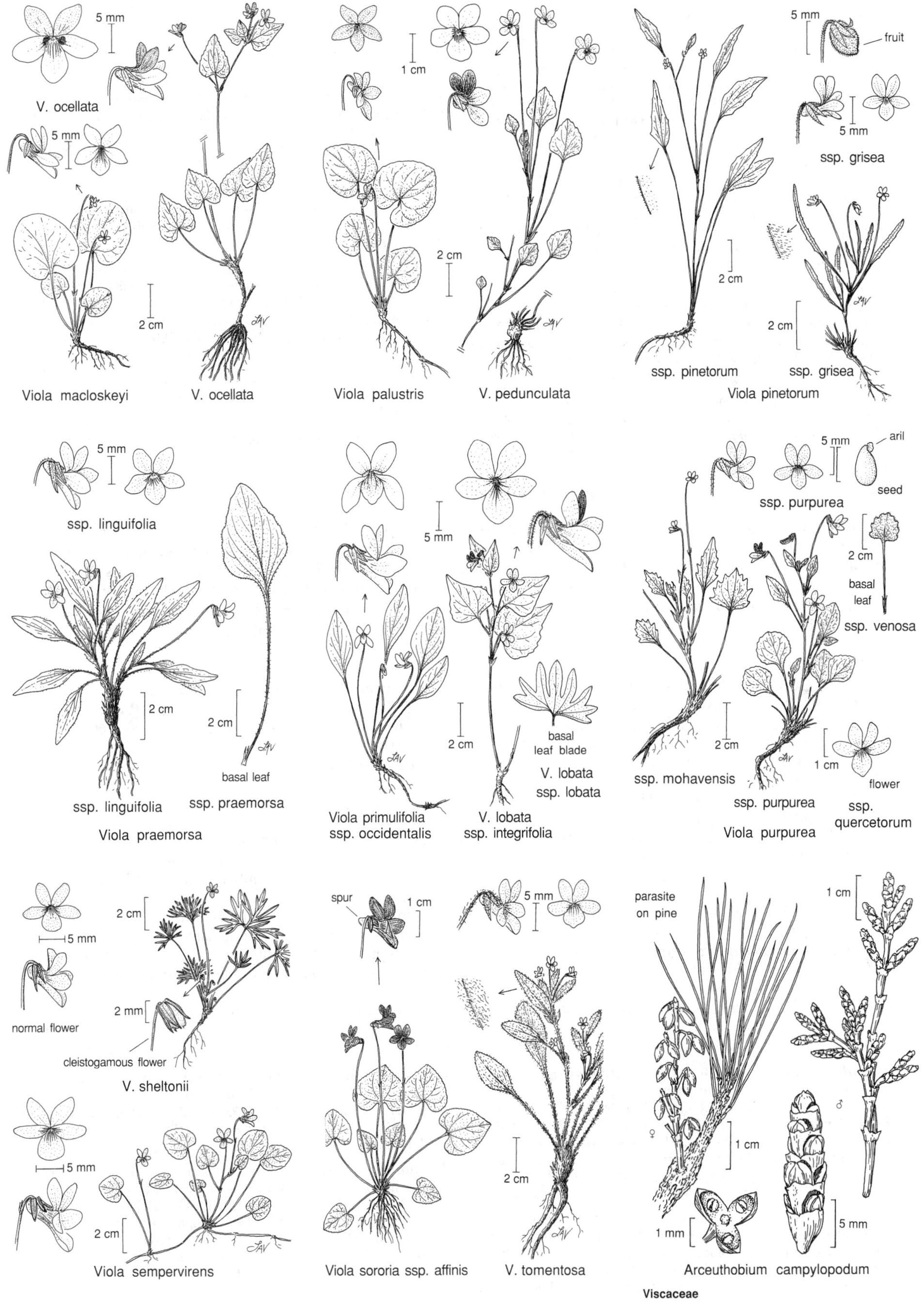
5 mm
V. ocellata
5 mm
2 cm
Viola macloskeyi
V. ocellata
1 cm
2 cm
Viola palustris
V. pedunculata
5 mm
fruit
5 mm
ssp. grisea
2 cm
2 cm
ssp. pinetorum
ssp. grisea
Viola pinetorum
5 mm
ssp. linguifolia
2 cm
2 cm
basal leaf
ssp. linguifolia
ssp. praemorsa
Viola praemorsa
5 mm
2 cm
basal
leaf blade
V. lobata
ssp. lobata
Viola primulifolia
ssp. occidentalis
V. lobata
ssp. integrifolia
5 mm
aril
seed
ssp. purpurea
2 cm
basal
leaf
ssp. venosa
2 cm
1 cm
flower
ssp. mohavensis
ssp. purpurea
ssp.
quercetorum
Viola purpurea
5 mm
2 cm
2 mm
normal flower
cleistogamous flower
V. sheltonii
5 mm
2 cm
Viola sempervirens
spur
1 cm
5 mm
2 cm
Viola sororia ssp. affinis
V. tomentosa
parasite
on pine
1 cm
1 cm
1 mm
5 mm
Arceuthobium campylopodum
Viscaceae

A. cyanocarpum J. Coulter & Nelson LIMBER-PINE DWARF MISTLETOE **ST** 3–4 cm, 1–2 mm wide at base, yellow-green. **FL** August–September. **SEED** mature August–September. *n*=14. Very uncommon. Pine forests, on *Pinus albicaulis, P. flexilis*, rarely *P. balfouriana* ssp. *b., P. monticola*; 2100–2800 m. KR, NCoR, c&s SNH, SnBr; to OR, MT, Colorado, NM. [*A. campylopodum f. cyanocarpum* (J. Coulter & Nelson) Gill] Causes high mortality in *P. albicaulis* forests, ne slope Mount Shastina.

A. divaricatum Engelm. PINYON DWARF MISTLETOE **ST** 7–12 cm, 2–4 mm wide at base, olive-green to brown. **FL** August–September. **SEED** mature September–October. *n*=14. Woodlands, on *Pinus edulis, P. monophylla* (on *P. quadrifolia* in n Baja CA, expectedly so in s CA); 1400–2300 m. c&s SNH, TR, PR, SNE, DMtns; to Colorado, TX, Baja CA. [*A. campylopodum* f. *d.* (Engelm.) Gill]

A. douglasii Engelm. DOUGLAS-FIR DWARF MISTLETOE **ST** gen < 2 cm, 1 mm wide at base, green; 2° branches rare. **FL** May–June. **SEED** mature September–October. *n*=14. Uncommon. Mixed conifer woodlands, on *Pseudotsuga menziesii*, rarely associated *Abies*; 1500–2000 m. KR, CaRH; to B.C., Colorado.

A. littorum Hawksw. & D. Nickrent COASTAL DWARF MISTLETOE **ST** 10–20 cm, 2–5 mm wide at base, dark brown to olive-green. **FL** September–October. **SEED** mature October–December. Within 10 km of coast, on *Pinus muricata, P. radiata*; 0–250 m. NCo, CCo.

A. monticola Hawksw., Wiens, & D. Nickrent **ST** 5–10 cm, 2–4 mm wide at base, dark brown to reddish. **FL** July–August. **SEED** mature October–November. On *P. monticola*; 700–1600 m. KR; to sw OR.

A. occidentale Engelm. FOOTHILL-PINE DWARF MISTLETOE **ST** 6–12 cm, 1.5–4 mm wide at base, light green to straw. **FL** August–October. **SEED** mature September–November. *n*=14. On *Pinus sabiniana*, rarely *P. coulteri*; 100–1200 m. NW, CaR, SN, CW, WTR.

A. siskiyouense Hawksw., Wiens, & D. Nickrent KNOBCONE-PINE DWARF MISTLETOE **ST** 6–10 cm, 2–2.5 mm wide at base, brown. **FL** September. **SEED** mature September–October. On *Pinus attenuata*; 400–1200 m. KR; sw OR.

A. tsugense (C. Rosend.) G. Jones HEMLOCK DWARF MISTLETOE **ST** 3–13 cm, 2–3 mm wide at base, green to reddish green. **FL** August–September. **SEED** mature September–October. *n*=14. On *Tsuga heterophylla, T. mertensiana*, associated *Pinus monticola*, rarely *Picea breweriana*; 0–1000, 1300–2500 m. NW, CaRH; to se AK, s B.C. [*A. campylopodum* f. *t.* (C. Rosend.) Gill]

ssp. ***mertensianae*** Hawksw. & D. Nickrent **ST** 3–9 cm. On *Tsuga mertensiana, Pinus monticola*; 1300–2500 m. KR, NCoR, CaRH; to s B.C.

ssp. ***tsugense*** **ST** 3–13 cm. On *Tsuga heterophylla*; 0–1000 m. NCo; to se AK.

PHORADENDRON MISTLETOE

Shrub, woody at least at base, glabrous or hairy, dioecious in CA. **ST** gen > 20 cm, rounded, green, less often reddish. **LF** with blade or lf scale-like (then each pair fused). **INFL**: spikes, many-fld, open or ± interrupted, short-peduncled; fls sunken into axis. **FL**: perianth parts gen 3. **STAMINATE FL**: anthers 2-chambered. **PISTILLATE FL**: perianth parts persistent. **FR** ± 3–6 mm, ± spheric, 1-colored, white, pink, or reddish, maturing (in temp) in 2 seasons, dispersed by consumption (by birds); pedicel 0. ± 200 spp.: temp, trop Am. (Greek: tree thief) *P. tomentosum* (DC.) Engl. collected in Texas for sale nationally in Christmas trade; other spp. similarly important locally. [Wiens 1974 Brittonia 16:11–54]

1. Lf scale-like, << 1 mm
  2. St canescent, esp at tip, reddish to green; spike with 2–3(7) fertile internodes; on *Acacia, Cercidium, Larrea* (rarely), *Olnyea, Prosopis*; < 1200 m .... ***P. californicum***
  2′ St glabrous, ± green or yellow-green; spike with 1–2 fertile internodes; on *Calocedrus, Juniperus*; 1300–2600 m
    3. St gen erect, internodes 5–10(12) mm; on *Juniperus*; 1700–2600 m .... ***P. juniperinum***
    3′ St often ± pendent in age, internodes (6)10–20 mm; on *Calocedrus*; 1300–1900 m .... ***P. libocedri***

1′ Lf with blade, gen 5–47 mm
  4. Lf length gen < 1.5 × width, 15–47 mm, gen > 20 mm; st gen short-hairy, at least at tip; spike with 2–5(7) fertile internodes, pistillate 6–15(20)-fld; on woody dicots
    5. Lf 30–42 mm, 15–23 mm wide, gen glabrous, if hairy not very densely so; fr glabrous; staminate fertile internodes 2–5(7), gen 30–35-fld, pistillate 2–4(5), 6–10(20)-fld; on woody dicots other than *Quercus* (esp *Alnus, Fraxinus, Juglans, Platanus, Populus, Prosopis, Robinia, Salix*); fl December–March; < 1200 m .... ***P. macrophyllum***
    5′ Lf 15–47 mm, 10–25 mm wide, very densely short-hairy; fr short-hairy near tip; staminate fertile internodes 2–4, gen 25–30-fld, pistillate 2(–3), gen 10–15-fld; on gen *Quercus*, less often other associated woody dicots (e.g. *Adenostoma, Arctostaphylos, Rhus, Umbellularia*); fl July-September; 60–2100 m .... ***P. villosum***
  4′ Lf length gen > 3 × width, 5–25 mm (gen < 20 mm); st glabrous; spike with gen 1 fertile internode, pistillate 2-fld; on conifers
    6. Lf gen 10–15 mm, 2–5 mm wide; on *Cupressus, Juniperus* (reported on *Pinus monophylla*); 200–2300 m .... ***P. densum***
    6′ Lf gen 5–25 mm, 5–8 mm wide; on *Abies concolor*; 1400–2500 m .... ***P. pauciflorum***

P. californicum Nutt. (p. 1103) DESERT MISTLETOE **ST** 4–10 dm, gen pendent in age, reddish to green, canescent, esp at tip, ± glabrous in age; internodes 13–28 mm. **LF** scale-like. **STAMINATE INFL**: fertile internodes 2–3(6), 4–10-fld. **PISTILLATE INFL**: fertile internodes 2–3(7), 2–3-fld. **FL** January–March. **FR** ± 3 mm, white to reddish pink, ± glabrous. *n*=14. Deserts, on *Acacia, Cercidium, Larrea* (rarely), *Olneya, Prosopis*; < 1200 m. D; NV, AZ, Baja CA. [vars. *distans* Trel., *leucocarpum* (Trel.) Jepson]

P. densum Trel. DENSE MISTLETOE **ST** 3–5 dm, gen ± erect, green, glabrous; internodes 6–17 mm. **LF** gen 10–15 mm, 2–5 mm wide, gen oblanceolate-oblong, ± sessile. **STAMINATE INFL**: fertile internodes 1(–2), 6–13-fld. **PISTILLATE INFL**: fertile internodes 1(–2), 2-fld. **FL** June–August. **FR** ± 4 mm, white to straw or pinkish, glabrous. *n*=14,27. Juniper/pinyon woodlands, on *Cupressus, Juniperus* (reported on *Pinus monophylla*, in Mount Pinos area, Ventura Co.); 200–2300 m. KR, NCoR, CaR, CW, TR, PR, GB, DMtns; to s OR, c AZ, nw Mex. [*P. bolleanum* (Seemann) Eichler var. *d.* (Trel.) Fosb.; *P. b.* ssp. *d.* (Trel.) Wiens]

P. juniperinum A. Gray JUNIPER MISTLETOE **ST** 2–4 dm, gen erect, gen woody only at base, yellow-green, glabrous; internodes 5–10(12) mm. **LF** scale-like. **STAMINATE INFL**: fertile internodes 1(2), 5–9-fld. **PISTILLATE INFL**: fertile internode 1, 2-fld. **FL** June–July. **FR** ± 4 mm, pinkish white, glabrous. *n*=14. Juniper/pinyon woodlands, on *Juniperus*; 1700–2600 m. CaRH, SNH, SnBr, MP, DMtns; to OR, Colorado, TX, Mex. [var. *ligatum* (Trel.) Fosb.]

P. libocedri (Engelm.) Howell INCENSE-CEDAR MISTLETOE **ST** 4–8 dm, often ± pendent in age, gen woody only at base, ± green, glabrous; internodes (6–)10–20 mm. **LF** scale-like. **STAMINATE INFL**: fertile internodes 1(–2), ± 7-fld. **PISTILLATE INFL**: fertile internode 1, 2–9-fld. **FL** July–September. **FR** ± 3–4 mm, pinkish white or straw, glabrous. *n*=14. Ponderosa-pine forests, on *Calocedrus decurrens*; 1300–1900 m. KR, NCoR, CaRH, SNH, SCoRI, TR, PR; to s OR, Baja CA. [*P. juniperinum* var. *l.* Engelm.; *P. j.* ssp. *l.* (Engelm.) Wiens]

P. macrophyllum (Engelm.) Cockerell BIG LEAF MISTLETOE **ST** > 1 m, ± erect, green, short-hairy, esp near tip, ± glabrous in age; internodes 22–59 mm. **LF** 30–42 mm, 15–23 mm wide, obovate to elliptic-round, ± petioled to ± not, gen glabrous, if hairy not very densely so. **STAMINATE INFL**: fertile internodes 2–5(7), gen 30–35-fld. **PISTILLATE INFL**: fertile internodes 2–4(5), 6–10(20)-fld. **FL** December–March. **FR** ± 4–5 mm, white, pink-tinged or not, glabrous. *n*=14. On woody dicots other than *Quercus* (esp *Alnus, Fraxinus, Juglans, Platanus, Populus, Prosopis, Robinia, Salix*); < 1200 m. NCoRO, NCoRI, SNF, GV, CW, SCo, TR, PR, D; to Colorado, w TX, Baja CA. [*P. flavescens* (Pursh) Nutt. var. *m.* Engelm.; *P. tomentosum* ssp. *m.* (Engelm.) Wiens]

P. pauciflorum Torrey FIR MISTLETOE **ST** 3–6 dm, ± erect, green, glabrous; internodes 10–21 mm. **LF** gen 5–25 mm, 5–8 mm wide, gen oblanceolate, ± petioled. **STAMINATE INFL**: fertile internodes 1(2), 6–14-fld. **PISTILLATE INFL**: fertile internode 1, 2-fld. **FL** July–August. **FR** 4 mm, pinkish white to straw, glabrous. *n*=14. Ponderosa pine forests, on *Abies concolor*; 1400–2500 m. c&s SN, TR; to s AZ, n Mex. [*P. bolleanum* var. *p.* (Torrey) Fosb.; *P. b.* ssp. *p.* (Torrey) Wiens]

P. villosum (Nutt.) Nutt. (p. 1103) OAK MISTLETOE **ST** ± 1 m, ± erect, gray-green, gen densely short-hairy when young, ± glabrous in age; internodes 15–38 mm. **LF** 15–47 mm, 10–25 mm wide, obovate-elliptic, ± petioled or not, very densely short-hairy. **STAMINATE INFL**: fertile internodes 2–4, gen 25–30-fld. **PISTILLATE INFL**: fertile internodes 2(–3), gen 10–15-fld. **FL** July–September. **FR** ± 3–4 mm, pinkish white, short-hairy near tip. *n*=14. Oak woodlands, on gen *Quercus*, less often other associated woody dicots (e.g. *Adenostoma, Arctostaphylos, Rhus, Umbellularia*); 60–2100 m. KR, NCoR, SNF, ScV, SCoR, SCo, TR, PR; to n OR, TX, Mex. [*P. flavescens* var. *v.* (Nutt.) Engelm.]

VISCUM MISTLETOE

Shrub, glabrous, evergreen, dioecious in CA. **ST** gen < 20 cm, rounded, green, less often reddish; 2° branches opposite, sometimes whorled and in > 1 plane. **LF**: blade well developed. **INFL**: cymes, few-fld, dense, short-peduncled or sessile, subtended by pair of fused bracts. **FL**: perianth parts gen 4. **STAMINATE FL**: anthers several-chambered. **PISTILLATE FL**: perianth parts gen deciduous. **FR** 6–10 mm, ± spheric, white in CA, maturing (in temp) in 2 seasons, dispersed by consumption (by birds); pedicel short, ± straight, or 0. ± 125 spp.: temp, trop, Old World. (Latin: mistletoe) [Hawksworth & Scharpf 1987 Eur J Forest Path 16:1–5]

V. album L. EUROPEAN MISTLETOE **ST**: internodes ± 5 cm. **LF** gen 5–8 cm, ± 1.5 cm wide, narrow-obovate; petiole short. **INFL** 3–5-fld. **FL** February–March. **FR** glabrous. *n*=10. On *Acer, Alnus, Betula, Crataegus, Malus, Populus, Robinia, Salix, Ulmus*, other deciduous trees; 60–100 m. NCoRO (Sebastopol, Santa Rosa, Sonoma Co.); native to Eurasia. Introduced to CA by Luther Burbank ± 1900, sometimes sold locally in Christmas trade.

VITACEAE GRAPE FAMILY

Michael O. Moore

Woody vines; tendrils opposite lvs; fls sometimes unisexual. **LVS** gen many, cauline, simple or compound, alternate, petioled, deciduous; stipules gen deciduous. **INFL** cyme or panicle, gen opposite lf, peduncled. **FL** radial; sepals gen reduced, gen fused, lobes 5 or 0; petals gen 5, free, reflexed and falling individually (or adherent at tips, ± erect, and falling as unit), reddish or yellowish; stamens gen 5, opposite petals; nectaries 0 or between stamens as ± free glands; ovary 1, superior, chambers gen 2(–4), style 1 or 0, stigma inconspicuous or head-like. **FR**: berry. **SEEDS** 1–6, large. 15 genera, ± 800 spp.: esp warm regions; some cult (*Cissus*, grape ivy; *Parthenocissus*; *Vitis*). [Moore 1991 Sida 14:339–367]

1. Lf palmately compound; st center white, partitions 0; petals free; fl gen bisexual; infl a cyme **PARTHENOCISSUS**
1' Lf simple; st center brown, partitioned at nodes; petals adherent at tips; fl unisexual or bisexual; infl a panicle .... **VITIS**

PARTHENOCISSUS VIRGINIA CREEPER, WOODBINE

ST: bark not peeling; st center white, nodal partitions 0; tendrils gen tipped with adhering disks. **LF** palmately compound; lflets 3–7, coarsely serrate. **INFL**: cyme. **FL** gen bisexual; calyx red, lobes shallow; petals free, reddish, margins greenish; nectaries obscure or 0. **FR** obovoid. **SEEDS** 1–4, obovoid. 15 spp.: temp, trop. (Greek: virgin ivy)

P. vitacea (Knerr) Hitchc. WOODBINE **ST**: tendril branches few, adherent disks gen 0. **LF**: lflet upper surface glossy, glabrous, lower surface ± dull, glabrous to hairy. **INFL**: forked at peduncle tip, sometimes again above. **FR** 9–12 mm wide, dark blue to black. 2*n*=40. Uncommon. Hillsides, thickets, ravines, open woodlands; < 1000 m. GV, CW, SW; to TX, ne US. [*P. inserta* (Kerner) K. Fritsch misapplied] Possibly alien in CA. ❀4,5,**6**;IRR:1–3,**7–10**,11,12,**14–16**,17,**18–21**,22–24.

VITIS GRAPE

ST: bark peeling; st center brown, clearly partitioned at nodes; tendrils not tipped with adhering disks. **LF** simple; blade crenate to serrate. **INFL**: panicle. **FL** unisexual or bisexual; calyx greenish, lobes 0 or shallow; petals adherent at tips, yellowish; stamens 3–9, gen erect; nectaries ± free glands. **PISTILLATE FL**: stamens reflexed or 0. **FR** 4–20 mm wide, spheric to ovoid, glaucous or not. **SEEDS** 1–4, obovoid, with a round structure opposite attachment scar that is either raised or sunken. 65 spp.; temp, subtrop. (Latin: vine) [Olmo & Koyama 1980 Proc 3rd Intl Symp Grape Breeding 33–41]

1. Fl bisexual; fr ± ovoid, skin adherent to pulp; st hairy when young, gen becoming glabrous .... ***V. vinifera***
1′ Fl unisexual; fr spheric, skin separating from pulp; st ± tomentose when young, sometimes becoming sparsely so
2. Fr purple, very glaucous, when 3–4-seeded gen > 8 mm wide; seed structure opposite attachment scar gen raised; stipules gen < 3.5 mm; st tomentose when young, becoming less so .... ***V. californica***
2′ Fr black, slightly to not glaucous, when 3–4-seeded gen < 8 mm wide; seed structure opposite attachment scar gen sunken; stipules gen > 3.5 mm; st densely tomentose when young, ± remaining so .... ***V. girdiana***

V. californica Benth. (p. 1103) CALIFORNIA WILD GRAPE **ST** tomentose when young, becoming less so; nodal partitions gen 3–4 mm thick. **LF**: stipules gen < 3.5 mm; blade lobes 0–3, shallow, margin crenate to slightly serrate, lower surface ± tomentose. **FL** unisexual. **FR** gen > 8 mm wide, spheric, purple, glaucous; skin separating from pulp. **SEED**: round structure opposite attachment scar gen raised. $2n$=38. Streamsides, springs, canyons; < 1000 m. NW, CaRF, SNF, GV, CW, SNE; OR. ❀4,5,**6**;IRR:**7–9**,10–12,**14–24**;CVS.

V. girdiana Munson (p. 1103) DESERT WILD GRAPE **ST** ± densely tomentose; nodal partitions gen 2–3 mm thick. **LF**: stipules gen > 3.5 mm; blade lobes 0 or 3–5 and shallow, margin gen serrate, lower surface tomentose to densely so. **FL** unisexual. **FR** gen < 8 mm wide, spheric, black, not or slightly glaucous; skin separating from pulp. **SEED**: round structure opposite attachment scar gen sunken. $2n$=38. Streamsides, canyon bottoms; < 1250 m. SW, s SNE, D; Baja Ca. Intergrades with (perhaps best a var. of) *V. californica.* ❀4–6;IRR:**7–16**,17,**18–24**.

V. vinifera L. CULTIVATED GRAPE, WINE GRAPE **ST** hairy when young, gen becoming glabrous; nodal partitions gen 3–5 mm thick. **LF**: stipules gen < 3.5 mm; blade lobes 0 to 3–5 and deep, margin gen serrate, lower surface glabrous or hairy. **FL** bisexual. **FR** gen > 8 mm wide, ± ovoid, purple to bluish black, densely to not glaucous; skin adhering to pulp. **SEED**: round structure opposite attachment scar sunken or raised. $2n$=38,57,76. Abandoned fields, roadsides; < 1000 m. GV, CW; native to Eur. Hybridizes with native spp.

ZYGOPHYLLACEAE CALTROP FAMILY

Duncan M. Porter

Herb, shrub, often armed; caudex present or not. **ST** branched; nodes often angled, swollen. **LVS** 1-compound, opposite; stipules persistent or deciduous; lflets entire. **INFL**: fls 1–2 in axils. **FL** bisexual; sepals 5, free, persistent or deciduous; petals 5, free, gen spreading, sometimes twisted and appearing propeller-like; stamens 10, sometimes appendaged on inside base; ovary superior, chambers 5–10, ovules 1–several per chamber, placentas axile. **FR**: capsule or splitting into 5–10 nutlets. 26 genera, ± 250 spp.: widespread esp in warm, dry regions; some cult (*Guaiacum*, lignum vitae; *Peganum*, harmal (NOXIOUS and illegal); *Tribulus*, caltrop (pernicious)). [Porter 1972 J Arnold Arbor 53:531–552]

1. Lflets 2
2. Lflets fused at base .... **LARREA**
2′ Lflets free at base .... **ZYGOPHYLLUM**
1′ Lflets 3 or more
3. Lflets 3, palmate, spine-tipped; stipules spine-tipped .... **FAGONIA**
3′ Lflets 6–18, pinnate, not spine-tipped; stipules not spine-tipped
4. Fr tubercled, nutlets 10 .... **KALLSTROEMIA**
4′ Fr spiny, nutlets 5 .... **TRIBULUS**

FAGONIA

Per, shrub. **ST** < 1 m, spreading, angled or ridged. **LF** palmately compound; stipules stiff, spine-tipped; lflets 3, spine-tipped, terminal largest. **INFL**: fls solitary in axils. **FL**: sepals deciduous; petals clawed, twisted, propeller-like, purple to pink, deciduous. **FR**: capsule, deeply 5-lobed, obovoid, ± septicidal; style persistent; peduncle reflexed. **SEED** 1 per chamber. ± 18 spp.; sw N.Am, Chile, Medit, sw Afr. (G.C. Fagon, French physician to Louis XIV, 1638–1718)

1. St ascending to erect, scabrous; glands only on youngest herbage, << 0.1 mm wide; stipules curved; lflets lanceolate .... ***F. laevis***
1′ St prostrate, not scabrous; glands also on older herbage, ± 0.15 mm wide; stipules straight; lflets elliptic to ovate .... ***F. pachyacantha***

F. laevis Standley (p. 1103) Shrub < 1 m, intricately branched. **LF**: lflets 3–9 mm, gen < petiole, 1–4 mm wide. **FL** ± 1 cm wide. **FR** 4–5 mm wide, minutely strigose or hairy, gen with some glands, rarely glabrous; style 1 mm, wider at base. Rocky hillsides to sandy washes; 0–700 m. s DMoj, DSon; to sw UT, nw Mex. [*F. californica* Benth. ssp. *l.* (Standley) Wiggins] Pls with minute, glandular hairs on frs have been called *F. longipes* Standley. ❀TRY.

F. pachyacantha Rydb. (p. 1103) Per; caudex woody. **LF**: lflets < 25 mm, ± = or > petiole, < 9 mm wide. **FL** ± 1.5 cm wide. **FR** 5 mm wide, hairy, gen with some glands; style 2–3 mm, not or barely wider at base. Flat, sandy or rocky habitats; 0–500 m. DSon; to sw AZ, Mex. [*F. californica* Benth. var. *glutinosa* Vail] ❀TRY.

KALLSTROEMIA

Ann, unarmed. **ST** spreading radially, prostrate to ascending, < 1 m. **LF** even-1-pinnate; stipules narrow, green. **INFL**: fls solitary in axils. **FL**: sepals deciduous or persistent; petals yellow to orange, withering but not deciduous. **FR** 10-lobed, splitting into 10 tubercled nutlets; style persistent; peduncle reflexed. **SEED** 1 per chamber. 17 spp.: warm and trop Am. (A. Kallstroem, obscure contemporary of Scopoli, author of genus) [Porter 1969 Contr Gray Herb 198:41–153]

1. Peduncle < subtending lf; petal 3–5 mm; sepals deciduous; style < fr body .... ***K. californica***
1′ Peduncle > subtending lf; petal 6–30 mm; sepals persistent; style > fr body
2. Sepals >> fr body, spreading; petal 15–30 mm, yellow to orange, darker at base .... ***K. grandiflora***
2′ Sepals ± = to slightly > fr body, appressed; petals 6–12 mm, orange .... ***K. parviflora***

K. californica (S. Watson) Vail (p. 1103) **ST** prostrate to decumbent, < 0.7 m, glabrous to strigose. **LF**: stipules 1.5–5 mm; lflets 6–12. **FL**: peduncle < subtending lf; sepals deciduous; petals 3–5 mm, yellow. **FR** ovoid; body > style, 3–5 mm wide. RARE in CA. Flat, sandy or disturbed areas; 0–600 m. D; to TX, Mex. Last collected in CA in 1948.

K. grandiflora A. Gray **ST** decumbent to ascending, < 1 m; some hairs densely silky, others sharply bristly. **LF**: stipules 4–10 mm; lflets 8–16. **FL**: peduncle > subtending lf; sepals persistent, margins strongly inrolled, >> fr body, spreading; petals 15–30 mm, yellow to orange, darker at base. **FR** 2–3 mm wide, ovoid; style 3 × fr body. Sandy roadsides; 300–900 m. Uncommon. DSon (near Desert Center, Riverside Co.; Jacumba, San Diego Co; expected elsewhere); native AZ to TX, w Mex. Waif in CA.

K. parviflora Norton (p. 1103) **ST** prostrate to decumbent, < 1 m, strigose to glabrous. **LF**: stipules 5–7 mm; lflets 6–10. **FL**: peduncle > subtending lf; sepals persistent, ± = to slightly > fr body, appressed; petals 6–12 mm, orange. **FR** 4–6 mm wide, ovoid; style to 3 × fr body. Uncommon. Sandy roadsides, slopes; 1500–2000 m. c PR (near Warner Hot Springs, San Diego Co.), e DMtns (Clark Mtns, San Bernardino Co.), expected elsewhere; native AZ to c US, c Mex. Waif in CA.

LARREA CREOSOTE BUSH

Shrub, unarmed. **ST** branched, erect to prostrate, < 4 m, reddish becoming gray; nodes swollen, darker; hairs 0 or appressed. **LF**: stipules persistent; lflets 2, fused at base. **INFL**: fls solitary in axils. **FL**: sepals deciduous, unequal, overlapping; petals clawed, twisted, propeller-like, yellow, deciduous; stamen appendages bract-like, coarsely toothed. **FR** 5-lobed, spheric, short-stalked, hairy, splitting into 5 hairy, 1-seeded nutlets. 5 spp.: warm, dry Am. (J.A. Hernandez de Larrea, Spanish clergyman)

L. tridentata (DC.) Cov. (p.1103) **LF**: lflets < 18 mm, < 8.5 mm wide, obliquely lanceolate to curved; deciduous awn between lflets < 2 mm. **FL** < 2.5 cm wide; sepals ovoid, appressed-hairy; petal claw brownish; stamens > appendages; ovary hairs dense, straight, stiff, silvery, (reddish brown in fr); style 4–6 mm, persistent on young fr. **FR** 4.5 mm wide. Common. Desert scrub; < 1000 m. SNE, D, (uncommon in Teh, SCo, SnJt); to sw UT, TX, c Mex. [var. *arenaria* L. Benson, Algodones creosote bush; *L. divaricata* ssp. *t.* (DC.) Lowe & Felger] Closely related to s S.Am *L. d.*. Clones may live 10,000 years, longer than any other living pls known; resinous odor characteristic; dominant shrub over vast areas of desert. ❀DRN,SUN:7,**8–10**,**14**,18,**19–21**,22,23&IRR:**11–13**; also STBL.

TRIBULUS PUNCTURE VINE, CALTROP

Ann. **ST** prostrate, spreading radially, < 1 m. **LF** even-1-pinnate; stipules ± lf-like. **INFL**: fls solitary in axils. **FL**: sepals deciduous; petals yellow, deciduous. **FR** 5-lobed, splitting into 5 nutlets, each with 2–4 stout spines; style deciduous; peduncle reflexed. **SEEDS** 3–5 per chamber. ± 12 spp.: esp dry Afr. (Greek: caltrop, weapon used to impede cavalry, from armed fr)

T. terrestris L. (p. 1103) **ST** ± silky or appressed-hairy, sharply bristly to glabrous. **LF**: stipules 1–5 mm; lflets 6–12. **FL** < 5 mm wide; peduncle < subtending lf. **FR** 5 mm, < 1 cm wide, ± flat, hairy, gray or yellowish; spines 4–7 mm, spreading, hairy to glabrous. Roadsides, railways, vacant lots, other dry, disturbed areas; gen < 1000 m. CA; to WY, e US, c Mex; native to Medit. First collected in CA in 1902; long a pernicious weed, now controlled by introduced weevils. TOXIC to livestock in vegetative condition, frs cause mechanical injury.

ZYGOPHYLLUM BEAN-CAPER

Per, unarmed; caudex woody. **ST** branched, fleshy. **LF**: stipules ± lf-like or membranaceous; lflets 2 or more, free, fleshy. **INFL**: fls 1–2 in axils. **FL**: sepals often unequal, deciduous or not; petals ± erect, white, yellow, or orange, base sometimes orange or red. **FR**: capsule, cylindric to spheric, 5-angled, septicidal. **SEEDS** 1–few per chamber. ± 70 spp.: Eurasia, s Afr, Australia. (Greek: yoke lf, from sometimes oblique lflets)

Z. fabago L. (p. 1103) **ST** erect, glabrous. **LF**: lflets < 4.5 cm, 3 cm wide, obliquely ovate; awn between lflets 1 mm. **FL** 6–7 mm; sepals ± = petals; stamens exserted, appendages < stamens, ± linear, divided at tip. **FR** 25–35 mm, oblong-cylindric; style persistent, thread-like, < 7 mm; peduncle reflexed. Uncommon. Disturbed areas; < 1000 m. NCoRO, SnJV, D; native to Medit, c Asia. [var. *brachycarpum* Boiss.] Forms large colonies.

ALISMATACEAE WATER-PLANTAIN FAMILY

Charles E. Turner

Ann, per from corms, stolons, rhizomes, or tubers, aquatic (± emergent or on mud), gen bisexual; roots fibrous. **ST**: caudex short. **LVS**: basal, simple, palmately veined, sometimes floating; submersed blades gen linear to ovate; emergent blades linear to sagittate. **INFL** gen scapose, umbel- to panicle-like; fls whorled, in interrupted clusters. **FLS** bisexual or unisexual, radial; sepals 3, gen green, gen persistent; petals 3, gen > sepals, white or pink; stamens 6–many; pistils 6–many, gen simple. **FR**: achene, gen compressed, beaked. ± 12 genera, 75–100 spp.: esp n hemisphere. [Rogers 1983 J Arnold Arbor 64: 383–420]

1. Pistils (and frs) in 1 whorl; fls bisexual, stamens gen 6
 2. Frs erect, beak < 1 mm; petals ± entire or slightly cut; pistils free at base .... **ALISMA**
 2′ Frs spreading, beak 3–6 mm; petals deeply cut; pistils ± fused at base .... **DAMASONIUM**
1′ Pistils (and frs) in spheric cluster; fls bisexual or unisexual, stamens 0 or > 6
 3. Infl axes and petioles gen angled; fls bisexual, gen 3+ per node; fr cluster bur-like .... **ECHINODORUS**
 3′ Infl axes and petioles gen smooth; fls gen unisexual, staminate above, pistillate below, gen 3 per node; fr cluster not bur-like .... **SAGITTARIA**

ALISMA WATER PLANTAIN

Per; caudex gen corm-like. **LF**: blade linear to ovate, tapered to base or petioled. **INFL**: peduncle gen smooth; pedicels < 45 mm in fr. **FL** bisexual; receptacle gen flat; sepals gen 1–4 mm; petals ± entire to slightly cut, white or pink; stamens gen 6; pistils many, free, in 1 whorl. **FR**: body gen 1.5–3 mm, erect, gen strongly compressed, lateral walls opaque to translucent, back thin-ridged; beak < body, gen lateral. ± 9 spp.: gen n temp. (Greek: ancient name) [Bjorkqvist 1968 Opera Bot 19:1–138] N.Am spp. need careful study.

1. Infl < to barely > lvs; style ± curved; lf blade linear to narrowly lanceolate, base tapered; pedicels gen spreading to recurved .... ***A. gramineum***
1′ Infl gen >> lvs; style ± straight, erect; lf blade gen lanceolate to ovate, base tapered, truncate, or slightly lobed; pedicels gen ascending to erect
 2. Petals pink; lf blade gen lanceolate, base tapered; lateral walls of fr thin, translucent .... ***A. lanceolatum***
 2′ Petals gen white, sometimes pink; lf blade lanceolate to ovate, base truncate to slightly lobed; lateral walls of fr thick, opaque .... ***A. plantago-aquatica***

A. gramineum Lej. **LF** 6–30 cm; blade 3–7 cm, 0.8–2 cm wide, linear to narrowly lanceolate, base tapered. **INFL** < to barely > lvs; pedicels gen spreading to recurved. **FL**: petals white or pink; style ± curved. **FR**: lateral walls gen thick, opaque. 2*n*=14. Ponds, ditches; 1200–1800 m. MP; to Can, e US, Eurasia. [*A. geyeri* Nicollet] ❀TRY.

A. lanceolatum With. **LF** 12–40 cm; blade 6–12 cm, 1–3 cm wide, gen lanceolate, base tapered. **INFL** gen >> lvs; pedicels gen ascending to erect. **FL**: petals pink; style ± straight, erect. **FR**: lateral walls gen thin, translucent. 2*n*=26,28. Ponds, rice fields, ditches, slow streams; < 500 m. NW, n SNF, ScV; OR, also Chile, Australia; native to Eurasia, n Afr.

A. plantago-aquatica L. (p. 1103) **LF** 7–45 cm; blade 5.5–15 cm, 1.5–10 cm wide, lanceolate to ovate, base truncate to slightly lobed. **INFL** gen >> lvs; pedicels gen ascending to erect. **FL**: petals gen white; style ± straight, erect. **FR**: lateral walls gen thick, opaque. 2*n*=14,28. Ponds; < 1600 m. CA-FP; to Can, se US, Eurasia, e Afr, perhaps also Australia. [*A. triviale* Pursh; *A. subcordatum* Raf.] ❀WET, fresh water margins and mud, SUN:**4–9**,10–12,**14–24**.

DAMASONIUM

Per; caudex gen corm-like. **LF**: blade linear to ovate, tapered to base or clearly petioled. **INFL**: peduncle gen smooth. **FL** bisexual; receptacle ± flat; sepals gen 3.5–6 mm; petals white or pink; stamens 6; pistils 6–15 in 1 whorl, fused at base. **FR**: body gen spreading, laterally compressed, sides gen opaque, back ± rounded, ribbed; beak = or > body, gen terminal. ± 5 spp.: N.Am, Eur, Australia. (Greek: ancient name) [Vuille 1987 Plant Syst & Evol 157:63–71]

D. californicum Benth. (p. 1103) **LF** 5–20 cm; blade 3–9 cm, 0.5–3 cm wide, gen < petiole, linear to narrowly ovate. **INFL** gen > lvs; pedicels 15–65 mm in fr, spreading to ascending. **FL**: petals 6–10 mm, deeply cut, white to pink, basal spot yellow; anthers reddish; ovule 1 per chamber. **FR**: body 3–5.5 mm; beak 3–6 mm. Ponds, vernal pools, streams, ditches; < 1700 m. NCoRI, n SNF, ScV, MP; to OR, ID, NV. [*Machaerocarpus c.* (Benth.) Small] ❀TRY;DFCLT.

ECHINODORUS BURHEAD

Ann or short-lived per; roots not partitioned. **LF**: petiole angled; blade linear to cordate. **INFL**: peduncle gen smooth; fls > 3 per node. **FL** bisexual; receptacle convex; sepals gen 2–6 mm, dark green; petals gen entire, white; stamens 9–many; pistils many, free, in spheric cluster. **FR**: cluster bur-like; body ± compressed, gen ribbed, tapered to beak. ± 50 spp.: Am, esp tropics. (Greek: spiny, leathery container, from fr) [Haynes & Holm-Nielsen 1986 Brittonia 38:325–332]

E. berteroi (Sprengel) Fassett (p. 1123) Gen ann in CA. **LVS** 8–30 cm; blade with transparent lines, coarsely veined; submerged blades linear, wavy; floating and emergent blades 6–14 cm, 3–15 cm wide, elliptic to cordate. **INFL** gen > lvs; peduncle angled; pedicels < 20 mm, gen ascending. **FL**: petals 6–9 mm; stamens ± 12. **FR**: body 1.5–3 mm, ribs gen 5; beak < 2 mm. 2*n*=22. Ponds, ditches; < 300 m. NCoRI, GV, CW, SW; to se US, S.Am. [var. *lanceolatus* (Engelm.) Fassett] ❀TRY.

SAGITTARIA ARROWHEAD, TULE POTATO, WAPATO

Ann, per, gen monoecious; roots partitioned. **LVS**: petiole unangled; submerged blades tapered to base; floating or emergent blades gen sagittate (or linear to ovate). **INFL**: lowest node gen with 3 pistillate fls; staminate fls above. **FLS** gen unisexual; sepals 3–10 mm, reflexed to appressed in fr; petals gen entire. **PISTILLATE FL**: receptacle convex; pistils many, in spheric cluster. **STAMINATE FL**: stamens many. **FR**: body gen 2–3.5 mm, strongly compressed, back winged or ridged; beak gen lateral, spreading or erect. ± 20 spp.: worldwide, esp Am. (Latin: arrow, from lf shape) [Bogin 1955 Mem NY Bot Gard 9:179–233] Some spp. weedy; tubers of some used for human and wildlife food.

1. Ann; stolons and tubers 0; lowest infl node with 2 bisexual fls .... ***S. montevidensis*** ssp. ***calycina***
1' Per from stolons and tubers; lowest infl node with 3 pistillate fls
  2. Emergent lf blades linear and 3-angled to lanceolate; pedicels of pistillate fls recurved, thickened in fr ***S. sanfordii***
  2' Emergent lf blades gen sagittate; pedicels of pistillate fls ascending in fr, not thickened
    3. Lower lobes of emergent lf blades gen 2 × terminal lobe; tubers spheric, tan .... ***S. longiloba***
    3' Lower lobes of emergent lf blades < or = terminal lobe; tubers oblong, ± white or bluish
      4. Fr beak ascending to ± erect, < 0.5 mm; lower lobes of emergent lf blades < terminal lobe .... ***S. cuneata***
      4' Fr beak spreading, 1–2 mm; lower lobes of emergent lf blade ± = terminal lobe .... ***S. latifolia***

S. cuneata E. Sheldon (p. 1123) Per; tuber oblong, ± white or bluish. **LVS**: emergent blades 5–15 cm, sagittate, lower lobes < terminal lobe. **PISTILLATE FL**: pedicel ascending in fr; sepals reflexed in fr. **STAMINATE FL**: filaments glabrous. **FR**: beak < 0.5 mm, ascending to ± erect. $2n$=22. Ponds, slow streams, ditches; < 2500 m. NW, CaR, SN, SnBr, GB; to s Can, n US, TX. ❀TRY.

S. latifolia Willd. (p. 1123) Per, sometimes dioecious; tubers oblong, ± white or bluish. **LVS**: emergent blades 6–30 cm, sagittate, lower lobes ± = terminal lobe. **PISTILLATE FL**: pedicel ascending in fr; sepals reflexed in fr. **STAMINATE FL**: filaments glabrous. **FR**: beak 1–2 mm, spreading. $2n$=22. Ponds, slow streams, ditches; < 1500 m. CA-FP, GB; to s Can, e US, n S.Am. ❀SUN, WET (fresh water):1–5,**6–9**,10–12,**14–24**.

S. longiloba Engelm. (p. 1123) Per; tubers spheric, tan. **LVS**: emergent blades 13–27 cm, sagittate, lower lobes gen 2 × terminal lobe. **PISTILLATE FL**: pedicel ascending in fr; sepals reflexed in fr. **STAMINATE FL**: filaments glabrous. **FR**: beak < 0.5 mm, ascending to ± erect. Ponds, ditches, rice fields; < 300 m. GV; to c US, n Mex. ❀TRY.

S. montevidensis Cham. & Schldl. ssp. ***calycina*** (Engelm.) C. Bogin (p. 1123) Ann; fls bisexual and staminate on same pl. **LVS**: emergent blades 5–15 cm, sagittate, lower lobes ± = terminal lobe. **INFL**: lowest node with 2 bisexual fls. **BISEXUAL FL**: pedicel recurved, thickened in fr; sepals appressed in fr; petals white, with green-yellow spot at base. **STAMINATE FL**: filaments papillate. **FR**: side oil-streaked when fresh; beak 0.5–1.5 mm, spreading. Ponds, ditches, rice fields; < 300 m. GV, SnFrB (Sonoma Co.), c SCo (Los Angeles Co.); to e US, n Mex. Ssp. *m.* in S.Am. ❀TRY.

S. sanfordii E. Greene (p. 1123) SANFORD'S ARROWHEAD Per; tubers spheric. **LVS**: emergent blades 14–25 cm, linear and 3-angled to narrowly ovate. **PISTILLATE FL**: pedicel recurved, thickened in fr; sepals appressed in fr. **STAMINATE FL**: filaments papillate. **FR**: side oil-streaked when fresh; beak < 0.5 mm, ascending to ± erect. UNCOMMON. Ponds, ditches; < 300 m. n NCo (Del Norte Co.), GV (where mostly extirpated), n SCo (Ventura Co.). Threatened by development. ❀SUN,WET(fresh water):4, **5**,7,**8**,**9**,**14–17**,**19–24**.

APONOGETONACEAE CAPE-PONDWEED FAMILY

Robert F. Thorne

Per from corm or short, thick rhizome, aquatic, glabrous. **LVS** simple, basal, gen submersed, petioled; blade expanded or not (if floating, blade elliptic to lanceolate). **INFL**: gen spike, terminal; branches 2–3(10), scapose above water, subtended by conspicuous, deciduous bract. **FL** gen bisexual; perianth parts 0–6, gen petal-like; stamens 6 in 2 whorls or many; pistils gen 3–6, fused ± 1/2–2/3 length, simple above, separating in fr, ovary superior, chamber 1, ovules 1–8, style short, stigmatic surface grooved. **FR**: follicles, leathery. **SEEDS** several; embryo straight. 1 genus, ± 40 spp.: trop Asia to Australia, s Afr.

APONOGETON

1 genus, ± 40 spp.: old world trop (Asia, Australia), s Afr. (Greek: from aquatic habitat)

A. distachyon L.f. CAPE-PONDWEED **LF**: petiole long; blade elliptic, floating. **FL**: perianth segment 1, base wide, white; stamens many. Ponds; < 150 m. c-w SnFrB, s SCo, expected elsewhere; native to s Afr. Widely cult for aquaria, often escaping but gen a waif.

ARACEAE ARUM FAMILY

Elizabeth McClintock

In CA per, terrestrial or aquatic, from short, gen erect caudex, often monoecious; elsewhere shrub, vine, or growing on other pl. **STS** sometimes above ground in addition to caudex. **LVS** simple or compound, basal (or cauline and 2-ranked). **INFL**: gen spike, fleshy, gen ill-smelling; fls gen many, bisexual or pistillate below, staminate above; bract subtending spike 1, gen showy (petal-like), gen > spike, sometimes sheathing. **FL**: perianth parts 4 or 6, free or fused; stamens gen 0, 4, or 6, free or fused; ovary superior to half-inferior and sunken in infl axis, chambers 1–3, stigma ± sessile. **FR**: berry. **SEEDS** 1–many. ± 110 genera, 1800 spp.: gen trop, subtrop. Some cult for food (*Colocasia*, taro) or orn (*Philodendron, Anthurium*). [Wilson 1960 J Arnold Arbor 41:47–63] Needle-like crystals in most tissues cause intense irritation when chewed; those of *Dieffenbachia*, dumb-cane, may induce temporary speechlessness.

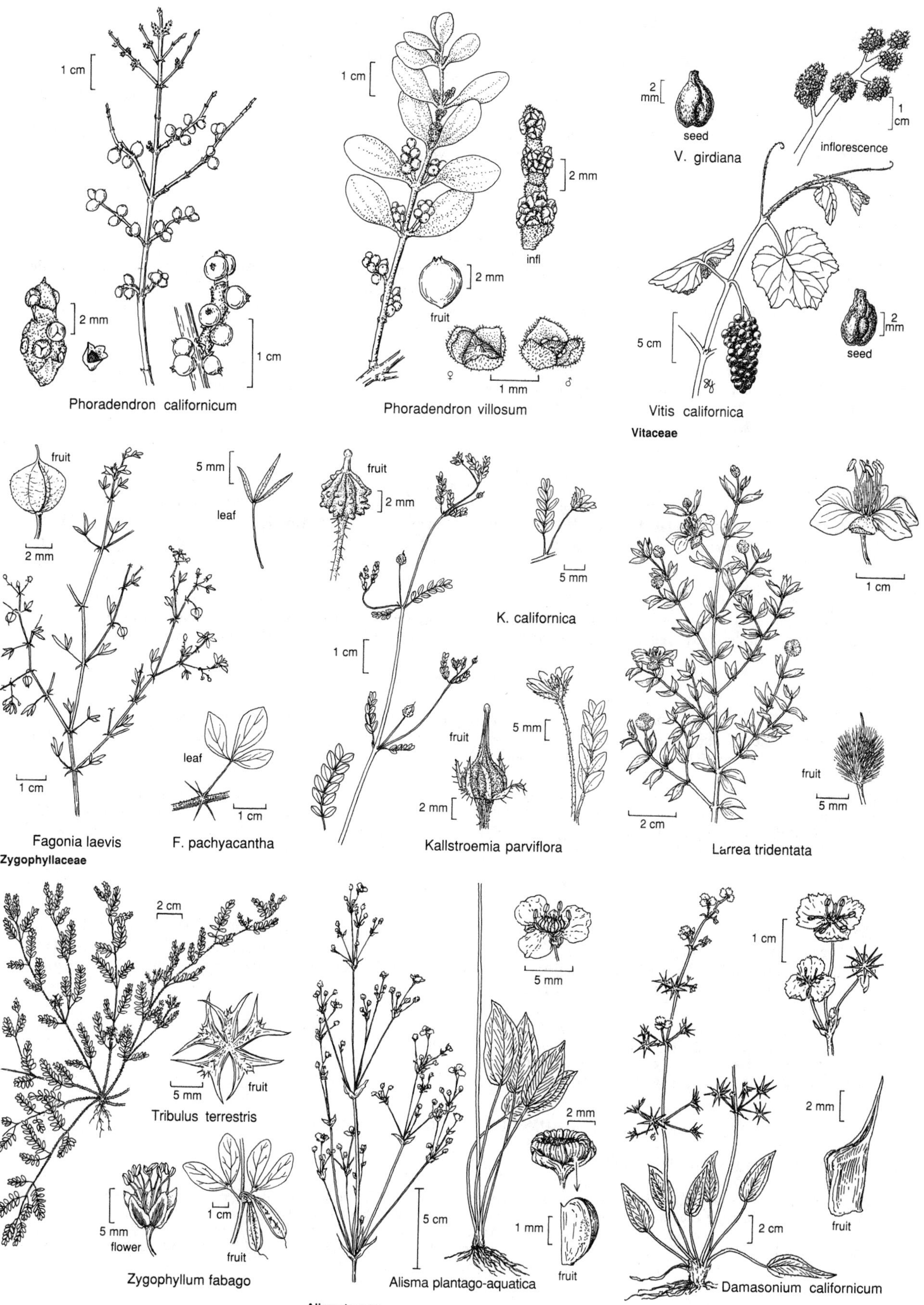
1 cm
2 mm
1 cm
Phoradendron californicum
1 cm
2 mm
infl
2 mm
fruit
♀
1 mm
♂
Phoradendron villosum
2 mm
seed
V. girdiana
inflorescence
1 cm
5 cm
2 mm
seed
Vitis californica
Vitaceae
fruit
2 mm
5 mm
leaf
1 cm
leaf
1 cm
Fagonia laevis
F. pachyacantha
Zygophyllaceae
fruit
2 mm
5 mm
K. californica
1 cm
fruit
5 mm
2 mm
Kallstroemia parviflora
1 cm
fruit
5 mm
2 cm
Larrea tridentata
2 cm
5 mm
fruit
Tribulus terrestris
5 mm
flower
1 cm
fruit
Zygophyllum fabago
5 mm
5 cm
2 mm
1 mm
fruit
Alisma plantago-aquatica
Alismataceae
1 cm
2 mm
fruit
2 cm
Damasonium californicum

1. Pl aquatic, free floating; infl ± sessile, inconspicuous among basal lvs . . . . **PISTIA**
1′ Pl terrestrial or near edge of water; peduncle short to long
2. Lvs 2-ranked; blade folded along midrib, linear, sessile; infl bract lf-like, narrow, sessile . . . . . . **ACORUS**
2′ Lvs ± whorled; blade ± flat, wider, petioled; infl bract wide, ± stalked
3. Infl bract lemon-yellow; petiole < lf blade . . . . **LYSICHITON**
3′ Infl bract green or white (or light yellow in *Arum*); petiole ± = or > blade
4. Peduncle gen = or > lf; infl bract white . . . . **ZANTEDESCHIA**
4′ Peduncle gen < lf; infl bract green or greenish white to yellow
5. Infl sterile in ± upper half, fls only in lower half; infl bract greenish white to yellow . . . . . . **ARUM**
5′ Infl fertile throughout; infl bract green . . . . **PELTANDRA**

ACORUS

Per, terrestrial or near edge of water. **LVS** 2-ranked; blade folded along midrib, linear to sword-shaped, sheathing, sessile. **INFL** erect to curved; peduncle lf-like, fused to bract; bract erect, lf-like, narrow, green, sessile. **FL** bisexual; perianth parts 6, greenish; stamens 6; ovary superior, chambers 2–3. 2–6 spp.: gen n temp. (Latin: aromatic plant) [Grayum 1987 Taxon 36:723–729] Sometimes placed in Acoraceae, because needle-like crystals 0, lf terminal, lvs and infl bract sessile.

A. calamus L. SWEET FLAG **LF** 2–7.5 dm, 7–20 mm wide. **INFL** 5–10 cm, ± 2 cm wide; peduncle ± = lvs; bract > peduncle, green. **FL** greenish yellow. Uncommon. Moist ground; < 100 m. n NCo (Humboldt Co.), NCoRI (Lake Co.), expected elsewhere; native to e N.Am, Eurasia. Sometimes cult for fragrant oil in rhizomes, used medicinally.

ARUM

Per, terrestrial, monoecious. **LVS** basal; blade flat, hastate to sagittate; petiole ± = or > blade. **INFL** < bract, tip with cylindric appendage; peduncle < petiole; bract ± tubular, enclosing infl at base, withering before fr; fls pistillate near base, staminate above, gen sterile between. **STAMINATE FL**: perianth 0; stamens 3–4. **PISTILLATE FL**: ovary chamber 1. ± 15 spp.: Eurasia, n Afr. (Greek: ancient name) Some cult as orn or for food (*A. maculatum* rhizomes edible when thoroughly cooked).

A. italicum Miller **LVS** appearing fall to early winter, after fl; blade 15–35 cm, veins whitish; petiole < 40 cm. **INFL** < bract; bract < 40 cm, greenish white to yellow outside, pale yellow-green inside, drooping at tip, withering. **FR** orange-red. Uncommon. Disturbed, gen shaded areas; < 500 m. NCo, NCoRO, ScV, SnFrB; native to Eur, w Asia, n Afr. Sometimes cult as orn.

LYSICHITON

Per, terrestrial. **LVS** basal; petiole < blade, stout; blade flat, ovate to oblanceolate. **INFL** < bract; peduncle < lvs, stout; bract boat-shaped, base sheathing peduncle, partly enclosing infl, deciduous. **FL** bisexual; perianth parts 4, fused at base; ovary partly embedded in infl axis, chambers 2. 2 spp.: ne Asia, w N.Am. (Greek: loosening tunic, from deciduous bract)

L. americanum Hultén & St. John (p. 1123) YELLOW SKUNK CABBAGE **LVS** appearing gen in spring after fl; blade 30–150 cm, ± fleshy, midvein thick, graduated into thick, wide petiole. **INFL** < 12 cm, ± 2.5 cm thick; peduncle 3–5 dm; bract < 20 cm, lemon-yellow. **FL** yellow-green, ill-smelling. **FR** greenish white. Uncommon. Marshy areas, stream edges, gen in coniferous forest; < 1300 m. NCo, SnFrB; to AK, MT, ID. ❀WET(fresh water)**4,5**&SHD: 1–3,**6**,7,14,**15–17**,18,19.

PELTANDRA

Per from thick, fibrous, clumped rhizomes, monoecious. **LVS** basal; petiole long; blade sagittate to hastate, pinnately veined. **INFL** gen > bract; bract oblong, enclosing pistillate fls, opening slightly above, exposing staminate fls. **FLS** embedded in infl axis; perianth 0. **STAMINATE FL**: stamens 4–5. **PISTILLATE FL**: ovary chamber 1, style stout. **FR** red or green. **SEEDS** 1–3. 3–4 spp.; native to e N.Am. (Greek: hidden anthers)

P. virginica (L.) Schott & Endl. GREEN ARROW ARUM, TUCKAHOE Pl < 1 m. **LF**: blade 10–30 cm, bright green, lower lobes long, acute. **INFL** < bract; bract 10–20 cm, green, margin pale green or white, upper part (surrounding staminate fls) decaying, lower part persistent in fr. **FR** brown. **SEEDS** 1–3, enclosed by gelatinous, translucent pulp. Uncommon. Ponds, reservoirs; < 400 m. sw SnJV (ne San Luis Obispo Co.); native to e N.Am.

PISTIA WATER-LETTUCE

1 sp. (Greek: liquid, from aquatic habitat)

P. stratiotes L. Per, aquatic, floating, monoecious, spreading from stolons; roots many. **ST** short, corm-like. **LVS** basal, whorled, sessile; blade < 20 cm, < 7 cm wide, widely wedge-shaped, tip truncate or notched, velvety-hairy. **INFL**: spike < bract, fused to bract (exc tip); bract ± 1.5 cm, hairy outside, upper half open, ± flared, reflexed, lower half closed. **FLS**: perianth 0. **STAMINATE FLS** several, whorled, sessile; stamens 2–8. **PISTILLATE FL** 1; ovary ± superior, chamber 1, style stout, curved, persistent. **FR** green. **SEEDS** many. Ditches; < 50 m. e DSon (Colorado River), expected elsewhere; to se US, trop worldwide.

ZANTEDESCHIA

Per from rhizomes, terrestrial, monoecious, gen fragrant. **LVS** basal, clumped; petiole ± = to > blade, spongy; blade hastate or sagittate to ovate, base deeply lobed. **INFL** < bract; peduncle ± = or > lvs, spongy; bract white or brightly colored. **FLS**: perianth 0. **STAMINATE FL**: stamens 2–3, free. **PISTILLATE FL**: ovary chambers 3. ± 6 spp.: s Afr. (F. Zantedeschi, Italian botanist, 1773–1846) [Letty 1973 Bothalia 11:5–26] Cult for showy infl, bract; cut infl long-lasting.

Z. aethiopica (L.) Sprengel CALLA LILY **LVS** appearing with infl; petiole < 9 dm; blade 15–45 cm, 10–25 cm wide. **INFL**: spike ± = bract, yellow; bract < 25 cm, funnel-shaped, white, tip linear, recurved, green. Uncommon. Waste ground near former habitations; < 300 m. NCo, NCoRO, SnFrB, SCoRO, SCo; native to s Afr. Commonly cult.

ARECACEAE [Palmae] PALM FAMILY

Elizabeth McClintock

Shrub, tree, evergreen, monoecious, dioecious, or fls bisexual. **ST**: trunk gen ± erect, unbranched. **LVS** splitting to be palmately or pinnately dissected or compound, alternate, forming a terminal crown, large; base sheathing; petiole often long. **INFL**: gen large panicle, axillary; peduncle sheathed by 1 or more large bracts; fls many, gen ± sessile. **FL** gen small, ± radial; sepals and petals gen 3, sometimes similar, fused at base or free; stamens gen 6; pistils 1 or 3, ovaries superior, gen 3, (if 1, chambers gen 3), styles free or fused. **FR**: often a drupe. **SEED** 1. ± 200 genera, 3,000 spp.: trop, subtrop; many cult, esp for orn. [Uhl & Dransfield 1987 Genera Palmarum] Used for food (fats, oils, frs, seeds) and building materials.

1. Lf blade pinnately compound, ± elongate; fl unisexual . . . . **PHOENIX**
1′ Lf blade palmately divided, ± round; fl bisexual . . . . **WASHINGTONIA**

PHOENIX DATE PALM

Tree, dioecious. **LVS** pinnately compound; bases persistent on trunk; lflets folded longitudinally with margins upward, lower sometimes smaller, spine-like. **INFL** within crown, < lvs. **FL**: perianth yellowish; calyx 3-lobed; petals gen free; ovaries 3, free, simple. ± 12 spp.: Afr, Asia. (Greek: name for date palm, of uncertain meaning)

1. Trunk thick, < 20 m; lvs ± 50–100, in dense crown, all ± arching; basal sprouts 0 (trunk 1) . . . . ***P. canariensis***
1′ Trunk slender, gen < 30 m; lvs 20–40, in ± open crown, uppermost erect, others ± stiffly drooping; basal sprouts present when young (trunks several if pl unpruned) . . . . ***P. dactylifera***

P. canariensis Chabaud CANARY ISLAND DATE PALM **LF** gen 5–7 m. **FR** ± 2 cm, rounded to ovate, brown, pulp thin. Uncommon. Near habitations, other disturbed areas; < 1000 m. SnFrB, SCo; native to Canary Islands. Abundantly cult; fr pulp sweet, edible.

P. dactylifera L. DATE, DATE PALM **LF** gen < 7 m. **FR** 2.5–5 cm, oblong-ovate, brown, pulp thick. Uncommon. Near habitations, adjacent moist areas; < 200 m. SCo, DSon; native to n Afr. Abundantly cult; fr (commercial date) pulp sweet, edible.

WASHINGTONIA FAN PALM

Tree. **LVS** persistent as a "skirt," palmately divided; segments folded longitudinally with margins upward; petiole margins and lower blade gen spiny. **INFL** within crown, > lvs. **FL** bisexual; perianth white; calyx lobes 3, ± erect; corolla lobes 3, reflexed; ovary 3-lobed, chambers 3. 2 spp.: deserts of s CA, AZ, n Mex. (George Washington, 1st president, US, 1732–1799) [Bailey 1936 Gentes Herb 4: 52–82] Widely cult as orn, CA to Florida.

W. filifera (L. Linden) H.A. Wendl. (p. 1123) CALIFORNIA FAN PALM **ST**: trunk thick, robust, < 20 m; base gen not swollen. **LF** gray-green; petiole 1–2 m; blade 1–2 m, divided nearly to middle, segments 40–60, margins with thread-like fibers, tips ± reflexed. **FR** oblong or ovate, black. UNCOMMON (but locally abundant). Groves, moist places, seeps, springs, stream sides; < 1200 m. DSon; se AZ, n Baja CA. Reportedly naturalized along Kern River (Kern Co.); expected elsewhere. ❀SUN:**14–16**,17,**22**,**23**,24&IRR:**8**,**9**,10,**11–13**,**18–21**.

COMMELINACEAE SPIDERWORT FAMILY

Elizabeth McClintock

Ann, per, gen glabrous. **ST** prostrate to erect or climbing; nodes often rooting. **LVS** alternate, entire, simple, linear to ovate, with closed basal sheath or lower lf clasping st. **INFL**: gen cyme or umbel, axillary. **FL** gen bisexual, bilateral or radial, gen insect-pollinated; sepals 3, gen green; petals 3, blue, white, or sometimes rose or purple, gen ephemeral; stamens 6 (3 sometimes sterile), filaments gen slender, often hairy; ovary superior, chambers 3. **FR**: gen capsule. **SEEDS** few. ± 50 genera, 700 spp.: esp trop, subtrop; some cult as orn.

1. Petals unequal; fertile stamens 3, sterile stamens 3 . . . . **COMMELINA**
1′ Petals equal; fertile stamens 6 . . . . **TRADESCANTIA**

COMMELINA DAYFLOWER

Per. **INFL**: cyme, subtended by boat-shaped bract. **FL** bilateral, gen blue, ephemeral; petals unequal, 2 large and showy; fertile stamens 3, sterile stamens 3, filaments glabrous. ± 150 spp.: trop, warm temp regions. (Johan, 1629–1692, and nephew Caspar, 1667–1731, Commelin, Holland) [Maheshwari & Maheshwari 1955 Phytomorphology 5:413–422]

C. benghalensis L. **ST** straggling, rooting at nodes. **LF** 30–60 mm, ± widely lanceolate; margins gen wavy and hairy. **FLS** of 2 kinds; those of upper st opening, ± 1 cm diam, petals blue, filaments glabrous; cleistogamous fls on underground sts. Disturbed, moist, urban areas; < 500 m. SCo; trop weed; native to trop Asia, Afr.

TRADESCANTIA SPIDERWORT

Per. **INFL**: gen umbel, subtended by 2–3 bracts. **FL** radial; petals equal, blue, rose, purple, or white; fertile stamens 6, filaments gen hairy. ± 25–30 spp.: N.Am, S.Am. (John Tradescant, gardener to King Charles I of England, 1608–1662)

T. fluminensis Vell. Conc. (p. 1123) Per. **ST** prostrate or decumbent, rooting at nodes. **LF** 25–60 mm, oblong to ovate, glabrous. **INFL**: fls few–many, none cleistogamous; petals 7–8 mm, white; filaments hairy, tissue between anthers triangular. Disturbed, shaded, urban areas; 500 m. SnFrB, SCo; uncommonly naturalized; native to S.Am. Streaming protoplasm was discovered in the filament hairs by Robert Brown in England, 1828.

CYMODOCEACEAE MANATEE-GRASS FAMILY

Robert F. Thorne

Per from long, slender, jointed rhizomes, dioecious, marine aquatic, glabrous. **ST** erect, sometimes short. **LVS** simple, cauline, alternate or opposite, 2-ranked; sheath open; ligule present; blade 0.5–15 mm wide, ± linear, ± flat or cylindric, margin entire. **INFL**: fls 1–2 (or bracted cyme), axillary. **FLS** minute; perianth 0. **STAMINATE FL**: stamens 2, filaments 0 or short and ± fused; pollen filament-like, sticking together in string-like bodies < 1 mm wide. **PISTILLATE FL**: pistils 1–2, simple, ovary superior, chamber 1, ovule 1, style 1, lobes 0–3. **FR**: utricle. **SEED** 1. 5 genera, ± 13 spp.: trop, subtrop worldwide.

HALODULE

STS erect, several, short. **LVS** alternate to ± opposite, narrowly linear; tip gen 3-toothed. **FLS** 1 per node, ± enclosed by lf sheaths. **STAMINATE FL**: anther sacs attached to filament at different places. **PISTILLATE FL** short-stalked; pistils 2, simple, attached to receptacle at slightly different places, ovary asymmetric, style 1, simple, thread-like, from side of ovary. 1–2 spp.: trop, subtrop marine. (Greek: perhaps from ancient name meaning under salty water) Poorly known exc to students of algae. Sometimes treated in Zannichelliaceae.

H. wrightii (Asch.) Asch. SHOAL-GRASS **LF** gen 10–20 mm, 0.5–3 mm wide. Aquatic; –70 m. DSon (Salton Sea, where presumed extinct), expected in warm, coastal, marine habitats; native to coastal Mex and Caribbean Sea.

CYPERACEAE SEDGE FAMILY

Raymond Cranfill, except as specified

Ann or per, often rhizomed, often of wet open places, gen monoecious; roots fibrous, hairy. **ST** gen 3-sided. **LVS** often 3-ranked; sheath gen closed; ligule gen 0; blade (0) various, parallel-veined. **INFL**: spikelets variously clustered; fls gen sessile in axil of fl bract. **FL** small, gen wind-pollinated; perianth 0 or bristle-like; stamens gen 3, anthers attached at base, 4-chambered; ovary superior, 1-chambered, 1-ovuled, style 2–3-branched. **FR**: achene, gen 3-sided. ± 110 genera, 3600 spp.: worldwide, esp temp. [Tucker 1987 J Arnold Arbor 68:361–445] Difficult: taxa differ in technical characters of infl and fr.

1\. Fl and fr enveloped in ± closed bract (perigynium)
  2\. Perigynium open only at tip .... **CAREX**
  2′ Perigynium open along 1 side .... **KOBRESIA**
1′ Fl and fr not enveloped in a perigynium
  3\. Gen only uppermost fls bisexual and fruiting (lower staminate)
    4\. Style base much expanded, persistent as tubercle on fr .... **RHYNCHOSPORA**
    4′ Style base little expanded, if persistent on fr, not tubercle-like
      5\. Perianth bristles 0; fr not whitish and bony .... **CLADIUM**
      5′ Perianth bristles present; fr ± white, bony .... **SCHOENUS**
  3′ Fls gen bisexual only, fruiting (exc sometimes lowest 1–3)
    6\. Fl bracts 2-ranked
      7\. Perianth bristles 5–8; spikelet clusters axillary, emerging from sheaths of st lvs; fr beak long ... **DULICHIUM**
      7′ Perianth bristles 0; spikelet clusters gen terminal, not emerging from sheaths of st lvs; fr beak 0
        8\. Spikelet 1–many-fld, gen not shed as unit .... **CYPERUS**
        8′ Spikelet 1–2-fld, shed as unit .... **KYLLINGA**

6′ Fl bracts spiraled
9. Perianth bristles gen present
10. Perianth bristles conspicuous, >> fl bract .... **ERIOPHORUM**
10′ Perianth bristles inconspicuous, hidden by fl bract
11. Fr top with tubercle of different color and texture; st lvs 0 .... **ELEOCHARIS**
11′ Fr top gen beaked, tubercle 0; st lvs gen 1–several .... [2]**SCIRPUS**
9′ Perianth bristles 0
12. Fl subtended by bractlets and fl bract .... **LIPOCARPHA**
12′ Fl subtended only by fl bract
13. Style ± equally wide throughout .... [2]**SCIRPUS**
13′ Style widest at base
14. Fr 3-sided, cells vertically elongate, top tubercled; style 3-branched, not fringed .... **BULBOSTYLIS**
14′ Fr 2–3-sided or round, cells ± square or horizontally elongate, tubercle 0; style 2–3-branched, gen fringed below branches .... **FIMBRISTYLIS**

BULBOSTYLIS

Ann or per, cespitose. **ST** slender, round, ridged and grooved, solid. **LVS** several, basal, thread-like to linear, folded or upcurled. **INFL**: spikelet clusters terminal, panicle-like; fl bracts spiraled, deciduous. **FLS** bisexual; perianth bristles 0; stamens gen 3; style 3-branched at tip, base bulb-like, persistent. **FR** 3-sided; cells gen vertically elongate; top tubercled. ± 120 spp.: worldwide. (Latin: bulb-like style) [Kral 1971 Sida 4:57–227]

B. capillaris (L.) C.B. Clarke (p. 1123) Ann < 3 dm. **ST** gen >> lf. **LF**: sheath margin wide, tan, gen fringed near tip; blade < 0.5 mm wide. **INFL**: spikelets gen long-peduncled, fusiform, gen few-fld; fruiting fl bracts dark reddish brown, midrib prominent, greenish. **FR** ± 1 mm, ± pale brown; angles ridged; tubercle ± round. $2n=72$. Meadows, grassy clearings; > 1000 m. c SNH (Mariposa, Tuolumne cos.); to e N.Am, Caribbean, C.Am; also e Asia. [*Fimbristylis c.* (L.) A. Gray]

CAREX SEDGE

Joy Mastrogiuseppe

Per, cespitose or from rhizomes, gen monoecious. **ST** gen sharply 3-angled, gen solid. **LVS** 3-ranked, gen glabrous exc gen scabrous on midrib, margin; sheath closed, back (blade side of st) green, ribbed, front gen thin, translucent, forming gen U-shaped mouth at top. **INFL**: spikelets gen several–many, arrayed in raceme, panicle, or head-like cluster, each 1–many-fld, gen subtended by a spikelet bract. **FLS** unisexual, each subtended by 1 fl bract; perianth 0. **STAMINATE FL**: stamens gen 3. **PISTILLATE FL** enclosed by perigynium (sac-like bract); perigynium body 2–3-sided or round, wall gen delicate; perigynium beak tip open, often notched; style 1, gen deciduous, stigmas 2–4, exserted. **FR** 2–4-sided. (Latin: cutter, from sharp lf and st edges) [Standley 1985 Syst Bot Monogr 7:1–106] Fully mature perigynia needed for identification, so are described under "**FR**" (long-persistent perigynia are often atypical); perigynium "front" faces spikelet axis; "fr" refers to achene body (excluding beak). "Shredding" lower lf sheath fronts become a network or fringe of veins; some others shred longitudinally only. Difficult because of many spp. and minute key characters; longer key statements and descriptions are designed to enhance both ease and probability of correct identification. Group descriptions are assumed in specific descriptions. ❀Many ssp. especially those with rhizomes are INVASIVE. This is one of the most effective genera for knitting moist or wet soil.

Key to Groups

1. Spikelet 1 per infl; staminate fls above pistillate if both in a spikelet; bristle-like axis sometimes in perigynium .... **Group 1**
1′ Spikelets > 1 per infl (sometimes in a dense, head–like cluster; staminate fls often below pistillate if both in a spikelet; bristle-like axis in perigynium 0
2. Dioecious
3. Spikelets > 3, ovate to round; perigynium faces glabrous; stigmas 2 .... [2]**Group 7**
3′ Spikelets 2(3), narrowly oblong; perigynium faces minutely hairy at least near beak; stigmas 3(4)
4. Pistillate fl bract obtuse or with a small point, ciliate or not, surface often hairy, width > or = perigynium, white margin wide; perigynium 3-sided, purplish .... ***C. scirpoidea*** var. ***pseudoscirpoidea***
4′ Pistillate fl bract acute to awned, ciliate at least at tip, surface glabrous, width < perigynium, white margin 0 or narrow; perigynium ± flat, brown to blackish .... ***C. gigas***
2′ Monoecious
5. Perigynium faces hairy, at least near tip; stigmas 3–4; fr 3–4-sided
6. Mouth of lf sheath finely toothed; infl < 4 cm; lateral, pistillate spikelets 1–2, < 1.5 cm .... **Group 2**
6′ Mouth of lf sheath entire; infl > 4 cm; lateral, pistillate spikelets gen > 2, > 1.5 cm .... **Group 3**
5′ Perigynium faces glabrous; stigmas 2–3; fr 2–3-sided
7. Stigmas 3; fr 3-sided
8. Style tough, persistent, not jointed to fr; perigynium veins many, each gen rib-like; staminate spikelets 1–5 .... **Group 4**
8′ Style delicate, deciduous, breaking easily at joint to fr; perigynium veins 0–many, gen 2 rib-like; staminate spikelet 0–1 .... **Group 5**
9. Perigynia 2–3 per spikelet, bristle-like axis present in perigynium; pistillate fl bracts lf–like, 1–15 cm .... ***C. tompkinsii***
9′ Perigynia > 3 per spikelet, bristle-like axis in perigynium 0; pistillate fl bracts not lf–like, < 1 cm .. **Group 5**

7′ Stigmas 2; fr 2-sided
10. Spikelets gen > 1.5 cm, sessile or not, ± cylindric, lower pistillate (or tip staminate), upper gen staminate .... **Group 6**
10′ Spikelets gen < 1.5 cm, sessile, ovoid to elliptic, each gen both staminate and pistillate
11. Perigynia 1–3 per spikelet, 2.2–3 mm; infl 3–4 mm wide; terminal spikelet often with a conspicuous, narrow, staminate tip .... ***C. disperma***
11′ Perigynia > 3 per spikelet, 2–8 mm; infl > 5 mm wide; terminal spikelet tip, if staminate, not conspicuous and narrow
12. Spikelets staminate at tip, pistillate at base; perigynium thick, ribs 0 or 2 .... [2]**Group 7**
12′ Spikelets staminate at base, pistillate at tip; perigynium thin to thick, ribs 2
13. Perigynium often not ribbed to base, ± thin-edged above but unwinged, entire exc for sometimes on beak, body strongly planoconvex, lower wall often filled with pithy tissue; st solid .... **Group 8**
13′ Perigynium ribbed to base, with distinct, gen serrate wings at least on upper body and beak, body planoconvex or flat, lower wall without pithy tissue; st hollow .... **Group 9**

Group 1

Some dioecious. **LF**: blade flat or folded, less often rolled. **INFL**: spikelets 1(2) per infl, each staminate at tip, pistillate below (or staminate on separate spikelet); spikelet bract gen 0; pistillate fl bract gen not lf-like, < 1 cm (exc awn, if present). **PISTILLATE FL**: stigmas gen 3. **FR**: perigynium 2–7 mm, gen glabrous, beak tip narrowly white-margined; bristle-like axis gen present within perigynium; fr gen 3-sided.

1. Lowest pistillate fl bracts lf-like, 1–15 cm, base clasping perigynium; pl cespitose; bristle-like axis in perigynium present
2. Upper st round or bluntly 3-angled; longest lvs << st, blade 1–1.5 mm wide; perigynia 1–6 per spikelet, 5–7.2 mm .... ***C. multicaulis***
2′ Upper st sharply 3-angled; longest lvs > or ± = st, blade 1.5–3 mm wide; perigynia gen 2–3 per spikelet, 4.5–6 mm .... ***C. tompkinsii***
1′ Lowest pistillate fl bracts not lf-like, < 1 cm, exc awns, if present, base clasping perigynium or not; pl cespitose or rhizomed; bristle-like axis in perigynium present or 0
3. Spikelets unisexual, ± linear, pistillate dark, staminate pale; spikelet bract often present; perigynium faces gen hairy, at least near beak
4. Pistillate fl bract obtuse or with a small point, ciliate or not, surface often hairy, width > or = perigynium, white margin wide; perigynium 3-sided, purplish .... ***C. scirpoidea*** var. ***pseudoscirpoidea***
4′ Pistillate fl bract acute to awned, ciliate at least at tip, surface glabrous, width < perigynium, white margin 0 or narrow; perigynium ± flat, brown to blackish .... ***C. gigas***
3′ Spikelets bisexual, oblong to triangular, pale or dark; spikelet bract 0 (exc sometimes *C. nigricans*); perigynium faces glabrous or hairy
5. Lf blade flat to folded, not rolled, not quill-like, < or > 1 mm wide
6. Perigynium 4.9–7 mm, 1.7–2.5 mm wide, bristle-like axis within present; moist to dry places .... ***C. geyeri***
6′ Perigynium 2.5–4.6 mm, 0.8–1.5 mm wide, bristle-like axis within 0; wet places
7. Pistillate fl bracts early deciduous; spikelet 6–10 mm wide; lowest perigynia spreading to reflexed, beaks conspicuous in infl, faces unveined; 2300–3700 m .... ***C. nigricans***
7′ Pistillate fl bracts persistent; spikelet 2–3 mm wide; perigynia ascending, beaks 0, faces veined; 0–700 m .... ***C. leptalea***
5′ Lf blade rolled, quill-like, < or ± = 1 mm wide
8. Stigmas 2; fr 2-sided; bristle-like axis in perigynium present; wet places .... ***C. capitata***
8′ Stigmas 3; fr 3-sided; bristle-like axis in perigynium present or 0; moist to dry places
9. Perigynium flat, thin, > fr, 2.5–7 mm, 0.9–4.8 mm wide, faces glabrous; style base not black, not persistent, style not exserted
10. Perigynium 2.5–4.1 mm, 0.9–2 mm wide, fr width > distance from fr to perigynium side; spikelet elliptic to narrowly conic, 3–6 mm wide .... ***C. subnigricans***
10′ Perigynium 4–7 mm, 2.1–4.8 mm wide, fr width < distance from fr to perigynium side; spikelet widely conic, 6–10 mm wide .... ***C. breweri*** var. ***breweri***
9′ Perigynium 3-sided to round, thick, ± = fr, 1.9–3.7 mm, 1.3–2.1 mm wide, faces gen hairy at least just below beak; style base black, persistent, style exserted .... ***C. filifolia***
11. Perigynium 1.9–3 mm, body tip tapered, beak 0–0.3 mm; fr 1.6–2.2 mm; longest pistillate fl bracts (exc awns) gen < 2.5 mm, gen < perigynia .... var. ***erostrata***
11′ Perigynium 2.9–3.7 mm, body tip gen abruptly narrowed, beak 0.1–0.8 mm; fr 2.2–3 mm; longest pistillate fl bracts (exc awns) gen > 3 mm, gen = or > perigynia .... var. ***filifolia***

Group 2

Monoecious. **ST** gen < lvs, base gen brown to red. **LVS**: sheath mouth finely toothed, gen U-shaped; basal blades gen not minute. **INFL** < 4 cm; terminal spikelet staminate; lateral pistillate spikelets 1–2, < 1.5 cm, sessile; solitary pistillate spikelets on stalks from pl base gen present; spikelet bracts gen green, lowest gen lf-like, gen sheathless; pistillate fl bract white-margined, gen not ciliate. **PISTILLATE FL**: stigmas gen 3. **FR**: perigynium ascending, gen ± unveined, gen green, stalked, faces hairy at least near tip, body ± = fr, beak gen > 0.4 mm; fr 3-sided, gen brown.

1. Spikelet bract purple; stigmas gen 4; fr 4-sided; lf sheath mouth V-shaped, blade sickle-shaped; pistillate fl bract ciliate; solitary pistillate spikelets 0 .... ***C. concinnoides***
1′ Spikelet bract green; stigmas 3; fr 3-sided; lf sheath mouth U-shaped, blade gen not sickle-shaped (exc *C. brevicaulis*); pistillate fl bract not ciliate; solitary pistillate spikelets present or 0
2. Staminate spikelet of taller infls 5–10 mm; faces of most perigynia unveined or with a few short veins
3. St scabrous at least on 1 angle exc near base; lf blade gen stiff, folded, gen ± sickle-shaped, bright green; pls in ± continuous stands; perigynium 3.8–4.5 mm, 1.3–2.3 mm wide, beak tip notched in front < 0.1 mm; coastal .... [2]***C. brevicaulis***
3′ St glabrous or scabrous on angles only just below infl, sometimes scabrous nearly to base; lf blade stiff or not, gen not sickle-shaped, ± flat, bright or gray-green; pls cespitose; perigynium 2.4–5 mm, 1–1.7 mm wide, beak tip notched in front > 0.1 mm (if < 0.1 mm, not near coast); coastal or inland ..... ***C. rossii***
2′ Staminate spikelet of taller infls 10–30 mm (exc sometimes *C. brainerdii*); faces of most perigynia veined or several-ribbed at least below (exc *C. brevicaulis*, sometimes *C. inops*)
4. Solitary pistillate spikelets gen 0, sometimes on long stalks; perigynia gen > 5 per spikelet; staminate spikelet 12–30 mm; pistillate fl bract tip gen acuminate or with a minute point; CaRH (Mount Shasta area) .... ***C. inops*** ssp. ***inops***
4′ Solitary pistillate spikelets present, on long or short stalks; perigynia gen < 6 per spikelet; staminate spikelet gen 10–17 mm; pistillate fl bract tip obtuse to awned; most of CA-FP
5. Perigynium face veins 0 or few, short; lf blade gen sickle-shaped; st gen < 15 cm — pls in ± continuous stands .... [2]***C. brevicaulis***
5′ Perigynium faces gen veined or several-ribbed at least below; lf blade gen not sickle-shaped; st gen > 10 cm
6. Perigynium beak tip to fr top 1–2 mm; basal lf sheaths dark purple-red; fr gen tan or brown; fresh lvs gray-green; lowest spikelet bract sheath of bisexual infl > or = 2 mm; bisexual infl gen < longer lvs .... ***C. brainerdii***
6′ Perigynium beak tip to fr top 0.4–1.3(1.6) mm; basal lf sheaths brown to red; fr gen white; fresh lvs ± bright green; lowest spikelet bract sheath of bisexual infl gen < 2 mm; bisexual infl < or > longer lvs .... ***C. globosa***

Group 3

Monoecious, gen cespitose. **LF**: sheath mouth entire; blades gen flat, hairy or not, basal gen not minute. **INFL** > 4 cm; top 1–3 spikelets staminate; pistillate spikelets gen > 2, > 1.5 cm, gen long-stalked; lowest spikelet bract lf-like, sheath gen < 5 mm; pistillate fl bract gen awned, surface gen glabrous, margin ciliate or not. **PISTILLATE FL**: style gen fragile, deciduous, jointed to fr, stigmas 3. **FR**: perigynium gen ascending, gen green to brown, 3-sided, gen abruptly narrowed above, beaked, faces hairy; fr 3-sided.

1. Style tough, persistent, not jointed to fr, bent; lowest spikelet bract > infl, sheathless
2. Lf blade hairy; perigynium many-ribbed, body ovate, 5–8 mm, beak teeth 0.7–1.5 mm, outcurved; lateral spikelets with < 100 ascending perigynia .... ***C. sheldonii***
2′ Lf blade glabrous; perigynium 2-ribbed, with few or 0 veins on faces, body obovate, 3–5 mm, beak teeth < 0.5 mm, straight; lateral spikelets with 150–300 spreading perigynia (see Group 4) .... ***C. spissa***
1′ Style fragile, deciduous, jointed to fr, not bent (base sometimes bent, persistent); lowest spikelet bract < to > infl, sheathed or not
3. Lowest spikelet bract sheath < 5 mm
4. Perigynium thin-walled, beak tip unnotched or notched only on back gen < 0.2 mm
5. Lf blade glabrous exc sometimes near sheath; spikelets dark; pistillate fl bract < perigynium; talus, 2600–3900 m .... ***C. congdonii***
5′ Lf blade conspicuously hairy; spikelets greenish; pistillate fl bract often > perigynium; moist to wet places, 1200–2600 m .... ***C. sartwelliana***
4′ Perigynium conspicuously thick-walled, beak tip notched on back and front > 0.2 mm
6. Lf blade strongly V-folded at least near base, basal blades not minute; pistillate fl bract not ciliate, acute, < or ± = perigynium; lowest spikelet stalk gen incl in bract; perigynium green to gold, ribs strong, sometimes obscured by hair; dry forest edges, often in pumice .... [2]***C. halliana***
6′ Lf blades flat or rolled, basal minute; pistillate fl bract ciliate, awned, > perigynium; lowest spikelet gen sessile; perigynium gen purplish, ribs weak; wet places
7. Upper lf ligules > 2 mm, thin, membranous; lf blade flat, not rigid, < or > infl, 1.5–5 mm wide, slender tip short, sheath mouth densely hairy or not but not ciliate; lowest infl internode 0.5–0.9 mm wide; perigynium beak teeth ± outcurved; style base gen straight; lower lf sheath fronts often not shredding; gen marshy places, in standing water or not .... ***C. lanuginosa***
7′ Upper lf ligules < 2 mm, thick, tough; lf blade rolled, rigid, >> infl, 0.5–2 mm wide, slender tip long, sheath mouth often ciliate; lowest infl internode gen < 0.8 mm wide; perigynium beak teeth gen ± erect; style base gen bent; lower lf sheath fronts shredding; lake, pond shores, gen in standing water .... ***C. lasiocarpa***
3′ Lowest spikelet bract sheath > 5 mm
8. Lf blade hairy; pistillate fl bract surface hairy, margin not ciliate
9. Perigynium 3.8–5.5 mm, base long-tapered, < upper 1/3 of body hairy, beak 0.5–1 mm; terminal spikelet staminate at least at tip .... ***C. gynodynama***
9′ Perigynium 3.5–4 mm, base short-tapered, at least upper 1/2 of body hairy, beak > 1 mm; terminal spikelet gen pistillate at least at tip .... ***C. hirtissima***
8′ Lf blade glabrous or very sparsely hairy; pistillate fl bract surface glabrous or midrib sometimes hairy, margin ciliate or not

10. Perigynium beak < 0.5 mm; lf sheath front often red or red-dotted (see Group 5)
 11. Perigynium hairs appressed, inconspicuous; fr 1.7–2.5 mm; pistillate fl bract ciliate .... ***C. mendocinensis***
 11′ Perigynium hairs spreading, conspicuous; fr 3.2–3.5 mm; pistillate fl bract not ciliate .... ***C. triquetra***
10′ Perigynium beak > 0.6 mm; lf sheath front not red or red-dotted
 12. Perigynium 5–8 mm; fr 2.7–3.3 mm .... ***C. obispoensis***
 12′ Perigynium < 5.5 mm; fr < 2.5 mm
  13. Lowest spikelet bract > infl; lvs stiff, ascending, at least lower part of blades sharply V-folded; spikelets stiffly erect, lowest spikelet stalk not or short-exserted; perigynium strongly many-ribbed ..... [2]***C. halliana***
  13′ Lowest spikelet bract << infl; lvs not stiff, spreading, blades not sharply V-folded; spikelets erect to nodding, lowest spikelet stalk short- to long-exserted; perigynium 2-ribbed, less often strongly veined as well
   14. Spikelet bract sheath mouth gen deeply Y- or V-shaped, often purple-banded at top, sheath ± expanded to mouth or not; perigynium ascending, gen primarily red-brown or purple, beak tip to fr top > 2 mm; pistillate fl bract tip acute or minutely pointed; lateral spikelets gen purplish to brown, tips tapered (see Group 5) .... ***C. fissuricola***
   14′ Spikelet bract sheath mouth gen shallowly U-shaped or truncate, gen not purple-banded at top, sheath linear; perigynium widely ascending to spreading, gen primarily green, beak tip to fr top < 2.5 mm; pistillate fl bract tip obtuse to acute; lateral spikelets gen 2-colored (perigynia vs. fl bracts), tips truncate or tapered (see Group 5) .... ***C. luzulina*** var. ***luzulina***

Group 4

Monoecious, gen long-rhizomed. **LF**: blade with small, raised, irregularly spaced cross-walls; lower sheath fronts gen shredding to network or fringe of veins. **INFL** > 4 cm; top 1–5 spikelets staminate; pistillate spikelets > 2, > 2.5 cm, cylindric; lowest spikelet bract lf-like, > infl, gen sheathless; lower pistillate fl bracts gen with long, serrate awn. **PISTILLATE FL**: style tough, persistent, not jointed to fr, gen bent, stigmas 3. **FR**: perigynium gen glabrous, gen many-ribbed, gen round in X-section, gen tapered above, body > fr, beak > 1 mm, tip deeply notched; fr 3-sided.

1. Lower spikelets staminate at tip; pl 1–2 m; perigynium beak 0.1–0.8 mm, often bent; style bent (keys also in Group 3) .... ***C. spissa***
1′ Lower spikelets pistillate; pl < 1.5 m; perigynium beak > 1 mm, straight; style straight or bent
 2. Lf sheath, blade lower surface soft-hairy; perigynium ascending, beak teeth 1.1–3 mm, outcurved ... ***C. atherodes***
 2′ Lf sheath, blade lower surface glabrous; perigynium ascending to reflexed, beak teeth 0.1–2.3 mm, outcurved or not
  3. Lower spikelets on long, nodding stalks; pistillate fl bract gen white or cream with pale reddish center, awn > body; fr 1.2–2 mm
   4. Perigynium stalked, beak notch 1.1–2.3 mm, outcurved, body ± angled; pistillate fl bract awn linear; ligule length < or = width .... ***C. comosa***
   4′ Perigynium ± sessile, beak notch 0.2–1 mm, erect, body ± not angled; pistillate fl bract awn lanceolate; ligule length >> width .... ***C. hystricina***
  3′ Lower spikelets ± erect, stalks 0 or short; pistillate fl bract gen mostly green to purplish, awn 0 or < body; fr gen 1.1–2.3 mm
   5. Perigynia in many dense rows, ascending-spreading to reflexed, 3.9–7 mm, abruptly narrowed above, beak 1–2 mm, teeth 0.1–0.8 mm; rhizomes long; st angles blunt; ligule length < or ± = width ..... ***C. utriculata***
   5′ Perigynia in a few ± open rows, ascending, 4–11 mm, tapered or ± abruptly narrowed above, beak 1–3 mm, teeth 0.2–1.5 mm; rhizomes short; st angles sharp; liqule length > width .... ***C. vesicaria***
    6. Perigynium gen 7–11 mm, tapered above, beak 2–3 mm, teeth 0.2–1.5 mm; lower lf sheath fronts gen shredding to network or fringe of veins .... var. ***major***
    6′ Perigynium gen 4–8 mm, + abruptly narrowed above, beak 1–2 mm, teeth 0.2–1 mm; loewr lf sheath sheath fronts gen not shredding .... var. ***vesicaria***

Group 5

Monoecious. **LF** gen < st; blades flat or folded, basal gen not minute; lower sheath fronts sometimes shredding to network or fringe of veins. **INFL**: terminal spikelet gen staminate, gen > 2 mm wide, gen oblong; lower spikelets pistillate or with staminate tips, gen erect, gen 1–4 cm, at least lowest ones gen stalked; lowest spikelet bract ± lf like, gen < or = infl, gen sheathless, gen ascending. **PISTILLATE FL**: stigmas 3. **FR**: perigynia gen 5–50 per spikelet, gen appressed to ascending, flat or 3-sided, faces gen glabrous, 0–many but gen 2-ribbed, beak tip gen notched, teeth gen glabrous; fr 3-sided.

1. Lf blade with long, soft hairs, sheath front brownish; fr ± = perigynium body; dry places .... ***C. whitneyi***
1′ Lf blade without long, soft hairs, sheath front white to red; fr < or ± = perigynium body; wet or dry places
 2. Lowest spikelet bract sheath > 6 mm; lower lf sheath fronts often shredding
  3. St ± 2–8 cm, < lvs; perigynia 3–8 per spikelet, 1.3–2 mm, 0.4–0.9 mm wide, spines few, curved, on upper body margins; pistillate spikelet < 5 mm, fl bracts early deciduous; lf blade margins and midrib at 10× coarsely serrate .... ***C. tiogana***
  3′ St > 10 cm, gen > lvs; perigynia > 5 per spikelet, > 2 mm, > 0.8 mm wide, spines 0; pistillate spikelet > 1 cm, fl bracts not early deciduous; lf blade margins and midrib at 10 × at most finely serrate
   4. Perigynium body tip wide conic, beak gen < 0.2 mm or indistinct, tip unnotched; fr > 2 mm
    5. Lf blade 6–14 mm wide, flat, not glaucous; perigynium 4.3–6.5 mm, not glaucous, with 1 fr; fr 2.7–3.3 mm; pl cespitose .... ***C. hendersonii***

5′ Lf blade 1–3.5 mm wide, flat or V folded, ± glaucous; perigynium 3.5–4.6 mm, green glaucous, with 1–2 frs; fr 2–3.5 mm; pl rhizomed .... ***C. livida***

4′ Perigynium body tip conic to rounded, beak > 0.2 mm, tip notched in front; fr gen < 2.5 mm

6. Lower spikelet bract blades ascending-spreading, >> infl; perigynia spreading to reflexed, crowded; upper spikelets densely clustered .... [2]***C. viridula*** var. ***viridula***

6′ Lower spikelet bract blades erect to ascending, < or ± = infl; perigynia appressed to spreading, crowded or not; upper spikelets sometimes densely clustered

7. Perigynium, pistillate fl bract midstripe, lf blade lower side glandular papillate; pl strongly rhizomed; lf blade gray green .... ***C. californica***

7′ Perigynium, pistillate fl bract midstripe, lf blade lower side not glandular, papillate or not; pl rhizomed or cespitose; lf blade gray or bright green

8. Lowest spikelet bract blade < infl, sheath mouth hairy; perigynium beak < 0.5 mm; lf sheath front often red or red dotted — lateral spikelets linear, 15–50 mm (keys also in Group 3) ...... ***C. mendocinensis***

8′ Lowest spikelet bract blade << infl, sheath mouth glabrous; perigynium beak > 0.5 mm; lf sheath front white

9. Perigynia dark purple, 1.7–2.5 mm wide, flat margin > 1/2 fr width, beak, pistillate fl bract (exc sometimes awn) glabrous; spikelet bract sheath conspicuously expanded to mouth, mouth > 2 mm wide, gen shallowly U-shaped, with wide purple band at top; lowest spikelet stalk long exserted from bract sheath, nodding; lf blade 5–20 mm wide — lateral spikelets > 1.2 cm, purple to dark brown .... ***C. luzulaifolia***

9′ Perigynia green to purple, < 1.7 mm wide (if wider and with flat margin around fr then beak, pistillate fl bract tip and midrib ciliate); spikelet bract sheath linear or slightly expanded to mouth, mouth < 2 mm wide, U-, V-, or Y-shaped, sometimes with narrow but without wide purple band at top; lowest spikelet stalk exserted or not from bract sheath, erect or nodding; lf blade gen 3–10 (1.5–15) mm wide

10. Perigynium 2.5–4.5 mm, gen mostly green, partly purple, beak tip gen ciliate, gen < 1.4 mm to fr top; lowest spikelet stalk incl in or exserted gen < 1 cm from bract sheath; lf blade 1.5–6 mm wide — lowest spikelet bract bladed

11. Infl > 15 cm, lowest internode 10–25 cm; pistillate fl bract gen white with green midrib; lf blade 3–6 mm wide; terminal spikelet gen pistillate below tip; longest lateral spikelets > 15 mm; larger perigynia 3.5–4.5 mm .... ***C. albida***

11′ Infl gen < 15 cm, lowest internode gen < 10 cm; pistillate fl bract gen at least partly red brown, sometimes all white; lf blade gen 1.5–3(4) mm wide; terminal spikelet gen staminate; longest lateral spikelets < 15 mm; larger perigynia 2.7–3.5(4) mm .... ***C. lemmonii***

10′ Perigynium 3–5.5 mm, green to purple, beak tip ciliate or not, > 1.5 mm to fr top; lowest spikelet stalk gen exserted > 1 cm from bract sheath; lf blade (1.5)3–10 mm wide

12. Spikelet bract sheath ± expanded to mouth or not, mouth gen deeply Y- or V-shaped, often with purple band at top; perigynium ascending, gen ± red brown or purple, beak tip to fr top > 2 mm; pistillate fl bract tip acute or minutely pointed (keys also in Group 3) .... ***C. fissuricola***

12′ Spikelet bract sheath linear, mouth gen shallowly U-shaped or truncate, gen without purple band at top; perigynium ascending to spreading, ± green or purple, beak tip to fr top < 2.5 mm; pistillate fl bract tip obtuse to acute .... ***C. luzulina***

13. Perigynium gen mostly purple, ± stalked, ascending, 3–4.5 mm, body ± = fr body; lateral spikelet tip tapered, sides ± deeply jagged; pistillate fl bract dark purple, midstripe gen < 0.2 mm wide, gen not to tip, tip acute; lf blade 1.5–6 mm wide .... var. ***ablata***

13′ Perigynium gen ± green to red-brown, sessile, ascending spreading to spreading, 3.5–5.5 mm, body ± > fr body; lateral spikelet tip truncate, rounded, or abruptly narrowed to staminate part, sides ± smooth or bristly, not deeply jagged; pistillate fl bract gen red-brown, midstripe > 0.2 mm wide, ± to tip, tip obtuse to acute; lf blade 4–10 mm wide (keys also in Group 3) .... var. ***luzulina***

2′ Lowest spikelet bract sheath < 5 mm; lower lf sheath fronts sometimes shredding

14. Perigynium body tapered to widely conic tip, gen beakless, papillate above; style base often persistent, black, exserted; lowest spikelet on nodding stalk; root conspicuously hairy; pl long-rhizomed; pistillate fl bract gen brown .... ***C. limosa***

14′ Perigynium body tapered or abruptly rounded to short, linear beak, papillate above or not; style base gen not persistent, brown, incl or not; lowest spikelet sessile or on erect or nodding stalk; root not conspicuously hairy; pl cespitose or rhizomed; pistillate fl bract ± brown or dark purple to black with pale midstripe

15. Pistillate fl bract < perigynium, awn hairy; perigynium beak teeth hairy .... ***C. serratodens***

15′ Pistillate fl bract < or > perigynium, awn 0 or glabrous; perigynium beak teeth glabrous or 0

16. Terminal spikelet staminate at tip or throughout

17. Perigynia > 60 per spikelet, spreading; lf blade 8–20 mm wide, with small, raised, irregularly spaced cross walls; st sharply 3-winged; lateral spikelets 3.5–14 cm .... ***C. amplifolia***

17′ Perigynia < 50 per spikelet, ascending or spreading; lf blade < 10 mm wide, without small raised cross walls; st sharply or obtusely 3-angled, ± unwinged; lateral spikelets < 4 cm

18. Perigynium appressed to ascending, gen ± purple at least above, body flat or flat around fr, ± unveined, width >> fr; pistillate fl bract with pale midstripe gen to tip, awned or tip long-tapered; lowest spikelet often on nodding stalk

19. Spikelets gen conspicuously 2-colored (perigynia vs. fl bracts), erect; perigynium not papillate, body elliptic to obovate, 3-sided with flat margin around fr; basal lf blades not minute .... [2]***C. heteroneura*** var. ***heteroneura***

19′ Spikelets purple, often nodding; perigynium papillate, body ovate, ± flat or 3-sided; basal lf blades gen minute . . . . *C. spectabilis*

18′ Perigynium spreading, gen green, body spheric or flat around fr above, veined or many ribbed, width ± = fr; pistillate fl bract with pale midstripe 0 or not to tip, awn 0; lowest spikelet sessile or on erect stalk

20. Staminate spikelet oblong, > 2 mm wide; lower spikelet bract blades ascending, < or > infl, not bright yellow-green; perigynium 1.6–2.1 mm wide, flat around fr above . . . . *C. raynoldsii*

20′ Staminate spikelet linear, 1–2 mm wide; lower spikelet bract blades ascending spreading, >> infl, bright yellow-green; perigynium 0.9–1.5 mm wide, not flat around fr . . . . [2]*C. viridula* var. *viridula*

16′ Terminal spikelet pistillate above, gen staminate at base

21. Lowest spikelet gen on long, nodding stalk; spikelets 6–10, gen > 2 cm; lowest spikelet bract > infl; perigynium 3.9–5.5 mm, very flat, > 3 × fr width . . . . *C. mertensii*

21′ Lowest spikelet sessile or on short, erect stalk; spikelets 1–6, < or > 2 cm; lowest spikelet bract < infl; perigynium < 4.5 mm, ± round to flat, < 3 × fr width (often > 3 × in *C. heteroneura* var. *epapillosa*)

22. Perigynium 1.6–3 mm, obovate, edge gen with a few stiff hairs on and just below beak, upper body gen papillate; terminal spikelet gen pistillate, or staminate at base or tip; pistillate fl bract white-margined . . . . *C. parryana* var. *hallii*

22′ Perigynium > 2.5 mm (exc sometimes *C. norvegica*), ovate to obovate, edge ± without hairs, upper body papillate or not; terminal spikelet pistillate above, gen staminate at base; pistillate fl bract white-margined or not

23. Upper body of perigynium conspicuously papillate

24. Pistillate fl bract awned, white margin or tip indistinct or 0; lower lf sheath fronts shredding — pl rhizomed . . . . *C. buxbaumii*

24′ Pistillate fl bract obtuse to with a small point, with distinct white margin and tip; lower lf sheath fronts not shredding

25. Perigynium ± inconspicuous in gen head-like infl, ± = pistillate fl bract, 1.3–2 mm wide, body at least 1/4 empty above fr, ± flat; lvs gray-green; pl cespitose, 10–30 cm; ± dry, rocky places, 3000–3900 m, often on summits . . . . *C. albonigra*

25′ Perigynium conspicuous in gen ± open infl, >> pistillate fl bract, 1–1.3 mm wide, body ± filled by fr, 3-sided; lvs green; pl short-rhizomed, 10–70 cm; wet places, 2900 m . . . . *C. norvegica*

23′ Upper body of perigynium not or inconspicuously papillate

26. Pistillate fl bract gen < 1 mm wide, red-brown to purple, length >, width <<, color ± = perigynium; infl gen head-like; st gen < 30 cm; 2400–4100 m . . . . *C. helleri*

26′ Pistillate fl bract gen > 1.2 mm wide, dark purple, length < or >, width < or >, color gen not = perigynium; infl not head-like, at least lowest spikelet not clustered with others; st 25–100 cm; 1800–3800 m . . . . *C. heteroneura*

27. Perigynium > 3.5 mm, flat margin (0.3(0.5–0.7 mm wide fr base to perigynium base 0.5–1.1 mm . . . . var. *epapillosa*

27′ Perigynium gen < 3.5 mm, flat margin < 0.4 mm wide;fr base to perigynium base 0–0.4 mm . . . . [2]var. *heteroneura*

Group 6

Monoecious, gen rhizomed. **LF**: blades gen flat, basal gen no minute; bladeless sheaths gen < 7 mm wide at midlength, gen no keeled. **INFL** > 2.5 cm; spikelets gen 1 per node, terminal 1–3 gen staminate, gen < 5 cm, lower 3 or more gen pistillate, gen erect, at least lowest 1 gen stalked; lowest spikelet bract gen lf-like, gen not sheathing; pistillate fl bract gen > 0.5 mm wide, rarely lf-like, gen acute, gen purple to black, gen glabrous. **PISTILLATE FL**: stigmas 2. **FR**: perigynium gen ascending, gen abruptly narrowed above, gen papillate, gen glabrous, 2-ribbed, gen green, beak gen ± purple or brown, tip gen notched < 0.1 mm, teeth gen glabrous; fr 2-sided, sometimes deeply indented on 1 or both sides.

1. Lowest spikelet bract sheath > 4.5 mm; lower spikelet stalks often from near pl base; longest pistillate spikelets < 2.5 cm; perigynium green or white to orange, papillate, ± fleshy

2. Perigynium 2.5–4 mm, body tapered to red-brown-margined tip, not translucent; lower pistillate fl bracts stout-awned, often lf-like; lower staminate fl bracts gen ± long-awned, like spikelet bract; spikelets often > 1 per lower branch . . . . *C. saliniformis*

2′ Perigynium 1.8–3.1 mm, body widely blunt-tipped or slightly tapered, tip not red-brown exc on beak (if present), ± translucent; lower pistillate fl bracts obtuse to narrow-awned, not lf-like; lower staminate fl bracts not long-awned, not like spikelet bract; spikelets 1 per node throughout infl

3. Pistillate fl bracts gen crowded, red-brown to purplish, often > perigynia, tip gen widely obtuse; spikelet bract sheath mouth V-shaped; spikelets, exc basal, gen densely clustered, terminal 1 gen pistillate at least at tip . . . . *C. garberi*

3′ Pistillate fl bracts gen not crowded, white to red-brown, < perigynia, tip obtuse to awned; spikelet bract sheath mouth U-shaped; spikelets not densely clustered, terminal one sometimes pistillate at tip

4. Lower pistillate fl bracts not appressed against, often falling before fully expanded perigynia; staminate portion of terminal spike 0.9–2 mm wide, lower fl bracts 2–3.5(4) mm; perigynium spheric in spike, gen sessile, body gen strongly ribbed, tip gen not conspicuously papillate at 10 ×, body when fresh orange, when dry orange or gold to purplish, often white above, ± squashed below . . . . *C. aurea*

4′ Lower pistillate fl bracts closely appressed against, falling after fully expanded perigynia; staminate portion of terminal spike (1.8)2–3.5 mm wide, lower fl bracts 3–6 mm; perigynium ± obovate, stalked, body veined, gen not ribbed, tip conspicuously papillate at 10 ×, body gold or white when fresh, when dry greenish white or pale gold, lower part sometimes gold, gen not squashed .... ***C. hassei***

1′ Lowest spikelet bract sheath < 4 mm; lower spikelet stalks not from near pl base; longest pistillate spikelets often > 3 cm; perigynium green, brown, or purple, papillate or not, not fleshy

5. Perigynium wall ± tough, resisting puncture, brown, sometimes red- or purple-dotted

6. Perigynium wall very thick, very tough, beak 0–0.3 mm, notch at tip 0–0.1 mm, tip glabrous; fr 1.4–2.5 mm, often deeply indented on 1 or both sides

7. Lowest spikelet ± truncate at base, stiff, straight, axis (1.8)2–8 cm, stalk long, nodding; perigynium dull, papillate, ± conspicuously veined; fr wide-stalked; lower lf sheath fronts not shredding; long-rhizomed .... ***C. lyngbyei***

7′ Lowest spikelet tapered at base, flexible, nodding, axis (2.5)3–15 cm, stalk 0 or ± short, gen erect; perigynium shiny, not papillate, inconspicuously veined; fr sessile; lower lf sheath fronts shredding; rhizomed, forming beds or large, dense, raised clumps .... ***C. obnupta***

6′ Perigynium wall thick, ± tough, beak 0.2–0.6 mm, notch at tip > 0.1 mm, tip hairy; fr 1.3–2 mm, not indented

8. Pistillate fl bract awned, tip or awn hairy; lower lf sheath fronts shredding, blades gen not tufted at st base, not blue-glaucous; fully mature spikelets uniformly brown, not pressing flat; perigynium spreading, 3–4.5 mm .... ***C. barbarae***

8′ Pistillate fl bract obtuse to awned, tip or awn glabrous; lower lf sheath fronts not shredding, blades gen tufted at st base, gen blue-glaucous; mature spikelets gen 2-colored, pressing ± flat; perigynium ascending, 2.6–4 mm .... [2]***C. nebrascensis***

5′ Perigynium wall thin, soft, easy to puncture, green to brown, often red or purple-dotted or -blotched

9. Perigynium beak 0.2–0.6 mm, tip notched > 0.1 mm, teeth minutely hairy; fresh lf blade gen blue-glaucous; lvs gen densely tufted at st base .... [2]***C. nebrascensis***

9′ Perigynium beak gen < 0.4 mm, tip notched < 0.1 mm, glabrous; fresh lf blade gen not blue-glaucous; lvs tufted at st base or not

10. Veins on perigynium faces 0 or indistinct; lower lf sheath fronts gen not shredding

11. Spikelets 4.5–11.5 cm, tips gen staminate, lower stalks gen nodding, < 11 cm; lowest spikelet bract gen > infl; lf sheath front purple-dotted, mouth purple-brown .... ***C. aquatilis*** var. ***dives***

11′ Spikelets < 10 cm, tips gen pistillate, lower stalks 0 or gen erect, < 4.2 cm; lowest spikelet bract gen < or ± = infl; lf sheath front red- or purple-dotted or not, mouth not or weakly purple

12. Lowest spikelet 10–100 mm, its bract gen ± = infl, its stalk 0–42 mm; infl gen open, lowest internode 20–180 mm; perigynium ± red-dotted or -blotched, weakly papillate; fr shiny .... ***C. aquatilis*** var. ***aquatilis***

12′ Lowest spikelet 8–30 mm, its bract gen < 1/2 infl, its stalk 0–15 mm; infl ± dense, lowest internode 5–45 mm; perigynium gen purple above, strongly papillate; fr dull .... ***C. scopulorum*** var. ***bracteosa***

10′ Veins on perigynium back face or on both faces gen distinct; lower lf sheath fronts shredding or not

13. Perigynium stalk gen > 0.1 mm, body ± slightly, abruptly narrowed near base, green or white over fr, purple-dotted or not, papillate; lower lf sheaths backs not scabrous, front not shredding (exc sometimes longitudinally); pistillate fl bract widely obtuse or acute, unawned, <, often << perigynium; lf blades 1–3.5 mm wide, basal no minute; lowest spikelet bract > or = infl .... ***C. lenticularis***

14. Lower spikelets >> internodes between, lowest 4–6 mm wide; perigynium > pistillate fl bract, stalk 0.4–0.7 mm .... var. ***limnophila***

14′ Lower spikelets < or less often > internodes between, lowest 3–4 mm wide; perigynium often >> pistillate fl bract, stalk < 0.5 mm

15. Perigynium veins gen 3 on back, 0 or < 3 on front, body often purple-dotted, beak dark throughout or with purple or brown ± triangle from tip, stalk < 0.2 mm .... var. ***impressa***

15′ Perigynium veins gen 5–7 on back and front, body green, beak purple or brown only at tip, stalk 0.2–0.5 mm .... var. ***lipocarpa***

13′ Perigynium ± sessile, body not even slightly abruptly narrowed near base, green to brown, often purple-dotted or -blotched, papillate or not; lower lf sheath backs scabrous, fronts shredding; pistillate fl bract obtuse to awned, < or > perigynium; lf blades 2–12 mm wide, basal minute or not; lowest spikelet bract < or > infl

16. Pl 1–1.5 m; lf blade 4–12 mm wide; bladeless sheath > 7 mm wide at midlength, keeled; perigynium (2.7)3–4.5 mm; staminate spikelets 2–7, terminal often > 5 cm; lateral spikelets gen 5–20 cm, 5–7 mm wide, stalks 0 or long, ± nodding; lowest spikelet bract lf-like, gen > infl .... ***C. schottii***

16′ Pl gen < 1 m; lf blade 2–9 mm wide; bladeless sheath < 7 mm wide at midlength, not keeled; perigynium 2–4 mm; staminate spikelets gen 1–2, terminal < 5 cm; lateral spikelets (1.5)2–9 cm, < 6 mm wide, stalks 0 or erect; lowest spikelet bract lf- or bristle-like, < or > infl

17. Pl forming dense stands but not large, raised clumps, long-rhizomed; lowest spikelet gen < 5 mm wide, its bract > or ± = infl; pistillate fl bract often > perigynium body, acute or tapered to gen pointed tip, often < 0.5 mm wide; perigynium 2.2–3.1 mm, ± weakly 1–3-veined both sides, green, purple-dotted above or not, weakly papillate; lf blade 2–7 mm wide .... ***C. angustata***

17′ Pl forming large, dense, raised clumps connected by rhizomes or not; lowest spikelet gen > or ± = 5 mm wide, its bract < or > infl; pistillate fl bract gen < perigynium body, obtuse- to acuminate-tipped, gen > 0.5 mm wide; perigynium 2.2–4 mm, 3–9-veined on back, green or above purple, papillate or not; lf blade 2–5 mm wide

18. Perigynium 1.2–1.6(1.8) mm wide, gen dark purple on beak, upper body; lf blades 2–4 mm wide, lower sheath backs gen without raised cross-walls; rocky or sandy streambeds below high-water mark; clumps not connected by rhizomes, not in ± continuous stands .... ***C. nudata***

18′ Perigynium 1.4–2.2 mm wide, sometimes dark purple on beak, upper body; lf blades 3–5 mm wide, lower sheath backs with small, raised, irregularly spaced cross-walls; streambanks, marshy areas, meadows; clumps connected by rhizomes, in ± continuous stands or not .... ***C. senta***

Group 7

Sometimes dioecious. **LF**: basal blades minute. **INFL** gen dense, > 5 mm wide; spikelets each staminate at tip, pistillate below, gen < 1.5 cm, ovate to oblong, sessile; lowest spikelet bract gen bristle-like, gen < infl. **PISTILLATE FL**: style gen < 1.5 mm, stigmas 2, gen < 4 mm. **FR**: perigynia > 3 per spikelet, planoconvex, gen widest above base, gen abruptly narrowed above, glabrous, gen stalked, margin of upper body, beak gen serrate, beak > 0.2 mm, tip gen notched; fr 2-sided.

1. Pl staminate

2. Pl cespitose; spikelets > 1 per lower node or branch; lf sheath front gen red-dotted

3. Anther > 1.9 mm; lf sheath front not cross-wrinkled, mouth gen with membranous, tongue-like extension, not copper, not purplish, without thick rim .... [2]***C. alma***

3′ Anther < 1.8 mm; lf sheath front often cross-wrinkled, mouth gen U-shaped, copper or purplish, often with thick rim .... [2]***C. cusickii***

2′ Pl rhizomed; spikelets gen 1 per node throughout infl; lf sheath front gen not red-dotted

4. Anther with narrow, < 0.1 mm tip; lf blade < 1.5 mm wide; rhizome < 2 mm thick; infl < 20 mm .. [2]***C. eleocharis***

4′ Anther with 0.1–1 mm awn; lf blade 1–5 mm wide; rhizome 1–5 mm thick; infl < 40 mm (exc some *C. praegracilis*)

5. Rhizome 1–2 mm thick, brown; anther awn 0.2–1 mm, glabrous; lf sheath mouth with thick rim; infl (7)10–25 mm wide .... [2]***C. douglasii***

5′ Rhizome 2–5 mm thick, dark brown; anther awn 0.1–0.4 mm, glabrous or hairy; lf sheath mouth fragile or with thick rim; infl < 15 mm wide exc in some *C. pansa*

6. Longest anther awns 0.2–0.4 mm, filaments gen incl; pl gen < 30 cm, on coast or islands — sand .. [3]***C. pansa***

6′ Longest anther awns 0.1–0.2 mm, filaments exserted; pl 20–75 cm, near coast or inland

7. Anther awn gen hairy at 20 ×, slender, tip tapered or obtuse; lf sheath mouth often with thick rim, without brown stripe; lf blade 1.5–3 mm wide .... [3]***C. praegracilis***

7′ Anther awn glabrous at 20 ×, stout, tip obtuse; lf sheath mouth without thick rim, often with brown stripe across; lf blade 2–5 mm wide .... [3]***C. simulata***

1′ Pl bisexual or pistillate

8. Perigynium wall filled with pithy tissue below, base often ± whitish; spikelet bracts often conspicuous

9. Perigynium long-tapered to poorly defined beak, beak tip to fr top > 1/2 total length, body widest ± at base

10. Perigynium pale to dark brown, veins on back few, on front 0; pistillate fl bract white or widely white-margined; fresh st not spongy .... [2]***C. alma***

10′ Perigynium green to medium brown, veins many both sides, each a rib; pistillate fl bract pale to black, white-margined or not; fresh st spongy (exc *C. jonesii*)

11. Lf sheath front with thick rim at mouth, not cross-wrinkled, back white-spotted; perigynium 3–4.5 mm; fresh st deeply concave between sharp angles or wings, spongy .... ***C. nervina***

11′ Lf sheath front without thick rim at mouth, often cross-wrinkled below mouth, back gen not white-spotted; perigynium 2.5–6 mm; fresh st deeply or shallowly concave between blunt or sharp angles or wings, spongy or firm

12. Spikelets > 1 per lower node or branch; infl 2–10 cm, 1–3 cm wide, spikelets >> 10; lf blade 5–11 mm wide; perigynium 3.6–6 mm, 1.5–1.8 mm wide .... ***C. stipata*** var. ***stipata***

12′ Spikelet 1 per node throughout infl; infl 1–2.5 cm, 0.6–1.5 cm wide, spikelets < 10; lf blade 1–5 mm wide; perigynium 2.5–4 mm, 0.8–1.5 mm wide

13. Margin of perigynium beak and upper body entire; lvs gen tufted near st base, sheath fronts gen ± not cross-wrinkled, mouth without tongue-like extension; pistillate fl bract gold to black; fresh st ± concave between ± sharp angles, ± firm; pl gen not forming large raised clump .... [2]***C. jonesii***

13′ Margin of perigynium beak and gen upper body serrate at least on 1 side; lvs gen not tufted near st base, sheath fronts often cross-wrinkled, mouth with tongue-like extension; pistillate fl bract brown, white-margined; fresh st deeply concave between sharp angles or wings, spongy; pl often forming large raised clump .... ***C. neurophora***

9′ Perigynium short-tapered or abruptly narrowed to conic or ± linear beak, beak tip to fr top < 1/2 total length, body widest at or above base

14. Perigynium back (near base) with lengthwise groove 0.1–0.3 mm wide, beak with white flap on back

15. Pl with rhizomes; perigynium beak 0.2–0.6 mm, gen < 1/4 body .... [3]***C. simulata***

15′ Pl cespitose (rhizomes 0); perigynium beak 0.7–1.5 mm, gen > 1/3 body

16. Lf sheath front near mouth copper or purplish, often cross-wrinkled, blade 2.5–6 mm wide; infl 4–8 cm, ± flexible, lower internodes >> others; perigynium 2–4 mm, 1–2 mm wide, lengthwise groove 0.1–0.3 mm wide near base .... [2]***C. cusickii***

16′ Lf sheath front near mouth white or at most with a narrow brown margin, not cross-wrinkled, blade 1–3 mm wide; infl 2–5 cm, stiff, lower internodes sometimes > but gen not >> than others; perigynium 2–3 mm, 0.9–1.4 mm wide, lengthwise groove 0.2–0.3 mm wide .... ***C. diandra***

14′ Perigynium back (near base) with lengthwise groove 0 or < 0.1 mm wide, beak without flap on back

17. Spikelets > 1 per lower node or branch; infl 2–15 cm, (0.8)1–3 cm wide, spikelets > (often >>) 10; pl cespitose
18. Perigynium gen < 3 mm, 0.4–0.7 mm thick, length gen < or ± 2 × width, body deltate, front wall filled with pithy tissue in whitish U-shaped area around fr, ± flat over fr, veins 0 or few, short, beak often entire; lower pistillate fl bract awns gen > 1 mm . . . . ***C. vulpinoidea***
18′ Perigynium gen > 3 mm (exc often *C. dudleyi*), 0.7–1.1 mm thick, length gen > 2 × width, body ovate to deltate, front wall filled with pithy tissue below, rounded over fr, veined or not, beak serrate or ciliate; lower pistillate fl bract awns < or > 1 mm
19. Pistillate fl bract primarily white or widely white-margined, ± = or > perigynium
20. Perigynium 3.5–4.3 mm, gold to dark brown, body ovate, widest ± at base; infl 2.5–15 cm; lower lf sheath front red-dotted, gen not cross-wrinkled, back not white-spotted . . . . [2]***C. alma***
20′ Perigynium 2–3.7(4) mm, gold, body diamond-shaped, widest ± at middle; infl 1.5–5(6.5) cm; lower lf sheath front sometimes red-dotted, gen cross-wrinkled, back often white-spotted . . . . [3]***C. dudleyi***
19′ Pistillate fl bract pale gold to brown, narrowly or not white-margined (base white or not), < perigynium — these 2 spp. intergrade
21. Perigynium gen 3–4.5 mm (if < 3 mm, length < 2 × width), body ovate, with fr below top, tapered to beak, with indistinct or wide flat margin around fr, beak conic, gen > 0.6 mm wide 0.2 mm above fr . . . . [2]***C. densa***
21′ Perigynium gen 2–3.7 mm, length > 2 × width, body ± diamond-shaped, with fr at top, gen abruptly narrowed above, with narrow flat margin around fr, beak conic to ± linear, gen < 0.6 mm wide 0.2 mm above fr . . . . [3]***C. dudleyi***
17′ Spikelet 1 per node throughout infl (if > 1 per lower node or branch, pl rhizomed); infl 0.8–5 cm, gen < 1 cm wide (–2.5 in *C. douglasii*, *C. pansa*), spikelets gen < 10; pl cespitose or rhizomed
22. Pistillate fl bract white to greenish; perigynium green to pale gold; lf sheath front not cross-wrinkled; pl cespitose; alien
23. Perigynium 1.2–2.3 mm wide, beak 0.4–0.8 mm, serrate; fr 1.1–1.7 mm wide; lf blade 1–3 mm wide . . . . ***C. leavenworthii***
23′ Perigynium 0.8–1.2 mm wide, beak 0.5–1 mm, entire; fr 0.8–1 mm wide; lf blade 0.7–1.5 mm wide . . . . ***C. retroflexa*** var. ***texensis***
22′ Pistillate fl bract gold to black, white-margined or not; perigynium gold to brown; lf sheath front cross-wrinkled or not; pl cespitose or rhizomed; native
24. Pl cespitose; perigynium beaks entire, conspicuous in infl, body shiny, entire above . . . . [2]***C. jonesii***
24′ Pl rhizomed; perigynium beaks serrate, gen ± inconspicuous in infl, body shiny or dull, serrate above
25. Perigynium < 2.6 mm, shiny, beak gen < 1/4 body, style exserted; lower pistillate fl bracts ± awned . . . . [2]***C. simulata***
25′ Perigynium > 2.6 mm, shiny or dull, beak > 1/4 body, style exserted or not; lower pistillate fl bracts minutely pointed to awned
26. Perigynium gen shiny, base narrowed for > 0.3 mm below fr; pistillate fl bract gen shiny, gen dark brown, less often pale; longest anther awns 0.2–0.4 mm, filaments gen incl; pl 10–30 cm; fr 1.4–2.1 mm; infl gen ovate; on coast or islands, sand . . . . [3]***C. pansa***
26′ Perigynium ± dull, base narrowed for < 0.3 mm below fr; pistillate fl bract ± dull, gold to medium brown; longest anther awns 0.1–0.2 mm, filaments exserted; pl gen > 20 cm; fr 1.2–1.9 mm; infl oblong-elliptic; near coast or inland, ± moist, often alkaline places, incl sand . . . . [3]***C. praegracilis***

8′ Perigynium wall not filled with pithy tissue, brown to blackish; spikelet bracts conspicuous or not
27. Spikelets > 1 per lower node or branch; infl 2–8 cm, 1–3 cm wide, spikelets >, often >> 10; lower lf sheath fronts often cross-wrinkled; pl cespitose
28. Perigynium gen 3–4.5 mm (if < 3 mm, length < 2 × width), body ovate, with fr below top, ± tapered above, with indistinct or wide flat margin around fr, beak wide-conic, > 0.6 mm wide 0.2 mm above fr . . . . [2]***C. densa***
28′ Perigynium gen 2–3.7 mm, length > 2 × width, body diamond-shaped, with fr at top, gen abruptly narrowed to beak, with narrow flat margin around fr, beak narrow-conic to linear, < 0.6 mm wide 0.2 mm above fr . . . . [3]***C. dudleyi***
27′ Spikelet 1 per node throughout infl (if > 1 per lower node or branch, pl rhizomed); infl 0.5–5 cm, 0.4–1.6 (–2.7 in *C. douglasii*, *C. pansa*) cm wide, spikelets gen < 10; lower lf sheath fronts not cross-wrinkled; pl cespitose or rhizomed
29. Perigynium beak, upper body entire, body ± shiny
30. Pl < 6 cm; rhizome long, gen wavy; lf > infl, blade 0.7–1.5 mm wide; perigynium 2.8–3.5 mm, > pistillate fl bract; fr round to square; open, dry, gravelly or rocky slopes, 3700–4000 m . . . . ***C. incurviformis*** var. ***danaensis***
30′ Pl 5–30 cm; rhizome short, straight; lf < infl, blade 2–4 mm wide; perigynium 3.3–4.8 mm, < pistillate fl bract; fr ovate; open, often rocky places wet from snow-melt, 1800–4000 m — some immature pls of *C. nervina*, *C. neurophora* will key here . . . . ***C. vernacula***
29′ Perigynium beak, upper body serrate, body shiny or dull
31. Perigynium conspicuously spreading, > pistillate fl bract (exc *C. occidentalis*)
32. Perigynium back bulged so that marginal ribs are on front, not sides, very shiny, beak < 1 mm; infl 4–8 mm wide; fr 1.6–2.5 mm . . . . ***C. vallicola***
32′ Perigynium back not bulged (marginal ribs are on sides), dull or ± shiny, beak < or > 1 mm; infl 5–15 mm wide; fr 1.2–2.1 mm
33. Infl ± open, 1.5–3 cm; perigynium ± = pistillate fl bract . . . . [2]***C. occidentalis***
33′ Infl dense (sometimes lowest spikelet separate), 0.7–2 cm; perigynium > pistillate fl bract

34. Perigynium 2–3.5 mm, pale green to brown; fr ± sessile; pistillate fl bract awned, white or greenish, midrib green; lower spikelet bracts > spikelets; alien, lawns .... ***C. cephalophora***

34′ Perigynium 3.4–5 mm, brown with green margin; fr long-stalked; pistillate fl bract acute to long-acuminate, red-brown, midrib green, margin white; lower spikelet bracts gen < spikelets; native, rocky-gravelly slopes, meadow edges .... ***C. hoodii***

31′ Perigynium appressed to ascending, < or = pistillate fl bract

35. Pl cespitose or short-rhizomed, bisexual; perigynium 1.5–2 mm wide, base narrowed for > 0.4 mm below fr

36. Perigynium ascending, beak 0.6–1.2 mm; fr 1.3–2 mm; lowest spikelet bract like pistillate fl bract, < infl .... [2]***C. occidentalis***

36′ Perigynium appressed, beak 1.2–3 mm; fr 1.8–2.3 mm; lowest spikelet bract bristle-like, often > infl .... ***C. tumulicola***

35′ Pl long-rhizomed, often unisexual; perigynium 1.3–2 mm wide, base narrowed for < or > 0.4 mm below fr

37. Rhizome < 2.1 mm thick; lf blade margin gen inrolled, (sometimes ± flat)

38. Stigmas 4–6 mm, persistent, style 1.8–3.5 mm, exserted; perigynium 3.5–4.6 mm, gold to medium brown, beak 0.9–1.8 mm, ± = body; lf blade 1–2.5 mm wide; infl on pistillate pl 1.5–3.5 cm, 1.3–2.7 cm wide; lf sheath mouth with thick rim; gen dioecious .... [2]***C. douglasii***

38′ Stigmas < 4 mm, deciduous, style < 1.5 mm, incl; perigynium 2.5–3.5 mm, white to black, beak 0.4–1(1.2) mm, < body; lf blade 0.5–1.5 mm wide; infl on pistillate or bisexual pl < 2 cm, < 1 cm wide; lf sheath mouth without thick rim; monoecious or dioecious .... [2]***C. eleocharis***

37′ Rhizome > 2 mm thick; lf blade flat or V-shaped

39. Perigynium gen shiny, base narrowed for > 0.3 mm below fr; pistillate fl bract gen shiny, gen dark brown, less often pale; longest anther awns 0.2–0.4 mm, filaments gen incl; pl 10–30 cm; fr body 1.4–2.1 mm; infl gen ovate; on coast or islands, in sand .... [3]***C. pansa***

39′ Perigynium ± dull, base narrowed for < 0.3 mm below fr; pistillate fl bract ± dull, gold to medium brown; longest anther awns 0.1–0.2 mm, filaments exserted; pl gen > 20 cm; fr 1.2–1.9 mm; infl oblong-elliptic; near coast or inland, ± moist, often alkaline places, incl sand .... [3]***C. praegracilis***

Group 8

Monoecious, gen cespitose. **LF**: blades gen flat, basal gen minute. **INFL**: spikelets each staminate at base, pistillate above, gen < 1.5 cm, ovate to oblong, sessile; lowest spikelet bract gen bristle-like. **PISTILLATE FL**: stigmas 2. **FR**: perigynia > 3 per spikelet, gen ascending, planoconvex, gen abruptly narrowed above, glabrous, gen veined both sides, ribs 2, ± thin-edged above but unwinged, gen serrate or ciliate above, often not reaching to base, beak > 0.2 mm, gen serrate, tip ± notched; fr 2-sided.

1. Infl < 15 mm, dark brown, lowest infl internode < 1 mm; perigynia spreading, entire, beaks conspicuous in infl .... ***C. illota***

1′ Infl > 15 mm, green to medium brown, lowest infl internode > 2 mm; perigynia appressed to spreading, entire or serrate, beaks conspicuous in infl or not

2. Perigynium wall filled at least in lower 1/2 of body with pithy tissue, upper body ± papillate

3. Perigynium 2–3.4 mm, beak 0.6–1.2 mm, > 1/4 body, conic; lower lf sheath fronts at least sparsely minutely red-dotted; infl dense, spikelets 7–15 .... [2]***C. arcta***

3′ Perigynium 1.5–3 mm, beak < 0.55 mm, < 1/4 body, wide-conic; lf sheath fronts not red-dotted; infl dense or open, spikelets < 10

4. Perigynium beak without brown stripe or flap; pistillate fl bract white, sometimes gold or pale brown; infl 2–15 cm, ± open, lowest spikelet gen < 2 × lowest internode; lf blade 1.5–4 mm wide; pl 10–80 cm; 1100–3200 m .... ***C. canescens***

4′ Perigynium beak with brown stripe or flap along middle of back; pistillate fl bract light brown, margin and tip white; infl (1)1.5–2.5 cm, often dense, lowest spikelet gen > 2 × lowest internode; lf blade 1.2–2.5 mm wide; pl 10–30 cm; 2400–3500 m .... ***C. praeceptorum***

2′ Perigynium wall not or filled only near base with pithy tissue, body not papillate

5. Perigynia spreading to reflexed, widest ± at base, wall filled with pithy tissue there; spikelets ± star-shaped; staminate fl bracts of terminal spikelet appressed, internodes between them 2/3–3/4 their length; terminal spikelet base narrowly tapered

6. Perigynium 1.9–3.3 mm, beak densely serrate, 1/4–1/3 body, tip to fr top 0.5–1 mm; st edges sharp; wet places, often calcareous, not sphagnum bogs .... ***C. interior***

6′ Perigynium (2.4)2.6–4.8 mm, beak sparsely serrate, > or ± = body, tip to fr top 0.8–2 mm; st edges ± blunt; acidic, gen in sphagnum bogs .... ***C. echinata***

7. Perigynium (2.4)2.6–4 mm, ± abruptly narrowed above, front gen unveined, with bulge of pithy tissue below fr, beak tip to fr top gen 1–1.5 mm; fr ± round, 1.1–1.6 mm, 0.8–1.3 mm wide; lf blades 1–3.3 mm at widest .... ssp. ***echinata***

7′ Perigynium 3.5–4.8 mm, tapered above, front 2–12-veined, without bulge of pithy tissue, beak tip to fr top 1.4–2.2 mm; fr ovate, 1.5–2.2 mm, 1–1.5 mm wide; lf blades 1.7–4 mm at widest .. ssp. ***phyllomanica***

5′ Perigynia appressed to spreading, widest near or above base, wall filled with pithy tissue there or not; spikelets not star-shaped; staminate fl bracts of terminal spikelet appressed to spreading, internodes between them < or > 1/2 their length; terminal spikelet base round to narrowly tapered

8. Perigynium beak conic, 0.5–1.3 mm; perigynium ascending, 2–4 mm, ovate; lower pistillate fl bracts obtuse to acute

9. Infl ± dense; lf blade 2–4 mm wide; perigynium beak conic, minutely serrate at 10×, not recurved . . . . [2]*C. arcta*
9′ Infl open; lf blade 1–2 mm wide; perigynium beak narrow-conic, often entire at 10×, sometimes recurved . . . . *C. laeviculmis*
8′ Perigynium beak long-tapered, (1)1.2–2.7 mm; perigynium appressed to spreading, 2.8–5 mm, lanceolate to narrowly elliptic; lower pistillate fl bracts awned
10. Spikelets linear to oblong, gen > 13 mm, gen not jagged-sided, base of lower ones staminate, narrowly tapered; perigynium 3.3–5 mm, beak teeth > 0.2 mm, often outcurved; pistillate fl bract gen golden-brown at least near center, gen covering fr, awn often > 1.2 mm . . . . *C. bolanderi*
10′ Spikelets round to oblong, 5–13 mm, jagged-sided, base of lower ones gen ± pistillate, gen rounded or short-tapered; perigynium 2.8–4.5 mm, beak teeth gen < 0.3 mm, erect to incurved; pistillate fl bract gen white, gen not covering fr top, awn < 1.2 mm . . . . *C. deweyana* ssp. *leptopoda*

Group 9

Monoecious, cespitose. **ST** hollow. **LF**: blades gen flat, basal minute; ligule gen < 2.5 mm. **INFL** dense or ± open; spikelets each staminate at base, pistillate above, gen erect, gen < 15 mm, ovate to fusiform, sessile; lowest spikelet bract gen ± like pistillate fl bract; pistillate fl bract gen < perigynium, in width < or ± = perigynium, gen brown, gen acute. **PISTILLATE FL**: stigmas 2. **FR**: perigynia > 3 per spikelet, gen ascending, gen ovate, planoconvex or flat, glabrous, back veins gen < 9, wings from base to near beak tip, gen serrate, beak tip notched; fr 2-sided, gen at base of perigynium body.

1. Larger perigynia 5.7–8.1 mm, lanceolate, appressed to ascending, planoconvex, width gen < 1/2 length, beak tip cylindric, ± entire; fr > 2.1 mm; infl open; spikelet base long-tapered
2. Perigynium front veins 0–few, < 1/2 fr length, body ± abruptly narrowed to beak, gen cream-white, gen translucent . . . . [2]*C. praticola*
2′ Perigynium front veins > 1/2 fr length, body tapered to beak, cream-white to brown, gen not translucent
3. Spikelets 1–3; pistillate fl bract reddish, not or narrowly white-margined, covering ± 3/4 perigynium; perigynium front veins 0–3 long, 1–3 short, beak tip red-brown; staminate fl bract not or narrowly white-margined; infl orange-brown, 2nd lowest internode 2–6.5 mm when present . . . . *C. davyi*
3′ Spikelets 3–8; pistillate fl bract gen white or widely white-margined, ± covering perigynium; perigynium front veins 5–8, long, beak tip red-brown, very tip white; staminate fl bract widely white-margined; infl often whitish, 2nd lowest internode gen 4–11 mm . . . . *C. petasata*
1′ Larger perigynia < 6 mm (if > 6 mm, then infl dense, perigynium flat, beak tip flat, or fr < 2.1 mm), lanceolate to elliptic, appressed to spreading, planoconvex or flat exc over fr, width < or > 1/2 length, beak tip cylindric-entire or flat-minutely serrate; fr < 2 mm; infl open or dense; spikelet base long-tapered or not
4. Perigynium wing entire throughout, < 0.1 mm wide, beak tip cylindric . . . . *C. integra*
4′ Perigynium wing minutely serrate at least on upper body and beak, < or > 0.1 mm wide, beak tip cylindric or flat
5. Lower 2–3 spikelet bracts (sometimes not present or damaged late in season) elongate, ± lf-like, at least lowest >> infl; infl dense, < 3 cm
6. Perigynium body planoconvex, ovate to elliptic, ascending to spreading, wall ± tough; fr 0.9–1.4 mm wide; coastal or on islands . . . . [2]*C. harfordii*
6′ Perigynium body flat around fr, lanceolate to narrowly ovate, appressed to ascending, wall delicate; fr 0.7–1.2 mm wide; coastal or inland
7. Most perigynium beak tips unwinged and entire for > 0.4 mm, gen green, very tip white; larger perigynia gen < 1.5 mm wide, lanceolate or narrowly ovate, ± stalked, fr near body base; infl < 1.5 cm exc on vigorous pls, axis ± erect; lowest spikelet bract ascending, gen < 1.8 mm wide . . . . *C. athrostachya*
7′ Most perigynium beak tips unwinged and entire for < 0.4 mm, end gen red-brown; larger perigynia > 1.2 mm wide, ovate, gen sessile, fr near body middle; infl gen > 1.5 cm, axis gen ascending; lowest spikelet bract gen ± erect, often > 2 mm wide . . . . *C. unilateralis*
5′ Lower 2–3 spikelet bracts gen like pistillate fl bracts or bristle-like, < infl, lowest sometimes lf-like and > infl; infl dense or open, < or > 3 cm
8. Ligule > 2.5 mm; spikelets sometimes > 1 per lower node or branch; lf sheath often with a membranous tongue-like extension above mouth — infl gen (2.5)3–8 cm
9. Spikelet sides smooth; perigynia ± appressed, 4–6 mm, (1.2)1.5–2.5 mm wide, flat exc over fr, beak tip to fr top 2.8–3.8 mm, beak tip cylindric and entire for < 0.4 mm; pistillate fl bract < perigynium, gold or brown; lf sheath front white, membranous; lower spikelet bracts gen like pistillate fl bracts; spikelets 1 per lower node or branch . . . . *C. scoparia*
9′ Spikelet sides rough; perigynia appressed to ascending, 2.5–5 mm, 1–2.1 mm wide, planoconvex or ± flat, beak tip to fr top 1.3–3 mm, beak tip cylindric and entire for < or > 0.4 mm; pistillate fl bract ± covering perigynium, gen white or pale gold, sometimes brownish; lf sheath front green, ribbed at least in lower 1/2; lower spikelet bracts lf- or bristle-like; spikelets often > 1 per lower node or branch
10. Lf sheath front green-ribbed exc for a ± triangular, delicate, membranous, white area extending < 6 mm below mouth and gen a membranous ± tongue-like brownish extension above mouth; flat margin of perigynium body incl wing 0.3–0.6 mm wide, beak tip reddish . . . . *C. feta*
10′ Lf sheath front sometimes green-ribbed above but with at least a narrow, delicate, membranous white strip extending > 10 mm below mouth and a membranous ± tongue-like white extension above mouth; flat margin of perigynium body incl wing 0.1–0.4 mm wide, beak tip pale

11. Larger perigynia > 4 mm, tapered above, beak tip to fr top (1.5)2–2.5 mm, beak tip cylindric and entire for < 0.4 mm; lowest spikelet bract base gen around st .... [2]***C. amplectens***
11′ Larger perigynia < 4(4.5) mm, ± abruptly narrowed above, beak tip to fr top gen < 2 mm, beak tip cylindric and entire for > 0.4 mm; lowest spikelet bract base often not around st .... ***C. fracta***
8′ Ligule gen < 2 mm; spikelets 1 per lower node or branch; lf sheath gen without tongue-like extension above mouth
12. Perigynium gen > 1.8 mm wide, body round to widely ovate, flat margin incl wing > 0.4 mm wide
13. Most pistillate fl bracts ± = perigynium, covering most of its beak; perigynia gen inconspicuous in infl; spikelet oblong or fusiform to rounded, 3–5 mm wide
14. Lf blade 1.5–4 mm wide, flat; perigynium beak tip red-brown, with white flap along middle of back or not; fr ± sessile
15. Perigynium front veins gen several to fr top, upper body margin gold, beak without white flap along middle of back; pistillate fl bract center gold, tip white .... [2]***C. ovalis***
15′ Perigynium front veins gen 0–2 to fr top, upper body margin green, beak with white flap along middle of back or not; pistillate fl bract center gen green, tip brown to gold .... [6]***C. preslii***
14′ Lf blade 0.5–2.5 mm wide, folded to rolled; perigynium beak tip brown to red-brown, white at very tip, with white flap along middle of back; fr often stalked
16. Perigynium body 1.2–2.5 mm wide, ± = pistillate fl bract width, length 2.2–4 × width .. [5]***C. phaeocephala***
16′ Perigynium body 1.8–3 mm wide, 2–3 × pistillate fl bract width, length 1.7–2.6 × width .... [2]***C. proposita***
13′ Most pistillate fl bracts < perigynium, covering part but not most of its beak; perigynia gen conspicuous in infl; spikelet ovate to spheric, gen > 4 mm wide
17. Perigynium planoconvex
18. Perigynium veins on back > 8, on front gen > 3 .... [2]***C. multicostata***
18′ Perigynium veins on back < 8, on front gen < 3
19. Spikelet ± spheric; infl green; flat perigynium margin incl wing 0.4–0.7 mm wide, beak tip 0.3–0.4 mm wide; pistillate fl bract white to pale green or pale gold; alien .... ***C. molesta***
19′ Spikelet ovate; infl gen 2-colored, brown and green; flat perigynium margin incl wing (0.1)0.2–0.5 mm wide, beak tip 0.1–0.3 mm wide; pistillate fl bract brown to gold, gen with green center; native .... [6]***C. preslii***
17′ Perigynium flat exc over fr
20. Perigynium margin crinkled at least above, beak tip cylindric and entire for < 0.4 mm, red-brown, very tip white; lf blade flat .... ***C. straminiformis***
20′ Perigynium margin gen not crinkled, beak tip cylindric and ± entire for > 0.3 mm, green to black; lf blade flat or folded or margin downcurved
21. Perigynium abruptly narrowed to short beak, body 2–3 × pistillate fl bract width, flat margin incl wing 0.4–0.9 mm; lf blade 0.5–2.5 mm wide, gen folded or margin downcurved; spikelets ± distinct, widely ovate to spheric; pl 10–30 cm; anthers ± persistent; 3000–4100 m ...... [2]***C. proposita***
21′ Perigynium ± tapered to long beak, body < 2 × pistillate fl bract width, flat margin incl wing 0.2–0.7 mm; lf blade 1.5–6 mm wide, flat; spikelets distinct or not, ovate; pl 5–100 cm; anthers deciduous; 1500–4200 m
22. Perigynium 4.2–6.5 mm, beak tip to fr top gen > 2.5 mm, flat margin incl wing 0.3–0.7 mm wide; fr 1.2–2.3 mm, stalk 0.4–0.7 mm; infl dark brown, less often greenish, spikelets indistinct; pl often < 20 cm; st often decumbent; 2400–4200 m .... [2]***C. haydeniana***
22′ Perigynium 2.9–5 mm, beak tip to fr top gen < 2.5 mm, flat margin incl wing 0.2–0.5 mm wide; fr 1–1.6 mm, stalk 0.3–0.5 mm; infl green to less often dark brown, spikelets ± indistinct; pl gen > 20 cm; st erect; 1500–3400 m .... [2]***C. microptera***
12′ Perigynium < 2.2 mm wide, body ovate, elliptic, lanceolate, or fusiform, flat margin incl wing < 0.4 mm wide
23. Perigynium veins on back > 9, on front gen > 3; pistillate fl bract often white or with wide white margin; ± dry places
24. Perigynium beak tip cylindric and entire for > 0.4 mm, brown to reddish, white at very tip, with minute white flap along middle of back; lf blade 0.5–2.5 mm wide; infl whitish to medium- or orange-brown
25. White margin of pistillate fl bract < 0.3 mm wide; fr 1.5–2.1 mm, 0.8–1.2 mm wide, 0.2–0.5 mm thick; perigynia ascending, body ± planoconvex, front veins often uneven or 0, < fr top, beak tip to fr top 1.5–2.6 mm; spikelets ascending, > 5 mm wide; infl > 10 mm wide .... [5]***C. phaeocephala***
25′ White margin of pistillate fl bract 0.3–0.6 mm wide; fr 1.8–2.3 mm, 1.1–1.4 mm wide, 0.5–0.7 mm thick; perigynia appressed, body strongly planoconvex, front veins even, gen 5–8 to fr top, beak tip to fr top 0.6–2 mm; spikelets appressed, < 5 mm wide; infl < 11 mm wide .... ***C. tahoensis***
24′ Perigynium beak tip cylindric and entire for < 0.4 mm, pale gold to reddish; lf blade 2–6.5 mm wide; infl white-green to light brown
26. Infl 4–6.5 cm, white-green, lowest internode 7–20 mm, next 4.7–7 mm; pistillate fl bract white, tinged pale gold or not, ± covering perigynium; lowest spikelet bract lf-like, base gen around st; perigynium gen > 4 mm, beak tip to fr top (1.5)2–2.5 mm .... [2]***C. amplectens***
26′ Infl 1–4.6 cm, gold, brown, or brown and green, lowest internode < 8 mm, next 2–5 mm; pistillate fl bract gold to brown, with wide white margin or not, < perigynium; lowest spikelet bract like pistillate fl bract or ± lf-like, base partly around st; perigynium 3.5–6.6 mm, beak tip to fr top 1.8–3.4 mm

27. Perigynium 3.5–6.3 mm, gen > 1.8 mm wide, ovate to widely so, beak tip to fr top gen < 2.5 mm, flat margin incl wing 0.2–0.5 mm wide, fr stalk 0.4–0.8 mm; lf sheath front membranous, not or weakly ribbed, blade flat, ligule < 2 mm; infl ± triangular, sometimes elongate, spikelets gen < 6 . . . . [2]***C. multicostata***

27′ Perigynium 4.8–6.6 mm, gen < 1.8 mm wide, lanceolate to fusiform, beak tip to fr top gen > 2.5 mm, flat margin incl wing 0.2–0.4 mm wide, fr stalk 0.6–1 mm; lf sheath front often ± firm, ribbed, blade flat or folded, ligule < 2.5 mm; infl gen oblong, spikelets gen > 6 . . . . ***C. specifica***

23′ Perigynium veins on back < 9, on front 0 to > 3; pistillate fl bract gold or brown, with white margin or not; wet or dry places

28. Infl open, lowest bract internode > 4 mm; spikelets distinct, at least the lowest < 2 × lowest internode

29. Perigynium boat-shaped, < 4.2 mm, < 1.2 mm wide, wings incurved to front, < 0.12 mm wide, beak tips inconspicuous in infl; most pistillate fl bracts covering perigynia; spikelet gen fusiform; lf blade 0.5–1.5(2) mm wide . . . . [2]***C. leporinella***

29′ Perigynium not boat-shaped, < or > 4.2 mm, < or > 1.2 mm wide, wings not incurved to front, < or > 0.12 mm wide, or, if perigynium boat-shaped and wings incurved to front, beak tips conspicuous in infl; most pistillate fl bracts < perigynia if wings incurved to front; spikelet ovate, elliptic, or fusiform; lf blade 0.5–4.5 mm wide

30. Perigynium front veins several, even, to above fr

31. Perigynium body ± or not planoconvex, flat margin incl wing 0.2–0.6 mm wide, beak tip with or without white flap along middle of back; pistillate fl bract gen covering perigynium

32. Perigynium length 1.8–2.6 × width, beak tip red-brown; lf blade 1.5–4 mm wide . . . . [2]***C. ovalis***

32′ Perigynium length 2.2–4 × width, beak tip brown to red-brown, white at very tip, with a white flap along middle of back; lf blade 0.5–2.5 mm wide . . . . [5]***C. phaeocephala***

31′ Perigynium body planoconvex, flat margin incl wing gen < 0.25 mm wide, beak tip without white flap along middle of back; pistillate fl bract gen not covering perigynium

33. Lowest 2 infl internodes together < 1/3, gen ± 1/5 total infl length . . . . [3]***C. abrupta***

33′ Lowest 2 infl internodes together < 1/2, gen ± 1/3 total infl length

34. Upper spikelets distinct, base of at least lowest spikelet often long-tapered; perigynium sometimes boat-shaped with wings incurved to front, wall delicate, beak < 2/5, gen ± 1/3 perigynium length; fr 0.8–1.3 mm wide; lowest spikelet bract > lowest spikelet or not, < infl, base gen not around st; pl self-supported; 1200–3200 m . . . . [3]***C. mariposana***

34′ Upper spikelets gen indistinct, base wedge-shaped or rounded; perigynium not boat-shaped, wings not incurved to front, wall ± tough, beak < 1/2, gen ± 2/5 perigynium length; fr 0.9–1.7 mm wide; lowest spikelet bract gen > lowest spikelet, < or > infl, base often around st; pl often supported by vegetation; < 900 m . . . . [3]***C. subbracteata***

30′ Perigynium front veins 0 or few, uneven, sometimes 1–2 to above fr

35. Longer pistillate fl bracts gen < 3.5 mm

36. Fr 1–1.6 mm, 0.7–1.2 mm wide; perigynium appressed to spreading, gen < 3.5 mm, flat margin incl wing gen < 0.2 mm wide, body whitish green, pale gold, or light brown, upper margin green or gold; spikelet fine-textured (parts small) . . . . [2]***C. subfusca***

36′ Fr gen 1.4–2.3 mm, 0.9–1.7 mm wide; perigynium ascending-spreading, gen > 3.5 mm, flat margin incl wing gen > 0.2 mm wide, body green to brown; spikelet coarse-textured (parts large)

37. Spikelets gen > 5, not or only lower distinct; pistillate fl bract gen reddish to brown ± throughout; perigynium beak tip to fr top 1.5–2.9 mm, back bulged, upper body margin widely dark brown or coppery (narrowly green to gold), ± = pistillate fl bract in color; fr gen 1.4–1.7(1.2–2.2) mm, ovate; infl 9–24 mm . . . . [3]***C. pachystachya***

37′ Spikelets 3–7, distinct; pistillate fl bract brown to gold, center gen green; perigynium beak tip to fr top (1.8)2.4–3.7 mm, front, back bulged, upper body margin widely green to gold, not = pistillate fl bract in color; fr gen 1.7–2.1(1.5–2.3) mm, ± squared; infl 17–25 mm . . . . . . [6]***C. preslii***

35′ Longer pistillate fl bracts gen > 3.5 mm

38. Perigynium beak tip with white flap along middle of back, very tip gen white; pistillate fl bract gen covering perigynium

39. Fr 0.2–0.5 mm thick; perigynium beak 0.5–1.2 mm, beak tip to fr top (1.5)1.8–2.6 mm, body gen gold to brown (cream-white), sometimes translucent, flat or ± planoconvex; lf blade 0.5–2.5 mm wide; infl stiff to ± flexible, erect, 10–35 mm, 2nd lowest internode 2–4.5 mm; st 5–40 cm; inland, 2700–3900 m . . . . [5]***C. phaeocephala***

39′ Fr 0.5–0.7 mm thick; perigynium beak 1–1.6 mm, beak tip to fr top 2–3.7 mm, body gen cream-white, translucent, planoconvex; lf blade 1.5–4 mm wide; infl flexible, often nodding, 15–50 mm, 2nd lowest internode (2.5)4–10 mm; st 25–80 cm; coastal or inland, < 3200 m . . . . [2]***C. praticola***

38′ Perigynium beak tip without white flap on back, very tip gen not white; pistillate fl bract gen not covering perigynium

40. Perigynium wall delicate, beak tip to fr top (1.8)2.4–3.7 mm; spikelets gen distinct; lowest spikelet bract < or > lowest spikelet, like pistillate fl bract or bristle-like; st self-supporting; 30–3000 m, ± inland . . . . [6]***C. preslii***

40′ Perigynium wall ± tough, beak tip to fr top 1.5–2.5 mm; upper spikelets gen indistinct; lowest spikelet bract gen > lowest spikelet, bristle- or lf-like; st often supported by vegetation; < 900 m, coastal

41. Lowest infl internode often > lowest spikelet, lowest 2 internodes together < 14 mm, < 1/2 infl length; lf blade 1.2–3 mm wide; perigynium wing often entire exc on and just below beak, front with 0–4 veins ± = fr; coastal or inland, 0–700 m .... ***C. gracilior***

41′ Lowest infl internode gen < lowest spikelet, lowest 2 internodes together < 9 (11) mm, < 2/5 infl length; lf blade 1.3–4.5 mm wide; perigynium wing serrate on upper 1/2 body, front with 0–7 veins ± = fr; coastal, < 900 m .... [3]***C. subbracteata***

28′ Infl dense, lowest bract internode < 3.5 mm; spikelets distinct or not, > 2 × lowest internode

42. Perigynia < 4.2 mm, < 1.2 mm wide, boat-shaped, wings incurved to front, gen ± 0.1 mm wide, beak tips inconspicuous in infl; most pistillate fl bracts covering perigynia; spikelet fusiform; lf blade 0.5–1.5(2) mm wide .... [2]***C. leporinella***

42′ Perigynia < or > 4.2 mm, < or > 1.2 mm wide, not boat-shaped, wings not incurved to front or, if boat-shaped and wings incurved to front, beak tips conspicuous in infl; most pistillate fl bracts < perigynia; spikelet ovate, elliptic, or fusiform; lf blade 0.5–4.5 mm wide

43. Perigynium body much wider than fr, flat exc over fr or planoconvex due to air within, beak tip to fr top ± > 1/2 perigynium length

44. Perigynium 4.2–6.5 mm, beak tip to fr top gen > 2.5 mm; fr stalk 0.4–0.7 mm, body (1.2)1.4–2.3 mm; infl gen dark brown, less often greenish; pl often < 20 cm, st often decumbent; 2400–4200 m .... [2]***C. haydeniana***

44′ Perigynium 2.9–5.4 mm, beak tip to fr top gen < 2.5 mm; fr stalk 0.3–0.5 mm, body 1–1.8 mm; infl green to dark brown; pl gen > 20 cm, st erect; 1400–3400 m

45. Perigynium 3.6–5.4 mm, beak gen < 1 mm, 1/4–1/3 body length, tip cylindric, unwinged for gen > 0.6 mm, entire, dark brown; fr 1.2–1.8 mm .... [3]***C. abrupta***

45′ Perigynium 2.9–5 mm, beak 1–1.5 mm, 1/3–1/2 body length, tip cylindric, unwinged for gen < 0.5 mm, ± serrate, green or brown; fr body 1–1.6 mm .... [2]***C. microptera***

43′ Perigynium body as wide as fr, planoconvex, not flat, beak tip to fr top < or > 1/2 perigynium length

46. Lowest 2 infl internodes together < 1/3 infl length

47. Fr filling < 1/2 perigynium body; perigynium front veins (2)4–8, gen > fr, beak > 2/5 perigynium length, perigynium sometimes boat-shaped, wings incurved to front; 1400–3300 m, inland .... [3]***C. abrupta***

47′ Fr filling > 1/2 perigynium body; perigynium front veins 0–4(6), gen < 1/2 fr length, sometimes > fr, beak < or > 2/5 perigynium length, perigynium not boat-shaped, wings not incurved to front; 0–3500 m, coastal or inland

48. Perigynium green or brown, margin green, flat margin incl wing < 0.3 mm wide; fr filling 1/2–2/3 perigynium body; lowest spikelet bract bristle- or lf-like .... [2]***C. harfordii***

48′ Perigynium dark gold to brown, metallic, upper margin sometimes green, flat margin incl wing 0.2–0.5 mm wide; fr filling 2/3–4/5 perigynium body; lowest spikelet bract gen like pistillate fl bract .... [2]***C. pachystachya***

46′ Lowest 2 infl internodes together > 1/3 infl length

49. Pistillate fl bract gen ± > 3.5 mm

50. Perigynium body flat or ± planoconvex, beak tip red with white flap along middle of back; lf blade 0.5–2.5 mm wide; 2700–3900 m, inland .... [5]***C. phaeocephala***

50′ Perigynium body planoconvex, beak tip green to red-brown; lf blade 1.3–4.5 mm wide; 0–3200 m, coastal or inland

51. Perigynium sometimes boat-shaped, wings incurved to front, front veins (2)4–8, thin, > fr; base of at least lowest spikelet gen long-tapered .... [3]***C. mariposana***

51′ Perigynium not boat-shaped, wings not incurved to front, front veins 0 or few, thick, gen < fr; spikelet base gen ± round to wedge-shaped

52. Perigynium wall delicate, beak tip to fr top (1.8)2.4–3.7 mm; spikelets gen distinct; lowest spikelet bract < or > lowest spikelet, like pistillate fl bract or bristle-like; st self-supported; 30–3000 m, ± inland .... [6]***C. preslii***

52′ Perigynium wall ± tough, beak tip to fr top 1.5–2.5 mm; upper spikelets gen indistinct; lowest spikelet bract > lowest spikelet, bristle- or lf-like; st often supported by vegetation; < 900 m, coastal .... [3]***C. subbracteata***

49′ Pistillate fl bract gen ± < 3.5 mm

53. Perigynium front veins (2)4–8, thin, > fr; base of at least lowest spikelet gen long-tapered .... [3]***C. mariposana***

53′ Perigynium front veins 0–few, thick, gen < fr; spikelet base gen ± round to wedge-shaped

54. Fr 1–1.6 mm, 0.7–1.2 mm wide; perigynia appressed to spreading, gen < 3.5 mm, flat margin incl wing gen < 0.2 mm wide, beak tip light to medium brown; spikelet fine-textured (parts small) .... [2]***C. subfusca***

54′ Fr 1.4–2.2 mm, 1–1.5 mm wide; perigynia ± spreading, often > 3.5 mm, flat margin incl wing gen > 0.2 mm wide, beak tip gen dark brown; spikelet coarse-textured (parts large)

55. Pistillate fl bract gen reddish to brown ± throughout; perigynium ± = pistillate fl bract in color, back only gen bulged, upper margin dark brown, coppery, or narrowly green to gold, beak tip to fr top 1.5–2.9 mm; fr body gen 1.4–1.7 (1.2–2.2) mm, ovate; spikelets gen > 5, not or only lower distinct; infl 9–24 mm .... [2]***C. pachystachya***

55′ Pistillate fl bract brown to gold, center green; perigynium not = pistillate fl bract in color, front and back gen bulged, upper margin widely green to gold, beak tip to fr top (1.8)2.4–3.7 mm; fr gen 1.7–2.1(1.5–2.3) mm, ± squared; spikelets 3–7, ± distinct; infl 17–25 mm . . . . [6]***C. preslii***

C. abrupta Mackenzie (p. 1123) ABRUPT-BEAKED SEDGE (Group 9) **ST** erect, gen > 20 cm. **LF**: blade 1.5–4.5 mm wide. **INFL** gen dense, 10–22 mm, brown; lowest 2 internodes together < 1/3, gen ± 1/5 total infl length; pistillate fl bract gen < perigynium, obtuse, brown or coppery. **FR**: perigynium 3.6–5.4 mm, 1–2.5 mm wide, sometimes boat-shaped with wings incurved to front, flat margin incl wing < 0.3 mm wide, body ± planoconvex, veined both sides, brown, upper margin green to gold, beak gen < 1 mm, tip cylindric, entire and unmargined for gen > 0.6 mm, gen dark brown, 1.6–2.2 mm to fr top; fr 1.2–1.8 mm, 0.7–1.2 mm wide, stalk 0.3–0.5 mm. Gen moist meadows, open forest; 1400–3300 m. KR, NCoR, CaRH, SN, TR, PR, MP, SNE (Sweetwater Mtns), DMtns; to OR, WY, NV. Intergrades with *C. mariposana*. ❀TRY.

C. albida L. Bailey (p. 1123) WHITE SEDGE (Group 5) Short-rhizomed. **ST** 40–60 cm. **LF**: blade 3–6 mm wide. **INFL** > 15 cm; lowest internode 10–25 cm; terminal spikelet staminate at least at tip; lateral spikelets pistillate at least below, longest > 15 mm, stalks gen exserted < 1 cm; lowest spikelet bract blade << infl, sheath long, linear, mouth < 2 mm wide, not purple-banded; pistillate fl bract gen white with green midrib. **FR**: perigynium spreading, 3.1–4.5 mm, 1–1.6 mm wide, green, beak 0.6–1.2 mm, white, often ciliate, tip to fr top < 1.4 mm; fr 1.2–1.9 mm, 0.9–1.5 mm wide. **ENDANGERED** CA. Sphagnum bogs; < 90 m. s NCoRO (Pitkin Marsh, Sonoma Co.). [*C. sonomensis* Stacey] Like *C. lemmonii*.

C. albonigra Mackenzie (p. 1123) (Group 5) Cespitose. **ST** 10–30 cm. **LF**: blade 2.5–5 mm wide, gray-green. **INFL** gen head-like; terminal spikelet pistillate above, staminate at base; pistillate fl bract ± = perigynium, obtuse to with a small point, dark purple, margin, tip ± white. **FR**: perigynium not conspicuous in infl, 2.5–3.4 mm, 1.3–2 mm wide, ± flat, gen papillate above, dark purple, empty space above fr at least 1/4 body, beak 0.1–0.5 mm; fr 1.3–2 mm, 0.7–1.3 mm wide, < perigynium body. $2n$=52. ± Dry, rocky soil, often summits; 3000–3900 m. c&s SNH, W&I; to w Can, Colorado. ❀TRY.

C. alma L. Bailey (p. 1123) (Group 7) Cespitose, rarely dioecious. **LF**: blade 3–6 mm wide; sheath front red-dotted. **INFL** open or dense, 2.5–15 cm, 1–2 cm wide; spikelets >> 10, > 1 per lower node or branch; spikelet bracts conspicuous; pistillate fl bract > or ± = perigynium, ± white or widely white-margined, awned. **STAMINATE FL**: anther > 1.9 mm, awn < 0.1 mm. **FR**: perigynium 3.5–4.3 mm, 1.5–2.3 mm wide, 0.7–1.1 mm thick, body ovate to deltate, tapered or abruptly narrowed above, gold to dark brown, veins on back few, on front 0–few, wall rounded over fr, filled with pithy tissue below, beak 0.6–1.4 mm, serrate or ciliate; fr 1.5–2.5 mm, 0.9–1.8 mm wide, ovate. Springs, streambanks; 120–2400 m. c&s SNH, SnJV, CCo, SCo, TR, PR, DMtns; NV, AZ, Baja CA. ❀TRY.

C. amplectens Mackenzie (p. 1123) (Group 9) **LF**: blade 3–6.5 mm wide; ligule gen < 2.5 mm. **INFL** open, 40–65 mm, whitish green; lowest internode 7–20 mm, next 4.7–7 mm; spikelets distinct; lowest spikelet bract lf-like, base gen around st; pistillate fl bract ± covering perigynium, white, tinged pale gold or not, acute to ± awned. **FR**: perigynium appressed, 3.5–5 mm, 1.5–1.8 mm wide, body lanceolate-ovate, planoconvex, tapered above, green to pale gold, flat margin incl wing 0.2–0.4 mm wide, veins on back > 9, long, on front gen > 3, > 2/3 fr length, beak tip cylindric and entire for < 0.4 mm, very tip green or gold, 1.5–2.5 mm to fr top; fr 1.8–2 mm, 1.1–1.3 mm wide. Moist to ± dry meadows, open forest; 1200–2100 m. CaRH, SNH. Like *C. specifica*, *C. fracta*; needs study. ❀TRY.

C. amplifolia Boott (p. 1123) (Group 5) Rhizomed. **ST** 50–100 cm, sharply 3-winged. **LF**: blade 8–20 mm wide, with small, raised, irregularly spaced cross-walls; sheath back ± hairy. **INFL**: lateral spikelets 3.5–14 cm; lower pistillate fl bracts purplish, tip white, glabrous-awned. **FR**: perigynia > 60 per spikelet, spreading, 2.5–3.5 mm, 1.2–2 mm wide, green and brown, beak 0.7–1.2 mm, curved, tip ± unnotched; fr 1.3–2 mm, 1–1.5 mm wide. At least seasonally wet places; < 2400 m. KR, NCoR, CaRH, SNH, CCo, SnFrB, MP; to B.C., ID. ❀IRR or WET:**4–6,17**&SHD:**1–3,7**,8,9,**14–16**,18–24&STBL.

C. angustata Boott (p. 1123) (Group 6) Long-rhizomed, forming dense stands but not large, raised clumps. **ST** < ± 1 m. **LF**: blades 2–7 mm wide, basal minute; sheath backs scabrous, fronts with reddish prickles, lower shredding to network or fringe of veins. **INFL**: staminate spikelets 1–2, terminal one < 5 cm; lateral spikelets 2–7 cm, 3–5 mm wide, pistillate exc tip of upper sometimes staminate; lowest spikelet bract > or ± = infl; pistillate fl bract often > perigynium body, acute or tapered to gen pointed tip, often < 0.5 mm wide, sometimes hairy-awned. **FR**: perigynium 2.2–3.1 mm, 1.2–2 mm wide, ± weakly 1–5-veined both sides, weakly papillate, green, purple-dotted above or not, upper margin sometimes with stiff, curved hairs, beak 0.2–0.5 mm, tip notched < 0.1 mm, glabrous; fr 1–1.5 mm, 0.7–1.3 mm wide. $2n$=66,68. Wet meadows; 300–2300 m. KR, NCoRO, NCoRH, CaRH, SNH, MP; to e WA, also in c ID. [*C. eurycarpa* Holm; *C. oxycarpa* Holm]. ❀TRY:STBL.

C. aquatilis Wahlenb. (Group 6) **INFL** gen open; lowest spikelet bract lf-like, > 1/2 infl; pistillate fl bract tip often white. **PISTILLATE FL**: style often exserted. **FR**: perigynium faces unveined, weakly papillate, beak tip notched < 0.1 mm, glabrous; fr 1.1–1.8 mm, 0.7–1.6 mm wide, shiny. $2n$=72,74,76,>80. Wet places; < 3200 m. NW, CaRH (Butte Co.), SNH, CCo, SnBr, Wrn, W&I; to AK, e Can; Eur.

var. ***aquatilis*** **LF**: blade 3–8 mm wide; sheath front not purple-dotted, mouth white or pale brown. **INFL**: lowest internode 2–18 cm; lateral spikelets 1–10 cm, 3–7 mm wide, stalks 0 or gen erect, < 4.2 cm, tip gen pistillate; lowest spikelet bract gen ± = infl. **FR**: perigynium 2–3.6 mm, 1.2–2.3 mm wide, ± red-dotted or -blotched, beak 0.1–0.2 mm, thickened. $2n$=72,76,78,80. Habitats of sp. KR, NCoR, SNH, SnBr, Wrn, W&I; to AK, e Can; n Eur. ❀TRY:STBL.

var. ***dives*** (Holm) L. Standley (p. 1123) **LF**: blade 5–18 mm wide; sheath front purple-dotted, mouth purple-brown. **INFL**: lateral spikelets 4.5–11.5 cm, 4–7 mm wide, lower stalks gen nodding, < 11 cm, tip gen staminate; lowest spikelet bract gen > infl. **FR**: perigynium 1.9–3.5 mm, 1–1.2 mm wide, ± yellow- or purple-dotted, beak 0.2–0.4 mm, gen not thickened. $2n$=72,76,78,80. Habitats of sp.; < 1100 m. NCo, KR, CaRH (Butte Co.), CCo; to AK. [*C. sitchensis* Prescott] Hybridizes with *C. lyngbyei*. ❀TRY.

C. arcta Boott (p. 1123) (Group 8) Cespitose. **LF**: blade 2–4 mm wide; lower sheath front at least sparsely minutely red-dotted. **INFL** ± dense, 1.5–3 cm, green to medium brown; spikelets 7–15, distinct, 5–10 mm, upper sometimes narrowly tapered at base, lower ± separate; lowest spikelet bract < or > infl; pistillate fl bract white, obtuse to minutely pointed. **FR**: perigynium ± spreading, 2–3.4 mm, 1.1–1.5 mm wide, ovate, green, upper body ± papillate, wall filled with pithy tissue at base and gen upper body, beak 0.6–1.2 mm, > 1/4 body, conic, minutely serrate at 10×, not recurved, tip reddish; fr 1.2–1.6 mm, 0.8–1.1 mm wide. $2n$=60. Wet places, esp sphagnum bogs; < 1400 m. NCoRO; to w Can.

C. atherodes Sprengel (p. 1123) (Group 4) **ST** 30–150 cm. **LF**: blade hairy on lower surface; sheath hairy at least near mouth. **INFL**: pistillate spikelets sessile or stalked; lowest spikelet bract long-sheathing; pistillate fl bract red-brown, white-margined. **PISTILLATE FL**: style straight. **FR**: perigynium ascending, 7–10 mm, 1.7–2.5 mm wide, stalked, green, beak > 1 mm, straight, teeth 1.1–3 mm, outcurved; fr 2.3–3.2 mm, 1.2–1.5 mm wide. Wet places, shallow water; 1300–1400 m. MP (Alturas); to n Can, ne US; Eurasia. ❀TRY:STBL.

C. athrostachya Olney (p. 1125) (Group 9) **LF**: blade 1.5–4 mm wide. **INFL** dense, 10–20 mm, green to light brown; axis ± erect; spikelets ± indistinct; lowest 2–3 spikelet bracts ascending, >> infl, gen < 1.8 mm wide, lf-like, base expanded ± around st; pistillate fl bract < perigynium, often short-awned. **FR**: perigynium appressed to ascending, 2.8–4.8 mm, 0.8–1.8 mm wide, lanceolate or narrowly ovate, ± stalked, green to light brown, flat margin incl wing 0.1–0.2 mm wide, veins weak, < 8 on back, 0–4 on front, beak tip gen cylindric, unwinged and entire for > 0.4 mm, gen green above, very tip white, 2.1–2.5 mm to fr top; fr 1.1–1.7 mm, 0.7–1.1 mm wide. Common. Seasonally moist meadows, marshes; 100–3200 m. KR, NCoR, CaR, SN, SnBr, MP; w N.Am. Intergrades with *C. unilateralis*. ❀TRY.

C. aurea Nutt. (p. 1125) (Group 6) **LF**: blade 2–4 mm wide. **INFL**: terminal spikelet sometimes pistillate above, 0.9–2 mm wide in staminate portion, at least lowest non-basal spikelet separate, lateral ones erect to nodding, 4–20 mm, 3–5 mm wide; lowest spikelet bract sheath > 4.5 mm, mouth U-shaped; lower staminate fl bracts 2–4 mm; pistillate fl bract <, not appressed against, often falling before fully expanded perigynium, acute to narrow-awned, white to red-brown. **FR**: perigynia gen 4–10 per spikelet, ± ascending to spreading, 1.8–3 mm, 1–2 mm wide, spheric in spikelet, gen round or wide-tapered at base, gen sessile, gen strongly-ribbed, green at full size but turning orange just before falling (rare in genus), translucent, fleshy, when dry orange or gold to purplish, upper part often white, lower ± squashed, body tip wide, blunt, gen not conspicuously papillate at 10×, beak ± 0, tip unnotched, often red-brown; fr 1.3–2 mm, 1–1.6 mm wide, beak < 0.1 mm. $2n$=52. Wet places; 1100–3300 m. NCoRH, CaRH, SNH, TR, MP, W&I; to B.C., ne N.Am, UT, NM. Dried, immature perigynia ± indistinguishable from *C. hassei*. ❀TRY:STBL.

C. barbarae Dewey (p. 1125) (Group 6) **LF**: blade 3.5–9 mm wide; sheath front purple, lower shredding to network or fringe of veins. **INFL**: lateral spikelets 2.5–8 cm, 5–8 mm wide, brown, not pressing flat when fully mature, tips staminate or not; lowest spikelet bract < or > infl; pistillate fl bract awned, awn or tip hairy. **FR**: perigynia 50–200 per spikelet, spreading, 3–4.5 mm, 1.9–2.5 mm wide, brown, red-dotted, wall thick, tough, beak 0.2–0.5 mm, tip notched > 0.1 mm in front, teeth minutely hairy; fr 1.7–2 mm, 1.2–1.7 mm wide. Seasonally wet places; < 900 m. KR, NCoR, CaRF, SNF, GV, SCo; s OR. ❀IRR or WET:**4–9,14–24**&STBL: may be INVASIVE; very important traditional basket fiber plant.

C. bolanderi Olney (p. 1125) (Group 8) Cespitose, rhizomed. **ST** 15–90 cm. **LF**: blade 2–5 mm wide. **INFL** 3–8 cm; spikelet gen > 13 mm, linear to oblong, gen not jagged-sided, base of lower spikelets staminate, narrowly tapered; lowest spikelet bract lf-like or not, gen < infl; pistillate fl bract gen covering fr, gen golden-brown at least near center, awn often > 1.2 mm. **FR**: perigynium appressed to ascending, 3.3–5 mm, 1–1.4 mm wide, gen > 3 × longer than wide, green, body lanceolate or narrowly elliptic, lower wall filled with pithy tissue, beak 1–2.7 mm, long-tapered, teeth > 0.2 mm, often outcurved, tip to fr top gen 1.2–2.7 mm; fr 1.5–1.8 mm, 1–1.3 mm wide. Moist meadows, forests; < 2500 m. NCo, KR, NCoR, CaRH, SNH, CCo, SnFrB, SW; to B.C., MT, NM. ❀IRR: **4–6,17**&SHD:**1–3,7**,8,9,**14–16**,18–24&STBL.

C. brainerdii Mackenzie (p. 1125) (Group 2) In clumps connected by rhizomes. **ST** gen 10–30 cm. **LF**: blade gen > bisexual infl, 1.5–3 mm wide, gray-green; basal sheaths dark purple-red. **INFL**: staminate spikelet 10–17 mm, 1.5–3.5 mm wide; pistillate spikelets 4–6, solitary ones on short or long, stout, erect stalks; lowest spikelet bract of bisexual infl with > or = 2 mm sheath tearing in age; pistillate fl bract reddish, at least some awned. **FR**: perigynia 1–6 per spikelet, 4–5.2 mm, 1.5–2.2 mm wide, faces veined or several-ribbed at least below, beak tip to fr top 1–2 mm; fr 2.1–2.5 mm, 1.5–2.1 mm wide, gen tan or brown. Dry rocky areas, open forests; 900–2800 m. KR, NCoRO, CaRH, n&c SNH, SCoRO (San Luis Obispo Co.); s OR. ❀TRY:STBL.

C. brevicaulis Mackenzie (p. 1125) (Group 2) Forming dense turf, short-rhizomed. **ST** 3–15 cm, < lvs, scabrous at least on 1 angle exc near base. **LF**: blade 1–3.5 mm wide, gen sickle-shaped, stiff, gen folded, bright green. **INFL**: staminate spikelet of taller infls 5–17 mm, 1.5–2 mm wide; pistillate spikelets 2–4; pistillate fl bract red-brown-tinged, obtuse to short-awned. **FR**: perigynia 1–6 per spikelet, 3.8–4.5 mm, 1.3–2.3 mm wide, beak tip notched in front < 0.1 mm; fr 1.5–2.6 mm, 1.5–2 mm wide. $2n$=28. Rocky or sandy soil; < 2500 m. NCo, CCo; to B.C. ❀IRR:**4–6,15–17**& SHD:**1–3,7**,8,9,**14**,18–24; turf or GRCVR&STBL.

C. breweri Boott var. ***breweri*** (p. 1125) (Group 1) Rhizomed. **ST** 10–25 cm, > lvs. **LF**: blade 1 mm wide, rolled, quill-like. **INFL**: spikelet 6–10 mm wide, wide-conic; pistillate fl bract golden-brown to black, 3-veined, very wide, margin wide, whitish. **FR**: perigynia 10–40 per spikelet, ascending to spreading, 4–7 mm, 2.1–4.8 mm wide, very flat, thin, golden-brown, a bristle-like axis within; fr 1.7–2.3 mm, 0.8–1 mm wide (width < distance from fr to perigynium side), << perigynium body, 3-sided. Dry gravelly, to sandy, open areas; 2300–3700 m. CaRH, SNH; to WA. ❀TRY: STBL.

C. buxbaumii Wahlenb. (p. 1125) (Group 5) Rhizomed. **ST** 25–100 cm. **LF**: blades 1.5–4 mm wide, basal minute; lower sheath front shredding to network or fringe of veins. **INFL**: terminal spikelet pistillate above; lowest spikelets erect, often sessile; pistillate fl bract purple or brown, glabrous-awned. **FR**: perigynium ascending to ± spreading, 2.5–4.3 mm, 1.4–2.1 mm wide, papillate, grayish green, beak < 0.3 mm; fr 1.4–2.1 mm, 1.1–1.5 mm wide. $2n$=±74,±100,±105. Uncommon. Wet places; < 3300 m. NCo, NCoR, c&s SNH, CCo; to AK, e N.Am. ❀TRY:STBL.

C. californica L. Bailey (p. 1125) CALIFORNIA SEDGE (Group 5) Rhizomed. **ST** 20–70 cm. **LF**: blade 2–5 mm wide, gray-green, lower sides glandular-papillate, basal blades minute; sheath front red or red-dotted. **INFL**: lowest spikelet bract long-sheathing; pistillate fl bract purplish brown, glandular-papillate on midstripe, awned or not. **FR**: perigynium appressed to ascending, 2.7–5.1 mm, 1.7–2.3 mm wide, glandular-papillate, green, beak 0.2–1 mm; fr 1.7–3.1 mm, 1.4–2.1 mm wide. RARE in CA. Meadows, drier areas of swamps; < 90 m. NCo; to WA; also in ID.

C. canescens L. (p. 1125) (Group 8) Cespitose. **ST** 10–80 cm. **LF**: blade 1.5–4 mm wide, glaucous. **INFL** ± open, 2–15 cm; lowest internode > 2 mm; spikelets < 10, 3–12 mm, green to light brown, upper narrowly tapered at base or not, lowest gen < 2 × lowest internode; lowest spikelet bract < infl; pistillate fl bract white, sometimes gold or pale brown, obtuse to acute. **FR**: perigynium ± ascending, 1.7–2.8 mm, 0.9–1.8 mm wide, green to gold or silvery, wall filled with pithy tissue to upper body, upper body papillate, beak 0.2–0.5 mm, < 1/4 body, wide-conic, ± entire, tip at most minutely notched, reddish; fr 1.2–1.5 mm, 0.7–1 mm wide. $2n$=54,56,62. Wet, open places; 1100–3200 m. NCoRO, n&c SNH; to AK, e N.Am, Eurasia; also in S.Am, Australia. ❀TRY.

C. capitata L. (p. 1125) (Group 1) Loosely cespitose. **ST** 10–35 cm, > lvs. **LF**: blade < 1 mm wide, rolled, quill-like. **INFL**: spikelet bisexual; spikelet bract 0; pistillate fl bracts brown, with wide, white margin. **PISTILLATE FL**: stigmas 2. **FR**: perigynia 6–25 per spikelet, ascending to spreading, 2–3.5 mm, 1.3–2 mm wide, flat around fr, finely veined on back, pale green, a bristle-like axis within, edge sharp to base, entire, faces glabrous, beak tip dark; fr 1–1.8 mm, 0.5–1.2 mm wide, < perigynium body, 2-sided. $2n$=50. Gen wet places; 1900–3900 m. CaRH, SNH; to AK, ne N.Am, S.Am, Eurasia. ❀TRY:STBL.

C. cephalophora Muhlenb. (Group 7) Cespitose. **LF**: blade 2–4.5 mm wide; sheath mouth with ± thick rim. **INFL** dense, 0.7–2 cm, 5–10 mm wide; spikelets 3–8, distinct, lower sometimes separate; lower spikelet bracts < or > infl, > spikelets; pistillate fl bract < perigynium, awned, white or greenish, midrib green. **FR**: perigynium ± spreading, 2–3.5 mm, 1.5–1.9 mm wide, pale green to brown, abruptly narrowed above, veins on front 0, on back 0 or light, beak 0.7–1.7 mm, tip narrowly white-margined; fr 1.2–1.8 mm, 1–1.5 mm wide, ± sessile. Lawns; < 100 m. c SCo (Pasadena, Los Angeles Co.); native to e N.Am.

C. comosa Boott (p. 1125) (Group 4) Short-rhizomed. **ST** 50–100 cm. **LF**: ligule length < or ± = width. **INFL**: lower spikelets on long, nodding stalks; pistillate fl bract gen white or cream with pale reddish center, awn linear, > body. **FR**: perigynium spreading, 5–7.5 mm, 1.1–1.6 mm wide, ± angled, stalked, shiny, green to gold, beak teeth 1.1–2.3 mm, outcurved; fr 1.5–2 mm, 0.7–1.1 mm wide, ± = perigynium body. Wet places; < 400 m. NCoRI, CaRH, GV, n CCo (Bodega Bay), SnFRB, SnBr, MP (Shasta Co.); to WA, e N.Am. ❀TRY:STBL.

C. concinnoides Mackenzie (p. 1125) (Group 2) Rhizomed. **ST** 15–35 cm, < or > lvs; base dark purple-red. **LF**: blade 2–5 mm

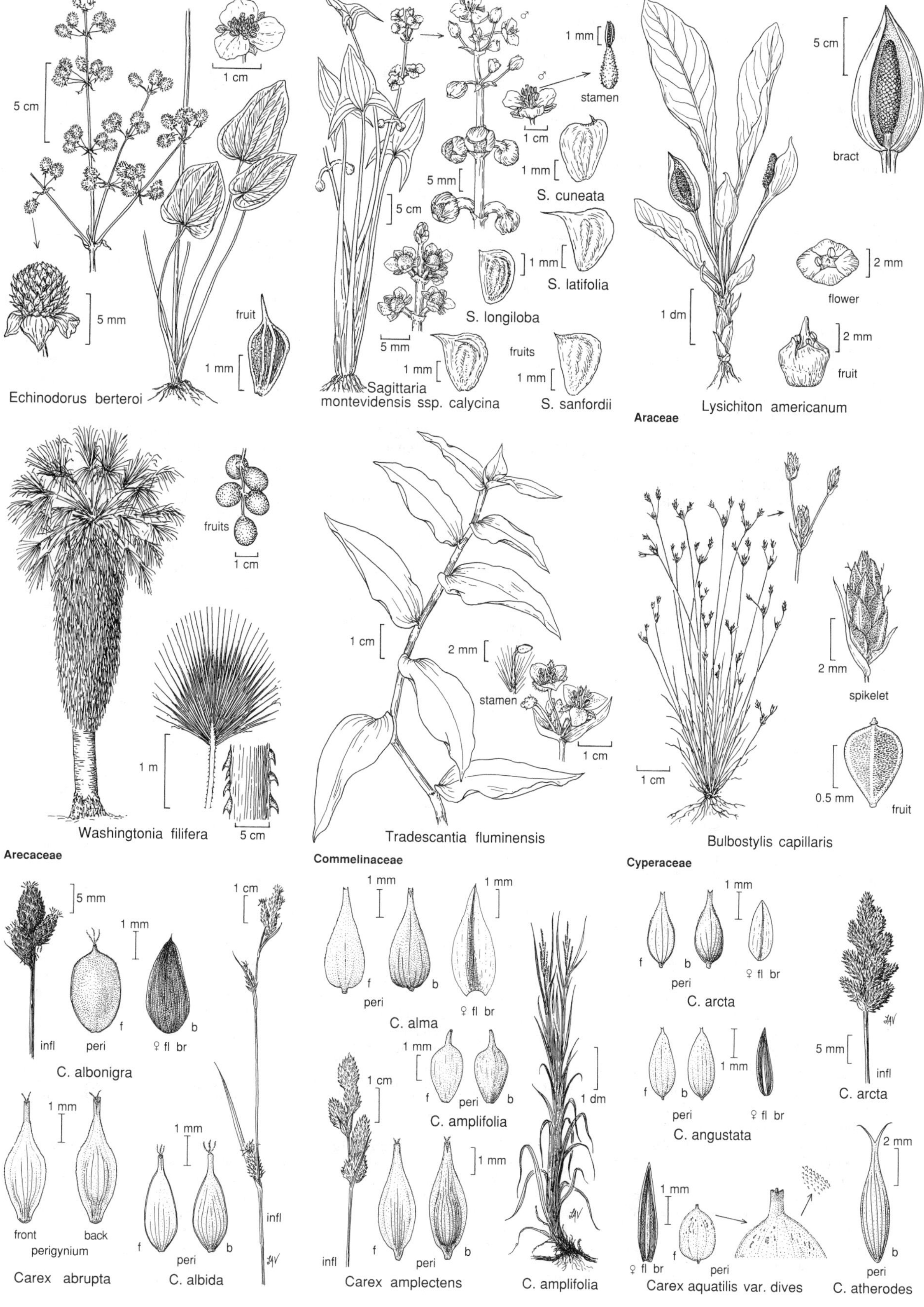

5 cm
1 cm
5 mm
fruit
1 mm
Echinodorus berteroi
5 cm
5 mm
1 mm
♂
stamen
1 cm
S. cuneata
1 mm
S. latifolia
S. longiloba
5 mm
fruits
1 mm
Sagittaria
montevidensis ssp. calycina
S. sanfordii
5 cm
bract
2 mm
flower
1 dm
2 mm
fruit
Lysichiton americanum
Araceae
fruits
1 cm
1 m
5 cm
Washingtonia filifera
Arecaceae
1 cm
2 mm
stamen
Tradescantia fluminensis
Commelinaceae
2 mm
spikelet
1 cm
0.5 mm
fruit
Bulbostylis capillaris
Cyperaceae
5 mm
1 mm
infl
peri
f
b
♀ fl br
C. albonigra
1 cm
1 mm
front
back
perigynium
Carex abrupta
1 mm
f
peri
b
C. albida
infl
f
peri
b
♀ fl br
C. alma
1 mm
f
peri
b
C. amplifolia
1 cm
1 mm
infl
f
peri
b
Carex amplectens
1 dm
C. amplifolia
1 mm
f
b
peri
♀ fl br
C. arcta
5 mm
infl
C. arcta
f
b
peri
♀ fl br
C. angustata
1 mm
♀ fl br
f
peri
Carex aquatilis var. dives
2 mm
b
peri
C. atherodes

wide, sickle-shaped, uppermost very short; sheath mouth V-shaped. **INFL**: staminate spikelet 2–4 mm wide; pistillate spikelets 1–2, linear to oblong, solitary ones 0; spikelet bracts purple, ± sheathing; pistillate fl bract purple, margin ciliate. **FL**: stigmas gen 4. **FR**: perigynia 5–10 per spikelet, ascending, 2.5–3.5 mm, 1.1–1.6 mm wide, 2-ribbed, white to light brown, beak 0.1–0.5 mm, tip often purple, very tip white; fr 1.9–2.8 mm, 0.8–1.7 mm wide, 4-sided. Clay, silt; 60–900 m. KR, NCoR; to w Can, MT. ❀IRR:**4–7,15–17**&SHD:**8,9,14,18–24**&STBL.

C. congdonii L. Bailey (p. 1125) CONGDON'S SEDGE (Group 3) Rhizomed. **ST** 40–90 cm. **LF**: blade 3–8 mm wide, glabrous or hairy just above sheaths, basal sometimes minute; sheath hairy, front red-dotted. **INFL**: spikelets dark; lowest spikelet bract ± = infl; pistillate fl bract < perigynium, brown, surface minutely hairy, margin not ciliate. **FR**: perigynia 40–200 per spikelet, 3.1–4 mm, 1.2–1.8 mm wide, 2-ribbed, thin–walled, purplish, tapered to 0.3–0.8 mm beak tip gen unnotched; fr 1.5–2.5 mm, 1.3–1.9 mm wide. UNCOMMON. Talus; 2600–3900 m. c&s SNH.

C. cusickii Mackenzie (p. 1125) (Group 7) Cespitose, sometimes dioecious. **LF**: blade 2.5–6 mm wide; sheath front often cross-wrinkled, red-dotted, mouth gen U-shaped, often thick-rimmed, copper or purplish; ligule length > 2 × width. **INFL** 4–8 cm, 1–2 cm wide, ± flexible; lower internodes >> others; spikelets > 1 per lower node or branch; spikelet bracts ± inconspicuous; pistillate fl bract brown, white-margined, acuminate, awned or not, often early deciduous. **STAMINATE FL**: anther < 1.8 mm, awn < 0.1 mm. **FR**: perigynium spreading, 2–4 mm, 1.1–2 mm wide, widest ± at base, blackish, shiny, back few-veined, with ± pale, central lengthwise groove 0.1–0.3 mm wide near base, lower wall pithy inside, front veins 0, beak 1–1.5 mm, with white flap along middle of back, tip unnotched; fr 1.1–1.8 mm, 0.8–1.2 mm wide. Wet places; < 2000 m. NCo, n&c SNH, CCo; to w Can, MT, WY. ❀TRY.

C. davyi Mackenzie (p. 1125) DAVY'S SEDGE (Group 9) **ST** 25–35 cm. **LF**: blade 1.5–2.5 mm wide. **INFL** open, 10–25 mm, orange-brown; 2nd lowest internode 2–6.5 mm when present; spikelets 1–3, distinct, base long-tapered; staminate fl bract not or narrowly white-margined; pistillate fl bract reddish, not or narrowly white-margined, obtuse, covering ± 3/4 perigynium. **FR**: perigynium appressed, 5.9–8 mm, 1.5–2.1 mm wide, lanceolate, planoconvex, red-gold, veins on back > 6, on front 0–3 long, 1–3 short, flat margin incl wing 0.3–0.6 mm wide, beak tip cylindric, ± entire, red-brown, 3–4.1 mm to fr top; fr 1.9–3 mm, 1.1–1.5 mm wide. UNCOMMON. Moist meadows; 1500–3200 m. n&c SNH; possibly related pls at Mount Adams, WA. Like *C. petasata*.

C. densa L. Bailey (p. 1125) (Group 7) Cespitose. **LF**: blade 3–7 mm wide; sheath front gen cross-wrinkled, red-dotted or not. **INFL** 1.5–8 cm, 0.8–1.5 cm wide; spikelets gen >> 10, > 1 per lower node or branch; spikelet bracts conspicuous; pistillate fl bract < perigynium, brown or red-brown, base sometimes white, lower awned or not, awns gen ciliate. **FR**: perigynium ascending to spreading, 2.5–4.5 mm (if < 3 mm, length < 2 × width), 1.3–2.1 mm wide, 0.7–1.1 mm thick, gold to brown, or green-margined, front veined or not, body ovate, with fr below top, with indistinct or wide flat margin around fr, lower wall gen filled with pithy tissue, beak 0.8–2.2 mm, conic, gen > 0.6 mm wide 0.2 mm above fr, serrate or ciliate, tip reddish; fr 1.3–1.9 mm, 1.1–1.6 mm wide. At least seasonally wet places; < 1500 m. NW, CaR, SN, GV, CCo, SnFrB, SCo, SnGb; to WA, NV. [*C. vicaria* L. Bailey in part, *C. breviligulata* L. Bailey] ❀TRY.

C. deweyana Schwein. ssp. ***leptopoda*** (Mackenzie) Calder & R. Taylor (p. 1125) (Group 8) Cespitose or rhizomed. **ST** 20–80 cm. **LF**: blade 2.5–5 mm wide. **INFL** 2–4 cm; spikelets 5–15 mm, round to oblong, jagged-sided, base of lower ones gen ± pistillate, gen rounded or short-tapered; lowest spikelet bract lf-like or not, gen < infl; pistillate fl bract gen not covering fr top, gen white, awn < 1.2 mm. **FR**: perigynium ascending to spreading, 2.8–4.5 mm, 1–1.5 mm wide, gen < 3 × longer than wide, green, body lanceolate or narrowly elliptic, lower wall filled with pithy tissue, beak long-tapered, 1–2.7 mm, teeth erect to incurved, gen < 0.3 mm, tip to fr top gen 1.2–1.8 mm; fr 1.5–1.9 mm, 1–1.3 mm wide. $2n$=54. Moist soil, wooded areas; < 2400 m. NW, CaRH, SNH, CCo, SnFrB, SW; w temp N.Am. [*C. leptopoda* Mackenzie] ❀TRY:STBL.

C. diandra Schrank (p. 1125) (Group 7) Cespitose. **LF**: blade 1–3 mm wide; sheath front red-dotted, mouth white or with at most a narrow brown margin; ligule length < 2 × width. **INFL** 2–5 cm, < 1.5 cm wide, stiff; lower internodes sometimes > but gen not >> others; spikelets sometimes > 1 per lower node or branch; spikelet bracts ± inconspicuous; pistillate fl bract acute, brown. **FR**: perigynium spreading, 2–3 mm, 0.9–1.4 mm wide, widest ± at base, brown, shiny, back few-veined, with ± pale, central lengthwise groove 0.2–0.3 mm wide, lower wall pithy inside, front veins 0, beak 1–1.5 mm, with white flap along middle of back, tip unnotched; fr 1–1.4 mm, 0.8–1 mm wide. $2n$=48,50,60. Marshy meadows; 150–2400 m. NCoR, n SNH (El Dorado Co.), CW, SW, MP; sporadic in w N.Am. ❀TRY.

C. disperma Dewey (p. 1125) Rhizomed. **LF**: blade 0.7–2 mm wide. **INFL** open, 1.5–2.5 cm, 3–4 mm wide; spikelets 2–4, lower << internodes between; staminate tip of terminal spikelet often very narrow, conspicuous; pistillate fl bract white, acuminate. **FR**: perigynia 1–3 per spikelet, 2.2–3 mm, 1.1–1.6 mm wide, light- to yellow-green, veined both sides, wall filled with pithy tissue to near beak, beak 0.2–0.3 mm, margin entire, tip white-margined; fr 1–2 mm, 0.9–1.3 mm wide. $2n$=70. Wet streamsides, lake margins; 1100–3400 m. SNH, W&I; to n Can, e N.Am, Eurasia. ❀TRY: STBL.

C. douglasii Boott (p. 1125) (Group 7) Gen dioecious; rhizome 1–2 mm thick, brown. **LF**: blade 1–2.5 mm wide, folded or inrolled at margin, thick; sheath mouth with thick rim. **STAMINATE INFL** dense, < 3 cm, 7–25 mm wide; lowest spikelet bract like pistillate fl bract. **PISTILLATE INFL** dense, 1.5–3.5 cm, 1.3–2.7 cm wide; spikelets gen < 10; pistillate fl bract > perigynium, gold, minutely pointed. **STAMINATE FL**: filament exserted, anther conspicuous, ± 3.5 mm, awn 0.2–1 mm, glabrous. **PISTILLATE FL**: style 1.8–3.5 mm, exserted; stigmas 4–6 mm, conspicuous, persistent. **FR**: perigynium appressed, 3.5–4.6 mm, 1.3–1.8 mm wide, many-veined, often obscurely, both sides, tapered to tip, gold to medium brown, beak 0.9–1.8 mm, ± = body, tip with narrow white margin; fr 1.4–1.9 mm, 1–1.5 mm wide. Dryish sandy, gravelly, or alkaline areas; 300–3800 m. KR, SNH, SnJV, SCoR, TR, PR, GB; to w&c Can, c US, NM. Distinctive. ❀TRY:STBL.

C. dudleyi L. Bailey. (p. 1125) (Group 7) Cespitose. **LF**: blade 3–6 mm wide; sheath back often white-spotted, front gen cross-wrinkled, red-dotted or not; ligule 0–9 mm. **INFL** 1.5–6.5 cm, 0.8–1.5 cm wide; spikelets >> 10, > 1 per lower node or branch; spikelet bracts conspicuous; pistillate fl bract gen red-brown, sometimes white, often awned. **FR**: perigynium ascending to spreading, 2–4 mm, 1.3–2.1 mm wide, pale gold, translucent, back veined, front veined or not, rounded over, with narrow flat margin around fr, body ± diamond-shaped, lower wall rarely filled with pithy tissue, beak 0.8–1.5 mm, conic to ± linear, gen < 0.6 mm wide 0.2 mm above fr, serrate or ciliate, tip reddish; fr 1.5–1.9 mm, 1.1–1.6 mm wide, 0.7–1.1 mm thick. At least seasonally wet places; 30–600 m. NCo, NCoR, GV, SN, CW, WTR, MP, DMtns; OR. ❀TRY.

C. echinata Murray (Group 8) **ST**: edges ± blunt. **LF**: blade 0.7–4 mm wide. **INFL**: spikelets 3–15.5 mm, ± star-shaped, base of terminal narrowly tapered; lowest spikelet bract << infl; staminate fl bracts of terminal spikelet appressed, internodes between them 2/3–3/4 their length; pistillate fl bract brown, white-margined, gen acute. **FR**: perigynium spreading to reflexed, 2.4–4.8 mm, 0.8–2 mm wide, widest ± at base, green to brown, lower wall filled with pithy tissue, beak 0.8–2 mm, > or ± = body, sparsely serrate, tip brownish, 0.8–2 mm to fr top. Wet places, esp sphagnum bogs; < 3200 m. NCo, KR, NCoRO, CaRH, SNH, CCo, SnBr; to B.C., e N.Am.

ssp. ***echinata*** (p. 1129) **LF**: blade 1–3.3 mm at widest. **INFL** gen open, 0.7–7.8 cm. **FR**: perigynium 2.4–4 mm, ± abruptly narrowed above, front gen unveined, with bulge of pithy tissue below fr, body margin gen serrate, beak tip to fr top gen 1–1.5 mm; fr 1.1–1.6 mm, 0.8–1.3 mm wide, ± round. $2n$=50,52,58. Habitats of sp.; < 3200 m. KR, NCoRO, CaRH, SNH, SnBr; to B.C., e N.Am. [*C. angustior* Mackenzie, *C. ormantha* (Fern.) Mackenzie; not *C. muricata* L.] ❀TRY.

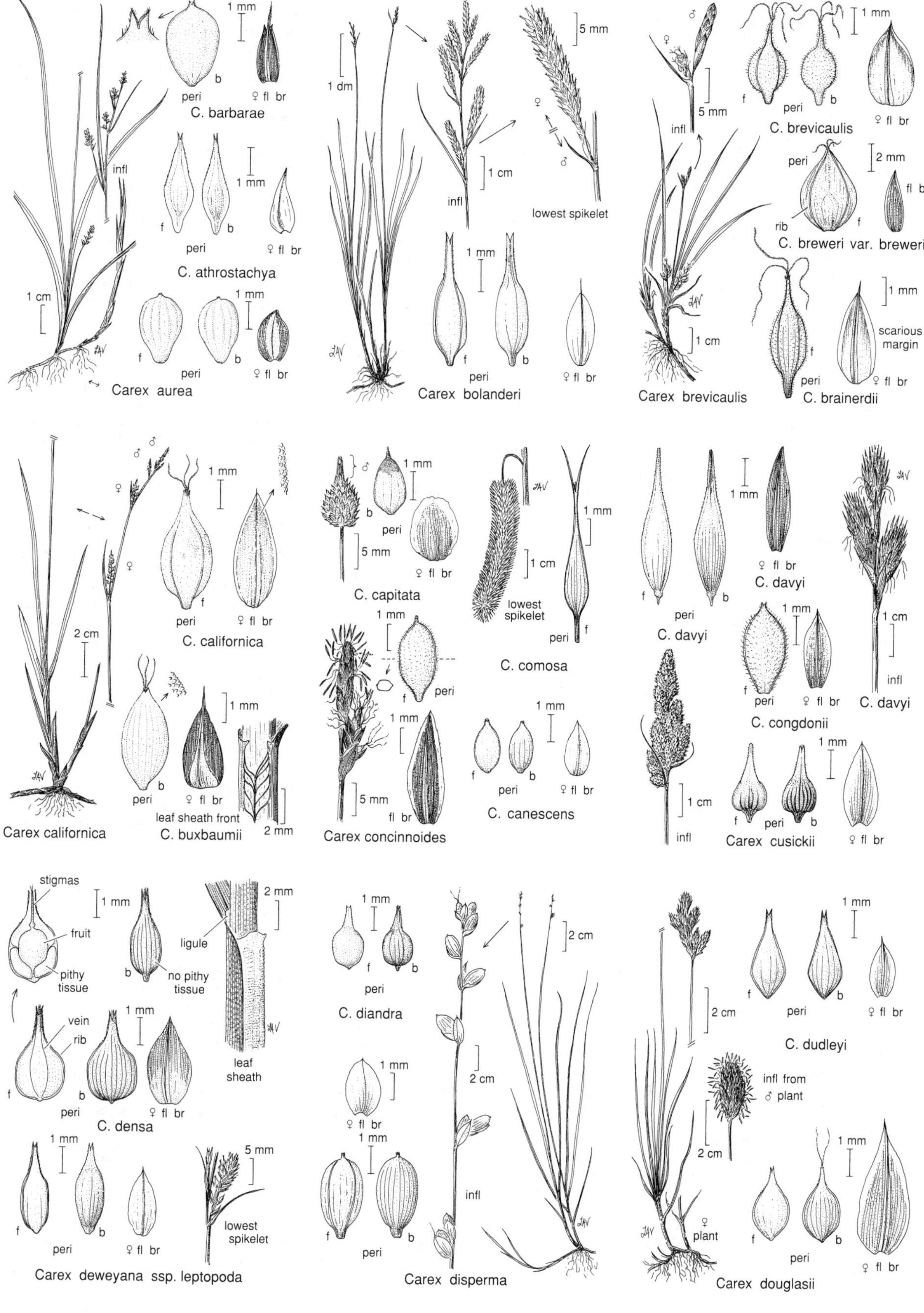

peri
b
♀ fl br
1 mm
C. barbarae
infl
f
b
peri
1 mm
♀ fl br
C. athrostachya
1 cm
1 mm
f
peri
b
♀ fl br
Carex aurea
1 dm
infl
1 cm
5 mm
lowest spikelet
1 mm
f
peri
b
♀ fl br
Carex bolanderi
infl
5 mm
1 mm
f
peri
b
♀ fl br
C. brevicaulis
peri
2 mm
fl br
rib
f
C. breweri var. breweri
1 cm
1 mm
scarious margin
f
peri
♀ fl br
Carex brevicaulis
C. brainerdii
1 mm
peri
f
♀ fl br
C. californica
2 cm
1 mm
b
peri
♀ fl br
leaf sheath front
2 mm
Carex californica
C. buxbaumii
1 mm
b
peri
♀ fl br
5 mm
C. capitata
1 mm
f
peri
1 mm
5 mm
fl br
Carex concinnoides
1 mm
1 cm
lowest spikelet
peri
f
C. comosa
1 mm
f
peri
b
♀ fl br
C. canescens
1 mm
♀ fl br
C. davyi
f
peri
b
C. davyi
1 mm
f
peri
♀ fl br
C. congdonii
1 cm
infl
C. davyi
1 cm
infl
1 mm
f
peri
b
♀ fl br
Carex cusickii
stigmas
1 mm
fruit
pithy tissue
b
no pithy tissue
ligule
2 mm
vein
rib
1 mm
f
b
peri
♀ fl br
C. densa
leaf sheath
1 mm
5 mm
f
peri
b
♀ fl br
lowest spikelet
Carex deweyana ssp. leptopoda
1 mm
f
b
peri
C. diandra
1 mm
♀ fl br
1 mm
f
peri
b
2 cm
2 cm
infl
Carex disperma
2 cm
1 mm
f
peri
b
♀ fl br
C. dudleyi
infl from ♂ plant
2 cm
1 mm
♀ plant
f
peri
b
♀ fl br
Carex douglasii

ssp. ***phyllomanica*** (W. Boott) A.A. Reznicek **LF**: blade 1.7–4 mm at widest. **INFL** dense, 1.2–4 cm. **FR**: perigynium 3.5–4.8 mm, tapered above, front 2–12-veined, without bulge of pithy tissue, body margin gen entire, beak tip to fr top 1.4–2.2 mm; fr 1.5–2.2 mm, 1–1.5 mm wide, ovate. $2n$=54,70. Sometimes brackish places; < 200 m. NCo, CCo; to AK. [*C. phyllomanica* W. Boott] ❀TRY:STBL.

C. eleocharis L. Bailey (p. 1129) SPIKERUSH SEDGE (Group 7) Sometimes dioecious; rhizome 1–2 mm thick, brown. **LF**: blade 0.5–1.5 mm wide, margin inrolled. **INFL** dense, 0.5–2 cm, 5–10 mm wide; spikelets < 10; lowest spikelet bract like pistillate fl bract to bristle-like; pistillate fl bract > perigynium, gold to brown, minutely pointed. **STAMINATE FL**: anther with narrow, < 0.1 mm tip. **PISTILLATE FL**: style < 1.5 mm, incl; stigmas < 4 mm, deciduous. **FR**: perigynium ascending, 2.5–3.5 mm, 1.5–1.8 mm wide, thick-walled, white to black, back veined, front unveined, beak 0.4–1.2 mm, < body, tip unnotched, white-margined; fr 1.5–2 mm, 1.25–1.7 mm wide. $2n$=60. RARE in CA. Dry areas in sagebrush shrubland, coniferous forest; 3500–4100 m. W&I; scattered w N.Am; also Eurasia. Sometimes considered *C. stenophylla* Wahlenb.

C. feta L. Bailey (p. 1129) (Group 9) **LF**: blade 2.5–5 mm wide; sheath front firm, green-ribbed ± to mouth, a delicate, ± triangular white area extending < 6 mm below mouth, gen a membranous, ± tongue-like, gen brownish extension above; ligule gen 3–8 mm. **INFL** open, 3–8 cm, whitish green to pale gold; spikelets distinct, sometimes > 1 per lower node or branch; lowest spikelet bract bristle- or lf-like, base around st or not; pistillate fl bract mostly white or pale gold ± covering perigynium. **FR**: perigynium ascending, 3–4.2 mm, 1.5–2.1 mm wide, green to pale gold, flat margin incl wing 0.3–0.6 mm wide, back veined, front ± unveined, beak tip cylindric and ± entire for < 0.4 mm, very tip reddish, 1.5–2.5 mm to fr top; fr 1.3–2 mm, 0.9–1.1 mm wide. Meadows, streambanks, wet soil; 30–2400 m. KR, NCoR, CaRH, SNH, SnFrB, SnBr, MP; to B.C. Like *C. fracta*. ❀TRY.

C. filifolia Nutt. (Group 1) Cespitose. **LF**: blade 0.2–0.7 mm wide, rolled, quill-like. **INFL**: lower pistillate fl bracts pale red-brown, white- or yellowish-brown-margined, obtuse to awned, clasping perigynium base. **PISTILLATE FL**: style exserted, base black, persistent. **FR**: perigynium appressed to ascending, 1.9–3.7 mm, 1.3–2.1 mm wide, 3-sided to round, a bristle-like axis within, faces hairy at least just below beak, unveined, whitish to gold; fr 1.6–3 mm, 1.2–1.8 mm wide, ± = perigynium body. Meadows, dryish areas with subsurface moisture; 1500–3700 m. SNH, SnBr, W&I; w N.Am.

var. ***erostrata*** Kük. (p. 1129) **ST** 5–25 cm, < or > lvs. **INFL**: longest pistillate fl bracts (exc awns) gen < 2.5 mm, gen < perigynia. **FR**: perigynium 1.9–3 mm, tapered to body tip, sometimes ruptured by fr, beak 0 or < 0.3 mm; fr 1.6–2.2 mm. Meadows; 1500–3700 m. SNH, SnBr, W&I; s OR, w NV. [*C. exserta* Mackenzie] ❀TRY.

var. ***filifolia*** **ST** 8–30 cm, > or = lvs. **INFL**: longest pistillate fl bracts (exc awns) gen > 3 mm, gen = or > perigynia. **FR**: perigynium 2.9–3.7 mm, body gen abruptly narrowed above, beak 0.1–0.8 mm; fr 2.2–3 mm. Dryish areas with subsurface moisture; ± 3200 m. c SNH (Inyo Co.); scattered w N.Am. Most SN pls called var. *filifolia* are vigorous pls of var. *erostrata*. ❀TRY.

C. fissuricola Mackenzie (p. 1129) (Groups 3,5) Cespitose or short-rhizomed. **ST** 50–80 cm. **LF**: blade 3–8 mm wide. **INFL**: terminal spikelet staminate at least at tip; lateral spikelets gen purplish to brown, tips tapered, stalks long-exserted, ± nodding; lowest spikelet bract blade << infl, sheath long, linear or ± expanded to mouth, mouth < 2 mm wide, gen Y- or V-shaped, purple bordered; pistillate fl bract brown, ± = perigynium in color, midstripe pale, ciliate, margin ciliate, tip acute or minutely pointed. **FR**: perigynium ascending, 3.2–5.5 mm, 0.9–2 mm wide, gen ± red-brown or purple, 2-ribbed, flat margin > 1/2 fr width, faces sparsely hairy when young, upper margin ciliate, beak 0.8–1.8 mm, ciliate, tip to fr top gen > 2 mm; fr 1.5–2 mm, 0.8–1 mm wide. Meadows, rocky streamsides; 1500–3500 m. c&s SNH; also in ID, ne NV, UT. Intermediate between *C. luzulaifolia*, *C. luzulina*. ❀IRR,DRN, SUN:1–6,7,14,**15–17**,18.

C. fracta Mackenzie (p. 1129) (Group 9) **LF**: blade 2.5–6 mm wide; sheath front ribbed at least in lower 1/2, with at least a narrow, delicate, membranous, white strip extending > 10 mm below mouth and gen a membranous, white, ± tongue-like extension above mouth; ligule gen 3–8 mm. **INFL** dense above or open, 2.5–8 cm, whitish green to pale gold; spikelets often > 1 per lower node or branch; lowest spikelet bract often lf-like, base around st or not; pistillate fl bract mostly white, sometimes gold-tinged, sometimes brownish, awned, ± covering perigynium. **FR**: perigynium ascending, 2.5–4.5 mm, 1–1.9 mm wide, planoconvex, green to pale gold, ± abruptly narrowed above, flat margin incl wing 0.1–0.4 mm wide, veins strong on back, sometimes 0 on front, beak tip cylindric and ± entire for > 0.4 mm, pale, tip to fr top 1.3–3 mm; fr 1.2–2 mm, 0.8–1.2 mm wide. Common. Montane meadows, open forests, moist soil; 800–3300 m. KR, NCoR, CaRH, SNH, TR, PR; to WA. Like *C. feta*. ❀TRY.

C. garberi Fern. (p. 1129) (Group 6) Loosely cespitose. **LF**: blade 2–3 mm wide. **INFL**: spikelets gen densely clustered exc basal ones, terminal one gen pistillate at least at tip, staminate part 1.2–2.5 mm wide; lateral spikelets 10–25 mm, 3.5–4.5 mm wide, stalks erect, long, lower gen from near pl base; lowest spikelet bract sheath > 4.5 mm, mouth V-shaped; pistillate fl bract appressed against, often > fully expanded perigynium, gen widely obtuse, red-brown to purplish. **FR**: perigynia gen > 10 per spikelet, 1.9–3 mm, 1–1.5 mm wide, gen veined, green to pale orange or white, fleshy, papillate, stalk 0.4–0.6 mm, base tapered, body tip wide, blunt, beak 0–0.1 mm, tip often red-brown, unnotched; fr 1.4–1.8 mm, 1–1.5 mm wide, beak < 0.1 mm. Wet places; < 100 m (much higher outside CA). NCo (Marin, Humboldt cos.), SnFrB (Santa Cruz Mtns); to B.C., e N.Am. Outside CA, perigynia sometimes finally become orange. ❀TRY:STBL.

C. geyeri Boott (p. 1129) GEYER'S or ELK SEDGE (Group 1) In clumps connected by rhizomes. **ST** 10–40 cm, < or > lvs. **LF**: blade 2–3.5 mm wide, flat or folded. **INFL**: lower pistillate fl bracts green, short-awned, > perigynium. **FR**: perigynia 1–3 per spikelet, appressed to ascending, 4.9–7 mm, 1.7–2.5 mm wide, planoconvex, green, a bristle-like axis within, body unveined, beak minute; fr 4.5–6.2 mm, 1.7–2.5 mm wide, ± = perigynium body. Uncommon. Open forests, slopes; 1500–2100 m. KR; to w Can, Colorado; apparently alien in Pennsylvania.

C. gigas (Holm) Mackenzie (p. 1129) SISKIYOU SEDGE (Group 1) Dioecious, short-rhizomed. **ST** 30–45 cm, >> lvs. **LF**: blade 2–3 mm wide. **PISTILLATE INFL**: spikelet bract like a bristle or pistillate fl bract, an additional sheathing, lf-like bract below often > infl, often with 1–2 spikelets in axil; pistillate fl bract acute to awned, in width < perigynium, margin narrowly white or not, ciliate at least at tip, surface glabrous, brown. **FL**: stigmas sometimes 4. **FR**: perigynia 20–40 per spikelet, ascending, 2.8–3.5 mm, 1.4–2 mm wide, ± flat, few-veined, brown to blackish, faces minutely hairy at least near beak, beak 0.3–0.5 mm; fr 1.5–1.8 mm, 0.7–1.2 mm wide, < perigynium body, sometimes 4-sided. UNCOMMON. Wet meadows; 850–1800 m. KR, n SNH (Plumas Co.). ❀IRR, DRN:4,5,**6**,**7**,8,9,**14–17**.

C. globosa Boott (p. 1129) (Group 2) In clumps connected by rhizomes. **ST** 15–40 cm. **LF**: blade 1.5–2.5 mm wide, ± bright green; basal sheaths brown to red. **INFL**: staminate spikelet of taller infls 10–20 mm, 1.5–4 mm wide; pistillate spikelets 2–3, solitary ones gen on long, slender, often arching stalks; lowest spikelet bract sheath of bisexual infl gen < 2 mm; pistillate fl bract tinged red-brown or purple, obtuse to awned. **FR**: perigynia 1–10 per spikelet, 3.3–5 mm, 1.8–2.3 mm wide, faces strongly veined at least below, beak tip to fr top 0.4–1.6 mm; fr 1.9–2.5 mm, 1.6–1.9 mm wide, gen white. Well drained soil of wooded areas; < 1300 m. NCo, NCoRO, CW, SW. ❀IRR,DRN:4–6,**17**&SHD:14,**15**,**16**&STBL.

C. gracilior Mackenzie (Group 9) **ST** self- supporting. **LF**: blade 1.2–3 mm wide. **INFL** ± open, 12–20 mm; lowest internode often > lowest spikelet, lowest 2 together < 14 mm, < 1/2 infl length; lower spikelets gen distinct, wedge-shaped or rounded; lowest spikelet bract often bristle-like, gen > lowest spikelets, < or > infl, base around st; pistillate bract gen > 3.4 mm, < or > perigynium, reddish. **FR**: perigynium ascending, 2.9–5.5 mm, ± 1.3–2.2 mm wide, flat, gen gold with brown margin, metallic, margin incl wing < 0.2 mm wide, often entire exc on or just below beak, wall ± tough, back veined, veins on front 0–4, to fr top, beak gen ± 2/5 perigynium length, tip cylindric and ± entire for > 0.4 mm, brown to red-brown, 1.5–2.5 mm to fr top; fr 1.5–2.1 mm, 0.9–1.7 mm wide. At least seasonally moist soil, grasslands to open forests; < 700 m. c&s NCo, c&s NCoR, SNF, CCo, SnFrB, SCoR.

C. gynodynama Olney (p. 1129) (Group 3) Loosely cespitose. **ST** 20–90 cm. **LF** hairy; blade 3–9 mm wide. **INFL**: terminal spikelet staminate at least at tip; lowest spikelet bract < infl, sheath > 5 mm; pistillate fl bract brown, obtuse to minutely pointed, surface hairy, margin not ciliate. **FR**: perigynia 20–40 per spikelet, 3.8–5.5 mm, 1.4–2 mm wide, many-veined, red-dotted, base long-tapered, beak 0.5–1.1 mm, conic, white at tip, teeth erect, < 0.5 mm; fr 2.3–2.5 mm, 1.3–1.8 mm wide. Moist meadows, open forests; < 600 m. NCo, NCoRO, n CCo (Marin, Santa Cruz cos.), SCoRO. Hybridizes with *C. hendersonii*, *C. mendocinensis*. ❀TRY:STBL.

C. halliana L. Bailey (p. 1129) HALL'S SEDGE (Group 3) Rhizomed. **ST** 10–50 cm. **LF**: blade ascending, 3–5 mm wide, strongly V-folded at least near base, glabrous; sheath hairy or not. **INFL**: spikelets erect; lowest spikelet bract > infl, gen incl lowest spikelet stalk, sheath < or > 0.5 mm; pistillate fl bract < or ± = perigynium, gen purplish, acute, margin white, not ciliate. **FR**: perigynia 20–40 per spikelet, 3.6–5 mm, 1.7–2.3 mm wide, thick-walled, strongly many-ribbed, green to gold, beak 1–1.7 mm, teeth erect, 0.2–0.5 mm; fr 1.9–2.5 mm, 1.3–1.8 mm wide. RARE in CA. Dry forest edges, often in pumice; ± 1800 m. KR (Medicine Lake, Siskiyou Co.); to WA. [*C. oregonensis* Olney]

C. harfordii Mackenzie (p. 1129) (Group 9) **LF**: blade gen 2–5 mm wide; basal sheath sometimes very long. **INFL** dense, 11–35 mm, greenish or brown; spikelets ± indistinct; lowest spikelet bract lf- or bristle-like, < to > infl; pistillate fl bract pale green to reddish, acute to short-awned, < or ± covering perigynium. **FR**: perigynium ascending to spreading, 2.5–4.5 mm, 1.1–2 mm wide, ovate to elliptic, green to brown, margin green, body planoconvex, veined both sides, wall ± tough, flat margin incl wing < 0.3 mm wide, beak tip cylindric and ± entire for > 0.4 mm, ± dark, 1.3–2.3 mm to fr top; fr 1.3–2 mm, 0.9–1.4 mm wide, filling 1/2–2/3 perigynium body. Marshy soil; < 600 m. NCoR, CW, ChI. [*C. montereyensis* Mackenzie] ❀TRY:STBL.

C. hassei L. Bailey (p. 1129) (Group 6) **LF**: blade 2–4 mm wide. **INFL**: terminal spikelet sometimes pistillate at tip, staminate part 1.8–3.5 mm wide, lowest non-basal spikelet separate, lateral ones erect to nodding, 3–5, 7–25 mm, 3.5–4.5 mm wide, long-stalked, lower gen from near pl base; lowest spikelet bract >> infl, sheath > 4.5 mm, mouth U-shaped; lower staminate fl bracts 3–6 mm; pistillate fl bract <, appressed against, falling after fully expanded perigynium, obtuse to awned, red-brown, margin, tip white. **FR**: perigynia gen 10–20 per spikelet, ascending to spreading, 2.1–3.1 mm, 1.2–1.6 mm wide, ± obovate, stalked, veined but gen not ribbed, when fresh white to gold, ± translucent, fleshy, when dry greenish white or pale gold, sometimes gold at base, gen not squashed, body wide-obovate or -elliptic, base tapered, tip wide, blunt, conspicuously papillate at 10 ×, beak < 0.2 mm, tip often red-brown, unnotched; fr 1.5–2 mm, 1–1.5 mm wide, beak < 0.1 mm. Wet places; < 2700 m. NCo, KR, SNH, CCo, SnFrB, SnGb, SnBr, SNE, DMtns; to B.C.; also in NV. [*C. aurea* var. *androgyna* Olney] See note under *C. aurea*. ❀TRY:STBL.

C. haydeniana Olney (p. 1129) (Group 9) **ST** often decumbent, often < 20 cm. **LF**: blade 1.5–4 mm wide, flat. **INFL** dense, 9–18 mm, ± spheric, gen dark brown, less often greenish; spikelets gen indistinct; pistillate fl bract < perigynium, dark brown to blackish. **FR**: perigynium 4.2–6.5 mm, 1.5–2.6 mm wide, dark brown, margin sometimes green, flat margin incl wing 0.3–0.7 mm wide, back ± veined, front gen unveined, beak tip cylindric for > 0.3 mm, ± minutely serrate, dark brown to black, 2.3–3.8 mm to fr top; fr 1.2–2.3 mm, 0.8–1.1 mm wide, stalk 0.4–0.7 mm. 2*n*=82. Uncommon. Rocky slopes, flats, moist soil; 2400–4200 m. c&s SNH, Wrn; OR; also in B.C., Rocky Mtns. [*C. nubicola* Mackenzie]

C. helleri Mackenzie (p. 1129) (Group 5) Cespitose. **ST** 15–50 cm. **LF**: blade 2–3.5 mm wide. **INFL** gen head-like; lowest spikelet sometimes separate; terminal spikelet pistillate above; lower spikelets sessile; pistillate fl bract gen < 1 mm wide, > but width << perigynium, red-brown to purple, tip long-tapered. **FR**: perigynium 2.5–3.8 mm, 1.5–2.8 mm wide, gen red-brown to purple at least above, beak 0.2–0.5 mm; fr 1.4–1.8 mm, 0.7–1.1 mm wide. Dry, rocky or gravelly slopes; 2400–4100 m. CaRH, SNH, SNE (Sweetwater Mtns), W&I; NV (Elko Co.). ❀TRY.

C. hendersonii L. Bailey (p. 1129) (Group 5) Cespitose. **ST** 40–100 cm, erect or decumbent. **LF**: blade 6–14 mm wide. **INFL**: lowest spikelet bract sheath > 6 mm; pistillate fl bract white, ± awned. **FR**: perigynium 4.3–6.5 mm, 1.3–2 mm wide, ± stalked, green, body tip wide-conic, beak gen < 0.2 mm or indistinct, tip unnotched; fr 2.7–3.3 mm, 1.6–2 mm wide. Coastal forests; < 900 m. NCo, CCo; to B.C.; also in ID. Hybridizes with *C. gynodynama*. ❀TRY.

C. heteroneura W. Boott (Group 5) Cespitose. **ST** 25–100 cm. **INFL** ± dense, not head-like; terminal spikelet gen pistillate above; lowest spikelets gen erect, gen conspicuously 2-colored; pistillate fl bract gen > 1.2 mm wide, dark purple, awned or not. **FR**: perigynium ascending, green or purplish above, 3-sided but flat around fr, body elliptic to obovate, ± rounded at tip, beak 0.2–0.5 mm; fr stalked. Wet meadows, forest openings, rocky slopes; 1800–3800 m. KR, CaRH, SNH, SnGb, SnBr, SnJt, Wrn, SNE (Sweetwater Mtns), W&I; to WA, Rocky Mtns.

var. ***epapillosa*** (Mackenzie) F. Herm. **LF**: blade 3.5–7 mm wide. **INFL**: pistillate fl bract midstripe inconspicuous. **FR**: perigynium 3.5–4.5 mm, 1.7–3.2 mm wide, flat margin (0.3)0.5–0.7 mm wide; fr 1.1–1.9 mm, 0.7–1.0 mm wide, base to perigynium base 0.5–1.1 mm. Habitats of sp. KR (Medicine Lake, Siskiyou Co.), CaRH, n&c SNH, Wrn, SNE (Sweetwater Mtns); to WA, Rocky Mtns. [*C. epapillosa* Mackenzie] ❀TRY.

var. ***heteroneura*** (p. 1129) **LF**: blade 2–4 mm wide. **INFL**: pistillate fl bract midstripe pale. **FR**: perigynium 2.5–3.9 mm, 1.4–2.5 mm wide, flat margin < 0.4 mm; fr 1–2 mm, 0.7–1.4 mm wide, base to perigynium base < 0.4 mm. Meadows, forest openings, rocky slopes; 1800–3500 m. KR, CaRH, SNH, SnGb, SnBr, SnJt, Wrn, SNE (Sweetwater Mtns), W&I; to WY. ❀TRY.

C. hirtissima W. Boott (p. 1129) (Group 3) Loosely cespitose. **ST** 30–60 cm, sparsely hairy. **LF** hairy; blades 3–7 mm wide, basal minute. **INFL**: terminal spikelet gen pistillate at least at tip; lowest spikelet bract < infl, sheath > 5 mm; pistillate fl bract ± hairy, gold, obtuse to minutely pointed, margin white, not ciliate. **FR**: perigynia 20–30 per spikelet, 3.5–4 mm, 1.4–1.8 mm wide, 2-ribbed, weakly veined, base short-tapered, beak > 1 mm, conic, white at tip, teeth erect, 0.3–0.8 mm; fr 2–2.4 mm, 1.3–1.6 mm wide, ± = perigynium body. Wet places; 60–1200 m. NCoR (Lake, Mendocino cos.), n&c SN. ❀TRY:STBL.

C. hoodii Boott (p. 1129) (Group 7) Cespitose. **LF**: blade 1.5–3.5 mm wide. **INFL** dense, 1–2 cm, 8–15 mm wide; spikelets 4–8, indistinct exc sometimes lowest; lower spikelet bracts gen < spikelets; pistillate fl bract < perigynium, acute to long-acuminate, red-brown, midrib green, margin white. **FR**: perigynium spreading, 3.4–5 mm, 1.4–2 mm wide, sessile, coppery brown, margin green, veins gen 0 both sides, beak 0.8–1.8 mm, tip notched; fr 1.7–2.1 mm, 1.3–1.7 mm wide, stalked. 2*n*=60. Rocky or gravelly slopes, meadow edges; 1200–3400 m. KR, NCoR, CaRH, SNH, SnBr, Wrn, W&I; to w Can, SD, Colorado. ❀TRY.

C. hystricina Muhlenb. (p. 1129) BOTTLEBRUSH SEDGE (Group 4) Rhizome short. **ST** 15–100 cm. **LF**: ligule length >> width. **INFL**: lower spikelets on long, nodding stalks; pistillate fl bract mostly white or cream, center pale reddish, awn > body, lanceolate. **FR**: perigynium ascending-spreading to spreading, 5–7 mm, 1.5–2.1 mm wide, round in X-section, ± sessile, green to gold, beak > 1 mm, teeth erect, 0.2–1 mm; fr 1.2–2 mm, 0.9–1.3 mm wide, << perigynium body. 2*n*=58. RARE in CA. Wet places; < 500 m. KR (Trinity Co.); to w Can, e N.Am.

C. illota L. Bailey (p. 1129) (Group 8) **LF**: blade 1–3 mm wide. **INFL** dense, 6–15 mm, dark brown; lowest internode < 1 mm; spikelets indistinct exc sometimes lower; pistillate fl bract < perigynium, blackish, shiny, ± obtuse. **FR**: perigynium spreading, 2.5–3.5 mm, 1–1.5 mm wide, planoconvex, dark brown, blackish above, unwinged but body with thin edge < 0.1 mm wide, entire, veins 0–6 on back, 0(–6) on front, lower wall often filled with pithy tissue, beaks conspicuous in infl, tip cylindric, ± entire, 1.2–1.6 mm to fr top; fr 1.2–1.5 mm, 0.7–1 mm wide. 2*n*=64. Marshes, bogs, wet meadows; 2100–3400 m. CaRH, SNH, SnBr, Wrn; w N.Am. ❀TRY:STBL.

C. incurviformis MacKenzie var. ***danaensis*** (Stacey) F. Herm. (p. 1129) (Group 7) Rhizome long, gen wavy. **ST** < 6 cm. **LF** > infl; blade 0.75–1.5 mm wide. **INFL** dense, 6–9 mm, 5–8 mm wide; spikelets < 10, indistinct; lowest spikelet bract like pistillate fl bract; pistillate fl bract acuminate, brown, margin white. **FR**: perigynium spreading, 2.8–3.5 mm, 1–1.6 mm wide, > pistillate fl bract, brown, darker near beak, shiny, sessile, veins many both sides, margin entire above, beak 0.7–1.5 mm, entire, tip unnotched, narrowly white-margined; fr 1.4–1.7 mm, 0.8–1.3 mm wide, round to square. Open, dry gravelly or rocky slopes; 3700–4000 m. c&s SNH, W&I; also in Colorado. [*C. danaensis* Stacey] ❀TRY: STBL.

C. inops L. Bailey ssp. ***inops*** (p. 1129) (Group 2) Long-rhizomed. **ST** 10–50 cm, < or = lvs. **LF**: blade 1–3 mm wide. **INFL**: staminate spikelets of taller infls 12–30 mm, 2–4 mm wide; pistillate spikelets 1–4, oblong to ± round, solitary ones gen 0, sometimes long-stalked; pistillate fl bract purple or dark red-brown. **FR**: perigynia 4–20 per spikelet, 2.6–4.8 mm, 1.4–2.2 mm wide, gen veined; fr 1.5–2.5 mm, 1.1–2 mm wide, gen white. Dry, rocky soil; 1000–2000 m. CaRH (Mount Shasta area); to B.C. [*C. pensylvanica* var. *vespertina* L. Bailey] ❀TRY:STBL.

C. integra Mackenzie (p. 1129) (Group 9) **LF**: blade 1–2 mm wide. **INFL** ± dense, 15–30 mm, gold to medium brown; lowest internode > 2 mm; spikelets ± distinct; pistillate fl bract < perigynium, gold to dark brown. **FR**: perigynium appressed to ascending, 2.2–3.3 mm, 0.7–1.2 mm wide, planoconvex, brown, wing < 0.1 mm wide, entire or beak sometimes ± serrate, veins on back weak, on front 0, beaks inconspicuous in infl, tip cylindric, entire, 1.2–1.6 mm to fr top; fr 1–2 mm, 0.6–0.9 mm wide. Seasonally moist soil; 900–3400 m. KR, NCoR, CaRH, SN, SnBr, PR, MP; OR, NV. Like *C. subfusca*, *C. illota*. ❀TRY.

C. interior L. Bailey (p. 1133) (Group 8) Cespitose. **LF**: blade 0.6–2.7 mm wide. **INFL** open, 1–3.7 cm; spikelets 3–20 mm, ± star-shaped, base of terminal one narrowly tapered; lowest spikelet bract << infl; staminate fl bracts of terminal spikelet appressed, internodes between them 2/3–3/4 their length; pistillate fl bract gen obtuse, gold, white-margined. **FR**: perigynium spreading to reflexed, 1.9–3.3 mm, widest ± at base, tapered above, green to brown, veined on back, gen not on front, lower wall filled with pithy tissue, beak 0.5–1 mm, 1/4–1/3 body length, densely serrate, tip reddish, 0.5–1 mm to fr top; fr 1.2–1.8 mm, 0.9–1.5 mm wide. Wet meadows, shores, often calcareous, not in sphagnum bogs; 1100–2100 m. KR, CaRH; to B.C., e N.Am. ❀TRY.

C. jonesii L. Bailey (p. 1133) (Group 7) Cespitose. **LVS** tufted near st base; blades 1–3 mm wide; sheath fronts gen not cross-wrinkled, gen hidden by other lvs, mouths U-shaped. **INFL** dense, 0.8–2 cm, 6–12 mm wide; spikelets < 10, indistinct; lowest spikelet bract like pistillate fl bract; pistillate fl bract gold to black, obtuse or with a small point. **FR**: perigynium spreading, 2.5–4 mm, 1–1.5 mm wide, gen widest ± at base, tapered above, gold to brown, shiny, gen at least back many-ribbed, lower wall often filled with pithy tissue, margin gen entire, beak ± poorly defined, 0.5–2 mm, conspicuous in infl, tip unnotched; fr 1–1.8 mm, 0.5–1 mm wide. Moist places; 900–3200 m. KR, NCoR, CaRH, SNH, SnBr, Wrn; to WA, Rocky Mtns. ❀TRY.

C. laeviculmis Meinsh. (p. 1133) (Group 8) Cespitose. **LF**: blade 1–2 mm wide. **INFL** open, 2–6 cm; spikelets 3–10 mm, base of terminal one round to ± narrowly tapered; lowest spikelet bract << infl; pistillate fl bract gold, white-margined, acute, midrib raised. **FR**: perigynium ascending, 2–4 mm, 0.9–1.5 mm wide, body ovate, veined on back, often on front, green to brown, beak 0.5–1.3 mm, narrow-conic, often entire at 10×, sometimes recurved; fr 1.3–1.5 mm, 0.7–1 mm wide. 2*n*=56. Moist soil in woodlands; 700–1800 m. KR, NCoRO, n&c SNH, Wrn; to AK, Rocky Mtns. ❀TRY.

C. lanuginosa Michaux (p. 1133) WOOLLY SEDGE (Group 3) Rhizomed. **ST** 30–100 cm. **LF**: blade 1.5–5 mm wide, flat, glabrous, slender tip short, basal blades minute; ligule of upper lvs > 2 mm, thin, membranous; sheath mouth densely hairy or not but not ciliate, lower fronts often not shredding. **INFL**: lowest internode 0.5–0.9 mm wide; pistillate spikelets gen sessile; pistillate fl bract > perigynium, ± ciliate, purplish or brown, with very narrow white margin, awned. **FR**: perigynia 25–75 per spikelet, 2.8–5 mm, 1.5–2 mm wide, thick-walled, weakly many-ribbed, gen purplish, beak 0.5–1.2 mm, teeth ± outcurved, 0.2–0.9 mm; fr 1.5–2.1 mm, 1–1.5 mm wide, persistent style base gen straight. Gen marshy places; 60–3300 m. NCo, KR, CaRH, SNH, CCo, SnFrB, SnGb, SnBr, GB, DMoj; to B.C., e N.Am. [*C. lasiocarpa* var. *lanuginosa* Kük.] ❀TRY:STBL.

C. lasiocarpa Ehrh. (p. 1133) SLENDER SEDGE (Group 3) Rhizomed. **ST** 30–120 cm. **LF**: blade >> infl, 0.5–2 mm wide, rolled, glabrous, slender tip very long; basal blades minute; upper ligules < 2 mm, thick, tough; sheath mouth often ciliate, lower fronts gen shredding. **INFL**: lowest internode gen < 0.8 mm wide; pistillate spikelets sessile; pistillate fl bract > perigynium, ± ciliate, purplish or brown, narrowly white-margined, awned. **FR**: perigynia 15–50 per spikelet, 2.8–5 mm, 1.5–2 mm wide, thick-walled, weakly many-ribbed, gen purplish, beak 0.5–1.2 mm, teeth gen ± erect, 0.2–0.9 mm; fr 1.5–2.1 mm, 1–1.5 mm wide, persistent style base gen bent. 2*n*=54. RARE in CA. Lake, pond shores, gen in standing water; 1800–2100 m. CaRH (nw Plumas Co.), n SNH (s Plumas Co.).

C. leavenworthii Dewey (Group 7) Cespitose. **LF**: blade 1–3 mm wide. **INFL** dense, 8–20 mm, 5–9 mm wide; spikelets 3–8; pistillate fl bract white to greenish, acute to minutely pointed. **FR**: perigynium ascending to spreading, 2–3.5 mm, 1.2–2.3 mm wide, widest above base, ± short-stalked, green to pale gold, veins on back few, on front 0, lower wall filled with pithy tissue, beak 0.4–0.8 mm, conic, serrate; fr 1–1.8 mm, 1.1–1.7 mm wide. Lawns; < 100 m. SnFrB (Alameda Co.), SCo (Santa Barbara, Los Angeles cos.); native to e N.Am.

C. lemmonii W. Boott (p. 1133) (Group 5) Cespitose or short-rhizomed. **ST** 20–80 cm. **LF**: blade 1.5–4 mm wide. **INFL** gen < 15 cm; lowest internode gen < 10 cm; lateral spikelets pistillate at least below, longest < 15 mm, stalks gen exserted < 1 cm, erect or spreading; lowest spikelet bract blade << infl, sheath long, linear, mouth not purple-banded; pistillate fl bract gen at least partly red-brown, with wide white margin, sometimes all white. **FR**: perigynium spreading, 2.7–4 mm, 1–1.7 mm wide, gen green, with a few purple blotches or not, beak 0.7–1.2 mm, gen ciliate, tip gen brown or purple, < 1.4 mm to fr top; fr 1.5–1.9 mm, 0.9–1.2 mm wide. Marshes, bogs; 700–3000 m. KR, NCoR, CaRH, SNH, SnBr, SnJt. Pls from Lassen, Plumas, and Butte cos. have the white pistillate fl bracts and perigynium beak of *C. albida*; some pls in SnBr approach *C. albida* in lf width and spikelet size, but have perigynia 3.8–4.6 mm, 1–1.5 mm wide, and fr 1.7–1.9 mm, 0.8–1.2 mm wide; pls of SN with few, appressed perigynia may be distinct. ❀TRY.

C. lenticularis Michaux (Group 6) Cespitose. **LF**: blade 1–3.5 mm wide, folded below; sheath front ± yellow-dotted. **INFL**: terminal spikelet rarely pistillate above; lateral spikelets 1.5–4.5 cm, 3–4 mm wide; lowest spikelet bract > or ± = infl, lf-like, sheath short; pistillate fl bract < or << perigynium, widely obtuse or acute. **FR**: perigynium 1.8–3.5 mm, 1–1.8 mm wide, body ± slightly, abruptly narrowed near base, stalk gen > 0.1 mm, gen veined, papillate, green, or white over fr, purple-dotted or not, early deciduous, beak 0.1–0.5 mm, tip notched < 0.1 mm. Wet places; < 3000 m. NW, CaRH, SNH, CCo, SnFrB, Wrn; to AK, Rocky Mtns. ❀TRY.

var. ***impressa*** (L. Bailey) L. Standley **INFL**: lower spikelets < or less often > internodes between, lowest spikelet 3–4 mm wide. **FR**: perigynium 1.8–3 mm, 1.1–1.5 mm wide, often >> pistillate fl bract, body often purple-dotted above, veins gen 3 on back, 0 or < 3, obscure on front, stalk < 0.2 mm, beak dark throughout or with purple or brown ± triangle from tip; fr 1–1.4 mm, 0.9–1.3 mm wide. 2*n*=92. Wet places; 1200–3000 m. KR, NCoR, CaRH, SNH, Wrn; to WA. [*C. paucicostata* Mackenzie]

var. ***limnophila*** (Holm) Cronq. **INFL**: lower spikelets >> internodes between, lowest spikelet 4–6 mm wide. **FR**: perigynium 2.5–3.5 mm, 1.3–1.6 mm wide, > pistillate fl bract, body green to gold, veins 5–7 both sides, stalk 0.4–0.7 mm, beak purple-tipped; fr 1.1–1.5 mm, 0.8–1.6 mm wide. Wet places; ± 0 m. NCo; to AK. [*C. hindsii* C.B. Clarke]

var. ***lipocarpa*** (Holm) L. Standley (p. 1133) **INFL**: lower spikelets < or less often > internodes between, lowest spikelet 3–4 mm wide. **FR**: perigynium 2–3.2 mm, 1–1.8 mm wide, often >> pistillate fl bract, body green, veins gen 5–7 both sides, stalk

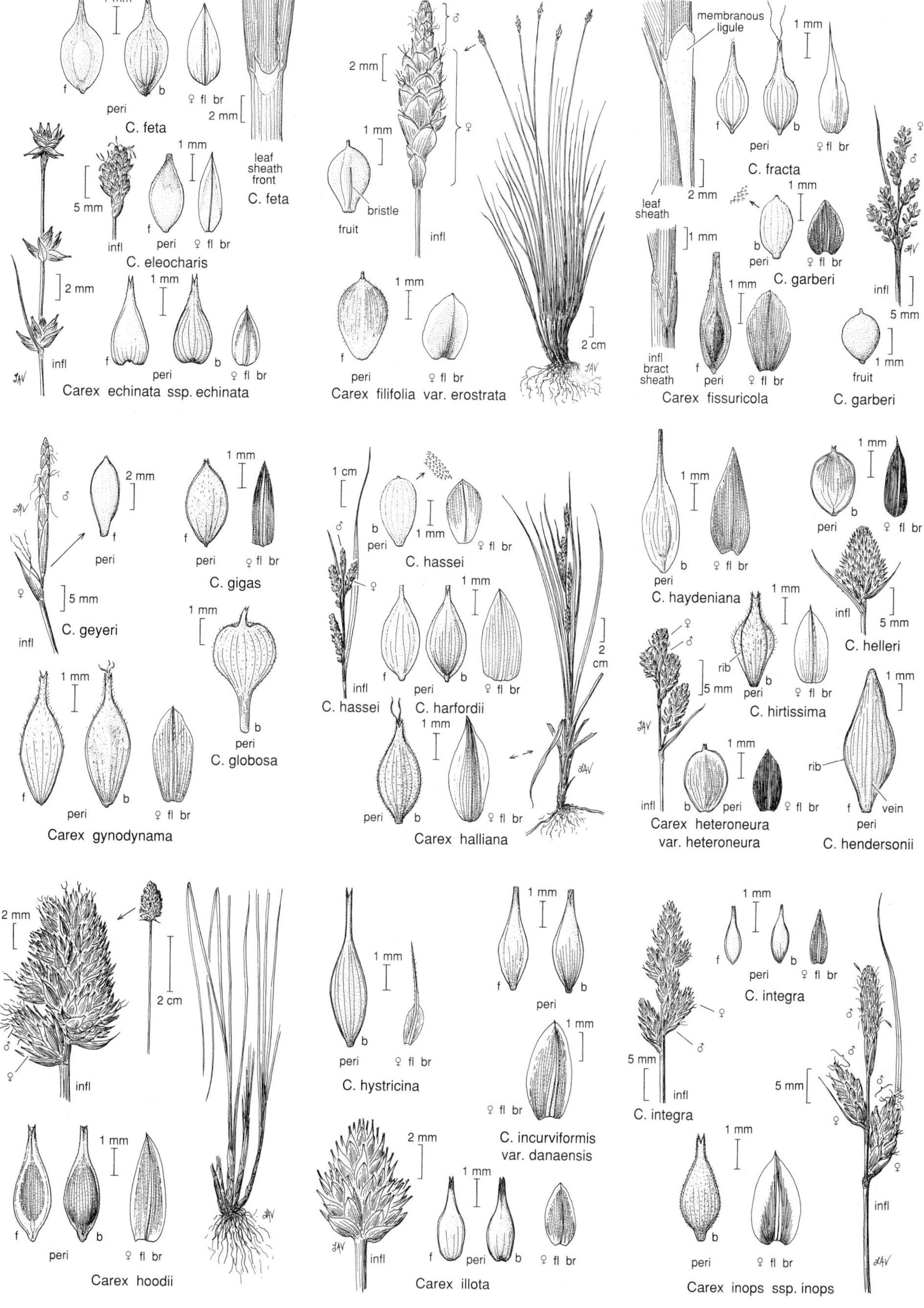

1 mm
f
peri
b
♀ fl br
2 mm
C. feta
leaf sheath front
C. feta
5 mm
infl
1 mm
f
peri
♀ fl br
C. eleocharis
2 mm
1 mm
infl
f
peri
b
♀ fl br
Carex echinata ssp. echinata
♂
2 mm
♀
1 mm
fruit
bristle
infl
1 mm
f
peri
♀ fl br
2 cm
Carex filifolia var. erostrata
membranous ligule
1 mm
f
peri
b
♀ fl br
C. fracta
leaf sheath
2 mm
1 mm
b
peri
♀ fl br
C. garberi
1 mm
1 mm
infl bract sheath
f
peri
♀ fl br
Carex fissuricola
♀
♂
infl
5 mm
fruit
1 mm
C. garberi
♂
2 mm
peri
f
♀
5 mm
C. geyeri
infl
1 mm
f
peri
♀ fl br
C. gigas
1 mm
b
peri
C. globosa
1 mm
f
peri
b
♀ fl br
Carex gynodynama
1 cm
♂
♀
b
peri
1 mm
♀ fl br
C. hassei
1 mm
infl
f
peri
b
♀ fl br
C. hassei
C. harfordii
1 mm
peri
b
♀ fl br
Carex halliana
2 cm
1 mm
peri
b
♀ fl br
C. haydeniana
1 mm
peri
b
♀ fl br
infl
5 mm
C. helleri
♀
♂
1 mm
rib
5 mm
peri
b
♀ fl br
C. hirtissima
1 mm
infl
b
peri
♀ fl br
Carex heteroneura var. heteroneura
1 mm
rib
f
vein
peri
C. hendersonii
2 mm
2 cm
♂
♀
infl
1 mm
f
peri
b
♀ fl br
Carex hoodii
1 mm
b
peri
♀ fl br
C. hystricina
1 mm
f
b
peri
1 mm
♀ fl br
C. incurviformis var. danaensis
2 mm
infl
1 mm
f
peri
b
♀ fl br
Carex illota
1 mm
f
peri
b
♀ fl br
C. integra
♀
♂
5 mm
infl
C. integra
♂
♂
5 mm
♂
♀
♀
infl
1 mm
b
peri
♀ fl br
Carex inops ssp. inops

0.2–0.5 mm, beak purple or brown only at tip; fr 1–1.7 mm, 0.8–1.5 mm wide. $2n$=92. Wet places; < 3000 m. KR, NCoR, CaRH, SNH, CCo, SnFrB; to AK, Rocky Mtns. [*C. kelloggii* W. Boott]

C. leporinella Mackenzie (p. 1133) (Group 9) **LF**: blade 0.5–2 mm wide, gen ± rolled. **INFL** gen open, 15–35 mm; spikelets distinct, gen fusiform; lowest spikelet bract sometimes bristle-like; pistillate fl bract gen covering perigynium, red-brown, white-margined. **FR**: perigynium appressed, 3.2–4.2 mm, 0.8–1.2 mm wide, planoconvex, gold, veined both sides, boat-shaped, wings incurved to front, flat margin incl wing ± 0.1 mm wide, beak tips inconspicuous in infl, cylindric and entire for > 0.4 mm, gold to brown, 1.5–2.2 mm to fr top; fr 1.4–1.9 mm, 0.7–1 mm wide. Moist meadows; 2100–4000 m. KR, CaRH, SNH, Wrn, SNE (Sweetwater Mtns), W&I; to WA, UT. ❀WETorIRR,SUN:1,**6**,**15**,**16**,17.

C. leptalea Wahlenb. (p. 1133) FLACCID SEDGE (Group 1) Rhizomed. **ST** 10–40 cm, = or > lvs. **LF**: blade 0.5–1 mm wide, flat or folded. **INFL**: spikelet bisexual, 2–3 mm wide; spikelet bract 0; pistillate fl bract < 1/2 × perigynium, white to brownish, red-dotted, persistent. **FR**: perigynia 3–8 per spikelet, ascending, 2.5–4.5 mm, 0.9–1.5 mm wide, green, faces veined, beak 0; fr 1.3–1.8 mm, 0.8–1 mm wide, < perigynium body. $2n$=50,52. RARE in CA. Wet meadows, swamps; < 700 m. NCo, NCoRO, CCo. [*C. polytrichoides* Muhlenb.; *C. microstachys* Michaux]

C. limosa L. (p. 1133) (Group 5) Long-rhizomed; root conspicuously hairy. **ST** 20–60 cm. **LF**: blades 1–3 mm wide, basal minute. **INFL**: lateral spikelets on long, nodding stalks; pistillate fl bract gen brown. **FR**: perigynium 2.3–4.5 mm, 1.6–2.6 mm wide, papillate above, brown-glaucous, body tapered to wide-conic tip, wall thick, tough, beak < 0.2 mm or gen 0; fr 1.5–2.7 mm, 1–1.8 mm wide; style exserted, base black, often persistent. $2n$=56,62,64. Sphagnum bogs; 1200–2700 m. SNH; to AK, MT, e Can, UT, Eurasia.

C. livida (Wahlenb.) Willd. (p. 1133) LIVID or PALE SEDGE (Group 5) Rhizomed. **ST** 15–60 cm. **LF**: blade 1–3.5 mm wide, ± glaucous. **INFL**: lowest spikelet bract sheath > 6 mm; pistillate fl bract red-brown. **FR**: perigynium 3.5–4.6 mm, 1.2–2.4 mm wide, papillate, green-glaucous, body tip wide-conic, beak < 0.15 mm, tip unnotched; fr 2–3.5 mm, 0.9–1.8 mm wide, sometimes 2 per perigynium. $2n$=32,50,52,76. EXTIRPATED in CA. Bogs, swamps; ± 0 m. NCo (Mendocino Co.); scattered to AK, e US. Last seen in CA in 1866.

C. luzulaifolia W. Boott (p. 1133) (Group 5) Cespitose or short-rhizomed. **ST**: 40–100 cm. **LF**: blade 5–20 mm wide. **INFL**: lateral spikelets pistillate, > 1.2 cm, purple to dark brown, lower stalks long-exserted, nodding; lowest spikelet bract nearly bladeless, sheath long, conspicuously gradually expanded to mouth, mouth > 2 mm wide, gen shallowly U-shaped with wide purple band at top; pistillate fl bract dark purple with narrow white margin, glabrous but sometimes ciliate-awned. **FR**: perigynium 4–5.3 mm, 1.7–2.5 mm wide, with flat margin > 1/2 fr width, short-stalked, dark purple, beak 1–2 mm, glabrous; fr 1.5–2.2 mm, 0.8–1.2 mm wide. Wet meadows; 2100–2800 m. CaRH (near Lassen Peak), n SNH, c SNH (n Tuolumne Co.); to MT. [misspelled *luzulaefolia*] ❀TRY.

C. luzulina Olney (Groups 3,5) Cespitose or short-rhizomed. **ST** 15–90 cm. **INFL**: lateral spikelets pistillate or tips staminate, lowest stalk exserted from bract sheath > 5 (gen > 10) mm; lowest spikelet bract sheath long, linear, mouth gen shallowly U-shaped or truncate, gen not purple, blade << infl; pistillate fl bract midstripe pale. **FR**: perigynium gen without flat margin, 2-ribbed, veins gen raised, beak 0.6–1.5 mm, dark purple, often ciliate, tip to fr top < 2.5 mm; fr 1.3–2 mm, 0.7–1.1 mm wide. Wet meadows, bogs; < 2400 m. NCo, KR, CaRH, n&c SNH, CCo, SnBr, Wrn; to B.C., MT, WY. Vars. intergrade. ❀TRY.

var. ***ablata*** (L. Bailey) F. Herm. (p. 1133) **LF**: blade 1.5–6 mm wide. **INFL**: upper spikelets sometimes clustered, lateral gen purple, sides ± deeply jagged, tip tapered; pistillate fl bract dark purple, midstripe gen < 0.2 mm wide, gen not to tip, tip acute. **FR**: perigynium ascending, 3–4.5 mm, 0.9–1.5 mm wide, gen mostly purple, ± stalked, body ± filled by fr. Habitat of sp.; 1200–2200 m. KR, CaRH, n&c SNH, Wrn; to B.C., MT, WY. [*C. ablata* L. Bailey.]

var. ***luzulina*** **LF**: blade 4–10 mm wide. **INFL**: upper spikelets gen densely clustered, lateral gen 2-colored, sides ± smooth or bristly, not deeply jagged, tip truncate, rounded, or abruptly narrowed to staminate part; pistillate fl bract gen red-brown, midstripe > 0.2 mm wide, ± to tip, tip obtuse to acute. **FR**: perigynium ascending-spreading to spreading, gen 3.5–5.5 mm, 0.9–1.6 mm wide, with flat margin < 1/2 fr width, mostly green to red-brown, less often purple, sessile, body ± not filled by fr. Habitat of sp.; < 2400 m. NW, CaRH, n SNH, CCo, SnBr; to WA, ID. Perigynia sometimes hairy in CaRH, n SNH; sometimes 1.6–1.8 mm wide, flat around fr (NCo).

C. lyngbyei Hornem. (p. 1133) (Group 6) Long-rhizomed. **LF**: blades 2–10 mm wide, basal minute. **INFL**: lateral spikelets 1.8–8 cm, 5–8 mm wide, stiff, straight on long, nodding stalks, tip gen staminate, base ± truncate; lowest spikelet bract > infl; pistillate fl bract tip often white, awned. **FR**: perigynium ascending to spreading, 2.2–4 mm, 1.5–2.8 mm wide, ± veined, papillate, brown, dull, wall very thick, very tough, beak 0–0.3 mm, tip glabrous, notch shallow or 0; fr indented 1.7–2.5 mm, 1.2–1.8 mm wide, often deeply indented on 1 or both sides, wide-stalked. $2n$= 68,70,72,76. Brackish areas; ± 0 m. NCo, CCo; to AK, e N.Am, Eurasia. Hybridizes with *C. aquatilis* var. *dives*. ❀TRY:STBL.

C. mariposana L. Bailey (p. 1133) (Group 9) **LF**: blade 1.5–4 mm wide. **INFL** ± open, 16–48 mm; spikelets gen distinct, base often long-tapered; lowest spikelet bract < infl, like pistillate fl bract or bristle-like; pistillate fl bract < or covering perigynium, brown or reddish. **FR**: perigynium appressed to ascending, 3.4–5.3 mm, 1.1–2.2 mm wide, green to gold, veined both sides, planoconvex, sometimes boat-shaped with wings incurved to front, flat margin incl wing gen < 0.2 mm wide, beak tip cylindric and ± entire for > 0.4 mm, brown to red-brown, white-margined or not, 1.8–2.5 mm to fr top; fr 1.5–2 mm, 0.8–1.3 mm wide. Meadows; 1200–3200 m. NW, CaRH, SN; NV (Washoe Co.). [*C. paucifructus* Mackenzie, Sierra Sedge] Variable. ❀TRY.

C. mendocinensis Olney (p. 1133) (Groups 3,5) Cespitose. **ST** 25–80 cm, nodding. **LF** ± glabrous; blade 2–5 mm wide; sheath ± hairy, at least near mouth, front often red or red-dotted. **INFL**: lateral spikelets nodding, 15–50 mm, linear; lowest spikelet bract blade < infl, sheath 2–3 cm, mouth minutely hairy; pistillate fl bract white or pale gold, red-dotted or not, ciliate, awned or not. **FR**: perigynium 3–5 mm, 1.2–1.7 mm wide, often minutely appressed-hairy near beak, green to gold, red-dotted or not, beak < 0.5 mm, gen ciliate; fr 1.7–2.5 mm, 1.2–1.5 mm wide. Moist areas, often serpentine; 150–1600 m. NCo, KR, n SNH (Butte, El Dorado cos.), CCo; s OR. Hybridizes with *C. gynodynama*. ❀IRR:**4–6**,**15–17**& SHD:**7**,**8**,**9**,**14**,18–24.

C. mertensii Prescott (p. 1133) (Group 5) Cespitose. **ST** 30–120 cm. **LF**: blades 4–7 mm wide, basal minute. **INFL**: spikelets 6–10, gen > 2 cm, terminal pistillate above, lowest gen on long, nodding stalk; lowest spikelet bract > infl; pistillate fl bract dark brown to dark purple, tip with small point or not. **FR**: perigynium 3.9–5.5 mm, 2.1–3.5 mm wide, > 3 × fr width, very flat, green, purple-dotted, beak 0.1–0.7 mm; fr 1.6–2.1 mm, 0.7–1.1 mm wide. $2n$=62. Wet open areas; ± 1500 m. KR; to nw Can, MT. ❀TRY.

C. microptera Mackenzie (p. 1133) (Group 9) **ST** gen > 20 cm, erect. **LF**: blade 2–6 mm wide. **INFL** dense, 12–25 mm, green to less often dark brown; spikelets ± indistinct; lowest spikelet bract like pistillate fl bract, << infl, or sometimes lf-like and > infl; pistillate fl bract, < perigynium, brown. **FR**: perigynium ascending-spreading, 2.9–5 mm, 1–2.2 mm wide, flat but sometimes planoconvex due to air within, veined both sides, front sometimes only at base, green to gold, center often brown, flat margin incl wing 0.2–0.5 mm wide, beak 1–1.5 mm, tip cylindric for gen 0.3–0.5 mm, ± serrate, green or brown, gen 1.6–2.5 mm to fr top; fr 1–1.6 mm, 0.7–1.4 mm wide, stalk 0.3–0.5 mm. $2n$=80,82,90. Common. Meadows; 1500–3400 m. KR, NCoRH, CaRH, SNH, TR, PR, Wrn, SNE (Sweetwater Mtns), W&I, DMtns; w N.Am. [*C. festivella* Mackenzie] Highly variable; intergrades with *C. pachystachya*, others.

C. molesta Mackenzie (Group 9) Cespitose with short rhizomes. **LF**: blade 2–3 mm wide; sheath back white-spotted, front with membranous tongue-like extension above mouth. **INFL** ± open, 20–30 mm, green; spikelets distinct, ± spheric; lowest spikelet bract like pistillate fl bract or bristle-like; pistillate fl bract white to pale green or pale gold, obtuse to acute, < perigynium. **FR**: perigynium 4.1–4.6 mm, 1.8–2.5 mm wide, body planoconvex, light green, veins on back weak, on front 0 or weak, flat margin incl wing 0.4–0.7 mm wide, beak tip 0.3–0.4 mm wide, flat, minutely serrate, gold, 1.5–2.5 mm to fr top; fr 1.6–1.9 mm, 1.1–1.3 mm wide. Disturbed ground, parking lots; < 100 m. SCo (Santa Barbara Co.); native to e&c US.

C. multicaulis L. Bailey (p. 1133) (Group 1) Cespitose. **STS** many, 20–60 cm, >> lvs, round or upper angles blunt. **LF**: blade 1–1.5 mm wide, short, folded or rolled. **INFL**: lower pistillate fl bracts lf-like, 1–15 cm, often > spikelet, green, margins white, base clasping perigynium. **FR**: perigynia 1–6 per spikelet, appressed to ascending, 5–7.2 mm, 2.2–2.5 mm wide, faintly many-veined, minutely ciliate near or on beak, green, a bristle-like axis within; fr 4.5–5.4 mm, 2.1–2.3 mm wide, ± = perigynium body. Forests; 15–2200 m. KR, NCoR, CaRH, SNH, SCoR, TR, PR, MP. Habit (± rush-like) distinctive. ❀IRR,DRN:**4–6**,**17**&SHD:**7**,8,9,**14–16.**

C. multicostata Mackenzie (p. 1135) (Group 9) **LF**: blade 2–6 mm wide. **INFL** dense, 10–40 mm, ± triangular, sometimes elongate; lowest internode < 8 mm; spikelets gen < 6, ± distinct; lowest spikelet bract bristle-like; pistillate fl bract light red-brown, often with wide white margin, < perigynium. **FR**: perigynium appressed-ascending, 3.5–6.3 mm, 1.5–2.5 mm wide, body ovate to wide-ovate, planoconvex, green to light brown, flat margin incl wing 0.2–0.5 mm wide, veins on back > 8, on front gen > 3, beak tip cylindric and entire for < 0.4 mm, gold, 1.8–3 mm to fr top; fr 1.7–2.7 mm, 1.1–1.7 mm wide, stalk 0.4–0.8 mm. Dry soil; 150–3500 m. KR, n&c NCoR, CaRH, SNH, SnFrB, TR, PR; to OR, ID, w NV. Sometimes like *C. specifica*. ❀TRY.

C. nebrascensis Dewey (p. 1135) (Group 6) **LVS** gen tufted at st base; blade 3–12 mm wide, thick, gen strongly blue-glaucous. **INFL**: lateral spikelets 3–6 mm, 5–9 mm wide, gen 2-colored, pressing ± flat; lowest spikelet bract gen > infl, base often dark-margined; pistillate fl bract obtuse to awned, tip or awn glabrous, margin often white. **FR**: perigynia 30–150 per spikelet, ascending, 2.6–4 mm, 1.6–2.5 mm wide, sessile, not papillate, veined, gen ± thick-walled, often red-dotted, beak 0.2–0.6 mm, tip brown or purple, notched > 0.1 mm, teeth minutely hairy; fr 1.3–2 mm, 0.9–1.8 mm wide. $2n$ =66,68. Meadows, swamps; < 2500 m. KR, CaRH, SNH, GV, WTR, SnJt, GB, DMtns; w US. ❀WET,SUN:1–3,**4–7**, 8,9,**14–18**&STBL.

C. nervina L. Bailey (p. 1135) (Group 7) Cespitose, often forming large raised clump. **ST** winged or not, spongy when fresh. **LVS** not tufted at st base; blades 3.5–5 mm wide; sheath back white-spotted, front not cross-wrinkled, mouth U-shaped, with thick rim. **INFL** dense, 1.5–3 cm, 6–18 mm wide; spikelets < 10, indistinct; pistillate fl bract acute, brown, margin white. **FR**: perigynium spreading, 3–4.5 mm, 1–1.8 mm wide, widest ± at base, body ± long-tapered above, many-ribbed, green to medium brown, shiny, lower wall filled with pithy tissue, beak 0.8–1.7 mm, poorly defined, tip notched; fr 1.5–1.8 mm, 1–1.3 mm wide. Moist to wet places; 1200–3000 m. KR, NCoR, CaRH, SNH, SnJV; s OR, w NV. ❀TRY.

C. neurophora Mackenzie (p. 1135) (Group 7) Cespitose, often forming large raised clump. **ST** winged or not, spongy when fresh. **LVS** not tufted at st base; blades 1.5–3 mm wide; sheath fronts often cross-wrinkled, mouth with tongue-like extension. **INFL** dense, 1–2.5 cm, 6–15 mm wide; spikelets < 10, indistinct; lowest spikelet bract like pistillate fl bract; pistillate fl bract brown, white-margined, obtuse to acute. **FR**: perigynium spreading, 2.9–3.9 mm, 0.8–1.5 mm wide, widest ± at base, body ± long-tapered above, many-ribbed, brown, shiny, lower wall filled with pithy tissue, beak poorly defined, serrate at least on 1 side; fr 1.1–1.6 mm, 0.8–1 mm wide. Moist to wet places; 1500–2100 m. NCoR; to WA, Rocky Mtns.

C. nigricans C. Meyer (p. 1135) (Group 1) Rhizomed. **ST** 5–30 cm, >> lvs. **LF**: blade 1.5–3 mm wide. **INFL**: spikelet bisexual, 6–10 mm wide; spikelet bract gen 0, very rarely well developed; pistillate fl bract dark brown, early deciduous. **FR**: perigynium spreading to, esp lowest, reflexed, 3.1–4.6 mm, 0.8–1.5 mm wide, shiny, unveined, dark brown, beak conspicuous in infl; fr 1.1–1.8 mm, 0.6–0.9 mm wide, < perigynium body. $2n$=±72. Wet, rocky slopes, meadows; 2300–3700 m. KR, CaRH, SNH; to AK, Colorado. ❀TRY:STBL.

C. norvegica Retz. (p. 1135) SCANDANAVIAN SEDGE (Group 5) Short-rhizomed. **ST** 10–70 cm. **LF**: blade 1.5–3.5 mm wide. **INFL** gen ± open; terminal spikelet 6–14 mm, pistillate above, lateral spikelets sessile or short-stalked; pistillate fl bract << perigynium, obtuse to with a small point, purple-black or brownish black, margin, tip white. **FR**: perigynium 2–3 mm, 1–1.3 mm wide, 3-sided, papillate, green or gold to dark purple, conspicuous in infl, beak 0.3–0.4 mm; fr 1.2–1.7 mm, 0.7–1 mm wide. RARE in CA. Wet meadows; 2900 m. W&I (White Mtns); to AK, ID, UT, NM, n-c US; circumboreal.

C. nudata W. Boott (p. 1135) (Group 6) Cespitose, in large, raised, dense clumps not connected by rhizomes, not in ± continuous stands. **ST** gen < ± 1 m. **LF**: blade 2–4 mm wide; sheath back scabrous, front purple-dotted or -blotched, gen with purple prickles, lower shredding. **INFL**: lateral spikelets 1.5–6 cm, 4–6 mm wide, stalk erect, tip sometimes staminate; lowest spikelet bract gen < infl; pistillate fl bract obtuse. **FR**: perigynium 2.2–4 mm, 1.2–1.8 mm wide, sessile, not papillate, green, gen dark purple above, veins 3–9 on back, sometimes on front, beak 0.1–0.3 mm, glabrous, tip notched < 0.1 mm; fr 1.3–2 mm, 1–1.5 mm wide. $2n$= 70,72. Rocky or sandy streambeds below high-water mark; < 1500 m. NCo, KR, NCoRO, SNH, ScV, CCo, SCoR; OR. ❀WET:**4–7, 17**&SHD:1–3,8,9,**14–16**,18–24.

C. obispoensis Stacey (p. 1135) SAN LUIS SEDGE (Group 3) Cespitose. **ST** 60–180 cm. **LF**: blade 2.5–8 mm wide, V-folded. **INFL**: pistillate spikelets nodding, lowest spikelet stalk long-exserted; lowest spikelet bract < infl, sheath > 9 mm; pistillate fl bract brown with wide white margin. **FR**: perigynia 40–60 per spikelet, 5–8 mm, 1.4–1.5 mm wide, unribbed, body tapered above, beak 1.4–2 mm, stout, teeth erect, 0.1–0.5 mm; fr 2.7–3.3 mm, 1.4–1.6 mm wide. RARE. Springs, streamsides, on serpentine; < 600 m. s CCo (San Luis Obispo Co.).

C. obnupta L. Bailey (p. 1135) (Group 6) Rhizomed, forming beds or large, dense, raised clumps. **LF**: blade 3–7 mm wide; sheath front with purplish prickles, thickened mouth, lower shredding to network or fringe of veins. **INFL**: lateral spikelets gen staminate at tip, 2.5–25 cm, 4.5–10 mm wide, tapered at base, flexible and nodding, stalk 0 or gen erect, ± short; lowest spikelet bract > infl; pistillate fl bract margin white, tip long-tapered. **FR**: perigynium ascending to spreading, 2–3.8 mm, 1.4–2.2 mm wide, not papillate, ± unveined, dark brown, purple-dotted, shiny, wall very thick, very tough, beak 0.1–0.3 mm, tip notch 0 or < 0.1 mm; fr 1.4–2.5 mm, 0.7–1.7 mm wide, often deeply indented on 1 or both sides, sessile. $2n$=70,72,74,±76. Moist to wet, often saline places; < 900 m. NCo, nw KR, CCo, SnFrB; to B.C.; also in n-c WA. ❀WET:**4,5,17**& SHD:**6**,7–9,**14–16**,24;INV,STBL.

C. occidentalis L. Bailey (p. 1135) (Group 7) Cespitose or short-rhizomed. **LF**: blade 1.5–2.5 mm wide. **INFL** ± open, 1.5–3 cm, 6–8 mm wide; spikelets gen < 10, ± distinct, lower ± separate; lowest spikelet bract < infl, like pistillate fl bract; pistillate fl bract brown, short-awned. **FR**: perigynium ascending to spreading, 2.5–3.5 mm, 1.5–1.9 mm wide, ± = pistillate fl bract, green to brown, veins ± 0, beak 0.6–1.2 mm, tip notched; fr 1.3–2 mm, 1.1–1.7 mm wide. Dry woodlands, meadows; ± 1900 m. SnBr; to WY, NM. ❀TRY.

C. ovalis Good (p. 1135) (Group 9) **LF**: blade 1.5–4 mm wide. **INFL** gen open, 15–40 mm; lowest internode > 4 mm; spikelets gen distinct; pistillate fl bract ± covering perigynium, red-gold to brown or greenish, center gold, tip white. **FR**: perigynium 3.4–5.2 mm, 1.3–2.5 mm wide, wide-ovate, length 1.8–2.6 × width, gold to light brown, flat margin incl wing 0.2–0.6 mm wide, veins gen both sides, beak tip cylindric, and ± entire for > 0.4 mm, red-brown, 1.4–2.9 mm to fr top; fr 1.2–2 mm, 0.9–1.2 mm wide, ± sessile. $2n$=62, 64,66,68. Seasonally wet soil; < 1100 m. NCoR, SnFrB, Wrn; to N.Am, Eurasia. [*C. leporina* L. misapplied; *C. tracyi* Mackenzie] ❀TRY:STBL.

C. pachystachya Cham. (p. 1135) (Group 9) **LF**: blade 1.2–6.5 mm wide. **INFL** 9–24 mm, ± dense; spikelets gen > 5, at most only lower distinct; lowest spikelet bract gen like pistillate fl bract; pistillate fl bract gen < 3.5 mm, gen < perigynium, gen reddish to brown ± throughout. **FR**: perigynium ascending-spreading, 2.8–5.1 mm, 1.1–2.3 mm wide, planoconvex, dark gold to brown, metallic, often copper, upper margin gen dark brown or coppery, less often narrowly green or gold, ± = in color to pistillate fl bract, flat margin incl wing 0.2–0.5 mm wide, minutely serrate, back veined, front veins 0, few to below fr middle, or 6 to beak, beak tip cylindric for > 0.4 mm, entire or serrate, dark brown to black, 1.5–2.9 mm to fr top; fr gen 1.2–2.2 mm, 1–1.5 mm wide, ovate, filling 2/3–4/5 perigynium body. $2n$=74,76,78,82. Dryish meadows, open forests; ± 0–3500 m. KR, NCoR, CaRH. [var. *gracilis* (Olney) Mackenzie; *C. olympica* Mackenzie] Variable; intergrades with *C. microptera*. ❀TRY.

C. pansa L. Bailey (p. 1135) (Group 7) Sometimes dioecious; rhizome 2–5 mm thick, dark brown. **LF**: blade 1–3 mm wide, flat or V-shaped; sheath mouth with thick rim or not. **INFL** 1.5–2.5 cm, 1–2.5 cm wide, ± head-like, gen ovate; spikelets gen < 10, sometimes > 1 per lower node or branch; lowest spikelet bract gen like pistillate fl bract, minutely pointed to bristle-like; pistillate fl bract gen dark brown, white-margined or not, gen shiny, minutely pointed to awned. **STAMINATE FL**: filament gen incl, longest anther awns 0.2–0.4 mm, glabrous or minutely hairy. **PISTILLATE FL**: style incl or not. **FR**: perigynium appressed to ascending, 3–4.5 mm, 1.3–2 mm wide, body tapered above, brown, gen shiny, veins 0 on front, many on back, lower wall sometimes filled with pithy tissue, base narrowed for > 0.3 mm below fr, beak 0.5–1.5 mm, serrate, tip unnotched; fr 1.4–2.1 mm, 1.1–1.6 mm wide. Coastal sand; < 10 m. NCo, CCo, SnFrB, ChI; to WA. ❀IRR,DRN:**4,5,17,24**;GRCVR&STBL.

C. parryana Dewey var. ***hallii*** (Olney) Kük. (p. 1135) PARRY'S SEDGE (Group 5) Short-rhizomed. **ST** 10–60 cm. **LF**: blade 2–4 mm wide. **INFL** ± open; spikelets pistillate or terminal one staminate at base or tip, terminal spikelet 10–30 mm, lateral sessile or short-stalked; pistillate fl bract < perigynium, gen dark purple or brown, margin white. **FR**: perigynium 1.6–3 mm, 1.1–1.6 mm wide, gen brown, upper body gen papillate, hairs on, just below beak few, stiff, beak 0.1–0.5 mm; fr 1.3–2 mm, 0.9–1.5 mm wide, ± = perigynium body. RARE in CA. Meadows; ± 3200 m. W&I (White Mtns); to n Can, ID, UT. [*C. hallii* Olney]

C. petasata Dewey (p. 1135) LIDDON'S SEDGE (Group 9) **LF**: blade 2–5 mm wide. **INFL** open, 2–6 cm, often whitish; 2nd lowest internode gen 4–11 mm; spikelets 3–8, distinct, base long-tapered; lowest spikelet bract sometimes bristle-like; staminate fl bract widely white-margined; pistillate fl bract ± covering perigynium, gen white or widely white-margined. **FR**: perigynium appressed, 5.7–8.1 mm, 1.5–2.5 mm wide, body planoconvex, tapered above, cream-white to brown, flat margin incl wing 0.2–0.5 mm wide, veins 5–8 both sides, beak tip cylindric, ± entire, red-brown, very tip white, 2.8–4.6 mm to fr top; fr 2.2–3 mm, 1.1–1.8 mm wide. Uncommon. Dry to wet meadows, grasslands; 600–3200 m. e KR, CaRH (Lassen Co.), n SNH (Alpine Co.), MP; to B.C., Rocky Mtns. [*C. liddonii* Boott]

C. phaeocephala Piper (p. 1135) (Group 9) **LF**: blade 0.5–2.5 mm wide, folded or margin downcurved. **INFL** dense or open, 10–35 mm, > 10 mm wide; 2nd lowest internode 2–4.5 mm; spikelets 2–7, distinct, ascending, club-shaped or fusiform; pistillate fl bract ± = perigynium in length, width, dark brown to orangish, white margin < 0.3 mm. **FR**: perigynium ascending, 3.5–5.8 mm, 1.2–2.5 mm wide, 2.2–4 × longer than wide, gen ± planoconvex, gen gold to brown, sometimes cream-white, green-margined, sometimes translucent, flat margin incl wing 0.2–0.6 mm wide, back veined, front veined or not, beak 0.5–1.2 mm, tip cylindric and ± entire for > 0.4 mm, gen red-brown, white at very tip, with white flap along middle of back, tip to fr top 1.5–2.6 mm; fr 1.5–2.1 mm, 0.8–1.2 mm wide, 0.2–0.5 mm thick, often long-stalked. $2n$=84. Often rocky soils; 2700–3900 m. CaRH, SN, Wrn, SNE (Sweetwater Mtns), W&I; to w Can, Colorado. [*C. eastwoodiana* Stacey; *C. phaeocephala* f. *eastwoodiana* (Stacey) F. Herm.] ❀TRY:STBL.

C. praeceptorum Mackenzie (p. 1135) (Group 8) Cespitose. **ST** 10–30 cm. **LF**: blade 1.2–2.5 mm wide. **INFL** often dense, 1– 2.5 cm; spikelets < 10, ± distinct, 4–7 mm, lowest 1 gen > 2 × lowest internode, terminal one narrowly tapered at base; lowest spikelet bract << infl; pistillate fl bract light brown with white margin and tip, obtuse to acute. **FR**: perigynium ascending to spreading, 1.5–2.5 mm, 1–1.2 mm wide, gold, upper body papillate, wall filled with pithy tissue to upper body, beak 0.2–0.4 mm, wide-conic, with a brown stripe or flap along middle of back; fr 1.2–1.5 mm, 0.8–1 mm wide. Wet places; 2400–3500 m. CaRH, c&s SNH; to WA, ID.

C. praegracilis W. Boott (p. 1135) (Group 7) Sometimes dioecious; rhizome 2–5 mm thick. **LF**: blade 1.5–3 mm wide, flat or V-shaped; sheath mouth with ± thick rim. **INFL** gen dense, 1–5 cm, 6–10 mm wide; spikelets gen < 10, sometimes > 1 per lower node or branch; pistillate fl bract ± dull, gold to brown, white-margined or not, minutely pointed to awned. **STAMINATE FL**: filament exserted; longest anther awns 0.1–0.2 mm, slender, gen minutely hairy at 20 ×. **PISTILLATE FL**: style gen incl. **FR**: perigynium ascending, 2.8–4 mm, 1.3–1.6 mm wide, widest above base, tapered above, dark brown, gen dull, veins several on back, 0 on front, lower wall sometimes filled with pithy tissue, base narrowed for < 0.3 mm below fr, beak 0.6–1.5 mm, serrate, tip unnotched, very tip white; fr 1.2–1.9 mm, 1–1.4 mm wide. Common. Often alkaline, ± moist places; < 2700 m. CA-FP; w&c N.Am, S.Am. ❀TRY:STBL.

C. praticola Rydb. (p. 1135) MEADOW SEDGE (Group 9) **LF**: blade 1.5–4 mm wide. **INFL** open, often nodding, 15–50 mm; lowest internode gen > 5 mm, second lowest 2.5–10 mm; spikelets 2–7, distinct, base often long-tapered; lowest spikelet bract gen bristle-like; pistillate fl bract gen covering perigynium, reddish or brown, white-margined. **FR**: perigynium appressed to ascending, 3.7–6.5 mm, 1.2–2.5 mm wide, body planoconvex, ± abruptly narrowed above, gen cream-white, translucent, flat margin incl wing 0.1–0.5 mm wide, veins on back weak, on front 0 or few, < 1/2 fr length, beak 1–1.6 mm, tip cylindric, ± entire, brown, with white flap along middle of back, very tip white, 2–3.7 mm to fr top; fr 1.5–2.7 mm, 1.2–1.7 mm wide, 0.5–0.7 mm thick. $2n$=±70. UNCOMMON. Moist to wet meadows; < 3200 m. NCo (Humboldt Co.), c&s SNH; N.Am. [*C. piperi* Mackenzie]

C. preslii Steudel (p. 1135) (Group 9) **LF**: blade 2–4 mm wide. **INFL** dense or open, 17–25 mm, green and brown; spikelets 3–7, gen distinct; lowest spikelet bract bristle-like or not; pistillate fl bract < or ± perigynium, brown to gold, center gen green, tip brown to gold. **FR**: perigynium ascending-spreading, 3–5 mm, 1.4–2.3 mm wide, gen bulged on front, back, green to gold, upper body margin green, not = pistillate fl bract in color, flat margin incl wing 0.1–0.5 mm wide, back veined, front veins gen 0 or few, irregular, beak tip flat or cylindric, serrate or entire, gen green to red-brown, 1.8–3.7 mm to fr top; fr 1.5–2.3 mm, 0.9–1.7 mm wide, ± squared, ± sessile. $2n$=±78. Meadows, open forest; 30–3000 m. KR, NCoR, SNH, Wrn; w N.Am. Sometimes incl in *C. pachystachya*. ❀TRY.

C. proposita Mackenzie (p. 1135) (Group 9)**LF**: blade 0.5–2.5 mm wide, folded or margin downcurved. **INFL** ± open, 15–25 mm; spikelets ± distinct, widely ovate to spheric; lowest spikelet bract bristle- or lf-like; pistillate fl bract < or > perigynium. **STAMINATE FL**: anthers ± persistent. **FR**: perigynium ascending-spreading, 4–6 mm, 1.8–3 mm wide, 1.7–2.6 × longer than wide, body round to ovate, 2–3 × pistillate fl bract in width, flat exc over fr, brown, green-margined, flat margin incl wing 0.4–0.9 mm wide, veins on back weak or not, on front 0 or few, beak tip cylindric and entire for > 0.3 mm, green to red-brown, white at very tip, with white flap along middle of back, tip to fr top 1.9–3.7 mm; fr 1.2–2.1 mm, 0.7–1.3 mm wide, often stalked. Uncommon. Rocky places; 3000–4100 m. c&s SNH; also in WA, c ID.

C. raynoldsii Dewey (p. 1135) (Group 5) Rhizomed. **ST** 20–75 cm. **LF**: blade 3–8 mm wide. **INFL**: pistillate fl bract purple or with pale midstripe not extending to tip, unawned. **FR**: perigynium spreading, 3–4.5 mm, 1.6–2.1 mm wide, with flat margin above, green, faces veined, beak 0.1–0.5 mm, purplish; fr 1.8–2.4 mm, 1.1–1.8 mm wide, ± = perigynium in width. $2n$=58. Meadows; 1800–3100 m. KR, CaRH, SNH, Wrn; to B.C., Rocky Mtns. ❀TRY:REVEG.

C. retroflexa Willd. var. *texensis* (Torrey) Fern. (Group 7) **LF**: blade 0.7–1.5 mm wide. **INFL** open, 1–3 cm, < 5 mm wide; spikelets gen < 10; pistillate fl bract white to greenish, acuminate to awned, deciduous before perigynium. **FR**: perigynium spreading to

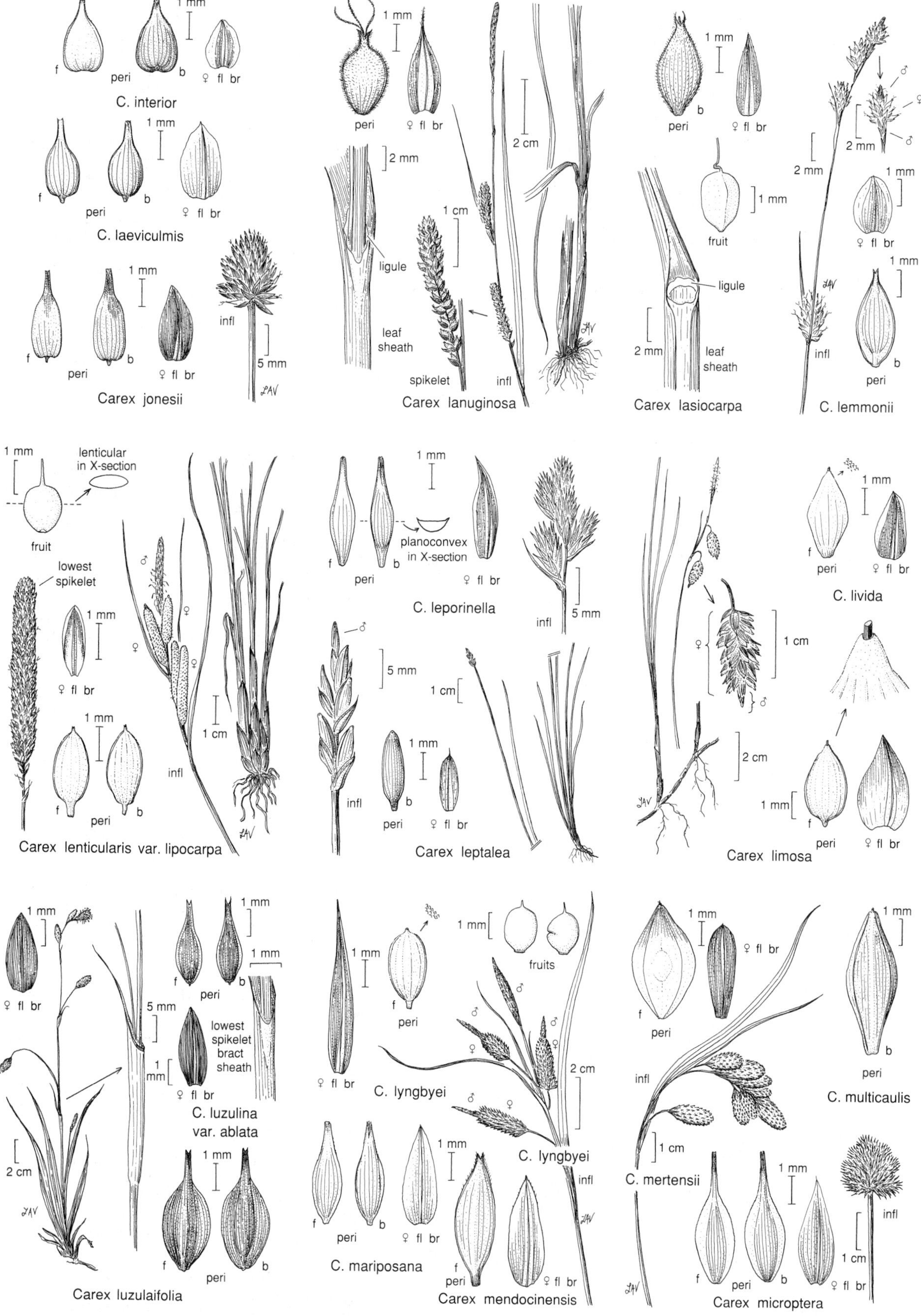

C. interior
C. laeviculmis
Carex jonesii
Carex lanuginosa
Carex lasiocarpa
C. lemmonii
Carex lenticularis var. lipocarpa
C. leporinella
Carex leptalea
C. livida
Carex limosa
C. luzulina var. ablata
Carex luzulaifolia
C. lyngbyei
C. mariposana
Carex mendocinensis
C. mertensii
C. multicaulis
Carex microptera
peri
fl br
fruit
ligule
leaf sheath
spikelet
infl
lenticular in X-section
lowest spikelet
planoconvex in X-section
lowest spikelet bract sheath
fruits
1 mm
2 mm
5 mm
1 cm
2 cm

reflexed, 2.4–3 mm, 0.8–1.2 mm wide, sessile, tapered above, unveined, green to pale gold, lower wall filled with pithy tissue, margin entire above, beak 0.5–1 mm, conic, entire, tip narrowly white-margined; fr 1.3–1.6 mm, 0.8–1 mm wide. Gardens, lawns, disturbed ground; < 100 m. ScV (Butte Co.), SCo (Santa Barbara, Los Angeles cos.); native to c&se US. [*C. texensis* (Torrey) L. Bailey]

C. rossii Boott (p. 1135) (Group 2) Cespitose. **ST** 3–40 cm, < or > lvs; glabrous or scabrous on angles only just below infl, sometimes scabrous nearly to base. **LF**: blade 1–4 mm wide, stiff or not, ± flat, bright or gray-green. **INFL**: staminate spikelet of taller infls 5–10 mm, 1–2.5 mm wide; pistillate spikelets 1–3; pistillate fl bract green or reddish tinged, awned or not. **FR**: perigynia 3–15 per spikelet, 2.4–5 mm, 1–1.7 mm wide, beak tip notched in front gen > 0.1 mm; fr 1.3–2.4 mm, 1.1–1.7 mm wide. Dry forests, meadows; < 3800 m. NW, CaRH, SNH, SnGb, SnBr, MP, W&I; to n Can, Colorado, AZ, scattered to n-c US. [*C. novae-angliae* Schwein. var. *r.* (Boott) L. Bailey; *C. brevipes* W. Boott] ❀TRY.

C. saliniformis Mackenzie (p. 1137) (Group 6) Loosely cespitose. **ST**: upper angles blunt, glabrous. **LF**: blade 2–5 mm wide. **INFL**: lateral spikelets 3–4, 6–15 mm, 3–5 mm wide, often > 1 per lower branch, stalk long; lowest spikelet bract gen > infl, sheath > 4.5 mm; lowest staminate fl bract often > 1/3 spikelet; lower pistillate fl bracts often lf-like, green, margin white or red-brown, awn stout, ciliate. **FR**: perigynia 8–20 per spikelet, 2.5–4 mm, 1.3–2.1 mm wide, sessile, papillate, body tapered to dark red-brown tip, beak 0–0.3 mm, tip unnotched; fr 1.4–1.9 mm, 0.8–1.5 mm wide, beak > 0.2 mm. Moist to wet, open areas; < 120 m. CCo. [misspelled *salinaeformis*] ❀TRY:STBL.

C. sartwelliana Olney (p. 1137) (Group 3) Cespitose. **ST** 30–90 cm. **LF** hairy; blade 3–7 mm wide. **INFL**: pistillate spikelets ± sessile; lowest spikelet bract ± = infl; pistillate fl bract often > perigynium, purplish, white-margined, ciliate. **FR**: perigynia 40–200 per spikelet, ascending, 2.3–3.9 mm, 1.2–1.8 mm wide, 2-ribbed, weakly veined, thin-walled, beak 0.4–1 mm, tip unnotched or notched on back < 0.2 mm; fr 1.5–2.1 mm, 1.2–1.5 mm wide. Meadows, standing water; 1200–2600 m. c&s SNH, SnJt. [*C. yosemitana* L. Bailey] ❀TRY.

C. schottii Dewey (p. 1137) (Group 6) Cespitose. **ST** 100–150 cm. **LF**: blades 4–12 mm wide, basal minute; sheath back scabrous, front red-dotted, shredding to network or fringe of veins; bladeless sheaths > 7 mm wide at midlength, keeled, scabrous. **INFL**: staminate spikelets 2–7, terminal one often > 5 cm; lateral spikelets staminate at least at tip, gen 5–20 cm, 5–7 mm wide, ± nodding, stalks 0 or long; lowest spikelet bract gen > infl, lf-like; pistillate fl bract obtuse to with a minute point. **FR**: perigynium 2.7–4.5 mm, 1.5–2.5 mm wide, sessile, veined, papillate or not, gold to brown, beak 0.1–0.5 mm, glabrous, tip notched < 0.5 mm; fr 1.6–2.3 mm, 1–1.6 mm wide. Streambanks, swamps; < 2500 m. NCoRO, SnFrB, SCo, WTR, PR. ❀TRY.

C. scirpoidea Michaux var. ***pseudoscirpoidea*** (Rydb.) Cronq. (p. 1137) (Group 1) Dioecious, long-rhizomed. **ST** 10–40 cm, > lvs. **LF**: blade 1.5–4 mm wide. **PISTILLATE INFL**: spikelets gen 1, sometimes 2, upper gen larger; spikelet bract 0 or < spike, sometimes lf-like and sheathing; fl bract obtuse or with a small point, = or > perigynium in width, ciliate or not, often minutely hairy on surface, brown to black with wide white margin. **FR**: perigynia 20–40 per spikelet, ascending, 2–3.5 mm, 1–1.7 mm wide, 3-sided, faintly veined, purplish, faces minutely hairy, at least near beak, beak 0.4–0.7 mm; fr 1.5–2.1 mm, 0.7–1.1 mm wide, ± = perigynium body. Rocky places, at least seasonally wet; 3500–3700 m. c SNH; to ID, MT, UT, NM. [*C. pseudoscirpoidea* Rydb.] ❀TRY: STBL.

C. scoparia Schk. (p. 1137) POINTED BROOM-SEDGE (Group 9) **LF**: blade gen 1.5–3 mm wide. **INFL** open, 15–70 mm; spikelets 3–8, distinct, fusiform, sides smooth; lowest spikelet bract gen like pistillate fl bract, less often lf-like; pistillate fl bract < perigynium, often short-awned, gold or brown. **FR**: perigynium ± appressed, 4–6 mm, 1.2–2.5 mm wide, body tapered above, green to gold, flat exc over fr, veins both sides, weak on front or not, flat margin incl wing 0.2–0.5 mm wide, beak tip cylindric and entire for < 0.4 mm, 2.8–3.8 mm to fr top; fr 1.3–1.9 mm, 0.7–1.2 mm wide. $2n$=60,64,68. RARE in CA. Wet, open places; gen < 1000 m. CaRF (Shasta Co.), n SNH (Plumas Co.); N.Am.

C. scopulorum Holm var. ***bracteosa*** (L. Bailey) F. Herm. (p. 1137) (Group 6) Loosely cespitose. **LF**: blade 3–6 mm wide; sheath front purple-dotted. **INFL** ± dense, gen dark purple; lowest internode 5–45 mm; lateral spikelet 8–30 mm, 4–6 mm wide, gen erect, stalk 0–15 mm, tip gen pistillate; lowest spikelet bract < 1/2 infl, ± like pistillate fl bract; pistillate fl bract obtuse to acute. **FR**: perigynia 8–50 per spikelet, ascending to spreading, 2–4 mm, 1.2–2.3 mm wide, unveined, papillate, gen purple above, beak 0.1–0.3 mm, glabrous, tip notched < 0.1 mm; fr 1.2–1.8 mm, 0.9–1.5 mm wide, dull. $2n$=72,76,78,80. Wet places; 1200–3400 m. KR, NCoRO, NCoRH, CaRH, SNH, Wrn, SNE (Sweetwater Mtns.); to B.C., Rocky Mtns [*C. gymnoclada* Holm] ❀TRY:STBL.

C. senta Boott (p. 1137) (Group 6) Forming large, dense, raised clumps connected by rhizomes, in ± continuous stands or not. **LF**: blade 3–5 mm wide; sheath backs scabrous, lower with small, raised, irregularly spaced cross-walls, lower fronts. **INFL**: lateral spikelet 2.5–5 cm, ± 5 mm wide, stalk short or 0, tip gen pistillate; lowest spikelet bract base dark-margined; pistillate fl bract < or ± = perigynium, obtuse to acuminate, gen dark purple. **FR**: perigynia 25–100 per spikelet, 2.9–4 mm, 1.4–2.2 mm wide, sessile, veined, gen not papillate, green, often purple-dotted or -blotched, beak 0.2–0.4 mm, glabrous, dark purple or not, tip notched < 0.1 mm; fr 1.5–2 mm, 1–1.5 mm wide. Streambanks, marshy areas, meadows; < 2900 m. n&c SNH, s ScV, CCo, SCo, ChI, TR, SnJt; Baja CA, AZ. Much like *C. nudata*. ❀WET:1–3,**7–9**,**14–16**,17,**18–24**&STBL.

C. serratodens W. Boott (p. 1137) (Group 5) Cespitose. **ST** 30–120 cm. **LF**: blades 1.7–4 mm wide, basal minute. **INFL**: terminal spikelet staminate at least at base; lower spikelets sessile; pistillate fl bract < perigynium, red-brown with pale midstripe, hairy-awned. **FR**: perigynium spreading, 3–5 mm, 1.2–2 mm wide, 3-sided, ± flattened around fr, green, purple-dotted, beak 0.3–1 mm, purplish, teeth minutely hairy; fr 1.8–2.5 mm, 1.1–1.8 mm wide. Moist places; < 1800 m. NCoR, SN, SnFrB, SCoR. [*C. bifida* Boott] ❀IRR:4,5,**6**,**15–17**,24&WET:**7**,8,9,**14**.

C. sheldonii Mackenzie (p. 1137) SHELDON'S SEDGE (Group 3) Rhizomed. **ST** 40–100 cm. **LF** hairy; blade 3–6 mm wide. **INFL**: pistillate spikelet stalks 0 or short; lowest spikelet bract > infl, sheathless; pistillate fl bract green to purplish, white-margined, short-awned or not. **FR**: perigynia 25–100 per spikelet, 5–8 mm, ± 2 mm wide, body ovate, many-ribbed, tapered to 2 mm beak with purple tip, teeth outcurved, 0.7–1.5 mm; fr 1.8–2.5 mm, 1–1.3 mm wide, << perigynium body; style bent, tough, not jointed to fr, persistent. RARE in CA. Wet places; 1200–1500 m. n SNH (Plumas, Placer cos.), MP; to OR, ID, UT.

C. simulata Mackenzie (p. 1137) (Group 7) Often dioecious; rhizome long, 1.5–3 mm thick, dark brown. **LF**: blade 2–5 mm wide; sheath mouth fragile, often with brown stripe across. **INFL** dense, 1.2–3.5 cm, 6–10 mm wide; spikelets < 10; lowest spikelet bract < infl, like pistillate fl bract; pistillate fl bract brown, white-margined, lower ± awned. **STAMINATE FL**: filaments exserted, longest anther awns 0.1–0.2 mm, stout, glabrous at 20 ×, tip obtuse. **PISTILLATE FL**: style exserted. **FR**: perigynium 1.7–2.6 mm, 1.3–1.6 mm wide, brown, shiny, veins few both sides, lower wall filled with pithy tissue, beak 0.2–0.6 mm, gen < 1/4 body, serrate, tip unnotched, white-margined; fr 1.1–2.3 mm, 0.7–1 mm wide. Moist soil; < 3300 m. KR, CaRH, SNH, CCo, SnFrB, GB; to WA, MT, UT, NM. ❀TRY:STBL.

C. specifica L. Bailey (p. 1137) (Group 9) **LF**: blade 2–5 mm wide, flat or folded; sheath front often ± firm and ribbed; ligule < 2.5 mm. **INFL** gen dense, 15–40 mm, gen oblong; lowest internode < 8 mm; spikelets gen > 6, ± indistinct, fusiform; lowest spikelet bract bristle-like or not; pistillate fl bract < perigynium, green or reddish. **FR**: perigynium appressed to ascending, 4.8–6.6 mm, 1.2–2.3 mm wide, body lanceolate to fusiform, planoconvex, veined both sides, gold, flat margin incl wing 0.2–0.4 mm wide, beak tip cylindric and entire for < 0.4 mm, very tip reddish, 2.5–3.4 mm to fr top; fr 1.7–2.1 mm, 1–1.5 mm wide, stalk 0.6–1 mm. Dry soil, meadows, open forests; 1200–3500 m. KR, CaRH, SNH; w NV. Perigynia sometimes narrow like those of *C. petasata* but smaller, and infl denser. ❀TRY.

C. spectabilis Dewey (p. 1137) (Group 5) Cespitose or short-rhizomed. **ST** 20–90 cm. **LF**: blades 2–5 mm wide, basal gen minute. **INFL**: lowest spikelets often on nodding stalks, purple;

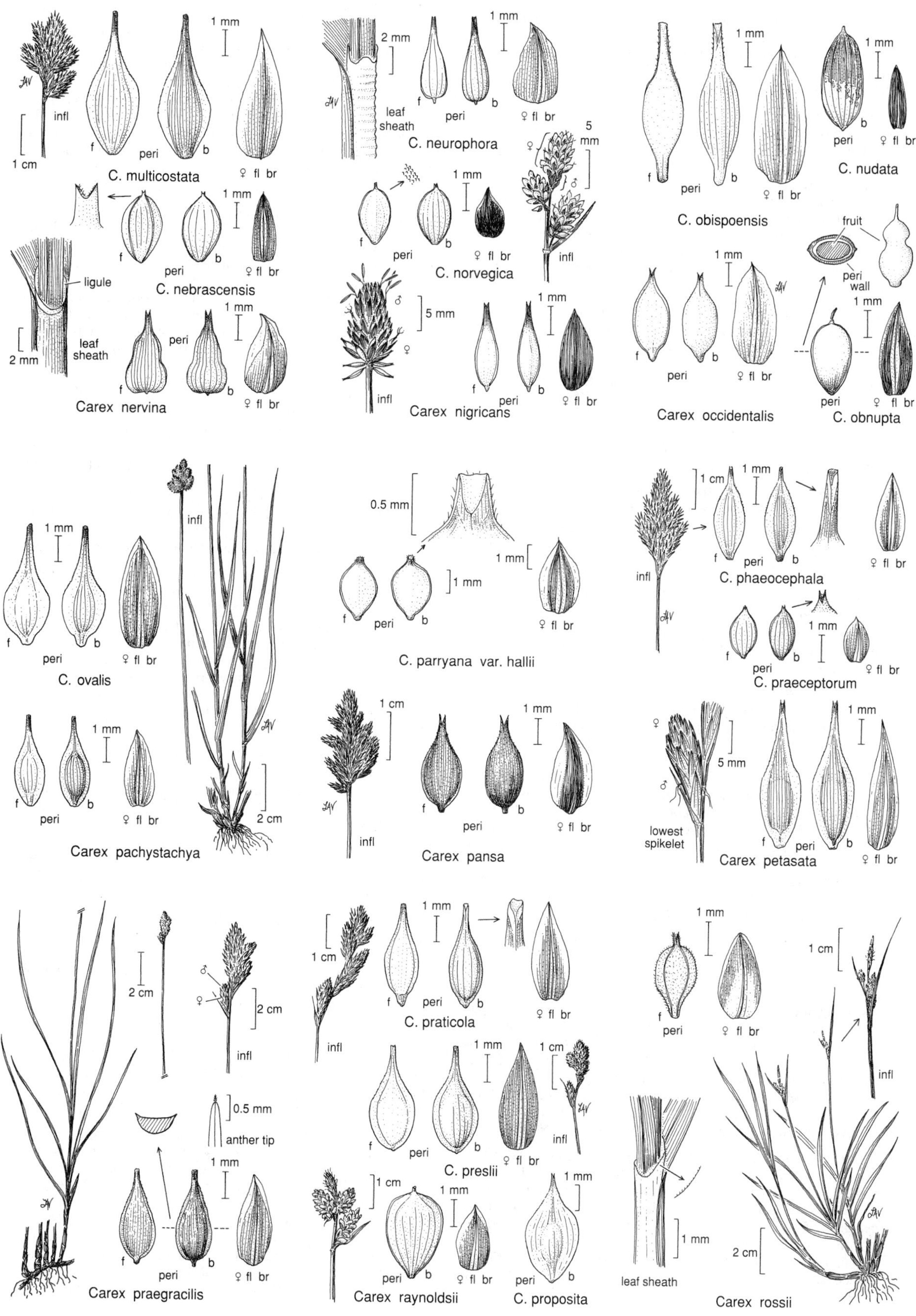

C. multicostata
C. nebrascensis
Carex nervina
C. neurophora
C. norvegica
Carex nigricans
C. obispoensis
C. nudata
Carex occidentalis
C. obnupta
C. ovalis
Carex pachystachya
C. parryana var. hallii
Carex pansa
C. phaeocephala
C. praeceptorum
Carex petasata
Carex praegracilis
C. praticola
C. preslii
Carex raynoldsii
C. proposita
Carex rossii

pistillate fl bract dark purple with pale midstripe, awned or tip long-tapered. **FR**: perigynium ascending, 2.6–5 mm, 1.6–2.3 mm wide, ± flat or 3-sided, papillate, purple above, body ovate, tapered to tip, beak 0.2–0.5 mm; fr 1.6–2 mm, 0.7–1 mm wide. $2n$=±42. Wet places; 1800–3700 m. KR, CaRH, SNH; to n&w Can, MT. ❀TRY.

C. spissa L. Bailey (p. 1137) (Groups 3,4) **ST** 1–2 m, stout. **INFL**: lower spikelets staminate at tip, stalked, erect; pistillate fl bract gold to red-brown, margin gen white, awn hairy. **PISTILLATE FL**: style bent, tough, not jointed to fr, persistent. **FR**: perigynia 150–300 per spikelet, spreading, 3–5 mm, 1.5–1.8 mm wide, body obovate, 2-ribbed, with few or 0 veins on faces, abruptly narrowed above, glabrous or hairy, glaucous, green, red-dotted, beak 0.1–0.8 mm, often bent, tip dark, teeth < 0.5 mm, straight; fr 2–2.6 mm, 1–1.3 mm wide. Waterways, serpentine or not; < 600 m. CCo, SCo; Baja CA. Resembles small cattails. ❀WET:7,8,9,**14–17**,19, **20–24**.

C. stipata Muhlenb. var. ***stipata*** (p. 1137) (Group 7) Cespitose. **ST** deeply concave between sharp or winged angles, spongy when fresh. **LF**: blade 5–11 mm wide; sheath front cross-wrinkled. **INFL** 2–10 cm, 1–3 cm wide; spikelets >> 10, > 1 per lower node or branch, indistinct; pistillate fl bract pale, often awned. **FR**: perigynium spreading, 3.6–6 mm, 1.5–1.8 mm wide, body widest ± at base, many-ribbed both sides, brown, long-tapered to poorly defined beak, lower wall filled with pithy tissue; fr 1.3–2.3 mm, 1.5–1.8 mm wide. Wet places; < 1700 m. NW, n&c SNH, SnFrB; N.Am; also in Japan. ❀TRY.

C. straminiformis L. Bailey (p. 1137) MOUNT SHASTA SEDGE (Group 9) **LF**: blade 2–4 mm wide. **INFL** dense, 15–25 mm; spikelets distinct, ± spheric; pistillate fl bract < perigynium, red-gold to brown. **FR**: perigynium ascending to spreading, 4–5.8 mm, 1.8–3.4 mm wide, body green or gold, flat exc over fr, flat margin incl wing 0.4–1 mm wide, edge minutely crinkled at least above, veins many at least on back, weak, beak tip cylindric and entire for < 0.3 mm, red-brown, very tip white, 2.2–3 mm to fr top; fr 1.3–2.3 mm, 1–1.6 mm wide. Common. Rocky or gravelly soils; 2000–3800 m. KR, NCoRI, CaRH, SNH, MP, SNE (Sweetwater Mtns), W&I; to WA; also in ID, UT. ❀TRY.

C. subbracteata Mackenzie (p. 1137) (Group 9) **ST** often supported by vegetation. **LF**: blade 1.3–4.5 mm wide. **INFL** ± dense, 13–35 mm; lowest internode gen < lowest spikelet, lowest 2 together < 11 mm, 2/5 infl length; lower spikelets gen distinct, wedge-shaped or rounded; lowest spikelet bract often bristle-like, gen > lowest spikelets, < or > infl, base around st; pistillate bract gen > 3.4 mm, < or > perigynium, reddish. **FR**: perigynium ascending, 2.9–5.5 mm, ± 1.3–2.2 mm wide, planoconvex, gen gold with brown margin, metallic, margin incl wing < 0.3 mm wide, serrate on upper 1/2 body, wall ± tough, back veined, veins on front 0–7, to fr top, beak gen ± 2/5 perigynium length, tip cylindric and ± entire for > 0.4 mm, brown to red-brown, 1.5–2.5 mm to fr top; fr 1.5–2.1 mm, 0.9–1.7 mm wide. At least seasonally moist soil, grasslands to open forests; < 900 m. NCo, NCoR, CCo, SnFrB, SCoR.

C. subfusca W. Boott. (p. 1137) (Group 9) **LF**: blade 1.2–3.7 mm wide. **INFL** open or dense, 11–30 mm; spikelets distinct; lowest spikelet bract bristle-like or not, > infl or not; lower pistillate fl bracts, < or ± = perigynium, gen ± < 3.5 mm, pale brown to reddish. **FR**: perigynium appressed to spreading, 2.5–4.3 mm, 0.9–1.9 mm wide, planoconvex, whitish green, pale gold, or light brown, upper margin green or gold, flat margin incl wing 0.1–0.3 mm wide, at least back veined, beak tip flat or cylindric, ± serrate, light to medium brown, very tip white, 1.1–2 mm to fr top; fr 1–1.6 mm, 0.7–1.2 mm wide. $2n$=84. Common. Seasonally wet meadows; 100–3500 m. KR, NCoR, CaRH, SN, SnFrB, SCoR, TR, PR, Wrn, DMtns; w N.Am. [*C. teneraeformis* Mackenzie] ❀TRY.

C. subnigricans Stacey (p. 1137) (Group 1) Rhizomed. **ST** 5–20 cm, > lvs. **LF**: blade 1 mm wide, rolled, quill-like. **INFL**: spikelet bisexual, 3–6 mm wide; spikelet bract 0; pistillate fl bract wide, gen 1-veined, light brown. **FR**: perigynia 10–40 per spikelet, ascending, 2.5–4.1 mm, 0.9–2 mm wide, flat, golden brown, a bristle-like axis within, faces glabrous; fr 1.3–1.8 mm, 0.7–1 mm wide (width > distance from fr to perigynium side), < perigynium body. Meadows, gen dry, rocky slopes; 2600–3800 m. SNH, SNE (Sweetwater Mtns), W&I; to ID, UT. ❀TRY:STBL.

C. tahoensis F.J. Smiley (p. 1137) (Group 9) **LF**: blade 0.7–2 mm wide, margin downcurved. **INFL** open, 15–30 mm, < 11 mm wide; spikelets distinct, appressed, < 5 mm wide; pistillate fl bract ± covering perigynium, white margin 0.3–0.6 mm wide. **FR**: perigynium appressed, 3.7–5 mm, 1.3–2 mm wide, planoconvex, gold to brown, flat margin incl wing 0.1–0.4 mm wide, back veined, front gen with 5–8 veins to fr top, beak tip cylindric and ± entire for > 0.4 mm, reddish, white at very tip, with white flap along middle of back, tip to fr top 0.6–2 mm; fr 1.8–2.3 mm, 1.1–1.4 mm wide, 0.5–0.7 mm thick. Uncommon. Open, rocky slopes; 2900–3700 m. SNH, SNE (Sweetwater Mtns).

C. tiogana D. Taylor & J. Mastrogiuseppe (p. 1137) TIOGA SEDGE (Group 5) Cespitose. **ST** 1.8–7.5 cm, < lvs. **LF**: blade 1.5–3 mm wide, ± sickle-shaped, margin, midrib densely, coarsely serrate at 10 ×. **INFL**: lateral spikelets erect to nodding; lowest spikelet bract sheath > 6 mm; pistillate fl bract white, early deciduous, midrib of at least some raised, spiny or serrate. **FR**: perigynia 3–8 per spikelet, ascending to spreading, 1.3–2 mm, 0.4–0.9 mm wide, with a few curved spines on upper body margin, shiny brown, beak 0.2–0.4 mm, ciliate; fr 0.8–1.2 mm, 0.4–0.8 mm wide. RARE. Coarse, wet, limey soil; 3100–3300 m. c SNH (Mono Co.).

C. tompkinsii J. Howell (p. 1137) TOMPKINS' SEDGE (Group 1) Cespitose. **STS** many, 10–60 cm, < or ± = longest lvs, upper angles sharp. **LF**: blade 1.5–3 mm wide. **INFL**: spikelets often 2; lower pistillate fl bracts 1–15 cm, > infl, lf-like, green, margins white, base clasping perigynium. **FR**: perigynia gen 2–3 per spikelet, appressed to ascending, 4.5–6 mm, 1.6–2.6 mm wide, unveined, minutely ciliate near or on beak, green, a bristle-like axis within; fr 3.1–4.9 mm, 1.2–2.5 mm wide, distending perigynium body. **RARE** CA. Open forests, slopes; 1200–1800 m. c SNF (near El Portal, Mariposa Co.), s SNH (near Kings Canyon, Fresno Co.).

C. triquetra Boott (p. 1137) (Group 3) Cespitose. **ST** 30–60 cm. **LF**: blade 2.5–6 mm wide, upper surface minutely papillate, sheath front red or red-dotted. **INFL**: lateral spikelets often staminate at base, nodding; lowest spikelet bract gen < infl, sheath > 5 mm; pistillate fl bract copper, acute to minutely pointed, midrib sometimes hairy. **FR**: perigynia 5–30 per spikelet, 3.3–5 mm, 1.6–2.8 mm wide, weakly veined, hairs spreading, beak < 0.5 mm, teeth erect, < 0.3 mm; fr 3.2–3.5 mm, 1.8–2.5 mm wide, ± = perigynium body. Clay, often rocky soil; 600–3000 m. CCo, SCo, s ChI, WTR, SnBr, PR; n Baja CA. ❀TRY.

C. tumulicola Mackenzie (p. 1137) (Group 7) Short-rhizomed. **LF**: blade 1–2.5 mm wide. **INFL** open, often flexible, 2–5 cm, 6–8 mm wide; lowest spikelet bract bristle-like, often > infl; pistillate fl bract ± = perigynium, red, white-margined, awned. **FR**: perigynium appressed, 3.5–5 mm, 1.5–2 mm wide, tapered above or not, light green to brown, veins several on back, 0 on front, beak 1.2–3 mm, teeth reddish, tip notched; fr 1.8–2.3 mm, 1.3–2 mm wide. Meadows, open woodlands; < 1200 m. NCo, NCoR, SN, CCo, SnFrB, ChI; to WA. ❀TRY:STBL.

C. unilateralis Mackenzie (p. 1137) (Group 9) **LF**: blade 2–5 mm wide. **INFL** dense, 10–30 mm; axis gen ascending; spikelets ± indistinct; lowest spikelet bract ± erect, >> infl, often > 2 mm wide, lf-like, base ± around st; pistillate fl bract < perigynium, reddish, often awned. **FR**: perigynium appressed to ascending, 3.5–5 mm, 1.2–1.8 mm wide, ovate, gen sessile, body green, flat margin incl wing 0.2–0.3 mm wide, at least back veined, beak tip flat, winged, and serrate throughout or less often cylindric, unwinged, and entire for < 0.4 mm, gen green, very tip gen red-brown, 1.4–2.5 mm to fr top; fr 1.3–2 mm, 0.7–1.2 mm wide, near middle of perigynium body. Seasonally wet places; < 1000 m. n&c NCoR; to B.C. Much like (possibly best a var. of) *C. athrostachya*. ❀TRY.

C. utriculata Boott (p. 1141) (Group 4) Long-rhizomed. **ST** 30–120 cm, ± = lvs, angles blunt. **LF**: ligule length < or ± = width. **INFL**: lateral spikelets erect, sessile or short-stalked; pistillate fl bract green with reddish margin, awn < body or 0. **FR**: perigynia 50–200 per spikelet, in many dense rows, ascending-spreading to reflexed, 3.9–7 mm, 1.3–3 mm wide, abruptly narrowed to 1–2 mm beak, shiny, gold to brown, teeth erect, 0.1–0.8 mm; fr 1.1–2 mm, 0.9–1.3 mm wide. $2n$=72,76,82. Common. Wet places, shallow water; < 3400 m. KR, NCoR, CaRH, SNH, CCo, SnFrB, SnBr, SNE; N.Am, Eurasia. [*C. rostrata* Stokes] ❀TRY:STBL.

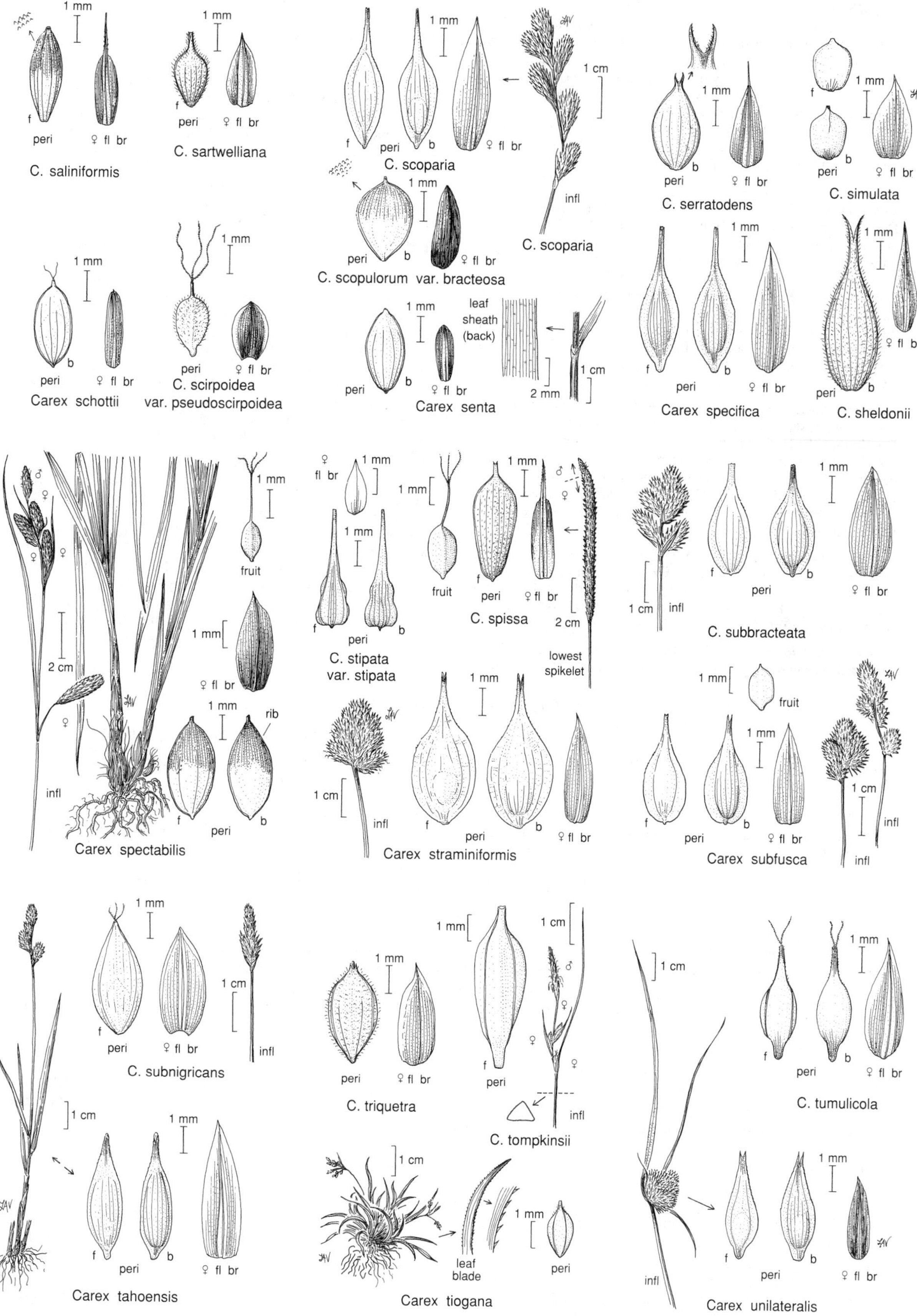

1 mm
f
peri
♀ fl br
C. saliniformis
1 mm
f
peri
♀ fl br
C. sartwelliana
1 mm
f
peri
b
♀ fl br
C. scoparia
1 cm
infl
C. scoparia
1 mm
peri
b
♀ fl br
C. scopulorum var. bracteosa
1 mm
b
peri
♀ fl br
C. serratodens
1 mm
f
b
peri
♀ fl br
C. simulata
1 mm
b
peri
♀ fl br
Carex schottii
1 mm
peri
♀ fl br
C. scirpoidea var. pseudoscirpoidea
1 mm
leaf sheath (back)
peri
b
♀ fl br
2 mm
1 cm
Carex senta
1 mm
f
peri
b
♀ fl br
Carex specifica
1 mm
♀ fl br
peri
b
C. sheldonii
♂
♀
2 cm
infl
1 mm
fruit
1 mm
♀ fl br
1 mm
rib
f
peri
b
Carex spectabilis
♀ fl br
1 mm
1 mm
f
peri
b
C. stipata var. stipata
1 mm
1 mm
fruit
f
peri
♀ fl br
C. spissa
2 cm
lowest spikelet
1 cm
infl
1 mm
f
peri
b
♀ fl br
Carex straminiformis
1 cm
infl
1 mm
f
peri
b
♀ fl br
C. subbracteata
1 mm
fruit
1 mm
f
peri
b
♀ fl br
Carex subfusca
1 cm
infl
infl
1 mm
f
peri
♀ fl br
1 cm
infl
C. subnigricans
1 cm
1 mm
f
peri
b
♀ fl br
Carex tahoensis
1 mm
peri
♀ fl br
C. triquetra
1 mm
1 cm
f
peri
infl
C. tompkinsii
1 cm
leaf blade
1 mm
peri
Carex tiogana
1 cm
1 mm
f
peri
b
♀ fl br
C. tumulicola
1 mm
infl
f
peri
b
♀ fl br
Carex unilateralis

C. vallicola Dewey (p. 1141) (Group 7) Cespitose, sometimes long-rhizomed. **ST** 15–60 cm. **LF**: blade 0.5–2 mm wide. **INFL** 1–3 cm, 4–8 mm wide; spikelets < 10; lowest spikelet bract like pistillate fl bract or bristle-like; pistillate fl bract < perigynium, white, minutely pointed. **FR**: perigynium spreading, 2.5–4 mm, 1.5–2.2 mm wide, sessile, green to brown, very shiny, back bulged so that marginal ribs on front rather than sides, ± veined, front unveined, beak 0.6–1 mm, teeth reddish; fr 1.6–2.5 mm, 1.4–2.1 mm wide. Moist to ± dry slopes, montane; < 2700 m. SNE (Sweetwater Mtns); to MT, SD, UT. ❀TRY:STBL.

C. vernacula L. Bailey (p. 1141) (Group 7) Cespitose; rhizomes short, straight. **ST** 5–30 cm. **LF** < infl; blade 2–4 mm wide. **INFL** dense, 8–16 mm, 8–16 mm wide; spikelets < 10, indistinct; lowest spikelet bract like pistillate fl bract; pistillate fl bract dark brown with narrow white tip, shiny, long-tapered or awned. **FR**: perigynium ascending to spreading, 3.3–4.8 mm, < pistillate fl bract, 1.2–1.8 mm wide, brown, often green-margined, ± shiny, back veined, front veined or not, margin entire above, beak 0.8–1.7 mm, entire; fr 1.2–1.6 mm, 0.7–1.1 mm wide, ovate. Open, often rocky areas, wet from snow melt; 1800–4000 m. CaRH, SNH, Wrn, SNE (Sweetwater Mtns), W&I; scattered in montane w US. ❀TRY.

C. vesicaria L. (Group 4) Rhizome short. **ST** 30–100 cm, gen >> lvs; angles sharp. **LF**: ligule length > width. **INFL**: lateral spikelets short-stalked, erect; pistillate fl bract gold to purplish, awned or not. **FR**: perigynia 20–100 per spikelet, in few ± open rows, ascending, 4–11 mm, 1.7–3 mm wide, green to brown, beak teeth erect to outcurved. $2n$=74,82,86. Wet places, shallow water; < 3300 m. NW, CaRH, SNH, CCo, SnFrB, MP; to AK, e N.Am; also in Eurasia. ❀TRY.

var. ***major*** Boott (p. 1141) **LF**: lower sheath fronts gen shredding to network or fringe of veins. **FR**: perigynium gen 7–11 mm, 1.9–3 mm wide, tapered above, beak 2–3 mm, teeth 0.2–1.5 mm; fr 1.7–3 mm, 1.1–1.8 mm wide. Wet places; < 1800 m. NCo, s KR, n SN (Plumas Co.), CCo; to AK, MT. [*C. exsiccata* L. Bailey]

var. ***vesicaria*** (p. 1141) **LF**: lower sheath fronts gen not shredding. **FR**: perigynium gen 4–8 mm, 1.7–2.6 mm wide, ± abruptly narrowed above, beak 1–2 mm, teeth 0.2–1 mm; fr 1.7– 2.3 mm, 1.1–1.5 mm wide. Wet places; < 3300 m. KR, NCoR, CaRH, SNH, SnFrB, MP; to B.C., e N.Am; also in Eurasia. [*C. inflata* Hudson]

C. viridula Michaux var. ***viridula*** (p. 1141) (Group 5) Cespitose. **ST** 5–40 cm. **LF**: blade 1.5–3 mm wide. **INFL** dense; staminate spikelet 1–2 mm wide, linear; lateral spikelets sessile or lowest 1 stalked, erect; lower spikelet bracts >> infl, gen long-sheathing, blade ascending-spreading; pistillate fl bract reddish, center green. **FR**: perigynia crowded, spreading to reflexed, 2–3.5 mm, 0.9–1.5 mm wide, body many-ribbed, yellow-green to brown, beak 0.6–1.2 mm, narrowly conic; fr 1.1–1.8 mm, 0.8–1.1 mm wide. $2n$=70,72. Sphagnum bogs; 0–1600 m. NCo, n SNH (1 site); to AK, e N.Am; also in Japan. [*C. oederi* Retz.]

C. vulpinoidea Michaux (p. 1141) FOX SEDGE (Group 7) Cespitose. **LF**: blade 2–5 mm wide; sheath front gen cross-wrinkled, minutely red-dotted. **INFL** dense, 5–10 cm, 1–1.5 cm wide; spikelets >> 10, > 1 per lower node or branch; pistillate fl bract brown to gold, base, margin white or not, awns often glabrous, lower gen > 1 mm. **FR**: perigynium spreading, gen 2–3 mm, 1–2 mm wide, < or = 2 × longer than wide, 0.4–0.7 mm thick, often entire above, beige, ± translucent, back veined, front veins 0 or 1–2, irregular, to fr middle, front wall moderately filled with pithy tissue in whitish U-shaped area around fr, ± flat over fr, beak 0.7–1.8 mm, teeth gold; fr 1.2–1.6 mm, 0.8–1 mm wide. $2n$=52,54. UNCOMMON. Wet places; < 1200 m. se KR, n CaRH, n ScV; to B.C., Colorado, AZ. ❀TRY.

C. whitneyi Olney (p. 1141) (Group 5) Cespitose. **ST** 25–100 cm. **LF**: blades 2–6 mm wide, with long, soft hairs, basal minute; sheath fronts brownish. **INFL**: lateral spikelets sessile; lowest spikelet bract sheath < 5 mm; pistillate fl bract spreading, ± ciliate above, white. **FR**: perigynium 3.7–5.5 mm, 1.5–2.5 mm wide, many-ribbed, green, beak 0.2–1 mm, tip with narrow white margin; fr 2.8–3.9 mm, 1.7–2.4 mm wide, ± = perigynium body. Dry, sandy or gravelly meadows, open forests; 1200–3400 m. KR, CaRH, n&c SNH, Wrn; s OR, w NV. [*C. jepsonii* J. Howell] ❀TRY.

CLADIUM TWIG RUSH

Per. **ST** rounded, leafy, hollow. **INFL**: spikelets in terminal and axillary, panicle-like clusters; fl bracts spiraled, basal sterile, reduced; uppermost fls bisexual, others staminate. **FL**: perianth bristles 0; stamens 2; style 2–3-branched. **FR** weakly 2-sided, truncate at base; style base persistent; tubercle 0. ± 40 spp.: worldwide, esp Australia. (Greek: branch, from infl)

C. californicum (S. Watson) O'Neill (p. 1141) SAW-GRASS Pl 10–20 dm, stout, coarse, from long rhizomes, forming dense bunches. **LF** 7–10 mm wide, flat; margin sharply serrate. **FR** purple-brown; tip 2–2.5 mm, buff. $2n$=36. Uncommon. Gen alkaline marshes, swamps; < 2000 m. s CCo, SCoRO, SCo, WTR, D; NV, AZ. [*C. mariscus* R.Br. var. *c.* S. Watson] ❀TRY;STBL.

CYPERUS NUTSEDGE, GALINGALE

Gordon C. Tucker

Ann or per, glabrous. **STS** gen > 1, erect, 2–200 cm, 3-angled or round. **LVS** basal; blades 0 or linear. **INFL**: bracts 1–22, lf-like, spreading or erect; rays < 10 cm; spikelets flat to ± cylindric; fl bracts 2-ranked, 1 per fl, 2–36 per spikelet. **FL** bisexual; perianth 0; stigmas gen 3. **FR** (ob)ovoid, gen 3-angled, brown, gen not beaked. (Greek: name for 1 sp.) [Tucker 1983 Syst Bot Monogr 2:1–85] Mature fr gen needed for identification. *C. unioloides* R.Br. & *C. virens* Michaux probably only waifs in CA.

1. St 1; stolons with tubers; fr seldom maturing
  2. Fl bracts yellow-brown to brown .... ***C. esculentus***
  2′ Fl bracts ± dark red to purple .... ***C. rotundus***
1′ Sts gen > 1, clumped or not; stolons, tubers 0; fr maturing
  3. Fr lenticular; stigmas 2
    4. Fr face next to spikelet axis .... ***C. laevigatus***
    4′ Fr angle next to spikelet axis
      5. Ann; rhizomes 0; spikelets in open spikes; fl bracts brown or ± red-brown; fr (ob)ovoid, black .... ***C. bipartitus***
      5′ Per; rhizomes short; spikelets gen in head-like spikes; fl bracts light brown to black; fr ovoid to elliptic, brown .... ***C. niger***
  3′ Fr 3-angled; stigmas 3
    6. Tip of fl bract outcurved, narrowly acute or bristle-like
      7. Fl bract 3-veined; fr widest ± near middle .... ***C. acuminatus***
      7′ Fl bract 5–11-veined; fr widest just below tip .... ***C. squarrosus***

6′ Tip of fl bract not outcurved, acute, obtuse, or rounded, sometimes with small point but not bristle-like
8. Basal lf blades 0; infl bracts (10)14–22 . . . . *C. involucratus*
8′ Basal lf blades present; infl bracts 1–12
9. Fl bract < 1.1 mm; ann; rhizomes 0
10. Spikelets 50–100+ in dense heads . . . . *C. difformis*
10′ Spikelets 3–15 in open heads . . . . *C. fuscus*
9′ Fl bract 1.3–4 mm; ann or per; rhizomes gen present
11. Fr body ± as long as wide, stalked . . . . *C. eragrostis*
11′ Fr body 1.5–3 × longer than wide, ± sessile
12. Spikelets flat; fl bract reddish . . . . *C. parishii*
12′ Spikelets slightly flat or subcylindric; fl bract brown, light reddish brown, or yellowish
13. St densely bumpy or warty; spikelets in dense, cylindric spikes . . . . *C. ligularis*
13′ St smooth; spikelets in open, ovoid to cylindric spikes
14. Fl bract 1.3–1.5 mm; fr 0.7–1 mm . . . . *C. erythrorhizos*
14′ Fl bract 2–4 mm; fr 1.5–2.4 mm
15. Ann; st base not corm-like; fl bract persistent, light brown splotched reddish, midvein green; spikelet axis jointed at each bract, spikelet falling apart . . . . *C. odoratus*
15′ Per; st base corm-like; fl bract yellowish, persistent or not; spikelet axis continuous, spikelet falling as unit . . . . *C. strigosus*

C. acuminatus Torrey & Hook. (p. 1141) Ann 5–40 cm. **ST** 0.4–1 mm thick. **INFL**: bracts 3–6; rays 0–5, < 20 mm; spikelets 4–7 mm, ovate, in ± spheric heads 5–18 mm wide; fl bract 1.3–2 mm, ovate-lanceolate, 3-veined, tip outcurved, narrowly acute. **FR** 0.7–1.1 mm, obovoid-elliptic, stalked, smooth, light brown; tip acute, beaked. Edges of temporary pools, ponds, streams, ditches; 0–400 m. KR, GV, WTR; to WA, n-c&e US.

C. bipartitus Torrey (p. 1141) Ann 3–25 cm. **INFL**: bracts 2–3, 1–12 cm; rays 1–4, 3–20 mm; spikelets 3–10, 8–20 mm, linear-oblong, flat, in open, ± ovoid spikes 8–15 mm wide; fl bract 1.8–2.3 mm, (red-)brown, lateral veins 0. **FL**: stigmas 2. **FR** 1–1.5 mm, (ob)ovoid, lenticular, bumpy or warty, black. Sandbars, pond shores, ditches; 100–1500 m. NW, c SNF, n SNH; to Can, S.Am. [*C. rivularis* Kunth]

C. difformis L. Ann 3–40 cm. **INFL**: bracts 2–3; rays 0–4, 6–30 mm; spikelets 50–100+ in dense, subspheric, yellowish or purplish, rayed or sessile heads; fl bracts 6–30 per spikelet, 0.6–0.8 mm, obovate, yellow to yellowish brown. **FR** 0.6–0.8 mm, obovoid-elliptic, glossy. Ditches, pond shores, rice fields; 100–500 m. GV, SW; native to Old World.

C. eragrostis Lam. (p. 1141) Per 10–90 cm. **INFL**: bracts 4–8, 3–50 cm; rays 0–10, 20–100 mm; spikelets 20–70, 5–20 mm, flat, oblong, in spheric heads 1.5–4 cm wide; fl bracts 6–12 per spikelet, 2–2.3 mm, ovate-lanceolate, beige; tip acute. **FR** 1.2–1.4 mm, stalked, finely reticulate, black or dark brown; tip beaked. Vernal pools, streambanks, ditches; 0–700 m. CA-FP; OR; also temp S.Am. ❀STBL;INV.

C. erythrorhizos Muhlenb. (p. 1141) Ann 5–100 cm; roots reddish. **INFL**: bracts 3–11, 5–70 cm; spikelets 20–150, 3–11 mm, linear, subcylindric, light (reddish) brown, in open, ovoid-cylindric spikes, these 1–6 per ray; fl bracts 6–30 per spikelet, 1.3–1.5 mm, overlapped, tips obtuse, lateral veins 0. **FR** 0.7–1 mm, light gray to brown, glossy; sides unequal. Ditches, riverbanks, shores; 0–500 m. CA; US, n Baja CA.

C. esculentus L. (p. 1141) Per 20–80 cm; stolons with tubers < 15 mm wide. **ST** 1. **INFL**: bracts 3–7; rays 5–10, 20–200 mm; spikelets 5–30 mm, linear, brown, in ± open, widely elliptic spikes; fl bracts 2–3 mm, overlapped, ovate to elliptic, (yellowish) brown. **FR** 1–1.6 mm, elliptic, seldom maturing. Croplands, disturbed places; 0–1000 m. CA; worldwide weed.

C. fuscus L. Ann 2–30 cm. **INFL**: bracts 2–3; rays 1–3, 2–15 mm (20 rays on robust pls < 5 mm); spikelets 3–15, 3–7 mm, narrowly elliptic, flat, in open, subspheric, yellowish to purplish brown heads; fl bracts 8–12(16) per spikelet, ± 1 mm, spheric, tip with small point, lateral veins 0, midveins 3. **FR** 0.7–0.9 mm, elliptic, ± sessile, smooth, light brown; base wedge-shaped; tip acute. Damp disturbed soil, receding shorelines, puddles; 0–50 m. SnJV; to e N.Am; native to temp Eurasia.

C. involucratus Rottb. Per 1–2 m. **LVS**: basal blades 0. **INFL**: bracts (10)14–22; rays 14–22, 20–200 mm; spikelets 5–15 mm, ± linear to ovate, in stellate clusters 15–30 mm wide; fl bract 1.6–2.4 mm, ovate-deltate, light brown, lateral veins 0. **FR** 0.6–0.8 mm, finely pitted. Ditches, shores; 0–200 m. SCo; native to e Afr, sometimes cult. [*C. alternifolius* L. misapplied]

C. laevigatus L. (p. 1141) Per 1–60 cm; rhizomes 1–2 mm thick, horizontal. **STS** clumped or not, 1–30 mm apart. **LF**: blades gen 0 (–7 cm). **INFL**: bracts 1–3, 1–10 cm; spikelets 1–14, 4–12 mm, elliptic to oblong-lanceolate, subcylindric, greenish white to reddish or dark brown, in open, ± ovoid heads; fl bracts 8–24 per spikelet, 1.5–2 mm, obovate to round, overlapped, lateral veins 0. **FL**: stigmas 2. **FR** 1.2–1.8 mm, oblong-elliptic to ovate, lenticular, dark brown or black. Alkaline or brackish, wet soils, hot springs, permanent pools in arroyos; 30–1000 m. SW, GB, D; to TX, Mex, scattered in trop, warm temp worldwide. ❀TRY;STBL.

C. ligularis L. Per 30–130 cm, cespitose; rhizome short, oblique. **ST** densely bumpy or warty. **LF**: blade stiff, margin harshly scabrous. **INFL**: bracts 3–12, 4–90 cm; rays 3–12, 10–160 mm; spikelets 3–7 mm, oblong-elliptic, subcylindric, reddish brown, in dense, cylindric spikes 10–35 mm, these in clusters of 3–7 at ray-ends; fl bracts 3–5 per spikelet, 2.5–3.3 mm, ovate, 9–11-veined, persistent. **FR** 1.5–1.6 mm, widely elliptic to obovoid, finely pitted. Uncommon. Wet soil, canyons, ditches near ocean; ± 0 m. w SnFrB (San Francisco), s SCo (San Diego); native to FL, Mex, Brazil, w Afr.

C. niger Ruiz López & Pavon (p. 1141) Per 1–50 cm; rhizomes short, 1 mm thick. **INFL**: bracts 2–3, 1–15 cm; rays 0–2, 0–5(40) mm; spikelets 3–25, 3–9 mm, linear-oblong, flat, black to (rarely light) brown, in gen dense, ovoid, head-like spikes 5–15 mm wide; fl bracts 3–9 per spikelet, 1.5–2.2 mm, ovate, light brown to black, lateral veins 0. **FL**: stigmas 2. **FR** 1.2–1.4 mm, elliptic to ovoid, lenticular, brown. Marshes, swamps, moist roadsides; 0–1500 m. NCoR, CaRF, SNF, n SNH, GV, CW, SCo, PR, SNE; to OK, S.Am. ❀TRY;STBL.

C. odoratus L. (p. 1141) Ann 10–50 cm. **INFL**: bracts 5–9, 5–24 cm; rays 6–12, 10–100 mm; spikelets linear, cylindric or slightly flat, in ± ovoid spikes; fl bracts 6–24 per spikelet, 2.1–3.5 mm, elliptic to ovate, light brown splotched reddish, midvein conspicuous. **FR** 1.5–1.9 mm, slightly flat front-to-back. Wet disturbed soils; 0–500 m. GV, SCo, D; trop and warm temp. [*C. ferax* Rich.]

C. parishii Britton (p. 1141) Ann 5–25 cm, cespitose. **INFL**: bracts 2–5, 3–20 cm; rays 1–6, 20–70 mm; spikelets 5–30, 6–22 mm, linear, flat, red, reddish purple, or reddish brown, in open, spheric spikes; fl bracts 8–10 per spikelet, gen ± 3 mm, elliptic, reddish, deciduous, 7–9-veined, midvein green, gen prominent. **FR** 1–1.3 mm, widely elliptic, (dark purplish) brown. Streambanks, roadsides; 0–800 m. SW, D; to Arkansas.

C. rotundus L. PURPLE NUTSEDGE Per 10–40 cm; stolons with tubers < 1 cm thick. **INFL**: bracts 2–5, 5–25 cm; rays 5–10, 10–150 mm; spikelets 3–12, 4–40 mm, linear to ± lanceolate, flat, reddish purple, in open, elliptic spikes 1–5 cm wide; fl bracts 6–36 per spikelet, 2–3.4 mm, ovate, 7–9-veined. **FR** 1.4–1.9 mm, elliptic, black, seldom maturing. Disturbed soils, croplands; 0–250 m. GV, SCo, DSon; trop and warm temp; native to Eurasia. Often considered world's worst weed.

C. squarrosus L. (p. 1141) Ann 1–10(20) cm. **INFL**: bracts 1–4, 1–15 cm; rays 0–6, 4–40 mm; spikelets 1–30, 2–20 mm, widely lanceolate to oblong, flat, greenish, yellowish, or reddish brown, in open, ovoid spikes 5–15 mm wide; fl bracts 2–8 per spikelet, 1.2–2 mm, oblong-lanceolate, 5–11-veined, bristle-like tip 0.5–1.3 mm. **FL**: stamen 1. **FR** 0.7–1.1 mm, ± obovoid, light brown to black. Moist, sunny, disturbed places, esp pond margins, riverbanks; 0–1500 m. CA; temp and trop ± worldwide. [*C. aristatus* Rottb.]

C. strigosus L. (p. 1141) FALSE NUTSEDGE Per 5–70 cm. **ST** corm-like at base. **INFL**: bracts 3–6, 5–30 cm; rays 0–7, 20–150 mm; spikelets 15–50, 6–30 mm, linear, subcylindric, in spikes 1–5 cm wide; fl bracts 3–11 per spikelet, 2.8–4 mm, lanceolate, yellowish, 7–11-veined. **FR** 1.8–2.4 mm, narrowly elliptic. Moist soils, pond margins, ditches, roadsides; 0–1000 m. CA; US, s Can.

DULICHIUM

1 sp. (Latin: for some sedge)

D. arundinaceum (L.) Britton (p. 1145) THREE-WAY SEDGE Pl < 10 dm, strongly rhizomed. **LVS** cauline, 2–8 cm, 4–8 mm wide, sheathing; uppermost 3-ranked. **INFL**: spikelets in axillary, raceme-like clusters that emerge from lf sheaths, gen 5–25 mm; fl bracts 2-ranked, 3–8 mm, strongly veined, tip acute or short-pointed, margin translucent. **FLS** bisexual; perianth bristles 5–8, ± 3–8 mm, persistent in fr, barbs reflexed; stamens 3; stigma 2-branched. **FR** linear-ellipsoid, ± flat, yellow; beak long. $2n=32$. Uncommon. Lake, pond margins, often in standing water; 1500–2400 m. KR, CaR, SN; to B.C.; also e N.Am. Known as fossil from Eur, e Asia. ❀IRR or WET,SUN:**1**,**2**3–5,**6**,**7**,**14**–**17**,18–24.

ELEOCHARIS SPIKERUSH

Ann or per. **ST** gen round, ridged and grooved, gen solid. **LVS** basal, 1(–4); base sheathing; blade gen 0. **INFL**: spikelet solitary, terminal, erect; spikelet bract 0; fl bracts gen spiraled. **FLS** bisexual; perianth bristles 0–6, persistent, barbs gen recurved; stamens (1–)3; style 2–3-branched, base bulb-like, persistent. **FR** 2–3-sided or round; top tubercled. ± 250 spp.: worldwide. (Greek: marsh grace) St shape best seen in fresh material (or just below spikelet); drying exaggerates ridges and grooves.

1. St 4-sided
 2. St gen < 1 mm wide, solid; spikelet much wider than st ..... [2]***E. pachycarpa***
 2′ St 3–5 mm wide, spongy; spikelet ± as wide as st ..... ***E. quadrangulata***
1′ St round (sometimes flattened in pressing)
 3. Fr brownish to black, shiny — tubercle hat-like, gen < 1/5 fr body
  4. Lf tip whitish, membranous ..... ***E. thermalis***
  4′ Lf tip green to reddish or becoming straw-colored, not membranous
   5. Fr ± 0.5 mm; tubercle base strongly narrowed ..... ***E. atropurpurea***
   5′ Fr ± 0.7–1 mm; tubercle base barely narrowed ..... ***E. geniculata***
 3. Fr whitish to brown, shiny or dull
  6. Style gen 2-branched; fr 2-sided
   7. Per, rhizomed; tubercle not flat, base narrowed; spikelet 5–25 mm, elongate, tip ± acute ..... ***E. macrostachya***
   7′ Ann, rhizome 0; tubercle flat, base not narrowed; spikelet 2–16 mm, oblong to ± round, tip ± obtuse ..... ***E. obtusa***
    8. Tubercle < 1/4 fr body; perianth bristles gen < fr ..... var. ***engelmannii***
    8′ Tubercle 1/3–1/2 fr body; perianth bristles > fr ..... var. ***obtusa***
  6′ Style (2)3-branched; fr (2)3-sided to round
   9. Fr longitudinal ridges gen > 3, crossbars many
    10. Ann, rhizome 0 ..... ***E. acicularis*** var. ***bella***
    10′ Per, rhizomed
     11. St solid, gen < 0.5 mm wide, not flattened in pressing; fl bract with (reddish) brown stripes ..... ***E. acicularis*** var. ***acicularis***
     11′ St spongy, gen > 0.5 mm wide, often flattened in pressing; fl bract without (reddish) brown stripes ..... ***E. radicans***
   9′ Fr longitudinal ridges 0–3, crossbars 0
    12. Spikelet 2–9-fld
     13. St gen 2–7 cm, not rhizomed; fr 1–1.3 mm, gen = perianth bristles; spikelet 2–3.5 mm ..... ***E. parvula***
     13′ St gen 10–40 cm, rhizomed; fr 2–3 mm, gen < perianth bristles; spikelet 4–7 mm ..... ***E. pauciflora***
    12′ Spikelet gen 10–many-fld
     14. Fl bracts weakly 2-ranked; fr tubercle base 3-lobed ..... [2]***E. pachycarpa***
     14′ Fl bracts spiraled; fr tubercle base ± round
      15. Fr ± 1.5 mm, tubercle flat; pl cespitose, rhizome short; spikelet 3–8 mm ..... ***E. bolanderi***
      15′ Fr ± 1–3 mm, tubercle not flat; pl loosely cespitose, rhizome long; spikelet 3–20 mm
       16. Fr tubercle base not narrowed; st tip often rooting to form new pl ..... ***E. rostellata***
       16′ Fr tubercle base slightly narrowed; st gen ascending, tip not rooting
        17. Spikelet oblong to ovate, ± as wide as st, tip obtuse to acute; perianth bristles 2–6, gen < fr ..... ***E. montevidensis***
        17′ Spikelet linear-lanceolate to fusiform, much wider than st, tip acute to acuminate; perianth bristles 6–7, gen > fr ..... ***E. parishii***

E. acicularis (L.) Roemer & Schultes Ann or per 2–15 cm. **LVS** several, gen spaced, needle-like (1 loosely sheathing, base reddish, tip abruptly expanded and scarious). **INFL**: spikelet 3–7 mm, wider than st, ± linear, 5–15-fld; fl bract straw-colored with (reddish)

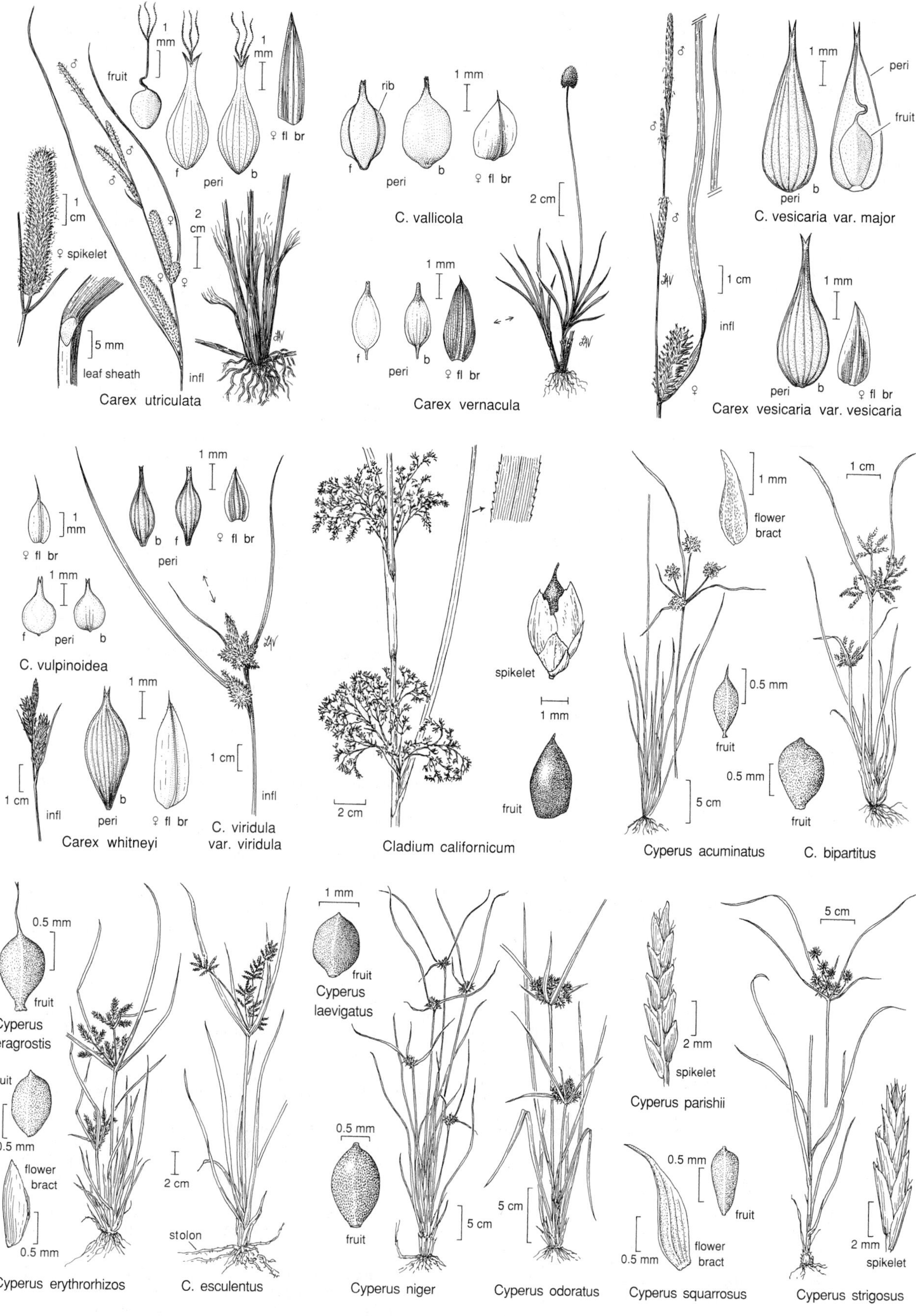
fruit
1 mm
1 mm
f
peri
b
♀ fl br
♂
1 cm
♀ spikelet
2 cm
5 mm
leaf sheath
infl
Carex utriculata
rib
1 mm
f
peri
b
♀ fl br
C. vallicola
2 cm
1 mm
f
peri
b
♀ fl br
Carex vernacula
1 cm
infl
1 mm
peri
fruit
peri
b
C. vesicaria var. major
1 mm
peri
b
♀ fl br
Carex vesicaria var. vesicaria
1 mm
♀ fl br
1 mm
b
f
peri
♀ fl br
1 mm
f
peri
b
C. vulpinoidea
1 mm
1 cm
infl
peri
b
♀ fl br
Carex whitneyi
1 cm
infl
C. viridula var. viridula
spikelet
1 mm
fruit
2 cm
Cladium californicum
1 mm
flower bract
0.5 mm
fruit
5 cm
Cyperus acuminatus
1 cm
0.5 mm
fruit
C. bipartitus
0.5 mm
fruit
Cyperus eragrostis
fruit
0.5 mm
flower bract
0.5 mm
Cyperus erythrorhizos
2 cm
stolon
C. esculentus
1 mm
fruit
Cyperus laevigatus
0.5 mm
fruit
5 cm
Cyperus niger
5 cm
Cyperus odoratus
2 mm
spikelet
Cyperus parishii
0.5 mm
fruit
0.5 mm
flower bract
Cyperus squarrosus
5 cm
2 mm
spikelet
Cyperus strigosus

brown stripes, tip ± obtuse. **FL**: style 3-branched. **FR**: perianth bristles 0 or 3; body 0.7–1 mm, elliptic to obovate, weakly 3-sided with several ridges and many crossbars, dull yellow-white; tubercle widely conic, flat, base narrowed. Marshes, meadows, riverbanks, vernal pools, in sagebrush scrub to lodgepole-pine forest; < 2500 m. CA-FP, MP; to se US, n Mex; circumboreal. Vars. intergrade in CA.

var. ***acicularis*** (p. 1145) Per, rhizomed. **FR**: crossbars 40–50 per row; perianth bristles gen 3, unequal. $2n$=20,30–38,50–58. Habitats and range of sp. (exc MP). ❀WET:1–3,**4–7**,8,9,**14–24**; also STBL.

var. ***bella*** Piper (p. 1145) Ann; rhizome 0. **FR**: crossbars ± 30–40 per row; perianth bristles gen 0 (or weak, appressed to fr). Marshes in sagebrush scrub to lodgepole-pine forest; 900–2500 m. NW, CaR, SN, MP; to WA, n-c US, n Mex. [*E. b.* (Piper) Svenson]

E. atropurpurea (Retz.) Kunth (p. 1145) Ann < 1.2 dm. **LF**: base deep brown; tip oblique, 1-toothed, green to reddish. **INFL**: spikelet 3–8 mm, much wider than st, oblong-ovate, gen 10–many-fld; fl bracts ovate-obtuse, greenish with dark brown sides, lower deciduous. **FL**: style 2-branched. **FR**: perianth bristles gen < fr; body ± 0.5 mm, obovate, sharply 2-sided, black, shiny; tubercle < 1/5 fr body, hat-like, base strongly narrowed. Rice fields, wet fallow ground; < 500 m. SCo, GV; to WA, c&se US; trop.

E. bolanderi A. Gray (p. 1145) Per 1–3 dm, short-rhizomed, ± glaucous. **LF**: base purplish, loosely sheathing upward, becoming straw-colored, spotted purple; tip truncate, often 1-toothed. **INFL**: spikelet 3–8 mm, wider than st, elliptic to ovate, 10–many-fld; fl bracts dark purplish brown, often with a short, scarious tips, lowest nearly round, margins wide, membranous. **FL**: style 3-branched. **FR**: perianth bristles 3–4, 1/2–3/4 fr; body ± 1.5 mm, obovate, weakly 3-sided, 3-ridged, yellowish brown; tubercle flat, tip with small point. Meadows, openings; 1000–2000 m. SN, Wrn; to ID. [*E. montevidensis* var. *b.* (A. Gray) V. Grant] ❀TRY.

E. geniculata (L.) Roemer & Schultes (p. 1145) Ann < 4 dm. **LF**: base brown, becoming straw-colored above; tip oblique, 1-toothed. **INFL**: spikelet 2–8 mm, much wider than st, ovate to ± round, 10–many-fld; fl bracts ovate-obtuse, greenish with dark brown sides, lower deciduous. **FL**: style 2-branched. **FR**: perianth bristles gen > fr; body ± 0.7–1 mm, obovate, strongly 2-sided, (purplish) black, shiny; tubercle < 1/5 fr body, wide, hat-like, base weakly narrowed. Wet soil, esp marshes, streambanks; < 1000 m. SW, D. [Wilson 1990 Cyper Newsl 7:6–7]

E. macrostachya Britton (p. 1145) Per 5–10 dm, rhizomed. **LF** loosely sheathing, base purplish, becoming straw-colored above; tip truncate, often 1-toothed. **INFL**: spikelet 5–25 mm, not much wider than st, gen 10–many-fld, elongate, tip acute; fl bract lanceolate-acute, brown to purplish. **FL**: style 2-branched. **FR**: perianth bristles unequal, < to > fr body; body ± 1.5–2.5 mm, obovate, strongly 2-sided, yellowish brown; tubercle ± conic, base narrowed. n=5,8,16,18,19,21,23. Marshes, pond margins, vernal pools, ditches; < 2500 m. CA; temp, montane w hemisphere. [*E. palustris* (L.) Roemer & Schultes, in part] Complex, polyploid, needs study. ❀TRY;also STBL.

E. montevidensis Kunth (p. 1145) Per 1–5 dm; rhizome long, reddish. **ST** not glaucous. **LF** purplish brown, becoming straw-colored above; tip truncate, 1-toothed. **INFL**: spikelet 3–8 mm, ± as wide as st, oblong to ovate, gen 10–many-fld, tip obtuse to acute; fl bract brownish to yellowish, margin translucent, tip obtuse. **FL**: style 3-branched. **FR**: perianth bristles 2–6, gen < fr; body ± 1 mm, obovate, weakly 3-sided, yellowish brown, shiny; tubercle short, conic, base slightly narrowed. Moist, often sandy openings; < 1000 m. NW, CW, SW; to TX, S.Am. [var. *decumbens* (C.B. Clarke) V. Grant] Reportedly intergrades with *E. parishii*. ❀IRR or WET:4–6,7,8–10,**14–24**;GRNCVR; also STBL.

E. obtusa (Willdenow) Schultes Ann < 5 dm. **ST** spongy, flattened in pressing. **LF** purplish below, becoming straw-colored above; tip oblique, entire. **INFL**: spikelet 2–16 mm, often wider than st, oblong to ± round, tip obtuse, gen 10–many-fld; fl bract ovate, brownish, tip obtuse to acute. **FL**: style 2-branched. **FR**: perianth bristles < to > fr; body ± 1–1.5 mm, ± obovoid, 2-sided, (pale) brown, shiny; tubercle flat, base not narrowed. $2n$=10. Wet soil of pond and lake margins, marshes, often yellow-pine forests; < 2600 m. NW, SN, MP; to B.C., ID. Also e N.Am, with other vars.

var. ***engelmannii*** (Steudel) Gilly (p. 1145) **INFL**: spikelet 5–16 mm, oblong to ovate. **FR**: perianth bristles < to = fr; tubercle < 1/4 fr body. Habitats of sp. s NW (Marin, Lake cos.), SN, n MP; to WA, e N.Am.

var. ***obtusa*** (p. 1145) **INFL**: spikelet 2–13 mm, ± round, tip ± obtuse. **FR**: perianth bristles > fr; tubercle 1/3–1/2 fr body. Habitats of sp.; < 1000 m. NW; to B.C., ID; also e N.Am.

E. pachycarpa Desv. (p. 1145) Per 1–4 dm, short-rhizomed. **ST** gen < 1 mm wide, 4-sided or round, tip sometimes arched and rooting to form new pl. **LF** (purplish) brown to straw-colored; tip rigid, acute, 1-toothed. **INFL**: spikelet 5–10 mm, wider than st, ovate, laterally compressed, gen 10–many-fld; fl bracts weakly 2-ranked, purplish brown to blackish, tips obtuse to acute. **FL**: style 3-branched. **FR**: perianth bristles 0–6, = fr; body ± 1.2–1.5 mm, elliptic to ± round, weakly 3-sided, yellow-white; tubercle deltate, strongly 3-lobed, base not narrowed. Saltmarshes, vernal pools; < 2000 m. n NCoRO (Humboldt Co.), SN (Amador, El Dorado cos.); NV; native to Chile.

E. parishii Britton (p. 1145) Per 1–3 dm, not glaucous; rhizome long, reddish. **LF** purplish brown, becoming straw-colored above; tip truncate, 1-toothed. **INFL**: spikelet 10–15 mm, much wider than st, ± linear-lanceolate, gen 10–many-fld, tip acute to acuminate; fl bract ± dark brown, tip short, translucent. **FL**: style 3-branched. **FR**: perianth bristles 6–7, unequal, = or > fr; body ± 1 mm, obovate, appearing 2-sided, yellowish to light brown, shiny; tubercle short, conic, base barely narrowed. n=5. Moist, often sandy openings; gen < 2000 m. CA (exc MP); to OR, NV, NM, n Mex. [*E. montevidensis* vars. *disciformis* (Parish) V. Grant and *p.* (Britton) V. Grant] ❀TRY;GRNCVR;also STBL.

E. parvula (Roemer & Schultes) Link (p. 1145) SMALL SPIKERUSH Per < 1 dm; rhizome 0. **ST** often spongy when fresh, flattened in pressing. **LF** membranous. **INFL**: spikelet 2–3.5 cm, wider than st, ovate, 2–9-fld; fl bract ovate, greenish to brownish, tip obtuse to acute. **FL**: style 3-branched. **FR**: perianth bristles gen = fr; body 1–1.3 mm, obovate, strongly 3-sided, straw-colored, shiny; tubercle base not narrowed. $2n$=8,10. UNCOMMON. Wet, gen saline flats, marshes; < 2500 m. NCo, GB, D; circumboreal. [var. *coloradoensis* (Britton) Beetle] Pls from interior N.Am sometimes separated.

E. pauciflora (Light.) Link (p. 1145) Per 1–4 dm, rhizomed. **LF** straw-colored to brownish; tip truncate. **INFL**: spikelet 4–7 mm, gen wider than st, ovate, 2–7-fld; fl bract ovate, dark brown to blackish, margin scarious, tip obtuse to acute. **FL**: style 3-branched. **FR**: perianth bristles gen > fr, barbs irregular; body 2–3 mm, obovate to fusiform, weakly 2–3-sided, yellow-brown, weakly netted; tubercle base not narrowed, tip dark. Uncommon. Meadows; 1500–3700 m. NW, CaR, SN, SnBr, SnJt, MP; circumboreal; also s Chile. [vars. *bernardina* (Munz & I.M. Johnston) Svenson and *suksdorfiana* (Beauverd) Svenson]

E. quadrangulata (Michaux) Roemer & Schultes (p. 1145) FOUR-ANGLED SPIKERUSH Per 5–10 dm, from caudex. **ST** 3–5 mm wide, sharply 4-sided, spongy, flattened in pressing. **LF**: tip rarely with short, blade-like structure. **INFL**: spikelet 15–35 mm, ± as wide as st, oblong, gen 10–many-fld; fl bract widely ovate, straw-colored, tip blunt. **FL**: style 3-branched. **FR**: perianth bristles gen < fr, barbs minute; body 2.7–4.2 mm, obovate, dark brown, shiny, ± netted; tubercle ± deltate, base narrowed. RARE in CA. Marshes and lake, pond margins; < 500 m. GV (Butte, Merced cos.); e N.Am. Formerly thought alien in CA.

E. radicans (Poiret) Kunth (p. 1145) Per < 8 dm, often floating, rhizomed. **ST** gen > 0.5 mm wide, spongy, often flattened in pressing. **LVS** several, gen spaced, needle-like (1 sheathing, ± appressed, base greenish, tip not expanded, not scarious). **INFL**: spikelet 3–4 mm, wider than st, linear to ovate, 4–12-fld; fl bract greenish to straw-colored, tip ± obtuse. **FL**: style 3-branched. **FR**: perianth bristles (0) gen > fr; body 0.7–0.8 mm, elliptic to obovate, weakly 3-sided, several-ridged, crossbars 30–40 per row, dull yellow-white; tubercle conic, ± flat, base narrowed. Marshes, lake edges; < 1500 m. NW, SnJV, CW, SW; s U.S. to S.Am. [*E. acicularis* var. *r.* (Poiret) Britton misapplied]

E. rostellata (Torrey) Torrey (p. 1145) Per 2–15 dm, rhizomed. **ST** spongy, flattened in pressing; tip often arched, rooting to form new pl. **LF** rigid, straw-colored to greenish; tip truncate to oblique.

INFL: spikelet 8–20 mm, wider than st, ovate, gen 10–many-fld; fl bracts light brown, often purple-spotted, lowest rounded, upper acute. **FL**: style 3-branched. **FR**: perianth bristles ± = fr; body 2–3 mm, obovate, weakly 2–3-sided, olive, shiny, weakly netted; tubercle deltate, not flat, base not narrowed. Alkaline marshes, sinks, springs; < 2000 m. SnFrB, SW, SNE, D; to Caribbean, Mex, also w S.Am. Indicator of saline, calcareous soils. ❀TRY;GRNCVR; also STBL.

E. thermalis Rydb. (p. 1145) Per < 4 dm, rhizomed. **LF**: base greenish; tip whitish, membranous. **INFL**: spikelet 2–6 mm, much wider than st, oblong to ± round, gen 10–many-fld; fl bract ovate, greenish with purplish brown sides, tip rounded. **FL**: style 2-branched. **FR**: perianth bristles > fr, whitish; body ± 1 mm, obovate, 2-sided, brownish black, shiny, minutely dotted; tubercle < 1/5 fr body, hat-like, base much narrowed. Stream margins, marshes; < 1000 m. SNF, GV; to WY, UT. [*E. flavescens* (Poiret) Urban misapplied] Formerly thought alien in CA. ❀TRY; also STBL.

ERIOPHORUM COTTON GRASS

Per. **ST** leafy, solid. **LVS** sheathing; uppermost often bladeless. **INFL** subtended by 1–several, lf- or scale-like bracts; spikelets terminal, solitary, or in head-like clusters; fl bracts spiraled. **FLS** bisexual; perianth bristles 6–many, >> fl bracts, persistent, often greatly elongating; stamens 3; style 3-branched. **FR** ± flat or 3-sided. ± 20 spp.: esp n temp (1 s Afr). (Greek: wool-bearing)

1. St 3-sided; perianth bristles 6–7, barbed, tawny ..... ***E. criniger***
1′ St rounded; perianth bristles many, not barbed, white ..... ***E. gracile***

E. criniger (A. Gray) Beetle (p. 1145) **ST** 2–10 dm, erect, 3-sided. **LVS** 4–10 cm, flat; basal present at fl. **INFL**: spikelets 5–10. **FL**: perianth bristles 6–7, barbed, tawny. Wet meadows, streambanks; gen > 2000. n&c NW, CaR, SNH; s OR. [*Scirpus c.* A. Gray] ❀WET or IRR,DRN:**1**,**2**,3,**4**–**7**,14,**15**–**17**.

E. gracile Koch (p. 1145) **ST** 30–60 cm, rounded, from spreading rhizome. **LVS** 2–30 cm, 3-sided; basal 0 at fl. **INFL**: spikelets 2–5. **FL**: perianth bristles many, not barbed, white. $2n$=30,38. Wet meadows, bogs; gen > 2000 m. NW, CaR, n&c SNH; circumboreal. Formerly in SnFrB. ❀TRY.

FIMBRISTYLIS

Ann or per, cespitose. **ST** round, slender, solid, ridged and grooved. **LVS** several, basal; bases sheathing, margins wide, scarious; blades 0 to linear, flat, folded, or upcurled. **INFL**: spikelets in terminal, panicle-like clusters; fl bracts spiraled, deciduous; **FLS** bisexual; perianth bristles 0; stamens gen 3; style 2–3-branched, gen fringed below branches, widest at base, deciduous. **FR** 2–3-sided or round, surface cells ± square or horizontally elongate. ± 200 spp.: worldwide. (Latin: fringed style) [Kral 1971 Sida 4:57–227]

1. Spikelet nearly round, obtuse; style 3-branched ..... ***F. miliacea***
1′ Spikelet elongate, acute; style 2-branched
 2. Peduncles gen long; fl bracts densely hairy; per ..... ***F. thermalis***
 2′ Spikelet subsessile; fl bracts glabrous; ann ..... ***F. vahlii***

F. miliacea (L.) Vahl Ann < 5 dm. **ST** = or > lvs, ± flat. **LVS** 2-ranked. **INFL**: spikelet long-peduncled, nearly round, obtuse; fl bracts glabrous, fruiting bracts widely ovate, gen dark brown with paler or greenish midribs, tips obtuse to notched. **FL**: style 3-branched, entire to weakly fringed below branches. **FR** ± 1 mm, obovate, ± 3-sided, 3-ridged, pale brown. $2n$=10. Rice fields; > 300 m. GV; native to Old World trop, widespread alien. Collected in SnFrB in 1866.

F. thermalis S. Watson (p. 1145) HOT-SPRINGS FIMBRISTYLIS Per < 15 dm, rhizomed. **ST** > lf, round below, flat above. **LVS** spiraled; blades linear, flat or ± upcurled. **INFL**: spikelet gen long-peduncled, elongate, tip acute; fl bracts densely hairy, fruiting bracts narrowly ovate, pale brown, midrib greenish, exserted, tip acute. **FL**: style 2-branched. **FR** ± 1.5 mm, obovate, 2-sided, dark brown, shiny. $2n$=20. RARE in CA. Wet, alkaline soils, gen near hot springs; > 500 m. DMoj; s NV, AZ. [*F. spadicea* (L.) M. Vahl misapplied] Like *F. caroliniana* (Lambert) Fern. of se US.

F. vahlii (Lambert) Link (p. 1145) Ann < 1 dm. **ST** > lf, round. **LVS** spiraled; blade ± thread-like, upcurled. **INFL**: spikelets in tight clusters, subsessile, elongate, tips acute; fl bracts glabrous, fruiting bracts lanceolate, straw-colored to pale green, midrib dark green, exserted, tip acute. **FL**: style 2-branched. **FR** ± 0.6 mm, obovoid to round, pale brown. $2n$=20. Wet soil, often mud flats, silty levees; < 500 m. SnJV, e DSon; to se US, C.Am. Perhaps alien.

KOBRESIA

Per. **LVS** basal, sheathing. **INFL**: spikelets in terminal clusters, spiraled, subtended by spikelet bract, gen 1–2-fld, lower fls pistillate, upper staminate. **FL**: perianth bristles 0. **PISTILLATE FL**: perigynium open along 1 side, fused at base or not at all; style 3-branched. **FR** 3-sided; tubercles 0. ± 35 spp.: n temp. (J.P. von Kobres, Austrian botanist, 1747–1823)

K. bellardii (All.) Degl. (p. 1145) Pl 1–5 dm, cespitose, slender, wiry. **LF** < st, thread-like. **INFL**: spikelet clusters 15–25 mm, gen dense. **FR** ± 2 mm, 1 mm wide. $2n$=52. RARE in CA. Rocky seeps; > 3000 m. c SNH (Convict Basin, Mono Co.); circumboreal. [*K. myosuroides* (Villars) Fiori & Paol.]

KYLLINGA

Gordon C. Tucker

Per, glabrous. **ST** erect, 2–100 cm. **LVS** basal, linear. **INFL**: bracts 2–4; spikelets 50–100+, 1–2-fld, slender-elliptic, breaking apart just above 2 basal fl bracts at maturity, in 1–3, sessile, dense, ovoid, light green spikes; fl bracts 2-ranked, 4–5 per spikelet, 2 at base << others, 3rd subtending a bisexual fl, 4th (and rarely 5th) ± = 3rd but sterile or subtending staminate fl. **FL**: perianth 0; stamens 1–3; stigmas 2. **FR** lenticular, ± smooth, brown. (P. Kylling, Danish botanist, died 1696) [Tucker 1984 Rhodora 86:507–538]

K. brevifolia Rottb. (p. 1149) Pl rhizomed. **STS** 1–6, 2–55 cm. **INFL**: longest bract erect; spikelets 2.2–3.2 mm, oblong-lanceolate, in 1–3 spikes 4–7 mm. **FR** 1–1.3 mm, ± elliptic, ± stalked. Lawns, ditches, croplands; 0–300 m. GV, SW, D; trop and warm temp; native to trop Am.

LIPOCARPHA

Gordon C. Tucker

Ann, glabrous. **ST** ± erect, 1–20 cm. **LVS** basal, 1–3. **INFL**: bracts 1–3, lf-like; spikelets 50–150, dense, spiralled, sessile, in dense, spheric to cylindric spikes; fl bracts spirally arranged, (0)1–2 per fl, 100–400 per spikelet, outer > inner, awned at tip, with 1 central green and 2 lateral whitish veins, margins reddish, inner bract translucent. **FL** bisexual; perianth 0; stamens 1–3, anthers 0.2–0.3 mm; styles 2-branched. **FR** 3-angled to subcylindric, abruptly soft-pointed, bumpy or warty, brown. (Greek: half chaff, from translucent inner fl bract) [Friedland 1941 Amer J Bot 28:855–861]

1. Awn of outer fl bract (0.5)1–1.5 mm, outcurved; fr light brown .... ***L. occidentalis***
1′ Awn of outer fl bract < 1 mm, ± straight to outcurved; fr light (reddish) brown to black
  2. Inner fl bract ± = fr, awn of outer fl bract 0.5–1 mm; fr light reddish brown to black .... ***L. aristulata***
  2′ Inner fl bract 0 or < 1/2 fr, awn of outer fl bract < 0.2 mm; fr light brown .... ***L. micrantha***

L. aristulata (Cov.) G.C. Tucker (p. 1149) **ST** 1–15 cm. **INFL**: spikes 1–3, 3–7 mm, gen ± open, oblong to gen cylindric; outer fl bract oblanceolate, body 0.8–1 mm, awn 0.5–1 mm; inner fl bract ± = fr. **FR** 0.7–0.9 mm, widest just below tip, bumpy or warty, light reddish brown to black; faces gen flat to concave. Wet soil; 100–400 m. NCoRO, ScV; to WA, se US. [*Hemicarpha micrantha* (Vahl) Pax var. *a.* Cov.]

L. micrantha (M. Vahl) G.C. Tucker (p. 1149) **ST** 1–10 cm. **INFL**: spikes 1–3, 3–6 mm, dense, cylindric; outer fl bract widely lanceolate, body 0.8–1 mm, awn 0.1–0.2 mm; inner fl bract 0 or < 1/2 fr. **FR** 0.7–0.8 mm, widest near middle, finely pitted, light brown; faces convex. Wet soil; 0–1500 m. NCoRO, SNF, GV; US, s Can, trop Am. [*Hemicarpha m.* (M. Vahl) Pax. incl var. *minor* (Schrader) Friedl.]

L. occidentalis (A. Gray) G.C. Tucker (p. 1149) **ST** 1–7 cm. **INFL**: spikes 1–2, 2–5 mm, dense, gen ± spheric; outer fl bract lanceolate, body 0.5–0.8 mm, awn (0.5)1–1.5 mm; inner fl bract ± = fr. **FR** 0.5–0.6 mm, widest at or just below tip, finely pitted, light brown; faces convex. Wet soil; 1200–1900 m. NCoRO, SN, PR; to WA. [*Hemicarpha o.* A. Gray]

RHYNCHOSPORA BEAKED-RUSH

Per. **ST** rounded or 3-sided, leafy, solid. **LVS** sheathing. **INFL** gen subtended by 1+ awn-like bracts; spikelets 1–many in gen terminal clusters; fl bracts spiraled, lower gen sterile, upper gen staminate, others bisexual. **FL**: perianth bristles 5–many, persistent; stamens 3; style 2-branched, base much expanded, persistent as tubercle on fr. **FR** 2-sided. ± 200 spp.: worldwide, esp trop. (Greek: beaked seed) [Gale 1944 Rhodora 46:89–134;159–197;207–249;255–278]

1. Perianth bristle barbs reflexed
  2. Fl bracts tawny to whitish; perianth bristles 10–12 .... ***R. alba***
  2′ Fl bracts ± dark brown; perianth bristles 5–7 .... ***R. capitellata***
1′ Perianth bristle barbs ascending
  3. Perianth bristles = or > fr; fr body obovoid, tubercle chalky-white .... ***R. californica***
  3′ Perianth bristles < 1/2 fr; fr body ovoid to ± spheric, tubercle not chalky-white .... ***R. globularis***

R. alba (L.) M. Vahl (p. 1149) WHITE BEAKED-RUSH **INFL**: fl bracts tawny to whitish. **FL**: perianth bristles 10–12, = or > fr, barbs reflexed. **FR** 1.6–2 mm; body ovoid, pale, faintly ridged. 2*n*=26,42. UNCOMMON. Marshes, bogs; < 2000 m. c&s NW (Mendocino, Sonoma cos.), se CaRH (Lassen Co.); circumboreal. ❀IRR or WET,SUN,DRN:**1**,**2**,3,**4**–**6**,7,14,**15**–**17**.

R. californica Gale (p. 1149) CALIFORNIA BEAKED-RUSH **INFL**: fl bracts reddish to dark brown. **FL**: perianth bristles 6–7, = or > fr, barbs ascending. **FR** ± 2 mm; body obovoid, dark, strongly ridged; tubercle chalky- white. RARE. Marshes, bogs, seeps; < 1000 m. s NW (Sonoma Co.), n&c SNF (Butte, Mariposa? cos.), n SnFrB. Mariposa Co. pls not recently collected, may be undescribed.

R. capitellata (Michaux) M. Vahl **INFL**: fl bracts ± dark brown. **FL**: perianth bristles 5–7, = or > fr, barbs reflexed. **FR** 1.3–1.8 mm, body obovoid, reddish brown, smooth. Marshes, seeps; < 2000 m. NW (Sonoma, Trinity cos.), n SNF (Butte Co.); also e N.Am. [*R. glomerata* (L.) M. Vahl vars. *c.* (Michaux) Kük. and *minor* Britton] ❀TRY.

R. globularis (Chapman) Small var. ***globularis*** (p. 1149) ROUND-HEADED BEAKED-RUSH **INFL**: fl bracts tawny to whitish. **FL**: perianth bristles 5–6, < 1/2 fr, barbs ascending. **FR** 1.2–1.6 mm, body ovoid to ± spheric, dark, strongly ridged; tubercle not chalky- white. RARE in CA. Marshes, seeps; < 500 m. s NCoRO (Pitkin Marsh, Sonoma Co.); also e N.Am. Remarkably separated from e N.Am range.

SCHOENUS

Per. **LVS** basal, erect, stiff, sheathing. **INFL** subtended by 2 unequal, lf-like bracts; spikelet bracts 1–2, sheathing; spikelets in terminal head-like clusters, flat; fl bracts 2-ranked, only uppermost fruiting. **FLS** bisexual and staminate; perianth bristles 3–6, gen < 1 mm, deciduous; stamens 3; style 3-branched. **FR** obovoid; tubercle gen 0. ± 80 spp.: worldwide, esp Australia. (Greek: a rush)

S. nigricans L. (p. 1149) BLACK SEDGE Pl 2–7 dm, densely cespitose; rhizomes 0. **LF** > 1/2 st, ± 1 mm wide, entire to minutely serrate, uprolled; ligule wide, very dark brown. **FR** 3-sided, with a ± white, bony covering. 2*n*=44,54. Uncommon. Marshes, swamps, springs, gen alkaline soils; < 2000 m. SnBr, DMoj; to Caribbean, Eurasia, n Afr. Used as roof thatch in Ireland. ❀TRY.

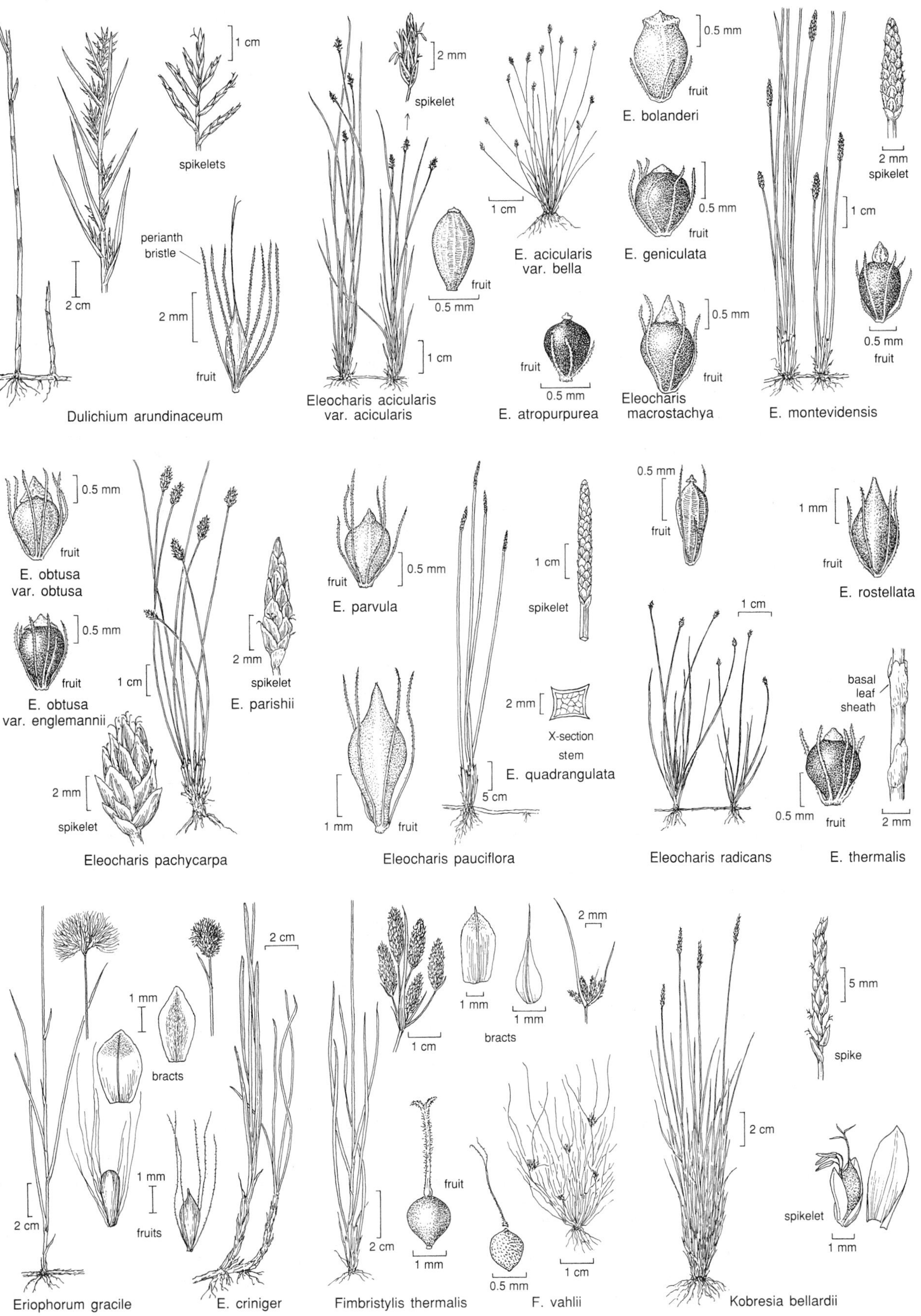
1 cm
spikelets
2 cm
perianth bristle
2 mm
fruit
Dulichium arundinaceum
2 mm
spikelet
fruit
0.5 mm
1 cm
Eleocharis acicularis var. acicularis
1 cm
E. acicularis var. bella
fruit
0.5 mm
E. atropurpurea
0.5 mm
fruit
E. bolanderi
0.5 mm
fruit
E. geniculata
0.5 mm
fruit
Eleocharis macrostachya
2 mm
spikelet
1 cm
0.5 mm
fruit
E. montevidensis
0.5 mm
fruit
E. obtusa var. obtusa
0.5 mm
fruit
E. obtusa var. englemannii
1 cm
2 mm
spikelet
2 mm
spikelet
E. parishii
Eleocharis pachycarpa
fruit
0.5 mm
E. parvula
1 mm
fruit
1 cm
spikelet
2 mm
X-section stem
E. quadrangulata
5 cm
Eleocharis pauciflora
0.5 mm
fruit
1 cm
Eleocharis radicans
1 mm
fruit
E. rostellata
basal leaf sheath
0.5 mm
fruit
2 mm
E. thermalis
1 mm
bracts
2 cm
1 mm
fruits
Eriophorum gracile
2 cm
E. criniger
1 cm
1 mm
1 mm
bracts
2 mm
fruit
2 cm
1 mm
Fimbristylis thermalis
0.5 mm
1 cm
F. vahlii
5 mm
spike
2 cm
spikelet
1 mm
Kobresia bellardii

SCIRPUS

S. Galen Smith, A.E. Schuyler, & William J. Crins

Ann or per, rhizomed or not; roots fibrous. **ST** gen erect, 3-angled or cylindric, solid. **LVS** basal or cauline, alternate, 3-ranked; sheaths closed; ligule 0 or present; blades linear, sometimes vestigial and scale-like. **INFL** panicle- or head-like; bracts 1–several, lf- or ± st-like; spikelets 1–many, gen many-fld; fl bracts spiraled, gen ovate, scale-like. **FL** bisexual; perianth bristles gen < and hidden by fl bracts, gen ± straight, gen slender, gen stiff, gen persistent on fr, gen finely spined or fringed, sometimes 0 or vestigial; stamens gen 3; style 1, ribbon- or thread-like, stigmas 2–3, gen exserted. **FR**: achene, gen obovoid, lenticular or ± 2–3-angled, gen beaked, not tubercled. ± 200 spp.: gen wet sites, worldwide. Some spp. mistaken for *Eleocharis*.

1. Spikelets gen 1(3), ± sessile; st and lf blade << 1.5 mm wide
2. Pl submersed exc infl, lf tip gen floating; spikelet 6–13 mm .... ***S. subterminalis***
2′ Pl terrestrial or emersed, lf tip not floating; spikelet 2–7 mm
3. Largest spikelets gen < 6-fld; fr smooth; SNH, > 2400 m
4. Rhizome 0; fr sharply 3-angled; infl bract > spikelet, tip with awn gen > 1 mm .... ***S. clementis***
4′ Rhizome slender; fr compressed, 2–3-angled; infl bract < spikelet, tip abruptly pointed .... ***S. pumilus***
3′ Largest spikelets gen > 8-fld (exc some *S. cernuus*); fr gen papillate, pitted, or ribbed; w of SNH, < 1500 m
5. Fl bract strongly keeled throughout, ± folded in fr; fr sides concave to ± flat, fine-papillate .... ***S. koilolepis***
5′ Fl bract keeled above middle, ± flat in fr; fr sides convex, ribbed or fine-papillate to -pitted
6. Fr sharply 3-angled, sides fine-papillate or -pitted; rhizome 0; spikelet 1 .... ***S. cernuus***
6′ Fr obscurely 3-angled, sides ribbed; rhizome thread-like; spikelets 1–3 .... ***S. setaceous***
1′ Spikelets gen 2–many, sessile or stalked; st and lf blade (0.6)2–15 mm wide (st gen < 1 mm wide in *S. nevadensis* and *S. saximontanus*)
7. Ann; rhizome 0; st gen < 1 mm wide; fr sharply transverse-ribbed .... ***S. saximontanus***
7′ Per; rhizome short to long; st (0.6)2–15 mm wide; fr smooth (low-ridged in *S. mucronatus*)
8. Well developed infl bracts gen 2+, thin, lf-like, gen spreading; lvs ± cauline, blades gen > sheaths, thin
9. Spikelet 1–3.5 mm wide; fr ± 0.5–2 mm; fl bract glabrous, tip acute to abruptly pointed; tubers 0
10. Spikelets not clearly clustered, ± stalked .... ***S. pendulus***
10′ Spikelets mostly in head-like clusters, sessile
11. Fr gen 2-angled; stigmas 2; perianth bristles straight in fr, fine-toothed throughout .... ***S. microcarpus***
11′ Fr 3-angled; stigmas gen 3; perianth bristles ± contorted to straight in fr, fine-toothed above middle
12. Fl bract dark brown to blackish in fr; perianth bristles < 4 mm, teeth sparse, ascending ...... ***S. congdonii***
12′ Fl bract brown in fr; perianth bristles gen < 2.5 mm, teeth dense, reflexed .... ***S. diffusus***
9′ Spikelet 4–10 mm wide; fr 2–4.5 mm; fl bract scabrous or sparse-puberulent, tip awned; tubers gen present
13. Perianth bristles persistent on fr; fr sharply 3-angled, sides ± concave; longest infl bract gen 4–15 mm wide when dry; lf blade gen 6–20 mm wide when dry; stigmas 3 .... ***S. fluviatilis***
13′ Perianth bristles gen not persistent on fr; fr ± weakly 2–3-angled, sides convex; longest infl bract gen 2–6 mm wide when dry; lf blade gen 2–12 mm wide when dry; stigmas 2–3
14. Spikelet 4–5 mm wide; fr 2–2.5(3) mm; stigmas 3 .... ***S. tuberosus***
14′ Spikelet 5–10 mm wide; fr (2.5)3–3.5 mm; stigmas 2–3
15. Fl bract not strongly appressed, ± pale grayish to orange-brown; spikelets often in 1 dense cluster; lf sheath top opposite blade with a triangular, veinless area .... ***S. maritimus***
15′ Fl bract strongly appressed, ± orange-brown; spikelets gen on open, elongate branches; lf sheath top opposite blade clearly veined .... ***S. robustus***
8′ Well developed infl bract gen 1, thick, ± st-like at least above, erect or bent to side; lvs ± basal; blades 0 or gen < sheaths (exc *S. nevadensis, S. pungens*), thin to st-like
16. St 3-angled, at least above
17. Infl panicle-like, dense to open, spikelets 20–many; perianth bristles on fr thick, flat, fringed with soft hairs; st often cylindric below, < 4 m .... ***S. californicus***
17′ Infl head-like, or rarely with 1–2 short branches, spikelets 1–12(20); perianth bristles on fr slender, minutely spiny; st 3-angled throughout, gen < 2.2 m
18. Lf blade vestigial or 0; infl bract gen bent to side; fl bract tip abruptly pointed, notch 0; fr low-ridged; rhizome short, obscured by crowded sts .... ***S. mucronatus***
18′ Lf blade < to > sheath; infl bract gen erect; fl bract tip awned from notch; fr smooth; rhizome long, not obscured
19. Fl bract tip with awn < 0.5 mm; uppermost lf blade gen < sheath; st sides deep-concave .... ***S. americanus***
19′ Fl bract tip with awn 0.5–1.5 mm; uppermost lf blade > sheath; st sides ± flat to shallow-concave .... ***S. pungens***
16′ St cylindric throughout
20. Infl head-like; spikelets 1–10; fl bract tip acute to obtuse, notch 0 .... ***S. nevadensis***
20′ Infl panicle-like; spikelets 3–100+; fl bract tip short-awned, notched (tip sometimes worn off in age)
21. Spikelets solitary at branch tips; fr 3-angled, 2.5–3 mm, beak 0.3–1 mm; stigmas 3 .... ***S. heterochaetus***
21′ Spikelets many, mostly clustered at branch tips; fr ± 2–3-angled, ± 2–2.5 mm, beak 0.1–0.4 mm; stigmas 2–3
22. Fl bract gen prominently spotted, awn gen contorted, 0.5–1 mm; spikelets gen 1–7 per cluster .... ***S. acutus*** var. ***occidentalis***
22′ Fl bract faintly spotted or not, awn ± straight, gen < 0.5 mm; spikelets gen solitary or 2–3 per cluster .... ***S. tabernaemontani***

S. acutus Bigelow var. ***occidentalis*** (S. Watson) Beetle (p. 1149) TULE Per 150–400 cm; rhizome long. **STS** erect, 4–12 mm wide in middle, cylindric. **LVS** ± basal; sheaths prominent; blades 1–2, << sheaths, flat, glabrous. **INFL** panicle-like; branches stiff to flexible, erect to arched; spikelets 3–many, in clusters of 1–7 at branch tips, 7–10(24) mm, 3–4.5 mm wide; main bract 1–4(11) cm, gen erect, stiff, st-like at least above; fl bract ± 4 mm, finely prickly to papillate above middle, ± colorless to straw-colored, or orange to dark red-brown above, gen prominently spotted, margin woolly, tip notched, awn 0.5–1 mm, gen contorted. **FL**: perianth bristles gen 6, < or = fr. **FR** 2–2.5 mm, obtusely 2–3-angled, smooth, gray-brown; beak 0.1–0.3 mm. Marshes, lakes, streambanks; < 2500 m. CA (exc e D); temp N.Am. Hybridizes with *S. californicus, S. tabernaemontani*; some pls in mtns with < 40 spikelets per infl, stigmas 2, fr 2-angled, perianth bristles ± = fr may be var. *acutus* (gen e N.Am). ❀STBL.

S. americanus Pers. (p. 1149) Per 50–220 cm; rhizome long. **STS** erect to arched, 3–7 mm wide in middle, 3-angled; sides deeply concave. **LVS** ± basal, < 1/2 st; blades 1–2, gen < sheaths, flat, folded below, subcylindric above, glabrous. **INFL** head-like; spikelets 2–12(20), 5–11 mm, 4–5 mm wide; bract 1, 1–4 cm, gen erect, stiff, less often bent to side, ± st-like; fl bract 3–3.5 mm, glabrous, orange- to red-brown, margin short-woolly, tip notch < 0.5 mm, awn < 0.5 mm, ± straight. **FL**: perianth bristles 2–7, 1/2 to > fr. **FR** ± 2–3 mm, obtusely 2-angled, or back low-ridged, smooth, dark brown. *n*=39. Marshes, ponds; < 2000 m. CA (exc SNH); to e N.Am, S.Am. [*S. olneyi* A. Gray misapplied] Hybridizes with *S. pungens*. ❀STBL.

S. californicus (C. Meyer) Steudel (p. 1149) Per 200–400 cm; rhizome long. **STS** erect, < 10 mm wide in middle, 3-angled throughout or cylindric below. **LVS** ± basal; sheaths < 1/6 st; blades 0 or 1–2, < sheaths, flat, glabrous. **INFL** panicle-like; spikelets 20–many, clustered at branch tips, 5–11 mm, ± 3 mm wide; branches flexible, arched; main bract 3–8 cm, erect, stiff, ± st-like (at least above); fl bract 2.5–3 mm, glabrous, orange-brown, midrib thick, margin woolly, tip notch short, awn < 0.5 mm, gen bent. **FL**: perianth bristles 2–4, ± = fr, flat, thick, fringed; stigmas 2. **FR** ± 2 mm, 2-angled, smooth, dark brown; beak ± 0.2 mm. Marshes; < 200 m. NCo, GV, CCo, SnFrB, SCo, e D (Colorado River); to s US, S.Am, oceanic islands. ❀STBL.

S. cernuus Vahl (p. 1149) Ann 4–40 cm. **STS** erect to strongly arched, dense-tufted, < 0.5 mm wide, subcylindric, grooved when dry. **LVS** ± basal; sheaths < 3 cm; upper blades 3–8 cm, flat, glabrous. **INFL**: spikelet 1, 3–5 mm, ± 2 mm wide; bract 1, 0.2–0.6(2) cm, erect, stiff, st-like; fl bract ± 1.5 mm, glabrous, greenish to pale or dark orange- to red-brown, keeled near abruptly pointed tip. **FL**: perianth bristles 0; stigmas (2)3. **FR** ± 1 mm, sharply 3-angled, brown; sides strongly convex, finely papillate or pitted in vertical rows; beak minute. Moist, gen sandy soils; < 800 m. NCo, deltaic GV, CCo, SnFrB, SCo, SnBr; to B.C., S.Am., Eurasia. [var. *californicus* (Torrey) Beetle] ❀IRR or WET:**4–7**,8,9,**14–24**.

S. clementis M.E. Jones (p. 1149) YOSEMITE BULRUSH Per, 3–12 cm, cespitose; rhizome 0. **STS** erect, tufted, gen 1–2 mm wide, subcylindric, grooved. **LVS** ± basal; sheaths short, not prominent, lower sheaths brown; blades >> sheaths, ± flat, thick, glabrous. **INFL**: spikelet 1, 3–7 mm, 1–3 mm wide, 2–5-fld; bract 1, 4–8 mm, tip erect, stiff, ± st-like, awn 1–6 mm; fl bract 2–3 mm, glabrous, orange-brown, midrib thick, tip rounded to abruptly pointed. **FL**: perianth bristles 0–6, < fr, flat, gen fragile, ± deciduous; stigmas 3. **FR** ± 1–1.5 mm, sharply 3-angled, ± smooth. UNCOMMON. Dry to wet meadows, streambanks; 2400–3600 m. c&s SNH.

S. congdonii Britton (p. 1149) Per 30–50 cm; rhizome ± short. **STS** erect, solitary or ± clustered, 2–4 mm wide in middle, ± 3-angled. **LVS** basal and cauline; sheaths prominent; blades gen < or = sheaths, flat, glabrous, midrib and margins scabrous. **INFL** panicle-like; spikelets 20–many, gen in head-like clusters of 5–15 at branch tips, 2.5–6 mm, 1–2.5 mm wide; longest bract 1.5–4 cm, ascending to erect, lf-like; fl bract 1.5–3 mm, glabrous, dark brown to blackish, midrib pale, tip acuminate to short-pointed. **FL**: perianth bristles 6, < 4 mm, contorted, fragile; teeth above middle sparse, ascending. **FR** ± 2 mm, 3-angled, smooth. Meadows, marshes, streambanks; 600–3000 m. KR, NCoR, CaR, MP; s OR. ❀IRR,DRN,SUN:1–7,14–24;DFCLT.

S. diffusus Schuyler (p. 1149) Per 40–100 cm; rhizome ± short. **STS** erect, solitary or clustered, 2–5 mm wide in middle, ± 3-angled. **LVS** basal and cauline; sheaths prominent; blades gen < or = sheaths, flat, glabrous, midrib and margin scabrous. **INFL** panicle-like; spikelets 25–many, gen in head-like clusters of 3–12 at branch tips, 3–6 mm, 1.5–2.5 mm wide; longest bract 2–5 cm, ascending to erect, lf-like; fl bract ± 1.5–3 mm, glabrous, brown, midrib pale, tip acuminate to short-pointed. **FL**: perianth bristles 6, gen < 2.5 mm, ± contorted to straight, teeth above middle dense, reflexed. **FR** 1.5–2 mm, 3-angled, smooth. Meadows, marshes, streambanks; 300–2300 m. KR, s CaR, SN. ❀TRY.

S. fluviatilis (Torrey) A. Gray (p. 1149) RIVER BULRUSH Per ± 100–200 cm; rhizome long with tubers < 3 cm wide. **STS** erect, 5–15 mm wide, sharply 3-angled, smooth or angles finely scabrous above. **LVS** cauline; sheaths clearly veined at top; blades >> sheaths, gen 6–20 mm wide when dry, flat, glabrous, midrib and margins finely scabrous. **INFL** panicle-like (sometimes head-like); spikelets ± 8–50, 15–28 mm, 6–10 mm wide, most gen in clusters of 1–4(8) at branch tips; 2–3 bracts >> infl, spreading, lf-like, longest gen 4–15 mm wide when dry; fl bract 7–10 mm, not strongly appressed, scabrous, light brown, tip notch < 1 mm, awn 2–3 mm, ± bent. **FL**: perianth bristles gen 6, 1/2 to > fr; stigmas 3. **FR** ± 3.5–4.5 mm, strongly 3-angled, not compressed, smooth, pale brown, ± shiny; sides ± concave; tip rounded, beak 0.3–0.8 mm. 2*n*=94. Marshes; < 1300 m. NCoR, n SNH, ScV, SnFrB, MP; temp N.Am, Asia. ❀STBL.

S. heterochaetus A. Chase (p. 1149) SLENDER BULRUSH Per 150–200 cm; rhizome long. **STS** erect, 4–5 mm wide in middle, cylindric. **LVS** ± basal; sheaths prominent, fragile, often lf-like above; blades 1–2, < or = sheaths, flat, glabrous. **INFL** panicle-like; branches flexible, arched; spikelets 10–30, 1(2) per branch tip, 8–10(15) mm, 3–4 mm wide; main bract 2.5–11 cm, erect, stiff, ± st-like at least above; fl bract 3–4 mm, straw to light orange-brown, midrib scabrous above middle, margin ± ciliate, tip notched, awn 0.5–1 mm, gen contorted. **FL**: perianth bristles 4(5), unequal, some << fr; stigmas 3. **FR** ± 2.5–3 mm, obtusely 3-angled, ± compressed, smooth, brown; beak 0.3–1 mm. *n*=19. RARE in CA. Lake margin; 1600 m. s CaRH (Wilson Lake); to WA, e N.Am.

S. koilolepis (Steudel) Gleason Ann 1–20 cm. **STS** gen erect, tufted, < 0.5 mm wide, subcylindric. **LVS** ± basal; sheaths < 1 cm; upper 2–3 blades < 5 cm, C-shaped or flat, thick near tip. **INFL** head-like; spikelets 1–3, 2–6 mm, 1–2 mm wide; bract 1, (5)10–25 mm, erect, stiff, ± st-like; fl bract 2–3 mm, strongly keeled throughout, ± folded in fr, glabrous, greenish to orange-brown, tip acute to short-awned. **FL**: perianth bristles 0; stigmas 3. **FR** ± 1–1.5 mm, sharply 3-angled, shiny, dark orange-brown; sides concave to flat, papillate in vertical rows; beak minute. Moist, open sites, woodlands; < 750 m. NCo, NCoRO, CCo, SnFrB; to e US.

S. maritimus L. (p. 1149) Per 80–150 cm; rhizome long with tubers < 2 cm wide. **STS** erect, ± 3–8 mm wide, sharply 3-angled; sides smooth, angles minute-scabrous. **LVS** ± cauline; sheath top opposite blade with V-shaped veinless area; upper blades >> sheaths, gen 3–12 mm wide when dry, flat to V-shaped, midrib and margin fine-scabrous. **INFL**: spikelets 4–many, 10–30 mm, (5)7–8 mm wide, often in 1 sessile, dense cluster or < 1/2 in clusters of 1–15 at branch tips; 2–3 bracts >> infl, spreading, lf-like, longest gen 2–5 mm wide when dry; fl bract not strongly appressed, 5–7 mm, pale grayish to orange-brown, ± translucent, scabrous, tip notch ± 0.5 mm, awn 1.5–3 mm, ± contorted. **FL**: perianth bristles gen 6, gen < 1/2 fr, gen deciduous. **FR** 2.5–3.5 mm, ± compressed, ± weakly 2–3-angled, smooth, shiny, dark brown; sides convex; tip truncate, beak 0.1–0.2 mm. 2*n*=90. Marshes; < 2500 m. NW, GV, CW, SW, MP, D; almost worldwide. [var. *paludosus* (Nelson) Kük.] Apparently hybridizes with *S. robustus*. ❀STBL.

S. microcarpus C. Presl (p. 1149) Per 70–160 cm; rhizome long. **STS** erect, clumped or not, 5–7 mm wide, ± 3-angled. **LVS** basal and cauline; sheaths prominent; blades < to = sheaths, flat, glabrous, margin scabrous. **INFL** ± panicle-like; spikelets 50+, gen in head-like clusters of 4–12 at branch tips, 2–8 mm, 1–3.5 mm wide; longest bract 1.2–4 cm, ascending to erect, lf-like; fl bract 1–3 mm, glabrous, dark brown, tip ± abruptly pointed. **FL**: perianth bristles gen 4, < to > fr, fine-toothed throughout; stigmas gen 2. **FR** ± 0.5–1.5 mm, gen 2-angled, smooth. Marshes, wet meadows, streambanks, pond margins; < 2900 m. CA (exc D); to AK, e US, Asia. ❀IRR or WET:1,**2–10,14–24**;also STBL;INV.

S. mucronatus L. Per < 80 cm, cespitose; rhizome short, obscured by sts. **STS** erect, crowded, 2–4 mm wide, sharply 3-angled. **LVS** ± basal; sheaths prominent; blades vestigial or 0. **INFL** head-like; spikelets 1–10(25), 6–12 mm, 4–5 mm wide; bract 2.5–13 cm, gen bent to side, stiff, st-like; fl bract 3–3.5 mm, glabrous, brown, gen with wide, green stripe along midrib, tip abruptly pointed. **FL**: perianth bristles 6, ± = fr; stigmas 3. **FR** ± 2 mm, ± compressed, obtusely 3-angled, with many, low, transverse ridges. Wet places, rice-fields; < 150 m. NCoRO, ScV, SnFrB, SCo; to c&e US; native to Eurasia.

S. nevadensis S. Watson (p. 1149) Per 10–45 cm; rhizomes long, some vertical. **STS** erect to arched, well spaced to clumped, < or = 1 mm wide, cylindric, stiff, wiry. **LVS** basal and cauline; lower sheaths coarsely fibrous; upper blades >> sheaths, ± strongly curved, C-shaped to ± cylindric, wiry. **INFL** head-like; spikelets 1–10, 10–20 mm, ± 5 mm wide; bract gen 1, 1–10 cm, erect to spreading, stiff, ± st-like; fl bract ± 4 mm, ± glabrous, brown, tip acute to obtuse. **FL**: perianth bristles 1–4, << to 1/2 fr; stigmas 2. **FR** ± 2.5 mm, 2-angled, shiny, brown, pitted; tip ± rounded, beak 0. Wet places, saline soils; 1000–2500 m. GB; to B.C., Great Plains, S.Am. ❀STBL.

S. pendulus Muhl. Per 50–100 cm; rhizome short to long, thick. **STS** erect, 1–few, tufted, 0.6–10 mm wide, 3-angled. **LVS** basal and cauline; sheaths prominent; blades < or = sheaths, flat, glabrous, margin scabrous. **INFL** panicle-like; spikelets 50+, ± stalked, 5–10 mm, 2–3 mm wide, cylindric; longest bract 2–5(8) cm, ascending to ± erect, lf-like; fl bract ± 1.5–2 mm, ± glabrous, brown, tip acuminate to abruptly pointed. **FL**: perianth bristles gen 6, ± unequal, ± 2 × fr, contorted; stigmas gen 3. **FR** 1–1.5 mm, 3-angled, smooth. *n*=20. Uncommon. Marshes, wet meadows, ditches; 800–1000 m. n CaRH (near Yreka); s OR, Can, e US, Mex.

S. pumilus Vahl DWARF BULRUSH Per 3–17 cm; rhizome long, **STS** erect, tufted, gen < 1.5 mm wide, subcylindric, ± grooved. **LVS** ± basal; sheaths short, not prominent; blades >> sheaths, ± flat, thick, glabrous. **INFL**: spikelet 1, 3–5 mm, 1–3 mm wide, 3–6-fld; bract 1, 1.5–3 mm, erect, stiff, ± st-like, tip abruptly pointed; fl bract 2–3 mm, glabrous, orange-brown, midrib thick, margin translucent, tip rounded to abruptly pointed. **FL**: perianth bristles 0; stigmas 3. **FR** ± 1.5–2 mm, compressed, 2–3-angled, smooth. RARE in CA. Wet sites, limestone soils; 3100–3250 m. c SNH (Convict Creek, Mono Co.); to Can, Colorado, Eurasia. [*S. rollandii* Fern., Rolland's bulrush, misapplied]

S. pungens Vahl COMMON THREESQUARE Per 20–200 cm; rhizome long. **STS** erect to strongly arched, 2–6 mm wide in middle, sharply 3-angled; sides flat to ± concave. **LVS** ± basal, ± 1/2 to = st; blades 2–6, gen > sheaths, ± flat to subcylindric below, subcylindric above, glabrous or minutely scabrous near tip. **INFL** head-like; spikelets 1–5(10), 7–13 mm, 4–5 mm wide; main bract 3–11(20) cm, gen erect, ± st-like, next bract 5–12 mm; fl bract 3.5–5 mm, glabrous, orange- to dark red-brown, margin short-woolly, tip notch ± 0.5–1 mm, awn 0.5–1.5 mm, ± straight or tip contorted. **FL**: perianth bristles gen 2–7, gen 1/2 to = fr; stigmas 3. **FR** ± 2–3.5 mm, obtusely 3-angled, ± compressed, smooth, dark brown. Marshes, lake shores; < 2000 m. NCo, SN, SnJV, CW, SCo, GB, DMoj; N.Am, S.Am, Eur, s Pacific. [*S. americanus* var. *longispicatus* Britton & ssp. *monophyllus* (C. Presl) T. Koyama] ❀STBL.

S. robustus Pursh (p. 1149) Per 50–150 cm; rhizome long with 1–2 cm wide tubers. **STS** erect, 4–6 mm wide, sharply 3-angled; sides smooth, angles papillate, scabrous. **LVS** ± cauline; sheaths clearly veined at top; blades >> sheaths, gen 4–12 mm wide when dry, flat to V-shaped, glabrous, midrib and margin finely scabrous. **INFL** panicle- or head-like; spikelets ± 6–30, 15–30 mm, (6)8–10 mm wide, gen < 1/2 to > 1/2 in clusters of 1–3 at branch tips; 2–3 bracts >> infl, spreading, lf-like, longest gen 2–6 mm wide when dry; fl bract strongly appressed, ± 6 mm, ± orange-brown, ± opaque, scabrous, tip notch ± 0.5 mm, awn ± 2–3 mm, ± contorted. **FL**: perianth bristles gen 6, < to = fr, gen deciduous. **FR** ± 3–3.5 mm, weakly 2–3-angled, smooth, dark brown; sides convex; tip truncate, beak 0.1–0.2 mm. Marshes; < 800 m. KR, SnFrB, SCo; to Mex, S.Am. Hybridizes often with *S. maritimus*. ❀STBL.

S. saximontanus Fern. Ann 5–30 cm. **STS** erect to arched, tufted, gen < 1 mm wide, subcylindric, ridged when dry. **LVS**: cauline on lower 1/2 st; sheaths prominent; lower blades << sheaths, upper > sheaths, flat, thick, glabrous. **INFL** ± head-like, sometimes short-branched; spikelets 3–12, 8–15 mm, 2.5–3 mm wide; main bract 3–12 cm, erect, stiff, ± st-like; fl bract ± 3 mm, glabrous, midrib green, thick, sides pale orange, translucent, margins minutely ciliate, tip acute, abruptly short-pointed to -awned. **FL**: perianth bristles 0; stigmas 3. **FR** ± 1.5 mm, sharply 3-angled, with many sharp, transverse ridges. Pond, lake margins; < 300 m. GV, WTR; to c US, Mex.

S. setaceous L. (p. 1149) Per (may fl 1st year), 3–20 cm; rhizome thread-like. **STS** erect to strongly arched, tufted, < 0.5 mm wide, subcylindric, grooved when dry. **LVS** ± basal; sheaths < 2 cm, gen reddish at base; upper blades < 6 cm, flat, thick, glabrous. **INFL** head-like; spikelets 1–3, 3–6 mm, ± 2 mm wide; bract 1, 3–10 mm, erect, ± stiff, ± st-like; fl bract ± 1.5 mm, glabrous, keeled near abruptly pointed tip, orange-to red-brown or blackish. **FL**: perianth bristles 0; stigmas (2)3. **FR** ± 1 mm, obscurely 3-angled, brown; sides convex, ribbed. Wet places, streambanks, pond margins; < 1500 m. NCo, n SN; to B.C.; native to Eurasia, Afr, Australia.

S. subterminalis Torrey (p. 1149) WATER BULRUSH Per 20–140 cm, submersed exc infls; lf tips gen floating; rhizome long, delicate. **STS** gen erect, < 1 mm wide, cylindric. **LVS** basal and cauline; sheaths < 1/2 st; blades >> sheaths, weak, slender. **INFL**: spikelet 1, 6–13 mm, 4–7 mm wide; bract 1, 1–6 cm, erect, stiff, ± st-like; fl bract 4–6 mm, glabrous, green, pale brown in age, tip abruptly pointed. **FL**: perianth bristles gen 6, < to = fr, ± straight to contorted. **FR** ± 2.5–4 mm, sharply 3-angled, smooth. RARE in CA. Lakes, ponds, marshes; 750–2250 m. KR, n SNH; to AK, e N.Am.

S. tabernaemontani C. Gmelin (p. 1149) Per 100–200(300) cm; rhizome long. **STS** erect, densely clustered, 4–8 mm wide in middle, cylindric. **LVS** ± basal; sheaths prominent; blades 1–2, << sheaths, flat, glabrous. **INFL** panicle-like; branches flexible, arched; spikelets 12–125, solitary or in clusters of 2–4 at branch tips, 5–10(13) mm, 3–4 mm wide; main bract 2–5 cm, gen erect, stiff, ± st-like (at least above); fl bract 2.5–3 mm, orange-brown, faintly or not spotted, midrib minutely spiny-papillate, margin woolly, tip ± rounded, notched, awn gen < 0.5 mm, ± straight. **FL**: perianth bristles 6, = fr; stigmas gen 2. **FR** ± 2–2.5 mm, ± obtusely 2-angled, smooth, brown; beak < 0.2 mm. *n*=21. Marshes, lakes, streambanks; < 200 m. NW, SnJV, SnFrB, SCo; N.Am, S.Am, Eurasia, South Pacific. [*S. validus* Vahl] ❀TRY.

S. tuberosus Desf. Per < 50 cm; rhizome long with 1–2 cm wide tubers. **STS** erect, gen densely clumped, ± 1.5–3 mm wide, sharply 3-angled. **LVS** ± cauline; sheath top opposite blade V-shaped, veinless; upper blades >> sheaths, gen 2–5 mm wide when dry, midrib and margin finely scabrous. **INFL** panicle- or head-like; spikelets ± 3–50, 10–25(40) mm, 4–5 mm wide, gen many in clusters at branch tips; 2–3 bracts >> infl, lf-like, gen < 3 mm wide when dry; fl bract ± 5 mm, not strongly appressed, sparsely scabrous, ± colorless to orange-brown, tip notch ± 0.5 mm, awn ± 1–1.5 mm, irregularly bent. **FL**: perianth bristles gen 6, < 1/2 to = fr, gen deciduous; stigmas 3. **FR** 2–3 mm, weakly 2–3-angled, smooth, brown; sides convex; tip truncate, beak 0.1–0.2 mm. Ditches, marshes, rice-fields; < 150 m. GV, SnFrB; to OR, e N.Am; native to Eur. [*S. maritimus* var. *t.* (Desf.) Roemer & Schultes] Cult for waterfowl food.

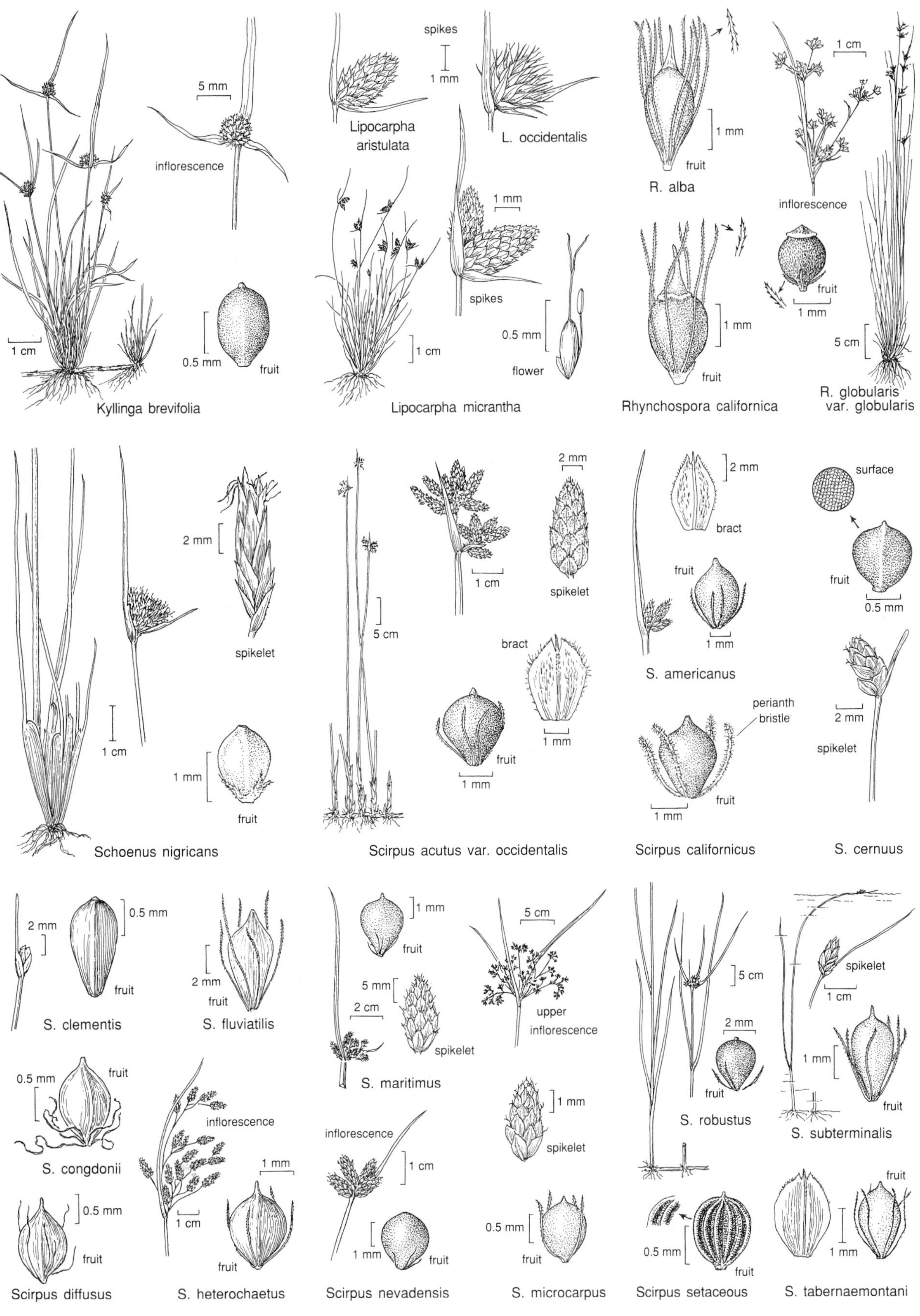

5 mm
inflorescence
1 cm
0.5 mm
fruit
Kyllinga brevifolia
spikes
1 mm
Lipocarpha aristulata
L. occidentalis
1 mm
spikes
1 cm
0.5 mm
flower
Lipocarpha micrantha
1 mm
fruit
R. alba
1 cm
inflorescence
1 mm
fruit
1 mm
fruit
Rhynchospora californica
5 cm
R. globularis var. globularis
2 mm
spikelet
1 cm
1 mm
fruit
Schoenus nigricans
2 mm
spikelet
1 cm
5 cm
bract
1 mm
fruit
1 mm
Scirpus acutus var. occidentalis
2 mm
bract
fruit
1 mm
S. americanus
perianth bristle
1 mm
fruit
Scirpus californicus
surface
fruit
0.5 mm
2 mm
spikelet
S. cernuus
2 mm
0.5 mm
fruit
S. clementis
2 mm
fruit
S. fluviatilis
1 mm
fruit
5 mm
2 cm
spikelet
S. maritimus
5 cm
upper inflorescence
5 cm
2 mm
fruit
S. robustus
spikelet
1 cm
1 mm
fruit
S. subterminalis
0.5 mm
fruit
S. congdonii
inflorescence
1 cm
1 mm
fruit
S. heterochaetus
0.5 mm
fruit
Scirpus diffusus
inflorescence
1 cm
1 mm
fruit
Scirpus nevadensis
1 mm
spikelet
0.5 mm
fruit
S. microcarpus
0.5 mm
fruit
Scirpus setaceous
fruit
1 mm
S. tabernaemontani

HYDROCHARITACEAE WATERWEED FAMILY

Robert F. Thorne

Ann, per, aquatic, freshwater or marine, bisexual, monoecious, or dioecious. **LVS** basal or cauline, alternate, opposite, or whorled, gen ± sheathing at base, glabrous. **INFL**: cyme; fls 1–few, subtended by ± sheathing, entire or lobed bract; staminate fls sometimes deciduous, free-floating. **FL** radial; perianth tube 0 or much elongated, peduncle-like in fl; sepals (0)3, green; petals (0)3, colored or white; stamens (1)3–many, gen in 1+ series; ovary inferior, chamber 1, placentas parietal or basal, ovules 1–many; style lobes gen 3, linear, lobed or notched. **FR**: achene or berry-like and dehiscing irregularly, linear to spheric, submersed. ± 17 genera, ± 130 spp.: worldwide; some cult for aquaria, others noxious weeds.

1. Lf petioled, blade ovate to reniform; fl bisexual, emergent .... **OTTELIA**
1' Lf sessile, blade thread-like to oblong; fls unisexual or bisexual, at water surface or submersed — staminate fls sometimes deciduous, free-floating
  2. Fl sessile, 1–few per axil; perianth 0; fr achene with net-like surface .... **NAJAS**
  2' Fl with perianth tube peduncle-like, sepals green, petals colored or white; fr berry-like and dehiscing irregularly, surface not net-like
    3. Lvs 2–4 cm, 2–5 mm wide, gen 3–6 per node; staminate fls 2–3, petals 8–10 mm .... **EGERIA**
    3' Lvs < 1.5(2) cm, < 3 mm wide, opposite or 2–3 per node; staminate fl 1, petals gen < 5 mm
      4. Lf margin minutely and finely serrate, midrib on lower surface ± smooth .... **ELODEA**
      4' Lf margin clearly and sharply serrate, midrib on lower surface ± rough .... **HYDRILLA**

EGERIA BRAZILIAN WATERWEED

Per, rooted in mud, submersed or floating, dioecious (CA pls staminate). **ST** slender, sometimes branched or not. **LVS** opposite below, crowded and whorled above, 3–6 per whorl, sessile. **STAMINATE INFL**: fls 2–3, clustered, peduncled; bract slender. **PISTILLATE INFL**: fl 1, axillary; bract slender, ± tubular. **STAMINATE FL** floating; perianth tube elongated, slender, peduncle-like, persistent on pl; stamens 9. **PISTILLATE FL** floating; perianth tube elongated, slender, peduncle-like, persistent on pl; stigmas 3, 3–4-lobed. 2 spp.: S.Am. (Latin: a mythical water nymph)

E. densa Planchon **ST** 2–3 mm thick. **LF** 2–4 cm, 2–5 mm wide, narrowly oblong; margin finely serrate. **STAMINATE FL**: perianth tube < 6 cm; petals 8–10 mm, white, ± round. **PISTILLATE FL**: perianth tube < 6 cm; petals 6–7 mm, white, ± round. $2n$=48. Streams, ponds, sloughs; < 2200 m. n&s SNF, SnJV, SnFrB, SnJt, SNE (expected elsewhere); to e US, Eur; native to S.Am. Staminate pls widely cult for aquaria. [*Elodea d.* (Planchon) Caspary]

ELODEA

Per, rooting at nodes, gen submersed, dioecious or some fls bisexual. **ST** slender, gen branched. **LVS** opposite or whorled, gen 3 per whorl, sessile; blade margin minutely, finely toothed. **INFL**: fls solitary, axillary, sessile or peduncled; bract gen notched. **STAMINATE FL** floating, deciduous or not; perianth tube elongated, slender, peduncle-like; stamens 3–9. **PISTILLATE FL** floating; perianth tube elongated, slender, peduncle-like; style slender, stigmas 3, simple or 2-lobed. **FR** cylindric to ovoid. **SEEDS** several. ± 12 spp.: temp and trop Am. (Greek: of marshes) [St. John, 1965 Rhodora 67:1–35,155–180]

1. Lf 1–4 mm wide, tip obtuse or abruptly pointed; staminate fl not deciduous, not free-floating; staminate sepals 3.5–5 mm, petals ± 5 mm; pistillate sepals ± 2–3 mm .... ***E. canadensis***
1' Lf 0.3–2 mm wide, tapered to acute tip; staminate fl deciduous, free-floating; staminate sepals ± 2 mm, petals < 1 mm or 0; pistillate sepals ± 1 mm .... ***E. nuttallii***

E. canadensis Rich. (p. 1161) COMMON WATERWEED **LVS** gen crowded at st tips; middle and upper 3 per whorl; blade gen 9–15 mm, 1.5–3 mm wide, linear, tip obtuse or abruptly pointed. **STAMINATE INFL**: bract ovoid to elliptic, swollen; fl not deciduous, not free-floating. **STAMINATE FL**: sepals 3.5–5 mm, ± 4 mm wide; petals ± 5 mm. **PISTILLATE FL**: sepals ± 2–3 mm. **SEED** glabrous. $2n$=24,48. Shallow water, ditches, sloughs, ponds, lakes; 300–2600 m. NCoRO, CaRH, SNH, GV, SnFrB, SnGb, SnBr, GB; to B.C., e US; naturalized in Eur.

E. nuttallii (Planchon) H. St. John (p. 1161) NUTTALL'S WATERWEED **LVS** gen not crowded at st tips; middle and upper 3(4) per whorl; blade gen 6–13 mm, 0.3–1.5(2) mm wide, linear to narrowly lanceolate, tapered to acute tip. **STAMINATE INFL**: bract ovoid, abruptly pointed; fl deciduous, free-floating. **STAMINATE FL**: sepals ± 2 mm, ± 1.5 mm wide; petals gen 0 or ± 0.5 mm. **PISTILLATE FL**: sepals ± 1 mm. **SEED** short-soft-hairy. $2n$=48. Shallow water, streams, lakes, ponds, ditches; 500–2800 m. KR, SNH, ScV, SnBr, GB; to WA, c US.

HYDRILLA

1 sp. (Greek: growing in water)

H. verticillata (L.f.) Caspary (p. 1161) Per, rooting at nodes, submersed, dioecious. **STS**: horizontal, tuber-like; erect sts elongate, much-branched. **LVS** whorled, gen 4–8 per whorl, sessile, 1–2 cm, 1.5–2 mm wide, oblong, margin clearly, sharply toothed, mid-rib ± keeled below, rough, with tooth-like, conic bumps, tip gen sharp-pointed. **STAMINATE INFL**: fl 1, bract at base of elongated perianth tube, spheric. **PISTILLATE INFL**: fl 1, bract at base of elongated perianth tube, ± cylindric. **STAMINATE FL** deciduous, free-floating; perianth parts 6 in 2 whorls, 3–5 mm; stamens 3. **PISTILLATE FL** persistent, floating; perianth parts 6 in 2 whorls, 3–5 mm; stigmas 3. **FR** 5–6 mm, ± fusiform. Ditches, canals, ponds, reservoirs, lakes; < 200 m. NCoRI, n&c SNF, ScV, SCo, D; to s US, C.Am.; native to Eurasia. NOXIOUS WEED.

NAJAS

Ann, aquatic, submersed, sometimes mat-like, monoecious or dioecious. **STS** several, much-branched, slender. **LVS** simple, cauline, opposite or appearing ± whorled; sheath gen wider than blade, expanded abruptly at junction with blade; blade gen linear, margin entire to spiny. **INFL** axillary; fls 1–few, clustered, inconspicuous. **STAMINATE FL**: perianth 0; stamen 1, anther opening irregularly, subsessile, subtended by 2 minute involucres, inner membranous, flask-shaped, outer cup-like, tubular, or with free scales. **PISTILLATE FL**: perianth part 0–1; ovary 1, chamber 1, ovule 1, style short, stigmas 2–4, linear. **FR**: achene, fusiform; outer wall thin, ± translucent. ± 50 spp.: ± worldwide. (Greek: water nymph) [Shaffer-Fehre 1991 Bot J Linn Soc 107:189–209]

1. Lf coarsely spine-toothed, stiff, 1–3 mm wide, bright green; internodes often minutely spiny; fr gen << 5 mm, 2–3 mm wide; dioecious ..... ***N. marina***
1′ Lf entire to minutely toothed (at 20× magnification), flexible, 0.1–2 mm wide, light to olive-green or reddish; internodes not spiny; fr gen < 3 mm, < 2 mm wide; monoecious
  2. Lf sheath appendages obvious, rounded to acuminate
    3. Lf thread-like, not recurved; fr linear-oblong, often slightly curved, net-like surface with angular spaces longer than wide ..... ***N. gracillima***
    3′ Lf linear, often recurved; fr oblong or elliptic in outline, gen straight, net-like surface with angular spaces as long as wide ..... ***N. graminea***
  2′ Lf sheath junction with blade abrupt, top truncate or rounded, not appendaged (following reliably distinguished only by fr)
    4. Fr surface smooth, shiny (finely net-like at 20 × magnification); anthers 1-chambered; stigmas gen 3 .. ***N. flexilis***
    4′ Fr surface dull, clearly net-like, pitted; anthers 4-chambered; stigmas 2–3 ..... ***N. guadalupensis***

N. flexilis (Willd.) Rostkov & Schmidt (p. 1161) SLENDER WATER-NYMPH Monoecious. **ST** < 20 dm, much-branched. **LVS** 1–2.5 cm, 1–2 mm wide, crowded on distal branches, flexible; sheath obtuse at junction with blade; blade margin minutely toothed, tip narrowly long-tapered. **STAMINATE FL** 2.5–3 mm; anther 1-chambered. **PISTILLATE FL** ± 3 mm; stigmas gen 3, sometimes 2–4. **FR**: surface smooth, shiny (finely net-like at 20× magnification). Uncommon. Ponds, lakes; < 1500 m. KR, NCoRI (expected elsewhere); to Can, ne US, n Eur.

N. gracillima (A. Braun) Magnus THREAD-LEAVED WATER-NYMPH Monoecious. **ST** slender, delicate. **LF** ± 3 cm, 0.1–0.3(0.5) mm wide, thread-like, not recurved; sheath appendaged, appendages 2, acuminate, margins fringed or minutely spined. **FR** ± 3 mm, ± 1 mm wide, linear-oblong, often slightly curved; surface net-like with angular spaces longer than wide. RARE in CA. Rivers; ± 100 m. n ScV, expected elsewhere; native to ne US.

N. graminea Del. RICE-FIELD WATER-NYMPH Monoecious. **ST** 0.5–6 dm; lateral branches short. **LVS** 1–2.5 cm, 0.5–1 mm wide, tufted on lateral branches; tip recurved; sheath appendaged, appendages 2, ± = 1/2 sheath length, acuminate, minutely spine-tipped; blade margin entire, tip with 1–several minute spines. **STAMINATE FL** < 2 mm; anther 4-chambered. **PISTILLATE FL** 1.5–3 mm; stigmas 2–3. **FR**: surface dull, clearly net-like. Very uncommon. Irrigation ditches, rice fields; < 150 m. ScV (Butte, Colusa cos.); native to trop Asia.

N. guadalupensis (Sprengel) Magnus (p. 1161) COMMON WATER-NYMPH Monoecious. **ST** < 6 dm, gen much-branched. **LVS** 1–2.5 cm, 0.5–1 mm wide, gen ± evenly spaced, flexible; sheath rounded to obtuse at junction with blade, few-toothed; blade margin minutely few-toothed, tip acute. **STAMINATE FL** 2–3 mm; anther 4-chambered. **PISTILLATE FL** 2–3 mm; stigmas 2–3. **FR**: surface dull, clearly net-like, pitted. $2n$=12,36,42,48,54,60. Ponds, lakes, irrigation ditches, reservoirs; < 1200 m. NCoRO, GV, CW, SCo, D; to OR, e US, S.Am.

N. marina L. (p. 1161) HOLLY-LEAVED WATER-NYMPH Dioecious. **ST** < 4 dm, branched from base; internodes often minutely spiny. **LVS** 1–4 cm, 1–3 mm wide, ± evenly spaced, stiff; sheath rounded at junction with blade, entire or minutely few-toothed; blade margin and lower surface coarsely spine-toothed. **STAMINATE FL** 3–4 mm; anther 4-chambered. **PISTILLATE FL** 3–4 mm; stigmas 3. **FR**: surface smooth (finely net-like at 20× magnification). Ponds, lakes, marshes, rivers; < 1000 m. NCoR, s SNF, CCo, SCoR, SCo, SnBr, PR, D; to n-c US, Baja CA, Eurasia, Pacific Islands.

OTTELIA

Per, submersed to emergent, rooted in mud. **STS** gen short, erect. **LVS** gen ± basal, petioled; submersed blades gen strap-like, floating or emergent ones wide. **INFL** terminal; fl solitary, persistent. **FL** bisexual, fragrant; sepals 3; petals 3; stamens 3, staminodes 6+; ovary oblong, stigmas 6–9, linear, notched. **FR** oblong, irregularly dehiscent, gen ribbed. **SEEDS** several. ± 40 spp.: trop, warm temp Eurasia. (Ancient name from Malabar, India)

O. alismoides (L.) Pers. **LVS** 10–25 cm; lower blades elliptic to ovate, upper blades cordate to reniform; veins parallel, curved. **FL**: sepals ± 1.5 cm, 4 mm wide; petals 2–3 cm, obovate, white to pinkish; anthers bright yellow. **FR** < 4 cm, oblong; ribs 2–10, scarcely winged; tip beaked. Ditches, rice-fields; ± 50 m. Formerly e ScV (Butte Co.), presumed extirpated; native to Afr, India, sw Pacific. Potentially noxious weed.

IRIDACEAE IRIS FAMILY

Elizabeth McClintock, except as specified

Per, bulbed, cormed, or rhizomed. **ST** gen erect. **LVS** gen basal (a few cauline), 2-ranked, ± linear, gen grass-like, gen sharply folded along midrib; bases overlapping, sheathing. **INFL**: spike, raceme, panicle, ± terminal, or fls solitary; bracts ± like lf bases, sheathing. **FL** gen bisexual, gen radial; hypanthium fused to ovary; perianth parts gen fused into tube above ovary, gen petal-like, in 2 series of 3, outer (sepals) gen ± like inner (petals); stamens 3, gen attached to sepals, filaments fused below into a tube or not; ovary inferior, 3-chambered, placentas gen axile, style 1, each of 3 branches entire or 2-branched or -lobed, petal-like or not, with stigma on under surface instead of at tip. **FR**: capsule, loculicidal. **SEEDS** few–many. 80 genera, ± 1500 spp.: worldwide, esp Afr; many cult (e.g., *Iris, Gladiolus, Crocus, Freesia*).

1. Style branches petal-like, conspicuous, covering stamens; sepals unlike petals in shape, size, position .... **IRIS**
1′ Style branches not petal-like, not conspicuous, not covering stamens; sepals ± like petals (exc *Libertia*)
2. Fls in umbel-like clusters
3. Perianth parts dissimilar, outer sepal-like, inner petal-like, larger, gen white .... **LIBERTIA**
3′ Perianth parts alike, blue or yellow — fls ephemeral, opening in sun .... **SISYRINCHIUM**
2′ Fls not in umbel-like clusters
4. Fls gen solitary on erect, basal peduncle; pl gen < 20 cm — lf gen compressed-cylindric .... **ROMULEA**
4′ Fls in spikes, racemes, or panicles of spikes; pl gen > 20 cm
5. Infl 1-sided, axis bent ± 90° near lowest fl .... **FREESIA**
5′ Infl not 1-sided, axis sometimes ± zig-zag but not bent ± 90°
6. Infl gen < 15-fld (*Watsonia marginata* > 30-fld)
7. Infl bracts papery, cut into lanceolate-linear segments in upper half or at tip .... **SPARAXIS**
7′ Infl bracts leathery, entire
8. Style branches each entire; perianth purplish pink to reddish or white to yellow; lf axils without bulblets .... **GLADIOLUS**
8′ Style branches each 2-lobed; perianth brick-red or rose-pink; lf axils with or without bulblets after fl .... **WATSONIA**
6′ Infl gen > 15-fld
9. Infl gen a panicle of spikes; perianth funnel-shaped, lobes spreading, ± equal .... **CROCOSMIA**
9′ Infl gen a spike; perianth funnel-shaped to salverform, lobes spreading, reflexed, or ± erect, ± equal or not
10. Perianth funnel-shaped, tube curved, lobes unequal .... **CHASMANTHE**
10′ Perianth salverform, tube straight, lobes ± equal .... **IXIA**

CHASMANTHE

Per; corms ± round, with thin, fibrous covering. **ST** gen unbranched. **INFL**: gen spike, terminal; fls each subtended by 2 ± equal, entire bracts. **FL** asymmetric; perianth funnel-shaped, tube curved, basal part ± cylindric, gen > bracts, lobes unequal, upper longest, ± curved, parallel to lateral 2, lowest 3 < upper, reflexed; stamens arched under upper perianth lobe, exserted; style parallel to stamens, exserted, branches each entire. **SEEDS** many, colored. ± 9 spp.: s Afr, several orn. (Greek: gaping fl)

C. floribunda (Salisb.) N.E. Br. **ST** gen < 1 m. **LF** 20–65 cm, 2.5–5 cm wide, sometimes wider on 1 side of midrib. **INFL**: fls 20–30, 2-ranked. **FL**: perianth < 4 cm, scarlet and yellow. Uncommon. Disturbed urban areas; < 50 m. CCo, SnFrB; native to s Afr. Long mistaken as *C. aethiopica* (L.) N.E. Br.; pls sometime persist from garden waste.

CROCOSMIA

Per; corms round to ± flat, with fibrous covering, often produced by rhizomes. **ST** branched or not. **INFL**: gen panicle of spikes, gen > lvs; fls each subtended by 2 bracts. **FL**: perianth funnel-shaped, orange to brick-red, tube gen curved, lobes spreading, ± equal; stamens free, exserted from tube; style ± > stamens, branches each entire. **SEEDS** gen 3–6, some aborted. ± 7 spp.: s Afr. (Greek: saffron, from smell of dried fls in warm water) [Kostelijk 1984 Plantsman 4:246–253; DeVos 1984 J S Afr Bot 50:463–502]

C. ×crocosmiiflora (Burb. & Dean) N.E. Br. **ST** < 1 m. **LF** 30–60 cm, 1–2.5 cm wide; midrib conspicuous. **FL**: perianth 2–5 cm, lobes < 5 cm wide. Common. Gen in disturbed coastal, urban areas, as roadsides, often from garden waste; < 50 m. NCo, SnFrB, SCo, expected elsewhere. Made in France from *C. pottsii* (Baker) N.E. Br. × *C. aurea* (Pappe) Planchon, both s Afr; highly variable, spreading by corms, rhizomes.

FREESIA

Per; corms with fibrous covering. **STS** branched or not. **INFL**: spike, 1-sided, axis bent ± 90° near lowest fl; fls ± erect, only on upper side of axis, each subtended by 2 membranous bracts. **FL** often fragrant; perianth funnel-shaped, tube narrow, upper-middle lobe unlike others; stamens free, gen incl in tube; style branches each 2-branched. **SEEDS** many. ± 20 spp.: s Afr. (Friedric H.T. Freese, German physician-botanist, 1795–1876) [Brown 1935 J S Afr Bot 1:1–31] Several spp. and hybrids cult as orn.

F. refracta (Jacq.) Klatt **ST** gen branched, < 20 cm. **LF** 5–15 cm, 5–10 mm wide. **FL**: perianth < 5 cm, < 3 cm wide at top, colors varied, lobes spreading. Uncommon. Disturbed urban, coastal areas; < 50 m. CCo, SnFrB, SCo; native to s Afr.

GLADIOLUS

Per; corm rounded, with fibrous covering. **ST** gen unbranched. **INFL**: gen spike; fls each subtended by 2 leathery, entire bracts. **FL** gen bisexual, bilateral; perianth funnel-shaped, tube ± curved, upper 3 lobes > lower 3; stamens attached to base of perianth tube, arching to upper lobes; style ± = stamens, branches each entire, arched. **SEEDS** many, winged or not. ± 180 spp.: esp Afr, also Madagascar, Eur. (Latin: little sword, from lf shape) [Lewis and others 1972 J S Afr Bot, Suppl vol 10] Many spp. and hybrids cult as orn.

1. Fl 4–5 cm; perianth purplish pink to reddish, tube < 1 cm; basal lf 5–15 mm wide, flat, not twisted ..... ***G. italicus***
1′ Fl 5–9 cm; perianth white to yellow, gen green-tinged, tube gen 4–6 cm; basal lf < 5 mm wide, 4-angled, ± twisted ..... ***G. tristis***

G. italicus Miller **ST** < 100 cm, > lvs. **INFL**: fls gen < 15. **FL** gen bisexual (some pls pistillate); stamens, style exerted from perianth tube. Uncommon. Disturbed ground, orchards, old fields; < 500 m. NCoRO, SCo; native to s Eur. [*G. seqetum* Ker Gawler]

G. tristis L. **ST** 50–70 cm, < or = lvs. **INFL**: fls gen 3–4(8). **FL** bisexual; stamens, style incl in perianth tube. Uncommon. Disturbed ground, along roadsides; < 800 m. SCo; native to s Afr.

IRIS IRIS

Douglass M. Henderson & Anita F. Cholewa

Per; rhizome creeping or ± tuber-like. **INFL**: fls 1–many. **FL**: perianth parts clawed, sepals (wider, spreading or reflexed) unlike petals (gen narrower, erect); style branches ± petal-like, arching over stamens, each with flat, scale-like stigma on surface facing stamen, just below gen 2-lobed tip (crest). **SEEDS** compressed, pitted. Perhaps 150 spp.: gen n temp. (Greek: rainbow, from fl colors) [Lenz 1958 Aliso 4:1–72; Clarkson 1959 Madroño 15:115–122] Hybrids are common, esp in *I. tenax* alliance; some authors lump taxa recognized here. ❀Pacific Iris hybrids; CVS.

1. Rhizome gen 10–40 mm diam; stigma 2-lobed or rounded; fr often >> 5 cm (sect. *Longipetalae*)
  2. Lf 10–25 mm wide; fl st 5–15 dm, branched; fls many ..... ***I. pseudacorus***
  2′ Lf 3–9 mm wide; fl st 2–6 dm, rarely branched; fls 1–8
    3. Lowest 2 bracts gen alternate (5–100 mm apart), not scarious or only so near tip, lower of these 7–15 cm; fls 3–6 ..... ***I. longipetala***
    3′ Lowest 2 bracts opposite, gen scarious, outer of these 4–8 cm; fls gen 2 ..... ***I. missouriensis***
1′ Rhizome gen 3–10 mm diam (exc *Iris tenax*); stigma triangular, truncate, tongue-shaped, or rounded; fr often < 5 cm (sect. *Californicae*)
  4. Perianth tube 5–12 mm, gen stout
    5. Lowest 2 bracts opposite, adherent at base, not spreading; lvs dark green on 1 side, pale yellow on the other, some bract-like — perianth yellow ..... ***I. bracteata***
    5′ Lowest 2 bracts gen alternate, free, spreading; lvs colored ± equally on both sides, none or few bract-like
      6. Fl st < 7 dm, gen 3-fld; lf 9–20 mm wide; lowest bract 8–14 mm wide; petals 5–9.5 cm ..... ***I. munzii***
      6′ Fl st seldom >> 4 dm, 1–3-fld; lf gen < 10 mm wide; lowest bract 3–9 mm wide; petals 3.5–6 cm ..... ***I. hartwegii*** in part
        7. Fls 3 ..... ssp.***columbiana***
        7′ Fls 1–2
          8. Lowest bract 6–9 mm wide; lf base gen with some red pigment ..... ssp. ***australis***
          8′ Lowest bract 4–7 mm wide; lf base gen without red pigment ..... ssp. ***hartwegii***
  4′ Perianth tube 11–120 mm, slender to ± stout
    9. Lowest 2 bracts alternate, spreading; fls 1–2; perianth tube (above ovary) 11–20 mm
      10. Perianth tube 11–15 mm, abruptly flaring, throat distinct; sepals narrowly oblanceolate, 1.2–2 cm wide; lf base with little or no red pigment; style crest very slender ..... ***I. hartwegii*** ssp. ***pinetorum***
      10′ Perianth tube 11–20 mm, gradually flaring, throat indistinct; sepals oblanceolate to widely obovate, 1.5–2.4 cm wide; lf base with obvious red pigment; style crest narrowly ovate ..... ***I. tenax*** ssp. ***klamathensis***
    9′ Lowest 2 bracts opposite (if alternate, fls 3); perianth tube almost always 15–120 mm
      11. Stigmas truncate or rounded, margins gen minutely and irregularly toothed ..... ***I. purdyi***
      11′ Stigmas triangular, tongue-shaped, or rounded, margins entire (rarely minutely but regularly toothed)
        12. Perianth tube 10–30 mm
          13. St sometimes branched; lowest bract lanceolate, 6–15.5 cm, gen >> 7 mm wide; perianth tube 10–28 mm; ovary tip with nipple-like projection; fls 2–3; lf < 22 mm wide, not grass-like .... ***I. douglasiana***
          13′ St unbranched; lowest bract widely lanceolate to ovate, 3.3–6 cm, 5–7 mm wide; perianth tube 15–30 mm; ovary tip rounded to ± acute; fls 1–2; lvs 2–4 mm wide, grass-like ..... ***I. innominata***
        12′ Perianth tube 30–120 mm
          14. Style crests << style branches; perianth parts spreading or not, cream-yellow to blue-purple, with or without dark veins
            15. Outermost bract 6–11 mm wide, widely lanceolate; perianth throat distinct; lf 4–8 mm wide, base often intensely reddish; perianth cream-yellow, veins sometimes darker ..... ***I. fernaldii***
            15′ Outermost bract 3–9 mm wide, linear-lanceolate; perianth throat indistinct (tube gen with a short, bowl-shaped enlargement near top); lf < 5 mm wide, base gen colorless; perianth golden yellow to cream, pale lavender, or deep blue-purple, veins gen darker ..... ***I. macrosiphon***
          14′ Style crests often = style branches; perianth parts spreading, whitish to cream, gen with dark veins
            16. Perianth throat indistinct (tube gen with a short, bowl-shaped enlargement near top) ...... ***I. chrysophylla***
            16′ Perianth throat distinct ..... ***I. tenuissima***
              17. St-lvs 3–4, free from st only at tips, gen flushed with pink; stigmas widely triangular to rounded ..... ssp. ***purdyiformis***
              17′ St-lvs 1–3, free from st exc near bases, green; stigmas triangular ..... ssp. ***tenuissima***

I. bracteata S. Watson (p. 1161) SISKIYOU IRIS Rhizome 6–9 mm diam. **ST** < 6 dm, with 3–6 bract-like lvs. **LF** 6–10 mm wide. **INFL**: fls 2; lowest 2 bracts opposite, outer of these 5–9 cm, 6–10 mm wide. **FL**: perianth yellow, veined with deep maroon or brown, tube 5–10 mm, funnel-shaped, sepals 4.3–7.8 cm, 17–30 mm wide; style branches 22–30 mm, crests 9–17 mm, stigmas triangular. $2n=40$. UNCOMMON. Partly shady places, gen in yellow-pine forest; ± 500–1000 m. KR (Del Norte Co.); sw OR. ❀DRN:**4,5**&IRR:**16**&SHD:1–3,**6,7,14–17**,18–21.

I. chrysophylla Howell YELLOW-FLOWERED IRIS Rhizome < 10 mm diam. **ST** < 2 dm. **LF** 3–5 mm wide. **INFL**: fls 2; lowest 2 bracts opposite, outer of these 5–8.6 cm, < inner, 6–10 mm wide. **FL**: perianth pale creamy yellow to ± white, sometimes with faint bluish tinge, gen with darker veins, tube 43–120 mm, throat indistinct (tube gen with a short, bowl-shaped enlargement near top), sepals 4.6–6.7 cm, 10–20 mm wide; style branches 17–25 mm, crests 15–25 mm, stigmas triangular. $2n$=40. UNCOMMON. Open coniferous forests; ± 600–1000 m. w&s KR; w OR. ❀DRN:4,5,**6**&IRR:**17**&SHD:1–3,**7,14–16**,18–21.

I. douglasiana Herbert (p. 1161) Rhizome < 9 mm diam. **ST** 1.5–8 dm, sometimes branched. **LF** < 22 mm wide. **INFL**: fls 2–3; lowest 2 bracts opposite, sometimes alternate, outer or lower of these 6–15.5 cm, 4–12 mm wide. **FL**: perianth pale cream to light, dark lavender or deep reddish purple, tube 10–28 mm, sepals 5–8.7 cm, 14–30 mm wide; ovary tip with nipple-like projection, style branches 17–35 mm, crests 10–20 mm, stigmas triangular. $2n$=40. Common. Grassy places, esp near coast; gen < 100(–1000) m. NW, CW, n SW; OR. [var. *altissima* Jepson; var. *oregonensis* R. Foster] Highly variable. Noxious weed in pastures, because lvs unpalatably bitter. ❀**4–6,16,17**&IRR:**15,24**&SHD:7,8,9,**14,18–23**;GRCVR.

I. fernaldii R. Foster (p. 1161) Rhizome averaging 6 mm diam. **ST** 2–4 dm. **LF** 4–8 mm wide. **INFL**: fls 2; lowest 2 bracts opposite, outer of these 5–9 cm, 6–11 mm wide. **FL**: perianth cream-yellow, sometimes with darker veins, tube 30–62 mm, with a distinct throat, sepals 4.7–6.8 cm, 12–21 mm wide; style branches 22–30 mm, crests 10–17 mm, stigmas triangular. Shady places; ± 50–2000 m. s NCo, s NCoRI, sw ScV, n CCo, SnFrB. Pls dry a peculiar gray-green. ❀DRN,DRY:**6**,17&SHD:**7,14–16**;DRN,IRR:1,2, 8,9,18.

I. hartwegii Baker Rhizome 5–9 mm diam. **ST** 0.5–4 dm. **LF** gen 2–10 mm wide. **INFL**: fls 1–3; lowest 2 bracts gen alternate, up to 9 cm apart, lowermost 5–12.5 cm, 3–9 mm wide. **FL**: perianth gen pale yellow or cream with darker veins, sometimes deep yellow, lavender, purple, or bluish violet, tube 5–15 mm, sepals 4–7 cm, 14–27 mm wide, petals 3.5–6 cm, 5–14 mm wide, narrowly oblanceolate; style branches 16–30 mm, crests 5–15 mm, stigmas triangular. Open or partly shaded slopes; 600–2300 m. s CaR, SN, SnGb, SnBr, SnJt.

ssp. ***australis*** (Parish) L. Lenz **LF**: base gen pinkish. **INFL**: fls 2; lowest bract 6–9 mm wide. **FL**: perianth purple to bluish violet. $2n$=40. Common. Partial shade in yellow-pine forests; 1600–2300 m. e SnGb, SnBr, SnJt. ❀DRN,SHD:**15,16**&IRR:1–3,**7**,14, **18**,19–21;DFCLT.

ssp. ***columbiana*** L. Lenz TUOLUMNE IRIS **LF**: base slightly red. **INFL**: fls 3; lowest bract 5–8 mm wide. **FL**: perianth pale creamy yellow, golden-veined. UNCOMMON. Dry slopes; 600–1400 m. c SNH (Tuolumne Co.). ❀DRN&IRR:**17**&SHD:1–3,**7**, 9,14,**15,16**,18.

ssp. ***hartwegii*** (p. 1161) **LF**: base gen green. **INFL**: fls 2; lowest bract 6–11 cm, 4–7 mm wide. **FL**: perianth pale to golden yellow or lavender, tube 5–10 mm. Common. Habitat of sp.; 600–2300 m. s CaR, SN. ❀DRN:**17**&IRR,SHD:1–3,**7**,9,14,**15,16**,18.

ssp. ***pinetorum*** (Eastw.) L. Lenz **LF** 3–5 mm wide; base faint pink. **INFL**: fls 2; lowest bract 5–7 cm, 3–6 mm wide. **FL**: perianth pale creamy yellow, tube 11–15 mm, sepals conspicuously golden veined. Common. Yellow-pine forests; 1000–1400 m. s CaR, n&c SN. ❀DRN:17&SHD,IRR1–3,**7**,9,14–18.

I. innominata L. Henderson (p. 1161) DEL NORTE COUNTY IRIS Rhizome 3–4 mm diam. **ST** < 2 dm, with 2–4 slightly sheathing lvs. **LF** 2–4 mm wide. **INFL**: fls 1–2; lowest 2 bracts opposite, outermost 3.3–6 cm, 5–7 mm wide. **FL**: perianth deep golden yellow with darker veins or sometimes yellow throughout, tube 15–30 mm, sepals 4.4–6.3 cm, 17–30 mm wide; style branches 19–26 mm, crests 9–14 mm, stigmas triangular. $2n$=40. UNCOMMON. Open or partly shaded slopes with well drained soil; ± 300–2000 m. NCo, KR (Del Norte Co.); sw OR. ❀DRN:**4–7**&IRR:**15–17, 24**&SHD:8,9,**14,18–23**.

I. longipetala Herbert Rhizome 10–25 mm diam. **ST** 3–6 dm, gen with 1–2 lvs. **LF** 4–9 mm wide. **INFL**: fls 3–6; lowest 2 bracts alternate, 0.5–10 cm apart, rarely opposite, lower or outer of these 7–15 cm. **FL**: perianth lilac-purple with darker veins, tube 5–13 mm, sepals 6–10 cm, gen 30–50 mm wide, petals notched at tip; style branches < 40 mm, crests < 15 mm, stigmas ± 2-lobed. $2n$= 86–88. Moist, open places; < 600 m. c&s NCo, s NCoRO, n&c CCo, SnFrB. ❀**4–6,17**;IRR:**24**&SHD:7,14,**15,16**,19–23.

I. macrosiphon Torrey Rhizome averaging 8 mm diam. **ST** < 2.5 dm or ± 0. **LF** 2–5 mm wide; base gen colorless. **INFL**: fls 1–2; lowest 2 bracts opposite, outermost 3.9–9.5 cm, 3–9 mm wide. **FL**: perianth golden yellow to cream or pale lavender to deep blue-purple, gen with darker veins, tube 36–86 mm, upper part with a bowl-shaped enlargement, sepals 3.9–6.8 cm, 12–16 mm wide; style branches 19–33 mm, crests 8–18 mm, stigmas triangular. Common. Open to partly shaded slopes; gen < 1000 m. CaRF, n&c SNF, NCoRI, SnFrB. Highly variable. ❀DRN,DRY:**6,17**&SHD: **7**,8,9,**14–16**&IRR:18.

I. missouriensis Nutt. (p. 1161) WESTERN BLUE FLAG Rhizome 20–30 mm diam. **ST** 2–5 dm, sometimes branched. **LF** 3–9 mm wide; base sometimes purplish. **INFL**: fls 1–2; lowest 2 bracts opposite, gen scarious, outermost 4–8 cm, 5–7 mm wide. **FL**: perianth pale lilac to whitish with lilac-purple veins, tube < 12 mm, sepals < 6 cm, 2 cm wide; style branches < 25 mm, crests averaging 8 mm, stigmas 2-lobed. $2n$=38. Moist, grassy places; 900–3400 m. NCoR, SN, SCoRI, TR, PR, GB; much of w N.Am, n Mex. NOXIOUS WEED in pastures, because lvs unpalatably bitter. ❀IRR:1, **2–7,9**,10,11,14,**15**,16,**18**,19–21.

I. munzii R. Foster (p. 1161) MUNZ'S IRIS Rhizome < 10 mm diam. **ST** < 7 dm. **LF** 9–20 mm wide; base gen evergreen, without red pigment. **INFL**: fls gen 3; lowest 2 bracts gen alternate, up to 19 cm apart, spreading, outermost 6.5–11 cm, 8–14 mm wide. **FL**: perianth pale lavender to bluish or reddish violet, tube 7–12 mm, sepals 6–9 cm, 18–37 mm wide; style branches gen 30 mm, crests 11–20 mm, stigmas triangular, sometimes with acuminate tips. $2n$= 40. UNCOMMON. Partly shaded slopes; 540–800 m. s SNF (Tule River, Tulare Co.). ❀DRN,SHD,IRR:**7,9**,14–16,**18–21**,23; DFCLT.

I. pseudacorus L. Rhizome 30–40 mm diam. **ST** 5–15 dm, branched. **LF** 10–25 mm wide, stiff. **INFL**: fls many; lowest 2 bracts gen opposite, outermost 4.9–5.1 cm, 7–10 mm wide. **FL**: perianth bright yellow or cream, crest area outlined by series of brown marks, tube 12–13 mm, sepals < 6 cm, petals narrowed near middle; style branches < 25 mm, crests 7–10 mm, stigmas rounded. **FR** 5–8 cm. $2n$=24,30,32,34. Irrigation ditches, pond margins; gen < 100 m. SnFrB, s SnJV; widespread e of Rocky Mtns; native to Eur.

I. purdyi Eastw. (p. 1161) Rhizome 4–6 mm diam. **ST** 0.6–3.5 dm. **LF** 4–8 mm wide. **INFL**: fls 2; lowest 2 bracts opposite, outermost 5–7.7 cm, 8–13 mm wide. **FL**: perianth pale cream-yellow with prominent, brownish purple veins or whitish with lavender tinge, tube 2.8–4.8 cm, sepals 5.5–8.4 cm, 1.6–2.7 cm wide; style branches 2–3.2 cm, crests 9–12 mm, stigmas truncate or rounded, margins gen minutely, irregularly toothed. $2n$=40. Common. ± open places; gen < 1200 m. NCo, s KR, NCoRO. ❀DRN:**4–6, 17**&IRR:**15,16,24**&SHD:7,8,9,**14,18–23**.

I. tenax Douglas ssp. ***klamathensis*** L. Lenz (p. 1161) ORLEANS IRIS Rhizome 30–50 mm diam. **ST** < 4 dm. **LF** 3–8 mm wide, gen >> st; base bright red or pink. **INFL**: fls 1–2; lowest 2 bracts alternate, < 18 mm apart, spreading, lower of these 4.7–9.5 cm, 3–6 mm wide. **FL**: perianth pale buff yellow, with deep maroon or brownish veins on sepals, tube 11–20 mm, sepals 5.3–7.3 cm, 15–24 mm wide; style branches 20–30 mm, crests 8–16 mm, narrowly ovate, stigmas triangular. UNCOMMON. ± shaded mixed-evergreen forests; ± 100–500 m. w KR (near Orleans, Humboldt Co.). ❀DRN:**4,5**&IRR:**15–17**&SHD:**6,7**,14.

I. tenuissima Dykes (p. 1161) Rhizome < 10 mm diam. **LF** < 6 mm wide. **INFL**: fls 2; lowest 2 bracts opposite, outermost 3.5–10 cm, 5–10 mm wide. **FL**: perianth tube 30–58 mm, throat distinct, sepals 4.7–7.5 cm, 11–18 mm wide; style branches 20–30 mm, crests 9–23 mm, stigmas triangular to rounded or tongue-shaped. Common. Dry, open woodlands; 100–2000 m. NW, CaR, n SNH, ScV.

ssp. ***purdyiformis*** (R. Foster) L. Lenz **ST**: lvs 3–4, bract-like, free only at tips, gen flushed with pink. **FL**: perianth pale yellow or cream with few or no dark veins; stigmas widely triangular to rounded. Shaded places, yellow-pine forests; ± 600–2000 m. se KR (Shasta Co.), CaR, n SNH. ❀DRN,DRY:**6**,17&SHD:**7**,**15**: DRN,IRR:1,2,8,9,14,16,18.

ssp. ***tenuissima*** ST-lvs 1–3, bract-like, free exc near bases. **FL**: perianth gen pale cream with distinct lavender, brownish or reddish brown veins; stigmas triangular. 2*n*=40. Dry, open woodlands; 100–1000 m. NW, ScV. ❀DRN,DRY:**6**,**17**&SHD:**7**,**15**& IRR:2,8,9,**14**,**16**,18.

IXIA

Per; corm gen round with fibrous covering. **STS** erect, gen unbranched. **INFL**: spike; fls few–many, each subtended by 2, gen papery bracts. **FL**: perianth ± salverform, > stamens, tube straight, slender, lobes spreading, ± equal; filaments gen free; style ± = stamens, branches each entire. **FR** thin-walled. **SEEDS** many, small. ± 45 spp.: s Afr. (Greek: bird lime, from sticky sap of some spp.)

I. maculata L. **ST** 20–50 cm, unbranched. **LVS**: basal 5–8, 10–35 cm, < 5 mm wide, often twisted, veins several. **INFL** gen 4–12-fld, dense. **FL**: perianth 5–10 mm, lobes > tube, 1.5–2 cm, < 1 cm wide, spreading, orange to yellow, dark basally; anthers 7–9 mm, > filaments. Uncommon. Urban areas, disturbed coastal dunes; < 50 m. SnFrB; native to s Afr.

LIBERTIA

Per; rhizomes creeping; roots fibrous. **ST** branched or not. **INFL**: fls in umbel-like clusters. **FL**: perianth parts free, spreading, dissimilar, outer gen sepal-like, inner petal-like, larger, gen white; filaments shortly fused at base; style branches each entire. **SEEDS** many. ± 20 spp.: esp Australasia, some temp S.AM. (Marie A. Libert, Belgium, 1782–1863)

L. formosa Graham **ST** 60–125 cm. **LVS**: basal 15–45 cm, 5–15 mm wide. **INFL**: pedicels 6–12 mm. **FL**: sepals < 10 mm, lanceolate, greenish brown, petals spreading, 12–18 mm, obovate, white. **FR** 6 mm wide, round. Uncommon. Disturbed urban, coastal areas; < 50 m. SnFrB; native to Chile.

ROMULEA

Per; corms rounded, membranous covering split at top into several pointed teeth. **INFL**: fl gen solitary on erect, basal peduncle, subtended by 2 bracts. **FL**: perianth gen funnel-shaped, tube gen < or = lobes; style gen ± = stamens, branches each 2-lobed. **SEEDS** many. ± 90 spp.: esp Afr, Medit. (Romulus, legendary founder of Rome) [DeVos 1983 J S Afr Bot, Suppl vol. 9]

R. rosea (L.) Ecklon var. *australis* (Ewart) De Vos Pl gen < 20 cm. **LF** 5–35 cm, 1–2.5 mm wide, compressed-cylindric, gen 4-grooved. **INFL**: peduncle 5–15 cm; outer bract ± 15 mm, inner ± 10 mm. **FL** 1.5–2 cm; perianth tube yellow, lobes pale lilac; style 7–10 mm, not exceeding stamens. **FR** 10–15 mm, thin-walled; bracts persistent. Uncommon. Disturbed areas, dry, sandy or often hard-packed soil; < 50 m. NCo, CCo, SnFrB; native to s Afr. [*R. lonqifolia* (Salisb.) Baker]

SISYRINCHIUM

Douglass M. Henderson & Anita F. Cholewa

Per; rhizomes compact. **STS** single or tufted, gen ± flat and winged or rounded, sometimes with lf-bearing nodes well above basal lvs, each with 1 or more fl-branches. **LF** narrow, grass-like. **INFL**: fls in umbel-like clusters; bracts 2, equal in length or not, margins translucent. **FL** ephemeral; perianth reddish purple, bluish, violet, yellow, rarely white, parts ± alike, but outer gen wider; filaments ± completely free to ± completely fused. **SEEDS** ovoid, smooth or pitted. ± 70 spp.: w hemisphere. (Name used by Theophrastus for Iris-like pl) [Henderson 1976 Brittonia 28:149–176]

1. Perianth reddish purple, gen >> 18 mm; filaments fused ± in basal half; st rounded, not winged (sect. *Eriphilema*) ..... ***S. douglasii*** var. ***douglasii***

1′ Perianth bluish, violet, yellow, or white, gen < 18 mm; filaments ± completely free or ± completely fused; st ± flat, winged

2. Perianth yellow; filaments ± completely free (sect. *Echthronema*)

3. St gen >> 2 mm wide; perianth gen 11–18 mm ..... ***S. californicum***

3′ St gen < 2 mm wide; perianth gen 7–11 mm ..... ***S. elmeri***

2′ Perianth bluish to violet, rarely white, each part with a basal yellow blotch; filaments ± completely fused (sect. *Sisyrinchium*)

4. Inner bract with ± uniformly narrow translucent margins; outer perianth parts gen elliptic to narrowly oblanceolate ..... ***S. idahoense***

5. Upper st margins gen with a few widely spaced, fine teeth; margins of outer bract fused in basal 4–7 mm; perianth gen >> 13 mm ..... var. ***idahoense***

5′ Upper st margins gen entire; margins of outer bract fused gen < 4.5 mm; perianth gen < 13 mm ..... var. ***occidentale***

4′ Inner bract with translucent margins wider at or just below tip; outer perianth parts gen wedge-shaped

6. St gen without lf-bearing nodes — translucent margins of inner bract extending above tip as 2 rounded teeth; perianth gen medium blue to blue-violet, 9–12 mm, outer parts gen widely wedged-shaped; mostly SNE ..... ***S. halophilum***

6′ Sts gen with 1 or more lf-bearing nodes

7. Translucent margins of inner bract not extending above tip; perianth deep bluish purple to blue-violet, or pale blue, rarely white, 10.5–17 mm, outer parts gen widely wedge-shaped; CA, OR .... ***S. bellum***
7′ Translucent margins of inner bract extending above tip as 2 rounded teeth; perianth gen pale blue, 10–15 mm, outer parts gen narrowly wedge-shaped; ne DMoj .... ***S. funereum***

S. bellum S. Watson (p. 1161) BLUE-EYED-GRASS **STS** gen tufted, < 64 cm, sometimes 5.3 mm wide, almost always with lf-bearing nodes. **INFL**: translucent margins of inner bract wider just below tip, not extending above tip. **FL**: perianth 10.5–17 mm, deep bluish purple to blue-violet, or pale blue, rarely white, tips truncate to notched, with a small point. *n*=16. Common. Open, gen moist, grassy areas, woodlands; gen < 2400 m. CA; OR. [*S. eastwoodiae* E. Bickn.; *S. greenei* E. Bickn.; *S. hesperium* E. Bickn.] Gen self-incompatible. ❀DRN:4,5,**6,7**,8,**9,14–18,20–24**&IRR:3,**19**&SHD:10,11;CVS.

S. californicum (Ker Gawler) Dryander (p. 1161) GOLDEN-EYED-GRASS **STS** tufted, < 62 cm, gen 2–6.2 mm wide, dull green, drying ± black. **FL**: perianth 11–18 mm, medium to bright yellow, veins dark brownish. Gen moist places near coast; gen < 600 m. *n*=17. NCo, n&c CCo, SnFrB; to B.C. [*S. lineatum* Torrey; *Hydastylus borealis* E. Bickn.; *H. brachypus* E. Bickn.] ❀IRR:**4–7,15–17,22–24**&SHD:8,9,**14,18–21**;INV.

S. douglasii A. Dietr. var. ***douglasii*** (p. 1161) PURPLE-EYED-GRASS, GRASS WIDOWS **STS** tufted, < 31 cm, round, not winged. **LVS** basal, bract-like. **FL**: perianth gen >> 18 mm, reddish purple; filaments fused in basal half, tube ± enlarged above base. *n*=16. Gen open, often rocky places where moist only in spring; < 2000 m. KR, n NCoRO, CaR, MP; to B.C., ID, UT. ❀DRN:1–3&DRY:4, 5,**6**&SHD:**7**,14,15;DFCLT.

S. elmeri E. Greene **STS** tufted, < 20 cm, gen < 2 mm wide, medium green, drying dark but not ± black. **FL**: perianth 7–11 mm, deep yellow to yellow-orange, veins dark brownish. Wet meadows, similar places; 1200–2580 m. s KR, s CaRH, SN, SnBr. ❀IRR, DRN:**6,7**,14–16&SUN1–3,**4,5,17**.

S. funereum E. Bickn. **STS** tufted, < 70 cm, almost always with lf-bearing nodes, pale green, glaucous. **INFL**: translucent margins of inner bract widest at tip, extending above tip as 2 rounded teeth. **FL**: perianth 10–15 mm, gen pale blue, tips truncate to sometimes notched, with a small point. *n*=16. Gen strongly alkaline margins of wet areas; < 800 m. ne DMoJ (Death Valley region); adjacent NV. Self-incompatible. ❀IRR,ALKALINE:**8–14,19–23**.

S. halophilum E. Greene (p. 1161) **STS** tufted, < 40 cm, gen without lf-bearing nodes, glaucous. **INFL**: translucent margins of inner bract widest at tip, extending above tip as 2 rounded teeth. **FL**: perianth gen 9–12 mm, gen medium blue to blue-violet, outer parts gen widely tapered at base, tips truncate to rarely notched, with a small point. *n*=16. Gen moist, alkaline meadows; < 2550 m. SNE, adjacent DMoj; to UT. [*S. leptocaulon* E. Bickn.] Self-incompatible. ❀IRR,ALKALINE:**8–14**,15–17,**18–23**.

S. idahoense E. Bickn. (p. 1161) **STS** tufted, < 45 cm, sometimes with 1 or more lf-bearing nodes, green to glaucous. **INFL**: translucent margins of inner bract ± uniformly narrow. **FL**: perianth 8–17 mm, gen blue to blue-violet, outer parts gen narrowly elliptic or oblanceolate, tips rounded to deeply notched, with a small point. Open, moist, grassy places; 900–3150 m. KR, CaR, SN, GB; to B.C., MT, Colorado. Highly variable, self- or cross-pollinating. Vars. intergrade.

var. ***idahoense*** **ST**: upper margin gen with a few widely spaced, fine teeth. **INFL**: margins of outer bract fused in basal 4–7 mm. **FL**: perianth gen >> 13 mm. *n*=32,48. Common. Habitat of sp. KR, CaR, MP; to B.C., MT. ❀IRR,DRN:**6,7**,14–16&SUN:1–3,**4,5**,17.

var. ***occidentale*** (E. Bickn.) D. Henderson **ST**: upper margin gen entire. **INFL**: margins of outer bract fused gen in basal 4.5 mm. **FL**: perianth gen < 13 mm. *n*=32. Habitat of sp. SN, SNE; to e WA, MT, Colorado. [*S. occidentale* E. Bickn.] ❀IRR,DRN:**6,7**,14–16 &SUN:1–3.**4,5**,17.

SPARAXIS HARLEQUIN FLOWER

Per; corms papery to fibrous. **STS** erect, < 45 cm, stiff, branched or not. **LVS** gen basal, 2-rowed, < 3 dm, lanceolate to ± sickle-shaped, veined, with false midrib. **INFL**: spike; fls few, scattered; bracts paired, papery, streaked with brown, cut into lanceolate-linear segments in upper half or irregularly at tip. **FL** bisexual, gen radial; perianth funnel-shaped, tube ± = lobes; stamens 3, free, < style, stigma 3-lobed. **FR**: capsule, spheric, firm. **SEEDS** several, spheric to angled, large, gen shiny. 6 spp.: s Afr. (Greek: to tear, from torn appearance of bracts) [Goldblatt 1969 J S Afr Bot 35:219–252]

1. Perianth tube 10–15 mm, lobes cream-white to yellow, with dark marks, or light purple; bracts deeply cut into linear-lanceolate segments in upper half .... ***S. grandiflora***
1′ Perianth tube < 10 mm, lobes light purple to orange, brown, purple-brown or cream-white to pinkish, with yellow center; bracts irregularly cut into lanceolate-linear segments at tip .... ***S. tricolor***

S. grandiflora (Delaroche) Ker Gawler **FL**: perianth tube 10–15 lobes cream-white to yellow, with dark marks, or light purple; bracts deeply cut into linear-lanceolate segments in upper half. Uncommon. Disturbed sites; < 50 m. CCo; native to s Afr.

S. tricolor (Schneev.) Ker Gawler **FL**: perianth tube < 10 mm, lobes light purple to orange, brown, purple-brown or cream-white to pinkish, with yellow center; bracts irregularly cut into lanceolate-linear segments at tip. Uncommon. Disturbed sites; < 50 m. CCo; native to s Afr.

WATSONIA

Per; corms rounded, coating fibrous. **INFL**: spike, branched below or not; fls facing in 2 opposite directions, appearing to be in 1 plane, each subtended by 2 leathery, entire bracts. **FL** asymmetric, sessile; perianth funnel-shaped, tube bent, lobes ± equal, oblong or lanceolate; stamens attached to perianth throat; style ± = stamens, branches each 2-lobed. **FR** oblong. ±60 spp.: s Afr. (Sir William Watson, English botanist-physician, 1715–1787)

1. Perianth brick-red, tube 4–5 cm, sharply bent; lf axils with bulblets after fl .... ***W. bulbillifera***
1′ Perianth rose-pink, tube < 2 cm, gradually bent; lf axils without bulblets .... ***W. marginata***

W. bulbillifera Matthews & L. Bolus **ST** 1–1.5 m. **LVS** ± 5–6, < 60 cm, < 6 cm wide; axils with bulblets after fl. **INFL** 10–15-fld, ± open. **FL**: perianth 6–7 cm, brick-red, tube 4–5 cm, sharply bent, lobes ± 2.5 cm, oblong to obovate. Uncommon, but may be locally abundant. Disturbed roadsides, fields, waste places; < 50 m. NCo; native to s Afr. Reproduces by bulblets, can be invasive.

W. marginata (Ecklon) Ker Gawler **ST** 1–1.5 m. **LVS** 3–4, 45–80 cm, < 5 cm wide; bulblets 0. **INFL** > 30-fld, gen dense. **FL**: perianth 3.5–4.5 cm, rose-pink, tube < 2 cm, gradually bent, lobes 1–1.5 cm, ovate-oblong. Disturbed ground, roadsides; < 100 m. SCo, native to s Afr. Sometimes persisting from garden waste.

JUNCACEAE RUSH FAMILY

Janice Coffey Swab

Ann, per, gen from rhizomes. **ST** round or flat. **LVS** gen mostly basal; sheath margins fused, or overlapping and gen with 2 ear-like extensions at blade junction; blade round, flat, or vestigial, glabrous or margin hairy. **INFL**: head-like clusters or single fls, variously arranged; bracts subtending infl 2, gen lf-like; bracts subtending infl branches 1–2, reduced; bractlets subtending fls gen 1–2, gen translucent. **FL** gen bisexual, radial; sepals and petals similar, persistent, green to brown or purplish black; stamens gen 3 or 6, anthers linear, persistent; pistil 1, ovary superior, chambers gen 1 or 3, placentas 1 and basal or 3 and axile or parietal, stigmas gen >> style. **FR**: capsule, loculicidal. **SEEDS** 3–many, often with white appendages on 1 or both ends. 9 genera, 325 spp.: temp, arctic, trop mtns. Fls late spring to early fall.

1. Lvs stiff, glabrous, sheath open; fr 3-chambered (rarely 1 in minute pls); seeds many .... **JUNCUS**
1′ Lvs flexible, margins gen hairy, sheath closed; fr 1-chambered; seeds 3 .... **LUZULA**

JUNCUS RUSH

Ann, per; rhizome (if any) gen with scale-like lvs. **ST** gen cylindric or flat. **LF**: blade well developed and cylindric or flat, or reduced to small point; crosswalls often present (pull fresh blade apart lengthwise to see or slide lf between fingers to feel); appendages often present at blade-sheath junction. **INFL** gen terminal (appearing lateral when pushed aside by lowest infl bract); bractlets 0–2. **FLS**: stamens gen 3 or 6 (2 in some very small ann taxa); pistil 1, ovary chambers 1 or 3, placentas axile or parietal, stigmas gen 3 (2). **SEEDS** many. 225 spp.: worldwide, esp n hemisphere. (Latin: to join or bind, from use of sts) [Ertter 1986 Mem NY Bot Gard 39:1–90]

1. Ann, gen < 10 cm; lvs narrow, often inrolled, gen < 1 mm wide
 2. Fls solitary at nodes (infl sometimes dense but not of head-like clusters), each fl subtended by bractlets; st leafy, gen branched; stamens 6
  3. Petals rounded to obtuse, < to = fr; fr tip truncate .... ***J. ambiguus***
  3′ Petals acuminate, > fr; fr tip acute .... ***J. bufonius***
   4. Perianth 3–4 mm .... var. ***occidentalis***
   4′ Perianth 4–7 mm
    5. Fls solitary .... var. ***bufonius***
    5′ Most fls in terminal clusters that appear slightly coiled .... var. ***congestus***
 2′ Fls terminal, 1–6 in clusters subtended by bractlets; st lfless, rarely branched; stamens 2–3
  6. Lower bracts lf-like, > infl; sepals >> and more acuminate than petals .... ***J. capitatus***
  6′ Lower bracts inconspicuous, membranous; sepals and petals similar in size and shape
   7. Pl gen > 7 cm; style and anthers > 0.7 mm (fls outcrossing)
    8. Seed striate at 10×; st hair-like, < 0.4 mm diam; fr < perianth .... ***J. triformis***
    8′ Seed not striate at 10×; st not very hair-like, often > 0.4 mm diam; fr = or > perianth .... ***J. leiospermus***
     9. Fls gen 1 per st .... var. ***ahartii***
     9′ Fls gen 2 or more per st .... var. ***leiospermus***
   7′ Pl gen < 6.5 cm; style and anthers < 0.7 mm (fls gen self-pollinating)
    10. Fls gen 2 or more (1–7) per st; bracts acute to acuminate; seed ± striate
     11. Fr < perianth and much lighter in color; seed 0.5–0.8 mm, not > 3 per row or 9 per capsule, ± obscurely striate .... ***J. capillaris***
     11′ Fr = to > perianth and similar in color; seed < 0.5 mm, often > 3 per row or 9 per capsule, obviously striate
      12. Fr gen > perianth; perianth parts gen 4 — fr and perianth pale green or pink .... ***J. tiehmii***
      12′ Fr ± = perianth; perianth parts gen 6
       13. Fls gen 3 (1–4) per st; fr and perianth gen becoming dark reddish, seed 0.4–0.5 mm .... ***J. kelloggii***
       13′ Fls rarely > 1 per st; fr and perianth gen remaining pale yellow-green; seed 0.3–0.4 mm ..... ***J. luciensis***
    10′ Fls 1 per st; bracts 0 or truncate to acute; seed not striate
     14. Fr < shiny perianth; st hair-like, 0.1–0.2 mm diam .... ***J. bryoides***
     14′ Fr ± = or > dull perianth; st not very hair-like, gen > 0.2 mm diam
      15. Infl bract 1, truncate, sheathing st; stamens gen 3 .... ***J. uncialis***
      15′ Infl bracts 0–2, ovate to rounded, not sheathing st; stamens gen 2 .... ***J. hemiendytus***
       16. Infl bracts 0; fr gen < perianth .... var. ***abjectus***
       16′ Infl bracts 1–2; fr > perianth .... var. ***hemiendytus***
1′ Per, gen larger in all respects
 17. Lower infl bract cylindric, resembling continuation of st; infl appearing lateral
  18. Infl of small clusters subtended by one bract; bractlets 0
   19. Sepals obtuse, thin; fr gen ± 2 × perianth .... ***J. acutus*** ssp. ***leopoldii***
   19′ Sepals acute, firm; fr ± = perianth .... ***J. cooperi***
  18′ Infl ± open; fls solitary, each subtended by 2 bractlets
   20. Fls gen 1–3; seed with conspicuous appendages
    21. Inner lf blades reduced to a bristle; fr tip indented .... ***J. drummondii***
    21′ Inner lf blades well developed; fr tip acute .... ***J. parryi***
   20′ Fls gen many; seed appendages 0 (or minute)

22. Anther gen < filament; perianth 1.8–4.2 mm
23. Stamens 6; fr ± spheric, with soft beak .... ***J. patens***
23′ Stamens 3; fr obovoid, without soft beak .... ***J. effusus***
24. Perianth 2.5–4.2 mm, firm when fresh, rigid when dry .... var. ***pacificus***
24′ Perianth 1.8–3.5 mm, soft when fresh, not rigid when dry
25. Perianth light brown, < 2 mm .... var. ***exiguus***
25′ Perianth dark brown, > 2 mm
26. Base of lf sheath dull brown, top much lighter .... var. ***brunneus***
26′ Base of lf sheath shiny, top dark .... var. ***gracilis***
22′ Anther gen > filament; perianth 3.5–8 mm
27. St gen flattened, often twisted
28. Rhizome stout; infl compact; lf blades 0 .... ***J. breweri***
28′ Rhizome slender; infl open; upper lf blades often present .... ***J. mexicanus***
27′ St cylindric, not twisted
29. Fr obtuse .... ***J. textilis***
29′ Fr acute
30. Perianth gen 3–6 mm .... ***J. balticus***
30′ Perianth 5–8 mm .... ***J. lesueurii***
17′ Lower infl bract not resembling continuation of st, or if so, channeled along inner side; infl appearing terminal
31. Lf blade cylindric (sometimes slightly flat) — bases not overlapping; crosswalls gen complete
32. Stamens 3, anthers < filaments
33. Pl growing submerged in fresh water — early lvs hair-like, later lvs erect .... ***J. supiniformis***
33′ Pl not submerged
34. Infl dense, clusters 1–5, > 20-fld .... ***J. bolanderi***
34′ Infl open, clusters gen 6–150, 2–20-fld
35. Fr ± = perianth .... ***J. acuminatus***
35′ Fr 2 × perianth length .... ***J. diffusissimus***
32′ Stamens 6, anthers sometimes > filaments
36. Infl gen a single cluster
37. Lower infl bract not sheathing, narrow; perianth brown, stiff .... ***J. duranii***
37′ Lower infl bract short, sheathing, wide; perianth purplish black, soft .... ***J. mertensianus***
36′ Infl of several to many clusters
38. Anthers gen > 2 × filament length
39. Perianth light; lf sheath appendages 4–6 mm .... ***J. chlorocephalus***
39′ Perianth dark; lf sheath appendages 2–3 mm .... ***J. nevadensis***
38′ Anthers < to > filaments
40. Lf, st surfaces conspicuously wrinkled .... ***J. rugulosus***
40′ Lf, st surfaces not wrinkled
41. Infl clusters many (<150) .... ***J. dubius***
41′ Infl clusters gen < 25
42. Sepals > petals .... ***J. torreyi***
42′ Sepals = or somewhat < petals
43. Infl clusters 3–12-fld; perianth 1.9–3.1 mm .... ***J. articulatus***
43′ Infl clusters 10–25-fld; perianth 2.5–4.1 mm .... ***J. nodosus***
31′ Lf blade flat
44. Lf blade oriented with edge toward st (at least from middle to tip), bases overlapping, crosswalls gen incomplete
45. Anthers inconspicuous, gen < or = filaments
46. Lf sheath margins membranous, prolonged into small ear-like appendages .... ***J. saximontanus***
46′ Lf sheath appendages 0 or inconspicuous
47. Infl gen of few, many-fld clusters; stamens gen 3; fr abruptly tapered to short beak .... ***J. ensifolius***
47′ Infl gen of many, few-fld clusters; stamens gen 6; fr gradually tapered to beak .... ***J. xiphioides***
45′ Anthers large, conspicuous, >> filaments
48. Perianth gen < 4 mm
49. Fr < perianth, abruptly pointed .... ***J. macrandrus***
49′ Fr > perianth, gradually tapered to sharp point .... ***J. oxymeris***
48′ Perianth gen > 4 mm .... ***J. phaeocephalus***
50. Infl 5–10 mm wide, of 10+ few-fld clusters .... var. ***paniculatus***
50′ Infl 10–15 mm wide, of 1–10 many-fld clusters .... var. ***phaeocephalus***
44′ Lf blade oriented with flat side toward st; crosswalls 0
51. Fls solitary, each subtended by 2 bractlets
52. Fr chambers 3
53. Perianth 4–5.5 mm; fls many; infl open .... ***J. brachyphyllus***
53′ Perianth 3.5–4 mm; fls few; infl gen compact .... ***J. confusus***
52′ Fr chamber 1 (walls incomplete)
54. Lf sheath appendages firm, rigid, glossy, 0.5–2 mm .... ***J. dudleyi***
54′ Lf sheath appendages membranous, 0.5–6 mm
55. Lvs gen < 1/2 st length; perianth gen > 4 mm .... ***J. occidentalis***

55′ Lvs gen > 1/2 st length; perianth gen < 4 mm .... ***J. tenuis***
51′ Fls in head-like clusters
56. Lvs evenly spaced on st, often overlapping
57. Stamens 6; anthers yellow .... ***J. cyperoides***
57′ Stamens 3; anthers purplish red .... ***J. marginatus***
56′ Lvs entirely or mostly basal
58. Seed appendages gen > 1/2 seed body length
59. Appendages < seed body .... ***J. howellii***
59′ Appendages > seed body .... ***J. regelii***
58′ Seed appendages 0 or short
60. Lf sheath appendages well developed
61. Sepals = to > petals .... ***J. longistylis***
61′ Sepals < petals .... ***J. macrophyllus***
60′ Lf sheath appendages 0 or obscure
62. Infl clusters 3–7-fld .... ***J. covillei***
63. Fl and fr dark brown; petals rounded .... var. ***covillei***
63′ Fl and fr pale brown; petals pointed .... var. ***obtusatus***
62′ Infl clusters 5–25-fld
64. Sandy places near ocean .... ***J. falcatus***
64′ Inland habitats .... ***J. orthophyllus***

J. acuminatus Michaux (p. 1161) Per, cespitose, 20–80 cm; rhizome short. **LVS** mostly basal; sheath appendages 1.5–5 mm, rounded; blade cylindric, crosswalls complete, obvious. **INFL**: lowest bract < infl; branches spreading; clusters gen 6–50, 5–20-fld. **FL**: perianth segments 2.5–3.5 mm, ± equal, narrowly acuminate, light brown to greenish; stamens gen 3, filaments slightly > anthers. **FR** ± = perianth, light brown; chamber 1. **SEED** 0.3–0.4 mm; appendages minute. $2n=40$. Uncommon. Sandy riverbanks, lake shores, ditches; gen < 500 m. NCo, KR, CaRF, GV, SnFrB; to B.C., e N.Am, Mex. All spp. with complete lf crosswalls may be enlarged and deformed by sucking insects.

J. acutus L. ssp. ***leopoldii*** (Parl.) Snog. (p. 1161) Per, cespitose, 50–140 cm; rhizome much-branched. **ST** rigid, hardened, cylindric. **LVS** basal, 40–120 cm, rigid; tip hard, sharp; sheath appendages firm. **INFL** appearing lateral, gen ± open; lowest bract cylindric, resembling st, < to = infl; branches uneven; clusters 2–4-fld, each subtended by 1 obvious, clasping bract; bractlets 0. **FL**: perianth 2–4 mm, margins membranous, sepals thin, obtuse, petals ± rounded; anthers 6, >> filaments, reddish brown. **FR** gen ± 2 × perianth, nearly spheric, shiny brown. **SEED** irregular; appendages small. $2n=48$. Moist saline places (salt marshes, alkaline seeps); gen < 300 m. CCo, SCo, s ChI, DSon; to AZ, Baja CA, S.Am, s Afr. [var. *sphaerocarpus* Engelm., spiny rush] ❀IRR or WET:5,7,**8,9**, **12–17,19–24**;also STBL;good barrier plant.

J. ambiguus Guss. (p. 1163) Ann, densely cespitose or solitary, 3–17 cm. **ST** 0.5–1 mm wide. **LVS** ± cauline, gen 1–2 per st. **INFL**: fls ± crowded near slightly coiled branch tips; lowest bract > infl; upper bracts 1/3 to ± = infl; bractlets 1.5 mm. **FL**: sepals 4–6.8 mm, acute; petals 3.3–5.3 mm, obtuse or rounded; stamens 6, filaments > anthers. **FR** = to > petals; tip rounded. **SEED** 0.3–0.4 mm, ovoid or barrel-shaped; appendages 0. $2n=34$. Uncommon. Saline places near coast; gen < 10 m. NCo, CCo; scattered in N.Am, Eurasia, n Afr. [*J. bufonius* L. var. *halophilus* Buchenau & Fern.]

J. articulatus L. Per 10–60 cm; rhizome short, branching. **ST** erect or ascending; nodes often rooting. **LVS**: basal lf blades reduced; cauline lvs 1–3, sheaths loose, blades 5–10 cm, cylindric, crosswalls complete, sheath appendages rounded, 1–1.5 mm, firm. **INFL** 2–15 cm; lowest bract inconspicuous; branches spreading; clusters gen < 25, 3–12-fld. **FL**: perianth segments 1.9–3.1 mm, ± equal, acute to acuminate, dark brown; stamens 6, filaments ± = anthers. **FR** > perianth, dark brown, 3-angled, tapering from near middle to obvious tip. **SEED** 0.5 mm, oblong; appendages minute. $2n=80$. Uncommon. Moist ground; > 500 m. NCo, s ScV; sporadic to B.C., e N.Am, Eurasia.

J. balticus Willd. (p. 1163) Per 35–110 cm; rhizome scaly, creeping, gen unbranched, slender to stout. **ST** 1–6 mm wide, gen cylindric. **LVS** basal; blades 0; sheaths variable, 2–15 cm. **INFL** appearing lateral, ± open; lowest bract cylindric, resembling st, gen >> infl; fls 5–50 or more; bractlets 2, membranous. **FL**: perianth segments 3–6 mm, sepals ± = to > petals, scarious margins of petals wider than those of sepals; stamens 6, filaments << anthers. **FR** ± = perianth; beak small but obvious. **SEED** 0.4–0.8 mm; appendages 0. $2n=40,80$. Moist to rather dry places; gen < 2200 m. CA; to AK, e N.Am, S.Am, Eurasia. [var. *montanus* Engelm.] Highly variable, intergrading complex needing study. ❀IRR,DRN:1–3,**4–7**,8–13, **14–24**;also STBL;INV.

J. bolanderi Engelm. (p. 1163) Per 30–80 cm; rhizome stout, creeping. **LVS**: basal scale-like; cauline 3–4, sheath appendages 3–5 mm, clear, blade 10–20 cm, 1–1.5 mm wide, cylindric, blade crosswalls complete. **INFL**: lowest bract gen > infl; clusters 1–4, gen > 20-fld. **FL**: perianth segments 3–3.5 mm, ± equal, narrowly acuminate, dark to light brown; stamens 3, filaments > anthers. **FR** ± = perianth, oblong, short-beaked. **SEED** ± 0.5 mm, obovoid; appendages minute. Swampy or sandy ground; < 1600 m. NW; to B.C. ❀WET or IRR:**4–7**,14,**15–17**,19–21,**22–24**.

J. brachyphyllus Wieg. Per, cespitose, 30–50 cm. **ST** stiff, stout, conspicuously grooved. **LVS** ± basal; sheath appendages conspicuous, membranous; blade with flat side toward st, << st, stiff, spreading, crosswalls 0. **INFL** open; lowest bract = to gen > infl; branches ascending; fls solitary, many. **FL**: perianth segments 4–5.5 mm (sepals > petals), long-acuminate, sharp; stamens 6, filaments > anthers. **FR** nearly = petals, widely cylindric, ± 3-angled; chambers 3. **SEEDS** slightly twisted; appendages minute. $2n=80$. Uncommon. Moist places; 900–1600 m. MP; to WA, ID, TX, e US. Typical of sandy prairies of c US; montane pls of nw US warrant further study. ❀TRY.

J. breweri Engelm. Per 10–130 cm; rhizome heavy, creeping, sometimes vertical. **ST** slender, twisted or arching, firm. **LVS** basal; sheaths very dark brown, shiny, loose; blades 0. **INFL** appearing lateral, compact; lowest bract = or > st, cylindric, resembling st; bractlets 2 per fl; fls gen few. **FL**: perianth segments 5–8 mm (sepals gen > petals), very dark, middle greenish, sides brown to purplish, margins membranous. **FR** < perianth, ovoid, very dark brown, shiny; tip acute. **SEED** 0.7–1.1 mm, ovoid, appendages 0. Coastal dunes and marshes; < 100 m. NCo, CCo; to B.C. [*J. lesueurii* var. *tracyi* Jepson] ❀STBL.

J. bryoides F.J. Herm. (p. 1163) Ann, densely cespitose, 0.1–2.2 cm, reddish brown in fr. **ST** hair-like (0.1–0.2 mm wide). **LVS** basal, 1/4 to = st; sheath appendages 0. **INFL**: fl 1 per st; bracts 1–2, < 1 mm. **FL**: perianth segments 4–8, 1–2.8 mm (sepals < to = petals), incurved over fr; stamens 2–4, filaments > anthers. **FR** < perianth, ovoid to elliptic; chambers 2–3. **SEED** 0.3–0.5 mm, not striate; appendages small. Common. Wet places, washes, meadows, granitic seeps; 600–3600 m. SNH, SCoRO, TR, PR, W&I; to OR, ID, Colorado, Baja CA. Gen self-pollinating.

J. bufonius L. TOAD RUSH Ann, gen branched from base, 2–30 cm. **ST** gen ± 1 mm wide. **LVS** ± cauline, 1–3 per st, 0.5–1.5 mm wide. **INFL**: fls 1–few in small clusters, ± throughout pl; lowest bracts lf-like; bractlets 1–2.5 mm. **FL**: perianth segments 2–7 mm (sepals gen > petals), acuminate (or petals obtuse); stamens 6, filaments < to > anthers. **FR** < perianth, oblong to obovoid; tip acute.

SEED 0.3–0.6 mm, ovoid to elliptic; appendages 0 or small. Moist (sometimes saline) open or disturbed places; < 3200 m. CA; ± worldwide. Detailed study of variation in N.Am pls needed.

var. ***bufonius*** (p. 1163) Pl relatively large in all features. **INFL**: fls solitary at nodes. **FL**: perianth 4–7 mm, petals acuminate. **FR** < petals. $2n$=108. Habitat and range of sp. ❀STBL.

var. ***congestus*** Wahlenb. **INFL**: most fls crowded near slightly coiled branch tips (some lower fls gen solitary). **FL**: perianth 4–7 mm, petals acuminate. **FR** < petals. Uncommon. Saline habitats; gen < 200 m. c GV, CW. [var. *congdonii* (S. Watson) J. Howell]

var. ***occidentalis*** F.J. Herm. Pl relatively small in all features. **INFL**: fls solitary at nodes. **FL**: perianth 3–4 mm, petals acuminate. **FR** < petals. Dry pools, streamsides; < 3200 m. CA; to OR, ID, TX. [gen misidentified as *J. sphaerocarpus* Nees of Eur]

J. capillaris F.J. Herm. (p. 1163) Ann, densely cespitose, 0.7–5.7 cm, reddish brown in fr. **ST** 0.1–0.3 mm wide. **LVS** basal, < 2.2 cm; sheath appendages 0. **INFL**: fls 1–3 per st; bracts 0.4–1.5 mm, acute to acuminate, membranous. **FL**: perianth segments 4–6, 1.8–2.8 mm (petals > sepals); stamens 2–3, filaments > anthers (anthers < 0.7 mm). **FR** < perianth and much lighter in color, spheric to elliptic. **SEEDS** 9 or fewer per fr (3 or fewer per row), 0.5–0.8 mm, ± obscurely striate; appendages small. $2n$=36. Moist, bare to grassy places, meadows, streambanks, granitic seeps; 1200–3200 m. SNH; s OR. Gen self-pollinating.

J. capitatus Weigel Ann < 10 cm. **ST** thread-like, channeled, angled, or flat. **LVS** basal; blade 1.5–3.5 cm, margins ± inrolled; sheath loose, appendages 0. **INFL**: cluster 1 per st; fls 2–6; lower bracts 5–15 mm, gen > infl, lf-like, ± keeled. **FL**: perianth segments as few as 2 under poor conditions, sepals 3–4 mm, acuminate, petals 2–2.5 mm, ovate, thin; stamens 3. **FR** < petals, ovoid. **SEED** 0.2–0.3 mm, asymmetric. $2n$=18. Uncommon. Vernal pools; < 1000 m. CaRF, n SNF, GV; scattered in s-c US; native to Eurasia, n Afr.

J. chlorocephalus Engelm. (p. 1163) Per 12–50 cm; rhizome ± stout, mat-forming. **LVS** mostly basal, purple-red, blade 0; cauline lvs 3–4, sheath appendages 4–6 mm, clear, blade ± 1/3 st length, < 1 mm wide, cylindric to ± flat, blade crosswalls complete but obscure. **INFL**: lowest bract inconspicuous, gen reaching lowest cluster; clusters 1–3, ± as wide as long; bractlets clear, ± 1/3 perianth length. **FL**: perianth segments ± 4 mm, ± equal, middle pale green to red, margins wide, clear, sepals pointed, petals rounded; stamens 6, filaments < 1/2 anthers; style > perianth, stigmas prominent, 2.7 mm. **FR** << perianth, 3-angled, brown. **SEED** 0.6 mm, asymmetric, light brown. Wet areas in montane conifer forest; 1200–3000 m. n&c SN; NV. ❀TRY;GRNCVR.

J. confusus Cov. (p. 1163) Per, cespitose, 30–50 cm. **ST** slender, light green. **LVS** basal; sheath appendages ± 1 mm, thin, whitish; blades with flat side toward st, nearly thread-like, < 2/3 st length, thick, crosswalls 0, margins gen inrolled. **INFL** gen compact; lowest bract 2–8 cm; fls solitary, few; bractlets 2 per fl, large, ovate. **FL**: perianth segments 3.5–4 mm, ± equal, middle straw-colored, lateral stripes dark, margins wide, scarious; stamens 6, filaments >> anthers. **FR** oblong, slightly < perianth, 3-angled; chambers 3. **SEED** 0.4–0.5 mm; ends oblique; appendages 0. $2n$=80. Conifer forest; 1200–2000 m. KR, NCoRI, CaRH, n&c SNH; to w Can, SD, Colorado.

J. cooperi Engelm. (p. 1163) Per, ± cespitose, 40–80 cm; rhizome short, thick, much-branched; roots large, spongy. **LVS** basal; blades short, stiff, cylindric, tips sharp. **INFL** appearing lateral; lowest bract cylindric, resembling st, > infl, tip sharp; branches unequal; clusters 2–10-fld; bracts within infl obvious, > cluster; bractlets white. **FL**: perianth segments 5–6 mm (sepals > petals), pale greenish straw, tips acute, firm; stamens 6, large, filaments < anthers. **FR** slightly > perianth, narrowly oblong, 3-angled. **SEED** with a conspicuous white ridge; appendages unequal. Uncommon. Alkaline places; < 600 m. D; NV.

J. covillei Piper Per, cespitose, 5–25 cm; rhizome creeping, heavy. **LVS** mostly basal; sheath appendages 0; blade with flat side toward st, often = infl, 2–4 mm wide; cauline lvs 0–2. **INFL**: lowest bract < to > infl; branches 1–5; clusters 10–15, 3–7-fld. **FL**: perianth segments 2–3.5 mm (sepals > petals), middle papillate, brown or green; stamens 6, filaments slightly < anthers. **FR** > perianth, widely cylindric, dark; tip ± notched, short-beaked. **SEED** 0.4 mm; appendages minute. $2n$=38. Moist places, esp in forests; < 3000 m. NW, CaR, SN, SnGb, SnBr; to B.C.

var. ***covillei*** (p. 1163) **FL**: perianth dark brown; petals rounded. **FR** obviously > perianth, dark brown. Sandy places, streamsides; gen < 300 m. NCo; to B.C. ❀STBL.

var. ***obtusatus*** (Engelm.) C. Hitchc. **FL**: perianth pale brown; petals pointed. **FR** barely > perianth, pale brown. Montane forests, other moist places; gen 500–3000 m. KR, NCoR, CaR, SN, SnGb, SnBr; to ID. ❀STBL.

J. cyperoides Laharpe Per 5–30 cm; rhizome grading into st. **ST** branched from near base; nodes conspicuous. **LVS** evenly spaced on st, sometimes overlapping; sheath appendages 0; blade with flat side toward st, crosswalls 0. **INFL**: lowest bract < infl, resembling cauline lvs; branches many; clusters 1–40, 3–10-fld; bracts membranous. **FL**: perianth segments 2.2–5 mm, ± equal, acuminate, light-colored, sepals concave, petals flat; stamens 6, filaments >> anthers. **FR** ± = to > perianth, 3-angled; beak ± 0.5 mm. **SEED** 0.4 mm; appendages minute. Yellow-pine forest; ± 1000 m. n SNH (Butte Co); native to S.Am. Only N.Am site, 5600 km from closest population in S.Am Andes.

J. diffusissimus Buckley Per, cespitose, 20–75 cm. **ST** thin. **LVS** few; sheath appendages purplish at base; blade cylindric, crosswalls complete. **INFL** 10–30 cm; lowest bract inconspicuous; branches spreading; clusters many, 2–5-fld. **FLS** gen not all developing; perianth segments 2.5–3.5 mm, ± equal; stamens 3, filaments gen > anthers. **FR** gen 2 × perianth, slender, 3-angled. **SEED** 0.3–0.4 mm; appendages minute. Muddy gravel in open places; < 100 m. e ScV; native to c & se US.

J. drummondii E. Meyer (p. 1163) Per, densely cespitose, 5–40 cm. **LVS** basal; sheaths 2–7 cm, pale; blades 0 or small, inner bristle-like. **INFL** appearing lateral; lowest bract cylindric, resembling st, > infl; fls gen 2–3. **FL**: perianth 5–7 mm, sepals wider, gen > petals; stamens 6, filaments << anthers. **FR** = to > perianth; tip indented. **SEED**: body 0.5 mm, narrow; appendages > body. Moist, rocky places in conifer forests; 2000–3500 m. KR, CaR, SN; to AK, WY, NM.

J. dubius Engelm. (p. 1163) Per, gen densely matted, 15–70 cm; rhizome stout. **LVS**: basal blades 0; cauline lvs few, sheath appendages 4–6 mm, prominent, clear, blade cylindric, blade crosswalls complete but obscure. **INFL**: lowest bract gen 1–2 cm, inconspicuous; branches spreading; clusters gen many, 4–10-fld; bracts and bractlets clear. **FL**: perianth segments 2.5–3 mm, ± equal, bristle-tipped; stamens 6, filaments << anthers. **FR** barely > perianth, 3-sided. **SEED** obovate; appendages minute. Wet places; < 2000 m. CaRF, c&s SNF, SnFrB, SCoRO, PR; OR. ❀TRY;GRNCVR.

J. dudleyi Wieg. (p. 1163) DUDLEY'S RUSH Per, cespitose, 20–60 cm. **LVS** ± basal; sheath margins scarious; sheath appendages < 2 mm, firm, yellowish to reddish, glossy; blade with flat side toward st, gen < 1/2 st length, 1–1.3 mm wide, soft, margins gen inrolled. **INFL**: lowest bracts lf-like, 1 gen >> infl; branches unequal; fls solitary, gen denser near st tips; bractlets 2 per fl, blunt. **FL**: perianth 4–5 mm, lanceolate, acuminate, conspicuously spreading in fr; stamens 6, filaments >> anthers. **FR** < perianth, widely ovoid; chamber 1 (septa incomplete). **SEED** < 0.4 mm, oblong; appendages minute. $2n$=80. UNCOMMON. Wet areas in montane conifer forest; < 2000 m. KR; scattered to e N.Am. [*J. tenuis* Willd. var. *d.* (Wieg.) F.J. Herm.]

J. duranii Ewan (p. 1163) DURAN'S RUSH Per, cespitose, 3–20 cm; rhizome vertical. **ST** slender, ± flat. **LVS** mostly cauline; sheath appendages prominent, rounded; blade cylindric, gen = st, tip long, crosswalls complete but obscure. **INFL**: lowest bract > infl, narrow, not sheathing, rolled tip > rest of bract blade (not gen present in fr); cluster gen 1; bractlets awned. **FL**: perianth segments ± 3 mm, ± equal, linear-lanceolate, stiff, brown; stamens 6, filaments < or = anthers. **FR** ± = perianth, obovoid. **SEED** 0.5 mm, narrowly lanceolate. UNCOMMON. Wet places in montane conifer forests; 1800–2750 m. SnGb, SnBr, SnJt. [*J. mertensianus* var. *d.* (Ewan) F.J. Herm.]

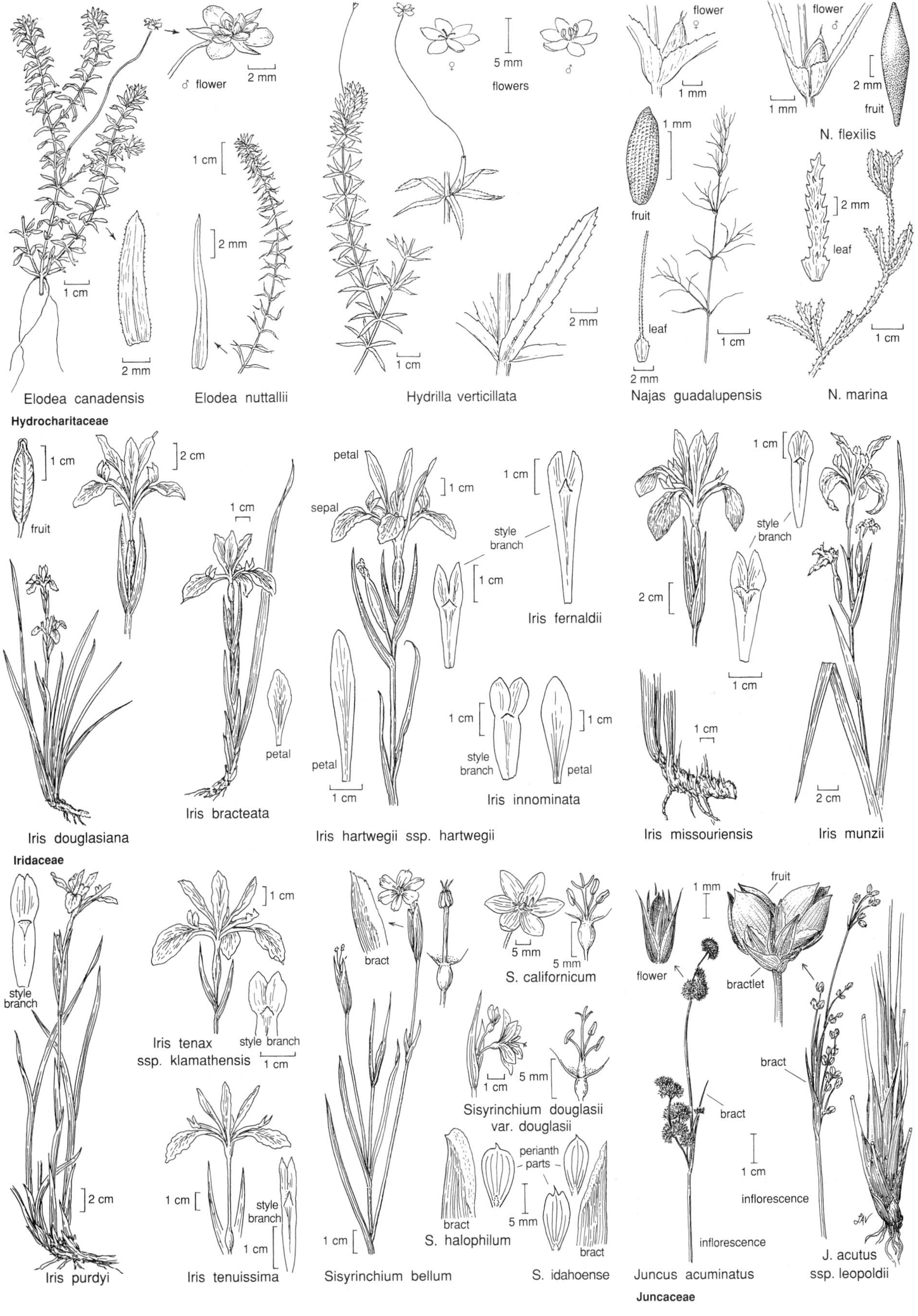
♂ flower
2 mm
1 cm
1 cm
2 mm
2 mm
Elodea canadensis
Elodea nuttallii
5 mm
♀
♂
flowers
2 mm
1 cm
Hydrilla verticillata
flower
♀
1 mm
1 mm
fruit
leaf
2 mm
1 cm
Najas guadalupensis
flower
♂
1 mm
2 mm
fruit
N. flexilis
2 mm
leaf
1 cm
N. marina
Hydrocharitaceae
1 cm
fruit
2 cm
Iris douglasiana
1 cm
petal
Iris bracteata
petal
sepal
1 cm
style
branch
1 cm
petal
1 cm
Iris hartwegii ssp. hartwegii
1 cm
Iris fernaldii
1 cm
style
branch
1 cm
petal
Iris innominata
2 cm
1 cm
style
branch
1 cm
1 cm
Iris missouriensis
2 cm
Iris munzii
Iridaceae
style
branch
2 cm
Iris purdyi
1 cm
Iris tenax
ssp. klamathensis
style branch
1 cm
1 cm
style
branch
1 cm
Iris tenuissima
bract
1 cm
Sisyrinchium bellum
5 mm
5 mm
S. californicum
1 cm
5 mm
Sisyrinchium douglasii
var. douglasii
bract
perianth
parts
5 mm
S. halophilum
bract
S. idahoense
1 mm
flower
fruit
bractlet
bract
bract
1 cm
inflorescence
inflorescence
Juncus acuminatus
J. acutus
ssp. leopoldii
Juncaceae

J. effusus L. Per, cespitose, 60–130 cm; rhizome stout, branched. **ST** 1–3.5 mm wide at base. **LVS** basal; sheaths 5–15 cm; blades 0. **INFL** appearing lateral, open; lowest bract cylindric, resembling st, >> infl; fls gen many, single; bractlets 2 per fl. **FL**: perianth segments 1.8–4.2 mm, ± equal, gen ascending, very pale to dark; stamens 3, filaments > anthers. **FR** ± = perianth, obovoid, ± truncate. **SEED** 0.5 mm; appendage 1, minute. 2*n*=40. Wet places; < 2500 m. CA exc D; to B.C., e US, Mex, Eurasia.

var. ***brunneus*** Engelm. (p. 1163) **LF**: sheath dull brown at base, becoming much lighter at top, edges not overlapping at top. **FL**: perianth 2–3.5 mm, soft when fresh, not rigid when dry, dark brown. Moist places incl salt marshes; gen < 100 m. NCo, CCo, SCo; to B.C. ❀IRR or WET:**4,5**,8,9,14,**15–17**,19–21,**22–24**; may be INV; also STBL.

var. ***exiguus*** Fern. & Wieg. (p. 1163) **LF**: sheath short, straw-colored to greenish, dull. **FL**: perianth 1.8–2 mm, ± spreading from base, soft when fresh, not rigid when dry, pale brown. Montane; 700–2300 m. KR, NCoR, CaR, n&c SN, SnBr; to WA. ❀TRY.

var. ***gracilis*** Hook. (p. 1163) **LF**: sheath dark at top, shiny. **FL**: perianth 2–3.5 mm, soft when fresh, not rigid when dry, dark brown. Moist places; < 2500 m. NW, CaR, SN; to B.C. ❀TRY.

var. ***pacificus*** Fern. & Wieg. (p. 1163) **LF**: sheath uniformly dark, edges overlapping at top. **FL**: perianth 2.5–4.2 mm, firm when fresh, rigid when dry, green to pale brown. Many moist habitats; < 2500 m. CA exc D; to B.C. A large, s CA variant with dark lf sheaths may warrant status as var. ❀IRR or WET:1–3,**4–7**,8,9, **14–24**.

J. ensifolius Wikström (p. 1163) Per 20–60 cm; rhizome slender, creeping. **LVS** mostly basal; bases overlapping; sheath appendages obscure; blade flat, with edge toward st, 2–5 mm wide, gen curved, crosswalls incomplete, tip long. **INFL**: lowest bract ± 1/2 infl length, gen curved; clusters 2–7, many-fld, hemispheric. **FL**: perianth segments ± 3 mm, ± equal, dark brown to black; stamens gen 3, ± 1/2 perianth length, filaments > anthers. **FR** barely > perianth, oblong, abruptly short-beaked. **SEED** ± 0.5 mm, widely fusiform; appendages 0. 2*n*=40. Common. Wet places; < 2800 m. NW, CaR, SN, CCo, SnBr; to AK, w Can, UT, Mex. ❀STBL.

J. falcatus E. Meyer var. ***falcatus*** (p. 1163) Per 3–30 cm; rhizome long, scaly. **LVS** mostly basal; sheath appendages 0 or small; blade with flat side toward st, gen = infl, stiff and somewhat curved; cauline lvs 1–2, < basal, stiff. **INFL**: lowest bract gen > infl; clusters 1–3, 5–25-fld, hemispheric. **FL**: perianth segments 5–6 mm, ± equal, papillate, margins wide, clear, very dark; stamens 6, filaments << anthers. **FR** ± = perianth, ± spheric; tip notched; beak short but obvious. **SEED** asymmetric; outer covering loose, white; appendages 0. 2*n*=38. Moist, sandy, coastal areas; gen < 100 m. NCo, CCo, SCo; to B.C., e Asia, Australia. ❀STBL.

J. hemiendytus F.J. Herm. Ann, densely cespitose, 0.1–3.2 cm. **ST** 0.1–0.5 mm wide. **LVS** basal, < 1.8 cm; sheath appendages 0. **INFL**: fl gen 1 per st; bracts 0–2, < 2 mm, ovate to rounded, not sheathing st. **FL**: perianth segments 4–6, 1.9–3.5 mm (sepals = or > petals), dull; stamens 2–3, filaments > anthers, (anthers < 0.7 mm). **FR** < to > perianth, obovoid to oblong; chambers 2–3. **SEED** 0.3–0.55 mm, not striate; appendages small. Damp open areas, esp vernally wet places; 400–3400 m. NW, SN, SnBr, GB; to WA, ID, UT. Gen self-pollinating.

var. ***abjectus*** (F.J. Herm.) B. Ertter CENTER BASIN RUSH **ST** 0.2–0.5 mm wide, wider below fl. **INFL**: bracts 0. **FR** gen < perianth. UNCOMMON. Habitat of sp.; 1400–3400 m. SNH; to OR, ID, NV.

var. ***hemiendytus*** (p. 1163) **ST** 0.1–0.3 mm wide, not wider below fl. **INFL**: bracts 1–2. **FR** > perianth. 2*n*=32. Habitat and range of sp.; 400–3200 m.

J. howellii F.J. Herm. (p. 1163) Per 15–60 cm; rhizome branched, stout. **LVS** mostly basal; sheath appendages 1–3 mm, acute; blade with flat side toward st, 2–4.5 mm wide, crosswalls 0, margins minutely scabrous; cauline lvs 1–2. **INFL**: lowest bract < 1.5 cm; clusters 2–10, 3–8-fld. **FL**: perianth segments 5–6.5 mm (sepals < petals), acuminate, middle dark brown or green and sometimes minutely papillate; stamens 6, filaments << anthers. **FR** < perianth, obovoid; beak small. **SEED** 0.5–0.7 mm; appendages gen > 1/2 body length. Meadows; 850–2380 m. CaRH, n SNH; OR. ❀TRY.

J. kelloggii Engelm. (p. 1163) Ann, densely cespitose, 0.1–6 cm; reddish in fr. **ST** 0.1–0.3 mm wide, hair-like. **LVS** basal, < 2.5 cm, 0.1–0.4 mm wide. **INFL**: fls gen 3 (1–4) per st; bracts 1–2.5 mm, acute to acuminate. **FL**: perianth segments 4–6, 1.5–3.2 mm, unequal; stamens 2–3, filaments > anthers, anthers < 0.7 mm. **FR** ± = perianth and similar in color, obovoid to elliptic. **SEED** 0.4–0.5 mm, ovoid, obviously striate; appendages minute. 2*n*=34. Uncommon. Damp sandy or clay soils, vernal pools, seeps, fields, meadows; < 800 m. NCo, NCoR, ScV, CCo; to B.C. Gen self-pollinating.

J. leiospermus F.J. Herm. DWARF RUSH Ann, densely cespitose, 1.9–11.6 cm, pale to reddish brown. **ST** gen > 0.4 mm wide. **LVS** basal, < 3/4 st length; sheath appendages 0. **INFL**: cluster 1, 1–7-fld; bracts 0.7–2.4 mm. **FL**: perianth segments 6–10, 1.5–4.6 mm (petals sometimes > sepals); stamens 3–5, filaments 1/4 to = anthers, anthers > 0.7 mm. **FR** spheric to oblong, gen > perianth, resembling perianth in color; chambers 3–5. **SEED** 0.3–0.4 mm, ovoid, not striate; appendages minute (or 1 obvious). 2*n*=32. Vernal pool margins, wet places in chaparral, woodland; 50–500 m. CaRF, ScV, ne SnJV.

var. ***ahartii*** B. Ertter AHART'S DWARF RUSH Pl features at small end of range; parts gen fewer. **FLS** 1–2 per st. RARE. Vernal pool margins; 50–100 m. e ScV, ne SnJV (Butte, Calaveras cos.).

var. ***leiospermus*** (p. 1163) RED BLUFF DWARF RUSH Pls at larger end of range in all features; parts gen more. **FLS** 2–7 per st. 2*n*=32. RARE. Habitat of sp.; 280–500 m. n ScV, CaRF (Shasta, Tehama, Butte cos.).

J. lesueurii Bolander Per 30–140 cm; rhizome creeping. **LVS** basal; sheaths light brown; blades 0. **INFL** appearing lateral; branches open; lowest bract cylindric, resembling st, gen < 1/2 st length; bractlets 2 per fl; fls many. **FL**: perianth 5–8 mm (sepals > petals), acuminate, middle greenish, margins dark; stamens 6, filaments << anthers. **FR** < perianth, light brown; tip acute, 3-angled. **SEED** 0.4–0.7 mm, ovoid; appendages 0. Salt or freshwater marshes; < 100 m. NCo, CCo; S.Am. ❀IRR or WET:4,5,8,9,**14–17**,19–21,**22–24**;INV;STBL.

J. longistylis Torrey (p. 1163) Per, loosely cespitose, 20–60 cm. **ST** slender. **LVS** mostly basal; sheath appendages obvious, 0.5–2 mm, often truncate; blade with flat side toward st, < 1/2 st length, 1–3 mm wide, crosswalls 0; cauline lvs 1–3. **INFL**: lowest bract 1–2 cm, membranous; clusters 1–9, 3–12-fld. **FL**: perianth segments 5–6 mm, ± equal (or sepals > petals), middle dark green, scarious margin wide; stamens 6, filaments < anthers. **FR** < perianth, 3-angled, brown; beak short. **SEED** 0.5 mm, narrow; appendages short. 2*n*=40. Moist places in conifer forests; 1800–2900 m. SNH, TR, PR, Wrn, W&I; to B.C., e N.Am, NM.

J. luciensis B. Ertter Ann, densely cespitose, 0.4–6.2 cm, remaining pale yellow-green. **ST** 0.1–0.3 mm wide, hair-like. **LVS** basal, < 1.5 cm; sheath appendages 0. **INFL**: fls 1–2 per st; bracts 0.4–1.6 mm, acute to acuminate. **FL**: perianth segments 4–6, 1.6–4.2 mm, gen ± equal; stamens 2–3, filaments > anthers, anthers < 0.7 mm. **FR** ± = perianth; chambers 2–3. **SEED** 0.3–0.4 mm, obviously striate; appendages 0 or small. 2*n*=32. Uncommon. Wet, sandy soils of seeps, meadows, vernal pools, streamsides; 300–1900 m. n SNH, SCoRO, TR, PR. Gen self-pollinating.

J. macrandrus Cov. (p. 1167) Per, cespitose, 30–50 cm; rhizome creeping. **LVS** mostly basal; bases overlapping; sheath appendages 0; blade flat, with edge toward st, 2–3 mm wide, crosswalls gen incomplete. **INFL**: lowest bract << infl; clusters 3–many, 3–5-fld. **FL**: perianth 3–4 mm, dark purple-brown; stamens 6, filaments << conspicuous anthers. **FR** < perianth, widely oblong; beak abruptly pointed. **SEED** squarish; appendages small. Wet places, montane conifer forests; 1200–2800 m. SNH, TR.

J. macrophyllus Cov. (p. 1167) Per 20–100 cm; rhizome spreading. **LVS** mostly basal; sheath appendages 1.5–3 mm, membranous; blade with flat side toward st, often = st, 1.5–3 mm wide, somewhat channeled, midrib prominent; cauline lvs 1–2, thick. **INFL**: lowest bract lf-like, < infl; branches ascending; clusters 8–25, 3–5-fld. **FL**: perianth segments 5–6 mm (sepals < petals), margins membranous;

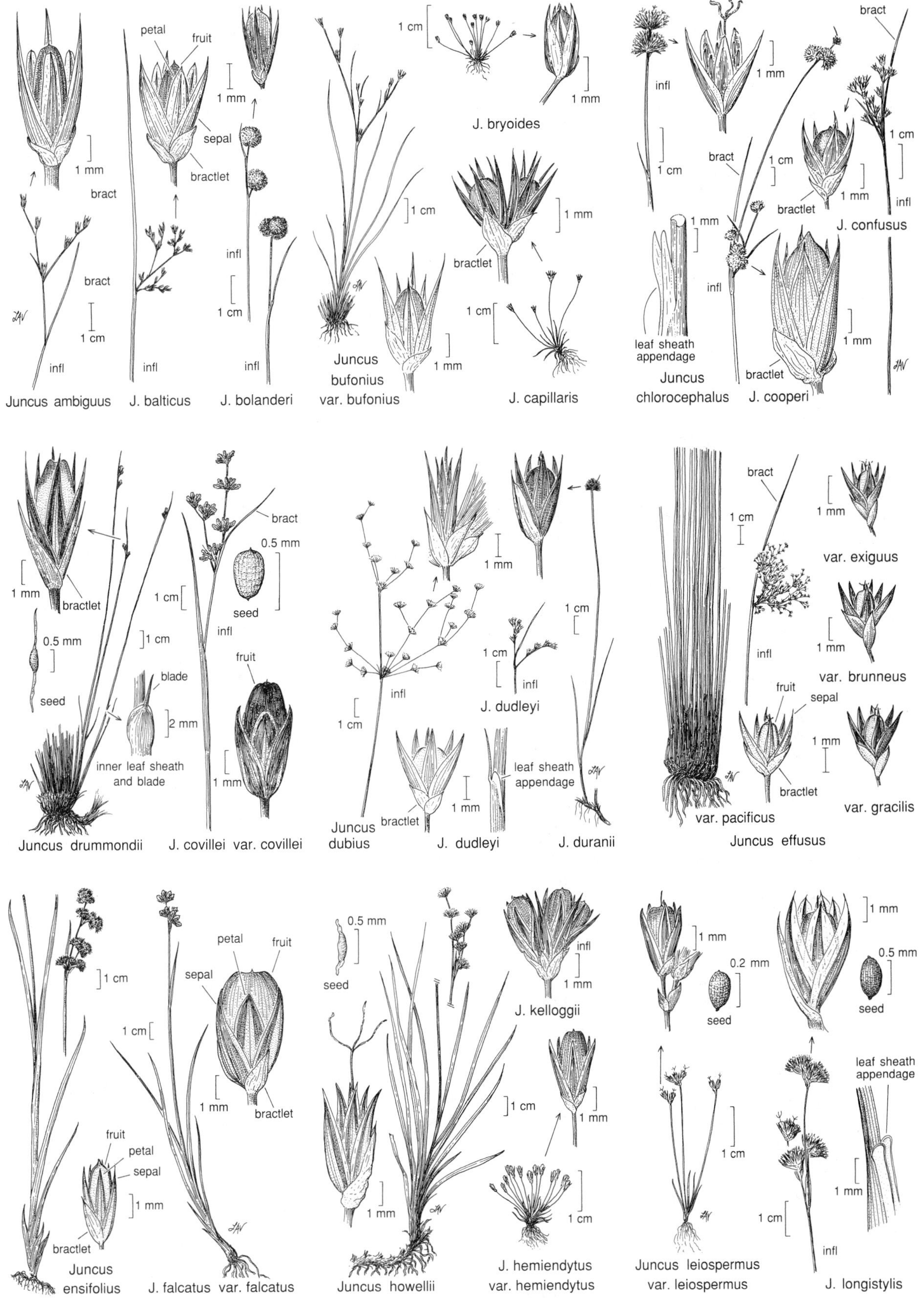

bract
infl
1 mm
1 cm
petal
fruit
sepal
bractlet
Juncus ambiguus
J. balticus
J. bolanderi
Juncus bufonius var. bufonius
J. bryoides
J. capillaris
leaf sheath appendage
Juncus chlorocephalus
J. cooperi
J. confusus
Juncus drummondii
0.5 mm
seed
2 mm
blade
inner leaf sheath and blade
J. covillei var. covillei
Juncus dubius
J. dudleyi
J. duranii
var. exiguus
var. brunneus
var. pacificus
var. gracilis
Juncus effusus
Juncus ensifolius
J. falcatus var. falcatus
Juncus howellii
J. kelloggii
J. hemiendytus var. hemiendytus
0.2 mm
Juncus leiospermus var. leiospermus
J. longistylis

stamens 6, filaments << anthers. **FR** gen < to ± = perianth, obovoid, short-beaked, shiny brown. **SEED** plump; appendages 0. Uncommon. Wet slopes; < 2600 m. SCoRO, TR, DMtns; to AZ, Baja CA.

J. marginatus Rostkov var. ***marginatus*** (p. 1167) Per 15–70 cm; rhizome short, thick, knotty. **ST** slender. **LVS** distributed along st; sheath appendages rounded, scarious; blade with flat side toward st, < infl, 1–6 mm wide, soft, crosswalls 0. **INFL** open or compact; lowest bract << infl; clusters 2–40, 5–12-fld. **FL**: perianth segments 2.5–3.5 mm (sepals < petals), sharply acute (petals blunter), reddish brown; stamens 3, filaments >> purplish red anthers. **FR** = perianth, spheric; beak 0. **SEED** 0.5 mm, ovoid, brown; appendages minute. $2n$=38,40. RARE in CA. Swampy places; < 1000 m. n SNF; scattered across US, s Can.

J. mertensianus Bong. (p. 1167) Per 15–45 cm; rhizome vertical, stout. **ST** flat. **LF**: cauline lvs 2–3; sheath appendages prominent, rounded to acute, opaque; blade ± flat with edge toward st, crosswalls complete but gen obscure. **INFL**: lowest bract short-sheathing, wide, narrow tip > infl; cluster gen 1, gen > 12-fld; bractlets short-awned, dark brown, opaque. **FL**: perianth segments 3–4 mm, soft, shiny brownish black, sepals narrowly acuminate, petals acute; stamens 6, filaments < to > anthers. **FR** ± = perianth, oblong; tip notched. **SEED** ± 0.5 mm, ovate-lanceolate; appendages minute. $2n$=40. Common. Alpine and subalpine meadows, streambanks, lake margins; 1200–3500 m. KR, NCoRI, CaRH, SNH, Wrn, W&I; to AK, SD, Colorado. Important forage for sheep. ❀WET: **1–3**,18.

J. mexicanus Willd. (p. 1167) Per 10–60 cm; rhizome heavy. **ST** erect or gen spirally twisted, flattened, slender. **LVS** basal; sheaths loose, appendages short, firm; upper sheaths gen bearing 5– 20 cm blades that resemble st (blades less common farther n). **INFL** appearing lateral, ± compact; lowest bract 1–25 cm, cylindric, resembling st; fls 2–many. **FL**: perianth segments 3–5.5 mm (sepals > petals), gen acuminate, color variable, midstripe varying in width and intensity, margins gen clear; stamens 6, filaments << anthers. **FR** < to > perianth, ovoid, 3-angled; beak 0.3 mm. **SEED** 0.5–0.7 mm; appendages 0. Common. Coast to montane meadows; < 3800 m. CA (exc GV); to WA, Colorado, TX, S. Am. ❀STBL.

J. nevadensis S. Watson (p. 1167) Per gen 10–50 cm; rhizome creeping. **ST** slender. **LF**: sheath appendages 2–3 mm, membranous; blade gen < 2 mm wide, cylindric, crosswalls complete, obvious or obscure. **INFL** variable; lowest bract inconspicuous; clusters gen 1–4 (many), < 10-fld. **FL**: perianth segments 3–3.5 mm, ± equal, dark brown; stamens 6, filaments << anthers. **FR** < to ± = perianth, oblong, shiny brown; tip abruptly beaked. **SEED** 0.6 mm, ± spheric, brown; appendages 0. Common. Mtn meadows, streambanks; 1200–3300 m. KR, NCoRH, NCoRI, CaRH, SNH, W&I, SnGb, SnBr; to B.C., WY. [*J. phaeocephalus* Engelm. var. *gracilis* Engelm.; *J. mertensianus* ssp. *gracilis* (Engelm.) F.J. Herm.] Important forage for cattle and horses.

J. nodosus L. (p. 1167) KNOTTED RUSH Per 15–60 cm; rhizome creeping, slender, tuber-bearing. **ST** slender. **LF**: sheath appendages 0.5–1 mm, rounded, firm; blade cylindric, upper > infl, crosswalls complete, prominent. **INFL**: lowest bract = to > infl; clusters 2–20, 10–25-fld, spreading, spheric. **FL**: perianth segments 2.5–4.1 mm, acuminate, ± equal; stamens 6, filaments > anthers. **FR** > perianth, slender, sharply 3-angled, long-tapered. **SEED** 0.5 mm; appendages small. $2n$=40. UNCOMMON. Streambanks, lake shores, wet meadows; < 1700 m. se SNH, n DMtns; scattered across US, s Can.

J. occidentalis (Cov.) Wieg. (p. 1167) Per, cespitose, stiff, 30–60 cm. **LVS** many, basal; sheath appendages < 1 mm, membranous; blade with flat side toward st, gen < 1/2 st, 1–1.5 mm wide, crosswalls 0. **INFL** gen open; fls solitary; lowest bract > infl; bractlets 2 per fl, conspicuous. **FL**: perianth segments 4–5 mm, ± equal, middle green, scarious margins wide; stamens 6, filaments >> anthers. **FR** < 3/4 perianth, oblong to ovate, 3-angled, dark brown; tip notched; chamber 1 (septa incomplete). **SEED** oblong; appendages obvious. Moist areas; < 2300 m. NCo, CCo, n&c SN; OR, NV, AZ. [*J. tenuis* var. *o.* Cov.; *J. t.* var. *congestus* Engelm.] ❀IRR or WET:1–3,**4–7,14–18**,19–24.

J. orthophyllus Cov. (p. 1167) Per 20–50 cm; rhizome scaly, hard, dark, creeping. **LVS** mostly basal; sheath appendages 0 or minute; blade with flat side toward st, << to ± = st, 2–6 mm wide, crosswalls 0; cauline lvs 1–3, above middle. **INFL** open; lowest bract inconspicuous; branches papillate; clusters 1–many, 6–10-fld, hemispheric; bractlets prominent, ± 1/2 perianth. **FL**: perianth segments 5–6 mm, minutely papillate, bristle-tipped, middle green, margins brown, sepals < petals; stamens 6, filaments < anthers. **FR** slightly < perianth; tip flat; beak small. **SEED** 0.5 mm, brown; appendages minute. Inland wet places, esp meadows, streambanks in forest; 1200–3500 m. NCoRH, CaRH, SNH, TR, DMtns; to WA, ID, NV.

J. oxymeris Engelm. (p. 1167) Per 30–60 cm; rhizome creeping. **LF**: bases overlapping; sheath appendages 0 or small; blade flat, with edge toward st, 3–7 mm wide, crosswalls gen incomplete. **INFL** large, open; lowest bract inconspicuous; branches ascending; clusters 10–70, few-fld. **FL**: perianth segments 2.5–4 mm, ± equal, sharp-pointed, light brown to green, margins scarious; stamens 6, filaments << conspicuous anthers. **FR** gen > perianth, 3-angled, gradually tapered to slender beak, brown. **SEED** gen 0.5 mm; appendages small, blunt. Uncommon. Montane meadows, shrubland, yellow-pine forest; < 2100 m. NCoRI, SN, GV, TR; to WA. Highly variable, possibly from hybridization with *J. phaeocephalus*.

J. parryi Engelm. (p. 1167) Per, densely cespitose, 6–30 cm. **LVS** basal; sheaths many, appendages 0 (or obscure, membranous); blades on inner lvs only, 3–8 cm, thread-like. **INFL** appearing lateral; lowest bract gen >> infl, cylindric; fls 1–3, sometimes sessile; bractlets 2 per fl, dissimilar. **FL**: perianth segments 5–7 mm (sepals > petals), pale brown, tips acuminate, veins obvious in lower part; stamens 6, filaments << anthers. **FR** gen > perianth, 3-angled; tip acute. **SEED** 2 mm, ovate; appendages > seed body. Dry granitic slopes; 2000–3800 m. KR, SN, SnBr; to sw Can, Colorado.

J. patens E. Meyer (p. 1167) Per, densely cespitose, 30–90 cm; rhizome stout, branched. **ST** bluish gray-green when fresh, distinctly grooved. **LVS** basal; sheaths dark brown; blades 0 or rudimentary. **INFL** appearing lateral, open to compact; lowest bract gen >> infl, cylindric, resembling st; fls gen many. **FL**: perianth segments 2.5–3 mm (sepals > petals), narrow, acuminate, spreading in fr; stamens 6, filaments >> anthers. **FR** < to = perianth, ± spheric, soft-beaked. **SEED** 0.4–0.5 mm, asymmetric; appendages minute. Marshy places; < 1600 m. KR, NCoR, CW, SCo, ChI, WTR; OR. ❀**4,5,17** IRR or WET:7,8,9,**14–16,18–24**; also STBL.

J. phaeocephalus Engelm. Per 10–50 cm; rhizome stout, creeping. **ST** flat. **LF**: bases overlapping; sheath appendages indistinct; blade flat, with edge toward st, often > st, 1.5–4 mm wide, crosswalls gen incomplete, tip fine-pointed. **INFL**: lowest bract slightly > lowest cluster; clusters 1–many, few–many-fld. **FL**: perianth segments 4–5 mm, ± equal, widely lanceolate, dark brown; stamens 6, filaments << conspicuous anthers; stigmas long-exserted. **FR** < to = perianth; beak long-tapered. **SEED** 0.6 mm, ovoid. Moist places; < 2200 m. NCo, n SNH, CCo, SnFrB, SCo, n ChI, SnBr, PR.

var. ***paniculatus*** Engelm. **STS** in large clumps. **INFL**: clusters > 10, 5–10 mm wide, few-fld. Wet places, coastal and inland; 2200 m. NCo, n SNH, SnFrB, SnBr, PR. ❀TRY.

var. ***phaeocephalus*** (p. 1167) **STS** separated along rhizome. **INFL**: clusters 1–10, 10–15 mm wide, many-fld. Coastal meadows and borders of marshes; gen < 100 m. NCo, CCo, SCo, n ChI. ❀IRR:4,5,14,**15–17**,19–21,**22–24**;INV;also STBL.

J. regelii Buchenau (p. 1167) REGEL'S RUSH Per, sometimes cespitose, 30–60 cm; rhizome stout. **ST** slender, ± flat. **LVS** mostly basal; sheath appendages 0 or minute; blade with flat side toward st, 1–3 mm wide, crosswalls 0; cauline lvs 1–3, < infl. **INFL**: lowest bract 1–4 cm, papillate; branches few, straight; clusters 1–5, 10–30-fld. **FL**: perianth segments 4–6 mm, ± equal, papillate, bristle arising below tip, sepals pointed, petals more rounded; stamens 6, filaments < to = anthers. **FR** ± = perianth, spheric, dark purplish brown; beak ± 0.5 mm. **SEED** 1.2–1.8 mm; appendages > seed body. UNCOMMON. Mtn meadows; 800–1900 m. KR; to B.C., WY, UT.

J. rugulosus Engelm. (p. 1167) WRINKLED RUSH Per, often densely matted, 15–70 cm; rhizome stout, horizontal; surfaces prominently wrinkled. **LF**: basal blades 0; sheath appendages 4–6 mm, prominent, clear; cauline blades cylindric, crosswalls complete but obscure. **INFL**: lowest bract gen inconspicuous; branches spreading; clusters gen many (<150), 4–8-fld; bracts and bractlets clear. **FL**: perianth segments brown to brownish red, clear margins wide, sepals 2.5 mm, ± bristle-tipped, petals 2.5–3 mm, petal veins prominent at base; stamens 6, filaments ± = anthers. **FR** > perianth, 3-angled, long-beaked, bright brownish red. **SEED** 0.4 mm, plump, brown; appendages minute. 2*n*=40. Common. Wet places; < 2100 m. s SNF, CCo, TR, PR, DMtns. ❀TRY;STBL.

J. saximontanus Nelson (p. 1167) Per 30–60 cm; rhizome stout, creeping. **LF**: bases overlapping; sheath margins membranous, appendages small but distinct; blade flat, with edge toward st, 2–5 mm wide, crosswalls gen incomplete. **INFL**: lowest bract short; clusters few–many, 4–25-fld. **FL**: perianth segments 2.5–3 mm (sepals > petals), lanceolate, pale to dark brown; stamens 6 (3), ± 2/3 perianth length, filaments > inconspicuous anthers. **FR** < to ± = perianth, oblong. **SEED** 0.4–1 mm; appendages minute to 0.3 mm. Wet places, montane conifer forest; 1500–2900 m. KR, CaRH, SNH, SnGb, SnBr, W&I; to AK, TX, Mex. [forma *brunnescens* (Rydb.) F. J. Herm]

J. supiniformis Engelm. (p. 1167) HAIR-LEAVED RUSH Per, submerged when young, cespitose, 8–40 cm; rhizome matted, slender, spreading. **ST**: nodes often rooting, forming new plantlets; erect fl-sts appear as water recedes. **LVS**: submerged lvs < 30 cm, hair-like; sheath appendages 1–2 mm, membranous; cauline blades > st, cylindric, crosswalls complete but obscure. **INFL**: lowest bract > infl; clusters 2–6, 3–9-fld. **FL**: perianth segments ± 4 mm, ± equal, narrowly lanceolate, 3–4-veined; stamens gen 3 (6), filaments > anthers. **FR** > perianth, oblong, short-beaked. **SEED** 0.6–0.7 mm, obovoid; ends pointed; appendages minute. UNCOMMON. Marshes, ponds, ditches; gen < 100 m. n&c NCo; to AK. [*J. oreganus* S. Watson]

J. tenuis Willd. (p. 1167) Per, cespitose, 20–60 cm. **LF**: sheath margins membranous, appendages 1–6 mm; blade with flat side toward st, 1/2 to = st, 1–1.3 mm wide, soft, crosswalls 0. **INFL** ± open; lower bracts rarely 3, lf-like, >> infl; branches unequal; fls solitary, somewhat denser near branch tips; bractlets 2 per fl, acute. **FL**: perianth segments 3–4.5 mm, ± equal, lanceolate, acuminate, spreading in fr, pale brown; stamens 6, filaments >> anthers. **FR** < perianth, widely ovoid; chamber 1 (septa incomplete). **SEED** < 0.4 mm, oblong; appendages minute, blunt. 2*n*=80. Uncommon. Damp places; gen < 1500 m. NW, CW; ± worldwide. ❀IRR or WET:**4–7**,8,9,**14–17**,18–21,**22–24**.

J. textilis Buchenau (p. 1167) Per 100–200 cm; rhizome heavy, creeping. **ST** cylindric, striate; base 3–5 mm wide. **LVS** basal; sheaths < 18 cm, brown; blades 0. **INFL** lateral, open; lowest bract = to > infl, cylindric, resembling st; branches 5–10 cm, extremely variable; fls many, solitary; bractlets 2 per fl, tips rounded. **FL**: perianth segments 3.5–5 mm (sepals barely > petals, acute), green to light brown; stamens 6, filaments << anthers. **FR** < to ± = perianth, obovate-obtuse, dark brown, shiny, often aborted. **SEED** 0.5–0.8 mm, plump; appendages 0. Dry or moist soils; < 1800 m. s SCoRO, n&c SW. ❀STBL.

J. tiehmii B. Ertter (p. 1167) Ann, densely cespitose, 0.5–6 cm. **ST** 0.1–0.2 mm wide. **LVS** basal, < 2.5 cm; sheath appendages 0. **INFL**: fls 1–7 per st; bracts 0.6–1.5 mm, acute or acuminate. **FL**: perianth segments 4–6, 1–2.9 mm, ± equal, pale green or pink; stamens 2–3, filaments > anthers, anthers < 0.7 mm. **FR** gen > perianth, obovoid to oblong, color of perianth. **SEED** 0.3–0.5 mm, obviously striate; appendages small. 2*n*= 34. Bare, moist granitic sand of seeps, streamsides, meadows; 300–3100 m. SN, SCoR, TR, PR; to OR, ID, NV, Baja CA.

J. torreyi Cov. (p. 1167) Per 30–100 cm; rhizome thin, creeping, bearing narrow tubers. **ST** stout. **LF**: sheath appendages prominent, thin; blade cylindric, crosswalls complete. **INFL**: lowest bract > infl; clusters 1–20, 25–80-fld, crowded. **FL**: perianth segments slender, tapered to rigid points, sepals 4–5.6 mm, petals 3.5–5 mm; stamens 6, filaments > anthers. **FR** ± > perianth, thin, 3-angled, tapered to point, with seeds only below middle. **SEED** 0.4–0.5 mm, oblong; appendages 0. 2*n*=40. Meadows, moist woodlands; 0–1800 m. SW, GB, D; to s Can, much of US.

J. triformis Engelm. (p. 1173) Ann, densely cespitose, 1.8–16.5 cm; often reddish in fr. **ST** < 0.4 mm wide, hair-like. **LVS** basal, < 1/3 st length; sheath appendages 0. **INFL**: fls 1–8 per st; bracts 0.7–2.4 mm. **FL**: perianth segments 4–6, 1.3–4.5 mm (sepals gen < petals), middle green to red; stamens 2–3, filaments < anthers, anthers > 0.7 mm. **FR** gen < perianth, spheric to oblong, color of perianth; chambers 2–3. **SEED** 0.3–0.7 mm, striate at 10×. 2*n*=36. Uncommon. Vernal pools, granitic seeps; 50–2500 m. SN, SnJV, SCo, PR. [*J. megaspermus* F.J. Herm.]

J. uncialis E. Greene (p. 1173) Ann, densely cespitose, 0.3–3.5 cm, not reddish in fr. **ST** 0.1–0.5 mm wide. **LVS** basal, < 3/4 st length; sheath appendages 0. **INFL**: fl 1 per st; bract 1, sheathing st, 0.2–0.9 mm, truncate. **FL**: perianth segments 4–8, 1.7–5.3 mm (sepals gen > petals), dull; stamens 2–3, filaments > anthers, anthers < 0.7 mm. **FR** ± = to > perianth, ovoid or elliptic; chambers 2–4. **SEED** 0.3–0.4 mm, ovoid, not striate; appendages 0 or small. 2*n*=32. Uncommon. Vernal pool margins, other drying places; 45–1700 m. KR, NCoRI, GV, SCoRO, SCo, MP; OR, NV. Gen self-pollinating.

J. xiphioides E. Meyer (p. 1173) Per 40–80 cm; rhizome stout, creeping. **LF**: sheath appendages obscure; blade 3–12 mm wide, flat, with edge toward st, curved, crosswalls gen incomplete, tip long. **INFL**: lowest bract < 1/2 infl length; clusters many, 3–10-fld. **FL**: perianth segments 3–3.5 mm, very narrow (revealing fr between), green to brown; stamens gen 6, filaments = to > inconspicuous anthers. **FR** = to > perianth, oblong, gradually tapered to beak. **SEED** ± 0.5 mm; appendages minute. 2*n*=40. Wet places; 0–2100 m. CA; OR, AZ, Baja CA. ❀STBL;INV.

LUZULA HAIRY WOOD RUSH

Per; rhizome often short, vertical. **ST** round. **LVS** mostly basal, reduced upward; sheath closed; margin and sheath opening gen with long, soft hairs; blades flat or channeled, veins indistinct, tips often thick. **INFL**: head-like clusters or panicles of separate fls; bractlets 1–3, margins often hairy. **FL**: stamens 6; pistil 1, chamber 1, placenta basal. **SEEDS** 3, plump, elliptic, often with a distinct ridge, sometimes attached to placenta by tuft of hairs. 80 spp.: worldwide, esp n hemisphere. (Latin: light; Italian: glowworm) When present, fleshy seed appendage (outer seed coat) adapts large seeds to ant dispersal.

1. Fls solitary or 2–3 at ends of infl branches
  2. Infl branches widely spreading, stiff; lf-sheath opening glabrous .... ***L. divaricata***
  2′ Infl branches erect or arching; lf-sheath opening hairy .... ***L. parviflora***

1′ Fls in 1 or more clusters, often head-like
  3. Lvs channeled; infl nodding — infl dense .... ***L. spicata***
  3′ Lvs flat; infl not nodding
    4. Infl not distinctly head-like; perianth 1.5–2 mm .... ***L. subcongesta***
    4′ Infl ± head-like; perianth 2–5 mm
      5. St and infl gen greenish, sometimes brown; infl gen at least slightly open, with spreading branches; widespread .... ***L. comosa***
      5′ St and infl blackish; infl very compact; alpine .... ***L. orestera***

L. comosa E. Meyer (p. 1173) Pl cespitose, 10–40 cm. **LF**: sheath opening and blade margin with long, soft, wavy hairs; blade 5–15 cm, 3–7 mm wide, tip thickened. **INFL** umbel-like; clusters 1–6, spike-like, 5–15 mm, 5–7 mm diam, spheric to cylindric; lowest bract < to >> infl; bractlets clear, margins ciliate, esp in upper 1/2. **FL**: perianth segments 2–5 mm (sepals often > petals), dark brown to pale, margins wide, translucent; filaments << to > anthers. **FR** spheric, = to << perianth, greenish to dark brown. **SEED** 1–1.5 mm, cylindric, red-brown to brown; appendage 1/4–1/2 seed-body length. $2n$=12,24. Common. Meadows, open woodlands, coniferous forests; 50–3200 m. NW, CaRH, SN (exc Teh), CCo, SnFrB, ChI, SnGb, SnBr, SnJt, MP; to AK, Rocky Mtns. [vars. *congesta* S. Watson, *macrantha* S. Watson, *subsessilis* (Buchenau) S. Watson; *L. campestris* var. *columbiana* H. St. John] Highly variable, the intergrading and poorly defined forms often treated under *L. campestris* (L.) DC. or *L. multiflora* (Ehrh.) Lej. ❀DRN:**4–6**&SHD: **17**&IRR:1–3,7,14,**15,16**,18–24.

L. divaricata S. Watson (p. 1173) Pl densely cespitose, 6–30 cm. **ST** 2 mm wide, reddish. **LVS** many; blade < 20 cm (sometimes > st), 4–6 mm wide, ± glabrous, tip < 12 mm, long-pointed, not thickened. **INFL** cyme-like, open, 5–15 cm, width 1/2 to = length; branching mostly at right angles; fls single on long pedicels; lowest bract < 2 cm, inconspicuous, bracts and bractlets clear. **FL**: perianth segments 1.8–2.4 mm (sepals slightly > petals), narrowly acuminate, pale brown with reddish tint, midrib distinct; filaments ± = anthers. **FR** < to slightly > perianth, deep reddish brown. **SEED** 1.2 mm, light brown, with hair tuft. Subalpine forests to alpine granitic slopes; 2100–3700 m. CaRH, SNH; OR, NV.

L. orestera Sharsm. (p. 1173) Pl densely cespitose, stiffly erect, 3–26 cm. **ST** reddish brown. **LVS** many, overlapping; blade 2.5–7 cm, 2–5 mm wide, firm, mostly glabrous, tip thickened, reddish. **INFL**: 1–5 clusters, appearing as single pyramidal infl, 5–10 mm wide, sometimes with 1–2 smaller clusters on short peduncles; lowest bract < to > infl, gen stiff, reddish; bractlets clear. **FL**: perianth segments 2–3 mm (sepals slightly > petals), very dark, margins clear; filaments gen = anthers. **FR** << perianth, dark brown to black in upper portion. **SEED** 0.8 mm, oval; appendage ± 0.2 mm. $2n$= 20,22. Alpine and subalpine meadows, fell-fields; 2700–3600 m. SNH. ❀IRR:1&SHD:2,3,7,15,16,18;DFCLT.

L. parviflora (Ehrh.) Desv. (p. 1173) Pl loosely cespitose, 20–50 cm. **LVS** many; sheath opening with long, soft hairs; blade 12–17 cm, 5–8 mm wide, tip thickened; cauline lvs 3–5, 7–9 cm, 3–5 mm wide. **INFL** cyme-like, open; fls gen many; major branches long, nodding; terminal branches short; fls gen 2–4 and crowded (or solitary); lowest bract inconspicuous; bracts and bractlets sparsely fringed, clear or brown. **FL**: perianth segments 2–2.5 mm (sepals = petals), widely lanceolate, tips acute, pale to brown; filaments ± = anthers. **FR** gen > perianth, spheric, greenish to brown. **SEED** 1.1–1.3 mm, brown to brownish red or purple, with hair tuft. $2n$=22,24. Moist places in coniferous woodlands; 1000–3300 m. NW (exc NCoRI), SN; to AK e N.Am, Eurasia. [var. *fastigiata* Buchenau; var. *melanocarpa* (Michaux) Buchenau]

L. spicata (L.) DC. (p. 1173) Pl densely cespitose, 3–33 cm. **ST**: bases thick, extending several cm into soil, reddish. **LF**: sheath opening densely hairy; blade erect, 2–15 cm, 1–4 mm wide, linear, channeled, tip not thickened. **INFL**: panicle of dense, nodding, spike-like clusters, each 1–25 mm, often interrupted by 10–70 mm; lowest bract gen > infl; bracts and bractlets = to > fls, clear, margin hairy, tip narrow and extended. **FL**: perianth segments 2–2.5 mm (sepals > petals), bristle-pointed, brown with clear margins or very pale throughout. **FR** gen < perianth, round with ± acute tip, pale to dark brown. **SEED** 1 mm, brown; appendage 0.2 mm. $2n$=24. Alpine slopes, subalpine forests; 2900–3700 m. SNH, Wrn; circumpolar, to AK, high mtns of N.Am, Eurasia. [var. *nova* Smiley; ssp. *saximontana* A. Löve & D. Löve]

L. subcongesta (S. Watson) Jepson (p. 1173) Pl from stout horizontal rhizome, 15–48 cm. **LF**: blade < 19 cm, 3–5 mm wide, bluish green, tapered abruptly to long tip. **INFL**: clusters 3–12-fld, crowded at ends of uneven, often arching peduncles; lowest bract 2–3 cm, inconspicuous; bractlets clear, margin hairy. **FL**: perianth segments 1.5–2 mm (sepals = petals), pale to dark brown, tips acute; filaments ± = anthers. **FR** < perianth, pale to dark brown-purple, short-beaked. **SEED** 1.1–1.2 mm, 0.6 mm wide, brown; appendage 0. $2n$=24. Alpine to subalpine moist or wet places; 2000–3500 m. KR, CaRH, SNH. ❀STBL.

JUNCAGINACEAE ARROW-GRASS FAMILY

Robert F. Thorne

Ann or per from rhizomes, submersed or emergent, sometimes dioecious or pls with some unisexual fls. **ST** short, erect, ± scapose. **LVS** basal or cauline, alternate, gen narrowly cylindric; sheath open, gen liguled. **INFL**: spike or raceme (terminal) or fls solitary in axils; bracts 0. **FL**: perianth parts gen 6 in 2 whorls (exc 0–1 in *Lilaea*), free, greenish; stamens gen 1, 3, or 6, filament short, ± fused to inner perianth parts, anthers elongate, dehiscing outward; pistil 1 (simple) or seemingly so (ovaries 3,4 or 6, fused to central axis, each with 1 chamber and 1 style), ovule 1 per chamber, style short and plumose or long and thread-like. **FR**: follicle or nutlet. 5 genera, ± 20 spp.: temp and circumboreal.

1. Perianth parts 0 or 1; stamen 1; ovary chambers (and seed) 1; fls bisexual and staminate (many on emergent spike) and pistillate (few, axillary, submersed) .... **LILAEA**

1′ Perianth parts 3–6; stamens 3 or 6; ovary chambers (and seeds) 3–6; fls bisexual (on aerial raceme or spike) .... **TRIGLOCHIN**

LILAEA

1 sp. (A. Raffeneau-Delile, French botanist, 1778–1850)

L. scilloides (Poiret) Hauman (p. 1173) FLOWERING-QUILLWORT Ann from fibrous roots, aquatic, gen emergent. **ST** erect, short, obscure. **LVS** basal, tufted, 5–20(45) cm, 1–5 mm wide; sheath 3–10 cm, translucent; blade tip acute to pointed. **INFL**: bisexual and staminate fls in erect, gen emergent spike, 6–20 cm; pistillate fls solitary in submersed axils, enclosed by lf sheaths. **BISEXUAL FL**: perianth part 1, 2–3 mm; stamen 1; ovary 1, chamber 1, ovule 1, style short. **STAMINATE FL**: perianth part 1, 2–3 mm; stamen 1. **PISTILLATE FL**: perianth 0; ovary 1, chamber 1, ovule 1, style 6–20 cm, thread-like, stigma floating. **FR**: nutlet, 2–10 mm, ribs 25–30; tip beaked. $2n$=12. Vernal pools, ditches, streams, ponds, lake margins: < 1700 m. NCo, NCoRI, SN, GV, CW, SCo, PR, GB; to w Can, MT, Mex, Chile; naturalized in Australia. Sometimes treated in Lilaeaceae. ❀WET,fresh water margins & mud;SUN: **1–3**,4,5,**6–9**,10–12,**14–24**.

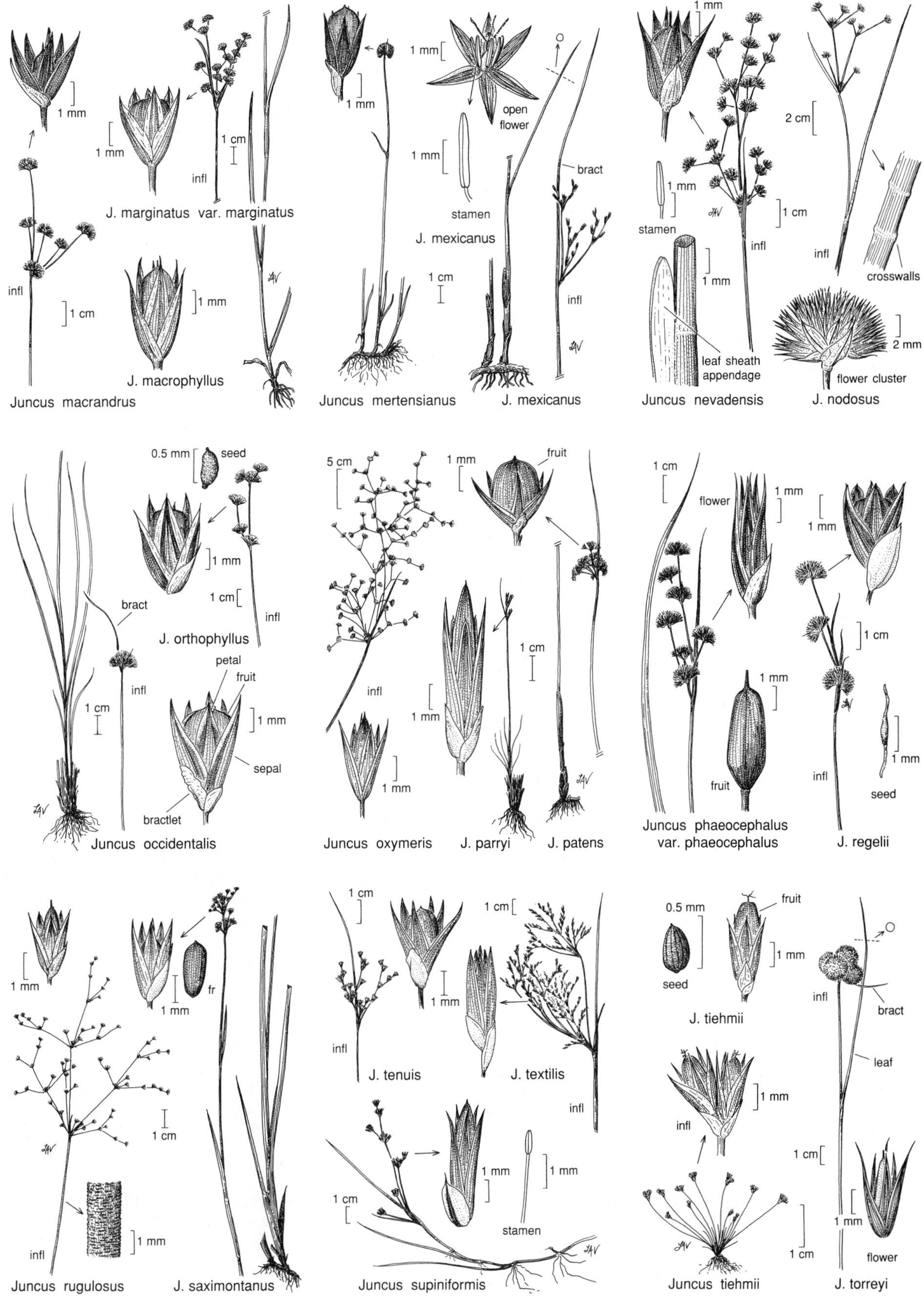
1 mm
1 mm
1 cm
infl
1 mm
1 cm
infl
J. marginatus var. marginatus
1 mm
J. macrophyllus
Juncus macrandrus
1 mm
1 mm
open flower
1 mm
stamen
J. mexicanus
1 cm
bract
infl
Juncus mertensianus
J. mexicanus
1 mm
1 mm
stamen
1 cm
infl
1 mm
leaf sheath appendage
Juncus nevadensis
2 cm
infl
crosswalls
2 mm
flower cluster
J. nodosus
0.5 mm
seed
1 mm
1 cm
infl
bract
J. orthophyllus
1 cm
infl
petal
fruit
1 mm
sepal
bractlet
Juncus occidentalis
5 cm
1 mm
fruit
infl
1 mm
1 cm
1 mm
Juncus oxymeris
J. parryi
J. patens
1 cm
flower
1 mm
1 mm
1 cm
1 mm
fruit
infl
1 mm
seed
Juncus phaeocephalus var. phaeocephalus
J. regelii
1 mm
1 mm
fr
1 cm
infl
1 mm
Juncus rugulosus
J. saximontanus
1 cm
1 cm
1 mm
infl
J. tenuis
J. textilis
infl
1 mm
1 mm
1 cm
stamen
Juncus supiniformis
0.5 mm
fruit
seed
1 mm
J. tiehmii
infl
bract
leaf
1 mm
infl
1 cm
1 cm
1 mm
flower
Juncus tiehmii
J. torreyi

TRIGLOCHIN ARROW-GRASS

Per from rhizomes in CA, terrestrial or aquatic. **ST** erect, short, obscure. **LVS** basal, ± tufted; sheath membranous; ligule entire to 2-lobed. **INFL**: raceme, scapose, narrowly cylindric, glabrous; pedicels short. **FL** bisexual; perianth parts 3–6, gen green, inner surface concave; stamens (1)3–6, subsessile, anthers wide; ovaries 3 or 6 (if 6, 3 sometimes sterile), 1-chambered, ± fused to central axis, style short, stigma papillate. **FR**: follicles 3–6, separating from axis. **SEED** 1, linear, ± flat or angled. ± 12 spp.: temp and circumboreal. (Greek: 3-pointed, from frs of some) TOXIC when fresh from cyanogenic compounds.

1. Fertile ovaries 3; fr linear or subspheric; pl gen with 1–2 racemes
  2. Fr linear, tapered to base .... *T. palustris*
  2′ Fr subspheric, not tapered to base .... *T. striata*
1′ Fertile ovaries 6; fr oblong to ovoid; pl gen with 2 or more racemes
  3. Ligule entire to slightly notched, (1)1.5–5 mm; lf blade somewhat compressed, gen 1.5–4 mm wide; pl gen > 3 dm .... *T. maritima*
  3′ Ligule deeply 2-lobed, 0.5–1.5 mm; lf blade ± cylindric, gen 1–1.5 mm wide; pl gen < 3(–6) dm ..... *T. concinna*
    4. Pl gen < 3(4.5) dm; coastal salt marshes; CCo, SCo .... var. ***concinna***
    4′ Pl gen > 3(6) dm; inland alkaline marshes, meadows; s SNH, GB, DMoj .... var. ***debilis***

T. concinna Burtt Davy ARROW-GRASS Per from spreading to ascending rhizomes, < 3(4.5–6) dm. **LF** < 20 cm, ± 1 mm wide, subcylindric, ± fleshy; ligule 0.5–1.5 mm, 2-lobed. **INFLS** 2 or more per pl, > lvs; pedicels < 5 mm, ascending. **FL**: perianth parts gen 6, ± 1.5 mm; stamens gen 6; fertile ovaries 6. **FRS** 6 per fl, < 5 mm, oblong-ovoid, falling from axis. $2n=48,96$. Salt marshes, alkaline meadows, seeps, mudflats, stream and lake margins; < 2500 m. NCo, s SNH, CCo, SCo, GB, DMoj; to Can, Baja CA.

var. ***concinna*** (p. 1173) Pl < 3(4.5) dm. Uncommon. Salt marshes; < 5 m. NCo, CCo, SCo; to B.C., Baja CA. ❀WET,saline to fresh water margins & mud;SUN:**4,5**,15,16,**17,24**.

var. ***debilis*** (M.E. Jones) J. Howell Pl 3–6 dm. Alkaline meadows, seeps, mudflats, stream and lake margins; < 2500 m. s SNH, GB, DMoj; to Can, Colorado. ❀TRY.

T. maritima L. (p. 1173) SEASIDE ARROW-GRASS Per from short, thick rhizomes, gen > 3 dm. **LF** < 70 cm, 1.5–5 mm wide, flat; ligule 1–5 mm, entire to slightly notched. **INFLS** 2 or more per pl, gen > lvs; pedicels 1–3(5) mm, ascending. **FL**: perianth parts 6, ± 2 mm; stamens gen 6; fertile ovaries 6. **FRS** 6 per fl, 3–5(7) mm, oblong to ovoid, falling from axis. $2n=12,24,30,36,48,60,96,120,144$. Uncommon. Saline, brackish, or alkaline marshes; < 2800 m. NCo, SN, CCo, SCo, SnBr, GB; circumboreal. ❀WET or IRR, alkaline to fresh water margins & mud; SUN:**1–9,14–24**; STBL.

T. palustris L. (p. 1173) MARSH ARROW-GRASS Per from short, ascending rhizomes, 1–3(5.5) dm. **LF** gen < 30 cm, 0.5–2 mm wide, ± flat; ligule 0.5–1.5 mm, gen 2-lobed. **INFLS** gen 1–2 per pl, gen > lvs; pedicels 2–4 mm in fr, slender, erect. **FL**: perianth parts gen 6, 1.5–2 mm; stamens 6; fertile ovaries 3. **FRS** 3 per fl, 6–7(9) mm, linear to narrowly club-shaped, separating from axis base, suspended from axis tip before falling. $2n=24$. Uncommon. Wet meadows, wet flats, stream and lake margins; 2400–3700 m. c&s SNH; circumboreal. ❀TRY.

T. striata (p. 1173) Ruiz Lopez & Pavon THREE-RIBBED ARROW-GRASS Per from slender, spreading to ascending rhizomes, 1–2(4.5) dm. **LF** 5–20(45) cm, 1–2 mm wide, flat; ligule 1–2.5 mm, gen entire, tip rounded to acuminate. **INFLS** gen 1–2 per pl, 5–15(32) cm; pedicels 1–1.5(3.5) mm in fr, gen ascending. **FL**: perianth parts 3–5, greenish yellow; stamens 1–2(3), often unequal; ovaries 6, sterile 3 alternate with fertile 3, wing-like and persistent in fr. **FRS** 3 per fl, subspheric, 3-angled, falling from axis. Uncommon. Saline or brackish marshes; < 10 m. NCo, CCo, SCo; to WA, e US; also Chile, Australasia. ❀TRY.

LEMNACEAE DUCKWEED FAMILY

Wayne P. Armstrong

Per, floating aquatics, small, clonal, in dense populations; new pls produced in budding pouch at base or along margins; may overwinter on bottom as dense, rootless, starch-filled daughter pl (winter bud); roots 0–many. **PL BODY** 0.4–10 mm, flat and tongue-shaped to spheric. **FL** 1, rarely seen, minute, appearing like 2–3 unisexual fls, often sheathed by minute membrane; perianth 0; stamens 1–2; pistil 1, simple, maturing before stamens. **FR** achene-like, sometimes winged. **SEEDS** 1–3, smooth or ribbed. Spp. best separated by chemistry and fr; clones vary; magnification, backlighting gen needed to identify vegetative pls. 4 genera, 34 spp.: worldwide; orn in pools, aquaria. [Landolt 1986 Veröff Geobot Inst ETH Stiftung Rübel Zürich 71] ❀May be used in still ponds, 0–2500 m, depending on individual tolerances; very invasive.

1. Roots present
  2. Root 1 .... **LEMNA**
  2′ Roots gen 2–16 .... **SPIRODELA**
1′ Roots 0 (see also *Lemna trisulca*)
  3. Pl body ± spheric, ovoid, or cylindric, gen ± 1 mm .... **WOLFFIA**
  3′ Pl body flat, 3–10 mm .... **WOLFFIELLA**

LEMNA DUCKWEED

Pls gen in clusters of 2–many; root gen 1, gen 2–6 mm, sheath near base gen not winged. **PL BODY** gen 2–5 mm, flat, gen widely elliptic to oblong, pale to dark green, often ± red; veins 1–5 (visible in backlight); winter buds gen 0. **FLS** in 2 lateral pouches, sheathed by minute membrane. **FR** gen unwinged. **SEED** ribbed, gen smooth between ribs. 13 spp.: worldwide. (Greek: lake or swamp)

1. Pl body 6–10 mm, on a tapered stalk, often connected in branched chains of 10–30 .... *L. trisulca*
1′ Pl body 1–5 mm, appearing sessile, single or in clusters of 2–8
  2. Vein 1 per pl body (sometimes obscure)

3. Pl body 1–2.5 mm, widely elliptic, gen thicker in middle, gen in pairs; vein < 2/3 distance from root node to tip of pl body . . . . . *L. minuscula*
3′ Pl body 2–4 mm, ± oblong, very thin, transparent green, gen 4–8 attached; vein up to 3/4 distance from root node to tip of pl body . . . . . *L. valdiviana*
2′ Veins 3–5 per pl body
4. Root sheath with 2 obvious, wing-like appendages at base . . . . . *L. aequinoctialis*
4′ Root sheath not winged
5. Pl body widely elliptic to round, tip asymmetric, lower surface gen much swollen or rounded . . . . . *L. gibba*
5′ Pl body widely elliptic, tip symmetric, lower surface flat
6. Upper surface gen smooth; lower surface green; winter buds 0 . . . . . *L. minor*
6′ Upper surface gen with midline row of minute bumps; lower surface becoming ± red; winter buds dark green or brown in autumn . . . . . *L. turionifera*

L. aequinoctialis Welw. Root-sheath winged. **PL BODIES** in pairs, 2–3.5 cm, obovate-elliptic, light green; base asymmetric; tip symmetric; upper surface tip and root node with minute bumps; lower surface light green, smooth. **SEED** cross-lined between ribs. Freshwater in hot regions; < 200 m. GV, CCo, SCo, DSon; worldwide. [*L. paucicostata* Hegelm.; not *L. perpusilla* Torrey]

L. gibba L. (p. 1173) **PL BODIES** gen in pairs or 3's, 3–6 mm, widely elliptic to round, glossy green or yellow-green mottled red; base symmetric; tip asymmetric; upper surface barely ridged, midline bumps gen 0; lower surface gen swollen, with enlarged air spaces often bordered in red. **FR** strongly winged. Common. Fresh or brackish water; gen < 1500 m. CA; worldwide. Variable: vegetative, non-swollen forms appear much like *L. minor*. Gen replaced > 1500 m by *L. turionifera*.

L. minor L. **PL BODIES** in pairs or 3's, 2–5 mm, obovate-elliptic, glossy-green; base and tip gen symmetric; upper surface smooth, midline bumps gen 0; lower surface air spaces ± obscure. Common. Freshwater; < 2000 m. CA; worldwide. [*L. m.* var. *minima* Chevall.]

L. minuscula Herter **PL BODIES** gen in pairs (esp when crowded, in full sunlight), 1–2.5 mm, widely elliptic to oblong, pale green; base and tip gen symmetric; margin clearly thinner than center; vein 1, obscure, < 2/3 distance from root node to tip of pl body (gen < region of visible air spaces between cells). **SEED** cross-lined between ribs. Common. Freshwater; < 2200 m. CA except MP; w US, S.Am, Eur, n Asia. [*L. minuta* Kunth; *L. minima* (Hegelm.) Hegelm.] Transparent green pls, 2–4 mm, in 4's, are much like *L. valdiviana*. Needs study.

L. trisulca L. Root often 0. **PL BODY** 6–10 mm, lanceolate to oblong, transparent green, on a long, tapered stalk, connected in branched chains of 8–30; base and tip symmetric; surfaces smooth. Meadows, mtn streams; < 3000 m. CaR, SN, SCo?, SnBr, GB, DMoj?; nearly worldwide. Gen forming dense, tangled masses below water surface.

L. turionifera Landolt **PL BODIES** gen in pairs or 3's, 2–5 mm, widely elliptic to obovate, glossy green; base and tip gen symmetric; upper surface shiny green with row of minute bumps at midline; lower surface often ± red; winter buds 1–2 mm, dark green or brown. Freshwater; < 3000 m. Most of CA; N.Am, Asia. Like *L. minor*, exc for winter buds.

L. valdiviana Philippi **PL BODIES** often in clusters of 4–8, 2–4 mm, thin, elliptic to narrowly oblong, uniformly transparent green; base often asymmetric; tip symmetric; surfaces smooth; vein up to 3/4 distance from root node to tip of pl body (gen > region of visible air spaces between cells). **SEED** cross-lined between ribs. Freshwater; < 1500 m. NCo, NCoR, CCo, SCoR, SCo; to e US, S.Am. [*L. cyclostasa* (Elliott) C. Thompson, not Chevall.] Variable. Apparently uncommon in CA, esp in mtns. Forms tangled masses under other pls.

SPIRODELA DUCKMEAT

Pls gen in clusters of 2–5; roots 1–16, some passing through minute scale on lower surface. **PL BODY** 3–10 mm, oblong to round, flat; upper surface dark shiny green, lower gen red-purple; veins 3–12 (visible in backlight); young pls with minute scale-like lf on each side at base; winter buds often produced. **FLS** in 2 lateral budding pouches, sheathed by minute membrane. **SEED** ribbed. 3 spp.: worldwide. (Greek: visible thread, from roots)

1. Roots gen 7–16; pl body 5–10 mm, ± round . . . . . *S. polyrrhiza*
1′ Roots gen 2–5; pl body 3–5 mm, oblong-obovate . . . . . *S. punctata*

S. polyrrhiza (L.) Schleiden (p. 1173) Roots 5–16, 1–2 passing through minute scale. **PL BODY** 5–10 mm, round-ovate, symmetric; veins 7–12; upper surface smooth; winter buds produced in autumn. Freshwater; 0–2500 m. CA; nearly worldwide.

S. punctata (G. Meyer) C. Thompson Roots 1–6, all passing through minute scale. **PL BODY** 3–5 mm, oblong-obovate, asymmetric; veins 3–5; upper surface often with row of minute bumps at midline. Still water, coastal and interior valleys; < 500 m. NCo, GV, CCo, SnFrB, SCo; nearly worldwide. [*S. oligorrhiza* (Kurz) Hegelm.]

WOLFFIA WATER-MEAL

Pls gen in unequal pairs; roots 0. **PL BODY** 0.4–1.3 mm, nearly spheric to ± cylindric, floating on or partially below water surface; veins 0; budding pouch funnel-shaped; winter buds often produced. **FL** produced in cavity on upper surface; sheathing membrane 0; stamen 1; pistil 1. **SEED** smooth. 9 spp.: worldwide, esp temp, trop. (J.F. Wolff, German botanist and physician, 1778–1806) [Armstrong & Thorne 1984 Madroño 31:172–179] CA spp. may be distinguished by size and shape with > 10× magnification.

1. Tip of pl body clearly pointed, upper surface distinctly flattened — dead pls dotted with brown pigment cells . . . . . *W. borealis*
1′ Tip of pl body rounded, upper surface ± rounded (exc *W. arrhiza*)
2. Upper surface of mature vegetative pl body with a prominent bump; dead pls dotted with brown pigment cells . . . . . *W. brasiliensis*
2′ Upper surface of mature vegetative pl body smooth; brown pigment cells 0

3. Upper surface dark green, distinctly flattened; stomates gen 15–100 .... ***W. arrhiza***
3′ Upper surface transparent green, ± rounded; stomates 1–10(30)
  4. Pl body 0.8–1.3 mm, almost spheric, most of upper surface clearly rounded, the top flat .... ***W. columbiana***
  4′ Pl body 0.4–0.8 mm, longer than wide, upper surface barely rounded throughout .... ***W. globosa***

W. arrhiza (L.) Wimmer **PL BODY** 0.8–1.3 mm, nearly spheric; upper surface distinctly flattened, with 15–100 stomates, dark green (transparent below), only central portion floating above water. Uncommon. Freshwater; < 200 m. SCo (San Diego Co.); to Eur, sw Asia, Afr. In backlight, pls appear slightly darker than other transparent green spp. Easily confused with *W. columbiana*.

W. borealis (Hegelm.) Landolt & O. Wildi (p. 1173) **PL BODY** 0.7–1.2 mm, longer than wide; upper surface slightly concave or flat, with 50–100 stomates, dark green (transparent below), floating above water; tip acutely pointed, slightly upturned when viewed from side; dead pl brown-dotted. Uncommon. Freshwater; < 1000 m. NCoR, CaRH, SCo; to n US, Can. [not *W. punctata* Griseb.] Widely scattered localities. In backlight, appears darker than other transparent green spp.

W. brasiliensis Wedd. **PL BODY** 0.7–1.2 mm, longer than wide; upper surface flat near margin, gradually elevated to a prominent, central, conical bump, with 50–100 stomates, dark green (transparent below), floating above water; tip round; dead pl brown-dotted. Uncommon. Ponds; < 100 m. ScV (near Sacramento River); to e US, S.Am. In backlight, appears darker than other transparent green spp.

W. columbiana Karsten (p. 1173) **PL BODY** 0.8–1.3 mm, nearly spheric, transparent green; upper surface with 1–10(30) stomates, most of it round, central portion flat and floating above water; tip round. Freshwater; < 200 m. SnJV, CCo, SCoRO, SCo; to Can, S.Am. Small pls much like *W. globosa*.

W. globosa (Roxb.) Hartog & Plas (p. 1173) **PL BODY** 0.4–0.8 mm, longer than wide, sometimes ± cylindric, transparent green; upper surface with 1–10 stomates, barely rounded throughout, sometimes darker green, only central portion floating above water; tip round; budding pouch gen with collar of long cells at junction with daughter pl. Ponds of hot interior valleys; < 200 m. SNF, SnJV, SCo; worldwide, esp trop. [*W. cylindracea* Hegelm.] Smallest of all known angiosperms.

WOLFFIELLA MUD-MIDGET

Pls gen in unequal pairs; roots 0. **PL BODY** 3–10 cm, thin-membranous, flat, linear to oblong, floating just below water surface; free ends gen recurved; budding pouch triangular, with track of long cells on lower surface between midline and margin; winter buds 0. **FL** produced in cavity on upper surface; sheathing membrane 0; stamen 1; pistil 1. **SEED** smooth. 9 spp.: worldwide, esp warm temp, trop. (Diminutive of Wolffia) [Landolt 1984 Veröff Geobot Inst ETH Stiftung Rübel Zürich 51:164–172]

1. Pl body 4–10 mm, widely oblong, ends strongly recurved, budding pouch angle 80–120° .... ***W. lingulata***
1′ Pl body 3–5 mm, linear to narrowly oblong, ends gen not to slightly recurved, budding pouch angle gen 40–70° .... ***W. oblonga***

W. lingulata (Hegelm.) Hegelm. (p. 1173) **PL BODY**: ends recurved to form a semicircle when viewed from side; upper surface gen concave; budding pouch with track of long cells between middle and edge of lower wall. Coastal and interior valleys; < 200 m. NCo, SNF, GV, CCo, SnFrB, SCo?; to se US, S.Am. Variable. Small pls much like *W. oblonga*.

W. oblonga (Philippi) Hegelm. (p. 1173) **PL BODY** not forming semicircle when viewed from side; upper surface gen flat; budding pouch with track of long cells along edge of lower wall. Uncommon. Habitat and range of *W. lingulata*. Daughter pl often angled from axis of parent, so pair appears boomerang-like.

LILIACEAE LILY FAMILY

Dale W. McNeal, except as specified

Per to trees, from membranous bulb, fibrous corm, scaly rhizome, or erect caudex. **ST** gen underground. **LVS** gen basal, often withering early, alternate, gen ± linear. **INFL** various, gen bracted. **FL** gen bisexual, gen radial; perianth often showy, segments gen 6 in two petal-like whorls (outer sometimes sepal-like), free or fused at base; stamens 6 (or 3 + gen 3 ± petal-like staminodes), filaments sometimes attached to perianth or fused into a tube or crown; ovary superior or inferior, chambers 3, placentas gen axile, style gen 1, stigmas gen 3. **FR**: gen capsule, loculicidal or septicidal (berry or nut). ± 300 genera, 4600 spp.: esp ± dry temp and subtrop; many cult for orn or food; some TOXIC. Here incls genera sometimes treated in Agavaceae, Amaryllidaceae, and other families.

1. Shrub- or tree-like, coarse; lvs large, ± sword-like, persistent
  2. Ovary inferior; perianth (greenish) yellow .... **AGAVE**
  2′ Ovary superior; perianth white or purplish
    3. Fls bisexual and unisexual, < 1 cm; lvs not spine-tipped .... **NOLINA**
    3′ Fls bisexual, > 3 cm; lvs spine-tipped .... **YUCCA**
1′ Per (± fleshy if coarse); lvs gen withering (exc *Aloe, Xerophyllum*)
  4. Perianth parts 4, petal-like; stamens 4; ovary 2-chambered .... **MAIANTHEMUM**
  4′ Perianth parts 6 and petal-like (or 3 sepal-like)
    5. Infl a scapose, bracted umbel; pl from scaly bulb or corm
      6. Umbel 1-fld; pl with onion odor .... [2]**IPHEION**
      6′ Umbel 3–many-fld; pl with onion odor or not
        7. Perianth ± free at base, not forming an obvious tube
          8. Filaments with cup-like, basal appendage; pedicels jointed .... **BLOOMERIA**
          8′ Filaments sometimes winged or fused but not appendaged; pedicels not jointed
            9. Pls with onion odor; pedicels not each subtended by scarious bractlets .... **ALLIUM**
            9′ Onion odor 0; pedicels subtended by scarious bractlets above main bracts .... **MUILLA**

7′ Perianth parts fused at base, forming an obvious tube
10. Fertile stamens 3 (or 6 but 3 much smaller)
11. Infl axis gen straight; umbel open; staminodes gen 3, much unlike and alternate 3 fertile stamens; crown-like filament appendages 0 .... **BRODIAEA**
11′ Infl axis gen curved or twisted; umbel gen dense; filaments forming crown-like tube outside anthers .... **DICHELOSTEMMA**
10′ Fertile stamens 6
12. Filaments fused into crown-like tube .... **ANDROSTEPHIUM**
12′ Filaments free
13. Bulb with membranous outer scales; noxious weed of disturbed areas .... **NOTHOSCORDUM**
13′ Corm with ± fibrous outer scales; common native .... **TRITELEIA**
5′ Infl various (not a scapose, bracted umbel); pl gen from bulbs or rhizomes
14. Lvs very fleshy, margin gen horny-toothed .... **ALOE**
14′ Lvs not or barely fleshy, margins not horny-toothed
15. Pl from caudex or rhizome (bulb 0)
16. St not evident (underground, short, vertical)
17. Lvs many, linear; perianth white .... **LEUCOCRINUM**
17′ Lvs 2, wider than linear; perianth greenish or yellowish, dark-mottled .... **SCOLIOPUS**
16′ St obvious above ground
18. Lvs ± grass-like, very tough, long-persistent — infl a club-like raceme .... **XEROPHYLLUM**
18′ Lvs not very tough, persistent for 1 year or less
19. Lvs basal (or cauline much reduced)
20. Lf lanceolate to ovate; fls > 10 mm
21. Fr a berry; perianth cylindric to bell-shaped .... **CLINTONIA**
21′ Fr a capsule; perianth parts ± reflexed .... [2]**ERYTHRONIUM**
20′ Lvs narrow, grass-like; fls < 10 mm
22. Infl head-like < 7 cm; pl glandular above; fls 3–6 mm .... **TOFIELDIA**
22′ Infl raceme or panicle > 7 cm; pl glabrous; fls 5–12 mm
23. Infl a panicle; fls white to pinkish; filaments glabrous; seed tails 0 .... **ASPHODELUS**
23′ Infl a raceme; fls greenish yellow; filaments woolly; seeds bristle-tailed at both ends .... **NARTHECIUM**
19′ Lvs cauline, well distributed on st
24. Lvs scale-like, papery — st much-branched, green, branchlets lf-like, flat or thread-like **ASPARAGUS**
24′ Lvs expanded, green
25. Dioecious; sts climbing or trailing, 2–5 m; petiole gen with 2 tendrils .... **SMILAX**
25′ Monoecious or fls bisexual; st erect, < 2 m; tendrils 0
26. Lvs 3, in 1 whorl subtending fl .... **TRILLIUM**
26′ Lvs > 3, alternate
27. Infl a raceme or panicle; fls many, obvious
28. Pl < 1 m; lvs ± finely veined; infl < 1.5 dm; fr a berry .... **SMILACINA**
28′ Pl 1–2 m; lvs coarsely veined; infl 2–7 dm; fr a capsule .... **VERATRUM**
27′ Fls 1–3 per branch or axil, ± hidden by lvs
29. Fls 1–7 per terminal branch; peduncle not twisted; lf base oblique .... **DISPORUM**
29′ Fls 1–2 per axil; peduncle twisted, fused to st; lf base cordate .... **STREPTOPUS**
15′ Pl from a scaly bulb or corm
30. Some lvs obviously cauline
31. Distinct glandular area on petals 0
32. Perianth > 3 cm; style 1; stigma 3-lobed .... **LILIUM**
32′ Perianth < 2 cm; styles 3 .... [2]**STENANTHIUM**
31′ Petals with distinct glandular area in lower half
33. Outer perianth parts narrower, ± sepal-like .... [2]**CALOCHORTUS**
33′ Outer and inner perianth parts similar
34. Lvs mostly cauline, sometimes whorled; fls gen nodding; style 1, sometimes 3-branched .... **FRITILLARIA**
34′ Lvs mostly basal (cauline ± reduced, alternate); fls gen erect; styles 3 .... [2]**ZIGADENUS**
30′ Lvs ± basal (infl scapose or cauline lvs much reduced)
35. Styles 3, ± unfused
36. Perianth parts glandless; corolla greenish yellow to purplish .... [2]**STENANTHIUM**
36′ Perianth parts with 1–2 prominent glands near base; corolla white to yellowish .... [2]**ZIGADENUS**
35′ Style 1 (sometimes 3-branched above middle)
37. Perianth parts fused into a tube
38. Infl dense, spike-like; perianth deep blue exc very short white lobes .... **MUSCARI**
38′ Infl an open raceme or panicle; perianth white to yellowish, lobes = or > tube
39. Perianth 4–6 cm, funnel-shaped .... **HESPEROCALLIS**
39′ Perianth ± 1 cm, lobes spreading or reflexed .... **ODONTOSTOMUM**
37′ Perianth parts free
40. Outer perianth parts narrower, ± sepal-like .... [2]**CALOCHORTUS**
40′ Outer and inner perianth parts similar
41. Lvs 2 (1 in non-fl pls); infl unbracted .... [2]**ERYTHRONIUM**

41′ Lvs > 2; infl bracts scarious
42. Infl 1-fld; pl with onion odor .... [2]**IPHEION**
42′ Infl 2+ fld
43. Infl a raceme
44. Perianth parts 1.5–4 cm, blue to purplish, (white) .... **CAMASSIA**
44′ Perianth parts < 1 cm, white to yellowish or purplish .... [2]**HASTINGSIA**
43′ Infl a panicle
45. Style thread-like, deciduous; perianth twisted above ovary in fr .... **CHLOROGALUM**
45′ Style ± stout, persistent; perianth not twisted together in fr .... [2]**HASTINGSIA**

AGAVE

Katy K. McKinney

Shrub-like, often producing rosettes without seeds, blooms once and dies. **LVS** in basal rosette, long-lived, < 50 cm, sessile, linear to ovate, fleshy, glabrous; lateral teeth and tip spin-like. **INFL**: panicle or raceme-like, scapose, gen bracted, gen 2–4 m. **FL**: perianth segments 6 in 2 petal-like whorls, 3–10 cm; stamens 6; ovary inferior, chambers 3. **FR**: capsule, ± ovoid, loculicidal. **SEEDS** many, flat, black. ± 300 spp.: warm and trop Am. (Greek: noble, from imposing stature) [Gentry 1982 Agaves of N.Am, Univ AZ Press]

1. Infl ± raceme-like; bracts on peduncle 0; n&e DMtns .... ***A. utahensis***
1′ Infl a panicle; bracts on peduncle obvious; more widespread
2. Lf lanceolate, gray; bracts on peduncle scarious; s DMtns, DSon .... ***A. deserti***
2′ Lf ovate, dark green; bracts on peduncle large, fleshy; s SCo .... ***A. shawii***

A. deserti (Engelm.) Gentry (p. 1173) Pl gray-glaucous; caudex 30–50 cm, unbranched. **LF** 25–40 cm, gen lanceolate; marginal teeth 2–8 mm, slender, widely spaced. **INFL**: panicle; peduncle bracts triangular, scarious; fls pedicelled in small umbels. **FL**: perianth 40–60 mm, light yellow, tube 4–6 mm, lobes equal; nectary disk thick; filaments attached near base of tube, 25–35 mm; ovary 22–40 mm, neck slightly narrowed. **FR** 35–50 mm, short-stalked, beaked. Rocky slopes, washes in desert scrub; < 1500 m. s DMtns, DSon; AZ, Baja CA. ❀DRN:3,7,**8–10**,11,**12**,13,**14,19–21**&SUN, DRY:15,16,18,22,23.

A. shawii (Engelm.) Gentry (p. 1173) SHAW'S AGAVE Pl dark green; caudex 10–30 cm, branched. **LF** 20–50 cm, gen ovate; marginal teeth variable. **INFL**: panicle; peduncle bracts purple, fleshy; fls sessile in tight clusters. **FL**: perianth 75–100 mm, yellow to reddish, tube 12–16 mm, lobes unequal; filaments attached near middle of tube, 60–70 mm; ovary 35–50 mm, fleshy. **FR** 50–70 mm, beaked. RARE in CA. Coastal bluffs, slopes, < 500 m. s SCo (San Diego Co.); Baja CA. In cult.

A. utahensis (Engelm.) Gentry UTAH AGAVE Pl blue-glaucous; caudex 18–30 cm, branched. **LF** 15–30 cm, gen narrowly lanceolate; lateral teeth 2–4 mm, blunt, detachable. **INFL** raceme-like; peduncle bracts 0; fls 2–8 per cluster. **FL**: perianth 25–31 mm, yellow, tube 2.5–4 mm, lobes equal; filaments attached near base of tube, 18–20 mm; ovary 10–25 mm. **FR** 15–35 mm, short-beaked. UNCOMMON. Shadscale scrub, Joshua-tree woodland; 900–1500 m. n&e DMtns; to UT, AZ. [var. *eborispina* (Hester) Breitung, ivory-spined agave; var. *nevadensis* Engelm., Clark Mtn agave] ❀DRN,DRY,SUN:3,7–9,**10–12**,13,14,**18**,19–21.

ALOE

Reid Moran

Per, sometimes tree; sap bitter. **ST** branched or not. **LVS**: basal >> cauline, fleshy; base clasping, margin gen horny-toothed. **INFL**: panicle, gen axillary; branches 1–many, raceme-like. **FL** tubular, odorless, colorful; perianth parts in 2 sets of 3, ± fused; stamens 6, unequal; ovary chambers 3. **FR**: capsule, loculicidal; valves 3. **SEEDS** many, flat or angled. ± 250 spp.: Medit Eur, s Asia, esp Afr. (Greek: ancient name) [Reynolds 1950 Aloes of S Afr, Cape Times Ltd]

A. saponaria (Aiton) Haw. × *A. striata* Haw. **ST** decumbent, < 4 dm, covered with dead lvs; basal lf rosettes 3–5 dm wide, loosely clustered. **LVS**: basal 15–30, 2–2.5 dm, triangular-ovate, white-spotted, tip withering, marginal teeth 1–2 mm. **INFL** 4–9 dm; racemes 5–10, dense. **FL** 25–32 mm, red. **FR** 2–3 cm, oblong. **SEED** 5 mm, 3-winged. Coastal-sage scrub; < 100 m. s SCo (La Jolla, San Diego Co.); native to s Afr. Apparently sterile but spreading vegetatively.

ALLIUM ONION, GARLIC

Per with onion odor, taste; bulb solitary or on rhizomes, reforming each year, divides at base into daughter bulbs, outer bulb coats gen brown or gray, cell sculpture gen important to identification, inner bulb coats gen white. **ST** scapose, cylindric or flat. **LVS** basal, 1–5, linear, cylindric, channeled, or flat, gen withering from tip before fl. **INFL**: umbel; bracts gen 2–4, conspicuous, ± fused, scarious. **FL**: perianth segments 6, in 2 petal-like whorls; stamens 6, epipetalous, filaments wide at base, fused into a ring; ovary superior, sometimes with 3 or 6 crests, chambers 3, ovules gen 2 per chamber, style 1, stigma entire or 3-lobed. **FR**: capsule, loculicidal. **SEED** obovoid, black, sculpture net-like, smooth, or granular. 500 spp.: worldwide, esp CA. (Latin: garlic)

1. Fls gen replaced by bulblets; weedy .... ***A. vineale***
1′ Fls never replaced by bulblets
2. Lf 1 per st, cylindric; ovary crests 6, prominent, ± triangular
3. Stigma ± entire
4. Stamens exserted; perianth parts 5–9 mm, inner > outer, margins irregular to jagged; st 20–50 cm .... ***A. sanbornii*** var. ***sanbornii***
4′ Stamens incl; perianth parts 8–20 mm, ± equal, entire; st gen < 20 cm

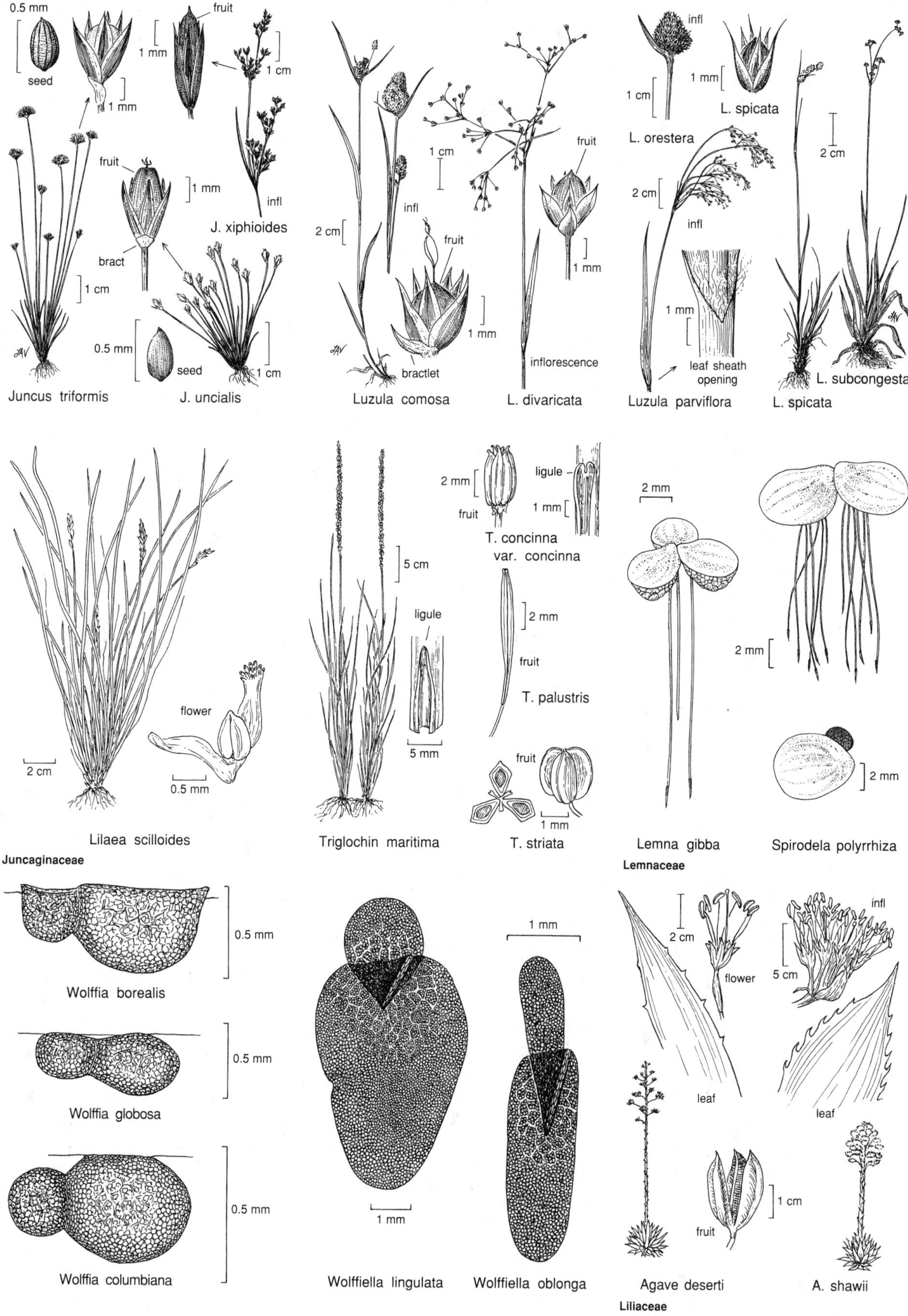
0.5 mm
seed
1 mm
fruit
1 mm
1 cm
fruit
1 mm
infl
J. xiphioides
bract
1 cm
0.5 mm
seed
1 cm
Juncus triformis
J. uncialis
1 cm
infl
2 cm
fruit
1 mm
bractlet
Luzula comosa
fruit
1 mm
inflorescence
L. divaricata
infl
1 cm
1 mm
L. spicata
L. orestera
2 cm
2 cm
infl
1 mm
leaf sheath opening
Luzula parviflora
L. spicata
L. subcongesta
2 cm
flower
0.5 mm
Lilaea scilloides
Juncaginaceae
5 cm
ligule
5 mm
Triglochin maritima
2 mm
fruit
ligule
1 mm
T. concinna var. concinna
2 mm
fruit
T. palustris
fruit
1 mm
T. striata
2 mm
Lemna gibba
Lemnaceae
2 mm
2 mm
Spirodela polyrrhiza
0.5 mm
Wolffia borealis
0.5 mm
Wolffia globosa
0.5 mm
Wolffia columbiana
1 mm
Wolffiella lingulata
1 mm
Wolffiella oblonga
2 cm
flower
leaf
fruit
1 cm
Agave deserti
Liliaceae
infl
5 cm
leaf
A. shawii

5. Outer bulb coat sculpture clearly net-like with ± long, twisted meshes .... ***A. nevadense***
5′ Outer bulb coat sculpture 0, only 2–3 vertical rows of cells above root pad, or indistinctly polygonal
6. Perianth parts 8–12 mm; pedicels slender, > fls .... ***A. atrorubens***
7. Perianth gen deep red-purple (white), lanceolate, long-acuminate .... var. ***atrorubens***
7′ Perianth pale pink, dark-veined, ± ovate, acute to acuminate .... var. ***cristatum***
6′ Perianth parts 12–20 mm; pedicels stout, gen < fls
8. Perianth parts linear-lanceolate, long-tapered; daughter bulbs 1–2, stalked, basal; TR, w PR ... ***A. monticola***
8′ Perianth parts lanceolate, acute; daughter bulbs 0; n SnBr, DMtns .... ***A. parishii***
3′ Stigma lobes 3, often slender and recurved
9. Stamens exserted or = perianth
10. Inner perianth parts 1.5 × outer; ovary crests entire or notched .... ***A. sanbornii*** var. ***congdonii***
10′ Perianth parts ± equal; ovary crests dentate to deeply cut .... ***A. howellii***
11. St gen ± 20(10–35) cm, slender; fls gen 10–30; perianth gen pink to lavender; ovary crests purple; fls spring — s SnJV and adjacent mtns .... var. ***howellii***
11′ St 20–60 cm, stout; fls gen > 50; perianth gen white; ovary crest white or green; fls ± early summer
12. Stamens exserted 0–2 mm; n WTR .... var. ***clokeyi***
12′ Stamens exserted 2–4 mm; s SCoRI .... var. ***sanbenitense***
9′ Stamens incl
13. Perianth parts (at least inner) dentate to jagged
14. St 25–40 cm .... ***A. jepsonii***
14′ St 5–20 cm
15. Bulb coats brown to gray; perianth ± spreading, tips reflexed, inner parts jagged .... ***A. abramsii***
15′ Bulb coats red-brown; perianth erect, ± straight, inner parts dentate .... ***A. denticulatum***
13′ Perianth parts ± entire
16. Ovary crests entire, notched, or irregular (not dentate)
17. Perianth 10–18 mm, maroon or deep red-purple
18. Perianth deep red-purple, tips reflexed; serpentine, SnFrB .... ***A. sharsmithae***
18′ Perianth maroon, outer tips recurved, inner tips spreading to reflexed; s SNH .... ***A. shevockii***
17′ Perianth 6–9 mm, white to pink, becoming darker
19. Infl compact, pedicels straight; perianth parts elliptic to ovate, obtuse to acute .... ***A. munzii***
19′ Infl loose, pedicels curved in fr; perianth parts lanceolate, acuminate .... ***A. parryi***
16′ Ovary crests dentate to deeply cut
20. Perianth deep red-purple, tips gen recurved-spreading .... [2]***A. fimbriatum*** var. ***fimbriatum***
20′ Perianth white to lavender, midveins darker, tips erect
21. Fls gen 8–12 mm .... [2]***A. fimbriatum***
22. St slender, 10–25 cm; SNE, DMoj .... var. ***mohavense***
22′ St stout, 10–37 cm; serpentine, s NCoR .... var. ***purdyi***
21′ Fls gen 6–9 mm
23. St slender, 7–20 cm; perianth erect; SCoR, WTR .... ***A. diabloense***
23′ St stout, 25–50 cm; perianth spreading from base; serpentine, c SNF .... ***A. tuolumnense***
2′ Lvs 2+ per st (if 1, lf flat or channeled); ovary crests gen ± 0
24. Bulb oblong, clustered on stout rhizome — bulb coats finely lined with narrow, vertically oriented cells
25. St 10–40 cm; perianth parts narrowly ovate, acute; stamens incl .... ***A. haematochiton***
25′ St 50–100 cm; perianth parts narrowly lanceolate, acuminate; stamens exserted .... ***A. validum***
24′ Bulb ovoid to ± spheric; rhizomes 0 or slender, with 1 new bulb at tip
26. Ovary crests 6, prominent, triangular; daughter bulbs gen clustered at bulb base or on slender rhizome
27. Lvs gen withered in fl; perianth rigid-keeled, ± shiny in fr, tip margins rolled inward ...... ***A. campanulatum***
27′ Lvs gen green in fl; perianth not keeled, papery in fr, tip margins ± flat
28. Perianth parts lanceolate, acuminate; s SNH, GB .... ***A. bisceptrum*** var. ***bisceptrum***
28′ Perianth parts ovate to elliptic, acute; n&c CA .... ***A. membranaceum***
26′ Ovary crests 0 or inconspicuous, not triangular; daughter bulbs 1–3, ± = parent bulb and formed within its coats or on stout rhizomes
29. Lvs flat or widely channeled, ± sickle-shaped; (dry) pl in fr breaking at soil surface
30. Outer bulb coat sculpture clearly net-like throughout
31. Bulb coat sculpture laterally elongate, rectangular, thin-walled, in ± regular rows
32. Perianth parts linear-lanceolate .... ***A. anceps***
32′ Perianth parts oblanceolate to ovate
33. St 15–20 cm, infl held well above soil surface; pedicel > fl .... ***A. lemmonii***
33′ St 3–10 cm, infl held close to soil surface; pedicel ± = fl .... ***A. punctum***
31′ Bulb coat sculpture ± square or polygonal, curved and irregular if laterally elongate
34. Bulb coat sculpture ± laterally elongate, curved, irregularly arranged .... ***A. tribracteatum***
34′ Bulb coat sculpture ± rectangular, square, or polygonal .... ***A. obtusum***
35. Lf 1 per st; perianth parts pink, lanceolate, acute .... var. ***conspicuum***
35′ Lvs 1 or 2 per st; perianth parts white, oblong-elliptic, obtuse .... var. ***obtusum***
30′ Outer bulb coat sculpture 0 or only 2–3 rows of vertical cells above root pad
36. St cylindric or slightly compressed, not winged
37. Stamens well incl
38. Lvs straight or barely sickle-shaped; fls > 10; CA-FP .... ***A. cratericola***
38′ Lvs strongly sickle-shaped; fls gen 5–10; CaRH, SNH, GB .... [2]***A. parvum***

37′ Stamens exserted or ± = perianth
39. Lvs 2 per st .... ***A. yosemitense***
39′ Lf 1 per st
40. Filaments smooth .... ***A. burlewii***
40′ Filaments bumpy or warty near base .... ***A. hoffmanii***
36′ St flattened, gen winged above
41. Perianth parts narrowly lanceolate, long-acuminate; stamens exserted .... ***A. platycaule***
41′ Perianth parts lanceolate to ovate, obtuse to long-acuminate; stamens incl
42. Lvs gen 6–8 mm wide; perianth parts 9–15 mm, long-acuminate .... ***A. falcifolium***
42′ Lvs < 6 mm wide; perianth parts 6–10 mm, obtuse to acute
43. Perianth parts lanceolate, acute (tip margins inrolled, so appearing acuminate) ... ***A. tolmiei*** var. ***tolmiei***
43′ Perianth parts widely lanceolate to ovate, obtuse, ± flat
44. Inner bulb coats white; perianth parts not keeled .... [2]***A. parvum***
44′ Inner bulb coats gen pink to red; perianth parts keeled in fr .... ***A. siskiyouense***
29′ Lvs narrowly channeled to ± cylindric, if flat, not sickle-shaped; pl persistent (sometimes not dry) in fr
45. Bulb coat sculpture 0 or obscure
46. Bulb symmetric, present at fl; rhizome 0
47. Longest infl bracts >> pedicels; perianth parts 4.5–7 mm .... ***A. paniculatum*** var. ***paniculatum***
47′ Longest infl bracts < to = pedicels; perianth parts 7–18 mm
48. Infl ± hemispheric; fls erect; perianth parts widely elliptic, obtuse .... ***A. neapolitanum***
48′ Infl open, ± 1-sided; fls drooping; perianth parts lanceolate, acute .... ***A. triquetrum***
46′ Bulb oblique, gone by fl; new bulbs borne on rhizome tips originating from old bulb root pad
49. Rhizomes conspicuous, gen > 3 cm; lvs widely channeled, flat, or keeled; perianth parts ovate, spreading, entire .... ***A. unifolium***
49′ Rhizomes inconspicuous, < 2 cm; lvs ± cylindric; perianth parts ± lanceolate, erect, at least inner finely serrate .... ***A. bolanderi***
50. Bulb oblique-ovoid, coats breaking with a serrate edge; perianth parts 8–12 mm, narrowly ovate .... var. ***bolanderi***
50′ Bulb oblique-oblong, tuber-like, coats breaking with an irregular edge; perianth parts 9–14 mm, narrowly lanceolate .... var. ***mirabile***
45′ Bulb coat sculpture obvious, net-like (with hand lens), meshes prominent, thick-walled
51. Bulb coat sculpture square or polygonal
52. Perianth parts 8–15 mm, spreading or tips reflexed, inner dentate .... ***A. acuminatum***
52′ Perianth parts 4–9 mm, erect or spreading from base, entire .... ***A. lacunosum***
53. St 10–25 cm; infl dense, pedicel 0.7–1.5 × fl
54. Lvs gen < st; st 15–25 cm; fls gen < 7 mm; s SNF, Teh, w DMoj .... var. ***kernensis***
54′ Lvs > st; st 10–20 cm; fls 7–9 mm; coast, coast ranges .... var. ***lacunosum***
53′ St 15–35 cm; infl open, pedicel 1.5–3.5 × fl
55. Bracts 2; fl 6–8 mm; TR, SNE, DMoj .... var. ***davisiae***
55′ Bracts 3; fl 4–6 mm; SCoRI .... var. ***micranthum***
51′ Bulb coat sculpture laterally elongate, in ± vertical herringbone pattern, or ± twisted
56. Bulb coat sculpture in wavy lateral rows (herringbone pattern indistinct); perianth parts spreading, outer ± = inner
57. St 5–17 cm; perianth parts erect in fr; fls persistent — CCo .... ***A. hickmanii***
57′ St 15–50 cm; perianth parts folding over ovary in fr; fls and pedicels deciduous
58. Ovary crests 6, lateral, ± rectangular; infl dense, pedicels 0.7–2 × fl .... ***A. amplectens***
58′ Ovary crests 3, central, minute or 0; infl loose, pedicels 1.5–4 × fl length
59. Lvs 1–3(4) mm wide, widely channeled or cylindric; inner bulb coats light yellow; perianth translucent in fr .... ***A. hyalinum***
59′ Lvs (3)4–6 mm wide, keeled, widely channeled or flat; inner bulb coats white; perianth ± opaque, papery in fr .... ***A. praecox***
56′ Bulb coat sculpture in sharply serrate lateral rows, herringbone pattern distinct; perianth parts erect, outer > inner
60. Perianth parts folded over ovary in fr; fls and pedicels deciduous .... ***A. serra***
60′ Perianth parts erect, rigid in fr; fls persistent
61. Lvs 3–6, curved to curled; infl dense, pedicels erect, 0.7–2 × fl; sea cliffs .... ***A. dichlamydeum***
61′ Lvs 2–3, straight to ± curved; infl open, pedicels spreading, 1–3 × fl; not sea cliffs
62. Inner perianth parts dentate, curled .... ***A. crispum***
62′ Inner perianth parts entire to dentate, not curled .... ***A. peninsulare***
63. Lvs curved; fls 8–12 mm, stigma minute, entire; CCo, SnFrB .... var. ***franciscanum***
63′ Lvs straight; fls 10–15 mm, stigma 3-lobed or head-like; widespread .... var. ***peninsulare***

A. abramsii (Traub) D. McNeal Bulb 10–15 mm, ovoid to ± spheric; outer coat red-brown, sculpture 0 or 2–3 rows of vertical cells above root pad. **ST** 5–15 cm. **LF** < 3 × st, cylindric. **INFL**: fls 6–40; pedicels 6–15 mm. **FL** 8–15 mm; perianth parts linear to narrowly lanceolate, rose-purple, inner < outer, jagged; ovary tip entire or notched, crests 6, prominent, triangular. *n*=7. Uncommon. Granitic sand; 1400–2000 m. c&s SNH. [*A. fimbriatum* var. *a.* Traub]

A. acuminatum Hook. (p. 1181) Bulb 8–16 mm, ovoid to ± spheric; outer coat yellow-brown, sculpture prominent, thick-walled, square or polygonal. **ST** 10–35 cm. **LVS** 2–3, ± = st, ± cylindric. **INFL**: fls 10–40; pedicels 6–25 mm. **FL** 8–15 mm; perianth parts lanceolate to narrowly ovate, white to rose-purple, inner finely dentate; ovary crests 3, minute, 2-lobed, central. *n*=7. Hills and plains; < 1900 m. NW, CaR, ne SnFrB (Mount Diablo), MP; to B.C., Rocky Mtns, AZ. ❀DRN,DRY,SUN:6,7,14–18;often DFCLT.

A. amplectens Torrey (p. 1181) Bulb 6–15 mm, ovoid to ± spheric; outer coat brown to gray, sculpture laterally elongate, wavy, in irregular herringbone pattern; inner coats gen red. **ST** 15–50 cm. **LVS** 2–4, < st, ± cylindric. **INFL**: fls 10–50; pedicels 4–16 mm. **FL** 5–9 mm; perianth parts spreading, lanceolate, white to pink, in fr folded over ovary; ovary crests 6, ± prominent, lateral. *n*=7,14;2*n*=21. Clay soils incl serpentine, open or wooded places; < 1800 m. CA-FP (exc SNH); to B.C. ❀DRY,SUN:4–6,**7**,8–10, **14–18**,19–21;may be DFCLT.

A. anceps S. Watson Bulb 15–20 mm, ovoid; outer coat ± yellow-brown, sculpture delicate, narrowly rectangular, laterally elongate. **ST** 10–15 cm. **LVS** 2, 2 × st, flat, sickle-shaped. **INFL**: fls 15–35; pedicels 15–30 mm. **FL** 6–12 mm; perianth parts linear-lanceolate, entire, ± pink, green-veined; ovary crests 6, minute, central. *n*=7. Barren clay, rocky slopes; 1200–1550 m. MP; s OR, w NV.

A. atrorubens S. Watson Bulb 10–16 mm, ovoid to ± spheric; outer coat red-brown, sculpture 0 or 2–3 rows of vertical cells above root pad; inner coats pink to white. **ST** 5–17 cm. **LF** < 2 × st, cylindric, tip tightly coiled when fresh. **INFL**: fls 5–50; pedicels 6–20 mm. **FL** 8–12 mm; perianth parts lanceolate to ovate, entire, red-purple or pink (rarely white), inner narrower, = or > outer; ovary crests 6, prominent, triangular, tip entire or notched. Rocky or sandy soil; 1200–2100 m. GB; to UT, AZ.

var. ***atrorubens*** **FL**: perianth parts ± narrowly lanceolate, ± acute, red-purple (rarely white), tip margins rolled inward, so appearing long-acuminate. *n*=7. Common. Habitat and range of sp.

var. ***cristatum*** (S. Watson) D. McNeal **FL**: perianth parts ± ovate, acute, pale pink with deep pink midveins, tip margins flat. *n*=7. Uncommon. Sandy soils; 1200–2100 m. SNE; w NV. [var. *inyonis* (M.E. Jones) F. Ownbey & Aase; *A. nevadense* S. Watson var. *c.* (S. Watson) F. Ownbey]

A. bisceptrum S. Watson var. ***bisceptrum*** (p. 1181) Bulb 10–15 mm, ovoid to ± spheric; daughter bulbs gen clustered at bulb base; outer coat brown to gray, sculpture ± square, walls very wavy. **STS** 1–3, 10–35 cm. **LF** ± = st, widely channeled. **INFL**: fls 15–40; pedicels 10–25 mm. **FL** 6–10 mm; perianth parts lanceolate, acuminate, entire, rose-purple, papery in fr; ovary crests 6, low, ± triangular, minutely dentate. *n*=7. Meadows, aspen groves; 2000–2900 m. GB; to ID, UT.

A. bolanderi S. Watson (p. 1181) Bulb 7–20 mm ovoid to oblong, outer coat sculpture obscure, ± V-shaped; rhizomes 1–2, short, with new bulbs at tips. **ST** 10–35 cm. **LVS** 2–3, ± = st, cylindric. **INFL**: fls 10–20; pedicels 10–20 mm. **FL** 8–14 mm; perianth parts finely dentate, obtuse to acute, rolled inward so appearing narrower, red-purple (rarely white); ovary crests 3, minute, 2-lobed, central. Rocky clay soils; < 1000 m. NW, se SnFrB (Mount Hamilton); sw OR.

var. ***bolanderi*** Bulb 7–12 mm, oblique-ovoid; outer coats pale brown to gray, breaking with a serrate edge. **FL**: perianth parts narrowly ovate. *n*=7. Uncommon. Rocky clay soils, incl serpentine; < 1000 m. Range of sp. ❀DRN,DRY,SUN:7,14–16;DFCLT.

var. ***mirabile*** (L. Henderson) D. McNeal Bulb 10–20 mm, irregularly oblique-oblong, tuber-like; outer coats pale brown, breaking with an irregular edge. **FL**: perianth parts narrowly lanceolate. *n*=7. Uncommon. Rocky clay soils, incl serpentine; < 1000 m. KR, n NCoR; sw OR.

A. burlewii Davidson Bulb 15–25 mm, ovoid; outer coat brown to gray, sculpture 0 or 2–3 rows of vertical cells above root pad. **ST** 2–8 cm. **LF** < 2 × st, widely channeled. **INFL**: fls 8–20; pedicels 6–10 mm. **FL** 7–10 mm; perianth parts ovate, strictly erect, entire, dull purple, midveins dark; ovary crests 0 or 6, minute, central. *n*=7. Granitic sands on dry slopes, ridges; 1800–2800 m. s SNH, Teh, SCoRI, SW. ❀very DFCLT.

A. campanulatum S. Watson (p. 1181) Bulb 1–2 cm, ovoid; daughter bulbs gen clustered at bulb base or on long thread-like rhizomes; outer coat brown to gray, sculpture ± square, walls very wavy; inner coats pink to white. **ST** 10–30 cm. **LVS** 2, ± = st, widely channeled. **INFL**: fls 10–50; pedicels 1–2 cm. **FL** 5–8 mm; perianth parts lanceolate to ovate, acuminate, entire, rose to purple (rarely white) with a darker crescent at base, in fr rigid, ± keeled; ovary crests 6, low, triangular, minutely dentate. *n*=7,14. Dry mtn slopes; 600–2600 m. CA-FP (exc SCoR); NV. ❀DRY,SUN, DRN:1–3,7–9,14–18;DFCLT.

A. cratericola Eastw. Bulb 15–25 mm, ovoid; outer coat (red-) brown, sculpture 0 or 2–3 rows of vertical cells above root pad. **ST** 3–10 cm. **LVS** 1–2, 1.5–4 × st, ± flat to widely channeled. **INFL**: fls 20–30; pedicels 5–18 mm. **FL** 7–14 mm; perianth parts ± oblong to elliptic, entire, ± pink; ovary crests 3, minute, central. *n*=7, 14. Uncommon. Open, serpentine, volcanic, or granitic places; 300–1800 m. KR, NCoR, n&c SNF, s SNH, Teh, WTR, SnJt. [*A. tribracteatum* Torrey var. *jacintense* Munz] Most CA pls have 1 lf; s CA pls and 2 populations in Mariposa Co. have 2 lvs; some NCoR populations are mixed. ❀DRY,SUN:**7**,8,9,14–21;DFCLT.

A. crispum E. Greene (p. 1181) Bulb 9–15 mm, ovoid to ± spheric; outer coat brown to gray, sculpture laterally elongate, V-shaped, herringbone-like. **ST** 15–35 cm. **LVS** 2–3, < st, ± cylindric. **INFL**: fls gen 10–40; pedicels 10–35 mm. **FL** 8–13 mm; perianth parts rose-purple, outer ovate, entire, inner narrower, finely dentate, curled, tip ± recurved; ovary crests 3, minute, 2-lobed, central. *n*=7. Clay slopes, incl serpentine; < 800 m. CW. ❀DRY, SUN:7,14–16;DFCLT.

A. denticulatum (Traub) D. McNeal Bulb 10–14 mm, ovoid to ± spheric; outer coats red-brown, sculpture 0 or 2–3 rows of vertical cells above root pad; inner coats pale brown to white. **ST** 5–18 cm. **LF** 1.5–2 × st, cylindric. **INFL**: fls 5–30; pedicels 5–20 mm. **FL** 9–17 mm; perianth parts minutely dentate near tip, rose-purple; ovary crests 6, prominent, entire to finely and irregularly dentate. *n*=7. Dry slopes; 900–1600 m. s SN, Teh, w DMoj. [*A. fimbriatum* var. *d.* Traub]

A. diabloense (Traub) D. McNeal Bulb 10–16 mm, ovoid to ± spheric; outer coats red-brown, sculpture 0 or 2–3 rows of vertical cells above root pad; inner coats pale brown to white. **ST** gen 7–20 cm. **LF** 1.5–3 × st, cylindric. **INFL**: fls 10–50; pedicels 7–20 mm. **FL** 6–10 mm; perianth parts entire, white, midveins or tips pink; ovary crests 6, prominent, jagged to deeply cut. *n*=7. Dry serpentine; 500–1500 m. SCoR, WTR. [*A. fimbriatum* var. *d.* Traub] ❀DRY,SUN:7,14,19–23;DFCLT.

A. dichlamydeum E. Greene Bulb 10–15 mm, ovoid to ± spheric; outer coat brown to gray, sculpture laterally elongate, V-shaped, herringbone-like. **ST** 10–30 cm, stout. **LVS** 3–6, = or > st, channeled to ± cylindric. **INFL**: fls gen 5–30; pedicels 5–20 mm. **FL** 9–12 mm; perianth parts ovate, entire or inner minutely dentate, deep red-purple; ovary crests 3, minute, 2-lobed, central. *n*=7. Dry clay soil on or near sea cliffs; 50–150 m. NCo, n CCo. ❀DRY,SUN, DRN:5,14–16,**17**,19–24;often DFCLT.

A. falcifolium Hook. & Arn. (p. 1181) Bulb 15–25 mm, ovoid; outer coats ± red-brown, sculpture 0 or 2–3 rows of vertical cells above root pad; inner coats white or pink. **ST** 5–20 cm, flat, winged. **LVS** 2, 1.5–3 × st, flat, sickle-shaped. **INFL**: fls 10–30; pedicels 8–15 mm. **FL** 9–15 mm; perianth parts lanceolate, at least inner gen dentate with minute glands, rose-purple or dingy white; ovary crests 3, low, wide. *n*=7. Common. Heavy soils, incl serpentine; 100–2100 m. NW, SnFrB; sw OR. ❀DRY,DRN,SUN:5,**7**,8,9,14–17; usually DFCLT.

A. fimbriatum S. Watson Bulb 10–17 mm, ovoid to ± spheric; outer coats red-brown, sculpture 0 or 2–3 rows of vertical cells above root pad; inner coats pale brown to white. **ST** 10–37 cm. **LF** 1.5–2 × st, cylindric. **INFL**: fls 6–75; pedicels 6–20 mm. **FL** 6–12 mm; perianth parts lanceolate to ovate, entire, dark red-purple to white; ovary crests 6, prominent, finely dentate to deeply cut. Dry slopes and flats; 300–2700 m. s NCoR, s SNF, Teh, CW, SW, SNE, w D; n Baja CA.

var. ***fimbriatum*** (p. 1181) **ST** 10–20 cm. **INFL**: fls 6–35. **FL**: perianth parts dark red-purple; ovary crests finely dentate to deeply cut. *n*=7. Common. Habitats and range of sp. (exc SNE, D). [*A. anserinum* Jepson] ❀DRY,DRN,SUN:3,7–9,11,14–24; DFCLT.

var. ***mohavense*** Jepson **ST** 10–25 cm. **INFL**: fls 12–60. **FL**: perianth parts white, pink, or light lavender; ovary crests deeply cut, sometimes with additional outgrowths on ovary. *n*=7. Common. Habitats of sp.; 700–1400 m. SNE, w DMoj. Pls from n base SnBr intermediate to var. *f.* are placed here provisionally.

var. ***purdyi*** (Eastw.) D. McNeal PURDY'S ONION **ST** 10–37 cm. **INFL**: fls 20–75. **FL**: perianth parts white to lavender; ovary crests finely, irregularly dentate to deeply cut, rarely 0. *n*=7. UNCOMMON. Serpentine outcrops; 300–600 m. c NCoRI (w Colusa, e Lake cos.) ❀DRN,DRY,SUN:7,9,14–17,19–21;DFCLT.

A. haematochiton S. Watson (p. 1181) Bulbs clustered, 2–3 cm, oblong, on short rhizome; outer coats ± red-brown; inner coats deep red to white. **ST** 10–40 cm. **LVS** 4–6, < or ± = st, flat. **INFL**: fls 10–30; pedicels 7–15 mm. **FL** 6–8 mm; perianth parts narrowly ovate, entire, white to rose, midveins darker; ovary crests 6, short, round, densely papillate. *n*=7. Common. Dry slopes, ridges; < 800 m. SCoRO, SCo, WTR, PR; n Baja CA. ❀DRN,SUN:7,**14–17,19–24**.

A. hickmanii Eastw. HICKMAN'S ONION Bulb 8–12 mm, ovoid to ± spheric; outer coat pale brown to gray, sculpture laterally elongate, wavy, irregularly herringbone-like. **ST** 5–17 cm. **LVS** 2, 1.5 × st, ± cylindric. **INFL**: fls 4–15; pedicels 4–12 mm. **FL** 5–7 mm; perianth parts lanceolate to narrowly ovate, entire, white to pale pink; ovary crests 0 or 3, minute, central. *n*=7. RARE. Grassy, wooded slopes; ± 50 m. c CCo (Monterey Peninsula; Arroyo de la Cruz, San Luis Obispo Co.) In cult.

A. hoffmanii Traub Bulb 15–25 mm, ovoid; outer coat red-brown to brown, sculpture 0 or 2–3 rows of vertical cells above root pad. **ST** 5–10 cm. **LF** < 2 × st, widely channeled. **INFL**: fls gen 10–40; pedicels 8–15 mm. **FL** 8–10 mm, ± narrowed above ovary; perianth parts erect, linear-lanceolate, long-tapered, pink to purple, midveins prominent, ± green; stamens exserted, distinctly bumpy or warty near base; ovary crests 6, obscure, central. *n*=7. Serpentine outcrops; 1100–1800 m. s KR, n NCoRH. Locally common. ❀DRN,SUN:7,16;DFCLT.

A. howellii Eastw. Bulb 9–17 mm, ovoid to ± spheric; outer coats red-brown, sculpture 0 or 2–3 rows of vertical cells above root pad; inner coats pale brown to white. **ST** 15–60 cm. **LF** 0.7–1.5 × st, cylindric. **INFL**: fls gen 10–100; pedicels 6–25 mm. **FL** 5–8 mm; perianth parts ovate, entire, white to pale lavender, midveins darker; ovary crests 6, prominent, finely and irregularly dentate to deeply cut. Common. Serpentine and granitic soils; 200–1850 m. s SNF, Teh, SnJV, SnFrB, SCoR, WTR, SnBr.

var. ***clokeyi*** Traub **ST** 20–40 cm, stout. **INFL**: fls gen 50–100; pedicels 15–25 mm. **FL**: perianth parts white, midveins sometimes pink; stamens exserted 0–2 mm. *n*=7. Open slopes, sagebrush scrub; 1300–1850 m. n WTR (Mount Pinos region). Locally common.

var. ***howellii*** **ST** 15–30 mm, slender. **INFL**: fls gen 15–40; pedicels 7–15 mm. **FL**: perianth parts gen lavender (or white); stamens exserted 0–2 mm. *n*=7. Common. Grassy slopes; 200–900 m. s SNF, Teh, SnJV, SnFrB, SCoR, WTR. ❀DRY,SUN:7–9,14–16;DFCLT.

var. ***sanbenitense*** (Traub) F. Ownbey & Aase **ST** 25–60 cm, stout. **INFL**: fls gen 50–90; pedicels 10–20 mm. **FL**: perianth parts white or pale pink; stamens exserted 2–4 mm. *n*=7. Uncommon. Grassy slopes; 300–1000 m. SCoRI (se San Benito, w Fresno cos.).

A. hyalinum Curran (p. 1181) Bulb 5–12 mm, ovoid to ± spheric, gen clustered; outer coat brown to gray, sculpture laterally elongate, wavy, irregularly herringbone-like or twisted; inner coats light yellow. **ST** 15–45 cm. **LVS** 2–3, 0.7–1.5 × st, channeled to ± cylindric. **INFL**: fls gen 5–25; pedicels 10–35 mm. **FL** 6–10 mm; perianth parts spreading, white or pale pink, in fr translucent, folded over ovary; ovary crest 0. *n*=7. Common. Grassy slopes, outcrops; 50–1500 m. SNF, SnJV. ❀SUN,DRN,DRY:**7–9,14–16**,17–24;INV.

A. jepsonii (Traub) S. Denison & D. McNeal JEPSON'S ONION Bulb 15–25 mm, ovoid; outer coat red-brown to gray, sculpture 0 or 2–3 rows of vertical cells above root pad. **ST** 25–40 cm. **LF** ± = st, cylindric. **INFL**: fls gen 20–60; pedicels 7–20 mm. **FL** 7–8.5 mm; perianth parts erect, ovate-elliptic, jagged, white, midveins deep pink, tip reflexed; ovary crests 6, prominent, ± jagged. *n*=7. RARE. Open, serpentine or volcanic slopes, flats; 300–600 m. n&c SNF (Butte, Tuolumne cos.). [*A. sanbornii* var. *j.* Traub]

A. lacunosum S. Watson (p. 1181) Bulb 1–2 cm, ovoid; outer coats often many, thickly surrounding bulb, yellow brown, sculpture ± square, walls obscurely wavy. **ST** 10–35 cm. **LVS** 2, 0.7–2 × st, ± cylindric or flat. **INFL**: fls 5–45; pedicels 5–25 mm. **FL** 4–9 mm; perianth parts oblanceolate to narrowly ovate, entire, white or pale pink, midveins darker; ovary crests 3, minute, 2-lobed, central, crests and upper ovary densely papillate. Common. Dry, open hillsides; 50–2100 m. s SNF, Teh, SnFrB, SCoR, SCo, n ChI, WTR, SnBr, SNE, DMoj.

var. ***davisiae*** (M.E. Jones) D. McNeal & F. Ownbey **INFL** open; bracts 2; fls 10–35; pedicels 10–25 mm. **FL** 6–8 mm. *n*=7. Uncommon. Open, sandy slopes, ridges; 600–2100 m. WTR, SnBr, SNE, DMoj. [*A. d.* M.E. Jones] ❀SUN,DRN,DRY:3,7–9,11,14–16,18–21;DFCLT.

var. ***kernensis*** D. McNeal & F. Ownbey **INFL** dense; bracts 2; fls 10–45; pedicels 10–15 mm. **FL** 5–7 mm. *n*=7. Uncommon. Open sandy slopes; 700–1300 m. s SNF, Teh, w DMoj.

var. ***lacunosum*** **INFL** dense; bracts 2; fls 5–25; pedicels 5–12 mm. **FL** 6–9 mm. *n*=7. Serpentine outcrops; 50–1000 m. SnFrB, SCoR, SCo, n ChI. ❀SUN,DRN,DRY:7,14–16,19–24;DFCLT.

var. ***micranthum*** Eastw. **INFL** open; bracts 3; fls gen 10–30; pedicels 6–13 mm. **FL** 4–6 mm. *n*=7. Uncommon. Serpentine outcrops; 300–600 m. SCoRI. ❀SUN,DRN,DRY:7,9,14–16,19–21;DFCLT.

A. lemmonii S. Watson (p. 1181) Bulb 15–22 mm, ovoid; outer coat sculpture delicate, narrowly rectangular, laterally elongate. **ST** 15–20 cm, flat, narrowly winged. **LVS** 2, ± = st, flat, sickle-shaped. **INFL**: fls 10–40; pedicels 8–16 mm. **FL** 6–9 mm; perianth parts narrowly ovate, entire, white to pink; ovary crests 6, obscure, ridge-like. *n*=7. Common. Drying clay soils; 1200–1900 m. n&c SNH, MP; to OR, ID, NV. ❀SUN,DRY:1–3,7,14–18;DFCLT.

A. membranaceum Traub Bulb 10–16 mm, ovoid; daughter bulbs sometimes in tight basal cluster; outer coat sculpture ± square, walls very wavy. **ST** 15–40 cm. **LVS** 2–3, ± = st, flat. **INFL**: fls 15–35; pedicels 10–20 mm, slender. **FL** 7–12 mm; perianth parts spreading, elliptic to ovate, entire, white to pale pink, papery in fr; ovary crests 6, short, triangular. *n*=7. Uncommon. Wooded slopes; 150–1400 m. KR, CaRF, n&c SNF.

A. monticola Davidson Bulb 10–22 mm, ovoid; daughter bulbs gen 1–2, large, stalked, at bulb base; outer coat gray, sculpture 0, very obscure, or ± square; inner coats white or pink. **ST** 6–25 cm, glaucous. **LF** < 2 × st, cylindric, glaucous. **INFL**: fls 8–25; pedicels 5–12 mm. **FL** 12–19 mm; perianth parts erect, lanceolate, entire, pink to rose-purple near tip, often white below; ovary crests 6, prominent, ± linear to narrowly triangular, entire. *n*=7. Uncommon. Rocky ridges, talus slopes; 1400–3200 m. TR, nw PR (Orange Co.). [ssp. *keckii* (Munz) F. Ownbey & Aase, mountain onion]

A. munzii (Traub) D. McNeal MUNZ'S ONION Bulb 10–15 mm, ovoid; outer coats red-brown, sculpture 0 or 2–3 rows of vertical cells above root pad; inner coats pale brown, white, or pink. **ST** 15–35 cm. **LF** ± 1.5 × st, cylindric. **INFL**: fls 10–35; pedicels 7–12 mm. **FL** 6–8 mm; perianth parts erect, ovate, entire, white, red in fr, midveins sometimes pink or green; ovary crests 6, prominent, finely and irregularly dentate. *n*=7. RARE. Grassy openings in coastal-sage scrub; 300–900 m. e SCo, nw PR (w Riverside Co). [*A. fimbriatum* var. *m.* Traub] Threatened by urbaniztion, citrus culture.

A. neapolitanum Cirillo Bulb 1–2 cm, ± spheric; outer coat sculpture ± square, walls thick, appearing crusted. **ST** 20–60 cm, 3-angled, 2 angles slightly winged. **LVS** 2–3, < or = st, flat, linear to narrowly lanceolate. **INFL**: fls 10–25; pedicels 15–35 mm. **FL** 7–12 mm; perianth parts ± spreading, elliptic, entire, white; ovary crest 0. Disturbed ± urban places; < 100 m. NCo, NCoR, ScV, CCo, SnFrB, SCo; native to Medit. Cult as orn.

A. nevadense S. Watson (p. 1181) Bulb 9–15 mm, ovoid; daughter bulbs gen 1–2, stalked, at bulb base; outer coat brown to gray, sculpture ± elongate, intricately twisted; inner coats white or pink. **ST** 5–15 cm. **LF** 1.5–2 × st, cylindric, tip tightly coiled before withering. **INFL**: fls gen 5–25; pedicels 6–17 mm. **FL** 7–12 mm; perianth parts spreading, lanceolate to ovate, ± recurved at tip, entire, white or pink, midveins dark pink; ovary crests 6, prominent, triangular, entire or tip notched. *n*=7,14. Uncommon. Sandy or gravelly slopes; 1300–1700 m. DMtns; to OR, ID, AZ.

A. obtusum Lemmon (p. 1181) Bulb 1–2 cm, ovoid; outer coat gen gray, sculpture ± square, polygonal, or rectangular. **ST** 1.5–17 cm. **LVS** 1–2, 1–4 × st. **INFL**: fls 6–60; pedicels 2–15 mm. **FL** 4–12 mm; perianth parts erect, lanceolate to oblong-elliptic, entire, white or pink, midveins dark purple; ovary crests 3, obscure to ± prominent, 2-lobed, central. Common. Granitic sands; 800–3500 m. CaRH, n&c SNF, SNH; NV.

var. ***conspicuum*** W. Mortola & D. McNeal **LF** 1, 1–2.5 × st, widely channeled. **INFL**: fls 10–60. **FL**: perianth parts lanceolate, acute, pink. *n*=7. Uncommon. Habitat of sp.; 800–3000 m. n&c SNF, n&c SNH.

var. ***obtusum*** **LVS** 1–2, 1.5–4 st length, channeled. **INFL**: fls 6–30. **FL**: perianth parts oblong-elliptic, obtuse, white. *n*=7. Common. Habitat and range of sp.; 1500–3500 m. Small, easily overlooked.

A. paniculatum L. var. *paniculatum* Bulb 10–15 mm, ovoid; outer coats dark brown, sculpture obscure, square; inner coats light brown. **ST** 30–70 cm. **LVS** 3–5, < or = st, channeled, sheathing st in lower 30–50%. **INFL**: fls 25–100; bracts 2, long-tapered, unequal, longer 5–14 cm, >> fls; pedicels unequal, 10–45 mm. **FL** 5–7 mm, bell-shaped; perianth parts ovate, entire, white to lilac-pink; ovary crest 0. *n*=8,16. Disturbed areas; ± 50 m. s NW, n CW; native to s Eur. NOXIOUS WEED.

A. parishii S. Watson PARISH'S ONION Bulb 10–15 mm, ovoid; outer coats red-brown, sculpture 0, very obscure, or square; inner coats pink. **ST** 5–25 cm. **LF** 1, < 2 × st, cylindric. **INFL**: fls 6–25; pedicels 5–15 mm. **FL** 12–18 mm; perianth parts spreading, lanceolate, entire, pale pink, midveins darker; ovary crests 6, entire or finely and irregularly dentate. *n*=7. UNCOMMON. Open rocky slopes; 900–1400 m. DMoj; w AZ. Populations scattered.

A. parryi S. Watson Bulb 8–14 mm, ovoid to ± spheric; outer coats ± red-brown, sculpture 0 or 2–3 rows of vertical cells above root pad; inner coats lighter. **ST** 5–20 cm. **LF** < 1.5 × st, cylindric. **INFL**: fls 8–50; pedicels 6–20 mm. **FL** 6–9 mm; perianth parts lanceolate to narrowly ovate, entire, white or midveins pink, in fr ± red; ovary crests 6, prominent, entire or finely and irregularly dentate. *n*=7. Dry slopes, flats; 900–2200 m. s SNH, SnBr, PR; n Baja CA. [*A. fimbriatum* var. *p.* (S. Watson) F. Ownbey & Aase]

A. parvum Kellogg Bulb 10–25 mm, oblique to ovoid; outer coat brown to gray, sculpture 0, obscure, or ± square. **ST** 3–12 cm. **LVS** 2, 1.5–3 × st, flat, sickle-shaped. **INFL**: fls 5–30; pedicels 3–12 mm. **FL** 6–12 mm; perianth parts erect, oblong to elliptic, entire, white or pink, midveins wide, dark; ovary crests 3, minute, round, central. *n*=7. Common. Stony clay slopes, talus; 1200–2800 m. KR, CaRH, n&c SNH, MP; to e OR, ID, UT.

A. peninsulare Lemmon (p. 1181) Bulb 8–15 mm, ovoid to ± spheric; outer coat brown to gray, sculpture laterally elongate, V-shaped, herringbone-like. **ST** 12–45 cm. **LVS** 2–3, 0.7–1.5 × st, channeled to ± cylindric. **INFL**: fls gen 5–35; pedicels 1–4 cm. **FL** 8–15 mm; perianth parts erect (tips ± recurved), lanceolate to elliptic, entire or inner minutely dentate, red-purple; ovary crests 3, minute, 2-lobed, central. Common. Dry slopes and flats; 300–1100 m. SNF, Teh, ScV, CCo, SnFrB, SCoR, SW; s OR, n Baja CA.

var. ***franciscanum*** D. McNeal & F. Ownbey **LF** curved. **INFL**: pedicels 8–20 mm. **FL** 8–12 mm; stigma barely thickened. *n*=7. Uncommon. Dry hillsides; < 300 m. CCo, SnFrB.

var. ***peninsulare*** **LF** ± straight. **INFL**: pedicels 10–40 mm. **FL** 10–15 mm; stigma 3-lobed, sometimes ± head-like. *n*=7. Common. Habitat and range of sp. ❀SUN,DRY:7,8,9,**14–24**;may be DFCLT.

A. platycaule S. Watson Bulb 2–3 cm, ovoid; outer coat gen gray, sculpture 0. **ST** 7–25 cm, stout, flat, winged. **LVS** 2, ± 2 × st, flat, sickle-shaped. **INFL**: fls 30–90; pedicels 10–25 mm. **FL** 8–15 mm; perianth parts spreading, narrowly lanceolate, long-acuminate, entire, bright pink to rose, in fr erect, narrowed above ovary; ovary crest 0. *n*=7. Common. Rocky or sandy slopes; 1500–2500 m. n SNH, MP; s OR, w NV. ❀DRN,SUN:1–3,7,14–18;DFCLT.

A. praecox Brandegee Bulb 10–18 mm, ovoid to ± spheric; outer coat brown to gray, sculpture laterally elongate, wavy, irregularly herringbone-like or twisted. **ST** 20–60 cm. **LVS** 2–3, 0.7–1.5 × st, widely channeled, keeled. **INFL**: fls gen 5–40; pedicels 15–40 mm. **FL** 8–12 mm; perianth parts spreading, ± ovate, entire, pale pink, midveins purple, in fr papery, folded over ovary, dull purple; ovary crests 3, minute, 2-lobed, central. *n*=7. Uncommon. Shaded, grassy slopes; < 800 m. SW (exc SnGb, SnBr); n Baja CA. ❀DRN,DRY,part SHD:7,9,14–17,19–24;DFCLT.

A. punctum L. Henderson Bulb 1–2 cm, ovoid; outer coats yellow-brown to gray, sculpture delicate, narrowly rectangular, laterally elongate. **ST** 3–10 cm. **LVS** 2, 1.5–2 × st, flat, sickle-shaped. **INFL**: fls 6–20; pedicels 5–11 mm. **FL** 6–13 mm; perianth parts erect, oblong-lanceolate, entire, white to pink, midveins purple; ovary crests 3, wide, low, central. *n*=7. Uncommon. Rocky flats; 1200–1600 m. MP; s OR, w NV.

A. sanbornii Alph. Wood Bulb 15–25 mm, ovoid; outer coats red-brown, sculpture 0 or 2–3 rows of vertical cells above root pad; inner coats light brown or white. **ST** 18–60 cm. **LF** = st, cylindric. **INFL**: fls gen 18–150; pedicels 5–22 mm. **FL** 4–9 mm; perianth parts erect, ± ovate, entire to jagged, white to deep pink, inner whorl longer; ovary crests 6, prominent, entire. UNCOMMON. Serpentine outcrops; 300–1400 m. CaRF, n&c SNF; s OR.

var. ***congdonii*** Jepson CONGDON'S ONION **FL**: perianth parts acuminate, ± jagged, inner 1.5 × outer; stigma distinctly 3-lobed. *n*=7. UNCOMMON. Habitat of sp.; 300–700 mm. n&c SNF.

var. ***sanbornii*** SANBORN'S ONION **FL**: perianth parts acute, entire to ± irregular, inner 1.3 × outer; stigma head-like to obscurely 3-lobed. *n*=7. UNCOMMON. Habitat and range of sp. ❀DRY, SUN,DRN:7,14–17;may be DFCLT.

A. serra D. McNeal & F. Ownbey Bulb 8–12 mm, ovoid to ± spheric; outer coat gen brown, sculpture laterally elongate, V-shaped, herringbone-like. **ST** 15–40 cm. **LVS** 2–3, < or = st, ± cylindric. **INFL**: fls 10–40; pedicels 7–15 cm. **FL** 8–11 mm; perianth parts spreading, ± lanceolate, pink to rose, in fr papery, folded over ovary; ovary crests 3, minute, 2-lobed, central. *n*=7. Common. Grassy slopes; 30–1200 m. NCoR, SnFrB, SCoRI. [*A. serratum* S. Watson misapplied] ❀TRY.

A. sharsmithae (Traub) D. McNeal SHARSMITH'S ONION Bulb 10–18 mm, ovoid to ± spheric; outer coats red-brown, sculpture 0 or 2–3 rows of vertical cells above root pad. **ST** 4–17 cm. **LF** ± 2 × st, cylindric. **INFL**: fls 5–50; pedicels 6–19. **FL** 10–18 mm; perianth parts erect, lanceolate, entire, purple, tips ± recurved; ovary crests 6, prominent, entire, gen papillate. *n*=7. UNCOMMON. Rocky serpentine slopes; 400–1200 m. se SnFrB (Mount Hamilton Range). [*A. fimbriatum* var. *s.* Traub] ❀DFCLT.

A. shevockii D. McNeal (p. 1181) SPANISH NEEDLE ONION Bulb 10–15 mm, ± spheric; daughter bulbs 1–2, large, stalked, near bulb base, forming thread-like rhizomes with 1 terminal bulb; outer coat brown to gray, sculpture 0 or 2–3 rows of vertical cells above root pad; inner bulb coats light yellow, becoming red. **ST** 10–20 cm. **LF** 1.5–2.5 × st, cylindric. **INFL**: fls gen 12–30; pedicels 10–16 mm. **FL** 12–14 mm; perianth parts oblanceolate to ovate, maroon near tip, white or green below, outer tips reflexed and curled, inner ± flaring; ovary crests 6, prominent, margins irregular, tip gen notched. RARE. Metamorphic outcrops, talus; 2000–2500 m. s SNH (Spanish Needle Peak, ne Kern Co.). In cult.

A. siskiyouense Traub SISKIYOU ONION Bulb 8–20 mm, ovoid; outer coat red-brown, sculpture 0; inner coat pink to red. **ST** 3–8 cm. **LVS** 2, ± 2 × st, flat, sickle-shaped. **INFL**: fls 10–35; pedicels 5–16 mm. **FL** 8–11 mm; perianth parts erect, lanceolate to elliptic, entire or minutely dentate near tip, pink, midveins often darker; ovary crests 3, minute, 2-lobed, central. *n*=7. UNCOMMON. Rocky slopes, incl serpentine; 900–2500 m. KR, NCoR; sw OR. [*A. falcifolium* var. *demissum* Jepson] ❀DRN,DRY:1–3,7,14–16; DFCLT.

A. tolmiei Baker & S. Watson var. ***tolmiei*** Bulb 1–2 cm, ovoid; outer coat brown to gray, sculpture 0. **ST** 5–15 cm, flat, winged.

LVS 2, < 2 × st, flat, sickle-shaped. **INFL**: fls gen 10–40; pedicels 10–25 mm. **FL** 6–10 mm; perianth parts erect, lanceolate, entire, white to pink, midveins gen dark pink; ovary crests 3, obscure, 2-lobed, central. *n*=7. Uncommon. Rocky clay flats; 1500–2200 m. MP; to WA, ID, NV.

A. tribracteatum Torrey (p. 1181) THREE-BRACTED ONION Bulb 1–2 cm, ovoid; outer coat sculpture laterally elongate, irregularly curved. **ST** 2–7 cm. **LVS** 2, 1.5–3 × st, channeled. **INFL**: fls 10–30; pedicels 6–10 mm. **FL** 6–8 mm; perianth parts erect, lanceolate to elliptic, entire, white to purple; ovary crests 3, minute, ± lateral. *n*=7. RARE. Volcanic slopes; 1300–3000 m. c SNH (Tuolumne Co).

A. triquetrum L. Bulb 1–2 cm, ovoid; outer bulb coats thin, ± translucent, yellow-brown, sculpture obscure, vertically elongate, twisted. **ST** 10–40 cm; sharply 3-angled. **LVS** 2–3, ± = st, flat. **INFL** gen 1-sided, open; fls 3–15; pedicels 15–25 mm. **FL** 10–18 mm, pendent, bell-shaped; perianth parts lanceolate, entire, white, midveins green; ovary crest 0. **SEEDS** appendaged. Shady ± disturbed places; < 100 m. NCo, CCo; native to w Medit. Cult as orn; locally common.

A. tuolumnense (Traub) S. Denison & D. McNeal RAWHIDE HILL ONION Bulb 10–25 mm, ovoid; outer coat dark red-brown, sculpture 0 or 2–3 rows of vertical cells above root pad; inner coats light brown. **ST** 25–50 cm. **LF** ± = st. **INFL**: fls gen 20–60; pedicels 7–20 mm. **FL** 6–8 mm; perianth parts spreading, ovate, entire, white or pink; ovary crests 6, prominent, deeply cut. *n*=7. RARE. Serpentine slopes; 300–600 m. c SNF (Rawhide Hill, Redhills, Tuolumne Co.). [*A. sanbornii* var. *t.* Traub]

A. unifolium Kellogg (p. 1181) Bulb 1–2 cm, ovoid to oblique; rhizomes 1–3, < 5 cm, conspicuous, bearing new bulbs at tips; outer bulb coat pale brown, sculpture obscure, ± rectangular. **ST** 30–80 cm. **LVS** 2–3, < st, widely channeled, keeled. **INFL**: fls gen 15–35; pedicels 15–40 mm. **FL** 11–15 mm; perianth parts spreading, ovate to obovate, entire, pink or white; ovary crests 6 longitudinal ridges. *n*=7. Uncommon. Moist clay or serpentine, esp grassy streambanks; < 1100 m. w NW, w CW; OR. ❀4–6;IRR:**7**,8,**9**,**14–17**,18, **19–24**;may be INV.

A. validum S. Watson (p. 1181) Bulb 3–5 cm, oblong-ovoid, clustered on short, stout rhizome; outer coat brown to gray, fibrous, vertically lined. **ST** 50–100 cm, angled. **LVS** 3–6, ± = st, flat or ± keeled. **INFL**: fls 15–40; pedicels 7–12 mm. **FL** 6–10 mm; perianth parts ± erect, narrowly lanceolate, acuminate, entire, rose to white; stamens exserted; ovary crest 0. *n*=14,28. Common. Wet meadows; 1200–3400 m. NW, CaRH, SNH, Wrn; to B.C., ID, NV. ❀IRR or WET:**1**,**2**,3–5,**6**,7,14–16,**17**,18,24;DFCLT.

A. vineale L. Bulb 1–2 cm, ovoid; daughter bulbs several, stalked, hard-shelled, carried inside bulb coats; outer coat ± yellow-brown, vertically lined, splitting into strips. **ST** 30–100 cm. **LVS** 2–4, << st, sheathing lower st. **INFL**: fls gen few, gen replaced by bulblets; pedicels 12–15 mm. **FL** 3–4 mm; perianth parts ovate, entire, green-white to purple; inner filaments with 2 prominent, lateral appendages; ovary crest 0. Uncommon. Disturbed places; < 100 m. s NW, n SNF, ScV, n CW; native to Eur. NOXIOUS WEED.

A. yosemitense Eastw. (p. 1187) YOSEMITE ONION Bulbs 2–3 cm, ovoid, in clusters of 2–12 or more. **ST** 6–23 cm. **LVS** 2, 1–3 × st, widely channeled to ± flat. **INFL**: fls 10–50; pedicels 7–34 mm. **FL** 7–15 mm; perianth parts linear-oblong, entire, white to pink; ovary crests 3, minute, 2-lobed, central. *n*=7. **RARE**. Open, rocky slopes; 800–2200 m. c SN (Tuolumne, Mariposa cos.).

ANDROSTEPHIUM

Glenn Keator

Per from spheric, fibrous-coated corm. **LVS** basal, linear, channeled. **INFL** umbel-like, scapose, straight, cylindric; bracts papery; pedicels unjointed, erect. **FL**: perianth segments 6 in 2 petal-like whorls, tube funnel-shaped, lobes narrowly oblong; stamens 6, filaments fused into a crown-like tube with toothed lobes between anthers, anthers attached near middle; ovary superior, sessile, chambers 3, style persistent. **FR**: capsule, subspheric, obtusely 3-angled, loculicidal. **SEEDS** several per chamber, flat, black. 3 spp.: sw US. (Greek: stamen crown, from fused filaments)

A. breviflorum S. Watson **LF** 10–30 cm, 2 mm wide, scabrous. **INFL** 10–30 cm, scabrous near base; bracts lanceolate; pedicels 15–30 mm; fls 3–12. **FL**: perianth white to light violet drying yellow-brown, tube 5–7 mm, lobes 10–14 mm; filaments 8–10 mm, tube ± funnel-shaped, appendages ± 2 mm, anthers ± 3 mm. **FR** 10–15 mm, deeply 3-lobed. Open desert scrub; 700–1600 m. e D?; to w Colorado. Documentation of occurrence in CA needed.

ASPARAGUS

Per, ± dioecious; rhizome gen with fleshy tubers. **ST** erect to ± climbing, much-branched; branchlets flat or thread-like, functioning as lvs. **LVS** scale-like, papery, gen with a spiny spur at base. **INFL**: raceme or umbel; pedicels slender, jointed. **FL** small, white to greenish yellow; perianth segments 6 in 2 petal-like whorls, free or ± fused; stamens 6, epipetalous, ± = perianth; ovary superior, chambers 3, style 1, short, slender, stigmas 3. **FR**: berry, spheric. **SEEDS** 1–6, black. ± 300 spp.: esp n temp. (Greek: ancient name)

1. Branchlets flat, lf-like, 6–20 mm wide .... ***A. asparagoides***
1′ Branchlets thread-like, < 1 mm wide .... ***A. officinalis*** ssp. ***officinalis***

A. asparagoides (L.) Druce Roots with fleshy tubers. **ST** 1–5 m, twining; branchlets spreading to reflexed, lf-like, 6–20 mm wide; young sts simple, slender. **LVS**: spurs at base 0. **INFL**: pedicels narrow, jointed above. **FL** 5–7 mm, drooping, perianth spreading to reflexed, greenish white. **FR** 6–8 mm, red or blue. Disturbed places, fields; < 200 m. CCo, n SCo; native to s Afr.

A. officinalis L. ssp. *officinalis* Rhizome thick, matted, tubers 0. **ST** 1–3 m, erect; branchlets thread-like; young sts simple, stout, edible. **INFL**: pedicels 1–2 cm, thread-like, jointed near middle. **FL** 3–7 mm, drooping, bell-shaped, greenish white. **FR** 6–8 mm, red. *n*=10. Disturbed places, roadsides, fields; gen < 200 m. CA-FP; native to Eur. Cult for food. Naturalized populations scattered.

ASPHODELUS ASPHODEL

Ann or per from rhizomes. **LVS** basal, spiral, linear. **INFL**: raceme or panicle, dense; bracts persistent, narrowly lanceolate, scarious; pedicels jointed. **FL** small; perianth parts 6, ± free, in 2 petal-like whorls, white or pinkish; stamens 6, filaments wider at base; ovary superior, chambers 3, stigma head-like, slightly 3-lobed. **FR**: capsule, loculicidal. **SEEDS** 6, black. 12 spp.: s Eur. (Greek: ancient name)

A. fistulosus L. Ann or short-lived per; roots many from ± tuber-like st bases. **ST** 15–70 cm, branched, hollow. **LVS** many, 10–30 cm, ± 4 mm diam, subcylindric, hollow; bases wide, membranous. **INFL**: panicle, open. **FL** 5–12 mm; perianth parts oblong, obtuse; stamens incl, of 2 lengths, longest = style. **FR** 5–7 mm, ± spheric. **SEEDS** 3–4 mm. 2*n*=28,56. Disturbed areas, fields; < 50 m. s SnJV, SCo (expected elsewhere); native to s Eur. NOXIOUS WEED.

BLOOMERIA GOLDENSTAR

Glenn Keator

Per from subspheric, fibrous-coated corm with daughter corms. **LVS** basal, 1–2, linear-lanceolate, keeled, margin entire, withered in fl. **INFL** umbel-like, stiff, straight, cylindric; bracts 2–4, scarious in fl; pedicels long, ray-like, erect; fls 10–35+. **FL**: perianth segments 6 in 2 petal-like whorls, barely fused at base, golden-yellow, striped brownish, lobes widely spreading; stamens 6, filaments ± 6 mm, thread-like, with cup-like, winged, basal appendage 3 mm, anthers attached near middle; style 5 mm, thread- or club-like, persistent, stigma 3-lobed. **FR**: capsule, 5–6 mm, subspheric, 3-angled, sessile, loculicidal. **SEEDS** angled, black; coat wrinkled. 2 spp.: c&s CA, n Baja CA. (H.G. Bloomer, early San Francisco botanist) [Hoover 1955 Herbertia 11:13–23]

1. Lf gen 1; infl 15–60 cm; perianth lobes abruptly spreading .... ***B. crocea***
1′ Lvs 1–2; infl gen < 8 cm; perianth lobes ascending at base, then spreading .... ***B. humilis***

B. crocea (Torrey) Cov. (p. 1187) COMMON GOLDENSTAR **LF** gen 1, < 30 cm. **INFL** 15–60 cm; pedicels 2–6 cm. **FL**: perianth lobes abruptly spreading, 8–12 mm. *n*=9. Grassland, open woodlands, chaparral edges; < 1700 m. SCoR, TR, PR; n Baja CA. [vars. *aurea* (Kellogg) Ingram & *montana* (E. Greene) Ingram] In cult.

B. humilis Hoover (p. 1187) DWARF GOLDENSTAR **LVS** 1–2, 5–10 cm. **INFL** < 8 cm; pedicels 1–5 cm. **FL**: perianth lobes ascending at base, then spreading, 7–10 mm. **RARE** CA. Grassland chaparral edges; < 30 m. c&s CCo, adjacent c SCoR (Monterey, n San Luis Obispo cos.).

BRODIAEA

Glenn Keator

Per from subspheric, dark-brown-fibrous-coated corm with daughter corms. **LVS** basal, gen 3–5, linear (gen crescent-shaped in X-section), entire, glabrous. **INFL** umbel-like, scapose, open (pedicels > fls); axis slender, straight, cylindric; bracts scarious. **FL** erect; perianth outer surface shiny, tube narrowly bell- to funnel-shaped, lobes 6 in 2 petal-like whorls, inner wider; staminodes (0)3, gen much unlike stamens, often ± inrolled as seen from above; stamens 3, epipetalous, opposite inner perianth, ± = style, anthers attached at base; style 1, stigma with 3 decurrent wings. **FR**: capsule, sessile, ovoid, loculidical. **SEEDS** black, lined. 13 spp.: coastal w US, esp n CA. (Brodie, Scottish plantsman) [Niehaus 1971 Univ Calif Publ Bot 60:1–66]

1. Staminodes 0 .... ***B. orcuttii***
1′ Staminodes present
 2. Perianth lobes widely spreading
  3. Staminodes = perianth lobe width; serpentine, NCoR .... [2]***B. stellaris***
  3′ Staminodes < perianth lobe width; if on serpentine not in NCoR
   4. Perianth tube not narrowed above ovary
    5. Staminodes strongly inrolled .... ***B. insignis***
    5′ Staminodes not inrolled
     6. Staminodes tapered, reflexed against perianth .... ***B. filifolia***
     6′ Staminodes abruptly pointed, erect .... ***B. kinkiensis***
   4′ Perianth tube narrowed above ovary
    7. Anthers obcordate, axis tissue on outside with obvious enlarged cells .... [2]***B. pallida***
    7′ Anthers linear, axis tissue without enlarged cells on outside
     8. Infl 2–10 cm; inner perianth lobes 5–7 mm wide .... ***B. minor***
     8′ Infl 10–25 cm; inner perianth lobes 4–5 mm wide .... ***B. purdyi***
 2′ Perianth lobes ascending
  9. Perianth tube narrowed above ovary or staminodes = perianth width
   10. Perianth tube narrowed above ovary; staminodes narrowed at tip, 2-lobed .... [2]***B. pallida***
   10′ Perianth tube not narrowed above ovary; staminodes wide throughout, unlobed .... [2]***B. stellaris***
  9′ Perianth tube not narrowed above ovary; staminodes narrower than perianth
   11. Staminodes narrowly linear, wavy
    12. Filaments with 2 thread-like appendages; perianth lobes gen < 2 × tube .... ***B. appendiculata***
    12′ Filaments appendages gen 0; perianth lobes > 2 × tube .... ***B. californica***
     13. Perianth lobes 6–10 mm wide; ovary 7–11 mm .... var. ***californica***
     13′ Perianth lobes 4–7 mm wide; ovary 5–7 mm .... var. ***leptandra***
   11′ Staminodes wider at base, not wavy
    14. Staminodes standing away from stamens; perianth tube funnel-shaped .... ***B. elegans***
     15. Staminodes flat, straight; NW, SN .... ssp. ***elegans***
     15′ Staminodes inrolled, tips reflexed; nw KR? .... ssp. ***hooveri***
    14′ Staminodes leaning in toward stamens; perianth tube bell-shaped
     16. Fl 24–38 mm; staminodes strongly inrolled .... ***B. coronaria***

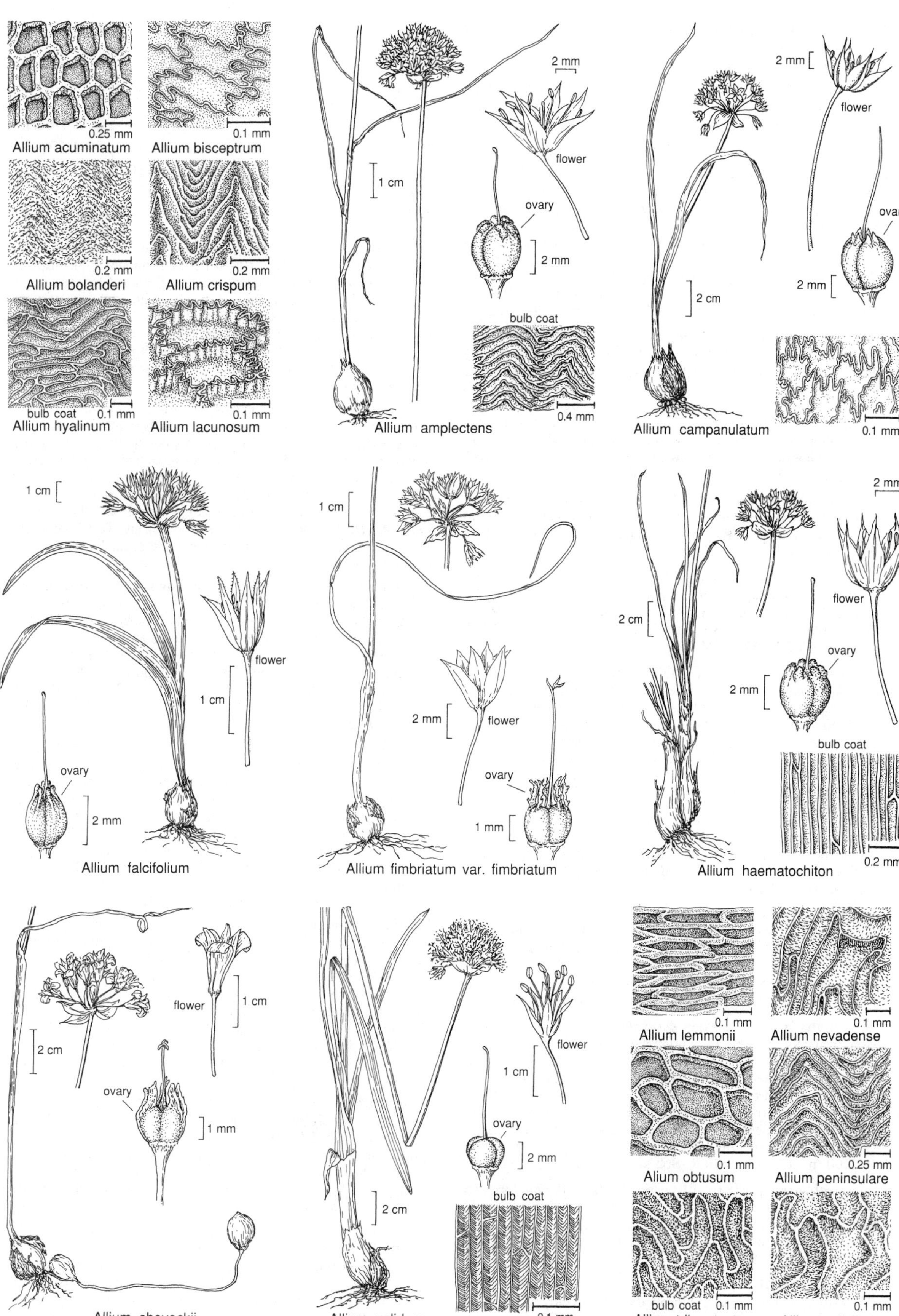
0.25 mm
Allium acuminatum
0.1 mm
Allium bisceptrum
0.2 mm
Allium bolanderi
0.2 mm
Allium crispum
bulb coat
0.1 mm
Allium hyalinum
0.1 mm
Allium lacunosum
1 cm
2 mm
flower
ovary
2 mm
bulb coat
0.4 mm
Allium amplectens
2 mm
flower
ovary
2 mm
2 cm
0.1 mm
Allium campanulatum
1 cm
flower
1 cm
ovary
2 mm
Allium falcifolium
1 cm
2 mm
flower
ovary
1 mm
Allium fimbriatum var. fimbriatum
2 mm
flower
2 cm
ovary
2 mm
bulb coat
0.2 mm
Allium haematochiton
flower
1 cm
2 cm
ovary
1 mm
Allium shevockii
flower
1 cm
ovary
2 mm
2 cm
bulb coat
0.1 mm
Allium validum
0.1 mm
Allium lemmonii
0.1 mm
Allium nevadense
0.1 mm
Alium obtusum
0.25 mm
Allium peninsulare
bulb coat
0.1 mm
Allium tribracteatum
0.1 mm
Allium unifolium

17. Perianth blue-purple; widespread ..... ssp. ***coronaria***
17′ Perianth rose-purple; NCoRI ..... ssp. ***rosea***
16′ Fl 14–24 mm; staminodes little inrolled
18. Anther axis tissue entire; ovary purple ..... ***B. jolonensis***
18′ Anther axis tissue dentate; ovary green ..... ***B. terrestris***
19. Infl 2–20 cm; fls 26–33 mm ..... ssp. ***kernensis***
19′ Infl 0.5–7 cm; fls 16–25 mm ..... ssp. ***terrestris***

B. appendiculata Hoover (p. 1187) Corm coat coarsely fibrous. **INFL**: axis 10–45 cm, stout; pedicels 4–10 cm. **FL**: perianth violet-purple, tube 8–12 mm, cylindric (translucent, splitting in fr), lobes ascending, 15–22 mm, recurved; staminodes erect, 8–15 mm, linear, gen white, margins inrolled 1/2, wavy; filaments 4–7 mm, wider at base, tip appendage forked, anthers 3–6 mm, tips of pollen chambers hooked; ovary 5–6 mm, style 9–12 mm. *n*=6. Uncommon. Grassland, open woodland, gravelly clay soil; < 600 m. s NCoRI, c SNF, SnFrB. ❀DRY,SUN:7,**14–16**,17,**19–21**,22–24.

B. californica Lindley Corm coat coarsely fibrous. **INFL**: axis 20–70 cm, stout; pedicels 2–10 cm. **FL**: perianth pale lilac (white), tube 9–12 mm, cylindric (translucent, splitting in fr), lobes ascending, 20–30 mm, 4–10 mm wide, tips recurved; staminodes 16–27 mm, erect, linear, gen flat, white to pale lilac, margins inrolled 1/4–1/2; filaments 7–10 mm, wider at base, rarely with a forked appendage, anthers 9–12 mm, linear; ovary 5–11 mm, style 15–23 mm. *n*=6,12. Grassland, open woodland, chaparral, gravelly clay soils or serpentine; 60–900 m. NCoR, n SN.

var. ***californica*** (p. 1187) **FL**: perianth lobes 6–10 mm wide; staminode margins inrolled ± 1/4; ovary 7–11 mm. *n*=6. Grassland, open woodland, gravelly clay soils; 60–800 m. Range of sp. ❀DRY,SUN:**7**,8,9,**14–17**,19–24.

var. ***leptandra*** (E. Greene) Hoover (p. 1187) **FL**: perianth lobes 4–7 mm wide; staminode margins inrolled 1/2; ovary 5–7 mm. *n*=6,12. Uncommon. Open forest, chaparral, often on serpentine; < 900 m. c NCoRI (Napa, Lake, Sonoma cos.). ❀DRY,SUN or part SHD:**7,14–17**,19–24.

B. coronaria (Salisb.) Engl. **INFL**: axis 4–25 cm; pedicels 1–5 cm. **FL**: perianth blue-violet, blue-purple, pink-purple, or rose, tube bell-shaped (opaque, not splitting in fr), 6–13 cm, lobes ascending, 12–25 mm, tips recurved; staminodes held close to stamens, 10–11 mm, white or pink, margins inrolled 3/4; filaments 3–4 mm, free part ± triangular, anthers 5–7 mm; ovary 6–9 mm, style 6–11 mm. *n*=6,12. Grassland, volcanic mesas; 0–1600 m. NW, CaR, n&c SN; to B.C.

ssp. ***coronaria*** Corm coat thick. **INFL**: axis 5–25 cm. **FL**: perianth blue-violet or blue-purple, tube 6–13 cm, lobes 15–25 cm; staminodes white; tips of pollen chambers hooked; ovary 8–9 mm, style 9–11 mm. *n*=6,12. Habitat and range of sp. ❀DRY,SUN:4,**5**,6,7,9,**14–17**,18–24.

ssp. ***rosea*** (E. Greene) T. Niehaus (p. 1187) INDIAN VALLEY BRODIAEA Corm coat thin. **INFL**: axis 4–7 cm. **FL**: perianth rose to pink-purple, tube 7–9 mm, lobes 12–18 mm; staminodes white to pink; tips of pollen chambers rounded; ovary 6–8 mm, style 6–8 mm. *n*=6. **ENDANGERED** CA. Grassland, often on serpentine; ± 30–100 m. NCoRI (Tehama, Glenn, Lake cos.). In cult.

B. elegans Hoover HARVEST BRODIAEA **INFL**: axis 10–50 cm, stout; pedicels 5–10 cm. **FL**: perianth blue-purple to violet, tube 8–19 mm, funnel-shaped (opaque, not splitting in fr), lobes ascending, 15–30 mm, tips recurved; staminodes 6–9 mm, = stamens, erect, flat (or margin inrolled 1/4), acute or 3-lobed, white to pale lilac, standing away from stamens, tip erect or slightly recurved; filaments 4–6 mm, narrowly winged, anthers 4–10 mm, linear; ovary 9–15 mm, style 7–15 mm. Grassland, meadows, open woodland; 0–2200 m. KR, n&c NCoR, SN; sw OR.

ssp. ***elegans*** (p. 1187) **FL**: perianth tube 8–19 mm, lobes 15–30 mm; staminode margin flat, tip erect; anthers 4–10 mm; ovary 9–15 mm, style 7–15 mm. *n*=6,8,12,16. Habitat and range of sp. [vars. *australis* Hoover & *mundula* (Jepson) Hoover] ❀DRN:1,6,**7**,14,**15–17**,18–24&SUN:**4,5**.

ssp. ***hooveri*** T. Niehaus **FL**: perianth tube 13–17 mm, lobes 22–30 mm; staminode margin inrolled 1/4, tip slightly recurved; anthers 6–8 mm; ovary 9–11 mm, style 12–15 mm. *n*=20. Grasslands; 0–100 m. nw KR?; w OR. Documentation of occurrence in CA needed.

B. filifolia S. Watson (p. 1187) THREAD-LEAVED BRODIAEA **INFL**: axis 20–30 cm; pedicels 1–4 cm. **FL**: perianth violet-red-purple, tube 6–8 mm, narrowly cylindric (transparent, splitting in fr), lobes widely spreading, 10–14 mm; staminodes ± inconspicuous, 2–4 mm, tapered, reflexed against perianth; filaments 0.5–1 mm, triangular-winged, anthers 3–5 mm, with a wide notch between pollen chambers; ovary 4–5 mm, style 6–7 mm. *n*=12. **ENDANGERED** CA. Grassland, vernal pools; 60–300 m. s SCo (w San Diego Co.), w PR (Riverside, San Diego cos.); extirpated from Los Angeles, San Benardino cos. Threatened by development, vehicles. In cult.

B. insignis (Jepson) T. Niehaus (p. 1187) KAWEAH BRODIAEA **INFL**: axis 7–25 cm; pedicels 2–9 cm. **FL**: perianth rose- to pink-purple, tube 6–9 mm, cylindric (opaque, not splitting in fr), lobes widely spreading, 11–15 mm; staminodes 6–8 mm, erect, held close to stamens, white, margin inrolled 3/4, tip ± 2-lobed; filaments 1–2 mm, linear, triangular- winged, tips of pollen chambers hooked; ovary 4–5 mm, style 4–5 mm. *n*=16. Foothill woodland; 250–500 m. **ENDANGERED** CA. s SNF (Kaweah and Tule river drainages, Tulare Co.). Threatened by development, road maintenance, grazing. In cult.

B. jolonensis Eastw. (p. 1187) **INFL**: axis 5–15 cm; pedicels 1–4 cm. **FL**: perianth blue-violet, tube 7–9 mm, cylindric or narrowly bell-shaped (thick, opaque, not splitting in fr), lobes ascending, 11–18 mm, tips recurved; staminodes held close to stamens, 5–6 mm, violet, margin inrolled 1/4, tip hooded and ± notched; filaments 1–2 mm, anthers 5–6 mm; ovary 5–6 mm, light purple, style 5–7 mm. *n*=6. Grassland, foothill woodland on clay; 0–300 m. SCoRO, w SW (incl Santa Cruz Island); n Baja CA. ❀DRY:14,**19–23**&SUN:**15–17,24**.

B. kinkiensis T. Niehaus (p. 1187) SAN CLEMENTE ISLAND BRODIAEA **LF** 1, ± cylindric (grooved in X-section). **INFL**: axis 2–3 cm; pedicels 3–8 cm. **FL**: perianth purple or violet, tube 9–12 mm, narrowly bell-shaped (opaque, not splitting in fr), lobes widely spreading, 12–18 mm, inner ± round, tips of inner lobes widely rounded; staminodes erect, held away from stamens, 3–7 mm, tip abruptly pointed; filaments 1.5–2 mm, free part triangular-winged, anthers 4–7 mm, with a wide notch between pollen chambers; ovary 6–9 mm, style 5–7 mm. *n*=16. RARE. Grassland on clay flats; ± 150–200 m. s ChI (San Clemente Island). In cult.

B. minor (Benth.) S. Watson (p. 1187) DWARF BRODIAEA **INFL**: axis 2–10 cm; pedicels 1–3 cm. **FL**: perianth pale bluish to lilac, tube 6–10 mm, urn-shaped (narrowed above ovary) (tough, opaque, not splitting in fr), lobes widely spreading, 11–17 mm; staminodes erect, held close to stamens, 5–8 mm, white, margin inrolled 1/2 at midlength, tip notched; filaments 0.5–4 mm, anthers 1–5 mm, tips of pollen chambers hooked; ovary 3–4 mm, style 5–6 mm. *n*=6. Grassland, gravelly clay soils; 30–500 m. n&c SNF, e ScV, ne SnJV. [var. *nana* (Hoover) Hoover] ❀DRY,SUN:7,8,**9,14–17**,**19–24**.

B. orcuttii (E. Greene) Baker (p. 1187) ORCUTT'S BRODIAEA **LF** ± cylindric (widely channeled in X-section). **INFL**: axis 8–25 cm; pedicels 1–5 cm. **FL**: perianth violet, tube 3–5 mm (transparent, splitting in fr), lobes widely spreading, 12–19 mm; staminodes 0; filaments 4–6 mm, anthers 4–6 mm, linear; ovary 4–6 mm, style 7–11 mm. *n*=12. RARE. Grassland near streams, vernal pools; 0–1600 m. PR (s Riverside, San Diego cos.).

B. pallida Hoover (p. 1187) CHINESE CAMP BRODIAEA **INFL**: axis 10–20 cm; pedicels 5–30 mm. **FL**: perianth pale purple or lilac, tube 9–11 mm, ± urn-shaped (slightly narrowed above ovary) (thin, opaque, not splitting in fr), lobes ascending to strongly recurved, 9–11 mm; staminodes erect, held close to stamens, 8–11 mm, white, margin inrolled 1/2 at midlength, tip deeply notched;

filaments 4–5 mm, linear-winged, anthers 2–3 mm, ± obcordate, widely V-notched between pollen chambers; ovary 4–5 mm, style 8–11 mm. *n*=6. **ENDANGERED** CA. Vernal streambed on serpentine; ± 380 m. c SNF (near Chinese Camp, Tuolumne Co.). 1 population; hybridizes with *B. elegans*. Threatened by development. In cult.

B. purdyi Eastw. **INFL**: axis 10–25 cm; pedicels 1–3 cm. **FL**: perianth blue-violet, tube 6–8 mm, urn-like (strongly narrowed above ovary) (tough, opaque, not splitting in fr), lobes spreading, 9–19 mm; staminodes erect, held close to stamens, 6–8 mm, white, margin inrolled 3/4, tip notched; filaments 1–2 mm, narrowly triangular, anthers 3–4 mm, V-notched between pollen chambers; ovary 4–5 mm, style 7–9 mm. *n*=12,24. Open woodland, often on serpentine; 100–600 m. s CaRF, n SNF. Much like *B. minor*; might be made a ssp. of it. ❀DRY,SUN:**7,14–16**,17,19–24.

B. stellaris S. Watson (p. 1187) **INFL**: axis 2–6 cm; pedicels 1–5 cm. **FL**: perianth blue-purple, tube 7–10 mm, bell-shaped (transparent, not splitting in fr), lobes ascending, 7–15 mm, tips recurved; staminodes erect, held close to stamens, 4–8 mm, wide, white, margin inrolled 1/4, tip widely notched; filaments 1–3 mm, with a forked appendage at tip, anthers 4–6 mm, pollen chambers separated; ovary 6–9 mm, style 4–5 mm. *n*=6. Openings in coastal forest, on serpentine; 0–900 m. c&s NCo, NCoRO. ❀DRN, SUN:5,15,16,**17**;DFCLT.

B. terrestris Kellogg **INFL**: axis 0.5–20 cm; pedicels 1–15 cm. **FL**: perianth blue-violet, tube 6–13 mm, narrowly bell-shaped (opaque, not splitting in fr), lobes ascending, 10–20 mm, tips recurved; staminodes leaning inward slightly but held away from stamens, 4–5 mm, violet to whitish, margin inrolled 1/4, tip slightly hooded, ± 2-notched; filaments 2–3 mm, narrowly triangular-winged, anthers 3–6 mm, pollen chambers ± separated; ovary 7–9 mm, style 4–9 mm. Grassland, open woodlands; 0–1500 m. NCo, w NCoRO, s SNF, Teh, CW, TR, PR; s OR.

ssp. ***kernensis*** (Hoover) T. Niehaus (p. 1187) **LF** not very crescent-shaped in X-section. **INFL**: axis 2–20 cm; pedicels 1–7 cm. **FL**: perianth tube 11–13 mm, lobes 15–20 mm, inner ± ovate; anthers 5–6 mm; ovary 8–9 mm, style 7–9 mm. *n*=24. Habitats of sp. s SNF, Teh, SCoR, TR, PR. [*B. coronaria* var. *k.* Hoover; *B. elegans* var. *australis* Hoover] ❀DRY,SUN:**7**,8,9,**14–17**,18, **19–24**.

ssp. ***terrestris*** (p. 1187) **LF** crescent-shaped in X-section. **INFL**: axis 0.5–7 cm; pedicels 3–15 cm. **FL**: perianth tube 6–9 mm, lobes 10–16 mm, inner ± lanceolate; anthers 3–4 mm; ovary 7–9 mm, style 4–5 mm. *n*=6,18. Habitats of sp.: 0–450 m. NCo, w NCoRO, CW (exc SCoRI); s OR. [*B. coronaria* var. *macropoda* (Torrey) Hoover] ❀SUN:5,7,14,**15–17**.

CALOCHORTUS

Peggy Fiedler & Bryan Ness

Bulb coat gen membranous. **LVS** gen linear to lanceolate; basal lf 1; cauline lvs 0–several, smaller upwards. **INFL** often ± umbel-like; fls 2–many. **FL** spheric and closed to nearly rotate; sepals gen < petals, gen ± lanceolate (ovate), gen nearly glabrous; petals gen widely wedge-shaped, gen hairy inside, nectary near base; filaments ± flat, anthers gen attached at base; style 1, stigmas 3. **FR**: capsule, septicidal, gen ± oblong, gen 3-angled or -winged; chambers 3. **SEEDS** many in 2 rows per chamber, gen flat, gen netted, gen ± yellow. ± 65 spp.: w N.Am, C.Am; many cult. Bulbs of some eaten by native Americans. Nectary shape and hairs important to identification. (Greek: beautiful grass) [Ness 1989 Syst Bot 14:495–505] Sect. *Calochortus* by Bryan Ness.

1. Fl nodding; perianth closed, oblong to spheric (sect. *Calochortus* in part)
 2. Petals white, pink, or rose
  3. Sepals appressed to petals; lower nectary membranes 1/3–2/3 petal width; petals gen white to pink; n&c SNF, CCo, n ChI .... ***C. albus***
  3′ Sepals spreading; lower nectary membranes ± = petal width; petals rose; c&s SNF .... ***C. amoenus***
 2′ Petals yellow
  4. St gen simple above base; petals > sepals, hairy; pl gen > 30 cm; pl very glaucous .... ***C. raichei***
  4′ St gen branched above base; petals gen < or = sepals, glabrous to sparsely hairy; pl gen < 30 cm; pl not very glaucous
   5. Pl slightly glaucous; petals deep yellow; NCoRO, nw SnFrB .... ***C. amabilis***
   5′ Pl green; petals light yellow; ne SnFrB .... ***C. pulchellus***
1′ Fl erect to spreading; perianth open, bell-shaped
 6. Fr nodding; petal not much spotted or striped (if patterned, above-ground st < 10 cm) (sect. *Calochortus* in part)
  7. Petals gen > 30 mm, pink to purplish; ne KR .... ***C. persistens***
  7′ Petals gen < 25 mm, color various; widespread (not s CA)
   8. Petal ± (sub)glabrous inside
    9. Cauline lvs 0; SN .... ***C. minimus***
    9′ Cauline lvs present; NW, CW
     10. St 8–25 cm, branched, bulblets 0 .... ***C. umbellatus***
     10′ St gen < 5 cm, gen simple, bulblets present .... ***C. uniflorus***
   8′ Petal ± conspicuously hairy inside
    11. Petal deep yellow; CaRF,n&c SNF .... ***C. monophyllus***
    11′ Petal white, purple, or blue
     12. Petal lanceolate to obovate, ± ciliate on sides but not around tip; NW, ScV, CW, s SN
      13. Petal obovate; st gen branched; NW, ScV, CW .... ***C. tolmiei***
      13′ Petal lanceolate; st simple; s SN .... ***C. westonii***
     12′ Petal obovate, conspicuously ciliate around entire margin; NW, CaR, SNH
      14. Petal bluish (purple crescent 0), ± smooth inside; NW, CaR, SNH .... ***C. coeruleus***
      14′ Petal greenish white, base dark purplish, minutely bumpy inside; e KR .... ***C. elegans***
 6′ Fr erect; petal obviously spotted or striped (exc *C. nudus*)
  15. Basal lf ± persistent at fl (sect. *Calochortus* in part)
   16. Petal flecked yellow-green, streaked purple-brown; nw SnFrB .... ***C. tiburonensis***
   16′ Petal pink to purple; KR, CaR, MP

17. Petal with purple crescent above nectary .... ***C. longebarbatus*** var. ***longebarbatus***
17′ Petal crescent 0 inside
18. Petal 30–40 mm with dark purple crescent outside; nectary deep; dry places; < 1100 m .... ***C. greenei***
18′ Petal 14–16 mm, crescent 0 outside; nectary ± shallow; moist places; > 1200 m .... ***C. nudus***
15′ Basal lf ± withered by fl (exc *C. excavatus*)
19. Nectary surface ± glabrous, ± hidden by dense bordering hairs; bulb coat fibrous — s CW, SW (sect. *Cyclobothra*)
20. Petal ± pink, gen toothed, not ciliate .... ***C. plummerae***
20′ Petal yellow to red-brown, gen ± ciliate
21. Petal tip dark-hair-tufted .... ***C. obispoensis***
21′ Petal tip fringed, not hair-tufted .... ***C. weedii***
22. Petal deep yellow .... var. ***weedii***
22′ Petal cream to purplish or red-brown
23. Petal purplish, rounded; anther rounded; c SCo, n PR .... var. ***intermedius***
23′ Petal sometimes cream or red-brown, squarish; anther abruptly pointed; SCoRO, WTR ...... var. ***vestus***
19′ Nectary surface hairy; bulb coat membranous (sect. *Mariposa*)
24. Nectary ± round, clearly depressed, encircled by fringed membrane
25. Petal gen orange to red (yellow); st sometimes twisted .... ***C. kennedyi***
26. Petal orange (e DMoj) to red (n TR, w DMoj) .... var. ***kennedyi***
26′ Petal ± yellow; n&e DMtns .... var. ***munzii***
25′ Petal yellow or white gen tinged lilac; st sometimes zigzag but not twisted
27. Petal white, tinged lilac, or lavender, green-striped outside; sepals sometimes dark-spotted
28. Petal base marked dark purple; bracts gen 3–8 cm; anthers red-brown — sw SNE .... ***C. excavatus***
28′ Petal white or tinged lilac; bracts gen 2–5 cm; anthers yellow, maroon, blue, purple, or red
29. Sepal base not spotted; petals ± widely wedge-shaped .... ***C. invenustus***
29′ Sepal base dark-spotted; petals narrowly obovate — SNE, DMtns
30. Petal with red or purple arch above yellow-encircled nectary; SNE .... ***C. bruneaunis***
30′ Petal without dark arch above red- or purple-encircled nectary; n DMtns .... ***C. panamintensis***
27′ Petal yellow to gold, not green-striped outside; sepal base gen dark-spotted
31. Perianth bell-shaped; st straight; nectary densely slender-hairy, encircled by slender hairs; SW .... ***C. concolor***
31′ Perianth cup-shaped; st sometimes zigzag; nectary densely knobby-hairy, encircled by club-shaped hairs; SN, CW .... ***C. clavatus***
32. Petal < sepal; nectary very deep; n&c SNF .... var. ***avius***
32′ Petal > sepal; nectary shallow to moderately deep; w GV, CW, n SW
33. St gen < 30 cm, slender, straight
34. Petal 30–40 mm, sparsely hairy; lvs not recurved; SnGb .... var. ***gracilis***
34′ Petal 40–50 mm, hairy; lvs strongly recurved; s-c CCo .... ssp. ***recurvifolius***
33′ St gen 50–100 cm, coarse, zigzag
35. Petal deep yellow, hairs very knobby; anther deep purple; SCoR .... ssp. ***clavatus***
35′ Petal light yellow, hairs not very knobby; anther yellow to medium purple; esp w SnJV ssp. ***pallidus***
24′ Nectary not simultaneously round, depressed, and surrounded by fringed membrane (but sometimes 1 of these)
36. St base not bulblet-bearing (exc some *C. splendens*)
37. Petal purple-veined, not spotted .... ***C. striatus***
37′ Petal spotted, not obviously veined or striped
38. St twining or sprawling; perianth yellow-banded .... ***C. flexuosus***
38′ St ± erect; perianth not yellow-banded
39. Petal white, gen flushed pink, red-spotted above nectary, yellow-hairy near nectary; PR ...... ***C. dunnii***
39′ Petal lavender to deep purple, base gen purple-spotted, sparsely (not yellow) hairy; NCoRI, CW, SW .... ***C. splendens***
36′ St base gen bulblet-bearing (exc *C. palmeri* var. *munzii*)
40. Petal bright or deep yellow
41. Nectary crescent-shaped to oblong .... ***C. luteus***
41′ Nectary gen ± square (ovate) .... [2]***C. venustus***
40′ Petal white to pink or purple (sometimes yellowish)
42. Nectary gen ± ovate or oblong (sometimes ± round)
43. Anthers sagittate; SN, MP .... ***C. leichtlinii***
43′ Anthers not sagittate; se KR, CW, SW
44. Fr narrowly oblong, not angled .... ***C. catalinae***
44′ Fr linear, angled
45. Perianth gen 40–50 mm; nectary hairs slender, simple; ne KR .... ***C. monanthus***
45′ Perianth gen 20–30 mm; nectary hairs thick, ± branched, knobby; s CA-FP .... ***C. palmeri***
46. Nectary glabrous or purple-hairy; bulblets 0; bracts alternate; SnJt .... var. ***munzii***
46′ Nectary yellow-hairy; bulblets present; bracts opposite; Teh, s CW, TR, SnJt .... var. ***palmeri***
42′ Nectary 1–2 crescents or chevrons, ± triangular, or ± square
47. Nectary ± triangular-sagittate; sepals gen > petals .... ***C. macrocarpus***
47′ Nectary 1–2 crescents or chevrons, or ± square; sepals < or = petals

48. Nectary ± square; petal sometimes with 2nd, distal, paler spot
 49. Nectary in red spot; 2nd petal spot 0; seed 5–6 mm, 3 mm wide .... ***C. simulans***
 49′ Nectary not in red spot; 2nd petal spot gen present; seed 4–5 mm, 1–2 mm wide .... [2]***C. venustus***
48′ Nectary 1–2 crescents or chevrons; petal spot 1, gen in yellow zone
 50. Nectary 2 crescents or chevrons .... ***C. vestae***
 50′ Nectary 1 crescent or chevron
  51. Fr lanceolate; CW .... ***C. argillosus***
  51′ Fr linear; widespread .... ***C. superbus***

C. albus Benth. (p. 1187) WHITE GLOBE LILY, FAIRY LANTERN **ST** 20–80 cm, branched. **LVS**: basal 30–70 cm, persistent. **INFL**: fls 2–many, nodding. **FL**: perianth ± closed at tip, ± oblong; sepals 10–15 mm, appressed to petals; petals 20–25 mm, white to pink, ± elliptic, sparsely ciliate, hairs above nectary slender, nectary ± depressed, with several fringed membranes 1/3–2/3 petal width. **FR** nodding, 25–40 mm, winged. **SEED** irregular, dark brown. 2*n*=20. Common. Shady to open woodlands, shrubland; 0–2000 m. n&c SNF, CW (exc CCo), WTR, n ChI. Pls with deep rose fls from sw SnFrB and n SCoRO have been called var. *rubellus* E. Greene. Hybridizes with *C. monophyllus*. ❀DRN,DRY,part SHD:7,9,**14**, 15–24;DFCLT.

C. amabilis Purdy DIOGENES' LANTERN Pl slightly glaucous. **ST** 10–50 cm, gen branched. **LVS**: basal 20–50 cm, persistent. **INFL**: fls 2–many, nodding. **FL**: perianth closed at tip, spheric; sepals 15–20 mm, spreading; petals 16–20 mm, deep yellow, brown-spotted outside, widely lanceolate, densely ciliate throughout, surface glabrous, nectary deep, bordered above by band of hairs. **FR** nodding, 20–30 mm, winged. **SEED** irregular, dark brown. 2*n*=20. Common. Shady or open woodlands, shrubland; 100–1000 m. NCoRI, nw SnFrB. ❀DRN,DRY,part SHD: **7**,14–21;DFCLT.

C. amoenus E. Greene **ST** 20–50 cm, branched. **LVS**: basal 20–50 cm, persistent; cauline 2–5, acuminate. **INFL**: fls 2–many, nodding. **FL**: perianth closed at tip, spheric; sepals 10–15 mm, spreading; petals 16–25 mm, rose, ± widely elliptic, sparsely ciliate, sparsely hairy, nectary not very deep, with 4–5 ciliate membranes, lowest ± = petal width. **FR** nodding, 20–30 mm, ± winged. **SEED** irregular, dark brown. 2*n*=20. Grassy slopes in partial shade. 500–1500 m. c SNF (Madera, Fresno cos.), s SNF. ❀DRN,DRY,part SHD or SUN:**7**,14–21;DFCLT.

C. argillosus (Hoover) R. Zebell & P. Fiedler (p. 1187) CLAY MARIPOSA LILY **ST** 40–60 cm, simple; base bulblet-bearing. **LVS**: basal 20–30 cm, withering; cauline reduced upwards. **INFL** ± umbel-like; fls 1–4, erect; bracts 2–8 cm. **FL**: perianth bell-shaped; sepals 20–40 mm; petals 20–40 mm, ± rounded, white to purple, central red spot within pale yellow, sparsely hairy; nectary 1 crescent or chevron, not depressed, densely short-hairy. **FR** erect, 4–6 cm, lanceolate. **SEED** unknown. RARE. Hard clay from volcanic or metamorphic rocks; < 800 m. w CW. [*C. superbus* Munz misapplied] In cult.

C. bruneaunis Nelson & J.F. Macbr. (p. 1187) **ST** 10–40 cm, gen simple; base bulblet-bearing. **LVS**: basal 10–20 cm, withering; upper cauline inrolled. **INFL**: fls 1–4, erect; bracts 2–4 cm. **FL**: perianth bell-shaped; sepals 10–40 mm, dark-spotted near base; petals 20–40 mm, narrowly obovate, white, tinged lilac, striped green outside, with dark red or purple arch above nectary, ± glabrous, nectary surrounded by yellow spot, round, depressed, densely short-hairy, encircled by fringed membrane. **FR** 3–7 cm, erect, linear-lanceolate, angled. *n*=7. Dry shrubs or grass in pinyon/juniper woodland; 1700–3000 m. s SNE; to OR, MT, NV, UT. [*C. nuttallii* Torrey var. *b.* (Nelson & J.F. Macbr.) F. Ownbey] ❀DFCLT.

C. catalinae S. Watson (p. 1187) CATALINA MARIPOSA LILY **ST** 20–60 cm, gen branched above; base bulblet-bearing. **LVS**: basal 10–30 cm, withering. **INFL**: fls 1–4, erect; bracts 2–10 cm, opposite. **FL**: perianth bowl-shaped; sepals 20–30 mm, purple-spotted near base; petals 20–50 mm, white, tinged lilac, purple-spotted near base, nearly glabrous, nectary not depressed, oblong, densely branched-hairy. **FR** erect, 2–5 cm, not angled. *n*=7. UNCOMMON. Heavy soil, open grassland or shrubland; < 700 m. s CCo, w SCo, esp ChI. Threatened by development. In cult.

C. clavatus S. Watson **ST** 20–100 cm; bulblets gen 0. **LVS**: basal 10–20 cm, withering. **INFL** ± umbel-like; fls 1–6, erect; bracts 4–8 cm, bases widest. **FL**: perianth cup-shaped; sepals 20–40 mm, gen marked red-brown near base; petals 30–50 mm, yellow, gen banded darker above nectary, hairs near nectary club-shaped, nectary round, ± depressed, surrounded by fringed membrane, surface densely short-knobby-hairy. **FR** erect, 6–9 cm, narrowly lanceolate, angled. *n*=8. Rocky slopes, chaparral, open forest, often on serpentine; < 1800 m. n&c SNF, w GV, e SnFrB, SCoR, n WTR, SnGb. Vars. and sspp. here considered equivalent.

var. ***avius*** Jepson (p. 1187) PLEASANT VALLEY MARIPOSA LILY **ST** 50–100 cm, coarse, zigzag. **FL**: petals gen < sepals, nectary very deep. RARE. Open oak/pine forest; 900–1800 m. n&c SNF (El Dorado, Amador cos.).

ssp. ***clavatus*** (p. 1187) CLUB-HAIRED MARIPOSA LILY **ST** 50–100 cm, coarse, zigzag. **FL**: petals 40–50 mm, deep yellow; anthers 8–10 mm, deep purple. UNCOMMON. Gen on serpentine; < 1300 m. s SCoRO, n SCoRI. ❀DRN,DRY,SUN:7,14–24;DFCLT.

var. ***gracilis*** F. Ownbey SLENDER MARIPOSA LILY **ST** 20–30 cm, slender. **FL**: petals 30–40 mm, sparsely hairy, with reddish brown line above small, shallow, nectary; anthers 4–7 mm. UNCOMMON. Shaded foothill canyons; < 1000 m. SnGb.

ssp. ***pallidus*** (Hoover) Munz **ST** 50–100 cm, coarse, zigzag. **FL**: petals light yellow, hairs not very knobby; anthers yellow to medium purple. Habitats of sp. w SnJV (and adjacent e SnFrB, SCoRI), n WTR. ❀DFCLT.

ssp. ***recurvifolius*** (Hoover) Munz (p. 1187) ARROYO DE LA CRUZ MARIPOSA LILY **ST** 8–20 cm, slender. **LF** strongly recurved. **FL**: petals 40–50 mm, deep yellow; anthers 8–10 mm, deep purple. RARE. Ocean bluffs; ± 50 m. s-c CCo (nw San Luis Obispo Co.).

C. coeruleus (Kellogg) S. Watson BEAVERTAIL-GRASS **ST** 3–15 cm, simple, sometimes slender. **LVS**: basal 10–20 cm, persistent. **INFL**: fls 1–10, erect to spreading. **FL**: perianth bell-shaped; sepals ± 10 mm; petals 8–12 mm, obovate, bluish, ciliate throughout, hairy inside, nectary slightly depressed, bordered below by ciliate membranes, above by short hairs. **FR** nodding, 10–15 mm, not angled. **SEED** irregular. Common. Gravelly openings in woodlands; 600–2500 m. NW, CaR, SNH. [var. *fimbriatus* F. Ownbey] Intermediates to *C. tolmiei* scattered but common in NW, CaR. ❀DFCLT.

C. concolor (Baker) Purdy **ST** 30–60 cm, straight, gen stout; bulblets gen 0. **LVS**: basal 10–20 cm, glaucous, withering; upper cauline inrolled. **INFL** ± umbel-like; fls 1–7, erect; bracts 4–8 cm, opposite. **FL**: perianth bell-shaped; sepals 20–30 mm, gen dark-red-blotched near base; petals 30–50 mm, yellow (often drying purplish), sparsely long-yellow-hairy near nectary, nectary gen small, ± round, depressed, encircled by fringed membrane, densely slender-hairy. **FR** erect, 5–8 cm, narrowly lanceolate, angled. 2*n*=±14. Dry, often granitic slopes in chaparral, yellow-pine forest; 600–2500 m. SnBr (s slope), PR; n Baja CA. ❀DFCLT.

C. dunnii Purdy (p. 1187) DUNN'S MARIPOSA LILY **ST** 20–60 cm, slender, gen branched; bulblets 0. **LVS**: basal 10–20 cm, channeled, withering. **INFL**: fls 2–6, erect; bracts 1–2 cm. **FL**: perianth widely bell-shaped; sepals 10–20 mm; petals 20–30 mm, gen rounded, white or flushed pink, red-brown-spotted above nectary, yellow-hairy near and in nectary, nectary round, not depressed. **FR** erect, 2–3 cm, linear, angled. *n*=7. **RARE** CA. Dry, stony ridges in chaparral and yellow-pine forest; 1500–1700 m. s PR (s San Diego Co.); n Baja CA.

C. elegans Pursh CAT'S EAR **ST** 5–15 cm, gen simple, ± slender. **LVS**: basal 10–20 cm, persistent; cauline gen 0. **INFL**: fls 1–7, erect to spreading. **FL**: perianth bell-shaped; sepals 12–16 mm, ± papillate inside; petals 12–20 mm, ± oblanceolate, greenish white, base dark purplish, ciliate, hairy, nectary shallow, bordered below with ± ciliate membranes, above with short hairs. **FR** nodding, 15–20 mm, winged. **SEED** irregular, light brown. 2*n*=20. Uncommon. Open woodlands; 1500–2500 m. e KR; to WA, MT. [var. *nanus* Alph. Wood] ❀DFCLT.

C. excavatus E. Greene (p. 1187) INYO COUNTY STAR-TULIP **ST** 10–30 cm, slender, simple; base gen bulblet-bearing. **LVS**: basal 10–20 cm, gen persistent. **INFL** ± umbel-like; fls 1–6, erect; bracts 3–8 cm, paired. **FL**: perianth widely bell-shaped; sepals 20–30 mm, not spotted; petals 30–40 mm, white, base marked dark purple, green-striped outside, sparsely short-hairy near nectary, nectary round, depressed, encircled by fringed membrane, densely short-branched-hairy; anthers red-brown. **FR** erect, narrowly lanceolate, angled. RARE. Grassy meadows in shadscale scrub; 1300–2000 m. w SNE (Mono, Inyo cos.). Threatened by groundwater development.

C. flexuosus S. Watson (p. 1187) **ST** 10–20 cm, branched, gen wavy, ± sprawling; bulblets 0. **LVS**: basal 1–2, withering. **INFL**: fls 1–6, erect; bracts 1–3 cm. **FL**: perianth bell-shaped; segments purple-spotted, yellow-banded; sepals 20–30 mm; petals 30–40 mm, white, lilac-tinged, sparsely short-hairy near nectary, nectary 1 crescent, not depressed, densely short-hairy. **FR** erect, 3–4 cm, stout, angled. n=7. Dry, rocky sites; 600–1700 m. Creosote-bush or sagebrush scrub. e D; to Colorado. ❀DFCLT.

C. greenei S. Watson (p. 1187) GREENE'S MARIPOSA LILY **ST** 10–30 cm, gen branched. **LVS**: basal ± 20 cm, persistent; cauline 1–2. **INFL**: fls 1–5, erect. **FL**: perianth bell-shaped; sepals 25–30 mm; petals 30–40 mm, purplish with darker crescent above nectary, not ciliate, ± hairy, nectary deeply depressed, bordered below by wide ciliate membranes, above by short hairs. **FR** erect, 20–25 mm, winged. **SEED** irregular. RARE. Shrubby hillsides, open woodlands; ± 700–1100 m. e KR, nw CaR, w MP; s OR.

C. invenustus E. Greene (p. 1191) **ST** 20–50 cm, slender, gen simple; base bulblet-bearing. **LVS**: basal 10–20 cm, withering; upper cauline inrolled. **INFL** ± umbel-like; fls 1–6, erect; bracts 2–5 cm. **FL**: perianth bell-shaped; sepals 20–30 mm; petals 20–40 mm, white or tinged lilac, sometimes purple-spotted below nectary, green-striped outside, sparsely short-hairy near nectary, nectary small, round, slightly depressed, encircled by fringed membrane, densely short-branched-hairy. **FR** erect, 5–7 cm, angled. n=7. Dry soil, gen granitic, gen in montane coniferous forest; 1500–3000 m. c SNH, Teh, se SnFrB, SCoR, SW. ❀DFCLT.

C. kennedyi Porter (p. 1191) **ST** 10–20(50) cm, gen simple, sometimes twisted; bulblets gen 0. **LVS**: basal 10–20 cm, glaucous, channeled, withering. **INFL** ± umbel-like; fls 1–6, erect; bracts 2–4 cm, base widest. **FL**: perianth bell-shaped; segments often dark-spotted near base; sepals 20–30 mm; petals 30–50 mm, yellow to red, sparsely club-like-hairy near nectary, nectary round, depressed, encircled by fringed membrane, densely simple- or forked-hairy. **FR** erect, 4–6 cm, ± lanceolate, angled, striped. n=8. Heavy or rocky soil in creosote-bush scrub, pinyon/juniper woodland; 600–2200 m. n TR, DMoj; NV, AZ.

var. ***kennedyi*** **FL**: petal orange (esp e DMoj) or red (esp n TR, w DMoj). Habitats and range of sp. ❀DFCLT.

var. ***munzii*** Jepson **FL**: petal yellow. n&e DMtns (Panamint, Clark, Providence mtns). ❀DFCLT.

C. leichtlinii Hook. f. (p. 1191) **ST** 20–60 cm, simple; base bulblet-bearing. **LVS**: basal 10–15 cm, withering. **INFL** ± umbel-like; fls 1–5, erect; bracts paired. **FL**: perianth bell-shaped; sepals 10–20 mm; petals 10–40 mm, white to smoky-blue, often tinged pink, red- to black-spotted above nectary, sparsely short-hairy near nectary, nectary slightly depressed, ± ovate, densely short-hairy; anthers sagittate. **FR** erect, 3–6 cm, narrowly lanceolate. n=7. Common. Open, gravelly places in chaparral or montane coniferous forest; 1300–4000 m. SNH, MP; w NV. ❀DFCLT.

C. longebarbatus S. Watson var. ***longebarbatus*** (p. 1191) LONG-HAIRED STAR-TULIP **ST** 10–30 cm, gen simple; base bulblet-bearing. **LVS**: basal 20–30 cm, persistent; cauline 1–2. **INFL**: fls 1–4, erect. **FL**: perianth bell-shaped; sepals 15–20 mm; petals 20–30 mm, pink to lavender, with central red-purple crescent, not ciliate, sparsely long-hairy, nectary bordered below by ciliate membrane, above by short hairs; anther 1/2 filament length. **FR** erect, 20–25 mm, sometimes spheric, winged. **SEED** irregular, light brown. $2n$=20. RARE. Vernal meadows, heavy clay soil; 1200–1900 m. CaRH, MP; to s-c WA. Threatened by grazing. Var. *peckii* F. Ownbey in c OR.

C. luteus Lindley (p. 1191) **ST** 20–50 cm, slender; base bulblet-bearing. **LVS**: basal 10–20 cm, persistent. **INFL** ± umbel-like; fls 1–7, erect; bracts 1–8 cm. **FL**: perianth bell-shaped; sepals 20–30 mm, long-tapered; petals 20–40 mm, deep yellow, gen lined red-brown inside, often with central red-brown blotch inside, sparsely slender-hairy near nectary, nectary not depressed, ± crescent-shaped to oblong, matted-short-hairy. **FR** erect, 3–6 cm, narrowly lanceolate, angled. n=7,10. Heavy soils in grassland, woodland, mixed-evergreen forest; < 700 m. c&s NW, SNF, ScV, CW, n ChI. Interior pls gen n=7; coastal pls gen n=10. ❀DRN,DRY,SUN:**7**,8, 9,**14–17,19–24**;often DFCLT.

C. macrocarpus Douglas (p. 1191) **ST** 20–50 cm, stout, gen simple; base gen bulblet-bearing. **LVS**: basal 5–10 cm, withering; cauline inrolled, tips curled. **INFL** ± umbel-like; fls 1–6, erect; bracts 3–5 cm. **FL**: perianth bell-shaped; sepals 40–50 mm, gen > petals; petals 35–60 mm, narrowly obovate, purple, green-striped outside, gen purple-banded and bearded above nectary, nectary slightly depressed, ± triangular-sagittate, densely slender-hairy. **FR** erect, 4–5 cm, ± lanceolate, angled. n=7. Common. Sagebrush scrub, yellow-pine forest, gen in volcanic soil; 1300–2000 m. n CaR, GB; to B.C. ❀DFCLT.

C. minimus F. Ownbey **ST** < 10 cm, simple. **LVS**: basal 10–20 cm, persistent; cauline ± 0. **INFL**: fls 1–5(10), ± erect. **FL**: perianth bell-shaped; sepals 8–10 mm; petals 10–14 mm, white, not ciliate, ± glabrous exc near nectary, nectary bordered below by ± ciliate membrane. **FR** nodding, 15–20 mm, winged. **SEED** irregular, yellow to light brown. $2n$=20. Common. Moist, open woodlands, lake margins; 1200–3000 m. SNH. Derivatives of hybridization with *C. nudus* (pink- to lavender-tinged petals with rounded tips) occur in n SNH [Ness 1990 Amer J Bot 77:1519–1531].

C. monanthus F. Ownbey SINGLE-FLOWERED MARIPOSA LILY **ST** simple, straight; base bulblet-bearing. **LVS**: basal withering. **INFL**: fl 1, erect; bracts opposite; peduncle long. **FL**: perianth ± narrowly bell-shaped; sepals ± 40 mm; petals 40–50 mm, irregularly toothed distally, pinkish with a chevron-shaped, dark red spot above each nectary, irregularly toothed above, sparsely slender-hairy near nectary, nectary not depressed, oblong, densely slender-hairy. **FR** erect, linear, angled. **SEED** unknown. **PRESUMED EXTINCT**. Vernal meadow; ± 800 m. ne KR (near Yreka, Siskiyou Co.). Collected once in 1876.

C. monophyllus (Lindley) Lemaire **ST** 8–20 cm, simple or branched, slender. **LVS**: basal 10–30 cm, persistent; cauline 0–3. **INFL**: fls 1–6, ± erect. **FL**: perianth bell-shaped; sepals 16–20 mm; petals ± = sepals, narrowly obovate, deep yellow, often red-brown-spotted above base, ciliate, densely branched-hairy, nectary bordered below by ciliate membrane, above by short hairs. **FR** nodding, 12–20 mm, winged. **SEED** irregular, dark brown. $2n$=20. Wooded slopes, clay-loam soils; 400–1200 m. CaRF, n&c SNF. ❀DRN,DRY,part SHD:**7**,9,14–21;DFCLT.

C. nudus S. Watson (p. 1191) **ST** 10–25 cm, gen simple. **LVS**: basal, 5–15 cm, persistent; cauline 0. **INFL**: fls 1–several, erect. **FL**: perianth bell-shaped; sepals 10–12 mm; petals 14–16 mm, pink to lavender, not ciliate, ± glabrous, nectary shallow, bordered below by wide, ciliate membrane. **FR** erect, 15–20 mm, winged. **SEED** irregular, light brown. $2n$=20. Moist, grassy areas, lake and bog margins; 1200–2500 m. KR, CaRH; sw OR. [var. *shastensis* (Purdy) Jepson, Shasta star-tulip] Hybridizes with *C. minimus*. ❀DFCLT.

C. obispoensis Lemmon (p. 1191) SAN LUIS MARIPOSA LILY Bulb coat fibrous. **ST** 30–60 cm, slender, branched. **LVS**: basal 20–30 cm, withering; cauline 4–12 cm, upper inrolled. **INFL**: fls 2–6, erect. **FL**: perianth ± rotate; sepals 10–30 mm, reflexed; petals 10–20 mm, yellow to deep orange, ± ovate, coarsely hairy inside, fringed, tip dark-hair-tufted, nectary round, glabrous, slightly depressed, ± hidden by dense, slender, bordering hairs with fused bases. **FR** erect, 3–4 cm, linear, angled. n=9. RARE. Dry serpentine gen in chaparral; 100–500 m. s SCoRO (San Luis Obispo Co.)

C. palmeri S. Watson **ST** 30–60 cm, straight, gen branched; base gen bulblet-bearing. **LVS**: basal 10–20 cm, withering. **INFL**: fls 1–6, erect; bracts 1–2 cm. **FL**: perianth widely bell-shaped; sepals ± 30 mm, gen brown-spotted near base; petals 20–30 mm,

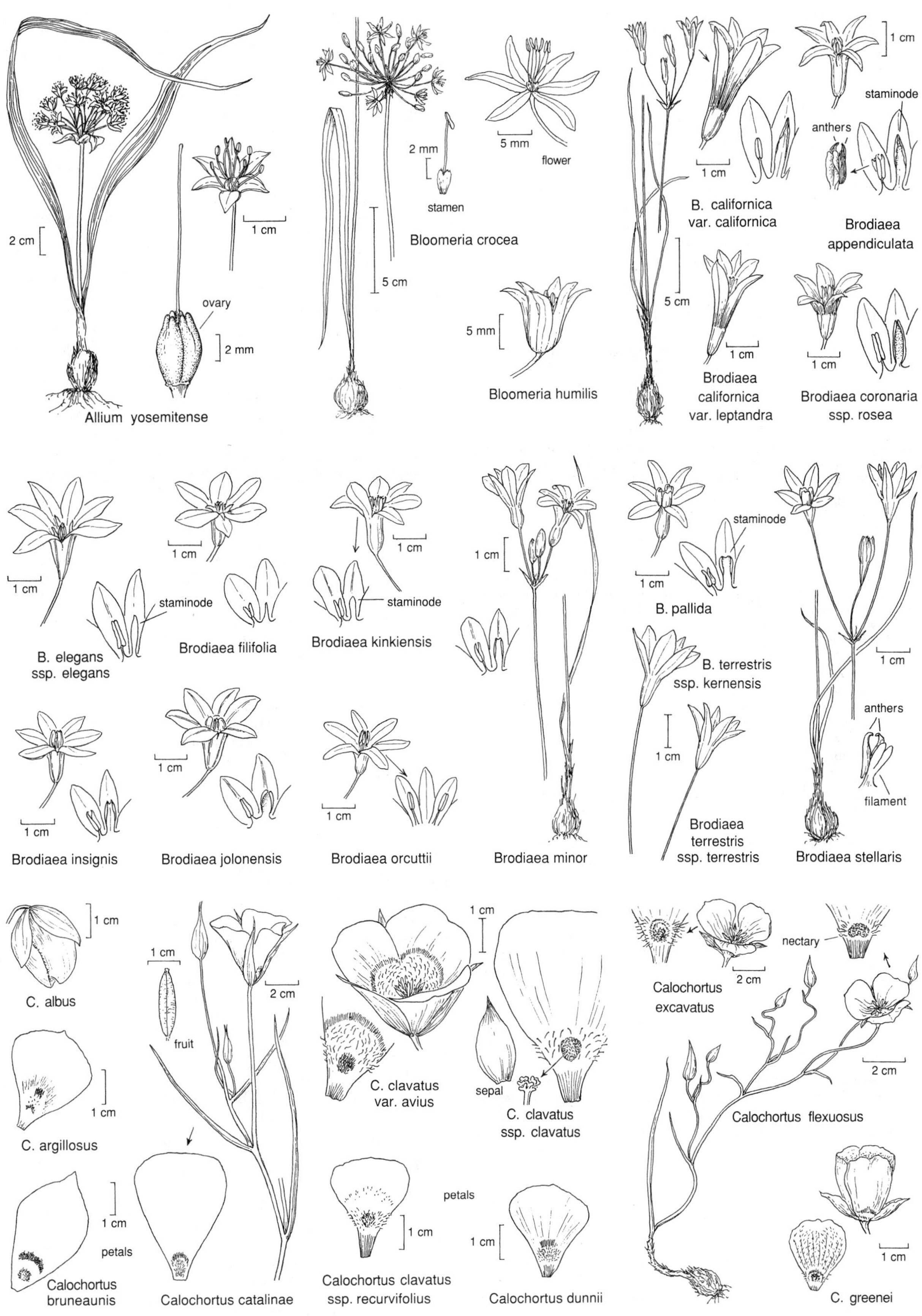
2 cm
1 cm
ovary
2 mm
Allium yosemitense
2 mm
stamen
5 mm
flower
5 cm
Bloomeria crocea
5 mm
Bloomeria humilis
1 cm
B. californica
var. californica
5 cm
1 cm
Brodiaea
californica
var. leptandra
1 cm
staminode
anthers
Brodiaea
appendiculata
1 cm
Brodiaea coronaria
ssp. rosea
1 cm
staminode
B. elegans
ssp. elegans
1 cm
Brodiaea filifolia
1 cm
staminode
Brodiaea kinkiensis
1 cm
staminode
1 cm
B. pallida
B. terrestris
ssp. kernensis
1 cm
Brodiaea
terrestris
ssp. terrestris
1 cm
anthers
filament
Brodiaea stellaris
1 cm
Brodiaea insignis
1 cm
Brodiaea jolonensis
1 cm
Brodiaea orcuttii
Brodiaea minor
1 cm
C. albus
1 cm
C. argillosus
1 cm
petals
Calochortus
bruneaunis
1 cm
fruit
2 cm
Calochortus catalinae
1 cm
C. clavatus
var. avius
sepal
C. clavatus
ssp. clavatus
petals
1 cm
Calochortus clavatus
ssp. recurvifolius
1 cm
Calochortus dunnii
2 cm
Calochortus
excavatus
nectary
2 cm
Calochortus flexuosus
1 cm
C. greenei

white to lavender, sometimes brown-spotted above nectary, gen yellow-hairy near nectary, nectary not depressed, ± round, gen densely thick-knobby-yellow- (or purple-) hairy. **FR** erect, ± 5 cm, linear, angled. *n*=7. Meadows, vernally moist places in yellow-pine forest, chaparral; 1200–2200 m. Teh, s CW, TR, SnJt.

var. ***munzii*** F. Ownbey (p. 1191) MUNZ'S MARIPOSA LILY **ST**: bulblets 0. **INFL:** pedicels paired; bracts opposite. **FL**: nectary glabrous or purple-hairy. RARE Yellow-pine forest; 1200–2200 m. SnJt.

var. ***palmeri*** PALMER'S MARIPOSA LILY **ST**: base bulblet-bearing. **INFL**: bracts alternate. **FL**: nectary yellow-hairy. RARE. Habitats and range of sp.

C. panamintensis (F. Ownbey) Rev. PANAMINT MARIPOSA LILY **ST** 40–60 cm, gen simple; base bulblet-bearing. **LVS**: basal 10–20 cm, withering; upper cauline inrolled. **INFL**: fls 1–4, erect; bracts 2–4 cm. **FL**: perianth bell-shaped; sepals 10–40 mm, dark-spotted near base; petals 20–40 mm, narrowly obovate, white tinged lilac, green-striped outside, not spotted, ± glabrous, nectary in red or purple spot, round, depressed, encircled by fringed membrane, densely short-hairy. **FR** erect, ± 7 cm, ± linear, angled. *n*=7. RARE. Dry pinyon/juniper woodland; 2500–3200 m. n DMtns (Panamint Mtns). [*C. nuttallii* Torrey var. *p.* F. Ownbey]

C. persistens F. Ownbey (p. 1191) SISKIYOU MARIPOSA LILY **ST** ± 10 cm, simple. **LVS**: basal ± 20 cm, persistent; cauline 1, small. **INFL** ± umbel-like; fls 2, erect. **FL**: perianth bell-shaped; sepals 35–40 mm; petals ± = sepals, pink to purplish, obovate, yellow-ciliate, hairy only above nectary, nectary bordered below by wide, ciliate membrane, above by short hairs. **FR** nodding, ± 1 cm, winged. 2*n*=20. **RARE** CA. Open, rocky areas; 1000–1500 m. ne KR (Siskiyou Co.).

C. plummerae E. Greene (p. 1191) PLUMMER'S MARIPOSA LILY Bulb coat fibrous. **ST** 30–60 cm, slender, gen branched. **LVS**: basal 20–40 cm, withering; cauline 4–17 cm, upper inrolled. **INFL**: fls 2–6, erect; bracts lf-like. **FL**: perianth widely bell-shaped; sepals 30–50 mm, long-tapered; petals 30–40 mm, pale pink to rose, drying purplish, margin toothed (rarely fringed), long-yellow-hairy in wide central band, nectary round, slightly depressed, ± glabrous, ± hidden by dense, orange bordering hairs. **FR** erect, 4–8 cm, linear, angled. *n*=9. RARE. Dry, rocky chaparral, yellow-pine forest; < 1700 m. SCo, PR.

C. pulchellus Benth. (p. 1191) MOUNT DIABLO FAIRY LANTERN Herbage green. **ST** 10–30 cm, branched. **LVS**: basal 10–40 cm, persistent; cauline 2–3. **INFL**: fls 1–many, nodding. **FL**: perianth closed at tip, spheric; sepals 20–30 mm; petals 25–33 mm, narrowly ovate, light yellow, ciliate with thick hairs, sparsely hairy, nectary deep, bordered above by slender hairs. **FR** nodding, 20–30 mm, winged. **SEED** irregular, dark brown. 2*n*=20. RARE. Wooded slopes, chaparral; 200–800 m. ne SnFrB (Mount Diablo, Contra Costa Co.). In cult.

C. raichei S. Farwig & V. Girard (p. 1191) CEDARS FAIRY LANTERN Pl very glaucous. **ST** 1–100 cm, gen simple. **LVS**: basal 10–40 cm, persistent; cauline 1–4. **INFL**: fls 1–2, nodding. **FL**: perianth closed at tip, ± spheric; sepals 15–25 mm; petals 35–45 mm, pale yellow, obovate, ciliate throughout, surface hairy, nectary ± depressed, bordered above by long, slender hairs. **FR** nodding, 25–35 mm, angled. **SEED** irregular, dark brown. RARE. Open serpentine in woodlands; 200–300 m. s NCoRO (Big Austin Creek, Sonoma Co.). Potentially threatened by mining.

C. simulans (Hoover) Munz (p. 1191) SAN LUIS OBISPO MARIPOSA LILY **ST** 10–60 cm, gen branched; base bulblet-bearing. **LVS**: basal 10–20 cm, withering. **INFL** ± umbel-like; fls 1–3, erect; bracts 2–8 cm. **FL**: perianth bell-shaped; sepals 20–30 mm, tips recurved; petals 30–50 mm, not clawed, white to yellow, sparsely hairy near base, nectary in dark red spot (2nd spot 0), not depressed, ± square, short-hairy. **FR** erect, 5–6 cm, linear, angled. *n*=7. UNCOMMON. Sand (often granitic), grassland to yellow-pine forest; < 1100 m. se SCoRO (c San Luis Obispo Co.). ❀DFCLT.

C. splendens Benth. (p. 1191) **ST** 20–60 cm, branched; bulblets gen 0. **LVS**: basal 10–20 cm, withering. **INFL**: central axis distinct; fls 1–4, erect; bracts 2–5 cm. **FL**: perianth bell-shaped, base narrowed; sepals 20–30 mm, deep lilac, often purple-spotted; petals 30–50 mm, toothed above, gen purple-spotted and sparsely white-hairy near base, nectary not depressed, ± square, gen densely branched-hairy. **FR** erect, 5–7 cm, linear, angled. *n*=7,14. Dry, often granitic soils, grassland, chaparral, yellow-pine forest; < 2800 m. c&s NCoRI, CW, w SW; n Baja CA. ❀DRN,DRY,SUN:1–3, 7,9,14–24;DFCLT.

C. striatus Parish (p. 1191) ALKALI MARIPOSA LILY **ST** 1–5 cm; bulblets 0. **LVS**: basal 10–20 cm, gen withering. **INFL** ± umbel-like; fls 1–5, erect; bracts 1–3 cm. **FL**: perianth bell-shaped, base narrowed; sepals 10–20 mm; petals 20–30 mm, irregularly toothed above, white to lavender, purple-veined, sparsely hairy near nectary, nectary not depressed, oblong, densely simple-hairy. **FR** erect, 4–5 cm, linear, angled. RARE. Alkaline meadows, moist creosote-bush scrub; 800–1400 m. w DMoj; w NV. Threatened by grazing, urbanization. In cult.

C. superbus J. Howell (p. 1191) **ST** 40–60 cm; base bulblet-bearing. **LVS**: basal 20–30 cm, withering. **INFL** ± umbel-like; fls 1–3, erect; bracts 2–8 cm. **FL**: perianth bell-shaped; segments centrally dark-blotched in bright yellow zone; sepals 20–40 mm; petals 20–40 mm, white to yellowish or lavender, gen purple-lined near base, sparsely short-hairy near nectary, nectary not depressed, 1 crescent or chevron, densely short-hairy. **FR** erect, 5–6 cm, linear, angled. *n*=6,7. Common. Open grassland, woodland, dry meadows, yellow-pine forest; < 1700 m. NW, CaRF, SNF, CW, SW. ❀DRN,SUN,DRY:7,14–24;DFCLT.

C. tiburonensis A.J. Hill (p. 1191) TIBURON MARIPOSA LILY **ST** 10–60 cm, gen branched. **LVS**: basal 10–70 cm, persistent; cauline 1–several. **INFL**: fls 2–7, erect. **FL**: perianth bell-shaped; sepals 20–35 mm; petals ± = sepals, light yellow-green, flecked purplish brown, oblanceolate, ciliate to near tip, hairy, nectary depressed, bordered below by ciliate membrane, above by 2+ rows of short hairs. **FR** erect, 20–30 mm, angled. **SEED** irregular, dark brown to black. 2*n*=20 **THREATENED** CA. Serpentine grassland; 50–150 m. nw SnFrB (Ring Mtn, Marin Co.). In cult.

C. tolmiei Hook. & Arn. (p. 1191) PUSSY EARS **ST** 10–40 cm, simple or branched, ± slender. **LVS**: basal 10–40 cm, persistent; cauline gen 1. **INFL** ± umbel-like; fls 1–several, erect or spreading. **FL**: perianth bell-shaped; sepals 10–15 mm; petals 12–25 mm, obovate, white to pinkish or purplish, ± ciliate, densely hairy, nectary bordered below by ciliate membrane, above by short hairs. **FR** nodding, 20–30 mm, winged. **SEED** irregular, dark brown. 2*n*=20. Common. Dry, grassy slopes, woodlands, often in poor soil. 50–2000 m. NW, n ScV, n CCo, SnFrB; to WA, ID. ❀DRN,DRY:1, 4–6,**7**,15–17;DFCLT.

C. umbellatus Alph. Wood OAKLAND STAR-TULIP **ST** 8–25 cm, gen branched; bulblets 0. **LVS**: basal 20–40 cm, persistent; cauline 1. **INFL** ± umbel-like; fls 3–12, ± erect. **FL**: perianth bell-shaped; sepals 10–14 mm; petals 12–18 mm, ± toothed, white to pale pink, often purple-spotted near nectary, not ciliate, ± glabrous, nectary rounded below, truncate above, covered by ciliate membrane, bordered above by short hairs. **FR** nodding, 10–14 mm, winged. **SEED** irregular, dark brown. 2*n*=20. RARE. Open chaparral or woodlands, gen on serpentine; 100–700 m. s NCoRO, SnFrB.

C. uniflorus Hook. & Arn. (p. 1191) LARGE-FLOWERED STAR-TULIP **ST** gen < 5 cm, gen simple; base bulblet-bearing. **LVS**: basal 10–40 cm, persistent; cauline 1–3. **INFL** ± umbel-like; fls 1–5, erect. **FL**: perianth bell-shaped; sepals 12–16 mm; petals 15–28 mm, ± white to pink, often purple-spotted above nectary, not ciliate, sparsely hairy near nectary, nectary covered by ciliate membrane, bordered above by short hairs. **FR** nodding, 15–25 mm, narrowly winged. **SEED** irregular, brown. 2*n*=20,40. UNCOMMON. Moist meadows; < 200 m. NW, SnFrB, n&c CCo; OR. ❀DRN,SUN, IRR:4,**5**,6,**7**,9,**14–17**,18,**19–24**.

C. venustus Benth. (p. 1191) **ST** 10–60 cm, gen branched; base bulblet-bearing. **LVS**: basal 10–20 cm, withering. **INFL** ± umbel-like; fls 1–6, erect; bracts 2–8 cm. **FL**: perianth bell-shaped; sepals 20–30 mm, tips recurved; petals 30–50 mm, ± obovate, ± clawed, white, yellow, purple, or dark red, centrally dark-blotched, gen with 2nd paler blotch above, distally dark-margined, sparsely hairy toward base, nectary not in red spot, not depressed, ± square, short-hairy. **FR** erect, 5–6 cm, linear, angled. *n*=7. Sandy (often granitic) soil in grassland, woodland, yellow-pine forest. 300–2700 m. SNF, CW, WTR, SnGb. Fls highly variable, gen very showy. ❀DRN, DRY:1–3,**7**,9,**14**,**18**,19–24&SUN:**15–17**;may be DFCLT.

C. vestae Purdy (p. 1191) **ST** 30–50 cm, gen branched; base strongly bulblet-bearing. **LVS**: basal 10–20 cm, withering. **INFL** ± umbel-like; fls 1–6, erect; bracts 2–5 cm. **FL**: perianth bell-shaped; sepals 20–30 mm; petals 30–40 mm, white to purplish, ± red-lined below nectary, centrally red-brown-blotched in pale yellow zone, sparsely short-hairy near nectary, nectary not depressed, 2 side-by-side crescents or chevrons, densely short-hairy. **FR** erect, linear, angled. n=14. Clay soil in mixed-evergreen or yellow-pine forests; 500–900 m. s KR, NCoR. ❀DRN,DRY,SUN:**7**,14–21;may be DFCLT.

C. weedii Alph. Wood Bulb coat fibrous. **ST** 30–90 cm, slender, gen branched. **LVS**: basal 20–40 cm, withering; upper cauline inrolled. **INFL**: fls 2–6, erect. **FL**: perianth widely bell-shaped; sepals 20–30 mm; petals < or = sepals, ovate to wedge-shaped, cream to deep yellow, purplish, or red-brown, flecked and often margined red-brown, ± fringed, long-hairy, nectary round, slightly depressed, ± glabrous, ± hidden by dense, long, yellow bordering hairs. **FR** erect, 4–5 cm, linear, angled. n=9. Dry chaparral, often in heavy or rocky soil; < 1900 m. SCoRO, SCo, WTR, PR.

var. ***intermedius*** F. Ownbey **FL**: petals rounded, purplish, dark- or yellow-hairy; anthers rounded. Dry, rocky, open slopes; < 680 m. c SCo & n PR (Orange Co.). ❀DFCLT.

var. ***vestus*** Purdy (p. 1191) **FL**: petals squarish, pale cream, purplish, or red-brown, dark-hairy; anthers abruptly pointed. n=9. Dry, open coastal woodland, chaparral; < 900 m. SCoRO, WTR. ❀DRN,DRY,SUN:**7**,14–24;DFCLT.

var. ***weedii*** (p. 1191) **FL**: petals squarish, deep yellow, dark- or yellow-hairy; anthers acute. n=9. Habitats and range of sp. ❀DFCLT.

C. westonii Eastw. (p. 1191) SHIRLEY MEADOWS STAR-TULIP **ST** 3–15 cm, simple, slender. **LVS**: basal, 10–20 cm, persistent; cauline 0. **INFL**: fls 1–10, erect to spreading. **FL**: perianth bell-shaped; sepals ± 10 mm; petals 8–12 mm, bluish, lanceolate, not ciliate near tip, sparsely hairy, nectary slightly depressed, bordered below by ciliate membranes, above by short hairs. **FR** nodding, 10–15 mm, angled. **SEED** irregular. RARE. Meadows, open woodlands; ± 1500–2000 m. s SNF (Greenhorn Mtns). [*C. coeruleus* var. *w.* (Eastw.) F. Ownbey]

CAMASSIA CAMAS

Susan M. D'Alcamo

Per; bulbs solitary or clustered. **LVS** basal, 15–60 cm, linear-lanceolate, glabrous. **INFL**: raceme, scapose, 2–13 dm; fls 3–many; bracts 1–6 cm, lanceolate, becoming scarious; pedicels 1–5 cm, spreading or incurved in fr. **FL** ± radial; perianth parts 6, petal-like, 15–40 mm, lanceolate, purplish blue to white, 3–9-veined, sometimes twisting over fr; ovary superior, chambers 3, style 1, stigma lobes 3. **FR**: capsule, ± spheric to oblong, cross-ridged, loculidical. **SEEDS** 3–many, ± shiny black. ± 4 spp.: N.Am. (Native American name: camas or quamash) [Gould 1942 Amer Midl Naturalist 28:712–742] Spp. differentiated mostly by geog.

C. quamash (Pursh) E. Greene Bulb gen solitary. **FL**: perianth gen (purplish) blue (white). **FR** ovate to oblong. Damp forests, meadows, streamsides; < 3300 m. NW, CaR, SNH, n SnFrB, MP; to sw Can, WY, UT. Highly variable among populations and with habitat. Probably only 2 forms warrant recognition; needs study. Bulbs were used widely for food and traded among native tribes, perhaps blurring distinctions or creating local forms.

1. Perianth segments 15–20 mm; anthers bright yellow; infl dense; pl appearing ± stout .... ssp. ***breviflora***
1′ Perianth segments 20–40 mm; anthers dull yellow to violet; infl ± open; pl appearing ± slender ... ssp. ***quamash***

ssp. ***breviflora*** Gould (p. 1191) Pl appearing ± stout. **LVS** relatively wide, gen ± glaucous. **INFL** dense. **FL**: perianth segments 15–20 mm; anthers bright yellow. Wet meadows in full sun; 1500–2200 m. MP, adjacent n SNH; to WA. ❀IRR or WET:1,4–7;DFCLT.

ssp. ***quamash*** (p. 1191) Pl appearing ± slender. **LVS** relatively narrow, gen not very glaucous. **INFL** ± open. **FL**: perianth segments 20–40 mm; anthers dull yellow to violet. Habitats and range of sp. [ssp. *linearis* Gould; *C. leichtlinii* (Baker) S. Watson ssp. *suksdorfii* (Greenman) Gould] ❀IRR:**1,4–6**&SHD:2,3,7,14–18; DFCLT.

CHLOROGALUM SOAP PLANT, AMOLE

Judith A. Jernstedt

Per; bulb ovoid to ± elongate, outer coats white to brown, often very fibrous. **LVS** basal, linear, reduced to scarious bracts in infl. **INFL**: panicle; fls 1–several per node. **FL**: perianth segments 6 in 2 petal-like whorls, free, white, purple, or pinkish, persistent in fr and twisted together above capsule; stamens 6, attached to bases of perianth parts, anthers attached at middle; ovary superior, chambers 3, style slender, stigma slightly 3-lobed. **FR**: capsule, stalked, loculicidal. **SEEDS** ovoid, black, 1–2 per chamber. 5 spp.: w N.Am, esp CA. (Greek: green milk or juice) [Hoover 1940 Madroño 5:137–147]

1. Fls open during the day (not opening in the evening)
  2. Perianth white or pinkish; fls or buds 2 or more per node .... ***C. parviflorum***
  2′ Perianth deep blue to purple; fls and buds 1 per node .... ***C. purpureum***
    3. Infl 25–40 cm .... var. ***purpureum***
    3′ Infl 10–20 cm .... var. ***reductum***
1′ Fls opening in the evening
  4. Lf 2–5 mm wide, flat or nearly so .... ***C. angustifolium***
  4′ Lf 4–25 mm wide, margin strongly wavy
    5. Pedicels 2–5 mm, << perianth .... ***C. grandiflorum***
    5′ Pedicels 5–35 mm, ± = or > perianth .... ***C. pomeridianum***
      6. Bulb coats membranous or with few coarse fibers .... var. ***minus***
      6′ Bulb coats with many coarse fibers
        7. Infl nearly prostrate or branches widely spreading from base, < 40 cm high .... var. ***divaricatum***
        7′ Infl erect, 50–250 cm high .... var. ***pomeridianum***

C. angustifolium Kellogg (p. 1191) Heterostylous; bulb 3–5 cm; coats reddish brown, membranous, with delicate fibers. **LF** 2–5 mm wide, not strongly wavy. **INFL** (15–20) 30–70 cm; branches ascending; pedicels 2–3 mm, slender. **FL** opening in evening, closing by following morning; perianth segments 8–12 mm in fl, oblong, white with greenish yellow midvein, spreading but not strongly recurved; stamens 8–12 mm, < or = perianth, anthers 1.5–3 mm, yellow; style 6–8 mm or 2–4 mm. **FR** 1.5–3 mm. *n*=17,17+. Heavy soils of grassland or woodland; < 500 m. n&c NCoRI, SNF (esp n&c); s OR. ❀SUN,DRY:**7**,8,9,**14–24**.

C. grandiflorum Hoover RED HILLS SOAPROOT Bulb 5–7 cm; coats reddish to brown, membranous; outer bulb scales with few delicate fibers. **LVS** basal, 4–12 mm wide, strongly wavy. **INFL** 30–100 cm; branches ascending; pedicels 2–5 mm, << perianth, stout. **FL** opening in evening, closing by next morning; perianth segments 15–30 mm, linear, white with purplish midvein, recurved at fl; stamens gen slightly < perianth, anthers 3 mm, yellow; style 12–28 mm, = or > perianth. **FR** 5–8 mm. RARE. Serpentine outcrops, open shrubby or wooded hills; 300–500 m. n SNF (Placer, El Dorado cos.), c SNF (Tuolumne Co.). Differs from *C. pomeridianum* var. *minus* by short, stout pedicels, larger fls. In cult.

C. parviflorum S. Watson Bulb 4–7 cm; coats dark brown, membranous. **LF** 3–9 mm wide, wavy. **INFL** 30–90 cm; pedicels gen 2 or more per node, 2–8 mm (gen < fls), slender. **FL** open during daylight, closing after 1 day; perianth 7–8 mm, white or pinkish (midvein darker), spreading from above base; stamens 3–4 mm, < perianth, anthers 1.5–2 mm, yellow; style 7–9 mm, > perianth segments. **FR** 3–4 mm. *n*=30. Dry, open coastal-sage scrub; < 750 m. c&s SCo; n Baja CA. ❀DRN,SUN,DRY:7–9,**14–24**.

C. pomeridianum (DC.) Kunth Bulb 7–15 cm; coat thick, fibers coarse (exc var. *minus*), brown. **LF** 20–70 cm, 6–25 mm wide; margins gen wavy. **INFL** highly branched; pedicels 5–25(35) mm, slender, ± = perianth. **FL** opening in evening, closing by next morning; perianth segments spreading, 15–25 mm, linear, white with green or purple midvein, recurved at fl; stamens < perianth, anthers ± 2 mm, purple or yellow, pollen yellow or cream; style 10–15 mm, < to > perianth. **FR** 5–7 mm, spheric to slightly lobed. Bluffs, grassland, chaparral, dry open woodland; < 1500 m. w CA-FP; sw OR.

var. ***divaricatum*** (Lindley) Hoover **INFL** nearly prostrate or branches < 40 cm, spreading widely from base. *n*=18. Coastal bluffs, hills; gen < 100 m. s NCo, CCo. ❀DRN,SUN,DRY:**5,14–17**,18–21,**22–24**.

var. ***minus*** Hoover Bulb coats membranous or with few coarse fibers. **INFL** erect, 30–40 cm; branches ascending. *n*=18. Serpentine outcrops in chaparral; probably gen < 750 m. NCoRI, SnFrB, SCoRO. See note under *C. grandiflorum*. ❀SUN,DRY:**7**, 8,9,**14–24**.

var. ***pomeridianum*** (p. 1191) **INFL** erect, 50–250 cm. *n*= 15,18,18+. Common. Open grassland, chaparral, woodland; < 1500 m. NW, SNF, w GV, CW, SW; sw OR. Bulb juices produce lather in water; roasted bulbs are a native food. ❀DRN,DRY, SUN:5,**7**,8,**9,14–24**.

C. purpureum Brandegee Bulb 2.5–3 cm; scales white to brown, membranous. **LVS** basal, 2–5 mm wide, linear, wavy. **INFL** 10–40 cm; branches few; pedicels 4–10 mm, slender, gen > fls. **FL** open during day; perianth deep blue to purple, segments 5–7 mm, recurved; stamens ± = perianth, anthers 1 mm, yellow; style exserted, 5–6 mm. **FR** 3 mm. Open woodland, sometimes on serpentine; ± 300 or ± 600 m. e SCoRO. Vars. intergrade but apparently retain distinctions in a common garden, produce fertile seeds when artificially crossed.

var. ***purpureum*** PURPLE AMOLE **INFL** 25–40 cm. *n*=30. RARE. Woodland; ± 300 m. ne SCoRO (e side Santa Lucia Mtns, Monterey Co.). Threatened by trampling, grazing, vehicles. In cult.

var. ***reductum*** Hoover (p. 1191) CAMATTA CANYON AMOLE **INFL** 10–20 cm. **RARE** CA. Serpentine woodland; ± 600 m. se SCoRO (ne La Panza Range, San Luis Obispo Co.). In cult.

CLINTONIA

Per; rhizome slender, spreading. **LVS** ± basal, wide; petioles sheathing. **INFL**: umbel or fls solitary on scapose peduncles. **FL**: perianth part 6 in 2 petal-like whorls, free; stamens 6, attached to perianth, hairy near base in CA; ovary superior, chambers 2–3, stigma obscurely 2–3-lobed. **FR**: berry, ovoid, blue in CA. **SEEDS** 2–several, black. 6 spp.: w N.Am, e N.Am, Asia. (De Witt Clinton, naturalist, governor of New York, 1769–1828)

1. Infl an umbel; fls several–many; perianth pink to rose-purple, peduncle >> lvs .... ***C. andrewsiana***
1′ Fls solitary (2), white; peduncle gen < lvs .... ***C. uniflora***

C. andrewsiana Torrey (p. 1197) Rhizome 4–10 mm diam. **LVS** 5(6), 15–30 cm, 5–12 cm wide, elliptic, ± hairy. **INFL**: umbels gen 1–4, lateral; peduncle 25–50 cm, hairy; bracts 1–3; pedicels 1–3 cm, unequal, hairy. **FL** 10–18 mm; perianth pink to rose-purple, hairy, segments pouch-like at base. **FR** 8–12 mm. *n*=14. Shaded, damp redwood forest; < 400 m. NCo, KR, NCoR, CCo, w SnFrB, n SCoRO; sw OR. ❀SHD,DRN,IRR;DRY when dormant:4,**5**,7, 14–16,**17**;DFCLT.

C. uniflora (Schultes) Kunth (p. 1197) Rhizome 1–3 mm diam. **LVS** 2–3, 7–15 cm, 2–6 cm wide, oblanceolate to oblong or elliptic, ± hairy. **INFL**: fl 1(2); peduncle 4–8 cm, hairy; bracts 1–2. **FL** 18–22 mm; perianth white, hairy, segments not pouched at base. **FR** 6–10 mm. *n*=14. Shaded coniferous forest; 1000–1900 m. KR, NCoRO, CaR, SN; to sw Can. ❀SHD,IRR,DRN:1,**4–6**,7,14–17; DFCLT.

DICHELOSTEMMA

Glenn Keator

Per from spheric, fibrous-coated corm; cormlets gen sessile. **LVS** basal, 2–5, narrowly lanceolate, gen keeled, entire, glabrous, sometimes withered by fl. **INFL** umbel- or raceme-like, gen dense (pedicels < fl); axis gen curved to twining, cylindric; bracts 2–4, ± papery. **FL**: perianth tube cylindric to bell-shaped, lobes 6 in 2 petal-like whorls; staminodes gen 0 (stamen-like in 1 sp.); stamens 3 (6 in 1 sp.), filaments fused to perianth and into a crown-like tube, free filaments gen ± 0, anthers attached at base; style 1, stigma 3-lobed. **FR**: capsule, gen not stalked, gen ovoid, 3-angled, loculicidal. **SEEDS** sharply angled, black. *n*=9. 5 spp.: w US, esp n CA (Greek: toothed crown, from stamen appendages) [Keator 1992 Four Seasons 9:24–39]

1. Fl long-cylindric; umbel open; perianth red and green .... ***D. ida-maia***
1′ Fl bell-shaped to salverform; umbel dense; perianth not red and green
 2. Infl weak, twining; perianth pink; stamen-like staminodes alternate fertile stamens .... ***D. volubile***
 2′ Infl self-supporting; perianth blue, purple, or white; staminodes 0
  3. Fertile stamens 6 (3 smaller); perianth tube not narrowed above ovary .... ***D. capitatum***
   4. Bracts gen dark purple; fls gen > 6; pedicels gen < bracts .... ssp. ***capitatum***
   4′ Bracts whitish or streaked purple; fls 2–5; pedicels > bracts .... ssp. ***pauciflorum***

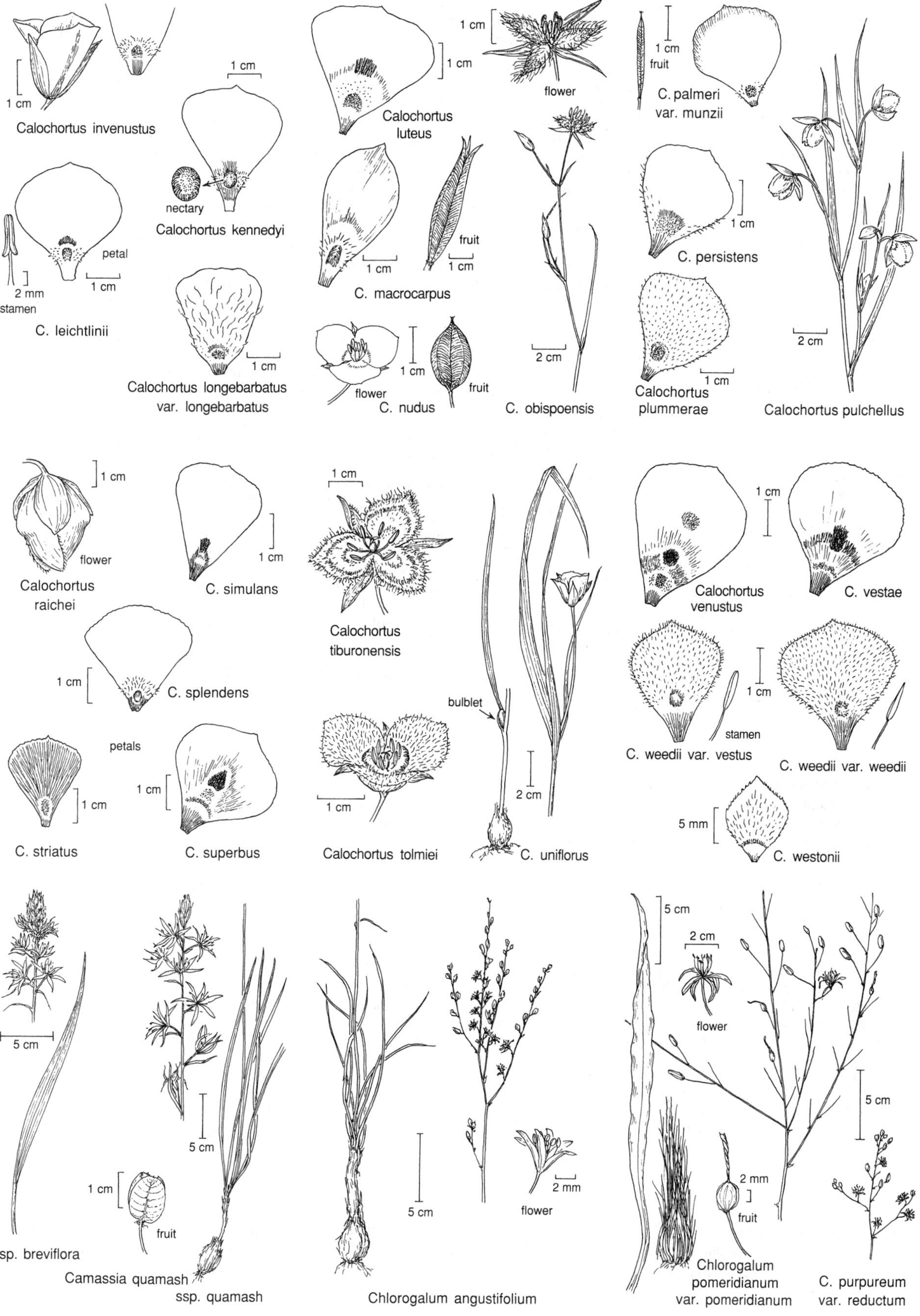
1 cm
Calochortus invenustus
1 cm
nectary
Calochortus kennedyi
petal
2 mm
stamen
1 cm
C. leichtlinii
1 cm
Calochortus longebarbatus var. longebarbatus
1 cm
Calochortus luteus
1 cm
flower
1 cm
fruit
1 cm
1 cm
C. macrocarpus
1 cm
flower
fruit
C. nudus
2 cm
C. obispoensis
1 cm
fruit
C. palmeri var. munzii
1 cm
C. persistens
1 cm
Calochortus plummerae
2 cm
Calochortus pulchellus
1 cm
flower
Calochortus raichei
1 cm
C. simulans
1 cm
C. splendens
petals
1 cm
C. striatus
1 cm
C. superbus
1 cm
Calochortus tiburonensis
1 cm
Calochortus tolmiei
bulblet
2 cm
C. uniflorus
1 cm
Calochortus venustus
C. vestae
1 cm
stamen
C. weedii var. vestus
C. weedii var. weedii
5 mm
C. westonii
5 cm
ssp. breviflora
5 cm
1 cm
fruit
Camassia quamash
ssp. quamash
5 cm
2 mm
flower
Chlorogalum angustifolium
5 cm
2 cm
flower
2 mm
fruit
Chlorogalum pomeridianum var. pomeridianum
5 cm
C. purpureum var. reductum

3′ Fertile stamens 3, equal; perianth tube narrowed above ovary
5. Infl a dense raceme; anther appendages deeply notched .... ***D. congestum***
5′ Infl an umbel; anther appendages truncate or rounded .... ***D. multiflorum***

D. capitatum Alph. Wood BLUE DICKS Cormlets sessile and on stolons. **LVS** 2–3, 10–40 cm, barely keeled. **INFL** head- or umbel-like, dense; axis < 65 cm; bracts widely lanceolate, whitish to dark purple; fls 2–15. **FL**: perianth blue, blue-purple, pink-purple, or white, tube 3–12 mm, narrowly cylindric to short-bell-shaped, lobes gen ascending, 7–12 mm; stamens 6, crown segments 4–6 mm, deeply notched, lanceolate, white, angled inward, slightly reflexed at tip, outer filaments wider at base, outer anthers 2–3 mm, inner free filaments ± 0, inner anthers 3–4 mm; ovary sessile, style 4–6 mm. Open woodlands, scrub, desert, grassland; 0–2300 m. CA; to OR, UT, NM, n Mex. [*Brodiaea pulchella* (Salisb.) E. Greene; *D. p.* (Salisb.) A.A. Heller] Variable; sspp. intergrade.

ssp. ***capitatum*** (p. 1197) **INFL**: bracts gen dark purple (or paler and striped dark purple); pedicels 1–15 mm, gen < bracts; fls 6–15. **FL**: perianth lobes gen ascending. *n*=9,18,27,36,45. Habitats and range of sp. [*D. lacuna-vernalis* L. Lenz, vernal pool brodiaea] ❀DRN,DRY,SUN:1–3,5,6,**7–10**,11,12,**14–24**.

ssp. ***pauciflorum*** (p. 1197) (Torrey) Keator **INFL**: bracts whitish; pedicels 6–35 mm, > bracts; fls 2–5. **FL**: perianth lobes widely spreading. *n*=9,27. Deserts, open scrub: 300–2100 m. SNE, D; to UT, NM, n Mex. [*Brodiaea pulchella* var. *pauciflora* (Torrey) C. Morton] ❀DRN,DRY,SUN:2,3,**7–10**,11,**12**,**14–24**.

D. congestum (Sm.) Kunth (p. 1197) OOKOW **LVS** 3–4, 4–10 cm, ± glaucous, strongly keeled. **INFL** raceme-like, very dense; axis < 80 cm, ± scabrous; bracts widely lanceolate, pale purple to green; pedicels 1–6 mm; fls 6–15. **FL**: perianth blue-purple, tube 8–10 mm, narrowly ovoid, slightly narrowed above ovary, lobes ascending, 8–10 mm; filament crown segments erect, 5–6 mm, narrowly lanceolate, deeply notched, purplish, outer filaments minute stubs, inner anthers 4–5 mm; ovary sessile, style 5–6 mm. *n*=9,18. Open woodland, grassland near coast; 0–2000 m. NW, CaR, n SN; to B.C. [*Brodiaea c.* Sm.; *B. pulchella* (Salisb.) E. Greene misapplied] ❀DRN,DRY,SUN:4–6,**7**,9,14,**15–17**,18–24.

D. ida-maia (Alph. Wood) E. Greene (p. 1197) FIRECRACKER FLOWER **LVS** 3–5, 30–50 cm, glaucous, keeled. **INFL** umbel-like, open; axis < 100 cm; bracts lanceolate, reddish; pedicels 1–5 cm; fls 6–20 (nodding in fl, erect in fr). **FL**: perianth tube 20–27 mm, cylindric, red, with 6 sac-like swellings on lower 1/3, lobes recurved (erect in fr), 7–9 mm, green; crown segments erect, 3–4 mm, entire, rounded, inrolled, whitish, anthers 6–8 mm; ovary narrowly ovoid, stalk 6 mm, style 15–20 mm. *n*=24. Forest edges, grassland near coast; 30–2000 m. NCo, KR, n&c NCoR; s OR. [*Brodiaea i.-m.* (Alph. Wood) E. Greene] Sporadic hybrids with *D. congestum* or *D. multiflorum*, with fls ± horizontal, perianth tube 14–17 mm and rose-purple have been called *D. venustum* (E. Greene) Hoover, rose firecracker flower, or *Brodiaea v.* (E. Greene) E. Greene. ❀DRN,DRY,part SHD:4–6,**7**,**14–17**,18–24.

D. multiflorum (Benth.) A.A. Heller (p. 1197) WILD HYACINTH **LVS** 3–4, 30–85 cm, glaucous, unkeeled. **INFL** umbel-like, dense, spheric; axis < 80 cm, scabrous; bracts 7–12 mm, ovate, acuminate, streaked purple; pedicels 3–15 mm; fls 10–35, erect. **FL**: perianth pink- to blue-purple, tube 8–10 mm, ± cylindric, strongly narrowed above ovary, tough and persistent in fr, lobes widely spreading, 8–10 mm; crown segments 5–6 mm, truncate to rounded, entire to minutely dentate, strongly inrolled, arching slightly inward, white or pale purple, anthers 4–5 mm; ovary sessile, spheric, style 7–8 mm. *n*=9,18,27; 2n=45. Foothill grassland, open woodland, scrub; 15–2000 m. NW, n SN, uncommon in n&w SnFrB; s OR. [*Brodiaea m.* Benth.] ❀DRN,DRY:5,6,7,8,9,**14–24**.

D. volubile (Kellogg) A.A. Heller (p. 1197) TWINING BRODIAEA, SNAKE LILY **LVS** 3–4, 30–70 cm, green, strongly keeled. **INFL** umbel-like, dense, spheric; axis < 150 cm, scabrous; bracts 12–15 mm, widely ovate, acute, pinkish; pedicels 10–40 mm; fls 10–30 (sometimes horizontal in fl, stiffy erect in fr). **FL**: perianth pink, tube 5–7 mm, ± spheric, narrowed above ovary, with 6 sac-like angles, lobes ± widely spreading (ascending in fr), 5–7 mm; staminodes 3, ± 3 mm, linear-oblong, gen notched, cream, ciliate-dentate; crown segments 2 per stamen, 3–4 mm, narrowly lanceolate, white, folded inward, anthers 3–4 mm; ovary subspheric, sessile or short-stalked, style 3–4 mm. *n*=9,18. Foothill woodland, scrub; 100–1600 m. NCoRI, CaRF, SNF. [*Brodiaea v.* (Kellogg) Baker] ❀DRN,DRY:5,6,**7**,8,9,**14–16**,17,**18–24**.

DISPORUM

Per; rhizome slender, creeping. **ST** erect, branched, scaly below, leafy above. **LVS** alternate, sessile to clasping, ± oblique at base, strongly veined. **INFL** ± umbel-like, terminal; fls 1–several, drooping, ± hidden by lvs; perianth segments 6 in 2 petal-like whorls, free, white to greenish in CA; stamens 6, free; ovary superior, chambers 3, style slender, stigma entire or 3-lobed. **FR**: berry, ovoid to spheric, yellow to red. **SEEDS** black. ± 15 spp.; temp N.Am, e Asia. (Greek: double seed)

1. Fl 8–15 mm; perianth parts narrowly elliptic, narrowed at base, spreading from middle or lower; stamens = or > perianth .... ***D. hookeri***
1′ Fl 15–25 mm; perianth parts oblanceolate, scarcely narrowed at base, spreading only at tip; stamens < perianth .... ***D. smithii***

D. hookeri (Torrey) Nicholson (p. 1197) **ST** 30–80 cm, sparingly branched, hairy at least above. **LF** 3–15 cm, (ob)ovate, acute to acuminate, cordate, oblique at base, ciliate, glabrous to densely hairy, esp on veins below. **INFL**: fls 1–3. **FL** narrowly bell-shaped; perianth parts 8–15 mm, narrowly elliptic, spreading from middle or lower, greenish; stamens = or > perianth, anthers glabrous or minutely sharp-bristly; ovary and style glabrous or ± hairy. **FR** 7–9 mm, spheric, red. Shady woodlands; < 1600 m. NW, SN, CW; to w Can, MT. [var. *parvifolium* (S. Watson) Britton; var. *trachyandrum* (Torrey) Q. Jones] ❀SHD,DRN:**4–6**&IRR:1,2,7,14–16,**17**.

D. smithii (Hook.) Piper (p. 1197) **ST** 30–90 cm, ± freely branched, ± glabrous to hairy. **LF** 5–12 cm, ± ovate, acute to abruptly acuminate; base rounded to subcordate; margin not ciliate (or sparsely so near tip); veins ± glabrous below. **INFL**: fls 1–7. **FL** cylindric; perianth parts 15–28 mm, oblanceolate, spreading slightly at tips, whitish; stamens < perianth, anthers glabrous; ovary glabrous, style hairy to tip. **FR** 12–15 mm, obovoid, orange to red. Moist, shady woodlands, esp redwood forest; < 800 m. w NW, SnFrB; to B.C. ❀SHD,DRN:**4**,**5**,6,&IRR:7,14,**15–17**.

ERYTHRONIUM FAWN LILY

Geraldine A. Allen

Per from elongate bulb of 1 fleshy scale, gen with small, bead-like segments of persistent rhizome. **LVS** 2 (1 in non-fl pl), basal, 6–35 cm, lanceolate to ovate (solitary basal lf wider), narrowed to petiole, glabrous; margin entire to wavy. **INFL**: raceme; fls 1–10; bracts 0. **FL** showy, nodding; perianth segments 6, similar, free, ± lanceolate, ± strongly recurved; stamens 6;

style 1, stigma entire to 3-lobed. **FR** ovoid to oblong. **SEED** ± ovoid, ± angular, brown. n=12. ± 25 spp.: esp temp N.Am. (Greek: red, from fls of some) [Applegate 1935 Madroño 3:58–113; Shevock et al. 1990 Madroño 37:261–273] Attractive in woodland gardens. Pressed specimens fade, so lf and fl markings should be recorded fresh.

1. Lvs uniformly green, not mottled
 2. Perianth white with yellow base; stigma ± entire
  3. Perianth segments 10–20 mm, without sac-like folds at base ..... ***E. purpurascens***
  3′ Perianth segments 18–35 mm, inner with small sac-like folds at base
   4. Perianth segments < 1/3 yellow; lvs 6–17 cm; fls 1–3; KR, CaRH ..... ***E. klamathense***
   4′ Perianth segments 1/3–1/2 yellow; lvs 10–35 cm; fls 1–8; s SNH ..... ***E. pusaterii***
 2′ Perianth yellow; stigma entire or lobed
  5. Perianth segments 15–28 mm; style and filaments yellow; stigma entire or with lobes < 1 mm; fls 1–10 ..... ***E. pluriflorum***
  5′ Perianth segments 20–35 mm; style and filaments white; stigma entire or with slender, recurved, 2–4 mm lobes; fls 1–5
   6. Fresh perianth segments with narrow, pale zone at base; stigma lobes 2–4 mm; KR, NCoR ... ***E. grandiflorum***
   6′ Fresh perianth segments lacking pale zone at base; stigma entire or with lobes < 1 mm; c SNF . ***E. tuolumnense***
1′ Lvs irregularly mottled with brown or white
 7. Filaments flat at base, gen > 1.5 mm wide; stigma lobes 4–6 mm
  8. Perianth segments pink with yellow base ..... ***E. revolutum***
  8′ Perianth segments white with yellow base ..... ***E. oregonum***
 7′ Filaments < 0.8 mm wide; stigma entire to lobed
  9. Perianth violet to pink with dark purple base ..... ***E. hendersonii***
  9′ Perianth ± white with yellow base
   10. Style 4–10 mm; anthers yellow, white, or cream
    11. Anthers white to cream; style ± straight; bulblets 0 ..... ***E. citrinum***
    11′ Anthers yellow; style ± bent; bulblets 0 or sessile ..... ***E. helenae***
   10′ Style 10–15 mm; anthers white to cream
    12. Fl-stalk (when fls > 1) branched well above lvs; bulblets 0 ..... ***E. californicum***
    12′ Fl-stalk (when fls > 1) branched near ground level; bulblets produced from long, slender rhizomes ..... ***E. multiscapoideum***

E. californicum Purdy (p. 1197) Bulb 35–60 mm, ovoid. **LF** 7–15 cm, oblong to narrowly ovate, mottled with brown or white. **INFL**: stalk 10–30 cm, ± reddish; fls 1–3. **FL**: perianth segments 25–40 mm, ± narrowly ovate, white to cream with yellow base, ± banded with brown to red, inner with small sac-like folds at base; stamens 12–25 mm, filaments slender, ± white, anthers ± white to cream; style 10–14 mm, ± white, stigma lobes < 2 mm, spreading. Dry woodlands, openings; 0–1900 m. KR, NCoR. May intergrade with *E. citrinum*, *E. multiscapoideum*. ❀SHD,DRN,IRR or DRY:6,7, 14–16,**17**;DFCLT.

E. citrinum S. Watson LEMON-COLORED FAWN LILY Bulb 40–50 mm, slender. **LF** 9–15 cm, lanceolate to narrowly ovate, ± wavy-margined, mottled with brown or white. **INFL**: stalk 12–35 cm; fls 1–3. **FL**: perianth segments 25–45 mm, lanceolate to narrowly elliptic, white with yellow base, ± pinkish in age, inner ± with small sac-like folds at base; stamens gen 11–17 mm, filaments slender, white, anthers white to cream; style 6–10 mm, white, stigma entire or with lobes < 1 mm. UNCOMMON. Dry woodlands, shrubby slopes (± on serpentine); 100–1100 m. KR; sw OR. Pls lacking sac-like folds on inner perianth segments have been called *E. howellii* S. Watson, Howell's fawn-lily. ❀SHD,DRN,DRY:5,17;DFCLT.

E. grandiflorum Pursh GLACIER LILY Bulb 30–50 mm, slender. **LF** 5–20 cm, lanceolate, ± wavy-margined, green. **INFL**: stalk 5–30 cm; fls 1–3. **FL**: perianth segments 20–35 mm, lanceolate, yellow with narrow, pale zone at base, inner ± with sac-like folds at base; stamens 11–18 mm, filaments ± slender, white, anthers cream to yellow (sometimes dark red elsewhere); style 10–15 mm, white, stigma lobes 2–4 mm. Subalpine meadows; 1400–2300 m. KR, NCoR; to sw Can, Colorado. Pls with pale anthers (all CA material) have been called var. *pallidum* H. St. John. ❀IRR,DRN:1–3; DFCLT.

E. helenae Appleg. Bulb 30–55 mm, ovoid, sometimes producing sessile bulblets. **LF** 10–20 cm, widely lanceolate to ovate, ± wavy-margined, mottled with brown or white. **INFL**: stalk 15–30 cm; fls 1–3. **FL**: perianth segments 25–40 mm, lanceolate to ovate, white with yellow base, ± pinkish in age, inner with small sac-like folds at base; stamens 8–13 mm, filaments ± slender, ± yellow, anthers yellow; style 5–8 mm, ± bent, ± white, stigma lobes < 1 mm. Dry woodlands, on serpentine; 500–1200 m. s NCoRI (near Mt. St. Helena). ❀SHD,DRN,DRY or IRR:6,7,14–17;DFCLT.

E. hendersonii S. Watson HENDERSON'S FAWN LILY Bulb 40–55 mm, slender. **LF** 10–25 cm, oblong to ovate, entire to ± wavy-margined, mottled with brown or white. **INFL**: stalk 12–30 cm, ± reddish; fls 1–4. **FL**: perianth segments 18–35 mm, widely lanceolate, violet to pink (darker toward tip) with dark purple zone at base, inner with small sac-like folds at base; stamens 10–14 mm, filaments slender, purple, anthers pale brown to purple; style 6–8 mm, violet, stigma entire or with lobes < 1 mm. UNCOMMON. Dry woodlands, openings; 300–1600 m. KR; sw OR. ❀SHD, DRN,IRR;5–7,15–17;DFCLT.

E. klamathense Appleg. KLAMATH FAWN LILY Bulb 25–40 mm, slender. **LF** 6–17 cm, lanceolate to narrowly elliptic, ± folded along midvein, entire to wavy-margined, green. **INFL**: stalk 6–20 cm; fls 1–3. **FL**: perianth segments 20–35 mm, widely lanceolate, white with yellow base, ± pinkish in age, inner with small sac-like folds at base; stamens 8–14 mm, filaments slender, white, anthers ± yellow; style 4–9 mm, white, stigma ± entire. UNCOMMON. Montane meadows, forest openings; 1200–1850 m. KR, CaRH; s OR. ❀IRR,DRN:1;DFCLT.

E. multiscapoideum (Kellogg) Nelson & Kenn. Bulb 20–50 mm, ovoid, producing bulblets from slender rhizomes. **LF** 4–15 cm, ± elliptic, entire to wavy-margined, mottled with brown or white. **INFL**: stalk 8–20 cm, ± branched just above lvs; fls 1–4. **FL**: perianth segments 16–35 mm, widely lanceolate to elliptic, ± white with yellow base; stamens 10–15 mm, filaments slender, white, anthers white to cream; style 10–13 mm, white, stigma lobes 1–4 mm, recurved. Open woodlands, shrubby slopes; 400–1000 m. CaRF, n&c SNF. May intergrade with *E. californicum*. ❀SHD, DRN,DRY or IRR:**3,6,7,15–17**;CVS.

E. oregonum Appleg. Bulb 35–50 mm, narrowly ovoid. **LF** 10–22 cm, widely lanceolate to ovate, entire to ± wavy-margined, mottled with brown or white. **INFL**: stalk 15–45 cm; fls 1–3. **FL**: perianth segments 30–45 mm, lanceolate to narrowly elliptic, white with yellow base, sometimes red-banded, inner with sac-like folds at base; stamens 12–22 mm, filaments 2–3 mm wide, white, anthers cream to yellow; style 12–18 mm, white, stigma lobes 4–6 mm, slender, recurved. Openings in woodlands; 100–500 m. NCoRO; s OR to B.C. [ssp. *leucandrum* Appleg.] CA populations are geographically separate, may be a white form of *E. revolutum*. ❀SHD,DRN,IRR or DRY:4–6,15–17;DFCLT.

E. pluriflorum J.R. Shevock, J. Bartel & G. Allen Bulb 40–75 mm, ± ovoid. **LF** 7–30 cm, oblanceolate to elliptic, ± wavy-margined, green. **INFL**: stalk 8–35 cm; fls 1–10. **FL**: perianth segments 15–28 mm, lanceolate, yellow, aging bronze, lacking sac-like folds at base; stamens 8–12 mm, filaments slender, yellow, anthers yellow; style 6–8 mm, yellow, stigma entire or with lobes < 1 mm. Open, rocky places; 2300–2550 m. c SNH (Madera Co.). ❀DRN:1,2;DFCLT.

E. purpurascens S. Watson (p. 1197) Bulb 25–40 mm, slender. **LF** 6–15 cm, lanceolate to narrowly ovate, ± wavy-margined, green. **INFL**: stalk 7–20 cm; fls 1–6. **FL**: perianth segments 10–20 mm, lanceolate, white with yellow base, pinkish purple in age, lacking sac-like folds at base; stamens 8–12 mm, filaments slender, yellow, anthers ± yellow; style 4–5 mm, yellow, stigma ± entire. Open forests, meadows, rocky places; 1500–2700 m. CaRH, n&c SNH. ❀SHD,DRN,DRY:1–3;DFCLT.

E. pusaterii (Munz & J. Howell) J.R. Shevock, J. Bartel, & G. Allen KAWEAH LAKES FAWN LILY Bulb 40–60 mm, narrowly ovoid. **LF** 10–35 cm, lanceolate, ± wavy-margined, green. **INFL**: stalk 12–40 cm; fls 1–8. **FL**: perianth segments 25–45 mm, lanceolate, white, lower 1/3–1/2 bright yellow, aging ± pink, inner with sac-like folds at base; stamens 8–15 mm, filaments slender, ± white, anthers yellow; style 7–10 mm, ± white, stigma lobes gen < 1 mm. Meadows, rocky ledges; 2100–2500 m. RARE. s SNH (Tulare Co.). [*E. grandiflorum* ssp. *p.* Munz & J. Howell] In cult.

E. revolutum Smith (p. 1197) Bulb 35–50 mm, narrowly ovoid. **LF** 10–25 cm, widely lanceolate to ovate, entire to ± wavy-margined, mottled with brown or white. **INFL**: stalk 15–40 cm; fls 1–3. **FL**: perianth segments 25–40 mm, lanceolate to narrowly elliptic, pink with yellow bands at base, inner with small sac-like folds at base; stamens 12–22 mm, filaments 2–3 mm wide, ± lanceolate, white to pink, anthers yellow; style 12–18 mm, white to pink, stigma lobes 4–6 mm, slender, recurved. Streambanks and wet places in woodlands; 0–1000 m. NCo, NCoRO; to B.C. ❀SHD,DRN, IRR;**16**,17;CVS.

E. tuolumnense Appleg. (p. 1197) TUOLUMNE FAWN LILY Bulb 50–100 mm, ovoid, readily producing sessile bulblets. **LF** 15–35 cm, elliptic to ovate, entire to ± wavy-margined, green. **INFL**: stalk 15–35 cm, ± reddish; fls 1–5. **FL**: perianth segments 20–35 mm, ± narrowly ovate, yellow, inner with small sac-like folds at base; stamens 12–16 mm, filaments slightly widened, 0.4–0.6 mm wide at base, ± white, anthers yellow; style 8–10 mm, ± white, stigma ± entire. RARE. Open woodlands; 600–950 m. c SNF (Tuolumne Co.). In cult.

FRITILLARIA FRITILLARY

Bryan D. Ness

Per; bulb with 1–several large fleshy scales, 0–many small scales. **ST** erect, simple (0 in non-fl pls). **LVS** alternate (or whorled below), sessile, linear to ± ovate (1 "bulb-lf" in non-fl pls). **INFL**: raceme; bracts lf-like. **FL** gen nodding, bell- or cup-shaped; perianth segments 6, of 2 similar whorls; nectaries 6, on perianth parts; stamens 6, incl, inserted at perianth base, anthers attached ± near middle; ovary ± sessile, style 1, entire or 3-branched. **FR**: capsule, loculicidal, thin walled, ± rounded, 6-angled, or winged, chambers 3. **SEEDS** many, 2 rows per chamber, flat, brownish. ± 100 spp.: n temp. (Latin: dicebox, from fr shape) [Turrill & Sealy 1980 Hooker's Icones Plantarum 34:1–275] Bulbs of some eaten by Indians of N.Am. ❀DRN: for pots or rock gardens; DRY when dormant. Most are very DFCLT.

1. Style entire or branches < 1.5 mm; perianth never scarlet
  2. Perianth either < 20 mm or yellow to orange
    3. Lvs whorled; perianth pinkish to purplish; s SN, Teh ..... ***F. brandegei***
    3′ Lvs ± opposite to scattered; perianth yellow to orange; n CA ..... ***F. pudica***
  2′ Perianth > 20 mm, not yellow to orange
    4. Perianth segments pinkish purple, obovate, tips rounded to acute, not recurved; fls not noticeably scented; NCoRI, n SNF ..... ***F. pluriflora***
    4′ Perianth segments white to pink, oblanceolate, tips acute to short-tapered, gen recurved; fls fragrant; s SNF ..... ***F. striata***
1′ Style branches obvious, > 1.5 mm; perianth sometimes scarlet
  5. Perianth scarlet, clearly mottled, tips gen recurved ..... ***F. recurva***
  5′ Perianth not both scarlet and mottled, tips gen not recurved (sometimes slightly recurved in *F. eastwoodiae*)
    6. Lvs > 10 or in whorls of 3–6 on lower st, not sickle-shaped
      7. Fls ± erect; upper lvs gen < 1/2 lowest lf; lvs often > infl; c SNH (Mono Co.), Teh, TR ..... [2]***F. pinetorum***
      7′ Fls nodding; upper lvs gen ± = lowest lf; lvs gen < infl; more widespread
        8. Perianth purplish brown, spotted yellow or white; lvs 2–3 per node; 1000–3200 m, esp inland mtns ..... [2]***F. atropurpurea***
        8′ Perianth nearly black, purplish, greenish white, greenish yellow, or red; lvs gen > 3 per node; 200–1800 m, esp coastal mtns
          9. Perianth dull greenish yellow, dark-dotted; nectary widely elliptic to ± diamond-shaped, paler than rest of perianth segment; SCoRO, WTR ..... ***F. ojaiensis***
          9′ Perianth greenish white, pale green to red, or almost black, mottled in some (esp *F. affinis* var. *a.*); nectary lanceolate or narrower; NW, CaR, SN, CW
            10. Perianth pale green to almost black, not mottled; nectary ± 1/2 perianth length, green; small bulb scales 0–4; SCoR ..... ***F. viridea***
            10′ Perianth greenish white or greenish yellow to red or purplish, sometimes mottled; nectary < 1/2 perianth length, color various; small bulb scales gen > 10 (sometimes < 10 in *F. a.* var. *affinis*); NW, CaR, SN, CW
              11. Perianth gen > 2 cm, gen clearly purple-mottled — NW, CaR, n SNF, CCo, SnFrB ..... ***F. affinis***
                12. Perianth clearly mottled; small bulb scales gen 2–20; KR, NCoR, CaR, n SNF, SnFrB ..... var. ***affinis***
                12′ Perianth scarcely mottled; small bulb scales > 50; NCo, CCo ..... var. ***tristulis***
              11′ Perianth gen < 2 cm, not (or faintly) mottled
                13. Style branches barely recurved; perianth greenish yellow to red, tips gen recurved (or flared); nectaries green, gold, or yellow; CaR ..... ***F. eastwoodiae***

13′ Style branches distinctly recurved; perianth parts purplish to greenish white, tips not flared or recurved; nectary greenish, dotted purple; SN ..... ***F. micrantha***
6′ Lvs < 10 or alternate (sometimes crowded), sometimes sickle-shaped
14. Perianth clearly mottled; small bulb scales gen > 10 (exc *F. purdyi*)
15. Lvs sickle-shaped; fls erect; SnFrB, SCoRI ..... ***F. falcata***
15′ Lvs gen not sickle-shaped; fls nodding or horizontal; not SnFrB or SCoRI
16. Lvs 2–10, ovate, some sometimes sickle-shaped; perianth white, spotted purple; small bulb scales 0–3 ..... ***F. purdyi***
16′ Lvs gen > 4, not sickle-shaped, perianth purplish brown, spotted yellow and white; small bulb scales 45–50
17. Fls nodding; upper lvs slightly < or = lowest cauline lf; cauline lvs gen < infl; KR, CaR, SN ..... [2]***F. atropurpurea***
17′ Fls ± erect; upper lvs gen < 1/2 lowest cauline lf; lvs > infl; c SNH (Mono Co.), Teh, TR ..... [2]***F. pinetorum***
14′ Perianth not clearly mottled; small bulb scales < 10
18. Lvs 2–4, sickle-shaped ..... ***F. glauca***
18′ Lvs gen > 4, not sickle-shaped
19. Nectaries obscure (narrow band 1/2–2/3 perianth length); perianth white, striped green, odorless or faintly fragrant; coastal valleys ..... ***F. liliacea***
19′ Nectaries prominent (narrow band 2/3 to = perianth); perianth brown, purplish brown, or greenish purple at least inside; inland valleys and hills
20. Perianth greenish white or yellow outside, purplish brown inside; fls odor definitely unpleasant; gen in clay depressions ..... ***F. agrestis***
20′ Perianth dark brown to greenish purple or yellowish green; fls odorless (exc var. *ineziana*); gen on hillsides, mesas ..... ***F. biflora***
21. Lvs widely lanceolate to oblanceolate ..... var. ***biflora***
21′ Lvs linear to narrowly lanceolate ..... var. ***ineziana***

F. affinis (Schultes) Sealy CHECKER LILY Bulb; large scales 2–5; small scales 2–20. **ST** 1–12 dm. **LVS** in 1–4 whorls of 2–8 below, alternate above, 4–16 cm, linear-lanceolate to ovate. **FL** nodding; perianth parts 1–4 cm, oblong to ovate, brownish purple mottled yellow to pale yellowish green mottled purple; nectary 1/2–2/3 perianth length, lanceolate, yellow with purple dots; style divided 1/2. **FR** widely winged. 2*n*=24,36,48. Common. Oak or pine scrub, grasslands; < 1800 m. KR, NCoR, CaR, n SNF, SnFrB; to B.C., ID. [*F. lanceolata* Pursh (illegitimate name); *F. phaeanthera* Purdy] Highly variable; needs further study; hybridizes with *F. recurva*.

var. ***affinis*** (p. 1197) Bulb with 2–20 small scales. **FL**: perianth parts clearly mottled. Habitats and range of sp. ❀DRN, DRY,SHD:5,6,15–17&IRR:7,14,18;DFCLT.

var. ***tristulis*** (A.L. Grant) B. Ness Bulb with 60–100+ small scales. **FL**: perianth parts scarcely mottled. Very uncommon. Coastal grassland; < 100 m. n CCo (w Marin Co.)

F. agrestis E. Greene (p. 1197) STINKBELLS Bulb; large scales 2–9; small scales 0–2. **ST**: 3–6 dm. **LVS** 5–12, alternate, crowded near lower center of st, 5–15 cm, linear to lance-oblong. **FL** nodding, odor unpleasant; perianth parts 1.8–3.5 cm, ovate, greenish white outside, purplish brown inside; nectary prominent, 2/3 perianth length, narrowly linear, green; style divided 1/2. **FR** ± angled. 2*n*=24. UNCOMMON. Clay depressions or other low heavy soils; < 500 m. NCoRO (Mendocino Co.), SNF, GV, CW. [*F. biflora* var. *a.* E. Greene] ❀DRY,SUN:14;DFCLT.

F. atropurpurea Nutt. (p. 1197) Bulb; large scales 2–5; small scales 45–50. **ST** 1–6 dm. **LVS** 2–3 per node, 4–12 cm, linear to lanceolate. **FL** nodding; perianth widely open, parts 1–2.5 cm, oblong to ± diamond shaped, purplish brown mottled yellow or white; nectary indistinct, covering most of perianth part, elliptic, yellow with dark reddish dots; style divided > 1/2. **FR** acutely angled. Common. Lf mold under trees; 1000–3200 m. KR, CaR, SN, GB, DMtns; to OR, MT, NM. Easily confused with *F. pinetorum*

F. biflora Lindley CHOCOLATE LILY, MISSION BELLS Bulb; large scales 2–8; small scales 0–4. **ST** 1–4.5 dm. **LVS** 3–7, alternate, often ± crowded just above ground level, 5–19 cm, linear or oblong to narrowly ovate. **FL** nodding; perianth parts 1.8–4 cm, narrowly ovoid, dark brown, greenish purple, or yellowish green; nectary prominent, 2/3 perianth length, narrowly linear, purplish to greenish; style divided 1/2–2/3. **FR** angled. Uncommon. Grassy slopes, mesas, serpentine barrens; < 1200 m. NCoR (Mendocino, Napa cos.), CW, SW. Sometimes confused with *F. agrestis*, which has unpleasant odor, grows in heavier soil.

var. ***biflora*** (p. 1197) **LF** 8–40 mm wide, oblong to narrowly ovate. **FL** 1.8–4 cm, without unpleasant odor. 2*n*=24. Habitat and range of sp. [*F. roderickii* W. Knight] ❀DRN,SUN:14–17,**20**,**21**, 24;CVS;DFCLT.

var. ***ineziana*** Jepson **LF** 3–6 mm wide, linear. **FL** 1.5–2 cm, with unpleasant odor. Very uncommon. Probably on serpentine; ± 150 m. SnFrB (Hillsborough, San Mateo Co).

F. brandegei Eastw. (p. 1197) GREENHORN FRITILLARY Bulb; large scales 8–12; small scales 60–200+. **ST** 4–10 dm. **LVS** in 1–2 whorls of 4–8 below, alternate above, 4–11 cm, lanceolate. **FL** nodding; perianth parts 1.2–2 cm, oblong-lanceolate, pink to purplish; nectary 1/3 perianth length, lanceolate, green with reddish edges; style entire. **FR** winged. 2*n*=24. RARE. Granitic soils, open forests; 1500–2100 m. s SN (esp Greenhorn Mtns), Teh.

F. eastwoodiae MacFarl. (p. 1197) BUTTE COUNTY FRITILLARY Bulb; large scales 2–5; small scales 10–60. **ST** 2–8 dm. **LVS** in 1–2 whorls of 3–5 below, alternate above, 5–10 cm, linear to narrowly lanceolate, ± glaucous. **FL** nodding; perianth parts 1–1.7 cm, flared to barely recurved at tips, narrowly elliptic, pale greenish yellow to ± red; nectary < 1/3 perianth length, lanceolate, green, gold, or yellow; style divided < 1/2, branches barely recurved. **FR** angled. 2*n*=24,34,36. UNCOMMON. Dry benches and slopes; gen 500–1500 m. CaR (Shasta, Tehama, Butte cos.). [*F. phaeanthera* Eastw., not Purdy] [MacFarlane 1978 Madroño 25:93–100] ❀DRN,DRY, SHD:**3**,6,14–17;DFCLT.

F. falcata (Jepson) D. Beetle (p. 1197) TALUS FRITILLARY Bulb; large scales 2–4; small scales 8–32. **ST** 0.7–2 dm. **LVS** 2–6, alternate, ± fleshy near st, 3.5–8.5 cm, widely linear, folded, sickle-shaped. **FL** erect; perianth parts 1.5–2.2 cm, obovate, greenish outside, mottled rusty-brown and yellow inside; nectary indistinct, 1/2–2/3 perianth length, widely lanceolate to ± diamond-shaped, yellowish green spotted brownish; style divided ± 2/3. **FR** acutely angled. 2*n*=24. RARE. Serpentine talus; 300–1200 m. SnFrB, SCoRI (San Benito Co.) In cult.

F. glauca E. Greene (p. 1197) Bulb; large scales 3–9; small scales 1–9. **ST** 0.8–2 dm. **LVS** 2–4, alternate, 3.5–9 cm, lanceolate-oblong, sickle-shaped, glaucous. **FL** nodding; perianth parts 1.5–2.5 cm, lance-oblong, purplish or greenish marked with yellow; nectary < 1/2 perianth length, widely lanceolate, green with maroon dots; style divided 1/2. **FR** widely winged. 2*n*=24. Uncommon. Talus slopes, on serpentine; 600–2100 m. NW; s OR. ❀DRN, DRY,SUN:1–3,14–15;DFCLT.

F. liliacea Lindley (p. 1197) FRAGRANT FRITILLARY Bulb; large scales 2–7; small scales 1–2. **ST** 1–3.5 dm. **LVS** 2–20, alternate, 3.5–12 cm, linear to ovate. **FL** nodding, odorless or sweet-scented; perianth parts 1–1.6 cm, white, striped green; nectary 1/2–2/3 perianth length, narrowly linear, purplish to greenish; style divided 1/2. **FR** obtusely angled. 2*n*=24. RARE. Heavy soil, open hills and fields near coast; gen < 200 m. CW. Sometimes confused with *F. agrestis*, which has an obnoxious odor. In cult.

F. micrantha A.A. Heller (p. 1197) BROWN BELLS Bulb; large scales 1–4; small scales 12–60. **ST** 4–9 dm. **LVS** in 1–3 whorls of 4–6 below, alternate above, 4.5–15 cm, linear to narrowly lanceolate. **FL** nodding; perianth parts 1–2 cm, purplish or greenish white, sometimes faintly mottled; nectary > 1/3 perianth length, narrowly lanceolate, greenish white dotted purple; style divided 1/3–2/3, branches strongly recurved. **FR** widely winged. 2*n*=24. Common. Dry benches, slopes; 300–1800 m. SN. ❀DRN,DRY,SHD:3,14–17;DFCLT.

F. ojaiensis David (p. 1197) OJAI FRITILLARY Bulb; large scales 3–5; small scales 1–3. **ST** 4–7 dm. **LVS** in 1–3 whorls of 3–5 below, alternate or opposite above, 4–13 cm, linear to narrowly lanceolate. **FL** nodding; perianth parts 1.5–3 cm, widely lanceolate, dull greenish yellow below with scattered to profuse dark dots; nectary distinct or not, 1/3 perianth length, ± diamond-shaped to ovate, paler than rest of perianth part; style divided 1/2–2/3. **FR** winged. 2*n*=24. RARE. Rocky slopes, river basins; 300–500 m. SCoRO (San Luis Obispo, Santa Barbara cos.), WTR (Ventura Co.). Closely related to *F. affinis*.

F. pinetorum A. Davids (p. 1201) Bulb; large scales 2–5; small scales 45–50. **ST** 1–4 dm, ± glaucous. **LVS** 4–20, 2–3 per node below, alternate above, 5–15 cm, glaucous, linear; upper cauline lvs ± 1/3–1/2 lowest cauline lf, lvs often = or > infl. **FL** ± erect; perianth parts 1.4–1.9 cm, purplish, mottled greenish yellow; nectary indistinct, 2/3 perianth length, widely ovate, yellow, dotted brown; styles divided to near base. **FR** angled. 2*n*=26. Uncommon. ± shaded granitic slopes; 1800–3200 m. c SNH (Mono Co.), Teh, TR. Easily confused with *F. atropurpurea*, which has upper lvs ± = lowest cauline lf.

F. pluriflora Benth. (p. 1201) ADOBE-LILY Bulb; large scales 1–12; small scales 0–2. **ST** 1.5–4.5 dm. **LVS** 3–10, alternate, clustered near ground, 6–15 cm, elliptic to obovate-oblong. **FL** nodding; perianth parts 2–3.5 cm, obovate, pinkish purple, tips rounded to acute; nectary 2/3 perianth length, narrowly linear, lavender; style entire. **FR** obtusely angled. 2*n*=24. RARE. Adobe soil of interior foothills; < 500 m. NCoRI, n SNF, edges of ScV; s OR. In cult.

F. pudica (Pursh) Sprengel (p. 1201) YELLOW FRITILLARY Bulb; large scales 4–5; small scales 85–125. **ST** 0.7–3 dm. **LVS** 2–8, alternate, 3–20 cm, linear to lanceolate. **FL** nodding; perianth parts 0.8–2.2 cm, yellow to orange, some lined brown, aging brick-red; nectary near base of perianth part, elliptic to round, green; style entire. **FR** angled. 2*n*=24,26. Common. Grassy, shrubby, or wooded slopes; < 2100 m. KR, CaR, n SN, MP; to B.C., MT, WY. Highly variable. ❀DRY,SUN:1–3,7,14,15;DFCLT.

F. purdyi Eastw. (p. 1201) PURDY'S FRITILLARY Bulb; large scales 2–8; small scales 0–3. **ST** 1–4 dm. **LVS** 2–10, alternate, ± crowded near ground, 2.5–10 cm, ovate. **FL** horizontal or nodding; perianth parts 1.5–3 cm, white with purple spots or lines and pink shading, tips often recurved; nectary obscure, ± = perianth length, widely linear, colored like perianth; style divided 1/2. **FR** acutely angled. 2*n*=24. UNCOMMON. Dry ridges, gen on serpentine; 400–2100 m. NW. ❀DRY:14–17;DFCLT.

F. recurva Benth. (p. 1201) SCARLET FRITILLARY Bulb; large scales 4–6; small scales 20–30. **ST** 3–9 dm. **LVS** in 1–3 whorls of 2–5 below, alternate above, 3–15 cm, linear to narrowly lanceolate. **FL** nodding; perianth parts 1.5–3.7 cm, scarlet, checkered with yellow inside and purple outside, tips gen recurved; nectary 1/5 perianth part, narrowly lanceolate, yellow; style divided 1/4–1/2. **FR** winged. 2*n*=24,36. Common. Dry hillsides in shrubland or woodlands; 300–2200 m. NW, CaR, SN; w NV. [var. *coccinea* E. Greene] May hybridize with *F. affinis*. ❀DRN,DRY,SHD:3–7, 14–17;DFCLT.

F. striata Eastw. (p. 1201) STRIPED ADOBE-LILY Bulb; large scales 2–7; small scales 0–1. **ST** 2.5–3.8 dm. **LVS** 3–10, alternate, 6–7 cm, oblong-ovate, ± glaucous. **FLS** nodding, fragrant; perianth parts 2–3.5 cm, oblanceolate, white to pink, often striped red, tips acute to short-tapered, gen recurved; nectary at base of perianth part, linear, green, outlined with lavender; style barely divided. **FR** angled. 2*n*=24. **THREATENED** CA. Adobe soil; < 1000 m. s SN (Tulare, Kern cos., esp Greenhorn Mtns).

F. viridea Kellogg (p. 1201) SAN BENITO FRITILLARY Bulb; large scales 3–5; small scales 2–4. **ST** 3–6.5 dm. **LVS** in 1–2 whorls of 3–4 below, alternate above, 4–10 cm, narrowly lanceolate. **FL** nodding; perianth parts 0.9–1.8 cm, lanceolate, pale green to almost black; nectary ± 1/2 perianth length, lanceolate, green; style divided 1/2. **FR** winged. RARE. Serpentine slopes; 200–1500 m. SCoR (San Benito, San Luis Obispo cos.). Closely related to *F. affinis*.

HASTINGSIA

Per; bulb ovoid to ± elongate, outer coats black. **ST**: 25–90 cm, slender, ± scapose. **LVS** basal, 19–55 mm, 4–14 mm wide, grass-like, keeled, ± glaucous. **INFL**: raceme or panicle, terminal, bracted; fls 20–70+. **FL**: perianth segments 6 in 2 petal-like whorls, free; stamens 6, epipetalous; ovary superior, chambers 3, style 1, stigma 3-lobed. **FR**: capsule, short-stalked, loculicidal. **SEEDS** ovoid, flat, black. 4 spp.: CA, OR. (Namesake unknown) [Sherman & Becking 1991 Madroño 38:130–138] Segregated from *Schoenolirion* of se US.

1. Perianth parts ± spreading in outer 1/2; stamens incl or barely exserted .... ***H. alba***
1′ Perianth parts reflexed > 2/3 their length; stamens exserted .... ***H. serpentinicola***

H. alba (Durand) S. Watson (p. 1201) Bulb 26–56 mm, 17–31 mm wide. **ST** 40–90 cm. **LF** 20–50 cm. **INFL** dense; branches gen 2–3. **FL** 6–8 mm; perianth parts elongating as anthers mature, ± spreading above middle, equal, white to yellowish, outer parts ± 1 mm wide, linear, blunt, inner parts ± 2 mm wide, ovate, acute; stamens ± = perianth (outer 3 at first > perianth, opening before inner 3). **FR** 6–9 mm, oblong. *n*=26. Wet meadows, bogs, rocky seeps; 500–2300 m. NW, CaR, n SNH; sw OR. [*Schoenolirion a.* Durand] ❀IRR to WET,SUN:1,2,**4–6**,7,14,**15,16**,17.

H. serpentinicola R. Becking (p. 1201) Bulb 23–40 mm, 14–21 mm wide. **ST** 28–52 cm. **LF** 19–35 cm. **INFL** open; branches gen 0(–3). **FL** 5–6 mm; perianth parts 1–2 mm, linear, reflexed > 2/3 their length, greenish to yellowish, midvein greenish yellow to light purple; stamens ± equal, exserted. **FR** 5–8 mm, oblong. *n*=26, 27?. Well drained, exposed serpentine; 1800–2200 m. n&c NW; sw OR. ❀DRN,SUN,IRR:1,2,**4–7**,14,**15,16**.

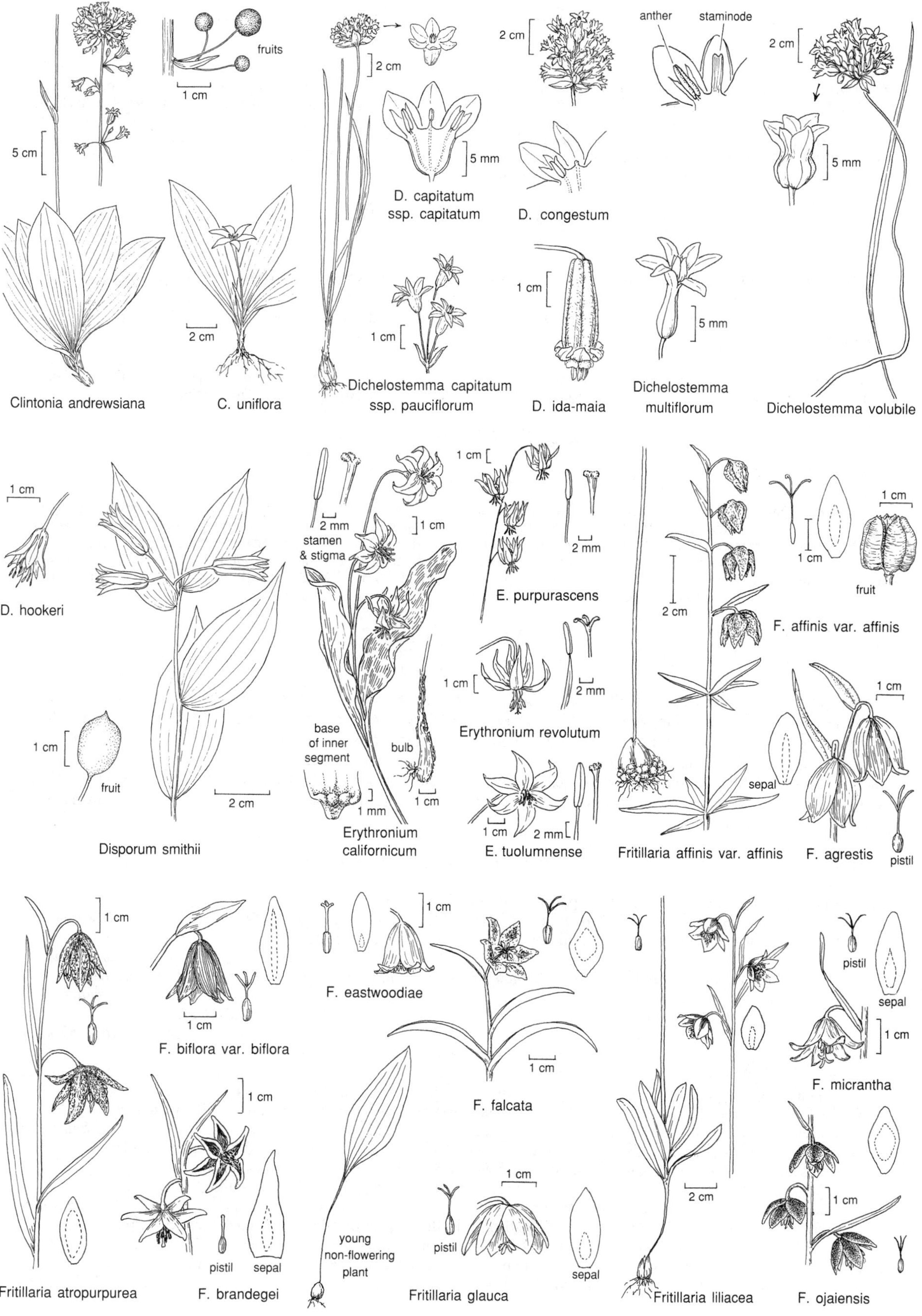

fruits
1 cm
5 cm
2 cm
Clintonia andrewsiana
C. uniflora
2 cm
5 mm
D. capitatum ssp. capitatum
1 cm
Dichelostemma capitatum ssp. pauciflorum
2 cm
D. congestum
anther
staminode
1 cm
D. ida-maia
5 mm
Dichelostemma multiflorum
2 cm
5 mm
Dichelostemma volubile
1 cm
D. hookeri
1 cm
fruit
2 cm
Disporum smithii
2 mm
stamen & stigma
1 cm
base of inner segment
1 mm
bulb
1 cm
Erythronium californicum
1 cm
2 mm
E. purpurascens
1 cm
2 mm
Erythronium revolutum
1 cm
2 mm
E. tuolumnense
2 cm
Fritillaria affinis var. affinis
1 cm
1 cm
fruit
F. affinis var. affinis
1 cm
sepal
F. agrestis
pistil
1 cm
Fritillaria atropurpurea
1 cm
F. biflora var. biflora
1 cm
pistil
sepal
F. brandegei
1 cm
F. eastwoodiae
1 cm
F. falcata
young non-flowering plant
1 cm
pistil
sepal
Fritillaria glauca
2 cm
Fritillaria liliacea
pistil
sepal
1 cm
F. micrantha
1 cm
F. ojaiensis

HESPEROCALLIS DESERT LILY

1 sp. (Greek: western beauty)

H. undulata A. Gray (p. 1201) Per; bulb 4–6 cm, ovoid, deep. **ST** 30–180 cm, gen simple, ± scapose. **LVS** mostly basal (cauline reduced), 20–50 cm, 8–15 mm wide, blue-green; margins wavy, white. **INFL**: raceme; bracts conspicuous, ± ovate, papery; pedicels ± 1 cm. **FL** jointed to pedicel, 4.5–6 cm, funnel-shaped; perianth parts 6, fused below, petal-like, white with silver-green midstripe, lobes 3–4 cm, ± oblanceolate, spreading; stamens 6, attached to perianth; ovary superior, chambers 3, style slender, stigma slightly 3-lobed. **FR**: capsule, 12–16 mm, 3-lobed, loculicidal. **SEED** ± 5 mm, flat, black. *n*=24. Sandy flats; < 800 m. D; w AZ. ❀DRN, sand,SUN:12,13;DFCLT.

IPHEION STAR FLOWER

Glenn Keator

Per from deep, membrane-coated bulb. **LVS** several, basal, unkeeled, with onion odor. **INFL** scapose; fl solitary; bracts 2, fused at base. **FL** showy; perianth tube short, lobes 6 in 2 petal-like whorls, widely spreading, ± equal; stamens 6, anthers attached at middle; ovary superior, style thread-like, stigma barely 3-lobed. **FR**: capsule. ± 6 spp.: s S.Am. (Greek: origin obscure)

I. uniflorum Raf. **LF** < 30 cm, nearly flat, glaucous. **INFL** < 20 cm. **FL**: perianth white flushed bluish purple; perianth lobes 3–3.5 cm wide. Uncommon. Disturbed places near gardens; gen < 300 m. SW; native to Argentina. [*Brodiaea u.* (Raf.) Engl.] Perhaps a waif.

LEUCOCRINUM SAND LILY

1 sp. (Greek: white lily)

L. montanum Nutt. (p. 1201) Per from short, often deep caudex; roots fleshy. **ST** ± 0. **LVS** basal, 10–20 cm, linear, spreading, base sheathed by membranous bracts. **INFL**: pedicels underground. **FL** 5–10 cm; perianth ± salverform, parts 6, fused into slender tube, petal-like, white, showy, persistent, lobes 2–2.5 cm, ± oblong, spreading; stamens 6, attached near top of perianth tube, anthers 4–6 mm, linear; ovary superior, underground, chambers 3, style 1, persistent, stigma slightly 3-lobed. **FR**: capsule, 5–8 mm, 3-angled, ± wrinkled, loculicidal. **SEED** 3–4 mm, angled, black. 2*n*=28. Sandy flats, sagebrush scrub, juniper woodland, montane forest; 1000–1500 m. KR, CaRH, MP; to OR, NE, UT. ❀DFCLT.

LILIUM LILY

Mark W. Skinner

Per from bulb-like, scaly rhizomes (called bulbs), gen not clonal, ± glabrous. **ST** erect. **LVS** ± whorled (often some scattered), sessile, gen ± elliptic; veins gen 3; stipule 0. **INFL**: fls axillary, 1–40+; bracts gen 2 per fl. **FL** gen radial, gen bell- or trumpet-shaped; perianth segments 6 in 2 petal-like whorls, outer gen ± clawed, gen red-purple-spotted on inner base; stamens 6, anthers attached near middle (measures are after dehiscence); style 1, stigma 3-lobed. **FR**: capsule, erect, gen ± smooth, loculicidal. **SEEDS** many, flat, in 6 stacks. ± 90 spp.: n temp, trop mtns of e Asia. (Greek: lily) [Skinner 1988 PhD Harvard Univ] Variable; hybridization common. Many spp. declining from habitat destruction and collecting; few thrive in gardens. Gen fls May to August.

1. Fl nodding (between horizontal and hanging vertically) to erect; perianth segments gen weakly recurved (exc *L. maritimum*)
  2. Perianth ± white (often aging purple or pink); fls horizontal to erect
    3. Outer perianth segments 4–7 cm; fl ascending to erect; bulb scales unsegmented; NW, SnFrB ..... ***L. rubescens***
    3′ Outer perianth segments 6–12 cm; fl nodding to ascending; bulb scales sometimes crimped or segmented; KR, SN, CaR ..... ***L. washingtonianum***
      4. Perianth aging deep pink or lavender; some bulb scales gen crimped or segmented; perianth 6–10 cm; KR ..... ssp. ***purpurascens***
      4′ Perianth aging ± white; crimped or segmented bulb scales gen 0; perianth 8–12 cm; SN, CaR ..... ssp. ***washingtonianum***
  2′ Perianth yellow to red (rarely pink); fls nodding to ascending
    5. Perianth bright yellow, segments 7–11 cm; TR, PR ..... ***L. parryi***
    5′ Perianth red, orange, or pink (rarely pale off-yellow), segments < 5 cm; NCo, KR, SN
      6. Lf 1.8–7 cm, glaucous, margins gen wavy; longest bulb scales 3–6 cm, unsegmented; bulb ovoid; dry places; KR ..... ***L. bolanderi***
      6′ Lf 3–18 cm, not obviously glaucous, margins gen not wavy; longest bulb scales < 3 cm (exc large pls), gen segmented; bulb horizontal, rhizome-like; ± wet or seasonally inundated places; NCo, SN
        7. Perianth red (rarely red-orange) on both surfaces (exc base green), tips strongly recurved; fl nodding to horizontal; fr 2.4–4 cm; bulb scales 0(2)-segmented; NCo ..... ***L. maritimum***
        7′ Perianth yellowish, yellow-orange, or pinkish (if reddish, only on distal 40% of inner surfaces), tips weakly recurved; fl nodding to ascending; fr 1.5–3 cm; bulb scales gen 2–3-segmented; SN ..... ***L. parvum***
1′ Fl pendent (hanging vertically) (sometimes some nodding); perianth segments strongly recurved
  8. Perianth white to pink (gen aging pink), midribs yellow; KR, n NCoRO ..... ***L. kelloggii***

8′ Perianth yellow to red; widespread

9. Pl of ± dry places; perianth yellow or orange; longest bulb scale > 3 cm, segmented or not; bulb ± ovoid, erect

10. Pistil 2.4–4 cm; perianth segments 3–7 cm; anthers 6–13 mm; fr 1–2 cm diam; NW .... ***L. columbianum***

10′ Pistil 4.5–7 cm; perianth segments 5–10 mm; anthers 11–19 mm; fr 1.8–3.5 cm diam; s CaRH, SNH, SW .... ***L. humboldtii***

11. Perianth uniformly orange, red spots unmargined; bulb scales unsegmented, off-white; s CaRH, SNH .... ssp. ***humboldtii***

11′ Perianth yellow-orange, red spots near tip margined with lighter red; bulb scales crimped or segmented, often purplish; SW, s SCoRO .... ssp. ***ocellatum***

9′ Pl of ± moist places; perianth yellow to reddish; longest bulb scales < 3 cm (exc large pls), gen segmented; bulb horizontal, rhizome-like

12. Perianth ± uniformly yellow or yellow-orange (green at base)

13. Fl mildly fragrant; anthers 3–6 mm, magenta or dull red; pistil 2.5–3.5 cm; fr 1.5–3 cm; not clonal; c&s SNH .... ***L. kelleyanum***

13′ Fl not fragrant; anthers 5–13 mm, yellow; pistil 3–5 cm; fr 2.5–4 cm; weakly clonal; KR .... ***L. pardalinum*** ssp. ***wigginsii***

12′ Perianth ± 2-toned, with green, yellow, or orange centers, darker orange or red tips

14. Perianth red to maroon (rarely orange), center green to yellow, outer surface green on basal 40–50%, segments 4–8 cm; filaments ± parallel; rhizome unbranched; NCo .... ***L. occidentale***

14′ Perianth yellow to orange, tips redder, outer surface green only on basal ± 20%, segments 3–11 cm; filaments spreading; rhizome often branched; CA-FP .... most ***L. pardalinum***

15. Perianth segments 5–11 cm; gen in large clones; anthers 10–17 mm; CA-FP .... ssp. ***pardalinum***

15′ Perianth segments 3.5–9 cm; gen in small clones; anthers 5–14 mm; KR, CaR, n SNH

16. Lf ± linear (length 7–35 × width); KR .... spp. ***vollmeri***

16′ Lf ± elliptic (length 2–13 × width); KR and elsewhere

17. Pollen red- or brown-orange; anthers magenta; bulb scales gen 2-segmented; s NCoRO ... ssp. ***pitkinense***

17′ Pollen gen yellow to bright orange; anthers gen orange (magenta); bulb scales gen 2–4-segmented; KR, CaR, n SNH .... spp. ***shastense***

L. bolanderi S. Watson (p. 1201) BOLANDER'S LILY Pl < 1 m, glaucous; bulb erect, ± ovoid, longest scales 3–6 cm, unsegmented. **LVS** in 2–6 whorls, 1.8-7 cm, gen ± obovate, gen distinctly glaucous; margin gen wavy. **INFL**: fls 1–9, nodding to horizontal. **FL** narrowly bell-shaped, not fragrant; perianth segments 3–5 cm, ± elliptic, red or magenta (rarely salmon or pale yellow), inside basal 30–50% often yellowish, 25% recurved; stamens ± = perianth, filaments ± parallel, anthers 3–8 mm, reddish or magenta, pollen red-brown, orange, or yellow; pistil 2–4 cm, ovary 1–2 cm. **FR** 2–4 cm. $2n$=24. UNCOMMON. Serpentine soil in chaparral, conifer forest, gen with *Xerophyllum*; 150–1600 m. KR; sw OR. Hybridizes with *L. pardalinum, L. washingtonianum* ssp. *purpurascens*, & *L. rubescens*. ❀IRR,DRN,part SHD:4–6,**7**,14–17;DFCLT.

L. columbianum Baker COLUMBIA LILY Pl < 2.5 m, glaucous or not; bulb erect to ascending, oblique or ± ovoid, longest scales 3–7 cm, (0)2–3(4–5)-segmented. **LVS** in 2–9 whorls, gen ascending, 1.5–16 cm, gen ± obovate; margin gen wavy. **INFL**: fls 1–18, pendent to nodding. **FL** widely bell-shaped, not fragrant; perianth segments 3–7 cm, 50% strongly recurved, orange (often reddish outside); stamens > perianth, filaments spreading, anthers 6–13 mm, ± yellow, pollen orange or yellow; pistil 2.4–4 cm. **FR** 2–6 cm, 1–2 cm diam. $2n$=24. Dry scrub, coastal prairie, gaps and roadsides in conifer forest, esp along coast; < 1300 m. NW; to B.C., ID. May intergrade with *L. kelloggii* near OR border. Hybridizes with *L. pardalinum* sspp. *vollmeri* & *wigginsii*, often with *L. occidentale* in OR. ❀SHD,DRN,DRY July to October:3–7,14–19;may be DFCLT.

L. humboldtii Roezl & Leichtlin Pl < 3 m; bulb erect, oblique, or ± ovoid, longest scales 3–12 cm, segmented or not. **ST** sometimes brown-purple. **LVS** in 4–9 whorls, gen ascending, 4–15 cm, gen ± oblanceolate; margin gen wavy. **INFL**: fls 1–40, pendent. **FL** ± widely bell-shaped, not fragrant; perianth segments 5–10 cm, 75% strongly recurved, orange or yellow, inner surface ridged basally; stamens >> perianth, filaments spreading, anthers 11–19 mm, purple, pollen red-brown to tan-yellow; pistil 4.5–7 cm. **FR** 2–6 cm, 1.8–3.5 cm diam, ribbed. Yellow-pine forest; < 1800 m. s CaRH, SNH, SW. Declining.

ssp. ***humboldtii*** HUMBOLDT LILY Bulb scales unsegmented, off-white (sometimes purple-speckled). **FL**: perianth orange, magenta spots unmargined; pollen red- or rusty-brown, rusty-orange. UNCOMMON. Yellow-pine forest; 600–1100 m. s CaRH, SNH. ❀DRN,DRY July to September,SHD:1,2,**7**,14–24;DFCLT.

ssp. ***ocellatum*** (Kellogg) Thorne (p. 1201) OCELLATED LILY Bulb scales obscurely (0)2–5–segmented, often purple at tip. **FL**: perianth yellow or light orange, spots margined lighter red (toward tip larger, with wider margins); pollen ± tan or tan-yellow. UNCOMMON. Yellow-pine forest and openings, oak canyons; < 1800 m. SCoRO, SW. [var. *o.* (Kellogg) Elwes; var. *bloomerianum* (Kellogg) Jepson] ❀DRN,DRY mid summer,SHD:**7**,14–24;DFCLT.

L. kelleyanum Lemmon Pl < 2 m; bulb horizontal, rhizome-like, longest scales 1–3 cm, (0)2–3-segmented. **LVS** in 2–5 whorls, 7–16 cm, gen drooping at tips; margin not wavy. **INFL**: fls 1–25, pendent. **FL** ± widely bell-shaped, mildly fragrant; perianth segments 4–6 cm, 60% strongly recurved, yellow or yellow-orange; stamens >> perianth, filaments spreading, anthers 3–6 mm, magenta or dull red, pollen orange, tan-orange, or red-brown; pistil 2.5–3.5 cm. **FR** 1.5–3 cm. $2n$=24. Hillside seeps, wet thickets, streamsides in subalpine forest; 2200–3000 m. c&s SNH. Intergrades with *L. parvum* in s Mono Co. ❀IRR or WET,DRN,SHD:1–7,14–17;DFCLT.

L. kelloggii Purdy Pl < 2 m; bulb erect, ± ovoid, longest scales 3–6.5 cm, unsegmented. **LVS** in 2–7 whorls, 6–16 cm, gen ± oblanceolate; margin gen not wavy. **INFL**: fls 1–27, pendent. **FL** ± widely bell-shaped, fragrant; perianth segments 3–8 cm, 67% strongly recurved, pink or white (gen aging pink), midrib yellow; stamens >> perianth, filaments spreading, anthers 5–14 mm, pale red-orange or magenta, pollen orange; pistil ± 3–4 cm. **FR** 3–6 cm. $2n$=24. Gaps and roadsides in conifer forest; 175–1300 m. KR, n NCoRO; sw OR. ❀part SHD,DRY July to September,DRN:5,7, 14–17;may be DFCLT.

L. maritimum Kellogg (p. 1201) COAST LILY Pl < 2.5 m (< 0.3 m on coastal bluffs); bulb horizontal, ± rhizome-like, lumpy, longest scales 1.5–4 cm, 0(2)-segmented. **LVS** basal, scattered, or in 1–4 whorls, 3–18 cm, gen ± oblanceolate; margin not wavy. **INFL**: fls 1–13, nodding (rarely horizontal). **FL** bell-shaped, not fragrant; perianth segments 3–5 cm, 33% strongly recurved or rolled, elliptic, red or red-orange, darker spots concentrated mid-basally, surrounded by light orange (or yellow-green); stamens ± = perianth, filaments ± parallel, anthers 4–12 mm, light magenta, pollen orange; pistil ± 2–3 cm. **FR** 2.4–4 cm. $2n$=24. RARE. Coastal prairie or scrub, bogs, gaps in closed-cone-pine forest; < 150 m. s NCo (extirpated in n CCo). Hybridizes with *L. pardalinum*. In cult.

L. occidentale Purdy (p. 1201) WESTERN LILY Pl < 2.5 m; bulb horizontal, rhizome-like, unbranched, longest scales 1–2.5 cm, 1–2(3)-segmented. **LVS** in 1–9 whorls (rarely scattered), 4–19 cm, sometimes ± linear; margin not wavy. **INFL**: fls 1–35, pendent. **FL** ± widely bell-shaped, not fragrant; perianth segments 4–8 cm, 60% strongly recurved, ± oblanceolate or narrowly elliptic, ± 2-toned, inner surface red to maroon (rarely orange) on distal 50–60%, base yellow to green, sometimes with an intervening band of orange or yellow, strongly green outside on basal 40–50%; stamens >> perianth, filaments ± parallel, anthers 5–14 mm, dull red or magenta, pollen gen red-brown (± orange); pistil 3–5 cm. **FR** 2–5 cm. $2n$= 24. **ENDANGERED** CA. Coastal scrub or prairie, gaps in coniferous forest; < 100 m. n NCo (Humboldt Bay, Humboldt Co.); sw OR. Description is of CA pls. Central "green star" of ± all new fls ages yellow or yellow-orange. In cult.

L. pardalinum Kellogg Pl < 3 m, ± clonal; bulb horizontal, rhizome-like, often branched, scales (0)2–4-segmented, longest 10– 33 mm. **LVS** scattered or in 1–8 whorls, 4–27 cm, gen ± elliptic; margin gen not wavy. **INFL**: fls 1–35, pendent. **FL** ± widely bell-shaped, gen not fragrant; perianth segments 3-11 cm, 60–75% strongly recurved, gen ± 2-toned, inner surface gen pale orange to red on distal 25–60%, lighter near base, maroon spots near tip margined yellow or orange; stamens >> perianth, filaments spreading ± widely, anthers 6–17 mm, ± magenta to orange or yellow, becoming darker, pollen red-brown to yellow, becoming lighter; pistil 3–8 cm. **FR** 2–6 cm. $2n$=24. Moist places, streambanks, along coast; < 2000 m. CA-FP. Intergrading complex.

ssp. ***pardalinum*** LEOPARD LILY Strongly clonal; bulb scales (0)2(3)-segmented. **LVS** whorled. **FL** rarely fragrant; perianth segments 5–11 cm, 2-toned, tips darker; anthers 10–17 mm, pale magenta or magenta, becoming purple (or yellowish), pollen red- to brown-orange; pistil 5–8 cm, ovary ± 1–2 cm. **FR** 3–6 cm. $2n$=24. Moist places (drier along coast); < 1700 m. CA-FP. Variable. Hybridizes with *L. parvum, L. maritimum*, others. ± replaced by *L. parryi* in SW mtns. ❀IRR,DRN,SHD:1–6,**7,14–17**,18–24.

ssp. ***pitkinense*** (Beane & Vollmer) M. Skinner PITKIN MARSH LILY Weakly clonal; bulb scales (0)2-segmented. **LVS** whorled. **FL**: perianth segments 5–7 cm, 2-toned, tips darker; anthers 6–11 mm, magenta, pollen red- or brown-orange; pistil 3–5 cm. **ENDANGERED** CA. Marshes, valley-oak scrub; 35–60 m. s NCo-RO (Pitkin Marsh, Sonoma Co.). [*L. p.* Beane & Vollmer] Barely distinct from ssp. *pardalinum*. Threatened by habitat loss, grazing, competition, collecting. In cult.

ssp. ***shastense*** (Eastw.) M. Skinner Weakly clonal; bulb scales (0)2–4-segmented. **LVS** whorled (or scattered in young pls). **FL**: perianth segments 3–8 cm, 2-toned, tips darker; anthers 5–14 mm, orange to magenta, pollen yellow to red-orange; pistil 3–5 cm. **FR** 2–5 cm. Wet meadows, streamsides in coniferous forest; 1000–1800 m. KR, CaR, n SNH. [*L. nevadense* Eastw. var. *s.* Eastw.] Pls in Plumas Co. probably intergrade with ssp. *pardalinum*. May intergrade with ssp. *wigginsii* in KR; needs study. ❀IRR or WET,DRN,SHD or part SUN:1–7,14–17;DFCLT.

ssp. ***vollmeri*** (Eastw.) M. Skinner (p. 1201) VOLLMER'S LILY Weakly clonal; bulb scales 1–2-segmented. **LVS** whorled (or scattered esp in small pls), ± linear. **FL**: perianth segments 5–9 cm, 2-toned, tips darker; anthers 6–14 mm, magenta or purple, pollen red-orange or orange; pistil 3–6 cm. **FR** 2.5–5 cm. UNCOMMON. Bogs, streams, springs; 150–1300 m. KR; sw OR. [*L. v.* Eastw.] Forms hybrid swarms with ssp. *wigginsii* in w Siskiyou Co. Pls in deep shade gen have wider lvs, are much like ssp. *pardalinum*. ❀WET or IRR,DRN,SHD:1,4–7,14–17;DFCLT.

ssp. ***wigginsii*** (Beane & Vollmer) M. Skinner (p. 1201) WIGGIN'S LILY Weakly clonal; bulb scales 2–3(4)-segmented. **LVS** whorled (or scattered esp in small pls). **FL**: perianth segments 3–7 cm, not 2-toned (gen uniformly orange or yellow-orange); stamens often malformed or shrunken, anthers 5–13 mm, pale yellow, pollen yellow or orange; pistil 3–5 cm. **FR** ± 2–4 cm. UNCOMMON. Wet thickets, meadows, streams among conifers; 800–2000 m. KR; sw OR. [*L. w.* Beane & Vollmer] ❀WET or IRR,SHD,DRN: 1,4–7,14–17;DFCLT.

L. parryi S. Watson (p. 1205) LEMON LILY Pl < 2.5 m; bulb horizontal, rhizome-like, scales (0)2(4)-segmented, longest 9–37 mm. **LVS** in 1–8 whorls (or scattered in young pls), 7–29 cm, sometimes narrowly linear; margin not wavy. **INFL**: fls 1–31, horizontal or ± nodding. **FL** gen slightly bilateral, trumpet-shaped, strongly fragrant; perianth segments 7–11 cm, 40% recurved (outer < inner), ± oblanceolate, bright yellow, maroon spots gen sparse, minute; stamens ± = perianth, filaments ± parallel, anthers 8–14 mm, pale magenta-brown, pollen rusty- or brown-orange; pistil 5–10 cm. **FR** 4–6 cm. $2n$=24. RARE. Meadows, streams in montane coniferous forest; 1300–2600 m. TR, PR. In cult.

L. parvum Kellogg ALPINE LILY Pl < 2 m; bulb horizontal, rhizome-like, longest scales 1–3.5 cm, (0)2–3(4)-segmented. **LVS** in 2–6 whorls, 4–15 cm; margin not wavy. **INFL**: fls 1–41, nodding to ascending. **FL** sometimes slightly bilateral, ± trumpet-shaped, not fragrant; perianth segments 3–4 cm, weakly recurved (outer sometimes < inner), oblanceolate, inner surface light orange to red (purplish) on distal 25–50%, lighter near base (sometimes uniformly light orange); stamens ± > perianth, filaments ± spreading, anthers 3–8 mm, pale yellow, orangish, or magenta, pollen yellowish to red-orange; pistil 2–4 cm. **FR** 1.5–3 cm. $2n$=24. Wet meadows, willow thickets, streams in coniferous forest; 1400–2900 m. n&c SNH. [var. *luteum* Purdy] ❀SHD,IRR or WET,DRN:1–7,14–17;DFCLT.

L. rubescens S. Watson (p. 1205) REDWOOD LILY Pl < 2 m, often glaucous; bulb erect, ± ovoid, longest scales 4–9 cm, unsegmented. **LVS** in 3–13 whorls, gen ± ascending, 3–13 cm, gen oblanceolate; margin gen wavy. **INFL**: fls 1–40, ascending to erect. **FL** trumpet-shaped, fragrant; perianth segments 4–7 cm, gen weakly recurved, strongly oblanceolate (inner segments wider), inner surface white becoming pink-purple, magenta spots minute, outer surface often reddish or purplish; stamens = perianth, filaments ± parallel exc distally, anthers 4–8 mm, pale yellow, pollen yellow; pistil 2–4 cm, ovary 1–2 cm. **FR** 2–4 cm, gen ribbed. $2n$=24. UNCOMMON. Dry soils in chaparral, gaps in conifer forest; 30–1500 m. NW, SnFrB. Has unique chromosomes for the genus. ❀DRN, DRY,part SHD:7,14–17;DFCLT.

L. washingtonianum Kellogg WASHINGTON LILY Pl < 3 m, often glaucous; bulb ± horizontal to erect, ± ovoid, oblique, or ± rhizome-like, scales 0–2-segmented, longest 3–12 cm. **LVS** in 1–9 whorls, horizontal to nearly clasping st, 3–13 cm, gen oblanceolate; margin wavy or not. **INFL**: fls 1–26, nodding to ascending. **FL** gen slightly bilateral, ± trumpet-shaped, strongly fragrant; perianth segments 6–12 cm, gen weakly recurved (outer < inner), strongly oblanceolate (inner wider), inner surface white or becoming deep pink, magenta spots minute; stamens < or > perianth, filaments ± parallel, anthers 8–15 mm, off-white or cream, pollen yellow or cream; pistil 2–11 cm. **FR** 2.5–6 cm. $2n$=24. Conifer forest, esp gaps, burned clearcuts; 400–2200 m. KR, CaR, SN; OR.

ssp. ***purpurascens*** (Stearn) M. Skinner (p. 1205) PURPLE-FLOWERED SHASTA LILY Bulb ascending to erect; most scales indistinctly 2(3)-segmented. **FL**: perianth segments 6–10 cm, 33–50% recurved, becoming deep pink, gen purplish (often faintly so) outside, yellow midstripe gen 0; anthers cream, becoming yellow, pollen pale (bright) yellow. **FR** gen ribbed. UNCOMMON. Douglas-fir forest; 400–1800 m. KR; OR. [var. *p.* Stearn] ❀DRN,SHD, DRY in summer:1–7,14–17;DFCLT.

ssp. ***washingtonianum*** Bulb ± horizontal or ascending; most scales unsegmented. **FL**: perianth segments 8–12 cm, 25–33% recurved, rarely becoming pink, gen white outside, with yellow midstripe in basal 50%; anthers off-white or cream, becoming pale pink or dirty yellow, pollen very pale yellow or cream. **FR** sometimes ribbed. Coniferous forest; 1200–2200 m. CaR, SN. [var. *minus* Purdy] ❀DRN,IRR,SHD:1,2,6,7,14–17;DFCLT.

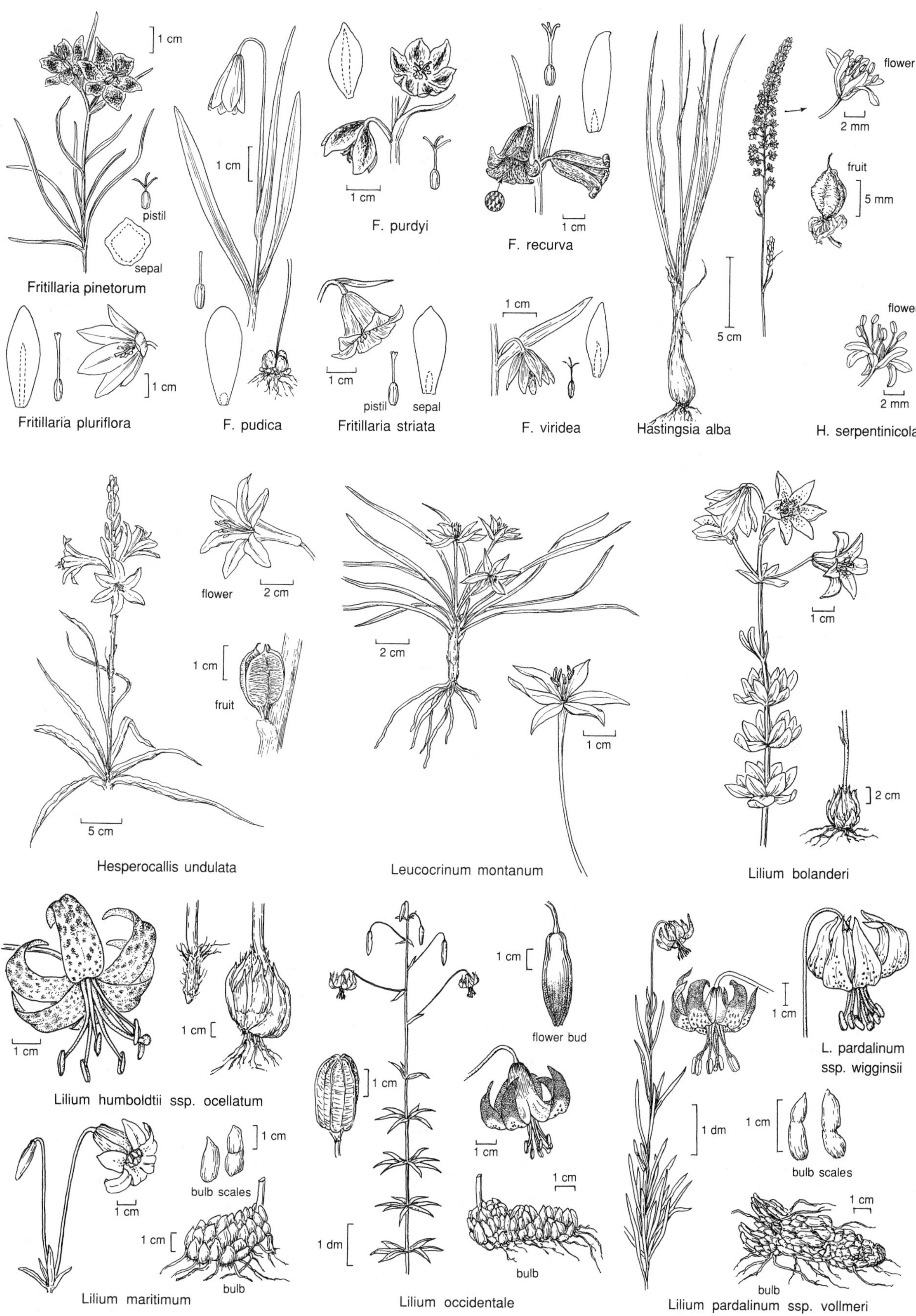
1 cm
pistil
sepal
Fritillaria pinetorum
1 cm
1 cm
F. purdyi
1 cm
F. recurva
flower
2 mm
fruit
5 mm
5 cm
flower
2 mm
1 cm
Fritillaria pluriflora
F. pudica
1 cm
pistil
sepal
Fritillaria striata
1 cm
F. viridea
Hastingsia alba
H. serpentinicola
flower
2 cm
1 cm
fruit
5 cm
Hesperocallis undulata
2 cm
1 cm
Leucocrinum montanum
1 cm
2 cm
Lilium bolanderi
1 cm
1 cm
Lilium humboldtii ssp. ocellatum
1 cm
flower bud
1 cm
1 cm
1 dm
1 cm
bulb
Lilium occidentale
1 cm
L. pardalinum
ssp. wigginsii
1 dm
1 cm
bulb scales
1 cm
bulb
Lilium pardalinum ssp. vollmeri
1 cm
bulb scales
1 cm
1 cm
bulb
Lilium maritimum

MAIANTHEMUM FALSE LILY-OF-THE-VALLEY

Per; rhizome slender, creeping. **ST** erect, simple; base sheathed by 2 membranous scales. **LVS** cauline, alternate, gen 2 (if 3, upper ± bract-like). **INFL**: raceme. **FL** small; perianth parts 4 in 2 petal-like pairs, white; stamens 4, attached to perianth; ovary superior, chambers 2, style 1, stigma 2-lobed. **FR**: berry. **SEEDS** 1–3. 3 spp.: n temp. (Greek: May fl, from fl season)

M. dilatatum (Alph. Wood) Nelson & J.F. Macbr. (p. 1205) **ST** 10–40 cm. **LF** 5–20 cm, 5–10 cm wide, cordate to sagittate; petioles 5–15 cm. **INFL** 2–8 cm. **FL** ± 2.5 mm; stamens < perianth. **FR** ± 6 mm, spheric, red. **SEED** ± 3 mm, brown. *n*=18. Moist, shady, coniferous forest; < 500 m. w NW; to AK, ID. ❀**4,5,17**&SHD:**6**&IRR:7,14,**15,16**,24;deciduous GRNCVR;may be INV.

MUILLA

Glenn Keator

Per from small, fibrous-coated corm. **LVS** few–several, basal, narrow, gen ± cylindric. **INFL** umbel-like; axis stiff, straight, cylindric; bracts gen 3, papery, acuminate; pedicels slender, unjointed, erect. **FL**: perianth segments 6 in 2 petal-like whorls, barely fused at base, lobes gen equal, lanceolate or oblong, widely spreading, midribs 2-veined; stamens 6, filaments thread-like to winged, anthers attached at middle; style short, club-shaped, persistent, stigma 3-lobed. **FR**: capsule, subspheric, 3-angled, loculicidal. **SEEDS** irregularly angled, black. ± 6 spp.: sw US, n Mex. (Anagram of *Allium*, from superficial resemblance) [Shevock 1984 Aliso 10:621–627]

1. Perianth yellow .... ***M. clevelandii***
1′ Perianth white, tinged greenish to lilac
  2. Filaments thread-like or slightly wider at base .... ***M. maritima***
  2′ Filaments much wider at base, forming a crown
    3. Infl 3–15 cm; perianth 3–4 mm .... ***M. coronata***
    3′ Infl 15–50 cm; perianth 6–8 mm .... ***M. transmontana***

M. clevelandii (S. Watson) Hoover (p. 1205) SAN DIEGO GOLDENSTAR **LF** 6–15 cm. **INFL** < 30 cm; pedicels 20–35 mm; fls 20–30. **FL**: perianth lobes 6–10 mm, yellow, striped green; filaments 3–5 mm, appendages 3 mm, entire, anthers white or pale yellow. **FR** 4–5 mm. *n*=7. RARE. Mesa grasslands, scrub edges; < 50 m. s SCo (sw San Diego Co.). [*Bloomeria c.* S. Watson] Threatened by urbanization, vehicles. In cult.

M. coronata E. Greene (p. 1205) CROWNED MUILLA **LF** < 18 cm. **INFL** 3–15 cm; pedicels 10–30 mm; fls 3–10. **FL**: perianth lobes 3–4 mm, greenish outside, white or faded blue inside; stamens 2–4 mm, filaments petal-like, ± translucent, much wider at base and fused into cylindric crown, anthers yellow. **FR** 3–7 mm. *n*=7. UNCOMMON. Open desert scrub, woodland, in heavy soils; 1000–1600 m. s SNH (e slope), SNE, n&w DMoj; w NV.

M. maritima (Torrey) S. Watson (p. 1205) COMMON MUILLA **LF** 10–60 cm. **INFL** 10–50 cm; pedicels 10–50 mm; fls 4–20. **FL**: perianth lobes 3–6 mm (inner slightly wider), greenish white, striped brown; stamens 2–5 mm, filaments thread-like or slightly wider at base, anthers blue, green, or purple. **FR** 5–8 mm. *n*=7,8, 10. Grassland, open scrub, woodland, in alkaline, granitic, or serpentine soils; 0–2300 m. c&s NW, CW, SW, uncommon in c SNF, GV, w D; Baja CA. ❀DRN,DRY,SUN:3,7,14–14;may be DFCLT.

M. transmontana E. Greene **LF** 20–60 cm. **INFL** 15–50 cm; pedicels 2–3 cm; fls 12–30. **FL**: perianth lobes 6–8 mm, white aging lilac; stamens 4–6 mm, filaments wider and fused at base into cup, anthers yellow. **FR** 8–10 mm. *n*=10. High desert scrub, open forest; 1200–2300 m. GB; w NV. ❀DFCLT.

MUSCARI GRAPE HYACINTH

Per; bulb ovoid to ± spheric. **ST** scapose, cylindric. **LVS** 2–7, basal, linear. **INFL** spike-like, unbracted; terminal fls often sterile and modified. **FL**: fertile fls 2–12 mm, spheric to bell-shaped; perianth segments 6, petal-like, fused below, blue, purple, brownish, or yellow, lobes minute, often different color from tube; stamens attached to perianth, incl; ovary superior, chambers 3, style 1, stigma head-like. **FR**: capsule, sharply angled or winged, loculicidal. **SEEDS** 6, angled, black. 50 spp.: Eur, Medit, w Asia. (Latin: from musky odor of some spp.)

M. botryoides (L.) Miller **ST**: scape 5–30 cm. **LVS** 2–3(4), 10–25 cm, linear, ± channeled, abruptly narrower at tip. **INFL** dense, elongating in fr; pedicels 2–4 mm. **FLS**: fertile fls nodding, 3–5 mm, spheric; perianth narrowed below lobes, blue, lobes white; upper sterile fls few, smaller, paler. **FR** 4–6 mm, spheric. 2*n*=18, 36,48,63. Disturbed urban areas; gen < 200 m. CCo (expected elsewhere); native to s Eur.

NARTHECIUM BOG ASPHODEL

Per; rhizome creeping. **ST** erect, ± scapose, 10–70 cm, simple. **LVS** mostly basal (astride rhizome), linear, grass-like, densely overlapping; upper reduced. **INFL**: raceme; pedicels with bractlet near middle. **FL** small; perianth parts 6, petal-like, free, greenish yellow; stamens 6, filaments woolly; ovary superior, chambers 3, style ± 0, stigma slightly 3-lobed. **FR**: capsule, loculicidal. **SEEDS** many, bristle-tailed on both ends, brown. 4 spp.: n temp. (Greek: ancient name)

N. californicum Baker (p. 1205) **ST** 20–60 cm. **LF** 10–30 cm. **INFL** 8–15 cm; pedicels 5–12 mm. **FL** 5–10 mm; perianth parts linear-lanceolate; anthers 3–4 mm. **FR** ± 2 × perianth, lanceolate. **SEED** 6–9 mm (incl bristle-tails). 2*n*=26. Wet meadows, streambanks; 700–2600 m. KR, NCoRO, CaRH, n&c SNH. ❀IRR or WET,DRN:1,4–6&SHD:2,7,15–17;DFCLT.

NOLINA BEARGRASS

James C. Dice

Tree-like subshrubs, dioecious or some fls bisexual. **STS** thick and woody or ± underground. **LVS** densely rosetted, 6–20 dm, stiff, sword-like; bases much expanded, white, fleshy. **INFL**: panicle or raceme-like, scapose, bracted, < 4 m. **FL**: perianth parts 6 in 2 petal-like whorls, ± white, < 6 mm; stamens 6, filaments slender; ovary superior, 3-chambered, style and 3 stigmas short, ovules 2 per chamber. **FR**: capsule, papery. **SEEDS** 1–3 per fr, ovoid. ± 25 spp.: s US, Mex. (P.C. Nolin, 18th century French agriculturist) [Munz & Roos 1950 Aliso 2:217–238]

1. Old lf margins fibrous-shredding; bracts deciduous; seeds gray — s DMoj, DSon .... *N. bigelovii*
1′ Lf margins minutely and persistently serrate; bracts persistent; seeds reddish brown
 2. Aboveground st ± 0; mature rosettes with < 45 lvs; lf bases barely expanded; infl < 1.6 m; bracts ± small — sw PR .... *N. interrata*
 2′ Aboveground st obvious; mature rosettes with 45–200 lvs; lf bases strongly expanded; infl 1.6–4 m; bracts large, papery .... *N. parryi*

N. bigelovii (Torrey) S. Watson (p. 1205) **ST** 10–25 dm, simple to several-branched aboveground. **LVS** 50–150 per rosette, 15–45 mm wide just above expanded base, ± glaucous; base 7–15 cm, 5–11 cm wide; margin minutely serrate when young, shredding-fibrous when mature. **INFL** 15–35 dm; axis 20–40 mm diam at base; bracts ± lanceolate, gen early deciduous. **SEEDS** 2.5–3.5 mm, grayish white. *n*=19. Rocky slopes and ridges; 300–1500 m. DMtns, DSon; s NV, w AZ, nw Mex. Scattered and local. ❀DRN,SUN:2,3,**7,10,14**,15,16,**18–24**&IRR:**8,9**,11–13.

N. interrata H. Gentry (p. 1205) DEHESA NOLINA **ST** branching underground, forming a platform of rosettes. **LVS** 10–45 per rosette, 12–30 mm wide just above expanded base, gen glaucous; base 5–10 cm, 2–6 cm wide; margins minutely serrate. **INFL** 5–16 dm; axis 5–16 mm diam at base; bracts gen inconspicuous and persistent. **SEEDS** 4–6 mm, reddish brown. **ENDANGERED** CA. Foothills on gabbro soils; 200–700 m. sw PR (sw San Diego Co.); nw Baja CA.

N. parryi S. Watson (p. 1205) **ST** < 20 dm, simple to several-branched aboveground (sometimes belowground). **LVS** 40–200 per rosette, 12–40 mm wide just above expanded base, gen green; base 4–21 cm, 3–15 cm wide; margins minutely serrate. **INFL** 16–38 dm; axis 17–80 mm diam at base; bracts large, papery, persistent. **SEEDS** 3–5 mm, reddish brown. *n*=19,20. Dry slopes and ridges; 900–1200 m. s SNH (Kern Plateau), WTR, e SnBr, PR, DMtns, DSon. [ssp. *wolfii* Munz] Scattered, local. ± coastal pls from WTR (Ventura Co.) and n&w PR that are smaller throughout (exc seeds) with *n*=19 are an undescribed sp. ❀SUN,DRN:1–3,**7,12,14–16**, 17,**18–24**&IRR:**8,9,11,13**.

NOTHOSCORDUM FALSE GARLIC

Per; bulb sometimes with basal bulblets; outer coats membranous, brown. **ST** scapose, cylindric. **LVS** basal, linear, flat or channeled, withering from tip before fl; bases sheathing. **INFL**: umbel; bracts conspicuous, gen 2, ± fused, scarious. **FL**: perianth parts 6, petal-like, ± spreading; stamens 6, attached to perianth, filaments wider at base; ovary superior, chambers 3, style 1, stigma head-like. **FR**: capsule, loculicidal. **SEED** angled or flat, black. ± 35 spp.: Am. Like *Allium*, but onion odor 0. (Greek: origin unknown)

N. inodorum (Aiton) Nicholson Bulb ± 1.5 cm. **LVS** 2–several, 25–30 cm, 4–10 mm wide. **ST** 20–40 cm. **INFL** < 4 cm diam; bracts < umbel, persistent; pedicels erect to spreading, unequal. **FL** 8–14 mm; perianth white with greenish bases, reddish midveins, lobes oblanceolate, obtuse; stamens incl; style persistent in fr. **FR** 6–7 mm, obovoid. Disturbed areas; < 100 m. GV, CW, SCo; native to S.Am. NOXIOUS WEED.

ODONTOSTOMUM

1 sp. (Greek: tooth mouth, from staminodes)

O. hartwegii Torrey (p. 1205) Per; corm 2.5–3 cm, ovoid, deep. **ST** ± scapose, 12–50 cm, ± curved, gen branched. **LVS** mostly basal, 10–30 cm, linear, sheathing; upper reduced. **INFL**: panicle or raceme; bracts 3–10 mm, linear, scarious; pedicels 3–30 mm, with scarious bractlet above middle. **FL**: perianth parts 6, fused below, petal-like, white to yellowish; tube 4–5 mm, strongly veined; lobes ± = tube, spreading (becoming reflexed), 3–5-veined; stamens 6, attached to perianth, filaments 1–2 mm, alternate 6 staminodes 0.5–1 mm, anthers dehiscing by pores at tip; ovary superior, chambers 3, style 4–5 mm, thread-like, stigma slightly 3-lobed. **FR**: capsule, loculicidal. **SEED** obovoid, brown. *n*=10. Clay, often serpentine soils; < 600 m. NCoRI, n&c SNF. ❀SUN,IRR until fls. open, then DRY:7,8,9,**14–17**.

SCOLIOPUS

Per; rhizome short, slender. **ST** short, underground. **LVS** 2, basal, sheathing, 5–20 cm, 5–10 cm wide, elliptic to oblong, many-veined, mottled. **INFL**: umbel, subsessile; peduncle underground; pedicels 2–12, 8–20 cm, appearing scapose, 3-angled, twisting and recurving so fr ± touches soil. **FL** ill-scented when fresh, greenish or yellowish, mottled, heavily lined purple or dark brown; perianth parts 6 in 2 very different whorls; sepals much wider, spreading, deeply concave near base; petals narrow, erect or arching over ovary; stamens 3, attached to sepals; ovary 3-angled, chamber 1, style branches 3, linear, spreading or ± recurved. **FR** capsule-like, thin-walled, irregularly dehiscent or indehiscent. 2 spp.: w coast US. (Greek: crooked foot, from curving pedicel)

S. bigelovii Torrey (p. 1205) **INFL**: pedicels 3–12, 10–20 cm. **FL**: sepals 14–17 mm, narrowly ovate, abruptly spreading; petals ± = sepals, linear; stamens 5–6 mm; style branches 5–6 mm. **FR** 15–18 mm, elliptic. **SEED** ± 3 mm, oblong. *n*=8. Moist, shady redwood forest; < 1100 m. NCoRO, SnFrB. ❀SHD,IRR,DRN:**5**, 7,14,**15–17**.

SMILACINA FALSE SOLOMON'S SEAL

Per; rhizome creeping. **ST** erect, scaly below, leafy above. **LVS** alternate, > 5. **INFL**: panicle or raceme, terminal. **FL** small; perianth parts 6, petal-like, white; stamens 6, attached to perianth; ovary superior, chambers 3, style 1, stigma ± 3-lobed. **FR**: berry. **SEEDS** 1–3. ± 25 spp.: n temp. (Greek: little smilax) CA spp. sometimes considered part of *Maianthemum*.

1. Infl a well developed panicle; perianth parts 1–2 mm, << stamens .... ***S. racemosa***
1′ Infl a raceme (or with a few branches near base); perianth parts 4–7 mm, > stamens .... ***S. stellata***

S. racemosa (L.) Link (p. 1205) Rhizome thick. **ST** 30–90 cm, glabrous to finely hairy above. **LF** 7–20 cm, ovate to oblong-elliptic, acute to acuminate, glabrous to finely hairy below, (sub)-sessile; bases ± clasping. **INFL**: panicle, 5–12 cm; fls > 20. **FL** 1–2 mm; perianth parts ± erect, narrowly oblong; stamens > perianth. **FR** 5–7 mm, spheric, red, dotted purple. **SEED** black. n=±18. Moist woodlands, streambanks, open woodlands; < 2000 m. NW, CaR, SN, CW, TR; to AK, e N.Am. [var. *amplexicaulis* (Nutt.) S. Watson; var. *glabra* J.F. Macbr.] ❀DRN,almost DRY when dormant:**4**& SHD:**5**,**6**,**17**&IRR:1,2,7,8,9,**14–16**,18–24.

S. stellata (L.) Desf. (p. 1205) Rhizome slender. **ST** 30–70 cm, straight or ± zigzag above, glabrous to puberulent. **LF** 5–17 cm, (ob)lanceolate to elliptic, acuminate, puberulent below, sessile, clasping. **INFL**: raceme (rarely branched at lowest node), gen 2–8 cm; fls 5–15. **FL** 4–6 mm; perianth parts oblong to lanceolate, spreading; stamens < perianth. **FR** 7–10 mm, spheric, reddish purple to black. **SEED** brown. n=18. Moist woodlands, streambanks, open slopes; < 2400 m. CA-FP; to B.C., e N.Am. [var. *sessilifolia* (Baker) L. Henderson] ❀DRN:**4**&IRR:1,2,**5**,**6**&SHD:3,**7**,8–10, 14,**15–17**,18–24.

SMILAX

Per, dioecious; caudex gen large, tuber-like. **ST** climbing or trailing, sometimes woody below, often prickly, scaly below, leafy above. **LVS** alternate, simple, deciduous or persistent, strongly veined; base rounded to sagittate; petiole gen with 2 tendrils near base. **INFL**: umbels or clusters, axillary. **FL**: perianth parts 6, petal-like, free, white to greenish or yellowish; stamens 6, free; ovary superior, chambers gen 3, style ± 0, stigmas 3, spreading. **FR**: berry, spheric, black, red, or purple (white). **SEEDS** 1–6, black. ± 350 spp.; esp trop, also temp N.Am, e Asia. (Greek: origin uncertain)

1. St not shiny, gen armed with needle-like prickles; staminate fls 5–6 mm; pistillate fls 3–3.5 mm .... ***S. californica***
1′ St smooth, shiny, unarmed; staminate and pistillate fls ± 2.5 mm .... ***S. jamesii***

S. californica (A. DC.) A. Gray (p. 1209) Vine; caudex short, knotty. **ST** 2–5 m, ± woody, gen armed with needle-like prickles; central pith 0. **LF** 5–10 cm, ovate, acuminate, dull green; base ± cordate; lower surface not glaucous. **INFLS**: staminate gen 15–30-fld; pistillate gen 15–35-fld; peduncles 2–5 cm, drooping. **FLS**: staminate 5–6 mm; pistillate 3–3.5 mm; perianth parts ± 1.5 mm wide, linear, reflexed from middle. **FR** ± 6 mm, black, drying black. **SEEDS** 3. n=16. Streambanks in coniferous forest; < 1600 m. NW, CaRF, n SNF; sw OR. ❀IRR,SHD:1,**2**,3,6,**7**,14,**15**,**16**,17–24.

S. jamesii G.A. Wallace (p. 1209) ENGLISH PEAK GREENBRIAR Per vine with central pith; rhizome long, ± zigzag. **ST** 2–3 m, smooth, shiny, unarmed. **LF** 5–8 cm, triangular to ± ovate, acute or rigid-tipped; base truncate to ± cordate; upper surface dark green; lower surface glaucous, minutely papillate. **INFL**: staminate fls gen < 20; pistillate fls gen < 40; peduncles stout, ± erect, staminate 5–8 cm, pistillate 5–13.5 cm. **FL** ± 2.5 mm; perianth parts ± 1.5 mm wide, spreading or reflexed, elliptic to oblong. **FR** 6–8 mm, probably dark blue (drying dull maroon). **SEEDS** 6. RARE. Lakesides, streambanks, alder thickets in montane coniferous forest; gen 1500–2500 m. KR, CaR?. Member of *S. herbacea* L. complex. In cult.

STENANTHIUM

Per; bulbs small. **ST** ± scapose, simple or branched. **LVS** mostly basal, grass-like; cauline lvs 1–2, bract-like. **INFL**: raceme or panicle. **FL** nodding, sometimes unisexual; perianth parts 6, petal-like, fused below; stamens 6, attached to perianth, filaments wider at base, anthers 1-chambered; ovary partly inferior (at very base), ovoid, chambers 3, styles 3. **FR**: capsule, beaked by 3 persistent styles, septicidal. **SEEDS** oblong, coat loose, winged at both ends. 5 spp.: N.Am, Asia. (Greek: narrow fl)

S. occidentale A. Gray (p. 1209) Bulb 2–4 cm. **ST** 20–50 cm. **LVS** 10–30 cm, 3–15 mm wide. **INFL** 10–20 cm; bracts linear-lanceolate, scarious; pedicels 5–30 mm, ascending. **FL** nodding, 1–2 cm, bell-shaped; perianth parts oblong-lanceolate, pale greenish yellow to purplish. **FR** 15–20 mm (incl persistent styles). **SEED** ± 3 mm. n=8. Moist banks, thickets, meadows; 1500–1900 m. KR; to w Can, MT. ❀SHD,WET or IRR,DRN:1,4,5,**6**,7,15–17;DFCLT.

STREPTOPUS TWISTED-STALK

Per; rhizomes long, horizontal. **ST** simple or branched. **LVS** alternate, sessile or clasping, oblong to ovate, acute to acuminate. **INFL**: fls 1–2 on slender, twisted peduncles that are fused to st for 1 internode (and thus appear to originate opposite the lf above their actual origin). **FL** bell- to saucer-shaped; perianth parts 6, petal-like, outer flat, inner ± keeled, slightly narrower; stamens 6, filaments flat, anthers abruptly tipped; ovary superior, chambers 3, style 1, entire or 3-lobed. **FR**: berry, greenish to dark red. **SEEDS** 3–15, elliptic, grooved or wrinkled. 7 spp.: N.Am, Eurasia. (Greek: twisted foot, from peduncles)

S. amplexifolius (L.) DC. var. ***americanus*** Schultes (p. 1209) Rhizome thick; roots thick, fibrous. **ST** 30–100 cm, repeatedly branched, glabrous to densely hairy. **LF** 5–15 cm, 2–5 cm wide, entire or minutely dentate, cordate-clasping, acuminate, ± glaucous below. **FL** 9–15 mm, bell-shaped; perianth parts narrowly oblong-lanceolate, spreading to ± recurved at tips, white, tinged (yellow-) green, inner clasping stamens; stamens 6, outer filaments ± 1 mm, inner 2–3 mm; style 4–5 mm, thick, ± entire. **FR** 10–15 mm, yellow or red. **SEEDS** many, ± 3 mm, grooved. n=16. Moist, shaded areas; 250–1700 m. KR, NCoRH, CaRH; to AK, c N.Am. [var. *denticulatus* Fassett] ❀SHD,WET,DRN:**4–6**,7,15–17;DFCLT.

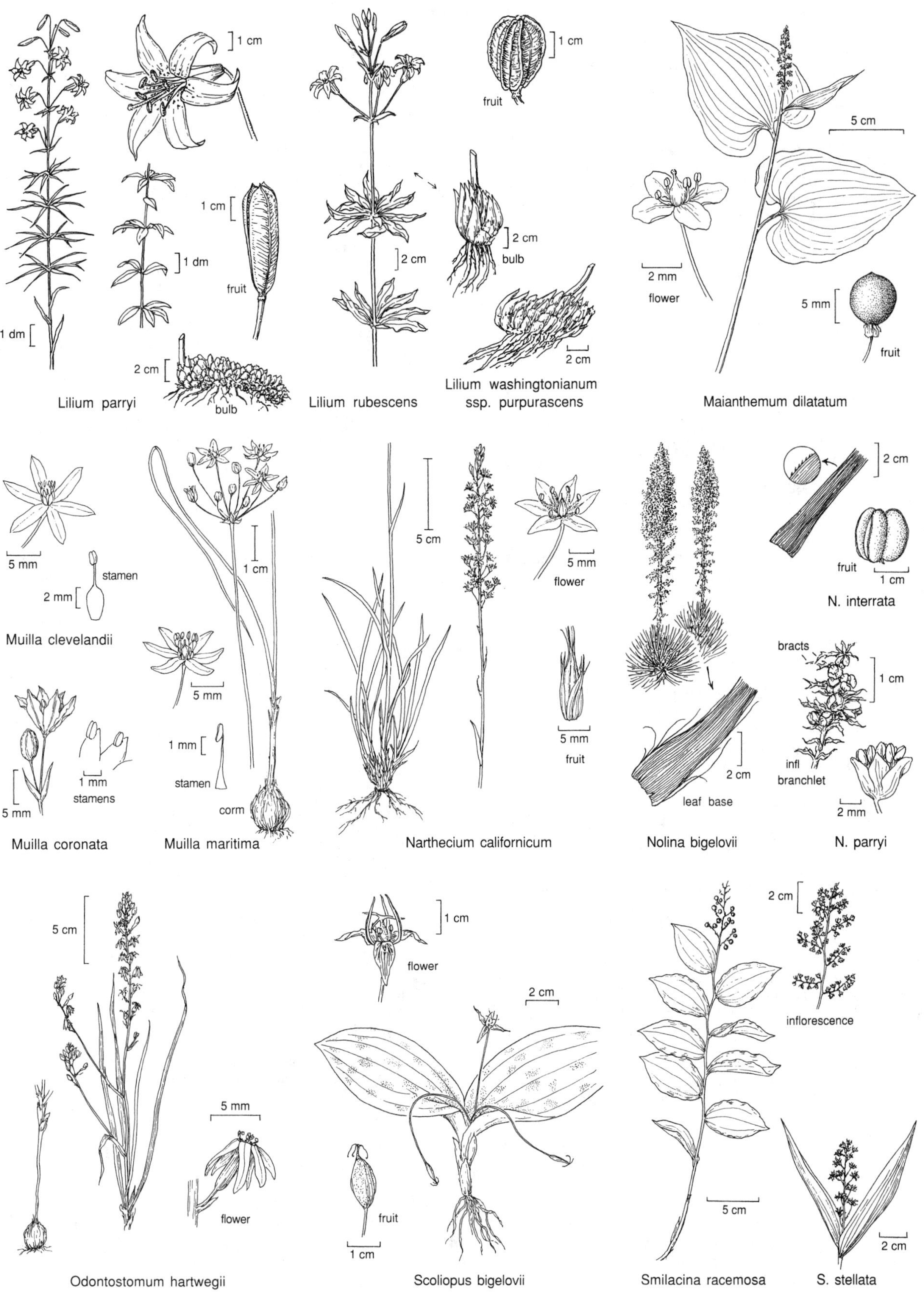
1 cm
1 cm
1 dm
fruit
1 dm
2 cm
bulb
Lilium parryi
2 cm
Lilium rubescens
1 cm
fruit
2 cm
bulb
2 cm
Lilium washingtonianum
ssp. purpurascens
5 cm
2 mm
flower
5 mm
fruit
Maianthemum dilatatum
5 mm
stamen
2 mm
Muilla clevelandii
1 mm
stamens
5 mm
Muilla coronata
1 cm
5 mm
1 mm
stamen
corm
Muilla maritima
5 cm
5 mm
flower
5 mm
fruit
Narthecium californicum
2 cm
leaf base
Nolina bigelovii
2 cm
fruit
1 cm
N. interrata
bracts
1 cm
infl
branchlet
2 mm
N. parryi
5 cm
5 mm
flower
Odontostomum hartwegii
1 cm
flower
2 cm
fruit
1 cm
Scoliopus bigelovii
2 cm
inflorescence
5 cm
Smilacina racemosa
2 cm
S. stellata

TOFIELDIA

Per; rhizomes short, spreading. **ST** erect, ± scapose. **LVS** mostly basal, linear, grass-like, ± 2-ranked; cauline lvs 2–3, on lower 1/3 of st, sheathing. **INFL**: raceme, terminal, short. **FL** small; perianth parts 6, petal-like, white to greenish or yellowish; stamens 6; ovary superior, 3-lobed at tip, chambers 3, styles 3. **FR**: capsule, 3-beaked from persistent styles, septicidal. **SEEDS** many; coat often ± loose. ± 15 spp.: n temp, w S.Am. (Thomas Tofield, British botanist, 1730–1799)

T. occidentalis S. Watson ssp. ***occidentalis*** (p. 1209) **ST** 30–80 cm, ± glabrous below, densely glandular above. **LF** 5–20 cm, 3–8 mm wide. **INFL** 1–3 cm (–7 cm in fr), ± head-like; pedicels 1–6 mm, tipped with 3 free to ± fused bracts. **FL** 3–6 mm; perianth parts oblong-ovate, inner narrower and > outer; stamens ± = perianth. **FR** 5–9 mm. **SEED** 0.5–1.5 mm, brown; coat white, spongy, appendaged at 1 end. n=15. Wet meadows, bogs; < 3100 m. NW, SNH; sw OR. Other sspp. OR to AK, w Can. *T. glutinosa* Pers. distinct (Can, e N.Am). ❀WET:4,6,17&SHD:1,2,15,16;DFCLT.

TRILLIUM WAKEROBIN, TRILLIUM

Bryan D. Ness

Per; rhizome short, thick, horizontal to erect. **ST** erect, 1 or more per pl. **LVS** 3 in a single whorl, subtending fl, ± ovate. **FL** bisexual, 1 per st, erect to nodding; sepals 3, free, persistent, greenish; petals 3, free, withering, white, pinkish, yellowish, or purple; stamens 6; ovary chambers 3, styles 3. **FR**: capsule, berry-like. **SEEDS** many, ovoid. ± 30–40 spp.: N.Am, Asia. (Latin: three, from lvs) [Freeman 1975 Brittonia 27:1–62]

1. Fl sessile
 2. Stamens < 1.25 × ovary length; fl gen with musty or fetid odor; w KR, SN, ScoRO .... ***T. angustipetalum***
 2′ Stamens ± 2 × ovary length; fl gen with sweet rose-like or ± spicy scent; NW, CaR, n SN, SnFrB
  3. Ovary and tissue between anther sacs greenish; petals white to pink (base sometimes purplish); NW, CaR, n SN, SnFrB .... ***T. albidum***
  3′ Ovary and tissue between anther sacs purple; petals yellow to purple (sometimes white); NCoRI, SnFrB .... ***T. chloropetalum***

1′ Fl stalked
 4. Lvs gen petioled; fl recurved to nodding; petals ovate-cordate, white, gen purple-spotted; KR .... ***T. rivale***
 4′ Lvs ± sessile; fl erect to nodding; petals linear to ovate, white, aging pink; NW, CW .... ***T. ovatum***
  5. Petals < 20 mm (often < 15 mm), < 6 mm wide; fl ± nodding; > 1200 m; KR .... ssp. ***oettingeri***
  5′ Petals gen > 15 mm, > 6 mm wide; fl gen ± erect; < 1600 m; NW, CW .... ssp. ***ovatum***

T. albidum Freeman (p. 1209) **ST** 2–7 dm. **LF** sessile, 7–20 cm, rounded to obtuse, gen ± brown-spotted. **FL** gen with a sweet rose-like or ± spicy scent; sepals 3–8 cm, lanceolate; petals 4–11 cm, oblanceolate to obovate, white to pinkish, sometimes purplish near base; stamens 14–31 mm, tissue between anther sacs greenish; ovary greenish or rarely purplish. n=5. Common. Edges of redwood or mixed-evergreen forest, coastal scrub and chaparral, moist canyon slopes, ravine banks; < 2000 m. NW, CaR, n SN, SnFrB; to WA.

T. angustipetalum (Torrey) Freeman (p. 1209) **ST** 2–7 dm. **LF** subsessile, 9–25 cm, rounded to obtuse, sometimes lightly brown- or dark green-spotted. **FL** gen with a musty or fetid odor; sepals 3–6.5 cm, oblong-lanceolate to linear-lanceolate; petals 5–11 cm, linear, dark purple; stamens 11–23 mm, tissue between anther sacs purple; ovary dark purple. n=5. Montane coniferous forest, foothill woodland, chaparral; 100–2000 m. w KR, SN (exc Teh), SCoRO (Santa Lucia Mtns); s OR. [*T. chloropetalum* (Torrey) Howell var. *a.* (Torrey) Munz; *T. kurabayashii* Freeman, Kurabayashi's wakerobin]

T. chloropetalum (Torrey) Howell (p. 1209) GIANT TRILLIUM **ST** 2–7 dm. **LF** sessile, 7–21 cm, rounded to obtuse, gen ± brown-spotted. **FL** gen with a sweet rose-like or ± spicy scent; sepals 3–7 cm, lanceolate; petals 5–11 cm, linear-oblanceolate to obovate, yellow to dark purple (sometimes white); stamens 15–35 mm, tissue between anther sacs purple; ovary purple. n=5. Edges of redwood forests and chaparral, gen on moist slopes and canyon banks in alluvial soils; 100–2000 m. NCoRI, SnFrB. [var. *giganteum* (Hook. & Arn.) Munz] Highly variable; a population may contain all color variants.

T. ovatum Pursh WHITE or WESTERN TRILLIUM **ST** 1–7 dm. **LF** ± sessile, 4–20 cm, acute to acuminate. **FL** erect to nodding; sepals 0.5–6 cm, 1.5–45 mm wide, lanceolate; petals 0.5–7 cm, 1.8–45 mm wide, linear to widely obovate, white, aging pink; stamens 1.7–15 mm. n=5. Common. Redwood and mixed-evergreen forest on moist wooded slopes; 10–2000 m. NW, CW; to w Can, Colorado. Highly variable. ❀DRN,SHD:14–17;DFCLT.

ssp. ***oettingeri*** Munz & Thorne (p. 1209) SALMON MOUNTAINS WAKEROBIN **ST** 1–3 dm. **LF** subsessile. **FL** ± nodding; sepals 5–25 mm, 1.5–5.3 mm wide, linear to linear-lanceolate; petals 5–23 mm, 1.8–5.5 mm wide, linear to linear-lanceolate. UNCOMMON. Mixed-evergreen forest on moist wooded slopes; 1200–2000 m. c KR (esp Marble Mtns, Siskiyou Co.).

ssp. ***ovatum*** (p. 1209) **ST** 1.5–7 dm. **LF** gen sessile. **FL** gen ± erect; sepals 1–6 cm, 2–24 mm wide, lanceolate; petals 1.5–7 cm, 5–45 mm wide, oblong-lanceolate to widely obovate. Habitat of sp.; 10–1600 m. Range of sp.

T. rivale S. Watson (p. 1209) BROOK WAKEROBIN **ST** 0.5–2 dm. **LF** petioled (rarely subsessile), 1.7–11 cm, sometimes lanceolate, acute to acuminate. **FL** gen ± nodding; sepals 1–2 cm, 4–11 mm wide, lanceolate or ovate to oblong; petals 1.5–3 cm, 8–19 mm wide, ovate-cordate, white, gen purple-spotted; stamens 6–15 mm. n=5. UNCOMMON. Yellow-pine forest along rocky streambanks; 40–1500 m. KR; sw OR.

TRITELEIA

Glenn Keator

Per from spheric, tan, fibrous-coated corm; cormlets sessile. **LVS** basal, 2–3, narrowly lanceolate, keeled, glabrous, entire, often withered at fl. **INFL** umbel-like, open; axis 2–100 cm, stiff, erect, cylindric; bracts ± lanceolate, ± scarious; pedicels ± erect, gen > perianth; fls gen many. **FL**: perianth tube gen funnel-shaped, lobes 6 in 2 petal-like whorls, gen ascending to spreading; stamens 6, filaments appendages gen 0, anthers attached at middle, gen angled away from stigma; style 1, stigma weakly 3-lobed. **FR**: capsule, gen stalked, ovoid, loculicidal. **SEEDS** subspheric, black. 14 spp.: w N.Am, esp n&c CA. (Greek: three complete, because all fl parts in 3's) [Hoover 1941 Amer Midl Naturalist 25:73–100]

1. Perianth evenly purple or blue (exc midveins gen darker)
 2. Perianth tube rounded at base; ovary stalk < ovary ..... ***T. grandiflora*** ssp. ***howellii***
 2′ Perianth tube tapered at base; ovary stalk > ovary
  3. Stamens attached to perianth tube at 1 level; perianth throat translucent, shiny ..... ***T. bridgesii***
  3′ Stamens attached to perianth tube at 2 levels; throat gen not translucent and shiny
   4. Perianth 14–19 mm; filaments ± linear — c KR ..... ***T. crocea*** var. ***modesta***
   4′ Perianth gen 20–45 mm; filaments sometimes wider at base
    5. Anthers ± 1.5 mm; filaments wider at base; s ChI ..... ***T. clementina***
    5′ Anthers 2–5 mm; filaments linear; widespread ..... ***T. laxa***
1′ Perianth white (sometimes flushed purple) to golden-yellow
 6. Perianth white, sometimes flushed purple
  7. Pedicel > 2 × perianth; ovary yellow or green
   8. Ovary green; perianth lobes reflexed ..... ***T. ixioides*** ssp. ***cookii***
   8′ Ovary bright yellow; perianth lobes ascending ..... ***T. peduncularis***
  7′ Pedicel < 2 × perianth; ovary green
   9. Filaments wider at base; anthers yellow ..... ***T. hyacinthina***
   9′ Filaments linear; anthers lilac ..... ***T. lilacina***
 6′ Perianth straw-colored to bright golden-yellow
  10. Stamens attached to perianth tube at 2 levels ..... ***T. crocea*** var. ***crocea***
  10′ Stamens attached to perianth tube at 1 level
   11. Forked appendages outside of anthers 0
    12. Filaments of 2 lengths ..... ***T. lugens***
    12′ Filaments ± equal ..... ***T. montana***
   11′ Forked appendages present outside of anthers
    13. Appendages blunt, short; perianth tube gen = lobes ..... ***T. dudleyi***
    13′ Appendages pointed, conspicuous; perianth tube << lobes ..... ***T. ixioides***
     14. Perianth tube 7–10 mm; perianth often golden-yellow; c&s CW ..... ssp. ***ixioides***
     14′ Perianth tube 3–7 mm; perianth gen straw-colored to pale yellow; KR, SN
      15. Anthers gen blue (cream); filament appendages straight or curved in ..... ssp. ***anilina***
      15′ Anthers cream or yellow (blue); filament appendages gen curved out ..... ssp. ***scabra***

T. bridgesii (S. Watson) E. Greene (p. 1209) **LF** 20–55 cm, 3–10 mm wide. **INFL** 10–50 cm, smooth exc weakly scabrous near base; pedicels 20–90 mm. **FL**: perianth lilac, blue-purple, or red-purple, tube 17–25 mm, throat translucent and shiny, lobes 10–20 mm; stamens attached at 1 level, filaments 3–4 mm, triangular, anthers 3.5–4.5 mm, bluish; ovary stalk 3–4 × ovary. *n*=8. Forest edges, often on rocks; 0–100 m. KR, NCoR, CaRF, n&c SNF. [*Brodiaea b.* S. Watson] ❀DRN:5,14,**15–17**,24.

T. clementina Hoover (p. 1209) SAN CLEMENTE ISLAND TRITELEIA **LF** 30–50 cm, 4–10 mm wide. **INFL** 30–40 cm, smooth; pedicels 30–80 mm. **FL**: perianth light blue, tube 7–10 mm, lobes 10–15 mm; filaments 2 mm, attached at 2 levels, triangular, anthers 1.5 mm, purple; ovary stalk = ovary. *n*=8. RARE. Damp clefts, rocky walls; 0–150 m. s ChI (San Clemente Island). [*Brodiaea c.* (Hoover) Munz] In cult.

T. crocea (Alph. Wood) E. Greene **LF** 15–25 cm, 4–10 mm wide. **INFL** 10–30 cm, smooth exc weakly scabrous near base; pedicels 7–20 mm. **FL**: perianth bright yellow or pale blue, tube 5–8 mm, lobes widely spreading, ± = tube, striped greenish; filaments attached at 2 levels, unequal (1.5 or 3 mm), linear or base barely wider, anthers 1–2 mm; ovary stalk ± = ovary. Open conifer forest, dry slopes; 1200–2200 m. KR; sw OR. [*Brodiaea c.* (Alph. Wood) S. Watson] Vars. need study.

var. ***crocea*** (p. 1209) YELLOW TRITELEIA **FL**: perianth bright yellow, lobes entire. *n*=8. UNCOMMON. Habitat and range of sp. ❀DRN,DRY:1,7,14–17;DFCLT.

var. ***modesta*** (H.M. Hall) Hoover (p. 1209) TRINITY MOUNTAINS TRITELEIA **FL**: perianth pale blue, lobes slightly fringed toward tips. UNCOMMON. Habitat of sp.; 1200–2000 m. c KR (Trinity Mtns). [*Brodiaea c.* var. *m.* (H.M. Hall) Munz] ❀DRN, DRY,SUN:1,7,15,16;DFCLT.

T. dudleyi Hoover (p. 1209) **LF** 10–30 cm, 4–8 mm wide. **INFL** 10–30 cm, smooth; pedicels 15–35 mm, slender. **FL**: perianth pale yellow, drying purplish, tube 8–12 mm, ± cylindric to narrowly funnel-shaped, lobes 10–12 mm; filaments attached at 1 level, unequal (2 or 3.5 mm), narrowly triangular, sometimes with blunt appendages, anthers 1 mm, lavender; ovary stalk = ovary. *n*=8. Subalpine forest; 3000–3500 m. s SNH, TR. [*Brodiaea d.* (Hoover) Munz]

T. grandiflora Lindley ssp. ***howellii*** (S. Watson) Hoover (p. 1209) HOWELL'S TRITELEIA **LF** 20–40 cm, 4–10 mm wide. **INFL** 20–70 cm, smooth; pedicels 10–40 mm, ascending. **FL**: perianth white to blue-purple, tube 8–18 mm, becoming swollen, base rounded, lobes 9–13 mm; filaments attached at 2 levels, unequal (± 2 mm or 3–4 mm), widely triangular, anthers 3–4 mm; ovary stalk 1/2 ovary length. *n*=16. RARE in CA. Sage grassland; ± 700 m. n CaRH (Hornbrook, n-c Siskiyou Co.); to B.C. [*Brodiaea douglasii* S. Watson var. *h.* (S. Watson) M. Peck]

T. hyacinthina (Lindley) E. Greene (p. 1209) WHITE BRODIAEA **LF** 10–40 cm, 4–22 mm wide. **INFL** 30–60 cm, smooth to scabrous; pedicels 5–50 mm. **FL**: perianth white, sometimes flushed purple outside, tube 2–4 mm, shallowly bowl-shaped, lobes ascending to spreading, 7–12 mm; filaments 2–4 mm, bases wider, anthers 1–2 mm, whitish; ovary stalk 1/2 ovary length. *n*=14,28,35. Grassland, vernally wet meadows; 0–2000 m. NW, CaR, SN, GV, n&c CW; to B.C., ID. [*Brodiaea h.* (Lindley) Baker] Variable. ❀DRN,SUN,DRY in summer:1–6,**7**,8,9,14,**15–17**,18–24.

T. ixioides (S. Watson) E. Greene **LF** 20–40 cm, 3–14 mm wide. **INFL** 20–80 cm, gen smooth (exc base); pedicels 10–70 mm (> 2 × fl), curved upward. **FL**: perianth straw-colored or golden-yellow striped dark (or white flushed purple), tube 4–10 mm, lobes ascending to reflexed, 7–16 mm; stamens held close to pistil, filaments attached at 1 level, unequal (2.5–3 or 4–5 mm), flat, tip forked into wing-like appendages, anther 1.5–2 mm, gen yellow; ovary stalk < ovary. Edges of coastal coniferous and mixed forests, often in sandy soils; 0–3000 m. KR, CaRF, SN, cw CW; sw OR. [*Brodiaea lutea* (Lindley) C. Morton]

ssp. ***anilina*** (E. Greene) L. Lenz (p. 1209) PRETTY FACE **FL**: perianth dull yellow (drying bluish), tube 4–7 mm, lobes widely spreading; filament appendages gen short, curved toward anther, anther gen blue. *n*=5,7,20,21,25. Coniferous forest edge, often in gravel or sand; 600–3000 m. n KR, SN; sw OR. [*Brodiaea lutea* var. *a.* (E. Greene) Munz; misspelled *analina* by Munz] ❀DRN, SHD,IRR:1,2,5–7,15–17;DFCLT.

ssp. ***cookii*** (Hoover) L. Lenz (p. 1209) COOK'S TRITELEIA **FL**: perianth white to pale straw-colored, tube = lobes, lobes strongly reflexed; filament appendages curled away from anther or reflexed.

n=7. UNCOMMON. Streamsides, wet ravines on serpentine near cypresses; < 500 m. w-c SCoRO (s Santa Lucia Mtns). [*Brodiaea lutea* var. *c.* (Hoover) Munz]

ssp. ***ixioides*** (p. 1209) GOLDEN BRODIAEA **FL**: perianth gen golden-yellow, tube 7–10 mm, lobes ascending to spreading; filament appendages ± straight to recurved. *n*=7. Closed-cone-pine forest, coastal foothill woodland; gen < 300 m. c&s CCo, sw SnFrB, w SCoRO. Sterile hybrids with *T. hyacinthina* from c CCo (Point Lobos, Monterey Co.) have been called *T.* × *versicolor* Hoover; possibly extinct due to successional change. ❀DRN,DRY, SUN:5,14,**15–17**,24.

ssp. ***scabra*** (E. Greene) L. Lenz (p. 1209) **INFL**: axis downwardly scabrous. **FL**: perianth gen straw-colored to pale yellow, tube 4–6 mm, lobes ascending to reflexed; filament appendages ± straight to recurved. *n*=5,6,8. Scrub edges, mixed or coniferous forest, in clay and granite soils; 150–2200 m. s CaRF, SNF. [*Brodiaea lutea* var. *s.* (E. Greene) Munz] ❀DRN,DRY,SUN:**7,14–16**, 17–24.

T. laxa Benth. (p. 1213) ITHURIEL'S SPEAR **LF** 20–40 cm, 4–25 mm wide. **INFL** 10–70 cm, smooth or scabrous; pedicels 20–90 mm. **FL**: perianth blue, blue-purple, or white, tube 12–25 mm, lobes 8–20 mm; filaments attached at 2 levels, 3–6 mm, ± equal, linear, anthers 2–5 mm, gen white to bluish; ovary stalk 2–3 × ovary. *n*=7,8,14,16,21,24. Common. Open forest, woodland, grassland on clay soil; 0–1500 m. NW, CaR, SN, CW, TR; sw OR. Highly variable; more study needed. [*Brodiaea l.* (Benth.) S. Watson] ❀DRN,SUN,DRY in summer:5,7,8,9,**14–17**,**19–24**.

T. lilacina E. Greene **LF** 10–40 cm, 4–22 mm wide. **INFL** 30–60 cm, smooth to scabrous; pedicels 5–50 mm. **FL**: perianth 7–10 mm, white, unstriped, tube 2–4 mm, shallowly bowl-shaped, glassy-beaded when fresh (translucent-shiny when dry), lobes 7–12 mm; filaments 2–4 mm, linear, anthers 1–2 mm, lilac or lilac-dotted; ovary stalk 1/2 ovary length. *n*=8. Volcanic hills and mesas; 70–150 m. s CaRF, n&c SNF, e ScV, ne SnJV. [*Brodiaea hyacinthina* var. *greenei* (Hoover) Munz] ❀DRN,DRY,SUN:7–9,14–16, 19–24;DFCLT.

T. lugens E. Greene (p. 1213) **LF** 10–40 cm, 3–10 mm wide. **INFL** 10–40 cm, ± smooth; pedicels 10–25 mm. **FL**: perianth yellow (pale to golden) striped dark, tube 4–5 mm, lobes 8–9 mm, spreading; filaments attached at 1 level, unequal (1–2 or 2–3 mm), bases wide (bases of inner ± = perianth lobe width), anthers 1.5–2 mm, yellow or blue; ovary stalk ± = ovary. *n*=8. Uncommon. Edges of chaparral, mixed forests; 100–1000 m. s NCoR, SnGb?. [*Brodiaea l.* (E. Greene) Baker] Related to *T. ixioides*; scattered occurrences suggest further study. ❀DRY,SUN:7–14–21;DFCLT.

T. montana Hoover (p. 1213) **LF** 10–30 cm, 2–5 mm wide. **INFL** 5–25 cm, ± scabrous; pedicels 5–30 mm. **FL**: perianth yellow, striped brown, tube 4–5 mm, narrowly funnel-shaped, lobes 8–10 mm; filaments attached at 1 level, 5–6 mm, thread-like, anthers 1–1.5 mm, cream to blue; ovary stalk = ovary. *n*=8. Open montane coniferous forests, esp on granite; 1200–3000 m. n&c SNH. [*Brodiaea gracilis* S. Watson] ❀DFCLT.

T. peduncularis Lindley (p. 1213) LONG-RAYED BRODIAEA **LF** 20–40 cm, 5–15 mm wide. **INFL** 10–80 cm, smooth; pedicels 20–160 mm, straight. **FL**: perianth white (outside often flushed violet), tube 7–10 mm, lobes 8–16 mm; filaments attached at 2 levels, unequal (1–1.5 mm or 2–3 mm), nearly linear, anthers 2–4 mm, white; ovary stalk = ovary, ovary bright yellow in fl. *n*=7,14. Wet grassland, vernal streams and pools, often serpentine; 0–800 m. NW, n&c CW. [*Brodiaea p.* (Lindley) S. Watson] ❀WET in spring;DRY in summer,SUN:**7,14–17**.

VERATRUM CORN LILY, FALSE HELLEBORE

Per, coarse, leafy; rhizome thick. **ST** erect, 1–2 m, simple, hollow. **LVS** many, alternate, lanceolate to widely ovate, gen acute, clasping, coarsely veined, reduced upward. **INFL**: panicle; fls many. **FL** bisexual or staminate; perianth parts 6, petal-like, free, widely spreading, white or greenish to red-brown, nectary glands 1–2 near base; stamens 6, attached to perianth; ovary slightly inferior, chambers 3, styles 3, short, stigmas long. **FR**: capsule, septicidal. ± 25 spp.: n temp. (Latin: dark roots) Alkaloids used medicinally and TOXIC to both livestock and humans.

1. Ovary and fr ± densely tomentose; perianth parts (at least inner) irregularly and shallowly fringed ..... ***V. insolitum***
1′ Ovary and fr glabrous or with only a few straight hairs; perianth parts entire to deeply fringed
  2. Perianth deeply fringed ..... ***V. fimbriatum***
  2′ Perianth entire or finely dentate
    3. Infl dense, branches crowded, spreading or ascending ..... ***V. californicum*** var. ***californicum***
    3′ Infl loose, branches ± separated, lowest drooping ..... ***V. viride***

V. californicum Durand var. ***californicum*** (p. 1213) **LVS** ovate; lower 20–40 cm, tomentose-ciliate, lower surface curly-hairy, upper surface glabrous or veins sparsely short-hairy. **INFL** erect, gen 30–60 cm, tomentose; branches ascending or spreading; pedicels 1–6 mm. **FL** 10–15 mm; perianth parts elliptic to obovate, white or greenish, glabrous to sparsely woolly below, entire to ± dentate, glands 1–2, Y-shaped, green; stamens 1/2–2/3 perianth length; ovary glabrous. **FR** 2–3 cm, narrowly ovoid. **SEED** 10–12 mm, ± winged. *n*=16. Streambanks, moist meadows, forest edges; ± 1000–3500 m. CA-FP, MP; to WA, MT, Colorado, Mex. ❀WET:1,**2**,6,7.

V. fimbriatum A. Gray (p. 1213) FRINGED FALSE HELLEBORE **LVS** lanceolate; lower 20–50 cm, glabrous or sparsely hairy. **INFL** erect, gen 15–50 cm, tomentose; branches spreading; pedicels 6–12 mm. **FL** 6–10 mm; perianth parts diamond-shaped to ovate, white, glabrous, margins conspicuously fringed, glands 2, elliptic, yellow; stamens ± 1/2 perianth length; ovary glabrous. **FR** ± 8 mm, obovoid. **SEED** ± 6 mm, scarcely margined. *n*=16. UNCOMMON. Wet meadows in coastal scrub; < ± 100 m. c NCo (Mendocino, Sonoma cos.). ❀IRR or WET,SHD,DRN:5,15–17;DFCLT.

V. insolitum Jepson (p. 1213) SISKIYOU FALSE HELLEBORE **LVS** elliptic; lower 10–23 cm, hairy below and on sheaths. **INFL** erect, gen 20–50 cm, grayish woolly; branches ascending; pedicels 6–15 mm. **FL** 6–9 mm; perianth parts widely ovate, white to yellowish, hairy, outer irregularly ciliate, inner ± fringed, glands 2, elliptic, greenish; stamens ± = perianth; ovary woolly. **FR** 2–3 cm, oblong-ovoid, ± woolly. **SEED** 10–15 mm, widely winged. *n*=16. RARE. Openings in thickets and mixed-evergreen forest on red clay; > 900 m. KR; to WA.

V. viride Aiton **LVS** elliptic to widely ovate; lower 15–35 cm, sparsely curly-ciliate, lower surface hairy, upper surface veins sparsely short-hairy. **INFL** gen 30–70 cm, woolly; main axis erect; lower branches drooping; pedicels 2–5 mm. **FL** 6–10 mm; perianth parts oblanceolate to oblong, green to yellowish, ± irregularly toothed, hairy above, glands 2, elliptic, darker green or yellowish; stamens ± 1/2 perianth length; ovary glabrous. **FR** 2–3 cm, oblong-ovoid. **SEED** 7–10 mm, widely winged. *n*=16. Wet subalpine meadows; 1500–2000 m. KR; to AK, e N.Am. ❀DFCLT.

XEROPHYLLUM BEAR-GRASS, INDIAN BASKET-GRASS

Per, non-fl for several years, stalk dying after fr; rhizome woody, tuber-like. **ST** stout, simple, leafy. **LVS** persistent, tough, wiry, grass-like, strongly scabrous. **INFL**: raceme, dense, ± club-shaped, longer (< 50 cm) in fr. **FL** small; perianth parts 6,

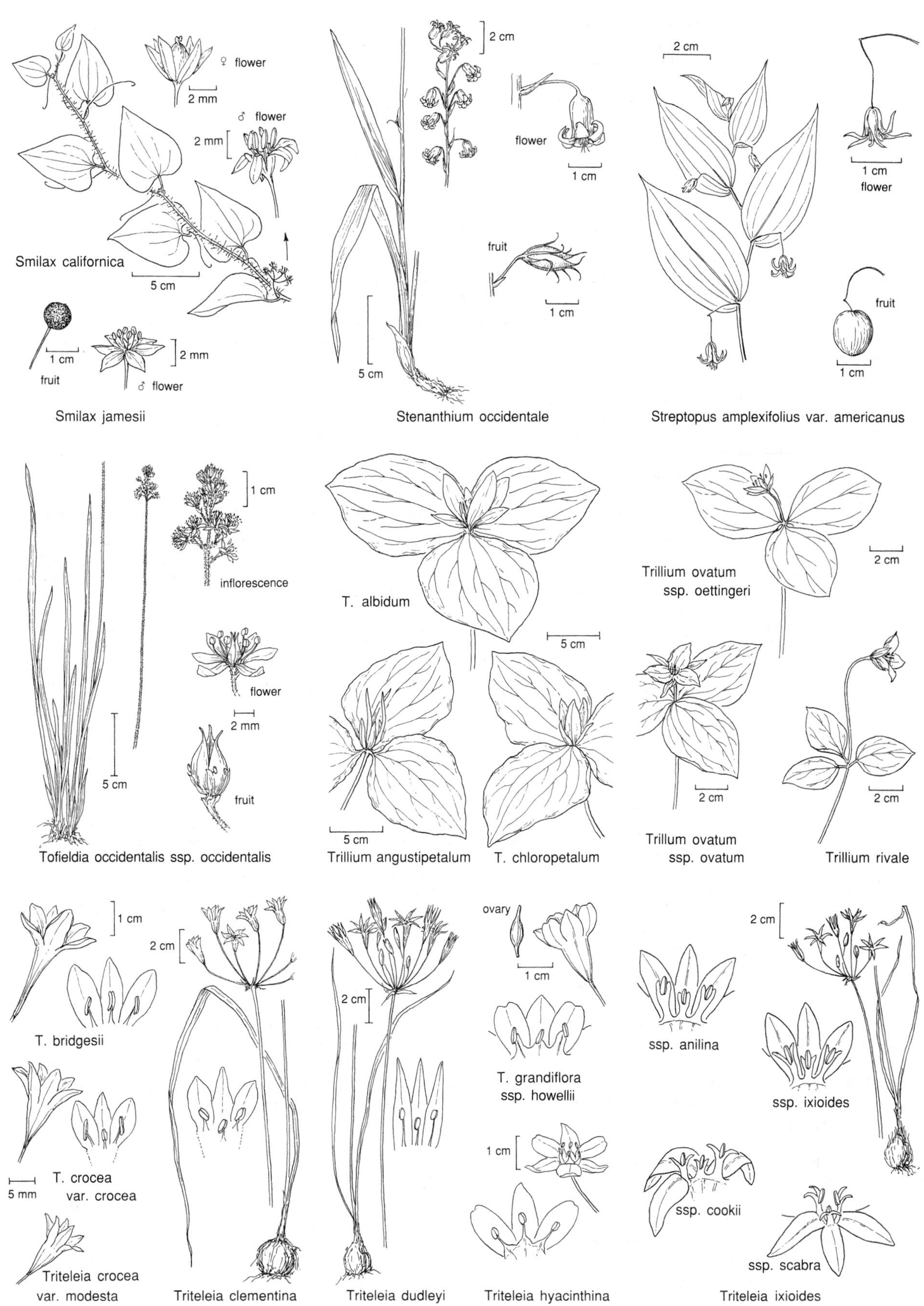
♀ flower
2 mm
♂ flower
2 mm
Smilax californica
5 cm
1 cm
fruit
2 mm
♂ flower
Smilax jamesii
2 cm
flower
1 cm
fruit
1 cm
5 cm
Stenanthium occidentale
2 cm
1 cm
flower
fruit
1 cm
Streptopus amplexifolius var. americanus
1 cm
inflorescence
flower
2 mm
5 cm
fruit
Tofieldia occidentalis ssp. occidentalis
T. albidum
5 cm
5 cm
Trillium angustipetalum
T. chloropetalum
Trillium ovatum
ssp. oettingeri
2 cm
2 cm
2 cm
Trillum ovatum
ssp. ovatum
Trillium rivale
1 cm
T. bridgesii
5 mm
T. crocea
var. crocea
Triteleia crocea
var. modesta
2 cm
Triteleia clementina
2 cm
Triteleia dudleyi
ovary
1 cm
T. grandiflora
ssp. howellii
1 cm
Triteleia hyacinthina
ssp. anilina
2 cm
ssp. ixioides
ssp. cookii
ssp. scabra
Triteleia ixioides

petal-like, free, oblong to ovate, persistent, whitish to cream, 5–7-veined; stamens 6, filaments wider at base; ovary chambers 3, style short, stigmas 3, thread-like. **FR**: capsule, 3-angled, loculicidal. **SEEDS** 9–18, oblong, 3-angled, black. 2 spp.: N.Am. (Greek: dry lf, from tough, persistent lvs)

X. tenax (Pursh) Nutt. (p. 1213) **ST** 15–150 cm. **LVS**: basal in dense clump, 30–100 cm, 2–6 mm wide; cauline becoming bract-like upward. **INFL**: pedicels 2–5 cm, spreading to ± erect. **FL** 5–10 mm; perianth parts ± oblong; style ± 1 mm, stigma branches 2–4 mm. **FR** 5–7 mm, strongly 3-lobed. n=15. Dry open slopes, ridges, montane coniferous forest; < 2300 m. KR, NCoRO, CaRH, n SNH, CCo, SnFrB; to B.C., MT, WY. ❀DRN:**4–6**,17&SHD, IRR:1,2,7,15,16;DFCLT.

YUCCA SPANISH BAYONET

Katy K. McKinney & James C. Hickman

Subshrub or tree-like, sometimes dying after fr. **LVS** rosetted (basal or elevated on branches), 2–15 dm, linear, stiff, sword-like, stoutly spine-tipped; bases ± expanded; edges gen curved up. **INFL**: panicle, dense; fls pendent. **FL** 2–13 cm; perianth parts 6 in 2 petal-like whorls, gen ± fused, ± white, fleshy, waxy; stamens 6, filaments ± thick, fleshy; ovary superior, style short, stigma 3-lobed, concave or dome-like. **FR**: gen capsule. **SEEDS** ± many in 2 rows per chamber, black, often flat. ± 40 spp.: esp dry sw N.Am. (Haitian: yuca, or manihot, because young infls sometimes roasted for food) Pollinated at night by small moths that simultaneously lay eggs in ovary.

1. Lf margins strongly fibrous-shredding; fr ± pendent
 2. Aboveground st ± 0; perianth 5–13 cm; young fr fleshy; e DMtns .... ***Y. baccata***
 2′ Aboveground st conspicuous, gen few-branched; perianth 3–5 cm; young fr ± dry; s SW, s DMoj, nw DSon .... ***Y. schidigera***
1′ Lf margins minutely serrate (not at all fibrous-shredding); fr spreading to erect
 3. Pl tree-like (rosettes aerial); lvs 20–35 cm, ± dark green; DMoj .... ***Y. brevifolia***
 3′ Aboveground st ± 0 (rosettes ± basal); lvs gen 40–100 cm, ± gray-green; s SN, s SCoR, SW .... ***Y. whipplei***

Y. baccata Torrey (p. 1213) **STS** ± 0 aboveground; rosettes ± open, solitary or in small clumps. **LF** 50–75 cm, ± dark green; expanded base ± 10 cm, 5 cm wide, reddish; margins strongly fibrous-shredding. **INFL** 6–8 dm, heavy, purple-tinged; peduncle < 2 dm. **FL**: perianth 5–13 cm, bell-shaped, reddish brown outside, ± white inside, segments lanceolate (outer narrower), fused below; pistil 5–8 cm. **FR** 15–17 cm, fleshy when young, eventually pendent. Uncommon. Dry Joshua-tree woodland; 800–1300 m. e DMtns (Clark, New York, Providence mtns); to UT, TX. [var. *vespertina* McKelvey] Reported to hybridize with *Y. schidigera*. ❀DRN, SUN,DRY:2,3,**7**,8,9,**10**,11,14–16,**18–21**,22,23.

Y. brevifolia Engelm. (p. 1213) JOSHUA TREE Tree-like, 1–15 m, gen openly branched (sometimes clumped or only low-branching); rosettes at st tips, dense. **LF** 20–35 cm, dark green; expanded base 2–4 cm, 4–5 cm wide, ± white; margins minutely serrate. **INFL** 3–5 dm, heavy; peduncle gen ± 1 dm. **FL**: perianth 4–7 cm, ± bell-shaped, cream to greenish, segments ± widely lanceolate, fused ± to middle; filaments thick throughout; pistil ± 3 cm, stigma cavity surrounded by lobes. **FR** spreading to erect. Desert flats, slopes; 500–2000 m. DMoj; to sw UT, w AZ. [vars. *herbertii* (J.M. Webber) Munz & *jaegeriana* McKelvey] Growth form variable. ❀DRN,SUN,DRY:2,**3**,7,9,**10**,11,12,14–16,**18–21**,22,23.

Y. schidigera K.E. Ortgies (p. 1213) MOHAVE YUCCA Shrub- or tree-like 1–5 m, 1–few-branched; rosettes at st tips, not very dense. **LF** 30–150 cm, yellowish or bluish green; expanded base 2–8 cm, 4–11 cm wide, ± white; margins coarsely fibrous-shredding. **INFL** 6–12 dm, heavy; peduncle 1–5 dm. **FL**: perianth 3–5 cm, narrowed at base, gen cream; pistil 2–3 cm, stigma cavity suronded by lobes. **FR**: capsule, ± pendent. Chaparral, creosote-bush scrub; < 2500 m. s SW (San Diego Co.), s DMoj, nw DSon; NV, AZ, n Baja CA. ❀DRN,SUN,DRY:2,3,7–9,**10**,11,12,**14**,15,16,**18–23**,24.

Y. whipplei Torrey (p. 1213) OUR LORD'S CANDLE Pl dies after fr. **STS** ± 0 aboveground; rosettes 1–many, very dense. **LF** gen 40–100 cm, flat or ± 3-angled, ± gray-green; expanded base ± 4–7 cm, 4–7 cm wide, ± white to greenish; margins minutely serrate. **INFL** 2–40 dm, not appearing heavy; peduncle 15–35 dm; branches and fls very many. **FL**: perianth ± 3 cm, ± spheric, white, gen purple-tipped; filaments linear below, tip angled, club-like; pistil 1–2 cm, stigmas domed, clear-papillate. **FR** spreading to erect. Chaparral, coastal or desert scrub; < 2500 m. s SN (esp e slope), s SCoR, SW; n Baja CA. Growth form highly variable; branched pls from desert edge have been called ssp. *cespitosa* (M.E. Jones) A.L. Haines; late-branching coastal pls have been called ssp. *intermedia* A.L. Haines; large, unbranched pls from s slope SnGb and SnBr have been called ssp. *parishii* (M.E. Jones) A.L. Haines; rhizomed colonial pls from s SCoR have been called ssp. *percursa* A.L. Haines. ❀DRN,SUN,DRY:1–3,**7**,8–12,**14–16**,17,**18–23**,24.

ZIGADENUS DEATH CAMAS

Per from bulb or rhizome. **ST** ± scapose. **LVS** many, ± basal (reduced upward), linear, gen folded, ± curved. **INFL**: raceme or panicle. **FL** bisexual, staminate, or sterile; perianth parts 6, petal-like, free or ± fused to ovary base, white to yellowish in CA, glands 1–2 near base; stamens 6, free to ± attached to perianth; ovary chambers 3, styles 3. **FR**: capsule, septicidal. **SEEDS** many. ± 15 spp.: temp N.Am, Asia. (Greek: yoke-gland, from gland shape of some) All taxa should be considered highly TOXIC to livestock (gen unpalatable) and humans from alkaloids (esp in bulbs); caused serious illness to some members of Lewis & Clark Expedition.

1. Perianth parts 3–6 mm, stamens = or > perianth
 2. Infl a well developed panicle; fls of main axis bisexual; fls of branches often staminate; stamens 1–2 mm > perianth .... ***Z. paniculatus***
 2′ Infl raceme-like (rarely branched below); fls all bisexual; stamens ± = perianth
  3. Pl scabrous; infl widely conic, pedicels 15–40 mm, spreading; perianth parts short-clawed .... ***Z. micranthus*** var. ***micranthus***
  3′ Pl glabrous; infl narrowly conic, pedicels 6–12 mm, ascending; perianth parts long-clawed .... ***Z. venenosus*** var. ***venenosus***

1′ Perianth parts 5–15 mm, stamens < perianth
4. Perianth parts 5–8 mm — pedicels spreading
5. Pedicels not upturned at tip; styles widely spreading or recurved; DMoj, Teh, WTR .... ***Z. brevibracteatus***
5′ Pedicels upturned at tip; styles erect; wet soils, NCoR, CW .... ***Z. micranthus*** var. ***fontanus***
4′ Perianth parts 8–15 mm
6. Fls of lower infl branches staminate; perianth parts 8–12 mm, ovate, obtuse, stamens 2/3–3/4 perianth length; SN .... ***Z. exaltatus***
6′ Fls bisexual; perianth parts 8–15 mm, lanceolate to ovate, acute, stamens ± 1/2 length parts; w CA .. ***Z. fremontii***

Z. brevibracteatus (M.E. Jones) H.M. Hall (p. 1213) Bulb 25–40 cm diam, ± spheric; outer coats dark brown to black. **ST** 30–50 cm, glabrous. **LF** 15–30 cm, 5–10 mm wide, scabrous-ciliate. **INFL**: gen panicle, 10–35 cm, open; fls of branches mostly bisexual or staminate; pedicels spreading, 10–35 mm. **FL** 5–8 mm; perianth parts ovate to elliptic, ± obtuse, short-clawed, glands greenish, indistinct exc ridge at lower edge; stamens ± 2/3 perianth; styles spreading or recurved. **FR** 12–20 mm, oblong. *n*=11. Sandy desert; 600–1800 m. Teh, WTR, DMoj. ❀DRN,DRY,SUN:2,3,7,9,**11**,12,**14**,15,16.

Z. exaltatus Eastw. Bulb 60–100 mm, 30–50 mm diam, ± oblong; outer coats dark brown. **ST** 60–100 cm, glabrous. **LF** 50–80 cm, 15–30 mm wide, scabrous-ciliate. **INFL**: panicle, 20–40 cm, open; fls of lower branches staminate (or with stunted ovary, 3 long styles); pedicels ascending, 10–40 mm. **FL** 8–12 mm; perianth parts ovate to elliptic, obtuse, outer subsessile, inner sharply infolded to form a 1–2 mm claw, glands greenish yellow with thick teeth along veins; stamens ± 2/3 perianth; styles erect to ± spreading. **FR** 15–25 mm, oblong. *n*=11. Meadows, wooded slopes; 600–1500 m. SNF. ❀DRN,SUN,DRY:2,**7**,9,14–16,19–21.

Z. fremontii (Torrey) S. Watson (p. 1213) Bulb 20–35 mm diam, ± spheric; outer coats black. **ST** 40–90 cm, glabrous. **LF** 20–50 cm, 8–30 mm wide, curved, scabrous-ciliate. **INFL**: panicle or raceme, 5–40 cm, open; fls bisexual; pedicels spreading to ascending, 10–45 mm. **FL** 8–15 mm; perianth parts widely ovate, obtuse, outer very short-clawed, inner with wide, 2–3 mm claw, glands greenish yellow with thick teeth along veins; stamens ± 1/2 perianth; styles erect to ± spreading. **FR** 15–35 mm, cylindric. *n*=11. Grassy or wooded slopes, outcrops; < 1000 m. NW, ScV, CW, SW; OR, n Baja CA. [vars. *inezianus* Jepson, *minor* (Hook.& Arn.) Jepson, & *salsus* Jepson] ❀DRN,SUN,mid summerDRY:2,7,8,9,**14–24**.

Z. micranthus Eastw. Bulb 10–45 mm diam, spheric; outer coats dark brown to black. **ST** 15–80 cm, glabrous to ± scabrous. **LF** 15–80 cm, 4–25 mm wide, scabrous-ciliate. **INFL**: raceme or panicle, 5–40 cm, open; fls bisexual; pedicels spreading, 15–40 mm, tips sometimes upturned in fr. **FL** 3–7 mm; perianth parts ovate to elliptic, obtuse, clawed, glands yellow, with thick teeth along veins; stamens ± = perianth; styles erect. **FR** 8–25 mm, cylindric. *n*=11. Dry slopes, flats, seeps; < 1000 m. NW, SnFrB, SCoRI; OR.

var. ***fontanus*** (Eastw.) D. McNeal (p. 1213) Bulb 25–45 mm thick; outer coats brown. **ST** 60–80 cm, glabrous. **LF** erect, often > st, 10–25 mm wide. **INFL**: panicle; pedicels 25–45 mm, tips upturned in fr. **FL** 5–7 mm; perianth parts cordate at base, glands not or slightly ridged on lower edge. Vernally moist or marshy areas, often on serpentine; < 500 m. NCoR, SnFrB, SCoRI. [*Z. f.* Eastw.] ❀SUN,mid summerDRY:7–9,**14–17**,19–24.

var. ***micranthus*** Bulb 10–25 mm thick; outer coat dark brown to black. **ST** 15–50 cm, scabrous at least below. **LF** < st, 4–10 mm wide. **INFL** raceme-like (sometimes with basal branches); pedicels 15–40 mm, spreading, tips only slightly upturned in fr. **FL** 3–5 mm; perianth parts not cordate, glands with heavy, infolded ridge on lower edge. Dry slopes, flats; < 1000 m. NW; sw OR. ❀SUN:4–6,**17**&IRR:**7,14–16**,19–24.

Z. paniculatus (Nutt.) S. Watson (p. 1213) Bulb 30–50 mm, 8–30 mm wide, ovoid; outer coats dark brown to black. **ST** 20–70 cm, glabrous. **LF** 20–50 cm, 6–16 mm wide, scabrous-ciliate. **INFL**: panicle, 10–30 cm, densely fld at tip; fls bisexual (or sterile or staminate on branches); pedicels ascending, 5–25 mm. **FL** 3–6 mm; perianth parts unequal, outer 3–5 mm, ovate, acute to subacuminate, short-clawed, inner 3.5–6 mm, triangular-ovate, acute, longer-clawed, glands yellowish green, indistinct; stamens 1–2 mm > perianth; styles erect or ± spreading. **FR** 8–20 mm, cylindric. *n*=11. Dry sagebrush scrub to coniferous forest; 1200–2300 m. NW, CaR, n SNH, MP; to WA, MT, Colorado, NM. ❀DRN,SUN,DRY:1,**2**,3,**7**,10,14–16,18–21.

Z. venenosus S. Watson var. ***venenosus*** (p. 1213) Bulb 12–25 mm diam, widely ovate; outer coats ± brown. **ST** 15–70 cm, glabrous. **LF** 10–40 cm, 4–10 mm wide, scabrous-ciliate. **INFL** raceme-like (sometimes with basal branches), 5–25 cm; pedicels erect or ascending, 10–20 mm. **FLS** bisexual, 4–6 mm; perianth parts unequal, ± ovate, obtuse, outer 4–5.5 mm, gen subsessile, inner ± 4.5–6 mm, claw ± 1 mm, glands yellowish green; upper margin a heavy, irregular ridge; stamens = or > perianth; styles erect. **FR** 8–14 mm, cylindric. *n*=11. Moist meadows to dry rocky hillsides; < 2600 m. NW, CaR, SN, CW, WTR, PR, MP, DMtns; to B.C., c US, AZ, n Baja CA. ❀SUN:**4–6**&IRR:1,**2,3,7,14–21**,22–24.

ORCHIDACEAE ORCHID FAMILY

Dieter H. Wilken & William F. Jennings

Per, terrestrial in CA, some nongreen, gen from rhizomes. **LVS** linear to ± round or scale-like, gen sessile. **INFL**: gen raceme or spike, bracted. **FL** bisexual, bilateral, sometimes spurred; sepals gen 3, gen petal-like, gen free, uppermost gen erect; petals 3, lowest different ("lip"); stamen gen 1, fused with style and stigma into column, pollen gen sticky, gen removed as sessile anther sacs; ovary inferior, gen twisted 180° (so lip appears to be lowest perianth segment), 1-chambered, placentas 3, parietal; stigmas 3, gen under column tip. **FR**: capsule. **SEEDS** very many, minute. ± 800 genera, ± 18,000 spp.: esp trop (worldwide exc deserts). Many cult for orn, esp *Cattelya, Cymbidium, Epidendrum, Oncidium, Paphiopedalum; Vanilla planifolia* frs used as source of food flavoring. [Luer 1975 Orchids US and Can, NY Bot Garden] Nongreen pls derive nutrition through fungal intermediates.

1. Pl nongreen; lvs scale- or bract-like, blade 0 or << sheath
2. Sepals 12–20 mm; pl white; lip folded lengthwise below middle, lobes erect .... **CEPHALANTHERA**
2′ Sepals gen < 10 mm; pl yellow-green to purplish; lip ± flat to concave, lobes ± spreading .... **CORALLORHIZA**
1′ Pl green; lf blades 1–many, lower gen > sheath
3. Lip ± balloon-like, mouth puckered, margin inrolled; lower sepals ± fused, pendent behind lip; stamens 2 .... **CYPRIPEDIUM**
3′ Lip not ballon-like, flat to pouch-like; lower sepals free; stamen 1 (pollen sacs 2, sometimes separated by column tip)

4. Lip clearly spurred at back
 5. St ± scapose; lvs gen basal, ± withered in fl; sepals gen 1-veined; caudex tuber-like .... **PIPERIA**
 5′ St leafy; lvs basal and cauline, persistent in fl; sepals gen 3-veined; caudex cylindric to fusiform .... **PLATANTHERA**
4′ Lip not spurred
 6. Lvs 1–2, basal or opposite (sheath-like bracts may also be present)
  7. Lvs 2, opposite, ± sessile .... **LISTERA**
  7′ Lf 1, basal, petioled or tapered to sheathing base
   8. Fl gen 1, perianth parts > 15 mm, pink to purple; blade clearly petioled .... **CALYPSO**
   8′ Fls 10+, perianth parts < 3 mm, green to yellowish; blade tapered to sheathing base .... **MALAXIS**
 6′ Lvs 3–many, basal or alternate
  9. Basal lvs in rosettes, white-mottled or -striped; pl of dry, shaded places .... **GOODYERA**
  9′ Basal lvs not in rosette, not mottled or striped; pl gen of wet, open places
   10. Lvs cauline; infl raceme, open, ± 1-sided; perianth gen reddish tinged or veined .... **EPIPACTIS**
   10′ Lvs mostly basal; infl spike, dense, ranks of fls helical; perianth gen ± white .... **SPIRANTHES**

CALYPSO FAIRY SLIPPER

1 sp. (Greek: Kalypso, sea nymph in Homer's *Odyssey*, from her beauty and secretive behavior)

C. bulbosa (L.) Oakes (p. 1213) Pl 7–18 cm; caudex corm-like, subspheric; rhizome 0. **ST** ± scapose. **LVS**: basal 1, petioled; blade 3–6.5 cm, elliptic to ovate; cauline 2–4, sheath-like. **INFL**: fl gen 1; fl bract 1, pinkish to lavender. **FL**: sepals free, 15–25 mm, gen pink; lateral petals sepal-like; lip ± pendent, pouch-like in lower 2/3, purplish outside, purple-striate inside, distal 1/3 ± concave, white to pink with white to yellow, hairy, ± red-spotted base; column 8–11 mm, widely ovate, hood-like, arched over pouch, gen pink. **FR** erect. $2n$=28. Moist, gen shaded coniferous forest; < 1800 m. NW (exc NCoRI), sw SnFrB; to AK, ne N.Am, NM, Eurasia. [ssp. *occidentalis* (Holz.) Calder & Roy Taylor] ❀SHD,IRR acidic soil:1, 4–7,15–17;very DFCLT.

CEPHALANTHERA

Pls gen green (white in N.Am sp.). **ST** ± scapose. **LVS** mostly scale- or bract-like. **INFL**: raceme; fl bract < fl, scale-like. **FL**: sepals free, lower gen ascending, ± curved over column and lip; lateral petals gen > lip; lip gen narrowed at middle, concave or folded below middle, tip ± spreading to reflexed; column subcylindric, anther head-like, short-stalked. **FR** gen erect. ± 15 spp.: esp Eurasia. (Greek: head-like anther) Some spp. ± subterranean; closely related to *Epipactis*.

C. austiniae (A. Gray) A.A. Heller (p. 1213) PHANTOM ORCHID Pl 20–55 cm, white, becoming yellowish or brown. **LVS** 3–6 cm, sheathing. **FL** white; sepals 12–20 mm, elliptic to oblanceolate, acute; lip folded lengthwise and lobed below middle, reflexed above middle, with a yellow, papillate spot; column 6–9 mm; anther stalk ± hinge-like. Decomposed litter of rich soil in mixed-evergreen or coniferous forest; 300–2200 m. NW, CaR, SNH, SnFrB, n SCoRO, SnBr; to B.C., ID. [*Eburophyton a.* (A. Gray) A.A. Heller]

CORALLORHIZA CORALROOT

Pls yellowish green to purplish; rhizome branches many, short, scaly, together coral-like. **ST** ± scapose. **LVS** bract-like, ± sheathing. **INFL**: raceme; fl bract << fl, often scale-like. **FL**: sepals ± alike, oblong to (ob)lanceolate, gen curved over column and lip, gen 3-veined, lower gen fused at base; lateral petals = or > lip, spreading or curved toward lip; lip simple to 3-lobed, spreading to reflexed, sometimes short-spurred; column gen convex above, concave below, curved over lip. **FR** pendent. ± 10 spp.: N.Am, C.Am. (Greek: coral root) Albino pls require careful comparison of sepals and lips for identification.

1. Perianth with 3–5 (faint) reddish to purplish stripes .... *C. striata*
1′ Perianth not clearly striped
 2. St (greenish) yellow; sepals white to yellowish, 4.5–6 mm, 1-veined — n SNH .... *C. trifida*
 2′ St gen reddish to brown or purple; sepals gen yellowish green to reddish brown, 5.5–10 mm, 3-veined
  3. Lip spots 4–8; column 3–5 mm .... *C. maculata*
  3′ Lip spots 0–2; column 5.5–8 mm .... *C. mertensiana*

C. maculata Raf. (p. 1213) SPOTTED CORALROOT Pl 17–55 cm. **ST** red to yellow-brown. **FL**: sepals 5.5–10 mm, lower spreading, gen colored like sts; lateral petals gen deep pink to red, sometimes dark-spotted; lip 5–7 mm, white, clearly red- to purple-spotted, base with 2 lateral lobes, tip finely crenate or toothed; spur < 2.5 mm; column 3–5 mm, yellowish, purple-spotted. **FR** 15–20 mm. $2n$=42. Shaded coniferous forest, in decomposing lf litter; < 2800 m. NW, CaR, SN (exc Teh), sw SnFrB, SnGb, SnBr, PR; to B.C., e US, NM.

C. mertensiana Bong. (p. 1213) WESTERN CORALROOT Pl 15–45 cm. **ST** gen reddish. **FL**: sepals 7–10 mm, gen pink, lower spreading; lateral petals deep pink to red, veins gen yellow or dark red; lip 6–9 mm, deep pink to red, gen with 3 dark red veins, base with 2 lateral, short teeth, tip irregularly crenate or toothed; spur 0.5–2.5 mm; column 5.5–8 mm. **FR** 15–25 mm. Shaded to open mixed-evergreen or coniferous forest, in decomposing lf litter; < 2200 m. NW (exc NCoRI); to AK, MT, WY.

C. striata Lindley (p. 1213) STRIPED CORALROOT Pl 15–50 cm. **ST** gen red-brown to purplish (yellowish). **FL**: sepals 6–17 mm, gen yellowish pink to pale brown, with 3–5 longitudinal, reddish to purplish stripes, lower curved forward; lateral petals ± sepal-like; lip 8–15 mm, entire, pale yellow-brown to reddish; column 4–7 mm, purple-spotted. **FR** 12–25 mm. $2n$=42. Open to shaded coniferous forest, in decomposing lf litter; 100–2200 m. NW (exc NCoRI), CaR, SN (exc Teh), sw SnFrB; to B.C., e Can, n Mex. Pls from SN with sepals 6–10 mm have been called var. *vreelandii* (Rydb.) L.O. Williams.

C. trifida Chatel. (p. 1213) NORTHERN CORALROOT Pl 8–30 cm. **ST** (greenish) yellow. **FL**: sepals 4.5–6 mm, gen 1-veined, white or

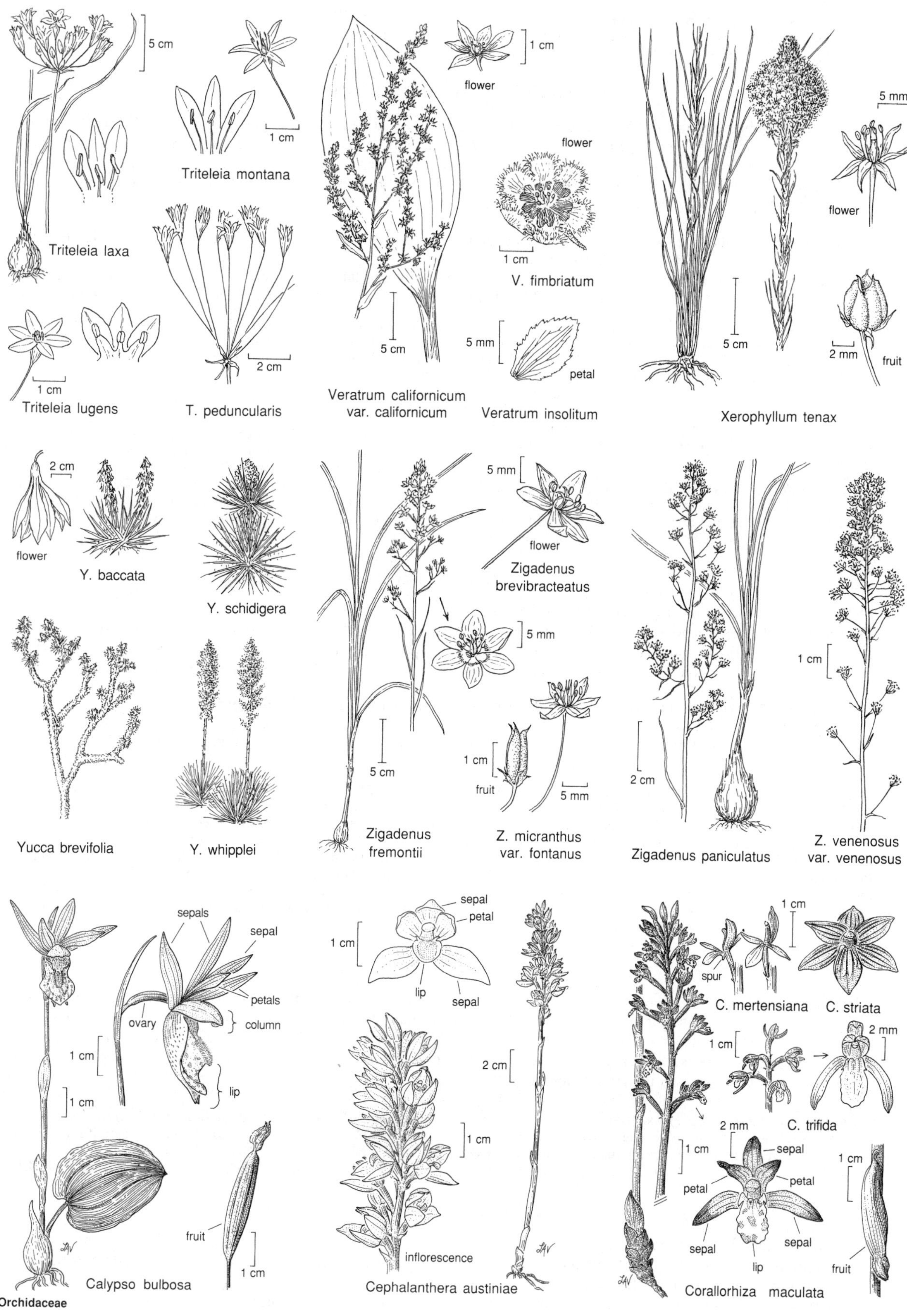
5 cm
1 cm
Triteleia montana
Triteleia laxa
1 cm
Triteleia lugens
2 cm
T. peduncularis
flower
1 cm
5 cm
Veratrum californicum
var. californicum
flower
1 cm
V. fimbriatum
5 mm
petal
Veratrum insolitum
5 mm
flower
5 cm
2 mm
fruit
Xerophyllum tenax
2 cm
flower
Y. baccata
Y. schidigera
Yucca brevifolia
Y. whipplei
5 mm
flower
Zigadenus
brevibracteatus
5 mm
5 cm
1 cm
fruit
5 mm
Zigadenus
fremontii
Z. micranthus
var. fontanus
2 cm
Zigadenus paniculatus
1 cm
Z. venenosus
var. venenosus
sepals
sepal
petals
ovary
column
1 cm
1 cm
lip
fruit
1 cm
Calypso bulbosa
sepal
petal
1 cm
lip
sepal
2 cm
1 cm
inflorescence
Cephalanthera austiniae
1 cm
spur
C. mertensiana
C. striata
1 cm
2 mm
C. trifida
2 mm
1 cm
sepal
petal
petal
sepal
sepal
lip
1 cm
fruit
Corallorhiza maculata
Orchidaceae

yellowish, lower curved forward; lateral petals ± sepal-like; lip 3–4.5 mm, white, sometimes faintly red-spotted, base gen short-lobed or toothed, finely crenate above middle; column 2.5–6 mm, white or yellowish. **FR** 8–14 mm. $2n$=42. RARE in CA. Wet, open to shaded, gen coniferous forest; 1400–1700 m. n SNH (Plumas Co.); to AK, e US, NM.

CYPRIPEDIUM LADY'S-SLIPPER

ST leafy, often puberulent above. **LVS**: basal 1–2, < cauline; cauline lanceolate to ovate, often ribbed. **INFL**: fls 1–several; bracts ± lf-like. **FL**: lower sepals ± fused, pendent behind lip; lateral petals descending to spreading, ± like upper sepal, sometimes twisted; lip balloon-like, gen obovoid, mouth ± puckered, margin inrolled; column curved over lip mouth; stamens 2, separated by petal-like staminode. **FR** pendent to spreading, ribbed. ± 35 spp.: temp N.Am, Eurasia. (Greek: Venus foot, from lip shape) Some cult for orn.

1. Lvs 2, ± basal, ± opposite; lip yellow-green below, purple above .... ***C. fasciculatum***
1′ Lvs 4–10+, cauline, alternate; lip mostly white
  2. Upper sepal and petals yellowish green; petals 12–18 mm, oblong to lanceolate, flat .... ***C. californicum***
  2′ Upper sepal and petals gen purplish; petals 25–60 mm, narrowly lanceolate, twisted .... ***C. montanum***

C. californicum A. Gray (p. 1217) CALIFORNIA LADY'S-SLIPPER Pl 18–110 cm. **LVS** 5–12, cauline, alternate, 5–15 cm, lower elliptic, upper ± lanceolate. **INFL** open; fls (1)3–12. **FL**: upper sepal 15–20 mm, ± elliptic to ovate, yellowish green; lateral petals 12–18 mm, lanceolate to oblong, spreading; lip 15–20 mm, white; staminode ± round, reflexed, white, gen with green stripe. UNCOMMON. Moist slopes, streambanks, mixed-evergreen or coniferous forest; 50–2200 m. KR, n NCoRO, w CaR, n SN, nw SnFrB; sw OR. ❀very DFCLT.

C. fasciculatum S. Watson (p. 1217) CLUSTERED LADY'S-SLIPPER Pl 12–20 cm. **LVS** 2, ± basal, ± opposite, 5–12 cm, widely elliptic. **INFL**: fls 1–4, ± clustered, ± nodding. **FL**: upper sepal 1.5–2.5 cm, lanceolate, greenish to brown, veins dark brown; lateral petals 15–25 mm, lanceolate, descending; lip 8–15 mm, yellow-green below, purple above; staminode < 5 mm, pale (greenish) yellow. RARE. Open coniferous forest; 100–1700 m. NW, CaR, n SN, sw SnFrB; to B.C., MT, Colorado. In cult.

C. montanum Lindley (p. 1217) MOUNTAIN LADY'S SLIPPER Pl 25–70 cm. **LVS** 4–6, cauline, alternate, 5–15 cm, elliptic to ovate. **INFL** open; fls gen 1–3. **FL**: upper sepal 3–6 cm, lanceolate, twisted or wavy, gen purplish; lateral petals 25–60 mm, narrowly lanceolate, descending, twisted; lip 20–30 mm, white; staminode 8–12 mm, yellow, red- to purple-spotted. UNCOMMON. Moist areas, dry slopes, mixed-evergreen or coniferous forest; 200–2200 m. NW, CaR, n&c SN, sw SnFrB, MP; to AK, MT, WY. ❀very DFCLT.

EPIPACTIS

LVS gradually reduced upward, lanceolate to widely ovate, often ± ribbed. **INFL** ± 1-sided, open; fls 4+; fl bract ± lf-like. **FL**: sepals ± alike, lanceolate to ovate, lower spreading to descending; lateral petals ascending or curved forward, shape and color ± like sepals; lip abruptly narrowed at ± middle, concave to pouch-like below middle, ± flat or grooved above middle; column curved over lip, convex above. **FR** spreading to pendent. ± 25 spp.: N.Am, Eurasia, n Afr. (Greek: ancient name)

1. Sepals and petals 12–20 mm, yellowish to greenish, veined red-purple; lip 14–20 mm, grooved to tip; wet places .... ***E. gigantea***
1′ Sepals and petals 8–13 mm, white to greenish, tinged pink-purple; lip 10–12 mm, ± flat above middle; ± dry places .... ***E. helleborine***

E. gigantea Hook. (p. 1217) STREAM ORCHID Pl 30–70(100) cm. **LVS** 5–15 cm, lanceolate to widely elliptic. **INFL**: fl bract lanceolate to oblong. **FL**: sepals 12–20 mm, green, purple-veined; lateral petals 13–15 mm; lip 14–20 mm, lower half deeply concave, greenish to yellowish, veined red-purple, upper half grooved, yellow, reddish tinged or veined below; column 5–9 mm. **FR** 20–28 mm. n=20. Seeps, wet meadows, streambanks; < 2600 m. CA-FP (exc GV, s ChI), SNE, D; to B.C., SD, TX, Mex. ❀IRR or WET, DRN:**4–6**,**15–17**,**24**&SHD:1–3,7,8–10,**14**,18–21,**22**,**23**;CV.

E. helleborine (L.) Crantz Pl 40–100 cm. **LVS** 6–10 cm, lanceolate to ovate. **INFL**: fl bract linear to narrowly lanceolate. **FL**: sepals 10–13 mm, green, purple-tinged (esp outside); lateral petals 8–11 mm; lip 10–12 mm, lower half pouch-like, white to pinkish outside, brown to purplish inside, upper half ± flat, white to pinkish, with 2 thick bumps near base; column 3–5 mm. **FR** 1–1.5 cm. $2n$=38. Gen dry slopes, road-cuts; < 1300 m. s NCoRO, c SNH, CCo, SnFrB (expected elsewhere); to e Can, c&ne US; native to Eur.

GOODYERA

Rhizomes slender. **ST** ± scapose. **LVS**: basal in rosettes, blades tapered to base, white-veined to -mottled; cauline bract-like. **INFL**: fls ± 1-sided or in spiral ranks; fl bract ± = fl. **FL**: sepals ± equal, ± enclosing column and lip, upper adherent to lateral petals, forming a hood, lower sepals spreading to reflexed; lateral petals fused; lip gen pouch-like below middle, deeply grooved to tip; column beaked. **FR** ascending to erect. ± 25 spp.: esp temp n hemisphere, also trop. (John Goodyer, English botanist, 1592–1664) [Ackerman 1975 Madroño 23:191–198]

G. oblongifolia Raf. (p. 1217) RATTLESNAKE PLANTAIN Pl 18–35 cm. **LVS**: basal 4–9 cm, lanceolate to widely elliptic, midrib white-striped, veins net-like, white; cauline 2–5 on lower st. **INFL** dense; bracts 7–11 mm, gen < fls. **FL** white; upper sepal 6–11 mm, lower sepals 5–9 mm, ± round; lip 6–10 mm, beaked; column 3–5 mm. $2n$=22. Dry coniferous forest, in decomposing lf litter; 500–2200 m. NW, CaR, n&c SNH, sw SnFrB, MP; to AK, e N.Am, Mex. ❀SHD,IRR,DRN,acidic soil:1–6,14–16,**17**;usually DFCLT.

LISTERA TWAYBLADE

Rhizomes slender. **LVS** cauline, 2, ± opposite, gen ovate to ± round. **INFL**: raceme, ± open; fl bract < fl. **FL**: sepals ± equal, green to purplish, lower spreading; lateral petals ± like sepals, ascending to erect; lip gen >> sepals and petals, spreading to descending, gen wedge-shaped, flat, tip entire to deeply lobed; column ± subcylindric, straight to curved, anther at tip. **FR** gen spreading. ± 25 spp.: temp and arctic N.Am, Eurasia. (Martin Lister, English naturalist, 1638–1711)

1. Lip deeply 2-lobed, lobes ± linear; lf base cordate .... ***L. cordata***
1′ Lip ± oblanceolate, tip rounded or notched; lf base abruptly tapered or rounded
 2. Lip 4.5–7 mm, tapered to base, base wider than column, gen with 2 short, thread-like teeth, tip rounded .... ***L. caurina***
 2′ Lip 8–13 mm, short-clawed, claw width ± = column width, teeth 0, tip notched .... ***L. convallarioides***

L. caurina Piper (p. 1217) NORTHWEST TWAYBLADE Pl 10–30 cm. **LF**: blade 2.5–7 cm, abruptly tapered to base. **INFL** 2–5 cm. **FL**: sepals 3–4 mm, narrowly lanceolate; lip 4.5–7 mm, ± oblanceolate, tapered to base, gen with 2 lateral, thread-like teeth, tip rounded; column 1.5–2.5 mm. Moist, shady coniferous forest; 100–1800 m. KR; to AK, MT, WY. ❀very DFCLT.

L. convallarioides (Sw.) Nutt. (p. 1217) BROAD-LEAVED TWAYBLADE Pl 10–35 cm. **LF**: blade 2.5–7 cm, abruptly tapered to base. **INFL** 3–4 cm. **FL**: sepals 4–5.5 mm, ± lanceolate; lip 8–13 mm, ± oblanceolate, short-clawed at base, teeth 0, tip notched; column 2.5–3.5 mm. 2*n*=36. Moist, shady coniferous forest; 1000–2900 m. KR, NCoRH, CaR, SNH, SnBr, SnJt, MP; to AK, e N.Am, AZ, Asia. ❀very DFCLT.

L. cordata (L.) R.Br. (p. 1217) HEART-LEAVED TWAYBLADE Pl 6–25 cm. **LF**: blade 1–4 cm, base cordate. **INFL** 2.5–4 cm. **FL**: sepals 2–4 mm, oblong to narrowly elliptic; lip 8–10 mm, oblong below middle, with 2 short teeth at base, deeply 2-lobed above middle, lobes linear; column 0.5–1.5 mm. 2*n*=36,38,40,42. RARE in CA. Shady, mixed-evergreen or coniferous forest; 100–1300 m. KR; to AK, e N.Am, Eurasia. In cult.

MALAXIS

Caudex short, slender to bulb-like; rhizome 0. **LVS** ± basal, 1–5, sheathing. **INFL**: spike or raceme; bracts << fls. **FL**: lower sepals free, spreading to ascending, gen green; lateral petals gen < and narrower than sepals, spreading to erect; lip spreading to ± pendent; column gen > petals, oblong to triangular, gen concave. **FR** spreading to erect. ± 100 spp.; worldwide (exc Afr). (Greek: soft, from lvs)

M. monophyllos (L.) Sw. ssp. ***brachypoda*** (A. Gray) A. Löve & D. Löve (p. 1217) ADDER'S-MOUTH Pl 10–15 cm; caudex bulb-like. **LF** 1, 4–5.5 cm, widely lanceolate to ovate, sheath 2–3 cm. **INFL**: raceme, 2–6 cm, open; pedicels < 1.5 mm. **FL** green to yellow; sepals 1–2 mm, narrowly lanceolate; lateral petals ± = lip, linear, ± curved behind fl; lip 1.5–3 mm, triangular, base cordate, tip beak-like; column < 1 mm. **FR** 3.5–5.5 mm. 2*n*=28. RARE in CA. Wet meadows, shaded places, coniferous forest; 2200– 2700 m. SnBr, SnJt (where presumed extinct); to AK, ne N.Am, Colorado. [*M. brachypoda* (A. Gray) Fern.]

PIPERIA PIPERIA

Caudex tuber- or bulb-like, 1–4 cm, gen ovoid. **LVS**: basal 2–5, linear to widely oblanceolate; cauline bract-like. **INFL**: spike or raceme, gen cylindric; fl bract gen < fl. **FL**: perianth white to green; sepals gen 2–5.5 mm, 1–2 mm wide, 1-veined, upper pointed forward to erect, lower free, spreading to reflexed; lateral petals ± = sepals, spreading to erect; lip pointed forward or down, spurred; column < lip. **FR** ascending to erect. ± 8 spp.: w N.Am. (Charles V. Piper, Am botanist, 1867–1926) [Morgan & Ackerman 1990 Lindleyana 5:205–211] Some spp. difficult to separate; green-fld spp. doubtfully distinct from *P. unalascensis*.

1. Spur 6–15 mm
 2. Spur ± straight, perpendicular to infl axis; upper sepal pointed forward; st gen < 3 mm diam .... ***P. transversa***
 2′ Spur ± curved, parallel to infl axis; upper sepal erect, lip curved down or pointed forward; st gen > 3 mm diam
  3. Sepals white with dark green midvein, petals white to pale green .... ***P. elegans***
  3′ Perianth green to yellow-green throughout
   4. Lip triangular; lateral petals ± sickle-shaped, flat; lower sepals ± reflexed .... ***P. elongata***
   4′ Lip ± ovate-deltate; lateral petals ± straight, ± concave; lower sepals spreading .... ***P. michaelii***
1′ Spur 1.5–6 mm
 5. Upper sepal and lateral petals mostly white or with white margins; lower sepals and lip mostly white; lip recurved toward spur
  6. Infl gen 1-sided; lateral petals straight, white, midvein green; lip ± obtuse; NW, sw SnFrB .... ***P. candida***
  6′ Infl ± cylindric; lateral petals ± curved, green with wide white margin; lip ± acute; c CCo .... ***P. yadonii***
 5′ Perianth green throughout; lip ± straight to upcurved above middle
  7. Lip narrowly lanceolate; lateral petals ± recurved; spur curved, pointed down .... ***P. leptopetala***
  7′ Lip oblong to ± deltate; lateral petals spreading to ± pointed forward; spur straight or curved, pointed back or down
   8. Lip flat; seeds dark brown; SW .... ***P. cooperi***
   8′ Lip (esp tip) upcurved; seeds tan to reddish brown; CA-FP (exc SW), MP .... ***P. unalascensis***

P. candida R. Morgan & J. Ackerman (p. 1217) WHITE FLOWERED PIPERIA Pl 10–55 cm. **LVS**: basal 7–18 cm, 12–30 mm wide. **INFL** 7–30 cm, ± 1-sided, ± open. **FL**: upper sepal ± pointed forward, white or green with white margins, lower sepals white, midvein green; lateral petals ± pointed forward to slightly ascending, white, midvein green; lip 1.5–3.5 mm, narrowly triangular, recurved toward spur, white; spur 1.5–3.5 mm, pointed down. *n*=21. RARE. Open to shaded sites, gen coniferous forest; < 1300 m. NW, sw SnFrB; to AK. In cult.

P. cooperi (S. Watson) Rydb. Pl 15–85 cm. **LVS**: basal 8–20 cm, 8–30 mm wide. **INFL** 10–45 cm, open. **FL**: perianth green; upper sepal ascending to pointed forward; lateral petals spreading to ascending; lip 2–4 mm, ovate-deltate, flat; spur 2.5–6 mm, pointed back or down. *n*=21. Uncommon. Gen dry sites, shrubland, woodland, forest; < 1500 m. SCo, s ChI (Santa Catalina Island), SnGb, PR; Baja CA. ❀very DFCLT.

P. elegans (Lindley) Rydb. (p. 1217) Pl 15–70 cm. **LVS**: basal 6–30 cm, 10–75 mm wide. **INFL** 3–40 cm, gen dense. **FL**: sepals white, midvein dark green, upper ± erect; lateral petals spreading to erect, white to pale green, base or midvein green; lip 3.5–6.5 mm, ± lanceolate, ± curved downward, white to pale green; spur 6–14 mm, pointed down. *n*=21. Gen dry, open sites, shrubland, coniferous forest; < 500 m. NCo, w KR, NCoRO, CCo, SnFrB; to B.C., MT. [*Habenaria e.* (Lindley) Bolander incl var. *maritima* (E. Greene) Ames] ❀SHD,DRN,IRR,acidic soil:1,4–6,14–17;very DFCLT.

P. elongata Rydb. (p. 1217) Pl 20–130 cm. **LVS**: basal 8–30 cm, 10–65 mm wide. **INFL** 15–30 cm, 10–30 mm wide, open to dense.

FL: perianth green; upper sepal ± erect, lower sepals ± reflexed; lateral petals ± sickle-shaped, gen erect, flat; lip 2–5.5 mm, triangular, reflexed; spur 6.5–15 mm, slender, gen pointed down. *n*=21. Gen dry sites, shrubland, mixed-evergreen or coniferous forest; < 2200 m. CA-FP (exc GV, s ChI); to B.C., MT.

P. leptopetala Rydb. Pl 15–70 cm. **LVS**: basal 7–15 cm, 15–30 mm wide. **INFL**: 4–40 cm, ± open. **FL**: perianth green; upper sepal erect to recurved, lower sepals reflexed; lateral petals ± linear, spreading or erect, ± recurved; lip 2.5–5 mm, narrowly lanceolate, pointed forward or down; spur 4–7 mm, curved, pointed down. Uncommon. Gen dry sites, shrubland, woodland; < 2200 m. NCoR, CaR, SN, SCoR, TR, PR; to WA.

P. michaelii (E. Greene) Rydb. (p. 1217) MICHAEL'S PIPERIA Pl 15–70 cm. **LVS**: basal 7–24 cm, 10–50 mm wide. **INFL** 5–25 cm, dense to open. **FL**: perianth green to yellow-green; upper sepal ascending, lower sepals spreading; lateral petals ± ascending, ± concave; lip 3–6 mm, ovate-deltate; spur 8–12 mm, gen pointed down. UNCOMMON. Gen dry sites, coastal shrubland, woodland, mixed-evergreen or closed-cone-pine forest; < 700 m. NCo, SNF, CCo, SnFrB, n SCo, n ChI (Santa Cruz Island), WTR. ❀very DFCLT.

P. transversa Suksd. (p. 1217) Pl 15–55 cm. **LVS**: basal 6–18 cm, 10–40 mm wide. **INFL** 7–26 cm, open to dense. **FL**: sepals and lateral petals white to yellowish with green midvein; upper sepal pointed forward; lateral petals spreading and ± curved back; lip 2.5–5 mm, oblong to ± ovate, pointed forward or down, white; spur 6–10 mm, gen straight, pointed back. *n*=21. Gen dry sites, shrubland, mixed-evergreen or coniferous forest; < 2600 m. NW, SN, CW, TR, PR; to B.C.

P. unalascensis (Sprengel) Rydb. (p. 1217) Pl 15–70 cm. **LVS**: basal 7–15 cm, 5–40 mm wide. **INFL** 5–40 cm, gen open, sometimes dense above. **FL**: perianth green; upper sepal ascending or pointed forward; lateral petals ± erect to pointed forward; lip 2–5 mm, oblong to deltate, gen pointed down, tip upcurved; spur 1.5–5 mm, pointed back or down. *n*=21. Gen dry sites, shrubland, woodland, forest; < 3000 m. CA-FP (exc SW), MP; to AK, ne Can, Colorado. [*Habenaria u.* (Sprengel) S. Watson] ❀very DFCLT.

P. yadonii R. Morgan & J. Ackerman (p. 1217) YADON'S PIPERIA Pl 10–50 cm. **LVS**: basal gen 10–15 cm, 20–35 mm wide. **INFL** (2)5–15(30) cm, dense. **FL**: upper sepal green with white margin, ± erect; lower sepals white, midvein faintly green; lateral petals ± erect, ± curved, green with wide white margin; lip 2.5–5 mm, narrowly triangular, recurved toward spur, white; spur 2.5–4 mm, pointed down. RARE. Gen sandy soil or sandstone, coastal shrubland, Monterey-pine forest; < 150 m. c CCo (n Monterey Co.).

PLATANTHERA BOG-ORCHID

Caudex cylindric to fusiform. **LVS** basal and cauline; basal 1–2 gen < lower cauline; cauline linear to elliptic, gradually reduced upward. **INFL** gen spike; fl bracts lf-like. **FL**: perianth white to greenish; sepals ± equal, upper gen hood-like, lower free, gen spreading; lateral petals gen erect; lip pendent to upcurved, spurred from back; column ± erect, tip (often the stigma) separating pollen sacs. **FR** ascending to erect. ± 85 spp.: temp N.Am, Eurasia. (Greek: wide anther)

1. Spur club-shaped, ± straight; lip gen oblong .... *P. stricta*
1′ Spur gen cylindric, ± curved; lip base wider than tip
  2. Perianth white to cream; lip abruptly wider below middle, ± oblong above middle .... *P. leucostachys*
  2′ Perianth green to yellow-green; lip ± gradually tapered to tip
    3. Column gen 1.5–2 mm; lip gen 4–7 mm; spur tip blunt; infl gen 5–15 cm, dense .... *P. hyperborea*
    3′ Column gen 2.5–4 mm; lip gen 6–10 mm; spur tip acute; infl gen ± 15–40 cm, open (esp below) ... *P. sparsiflora*

P. hyperborea (L.) Lindey (p. 1217) GREEN-FLOWERED BOG-ORCHID Pl 15–100 cm. **LVS**: cauline 4–12 cm, 6–30 mm wide. **INFL** gen 5–15 cm, ± dense; lower bracts 18–30 mm. **FL**: perianth green to yellowish green; sepals 3–6 mm; lip 4–7 mm, narrowly lanceolate, descending to slightly upcurved; spur slightly < lip, ± cylindric, slightly curved, tip blunt; column 1.5–2 mm. 2*n*=42,84. Wet coniferous or subalpine forest; 1800–3200 m. KR, SNH (e slope), n W&I; to AK, ne N.Am, Colorado, Iceland, Japan. [*Habenaria h.* (L.) R. Br.]

P. leucostachys Lindley (p. 1217) WHITE-FLOWERED BOG-ORCHID Pl 15–100 cm. **LVS**: cauline 5–25 cm, 9–30 mm wide. **INFL** 5–35 cm, gen dense; lower bracts 9–25 mm. **FL**: perianth white to cream; sepals 4–8 mm; lip 5–10 mm, abruptly wider below middle, pendent; spur 5–15 mm, 1–2 × lip in CA, gen cylindric, ± curved; column gen < 1/2 upper sepal. 2*n*=42. Wet, gen open places, meadows; < 3400 m. CA-FP (exc GV), GB; to AK, MT, UT. [*Habenaria dilatata* (Pursh) Hook. var. *l.* (Lindley) Ames] Pls with spur ± = lip and barely curved have been incorrectly called *P. dilatata* (Pursh) Hook., of n N.Am. ❀IRR,DRN:1,4–7,14–17;DFCLT.

P. sparsiflora (S. Watson) Schltr. (p. 1217) SPARSE-FLOWERED BOG-ORCHID Pl 25–55 cm. **LVS**: cauline 4–15 cm, 5–30 mm wide. **INFL** gen 15–40 cm, ± open; lowest fls gen not overlapping. **FL**: perianth green; sepals 5–9 mm; lip 6–10 mm, ± linear; spur ± = lip, ± cylindric, slightly curved, tip acute; column 2.5–4 mm. Wet meadows, streambanks, coniferous forest; 300–3400 m. NW (exc NCoRI), SNH, TR, PR, SNE; to WA, Colorado, NM. [*Habenaria s.* S. Watson] ❀DFCLT.

P. stricta Lindley (p. 1217) SLENDER BOG-ORCHID Pl 20–80 cm. **LVS**: cauline 4–11 cm, 5–25 mm wide. **INFL** 5–22 cm, gen open below; lower bracts 5–35 mm. **FL**: perianth green, sometimes tinged red-purplish; sepals 3–6 mm; lip 5–7 mm, gen oblong, ± pendent; spur 0.5–1 × lip, club-shaped, pendent; column ± 2 mm. 2*n*=42. Wet meadows, coniferous forest; 1000–2300 m. KR, CaR, MP; to AK, MT, NM. [*Habenaria saccata* E. Greene]

SPIRANTHES LADIES TRESSES

LVS basal and cauline, reduced upward, linear to oblong. **INFL**: spike, gen dense, fls in spiral ranks; bracts < to > fls, gradually reduced upward. **FL**: sepals and lateral petals narrowly lanceolate; upper sepal ± fused to lateral petals, together hood-like, enclosing column; lower sepals ± free, ± = lip, adherent to hood; lip deeply grooved below middle, concave above middle; column < lip, tip with anther on back. **FR** spreading to ascending. ± 40 spp.: esp Am, also Japan, Australia, New Zealand. (Greek: coiled fls) [Sheviak 1990 Rhodora 92:213–231]

1. Lip lanceolate to ± ovate, tip puberulent above; perianth gen yellowish .... *S. porrifolia*
1′ Lip ± violin-shaped, tip glabrous above; perianth gen white .... *S. romanzoffiana*

S. porrifolia Lindley (p. 1231) Pl 18–56 cm. **LVS**: basal 8–14 cm; cauline 1.5–10 cm. **INFL** 5–14 cm; bracts 7–20 mm. **FL**: perianth 7–12 mm, gen yellowish, sometimes cream; upper sepal and lateral petals fused ± < 1/2, fused portion cylindric, lobes spreading to recurved; lip ± ovate to lanceolate, tip puberulent above; column 2–4 mm. Wet meadows, freshwater marshes, seeps; < 2500 m. NW (exc NCoRI), CaR, SN, SnFrB, SnGb, PR; to WA. ❀very DFCLT.

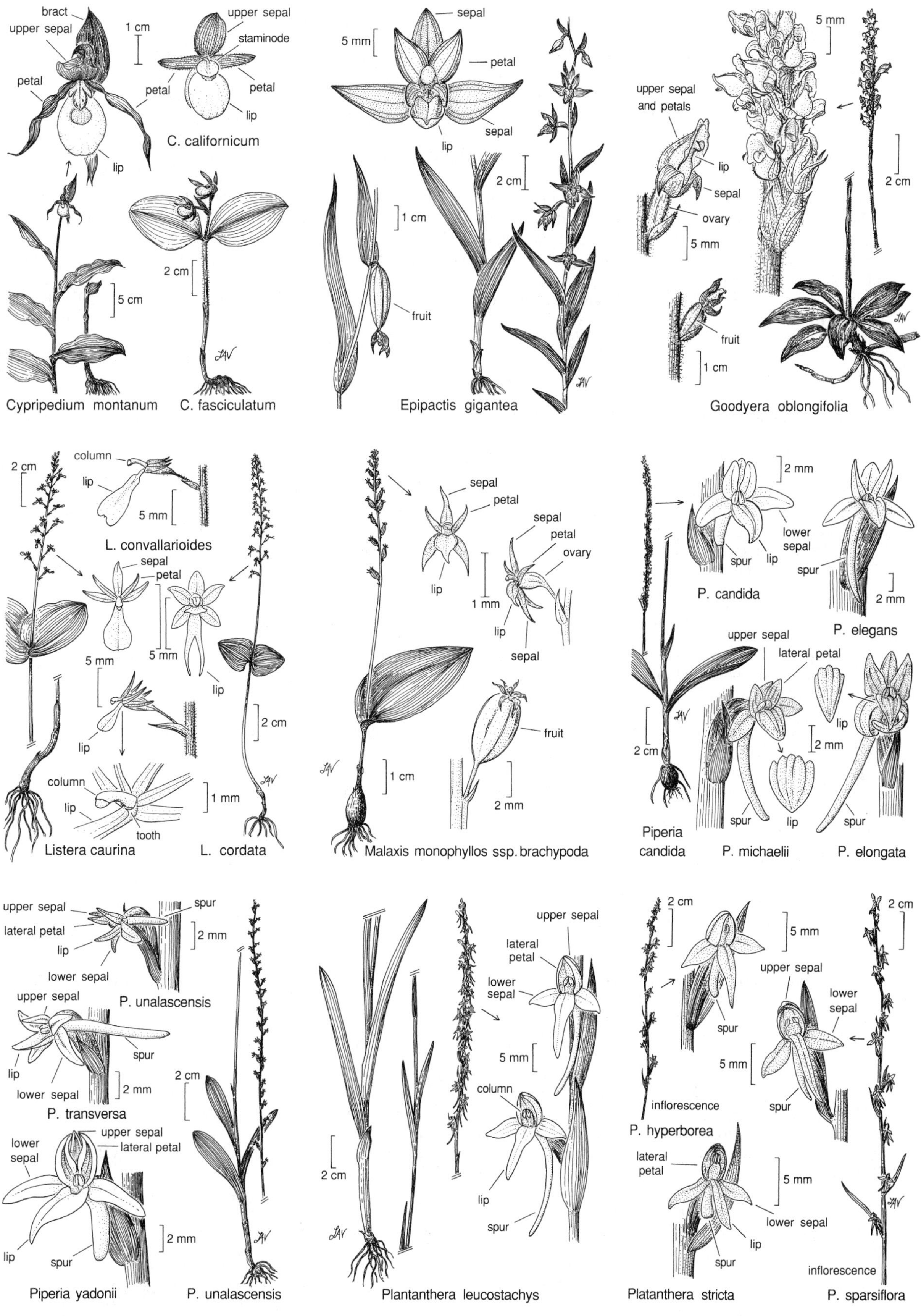
bract
upper sepal
petal
lip
1 cm
upper sepal
staminode
petal
petal
lip
C. californicum
5 cm
2 cm
Cypripedium montanum
C. fasciculatum
sepal
5 mm
petal
sepal
lip
1 cm
fruit
2 cm
Epipactis gigantea
5 mm
upper sepal
and petals
lip
sepal
ovary
5 mm
2 cm
fruit
1 cm
Goodyera oblongifolia
2 cm
column
lip
5 mm
L. convallarioides
sepal
petal
5 mm
5 mm
lip
2 cm
lip
column
lip
tooth
1 mm
Listera caurina
L. cordata
sepal
petal
sepal
petal
ovary
lip
1 mm
lip
sepal
fruit
1 cm
2 mm
Malaxis monophyllos ssp. brachypoda
2 mm
lower
sepal
spur
lip
spur
P. candida
2 mm
P. elegans
upper sepal
lateral petal
lip
2 mm
2 cm
spur
lip
spur
Piperia
candida
P. michaelii
P. elongata
upper sepal
lateral petal
lip
lower sepal
spur
2 mm
P. unalascensis
upper sepal
spur
lip
lower sepal
2 mm
2 cm
P. transversa
lower
sepal
upper sepal
lateral petal
2 mm
lip
spur
Piperia yadonii
P. unalascensis
upper sepal
lateral
petal
lower
sepal
5 mm
column
2 cm
lip
spur
Plantanthera leucostachys
2 cm
5 mm
upper sepal
lower
sepal
spur
5 mm
spur
inflorescence
P. hyperborea
lateral
petal
5 mm
lower sepal
lip
spur
2 cm
inflorescence
Platanthera stricta
P. sparsiflora

S. romanzoffiana Cham. (p. 1231) Pl 7–30 cm. **LVS**: basal 3–13 cm; cauline 1.5–7 cm. **INFL** 2–14 cm; bracts 6–14 mm. **FL**: perianth 8–12 mm, gen white, sometimes cream; upper sepal and lateral petals fused ± > 1/2, fused portion ± inflated at base; lip ± violin-shaped, tip glabrous above; column 1.5–4 mm. $2n=44$. Wet meadows, freshwater marshes, seeps; < 3300 m. NW, CaR, SN, CCo, SnFrB, MP; to AK, ne N.Am, NM. ❀very DFCLT.

POACEAE [Gramineae] GRASS FAMILY

James P. Smith, Jr., except as specified

Ann to bamboo-like; roots gen fibrous. **ST** gen round, hollow; nodes swollen, solid. **LVS** alternate, 2-ranked, gen linear; sheath gen open; ligule membranous or hairy, at blade base. **INFL** various (of gen many spikelets). **SPIKELET**: glumes gen 2; florets (lemma, palea, fl) 1–many; lemma gen membranous, sometimes glume-like; palea gen ± transparent, ± enclosed by lemma. **FL** gen bisexual, minute; stamens gen 3; stigmas gen 2, gen plumose. **FR**: achene-like grain. 650–900 genera; ± 10,000 spp.: worldwide; greatest economic importance of any family (wheat, rice, maize, millet, sorghum, sugar cane, forage crops, orn, weeds; thatching, weaving, building materials). [Hitchcock 1951 Manual grasses US, USDA Misc Publ 200; Clayton & Renvoise 1986 Kew Bull Add Series 13] See Glossary p. 26 for illustrations of general family characteristics. Gen wind-pollinated.

Key to Groups

1. Pl very tough or woody, gen 2–7 m, gen > 10 mm diam; infl often plume-like .... **Group 1**
1′ Ann or per gen < 1.5 m, gen < 5 mm diam; infl various
 2. Infl branches not easily seen at arm's length (infl seeming 1 unit: 1 spikelet, head-like, or densely panicle-like) .... **Group 2**
 2′ Infl branches easily seen at arm's length (infl obviously > 1 unit: umbel- to ± openly panicle-like)
  3. Infl gen umbel-like or panicle-like with spike-like branches, gen dense, spikelets often paired .... **Group 3**
  3′ Infl clearly panicle-like, 1° branches further branched 2+ ×, gen open; spikelets gen not paired
   4. Floret 1 per spikelet .... **Group 4**
   4′ Florets 2+ per spikelet (1+ sometimes sterile or minute) .... **Group 5**

Group 1: Tough or woody, very tall; infl often plume-like

1. St woody, branched near middle; infl axillary, panicle, or cluster [cult bamboos, often spreading by rhizomes but not naturalized, not treated further]
1′ St tough, ± inflexible, not woody, not branched near middle; infl terminal, often plume-like
 2. Spikelets without silky hairs
  3. Infl open, lower branches spreading to ascending; spikelets unisexual, lower staminate, drooping, stamens 6, upper pistillate, ascending to erect; wet places .... **ZIZANIA**
  3′ Infl gen dense, branches ascending to erect; spikelets bisexual, stamens gen 3; dry places
   4. Glumes 2–5-veined at middle; lemma awns 7–35 mm .... **ELYMUS**
   4′ Glumes 1-veined at middle; lemmas acute (or awn < 4 mm) .... **LEYMUS**
 2′ Spikelets with silky hairs
  5. Florets 2, lower vestigial, obscure, upper fertile, gen plump; spikelets paired, lower sessile to short-stalked
   6. Lower spikelet stalk 1–2 mm; spikelets of pair falling separately; lemma awn 8–12 mm; infl fan-shaped .... **MISCANTHUS**
   6′ Lower spikelet sessile; spikelet pair gen falling together; lemma awn 3–5 mm; infl plume-like **SACCHARUM**
  5′ Florets 2–10, gen all fertile; spikelets not clearly paired, gen all stalked
   7. Lemma glabrous (spikelet axis long-soft-hairy) .... **PHRAGMITES**
   7′ Lemma long-hairy, esp below middle
    8. St solid; infl ± 1-sided; lemmas 5–7-veined .... **AMPELODESMOS**
    8′ St hollow; infl not 1-sided; lemmas gen 1–3-veined
     9. Lvs cauline; glumes 3–5-veined; lemma acute, awn ± 0; wet places .... **ARUNDO**
     9′ Lvs basal, clumped; glumes 1-veined; lemma acute to acuminate, short-awned; gen dry places **CORTADERIA**

Group 2: Infl branches obscure

1. Spikelet 1 (per infl)
 2. Lvs 3–8 cm, flat to ± inrolled; glumes 2; lemmas awned; mtns .... ***Danthonia unispicata***
 2′ Lvs ± 1 cm, needle-like; glumes 0; lemmas not awned; coastal salt marshes .... [2]**MONANTHOCHLOË**
1′ Spikelets 2–many
 3. Spikelets enclosed in ± spiny, bur-like involucre .... **CENCHRUS**
 3′ Spikelets not enclosed by bur-like involucre
  4. Pl gen < 10 cm, often tufted or matted; mature infl gen partly hidden in lvs
   5. Glumes 0
    6. Lf blade < 2 mm wide, ± needle-like; spikelets 1–3; lemma boat-shaped; salt marshes [2]**MONANTHOCHLOË**
    6′ Lf blade 5–12 mm wide, flat; spikelets many, infl spike-like, cylindric; lemma fan-shaped; vernal pools .... **NEOSTAPFIA**
   5′ Glumes 1–2

7. Floret 1; lemma strongly 1-veined, acute or short-pointed, not awned .... **CRYPSIS**
7′ Florets 2–20+; lemmas 3–13-veined, gen awned (exc *Sclerochloa, Pennisetum*)
8. Lemma tips entire, margin glabrous, awn 0
9. Infl raceme-like, subtended by 5–10 bristles, spikelet 1–2 mm; florets 2 .... ***Pennisetum clandestinum***
9′ Infl spike-like, ± 1-sided, not subtended by bristles, spikelet 6–10 mm; florets 3–5 ..... **SCLEROCHLOA**
8′ Lemma tips 2–3-toothed or -lobed, margin ciliate to soft-hairy, awns 1 or 3
10. Lemmas 3-lobed, awns 3 (central largest); florets 4, falling together, leaving glumes **BLEPHARIDACHNE**
10′ Lemmas 2-lobed to -toothed, awn 1; florets 4–20, falling separately
11. Glumes 3–9 mm, gen acuminate; lemma back densely soft-hairy below middle .... **ERIONEURON**
11′ Glumes 2–4 mm, gen acute; lemma back short-stiff-hairy on outer veins .... **MUNROA**
4′ Pl gen > 10 cm; mature infl well exserted
12. Lemmas with 9 awns or 5 awn-tipped teeth
13. Awns plumose; glumes entire; dry places, e DMoj .... **ENNEAPOGON**
13′ Awns glabrous to scabrous; glumes irregularly toothed; vernal pools, CA-FP .... **ORCUTTIA**
12′ Lemmas with 0–3 awns, lobes, or teeth
14. Spikelets a mix of fertile and sterile, gen very different in shape and size
15. Glume keel wing-like; fertile spikelet 1, sterile 6 .... ***Phalaris paradoxa***
15′ Glumes not winged; spikelets in 2's, 3's, or clusters
16. Infl panicle-like, dense, ± 1-sided; spikelet clusters spreading to drooping; fertile spikelet 1, sterile 1–3 per cluster .... **LAMARCKIA**
16′ Infl spike- or head-like (± 1-sided in *Cynosurus*)
17. Spikelets 2 per node, lower sessile, with 2–3 fertile florets, upper sterile, short stalked ... **CYNOSURUS**
17′ Spikelets 3 per node, ± sessile, central gen with 1(2) fertile florets, lateral staminate or sterile
18. Base of spikelet cluster gen not hairy-tufted; glumes gen awn-like, tip gen glabrous; awns gen 5–90 mm .... [4]**HORDEUM**
18′ Base of spikelet cluster gen hairy-tufted; glumes oblanceolate, tip ciliate to fringed; awns < 5.5 mm .... [2]**PLEURAPHIS**
14′ Spikelets fertile, alike
19. Spikelets on 1 side of infl axis or in dense, 1-sided, spike-like clusters
20. Infl of lateral, 1-sided, spike-like branches; floret 1; wet places .... **SPARTINA**
20′ Infl spike-like; florets 2–3; dry places
21. Upper floret(s) gen sterile, vestigial, lemmas 3-veined, 3-awned; infl axis cylindric ...... **BOUTELOUA**
21′ Upper floret bisexual, lemma 7–9-veined, not awned; infl axis flat, thick — spikelet subtended by 1 bristle .... **STENOTAPHRUM**
19′ Infl not clearly 1-sided, dense, gen cylindric or spike-like (exc *Schizachyrium*)
22. Spikelet clearly subtended by bristles
23. Bristles and spikelets (or spikelet clusters) falling together .... **PENNISETUM**
23′ Bristles persistent, spikelets falling separately .... **SETARIA**
22′ Spikelet not subtended by bristles
24. Infl axis thickened; lower spikelets strongly appressed or sunken into axis
25. Glumes or lemmas awned
26. Glumes 1–5-awned; florets 2–5 .... **AEGILOPS**
26′ Glumes not awned; floret 1 .... **SCRIBNERIA**
25′ Glumes and lemmas not awned
27. Glume 1; lemmas 3-veined .... **HAINARDIA**
27′ Glumes 2 (to 1 side of spikelet or overlapping); lemmas 1-veined .... **PARAPHOLIS**
24′ Infl axis not clearly thickened; spikelets neither strongly appressed nor sunken into axis
28. Spikelets 2–4 per node
29. St with axillary infls .... **SCHIZACHYRIUM**
29′ St without axillary infls
30. Glumes awnless (0 in *Elymus californicus*)
31. Fertile florets gen 2–7
32. Glumes narrowly elliptic to lanceolate; lemma awns 16–30 mm .... [4]**ELYMUS**
32′ Glumes gen awl-like (exc *L. mollis*); lemma awns gen 0–5 mm .... [3]**LEYMUS**
31′ Fertile floret 0–1
33. Spikelets 3 per node, central sessile, lateral short-stalked; infl axis breaking at nodes [4]**HORDEUM**
33′ Spikelets 2 per node, short-stalked or lower sessile; infl axis not breaking, spikelets falling in pairs
34. Lower spikelet sessile; awns > 5 cm, red-brown, puberulent .... **HETEROPOGON**
34′ Lower spikelet stalked; awns 0 .... **IMPERATA**
30′ Glumes awned (or awn-like throughout)
35. Some glumes awned from middle and tip .... **PLEURAPHIS**
35′ Glumes gen awned only from tip
36. Infl axis breaking apart in fr
37. Spikelets 3 per node, central sessile, with 1 fertile floret, lateral spikelets sterile, short-stalked .... [4]**HORDEUM**
37′ Spikelets 2 per node, gen sessile, each with 2–7 fertile florets
38. Glumes ± elliptic to lanceolate; lemma awns 16–30 mm .... [4]**ELYMUS**
38′ Glumes gen awl-like (exc *L. mollis*); lemma awns gen 0–5 mm .... [3]**LEYMUS**
36′ Infl axis not breaking apart in fr

39. Lemmas 3-veined
40. Lemma awn 2–7 mm; spikelet pair falling together; lower glume 2–3-awned ..... [2]**LYCURUS**
40′ Lemma awn 30–70 mm; spikelets not falling together, spikelet axis breaking at nodes, leaving glumes; lower glume 1-awned .... **TAENIATHERUM**
39′ Lemmas 5-veined
41. Fertile floret 1 per spikelet; ann or per; rhizomes gen 0 .... [4]**HORDEUM**
41′ Fertile florets 2–7 per spikelet; per, often rhizomed
42. Glumes ± elliptic to lanceolate; lemma awns 16–30 mm .... [4]**ELYMUS**
42′ Glumes gen awl-like (exc *L. mollis*); lemma awns gen 0–5 mm .... [3]**LEYMUS**
28′ Spikelets 1 per node
43. Floret 1 per spikelet, fertile
44. Glumes plumose or with keel clearly stiff- to soft-hairy (exc *Alopecurus myosuroides*)
45. Glumes not awned; spikelet gen breaking below glumes, falling as 1 unit .... **ALOPECURUS**
45′ Glumes pointed to awned; spikelet axis breaking above glumes, floret falling as 1 unit
46. Lemmas 3-awned; glumes ± soft-hairy, gen tapered to tip .... **LAGURUS**
46′ Lemmas not awned; glumes stiff-ciliate, gen truncate .... **PHLEUM**
44′ Glumes glabrous to sparsely hairy, keel neither plumose nor clearly stiff- to soft-hairy
47. Glumes clearly awned
48. Lower glume 2–3-veined, 2–3-awned .... [2]**LYCURUS**
48′ Lower glume gen 1-veined, short-pointed or 1-awned
49. Spikelet axis breaking above glumes, floret falling as 1 unit; lemma 3-veined; awn 0 or gen < lemma body .... [2]**MUHLENBERGIA**
49′ Spikelet breaking below glumes and part of stalk, falling as 1 unit; lemma 5-veined; awn 0 or gen > body .... **POLYPOGON**
47′ Glumes long-tapered to short-toothed, not clearly awned
50. Glumes swollen at base, long-tapered; upper glume 3–5 × > lemma body .... **GASTRIDIUM**
50′ Glumes not swollen at base, tip obtuse to acuminate; upper glume 1–2 × > lemma body
51. Palea < 1/3 lemma .... **AGROSTIS**
51′ Palea ± = lemma
52. Awn from lemma middle or below, slender, often twisted and bent ...... **CALAMAGROSTIS**
52′ Awn 0 or from lemma tip
53. Spikelet > 10 mm; lemma 5–7-veined .... **AMMOPHILA**
53′ Spikelet < 6 mm; lemma 1–3-veined
54. Lemma gen 3-veined; top of lf sheath gen glabrous .... [2]**MUHLENBERGIA**
54′ Lemma gen 1-veined; top of lf sheath gen densely hairy .... **SPOROBOLUS**
43′ Florets 2–20+ per spikelet
55. Fertile floret 1, gen subtended by 2 sterile florets (often obscure in *Phalaris*)
56. Sterile florets gen > fertile floret; sterile lemmas ± like fertile one, awned .... **ANTHOXANTHUM**
56′ Sterile florets gen < fertile floret; sterile lemmas awl- or chaff-like, not awned .... **PHALARIS**
55′ Fertile florets 2+; sterile florets, if any, above fertile floret(s)
57. Lf sheath and blade continuous; lemmas clearly 11–15-veined .... **TUCTORIA**
57′ Lf sheath separated from blade by line, fold, or crease; lemmas 3–9-veined
58. Lemmas 3-veined
59. Pl from short stolons; spikelets unisexual; staminate lemmas with 1 short awn, pistillate lemmas with 3 long awns .... **SCLEROPOGON**
59′ Pl tufted; spikelets bisexual; lemmas entire to awn-tipped
60. Lemma tip minutely 2-toothed, awn < 2 mm, back hairy throughout .... ***Erioneuron pilosum***
60′ Lemma tip entire to minutely notched, veins of back densely hairy .... ***Tridens muticus***
58′ Lemmas 5–9-veined
61. Lemma veins parallel, not converging at tip .... **PLEUROPOGON**
61′ Lemma veins converging at tip
62. Glume gen 1, on outside edge of spikelet .... **LOLIUM**
62′ Glumes 2
63. Spikelets sessile
64. Lemma keel and margin ciliate .... **SECALE**
64′ Lemmas not ciliate
65. Palea gen ciliate above middle; spikelets not clearly 2-ranked .... **BRACHYPODIUM**
65′ Palea entire; spikelets gen 2-ranked (exc *L. triticoides*)
66. Spikelets strongly overlapping; infl axis internodes gen < 3 mm .... **AGROPYRON**
66′ Spikelets not strongly overlapping; infl axis internodes 4–15 + mm
67. Glumes thick, truncate to acute, awn 0, scabrous distally, midvein prominent; lemma awn gen 0 .... **ELYTRIGIA**
67′ Glumes ± thin, ± acuminate, awned or not, gen smooth, midvein not prominent; lemmas gen awned
68. Glume widest at middle, ± curved to side, tip acute; infl gen strongly glaucous .... **PASCOPYRUM**
68′ Glume widest below middle, straight, tip obtuse to acute; infl not strongly glaucous
69. Lemma awns (5)10–35 mm .... [4]**ELYMUS**
69′ Lemma awns gen < 5 mm

70. Pl rhizomed; glumes awl-like; moist, often saline places .... ***Leymus triticoides***
70′ Rhizomes 0; glumes elliptic to lanceolate; dry places .... **PSEUDOROEGNERIA**
63′ Spikelets short-stalked
71. Rhizomes thick, scaly; st stiff, ridged, base ± woody — n-most DMoj .... **SWALLENIA**
71′ Rhizomes 0 or thin; st not stiff and ridged, base not woody
72. Upper glume > and wider than lower when flat .... **KOELERIA**
72′ Glumes ± alike (upper gen < lower when flat)
73. Lemma not awned
74. Glumes 5–7-veined, gen > lower floret; lemma gen 9-veined .... **SCHISMUS**
74′ Glumes 1–3-veined, gen < lower floret; lemma gen 5-veined
75. Spikelet axis not prolonged, not bristly; lf blade prow-tipped .... **POA**
75′ Spikelet axis prolonged behind upper floret, bristly; lf tip ± flat .... ***Trisetum wolfii***
73′ Lemma awned from tip or back
76. Glumes < florets; lemma awned from tip .... **VULPIA**
76′ 1–2 glumes = or > florets; lemma awned from back
77. Glumes gen > lower floret; florets 3–8; lemma awn flat .... **DANTHONIA**
77′ Glumes gen < or = lower floret; florets 2–3; lemma awn ± cylindric .... **TRISETUM**

Group 3: Infl umbel-like or branches spike-like, not branched again

1. Spikelets paired (gen 1 sessile, 1 stalked), gen falling together; lower glume firm to hard, enclosing upper glume; fertile lemma thin
2. Spikelets bisexual
3. Infl axis not breaking in fr; spikelets breaking below glumes, falling separately .... **MISCANTHUS**
3′ Infl axis breaking apart in fr; spikelet pair falling with axis segment (or stalked spikelet falling alone) .... **SACCHARUM**
2′ Sessile spikelet bisexual, stalked spikelet staminate or sterile (vestigial in *Andropogon*)
4. Upper infl axis and upper stalks flat, with ± membranous groove .... **BOTHRIOCHLOA**
4′ Upper infl axis and upper stalks round to flat, not grooved
5. Awns < 5 mm; sessile spikelet bisexual .... **ANDROPOGON**
5′ Awns 10–35 mm; lowest spikelet pairs gen staminate, upper sessile spikelets bisexual ...... **HYPARRHENIA**
1′ Spikelets 1–3 per node (if 2, glumes and lemmas ± alike or fertile lemma thicker than glumes)
6. Fertile floret 1; sterile floret 0 or reduced and below fertile floret (exc some *Cynodon*)
7. Spikelet axis minutely prolonged above fertile floret, bristle-like at 20× magnification .... **CYNODON**
7′ Spikelet axis not prolonged above fertile floret
8. Fertile lemma firm to rigid, gen thicker than upper glume
9. Fertile lemma gen abruptly pointed to short-awned (exc *E. acuminata*); lower glume forming disk- or cup-like ring below stalk and upper glume .... [2]**ERIOCHLOA**
9′ Fertile lemma not awned or abruptly pointed; lower glume 0 or minute
10. Spikelet narrowly oblong; fertile lemma back facing away from infl axis, minutely bristle-tipped .... [2]**AXONOPUS**
10′ Spikelet elliptic to widely ovate; fertile lemma back facing infl axis, not bristle-tipped ...... [3]**PASPALUM**
8′ Fertile lemma thin, flexible, gen like upper glume(s)
11. Glumes clearly 3–5-veined, veins minutely ridged .... **DIGITARIA**
11′ Glumes ± faintly 1–3(5)-veined
12. Ann; spikelets round; glumes wrinkled, back winged; lemmas 5-veined .... [2]**BECKMANNIA**
12′ Per, gen rhizomed; spikelets narrow; glumes not wrinkled, back keeled; lemmas 1–3-veined ... **SPARTINA**
6′ Fertile florets gen 2 (if 1, subtended by 1–2 sterile florets or lemmas; sterile lemmas sometimes mistaken for glumes)
13. Fertile florets 2–12
14. Infl gen panicle-like; infl branches not clearly 1-sided (exc *Sclerochloa*)
15. Lemmas (1)3-veined; gen wet places .... **LEPTOCHLOA**
15′ Lemmas (3)5–9-veined; gen dry places
16. Upper glume 7–9-veined, tip blunt; spikelet falling as 1 unit .... **SCLEROCHLOA**
16′ Upper glume 3-veined, tip acute to short-pointed; spikelet axis breaking above glumes and between florets
17. Lemma unawned .... **DESMAZERIA**
17′ Lemma awn gen 1–22 mm .... **VULPIA**
14′ Infl umbel-like (branches in 2–4 whorls in *C. verticillata*), spike-like; spikelets ± on 1 side of infl branches
18. Glumes and lemmas unawned .... **ELEUSINE**
18′ Glumes or lemmas awned
19. Infl branch axis with spikelets to tip; both glumes persistent, awns gen straight .... [2]**CHLORIS**
19′ Infl branch axis with naked, bristle-like tip; lower glume persistent, upper deciduous, awns ± curved .... **DACTYLOCTENIUM**
13′ Fertile floret 1
20. Glumes ± semicircular, wrinkled, winged .... [2]**BECKMANNIA**
20′ Glumes not semicircular, not wrinkled, not winged
21. Infl panicle-like, branches falling with attached spikelets .... [2]**BOUTELOUA**
21′ Infl gen panicle- to umbel-like, branches persistent, spikelets or florets falling separately

22. Spikelets gen 2–3 per node
23. Glumes and lemmas gen awn-tipped; ligule 0 .... **ECHINOCHLOA**
23′ Glumes and lemmas rounded to acute; ligule present
24. Fertile lemma flexible, thinner than sterile lemma and upper glume, back facing away from infl axis .... **DIGITARIA**
24′ Fertile lemma firm to hard, thick, gen like sterile lemma and upper glume, back facing infl axis .... [3]**PASPALUM**
22′ Spikelets 1 per node
25. Fertile lemma membranous, gen like upper glume; glumes 2 (lower glume << upper glume in *Digitaria*)
26. Infl panicle-like, branches 1 per node .... [2]**BOUTELOUA**
26′ Infl umbel-like or in 2–4 whorls .... [2]**CHLORIS**
25′ Fertile lemma firm to hard, thicker than upper glume; lower glume 0 or << upper glume
27. Fertile lemma gen abruptly pointed to short-awned (exc *E. acuminata*); lower glume forming a disk- or cup-like ring below stalk and upper glume .... [2]**ERIOCHLOA**
27′ Fertile not awned or abruptly pointed; lower glume 0 or minute
28. Spikelet narrowly oblong; fertile lemma back facing away from infl axis, minutely bristle-tipped .... [2]**AXONOPUS**
28′ Spikelet elliptic to widely ovate; fertile lemma back facing infl axis, not bristle-tipped ... [3]**PASPALUM**

Group 4: Infl panicle-like; floret 1 per spikelet

1. Glumes 0 or vestigial
2. Monoecious; lower infl branches with staminate spikelets, upper branches with pistillate spikelets ..... **ZIZANIA**
2′ Spikelets bisexual
3. Per from long rhizomes; lemmas awnless .... **LEERSIA**
3′ Ann; lemmas abruptly short-pointed to rigidly awned .... [3]**ORYZA**
1′ Glumes 2, well developed
4. Glumes ± semicircular, winged, wrinkled .... **BECKMANNIA**
4′ Glumes not semicircular, winged (exc *Phalaris*), or wrinkled
5. Glumes and lemmas awnless (or awns early deciduous)
6. Lemma base clearly hairy-tufted
7. Spikelets 10–13 mm, subsessile; infl ± dense; coastal dunes .... **AMMOPHILA**
7′ Spikelets 3–10 mm, gen clearly stalked; infl gen open; ± dry places away from coast
8. Lemma clearly 3-veined; spikelet compressed .... [6]**MUHLENBERGIA**
8′ Lemma faintly 5–7-veined; spikelets ± round in X-section
9. Glumes 5–10 mm; floret 3–6 mm .... [2]**ACHNATHERUM**
9′ Glumes 2.5–3.5 mm; floret 1.5–2.5 mm .... [2]**PIPTATHERUM**
6′ Lemma base not hairy-tufted (exc some *Agrostis*)
10. Lower glume (or both glumes) < lemma
11. Lemma 5-veined; glumes scale-like; stamens 6; wet places .... [3]**ORYZA**
11′ Lemma 1–3-veined; glumes not scale-like; stamens 2–3; ± dry places
12. Lemma 3-veined; fr falling with lemma and palea .... [6]**MUHLENBERGIA**
12′ Lemma 1-veined; fr falling from floret .... **SPOROBOLUS**
10′ Glumes = or sometimes 1 > lemma
13. Glumes with thin, membranous wing .... **PHALARIS**
13′ Glumes not winged
14. Spikelet axis prolonged as a short bristle behind floret; spikelet breaking below glumes, falling as 1 unit .... [2]**CINNA**
14′ Spikelet axis gen not prolonged; floret or fr gen breaking above persistent glumes
15. Lemma gen 5-veined; palea 0 or << lemma .... [3]**AGROSTIS**
15′ Lemma 3-veined; palea ± = lemma .... [6]**MUHLENBERGIA**
5′ Glumes or lemmas awned
16. Lemma 3-awned, central sometimes >> lateral .... **ARISTIDA**
16′ Lemma gen 1-awned (sometimes awnless in *Cinna*)
17. Upper or both glumes abruptly short-pointed to awned from tip or from between 2 teeth
18. Glumes rosy (but fading), upper silky .... **RHYNCHELYTRUM**
18′ Glumes not rosy, upper gen glabrous or scabrous
19. Glume awn gen >> body (exc *P. elongatus*) .... **POLYPOGON**
19′ Glume awn < glume body
20. Glumes 12–19 mm; lemma body > 7.5 mm .... [2]**NASELLA**
20′ Glumes < 6 mm; lemma body < 7 mm
21. Palea < 2/3 lemma; lemma gen 5-veined .... [3]**AGROSTIS**
21′ Palea gen = lemma; lemma gen 3-veined .... [6]**MUHLENBERGIA**
17′ Glumes obtuse, acute, or tapered to tip
22. Glumes minute, scale-like; stamens 6 .... [3]**ORYZA**
22′ Glumes neither minute nor scale-like; stamens gen 2–3
23. Lemma stiffly membranous to hard, ± tightly enfolding palea and fr
24. Lemma margins inrolled, clasping palea; palea grooved, tip exposed .... **PIPTOCHAETIUM**
24′ Lemma margins flat, either overlapping or leaving back of palea exposed; palea not grooved

25. Infl wide, branches widely spreading to ± drooping; palea < 1/3 lemma ... **[2]NASELLA**
25′ Infl narrow, branches gen ascending; palea > 1/2 to = lemma
26. Glumes 2.5–4.5 mm; lemma awn 3–14 mm; floret base blunt
27. Lemma awn deciduous, 3–11 mm; floret 1–2.5 mm; dry places ... **[2]PIPTATHERUM**
27′ Lemma awn persistent, 10–14 mm; floret gen ± 3–4 mm; wet places ... **PTILAGROSTIS**
26′ Glumes 5–22 mm; lemma awn gen 12–200 mm; floret base sharp (exc some *Achnatherum*)
28. Floret base < 2 mm, hairs gen < or = hairs on lemma body; palea < lemma, gen flexible ... **[2]ACHNATHERUM**
28′ Floret base 2–3 mm, hairs clearly > hairs on lemma body; palea = lemma, hard or stiff ... **HESPEROSTIPA**
23′ Lemma thin, flexible, ± not enfolding palea and fr
29. Lemma awned ± below middle
30. Floret axis not prolonged; palea 0 or < lemma; floret base glabrous or hairs < 1 mm ... **[3]AGROSTIS**
30′ Floret axis prolonged beyond floret, hairy; palea ± = lemma; floret base hairs gen 1–4 mm ... **CALAMAGROSTIS**
29′ Lemma abruptly short-pointed or awned at (near) tip
31. Lemma awn 3+ × > lemma body
32. Glumes > lemma body; lemma awns 4–9 mm ... **APERA**
32′ 1–2 glumes gen < lemma body; lemma awns 10–30 mm ... **[6]MUHLENBERGIA**
31′ Lemma awn < or = lemma body
33. Floret axis prolonged behind floret, short-bristly; spikelet falling as 1 unit; palea ± = lemma ... **[2]CINNA**
33′ Floret axis not prolonged; floret breaking above glumes; palea gen << lemma ... **[6]MUHLENBERGIA**

Group 5: Infl panicle-like; florets 2+ per spikelet

1. Spikelets enclosed in stiff-bristly or bur-like involucre ... **CENCHRUS**
1′ Spikelets not enclosed by involucre
2. Lemma with 3–9 awns, awn-like lobes, or teeth
3. Lemma 3-awned
4. Lemma ciliate; awns ± 6 mm; fls bisexual ... **BLEPHARIDACHNE**
4′ Lemma glabrous; awns 60–120 mm; fls pistillate ... **[2]SCLEROPOGON**
3′ Lemma with 5–9 awns or teeth
5. Lemma with 9 plumose awns; rocky slopes, e D ... **ENNEAPOGON**
5′ Lemma with 5 awn-like teeth; vernal pools, CA-FP ... **ORCUTTIA**
2′ Lemma abruptly short-pointed; awn 0–1
6. Lemma of fertile floret(s) firmer or thicker than glumes
7. Spikelets (or clusters) subtended by 1–many bristles
8. Upper glume and fertile lemma abruptly pointed to awned; ligule 0 ... **ECHINOCHLOA**
8′ Glumes and fertile lemma acute to blunt, awn 0; ligule present
9. Spikelet (or cluster) and subtending bristles falling as 1 unit; bristles ± plumose, esp below **PENNISETUM**
9′ Spikelet falling as 1 unit; bristles persistent, rough or short-bristly, not plumose ... **SETARIA**
7′ Spikelets not subtended by bristles
10. Upper glume and sterile lemma silky or appressed-hairy
11. Infl dense, spikelets 2–3 mm, 2–3 per node, subsessile to short-stiff-stalked ... **DIGITARIA**
11′ Infl ± open, spikelets 3–5.5 mm, gen 1 per node, on short, wiry stalks ... **RHYNCHELYTRUM**
10′ Upper glume and sterile lemma glabrous to scabrous
12. Upper glume gen clearly awned, fertile lemma abruptly pointed; ligule 0; palea free from lemma ... **[2]ECHINOCHLOA**
12′ Upper glume and fertile lemma gen awnless; ligule present; palea enfolded by lemma ... **PANICUM**
6′ Lemma of fertile floret(s) and glumes ± alike in texture (or glumes thicker than fertile lemma)
13. Glumes (esp of sessile spikelet) leathery, shiny, thicker than fertile lemma ... **SORGHUM**
13′ Glumes and lemmas ± alike in texture
14. Spikelets 2–many per node
15. Spikelets subsessile to stalked; lower glume 0 or < 0.5 mm, << upper glume ... **[2]DIGITARIA**
15′ Spikelets sessile to sunken into infl axis; lower glume ± = lanceolate to awl-like upper glume
16. Glume body narrowly elliptic to lanceolate; lemma awn 16–30 mm ... **ELYMUS**
16′ Glume body awl-like; lemma acute or awn < 5 mm ... **LEYMUS**
14′ Spikelets gen 1 per node (± paired in *Cynosurus*)
17. Spikelets both fertile and sterile, unlike
18. Spikelets ± paired, ascending; fertile spikelet gen < sterile spikelet, sessile, breaking apart above glumes; sterile spikelet persistent ... **CYNOSURUS**
18′ Spikelets gen 2–4 per cluster, spreading or drooping; fertile spikelet 1, sterile spikelets 1–3, cluster falling as 1 unit ... **LAMARCKIA**
17′ Spikelets gen all fertile (exc *Poa bulbosa*), ± alike
19. Spikelets unisexual or florets forming leafy bulblets
20. Florets forming leafy bulblets, st base bulb-like ... ***Poa bulbosa***
20′ Florets staminate or pistillate, st base narrow
21. Lemma base often cobwebby-tufted; lf tips folded, prow-like ... **[3]POA**
21′ Lemma base not hairy-tufted; lf tip flat, tapered

22. Pl densely tufted; sts hollow; dry mtn slopes .... *Festuca kingii*
22′ Pl from scaly rhizomes; sts solid; alkaline or saline places .... **DISTICHLIS**
19′ Bisexual florets 1–many; unisexual or sterile florets gen 0–2
23. Bisexual floret 1, gen subtended by 1–2 staminate or sterile florets (exc *Holcus*)
24. Sterile or staminate florets above bisexual floret — uppermost floret with straight to curved awn .... **HOLCUS**
24′ Sterile or staminate florets below bisexual floret
25. Bisexual floret awned (exc *Arrhenatherum*)
26. Bisexual floret subtended by 1 staminate floret; lower floret awn 10–17 mm; upper floret awn (0) < 4 mm .... **ARRHENATHERUM**
26′ Bisexual floret subtended by 2 staminate florets; lower floret awn < 5 mm; upper floret awn 10–15 mm .... **VENTENATA**
25′ Bisexual floret awnless
27. Lower florets gen << bisexual floret, awl- or knob-like .... **PHALARIS**
27′ Lower florets gen = and like bisexual floret in texture
28. Lower glume ± 1/2 upper glume .... **ANTHOXANTHUM**
28′ Glumes ± equal
29. Sterile lemmas pointed to short-awned; glumes clearly 3-veined .... *Ehrharta calycina*
29′ Sterile lemmas rounded to acute; glumes faintly 3–7-veined
30. Lemma ± glabrous; glumes dull; sterile lemma cross-wrinkled near tip .... *Ehrharta erecta*
30′ Lemma ± hairy or ciliate; glumes shiny; sterile lemmas smooth .... **HIEROCHLOË**
23′ Bisexual (or pistillate) florets 2–many; staminate or sterile florets, if any, above pistillate
31. Lemma awns 3 .... [2]**SCLEROPOGON**
31′ Lemma awn 0 or 1
32. Lemma gen clearly 3-veined
33. Lemma back glabrous (floret base sometimes puberulent or short-hairy)
34. Florets (3)5–40 per spikelet; glumes and lemmas obtuse to acuminate .... **ERAGROSTIS**
34′ Florets 1–2 per spikelet; glumes and lemmas gen short-pointed to awned .... **MUHLENBERGIA**
33′ Lemma back, esp veins, puberulent throughout or hairy below middle
35. Florets 2 per spikelet; glumes > florets; s CHI (presumed extinct) .... **DISSANTHELIUM**
35′ Florets 4–20 per spikelet; glumes < or = lower florets; SNE, DMtns
36. Glumes 2–4 mm; infl ± hidden by upper lf clusters .... **MUNROA**
36′ Glumes 3–9 mm; infl not hidden by lvs
37. Lf blade white-margined; lemmas (and some glumes) short-awned; stigmas white **ERIONEURON**
37′ Lf blades not white-margined; lemmas and glumes awnless; stigmas purple .... **TRIDENS**
32′ Lemmas ± 5–many-veined
38. Some lemmas awned
39. Lemma awns 15–90 mm
40. Glumes > lemmas (exc awns); lemmas gen awned from back .... **AVENA**
40′ Glumes < lemmas (exc awns); lemma awned from 2-lobed or -toothed tip .... [3]**BROMUS**
39′ Lemma awns gen < 12(15) mm
41. Lemmas awned at or below middle
42. Lemmas 1–2 mm, tips with 2 slender teeth; spikelet axis not prolonged, bristly; floret base short-bristly .... **AIRA**
42′ Lemmas 2–4 mm, tips truncate, with 2–4 short teeth; spikelet axis prolonged beyond upper floret; floret base soft-hairy .... **DESCHAMPSIA**
41′ Lemmas awned from or near tip
43. Lower glume > lowest floret
44. Spikelet axis ± glabrous, not prolonged .... **DANTHONIA**
44′ Spikelet axis stiff- to short-hairy, prolonged behind upper floret .... [2]**TRISETUM**
43′ Lower glume < or = lowest floret
45. Upper glume wider (when flattened) than lower glume
46. Upper glume widest below middle; floret axis gen prolonged as slender bristle or vestigial floret; dry places .... [2]**KOELERIA**
46′ Upper glume widest above middle; floret axis not prolonged; wet places .. [2]**SPHENOPHOLIS**
45′ Glumes ± alike in shape
47. Glumes (and lemmas) papery; lemmas clearly 5–7-veined, upper margin gen translucent; st base often bulb-like .... [2]**MELICA**
47′ Glumes not papery; lemmas obscurely veined, margin ± opaque; st base narrow
48. Lemma tip 2-lobed; spikelets gen > 1.5 cm; sheath closed .... [3]**BROMUS**
48′ Lemma tip entire; spikelets gen < 1 cm; sheath open (exc *Dactylis*)
49. Upper glume = or > lower floret; lemma awned from below tip (awn 0 in *T. wolfii*) .... [2]**TRISETUM**
49′ Upper glume < lower floret; lemma awned from tip
50. Spikelets gen on 1 side of branch tips, crowded; sheath closed, back keeled ... **DACTYLIS**
50′ Infl not 1-sided; sheath open, back ± rounded
51. Per; stamens 3 .... [3]**FESTUCA**
51′ Ann; stamen gen 1 .... **VULPIA**
38′ Lemmas awnless

52. Glumes and lemmas papery; lemma wider than long .... **BRIZA**
52′ Glumes and lemmas membranous; lemma longer than wide
53. Lower glume clearly > lowest lemma .... **SCHISMUS**
53′ Lower glume < or = lowest lemma
54. Lemma veins ± equally spaced, parallel, not converging at lemma tip — gen wet places
55. Sheath closed to near top; upper glume gen 1-veined .... **GLYCERIA**
55′ Sheath open to base; upper glume gen 3-veined
56. Lemmas faintly 5-veined; ann or per, rhizomes 0; open, saline or alkaline soils **PUCCINELLIA**
56′ Lemmas clearly 7–9-veined; per from rhizomes; near freshwater .... **TORREYOCHLOA**
54′ Lemma veins not equally spaced, gen curved, gen converging at tip
57. Upper glume wider (when flattened) than lower — lower glume 1-veined, upper 3–5-veined
58. Upper glume widest below middle; floret axis prolonged as slender bristle or vestigial floret; dry places .... [2]**KOELERIA**
58′ Upper glume widest above middle; floret axis not prolonged; wet places .. [2]**SPHENOPHOLIS**
57′ Glumes ± alike in shape
59. Floret base cobwebby-tufted .... [3]**POA**
59′ Floret base glabrous or puberulent
60. Lemmas 5-veined (2 may be faint)
61. Spikelet stalk thick, flat, rigid; central infl axis ± wavy or zigzag; lemma ± circular in X-section .... **DESMAZERIA**
61′ Spikelet stalk slender, round, flexible, not flat; central infl axis not wavy or zigzag; lemma V-shaped or ± flat in X-section
62. Lf tip flat, tapered; lemma gen short-awned from tip (or 0) .... [3]**FESTUCA**
62′ Lf tip folded, prow-like; lemmas awnless .... [3] **POA**
60′ Lemmas 7–9-veined
63. Glumes ± papery-translucent; lemma clearly veined, margin thin, not green; st base often bulb-like .... [2]**MELICA**
63′ Glumes not papery-translucent; lemma faintly veined, margin gen green; st base narrow
64. Lemma tip 2-lobed; lf sheath closed; spikelets (exc awns) gen > 1.5 cm .... [3]**BROMUS**
64′ Lemmas tapered to tip; lf sheath gen open; spikelets (exc awns) gen < 1 cm ... [3]**FESTUCA**

ACHNATHERUM NEEDLEGRASS

Mary E. Barkworth

Per, tufted. **ST** gen erect. **LF**: ligule membranous, sometimes long-ciliate, blade gen flat. **INFL** panicle-like, gen narrow; branches gen ascending. **SPIKELET**: glumes > floret (exc awn), tapered below midpoint; axis breaking above glumes; floret 1, gen cylindric; callus blunt or sharp, hairs stiff; lemma stiffly membranous to hard, evenly hairy or glabrous above, awned from tip; awn > 10 mm, persistent, with 1–2 bends, or < 10 mm, readily deciduous, ± straight; palea < lemma, hairy, veined. ± 75 spp.: temp worldwide. (Greek: awned scale, from lemma) Segregated mostly from *Stipa*; see also *Hesperostipa, Nassella*.

1. Basal segment of awn hairy, some hairs > 1 mm; callus sharp
2. Lowest hairs on awn > upper lemma hairs
3. Awn with 1 bend; uppermost ligule 0.3–1 mm, long ciliate .... ***A. speciosum***
3′ Awn with 2 bends; uppermost ligule 2–7 mm, glabrous
4. Infl 15–30 cm; uppermost ligule 2–4 mm, opaque .... ***A. latiglumis***
4′ Infl 7–13 cm; uppermost ligule 3–7 mm, translucent .... ***A. thurberianum***
2′ Lowest hairs on awn < upper lemma hairs
5. Palea 2/3–3/4 lemma length .... ***A. nevadensis***
5′ Palea 1/3–1/2 lemma length .... ***A. occidentalis***
6. All awn segments hairy .... ssp. ***occidentalis***
6′ Lower awn segments hairy, upper glabrous
7. Hairs of lemma tip = hairs at awn base .... ssp. ***californicum***
7′ Hairs of lemma tip > hairs at awn base .... ssp. ***pubescens***
1′ Basal segment of awn scabrous; callus sharp or blunt
8. Hairs of lemma tip 2–7 mm
9. Awn 3–11 mm, straight, readily deciduous
10. Infl open, branches and pedicels widely spreading .... ***A. hymenoides***
10′ Infl dense, branches and pedicels ascending .... ***A. webberi***
9′ Awn 13–44 mm, with 1 or 2 bends, persistent
11. Hairs of lemma tip = hairs at lemma midpoint .... ***A. pinetorum***
11′ Hairs of lemma tip 1–7 mm, > hairs at midpoint; awn with 1 or 2 bends
12. Awn bent twice; basal lf blade 2.5–7 mm .... ***A. coronatum***
12′ Awn bent once; basal lf blade 2–3 mm .... ***A. parishii***
8′ Hairs of lemma tip < 2 mm
13. Distal segment of awn wavy .... ***A. aridum***
13′ Distal segment of awn straight

14. Palea 1/3–2/3 lemma length
 15. Internodes above middle with reflexed hairs .... ***A. diegoensis***
 15′ Internodes gen glabrous or inconspicuously hairy at nodes .... ***A. nelsonii***
14′ Palea > 3/4 lemma length
 16. Lemma tip 2-toothed, awn-like teeth 1–2 mm; sts 6–15 dm .... ***A. stillmanii***
 16′ Lemma tip 2-toothed or not, teeth < 1 mm; sts 1.5–9 dm
  17. Lemma tip with thick lobe ± 0.1 mm; floret 5.5–7 mm, somewhat compressed .... ***A. lemmonii***
  17′ Lemma tip with membranous lobe 0.2–1 mm; floret 4.5–6 mm, cylindric
   18. Palea thick, stiff, veins evident .... ***A. brachychaetum***
   18′ Palea thin, flexible, veins obscure .... ***A. lettermanii***

A. aridum (M.E. Jones) Barkworth MORMON NEEDLEGRASS **ST** 3.5–8.5 dm. **LF**: blade 0.9–3 mm wide. **INFL** 5–17 cm, often partly enclosed by uppermost lf sheath. **SPIKELET**: glumes 8–15 mm, lower > upper; floret 4–6.5 mm; callus sharp; lemma 1.4–2.4 × palea length, hairs short, awn 40–80 mm, distal segment wavy; palea 2–3.2 mm. RARE in CA. Outcrops, shrub-steppe, pinyon/juniper; 1200–1550 m. e DMoj; to Colorado, AZ. [*Stipa a.* M.E. Jones]

A. brachychaetum (Godron) Barkworth **ST** 4–10 dm. **LF**: sheath glabrous; blade 1–2 mm wide. **INFL** 20–30 cm. **SPIKELET**: glumes 6–8 mm, ± equal; floret 4.5–5.5 mm; callus blunt; lemma tip hairs 0.8–1.7 mm, > body hairs, awn 15–20 mm, bent twice. Disturbed sites; < 300 m. nw SnJV (eradicated from SnJV in Fresno Co.), n SCo; native to Argentina. [*Stipa b.* Godron] NOXIOUS WEED.

A. coronatum (Thurber) Barkworth (p. 1231) **ST** 5–21 dm, glabrous to hairy. **LF**: sheath ciliate at top, hairs 1–2.5 mm; blade 2.5–7 mm wide. **INFL** 15–60 cm; branches widely spreading to ascending. **SPIKELET**: glumes unequal, lower 16–21 mm, upper 11–18 mm; floret 6.5–10 mm; callus blunt; lemma 1.2–1.7 × palea length, hairs 1.5–5 mm, awn 25–45 mm, bent twice. Gravelly, rocky slopes, chaparral; < 1550 m. SCoR, SW; Baja CA. [*Stipa c.* Thurber] ❀DRN,DRY,SUN:7,11,**14–16**,17,19–21,**22,23**;STBL.

A. diegoensis (Swallen) Barkworth SAN DIEGO COUNTY NEEDLEGRASS **ST** 11–14 dm; lowest internode hairy throughout, others less so. **LF**: basal sheaths hairy. **INFL** 21–25 cm. **SPIKELET**: glumes 8–11.5 mm, ± equal; floret 5–7.5 mm; callus blunt; lemma 1.6–2 × palea length, hairs 0.5–1 mm, awn 26–36 mm, bent twice. UNCOMMON. Rocky soil, chaparral, coastal-sage scrub, often near streams on mainland; < 350 m. ChI, PR; Baja CA. [*Stipa d.* Swallen]

A. hymenoides (Roemer & Schultes) Barkworth (p. 1231) INDIAN RICEGRASS **ST** 2.5–7 dm. **LF**: blade rolled, < 1 mm diam. **INFL** ± open; branches, spikelet stalks widely spreading. **SPIKELET**: glumes 5–9 mm, ± equal, base swollen; floret 3–4.5 mm; callus sharp; lemma ± spheric, hairs 2.5–6 mm, awn 3–6 mm, deciduous. $2n$=46,48. Dry, well drained, often sandy soil, desert-shrub, sagebrush shrubland, pinyon/juniper; < 3400 m. CaR, SN, SW, GB, D; to B.C., Great Plains, Mex. [*Oryzopsis h.* (Roemer & Schultes) Ricker] Hybrids with other spp., incl *O.* ×*bloomeri* (Bolander) Ricker, have narrower florets and awns 12–16 mm. Used by Native Americans for food; highly palatable to livestock. ❀DRN,DRY,SUN:1,**2,3,7**,8–11,**14**,15,16,**18–21**,22–24;STBL.

A. latiglumis (Swallen) Barkworth **ST** 5–11 dm, hairy below. **LF**: basal sheaths hairy; ligule 1–4 mm; blade < 3 mm wide. **INFL** 15–30 cm. **SPIKELET**: glumes 12–15 mm, subequal, abruptly acuminate; floret 8–9 mm; callus 1 mm, sharp; lemma awn 33–45 mm, bent twice, lower 2 segments long-hairy, distal segment scabrous. $2n$=70. Dry slopes, coniferous forest; 1300–1250 m. c&s SN, TR. [*Stipa l.* Swallen] Polyploid derivative of *A. nelsonii* and *A. lemmonii*.

A. lemmonii (Vasey) Barkworth (p. 1231) **ST** 1.5–9 dm. **LF**: basal ligules 0.5–1.2 mm; blade 0.5–1.5 mm wide. **INFL** 7–21 cm. **SPIKELET** somewhat laterally compressed; glumes 7–11.5 mm, ± equal; floret 5.5–7 mm; callus blunt; lemma 1.1–1.3 × palea length, hairs short, tip lobe ± 0.1 mm, thick, awn 16–30 mm. $2n$=34,36. Sagebrush shrubland, coniferous forest; < 2300 m. KR, NCoR, CaR, SN, Teh, GB; to B.C., ID, UT. [*Stipa l.* (Vasey) Scribner, incl var. *pubescens* Crampton, pubescent needlegrass; *S. columbiana* Macoun] ❀DRN,SUN:1,**2**,3–6,**7**,9,**14–18**,19–24;STBL.

A. lettermanii (Vasey) Barkworth LETTERMAN'S NEEDLEGRASS **ST** 2.5–8 dm. **LF**: blade 0.5–2 mm wide, gen curled. **INFL** 7–19 cm. **SPIKELET**: glumes 6.5–9 mm; floret 4.5–6 mm; callus 0.4–1 mm, blunt; lemma 1–1.2 × palea length, hairs < 1.5 mm below, < 2 mm at top, awn 12–25 mm, bent twice. $2n$=32. UNCOMMON. Meadows, dry slopes, sagebrush shrubland, coniferous forest; 1700–3400 m. n&c SN, SnBr, MP; to ID, NM. [*Stipa l.* Vasey]

A. nelsonii (Scribner) Barkworth ssp. ***dorei*** (Barkworth & J. Maze) Barkworth **ST** 4–17.5 dm. **LF**: blade 0.5–4.5 mm wide. **INFL** 9–36 cm. **SPIKELET**: glumes 6–12.5 mm, subequal; floret 4.5–7 mm; callus blunt; lemma ± 2 × palea length, hairs 1–2 mm; awn 13–45 mm. $2n$=36,44. Clearings, sagebrush shrubland steppe, meadows; 500–3500 m. CaR, SN, MP; to Yukon, WY, UT. [*Stipa columbiana* Macoun misapplied] ❀TRY.

A. nevadensis (B. Johnson) Barkworth **ST** 2–9 dm, hairy below nodes. **LF**: blade 1–3 mm wide, gen inrolled. **INFL** 6–20 cm. **SPIKELET**: glumes 8–14 mm; floret 5.5–8.5 mm; callus sharp; lemma ± 1.3 × palea length, tip hairs < 1.5 mm, > lower awn hairs; awn 20–35 mm, bent twice, lower 2 segments hairy. $2n$=68. Sagebrush shrubland, open woodlands; < 3100 m. SNE; NV. [*Stipa n.* B. Johnson] Polyploid derivative of *A. nelsonii* and *A. lettermanii*. ❀DFCLT.

A. occidentalis (Thurber) Barkworth **ST** 2–12 dm. **LF** glabrous to hairy; sheath gen ciliate at top; blade 0.5–1 mm wide, gen rolled. **INFL** 8–30 cm. **SPIKELET**: glumes 9–15 mm, ± equal; floret 5.5–7.5 mm; callus sharp; lemma 1.6–3 × palea length, hairs short; awn 15–40 mm, bent twice, lower 2 segments hairy. $2n$=36. Open, dry sites, sagebrush shrubland, coniferous forest, alpine; 150–3400 m. CA-FP (exc GV, CW, PR); to WA, ID, UT.

ssp. ***californicum*** (Merr. & Burtt Davy) Barkworth (p. 1231) **SPIKELET**: lemma tip hairs < awn hairs; upper awn segment smooth to rough, not hairy. Sagebrush shrubland, coniferous forest; 150–3100 m. KR, CaR, SN, SnBr, SnJt, GB; to WA, ID, NV. Intergrades with ssp. *pubescens*. [*Stipa c.* Merr. & Burtt Davy] ❀DRN:**1**,15,16,24SHD:**2**,3,7,**14,18**,19–23;STBL.

ssp. ***occidentalis*** (p. 1231) **SPIKELET**: lemma tip hairs gen = awn hairs; all awn segments hairy. Coniferous forest, alpine; gen 3000–3400 m. SN. [*Stipa o.* Thurber] ❀DRN,SHD:**1–3**;STBL.

ssp. ***pubescens*** (Vasey) Barkworth **SPIKELET**: lemma tip hairs gen = awn hairs, distal awn segment smooth to slightly rough. Sagebrush shrubland, coniferous forest; 150–3100 m. KR, CaR, SN, SnBr; to WA, ID, UT. [*Stipa elmeri* Scribner] ❀DRN:**1,15, 16,24**&SHD:**2**,3,7,**14,18**,19–23;STBL.

A. parishii (Vasey) Barkworth (p. 1231) **ST** 3–8 dm. **LF**: basal sheath ciliate at top, hairs 2–3 mm; blade 1.5–3 mm wide. **INFL** 11–15 cm. **SPIKELET**: glumes unequal, lower 11–21 mm, upper 2–3 mm; floret 4–8 mm; callus blunt; lemma 1.4–2 × palea length, hairs < 5 mm, tip hairs longest, awn 15–35 mm, bent once. Dry rocky slopes, shrubland, pinyon/juniper woodland; 900–2700 m. s SN, Teh, TR, PR, W&I; to UT, AZ, Baja CA. [*Stipa coronata* var. *depauperata* (M.E. Jones) A. Hitchc.] ❀DRN,SUN:1,**2,3,7**,8–10, **14–18**,19–24;STBL.

A. pinetorum (M.E. Jones) Barkworth **ST** 1–5 dm. **LF**: blade 0.5–1 mm wide. **INFL** 4.5–20 cm; branches appressed. **SPIKELET**: glumes 7–11 mm, ± equal; floret 3.5–5.5 mm; callus blunt to sharp; lemma 1–1.5 × palea length, hairs at midlength 1.5–3.5 mm, at tip < 5 mm, awn 13–25 mm, bent twice. $2n$=32. Rocky soil, pinyon/juniper woodland, coniferous forest; 1900–3810 m. SN, W&I, DMtns; to ID, NM. [*Stipa p.* M.E. Jones] ❀TRY.

A. speciosum (Trin. & Rupr.) Barkworth (p. 1231) DESERT NEEDLEGRASS **ST** 3–6 dm. **LF**: basal sheath hairy; blade gen rolled, < 1 mm diam. **INFL** 10–15 cm, gen partly enclosed by uppermost lf sheath. **SPIKELET**: glumes 14–20 mm, ± equal; floret 8–9 mm; callus sharp; lemma 1.2–2.1 × palea length, densely short-hairy near base, glabrous near tip, awn 35–40 mm, bent once, hairs of lower segment < 8 mm. 2*n*=66,68,±74. Rocky slopes, canyons, washes; < 2200 m. s SN, Teh, SCoRO, SW, s SNE, D; to Colorado, Mex, S.Am. [*Stipa s.* Trin. & Rupr.] ❀DRN,SUN:1,**2**,**3**,**7**,8,9,**10**, 11,**14–18**,19–21,**22–24**;also STBL.

A. stillmanii (Bolander) Barkworth (p. 1231) **ST** 6–15 dm. **LF**: basal sheath smooth, hairy above, at collar; blade 4–6 mm wide, folded or rolled. **INFL** 10–20 cm. **SPIKELET**: glumes 14–22 mm, ± equal; floret 8–12 mm; callus 1–1.5 mm, blunt; lemma short-hairy, tip 2-toothed, teeth 1–2 mm, awn-like, awn 19–32 mm, bent once, sometimes twice; palea ± = lemma. Uncommon. Rocky slopes, coniferous forest; 900–1550 m. n SNH [*Stipa s.* Bolander] ❀DRN,IRR,part SHD:1,2,6,7,15,16;DFCLT.

A. thurberianum (Piper) Barkworth **ST** 3–7.5 dm, hairy. **LF**: ligule 1.5–8 mm; blade 0.5–2 mm wide. **INFL** 7–15 cm. **SPIKELET**: glumes 10–15 mm, 2.5–4 mm wide, lower > upper; floret 6–9 mm; callus sharp; lemma 1.1–1.4 × palea length, tip hairs 2 × lower hairs; awn 32–56 mm, bent twice, lower segments long-hairy. 2*n*=34. Canyons, foothills, sagebrush shrubland, juniper woodland; 1500–2600 m. SN, Teh, WTR, GB; to WY, Colorado. [*Stipa t.* Piper] ❀DRN:1,**2**,**3**,**7**,**10**,11,12,**14–24**;STBL.

A. webberi (Thurber) Barkworth **ST** 1–3.5 dm. **LF**: blade gen rolled, < 0.5 mm diam, stiff. **INFL** 2.5–7 cm, dense, often partly enclosed by upper lf sheath. **SPIKELETS**: glumes 6–10 mm, ± equal; floret 4.5–6 mm, callus blunt; lemma evenly, densely long-hairy, hairs 2.5–3.5 mm, awn 5–11 mm, straight, readily deciduous; palea ± = lemma. 2*n*=32. Dry, open flats, rocky slopes, gen with sagebrush; 1500–3000 m. SN, MP, W&I; to OR, Colorado. [*Oryzopsis w.* (Thurber) Benth.] ❀DRN,SUN:1,**2**,**3**,**7**,8,9,**10**,11,**14–24**; STBL.

AEGILOPS GOATGRASS

Mary E. Barkworth

Ann. **ST** gen erect to abruptly bent at base, gen glabrous. **LF**: sheath appendaged; ligule membranous; blade flat, spreading. **INFL** spike-like, gen cylindric, ± open to dense; spikelets 2-ranked, 1 per node, lower gen vestigial. **SPIKELET**: glumes, lemmas toothed, gen awned; glumes thick, hard, 3+ -veined; axis breaking above glumes and between florets; lemma firm, flat or rounded, veins parallel, not converging; anthers 1.5–4 mm. 27 spp.: Medit, sw&c Asia. (Latin: ancient name for wheat) [Kimber & Feldman 1987 Univ of Missouri Spec Rep 353] Sometimes included in *Triticum*, wheat.

1. Glumes of terminal spikelet 1-awned, glumes of lower spikelets 1-awned or with long tooth .... ***A. cylindrica***
1′ Glumes of terminal spikelet 3-awned, glumes of lower spikelets 2–5-awned
 2. Glumes of lowermost spikelet with 2–5 awns; upper spikelets gen sterile .... ***A. ovata***
 2′ Glumes of lowermost spikelet with 2–3 awns; upper spikelets fertile .... ***A. triuncialis***

A. cylindrica Host (p. 1231) JOINTED GOATGRASS **ST** 14–50 cm. **INFL** 2–12 cm; axis breaking apart in fr; spikelets partly sunken in axis. **SPIKELET**: glumes of lower fertile spikelets acute, tapered, or short-awned, glumes of terminal spikelet awned; florets 2–5; lemma of lower spikelets abruptly pointed or with awn 1–5 mm; lemma of terminal spikelets with awn 4–5 cm. 2*n*=56. Disturbed, dry sites, cult fields; < 1500 m. CaR, ScV, SW, MP; to WA, Great Plains, Mex; native to Medit Eur, w Asia. NOXIOUS WEED.

A. ovata L. **ST** 10–24 cm. **INFL** 1–3 cm; axis breaking apart in fr; spikelets not sunken in axis. **SPIKELET**: glumes of lowermost spikelet 2–5-awned, awns 2–3.5 cm; florets gen 5, upper 3 gen sterile; lemma of lower spikelets gen 2-awned; lemma of upper spikelets 3-awned; lemma awns < glume awns. 2*n*=28. Disturbed fields, roadsides; < 500 m. NCoR (Mendocino Co.), expected elsewhere; native to Medit Eur, w Asia. NOXIOUS WEED.

A. triuncialis L. (p. 1231) BARBED GOATGRASS **ST** 17–45 cm. **INFL** 2–5.5 cm; axis breaking apart in fr; spikelets gen not sunken in axis. **SPIKELET**: glumes of lowermost spikelet 2–3 awned; glumes of upper spikelets 3-awned, awns gen 4–8 cm; florets gen 4, upper 2 gen sterile; lemmas short-awned or awn 0. 2*n*=28. Disturbed sites, cult fields, roadsides; < 1000 m. s NCoR, CaRF, n&c SNF, ScV, n CW; native to Medit Eur, w Asia. NOXIOUS WEED.

AGROPYRON CRESTED WHEATGRASS

Mary E. Barkworth

Per. **ST** erect, gen tufted. **LF**: sheath sometimes appendaged; ligule membranous; blade flat or rolled. **INFL** spike-like, dense; axis not breaking apart in fr; spikelets 2-ranked, strongly overlapping, 1 per node, gen spreading. **SPIKELET**: glumes gen ± equal, = or < lower floret, lanceolate, keeled, acute to short-awned; axis breaking above glumes and between florets; lemma firm, acute to awn-tipped; anthers 2–5 mm. 12 spp.: Medit, e Eur, c Asia. (Greek: ancient name for wild wheat) [Barkworth & Dewey 1985 Amer J Bot 72:767–776] See also *Elymus, Elytrigia, Pascopyrum, Pseudoroegneria* for spp. sometimes treated here.

A. desertorum (Fischer) Schultes (p. 1231) DESERT CRESTED WHEATGRASS **ST** 2.5–10 dm. **LF**: sheath, blade gen glabrous, sometimes soft-hairy or scabrous. **INFL** 3–10 cm; internodes 1–3 mm; spikelets spreading to ascending. **SPIKELET** 7–11 mm; glumes < lower floret, translucent margin ± 0.4 mm wide, gen 3-veined; florets 3–8; lemma 5–9 mm, gen 5-veined, awn < 3.5 mm. 2*n*=28. Disturbed areas, roadsides; 600–1500 m. CaRF, n SNH, GB; to Great Plains; native to e Eur. Pls with infl internodes < 1 mm, spikelets spreading are called *A. cristatum* (L.) Gaertner. Pls with spikelets appressed, lemma awn < 1 mm, are called *A. fragile* Roth [*A. sibiricum* Willd.] All forms (and hybrids) are planted for erosion control.

AGROSTIS BENT GRASS

M. J. Harvey

Ann or per, gen tufted, sometimes from rhizomes or stolons. **STS** gen erect. **LF**: sheath gen smooth, glabrous; ligule membranous; blade flat to rolled. **INFL** panicle-like, densely cylindric to openly ovate. **SPIKELET**: glumes gen subequal, back gen glabrous, vein gen finely scabrous, 1-veined, gen acute; floret 1, < glumes, gen breaking above glumes; callus glabrous to densely hairy; lemma gen 5-veined, veins not converging, sometimes extended as short teeth, awned from back or not; palea 0 to +/– = lemma, translucent; anthers gen 3. +/– 200 spp.: esp temp Am, Eurasia. (Greek: pasture) [Carlbom 1967 PhD OR State Univ] Some cult in pastures, lawns.

1. Rhizomes or stolons well developed, clearly present; per
2. Pl from stolons
3. Glume back finely scabrous throughout; spikelet falling as 1 unit; infl axes stiff .... [2]***A. viridis***
3′ Glume back +/– glabrous, finely scabrous only on keel; floret axis breaking above persistent glumes; infl axes thread-like, flexible
4. Ligule 0.5–2 mm, gen wider than long; stolons < 5 cm; infl widely ovate in outline, 1° branches spreading, spikelets not crowded .... [3]***A. capillaris***
4′ Ligule 2–5 mm, longer than wide; stolons 5–100 cm; infl elliptic to lanceolate in outline, 1° branches gen all ascending, spikelets overlapping, crowded .... [2]***A. stolonifera***
2′ Pl from rhizomes
5. Floret callus hairs 1.5–2 mm, gen > 1/2 lemma; ligule 4–7 mm .... [3]***A. hallii***
5′ Floret callus hairs gen minute, sparse, or 0; ligule gen < 3 mm (exc *A. gigantea*)
6. Palea 0 or minute and << lemma
7. Lf blade 1–6 mm wide; lemma tip minutely toothed; rhizome < 10 cm; anthers gen > 1 mm .... [3]***A. pallens***
7′ Lf blade < 1 mm wide; lemma tip +/– acute; rhizome < 5 cm; anthers < 1 mm .... [3]***A. variabilis***
6′ Palea well developed, +/– 1/2 to slightly < lemma
8. Infl +/– oblong in outline, somewhat dense, branches gen all ascending to erect; montane to alpine, gen > 1500 m .... [2]***A. thurberiana***
8′ Infl gen ovate in outline, open, most branches spreading; open, gen disturbed places, gen < 1000 m
9. Rhizomes < 5 cm, slender, not clearly scaly; ligule wider than long; infl branches with spikelets on distal half .... [3]***A. capillaris***
9′ Rhizomes < 25 cm, +/– thick, scaly; ligule longer than wide; infl branches with spikelets +/– throughout .... [2]***A. gigantea***
1′ Rhizomes or stolons 0; ann or per
10. Lemma awned from back or near tip
11. Infl open, gen oblong to ovate in outline; spikelets not crowded, infl axes clearly visible
12. Infl gen oblong to lanceolate in outline, 1° branches gen ascending
13. Lemma awned at or above middle; anthers < 0.6 mm; lower lf sheaths gen glabrous .... [3]***A. exarata***
13′ Lemma awned below middle; anthers 1–1.5 mm; lower lf sheaths finely tomentose .... ***A. hooveri***
12′ Infl gen ovate in outline, lowest 1° branches spreading, upper branches gen ascending
14. Lemma back puberulent below middle; floret axis prolonged beyond lemma +/– 1 mm, short-hairy-tufted; palea > 1 mm .... ***A. avenacea***
14′ Lemma back glabrous or fine-scabrous; floret axis not prolonged beyond lemma; palea 0 or minute
15. Lemma awned near tip, awn 3–10 mm, wavy; callus hairs < 1/2 lemma, dense; anther 1, persistent in fr; ann .... ***A. elliottiana***
15′ Lemma awned below middle, awn < 2 mm, +/– straight; callus hairs << lemma, sparse; anthers 3, deciduous; per .... [2]***A. scabra***
11′ Infl dense, gen cylindric; spikelets crowded, overlapping, infl axes not clearly visible
16. Primary infl branches gen > 0.5 cm, often evident at base
17. Lf blade gen < 1 mm wide, +/– inrolled; floret callus glabrous; infl base often partly enclosed by upper lf; anthers 1–2 mm .... [2]***A. blasdalei***
17′ Lf blade gen 2–10 mm wide, flat; floret callus minutely hairy; infl clearly stalked; anthers +/– 0.5 mm
18. Back of glume fine-scabrous throughout; palea 0.5–0.7 mm; ligule 1.5–2 mm .... [3]***A. densiflora***
18′ Back of glume +/– glabrous (keel fine-scabrous); palea +/– 0.3 mm; ligule 2.5–4 mm .... [3]***A. exarata***
16′ Primary infl branches gen < 0.5 cm, obscured by densely clustered spikelets
19. Lemma teeth 4, unequal (two < 1 mm, other two 1–1.5 mm); lemma awned below middle; callus densely short-hairy .... ***A. tandilensis***
19′ Lemma teeth 0 or 2, equal; lemma awned at or above middle; callus gen sparsely hairy, hairs minute
20. Lemma awned above middle, awn < 3.5 mm, straight; glume tips acute; per .... [3]***A. densiflora***
20′ Lemma awned +/– at middle, awn 3.5–10 mm, gen bent; glume tips narrowly acuminate to awn-like; ann
21. Lemma body 2–4 mm, awn 8–10 mm .... ***A. hendersonii***
21′ Lemma body 1.5–2 mm, awn 4–8 mm .... ***A. microphylla***
10′ Lemma awnless
22. Spikelets crowded and often overlapping on same branch, spikelet stalks and 2° axes not clearly visible
23. Palea slightly < lemma (or +/– 1/3 × lemma in *A. densiflora*)
24. Spikelet breaking below glumes, falling as 1 unit; sts often decumbent to long-trailing .... [2]***A. viridis***
24′ Floret breaking above persistent glumes; sts gen erect
25. Lower infl branches < 1.5 cm; ligule 1.5–2 mm; anthers ± 0.5 mm .... [3]***A. densiflora***
25′ Lower infl branches > 2 cm; ligule 2–5 mm; anthers > 1 mm .... [2]***A. stolonifera***
23′ Palea 0 or < 1/3 lemma
26. Lf blade < 1 mm wide, gen inrolled or folded
27. Anthers > 1 mm; infl branches gen < 0.5 cm; coastal habitats < 100 m .... [2]***A. blasdalei***
27′ Anthers < 1 mm; infl branches 0.5–1.5 cm; inland mtns, 1600–4000 m .... [3]***A. variabilis***
26′ Lf blade gen 2–7 mm wide, gen flat
28. Floret callus with hairs 1.5–2 mm, gen slightly > 1/2 lemma .... [3]***A. hallii***
28′ Floret callus with hairs 0 or < 0.5 mm, < 1/2 lemma
29. Lower infl branches 1–2 cm; anthers 0.3–0.6 mm .... [3]***A. exarata***
29′ Lower infl branches 2–5 cm; anthers 0.7–1.8 mm .... [3]***A. pallens***

22′ Spikelets not crowded, gen well spaced on same branch, axes clearly visible
30. Primary infl axes branched 1–2 × above middle, spikelets 0 on lower 1/2
31. Lf blades 2–4 mm wide; glumes 2–3 mm; palea 1/4–1/5 lemma; anthers 0.6–0.8 mm; moist areas . . . . . ***A. oregonensis***
31′ Lf blades 0.7–2 mm wide; glumes 1.5–2.7 mm; palea 0 or minute, << lemma; anthers < 0.7 mm; well drained areas
32. Lvs basal and cauline; infl +/– 2 × longer than wide; 1° axes +/– stiff . . . . . ***A. idahoensis***
32′ Lvs mostly basal; infl +/– long as wide; 1° axes flexible, lower +/– arched . . . . . [2]***A. scabra***
30′ Primary infl axes branched 1–2 × from base upwards, spikelets distributed throughout
33. Floret callus with hairs 1.5–2 mm, gen slightly > 1/2 lemma; anthers > 1.5 mm . . . . . [3]***A. hallii***
33′ Floret callus glabrous or with minute hairs << lemma; anthers gen < 1.5 mm (exc some *A. pallens*)
34. Palea 0 or minute, << lemma
35. Lf blades 1–6 mm wide; pls gen 10–70 cm; anthers 0.7–1.8 mm . . . . . [3]***A. pallens***
35′ Lf blades gen < 1 mm wide; pls 4–30 cm; anthers 0.5–0.7 mm . . . . . [3]***A. variabilis***
34′ Palea > 1/2 lemma
36. Infl branches spreading to ascending; anthers 1–1.5 mm; wet or disturbed areas, gen < 1500 m
37. Ligule 0.5–2 mm, gen wider than long; infl branches with spikelets on distal half . . . . . [3]***A. capillaris***
37′ Ligule 2–6 mm, longer than wide; infl branches with spikelets +/– throughout . . . . . [2]***A. gigantea***
36′ Infl branches ascending to erect; anthers < 1 mm; moist to +/– dry areas, gen > 1500 m
38. Infl 2–5 cm, slightly flattened; pls gen < 15 cm; alpine . . . . . ***A. humilis***
38′ Infl gen 4–11 cm, not clearly flattened; pls 8–50 cm; < subalpine . . . . . [2]***A. thurberiana***

A. avenacea Gmelin Per 15–65 cm. **LVS**: ligule 3–5 mm; lower blades 8–20 cm, 1–3 mm wide, gen flat, fine-scabrous. **INFL** 7–30 cm, widely ovate, open; lower 1° branches spreading, lower 5–15 cm, axes branched above middle, thread-like. **SPIKELET**: glumes 2.8–3.6 mm, back puberulent below middle; floret axis prolonged beyond floret +/– 1 mm, short-hairy-tufted; callus hairs < 0.7 mm; lemma +/– 2 mm, back puberulent below middle, tip 2-toothed, awned from middle, awn 4–7.5 mm, bent; palea > 1 mm, +/– 1/2 × lemma; anther +/– 0.5 mm. 2*n*=56. Open, often disturbed places; < 300 m. s NCo, s NCoR, SNF, GV, CW, n SCo; to e US, TX; native to s Pacific Islands.

A. blasdalei A. Hitchc. (p. 1231) BLASDALE'S BENT GRASS Per 6–30 cm, decumbent to erect. **LVS**: ligule gen 1–1.5 mm; lower blades 2–5 cm, gen < 1 mm wide, inrolled. **INFL** 2–8 cm, cylindric, dense; base often partly enclosed by upper lf; 1° branches ascending to appressed, lower gen < 0.5 cm. **SPIKELET**: glumes 1.8–4 mm; callus glabrous; lemma 1.5–3 mm, sometimes awned above middle, awn < 0.7 mm, straight; palea +/– 0.3 mm, < 1/3 × lemma; anthers 1–2 mm. 2*n*=42. RARE. Dunes, gravelly soils, coastal bluffs, shrubland; < 100 m. s NCo, n CCo, n SnFrB. [var. *marinensis* B. Crampton, marin bent grass, **RARE** CA]. May intergrade locally with *A. densiflora*; needs study.

A. capillaris L. COLONIAL BENT Per 10–75 cm; stolons or rhizomes < 5 cm, slender. **LVS**: ligule 0.5–2 mm, gen wider than long; lower blades 3–10 cm, 1–5 mm wide, gen flat. **INFL** 3–20 cm, widely ovate in outline, open; 1° branches mostly spreading, lower 1.5–4 cm, axes thread-like. **SPIKELET**: glumes 2–3 mm; callus glabrous or minutely hairy; lemma 1.5–2.5 mm, sometimes short-awned near tip; palea 1/2–2/3 × lemma; anthers 0.8–1.3 mm. 2*n*= 28. Roadsides, open, disturbed places; < 1000 m. KR, CaR, n&c SN, CCo, SnFrB, SCo; to AK, e US; native to Eur. [*A. tenuis* Sibth.]

A. densiflora Vasey (p. 1231) Per 9–85 cm. **LVS**: ligule 1.5–2 mm; lower blades 2–12 cm, 2–10 mm wide, flat. **INFL** 2–10 cm, +/– cylindric, dense; 1° branches +/– appressed, < 1.5 cm. **SPIKELET**: glumes 2–3 mm, back finely scabrous; callus minutely hairy; lemma 1.5–2 mm, sometimes awned above middle, awn < 3.5 mm, straight; palea +/– 1/3 × lemma; anthers +/– 0.5 mm. 2*n*=42. Coastal bluffs, sandy soils; < 200 m. NCo, CCo, w SnFrB. [*A. californica* Trin.; *A. clivicola* B. Crampton (coastal bluff bent grass) incl var. *punta-reyensis* B. Crampton (Point Reyes bent grass)]

A. elliottiana Schultes (p. 1231) Ann 5–45 cm. **LVS**: ligule 2–3 mm; lower blades 0.5–4 cm, < 1 mm wide, flat to inrolled. **INFL** 5–20 cm, gen widely ovate in outline, open; lower 1° branches 1–4.5 cm, spreading, upper ascending, axes thread-like with spikelets clustered near tip. **SPIKELET**: glumes 1.5–2 mm; callus hairs < 0.6 mm, dense; lemma 1–2 mm, gen awned from near tip, awn 3–10 mm, wavy; palea 0; anther 1, persistent in fr. 2*n*=14. Vernal pool margins; < 500 m. NCoRI, CaRF, n SNF, n ScV; to s US. [*A. exigua* Thurber]

A. exarata Trin. (p. 1231) Per 8–100 cm. **LVS**: ligule 2.5–4 mm; lower blades 4–15 cm, 2–7 mm wide, flat. **INFL** 5–30 cm, oblong to +/– ovate in outline, +/– open to dense, sometimes interrupted near base; 1° branches 1–2 cm, ascending to +/– appressed. **SPIKELET**: glumes 1.5–3.5 mm, acute to narrowly acuminate; callus hairs < 0.5 mm; lemma 1–2 mm, sometimes awned above middle, awn < 3.5 mm, straight to bent; palea < 1/3 × lemma; anthers 0.3–0.6 mm. 2*n*=28,42,56. Common. Moist or disturbed areas, open woodland, coniferous forest; < 2000 m. CA-FP; to AK, SD, Mex. [*A. ampla* A. Hitchc.; *A. longiligula* Vasey vars. *l.* and *australis* J. Howell] Pls with dense, narrowly cylindric infl and awned lemmas have been called var. *monolepis* (Torrey) A. Hitchc. ❀STBL.

A. gigantea Roth Per 20–100 cm; rhizomes < 25 cm, +/– scaly. **LVS**: ligule 2–6 mm; lower blades 4–10 cm, 3–8 mm wide, flat. **INFL** 8–25 cm, widely ovate in outline, open; 1° branches mostly spreading, lower 4–7 cm, axes thread-like. **SPIKELET**: glumes 2–3 mm; callus hairs 0 or minute; lemma 1.5–2 mm, awn 0; palea 0.7–1.4 mm; anthers 1–1.4 mm. 2*n*=42. Roadsides, fields, ditches, disturbed areas; < 2000 m. CA-FP; e N.Am; native to Eur. [*A. alba* L. misapplied, in part] Difficult to separate from *A. stolonifera*.

A. hallii Vasey (p. 1231) Per 17–100 cm; rhizomes < 50 cm. **LVS**: ligule 4–7 mm; lower blades 7–20 cm, 2–5 mm wide, flat. **INFL** 7–22 cm, lanceolate to narrowly ovate in outline, +/– open to dense; 1° branches ascending to +/– appressed, lower 1–5 cm. **SPIKELET**: glumes 2.5–4 mm; callus hairs 1.5–2 mm; lemma 2–3 mm, awn 0; palea minute, << lemma; anthers 1.5–2.3 mm. 2*n*=42. Open oak woodland, coniferous forest; < 1500 m. w NW, CCo, SnFrB, n SCo, WTR; OR. ❀TRY;GRNCVR.

A. hendersonii A. Hitchc. (p. 1231) HENDERSON'S BENT GRASS Ann 6–70 cm. **LVS**: ligule 1–4 mm; lower blades 1–4 cm, +/– 1 mm wide, flat to weakly inrolled. **INFL** 1–5 cm, cylindric, dense; 1° branches < 0.5 cm, ascending to +/– appressed. **SPIKELET**: glumes 5–7 mm, tip narrowly acuminate to awn-like; callus hairs +/– 0.7 mm; lemma 2–4 mm, awned +/– at middle, awn 8–10 mm, +/– bent; palea 0; anthers +/– 0.5 mm. 2*n*=42. UNCOMMON. Vernal pools; < 300 m. CaRF, n SNF, ScV, n SnJV; OR. [*A. microphylla* var. *h.* (C. Hitchc.) Beetle] Perhaps = *A. microphylla*.

A. hooveri Swallen (p. 1231) HOOVER'S BENT GRASS Per 30–80 cm. **LVS**: lower lf sheaths finely tomentose; ligule 4–6 mm; lower blades 10–16 cm, 1–2 mm wide, flat, becoming inrolled. **INFL** (4)10–17 cm, gen lanceolate in outline, open; 1° branches +/– ascending, 15–40 cm, axes thread-like. **SPIKELET**: glumes 2–3 mm; callus hairs < 0.3 mm, dense; lemma 1.5–2 mm, awned below middle, awn < 2.5 mm, bent; palea 0; anthers 1–1.5 mm. UNCOMMON. Dry sandy soils, open chaparral, oak woodland; < 600 m. s CCo, s SCoRO (San Luis Obispo, Santa Barbara cos.).

A. humilis Vasey (p. 1231) MOUNTAIN BENT GRASS Per gen < 15 cm. **LVS** mostly basal; ligule 0.5–2 mm; lower blades 1–3 cm,

0.5–1 mm wide, flat to +/– folded. **INFL** 2–5 cm, narrowly oblong in outline, +/– open; 1° branches gen erect, gen < 0.5 cm. **SPIKELET**: glumes 1.5–2 mm; callus hairs minute, sparse; lemma 1.5–2 mm, awn 0; palea 1–1.5 mm; anthers 0.5–0.7 mm. RARE in CA. Subalpine or alpine meadows, slopes; 2700–3200 m. c&s SNH; to B.C., MT, NM. More common than previously assumed, expected elsewhere; intergrades with *A. thurberiana*.

A. idahoensis Nash (p. 1231) Per 8–30 cm. **LVS**: ligule 1–3 mm; lower blades 1–5 cm, 0.5–2 mm wide, flat, often inrolled with age. **INFL** 3–13 cm, lanceolate to ovate in outline, +/– open; 1° branches gen ascending, lower 1–4 cm, axes thread-like. **SPIKELET**: glumes 1.5–2.5 mm; callus glabrous or hairs < 0.3 mm; lemma 1–2 mm, awn 0; palea minute, << lemma; anthers 0.3–0.5 mm. Common. Open, wet meadows, coniferous forest; < 3500 m. NW, CaR, SN, n SnFrB, SnBr, SnJt; to AK, MT, NM. [*A. tenuis* Vasey]

A. microphylla Steudel (p. 1231) Ann 8–45 cm. **LVS**: ligule 1.5–4 mm; lower blades 3–15 cm, 0.7–2.5 mm wide, fine-scabrous, flat, becoming inrolled. **INFL** 2–12 cm, +/– cylindric, dense; 1° branches ascending to +/– appressed, lower 0.3–1.5 cm. **SPIKELET**: glumes 2.5–5 mm, tips narrowly acuminate to awn-like; callus hairs < 0.5 mm; lemma 1.5–2 mm, awned from middle, awn 3.5–8 mm, slightly bent; palea 0; anther +/– 0.5 mm. Thin, rocky soils, cliffs, sometimes on serpentine, vernal pools; < 200 m. NCo, s NCoR, CCo, SCo; to B.C., Baja CA. [var. *intermedia* Beetle; *A. aristiglumis* Swallen, awned bent grass]

A. oregonensis Vasey Per 12–75 cm. **LVS**: ligule 2–4.5 mm; lower blades 10–30 cm, 2–4 mm wide, gen flat. **INFL** 8–35 cm, lanceolate to ovate in outline, open; 1° branches ascending, lower 2–6 cm, axes thread-like. **SPIKELET**: glumes 2–3 mm; callus hairs 0 or minute, sparse; lemma 1.5–2.5 mm, sometimes awned above middle, awn < 2 mm, straight; palea 1/5–1/4 × lemma; anthers 0.6–0.8 mm. Moist areas, meadows, streambanks; < 2100 m. KR, NCoR, CaR, SN, SnBr?, SnJt?; to AK, WY.

A. pallens Trin. Per 10–70 cm, sometimes from rhizomes < 10 cm. **LVS**: ligule 1.5–3 mm; lower blades 1.5–5 cm, 1–6 mm wide, flat to inrolled. **INFL** 5–20 cm, lanceolate to narrowly ovate in outline, +/– open; 1° branches gen ascending, lower 2–5 cm. **SPIKELET**: glumes 2–3 mm; callus hairs minute; lemma 1.5–2.5 mm, sometimes awned near tip, awn < 0.5 mm, straight; palea 0 or minute, << lemma; anthers 0.7–1.8 mm. 2*n*=42,56. Common. Open meadows, woodland, forest, subalpine; 200–3500 m. CA-FP, GB; to B.C., MT. [*A. lepida* A. Hitchc.; *A. diegoensis* Vasey] Geog and ecological variation need study. ❀DRN,part SHD:1,2,3,**4–7**,8,9,**14–24**;decid.GRNCVR.

A. scabra Willd. (p. 1231) Per 20–75 cm. **STS** ascending to erect. **LVS** mostly basal; ligule 2–5 mm; lower blades 4–14 cm, 1–3 mm wide, flat, finely scabrous. **INFL** 8–25 cm, ovate in outline, open; 1° branches spreading below, ascending above, lower 4–11 cm, axes thread-like, branched 1–2 × above middle, often breaking at base in fr. **SPIKELET**: glumes 1.5–3 mm; callus hairs minute, sparse; lemma 1.5–2 mm, sometimes awned from below middle, awn < 2 mm, +/– straight; palea 0 or minute, << lemma; anthers 0.4–0.7 mm. 2*n*=42. Open roadsides, meadows, coniferous forest; 1000–3100 m. KR, NCoR, SN, TR, SnJt, SNE; to AK, ne US. Stunted alpine pls have been called var. *geminata* (Trin.) Swallen. ❀SUN:4,5,**6**&IRR:1–3,**7**,14,**15–17**,18–24;may be INV.

A. stolonifera L. (p. 1237) CREEPING BENT Per 8–60 cm, decumbent to erect, often mat-like; stolons 5–100 cm. **LVS**: ligule 2–5 mm; lower blades 2–10 cm, 2–5 mm wide, flat. **INFL** 3–15 cm, elliptic to lanceolate in outline, +/– dense; 1° branches ascending to +/– erect, lower gen 2–6 cm. **SPIKELET**: glumes 1.5–3 mm; callus hairs minute, sparse; lemma 1.5–2 mm, awn 0; palea slightly < lemma; anthers 1–1.5 mm. 2*n*=28. Ditches, lake margins, marshes; < 1000 m. NW, CaR, n SN, CW, SW (exc ChI); to s Can, s US; native to Eur. [*A. alba* L. vars. *major* (Gaudin) Farw. and *palustris* (Hudson) Pers.; *A. alba* var. *a.* misapplied in part] Difficult to separate from *A. gigantea*.

A. tandilensis (Kuntze) L. Parodi Ann 9–21 cm. **LVS**: ligule 2–2.5 mm; lower blades 2–5 cm, < 1 mm wide, flat, inrolled with age. **INFL** 2–5 cm, cylindric, dense; 1° branches erect to appressed, gen < 0.5 cm. **SPIKELET**: glumes 3–3.5 mm; callus densely short-hairy; lemma +/– 1.5 mm, back densely puberulent below middle, awned below middle, awns < 6 mm, bent, lemma tip 4-toothed, 2 teeth < 1 mm, other teeth 1–1.5 mm; palea 0; anther 1, < 0.2 mm. Vernal pools; < 100 m. Deltaic GV (Solano Co.), s SCo (San Diego Co.), expected elsewhere; native to Argentina. [*A. kennedyana* Beetle]

A. thurberiana A. Hitchc. Per 8–50 cm, sometimes from rhizomes < 3 cm. **LVS**: basal withering in fl; ligule 0.8–3 mm; lower blades 8–15 cm, 1–3 mm wide, flat. **INFL** 4–11 cm, oblong in outline, +/– open; 1° branches ascending to erect, gen 1–2 cm. **SPIKELET**: glumes 1.5–2.5 mm; axis gen prolonged beyond floret, 0.5 mm, short-hairy-tufted; callus hairs minute, sparse; lemma 1–2 mm, awn 0; palea +/– 1.5 mm; anthers 0.6–0.8 mm. 2*n*=14. Moist, often heavy soils, coniferous forest; 1300–3500 m. KR, NCoRH, CaRH, SNH; to AK, MT, Colorado. Intergrades with *A. humilus*. ❀TRY.

A. variabilis Rydb. (p. 1237) Per 4–30 cm, sometimes from rhizomes < 5 cm. **LVS** mostly basal; ligule 1–2.5 mm; lower blades 3–7 cm, < 1 mm wide, flat, becoming folded. **INFL** 2.5–6 cm, +/– cylindric, +/– dense; 1° branches ascending to erect, lower 0.5–1.5 cm. **SPIKELET**: glumes 2–2.5 mm; callus hairs minute; lemma 1.5–2 mm, awn gen 0; palea 0; anthers 0.5–0.7 mm. Meadows, subalpine forest, talus, alpine; 1600–4000 m. KR, NCoRH, CaRH, SN, Wrn; to AK, w Can, Colorado.

A. viridis Gouan (p. 1237) Gen per 10–75 cm, decumbent to long-trailing, rooting at nodes. **LVS**: ligule 1–6 mm; lower blades 3–18 cm, 2–10 mm wide, flat. **INFL** 2–15 cm, narrowly lanceolate to elliptic in outline, dense, often interrupted; 1° branches erect to +/– appressed, lower 1–4 cm. **SPIKELET** breaking below glumes, falling as 1 unit; glumes 1.5–2.5 mm, back finely scabrous; callus glabrous; lemma 1–1.5 mm, awn 0; palea slightly < lemma; anthers 0.5–0.7 mm. 2*n*=28. Common. Disturbed areas, wet areas, ponds, ditches, streambanks; < 2000 m. CA; to s US; native to Eur. [*A. semiverticillata* (Forsskal) C. Chr.]

AIRA EUROPEAN HAIRGRASS

Dieter H. Wilken

Ann. **STS** 1–15, gen solitary, sometimes tufted, glabrous to puberulent. **LVS** ± basal; collar glabrous to puberulent; ligule < 1 mm, membranous; blade thread-like, flat to rolled. **INFL** panicle-like. **SPIKELET** bisexual, 3–5 mm; glumes > lower floret, translucent, keel scabrous; callus short-bristly; axis breaking above glumes and between florets; florets 2; lemma faintly 3–5-veined, gen glabrous, tip with 2 slender teeth, slightly scabrous, awned at or below the middle, awn bent once, exserted (sometimes reduced or 0 in lower floret); palea slightly < lemma. ± 10 spp.: s Eur. (Greek: a grass)

1. Infl narrow, pedicels < spikelets ..... ***A. praecox***

1′ Infl wide, pedicels > spikelets

  2. Lower lemma awn present ..... ***A. caryophyllea***

  2′ Lower lemma awn gen 0 ..... ***A. elegantissima***

A. caryophyllea L. (p. 1237) SILVER EUROPEAN HAIRGRASS **STS** 1–10, 6–45 cm, gen glabrous (sometimes puberulent just below nodes). **LF**: sheath slightly scabrous; ligule minutely scabrous. **INFL** > 2 cm wide, open. **SPIKELET** 3–4 mm; glumes 2–3 mm; lemmas 1.5–2 mm; awns ± 3 mm. 2*n*=14,28. Sandy soils, open or disturbed sites; < 1900 m. NW, w CaR, SNF, GV, CW, e SW; to B.C., e US, Baja CA; native to Eur.

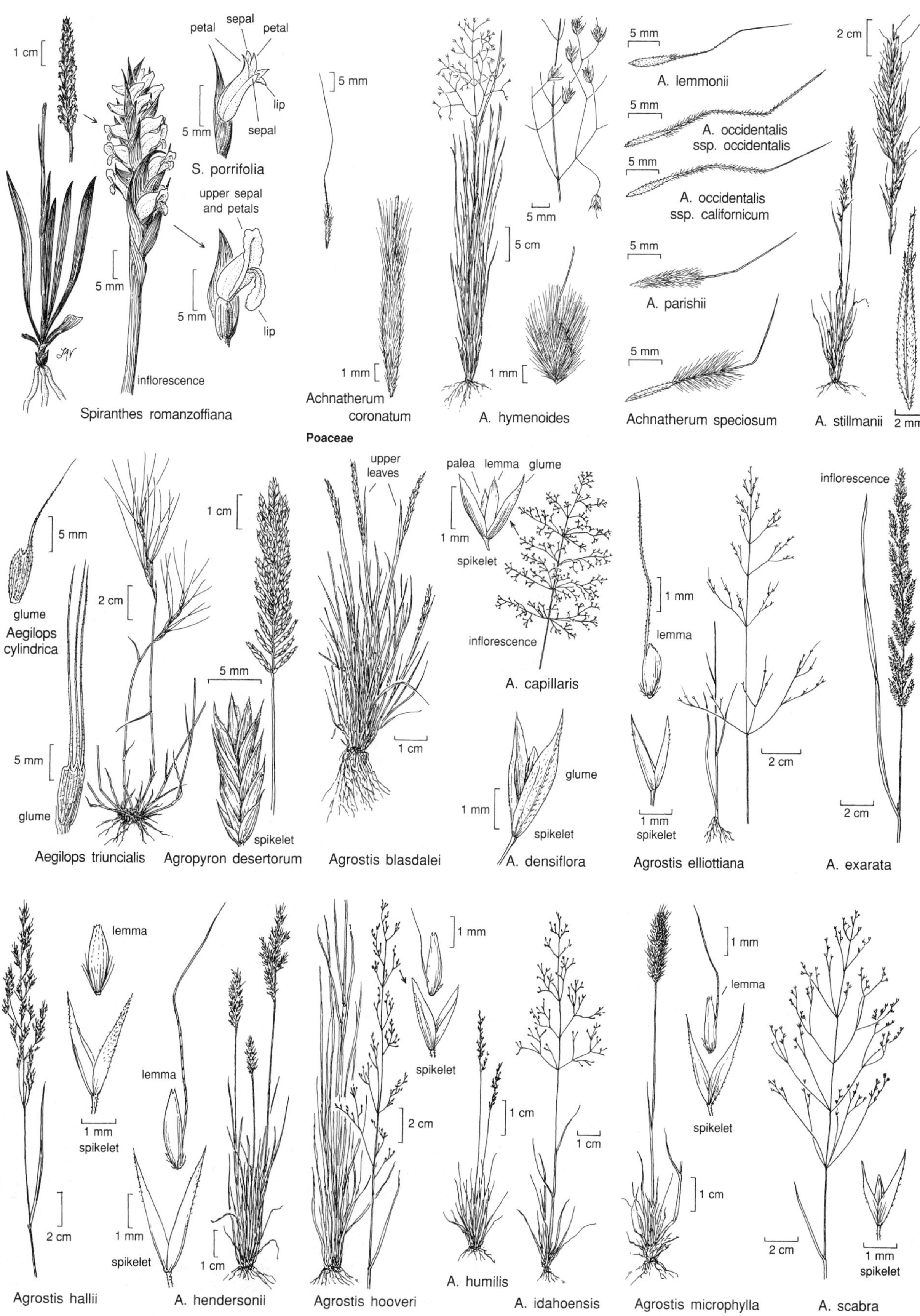
1 cm
5 mm
petal
sepal
petal
lip
sepal
5 mm
S. porrifolia
upper sepal and petals
5 mm
lip
inflorescence
Spiranthes romanzoffiana
5 mm
1 mm
Achnatherum coronatum
5 mm
5 cm
1 mm
A. hymenoides
5 mm
A. lemmonii
5 mm
A. occidentalis ssp. occidentalis
5 mm
A. occidentalis ssp. californicum
5 mm
A. parishii
5 mm
Achnatherum speciosum
2 cm
A. stillmanii
2 mm
Poaceae
5 mm
glume
Aegilops cylindrica
5 mm
glume
2 cm
Aegilops triuncialis
1 cm
5 mm
spikelet
Agropyron desertorum
upper leaves
1 cm
Agrostis blasdalei
palea
lemma
glume
1 mm
spikelet
inflorescence
A. capillaris
glume
1 mm
spikelet
A. densiflora
1 mm
lemma
1 mm
spikelet
2 cm
Agrostis elliottiana
inflorescence
2 cm
A. exarata
lemma
1 mm
spikelet
2 cm
Agrostis hallii
lemma
1 mm
spikelet
1 cm
A. hendersonii
1 mm
spikelet
2 cm
Agrostis hooveri
1 cm
A. humilis
1 cm
A. idahoensis
1 mm
lemma
spikelet
1 cm
Agrostis microphylla
2 cm
1 mm
spikelet
A. scabra

A. elegantissima Schur (p. 1237) ELEGANT EUROPEAN HAIRGRASS **STS** 1–8, 9–35 cm, gen glabrous. **LF**: sheath and ligule glabrous to slightly scabrous. **INFL** > 2 cm wide, open. **SPIKELET** 3–4 mm; glumes 2–2.5 mm; lemmas ± 2 mm; awns ± 3 mm. 2*n*=14. Sandy to clay soils, open sites; < 300 m. c&s NCoRO, ScV, SnFrB; to WA, e&s US; native to s Eur. [formerly known as *A. elegans* Gaudin, an illegitimate name]

A. praecox L. (p. 1237) **STS** 1–12, 5–25 cm, gen glabrous. **LF**: sheath glabrous to slightly scabrous; ligule slightly scabrous. **INFL** ≤ 1 cm wide, narrow. **SPIKELET** gen 4 mm; glumes 2–3 mm; lemmas ± 2 mm; awns 3–4 mm. 2*n*=14. Sandy soils, open sites; < 100 m. NCo, s NCoRO; to B.C., e US; native to s Eur.

ALOPECURUS FOXTAIL

William J. Crins

Ann, per, cespitose or from stolons. **ST** decumbent to erect, 1–8 dm; nodes visible, brown. **LVS**: ligule 1–6 mm, membranous, truncate to acute, gen scabrous; blade flat, glabrous or scabrous. **INFL** panicle-like, gen cylindric, dense; branches short. **SPIKELET** ± compressed, breaking below glumes, falling as 1 unit; glumes ± equal, gen = spikelet, membranous, gen keeled, keel and lateral veins gen stiff- or appressed-hairy, margins free or fused near base, tip obtuse, acute, or short-awned, 3-veined; floret 1; lemma membranous, margins keeled, sometimes fused near base, truncate to acute, 3–5-veined, awned on back below middle, awn straight or abruptly bent gen at lemma tip; palea gen 0; anthers 0.5–4 mm. **FR** glabrous. ± 35 spp.: temp N.Am, Eurasia. (Greek: fox tail) [Rubtzoff 1961 Leafl West Bot 9:165–180]

1. Spikelets exc awns > 4 mm
  2. Per; st gen > 4.5 dm; glumes, lemma acute ..... ***A. pratensis***
  2′ Ann; st gen < 4.5 dm; glumes, lemma acute or obtuse
    3. Glumes ciliate on keel near base, glabrous on lateral veins, winged and appearing asymmetrical; anthers > 3.5 mm; infl gen > 7 cm, tapered at both ends; lf sheaths not inflated ..... ***A. myosuroides***
    3′ Glumes ciliate on all nerves, not winged; anthers < 1.5 mm; infl < 6.5 cm, not tapered; upper lf sheaths conspicuously inflated ..... [2]***A. saccatus***

1′ Spikelets exc awns < 4 mm
  4. Awn straight, > lemma by 0–2 mm; per ..... ***A. aequalis***
  4′ Awn bent, exceeding lemma by 1.5–5 mm; per or ann
    5. Spikelets > 3 mm; upper lf sheath conspicuously inflated; awn exceeding lemma by 3–5 mm ..... [2]***A. saccatus***
    5′ Spikelets gen < 3 mm; upper lf sheath not inflated; awn exceeding lemma by 1.5–3.5 mm
      6. Ann; anther < 1 mm ..... ***A. carolinianus***
      6′ Per; anther 1.2–2 mm ..... ***A. geniculatus***

A. aequalis Sobol. (p. 1237) SHORT-AWN FOXTAIL Per. **ST** 0.9–4.7 dm. **LF**: ligule 2–5.5 mm; blade 2.5–10 cm, 1–5(8) mm wide. **INFL** 1–7.5 cm, 3–6(9) mm wide. **SPIKELET**: glumes 2–3(3.5) mm; lemma awn straight, exceeding lemma body by 0–2 mm; anthers 0.5–1 mm. 2*n*=14. Common. Wet meadows, shores; 100–3500 m. CA-FP, GB; to subarctic, e N.Am, Eurasia. Pls in CCo with spikelet < 3.4 mm, lemma awn exceeding lemma body by 0–2, have been called var. *sonomensis* Rubtzoff, Sonoma alopecurus; pls from high elevations in SNH also may have long awns.

A. carolinianus Walter Ann. **ST** 0.6–4.5 dm. **LF**: ligule 3–4.5 mm; blade 1–8 cm, 1–3 mm wide. **INFL** 1–7 cm, 4–6 mm wide. **SPIKELET**: glumes 2–3 mm; lemma awn bent, exceeding lemma body by 1.5–3.5 mm; anthers 0.5–1 mm. 2*n*=14. Uncommon. Vernal pools, open, disturbed ground; 50–1400 m. NW, GV, SW; to B.C., e N.Am. Occurs in natural habitats in s SCo, weedy elsewhere.

A. geniculatus L. WATER FOXTAIL Per. **ST** 1.4–5.5 dm. **LF**: ligule 2.5–5 mm; blade 2–8 cm, 1–4(7) mm wide. **INFL** 1.5–6 cm, 4–7.5 mm wide. **SPIKELET**: glumes 2.5–3 mm; lemma awn bent, exceeding lemma body by (1.5)2–3.5 mm; anthers (1)1.5–2 mm. 2*n*=28. Open, wet meadows, pools, shores, streambanks; < 1550 m. NW, CaR, SN, CW, MP; to B.C., e N.Am., Eurasia. [*A. pallescens* Piper] Variable. Pls from NW, CW probably naturalized.

A. myosuroides Hudson Ann. **ST** 4–4.5 dm. **LF**: ligule 4.5–6 mm; blade 8–11 cm, 3.5–6 mm wide. **INFL** 7–10 cm, 4–7 mm wide. **SPIKELET**: glumes 4.5–5.5 mm, glabrous exc keel ciliate at base; lemma awn bent, exceeding lemma body by 3–4 mm; anthers ± 4 mm. 2*n*=14,28. Pastures; < 100 m. SnJV, SCo; native to Eurasia. Cult for forage but not very successful in CA.

A. pratensis L. MEADOW FOXTAIL Per. **ST** 4.8–7.7 dm. **LF**: ligule 1.5–3 mm; blade 2.5–16 cm, 2–6 mm wide. **INFL** 3.5–7.5 cm, 6–10 mm wide. **SPIKELET**: glumes 4–5 mm; awn bent, exceeding lemma body by 2–5.5 mm; anthers 2–3.5 mm. 2*n*=28,42. Open, damp meadows; 100–1700 m. NW, CaR, SN, CW; to AK, e N.Am.; native to Eurasia. Cult for forage.

A. saccatus Vasey (p. 1237) Ann. **ST** 1.2–4.5 dm. **LF**: upper sheath inflated; ligule 1.5–5.5 mm; blade 1–8(13) cm, 1–4 mm wide. **INFL** 1.5–6.5 cm, 5.5–10 mm wide. **SPIKELET**: glumes 3–5 mm; lemma awn bent, exceeding lemma body by 3–5 mm; anthers ± 1 mm. Vernal pools, moist, open meadows; < 700 m. NW, c SN, CW, GV, SW; to WA. [*A. howellii* Vasey]

AMMOPHILA BEACHGRASS

Dieter H. Wilken

Per with long, thick rhizomes. **STS** clumped, stiff, erect. **LVS** basal and cauline, erect; ligule membranous, acute; blade inrolled. **INFL** panicle-like. **SPIKELETS** subsessile, laterally compressed, strongly keeled; glumes ± > floret, firmly membranous, obtuse to acuminate, lower gen 1-veined, upper 3-veined; callus hairs long, tufted; floret bisexual, breaking above glumes; lemma firmly membranous, 5–7-veined; palea = lemma, membranous. 4 spp.: coastal e N.Am, Eur. (Greek: sand loving)

A. arenaria (L.) Link (p. 1237) EUROPEAN BEACHGRASS **STS** 5–12 dm. **LVS** 4–11 dm; ligules 1–3 cm, acute; blades 2–5 mm wide. **INFL** 15–30 cm, ± 2 cm wide, cylindric. **SPIKELET** 10–13 mm; callus hairs 2–4 mm; lemma 8–10 mm. 2*n*=28. Sand dunes; < 20 m. NCo, CCo, SCo; native to n Eur. Cult for dune stabilization, baskets, brooms. Pls from c SCo (Orange Co.) and n CCo (Angel Island), with ligules 1–5 mm, firm, minutely ciliate belong to *A. breviligulata* Fern., native to coastal e US.

AMPELODESMOS DIS GRASS

1 sp. (Greek: to tie grapevines together, from early use of lvs)

A. mauritanicus (Poiret) Durand & Schinz Per. **STS** clumped, < 3 m; solid in X-section. **LF**: blade < 1 m, ± 7 mm wide; ligule membranous, ciliate; blade margins serrate, tip long tapered. **INFL**: panicle-like, ± 1-sided, 2–6 dm, terminal, dense; branches slender, drooping. **SPIKELET** bisexual, 10–16 mm; glumes membranous, acuminate, 3–5-veined; axis breaking apart above glumes and between florets; florets 2–6; callus long-hairy; lemma back rounded, tip 2-toothed, long-hairy near base, 5–7-veined, short-awned at tip between teeth; palea veins extending as 2 short awns. Oak woodland; < 1000 m. NCoRO (Napa Co.); native to Medit. [Decker 1964 Brittonia 16: 76–79] Cult as orn.

ANDROPOGON BLUESTEM

Kelly W. Allred

Per, cespitose. **STS** erect, branched; nodes gen hairy. **LVS** cauline; ligule membranous, minutely ciliate; blade flat or folded. **INFL** panicle-like with 2 or more spike-like branches, solitary or compactly clustered, partly enclosed in lf sheaths; axes breaking apart with age; spikelet sessile, subtended by hairy, naked stalk and axis segment, falling with stalk and axis segment as 1 unit. **SPIKELET**: glumes ± = florets, lanceolate, ± translucent; florets 2, lower vestigial, obscure, upper fertile, lemma translucent, awned, palea << lemma or 0; stamens 1–3. **FR** oblong, brownish or purplish. ± 100 spp.: warm temp, trop. Some spp. cult for forage, revegetation. (Greek: man beard, from hairy staminate spikelets) [Campbell 1983 J Arn Arbor 64:171–254]

1. Ligule > 1 mm; lower glume keel scabrous in lower half; lf sheath minutely scabrous .... ***A. glomeratus*** var. ***scabriglumis***
1′ Ligule < 1 mm; lower glume keel glabrous in lower half; lf sheath glabrous .... ***A. virginicus*** var. ***virginicus***

A. glomeratus (Walter) Britton, Sterns, & Pogg. var. ***scabriglumis*** C.S. Campbell (p. 1237) SOUTHWESTERN BUSHY BLUESTEM **ST** 0.8–1.5 m. **LF**: sheath scabrous; ligule 1–2.2 mm; lower blades 3–6 dm, 3.5–6 mm wide. **INFLS** many, compactly clustered, plume-like; branches gen 2–4. **SPIKELET** 4–4.5 mm; lower glume keel gen scabrous at base; callus hairs 1–2 mm; awn 0.5–2 mm. $2n$=20,40. Moist, open, disturbed areas, seeps; < 600 m. TR, DMoj, naturalized in NCo, NCoRO, n SNF, SNH, ScV, SCo; to NM, Baja CA. ❀IRR or WET,SUN:5,**7–11**,12,**14–16**,17,18,**19–24**.

A. virginicus L. var. *virginicus* (p. 1237) BROOMSEDGE BLUESTEM **ST** 0.5–2 m. **LF**: sheath glabrous; ligule < 1 mm; lower blades 1–5 dm, 1.5–5 mm wide. **INFLS** several, compactly clustered; branches 2–7. **SPIKELET** 3.5–4 mm; lower glume keel gen glabrous at base; callus hairs 1–2.5 mm; awn 0.5–2 mm. $2n$=20. Moist, open, disturbed areas, seeps; < 300 m. NCoRO, CaRF, n&c SNF, ScV; native to e U.S.

ANTHOXANTHUM VERNAL GRASS

Dieter H. Wilken

Ann, per. **STS** ascending to erect. **LVS** ± basal, fragrant; appendages < 1 mm, rounded (sometimes obscure); ligule membranous; blade flat. **INFL** gen panicle-like. **SPIKELETS** subsessile, laterally compressed; glumes > florets, lower glume 1/2 length upper, tip acute, lower 1-veined, upper 3-veined; florets 3, lower 2 sterile, on opposite sides of upper, fertile floret, lower and upper florets breaking above glumes, falling as 1 unit; lemma of lower florets > upper floret, awned at or below middle, tip 2-forked or lobed, hairy, 3-veined, lemma of upper floret awnless, tip rounded, ± glabrous, faintly 5–7-veined; palea 0 in lower florets, present and < lemma in fertile floret, 1 veined. 18–20 spp.: Eur, w Asia, montane n Afr. (Greek: yellow flower)

1. Ann, lf blade < 2 mm wide .... ***A. aristatum***
1′ Per, lf blade 3–7 mm wide .... ***A. odoratum***

A. aristatum Boiss. (p. 1237) **STS** 1–3 dm. **LVS**: upper sheaths 1–6 cm; ligules < 2 mm; upper blades 1–3 cm, ciliate, soft-hairy. **INFL** 0.8–3 cm, 3–9 mm wide. **SPIKELETS** 4–8 mm; lemma hairs ± stiff, hairs at lemma base brown. $2n$=10. Disturbed sites; < 200 m. NCo, NCoR, n CaRF, CCo, SnFrB; to B.C., c&e US; native to Eur.

A. odoratum L. (p. 1237) SWEET VERNAL GRASS **STS** 3–6 dm. **LVS**: upper sheaths 4.5–9 cm; ligules 1–3 mm; upper blades 3–6 cm, slightly ciliate and soft-hairy at base. **INFL** 2–8 cm, 5–15 mm wide. **SPIKELETS** 7–9 mm; lemma hairs ± soft, hairs at lemma base colorless. $2n$=10,20,28. Disturbed sites; < 1100 m. NCo, NCoRO, NCoRI, CCo, SnFrB, n&c SN; widespread US, temp; native to Eur.

APERA

Dieter H. Wilken

Ann. **STS** erect, solitary to tufted, gen glabrous. **LVS** basal and cauline; ligule membranous, acute to toothed at tip; blade flat to rolled, gen ridged. **INFL** panicle-like; branches spike-like, minutely scabrous. **SPIKELET**: glumes subequal, > floret, membranous, lower 1-veined, upper 1–3-veined; floret 1, bisexual, breaking above glumes; lemma rounded on back, membranous, short bristly at base, minutely scabrous and awned near tip, obscurely 3–5-veined, awn straight, flexible; palea ± = lemma. ± 3 spp.: temp Eur, Asia. Cult for revegetation. (Greek: not maimed, alluding to occasional presence of vestigial florets)

1. Infl branches ascending .... ***A. interrupta***
1′ Infl branches spreading .... ***A. spica-venti***

A. interrupta (L.) Beauv. **STS** gen < 3, 1–6 dm, sometimes tufted. **LF**: ligule 1–4 mm; blade 1–3 mm wide, minutely scabrous above. **INFL** 5–15 cm; branches ascending; glumes 1.5–3 mm; lemma 1.5–2 mm, minutely scabrous at tip, awn 5–8 mm. 2*n*=14, 28. Disturbed sites; 1220–1700 m. MP (w of Alturas, Modoc, Co.), DMtns (Furnace Creek, e Inyo Co.); reported elsewhere in w US; native to Eur, Asia.

A. spica-venti (L.) Beauv. **STS** gen < 4, 2–12 dm, sometimes tufted. **LF**: ligule 3–9 mm; blade 3–5 mm wide, ± glabrous to minutely scabrous above. **INFL** 8–25 cm; branches spreading; glumes 1.5–3 mm; lemma 1.5–2.5 mm, minutely scabrous at tip, awn 4–9 mm. 2*n*=14. Disturbed and open sites; 90–430 m. n SnJV (near La Grange, e Stanislaus Co.), s SCoRO (Cuyama Valley, Santa Barbara Co.); reported elsewhere in w US; native to temp Eur.

ARISTIDA THREE-AWN

Kelly W. Allred

Ann, per, cespitose. **ST** ascending to erect. **LVS** basal and cauline; basal often tufted; ligule hairy; blades flat or inrolled. **INFL** raceme-like or panicle-like; branches spike-like. **SPIKELET**: glumes narrowly lanceolate, thin, 1-veined, awn gen 0; floret 1, breaking above glumes; lemma ± fusiform, hard when mature, 3-veined, tip beak-like or not, awned at tip, awns 3, equal or unequal; palea < lemma, enclosed by lemma, transparent. **FR** narrowly fusiform. ± 300 spp.: worldwide, arid warm temp. Some spp. noxious. (Latin: awn) [Allred 1992 Great Basin Nat 52:41–52]

1. Lower sts densely hairy ..... ***A. californica***
1' Lower sts glabrous
 2. Ann
  3. Lemma awns < 2.5 cm; glumes 5–10 mm ..... ***A. adscensionis***
  3' Lemma awns > 3.5 cm; glumes 20–34 mm ..... ***A. oligantha***
 2' Per
  4. Base of 1° infl branches abruptly spreading or ascending from main axis
   5. Lower infl branches ascending, upper appressed ..... ***A. purpurea*** var. ***parishii***
   5' All infl branches spreading
    6. Lf blade base glabrous; lemma tip much twisted; anthers ± 1 mm ..... ***A. divaricata***
    6' Lf blade base sparsely long-hairy; lemma tip slightly twisted or straight; anthers 1–2 mm ..... ***A. ternipes*** var. ***hamulosa***
  4' Base of 1° infl branches branches appressed to main axis, tips sometimes spreading or ascending
   7. Most awns 4–10 cm ..... ***A. purpurea***
    8. Upper glume gen > 16 mm; lemma ± linear ..... var. ***longiseta***
    8' Upper glume gen < 16 mm; lemma gradually tapered to tip ..... [2]var. ***purpurea***
   7' Most awns < 3.5 cm
    9. Mature infl branches and spikelet stalks slender, curving to drooping, appearing S- or U-shaped ..... [2]var. ***purpurea***
    9' Mature infl branches and spikelet stalks gen stiff, straight, erect to spreading
     10. Tip of lemma gen < 0.2 mm wide; awns delicate, gen < 0.2 mm wide at base ..... var. ***nealleyi***
     10' Tip of lemma gen > 0.2 mm wide; awns stiff, gen > 0.2 mm wide at base
      11. Infl gen 3–14 cm; lf blade gen < 10 cm ..... var. ***fendleriana***
      11' Infl gen 15–30 cm; lf blade gen > 10 cm
       12. Infl dense, reddish, fading to straw-colored; lower branches with 8–18 spikelets ..... var. ***parishii***
       12' Infl somewhat loose, tan to brown, fading to straw-colored; lower branches with 2–10 spikelets ..... var. ***wrightii***

A. adscensionis L. (p. 1237) SIX-WEEKS THREE-AWN Ann. **ST** branched below, 0.5–8 dm. **LF**: blade < 10 cm, inrolled. **INFL** 2–22 cm, narrow. **SPIKELET**: lower glume 5–7 mm, upper 8–10 mm; lemma 6–13 mm, awns ± equal, 7–23 mm. 2*n*=22. Dry, open places, rocky sites, shrubland ; < 1400 m. CCo, SCo, s ChI, WTR, PR, D; to MO, TX, S.Am. Some pls (Riverside Co.) have very short lateral awns.

A. californica Thurber var. ***californica*** CALIFORNIA THREE-AWN Per, ± bushy. **ST** much-branched, gen 1–4 dm, densely hairy. **LF**: sheath << internodes; blade < 5 cm, gen inrolled. **INFL** 2–6 cm. **SPIKELET**: lower glume 4–8 mm, upper 9–12 mm; lemma 5–7 mm, narrow beak at tip 8–26 mm, awns 2–4.5 cm, beak and awns breaking from lemma. 2*n*=22. Dry sandy sites, shrubland; < 150 m. D; AZ, nw Mex. ❀TRY.

A. divaricata Willd. POVERTY THREE-AWN Per. **ST** erect, 2.5–7 dm. **LF**: blade 5–20 cm, loosely inrolled, glabrous. **INFL** 10–30 cm, 6–25 cm wide, open; stalk flattened, easily broken; primary branches stiffly spreading. **SPIKELET**: glumes ± equal, 8–12 mm; lemma 8–13 mm, twisted 4 or more times at tip, awns gen 10–22 mm, lateral < or = central; anthers 0.8–1 mm. 2*n*=22. Uncommon. Dry slopes, shrubland, grassland; < 1000 m. s SnJV, SCo, s PR; to Great Plains, C.Am. Pls with lateral awns much < central or absent are *A. orcuttiana* Vasey, once collected near San Diego.

A. oligantha Michaux (p. 1237) OLDFIELD THREE-AWN Ann. **ST** much-branched, 3–7 dm. **LF**: blade 3–22 cm, 1–2 mm wide, gen inrolled. **INFL** 12–25 cm, open, raceme-like. **SPIKELETS** ± sessile, spreading; glumes subequal, 20–34 mm, short-awned, lower 3-5-veined; lemma 13–20 mm, central awn gen 3.5–7 cm, lateral awns slightly shorter. 2*n*=22. Dry slopes, fields, grassland, shrubland, woodland; < 1000 m. NW, CaRF, SNF, GV, MP; to OR, e US.

A. purpurea Nutt. Per. **ST** gen erect, unbranched, 10–80 cm. **LF**: blade 5–25 mm, 1–2 mm wide, gen inrolled. **SPIKELET**: glumes thin, lower 4–12 mm, upper 7–25 mm; awns 0.5–6 cm, equal or central slightly longer. 2*n*=22,44,66,88. Sandy to rocky soils, slopes, plains; < 2000 m. SW, D; to sw can, Great Plains, n Mex.

var. ***fendleriana*** (Steudel) Vasey FENDLER THREE-AWN **LVS** gen basal; blade gen < 10 cm. **INFL** 3–14 cm, narrow. **SPIKELET**: lower glume 5–8 mm, upper 10–15 mm; lemma 8–14 mm, awn gen 1.8–4 cm, 0.2–0.3 mm wide at base. 2*n*=22,44. Dry, rocky slopes, shrubland; 1000–2000 m. SnBr, PR, SNE, DMoj; to MT, Great Plains, n Mex. [*A. f.* Steudel] ❀STBL.

var. ***longiseta*** (Steudel) Vasey (p. 1237) RED THREE-AWN **LVS** basal or cauline; blade 4–16 cm. **INFL** 5–15 cm; branches stiff and erect to delicate and drooping. **SPIKELET**: lower glumes 8–12 mm, upper gen 16–25 mm; lemma 12–16 mm, 0.4–0.8 mm wide just below awns, awn 4–10 cm, 0.2–0.5 mm wide at base, stiff. $2n$=22,44,66,88. Dry slopes, plains, shrubland; 300–1600 m. SnBr, D; to sw Can, n Mex. [*A. l.* Steudel] ❀TRY.

var. ***nealleyi*** (Vasey) K.W. Allred (p. 1237) NEALLEY THREE-AWN **LVS** gen basal; blade 5–25 cm. **INFL** 8–18 cm, narrow, light brown; branches gen erect. **SPIKELET**: lower glume 4–7 mm, upper 8–14 mm; lemma 7–13 mm, 0.1–0.2 mm wide just below awns; awns 1.5–2.5 cm, gen 0.1 mm wide at base, delicate. $2n$=22,44. Dry slopes, plains, shrubland; 200–2000 m. SCo, SnBr, PR, D; to s UT, OK, n Mex. [*A. glauca* (Nees) Walp.]

var. ***parishii*** (A. Hitchc.) K.W. Allred (p. 1237) PARISH THREE-AWN **LVS** cauline; blade > 10 cm, gen flat. **INFL** 15–24 cm, narrow; lower branches sometimes stiffly spreading, reddish when young. **SPIKELET**: lower glume 7–11 mm, upper 10–15 mm; lemma 10–13 mm, 0.2–0.3 mm wide just below awns, awns 2–3 cm, 0.2–0.3 wide at base. Dry slopes, plains, chaparral, shrubland; 300–1300 m. SCo, SnBr, D; s NV, Baja CA. [*A. parishii* A. Hitchc.]

var. ***purpurea*** PURPLE THREE-AWN **LVS** gen cauline; blade 5–25 cm. **INFL** 5–20 cm, gen nodding, purplish; branches gen delicate, wavy or drooping. **SPIKELET**: lower glume 4–9 mm, upper 7–16 mm; lemma 6–12 mm, 0.1–0.3 mm wide just below awns, awns gen 2–6 cm, 0.1–0.3 mm wide at base. $2n$=22,44,66,88. Dry slopes, shrubland; 250–800 m. SCo, SnBr, PR, DMtns; to Arkansas, n Mex. ❀TRY.

var. ***wrightii*** (Nash) K.W. Allred (p. 1237) WRIGHT THREE-AWN **LVS** cauline, 10–25 cm, flat to inrolled. **INFL** gen 14–30 cm; branches gen erect. **SPIKELET**: lower glume 5–10 mm, upper 10–16 mm; lemma 8–14 mm, 0.2–0.3 mm wide just below awns, awns gen 2–3.5 cm, 0.2–0.3 wide at base. $2n$=22,44,66. Sandy to rocky slopes, plains, shrubland; 500–1800 m. PR, DMoj; to s UT, OK, n Mex. [*A. w.* Nash] ❀DRN,SUN,DRY:3,**7**,8,**9**,**10**,11,12,**14–24**.

A. ternipes Cav. var. ***hamulosa*** (Henrard) J.S. Trent (p. 1237) HOOK THREE-AWN Per, sometimes bushy. **STS** few, prostrate to erect, 25–80 cm. **LF**: blade 5–40 cm, flat to inrolled, base sparsely long-hairy. **INFL** 15–40 cm, open; primary branches spreading. **SPIKELET**: glumes ± equal, 9–15 mm; lemma 10–15 mm, gen straight at tip, awns equal to unequal, central awn 10–25 mm, lateral 6–23 mm. $2n$=44. Dry hills and slopes; 100–1350 m. GV, SCo, TR, PR, DMoj; to TX, C.Am. [*A. h.* Henrard] ❀DRN,SUN:2,3,**7**, 8,9,**10**,11,12,**14–16**,17,**18–24**;also STBL.

ARRHENATHERUM

Dieter H. Wilken

Per. **STS** erect. **LVS** basal and cauline; ligule membranous, obtuse, minutely soft hairy; blade flat, soft hairy. **INFL** panicle-like, narrow. **SPIKELET**: glumes unequal, keeled, acute, lower 1-veined, upper 3-veined; florets 2, lower > upper, lower gen staminate, upper bisexual, breaking above glumes, falling as 1 unit; callus hairy; lemmas keeled, 5–7-veined, lemma of lower floret awned below middle, awn bent, stiff, upper lemma awnless or awned at tip, awn straight; palea < lemma. 6 spp.: temp Eur, Asia. (Greek: masculine awn, referring to awned staminate floret)

A. elatius (L.) J.S. Presl & C. Presl (p. 1241) **ST** ascending to erect, 7–15 dm, sometimes bulbous at base. **LF**: ligule 1–3 mm, obtuse, minutely ciliate; blade 3–8 mm wide, flat, glabrous to minutely scabrous. **INFL** 9–25 cm. **SPIKELET** subsessile to stalked; lower glume 4–7 mm, upper 7–10 mm, glume margins translucent, keel soft-hairy; lemmas 6–9 mm, awn of lower lemma 10–17 mm, awn of upper < 4 mm (sometimes 0); palea translucent. $2n$=14,28, 42. Disturbed and open sites; 30–1800 m. NW, CaR, n&c SNF, c SNH, CCo, SnFrB; to B.C., w US; native to temp Eur. Cult for stabilizing bare soil. Pls with sts swollen at base have been called ssp. *bulbosum* (Willd.) Schuebler & Martens.

ARUNDO

Kelly W. Allred

Per; rhizomes thick. **ST** erect, cane-like. **LVS** cauline; sheaths > internodes, glabrous; ligule thinly membranous, fringed; blade flat or folded, glabrous, margin scabrous. **INFL** panicle-like, dense, silvery to purplish. **SPIKELET** laterally compressed; glumes > florets, 3–5-veined; axis glabrous; florets 4–5, breaking above glumes and between florets; lemmas 3–7-veined, hairy, awn ± 0; palea < lemma; 3 spp.: warm temp, trop. (Latin: a reed grass)

A. donax L. (p. 1241) GIANT REED **ST** < 8 m; nodes glabrous; internodes < 4 cm thick. **LF**: blade < 1 m, 2–6 cm wide. **INFL** 3–6 dm, plume-like; branches ascending. **SPIKELETS** 10–14 mm; glumes 10–13 mm, thin, brownish or purplish; lemmas 8–12 mm, tip 2-toothed, hairs < 8 mm, silky; palea 3–5 mm, hairy at base; anthers 2.5–3 mm. $2n$=110. Moist places, seeps, ditchbanks; < 500 m. c SNF, CCo, SCo, SnGb, D; native to Eur.

AVENA OATS

Dieter H. Wilken

Ann. **STS** erect, 1–6, ± glabrous. **LVS** basal and cauline; ligules 2–5 mm, membranous, rounded at tip; blade flat. **INFL** panicle-like, open. **SPIKELETS** gen stalked, ± pendent; glumes subequal, gen > florets, membranous, 5–7-veined, gen glabrous; axis prolonged behind upper floret, vestigial floret at tip; florets 2–3, upper floret > lower, bisexual, breaking above glumes and between florets or not breaking; lemmas hard, glabrous to hairy below awn, awned at or slightly below middle, tip 2-forked, 5–7-veined, awn stiff, bent to straight, slightly to strongly coiled below bend; palea ± < lemma. ± 10–15 spp.: temp Eur, Asia. Cult for grain, hay. (Latin: oats)

1. Awns short-hairy below midpoint .... ***A. sterilis***
1′ Awns glabrous to minutely scabrous
  2. Forks at lemma tip bristly, bristles > 2 mm
    3. Lemma gen soft-hairy below awn; spikelet breaking at maturity .... ***A. barbata***
    3′ Lemma ± glabrous to soft-hairy at base; spikelet axis not breaking at maturity .... ***A. strigosa***
  2′ Forks at lemma tip < 1 mm or 0

4. Upper and lower floret awns gen subequal, awn bent, twisted below bend .... ***A. fatua***
4′ Upper floret awn << lower floret awn or 0, awn gen straight, slightly twisted .... ***A. sativa***

A. barbata Link (p. 1241) SLENDER WILD OAT **ST** 3–6 dm. **LF**: blade 2–6 mm wide, glabrous to minutely scabrous (sometimes ciliate). **SPIKELET**: glumes 18–30 mm, 5–7-veined; lemmas 12–18 mm, gen soft-hairy below awn, forks at tip 2–6 mm, bristly, awn 20–40 mm, bent, twisted below bend. 2*n*=28. Disturbed sites; 40–1200 m. CA-FP, MP, DMoj; reported elsewhere in US; native to s Eur.

A. fatua L. (p. 1241) WILD OAT **ST** 3–12 dm. **LF**: blade 4–12 mm wide, minutely scabrous. **SPIKELET**: glumes 18–25 mm, 7–9-veined; lemmas 14–20 mm, gen glabrous on back to soft-hairy in lower third, forks at tip < 1 mm, awn 25–40 mm, bent, twisted below bend. 2*n*=42. Disturbed sites; 25–1220 m. CA-FP, MP, DMoj; to AK, e Can, most of US; native to Eur. *A. occidentalis* Durand, with 4 florets, has been reported from CA. Hybridizes with *A. sativa, A. sterilis.*

A. sativa L. (p. 1241) CULTIVATED OAT **ST** 4–9 dm. **LF**: blade 3–16 mm wide, minutely scabrous. **SPIKELET**: glumes 15–30 mm, 7–9-veined; lemmas 12–25 mm, ± glabrous on back (sometimes stiff-hairy at base), tip acute to minutely 2-forked, forks < 1 mm, awn of upper floret 15–30 mm, straight, slightly twisted, awn of lower floret < 15 mm or 0. 2*n*=42. Disturbed sites; 20–1100 m. CA-FP; to AK, e Can, most of US; origin in Eur. Cult for grain; believed to be derived from wild *A. fatua* by early humans. Hybridizes with *A. fatua*.

A. strigosa Schreber **ST** 3–8 dm. **LF**: blade 3–10 mm wide, minutely scabrous. **SPIKELET**: glumes 12–20 mm, 7–9-veined; lemmas 13–16 mm, ± glabrous to soft-hairy at base, forks at tip bristly, bristles 5–8 mm, awn 20–30 mm, bent, slightly twisted below bend. Disturbed sites; 10–30 m. s NCo (Pt. Reyes, Marin Co.); native to n&c Eur.

A. sterilis L. **ST** 5–12 dm. **LF**: blade 4–8 mm wide, minutely scabrous. **SPIKELET**: glumes 25–45 mm, 7–11-veined; lemmas 18–30 mm, stiff-hairy below awn, forks at tip ± 2 mm, awn 35–90 mm, bent, twisted and short-hairy below bend. Disturbed sites; 30–120 m. SnFrB; native to s Eur. Hybridizes with *A. fatua*.

AXONOPUS

Robert Webster

Ann, per. **STS** gen erect; internode solid or spongy inside. **LVS** gen basal; ligule short-hairy or membranous, ciliate; blade flat or folded. **INFL**: branches ± spike-like, ascending to spreading; spikelets gen many, gen 1 per node, subsessile, on one side of axis. **SPIKELET** compressed, falling as 1 unit; glume 1; florets 2, lower floret sterile or unisexual, lemma ± = glume, palea 0, upper floret fertile, lemma thick, ± rigid, back rounded, facing away from infl axis, margin gen inrolled, tip gen obtuse, palea ± = lemma. ± 100 spp.: subtrop, trop Am. (Greek: axis foot, from spikelet at branch tip) [Black 1963 Advancing Frontiers Pl Sci 5:1–186]

A. affinis Chase (p. 1241) Per. **ST** 1.5–7 dm; nodes 2–5. **LF**: sheath glabrous; ligule < 0.5 mm; blade 2.5–25 cm, 3–8 mm wide, upper surface glabrous. **INFL**: main axis < 3 cm, ± 0.5 mm wide; branches 2–5, 3–8 cm, axis glabrous or hairy. **SPIKELET** ± 2 mm, ± 1 mm wide, ± oblong; lower floret pistillate; lemma of upper floret 4-veined, minutely bristle-tipped. Disturbed places, lawns; < 200 m. SCo, expected elsewhere; native to s US, subtrop Am.

BECKMANNIA

Ann, per. **STS** gen erect. **LVS** cauline; sheath glabrous; ligule membranous, acute, entire or irregularly cut; blades flat. **INFL** panicle-like; branches short, spike-like, ascending to appressed; spikelets 2-ranked, overlapping. **SPIKELET** ± round in outline; glumes ± equal, strongly compressed, wrinkled, 3-veined, winged; axis breaking below glumes, falling as 1 unit; florets 1–2, bisexual; lemma 5-veined, acuminate; palea narrow, < lemma. 2 spp.: N.Am, Eur. (J. Beckmann, German botanist, 1739–1811) [Reeder 1953 Bull Torrey Bot Club 80:187–196]

B. syzigachne (Steudel) Fern. (p. 1241) **ST** 2.5–10 dm. **LF**: ligule 4–8 mm; blades 6–21 cm, 4–10 mm wide. **INFL** 9–28 cm, narrow, ± dense; branches 0.5–4 cm, ascending to appressed. **SPIKELET** 1.5–3 mm; glumes glabrous to finely scabrous, tip short-pointed; floret 1; lemma 2–3 mm, lanceolate, acuminate; palea < lemma, 2-toothed. 2*n*=14. Pond margins, streams, marshes; < 1500 m. NW, CaR, n SN, n CCo, SnFrB, MP, n SNE; to AK, e US.

BLEPHARIDACHNE EYELASH GRASS

Ann, per. **LVS** short, tufted. **INFL** panicle-like, dense, short, barely elevated above lvs. **SPIKELET** bisexual; glumes > florets, thin, 1-veined; axis breaking apart above glumes, falling as 1 unit; florets 4, lower 2 sterile or staminate, 3rd bisexual, uppermost sterile, ± vestigial; lemma ciliate, back rounded, 3-veined, deeply 2-lobed to middle, lobes short-awned, tip awned between lobes, awn = lobes; palea ciliate or glabrous. 4 spp.: sw N.Am, Argentina. (Greek: eyelash + chaff, from ciliate lemma)

B. kingii (S. Watson) Hackel (p. 1241) KING'S EYELASH GRASS Per, tufted. **ST** gen < 10 cm. **LF**: blade 1–3 cm, < ± 1 mm wide; sheath margins translucent; blade curved, ± stiff, sharp-pointed. **INFL** < 2 cm, head-like, terminal, exceeded by upper blades, straw-colored or purplish. **SPIKELET** compressed; glumes 7–8 mm, ± equal, acuminate; callus soft-hairy; lemmas ± 6 mm, margins ciliate, base soft-hairy; sterile palea narrower and < lemma; fertile palea = lemma. 2*n*=14. RARE in CA. Pinyon/juniper woodland; 1350–1600 m. SNE (Inyo, Mono cos.); to s UT. [Hunziker & Anton 1979 Brittonia 31:446–453]

BOTHRIOCHLOA

Kelly W. Allred

Per, cespitose. **ST** gen erect; nodes gen short-hairy. **LVS** basal and cauline; ligule membranous; blade flat or folded. **INFL** panicle-like, branching 1–several ×; branches spike-like, long-soft-hairy; axes grooved, breaking apart with age. **SPIKELETS** paired; lower sessile, bisexual; upper stalked, staminate or sterile; pair with subtending axis segment falling as 1 unit.

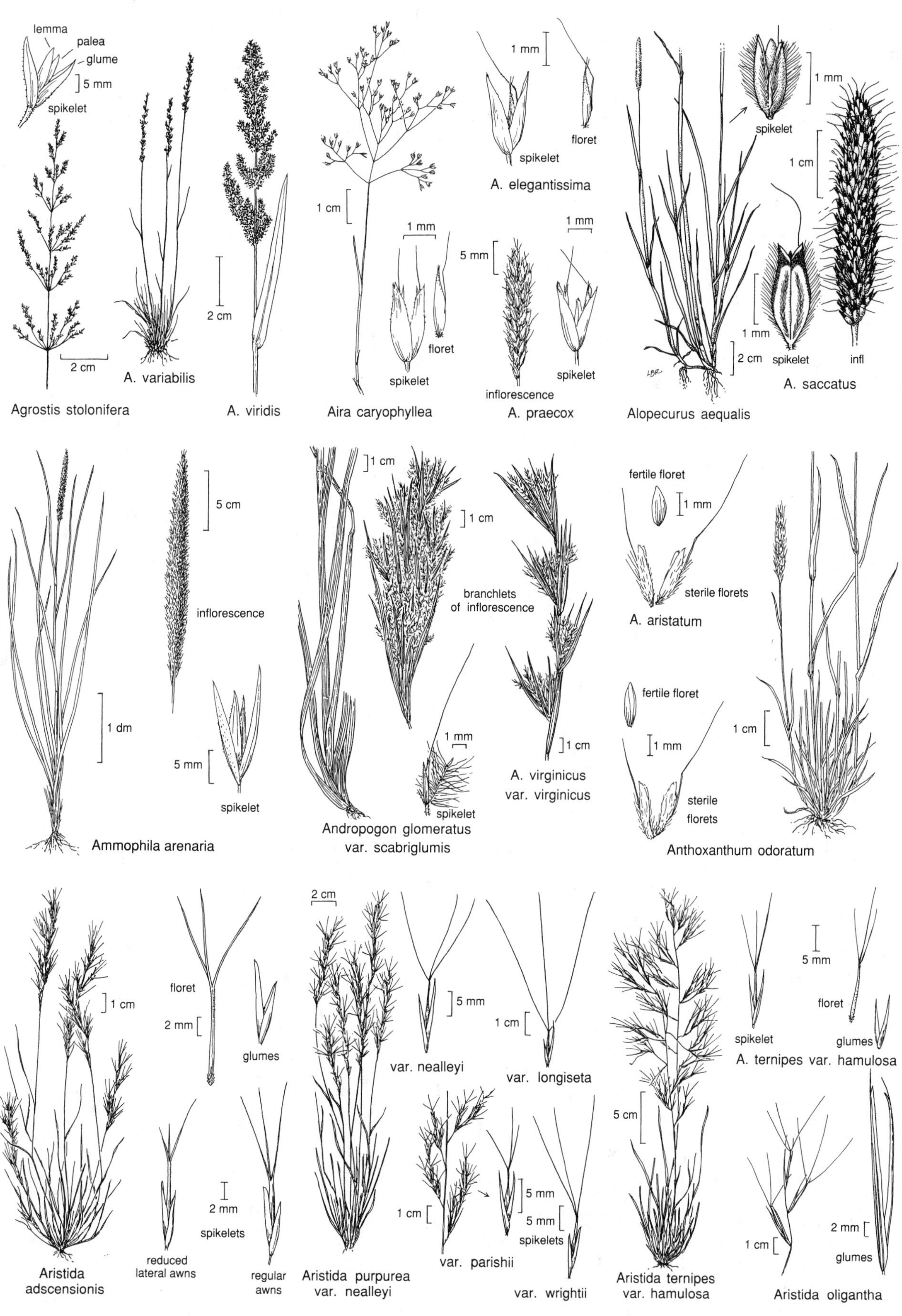
lemma
palea
glume
5 mm
spikelet
2 cm
Agrostis stolonifera
A. variabilis
2 cm
A. viridis
1 cm
1 mm
floret
spikelet
Aira caryophyllea
1 mm
spikelet
floret
A. elegantissima
5 mm
1 mm
inflorescence
spikelet
A. praecox
2 cm
Alopecurus aequalis
spikelet
1 mm
1 cm
1 mm
spikelet
infl
A. saccatus
5 cm
inflorescence
1 dm
5 mm
spikelet
Ammophila arenaria
1 cm
1 cm
branchlets of inflorescence
1 mm
spikelet
Andropogon glomeratus var. scabriglumis
1 cm
A. virginicus var. virginicus
fertile floret
1 mm
sterile florets
A. aristatum
fertile floret
1 mm
sterile florets
1 cm
Anthoxanthum odoratum
1 cm
floret
2 mm
glumes
reduced lateral awns
2 mm
spikelets
regular awns
Aristida adscensionis
2 cm
Aristida purpurea var. nealleyi
5 mm
var. nealleyi
1 cm
var. longiseta
1 cm
var. parishii
5 mm
5 mm
spikelets
var. wrightii
5 cm
Aristida ternipes var. hamulosa
5 mm
spikelet
floret
glumes
A. ternipes var. hamulosa
1 cm
2 mm
glumes
Aristida oligantha

SESSILE SPIKELET: glumes > florets, membranous; florets 2, lower vestigial, obscure, upper fertile; lemma translucent, tip awned. **FR** oblong to fusiform. ± 35 spp.: worldwide, warm temp, trop. (Greek: pit, from pitted glumes of some spp.) [Allred & Gould 1983 Syst Bot 8:168–184] Cult for forage, revegetation.

1. Sessile spikelet ± = stalked spikelet; infl reddish to purplish .... ***B. ischaemum*** var. ***songarica***
1′ Sessile spikelet < and narrower than stalked spikelet; infl tan to silvery
  2. Sessile spikelet > 4.5 mm; lemma awn > 18 mm .... ***B. barbinodis***
  2′ Sessile spikelet < 4.5 mm; lemma awn < 18 mm .... ***B. laguroides*** ssp. ***torreyana***

B. barbinodis (Lagasca) Herter (p. 1241) CANE BLUESTEM **STS** 6–12 dm, clumped. **LVS** basal and cauline; blade 20–30 cm. **INFL** 7–14 cm, gen branching 2 ×, primary branches 4–9 cm. **SPIKELET**: upper spikelet stalk 3–4 mm. **SESSILE SPIKELET** > stalked spikelet, 4.5–7.5 mm, gen tan; lower glume sometimes pitted on back; awn 2–3 cm. $2n$=180. Dry slopes; < 1200 m. SCo, s ChI, WTR, SnGb, PR; w to OK, TX, Mex. [*Andropogon b.* Lagasca] ❀TRY.

B. ischaemum (L.) Keng var. *songarica* (Fischer & C. Meyer) Celarier & Harlan **ST** gen solitary, 4–10 dm. **LVS** gen basal; blade 10–20 cm. **INFL** 5–12 cm, branching 1 ×; branches 3–8, 4–8 cm. **SPIKELET**: upper spikelet stalk 1–3 mm. **SESSILE SPIKELET** ± = stalked spikelet, 3–4.5 mm, reddish to purplish; lower glume pit 0; awn 1–1.5 cm. $2n$=40,50,60. Disturbed ground; ± 1000 m. CaRF (Shasta Co.), expected more widely; native to c Eur, Asia.

B. laguroides (DC.) Herter ssp. *torreyana* (Steudel) Allred & Gould (p. 1241) **STS** 4–13 dm, gen clumped. **LVS** gen basal; blade 5–21 cm. **INFL** 4–12 cm, branching 1–2 ×, 1–3 cm. **SPIKELET**: upper spikelet stalk 1–3 mm. **SESSILE SPIKELET** > stalked spikelet, 3.5–4 mm, tan; lower glume pit 0; awn 1–2 cm. Rocky slopes, disturbed areas; ± 900 m. c SNF (Mariposa Co.), SnFrB (Alameda Co.), expected elsewhere; native to c&s US, N Mex, S.Am. Cult for forage.

BOUTELOUA GRAMA

J. Travis Columbus

Ann, per, gen cespitose. **ST** solid, gen glabrous. **LVS** gen basal; ligule gen < 1 mm, gen hairy; blade flat to inrolled, upper surface gen ± short-hairy, often ciliate near ligule, hairs long, bulbous-based. **INFL** gen panicle-like; branches spike-like, 1 per node, persistent or deciduous in fr; spikelets 2-rowed on 1 side of axis, overlapping. **SPIKELET** sessile or short-stalked, ± cylindric to laterally compressed; glumes gen unequal, gen lanceolate, 1-veined, upper glume firmer than lower; axis (if infl branch persistent) breaking between glumes and lower floret; florets gen 2–3, lower floret bisexual, > upper, upper florets gen vestigial, sterile; lemmas 3-veined, gen 3-awned, awns straight, scabrous; palea ± = lemma. ± 40 spp.; Am. (Claudio (born 1774) and Esteban (born 1776) Boutelou, Spanish botanists, horticulturists) [Gould 1979 Ann Missouri Bot Gard 66:348–416] Many spp. important for forage.

1. Ann
  2. Spikelets 1–4 per infl branch; branch axis prolonged beyond terminal spikelet node 5+ mm; infl branches deciduous in fr, spikelets falling with branch .... ***B. aristidoides***
  2′ Spikelets gen > 7 per infl branch; branch axis terminated by spikelet; infl branches persistent, spikelet axis breaking between glumes and lower floret
    3. Tip of lower floret lemma 2-lobed, central awn from sinus; base of awned upper floret hairy-tufted ... ***B. barbata***
    3′ Tip of lower floret lemma tapered to central awn; base of upper floret glabrous .... [2]***B. trifida***
1′ Per, rhizomes or stolons sometimes present
  4. Infl branches 13–60, deciduous in fr, spikelets falling with branch .... ***B. curtipendula***
  4′ Infl branches < 9, persistent in fr, spikelet axis breaking between glumes and lower floret
    5. Lower st internodes hairy; infl branch axis slightly prolonged beyond terminal spikelet node .... ***B. eriopoda***
    5′ Lower st internodes glabrous or minutely scabrous; infl branch axis terminated by fertile or vestigial spikelet
      6. Base of awned upper floret hairy-tufted; tip of lower floret lemma 2-lobed, central awn from sinus .. ***B. gracilis***
      6′ Base of upper floret glabrous; tip of lower floret lemma tapered to central awn .... [2]***B. trifida***

B. aristidoides (Kunth) Griseb. var. ***aristidoides*** (p. 1241) NEEDLE GRAMA Ann. **ST** decumbent to erect, 0.5–3.5 dm. **LF**: blade < 7 cm, < 2 mm wide. **INFL**: branches 4–16, 8–25 mm, appressed to reflexed, deciduous in fr; branch axis exceeding terminal spikelet node > 5 mm, base densely short-hairy; spikelets 1–4 per branch, gen appressed, falling with branch. **SPIKELET**: upper glume 5–7 mm, glabrous or hairy, acute; florets (1)2; lower floret lemma ± = upper glume, glabrous or hairy, lobes 0, awns 0 or 2, < 1 mm; upper floret base hairy-tufted, lobes gen 0 between awn bases, awns 2–7 mm, gen unequal. $2n$=40. Dry, open, sandy to rocky slopes, flats, washes, disturbed sites, scrub, woodland; < 1800 m. e PR, e&s DMoj, DSon; to UT, TX, s Mex, S.Am. Another var. in AZ, NM, n Mex. ❀STBL.

B. barbata Lagasca var. ***barbata*** (p. 1241) SIX WEEKS GRAMA Ann. **ST** prostrate to erect, 0.5–3 dm. **LF**: blade < 6 cm, < 2 mm wide. **INFL**: branches 2–8, 6–25 mm, spreading to appressed, persistent in fr; branch axis terminated by spikelet, base glabrous or puberulent; spikelets 7–40 per branch, spreading to ascending, breaking apart between glumes and lower floret. **SPIKELET**: upper glume 1–3.5 mm, glabrous or hairy, tip gen notched, gen awned from sinus < 1 mm; florets 2–3, lower floret lemma ± = upper glume, hairy below middle, tip 2-lobed, awns < 3.5 mm, ± equal, central awn from sinus; base of middle or, if only 2 florets, upper floret hairy-tufted, lobed between awn bases, awns 1–3.5 mm, ± equal; uppermost floret (if present) < 1 mm, awn 0. $2n$=20,40. Gen open, sandy to rocky slopes, flats, washes, roadsides, disturbed sites, scrub, woodland, pine forest; < 1700 m. SnJV, e PR, D; to Colorado, TX, s Mex. Other vars. in AZ, NM, n Mex. ❀STBL.

B. curtipendula (Michaux) Torrey (p. 1241) SIDE-OATS GRAMA Per, sometimes rhizomed. **ST** gen erect, 2–9 dm. **LF**: blade < 25 cm, < 4 mm wide. **INFL**: branches 13–60, 5–20 mm, gen pendent, deciduous in fr; branch axis slightly exceeding terminal spikelet node, base puberulent; spikelets 1–13 per branch, ascending to appressed, falling with branch. **SPIKELET**: upper glume 3–10 mm, glabrous or scabrous, acute or awned < 0.5 mm; florets 1–2; lower floret lemma < to ± = upper glume, glabrous or sparsely hairy, lobes 0, awns 2, < 2 mm; base of upper floret (if present) glabrous, tip 2-lobed, central awn from sinus, < 6 mm, lateral awns < 4 mm. $2n$=20,28,35,40–103. Dry, rocky slopes, crevices, sandy to rocky drainages, scrub, woodland; < 1900 m. s ScV (Yolo Co. as roadside waif), PR (Santa Rosa, Cuyamaca mtns), e&s DMtns; to s Can, e US, S.Am. Pls with erect sts and rhizomes 0 have been called var. *caespitosa* Gould & Kapadia. ❀DRN,SUN:2,3,**7**,**14–16**,17,**18**,**24**&IRR:**8–12**,**19–23**;also STBL.

B. eriopoda (Torrey) Torrey (p. 1241) BLACK GRAMA Per, sometimes stoloned. **ST** decumbent to erect, 1.5–6 dm; internodes, esp lower, hairy. **LF**: blade < 10(15) cm, < 2 mm wide. **INFL**: branches 2–7, 10–40 mm, spreading to appressed, persistent in fr; branch axis slightly exceeding terminal spikelet node, base densely hairy; spikelets 6–18 per branch, ascending to appressed, breaking apart between glumes and lower floret. **SPIKELET**: upper glume 4–8 mm, gen glabrous, sharply acute; florets 2; lower floret lemma ± = to > upper glume, base hairy-tufted, glabrous or sparsely hairy above, lobes 0, central awn 1.5–4 mm, lateral awns < 2 mm; upper floret base hairy-tufted, lobes gen 0 between awn bases, awns 4–8 mm, ± equal. $2n$=20,21,28. Dry, open, sandy to rocky slopes, flats, washes, scrub, woodland; 900–1900 m. e DMtns; to WY, OK, n Mex.

B. gracilis (Kunth) Griffiths (p. 1241) BLUE GRAMA Per, gen short-rhizomed. **ST** decumbent to erect, 1–6 dm. **LF**: blade < 15 cm, < 2 mm wide. **INFL**: branches 1–4(6), 10–50 mm, spreading to appressed, persistent in fr; branch axis terminated by vestigial spikelet, base hairy; spikelets 20–80 per branch, spreading to ascending, breaking apart between glumes and lower floret. **SPIKELET**: upper glume 3.5–6 mm, vein gen with long, bulbous-based hairs, acute or awned < 0.5 mm; florets 2–3; lower floret lemma ± = upper glume, base hairy-tufted, back hairy, tip 2-lobed, awns 1–3 mm, unequal, central awn from sinus; base of middle or, if only 2 florets, upper floret hairy-tufted, lobed between awn bases, awns 2.5–6 mm, ± equal; uppermost floret (if present) < 2 mm, awn 0. $2n$=20, 21,28,35,40,42,60,61,77,84. Sandy to rocky slopes, flats, drainages, scrub, woodland, pine forest; < 2300 m. SnBr, e DMtns (Ivanpah, New York, Clark mtns), waif elsewhere; to s Can, e US, s Mex, S.Am. ❀DRN,SUN:**7,15,16**,17&IRR:2,3,**8,9,14,18–24**; GRCVR;also STBL.

B. trifida Thurber (p. 1241) RED GRAMA Per, sometimes fl 1st year, sometimes short-rhizomed. **ST** ascending to erect, 1–3 dm. **LF**: blade < 5 cm, < 1.5 mm wide. **INFL**: branches 1–7, 10–35 mm, ascending to appressed, persistent in fr; branch axis terminated by spikelet, base puberulent to hairy; spikelets 8–32 per branch, ascending, breaking between glumes and lower floret. **SPIKELET**: upper glume 2–5 mm, glabrous, acute or notched, awned from sinus < 1 mm; florets 2(3), lower floret lemma < upper glume, glabrous or hairy, tip tapered to central awn, awns 2–8 mm, ± equal; upper floret base glabrous, lobes 0 between awn bases, awns 2–8 mm, ± equal. $2n$=20,28. RARE in CA. Dry, rocky, often calcareous slopes, crevices, scrub; 700–2000 m. e DMtns; to UT, TX, c Mex.

BRACHYPODIUM

Dieter H. Wilken

Ann, per from rhizomes. **STS** decumbent to erect; nodes often hairy. **LVS** gen cauline; blade flat to inrolled, glabrous to short-hairy; ligule membranous. **INFL** spike-like to raceme-like; spikelets gen 1 per node, subcylindric, ascending to appressed, sessile to short-stalked. **SPIKELET**: glumes unequal, < lowest floret, 5–9-veined, acute to awned; florets 7–20+, bisexual; axis breaking above glumes and between florets; lemma backs rounded, 7–9-veined, acute to awned from tip; palea slightly < lemma, clearly ciliate or toothed. ± 15 spp.: temp, subtrop worldwide. (Greek: short foot, from short, thick spikelet stalk in some spp.)

1. Ann; lf blades 1.5–8 cm; lemma awn 4–11 mm .... ***B. distachyon***
1′ Per from rhizomes; lf blade (7)10–30 cm; lemma acute to short-awned, awn < 3 mm .... ***B. pinnatum***

B. distachyon (L.) Beauv. (p. 1245) Ann 15–40 cm. **STS** decumbent to erect. **LF**: blades 1.5–8 cm, 3–5 mm wide, flat. **INFL** 1–8 cm; spikelets 1–6 per st. **SPIKELET**: glumes 5–8 mm; florets 6–20; lemma 7–9 mm, awn 4–11 mm; palea stiff-ciliate to minutely toothed above middle. Disturbed areas, dry slopes; < 600 m. s NCoR, s CaRF, SNF, GV, CW, SCo, s ChI (Santa Catalina Island); native to s Eur.

B. pinnatum (L.) Baeuv. (p. 1245) Per from rhizomes, gen > 50 cm. **ST** gen erect. **LF**: blades (7)10–30 cm, 1.5–4 mm wide, ± inrolled when dry. **INFL** 9–20 cm; spikelets 6–15(20) per st. **SPIKELET**: glumes 5–8 mm; florets 7–24; lemma 7–9 mm, acute to short-awned, awn < 3 mm; palea finely ciliate above middle. Dunes, disturbed places; < 30 m. s NCo, expected elsewhere; native to Eur. Pls with lf blades tightly inrolled, rhizomes many-branched, have been called *B. phoenicoides* (L.) Roemer & Shultes.

BRIZA QUAKING GRASS

Dieter H. Wilken

Ann, per. **STS** ascending to erect, 1(–6). **LVS** basal to cauline; ligules membranous to translucent; blades flat. **INFL** panicle-like, open. **SPIKELETS** erect to pendent, ± compressed, subconic to ovoid; glumes subequal, papery, rounded at tip, 3–9-veined; florets 6–20, breaking above glumes and between florets; lemma width > length, papery to translucent, rounded at tip, 7–9-veined; palea ± = lemma. 10–12 spp.: Eur, n Afr. (Greek: a kind of grain)

1. Spikelets 10–19 mm, ovoid, pendent .... ***B. maxima***
1′ Spikelets 2–5 mm, subconic, erect .... ***B. minor***

B. maxima L. (p. 1245) Ann. **STS** 1–4, 3–60 cm. **LF**: ligule 1–4 mm; blade 1–7 mm wide. **INFL** 2–10 cm. **SPIKELETS** 1–14 per infl, 10–19 mm, pendent, ovoid, obtuse at base; glumes 4–7 mm, 5–9-veined; florets 12–19; lemmas 6–8 mm. $2n$=14. Shaded sites; 20–800 m. NCo, NCoRO, n SNF, n CCo, n&c SCoRO, n SCo; to B.C.; native to s Eur. Cult for orn.

B. minor L. (p. 1245) Ann. **STS** 1–8, 10–50 cm. **LF**: ligule 3–6 mm, blade 3–10 mm wide. **INFL** 3–20 cm. **SPIKELETS** gen > 15 per infl, 2–5 mm, erect, subconic, truncate at base; glumes 1.5–4 mm, 3–5-veined; florets 4–6; lemmas 1–2 mm. $2n$=10. Shaded or moist, open sites; 20–600 m. NCo, NCoRO, CCo, n SCoRO, n&c SNF, s ScV, n SnJV, DSon (Rancho Mirage); to OR, e US; native to s&w Eur.

BROMUS BROME

Dieter H. Wilken & Elizabeth L. Painter

Ann to per. **LVS** basal and cauline; sheath closed, gen hairy; ligule gen < 5 mm, membranous, entire to fringed; blade flat to inrolled. **INFL** gen panicle-like, open to dense; spikelet stalk gen stiff, rigid. **SPIKELET** strongly compressed to cylindric; axis breaking above glumes and between florets; glumes unequal, gen < lower floret, lower gen 1–3-veined, upper

3–7-veined, back rounded to keeled, tip acute; lemmas faintly 5–9-veined, tip gen 2-toothed, short-pointed to straight-awned from between teeth; palea gen < lemma. ± 150 spp.: temp worldwide. (Greek: ancient name) [Stebbins 1947 Contr Gray Herb 165:42–55; Wagnon 1952 Brittonia 7:415–480] Native spp. need careful study.

1. Spikelet strongly compressed, glumes and lemmas clearly keeled
 2. Lemma awn 0–3 mm; lower glume gen 5–7-veined ..... ***B. catharticus***
 2′ Lemma awn 3–15 mm; lower glume 3–7-veined
  3. Ann; lemma body ± = upper glume, back glabrous to scabrous, short-hairy between outer 1–2 veins and margin ..... ***B. arizonicus***
  3′ Per; lemma body gen > upper glume, back evenly scabrous to short-hairy
   4. Fr compressed; lemma wing below middle and between outer vein and margin ± 1 mm wide; disturbed places ..... ***B. stamineus***
   4′ Fr cylindric; lemma wing gen < 0.5 mm wide; ± undisturbed places ..... ***B. carinatus***
    5. Spikelets not crowded or overlapping; branches and stalks gen > spikelets; gen inland ..... var. ***carinatus***
    5′ Spikelets crowded, overlapping; branches and stalks gen < spikelets; coastal ..... var. ***maritimus***
1′ Spikelet subcylindric to compressed, lemmas (sometimes glumes) slightly compressed or with back rounded
 6. Lemma awns gen 15–55 mm, teeth gen 3–7 mm (exc *B. sterilis*); lower glume gen 1-veined (exc *B. alopecurus*)
  7. Lemma awn bent, gen twisted 1+ ×
   8. Infl dense; spikelets > lower branches or stalks; lemma teeth acute to rounded, not thread-like ..... [2]***B. alopecurus***
   8′ Infl ± open; spikelets gen < or = lower branches or stalks; lemma teeth thread-like ..... ***B. trinii***
  7′ Lemma awn gen straight, not bent, not twisted
   9. Infl dense; spikelets all ascending to erect, > lower branches or stalks ..... [2]***B. madritensis***
    10. St, lf sheath gen glabrous; infl narrowly ovoid to oblong; most spikelet stalks visible ..... ssp. ***madritensis***
    10′ St, lf sheath gen puberulent; infl ovoid; most spikelet stalks obscure ..... ssp. ***rubens***
   9′ Infl gen open; lower spikelets (or branches) spreading to drooping, spikelets gen < lower branches or stalks
    11. Lemma body 18–30 mm, teeth 3–7 mm; awn 30–55 mm ..... ***B. diandrus***
    11′ Lemma body 9–15 mm, teeth 0–3 mm; awn 8–30 mm
     12. Lemma awn 18–30 mm; 1° infl branches gen with 1(2) spikelets ..... ***B. sterilis***
     12′ Lemma awn 8–18 mm; 1° infl branches gen with 2+ spikelets ..... [2]***B. tectorum***
 6′ Lemma awns gen 0–15 mm, teeth gen 0–3 mm; lower glume 1–3(5)-veined
  13. Lemma awn bent to curved, teeth 3–4.5 mm ..... [2]***B. alopecurus***
  13′ Lemma awn straight, gen erect, teeth 0–3(4) mm
   14. Lemma awns gen < 3 mm
    15. Glumes obtuse; lemma 2–3 mm wide (midvein to margin), body 6–9.5 mm; ann ..... ***B. briziformis***
    15′ Glumes acute to long-tapered; lemma 1–2 mm wide (midvein to margin), body 7–14 mm; per
     16. Infl gen nodding, lower branches drooping; upper glume 5–7-veined; glumes, lemmas puberulent ..... [3]***B. anomalus***
     16′ Infl erect, branches ascending to erect; upper glume 3-veined; glumes, lemmas gen glabrous
      17. Pls from rhizomes; upper glume 6–10 mm, 1–2 mm > lower glume; lemma body 8–12 mm, awn 0–2.5 mm ..... ***B. inermis***
      17′ Rhizomes 0; upper glume10–13 mm, 2–3 mm > lower glume; lemma body 10–14 mm, awn 2–5 mm ..... [2]***B. suksdorfii***
   14′ Lemma awns gen 3–25 mm
    18. Ann; lemma teeth gen 0.5–4 mm, acute
     19. Lower glume 1-veined, upper glume 3-veined
      20. Infl dense; spikelets all ascending to erect, stalks gen < 10 mm ..... [2]***B. madritensis*** (see 10. for sspp.)
      20′ Infl gen open, lower branches or spikelets spreading to nodding, stalks gen > 10 mm ..... [2]***B. tectorum***
     19′ Lower glume 3–5-veined, upper glume 5–7-veined
      21. Infl ± dense; spikelets all ascending to erect, stalks gen 0–5 mm ..... ***B. hordeaceus***
      21′ Infl open; lower spikelets spreading to nodding, some stalks or branches > 10 mm
       22. Lemma body becoming hard, margins below middle inrolled, enclosing fr; spikelet axis exposed in fr ..... ***B. secalinus***
       22′ Lemma body ± membranous, margins touching, not inrolled, not enclosing fr; spikelet axis concealed in fr
        23. Spikelet gen soft-hairy; lower glume 7–10 mm, upper glume 8–12 mm ..... ***B. arenarius***
        23′ Spikelet gen glabrous; lower glume 4.5–7 mm, upper glume 5–8 mm ..... ***B. japonicus***
    18′ Per; lemma tip gen rounded to minutely lobed
     24. Infl narrow, ± cylindric, interrupted, branches ascending to erect, gen < spikelets — lower glume 1-veined, glabrous ..... [2]***B. suksdorfii***
     24′ Infl pyramidal, lower branches nodding to spreading, gen > spikelets
      25. Lemma unevenly hairy, back gen glabrous above middle, puberulent below middle or between outer veins and margin
       26. Lower glume 1(3)-veined, upper glume 3-veined; anthers gen 1–3 mm; ligule gen < 1.5 mm ..... ***B. ciliatus***
       26′ Lower glume 3-veined, upper glume 5–7-veined; anthers gen 3–4(5) mm; ligule gen 2–4 mm ..... [2]***B. laevipes***
      25′ Lemma glabrous or evenly hairy throughout
       27. Lower glume 1-veined
        28. Upper glume 10–15 mm; cauline lf ligule gen 3–6 mm ..... ***B. vulgaris***
        28′ Upper glume 6–11 mm; cauline lf ligule gen 1–3(4) mm

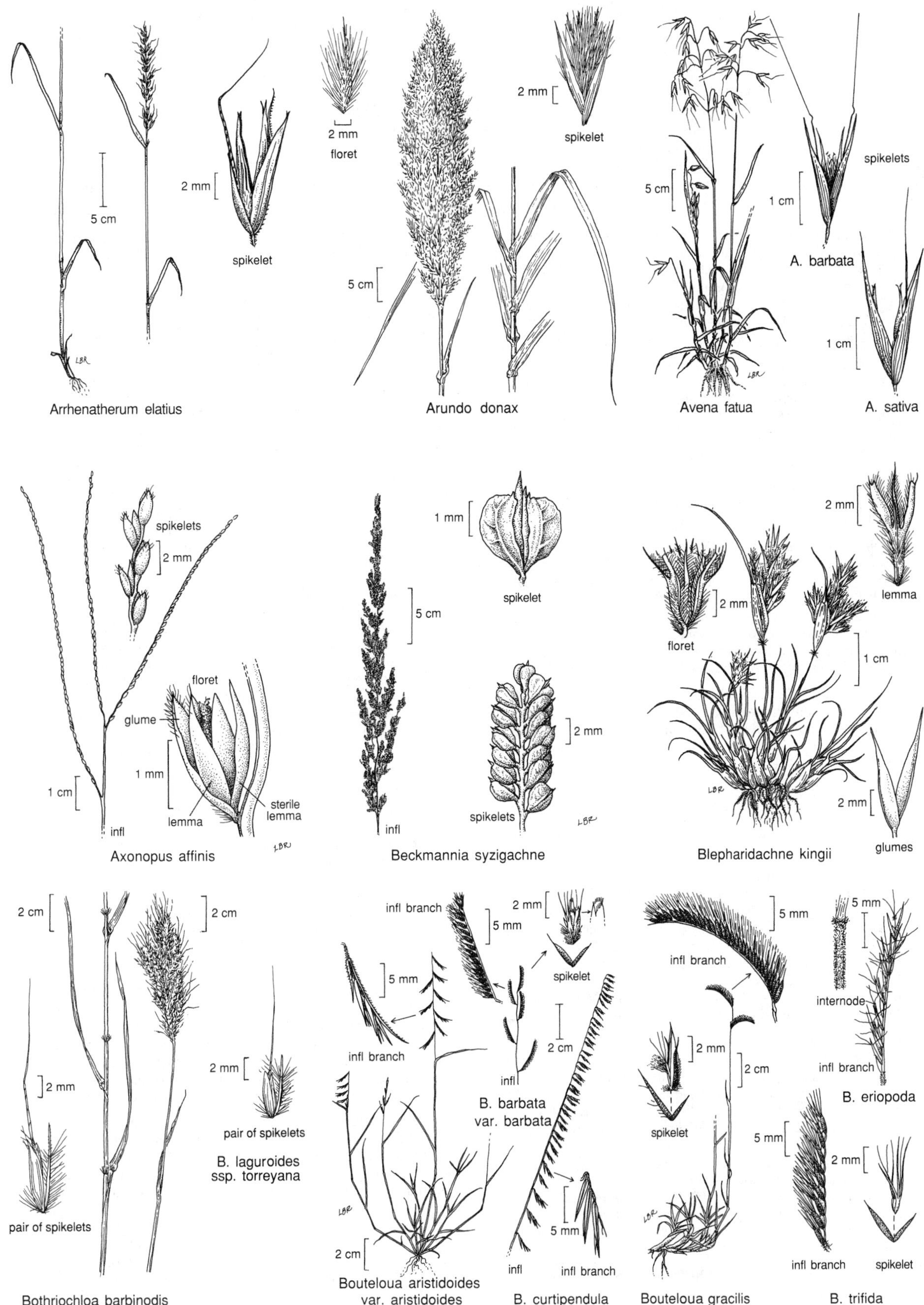
5 cm
2 mm
spikelet
Arrhenatherum elatius
2 mm
floret
5 cm
2 mm
spikelet
Arundo donax
5 cm
spikelets
1 cm
A. barbata
1 cm
Avena fatua
A. sativa
spikelets
2 mm
floret
glume
1 mm
1 cm
lemma
sterile lemma
infl
Axonopus affinis
1 mm
spikelet
5 cm
2 mm
spikelets
infl
Beckmannia syzigachne
2 mm
lemma
2 mm
floret
1 cm
2 mm
glumes
Blepharidachne kingii
2 cm
2 cm
2 mm
2 mm
pair of spikelets
pair of spikelets
B. laguroides ssp. torreyana
Bothriochloa barbinodis
infl branch
5 mm
2 mm
spikelet
5 mm
infl branch
2 cm
infl
B. barbata var. barbata
5 mm
2 cm
Bouteloua aristidoides var. aristidoides
infl
infl branch
B. curtipendula
5 mm
infl branch
2 mm
2 cm
spikelet
Bouteloua gracilis
5 mm
internode
infl branch
B. eriopoda
5 mm
2 mm
infl branch
spikelet
B. trifida

29. Lemma hairs gen > glume hairs, awn 1–3.5 mm; cauline lf blades gen inrolled; sheath glabrous . . . . [3]*B. anomalus*
29′ Lemma glabrous or hairs = glume hairs, awn 4–8 mm; cauline lf blades gen flat; sheath glabrous to soft-hairy . . . . [2]*B. orcuttianus*
27′ Lower glume 3-veined
30. Lemma hairs gen > glume hairs, awn 1–3.5 mm; cauline lf blade gen inrolled . . . . [3]*B. anomalus*
30′ Lemma and glume hairs gen ± equal, awn 3–8 mm; cauline lf blade gen flat
31. Upper glume gen 5-veined . . . . [2]*B. laevipes*
31′ Upper glume gen 3-veined
32. Infl 14–21 cm, lower branches nodding to spreading . . . . *B. grandis*
32′ Infl 6–17 cm, lower branches spreading to ascending . . . . [2]*B. orcuttianus*

B. alopecurus Poiret Ann 15–45 cm. **LF** soft-hairy; blade 1–4 mm wide. **INFL** 4–10 cm, dense; branches or spikelets erect. **SPIKELET** subcylindric, glabrous to puberulent; lower glume 8–10 mm, 3-veined; upper glume 9.5–12 mm, 3–5-veined; florets 7–13; lemma body 6–11 mm, back rounded, gen 5-veined, teeth 3–4.5 mm, awn 10–18 mm. Open, disturbed places; < 100 m. e ScV; also reported from ne US; native to Medit Eur.

B. anomalus Fourn. (p. 1245) Per 30–90 cm, often tufted. **LF** glabrous to soft-hairy; blade 1.5–5 mm wide. **INFL** 6–20 cm, open, gen nodding. **SPIKELET** ± compressed; glumes rounded, gen puberulent, lower 5–8 mm, 1–3-veined, upper 6–11 mm, 3-veined; florets 6–11; lemma body 7–13 mm, back rounded, 5–7-veined, puberulent to densely short-hairy, tip rounded to minutely lobed, awn 1–3.5 mm. $2n$=14. Uncommon. Dry, open places, slopes, coniferous forest; 3000–3500 m. W&I (reported from n SNH, SnBr); to w Can, TX, Mex. [*B. porteri* (J. Coulter) Nash] CA pls need study.

B. arenarius Labill. (p. 1245) Ann 15–60 cm. **LF** soft-hairy; blade 2–5 mm wide. **INFL** 4–19 cm, open; branches nodding to spreading, threadlike; spikelet stalks wavy to S-shaped. **SPIKELET** ± compressed, gen short-soft-hairy; glumes with flat to rounded back, lower 7–10 mm, gen 3-veined, upper 8–12 mm, 5–7-veined; florets (3)6–9; lemma body 7–11 mm, back rounded, 5–7-veined, teeth 0.5–2 mm, awn (6)9–14 mm. Open, disturbed places; < 2000 m. CA-FP, D; to OR, AZ; native to Australia.

B. arizonicus (Shear) Stebb. (p. 1245) Ann 40–90 cm. **LF**: glabrous to sparsely short-hairy; blade 3–9 mm wide, margin clearly ciliate below middle. **INFL** 5–30 cm, ± open; branches ascending to erect, lower ± spreading in fr. **SPIKELET** ± strongly compressed; glumes keel-like, glabrous, lower 8–13 mm, 3–5-veined, upper 9–15 mm, 5–9-veined; florets 5–9; lemma body 9–13 mm, keel-like, 7–9-veined, back glabrous to scabrous, short-hairy between margin and outer 1–2-veins, awn 7–15 mm. $2n$=84. Open places, grassland, shrubland; < 1000 m. SnJV, s SCoR, SCo, ChI, D; AZ, Baja CA.

B. briziformis Fischer & C. Meyer RATTLESNAKE BROME Ann 17–70 cm. **LF**: sheath densely soft-hairy; blade 1.5–4.5 mm wide, scabrous to gen soft-hairy. **INFL** 2.5–15 cm, open, nodding. **SPIKELET** compressed, gen glabrous; glumes membranous, rounded, lower 4.6–6 mm, faintly 3–5-veined, upper 5–8 mm, faintly 5–9-veined; florets 7–15; lemma body 6–9.5 mm, rounded, faintly 5–7-veined, tip acute to obtuse, awn 0–1 mm. $2n$=14. Open places, ditches; 1400–1700 m. KR, CaR, n SN, MP; to AK, ne US, Mex; native to Eurasia. Often confused with *Briza maxima*.

B. carinatus Hook. & Arn. CALIFORNIA BROME Per 45–150 cm, sometimes fl 1st yr. **LF**: sheath glabrous to soft-hairy; blade 3–12 mm wide, glabrous, scabrous, or soft-hairy. **INFL** 5–20 cm, ± dense, gen becoming open in fr; lowest branches gen spreading to ascending; upper branches ascending to erect. **SPIKELET** strongly compressed; glumes keel-like, glabrous to short-soft-hairy, lower 6.5–12 mm, gen 3-veined, upper 9–15 mm, 5–7(9)-veined; florets 7–11; lemma body 12–17 mm, 7–9-veined, keel-like, glabrous to densely short-hairy, awn 3–15 mm. $2n$=56. Open shrubland, woodland, coniferous forest; < 3500 m. CA (exc GV, DSon); to AK, TX, n Mex. Pls gen self-pollinating, florets often cleistogamous; forms formerly recognized as spp. widespread, often occurring together.

var. ***carinatus*** (p. 1245) **INFL**: spikelets not crowded or overlapping; branches and stalks gen > spikelets. Habitat and range of sp. [*B. breviaristatus* Buckley; *B. marginatus* Steudel; *B. polyanthus* Scribner, Great Basin brome] Pls from NCo with lf blade gen > 10 mm wide, spikelets 1–2 per branch, may be *B. sitchensis* Trin. ❀STBL;CV.

var. ***maritimus*** (Piper) C. Hitchc. (p. 1245) **INFL**: spikelets crowded, gen overlapping; branches and stalks gen < spikelets. Dunes, meadows; < 100 m. NCo, CCo, SnFrB, n SCo, ChI; OR. [*B. maritimus* (Piper) A. Hitchc.] ❀STBL.

B. catharticus Vahl RESCUE GRASS Ann or short-lived per, 30–100 cm. **LF**: sheath gen short-soft-hairy; blade 3–10 mm wide, glabrous to scabrous or short-soft-hairy. **INFL** 10–30 cm, ± open, gen erect; lower branches spreading to nodding; upper branches ascending. **SPIKELET** strongly compressed, glabrous to minutely scabrous; glumes keel-like, lower 7–12 mm, 5–7(9) veined, upper 8–13 mm, 7–9-veined; florets 5–13; lemma body 9–12 mm, keel-like, 7–11-veined, keeled, tip acute, awn 0–2.5 mm. $2n$=42. Open, gen disturbed places, fields; < 1500 m. CA; to s Can, e US, Eur; native to S.Am. [*B. haenkeanus* (C. Presl) Kunth, *B. unioloides* Kunth, *B. willdenovii* Kunth; Pinto-Escobar 1981 Bot Jahrb Syst 102:445–457]

B. ciliatus L. (p. 1245) FRINGED BROME Per 35–90 cm, gen tufted. **LF** glabrous to soft-hairy; blade 4–11 mm wide. **INFL** 8–21 cm, open; lower branches gen nodding; upper branches spreading to ascending. **SPIKELET** compressed; glumes rounded, glabrous, lower 5–8 mm, 1(3)-veined, upper 6–11 mm, gen 3-veined; florets 5–10; lemma body 7.5–14 mm, back rounded, gen 5-veined, glabrous, puberulent between outer veins and margin, tip minutely lobed, awn 2–6 mm; anthers 1–2.5 mm. $2n$=14,28. Meadows, coniferous forest; 1100–3200 m. SNH, SnBr, n W&I; to AK, ne N.Am, n Mex. Pls from s SNH, SnBr, with anthers 2–3 mm, $2n$=28, have been called *B. richardsonii* Link.

B. diandrus Roth RIPGUT GRASS Ann 15–80 cm. **LF** gen soft-hairy; blade 2–7 mm wide, margin scabrous. **INFL** 6–25 cm, ± open; lower branches gen nodding; upper spreading to ascending. **SPIKELET** ± compressed, glabrous to scabrous; lower glume 12–25 mm, gen 1-veined; upper glume 18–30 mm, gen 3-veined; florets 5–8; lemma body 18–30 mm, 5–7-veined, back rounded, tip with teeth 3–7 mm, awn 30–55 mm. $2n$=28,42,56. Open, gen disturbed places, fields; < 2000 m. CA; to B.C., S.Am; native to Eur. [*B. rigidus* Roth incl var. *gussonei* (Parl.) Cosson & Durieu misapplied]

B. grandis (Shear) A. Hitchc. Per 60–150 cm. **LF**: sparsely to densely short-soft-hairy; blade 6–12 mm wide. **INFL** 14–21 cm, open; branches spreading to nodding. **SPIKELET** compressed, puberulent; lower glume 4–7 mm, gen 3-veined; upper glume 7–9 mm, 3(5)-veined; florets 6–9 mm; lemma body 7–11 mm, back rounded, 5–7-veined, tip minutely lobed, awn 3–6 mm. $2n$=14. Dry, open places, shrubland, oak woodland, coniferous forest; 400–2300 m. c&s SN, CW, TR, PR; Baja CA. Related to *B. orcuttianus*.

B. hordeaceus L. (p. 1245) Ann 11–65 cm. **LF** gen soft-hairy; blade 1.5–5 mm wide. **INFL** 2.5–13 cm, gen ± dense; branches ascending to erect; spikelet > stalk. **SPIKELET** ± compressed, glabrous to short-soft-hairy; lower glume 5–8 mm, gen 3-veined, upper glume 6–9 mm, 5–7-veined; florets 5–10; lemma 6.5–10 mm, back rounded, 5–7-veined, teeth 0.5–1.5 mm, awn 4–10 mm; palea gen < lemma; anthers gen < 3 mm. $2n$=14,28. Open, often disturbed places; < 1000 (2100) m. CA (uncommon D); Am; native to Eurasia. [*B. mollis* L., *B. racemosus* L., & *B. scoparius* L. misapplied] Pls with awn outcurved in fr, ± flat near base, have been called ssp. *molliformis* (Godron) Maire [*B. molliformis* Godron]; pls with anthers 3–5 mm have been called *B. arvensis* L.

B. inermis Leysser ssp. *inermis* SMOOTH BROME Per 45–100+ cm, rhizomed. **LF** glabrous to sparsely scabrous; blade 5–12 mm wide, margin ± scabrous. **INFL** 9–20 cm, ± dense; branches gen ascending to erect. **SPIKELET** ± cylindric, glabrous to minutely scabrous; lower glume 5–9 mm, gen 1-veined; upper glume 6–10 mm, gen 3-veined; florets 5–10; lemma body 8–12 mm, 5–7-veined, back rounded, tip obtuse to minutely lobed, awn 0–2.5 mm. 2*n*=28,42,± 56. Meadows, ditches, fields; < 2700 m. SNH (e slope), SCoRO, SCo, PR, GB; to e US; native to Eur. Cult widely for forage, revegetation after fire. Ssp. *pumpellianus* (Scribner) Wagnon with puberulent lemmas is native to nw N.Am, Rocky Mtns.

B. japonicus Murr Ann 17–85 cm. **LF**: sheath soft-hairy; blade 1.5–6 mm wide, glabrous to sparsely soft-hairy. **INFL** 3–26 cm, ± open; branches spreading to ascending, lower sometimes nodding; spikelet stalks thread-like, flexible. **SPIKELET** subcylindric to slightly compressed, gen glabrous; glumes with rounded back, lower 4.5–7 mm, 3–5-veined, upper 5–8 mm, 5–7-veined; florets 3–9; lemma 7–10 mm, gen glabrous, back rounded, faintly 7–9-veined, teeth fused or free, < 1 mm, awn 5–11 mm, straight to slightly curved. Open, disturbed places; < 2300 m. CA; Am; native to Eurasia. [*B. commutatus* Schrad.] Pls with infl branches nodding and awns slightly curved are uncommon.

B. laevipes Shear Per 55–140+ cm. **LF** glabrous to short-hairy; blade 4–17 mm wide. **INFL** 7–27 cm, open; branches nodding to spreading, upper ± ascending. **SPIKELET** compressed; glumes glabrous to puberulent, lower 4–9 mm, 3-veined, upper 6–11 mm, 5–7-veined; florets 5–11; lemma body 9–15 mm, back rounded, 5–7-veined, puberulent or glabrous above middle, densely puberulent below middle near margin, tip obtuse to minutely lobed, awn 3.5–7 mm; anthers 3–5 mm. 2*n*=14. Shrubland, coniferous forest; 100–2500 m. NW, CaR, SN, CW, SCo, ChI, WTR, PR; to WA, Baja CA. Pls from CW and SW < 900 m, with glumes and lemmas evenly puberulent, have been called *B. pseudolaevipes* Wagnon.

B. madritensis L. FOXTAIL CHESS Ann 10–50 cm. **LF** glabrous to short-soft-hairy; blade 1–4 mm wide. **INFL** 3–11 cm, ± dense; branches ascending to appressed. **SPIKELET** cylindric to slightly compressed, glabrous to puberulent; lower glume 5–11 mm, 1-veined, upper 9–15 mm, 3-veined; florets 3–9; lemma body 10–18 mm, back gen rounded, 3–5-veined, teeth 2–4 mm, awn 10–25 mm. 2*n*=28. Open, gen disturbed places; < 2200 m. CA; to B.C., e US, n Mex; native to Eur.

ssp. *madritensis* (p. 1245) **ST** gen glabrous. **LF**: sheath gen glabrous. **INFL** 4–11 cm, narrowly ovoid; most branches visible. Habitats of sp. CA-FP, D (uncommon); s OR, Baja CA; native to Eur.

ssp. *rubens* (L.) Husnot (p. 1245) **ST** gen puberulent. **LF**: sheath gen puberulent. **INFL** 3–8 cm, ovoid, dense; branches (exc lowest) obscure. Habitats and range of sp. [*B. rubens* L.]

B. orcuttianus (Shear) A. Hitchc. Per 60–160 cm. **LF** glabrous to long-soft-hairy; blade 5–11 mm wide, glabrous to sparsely soft-hairy. **INFL** gen 6–17 cm, open; branches spreading to ascending. **SPIKELET** compressed, glabrous to puberulent; lower glume 5.5–8 mm, 1–3-veined; upper glume 8–10 mm, gen 3-veined; florets gen 5–8; lemma body 8–12 mm, back rounded, 3–7-veined, tip minutely lobed, awn 4–8 mm. Dry places, meadows, shrubland, coniferous forest; 900–2900 m. NW, CaR, SN, SCoRO, TR, PR, MP; to WA, w NV, Baja CA. [var. *hallii* A. Hitchc.] Related to *B. grandis*.

B. secalinus L. (p. 1245) Ann 45–100 cm. **LF** gen short-soft-hairy; blade 2–7 mm wide. **INFL** 8–17 cm, ± open; branches spreading to ascending, nodding in fr. **SPIKELET** ± compressed, glabrous to sparsely puberulent; glumes rounded, lower 4–5.5 mm, 3–5-veined, upper 5–8 mm, 5–7-veined; florets 5–9; lemma body 6.5–8.5 mm, hard in fr, back rounded, 5–7-veined, margins below middle inrolled, enclosing fr, teeth fused or free, 0.5–1 mm, awn (0)3–8 mm. 2*n*=14,28. Uncommon. Open disturbed places, fields; < 1500 m. NW, CaR, n&c SN, MP; to e N.Am; native to Eurasia.

B. stamineus Desv. Per 30–95+ cm. **LF**: sheath gen soft-hairy; blade 4–7 mm wide, gen glabrous to scabrous. **INFL** 8–17 cm, ± open; branches stiffly ascending. **SPIKELET** strongly compressed, glabrous to minutely scabrous; glumes keel-like, lower 8–12 mm, 5–7-veined, upper 12–20 mm, 7–9-veined; florets 5–8; lemma body 12–16 mm, keel-like, 7–13-veined, wing between outer vein and margin ± 1 mm wide, tip acute, awn 6–10 mm. Open, disturbed places, fields; < 200 m. s NCo, s NCoR, n&c SNF, deltaic GV, CCo, SnFrB, n ChI; native to Chile.

B. sterilis L. (p. 1245) POVERTY BROME Ann 25–85 cm. **LF**: sheath puberulent; blade 2–5 mm wide, puberulent to sparsely soft-hairy. **INFL** 10–25 cm, ± open; lower branches spreading to drooping; upper branches ascending. **SPIKELET** subcylindric to ± compressed; glumes ± rounded, glabrous, lower 7–13 mm, 1-veined, upper 10–18 mm, 3-veined; florets 6–11; lemma body 9–15+ mm, back rounded, gen 5-veined, glabrous to sparsely scabrous, teeth 0–2 mm, awn 18–30 mm. 2*n*=14,28. Open, often disturbed places; < 1100 m. NW, SNF, SnFrB, SCoRO, s ChI; to e N.Am; native to Eurasia.

B. suksdorfii Vasey (p. 1245) Per 57–90 cm. **LF** gen glabrous; blade 5–10 mm wide. **INFL**: branches ascending to erect. **SPIKELET** compressed; glumes gen glabrous, lower 6–11 mm, 1(3)-veined, upper 10–13 mm, gen 3-veined; florets 5–8; lemma body 10–14 mm, back rounded, 3–7-veined, glabrous to puberulent, tip rounded to minutely lobed, awn 2–5 mm. 2*n*=14. Slopes, meadows, coniferous forest; 1200–3200 m. KR, NCoRH, CaR, SNH, SCoRO, MP; to WA.

B. tectorum L. (p. 1245) CHEAT GRASS, DOWNY BROME Ann 5–40 cm. **LF**: sheath gen densely soft-hairy; blade 1–5 mm wide, ± glabrous to densely soft-hairy, gen long-ciliate near base. **INFL** 6–22 cm, open to ± dense; branches spreading to nodding. **SPIKELET** subcylindric to slightly compressed; glumes glabrous to short-hairy, lower 5–8 mm, 1-veined, upper 7–12 mm, 3-veined; florets 3–7; lemma body 9–13 mm, 5–7-veined, glabrous to short hairy, tip with 2 teeth 1–3 mm, awn 8–18 mm. Open, disturbed places; < 2200 m. CA; Am; native to Eurasia. [var. *glabratus* Shear]

B. trinii Desv. (p. 1245) CHILEAN CHESS Ann 20–90 cm, often tufted. **LF**: sparsely to densely long-soft-hairy; blade 2–9 mm wide. **INFL** 8–30 cm, ± open; branches gen ascending. **SPIKELET** ± compressed; glumes rounded, gen glabrous, lower 8–16 mm, 1-veined, upper 10–18 mm, 3-veined; florets 3–9; lemma body 6–15 mm, back rounded, densely silky-hairy, teeth 3–5 mm, thread-like, awn 14–22 mm, bent, twisted below middle. 2*n*=42. Open, sandy or gravelly soils; < 1700 m. NCoRI, CaRF, SN, SnJV, CW, SW, SNE, D; to Colorado, n Mex; probably native to w S.Am. [var. *excelsus* Shear]

B. vulgaris (Hook.) Shear (p. 1245) Per 45–110 cm. **LF** glabrous to soft-hairy; blade 5–12 mm wide. **INFL** 8–22 cm, open; lower branches spreading to ascending; upper branches ascending. **SPIKELET** compressed, glabrous to soft-hairy; lower glume 6–8.5 mm, gen 1-veined; upper glume 10–15 mm, 3-veined; florets 4–8; lemma body 10–15 mm, back rounded, tip rounded to minutely lobed, awn 5–10 mm. Shaded streambanks, coniferous forest; 50–1900 m. NW, CaR, n&c SN, SnFrB; to WA. ❀TRY.

CALAMAGROSTIS REED GRASS

Craig W. Greene

Per, gen from rhizomes. **STS** 1–15 dm, gen not branched, ± smooth; nodes gen 2–4. **LVS** gen basal and cauline; sheath smooth or scabrous; ligule membranous; blade flat to inrolled. **INFL** panicle-like, open to dense; branches ± drooping to appressed; spikelets ascending to appressed. **SPIKELET**: glumes subequal, gen lanceolate, acute to acuminate, lower gen 1-veined, upper 3-veined; floret 1, breaking above glumes; axis prolonged beyond floret, hairy; callus hairy; lemma < glumes, awned from below middle to near base, tip gen 4-toothed, veins 3–5, awn straight to twisted, bent; palea ± = lemma, thin. ± 100 spp.: cool temp (esp moist montane); some forage value. (Greek: reed grass) [Greene 1980 Ph.D. Thesis Harvard University] Hybridization, polyploidy, and asexual seed set contribute to taxonomic difficulty.

1. Awn exserted beyond glume tips 1–10 mm, twisted and bent
 2. Panicle open, branches spreading to ascending
  3. Pl from rhizomes; st > 4 dm; lf blades gen cauline, 3–10 mm wide; coastal, < 100 m .... ***C. bolanderi***
  3′ Pl cespitose, rhizomes 0; st gen < 3 dm; lf blades gen basal, 1–2 mm wide; subalpine to alpine, > 1300 m .... ***C. breweri***
 2′ Panicle dense, narrow, branches ascending to appressed
  4. Awn exserted 2–5+ mm beyond glume tips
   5. Glumes 8–10 mm, long-acuminate; awn exserted 5+ mm beyond glume tips .... ***C. foliosa***
   5′ Glumes 4.5–8 mm, acute to acuminate; awn exserted 2–4 mm beyond glume tips .... ***C. purpurascens***
  4′ Awn exserted < 1 mm beyond glume tips or sides
   6. Glumes 6.5–8 mm — serpentine soils .... [2]***C. ophiditis***
   6′ Glumes 4–6 mm
    7. Collar glabrous .... [2]***C. koelerioides***
    7′ Collar puberulent to short-hairy .... [2]***C. rubescens***
1′ Awn gen < glumes, sometimes prolonged beyond glume tips < 1 mm, straight or twisted and bent
 8. Infl ± open, lower branches spreading to ascending
  9. Callus hairs dense, length > 1/2 lemma; glumes 3–4.5 mm; awn gen slender, ± flexible, straight or slightly twisted .... ***C. canadensis***
  9′ Callus hairs sparse, length < 1/4 lemma; glumes 5–7+ mm; awn stiff, gen twisted .... ***C. nutkaensis***
 8′ Infl ± dense, narrow, lower branches appressed (sometimes ascending in *C. stricta*)
  10. Callus hair length > 1/2 lemma; awn gen straight; glumes 2–6 mm .... ***C. stricta***
   11. Glumes thick, margin gen opaque; blade 3–6 mm wide; ligule 2–6 mm .... ssp. ***inexpansa***
   11′ Glumes thin, margin ± translucent; blade 2–4 mm wide, gen inrolled; ligule 1–3 mm .... ssp. ***stricta***
  10′ Callus hair length < 1/2 lemma; awn gen twisted, bent; glumes > 4 mm
   12. Glumes 6.5–8 mm — serpentine soils .... [2]***C. ophiditis***
   12′ Glumes 4–6 mm
    13. Collar glabrous .... [2]***C. koelerioides***
    13′ Collar puberulent to short-hairy .... [2]***C. rubescens***

C. bolanderi Thurber (p. 1245) BOLANDER'S REED GRASS **ST** 5–15 dm; nodes gen 4. **LF**: ligule 3–5 mm; blade 3–10 mm wide, flat. **INFL** 10–25 cm, ± open; lower branches < 6–8 cm, spreading. **SPIKELET**: glumes 3–4(5) mm, smooth exc keel, tip scabrous; axis ± 1 mm, hairs 1 mm; callus hairs ± 1 mm, tufted; lemma ± = glumes, scabrous, awned near base; awn 5–6 mm, strongly twisted, bent. $2n$=56. UNCOMMON. Bogs, moist meadows, open woodlands; < 100 m. NCo. ❀IRR,DRN,SHD:4,5,15–17,24;DFCLT.

C. breweri Thurber (p. 1245) Cespitose; rhizomes 0. **ST** 1.5–3(5) dm; nodes 2–3. **LVS** gen basal and lower st; ligule 0.5–1.5 mm; blade 1 mm wide, flat or inrolled. **INFL** 3–8 cm, open; lower branches < 2 cm, ± spreading. **SPIKELET**: glumes 3–4 mm, smooth or keel scabrous; axis 1–2 mm, hairs ± 1 mm; callus hairs < 1 mm, sparse; lemma ± = glumes, awned near base; awn 4–5 mm, twisted, bent. $2n$=28,42. Moist sub-alpine and alpine meadows, lake margins, streambanks; 1300–3800 m. KR, SNH; OR. ❀DRN,IRR:1,2,6,7,15,16;DFCLT.

C. canadensis (Michaux) Beauv. (p. 1245) **ST** 6–15 dm, gen branched; nodes 3–8. **LF**: sheath glabrous to hairy near collar; ligule 3–8 mm, fringed; blade 3–8 mm wide, ± drooping, flat, scabrous. **INFL** 10–25 cm, open to dense; lower branches < 3–8 cm, ± spreading. **SPIKELET**: glumes 3–4.5 mm, scabrous, keel long scabrous, gen purplish; axis sparsely long-hairy; callus hairs ± = lemma, dense; lemma 2.5–4.5 mm, thin, awned just below middle; awn extending ± to lemma tip, straight. $2n$=42–66. Moist meadows, bogs, open woodlands; 1500–3400 m. KR, CaRH, SNH; to AK, e US. Pls with callus hairs 1/2 lemma length may be hybrids with *C. nutkaensis* and have been called *C. lactea* Beal. Some pls set seed asexually. ❀TRY.

C. foliosa Kearney (p. 1245) LEAFY REED GRASS **STS** 3–6 dm, tufted; nodes 2–3. **LVS** gen basal; ligule 5–6 mm; blade ± = st, 1–2 mm wide, inrolled, upper surface scabrous. **INFL** 5–12 cm, dense, narrow; lower branches < 3.5 cm, ascending to appressed. **SPIKELET**: glumes 8–10 mm, scabrous; axis ± 4 mm, hairs < 3 mm; callus hairs 2–3 mm; lemma 5–7 mm, awned near base; awn exserted 5+ mm beyond glume tips, bent, twisted. $2n$=28. UNCOMMON. Bluffs, cliffs, coastal scrub, forest; < 1200 m. NCo, NCoRO. ❀DRN,IRR:7,14,19–23&SUN:4,**5,15–17**,24.

C. koelerioides Vasey (p. 1245) **STS** 6–10+ dm, gen tufted; nodes 3–5. **LF**: sheath ± scabrous; ligule 3–7 mm; blade 3–7 mm wide, flat or inrolled. **INFL** 5–16 cm, gen dense; branches gen < 3 cm, appressed. **SPIKELET**: glumes 4–6 mm, scabrous esp on keel; axis ± 1 mm, hairs 1–2 mm; callus hairs < 2 mm, sparse, tufted; lemma 3.5–5 mm, awned near base; awn = lemma or extending beyond glume tips, stiff, twisted, bent. $2n$=28. Meadows, slopes, dry hills, ridges; < 2300 m. NW, CW, PR; to ID, WY. [*C. densa* Vasey, dense reed grass] Larger pls like *C. nutkaensis*. ❀DRN, SUN:1–3,6,**7**,8–10,**14–18**,19–24.

C. nutkaensis (C. Presl) Steudel (p. 1251) **STS** 6–10+ dm, densely tufted; nodes 2–3. **LF**: sheath loosely open at st base; ligule 1–4 mm, gen hidden by expanded collar; blade 4–7(10) mm wide, flat. **INFL** 12–30 cm, ± open below, narrow above; branches < 5–7+ cm, ascending. **SPIKELET**: glumes 5–7+ mm, smooth to scabrous esp on keel; axis 0.5–1 mm, hairs < 1 mm, sparse; callus hairs 1.5–2 mm; lemma 4–5 mm, awned ± near middle; awn ± = lemma, straight or slightly twisted, bent. $2n$=28. Wet areas, beaches, dunes, coastal woodland; < 1000 m. NCo, CCo, SnFrB; to AK. ❀IRR,DRN:**4–6**,7–9,14–17,19–24.

C. ophiditis (J. Howell) Nygren (p. 1251) SERPENTINE REED GRASS **STS** 6–10 dm, clumped; nodes 3–5. **LF**: ligule 2–5 mm; blade 2–4 mm wide, gen inrolled, scabrous. **INFL** 8–15 cm, ± dense; branches < 4 cm, gen appressed. **SPIKELET**: glumes (5)6.5–8 mm, scabrous esp on keel, gen pale; axis ± 2 mm, hairs ± 1 mm; callus hairs < 2 mm; lemma 5–6 mm, awned near base; awn gen > lemma, distal part sometimes exserted ± 1 mm, twisted, bent. $2n$=28. UNCOMMON. Serpentine soils; < 450 m. s NCoRO, n CCo, n SnFrB. ❀DRN,IRR,part SHD:7–9,**14,19–23**&SUN:4,5, **15–17,24**;"beautiful".

C. purpurascens R.Br. (p. 1251) Cespitose; rhizomes short. **ST** 1–8 dm; nodes 2–3. **LF**: collar glabrous to short-hairy; ligule 2–6 mm; blade 2–5 mm wide, flat, lower surface smooth, upper gen soft-hairy. **INFL** 4–15 cm, dense, narrow; branches < 3.5 cm. **SPIKELET**: glumes 4.5–8 mm, scabrous; axis 1–2 mm, hairs 1–2 mm; callus hairs < 1/4 lemma length; lemma ± = glumes, awned near base; awn > lemma, exserted 2–4 mm beyond glume tips, twisted, bent. $2n$=28,40–58,84. Subalpine, alpine rocky slopes, sandy soils; 1300–4000 m. CaRH, SN, W&I; to AK, Siberia, Greenland. Some pls set seed asexually. ❀DRN,IRR:**1**,2,3,7,14, 18–24&SUN:4–6,**15,16**,17.

C. rubescens Buckley (p. 1251) PINE GRASS **ST** 6–10 dm; nodes 2–3. **LF**: sheath smooth, gen puberulent or hairy-tufted near collar; ligule 3–5 mm; blade 2–5 mm wide, flat, lower surface smooth or scabrous, upper surface scabrous, sometimes short-hairy. **INFL** 6–15(25) cm, dense to ± open; branches < 2–4 cm. **SPIKELET**: glumes 4–5 mm, smooth to scabrous esp on keel; axis ± 1 mm, hairs < 2 mm; callus hairs < 1 mm; lemma 3–4 mm, awned

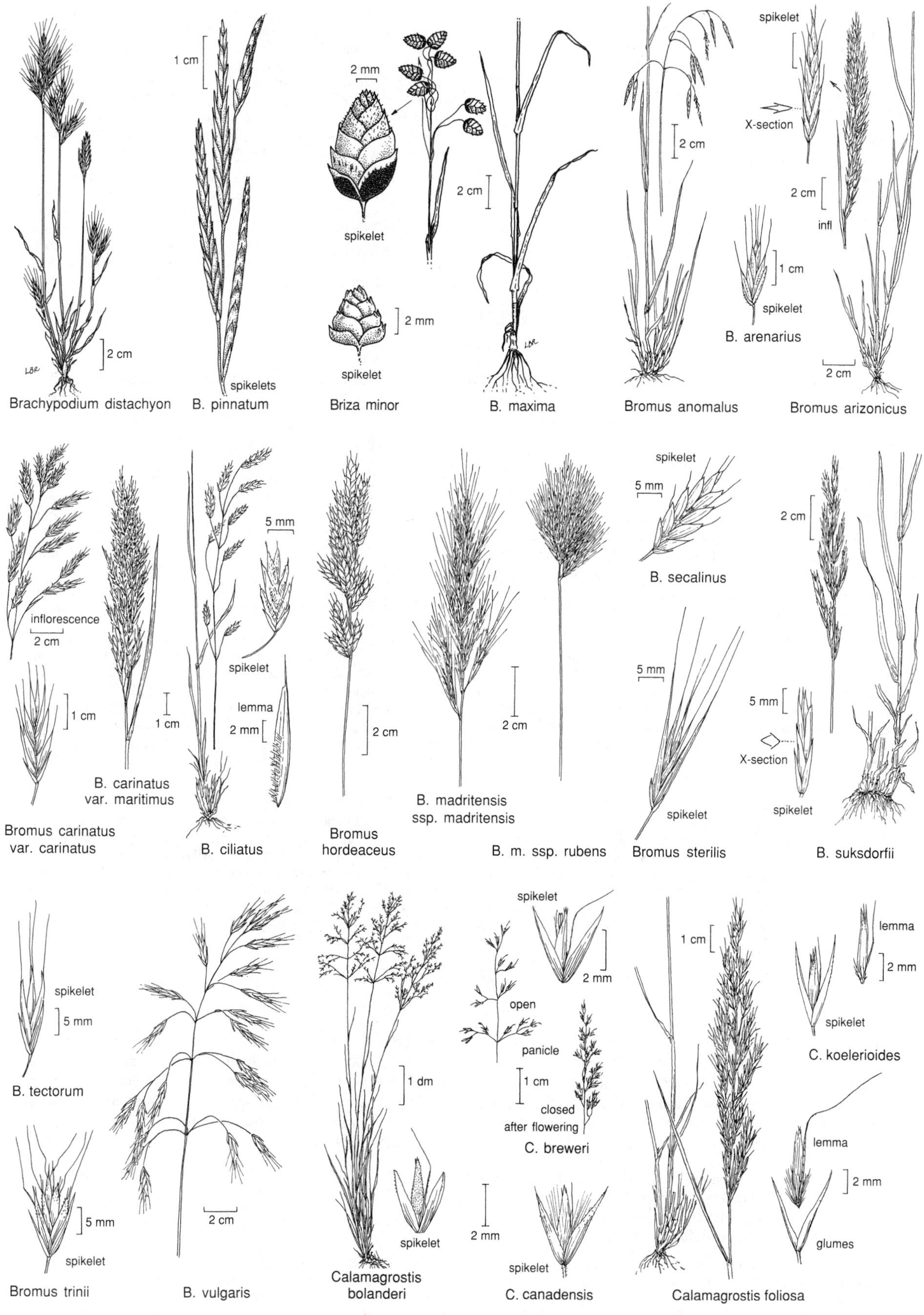
2 cm
Brachypodium distachyon
1 cm
spikelets
B. pinnatum
2 mm
spikelet
2 mm
spikelet
Briza minor
2 cm
B. maxima
2 cm
Bromus anomalus
spikelet
X-section
2 cm
infl
1 cm
spikelet
B. arenarius
2 cm
Bromus arizonicus
inflorescence
2 cm
1 cm
1 cm
B. carinatus
var. maritimus
Bromus carinatus
var. carinatus
5 mm
spikelet
lemma
2 mm
B. ciliatus
2 cm
Bromus
hordeaceus
B. madritensis
ssp. madritensis
2 cm
B. m. ssp. rubens
spikelet
5 mm
B. secalinus
5 mm
spikelet
Bromus sterilis
2 cm
5 mm
X-section
spikelet
B. suksdorfii
spikelet
5 mm
B. tectorum
5 mm
spikelet
Bromus trinii
2 cm
B. vulgaris
1 dm
spikelet
Calamagrostis
bolanderi
spikelet
2 mm
open
panicle
1 cm
closed
after flowering
C. breweri
2 mm
spikelet
C. canadensis
1 cm
Calamagrostis foliosa
lemma
2 mm
spikelet
C. koelerioides
lemma
2 mm
glumes

near base; awn 3.5–4.5 mm, exserted slightly beyond glume sides, strongly twisted, bent. 2*n*=28,42,56. Wooded slopes, montane forests; < 900 m. NCo, NCoRO, CW, n ChI (Santa Cruz Island); to sw Can, Rocky Mtns. ❀IRR,DRN:1,2,**7**,9,10,**14–23**&SUN:**4–6**,**24**.

C. stricta (Timm) Koeler Loosely cespitose. **ST** 2–24 dm. **LF**: sheath smooth; ligule 1–5.5 mm; blade 2–5 mm wide, gen inrolled, lower surface gen smooth, upper surface smooth to scabrous. **INFL** 5–20 cm, dense, narrow; branches < 1.5–5+ cm, ascending to appressed. **SPIKELET**: glumes 2–6 mm, smooth to scabrous; axis ± 1 mm, hairs = callus hairs; callus hairs 1/2 to = lemma length; lemma 2–5 mm, fine-scabrous, awned at or below middle; awn ± = glume tip, gen straight. Mtn slopes, meadows, coastal marshes; < 3400 m. NW, CaR, SNH, CCo; to AK, ne N.Am, Eurasia. Sspp. intergrade.

ssp. ***inexpansa*** (A. Gray) C.W. Greene (p. 1251) **ST** 4–12 dm. **LF**: ligule 2–5.5 mm; blade flat, strongly scabrous, upper surface gen glaucous. **INFL** 6–20 cm; longest branches < 5+ cm. **SPIKELET**: glumes 3–6 mm, gen thick, opaque; callus hairs 2/3 to = lemma; lemma 2.5–5 mm; awn sometimes twisted, bent, stiff; anthers gen sterile. 2*n*=28,56,58,70,84–±120. Habitats of sp. NW, CaR, n&c SNH, CCo; to AK, ne N.Am, ne Asia. [*C. i.* A. Gray] Some pls set seed asexually. Pls from CCo with hard glumes have been called *C. crassiglumis* Thurber, Thurber's reed grass.

ssp. ***stricta*** **ST** 2–9 dm. **LF**: ligule 1–3.5 mm; blade gen inrolled, upper surface smooth to scabrous. **INFL** 5–12 cm; longest branches < 4 cm. **SPIKELET**: glumes 2–4.5 mm, gen thin, margin sometimes translucent; callus hairs 1/2–3/4 lemma; lemma 2–4 mm; awn straight, slender; anthers gen fertile. 2*n*=28,42,56,±70. Coniferous forest, meadows, slopes; 1500–3350 m. SNH; to AK, ne N.Am, Eurasia.

CENCHRUS SANDBUR

Robert Webster

Ann, per. **STS**: internode solid to spongy inside. **LVS** basal and cauline; sheath gen smooth; ligule short-hairy or membranous, ciliate; blade flat or folded. **INFL**: main axis straight or wavy; spikelets in groups of 1(–8), gen enclosed by bur-like involucre, bracts bristle- or spine-like, fused; involucre and enclosed spikelets falling as 1 unit; spikelet sessile to ± embedded in short axis. **SPIKELET** ± compressed; glumes strongly unequal, lower 1-veined, upper ± = florets; florets 2, lower floret sterile or staminate, lemma gen 5-veined, palea gen present, upper floret fertile, lemma thick, ± hard, palea ± = lemma. ± 20 spp.: warm temp Am, Afr, s Asia. (Greek: ancient name) [Delisle 1963 Iowa State Coll J Sci 37:259–351]

1. Involucre bracts free to near base, ± stiff-bristly; per; central infl axis straight .... ***C. ciliaris***
1' Involcure bracts fused to form a bur-like involucre, gen spiny; ann; central infl axis wavy
  2. Lower involucre bracts whorled, ± flexible, upper bracts spiny .... ***C. echinatus***
  2' Lower involucre bracts not clearly whorled, all bracts spiny
    3. Involucre bracts 8–40; longest bract < 5 mm .... ***C. incertus***
    3' Involucre bracts > 50; longest bract > 5 mm .... ***C. longispinus***

C. ciliaris L. Per, sometimes from stolons. **ST** 2–15 dm. **LF**: sheath 2–7 cm, glabrous or hairy; ligule 0.5–1.5 mm; blade 5–30 cm, 2.5–11 mm wide, upper surface soft-hairy. **INFL** 2–13 cm; main axis straight; involucre bracts 30–50, ± stiff-bristly, free to near base. **SPIKELET** 2.5–4.5 mm, ± 1–1.5 mm wide, lanceolate to ovate, gray to green; lower glume ± 1–2.5 mm, upper ± 1.5–3.5 mm; lower floret staminate or sterile, lemma acute to acuminate, palea present or 0. Disturbed places, fields; < 100 m. SCo (Los Angeles Co.); to TX, n Mex; native to Afr, India.

C. echinatus L. SOUTHERN SANDBUR Ann. **ST** 1–5 dm. **LF**: sheath 3–7 cm; ligule ± 1–1.5 mm; blade 6–20 cm, 3.5–11 mm wide, upper surface glabrous or hairy. **INFL** 3.5–8 cm; main axis wavy; involucre bracts 40–60, fused. **SPIKELET** 5–6.5 mm, ± 1.5–2 mm wide, lanceolate to ovate, green; lower glume ± 1–3 mm, upper ± 4–6 mm; lower floret sterile, lemma acute. 2*n*=68,70. Disturbed places, fields; < 150 m. s ScV (Solano Co.), SCo (San Diego), DSon; native to s US, Mex, C. and S.Am. NOXIOUS WEED.

C. incertus M. Curtis COAST SANDBUR Ann. **ST** decumbent, 1–5 dm. **LF**: sheath 2.5–7 cm; ligule 0.5–1.5 mm; blade 4–12 cm, 2.5–5 mm wide, upper surface glabrous. **INFL** 2–5 cm; main axis wavy; involucre bracts 8–40, fused. **SPIKELET** 3.5–5.5 mm, ± 1–2.5 mm wide, ovate, green; lower glume 1–3.5 mm, upper glume 3–5 mm; lower floret staminate, lemma 5–7-veined, acute. 2*n*=34. Disturbed places, fields; < 800 m. GV, SCo, DMoj (near Daggett); OR; native to s US, Mex, C. and S. Am. NOXIOUS WEED.

C. longispinus (Hackel) Fern. (p. 1251) MAT SANDBUR Ann. **ST** 1–6 dm. **LF**: sheath 2.5–8 cm; ligule ± 1 mm; blade 4–12 cm, 2.5 mm wide, upper surface glabrous. **INFL** 3–5 cm; main axis wavy; involucre bracts 50–75, fused. **SPIKELET** ± 6–7.5 mm, 2–3 mm wide, lanceolate to ovate, green; lower glume 1.5–4 mm, upper 4.5–6 mm; lower floret staminate, lemma acute. 2*n*=34. Disturbed places, fields; < 900 m. s ScV, SnJV, SCo, MP (Lassen Co.), DMoj; native to c&e US. [*C. pauciflorus* Bentham misapplied]. NOXIOUS WEED.

CHLORIS

Dennis Anderson

Ann, per, cespitose or from rhizomes. **ST** decumbent to erect, 2–30 dm. **LF**: ligule membranous or hairy-tufted; blade gen 10–40 cm, 0.5–1.5 cm wide, flat. **INFL** gen umbel-like; branches 2–30, sometimes in distinct whorls, each raceme- or spike-like branch with 2 rows of overlapping spikelets on 1 side of axis. **SPIKELET**: glumes unequal, < florets, 1–3-veined; axis breaking above glumes; lower 1–2 florets fertile, upper 1–3 sterile, < 1/2 lower floret length; fertile floret lemma ovate to lanceolate, back glabrous, midvein hairy, 3-veined, awn 1; palea < lemma, translucent, obscure. **FR** ± fusiform, 3-angled. ± 50 spp.: warm temp, trop worldwide. (Greek: mother of Nestor, goddess of flowers)

1. Infl panicle-like; branches in 2–4 distinct whorls, spreading .... ***C. verticillata***
1' Infl umbel-like; branches ± erect
  2. Per, gen from stolons; vestigial florets gen 2+ .... ***C. gayana***
  2' Ann, gen < 5 dm; vestigial florets gen 1 .... ***C. virgata***

C. gayana Kunth RHODES GRASS Per, gen from stolons. **ST** 5–30 dm. **LF**: sheath glabrous to scabrous, hairy near collar; ligule hairy; blade < 30 cm, 1.5 cm wide. **INFL** umbel-like; branches erect, 8–15 cm. **SPIKELET** 3–5 mm; lower glume 1.5–3 mm, 0.3–0.5 mm wide, acute to short-awned; upper glume 2–3.5 mm, 0.3–0.5 mm wide, acute to short-awned; fertile florets 1 or 2, 2.5–4 mm, < 1 mm wide, ovate, obovate, or elliptic, margin gen hairy-tufted near tip, awn 1.5–6.5 mm; sterile florets (1)2–4, lowest

2.2–3.2 mm, 0.5–0.8 mm wide, awn 1–3 mm; upper < lower. **FR** 1–1.5 mm. 2*n*= 20,30,40. Disturbed areas; < 300 m. SnJV, SCo, expected elsewhere; native to Afr. Cult for forage.

C. verticillata Nutt. (p. 1251) Per. **ST** gen 1–4 dm. **LF**: sheath glabrous; ligule hairy; blade < 15 cm, ± 0.3 cm wide. **INFL** panicle-like; branches in 2–4 whorls, spreading, 5–15 cm. **SPIKELET** 2–3.5 mm; lower glume 2–3 mm, ± 0.3 mm wide, lanceolate; upper glume 3–3.5 mm, 0.3–0.5 mm wide, lanceolate; fertile floret 1, 2–3.5 mm, 1.5–2 mm wide, lanceolate to elliptic, glabrous to hairy, awn 5–9 mm; sterile floret 1, 1–2.5 mm, 0.5–1 mm wide, awn 3–7 mm. **FR** ± 1.5 mm. 2*n*=28,40,63. Uncommon. Disturbed areas; < 300 m. SnFrB, expected elsewhere; native to s Great Plains.

C. virgata Sw. Ann. **ST** gen 1–5 dm. **LF**: sheath glabrous to hairy near collar; ligule glabrous to hairy; blade < 30 cm, 1.5 cm wide. **INFL** umbel-like; branches ± erect, 5–10 cm. **SPIKELET** 2.5–4.5 mm; lower glume 1.5–2.5 mm, 0.2–0.5 mm wide, acute; upper glume 2.5–4.5 mm, 0.3–0.5 mm wide, acute; fertile floret 1, 2.5–4 mm, 0.5–1.5 mm wide, ovate, obovate, or elliptic, keel hairy near tip, margin gen hairy-tufted, awn 2.5–15 mm; sterile floret 1, 1.5–3 mm, 0.5–0.8 mm wide, awn 3–9.5 mm. **FR** 1.5–2 mm. 2*n*=20,26, 30,40. Disturbed areas; < 200 m. GV, SCo, D; to s Great Plains, n Mex; native to warm temp regions worldwide.

CINNA WOODREED

David M. Brandenburg

Per. **STS** erect. **LVS** gen cauline; sheath glabrous; blade flat, margin scabrous. **INFL** panicle-like; branches spreading to ascending. **SPIKELET** ± sessile to stalked, breaking below glumes, falling as 1 unit; glumes ± equal or lower < upper, lower 1-veined, upper 1–3-veined; floret 1, slightly < or > glumes, bisexual; axis prolonged behind palea, short-bristly; lemma faintly 3–5-veined, short-awned just below acute tip, or awnless; palea ± < lemma; stamens 1–2. 4 spp.: temp N.Am., n S.Am, Eurasia. (Greek: a grass) [Brandenburg et al. 1991 Sida 14:581–596]

1. Spikelet 4–5.5 mm; floret ± sessile; lemma faintly 5-veined; stamens 2, anther gen 1–2.5 mm .... ***C. bolanderi***
1′ Spikelet 2.5–4 mm; floret pedicel < 0.5 mm; lemma gen 3-veined; stamen 1, anther gen < 1 mm .... ***C. latifolia***

C. bolanderi Scribner SCRIBNER WOODREED **ST** 8.5–20 dm. **LF**: ligule 3.5–7 mm; blade < 40 cm, 2–19 mm wide. **INFL** 7.5–43 cm, 3–18 cm wide, green to golden-green. **SPIKELET** 4–5.5 mm; lower glume gen 3.5–5 mm, upper gen 4.5–5.5 mm; floret ± sessile; lemma 3–4.5 mm, faintly 5-veined, awn < 1.5 mm or 0; palea 3–3.5 mm; stamens 2, anthers 1–2.5 mm. Streambanks, wet meadows, moist sites in coniferous forest; 1850–2400 m. c&s SNH.

C. latifolia (Goeppert) Griseb. (p. 1251) DROOPING WOODREED **ST** 2–19 dm. **LF**: ligule 2–8 mm; blade < 28 cm, 1–20 mm wide. **INFL** 3–46 cm, 0.5–20 cm wide, green or purplish. **SPIKELET** 2.5–4 mm; glumes gen 2.5–4 mm; floret short-pedicelled; lemma 2–4 mm, gen 3-veined, awn < 2.5 mm or 0; palea 2–3.5 mm; stamen 1, anther < 1 mm. 2*n*=28. Streambanks, wet meadows, moist sites in coniferous forest; 1350–2800 m. KR, NCoR, CaRH, n&c SNH; circumboreal. ❀STBL.

CORTADERIA

Kelly W. Allred

Per, dioecious. **STS** densely clumped, erect. **LVS** gen basal; sheath glabrous to hairy; ligule short, hairy; blades flat or folded. **INFL** panicle-like, plume-like. **SPIKELETS** ± laterally compressed; glumes unequal, 1-veined; florets 2–8, breaking above glumes and between florets; lemmas silky-hairy, 3–5-veined, tip awned. 24 spp.: S.Am, New Zealand, New Guinea. (Argentine term for cutting) [Costas Lippmann 1977 Fremontia 4:25–27]

1. Sheaths densely hairy; st 4–5 × infl length .... ***C. jubata***
1′ Sheaths glabrous to sparsely hairy; st 2–3 × infl length .... ***C. selloana***

C. jubata (Lemoine) Stapf (p. 1251) Pls pistillate only, producing fr asexually. **ST** 2–7 m. **LF**: sheath densely hairy; blade 2–10 cm wide, deep green, distal half straight, upper surface hairy at base. **INFL** 3–10 dm. **SPIKELETS** 14–16 mm; florets 3–5; lemma hairy, tip acuminate, awn < 1 mm; stigmas incl. 2*n*=108. Disturbed sites, many habitats, esp coastal; < 800 m. NCo, NCoRO, CCo, SnFrB, SCo, WTR; native to montane w S.Am; invasive.

C. selloana (Schultes) Asch. & Graebner PAMPAS GRASS **ST** 2–4 m. **LF**: sheath glabrous to sparsely hairy; blade 3–8 cm wide, ascending, bluish to pale green, distal half curled, upper surface glabrous at base. **INFL** 3–13 dm. **SPIKELETS** 15–17 mm; florets 4–8; pistillate lemma hairy, staminate lemma glabrous, tip acute, awn 2.5–5 mm; stigmas exserted. 2*n*=72. Disturbed sites; < 300 m. SnFrB, SCo (Orange Co.), cult elsewhere; s US; native to e S.Am.

CRYPSIS PRICKLE GRASS

Ann. **STS** prostrate, ascending, or erect. **LF**: ligule hairy; blade gen short, linear to narrowly lanceolate. **INFL** panicle-like, dense, cylindric and exserted or head-like and enclosed by enlarged sheaths. **SPIKELET** bisexual, strongly compressed, falling as 1 unit; glumes, lemma acute or short-pointed; glumes < or = floret, gen lanceolate, keeled, strongly 1-veined; floret 1; lemma membranous, 1-veined; palea gen 2-veined, gen splitting with age; stamens gen 3. 8 spp.: Medit Eur, Asia, c Afr. (Greek: concealment, from partly hidden infl) [Hammel & Reeder 1979 Syst Bot 4:267–280]

1. Sheath margins hairy .... ***C. vaginiflora***
1′ Sheath margins glabrous
  2. Infl length 7–8 × > width; infl stalk > 1 cm, exserted from subtending sheath .... ***C. alopecuroides***
  2′ Infl length < 5 × > width; infl stalk < 1 cm, infl partly enclosed by subtending sheath .... ***C. schoenoides***

C. alopecuroides (Piller & Mitterp.) Schrader (p. 1251) Pl gen purple to black. **STS** ascending, 5–75 cm; branches few. **LF**: blade 5–12 cm, green or ± glaucous; sheath glabrous. **INFL** 1.5–6.5 cm, 4–6 mm wide, cylindric. **SPIKELET** 2–3 mm; glume keel hairy, lower glume < upper; lemma gen > glumes; palea 2-veined. 2*n*=16. Bottom-lands, reservoir and river margins; < 800 m. KR, NCoRI, CaRH, CCo (Marin Co.), SNE; to WA, e US; native to Eur.

C. schoenoides (L.) Lam. (p. 1251) SWAMP GRASS Pl gen pink to purple, gen mat-like. **STS** decumbent, 5–75 cm; branches few.

LF: blade 2–10 cm; sheath wide, glabrous. **INFL** 3–75 mm, 5–15 mm wide, ovoid to cylindric, enclosed by subtending sheath. **SPIKELET** ± 3 mm; glume margin glabrous, lower glume < upper; lemma gen > glumes. $2n$=32. Wet places; < 500 m. CA-FP; c&e US; native to Eur. [*Heleochloa s.* (L.) Host.] Used for wildfowl, livestock forage.

C. vaginiflora (Forsskal) Opiz Pl gen green, gen mat-like. **STS** prostrate, < 30 cm; branches many. **LF**: sheath hairy; blade 1–5 cm, ± rigid, breaking easily from sheaths. **INFL** 3–15 mm, 3–6 mm wide, ± ovoid, gen enclosed by subtending sheath. **SPIKELET** gen 2.5–3.2 mm; glumes ± equal, lower glume margin hairy; lemma ± = glumes. $2n$=48. Wet soils, lake margins, vernal pools; < 500 m. NCoR, CaR, GV, SW (Los Angeles Co.); ID; native to Eurasia. [*C. niliaca* Figari]

CYNODON

Per, mat-like, from rhizomes or stolons. **STS** ± branched. **LF**: blade short, flat, narrow, fleshy. **INFL** umbel-like; branches 2–20, spike-like, with 2 rows of overlapping spikelets along 1 side of axis; spikelets sessile. **SPIKELET** bisexual, strongly compressed; glumes ± equal, 1-veined, awn 0; floret 1, rarely 2, upper floret vestigial, breaking above glumes; lemma keeled, 3-veined, awn 0; palea = lemma, 2-veined. 8–10 spp.; trop, warm temp Eurasia, Afr. (Greek: dog tooth, from hard scales on rhizomes) [Harlan et al. 1970 Okla State Univ Agric Exp Sta Bull B–673]

1. Infl branches 2, gen 1–2 cm .... ***C. transvalensis***
1′ Infl branches 3–20, gen 3–8 cm
  2. Rhizomes present; glumes lanceolate, 1/4–1/2 × spikelet length; palea keels glabrous .... ***C. dactylon***
  2′ Rhizomes 0: glumes deltate, gen < 1/5–1/4 × spikelet length; palea keels hairy .... ***C. plectostachyus***

C. dactylon (L.) Pers. (p. 1251) BERMUDA GRASS Per from rhizomes or stolons. **ST** gen erect, 1–4 dm, flat. **LF**: ligule white-hairy; blade < 6 cm, glabrous or upper surface hairy. **INFL**: branches 2.5–5 cm, gen 4–7. **SPIKELET** ± 2 mm; glumes ± 1.5 mm, gen purplish; lemma ± 2 mm, boat-shaped, acute, keel and margins hairy; palea keels glabrous. $2n$=36. Disturbed sites; < 900 m. CA-FP, D; warm temp, trop; native to Afr. Cult for lawns, forage. TOXIC: important pollen source in hay fever; may produce contact dermatitis.

C. plectostachyus (Schumann) Englem. STAR GRASS Per from stolons. **ST** erect, 6–10 dm. **LF**: ligule thin, transparent; blade 10–30 cm, 4–7 mm wide, hairy. **INFL**: branches 3–7 cm, gen 7–20, curled with age. **SPIKELET** 2.5–3 mm; glumes < 0.3 mm, scale-like, deltate; lemma 2–3 mm, elliptic in outline, keel and margins hairy; palea keels hairy. $2n$=18. Uncommon. Roadsides; < 100 m. SnJV (Merced Co.); native to Afr.

C. transvalensis Burtt Davy AFRICAN BERMUDA GRASS Per from rhizomes and stolons. **ST** prostrate to ascending, gen < 10 cm. **LF**: ligule membranous, short; blade gen 10–30 cm, ± 1.5 mm wide, gen yellow-green, hairy. **INFL**: branches 1–2 cm, gen 2, sometimes 1 or 3. **SPIKELET** 2–4 mm; glumes unequal, < 3/4 spikelet length; lemma 2–4 mm, elliptic, keel ± hairy; palea keels glabrous. $2n$=18. Uncommon. Roadsides; < 100 m. DSon (Imperial Co.); native to Afr.

CYNOSURUS

Ann, per. **LVS** gen basal; ligule membranous; blade flat. **INFL** panicle-like, cylindric to head-like, dense; spikelets compressed, gen paired, pairs short-stalked; lower spikelet gen < upper, gen fertile, sessile; upper spikelet sterile, short-stalked. **FERTILE SPIKELET** bisexual; glumes lanceolate, 1-veined, awn 0; axis breaking above glumes and between florets; florets 2–3; lemma back rounded, faintly 5-veined, awned. **STERILE SPIKELET** persistent; axis not breaking apart; glumes, lemmas linear-lanceolate, rigid, 1-veined. $2n$=14,16. 8 spp.: Eur, w Asia, n Afr. (Greek: dog tail, from shape of infl) [Jirasek & Chrtek 1964 Novit Bot Inst Prag 1964:23–27]

1. Infl cylindric; fertile lemma awn < 1 mm; per .... ***C. cristatus***
1′ Infl ± head-like; fertile lemma awn 3–10 mm; ann .... ***C. echinatus***

C. cristatus L. (p. 1251) CRESTED DOGTAIL Per, ± puberulent. **STS** erect, 3–8 dm; base sometimes abruptly bent. **LF**: blade 6–10 cm, 0.5 mm wide. **INFL** 3–8 cm, cylindric, curved; central axis flat, ± zig-zag. **FERTILE SPIKELET** ± 3 mm; glumes keeled; lemma back rounded, awn < 1 mm. **STERILE SPIKELET** 4–5 mm; glumes ± flat; florets 7–9; lemmas ± = glumes. $2n$=14. Fields, disturbed places; < 300 m. NCo, CCo, SCo; to WA, ne US; native to Eur. Sometimes cult for forage.

C. echinatus L. (p. 1251) HEDGEHOG DOGTAIL Ann, glabrous. **STS** erect, 1–5 dm. **LF**: blade 4–10 cm, 2–5 mm wide. **INFL** 1–4 cm, head-like, sometimes interrupted; central axis ± cylindric; spikelets ± 1-sided. **FERTILE SPIKELET** 5–6 mm; glumes keeled, short-awned; lemma ± = glumes, back rounded, awn 3–10 mm. **STERILE SPIKELET**: glumes keeled, ciliate, long-pointed; lemma ± = glumes. $2n$=14,16. Open, disturbed sites; gen < 1000 m. NW, SNF, CW, WTR; to B.C., e US; native to Eur.

DACTYLIS ORCHARD GRASS

Dieter H. Wilken

1 sp. (Greek: finger, from crowded spikelets at infl tips) [Stebbins & Zohary 1959 Univ Calif Publ Bot 31:1–40]

D. glomerata L. (p. 1251) Per from short rhizomes. **STS** (1)2–5, 3–11 dm. **LVS** basal and cauline; sheath closed, keeled; appendages 0; ligules 4–9 mm, ± translucent, fringed to toothed at obtuse tip, glabrous to minutely soft-hairy; blades 3–6 mm wide, slightly scabrous. **INFL** 1–1.5 dm, panicle-like; branches spike-like, stiffly erect to spreading. **SPIKELETS** crowded, gen on 1 side of branch tips, short-stalked to subsessile, laterally compressed; florets 2–4, gen bisexual (upper sometimes staminate), breaking above glumes and between florets; glumes and lemmas short awned at tip, minutely ciliate to short-hairy on back; glumes 3–6 mm, subequal; lemmas 4–7 mm, margin translucent, 5-veined; palea slightly < lemma, 2-forked at tip. $2n$=14,21,27–31,42. Disturbed, often moist sites; < 2500 m. CA-FP, GB; to Can, US; native to Eurasia. Cult for forage, hay.

DACTYLOCTENIUM

Ann, per. **STS** decumbent to erect. **LF**: blade flat. **INFL** umbel-like; branches 2–11, spike-like, with 2 rows of overlapping, sessile spikelets gen along 1 side of axis; axis tip ± naked. **SPIKELET** bisexual, compressed; glumes unequal, wide, 1-veined, lower persistent, acute, upper deciduous, short-awned, awn gen curved; axis breaking apart above glumes and between florets; florets 3–5; lemma membranous, keeled, 3-veined, lateral veins gen faint, tip acuminate or short-awned, awn gen curved; palea ± = lemma. 13 spp.: Afr. (Greek: finger + a small comb, from spikelet arrangement) [Fisher & Schweickerdt 1941 Ann Natal Mus 10:47–77]

D. aegyptium (L.) Willd. (p. 1251) CROWFOOT GRASS Ann, gen mat-like. **ST** 2–4(10) dm, rooting at lower nodes. **LF**: blade < 15 cm, 4–5 mm wide, margin ciliate, hairs swollen at base. **INFL**: branches gen 4–5, 1–5 cm. **SPIKELET** ± 4 mm; upper glume keeled, 1.5–2 mm; florets 3; lemma widely-ovate. $2n$=20,36,40,45,48. Disturbed places; < 300 m. SnJV (Kern Co.), SCo, PR, DSon; e&s US, trop Am, Asia; native to Afr.

DANTHONIA OATGRASS

Kelly W. Allred

Per, cespitose. **STS** erect. **LVS** gen basal and cauline; sheaths < internodes; ligule short, densely ciliate; blades narrow, flat to folded. **INFL** gen raceme-like (sometimes panicle-like or spikelet 1). **SPIKELETS** ± laterally compressed; glumes ± equal, > florets, papery, 1–5-veined; florets 3–8, breaking above glumes and between florets; callus short-hairy; lemma rounded, 7–9-veined, tip 2-toothed, awned on back below teeth; awn gen bent, flat, coiled, below bend, straight, ± cylindric above bend, palea = lemma. **FR** elliptic. 20 spp.: warm temp, trop, Am, Eur, s Afr. (E. Danthione, France, early 19th century) Variation, esp in *D. californica, D. unispicata*, complex, needs critical study.

1. Upper lf blades abruptly spreading to reflexed; mature spikelets gen spreading .... ***D. californica***
 2. Lvs, esp sheaths, densely hairy, hairs papillate at base; florets gen 3–5 .... var. ***americana***
 2′ Lvs gen glabrous, hairy near ligule but not papillate; florets gen 4–8 .... var. ***californica***
1′ Upper lf blades erect to appressed; mature spikelets ascending to erect
 3. Lvs gen glabrous, not papillate .... ***D. intermedia***
 3′ Lvs gen densely hairy, hairs papillate at base or papillate without hairs
  4. Spikelets 6–18 per infl .... ***D. pilosa***
  4′ Spikelets 1–4 per infl .... ***D. unispicata***

D. californica Bolander CALIFORNIA OATGRASS **ST** 1–10 dm. **LVS** basal and cauline, glabrous to densely hairy; upper blades 8–25 cm, flat, spreading to reflexed. **INFL** 2–6 cm, open. **SPIKELETS** 1–5; stalk gen spreading, puberulent; glumes 10–23 mm; florets 3–8; lemma 8–15 mm, base, lower margin hairy, back ± glabrous; teeth 2–5 mm, awn 4–12 mm. $2n$=36. Gen moist, open sites, meadows, forests; < 2200 m. NW, CaR, SN, CW, SnBr, s PR, MP; w Can, w US, S.Am (Chile).

var. ***americana*** (Scribner) A. Hitchc. Upper lvs, esp sheath, densely hairy; hair bases papillate. **SPIKELET**: florets 3–5. Habitat of sp.; 150–2200 m. NCo, NCoR, CaRH, SN, CCo, SnBr, s PR, MP. ❀DRN,SUN:**4–6,15–17**&IRR:1–3,**7,14,18–24**;also STBL.

var. ***californica*** (p. 1255) Upper lvs gen glabrous, hairy near ligule, not papillate. **SPIKELET**: florets 4–8. Habitat of sp.; 100–1900 m. NW, CaR, SN, CW. ❀DRN,SUN:**4–6,15–17**&IRR:1–3, **7,14,18**,19–21,**22–24**; also STBL.

D. intermedia Vasey (p. 1255) INTERMEDIATE OATGRASS **ST** 1–5 dm. **LVS** gen basal, glabrous exc near ligule; upper blades 5–10 cm, ± inrolled, ascending. **INFL** 2–5 cm, narrow, compact; spikelets 4–10, stalk erect or appressed, glabrous. **SPIKELET**: glumes 9–14 mm; florets 3–6; lemma 6–10 mm, margins, base hairy, teeth 1–2 mm, awn 5–9 mm. $2n$=18,36. Meadows, bogs, damp banks, moist forests; 1500–3400 m. KR, NCoRH, CaRH, SNH, MP; to AK, e Can, w US. ❀TRY.

D. pilosa R.Br. (p. 1255) **ST** 3–6 dm. **LVS** basal and cauline, densely hairy, papillate at base; upper blades 2–4 cm, gen flat, ascending. **INFL** 3–10 cm, narrow, ± compact; spikelets 6–18, stalk ascending, puberulent. **SPIKELET**: glumes 10–14 mm; florets 4–8; callus ± 1 mm, sharp, hairy; lemma 5–6 mm, teeth 4–6 mm, awn 8–12 mm. $2n$=24,48. Disturbed, open sites, meadows, coniferous forest; < 800 m. NCo, KR, NCoRO, CCo, SnFrB; native to Australia.

D. unispicata (Thurber) Macoun ONE-SPIKE OATGRASS **ST** 1–3 dm. **LVS** basal and cauline, densely hairy, papillate at base; upper blades 3–8 cm, flat to ± inrolled, ascending. **SPIKELETS** gen 1 sometimes 2–4; stalk erect, puberulent; glumes 9–25 mm; florets 3–6; lemmas 6–14 mm, margin hairy, teeth 1–5 mm, awn 4–9 mm. $2n$=36. Dry meadows, rocky slopes, open sites in coniferous forest; 900–3200 m. KR, NCoR, CaRH, SNH, MP; to w Can, w US. ❀DRN,SUN,IRR:1,**2**,3–5,**6,7**,8–10,**14–24**.

DESCHAMPSIA HAIRGRASS

Dieter H. Wilken

Ann, per. **STS** erect, solitary to densely clumped. **LVS** basal to cauline; ligule narrow, decurrent to sheath, glabrous to minutely hairy; blades flat to inrolled. **INFL** panicle- to spike-like, open to narrow. **SPIKELET**: glumes and lemmas shiny; glumes equal to ± subequal, > lower floret; axis prolonged beyond upper floret, bristly (sometimes with vestigial floret at tip); florets 1–3, bisexual, breaking above glumes and between florets; callus soft-hairy; lemmas rounded, 2–4-toothed at truncate tip, faintly 3–7-veined, awned at or below middle, awn straight to bent; palea ± = lemma. 30–40 spp.: temp Am, Eurasia, New Zealand, Antarctica. (J. L-Deslongchamps, France, born 1774)

1. Ann .... ***D. danthonioides***
1′ Per
 2. Infl < 1 cm wide; basal lf blades ± 1 mm wide .... ***D. elongata***
 2′ Infl > 1 cm wide; basal lf blades 1–7 mm wide

3. Lemma awned at middle; cauline lf blades 3–7 mm wide .... ***D. atropurpurea***
3′ Lemma awned below middle; cauline lf blades 1–4 mm wide .... ***D. cespitosa***
4. Infl open, lower branches spreading to drooping .... ssp. ***cespitosa***
4′ Infl compact, lower branches ascending to erect .... ssp. ***holciformis***

D. atropurpurea (Wahlenb.) Scheele (p. 1255) MOUNTAIN HAIRGRASS Per. **STS** densely clumped, 1.5–6 dm. **LVS** basal and cauline; basal tufted, gen glabrous; ligule 1–3 mm, obtuse to truncate, minutely ciliate at tip; blades 4–9 cm, 3–7 mm wide, flat. **INFL** open; lower branches spreading to drooping. **SPIKELET** purplish; glumes 4–6 mm, equal, elliptic, acute, lower 1-veined, upper 3-veined; florets gen 2; callus hairs ± 1/2 lemma length; lemmas 2.5–4 mm, gen 2-toothed at tip, faintly 3-veined, awned near middle, awn 2–3 mm, straight to slightly bent. 2*n*=14. UNCOMMON. Wet sites, meadows, streambanks, in coniferous forest; 2000–2300 m. KR (Marble Mtns, Trinity Alps), CaRH (Mt Shasta); to B.C., e Can, ne US, n Eurasia.

D. cespitosa (L.) Beauv. TUFTED HAIRGRASS Per. **STS** densely clumped, 2–10 dm. **LVS** gen basal, tufted, glabrous to scabrous; ligule 3–8 mm, acute to obtuse, entire to toothed at tip; blades gen 8–20 cm, 1–4 mm wide, flat to inrolled. **INFL** narrow to open; lower branches erect to drooping. **SPIKELET**: glumes and tips of lemmas purplish; glumes 3–7 mm, ± subequal, lanceolate, acute, lower 1-veined, upper 3-veined; florets gen 2; callus hairs gen < 1/3 lemma length; lemmas 2–4 mm, gen 4-toothed at tip, faintly 5-veined, awned below middle, awn 2–6 mm, straight to slightly bent. 2*n*=26–28. Wet sites, meadows, streambanks, coastal marshes, forests, alpine; < 3900 m. NW, CaR, SN, CCo, SnFrB, TR, Wrn, W&I; to AK, e N.Am, S.Am, Eurasia, New Zealand. Sspp. intergrade.

ssp. ***cespitosa*** (p. 1255) **INFL** open; lower branches spreading to drooping. **SPIKELETS** solitary to clustered on exposed branchlets. Meadows, streambanks, coastal marsh, forests, alpine; < 3820 m. NW, CaR, SN, CCo, SnFrB, TR, Wrn, W&I; to AK, Can, e US, Eurasia. Variable in pl and spikelet size, awn length. Pls of NCo with glumes > 5 mm have been called ssp. *beringensis* (Hultén) W.E. Lawr. ❀SUN or part SHD:**4–6,17**&IRR:1–3,7,8–10,**14–16,18–24**; also STBL.

ssp. ***holciformis*** (C. Presl) W.E. Lawr. (p. 1255) **INFL** narrow, compact; lower branches ascending to erect. **SPIKELETS** densely clustered on short, obscure branchlets. Coastal marshes and meadows; < 850 m. NCo, NCoRO, CCo; to B.C. ❀IRR or WET,SUN:**4,5**,6–9,**14–17**,18–21,**22–24**;also STBL;tolerates saline water.

D. danthonioides (Trin.) Benth. (p. 1255) ANNUAL HAIRGRASS Ann. **STS** gen solitary to loosely clumped, 1.2–6 dm. **LVS** gen basal, glabrous; ligule 2–4 mm, acute to acuminate, entire; blades 1–9 cm, 1–2 mm wide, gen inrolled. **INFL** narrow to open; lower branches gen ascending. **SPIKELET**: glume and lemma tips sometimes purplish; glumes 4–9 mm, equal, lanceolate, acute to acuminate, 3-veined; florets gen 2; callus hairs < 1/2 lemma length; lemmas 2–3 mm, gen 2–3-toothed at tip, faintly 1–3-veined, awned below middle, awn 4–9 mm, slightly bent. 2*n*=26. Moist to drying, open sites, meadows, streambanks, temporary ponds; < 2700 m. CA-FP, MP; to AK, AZ, Baja CA, S.Am. ❀SUN,IRR:1–3,**4–10**,11,12,**14–24**.

D. elongata (Hook.) Benth. (p. 1255) SLENDER HAIRGRASS Per. **STS** densely clumped, 1–7 dm. **LVS** gen basal, tufted, glabrous; ligule 2–8 mm, acute to acuminate, entire; blades 4–8 cm, ± 1 mm wide, flat to inrolled. **INFL** narrow; branches spike-like, ascending. **SPIKELET**: glumes and lemmas green to tan, tips sometimes purplish; glumes 3–5 mm, ± subequal, narrow-lanceolate, acute to acuminate, 3-veined; florets gen 2; callus hairs ± 1/2 lemma length; lemmas 2–3 mm, 2–4-toothed at tip, faintly 5-veined, awned near middle, awn 1–5 mm, gen straight. 2*n*=26. Wet sites, meadows, lakeshores, shaded slopes; 100–3100 m. CA-FP, GB; to AK, WY, Baja CA, S.Am. ❀IRR or WET:1–3,**4–6,17**&part SHD:**7**,8,9,**14–16,18–24**.

DESMAZERIA

Dieter H. Wilken

Ann. **STS** prostrate to erect, solitary to clumped. **LVS** basal and cauline; sheath open; appendages 0; ligule membranous, toothed at tip; blade flat to inrolled. **INFL** raceme- or panicle-like. **SPIKELETS** subsessile to short-stalked; florets 5–25, bisexual, breaking above glumes and between florets; glumes equal to subequal, lower lanceolate, 1–3-veined, upper ± elliptic, 3-veined; lemma firmly membranous, acute to obtuse, 1–5-veined; palea < lemma. 3 spp.: Eur. (J.B. Desmazieres, French botanist) [see discussion in Clapham, Tutin, & Warburg 1952 Fl Brit Isles 1434].

D. rigida (L.) Tutin **ST** gen erect, 1–3 dm, glabrous. **LF**: blade 2–25 cm, 1–3 mm wide, flat, glabrous. **INFL** panicle-like, ovoid, 4–8 cm; branches and pedicels rigid, 3-angled; branches widely spreading. **SPIKELETS** 5–7 mm, 2–5 on lower branches, lowest in axil; florets 5–12; glumes and lemmas glabrous; glumes 1–2.5 mm; lemmas spreading, 2–3 mm, translucent at acute to abruptly soft-pointed tip. 2*n*=14. Open, often sandy sites; < 700 m. s NCo, CCo, SnFrB, SCo, s ChI (Santa Catalina Island); to Am; native to s&w Eur. [*Scleropoa r.* (L.) Griseb.]

DIGITARIA

Robert Webster

Ann, per. **STS** decumbent to erect. **LVS** basal and cauline; ligule membranous, ciliate or not; blade gen flat. **INFL** umbel- to panicle-like; 1° branches ± spike-like, spreading to ascending; spikelets gen many per branch, 2–3 per node, short-stalked to subsessile, on one side of axis. **SPIKELET** compressed, falling as 1 unit; glumes unequal, upper glume < or = spikelet, appressed-hairy, clearly 3–5-veined, veins minutely ridge-like; florets 2, lower floret sterile, lemma texture like upper glume, upper floret fertile, lemma ± thin, flexible, back facing away from infl axis, margin flat, tip gen obtuse; palea ± = lemma. ± 200 spp.: warm temp, trop, worldwide. (Latin: finger, from infl branch arrangement) [Webster 1987 Sida 12:209–222]

1. Spikelets gen 3 per node; tip of spikelet stalk disc-like; lower glume a translucent scale or 0; lf blade glabrous .... ***D. ischaemum***
1′ Spikelets gen 2 per node; tip of spikelet stalk truncate; lower glume membranous; lf blade short-soft-hairy .... ***D. sanguinalis***

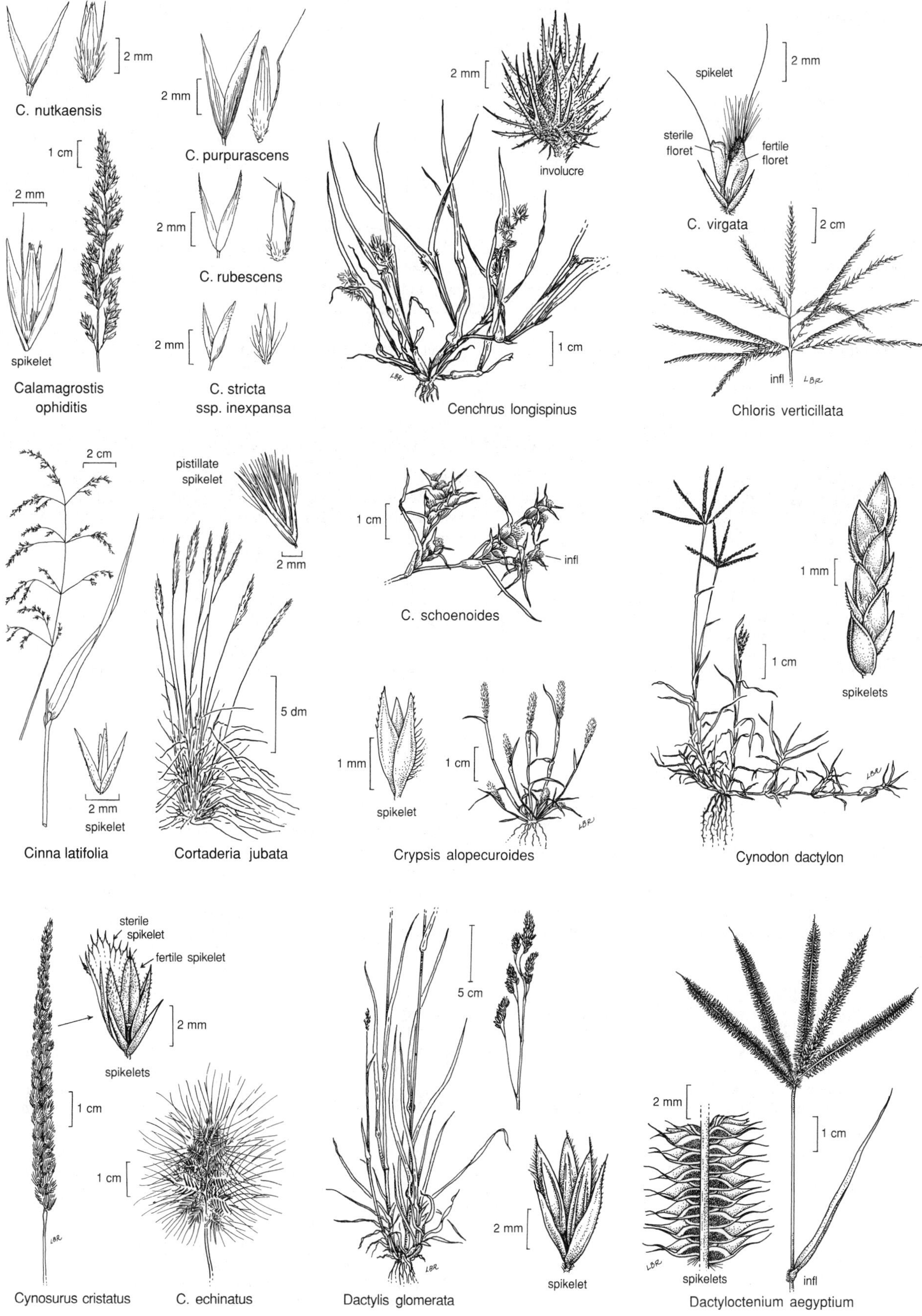
2 mm
C. nutkaensis
1 cm
2 mm
spikelet
Calamagrostis ophiditis
2 mm
C. purpurascens
2 mm
C. rubescens
2 mm
C. stricta ssp. inexpansa
2 mm
involucre
1 cm
Cenchrus longispinus
spikelet
2 mm
sterile floret
fertile floret
C. virgata
2 cm
infl
Chloris verticillata
2 cm
2 mm
spikelet
Cinna latifolia
pistillate spikelet
2 mm
5 dm
Cortaderia jubata
1 cm
infl
C. schoenoides
1 mm
spikelet
1 cm
Crypsis alopecuroides
1 mm
1 cm
spikelets
Cynodon dactylon
sterile spikelet
fertile spikelet
2 mm
spikelets
1 cm
1 cm
Cynosurus cristatus
C. echinatus
5 cm
2 mm
spikelet
Dactylis glomerata
2 mm
1 cm
spikelets
infl
Dactyloctenium aegyptium

D. ischaemum (Schreber) Muhlenb. Ann. **ST** 1.5–5 dm; nodes 2–5. **LF**: sheath 1–12 cm, glabrous; ligule ± 1–2 mm; blade 1.5–9 cm, 2.5–5 mm wide, upper surface glabrous. **INFL**: branches 1.5–6 cm; spikelets many, gen 3 per node, stalk 0.5–3 mm. **SPIKELET** ± 2–2.5 mm, ± 1 mm wide, elliptic, purple; lower glume << 0.5 mm, upper glume 3/4 to = spikelet length; lower floret lemma 7-veined, acute. $2n$=36. Disturbed places, fields; < 300 m. SnJV, CCo, SCo, expected elsewhere; to e US; native to Eur.

D. sanguinalis (L.) Scop. (p. 1255) Ann. **ST** 2–7 dm; nodes 2–5. **LF**: sheath 2.5–15 cm, hairy; ligule 1–3 mm; blade 3–17 cm, 2–14 mm wide, upper surface soft-hairy. **INFL**: branches 3–9 cm; spikelets many, gen 2 per node, stalk ± 0.5 mm. **SPIKELET** ± 2.5–3 mm, ± 1 mm wide, lanceolate to ovate, purple in fr; lower glume < 0.5 mm, upper glume ± 1/2–3/4 spikelet length; lemma of lower floret 5–7-veined, acuminate to acute. $2n$=36. Disturbed places, fields, roadsides; < 500 m. NCo, s CaRF, SNF, CW, SCo, expected elsewhere; OR, e US; native to Eur.

DISSANTHELIUM

Ann, per, sometimes from rhizomes or stolons. **STS** spreading to decumbent. **LF**: blade flat. **INFL** panicle-like, ± narrow; branches ± clustered, ascending to erect, ± dense. **SPIKELET** bisexual; glumes equal or subequal, gen > lower floret, ± membranous, acuminate, awn 0, lower glume ± 1-veined, upper 3-veined; axis breaking apart above glumes and between florets; florets 2; lemma membranous, keeled, 3-veined; palea < and enclosed by lemma. 16 spp.: sw CA (s ChI), Mex, w S.Am. (Greek: double small flower, from 2 small florets) [Swallen & Tovar 1965 Phytologia 11:361–376]

D. californicum (Nutt.) Hook. (p. 1255) CALIFORNIA DISSANTHELIUM Ann, ± fleshy. **ST** < ± 3 dm. **LF**: blade 1–1.5 dm, 2–4 mm wide. **INFL** < 1.5 dm; branches ascending or curving. **SPIKELET** 3–4 mm, equal; glumes > florets; lemma ± 2 mm, hairy, obtuse to acute. **PRESUMED EXTINCT** (last seen in 1912). Coastal-sage scrub; < 500 m. s ChI (Santa Catalina, San Clemente islands); Baja CA (Guadalupe Island).

DISTICHLIS SALTGRASS

Per from scaly rhizomes, gen dioecious. **STS** ascending to erect, stiff, glabrous, solid in X-section. **LVS** 2-ranked; ligule membranous, fringed; blade flat or ± rolled, pointed, glabrous, gen deciduous at collar. **INFL** panicle- or raceme-like; spikelets stalked. **SPIKELET** unisexual; pistillate gen = staminate, compressed; glumes unequal, firm, awns 0, lower 3(5)-veined, upper 5–7(9)-veined; axis breaking above glumes and between florets; florets 3–20; lemma wide, 9–11-veined, awn 0; palea < or > lemma; keels minutely hairy. 3–6 spp.: Am, 1 sp. in Australia. (Greek: 2-ranked, from lf arrangement) [Beetle 1943 Bull Torr Bot Club 70:638–650]

D. spicata (L.) E. Greene (p. 1255) SALTGRASS Per; rhizomes stout, yellowish; stolons sometimes present. **STS** erect, 1–5 dm, **LF**: blade gen 2–10 cm, 1–4 mm wide, stiff. **INFL** panicle-like, 2–8 cm, narrowly ± cylindric to elliptic in outline, ± dense. **SPIKELET** 6–20(25) mm, straw-colored to purplish; glumes 1.5–4 mm; florets 5–15; lemmas 3–6 mm; palea ± = lemma, keel ciliate in pistillate florets. $2n$=40. Salt marshes, moist, alkaline areas; < 1000 m. CA; s Can, US. [vars. *divaricata* Beetle, *nana* Beetle, *stolonifera* Beetle, *stricta* (Torrey) Beetle] ❀SUN:**4–7,14–17, 21–24**&IRR:3,**8,9**,10,11,**12,13,18–20**;GRNCVR;INV;STBL;tolerates salt & alkali.

ECHINOCHLOA

Robert Webster

Ann, sometimes per. **STS** decumbent to erect; internode spongy inside. **LVS** basal and cauline; sheath gen glabrous; ligule gen 0; blade gen flat, upper surface gen glabrous. **INFL** panicle-like, ± dense; branches gen ascending to appressed, axis gen glabrous; spikelets gen many, 1–2 per node, short-stalked to subsessile, ± on 1 side of axis. **SPIKELET** falling as one unit, ovoid to compressed; glumes unequal, lower < upper, short-bristly to hairy, gen green to purplish, upper glume gen awned; florets 2, lower floret sterile or staminate, lemma gen like glumes, upper floret fertile, lemma leathery or hard, shiny to dull, margin inrolled, tip abruptly pointed, palea free from lemma. ± 35 spp.: warm temp, subtrop, worldwide. (Greek: hedgehog grass, from bristly spikelet) [Gould, Ali, & Fairbrothers 1972 Amer Midl Nat 87:36–59]

1. 2° infl branches spreading ..... ***E. crus-pavonis***
1' 2° infl branches ascending to appressed
  2. Spikelets 4–5.5 mm, ± 2–2.5 mm wide; back of upper glume slightly keeled — weed in rice fields .. ***E. oryzoides***
  2' Spikelets 2–4 mm, ± 1–2 mm wide; back of upper glume rounded
    3. 1° infl branches gen 1–3 cm; spikelet 2–3 mm, ± 1–1.5 mm wide; lemma of lower floret acute ..... ***E. colona***
    3' 1° infl branches gen 3–7 cm; spikelet 3–4 mm, ± 1.5–2 mm wide; lemma of lower floret acuminate
      4. Lemma of upper floret abruptly pointed ..... ***E. crus-galli***
      4' Lemma of upper floret tapered to a point ..... ***E. muricata***

E. colona (L.) Link. Ann. **ST** decumbent to erect, 1–9 dm. **LF**: sheath 4–9 cm; blade 3–22 cm, 3–7.5 mm wide. **INFL** 3–13 cm; 1° branches 1–3 cm, axis glabrous; spikelets 2 per node, stalk < 1 mm. **SPIKELET** 2–3 mm, 1–1.5 mm wide, ovate to elliptic; lower glume 1–1.5 mm, 3–5-veined, upper glume ± = spikelet; florets ± equal; lower floret staminate or sterile, lemma 5-veined, acute. Wet places, fields; < 100 m. s SnFrB, DSon, expected elsewhere; to e US; native to Eurasia.

E. crus-galli (L.) P. Beauv. (p. 1255) Ann. **ST** erect, 2.5–15 dm. **LF**: sheath 3–7 cm; blade 0.5–30 cm, 6–20 mm wide. **INFL** 6–10 cm; 1° branches 3–7 cm, axis glabrous or hairy; spikelet 1 per node, stalk 0.5–1 mm. **SPIKELET** 3–4 mm, ± 1.5–2 mm wide, lanceolate to elliptic; lower glume ± 1–1.5 mm, 3–5-veined, upper glume ± = spikelet; florets subequal, lower sterile, lemma 5–7-veined, acuminate. Waste places, often wet sites, fields, roadsides; < 1500 m. CA (esp CA-FP); worldwide; native to Eurasia, Afr. [*E. c. var. zelayensis* (Kunth) A. Hitchc.]

E. crus-pavonis (Kunth) Schultes Ann. **ST** decumbent to erect, 6–18 dm. **LF**: sheath 7–20 cm; blade 12–60 cm, 6–25 mm wide. **INFL** 10–26 cm; 1° branches 3.5–7 cm; spikelets 1–2 per node, stalk < or = 1 mm. **SPIKELET** ± 3 mm, ± 1.5 mm wide, elliptic, green; lower glume 1–1.5 mm, 3-veined, upper glume ± = spikelet;

lower floret ± = upper, sterile or staminate, lemma 5–7-veined, acuminate. Disturbed places, fields; < 300 m. GV, SCo, expected elsewhere; to e US; native to Eurasia, Afr.

E. muricata (P. Beauv.) Fern. Ann, sometimes rooting at lower nodes. **ST** erect, 5–10 dm. **LF**: sheath 6–20 cm; blade 7–20 cm, 4–11 mm wide. **INFL** 6–25 cm; 1° branches 3–6 cm; spikelets 1–2 per node, stalk < 1 mm. **SPIKELET** ± 3–4 mm, ± 1.5–2 mm wide, elliptic; lower glume ± 1 mm, 3–5-veined, upper glume ± = spikelet; lower floret slightly < upper, sterile, lemma 5-veined, acuminate. Waste places, often wet sites, fields; < 1000 m. CA (esp CA-FP); to e US; native to Eurasia.

E. oryzoides (Ard.) Fritsch Ann. **ST** erect, 4–12 dm. **LF**: sheath 8–25 cm; blade 7–20 cm, 4–12 mm wide. **INFL** 8–15 cm; 1° branches 3.1–6 cm, axis glabrous to short-hairy; spikelets 2 per node, stalk < 2 mm. **SPIKELET** ± 4–5.5 mm, 2–2.5 mm wide, elliptic, green; lower glume 1.5–2 mm, 3-veined, upper glume ± = spikelet; lower floret slightly < upper, sterile, lemma 5-veined, acuminate. Wet places, esp rice fields; < 100 m. s ScV; native to Eurasia. [*E. oryzicola* (Vasinger) Vasinger var. *mutica* Vasinger misapplied]

EHRHARTA

Ann, per. **STS** gen erect. **LF**: ligule gen membranous, toothed; blade flat. **INFL** gen panicle-like; branches spreading to erect; spikelets sessile to stalked. **SPIKELET** bisexual, compressed; glumes ± equal to unequal, < to > florets, gen ovate; axis breaking above glumes, falling as 1 unit; florets 3, lower 2 sterile, palea 0, upper floret fertile, palea present; lemmas membranous, becoming hard, short-awned or not; palea 2(3–5)-veined or not; stamens 1–4 or 6. ± 35 spp.: s Afr, New Zealand. (J. Friedrich Ehrhart, German botanist, student of Linnaeus, 1742–1795) [Gibbs Russell & Ellis 1987 Bothalia 17:51–65]

1. Spikelets 5–8 mm; sterile lemmas soft-hairy, short-awned to pointed .... ***E. calycina***
1′ Spikelets 3–3.5 mm; sterile lemmas ± glabrous, awns 0 .... ***E. erecta***

E. calycina Smith (p. 1255) VELDT GRASS Per. **STS** 3–7.5 dm. **LF**: blade 7–20 cm, 2–7 mm wide. **INFL** 10–15 cm, ± open; spikelets subsessile to stalked, stalk < 5 mm, ± thread-like. **SPIKELET** 5–8 mm; glumes 5–7 mm, ± equal, > sterile florets, becoming purplish; sterile lemmas awned or pointed, soft-hairy; fertile lemma awnless, veins hairy. $2n$=24. Sandy soils;< 200 m. n CCo (Bodega Bay), SCoRO (Nipomo Mesa), WTR (Casitas Pass), expected elsewhere; native to s Afr. Cult for forage, erosion control.

E. erecta Lam. Per. **STS** 3–6 dm. **LF**: blade 5–12 cm, 4–9 mm wide. **INFL** 6–15 cm, ± open; spikelets sessile to subsessile, stalk gen < 1 mm, stiff. **SPIKELET** 3–3.5 mm; glumes 1.5–3 mm, ± equal, > sterile florets; sterile lemmas awnless, ± glabrous; fertile lemma awnless, ± glabrous. Disturbed places; < 200 m. e SnFrB, n SCo (Santa Barbara, Ventura cos.); native to s Afr.

ELEUSINE

Ann, per. **STS** decumbent to erect, flat, gen ± fleshy. **LF**: sheath compressed, strongly keeled; ligule membranous, ciliate; blade flat or folded. **INFL** umbel-like; branches 1–16, spike-like, with 2 rows of overlapping spikelets along 1 side of axis; spikelets sessile. **SPIKELET** bisexual, compressed; glumes unequal, lower 1-veined, upper 1–7-veined; axis breaking apart above glumes and between florets; florets 2–15; lemma keeled, acute, 1- or 3-veined, short-awned or not; palea < lemma. 9 spp.: Afr, 1 sp. in S.Am. (Greek: Eleusis, ancient town) [Phillips 1972 Kew Bull 27:251–270] Cult for grain in Afr, India.

1. Infl branches 4–15 cm, 3–5 mm wide; ann .... ***E. indica***
1′ Infl branches gen 1–2.5 cm, 7–15 mm wide; per .... ***E. tristachya***

E. indica (L.) Gaertner (p. 1255) GOOSE GRASS Ann. **STS** prostrate to erect, 1–5(8) dm. **LF**: blade 5–25 cm, 3–8 mm wide. **INFL**: branches 2–6(12), 4–15 cm. **SPIKELET** 5–8 mm; lower glume 1.5–3 mm; upper glume 2.5–4.5 mm; lemma 2.5–4 mm. $2n$=18. Roadsides, disturbed areas; < 200 m. CA-FP; warm temp, subtrop; native to s Eurasia.

E. tristachya (Lam.) Lam. Per. **STS** ascending, 1–3 dm. **LF**: blade 2–15 cm, 7–15 mm wide. **INFL**: branches 1–3(4), 1–2.5 cm, 7–15 mm wide. **SPIKELET** 3–4.5 mm; lower glume ± 1.5 mm; upper glume ± 2.5 mm; lemma 3–3.5 mm. $2n$=18. Disturbed areas; < 200 m. GV; native to Afr.

ELYMUS

Mary E. Barkworth

Per, sometimes from rhizomes. **ST** gen bent at base or erect, gen tufted. **LF**: sheath appendaged, appendages sometimes small, fragile; ligule membranous, truncate to obtuse; blades flat, folded, or rolled. **INFL** spike-like, open to dense; axis gen not breaking apart in fr; spikelets ± 2-ranked or not, 1–4 per node, gen ascending. **SPIKELET**: glumes lanceolate to awl-like, sometimes 0, awned from tip or not; florets 1–7; lemma gen > glumes, gen rounded, tip acute to awned, awn straight or curved outward; anthers 1–6 mm. 150 spp.: temp worldwide. (Greek: ancient name for millet). [Barkworth & Dewey 1985 Amer J Bot 72:767–776; Wilson 1963 Brittonia 15:303–323] See *Agropyron, Elytrigia, Leymus, Pseudoroegneria* for spp. sometimes treated here. Some spp. hybridize; hybrids with *Hordeum, Leymus, Pseudoroegneria* also occur.

1. Glumes 0; top of stalk below lower floret ± knot-like, minutely hairy .... ***E. californicus***
1′ Glumes present, lanceolate to awl-like
 2. Glumes awn- to awl-like, base gen < 1.5 mm wide, awn 25–100 mm
  3. Spikelets 3 per node; central spikelet fertile; lateral spikelets sterile, ± glume-like, florets vestigial .... ***E. elymoides*** ssp. ***hordeoides***
  3′ Spikelets gen 2(3) per node; lateral spikelets with some fertile florets
   4. Lower spikelets with 2 glumes; glumes entire .... ***E. elymoides*** ssp. ***brevifolius***
   4′ Lower spikelets with 2 glumes and 1+ glume-like, reduced florets; glumes entire to divided into 2+ awns

5. Glumes divided into 3+ outward-curving awns; sheath appendages evident, 0.5–1.5 mm .... ***E. multisetus***
5′ Glume entire or divided into 2 outward-curving awns; sheath appendages obscure, < 0.5(1) mm ***E. elymoides***
6. All glumes entire, awn gen < lemma awn .... ssp. ***californicus***
6′ Some glumes divided, awn(s) gen > lemma awn .... ssp. ***elymoides***
2′ Glumes flat, base gen > 1.5 mm wide, awn < 25 mm
7. Infl with 2–4 spikelets at all or most nodes
8. Infl nodding; glumes curving outward from base; base of lower florets exposed .... ***E. canadensis***
8′ Infl erect; glumes not curving outward at base, base of lower florets concealed — awns, if present, gen straight .... ***E. glaucus***
9. Lemma awn < 5 mm .... ssp. ***virescens***
9′ Lemma awn 10–30 mm
10. Lf sheaths smooth or rough but not hairy .... ssp. ***glaucus***
10′ Lf sheaths, blades sparsely to densely hairy, hairs spreading or directed downwards .... ssp. ***jepsonii***
7′ Infl with 1 spikelet at all or most nodes
11. Lemmas awned, awns curving out at maturity
12. Spikelets 9–15 mm; infl breaking apart at maturity .... ***E. scribneri***
12′ Spikelets 10–18 mm; infl not breaking apart at maturity
13. St 6–10 dm; lf blade gen > 10 cm .... ***E. arizonicus***
13′ St 3–5 dm; lf blade gen < 10 cm .... ***E. sierrae***
11′ Lemma awns straight or 0
14. Pls with rhizomes .... ***E. lanceolatus***
14′ Pls with tufted sts, rhizomes gen 0
15. Anthers 3–6 mm; most infl internodes > 15 mm .... ***E. stebbinsii***
15′ Anthers 1–2.5 mm; most infl internodes < 15 mm .... ***E. trachycaulus***
16. Lemma awn > lemma body, 8–30 mm .... ssp. ***subsecundus***
16′ Lemma awn < lemma body, < 7 mm .... ssp. ***trachycaulus***

E. arizonicus (Scribner & J.G. Smith) Gould ARIZONA WHEATGRASS **ST** 6–10 dm. **LF**: appendages slender; blade 2.5–6 mm wide, flat or folded, glabrous or hairy, gen glaucous. **INFL** 10–25 cm, nodding; internodes 9–17 mm; spikelet 1 per node. **SPIKELET** 15–18 mm; glumes ± equal, acute, acuminate, or with awn < 8 mm; lemma awn 15–35 mm, curving outward; anthers 3–5 mm. $2n=28$. Uncommon. Open, dry slopes; ± 600 m. NCoRO (Mendocino Co.); to TX, n Mex. [*Agropyron a.* Scribner & J.G. Smith] Not collected in many years; may be waif.

E. californicus (Bolander) Gould (p. 1261) CALIFORNIA BOTTLE-BRUSH GRASS **ST** 10–20 dm, rigid. **LF**: basal sheath hairs stiff, appendages 1–3 mm, slender; blade 10–20 mm wide, flat. **INFL** 10–25 cm, erect; internodes 7–12 mm; spikelets 3–4 per node. **SPIKELET** 12–15 mm; glumes 0; lemma awn ± 20 mm, straight. $2n=28$. UNCOMMON. Coniferous forest; < 300 m. NCo, NCoRO, n CCo, SnFrB (Santa Cruz Mtns). [*Hystrix c.* (Bolander) Kuntze] ❀part SHD:**5**&IRR:7,**14–17**,19–21,**22–24**.

E. canadensis L. CANADIAN WILDRYE **ST** 8–15 dm. **LF**: sheath gen glabrous, appendages 1–2 mm; blade 3–15 mm wide, flat. **INFL** 7–25 cm, nodding to drooping; internodes 5–10 mm; spikelets 2–4 per node. **SPIKELET** 12–20 mm; glumes 10–25 mm, narrowly elliptic, ascending, thick, membranous distally, smooth, whitish, awn ± 10 mm; base of lower florets exposed; lemma awn 16–35 mm, curving outward; anthers 2.5–3 mm. $2n=28$. Disturbed places; < 200 m. NCoRO (Sonoma Co.), expected elsewhere; to B.C., e US; native to c&e US.

E. elymoides (Raf.) Swezey (p. 1261) SQUIRRELTAIL **ST** 1–6.5 dm. **LF**: sheath glabrous to long-hairy, appendages < 1 mm; blade 1–6 mm wide, flat, folded, or rolled. **INFL** 2.5–15 cm (exc awns), breaking apart with age; internodes 3–10 mm; spikelets gen 2 per node. **SPIKELET** 12–20 mm; glumes 35–85 mm, awn-like, base narrow, thick, gen spreading, sometimes with 1–2 short awns at base; lemma awn 30–90 mm, spreading; anthers ± 2 mm. $2n=28$. Dry, open areas; 600–4200 m. KR, CaR, SN, TR, PR, GB, D; to Great Plains, TX, n Mex. [*Sitanion hystrix* (Nutt.) J.G. Smith] Hybrids with *E. trachycaulus* have been called *E. macounii* Vasey or (from c SNH, MP), *E. saundersii* Vasey [*Agropyron s.* (Vasey) A. Hitchc.] Hybrids with *E. glaucus* have been called *E.* ×*hansenii* Scribner [*Sitanion h.* (Scribner) J.G. Smith] Hybrids with *Pseudoroegneria spicata* have been called *Agropyron saxicola* (Scribner & J.G. Smith) Piper. (see also *E. multisetus*).

ssp. ***brevifolius*** (J.G. Smith) Barkworth **INFL**: spikelets gen 2 per node. **SPIKELET**: glumes entire, awn 50–110(125) mm; lowest floret fertile, not glume-like; fertile florets 1+; lemma awn 50–105 mm. $2n=28$. Habitat of sp.; 600–3000 m. SnBr, PR, MP, DMoj; to OR, Great Plains, n Mex. [*Sitanion longifolium* J.G. Smith] ❀SUN,DRN:1–3,**7**,8,9,**10**,11,**14–24**.

ssp. ***californicus*** (J.G. Smith) Barkworth **INFL**: spikelets gen 2 per node. **SPIKELET**: glumes entire, awn 16–60 mm; lowest floret gen sterile, glume-like, fertile florets 1+; lemma awn 25–70 mm. $2n=28$. Habitat of sp.; 800–4200 m. KR, CaR, SN, SnGb, SnBr, SnJt, SNE; to WA, MT, UT. [*Sitanion hystrix* (Nutt.) J.G. Smith var. *c.* (J. G. Smith) F.D. Wilson] ❀SUN,DRN:**1–3**,7,10, 14,**15–17**,18–24.

ssp. ***elymoides*** **INFL**: spikelets gen 2 per node. **SPIKELET**: some glumes divided, awns 2, 35–85 mm; lowest floret gen sterile, glume-like, fertile florets 1+; lemma awn 25–75 mm. $2n=28$. Habitat of sp.; 800–4000 m. TR, SnJt, GB, D; to WA, WY, Colorado. [*Sitanion hystrix* (Nutt.) J. G. Smith var. *h.*] ❀SUN,DRN:1,**2**,**3**,7–11,14–24.

ssp. ***hordeoides*** (Suksd.) Barkworth **INFL**: spikelets 3 per node. **SPIKELET**: glumes gen entire, awn 15–50 mm; central floret fertile, lateral florets sterile, ± glume-like; lemma awn < 10 mm. $2n=28$. Habitat of sp.; 800–1000 m. KR (Siskiyou Co.); to WA, NV. ❀SUN,DRN:1,2,4,6,7,14–24.

E. glaucus Buckley (p. 1261) BLUE WILDRYE Sometimes with short stolons. **ST** 6–14 dm. **LF**: sheath glabrous or hairy, appendages ± 2 mm; blade 4–12 mm wide, gen flat. **INFL** 6–16 cm (exc awns), not breaking apart with age; internodes 4–8 mm; spikelets gen 2 per node. **SPIKELET** 8–16 mm; glumes 6.5–19 mm, short-awned; lower florets concealed; lemma 8.5–14 mm, awn < 30 mm, gen straight; anthers 1.5–3 mm. $2n=28$. Open areas, chaparral, woodland, forest; < 2500 m. CA; to AK, Great Plains, n Mex. Hybridizes with *E. elymoides*, *E. stebbinsii*, *E. trachycaulus*. ❀DRN,SUN:**4–6**,**15–17**&IRR:1,**2**,**3**,**7**,8,9,**10**,11,**14**,**18–24**.

ssp. ***glaucus*** **LF**: sheath, blade ± glabrous or scabrous. **SPIKELET**: lemma awn 10–30 mm. $2n=28$. Habitats and range of sp.

ssp. ***jepsonii*** (Burtt Davy) Gould **LF**: sheath, blade sparsely to densely hairy. **SPIKELET**: lemma awn 10–20 mm. $2n=28$. Coniferous forest, woodland; 900–2200 m. KR, NCoR, CaR, n SN, SCoR, TR, PR; to Rocky Mtns, Baja CA. ❀part SHD,DRN, IRR:1,**2**,**3**,**6**,7,14,**15**,**16**,17.

ssp. ***virescens*** (Piper) Gould **LF**: sheath, blade ± glabrous or scabrous. **SPIKELET**: lemma awns < 5 mm. $2n=28$. Coniferous forest, chaparral; < 300 m. NCo, NCoRO, CW; to AK. ❀DRN:**4–6**&IRR,SHD:7,14,**15–17**.

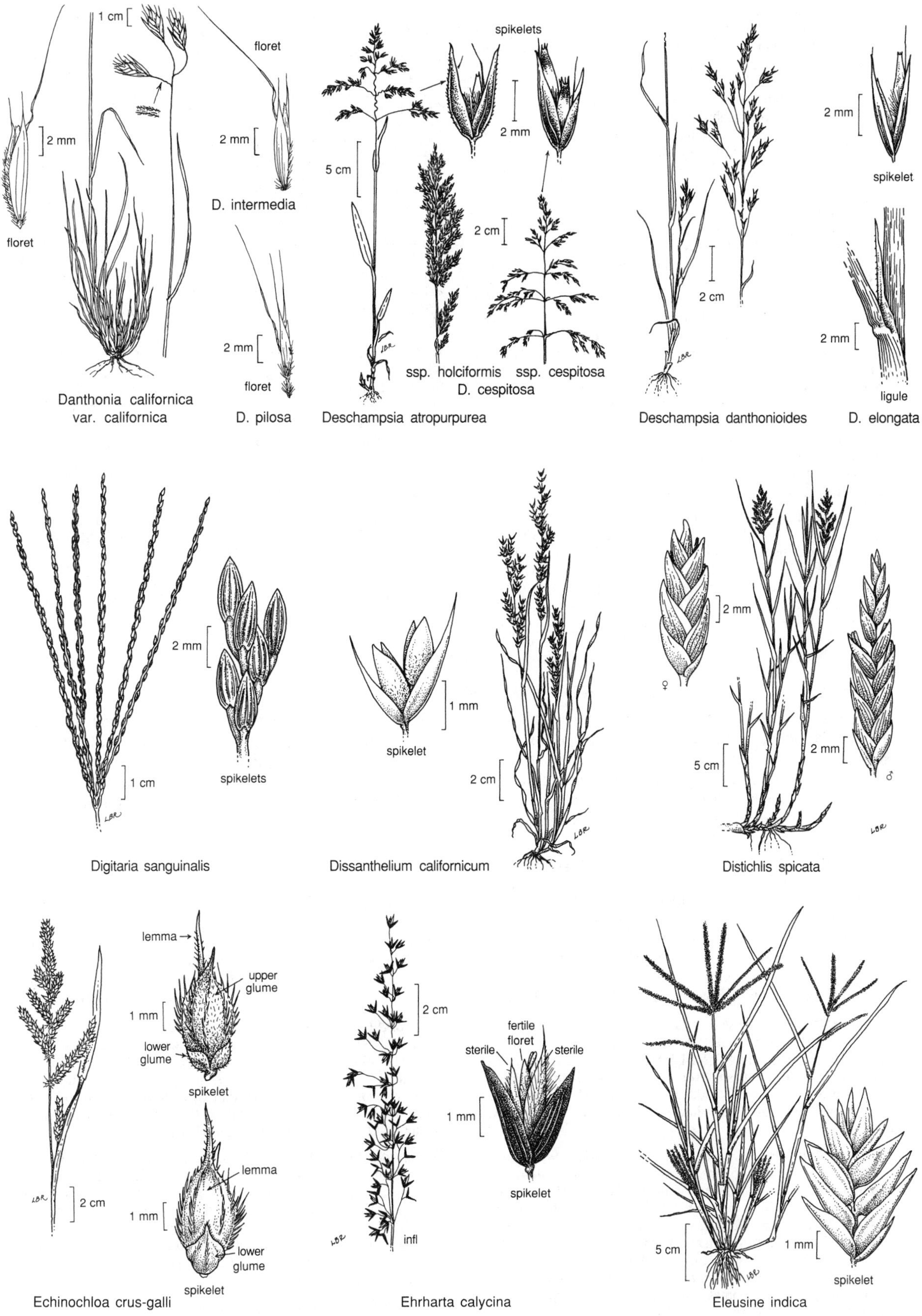

1 cm
floret
2 mm
2 mm
floret
D. intermedia
2 mm
floret
Danthonia californica
var. californica
D. pilosa
spikelets
5 cm
2 mm
2 cm
ssp. holciformis ssp. cespitosa
D. cespitosa
Deschampsia atropurpurea
2 cm
Deschampsia danthonioides
2 mm
spikelet
ligule
2 mm
D. elongata
2 mm
1 cm
spikelets
Digitaria sanguinalis
1 mm
spikelet
2 cm
Dissanthelium californicum
2 mm
♀
5 cm
2 mm
♂
Distichlis spicata
lemma
upper
glume
1 mm
lower
glume
spikelet
lemma
2 cm
1 mm
lower
glume
spikelet
Echinochloa crus-galli
2 cm
fertile
floret
sterile
sterile
1 mm
spikelet
infl
Ehrharta calycina
5 cm
1 mm
spikelet
Eleusine indica

E. lanceolatus (Scribner & J.G. Smith) Gould ssp. ***lanceolatus*** THICKSPIKE WHEATGRASS From rhizomes. **ST** 4–12 dm. **LF**: sheath glabrous to hairy, appendages 1–2 mm; blade 1–5 mm wide, rolled to flat. **INFL** 6–22 cm, erect; internodes 5–18 mm; spikelets gen 1 per node. **SPIKELET** 11–16 mm; lower glume 4–8.5 mm, ± 1/2 spikelet length; upper glume 5–11 mm, awn 0; lemma 7–11 mm, gen ± hairy, sometimes glabrous, acute to awn-tipped; anthers 3–5 mm. 2*n*=28. Open sites, woodland, coniferous forest; 500–1200 m. CaR, n SN, MP; to AK, Rocky Mtns. [*Agropyron dasystachyum* (Hook.) Vasey; *A. riparium* Scribner & J.G. Smith] Other ssp. in se Can, ne US. ❀STBL.

E. multisetus (J.G. Smith) Burtt Davy BIG SQUIRRELTAIL **ST** 1.5–6 dm. **LF**: sheath glabrous, appendages gen 0.5–1.5 mm; blade 1.5–5 mm wide, flat or rolled. **INFL** 3–17 cm (exc awns), breaking apart with age; internodes 4–8 mm; spikelets gen 2 per node. **SPIKELET** 8–11 mm; glumes divided near base into 3–5 awns, 25– 200 mm, distal half curving outward with age; lowest florets like glumes, vestigial; lemma 8–10 mm, tip gen 2-lobed, lobes awn-like, < 20 mm, awn between lobes 25–100 mm; anthers 1–2 mm. 2*n*=28. Open, sandy to rocky areas; < 3200 m. CA; to WA, Rocky Mtns. [*Sitanion jubatum* J.G. Smith] Hybrids with *Pseudoroegneria spicata* have been called *Agropyron saxicola* (Scribner & J.G. Smith) Piper. Hybrids with *E. trachycaulus* have been called *Agropyron saundersii* (Vasey) A. Hitchc. or *E. aristatus* Merr. Hybrids with *E. glaucus* have been called *Sitanion hansenii* (Scribner) J.G. Smith. (see also *E. elymoides*). ❀SUN,DRN:**1**,**2**,3–6,**7**,**14–24**; also STBL;may be INV.

E. scribneri (Vasey) M.E. Jones (p. 1261) SCRIBNER'S WHEATGRASS **ST** 1.5–5.5 dm, becoming prostrate. **LF**: sheath short-hairy, appendages < 1 mm; blade 1.5–3.5 mm wide, rolled or flat. **INFL** 3.5–8 cm, breaking apart with age; internodes 3.5–8 mm; spikelet 1 per node. **SPIKELET** 9–15 mm; glumes 4–7 mm, awn 10–25 mm, curving strongly outward; lemma 7–10 mm, awn curving outward; anthers 1–1.6 mm. 2*n*=28. RARE in CA. Rocky areas, alpine; 2900–4200 m. W&I; to B.C., Rocky Mtns. [*Agropyron scribneri* Vasey] Hybridizes with *E. trachycaulus* & *E. elymoides*.

E. sierrae Gould (p. 1261) **ST** 3–5 dm, decumbent. **LF**: sheath gen glabrous, appendages < 1 mm; blade 1–3 mm wide. **INFL** 4–7 cm, flexible; middle internodes 8–10 mm; spikelet 1 per node. **SPIKELET** 10–15 mm; glumes 7–8 mm, awn 5 mm, straight; lemma 7–9 mm, awn 15–25 mm, curving strongly outward; anthers 2–3 mm. 2*n*=28. Rocky slopes, coniferous forest; 2200–3400 m. SNH. [*Agropyron pringlei* Scribner & J.G. Smith) A. Hitchc.]

E. stebbinsii Gould Sometimes from rhizomes. **ST** 7–12 dm. **LF**: sheath gen glabrous, appendages 1–2 mm, persistent; ligule 1–2 mm, membranous; blade 2–6 mm wide, flat or rolled. **INFL** 10–25 cm; internodes 12–25 mm; spikelet 1 per node. **SPIKELET** 12–20 mm; glumes 10–15 mm; lemma 8–12 mm, awn 1–5 mm, straight; anthers 3–6 mm. Dry slopes, chaparral, coniferous forest; < 1600 m. NCoRI, SN, SCoR, TR, PR. 2*n*=28. [*Agropyron parishii* Scribner & J.G. Smith] Hybridizes with *E. glaucus* and *E. trachycaulus*. ❀TRY.

E. trachycaulus (Link) Shinn. (p. 1261) SLENDER WHEATGRASS **ST** 3–15 dm. **LF**: sheath glabrous to hairy, appendages < 1 mm; blade 2–8 mm wide, gen flat. **INFL** 4–25 cm, not breaking apart with age; internodes 4–15 mm; spikelet 1 per node. **SPIKELET** 10–20 mm; glumes 6–12 mm, acute to short-awned; lemma 6–13 mm, glabrous to hairy, awn 1–30 mm; anthers 1–2.5 mm. 2*n*=28. Dry to moist, open areas, forest, woodland; < 3400 m. CA (exc GV); to AK, e US. Hybrids with *Hordeum jubatum* have been called *E. macounii* Vasey.

ssp. ***subsecundus*** (Link) Gould **SPIKELET**: lemma awns 8–30 mm. 2*n*=28. Habitat of sp.; 1000–3400 m. SN, MP; to Rocky Mtns. [*Agropyron s.* (Link) A. Hitchc.]

ssp. ***trachycaulus*** **SPIKELET**: lemma awns < 7 mm. 2*n*=28. Habitat of sp.: < 3300 m. CA (exc GV); to AK, e US. [*Agropyron t.* (Link) Malte] ❀TRY.

ELYTRIGIA

James K. Jarvie & Mary E. Barkworth

Per, gen from rhizomes. **STS** erect, sometimes tufted. **LF**: sheath appendaged; ligule membranous; blade flat or rolled. **INFL** spike-like; axis gen not breaking apart in fr; spikelets 2-ranked, strongly overlapping, ± appressed to axis, 1 per node. **SPIKELET**: glumes thick, midvein gen prominent and scabrous at least above middle, tip truncate, obtuse, acute or short-awned; axis breaking above glumes and between florets; lemma gen awnless. 25 spp.: Medit Eur, Asia. (Greek: from combination of *Elymus* and *Triticum*) [Jarvie 1990 PhD dissertation UT State Univ] Some spp. cult for forage, erosion control; some serious weeds. See *Agropyron, Elymus, Pseudoroegneria*.

1. Pls without rhizomes; glume tip truncate; glumes weakly keeled
  2. Lowest spikelets > or = internodes .... ***E. elongata***
  2′ Lowest spikelets < internodes .... ***E. pontica*** ssp. ***pontica***
1′ Pls strongly rhizomed; glume tip gen acute to obtuse; glumes keeled
  3. Glumes acute to awn-tipped; blades drooping, distance between veins >> vein widths .... ***E. repens***
  3′ Glumes truncate to obtuse, not awn-tipped; blades stiff, distance between veins < or = vein width
    4. Infl axis not breaking apart in fr; blade veins > 8, not strongly ribbed .... ***E. intermedia*** ssp. ***intermedia***
    4′ Infl axis breaking apart in fr; blade veins < 8, strongly ribbed .... ***E. juncea*** ssp. ***boreali-atlantica***

E. elongata (Host) Nevski TALL WHEATGRASS **ST** 3.5–13 dm. **LF**: sheath glabrous; blade 2–5 mm wide, veins < 8, strongly ribbed. **INFL** 10–26 cm. **SPIKELET** 10–18 mm; glume tips truncate; florets 4–11; lemmas 6–10 mm; anthers 4–6 mm. 2*n*=14,28. Disturbed areas, slopes; < 1500 m. s SNF, n SNH, Teh, MP; to Great Basin; native to Medit Eur. [*Agropyron e.* (Host) Beauv. in part]

E. intermedia (Host) Nevski ssp. *intermedia* INTERMEDIATE WHEATGRASS **ST** 5–11.5 dm. **LF**: sheath glabrous or ciliate; blade 2–8 mm wide; veins many, weakly ribbed. **INFL** 8–21 cm. **SPIKELET** 11–18 mm; glume tips truncate to obtuse; florets 3–10; lemmas 7.5–10 mm, gen glabrous, sometimes rough-hairy, awn gen 0; anthers 5–7 mm. 2*n*=42. Open areas, slopes; < 2100 m. KR, NCoRI (Yolo Co.), CaR, SN, WTR, SnBr, DMtns; to B.C., Great Plains; native to Eurasia. [*Agropyron i.* (Host) Beauv.] Pls with hairy lf blades, rough-hairy lemmas have been called ssp. *barbulata* (Schur) A. Löve. [*Agropyron trichophorum* (Link) Richter]

E. juncea (L.) Nevski ssp. *boreali-atlantica* (Simonet & Guin.) Hylander **STS** 2.7–5 dm, tufted. **LF**: sheath glabrous; blade 2–4 mm wide, rolled; veins < 8, strongly ribbed. **INFL** 4–6 cm, breaking apart in fr. **SPIKELETS** 14–30 mm; florets 4–8; glume tips obtuse to acute; lemmas 10–16 mm with short awn; anthers 6–12 mm. 2*n*=28. Sand dunes; < 100 m. n CCo (San Francisco); native to Eur, n Afr. [*Agropyron junceum* (L.) Beauv.]

E. pontica (Podp.) Holub ssp. *pontica* (p. 1261) TALL WHEATGRASS Rhizomes 0. **STS** 5–22 dm, tufted. **LF**: sheath ciliate below middle; blade 2–6.5 mm wide, gen rolled; veins < or = 8, strongly ribbed. **INFL** 10–42 cm. **SPIKELETS** 13–30 mm; florets 6–12; glume tips truncate; lemmas 9–12 mm, awn 0; anthers 2.5–6 mm. 2*n*=70. Disturbed, often alkaline areas; < 1600 m. CA (exc NW); to B.C., e US; native to se Eur, w Asia. [*Agropyron elongatum* (Host) Beauv. in part]

E. repens (L.) Nevski (p. 1261) QUACKGRASS **ST** 5–10 dm. **LF**: sheath gen glabrous or lowermost soft-hairy; blade 2–14 mm wide, veins of 2 kinds, some faint, others strongly ribbed, widely spaced. **INFL** 8–20 cm. **SPIKELETS** 9–16 mm; florets 3–8; glumes gen with awn 0.5–4 mm, otherwise acute; lemmas 6–12 mm, tapering to point or awn 0.5–10 mm; anthers 4–5.5 mm. $2n$=42. Weed in disturbed areas, cult fields; < 1800 m. CA (exc D); to e US; native to Eurasia. [*Agropyron r.* (L.) Beauv.] NOXIOUS WEED.

ENNEAPOGON PAPPUS GRASS

Dieter H. Wilken

Per, gen cespitose. **STS** ascending to erect. **LVS** basal and cauline; ligule hairy; blade flat to inrolled. **INFL** panicle- to spike-like, narrow, gen compact. **SPIKELET**: glumes ± subequal, 3–9-veined; florets 3–6, breaking above glumes and weakly between florets, lower 1–3 florets fertile, bisexual, upper gen sterile, gradually reduced; lemmas < glumes, elliptic to ovate, firmly membranous, rounded on back, gen 9-veined, awned at truncate tip, awns 9, plumose; fertile palea slightly > lemma. ± 30 spp.: warm temp N.Am, Afr, Asia, Australia. (Greek: nine beards, from 9 plumose awns) [Renvoise 1968 Kew Bull 22: 393–401]

E. desvauxii Beauv. (p. 1261) NINE-AWNED PAPPUS GRASS **ST** 1–4 dm; nodes dense, short-hairy; internodes 2–5 cm. **LF** soft-hairy; sheath ciliate; ligule hairs < 1 mm; blade > 2 × internode, < 2 mm wide, ± inrolled. **INFL** spike-like, 3–6 cm, grayish. **SPIKELET**: glumes minutely soft-hairy on back, strongly veined, lower 3–5 mm, upper 4–6 mm; florets 3, lower 1 fertile, upper 2 sterile; lemmas 1–3 mm, awns 2–5 mm, exserted. $2n$=20. RARE in CA. Rocky slopes, crevices, calcareous soils, desert woodland; 1275–1825 m. e DMoj (DMtns); to Colorado, w TX, n Mex, S.Am. Lower sheaths sometimes enclose cleistogamous spikelets that disperse with st parts.

ERAGROSTIS LOVEGRASS

John R. Reeder

Ann, per, often glandular; glands often wart-like, circular, pitted. **LF**: sheath margin hairy on sides just below collar; ligules ciliate. **INFL** gen panicle-like, open or dense, sometimes spike-like, often glandular. **SPIKELET** laterally compressed; glumes ± unequal, acute or acuminate, 1(3)-veined; florets 3–many, axis breaking above glumes and between florets (or persistent with glumes and lemmas deciduous, paleas remaining attached or not); lemma keeled or rounded, acute or obtuse, 3-veined, veins gen obvious; palea ± = lemma. **FR** lens-shaped or elliptic, sometimes grooved, gen red-brown. ± 300 spp.: trop, warm temp. (Greek: eros, love, agrostis, a kind of grass) [Koch 1974 Ill Biol Monogr 48:1–74]

1. Per; anthers 0.5–1+ mm; spikelet axis breaking tardily above glumes and between florets
  2. Sts simple, erect, not weak or trailing; spikelets 1.5–2 mm wide; fr > 1 mm .... ***E. curvula***
  2′ Sts abruptly bent, branching, some becoming decumbent and rooting at nodes; spikelets ± 1 mm wide; fr gen < 0.8 mm .... ***E. lehmanniana***
1′ Ann; anthers gen 0.2–0.4 mm; spikelet axis not breaking apart, persistent, glumes and lemmas deciduous, paleas deciduous or remaining attached
  3. Sts widely creeping, rooting at nodes, forming mats .... ***E. hypnoides***
  3′ Sts often decumbent, not creeping, not mat-forming
    4. Fr with one or both ends truncate; surface checkered, with an evident groove on side opposite embryo .... ***E. mexicana***
      5. Spikelet ovoid to oblong, 1.5 mm + wide; fr ± rectangular; infl branches often with scattered glands .... ssp. ***mexicana***
      5′ Spikelet linear to linear-lanceolate, ± 1 mm wide; fr pear-shaped; infl not glandular .... ssp. ***virescens***
    4′ Fr elliptic, ovate, or sometimes ± spheric, tip rounded; surface smooth, not grooved
      6. Pls without conspicuous glands or glandular areas
        7. Paleas deciduous; lower glume < 1/2 length of lowest floret; lowest infl branches gen whorled ...... ***E. pilosa***
        7′ Paleas persistent; lower glume > 1/2 length of lowest floret; infl branches gen solitary or paired . ***E. pectinacea***
          8. Spikelet stalks spreading .... var. ***miserrima***
          8′ Spikelet stalks appressed to branches, diverging < 20° .... var. ***pectinacea***
      6′ Pls with conspicuous glands or glandular areas on lf sheaths, blade margins, infl axis and branches, spikelet stalks, or lemma keel
        9. Lf blade margin with conspicuous glands; glume keel, lemmas, other pl parts also often glandular
          10. Spikelets 2.5–3 mm wide; spikelet stalk gen not glandular .... ***E. cilianensis***
          10′ Spikelets < 2 mm wide; spikelet stalks gen with 1 or 2 glands .... ***E. minor***
        9′ Lf blade margin without glands; glume keel and lemmas not glandular
          11. Infl axis with conspicuous glandular bands or patches esp below lowest branches; lvs without glands .... ***E. barrelieri***
          11′ Infl axis without conspicuous glandular bands or patches, sometimes glandular below st nodes; lf sheath, blade with many circular glands .... ***E. lutescens***

E. barrelieri Daveau (p. 1261) Ann. **ST** decumbent or erect, tufted, sometimes prostrate-spreading, branching at base, (1)2.5–3(5) dm, sometimes glandular below nodes. **LF**: sheath glabrous, soft-hairy near collar; ligule ± 0.5 mm; blade 2–10(15) cm, 2–5 mm wide, gen glabrous, flat to inrolled at tip. **INFL** 3–15 cm, 2–6(8) cm wide, open; axis with glandular patches below branches; branches, esp lowest, spreading to stiffly ascending. **SPIKELET** ± 1 cm, 1–1.5 mm wide, slightly compressed, linear; lower glume 1–1.5 mm, slightly < upper; axis not breaking apart, paleas persistent; florets 10–15(20); lemma ± 2 mm, grayish green or with reddish tinge, lateral veins conspicuous; anthers ± 0.2 mm. **FR** ± 0.8 mm, elliptic-ovoid. $2n$=60. Disturbed soils; gen < 1000 m. GV, SCoRO, SCo, WTR, SnBr, PR; to KS, TX; native to s Eur.

E. cilianensis (All.) Janchen (p. 1261) STINKGRASS Ann. **STS** spreading or decumbent, sometimes abruptly bent, often branching,

< 6 dm; glands gen present below st nodes. **LF**: sheath glabrous, long-hairy below collar, keel glandular; blade 10–20 cm, 2–8 mm wide, flat or inrolled, margin with wart-like glands. **INFL** < 20 cm, 5–6+ cm wide, ± compact, gen gray-green. **SPIKELET** 2.5–3 mm wide, linear to ovate; glumes 1.5–2 mm, ± equal, midvein often glandular; axis not breaking apart, paleas persistent; florets (5)10–45; lemma 2–2.5 mm, lateral veins prominent, midvein gen glandular; anthers ± 0.2 mm. **FR** ± 0.6 mm, ± spheric. $2n$=20. Disturbed soils; < 1500 m. CA; to e US; native to Eur. [*E. megastachya* (Koeler) Link]

E. curvula (Schrader) Nees var. *curvula* (p. 1261) WEEPING LOVEGRASS Per. **ST** erect, densely tufted, unbranched, 4–12 dm, glabrous. **LF**: sheaths < internodes, glabrous or hairy, gen with long hairs on collar and inside upper sheath margin; ligule ± 1 mm; blade (1)2–3(5) dm, inrolled, long-tapered, distally thread-like, scabrous. **INFL** < 35 cm, 15 cm wide, open, gen nodding; branches flexible; lower axils long-hairy; spikelets short-stalked. **SPIKELET** 5–8 mm, gen gray-green; glumes acute, 1-veined, lower 1.5–2 mm, upper 2.5–3 mm; axis breaking apart tardily; lemmas obtuse to ± acute, 3-veined, lemma of lowest floret 2–3 mm; anthers 1–1.5 mm, purple. **FR** ± 1.5 mm, ± ovoid or oblong, light brown; embryo ± 1/2 fr length, dark brown to ± black. $2n$=20,40,50,60. Roadsides, near gardens; < 500 m. CaRF, GV, SCoRO, SnBr, PR, DMoj; to s US; native to s Afr. Orn, cult for erosion control.

E. hypnoides (Lam.) Britton, Sterns, & Pogg. (p. 1261) Ann, gen matted. **ST** creeping, branching; fl branches ± erect, 2–25 cm. **LF**: blade 0.5–5 cm, 0.8–5.5 mm wide, flat or inrolled, glabrous or ± hairy esp on upper surface. **INFL** 1–5 cm, elliptic, compact to ± open. **SPIKELET** 5–10(20) mm; axis not breaking apart, paleas persistent; lemma 1.5–2 mm, acute, translucent, 3-veined; anthers 0.2–0.3 mm. **FR** ± 0.5 mm, disc-like, slightly flattened laterally. $2n$=20. Sand or mud near streams, lakes; < 500 m. KR, NCoR, c SNF, GV, SnFrB; to e US, Mex, S.Am.

E. lehmanniana Nees (p. 1261) LEHMANN LOVEGRASS Per, glabrous. **STS** decumbent to erect, often abruptly bent at lower nodes, often stolon-like, branched, 3–6 dm; nodes glabrous. **LF**: sheaths gen < internodes, glabrous or sparsely hairy near collar; ligule ± 0.5 mm; blade (5)8–15 cm, 1–3 mm wide, flat to inrolled, tapered to a rigid point. **INFL** 6–20 cm, < 10 cm wide, open; lower branches loosely spreading. **SPIKELET** 4–8 mm, ± 1 mm wide; glumes unequal, lower ± 1 mm, acute, upper 1–2 mm, obtuse; axis breaking apart tardily; florets 6–10, linear, gray-green; lemmas membranous, obtuse, lowest floret lemma ± 1.5 mm; anthers ± 0.8 mm. **FR** 0.6–0.8 mm, oblong, pale; embryo ± 1/2 fr length, dark brown or black. $2n$=40,60. Roadsides; < 1000 m. SCo, DMoj; to TX, n Mex; native to s Afr. Cult for erosion control.

E. lutescens Scribner (p. 1261) Ann. **ST** ascending or erect, 5–25(70) cm, with ring of glands below nodes; glands circular, pitted. **LF**: sheath gen glabrous, ± soft-hairy near collar, veins ± glandular; ligule ± 0.5 mm; blade 2–10 cm, 1–3 mm wide, flat, folded, or inrolled, lower surface veins gen with many glands. **INFL** 2–10(25) cm, erect, narrow, sometimes ± open; branches ascending or appressed, glandular. **SPIKELET** 2.5–7(10) mm, yellowish, purple tinged, esp at tip; glumes acute, lower 1–1.5 mm, upper 1.5–2 mm; axis not breaking apart; florets 8–14; lemmas ± 2 mm, veins prominent; anthers 0.2 mm, purple; palea persistent. **FR** < 1 mm, elliptic, brown, not grooved. Sandy margins of streams, lakes; < 1050 m. c SNF, SnJV; to WA, Colorado, AZ.

E. mexicana (Hornem.) Link Ann. **ST** widely spreading to erect, gen 1.5–10 dm, gen with ring of glandular depressions below nodes. **LF**: sheath glabrous or papillate-soft-hairy on upper margins, often with glandular depressions on veins; blade 5–25 cm, 4–7 mm wide, flat, sometimes hairy below, midvein rarely glandular. **INFL** (5)10–35 cm, open; axis below nodes, branches, spikelet stalks gen sparsely glandular. **SPIKELET** ovoid to linear, gray-green to reddish; glumes ± 2 mm, upper slightly > lower, lanceolate; axis not breaking apart; florets 5–15; lemmas 1.5–2.5 mm, ovate, acute, gen glabrous; palea slightly < lemma, persistent. **FR** pear-shaped, elliptic, or ± rectangular; surface checkered, with shallow to deep groove on side opposite embryo. Disturbed, gen open sites; < 2000 m. CA; to OK, TX, S.Am.

ssp. ***mexicana*** (p. 1261) **INFL**: branches often with scattered glands. **SPIKELET** 1.5+ mm wide, ovoid to oblong. **FR** ± rectangular. $2n$=60. Disturbed soils in fields, forest margins, and waste places; < 1500 m. SnJV, SCo, WTR, SnBr; to OK, TX, n S.Am. [*E. neomexicana* Vasey]

ssp. ***virescens*** (C. Presl) Koch & E. Sánchez (p. 1261) **INFL** not glandular. **SPIKELET** ± 1 mm wide, linear to linear-lanceolate. **FR** elliptic or pear-shaped. Disturbed soils in fields, sandy river banks, etc.; < 2000 m. CA; NV, S.Am. [*E. orcuttiana* Vasey]

E. minor Host (p. 1261) Ann. **STS** erect, becoming prostrate, often branching at base, < 6 dm. **LF** gen glabrous; sheath long-soft-hairy near collar; blades like *E. cilianensis*. **INFL** gen 3.5–15 cm, 2.5–6 cm wide, gen gray-green; spikelet stalk with 1 or 2 glands near middle. **SPIKELET** ± 2 mm wide, linear to ovate; glumes not glandular; axis not breaking apart, paleas persistent; florets 8–12; lemma not glandular. **FR** ± 0.5 mm, subspheric to elliptic. $2n$=40. Disturbed soils; < 1400 m. NCoRO, n&c SNF, GV, SCo, MP, DMoj; e US, native to Eur. [*E. poaeoides* Roemer & Schultes]

E. pectinacea (Michaux) Nees Ann. **ST** erect, sometimes abruptly bent at base, 1.5–6(7.5) dm. **LF**: sheaths gen < internodes, glabrous exc white- hairy-tufted at collar margins; blade 2–15(20) cm, 1.5–3(5) mm wide, flat, glabrous. **INFL** (5)10–25 cm, 3–12(15) cm wide, open; primary branches spreading or ascending, straight, alternate or opposite; spikelet stalks spreading or appressed. **SPIKELET** 5–8(10) mm, 1.2–2 mm wide, gen linear; glumes thin, lower 0.5–1 mm, upper 1–1.5 mm; axis not breaking apart, paleas persistent; florets 5–15(20); lemma 1.5–2 mm, membranous, gray-green or red-tinged near tip, veins prominent. **FR** ± 1 mm, oblong to slightly pear-shaped, not grooved. Open disturbed sites, fields; < 1400 m. CA; widespread US, to Caribbean, C.Am.

var. ***miserrima*** (Fourn.) Reeder (p. 1261) **INFL**: spikelet stalks spreading. $2n$=60. Habitat of sp.; < 200 m. n SNF, GV; to se US, Caribbean, C.Am. [*E. arida* A. Hitchc.; *E. tephrosanthos* Schultes]

var. ***pectinacea*** (p. 1261) **INFL**: spikelet stalks appressed to branches, diverging < 20°. $2n$=60. Habitat and range of sp. [*E. diffusa* Buckley]

E. pilosa (L.) P. Beauv. var. ***pilosa*** (p. 1261) Ann. **ST** ascending to erect, 1–6.5 dm; axis below nodes rarely glandular. **LF**: sheath glabrous, margin sparsely hairy near collar, not glandular; ligule < 0.5 mm; blade 2–20 cm, 1–3.5 mm wide, flat to ± inrolled. **INFL** 4–20 cm, < 15 cm wide, ± open; lower 1–2 nodes with whorled branches; primary branches slender, spreading or ascending, rarely reflexed; spikelet stalks appressed to spreading. **SPIKELET** 3.5–10 mm, 1–2 mm wide; lower glume 0.5–1 mm, gen < 1/2 lowest lemma length, upper slightly > lower; axis not breaking apart, paleas deciduous; florets 5–16; lemma ± 1.5 mm, gray-green with purple or reddish tip, lateral veins obscure. **FR** 0.5–0.9 mm, light to dark brown, ovoid, smooth. $2n$=40. Disturbed sandy soils; < 200 m. KR, GV, MP; gen e US to TX, Caribbean, Mex.

ERIOCHLOA

Robert Webster

Ann, per. **STS** decumbent to erect. **LVS** basal and cauline; sheath glabrous or hairy; ligule gen < 1 mm, membrane hairy-fringed; blade gen flat. **INFL** panicle-like, ± dense; 1° branches spreading to appressed; 2° branches appressed; spikelets many, 1–2 per node, short-stalked to subsessile, on one side of axis. **SPIKELET** lanceolate, ± compressed, gen green, falling as 1 unit; glumes strongly unequal, lower glume gen 0, fused to spikelet base to form a disc- or cup-like ring between stalk and upper glume, upper glume ± = spikelet; florets 2, lower floret sterile, gen acuminate, palea 0, upper floret fertile, lemma firm or hard, gen wrinkled, margin inrolled, tip short-pointed to awned. ± 30 spp.; warm temp, trop, worldwide. (Latin: woolly grass) [Shaw & Webster 1987 Sida 12:165–207]

1. Upper floret awnless ..... ***E. acuminata***
1′ Upper floret tapered to short awn, awn ± 1–2.5 mm
 2. Spikelets 2 per node at middle of branch axis; upper lf blade surface glabrous; glume awn ± 1–3 mm; upper floret lemma abruptly pointed, point < 0.5 mm ..... ***E. aristata***
 2′ Spikelets gen 1 per node; upper lf blade surface short-hairy; glume awn ± 1 mm; upper floret lemma awn slightly < 1 mm ..... ***E. contracta***

E. acuminata (C. Presl) A. Hitchc. var. ***acuminata*** (p. 1261) Ann. **ST** 3–12 dm; nodes 2–5. **LF**: sheath 4–8 cm, glabrous or short-hairy; blade gen 5–12 cm, 5–12 mm wide, upper surface glabrous or short-hairy. **INFL**: main axis 7–16 cm; 1° branches 1–5 cm; spikelets gen 2 per node, 1 per node near the axis tip, stalk = or < 1 mm. **SPIKELET** 4–6 mm, ± 1–1.5 mm; lower floret lemma gen 5-veined; upper floret ± 0.8 × lower floret length, lemma acuminate. Seasonal streams, irrigated fields, orchards; < 200 m. SW, DSon; to s US, Baja CA. [*E. gracilis* (Fourn.) A. Hitchc.] Other var. native to s AZ, NM, TX, n Mex.

E. aristata Vasey var. ***aristata*** AWNED CUP GRASS **ST** 4–10 dm; nodes 3–10. **LF**: sheath 4–13 cm, glabrous; blade gen 6–20 cm, 6–20 mm wide, upper surface gen glabrous. **INFL**: main axis 5–20 cm; 1° branches 2–3.5 cm; spikelets 1–2 per node; stalk 0.5–2 mm. **SPIKELET** 4–7 mm (exc awn), ± 1–1.5 mm wide; glume awn ± 1–3 mm; lower floret lemma 3–7-veined; upper floret ± 0.6 × lower floret length. RARE in CA. Seasonal streams, riverbanks; < 100 m. DSon (Imperial Co.); to s US, n Mex. Other var. in C.Am.

E. contracta A. Hitchc. (p. 1261) Ann. **ST** 2–10 dm; nodes 2–5. **LF**: sheath 4–8 cm, glabrous to short-hairy; blade gen 8–12 cm, 2–8 mm wide, upper surface gen short-hairy. **INFL**: main axis 6–20 cm; 1° branches 1.5–4.5(6) cm; spikelets gen 1 per node; stalk = or < 1 mm. **SPIKELET** 3.5–4.5 mm, ± 1–2 mm wide, becoming purple; lower glume ± = spikelet; lower floret lemma 3–7-veined; upper floret ± 0.7 × lower floret length. Seasonal streams, ditches, irrigated fields; < 100 m. Deltaic GV (Solano Co.), SW, DSon; native to c&s US.

ERIONEURON

Per, tufted or mat-like, sometimes from short stolons. **STS** spreading to erect. **LVS** gen basal; ligule short-hairy; blade ± stiff, pointed. **INFL** raceme- or panicle-like, head-like, ± dense. **SPIKELET** compressed; glumes ± equal, membranous, gen acuminate, glabrous, 1-veined; axis breaking apart above glumes and between florets; florets 4–20, lower bisexual, upper staminate or sterile; lemma wide, back rounded, gen densely soft-hairy below middle, 3-veined, tip 2-lobed or not, short-awned from tip or not; palea densely hairy below middle; stamens 1 or 3. 5 spp.: sw US, Mex, S.Am. (Greek: woolly nerve, from lemma, palea hairs) [Tateoka 1961 Amer J Bot 48:565–573]

1. Infl stalked, > terminal lvs; lemma minutely 2-toothed, awn < 2 mm ..... ***E. pilosum***
1′ Infl ± subsessile, < some terminal lvs; lemma deeply 2-lobed, awn 2.5–4 mm ..... ***E. pulchellum***

E. pilosum (Buckley) Nash (p. 1265) HAIRY ERIONEURON Per, cespitose. **STS** erect, 1–3(4) dm. **LF**: sheath margin long-soft-hairy at collar; blade 3–6 cm, 1–1.5 mm wide, flat or folded, margin white. **INFL** raceme- or panicle-like, 1.5–4 cm, 1–2 cm wide, stalked and elevated above terminal lf cluster. **SPIKELET** 1–1.5 cm; glumes 3–7 mm, tan or purplish, awn 0; lemma 4–7 mm, tip minutely 2-toothed, awn < 2 mm. $2n$=16. RARE in CA. Rocky slopes, ridges, pinyon/juniper woodland; 1500–2000 m. SNE, e DMtns; to KS, TX, Mex. [*Tridens p.* (Buckley) A. Hitchc.]

E. pulchellum (Kunth) Tateoka (p. 1265) FLUFF GRASS Per, mat-like, with clusters of lvs from short stolon-like sts. **STS** spreading to erect, < 1 dm. **LF**: sheath margin short-soft-hairy at collar; blade 2–6 cm, inrolled. **INFL** ± panicle-like, 1–2.5 cm, 1–1.5 cm wide, short-stalked to subsessile and gen < terminal lvs. **SPIKELET** 6–9 mm; glumes 6–9 mm, tan or purplish, short-awned; lemma 2.5–5 mm, 2-lobed from near middle, awn 2.5–4 mm. $2n$=16. Sandy to rocky slopes, flats, desert shrubland, woodland; 300–1700 m. D; to Colorado, TX, c Mex. [*Tridens p.* (Kunth) A. Hitchc.]

FESTUCA FESCUE

Susan G. Aiken

Per, gen cespitose, gen ± glabrous; bisexual, dioecious in *F. kingii*. **ST** erect. **LVS** ± basal; sheath gen persisting; collar gen glabrous; ligule gen < 1 mm, membranous, truncate, minutely fringed; blade flat or rolled, basal lobes gen 0. **INFL** panicle-like; branches dense and appressed to open and spreading. **SPIKELET**: glumes < lowest floret, unequal, lower 1–3-veined, upper 3–5-veined; axis breaking above glumes and between florets, florets 2–10, gen bisexual; lemma base gen glabrous, 5-veined (rarely 3- or 7-veined), not converging at tip; awn gen terminal, straight, glabrous; palea ± = lemma; stamens 3. **FR** free from palea, gen beakless. (Latin: ancient name) [Frederiksen 1982 Nord J Bot 2:525–536]

1. Dioecious; spikelets staminate or pistillate; reduced, sterile pistils sometimes present (subg. *Leucopoa*) — rhizomes short; lemma awnless ..... ***F. kingii***
1′ Bisexual; florets gen bisexual (subg. *Festuca*)
 2. Basal lobes of lf blade prominent, ± clasping st
  3. Basal lobes hairy; lemma awn, gen 0.5–1.5 mm ..... ***F. arundinacea***
  3′ Basal lobes glabrous; lemma awn << 0.5 mm or 0 ..... ***F. pratensis***
 2′ Basal lobes of lf blade, if any, not clasping st
  4. Lf sheath closed, ± reddish, gen with downward-pointing hairs; rhizomes often present ..... ***F. rubra***
  4′ Lf sheath open at least half its length, gen green, glabrous; rhizomes 0
   5. Pl > 3 dm; sts densely clumped with conspicuous dead lf sheaths at pl base
    6. Lf collar gen hairy; ovary tip hairy ..... ***F. californica***
    6′ Lf collar and ovary tip glabrous ..... ***F. idahoensis***
   5′ Pls < 3 dm (> 3.5 dm in *F. elmeri*, *F. subulata*, *F. trachyphylla*, *F. viridula*); if sts densely clumped, dead lf sheaths not conspicuous
    7. Lf blade flat or loosely rolled, > 2 mm wide

8. Lemma base long, with a tuft of hairs .... *F. subuliflora*
8′ Lemma base very short, glabrous
9. Lemma prominently scabrous; awn subterminal from between 2 short teeth .... *F. elmeri*
9′ Lemma sparsely scabrous; awn terminal .... *F. subulata*
7′ Lf blade folded (exc *F. viridula*), < 2 mm wide
10. Lemma awns 3–12 mm .... *F. occidentalis*
10′ Lemma awns 0.5–2.5 mm
11. Spikelets 9–12 mm .... *F. viridula*
11′ Spikelets 2.5–9 mm (*F. ovina* complex; separations may prove difficult)
12. Anthers ± 1.5–3 mm
13. Infl 1–5 cm; anthers < 2 mm; high montane, native .... *F. saximontana* var. *purpusiana*
13′ Infl 3–13 cm; anthers 2–3 mm; disturbed places, naturalized .... *F. trachyphylla*
12′ Anthers ± 1–1.5 mm
14. Infl gen unbranched; ovary glabrous .... *F. brachyphylla* ssp. *breviculmis*
14′ Infl short-branched at lowest node; ovary with a few stiff hairs .... *F. minutiflora*

F. arundinacea Schreber (p. 1265) TALL FESCUE **ST** 8–20 dm, robust; nodes visible. **LF**: sheath shredding with age; ligule 0.5–1 mm; blade 25–70 cm, 4–10 mm wide, flat or loosely rolled, hairy, ± rigid, prominently ribbed above, basal lobes ± clasping st, hairy. **INFL** 15–35 cm; branches many, spreading. **SPIKELET** 8–16 mm; lower glume 3–6 mm, upper 4–9 mm; florets 3–8; lemma 6–12 mm, scabrous near tip, often tinged purple, awn 0.5–2 mm; anthers 3–4 mm; ovary tip glabrous. $2n$=28,42,70. Disturbed places; < 2700 (gen < 1000) m. CA-FP; to e N.Am; native to Eur.

F. brachyphylla Schultes & Schultes f. ssp. ***breviculmis*** S. Frederiksen (p. 1265) **STS** 0.4–2 dm, densely clumped; nodes ± concealed. **LF**: ligule < 0.5 mm; blade 1–6 cm, < 0.5 mm wide, folded. **INFL** 0.8–2.5 cm, gen unbranched, narrow. **SPIKELET** 3.5–5.5 mm; lower glume 2–3 mm, upper 2.5–4.5 mm; florets 2–4; lemma 3–4.5 mm, slightly scabrous near tip, awn ± 1–1.5 mm; anthers ± 1–1.5 mm; ovary tip glabrous. Rocky places, subalpine or alpine; 2800–4300 m. c&s SNH, W&I. Other sspp. to Arctic, e N.Am. ❀DRN,IRR,SUN:1,2,16,17;DFCLT.

F. californica Vasey (p. 1265) CALIFORNIA FESCUE **STS** 4.5–12 dm, clumped; nodes visible. **LF**: sheath ± scabrous, conspicuously persisting; collar glabrous to long-hairy; ligule 0.5–1.5(4) mm; blade 10–100 cm, 1.8–3.5 mm wide, flat or rolled, scabrous. **INFL** 10–27 cm, open; spikelets borne near branch tips. **SPIKELET** 13–18 mm; lower glume 4.5–7 mm, upper 6–8 mm; florets 4–6; lemma 7.5–11 mm, glabrous, awn 1.5–2.5 mm; anthers 4.5–5 mm; ovary tip hairy. $2n$=56. Open forest, chaparral; < 1800 m. NW, CaR, n&c SN, CW, SnBr; OR. Smaller pls with ± glabrous collars from SnBr have been called var. *parishii* (Piper) A. Hitchc. ❀4–6&IRR:**17**,22–24&part SHD:**7**,**8**,**9**,**14–16**,18–21;CVS;also STBL.

F. elmeri Scribner & Merr. (p. 1265) **STS** 4–10 dm, loosely tufted; nodes visible. **LF**: ligule < 0.5 mm; blade 10–40 cm, ± 2–6 mm wide, flat or loosely rolled, scabrous or hairy above. **INFL** 10–20 cm; branches slender, ± drooping. **SPIKELET** 7–11 mm; lower glume 2–4 mm, upper 3–4.5 mm; florets 2–6; lemma 5.5–7 mm, very scabrous, awn 2–5 mm, subterminal from between 2 short teeth, scabrous; anthers 3.5–4 mm; ovary tip hairy. Moist, wooded slopes, under trees in rich soil; gen < 300 m. NW, CW; OR. [ssp. *luxurians* Piper] ❀TRY.

F. idahoensis Elmer (p. 1265) IDAHO FESCUE, BLUE BUNCHGRASS **STS** 3–10 dm, gen densely clumped; nodes visible. **LF**: sheath clearly persistent; ligule < 0.5 mm; blade 5–35 cm, < 2 mm wide, rolled, scabrous, ± stiff. **INFL** 6–20 cm; branches ± appressed, scabrous. **SPIKELET** 7–17 mm; lower glume ± 2.5–6 mm, upper 4–8 mm; axis gen visible, gen zig-zag; florets 3–9; lemma ± 6–10 mm, scabrous near tip, awn 1–6 mm; anthers 3–4.5 mm; ovary tip glabrous. $2n$=28. Dry, open or shady places; gen < 1800 m. NW, CaR, n&c SN, n&c CW, MP; to Can, Colorado. ❀DRN,IRR:1–5, **6**,**7**,8–10,**14–24**;CVS;also STBL.

F. kingii (S. Watson) Cassidy (p. 1265) Dioecious; rhizomes short. **STS** 3–8 dm, tufted; nodes visible. **LF**: sheath glabrous to densely hairy, conspicuously persisting, gen reddish brown with age; ligule 0.5–3.5 mm, back puberulent; blade 3–30 cm, 1.5–7 mm wide, ± flat, stiffly erect, glaucous. **INFL**: branches appressed, some with spikelets near base. **STAMINATE SPIKELET** 5–10(12) mm; lower glume 3–5.5 mm, upper 4–6.5 mm; florets 2–5; lemma 5–8 mm, finely scabrous, awn 0; anthers 3.5–5 mm; pistil reduced or 0. **PISTILLATE SPIKELET** 5–8 mm; lower glume 3–5 mm, upper 3.5–6 mm; florets 2–5; lemma 5–8 mm, finely scabrous, awn 0; stamens sterile or 0; ovary tip hairy. **FR** plump, minutely beaked. $2n$=56. Dry, sandy places, sagebrush plains to subalpine forest; > 2000 m. s SNH, SnBr, GB; to OR, Great Plains. [*Hesperochloa kingii* (S. Watson) Rydb.; *Leucopoa k.* (S. Watson) W.A. Weber] ❀TRY.

F. minutiflora Rydb. **STS** 0.4–3 dm, loosely tufted; nodes ± concealed. **LF**: sheath open 1/2 length, shredding with age; ligule < 0.5 mm; blade 2–12 cm, << 0.5 mm wide, folded, often V-shaped in section, soft. **INFL** 1–4 cm, narrow, gen branched only at lowest node. **SPIKELET** 2.5–5 mm; lower glume ± 1–2.5 mm, upper 2–3.5 mm; florets 2; lemma ± 2–3.5 mm, sparsely scabrous near sharp tip, awn 0.5–1.5 mm; anthers ± 1 mm; ovary tip sparsely short-hairy. $2n$=28. Uncommon. Moist, shady banks; > 3000 m. c SNH (Mt. Dana, Tuolumne-Mono Co. line); to Yukon, w Can, Colorado. Sometimes confused with *F. brachyphylla*.

F. occidentalis Hook. (p. 1265) WESTERN FESCUE **STS** 4–11 dm, slender, loosely clumped; nodes visible. **LF**: sheath shredding with age; ligule << 0.5 mm; blade 5–25 cm, 0.5–1 mm wide, folded, often V-shaped in section, soft. **INFL** 5–20 cm; branches 1–2, often drooping. **SPIKELET** 6–12 mm; lower glume 2–5 mm, upper 3–6 mm; florets 2–6; lemma 4–8 mm, tip sparsely scabrous, awn 3–12 mm; anthers ± 2–3 mm; ovary tip hairy. $2n$=14,42,46,56,64,70. Open pine/oak woodland, redwood forest; gen < 1900 m. NW, CaR, SN, n&c CW, PR, MP; to B.C., e N.Am. ❀IRR,DRN:**4–6**, **15–17**&SHD:**7**,**14**,21–24.

F. pratensis Hudson (p. 1265) MEADOW FESCUE **STS** 3–13 dm, loosely clumped; nodes visible. **LF**: sheath shredding with age; ligule < 0.5 mm; blade 10–30 cm, 2–7 mm wide, flat or loosely rolled, basal lobes ± clasping st, glabrous. **INFL** 10–25 cm, narrow, branched only at lowest node. **SPIKELET** 12–15.5 mm; lower glume 2.5–4 mm, upper 3.5–5 mm; florets 4–10; lemma 6–8 mm, awn gen 0; anthers 2–4.5 mm; ovary tip glabrous. Disturbed places; gen < 2000 m. CA (exc D), less common in SW; to e N.Am; native to Eur. Grown for forage. [*F. elatior* L.]

F. rubra L. (p. 1265) RED FESCUE Rhizomes gen present, sometimes very short. **STS** 3–8 dm, ± clumped, decumbent at base; nodes visible. **LF**: sheath closed, ± reddish, shredding with age, hairs ± downward-pointing; ligule < 0.5 mm; blade 5–30 cm, < 3 mm wide, ± folded. **INFL** 5–20 cm, ± open; branches ± ascending. **SPIKELET** 9–12 mm; lower glume 2.5–3.5 mm, upper 3.5–5.5 mm; florets 3–10; lemma 5–7 mm, sometimes scabrous near tip, awn < 4 mm, scabrous; anthers 2.5–4 mm; ovary tip glabrous. $2n$=14,28,42,56,70,128. Sand dunes, grassland, subalpine forest; gen < 2500 m. NW, CaR, n&c SN, CW, TR; worldwide. Some commercial cultivars much naturalized in CA. Distinct forms occur in NCo on sand dunes, ocean bluffs, in salt marshes. ❀IRR:1, **4–6**,**15–17**,**24**&part SHD:2,3,**7**,**8**,**9**,**14**,**18–23**;CVS:GRNCVR.

F. saximontana Rydb. var. ***purpusiana*** (St.-Yves) S. Frederiksen & L.E. Pavlick **STS** 0.5–2.5 dm, clumped; nodes ± concealed. **LF**: ligule < 0.5 mm; blade 3–10 cm, < 0.5 mm wide, folded, scabrous near tip. **INFL** 1–5 cm, very narrow, branched only at lowest node. **SPIKELET** ± 5–7.5 mm; lower glume 2–3.5 mm, upper 3.5–5 mm; florets 2–5; lemma ± 3.5–5.5 mm, scabrous near awn,

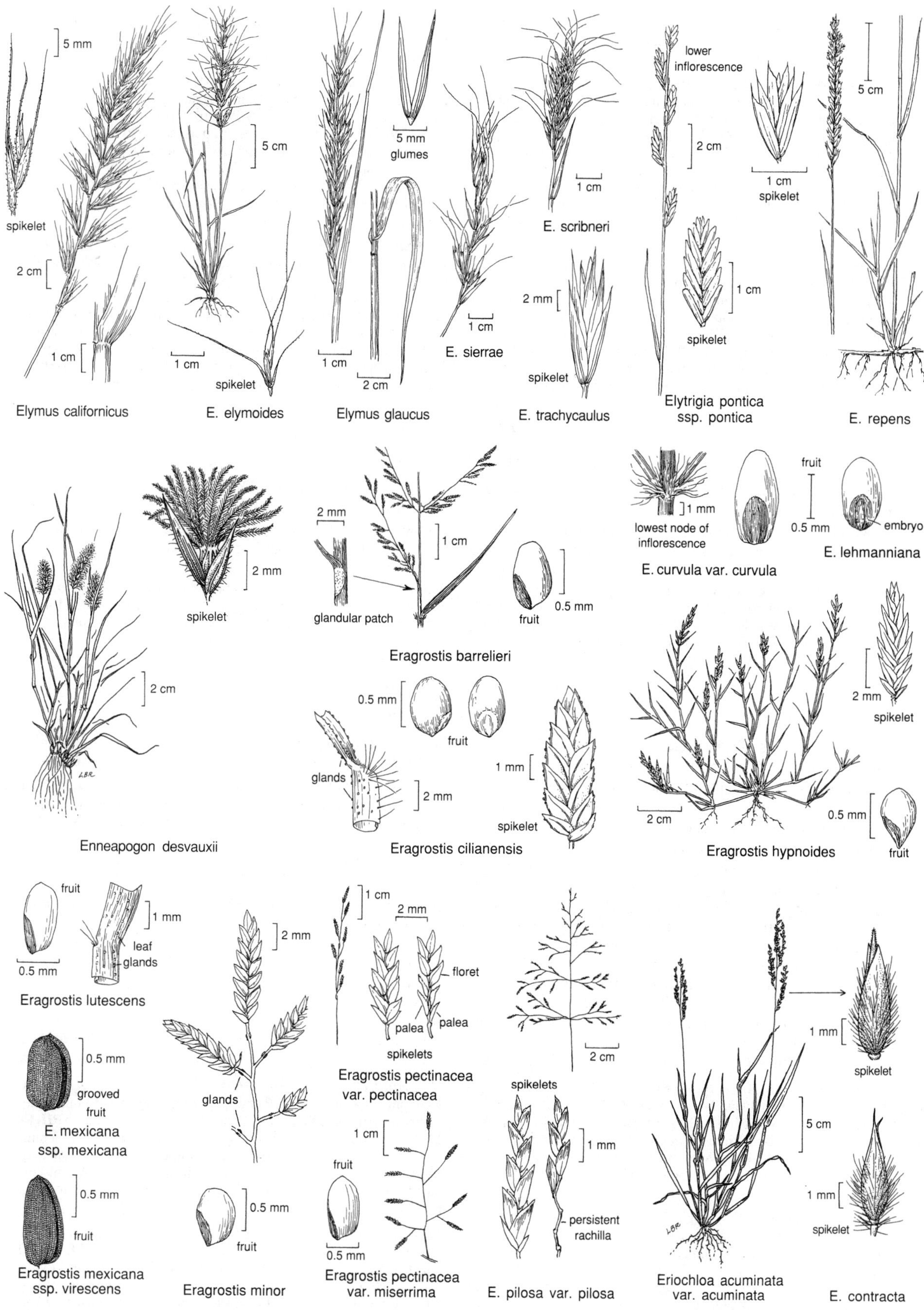
5 mm
spikelet
2 cm
1 cm
Elymus californicus
5 cm
1 cm
spikelet
E. elymoides
5 mm
glumes
1 cm
2 cm
Elymus glaucus
1 cm
E. sierrae
1 cm
E. scribneri
2 mm
spikelet
E. trachycaulus
lower
inflorescence
2 cm
1 cm
spikelet
1 cm
spikelet
Elytrigia pontica
ssp. pontica
5 cm
E. repens
2 mm
spikelet
2 cm
Enneapogon desvauxii
2 mm
1 cm
glandular patch
fruit
0.5 mm
Eragrostis barrelieri
0.5 mm
fruit
glands
2 mm
1 mm
spikelet
Eragrostis cilianensis
1 mm
lowest node of
inflorescence
E. curvula var. curvula
fruit
0.5 mm
embryo
E. lehmanniana
2 mm
spikelet
2 cm
0.5 mm
fruit
Eragrostis hypnoides
fruit
1 mm
leaf
glands
0.5 mm
Eragrostis lutescens
0.5 mm
grooved
fruit
E. mexicana
ssp. mexicana
0.5 mm
fruit
Eragrostis mexicana
ssp. virescens
2 mm
glands
0.5 mm
fruit
Eragrostis minor
1 cm
2 mm
floret
palea
palea
spikelets
Eragrostis pectinacea
var. pectinacea
1 cm
fruit
0.5 mm
Eragrostis pectinacea
var. miserrima
2 cm
spikelets
1 mm
persistent
rachilla
E. pilosa var. pilosa
5 cm
Eriochloa acuminata
var. acuminata
1 mm
spikelet
1 mm
spikelet
E. contracta

awn ± 1–2 mm; anthers 1.5–2 mm; ovary tip glabrous. $2n$=42. Alpine, subalpine summits, dry granitic gravel, talus fields, sagebrush scrub; gen > 3000 m. c&s SNH, SnBr.

F. subulata Trin. (p. 1265) BEARDED FESCUE Rhizomes short. **STS** 3.5–12 dm, loosely clumped; nodes visible. **LF**: sheath ± hairy; ligule < 0.5 mm; blade 10–30 cm, 3–10 mm wide, flat, glabrous or upper surface minutely scabrous. **INFL** 10–40 cm, open; branches drooping. **SPIKELET** 8.5–12 mm; lower glume ± 2–3 mm, upper 3–6 mm; florets 3–5; lemma 6–9 mm, glabrous to sparsely scabrous, awn 5–20 mm; anthers 1.5–3 mm; ovary tip hairy. $2n$=14,28. Open places, moist banks, forest; gen < 2500 m. NW, CaR, n&c SN; to AK, WY, UT. ❀DRN:**4**&SHD,IRR:1,2,**5–7**,**14–17**,18–21.

F. subuliflora Scribner (p. 1265) CRINKLE-AWN FESCUE **STS** 6–10 dm, loosely clumped, leafy to near panicle; nodes visible. **LF**: sheath and collar ± hairy; ligule < 0.5 mm; blade 15–30 cm, 2.5–8 mm wide, ± flat, ± hairy, soft. **INFL** 10–20 cm, open; branches gen 1 per node, drooping. **SPIKELET** 8–12.5 mm; lower glume 2–4 mm, upper 3.5–5.5 mm; florets 2–5; lemma 6–9 mm, ± strongly veined, sparsely scabrous, base hairy-tufted, tip slightly forked, awn 10–15 mm, crinkled; anthers 2.5–4 mm; ovary tip hairy. $2n$=28. Near streams, redwood, oak/pine forest; gen < 700 m. NCo, KR, NCoRO, n SNH, SnFrB; to B.C., esp near coast. ❀SHD, DRN,IRR:**4**,**5**,6,7,14,**15–17**,24.

F. trachyphylla (Hackel) Kraj. HARD or SHEEP FESCUE **STS** 2–7.5 dm, clumped, sparsely hairy near infl; nodes visible. **LF**: sheath ± hairy; ligule < 0.5 mm; blade 8–30 cm, < 1 mm wide, ± folded, ± hairy. **INFL** 3–13 cm; branches ± appressed. **SPIKELET** 5.5–9 mm; lower glume 2–4 mm, upper 3–5.5 mm; axis gen visible; florets 3–8; lemma 4–5 mm, awn 0.5–2.5 mm; anthers 2–3 mm; ovary tip glabrous. $2n$=28,42. Open places, slopes; 50–3000 m. NCoRO, n SNH, expected elsewhere; to e N.Am; native to Eur. Used for erosion control, esp on ski slopes.

F. viridula Vasey (p. 1265) MOUNTAIN BUNCHGRASS, GREEN FESCUE **STS** 5–10 dm, clumped; nodes visible. **LF**: sheath strongly veined, shredding with age; ligule < 0.5 mm; blade 10–30 cm, 1–2.5 mm wide, ± rolled. **INFL** 4–15 cm, open; branches gen 2 per node, ascending. **SPIKELET** 9–12 mm; lower glume 3.5–5 mm, upper 4.5–7 mm; florets 3–6 mm; lemma 5–8 mm, folded near tip, awn 0–1 mm; anthers 3–4 mm; ovary tip hairy. $2n$=28. Subalpine meadows, open forests, rocky slopes; gen > 2000 m. KR, n SNH; to B.C., ID, Colorado. Important forage grass. ❀DFCLT.

GASTRIDIUM

Dieter H. Wilken

Ann. **STS** ascending to erect. **LVS** basal and lower cauline; ligule ± translucent; blade flat. **INFL** panicle-like, narrow, compact; branches gen erect. **SPIKELETS** stalked; glumes unequal, membranous at swollen base, 1-veined; floret 1, breaking above glumes; axis prolonged as a minute bristle; lemma << glumes, translucent, truncate to obtuse, awnless to awned, 5-veined; palea ± = lemma. 2 spp.: warm temp Eur, w Asia. (Greek: small pouch, from swollen spikelet base)

G. ventricosum (Gouan) Schinz & Thell. (p. 1265) NIT GRASS Pls 1–4 dm, gen glabrous. **LF**: ligule 2–4 mm; blade 1–4 cm. **INFL** 1.5–9 cm, 4–10 mm wide. **SPIKELET**: glumes translucent between veins, keel minutely scabrous, lower glume 6–7 mm, upper 3–5 mm; lemma ± 1 mm, awned below truncate, toothed tip, awn 3–5 mm, straight to curved. $2n$=14. Open, gen dry, disturbed sites; 1200 m. CA-FP, MP; to OR, Can, ne US; native to Eur.

GLYCERIA MANNAGRASS

Mary E. Barkworth

Per from rhizomes. **ST**: base decumbent, erect above, gen rooting at lower nodes. **LF**: sheath closed to near top; ligule thin, membranous, acute; blade flat or folded, only midrib prominent. **INFL** panicle-like, 15–50 cm, nodding. **SPIKELET**: glumes < lowest floret, margin translucent, 1-veined; axis breaking above glumes and between florets; florets 8–15; lemma strongly 5–11-veined, veins not converging, ending short of tip, tip margin translucent; palea ± = lemma, keels prominent, ± curved. ± 40 spp.: temp worldwide. (Greek: sweet, from taste of grain) [Church 1949 Amer J Bot 36:155–165] See also *Catabrosa, Torreyochloa, Puccinellia*. ❀STBL.

1. Spikelet ovoid; infl open, branches spreading
  2. Palea tip jagged or widely V-notched; lemma widest at middle or below (sect. *Hydropoa*) .... ***G. grandis***
  2′ Palea tip narrowly notched; lemma widest above middle (sect. *Striatae*)
    3. Lf blade 6–12 mm wide, thin; lower glume 0.8–1.2 mm .... ***G. elata***
    3′ Lf blade 2–8 mm wide, firm; lower glume 0.4–0.8 mm .... ***G. striata***
1′ Spikelet cylindric; infl narrow, branches appressed (sect. *Glyceria*)
  4. Lemma glabrous between barely scabrous veins .... ***G. borealis***
  4′ Lemma minutely scabrous between distinctly scabrous veins
    5. Lemma tip gen rounded .... ***G. leptostachya***
    5′ Lemma tip jagged, longest between veins .... ***G. occidentalis***

G. borealis (Nash) Batch. (p. 1265) **ST** 8–15 dm, ± 2 mm diam. **LF**: ligule 4–12 mm; blade 2–7 mm wide. **INFL** 18–50 cm, narrow; spikelets appressed. **SPIKELET** 9–18 mm, cylindric; lower glume 1–2 mm, upper 2–3.5 mm; florets 6–11; lemma widest below middle. n=10. Shallow water, muddy shores, freshwater ponds, lakes, coniferous forest; 800–1250 m. CaR, SN; to AK, e US, NM.

G. elata (Lam.) A. Hitchc. (p. 1265) FOWL MANNAGRASS **ST** 6–15 dm, 3–5 mm diam. **LF**: ligule 2–6 mm; blade 4–12 mm wide, thin. **INFL** 15–25 cm; branches spreading. **SPIKELET** 2.5–5 mm, ovoid; lower glume 0.8–1.2 mm, upper 1–1.5 mm; florets 4–7; lemma widest above middle; palea tip narrowly notched. n=10. Common. Wet places, coniferous forest; < 2600 m. NCo, SN, SnBr, SnJt; to B.C., NM. Intergrades with *G. grandis*, *G. striata*.

G. grandis S. Watson (p. 1265) AMERICAN MANNAGRASS **ST** 9–20 dm, 3–6 mm diam. **LF**: ligule 2–7 mm; blade 4–15 mm wide, moderately thick. **INFL** 16–40 cm; branches spreading. **SPIKELET** 4–6.5 mm, ovoid; lower glume 1–1.8 mm, upper 1.5–2.5 mm; florets 4–7; lemma 2–2.8 mm, widest at middle or below; palea tip irregular or widely V-notched. n=10. RARE in CA. Wet places, meadows, lake and stream margins; < 500 m. NCo, NCoR; to B.C., e US.

G. leptostachya Buckley **ST** 10–15 dm, 2–4 mm diam. **LF**: ligule 4–9 mm; blade 4–10 mm wide, flat to rolled. **INFL** 20–40 cm, narrow; spikelets appressed. **SPIKELET** 10–20 mm, cylindric; florets 8–14; lower glume ± 1.5 mm, upper ± 3 mm; lemma 3–4 mm, tip rounded. n=20. Freshwater marshes and lakes; < 400 m. NW, CCo, SnFrB; to s AK.

G. occidentalis (Piper) J.C. Nelson **ST** 7–15 dm, 1.5–2 mm diam. **LF**: ligule 5–12 mm; blade 4–13 mm wide. **INFL** 15–40 cm, narrow; spikelets appressed. **SPIKELET** 15–20 mm, cylindric; lower glume 1.5–3.5 mm, upper 2.5–5 mm; florets 6–13; lemma 3.5–4.5 mm, tip margin irregular, longest between veins. *n*=20. Uncommon. Wet places; < 1300 m. NCo, NCoRO, n SNH, ScV, CCo; to B.C., ID.

G. striata (Lam.) A. Hitchc. **ST** 3–13 dm, 2–5 mm diam. **LF**: ligule 1–3.5 mm; blade 2–8 mm wide, firm. **INFL** 6–25 cm; branches spreading. **SPIKELET** 2.5–4.5 mm, ovoid; lower glume 0.4–0.8 mm, upper 0.7–1.2 mm; florets 3–6; lemma widest above middle; palea tip narrowly notched. *n*=20. Common. Wet meadows, stream margins, coniferous forest; 1500–2500 m. KR, NCoR, CaR, n&c SNH, MP; to B.C., Rocky Mtns.

HAINARDIA

Thomas Worley

1 sp. (P. Hainardi, Swiss phytogeographer). [Greuter & Rechinger 1967 Boissiera 13:22–196] See *Parapholis*.

H. cylindrica (Willd.) Greuter (p. 1265) Ann. **ST** ascending to erect, 2–5 dm, branched, glabrous. **LF**: sheath 1–6 cm; ligule 1 mm, membranous; blade gen flat, ribbed, upper surface scabrous. **INFL** spike-like, 8–20 cm, cylindric, stiff, straight, breaking at nodes; spikelets alternate, 2-ranked, appressed, embedded in axis, falling with axis segment, lowest incl in sheath. **SPIKELET**: glume 1, gen 5–7 mm, thick, rigid, margin sometimes inrolled, 9-veined; floret 1, bisexual; lemma 4–6 mm, translucent, back facing infl axis, 3-veined; palea ± = lemma, translucent; anther 2–3.5 mm. **FR** 2.5–3.5 mm. 2*n*=26,52. Coastal salt marshes, alkaline soils; < 300 m. n SNF (Amador Co.), deltaic GV, CCo, SnFrB, SCo; Baja CA; native to Eur. [*Monerma c.* (Willd.) Cosson & Durand]

HESPEROSTIPA

Mary E. Barkworth

Per, cespitose. **ST** erect, unbranched. **LF**: blade upper surface conspicuously ridged, gen inrolled. **INFL** panicle-like, narrow. **SPIKELET**: glumes tapered from near base to acute tip, awn 0; axis breaking above glumes; floret 1, 7–25 mm, narrowly cylindric; callus 2–5 mm, sharp, densely stiff-hairy; lemma hard, margins overlapping at maturity, upper portion fused, awn 6.5–18 cm, bent twice, lower segments twisted, last segment not twisted; palea = lemma, hard, 2-veined, veins terminating at tip. 4 spp.: N.Am. (Greek: western stipa) Segregated from *Stipa*; most closely related to *Piptochaetium, Nassella*.

H. comata (Trin. & Rupr.) Barkworth NEEDLE-AND-THREAD **ST** 1–11 dm. **INFL** 10–28 cm. **SPIKELET**: lower glume 18–35 mm, upper 1–3 mm shorter; floret 7–13 mm; lemma evenly hairy, hairs ± 1 mm, white; awn 65–195 mm. Well-drained soils; 200–3500 m. SN, TR, PR, GB, DMtns; to Yukon, Great Plains, Mex. [*Stipa c.* Trin. & Rupr.].

1. Awn 75–195 mm, distal segment wavy to curly . ssp. ***comata***
1′ Awn 65–130 mm, distal segment straight . . . ssp. ***intermedia***

ssp. ***comata*** (p. 1269) **SPIKELET**: awn 75–195 mm, distal segment wavy or curly. 2*n*=38,44,46. Grassland, sagebrush shrubland; 200–3500 m. SN, TR, PR, GB, DMtns, introduced elsewhere; to Yukon, c US, Mex. ❀DRN,SUN:1,**2,15–17**&IRR:**3,7**,14,18–24; also STBL.

ssp. ***intermedia*** (Scribner & Tweedy) Barkworth (p. 1269) **SPIKELET**: awn 65–130 mm, distal segment straight. 2*n*=44,46. Pinyon/juniper woodland, coniferous forest; 1500–3500 m. n SNH; to s Can, Rocky Mtns. ❀STBL.

HETEROPOGON

Kelly W. Allred

Ann, per, cespitose. **ST** erect, densely clumped. **LVS** cauline; sheaths flattened, gen < internodes; ligule truncate, thinly membranous; blade flat or folded. **INFL** ± spike-like. **SPIKELETS** on 1 side of axis, paired; spikelets of lower 1–4 pairs ± equal, sterile or staminate; spikelets of upper pairs strongly unequal, lower spikelet sessile, bisexual, awned, upper stalked, staminate or sterile, awn 0, pair breaking below sessile spikelet. **SESSILE SPIKELET**: glumes hard, short-hairy, tightly enclosing florets; florets 2, lower vestigial, obscure, upper fertile; lemma transparent, fragile. **STALKED SPIKELET**: glumes, lemmas thinly membranous. 6 spp.: warm temp, trop. (Greek: different beard, from awned and awnless spikelets)

H. contortus (L.) Roemer & Schultes (p. 1269) Per. **ST** 2–8 dm. **LF**: blade 6–24 cm, 4–8 mm wide. **INFL** 4–8 cm. **SESSILE SPIKELET**: glume 5–8 mm, dark brown, sharp-pointed, callus hairs red-brown, awn 5–10 cm, puberulent, red-brown. **STALKED SPIKELET** staminate. 2*n*=60. Uncommon. Rocky slopes, washes, open areas; < 800 m. DSon, naturalized in SCo (San Diego Co.); Baja CA, to s US, worldwide. NOXIOUS.

HIEROCHLOË SWEETGRASS

Dieter H. Wilken

Per with rhizomes. **STS** ascending to erect. **LVS** gen basal to lower cauline, fragrant; ligule membranous, fringed at acute to obtuse tip; blade flat to inrolled. **INFL** panicle-like, open to compact; lower branches ascending to drooping. **SPIKELETS** stalked; florets 3, breaking above glumes, falling as 1 unit, lower 2 florets staminate, with 3 stamens, upper bisexual, with 2 stamens; glumes ± equal, membranous, shiny, 1–3-veined, awnless; lemmas gen < glumes, elliptic to ovate, rounded on back, ciliate, awned to awnless, 3–5-veined; paleas 1–3-veined. ± 30 spp.: temp N.Am, Eurasia. (Greek: sacred grass) [Weimarck 1971 Bot Not 124: 129–175] Fresh lvs used for fragrance in churches on Saints' Days and as incense by native Americans.

1. Lf blades > 5 mm wide, upper blade spreading; lemmas rounded . . . . ***H. occidentalis***
1′ Lf blades gen < 5 mm wide, upper blade ± appressed; lemmas acute . . . . ***H. odorata***

H. occidentalis Buckley (p. 1269) Pls 3–10 dm. **LF**: sheath minutely scabrous; ligule 2–4 mm; blade 7–30 cm, 5–15 mm wide. **INFL** ± open, 7–10 cm; lower branches often drooping. **SPIKELET** 5–6 mm; lemmas rounded at slightly lobed tip, lemmas of lower florets glabrous to short hairy at base, hairs clear, upper lemma ± firmly membranous at base. $2n$=42. Moist to dry, coniferous forest; < 750 m. NCo, NCoRO, CCo, SnFrB, SCoRO; to WA. ❀IRR,DRN:**4,5,17**&SHD:**6**,14–16,19–24;rather DFCLT.

H. odorata (L.) Beauv. (p. 1269) VANILLA GRASS Pls 3–5 dm. **LF**: sheath glabrous to puberulent; ligule 3–5 mm; blade 3–5 cm, 2–5 mm wide. **INFL** gen open, 5–8 cm; lower branches drooping to spreading. **SPIKELET** 3–6 mm; lemmas acute, lemmas of lower florets minutely scabrous at base, short-hairy throughout, hairs tan to golden brown, upper lemma membranous at base. $2n$=28,42,56. RARE in CA. Wet sites, meadows; 1830 m. n CaRH (Ball Mtn, Siskiyou Co.); to AK, Can, e US, Eurasia.

HOLCUS VELVET GRASS

Dieter H. Wilken

Per, cespitose or from rhizomes, glabrous to velvety soft-hairy. **LVS** gen basal; ligule membranous, truncate, puberulent; blade flat. **INFL** panicle-like, ± congested. **SPIKELETS** laterally compressed, breaking below glumes, falling as 1 unit; glumes ± subequal, lower 1-veined, upper 3-veined; florets gen 2, lower floret bisexual, upper staminate or sterile; callus hairy; lemmas ± 2 mm, shiny, membranous, faintly 3–5-veined, lemma of lower floret awnless, lemma of upper floret awned near 2-lobed tip; palea ± = lemma. ± 8 spp.: temp Eurasia, Afr. (Latin: a grass)

1. Sts tufted, internodes soft-hairy; lemma awn ± 1 mm, twisted to recurved .... ***H. lanatus***
1′ Sts solitary to few, from rhizomes, internode glabrous; lemma awn 3–4 mm, bent to straight .... ***H. mollis***

H. lanatus L. (p. 1269) COMMON VELVET GRASS **STS** clumped, ascending to erect, 6–20 dm; nodes and internodes soft-hairy. **LF**: ligule 1–3 mm; blade 5–18 cm, 4–9 mm wide. **INFL** 7–15 cm. **SPIKELET**: glumes 3–6 mm, purplish, short-hairy on back, keel long-hairy; lemmas 3–4 mm, awn twisted to recurved, ± 1 mm. $2n$=14. Moist sites, roadbanks, cult fields, meadows; 100–2300 m. CA exc DSon; to AK, Can, widespread US, S.Am; native to Eur. Cult for forage, hay.

H. mollis L. (p. 1269) CREEPING VELVET GRASS **STS** solitary to few, from rhizomes, ascending to decumbent, 3–8 dm; nodes soft-hairy, hairs pointed downward, internodes gen glabrous. **LF**: ligule 2–3 mm; blade 3–15 cm, 4–9 mm wide. **SPIKELET**: glumes 4–5 mm, tan, tinged purplish, minutely scabrous on back, keel and veins short-hairy; lemmas 2–3 mm, awn bent to straight, 3–4 mm. $2n$= 28,35,42,49. Moist sites, ditches, lawns; < 120 m. NCo; to Can, scattered in e US; native to temp n Eur.

HORDEUM BARLEY

Mary E. Barkworth

Ann, per, sometimes from short rhizomes. **ST** decumbent to erect, gen abruptly bent at base. **LF**: sheath glabrous or hairy, short-appendaged or not; blade flat or ± rolled. **INFL** spike-like, dense; axis breaking apart at nodes in fr; spikelets 2-ranked, strongly overlapping, 3 per node, spikelets of 2 kinds. **CENTRAL SPIKELET** bisexual, gen sessile; with 1 stalked or sessile floret; glumes awn-like, gen > floret; lemma awned. **LATERAL SPIKELETS** 2, sterile or staminate, gen short-stalked; with 1 sessile floret; glumes awn-like, > floret, lemma gen awned. 32 spp.: temp worldwide exc Australia. (Latin: ancient name for Barley) [Baum & Bailey 1990 Canad J Bot 68:2433–2442]

1. Per
  2. Central spikelet glumes 20–80 mm, spreading with age
    3. Glumes 20–26 mm, flat near base; lemma awn of central spikelet 17–20 mm .... [2]***H. arizonicum***
    3′ Glumes 35–80 mm, not flat; lemma awn of central spikelet 25–90 mm .... [2]***H. jubatum***
  2′ Central spikelet glumes 7–19 mm, spreading or straight with age .... ***H. brachyantherum***
    4. Basal lf sheaths glabrous to short-hairy; lemma awn of central spikelet < 4.5 mm; glumes gen straight .... ssp. ***brachyantherum***
    4′ Basal lf sheaths short- and long-hairy; lemma awn of central spikelet < 7.5 mm; glumes gen spreading .... ssp. ***californicum***
1′ Ann
  5. Lf sheath appendages well-developed, 1–4 mm; lateral florets exc awns 8–14 mm .... ***H. murinum***
    6. Central spikelet sessile to subsessile, stalk < 0.5 mm; central floret (exc awn) gen = lateral spikelet floret; palea of lateral floret ± glabrous .... ssp. ***murinum***
    6′ Central spikelet stalk 1–2 mm; central floret (exc awn) < lateral spikelet floret; palea of lateral floret hairy
      7. Central floret slightly < lateral floret; anthers of lateral florets > 2 × anthers of central floret ...... ssp. ***glaucum***
      7′ Central floret << lateral florets; anthers of central and lateral florets ± equal .... ssp. ***leporinum***
  5′ Lf sheath appendages < 2 mm or 0; lateral florets exc awns < 5 mm
    8. Lemma of central spikelet with awn 10–90 mm
      9. Glumes straight with age .... ***H. marinum*** ssp. ***gussoneanum***
      9′ Glumes spreading with age
        10. Central spikelet glumes flat near base .... [2]***H. arizonicum***
        10′ Central spikelet glumes not flat .... [2]***H. jubatum***
    8′ Lemma of central spikelet with awn 3–12 mm
      11. Lemma of lateral spikelet with awn 3–8 mm .... ***H. marinum*** ssp. ***gussoneanum***
      11′ Lemma of lateral spikelet with awn < 2 mm
        12. Glumes of central spikelet slightly flat or rounded near base .... ***H. depressum***
        12′ Glumes of central spikelet clearly flat near base .... ***H. intercedens***

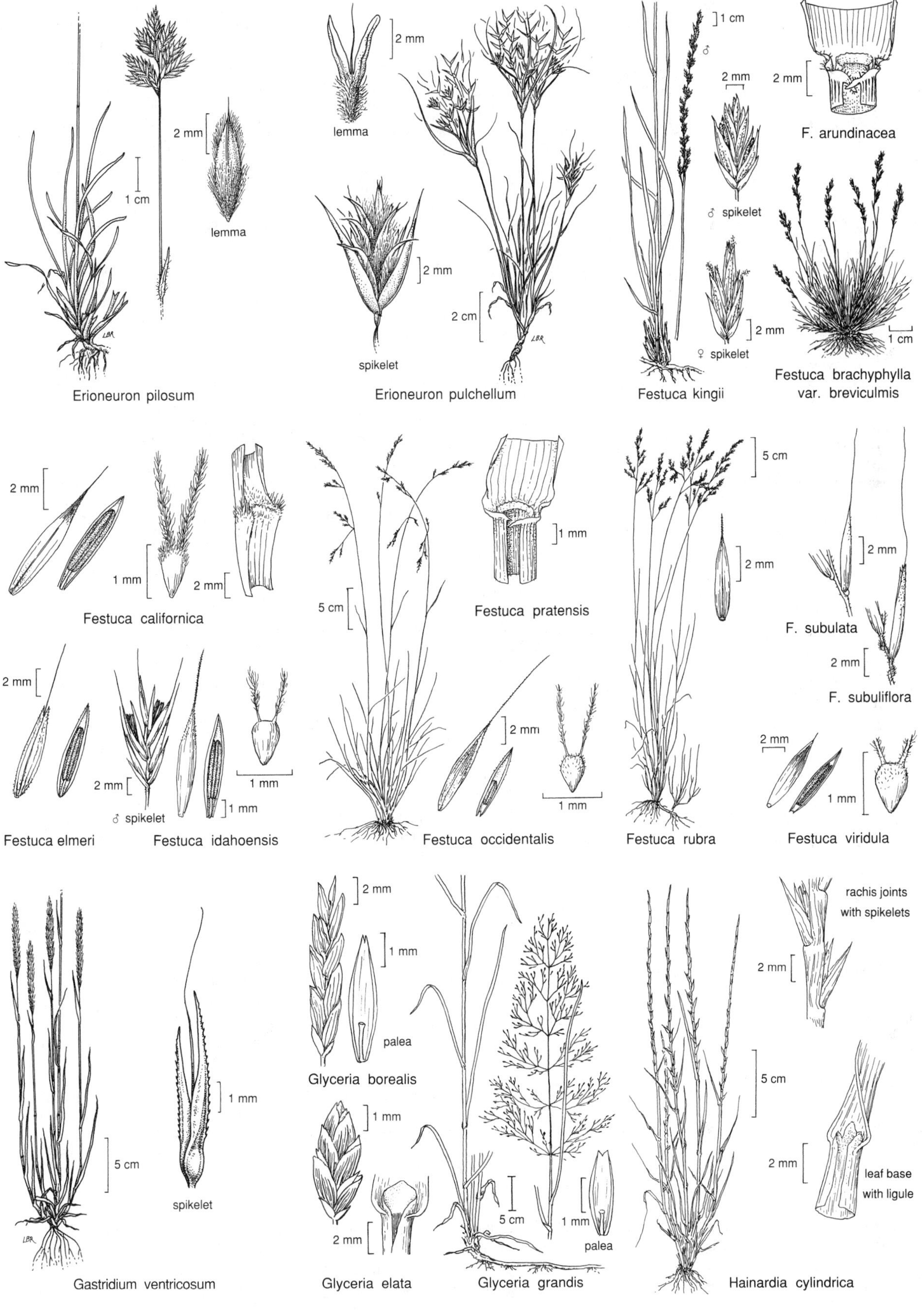
1 cm
2 mm
lemma
Erioneuron pilosum
2 mm
lemma
2 mm
spikelet
2 cm
Erioneuron pulchellum
1 cm
♂
2 mm
♂ spikelet
2 mm
♀ spikelet
Festuca kingii
2 mm
F. arundinacea
1 cm
Festuca brachyphylla
var. breviculmis
2 mm
1 mm
2 mm
Festuca californica
2 mm
2 mm
♂ spikelet
1 mm
1 mm
Festuca elmeri
Festuca idahoensis
5 cm
1 mm
Festuca pratensis
2 mm
1 mm
Festuca occidentalis
5 cm
2 mm
Festuca rubra
2 mm
F. subulata
2 mm
F. subuliflora
2 mm
1 mm
Festuca viridula
5 cm
1 mm
spikelet
Gastridium ventricosum
2 mm
1 mm
palea
Glyceria borealis
1 mm
2 mm
Glyceria elata
5 cm
1 mm
palea
Glyceria grandis
5 cm
rachis joints
with spikelets
2 mm
2 mm
leaf base
with ligule
Hainardia cylindrica

H. arizonicum Covas Ann, bien, sometimes per. **STS** 2.1–7.5 dm, gen erect, tufted; nodes glabrous. **LF**: lower sheaths hairy, upper sheaths glabrous, appendages < 2 mm or 0; blade < 4 mm wide, scabrous. **INFL** 5–12 cm, 6–10 mm wide, pale green. **CENTRAL SPIKELET**: glumes 22–26 mm, flat near base, spreading with age; lemma awn 17–20 mm. **LATERAL SPIKELET**: stalks curved; glumes 20–26 mm; floret 6–7.5 mm, narrow; lemma tip tapered. 2*n*=42. Uncommon. Wet places, irrigated fields; < 300 m. DSon; AZ.

H. brachyantherum Nevski (p. 1269) Per. **STS** < 9 dm, loosely to densely tufted; nodes glabrous. **LF**: sheath glabrous to sparsely hairy; appendages 0; blade 1.5–9 mm wide, glabrous or hairy. **INFL** 8–10 cm, green to ± purplish. **CENTRAL SPIKELET**: glumes 7–19 mm, sometimes flat at base, gen straight with age; floret 5.5–10 mm; lemma awn 3.5–22 mm. **LATERAL SPIKELETS**: glumes 6.5–19 mm, straight or spreading with age, lower glume sometimes flat at base; floret < 7 mm or vestigial; lemma awn < 7.5 mm; anthers 0.9–5.5 mm. 2*n*=14,28,42. Meadows, pastures, streambanks; < 3400 m. CA (exc D); to AK, Rocky Mtns, Mex, Eurasia.

ssp. ***brachyantherum*** **ST** gen robust. **LF**: blade < 19 cm, < 9 mm wide, glabrous or sparsely short-hairy. **CENTRAL SPIKELET**: glumes 7–17 mm, gen straight with age; lemma awn < 4.5 mm; anthers 1–3.5 mm. 2*n*=28,42. Habitat and range of sp.

ssp. ***californicum*** (Covas & Stebb.) v. Bothmer, N. Jacobsen & O. Seberg **ST** slender. **LF**: blade < 11.5 cm, gen < 3.5 mm wide, gen short- and long-hairy, rarely glabrous. **CENTRAL SPIKELET**: glumes 9–19 mm, gen spreading with age; lemma awn to 7.5 mm; anthers 1.5–5.5 mm. 2*n*=14. Habitat of sp.; < 500 m. CA-FP; OR. [*H. californicum* Covas]

H. depressum (Scribner & J.G. Smith) Rydb. (p. 1269) LOW BARLEY Ann. **STS** 1–5.5 dm, erect, loosely tufted; nodes glabrous. **LF**: basal sheaths hairy, appendages 0; blade < 4.5 mm wide, sparsely to densely hairy. **INFL** 2.2–7 cm, 4.8 mm wide, pale green or reddish. **CENTRAL SPIKELET**: glumes 5.5–20 mm, sometimes flattened to 0.5 mm wide at base; floret 5–9 mm; lemma awn 3–12 mm. **LATERAL SPIKELETS** staminate or sterile, sometimes bisexual; glumes 5–20 mm, lower glume ± flat at base; floret vestigial; lemma awn < 1 mm; anthers 0.5–1.5 mm. 2*n*=28. Moist sites, vernal pools, gen alkaline soils; < 1800 m. CA (exc Mtns); to WA, ID.

H. intercedens Nevski (p. 1269) Ann. **STS** 0.5–4 dm, bent at base or erect, loosely tufted; nodes gen hairy. **LF**: sheath hairs in vertical lines, appendages < 2 mm or 0; blade < 2 mm wide, sparsely to densely long-spreading-hairy. **INFL** 2.5–6.5 cm, pale green. **CENTRAL SPIKELET**: glumes 9–17 mm, flat at base; floret 4.5–7.5 mm; lemma awn 5.5–10 mm, gen spreading with age. **LATERAL SPIKELETS** gen sterile; glumes < 18 mm, flat at base; lemma 1.7–4.4 mm, tip obtuse, acute, or with awn < 1.2 mm. Vernal pools, dry, saline streambeds, alkaline flats; < 1000 m. 2*n*=14. SW; nw Baja CA. [*H. pusillum* Nutt. in part]

H. jubatum L. (p. 1269) FOXTAIL BARLEY Per or ann. **STS** 2–6 dm, bent at base or erect, densely tufted. **LF**: sheath glabrous to hairy, appendages 0; blade < 5 mm wide, scabrous to short-hairy. **INFL** 3–10 cm, breaking apart in fr, whitish green to light purple. **CENTRAL SPIKELET**: glumes (10)35–80, 35–80 mm, not flat at base, strongly spreading with age; floret 5.5–8 mm; lemma awn 25–90 mm. **LATERAL SPIKELETS** staminate or sterile; glumes 35–80 mm, not flat at base; floret 4–5 mm; lemma awn 2–7 mm. 2*n*=28. CA; to AK, e US, Mex. [var. *caespitosum* (Scribner) A. Hitchc.] Scabrous spikelet clusters can cause mechanical injury to animals.

H. marinum Hudson ssp. *gussoneanum* (Parl.) Thell. MEDITERRANEAN BARLEY Ann. **ST** 1–5 dm, bent at base or erect; nodes glabrous. **LF**: basal sheaths ± hairy, appendages < 2 mm or 0; blade 1–6 mm wide. **INFL** 1.5–7 cm, 5–20 mm wide, green to purple, breaking apart at nodes in fr; central and lateral spikelets falling together. **CENTRAL SPIKELET**: glumes 14–26 mm, not flat at base, ± straight with age; floret 5–8 mm; lemma awn 6–18 mm. **LATERAL SPIKELETS** sterile; glumes 10–24 mm, rounded to slightly flat; floret < 5 mm; lemma awn 3–8 mm. Dry to moist, disturbed sites; < 1500 m. CA (exc mtns); to B.C., ID, AZ; native to Eur. [*H. hystrix* Roth; *H. geniculatum* All.]

H. murinum L. Ann. **ST** 1–11 dm, erect, sometimes ± prostrate; nodes glabrous. **LF**: basal sheaths glabrous to ± hairy, appendages 1–4 mm; blade 2–5 mm wide, glabrous, scabrous or sparsely long-hairy. **INFL** 3–8 cm, 7–16 mm wide, green to glaucous, sometimes reddish or brown in fr. **CENTRAL SPIKELET**: glumes 11–25 mm, flattened at base, gen ciliate; floret 8–14 mm; lemma awn 20–40 mm. **LATERAL SPIKELETS**: glumes 11–35 cm, base flattened; floret 8–15 mm; lemma awn 20–50 mm. Moist, gen disturbed sites; gen < 1000 m. CA; to B.C., e US, n Mex; native to Eur.

ssp. *glaucum* (Steudel) Tzvelev (p. 1269) Summer ann. **ST** 1.5–4 dm. **INFL** green to glaucous, gen brown in fr. **CENTRAL SPIKELET** stalked; floret ± < lateral florets; lemma awn < awn of lateral floret. 2*n*=14. Habitat and range of sp. [*H. g.* Steudel, *H. stebbinsii* Covas]

ssp. *leporinum* (Link) Arcang. Winter ann. **ST** 3–11 dm. **INFL** gen green. **CENTRAL SPIKELET** stalked; floret << lateral florets; lemma awn slightly < awn of lateral floret. 2*n*=28,42. Habitat and range of sp. [*H. leporinum* Link]

ssp. *murinum* Winter ann. **ST** 3–6 dm. **INFL** green. **CENTRAL SPIKELET** sessile or subsessile; stalk < 0.5 mm; floret gen = lateral florets; lemma awn slightly < awn of lateral floret. 2*n*=28. Habitat and range of sp.

HYPARRHENIA

Kelly W. Allred

Ann, per, gen cespitose. **STS** erect, 1–3. **LVS** cauline; sheaths gen < internodes; ligule membranous; blade flat or folded. **INFL** long-stalked from upper axils, with 2 spike-like branches, branches ± equal, hairy. **SPIKELETS** paired; lowest pairs sessile, staminate; upper pairs with lower spikelet sessile, bisexual, upper stalked, staminate or sterile, pairs falling as 1 unit or spikelets breaking below glumes; glumes > lemmas, lanceolate, thinly membranous, hairy; florets 2, lower vestigial, obscure, upper fertile; lemma translucent, awned. 55 spp.: trop Afr, Asia. (Greek: below masculine, from basal staminate spikelets)

H. hirta (L.) Stapf (p. 1269) Per. **STS** ± clumped, 3–10 dm. **LF**: blade 5–30 cm, < 3 mm wide, ± folded or inrolled. **INFL**: stalk long-soft-hairy, subtended by expanded sheath, blade; branches 2–4 cm. **SPIKELET** 4–6 mm; glumes densely hairy; lemma hairy in upper half, awn 1–3.5 cm. Disturbed sites; < 300 m. SCo (Los Angeles), also cult; native to warm temp Eurasia, Afr.

IMPERATA

Kelly W. Allred

Per with rhizomes. **ST** erect, solid in X-section. **LVS** cauline; ligule membranous, truncate; blade flat. **INFL** panicle-like, ± cylindric; branches appressed, many, short, spike-like, densely silky. **SPIKELETS** in pairs, stalked, ± round in X-section, breaking below glumes, falling as 1 unit; glumes unequal, lower < upper, thinly membranous; florets 2, lower vestigial, obscure, upper bisexual, < glumes; lemma reduced, transparent or 0, awn 0; palea << lemma, ± vestigial; stamens 1–2; style exserted, stigmas plumose. 8 spp.: warm temp, trop. (F. Imperato, Italian naturalist, 1500's)

I. brevifolia Vasey (p. 1269) SATINTAIL Rhizomes hard, scaly. **ST** 0.7–1.5 m. **LF**: ligule densely ciliate; blade 15–50 cm, 4–15 mm wide, narrow at collar. **INFL** 1–3 dm, plume-like, densely white-silky-hairy, appearing speckled from adherent brown anthers and stigmas; hairs 8–15 mm. **SPIKELET**: glumes 2.5–5 mm, faintly 5-veined. Wet springs, meadows, streamsides, flood plains; < 500 m. SnJV, SCo, SnGb, SnBr, DMoj, cult elsewhere; to TX, n MEX. NOXIOUS WEED.

KOELERIA

Dieter H. Wilken

Ann, per. **STS** erect. **LVS** basal to cauline; ligule membranous, glabrous to minutely ciliate, toothed at obtuse to truncate tip; blade narrow, flat to inrolled. **INFL** panicle-like, gen compact, narrow. **SPIKELET** laterally compressed; glumes unequal, upper > and wider than lower, keeled, acute, lower 1-veined, upper faintly 3–5-veined; axis prolonged beyond fertile floret, bristly (sometimes with vestigial floret at tip); florets 2–5, bisexual, breaking above glumes and between florets; lower lemmas gen > glumes, awned or not, 5-veined; palea ± < lemma, tip minutely 2-forked. ± 30 spp.: temp N.Am, Eurasia. (G.L. Koeler, Germany, born 1765)

1. Per, infl axis puberulent, glumes and lower lemmas glabrous to minutely scabrous on back .... ***K. macrantha***
1' Ann, infl axis glabrous, glumes and lower lemmas papillate to coarse-hairy on back .... ***K. phleoides***

K. macrantha (Ledeb.) J. A. Shultes (p. 1269) JUNEGRASS Per, cespitose. **STS** 2–7 dm, glabrous to puberulent. **LVS** gen basal, tufted, glabrous to puberulent; ligule 1–2 mm; blade 3–20 cm, 1–2(3) mm wide, gen ridged. **INFL** 2–15 cm, 1–2 cm wide, cylindric to narrowly conic, axis and branches puberulent. **SPIKELET** 4–6 mm, ± shiny, tan (sometimes purplish); florets 2–3(4); glumes and lower lemmas minutely scabrous on back; lower glume ± 3 mm, upper ± 5 mm; lemmas 3–5 mm, acute to small-pointed at tip. $2n$=14. Dry, open sites, clay to rocky soils, shrubland, woodland, coniferous forest, alpine; < 3500 m. NW, CaR, SN, CW, TR, PR, MP, W&I; to AK, e Can, c&e US, n Mex. [*K. cristata* (L.) Pers., an illegitimate name; *K. pyramidata* (Lam.) Beauv. misapplied] ❀DRN:**4–6,14–17**&IRR,part SHD:1–3,**7**,8–10,**18–24**;also STBL.

K. phleoides (Villars) Pers. (p. 1269) Ann. **STS** 1–6(15), 5–40 cm, glabrous. **LVS** basal and cauline; basal loosely tufted, with soft to coarse hairs on back and margins; ligule < 1 mm; blade 1–5 cm, 1–2 mm wide, smooth to ridged. **INFL** 8–30 cm, 4–10 mm wide, cylindric, axis and branches glabrous. **SPIKELET** 3–4 mm, green to tan; glumes and lower lemmas papillate to coarse-hairy on back; lower glume ± 2 mm, upper ± 3 mm; lemmas 2–3 mm, lemmas awned at tip, lower lemma awn 1–2 mm. $2n$=26. Open, disturbed sites; < 350 m. NCoRI, CaRF, SNF, Teh, GV, s SCoRO, SCoRI; to WA, s US; native to Medit.

LAGURUS HARE'S TAIL

1 sp. (Greek: hare tail, from densely hairy infl)

L. ovatus L. (p. 1269) Ann, gen soft-hairy. **STS** ascending to erect, 1–6 dm. **LVS** basal and cauline; sheath loosely surrounding st; ligule membranous, truncate; blade 2–10 cm, 3–10 mm wide, flat, spreading to curving away. **INFL** spike-like, 1.5–3 cm, ovoid to ± subcylindric, dense; spikelets gen subsessile. **SPIKELET** 7–10 mm (exc awns), compressed; glumes equal, > floret, narrowly lanceolate, soft-hairy to plumose, long-tapered; axis breaking apart above glumes, prolonged beyond floret; floret 1; lemma back glabrous, base puberulent, faintly 5-veined, awns 3, 2 from tip < 2 mm, slender, 1 from lemma back 8–20 mm, ± stiff, bent. $2n$=14. Disturbed places; < 200 m. s NCo, CCo, SnFrB; ne US; native to s Eur. Cult for orn.

LAMARCKIA GOLDENTOP

Lynn G. Clark

1 sp.: native to Medit, naturalized in similar climates worldwide (J.B. Lamarck, French botanist, 1744–1829)

L. aurea (L.) Moench (p. 1269) Ann, cespitose, glabrous. **ST** gen erect, 7–40 cm. **LVS** cauline, ± evenly distributed; ligule 3–7 mm, membranous, glabrous, tip ± irregularly cut; blade 2.5–9 cm, 2.5–7 mm wide, flat. **INFL** panicle-like, terminal, 2–8 cm, dense, golden yellow to purplish; axis short-white-hairy in branch axils; spikelets short-stalked, with 1 fertile and 1–3 sterile spikelets in spreading to drooping clusters, each cluster gen falling as 1 unit. **FERTILE SPIKELET**: glumes 2.5–4 mm, ± equal, gen = spikelet; florets 2; lower floret fertile, 2.5–3 mm, lemma awned from near tip, awn 6–7 mm, straight; upper floret sterile, ± 0.5 mm, awn 4–5 mm. **STERILE SPIKELET** 6–9 mm, >> glumes, linear; glumes > lower floret; florets 5–8; lemmas ± overlapping, 1.5–2 mm, obtuse, tip ± fringed, awn 0 (stalk base sometimes with a reduced, sterile spikelet that is like the fertile in size and shape). n=7. Open ground, moist seeps, rocky hillsides, sandy soil; < 660 m. CA; AZ; native to Medit. Somewhat weedy.

LEERSIA CUTGRASS

Dieter H. Wilken

Per from long rhizomes. **STS** gen solitary, decumbent to erect. **LVS** cauline; ligule membranous; blade flat to folded. **INFL** panicle-like, open; lateral branchlets, spikelet stalks arched to wavy. **SPIKELET** laterally compressed; glumes 0; floret 1, bisexual, falling as 1 unit; lemma, palea firmly membranous; lemma strigose on back, awnless, 5-veined; palea ± = lemma. 17 spp.: trop, warm temp Am, Eurasia. (J.D. Leers, Germany, born 1727) [Pyrah 1969 Iowa State Coll J Sci 44:215–270]

L. oryzoides (L.) Sw. (p. 1275) RICE CUTGRASS **STS** 1–1.5 m; nodes short, soft-hairy. **LF**: sheath glabrous to minutely scabrous; ligule ± 1 mm, truncate; blade 10–28 cm, 8–14 mm wide, margin strongly scabrous, with downward-pointing teeth. **INFL** 12–20 cm; lower branches ± spreading. **SPIKELET** 4–5 mm, oblong to narrowly elliptic; lemma 4–5 mm, width 3–5 × palea width, strigose on back. $2n$=48. Marshes, streams, ponds; < 700 m. NW, ne SCo, SNE (Owens Valley); to B.C., e N.Am, Eurasia. Lateral infls enclosed by sheath, gen cleistogamous. ❀STBL.

LEPTOCHLOA SPRANGLETOP

Ann, per. **STS** spreading to erect. **LVS** gen cauline; ligule membranous, ± entire to jagged, sometimes ciliate; blade flat. **INFL** panicle-like; branches spike-like; spikelets short-stalked or sessile. **SPIKELETS** compressed or ± cylindrical; glumes equal or unequal, 1(3)-veined, short-awned or not; axis breaking apart above glumes and between florets; florets gen 2–12; lemma back rounded or keeled, glabrous or hairy, 3-veined, tip obtuse or minutely 2-lobed, awn gen 0; stamens 2 or 3. 40 spp.: warm temp, trop. (Greek: slender grass, from slender infl) [McNeill 1979 Brittonia 31:399–404]

1. Spikelets 1–3 mm, compressed; florets 2–4; lemma strongly keeled ..... ***L. filiformis***
1′ Spikelets 4–12 mm, ± cylindric; florets (4)5–12; lemma rounded
 2. Lemma obtuse to truncate, abruptly pointed ..... ***L. uninervia***
 2′ Lemma gen acute, awned, awn 0.5–3(5) mm
  3. Spikelets 6–12 mm ..... ***L. fascicularis***
  3′ Spikelets 4–6 mm ..... ***L. viscida***

L. fascicularis (Lam.) A. Gray (p. 1275) BEARDED SPRANGLETOP Ann. **STS** spreading to erect, 3–10 dm. **LF**: sheath gen glabrous; ligule 2–5 mm, membranous, jagged; blade 10–50 cm, 1–5 mm wide. **INFL** 1–3 dm; branches spreading to ascending, lower 8–15 cm. **SPIKELET** 6–12 mm, ± cylindric in X-section; glumes 2–4 mm, upper > and wider than lower; florets 6–12; lemma 3.5–5 mm, back rounded, densely hairy below middle, awn at tip 1–3(5) mm. $2n$=20. Marshes, wetlands, sometimes wet disturbed areas; < 1200 m. GV, GB, D; to e US, Mex, S.Am. Fls gen self-pollinating.

L. filiformis (Lam.) Beauv. (p. 1275) RED SPRANGLETOP Ann. **STS** decumbent to erect, 1–10 dm, sometimes reddish or purple. **LF**: sheath papillate; ligule 1–2 mm, entire to jagged and hairy; blade 5–30 cm, 3–10 mm wide. **INFL** 3–20 cm; branches gen ascending, lower 5–10 cm. **SPIKELET** 1–2 mm, compressed; glumes 1–2 mm, ± equal, purplish; florets (2)3–4; lemma 1–2 mm, keeled, veins hairy, tip minutely 2-lobed, awn 0. $2n$=20. Wet sites, drying ponds; < 100 m. DSon (Imperial Co.); to s US, Mex, S.Am.

L. uninervia (C. Presl) A. Hitchc. & Chase MEXICAN SPRANGLETOP Ann. **STS** erect, sometimes few-branched, 3–10 dm. **LF**: sheath glabrous or scabrous; ligule 2–6 mm, entire to jagged; blade 10–45 cm, 1–4 mm wide. **INFL** 1–3 dm; branches gen ascending, lower 3–6 cm. **SPIKELET** 5–7 mm, ± cylindric in X-section; glumes 1–2 mm, upper > and wider than lower; florets 6–9; lemma 2–3 mm, back rounded, tip obtuse to truncate, abruptly short-pointed, awn 0. $2n$=20. Ditches, drying ponds, disturbed wet places; gen < 1000 m. s SNF, SnJV, SW, GB, D; to s US, Mex, S.Am.

L. viscida (Scribner) Beal STICKY SPRANGLETOP Ann. **STS** spreading to erect, 1–5 dm. **LF**: sheath scabrous; ligule 1–3 mm, minutely jagged; blade 2–15 cm, 2–6 mm wide. **INFL** 3–8 cm; branches ascending to ± appressed, lower 1–2 cm. **SPIKELET** gen 4–6 mm, ± cylindric in X-section; glumes 2–3 mm, upper > and wider than lower; florets 4–7(8); lemma 1.5–3 mm, back rounded, ± sticky, veins hairy, tip minutely 2-lobed, awn 0.5–1 mm, slender. Wet places; < 100 m. SnJV (Kern Co.); to TX, Mex.

LEYMUS

Mary E. Barkworth

Per, gen from rhizomes. **LF**: ligule membranous; blade flat or rolled, strongly ribbed above, glabrous or hairy. **INFL** gen spike-like (panicle-like in *L. condensatus*), dense, gen > upper cauline lvs; some nodes with 2+, gen sessile spikelets. **SPIKELET**: glumes < spikelet, lanceolate, membranous, flexible or narrowly lanceolate to awl-like, stiff; axis breaking above glumes and between florets; florets 2–7; lemma acute to short-awned; palea < lemma; anthers 2.5–7 mm. 31 spp.: N.Am., Eurasia. (Anagram of *Elymus*) [Barkworth & Atkins 1984 Amer J Bot 71:609–625] Some spp. important in revegetation, often on saline soils. Sometimes treated in *Elymus*.

1. Glumes lanceolate, > 1 mm wide, clearly veined at midlength; pls from rhizomes
 2. Glumes 3–5 mm wide at midlength, back rounded ..... ***L. mollis***
 2′ Glumes 1–2 mm wide at midlength, back keeled ..... ***L. ×vancouverensis***
1′ Glumes awl-like, not clearly veined at midlength; pls cespitose or from short rhizomes (rhizomes well-developed in *L. pacificus, L. triticoides*)
 3. Pls gen > 12 dm; blades often > 10 mm wide
  4. Infl unbranched; spikelets sessile on central axis ..... ***L. cinereus***
  4′ Infl often branched at lowest nodes; some spikelets stalked, stalks 1–5 mm
   5. Anthers fertile; infl branched at lowest nodes ..... ***L. condensatus***
   5′ Anthers gen sterile (not dehiscent); infl gen unbranched ..... ***L. ×multiflorus***
 3′ Pls gen < 10 dm; blades < 10 mm wide
  6. Pl cespitose, rhizomes 0 ..... ***L. salinus*** ssp. ***mojavensis***
  6′ Pl from rhizomes
   7. St 1–6 dm; lf blade often > infl ..... ***L. pacificus***
   7′ St 4.5–13 dm; lf blade < infl ..... ***L. triticoides***

L. cinereus (Scribner & Merr.) A. Löve (p. 1275) Pl cespitose; rhizomes 0 or short. **ST** 7–21 dm, gen glabrous; lowest node often hairy. **LF**: ligule 2.5–6.5 mm; blade 3–12 mm wide, upper surface scabrous. **INFL** 9–19 cm; spikelets 2–17 per node. **SPIKELET**: glumes 8–18 mm, awl-like; lemma 6.5–12 mm, acute to awn-tipped, glabrous to short-hairy. $2n$=28,56. Streamsides, canyons, roadsides, sagebrush shrubland, open woodlands; < 3000 m. CaR, SN, Teh, ScV, TR, GB, e DMtns (Kingston Mtns); to Can, Colorado. [*Elymus c.* Scribner & Merr.] Hybridizes with *L. triticoides*. ❀TRY.

L. condensatus (C. Presl) A. Löve Pl cespitose; rhizomes 0 or short. **ST** 11–30 dm, glabrous. **LF**: blade 10–28 mm wide. **INFL** 17–44 cm, panicle-like; lower nodes short-branched; spikelets sessile or stalked. **SPIKELETS**: glumes 6–16 mm, awl-like; lemmas 7–14 mm, glabrous to hairy, acute to awn-tipped, awn < 4 mm. $2n$=28,56. Dry slopes, open woodland; < 1500 m. CW, SW, DMoj; Mex. [*Elymus c.* C. Presl] Hybridizes with *L. triticoides* (see *L. ×multiflorus*). ❀DRN:**7,14–24**&IRR:**8**·**12**;CV.

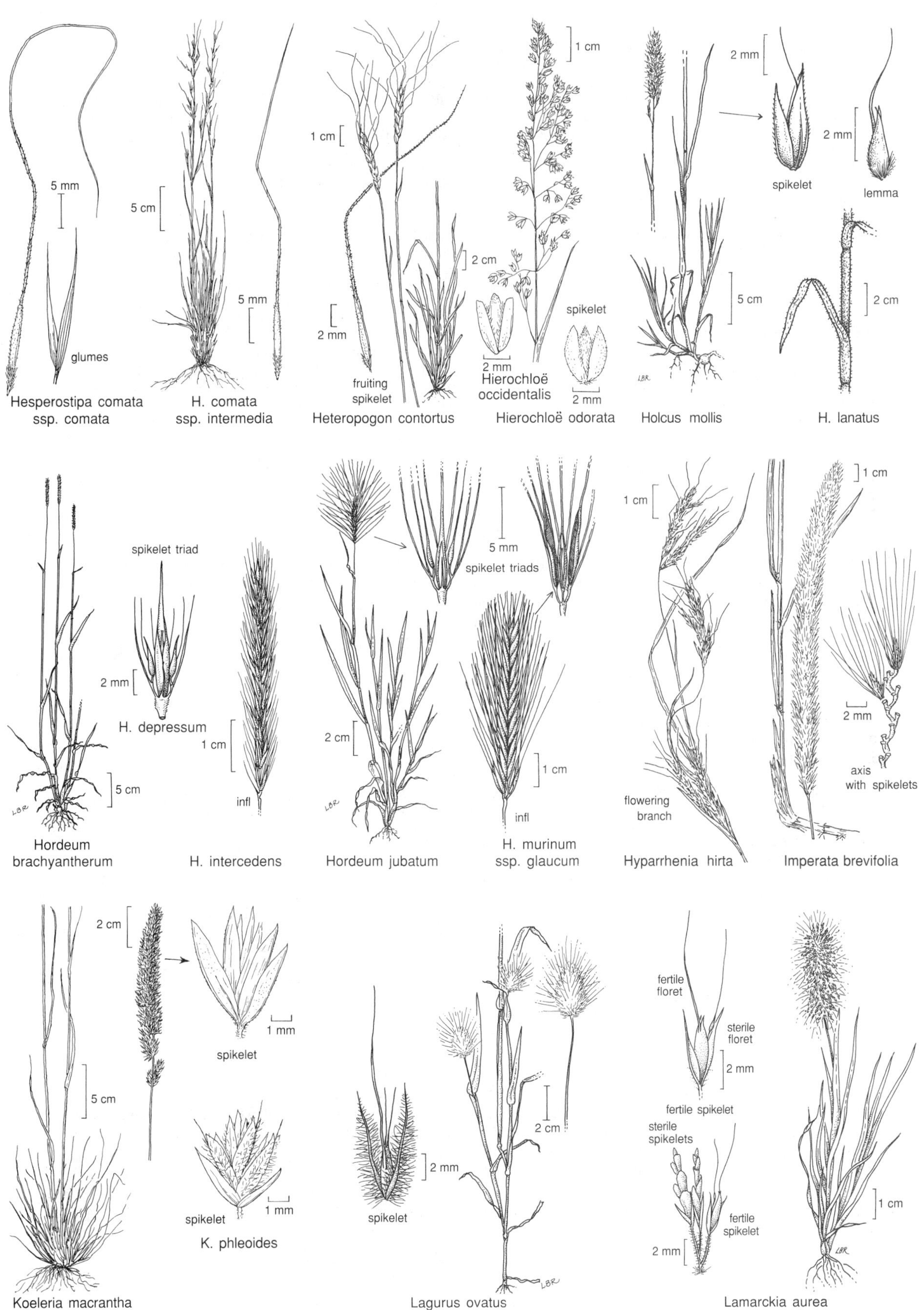
5 mm
glumes
Hesperostipa comata ssp. comata
5 cm
H. comata ssp. intermedia
5 mm
1 cm
2 mm
fruiting spikelet
2 cm
Heteropogon contortus
1 cm
2 mm
Hierochloë occidentalis
spikelet
2 mm
Hierochloë odorata
2 mm
spikelet
2 mm
lemma
5 cm
Holcus mollis
2 cm
H. lanatus
spikelet triad
2 mm
H. depressum
1 cm
infl
5 cm
Hordeum brachyantherum
H. intercedens
5 mm
spikelet triads
2 cm
Hordeum jubatum
1 cm
infl
H. murinum ssp. glaucum
1 cm
flowering branch
Hyparrhenia hirta
1 cm
2 mm
axis with spikelets
Imperata brevifolia
2 cm
1 mm
spikelet
5 cm
spikelet
1 mm
K. phleoides
Koeleria macrantha
2 mm
spikelet
2 cm
Lagurus ovatus
fertile floret
sterile floret
2 mm
fertile spikelet
sterile spikelets
fertile spikelet
2 mm
1 cm
Lamarckia aurea

L. mollis (Trin.) Pilger ssp. ***mollis*** (p. 1275) From rhizomes. **ST** 5–13 dm, densely hairy below infl. **LF**: ligule 0.5–1 mm; blade 5–15 mm wide, upper surface ± scabrous. **INFL** 12–30 cm; spikelets 3+ per node, sessile or stalked. **SPIKELET**: glumes 13–30 mm, lanceolate, flat, slightly to densely hairy; lemma 12–20 mm, densely hairy, acute to awn-tipped. 2*n*=28. Sandy beaches; gen < 10 m. NCo, CCo; to AK, Asia. [*Elymus m.* Trin.] Hybridizes with *L. triticoides* (see *L.* ×*vancouverensis*). Other ssp. arctic. ❀DRN:**4–5,15–17**&IRR:7–9,**14**,19–21,**22–24**;also STBL.

L. ×***multiflorus*** (Gould) Barkworth & D.R. Dewey Pl cespitose; rhizomes short. **ST** 6.5–20 dm, 3.5–5 mm diam at base, gen glabrous. **LF**: blade 6–15 mm wide. **INFL** 12–40 cm; spikelets sessile or stalked. **SPIKELET**: glumes > lower floret, awl-like; lemma glabrous. 2*n*=42. Open areas, often saline soils; < < 300 m. SnJV, CW. [*Elymus triticoides* Buckley ssp. *m.* Gould] Hybrid between *L. triticoides* and *L. condensatus.* ❀STBL.

L. pacificus (Gould) D.R. Dewey Pl from rhizomes. **ST** 1–6 dm, glabrous or sparsely hairy near nodes. **LF**: blade 10–30 cm, 2–4 mm wide, often > infl. **INFL** 2–8 cm; spikelets often 1 per node. **SPIKELET**: glumes 5–10 mm, awl-like; lemmas ± 10 mm, gen glabrous, pointed or short-awned. 2*n*=28. Coastal bluffs; < 100 m. s NCo, CCo, n ChI. [*Elymus p.* Gould] Fls infrequently, seldom collected. ❀STBL.

L. salinus (M.E. Jones) A. Löve ssp. ***mojavensis*** Barkworth & R.J. Atkins Pl cespitose. **ST** 4–14 dm, gen glabrous. **LF**: ligule 0.5–1 mm; blade 1–5 mm wide, upper surface evenly hairy. **INFL** 4–14 cm; spikelets gen 1 at lower, upper nodes, 2 at central nodes. **SPIKELET**: glumes 0 or < 13 mm, awl-like; lemma 7–13 mm, gen glabrous, acute or awn-tipped, awn < 2.5 mm. Hillsides; 1350–2000 m. DMtns; to ID, w Colorado, n AZ. [*Elymus s.* M.E. Jones] Other ssp. Great Basin. ❀TRY.

L. triticoides (Buckley) Pilger (p. 1275) Pl from rhizomes. **ST** 4.5–13 dm, glabrous to hairy. **LF**: ligule 0.2–1.3 mm; blade 2.5–4 mm wide, upper surface finely scabrous. **INFL** 5–20 cm, narrow; spikelets 1–3 per node. **SPIKELET**: glumes 5–16 mm, awl-like; lemmas 5–12 mm, glabrous to puberulent, gen awn-tipped, awn gen ± 3 mm. 2*n*=28. Moist, often saline, meadows; < 2300 m. CA (exc D); to WA, Rocky Mtns, TX. [*Elymus t.* Buckley] Hybridizes with *L. condensatus, L. mollis, L. cinereus.* ❀4,**5,6,15–17**&IRR: 1,**2,3,7–10**,11,**14,18–24**;GRNCVR;also INV;CVS.

L. ×***vancouverensis*** (Vasey) Pilger Pl from rhizomes. **ST** 6.5–13 dm, sparsely to densely hairy below infl. **LF**: blade < 9 mm wide. **INFL** 7–27 cm; spikelets 1–2 per node, sometimes short-stalked. **SPIKELET**: glumes 8–14 mm, base flat, tapered to 1–4 mm awn; lemma base glabrous, scabrous to hairy near tip. 2*n*=42. Sandy beaches; < 100 m. NCo, n CCo; to B.C. [*Elymus v.* Vasey] Hybrid between *L. mollis* and *L. triticoides.* ❀STBL.

LOLIUM RYEGRASS

Dieter H. Wilken

Ann, per. **STS** solitary to loosely clumped, ascending to erect. **LVS** basal and cauline; appendages acute; ligules membranous, obtuse to truncate; blade flat to folded. **INFL** spike-like (sometimes panicle-like). **SPIKELETS** gen 2-ranked, ± laterally compressed, narrow edge facing infl axis, sessile; glume 1, on outside edge of spikelet, > lower floret, 3–9-veined; florets 5–20, bisexual, breaking above glume and between florets; lemma ± membranous, rounded on back, awnless to awned at acute to obtuse tip, 5-veined; palea ± = lemma. 8 spp.: Eurasia. (Latin: ancient common name for ryegrass) [Terrell 1966 Bot Rev 32:138–164]

1. Glume gen = or > rest of spikelet exc awns, lower lemmas hard, ± swollen at base .... ***L. temulentum***
1' Glume < rest of spikelet, lower lemmas firmly membranous, flat to rounded at base
  2. Lemmas awned, pls without sterile shoots at base .... ***L. multiflorum***
  2' Lemmas awnless, pls with sterile shoots at base .... ***L. perenne***

L. multiflorum Lam. (p. 1275) ITALIAN RYEGRASS Ann, bien. **STS** ascending to erect, 4–8 dm, gen glabrous. **LF**: ligule 1–3 mm; blade 5–20 cm, 3–7 mm wide, rolled in bud. **INFL** 10–30 cm, spike-like. **SPIKELET** 10–15 mm; glume < rest of spikelet, 5–9 mm, linear-lanceolate, lemmas 5–7 mm, ± lanceolate, awn 1–8 mm. 2*n*=14. Disturbed sites, abandoned fields; < 1000 m. CA–FP; to AK, e N.Am; native to Eur. Hybridizes with *L. perenne.* Pls with infl axis flattened and angled are *L. rigidum* Gaudin [*L. strictum* Presl.], reported from GV.

L. perenne L. (p. 1275) PERENNIAL RYEGRASS Per. **STS** ascending to erect, (3)5–8 dm, gen glabrous. **LF**: ligule 1–3 mm; blade 4–25 cm, 2–5 mm wide, folded in bud. **INFL** (8)12–30 cm, spike-like (sometimes panicle-like, branches spike-like, spreading). **SPIKELET** 10–15 mm; glume < rest of spikelet, 5–10 mm, linear-lanceolate; lemmas 5–7 mm, ± lanceolate, awnless. 2*n*=14,28. Disturbed sites, abandoned fields, lawns; < 1000 m. CA–FP; to AK, e N.Am; native to Eur.

L. temulentum L. (p. 1275) DARNEL Ann. **STS** ascending to erect, 4–9 dm, glabrous to scabrous. **LF**: ligule < 2 mm; blade 5–20 cm, 2–7 mm wide, folded in bud. **INFL** 7–28 cm, spike-like. **SPIKELET** (15)20–30 mm; glume = or > rest of spikelet exc awns, 10–17(20) mm, lanceolate to oblong; lemmas 5–8 mm, awnless to awned at tip, awn 8–18 mm. 2*n*=14. Open, disturbed sites; 150–1750 m. CA-FP, MP; to B.C., e US; native to Medit. TOXIC: grain can contaminate flour or birdseed; severe poisoning rare.

LYCURUS

Per. **STS** erect, ± solid in X-section. **LVS** gen basal, gray-green; sheath compressed, keeled; ligule membranous, long tapered, entire; blade flat to folded. **INFL** panicle-like; branches short, ± appressed, dense. **SPIKELETS** stalked, paired; pair with subtending axis segment falling as 1 unit; lower spikelet staminate or sterile, upper bisexual; glumes unequal, < lemma, 2–3-veined, awned; floret 1; lemma firm, long-tapered, awned; palea ± = lemma. 3 spp.: sw US, Mex, S.Am. (Greek: wolf tail, from infl shape) [Reeder 1985 Phytologia 57: 283–291]

L. phleoides Kunth var. ***phleoides*** (p. 1275) WOLFTAIL Per, tufted. **ST** decumbent to erect, 2–6 dm, compressed, puberulent. **LF**: ligule 3–7 mm, decurrent to sheath; blade 3–10 cm, 1–2 mm wide, margin and midvein ± white. **INFL** 3–6 cm, ± 5 mm wide, cylindric. **SPIKELET** compressed; glumes 1–2 mm, ± equal, < floret, awns 3–7 mm, lower glume 2-veined, awns 2–3, upper glume 1-awned; floret 1; lemma 3–4 mm, 3-veined, awn 2–7 mm. 2*n*=40. RARE in CA. Joshua tree woodland, pinyon/juniper woodland; ± 500 m. e DMtns (New York Mtns); to OK, Mex.

MELICA MELIC, ONIONGRASS

Mary E. Barkworth

Per; rhizomes, corms gen 0. **STS** gen erect, gen densely clumped. **LVS** ± basal; sheath closed to near top, glabrous to short-hairy; ligule thin, membranous, tip obtuse to truncate, gen jagged; blade gen 2–5 mm wide, flat, veins inconspicuous. **INFL** raceme- or panicle-like, gen narrow. **SPIKELET**: glumes papery, back rounded, tip rounded, translucent, lower glume 3–5-veined, upper 1–3-veined; axis gen breaking above glumes; lower florets fertile, 1–7, uppermost florets sterile, ± densely clustered at axis tip; lemma ± like glumes, prominently 5–7-veined, veins not converging, base ± red; palea < lemma. ± 80 spp.: gen temp, exc Australia. (Latin: honey, or old Italian name for pl with sweet sap) [Boyle 1945 Madroño 8:1–26]

1. Glumes deciduous; spikelets falling as 1 unit, 1–2 per infl branch (sect. *Melica*) ..... ***M. stricta***
1′ Glumes persistent; spikelet axis breaking above glumes, spikelets > 2 per lower infl branch (sect. *Bromelica*)
 2. Lemma acuminate or awned; hairs (if any) near base
  3. Awn 5–12 mm; lemma surface glabrous, margin often hairy near base ..... ***M. aristata***
  3′ Awn 0–4 mm; lemma surface hairy near base
   4. Lemma obtuse to ± tapered, awn 1–4 mm, ciliate; corms 0 ..... ***M. harfordii***
   4′ Lemma strongly tapered, awn 0; corms present ..... ***M. subulata***
 2′ Lemma obtuse to acute, awn 0, hairs (if any) near tip
  5. Fertile florets 1–2 in all spikelets
   6. Sterile cluster at axis tip 0.5–4 mm, widest at middle or below; lemma glabrous or minutely scabrous ..... ***M. imperfecta***
   6′ Sterile cluster at axis tip 0.5–1.5 mm, truncate or widest above middle; lemma margin gen hairy .. ***M. torreyana***
  5′ Fertile florets 3–7 in some or all spikelets
   7. St 10–60 cm; floret stalks swollen when fresh, ± wrinkled and yellow-brown when dry ..... ***M. fugax***
   7′ St 45–200 cm; floret stalks not swollen nor wrinkled when dry
    8. Palea 1/2–3/4 × lemma length; lvs 3–5 per st, blades 3–9 cm ..... ***M. frutescens***
    8′ Palea ± 7/8 × lemma length; lvs 2–3 per st, blades gen >> 9 cm
     9. Sterile cluster at axis tip widest above middle, tip truncate ..... ***M. californica***
     9′ Sterile cluster at axis tip widest below middle, tip acute to acuminate
      10. Both glumes ± 1/2 lowest floret length; corms connected to rhizome by short stalk ..... ***M. spectabilis***
      10′ 1 or both glumes ± 3/4 lowest floret length; corms sessile on rhizome (or rhizome 0)
       11. Sheath of basal lf remaining intact; infl ± narrow, branches ± ascending ..... ***M. bulbosa***
       11′ Sheath of basal lf becoming fibrous; infl gen wide, branches spreading to reflexed ..... ***M. geyeri***

M. aristata Bolander AWNED MELIC **ST** 5–12 dm. **LF**: ligule 3–5 mm; blade 3–6 mm wide. **INFL** 10–23 cm, narrow; spikelets 2–10 per branch. **SPIKELET** 11–20 mm, cylindric; glumes 7–11 mm, translucent margin 1 mm wide, sterile cluster 2.5–6 mm; lemma 8–13 mm, awn 5–12 mm. *n*=9. Dry open sites, coniferous forest; 1000–3000 m. KR, NCoR, CaR, SN, SCoR, SnBr; to s WA.

M. bulbosa Geyer (p. 1275) ONIONGRASS Rhizomes short; corms clustered. **ST** 5–8 dm. **LF**: ligule 2–6 mm wide. **INFL** 7–20 cm, narrow to wide. **SPIKELET** 6–24 mm; lower glume 5–9 mm, upper 6–10 mm, ± = 3/4 lowest floret length; fertile florets 2–5, sterile cluster 1.5–5 mm, narrow, tapered, exserted; lemma 6–11 mm, gen obtuse. *n*=9. Dry rocky slopes, coniferous forest; < 3400 m. CA-FP (exc GV, SW); to B.C., Rocky Mtns, TX. Pls with spreading infl, large spikelets have been called var. *inflata* (Bolander) Boyle. ❀TRY.

M. californica Scribner CALIFORNIA MELIC Rhizomes short or 0; corms gen 0. **ST** 5–13 dm; base gen swollen; lowest 1–2 internodes short; nodes ± swollen. **INFL** 4–30 cm, narrow. **SPIKELET** 5–15 mm; glumes subequal, 3.5–13 mm; fertile florets 2–5, sterile cluster 1–3 mm, truncate; lemma 5–9 mm, tip obtuse or notched. *n*=9. Open hillsides, oak woodland, coniferous forest; < 2100 m. NW, SNF, Teh, CW, WTR. Pls with glumes < spikelet, glume tips obtuse are var. *californica*; pls with glumes ± = spikelet, glume tips acute have been called var. *nevadensis* Boyle. ❀DRN:2,3,**7**,8,9, **14**,18–23&SUN:5,**15–17**,24;also STBL.

M. frutescens Scribner Rhizomes short to long; corms 0. **ST** 4–20 dm, gen branching at basal nodes. **LF**: blade 5–9 cm, 2–4 mm wide. **INFL** 12–40 cm, narrow. **SPIKELET** 12–18 mm; lower glume 7–12 mm, upper 9–15 mm, translucent margin 1–2 mm wide; fertile florets 3–6, sterile cluster tapered, concealed by uppermost fertile floret; lemma 8–11 mm, acute; palea 1/2–3/4 lemma length. *n*=9. Dry slopes, chaparral, woodland; 300–1500 m. s SCoRI, TR, PR, D; AZ, Baja CA. ❀TRY.

M. fugax Bolander Rhizomes short; corms clustered. **ST** 1–6 dm. **LF**: ligule 1–3 mm; blade 2–4 mm wide. **INFL** 8–18 cm, narrow or wide. **SPIKELETS** 4–17 mm; lower glume 3–5 mm, upper 3.5–7 mm; fertile florets 2–5, floret stalk swollen when fresh, ± wrinkled, brown when dry, sterile cluster 2–3.5 mm, tapered; lemma 4–7 mm, obtuse, sometimes notched. *n*=9. Dry volcanic flats, hillsides, coniferous forest; 1200–2200 m. KR, NCoR, CaR, n&c SN, MP; to B.C. ❀TRY.

M. geyeri Bolander Rhizomes short to long; corms sessile. **ST** 8–20 dm. **LF**: ligule 2–3.5 mm; blade 2–8 mm wide. **INFL** 11–17 cm, wide; branches spreading to reflexed. **SPIKELET** 8–24 mm; lower glume 3.5–7 mm, upper 5.5–11 mm, tips extending to ± 3/4 length of lowest lemma; fertile florets 2–6, sterile cluster 3–7 mm, tapered; lemma 8–11 mm. *n*=9. Dry open sides, oak woodland, coniferous forest; < 2000 m. NW, CaR, n&c SN, n SCoRO; OR. Pls from SnFrB (Marin Co.) with lemma awns 0.5–2 mm have been called var. *aristulata* J. Howell. ❀DRN:4,**5,6,17**&SHD,IRR:**7**, 14–16,19–24.

M. harfordii Bolander **ST** 5–12 dm. **LF**: ligule 1–4 mm; blade 2–6 mm wide. **INFL** 6–23 cm, narrow. **SPIKELET** 7–20 mm; glumes 5–11, obtuse to ± acute; fertile florets 2–6, sterile cluster 3–5 mm, lanceolate in outline; lemma 6–16 mm, lower margins ciliate, tip notched, awn 1–4 mm. *n*=9. Dry slopes, coniferous forest; < 2150 m. NW, CaRF, n&c SN, SnFrB, n SCoRO; to B.C. ❀DRN:**4–6**&IRR:**15–17**&SHD:7,8,9,14,19–24.

M. imperfecta Trin. (p. 1275) **ST** 5–11 dm. **LF**: ligule 3–6 mm; blade 1–6 mm wide. **INFL** 5–36 cm, narrow to wide. **SPIKELET** 3.5–7 mm; glumes 2–6 mm, ± equal; fertile florets 1–2, sterile cluster 0.5–4 mm, acute to obtuse; lemma 3–7 mm, acute or obtuse, gen glabrous or minutely scabrous. *n*=9. Dry rocky hillsides, chaparral, woodland; < 1500 m. c&s SN, SnFrB (Santa Cruz Mtns), SCoR, SW, w DMoj; Baja CA. ❀**15–17**;part SHD:7,11&IRR:**14, 18–24**;also STBL.

M. spectabilis Scribner (p. 1275) PURPLE ONIONGRASS Rhizomes short to long; corms short-stalked, scattered. **ST** 0.5–10 dm. **LF**: ligule 1–3 mm. **INFL** 5–26 cm, gen narrow. **SPIKELET** 7–19 mm; glumes 3.5–7 mm, ± equal, < 1/2 spikelet length, obtuse; fertile florets 3–7, sterile cluster ± 2.5 mm, tapered, gen concealed; lemma 6–9 mm, gen widest above middle, acute to obtuse. UNCOMMON. Wet sites, meadows, coniferous forest; 1200–2600 m. KR, n NCoR; to B.C., Rocky Mtns.

M. stricta Bolander (p. 1275) **ST** 1–9 dm. **INFL** 3–30 cm, very narrow; spikelets 1–2 per branch. **SPIKELET** open, appearing V-shaped; glumes 6–18 mm, ± equal, spreading, upper half translucent; axis falling as 1 unit; fertile florets 2–5, sterile cluster 2–7 mm; lemma 8–16 mm, tip obtuse to acute, awn 0; palea 1/2–3/4 lemma length; anthers 1–3 mm. *n*=9. Open sites, coniferous forest, rocky areas in alpine; 1200–3350 m. KR, NCoRO, SN, Teh, TR, W&I; to OR, UT. Pls with straw-colored lf sheaths, long paleas, anthers 2–3 mm from WTR have been called var. *albicaulis* Boyle. ❀DRN,SUN:1–3,6,7,10,14–24;DFCLT.

M. subulata (Gris) Scribner Rhizomes short; corms sessile, clustered. **ST** 8–13 dm. **LF**: ligule 1–5 mm; blade 2–10 mm wide. **INFL** 8–25 cm, narrow to wide. **SPIKELET** 10–28 mm; lower glume 4–7 mm, upper 6–9 mm, acute; florets 2–5, sterile cluster 4–9 mm, tapered; lemma 8–15 mm, gen hairy on lower back, tip acuminate, awn 0. *n*= 9. Moist sites, streambanks, coniferous forest; < 2300 m. NW, CaR, n&c SN, SnFrB (Santa Cruz Mtns), n SCoRO, MP; to AK, Rocky Mtns; also in S.Am. ❀STBL.

M. torreyana Scribner (p. 1275) **ST** decumbent to erect, 3–10 dm. **LF**: ligule 2–5 mm; blade 1–4 mm wide. **INFL** 8–25 cm, gen narrow. **SPIKELET** 4–7 mm; glumes 3.5–7 mm, ± equal; florets 1–2; sterile cluster 0.5–1 mm, widest above middle; lemma 4–6 mm, back, margin hairy. *n*=9. Chaparral, coniferous forest; < 1200 m. NW, SN, CW. ❀DRN,DRY,part SHD:7,9,**14–24**&SUN:5.

MISCANTHUS

Kelly W. Allred

Per, cespitose, gen with rhizomes. **STS** tall, erect, solid in X-section. **LVS** cauline; sheaths > internodes; ligule membranous, truncate; blade flat. **INFL** panicle-like; branches ± spike-like, silky-hairy. **SPIKELETS** paired, unequally stalked, breaking below glumes, falling as 1 unit; stalks long-hairy at tip; glumes ± equal, > florets, membranous; florets 2, lower vestigial, obscure, upper bisexual; lemma translucent, tip awned; stamens 2–3. ± 20 spp.: trop Asia. (Greek: stalked flower, from spikelets) Cult.

M. sinensis Andersson **STS** densely clumped, < 3 m. **LF**: blade 4–10 dm, 6–10 cm wide, margin serrate to smooth. **INFL** 1–3 dm, fan-shaped. **SPIKELETS** 3–4 mm; lower spikelet stalk 1–2 mm, upper spikelet stalk 3–6 mm, hairs > spikelet; lower glume leathery, upper papery; awn 8–12 mm. Irrigation ditches; < 200 m. n SNF (El Dorado Co.); native to se Asia.

MONANTHOCHLOË SHOREGRASS

Dieter H. Wilken

Per, mat-like, dioecious. **STS**: central prostrate to decumbent, widely creeping, rooting at nodes; lateral ascending to erect. **LVS** cauline, tufted on prostrate sts, densely clustered on erect branches; sheath open, persistent, fibrous with age; ligule ring-like, ciliate; blade gen oblong, thick, obtuse to pointed. **SPIKELET** gen 1, sessile to short-stalked, concealed by upper lf blades, unisexual; glumes 0; florets 3–5, lower 1–2 fertile, upper sterile; florets of staminate spikelet breaking apart, lemmas firmly membranous; florets of pistillate spikelet weakly breaking apart or not, lower lemmas hard, strongly enfolding fl and fr; palea ± = lemma, membranous. 2 spp.: warm temp N.Am, Cuba, Mex; also Argentina. (Greek: one-flower grass)

M. littoralis Engelm. (p. 1275) **STS**: prostrate 3–8 dm; lateral 5–23 cm. **LVS**: upper ± 2-ranked; sheath 5–12 mm, scarious; blade 4–12 mm. **SPIKELET** 5–11 mm; staminate floret sessile to short-pedicelled; pistillate floret sessile, gen concealed by lvs. Salt marshes; ± 0 m. SCo, ChI; to s US, Cuba, Mex. ❀STBL.

MUHLENBERGIA MUHLY

Paul M. Peterson

Ann, per, sometimes mat-like, often rhizomed. **ST** decumbent to erect, ± clumped. **LVS** basal and cauline; sheath open; ligule membranous, entire to irregularly toothed, sometimes with 1 large tooth on each side; blade flat to rolled. **INFL** panicle-like, narrow to open; branches spreading to appressed. **SPIKELET**: glumes subequal, gen 1-veined, short-pointed to awned, upper glume sometimes 3-veined; florets 1, sometimes 2, breaking above glumes; lemma short-pointed to awned, glabrous to hairy, 3-veined; palea < to = lemma. **FR** ± fusiform, reddish brown, gen falling with lemma and palea. ± 160 spp.: temp Am, s Asia. (H.L.E. Muhlenberg, Pennsylvania botanist, 1753–1815) [Reeder 1981 in Gould and Moran 1981 San Diego Soc Nat Hist Memoir 12:67–78] ❀STBL.

1. Ann
 2. Lemma awned, awn 1–3 cm, lemma 2.5–6 mm
  3. Lemma gen 4.5–6 mm; infl branches closely appressed, < 1.5 cm wide; glumes gen 1–2 mm ...... ***M. appressa***
  3′ Lemma gen 2.5–4.5 mm; infl branches spreading to ascending, 1–5 cm wide; glumes gen < 1 mm .... [2]***M. microsperma***
 2′ Lemma abruptly pointed, point < 1 mm, or awn 0; lemma < 2.5 mm
  4. Infl narrow, < 1 cm wide; branches closely appressed; sts often rooting at lower nodes .... [2]***M. filiformis***
  4′ Infl open, 1.5–8 cm wide; branches reflexed to ascending; sts not rooting at lower nodes
   5. Infl branches stiffly reflexed ± 90° from axis; glumes glabrous; ligule teeth 2, 1 on each side .... ***M. fragilis***
   5′ Infl branches ascending < 80° from axis; glume tip short-hairy; ligule truncate to obtuse, irregularly short-toothed .... ***M. minutissima***
1′ Per
 6. Lemma awn 0 or short-pointed, point < 1 mm
  7. Infl 5–14 cm wide, open, branches spreading .... ***M. asperifolia***
  7′ Infl < 4 cm wide, narrow, branches ascending to appressed
   8. Rhizomes creeping, ± scaly; sts ± decumbent

9. Glumes 2.5–4 mm, ± = lemma; lower half lemma short-soft-hairy .... [2]*M. californica*
9′ Glumes 0.5–1.8 mm, ± 1/2 lemma length; lemma glabrous to ± scabrous
10. Ligule gen 1–2.5 mm; infl axis gen obscured by branches, spikelets .... *M. richardsonis*
10′ Ligule gen < 1 mm; infl axis gen visible between branches .... *M. utilis*
8′ Rhizomes 0; sts clumped or decumbent and ± rooting at the nodes
11. Blade < 4 cm; lemma gen < 2 mm; sts gen < 3 dm
12. Lemma short-awned, awn < 1 mm; glumes 0.5–1.2 mm; ligule 1–2.5 mm, obtuse to acute ... [2]*M. filiformis*
12′ Lemma awn 1.5–5 mm; glumes < 0.5 mm; ligule < 0.5 mm, truncate .... [2]*M. schreberi*
11′ Blade > 5 cm; lemma gen > 2.5 mm; sts 2–15 dm
13. Infl < 15 cm, 1.5–4 cm wide; branches loosely fld .... *M. jonesii*
13′ Infl 15–60 cm, < 1.2 cm wide; branches densely fld .... *M. rigens*
6′ Lemma awn 1–30 mm
14. Infl open, 6–15 cm wide, branches spreading .... *M. porteri*
14′ Infl < 5 cm wide, branches ascending to appressed
15. Glumes 1 or 2, gen < 0.5 mm, vein 0 .... [2]*M. schreberi*
15′ Glumes 2, gen > 0.5 mm, vein evident
16. Rhizome gen scaly, creeping
17. Blade gen rolled, < 2 mm wide; anther purple, 1.5–3 mm
18. Lemma, palea short-soft-hairy on lower half; sts loosely clumped, decumbent .... [2]*M. arsenei*
18′ Lemma base sparsely short-hairy; palea glabrous to ± scabrous; sts erect, ± rooting at lower nodes .... [2]*M. pauciflora*
17′ Blade gen flat, > 2 mm wide; anther yellow, gen < 1.5 mm
19. Hairs at lemma base 2–3.5 mm .... *M. andina*
19′ Hairs at lemma base < 1.5 mm
20. Lemma awn gen < 2.2 mm; anther 1–1.5 mm .... [2]*M. californica*
20′ Lemma awn 2.5–9 mm; anther < 0.5 mm .... *M. mexicana*
16′ Rhizomes 0
21. Upper glume 3-veined, 3-toothed .... *M. montana*
21′ Upper glume gen 1-veined, obtuse, acute, or awned
22. Glumes gen < 1.5 mm, obtuse; cleistogamous spikelets gen present in lowermost st axils [2]*M. microsperma*
22′ Glumes gen 1.5–3.5 mm, acute, acuminate, or awned; cleistogamous spikelets 0
23. Lemma, palea short-soft-hairy on lower half; sts loosely clumped, decumbent .... [2]*M. arsenei*
23′ Lemma base sparsely short-hairy; palea glabrous to ± scabrous; sts erect, ± rooting at lower nodes .... [2]*M. pauciflora*

M. andina (Nutt.) A. Hitchc. FOXTAIL MUHLY Per; rhizome scaly, creeping. **ST** 2.5–8.5 dm. **LF**: ligule 0.5–1.5 mm, truncate, ciliate; blade 4–16 cm, 2–4 mm wide, flat. **INFL** 2–15 cm, 5–15 mm wide, narrow; branches appressed, loosely fld. **SPIKELET**: glumes 2–4 mm, acuminate or short-awned; lemma 2–3.5 mm, hairs at base = lemma, awn 1–7 mm; anthers 0.5–1.5 mm, yellow. 2*n*=20. Canyons, streambanks, wet meadows; < 3100 m. KR, NCoRI, SN, SCoRI, SnBr, SNE, DMtns; to Can, Colorado, w TX.

M. appressa C.O. Goodd. Ann. **ST** 1–4 dm. **LF**: ligule 1.5–3 mm, truncate to obtuse, decurrent to sheath, toothed; blade 1–5 mm, 1–2 mm wide, flat or folded. **INFL** 4–14 cm, 0.5–1.5 cm wide, narrow, loosely fld; branches 0.5–1.5 cm, appressed. **SPIKELETS** in lower branch axils cleistogamous, enclosed by tightly rolled sheath; glumes 1–2 mm, obtuse to acute; lemma 4.5–6 mm, hairs at base short-appressed between veins, awn 1–3 cm; anther 0.5–1 mm, purple. Uncommon. Open canyon bottoms and rocky slopes; 20–1600 m. s ChI (San Clemente Island), DMtns (Providence Mtns); s AZ, Baja CA.

M. arsenei A. Hitchc. (p. 1275) TOUGH MUHLY Per; rhizomes ± short. **ST** 1.5–4 dm, decumbent at base. **LF**: ligule 1–2 mm, acuminate, toothed, ± decurrent to sheath, with 1 large tooth on each side; blade 1–5 cm, < 2 mm wide, rolled. **INFL** 4–12 cm, < 3 cm wide, narrow; branches ascending to appressed, loosely fld. **SPIKELET**: glumes 2–3 mm, acute, ± short-awned, awn < 1 mm; lemma 3.5–5 mm, short-soft-hairy on lower half, awn 4–12 mm; palea short-soft-hairy on lower half; anther 1.6–3 mm, purple. RARE in CA. Limestone rock outcrops, slopes; 1400–1860 m. DMtns (Clark Mtns); to se UT, n NM, n Baja CA.

M. asperifolia (Nees & Meyen) L. Parodi (p. 1275) SCRATCH-GRASS Per; rhizomes shiny, scaly; ± stoloned. **ST** decumbent to erect, 1–6 dm. **LF**: ligule 0.2–1 mm, truncate, minutely ciliate; blade 2–6 cm, 1–2.8 mm wide, flat or folded. **INFL** 6–17 cm, ovoid, open; branches 5–14 cm, spreading. **SPIKELET**: glumes 0.5–1.5 mm, acute; florets 1–2; lemma 1–2 mm, glabrous ± short-awned; anther 1–1.2 mm, purple. 2*n*=20,22,28. Moist, often alkaline meadows, seeps, hot springs; 120–2150 m. CA; w N.Am, also in s S.Am.

M. californica Vasey (p. 1279) CALIFORNIA MUHLY Per; rhizomes short, scaly, creeping. **ST** 3–7 dm. **LF**: ligule 0.8–2 mm, truncate, irregularly toothed, minutely ciliate; blade 4–16 cm, 2–6 mm wide, flat. **INFL** 5–13 cm, < 2 cm wide, narrow; branches ascending to erect, short, densely fld. **SPIKELET**: glumes 2.5–4 mm, acuminate, ± long-tapered to awned, awn < 1.2 mm; lemma 2.8–4 mm, short-soft-hairy on lower half, hairs < 1.5 mm, awn < 2.2 mm; anther 1–1.5 mm, yellow. 2*n*=80. UNCOMMON. Streambanks, canyons, moist ditches; 100–2000 m. SCo, SnGb, SnBr, SnJt.

M. filiformis (Thurber) Rydb. PULL-UP MUHLY Ann, often short stoloned. **ST** decumbent, loosely clumped, rooting at lower nodes, 0.2–3 dm. **LF**: ligule 1–2.5 mm, obtuse to acute, margin serrate; blade 1–4 cm, 1–2 mm wide, flat or rolled. **INFL** 1–6 cm, < 1 cm wide, cylindric, narrow; branches closely appressed. **SPIKELET**: glumes 0.5–1.2 mm, obtuse, ± toothed at 10×; lemma 1.5–2 mm, short-awned, awn < 1 mm; anther 0.5–1 mm, purple. 2*n*=18. Moist meadows, seeps, streambanks; 150–3350 m. NW, SN, SnBr, SnJt, MP; to B.C., Great Plains, NM, Mex. ❀WET:**1–3**&SUN:**6,7,14–16,18**.

M. fragilis Swallen (p. 1279) DELICATE MUHLY Ann. **ST** erect or spreading, 1–3.5 dm. **LF**: ligule 1–3 mm, decurrent to sheath, with 1 large tooth at each side; blade 1–6 cm, 1–2 mm wide, flat, margin, midvein strongly white-thickened. **INFL** 10–30 cm, 3.5–8 cm wide, ovoid, open; branches thread-like, stiffly spreading to reflexed ± 90° from central axis. **SPIKELET**: glumes 0.5–1 mm, obtuse to acute, glabrous; lemma ± 1 mm, gen glabrous, margin, midvein sometimes short-hairy; anther < 0.5 mm, purple. 2*n*=20. RARE in CA. Open, ± disturbed, limestone gravelly wash; ± 1600 m. e DMtns (Clark, New York mtns); to w TX, Mex. In cult.

M. jonesii (Vasey) A. Hitchc. Per. **ST** densely clumped, 2–5 dm. **LVS**: basal gen tufted; ligule 2–4.5 mm, acute; blade 5–12 cm, 1–2.5 mm wide, flat to ± folded. **INFL** 4–15 cm, 1.5–4 cm wide; branches ascending to loosely spreading, loosely fld. **SPIKELET**: glumes 0.6–1.8 mm, obtuse, upper irregularly toothed, ± 3-veined; lemma 2.8–3.5 mm, short-soft-hairy on lower third, short-awned, awn < 1 mm; anther 1.4–2.2 mm, purple. 2*n*=20. Open slopes; 1130–2130 m. KR, CaRH, n SNH.

M. mexicana (L.) Trin. Per; rhizomes scaly, creeping. **ST** 3–7 dm. **LF**: ligule 0.4–1 mm, truncate, irregularly toothed; blade 4–12 cm, 2–5 mm wide, flat. **INFL** 2–15 cm, 5–15 mm wide, narrow; branches stiffly ascending, densely fld. **SPIKELET**: glumes 1.5–3.5 mm, acuminate to short-awned, awn < 2 mm; lemma 1.8–3.4 mm, hairs at base < 1.5 mm, awn 2.5–9 mm; anther < 0.5 mm, yellow. Uncommon. 2*n*=40. Riverbanks and canyons; 60–1530 m. KR, NCoRO, n SN; to B.C., e US.

M. microsperma (DC.) Kunth (p. 1279) LITTLESEED MUHLY Ann, short-lived per. **ST** 1–6 dm. **LF**: ligule 1–2 mm, decurrent to sheath, truncate to obtuse, toothed; blade 2–6 cm, 1–2.5 mm wide, flat or loosely rolled. **INFL** 5–20 cm, 1–5 cm wide; branches spreading to ascending, loosely to densely fld. **SPIKELETS** in lower branch axils cleistogamous, enclosed by tightly rolled sheath; glumes 0.5–1 mm, obtuse; lemma 2.5–4.5 mm, short-soft-hairy at base, awn 1–3 cm; anther 0.5–1 mm, purple. 2*n*=20,40,60. Open ± disturbed sites; < 1650 m. CCo, SCoRO, SW, D; to sw UT, AZ, Mex, also in S.Am.

M. minutissima (Steudel) Swallen Ann. **ST** ascending to erect, 0.2–3 dm. **LF**: ligule 1–2 mm, truncate to obtuse, short-toothed; blade 0.5–4 cm, 1–2 mm wide, flat. **INFL** 1–20 cm, narrowly ovoid, open; branches 1.5–5 cm, ascending < 80° from central axis. **SPIKELET**: glumes 0.5–1 mm, obtuse, tip short-hairy; lemma 1–1.5 mm, margins, midvein short-hairy; anther 0.5–1 mm, purple. 2*n*=60,80. Open, ± disturbed, sandy slopes, seeps; 400–2300 m. KR, n&c SNH, SnBr, SnJt, SNE; to WA, MT, w TX, Mex.

M. montana (Nutt.) A. Hitchc. (p. 1279) MOUNTAIN MUHLY Per. **ST** 1–4 dm, densely tufted. **LF**: ligule 4–10 mm, acute; blade 5–12 cm, 1–2.5 mm wide, flat, ± rolled. **INFL** 4–15 cm, 2–6 cm wide, oblong in outline; branches spreading to ascending, loosely fld. **SPIKELET**: glumes 1.5–3 mm, upper 3-toothed, 3-veined; lemma 3–4.2 mm, margin, midvein short-soft-hairy on lower half, awn 6–18 mm; anther 1.5–2.2 mm, purple. 2*n*=20,40. Open slopes, granitic rock outcrops, dry meadows; 1640–3420 m. KR, SNH; w US to Mex, Guatemala. ❀DRN,IRR:1,2&SUN:**15,16**.

M. pauciflora Buckley (p. 1279) FEW-FLOWERED MUHLY Per; rhizomes ± short, knot-like. **ST** erect, 3–5 dm, wiry, rooting at lower nodes; lower nodes knot-like. **LF**: ligule 1–2.5 mm, decurrent to sheath, with 1 large tooth on each side; blade 5–8 cm, 0.5–1.5 mm wide, flat to ± folded. **INFL** 5–12 cm, < 3 cm wide, narrow; branches ascending to appressed, loosely fld. **SPIKELET**: glumes 1.4–3.2 mm, acuminate to ± short-awned, awn < 1 mm; lemma 4–5 mm, hairs at base short-appressed, awn 5–20 mm; palea glabrous to ± scabrous; anther 1.8–2 mm, purple. RARE in CA. Rocky slopes, ledges, canyons; 1755 m. e DMtns (New York Mtns); to s Colorado, w TX, Mex. In cult.

M. porteri Beal (p. 1279) Per. **ST** 2.5–8 dm, wiry; lower nodes knot-like. **LF**: ligule 1–2.5 mm, truncate, decurrent to sheath, toothed; blade 2–8 cm, 1–2 mm wide, flat to ± folded. **INFL** 4–15 cm, 6–15 cm wide, ovoid, open; branches thread-like, spreading. **SPIKELET**: glumes 2–3 mm, acuminate; lemma 3–4.2 mm, hairy below middle, awn 2–10 mm; anther 1.5–2.3 mm, purple to yellow. 2*n*=20,23,24,40. Among boulders or shrubs, rocky slopes, cliffs; 610–1680 m. SnBr, SnJt, SNE, DMoj; to Colorado, w TX, Mex.

M. richardsonis (Trin.) Rydb. (p. 1279) MAT MUHLY Per; rhizome scaly, matted. **ST** decumbent to erect, 0.5–4 dm; lower nodes often swollen or knot-like. **LF**: ligule 1–2.5 mm, acute to truncate, decurrent to sheath; blade 1–5 cm, 1–2 mm wide, flat to ± rolled. **INFL** 1–12 cm, 1–4 mm wide, cylindric, narrow; axis gen obscured by appressed branches. **SPIKELET**: glumes 0.8–1.8 mm, acute to ± short-awned; lemma 2–3 mm, glabrous, ± scabrous at tip, ± short-awned, awn < 0.5 mm; anther 1.2–1.5, yellow to purple. 2*n*=40. Open sites, ± moist meadows, talus slopes, along streams; 1220–3670 m. KR, CaRH, SNH, SCoRO, TR, SnJt, GB, DMtns; to Can, ne US, Mex. ❀IRR,SUN:**1–3,6,11,14–16,18,21**;GRCVR.

M. rigens (Benth.) A. Hitchc. (p. 1279) DEERGRASS Per. **ST** densely clumped, 5–15 dm. **LF**: ligule 0.5–2 mm, truncate, ± ciliate; blade 10–50 cm, 1.5–6 mm wide, flat. **INFL** 15–60 cm, 5–12 mm wide, cylindric, narrow; branches appressed, densely fld. **SPIKELET**: glumes 1.8–3 mm, acute or obtuse, ± scabrous; lemma 2.5–3.5 mm, base sparsely short-hairy, ± abruptly pointed; anther 1.3–1.7, yellow to purple. 2*n*=40. Sandy to gravelly places, canyons, stream bottoms; < 2150 m. CaRH, SN, GV, SCoRO, SCo, TR, SnJt, SNE, DMoj; to TX, Mex. ❀**6,14–17,22–24**;IRR:1–3,**7–11,18–21**.

M. schreberi S. Gmelin NIMBLEWELL Per. **ST** decumbent, 1–4 dm, rooting at lower nodes. **LF**: ligule < 0.5 mm, truncate, minutely toothed, ciliate; blade 3–9 mm, 1.5–5 mm wide, flat, sparsely long-hairy near sheath. **INFL** 3–15 cm, 2–10 mm wide, narrow; branches ascending to appressed. **SPIKELET**: glumes < 0.5 mm, obtuse, veins 0, lower ± vestigial, obscure or 0; lemma 2–2.5 mm, sparsely short-stiff-hairy on lower third, awn 1.5–5 mm; anther < 0.5 mm, purple. 2*n*=40,42. Disturbed sites, cult fields; < 1200 m. n&c SNF, ScV; native to e US, TX, e Mex, S.Am. NOXIOUS WEED.

M. utilis (Torrey) A. Hitchc. APAREJOGRASS Per; rhizome scaly. **ST** decumbent, often creeping, 0.5–3 dm. **LF**: ligule 0.3–0.8, truncate, decurrent to sheath; blade 1–3.5 cm, 0.6–1.2 mm wide, flat to ± rolled. **INFL** 1–5 cm, 1–3 mm wide, narrow; branches short, appressed; axis gen visible between branches. **SPIKELET**: glumes 0.5–1.5 mm, acute; lemma 1.5–2.5 mm, acute, glabrous; anther 1–1.2 mm, yellow to purple. 2*n*=20. Wet sites along streams, ponds; 250–1000 m. SCoRO, SCo, WTR; to TX, C.Am.

MUNROA

Dieter H. Wilken

Ann, multi-branched. **STS** spreading to erect. **LVS** tightly clustered; ligule short-hairy; blade stiff, flat or folded, margin white. **INFL** ± panicle-like; spikelets subsessile, clustered, ± concealed by terminal lf clusters. **SPIKELET**: glumes gen narrow, acute, 1-veined; axis gen breaking above glumes and between florets; florets bisexual; lemmas gen lanceolate, thick, pointed or short-awned, 3-veined; palea ± = lemma; lower spikelets with glumes subequal, slightly < lower floret, florets 3–4; upper spikelets with glumes strongly unequal, << florets, lower glume sometimes 0, florets 2–3. 3 spp.: temp N.Am, s S.Am. (W. Munro, English agrostologist, 1818–1880) [Parodi 1934 Revista Mus La Plata Secc. Bot 34:171–193]

M. squarrosa (Nutt.) Torrey (p. 1279) FALSE BUFFALOGRASS Pl mat-like, gen < 20 cm wide. **ST** gen 3–10 cm; internodes ± scabrous, sometimes puberulent; lower internodes evident, upper crowded, concealed by lvs. **LF**: sheath stiff-hairy near collar, margins ± ciliate; ligule hairs < 1 mm; blades 1–2 cm, 1–3 mm wide. **SPIKELETS** 6–8 mm; lower gen sessile; upper short-stalked; glumes 2–4 mm; lemmas 3–5 mm, outer veins short-stiff-hairy below middle, awn < 2 mm; anthers 1–1.5 mm. 2*n*=16. RARE in CA. Open, gravelly or rocky places; 1500–1800 m. DMtns (Clark Mtn.); to Great Plains, TX, n Mex.

NASSELLA

Mary E. Barkworth

Per. **ST** gen unbranched. **INFL** panicle-like; branches gen widely spreading. **SPIKELET**: glumes tapered from near base to tip, tip narrowly acute, gen reddish; axis breaking above glumes; floret 1, gen round in X-section; callus 0.5–4 mm, gen sharp; lemma papillate or minutely tubercled at tip, sometimes hairy, margins strongly overlapping at maturity, tip neck-like,

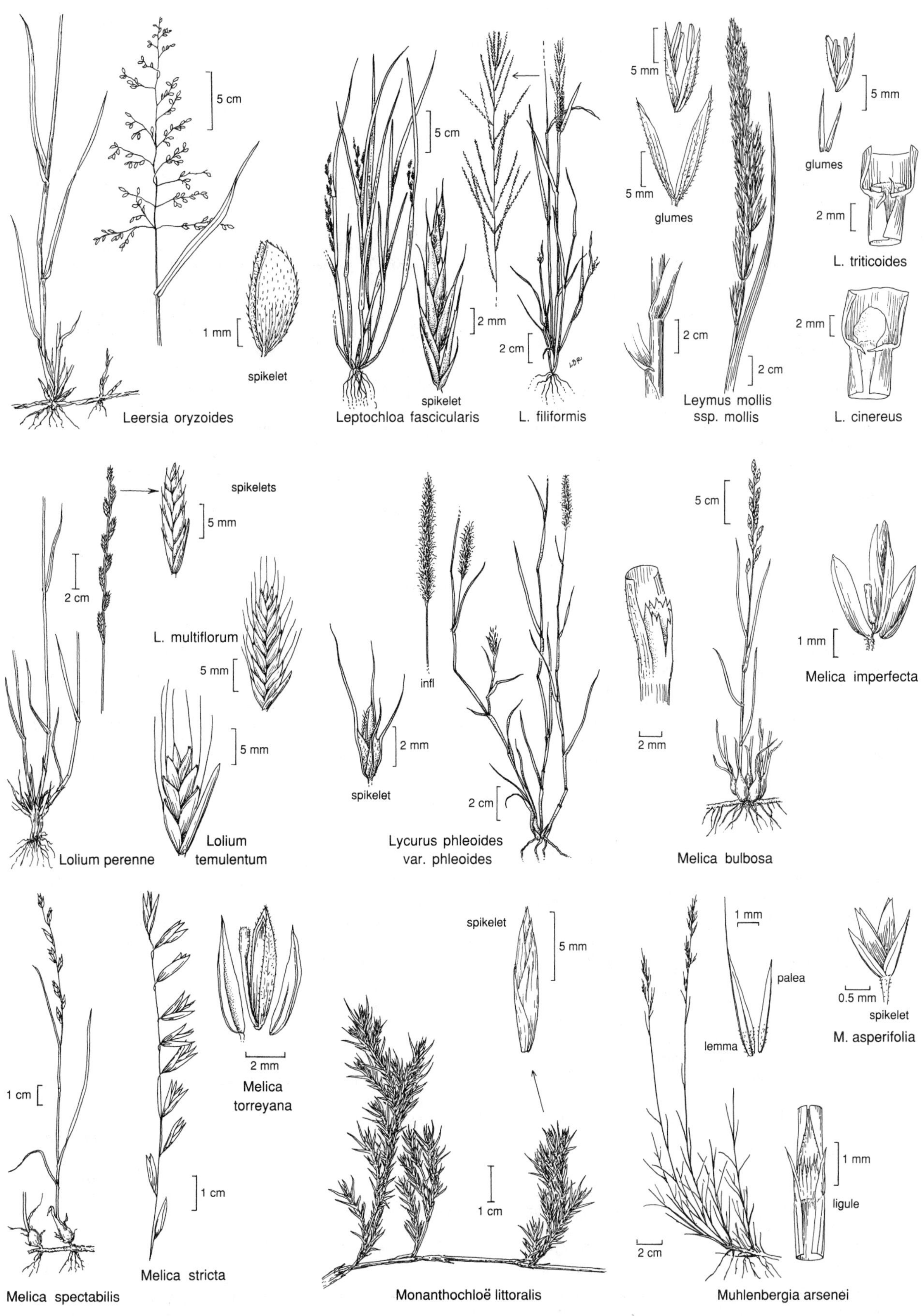
5 cm
1 mm
spikelet
Leersia oryzoides
5 cm
2 mm
2 cm
spikelet
Leptochloa fascicularis
L. filiformis
5 mm
5 mm
glumes
2 cm
2 cm
Leymus mollis
ssp. mollis
5 mm
glumes
2 mm
L. triticoides
2 mm
L. cinereus
spikelets
5 mm
2 cm
L. multiflorum
5 mm
5 mm
Lolium perenne
Lolium
temulentum
infl
2 mm
spikelet
2 cm
Lycurus phleoides
var. phleoides
5 cm
2 mm
Melica bulbosa
1 mm
Melica imperfecta
1 cm
Melica spectabilis
1 cm
Melica stricta
2 mm
Melica
torreyana
spikelet
5 mm
1 cm
Monanthochloë littoralis
1 mm
palea
lemma
0.5 mm
spikelet
M. asperifolia
1 mm
ligule
2 cm
Muhlenbergia arsenei

ciliate, gen pale colored; palea < 1/3 lemma length, glabrous, veins 0; anthers ciliate, often of 2 lengths. ± 80 spp.: Am. (Latin: nassa, a basket with a narrow neck) [Barkworth 1990 Taxon 39:597–614] Segregated from *Stipa*; see also *Achnatherum, Hesperostipa*.

1. Distal segment of awn wavy
  2. Glumes 12–21 mm; awn 45–110 mm .... ***N. cernua***
  2′ Glumes 4.5–12 mm; awn 20–46 mm .... ***N. lepida***
1′ Distal segment of awn straight
  3. Lemma glabrous between veins .... ***N. formicarum***
  3′ Lemma hairy between veins .... ***N. pulchra***

N. cernua (Stebb. & Löve) Barkworth NODDING NEEDLEGRASS **ST** 3–10 dm. **LF**: blade 0.4–1.5 mm wide. **INFL** 15–80 cm. **SPIKELET**: glumes unequal, lower 12–21 mm, upper 3–4 mm shorter; floret 4–9 mm; lemma evenly hairy on lower 1/4, then only over veins, tip neck-like, inconspicuous, awn 45–110 mm, distal segment wavy. 2*n*=70. Grassland, chaparral, juniper woodland; < 1400 m. NCoRI, e SnFrB, SCoR, TR, PR; Baja CA. [*Stipa c.* Stebb. & Löve; *S. lepida* var. *andersonii* (Vasey) A. Hitchc.] ❀DRY,SUN:**7**,8,9,11,**14–24**;also STBL.

N. formicarum (Del.) Barkworth **ST** 4–8 dm. **LF**: blade 1–2.5 mm wide. **INFL** 20–25 cm. **SPIKELET**: glumes 10–11 mm, ± equal; floret 5.5–7 mm; lemma hairy on lower third of midvein and marginal veins, neck 1–1.3 mm, conspicuous, white, awn 30–40 mm, distal segment straight. Open, disturbed areas, roadsides, fields; < 200 m. s NCoRO (Sonoma Co.), s ScV, n SnFrB; native to S.Am. [*Stipa f.* Del.]

N. lepida (A. Hitchc.) Barkworth (p. 1279) FOOTHILL NEEDLEGRASS **ST** 3–10 dm. **LF**: blade 1–3.5 mm wide. **INFL** 9–55 cm. **SPIKELET**: glumes unequal, lower 5.5–12 mm, upper 1–4 mm shorter; floret 4–7 mm; lemma hairy all over, becoming glabrous between veins, awn 20–46 mm, distal segment wavy. 2*n*=34. Dry slopes, chaparral, oak grassland; < 1700 m. NW, CW, SCo, ChI, WTR, w PR; Baja CA. [*Stipa l.* A. Hitchc. var. *l.*] ❀SUN or part SHD:**7**,8,**9**,11,**14–24**;also STBL.

N. pulchra (A. Hitchc.) Barkworth (p. 1279) PURPLE NEEDLEGRASS **ST** 3–10 dm. **LF**: blade 0.8–3.5 mm wide. **INFL** 18–60 cm. **SPIKELET**: glumes subequal, 12–19 mm; floret 7.5–11.5 mm; lemma evenly hairy, becoming glabrous on veins, neck conspicuous, not strongly narrowed at base, awn 38–100 mm, strongly bent twice, distal segment straight. 2*n*=64. Oak woodland, chaparral, grassland; < 1300 m. NW, n&c SNF, s ScV, CW, SCo, ChI, WTR, w PR; Baja CA. [*Stipa p.* A. Hitchc.] ❀DRN,SUN:5,**7–9**, 11,**14–24**;also STBL.

NEOSTAPFIA

John R. Reeder

1 sp. (O. Stapf, British botanist, 1857–1933)

N. colusana (Burtt Davy) Burtt Davy (p. 1279) COLUSA GRASS Ann, cespitose, brown-sticky-glandular with age. **ST** ascending, 1–3 dm. **LF**: collar 0; sheath and blade continuous, base loosely enclosing st; blade 5–12 mm wide. **INFL** 2–8 cm, 8–12 mm wide, spike-like, cylindric, dense; spikelets spirally arranged along axis; axis gen prolonged above spikelets, naked or with small scales. **SPIKELET**: glumes 0; lemma flat, ± fan-shaped, strongly 7–11-veined; florets gen 5, breaking between florets; lower lemmas ± 5 mm, ± translucent between veins, finely ciliate; palea ± = lemma; anthers ± 2.5 mm. **FR** gen 2.5 mm. 2*n*=40. **ENDANGERED** CA. Vernal pools; < 200 m. GV (Colusa, Merced, Solano, Stanislaus cos.). [Reeder 1982 Amer J Bot 69:1082–1095] Infls resemble miniature ears of corn.

ORCUTTIA

John R. Reeder

Ann, ± hairy. **STS** erect, becoming prostrate. **LVS** basal and cauline; collar 0; sheath and blade continuous; basal lvs long, floating when young, becoming dry and non-functional as aerial parts grow; cauline lf blade flat or inrolled when dry. **INFL** spike-like, exserted; spikelets 2-ranked. **SPIKELET** laterally compressed; glumes irregularly 2–5-toothed; florets 4–40, breaking above glumes and between florets with age; lemma strongly 5-toothed, each tooth 1/3–1/2+ lemma length, prominent central vein flanked by 2 weaker veins; palea < lemma; anthers white or pinkish, exserted, filaments slender, ribbon-like; stigma 1/3–1/2 style length, ± sparsely short-hairy. 5 spp.: CA, n Baja CA. (C.R. Orcutt, CA botanist, 1864–1929) [Reeder 1982 Amer J Bot 69:1082–1095]

1. Lemma teeth unequal, central tooth gen longest
  2. Lemma 6–7 mm, teeth awned, awns > 1 mm; fr ± 2.5 mm .... ***O. viscida***
  2′ Lemma < 5 mm, teeth sharp-pointed or awned, awns < 0.5 mm; fr < 2 mm
    3. Pl sparsely hairy; st gen prostrate; infl with spikelets well separated on axis below, crowded toward tip; SW .... ***O. californica***
    3′ Pl conspicuously hairy, grayish; st erect or ± decumbent; spikelets crowded, infl ± head-like; SnJV .... ***O. inaequalis***
1′ Lemma teeth ± equal
  4. St gen prostrate; fr 1.5–2 mm .... ***O. californica***
  4′ St erect, ascending, or decumbent; fr 2.5–3 mm
    5. St branching only at lower nodes; infl with upper spikelets densely crowded, lower well separated; pl gen densely hairy .... ***O. pilosa***
    5′ St ± slender, often branching from upper nodes; spikelets evenly spaced, not densely crowded; pl sparsely hairy .... ***O. tenuis***

O. californica Vasey (p. 1279) CALIFORNIA ORCUTT GRASS **ST** gen prostrate, sometimes forming mats, 5–15(20) cm. **LF**: sheath and blade separated by faint line when dry; blade 1–2 cm, 2–3 mm wide. **INFL** 3–6 cm, exserted with age; spikelets crowded at tip.

SPIKELET: glumes 2–3(4) mm, subequal, irregularly toothed; florets 5–15(25); lemma ± 5 mm, teeth equal or central tooth > lateral, short awn-tipped; palea ± = lemma; anthers ± 2 mm. **FR** 1.5–2 mm, narrowly elliptic. 2*n*=32. **ENDANGERED** CA. Vernal pools; < 625 m. SW (Los Angeles, Riverside, San Diego cos.); n Baja CA.

O. inaequalis Hoover (p. 1279) SAN JOAQUIN VALLEY ORCUTT GRASS Pl cespitose, hairy, grayish. **ST** gen erect, sometimes spreading, forming mats, 5–15(25) cm. **LF**: blade 2–4 mm wide. **INFL** 2–3.5(5) cm, ± head-like; spikelets densely crowded. **SPIKELET**: glumes ± 3 mm, subequal, irregularly toothed; florets 4–20(30); lemma 4–5 mm, teeth ± 1/2 lemma body length, awn-tipped, central tooth >> lateral; palea ± = lemma; anthers ± 2 mm. **FR** ± 1.5 mm, widely elliptic. 2*n*=24. **ENDANGERED** CA. Vernal pools; < 200 m. SnJV (Fresno, Madera, Merced, Stanislaus, Tulare cos.). [*O. californica* var. *i.* (Hoover) Hoover]

O. pilosa Hoover (p. 1279) HAIRY ORCUTT GRASS Pl cespitose, gen densely hairy. **ST** decumbent to erect, 5–20(35) cm. **LF**: sheath and blade gen separated by line when dry; blade 3–5(8) mm wide. **INFL** < 10 cm, exserted with age; spikelets crowded at tip, ± overlapping below. **SPIKELET**: glumes ± 3 mm, irregularly 3-toothed; florets 10–40; lemma 4–5 mm, acute or with awn-tipped teeth, teeth ± equal, 1/3–1/2 lemma body length; anthers 2.5–3 mm. **FR** ± 2 mm, elliptic. 2*n*=30. **ENDANGERED** CA. Vernal pools; < 200 m. GV (Madera, Merced, Stanislaus, Tehama cos.).

O. tenuis A. Hitchc. (p. 1279) SLENDER ORCUTT GRASS Pl gen with 1 main st, sometimes weakly cespitose, sparsely hairy. **ST** gen erect, sometimes becoming decumbent, 5–15(25) cm, often branching in upper 2–10 cm. **LF**: blade 1.5–2 mm wide. **INFL** 5–10 cm, exserted with age; spikelets ± evenly spaced on axis. **SPIKELET**: glumes 3–6 mm, subequal, 3–5-toothed, teeth < 1 mm; florets 5–20; lemma 4.5–6 mm, acute or with awn-tipped teeth, teeth ± equal, ± 1/2 lemma body length, spreading or slightly recurved; palea slightly < lemma; anthers ± 3 mm. **FR** ± 3 mm, narrowly oblong. 2*n*=26. **ENDANGERED** CA. Vernal pools; 200–1100 m. NCoRI (Lake Co), CaRF (Shasta, Tehama cos.).

O. viscida (Hoover) Reeder (p. 1279) SACRAMENTO ORCUTT GRASS Pl hairy, strongly aromatic, sticky. **ST** erect to spreading with age, 3–10(15) cm. **LF**: blade 2–4 mm wide. **INFL** 3–5 cm, exserted with age; spikelets ± crowded. **SPIKELET**: glumes 5–6 mm, subequal, unequally 3-toothed, teeth = glume body, awn-tipped; florets 6–20(30); lemma 6–7 mm, teeth = lemma body, awn-tipped, central tooth >> lateral; palea 3/4 to > lemma; anthers ± 2 mm. **FR** ± 2.5 mm, widely elliptic. 2*n*=28. **ENDANGERED** CA. Vernal pools; < 100 m. ScV (Sacramento Co.). [*O. californica* var. *v.* Hoover]

ORYZA RICE

Dieter H. Wilken

Ann, per. **STS** gen solitary, erect. **LVS** cauline; ligule membranous; blade flat to folded. **INFL** panicle-like. **SPIKELET** laterally compressed; glumes << floret, scale-like; floret 1, bisexual, breaking above glumes or not; lemma, palea firmly membranous; lemma 5-veined, small-pointed or rigidly awned at tip; palea ± = lemma; stamens 6. ± 20 spp.: trop worldwide. (Greek: rice) Cult for food. Spikelet also has been interpreted as 1 floret subtended by 2 scale-like, sterile lemmas and 2 minute, flap-like glumes.

O. sativa L. Ann. **ST** 4–15(20) dm. **LF**: sheath gen glabrous; ligule 4–10 mm, acute; blade 15–35 cm, 3–11 mm wide, glabrous to minutely scabrous. **INFL** 12–25 cm; branches ascending. **SPIKELET** elliptic; glumes equal, ± 3 mm, linear-lanceolate; lemma, palea 5–7 mm, distally sharp-bristly; lemma small-pointed to awned, awn (3)7–20 mm. Wet sites, ditches; < 70 m. GV; to s US, Medit; native to se Asia.

PANICUM MILLET, PANICGRASS

Robert Webster

Ann, per. **STS** gen erect; internode solid to hollow inside. **LVS** basal and cauline; sheath glabrous or hairy; ligule short-hairy or membranous, ciliate, hairs gen > membrane. **INFL** panicle-like, gen open; 1° branches spreading to ascending; 2° branches spreading to appressed; spikelets many, 1–2 per node, gen stalked, on one side of axis or not, stalk tip expanded, one side concave. **SPIKELET** falling as 1 unit, ± compressed, gen green to purplish; glumes gen unequal, lower gen < upper, free, clasping, upper glume ± = spikelet, membranous, ± thin; florets 2, lower sterile or staminate, lemma texture like glumes, upper floret fertile, lemma leathery to hard, firm, gen shiny, smooth to rough, margin inrolled, tip blunt, palea enclosed by lemma margin. ± 450 spp.: trop to warm temp, worldwide. (Latin: ancient name for millet) [Spellenberg 1975 Brittonia 27:87–95] Some spp. cult for food.

1. Lvs of 2 forms, basal < and gen wider than cauline, rosette well developed, gen persistent
  2. Spikelet 3–3.5 mm, 1–2 mm wide; upper glume 9-veined, lower glume acuminate .... ***P. oligosanthes*** var. ***scribnerianum***
  2′ Spikelet 1–2.5 mm, slightly < 1 mm wide; upper glume 7-veined, lower glume acute .... ***P. acuminatum***
    3. Lf blade and sheath hairy .... var. ***acuminatum***
    3′ Lf blade and sheath glabrous .... var. ***lindheimeri***
1′ Lvs gradually reduced upward, basal rosette not well developed
  4. Upper floret lemma with crescent-shaped scar at base .... ***P. hillmanii***
  4′ Upper floret lemma not scarred at base
    5. Per
      6. Spikelet 2.5–3 mm, glabrous; upper floret lemma with flat margin; ligule membranous, ciliate .... ***P. antidotale***
      6′ Spikelet 5–7.5 mm, short-hairy; upper floret lemma with inrolled margin; ligule short-hairy .... ***P. urvilleanum***
    5′ Ann
      7. Spikelet 4.5–5.5 mm; lf base minutely lobed .... ***P. miliaceum***
      7′ Spikelet 2–3.5 mm; lf base not lobed
        8. Spikelet axis visible, ± elongated between glumes and between florets; lower glume 1.5–2.5 mm .... ***P. hirticaule***
        8′ Spikelet axis not elongated, not clearly visible; lower glume 1–1.5 mm
          9. Lower glume ± 1/2 spikelet length; ultimate infl branches spreading; st nodes short-hairy .... ***P. capillare***
          9′ Lower glume < 1/3 spikelet length; ultimate infl branches appressed; st nodes glabrous .. ***P. dichotomiflorum***

P. acuminatum Sw. (p. 1279) Per. **ST** 2–8 dm. **LF**: sheath 2–7 cm, gen short-hairy; ligule 2–5 mm, hairy; blade 3–15 cm long, 2–13 mm wide, upper surface glabrous or short-soft-hairy. **INFL** 4–9 cm; 1° branches 2–4 cm, axis glabrous or short-hairy; spikelet 1 per node, stalk 0.5–3 mm. **SPIKELET** 1–2.5 mm, slightly < 1 mm wide, elliptic to slightly obovate; lower glume ± 0.5–1 mm, 1-veined, tip acute; lower floret sterile, lemma 7-veined, tip rounded, palea vestigial; upper floret slightly < lower floret. 2*n*=18. Moist places, marshes, streambanks; < 2600 m. NW, CaR, SN, GV, CCo, SnFrB, SCoRO, SW; to Can, e US, C. & n S.Am.

var. ***acuminatum*** **ST** 1–6 dm. **LF**: sheath hairy; ligule 2–4 mm; blade 5–10 cm, 5–12 mm wide, upper surface short-soft-hairy. **INFL** 5–8 cm. **SPIKELET** 1.5–2.5 mm, obovate to elliptic; lower glume 0.5–1 mm. Habitat and range of sp. [*P. huachucae* Ashe, *P. occidentale* Scribner, *P. pacificum* A. Hitchc. & Chase, *P. shastense* Scribner & Merr., *P. thermale* Bolander, Geyser's panicum, **ENDANGERED** CA]

var. ***lindheimeri*** (Nash) Fern. **ST** 3–9 dm. **LF**: sheath glabrous; ligule 2–5 mm; blade 3–15 cm, 3–11 mm wide, upper surface glabrous. **INFL** 4–12 cm. **SPIKELET** 1–2 mm, gen elliptic; lower glume gen < 0.5 mm. Moist places; < 100 m. GV; to e US, TX, n Mex. [*P. lindheimeri* Nash] ❀STBL.

P. antidotale Retz. BLUE PANICGRASS Per from rhizomes. **ST** erect, 1–25 dm; nodes 5–8. **LF**: sheath 4–8 cm, glabrous; ligule 1.5–2.5 mm, densely hairy; blade 15–30 cm, 4–12 mm wide, upper surface glabrous. **INFL** 13–28 cm; 1° branches 9–18 cm, glabrous; spikelets 1–2 per node, stalk < 2.5 mm. **SPIKELET** 2.5–3 mm, ± 1 mm wide, ovate to elliptic, brown, green, or purple; lower glume 1.5–2.5, 5-veined; lower floret staminate, lemma 7-veined, tip acute, palea ± = lemma; upper floret ± = lower floret. 2*n*=18. Open, gen disturbed areas, fields; < 100 m. s DMoj, DSon; to TX, n Mex; native to India. NOXIOUS WEED.

P. capillare L. (p. 1279) WITCHGRASS Ann. **ST** 2–9 dm. **LF**: sheath 4–8 cm, short-hairy; ligule 0.5–1.5 mm, tip hairy; blade 6–22 cm, 6–20 mm wide, upper surface short-soft-hairy to roughly hairy. **INFL** 15–40 cm; 1° branches 12–30 cm, axis sparsely short-hairy; spikelet 1 per node; stalk 2–6 mm. **SPIKELET** 2–3.5 mm, ± 1 mm wide, lanceolate to elliptic; lower glume ± 1–1.5 mm, 3-veined; lower floret sterile, lemma 7-veined, tapered to a tail-like tip, palea gen 0; upper floret ± 0.8 × lower floret length. 2*n*=18. Open places, fields; < 1500 m. CA; to Can, e US.

P. dichotomiflorum Michaux Ann. **ST** decumbent to erect, 2–10 dm. **LF**: sheath 4–9 cm, glabrous; ligule 1–2 mm, membranous; blade 8–22 cm, 4–6 mm wide, upper surface glabrous. **INFL** 7–20 cm; 1° branches 4–16 cm, axis glabrous; spikelets 1–2 per node, stalk 0.5–1 mm. **SPIKELET** 2.5–3 mm, ± 1 mm wide, ovate; lower glume ± 1 mm, 1–3-veined; lower floret sterile, lemma 7-veined, acute, palea ± = lemma; upper floret ± 0.7–0.8 × lower floret length. 2*n*=36,54. Waste places, fields; < 500 m. SnJV, SCo; native to e US.

P. hillmanii Chase Ann. **ST** 2–7 dm. **LF**: sheath 2.5–7 cm, roughly hairy; ligule 1–2.5 mm, membranous; blade 8–15 cm, 5–12 mm wide, upper surface short-soft-hairy. **INFL** 10–25 cm; 1° branches 5–13 cm, axis glabrous to short- hairy; spikelet 1 per node, stalk 1–6 mm. **SPIKELET** 2.5–3 mm, ± 1 mm wide, elliptic; lower glume 1–1.5 mm, 3-veined; lower floret sterile, lemma 7-veined, acute, palea gen < lemma; upper floret 0.8–0.9 × lower floret length. 2*n*=18. Waste places, fields; < 200 m. s ScV; native to Great Plains.

P. hirticaule C. Presl Per. **ST** 1–8 dm. **LF**: sheath 2–6 cm, axis glabrous to short-hairy; ligule ± 0.5–1.5 mm; blade 7–15 cm, 3–12 mm wide, upper surface gen sparsely short-hairy. **INFL** 5–12 cm; 1° branches 3–8 cm, glabrous; spikelets 1–2 per node, stalk 0.5–1 mm. **SPIKELET** ± 2.5–3 mm, ± 1 mm wide, lanceolate to ovate, green; axis between glumes and florets visible; lower glume ± 1.5–2.5 mm, gen 5-veined; lower floret sterile, lemma 7-veined, acuminate to acute, palea vestigial or 0; upper floret 0.7–0.8 × lower floret length. Sandy soils, open sites, creosote-bush scrub; < 1400 m. DSon; to TX, n Mex. ❀STBL.

P. miliaceum L. (p. 1279) BROOM CORN MILLET Ann. **ST** 2–10 dm. **LF**: sheath 3.5–8 cm, short-hairy; ligule 1–3 mm, membranous; blade 10–20 cm, 6–25 mm wide, upper surface velvety to roughly hairy. **INFL** 10–40 cm; 1° branches 4–18 cm, axis glabrous; spikelet 1 per node, stalk 2–10 mm. **SPIKELET** 4.5–5.5 mm, ± 2–2.5 mm wide, elliptic, green to brown; lower glume 2.5–4.5 mm, 3–5-veined; lower floret sterile, lemma 11-veined, acuminate to acute, palea vestigial; upper floret 0.8 × lower floret length. 2*n*=36. Disturbed places, fields, roadsides; < 1000 m. GV, SnFrB, SW, DSon; to e US; native to Eurasia. Cult for seed, food.

P. oligosanthes Schultes var. ***scribnerianum*** (Nash) Fern. Per. **ST** 3–6 dm. **LF**: sheaths 2–8 cm, glabrous or short-hairy; ligule 1–4 mm, hairy; blade 3–14 cm, 3–15 mm, upper surface glabrous or short-hairy. **INFL** 5–8 cm; 1° branches 2–4.5 cm, axis glabrous; spikelets 1–2 per node, stalk 2–5 mm. **SPIKELET** ± 2.7–3.5 mm, ± 1–2 mm wide, elliptic, green; lower glume ± 1–1.5 mm, 1-veined, tip acuminate; lower floret sterile, lemma 9–11-veined, acute to rounded, palea ± = lemma to vestigial; upper floret slightly < lower floret. 2*n*=18. Meadows, open sites in forest; < 1400 m. NW, n CaR; to Can, e US. Typical var. in e&s US.

P. urvilleanum Kunth Per from stolons. **ST** 4–10 dm. **LF**: sheath 12–35 cm, hairy; ligule 1–1.5 mm; blade 20–45 cm, 4–7 mm wide, upper surface short-hairy. **INFL** 20–35 cm; 1° branches 8–13 cm, glabrous; spikelet 1 per node, stalk 1.5–10 mm. **SPIKELET** 5–7.5 mm, 2–2.5 mm wide, elliptic, green; lower glume 4.5–6.5 mm, 5–7-veined; lower floret staminate, lemma 9-veined, acute, palea ± = lemma; upper floret 0.8–0.9 × lower floret length, lemma margin inrolled. 2*n*=36. Sandy soils, dunes; < 900 m. e SCo, D; AZ, N Mex, S.Am. ❀STBL.

PARAPHOLIS

Thomas Worley

Ann. **ST** decumbent to erect, 5–45 cm, branched, glabrous; nodes gen purple. **LF**: sheath 1–6 cm; ligule ± 1.5 mm, membranous; blade flat to ± inrolled, ribbed, upper surface scabrous. **INFL** spike-like, 3–18 cm, rigid, cylindric, straight to strongly curved, breaking at nodes; spikelets alternate, 2-ranked, appressed, embedded in axis, falling with axis segment. **SPIKELET**: glumes 2, edge-to-edge or slightly overlapping, 4–7 mm, lanceolate, keel to 1 side near base or obscure, margin translucent, ± enclosing floret, 5-veined; florets 1–2, bisexual; lemma 3.5–5.5 mm, translucent, 3-veined; palea ± = lemma, translucent. **FR** ± 3.5 mm. 5 spp.: native to Eur. (Greek: scales beside, from glume orientation) See *Hainardia*.

1. Infl gen curved, gen twisted; upper lf sheath margins expanded, enclosing lowest spikelets; anther 0.5–1 mm .... ***P. incurva***

1′ Infl gen straight; lf sheath margins all alike, all spikelets exserted from sheath; anther 2–4 mm .... ***P. strigosa***

P. incurva (L.) C.E. Hubb. (p. 1285) SICKLE GRASS **ST** decumbent to erect, 3–35 cm, branched throughout or not. **LF**: sheath 1–4 cm, margin of upper expanded, enclosing lower 1–4 spikelets; blade 1–10 cm, 1–3 mm wide, gen inrolled. **INFL** 2–15 cm, gen curved, gen twisted; spikelets 3–20. **SPIKELET**: glume keel ± 0.5 mm; floret sometimes cleistogamous; anther 0.5–1 mm. 2*n*=32,36, 38,42. Disturbed, well drained soils of salt marshes, gen above highest tide level; < 100 m. NCo, CCo, SCo, ChI; to B.C., Mex; native to Eur.

P. strigosa (Dumort.) C.E. Hubb. **ST** gen ascending to erect, 12–45 cm, branching at lower nodes. **LF**: sheath 2–6 cm, margins all alike, not enclosing lower spikelets; blade 1–10 cm, 1–3 mm wide, gen flat. **INFL** 7–18 cm, gen straight; spikelets 10–25. **SPIKELET**: glume keel obscure; floret never cleistogamous; anther 2–4 mm. 2*n*=14. Moist, salt marsh soils, gen below highest tide level; < 5 m. NCo (esp n Humboldt Bay); native to Eur.

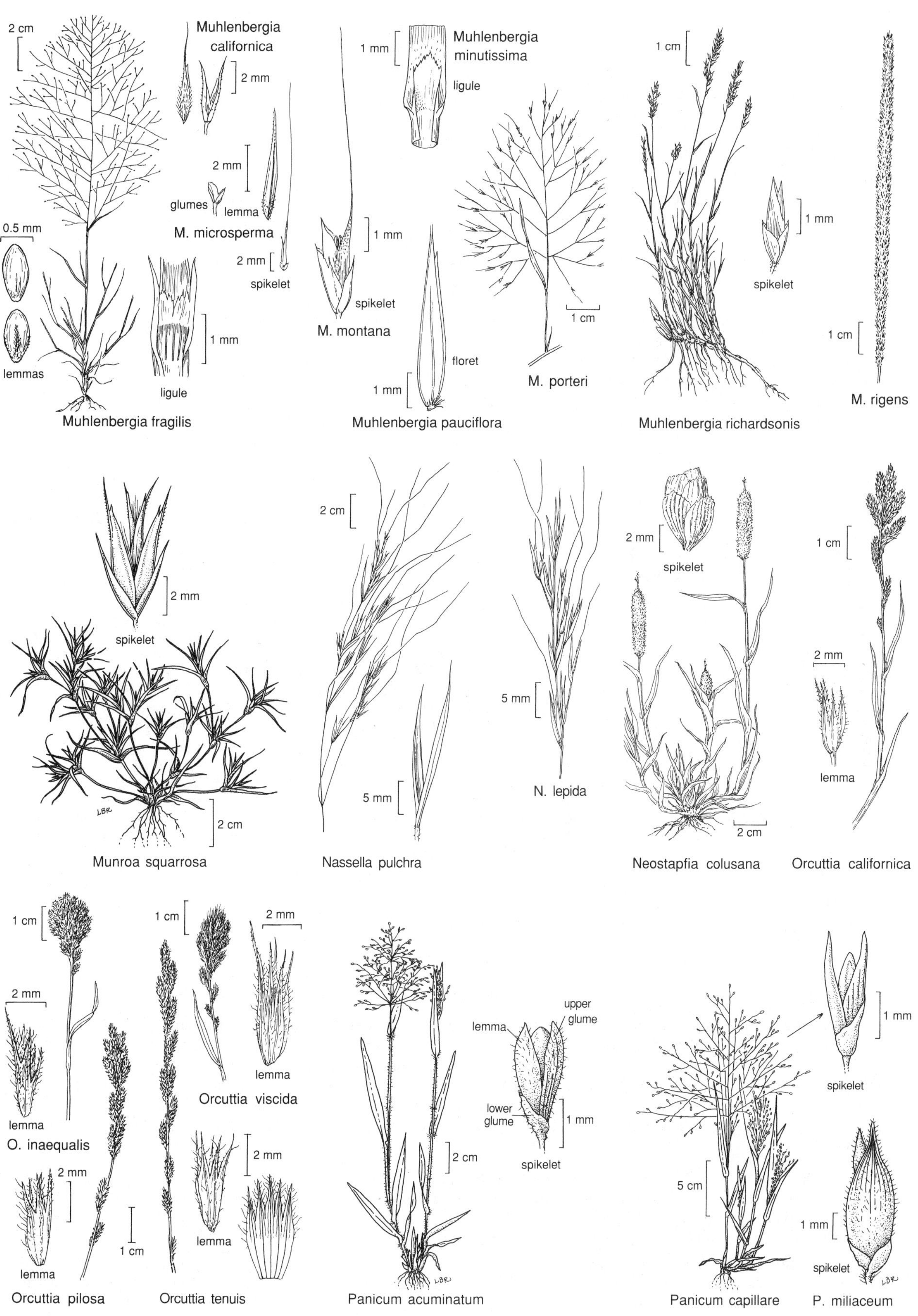
2 cm
Muhlenbergia californica
2 mm
2 mm
glumes
lemma
M. microsperma
2 mm
spikelet
0.5 mm
lemmas
1 mm
ligule
Muhlenbergia fragilis
1 mm
Muhlenbergia minutissima
ligule
1 mm
spikelet
M. montana
floret
1 mm
Muhlenbergia pauciflora
1 cm
M. porteri
1 cm
1 mm
spikelet
Muhlenbergia richardsonis
1 cm
M. rigens
2 mm
spikelet
2 cm
Munroa squarrosa
2 cm
5 mm
Nassella pulchra
5 mm
N. lepida
2 mm
spikelet
2 cm
Neostapfia colusana
1 cm
2 mm
lemma
Orcuttia californica
1 cm
2 mm
lemma
O. inaequalis
2 mm
lemma
1 cm
Orcuttia pilosa
1 cm
2 mm
lemma
Orcuttia viscida
2 mm
lemma
Orcuttia tenuis
upper glume
lemma
lower glume
1 mm
spikelet
2 cm
Panicum acuminatum
1 mm
spikelet
5 cm
1 mm
spikelet
Panicum capillare
P. miliaceum

PASCOPYRUM

Mary Barkworth

1 sp. (Latin & Greek: pasture wheat) [Barkworth & Dewey 1985 Amer J Bot 72:767–776]

P. smithii (Rydb.) A. Löve (p. 1285) Per, from rhizomes. **ST** erect, (2)4–9 cm, glaucous. **LVS** basal and cauline, glaucous; ligule membranous; sheath appendages 1–2 mm, clasping; blade flat, rolled when dry. **INFL** 4–15 cm, spike-like; spikelets 2-ranked, overlapping, 1(2) per node. **SPIKELET**: glumes subequal, < or = lower floret, glabrous to rough, tapered from middle to acute tip, 3-veined in lower half, midvein curving slightly to side; axis breaking above glumes and between florets; florets 4–11; lemma 8–14 mm, glabrous to hairy, awn < 5(15) mm; anthers 2–4.5 mm. $2n$=56. Uncommon. Dry, alkaline soils, flats; 1500–2000 m. GB; to s Can, Great Plains. [*Agropyron s.* Rydb.]Polyploid derived from hybrid between *Elymus* and *Leymus*.

PASPALUM

Robert Webster

Per in CA, gen from rhizomes or stolons. **STS** decumbent to erect; internode solid to hollow inside. **LVS** basal and cauline; sheath glabrous or hairy; ligule gen membranous. **INFL** panicle-like; axis gen glabrous; 1° branches raceme- to spike-like, spreading to appressed; spikelets many, 1–2 per node, gen short-stalked, on one side of axis. **SPIKELET** falling as 1 unit, compressed, gen green; glumes 1–2, lower glume minute or 0, upper ± = spikelet; florets 2, lower floret gen sterile, palea vestigial or 0, upper floret fertile, lemma firm, thick, sometimes hard, back facing infl axis, smooth or striate, margin inrolled, tip blunt. ± 300 spp.: trop, warm temp worldwide. (Greek: ancient name) [Gould 1975 The Grasses of Texas pp. 500–527, Texas A&M Press]

1. Margins of upper glume and lower floret lemma glabrous
  2. Upper glume puberulent on back; lvs clearly 2-ranked ..... ***P. distichum***
  2′ Upper glume glabrous on back; lvs not clearly 2-ranked ..... ***P. notatum***
1′ Margins of upper glume and lower floret lemma ciliate
  3. Main infl axis with 3–6 branches; spikelet ± 3–4 mm ..... ***P. dilatatum***
  3′ Main infl axis with 12–20 branches; spikelet ± 1.5–3 mm ..... ***P. urvillei***

P. dilatatum Poiret (p. 1285) DALLIS GRASS Per from short rhizomes. **ST** decumbent to erect, 2.5–14 dm; nodes 2–6. **LF**: sheath 6–30 cm, glabrous to hairy; ligule 2–8 mm; blade 9–35 cm, 4–10 mm wide, upper surface glabrous, sometimes sparsely hairy at base. **INFL**: main axis 3–20 cm; 1° branches 3–6, 4–12.5 cm; spikelets many, 2 per node, stalk 1–1.5 mm. **SPIKELET** 3–4 mm, 2–2.5 mm wide, elliptic, green to purple; lower glume 0; lower floret lemma 5–9-veined, tip acute to rounded; upper floret 0.7–0.9 × lower floret length. $2n$=40,50. Moist places, ditches, roadsides; < 400 m. CA-FP, DMoj; to WA, e US, Eur; native to S.Am.

P. distichum L. (p. 1285) Per from stolons and rhizomes. **ST** decumbent to erect, 0.8–6 dm; nodes 5–15. **LF**: sheath 3–20 cm, glabrous; ligule < 1.5 mm; blade 2–22 cm, 2–7 mm wide, upper surface glabrous. **INFL**: main axis < 1.5 cm; 1° branches 2–3, 1.5–5.5 cm; spikelets many, 1–2 per node, stalk < 0.5 mm. **SPIKELET** 2.5–3.5 mm, ± 1.5 mm wide, obovate to elliptic; lower glume < 2.5 mm, 1-veined; lower floret lemma 3–5-veined, tip acute; upper floret slightly < lower floret. $2n$=40,60. Moist places, marshes, ditches; < 1650 m. CA-FP (exc mtns), GB, n DMoj; to WA, e US, S.Am. ❀STBL.

P. notatum J. Fleugge Per from short rhizomes. **ST** erect, 2–10 dm; nodes 2–5. **LF**: sheath 4–20 cm, glabrous; ligule < 1.5 mm; blade 6–25 cm, 4–10 mm wide, upper surface glabrous. **INFL**: main axis < 2 cm; 1° branches 2–3, 2.5–13 cm; spikelets many, 1 per node, stalk < 1 mm. **SPIKELET** ± 3–4 mm, 1–2.5 mm wide, ovate to elliptic; lower glume 0; lower floret lemma 5-veined, tip rounded; upper floret slightly < lower floret. Uncommon. Moist places, fields; < 200 m. SnJV, expected elsewhere; to TX; native to Mex, C. & S. Am.

P. urvillei Steudel Per from short rhizomes. **ST** erect, 8–20 dm; nodes 3–6. **LF**: sheath 10–30 cm, glabrous to hairy; ligule 2–9 mm; blade 25–60 cm, 4–25 mm wide, upper surface glabrous. **INFL**: main axis 12–30 cm; 1° branches 12–20, 7–14 cm; spikelets many, 2 per node, stalk ± 1–1.5 mm. **SPIKELET** ± 1.5–3 mm, ± 1–1.5 mm wide, elliptic, green; lower glume 0; lower floret lemma 3–5-veined, tip acute; upper floret slightly < lower floret. $2n$=40. Moist places, fields; < 100 m. ScV, SCo, expected elsewhere; native to S.Am.

PENNISETUM

Robert Webster

Ann, per. **STS** prostrate to erect and tufted; internode spongy inside, sometimes hollow. **LVS** basal and cauline; sheath gen glabrous; ligule short-hairy or membranous, ciliate. **INFL** gen panicle-like, dense, ± cylindric (raceme-like, spikelets few in *P. clandestinum*), spikelets many, short-stalked to sessile, clustered, gen 1–4 per cluster, subtended by 5–50 flexible bristles; spikelet cluster and bristles gen falling as 1 unit. **SPIKELET** compressed; glumes 1–2, lower glume < upper or 0, upper ± = spikelet; florets 2, lower floret gen sterile, lemma like glumes, upper floret fertile, lemma firm, ± thick or hard, smooth or scabrous, gen dull, margin flat to inrolled, tip blunt. ± 80 spp.: warm temp, trop Eurasia, Afr. (Latin: feather bristle) Some spp. cult for orn, food.

1. Pls with evident stolons; infl raceme-like, ± enclosed by sheath; spikelets few ..... ***P. clandestinum***
1′ Pls cespitose, sts erect, stolons gen 0; infl panicle-like, dense, cylindric, exserted from sheath; spikelets many
  2. Main infl axis short-hairy; spikelets < 7 mm ..... ***P. setaceum***
  2′ Main infl axis smooth or scabrous; spikelets > 7 mm ..... ***P. villosum***

P. clandestinum Chiov. (p. 1285) KIKUYU GRASS Per from stolons. **STS**: vegetative st spreading; fl st decumbent, 0.5–4.5 dm. **LF**: sheath 1–10 cm, glabrous or hairy; ligule ± 1.5–2 mm; blade 1.5–3 cm, 2–6 mm wide, upper surface glabrous to short-hairy. **INFL** < 1 cm; 1° branches < 0.5 cm, glabrous; bristles subtending cluster 5–10. **SPIKELET** 10–20 mm, ± 1 mm wide, lanceolate, grayish green; lower glume ± 1–2 mm, 0–1-veined; upper glume ± = spikelet length; lower floret lemma 9–13-veined, tip acuminate, palea 0; upper floret slightly < lower floret. 2*n*=36. Disturbed places, roadsides; < 100 m. NCo, CCo, SnFrB, SCo, n ChI (Santa Cruz Island); to S.Am; native to Afr. NOXIOUS WEED.

P. setaceum Forsskal Per, cespitose. **ST** erect, 4–15 dm. **LF**: sheath 4–8 cm, glabrous; ligule 0.5–1 mm; blade 20–65 cm, 2–3.5 mm wide, upper surface glabrous. **INFL** 8–30 cm; 1° branches (incl bristles) 20–30 mm, short-hairy; bristles subtending cluster 15–40. **SPIKELET** 4.5–6.5 mm, ± 1 mm wide, lanceolate, purplish; lower glume minute or 0; upper glume gen < 0.6 × spikelet length; lower floret lemma 3-veined, tip acuminate, palea 0; upper floret ± = lower floret. Waste places, urban roadsides; < 100 m. Deltaic GV, CCo, SnFrB, SCo, expected elsewhere; Baja CA; native to Afr.

P. villosum R.Br. (p. 1285) Per, cespitose. **ST** erect, 1.6–7.5 dm. **LF**: sheath 4–10 cm, glabrous; ligule ± 1–1.5 mm; blade 5–40 cm, 2–4.5 mm wide, upper surface glabrous to hairy. **INFL** 4–8 cm; 1° branches (incl bristles) 3–5 cm, glabrous; bristles subtending cluster 30–50. **SPIKELET** 9–11 mm, ± 1.5 mm wide, lanceolate, green to white; lower glume < 1 mm, veins not visible; upper glume ± 0.5 × spikelet length; lower floret sterile or staminate, lemma 7–9-veined, tip acuminate, palea ± = lemma or 0; upper floret ± = to slightly > lower floret. Waste places, urban roadsides; < 100 m. SW, DSon, expected elsewhere; to TX; native to Afr.

PHALARIS

Dennis Anderson

Ann, per, cespitose or from rhizomes. **ST** gen erect, 2–20 dm. **LF**: sheath open; ligule membranous, truncate; blade gen 2–5 dm, 1–2 cm wide, flat. **INFL** panicle-like, gen cylindric to ovoid, dense; branches ascending to appressed, obscure. **SPIKELET** gen fertile, sometimes also sterile in *P. paradoxa*, compressed; glumes equal, > florets, sometimes with wing-like keel, 3–5-veined; axis gen breaking above glumes, gen falling as 1 unit; florets 2–3, lower 1–2 vestigial or 0, upper 1 fertile; upper floret lemma gen ovoid, glabrous or appressed-hairy, shiny, faintly 5-veined, awn 0; palea < lemma, translucent. **FR** ± fusiform. ± 15 spp.: temp N.Am, Eurasia. (Greek: ancient name for grass with shiny spikelets)

1\. Fertile spikelet surrounded by 5–6 sterile spikelets, falling together; sterile spikelets = fertile spikelet or club-shaped; fertile floret lemma ± glabrous; lower florets < 0.2 mm ..... ***P. paradoxa***
1′ Spikelets ± all fertile, gen breaking apart separately; fertile floret lemma gen hairy; lower florets gen > 0.2 mm, gen > 1/3 upper floret length
2\. Lower floret gen 1
3\. Glume wing not toothed; per ..... [2]***P. aquatica***
3′ Glume wing gen toothed; ann ..... ***P. minor***
2′ Lower florets 2
4\. Lower florets unequal, longer 1–2 mm, other < 0.5 mm ..... [2]***P. aquatica***
4′ Lower florets ± equal
5\. Lower florets fleshy to corky, < 1 mm ..... ***P. brachystachys***
5′ Lower florets awl-like or chaff-like
6\. Lower florets wide, chaff-like — gen < 1/2 upper floret ..... ***P. canariensis***
6′ Lower florets awl-like
7\. Glumes not winged (narrowly so in *P. californica*)
8\. Infl ± cylindric, ± interrupted at base, length > 2 × width; rhizomes evident ..... ***P. arundinacea***
8′ Infl widely ovoid with truncate base, length gen not 2 × width; rhizomes 0 ..... [2]***P. californica***
7′ Glumes winged
9\. Glume midvein scabrous with short delicate barbs; lateral veins ± glabrous or with < 5 stiff hairs
10\. Fertile floret lemma ± flat, ± appressed-hairy, tip acute ..... ***P. caroliniana***
10′ Fertile floret lemma ± swollen, hairs spreading, tip glabrous, acuminate ..... [2]***P. lemmonii***
9′ Glume midvein, lateral veins sharply scabrous with 9+ stiff hairs
11\. Fertile floret lemma ± swollen, tip glabrous, acuminate ..... [2]***P. lemmonii***
11′ Fertile floret lemma ± flat, tip hairy, acute
12\. Infl cylindric, base ± rounded, length > 3 × width; glume wing wide, evident; fr < 1.5 mm; ann ..... ***P. angusta***
12′ Infl widely ovoid, base truncate, length gen < 2 × width; glume wing very narrow, obscure; fr ± 2.5 mm; per, cespitose ..... [2]***P. californica***

P. angusta Nees Ann. **ST** 3–15 dm. **INFL** 2.5–17 cm, 0.6–1.5 cm wide, cylindric, narrow. **SPIKELET**: glumes 3–5.5 mm, 1–2.5 mm wide, wing wide, back scabrous near base, veins scabrous, base ± oblique, tip acute; lower florets 2, 0.5–1.5 mm, awl-like, base hairy-tufted; upper lemma 2–4 mm, 1–1.5 mm wide, lanceolate, densely hairy. **FR** < 1.5 mm, 0.7–0.9 mm wide. 2*n*=14. Uncommon. Gen wet areas, marshes, sloughs, ditches; < 1000 m. CA-FP; to se US, S.Am.

P. aquatica L. HARDING GRASS Per, cespitose, sometimes from short rhizomes. **ST** 4–15 dm. **INFL** 1.5–11 cm, 1–2.5 cm wide, gen cylindric, sometimes interrupted in lower 1/3, or ovoid. **SPIKELET**: glumes 4.5–7.5 mm, 1–2 mm wide, gen glabrous, sometimes minutely toothed, keel wing-like; lower florets 1–2, gen 0.2–2 mm, awl-like; upper lemma 3–4.5 mm, 1–1.5 mm wide, lanceolate-ovoid, densely hairy. **FR** 2–2.5 mm, 1–1.5 mm wide. 2*n*=28. Wet areas, ditches; < 1200 m. NW, c SN, CCo, SCo; OR; native to Medit Eur. [*P. tuberosa* L. var. *stenoptera* (Hackel) A. Hitchc.]

P. arundinacea L. REED CANARY GRASS Per from distinct rhizomes. **ST** 5–15 dm. **INFL** 7–40 cm, 2–11 cm wide, cylindric, interrupted near base; branches spreading in fl, appressed in fr. **SPIKELET**: glumes 3.5–7.5 mm, midvein scabrous, wing 0, tip acute; lower florets 2, 1–2.5 mm, awl-like, hairy; upper lemma 3–4.5 mm, ± 1.5 mm wide, narrowly lanceolate, glabrous to sparsely hairy. **FR** 1.5–2 mm, < 1 mm wide. 2*n*=14,27–31,35,42. Wet streambanks, moist areas, grassland, woodland; < 1600 m. CA-FP; temp N.Am, Eurasia. Cult for forage.

P. brachystachys Link Ann. **ST** 3–9 dm. **INFL** 1.5–4 cm, 0.8–1.8 cm wide, ovoid. **SPIKELET**: glume 6.5–8.5 mm, 1.5–2.5 mm wide, back glabrous or hairy, keel wing-like, tip acute; lower florets 2, 0.5–1 mm, ± fleshy, becoming corky; upper lemma 4.5–5.5 mm, 1.5–2 mm wide, ovoid, densely hairy. **FR** 3.5–4 mm, ± 1.5 mm wide. 2*n*=12. Wet, disturbed areas; < 200 m. ScV (Butte Co.), SnFrB, SCoRO (San Luis Obispo Co.); native to Medit Eur.

P. californica Hook. & Arn. (p. 1285) Per, cespitose. **ST** 5–15 dm. **INFL** 1.5–5 cm, 1–3 cm wide, widely ovoid; base ± truncate. **SPIKELET**: glumes 5–8 mm, 1–1.5 mm wide, gen glabrous, narrowly wing-like at tip or not, tip acute, purplish; lower florets 2, 2–3.5 mm, awl-like, densely hairy; upper lemma 3.5–5 mm, 1–1.5 mm wide, lanceolate, sparsely hairy. **FR** ± 2.5 mm, ± 1 mm wide. $2n$=28. Uncommon. Moist areas, meadows, woodlands; < 700 m. NCo, NCoRO, CCo, SnFrB, SCoRO; s OR. ❀IRR:**4**,**5**,6,**17**& SHD:**7**,8,9,**14–16**,19–23,**24**.

P. canariensis L. CANARY GRASS (p. 1285) Ann. **ST** 3–10 dm. **INFL** 1.5– 4 cm, 1.5–2 cm wide, ovoid to oblong. **SPIKELET**: glumes 7–10 mm, 1.5–2.5 mm wide, wing-like above middle, tip widely acute; lower florets 2, 2.5–4.5 mm, 0.5–1 mm wide, chaff-like; upper lemma 5–7 mm, 1.5–2.5 mm wide, ovoid, densely hairy. **FR** ± 4 mm, ± 1.5 mm wide. $2n$=12. Disturbed areas; < 300 m. NCo, CCo, SnFrB, SCo; to Great Plains; native to Medit Eur.

P. caroliniana Walter Ann. **ST** 5–10 dm. **INFL** 1–7 cm, 0.8–2 cm wide, ovoid to subovoid. **SPIKELET**: glumes 4–5.5 mm, 1–1.5 mm wide, narrowly winged, wing finely scabrous, lateral veins glabrous to slightly scabrous, tip acute to acuminate; lower florets 2, 1.5–2.5 mm, awl-like, appressed-hairy; upper lemma 3–4.5 mm, 1–2 mm wide, lanceolate, densely hairy. **FR** ± 2 mm, ± 1 mm wide. $2n$=14. Uncommon. Disturbed areas; < 650 mm. NCo, c SNF, CCo, SCoRO, SW; native to e US.

P. lemmonii Vasey (p. 1285) Ann. **ST** 5–13 dm. **INFL** 4–12 cm, 0.7–1.5 cm wide, ± cylindric, interrupted at base; lower branches < 2 cm. **SPIKELET**: glumes 5–7 mm, ± 1 mm wide, wing 0, veins scabrous, tip acuminate; lower florets 2, 1–1.5 mm wide, awl-like, appressed-hairy; upper lemma 4–5 mm, 1–1.5 mm wide, lanceolate-ovoid, ± swollen, hairs spreading exc tip glabrous. **FR** ± 2 mm, ± 1 mm wide. $2n$=14. Gen moist areas, shrubland, woodland; < 750 m. NCoRO, c SN, GV, SnFrB, SCoR, SCo.

P. minor Retz. (p. 1285) Ann. **ST** 2–10 dm. **INFL** 1–6 cm, 1–2 cm wide, oblong-ovoid. **SPIKELET**: glumes 4–6.5 mm, 1–1.5 mm wide, glabrous, wing entire or toothed; lower floret 1, 1–2 mm, gen awl-like; upper lemma 3–4 mm, 1–2 mm wide, widely ovate, hairy. **FR** ± 2–2.5 mm, ± 1.5 mm wide. $2n$=28,29. Disturbed areas; < 300 m. CA-FP; native to Medit.

P. paradoxa L. (p. 1285) Ann. **ST** 2–10 dm. **INFL** 3–9 cm, 1–2 cm wide, oblong; base tapered; tip ± truncate to acuminate; spikelets in clusters, cluster falling as 1 unit; fertile spikelet gen surrounded by 5–6 sterile, short-stalked spikelets; sterile spikelet vestigial or = fertile. **FERTILE SPIKELET**: glumes 5.5–8 mm, ± 1 mm wide, gen glabrous, wing lobed to toothed, tip acute to acuminate; lower florets vestigial, knob-like; upper lemma 2.5–3.5 mm, 1–1.5 mm wide, narrowly ovoid, gen glabrous or sparsely hairy near tip. **FR** ± 2.5 mm, ± 1 mm wide. $2n$=14. Disturbed areas; < 250 m. GV, SW; native to Medit Eur. Small pls gen have many, vestigial sterile spikelets.

PHLEUM TIMOTHY

Dieter H. Wilken

Ann, per. **STS** ascending to erect. **LVS** basal and cauline; appendages 0 or small, acute to obtuse; ligule membranous to translucent, obtuse to truncate; blade gen flat, margin minutely scabrous. **INFL** panicle-like, cylindric to ovoid, dense; branches spike-like, short. **SPIKELET** ± sessile, strongly laterally compressed; glumes subequal, membranous, keel gen stiff-ciliate (comb-like), pointed to awned at obtuse to truncate tip, 3-veined; floret 1, breaking above glumes, bisexual; lemma gen awnless at wide, truncate tip, 3–7-veined; palea ± = lemma. 15 spp.: temp Am, Eurasia. (Greek: a marsh reed)

1. Infl 1–6 cm, ovoid to cylindric, awn 2–3 mm .... ***P. alpinum***
1′ Infl 4–18 cm, cylindric, awn < 2 mm .... ***P. pratense***

P. alpinum L. (p. 1285) MOUNTAIN TIMOTHY Per. **STS** gen clumped, 2–6 dm. **LF**: basal loosely tufted; cauline blade 2–12 cm, 3–8 mm wide. **INFL** 1–6 cm, 7–12 mm wide, ovoid (often cylindric at low elevations). **SPIKELET**: glumes 2–5 mm, scabrous on back, awned at wide acuminate tip, awn 2–3 mm; lemma 2–3 mm, puberulent on back. $2n$=14,28. Wet meadows, streambanks, coniferous forest, alpine; 1500–3350 m (NCo, NCoR: < 1500 m). NCo, KR, n&c NCoR, CaRH, SNH, SnBr, SnJt, W&I; to AK, e Can, ne US, Mex; also in S.Am, Eur. ❀STBL.

P. pratense L. CULTIVATED TIMOTHY Per. **STS** solitary to loosely clumped, 5–10 dm; base gen swollen. **LF**: basal few, gen spreading; cauline blade 4–20 cm, 3–6 mm wide. **INFL** 4–18 cm, 5–8 mm wide, cylindric. **SPIKELET**: glumes 2–3 mm, lower 1/2 scabrous on back, awned at acuminate tip, awn gen < 2 mm; lemma 1–2.5 mm, veins puberulent. $2n$=14,21,28. Disturbed sites, roadsides, cult fields; < 2100 m. CA-FP, GB; N.Am, Mex; native to Eurasia. Widely cult for forage, hay.

PHRAGMITES

Kelly W. Allred

Per with thick rhizomes or stolons, forming dense stands. **ST** tall, erect. **LVS** cauline; sheaths open; ligule short, membranous, truncate or hairy; blade flat or folded, gen deciduous. **INFL** panicle-like. **SPIKELET**: glumes unequal, lower >> upper, 1–3-veined; axis long-soft-hairy; florets 1–10, breaking above glumes and between florets; lower florets sterile or staminate, upper bisexual; lemmas lanceolate, glabrous, gen 3–5-veined; palea << lemma; stamens gen 2–3. 3 spp.: temp, trop. (Greek: growing in hedges) [Clayton 1968 Taxon 17:168–169]

P. australis (Cav.) Steudel (p. 1285) COMMON REED **ST** 2–4 m. **LF**: blade gen 20–45 cm, 1–5 cm wide, margins scabrous, gen breaking at collar. **INFL** 15–50 cm, plume-like, oblong to obovoid, purplish to whitish. **SPIKELETS** 10–16 mm; lower glume 3–7 mm, upper glume 5–10 mm; florets 2–10. $2n$=36,44,46,48,49–52, 54,72,84,96. Pond and lake margins, sloughs, marshes; < 1600 m. CA; worldwide. [*P. communis* Trin.] Perhaps most widely distributed of all vascular seed pls. ❀IRR or WET,SUN:1,**2**,**3**,4,5,**6–11**, 12,**14**,15–17,**18–23**,24;STBL;INV.

PIPTATHERUM

Mary E. Barkworth

Per, cespitose. **ST** erect, unbranched. **LF**: ligule membranous, gen truncate; blade flat to inrolled, often tapered. **SPIKELET**: glumes ± equal, obtuse to acute, veins evident; axis breaking above glumes; floret 1, gen ± compressed; callus blunt; lemma margins widely separated with age, awn gen readily deciduous; palea ± = lemma, texture like lemma. 30 spp.: arid temp, subtrop Eurasia, Afr. (Greek: falling awn) Segregated from *Oryzopsis*.

1. Lf blade 0.5–1.5 mm wide; infl 8–13 cm, branches 1–3 per node ..... ***P. micranthum***
1′ Lf blade 2–10 mm wide; infl 15–40 cm, branches 3–7 per node ..... ***P. miliaceum***

P. micranthum (Trin. & Rupr.) Barkworth (p. 1285) SMALL-FLOWERED RICEGRASS **ST** 3–6.5 dm. **LF**: basal ligule 0.4–1.5 mm. **INFL** 8–13 cm; lowest branches becoming reflexed. **SPIKELET**: glumes 2.5–3.5 mm; floret 1.5–2.5 mm; lemma gen glabrous, sometimes sparsely but evenly short-hairy, awn 4–8 mm, weakly bent. $2n$=22. RARE in CA. Gravel benches, rocky slopes, creek banks; 700–2950 m. W&I, DMtns; to B.C., Rocky Mtns. [*Oryzopsis m.* (Trin. & Rupr.) Thurb.]

P. miliaceum (L.) Cosson SMILO GRASS **ST** 4–15 dm, becoming branched at lower nodes. **LF**: basal ligule 0.5–1.5 mm. **INFL** 15–40 cm; branches whorled. **SPIKELET**: glumes 3–3.5 mm; floret 1.5–2 mm; lemma glabrous, awn 3–4 mm, weakly bent. $2n$=24. Disturbed places; < 300 m. NCo, GV, CW, SCo; native to Eurasia. [*Oryzopsis m.* (L.) Asch. & Schweinf.]

PIPTOCHAETIUM

Mary E. Barkworth

Per, cespitose. **ST** prostrate or erect. **LVS** gen basal; blade gen inrolled, with translucent line on both sides of midvein, gen wavy toward tip. **INFL** panicle-like; spikelets gen only on distal half of branches. **SPIKELET**: glumes ± equal, gen > floret; axis breaking above glumes; floret 1, gen ovoid; lemma finely striate, hairs brown, tubercled near neck-like tip, margins inrolled, fitting into grooved palea, awned; palea slightly > lemma, grooved longitudinally. ± 35 spp.: Am. (Greek: falling hair) [Parodi 1944 Revista Mus La Plata, Secc Bot 6:213–310]

P. setosum (Trin.) Arechav. **ST** 2–4 dm. **LF**: blade ± 0.5 mm diam, thread-like. **INFL** 5–10 cm. **SPIKELET**: glumes ± equal, 7–8 mm; floret 3–3.5 mm, ovoid; callus blunt; lemma tip 0.5–1 mm wide, awn < 15 mm, bent, deciduous. Disturbed sites; < 200 m. n SnFrB (Marin Co.); native to Chile.

PLEURAPHIS

J. Travis Columbus

Per, cespitose, rhizomed. **ST** ascending to erect, solid; nodes gen hairy. **LF**: ligule membranous, ciliate-fringed; blade firm, flat to inrolled, sharply acute. **INFL** spike-like, gen cylindric; spikelets in clusters ± equal, 3 per node; clusters wedge-shaped, overlapping, appressed to ascending, hairy-tufted at base, falling as 1 unit; axis wavy; glumes of cluster together involucre-like. **CENTRAL SPIKELET** subsessile, appressed to or nearest infl axis; glumes equal, < floret, oblanceolate, keeled, ciliate, tip deeply 2-lobed, lobes lanceolate, awns 3–9, 1 from ± mid-keel, others terminal; florets gen 1–2, lower floret bisexual, upper floret (if present) bisexual or staminate; lemma lanceolate, 3-veined, gen ciliate, tip gen 2-lobed, gen 1-awned ± from sinus; palea ± = lemma. **LATERAL SPIKELETS** sessile; glumes < to ± = florets, ciliate, lower glume asymmetric with 1 awn from ± middle near margin, gen 2-lobed, lobes unequal; florets 1–4, gen staminate; lemma 3-veined, tip gen ciliate; palea ± = lemma. 3 spp.: w US, n Mex. (Greek: side needle, from awn position on lower glume of lateral spikelets) [Reeder & Reeder 1988 Madroño 35:6–9]

1. St internodes glabrous to puberulent ..... ***P. jamesii***
1′ St internodes sparsely to densely hairy ..... ***P. rigida***

P. jamesii Torrey (p. 1285) GALLETA **Pl** 1.5–4(6.5) dm, unbranched above base. **ST** ± 1 mm diam; internodes glabrous or puberulent; node hairs ± straight. **LVS** gen basal; glabrous or scabrous, long-ciliate near ligule; ligule membrane 1–3 mm, gen appendaged; blade < 13(21) cm, 2–3 mm wide, upper surface sometimes with short, ± straight, hairs. **INFL** 3–7 cm; spikelet clusters 6–9 mm. **CENTRAL SPIKELET**: glume margin hairs < 0.5 mm; lower lemma awn < 2.5 mm. **LATERAL SPIKELETS**: lower glume 1-awned; upper glume tip unlobed or 2-lobed, awn 0–0.4 mm, margin hairs < 0.5 mm; lemma tip unlobed or 2-lobed, awn 0–0.4 mm, margin hairs < 0.2 mm. $2n$=18,36,38,72. Dry, sandy to rocky slopes, flats, scrub, woodland; 1000–2500 m. SNE, n&e DMtns; to WY, TX. [*Hilaria j.* (Torrey) Benth.] ❀SUN,DRN:1,2,**3**,7,**8–10**, **14–16**,17,**18–23**,24&IRR:**11–13**;also STBL.

P. rigida Thurber (p. 1285) BIG GALLETA **Pl** 3.5–10 dm, branched above base, gen bush-like. **ST** 1.5–3.5 mm diam at base; internodes sparsely to densely curly-hairy; node hairs curly. **LVS** gen cauline, gen sparsely to densely curly-hairy, esp near and sometimes overlapping ligule; ligule membrane < 1 mm, appendages 0; blade < 10 cm, 2–4 mm wide. **INFL** 4–10 cm; spikelet clusters 7–11 mm. **CENTRAL SPIKELET**: glume margin hairs 0.5–3 mm; lower lemma tip sometimes 4-lobed, awns 3 ± from sinuses, central awn 2–5.5 mm. **LATERAL SPIKELETS**: lower glume with 1+ subsidiary lobes, or larger lobe tip fringed, awns 2–4; upper glume tip with 2+ lobes or fringed, awns 1–3, 0.4–2.5 mm, margin hairs 0.5–2 mm; lemma tip 2-lobed, awn 1 ± from sinus, 0.4–2 mm, margin hairs 0.2–1 mm. $2n$=18,36,±108. Common. Dry, open, sandy to rocky slopes, flats, and washes, sand dunes, scrub, woodland; < 1600 m. PR, e&s DMoj, DSon; to UT, n Mex. [*Hilaria r.* (Thurber) Scribner] Important forage; some pls from e DMoj and Ord Mtn. (San Bernardino Co.), with ± straight internode hairs, intermediate to *P. jamesii*. ❀SUN,DRN:2,3,7,**8–10**,**14–16**,17,18,**19–21**,22–24&IRR:**11–13**;also STBL.

PLEUROPOGON SEMAPHORE GRASS

Mary E. Barkworth

Ann, per. **ST** gen erect. **LVS** gen cauline. **LF**: sheath margins fused > 1/2 length; ligule membranous; blade flat or folded, drooping. **INFL** raceme-like. **SPIKELET** 1.5–8 cm; glumes << lowest floret, translucent, lower 1-veined, upper 1–3-veined; florets 5–20; lemma 7–9-veined, veins prominent, not converging, extending to tip, awn 1–5 mm; palea ± = lemma, veins with appendages. 5 spp.: temp w N.Am., 1 sp. in e Asia. (Greek: side beard, from awn at palea base in some spp.)

1. Lowest floret lemma 4.5–7.5 mm; st 1.5–9.5 dm; rhizome 0 ..... ***P. californicus***
1′ Lowest floret lemma 8–10 mm; st 8.5–16 dm; rhizome evident
 2. Awn 1–4 mm; spikelets ascending ..... ***P. hooverianus***
 2′ Awn 5–20 mm; spikelets reflexed ..... ***P. refractus***

P. californicus (Nees) Vasey Ann, per. **ST** decumbent to erect, clumped, 1.5–9.5 dm, often rooting at lower nodes. **LF**: ligule 2–6 mm; blade 3–6 mm wide. **INFL** 8–30 cm. **SPIKELET** 15–50 mm, spreading to erect; lower glume 1–4.5 mm, upper 2–6.5 mm; lemma 4.5–7.5 mm, awn < 11 mm. 2*n*=18. Wet places, redwood, oak forests; < 650 m. NW, CaRF, n&c SNF. Per, short-awned pls from NCoRO have been called *P. davyi* L. Benson.

P. hooverianus (L. Benson) J. Howell NORTH COAST SEMAPHORE GRASS Per. **ST** erect, 10–16 dm. **LF**: ligule 3–7 mm; blade 4–10 mm wide. **INFL** 20–35 cm. **SPIKELET** 2.8–4.5 cm, ascending; lower glume 3–6 mm, upper 4.5–7.5 mm; lemma 8–9 mm, awn 1–4 mm. 2*n*=36. **RARE** CA. Marshy areas, redwood groves; < 500 m. s NCo, n CCo.

P. refractus (A. Gray) Benth. (p. 1291) NODDING SEMAPHORE GRASS Per. **ST** erect, 8.5–15 dm. **LF**: ligule 2–7 mm; blade 5–14 mm wide. **SPIKELET** 2.5–8 cm, reflexed at maturity; lower glume 3–6 mm, upper 4–8.5 mm; lemma 8–10 mm, awn 5–20 mm. 2*n*= 36. UNCOMMON. Wet meadows, shady banks; < 1600 m. NCo, KR, NCoRO; to B.C.

POA BLUEGRASS

Robert J. Soreng

Ann, per, some ± dioecious. **ST** 0.3–12 dm. **LF**: sheath open to closed (best observed on upper st lf); ligule thin, flexible; blade grooved above on both sides of midvein, flat, folded, or inrolled, gen smooth or scabrous on veins, gen prow-tipped. **INFL** panicle-like; branches appressed to drooping. **SPIKELET** gen compressed, breaking between florets; glumes 2, similar, gen < lowest lemma, awnless; florets gen 2–6; callus indistinct, often with obvious tuft of long cobwebby hairs; lemma gen keeled to base, of same texture as glumes, awnless, veins gen 5, ± converging near tip; palea well developed, keels gen scabrous; fertile anthers 0.2–4.5 mm; ovary glabrous. ± 500 spp.: temp and cool regions. (Greek: ancient name) [Soreng 1991 Syst Bot 16:507–528] CA is center of diversity in N.Am. Spikelet features best observed on lowest florets of spikelet.

1. Ann (rarely bien); anthers gen 0.2–1.0 mm (–1.8 mm in *P. bolanderi*); fl gen bisexual
  2. Callus glabrous; upper fls of spikelet often pistillate (sect. *Ochlopoa*)
    3. Spikelet axis gen hidden, terminal internode < 1/2 terminal floret length; anthers 0.6–1 mm; widespread ..... ***P. annua***
    3′ Spikelet axis visible, terminal internode > 1/2 terminal floret length; anthers 0.2–0.5 mm; esp s CA-FP ..... ***P. infirma***
  2′ Callus with cobwebby hairs (sect. *Diversipoa*)
    4. Infl branches appressed; lf blade abruptly prow-tipped; lemma veins hairy ..... ***P. bigelovii***
    4′ Lower infl branches eventually spreading; lf blade not prow-tipped or lemma glabrous or scabrous
      5. Lemma glabrous or scabrous; lf blade abruptly prow-tipped; some spikelet internodes gen > 1 mm . ***P. bolanderi***
      5′ Lemma puberulent; lf blade tapered, slightly prow-tipped; some spikelet internodes < 1 mm ..... ***P. howellii***
1′ Per; anthers often > 1.2 mm; fl sometimes unisexual
  6. Rhizomes or stolons present; anthers 1–4 mm
    7. St and nodes distinctly compressed, wiry; lf sheath open nearly to base; infl branches scabrous (sect. *Tichopoa*) ..... ***P. compressa***
    7′ St and nodes little or not compressed or wiry; lf sheath sometimes not open so far; infl branches sometimes smooth
      8. Lemma glabrous or minutely scabrous; callus long-cobwebby
        9. Lf blade thin, gen flat; upper surface glabrous or sparsely scabrous; fls bisexual ..... ***P. kelloggii***
        9′ Lf blade thick, folded and inrolled, upper surface gen finely hairy; fls gen unisexual (sect. *Madropoa*) ..... ***P. piperi***
      8′ Lemma hairy at least on keel or callus glabrous
        10. Lemma gen < 3.5 mm, lemma and callus smooth, glabrous; st nodes gen hidden in basal tuft; infl < 8 cm, branches appressed, gen smooth; fls unisexual (sect. *Madropoa*) ..... ***P. atropurpurea***
        10′ Lemma gen > 3.5 mm or ± scabrous or hairy (if < 4 mm and smooth and glabrous, then some st nodes well exposed and infl > 8 cm); infl 2–30 cm, branches sometimes ± spreading, sometimes scabrous; fls sometimes bisexual
          11. Lf blade firm, inrolled, upper surface gen densely and finely hairy or scabrous; fls gen unisexual; coastal sand (sect. *Madropoa*)
            12. Lemma < 4.5 mm, sparsely hairy; pl < 30 cm; infl ± loose ..... ***P. confinis***
            12′ Lemma >> 4.5 mm, glabrous to densely hairy; pl sometimes > 30 cm; infl dense, branches appressed
              13. Upper glume < 7 mm; lowest lemma < 7.5 mm; st very scabrous or coarsely hairy below infl; s NCo, CCo, n ChI ..... ***P. douglasii***
              13′ Upper glume > 7 mm; lowest lemma > 7.5 mm; st smooth below infl; n&c NCo ..... ***P. macrantha***
          11′ Lf blade gen not firm or inrolled, upper surface glabrous or sparsely soft-hairy; fls often bisexual; widespread (if pl otherwise as in 11., then montane)
            14. Sheath of upper st lf closed nearly to top, blade > sheath (*P. nervosa* complex) ..... ***P. sierrae***
            14′ Sheath of upper st lf open at least 1/4 length, blade < or = sheath
              15. Callus glabrous (rarely short-cobwebby); lemma glabrous or hairy; fls often pistillate
                16. Lemma gen < 3.5 mm, sometimes glabrous; infl linear in outline, short-branched, interrupted, many-fld; upper st-lf blade > 1/5 sheath length; fls appearing bisexual but anthers gen sterile; tall leafy-st pls of moist, low-elevation meadows (see *P. secunda* ssp. *juncifolia*) ..... ***P. pratensis* × *P. secunda***
                16′ Lemma gen > 3.5 mm; infl lanceolate to ovate in outline (otherwise sometimes as above); upper st-lf blade sometimes < 1/6 sheath length; fls gen pistillate (anthers minute); cespitose or tall pls of well drained soils of slopes of forests

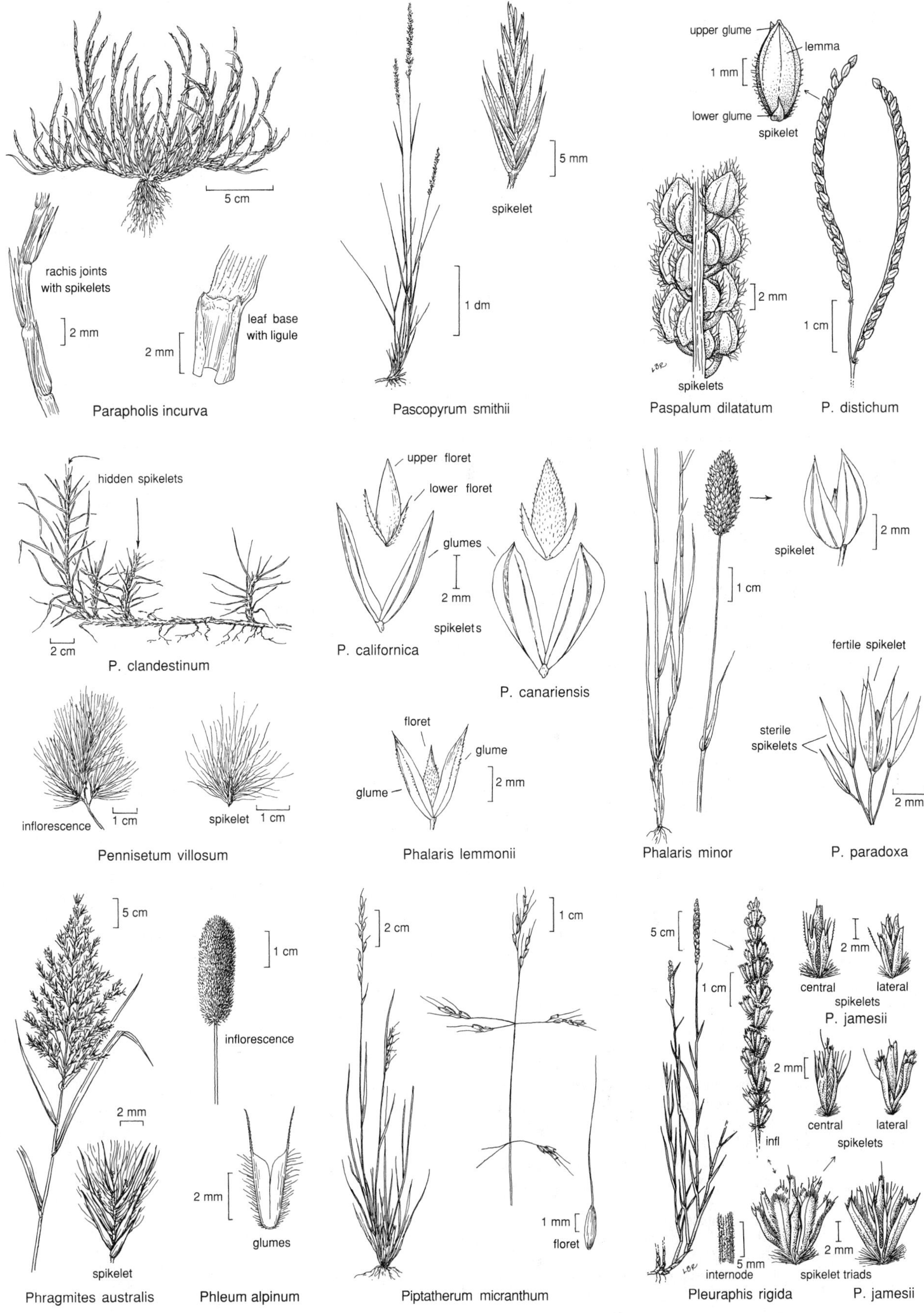
5 cm
rachis joints
with spikelets
2 mm
leaf base
with ligule
2 mm
Parapholis incurva
spikelet
5 mm
1 dm
Pascopyrum smithii
upper glume
lemma
1 mm
lower glume
spikelet
2 mm
1 cm
spikelets
Paspalum dilatatum
P. distichum
hidden spikelets
2 cm
P. clandestinum
upper floret
lower floret
glumes
2 mm
spikelets
P. californica
P. canariensis
2 mm
1 cm
spikelet
fertile spikelet
sterile
spikelets
2 mm
inflorescence
1 cm
spikelet
1 cm
Pennisetum villosum
floret
glume
glume
2 mm
Phalaris lemmonii
Phalaris minor
P. paradoxa
5 cm
1 cm
inflorescence
2 mm
2 mm
glumes
spikelet
Phragmites australis
Phleum alpinum
2 cm
1 cm
1 mm
floret
Piptatherum micranthum
5 cm
1 cm
2 mm
central
lateral
spikelets
P. jamesii
2 mm
central
lateral
spikelets
infl
5 mm
internode
2 mm
spikelet triads
Pleuraphis rigida
P. jamesii

17. St-lf blades strongly reduced upward, uppermost < 1/8 sheath length; lemma keel and marginal veins obviously hairy; pl densely tufted (sect. *Madropoa*) .... [2]***P. fendleriana*** ssp. ***longiligula***
17′ St-lf blades gen longest at middle of st, uppermost gen > 1/5 sheath length; lemma scabrous or keel and marginal veins sparsely hairy; pl loosely tufted .... ***P. wheeleri***
15′ Callus long-cobwebby (hairs > 1/2 lemma length); lemma keel and marginal veins hairy, glabrous between veins; fls bisexual (exc *P. rhizomata*)
18. Sheath gen open > 3/4 length; stolons gen present; st sometimes branching above base; lemma sometimes hairy only on keel
19. First glume gen 3-veined, not arched; lemma obviously hairy on keel and marginal veins (sect. *Stenopoa*) .... [2]***P. palustris***
19′ First glume 1-veined, ± sickle-shaped; lemma sparsely hairy on keel and rarely on marginal veins (sect. *Coenopoa*) .... [2]***P. trivialis***
18′ Sheath open < 3/4 length; pl with obvious rhizomes; st not branching above base
20. Pl often forming a dense sod; ligule truncate to rounded; infl densely fld, branches at lower nodes gen 4 or more (some may be small); lemma gen < 4 mm; fls rarely pistillate (sect. *Poa*) .... [2]***P. pratensis***
20′ Pl not sod-forming; ligule acute; infl sparsely fld, branches at lower nodes gen 1–2; lemma gen > 4 mm; fls sometimes pistillate (*P. nervosa* complex) .... ***P. rhizomata***
6′ Rhizomes and stolons 0; anthers 0.2–4.5 mm
21. Most florets replaced by leafy bulblets; st base bulbous .... ***P. bulbosa***
21′ Florets not mostly replaced by bulblets; st base not bulbous
22. Callus cobwebby; lemma hairy on keel and gen on marginal veins but not between veins
23. Infl < 10 cm, ± dense; pl densely cespitose
24. Pl > 20 cm; sheaths open < 1/2 length; fls pistillate, anthers vestigial .... [2]***P. cusickii*** ssp. ***purpurascens***
24′ Pl << 20 cm; sheaths open > 3/4 length; fl gen bisexual, anthers 0.8–1.2 mm (sect. *Abbreviatae*) .... ***P. pattersonii***
23′ Infl gen > 10 cm or open; pl loosely cespitose
25. Lower glume gen 3-veined; sheaths open 4/5 length or more (sect. *Stenopoa*)
26. Ligule of upper st lf truncate, gen < 0.8 mm; gen forest .... ***P. nemoralis***
26′ Ligule of upper st lf acute, gen > 1 mm; gen moist meadow or streamside .... [2]***P. palustris***
25′ Lower glume gen 1-veined; sheaths open < 3/4 length
27. Anthers < 1 mm; infl branches with fls in top 1/3; high montane to alpine (sect. *Oreinos*) .... ***P. leptocoma***
27′ Anthers > 1.3 mm; some infl branches with fls from midlength; low elevation (sect. *Coenopoa*) .... [2]***P. trivialis***
22′ Callus glabrous or not cobwebby (ring of short callus hairs present in some *P. secunda*, *P. unilateralis*); lemma often hairy between veins
28. Lemma obviously hairy at least on keel and marginal veins; fls unisexual; sheath open ± 1/3 length; uppermost st-lf blade firm, often highly reduced; callus glabrous (sect. *Madropoa*) .... [2]***P. fendleriana*** ssp. ***longiligula***
28′ Lemma often glabrous (if obviously hairy on keel and marginal veins or between, then fls bisexual, sheaths open > 3/4 or < 1/2 length, uppermost st-lf blade often thin and withering or not highly reduced); callus sometimes with a crown of short hairs
29. Spikelet lanceolate to narrowly ovate in outline, little compressed; lemma weakly keeled; sheath open > 3/4 length; fls bisexual (sect. *Secundae*)
30. Infl branches widely spreading in fl and fr; pl delicate, spring-active; 500–700 m, often on serpentine .... ***P. tenerrima***
30′ Infl branches gen ascending to appressed in fr (if widely spreading in fr, pl gen more robust, summer-active, of high elevation, not on serpentine) .... ***P. secunda***
31. Lemma glabrous (sometimes scabrous, rarely sparsely short-hairy); ligule of sterile shoots and often st < 2 mm, ± truncate, scabrous; lf blade ± firm, retaining shape; infl branches appressed; esp GB .... ssp. ***juncifolia***
31′ Lemma ± evenly short-hairy on keel and sides across base; ligule of sterile shoots and st gen > 2 mm, acute or acuminate, smooth or sparsely scabrous; lf blade gen ± soft, soon withering; infl branches appressed or spreading; CA .... ssp. ***secunda***
29′ Spikelet ± ovate in outline, obviously compressed; lemma obviously keeled to base; sheaths sometimes open < 3/4 length; fls sometimes unisexual
32. Uppermost sheath open < 1/2 length; lemma > 4.5 mm, sparsely hairy on keel, base, and sometimes marginal veins; fls pistillate (sect. *Madropoa*) .... [2]***P. cusickii*** ssp. ***purpurascens***
32′ Uppermost sheath gen open > 1/2 length; lemma gen < 4.5 mm or glabrous (if hairy, sheath open > 3/4 length, fls bisexual)
33. Lemma obviously hairy, < 3.6 mm; subalpine or higher (sect. *Stenopoa*) .... ***P. glauca*** ssp. ***rupicola***
33′ Lemma often glabrous, length various (if lemma obviously hairy, pl coastal)
34. Sheath open ± 9/10 length; fls bisexual; low salty places; coast or w CA (sect. *Secundae* subsect. *Halophytae*)
35. Infl branches ascending, very scabrous, with spikelets only in top 1/2; near hot springs, NCoRO, Napa Co. .... ***P. napensis***
35′ Infl branches appressed, smooth to sparsely scabrous, with spikelets from near base; seabluffs, NCo, n&c CCo .... ***P. unilateralis***
34′ Sheath open < 9/10 length; fls sometimes unisexual; gen ± salt-free places; montane or GB

36. Pl gen < 1 dm; longest anthers 0.2–2 mm; upper glume gen = or > first lemma; lemma sometimes sparsely short-hairy; fls bisexual (sect. *Abbreviatae*)
37. Lf firm; lemma 3–5 mm; anthers 0.7–2 mm .... ***P. keckii***
37′ Lf soft; lemma < 3 mm; anthers 0.2–0.7 mm .... ***P. lettermanii***
36′ Pl gen 1–7 dm; longest fertile anthers 2–4.5 mm; upper glume gen < first lemma; lemma glabrous; fls sometimes unisexual (sect. *Madropoa* subsect. *Epiles*)
38. Ligule of uppermost lf of sterile sts 2.5–6 mm, acute, smooth; lemma < 5.5 mm; lowest true lvs lacking blades; fls bisexual or pistillate; SNH .... ***P. stebbinsii***
38′ Ligule of uppermost lf of sterile sts gen < 2 mm (if 2–3 mm, lemma > 5.5 mm), truncate to obtuse, scabrous; lowest true lvs sometimes with blades; all fls gen bisexual; more widespread
39. Pl gen < 30 cm, base often decumbent; lf blades of sterile sts gen > 1.5 mm wide, arched, their ligules sometimes > 2 mm; upper glume sometimes = first lemma; KR, CaRH, n SNH .... ***P. pringlei***
39′ Pl 25–70 cm, base rarely decumbent; lf blades of sterile sts < 1.5 mm wide, erect, their ligules < 2 mm; upper glume obviously < first lemma; widespread, esp GB .... ***P. cusickii*** in part
40. Basal tuft of lvs sparse; infl branches ± smooth; 1–2 st nodes exposed; fls pistillate; moist subalpine .... ssp. ***epilis***
40′ Basal tuft of lvs dense; infl branches obviously scabrous; 0–1 st nodes exposed; fls sometimes staminate (rarely bisexual); moist or dry places
41. Infl branches slender, longest gen > 17 mm; moist sagebrush to dry montane slopes .. ssp. ***cusickii***
41′ Infl branches stout, longest gen < 15 mm; dry ± alpine ridges — uncommon in CA .... ssp. ***pallida***

P. annua L. (p. 1291) ANNUAL BLUEGRASS Ann, bien, cespitose or with stolons, 0.2–2 dm. **LF**: sheath open ± 2/3 length; ligule 1–5 mm, rounded to obtuse; blade gen 1–3 mm wide, soft, gen flat, bright or yellowish green. **INFL** 1–10 cm, triangular, 1.2–1.6 × longer than wide, open in fr; branches spreading, smooth, with spikelets only in top 1/2. **SPIKELET**: axis ± hidden; callus glabrous; lemma 2.5–4 mm, smooth, veins soft-hairy or glabrous; palea keels hairy. **FL**: bisexual or upper 1–2 pistillate; anther 0.6–1 mm. 2*n*=28. Abundant. Disturbed moist ground, lawns, etc.; gen < 2000 m. CA (esp near coast); ± worldwide, native to Eur. Fl winter and spring, continuously when moisture permits.

P. atropurpurea Scribner SAN BERNARDINO BLUEGRASS Per from rhizomes, tufted, 1–5.5 dm, dioecious. **LF**: sheath open ± 2/3 length; ligule 1–2 mm, truncate to rounded, smooth; blade 1.5–3 mm wide, ± firm, folded and inrolled. **INFL** 3–7 cm, lanceolate; branches appressed, smooth. **SPIKELET**: callus glabrous; lemma 2.5–3.5 mm, glabrous, smooth. **FL** unisexual; anthers 1.5–2 mm. 2*n*=28. RARE. Moist meadows; 1500–2200 m. SnBr, PR. Threatened by development, grazing, vehicles. PR pls pistillate; fls early summer. In cult.

P. bigelovii Vasey & Scribner Ann, cespitose, gen 1.5–4 dm. **LF**: sheath open 1/2–3/4 length; ligule 1.5–6 mm, truncate to obtuse, minutely scabrous; blade 1.5–5 mm wide, soft, gen flat, abruptly prow-tipped. **INFL** 5–15 cm, ± linear; branches appressed, with spikelets from near base. **SPIKELET**: callus cobwebby; lemma 2.5–4 mm, veins (sometimes between) hairy; palea keels hairy. **FL**: anthers 0.2–1 mm. 2*n*=28. Uncommon. Shady places in desert scrub, yellow-pine forests; < 1500 m. SW, D; to TX, nw Mex. Fls early spring.

P. bolanderi Vasey Ann, cespitose, 2–6 dm. **LF**: sheath open 1/4–1/2 length; ligule 2–7 mm, rounded to obtuse; blade 1.5–5 mm wide, short, soft, gen flat, abruptly prow-tipped. **INFL** gen 10–15 cm; branches appressed in fl, spreading in fr, with spikelets in tip 1/2. **SPIKELET**: internodes elongated, glabrous; callus cobwebby; lemma 2.5–3 mm, minutely scabrous to nearly smooth; palea keels scabrous. **FL**: anthers 0.5–1.8 mm. 2*n*=28. Mtns, esp in open pine forest; gen 1500–3000 m. NW, SNH, SnJt; to WA, ID, UT. Fls late spring. ❀SUN:**1–3**;STBL.

P. bulbosa L. (p. 1291) BULBOUS BLUEGRASS Per, densely cespitose, 1.5–6 dm. **ST**: bases ± bulbous. **LF**: sheath open to near base; ligule 2–4 mm, obtuse; blade 1–2 mm wide, soft, flat or folded, soon withering. **INFL** 3–10 cm, ovate to lanceolate; branches gen ascending, smooth. **SPIKELET**: florets gen formed into leafy bulblets. **FL**: anthers 1.2–1.5 mm in fertile lemmas. 2*n*= 21–42. Disturbed places; gen < 2000 m. CA; ± worldwide temp; native to Eur. Sporadic in CA. Fls spring.

P. compressa L. CANADIAN BLUEGRASS Per from long, stout rhizomes, 1.5–6 dm. **ST** (incl nodes) ± flattened, keeled, wiry; nodes obviously exposed. **LF**: sheath open 3/4 length to near base; ligule 1–3 mm, rounded; blade 1.5–4 mm wide, soft to ± firm, flat or folded. **INFL** 2–9 cm, lanceolate to ovate, dense (or sparse and interrupted); branches ± ascending, short, densely scabrous on angles. **SPIKELET**: callus glabrous or ± cobwebby; lemma 2.3–3.5 mm, keel and marginal veins hairy. **FL**: anthers 1.3–1.8 mm. 2*n*=42 most often. Moist, often disturbed low ground; < 1800 m. CA; ± worldwide temp; native to Eur. Fls summer.

P. confinis Vasey BEACH BLUEGRASS Per from slender rhizomes or stolons, delicate, tufted, 0.7–3 dm, ± dioecious. **LF**: sheath open 1/2–2/3; ligule < 2 mm, obtuse to acute, scabrous; blade 1–2 mm wide, ± firm, folded, inrolled, upper blade surface on sterile sts gen finely hairy. **INFL** gen 1–5 cm, ovate, ± tawny; branches ± appressed, ± scabrous. **SPIKELET**: callus glabrous or diffusely short-cobwebby; lemma 2–4.5 mm, glabrous or sparsely hairy. **FL**: fertile anthers 1.5–2 mm. 2*n*=42. Ocean beaches, stabilized dunes; < 100 m. NCo; to B.C. Fls spring. ❀DUNE:**5,17,24**; STBL.

P. cusickii Vasey Per, ± densely cespitose, 1–6 dm, ± dioecious. **LF**: sheath open 1/5–3/4 length; ligule gen 1–6 mm (on sterile sts < 2 mm, truncate, scabrous); blade longest at mid-st, on sterile sts gen 0.5–1 mm wide, ± firm, inrolled (sometime also folded), upper surface finely hairy. **INFL** 2–12 cm, lanceolate to ovate, gen dense; branches ascending to appressed, slender, smooth or scabrous. **SPIKELET**: callus gen glabrous; lemma keeled, gen glabrous (rarely keel sparsely hairy), smooth or scabrous; palea keels scabrous. **FL**: fertile anthers 2–3.5 mm. Moist to dry meadows, sagebrush scrub, montane forest; 1500–3600 m. KR, SNH, GB; to s Can, ND, Colorado.

ssp. ***cusickii*** **LVS**: basal tuft dense; sheath open 1/2–3/4 length; 0–1 nodes barely exposed. **INFL** 3–12 cm; branches slender, obviously scabrous, longest gen > 17 mm. **SPIKELET**: lemma 3.5–7.5 mm, glabrous or scabrous. **FL** gen staminate or pistillate. 2*n*=28. Moist meadows, dry slopes in sagebrush scrub or montane forest; 1500–2500 m. e KR, n SNH, GB; to WA, ID, n NV. [*P. hansenii* Scribner] Fls spring to early summer. ❀IRR:**1–3**.

ssp. ***epilis*** (Scribner) W.A. Weber (p. 1291) SKYLINE BLUEGRASS **LVS**: basal tuft sparse; sheath open 1/3–2/3 length; 1–2 nodes well exposed. **INFL** 3–6 cm, dense; branches appressed, ± smooth, longest gen 10–22 mm. **SPIKELET**: lemma 4–5.8 mm, glabrous or scabrous. **FL** pistillate. 2*n*=56. Moist subalpine, esp snowbeds; 2400–3600 m. SNH; to B.C., MT, Colorado. [*P. epilis* Scribner] ❀IRR:**1–3**.

ssp. ***pallida*** R. Soreng **LVS**: basal tuft dense; sheath open 1/2–4/5 length; 0–1 nodes barely exposed. **INFL** 3–5 cm, ± dense; branches appressed, obviously scabrous, stout, longest gen < 15 mm. **SPIKELET**: lemma 5–6.5 mm, glabrous or scabrous. **FL** pistillate in CA. 2*n*=56,59. Uncommon in CA. High montane to lower alpine dry meadows, ridges; 2000–3500 m. c&s SNH, W&I; w Can, ND, Colorado. [*P. subaristata* Beal, not Philippi] Fls early summer. ❀DRY:**1–3**.

ssp. ***purpurascens*** (Vasey) R. Soreng **LVS**: basal tuft sparse; sheath open 1/4–1/2 length; 1–2 nodes exposed. **INFL** 4–8 cm, sparse; branches ascending, slender, sparsely scabrous, 17–30 cm, few-fld. **SPIKELET**: callus sometimes sparsely short-cobwebby; lemma 4.5–7 mm, keel and marginal veins gen sparsely short-hairy near base. **FL** pistillate. $2n$=28. Uncommon in CA. Moist subalpine meadows, ledges; 2500–3500 m. KR, n SNH; to B.C. [*P. p.* Vasey, not Sprengel] Fls summer. ❀IRR:**1–3**.

P. douglasii Nees (p. 1291) SAND-DUNE BLUEGRASS Per from long rhizomes or stolons, tufted, 1–3 dm, dioecious. **LF**: sheath open ± 1/2 length; ligule 1–2 mm, finely scabrous; blade 1–2.5 mm wide, firm, slender, folded, inrolled, upper surface finely hairy. **INFL** 2–6 cm, narrowly ovate, dense; peduncle and branches densely scabrous. **SPIKELET**: callus glabrous or with a ring of hairs; lemma 5–7.5 mm, glabrous or veins hairy. **FL**: fertile anthers 1.5–2 mm. $2n$=28. Shifting coastal dunes; gen < 100 m. s NCo, CCo, n ChI. Threatened by alien spp. Fls early summer. ❀IRR, DRN,SUN:**16,17,24**.

P. fendleriana (Steudel) Vasey ssp. ***longiligula*** (Scribner & Williams) R. Soreng (p. 1291) LONGTONGUE MUTTON GRASS Per, densely cespitose, gen with short rhizomes, 1.5–7 dm, dioecious. **LF**: sheath open ± 1/3 length; ligule 1–18 mm, truncate to acuminate; blade 1.5–4 mm wide, firm, folded, inrolled, uppermost st-lf blade gen ± vestigial, sterile st blade upper surface gen finely hairy. **INFL** gen 2–12 cm, lanceolate to ovate, dense; branches scabrous. **SPIKELET**: callus glabrous; lemma 3.5–6 mm, keel and marginal veins (sometimes between) hairy. **FL** gen pistillate; fertile anthers 2–4 mm. $2n$=56. Mtn slopes, sagebrush scrub to subalpine; 2000–3200 m. s SNH, GB, DMtns; to B.C., SD, n Mex. [*P. longiligula* Scribner & Williams] Fls spring to early summer. ❀DRY:**1–3**; STBL.

P. glauca M. Vahl. ssp. ***rupicola*** (Nash) W.A. Weber TIMBERLINE BLUEGRASS Per, densely cespitose, all current shoots flowering, gen 0.5–1.5 dm. **LF**: sheath open > 4/5 length; ligule 0.5–3 mm, truncate and finely scabrous at margin to acute and smooth; blade 1–2 mm wide, soft, flat or folded, abruptly ascending or spreading. **INFL** 1–5 cm, lanceolate to ovate; branches ascending to appressed, gen < 1.5 cm, gen scabrous on angles. **SPIKELET**: upper internodes < 1 mm, not elongated; glumes 3-veined, upper 2.5–3.5 mm (< 3/4 length of lower); callus glabrous; lemma 2.5–3.5 mm, veins and base hairy. **FL**: anthers 1.2–1.8 mm. $2n$=42–70. Dry alpine slopes, ridges; 3300–4000 m. c&s SNH, W&I; to sw Can, NM. [*P. rupicola* Nash] Fls summer.

P. howellii Vasey & Scribner Ann, cespitose, 3–8 dm. **LF**: sheath open 1/8–1/2 length; ligule 2.5–5 mm, truncate to acute; blade gen 1–6 mm wide, soft, gen flat, barely prow-tipped. **INFL** gen 20–25 cm; branches ascending in fl, spreading to reflexed in fr, with spikelets only in tip 1/2, densely scabrous. **SPIKELET**: internodes gen < 1 mm, some hairy; callus cobwebby; lemma ± 3 mm, short-hairy over body; palea keels scabrous or short-hairy. **FL**: anthers 0.2–1 mm. Rocky banks, shaded slopes of woodlands, chaparral, disturbed places; gen < 1000 m. CA-FP; to B.C. [*P. bolanderi* ssp. *h.* (Vasey & Scribner) Keck] Fls spring.

P. infirma Kunth WEAK BLUEGRASS Ann, cespitose, 0.2–1.5 dm. **LF**: sheath open ± 2/3 length; ligule 1–5 mm; blade 1–4 mm wide, gen flat, yellowish green. **INFL** 1–6 cm, 1.5–3 5 longer than wide; branches ascending, smooth, spikelets only in tip 1/2. **SPIKELET**: axis plainly visible, last internode 1/2–3/4 length of last lemma; callus glabrous; lemma 2–2.5 mm, veins curly-hairy; palea keels hairy. **FL**: upper 1–2 pistillate; fertile anthers 0.2–0.5 mm. $2n$=14. Disturbed low ground; < 500 m. s GV, s CW, SW; native to Medit. Mostly confined to s CA-FP in US. Like *P. annua*. Fls early spring.

P. keckii R. Soreng Per, densely cespitose, gen 0.3–1 dm. **LF**: sheath open 4/5–9/10 length; ligule of uppermost sterile st lf < 2.5 mm, smooth or sparsely scabrous; blade 1–2 mm wide, firm, folded, inrolled. **INFL** 1.5–6 cm, lanceolate to narrowly ovate. **SPIKELET**: callus glabrous; lemma 3–5 mm, glabrous or sparsely short-hairy. **FL**: anthers 0.7–2 mm. High alpine, often in open ground; > 3000 m. SNH, SNE (White, Sweetwater Mtns). Formerly mistaken for *P. suksdorfii* (Beal) Piper of WA. Fls summer.

P. kelloggii Vasey Per from long rhizomes, 2.5–8.5 dm. **LF**: sheath open to near base; ligule 1–4 mm, finely scabrous, truncate to obtuse; blade 2–5 mm wide, soft, gen flat. **INFL** 10–20 cm, ovate to triangular, open, sparse; branches ascending in fl, ± drooping in fr, scabrous, with spikelets only in tip 1/3. **SPIKELET**: callus cobwebby; lemma 4–5 mm, glabrous. **FL**: anthers ± 2 mm. $2n$=56. Shady openings in mixed-conifer and redwood forests; < 500 m. NCo, n CCo; ?sw OR. Fls spring. ❀TRY.

P. leptocoma Trin. ssp. ***leptocoma*** BOG BLUEGRASS Per, loosely cespitose, 1–7 dm. **LF**: sheath open 1/2–3/4 length; ligule 1.5–4 mm, truncate to obtuse, smooth; blade 1–4 mm wide, soft, gen flat. **INFL** 4–15 cm, open; branches ascending in fl, ± drooping in fr, with spikelets in top 1/3. **SPIKELET**: lower glume 1-veined; callus cobwebby; lemma 3–4 mm, keel and marginal veins sparsely hairy. **FL**: anthers 0.2–1 mm. $2n$=42. Moist subalpine, lower alpine meadows; 1800–3200 m. KR, CaRH, n&c SNH, GB; to AK, MT, NM. Fls summer.

P. lettermanii Vasey Per, cespitose, delicate, gen 0.2–0.9 dm. **LF**: sheath open 3/4–9/10 length; ligule 1–4 mm, truncate to acute, smooth; blade 0.5–1.5 mm wide, soft, flat or folded. **INFL** gen 1–3 cm, narrowly lanceolate. **SPIKELET**: glumes (at least upper) > lowest lemma; callus glabrous; lemma 2.5–3 mm, glabrous or very sparsely short-hairy. **FL**: anthers 0.2–0.7 mm. $2n$=14. High alpine, in sandy soil around boulders; > 3500 m. s SNH, W&I; to sw Can, Colorado. Fls summer. ❀IRR:**1**.

P. macrantha Vasey LARGE-FLOWERED SAND-DUNE BLUEGRASS Per from very long, stout rhizomes or stolons, tufted, 1–6 dm, dioecious. **LF**: sheath open ± 1/2 length; ligule < 5 mm; blade 2–4 mm wide, firm, folded and inrolled, upper surface finely hairy. **INFL** 3–15 cm, narrowly ovate, dense, ± interrupted; branches smooth to ± scabrous. **SPIKELET**: callus glabrous or with a ring of hairs; lemma 6–11 mm, glabrous or keel and marginal veins hairy. **FL**: fertile anthers 2–5 mm. $2n$=28. Shifting coastal dunes; < 100 m. n&c NCo; to B.C. [*P. douglasii* Nees ssp. *m.* (Vasey) Keck] Fls late spring, early summer. ❀DUNE:**5,17**;STBL.

P. napensis Beetle NAPA BLUEGRASS Per, densely cespitose, gen 3–10 dm. **LF**: sheath open to near base; ligule 4–6 mm, obtuse to acute, scabrous; blade 1–3 mm wide, ± firm, folded to inrolled, scabrous. **INFL** gen 5–15 cm, ± open; branches ascending, with dense spikelets in tip 1/2, densely scabrous. **SPIKELET**: callus glabrous; lemma 3–4 mm, glabrous or very sparsely short-hairy, scabrous. **FL**: anthers 1.2–1.8 mm. $2n$=42. **ENDANGERED** CA. Low sterile ground near hot springs; 100–200 m. se NCoRO (near Calistoga, Napa Co.). Fls spring. In cult.

P. nemoralis L. WOOD BLUEGRASS Per, cespitose, all current shoots flowering, 3–7 dm. **LF**: sheath open 3/4 length to near base; ligule < 0.5 mm, minutely scabrous, truncate; blade 1.5–4 mm wide, soft, gen flat, abruptly spreading, barely prow-tipped. **INFL** gen 5–15 cm, open, lanceolate to narrowly ovate; branches ascending, scabrous on angles. **SPIKELET**: lower glume gen 3-veined; callus sparsely cobwebby; lemma 3–3.5 mm, keel and marginal veins hairy. **FL**: anthers 0.3–1.4 mm. $2n$=28,42. Disturbed moist places in forest, along streams, coastal to mid-montane; gen < 2000 m. NW, SN; cool temp; native to Eur. Fls spring.

P. palustris L. FOWL BLUEGRASS Per, cespitose or with stolons, gen 2.5–12 dm. **LF**: sheath open 3/4 length to near base; ligule 1–3 mm, acute to rounded; blade 1.5–6 mm wide, soft, gen flat, often > sheath, slenderly prow-tipped, base closely ascending. **INFL** gen 10–30 cm, eventually open, lanceolate to narrowly triangular, many-fld; branches ascending to spreading in fr, scabrous on angles. **SPIKELET**: lower glume gen 3-nerved; callus cobwebby; lemma gen 2–3 mm, keel and marginal veins hairy. **FL**: anthers 0.8–1.4 mm. $2n$=28. Disturbed ground in moist forests or sagebrush scrub, meadows, along streams; 1500–2000 m. NW, CaRH, SnGb, SnBr, SNE; cool temp; native to Eur. Fls spring to early summer.

P. pattersonii Vasey PATTERSON'S BLUEGRASS Per, densely cespitose, gen 0.5–1.5 dm; sterile shoots many. **LF**: sheath open 4/5 length to near base, basal sheaths persisting; ligule 1–3.5 mm, acute, smooth; blade 1.5–2.5 mm wide, folded, ± inrolled, closely ascending. **INFL** 2–6 cm, lanceolate, dense; branches appressed, smooth or sparsely scabrous. **SPIKELET**: glumes ± equal, upper

3.5–4.2 mm, gen = first lemma; callus gen cobwebby; lemma 3.5–4 mm, keel and marginal veins gen hairy; palea keels sparsely scabrous above, some sparsely short-hairy below. **FL**: anthers gen 0.7–1.2 mm. 2*n*=42,84. RARE in CA. High alpine open ground; gen > 3300 m. SNH; to Colorado, W&I. Fls mid-summer.

P. piperi A. Hitchc. (p. 1291) PIPER'S BLUEGRASS Per from rhizomes, tufted, 2–5.5 dm, dioecious. **LF**: sheath open 1/3–2/3 length; ligule 1–2 mm, scabrous; blade 1.5–3 mm wide, firm, folded, inrolled, collar and upper surface of sterile st blades finely hairy. **INFL** gen 4–8 cm, sparse; branches ascending. **SPIKELETS** few; callus diffusely cobwebby; lemma 4–7 mm, glabrous, smooth to scabrous. **FL**: fertile anthers 2–3 mm. 2*n*=28. UNCOMMON. Serpentine, talus, chaparral, forest openings; 100–500 m. nw KR (Del Norte Co.); sw OR. Sometimes mistaken for *P. rhizomata* A. Hitchc. ❀DRY:5,**17**.

P. pratensis L. ssp. *pratensis* (p. 1291) KENTUCKY BLUEGRASS Per from long, stout rhizomes, tufted or loose, gen 2–7 dm. **LF**: sheath open 1/2–3/4 length; ligule 1–4 mm, truncate, smooth to minutely scabrous at margin; blade gen 2–4 mm wide, soft to ± firm, flat or folded. **INFL** gen 6–15 cm, ovate to triangular; branches gen spreading, smooth or scabrous. **SPIKELET**: callus densely long-cobwebby; lemma 3–4 mm, keel and marginal veins hairy. **FL**: anthers 1.2–2 mm. 2*n*=21–117. Common. Many disturbed or stable habitats, incl saline or alkaline soils; 0–3500 m. CA; n temp; native to Eur. Widely planted as lawn or pasture grass. Fls spring to early summer. ssp. agassizensis (Boivin & D. Löve) Taylor & McBride is widespread and possibly native in CA, has dense infl < 6 cm with smooth branches, lf firm and folded, sterile shoot blade upper surface often sparsely soft-hairy; ssp. *angustifolia* (L.) Arcang. is probably introduced in CA, has long, folded sterile shoots gen < 0.5 mm wide, sometimes hairy as in ssp. *agassizensis*, narrowly triangular open infl with with smooth ascending branches.

P. pringlei Scribner Per, densely cespitose, 0.5–3.5 dm, dioecious. **LF**: sheath open 2/3–6/7 length; ligule 1–6 mm, truncate to acute, scabrous or smooth; blade 1.5–3 mm wide, ± firm, folded, inrolled, upper surface of sterile st blades finely hairy. **INFL** 1–6 cm, gen ± lanceolate, dense; branches steeply ascending or appressed. **SPIKELET**: glumes nearly = first lemma, thin, shiny; callus glabrous; lemma 5–8 mm, glabrous. **FLS** pistillate in n SNH; fertile anthers (KR) 2–4 mm. Open places, esp snowbeds, high montane to subalpine; 2000–3000 m. KR, CaRH, n SNH; sw OR. Fls early summer. ❀IRR:**1**.

P. rhizomata A. Hitchc. TIMBER BLUEGRASS Per from rhizomes, 2.5–6 dm, ± dioecious. **LF**: sheath open 1/2–2/3 length; ligule 2–8 mm, acute, smooth; blade 1.5–3.5 mm wide, soft, gen flat. **INFL** 4–10 cm, ovate, sparse; branches gen 1–2 per node, with few spikelets; tip ± nodding. **SPIKELET**: callus cobwebby; lemma 4–6.5 mm, keel and marginal veins hairy. **FL** pistillate or bisexual (rarely staminate); fertile anthers 2.5–4 mm. 2*n*=28. UNCOMMON. Shady moist slopes in forest, in rich loose soils, over granitics; 400–1000 m. KR (Siskiyou, n Trinity cos.); sw OR. Fls late spring to early summer. ❀SHD,IRR:**5,17**.

P. secunda J.S. Presl (p. 1291) Per, densely cespitose, 1.5–10 dm. **LF**: sheath open 3/4 length to near base; ligule 0.5–10 mm, truncate to acuminate, sometimes scabrous; blade gen 0.5–3 mm wide, soft to firm, flat to folded or inrolled. **INFL** 2–25 cm, often ± 1-sided, gen linear to lanceolate, gen dense; branches gen appressed to ascending (gen spreading only in fl), ± scabrous. **SPIKELET** ± cylindric or little compressed; upper internodes gen > 1.2 mm; callus glabrous or with a ring of short hairs; lemma 3.5–5 mm, weakly keeled to rounded across lower back, glabrous to ± evenly short-hairy across body (rarely soft-hairy only on veins), smooth to scabrous. **FL**: anther 1.5–3 mm. 2*n*=42–106 (mostly high polyploids). Common. Many habitats; 0–3800 m. CA; to AK, Rocky Mtns, nw Mex, also in S.Am. Many ecological forms; sspp. tend to intergrade.

ssp. ***juncifolia*** (Scribner) R. Soreng Pl 3–12 dm. **LF**: ligule 0.5–6 mm, truncate to acuminate, scabrous (those on lateral shoots 0.5–2 mm); blade gen < 1.5 mm wide, ± firm, tightly folded and inrolled, retaining shape, often glaucous. **INFL** gen 6–25 cm. **SPIKELET**: lemma 3.5–6 mm, glabrous (rarely sparsely short-hairy on keel and marginal veins near base); palea keels scabrous. 2*n*=±63 most often. Sagebrush shrubland to lower montane forest, often in alkaline depressions; 900–3000 m. CaRH, SNH (esp e slope), GB; to s Can, ND, NM. Fls early summer. Several important ecotypes: 1) "*P. ampla* Merr." (big bluegrass: pl 6–12 dm; lf blade flat, gen glaucous; non-alkaline); 2) "*P. juncifolia* Scribner" (alkali bluegrass: lf blade folded and inrolled, ligule truncate; often alkaline); 3) "*P. nevadensis* Scribner" (Nevada bluegrass: lf blade folded and inrolled, ligule long, acute; often alkaline). "*P. pratensis* × *P. secunda*" [*P. fibrata* Scribner; *P.* × *limosa* Scribner & Williams (Lassen Co. bluegrass, see key: 16.)] is a set of hybrids between *P. s.* ssp. *j.* and *P. pratensis*; 2*n*=63–64; low, often saline meadows; 800–2000 m; range of ssp. j. ❀DRY:**1–9,14–24**;salt tolerant.

ssp. ***secunda*** ONE-SIDED BLUEGRASS Pl 1.5–10 dm. **LF**: ligule 2–10 mm, acute to acuminate, smooth or sparsely scabrous; blade 0.5–3 mm wide, soft, gen flat, soon withering, sometimes glaucous, basal often thread-like. **INFL** gen 2–15 cm. **SPIKELET**: lemma 4–5 mm, base ± evenly short-hairy (sometimes nearly glabrous); palea keels and between them gen hairy in lower 1/2. 2*n*=±84 most often. Common. Habitat and range of sp. Fls spring and early summer (midsummer in subalpine). [*P. canbyi* (Scribner) Howell; *P. gracillima* Vasey; *P. incurva* Scribner & Williams; *P. sandbergii* Vasey; *P. scabrella* (Thurb.) Vasey] Many ecological forms have been named; all intergrade completely, probably do not warrant taxonomic recognition. ❀DRY:**1–9,14–24**;some forms tolerate IRR.

P. sierrae J. Howell (p. 1291) Per from rhizomes, 2–6 dm, slender, dioecious, gen with purplish, scaly, axillary buds. **LF**: sheath open < 1/10 length; ligule 3–6 mm, acute to acuminate, minutely scabrous; blade 1.5–2.5 mm wide, soft, gen flat, uppermost > 1.5 × sheath length. **INFL** 5–15 cm, narrowly ovate, open, sparse. **SPIKELET**: callus glabrous or sparsely cobwebby; lemma 4–7 mm, glabrous to sparsely hairy on base, keel, marginal veins. **FL**: fertile anthers 2–4 mm. Uncommon. Shady moist slopes, often on mossy rocks, in canyons, forests; 350–1500 m. n SNF. Fls early spring. ❀SHD,IRR:**7**.

P. stebbinsii R. Soreng Per, cespitose, gen 1–3.5 dm, slender, ± dioecious. **LF**: sheath open 3/5–4/5 length; ligule of uppermost lf of sterile st 2.5–6 mm, clear, smooth; blade ± firm, folded, inrolled, upper blade surface on sterile sts ± finely hairy. **INFL** 3–7 cm, narrowly lanceolate, dense to sparse and interrupted. **SPIKELET**: callus glabrous; lemma 3.7–5.5 mm, glabrous, smooth or barely scabrous, thin. **FL** bisexual or pistillate (rarely staminate); fertile anthers 2–4.5 mm. 2*n*=42,81. Subalpine to lower alpine meadows; 2700–3700 m. c&s SNH. Formerly mistaken for *P. hansenii* Scribner or *P. leibergii* Scribner. Hybridizes with *P. keckii*, *P. secunda* ssp. *s.* Fls summer.

P. tenerrima Scribner (p. 1291) DELICATE BLUEGRASS Per, cespitose, delicate, 1.5–4 dm; basal lf tuft < 8 cm. **LF**: sheath open 3/4 length to near base; blade 0.5–1.5 mm wide, soft, gen flat, fine. **INFL** 7–9 cm, open; branches widely spreading in fr, scabrous. **SPIKELET**: callus glabrous or with a ring of short hairs; lemma 3–4.2 mm, weakly keeled, short-hairy near base to sparsely bristly. **FL**: anthers 1.6–2.1 mm. 2*n*=42. Uncommon. Thin, drying soils, often on serpentine; < 700 m. SNF, se SnFrB, SCoR, WTR. Sometimes mistaken for *P. secunda* ssp. *s.* Fls early spring. ❀DRN:**7,9,14–18**,19–24.

P. trivialis L. ROUGH BLUEGRASS Per, short-lived, cespitose or with short stolons from a dense basal tuft, 3–10 dm. **LF**: sheath open gen 3/4, scabrous; ligule 3–7 mm, acute, scabrous; blade 1–5 mm wide, soft, gen flat, barely prow-tipped. **INFL** 8–25 cm, open, lanceolate to ovate; branches ascending to spreading, densely scabrous, many-fld from lower 1/2. **SPIKELET**: lower glume 1-veined, ± sickle-shaped; callus sparsely long-cobwebby; lemma ± 2–3 mm, keel (rarely marginal veins) hairy near base. **FL**: anthers 1.4–2 mm. 2*n*=14. Disturbed moist places; 0–700 m. NCo, NCoRO, probably elsewhere; to AK, e N.Am; native to Eur. Fls spring and early summer.

P. unilateralis Vasey OCEAN-BLUFF BLUEGRASS Per, densely cespitose, 0.5–4 dm. **LF**: sheath open to near base; ligule 1.5–5 mm, smooth; blade 1–5 mm wide, soft to moderately thick, gen flat. **INFL** gen 3–7 cm, dense, cylindric or ± lanceolate; branches short, smooth or sparsely scabrous, appressed, with spikelets from base.

SPIKELET: callus glabrous or with a ring of hairs; lemma 3–4.5 mm, glabrous or with curved hairs on keel and veins (esp marginal). **FL**: anthers 1.5–3 mm. Coastal bluffs, in ± saline soils; ≲ 200 m. NCo, n&c CCo; scattered to WA. [Incl *P. pachypholis* Piper of WA] Fls spring. ❀SUN,DRN:**17,24**.

P. wheeleri Vasey (p. 1291) Per from short rhizomes, ± tufted, gen 3.5–8 dm, dioecious. **LF**: sheaths open 1/3–2/3 length, lower often finely reflexed-scabrous or short-hairy; lower ligules 0.5–2 mm, truncate to rounded, scabrous; blade 1.5–3.5 mm wide, soft to ± firm, flat or folded; sterile st blades folded and ± inrolled, upper surface gen ± finely hairy. **INFL** gen 5–12 cm, ovate, sparse; branches gen 1–4, ± ascending to spreading, ± scabrous. **SPIKELET**: callus glabrous; lemma 3–6 mm, glabrous to sparsely hairy on keel and marginal veins, gen sparsely scabrous. **FL** pistillate (anthers ± vestigial). 2*n*=56,63,70–91. Common (esp SN). Mtns, open forest in rich soil; 1300–3800 m. KR, NCoRH, CaRH, SNH, GB; to s Can, Colorado. Formerly mistaken for *P. nervosa* (Hooker) Vasey. Fls summer. ❀DRY:**1–3**;REVEG.

POLYPOGON BEARD GRASS

Steven A. Conley

Ann, per. **STS** decumbent to erect, simple. **LF**: sheath open, loosely enclosing st, glabrous; ligule thinly membranous, ± ovate to oblong, obtuse to truncate, minutely ciliate to toothed; blades ± cauline, flat, scabrous, veins minutely prickly at 10×. **INFL** panicle-like, oblong to narrowly ovoid in outline, interrupted to compact, dense. **SPIKELETS** breaking below glumes and part of stalk, falling as 1 unit; glumes ± equal, 1-veined, entire or 2-lobed, awned at tip or between lobes, awn straight; floret 1; lemma ± 0.5 × glumes, translucent, 5-veined, tip toothed, awn < glume awn or 0; palea slightly < lemma, transparent; anthers tightly enclosed by lemma and palea. **FR** oblong, smooth, enclosed by lemma, palea. 18 spp.: warm temp Eurasia, Afr, S.Am. Some spp. orn. (Greek: much bearded)

1. Glume awn gen > 3.5 mm
 2. Lemma awn 0; glume lobes > 0.5 mm, ciliate-fringed .... ***P. maritimus***
 2′ Lemma awn 0.5–5 mm; glume lobes < 0.5 mm or 0, not ciliate-fringed
  3. Glumes ± truncate at 10×; ligules 2–4 mm; lemma awn gen > 2.5 mm .... ***P. australis***
  3′ Glumes minutely lobed at 10×; ligules 5–6 mm; lemma awn gen < 2.5 mm .... ***P. monspeliensis***
1′ Glume awn gen < 3.5 mm
 4. Spikelet stalk gen > 2 mm; glumes > 2.5 mm; ligule 6–8 mm .... ***P. elongatus***
 4′ Spikelet stalk gen < 1 mm; glumes < 2.5 mm; ligule gen 4–5 mm .... ***P. interruptus***

P. australis Brongn. CHILEAN BEARD GRASS Per. **ST** erect, 1–10 dm. **LF**: ligule 2–4 mm, truncate, ± entire; blade 3–14 cm, 5–7 mm wide. **INFL** 6–16.5 cm, 1–3.5 cm wide. **SPIKELET** ± subsessile; stalk < 0.5 mm; glumes 2–3 mm, minutely bristly, awn purplish in fl, 2–7(11) mm; lemma 1–2 mm, awn 1.2–5(7) mm. Edges of streams, ditches; < 1000 m. NCoRO, SN, GV, SW; Baja CA; native to S.Am.

P. elongatus Kunth Per. **ST** decumbent to erect, 6–10 dm. **LF**: ligule 6–8 mm, toothed; blade 5.5–11.5 cm, 1 cm wide. **INFL** 19–30 cm, 2 cm wide; lower branches stiffly erect. **SPIKELET**: stalk > 2 mm; glumes 2.5–4 mm, acute, sparse bristly below middle, awn 1–2.5 mm; lemma 1.5–2.5 mm, awn < 2 mm or 0. 2*n*=28,56. Sand dunes, salt marshes; < 100 m. SnFrB, SCoR; native to S.Am.

P. interruptus Kunth DITCH BEARD GRASS Per. **ST** 5–9 dm, clumped. **LF**: ligule 2–8(13) mm, obtuse to truncate, minutely hairy; blade 0.5–19.5 cm, 3–6 mm wide. **INFL** 1.5–18 cm. **SPIKELET**: stalk < 1 mm; glumes 1.5–3 mm, scabrous; awns 1.5–4.5 mm, purplish; lemma 1–2 mm, awn 0.5–3 mm. 2*n*=42. Common. Streambanks, ditches; < 1300 m. CA; to Great Plains, sc US; native to S.Am.

P. maritimus Willd. (p. 1295) MEDITERRANEAN BEARD GRASS Ann. **ST** 0.5–5 dm. **LF**: ligule < 6 mm, irregularly toothed, minutely hairy; blade 1–14 cm, 2–4 mm wide. **INFL** 1–8.5(15) cm, plume-like, densely fld. **SPIKELET**: stalk < 1 mm; glumes 1–2 mm, minutely bristly below middle, ciliate-fringed, tip lobes 0.5–1 mm, awn 4.5–11.5 mm; lemma 0.5–1.5 mm, awn 0. 2*n*=14,42. Common. Moist places; < 600 m. NW, SNF, GV; native to Medit Eur, Afr.

P. monspeliensis (L.) Desf. (p. 1295) ANNUAL BEARD GRASS Ann. **ST** 2–10 dm. **LF**: ligule 5–6 mm, irregularly toothed, minutely hairy; blade 1–20.5 cm, 4–6 mm wide. **INFL** 1–17 cm, plume-like, densely fld. **SPIKELET**: glumes 1–2.5 mm, minutely bristly, awn 2–10 mm; lemma 0.5–1.5 mm, awn 0.5–4.5 mm. 2*n*=28,35. Common. Moist places, along streams, ditches; < 2100 m. CA; N.Am; native to s&w Eur.

PSEUDOROEGNERIA

Mary E. Barkworth

Per, cespitose or from rhizomes. **ST** erect. **LVS** basal and cauline; ligule membranous; uppermost blade gen widely spreading. **INFL** spike-like; spikelets ± 2-ranked, 1 per node, each spikelet partly overlapping spikelet on opposite side of axis, not overlapping spikelets on same side but not adjacent. **SPIKELET**: glumes < or = lower floret, lanceolate, stiffly membranous, acute to obtuse, awn 0; axis breaking above glumes and between florets; lemma awn terminal, strongly spreading or 0; anthers 4–8 mm. 19 spp.: arid temp Eurasia, N.Am. (Greek: false roegneria, earlier name for *Elymus*) [Carlson 1986 MS Thesis, OR State Univ] Segregated from *Agropyron*.

P. spicata (Pursh) A. Löve ssp. ***spicata*** (p. 1295) **ST** 6–10 dm, slender, green or glaucous. **LF**: blade 1–4 mm wide, flat to loosely rolled. **INFL** 8–16 cm, narrow; middle internodes 0.8–2.5 cm. **SPIKELET**: glumes 6–13 mm, 0.9–2.2 mm wide, ± 1/2 spikelet length, veins evenly glabrous or scabrous; florets 6–8; lemmas 9–14 mm, awned or not. Sagebrush steppe, open woodland; 800–1650 m. NCoR, CaR, n&c SN, MP; to s Can, Colorado. [*Agropyron s.* (Pursh) Scribner & J. Smith] Other ssp. in nw US. ❀DRN,SUN:1,**2,3**,6,**7,14–16**,17&IRR:8–10,**19–24**.

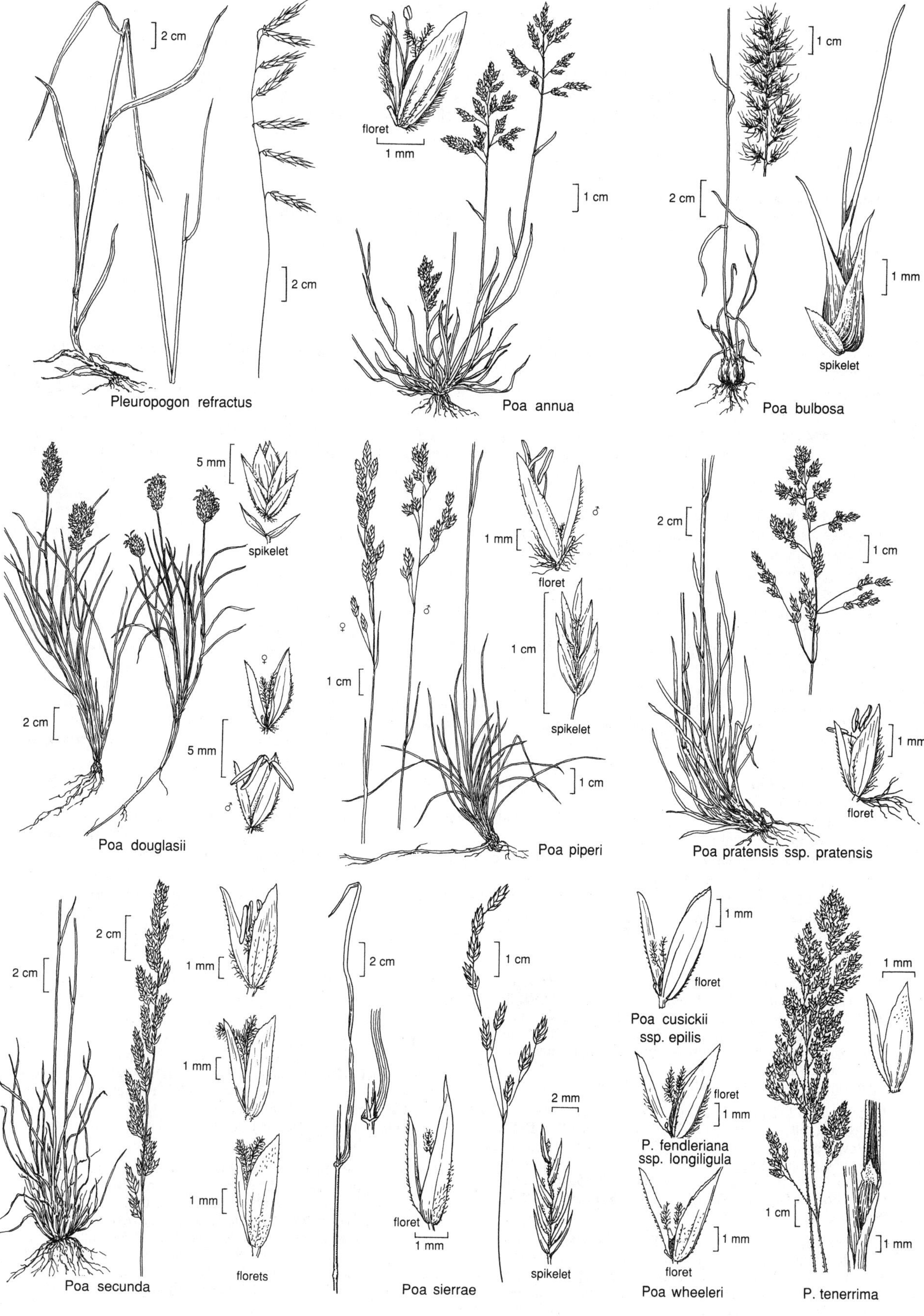
2 cm
2 cm
Pleuropogon refractus
floret
1 mm
1 cm
Poa annua
1 cm
2 cm
1 mm
spikelet
Poa bulbosa
5 mm
spikelet
2 cm
5 mm
Poa douglasii
1 mm
floret
1 cm
spikelet
1 cm
1 cm
Poa piperi
2 cm
1 cm
1 mm
floret
Poa pratensis ssp. pratensis
2 cm
2 cm
1 mm
1 mm
1 mm
florets
Poa secunda
2 cm
1 cm
2 mm
floret
1 mm
spikelet
Poa sierrae
1 mm
floret
Poa cusickii
ssp. epilis
floret
1 mm
P. fendleriana
ssp. longiligula
1 mm
floret
Poa wheeleri
1 mm
1 cm
1 mm
P. tenerrima

PTILAGROSTIS

Mary E. Barkworth

Per, cespitose. **ST** < 1 mm diam. **LF**: sheath glabrous; ligule membranous. **INFL** panicle-like; axis glabrous, branches glabrous to sparsely puberulent. **SPIKELET**: glumes ± equal, slightly > floret (exc awn), purplish, veins not evident, tip rounded to acute; axis breaking above glumes; floret 1; callus 0.1–0.8 mm, blunt; lemma thick, membranous, evenly hairy, margins open with age, awn 12–30 mm, persistent. ± 10 spp.: temp N.Am, Asia. (Greek: feather grass) [Barkworth 1983 Syst Bot 8:395–419]

P. kingii (Bolander) Barkworth (p. 1295) KING'S RICEGRASS **ST** 2–4 dm. **LF**: blade ± 0.3 mm diam, inrolled. **INFL** 6–10 cm; branches flexible, ascending. **SPIKELET**: glumes 3–4.5 mm; floret 2.8–4.2 mm; lemma awn 10–14 mm, weakly bent once or twice. 2*n*=22. Subalpine, alpine streambanks, meadows; 2700–3500 m. c&s SNH. [*Oryzopsis k.* (Bolander) Beal] ❀IRR,DRN,SUN:1–3, **6**,**7**,14,**15–17**,18–21.

PUCCINELLIA ALKALI GRASS

Jerrold I Davis

Ann, per; stolons and rhizomes gen 0. **STS** decumbent to erect. **LVS** basal and cauline; sheath open ± to base; ligule thinly membranous, acute to truncate, sometimes toothed. **INFL** panicle-like; lower branches reflexed to erect; spikelets stalked. **SPIKELET** bisexual; glumes < lowest floret, lower glume gen 1-veined, upper 3-veined; florets 2–9; lemma gen firm, back gen rounded, sometimes weakly keeled near tip, margin entire to scabrous-serrate near tip at 10×, glabrous to weakly puberulent at base, gen faintly 5-veined; awn 0; palea ± = lemma. 50 spp.: temp to arctic, N.Am, Eurasia. (B. Puccinelli, Italy, 1808–1850) [Davis 1983 Syst Bot 8:341–353] Gen on wet saline or alkaline soils; some spp. difficult to separate without hand lens.

1. Ann; remains of previous years growth not present
  2. Lowest lemma ± 2 mm; lemma tip obtuse to truncate .... ***P. parishii***
  2′ Lowest lemma 3.5–4 mm; lemma tip acute .... ***P. simplex***
1′ Per; previous years lvs persistent
  3. Spikelet stalk glabrous or sparsely fine scabrous
    4. Lemma margin near tip uniformly scabrous-serrate at 10×; lowest floret anthers 1.5–2 mm; KR (Shasta Co.) .... ***P. howellii***
    4′ Lemma margin near tip entire or sparsely fine-scabrous at 10×; lowest floret anthers < 1 mm; NCo .... ***P. pumila***
  3′ Spikelet stalk uniformly scabrous
    5. Lemma tip obtuse to truncate
      6. Lowest floret lemma 1.5–2 mm; lower infl branches spreading to reflexed in fr .... ***P. distans***
      6′ Lowest floret lemma 2–4 mm; lower infl branches erect to reflexed in fr
        7. Lowest floret lemma 3.5–4 mm; lower infl branches erect in fr; coastal habitats .... [2]***P. nutkaensis***
        7′ Lowest floret lemma 2–3.5 mm; lower infl branches erect to reflexed in fr; CA-FP, GB, gen inland .... [2]***P. nuttalliana***
    5′ Lemma tip acute
      8. Lowest floret lemma 3.5–4 mm; cauline lf blade 2–3 mm wide when flat; lower infl branches erect in fr; coastal habitats .... [2]***P. nutkaensis***
      8′ Lowest floret lemma 2–3.5 mm; cauline lf blade 1–4 mm wide when flat; lower infl branches erect to reflexed in fr; inland habitats
        9. Cauline lf blade inrolled, 1–2 mm wide when flat; lemma margin near tip entire to uniformly scabrous-serrate at 10×; infl gen < 10 cm .... ***P. lemmonii***
        9′ Cauline lf blade flat to inrolled, 1–4 mm wide when flat; lemma margin near tip scabrous-serrate at 10×; infl gen > 10 cm .... [2]***P. nuttalliana***

P. distans (Jacq.) Parl. EUROPEAN ALKALI GRASS Per. **LF**: cauline blade flat to inrolled, 1–7 mm wide when flat. **INFL** 2.5–22 cm; lower branches spreading to reflexed in fr; spikelet stalk scabrous.**SPIKELET**: lemma tip widely obtuse to truncate, margin near tip scabrous-serrate, lowest lemma 1.5–2 mm; anthers of lowest floret 0.5–1 mm. 2*n*=14,28,42. Saline meadows and flats; < 2700 m. CA-FP, GB; to AK, ne N.Am; native to Eurasia.

P. howellii J.I. Davis (p. 1295) HOWELL'S ALKALI GRASS Per. **LF**: cauline blade inrolled, 1.5–2 mm wide when flat. **INFL** 2–7 cm; lower branches erect to reflexed in fr; spikelet stalk glabrous to sparsely scabrous. **SPIKELET**: lemma tip acute to obtuse, margin near tip scabrous-serrate, lowest lemma 2.5–3.5 mm; anthers of lowest floret 1.5–2 mm. RARE. Mineral springs; ± 500 m. KR (Shasta Co.).

P. lemmonii (Vasey) Scribner (p. 1295) LEMMON'S ALKALI GRASS Per. **LF**: cauline blade inrolled, 1–2 mm wide when flat. **INFL** 2–10(18) cm; lower branches ascending to reflexed in fr; spikelet stalk scabrous. **SPIKELET**: lemma tip acute, margin near tip entire to scabrous-serrate, lowest lemma 2.5–3.5 mm; anthers of lowest floret 1–2 mm. 2*n*=14. Saline meadows and flats; 700–2000 m. KR, CaR, GB; to OR, ID, WY. ❀TRY;also STBL.

P. nutkaensis (J.S. Presl) Fern. & Weath. ALASKA ALKALI GRASS Per. **LF**: cauline blade flat to inrolled, 2–3 mm wide when flat. **INFL** 5.5–15 cm; lower branches erect in fr; spikelet stalk scabrous. **SPIKELET**: lemma tip acute to obtuse, margin near tip uniformly scabrous-serrate, lowest lemma 3.5–4 mm; anthers of lower floret 0.5–1 mm. 2*n*=42,56. Marshes, wet sites; < 10 m. CCo; to AK. Robust pls from CCo (Pt. Reyes) have been called *P. grandis* Swallen. ❀STBL.

P. nuttalliana (Schultes) A. Hitchc. NUTTALL'S ALKALI GRASS Per. **LF**: cauline blade flat to inrolled, 1–4 mm wide when flat. **INFL** (3.5)10–22 cm; lower branches erect to reflexed in fr; spikelet stalk scabrous. **SPIKELET**: lemma tip acute to obtuse, margin near tip scabrous-serrate, lowest lemma (2)2.5–3.5 mm; anthers of lowest floret 0.5–1.5 mm. 2*n*=28,42,56. Saline meadows and flats; < 2300 m. CA-FP, GB; to AK, e N.Am. [*P. airoides* (Nutt.) S. Watson & J. Coulter] ❀STBL;valuable forage grass.

P. parishii A. Hitchc. PARISH'S ALKALI GRASS Ann. **LF**: cauline blade gen inrolled, < 1 mm wide when flat. **INFL** 1–8 cm; lower branches erect to reflexed in fr; spikelet stalk scabrous (sometimes sparsely scabrous). **SPIKELET**: lemma veins hairy in lower half, tip obtuse to truncate, margin near tip scabrous-serrate, lowest lemma ± 2 mm; anthers of lowest floret ± 0.5 mm. 2*n*=14. RARE. Mineral springs; ± 1000 m. w DMoj (sw San Bernardino Co.); to NM.

P. pumila (Vasey) A. Hitchc. DWARF ALKALI GRASS Per. **LF**: cauline blade inrolled, 2–3 mm wide when flat. **INFL** 3–7 cm; lower branches reflexed to ascending in fr; spikelet stalk glabrous to sparsely scabrous. **SPIKELET**: lemma tip acute to obtuse, margin near tip entire to sparsely scabrous, lower lemma 2.5–3 mm; anthers of lowest floret 0.5–1 mm. 2*n*=42,56. UNCOMMON. Marshes and flats; < 10 m. NCo; to AK, ne N.Am.

P. simplex Scribner (p. 1295) Ann. **LF**: cauline blade gen inrolled, 1–2 mm wide when flat. **INFL** 1–18 cm; lower branches erect in fr; spikelet stalk ± glabrous to scabrous, often faintly short-soft-hairy. **SPIKELET**: lemma back short-hairy, rounded to weakly keeled at tip, acute, margin near tip entire to sparsely scabrous, lowest lemma 3.5–4 mm; anthers of lowest floret < 0.5 mm. 2*n*=56. Saline flats and mineral springs; < 800 m. Teh, GV, SnFrB, w DMoj; to UT.

RHYNCHELYTRUM

Robert Webster

Ann, per. **ST** gen erect; internode solid to hollow inside. **LVS** gen cauline; ligule hairy. **INFL** gen panicle-like, open; branches spreading to ascending; spikelets subsessile to stalked. **SPIKELET** falling as 1 unit; glumes strongly unequal, lower << upper, upper glume silky-hairy; florets 2, lower floret sterile or staminate, lemma ± = upper glume, upper floret fertile, lemma membranous to thick, firm, smooth, ± white in fr, margin flat, tip blunt, palea ± = lemma. ± 15 spp.: warm temp, subtrop, se Asia, Afr. (Latin: beaked scale, from beaked upper glume in some spp.)

R. repens (Willd.) C.E. Hubb. NATAL GRASS Per, sometimes cespitose. **ST** decumbent to erect, 3–10 dm; nodes 4–5. **LF**: sheath 3–9 cm, glabrous; ligule hairs 0.5–1.5 mm; blade 3–20 cm, 3–6 mm wide, upper surface glabrous. **INFL** 8–17 cm; 1° branches 2.5–6 cm, glabrous to puberulent; spikelet stalk 0.5–5 mm, ± wiry. **SPIKELET** ± 3–5.5 mm, ± 1–2 mm wide, ovate to elliptic; lower glume < 1.5 mm, 0–1-veined, upper glume ± = spikelet, silky hairs white to purplish; lower floret staminate, lemma ± like upper glume, 5-veined, tip minutely lobed, palea ± = lemma; upper floret ± 2/3 length lower floret, lemma firm, ± white, shiny. Disturbed places, fields, slopes; < 300 m. NCoRO, SnJV, s CCo, SCoRO, SCo; to s US; native to S.Afr. [*Tricholaena rosea* Nees] Used for soil stabilization.

SACCHARUM

Kelly W. Allred

Per with rhizomes. **STS** tall, erect. **LVS** cauline; sheaths > internodes; ligule short, membranous or hairy; blades flat or folded. **INFL** panicle-like, silky-hairy. **SPIKELET** bisexual, in pairs; lower spikelet sessile; upper spikelet stalked; pair falling with subtending axis segment as 1 unit or stalked spikelet deciduous; glumes, lemmas long-tapered; glumes > florets, 3–5-veined; florets 2, lower vestigial, obscure, upper fertile; lemma awned or awn 0; palea < lemma or 0. 35–40 spp.: trop, subtrop, se Asia. Some orn, sugarcane (*S. officinarum*) widely cult for sugar. (Latin: sugar)

S. ravennae (L.) Murray RAVENNAGRASS **STS** densely tufted, gen 2–4 m. **LF**: ligule < 1 mm, thin; blade < 12 mm wide, gen densely hairy near ligule, strongly serrate. **INFL** plume-like, 2.5–6 dm. **SPIKELET** 3.5–7 mm; stalked spikelet deciduous; glumes lanceolate, base densely silky-hairy; lemma awned, awn 3–5 mm. 2*n*=20. Ditchbanks, marshes; < 300 m. s DSon (Imperial Co.); native to Eurasia. [*Erianthus r.* (L.) P. Beauv.]

SCHISMUS MEDITERRANEAN GRASS

Kelly W. Allred

Ann, bien, cespitose. **ST** erect to prostrate. **LVS** gen basal, tufted, glabrous; ligule short-hairy; blades flat or inrolled. **INFL** panicle-like, dense. **SPIKELET** ± laterally compressed; glumes lanceolate, membranous, 5–7-veined; florets 3–8, bisexual, breaking above glumes and between florets; lemma 9-veined, ciliate proximally, toothed to notched, awn ± 0; palea < to > lemma. 5 spp.: warm temp, subtrop, Eurasia, Afr. (Greek: split, from notched lemma) [Conert and Türpe 1974 Abh. Senckenberg Naturf. Ges. 532:1–81]

1. Lower lemmas gen > 2.5 mm, teeth narrowly triangular; palea of lowest floret acute, < lemma .... ***S. arabicus***
1' Lower lemmas gen < 2.5 mm, teeth obtuse to widely triangular, palea of lowest floret obtuse, > lemma . ***S. barbatus***

S. arabicus Nees (p. 1295) Ann. **ST** gen 5–20 cm. **LF**: blade < 1 mm wide, thread-like. **SPIKELET**: glumes 4.5–6.5 mm; lemma 2.5–4 mm, teeth ± 0.3 × lemma, narrowly triangular; palea 2–3 mm, gen << lemma. 2*n*=12. Dry, open, gen disturbed areas; < 1300 m. SnJV, CW, s ChI, D; to TX, AZ; native to Eurasia.

S. barbatus (L.) Thell. (p. 1295) Ann. **ST** gen 2–16 cm. **LF**: blade < 2 mm wide, thread-like. **SPIKELET**: glume 4–5 mm; lemma 2–2.5 mm, teeth ± < 0.2 × lemma, obtuse to widely triangular; palea 1.5–2.5 mm, gen = > lemma. 2*n*=12. Dry, open, gen disturbed areas; < 1200 m. Teh, SnJV, SW, D; to TX, n Mex; native to s Eur, Afr.

SCHIZACHYRIUM

Kelly W. Allred

Ann, per. **STS** gen erect. **LVS** basal and cauline; sheaths gen keeled, < internodes; ligule membranous, minutely ciliate; blade flat or folded. **INFL** raceme-like, several often clustered together; branches 2–3, spike-like; axes breaking apart with age. **SPIKELETS** paired; pair with subtending axis segment falling as 1 unit; lower spikelet sessile, bisexual; upper spikelet stalked, staminate. **SESSILE SPIKELET**: glumes > florets, lanceolate, leathery; florets 2, lower vestigial, obscure, upper fertile, hyaline, awned at 2-toothed tip. ± 60 spp.: warm temp, trop. (Greek: split chaff, from toothed lemma)

S. scoparium (Michaux) Nash (p. 1295) LITTLE BLUESTEM Per. **ST** 5–15 dm, branched. **LF** glabrous to sparsely hairy; ligule 0.5–2.5 mm; blade gen 6–25 cm, 1.5–5 mm wide, green to glaucous. **INFL** spike-like; branches 2–3, 2–6 cm, axes zig-zag with age; stalks flat, ciliate on upper 2/3, spreading with age. **SESSILE SPIKELET**: glumes 6–9 mm, lower enclosing upper; lemma 4–6.5 m, awn 8–16 mm, bent, twisted. $2n$=40. Disturbed sites, roadsides; < 150 m. ScV (Solano Co.); s Can, c&e US, n Mex. Cult for forage, soil stabilization. Pls with straight infl axis, spikelet stalks ciliate at tip are *S. cirratum* (Hackel) Wooton & Standley, reported once from s San Diego Co.

SCLEROCHLOA HARDGRASS

Dieter H. Wilken

1 sp. (Greek: hard grass, from thick glumes)

S. dura (L.) Beauv. (p. 1295) Ann. **STS** gen clumped, prostrate to erect on same pl, 2–8 cm. **LVS** basal and cauline; sheath gen open; ligule < 2 mm, membranous, obtuse, tip toothed; blade 2–3 cm, flat. **INFL** spike-like, 1–2 cm. **SPIKELETS** 6–10 mm, subsessile, crowded, 2-ranked on 1 side of axis, breaking below glumes, falling as 1 unit (sometimes weakly breaking between florets); glumes and lemmas firmly membranous at base, distal margin translucent; glumes << lower lemmas, unequal, rounded, lower ± 2 mm, gen 3-veined, upper 4–6 mm, 7–9-veined; florets 3–5, lower 2–3 bisexual, upper staminate or sterile; lemmas 3–7 mm, strongly keeled, obtuse, 5-veined, awn 0; palea < lemma. $2n$=14. Disturbed sites; 700–800 m. w CaRH (Shasta Valley); to WA, Colorado, TX; native to s Eur.

SCLEROPOGON

1 sp. (Greek: hard beard, from firm awns) [Reeder & Toolin 1987 Phytologia 62:267–275]

S. brevifolius Philippi (p. 1295) BURRO GRASS Per, from short stolons, mat-like, gen dioecious or monoecious, rarely bisexual. **ST** erect, 1–2 dm. **LVS** gen basal, densely tufted; sheath smooth; ligule short-hairy; blade 2–8 cm, 1–3 mm wide, flat, firm, sharp-pointed. **STAMINATE INFL** 3–7 cm, raceme- or panicle-like. **PISTILLATE INFL** 10–20 cm, spike-like. **STAMINATE SPIKELET** 1–2.5 cm, compressed; glumes 3–8 mm, ± equal, lanceolate, 1(3)-veined, awn 0; axis breaking above glumes and falling as 1 unit; florets 5–10(20); lemma ± = glumes, 3(5)-veined, short-awned. **PISTILLATE SPIKELET** 8–15 cm incl awns; body exc awns ± cylindric, gen subtended by 1 bract; bract ± = lower glume; glumes 3-veined, acute, awn 0, lower glume 1–2 cm, upper 1.5–3 cm; axis breaking above glumes and ± between florets; florets 3–5, lower with sharp-pointed callus, upper reduced to awns; lemma 7– 11 mm, narrow, awns 3, 6–12 cm, ± spreading to ascending. $2n$=40. RARE in CA. Open creosote-bush shrubland; ± 1600 m. e DMtns (New York Mtns); to Colorado, TX, Mex, w S.Am.

SCRIBNERIA

Thomas Worley

1 sp. (Frank L. Scribner, American agrostologist, 1851–1938)

S. bolanderi (Thurber) Hackel (p. 1295) Ann. **ST** ascending to erect, 5–30 cm, branched, glabrous; nodes purple. **LF**: sheath 1–5 cm; ligule 2–4 mm, translucent; blade midrib, margins scabrous. **INFL** 4–11 cm, straight, ± purple; spikelets 1–2 per node, appressed, lower ± embedded in axis, upper (if present) with stalk 1–3 mm, overlapping node above. **SPIKELET**: glumes 2, 5–7 mm, edge-to-edge or overlapping, lower 2-veined, upper 4-veined; floret 1, breaking above glumes; callus short-bristly; lemma 4–6 mm, translucent, scabrous on upper 1/3, tip notched, awned, 5-veined; awn 2–4 mm, ± straight; palea gen < lemma, notched; stamen 1, anther gen > 1 mm. **FR** ± 2.5 mm. $2n$=26. Dry, disturbed areas; 500–2500 m. CA-FP (exc most SW); to WA. Reported from San Diego Co.

SECALE RYE

Mary E. Barkworth

Ann, per. **ST** gen erect. **LF**: sheath appendaged; ligule membranous; blade gen flat. **INFL** spike-like, dense, ± flat; axis sometimes breaking at nodes in fr; spikelets 2-ranked, 1 per node, sessile, not sunken. **SPIKELET**: glumes narrow, rigid, keeled, vein gen 1; florets 2, fertile, sessile and side-to-side, sometimes with vestigial floret between; lemma with keel near margin, keel and margins ciliate, veins 5, tip tapered, awn straight, scabrous. 5 spp.: Eurasia. (Latin: ancient name for rye)

S. cereale L. (p. 1295) Ann, sometimes bien. **ST** 6–12.5 dm, glabrous exc below infl. **LF**: sheath glabrous, appendages ± 1 mm; blade 3–10 mm wide. **INFL** 8–17 cm, gen not breaking apart. **SPIKELET**: glumes 6–17 mm, keeled; lemma 10–16 mm, awn 2–7 cm; anthers 7.5–8.5 mm. $2n$=14,16,27–29. Disturbed slopes, roadsides; < 1800 m. n SNH, sw SnFrB, s MP, expected elsewhere; native to sw Asia.

SETARIA

Robert Webster

Ann, per. **STS** gen erect; internode solid to hollow inside. **LVS** basal and cauline; ligule short-hairy or membranous, ciliate. **INFL** panicle-like, dense, gen cylindric; 1° branches spreading to appressed; spikelets many, gen clustered on one side of short 2° branch, short-stalked to subsessile, subtended by 1–15 bristles, bristles gen scabrous. **SPIKELET** falling as 1 unit,

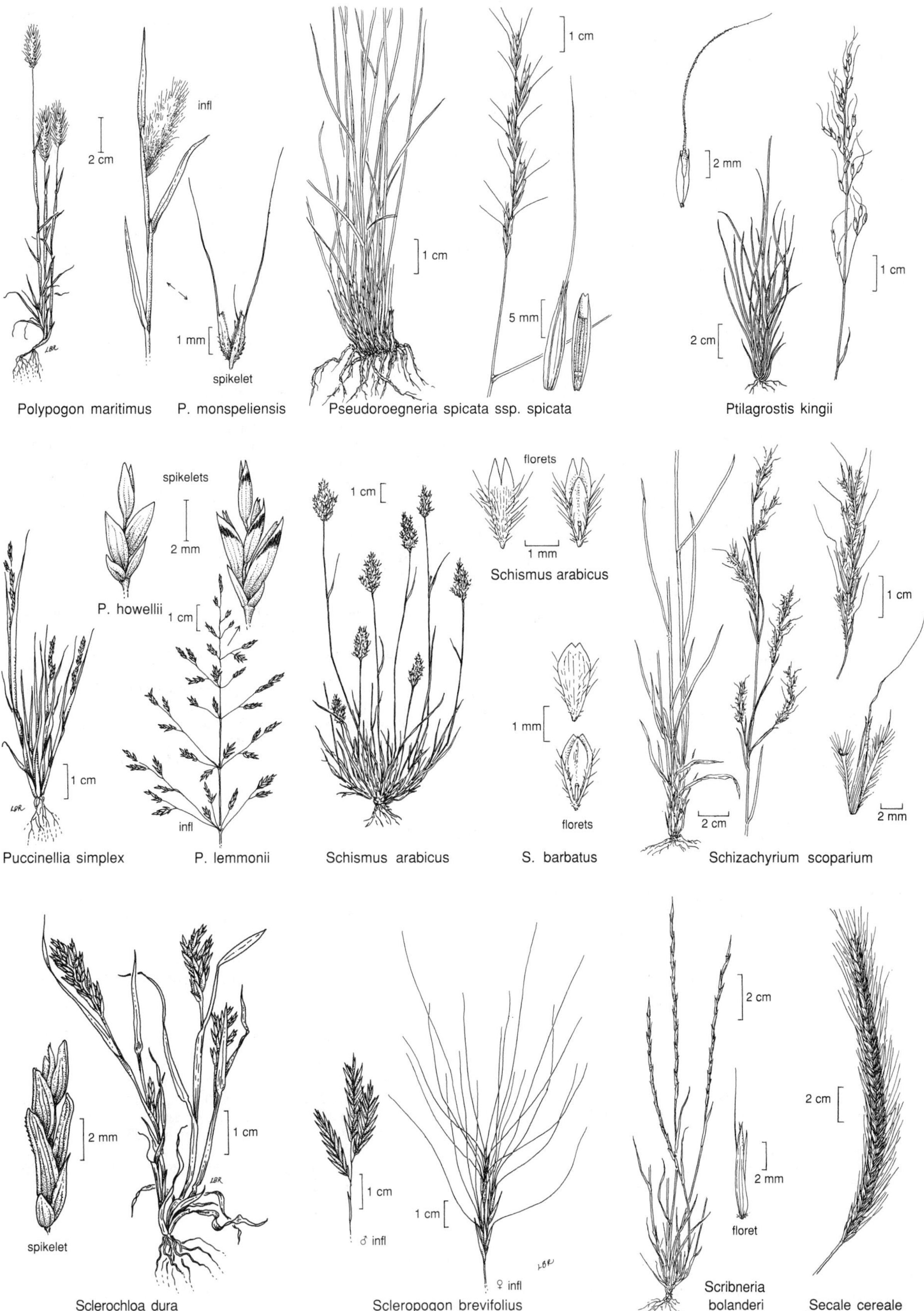

Polypogon maritimus P. monspeliensis Pseudoroegneria spicata ssp. spicata Ptilagrostis kingii

Puccinellia simplex P. lemmonii Schismus arabicus S. barbatus Schizachyrium scoparium

Sclerochloa dura Scleropogon brevifolius Scribneria bolanderi Secale cereale

gen elliptic; glumes unequal; florets gen 2, ± equal, lower floret sterile or staminate, palea gen < lemma, upper floret fertile, firm, gen hard, rough, margin inrolled, tip blunt. ± 100 spp.; warm temp, trop Eurasia, Afr. (Latin: bristly) [Rominger 1962 Illinois Biol Monogr 29:1–132] Some spp. cult for food.

1. Spikelet or spikelet cluster subtended by 1–3 bristles; lower floret palea gen < 2/3 × lemma
  2. Short stiff hairs on bristles pointed downward; main infl axis scabrous .... ***S. verticillata***
  2′ Short stiff hairs on bristles pointed toward tip; main infl axis short-hairy
    3. Spikelet 2.5–3 mm; upper lf blade surface short-soft- hairy .... ***S. faberi***
    3′ Spikelet ± 2 mm; upper lf blade surface glabrous .... ***S. viridis***
1′ Spikelet or spikelet cluster subtended by 4–15 bristles; lower floret palea gen well developed
  4. Spikelet ± 2.8–3.5 mm, gen 1.5–2 mm wide; ann .... ***S. pumila***
  4′ Spikelet gen < 2.8 mm, width gen = or < 1.5 mm; per
    5. Infl axis < 8 cm; base of fl sts with hard, knot-like swellings .... ***S. gracilis***
    5′ Infl axis > 8 cm; base of fl sts not swollen .... ***S. sphacelata***

S. faberi R. Herrm. Ann. **ST** erect, 1–20 dm; nodes 4–7. **LF**: sheath 6–11 cm, glabrous; ligule ± 1.5–2.5 mm; blade 15–30 cm, 1–2 cm wide, upper surface short-soft-hairy. **INFL** 7–18 cm; 1° branches 0.5–4 mm; bristles 2–3; spikelet stalk ± 0.5–1 mm. **SPIKELET** 2.5–3 mm, 1.4–1.8 mm wide; lower glume 1–1.5 mm, 3–5-veined, upper glume 0.7–0.9 × spikelet length; lower floret sterile, lemma 7-veined, tip acute, palea ± 2/3 lemma length. $2n$=36. Disturbed areas, fields; < 100 m. Deltaic GV, CCo, SCo; to e US; native to Asia. NOXIOUS WEED.

S. gracilis Kunth (p. 1301) Per, cespitose, from short rhizomes. **ST** erect, 7–12 dm; base with hard, knot-like swellings. **LF**: sheath 4–9 cm, glabrous; ligule < 1 mm; blade < 25 cm, 2–8 mm wide, upper surface glabrous. **INFL** 3–8 cm; 1° branches 3–8 mm; axis glabrous; bristles 4–12; spikelet stalk << 0.5 mm. **SPIKELET** 2–3 mm, ± 1–1.5 mm wide; lower glume 1–1.5 mm, 3- veined, upper glume 0.5–0.8 × spikelet length; lower floret sterile or staminate, lemma 5–7-veined, tip acute, palea ± = lemma. Open areas, grassland, chaparral; < 400 m. GV, CW, SCo, SNE (very uncommon), DMoj; to e US, C. & S.Am. [*S. geniculata* (Lam.) P. Beauv. misapplied] ❀STBL.

S. pumila (Poiret) Roemer & Schultes Ann. **ST** 2–13 dm. **LF**: sheath 4–9 cm, glabrous; ligule ± 0.5–1 mm; blade 5–30 cm, 3–10 mm wide, upper surface glabrous. **INFL** 2–6 cm; 1° branches 5–10 mm, axis short-hairy; bristles 4–12; spikelet stalk << 0.5 mm. **SPIKELET** 3–3.5 mm, ± 1.5–2 mm wide; lower glume 1–1.5 mm, 3–5-veined, upper glume 0.5–0.7 × spikelet length; lower floret staminate, lemma 5-veined, tip acute, palea = lemma. Gen moist sites, fields; < 1200 m. e KR, CaRF, SNF, GV, CW, SCo, SNE, DMoj; to s Can, e US, Mex; native to Eur. [*S. lutescens* (Weigel) Hubb., *S. glauca* (L.) P. Beauv.]

S. sphacelata (Schum.) Moss Per, cespitose or not. **ST** 6–30 dm. **LF**: sheath 4–10 cm, glabrous or hairy; ligule 1–2 mm; blade 10–50 cm, 3–15 mm wide, upper surface glabrous. **INFL** 9–40 cm; 1° branches 5–10 mm, axis glabrous; bristles 5–15; spikelet stalk < 0.5 mm. **SPIKELET** ± 2–3 mm, 1–1.5 mm wide; lower glume ± 1–1.5 mm, 2-veined, tip acute, upper glume 0.6–0.8 × spikelet length; lower floret staminate, lemma 5-veined, tip acute, palea = lemma. $2n$=18,36. Disturbed places, fields; < 200 m. GV; native to Afr.

S. verticillata (L.) P. Beauv. Ann. **ST** decumbent to erect, 1.6–10 dm. **LF**: sheath 2–10 cm, glabrous; ligule 1–2 mm; blade 5–25 cm, 3–12 mm wide, upper surface glabrous to hairy. **INFL** 3–13 cm; 1° branches 1.1–2 cm, axis glabrous; bristles 1–2; spikelet stalk << 0.5 mm. **SPIKELET** ± 1.7–2.5 mm, 1–1.5 mm wide; lower glume ± 0.5–1.5 mm, 1–3-veined, upper glume ± = spikelet length; lower floret sterile, lemma 5–7-veined, tip acute to rounded, palea vestigial. Waste places, fields, roadsides; < 200 m. SnJV, SW; native to Eur. [*S. carnei* A. Hitchc. misapplied]

S. viridis (L.) P. Beauv. (p. 1301) Ann. **ST** decumbent to erect, 2–10 dm. **LF**: sheath 5–15 cm, scabrous; ligule 1–2 mm; blade 8–20 cm, 3–12 mm wide, upper surface glabrous. **INFL** 2–15 cm; 1° branches 3–10 mm, glabrous or hairy; bristles 1–3; spikelet stalk < 0.5 mm. **SPIKELET** ± 2 mm, ± 1 mm wide; lower glume ± 1 mm, 3-veined, upper glume = spikelet length; lower floret sterile, lemma 5-veined, tip acute, palea vestigial. $2n$=18. Waste places, fields, roadsides; < 300 m. CA-FP, DSon; widespread N.Am; native to Eurasia.

SORGHUM

Kelly W. Allred

Ann, per, cespitose or with rhizomes. **ST** erect; internodes gen solid. **LVS** cauline; sheaths gen < internodes; ligule membranous; blades flat or folded. **INFL** panicle-like, open to compact. **SPIKELETS** in pairs (trios at branch tips); lower sessile, bisexual; upper 1(2) stalked, staminate; pair with subtending axis segment falling as 1 unit. **SESSILE SPIKELET** ovoid; lower glume leathery, shiny, glabrous to puberulent, enclosing upper glume, florets; florets 2, lower sterile, upper fertile; lemma membranous, fertile lemma awned, awn bent, twisted; palea < lemma. ± 20 spp.: trop, subtrop, Afr. Cult for food, forage, sugar. (Italian: Sorgho) [de Wet 1978 Amer J Bot 65:477–484]

1. Ann .... ***S. bicolor***
1′ Per with rhizomes .... ***S. halepense***

S. bicolor (L.) Moench SORGHUM, MILO, SUDAN GRASS Ann. **ST** erect, 1–2 m. **LF**: blade 3–10 cm wide. **INFL** 1–4 dm, open to compact; branches ± spreading to stiffly erect. **SESSILE SPIKELET** 4–9 mm; lemma 4–5 mm, awn 5–10 mm or 0. $2n$=20. Disturbed areas, roadsides, fallow fields; < 600 m. NCo, NCoR, GV, CCo, SCo, D; widely cult, native to Afr. [*S. lanceolatum* Stapf, *S. sudanense* (L.) Pers., *S. virgatum* (Hackel) Stapf]

S. halepense (L.) Pers. (p. 1301) JOHNSONGRASS Per with rhizomes. **ST** erect, 0.5–2 m. **LF**: blade 0.5–2 cm wide. **INFL** 1–5 dm, gen open; lower branches spreading to ascending. **SPIKELETS** 4–6.5 mm; lemma 4–5 mm, awn 9–15 mm or 0. **SESSILE SPIKELET** 4–6.5 mm, gen open, 10–60 cm, 5–25 cm wide. $2n$=40. Disturbed areas, ditchbanks, roadsides; < 800 m. NW, CaRF, SNF, GV, CW, SW; native to Medit.

SPARTINA CORD GRASS

John R. Baird & John W. Thieret

Per, gen with rhizomes. **ST** erect, unbranched. **LVS** basal and cauline; sheath open, > internodes, glabrous, margin of sheath opening sometimes with long, shaggy hairs; ligule a fringe of hairs, 0.5–2 mm; blade flat to inrolled, long tapered, upper surface ridged. **INFL** panicle-like; each spike-like branch with 2 rows of overlapping spikelets on lower side of axis.

SPIKELET laterally compressed, sessile, breaking below glumes, falling as 1 unit; glumes firmly membranous, obtuse to acuminate or with a small, sharp point, unequal, upper 1–3(5)-veined, gen > floret, lower 1-veined, < floret; floret 1; lemma 1(3)-veined, acute, awned or not, firmly membranous; palea ± = lemma, 2-veined. 15 spp.: Am, Eur, Afr. (Greek: a cord) [Spicher & Josselyn 1985 Madroño 32:158–167]

1. St 5–14 mm wide at base, internodes fleshy; middle lf blades gen flat when fresh or inrolled near tip, 4–25 mm wide at base; rhizome present; coastal
  2. Infl ± open, spikes loosely overlapping, loosely appressed or spreading (to 10°–20°) .... ***S. alterniflora***
  2′ Infl dense, cylindric, spikes closely overlapping, closely appressed .... ***S. foliosa***
1′ St 1.5–6 mm wide at base, internodes firm; middle lf blades gen inrolled when fresh, 1–8 mm wide at base; rhizome sometimes 0; coastal or interior
  3. Infl open, spikes ascending to spreading (to ± 60°), lowest often not overlapping; coastal .... ***S. patens***
  3′ Infl compact, spikes appressed, overlapping
    4. Rhizome gen 0 (rarely present, short, stout); coastal .... ***S. densiflora***
    4′ Rhizome present, elongate, slender; interior .... ***S. gracilis***

S. alterniflora Lois. SALT-WATER CORD GRASS Rhizome 4–7 mm wide. **ST** thick, solitary or in small clumps, 6–25 dm, 5–14 mm wide at base, fleshy in internodes. **LF**: blade 20–55 cm, 4–25 mm wide at base, flat when fresh or inrolled near tip; ridges on upper surface ± 6 per mm. **INFL** ± open, 10–40 cm, 7–22 mm wide; branches 5–30, loosely overlapping, loosely appressed or spreading (to 10°–20°), 4.5–13 cm, 3–5 mm wide. **SPIKELET** 8–15 mm; glume and lemma keels glabrous to softly shaggy-hairy; lower glume 4–10 mm; upper glume 8–15 mm; lemma 7.5–11 mm. $2n$=56,70. Coastal salt marshes; < 10 m. CCo (San Francisco Bay); native to e N.Am. Will likely spread unless eradicated.

S. densiflora Brongn. (p. 1301) DENSE-FLOWERED CORD GRASS Rhizome gen 0 (if present, short, stout, < 10 mm wide). **ST** cespitose, 2.7–15 dm, 3–6 mm wide at base, slender; internodes firm. **LF**: blade 12–43 cm, 4–8 mm wide at base, gen inrolled when fresh, ridges on upper surface ± 2 per mm. **INFL** 6–30 cm, 4–12 mm wide, compact; branches 2–20, overlapping, appressed, 1–11 cm, 2.5–6 mm wide. **SPIKELET** 8–14 mm; glume and lemma keels with short, sharp bristles, at least near tip; lower glume 4–7 mm; upper glume 8–14 mm; lemma 7–9 mm. Coastal salt marshes; < 10 m. NCo, CCo; native to s S.Am.

S. foliosa Trin. (p. 1301) CALIFORNIA CORD GRASS Rhizome 2–6 mm wide. **ST** solitary or in small clumps, 6–15 dm, 7–12 mm wide at base, thick; internodes fleshy. **LF**: blade 15–45 cm, 5–17 mm wide at base, flat when fresh or inrolled near tip, ridges on upper surfaces ± 5 per mm. **INFL** 9–25 cm, 5–13 mm wide, dense, cylindric; branches 3–25, closely overlapping, gen closely appressed, 2–8 cm, 3–7 mm wide. **SPIKELET** 10–25 mm; glume keels glabrous or with soft, shaggy hairs; lower glume 7–8 mm; upper glume 12–25 mm; lemma 10–12 mm, keel glabrous. $2n$=112. Salt marshes, mudflats, shores; < 10 m. NCo, CCo, SCo; Mex. ❀WET, SUN:**17,22–24**;STBL, salt tolerant.

S. gracilis Trin. (p. 1301) ALKALI CORD GRASS Rhizome 3–5 mm wide. **ST** gen solitary, 1.8–10 dm, 2–5 mm wide at base, slender, internodes firm. **LF**: blade 15–27 cm, 2.5–6 mm wide at base, gen inrolled when fresh, ridges on upper surface ± 5 per mm. **INFL** 4–25 cm, 5–12 mm wide, compact; branches 2–12, overlapping (often for only 1/2 their length, or lowest spike rarely separated), appressed, 1.5–8 cm, 2–6 mm wide. **SPIKELET** 6–11 mm; glume and lemma keels ciliate (hairs gen 0.3–1 mm) at least near tip; lower glume 3–7 mm; upper glume 5–11 mm; lemma 6.5–10 mm. $2n$= 40,42. UNCOMMON. Alkaline lake shores, stream banks, meadows, marshes; 1000–2100 m. SNE, n DMoj; to n&e Can, KS, NM. ❀WET,SUN:1,**2,3**;STBL, salt tolerant.

S. patens (Aiton) Muhlenb. SALT-MEADOW CORD GRASS Rhizome 2–4 mm wide or 0. **ST** slender, solitary or in small clumps, 3–12 dm, 1.5–4 mm wide at base, internodes firm. **LF**: blade 10–50 cm, 1–4 mm wide at base, gen inrolled when fresh, ridges on upper surface ± 3 per mm. **INFL** 5–22 cm, 10–90 mm wide, open; branches 2–13, often separated at least below, ascending to spreading (to ± 60°), 1–8 cm, 2–4 mm wide. **SPIKELET** 7–12 mm; glume and lemma keels scabrous at least near tip; lower glume 2–8 mm; upper glume 6–12 mm; lemma 5–8 mm. $2n$=28. Coastal salt marshes; < 10 m. CCo; native to se US.

SPHENOPHOLIS WEDGEGRASS

Dieter H. Wilken

Ann, per. **STS** solitary to clumped, erect. **LVS** cauline, glabrous to hairy; ligule ± membranous, obtuse to truncate; blade flat to inrolled. **INFL** panicle-like, open to compact; lower branches spreading to erect. **SPIKELET** subsessile to stalked, ± laterally compressed; glumes unequal, lower < upper, lower linear to lanceolate, 1-veined, upper obovate, 3–5-veined; florets 2–3, bisexual; lemmas membranous, rounded on back, compressed at acute to obtuse tip, weakly 5-veined, awned or not; palea < lemma. 4 spp.: N.Am, Caribbean. (Greek: wedge scale, from upper glume shape) [Erdman 1965 Iowa State J Sci 39:289–336]

S. obtusata (Michaux) Scribner (p. 1301) PRAIRIE WEDGEGRASS Per (sometimes fl first year). **ST** 2–8 dm. **LF**: sheath glabrous to scabrous; ligule 1–4 mm, tip jagged; blade 5–8 cm, flat. **INFL** 4–12 cm, gen erect, compact; lower branches ± ascending. **SPIKELET** 2–5 mm; lower glume ± 2 mm, ± linear, keel minutely scabrous, acute; upper glume 2–3 mm, widely obovate, obtuse; lemmas 2–3 mm, lower > glumes, acute to obtuse. $2n$=14. UNCOMMON. Wet meadows, streambanks, ponds; 300–2000 m. n SNH (Amador Co.), s SNH (Fresno Co.), ne SCo (Santa Ana River), sc PR (Cuyamaca Mtns); N.Am, Caribbean.

SPOROBOLUS DROPSEED, SACATON

Michael Curto

Ann, per. **STS** gen ascending to erect, 2–20 dm, gen tufted, ± solid in X-section. **LVS** gen basal; cauline few, ascending or curving away; distal sheath margin and collar glabrous or hairy; ligule < 1 mm, hairy or membranous, fringed; blade flat to inrolled, gen glabrous or scabrous, sometimes short-soft-hairy. **INFL** terminal, also sometimes axillary, panicle- or spike-like, gen partly enclosed by sheath; branches spreading or appressed. **SPIKELET** < 6 mm, gen pale to gray-green or purplish; glumes gen unequal, upper < or > lemma, membranous to translucent, 1-veined; floret bisexual, gen breaking above glumes; lemma texture gen like glumes, 1(3)-veined; palea < or > lemma. **FR** 1–3 mm, gen falling from floret, gen gelatinous when wet. ± 150 spp.: Am, Eurasia, Afr. (Greek: to throw seed, from deciduous seeds)

1. Ann; infl 1–5 cm, terminal and axillary at ± most st nodes; lemma leathery, tip beaked .... ***S. vaginiflorus***
1′ Per; infl 8–80 cm, gen terminal, sometimes axillary at some st nodes; lemma membranous, tip obtuse to acute
 2. Infl ± spike-like, cylindric, dense; 1° branches, gen < 3 cm, appressed, obscure
  3. Upper glume length > 0.5 × lemma, lanceolate, tip entire; glume and lemma midveins ± scabrous at 20× .... ***S. contractus***
  3′ Upper glume length < 0.5 × lemma, ovate, tip gen irregularly toothed; glumes and lemma midveins ± glabrous at 20× .... ***S. indicus***
 2′ Infl oblong to triangular in outline, ± open, sometimes enclosed by sheath; 1° branches gen > 3 cm, gen spreading
  4. Anthers 1–1.8 mm; glume and lemma backs ± rounded, midvein ± glabrous at 20×; lowest infl branches gen > 6 cm; old sheath bases straw-colored, glossy .... ***S. airoides***
  4′ Anthers < 1 mm; glume and lemma backs ± keeled, midvein ± scabrous at 20×; lowest infl branches gen < 6 cm; old sheath bases brown, dull
   5. 1° infl branches ± spike-like; 2°, 3° branches ± appressed, ± obscure .... ***S. cryptandrus***
   5′ 1° infl branches ± raceme-like; 2°, 3° branches spreading, evident .... ***S. flexuosus***

S. airoides (Torrey) Torrey (p. 1301) ALKALI SACATON Per. **STS** tufted, 3–20 dm. **LF**: distal sheath margin glabrous to short-hairy; collar margin glabrous or sparsely long-hairy; ligule < 0.5 mm, fringed; blade 12–40 cm, 2–4(6) mm wide. **INFL** gen terminal, 1–6 dm; base 4–25 cm wide, panicle-like, pyramid-shaped, open to dense; branches ascending to spreading, lowest gen > 6 cm, 2° gen evident, spreading. **SPIKELET** 1–3 mm; glume, lemma backs rounded; base purplish; glumes unequal, narrowly lanceolate, tip acute to obtuse, lower 0.5–2 mm, upper > 0.5 × lemma; floret breaking above glumes; lemma 2–3 mm, ovate to narrowly lanceolate; palea ± = lemma; anthers 1–1.8 mm. **FR** ± 1 mm. 2*n*=80,90,108,126. Seasonally moist, alkaline areas; < 2100 m. SNF, Teh, s ScV, SnJV, s SCoRO, SW, s SNE, D; to e WA, c&s US, Mex. Pls from s SCo, s PR, s DSon with sts 1–2 m, infl base 12–25 cm wide, 1° branches ascending, 2° and 3° branches appressed, spikelets dense have been called var. *wrightii* (Scribner) Gould. ❀SUN,IRR:1–3,**7–24**;also STBL.

S. contractus A. Hitchc. (p. 1301) SPIKE DROPSEED Per. **STS** tufted, 4–12 dm. **LF**: distal sheath margin glabrous to short-hairy; collar margin glabrous to densely long-hairy; ligule < 1 mm, hairy; blade 4–30 cm, 2–8 mm wide. **INFL** gen terminal, 1.5–5 dm; base < 1 cm wide, spike-like, cylindric, dense; branches gen < 3 cm. **SPIKELET** 2–3 mm; glumes, lemma keeled; glumes unequal, lower ± 1 mm, ± awl-like, upper > 0.5 × lemma, narrowly lanceolate, acute; lemma 2–2.5 mm, ovate to narrowly lanceolate, acute; palea ± = lemma; anthers 0.5–1 mm. **FR** ± 1 mm. 2*n*=36. Washes, rocky slopes, shrubland, pinyon/juniper woodland; < 1900 m. W&I, e D; to Colorado, TX, Mex. ❀SUN,DRN:1,**2**,3,**7**,**14–16**, **18–23**.

S. cryptandrus (Torrey) A. Gray (p. 1301) SAND DROPSEED Per. **STS** tufted, 3–10 dm. **LF**: distal sheath margin glabrous to short-hairy; collar margin glabrous to densely long-hairy; ligule < 0.5 mm, hairy; blade 3–15 cm, 1–5 mm wide. **INFL** gen terminal, 0.8–3(4) dm; base 2–6(10) cm wide, panicle-like, pyramid-shaped or oblong in outline, ± open, sometimes completely enclosed by sheath; lower 1° branches gen > 3 cm, ± spike-like; 2° and 3° branches appressed. **SPIKELET** 1.5–2.5 mm; glumes, lemma keeled; glumes unequal, lower 0.5–1 mm, ± awl-like, upper > 0.5 × lemma, narrowly lanceolate, acute; lemma 1.5–2.5 mm, ovate to narrowly lanceolate, acute; palea ± = lemma; anthers < 1 mm. **FR** ± 1 mm. 2*n*=18,36,38,72. Rocky to sandy washes, slopes, shrubland, woodland; 350–2800 m. e-c SNH, SnBr, PR, SNE, D (used for revegetation elsewhere); widespread s Can, US, n Mex. ❀SUN, DRN:1–3,**7**,9,**10**,**11**,12–24;also STBL.

S. flexuosus (Vasey) Rydb. MESA DROPSEED Per. **STS** tufted, 3–10 dm. **LF**: distal sheath margin short-hairy; collar margin densely long-hairy; ligule < 0.5 mm, fringed; blade 5–20 cm, 1–6 mm wide. **INFL** terminal, 1–3 dm; base 4–14 cm wide, pyramid-shaped or oblong in outline, ± open; lower 1° branches gen > 3 cm; 2° and 3° branches evident, spreading. **SPIKELET** 1.5–2.5 mm; glumes, lemma keeled; glumes unequal, lower 0.5–1 mm, ± awl-like, upper > 0.5 × lemma, narrowly lanceolate, acute; lemma 1.5– 2.5 mm, ovate to narrowly lanceolate, acute; palea ± = lemma; anthers < 1 mm. **FR** ± 1 mm. 2*n*=36, 38. Rocky to sandy washes, slopes, shrubland; < 1200 m. SNE, D; to s UT, w TX, N Mex. ❀SUN,DRN:2,3,7,9,**10–12**,13–24.

S. indicus (L.) R.Br. SMUTGRASS Per. **STS** 1–several, tufted, 3–6(10) dm. **LF**: sheath, collar margins glabrous to scabrous; sheath bases intact or becoming frayed, brown, dull; ligule < 0.5 mm, hairy; blade 8–30 cm, 3–5 mm wide. **INFL** gen terminal, 1–5(8) dm; base 0.5–1 cm wide; branches gen < 3 cm. **SPIKELET** 1.5–2.5 mm, ± green or brownish; glume, lemma backs rounded; glumes ± unequal, lower 0.5–1 mm, ± awl-like, upper < 0.5 × lemma, ovate, obtuse to acute; lemma 2–2.5 mm, ovate, acute; palea < or = lemma; anthers 0.5–1 mm. **FR** ± 1 mm. 2*n*=36. Open, disturbed areas, roadsides, lawns; < 1200 m. s CCo, SCo, s PR; to se US, Caribbean, Mex; native to trop Am. [*S. poiretii* (Roemer & Schultes) A. Hitchc. misapplied] Spikelets, upper lvs gen covered by black fungus, hence common name.

S. vaginiflorus (A. Gray) Alph. Wood POVERTY GRASS Ann. **STS** 1–several, tufted, 2–8 dm; base bent abruptly. **LVS** equally basal and cauline; sheath becoming straw-colored or purplish; sheath, collar margins gen hairy, hairs < 5 mm, swollen at base; ligule < 0.5 mm, densely hairy; blade 2–10 cm, 1–3 mm wide, scabrous. **INFL** terminal, ± axillary, 1–5 cm, 2–3 mm wide, partly or completely enclosed by sheath; branches appressed. **SPIKELET** 3–6 mm; glumes, lemmas rounded to keeled, glumes 2–5 mm, subequal, narrowly lanceolate, acute; lemma 3–6 mm, narrowly lanceolate, tip beaked; palea = or 2 mm > lemma; anthers < 3 mm. **FR** ± 2–3 mm, not gelatinous when wet. 2*n*=54. Open, gen sandy, disturbed areas; < 600 m. e KR (nw Shasta Co.), n SNH (Grass Valley, Truckee Canyon); native to c&e US, Mex.

STENOTAPHRUM

Robert Webster

Ann, per. **STS** prostrate to ascending, gen compressed; internode solid to spongy inside. **LVS** cauline; ligule mostly hairy, membrane < 0.5 mm; blade folded. **INFL** raceme- to panicle-like; 1° branches appressed, axis straight or wavy, tip awl-like; spikelets short-stalked to embedded in axis, subtended by 1+ bristles. **SPIKELET** compressed, gen falling as 1 unit, sometimes with infl axis; glumes equal to unequal; florets 2, lower floret sterile or staminate, lemma ± = upper glume, upper floret bisexual, lemma thick, firm, smooth or minutely rough, margin flat or inrolled, tip blunt. 7 spp.; trop worldwide. (Greek: narrow trench, from spikelet scars on infl axis) [Sauer 1972 Brittonia 24:202–222]

S. secundatum (Walter) Kuntze SAINT AUGUSTINE GRASS Per from long stolons. **STS**: vegetative gen prostrate; fl st decumbent, 1–4 dm. **LF**: sheath 1.2–10 cm, glabrous; ligule ± 0.5 mm; blade 5–15 cm, 5–15 mm wide, upper surface glabrous. **INFL** 5–10 cm;

axis straight or wavy, thickened, flat; each spikelet subtended by 1 bristle; spikelet stalk < 0.5 mm. **SPIKELET** 4–5 mm, ± 1.5–2 mm wide, lanceolate to elliptic; lower glume 1–2 mm; upper glume ± = spikelet length; lower floret sterile, lemma 7–9-veined, tip acute to acuminate, palea ± = lemma; upper floret slightly < lower floret. Fields, roadsides; < 150 m. CCo, SCo; s US, Afr, Pacific islands; probably native to S.Am.

SWALLENIA

1 sp. (Jason Swallen, American agrostologist, 1903–) [Henry 1979 Fremontia 7(2):3–6]

S. alexandrae (Swallen) Söderstrom & Decker (p. 1301) EUREKA VALLEY DUNE GRASS Per, tufted, from thick, scaly rhizomes; nodes ± woolly. **STS** ascending to erect, branching, 1.5–3.5 dm, stiff, ridged, gen glabrous. **LVS** gen cauline; sheath margins with soft-shaggy hairs near collar; ligule ± 1 mm, densely hairy; blade 5–14 cm, 3–6 mm wide, stiff, sharp-pointed, strongly veined. **INFL** panicle-like, 4–10 cm; branches spike-like, short, appressed; axis hairy. **SPIKELET** < 1.5 cm, persistent; glumes 9–14 mm, ± equal, > florets, wide, glabrous, 7–11-veined, awn 0; florets 3–7; lemma 7–9 mm, back rounded, lower margin with soft-shaggy hairs, 5–7-veined, awn 0; palea ± = lemma, margin hairs like lemma. **FR** 4 mm, 2 mm wide, falling from floret. $2n=20$. **ENDANGERED** US; **RARE** CA. Sand dunes; 900–1200 m. W&I (Eureka Valley, ne Inyo Co.). [*Ectosperma a.* Swallen]

TAENIATHERUM

Mary E. Barkworth

Ann. **STS** ascending to erect. **LF**: sheath short-appendaged; ligule membranous; blade flat to inrolled. **INFL** spike-like, dense; spikelets 2-ranked, 2 per node. **SPIKELET**: glumes awl- to awn-like, firm; axis breaking above glumes; florets 2, lower fertile, upper vestigial on prolonged axis; lemma lanceolate, gen scabrous, awn >> lemma body, flat, straight to curving outward; anthers < 2 mm. ± 3 spp.: temp Eurasia. (Greek: ribboned awn)

T. caput-medusae (L.) Nevski (p. 1301) **ST** 2–6 dm, decumbent to ascending, slender. **LF**: sheath glabrous, appendages < 3 mm; ligule < 0.5 mm, truncate; blade 1–3 mm wide, ± inrolled, glabrous to puberulent, long-ciliate near collar. **INFL** 1.5–5 cm exc awns. **SPIKELET**: glumes 1.5–4 cm, fused at base, awn-like, stiff, gen glabrous; lemma body 5–8 mm, narrowly lanceolate, awn 3–7 cm, flat, straight to curving outward. $2n=14$. NCoR, CaR, SNF, GV, n SCo, expected elsewhere; to WA, Rocky Mtns; native to Eur. [*Elymus c.-m.* L.] NOXIOUS WEED.

TORREYOCHLOA

Jerrold I Davis

Per from rhizomes. **STS** decumbent to erect, sometimes rooting at nodes. **LVS** basal and cauline; sheath open ± to base; ligule thinly membranous, acute to obtuse, sometimes toothed. **INFL** panicle-like; lower branches reflexed to erect. **SPIKELET** bisexual; glumes < lowest floret, lower glume gen 1-veined, upper gen 3-veined; florets 3–8; lemma firm, back rounded, scabrous, margin near tip scabrous-serrate at 10×, prominently (5)7–9-veined, veins not converging at tip; awn 0; palea ± = lemma. ± 5 spp.: temp N.Am, e Asia. (J. Torrey, Amer botanist, 1796–1873) [Davis 1991 Phytologia 70:361–365] Spp. gen occurring in freshwater wet habitats.

1. Infl linear to narrowly elliptic in outline, 5–15 × width, < 1 cm wide; lf blade 3.5–7 mm wide .... ***T. erecta***

1′ Infl ovate to elliptic or obovate in outline, 1–6 × width, 1–12 cm wide; lf blade 3.5–17.5 mm wide .... ***T. pallida*** var. ***pauciflora***

T. erecta (A. Hitchc.) Church **ST** 2–6 dm. **LF**: cauline lf blade 3.5–7 mm wide. **INFL** 5.5–11 cm, < 1 cm wide, linear to narrowly elliptic in outline, 5–15 × width; lower branches erect to ascending. **SPIKELET**: florets 4–6; lowest lemma 2–3 mm; anthers of lowest floret 0.6–0.8 mm. $2n=14$. Stream, lake margins, coniferous forest; 2000–3500 m. CaRH, SNH; to OR, NV. [*Puccinellia e.* (A. Hitchc.) Munz] ❀STBL.

T. pallida (J.S. Presl) Church var. ***pauciflora*** (J.S. Presl) J. I Davis (p. 1301) WEAK MANNAGRASS **ST** 2–14 dm. **LF**: cauline lf blade 3.5–17.5 mm wide. **INFL** (3)5–25 cm, 1–12 cm wide, deltate to ovate, elliptic or oblanceolate in outline, 1–6 × width; lower branches weakly reflexed to spreading or erect. **SPIKELET**: florets 3–8; lowest lemma 2–3.5 mm; anthers of lowest floret 0.5–0.7 mm. $2n=14$. Wet areas in forests, stream or lake margins; < 3500 m. NW, CaR, SN, CCo, MP; to AK, Rocky Mtns. [*Puccinellia p.* (J.S. Presl) Munz] Pls from > 3000 m, with infl ovate, 3–6 cm, length < 5 × width, have been called *Puccinellia californica* (Beetle) Munz, Sierra Nevada alkali grass. ❀STBL.

TRIDENS

Per, ± tufted. **STS** gen erect. **LF**: ligule fringed or short-hairy; blade flat or inrolled. **INFL** panicle-like, open to dense, or spike-like. **SPIKELET** ± compressed; glumes ± equal, membranous, lower 1-veined, upper 1–3(5)-veined; axis breaking above glumes and between florets; lemma wide, thin, back rounded, veins hairy below middle, 3-veined, tip notched or 2-toothed, gen short-pointed; palea gen < lemma, glabrous or minutely hairy. 18 spp.: N.Am. (Latin: 3-toothed, from lemma tip) [Tateoka 1961 Amer J Bot 48:565–573]

T. muticus (Torrey) Nash (p. 1301) SLIM TRIDENS Pl densely tufted. **ST** 2–5 dm. **LF**: sheath hairy, esp near collar; ligule short-hairy; blade 3–25 cm, 1–3 mm wide, gen inrolled, ± fine scabrous, sometimes sparsely hairy. **INFL** 4–20 cm, 3–8 mm wide, narrow; branches short, appressed; spikelets subsessile to short-stalked. **SPIKELET** 8–11 mm, ± cylindric; glumes 5–6 mm, 1-veined; florets 6–8, strongly overlapping, pale to light purple; callus densely hairy; lemma ± 5 mm, veins densely hairy below middle, tip entire to minutely notched; palea keels densely hairy. $2n=40$. Dry, rocky, gen limestone soils, creosote-bush shrubland, pinyon/juniper woodland; 900–2000 m. SNE, DMtns; to Colorado, TX, Mex. ❀DRN,SUN:1,**2**,**3**,**7**,8,9,**10**,11,**14**–**23**,24.

TRISETUM

Dieter H. Wilken

Ann, per. **STS** ascending to erect. **LVS** basal and cauline; ligule membranous, obtuse to truncate, toothed, tip ciliate or not; blade flat to inrolled. **INFL** panicle- to spike-like, open to compact, cylindric to narrowly conic. **SPIKELET**: glumes ± unequal, gen = or < lower floret, keeled, acute, lower 1-veined, upper 3-veined; axis stiff- to soft-hairy, gen prolonged behind upper floret, bristly or with vestigial floret; florets 2–3, bisexual, breaking above glumes and between florets (sometimes below glumes); callus short-hairy; lemmas ± keeled, tip 2-bristled or not, awned on back near tip or not, awn straight to bent; palea = or < lemma. 50–70 spp.: temp, trop mtns. (Latin: three bristle) Some spp. intergrade; needs critical study in w N.Am.

1. Lemma ± awnless .... *T. wolfii*
1′ Lemma awn 4–12 mm
 2. Infl gen spike-like, dense, awns spreading, lvs gen basal, tufted .... *T. spicatum*
 2′ Infl gen panicle-like, ± dense to open, awns gen ascending to erect, lvs basal and cauline
  3. Lower infl branches gen spreading, lower glume < 2 mm, upper glume tip irregularly 1–2-toothed ... *T. cernuum*
  3′ Lower infl branches ascending to erect, lower glume 4–5 mm; upper glume acute
   4. Spikelets green to tan, ligules gen > 1 mm, obtuse to truncate; infl axis ± hairy .... *T. canescens*
   4′ Spikelets golden brown, ligules < 1 mm, ring-like; infl axis glabrous .... *T. flavescens*

T. canescens Buckley (p. 1307) Per, ± cespitose. **STS** 5–8 dm, clumped. **LVS** basal and cauline; ligule 1–4 mm; cauline blade 2–5(8) mm wide. **INFL** panicle-like, 6–20 cm, narrow, compact to open; lower branches ascending to erect; central axis gen exposed, sparsely hairy. **SPIKELETS** gen on distal 2/3 of lower branches; glumes lanceolate, lower 4–5 mm, upper 6–7 mm, acute; lemmas 4–6 mm, awn 6–11 mm. $2n$=28,42. Open to shaded sites, meadows, coniferous forest; < 2700 m. NW, CaR, SN, CW, MP; to AK, MT, UT. [*T. cernuum* var. *c.* (Buckley) Beal] Pls from SN and CaR with conspicuously interrupted infls have been called *T. cernuum* var. *projectum* (Louis-Marie) Beetle. ❀TRY.

T. cernuum Trin. Per, sometimes from short rhizomes. **STS** 4–10 dm, solitary to loosely clumped. **LVS** basal and cauline; ligule 1–3 mm; cauline blade gen (4)6–12 mm wide. **INFL** panicle-like, 10–30 cm, open, triangular in outline; lower branches ± wiry, gen spreading; axis mostly exposed, glabrous. **SPIKELETS** gen on distal half of lower branches; lower glume < 2 mm, linear-lanceolate to ovate, upper 3–4 mm, narrowly elliptic, tip 1–2-toothed; axis conspicuous, segments between florets 1–2 mm; upper floret ± exserted; lemmas 4–6 mm, awn 7–12 mm. $2n$=42. Moist, shaded sites, redwood, coniferous forest; < 1000 m. NCo, NCoRO; to AK, MT. ❀IRR:**4–6,15–17**&SHD:1,14,22–24.

T. flavescens (L.) Beauv. Per. **STS** 6–8 dm, solitary to clumped. **LVS** basal and cauline; ligule < 1 mm; cauline blade 3–6 mm wide. **INFL** panicle-like, 7–14 cm, ovoid to elliptic in outline; lower branches weakly ascending; central axis ± exposed, glabrous. **SPIKELETS** gen on distal 4/5 of lower branches (sometimes axillary); glumes lanceolate, acute, lower ± 4 mm, upper 5–6 mm; lemmas 4– 5 mm, awn 4–7 mm. $2n$=24,28. Meadows, redwood forest; < 100 m. NCo; to Can, e US; native to temp Eur.

T. spicatum (L.) Richter (p. 1307) Per, cespitose. **STS** 0.5–4 dm, densely clumped. **LVS** mostly basal, tufted; ligule 1–3 mm; cauline blade gen 1–4 mm wide. **INFL** spike-like, 2–8(10) cm, dense, cylindric to narrowly elliptic in outline; lower branches erect, ± appressed; axis hairy, hidden by spikelets. **SPIKELETS** on branches from base to tip; glumes lanceolate, acute, lower 5–6 mm, upper 6–7 mm; lemmas 4–5 mm, awn 4–8 mm. $2n$=14,28,42. Open, dry to moist sites, meadows, streambanks; 1400–4000 m (KR: 150–2000 m). KR, CaRH, SN, W&I; temp Am, Eurasia. ❀DRN,SUN:4,5,**6**&IRR:**1**,2,**7,14–18**,19–24.

T. wolfii Vasey (p. 1307) Per, ± cespitose. **STS** 3–8 dm, clumped. **LVS** basal and cauline; ligule 1–5 mm; cauline blade 2–7 mm wide. **INFL** gen panicle-like, 3–9 cm, compact, narrow; lower branches ascending to stiffly erect; axis exposed to ± hidden by spikelets. **SPIKELETS** gen on distal 4/5 of lower branches (sometimes at base); glumes lanceolate, acute, lower 4–6 mm, upper 5–7 mm; lemmas 5–6 mm, short-awned or not, awn gen << 1 mm. Open, gen dry sites, meadows, coniferous forest, alpine; 2000–3700 m. c&s SNH, CaRH; to sw Can, MT, UT. Some alpine pls doubtfully distinct from *T. spicatum*. ❀TRY.

TUCTORIA

John R. Reeder

Ann, ± hairy. **STS** ascending to erect, unbranched, often fragile, easily breaking apart at nodes. **LVS** basal and cauline; collar 0; sheath and blade continuous; blade flat or becoming inrolled when dry. **INFL** spike-like, partly enclosed by upper sheath or exserted with age; spikelets spirally arranged on axis. **SPIKELET** laterally compressed; glumes entire or irregularly short-toothed; florets 5–40, tardily breaking above glumes and between florets; lemmas entire or minutely toothed, gen with central sharp point, 11–17-veined; palea ± = lemma; anthers exserted, filaments slender, ribbon-like; stigmas 1/3–1/2 style length, sparsely short-hairy. **FR** obovoid or oblong. 3 spp.: CA, s Baja CA. (Anagram of *Orcuttia*) [Reeder 1982 Amer J Bot 69:1082–1095]

1. Infl exserted; lemmas ± truncate; fr 2 mm, minutely wrinkled .... *T. greenei*
1′ Infl partly enclosed by upper lf; lemmas tapered gradually to a sharp-pointed tip; fr 3 mm, smooth *T. mucronata*

T. greenei (Vasey) Reeder (p. 1307) GREENE'S TUCTORIA **ST** erect, becoming decumbent, 5–15(30) cm; nodes often purplish. **LF** 2–3 cm, < 5 mm wide, curved outward. **INFL** < 8 cm, exserted with age; upper spikelets crowded, lower ± separated. **SPIKELET**: glumes 3–5 mm, subequal, tip irregularly short-toothed, strongly veined; florets < 40; lemma gen 4–5(6) mm, 9–13-veined; palea slightly < lemma; anthers 3–3.5 mm, whitish. **FR** ± 2 mm, slightly flattened laterally, oblong, minutely wrinkled. $2n$=24. **RARE** CA. Vernal pools; < 200 m. GV (Butte, Fresno, Madera, Merced, San Joaquin, Stanislaus, Tehama, Tulare cos.). [*Orcuttia g.* Vasey]

T. mucronata (Crampton) Reeder (p. 1307) CRAMPTON'S TUCTORIA **ST** ascending, becoming decumbent, < 12 cm. **LF** 2–4 cm, curved outward, inrolled, tapered to fine point. **INFL** 1.5–6 cm, ± enclosed in upper sheath; spikelets crowded. **SPIKELET**: glumes 4–7 mm, subequal, short-pointed, sometimes with 1 or 2 short lateral teeth; florets 5–10; lemma 5–7 mm, tip with a few minute teeth, 11–15-veined, central vein ending in a sharp point < 1 mm; anthers ± 3 mm, yellow, pinkish when drying. **FR** ± 3 mm, laterally flattened, widely oblong, smooth. $2n$=40. **ENDANGERED** CA, US. Vernal pool, grassland; < 150 m. sw ScV (Solano Co.). [*Orcuttia m.* Crampton]

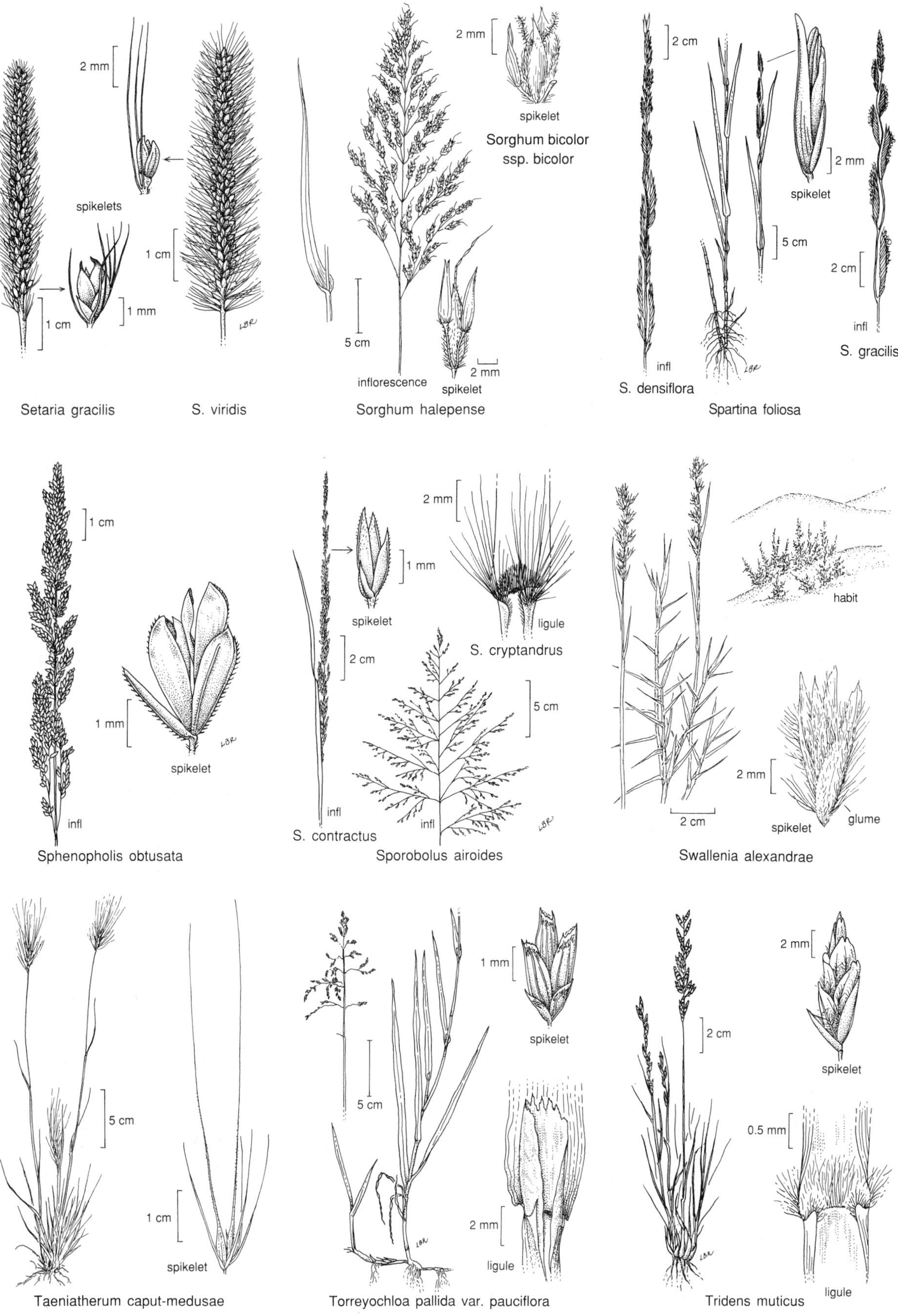

2 mm
spikelets
1 cm
1 mm
1 cm
Setaria gracilis
S. viridis
2 mm
spikelet
Sorghum bicolor
ssp. bicolor
5 cm
2 mm
inflorescence
spikelet
Sorghum halepense
2 cm
2 mm
spikelet
5 cm
2 cm
infl
S. gracilis
infl
S. densiflora
Spartina foliosa
1 cm
1 mm
spikelet
infl
Sphenopholis obtusata
2 mm
1 mm
spikelet
ligule
S. cryptandrus
2 cm
5 cm
infl
S. contractus
infl
Sporobolus airoides
habit
2 mm
2 cm
spikelet
glume
Swallenia alexandrae
5 cm
1 cm
spikelet
Taeniatherum caput-medusae
1 mm
spikelet
5 cm
2 mm
ligule
Torreyochloa pallida var. pauciflora
2 mm
2 cm
spikelet
0.5 mm
ligule
Tridens muticus

VENTENATA

Ann. **STS** ascending to erect, 1–5 dm. **LF**: sheath, blade glabrous to ± hairy; ligule membranous, gen acute. **INFL** panicle-like, open or dense. **SPIKELET** bisexual, compressed; glumes ± equal, < lemma, lower glume 3–7-veined, upper 3–9-veined; axis breaking above glumes and between florets; florets 2–7; lemma ± leathery, lemma of lower florets short-awned or not; uppermost awned from back, awn bent. 5 spp.: s Eur, w Asia. (Pierre Ventenat, French botanist, 1757–1805)

V. dubia (Leers) Durieu. **ST** erect, 1.5–7 dm, puberulent below nodes. **LF**: ligule 1–6 mm; blade 1–3 mm wide. **INFL** panicle-like, 3–10 cm, open; branches spreading to drooping; spikelets near branch tips, stalked. **SPIKELET** 10–15 mm; glumes lanceolate, acuminate, lower 5–7-veined, upper 7–9-veined; axis gen breaking apart above lowest floret; florets 2–3, lowest gen staminate, persistent to axis, awn straight; upper 1–2 florets bisexual, awn 1–1.5 cm. $2n$=14. Disturbed sites; 1200–1500 m. KR (Siskiyou Co.), n SNH (s Nevada Co.); to WA, ID; native to c&s Eur.

VULPIA

Susan G. Aiken & Robert I. Lonard

Ann. **STS** < 8 dm, solitary or loosely clumped, ascending to erect, unbranched. **LVS** gen cauline; sheath < internode; ligule < 1 mm, membranous, minutely fringed; blade < 15 cm, 0.5–2.5 mm wide, flat or rolled when dry; basal lobes 0. **INFL** panicle-like, narrow, dense or open. **SPIKELET**: glumes unequal, lanceolate, lower sometimes minute, upper 3-veined; axis breaking above glumes and between florets; florets 2–10, gen cleistogamous; lemma back round, 3–5-veined, awn < 22 mm, straight; palea ± = lemma, tip forked; stamen gen 1, < 1 mm; ovary tip glabrous. **FR** 3–6.5 mm, ± linear, ± sticking to palea. (J.S. Vulpius, pharmacist-botanist of Baden, Germany) [Lonard & Gould 1974 Madroño 22:217–230] ❀STBL.

1. Lower glume < 1/2 upper glume length, sometimes minute ..... ***V. myuros***
 2. Lemma margin ciliate near tip ..... var. ***hirsuta***
 2′ Lemma margin glabrous ..... var. ***myuros***

1′ Lower glume > 1/2 upper glume length
 3. Florets 5–10; closely overlapping; spikelet axis hidden; lemma awn < 4 mm ..... ***V. octoflora***
  4. Lemma back prominently long-scabrous near tip, margins long-scabrous ..... var. ***hirtella***
  4′ Lemma back glabrous or barely scabrous, margins often short-scabrous ..... var. ***octoflora***
 3′ Florets gen 1–5, loosely overlapping; spikelet axis visible, each internode > 1 mm; lemma gen awn 4–22 mm
  5. Lowest infl branches appressed to erect; branch axil not thickened ..... ***V. bromoides***
  5′ Lowest infl branches spreading or reflexed; branch axil thickened ..... ***V. microstachys***
   6. Spikelet glabrous or only scabrous ..... var. ***pauciflora***
   6′ Spikelet obviously hairy
    7. Glumes and lemmas hairy ..... var. ***ciliata***
    7′ Glumes glabrous and lemma hairy or glumes hairy and lemma glabrous
     8. Glumes hairy; lemma glabrous ..... var. ***confusa***
     8′ Glumes glabrous; lemma hairy ..... var. ***microstachys***

V. bromoides (L.) S.F. Gray (p. 1307) **ST** < 5 dm, glabrous or hairy. **INFL** 1.5–15 cm, dense, clearly > uppermost lf; branches 1–3 per node; spikelet stalk flat or winged. **SPIKELET** 5–10 mm; glumes 3.5–8 mm, ± equal; florets 4–7; lemma 5.5–8 mm, awn 2.5–12 mm, scabrous. **FR** ± 3.5–5 mm. $2n$=14. Uncommon. Dry, disturbed places, coastal-sage scrub, chaparral; gen < 1500 m. CA-FP; to e US; native to Eur. [*Festuca b.* L.; *F. dertonensis* (All.) Asch. & Graebner]

V. microstachys (Nutt.) Benth. (p. 1307) **ST** 1.5–7.5 dm, glabrous. **INFL** 2–24 cm, ± open, at least lower branches spreading or reflexed; branches 1 per node, < 7 cm; spikelet stalk angular. **SPIKELET** 5.5–10 mm; lower glume 2–5.5 mm, upper 3.5–7.5 mm; florets 2–4; lemma 4–9.5 mm, awn 3.5–12 mm. **FR** ± 4–6 mm. $2n$=42. Disturbed, open, gen sandy soils; gen < 1500 m. CA; to WA, ID, Baja CA.

var. ***ciliata*** (Beal) Lonard & Gould **SPIKELET**: glumes hairy; lemma hairy. Locally abundant. Open, gen disturbed places, sandy soils, hillsides, forest; gen < 1000 m. CA (esp CA-FP). [*Festuca eastwoodae* Piper, *F. grayi* (Abrams) Piper]

var. ***confusa*** (Piper) Lonard & Gould **SPIKELET**: glumes hairy; lemma glabrous. Uncommon. Coastal-sage scrub, yellow-pine forests, grassy plains, dry hillsides; gen < 1000 m. CA-FP; to WA. [*Festuca c.* Piper, *F. tracyi* C. Hitchc.]

var. ***microstachys*** **SPIKELET**: glumes glabrous; lemma hairy. Dry hillsides, coarse, sandy soils, crumbling serpentine or shale, open woodlands; gen < 1000 m. CA (esp CA-FP); to WA, ID. [*Festuca m.* Nutt., *F. arida* Elmer]

var. ***pauciflora*** (Beal) Lonard & Gould **SPIKELET**: glumes glabrous or scabrous; lemma glabrous or scabrous. Common. Dry, open, wooded hillsides, sandy, often disturbed places; gen < 1500 m. CA-FP, D; to WA, Baja CA. [*F. reflexa* Buckley; *F. pacifica* Piper; *F. m.* Nutt. var. *simulans* (Hoover) Hoover]

V. myuros (L.) C. Gmelin (p. 1307) **ST** < 7.5 dm, glabrous or scabrous only near infl. **INFL** 4–25 cm, < 2 cm wide, ± dense; base often enclosed in sheath at maturity; branches 1–3 per node; spikelet stalk < 1 mm, slender. **SPIKELET** 5–11.5 mm; lower glume gen < 1.5 mm, upper ± 2.5–5.5 mm; florets 3–6; lemma 4.5–6.5, awn 5–15 mm. **FR** 3.5–4.5 mm. $2n$=42. Common. Gen open places, sandy soils; < 2000 m. CA-FP, D; worldwide; probably native to Eur.

var. *hirsuta* (Hackel) Asch. & Graebner **SPIKELET**: lemma margin ciliate near tip, awn of lowermost floret 9.5–22 mm. Open places, hillsides, washes; gen < 1500 m. CA-FP, uncommon in D; worldwide; native to Eur. [*Festuca megalura* Nutt.]

var. *myuros* **SPIKELET**: lemma margin glabrous near tip, awn of lowermost floret 7.5–17 mm. Open places, hillsides, washes, or vernally moist sites in chaparral; gen < 1500 m. CA-FP; worldwide; native to Eur. [*Festuca m.* L.]

V. octoflora (Walter) Rydb. (p. 1307) **ST** < 6 dm, glabrous or hairy. **INFL** 0.4–16 cm, 0.5–2 cm wide, dense; branches 1 per node. **SPIKELET** 4.5–10 mm; lower glume ± 2–4.5 mm, upper ± 2.5–7 mm; florets 5–10; lemma ± 3–5 mm; awns 0.5–5 mm. **FR** 2–3.5 mm. $2n$=14. Sandy to rocky soils, open sites; < 2000 m. CA; widespread Am, Eur. [*Festuca o.* Walter]

var. ***hirtella*** (Piper) Henrard **SPIKELET**: lemma back prominently long-scabrous near tip. Sandy to rocky soils, open, gen disturbed areas, desert scrub to pinyon woodland, esp burned areas; gen < 1500 m. CA (esp s CA-FP); to B.C., Colorado, TX, Baja CA. [*Festuca o.* ssp. *hirtella* Piper]

var. ***octoflora*** **SPIKELET**: lemma back glabrous or barely scabrous, margins often scabrous. Sandy soils, washes, hills, chaparral; gen < 2000 m. s CA-FP, D; widespread Am, Eur.

ZIZANIA WILD RICE

Ann, per, monoecious, aquatic. **STS** erect, < 3 m, 1 cm wide. **LF**: ligule membranous, narrow; blade < 1.2 m, 2.5 cm wide. **INFL** panicle-like; lower branches spreading to ascending, bearing staminate spikelets; upper branches ascending to erect, bearing pistillate spikelets. **STAMINATE SPIKELET** gen drooping; glumes vestigial, forming a collar-like ridge; floret 1; lemma linear, 5-veined, acuminate or awned; palea 3-veined; stamens 6. **PISTILLATE SPIKELET** gen ascending to erect, cylindric, angled, firm; glumes gen 0; floret 1, lemma 3-veined, long-tapered, awned. **FR** 1–2 cm, cylindric, purplish black. 3 spp.: N.Am, e Asia. (Greek: ancient name for weed in grain fields) [Dore 1969 Canada Dept Agric Res Publ 1393]

Z. palustris L. var. *interior* (Fassett) W. Dore WILD RICE **ST** puberulent at nodes. **LF** 15–45 cm, 6–40 mm wide. **INFL** 2–5 dm. **STAMINATE SPIKELET** 6–8 mm. **PISTILLATE SPIKELET** 10–20 mm; awn 5–40 mm. $2n=30$. Wet meadows, shallow ponds, lake margins; < 1200 m. NCoRI, ScV, MP; native to Can, e US. Cult for food, waterfowl.

PONTEDERIACEAE PICKEREL-WEED FAMILY

Elizabeth McClintock

Per, ann, submersed, emersed, floating, or on wet ground. **LVS** simple, alternate or whorled, ± in basal rosette or not; blade linear (esp underwater) to round, sagittate or cordate or not, bases gen sheathing, veins parallel; petioles sometimes inflated or 0 (esp underwater). **INFL**: raceme, spike, panicle, or 1-fld, terminal but often appearing ± axillary, subtended by sheathing bract. **FL** bisexual, radial or ± bilateral; perianth lobes gen 6, in 2 series of 3, petal-like; stamens equal or not, attached to perianth tube at various levels, 3, 6, (1, 4), sometimes 3 sterile, modified; ovary superior, 1–3-chambered. **FR**: 1-seeded utricle or many-seeded, loculicidal capsule. **SEEDS** small, longitudinally ribbed or not. ± 6 genera, 30 spp.: most pantrop, some temp. Some cult as orns, some weeds, esp in rice fields. [Rosatti 1987 J Arnold Arb 68:35–71]

1. Pl gen free-floating; petiole inflated or not; infl gen many-fld .... **EICHHORNIA**
1′ Pl submersed, emersed, or on wet ground; petiole not inflated; infl 1–few-fld
  2. Infl 1-fld; stamens 3; perianth salverform, tube > 1 cm .... **HETERANTHERA**
  2′ Infl a few-fld raceme; stamens 6; perianth bell-shaped to rotate, tube < 1 cm .... **MONOCHORIA**

EICHHORNIA WATER HYACINTH

Per, gen free-floating. **STS** stout, erect, often connected by stolons. **LVS** ± in basal rosette; blade gen ovate to round; petiole inflated or not, gen > blade. **INFL**: spike, panicle, or 1-fld. **FL** ± bilateral; perianth funnel-shaped, lobes 6; stamens 6, unequal in length; ovary 3-chambered. **FR**: capsule. **SEEDS** many, longitudinally ribbed. ± 7 spp.: native to Am trop, perhaps Am subtrop, Afr; some widely naturalized. (J. A. F. Eichhorn, Germany, 1779–1856)

E. crassipes (C. Martius) Solms-Laubach (p. 1307) WATER HYACINTH **ST** > 30 cm. LF < 10 cm wide. **INFL** 5–15 cm, ± 3–35-fld. **FL**: perianth lilac or pale blue to white. Locally abundant. Ponds, sloughs, waterways; < 200 m. GV, SnFrB, SCo, PR; native to trop Am. Pls multiply, spread rapidly by vegetative means; perhaps the world's most troublesome aquatic weed.

HETERANTHERA MUD PLANTAIN

Per, ann, gen submersed, emersed, or on wet ground. **ST** ± erect, slender, elongate, ± spreading, rooting at nodes or stout. **LVS** alternate, linear, sessile, or ± in basal rosettes, ± ovate, long-petioled. **INFL** gen 1-fld. **FL** ± radial; perianth salverform, lobes 6; stamens 3, alike or of 2 sizes, shapes, anthers coiling or not; ovary 1-chambered. **FR**: capsule. **SEEDS** many, longitudinally ribbed. ± 12 spp.: trop, temp Am, Afr. (Greek: different, anther, from unequal anthers of most spp). Fls developed underwater gen cleistogamous.

1. Lvs linear, sessile, alternate on slender, ± spreading, gen much-branched, often tangled sts; fl sessile; perianth yellow .... ***H. dubia***
1′ Lvs ± ovate, long petioled, ± in basal rosette on gen stout, ± erect, often tufted, branched or unbranched, rarely tangled sts; fl peduncled; perianth white to blue-purple .... ***H. limosa***

H. dubia (Jacq.) MacMillan WATER STAR-GRASS (p. 1307) Per, gen submersed. **ST** often rooting at nodes. **LF** > 15 cm. **FL** appearing ± axillary; perianth yellow, tube ± 1.5(–7) cm; stamens 3, alike, anthers coiled. Uncommon. Still or moving water; < 1500 m. NCo, ScV, MP; widespread N.Am, C.Am. Vegetatively ± similar to spp. of *Potamogeton*, exc lvs lack distinct midrib. ❀Fresh water margins:1–5,**6–9**,10,11,**14–17**,18–24.

H. limosa (Schwartz) Willd. Ann, gen emersed or on wet ground (submersed as seedlings). **LF**: blade > 4 cm; petiole > 15 cm. **FL** appearing ± terminal; perianth white to blue-purple, tube > 1.5 cm; stamens 3, of 2 sizes, shapes, anthers not coiled. Uncommon. Rice fields; < 100 m. ScV; native to c&e US, trop Am.

MONOCHORIA

Per, submersed or emersed. **STS** ± erect, connected by rhizomes. **LF** 1 at st tip; blade ovate to narrowly lanceolate; petiole long. **INFL**: raceme, 1 at st tip, eventually reflexed. **FL** ± radial; perianth bell-shaped to rotate, tube < 1 cm, parts 6, nearly free; stamens 6, 1 gen > others; ovary 3-chambered. **FR**: capsule. **SEEDS** many, slightly ribbed longitudinally. 6 spp.: trop Afr, Asia, ne Australia. (Greek: 1 apart, from 1 stamen > others)

M. vaginalis (Burm. f.) Kunth **ST** > 12 cm. **LF**: blade 2.5–11 cm; petiole 4–70 cm. **INFL** > 5 cm, few-fld. **FL** gen opening under water; perianth blue, parts < 2 cm. Uncommon. Rice fields; > 100 m. ScV; native to trop Australia, Asia.

POTAMOGETONACEAE PONDWEED FAMILY

Robert F. Thorne

Ann, per, aquatic, (gen fresh to alkaline water), glabrous, from rhizomes or small, bulb-like, winter buds. **STS** erect, simple to branched, cylindric or flattened. **LVS** simple, cauline, alternate or in subopposite pairs; submersed thread-like to round, sessile or petioled; floating, if present, elliptic to ovate, petioled, leathery; sheath open, continuous with petiole or ± free from lf base, gen with stipules, stipules sometimes fused, ligule-like. **INFL**: spike or head-like, axillary or terminal, gen emergent, peduncled; bracts 0. **FL** bisexual; perianth parts 0 or 4, clawed; stamens 2 or 4, if 4, each fused to base of perianth part, sessile or filament short, wide, anthers open to outside; pistils 4, ovary 1-chambered, ovule 1, style short or stigmas sessile. **FR**: drupe. **SEED** 1. 3 genera, ± 95 spp.: worldwide. *Ruppia* sometimes treated in Ruppiaceae.

1. Perianth parts 4; stamens 4; fr sessile; ovule attached at base of chamber margin .... **POTAMOGETON**
1′ Perianth parts 0; stamens 2; fr long-stalked; ovule attached near top of chamber .... **RUPPIA**

POTAMOGETON PONDWEED

Ann, per, from rhizomes or small, bulb-like, winter buds. **ST** simple or branched, cylindric or flattened, rooting at lower nodes. **LVS** simple, cauline, gen alternate, gen flat, gen green, margin gen entire; submersed lvs sessile or petioled, linear to round, tip rounded to acuminate, veins 1–35; floating lvs, if any, elliptic to ovate, gen petioled, leathery; stipules free or fused, sheath-like below lf junction, free or fused (ligule-like) above lf junction. **INFL**: cylindric spike or head-like, axillary or terminal, floating to emergent. **FL**: inconspicuous; perianth parts 4, clawed, greenish; stamens 4, attached to base of perianth, anthers gen sessile; ovule attached at chamber base, style short or stigma sessile. **FR** gen obovate, sessile, floating. ± 90 spp.: mostly temp n hemisphere. (Greek: river neighbor, from aquatic habitat) [Haynes 1974 Rhodora 76:564–649; 1985 Sida 11:173–188; Wieglet 1988 Feddes Repert 99:249–266]

1. Pls with both submersed and floating lvs; floating lvs gen elliptic to ovate, gen leathery; submersed lvs linear to ovate, membranous
 2. Submersed lvs lanceolate to ovate, gen 1–7 cm wide, sometimes linear
  3. Submersed lvs gen lanceolate to ovate, curved, gen folded along midrib, sessile or short-petioled, blade veins 20–40; floating lvs elliptic to ovate, blade veins 30–50 .... ***P. amplifolius***
  3′ Submersed lvs linear to lanceolate, not curved, flat, upper long-petioled, lowermost sessile (all sessile in *P. gramineus*); all lf blades with < 30 veins
   4. Submersed lvs with 8+ veins, long-petioled .... ***P. nodosus***
   4′ Submersed lvs gen with 7 veins, all or lowermost sessile
    5. Margin of submersed lvs finely serrate near tip .... ***P. illinoensis***
    5′ Margin of submersed lvs entire
     6. Lvs reddish; submersed lf width = floating lf width; submersed lvs ± all on central st; floating lvs short-petioled .... ***P. alpinus*** ssp. ***tenuifolius***
     6′ Lvs green; submersed lvs narrower than floating lvs; submersed lvs on short, axillary branches; floating lvs long-petioled .... ***P. gramineus***
 2′ Submersed lvs linear, < 1 cm wide
  7. Stipules fused to lf base; infls of 2 kinds, lowermost spheric, submersed, short-peduncled, uppermost cylindric, emergent, long-peduncled; fr clearly compressed — lateral keels low .... ***P. diversifolius***
  7′ Stipules free from lf base; infls all emergent, ± cylindric; fr plump to slightly compressed
   8. Submersed lvs ribbon-like, 2–10 mm wide, with a clearly net-like band along the midrib .... ***P. epihydrus*** ssp. ***nuttallii***
   8′ Submersed lvs without a net-like band, cylindric, often petiole-like, < 2 mm wide, or linear to elliptic-lanceolate, 10–40 mm wide
    9. Floating lf blade oblong to widely elliptic, base slightly lobed, jointed to petiole; fr smooth or faintly keeled, sides concave .... ***P. natans***
    9′ Floating lf blade elliptic to ovate, base tapered to petiole, not jointed to petiole; fr 3-keeled, sides flat .... ***P. nodosus***
1′ Pls with all lvs submersed
 10. Lvs lanceolate, elliptic, oblong, or ovate, sessile, base tapered to clasping (short-petioled in *P. illinoensis*)
  11. Lf blade margin finely serrate; fr with long, slender beak 2–3 mm .... ***P. crispus***
  11′ Lf blade margin entire (finely serrate near tip in *P. illinoensis*); fr short-beaked
   12. Lf base tapered, not clasping

13. Upper lvs gen sessile; blade margins entire; lvs reddish .... ***P. alpinus*** ssp. ***tenuifolius***

13′ Upper lvs short-petioled; blade margin finely serrate near tip; lvs green .... ***P. illinoensis***

12′ Lf base clasping

14. Blades lanceolate to ovate, 5–20 cm, tip hood-like, splitting when dry; stipules 2–8 cm, persistent, whitish; sts zigzag; fr 4–5 mm, sharply 3-keeled .... ***P. praelongus***

14′ Blades narrowly ovate, 1–12 cm, tip not hood-like, flat, margin crinkly, not splitting when dry; stipules < 2 cm, becoming fibrous; sts straight; fr 2–4 mm, faintly 3-keeled .... ***P. richardsonii***

10′ Lvs linear

15. Stipules fused to lf base, sheath-like below, sheath 1+ cm

16. Lvs 2–8 mm wide; stipules fused above junction with blade, ligule-like

17. Lf margin entire; ligule entire; lf blade not appendaged at junction with sheath; fr smooth or faintly keeled .... ***P. latifolius***

17′ Lf margin finely serrate; ligule gen cut to fibrous; lf blade often appendaged at junction with sheath; fr clearly keeled .... ***P. robbinsii***

16′ Lvs linear to thread-like, < 3 mm wide; stipules free above junction with blade

18. Stigma sessile; fr 2–2.7 mm, not beaked; sheath partly closed; st branched below, few-branched above .... ***P. filiformis***

18′ Stigma raised on minute style, head-like, persisting in fr as minute beak; fr 2.5–5 mm; sheath open to base; st with many, forked branches above .... ***P. pectinatus***

15′ Stipules free from lf or fused to lf base for 1–2 mm

19. Lvs 2–5 mm wide, veins 9–35, somewhat rigid; sts clearly flattened, winged, width at least 1/2 lf width .... ***P. zosteriformis***

19′ Lvs < 3 mm wide, veins 1–7, weak; st not flattened or winged

20. Keel of fr back wing-like, wavy; glands at lf base 0 or faint .... ***P. foliosus***

21. Stipules coarsely veined and ciliate, rigid, breaking apart into fibers; infl gen interrupted; fr pale green, 1.4–1.7 mm, 1.1–1.2 mm wide, keel on back gen < 0.2 mm high, beak < or = 0.2 mm .... var. ***fibrillosus***

21′ Stipules finely veined, not coarsely ciliate, decaying early; infl gen not interrupted; fr olive to greenish brown, 1.5–2.7 mm, 1.2–2.2 mm wide, keel on back gen 0.2–0.4 mm high, beak 0.2–0.6 mm .... var. ***foliosus***

20′ Fr back rounded, not keeled; glands at base of some or all lvs 2, prominent .... ***P. pusillus***

22. Fr sides concave, tip beaked at top of central axis; infl clearly interrupted, with 2–4 whorls; stipules fused below middle .... var. ***pusillus***

22′ Fr sides rounded, tip oblique, beak to 1 side of central axis; infl with 1–2 whorls; stipules free, flat or margins inrolled .... var. ***tenuissimus***

P. alpinus Balbis ssp. ***tenuifolius*** (Raf.) Hultén Per, slender, from rhizomes. **ST** simple or few-branched, < 100 cm, cylindric. **LVS** reddish; submersed lvs sessile, 2–8 cm, 1–2 cm wide, lanceolate to elliptic-oblong, tapered to both ends, veins gen 7; floating lvs ± = submersed lvs, lanceolate, tapered to short petiole; stipules < 3 cm, free, membranous, early deciduous. **INFL**: spike, gen < 3 cm. **FR** ± 2.5 mm; back keeled. *n*=26. Uncommon. Ponds, lakes, marshes; gen 1300–2350 m. KR, SNH, CCo, MP; to AK, Greenland, Colorado, e Asia. [var. *t.* (Raf.) Ogden] Hybridizes with *P. nodosus, P. gramineus*. Ssp. *alpinus* in Eurasia.

P. amplifolius Tuckerman (p. 1307) BROAD-LEAVED PONDWEED Per from stout rhizome. **ST** simple or branched above, < 90 cm, cylindric. **LVS** often reddish brown; submersed lvs 8–20 cm, 2–7 cm wide, lanceolate to ovate, gen folded along midrib, curved, acute at both ends, veins 20–40; floating lvs 5–10 cm, 2.5–5 cm wide, elliptic to ovate, tapered or rounded at both ends; stipules gen 3–10 cm, free, becoming fibrous and stringy. **INFL**: spike, gen < 5 cm. **FR** 3–5 mm, clearly beaked; back rounded; sides flattened. *n*=26. Uncommon. Deep, clear-water lakes; 600–1850 m. NCoR, CaRH, SNH; to B.C., e N.Am, AZ. Hybridizes with *P. illinoensis, P. praelongus, P. richardsonii*.

P. crispus L. (p. 1307) CRISPATE-LEAVED PONDWEED Per from slender rhizomes. **ST** branched above, < 90 cm, somewhat flattened. **LVS** all submersed, sessile, 4–8 cm, 0.5–0.8 cm wide, oblong, thick, crisped or wavy; tip rounded; margin finely serrate; stipules gen < 1 cm, ± free, becoming fibrous. **INFL**: spike, < 2 cm. **FR** 4–6 mm; back clearly keeled; beak 2–3 mm, slender, erect or curved. Uncommon. 2*n*=36,42,50,52,72,78. Shallow water, ponds, reservoirs, streams; < 2100 m. NCoR, GV, CCo, SnFrB, SCo, ChI, SnGb, SnBr, DMoj; ± worldwide; native to Eurasia.

P. diversifolius Raf. (p. 1307) DIVERSE-LEAVED PONDWEED Per, delicate. **ST** much-branched above, < 50 cm, cylindric. **LVS**: submersed lvs sessile, < 6 cm, < 0.5 cm, linear, thin, tip acute, stipules < 1 cm, fused to blade, sheath-like below, ligule-like above; floating lvs with petiole < 2 × blade, blade 0.5–3 cm, 0.3–0.8 mm wide, elliptic, tapered at both ends, stipules < 2 cm, fused to lf, not sheathing, ± ligule-like. **INFL** < 1.5 cm; lowermost spheric, submersed, subsessile; upper spike-like, emergent, peduncled. **FR** 1.1–1.8 mm, subspheric, 3-keeled, compressed, with 2 low, lateral keels; back with 1 wavy or fine-toothed keel; sides flattened to concave. Uncommon. Shallow water, ditches, ponds, lakes; < 2500 m. NCoRI, c SNF, n SNH, GV, SCo, SnJt, MP; to B.C., e US, AZ, Mex. ❀fresh water,SUN:1–9,14–24.

P. epihydrus Raf. ssp. ***nuttallii*** (Cham. & Schldl.) Calder & Roy Taylor (p. 1307) NUTTALL'S PONDWEED Per from slender, matted rhizomes. **ST** simple or few-branched, < 170 cm, somewhat flattened. **LVS**: submersed lvs sessile, 5–25 cm, 0.2–1 cm wide, linear, ribbon-like, with median strip clearly net-like, tip acute; floating lvs gen opposite, petiole long, flattened, blade 2–8 cm, 0.4–3.5 cm wide, gen oblong to elliptic, tapered to petiole, tip rounded; stipules gen < 4 cm, fused, not sheathing, ligule-like. **INFL**: spike, 1–4 cm. **FR** 2–4 mm, spheric to obovate; back 3-keeled; sides concave. *n*=13. RARE in CA. Shallow water, ponds, lakes, streams, irrigation ditches; 400–1900 m. NCoRO, SNH, MP; to AK, e N.Am, Colorado. Ssp. *epihydrus* in e N.Am. Hybridizes with *P. nodosus*.

P. filiformis Pers. (p. 1307) SLENDER-LEAVED PONDWEED Per, slender, from rhizomes; tubers also present. **ST** branched below, < 65 cm, subcylindric. **LVS** all submersed, sessile, < 12 cm, < 0.3 cm wide, linear to bristle-like; tip rounded to acuminate; stipules < 3 cm, fused to lf, sheathing at base, tips free, scarious. **INFL**: spike, interrupted below, whorls 2–5, < 2.5 cm. **FR** 2–2.7 mm; back rounded. 2*n*=66. RARE in CA. Shallow, clear water of lakes and drainage channels; 300–2150 m. c SNH, SnJv, SnFrB, MP; to AK, Greenland, AZ, Eurasia. [var. *macounii* Morong misapplied to CA pls]

P. foliosus Raf. (p. 1307) LEAFY PONDWEED Per, from densely matted, slender rhizomes. **ST** many-branched, somewhat flattened. **LVS** all submersed, 1–10 cm, < 0.3 cm wide, linear; veins < 7; stipules fused, sheath-like. **INFL** subspheric to short spike, < 1 cm; peduncle club-shaped. **FR**: back keeled; sides rounded to slightly concave. Shallow water, ponds, lakes, streams, irrigation ditches; < 2300 m. CA; to AK, e Can, C.Am.

var. ***fibrillosus*** (Fern.) R. Haynes & Rev. FIBROUS PONDWEED **ST** < 60 cm. **LF** 2–4 cm, 1–2 mm wide; base gen with 2 glands; tip acute; stipules < 12 cm, persistent, clearly veined, becoming fibrous. **INFL** head-like or short spike, interrupted; peduncle 4–12 mm. **FR** 1.4–1.7 mm, 1.1–1.2 mm wide, pale green; back with keel < 0.2 mm high; beak ± 0.2 mm. RARE in CA. Shallow water, small streams; < 1300 m. n NCo (Crescent City), w MP, s SNE; to WA, ID, WY.

var. ***foliosus*** **ST** < 100 cm, flattened. **LVS** 1–10 cm, 0.3–2.5 mm wide; base without glands; tip acute to abruptly pointed; stipules < 2 cm, early deciduous, sometimes persistent as delicate fibers. **INFL** subspheric to head-like; peduncle 3–10 mm. **FR** 1.5–2.7 mm, 1.2–2.2 mm wide, olive to greenish brown; back with wavy, wing-like keel 0.2–0.4 mm high; beak 0.2–0.6 mm. 2*n*=28. Common. Habitats of sp.; < 2300 m. NW, CaRH, SNF, n SNH, GV, CW, SW (exc ChI), GB, D; to AK, e Can, C.Am. ❀TRY.

P. gramineus L. (p. 1307) GRASS-LEAVED PONDWEED Per from matted rhizomes. **ST** with many short branches, < 100 cm, cylindric. **LVS**: submersed lvs gen sessile, 2–11 cm, 1–12 mm wide, linear to oblanceolate, tip acute or long-tapered; floating lvs gen on short, axillary branches, with petiole > blade, blade 1.5–7 cm, 1–3 cm wide, elliptic to ovate, tip ± obtuse; stipules < 3 cm, free, persistent. **INFL**: spike, 1–4 cm. **FR** 1.5–3 mm; back faintly keeled; beak recurved. *n*=26. Uncommon. Shallow water, ponds, lakes, bogs; 900–2750 m. KR, NCoRI, CaRH, SNH, SnFrB, SnBr, GB; to AK, Greenland, AZ, Eurasia. Hybridizes with *P. alpinus, P. illinoensis, P. natans, P. nodosus, P. richardsonii.*

P. illinoensis Morong (p. 1307) SHINING PONDWEED Per from rhizomes. **ST** gen much-branched, < 150 cm, slender, cylindric. **LVS**: submersed lvs sessile or petioles < 2 cm, blade 6–20 cm, 15–50 mm wide, elliptic to oblanceolate, margin finely serrate near tip, tip acute to short-pointed; floating lvs, if any, with blades 4–12 cm, 2–6 cm wide, widely elliptic to oblong-elliptic, gen > petiole; stipules 2.5–7 cm, free, persistent. **INFL**: spike, gen < 6 cm. **FR** ± 4 mm; back keeled; beak slightly below tip. *n*=52. Lakes, ponds, streams; 400–2350 m. NCoR, CaR, SNH, SnJV, CCo, SnFrB, SCoRO, SnGb, SnBr, PR, GB; to B.C., e Can, TX, Baja CA; also C.Am. and Caribbean. Hybridizes with *P. gramineus, P. nodosus, P. richardsonii.* ❀TRY.

P. latifolius (Robb.) Morong (p. 1307) NEVADA PONDWEED Per from rhizome. **ST** much-branched, < 60 cm, subcylindric, whitish. **LVS** all submersed, sessile, 2–10 cm, 0.2–0.5 cm wide, linear; tip rounded or acute; veins 3–7; stipules < 3 cm, fused to blade, sheath-like below, ligule-like above, becoming deeply cut. **INFL**: spike, interrupted below, whorls 2–5, < 3 cm. **FR** < 3 mm, somewhat box-shaped, back faintly keeled. Very uncommon. Shallow, alkaline water, ponds, lakes; < 1450 m. c SnJV, GB, DMoj; to UT, TX, AZ. May be locally abundant.

P. natans L. (p. 1307) FLOATING-LEAVED PONDWEED Per from rhizome. **ST** gen simple, < 160 cm, subcylindric. **LVS**: submersed lvs 10–30 cm, < 2 mm wide, linear, blades linear or reduced to petiole; floating lvs 6–11 cm, < 60 mm wide, oblong to widely elliptic, base slightly lobed, jointed to petiole, tip rounded; stipules 6–8 cm, free, persistent. **INFL**: spike, < 5 cm. **FR** 3.5–5 mm; back rounded, smooth or faintly keeled; sides concave. 2*n*=52. Shallow, fresh or brackish water, lakes, ponds, bogs, marshes, lagoons, streams; < 2700 m. NCo, NCoRO, CaRH, SNH, SnJV, SnFrB, SnBr, PR, MP; to AK, e N.Am, Baja CA, Eurasia. Hybridizes with *P. gramineus, P. nodosus.* ❀water,SUN:1–9,14–24.

P. nodosus Poiret (p. 1307) LONG-LEAVED PONDWEED Per from rhizomes. **ST** simple to branched above, gen < 300 cm, subcylindric. **LVS**: submersed lvs long-petioled, 2–15 cm, 10–40 mm wide, lowermost sessile, blade linear to elliptic-lanceolate, tapered at both ends; floating lvs long-petioled, 5–10 cm, < 5 cm wide, elliptic to ovate; base tapered to rounded; tip rounded; stipules 3–9 cm, free, breaking apart early. **INFL**: spike, < 5 cm. **FR** 3–5 mm, clearly beaked; back rounded; sides flat. *n*=26. Shallow water, lakes, ponds, ditches, streams; 100–2750 m. NCoR, s SNF, n SNH, GV, CCo, SnFrB, SCo, SnBr, PR, GB, DMoj; worldwide (exc Australia). Hybridizes with *P. epihydrus, P. gramineus, P. illinoensis, P. natans, P. richardsonii.*

P. pectinatus L. (p. 1307) FENNEL-LEAF PONDWEED Per from matted rhizomes; tubers also present. **ST** many-branched, < 80 cm, ± cylindric. **LVS** all submersed, < 15(35) cm, < 1 mm wide, gen thread-like; tip acute to acuminate; stipules 2–5 cm, fused to blade, sheath-like below, sheath open, tips free. **INFL**: spike, interrupted in fr, whorls 2–6; peduncle 3–25 cm. **FR** 2.5–5 mm, plump; back rounded; beak very short. Common. Ponds, lakes, marshes, streams; < 2400 m. CA; worldwide (exc S.Am). Often weedy in reservoirs, irrigation canals; important food for waterfowl.

P. praelongus Wulfen (p. 1307) WHITE-STEMMED PONDWEED Per from stout rhizome. **ST** few-branched, often zig-zag, gen < 30 cm, subcylindric, whitish. **LVS** all submersed, sessile, 5–20 cm, 10–30 mm wide, lanceolate to ovate, often wavy; base gen lobed, clasping; tip hood-like, splitting when dry; stipules 3–10 cm, free, persistent, whitish. **INFL**: spike, 3–7.5 mm; peduncle 10–30 cm. **FR** 4–5 mm; back sharply 3-keeled; beak short. 2*n*=52. RARE in CA. Deep water, lakes; 1800–3000 m. CaRH, n&c SNH; to AK, Greenland, Mex, Eur, e Asia. Hybridizes with *P. richardsonii.*

P. pusillus L. (p. 1307) SMALL PONDWEED **ST** < 100 cm, cylindric. **LVS**: all submersed, thread-like to linear; base gen with 2 prominent glands, veins < 7; stipule tips gen becoming finely fibrous. **INFL** ± interrupted, < 1.5 cm. **FR** < 2.5 mm; back rounded; beak short. Shallow water, ponds, lakes, reservoirs, ditches, vernal pools, slow streams; < 2700 m. NCo, KR, n SNF, SNH, GV, SnFrB, SCoR, SW (exc ChI), GB, DMoj; circumboreal.

var. ***pusillus*** **ST** gen many-branched, < 100 cm. **LF** < 7 cm, gen (0.5)2–3 mm wide; tip acute; veins 3–5; stipules 6–15 mm, fused below middle, breaking apart early. **INFL**: spike, 6–15 mm, gen interrupted; whorls 2–4; peduncle 1.5–8 cm. **FR**: sides concave; tip beaked at top of central axis. Common. Habitat and distribution of sp.

var. ***tenuissimus*** Mert. & Koch **ST** simple to many-branched, < 80 cm. **LF** < 6 cm, gen < 1.5(2.5) mm wide; tip acute to obtuse; stipules 3–9 mm, free, flat or margins inrolled. **INFL** head-like to subcylndric spike; whorls 1–2; peduncle 0.5–4 cm. **FR**: sides round; tip beak to 1 side of central axis. *n*=13. Uncommon. Shallow water, mostly cold, acidic lakes, ponds; < 2100 m. NCo, KR, n SNF, SNH, SnJV, SCoRO, SnBr, GB; circumboreal. [*P. berchtoldii* Fieber]

P. richardsonii (A. Bennett) Rydb. (p. 1307) RICHARDSONS' PONDWEED Per from matted rhizomes. **ST** gen few-branched, < 70 cm, cylindric. **LVS** all submersed, sessile, 1–12 cm, gen 10–20 mm wide, gen narrowly ovate; margin crinkly near tip; base cordate; tip acute to rounded; stipules < 2 cm, free, becoming fibrous. **INFL**: spike, < 4 cm, slightly < peduncle. **FR** ± 4 mm; back rounded, smooth or faintly keeled. 2*n*=52. Uncommon. Shallow to deep water, ponds, lakes, lagoons, reservoirs, irrigation ditches, streams; < 2800 m. NCo, KR, CaRH, c SN, n SNH, MP; to AK, e N.Am, Colorado.

P. robbinsii Oakes (p. 1311) ROBBINS' PONDWEED Per from rhizomes. **ST** gen many-branched, < 200 cm, cylindric. **LVS** all submersed, clearly 2-ranked, sessile, < 12 cm, 3–4 mm wide, linear to lanceolate; base appendaged; margin gen finely serrate; stipules < 3 cm, fused to blade, sheathing below, free tips cut to fibrous, white. **INFL**: spike, < 2 cm; fls few, paired; peduncle club-shaped. **FR** (rarely collected), < 4 mm, back clearly keeled. RARE in CA. Deep water, lakes; 1600–3300 m. KR, c&s SNH; to AK, e US.

P. zosteriformis Fern. (p. 1311) EEL-GRASS PONDWEED **ST** gen few-branched, < 60 cm, 0.7–4 mm wide, flattened, winged. **LVS** all submersed, sessile, 5–20 cm, 2–5 mm wide, linear, ± rigid; tip obtuse to abruptly pointed; veins 9–35; stipules 2–6 cm, free, breaking apart. **INFL**: spike, < 3 cm; peduncle < 10 cm. **FR** ± 5 mm; back sharply keeled. RARE in CA. Ponds, lakes, streams; < 1300 m. s NCoRI (Clear Lake), GV, MP; to B.C., e N.Am.

RUPPIA DITCH-GRASS

Per from slender rhizomes. **STS** gen many-branched, thread-like, rooting at lower nodes. **LVS** cauline, alternate, < 10 cm, < 1 mm wide; tip acute; stipules < 15 mm, ± completely fused to lf base, sheath-like below lf junction, ± open. **INFL** terminal;

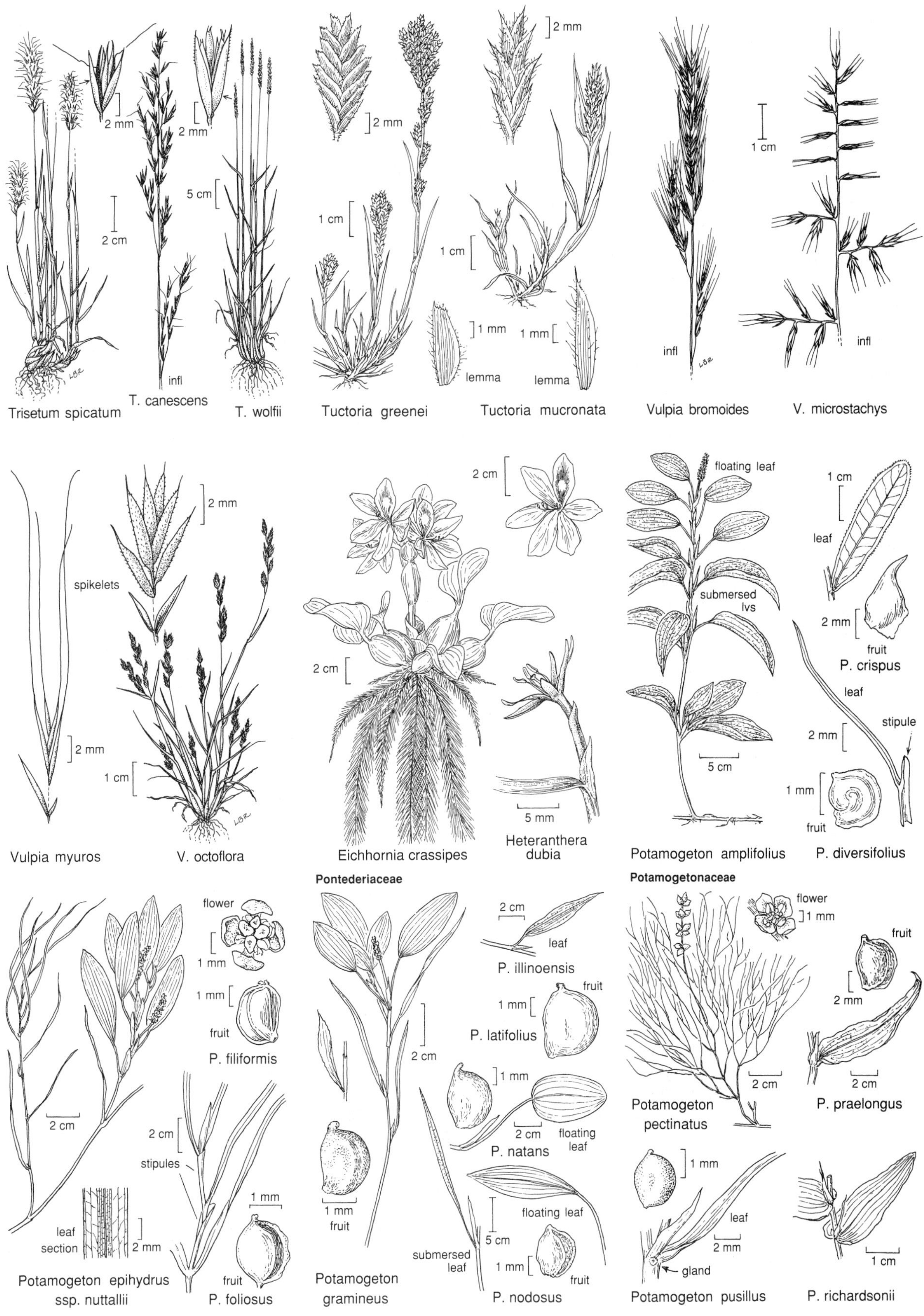

Trisetum spicatum · T. canescens · T. wolfii · Tuctoria greenei · Tuctoria mucronata · Vulpia bromoides · V. microstachys

Vulpia myuros · V. octoflora · Eichhornia crassipes · Heteranthera dubia · Potamogeton amplifolius · P. crispus · P. diversifolius

Pontederiaceae · **Potamogetonaceae**

Potamogeton epihydrus ssp. nuttallii · P. filiformis · P. foliosus · Potamogeton gramineus · P. illinoensis · P. latifolius · P. natans · P. nodosus · Potamogeton pectinatus · P. praelongus · Potamogeton pusillus · P. richardsonii

fls 2, sessile; peduncle elongated, straight or coiled in fr. **FL** minute; perianth 0; stamens 2, anthers sessile; pistils gen 4(2–8), simple, ovule attached at top of chamber, stigma sessile, peltate. **FR** ovoid, oblique, long-stalked. 2+ spp.; temp worldwide. (Heinrich Bernhard Ruppius, German botanist, 1688–1719) [Setchell 1946 Proc Calif Acad Sci 25:469–478]

1. Peduncle in fr 30–300 mm, coiled or flexible; lf tip acute ..... ***R. cirrhosa***
1′ Peduncle in fr 2–25 mm, straight; lf tip ± obtuse ..... ***R. maritima***

R. cirrhosa (Petagna) Grande (p. 1311) **LF** < 1 mm wide; tip acute. **INFL**: peduncle in fr 3–30 cm, coiled or flexible. $2n$=40. Marshes, ponds, sloughs; < 2045 m. NCo, NCoRI, Teh, SnJV, CW, SCo, SnBr, GB, D; ± worldwide. [*R. spiralis* Dumort.] Possibly indistinct from *R. maritima*; needs further study.

R. maritima L. **LF** < 0.5 mm wide; tip obtuse. **INFL**: peduncle in fr 2–25 mm, straight. $2n$=20. Marshes, ponds, sloughs; < 100 m. NCo, CW, SCo, ChI; ± worldwide.

SCHEUCHZERIACEAE SCHEUCHZERIA FAMILY

Robert F. Thorne

Per from rhizomes, aquatic (gen emergent) or terrestrial (sometimes on floating mats). **ST** erect, gen simple, slender. **LVS** basal and cauline, alternate, simple; sheath gen open; ligule prominent; blade erect, linear, subcylindric; tip with a large pore. **INFL**: raceme, terminal, bracted. **FL** bisexual, radial; perianth parts 6 in 2 whorls, free; stamens 6 in 2 whorls, free, anthers opening outward by slits; pistils 3(6), simple above, fused at base, ovary superior, chamber 1, ovules 1–3. **FR**: follicle. 1 genus, 1 sp.

SCHEUCHZERIA

The only genus. (Johann Jakob Scheuchzer, Swiss botanist, 1672–1733)

S. palustris L. ssp. ***americana*** (Fern.) Hulten (p. 1311) AMERICAN SCHEUCHZERIA **ST** in fl 3–4 dm. **LVS**: basal 1–2(3) dm; cauline shorter upward; ligule 2–10 mm; blade 1–3 mm wide. **INFL**: fls 3–12; peduncle in fr < 2.5 dm. **FL**: perianth parts ± 3 mm, 1-veined, greenish. **FR**: follicles 2–3, 5–10 mm, ± ascending. **SEEDS** 1–3, 4–5 mm. n=11. RARE. Floating mats, bogs, lake margins; 1400–2000 m. s CaRH (extant at Willow Lakes, extirpated at Lake Almanor, both nw Plumas Co.); to AK, e N.Am. Other ssp. in Eurasia.

TYPHACEAE CATTAIL FAMILY

Per from long rhizomes, colonial, glabrous, gen aquatic (submersed to emergent), monoecious. **ST** erect and stiff or submersed and floating above, cylindric, solid. **LVS** basal and cauline, alternate, ± 2-ranked, spongy or stiff; sheath open; blade linear, flat, keeled, or triangular in X-section, spongy. **INFL** spike-like (cylindric, dense) or head-like (spheric), terminal or axillary; staminate above pistillate, gen on same axis; fls subtended by 1, minute bract. **STAMINATE FL**: perianth parts 0 or 1–6 and scale-like; stamens 1–8. **PISTILLATE FL**: perianth parts 0 or 1–6 and flattened; ovary 1, chambers 1–2(3), ovules 1–2(3). **FR**: achene; wall thin, splitting in water. 2 genera, ± 25 spp.: worldwide. *Sparganium* formerly treated in Sparganiaceae. Family description and key to genera by R.F. Thorne.

1. Infl head-like, spheric; perianth parts 1–6, scale-like; fr 2–20 mm, sessile or pedicel short, thick, glabrous ..... **SPARGANIUM**
1′ Infl spike-like, cylindric; perianth 0; fr < 1 mm, pedicel long, thin, long-hairy ..... **TYPHA**

SPARGANIUM BUR-REED

Robert F. Thorne

Per from slender or corm-like rhizomes, glabrous, aquatic (submersed or emergent). **ST** slender, cylindric, solid; upper part erect or floating. **LF**: blade long, flat, keeled, or triangular, sometimes floating, spongy. **INFL** head-like, spheric, axillary and terminal, sessile or short-peduncled; bracts lf-like, gradually reduced upward; fls sessile, each gen subtended by 1 bractlet. **STAMINATE FL**: perianth parts 1–6, scale-like; stamens 1–8, filaments free or fused at base. **PISTILLATE FL**: perianth parts 1–6, oblanceolate to spoon-like, greenish, persistent in fr; ovary superior, chambers 1–2(3), ovule 1 per chamber, styles 1 or deeply 2(3)-lobed. **FR** fusiform to obconic, sessile or stalked; top tapered, truncate, or dome-like, beaked or not. 14 spp.: n temp, se Asia, sw Pacific. (Greek: swaddling band, from long, narrow lvs) [Cook & Nicholls 1986–7, Bot Helv 96:213–267;97:1–44]

1. Pl 10–26 dm, gen emergent; infl axis gen branched; stigmas 1–2; fr sessile, obconic, 6–10 mm, top truncate or dome-like, beaked
  2. Fr top dome-like; fr 4–7 mm, 2.5–5 mm wide; many fls with 1 stigma; many ovaries 1-chambered ..... ***S. erectum*** ssp. ***stoloniferum***
  2′ Fr top truncate to slightly rounded; fr 6–10 mm, 6–8 mm wide; many fls with 2 stigmas; many ovaries 2-chambered ..... ***S. eurycarpum*** ssp. ***eurycarpum***
1′ Pl 1–8 dm, submersed or emergent; infl axis simple; stigma 1; fr stalked, slender, fusiform to elliptic or oblanceolate, sometimes narrowed at middle, 3–5.5 mm; beak 0

3. Pl gen emergent; st ± stiff, erect; lvs and lower infl bracts 3–12 mm wide, keeled or triangular below middle; infl 12–25 mm diam in fr .... ***S. emersum*** ssp. ***emersum***

3′ Pl gen submersed; st ± flexible, floating above; lvs and lower infl bracts 1–5 mm wide, flat or sides convex, not keeled or triangular; infl 5–15 mm diam in fr

4. Lowest infl bract < 20(60) cm, >> infl; staminate infl gen 2+; pistillate infl gen 7–15 mm diam in fr, some pistillate infls on separate peduncle; fr 3.5–5 mm, 2–2.5 mm wide .... ***S. angustifolium***

4′ Lowest infl bract < 2 cm, ± = infl; staminate infl gen 1; pistillate infl 7–10 mm diam in fr; fr 2–4 mm, 1–2.5 mm wide .... ***S. natans***

S. angustifolium Michaux (p. 1311) NARROW-LEAVED BUR-REED Pl gen submersed (sometimes emergent and small). **ST** 3–10(17) cm, gen floating above, simple. **LF** 2–8(25) dm, 1.5–4(10) mm wide, ± flat, sides convex. **STAMINATE INFLS** gen 2–3, crowded on axis. **PISTILLATE INFLS** gen 2–4, 7–15(24) mm diam in fr; lowest gen on separate peduncle. **FR** 3–5.5 mm, 2–2.5 mm wide, elliptic to fusiform, gen narrowed at middle, short-stalked, brownish. 2*n*=30. Common. Shallow or deep water (1–2.5 m), lakes, ponds; 1200–3700 m. KR, SNH; to AK, ne N.Am, Colorado, circumboreal.

S. emersum Rehmann ssp. ***emersum*** (p. 1311) Pl gen emergent. **ST** 2–5(8) dm, erect, simple. **LF** 2–6(8) dm, 4–10 mm wide, keeled or triangular below middle (submersed lvs < 22 dm, < 18 mm wide, flat or slightly keeled). **STAMINATE INFLS** gen 4–7, well separated. **PISTILLATE INFLS** gen 3–4, 12–20(25) mm diam in fr; lowest on separate peduncle. **FR** 3.5–5.5 mm, ± 2–2.5 mm wide, fusiform, slightly narrowed at middle, brown, shiny. 2*n*=30. Lakes, ponds, marshes, streams; < 2600 m. NCo, KR, NCoRO, CaRH, SNH, SnBr, MP; to AK, Can. Another ssp. in Eurasia. [*S. multipedunculatum* (Morong) Rydb.; *S. simplex* Hudson]

S. erectum L. ssp. ***stoloniferum*** (Graebner) C. Cook & M.S. Nicholls (p. 1311) Pl gen emergent; rhizomes corm-like. **ST** 2–10(15) dm, erect, ± stiff; branches below infl 2–5. **LF** 5–15(35) dm, 10–20(28) mm wide, ± keeled. **STAMINATE INFLS** 6–9, well separated. **PISTILLATE INFLS** 1–3, gen 15–32 mm diam in fr. **FR** 4–7 mm, 2.5–5 mm wide, obconic, irregularly creased, light brown to straw-colored, sessile; top low-conic or dome-like; beak 2–3(4) mm. 2*n*=30. Lakes, ponds, marshes; < 850 m. NCo, NCoR, n SNH, GV, CCo, SnFrB, SCoRO, SCo, PR, MP; Baja CA; also in sw Asia, Australia. ssp. *erectum* in Eurasia. [*S. greenei* Morong]

S. eurycarpum Engelm. ssp. ***eurycarpum*** (p. 1311) Pl emergent (sometimes submersed); rhizomes corm-like. **ST** 5–15(26) dm, erect, ± stiff; branches below infl 2–5. **LF** 5–10(26) dm, 6–20 mm wide, ± keeled at base, flat above. **STAMINATE INFLS** 6–14, well separated. **PISTILLATE INFLS** 1–2, 20–35 mm diam in fr. **FR** 6–10 mm, 6–8 mm wide, wedge-shaped to obconic, ± abruptly narrowed and 3-angled below middle, light brown, sessile; top truncate to low-rounded; beak 2–4 mm. 2*n*=30. Pond and lake margins, marshes, streams; < 1400 m. NCoRI, s SNF, GV, CCo, SnFrB, SCoRI, MP; to B.C., e&s US. Another ssp. in Eurasia

S. natans L. (p. 1311) SMALL BUR-REED Pl gen submersed. **ST** 0.8–4 dm, floating above, flexible, simple. **LF** 0.6–4(6) dm, 1.5–6(10) mm wide, gen flat. **STAMINATE INFL** gen 1. **PISTILLATE INFLS** 1–3, well separated, 7–10 mm diam in fr. **FR** 2–4 mm, 1–2.5 mm wide, widely elliptic to obovoid-fusiform, slightly narrowed at middle or not, greenish or brownish, sessile or short-stalked. 2*n*=30. RARE in CA. Lakes, ponds; 2000–2500 m. CaRH, n&c SNH, MP; to AK, e N.Am; circumboreal. [*S. minimum* Fries]

TYPHA CATTAIL

S. Galen Smith

Per from tough rhizomes, emergent or terrestrial, colonial, glabrous, monoecious. **STS** erect, simple, hard. **LF**: sheath open; blade linear, C-shaped in X-section below, flat above. **INFL** spike-like, terminal, cylindric; staminate fls above, pistillate fls below; fls 1000+, staminate mixed with many papery scales; pistillate pedicels clustered on short, peg-like stalk. **STAMINATE FL**: perianth 0; stamens 2–7 on slender stalk; filaments slender, gen deciduous in fr. **PISTILLATE FLS** fertile and sterile; perianth 0; pedicel slender, long-hairy; ovary 1-chambered, ovule 1, style long, thread-like, stigma 1; sterile ovary truncate to rounded. **FR** minute, fusiform, falling with pedicel and hairs; wall thin, splitting in water. ± 8–13 spp.: worldwide. Rhizomes, pollen of some spp. used for food; lvs used for caning. (Greek: ancient name) [Smith 1987 Arch Hydrobiol Beih 27:129–138] All N.Am spp. hybridize.

1. Naked axis between staminate and pistillate fls gen 0; pistillate spike green when fresh, becoming brown, ± 28–36 mm wide in fr; pistillate stalk 1.5–3 mm, hair-like; stigmas widely lanceolate, gen persistent in fr; pistillate bractlets 0 .... ***T. latifolia***

1′ Naked axis between staminate and pistillate fls gen 1–8 cm; pistillate spike bright yellow-brown to dark brown in fl and fr, 15–25 cm wide in fr; pistillate stalk ± 0.5 mm, peg-like; stigmas linear, gen deciduous in fr; pistillate bractlets ± = or > pedicel hairs

2. Pistillate spike medium to dark brown; tip of pistillate bractlet dark brown, gen rounded; lf blade glands 0; sterile ovary green .... ***T. angustifolia***

2′ Pistillate spike bright yellow- to orange-brown; tip of pistillate bractlet straw-colored to orange-brown, acute to acuminate; base of lf blade gland-dotted on side facing st; sterile ovary pale brownish .... ***T. domingensis***

T. angustifolia L. (p. 1311) NARROW-LEAVED CATTAIL Pl 15–30 dm. **LF**: sheath top lobed, lobes ear-like, membranous, veiny, gen 0 with age; blade 4–12 mm wide when fresh, 3–8 mm wide when dry, glands 0. **INFL** gen < lvs; naked axis between staminate and pistillate fls gen 1–8 cm; staminate bractlets gen 2-lobed, some irregularly branched, brownish; pistillate stalk ± 0.5 mm, peg-like, spike (3.5)6–20 cm, 15–20 mm wide in fr, brown, bractlet = pedicel hairs, tip rounded, dark brown. **STAMINATE FL**: pollen grains single. **PISTILLATE FL**: stigma linear; sterile ovary truncate, ± = pedicel hairs, green when fresh, pale brown when dry, hair tips brownish. *n*=15. Marshes; < 2000 m. n SNH (Lake Tahoe), GV, CCo, SnFrB, SCo; to e N.Am, Eurasia. Possibly naturalized in CA. Hybrids with *T. latifolia* have been called *T.* ×*glauca* Godron.

T. domingensis Pers. SOUTHERN CATTAIL (p. 1311) Pl 15–40 dm. **LF**: sheath tapered to blade, lobes 0 to membranous, ear-like; blade 6–18 mm wide when fresh, 5–15 mm wide when dry, gland-dotted on inside near base. **INFL** ± = lvs; naked axis between staminate and pistillate fls (0)1–8 cm; staminate bractlets irregularly branched, straw-colored to cinnamon-brown; pistillate stalk ± 0.5 mm, peg-like, spike < 35 cm, 15–25 mm wide in fr, bright yellow- to orange-brown, bractlet > pedicel hairs, tip acute to acuminate. **STAMINATE FL**: pollen grains single. **PISTILLATE FL**: stigma linear; sterile ovary rounded to ± truncate, gen < pedicel hairs, pale brownish, hair tips straw-colored to orange-brown. *n*=15. Marshes; < 1500 m. NCo, NCoRO, GV, CW, SW, GB, D; warm temp, trop worldwide.

T. latifolia L. (p. 1311) BROAD-LEAVED CATTAIL Pls 15–30 dm. **LF**: sheath tapered to blade or shouldered; blade 10–29 mm wide when fresh, (5)10–20 mm wide when dry, glands 0. **INFL** ± = lvs; naked axis between staminate and pistillate fls 0 (sometimes 4–8 cm); staminate bractlets simple, hair-like, whitish; pistillate stalk 1.5–3 mm, hair-like, spike 5–25 cm, 28–36 mm wide in fr, green when fresh, brown in fr, bractlets 0. **STAMINATE FL**: pollen grains in groups of 4. **PISTILLATE FL**: stigma widely lanceolate; sterile ovary rounded, << pedicel hairs, hair tips whitish. n=15. Common. Marshes, ponds, lakes; < 2000 m. CA; temp N.Am, C.Am, Eurasia, Afr.

ZANNICHELLIACEAE HORNED-PONDWEED FAMILY

Robert F. Thorne

Per from slender, creeping rhizome, glabrous, aquatic, submersed, monoecious or dioecious. **ST** thread-like, weak. **LVS** alternate below to ± whorled and linear above; sheath fused to blade or free, stipule-like. **STAMINATE INFL** gen below pistillate infl, axillary; fls 1–3. **PISTILLATE INFL** terminal, short-stalked; fls 1–9, sessile, subtended by cup-like bract. **STAMINATE FL**: perianth parts 0 or 3 and scale-like; stamens 1–3, filaments free or fused; anther 1–2-chambered. **PISTILLATE FL**: perianth parts 0; ovary superior, chamber 1, ovule 1, style 1, simple or 3-lobed. **FR**: achene or nutlet. 4 genera; 9–10 spp.: ± worldwide.

ZANNICHELLIA HORNED-PONDWEED

Per, monoecious. **ST** gen few-branched. **LVS** ± opposite, clustered and ± whorled at upper nodes, stipuled. **STAMINATE FL**: stamen 1, filament slender, anther 2-chambered. **PISTILLATE FL**: style short, stigma peltate. **FR**: nutlet, stalked, compressed, tip beaked; back curved, ridged, toothed; front ± straight. 1–2 spp.: ± worldwide. (Gian G. Zannichelli, Venetian botanist, 1662–1729)

Z. palustris L. (p. 1311) **ST** 3–10 dm. **LF** 2–10 cm; blade < 1 mm wide, tip acute. **FR** 2–4 mm; beak 1–1.5 mm, flask-shaped. $2n$=12,24,28,32,36. Streams, ponds, ditches, lakes; < 2200 m. CA; ± worldwide.

ZOSTERACEAE EEL-GRASS FAMILY

Robert F. Thorne

Per, glabrous, submersed marine aquatic, rhizomed, monoecious or dioecious. **ST** flat, leafy, creeping or short with wide, thick base. **LVS** alternate, 2-ranked or tufted; sheath open or closed; ligule present; blade long, linear, ribbon-like, tip entire to notched. **INFL**: spike, axillary, flattened, enclosed by membranous, lf-like bract; fls many, 2-rowed on 1 side of axis, each enclosed by translucent, scale-like bractlets or not. **STAMINATE FL**: perianth 0; stamen 1, anther 2-chambered; pollen thread-like, < 2 mm. **PISTILLATE FL**: perianth 0; ovary 1, chamber 1, ovule 1, style 1, stigmas 2. **FR**: drupe or nutlet. 3 genera, ± 18 spp.: seacoasts, worldwide exc trop; some spp. source of food for marine animals, some aboriginal humans.

1. Dioecious; pls attached by short, tuber-like rhizome and roots to coastal, wave-swept rocks; lvs tufted at rhizome nodes, < 5 mm wide; fr base cordate .... **PHYLLOSPADIX**

1′ Monoecious; pls attached by long, slender, creeping rhizomes in bottom of bays, estuaries, and deep waters below wave action; lvs cauline, alternate, gen > 6 mm wide; fr base ± truncate .... **ZOSTERA**

PHYLLOSPADIX SURF-GRASS

Dioecious; rhizomes < 12 cm, thick, tuber-like. **ST** 10–60 cm, slender to thick, flat and winged below. **LVS** tufted at rhizome nodes, < 100(200) cm, 0.5–4 mm wide. **STAMINATE INFL** enclosed by membranous, lf-like bract; fls each enclosed by translucent, scale-like bractlets. **PISTILLATE INFL** enclosed by membranous, lf-like bract; fls alternating with vestigial anthers, each enclosed by translucent, scale-like bractlet. **STAMINATE FL**: anther sessile. **PISTILLATE FL**: ovary tapered to short style, style base persistent in fr. **FR**: nutlet, 2–3 mm, beaked, base with 2 downward projecting, horn-like lobes. 5 spp.: n Pacific coast; surf zones and rocky shores, just below low tide level. (Greek: lf-like infl)

1. Pistillate infl gen solitary, ± basal; peduncle gen 1–6 cm; lvs 1–4 mm wide, flat .... ***P. scouleri***

1′ Pistillate infls gen 3–7, cauline; peduncle 20–30 cm; lvs 0.5–2 mm wide, becoming folded or cylindric ... ***P. torreyi***

P. scouleri Hook. (p. 1311) **LF** 1–4 mm wide, flat. **PISTILLATE INFL** solitary (rarely 2); peduncle 1–6 cm. **FR** ± 3.5 mm. NCo, CCo, SCo; to AK, Baja CA.

P. torreyi S. Watson (p. 1311) **LF** 0.5–2 mm wide, flat, becoming folded or cylindric. **PISTILLATE INFLS** 3–7; peduncle gen 20–30 cm. **FR** ± 2–3 mm. NCo, CCo, SCo; to B.C., Baja CA.

ZOSTERA EEL-GRASS

Monoecious; rhizomes creeping, slender. **ST** < 300 cm, leafy. **LVS** 30–150 cm, alternate, cauline; blade with 3–7 parallel veins. **INFL** enclosed by sheath of subtending lf; staminate fls alternating with pistillate fls. **STAMINATE FL**: anther sessile. **PISTILLATE FL**: ovary tapered to short style. **FR**: nutlet, flask-like, wall translucent. **SEED** ± 3 mm, 1 mm diam, ± 20-ribbed or smooth, base truncate. 12 spp.: temp to cold marine waters worldwide. (Greek: from ribbon-like lvs) [Phillips & Echeverria 1990 Pacific Science 44:130–134] Important food source for marine animals, some aboriginal humans.

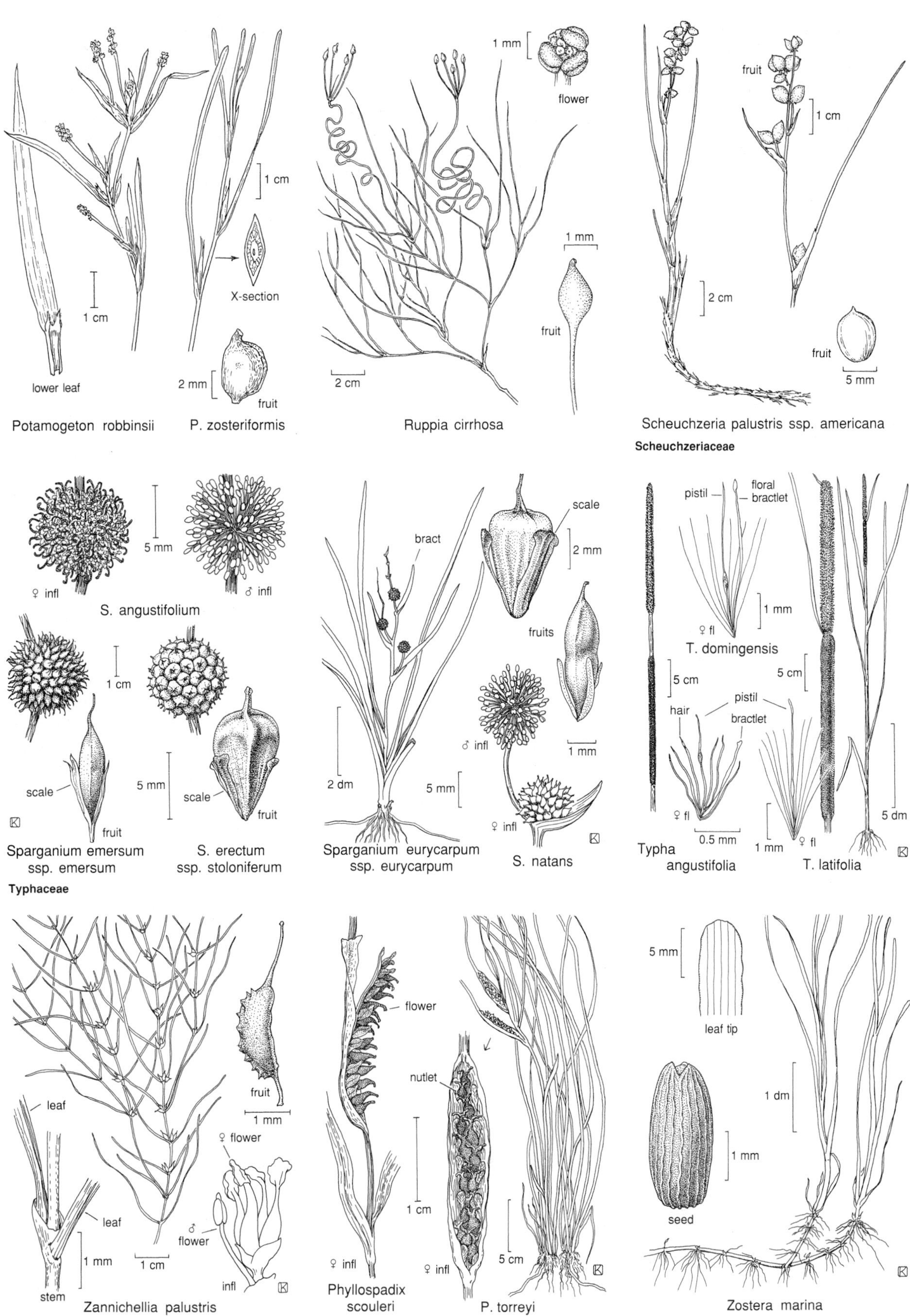

1 cm
lower leaf
X-section
1 cm
2 mm
fruit
Potamogeton robbinsii
P. zosteriformis
1 mm
flower
1 mm
fruit
2 cm
Ruppia cirrhosa
fruit
1 cm
2 cm
fruit
5 mm
Scheuchzeria palustris ssp. americana
Scheuchzeriaceae
5 mm
♀ infl
♂ infl
S. angustifolium
1 cm
scale
5 mm
scale
fruit
fruit
Sparganium emersum ssp. emersum
S. erectum ssp. stoloniferum
Typhaceae
bract
scale
2 mm
fruits
♂ infl
1 mm
2 dm
5 mm
♀ infl
Sparganium eurycarpum ssp. eurycarpum
S. natans
pistil
floral bractlet
1 mm
♀ fl
T. domingensis
5 cm
5 cm
hair
pistil
bractlet
♀ fl
0.5 mm
1 mm
♀ fl
5 dm
Typha angustifolia
T. latifolia
leaf
fruit
1 mm
♀ flower
leaf
♂ flower
1 mm
1 cm
stem
infl
Zannichellia palustris
Zannichelliaceae
flower
nutlet
1 cm
♀ infl
♀ infl
5 cm
Phyllospadix scouleri
P. torreyi
Zosteraceae
5 mm
leaf tip
1 dm
1 mm
seed
Zostera marina

1. Lvs 1.5–12 mm wide, tip gen obtuse; seed surface with ± 20–25 ridges; pls of shallow water, bays and estuaries; fl in March, fr in May-June .... ***Z. marina***
1′ Lvs 12–18.5 mm wide, tip gen notched; seed surface smooth or faintly ridged; pls of subtidal waters; fl in August (May), fr in September-October .... ***Z. pacifica***

Z. marina L. (p. 1311) **LF** 1.5–12 mm wide; tip gen obtuse. **SEED**: surface with ± 20–25 longitudinal ridges. *n*=6. Common. Shallow water, bays, estuaries; 0–2 m below mean low tide. NCo, CCo, SCo, ChI; to AK, Baja CA. [var. *latifolia* Morong]

Z. pacifica S. Watson **LF** 12–18.5 mm wide; tip gen notched. **SEED**: surface smooth or faintly ridged. Common. Subtidal waters; 5–17 m below mean low tide. NCo, CCo, SCo; to B.C. [*Z. asiatica* Miki]

APPENDIX I

FLORISTIC SUMMARY: NUMBERS OF TAXA IN TAXONOMIC GROUPS

The Jepson Manual includes descriptions of 173 families, 1222 genera, 5862 species, and 1169 subspecies or varieties known to occur outside of cultivation in California. Each of these taxonomic categories can be classified into one of three categories that relate to their geographic origin and natural distribution. Native taxa are those believed to have existed in California before initial visitation and colonization by European humans. The native taxa include a subset of endemics, taxa that do not occur naturally outside of California. Alien taxa are those believed to have been introduced from other parts of the world since the first European colonization and which have either become subsequently naturalized or are waifs.

Nineteen of the 173 plant families are composed entirely of naturalized genera and species. Among 1222 genera known to occur in California, 25 are endemic and 327 are composed exclusively of naturalized species. Among the 5862 species known to occur in California, 3423 species (58.4 %) are native, of which 1416 species (24.2 %) are endemic, and 1023 (17.4 %) are naturalized aliens. At least 26 of the endemic species are presumed extinct. At least 737 (63 %) of the additional 1169 subspecies and varieties are endemic. The proportion of endemics among all species, subspecies, and varieties is 30.6 %.

The Pteridaceae, Pinaceae, Asteraceae, and Poaceae are the largest families in each of the four major taxonomic groups respectively. Pteridaceae is composed of 10 genera, 33 native species (including 3 endemics and 3 aliens), and 4 subspecies or varieties (including 1 endemic). Pinaceae is composed of 5 genera, 33 species (including 4 endemics), and 3 varieties (including 2 endemics). Asteraceae, the largest family in California, is composed of 178 genera (including 11 endemics and 53 aliens), 748 species (including 187 endemics and 121 aliens), and 176 subspecies or varieties (including 107 endemics). Poaceae, the second largest family in California, is composed of 118 genera (including 2 endemics and 51 aliens), 428 species (including 37 endemics and 177 aliens), and 33 subspecies or varieties (including 3 endemics and 4 aliens).

MAJOR TAXONOMIC GROUPS

	Ferns and Fern Allies	Gymnosperms	Angiosperms		
			Dicots	Monocots	TOTAL
Families	14	5	133	21	173
Genera	32	14	948	228	1222
Species	103	60	4646	1058	5862
Endemic Species	5	7	1059	98	1169
Endemic Subspecies or Varieties	4	5	692	36	737
Alien Species	4	0	784	235	1023
Presumed Extinct	0	0	24	2	26

APPENDIX II

CLASSIFICATION OF CALIFORNIA PLANT FAMILIES

Ordering of Taxa

For *The Jepson Manual*, a hierarchical, alphabetical arrangement was chosen so that taxa would be easy to find with minimal reference to an index. However, an alphabetical ordering provides no information at all about relationships among families. These relationships are the subject of this appendix, which provides only a brief introduction to the complex subject of plant classification.

In condensing any classification into a linear array, most of the multi-dimensional relationships among groups are lost. Arthur Cronquist, our primary source for relationships among flowering plant families (reference below), presented families in the order of the classification that follows. This order has been adopted for some floras but may be confusing or misleading to beginners. Where, for example, would one look for willows? The willow family, Salicaceae, is in Subclass Dilleniidae, between Loasaceae of Violales and Capparaceae of Capparales, not in Subclass Hamamelidae with other superficially similar families that bear catkins. Even the two woody segregates of Saxifragaceae (Grossulariaceae and Philadelphaceae) are separated from their "parent" family by Crassulaceae (and, worldwide, by five other families as well).

The entire structure of and justification for a classification is important. Much less important is the linear ordering of its taxa. An extensive literature, and much controversy, exists on the subject of plant classification. For those wishing to pursue it, the works of Arthur Cronquist, Armen Takhtajan, and Robert F. Thorne are suggested as a beginning to classical approaches. Current thinking may be exemplified by the works of Michael Donoghue, Jeffrey Palmer, and Kare Bremer.

A Classification System for Plants

All true plants are considered to be members of the Kingdom Plantae (bacteria and fungi, which were also once included, are now considered to make up different kingdoms). *The Jepson Manual* does not cover the so-called "lower" true plants of California, which lack a vascular system (veins of xylem and phloem). Thus all algae, mosses, and liverworts are excluded, except in the Family Key, where some forms are distinguished from higher (or "vascular") plants.

Higher plants can be broken into three main groups that are treated sequentially in *The Jepson Manual*: 1) ferns and fern-allies, which reproduce by spores, like the lower plants; 2) gymnosperms, woody plants that reproduce by seeds not enclosed in an ovary (often seeds are borne in cones); 3) flowering plants, which produce true (albeit sometimes minute) flowers and develop their seeds enclosed in an ovary.

Flowering plants make up about 95% of Californian higher plants. Their two main groups — dicots (Class Dicotyledones) and monocots (Class Monocotyledones) — are treated separately. Thus, there are four sections of treatments: ferns & fern-allies (identified by "**F**" on page headings); gymnosperms ("**G**"), dicots ("**D**"), and monocots ("**M**").

Within these generally accepted major groups (the boundary between dicots and monocots is problematical), there is ongoing controversy about how to define and array less-inclusive groups — especially Subclasses, Orders, and Families. Given new techniques for analyzing current features, the predictable discovery of new features, and established differences of opinion about the relative importance of features in different groups, it is likely that all classification systems will continue to change over time.

For *The Jepson Manual*, classification within the ferns & fern-allies generally follows David Lellinger's 1985 *A Field Manual of the Ferns and Fern-Allies of the United States and Canada* (Smithsonian Institution, 389 pp.). Some changes suggested by Alan R. Smith were incorporated. Fern classification is in rapid flux; the version presented below is not likely to persist in its details for long. By contrast, there is general consensus about the much simpler classifications of fern-allies and gymnosperms.

Classification of flowering plants is currently a rich field for innovation. Much current work and thought may lead to a revolutionarily different classification that is capable of gaining reasonable consensus. Two kinds of advances are working toward such a result. First, new techniques are supplying more reliable information on genetics, chemistry, and anatomical details. Second, there is a growing emphasis on unique, derived features in the definitions of taxa.

The classification of Cronquist (1981, *An Integrated System of Classification of Flowering Plants*, Columbia Univ. Press, 1262 pp.) is presented below as it applies in California. Its components are also used in the text of *The Jepson Manual*, except where authors have argued persuasively for changes. Cronquist's system was chosen because it is widely known as a teaching tool and is relatively easy to understand and remember. It uses ten Subclasses — six for dicots — whereas other systems, such as those of Thorne and Takhtajan, have many more Subclasses. Because Cronquist's Subclasses are quite helpful in summarizing relationships, brief characterizations of them are given within the hierarchical scheme presented below.

The overall classification of higher plants used here is summarized as follows. In *The Jepson Manual* there are six Divisions (botanical equivalents of Phyla). Fern-allies make up two Divisions, ferns one, gymnosperms two, and flowering plants one that is divided into dicot and monocot Classes. The classification uses 16 Subclasses. The Divisions or Subclasses are further divided into 67 Orders, which in turn comprise 173 Families. (Orders are not specified for ferns; Subclasses are not specified for fern-allies or gymnosperms.)

Some Concepts of Classification

For their members to be recognizable, taxa at all levels must be defined by characteristics their members hold uniquely in common. Thus, they are not real entities, but concepts built around chosen features: any organism that shares the defining features of a taxon is automatically a member of that taxon. When this is recognized, it becomes clearer why classifications change as more is learned about organisms. The concepts that we call taxa can be equated (metaphorically and somewhat simplistically) to cross-referenced pigeon-holes arrayed in cabinets that are arrayed in rooms that are arrayed in buildings. There is nothing particularly special about any pigeon-hole — their contents may be combined, divided, or moved to another cabinet or room as new information (or a reinterpretation of old information) suggests.

Following the pigeon-hole analogy above, plant families are approximately comparable to rooms: the doors or corridors (cross-references) connecting them are sometimes altered or rooms are subdivided by new walls. Sometimes walls need to be torn down, or (parts of) the contents of a room moved to another building for better organization.

Much effort in systematics today is devoted not to the definition of taxa, but to the search for lineages — that is, sets of organismal relationship through parentage. Unlike taxa, which are necessarily conceptual, lineages are real things — they are composed of a definite (as opposed to an indefinite) number of entities, most of which are no longer extant. Most attention is given, by default, to the extant, contemporaneous members of the lineage, but they provide clues to the relationships of members no longer available for study.

Taxonomists generally wish to define taxa either so that their members share the greatest possible number of important features (the so-called "natural" approach) or so that they correspond as closely as possible to lineages (the so-called "phyletic" approach). However, the very act of breaking up a lineage (which is continuous through time) into separate taxa creates groups defined by unique common features and the "real thing" (the lineage) is lost. Philosophical deliberations about such issues will continue; however, it is helpful in understanding why there is so much change in taxonomy to recognize that taxa are never "real things" — they are simply the best concept we can define at the moment.

Examples of Changes

A few examples of differences between family definitions in *The Jepson Manual* from those in earlier manuals may help explain some of the reasons for changes.

Fumariaceae (bleeding-heart family) is easily recognized by its unique flower shape. It has long been known to be closely related to poppies (Papaveraceae). Recent detailed studies have shown clearly that "Fumariaceae" is more closely related to *Papaver* (the type genus of Papaveraceae) than *Papaver* is to other genera traditionally placed in the poppy family (such as *Eschscholzia*, our state flower). This new perception of overall pattern of relationship can be accommodated in only two ways — both require change. First, Fumariaceae could be retained as a family; logical consistency would require the shift into separate families of all those "poppies" that are at least as distantly related to *Papaver* as is Fumariaceae. In California, this would add at least Platystemonaceae (cream-cup family) and Eschscholziaceae (California-poppy family) to Papaveraceae and Fumariaceae. The alternative is to retain the traditional Papaveraceae, adding the genera of "Fumariaceae" to it. For simplicity and for the utility of keeping related groups together, the second alternative was chosen for *The Jepson Manual*. We viewed this as the conservative course, even though an attractive and popular family was "lost".

Other families have been split up elsewhere but remain unified in *The Jepson Manual*. It is easy to justify splitting Fabaceae (legume family) into three families, adding Mimosaceae (mimosa family) and Caesalpiniaceae (red-bud family). Ericaceae offers a similar situation. It is often divided into Monotropaceae (indian-pipe family), Pyrolaceae (pyrola family), Ericaceae proper (heath family), and Vacciniaceae (blueberry family). In these and other instances, broadly defined families were retained, recognizing the segregate groups as subfamilies. It may seem shocking that there is, in fact, no objective way to decide the rank at which to recognize a taxon — all such decisions are matters of tradition, utility, and taste and not scientific issues. We decided that utility (and most often tradition) were generally served by keeping related groups together.

The most extreme example of retaining a broadly defined family involves lilies and their relatives, which are very diverse worldwide. In this century, this lineage has been divided many different ways — in general, increasingly finely. The practical problem is that once any one of a large number of groups (such as the agave-like genera) is taken out, the justification for keeping other groups within the family is weakened. Cronquist was relatively conservative in his treatment of lilies, recognizing Agavaceae, Liliaceae, and Smilacaceae as separate (Californian) families on the basis of unique growth habits. *The Jepson Manual* recognizes only one family of lily relatives, (which includes both *Agave* and *Smilax*). If a modestly free hand in segregating lily-like families were allowed, in California we would have at least Agavaceae, Alliaceae, Aloeaceae, Asparagaceae, Calochortaceae, Convallariaceae, Liliaceae, Smilacaceae, and Trilliaceae. Again, for ease of comparing members of the lineage, these groups were kept together.

CLASSIFICATION OF THE FAMILIES OF HIGHER PLANTS FOUND IN CALIFORNIA

KINGDOM PLANTAE
- Algae: not covered
- Bryophytes [mosses & liverworts]: not covered

HIGHER (VASCULAR) PLANTS:

Ferns and Fern-allies [spore-bearers]

DIVISION LYCOPHYTA
- Lycopodiales
  - Lycopodiaceae
- Selaginellales
  - Selaginellaceae
- Isoetales
  - Isoetaceae

DIVISION SPHENOPHYTA
- Equisetales
  - Equisetaceae

DIVISION PTEROPHYTA
- SUBCLASS OPHIOGLOSSIDAE
  - Ophioglossaceae
- SUBCLASS SCHIZAEIDAE
  - Pteridaceae
- SUBCLASS GLEICHENIIDAE
  - Polypodiaceae
- SUBCLASS HYMENOPHYLLIDAE
  - Dennstaedtiaceae
  - Thelypteridaceae
  - Aspleniaceae
  - Dryopteridaceae
  - Blechnaceae
- SUBCLASS MARSILEIDAE
  - Marsiliaceae
- SUBCLASS SALVINIIDAE
  - Azollaceae

Gymnosperms [seeds not enclosed in ovary; gen with cones]

DIVISION GNETOPHYTA
- Ephedrales
  - Ephedraceae

DIVISION CONIFEROPHYTA
- Taxales
  - Taxaceae
- Coniferales
  - Pinaceae
  - Taxodiaceae
  - Cupressaceae

Flowering Plants [seeds enclosed in ovary; flowers present]

DIVISION ANTHOPHYTA
CLASS DICOTYLEDONES (DICOTS)
SUBCLASS MAGNOLIIDAE:
[Pistils generally simple; perianth parts and stamens free, generally many, spiralled.]
- Laurales
  - Calycanthaceae
  - Lauraceae
- Piperales
  - Saururaceae
- Aristolochiales
  - Aristolochiaceae
- Nymphaeales
  - Nymphaeaceae
  - Cabombaceae
  - Ceratophyllaceae
- Ranunculales
  - Ranunculaceae
  - Berberidaceae
- Papaverales
  - Papaveraceae

SUBCLASS HAMAMELIDAE:
[Mostly woody; flowers ± in unisexual catkins; perianth 0.]

Hamamelidales
  Platanaceae
Urticales
  Ulmaceae
  Cannabaceae
  Moraceae
  Urticaceae
Juglandales
  Juglandaceae
Myricales
  Myricaceae
Fagales
  Fagaceae
  Betulaceae

SUBCLASS CARYOPHYLLIDAE:
[Petals free (or 0 and sepals petal-like, sometimes fused); placentas basal or free-central.]

Caryophyllales
  Phytolaccaceae
  Nyctaginaceae
  Aizoaceae
  Cactaceae
  Chenopodiaceae
  Amaranthaceae
  Portulacaceae
  Basellaceae
  Molluginaceae
  Caryophyllaceae
Polygonales
  Polygonaceae
Plumbaginales
  Plumbaginaceae

SUBCLASS DILLENIIDAE:
[Petals ± free (if 0, sepals not petal-like); placentas generally parietal (or axile); stamens developing from inner to outer.]

Dilleniales
  Paeoniaceae
Theales
  Elatinaceae
  Hypericaceae
Malvales
  Sterculiaceae
  Malvaceae
Nepenthales
  Sarraceniaceae
  Droseraceae
Violales
  Cistaceae
  Violaceae
  Tamaricaceae
  Frankeniaceae
  Fouquieriaceae
  Cucurbitaceae
  Datiscaceae
  Loasaceae
Salicales
  Salicaceae
Capparales
  Capparaceae
  Koeberliniaceae
  Brassicaceae
  Resedaceae
Batales
  Bataceae
Ericales
  Empetraceae
  Ericaceae
Ebenales
  Styracaceae
Primulales
  Primulaceae

SUBCLASS ROSIDAE:
[Petals ± free; stamens more than petals or opposite them, developing from outer to inner; pistils sometimes simple; placentas most often axile.]

Rosales
  Pittosporaceae
  Philadephaceae
  Grossulariaceae
  Crassulaceae
  Saxifragaceae
  Rosaceae
  Crossosomataceae
Fabales
  Fabaceae
Proteales
  Elaeagnaceae
Haloragales
  Haloragaceae
  Gunneraceae
Myrtales
  Lythraceae
  Thymelaeaceae
  Myrtaceae
  Punicaceae
  Onagraceae
Cornales
  Cornaceae
  Garryaceae
Santalales
  Santalaceae
  Viscaceae
Rafflesiales
  Rafflesiaceae
Celastrales
  Celastraceae
  Aquifoliaceae
Euphorbiales
  Simmondsiaceae
  Euphorbiaceae
Rhamnales
  Rhamnaceae
  Vitaceae
Linales
  Linaceae

Polygalales
  Polygalaceae
  Krameriaceae
Sapindales
  Staphyleaceae
  Hippocastanaceae
  Aceraceae
  Burseraceae
  Anacardiaceae
  Simaroubaceae
  Meliaceae
  Rutaceae
  Zygophyllaceae
Geraniales
  Oxalidaceae
  Geraniaceae
  Limnanthaceae
  Tropaeolaceae
  Balsaminaceae
Apiales
  Araliaceae
  Apiaceae

SUBCLASS ASTERIDAE:
[Petals ± fused; stamens fewer than (or =) petals and alternate them; pistil compound, generally of 2 carpels.]

Gentianales
  Gentianaceae
  Apocynaceae
  Asclepiadaceae
Solanales
  Solanaceae
  Convolvulaceae
  Cuscutaceae
  Menyanthaceae
  Polemoniaceae
  Hydrophyllaceae
Lamiales
  Lennoaceae
  Boraginaceae
  Verbenaceae
  Lamiaceae
Callitrichales
  Hippuridaceae
  Callitrichaceae
Plantaginales
  Plantaginaceae
Scrophulariales
  Buddlejaceae
  Oleaceae
  Scrophulariaceae
  Myoporaceae
  Orobanchaceae
  Acanthaceae
  Martyniaceae
  Bignoniaceae
  Lentibulariaceae
Campanulales
  Campanulaceae
Rubiales
  Rubiaceae
Dipsacales
  Caprifoliaceae
  Valerianaceae
  Dipsacaceae
Asterales
  Asteraceae

CLASS MONOCOTYLEDONES
(MONOCOTS)

SUBCLASS ALISMATIDAE:
[Pistils simple; herbs, ± aquatic.]

Alismatales
  Alismataceae
Hydrocharitales
  Hydrocharitaceae
Potamogetonales
  Aponogetonaceae
  Scheuchzeriaceae
  Juncaginaceae
  Potamogetonaceae
  Zannichelliaceae
  Cymodoceaceae
  Zosteraceae

SUBCLASS ARECIDAE:
[Inflorescence often of many small flowers, enfolded or subtended by prominent bract(s); palm-like to minute aquatics.]

Arecales
  Arecaceae
Arales
  Araceae
  Lemnaceae

SUBCLASS COMMELINIDAE:
[Flowers small and subtended by chaffy bracts, or sepals and petals unlike; generally wind-pollinated.]

Commelinales
  Commelinaceae
Juncales
  Juncaceae
Cyperales
  Cyperaceae
  Poaceae
Typhales
  Typhaceae

SUBCLASS LILIIDAE:
[Flowers ± showy, insect-pollinated); sepals and petals general similar.]

Liliales
  Pontederiaceae
  Liliaceae
  Iridaceae
Orchidales
  Orchidaceae

APPENDIX III

NAME CHANGES FROM RECENT REFERENCES

This appendix lists changes in scientific names of California plants from those most recently accepted in certain statewide publications [viz., Munz's 1959 *A California Flora* as corrected by his 1968 *Supplement to A California Flora* and the *CNPS Inventory of Rare and Endangered Vascular Plants of California* (ed. 4, 1988)]. Taxa considered dubious by Munz (those not given in bold type) and names he listed as synonyms generally are not included.

Names not accepted in *The Jepson Manual* are listed alphabetically, in the order that they would have been found in *The Jepson Manual* had they been accepted. Taxa not accepted are generally dealt with in one of three ways in *The Jepson Manual*. First, names of unclearly differentiated or confusing taxa are here designated "indistinct (from)" in a note, generally corresponding to the phrase "have been called" in the text. This notation indicates uncertainty about whether the taxon should be accepted and implies that more study is needed.

The other two primary categories of names are generally found in brackets in *The Jepson Manual*. Names validly applicable in other areas but mistakenly used for California plants are designated "misapplied (to)". Names that are judged synonyms (and therefore not validly applicable anywhere) are simply listed in brackets in the main text. In this list, they are followed by "=" and the accepted name. Other situations are summarized as concisely as possible.

Also listed are the names of California taxa which have been described since Munz's *Supplement* was published or recognized since then to grow in California (both natives and naturalized aliens). [Ed. note: last-minute global alterations have produced minor inconsistencies]

FERNS & FERN-ALLIES

ASPLENIACEAE SPLEENWORT FAMILY

Asplenium septentrionale (L.) Hoffm. occurs in CA
A. trichomanes L. ssp. *trichomanes* occurs in CA
A. viride Hudson = *A. trichomanes-ramosum* L.

AZOLLACEAE MOSQUITO FERN FAMILY

Azolla mexicana C. Presl occurs in CA

BLECHNACEAE DEER FERN FAMILY: no changes

DENNSTAEDTIACEAE BRACKEN FAMILY: no changes

DRYOPTERIDACEAE WOOD FERN FAMILY

Athyrium filix-femina (L.) Roth vars. *californicum* F.K. Butters & *sitchense* Rupr. = var. *cyclosorum* Rupr.
Dryopteris dilatata (Hoffm.) A. Gray misapplied to *D. expansa* (C. Presl) Fraser-Jenkins & Jermy
Polystichum munitum (Kaulf.) C. Presl ssp. curtum Ewan = *P. imbricans* (D. Eaton) D.H. Wagner ssp. *curtum* (Ewan) D.H. Wagner
var. *imbricans* (D. Eaton) Maxon = *P. imbricans* (D. Eaton) D.H. Wagner ssp. imbricans
ssp. *nudatum* Ewan = *P. imbricans* (D. Eaton) D.H. Wagner ssp. *imbricans*

EQUISETACEAE HORSETAIL FAMILY: no changes

ISOETACEAE QUILLWORT FAMILY

Isoetes muricata Durieu = *I. echinospora* Durieu

LYCOPODIACEAE CLUB-MOSS FAMILY

Lycopodium inundatum L. = *Lycopodiella inundata* (L.) Holub

MARSILEACEAE MARSILEA FAMILY: no changes

OPHIOGLOSSACEAE ADDER'S-TONGUE FAMILY

Botrychium ascendens W.H. Wagner recently described
B. crenulatum W.H. Wagner recently described
B. lunaria (L.) Sw. var. *minganense* (Victorin) E.J. Dole = *B. minganense* Victorin
B. montanum W.H. Wagner recently described
B. multifidum (S. Gmelin) Rupr. sspp. *coulteri* (L. Underw.) R.T. Clausen & *silaifolium* (C. Presl) R.T. Clausen not recognized
B. pinnatum St. John occurs in CA
B. pumicola Cov. not known from CA
B. simplex Hitchc.: w N.Am forms may warrant sp. or ssp. status; var. *compositum* (Lasch) Milde indistinct
Ophioglossum lusitanicum L. ssp. *californicum* (Prantl) R.T. Clausen = *O. californicum* Prantl
O. vulgatum L. misapplied in N.Am to *O. pusillum* Raf.

POLYPODIACEAE POLYPODY FAMILY

Polypodium amorphum Suksd. occurs in CA
P. calirhiza S. Whitmore & A. Reed Smith recently described

PTERIDACEAE BRAKE FAMILY

Adiantum pedatum L. var. *aleuticum* Rupr. = *A. aleuticum* (Rupr.) C.A. Paris
Aleuritopteris cretacea (Liebm.) Fourn. misapplied to *Notholaena californica* D. Eaton
Argyrochosma limitanea (Maxon) M.D. Windham var. *limitanea* occurs in CA
Cheilanthes carlotta-halliae W. Wagner & Gilbert = *Aspidotis carlotta-halliae* (W. Wagner & Gilbert) Lellinger
C. cochisensis (Goodd.) Mickel = *Astrolepis cochisensis* (Goodd.) D.M. Benham & M.D. Windham

C. ×fibrillosa L. Underw. = *C. covillei* Maxon × *C. newberryi* (D. Eaton) Domin
C. jonesii (Maxon) Munz = *Argyrochosma jonesii* (Maxon) M.D. Windham
C. limitanea (Maxon) Mickel var. *limitanea* = *Argyrochosma limitanea* (Maxon) M.D. Windham var. *limitanea*
C. sinuata (Sw.) Domin var. *cochisensis* (Goodd.) Munz = *Astrolepis cochisensis* (Goodd.) D.M. Benham & M.D. Windham
Cryptogramma cascadensis E. Alverson recently described
Notholaena californica D. Eaton ssp. *nigrescens* Ewan not recognized
Onychium densum Brackenr. = *Aspidotis densa* (Brackenr.) Lellinger
Pellaea compacta (Davenp.) Mickel = *P. mucronata* (D. Eaton) D. Eaton var. *californica* (Lemmon) Munz & I.M. Johnston
P. longimucronata Hook. misapplied to *P. truncata* Goodd.
Pityrogramma triangularis (Kaulf.) Maxon var. *maxonii* Weath. = *Pentagramma triangularis* (Kaulf.) Yatskievych, Windham, & Wollenweber ssp. *maxonii* (Weath.) Yatskievych, Windham, & Wollenweber
var. *pallida* Weath. = *Pentagramma pallida* (Weath.) Yatskievych et al.
var. *triangularis* = *Pentagramma triangularis* (Kaulf.) Yatskievych, Windham, & Wollenweber ssp. *triangularis*
var. *viscosa* (D. Eaton) Weath. = *Pentagramma triangularis* (Kaulf.) Yatskievych, Windham, & Wollenweber ssp. *viscosa* (D. Eaton) Yatskievych, Windham, & Wollenweber
Pteris cretica L. reported from SnFrB
P. tremula R. Br. naturalized

SELAGINELLACEAE SPIKE-MOSS FAMILY

Selaginella douglasii (Hook. & Grev.) Spring undocumented in CA
S. engelmannii Hieron. var. *scopulorum* (Maxon) Reed = *S. densa* Rydb. var. *scopulorum* (Maxon) Tryon

THELYPTERIDACEAE THELYPTERIS FAMILY

Lastrea augescens (Link) J. Smith misapplied to *Thelypteris puberula* (Baker) C. Morton var. *sonorensis* A. Reed Smith
L. oregana (C. Chr.) Copel. = *Thelypteris nevadensis* (Baker) C. Morton

GYMNOSPERMS

CUPRESSACEAE CYPRESS FAMILY

Chamaecyparis lawsoniana (A. Murray) Parl. = *Cupressus lawsoniana* A. Murray
C. nootkatensis (D. Don) Spach = *Cupressus nootkatensis* D. Don
Cupressus bakeri Jepson ssp. *matthewsii* C. Wolf not recognized
C. nevadensis Abrams = *C. arizonica* E. Greene ssp. *nevadensis* (Abrams) E. Murray
C. pygmaea (Lemmon) Sarg. = *C. goveniana* Gordon ssp. *pigmaea* (Lemmon) J. Bartel
C. stephensonii C. Wolf = *C. arizonica* E. Greene ssp. *arizonica*
Juniperus communis L. var. *saxatilis* Pallas not recognized

EPHEDRACEAE EPHEDRA FAMILY: no changes

PINACEAE PINE FAMILY

Abies lasiocarpa (Hook.) Nutt. var. *lasiocarpa* occurs in CA
Pinus aristata Engelm. = *P. longaeva* D. Bailey
P. ×attenuradiata Stockw. & Right not recognized
P. balfouriana Grev. & Balf. ssp. austrina Mastrogiuseppe & Mastrogiuseppe recently described
P. murrayana Grev. & Balf. = *P. contorta* Loudon ssp. *murrayana* (Grev. & Balf.) Critchf.
P. remorata H. Mason = *P. muricata* D. Don

TAXACEAE YEW FAMILY: no changes

TAXODIACEAE BALD CYPRESS FAMILY: no changes

DICOTS

ACANTHACEAE ACANTHUS FAMILY

Beloperone californica Benth. = *Justicia californica* (Benth.) D. Gibson
Carlowrightia arizonica A. Gray occurs in CA

ACERACEAE MAPLE FAMILY: no changes

AIZOACEAE FIG-MARIGOLD FAMILY

Delosperma litorale (Kensit) L. Bolus naturalized
Glinus transferred to Molluginaceae
Mesembryanthemum chilensis Molina = *Carpobrotus chilensis* (Molina) N.E. Br.
M. cordifolium L.f. = *Aptenia cordifolia* (L.f.) N.E. Br.
M. crassifolium L. misapplied to *Malephora crocea* (Jacq.) Schwantes
M. edule L. = *Carpobrotus edulis* (L.) N.E. Br.
M. elongatum Haw. misapplied to *Conicosia pugioniformis* (L.) N.E. Br.
M. floribundum Haw. = *Drosanthemum* f. (Haw.) Schwantes
M. speciosum Haw. not naturalized [= *Drosanthemum* s. (Haw.) Schwantes]
Mollugo transferred to Molluginaceae

AMARANTHACEAE AMARANTH FAMILY

Alternanthera pungens Kunth = *A. caracasana* Kunth
Amaranthus torreyi (A. Gray) Benth. occurs in CA
Brayulinea densa (Willd.) Small = *Guilleminea densa* (Willd.) Moq. var. *aggregata* Uline & Bray
Tidestromia oblongifolia (S. Watson) Standley ssp. *cryptantha* (S. Watson) Wiggins not recognized

ANACARDIACEAE SUMAC or CASHEW FAMILY

Pistacia atlantica Desf. naturalized
Rhus diversiloba Torrey & A. Gray = *Toxicodendron diversilobum* (Torrey & A. Gray) E. Greene
R. glabra L. not naturalized
R. laurina Torrey & A. Gray = *Malosma laurina* (Torrey & A. Gray) Abrams
R. trilobata Torrey & A. Gray vars. *anisophylla* (E. Greene) Jepson, *malacophylla* (E. Greene) Jepson, & *quinata* Jepson not recognized
Schinus terebinthifolius Raddi naturalized

APIACEAE [UMBELLIFERAE] CARROT FAMILY

Angelica californica Jepson separate from *A. tomentosa* S. Watson
A. callii Mathias & Constance recently described
Anthriscus scandicina (G. Weber) Mansf. = *A. caucalis* M. Bieb.
Apium leptophyllum (Pers.) Benth. = *Ciclospermum leptophyllum* (Pers.) Britton & E. Wilson
A. nodiflorum (L.) Lagasca naturalized
Bupleurum subovatum Link misapplied to *B. lancifolium* Hornem.
Caucalis microcarpa Hook. & Arn. = *Yabea microcarpa* (Hook. & Arn.) Koso-Polj.

Cicuta bolanderi S. Watson = *C. maculata* L. var. *bolanderi* (S. Watson) Mulligan
Conioselinum chinense (L.) Britton, Sterns, & Poggenburg = *C. pacificum* (S. Watson) J. Coulter & Rose
Cymopterus ripleyi Barneby recently described
Eryngium aristulatum Jepson var. *hooveri* Y. Sheikh recently described
E. castrense Jepson separate from *E. vaseyi* J. Coulter & Rose
E. constancei Y. Sheikh recently described
E. mathiasiae Y. Sheikh recently described
E. spinosepalum Mathias separate from *E. vaseyi* J. Coulter & Rose
E. vaseyi J. Coulter & Rose var. *globosum* (Jepson) Mathias & Constance = *E. spinosepalum* Mathias
var. *vallicola* (Jepson) Munz not recognized
Glehnia leiocarpa Mathias = *G. littoralis* (A. Gray) Miq. ssp. *leiocarpa* (Mathias) Hultén
Hydrocotyle moschata Forster f. naturalized
Lilaeopsis masonii Mathias & Constance recently described
Lomatium hallii (S. Watson) J. Coulter & Rose occurs in CA
L. humile (J. Coulter & Rose) Mathias & Constance = *L. caruifolium* (Hook. & Arn.) J. Coulter & Rose var. *denticulatum* Jepson
L. leptocarpum Torrey & A. Gray = *L. bicolor* (S. Watson) J. Coulter & Rose var. *leptocarpum* (Torrey & A. Gray) M. Schlessman
L. macrocarpum (Hook. & Arn.) J. Coulter & Rose var. *ellipticum* Jepson not recognized
L. marginatum (Benth.) J. Coulter & Rose var. *purpureum* (Jepson) Jepson recognized
L. nevadense (S. Watson) J. Coulter & Rose var. *pseudorientale* Munz = var. *parishii* (J. Coulter & Rose) Jepson
L. parvifolium (Hook. & Arn.) Jepson var. *pallidum* (J. Coulter & Rose) Jepson not recognized
L. plummerae J. Coulter & Rose vars. *austiniae* (J. Coulter & Rose) Mathias & *sonnei* (J. Coulter & Rose) Jepson not recognized
L. shevockii R.L. Hartman & Constance recently described
L. stebbinsii M. Schlessman & Constance recently described
L. triternatum (Pursh) J. Coulter & Rose var. *anomalum* (M.E. Jones) Mathias not in CA
L. vaseyi J. Coulter & Rose = *L. utriculatum* (Torrey & A. Gray) J. Coulter & Rose
Oenanthe pimpinelloides L. naturalized
Oreonana purpurascens J.R. Shevock & Constance recently described
Perideridia bacigalupii Chuang & Constance recently described
P. bolanderi (A. Gray) Nelson & J.F. Macbr. ssp. *involucrata* Chuang & Constance recently described
P. gairdneri (Hook. & Arn.) Mathias ssp. *borealis* Chuang & Constance recently described
P. lemmonii (J. Coulter & Rose) Chuang & Constance separate from *P. parishii* (J. Coulter & Rose) Nelson & J.F. Macbr.
P. leptocarpa Chuang & Constance recently described
P. parishii (J. Coulter & Rose) Nelson & J.F. Macbr. ssp. latifolia (A. Gray) Chuang & Constance recognized
Petroselinum crispum (Miller) A.W. Hill naturalized
Spermolepis echinata (DC.) A.A. Heller occurs in CA
Torilis japonica (Houtt.) DC. misapplied to *T. arvensis* (Hudson) Link

APOCYNACEAE DOGBANE FAMILY

Amsonia brevifolia A. Gray = *A. tomentosa* Torrey & Frémont
Apocynum androsaemifolium L. var. *glabrum* Macoun not recognized
A. cannabinum L. var. *glaberrimum* A. DC. not recognized
A. medium E. Greene var. *floribundum* (E. Greene) Woodson = *A. androsaemifolium* L.
A. pumilum (A. Gray) E. Greene = *A. androsaemifolium* L.
A. sibiricum Jacq. var. *salignum* (E. Greene) Fern. = *A. cannabinum* L.
Catharanthus roseus (L.) G. Don. naturalized
Cycladenia humilis Benth. var. *tomentosa* (A. Gray) A. Gray = var. *humilis*

AQUIFOLIACEAE HOLLY FAMILY

Ilex aquifolium L. naturalized

ARALIACEAE GINSENG FAMILY: no changes

ARISTOLOCHIACEAE PIPEVINE FAMILY

Asarum marmoratum Piper occurs in CA

ASCLEPIADACEAE MILKWEED FAMILY

Asclepias cryptoceras S. Watson ssp. *davisii* (Woodson) Woodson indistinct (CA pls = ssp. *d.* if recognized)
A. linaria Cav. occurs in CA

ASTERACEAE [COMPOSITAE] SUNFLOWER FAMILY

Achillea borealis Bong. sspp. *arenicola* (A.A. Heller) Keck & *californica* (Pollard) Keck = *A. millefolium* L.
A. lanulosa Nutt. = *A. millefolium* L.
A. millefolium L. var. *gigantea* (Pollard) M. Nobs not recognized
Agoseris apargioides (Less.) E. Greene ssp. *maritima* (Sheldon) Q. Jones = var. *eastwoodiae* (Fedde) Munz
A. glauca (Pursh) E. Greene vars. *glauca*, *laciniata* (D. Eaton) F.J. Smiley, & *monticola* (E. Greene) Q. Jones, recognized
Ambrosia psilostachya DC. var. *californica* (Rydb.) S.F. Blake not recognized
A. pumila (Nutt.) A. Gray occurs in CA
Amphipappus fremontii Torrey & A. Gray ssp. *spinosus* (Nelson) Keck = var. *spinosus* (Nelson) C.L. Porter
Antennaria alpina (L.) Gaertner var. *media* (E. Greene) Jepson = *A. media* E. Greene var. *scabra* Jepson = *A. pulchella* E. Greene
A. flagellaris (A. Gray) A. Gray occurs in CA
A. microcephala A. Gray = *A. luzuloides* Torrey & A. Gray
A. neglecta E. Greene var. *howellii* (E. Greene) Cronq. = *A. howellii* E. Greene ssp. *howellii*
A. racemosa Hook. occurs in CA
A. rosea E. Greene ssp. *confinis* (E. Greene) R. Bayer recognized
Anthemis fuscata Brot. = *Chamaemelum fuscatum* (Brot.) Carv. Vasc.
Arctotis stoechadifolia P. Bergius var. *grandis* (Thunb.) Less. not recognized
Arnica chamissonis Less. vars. *bernardina* (E. Greene) Maguire, *incana* (A. Gray) Hultén, & *jepsoniana* Maguire not recognized
A. discoidea Benth. vars. *alata* (Rydb.) Cronq. & *eradiata* (A. Gray) Cronq not recognized
A. longifolia D. Eaton ssp. *myriadenia* (Piper) Maguire not recognized
A. parryi A. Gray ssp. *sonnei* (E. Greene) Maguire indistinct
A. rydbergii E. Greene misapplied in CA
A. spathulata E. Greene ssp. *eastwoodiae* (Rydb.) Maguire not recognized
Artemisia annua L. naturalized
A. arbuscula Nutt. ssp. *nova* (Nelson) G. Ward = *A. nova* Nelson ssp. *thermopola* Beetle recently described
A. cana Pursh: CA pls are ssp. *bolanderi* (A. Gray) G. Ward
A. spiciformis Osterh. separate from *A. rothrockii* A. Gray
A. tridentata Nutt. ssp. *vaseyana* (Rydb.) Beetle recognized
ssp. *wyomingensis* Beetle & A. Young recently described

Aster adscendens Lindley: spelling corrected to *A. ascendens*
A. brickelloides E. Greene var. *glabratus* E. Greene not recognized
A. campestris Nutt. var. *bloomeri* (A. Gray) A. Gray not recognized
A. chilensis Nees var. *invenustus* (E. Greene) Jepson not recognized
vars. *lentus* (E. Greene) Jepson & *sonomensis* (E. Greene) Jepson = *A. lentus* E. Greene
A. elatus (E. Greene) Cronq. = *A. alpigenus* (Torrey & A. Gray) A. Gray var. *andersonii* (A. Gray) M. Peck
A. exilis Elliott = *A. subulatus* Michaux var. *ligulatus* Shinn.
A. foliaceus Lindley var. *apricus* A. Gray indistinct from var. *parryi* (A. Eaton) A. Gray
A. greatai S.B. Parish: spelling corrected to *A. greatae*
A. hesperius A. Gray = *A. lanceolatus* Willd. ssp. *hesperius* (A. Gray) Semple & J. Chmielewski
A. intricatus (A. Gray) S.F. Blake = *Machaeranthera carnosa* (A. Gray) G. Nesom
A. occidentalis (Nutt.) Torrey & A. Gray var. *intermedius* A. Gray indistinct from var. *occidentalis*
vars. *delectabilis* (H.M. Hall) Ferris & *parishii* (A. Gray) Ferris = var. *occidentalis*
A. oregonensis (Nutt.) Cronq. ssp. *californicus* (Durand) Keck not recognized
A. paludicola Piper = *A. occidentalis* (Nutt.) Torrey & A. Gray var. *yosemitanus* (A. Gray) Cronq.
A. spinosus Benth. = *Chloracantha spinosa* (Benth.) G. Nesom var. *spinosa*
A. subspicatus Nees not known from CA
Baccharis glutinosa Pers. = *B. salicifolia* (Ruiz López & Pavón) Pers.
B. pilularis DC. var. *consanguinea* (DC.) Kuntze not recognized
B. plummerae A. Gray ssp. *glabrata* Hoover recognized
B. vanessae Beauch. recently described
B. viminea DC. = *B. salicifolia* (Ruiz López & Pavón) Pers.
Bahia neomexicana (A. Gray) A. Gray = *Schkuhria multiflora* Hook. & Arn. var. *multiflora*
Balsamorhiza hookeri Nutt.: CA pls are var. *lanata* W. Sharp
var. *neglecta* (W. Sharp) Cronq. = *B. hirsuta* Nutt.
B. sericea W.A. Weber recently described
Bebbia juncea (Benth.) E. Greene: CA pls are var. *aspera* E. Greene
Benitoa occidentalis (H.M. Hall) Keck = *Lessingia occidentalis* (H.M. Hall) M.A. Lane
Bidens connata Muhlenb. var. *petiolata* (Nutt.) Farw. = *B. tripartita* L.
Blepharipappus scaber Hook. ssp. *laevis* (A. Gray) Keck not recognized
Brickellia atractyloides A. Gray separate from *B. arguta* Robinson
Calycadenia ciliosa E. Greene = *C. fremontii* A. Gray
C. elegans E. Greene = *C. fremontii* A. Gray
C. hispida (E. Greene) E. Greene incl ssp. *reducta* Keck = *C. multiglandulosa* DC.
C. multiglandulosa DC. sspp. *bicolor* (E. Greene) Keck, *cephalotes* (DC.) Keck, & *robusta* Keck not recognized
C. tenella (Nutt.) Torrey & A. Gray = *Osmadenia tenella* Nutt.
C. truncata DC. sspp. *microcephala* Keck & *scabrella* (E. Drew) Keck not recognized
Carduus acanthoides L. naturalized
C. nutans L. var. *leiophyllus* (Petrovic) Arenes not recognized
Carthamus leucocaulos Sibth. & Smith naturalized
Centaurea cineraria L. not naturalized
C. eriophora L. not naturalized
C. jacea L. not naturalized
C. muricata L. not naturalized
C. ×pratensis Thuill. (*C. jacea* L. × *C. nigra* L.) naturalized
C. repens L. = *Acroptilon repens* (L.) DC.
C. salmantica L. not naturalized
C. virgata Lam. var. *squarrosa* (Willd.) Boiss. = *C. squarrosa* Willd.
Chaenactis alpina (A. Gray) M.E. Jones = *C. douglasii* (Hook.) Hook. & Arn var. *alpina* A. Gray
C. carphoclinia A. Gray var. *attenuata* (A. Gray) M.E. Jones = var. *carphoclinia*
C. douglasii (Hook.) Hook. & Arn. vars. *achilleifolia* (Hook. & Arn.) A. Gray, *montana* M.E. Jones, & *rubricaulis* (Rydb.) Ferris = var. *douglasii*
C. glabriuscula DC. var. *curta* (A. Gray) Jepson = var. *glabriuscula*
var. *denudata* (Nutt.) Munz = var. *lanosa* (DC.) H.M. Hall
var. *gracilenta* (E. Greene) Keck = var. *heterocarpha* (A. Gray) H.M. Hall
var. *tenuifolia* (Nutt.) H.M. Hall = var. *glabriuscula*
C. stevioides Hook. & Arn. var. *brachypappa* (A. Gray) H.M. Hall not recognized
C. tanacetifolia A. Gray = *C. glabriuscula* DC. var. *heterocarpha* (A. Gray) H.M. Hall
Chaetopappa alsinoides (E. Greene) Keck = *Pentachaeta alsinoides* E. Greene
C. aurea (Nutt.) Keck = *Pentachaeta aurea* Nutt.
C. bellidifolia (E. Greene) Keck = *Pentachaeta bellidifolia* E. Greene
C. exilis (A. Gray) Keck = *Pentachaeta exilis* (A. Gray) A. Gray
C. fragilis (Brandegee) Keck = *Pentachaeta fragilis* Brandegee
C. lyonii A. Gray (Keck) = *Pentachaeta lyonii* A. Gray
Chamaemelum nobile (L.) All. naturalized
Chrysanthemum foeniculaceum (Willd.) Desf. = *Argyranthemum foeniculaceum* (Willd.) Schultz-Bip.
C. leucanthemum L. = *Leucanthemum vulgare* Lam.
C. parthenium (L.) Bernardi = *Tanacetum parthenium* (L.) Schultz- Bip.
Chrysopsis breweri A. Gray = *Aster breweri* (A. Gray) Semple
C. oregona (Nutt.) A. Gray var. *oregona* = *Heterotheca oregona* (Nutt.) Schinn. var. *oregona*
var. *compacta* Keck = *Heterotheca oregona* (Nutt.) Schinn. var. *compacta* (Keck) Semple
var. *rudis* (E. Greene) Jepson = *Heterotheca oregona* (Nutt.) Schinn. var. *rudis* (E. Greene) Semple
var. *scaberrima* A. Gray = *Heterotheca oregona* (Nutt.) Schinn. var. *scaberrima* (A. Gray) Semple
C. villosa (Pursh) Nutt. var. *bolanderi* (A. Gray) Jepson = *Heterotheca sessiliflora* (Nutt.) Shinn. ssp. *bolanderi* (A. Gray) Semple
var. *echioides* (Benth.) A. Gray = *Heterotheca sessiliflora* (Nutt.) Shinn. ssp. *echioides* (Benth.) Semple
var. *fastigiata* (E. Greene) H.M. Hall = *Heterotheca sessiliflora* (Nutt.) Shinn. ssp. *fastigiata* (E. Greene) Semple
var. *hispida* (Hook.) D. Eaton = *Heterotheca villosa* (Pursh) Shinn. var. *hispida* (Hook.) Harms; also misapplied in part to *H. villosa* (Pursh) Shinn. vars. *cinerascens* (S.F. Blake) Semple & *shevockii* Semple
var. *sessiliflora* (Nutt.) A. Gray = *Heterotheca sessiliflora* (Nutt.) Shinn. ssp. *sessiliflora*
Chrysothamnus greenei (A. Gray) E. Greene occurs in CA
C. nauseosus (Pallas) Britton ssp. ×*viscosus* Keck = ssp. *hololeucus* (A. Gray) H.M. Hall & Clements × *Ericameria cuneata* (A. Gray) McClatchie
C. parryi (A. Gray) E. Greene ssp. ×*bolanderi* (A. Gray) H.M. Hall & Clements = *C. nauseosus* (Pallas) Britton ssp. *albicaulis* (Nutt.) H.M. Hall & Clements × *Ericameria discoidea*
C. viscidiflorus (Hook.) Nutt. ssp. *humilus* (E. Greene) H.M. Hall & Clements = *C. humilus* E. Greene
ssp. *lanceolatus* (Nutt.) H.M. Hall & Clements separate from ssp. *viscidiflorus*

sspp. *latifolius* (D. Eaton) H.M. Hall & Clements, *pumilus* (Nutt.) H.M. Hall & Clements, & *stenophyllus* (A. Gray) H.M. Hall & Clements indistinct from ssp. *viscidiflorus*

Cirsium breweri A. Gray = *C. douglasii* DC. var. *breweri* (A. Gray) Keil & C. Turner

C. californicum A. Gray incl var. *bernardinum* (E. Greene) Petrak = *C. occidentale* (Nutt.) Jepson var. *californicum* (A. Gray) Keil & C. Turner

C. callilepis (E. Greene) Jepson = *C. remotifolium* (Hook.) DC.

C. campylon H. Sharsmith = *C. fontinale* E. Greene var. *campylon* (H. Sharsm.) Keil & C. Turner

C. ciliolatum (L. Henderson) J. Howell occurs in CA

C. coulteri Harvey & A. Gray = *C. occidentale* (Nutt.) Jepson var. *occidentale*; also misapplied in part to *C. occidentale* (Nutt.) Jepson var. *venustum* (E. Greene) Jepson

C. drummondii Torrey & A. Gray misapplied to *C. scariosum* Nutt.

C. eatonii (A. Gray) Robinson misapplied to *C. arizonicum* (A. Gray) Petrak

C. foliosum (Hook.) DC. misapplied to *C. scariosum* Nutt.

C. nidulum (M.E. Jones) Petrak misapplied in part to *C. arizonicum* (A. Gray) Petrak

C. occidentale (Nutt.) Jepson var. *compactum* Hoover recognized

C. pastoris J. Howell = *C. occidentale* (Nutt.) Jepson var. *candidissimum* (E. Greene) J.F.Macbr.

C. proteanum J. Howell = *C. occidentale* (Nutt.) Jepson var. *venustum* (E. Greene) Jepson

C. tioganum Congdon = *C. scariosum* Nutt.

C. utahense Petrak misapplied in CA to *C. canovirens* Rydb.

C. vaseyi (A. Gray) Jepson = *C. hydrophilum* (E. Greene) Jepson var. *vaseyi* (A. Gray) J. Howell

C. walkerianum Petrak = *C. quercetorum* (A. Gray) Jepson

Conyza bilboana E.J. Remy separate from *C. canadensis* (L.) Cronq.

Coreopsis atkinsoniana Lindley = *C. tinctoria* Nutt.

C. basalis (C.F.E. Otto & D. Dietr.) S.F. Blake var. *wrightii* (A. Gray) S.F. Blake = *C. wrightii* (A. Gray) H. Parker

Corethrogyne californica DC. incl vars. *obovata* (Benth.) Kuntze & *lyonii* S.F. Blake = *Lessingia filaginifolia* (Hook. & Arn.) M.A. Lane var. *californica* (DC.) M.A. Lane

C. filaginifolia (Hook. & Arn.) Nutt. incl vars. *bernardina* (Abrams) H.M. Hall, *brevicula* (E. Greene) Canby, *hamiltonensis* Keck, *incana* (Nutt.) Canby, *latifolia* H.M. Hall, *linifolia* H.M. Hall, *peirsonii* Canby, *pinetorum* I.M. Johnston, *robusta* E. Greene, *sessilis* (E. Greene) Canby, *virgata* (Benth.) A. Gray, & *viscidula* (E. Greene) Keck = *Lessingia filaginifolia* (Hook. & Arn.) M.A. Lane var. *filaginifolia*

Crepis bakeri E. Greene ssp. *cusickii* (Eastw.) Babc. & Stebb. not recognized

C. bursifolia L. naturalized

C. modocensis E. Greene ssp. *subacaulis* (Kellogg) Babc. & Stebb. not recognized

C. nana Richards. ssp. *ramosa* Babc. not recognized

C. occidentalis Nutt. sspp. *conjuncta* (Jepson) Babc. & Stebb., *costata* (A. Gray) Babc. & Stebb., & *pumila* (Rydb.) Babc. & Stebb. indistinct & intergrading

Crupina vulgaris Cass. naturalized

Dicoria canescens A. Gray sspp. *clarkiae* (Kennedy) Keck & *hispidula* (Rydb.) Keck not recognized

Dimorphotheca ecklonis DC. = *Osteospermum ecklonis* (DC.) Norlindh

Dyssodia cooperi A. Gray = *Adenophyllym cooperi* (A. Gray) Strother

D. porophylloides A. Gray = *Adenophyllum porophylloides* (A. Gray) Strother

D. thurberi (A. Gray) Robinson = *Thymophylla pentachaeta* (DC.) Small var. *belenidium* (DC.) Strother

Eatonella congdonii A. Gray = *Lembertia congdonii* (A. Gray) E. Greene

Eclipta alba (L.) Hassk. = *E. prostrata* (L.) L.

Encelia farinosa Torrey & A. Gray var. *phenicodonta* (S.F. Blake) I.M. Johnston not recognized

E. virginensis Nelson ssp. *actoni* (Elmer) Keck = *E. actoni* Elmer

Enceliopsis argophylla (D.C. Eaton) Nelson var. *grandiflora* (M.E. Jones) Jepson = *E. covillei* (Nelson) S.F. Blake

Erechtites arguta (A. Richard) DC. = *E. glomerata* (Poiret) DC.

E. hieracifolia (L.) DC. var. *hieracifolia* naturalized

E. prenanthoides (A. Richard) DC. = *E. minima* (Poiret) DC.

Ericameria cuneata (A. Gray) McClatchie var. *macrocephala* Urbatsch recently described

Erigeron debilis (A. Gray) Rydb. = *Trimorpha acris* (L.) S.F. Gray var. *debilis* (A. Gray) G. Nesom

E. algidus Jepson separate from *E. petiolaris* E. Greene

E. angustatus E. Greene recognized

E. annuus (L.) Pers. occurs in CA

E. bloomeri A. Gray var. *pubens* Keck = var. bloomeri

E. breweri A. Gray var. *bisanctus* G. Nesom recently described

var. *elmeri* (E. Greene) Jepson = *E. elmeri* (E. Greene) E. Greene

var. *klamathensis* G. Nesom recently described

E. chrysopsidis A. Gray ssp. *austiniae* (E. Greene) Cronq. = *E. austiniae* E. Greene

E. compositus Pursh vars. *discoideus* A. Gray & *glabratus* Macoun not recognized

E. concinnus (Hook. & Arn.) Torrey & A. Gray var. *condensatus* D. Eaton occurs in CA

E. delicatus Cronq. = *E. cervinus* E. Greene

E. flexuosus Cronq. = *E. lassenianus* E. Greene var. *lassenianus*

E. foliosus Nutt. var. *blochmaniae* (E. Greene) H.M. Hall = *E. blochmaniae* E. Greene

var. *covillei* (E. Greene) Compton = *E. breweri* A. Gray var. *covillei* (E. Greene) G. Nesom

var. *mendocinus* E. Greene separate from var. *hartwegii* (E. Greene) Jepson

var. *stenophyllus* (Nutt.) A. Gray = var. *foliosus*

E. inornatus A. Gray var. *angustatus* A. Gray = *E. reductus* (Cronq.) G. Nesom var. *angustatus* (A. Gray) G. Nesom

var. *biolettii* (E. Greene) Jepson = *E. biolettii* E. Greene

var. *calidipetris* G. Nesom recently described

var. *keilii* G. Nesom recently described var. *reductus* Cronq. = *E. reductus* (Cronq.) G. Nesom

var. *reductus* var. *viscidulus* A. Gray = *E. petrophilus* E. Greene

var. *viscidulus* (A. Gray) G. Nesom *E. lassenianus* E. Greene var. *deficiens* Cronq. recognized

E. lonchophyllus Hook. = *Trimorpha lonchophylla* (Hook.) G. Nesom

E. mariposanus Congdon separate from *E. foliosus* Nutt.

E. nevadincola S.F. Blake = *E. eatonii* A. Gray var. *nevadincola* (S.F. Blake) G. Nesom

E. peregrinus (Pursh) E. Greene var. *angustifolius* (A. Gray) Cronq. = var. *callianthemus* (E. Greene) Cronq.

ssp. *callianthemus* (E. Greene) Cronq. = var. *callianthemus* (E. Greene) Cronq.

E. petiolaris E. Greene misapplied in CA to *E. algidus* Jepson

E. petrophilus E. Greene var. *sierrensis* G. Nesom recently described

E. pumilus Nutt. ssp. *concinnoides* Cronq. = *E. concinnus* (Hook. & Arn.) Torrey & A. Gray var. *concinnus*

E. serpentinus G. Nesom recently described

E. sonnei E. Greene = *E. eatonii* A. Gray var. *sonnei* (E. Greene) Cronq.

E. uncialis S.F. Blake: CA pls are var. *uncialis*

Eriophyllum confertiflorum (DC.) A. Gray var. *laxiflorum* A. Gray = var. *confertiflorum*

E. lanatum (Pursh) James Forbes var. *aphanactis* J. Howell indistinct from var. *achillaeoides* (DC.) Jepson
var. *leucophyllum* (DC.) W.R. Carter not in CA
var. *monoense* (Rydb.) Jepson = var. *integrifolium* (Hook.) F.J. Smiley
E. nubigenum E. Greene var. *congdonii* (Brandegee) Constance = *E. congdonii* Brandegee
E. staechadifolium Lagasca vars. *artemisiifolium* (Less.) J.F. Macbr. & *depressum* E. Greene not recognized
E. wallacei (A. Gray) A. Gray var. *rubellum* (A. Gray) A. Gray not recognized
Eupatorium adenophorum Sprengel = *Ageratina adenophora* (Sprengel) R. King & H. Robinson
E. herbaceum (A. Gray) E. Greene = *Ageratina herbacea* (A. Gray) R. King & H. Robinson
E. occidetale Hook. = *Ageratina occidentalis* (Hook.) R. King & H. Robinson
E. shastense D.W. Taylor & Stebb. recently described (CaR) = *Ageratina shastensis* (D.W. Taylor & Stebb.) R. King & H. Robinson
Evax acaulis (Kellogg) E. Greene = *Hesperevax acaulis* (Kellogg) E. Greene; also misapplied to *Hesperevax caulescens* (Benth.) A. Gray
E. acaulis E. Greene = *Hesperevax caulescens* (Benth.) A. Gray; also misapplied to *Hesperevax acaulis* (Kellogg) E. Greene
E. caulescens (Benth.) A. Gray = *Hesperevax caulescens* (Benth.) A. Gray
E. sparsiflora (A. Gray) E. Greene = *Hesperevax sparsiflora* (A. Gray) E. Greene
var. *brevifolia* (A. Gray) Jepson = *Hesperevax sparsiflora* (A. Gray) E. Greene; also misapplied in part to *Hesperevax acaulis* (Kellogg) E. Greene ssp. *robustior* J. Morefield
Filago vulgaris Lam. misapplied to *F. pyramidata* L. var. *pyramidata*
Galinsoga quadriradiata Ruiz López & Pavón naturalized
Gazania longiscapa DC. = *G. linearis* (Thunb.) Druce
Glyptopleura setulosa A. Gray = *G. marginata* D. Eaton
Gnaphalium beneolens Davidson = *G. canescens* DC. ssp. *beneolens* (Davidson) Stebb.
G. canescens DC. ssp. *thermale* (E. Nelson) Stebb. occurs in CA
G. chilense Sprengel = *G. stramineum* Kunth
G. leucocephalum A. Gray occurs in CA
G. microcephalum Nutt. = *G. canescens* DC. ssp. *microcephalum* (Nutt.) Stebb.
G. peregrinum Fern. = *G. purpureum* L.
Grindelia camporum E. Greene var. *parviflora* Steyerm. = var. *camporum*
G. fraxino-pratensis Rev. & Beatley recently described
G. hallii Steyerm. = *G. hirsutula* Hook. & Arn. var. *hallii* (Steyerm.) M.A. Lane
G. hirsutula Hook. & Arn. ssp. *rubricaulis* (DC.) Keck = ssp. *hirsutula*
G. humilis Hook. & Arn. = *G. hirsutula* Hook. & Arn. ssp. *hirsutula*; misapplied to *G. stricta* DC. var. *angustifolia* (A. Gray) M.A. Lane
G. ×*latifolia* Kellogg in part = *G. stricta* DC. var. *platyphylla* (E. Greene) M.A. Lane (+ hybrids)
G. maritima (E. Greene) Steyerm. = *E. hirsutula* Hook. & Arn. var. *maritima* (E. Greene) M.A. Lane
G. ×*paludosa* E. Greene not recognized
G. procera E. Greene = *G. camporum* E. Greene var. *camporum*
G. robusta Nutt. var. *bracteosa* (J. Howell) Keck = *G. camporum* E. Greene var. *bracteosa* (J. Howell) M.A. Lane
var. *robusta* = *G. camporum* E. Greene var. *bracteosa* (J. Howell) M.A. Lane in part and var. *camporum* in part (+ hybrids)
G. squarrosa (Pursh) Dunal: CA pls are var. *serrulata* (Rydb.) Steyerm.
G. stricta DC. ssp. *blakei* (Steyerm.) Keck = var. *stricta*
ssp. *venulosa* (Jepson) Keck = var. *platyphylla* (E. Greene) M.A. Lane
Gutierrezia bracteata Abrams = *G. californica* (DC.) Torrey & A. Gray
Haplopappus acaulis (Nutt.) A. Gray = *Stenotus acaulis* Nutt.
H. acradenius (E. Greene) S.F. Blake = *Isocoma acradenia* (E. Greene) E. Greene
ssp. *bracteosus* (E. Greene) H.M. Hall = *I. a.* var. *bracteosa* (E. Greene) G. Nesom
ssp. *eremophilus* (E. Greene) H.M. Hall = *I. a.* var. *eremophila* (E. Greene) G. Nesom
H. arborescens (A. Gray) H.M. Hall = *Ericameria arborescens* (A. Gray) E. Greene
H. bloomeri A. Gray = *Ericameria bloomeri* (A. Gray) J.F. Macbr.
H. brickellioides S.F. Blake = *Hazardia brickellioides* (S.F. Blake) W. Clark
H. canus (A. Gray) S.F. Blake = *Hazardia cana* (A. Gray) E. Greene
H. ciliatus (Nutt.) DC. = *Prionopsis ciliatus* (Nutt.) Nutt.
H. cooperi (A. Gray) H.M. Hall = *Ericameria cooperi* (A. Gray) H.M. Hall var. *cooperi*
H. cuneatus A. Gray = *Ericameria cuneata* (A. Gray) McClatchie var. *spathulatus* (A. Gray) S.F. Blake = *Ericameria cuneata* (A. Gray) McClatchie var. *spathulata* (A. Gray) H.M. Hall
H. eastwoodiae H.M. Hall = *Ericameria fasciculata* (Eastw.) J.F. Macbr.
H. ericoides (Less.) Hook. & Arn. incl ssp. *blakei* C. Wolf = *Ericameria ericoides* (Less.) Jepson
H. eximius H.M.Hall = *Tonestus eximius* (H.M. Hall) Nelson & J.F. Macbr.
H. gilmanii S.F. Blake = *Ericameria gilmanii* (S.F. Blake) G. Nesom
H. gooddingii (Nelson) Munz & I.M. Johnston = *Machaeranthera pinnatifida* (Hook.) Shinn. var. *gooddingii* (Nelson) B. Turner & R. Hartman
H. gracilis (Nutt.) A. Gray = *Machaeranthera gracilis* (Nutt.) Shinn.
H. greenei A. Gray = *Ericameria greenei* (A. Gray) G. Nesom
H. junceus E. Greene = *Machaeranthera juncea* (E. Greene) Shinn.
H. laricifolius A. Gray = *Ericameria laricifolia* (A. Gray) Shinn.
H. linearifolius DC. = *Ericameria linearifolia* (DC.) Urb. & G. Wussow
H. lyallii A. Gray = *Tonestus lyallii* (A. Gray) Nelson
H. macronema A. Gray = *Ericameria discoidea* (Nutt.) G. Nesom
H. nanus (Nutt.) D. Eaton = *Ericameria nana* Nutt.
H. ophitidis (J. Howell) Keck = *Ericameria ophitidis* (J. Howell) G. Nesom
H. palmeri A. Gray = *Ericameria palmeri* (A. Gray) H.M. Hall var. *pachylepis* H.M. Hall = *Ericameria palmeri* (A. Gray) H.M. Hall var. *pachylepis* (H.M. Hall) G. Nesom
H. parishii (E. Greene) S.F. Blake = *Ericameria parishii* (E. Greene) H.M. Hall: CA pls are var. *parishii*
H. peirsonii (Keck) J. Howell = *Tonestus peirsonii* (Keck) G. Nesom & R. Morgan
H. pinifolius (A. Gray) H.M. Hall = *Ericameria pinifolia* (A. Gray) H.M. Hall
H. propinquus S.F. Blake = *Ericameria brachylepis* (A. Gray) H.M. Hall
H. squarrosus Hook. & Arn. = *Hazardia squarrosa* (Hook. & Arn.) E. Greene
ssp. *grindelioides* (DC.) Keck = *Hazardia squarrosa* (Hook. & Arn.) E. Greene var. *grindelioides* (DC.) W. Clark
ssp. *obtusus* (E. Greene) H.M. Hall = *Hazardia squarrosa* (Hook. & Arn.) E. Greene var. *obtusa* (E. Greene) Jepson
ssp. *stenolepis* H.M. Hall = *Hazardia stenolepis* (H.M. Hall) Hoover

H. stenophyllus A. Gray = *Stenotus stenophyllus* (A. Gray) E. Greene
H. suffruticosus (Nutt.) A. Gray = *Ericameria suffruticosa* (Nutt.) G. Nesom
H. whitneyi A. Gray = *Hazardia whitneyi* (A. Gray) E. Greene
ssp. *discoideus* (J. Howell) Keck = *Hazardia whitneyi* (A. Gray) E. Greene var. *discoidea* (J. Howell) W. Clark
H. venetus (Kunth) S.F. Blake = *Isocoma menziesii* (Hook. & Arn.) G. Nesom
var. *argutus* (E. Greene) Keck = *Isocoma arguta* E. Greene
sspp. *furfuraceus* (E. Greene) H.M. Hall & *oxyphyllus* (E. Greene) H.M. Hall = *I. m.* var. *menziesii*
var. *sedoides* (E. Greene) Munz = *I. m.* var. *sedoides* (E. Greene) G. Nesom
ssp. *vernonioides* (Nutt.) H.M. Hall = *I. m.* var. *vernonioides* (Nutt.) G. Nesom
Hazardia detonsa (E. Greene) E. Greene separate from *H. cana* (A. Gray) E. Greene
H. orcuttii (A. Gray) E. Greene recognized
Helenium hoopesii A. Gray = *Dugaldia hoopesii* (A. Gray) Rydb.
Helianthus annuus L. ssp. jaegeri (Heiser) Heiser, ssp. *lenticularis* (Douglas) Cockerell, & var. *macrocarpus* (DC.) Cockerell not cognized
H. petiolaris Nutt. var. *canescens* A. Gray = *H. niveus* (Benth.) Brandegee ssp. *canescens* (A. Gray) Heiser
H. tephrodes A. Gray = *H. niveus* (Benth.) Brandegee ssp. *tephrodes* (A. Gray) Heiser
Hemizonia australis (Keck) Keck = *H. parryi* E. Greene ssp. *australis* Keck
H. calyculata (Babc. & H.M. Hall) Keck = *H. congesta* DC. ssp. *calyculata* Babc. & H.M. Hall
H. clevelandii E. Greene = *H. congesta* DC. ssp. *clevelandii* (E. Greene) Babc. & H.M. Hall
H. congesta DC. ssp. *leucocephala* (B.D. Tanowitz) Keil recently described
H. increscens (Keck) B.D. Tanowitz sspp. *foliosa* (Hoover) B.D. Tanowitz & *villosa* B.D. Tanowitz recently described
H. laevis (Keck) Keck = *H. pungens* (Hook. & Arn.) Torrey & A. Gray ssp. *laevis* Keck
H. lutescens (E. Greene) Keck = *H. congesta* DC. ssp. *congesta*
H. luzulaefolia DC. incl ssp. *rudis* (Benth.) Keck = *H. congesta* DC. ssp. *luzulifolia* (DC.) Babc. & H.M. Hall
H. multicaulis Hook. & Arn. incl ssp. *vernalis* Keck = *H. congesta* DC. ssp. *congesta*
H. paniculata A. Gray ssp. *increscens* Keck = *H. increscens* (Keck) B.D. Tanowitz ssp. *increscens*
H. ramosissima Benth. = *H. fasciculata* (DC.) Torrey & A. Gray
H. tracyi (Babc. & H.M. Hall) Keck = *H. congesta* DC. ssp. *tracyi* Babc. & H.M. Hall
Hesperevax acaulis (Kellogg) E. Greene vars. *ambusticola* J. Morefield & *robustior* J. Morefield recently described
Heteranthemis viscidehirta Schott naturalized; formerly mistaken for *Chrysanthemum segetum* L.
Heterotheca sessiliflora (Nutt.) Shinn. vars. *bolanderoides* Semple & *camphorata* (Eastw.) Semple indistinct from ssp. *echioides* (Benth.) Semple
var. *sanjacintensis* Semple indistinct from ssp. *fastigiata* (E. Greene) Semple
H. subaxillaris (Lam.) Britton & Rusby misapplied to *H. psammophila* B. Wagenkn.
Hieracium argutum Nutt. var. *parishii* (A. Gray) Jepson not recognized
H. cynoglossoides Arv.-Touv. misapplied in CA to *H. scouleri* Hook.
var. *nudicaule* A. Gray indistinct
Hofmeisteria pluriseta A. Gray = *Pleurocoronis pluriseta* (A. Gray) R. King & H. Robinson
Holocarpha obconica (J. Clausen & Keck) Keck ssp. *autumnalis* Keck indistinct
Hulsea callicarpha (H.M. Hall) Rydb. = *H. vestita* A. Gray ssp. *callicarpha* (H.M. Hall) Wilken
H. inyoensis (Keck) Munz = *H. vestita* A. Gray ssp. *inyoensis* (Keck) Wilken
H. mexicana Rydb. separate from *H. californica* Torrey & A. Gray
H. vestita A. Gray ssp. *gabrielensis* Wilken recently described
ssp. *parryi* (A. Gray) Wilken separate from ssp. *vestita*
var. *pygmaea* A. Gray = ssp. *pygmaea* (A. Gray) Wilken
Hymenoclea salsola A. Gray var. *patula* (Nelson) Peterson & Payne recognized
Hymenoxys cooperi (A. Gray) Cockerell var. *canescens* (D. Eaton) K. Parker = dwarfed subalpine form of *H. lemmonii* (E. Greene) Cockerell
Lactuca serriola L. var. *integra* Gren. & Godron not recognized
Lagophylla congesta E. Greene = *L. ramosissima* Nutt. ssp. *congesta* (E. Greene) Keck
L. glandulosa A. Gray ssp. *serrata* (E. Greene) Keck not recognized
Laphamia inyoensis Ferris = *Perityle inyoensis* (Ferris) A. Powell
L. megalocephala S. Watson = *Perityle megalocephala* (S. Watson) J.F. Macbr.
var. *intricata* (Brandegee) Keck misapplied to *Perityle megalocephala* (S. Watson) J.F. Macbr. var. *oligophylla* A. Powell
var. *megalocephala* = *Perityle megalocephala* (S. Watson) J.F. Macbr. var. *megalocephala*
L. villosa S.F. Blake = *Perityle villosa* (S.F. Blake) Shinn.
Lasthenia chrysostoma (Fisch. & C. Meyer) E. Greene = *L. californica* Lindley
L. glabrata Lindley ssp. *coulteri* (A. Gray) Ornd. recognized
L. macrantha (A. Gray) E. Greene ssp. *bakeri* (J. Howell) Ornd. recognized
L. minor (DC.) Ornduff ssp. *maritima* (A. Gray) Ornd. = *L. maritima* (A. Gray) M.C. Vasey
Layia chrysanthemoides (DC.) A. Gray ssp. *maritima* Keck not recognized
L. glandulosa (Hook.) Hook. & Arn. ssp. *lutea* Keck not recognized
L. paniculata Keck = *L. hieracioides* (DC.) Hook. & Arn.
L. platyglossa (Fischer & C. Meyer) A. Gray ssp. *campestris* Keck not recognized
L. ziegleri Munz = *L. platyglossa* (Fischer & C. Meyer) A. Gray
Leontodon leysseri (Wallr.) G. Beck = *L. taraxacoides* (Vill.) Merat
L. taraxacoides sspp. *longirostris* Finch & Sell & *taraxacoides* recognized
Lepidospartum squamatum (A. Gray) A. Gray var. *palmeri* (A. Gray) L. Wheeler not recognized
Lessingia germanorum Cham. var. *glandulifera* (A. Gray) J. Howell = *L. glandulifera* A. Gray var. *glandulifera*
vars. *parvula* (E. Greene) J. Howell & *tenuis* (A. Gray) J. Howell = *L. tenuis* (A. Gray) Cov.
var. *pectinata* (E. Greene) J. Howell = *L. glandulifera* A. Gray var. *pectinata* (E. Greene) Jepson
var. *tomentosa* (E. Greene) J. Howell = *L. glandulifera* A. Gray var. *tomentosa* (E. Greene) Ferris
L. hololeuca E. Greene var. *arachnoidea* (E. Greene) J. Howell = *L. arachnoidea* E. Greene
L. nemaclada E. Greene var. *albiflora* (Eastw.) J. Howell not recognized
L. ramulosa A. Gray var. *glabrata* Keck = *L. micradenia* E. Greene var. *glabrata* (Keck) Ferris
var. *micradenia* (E. Greene) J. Howell = *L. micradenia* E. Greene var. *micradenia*
Leucanthemum maximum Ramond naturalized
Leucelene ericoides (Torrey) E. Greene = *Chaetopappa ericoides* (Torrey) G. Nesom

Lygodesmia exigua A. Gray = *Prenanthella exigua* (A. Gray) Tomb
L. spinosa Nutt. = *Stephanomeria spinosa* (Nutt.) Tomb
Machaeranthera asteroides (Torrey) E. Greene var. *asteroides* occurs in CA
M. canescens (Pursh) A. Gray var. *incana* (Lindley) A. Gray occurs in CA
var. *ziegleri* (Munz) B. Turner recently described
M. cognata (H.M. Hall) Cronq. & Keck = *Xylorhiza cognata* (H.M. Hall) T.J. Watson
M. lagunensis Keck = *M. asteroides* (Torrey) E. Greene var. *lagunensis* (Keck) B. Turner
M. leucanthemifolia (E. Greene) E. Greene = *M. canescens* (Pursh) A. Gray var. *leucanthemifolia* (E. Greene) Welsh
M. orcuttii (Vasey & Rose) Cronq. & Keck = *Xylorhiza orcuttii* (Vasey & Rose) E. Greene
M. shastensis A. Gray vars. *eradiata* (A. Gray) Cronq. & *shastensis* = *M. canescens* (Pursh) A. Gray var. *shastensis* (A. Gray) B. Turner
vars. *glossophylla* (Piper) Cronq. & Keck & *montana* (E. Greene) Cronq. & Keck = *M. canescens* (Pursh) A. Gray var. *canescens*
M. tanacetifolia (Kunth) Nees occurs in CA
M. tephrodes (A. Gray) E. Greene = *M. canescens* (Pursh) A. Gray var. *canescens*
M. tortifolia (Torrey & A. Gray) Cronq. & Keck = *Xylorhiza tortifolia* (Torrey & A. Gray) E. Greene var. *tortifolia*
Madia capitata Nutt. = *M. sativa* Molina
M. doris-nilesiae T.W. Nelson & J.P. Nelson recently described
M. gracilis (Smith) Keck sspp. *collina* Keck & *pilosa* Keck not recognized
M. stebbinsii T.W. Nelson & J.P. Nelson recently described
Malacothrix blairii (Munz & I.M. Johnston) Munz = *Stephanomeria blairii* Munz & I.M. Johnston
M. coulteri A. Gray var. *cognata* Jepson not recognized
M. incana (Nutt.) Torrey & A. Gray var. *succulenta* (Elmer) Williams not recognized
M. indecora E. Greene separate from *M. foliosa* A. Gray
M. phaeocarpa W. Davis recently described
M. saxatilis (Nutt.) Torrey & A. Gray var. *altissima* (E. Greene) Ferris = var. *tenuifolia* (Nutt.) A. Gray
M. squalida E. Greene separate from *M. foliosa* A. Gray
M. torreyi A. Gray occurs in CA
Matricaria matricarioides (Less.) Porter = *Chamomilla suaveolens* (Pursh) Rydb.
M. occidentalis E. Greene = *Chamomilla occidentalis* (E. Greene) Rydb.
Micropus californicus Fischer & C. Meyer var. *subvestitus* A. Gray recognized
Munzothamnus blairii (Munz & I.M. Johnston) Raven = *Stephanomeria blairii* Munz & I.M. Johnston
Onopordum acanthium L.: CA pls are ssp. *acanthium*
O. illyricum L. naturalized
Palafoxia linearis (Cav.) Lagasca misapplied to *P. arida* B. Turner & M. Morris var. *gigantea* M.E. Jones = *P. arida* B. Turner & M. Morris
var. *gigantea* (M.E. Jones) B. Turner & M. Morris
var. *linearis* = *P. arida* B. Turner & M. Morris var. *arida*
Pentachaeta exilis (A. Gray) A. Gray ssp. *aeolica* Van Horn & Ornduff recently described
Perezia microcephala (DC.) A. Gray = *Acourtia microcephala* DC.
Petasites palmatus (Aiton) A. Gray = *P. frigidus* (L.) Fries var. *palmatus* (Aiton) Cronq.
Petradoria discoidea L. Anderson = *Chrysothamnus gramineus* H.M. Hall
Pluchea purpurascens (Sw.) DC. misapplied to *P. odorata* (L.) Cass.
Psilactis coulteri A. Gray misapplied to *Machaeranthera arida* B. Turner & D. Horne
Psilocarphus elatior (A. Gray) A. Gray occurs in CA; formerly confused with *P. brevissimus* Nutt.
P. tenellus Nutt. var. *tenuis* (Eastw.) Cronq. = var. *globiferus* (DC.) J. Morefield
Pulicaria hispanica (Boiss.) Boiss. = *P. paludosa* Link
Pyrrocoma lanceolata (Hook.) E. Greene var. *subviscosa* (E. Greene) G. Brown & Keil occurs in CA
Raillardella muirii A. Gray = *Raillardiopsis muirii* (A. Gray) Rydb.
R. scabrida Eastw. = *Raillardiopsis scabrida* (Eastw.) Rydb.
Rhagadiolus edulis Willd. misapplied to *R. stellatus* (L.) Gaertner
Senecio clevelandii E. Greene var. *heterophylla* Hoover not recognized
S. cruentus L'Hér. misapplied to complex incl florist's cineraria, *S. hybridus* Regel
S. douglasii DC. incl var. *tularensis* Munz = *S. flaccidus* Less. var. *douglasii* (DC.) B. Turner & T. Barkley
var. *monoensis* (E. Greene) Jepson = *S. flaccidus* Less. var. *monoensis* (E. Greene) B. Turner & T. Barkley
S. eurycephalus A. Gray var *lewisrosei* (J. Howell) T. Barkley recognized
S. foetidus Howell incl var. *hydrophiloides* (Rydb.) Cronq. = *S. hydrophiloides* Rydb.
S. ganderi T. Barkley & Beauch. recently described
S. ligulifolius E. Greene = *S. macounii* E. Greene
S. stygius E. Greene = *S. multilobatus* A. Gray
S. subnudus DC. misapplied to *S. cymbalarioides* J.N. Buek
Solidago gigantea Aiton occurs in CA; part of *S. canadensis* complex
S. missouriensis Nutt. misapplied in CA to *S. spectabilis* (D. Eaton) A. Gray
S. occidentalis (Nutt.) Torrey & A. Gray = *Euthamia occidentalis* Nutt.
S. sparsiflora A. Gray occurs in CA .
Soliva daucifolia Nutt. = *S. sessilis* Ruiz López & Pavón
S. pterosperma (A.L. Juss.) Less. = *S. sessilis* Ruiz López & Pavón
Sphaeromeria potentilloides (A. Gray) A.A. Heller var. *nitrophila* (Cronq.) A. Holmgren, L. Schultz, & T. Lowery recently described
Stenotus lanuginosus (A. Gray) E. Greene occurs in CA
Stephanomeria carotifera Hoover = *S. exigua* Nutt. ssp. *carotifera* (Hoover) Gottlieb
S. diegensis Gottlieb recently described
S. elata Nutt. occurs in CA; formerly confused with *S. exigua* Nutt.
S. exigua Nutt. var. *coronaria* (E. Greene) Jepson = ssp. *coronaria* (E. Greene) Gottlieb
var. *deanei* J.F. Macbr. = ssp. *deanei* (J.F. Macbr.) Gottlieb
ssp. *macrocarpa* Gottlieb recently described
S. myrioclada D. Eaton = *S. pauciflora* (Nutt.) Nelson var. *pauciflora*
S. virgata Benth. ssp. *pleurocarpa* (E. Greene) Gottlieb recognized
var. *tomentosa* (E. Greene) Munz = ssp. *virgata*
Stylocline amphibolus (A. Gray) J. Howell = *Micropus amphibolus* A. Gray
S. citroleum J. Morefield recently described
S. filaginea (A. Gray) A. Gray incl var. *depressa* Jepson = *Ancistrocarphus filagineus* A. Gray
S. intertexta J. Morefield recently described
S. masonii J. Morefield recently described
S. psilocarphoides M. Peck occurs in CA; formerly confused with other *Stylocline* or *Filago* spp.
S. sonorensis Wiggins occurs in CA
Tagetes patula L. = *T. erecta* L.
Tanacetum canum D. Eaton = *Sphaeromeria cana* (D. Eaton) A.A. Heller
T. douglasii DC. = *T. camphoratum* Less.
T. potentilloides A. Gray = *Sphaeromeria potentillloides* (A. Gray) A.A. Heller
Taraxacum laevigatum (Willd.) DC. = *T. officinale* Wigg.

Tetradymia axillaris Nelson var. *longispina* (M.E. Jones) Strother recognized
T. spinosa Hook. & Arn. separate from *T. axillaris*
T. tetrameres (S.F. Blake) Strother occurs in CA
Tolpis umbellata Bertol. = *T. barbata* (L.) Gaertner
Townsendia parryi D. Eaton occurs in CA
Viguiera ciliata (Robinson & Greenman) S.F. Blake = *Heliomeris hispida* (A. Gray) Cockerell
V. deltoidea A. Gray var. *parishii* (E. Greene) Vasey & Rose = *V. parishii* E. Greene
V. multiflora (Nutt.) S.F. Blake var. *nevadensis* (Nelson) S.F. Blake = *Heliomeris multiflora* Nutt. var. *nevadensis* (Nelson) Yates

BALSAMINACEAE TOUCH-ME-NOT FAMILY

Impatiens occidentalis Rydb. = *I. noli-tangere* L.

BASELLACEAE BASELLA FAMILY

Boussingaultia gracilis Miers var. *pseudobaselloides* L. Bailey = *Anredera cordifolia* (Ten.) Steenis

BATACEAE [BATIDACEAE] SALTWORT FAMILY: no changes

BERBERIDACEAE BARBERRY FAMILY

Achlys californica I. Fukuda & H.G. Baker recently described
Berberis amplectens (Eastw.) Wheeler = *B. aquifolium* Pursh var. *repens* (Lindley) H. Scoggan
B. dictyota Jepson = *B. aquifolium* Pursh var. *dictyota* (Jepson) Jepson
B. higginsiae Munz = *B. fremontii* Torrey
B. piperiana (Abrams) McMinn = *B. aquifolium* Pursh var. *aquifolium*
B. pumila E. Greene = *B. aquifolium* Pursh var. *repens* (Lindley) H. Scoggan
B. repens Lindley = *B. aquifolium* Pursh var. *repens* (Lindley) H. Scoggan
B. sonnei (Abrams) McMinn = *B. aquifolium* Pursh var. *repens* (Lindley) H. Scoggan
Mahonia higginsiae (Munz) Ahrendt = *Berberis fremontii* Torrey
M. nervosa (Pursh) Nutt. var. *mendocinensis* Roof not recognized
M. nevinii (A. Gray) Fedde = *Berberis nevinii* A. Gray
M. sonnei Abrams = *Berberis aquifolium* Pursh var. *repens* (Lindley) H. Scoggan

BETULACEAE BIRCH FAMILY

Alnus oregona Nutt. = *A. rubra* Bong.
A. sinuata (Regel) Rydb. = *A. viridis* (Chaix) DC. ssp. *sinuata* (Regel) A. Löve & D. Löve
A. tenuifolia Nutt. = *A. incana* (L.) Moench ssp. *tenuifolia* (Nutt.) Breitung

BIGNONIACEAE BIGNONIA FAMILY

Chilopsis linearis (Cav.) Sweet: CA pls are ssp. *arcuata* (Fosberg) Henrickson

BORAGINACEAE BORAGE FAMILY

Amsinckia furcata Suksd. = *A. vernicosa* Hook. & Arn. var. *furcata* (Suksd.) Hoover
A. gloriosa Suksd. = *A. tessellata* A. Gray var. *gloriosa* (Suksd.) Hoover
A. intermedia Fischer & C. Meyer incl var. *echinata* (A. Gray) Wiggins = *A. menziesii* (Lehm.) A. Nelson & J.F. Macbr. var. *intermedia* (Fischer & C. Meyer) Ganders
var. *eastwoodiae* (J.F. Macbr.) Jepson & Hoover = *A. eastwoodiae* J.F. Macbr.
A. spectabilis Fischer & C. Meyer var. *nicolai* (Jepson) Munz = var. *spectabilis*
Anchusa azurea Miller naturalized
A. procera Link = *A. officinalis* L.
Coldenia canescens DC. = *Tiquilia canescens* (DC.) A. Richardson incl vars. *canescens* & *pulchella* (I.M. Johnston) A. Richardson
C. nuttallii Hook. = *Tiquilia nuttallii* (Hook.) A. Richardson
C. palmeri A. Gray = *Tiquilia palmeri* (A. Gray) A. Richardson
C. plicata (Torrey) Cov. = *Tiquilia plicata* (Torrey) A. Richardson
Cryptantha circumscissa (Hook. & Arn.) I.M. Johnston vars. *hispida* (J.F. Macbr.) I.M. Johnston & *rosulata* J. Howell not recognized
C. clevelandii E. Greene var. *dissita* (I.M. Johnston) Jepson & Hoover indistinct
var. *florosa* I.M. Johnston not recognized
C. corollata (I.M. Johnston) I.M. Johnston indistinct from *C. decipiens* (M.E. Jones) A.A. Heller
C. hendersonii (Nelson) Piper = *C. intermedia* (A. Gray) E. Greene
C. hoffmannii I.M. Johnston indistinct from *C. virginensis* (M.E. Jones) Payson
C. inaequata I.M. Johnston = *C. holoptera* (A. Gray) J.F. Macbr.
C. jamesii (Torrey) Payson var. *abortiva* (E. Greene) Payson = *C. cinerea* (E. Greene) Cronq. var. *abortiva* (E. Greene) Cronq.
C. maritima (E. Greene) E. Greene var. *pilosa* I.M. Johnston not recognized
C. micrantha (Torrey) I.M. Johnston var. *lepida* (A. Gray) I.M. Johnston not recognized
C. muricata (Hook. & Arn.) Nelson & J.F. Macbr. var. *clokeyi* (I.M. Johnston) Jepson = *C. clokeyi* I.M. Johnston
vars. *denticulata* (E. Greene) I.M. Johnston & *jonesii* (A. Gray) I.M. Johnston not recognized
C. nana (Eastw.) Payson var. *ovina* Payson = *C. humilis* (E. Greene) Payson
C. nevadensis Nelson & Kenn. var. *rigida* I.M. Johnston not recognized
C. pterocarya (Torrey) E. Greene var. *purpusii* Jepson not recognized
C. rattanii E. Greene indistinct from *C. decipiens* (M.E. Jones) A.A. Heller
C. rostellata (E. Greene) E. Greene incl var. *spithamea* (I.M. Johnston) Jepson indistinct from *C. flaccida* (Lehm.) E. Greene
C. schoolcraftii Tiehm = *C. sobolifera* Payson
C. scoparia Nelson indistinct from *C. nevadensis* Nelson & Kennedy
C. sparsiflora (E. Greene) E. Greene indistinct from *C. flaccida* (Lehm.) E. Greene
C. subretusa I.M. Johnston = *C. sobolifera* Payson
C. torreyana (A. Gray) E. Greene var. *pumila* (A.A. Heller) I.M. Johnston not recognized
Echium fastuosum Aiton misapplied to *E. candicans* L.f. (*E. strictum* L.f., *E. pininana* Webb & Berth., hybrids, & others may be waifs)
Hackelia jessicae (MacGregor) Brand = *H. micrantha* (Eastw.) J. Gentry
H. longituba I.M. Johnston = *H. velutina* (Piper) I.M. Johnston
H. patens (Nutt.) I.M. Johnston misapplied to *H. brevicula* (Jepson) J. Gentry
Lappula echinata Gilib = *L. squarrosa* (Retz.) Dumort.
L. redowskii (Hornem.) E. Greene var. *desertorum* (E. Greene) I.M. Johnston = var. cupulata (A. Gray) M.E. Jones
Lithospermum incisum Lehm. occurs in CA
Lycopsis arvensis L. = *Anchusa arvensis* (L.) M. Bieb.
Mertensia ciliata var. *stomatechoides* (Kellogg) Jepson indistinct
M. cusickii Piper occurs in CA
Myosotis versicolor (Pers.) Smith = *M. discolor* Pers.
M. virginica (L.) Britton, Sterns & Pogg. = *M. verna* Nutt.

Pectocarya linearis DC. var. *ferocula* I.M. Johnston = ssp. *ferocula* (I.M. Johnston) Thorne
P. peninsularis I.M. Johnston occurs in CA
Pentaglottis sempervirens (L.) Tausch not clearly naturalized
Plagiobothrys californicus (A. Gray) E. Greene var. *californicus* = *P. collinus* (Philbr.) I.M. Johnston var. *californicus* (A. Gray) Higgins
var. *fulvescens* I.M. Johnston = *P. collinus* (Philbr.) I.M. Johnston var. *fulvescens* (I.M. Johnston) Higgins
var. *gracilis* I.M. Johnston = *P. collinus* (Philbr.) I.M. Johnston var. *gracilis* (I.M. Johnston) Higgins
var. *ursinus* (A. Gray) I.M. Johnston = *P. collinus* (Philbr.) I.M. Johnston var. *ursinus* (A. Gray) Higgins
P. diffusus (E. Greene) I.M. Johnston indistinct from *P. reticulatus* (Piper) I.M. Johnston var. *rossianorum* I.M. Johnston
P. fulvus (Hook. & Arn.) I.M. Johnston var. *campestris* (E. Greene) I.M. Johnston indistinct
P. salsus (Brandegee) I.M. Johnston occurs in CA
P. torreyi (A. Gray) A. Gray var. *diffusus* I.M. Johnston recognized
Symphytum officinale L. naturalized

BRASSICACEAE [CRUCIFERAE] MUSTARD FAMILY

Alyssum desertorum Stapf naturalized
A. minus (L.) Roth ssp. *micranthum* (C. Meyer) Dudley naturalized
A. strigosum Banks & Sol. naturalized
Arabis aculeolata E. Greene occurs in CA
A. bodiensis Rollins recently described
A. constancei Rollins recently described
A. ×*divaricarpa* Nelson var. *interposita* (E. Greene) Rollins not recognized
A. hirsuta (L.) Scop. var. *eschscholtziana* (Andrz.) Rollins not in CA (OR to AK)
var. *pycnocarpa* (M. Hopk.) Rollins recognized
A. koehleri Howell var. *stipitata* Rollins occurs in CA
A. lignifera Nelson undocumented in CA
A. pinzlae Rollins recently described
A. rollei Rollins recently described
A. suffrutescens S. Watson var. *horizontalis* (E. Greene) Rollins recognized
var. *perstylosa* Rollins = var. *suffrutescens*
A. tiehmii Rollins recently described
Aubrieta deltoidea (L.) DC. naturalized
Barbarea orthoceras Ledeb. var. *dolichocarpa* Fernald not recognized
Brassica campestris L. = *B. rapa* L.
B. fruticulosa Cyrillo not naturalized
B. geniculata (Desf.) Ball = *Hirschfeldia incana* (L.) Lagr.-Fossat
B. hirta Moench = *Sinapis alba* L.
B. kaber (DC.) Wheeler incl vars. *pinnatifida* (Stokes) Wheeler & *schkuhriana* (Rchb.) Wheeler = *Sinapis arvensis* L.
Camelina sativa (L.) Crantz misapplied to *C. microcarpa* Andrz.
Cardamine gambellii S. Watson = *Rorippa gambellii* (S. Watson) Rollins & Al-Shehbaz
C. gemmata E. Greene = *C. nuttallii* E. Greene var. *gemmata* (E. Greene) Rollins
C. hirsuta L. naturalized
C. lyallii S. Watson = *C. cordifolia* A. Gray var. *lyallii* (S. Watson) Nelson & J.F. Macbr.
C. occidentalis (Robinson) Howell recognized
Caulanthus crassicaulis (Torrey) S. Watson var. *glaber* M.E. Jones recognized
C. major (M.E. Jones) Payson var. *nevadensis* Rollins recently described
Cochlearia groenlandica L. misapplied to *C. officinalis* L. var. *arctica* (DC.) Gelert
Dentaria californica Torrey & A. Gray = *Cardamine californica* (Torrey & A. Gray) E. Greene
var. *cardiophylla* (E. Greene) Detl. = *Cardamine californica* (Torrey & A. Gray) E. Greene var. *cardiophylla* (E. Greene) Rollins
var. *cuneata* (E. Greene) Detl. = *Cardamine californica* (Torrey & A. Gray) E. Greene var. *cuneata* (E. Greene) Rollins
var. *integrifolia* (Torrey & A. Gray) Detl. = *Cardamine californica* (Torrey & A. Gray) E. Greene var. *integrifolia* (Torrey & A. Gray) Rollins
var. *sinuata* (E. Greene) Detl. = *Cardamine californica* (Torrey & A. Gray) E. Greene var. *sinuata* (E. Greene) O. Schulz
D. gemmata (E. Greene) Howell = *Cardamine nuttallii* E. Greene var. *gemmata* (E. Greene) Rollins
D. pachystigma S. Watson = *Cardamine pachystigma* (S. Watson) Rollins
var. *dissectifolia* Detl. = *Cardamine pachystigma* (S. Watson) Rollins var. *dissectifolia* (Detl.) Rollins
D. tenella Pursh var. *dissecta* O. Schulz = *Cardamine nuttallii* E. Greene var. *dissecta* (O. Schulz) Rollins
var. *palmata* Detl. = *Cardamine nuttallii* E. Greene var. *covilleana* (O. Schulz) Rollins
Descurainea pinnata (Walter) Britton ssp. *filipes* (A. Gray) Detl. = *D. incisa* (A. Gray) Britton ssp. *filipes* (A. Gray) Rollins
ssp. *paradisa* (Nelson & Kennedy) Detl. = *D. paradisa* (Nelson & Kennedy) O. Schulz
D. richardsonii (Sweet) O. Schulz ssp. *incisa* (A. Gray) Detl. = *D. incisa* (A. Gray) Britton ssp. *incisa*
ssp. *viscosa* (Rydb.) Detl. = *D. incana* (Fischer & C. Meyer) Dorn
Draba crassifolia Graham incl var. *nevadensis* C.Hitchc. misapplied in CA to *D. albertina* E. Greene
D. corrugata S. Watson forma *vestita* (Davidson) C. Hitchc. indistinct
D. cruciata Payson var. *integrifolia* C. Hitchc. & Sharsm. = *D. sharsmithii* Rollins & R.A. Price
D. cuneifolia Torrey & A. Gray var. *californica* Jepson = *D. californica* (Jepson) Rollins & R.A. Price
vars. *integrifolia* S. Watson & *sonorae* (E. Greene) Parish indistinct in CA
D. douglasii A. Gray var. *crockeri* (Lemmon) C. Hitchc. = *Cusickiella douglasii* (A. Gray) Rollins
D. fladnizensis Wulfen misapplied in CA to *D. monoensis* Rollins & R.A. Price
D. howellii S. Watson var. *carnosula* (O. Schulz) C. Hitchc. = *D. carnosula* O. Schulz
D. lanceolata Royle misapplied in CA to *D. cana* Rydb.
D. lemmonii S. Watson var. *incrassata* Rollins = *D. incrassata* (Rollins) Rollins & R.A. Price
D. monoensis Rollins & R.A. Price recently described
D. nivalis Lilj. var. *elongata* S. Watson = *D. lonchocarpa* Rydb. var. *lonchocarpa*
D. oligosperma Hook. var. *subsessilis* (S. Watson) O. Schulz indistinct
D. quadricostata Rollins = *Cusickiella quadricostata* (Rollins) Rollins
D. stenoloba Ledeb. vars. *nana* (O. Schulz) C. Hitchc. & *ramosa* C. Hitchc. = *D. albertina* E. Greene
D. subumbellata Rollins & R.A. Price recently described
D. verna L. var. *aestivalis* Lej. indistinct
Eruca sativa Miller = *E. vesicaria* (L.) Cav. ssp. *sativa* (Miller) Thell.
Erucastrum gallicum (Willd.) O. Schulz naturalized
Erysimum argillosum (E. Greene) Rydb. indistinct from *E. capitatum* (Douglas) E. Greene ssp. *capitatum*
E. asperum (Nutt.) DC. incl var. *stellatum* J. Howell = *E. capitatum* (Douglas) E. Greene ssp. *capitatum*

E. capitatum (Douglas) E. Greene var. *angustatum* (E. Greene) Rossbach = ssp. *angustatum* (E. Greene) R.A. Price
var. *bealianum* (Jepson) Rossbach indistinct from ssp. *capitatum*
E. concinnum Eastw. = *E. menziesii* (Hook.) Wettst. ssp. *concinnum* (Eastw.) R.A. Price
E. franciscanum Rossbach var. *crassifolium* Rossbach indistinct
E. menziesii (Hook.) Wettst. ssp. *eurekense* R.A. Price recently described
ssp. *yadonii* R.A. Price recently described
E. perenne (Cov.) Abrams = *E. capitatum* (Douglas) E. Greene ssp. *perenne* (Cov.) R.A. Price
E. suffrutescens (Abrams) Rossbach var. *grandifolium* Rossbach indistinct from *E. insulare* E. Greene ssp. *suffrutescens* (Abrams) R.A. Price
var. *lompocense* Rossbach = *E. capitatum* (Douglas) E. Greene ssp. *lompocense* (Rossbach) R.A. Price
var. *suffrutescens* = *E. insulare* E. Greene ssp. *suffrutescens* (Abrams) R.A. Price
Halimolobos diffusa (A. Gray) O. Schulz var. *jaegeri* (Munz) Rollins = *H. jaegeri* (Munz) Rollins
H. virgata (Nutt.) Schulz occurs in CA
Lepidium densiflorum Schrader var. *bourgeauanum* (Thell.) C. Hitchc. = *L. ramosissimum* Nelson var. *bourgeauanum* (Thell.) Rollins
var. *elongatum* (Rydb.) Thell. occurs in CA
L. fremontii S. Watson var. *stipitatum* Rollins recently described
L. heterophyllum Benth. naturalized
L. lasiocarpum Torrey & A. Gray var. *georginum* (Rydb.) C. Hitchc. misapplied in CA to var. *lasiocarpum*
L. latipes Hook. var. *heckardii* Rollins recently described
L. oblongum Small var. *insulare* C. Hitchc. recognized
L. ramosissimum Nelson: CA pls are var. *bourgeauanum* (Thell.) Rollins
Lesquerella kingii (S. Watson) S. Watson var. *cordiformis* (Rollins) Maguire & Holmgren misapplied in CA to ssp. *kingii*
ssp. *latifolia* (Nelson) Rollins & E. Shaw occurs in CA
L. palmeri S. Watson misapplied in CA to *L. tenella* Nelson
Malcolmia africana R. Br. naturalized
Phoenicaulis cheiranthoides Torrey & A. Gray ssp. *glabra* (Jepson) Abrams not recognized
P. eurycarpa (A. Gray) Abrams = *Anelsonia eurycarpa* (A. Gray) J.F. Macbr. & Payson
Polyctenium fremontii (S. Watson) E. Greene var. *confertum Rollins* recently described
Rorippa calycina (Engelm.) Rydb. var. *columbiae* (Robinson) Rollins = *R. columbiae* (Robinson) Howell
R. curvipes E. Greene occurs in CA, incl vars. *curvipes* & *truncata* (Jepson) Rollins
R. islandica (Oeder) Borbas = *R. palustris* (L.) Besser
vars. *fernaldiana* F.K. Butters & Abbe & *occidentalis* (S. Watson) F.K. Butters & Abbe = *R. palustris* (L.) Besser var. *occidentalis* (S. Watson) Rollins
var. *hispida* (Desv.) F.K. Butters & Abbe = *R. palustris* (L.) Besser
var. *hispida* (Desv.) Rydb.
R. microphylla (Boenn.) Hyland. undocumented in CA
R. obtusa (Nutt.) Britton misapplied in CA to *R. sphaerocarpa* (A. Bray) Britton
R. tenerrima E. Greene occurs in CA
Sisymbrium erysimoides Desf. naturalized
S. loeselii L. naturalized
Streptanthella longirostris (S. Watson) Rydb. var. *derelicta* Howell not recognized
Streptanthus cordatus Nutt. var. *duranii* Jepson recognized
var. *exiguus* Jepson = *S. oliganthus* Rollins
S. drepanoides Kruckeb. & J. Morrison recently described
S. glandulosus Hook. var. *hoffmannii* Kruckeb. indistinct from ssp. *secundus* (E. Greene) Kruckeb.
var. *niger* (E. Greene) Munz = *S. niger* E. Greene
var. *pulchellus* (E. Greene) Jepson = ssp. *pulchellus* (E. Greene) Kruckeb.
var. *sonomensis* Kruckeb. indistinct from ssp. *secundus* (E. Greene) Kruckeb.
S. heterophyllus Nutt. = *Caulanthus heterophyllus* (Nutt.) Payson
S. insignis Jepson ssp. *lyonii* Kruckeb. & J. Morrison recently described
S. tortuosus Kellogg var. *optatus* Jepson = (in part) var. *suffrutescens* (E. Greene) Jepson
var. *pallidus* Jepson = var. *tortuosus*
Teesdalia coronopifolia (Bergeret) Thell. naturalized
Thelypodium flavescens (Hook.) S. Watson = *Guillenia flavescens* (Hook.) E. Greene
T. integrifolium (Torrey & A. Gray) Endl. ssp. *affine* (E. Greene) Al-Shehbaz recognized
ssp. *complanatum* Al-Shehbaz recently described
T. jaegeri Rollins = *Caulostramina jaegeri* (Rollins) Rollins
T. laciniatum (Hook.) Endl. var. *millefolium* (Nelson) Payson = *T. millefolium* Nelson
T. lasiophyllum (Hook. & Arn.) E. Greene incl vars. *inalienum* Robinson, *rigidum* (E. Greene) Robinson, & *utahense* (Rydb.) Jepson = *Guillenia lasiophylla* (Hook. & Arn.) E. Greene
T. lemmonii E. Greene = *Guillenia lemmonii* (E. Greene) R. Buck
Thlaspi fendleri A. Gray var. *hesperium* (Payson) C. Hitchc. = *T. montanum* L. var. *montanum*
Thysanocarpus curvipes Hook. vars. *elegans* (Fischer & C. Meyer) Robinson, *eradiatus* Jepson, & *longistylis* Jepson indistinct
T. laciniatus Torrey & A. Gray var. *conchuliferus* (E. Greene) Jepson = *T. conchuliferus* E. Greene
vars. *crenatus* (Nutt.) Brewer, *hitchcockii* Munz, *ramosus* (E. Greene) Munz, & *rigidus* Munz indistinct
Tropidocarpum gracile Hook. var. *dubium* (Davidson) Jepson not recognized

BUDDLEJACEAE BUDDLEJA FAMILY

Chilianthus oleaceus Burchell = *Buddleja saligna* Willd.

BURSERACEAE TORCHWOOD FAMILY

Bursera hindsiana (Benth.) Engler not documented in CA

BUXACEAE: see **SIMMONDSIACEAE**

CABOMBACEAE WATER-SHIELD FAMILY

Brasenia shreberi J. Gmelin transferred from Nymphaeaceae

CACTACEAE CACTUS FAMILY

Cereus giganteus Engelm. = *Carnegiea gigantea* (Engelm.) Britton & Rose
C. emoryi Engelm. = *Bergerocactus emoryi* (Engelm.) Britton & Rose
Coryphantha vivipara (Nutt.) Britton & Rose = *Escobaria vivipara* (Nutt.) F. Buxb.
var. *alversonii* (J. Coulter) L. Benson = *E. v.* var. *alversonii* (J. Coulter) D. Hunt
var. *rosea* (Clokey) L. Benson = *E. v.* var. *rosea* (Clokey) D. Hunt
Echinocactus acanthodes Lemaire invalid name, = *Ferocactus cylindraceus* (Engelm.) Orcutt
E. johnsonii Engelm. = *Sclerocactus johnsonii* (Engelm.) N.P. Taylor
E. polyancistrus Engelm. & J. Bigelow = *Sclerocactus polyancistrus* (Engelm. & J. Bigelow) Britton & Rose
E. viridescens Torrey & A. Gray = *Ferocactus viridescens* (Torrey & A. Gray) Britton & Rose

Echinocereus mojavensis (Engelm. & J. Bigelow) Ruempler = *E. triglochidiatus* Engelm.
E. munzii (Parish) L. Benson = *E. engelmannii* (Engelm.) Lemaire
Mammillaria alversonii (J. Coulter) H. Zeissold = *Escobaria vivipara* (Nutt.) F. Buxb. var. *alversonii* (J. Coulter) D. Hunt
M. arizonica Engelm. = *Escobaria vivipara* var. *deserti* (Engelm.) D. Hunt
M. deserti Engelm. = *Escobaria vivipara* var. *deserti* (Engelm.) D. Hunt
M. dioica M. Brandegee var. *incerta* (Parish) Munz not recognized
M. microcarpa Engelm. invalid name, = *M. milleri* (Britton & Rose) Boed.
Opuntia acanthocarpa Engelm. & J. Bigelow = *O. a.* var. *coloradensis* L. Benson
ssp. ×*ganderi* C. Wolf not recognized
O. basilaris Engelm. & J. Bigelow vars. *ramosa* Parish & *whitneyana* (Baxter) Marshall & Bock = var. *basilaris*
O. bigelovii Engelm. var. *hoffmannii* Fosberg = *P.* ×*fosbergii* C. Wolf
O. echinocarpa Engelm. & J. Bigelow var. *parkeri* J. Coulter = *P. parryi* Engelm.
O. erinacea Engelm. & J. Bigelow var. *ursina* (A. Weber) Parish = var. *erinacea*
var. *xanthostemma* (K. Schumann) L. Benson = var. *utahensis* (Engelm.) L. Benson
O. littoralis (Engelm.) Cockerell var. *austrocalifornica* L. Benson & Walkington = *O.* ×*vaseyi* (J. Coulter) Britton & Rose
var. *martiniana* (L. Benson) L. Benson = *O. martiniana* of AZ (not known from CA)
var. *piercei* (Fosb.) L. Benson = *O. phaeacantha* Engelm.
O. megacantha Salm-Dyck misapplied in CA to *O. ficus-indica* (L.) Miller
O. mojavensis Engelm. & J. Bigelow = *O. phaeacantha* Engelm.
O. occidentalis Engelm. & J. Bigelow = *O.* ×*occidentalis* Engelm.
var. *covillei* (Britton & Rose) Parish = *O. vaseyi* (J. Coulter) Britton & Rose
var. *littoralis* (Engelm.) Parish = *O. littoralis* (Engelm.) Cockerell
var. *megacarpa* (Griffiths) Munz not recognized
var. *vaseyi* (J. Coulter) Benson & Walkington = *O.* ×*vaseyi* (J. Coulter) Britton & Rose
O. phaeacantha Engelm. vars. *major* Engelm. & *mojavensis* (Engelm. & J. Bigelow) Fosb. not recognized
O. pulchella Engelm. recently documented in CA
O. serpentina Engelm. = *O. parryi* Engelm.
O. treleasei J. Coulter = *O. basilaris* var. *treleasei* (J. Coulter) Toumey
O. ×*wigginsii* L. Benson not recognized
O. wrightiana (Baxter) Peebles [= *O. kunzei Rose*] not documented in CA

CALLITRICHACEAE WATER-STARWORT FAMILY

Callitriche heterophylla Pursh ssp. *bolanderi* (Hegelm.) Calder & Roy Taylor = var. *bolanderi* (Hegelm.) Fassett
C. longipedunculata Morong = *C. marginata* Torrey
C. stagnalis Scop. naturalized

CALYCANTHACEAE SWEET-SHRUB or CALYCANTHUS FAMILY: no changes

CAMPANULACEAE BELLFLOWER FAMILY

Campanula griffinii N. Morin recently described
C. sharsmithiae N. Morin recently described
C. shetleri Heckard recently described
Downingia bacigalupii Weiler recently described
D. elegans (Lindley) Torrey var. *brachypetala* (Gandg.) McVaugh not recognized
var. *corymbosa* (A. DC.) Nelson & J.F. Macbr. indistinct
D. humilis (E. Greene) E. Greene = *D. pusilla* (Don) Torrey
D. yina Applegate var. *major* McVaugh not recognized
Githopsis calycina Benth. = *G. specularioides* Nutt.
G. diffusa A. Gray ssp. *robusta* N. Morin recently described
G. filicaulis Ewan = *G. diffusa* A. Gray ssp. *filicaulis* (Ewan) N. Morin
G. latifolia Eastw. [= *Legousia speculum-veneris* A.DC.] not naturalized
G. pulchella Vatke ssp. *campestris* N. Morin recently described ssp. *serpentinicola* N. Morin recently described
G. specularioides Nutt. ssp. *candida* Ewan = *G. diffusa* A. Gray ssp. *candida* (Ewan) N. Morin
G. tenella N. Morin recently described
Howellia aquatilis A. Gray extinct in CA, not included in text
Lobelia cardinalis L.: CA pls are var. *pseudosplendens* McVaugh
Nemacladus rubescens E. Greene var. *tenuis* McVaugh indistinct

CANNABACEAE (formerly part of MORACEAE) HEMP FAMILY

Cannabis, Humulus: no changes

CAPPARACEAE [CAPPARIDACEAE] CAPER FAMILY

Cleomella hillmanii Nelson not documented in CA
C. plocasperma S. Watson var. *mojavensis* (Payson) Crum indistinct
Isomeris arborea Nutt. vars. *angustata* Parish, globosa Cov., & *insularis* Jepson indistinct
Polanisia trachysperma Torrey & A. Gray = *P. dodecandra* (L.) DC.
ssp. *trachysperma* (Torrey & A. Gray) Iltis
Wislizenia refracta Engelm. sspp. *californica* (E. Greene) C.S. Keller & *palmeri* (A. Gray) C.S. Keller recently recognized

CAPRIFOLIACEAE HONEYSUCKLE FAMILY

Lonicera involucrata (Richardson) Banks var. *ledebourii* (Eschsch.) Jepson recognized
L. subspicata Hook. & Arn. var. *johnstonii* Keck = var. *denudata* Rehder
Sambucus caerulea Raf. = *S. mexicana* C. Presl
S. callicarpa E. Greene = *S. racemosa* L. var. *racemosa*
S. microbotrys Rydb. = *S. racemosa* L. var. *microbotrys* (Rydb.) Kearney & Peebles
Symphoricarpos acutus (A. Gray) Dieck indistinct from *S. mollis* Nutt.
S. hesperius G. Jones indistinct from *S. mollis* Nutt.
S. parishii Rydb. = *S. rotundifolius* A. Gray var. *parishii* (Rydb.) Dempster
S. rivularis Suksd. = *S. albus* (L.) S.F. Blake var. *laevigatus* (Fern.) S. F. Blake
S. vaccinioides Rydb. = *S. rotundifolius* A. Gray

CARYOPHYLLACEAE PINK FAMILY

Arenaria californica (A. Gray) Brewer = *Minuartia californica* (A. Gray) Mattf.
A. confusa Rydb. = *A. lanuginosa* (Michaux) Rohrb. ssp. *saxosa* (A. Gray) Maguire
A. douglasii Torrey & A. Gray incl var. *emarginata* H. Sharsm. = *Minuartia douglasii* (Torrey & A. Gray) Mattf.

A. howellii S. Watson = *Minuartia howellii* (S. Watson) Mattf.
A. kingii (S. Watson) M.E. Jones ssp. *compacta* (Cov.) Maguire = var. *glabrescens* (S. Watson) Maguire
A. macradenia S. Watson var. *parishiorum* Robinson = var. *macradenia*
A. macrophylla Hook. = *Moehringia macrophylla* (Hook.) Fenzl
A. nuttallii Pax = *Minuartia nuttallii* (Pax) Briq.:
sspp. *fragilis* (Maguire & A. Holmgren) McNeill & *gracilis* (Robinson) Maguire not recognized
ssp. *gregaria* (A.A. Heller) Maguire = *Minuartia nuttallii* (Pax) Briq. ssp. *gregaria* (A.A. Heller) McNeill
A. obtusiloba (Rydb.) Fern. = *Minuartia obtusilobia* (Rydb.) House
A. pumicola Cov. & Leiberg var. *californica* Maguire = *A. aculeata* S. Watson
A. pusilla S. Watson var. *diffusa* Maguire = *Minuartia californica* (A. Gray) Mattf.
A. rosei Maguire & Barneby = *Minuartia rosei* (Maguire & Barneby) McNeill
A. rossii Richardson misapplied to *Minuartia stricta* (Sw.) Hiern
A. rubella (Wahlenb.) Smith = *Minuartia rubella* (Wahlenb.) Hiern
Cerastium beeringianum Cham. & Schldl.: CA pls are var. *capillare* Fern. & Wieg.
C. viscosum L. misapplied to *C. glomeratum* Thuill.
C. vulgatum L. misapplied to *C. fontanum* Baumg. ssp. *vulgare* (Hartman) Greuter & Burdet
Gypsophila elegans M. Bieb. var. *elegans* naturalized
G. scorzonerifolia Ser. naturalized
Herniaria cinerea DC. = *H. hirsuta* L. ssp. *cinerea* (DC.) Cout.
H. hirsuta L. ssp. *hirsuta* recognized
Kohlrauschia velutina (Guss.) Reichb. = *Petrorhagia dubia* (Raf.) G. Lopez & Romo
Loeflingia squarrosa Nutt. ssp. *artemisiarum* Barneby & Twisselm. = var. *artemisiarum* (Barneby & Twisselm.) R. Dorn
Lychnis alba Miller = *Silene latifolia* Poiret ssp. *alba* (Miller) Greuter & Burdet
Minuartia decumbens T.W. Nelson & J.P. Nelson recently described
M. stolonifera T.W. Nelson & J.P. Nelson recently described
Moenchia erecta (L.) P. Gaertner, Meyer, & Sherb. ssp. *erecta* naturalized
Paronychia ahartii B. Ertter recently described
P. echinulata Chater var. *echinulata* naturalized
Petrorhagia nanteuilii (Burnat) P. Ball & Heyw. naturalized
Sagina apetala Ard. var. *barbata* Fenzl not recognized
S. crassicaulis S. Watson = *S. maxima* A. Gray ssp. *crassicaulis* (S. Watson) G. Crow
S. occidentalis S. Watson = *S. decumbens* (Elliott) Torrey & A. Gray ssp. *occidentalis* (S. Watson) G. Crow
S. saginoides (L.) Karsten var. *hesperia* Fern. not recognized
Silene campanulata S. Watson ssp. *greenei* (S. Watson) C. Hitchc. & Maguire = ssp. *glandulosa* C. Hitchc. & Maguire
S. cucubalus Wibel = *S. vulgaris* (Moench) Garcke
S. douglasii Hook. var *monantha* (S. Watson) Robinson not recognized
S. hookeri Nutt. sspp. *bolanderi* (A. Gray) Abrams & *pulverulenta* (M.E. Jones) C. Hitchc. & Maguire not recognized
S. menziesii Hook. ssp. *dorrii* (Kellogg) C. Hitchc. & Maguire not recognized
S. montana S. Watson (incl vars. & sspp.) (invalid name), = *S. bernardina* S. Watson
S. nuda (S. Watson) C. Hitchc. & Maguire ssp. *insectivora* (L. Henderson) C. Hitchc. & Maguire not recognized
S. parishii S. Watson vars. *latifolia* C. Hitchc. & Maguire and *viscida* C. Hitchc. & Maguire not recognized
S. suksdorfii Robinson occurs in CA
Spergularia marina (L.) Griseb. var. *tenuis* (E. Greene) R. Rossbach not recognized
Stellaria calycantha (Ledeb.) Bong. ssp. *interior* Hultén & var. *simcoei* (J. Howell) Fern. not recognized
S. jamesii Torrey = *Pseudostellaria jamesii* (Torrey) W.A. Weber & R.L. Hartman
S. longifolia Willd. occurs in CA
S. longipes Goldie var. *laeta* S. Watson not recognized
S. pallida (Dumort.) Piré [= *S. media* (L.) Villars ssp. *pallida* (Dumort.) Asch. & Graebn.] naturalized
S. sitchana Steudel var. *bongardiana* (Fern.) Hultén = *S. borealis* Bigelow ssp. *sitchana* (Steudel) Piper
Tunica prolifera (L.) Scop. = *Petrorhagia prolifera* (L.) P. Ball & Heyw.
Vaccaria segetalis (Necker) Asch. = *V. hispanica* (Miller) Rauschert

CELASTRACEAE STAFF-TREE FAMILY

Forsellesia transferred (as *Glossopetalon*) to Crossosomataceae

CERATOPHYLLACEAE HORNWORT FAMILY: no changes

CHENOPODIACEAE GOOSEFOOT FAMILY

Atriplex argentea Nutt. ssp. *expansa* (S. Watson) H.M. Hall & Clements = var. *mohavensis* M.E. Jones
var. *hillmanii* M.E. Jones recognized
A. canescens (Pursh) Nutt. var. *laciniata* Parish not recognized
var. *macilenta* Jepson = ssp. *canescens*
A. coronata S. Watson var. *notatior* Jepson recognized
A. depressa Jepson recognized
A. elegans (Moq.) D. Dietr. var. *elegans* occurs in CA
A. heterosperma Bunge naturalized
A. lentiformis (Torrey) S. Watson ssp. *breweri* (S. Watson) H.M. Hall & Clements indistinct from ssp. *lentiformis*
A. minuscula Standley separate from *A. depressa* Jepson & *A. parishii* S. Watson
A. nuttallii S. Watson var. *falcata* M.E. Jones = *A. gardneri* (Moq.) D. Dietr. var. *falcata* (M.E. Jones) S. Welsh
A. patula L. ssp. *hastata* (L.) H.M. Hall & Clements misapplied to *A. triangularis* Willd.
ssp. *spicata* (S. Watson) H.M. Hall & Clements = *A. joaquiniana* Nelson
A. suberecta I. Verd. naturalized
A. subspicata (Nutt.) Rydb. occurs in CA
A. torreyi (S. Watson) S. Watson = *A. lentiformis* (Torrey) S. Watson ssp. *torreyi* (S. Watson) H.M. Hall & Clements
A. triangularis Willd. separate from *A. patula* L. ssp. *patula*
Chenopodium ambrosioides L. vars. *anthelminticum* (L.) A. Gray & *vagans* (Standley) J. Howell not recognized
C. berlandieri Moq. vars. *sinuatum* (Murray) Murray & *zschackei* (Murray) Murray not recognized
C. capitatum (L.) Asch. indistinct from *C. foliosum* (Moench) Asch.
C. desiccatum Nelson var. *leptophylloides* (Murray) Wahl = *C. pratericola* Rydb.
C. fremontii S. Watson var. *incanum* S. Watson = *C. incanum* (S. Watson) A.A. Heller var. *occidentale* D. Crawford
C. gigantospermum Aellen = *C. simplex* (Torrey) Raf.
C. glaucum L. ssp. *salinum* (Standley) Aellen not recognized
C. macrospermum Hook.f. var. *farinosum* (S. Watson) J. Howell = var. *halophilum* (Phillipi) Standley
C. opulifolium Schrader indistinct from *C. album* L.
C. overi Aellen = *C. foliosum* (Moench) Asch.

Corispermum hyssopifolium L. occurs in CA
Eurotia lanata (Pursh) Moq. = *Krascheninnikovia lanata* (Pursh) A.D.J. Meeuse & Smit
Salsola kali L. var. *tenuifolia* Tausch = *S. tragus* L.
S. soda L. occurs in CA
S. vermiculatus L. occurs in CA
Sarcobatus vermiculatus (Hook.) Torrey var. *baileyi* (Cov.) Jepson not recognized
Suaeda californica S. Watson vars. *pubescens* Jepson & *taxifolia* (Standley) Munz = *S. taxifolia* (Standley) Standley
S. depressa (Pursh) S. Watson var. *depressa* misapplied to *S. calceoliformis* (Hook.) Moq.
var. *erecta* S. Watson = *S. calceoliformis* (Hook.) Moq.
S. esteroa W. Ferren & S. Whitmore recently described
S. fruticosa (L.) Forsskal misapplied to *S. moquinii* (Torrey) E. Greene
S. occidentalis S. Watson indistinct from *S. calceoliformis* (Hook.) Moq.
S. torreyana S. Watson incl var. *ramosissima* (Standley) Munz = *S. moquinii* (Torrey) E. Greene

CISTACEAE ROCK-ROSE FAMILY

Cistus ladanifer L. naturalized
C. monspeliensis L. naturalized
C. salvifolius L. naturalized
C. villosus L. incl vars. *corsicus* (lois.) Grosser, *tauricus* Grosser, & *undulatus* Grosser = *C. incanus* L.
Helianthemum guttatum (L.) Miller = *Tuberaria guttata* (L.) Fourr.

COMPOSITAE: see **ASTERACEAE**

CONVOLVULACEAE MORNING GLORY FAMILY

Calystegia collina (E. Greene) Brummitt ssp. *oxyphylla* Brummitt recently described
ssp. *venusta* Brummitt recently described
C. fulcrata (A. Gray) Brummitt = *C. occidentalis* (A. Gray) Brummitt ssp. *fulcrata* (A. Gray) Brummitt
C. macrostegia (E. Greene) Brummitt ssp. *amplissima* Brummitt recently described
ssp. *longilobus* (Abrams) Brummitt = ssp. *intermedia* (Abrams) Brummitt
C. malacophylla ssp. *malacophylla* var. *berryi* (Eastw.) Brummitt indistinct
ssp. *tomentella* (E. Greene) Munz indistinct from *C. occidentalis* (A. Gray) Brummitt ssp. *occidentalis*
ssp. *tomentella* var. *deltoidea* (E. Greene) Munz = *C. occidentalis* ssp. *fulcrata* (A. Gray) Brummitt
C. polymorpha (E. Greene) Munz = *C. occidentalis* (A. Gray) Brummitt ssp. *occidentalis*
C. purpurata (E. Greene) Brummitt ssp. *solanensis* (Jepson) Brummitt not recognized
C. sepium (L.) R. Br. ssp. *americana* (Sims) Brummitt not naturalized
Ipomoea cairica (L.) Sweet naturalized

CORNACEAE DOGWOOD FAMILY

Cornus ×californica C.A. Meyer incl var. *nevadensis* Jepson = *C. sericea* L. ssp. *sericea*
C. occidentalis (Torrey & A. Gray) Cov. = *C. sericea* L. ssp. *occidentalis* (Torrey & A. Gray) Fosb.
C. stolonifera Michaux = *C. sericea* L. ssp. *sericea*

CRASSULACEAE STONECROP FAMILY

Aeonium arboreum (L.) Webb. & Berth. var. *arboreum* naturalized
A. haworthii Salm-Dyck naturalized
Cotyledon orbiculata L.: CA pls are var. *oblonga* (Haw.) DC.
Crassula solieri (C. Gay) F. Meigen occurs in CA
Dudleya abramsii Rose ssp. *affinis* K. Nakai recently described
D. alainae Reiser = *D. saxosa* (M.E. Jones) Britton & Rose ssp. *aloides* (Rose) Moran
D. arizonica Rose = *D. pulverulenta* (Nutt.) Britton & Rose ssp. *arizonica* (Rose) Moran
D. bettinae Hoover = *D. abramsii* Rose ssp. *bettinae* (Hoover) J. Bartel
D. blochmanae (Eastw.) Moran: spelling corrected to *D. blochmaniae*
D. brevifolia (Moran) Moran = *D. blochmaniae* (Eastw.) Moran ssp. *brevifolia* Moran
D. calcicola J. Bartel & J. Shevock recently described
D. cymosa (Lemaire) Britton & Rose ssp. *costafolia* J. Bartel & J. Shevock recently described
ssp. *crebrifolia* K. Nakai & Verity recently described
ssp. *gigantea* (Rose) Moran = ssp. *cymosa*
ssp. *minor* (Rose) Moran = *D. lanceolata* (Nutt.) Britton & Rose
ssp. *setchellii* (Jepson) Moran in part = *D. setchellii* (Jepson) Britton & Rose
D. parva Rose & Davidson = *D. abramsii* Rose ssp. *parva* (Rose & Davidson) J. Bartel
D. traskae (Rose) Moran: spelling corrected to *D. traskiae*
D. verityi K. Nakai recently described
Sedum albomarginatum R.T. Clausen recently described
S. album L. not naturalized
S. eastwoodiae (Britton) Berger separate from *S. laxum* (Britton) Berger
S. laxum (Britton) Berger ssp. *eastwoodiae* (Britton) R.T. Clausen = *S. eastwoodiae* (Britton) Berger
ssp. *flavidum* M. Denton recently described
ssp. *latifolium* R.T. Clausen = ssp. *laxum*
ssp. *perplexum* R.T. Clausen = ssp. *laxum*
ssp. *retusum* (Rose) R.T. Clausen = *S. obtusatum* A. Gray ssp. *retusum* (Rose) R.T. Clausen
S. obtusatum A. Gray ssp. *paradisum* M. Denton recently described = *S. paradisum* (M. Denton) M. Denton
S. purdyi Jepson = *S. spathulifolium* Hook.
S. radiatum S. Watson ssp. *depauperatum* not recognized
S. rosea (L.) Scop.: spelling corrected to *S. roseum*
S. spathulifolium Hook. sspp. *anomalum* (Britton) R.T. Clausen & Uhl & *pruinosum* (Britton) R.T. Clausen & Uhl not recognized
S. stenopetalum Pursh ssp. *radiatum* (S. Watson) R.T. Clausen = *S. radiatum* S. Watson
Tillaea aquatica L. = *Crassula aquatica* (L.) Schönl.
T. erecta Hook. & Arn. = *Crassula connata* (Ruiz López & Pavón) A. Berger
T. muscosa L. = *Crassula tillaea* Lester-Garl.

CROSSOSOMATACEAE CROSSOSOMA FAMILY

Forsellesia arida (M.E. Jones) A.A. Heller = *Glossopetalon spinescens* A. Gray
F. nevadensis (A. Gray) E. Greene = *Glossopetalon spinescens* A. Gray
F. pungens (Brandegee) A.A. Heller incl var. *glabra* Ensign = *Glossopetalon pungens* Brandegee
F. stipulifera (St. John) Ensign = *Glossopetalon spinescens* A. Gray

CRUCIFERAE: see **BRASSICACEAE**

CUCURBITACEAE GOURD FAMILY

Citrullus lanatus (Thunb.) Mansf. = *C. colocynthis* (L.) Schrader var. *lanatus* (Thunb.) Matsum. & Nakai
Marah fabaceus (Naudin) E. Greene var. *agrestis* (E. Greene) K.M. Stocking indistinct from var. *fabaceus*

CUSCUTACEAE DODDER FAMILY

Cuscuta approximata Bab. var. *urceolata* (Kunze) Yuncker misapplied
C. brachycalyx (Yuncker) Yuncker incl var. *apodanthera* (Yuncker) Yuncker = *C. californica* Hook. & Arn. var. *breviflora* Engelm.
C. campestris Yuncker = *C. pentagona* Engelm.
C. epithymum Murray = *C. approximata* Bab.
C. howelliana Rubtzoff recently described
C. indecora Choisy var. *neuropetala* (Engelm.) Hitchc. recognized
C. jepsonii Yuncker = *C. indecora* Choisy var. *indecora*
C. obtusiflora Kunth var. *glandulosa* Engelm. not known in CA
C. occidentalis Millsp. = *C. californica* Hook. & Arn. var. *breviflora* Engelm.
C. suaveolens Ser. misapplied to *C. indecora* Choisy var. *i.*
C. suksdorfii Yuncker incl var. *subpedicellata* Yuncker = *C. californica* Hook. & Arn. var. *breviflora* Engelm.

DATISCACEAE DATISCA FAMILY: no changes

DIPSACACEAE TEASEL FAMILY: no changes

DROSERACEAE SUNDEW FAMILY: no changes

ELAEAGNACEAE OLEASTER FAMILY: no changes

ELATINACEAE WATERWORT FAMILY

Elatine gracilis H. Mason indistinct from *E. chilensis* C. Gay
E. heterandra H. Mason separate from *E. brachysperma* A. Gray
E. obovata (Fassett) H. Mason = *E. brachysperma* A. Gray

EMPETRACEAE CROWBERRY FAMILY

Empetrum hermaphroditum (Lange) Hagerup [= *E. nigrum* L. ssp. *hermaphroditum* (Lange) Böcher] indistinct from *E. nigrum* L.

ERICACEAE HEATH FAMILY

Arctostaphylos acutifolia Eastw. = *A. patula* E. Greene
A. andersonii A. Gray var. *imbricata* (Eastw.) McMinn = *A. imbricata* Eastw.
var. *pallida* (Eastw.) McMinn = *A. pallida* Eastw.
A. bakeri Eastw. ssp. *sublaevis* P. Wells recently described
A. canescens Eastw. var. *candidissima* (Eastw.) Munz = ssp. *canescens*
ssp. *malloryi* W. Knight & R. Gankin = *A. malloryi* (W. Knight & R. Gankin) P. Wells
var. *sonomensis* (Eastw.) J. Adams = ssp. *sonomensis* (Eastw.) P. Wells
A. catalinae P. Wells recently described
A. ×*cinerea* Howell = *A. canescens* × *A. viscida*
A. columbiana Piper var. *tracyi* (Eastw.) J. Adams indistinct
A. crustacea Eastw. var. *crustacea* = *A. tomentosa* (Pursh) Lindley
ssp. *crustacea* (Eastw.) P. Wells
var. *rosei* McMinn = *A. tomentosa* (Pursh) Lindley ssp. *rosei* (Eastw.) P. Wells
A. edmundsii J. Howell var. ×*parvifolia* Howell = *A. glandulosa* × *A. nevadensis*
A. elegans Jepson = *A. manzanita* C. Parry ssp. *elegans* (Jepson) P. Wells
A. gabrielensis P. Wells recently described
A. glandulosa Eastw. var. adamsii Munz = ssp. *adamsii* (Munz) Munz
var. *campbelliae* (Eastw.) McMinn = *A. tomentosa* (Pursh) Lindley ssp. *crustacea* (Eastw.) P. Wells
var. *crassifolia* Jepson = ssp. *crassifolia* (Jepson) P. Wells
var. *cushingiana* (Eastw.) McMinn indistinct from ssp. *glandulosa*
ssp. *glaucomollis* P. Wells recently described
var. *howellii* (Eastw.) McMinn = ssp. *zacaensis* (Eastw.) P. Wells
var. *mollis* J. Adams = ssp. *mollis* (J. Adams) P. Wells
var. *zacaensis* (Eastw.) McMinn = ssp. *zacaensis* (Eastw.) P. Wells
A. glauca Lindley var. *puberula* J. Howell not recognized
A. hearstiorum Hoover & Roof = *A. hookeri* G. Don ssp. *hearstiorum* (Hoover & Roof) P. Wells
A. hookeri G. Don ssp. *ravenii* P. Wells recently described
A. insularis E. Greene var. *pubescens* Eastw. not recognized
A. intricata Howell = *A. glandulosa* Eastw. ssp. *glandulosa*
var. ×*oblongifolia* (Howell) Munz = *A. canescens* × *A. viscida*
A. klamathensis S. Edwards, T. Keeler-Wolf & W. Knight recently described
A. ×*knightii* R. Gankin & W. Hildreth = *A. glandulosa* × *A. nevadensis*
A. luciana P. Wells recently described
A. malloryi (W. Knight & R. Gankin) P. Wells recently described
A. manzanita C. Parry ssp. *glaucescens* P. Wells recently described
ssp. *roofii* (R. Gankin) P. Wells recently described
ssp. *wieslanderi* P. Wells recently described
A. mariposa Dudley = *A. viscida* C. Parry ssp. *mariposa* (Dudley) P. Wells
A. mendocinoensis P. Wells recently described
A. mewukka Merriam ssp. *truei* (W. Knight) P. Wells recently described
A. nortensis (P. Wells) P. Wells recently described
A. nummularia A. Gray var. *sensitiva* (Jepson) Munz not recognized
A. osoensis P. Wells recently described
A. ×*pacifica* Roof = *A. glandulosa* × *A. uva-ursi*
A. parryana Lemmon var. *pinetorum* (Rollins) Wiesl. & B. Schreiber = *A. patula* E. Greene
A. ×*parvifolia* Howell = *A. glandulosa* × *A. nevadensis*
A. pechoensis Abrams var. *viridissima* Eastw. = *A. viridissima* (Eastw.) McMinn
A. pilosula Jepson & Wiesl. ssp. *pismoensis* P. Wells not recognized
A. pseudopungens Roof not recognized
A. pungens Kunth ssp. *laevigata* (Eastw.) Roof = *A. manzanita* C. Parry ssp. *l.* (Eastw.) Munz
var. *montana* (Eastw.) Munz = *A. hookeri* G. Don ssp. *montana* (Eastw.) P. Wells
ssp. *chaloneorum* (Roof) Roof not recognized
A. purissima P. Wells recently described
A. regismontana Eastw. separate from *A. andersonii* A. Gray
A. stanfordiana C. Parry *ssp. bakeri* (Eastw.) J. Adams = *A. bakeri* Eastw. ssp. *bakeri*
ssp. *decumbens* (P. Wells) P. Wells recently described
ssp. *hispidula* (Howell) J. Adams = *A. hispidula* Howell
ssp. *raichei* W. Knight recently described
var. *repens* Roof = ssp. *decumbens* (P. Wells) P. Wells
A. subcordata Eastw. var. *confertiflora* (Eastw.) Munz = *A. confertiflora* Eastw.
var. *subcordata* = *A. tomentosa* (Pursh) Lindley ssp. *subcordata* (Eastw.) P. Wells
A. tomentosa (Pursh) Lindley ssp. *daciticola* P. Wells recently described
ssp. *eastwoodiana* P. Wells recently described
var. *hebeclada* (DC.) McMinn = ssp. *bracteosa* (DC.) J. Adams
ssp. *insulicola* P. Wells recently described
var. *tomentosiformis* (J. Adams) Munz = ssp. *crinita* (McMinn) R. Gankin
var. *trichoclada* (DC.) Munz = ssp. *bracteosa* (DC.) J. Adams

A. truei W. Knight = *A. mewukka* Merriam ssp. *truei* (W. Knight) P. Wells
A. uva-ursi (L.) Sprengel vars. *leobreweri* Roof, *marinensis* Roof, *suborbiculata* Roof, & ssp. *monoensis* Roof not recognized
A. viscida C. Parry ssp. *pulchella* (Howell) P. Wells recognized
A. wellsii W. Knight recently described
Chimaphila umbellata (L.) Bartram var. *occidentalis* (Rydb.) S.F. Blake not recognized
Kalmia polifolia Wangenh. ssp. *polifolia* also in CA
Ledum glandulosum Nutt. var. *californicum* (Kellogg) C. Hitchc., ssp. *columbianum* (Piper) C. Hitchc., & ssp. *olivaceum* C. Hitchc. not recognized
Moneses uniflora (L.) A. Gray var. *reticulata* (Nutt.) S.F. Blake not recognized
Monotropa hypopithys L.: spelling corrected to *M. hypopitys*
Ornithostaphylos oppositifolia (C. Parry) Small occurs in CA
Pyrola asarifolia Michaux var. *purpurea* (Bunge) Fern. = var. *asarifolia*
P. chlorantha Sw. occurs in CA
P. picta Smith forma *aphylla* (Smith) Campbell & ssp. *dentata* (Smith) Piper not recognized
P. secunda L. = *Orthilia secunda* (L.) House
Vaccinium arbuscula (A. Gray) Merriam = *V. cespitosum Michaux*
V. coccineum Piper indistinct from *V. membranaceum* Hook.
V. deliciosum Piper occurs in CA
V. nivictum Camp = *V. cespitosum* Michaux
V. occidentale A. Gray = *V. uliginosum* L. ssp. *occidentale* (A. Gray) Hultén
V. ovatum Pursh var. *saporosum* Jepson not recognized
V. uliginosum L. ssp. *uliginosum* not documented in CA

EUPHORBIACEAE SPURGE FAMILY

Bernardia incana C. Morton = *B. myricifolia* (Scheele) S. Watson
Croton californicus Muell. Arg. vars. *mohavensis* A. Ferg. & *tenuis* (S. Watson) A. Ferg. not recognized
Ditaxis adenophora (A. Gray) Pax & K. Hoffm. misapplied to *D. clariana* (Jepson) Webster
Euphorbia abramsiana Wheeler = *Chamaesyce abramsiana* (Wheeler) Koutnik
E. albomarginata Torrey & A. Gray = *Chamaesyce albomarginata* (Torrey & A. Gray) Small
E. arizonica Engelm. = *Chamaesyce arizonica* (Engelm.) J.C. Arthur
E. exigua L. not naturalized
E. exstipulata Engelm. var. *exstipulata* occurs in CA
E. fendleri Torrey & A. Gray = *Chamaesyce fendleri* (Torrey & A. Gray) Small
E. glyptosperma Engelm. = *Chamaesyce glyptosperma* (Engelm.) Small
E. hooveri Wheeler = *Chamaesyce hooveri* (Wheeler) Koutnik
E. maculata L. = *Chamaesyce maculata* (L.) Small
E. melanadenia Torrey = *Chamaesyce melanadenia* (Torrey) Millsp.
E. micromera Engelm. = *Chamaesyce micromera* (Engelm.) Wooton & Standley
E. nutans Lagasca = *Chamaesyce nutans* (Lagasca) Small
E. ocellata Durand & Hilg. = *Chamaesyce ocellata* (Durand & Hilg.) Millsp.
var. *arenicola* (Parish) Jepson = *C. o.* ssp. *arenicola* (Parish) Thorne
var. *ocellata* = *C. o.* ssp. *ocellata*
var. *rattanii* (S. Watson) Wheeler = *C. o.* ssp. *rattanii* (S. Watson) Koutnik
E. parishii E. Greene = *Chamaesyce parishii* (E. Greene) Millsp.
E. parryi Engelm. = *Chamaesyce parryi* (Engelm.) Rydb.
E. pediculifera Engelm. = *Chamaesyce pediculifera* (Engelm.) Rose & Standley
E. platysperma S. Watson = *Chamaesyce platysperma* (S. Watson) Shinn.
E. polycarpa Benth. = *Chamaesyce polycarpa* (Benth.) Millsp. var. *hirtella* (Boiss.) Parish indistinct
E. prostrata Aiton = *Chamaesyce prostrata* (Aiton) Small
E. revoluta Engelm. = *Chamaesyce revoluta* (Engelm.) Small
E. serpens Kunth = *Chamaesyce serpens* (Kunth) Small
E. serpyllifolia Pers. = *Chamaesyce serpyllifolia* (Pers.) Small
var. *hirtula* (S. Watson) Wheeler = ssp. *hirtula* (S. Watson) Koutnik
E. serrata L. naturalized
E. setiloba Torrey = *Chamaesyce setiloba* (Torrey) Millsp.
E. supina Raf. = *Chamaesyce maculata* (L.) Small
E. vallis-mortae (Millsp.) J. Howell = *Chamaesyce vallis-mortae* Millsp.
Tragia stylaris Muell. Arg. = *T. ramosa* Torrey

FABACEAE [LEGUMINOSAE] LEGUME FAMILY

Acacia baileyana F. Muell. naturalized
A. cyclops G. Don naturalized
A. decurrens (Wendl.) Willd. var. *dealbata* (Link) F. Muell. = *A. dealbata* Link
A. elata Benth. naturalized
A. mearnsii De Wild. probably naturalized but range not documented
A. minuta (M.E. Jones) Beauch. ssp. *minuta* misapplied to *A. farnesiana* (L.) Willd. var. *farnesiana*
A. paradoxa DC. naturalized
A. pycnantha Benth. naturalized
A. verticillata (L'Hér.) Willd. naturalized
Aeschynomene rudis Benth. naturalized
Albizia distachya (Vent.) Macbr. = *A. lophantha* (Willd.) Benth.
Alhagi camelorum Fischer = *A. pseudalhagi* (M. Bieb.) Desv.
Amorpha fruticosa L. var. *occidentalis* (Abrams) Kearney & Peebles not recognized
Astragalus antiselli A. Gray = *A. trichopodus* (Nutt.) A. Gray var. *phoxus* (M.E. Jones) Barneby
A. argophyllus Torrey & A. Gray var. *argophyllus* occurs in CA
A. didymocarpus Hook. & Arn. var. *daleoides* Barneby = var. *didymocarpus*
A. ertterae Barneby & J. Shevock recently described
A. kentrophyta A. Gray var. *elatus* S. Watson occurs in CA
var. *implexus* (Canby) Barneby = var. *tegetarius* (S. Watson) Dorn
A. leucopsis (Torrey) Torrey & A. Gray = *A. trichopodus* (Nutt.) A. Gray var. *lonchus* (M.E. Jones) Barneby
A. lentiginosus Hook. var. *carinatus* M.E. Jones = var. *lentiginosus*
var. *coulteri* (Benth.) M.E. Jones = var. *borreganus* M.E. Jones
var. *sesquimetralis* (Rydb.) Barneby occurs in CA
A. nuttallianus A. DC. var. *austrinus* (Small) F. Shreve & Wiggins probably occurs in CA
A. oophorus S. Watson var. *lavinii* Barneby recently described
A. purshii Hook. var. *longilobus* M.E. Jones = var. *tinctus* M.E. Jones
A. serenoi (Kuntze) E. Sheldon: CA pls are var. *shockleyi* (M.E. Jones) Barneby
A. shevockii Barneby recently described
A. tener A. Gray var. *ferrisiae* A. Liston recently described
A. vaseyi S. Watson = *A. palmeri* A. Gray
A. wootonii E. Sheldon = *A. allochrous* A. Gray var. *playanus* (M.E. Jones) Isely
Caesalpinia gilliesii (Hook.) D. Dietr. naturalized
C. spinosa (Molina) Kuntze naturalized
Caragana arborescens Lam. reportedly naturalized, not documented

Cassia armata S. Watson = *Senna armata* (S. Watson) H. Irwin & Barneby
C. covesii A. Gray = *Senna covesii* (A. Gray) H. Irwin & Barneby
C. tomentosa L. = *Senna multiglandulosa* (Jacq.) H. Irwin & Barneby
Colutea arborescens L. naturalized
Cytisus canariensis (L.) Kuntze = *Genista canariensis* L.
C. linifolius (L.) Lam. = *Genista linifolia* L.
C. maderensis (Webb & Berth.) Masferer = *Genista maderensis* (Webb & Berth.) Lowe
C. monspessulanus L. = *Genista monspessulana* (L.) L. Johnson
C. multiflorus (Aiton) Sweet naturalized
C. proliferus L. = *Chamaecytisus proliferus* (L.) Link
C. stenopetalus (Webb & Berth.) Christ = *Genista stenopetala* Webb & Berth.
C. striatus (Hill) Rothm. naturalized
Dalea arborescens (A. Gray) A.A. Heller = *Psorothamnus arborescens* (A. Gray) Barneby var. *arborescens*
D. californica S. Watson = *Psorothamnus arborescens* (A. Gray) Barneby ssp. *simplicifolius* (Parish) Barneby
D. emoryi A. Gray = *Psorothamnus emoryi* (A. Gray) Rydb.
D. fremontii A. Gray = *Psorothamnus fremontii* (A. Gray) Barneby
var. *minutifolia* (Parish) Benson = *Psorothamnus arborescens* (A. Gray) Barneby var. *minutifolius* (Parish) Barneby
var. *saundersii* (Parish) Munz = *Psorothamnus arborescens* (A. Gray) Barneby var. *arborescens*
D. ornata (Hook.) Eaton & J. Wright occurs in CA
D. parryi Torrey & A. Gray = *Marina parryi* (Torrey & A. Gray) Barneby
D. polydenia S. Watson (sometimes misspelled "polyadenia") = *Psorothamnus polydenius* (S. Watson) Rydb.
D. schottii Torrey = *Psorothamnus schottii* (Torrey) Barneby
D. spinosa A. Gray = *Psorothamnus spinosus* (A. Gray) Barneby
Halimodendron halodendron (L.) Voss naturalized
Hoffmannseggia densiflora A. Gray = *H. glauca* (Ortega) Eifert
H. microphylla Torrey = *Caesalpinia virgata* E.M. Fisher
Lathyrus angulatus L. formerly confused with *L. sphaericus* Retz.
L. biflorus T.W. Nelson & J.P. Nelson recently described
L. glandulosus Broich recently described
L. hitchcockianus Barneby & Rev. recently described, but not documented currently in CA
L. japonicus Willd. var. *glaber* (Ser.) Fern. not recognized
L. jepsonii E. Greene ssp. *californicus* (S. Watson) C. Hitchc. = var. *californicus* (S. Watson) Hoover
ssp. *jepsonii* = var. *jepsonii*
L. laetiflorus E. Greene ssp. *alefeldii* (T. White) Bradshaw = *L. vestitus* Nutt. var. *alefeldii* (T. White) Isely
ssp. *barbarae* (T. White) C. Hitchc. = *L. vestitus* Nutt. var. *vestitus*
ssp. *laetiflorus* = *L. vestitus* Nutt. var. *vestitus*
L. nevadensis S. Watson ssp. *lanceolatus* (J. Howell) C. Hitchc. not recognized
L. pauciflorus Fern. ssp. *brownii* (Eastw.) Piper = *L. brownii* Eastw.
L. sulphureus A. Gray var. *argillaceus* Jepson not recognized
L. tracyi Bradshaw = *L. lanszwertii* Kellogg var. *tracyi* (Bradshaw) Isely
L. vestitus Nutt. ssp. *bolanderi* (S. Watson) C. Hitchc. = var. *ochropetalus* (Piper) Isely in part; = var. *vestitus* in part
ssp. *puberulus* (E. Greene) C. Hitchc. = var. *vestitus*
Lotus aboriginum Jepson: corrected to *L. aboriginus* Jepson
L. argophyllus (A. Gray) E. Greene ssp. *adsurgens* (Dunkle) Raven = var. *adsurgens* (Dunkle) Isely
var. *decorus* (I.M. Johnston) Ottley = var. *argophyllus*
ssp. niveus (E. Greene) Munz = var. niveus (E. Greene) Ottley
var. *ornithopus* (E. Greene) Ottley (in part) = var. *argenteus* Dunkle
L. argyraeus (E. Greene) E. Greene ssp. *multicaulis* (Ottley) Munz = var. *multicaulis* (Ottley) Isely
var. *notitius* Isely recently described
L. crassifolius (Benth.) E. Greene var. *otayensis* Isely recently described
L. cupreus E. Greene = *L. oblongifolius* (Benth.) E. Greene var. *cupreus* (E. Greene) Ottley
L. davidsonii E. Greene = *L. nevadensis* (S. Watson) E. Greene var. *davidsonii* (E. Greene) Isely
L. douglasii E. Greene = *L. nevadensis* (S. Watson) E. Greene var. *nevadensis*
L. grandiflorus (Benth.) E. Greene var. *macranthus* (E. Greene) Isely occurs in CA
var. *mutabilis* Ottley = var. *grandiflorus*
L. heermannii (Durand & Hilg.) E. Greene var. *eriophorus* (E. Greene) Ottley = var. *orbicularis* (A. Gray) Isely
L. junceus (Benth.) E. Greene var. *biolettii* (E. Greene) Ottley recognized
L. neo-incanus Munz = *L. incanus* (Torrey) E. Greene
L. oblongifolius (Benth.) E. Greene var. *nevadensis* (A. Gray) Munz = var. *oblongifolius*
L. scoparius (Nutt.) Ottley var. *dendroideus* (E. Greene) Ottley = *L. dendroideus* (E. Greene) E. Greene var. *dendroideus*
var. *traskiae* (Noddin) Raven = *L. dendroideus* (E. Greene) E. Greene var. *traskiae* (Noddin) Isely
var. *veatchii* (E. Greene) Ottley = *L. dendroideus* (E. Greene) E. Greene var. *veatchii* (E. Greene) Isely
L. stipularis (Benth.) E. Greene var. *ottleyi* Isely recently described
L. strigosus (Nutt.) E. Greene var. *hirtellus* (E. Greene) Ottley not recognized
L. subpinnatus Lagasca misapplied to *L. wrangelianus* Fischer & C. Meyer
L. tomentellus E. Greene = *L. strigosus* (Nutt.) E. Greene
Lupinus abramsii C.P. Smith = *L. albifrons* Benth. var. *abramsii* (C.P. Smith) Hoover
L. adsurgens E. Drew var. *lilacinus* A.A. Heller indistinct in part; = *L. antoninus* Eastw. in part var. *undulatus* C.P. Smith indistinct in part; = *L. dalesiae* Eastw. in part
L. albicaulis Hook. var. *shastensis* (A.A. Heller) C.P. Smith indistinct
L. albifrons Benth. var. *eminens* (E. Greene) C.P. Smith = var. *albifrons*
var. *flumineus* C.P. Smith indistinct from var. *collinus* E. Greene
L. andersonii S. Watson var. *apertus* (A.A. Heller) C.P. Smith = *L. apertus* A.A. Heller
var. *christinae* (A.A. Heller) Munz = *L. angustiflorus* Eastw.
L. arboreus Sims var. *eximius* (Burtt Davy) C.P. Smith indistinct
L. arbustus Lindley vars. *calcaratus* (Kellogg) D. Dunn, *montanus* (Howell) D. Dunn, & *silvicola* (A.A. Heller) D. Dunn not recognized
L. argenteus Pursh *var. tenellus* (G. Don) D. Dunn = var. *argenteus*
L. bicolor Lindley sspp. *marginatus* D. Dunn, *microphyllus* (S. Watson) D. Dunn, pipersmithii (A.A. Heller) D. Dunn, *tridentatus* (C.P. Smith) D. Dunn, & *umbellatus* (E. Greene) D. Dunn not recognized
vars. *rostratus* (Eastw.) Jepson & *trifidus* (S. Watson) C.P. Smith not recognized
L. caespitosus Nutt. = *L. lepidus* Douglas var. *utahensis* (S. Watson) C. Hitchc.
L. caudatus Kellogg = *L. argenteus* Pursh var. *heteranthus* (S. Watson) Barneby

L. concinnus J. Agardh vars. *agardhianus* (A.A. Heller) C.P. Smith, *brevior* (Jepson) D. Dunn, *desertorum* (A.A. Heller) C.P. Smith, *optatus* C.P. Smith, *orcuttii* (S. Watson) C.P. Smith, & *pallidus* (Brandegee) C.P. Smith ± indistinct, not recognized
L. confertus Kellogg = *L. lepidus* Douglas var. *confertus* (Kellogg) C.P. Smith
L. congdonii (C.P. Smith) D. Dunn = *L. bicolor* Lindley
L. constancei T.W. Nelson & J.P. Nelson recently described
L. croceus Eastw. var. *pilosellus* (Eastw.) Munz indistinct
L. culbertsonii E. Greene = *L. lepidus* Douglas var. *culbertsonii* (E. Greene) C.P. Smith
L. deflexus Congdon = *L. citrinus* Kellogg var. *deflexus* (Congdon) Jepson
L. densiflorus Benth. vars. *aureus* (Kellogg) Munz, *densiflorus*, & *lacteus* (Kellogg) C.P. Smith = *L. microcarpus* Sims var. *densiflorus* (Benth.) Jepson
var. *glareosus* (Elmer) C.P. Smith = *L. microcarpus* Sims var. *horizontalis* (A.A. Heller) Jepson
vars. *austrocollium* C.P. Smith & *palustris* (Kellogg) C.P. Smith = *L. microcarpus* Sims var. *microcarpus*
L. elmeri E. Greene separate from *L. albicaulis* Hook.
L. formosus E. Greene var. *hyacinthinus* (E. Greene) C.P. Smith = *L. hyacinthinus* E. Greene
L. horizontalis A.A. Heller incl var. *platypetalus* C.P. Smith) = *L. microcarpus* Sims var. *horizontalis* (A.A. Heller) Jepson
L. hypolasius E. Greene = *L. lepidus* Douglas var. *ramosus* Jepson
L. inyoensis A.A. Heller = *L. argenteus* Pursh var. *heteranthus* (S. Watson) Barneby
L. leucophyllus Lindley var. *canescens* (Howell) C.P. Smith not recognized
L. lobbii E. Greene = *L. lepidus* Douglas var. *lobbii* (E. Greene) C. Hitchc.
L. lyallii A. Gray incl var. *danaus* (A. Gray) S. Watson = *L. lepidus* Douglas var. *lobbii* (E. Greene) C. Hitchc.
L. magnificus M.E. Jones vars. *glarecola* M.E. Jones & *hesperius* (A.A. Heller) C.P. Smith indistinct
L. meionanthus A. Gray = *L. argenteus* Pursh var. *meionanthus* (A. Gray) Barneby
L. micranthus Guss. misapplied to *L. bicolor* Lindley
L. milo-bakeri C.P. Smith indistinct from *L. luteolus* Kellogg
L. montigenus A.A. Heller = *L. argenteus* Pursh var. *montigenus* (A.A. Heller) Barneby
L. nanus Benth. sspp. *latifolius* (Benth.) D. Dunn & *menkerae* (C.P. Smith) D. Dunn not recognized
L. odoratus A.A. Heller var. *pilosellus* C.P. Smith indistinct
L. padre-crowleyi C.P. Smith separate from *L. inyoensis* A.A. Heller
L. palmeri S. Watson = *L. argenteus* Pursh var. *palmeri* (S. Watson) Barneby
L. polycarpus E. Greene = *L. bicolor* Lindley
L. polyphyllus Lindley sspp. *bernardinus* (Abrams) Munz & *superbus* (A.A. Heller) Munz = var. *burkei* (S. Watson) C. Hitchc.
L. ruber A.A. Heller = *L. microcarpus* Sims var. *microcarpus*
L. sellulus Kellogg incl var. *artulus* (Jepson) Eastw. & ssp. *ursinus* (Eastw.) Munz = *L. lepidus* Douglas var. *sellulus* (Kellogg) Barneby
L. sparsiflorus Benth. sspp. *inopinatus* (C.P. Smith) Dziekanowski & D. Dunn & *mohavensis* Dziekanowski & D. Dunn indistinct
L. sublanatus Eastw. = *L. argenteus* Pursh var. *argenteus*
L. subvexus C.P. Smith = *L. microcarpus* Sims var. *microcarpus*
L. tidestromii E. Greene var. *layneae* (Eastw.) Munz indistinct
L. uncialis S. Watson not documented in CA
L. vallicola A.A. Heller incl ssp. *apricus* (E. Greene) D. Dunn = *L. nanus* Benth.
Marina orcuttii (S. Watson) Barneby var. *orcuttii* occurs in CA
Medicago hispida Gaertner incl var. *confinis* (Koch) Burnat = *M. polymorpha* L.
M. lupulina L. var. *cupaniana* (Guss.) Boiss. not recognized
Onobrychis viciifolia Scopoli naturalized
Ornithopus roseus Dufour = *O. sativus* Brot.
Oxytropis viscida Nutt. = *O. borealis* DC. var. *viscida* (Nutt.) S. Welsh
Peteria thompsonae S. Watson occurs in CA
Phaseolus wrightii A. Gray = *P. filiformis* Benth. occurs in CA
Pisum sativum L. var. *arvense* (L.) Poiret not recognized
Psoralea bituminosa L. apparently not naturalized
P. californica S. Watson = *Pediomelum californicum* (S. Watson) Rydb.
P. castorea S. Watson = *Pediomelum castoreum* (S. Watson) Rydb.
P. lanceolata Pursh ssp. *scabra* (Nutt.) Piper = *Psoralidium lanceolatum* (Pursh) Rydb.
P. macrostachya DC. = *Hoita macrostachya* (DC.) Rydb.
P. orbicularis Lindley = *Hoita orbicularis* (Lindley) Rydb.
P. physodes Hook. = *Rupertia physodes* (Hook.) Grimes
P. rigida Parish = *Rupertia rigida* (Parish) Grimes
P. strobilina Hook. & Arn. = *Hoita strobilina* (Hook. & Arn.) Rydb.
Psorothamnus fremontii var. *attenuatus* Barneby recently described
Robinia neomexicana A. Gray occurs in CA
Rupertia hallii (Rydb.) Grimes separate from *Psoralea macrostachya* DC.
Senna didymobotrya (Fresen.) H. Irwin & Barneby naturalized
Thermopsis argentata E. Greene = *T. macrophylla* Hook. & Arn. var. *argentata* (E. Greene) Jepson
T. gracilis Howell = *T. macrophylla* Hook. & Arn. var. *venosa* (Eastw.) Isely
T. macrophylla Hook. & Arn. var. *agnina* J. Howell indistinct
Trifolium agrarium L. misappled to *T. aureum* Pollich
T. amplectens Torrey & A. Gray var. *amplectens* = *T. depauperatum* Desv. var. *amplectens* (Torrey & A. Gray) L.F. McDermott
var. *hydrophilum* (E. Greene) Jepson = *T. depauperatum* Desv. var. *hydrophilum* (E. Greene) Isely
var. *truncatum* (E. Greene) Jepson = *T. depauperatum* Desv. var. *truncatum* (E. Greene) Isely
T. appendiculatum Lojacono = phase of *T. variegatum* Nutt.
T. buckwestiorum Isely recently described
T. dedeckerae J.M. Gillett = *T. macilentum* E. Greene var. *dedeckerae* (J.M. Gillett) Barneby
T. dichotomum Hook. & Arn. = *T. albopurpureum* Torrey & A. Gray var. *dichotomum* (Hook. & Arn.) Isely
T. fucatum Lindley vars. *gambelii* (Nutt.) Jepson & *virescens* (E. Greene) Jepson not recognized
T. glomeratum L. naturalized
T. gracilentum Torrey & A. Gray var. *inconspicuum* Fern. not recognized
T. grayi Lojacono = *T. barbigerum* Torrey var. *andrewsii* A. Gray
T. longipes Nutt. vars. *elmeri* (E. Greene) L.F. McDermott, *nevadense* Jepson, var. *oreganum* (Howell) Isely, & *shastense* (House) Jepson recognized
var. *atrorubens* (E. Greene) Jepson indistinct from var. *nevadense* Jepson
T. microdon Hook. & Arn. var. *pilosum* Eastw. not recognized
T. monanthum A. Gray var. *eastwoodianum* J.S. Martin = var. *monanthum*
var. *parvum* (Kellogg) L.F. McDermott = var. *monanthum*
T. monoense E. Greene = *T. andersonii* A. Gray var. *beatleyae* (J.M. Gillett) Isely
T. olivaceum E. Greene = *T. albopurpureum* Torrey & A. Gray var. *olivaceum* (E. Greene) Isely
T. palmeri S. Watson = *T. gracilentum* Torrey & A. Gray var. *palmeri* (S. Watson) L.F. McDermott
T. polyodon E. Greene = phase of *T. variegatum* Nutt.
T. procumbens L. misapplied to *T. campestre* Shreber

T. productum E. Greene = *T. kingii* S. Watson var. *productum* (E. Greene) Jepson
T. tridentatum Lindley incl var. *aciculare* (Nutt.) L.F. McDermott = *T. willdenovii* Sprengel
T. variegatum Nutt.: 5 phases diagnosed without taxonomic rank
Vicia americana Willd.: CA pls are var. *americana*
V. angustifolia Reichard = *V. sativa* L. ssp. *nigra* (L.) Erhart
V. californica E. Greene = *V. americana* Willd. var. *americana*
V. dasycarpa Ten. = *V. villosa* Roth ssp. *varia* (Host) Corbiére
V. exigua Nutt. var. *exigua* = *V. ludoviciana* Nutt. var. *ludoviciana*
var. *hassei* (S. Watson) Jepson = *V. hassei* S. Watson
V. lathyroides L. naturalized

FAGACEAE OAK FAMILY

Quercus ×*acutidens* Torrey recognized, = *Q. cornelius-mulleri* × *Q. engelmannii*
Q. ×*alvordiana* Eastw. recognized, = *Q. douglasii* × *Q. john-tuckeri*
Q. berberidifolia Liebm. separate from *Q. dumosa* Nutt.
Q. chrysolepis Liebm. var. *nana* (Jepson) Jepson indistinct
Q. cornelius-mulleri K. Nixon & K. Steele recently described
Q. dumosa Nutt. misapplied in large part to *Q. berberidifolia* Liebm.
Q. dunnii Kellogg = *Q. palmeri* Engelm.
Q. durata Jepson var. *gabrielensis* K. Nixon & C.H. Muller recently described
Q. garryana Hook. var. *semota* Jepson = var. *breweri* (Engelm.) Jepson
Q. ×*macdonaldii* E. Greene recognized, = *Q. berberidifolia* × *Q. lobata*
Q. ×*morehus* Kellogg not recognized, = *Q. kelloggii* × *Q. wislizenii*
Q. parvula E. Greene formerly confused with *Q. wislizenii* A.DC var. *frutescens* Engelm.
var. *shrevei* (C.H. Muller) K. Nixon recognized
Q. turbinella E. Greene ssp. *californica* J. Tucker = *Q. john- tuckeri* K. Nixon & C.H. Muller
Q. wislizenii A.DC. var. *frutescens* Engelm. misapplied in part to *Q. parvula* E. Greene

FOUQUIERIACEAE OCOTILLO FAMILY

Fouquieria spendens Engelm.: CA pls are ssp. *splendens*

FRANKENIACEAE FRANKENIA FAMILY

Frankenia grandifolia Cham. & Schldl. incl var. *campestris* A. Gray = *F. salina* (Molina) I.M. Johnston

FUMARIACEAE: see **PAPAVERACEAE**

GARRYACEAE SILK TASSEL FAMILY

Garrya flavescens S. Watson var. *pallida* (Eastw.) Ewan not recognized

GENTIANACEAE GENTIAN FAMILY

Centaurium curvistamineum (Wittr.) Abrams = *C. muehlenbergii* (Griseb.) Piper
C. floribundum (Benth.) Robinson = *C. muehlenbergii* (Griseb.) Piper
C. namophilum Rev., R. Broome, & J. Beatley var. *namophilum* not documented in CA; var. *nevadense* R. Broome = *C. exaltatum* (Griseb.) Piper
C. umbellatum Gilib. = *C. erthyraea* Raf.
C. venustum (A. Gray) Robinson var. *abramsii* Munz not recognized
Frasera albicaulis Griseb. = *Swertia albicaulis* (Griseb.) Kuntze
ssp. *nitida* (Benth.) Post = *Swertia albicaulis* var. *nitida* (Benth.) Jepson
F. albomarginata S. Watson = *Swertia albomarginata* (S. Watson) Kuntze
F. neglecta H.M. Hall = *Swertia neglecta* (H.M. Hall) Jepson
F. parryi Torrey = *Swertia parryi* (Torrey) Kuntze
F. puberulenta Davidson = *Swertia puberulenta* (Davidson) Jepson
F. speciosa Griseb. = *Swertia radiata* (Kellogg) Kuntze
F. tubulosa Cov. = *Swertia tubulosa* (Cov.) H. St. John
F. umpquaensis M.E. Peck & Applegate = *Swertia fastigiata* Pursh
Gentiana affinis Griseb.: CA pls are var. *ovata* A. Gray (var. *parvidentata* Kusn. indistinct)
G. amarella L. = *Gentianella amarella* (L.) Boerner ssp. *acuta* (Michaux) J.M. Gillett
G. aquatica L. = *G. fremontii* Torrey
G. bisetaea Howell = *G. setigera* A. Gray
G. holopetala (A. Gray) Holm = *Gentianopsis holopetala* (A. Gray) Iltis
G. newberryi A. Gray var. *tiogana* (A.A. Heller) J. Pringle recognized
G. oregana A. Gray: CA records are *G. affinis* Griseb. var. *ovata* A. Gray
G. plurisetosa C.T. Mason recently described
G. simplex A. Gray = *Gentianopsis simplex* (A. Gray) Iltis
G. tenella Rottb. = *Gentianella tenella* (Rottb.) Boerner ssp. *tenella*
Menyanthes transferred to *Menyanthaceae*
Microcala quadrangularis (Lam.) Griseb. = *Cicendia quadrangularis* (Lam.) Griseb.

GERANIACEAE GERANIUM FAMILY

Erodium obtusiplicatum (Maire, Weiller, & Wilczek) J. Howell = *E. brachycarpum* (Godron) Thell.
Geranium anemonifolium L'Hér. naturalized
G. attenuilobum G. Jones & F. Jones = *G. viscosissimum* Fischer & C. Meyer
G. bicknellii Britton var. *longipes* (S. Watson) Fern. not recognized
G. concinnum G. Jones & F. Jones indistinct from *G. californicum* G. Jones & F. Jones
G. microphyllum Hook. f. = *G. potentilloides* DC.
G. nervosum Rydb. = *G. viscosissimum* Fischer & C. Meyer
G. pilosum Forster f. = *G. potentilloides* DC.
G. pyrenaicum Burm. f. not naturalized
Pelargonium capitatum (L.) Aiton not naturalized
P. graveolens L'Hér. misapplied to *P. quercifolium*
P. inodorum Willd. not naturalized
P. panduriforme Ecklon & Zeyher naturalized
P. quercifolium (L.) L'Hér. naturalized

GROSSULARIACEAE GOOSEBERRY FAMILY (segregated from SAXIFRAGACEAE)

Ribes amarum McClatchie var. *hoffmannii* Munz indistinct
R. divaricatum Douglas var. *divaricatum* not in CA
var. *inerme* (Rydb.) McMinn = *R. inerme* Rydb. var. *inerme*
var. *klamathense* (Cov. & Britton) McMinn = *R. inerme* Rydb. var. *klamathense* (Cov. & Britton) Jepson
var. *pubiflorum* Koehne occurs in CA
R. inebrians Lindley = *R. cereum* Douglas var. *inebrians* (Lindley) C. Hitchc.
R. malvaceum Sm. var. *clementinum* Dunkle = var. *malvaceum*
R. menziesii Pursh vars. *hystrix* (Eastw.) Jepson, *ixoderme* Quick, *leptosmum* (Cov.) Jepson, & *senile* (Cov.) Jepson indistinct
R. petiolare Douglas = *R. hudsonianum* A. Richards var. *petiolare* (Douglas) Jancz.

R. velutinum E. Greene var. *glanduliferum* (A.A. Heller) Jepson not recognized
R. viscosissimum Pursh var. *hallii* Jancz. indistinct

GUNNERACEAE GUNNERA FAMILY (segregated from HALORAGACEAE)

Gunnera chilensis Lam. = *G. tinctoria* (Molina) Mirbel

HALORAGACEAE WATER-MILFOIL FAMILY: for *Gunnera*, see **GUNNERACEAE**; for *Hippuris*, see **HIPPURIDACEAE**

Myriophyllum brasiliense Cambess. = *M. aquaticum* (Vell. Conc.) Verdc.
M. spicatum L. var. *exalbescens* (Fern.) Hultén misapplied to *M. sibiricum* V. Komarov

HIPPOCASTANACEAE BUCKEYE FAMILY: no changes

HIPPURIDACEAE MARE'S-TAIL FAMILY (segregated from HALORAGACEAE)

Hippuris: no changes

HYDRANGEACEAE: see **PHILADELPHACEAE**

HYDROPHYLLACEAE WATERLEAF FAMILY

Lemmonia californica A. Gray = *Nama californicum* (A. Gray) J. Bacon
Nama aretioides (Hook. & Arn.) Brand var. *aretioides* not in CA; var. *californicum* (Brand) Jepson recognized
N. demissum A. Gray var. *deserti* Brand = var. *demissum*
N. dichotomum (Ruiz Lόpez & Pavόn) Choisy var. *dichotomum* occurs in CA
N. hispidum A. Gray var. *revolutum* Jepson = var. *spathulatum* (Torrey) C. Hitchc.
Phacelia affinis A. Gray var. *patens* J. Howell indistinct
P. bicolor S. Watson: CA pls are var. *bicolor*
P. ciliata Benth. var. *opaca* J. Howell indistinct
P. cookei Constance & Heckard recently described
P. crenulata Torrey var. *funerea* J. Voss = var. *crenulata*
P. curvipes S. Watson var. *macrantha* (Parish) Munz = *P. davidsonii* A. Gray
P. divaricata (Benth.) A. Gray var. *congdonii* (E. Greene) Munz = *P. congdonii* E. Greene
var. *continentis* (J. Howell) Munz = *P. insularis* Munz var. *continentis* J. Howell
var. *insularis* (Munz) Munz = *P. insularis* Munz var. *insularis*
P. douglasii (Benth.) Torrey var. *cryptantha* Brand = *P. stellaris* Brand
var. *petrophila* Jepson indistinct
P. eisenii Brandegee var. *brandegeana* J. Howell indistinct
P. frigida E. Greene sspp. *dasyphylla* (J.F. Macbr.) Heckard & *frigida* = *P. hastata* Lehm. ssp. *compacta* (Brand) Heckard
P. heterophylla Pursh: CA pls are ssp. *virgata* (E. Greene) Heckard
P. imbricata E. Greene ssp. *bernardina* (E. Greene) Heckard = ssp. *imbricata*
P. ivesiana Torrey var. *pediculoides* J. Howell indistinct
P. ixodes Kellogg occurs in CA
P. minutiflora J. Voss = *P. crenulata* Torrey var. *minutiflora* (J. Voss) Jepson
P. mohavensis A. Gray var. *exilis* A. Gray = *P. exilis* (A. Gray) G.J. Lee
P. monoensis R. Halse recently described
P. oreopola Heckard ssp. *oreopola* indistinct from *P. imbricata* E. Greene ssp. *patula* (Brand) Heckard
ssp. *simulans* Heckard = *P. hastata* Lehm. ssp. *hastata*
P. ramosissima Lehm. var. *subglabra* M.E. Peck recognized
var. *suffrutescens* Parry = var. *latifolia* (Torrey) Cronq.
var. *valida* M.E. Peck = var. *eremophila* (E. Greene) J.F. Macbr.
P. sericea (Graham) A. Gray: CA pls are var. *ciliosa* Rydb.
P. stebbinsii Constance & Heckard recently described
P. umbrosa E. Greene separate from *P. distans* Benth.

HYPERICACEAE ST. JOHN'S WORT FAMILY

Hypericum canariense L. naturalized

JUGLANDACEAE WALNUT FAMILY

Juglans hindsii (Jepson) R.E. Smith = *J. californica* S. Watson var. *hindsii* Jepson
J. regia L. naturalized or long-persisting

KOEBERLINIACEAE JUNCO FAMILY

Koeberlinia spinosa Zucc.: CA pls are ssp. *tenuispina* (Kearney & Peebles) E. Murray

KRAMERIACEAE RHATANY FAMILY

Krameria parvifolia Benth. incl vars. *imparata* J.F. Macbr. & *glandulosa* (Rose & Painter) J.F. Macbr.) = *K. erecta* Schultes

LABIATAE see **LAMIACEAE**

LAMIACEAE [LABIATAE] MINT FAMILY

Acanthomintha obovata Jepson ssp. *cordata* J. Jokerst recently described
ssp. *duttonii* Abrams = *A. duttonii* (Abrams) J. Jokerst
Calamintha chandleri Brandegee = *Satureja chandleri* (Brandegee) Druce
Hedeoma drummondii Benth. occurs in CA
H. nana (Torrey) Briq. var. *californica* W.S. Stewart: spelling corrected to *H. nanum* var. *californicum*
Lycopus lucidus Benth. misapplied to *L. asper* E. Greene
Mentha arvensis L. var. *villosa* (Benth.) S.R. Stewart not recognized
M. citrata Ehrh. = *M. ×piperita* L.
M. rotundifolia (L.) Hudson misapplied to *M. suaveolens* Ehrh.
Molucella: spelling corrected to *Moluccella*
Monardella beneolens J. Shevock, B. Ertter, & J. Jokerst recently described
M. benitensis Hardham = *M. antonina* Hardham ssp. *benitensis* (Hardham) J. Jokerst
M. odoratissima Benth. ssp. *australis* (Abrams) Epling = *M. australis* Abrams
ssp. *glauca* (E. Greene) Epling = *M. glauca* E. Greene, in part; = *M. follettii* (Jepson) J. Jokerst, in part
ssp. *parvifolia* (E. Greene) Epling indistinct from *M. glauca* E. Greene
ssp. *×pinetorum* (A.A. Heller) Epling ssp. *pallida* (A.A. Heller) Epling × *M. villosa* Benth.
M. siskiyouensis Hardham recently described
M. stebbinsii Hardham & J. Bartel recently described
M. undulata Benth. var. *frutescens* Hoover = *M. frutescens* (Hoover) J. Jokerst
M. villosa Benth. var. *franciscana* (Elmer) Epling = ssp. *franciscana* (Elmer) J. Jokerst
ssp. *globosa* (Jepson) J. Jokerst separate from ssp. *villosa*
ssp. *neglecta* (E. Greene) Epling = *M. purpurea* Howell
var. *obispoensis* Hoover = ssp. *obispoensis* (Hoover) J. Jokerst
ssp. *sheltonii* (Torrey) Epling = *M. sheltonii* Torrey
var. *×subglabra* Hoover = *M. purpurea* Howell × *M. villosa* ssp. *villosa*
ssp. *subserrata* (E. Greene) Epling = ssp. *villosa*

Pogogyne clareana J. Howell recently described
P. douglasii Benth. ssp. *parviflora* (Benth.) J. Howell ± indistinct
P. serpylloides (Torrey) A. Gray ssp. *intermedia* J. Howell not recognized
Salvia ×*bernardina* E. Greene not recognized
S. columbariae Benth. var. *ziegleri* Munz not recognized
S. dorrii (Kellogg) Abrams ssp. *argentea* (Rydb.) Munz = var. *dorrii*
ssp. *carnosa* (E. Greene) Abrams & var. *carnosa* (E. Greene) Cronq. = var. *incana* (A. Gray) J.L. Strachan
ssp. *gilmanii* (Epling) Abrams = var. *dorrii*
var. *pilosa* (A. Gray) J.L. Strachan & Rev. recognized
S. grahamii Benth. not naturalized
S. longistyla Benth. not naturalized
S. pratensis L. not naturalized
Scutellaria austiniae Eastw. = *S. siphocampyloides* Vatke
S. holmgreniorum Cronq. = *S. nana* A. Gray
S. tuberosa Benth. ssp. *australis* Epling not recognized
Stachys acuminata E. Greene separate from *S. bullata* Benth.
S. emersonii Piper = *S. ajugoides* Benth. var. *rigida* Jepson & Hoover
S. mexicana Benth. = *S. ajugoides* Benth. var. *rigida* Jepson & Hoover
S. rigida Benth. sspp. *lanata* Epling, *quercetorum* (A.A. Heller) Epling, & *rivularis* (A.A. Heller) Epling not recognized

LAURACEAE LAUREL FAMILY: no changes

LEGUMINOSAE: see FABACEAE

LENNOACEAE LENNOA FAMILY

Ammobroma sonorae A. Gray = *Pholisma sonorae* (A. Gray) Yatskievych

LENTIBULARIACEAE BLADDERWORT FAMILY

Pinguicula macroceras Link = *P. vulgaris* L. ssp. *macroceras* (Link) Calder & Roy Taylor
Utricularia fibrosa Walter = *U. gibba* L.

LIMNANTHACEAE MEADOWFOAM FAMILY

Limnanthes floccosa Howell ssp. *bellingeriana* (M.E. Peck) Arroyo not documented for CA
L. vinculans Ornd. recently described

LINACEAE FLAX FAMILY

Hesperolinon serpentinum N. McCarten recently described
Linum lewisii Pursh vars. *alpicola* Jepson & *lewisii* recognized

LOASACEAE LOASA FAMILY

Mentzelia congesta Torrey & A. Gray var. *davidsoniana* (Abrams) J.F. Macbr. not recognized
M. desertorum (Davidson) H.J. Thompson & J. Roberts occurs in CA
M. eremophila (Jepson) H.J. Thompson & J. Roberts separate from *M. nitens* E. Greene
M. hirsutissima S. Watson var. *stenophylla* (Urb. & Gilg) I.M. Johnston not recognized
M. involucrata S. Watson var. *megalantha* I.M. Johnston indistinct
M. inyoensis B. Prigge recently described
M. leucophylla Brandegee misapplied to *M. oreophila* Darl.
M. montana (Davidson) Davidson separate from *M. congesta* Torrey & A. Gray and *M. veatchiana* Kellogg
M. multiflora (Nutt.) A. Gray: CA pls are ssp. *longiloba* (Darl.) Felger
M. nitens E. Greene var. *jonesii* (Urb. & Gilg) Darl. = *M. jonesii* (Urb. & Gilg) H.J. Thompson & J. Roberts
M. obscura H.J. Thompson & J. Roberts recently described
M. polita Nelson occurs in CA
M. pterosperma Eastw. occurs in CA
M. puberula Darl. indistinct from *M. oreophila* Darl.
M. tricuspis A. Gray var. *brevicornuta* I.M. Johnston = *M. tridentata* (Davidson) H.J. Thompson & J. Roberts
Sympetaleia rupestris (Baillon) A. Gray = *Eucnide rupestris* (Baillon) H.J. Thompson & W.R. Ernst

LOGANIACEAE: see BUDDLEJACEAE

LORANTHACEAE: see VISCACEAE

LYTHRACEAE LOOSESTRIFE FAMILY

Ammannia auriculata Willd. misapplied to *A. coccinea* Rottb.
Peplis portula L. = *Lythrum portula* (L.) D. Webb

MALVACEAE Mallow Family

Abutilon crispum (L.) Sweet = *Herrisantia crispa* (L.) Briz.
Althaea rosea (L.) Cav. = *Alcea rosea* L.
Eremalche kernensis C. Wolf = *E. parryi* (E. Greene) E. Greene ssp. *kernensis* (C. Wolf) D. Bates
Hibiscus californicus Kellogg = *H. lasiocarpus* Cav.
Lavatera assurgentiflora Kellogg ssp. *glabra* Philbr. indistinct
Malacothamnus arcuatus (E. Greene) E. Greene = *M. fasciculatus* (Torrey & A. Gray) E. Greene
M. densiflorus (S. Watson) E. Greene var. *viscidus* (Abrams) Kearney indistinct
M. fasciculatus (Torrey & A. Gray) E. Greene vars. *catalinensis* (Eastw.) Kearney, *laxiflorus* (A. Gray) Kearney, *nesioticus* (Robinson) Kearney, & *nuttallii* (Abrams) Kearney not recognized
M. fremontii A. Gray ssp. *cercophorus* (Robinson) Munz not recognized
M. gracilis (Eastw.) Kearney = *M. jonesii* (Munz) Kearney
M. hallii (Eastw.) Kearney = *M. fasciculatus* (Torrey & A. Gray) E. Greene
M. helleri (Eastw.) Kearney = *M. fremontii* A. Gray
M. howellii (Eastw.) Kearney = *M. fremontii* A. Gray
M. mendocinensis (Eastw.) Kearney = *M. fasciculatus* (Torrey & A. Gray) E. Greene
M. niveus (Eastw.) Kearney = *M. jonesii* (Munz) Kearney
M. orbiculatus (E. Greene) E. Greene = *M. fremontii* A. Gray
M. palmeri (S. Watson) E. Greene vars. *involucratus* (Robinson) Kearney & *lucianus* Kearney not recognized
M. parishii (Eastw.) Kearney = *M. fasciculatus* (Torrey & A. Gray) E. Greene
Malva rotundifolia L. misapplied to *M. neglecta* Wallr.
M. sylvestris ssp. *mauritiana* (L.) Boiss. not recognized
M. verticillata L. var. *crispa* L. misapplied to *M. neglecta* Wallr.
Malvastrum exile A. Gray = *Eremalche exilis* (A. Gray) E. Greene
M. kernense (C. Wolf) Munz = *Eremalche parryi* (E. Greene) E. Greene ssp. *kernensis* (C. Wolf) D. Bates
M. parryi E. Greene = *Eremalche parryi* (E. Greene) E. Greene ssp. *parryi*
M. rotundifolium A. Gray = *Eremalche rotundifolia* (A. Gray) E. Greene
Sida leprosa (Ortega) K. Schumann var. *hederacea* (Hook.) K. Schumann = *Malvella leprosa* (Ortega) Krapov
Sidalcea malvaeflora (DC.) Benth. sspp. *celata* (Jepson) C. Hitchc., *elegans* (E. Greene) C. Hitchc., & *nana* (Jepson) C. Hitchc. = ssp. *asprella* (E. Greene) C. Hitchc.
S. neomexicana A. Gray ssp. *thurberi* (A. Gray) C. Hitchc. not recognized

S. setosa C. Hitchc. = *S. oregana* (Torrey & A. Gray) A. Gray ssp. *oregana* (in part) & ssp. *spicata* (Regel) C. Hitchc. (in part, incl ssp. setosa)
S. stipularis J. Howell & True recently described
Sphaeralcea ambigua A. Gray var. *monticola* Kearney = var. *ambigua*
S. angustifolia (Cav.) G. Don var. *cuspidata* A. Gray not recognized
S. emoryi Torrey var. *arida* (Rose) Kearney = var. *emoryi* var. *nevadensis* Kearney = *S. angustifolia* (Cav.) G. Don var. *variabilis* (Cockerell) Kearney = var. *emoryi*
S. parvifolia Nelson misapplied to CA pls of *S. ambigua* A. Gray var. *a.*

MARTYNIACEAE UNICORN-PLANT FAMILY

Ibicella lutea (Lindley) Eselt. = *Proboscidea lutea* (Lindley) Stapf
Proboscidea altheaefolia (Benth.) Decne.: spelling corrected to *P. althaeifolia*

MELIACEAE MAHOGANY FAMILY: no changes

MENYANTHACEAE BUCKBEAN FAMILY (segregated from GENTIANACEAE)

Menyanthes: no changes
Nymphoides peltata (S. Gmelin) Kuntze occurs in CA

MOLLUGINACEAE CARPET-WEED FAMILY (segregated from AIZOACEAE)

Glinus, Mollugo: no changes

MORACEAE MULBERRY FAMILY (for *Cannabis, Humulus* see **CANNABACEAE**)

Ficus carica L. occurs as waif in CA
Maclura pomifera (Raf.) C. Schneider naturalized

MYOPORACEAE MYOPORUM FAMILY

Myoporum laetum Forster f. naturalized

MYRICACEAE WAX MYRTLE FAMILY: no changes

MYRTACEAE MYRTLE FAMILY

Eucalyptus citriodora Hook. naturalized
E. cladocalyx F. Muell. naturalized
E. pulverulenta Sims naturalized
E. sideroxylon Cunn. naturalized
E. viminalis Labill. naturalized
Eugenia apiculata DC. = *Luma apiculata* (DC.) Burret
Leptospermum laevigatum (Gaertner) F. Muell. naturalized

NYCTAGINACEAE FOUR O'CLOCK FAMILY

Abronia crux-maltae Kellogg = *Tripterocalyx crux-maltae* (Kellogg) Standley
A. micrantha Torrey = *Tripterocalyx micranthus* (Torrey) Hook.
Boerhavia annulata Cov. = *Anulocaulis annulatus* (Cov.) Standley
B. erecta L. var. *intermedia* (M.E. Jones) Kearney & Peebles = *B. intermedia* M.E. Jones
Hermidium alipes S. Watson = *Mirabilis alipes* (S. Watson) Pilz
Mirabilis froebelii (Behr) E. Greene incl var. *glabrata* (Standley) Jepson = *M. multiflora* (Torrey) A. Gray var. *pubescens* S. Watson
M. multiflora (Torrey) A. Gray var. *glandulosa* (Standley) J.F. Macbr. occurs in CA
Oxybaphus coccineus Torrey = *Mirabilis coccineus* (Torrey) Benth. & Hook.
O. comatus (Small) Weath. = *Mirabilis oblongifolia* (A. Gray) Heimerl
O. linearis (Pursh) Robinson = *Mirabilis linearis* (Pursh) Heimerl
O. nyctagineus (Michaux) Sweet = *Mirabilis nyctiginea* (Michaux) MacMillan
O. pumilus (Standley) Standley = *Mirabilis pumila* (Standley) Standley
Selinocarpus diffusus A. Gray not recognized
S. nevadensis (Standley) Fowler & B. Turner occurs in CA

NYMPHAEACEAE WATERLILY FAMILY (for *Brasenia* see **CABOMBACEAE**)

Nuphar polysepalum Engelm. = *N. luteum* (L.) Sibth. & Sm. ssp. *polysepalum* (Engelm.) E. Beal
Nymphaea mexicana Zucc. naturalized
N. odorata Aiton naturalized

OLEACEAE OLIVE FAMILY

Forestiera neomexicana A. Gray = *F. pubescens* Nutt.
Fraxinus velutina Torrey var. *coriacea* (S. Watson) Rehd. not recognized
Olea europaea L. naturalized

ONAGRACEAE EVENING-PRIMROSE FAMILY

Boisduvalia cleistogama Curran = *Epilobium cleistogamum* (Curran) P. Hoch & Raven
B. densiflora (Lindley) S. Watson incl vars. *pallescens* Suksd. & *salicina* (Torrey & A. Gray) Munz = *Epilobium densiflorum* (Lindley) P. Hoch & Raven
B. glabella (Nutt.) Walp. incl var. *campestris* (Jepson) Jepson = *Epilobium pygmaeum* (Speg.) P. Hoch & Raven
B. macrantha A.A. Heller = *Epilobium pallidum* (Eastw.) P. Hoch & Raven
B. pallida Eastw. = *Epilobium pallidum* (Eastw.) P. Hoch & Raven
B. stricta (A. Gray) E. Greene = *Epilobium torreyi* (S. Watson) P. Hoch & Raven
Camissonia benitensis Raven recently described
C. campestris (E. Greene) Raven ssp. *obispoensis* Raven recently described
C. confusa Raven recently described
C. hardhamiae Raven recently described
C. integrifolia Raven recently described
C. intermedia Raven recently described
C. lacustris Raven recently described
C. lewisii Raven recently described
C. luciae Raven recently described
C. pusilla Raven recently described
C. robusta Raven recently described
C. sierrae Raven recently described
ssp. *alticola* Raven recently described
C. tanacetifolia (Torrey & A. Gray) Raven ssp. *quadriperforata* Raven recently described
Clarkia australis E. Small recently described
C. borealis E. Small recently described
ssp. *arida* E. Small recently described
C. concinna (Fischer & C. Meyer) E. Greene ssp. *automixa* Bowman recently described
ssp. *raichei* Allen, Ford, & Gottlieb
C. cylindrica (Jepson) Harlan Lewis & M. Lewis ssp. *clavicarpa* W. Davis recently described
C. gracilis (Piper) Nelson & J.F. Macbr. ssp. *tracyi* (Jepson) Abdel-Hameed & S.R. Snow recognized
C. jolonensis D. Parnell recently described
C. lewisii Raven & D. Parnell recently described, separated from *C. bottae* (Spach) Harlan Lewis & M. Lewis
C. mosquinii E. Small recently described; both sspp. *mosquinii* & *xerophylla* E. Small presumed extinct
C. nitens Harlan Lewis & M. Lewis = *C. speciosa* Harlan Lewis & M. Lewis ssp. *nitens* (Harlan Lewis & M. Lewis) Harlan Lewis

C. rostrata W. Davis recently described
C. rubicunda (Lindley) Harlan Lewis & M. Lewis ssp. *blasdalei* (Jepson) Harlan Lewis & M. Lewis not recognized
C. tembloriensis Vasek ssp. *calientensis* (Vasek) Holsinger recently described
C. xantiana A. Gray ssp. *parviflora* (Eastw.) Harlan Lewis recognized
Epilobium adenocaulon Hausskn. incl vars. *holosericeum* (Trel.) Munz, occidentale Trel., & *parishii* (Trel.) Munz = *Epilobium ciliatum* Raf. ssp. *ciliatum*
E. angustifolium L. ssp. *angustifolium* not yet documented in CA
E. brevistylum Barbey var. *brevistylum* = *E. ciliatum* Raf. ssp. *glandulosum* (Lehm.) P. Hoch & Raven
var. *ursinum* (Trel.) Jepson = *E. halleanum* Hausskn.
E. canum (E. Greene) Raven ssp. *mexicanum* (C. Presl) Raven = ssp. *canum*
ssp. *septentrionale* (Keck) Raven = *E. septentrionale* (Keck) Raven
E. exaltatum E. Drew = *E. oreganum* E. Greene
E. glaberrimum Barbey var. *fastigiatum* (Nutt.) Trel. = ssp. *fastigiatum* (Nutt.) P. Hoch & Raven
E. glandulosum Lehm. = *E. ciliatum* Raf. ssp. *glandulosum* (Lehm.) P. Hoch & Raven
E. hornemannii Reichb.: CA pls are ssp. *hornemannii*
E. howellii P. Hoch recently described
E. leptophyllum Raf. naturalized
E. minutum Hook. = *E. minutum* Lehm.
E. obcordatum A. Gray var. *laxum* Jepson & ssp. *siskiyouense* Munz = *E. siskiyouense* (Munz) P. Hoch & Raven
E. palustre L. recently discovered to grow in CA
E. paniculatum Torrey & A. Gray incl vars. *laevicaule* (Rydb.) Munz, *jucundum* (Rydb.) Trel., & tracyi (Rydb.) Munz = *E. brachycarpum* C. Presl
E. pringleanum Hausskn. incl var. *tenue* (Trel.) Munz = *E. halleanum* Hausskn.
E. rigidum Hausskn. occurs in CA
E. saximontanum Hausskn. separate from *E. halleanum* Hausskn.
E. watsonii Barbey incl var. *franciscanum* (Barbey) Jepson = *E. ciliatum* Raf. ssp. *watsonii* (Barbey) P. Hoch & Raven
Gaura coccinea Pursh var. *glabra* (Lehm.) Torrey & A. Gray not recognized
G. odorata Lagasca misapplied to *G. drummondii* (Spach) Torrey & A. Gray
G. parviflora Douglas var. *lachnocarpa* Weath. not recognized
G. villosa Torrey incl var. *mckelveyae* Munz misapplied to *G. sinuata* Ser.
Heterogaura heterandra (Torrey) Cov. = *Clarkia heterandra* (Torrey) Harlan Lewis & Raven
Jussiaea repens L. misapplied to *Ludwigia peploides* (Kunth) Raven
var. *peploides* (Kunth) Griseb. misapplied to *Ludwigia peploides* (Kunth) Raven ssp. peploides
var. *montevidensis* (Spreng.) Munz misapplied to *Ludwigia peploides* (Kunth) Raven ssp. *montividensis* (Spreng.) Raven
J. uruguayensis Cambess. = *Ludwigia hexapetala* (Hook. & Arn.) Zardini, Gu, & Raven
Ludwigia natans Elliott var. *stipitata* Fern. & Griscom = *L. repens* Forster
L. palustris (L.) Elliott vars. *americana* (DC.) Fern. & Griscom & *pacifica* Fern. & Griscom not recognized
Oenothera abramsii J.F. Macbr. = *Camissonia pallida* (Abrams) Raven ssp. *pallida*
O. andina Nutt. = *Camissonia andina* (Nutt.) Raven
O. arenaria (Nelson) Raven = *Camissonia arenaria* (Nelson) Raven
O. avita (W. Klein) W. Klein = *O. californica* (S. Watson) S. Watson ssp. *avita* W. Klein
ssp. *eurekensis* (Munz & Roos) W. Klein = *O. californica* (S. Watson) S. Watson ssp. *eurekensis* (Munz & Roos) W. Klein
O. biennis L. naturalized
O. bistorta Torrey & A. Gray incl var. *veitchiana* Hook. = *Camissonia bistorta* (Torrey & A. Gray) Raven
O. boothii Douglas = *Camissonia boothii* (Douglas) Raven
ssp. *alyssoides* (Hook. & Arn.) Munz = *Camissonia boothii* (Douglas) Raven ssp. *alyssoides* (Hook. & Arn.) Raven
ssp. *condensata* (Munz) Munz = *Camissonia boothii* (Douglas) Raven ssp. *condensata* (Munz) Raven
ssp. *decorticans* (Hook. & Arn.) Munz = *Camissonia boothii* (Douglas) Raven ssp. *decorticans* (Hook. & Arn.) Raven
ssp. *desertorum* (Munz) Munz = *Camissonia boothii* (Douglas) Raven ssp. *desertorum* (Munz) Raven
ssp. *intermedia* Munz = *Camissonia boothii* (Douglas) Raven ssp. *intermedia* (Munz) Raven
ssp. *inyoensis* Munz = *Camissonia boothii* (Douglas) Raven ssp. *desertorum* (Munz) Raven
ssp. *rutila* (Davidson) Munz = *Camissonia boothii* (Douglas) Raven ssp. *decorticans* (Hook. & Arn.) Raven
O. breviflora Torrey & A. Gray = misapplied in CA to *Camissonia tanacetifolia* (Torrey & A. Gray) Raven ssp. *tanacetifolia*
O. brevipes A. Gray = *Camissonia brevipes* (A. Gray) Raven
ssp. *arizonica* Raven = *Camissonia brevipes* (A. Gray) Raven ssp. *arizonica* (Raven) Raven
ssp. *pallidula* (Munz) Raven = *Camissonia brevipes* (A. Gray) Raven ssp. *pallidula* (Munz) Raven
O. caespitosa Nutt. var. *crinita* (Rydb.) Munz = ssp. *crinita* (Rydb.) Munz
var. *marginata* (Hook. & Arn.) Munz = ssp. *marginata* (Hook. & Arn.) Munz
O. californica (S.Watson) S. Watson var. *glabrata* Munz = ssp. *californica*
O. campestris E. Greene = *Camissonia campestris* (E. Greene) Raven
ssp. *parishii* (Abrams) Munz = *Camissonia campestris* (E. Greene) Raven ssp. *campestris*
ssp. *obispoensis* Raven recently described
O. cardiophylla Torrey = *Camissonia cardiophylla* (Torrey) Raven
ssp. *robusta* Raven = *Camissonia cardiophylla* (Torrey) Raven ssp. *robusta* (Raven) Raven
var. *splendens* Munz & I.M. Johnston = *Camissonia arenaria* (Nelson) Raven
O. cavernae Munz not documented in CA
O. chamaenerioides A. Gray = *Camissonia chamaenerioides* (A. Gray) Raven
O. cheiranthifolia Sprengel = *Camissonia cheiranthifolia* (Sprengel) Raim.
var. *nitida* (E. Greene) Munz not recognized
ssp. *suffruticosa* (S. Watson) Munz = *Camissonia cheiranthifolia* (Sprengel) Raim. ssp. *suffruticosa* (S. Watson) Raven
O. claviformis Torrey & Frémont = *Camissonia claviformis* (Torrey & Frémont) Raven
ssp. *aurantiaca* (S. Watson) Raven = *Camissonia claviformis* (Torrey & Frémont) Raven ssp. *aurantiaca* (S. Watson) Raven
ssp. *cruciformis* (Kellogg) Raven = *Camissonia claviformis* (Torrey & Frémont) Raven ssp. *cruciformis* (Kellogg) Raven
ssp. *funerea* Raven = *Camissonia claviformis* (Torrey & Frémont) Raven ssp. *funerea* (Raven) Raven
ssp. *integrior* Raven = *Camissonia claviformis* (Torrey & Frémont) Raven ssp. *integrior* (Raven) Raven
ssp. *lancifolia* (A.A. Heller) Raven = *Camissonia claviformis* (Torrey & Frémont) Raven ssp. *lancifolia* (A.A. Heller) Raven

ssp. *peirsonii* (Munz) Raven = *Camissonia claviformis* (Torrey & Frémont) Raven ssp. *peirsonii* (Munz) Raven
var. *purpurascens* (S. Watson) Munz = *Camissonia claviformis* (Torrey & Frémont) Raven ssp. *integrior* (Raven) Raven
ssp. *yumae* Raven = *Camissonia claviformis* (Torrey & Frémont) Raven ssp. *yumae* (Raven) Raven
O. contorta Douglas = *Camissonia contorta* (Douglas) Raven
var. *flexuosa* (Nelson) Munz = *Camissonia parvula* (Torrey & A. Gray) Raven; also misapplied to C. pusilla Raven
O. cruciata (S. Watson) Munz = *Camissonia contorta* (Douglas) Raven
O. dentata Cav. misapplied to *Camissonia lacustris* Raven & *C. strigulosa* (Fischer & C. Meyer) Raven
O. ×erythrosepala Borbás = *O. glazioviana* Micheli
O. graciliflora Hook. & Arn. = *Camissonia graciliflora* (Hook. & Arn.) Raven
O. guadalupensis S. Watson ssp. *clementina* Raven = *Camissonia guadalupensis* (S. Watson) Raven ssp. *clementina* (Raven) Raven
O. hallii (Davidson) Munz = *Camissonia pallida* (Abrams) Raven ssp. *hallii* (Davidson) Raven
O. heterochroma S. Watson = *Camissonia heterochroma* (S. Watson) Raven ssp. *monoensis* (Munz) Raven not recognized
O. heterophylla Hook. & Arn. misapplied to *Camissonia bistorta* (Torrey & A. Gray) Raven
O. hirtella E. Greene = *Camissonia hirtella* (E. Greene) Raven
O. hookeri Torrey & A. Gray sspp. *angustifolia* (Gates) Munz, *grisea* (Bartlett) Munz, & *venusta* (Bartlett) Munz = O. *elata* Kunth ssp. *hirsutissima* (S. Watson) W. Dietr.
ssp. *hookeri* = *O. elata* ssp. *hookeri* (Torrey & A. Gray) W. Dietr. & W.L. Wagner
ssp. *montereyensis* Munz = *O. elata* Kunth ssp. *hookeri* (Torrey & A. Gray) W. Dietr. & W.L. Wagner
ssp. *wolfii* Munz = *O. wolfii* (Munz) Raven, W. Dietr., & Stubbe
O. ignota (Jepson) Munz = *Camissonia ignota* (Jepson) Raven
O. kernensis Munz = *Camissonia kernensis* (Munz) Raven
ssp. *gilmanii* (Munz) Munz = *Camissonia kernensis* (Munz) Raven
ssp. *gilmanii* (Munz) Raven
ssp. *mojavensis* Munz = *Camissonia kernensis* (Munz) Raven ssp. *gilmanii* (Munz) Raven
O. laciniata Hill ssp. *pubescens* (Sprengel) Munz = *O. pubescens* Sprengel
O. lamarckiana Ser. misapplied to *O. glazioviana* Micheli
O. leptocarpa E. Greene = *Camissonia californica* (Torrey & A. Gray) Raven
O. longissima Rydb. ssp. *clutei* (Nelson) Munz not recognized
O. micrantha Sprengel = *Camissonia micrantha* (Sprengel) Raven; also misapplied to *C. lewisii* Raven, *C. luciae* Raven, & *C. robusta* Raven
O. minor (Nelson) Munz = *Camissonia minor* (Nelson) Raven
O. multijuga S. Watson var. *parviflora* Munz misapplied to *Camissonia walkeri* (Nelson) Raven ssp. *tortilis* (Jepson) Raven
O. munzii Raven = *Camissonia munzii* (Raven) Raven
O. ovata Torrey & A. Gray = *Camissonia ovata* (Torrey & A. Gray) Raven
O. palmeri S. Watson = *Camissonia palmeri* (S. Watson) Raven
O. primiveris A. Gray ssp. *caulescens* (Munz) Munz = ssp. *primiveris*
O. pterosperma S. Watson = *Camissonia pterosperma* (S. Watson) Raven
O. pubens (S. Watson) Munz = *Camissonia pubens* (S. Watson) Raven
O. refracta S. Watson = *Camissonia refracta* (S. Watson) Raven
O. scapoidea Torrey & A. Gray var. *tortilis* Jepson = *Camissonia walkeri* (Nelson) Raven ssp. *t.* (Jepson) Raven
O. speciosa Nutt. var. *childsii* (Bailey) Munz not recognized
O. strigosa (Rydb.) Mackenzie & Bush = *O. villosa* Thunb. ssp. *strigosa* (Rydb.) W. Dietr. & Raven
O. subacaulis (Pursh) Garrett = *Camissonia subacaulis* (Pursh) Raven
O. tanacetifolia Torrey & A. Gray = *Camissonia tanacetifolia* (Torrey & A. Gray) Raven
O. walkeri (Nelson) Raven ssp. *tortilis* (Jepson) Raven = *Camissonia walkeri* (Nelson) Raven ssp. *tortilis* (Jepson) Raven
Zauschneria californica C. Presl = *E. canum* (E. Greene) Raven
sspp. *angustifolia* Keck, *californica*, & *mexicanum* (C. Presl) Raven = *E. c.* ssp. *canum*
ssp. *latifolia* (Hook.) Keck = *E. c.* ssp. *latifolium* (Hook.) Raven
Z. cana E. Greene = *Epilobium canum* (E. Greene) Raven ssp. *canum*
Z. garrettii Nelson = *Epilobium canum* (E. Greene) Raven ssp. *garrettii* (Nelson) Raven, not documented for CA
Z. septentrionalis Keck = *E. septentrionale* (Keck) Raven

OROBANCHACEAE BROOM-RAPE FAMILY

Orobanche californica Cham. & Schldl. vars. *californica* & *claremontensis* Munz misapplied to *O. vallicola* (Jepson) Heckard
var. *corymbosa* (Rydb.) Munz = *O. corymbosa* (Rydb.) Ferris
var. *parishii* Jepson = *O. parishii* (Jepson) Heckard
O. fasciculata Nutt. vars. *franciscana* Achey & *lutea* (Parry) Achey not recognized
O. grayana G. Beck var. *feudgei* Munz = *O. californica* Cham. & Schldl. ssp. *feudgei* (Munz) Heckard
var. *grayana* = *O. californica* Cham. & Schldl. ssp. *grayana* (G. Beck) Heckard
var. *jepsonii* Munz = *O. californica* Cham. & Schldl. ssp. *jepsonii* (Munz) Heckard
vars. *nelsonii* Munz & *violacea* (Eastw.) Munz = *O. californica* Cham. & Schldl. ssp. *californica*
O. ludoviciana Nutt.: probably not in CA [var. *arenosa* (Suksd.) Cronq. hard to separate from *O. parishii* (Jepson) Heckard var. *parishii* in GB]
var. *cooperi* (A. Gray) G. Beck = *O. cooperi* (A. Gray) Heller
var. *latiloba* Munz = *O. cooperi* (A. Gray) Heller
O. multiflora Nutt. var. *arenosa* (Suksd.) Munz = *O. ludoviciana* Nutt. var. *arenosa* (Suksd.) Cronq.
O. parishii (Jepson) Heckard ssp. *brachyloba* Heckard recently described
O. uniflora L. var. *minuta* (Suksd.) D.B. Achey, var. *sedi* (Suksd.) D.B. Achey, & ssp. *occidentalis* (E. Greene) Ferris not recognized
var. *purpurea* (Heller) Achey indistinct
O. valida Jepson ssp. *howellii* Heckard & L.T. Collins recently described

OXALIDACEAE OXALIS FAMILY

Oxalis martiana Zucc. = *O. latifolia* Kunth

PAEONIACEAE PEONY FAMILY: no changes

PAPAVERACEAE POPPY FAMILY (includes FUMARIACEAE)

Arctomecon californica Torrey & Frémont not documented in CA

Argemone munita Durand & Hilg. sspp. *argentea* Ownbey, *rotundata* (Rydb.) Ownbey, *robusta* Ownbey not recognized
Corydalis caseana A. Gray: CA pls are ssp. *caseana*
Dendromecon rigida Benth. ssp. *harfordii* (Kellogg) Raven = *D. harfordii* Kellogg
ssp. *rhamnoides* (E. Greene) Thorne [= *D. harfordii* Kellogg var. *rhamnoides* (E. Greene) Munz] indistinct
Dicentra formosa (Haw.) Walp.
ssp. *oregana* (Eastw.) Munz indistinct
Eschscholzia caespitosa Benth. ssp. *kernensis* Munz = *E. lemmonii* E. Greene ssp. *kernensis* (Munz) C. Clark
E. elegans E. Greene misapplied to *E. ramosa* E. Greene
E. hypecoides Benth. separate from *E. caespitosa* Benth.
E. rhombipetala E. Greene separate from *E. caespitosa* Benth.
Hesperomecon linearis (Benth.) E. Greene = *Meconella linearis* (Benth.) Nelson & J.F. Macbr.
Hunnemannia fumariifolia Sweet not naturalized
Meconella oregana Nutt. not documented in CA
var. *denticulata* (E. Greene) Jepson = *M. denticulata* E. Greene
Papaver apulum Ten. var. *micranthum* (Boreau) Fedde not naturalized
Platystemon californicus Benth. var. *ciliatus* Dunkle indistinct
vars. *crinitus* E. Greene, *horridulus* Jepson, *nutans* Brandegee, & *ornithopus* (E. Greene) Munz not recognized

PHILADELPHACEAE MOCK ORANGE FAMILY (segregated from SAXIFRAGACEAE and HYDRANGEACEAE)

Jamesia americana Torrey & A. Gray: CA pls are var. *rosea* C. Schneider
Philadelphus lewisii Pursh ssp. *californicus* (Benth.) Munz not recognized
ssp. *gordonianus* (Lindley) Munz indistinct
P. microphyllus A. Gray ssp. *stramineus* (Rydb.) C. Hitchc. not recognized
ssp. *pumilus* (Rydb.) C. Hitchc. indistinct

PHYTOLACCACEAE POKEWEED FAMILY: no changes

PITTOSPORACEAE PITTOSPORUM FAMILY

Pittosporum crassifolium Cunn. naturalized
P. tenuifolium Gaertner naturalized
P. tobira (Thunb.) Aiton naturalized
P. undulatum Vent. naturalized
Sollya heterophylla Lindley naturalized

PLANTAGINACEAE PLANTAIN FAMILY

Plantago bigelovii A. Gray var. *bigelovii* indistinct from *P. elongata* Pursh
var. *californica* (E. Greene) I.J. Bassett = *P. elongata* Pursh
P. elongata Pursh ssp. *pentasperma* I.J. Bassett indistinct
P. erecta Morris var. *rigidior* Pilger not recognized
P. hirtella Kunth var. *galeottiana* (Decne.) Pilger = *P. subnuda* Pilger
P. hookeriana Fischer & C. Meyer var. *californica* (E. Greene) I. Poe misapplied to *P. erecta* Morris
P. insularis Eastw. incl var. *fastigiata* (Morris) Jepson = *P. ovata* Forsskal
P. major L. vars. *pilgeri* Domin & *scopulorum* Fries & S.P. Broberg indistinct
P. maritima L. vars. *californica* (Fern.) Pilger & *juncoides* (Lam.) A. Gray indistinct
P. purshii Roemer & Schultes incl vars. *picta* Pilger & *oblonga* (Morris) Shinn. = *P. patagonica* Jacq.
P. rhodosperma Decne. indistinct from *P. virginica* L.

PLATANACEAE SYCAMORE FAMILY

Platanus racemosa Nutt. var. *wrightii* (S. Watson) L. Benson indistinct

PLUMBAGINACEAE LEADWORT FAMILY

Armeria maritima (Miller) Willd. var. *californica* (Boiss.) G. Lawr. = ssp. *californica* (Boiss.) Pors.
Limonium arborescens (Brouss.) Kuntze naturalized
L. californicum (Boiss.) A.A. Heller var. *mexicanum* (S.F. Blake) Munz not recognized
L. perfoliatum (Boiss.) Kuntze misapplied to *L. otolepis* (Schrenk) Kuntze

POLEMONIACEAE PHLOX FAMILY

Allophyllum violaceum (A.A. Heller) A.D. Grant & V. Grant = *A. gilioides* (Benth.) A.D. Grant & V. Grant ssp. *violaceum* (A.A. Heller) A. Day
Collomia debilis (S. Watson) E. Greene var. *larsenii* (A. Gray) Brand = *C. larsenii* (A. Gray) Payson
Eriastrum diffusum (A. Gray) H. Mason ssp. *harwoodii* (Craig) H. Mason = *E. sparsiflorum* (Eastw.) H. Mason
E. pluriflorum (A.A. Heller) H. Mason ssp. *sherman-hoytae* (Craig) H. Mason not recognized
E. sapphirinum (Eastw.) H. Mason sspp. *ambiguum* (M.E. Jones) H. Mason, *dasyanthum* (Brand) H. Mason, *gymnocephalum* (Brand) H. Mason not recognized
E. tracyi H. Mason = *E. brandegeae* H. Mason
Gilia brecciarum M.E. Jones ssp. *argusana* A.D. Grant & V. Grant = ssp. *neglecta* A.D. Grant & V. Grant
G. jacens A.D. Grant & V. Grant = *G. brecciarum* M.E. Jones ssp. *jacens* (A.D. Grant & V. Grant) A. Day
G. latiflora (A. Gray) A. Gray ssp. *cosana* A.D. Grant & V. Grant = *G. cana* (M.E. Jones) Heller ssp. *cana*
ssp. *excellens* (Brand) A.D. Grant & V. Grant = ssp. *davyi* (Milliken) A.D. Grant & V. Grant
G. leptalea (A. Gray) E. Greene ssp. *pinnatisecta* H. Mason & A.D. Grant = *G. sinistra* M.E. Jones ssp. *pinnatisecta* (H. Mason & A.D. Grant) A. Day
G. leptantha Parish ssp. *salticola* (Eastw.) A.D. Grant & V. Grant = *G. salticola* Eastw.
ssp. *vivida* A.D. Grant & V. Grant = *G. ochroleuca* M.E. Jones ssp. *vivida* (A.D. Grant & V. Grant) A.D. Grant & V. Grant
G. lottiae A. Day recently described
G. malior A.D. Grant & V. Grant recently described
G. sinistra M.E. Jones separate from *G. capillaris* Kellogg
G. tenerrima A. Gray occurs in CA
G. triodon Eastw. separate from *G. leptomeria* A. Gray
Ipomopsis aggregata (Pursh) V. Grant ssp. *arizonica* (E. Greene) V. Grant & A.D. Grant = *I. arizonica* (E. Greene) Wherry
ssp. *attenuata* (A. Gray) V. Grant & A.D. Grant misapplied to *I. tenuituba* (Rydb.) V. Grant
ssp. *formosissima* (E. Greene) Wherry separate from ssp. *bridgesii* (A. Gray) V. Grant & A.D. Grant
I. congesta (Hook.) V. Grant ssp. *palmifrons* (Brand) A. Day separate from var. *congesta*
I. effusa (A. Gray) Moran occurs in CA
Langloisia matthewsii (A. Gray) E. Greene = *Loeseliastrum matthewsii* (A. Gray) Timbrook
L. punctata (Coville) Goodding = *L. setosissima* (Torrey & A. Gray) E. Greene ssp. *punctata* (Coville) Timbrook
L. schottii (Torrey) E. Greene = *Loeseliastrum schottii* (Torrey) Timbrook
Leptodactylon californicum Hook. & Arn. ssp. *glandulosum* (Eastw.) H. Mason not recognized
L. pungens (Torrey) Rydb. sspp. *hallii* (Parish) H. Mason, *hookeri* (Douglas) Wherry, & *pulchriflorum* (Brand) H. Mason not recognized

Linanthus androsaceus (Benth.) E. Greene sspp. *croceus* (Milliken) H. Mason, *laetus* H. Mason, *luteus* (Benth.) H. Mason, & *micranthus* (Steud.) H. Mason = *L. parviflorus* (Benth.) E. Greene
L. bakeri H. Mason = *L. bolanderi* (A. Gray) E. Greene
L. dianthiflorus (Benth.) E. Greene ssp. *farinosus* (Brand) H. Mason not recognized
L. dichotomus Benth. ssp. *meridianus* (Eastw.) H. Mason not recognized
L. floribundus (A. Gray) Milliken sspp. *floribundus*, *glaber* R. Patterson, & *hallii* (Jepson) H. Mason recognized
L. harknessii (Curran) E. Greene ssp. *condensatus* H. Mason not recognized
L. liniflorus (Benth.) E. Greene ssp. *pharnaceoides* (Benth.) H. Mason not recognized
L. maculatus (Parish) Milliken = *Gilia maculata* Parish
L. nuttallii (A. Gray) Milliken ssp. *floribundus* (A. Gray) Munz = *L. floribundus* (A. Gray) Milliken
sspp. *howellii* T.W. Nelson & R. Patterson, *nuttallii*, & *pubescens* R. Patterson recognized
L. orcuttii (Parry & A. Gray) Jepson ssp. *pacificus* (Milliken) H. Mason not recognized
L. pachyphyllus R. Patterson recently described
L. parviflorus (Benth.) E. Greene separate from *L. androsaceus* (Benth.) E. Greene
Microsteris gracilis (Hook.) E. Greene incl ssp. *humilis* (E. Greene) V. Grant = *Phlox gracilis* E. Greene
Navarretia bakeri H. Mason = *N. leucocephala* Benth. ssp. *bakeri* (H. Mason) A. Day
N. cotulaefolia (Benth.) Hook. & Arn.: spelling corrected to *N. cotulifolia*
N. fossalis Moran recently described
N. heterodoxa (E. Greene) E. Greene ssp. *rosulata* (Brand) H. Mason = *N. rosulata* Brand
N. hirsutissima Brand separate from *N. atractyloides* (Benth.) Hook. & Arn.
N. minima Nutt. = *N. leucocephala* Benth. ssp. *minima* (Nutt.) A. Day
N. mitracarpa E. Greene ssp. *jaredii* (Eastw.) H. Mason = *N. jaredii* Eastw.
N. myersii P.S. Allen & A. Day recently described
N. nigellaeformis E. Greene: spelling corrected to *N. nigelliformis*
ssp. *radians* (J. Howell) A. Day recognized
N. pauciflora H. Mason = *N. leucocephala* Benth. ssp. *pauciflora* (H. Mason) A. Day
N. plieantha H. Mason = *N. leucocephala* Benth. ssp. *plieantha* (H. Mason) A. Day
N. propinqua Suksd. = *N. intertexta* (Benth.) Hook. ssp. *propinqua* (Suksd.) A. Day
N. viscidula Benth. ssp. *purpurea* (E. Greene) H. Mason indistinct
Phlox bryoides Nutt. = *P. muscoides* Nutt.
P. caespitosa Nutt. var. *pulvinata* Wherry (Cronq.) = *P. pulvinata* (Wherry) Cronq.
P. covillei E. Nelson = *P. condensata* (A. Gray) E. Nelson
P. diffusa Benth. ssp. *subcarinata* Wherry = *P. austromontana* Coville
P. hirsuta E. Nelson separate from *P. stansburyi* (Torrey) A.A. Heller
P. hoodii Richardson ssp. *lanata* (Piper) Munz = ssp. *canescens* (Torrey & A. Gray) Wherry
P. longifolia Nutt. var. *puberula* E. Nelson = *P. stansburyi* (Torrey) A.A. Heller
P. speciosa Pursh var. *nitida* Suksd. = ssp. *nitida* (Suksd.) Wherry
P. stansburyi (Torrey) A.A. Heller var. *brevifolia* (A. Gray) E. Nelson not recognized
var. *hirsuta* (E. Nelson) Jepson = *P. hirsuta* E. Nelson
Polemonium caeruleum L. ssp. *amygdalinum* (Wherry) Munz = *P. occidentale* E. Greene

POLYGALACEAE MILKWORT FAMILY

Polygala cornuta Kellogg var. *pollardii* Munz = var. *fishiae* (C. Parry) Jepson
P. intermontana T. Wendt occurs in CA
P. subspinosa S. Watson var. *heterorhyncha* Barneby = *P. heterorhyncha* (Barneby) T. Wendt

POLYGONACEAE BUCKWHEAT FAMILY

Centrostegia insignis (Curran) A.A. Heller = *Aristocapsa insignis* (Curran) Rev. & Hardham
C. leptoceras A. Gray = *Dodecahema leptoceras* (A. Gray) Rev. & Hardham
C. vortriedei (Brandegee) Goodman = *Systenotheca vortriedei* (Brandegee) Rev. & Hardham
Chorizanthe biloba Goodman var. *immemora* Rev. & Hardham recently described
C. californica (Benth.) A. Gray incl var. *suksdorfii* J.F. Macbr. = *Mucronea californica* Benth.
C. coriacea Goodman = *Lastarriaea coriacea* (Goodman) Hoover
C. cuspidata S. Watson var. *marginata* Goodman = var. *cuspidata*
C. cuspidata S. Watson var. *villosa* (Eastw.) Munz indistinct
C. insignis Curran = *Aristocapsa insignis* (Curran) Rev. & Hardham
C. leptoceras (A. Gray) S. Watson = *Dodecahema leptoceras* (A. Gray) Rev. & Hardham
C. perfoliata A. Gray = *Mucronea perfoliata* (A. Gray) A.A. Heller
C. procumbens Nutt. var. *albiflora* Goodman not recognized
C. pungens Benth. var. *hartwegii* (Benth.) Goodman = *C. robusta* C. Parry var. *hartwegii* (Benth.) Goodman [may also be called *C. pungens* Benth. var. *hartwegiana* Rev. & Hardham]
C. staticoides Benth. vars. *brevispina* Goodman, *elata* Goodman, & *latiloba* Goodman, & ssp. *chrysacantha* (Goodman) Munz not recognized
C. thurberi (Benth.) S. Watson = *Centrostegia thurberi* Benth.
C. vortriedei Brandegee = *Systenotheca vortriedei* (Brandegee) Rev. & Hardham
Dedeckera eurekensis Rev. & J. Howell recently described
Eriogonum anemophilum E. Greene misapplied to *E. rosense* Nelson & Kenn.
E. apricum J. Howell var. *prostratum* R. Myatt recently described
E. baileyi S. Watson var. *divaricatum* (Gand.) Rev. = *E. baileyi* var. *praebens* (Gand.) Rev.
E. bifurcatum Rev. recently described
E. breedlovei (J. Howell) Rev. var. *shevockii* J. Howell recently described
E. caninum (E. Greene) Munz = *E. luteolum* E. Greene var. *caninum* (E. Greene) Rev.
E. carneum Rev. = *E. glandulosum* (Nutt.) Benth.
E. contiguum (Rev.) Rev. recently described
E. deflexum Torrey var. *nevadense* Rev. recently described
E. diclinum Rev. recently described
E. ericifolium Torrey & A. Gray var. *thornei* Rev. & Henrickson recently described
E. gracile Benth. var. *cithariforme* (S. Watson) Munz = *E. cithariforme* S. Watson var. *cithariforme*
var. *incultum* Rev. recently described
var. *polygonoides* (S. Stokes) Munz = *E. cithariforme* S. Watson var. *agninum* (E. Greene) Rev.
E. grande E. Greene var. *timorum* Rev. recently described
E. heermannii Durand & Hilg. vars. *humilius* (S. Stokes) Rev. & *occidentale* S. Stokes recognized
E. heracleoides Nutt. var. *angustifolium* (Nutt.) Torrey & A. Gray = var. *heracleoides*
E. insigne S. Watson misapplied to *E. deflexum* Torrey var. *rectum* Rev.
E. kearneyi Tidestrom = *E. nummulare* M.E. Jones
E. libertini Rev. recently described
E. luteolum E. Greene separate from *E. vimineum* Benth. vars. *caninum* (E. Greene) Rev. & *luteolum* recognized

E. microthecum Nutt. vars. *alpinum* Rev., *corymbosoides* Rev., *johnstonii* Rev., & *lapidicola* Rev. recently described
var. *panamintense* S. Stokes separate from var. *laxiflorum* Hook.
var. *simpsonii* (A.DC.) Rev. recognized
E. molestum E. Greene recognized
var. *davidsonii* (E. Greene) Jepson = *E. davidsonii* E. Greene
E. nudum Benth. var. *decurrens* (S. Stokes) Bowerman separate from var. *auriculatum* (Benth.) Jepson
var. *murinum* Rev. recently described
var. *paralinum* Rev. recently described
var. *regirivum* Rev. & J. Stebb. recently described
var. *westonii* (S. Stokes) J. Howell separate from var. *pubiflorum* Benth.
E. nutans Torrey & A Gray separate from *E. collinum* M.E. Jones
var. *glabrum* Rev. not recognized
E. ochrocephalum S. Watson var. *alexanderae* Rev. recently described
E. ovalifolium Nutt. var. *eximium* (Tidestrom) J. Howell separate from var. *nivale* (Canby) M.E. Jones
var. *purpureum* (Nelson) Durand separate from var. *ovalifolium*
E. parvifolium Smith var. *lucidum* (S. Stokes) Rev. = var. *parvifolium*
var. *paynei* (Munz) Rev. = var. *parvifolium*
E. prattenianum Durand var. *avium* Rev. & J. Shevock recently described
E. prociduum Rev. recently described
E. puberulum S. Watson occurs in CA
E. racemosum Nutt. incl var. *desertorum* S. Stokes = *E. panamintense* C. Morton
E. rupinum Rev. recently described
E. strictum Benth. var. *greenei* (A. Gray) Rev. separate from var. *proliferum* (Torrey & A. Gray) Rev.
E. ternatum Howell var. *congdonii* (S. Stokes) J. Howell = *E. congdonii* (S. Stokes) Rev.
E. trichopes Torrey var. *hooveri* Rev. recently described
E. umbellatum Torrey var. *argus* Rev. recently described
var. *aureum* (Gand.) Rev. = var. *glaberrimum* (Gand.) Rev.
var. *furcosum* Rev. recently described (var. *stellatum* (Benth.) M.E. Jones misapplied)
var. *humistratum* Rev. recently described
var. *juniporinum* Rev. recently described
var. *munzii* Rev. recently described
var. *nevadense* Gand. recognized
var. *stellatum* (Benth.) M.E. Jones misapplied in CA to var. *furcosum* Rev.
var. *umbellatum* not in CA (misapplied in CA to var. *nevadense*)
E. ursinum S. Watson var. *nervulosum* S. Stokes = *E. nervulosum* (S. Stokes) Rev.
Fagopyrum sagittatum Gilib. = *F. esculentum* Moench
Muehlenbeckia complexa Meissner naturalized
M. hastatula (Smith) I.M. Johnston naturalized
Oxytheca luteola C. Parry = *Goodmania luteola* (C. Parry) Rev. & B. Ertter
O. parishii C. Parry vars. *cienegensis* B. Ertter & *goodmaniana* B. Ertter recently described
Polygonum aviculare L. misapplied in CA to *P. arenastrum* Boreau
P. coccineum Muhlenb. = *P. amphibium* L. var. *emersum* Michaux
P. confertiflorum Piper = *P. polygaloides* Meissner ssp. *confertiflorum* (Piper) J. Hickman
P. douglasii E. Greene var. *austiniae* (E. Greene) M.E. Jones = ssp. *austiniae* (E. Greene) E. Murray
var. *johnstonii* Munz = ssp. *johnstonii* (Munz) J. Hickman
var. *latifolium* (Engelm.) E. Greene = ssp. *douglasii*
P. esotericum Wheeler = *P. polygaloides* Meissner ssp. *esotericum* (Wheeler) J. Hickman
P. fusiforme E. Greene = *P. persicaria* L.
P. hydropiperoides Michaux var. *asperifloium* E. Stanford not recognized
P. kelloggii E. Greene = *P. polygaloides* Meissner ssp. *kelloggii* (E. Greene) J. Hickman
P. lapathifolium L. vars. *prostratum* Wimmer & *salicifolium* Sibth. not recognized
P. mexicanum Small not documented in CA
P. montereyense Brenckle = *P. arenastrum* Boreau
P. pensylvanicum L. naturalized
P. spergulariaeforme Meissner = *P. douglasii* E. Greene ssp. *spergulariiforme* (Meissner) J. Hickman
Rumex angiocarpus Murb. = *R. acetosella* L.
R. californicus Rech.f. = *R. salicifolius* J.A. Weinm. var. *denticulatus* Torrey
R. dentatus L. ssp. *klotzschianus* (Meissner) Rech.f. not recognized
R. fenestratus E. Greene = *R. occidentalis* S. Watson
R. fueginus Philippi = *R. maritimus* L.
R. lacustris E. Greene = *R. salicifolius* J.A. Weinm. var. *lacustris* (E. Greene) J. Hickman
R. obtusifolius L. ssp. *agrestis* (Fries) Danser not recognized
R. paucifolius S. Watson var. *gracilescens* Rech.f. not recognized
R. persicarioides L. = *R. maritimus* L.
R. transitorius Rech.f. = *R. salicifolius* J.A. Weinm. var. *transitorius* (Rech.f.) J. Hickman
R. triangulivalvis (Danser) Rech.f. = *R. salicifolius* J.A. Weinm. var. *triangulivalvis* (Danser) J. Hickman
R. utahensis Rech.f. = *R. salicifolius* J.A. Weinm. var. *denticulatus* Torrey

PORTULACACEAE PURSLANE FAMILY

Calandrinia ciliata (Ruiz López & Pavón) DC. var. *menziesii* (Hook.) J.F. Macbr. not recognized
Calyptridium umbellatum (Torrey) E. Greene var. *caudiciferum* (A. Gray) Jepson indistinct
Claytonia bellidifolia Rydb. = *C. megarhiza* (A. Gray) S. Watson
C. lanceolata Pursh vars. *peirsonii* Munz & I.M. Johnston & *sessilifolia* (Torrey) Nelson not recognized
C. megarhiza (A. Gray) S. Watson var. *bellidifolia* (Rydb.) C. Hitchc. not recognized
C. parviflora Hook. ssp. *grandiflora* John M. Miller & Chambers recently described
C. perfoliata Donn ssp. *mexicana* (Rydb.) John M. Miller & Chambers occurs in CA
C. washingtoniana (Suksd.) Suksd. separate from *C. sibirica* L.
Lewisia cantelowii J. Howell: spelling corrected to *L. cantelovii*
L. columbiana (Howell) Robinson included, but CA pls may be *L. cotyledon* (S. Watson) Robinson × *L. leana* (Porter) Robinson
L. pygmaea (A. Gray) Robinson ssp. *glandulosa* (Rydb.) Ferris = *L. glandulosa* (Rydb.) Dempster
ssp. *longipetala* (Piper) Ferris = *L. longipetala* (Piper) Clay
L. rediviva Pursh var. *minor* (Rydb.) Munz indistinct
L. serrata Heckard & Stebb. indistinct from *L. cantelovii* J. Howell
L. sierrae Ferris = *L. pygmaea* (A. Gray) Robinson
L. stebbinsii R. Gankin & W.R. Hildreth recently described
Montia cordifolia (S. Watson) Pax & K. Hoffm. = *Claytonia cordifolia* S. Watson
M. fontana L. sspp. *chondrosperma* (Fenzl) Walters, *amporitana* Sennen, & *variabilis* Walters indistinct
M. funstonii Rydb. indistinct from *M. fontana* L. ssp. *variabilis* Walters
M. gypsophiloides (Fischer & C. Meyer) Howell = *Claytonia gypsophiloides* Fischer & C. Meyer
M. hallii (A. Gray) E. Greene = *M. fontana* L.
M. perfoliata (Donn) Howell = *Claytonia perfoliata* Donn

forma *angustifolia* (E. Greene) J. Howell = *Claytonia perfoliata* Donn ssp. *perfoliata*
var. *depressa* (A. Gray) Jepson = *Claytonia rubra* (Howell) Tidestrom ssp. *depressa* (A. Gray) John M. Miller & Chambers; also misapplied to *Claytonia rubra* (Howell) Tidestrom ssp. *rubra*
forma *glauca* (Torrey & A. Gray) J. Howell = *Claytonia exigua* Torrey & A. Gray ssp. *glauca* (Torrey & A. Gray) John M. Miller & Chambers
var. *nubigena* (E. Greene) Jepson = *Claytonia gypsophiloides* Fischer & C. Meyer
forma *parviflora* (Hook.) J. Howell = *Claytonia parviflora* Hook. ssp. *parviflora*
var. *perfoliata* = *Claytonia perfoliata* Donn ssp. *perfoliata*
var. *utahensis* (Rydb.) Munz = *Claytonia parviflora* Hook. ssp. *parviflora*
M. saxosa (Brandegee) Robinson = *Claytonia saxosa* Brandegee
M. sibirica (L.) Howell var. *sibirica* = *Claytonia sibirica* L.
var. *heterophylla* (Torrey & A. Gray) Robinson misapplied to *Claytonia palustris* Swanson & Kelley
M. spathulata Howell (illegitimate name) incl vars. *exigua* (Torrey & A. Gray) Robinson, *rosulata* (Eastw.) J. Howell, & *tenuifolia* (Torrey & A. Gray) Munz = *Claytonia exigua* Torrey & A. Gray ssp. *exigua*
var. *viridis* Davidson = *Claytonia parviflora* Hook. ssp. *viridis* (Davidson) John M. Miller & Chambers
M. verna Necker = *M. fontana* L.
Portulaca mundula I.M. Johnston probably misapplied to *P. halimoides* L.; more study needed

PRIMULACEAE PRIMROSE FAMILY

Anagallis minima (L.) E.H. Krause = *Centunculus minimus* L.
Androsace filiformis Retz. occurs in CA
Dodecatheon alpinum (A. Gray) E. Greene ssp. *majus* H.J. Thompson not recognized
D. hansenii (E. Greene) H.J. Thompson = *D. hendersonii* A. Gray
D. hendersonii A. Gray sspp. *cruciatum* (E. Greene) H.J. Thompson & *parvifolium* (Knuth) H.J. Thompson not recognized
D. jeffreyi Van Houtte ssp. *pygmaeum* (H.M. Hall) H.J Thompson not recognized
D. pulchellum (Raf.) Merr. ssp. *monanthum* (E. Greene) H.J. Thompson not recognized
Trientalis arctica Hook. occurs in CA

PUNICACEAE POMEGRANATE FAMILY

Punica granatum L. naturalized

RAFFLESIACEAE RAFFLESIA FAMILY: no changes

RANUNCULACEAE BUTTERCUP FAMILY

Aconitum columbianum Nutt. var. *howellii* (Nelson & J.F. Macbr.) C. Hitchc. indistinct
A. geranioides E. Greene = *A. columbianum* Nutt.
A. hansenii E. Greene = *A. columbianum* Nutt. var. *howellii* (Nelson & J.F. Macbr.) C. Hitchc.
A. leibergii E. Greene = *A. columbianum* Nutt.
A. viviparum E. Greene = *A. columbianum* Nutt. var. *howellii* (Nelson & J.F. Macbr.) C. Hitchc.
Actaea rubra (Aiton) Willd. ssp. *arguta* (Nutt.) Hultén not recognized
Anemone quinquefolia L. vars. *grayi* (Behr & Kellogg) Jepson, *minor* (Eastw.) Munz, & *oregana* (A. Gray) Robinson = *A. oregana* A. Gray
Aquilegia formosa C. Fischer forma *anomala* J. Howell and vars. *hypolasia* (E. Greene) Munz, *pauciflora* (E. Greene) Boothman, & *truncata* (Fischer & C. Meyer) Baker not recognized
A. shockleyi Eastw. indistinct from *A. formosa* C. Fischer
Caltha howellii (Huth) E. Greene = *C. leptosepala* DC. var. *biflora* (DC.) G. Lawson
C. palustris L. not naturalized
Delphinium ajacis L. misapplied to *Consolida ambigua* (L.) Ball & Heywood
D. andersonii A. Gray ssp. *cognatum* (E. Greene) Ewan not recognized
D. diversifolium E. Greene ssp. *harneyense* Ewan = *D. depauperatum* Nutt.
D. hansenii (E. Greene) E. Greene ssp. *ewanianum* M. J. Warnock recently described
ssp. *kernense* (Davidson) Ewan recognized
D. kinkiense Munz = *D. variegatum* Torrey & A. Gray ssp. *kinkiense* (Munz) M.J. Warnock
D. menziesii DC. misapplied to *D. andersonii* A. Gray
D. parishii A. Gray ssp. *pallidum* (Munz) M.J. Warnock recognized
ssp. *purpureum* Harlan Lewis & Epling = *D. parryi* A. Gray ssp. *purpureum* (Harlan Lewis & Epling) M.J. Warnock
D. parryi A. Gray ssp. *eastwoodiae* Ewan recognized
ssp. *maritimum* (Davidson) M.J. Warnock recognized
ssp. *seditiosum* (Jepson) Ewan = ssp. *parryi*
D. patens Benth. ssp. *greenei* (Eastw.) Ewan = *D. gracilentum* E. Greene
D. pratense Eastw. = *D. gracilentum* E. Greene
D. sonnei E. Greene = *D. nuttallianum* Walp.
D. variegatum Torrey & A. Gray ssp. *thornei* Munz recently described
Myosurus aristatus Benth. = *M. apetalus* C. Gay
M. minimus L. var. *apus* (E. Greene) G.R. Campbell indistinct
var. *filiformis* E. Greene & ssp. *major* (E. Greene) G.R. Campbell not recognized
ssp. *montanus* G.R. Campbell = *M. apetalus* C. Gay
var. *sessiliflorus* (Huth) G.R. Campbell = *M. sessilis* S. Watson
Ranunculus alismifolius Benth. vars. *hartwegii* (E. Greene) Jepson & *lemmonii* (A. Gray) L. Benson = var. *alismifolius*
R. alveolatus Carter = *R. bonariensis* Poiret var. *trisepalus* (Gill) Lourt.
R. aquatilis L. var. *harrisii* L. Benson = var. *capillaceus* (Thuill.) DC.
R. californicus Benth. vars. *austromontanus* L. Benson, *cuneatus* E. Greene, *gratus* Jepson, & *rugulosus* (E. Greene) L. Benson not recognized
R. canus Benth. vars. *laetus* (E. Greene) L. Benson & *ludovicianus* (E. Greene) L. Benson not recognized
R. eschscholtzii Schldl. var. *e.* probably not in CA
R. flammula L. vars. *ovalis* (Jacob Bigelow) L. Benson & *samolifolius* (E. Greene) L. Benson not recognized
R. hystriculus A. Gray = *Kumlienia hystricula* (A. Gray) E. Greene
R. occidentalis Nutt. vars. *dissectus* L. Henderson, *eisenii* (Kellogg) A. Gray, *howellii* E. Greene, *rattanii* A. Gray, & *ultramontanus* E. Greene not recognized
R. orthorhynchus Hook. vars. *hallii* Jepson & *platyphyllus* A. Gray = var. *orthorhynchus*
R. repens L. var. *pleniflorus* Fern. indistinct
R. sceleratus L. var. *multifidus* Nutt. not recognized
R. subrigidus Drew = *R. aquatilis* L. var. *subrigidus* (Drew) Breitung
R. uncinatus D. Don var. *parviflorus* (Torrey) L. Benson indistinct

Thalictrum occidentale A. Gray var. *palousense* St. John not recognized
T. polycarpum (Torrey) S. Watson = *T. fendleri* A. Gray var. *polycarpum* Torrey
Trautvetteria grandis Torry & A. Gray = *T. caroliniensis* (Walter) Vail var. *occidentalis* (A. Gray) C. Hitchc.

RESEDACEAE MIGNONETTE FAMILY: no changes

RHAMNACEAE BUCKTHORN FAMILY

Ceanothus arboreus E. Greene var. *glabra* Jepson indistinct
C. ×arcuatus McMinn incl in *C. fresnensis* Abrams
C. crassifolius Torrey var. *planus* Abrams indistinct
C. cuneatus (Hook.) Nutt. vars. *dubius* J. Howell & *submontanus* (Rose) McMinn = var. *cuneatus*
C. dentatus Torrey & A. Gray var. *floribundus* (Hook.) Trel. not recognized
C. griseus (Trel.) McMinn var. *horizontalis* McMinn indistinct
C. impressus Trel. var. *nipomensis* McMinn indistinct
C. insularis Eastw. = *C. megacarpus* Nutt. var. *insularis* (Eastw.) Munz
C. integerrimus Hook. & Arn. vars. *californicus* (Kellogg) Benson, *macrothyrsus* (Torrey) Benson, & *puberulus* (E. Greene) Abrams indistinct
C. ×lobbianus Hook. incl in *C. dentatus* Torrey & A. Gray
C. ×lorenzenii (Jepson) McMinn indistinct from *C. velutinus* Hook. var. *velutinus*
C. oliganthus Nutt. var. *orcuttii* (C. Parry) Jepson indistinct from var. *oliganthus*
C. ×otayensis McMinn indistinct from *C. crassifolius* Torrey
C. prostratus Benth. vars. *laxus* Jepson & *occidentalis* McMinn indistinct from var. *prostratus*
C. ramulosus (E. Greene) McMinn var. *fascicularis* McMinn = *C. cuneatus* (Hook.) Nutt. var. *fascicularis* (McMinn) Hoover
var. *ramulosus* = *C. cuneatus* (Hook.) Nutt. var. *cuneatus*
C. ×regius (Jepson) McMinn incl in *C. papillosus* Torrey & A. Gray
C. rigidus Nutt. = *C. cuneatus* (Hook.) Nutt. var. *rigidus* (Nutt.) Hoover
C. roderickii W. Knight recently described
C. ×rugosus E. Greene incl in *C. prostratus* Benth.
C. ×serrulatus McMinn incl in *C. prostratus* Benth.
C. sorediatus Hook. & Arn. = *C. oliganthus* Nutt. var. *sorediatus* (Hook. & Arn.) Hoover
C. thyrsiflorus Eschsch. var. *repens* McMinn indistinct
C. tomentosus C. Parry var. *olivaceus* Jepson indistinct
C. ×veitchianus Hook. incl in *C. griseus* (Trel.) McMinn
Condalia lycioides (A. Gray) Weberb. var. *canescens* (A. Gray) Trel. = *Ziziphus obtusifolia* A. Gray var. *canescens* (A. Gray) M. Johnston
C. parryi (Torrey) Weberb. = *Ziziphus parryi* Torrey var. parryi
Rhamnus californica Eschsch. ssp. *crassifolia* (Jepson) C. Wolf = *R. tomentella* Benth. ssp. *crassifolia* (Jepson) J.O. Sawyer
ssp. *cuspidata* (E. Greene) C. Wolf = *R. tomentella* Benth. ssp. *cuspidata* (E. Greene) J.O. Sawyer
ssp. *tomentella* (Benth.) C. Wolf = *R. tomentella* Benth. ssp. *tomentella*
ssp. *ursina* (E. Greene) C. Wolf = *R. tomentella* Benth. ssp. *ursina* (E. Greene) J.O. Sawyer
R. crocea Nutt. ssp. *ilicifolia* (Kellogg) C. Wolf = *R. ilicifolia* Kellogg
ssp. *pilosa* (Trel.) C. Wolf = *R. pilosa* (Trel.) Abrams
ssp. *pirifolia* (E. Greene) C. Wolf = *R. pirifolia* E. Greene
R. rubra E. Greene vars. *modocensis* C. Wolf, *nevadensis* (Nelson) C. Wolf, *obtusissima* (E. Greene) C. Wolf, & *yosemitana* C. Wolf indistinct

ROSACEAE ROSE FAMILY

Acaena anserinifolia (Forster & Forster f.) Druce misapplied to *A. novae-zelandiae* Kirk
A. californica Bitter = *A. pinnatifida* Ruiz López & Pavón var. *californica* (Bitter) Jepson
Adenostoma fasciculatum Hook. & Arn. var. *obtusifolium* S. Watson indistinct
Alchemilla occidentalis Nutt. = *Aphanes occidentalis* (Nutt.) Rydb.
Amelanchier florida Lindley = *A. alnifolia* (Nutt.) Nutt. var. *semiintegrifolia* (Hook.) C. Hitchc.
A. pallida E. Greene indistinct from *A. utahensis* Koehne
A. pumila Nutt. = *A. alnifolia* (Nutt.) Nutt. var. *pumila* (Nutt.) Nelson
A. utahensis Koehne ssp. *covillei* (Standley) Clokey indistinct
Aruncus vulgaris Raf. = *A. dioicus* (Walter) Fern. var. *pubescens* (Rydb.) Fern.
Cercocarpus betuloides Torrey & A. Gray var. *traskiae* (Eastw.) Dunkle = *C. traskiae* Eastw.
C. ledifolius Nutt. var. *intercedens* C. Schneider = hybrids with *C. intricatus* C. Schneider
var. *intermontanus* N. Holmgren recently described
Cowania mexicana D. Don var. *stansburiana* (Torrey) Jepson = *Purshia mexicana* (D. Don) S. Welsh var. *stansburyana* (Torrey) S. Welsh
Crataegus suksdorfii (Sarg.) Kruschke separate from *C. douglasii* Lindley
Fragaria vesca L. ssp. *californica* (Cham. & Schldl.) Staudt not recognized
F. platypetala Rydb. = *F. virginiana* Duchesne
Geum canescens (E. Greene) Munz = *G. triflorum* Pursh
G. ciliatum Pursh = *G. triflorum* Pursh
G. strictum Aiton misapplied to *G. aleppicum* Jacq.
Heteromeles arbutifolia (Lindley) Roemer var. *macrocarpa* (Munz) Munz not recognized
Holodiscus boursieri (Carrière) Rehder = *H. discolor* (Pursh) Maxim.
H. discolor (Pursh) Maxim. vars. *delnortensis* Ley, *dumosus* (S. Watson) Dippel, *franciscanus* (Rydb.) Jepson not recognized
H. microphyllus Rydb. var. *sericeus* Ley = var. *microphyllus*
Horkelia bolanderi A. Gray ssp. *clevelandii* (E. Greene) Keck = *H. clevelandii* (E. Greene) Rydb.
ssp. *parryi* (S. Watson) Keck = *H. rydbergii* Elmer
H. congesta Hook. ssp. *nemorosa* Keck occurs in CA
H. daucifolia (E. Greene) Rydb. ssp. *latior* Keck not recognized
H. elata (E. Greene) Rydb. = *H. californica* Cham. & Schldl. ssp. *dissita* (Crum) B. Ertter
H. frondosa (E. Greene) Rydb. = *H. californica* Cham. & Schldl. ssp. *frondosa* (E. Greene) B. Ertter
H. fusca Lindley ssp. *pseudocapitata* (Rydb.) Keck = ssp. *parviflora* (Nutt.) Keck
H. hendersonii Howell occurs in CA
H. yadonii B. Ertter recently described
Ivesia aperta (J. Howell) Munz var. *canina* B. Ertter recently described
I. baileyi S. Watson vars. *baileyi* & *beneolens* (Nelson & J.F. Macbr.) B. Ertter occur in CA
I. kingii S. Watson: CA pls are var. *kingii*
I. longibracteata B. Ertter recently described
I. paniculata T.W. Nelson & J.P. Nelson recently described
I. purpurascens (S. Watson) Keck ssp. *congdonis* (Rydb.) Keck = *Horkeliella congdonis* (Rydb.) Rydb.
ssp. *purpurascens* = *Horkeliella purpurascens* (S. Watson) Rydb.
I. shockleyi S. Watson: CA pls are var. *shockleyi*
Lyonothamnus floribundus A. Gray var. *asplenifolius* (E. Greene) Brandegee = ssp. *asplenifolius* (E. Greene) Raven

Neviusia cliftonii J. Shevock, B. Ertter, & D.W. Taylor recently described
Osmaronia cerasiformis (Hook. & Arn.) E. Greene = *Oemleria cerasiformis* (Hook. & Arn.) J.W. Landon
Physocarpus alternans (M.E. Jones) J. Howell sspp. *annulatus* J. Howell & *panamintensis* J. Howell not recognized
Potentilla anglica Laicharding not naturalized
P. breweri S. Watson var. *viridis* Jepson misapplied to *P. drummondii* Lehm. ssp. *breweri* (S. Watson) B. Ertter
P. concinna Richardson occurs in CA
P. cristae W. Ferlatte & Strother recently described
P. diversifolia Lehm.: CA pls are var. *diversifolia*
P. egedii Wormsk. var. *grandis* (Torrey & A. Gray) J. Howell = *P. anserina* L. ssp. *pacifica* (Howell) Rousi
P. flabellifolia Hook. var. *grayi* (S. Watson) Jepson = *P. grayi* S. Watson
P. flabelliformis Lehm. = *P. gracilis* Hook. var. *flabelliformis* (Lehm.) Torrey & A. Gray
P. gracilis Hook. ssp. *nuttallii* (Lehm.) Keck = var. *fastigiata* (Nutt.) S. Watson
P. millefolia Rydb. var. *klamathensis* (Rydb.) Jepson indistinct
P. morefieldii B. Ertter recently described
P. norvegica L. ssp. *monspeliensis* (L.) Asch. & Graebn. not recognized
P. pectinisecta Rydb. = *P. gracilis* Hook. var. *elmeri* (Rydb.) Jepson
P. pensylvanica L.: CA pls may be var. *strigosa* Pursh
P. rivalis Nutt. var. *millegrana* (Engelm.) S. Watson not recognized
P. saxosa E. Greene incl ssp. *sierrae* Munz = *Ivesia saxosa* (E. Greene) B. Ertter
P. wheeleri S. Watson var. *rimicola* Munz & I.M. Johnston = *P. rimicola* (Munz & I.M. Johnston) B. Ertter
Prunus amygdalus Batsch. = *P. dulcis* (Miller) D.A. Webb
P. armeniaca L., *P. avium* L., *P. caroliniana* Aiton, *P. lusitanica* L., *P. mahaleb* L., & *P. persica* Batsch. not naturalized
P. cerasifera Ehrh. naturalized
P. lyonii (Eastw.) Sargent = *P. ilicifolia* (Nutt.) Walp. ssp. *lyonii* (Eastw.) Raven
P. virginiana L. var. *melanocarpa* (Nelson) Sargent = *P. v.* var. *demissa* (Nutt.) Torrey
Purshia glandulosa Curran = *P. tridentata* (Pursh) DC. var. *glandulosa* (Curran) M.E. Jones
Rosa gymnocarpa Nutt. var. *pubescens* S. Watson = *R. bridgesii* Crépin
R. minutifolia Engelm. occurs in CA
R. nutkana Presl vars. *muriculata* (E. Greene) G.N. Jones & *setosa* G.N. Jones not recognized
R. spithamea S. Watson var. *sonomensis* (E. Greene) Jepson not recognized
R. woodsii Lindley vars. *glabrata* (Parish) Cole & *gratissima* (E. Greene) Cole not recognized
Rubus allegheniensis Porter not naturalized
R. almus (L. Bailey) L. Bailey = *R. pensilvanicus* Poiret
R. glaucifolis Kellogg var. *ganderi* (L. Bailey) Munz indistinct
R. leucodermis Torrey & A. Gray vars. *bernardinus* (E. Greene) Jepson & *trinitatis* A. Berger indistinct
R. macropetalus Hook. = *R. ursinus* Cham. & Schldl.
R. parviflorus Nutt. var. *velutinus* (Hook. & Arn.) E. Greene indistinct
R. pedatus Smith misapplied in CA to *R. lasiococcus* A. Gray
R. procerus Mueller = *R. discolor* Weihe & Nees
R. spectabilis Pursh var. *franciscanus* (Rydb.) J. Howell indistinct
R. ursinus Cham. & Schldl. var. *sirbenus* (L. Bailey) J. Howell not recognized
R. vitifolius Cham. & Schldl. incl var. *eastwoodianus* (Rydb.) Munz = *R. ursinus* Cham. & Schldl.
Sanguisorba microcephala C. Presl = *S. officinalis* L. ssp. *microcephala* (C. Presl) Calder & Roy Taylor
S. minor Scop.: CA pls are ssp. *muricata* Briq.
Sorbus cascadensis G.N. Jones = *S. scopulina* E. Greene var. *cascadensis* (G.N. Jones) C. Hitchc.
S. sitchensis K.F. Roemer var. *grayi* (Wenzig) C. Hitchc. occurs in CA

RUBIACEAE MADDER FAMILY

Coprosma repens A. Rich. not naturalized
Crucianella angustifolia L. naturalized
Diodia teres Walter occurs in CA
Galium andrewsii A. Gray var. *gatense* Dempster = ssp. *gatense* (Dempster) Dempster & Stebb.
ssp. *intermedium* Dempster & Stebb. recently described
G. angustifolium Nutt. ssp. *borregoense* Dempster & Stebb. recently described
var. *foliosum* Hilend. & Howell = ssp. *foliosum* (Hilend. & Howell) Dempster & Stebb.
ssp. *gracillimum* Dempster & Stebb. recently described
ssp. *jacintinum* Dempster & Stebb. recently described
var. *pinetorum* Munz & I.M. Johnston = *G. johnstonii* Dempster & Stebb.
G. argense Dempster & Ehrend. recently described
G. boreale L.: CA pls are ssp. *septentrionale* (Roemer & Schultes) Iltis
G. californicum Hook. & Arn. ssp. *flaccidum* (E. Greene) Dempster & Stebb. recognized
ssp. *maritimum* Dempster & Stebb. recently described
var. *miguelense* (E. Greene) Jepson = ssp. *miguelense* (E. Greene) Dempster & Stebb.
ssp. *primum* Dempster & Stebb. recently described
ssp. *sierrae* Dempster & Stebb. recently described
G. catalinense A. Gray ssp. *acrispum* Dempster recently described
G. glabrescens (Ehrend.) Dempster & Ehrend. separate from *G. grayanum* Ehrend.
ssp. *modocense* Dempster & Ehrend. recently described
G. gabrielense Munz & I.M. Johnston = *G. angustifolium* Nutt. ssp. *gabrielense* (Munz & I.M. Johnston) Dempster & Stebb.
G. grayanum Ehrend. ssp. *nanum* Dempster & Ehrend. recently described
G. hypotrichium A. Gray ssp. *inyoense* Dempster & Ehrend. recently described
G. munzii Hilend & J. Howell var. *carneum* Hilend & J. Howell = *G. hilendiae* Dempster & Ehrend. ssp. *carneum* (Hilend & J. Howell) Dempster & Ehrend.
var. *kingstonense* Dempster = *G. hilendiae* Dempster & Ehrend. ssp. *kingstonense* (Dempster) Dempster & Ehrend.
G. nuttallii A. Gray var. *ovalifolium* Dempster = *G. porrigens* Dempster var. *porrigens*
var. *tenue* Dempster = *G. porrigens* Dempster var. *tenue* (Dempster) Dempster
G. pubens A. Gray = *G. bolanderi* A. Gray
G. saxatile L. naturalized
G. schultesii Vest naturalized
G. serpenticum Dempster sspp. *scotticum* Dempster & Ehrend. & *warnerense* Dempster & Ehrend. recently described
G. sparsiflorum Wight ssp. *glabrius* Dempster & Stebb. recently described
G. spurium L. indistinct from *G. aparine* L.
G. trifidum L. var. *subbiflorum* Wieg. = var. *pacificum* Wieg.
G. verum L. naturalized
G. watsonii (A. Gray) A.A. Heller = *G. multiflorum* Kellogg
Rubia tinctorum L. not naturalized

RUTACEAE RUE FAMILY: no changes

SALICACEAE WILLOW FAMILY

Populus ×*acuminata* Rydb. misapplied to *P. angustifolia* James
P. fremontii S. Watson ssp. *fremontii* vars. *arizonica* (Sarg.) Jepson & *macdougalii* (Rose) Jepson not recognized
P. trichocarpa Torrey & A. Gray = *P. balsamifera* L. ssp. *trichocarpa* (Torrey & A. Gray) Brayshaw
Salix alba L. naturalized
S. bebbiana Sarg. occurs in CA
S. commutata Bebb incl var. *denudata* Bebb misapplied to *S. eastwoodiae* A.A. Heller
S. caudata (Nutt.) A.A. Heller var. *bryantiana* C. Ball & Bracelin = *S. lucida* Muhlenb. ssp. *caudata* (Nutt.) E. Murray
S. coulteri Andersson = *S. sitchensis* Bong.
S. drummondiana Hook. var. *subcoerulea* (Piper) C. Ball not recognized
S. exigua Nutt. misapplied in part to *S. sessilifolia* Nutt. var. *stenophylla* (Rydb.) C. Schneider not recognized
S. geyeriana Andersson var. *argentea* (Bebb) C. Schneider not recognized
S. gooddingii C. Ball var. *variabilis* C. Ball not recognized
S. hindsiana Benth. incl vars. *leucodendroides* (Rowlee) C. Ball & *parishiana* (Rowlee) C. Ball indistinct from *S. exigua* Nutt.
S. laevigata Bebb incl var. *araquipa* (Jepson) C. Ball not recognized
S. lasiandra Benth. incl vars. *abramsii* C. Ball & *lancifolia* (Andersson) Bebb = *S. lucida* Muhlenb. ssp. *lasiandra* (Benth.) E. Murray
S. lasiolepis Benth. vars. *bracelinae* C. Ball & *sandbergii* (Rydb.) C. Ball not recognized
var. *bigelovii* (Torrey) Bebb indistinct
S. lutea Nutt. var. *watsonii* (Bebb) Jepson not recognized
S. mackenzieana (Hook.) Andersson = *S. prolixa* Andersson
S. melanopsis Nutt. vars. *bolanderiana* (Rowlee) C. Schneider, *gracilipes* C. Ball, & *tenerrima* (L. Henderson) C. Ball not recognized
S. nivalis Hook. = *S. reticulata* L. ssp. *nivalis* (Hook.) Löve et al.
S. parksiana C. Ball = *S. sessilifolia* Nutt.
S. phylicifolia L. var. *monica* (Bebb) Jepson = *S. planifolia* Cham. ssp. *planifolia*
S. piperi Bebb = *S. hookeriana* Hook.
S. pseudocordata Andersson = *S. boothii* Dorn
S. scouleriana Hook. var. *coetanea* C. Ball not recognized
S. tracyi C. Ball = *S. lasiolepis* Benth.

SANTALACEAE SANDALWOOD FAMILY

Comandra pallida A. DC. = *C. umbellata* (L.) Nutt. ssp. *californica* (Rydb.) Piehl

SARRACENIACEAE PITCHER-PLANT FAMILY

Sarracenia purpurea L. naturalized

SAURURACEAE LIZARD'S-TAIL FAMILY: no changes

SAXIFRAGACEAE SAXIFRAGE FAMILY: for *woody genera* see **GROSSULARIACEAE** or **PHILADELPHACEAE**

Boykinia elata (Nutt.) E. Greene = *B. occidentalis* Torrey & A. Gray
Heuchera leptomeria E. Greene var. *peninsularis* C. Rosend., F.K. Butters, & Lakela = *H. rubescens* Torrey var. *versicolor* (E. Greene) M.G. Stewart
H. micrantha Lindley vars. *erubescens* (A. Braun & C. Bouche) C. Rosend. & *pacifica* C. Rosend., F.K. Butters, & Lakela indistinct
var. *hartwegii* (W.E. Wheelock) C. Rosend. not recognized
H. parishii Rydb. separate from *H. rubescens* Torrey
H. pringlei Rydb. = *H. rubescens* Torrey
Jepsonia malvifolia (E. Greene) Small separate from *J. parryi* (Torrey) Small
J. parryi (Torrey) Small var. *heterandra* (Eastw.) Jepson = *J. heterandra* Eastw.
Lithophragma trifoliatum Eastw. = *L. parviflorum* (Hook.) Torrey & A. Gray var. *trifoliatum* (Eastw.) Jepson
Parnassia palustris L. var. *californica* A. Gray = *P. californica* (A. Gray) E. Greene
P. parviflora DC. separate from *P. californica* (A. Gray) E. Greene
Peltiphyllum peltatum (Torrey) Engler = *Darmera peltata* (Torrey) Voss
Saxifraga cespitosa L. occurs in CA
S. debilis Engelm. misapplied to *S. rivularis* L.
S. fragarioides E. Greene = *Saxifragopsis fragarioides* (E. Greene) Small
S. integrifolia Hook. recognized
S. nidifica E. Greene: CA pls are var. *nidifica*
S. nuttallii Small occurs in CA
S. sarmentosa L. = *S. stolonifera* Meerb.

SCROPHULARIACEAE FIGWORT FAMILY

Antirrhinum breweri A. Gray = *A. vexillo-calyculatum* Kellogg ssp. *breweri* (A. Gray) D. Thompson
A. cornutum Benth. var. *leptaleum* (A. Gray) Munz = *A. leptaleum* A. Gray
A. coulterianum Benth. forma *orcuttianum* (A. Gray) Munz not recognized
A. cyathiferum Benth. occurs in CA
A. nuttallianum Benth. var. *subsessile* (A. Gray) D. Thompson recognized
A. vexillo-calyculatum Kellogg ssp. *intermedium* D. Thompson recently described
Bacopa monnieri (L.) Wettst. naturalized
B. nobsiana H. Mason = *B. rotundifolia* (Michaux) Wettst.
B. repens (Sw.) Wettst. naturalized
Castilleja affinis Hook. & Arn. ssp. *insularis* (Eastw.) Munz & var. *contentiosa* (J.F. Macbr.) Bacigal. = ssp. *affinis*
ssp. *inflata* (Pennell) Munz indistinct from ssp. *affinis*
C. applegatei Fern. var. *applegatei* misapplied to *C. a.* Fern. var. *pinetorum* (Fern.) Chuang & Heckard
C. brevilobata Piper = *C. hispida* Benth. ssp. *brevilobata* (Piper) Chuang & Heckard
C. breweri Fern. incl var. *pallida* Eastw. = *C. applegatei* Fern. ssp. *pallida* (Eastw.) Chuang & Heckard
C. chromosa Nelson = *C. angustifolia* (Nutt.) G. Don
C. disticha Eastw. = *C. applegatei* Fern. ssp. *disticha* (Eastw.) Chuang & Heckard
C. exilis Nelson = *C. minor* (A. Gray) A. Gray ssp. *minor*
C. franciscana Pennell = *C. subinclusa* E. Greene ssp. *franciscana* (Pennell) Chuang & Heckard
C. gleasonii Elmer indistinct from *C. pruinosa* Fern.
C. hololeuca E. Greene = *C. lanata* A. Gray ssp. *hololeuca* (E. Greene) Chuang & Heckard
ssp. *grisea* (Dunkle) Munz = *C. grisea* Dunkle
C. inflata Pennell indistinct from *C. affinis* Hook. & Arn. ssp. *affinis*
C. jepsonii Bacigal. & Heckard = *C. subinclusa* E. Greene ssp. *subinclusa*
C. latifolia Hook. & Arn. ssp. *mendocinensis* Eastw. = *C. mendocinensis* (Eastw.) Pennell
C. leschkeana J. Howell misapplied to *C. chrymactis* Pennell
C. litoralis Pennell = *C. affinis* Hook. & Arn. ssp. *litoralis* (Pennell) Chuang & Heckard
C. martinii Abrams ssp. *ewanii* (Eastw.) Munz = *C. angustifolia* (Nutt.) G. Don
C. ×*montigena* Heckard = *C. applegatei* Fern. ssp. *ewanii* (Abrams) Chuang & Heckard × *C. angustifolia* (Nutt.) G. Don

C. neglecta E.M. Zeile = *C. affinis* Hook. & Arn. ssp. *neglecta* (E.M. Zeile) Chuang & Heckard
C. peirsonii Eastw. = *C. parviflora* Bong.
C. roseana Eastw. = *C. applegatei* Fern. ssp. *martinii* (Abrams) Chuang & Heckard
C. stenantha A. Gray incl ssp. *spiralis* (Jepson) Munz = *C. minor* (A. Gray) A. Gray ssp. *spiralis* (Jepson) Chuang & Heckard
C. uliginosa Eastw. indistinct from *C. miniata* Hook. ssp. *miniata*
C. wightii Elmer ssp. *inflata* (Pennell) Munz indistinct from *C. affinis* Hook. & Arn. ssp. affinis
ssp. *litoralis* (Pennell) Munz = *C. affinis* ssp. *litoralis* (Pennell) Chuang & Heckard
Collinsia antonina Hardham incl ssp. *purpurea* Hardham = *C. parryi* A. Gray
C. franciscana Bioletti = *C. multicolor* Lindley & Paxton
C. grandiflora Lindley var. *pusilla* A. Gray not recognized
C. heterophylla Buist var. *austromontana* Newsom indistinct
C. sparsiflora Fischer & C. Meyer var. *bruceae* (M.E. Jones) Newsom indistinct from var. *collina* (Jepson) Newsom
C. torreyi A. Gray var. *brevicarinata* Newsom recognized
Cordylanthus bernardinus Munz = *C. eremicus* (Cov. & C. Morton) Munz ssp. *eremicus*
C. canescens A. Gray = *C. maritimus* Benth. ssp. *canescens* (A. Gray) Chuang & Heckard
C. capitatus Benth. occurs in CA
C. ferrisianus Pennell = *C. rigidus* (Benth.) Jepson ssp. *rigidus*
C. filifolius Benth. misapplied to *C. rigidus* (Benth.) Jepson ssp. *setigerus* Chuang & Heckard
C. hansenii (Ferris) J.F. Macbr. = *C. pilosus* A. Gray ssp. *hansenii* (Ferris) Chuang & Heckard
C. helleri (Ferris) J.F. Macbr. = *C kingii* S. Watson ssp. *helleri* (Ferris) Chuang & Heckard
C. hispidus Pennell = *C. mollis* A. Gray ssp. *hispidus* (Pennell) Chuang & Heckard
C. littoralis (Ferris) J.F. Macbr. ssp. *littoralis* = *C. rigidus* (Benth.) Jepson ssp. *littoralis* (Ferris) Chuang & Heckard
ssp. *platycephalus* (Pennell) Munz = *C. rigidus* (Benth.) Jepson ssp. *rigidus*
C. maritimus Benth. ssp. *palustris* (Behr) Chuang & Heckard occurs in CA
C. pallescens Pennell = *C. tenuis* A. Gray ssp. *pallescens* (Pennell) Chuang & Heckard
C. palmatus (Ferris) J.F. Macbr. ssp. *carnulosus* (Pennell) Munz not recognized
C. pilosus A. Gray ssp. *bolanderi* (A. Gray) Munz = *C. tenuis* A. Gray ssp. *tenuis*
ssp. *diffusus* (Pennell) Munz = ssp. *pilosus*
ssp. *trifidus* (Robinson & Greenm.) Chuang & Heckard separate from ssp. *hansenii* (Ferris) Chuang & Heckard
C. ramosus Benth. ssp. *eremicus* (Cov. & C. Morton) Munz = *C. eremicus* (Cov. & C. Morton) Munz ssp. *eremicus*
ssp. *setosus* Pennell not recognized
C. tenuis A. Gray ssp. *barbatus* Chuang & Heckard recently described
ssp. *capillaris* (Pennell) Chuang & Heckard separate from ssp. *pallescens* (Pennell) Chuang & Heckard
C. viscidus (Howell) Pennell = *C. tenuis* A. Gray ssp. *viscidus* (Howell) Chuang & Heckard
Keckiella antirrhinoides (Benth.) Straw ssp. microphylla (A. Gray) Straw = var. *microphylla* (A. Gray) N. Holmgren
K. breviflora (Lindley) Straw ssp. *glabrisepala* (Keck) Straw = var. *glabrisepala* (Keck) N. Holmgren
K. rothrockii (A. Gray) Straw ssp. *jacintensis* (Abrams) Straw = var. *jacintensis* (Abrams) N. Holmgren
K. ternata (Torrey) Straw ssp. *septentrionalis* (Munz & I.M. Johnston) Straw = var. *septentrionalis* (Munz & I.M. Johnston) N. Holmgren
Lindernia anagallidea (Michaux) Pennell = *L. dubia* (L.) Pennell var. *anagallidea* (Michaux) Cooperrider
Mimulus acutidens E. Greene = *M. inconspicuous* A. Gray
M. arenarius A.L. Grant = *M. floribundus* Lindley
M. aridus (Abrams) A.L. Grant indistinct from *M. aurantiacus* Curtis
M. aurantiacus Curtis sspp. *australis* (McMinn) Munz & *lompocensis* (McMinn) Munz not recognized
M. barbatus E. Greene = *M. montioides* A. Gray
M. bifidus Pennell incl ssp. *fasciculatus* Pennell indistinct from *M. aurantiacus* Curtis
M. bigelovii (A. Gray) A. Gray var. *panamentensis* Munz = var. *cuspidatus* A.L. Grant
M. biolettii Eastw. = *M. filicaulis* S. Watson
M. brachiatus Pennell = *M. layneae* (E. Greene) Jepson
M. brandegei Pennell = *M. latifolius* A. Gray
M. coccineus Congdon = *M. mephiticus* E. Greene
M. densus A. L. Grant = *M. mephiticus* E. Greene
M. diffusus A. L. Grant = *M. palmeri* A. Gray
M. discolor A.L. Grant = *M. montioides* A. Gray
M. dudleyi A.L. Grant = *M. floribundus* Lindley
M. flemingii Munz indistinct from *M. aurantiacus* Curtis
M. floribundus Lindley ssp. *subulatus* A.L. Grant not recognized
M. glabratus Kunth ssp. *utahensis* Pennell = *M. guttatus* DC.
M. grayi A.L. Grant = *M. inconspicuous* A. Gray
M. guttatus DC. sspp. *arenicola* Pennell, *arvensis* (E. Greene) Munz, *litoralis* Pennell, & *micranthus* (A.A. Heller) Munz not recognized
M. longiflorus (Nutt.) A.L. Grant incl ssp. *calycinus* (Eastw.) Munz & var. *rutilus* A.L. Grant indistinct from *M. aurantiacus* Curtis
M. microphyllus Benth. = *M. guttatus* DC.
M. moschatus Lindley var. *moniliformis* (E. Greene) Munz not recognized
M. nasutus E. Greene = *M. guttatus* DC.
M. parryi A. Gray occurs in CA
M. parviflorus Lindley misapplied to *M. flemingii* Munz (which is indistinct from *M. aurantiacus* Curtis)
M. platylaemus Pennell = *M. bolanderi* A. Gray
M. primuloides Benth. var. *pilosellus* (E. Greene) Smiley indistinct from ssp. *primuloides*
M. puniceus (Nutt.) Steudel indistinct from *M. aurantiacus* Curtis
M. rattanii A. Gray ssp. *decurtatus* (A.L. Grant) Pennell not recognized
M. ringens L. naturalized
M. shevockii Heckard & Bacigal. recently described
M. subsecundus A. Gray = *M. fremontii* (Benth.) A. Gray
M. viscidus Congdon ssp. *constrictus* (A.L. Grant) Munz = *M. constrictus* (A.L. Grant) Pennell
M. whipplei A.L. Grant = *M. guttatus* DC.
Orthocarpus erianthus Benth. = *Triphysaria eriantha* (Benth.) Chuang & Heckard
var. *gratiosus* Jepson & Tracy = *Triphysaria eriantha* (Benth.) Chuang & Heckard ssp. *eriantha*
var. *micranthus* (A. Gray) Jepson = *Triphysaria micrantha* (A. Gray) Chuang & Heckard
var. *roseus* A. Gray = *Triphysaria eriantha* (Benth.) Chuang & Heckard ssp. *rosea* (A. Gray) Chuang & Heckard
O. faucibarbatus Benth. = *Triphysaria versicolor* Fischer & C. Meyer ssp. *faucibarbata* (A. Gray) Chuang & Heckard
var. *albidus* Keck = *Triphysaria versicolor* Fischer & C. Meyer ssp. *versicolor*
O. floribundus Benth. = *Triphysaria floribunda* (Benth.) Chuang & Heckard
O. pusillus Benth. = *Triphysaria pusilla* (Benth.) Chuang & Heckard

Parentucellia latifolia (L.) Caruel naturalized
Pedicularis centranthera A. Gray occurs in CA
P. densiflora Hook. ssp. *aurantiaca* E. Sprague not recognized
P. flavida Pennell = *P. bracteosa* Benth. var. *flavida* (Pennell) Cronq.
Penstemon barnebyi N. Holmgren recently described
P. bridgesii A. Gray misapplied in CA to *P. rostriflorus* Kellogg
P. cinereus Piper indistinct from *P. humilis* A. Gray var. *humilis*
P. clevelandii A. Gray ssp. *connatus* (Munz & I.M. Johnston) Keck = var. *connatus* Munz & I.M. Johnston
ssp. *mohavensis* Keck = var. *mohavensis* (Keck) McMinn
P. confusus M.E. Jones ssp. *patens* (M.E. Jones) Keck = *P. patens* (M.E. Jones) N. Holmgren
P. davidsonii E. Greene: CA pls are var. *davidsonii*
P. deustus Lindley ssp. *heterander* (Torrey & A. Gray) Pennell & Keck = var. *pedicellatus* M.E. Jones
ssp. *sudans* (M.E. Jones) Pennell & Keck = *P. sudans* M.E. Jones
var. *suffrutescens* L. Henderson occurs in CA
P. eatonii A. Gray ssp. *undosus* (M.E. Jones) Keck = var. *undosus* M.E. Jones
P. floridus Brandegee ssp. *austinii* (Eastw.) Keck = var. *austinii* (Eastw.) N. Holmgren
P. fruticiformis Cov. ssp. *amargosae* Keck = var. *amargosae* (Keck) N. Holmgren
P. grinnellii Eastw. ssp. *scrophularioides* (M.E. Jones) Munz = var. *scrophularioides* (M.E. Jones) N. Holmgren
P. heterodoxus A. Gray ssp. *cephalophorus* (E. Greene) Keck = var. *cephalophorus* (E. Greene) N. Holmgren
P. heterophyllus Lindley ssp. *australis* (Munz & I.M. Johnston) Keck = var. *australis* Munz & I.M. Johnston
ssp. *purdyi* Keck = var. *purdyi* (Keck) McMinn
P. humilis A. Gray: CA pls are var. *humilis*
P. janishiae N. Holmgren recently described
P. laetus A. Gray ssp. *leptosepalus* (A. Gray) Keck = var. *leptosepalus* A. Gray
ssp. *roezlii* (Regel) Keck = *P. roezlii* Regel
ssp. *sagittatus* Keck = var. *sagittatus* (Keck) McMinn
P. newberryi A. Gray ssp. *berryi* (Eastw.) Keck = var. *berryi* (Eastw.) N. Holmgren
ssp. *sonomensis* (E. Greene) Keck = var. *sonomensis* (E. Greene) Jepson
P. oreocharis E. Greene = *P. rydbergii* Nelson var. *oreocharis* (E. Greene) N. Holmgren
P. pahutensis N. Holmgren recently described
P. palmeri A. Gray: CA pls are var. *palmeri*
P. procerus Graham ssp. *brachyanthus* (Pennell) Keck = var. *brachyanthus* (Pennell) Cronq.
ssp. *formosus* (Nelson) Keck = var. *formosus* (Nelson) Cronq.
P. rattanii A. Gray ssp. *kleei* (E. Greene) Keck = var. *kleei* (E. Greene) A. Gray
P. shastensis Keck = *P. heterodoxus* A. Gray var. *shastensis* (Keck) N. Holmgren
P. speciosus Lindley ssp. *kennedyi* (Nelson) Keck not recognized
P. spectabilis Thurber ssp. *subviscosus* Keck = var. *subviscosus* (Keck) McMinn
P. venustus Lindley naturalized
Synthyris reniformis (Douglas) Benth. var. *cordata* A. Gray indistinct
Verbascum speciosum Schrader naturalized
Veronica alpina L. var. *alterniflora* Fern. = *V. wormskjoldii* Roemer & Schultes
V. comosa Richter = *V. catenata* Pennell

SIMAROUBACEAE QUASSIA or SIMAROUBA FAMILY

Holacantha emoryi A. Gray = *Castela emoryi* (A. Gray) Moran & Felger

SIMMONDSIACEAE JOJOBA FAMILY

(*Simmondisia* segregated from Buxaceae)

SOLANACEAE NIGHTSHADE FAMILY

Lycium andersonii A. Gray var. *deserticola* (C. Hitchc.) Jepson not recognized
L. chinense J.S. Miller not naturalized
Lycopersicon peruvianum (L.) Miller naturalized
Nicotiana bigelovii (Torrey) S. Watson var. *bigelovii* = *N. quadrivalvis* Pursh
var. *wallacei* A. Gray indistinct from *N. quadrivalvis* Pursh
N. trigonophylla Dunal = *N. obtusifolia* Martens & Galeotti
Petunia violacea Lindley not naturalized
Physalis angulata L. var. *lanceifolia* (Nees) Waterfall = *P. lancifolia* Nees
P. crassifolia Benth. var. *versicolor* (Rydb.) Waterfall not recognized
P. fendleri A. Gray = *P. hederifolia* A. Gray var. *fendleri* (A. Gray) Cronq.
P. greenei Vasey & Rose = *P. crassifolia* Benth.
P. hederifolia A. Gray var. *cordifolia* (A. Gray) Waterfall = var. *fendleri* (A. Gray) Cronq.
var. *palmeri* (A. Gray) C. Hitchc. recognized
P. ixocarpa Hornem. = *P. philadelphica* Lam.
P. lobata Torrey occurs in CA
P. mollis Nutt. = *P. viscosa* L.
P. neomexicana Rydb. misapplied in CA to *P. pubescens* L. var. *grisea* Waterfall
P. pruinosa L. = *P. pubescens* L. var. *grisea* Waterfall
P. subglabrata K.K. Mackenzie & Bush misapplied in CA to *P. longifolia* Nutt. (and perhaps other spp.)
Solanum clokeyi Munz = *S. wallacei* (A. Gray) Parish
S. gayanum (Remy) Phil.f. misapplied in CA to *S. furcatum* Dunal
S. laciniatum Aiton misapplied in CA to *S. aviculare* Forst.f.
S. nodiflorum Jacq. = *S. americanum* Miller
S. tenuilobatum Parish = *S. xanti* A. Gray
S. torreyi A. Gray = *S. dimidiatum* Raf.
S. umbelliferum Eschsch. vars. *glabrescens* Torrey & *incanum* Torrey not recognized
S. wallacei (A. Gray) Parish var. *clokeyi* (Munz) McMinn not recognized
S. xantii A. Gray: spelling corrected to *S. xanti*
vars. *hoffmannii* Munz, *intermedium* Parish, *montanum* Munz, & *obispoensis* (Eastw.) Wiggins not recognized

STAPHYLEACEAE BLADDERNUT FAMILY: no changes

STERCULIACEAE CACAO FAMILY

Fremontodendron californicum (Torrey) Cov. sspp. *crassifolium* (Eastw.) J.H. Thomas, *napense* (Eastw.) Munz, & *obispoense* (Eastw.) Munz not recognized

STYRACACEAE STORAX FAMILY

Styrax officinalis L. vars. *californicus* (Torrey) Rehd. & *fulvescens* (Eastw.) Munz & I.M. Johnston = var. *redivivus* (Torrey) H. Howard

TAMARICACEAE TAMARISK FAMILY

Tamarix africana Poiret indistinct from *T. gallica* L.
T. aralensis Bunge not naturalized

THYMELAEACEAE MEZEREUM FAMILY: no changes

TROPAEOLACEAE NASTURTIUM FAMILY: no changes

ULMACEAE ELM FAMILY

Ulmus carpinifolia Gleditsch not naturalized
U. procera Salisbury = *U. minor* Miller

UMBELLIFERAE: see APIACEAE

URTICACEAE NETTLE FAMILY

Boehmeria nivea (L.) Gaudich. naturalized
Parietaria floridana Nutt. misapplied to *P. hespera* B.D. Hinton
var. *californica* B.D. Hinton recently described
Soleirolia soleirolii (Req.) Dandy naturalized
Urtica californica E. Greene = *U. dioica* L. ssp. *gracilis* (Aiton) Selander
U. holosericea Nutt. = *U. dioica* L. ssp. *holosericea* (Nutt.) Thorne
U. lyallii S. Watson = *U. dioica* L. ssp. *gracilis* (Aiton) Selander
U. serra Blume misapplied to *U. dioica* L. ssp. *holosericea* (Nutt.) Thorne

VALERIANACEAE VALERIAN FAMILY

Kentranthus: spelling corrected to *Centranthus*
Plectritis anomala (A. Gray) Suksd. incl var. *gibbosa* (Suksd.) Dyal = *P. brachystemon* Fischer & C. Meyer
P. aphanoptera (A. Gray) Suksd. = *P. brachystemon* Fischer & C. Meyer
P. californica (Suksd.) Dyal var. *californica* = *P. ciliosa* (E. Greene) Jepson ssp. *ciliosa*
var. *rubens* (Suksd.) Dyal = *P. ciliosa* (E. Greene) Jepson ssp. *insignis* (Suksd.) D. Morey
P. ciliosa (E. Greene) Jepson var. *davyana* (Jepson) Dyal = ssp. *insignis* (Suksd.) D. Morey
P. congesta (Lindley) A. DC. var. *major* (Fischer & C. Meyer) Dyal = *P. brachystemon* Fischer & C. Meyer
P. eichleriana (Suksd.) A.A. Heller = *P. macrocera* Torrey & A. Gray
P. jepsonii (Suksd.) Burtt Davy = *P. macrocera* Torrey & A. Gray
P. macrocera Torrey & A. Gray vars. *collina* (A.A. Heller) Dyal, *grayi* (Suksd.) Dyal, & *mamillata* (Suksd.) Dyal not recognized
P. macroptera (Suksd.) Rydb. = *P. ciliosa* (E. Greene) Jepson
var. *macroptera* = *P. c.* ssp. *ciliosa*
var. *patelliformis* (Suksd.) Dyal = *P. c.* ssp. *insignis* (Suksd.) D. Morey
P. magna (E. Greene) Suksd. incl var. *nitida* (A.A. Heller) Dyal = *P. brachystemon* Fischer & C. Meyer
P. samolifolia (DC.) Hoeck incl var. *involuta* (Suksd.) Dyal misapplied to *P. brachystemon* Fischer & C. Meyer
Valeriana capitata Link ssp. *californica* (A.A. Heller) F. Meyer = *V. californica* A.A. Heller
V. pubicarpa Rydb. occurs in CA
Valerianella carinata Lois.: probably a waif in CA

VERBENACEAE VERVAIN FAMILY

Avicennia marina (Forsskal) Vierh. var. *resinifera* (Forster f.) Bakh. naturalized
Lantana camara L. waif in CA
Lippia incisa (Small) Tidestrom = *Phyla nodiflora* (L.) E. Greene var. *incisa* (Small) Mold.
L. lanceolata Michaux = *Phyla lanceolata* (Michaux) E. Greene
L. nodiflora (L.) Michaux = *Phyla nodiflora* (L.) E. Greene
vars. *canescens* (Kunth) Kuntze & *reptans* (Kunth) Kuntze = var. *nodiflora*
var. *rosea* (D. Don) Munz indistinct from var. *nodiflora*
Verbena brasiliensis Vell. = *V. litoralis* Kunth
V. ×clemensorum Mold. not recognized
V. gooddingii Briq. var. *nepetifolia* Tidestrom not recognized
V. hastata L. var. *scabra* Mold. not recognized
V. officinalis L. not recognized
V. robusta E. Greene = *V. lasiostachys* Link var. *scabrida* Mold.
V. tenera Sprengel = *V. tenuisecta* Briq.
Vitex agnus-castus L. not naturalized

VIOLACEAE VIOLET FAMILY

Viola adunca Smith var. *oxyceras* (S. Watson) Jepson not recognized
V. aurea Kellogg ssp. *mohavensis* M. Baker & J. Clausen = *V. purpurea* Kellogg ssp. *mohavensis* (M. Baker & J. Clausen) J. Clausen
V. bakeri E. Greene ssp. *grandis* M. Baker & J. Clausen illegitimate
ssp. *shastensis* M. Baker not recognized
V. beckwithii Torrey & A. Gray ssp. *glabrata* M. Baker not recognized
V. californica M. Baker = *M. glabella* Nutt.
V. lanceolata L. ssp. *occidentalis* (A. Gray) N. Russell = *V. primulifolia* L. ssp. *occidentalis* (A. Gray) McKinney & R.J. Little
V. lobata Benth. ssp. *psychodes* (E. Greene) Munz = ssp. *lobata*
V. nephrophylla E. Greene = *V. sororia* Willd. ssp. *affinis* (Le Conte) R.J. Little
V. pedunculata Torrey & A. Gray ssp. *tenuifolia* M. Baker & J. Clausen indistinct
V. praemorsa Douglas sspp. *arida* M. Baker, *major* (Hook.) M. Baker, & *oregona* M. Baker = ssp. *linguifolia* (Torrey & A. Gray) M. Baker & J. Clausen
V. purpurea Kellogg sspp. *atriplicifolia* (E. Greene) M. Baker & J. Clausen & *geophyta* M. Baker & J. Clausen = ssp. *venosa* (S. Watson) M. Baker & J. Clausen
ssp. *dimorpha* M. Baker & J. Clausen = ssp. *purpurea*
ssp. *mesophyta* M. Baker & J. Clausen = *V. pinetorum* E. Greene ssp. *pinetorum*
ssp. *xerophyta* M. Baker & J. Clausen = *V. pinetorum* E. Greene ssp. *grisea* (Jepson) R.J. Little, in part, and *V. p.* ssp. *pinetorum*, in part
V. quercetorum M. Baker & J. Clausen = *V. purpurea* Kellogg ssp. *quercetorum* (M. Baker & J. Clausen) R.J. Little

VISCACEAE MISTLETOE FAMILY (segregated from LORANTHACEAE)

Arceuthobium californicum Hawksw. & Wiens recently described
A. campylopodum Engelm. forma *abietinum* (Engelm.) Gill = *A. abietinum* (Engelm.) Hawksw. & Wiens
forma *blumeri* (Nelson) Gill not recognized
forma *cyanocarpum* (J. Coulter & Nelson) Gill = *A. cyanocarpum* J. Coulter & Nelson
forma *divaricatum* (Engelm.) Gill = *A. divaricatum* Engelm.
forma *microcarpum* (Engelm.) Gill not recognized
forma *tsugensis* (C. Rosend.) Gill = *A. tsugense* (C. Rosend.) G. Jones
A. littorum Hawksw. & Nickrent recently described
A. monticola Hawksw., Wiens, & Nickrent recently described

A. occidentale Engelm. separate from *A. campylopodum* Engelm.
A. siskiyouense Hawksw., Wiens, & Nickrent recently described
A. tsugense (C. Rosend.) G. Jones ssp. *mertensianae* Hawksw. & Nickrent recently described
Phoradendron bolleanum (Seemann) Eichler var. *densum* (Trel.) Fosb. = *P. densum* Trel.
var. *pauciflorum* (Torrey) Fosb. = *P. pauciflorum* Torrey
P. californicum Nutt. vars. *distans* Trel. & *leucocarpum* (Trel.) Jepson not recognized
P. flavescens (Pursh) Nutt. var. *macrophyllum* Engelm. = *P. macrophyllum* (Engelm.) Cockerell
var. *villosum* (Nutt.) Engelm. = *P. villosum* (Nutt.) Nutt.
P. juniperinum A. Gray var. *libocedri* Engelm. = *P. libocedri* (Engelm.) Howell
var. *ligatum* (Trel.) Fosb. not recognized

VITACEAE GRAPE FAMILY

Parthenocissus inserta (Kerner) K. Fritsch misapplied to *P. vitacea* (Knerr) Hitchc.

ZYGOPHYLLACEAE CALTROP FAMILY

Zygophyllum fabago L. var. *brachycarpum* Boiss. not recognized

MONOCOTS

ALISMATACEAE WATER-PLANTAIN FAMILY

Alisma geyeri Nicollet = *A. gramineum* Lej.
A. subcordatum Raf. = *A. plantago-aquatica* L.
A. triviale Pursh = *A. plantago-aquatica* L.
Machaerocarpus californicus (Benth.) Small = *Damasonium californicum* Benth.

APONOGETONACEAE CAPE-PONDWEED FAMILY: no changes

ARACEAE ARUM FAMILY

Peltandra virginica (L.) Schott & Endl. naturalized

ARECACEAE [PALMAE] PALM FAMILY

Phoenix canariensis Chabaud naturalized
P. dactylifera L. naturalized

COMMELINACEAE SPIDERWORT FAMILY

Commelina benghalensis L. naturalized

CYMODOCEACEAE MANATEE-GRASS FAMILY (*Halodule* segregated from ZANNICHELLIACEAE)

CYPERACEAE SEDGE FAMILY

Carex ablata L. Bailey = *C. luzulina* Olney var. *ablata* (L. Bailey) F. Herm.
C. angustior Mackenzie = *C. echinata* Murray ssp. *echinata*
C. breviligulata L. Bailey = *C. densa* L. Bailey
C. brevipes W. Boott = *C. rossii* W. Boott
C. breweri Boott: CA pls = var. *breweri*
C. epapillosa Mackenzie = *C. heteroneura* W. Boott var. *epapillosa* (Mackenzie) F. Herm.
C. exserta Mackenzie = *C. filifolia* Nutt. var. *erostrata* Kük.
C. exsiccata L. Bailey = *C. vesicaria* L. var. *major* Boott
C. eurycarpa Holm = *C. angustata* Boott
C. festivella Mackenzie = *C. microptera* Mackenzie
C. garberi Fern. occurs in CA
C. hindsii C.B. Clarke = *C. lenticularis* Michaux var. *limnophila* (Holm) Cronq.
C. inops L. Bailey: CA pls are ssp. *inops*
C. jepsonii J. Howell = *C. whitneyi* Olney
C. kelloggii W. Boott = *C. lenticularis* Michaux var. *lipocarpa* (Holm) L. Standley
C. leptopoda Mackenzie = *C. deweyana* Schwein. ssp. *leptopoda* (Mackenzie) Calder & Roy Taylor
C. luzulaefolia W. Boott: spelling corrected to *C. luzulaifolia*
C. montereyensis Mackenzie = *C. harfordii* Mackenzie
C. norvegica Retz. occurs in CA
C. ormantha (Fern.) Mackenzie = *C. echinata* Murray ssp. *echinata*
C. parryana Dewey var. *hallii* (Olney) Kük. occurs in CA
C. paucicostata Mackenzie = *C. lenticularis* Michaux var. *impressa* (L. Bailey) L. Standley
C. paucifructus Mackenzie = *C. mariposana* L. Bailey
C. phyllomanica W. Boott = *C. echinata* Murray ssp. *phyllomanica* (W. Boott) A.A. Reznicek
C. pseudoscirpoidea Rydb. = *C. scirpoidea* *Michaux* var. *pseudosciprpodea* (Rybd.) Cronq.
C. scoparia Schk. occurs in CA
C. scopulorum Holm: CA pls are var. *bracteosa* (L. Bailey) F. Herm.
C. sitchensis Prescott = *C. aquatilis* Wahlenb. var. *dives* (Holm) L. Standley
C. teneraeformis Mackenzie = *C. subfusca* W. Boott
C. texensis (Torrey) L. Bailey = *C. retroflexa* Willd. var. *texensis* (Torrey) Fern.
C. tiogana D.W. Taylor & J. Mastrogiuseppe recently described
C. tracyi Mackenzie = *C. ovalis* Good
C. vicaria L. Bailey: CA pls = *C. densa* L. Bailey
C. vulpinoidea Michaux occurs in CA
Cladium mariscus R. Br. var. *californicus* S. Watson = *C. californicum* (S. Watson) O'Neill
Cyperus alternifolius L. misapplied to *C. involucratus* Rottb.
C. aristatus Rottb. = *C. squarrosus* L.
C. ferax Rich. = *C. odoratus* L.
C. ligularis L. naturalized
C. niger Ruiz López & Pavón var. *capitatus* (Britton) O'Neill not recognized
var. *castaneus* (Pursh) Kük. = *C. bipartitus* Torrey
C. unioloides R. Br. not naturalized
C. virens Michaux not naturalized
Eleocharis acicularis (L.) Roemer & Schultes var. *radicans* (Poiret) Britton misapplied to *E. radicans* (Poiret) Kunth
E. flavescens (Poiret) Urban misapplied to *E. thermalis* Rydb.
E. montevidensis Kunth var. *bolanderi* (A. Gray) V. Grant = *E. bolanderi* A. Gray
var. *decumbens* (C.B. Clarke) V. Grant not recognized
vars. *disciformis* (Parish) V. Grant & *parishii* (Britton) V. Grant = *E. parishii* Britton
E. palustris (L.) Roemer & Schultes = *E. macrostachya* Britton, in part
E. parvula (Roemer & Schultes) Link var. *coloradoensis* (Britton) Beetle not recognized
E. pauciflora (Lightf.) Link vars. *bernardina* (Munz & I.M. Johnston) Svenson & *suksdorfiana* (Beauverd) Svenson not recognized
Fimbristylis capillaris (L.) A. Gray = *Bulbostylis capillaris* (L.) C.B. Clarke
F. miliacea (L.) Vahl waif in CA
F. spadicea (L.) Vahl misapplied to *F. thermalis* S. Watson
Hemicarpha micrantha (Vahl) Pax var. *aristulata* Cov. = *Lipocarpha aristulata* (Cov.) G.C. Tucker
vars. *micrantha* & *minor* (Schrader) Friedl. = *Lipocarpha micrantha* (Vahl) G.C. Tucker
H. occidentalis A. Gray = *Lipocarpha occidentalis* (A. Gray) G.C. Tucker

Kobresia myosuroides (Villars) Fiori & Paol. = *K. bellardii* (All.) Degl.
Rhynchospora glomerata (L.) Vahl vars. *capitellata* (Michaux) Kük. & *minor* Britton = *R. capitellata* (Michaux) M. Vahl
Scirpus acutus Bigelow: CA pls are all or nearly all var. *occidentalis* (S. Watson) Beetle
S. americanus Pers. vars. *longispicatus* Britton & *monophyllus* (C. Presl) T. Koyama = *S. pungens* Vahl
S. cernuus Vahl var. *californicus* (Torrey) Beetle not recognized
S. criniger A. Gray = *Eriophorum criniger* (A. Gray) Beetle
S. heterochaetus A. Chase occurs in CA
S. maritimus L. var. *paludosus* (Nelson) Kük. not recognized
var. *tuberosus* (Desf.) Roemer & Schultes = *S. tuberosus* Desf.
S. olneyi A. Gray misapplied to *S. americanus* Pers.
S. pendulus Muhl. occurs in CA
S. pumilus Vahl. occurs in CA
S. rollandii Fern. misapplied to *S. pumilus* Vahl
S. validus Vahl = *S. tabernaemonti* C. Gmelin

ERIOCAULACEAE PIPEWORT FAMILY
not naturalized

GRAMINEAE: see **POACEAE**

HYDROCHARITACEAE WATERWEED FAMILY

Elodea densa (Planchon) Caspary = *Egeria densa* Planchon
Hydrilla verticillata (L.f.) Caspary naturalized
Najas gracillima (A. Braun) Magnus naturalized
Ottelia alismoides (L.) Pers. presumed extirpated

IRIDACEAE IRIS FAMILY

Chasmanthe aethopica (L.) N.E. Br. misapplied to *C. floribunda* (Salisb.) N.E. Br.
Crocosmia crocosmiflora N.E. Br. = *C.* ×*crocosmiiflora* (Burb. & Dean) N.E. Br.
Gladiolus segetum Ker Gawler = *G. italicus* Miller
G. tristis L. naturalized
Ixia maculata L. naturalized
Sisyrinchium eastwoodiae E. Bickn. = *S. bellum* S. Watson
S. greenei E. Bickn. = *S. bellum* S. Watson
S. hesperium E. Bickn. = *S. bellum* S. Watson
S. idahoense E. Bickn. var. *occidentale* (E. Bickn.) D. Henderson separate from var. *idahoense*
Sparaxis grandiflora (Delaroche) Ker Gawler naturalized
S. tricolor (Schneev.) Ker Gawler naturalized
Watsonia marginata (Ecklon) Ker Gawler naturalized

JUNCACEAE RUSH FAMILY

Juncus abjectus F.J. Herm. = *J. hemiendytus* F.J. Herm. var. *abjectus* (F.J. Herm.) B. Ertter
J. acutus L. var. *sphaerocarpus* Engelm. = ssp. *leopoldii* (Parl.) Snog.
J. balticus Willd. var. *montanus* Engelm. not recognized
J. brachyphyllus Wieg. occurs in CA
J. bufonius L. var. *congdonii* (S. Watson) J. Howell = var. *congestus* Wahlenb.
var. *halophilus* Buchenau & Fern. = *J. ambiguus* Guss.
J. covillei Piper var. *obtusatus* C. Hitchc. recognized
J. cyperoides Laharpe occurs in CA
J. diffusissimus Buckley occurs in CA
J. leiospermus F.J. Herm. var. *ahartii* B. Ertter recently described
J. lesueurii Bolander var. *tracyi* Jepson misapplied to *J. breweri* Engelm.
J. luciensis B. Ertter recently described
J. marginatus Rostkov var. *marginatus* occurs in CA
J. megaspermus F.J. Herm. = *J. triformis* Engelm.
J. mertensianus Bong. var. *duranii* (Ewan) F.J. Herm. = *J. duranii* Ewan
J. saximontanus Nelson forma *brunnescens* (Rydb.) F.J. Herm. not recognized
J. sphaerocarpus Nees misapplied to *J. bufonius* L. var. *occidentalis* F.J. Herm.
J. tenuis Willd. var. *congestus* Engelm. = *J. occidentalis* (Cov.) Wieg.
var. *dudleyi* (Ewan) F.J. Herm. = *J. dudleyi* Wieg.
J. tiehmii B. Ertter recently described
Luzula glabrata (Hoppe) Desv. not in CA
L. subsessilis (Buchenau) S. Watson = *L. comosa* E. Meyer

JUNCAGINACEAE ARROW-GRASS FAMILY

Lilaea transferred from Lilaeaceae

LEMNACEAE DUCKWEED FAMILY

Lemna minima L. = *L. minuscula* Herter
L. perpusilla Torrey misapplied to *L. aequinoctialis* Welw.
L. turionifera Landolt recently described
L. valdiviana Philippi var. *abbreviata* Hegelm. not recognized
Spirodela oligorhiza (Kurz) Hegelm. = *S. punctata* (G. Meyer) C. Thompson
Wolffia brasiliensis Wedd. occurs in CA
W. cylindracea Hegelm. = *W. globosa* (Roxb.) Hartog & Plas
W. punctata Griseb. misapplied to *W. borealis* (Hegelm.) Landolt & O. Wildi

LILAEACEAE: see **JUNCAGINACEAE**

LILIACEAE LILY FAMILY

Agave utahensis (Engelm.) Gentry vars. *eborispina* (Hester) Breitung & *nevadensis* Engelm. not recognized
Allium atrorubens S. Watson var. *inyonis* (M.E. Jones) F. Ownbey & Aase = var. *cristatum* (S. Watson) D. McNeal
A. bolanderi S. Watson var. *mirabile* (L. Henderson) D. McNeal recognized
A. davisiae M.E. Jones = *A. lacunosum* S. Watson var. *davisiae* (M.E. Jones) D. McNeal & F. Ownbey
A. fimbriatum S. Watson var. *abramsii* Traub = *A. abramsii* (Traub) D. McNeal
var. *denticulatum* Traub = *A. denticulatum* (Traub) D. McNeal
var. *diabloense* Traub = *A. diabloense* (Traub) D. McNeal
var. *munzii* Traub = *A. munzii* (Traub) D. McNeal
var. *parryi* (S. Watson) F. Ownbey & Aase = *A. parryi* S. Watson
var. *sharsmithiae* Traub = *A. sharsmithiae* (Traub) D. McNeal
A. lacunosum S. Watson var. *kernensis* D. McNeal & F. Ownbey recently described
var. *micranthum* Eastw. recognized
A. monticola Davidson ssp. *keckii* (Munz) F. Ownbey & Aase not recognized
A. nevadense S. Watson var. *cristatum* (S. Watson) F. Ownbey = *A. atrorubens* S. Watson var. *cristatum* (S. Watson) D. McNeal
A. obtusum Lemmon var. *conspicuum* W. Mortola & D. McNeal recently described
A. paniculatum L. var. *paniculatum* naturalized
A. peninsulare Lemmon var. *franciscanum* D. McNeal & F. Ownbey recently described
A. punctum L. Henderson occurs in CA
A. sanbornii Alph. Wood
var. *jepsonii* Traub = *A. jepsonii* (Traub) S. Denison & D. McNeal
var. *tuolumnense* Traub = *A. tuolumnense* (Traub) S. Denison & D. McNeal

A. serratum S. Watson misapplied to *A. serra* D. McNeal & F. Ownbey
A. shevockii D. McNeal recently described
A. vineale L. naturalized
Aloe saponaria (Aiton) Haw. × *A. striata* Haw. naturalized
Asparagus asparagoides (L.) Druce naturalized
Brodiaea bridgesii S. Watson = *Triteleia bridgesii* (S. Watson) E. Greene
B. clementina (Hoover) Munz = *Triteleia clementina* Hoover
B. congesta Sm. = *Dichelostemma congestum* (Sm.) Kunth
B. coronaria (Salisb.) Engler var. *kernensis* Hoover = *B. terrestris* Kellogg ssp. *kernensis* (Hoover) T. Niehaus
var. *macropoda* (Torrey) Hoover = *B. terrestris* Kellogg ssp. *terrestris*
B. crocea (Alph. Wood) S. Watson = *Triteleia crocea* (Alph. Wood) E. Greene
var. *modesta* (H.M. Hall) Munz = *Triteleia crocea* (Alph. Wood) E. Greene var. *modesta* (H.M. Hall) Hoover
B. douglasii S. Watson var. *howellii* (S. Watson) M.E. Peck = *Triteleia grandiflora* Lindley ssp. *howellii* (S. Watson) Hoover
B. dudleyi (Hoover) Munz = *Triteleia dudleyi* Hoover
B. elegans Hoover var. *australis* Hoover = *B. terrestris* Kellogg ssp. *kernensis* (Hoover) T. Niehaus
var. *mundula* (Jepson) Hoover = *B. elegans* Hoover ssp. *elegans*
ssp. *hooveri* T. Niehaus recently described
B. gracilis S. Watson = *Triteleia montana* Hoover
B. hyacinthina (Lindley) Baker = *Triteleia hyacinthina* (Lindley) E. Greene var. *greenei* (Hoover) Munz = *Triteleia lilacina* E. Greene
B. ida-maia (Alph. Wood) E. Greene = *Dichelostemma ida-maia* (Alph. Wood) E. Greene
B. insignis (Jepson) T. Niehaus separate from *B. coronaria* (Salisb.) Engler
B. laxa (Benth.) S. Watson = *Triteleia laxa* Benth.
B. lugens (E. Greene) Baker = *Triteleia lugens* E. Greene
B. lutea (Lindley) C. Morton var. *lutea* = *Triteleia ixioides* (S. Watson) E. Greene
var. *analina* (E. Greene) Munz = *T. ixioides* ssp. *anilina* (E. Greene) L. Lenz
var. *cookii* (Hoover) Munz = *T. ixioides* ssp. *cookii* (Hoover) L. Lenz
var. *scabra* (E. Greene) Munz = *T. ixioides* ssp. *scabra* (E. Greene) L. Lenz
B. minor (Benth.) S. Watson var. *nana* (Hoover) Hoover not recognized
B. multiflora Benth. = *Dichelostemma multiflora* (Benth.) A.A. Heller
B. peduncularis (Lindley) S. Watson = *Triteleia peduncularis* Lindley
B. pulchella (Salisb.) A.A. Heller = *Dichelostemma capitatum* Alph. Wood
var. *pauciflora* (Torrey) C. Morton = *Dichelostemma capitatum* Alph. Wood var. *pauciflorum* (Torrey) Hoover
B. purdyi Eastw. separate from *B. minor* (Benth.) S. Watson
B. terrestris Kellogg separate from *B. coronaria* (Salisb.) Engler
B. uniflora (Raf.) Engler = *Ipheion uniflorum* Raf.
B. venusta (E. Greene) E. Greene = *Dichelostemma ida-maia* (Alph. Wood) E. Greene × *D. congestum* (Sm.) Kunth or *D. multiflorum* (Benth.) A.A. Heller
B. volubilis (Kellogg) Baker = *Dichelostemma volubile* (Kellogg) A.A. Heller
Calochortus coeruleus (Kellogg) S. Watson var. *fimbriatus* F. Ownbey not recognized
var. *westonii* (Eastw.) F. Ownbey = *C. westonii* Eastw.
C. elegans Pursh var. *nanus* Wood not recognized
C. longebarbatus S. Watson: CA pls are var. *longebarbatus*
C. nudus S. Watson var. *shastensis* (Purdy) Jepson not recognized
C. nuttallii Torrey var. *bruneaunis* (Nelson & J.F. Macbr.) F. Ownbey = *C. bruneaunis* Nelson & J.F. Macbr.
var. *panamintensis* F. Ownbey = *C. panamintensis* (F. Ownbey) Rev.
C. raichei S. Farwig & V. Girard recently described
C. superbus Munz misapplied in part to *Calochortus argillosus* (Hoover) R. Zebell & P. Fiedler
C. tiburonensis A.J. Hill recently described
Camassia leichtlinii (Baker) S. Watson ssp. *suksdorfii* (Greenman) Gould = *Camassia quamash* (Pursh) E. Greene ssp. *quamash*
C. quamash (Pursh) E. Greene ssp. *linearis* Gould = ssp. *quamash*
Dichelostemma lacuna-vernalis L. Lenz = *D. capitatum* Alph. Wood ssp. *capitatum*
D. ×*venustum* (E. Greene) Hoover = *D. ida-maia* (Alph. Wood) E. Greene × *D. congestum* (Sm.) Kunth or *D. multiflorum* (Benth.) E. Greene
Disporum hookeri (Torrey) Nicholson vars. *parvifolium* (S. Watson) Britton & *trachyandrum* (Torrey) Q. Jones not recognized
Erythronium grandiflorum Pursh var. *pallidum* St. John indistinct
ssp. *pusaterii* Munz & J. Howell = *E. pusaterii* (Munz & J. Howell) J. Shevock, J. Bartel, & G. Allen
E. howellii S. Watson indistinct from *E. citrinum* S. Watson
E. oregonum Applegate ssp. *leucandrum* Applegate not recognized
E. pluriflorum J. Shevock, J. Bartel, & G. Allen recently described
Fritillaria biflora Lindley var. *ineziana* Jepson separate from var. *biflora*
F. lanceolata Pursh = *F. affinis* (Schultes) Sealy
F. ojaiensis David separate from *F. affinis* (Schultes) Sealy
F. phaeanthera Eastw. = *F. eastwoodiae* MacFarl.
F. recurva Benth. var. *coccinea* E. Greene not recognized
F. roderickii W. Knight = *F. biflora* Lindley var. *biflora*
F. tristulis A.L. Grant in Jepson = *F. affinis* (Schultes) Sealy var. *tristulis* (A.L. Grant) B. Ness
F. viridea Kellogg separate from *F. affinis* (Schultes) Sealy
Hastingsia bracteosa S. Watson not documented in CA
H. serpentinicola R. Becking recently described
Lilium humboldtii Roezl & Leichtlin vars. *bloomerianum* (Kellogg) Jepson & *ocellatum* (Kellogg) Elwes = ssp. *ocellatum* (Kellogg) Thorne
L. pardalinum Kellogg ssp. *shastense* (Eastw.) M. Skinner recognized
L. pitkinense Beane & Vollmer = *L. pardalinum* Kellogg ssp. *pitkinense* (Beane & Vollmer) M. Skinner
L. vollmeri Eastw. = *L. pardalinum* Kellogg ssp. *vollmeri* (Eastw.) M. Skinner
L. washingtonianum Kellogg var. *minus* Purdy = ssp. *washingtonianum*
var. *purpurascens* Stearn = ssp. *purpurascens* (Stearn) M. Skinner
L. wigginsii Beane & Vollmer = *L. pardalinum* Kellogg ssp. *wigginsii* (Beane & Vollmer) M. Skinner
Muscari botryoides (L.) Miller naturalized
Nolina parryi S. Watson ssp. *wolfii* Munz not recognized
Schoenolirion album Durand = *Hastingsia alba* (Durand) S. Watson
S. bracteosa (S. Watson) Jepson = *Hastingsia bracteosa* S. Watson
Smilacina racemosa (L.) Link vars. *amplexicaulis* (Nutt.) S. Watson & *glabra* J.F. Macbr. not recognized
S. stellata (L.) Desf. var. *sessilifolia* (Baker) L. Henderson not recognized
Smilax jamesii G.A. Wallace recently described
Streptopus amplexifolius (L.) DC.: CA pls are var. *americanus* Schultes
Tofieldia glutinosa Pers. ssp. *occidentalis* (S. Watson) C. Hitchc. = *T. occidentalis* S. Watson ssp. *occidentalis*

Trillium albidum Freeman recently described
T. chloropetalum (Torrey) Howell var. *angustipetalum* (Torrey) Munz = *T. angustipetalum* (Torrey) Freeman
var. *giganteum* (Hook. & Arn.) Munz not recognized
T. ovatum Pursh ssp. *oettingeri* Munz & Thorne recently described
Yucca brevifolia Engelm. vars. *jaegeriana* McKelvey & *herbertii* (J.M. Webber) Munz not recognized
Y. whipplei Torrey sspp. *cespitosa* (M.E. Jones) A.L. Haines, *intermedia* A.L. Haines, *parishii* (M.E. Jones) A.L. Haines, & *percursa* A.L. Haines all indistinct
Zigadenus fontanus Eastw. = *Z. micranthus* Eastw. var. *fontanus* (Eastw.) D. McNeal
Z. fremontii (Torrey) S. Watson vars. *inezianus* Jepson, *minor* (Hook. & Arn.) Jepson, & *salsus* Jepson not recognized

NAJADACEAE: see **HYDROCHARITACEAE**

ORCHIDACEAE ORCHID FAMILY

Calypso bulbosa (L.) Oakes ssp. *occidentalis* (Holz.) Calder & Roy Taylor not recognized
Corallorhiza trifida Chatel. occurs in CA
Eburophyton austiniae (A. Gray) A.A. Heller = *Cephalanthera austiniae* (A. Gray) A.A. Heller
Habenaria dilatata (Pursh) Hook. var. *dilatata* not in CA
var. *leucostachys* (Lindley) Ames = *Platanthera leucostachys* Lindley
H. elegans (Lindley) Bolander incl var. *maritima* (E. Greene) Ames = *Piperia elegans* (Lindley) Rydb.
H. hyperborea (L.) R. Br. = *Platanthera hyperborea* (L.) Lindley
H. saccata E. Greene = *Platanthera stricta* Lindley
H. sparsiflora S. Watson = *Platanthera sparsiflora* (S. Watson) Schltr.
H. unalascensis (Sprengel) S. Watson = *Piperia unalascensis* (Sprengel) Rydb.
Malaxis brachypoda (A. Gray) Fern. = *M. monophyllos* (L.) Sw. ssp. *brachypoda* (A. Gray) A. Löve & D. Löve
Piperia candida R. Morgan & J. Ackerman recently described
P. cooperi (S. Watson) Rydb. separate from *P. unalascensis* (Sprengel) Rydb.
P. elongata Rydb. separate from *P. elegans* (Lindley) Rydb.
P. michaelii (E. Greene) Rydb. separate from *P. elegans* (Lindley) Rydb.
P. transversa Suksd. separate from *P. unalascensis* (Sprengel) Rydb.
P. yadonii R. Morgan & J. Ackerman recently described

PALMAE: see **ARECACEAE**

POACEAE [GRAMINEAE] GRASS FAMILY

Agropyron arizonicum Scribner & J.G. Smith = *Elymus arizonica* (Scribner & J.G. Smith) Gould
A. dasystachyum (Hook.) Vasey = *Elymus lanceolatus* (Scribner & J.G. Smith) Gould ssp. *lanceolatus*
A. elongatum (Host) Beauv. = *Elytrigia elongata* (Host) Nevski in part, *Elytrigia pontica* (Podp.) Holub ssp. *pontica* in part
A. junceum (L.) Beauv. = *Elytrigia juncea* (L.) Nevski ssp. *boreali-atlantica* (Simonet & Guin.) Hylander
A. parishii Scribner & J.G. Smith = *Elymus stebbinsii* Gould
A. pringlei (Scribner & J.G. Smith) A. Hitchc. = *Elymus sierrae* Gould
A. repens (L.) Beauv. = *Elytrigia repens* (L.) Nevski
A. riparium Scribner & J.G. Smith = *Elymus lanceolatus* (Scribner & J.G. Smith) Gould ssp. *lanceolatus*
A. ×saundersii (Vasey) A. Hitchc. = *Elymus trachycaulus* × *E. elymoides* or × *E. multisetus*
A. ×saxicola (Scribner & J.G. Smith) Piper = *Pseudoroegneria spicata* × *Elymus elymoides* or × *E. multisetus*
A. scribneri Vasey = *Elymus scribneri* (Vasey) M.E. Jones
A. smithii Rydb. = *Pascopyrum smithii* (Rydb.) A. Löve
A. spicatum (Pursh) Scribner & J. Smith = *Pseudoroegneria spicata* (Pursh) A. Löve var. *spicata*
A. subsecundum (Link) A. Hitchc. = *Elymus trachycaulus* (Link) Shinn. ssp. *subsecundum* (Link) Gould
A. trachycaulum (Link) Malte = *Elymus trachycaulus* (Link) Shinn. ssp. *trachycaulus*
A. trichophorum (Link) Richter = *Elytrigia intermedia* (Host) Nevski ssp. *intermedia*
Agrostis alba L. var. *alba* misapplied in part to *A. gigantea* Roth and *A. stolonifera* L.
vars. *major* (Gaudin) Farw. & *palustris* (Hudson) Pers. = *A. stolonifera* L.
A. ampla A. Hitchc. = *A. exarata* Trin.
A. aristiglumis Swallen = *A. microphylla* Steudel
A. blasdalei A. Hitchc. var. *marinensis* B. Crampton not recognized
A. clivicola B. Crampton incl var. *punta-reyensis* B. Crampton = *A. densiflora* Vasey
A. diegoensis Vasey = *A. pallens* Trin.
A. exarata Trin. var. *monolepis* (Torrey) A. Hitchc. indistinct
A. exigua Thurber = *A. elliottiana* Schultes
A. lepida A. Hitchc. = *A. pallens* Trin.
A. longiligula Vasey incl var. *australis* J. Howell = *A. exarata* Trin.
A. microphylla Steudel var. *hendersonii* (A. Hitchc.) Beetle = *A. hendersonii* A. Hitchc.
var. *intermedia* Beetle not recognized
A. scabra Willd. var. *geminata* (Trin.) Swallen indistinct
A. semiverticillata (Forsskal) C. Chr. = *A. viridis* Gouan
A. tenuis Sibth. = *A. capillaris* L.
A. tenuis Vasey = *A. idahoensis* Nash
Aira elegans Gaudin = *A. elegantissima* Schur
Alopecurus aequalis Sobol. var. *sonomensis* Rubtzoff indistinct
A. howellii Vasey = *A. saccatus* Vasey
A. pallescens Piper = *A. geniculatus* L.
Apera spica-venti (L.) Beauv. naturalized
Avena strigosa Schreber naturalized
A. sterilis L. naturalized
Axonopus compressus (Sw.) P. Beauv. misapplied to *A. affinis* Chase
Bouteloua hirsuta Lagasca not documented in CA
B. radicosa (Fourn.) Griffiths not documented in CA
B. rothrockii Vasey = *B. barbata* Lagasca var. *rothrockii* (Vasey) Gould not documented in CA
B. simplex Lagasca not documented in CA
Brachypodium pinnatum (L.) Beauv. naturalized
B. phoenicoides (L.) Roemer & Schultes indistinct from *B. pinnatum* (L.) Beauv.
Briza media L. not naturalized
Bromus alopecurus Poiret naturalized
B. arvensis L. indistinct from *B. hordeaceus* L.
B. breviaristatus Buckley = *B. carinatus* Hook. & Arn. var. *carinatus*
B. brizaeformis Fisch & C. Meyer: spelling corrected to *B. briziformis*
B. commutatus Schard. indistinct from *B. japonicus* Murr.
B. erectus Huds. not naturalized
B. haenkeanus (C. Presl) Kunth = *B. catharticus* Vahl
B. hordeaceus L. ssp. *molliformis* (Godron) Maire indistinct
B. inermis Leyss.: CA pls are ssp. *inermis*
B. marginatus Steudel = *B. carinatus* Hook. & Arn. var. *carinatus*
B. maritimus (Piper) A. Hitchc. = *B. carinatus* Hook. & Arn. var. *maritimus* (Piper) C. Hitchc.

B. mollis L. misapplied to *B. hordeaceus* L.
B. orcuttianus (Shear) A.Hitchc. var. *hallii* A. Hitchc. not recognized
B. polyanthus Scribner = *B. carinatus* Hook. & Arn. var. *carinatus*
B. porteri (J. Coulter) Nash = *B. anomalus* Fourn.
B. pseudolaevipes Wagnon indistinct from *B. laevipes* Shear
B. racemosus L. misapplied to *B. hordeaceus* L.
B. richardsonii Link indistinct from *B. ciliatus* L.
B. rigidus Roth incl var. *gussonei* (Parl.) Coss & Durieu misapplied to *B. diandrus* Roth
B. rubens L. = *B. madritensis* L. ssp. *rubens* (L.) Husnot
B. scoparius L. misapplied to *B. hordeaceus* L.
B. tectorum L. var. *glabratus* Shear not recognized
B. trinii Desv. var. *excelsus* Shear not recognized
B. unioloides Kunth = *B. catharticus* Vahl
B. willdenovii Kunth = *B. catharticus* Vahl
Calamagrostis crassiglumis Thurber indistinct from *C. stricta* (Timm) Koeler ssp. *inexpansa* (A. Gray) C.W. Greene
C. densa Vasey = *C. koelerioides* Vasey
C. inexpansa A. Gray = *C. stricta* (Timm) Koeler ssp. *inexpansa* (A. Gray) C.W. Greene
C. ×lactea Beal: probably *C. canadensis* (Michaux) Beauv. × *C. nutkaensis* (C. Presl) Steudel
C. stricta (Timm) Koeler ssp. *stricta* occurs in CA
Cenchrus ciliaris L. naturalized
C. incertus M. Curtis naturalized
C. pauciflorus Bentham misapplied to *C. longispinus* (Hackel) Fern.
Cinna bolanderi Scribner separate from *C. latifolia* (Goeppert) Griseb.
Crypsis niliaca Figari = *C. vaginiflora* (Forsskal) Opiz
Deschampsia cespitosa (L.) Beauv. ssp. *beringensis* (Hultén) W.E. Lawr. indistinct from ssp. *cespitosa*
Distichlis spicata (L.) E. Greene vars. *divaricata* Beetle, *nana* Beetle, *stolonifera* Beetle, & *stricta* (Torrey) Beetle not recognized
Echinochloa colonum (L.) Link: spelling corrected to *E. colona*
E. crus-pavonis (Kunth) Schultes naturalized
E. muricata (P. Beauv.) Fern. naturalized
E. oryzicola (Vasinger) Vasinger var. *mutica* Vasinger misapplied to *E. oryzoides* (Ard.) Fritsch
Ectosperma alexandrae Swallen = *Swallenia alexandrae* (Swallen) Söderstrom & Decker
Elymus ×aristatus Merr. = *E. multisetus* × *E. trachycaulus*
E. caput-medusae L. = *Taeniatherum caput-medusae* (L.) Nevski
E. cinereus Scribner & Merr. = *Leymus cinereus* (Scribner & Merr.) A. Löve
E. condensatus C. Presl = *Leymus condensatus* (C. Presl) A. Löve
E. elymoides (Raf.) Swezey ssp. *brevifolius* (J.G. Smith) Barkworth recognized
ssp. *californicus* (J.G. Smith) Barkworth recognized
ssp. *hordeoides* (Siksd.) Barkworth recognized
E. ×macounii Vasey = *E. trachycaulus* × *E. elymoides* or × *Hordeum jubatum*
E. mollis Trin. = *Leymus mollis* (Trin.) Pilger ssp. *mollis*
E. pacificus Gould = *Leymus pacificus* (Gould) D.R. Dewey
E. salinus M.E. Jones = *Leymus salinus* (M.E. Jones) A. Löve ssp. *mojavensis* Barkworth & R.J. Atkins
E. triticoides Buckley = *Leymus triticoides* (Buckley) Pilger
ssp. *multiflorus* Gould = *Leymus ×multiflorus* (Gould) Barkworth & D.R. Dewey
E. vancouverensis Vasey = *Leymus ×vancouverensis* (Vasey) Pilger
Eragrostis arida A. Hitchc. = *E. pectinacea* (Michaux) Nees var. *miserrima* (Fourn.) Reeder
E. diffusa Buckley = *E. pectinacea* (Michaux) Nees var. *pectinacea*
E. lehmanniana Nees naturalized
E. megastachya (Koeler) Link = *E. cilianensis* (All.) Janchen
E. neomexicana Vasey = *E. mexicana* (Hornem.) Link ssp. *mexicana*
E. orcuttiana Vasey = *E. mexicana* (Hornem.) Link ssp. *virescens*
E. oxylepis (Torrey) Torrey not naturalized
E. poaeoides Roemer & Schultes = *E. minor* Host
Eriochloa gracilis (Fourn.) A. Hitchc. = *E. acuminata* (C. Presl) A. Hitchc. var. *acuminata*
Festuca arida Elmer = *Vulpia microstachys* (Nutt.) Benth. var. *microstachys*
F. arizonica Vasey not documented in CA
F. brachyphylla Schultes & Schultes f.: CA pls are ssp. *breviculmis* S. Frederiksen
F. californica Vasey var. *parishii* (Piper) A. Hitchc. indistinct
F. confusa Piper = *Vulpia microstachys* (Nutt.) Benth. var. *confusa* (Piper) Lonard & Gould
F. dertonensis (All.) Asch. & Graebner = *Vulpia bromoides* (L.) S.F. Gray
F. eastwoodiae Piper = *Vulpia microstachys* (Nutt.) Benth. var. *ciliata* (Beal) Lonard & Gould
F. elatior L. = *F. pratensis* Hudson
F. elmeri Scribner & Merr. ssp. *luxurians* Piper not recognized
F. grayi (Abrams) Piper = *Vulpia microstachys* (Nutt.) Benth. var. *ciliata* (Beal) Lonard & Gould
F. megalura Nutt. = *Vulpia myuros* (L.) C. Gmelin var. *hirsuta* (Hackel) Asch. & Graebner
F. microstachys Nutt. = *Vulpia microstachys* (Nutt.) Benth. var. *microstachys*
F. minutiflora Rydb. separate from *F. brachyphylla* Schultes & Schultes f.
F. myuros L. = *Vulpia myuros* (L.) C. Gmelin var. *myuros*
F. octoflora Walter = *Vulpia octoflora* (Walter) Rydb. ssp. *hirtella* Piper = *V. o.* var. *hirtella* (Piper) Henrard
F. pacifica Piper = *Vulpia microstachys* (Nutt.) Benth. var. *pauciflora* (Beal) Lonard & Gould
F. reflexa Buckley = *Vulpia microstachys* (Nutt.) Benth. var. *pauciflora* (Beal) Lonard & Gould
F. saximontana Rydb. var. *purpusiana* (St.-Yves) S. Frederiksen & L.E. Pavlik separate from *F. brachyphylla* Schultes & Schultes f.
F. trachyphylla (Hackel) Kraj. naturalized
F. tracyi C. Hitchc. = *Vulpia microstachys* (Nutt.) Benth. var. *confusa* (Piper) Lonard & Gould
Heleochloa schoenoides (L.) Host. = *Crypsis schoenoides* (L.) Lam.
Hesperochloa kingii (S. Watson) Rydb. = *Festuca kingii* (S. Watson) Cassidy
Hesperostipa comata (Trin. & Rupr.) Barkworth ssp. *intermedia* (Scribner & Tweedy) Barkworth recognized
Hierochloe odorata (L.) Beauv. occurs in CA
Hilaria belangeri (Steud.) Nash not documented in CA
H. jamesii (Torrey) Benth. = *Pleuraphis jamesii* Torrey
H. rigida (Thurber) Scribner = *Pleuraphis rigida* Thurber
Hordeum californicum Covas & Stebb. = *H. brachyantherum* Nevski ssp. *californicum* (Covas & Stebb.) v. Bothmer, N. Jacobsen, & O. Seberg
H. geniculatum All. = *H. marinum* Hudson ssp. *gussoneanum* (Parl.) Thell.
H. glaucum Steudel = *H. murinum* L. ssp. *glaucum* (Steudel) Tzvelev
H. hystrix Roth = *H. marinum* Hudson ssp. *gussoneanum* (Parl.) Thell.
H. jubatum L. var. *caespitosum* (Scribner) A. Hitchc. not recognized
H. leporinum Link = *H. murinum* (L.) ssp. *leporinum* (Link) Arcang.
H. murinum L. sspp. *glaucum* (Steudel) Tzvelev, *leporinum* (Link) Arcang., & *murinum* recognized
H. pusillum Nutt. misapplied in part to *H. intercedens* Nevski
H. stebbinsii Covas = *H. murinum* L. ssp. *glaucum* (Steudel) Tzvelev

Hystrix californica (Bolander) Kuntze = *Elymus californicus* (Bolander) Gould
Koeleria cristata (L.) Pers. (illegitimate name) = *K. macrantha* (Ledeb.) J.A. Shultes
Lolium strictum C. Presl may not be naturalized, undocumented
Lycurus phleoides Kunth var. *phleoides* occurs in CA
Melica bulbosa Geyer var. *inflata* (Bolander) Boyle indistinct
M. californica Scribner var. *nevadensis* Boyle indistinct
M. geyeri Bolander var. *aristulata* J. Howell indistinct
Monerma cylindrica (Willd.) Cosson & Durand = *Hainardia cylindrica* (Willd.) Greuter
Muhlenbergia appressa C.O. Goodd. occurs in CA
M. fragilis Swallen occurs in CA
M. pauciflora Buckley occurs in CA
M. racemosa (Michaux) Britton, Sterns, & Poggenburg not documented in CA
Orcuttia californica Vasey var. *inaequalis* (Hoover) Hoover = *O. inaequalis* Hoover
var. *viscida* Hoover = *O. viscida* (Hoover) Reeder
O. greenei Vasey = *Tuctoria greenei* (Vasey) Reeder
O. mucronata Crampton = *Tuctoria mucronata* (Crampton) Reeder
Oryzopsis hymenoides (Roemer & Schultes) Ricker = *Achnatherum hymenoides* (Roemer & Schultes) Barkworth
O. kingii (Bolander) Beal = *Ptilagrostis kingii* (Bolander) Barkworth
O. micrantha (Trin. & Rupr.) Thurb. = *Piptatherum micranthum* (Trin. & Rupr.) Barkworth
O. miliacea (L.) Asch. & Schweinf. = *Piptatherum miliaceum* (L.) Cosson
O. webberi (Thurb.) Benth. = *Achnatherum webberi* (Thurb.) Barkworth
Panicum agrostoides Sprengel = *P. acuminatum* Sw. var. *lindheimeri* (Nash) Fern.
P. antidotale Retz. naturalized
P. arizonicum Scribner & Merr. not documented in CA
P. capillare L. var. *occidentale* Rydb. not recognized
P. huachucae Ashe = *P. acuminatum* Sw. var. *acuminatum*
P. lindheimeri Nash = *P. acuminatum* Sw. var. *lindheimeri* (Nash) Fern.
P. occidentale Scribner = *P. acuminatum* Sw. var. *acuminatum*
P. pacificum A. Hitchc. & Chase = *P. acuminatum* Sw. var. *acuminatum*
P. scribnerianum Nash = *P. oligosanthes* Schultes var. *scribnerianum* (Nash) Fern.
P. shastense Scribner & Merr. = *P. acuminatum* Sw. var. *acuminatum*
P. thermale Bolander = *P. acuminatum* Sw. var. *acuminatum*
Parapholis strigosa (Dumort.) C.E. Hubb. naturalized
Paspalum notatum J. Fleugge naturalized
Piptochaetium setosum (Trin.) Arechav. naturalized
Pleuropogon davyi L. Benson indistinct from *P. californicus* (Nees) Vasey
Poa ampla Merr. = *P. secunda* J. Presl ssp. *nevadensis* (Scribner) R. Soreng
P. bolanderi Vasey ssp. *howellii* (Vasey & Scribner) Keck = *P. howellii* Vasey & Scribner
P. bulbosa L. naturalized
P. canbyi (Scribner) Howell = *P. secunda* J. Presl ssp. *secunda*
P. cusickii Vasey ssp. *pallida* R. Soreng recently described
ssp. *purpurascens* (Vasey) R. Soreng recognized
P. douglasii Nees ssp. *macrantha* (Vasey) Keck = *P. macrantha* Vasey
P. epilis Scribner = *P. cusickii* Vasey ssp. *epilis* (Scribner) W.A. Weber
P. fendleriana (Steudel) Vasey: CA pls are ssp. *longiligula* (Scribner & Williams) R. Soreng
P. ×fibrata Scribner = *P. secunda* J. Presl ssp. *nevadensis* (Scribner) R. Soreng × *P. pratensis* L.
P. gracillima Vasey = *P. secunda* J. Presl ssp. *secunda*
P. hansenii Scribner misapplied in CA to *P. stebbinsii* R. Soreng
P. incurva Scribner & Williams = *P. secunda* J. Presl ssp. *secunda*
P. infirma Kunth naturalized
P. juncifolia Scribner = *P. secunda* J. Presl ssp. *nevadensis* (Scribner) R. Soreng
P. keckii R. Soreng recently described
P. nervosa (Hook.) Vasey misapplied to *P. wheeleri* Vasey
P. nevadensis Scribner = *P. secunda* J. Presl ssp. *nevadensis* (Scribner) R. Soreng
P. pattersonii Vasey occurs in CA
P. pratensis L. sspp. *agassizensis* (Boivin & D. Löve) Taylor & MacBryde & *angustifolia* (L.) Arcang. indistinct from ssp. *pratensis*
P. rupicola Nash = *P. glauca* M. Vahl ssp. *rupicola* (Nash) W.A. Weber
P. sandbergii Vasey = *P. secunda* J. Presl ssp. *secunda*
P. scabrella (Thurb.) Vasey = *P. secunda* J. Presl ssp. *secunda*
P. sierrae J. Howell recently described
P. suksdorfii (Beal) Piper misapplied in CA to *P. keckii* R. Soreng
P. wheeleri Vasey misapplied in CA to *P. nervosa* (Hook.) Vasey
Puccinellia airoides (Nutt.) S. Watson & J. Coulter = *P. nuttalliana* (Schultes) A. Hitchc.
P. californica (Beetle) Munz indistinct from *Torreyochloa pallida* (J. Presl) Church
P. erecta (A. Hitchc.) Munz = *Torreyochloa erecta* (A. Hitchc.) Church
P. grandis Swallen indistinct from *P. nutkaensis* (J. Presl) Fern. & Weath.
P. howellii J I Davis recently described
P. pauciflora (J. Presl) Munz = *Torreyochloa pallida* (J. Presl) Church var. *pauciflora* (J. Presl) J I Davis
P. pumila (Vasey) A. Hitchc. occurs in CA
Rhynchelytrum roseum (Nees) Stapf & C.E. Hubb. misapplied to *R. repens* (Willd.) C.E. Hubb.
Scleropoa rigida (L.) Griseb. = *Desmazeria rigida* (L.) Tutin
Scleropogon brevifolius Philippi occurs in CA
Setaria carnei A. Hitchc. misapplied to *S. verticillata* (L.) P. Beauv.
S. faberi R. Herrm. naturalized
S. geniculata (Lam.) P. Beauv. misapplied to *S. gracilis* Kunth
S. glauca (L.) P. Beauv. = *S. pumila* (Poiret) Roemer & Schultes
Sitanion ×hansenii (Scribner) J.G. Smith = *Elymus glaucus* Buckley × *E. elymoides* (Raf.) Swezey) or × *E. multisetus* (J.G. Smith) Burtt Davy
S. hordeoides Suksd. = *Elymus elymoides* (Raf.) Swezey ssp. *hordeoides* (Suksd.) Barkworth
S. hystrix (Nutt.) J.G. Smith = *Elymus elymoides* (Raf.) Swezey ssp. *elymoides*
S. jubatum J.G. Smith = *Elymus multisetus* (J.G. Smith) Burtt Davy
Spartina alterniflora Lois. naturalized
S. densiflora Brongn. naturalized
Sporobolus airoides (Torrey) Torrey var. *wrightii* (Scribner) Gould indistinct
S. poiretii (Roemer & Schultes) A. Hitchc. misapplied to *S. indicus* (L.) R. Br.
Stipa arida M.E. Jones = *Achnatherum aridum* (M.E. Jones) Barkworth
S. brachychaeta Godron = *Achnatherum brachychaetum* (Godron) Barkworth
S. californica Merr. & Burtt Davy = *Achnatherum occidentalis* (Thurber) Barkworth ssp. *californicum* (Merr. & Burtt Davy) Barkworth
S. cernua Stebb. & Löve = *Nassella cernua* (Stebb. & Löve) Barkworth

S. columbiana Macoun = *Achnatherum lemmonii* (Vasey) Barkworth; also misapplied to *Achnatherum nelsonii* (Scribner) Barkworth ssp. *dorei* (Barkworth & J. Maze) Barkworth
S. comata Trin. & Rupr. = *Hesperostipa comata* (Trin. & Rupr.) Barkworth
S. coronata Thurber var. *coronata* = *Achnatherum coronatum* (Thurber) Barkworth
var. *depauperata* (M.E. Jones) A. Hitchc. = *Achnatherum parishii* (Vasey) Barkworth
S. diegoensis Swallen = *Achnatherum diegoensis* (Swallen) Barkworth
S. elmeri Scribner = *Achnatherum occidentalis* (Thurber) Barkworth ssp. *pubescens* (Vasey) Barkworth
S. formicarum Del. = *Nassella formicara* (Del.) Barkworth naturalized
S. latiglumis Swallen = *Achnatherum latiglumis* (Swallen) Barkworth
S. lemmonii (Vasey) Scribner = *Achnatherum lemmonii* (Vasey) Barkworth
S. lepida A. Hitchc. var. *andersonii* (Vasey) A. Hitchc. = *Nassella cernua* (Stebb. & Löve) Barkworth
var. *lepida* = *Nassella lepida* (A. Hitchc.) Barkworth
S. lettermanii Vasey = *Achnatherum lettermanii* (Vasey) Barkworth
S. nevadensis B. Johnson = *Achnatherum nevadensis* (B. Johnson) Barkworth occurs in CA
S. occidentalis Thurber = *Achnatherum occidentalis* (Thurber) Barkworth ssp. *occidentalis*
S. pinetorum M.E. Jones = *Achnatherum pinetorum* (M.E. Jones) Barkworth
S. pulchra A. Hitchc. = *Nassella pulchra* (A. Hitchc.) Barkworth
S. speciosa Trin. & Rupr. = *Achnatherum speciosum* (Trin. & Rupr.) Barkworth
S. stillmanii Bolander = *Achnatherum stillmanii* (Bolander) Barkworth
S. thurberiana Piper = *Achnatherum thurberianum* (Piper) Barkworth
S. williamsii Scribner misapplied to *Achnatherum nelsonii* (Scribner) Barkworth ssp. *dorei* (Barkworth & J. Maze) Barkworth
Tridens pilosus (Buckley) A. Hitchc. = *Erioneuron pilosum* (Buckley) Nash
T. pulchellus (Kunth) A. Hitchc. = *Erioneruon pulchellum* (Kunth) Tateoka
Ventenata dubia (Leers) Durieu naturalized
Zizania palustris L. var. *interior* (Fassett) W. Dore naturalized

PONTEDERIACEAE PICKEREL-WEED FAMILY

Heteranthera limosa (Schwartz) Willd. naturalized

POTAMOGETONACEAE PONDWEED FAMILY

Potamogeton berchtoldii Fieber = *P. pusillus* L. var. *tenuissimus* Mert. & Koch
P. epihydrus Raf. ssp. *epihydrus* in e N.Am, not CA
var. *nuttallii* (Cham. & Schldl.) Fern. = ssp. *nuttallii* (Cham. & Schldl.) Calder & Roy Taylor
P. filiformis Pers. var. *macounii* Morong misapplied in CA
P. foliosus Raf. var. *fibrillosus* (Fern.) R. Haynes & Rev. occurs in CA
P. gramineus L. var. *maximus* Morong not recognized
Ruppia transferred from Ruppiaceae
R. spiralis Dumort. = *R. cirrhosa* (Petagna) Grande

RUPPIACEAE DITCH-GRASS FAMILY: see **POTAMOGETONACEAE**

SCHEUCHZERIACEAE SCHEUCHZERIA FAMILY: no changes

TYPHACEAE CATTAIL & BUR-REED FAMILY

Sparganium eurycarpum Engelm.: CA pls are ssp. *eurycarpum*
S. minimum Fries = *S. natans* L.
S. multipedunculatum (Morong) Rydb. = *S. emersum* Rehmann ssp. *emersum*
S. simplex Hudson = *S. emersum* Rehmann ssp. *emersum*
Typha ×*glauca* Godron = *T. angustifolia* × *T. latifolia*

ZANICHELLIACEAE HORNED-PONDWEED FAMILY

Halodule transferred to Cymodoceaceae

ZOSTERACEAE EEL-GRASS FAMILY

Zostera marina L. var. *latifolia* Morong not recognized
Z. pacifica S. Watson occurs in CA

INDEX

The following index includes all common names used in *The Jepson Manual*, scientific names of families and genera, and synonyms. It does not include scientific names of accepted species or infraspecific taxa, which are arrayed alphabetically in the text. Italics are not used for scientific names of genera or species in the index.

Page numbers for all common names, synonyms, and miscellaneous entries are given in regular type. Accepted scientific names of families are followed by the page number (in **bold type**) of the beginning of the family description. Scientific names of described genera are often followed by three or more page numbers. The first, in regular type, refer(s) to the page(s) on which the genus is keyed. Next, a single **bold type** entry indicates the page on which the generic description begins. Third, there may be one or more page numbers in *italics* that refer to illustration plate(s) for the genus.

Malvaceae, the Mallow Family, is listed for page **746**. Checkerbloom is listed for page 755, with various common names of species listed for pages 758–760. The entry for Sidalcea, the generic name for Checkerbloom, includes 747 (the page following the beginning of the family description, where the genus keys), **755** (the page where the generic description begins), and *761*, *767* (plate pages, which are always odd-numbered). In indented lines following the generic entry are several scientific names that have been accepted but are considered in *The Jepson Manual* to be synonyms.

A

C

D

E

F

G

M

P

S

T

U

V

W

Jacket design and production assistance by David Comstock
Text design and typesetting by William J. Stone
Illustration plate design by Susan M. D'Alcamo and Dr. Linda A. Vorobik
Dust jacket photo of *Calochortus pulchellus* by Jo-Ann Ordano

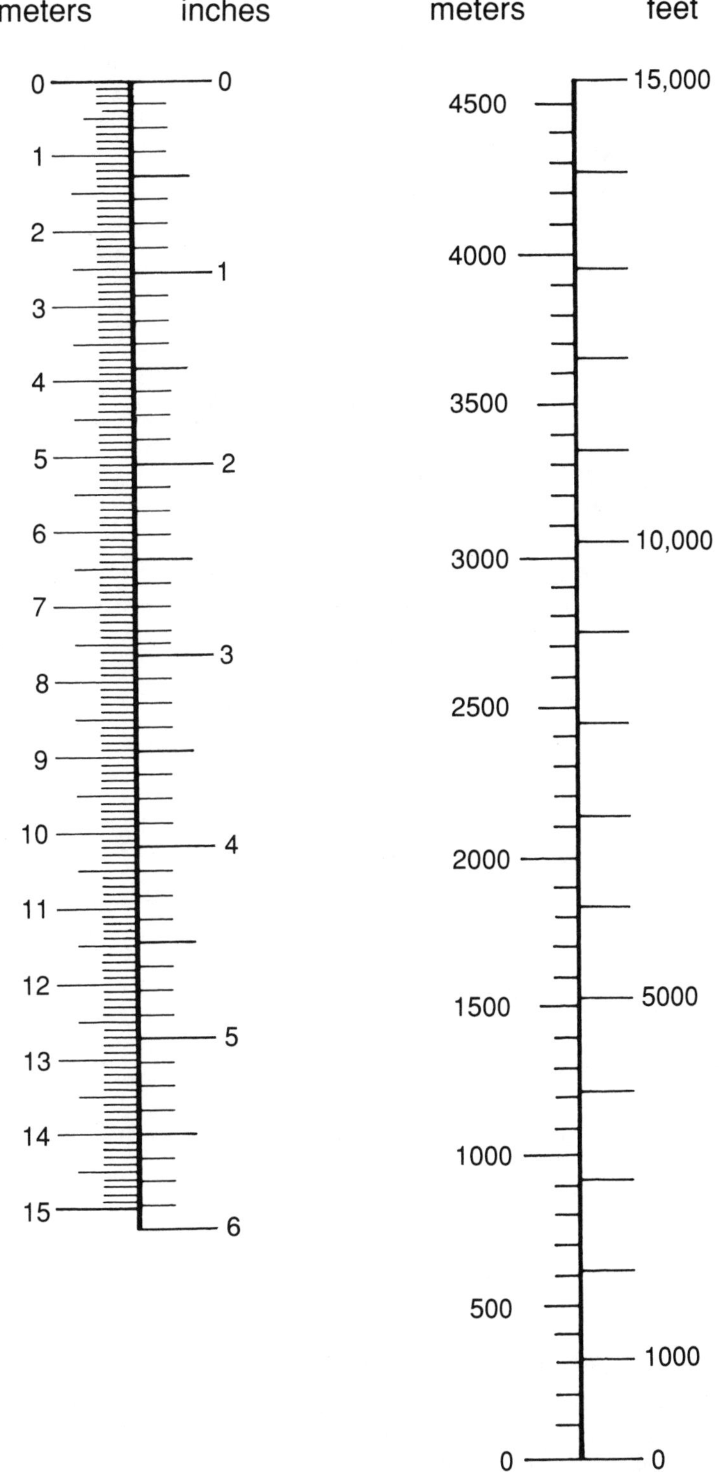
centimeters
inches
meters
feet
0
1
2
3
4
5
6
7
8
9
10
11
12
13
14
15
0
1
2
3
4
5
6
4500
4000
3500
3000
2500
2000
1500
1000
500
0
15,000
10,000
5000
1000
0